Compiled by B. N. Taylor, W. H. Parker,

Modern Physics, Vol. 41, p. 375, 1969)

in the last digits of the quoted value, computed on the basis of internal consistency.

Quantity	Symbol	Value	Error (ppm)	Units SI	Units cgs
Compton wavelength of the proton, $h/M_p c$	$\lambda_{C,p}$ $\lambda_{C,p}/2\pi$	1.3214409(90) 2.103139(14)	6.8 6.8	10^{-15} m 10^{-16} m	10^{-13} cm 10^{-14} cm
Compton wavelength of the neutron, $h/M_n c$	$\lambda_{C,n}$ $\lambda_{C,n}/2\pi$	1.3196217(90) 2.100243(14)	6.8 6.8	10^{-15} m 10^{-16} m	10^{-13} cm 10^{-14} cm
Gas constant	R_0	8.31434(35)	42	10^3 J kmole$^{-1} \cdot$ K^{-1}	10^7 erg mole$^{-1} \cdot$ K^{-1}
Boltzman's constant, R_0/N	k	1.380622(59)	43	10^{-23} J K^{-1}	10^{-16} erg K^{-1}
Stefan–Boltzman constant, $\pi^2 k^4 / 60 \hbar^3 c^2$	σ	5.66961(96)	170	10^{-8} W m^{-2} K^4	10^{-5} erg sec$^{-1} \cdot$ cm$^{-2} \cdot$ K^{-4}
First radiation constant, $8\pi hc$	c_1	4.992579(38)	7.6	10^{-24} J $\cdot$ m	10^{-15} erg $\cdot$ cm
Second radiation constant, hc/k	c_2	1.438833(61)	43	10^{-2} m $\cdot$ K	cm $\cdot$ K
Gravitational constant	G	6.6732(31)	460	10^{-11} N $\cdot$ m^2 kg^{-2}	10^{-8} dyn $\cdot$ cm^2 g^{-2}
kx-unit-to-angstrom conversion factor, $\Lambda = \lambda(\text{Å})/\lambda(\text{kxu}); \lambda(CuK\alpha_1) \equiv$ 1.537400 kxu	Λ	1.0020764(53)	5.3		
Å*-to-angstrom conversion factor, $\Lambda = \lambda(\text{Å})/\lambda(\text{Å}^*); \lambda(WK\alpha_1) \equiv$ 0.2090100 Å*	Λ^*	1.0000197(56)	5.6		

[a] Note that the unified atomic mass scale $^{12}C \equiv 12$ has been throughout, that amu = atomic mass unit, C = coulomb, G = gauss, Hz = hertz = cycles/sec, J = joule, K = kelvin (degrees kelvin), T = tesla (10^4 G), V = volt, and W = watt. In cases where formulas for constants are given (e.g., R_∞), the relations are written as the product of two factors. The second factor, in parentheses, is the expression to be used when all quantities are expressed in cgs units, with the electron charge in electrostatic units. The first factor, in brackets, is to be included only if all quantities are expressed in SI units. We remind the reader that with the exception of the auxiliary constants which have been taken to be exact, the uncertainties of these constants are correlated, and therefore the general law of error propagation must be used in calculating additional quantities requiring two or more of these constants.

Energy Conversion Factors

Quantity	Value	Unit	Error (ppm)
1 kg	5.609538(24)	10^{29} MeV	4.4
1 amu	931.4812(52)	MeV	5.5
Electron mass	0.5110041(16)	MeV	3.1
Proton mass	938.2592(52)	MeV	5.5
Neutron mass	939.5527(52)	MeV	5.5
1 electron volt	1.6021917(70)	10^{-19} J 10^{-12} erg	4.4
	2.4179659(81)	10^{14} Hz	3.3
	8.065465 (27)	10^5 m^{-1} 10^3 cm^{-1}	3.3
	1.160485(49)	10^4 K	42
Energy-wavelength conversion	1.2398541(41)	10^{-6} eV $\cdot$ m 10^{-4} eV $\cdot$ cm	3.3
Rydberg constant, R_∞	2.179914(17)	10^{-18} J 10^{-11} erg	7.6
	13.605826(45)	eV	3.3
	3.2898423(11)	10^{15} Hz	0.35
	1.578936(67)	10^5 K	43
Bohr magneton, μ_B	5.788381(18)	10^{-5} eV T^{-1}	3.1
	1.3996108(43)	10^{10} Hz T^{-1}	3.1
	46.68598(14)	m$^{-1} \cdot$ T^{-1} 10^{-2} cm$^{-1} \cdot$ T^{-1}	3.1
	0.671733(29)	K T^{-1}	43
Nuclear magneton, μ_n	3.152526(21)	10^{-8} eV T^{-1}	6.8
	7.622700(42)	10^6 Hz T^{-1}	5.5
	2.542659(14)	10^{-2} m$^{-1} \cdot$ T^{-1} 10^{-4} cm$^{-1} \cdot$ T^{-1}	5.5
	3.65846(16)	10^{-4} K T^{-1}	44
Gas constant, R_0	8.20562(35)	10^{-2} m$^3 \cdot$ atm kmole$^{-1} \cdot$ K^{-1}	42
Standard volume of ideal gas, V_0	22.4136	m^3 kmole^{-1}	

AMERICAN INSTITUTE OF PHYSICS HANDBOOK

OTHER McGRAW-HILL HANDBOOKS OF INTEREST

AMERICAN SOCIETY OF MECHANICAL ENGINEERS · ASME Handbooks:
    Engineering Tables   Metals Engineering—Processes
    Metals Engineering—Design   Metals Properties
BAUMEISTER AND MARKS · Standard Handbook for Mechanical Engineers
BERRY, BOLLAY, AND BEERS · Handbook of Meteorology
BLATZ · Radiation Hygiene Handbook
BRADY · Materials Handbook
BURINGTON · Handbook of Mathematical Tables and Formulas
BURINGTON AND MAY · Handbook of Probability and Statistics with Tables
CALLENDER · Time-Saver Standards
CHOW · Handbook of Applied Hydrology
CONDON AND ODISHAW · Handbook of Physics
CONSIDINE · Process Instruments and Controls Handbook
CONSIDINE AND ROSS · Handbook of Applied Instrumentation
ETHERINGTON · Nuclear Engineering Handbook
FINK AND CARROLL · Standard Handbook for Electrical Engineers
FLÜGGE · Handbook of Engineering Mechanics
GRANT · Hackh's Chemical Dictionary
HAMSHER · Communication System Engineering Handbook
HARRIS AND CREDE · Shock and Vibration Handbook
HENNEY · Radio Engineering Handbook
HICKS · Standard Handbook of Engineering Calculations
HUNTER · Handbook of Semiconductor Electronics
HUSKEY AND KORN · Computer Handbook
IRESON · Reliability Handbook
JURAN · Quality Control Handbook
KAELBLE · Handbook of X-rays
KALLEN · Handbook of Instrumentation and Controls
KING AND BRATER · Handbook of Hydraulics
KLERER AND KORN · Digital Computer User's Handbook
KOELLE · Handbook of Astronautical Engineering
KORN AND KORN · Mathematical Handbook for Scientists and Engineers
LANDEE, DAVIS, AND ALBRECHT · Electronic Designers' Handbook
LANGE · Handbook of Chemistry
MACHOL · System Engineering Handbook
MANTELL · Engineering Materials Handbook
MARKUS · Electronics and Nucleonics Dictionary
MEITES · Handbook of Analytical Chemistry
MERRITT · Standard Handbook for Civil Engineers
PERRY · Engineering Manual
PERRY, CHILTON, AND KIRKPATRICK · Chemical Engineers' Handbook
RICHEY · Agricultural Engineers' Handbook
ROTHBART · Mechanical Design and Systems Handbook
STREETER · Handbook of Fluid Dynamics
TERMAN · Radio Engineers' Handbook
TOULOUKIAN · Retrieval Guide to Thermophysical Properties Research Literature
TRUXAL · Control Engineers' Handbook
URQUHART · Civil Engineering Handbook
WOLMAN · Handbook of Clinical Psychology

American Institute of Physics Handbook

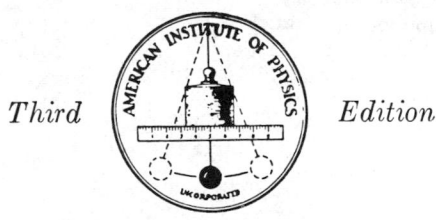

Third *Edition*

Section Editors

BRUCE H. BILLINGS, Ph.D.
Commissioner, Joint Commission
on Rural Reconstruction
Taipei, Taiwan

D. F. BLEIL, Ph.D.
Associate Technical Director
and Head, Research
U.S. Naval Ordnance Laboratory

RICHARD K. COOK, Ph.D.
Special Assistant for Sound Programs,
Office of Deputy Director
The National Bureau of Standards

H. M. CROSSWHITE, Ph.D.
Adjunct Professor of Spectroscopy,
Physics Department
The Johns Hopkins University

H. P. R. FREDERIKSE, Ph.D.
Chief, Solid State Physics Section
The National Bureau of Standards

R. BRUCE LINDSAY, Ph.D.
Professor of Physics, Emeritus,
Brown University

J. B. MARION, Ph.D.
Professor of Physics, Department
of Physics and Astronomy
University of Maryland

MARK W. ZEMANSKY, Ph.D.
Professor of Physics
The City College of the City
University of New York

Coordinating Editor

Dwight E. Gray, Ph.D.
American Institute of Physics

McGraw-Hill Book Company

*New York   St. Louis   San Francisco   Düsseldorf   Johannesburg
Kuala Lumpur   London   Mexico   Montreal   New Delhi
Panama   Rio de Janeiro   Singapore   Sydney   Toronto*

Library of Congress Cataloging in Publication Data

American Institute of Physics.
  American Institute of Physics handbook.

  Includes bibliographies.
  1. Physics—Handbooks, manuals etc.
I. Gray, Dwight E., ed.   II.   Title.
QC61.A5   1972   016.5301'5   72–3248
ISBN 07-001485-X

234567890   COCO   76543

The editors for this book were Daniel N. Fischel, Harold B. Crawford, Don A. Douglas,
and Winifred C. Eisler, and its production was supervised by George E. Oechsner.   It
was set in Modern 8A by The Maple Press Company.

It was printed and bound by The Colonial Press Inc.

Contents

Editor, *Dr. R. Bruce Lindsay*, Brown University

Fundamental concepts of mechanics. Units and conversion factors. Density of solids. Centers of mass and moments of inertia. Coefficients of friction. Elastic constants, hardness, strength, elastic limits, and diffusion coefficients of solids. Viscosity of solids. Astronomical data. Geodetic data. Seismological and related data. Oceanographic data. Meteorological information. Density and compressibility of liquids. Viscosity of liquids. Tensile strength and surface tension of liquid. Cavitation in flowing liquids. Diffusion in liquids. Liquid jets. Viscosity of gases. Molecular diffusion of gases. Compressible flow of gases. Laminar and turbulent flow of gases. Shock waves.

Editor, *Dr. Richard K. Cook*, The National Bureau of Standards

Acoustical definitions. Standard letter symbols and conversion factors for acoustical quantities. Propagation of sound in fluids. Acoustic properties of gases. Acoustic properties of liquids. Acoustic properties of solids. Properties of transducer materials.

v

The periodic system. The electronic structure of atoms. Energy-level diagrams of atoms. Persistent lines of the elements. Important atomic spectra. X-ray wavelengths and atomic energy levels. Constants of diatomic molecules. Constants of polyatomic molecules. Atomic transition probabilities.

Nuclear constants and calibrations. Properties of nuclides. Atomic mass formulas. Passage of charged particles through matter. Gamma rays. Neutrons. Nuclear fission. Elementary particles and interactions. Health physics. Particle accelerators.

Crystallographic properties. Structure, melting point, density, and energy gap of simple inorganic compounds. Electronic properties of solids. Properties of metals. Properties of semiconductors. Properties of ionic crystals. Properties of superconductors. Color centers and dislocations. Luminescence. Work function and secondary emission.

Index follows Section 9.

Contributors

J. R. Anderson, Ph.D., *University of Maryland*. Solid State Physics (9d)

Gordon Atkinson, Ph.D., *University of Maryland*. Electricity and Magnetism (5g)

Fred Ayres, Ph.D., *Reed College*. Mechanics (2n)

H. W. Babcock, Ph.D., *Hale Observatories*. Electricity and Magnetism (5i)

Julius Babiskin, Ph.D., *U.S. Naval Research Laboratory*. Solid State Physics (9d)

Stanley Ballard, Ph.D., *University of Florida*. Optics (6b, 6c)

Philip Baumeister, Ph.D., *University of Rochester*. Optics (6i)

J. A. Bearden, Ph.D., *The Johns Hopkins University*. Atomic and Molecular Physics (7f)

E. C. Beaty, Ph.D., *The National Bureau of Standards, Boulder*. Electricity and Magnetism (5e)

L. I. Beaubien, Ph.D., *U.S. Naval Research Laboratory*. Mechanics (2e)

Leo L. Beranek, Ph.D., *Bolt Beranek and Newman Inc.* Acoustics (3a, 3b, 3d, 3p)

Robert T. Beyer, Ph.D., *Brown University*. Acoustics (3o)

Hans Bichsel, Ph.D., *University of Washington*. Nuclear Physics (8d)

Bruce H. Billings, Ph.D., *Joint Commission on Rural Reconstruction, Taipei, Taiwan*. Optics (6c, 6f, 6h, 6l, 6o)

David T. Blackstock, Ph.D., *University of Texas*. Acoustics (3n)

E. Boldt, Ph.D., *NASA—Goddard Space Flight Center*. Electricity and Magnetism (5i)

R. M. Bozorth, Ph.D., *U.S. Naval Ordnance Laboratory*. Electricity and Magnetism (5f)

Willem Brouwer, Ph.D., *Diffraction Ltd. Inc.* Optics (6d)

James S. Browder, Ph.D., *University of Florida*. Optics (6b, 6c)

R. M. Burley, A.B., *Baird-Atomic, Inc.* Optics (6p)

Constance Carter, M.S., *Library of Congress*. Mathematics Bibliography; SI Units (1a, 1b)

Gregg E. Childs, Ph.D., *The National Bureau of Standards, Boulder*. Heat (4g)

R. J. Collins, Ph.D., *University of Minnesota*. Optics (6s)

W. R. Cook, Jr., M.A., *Gould, Inc.* Optics (6m)

H. M. Crosswhite, Ph.D., *The Johns Hopkins University*. Atomic and Molecular Physics (7a, 7b, 7c, 7d, 7e)

Evan A. Davis, Ph.D., *Westinghouse Research Laboratory*. Mechanics (2f)

R. DiPippo, Ph.D., *Southeastern Massachusetts University*. Mechanics (2r)

E. S. Domalski, Ph.D., *The National Bureau of Standards*. Heat (4j)

J. D. H. Donnay, Ph.D., *The Johns Hopkins University*. Solid State Physics (9a)

Thomas B. Douglas, Ph.D., *The National Bureau of Standards*. Heat (4e)

J. F. Ebersole, Ph.D., *University of Florida*. Optics (6b, 6c)

Phillip Eisenberg, Ph.D., *Hydronautics, Inc.* Mechanics (2o)

Eugene Epstein, Ph.D., *Aerospace Corporation*. Optics (6q)

John Evans, Ph.D., *Air Force Cambridge Research Laboratories*. Optics (6i)

Robley D. Evans, Ph.D., *Massachusetts Institute of Technology*. Nuclear Physics (8e)

H. P. R. Frederikse, Ph.D., *The National Bureau of Standards*. Solid State Physics (9b, 9c, 9e)

Eli Freedman, Ph.D., *Ballistic Research Laboratories*. Mechanics (2v)

R. J. Friauf, Ph.D., *University of Kansas*. Solid State Physics (9f)

Dudley Fuller, Ph.D., *Columbia University*. Mechanics (2d)

George T. Furakawa, Ph.D., *The National Bureau of Standards*. Heat (4e)

John S. Gallagher, A.B., *The National Bureau of Standards*. Heat (4i)

Murrey D. Goldberg, Ph.D., *Brookhaven National Laboratory*. Nuclear Physics (8f)

David T. Goldman, Ph.D., *The National Bureau of Standards*. Nuclear Physics (8b)

Edward F. Greene, Ph.D., *Brown University*. Mechanics (2v)

Martin Greenspan, B.S., *The National Bureau of Standards*. Acoustics (3e)

B. Gutenburg, Ph.D. (Deceased), *California Institute of Technology*. Mechanics (2i)

George A. Haas, Ph.D., *U.S. Naval Research Laboratory*. Solid State Physics (9j)

Lawrence Hadley, Ph.D., *Colorado State University*. Optics (6g)

Thomas A. Hahn, B.S., *The National Bureau of Standards*. Heat (4f)

William J. Hall, A.B., *The National Bureau of Standards, Boulder*. Heat (4a)

Cyril M. Harris, Ph.D., *Columbia University*. Acoustics (3j)

F. K. Harris, Ph.D., *The National Bureau of Standards*. Electricity and Magnetism (5c)

Miles F. Harris, Ph.D., *National Oceanic and Atmospheric Administration*. Mechanics (2k)

John A. Harvey, Ph.D., *Oak Ridge National Laboratory*. Nuclear Physics (8f)

Georg Hass, Ph.D., *U.S. Army Electronics Command*. Optics (6g)

J. P. Heppner, Ph.D., *NASA—Goddard Space Flight Center*. Electricity and Magnetism (5h, 5i)

G. Herzberg, Ph.D., *National Research Council of Canada*. Atomic and Molecular Physics (7h)

L. Herzberg, Ph.D. (Deceased), *National Research Council of Canada*. Atomic and Molecular Physics (7h)

D. B. Herrmann, Ph.D., *Bell Telephone Laboratories, Inc.* Electricity and Magnetism (5d)

Joseph Hilsenrath, M.A., *The National Bureau of Standards*. Heat (4h)

David L. Hogenboom, Ph.D., *Lafayette College*. Mechanics (2m)

Robert Howard, Ph.D., *Hale Observatories*. Electricity and Magnetism (5i)

K. P. Huber, Ph.D., *National Research Council of Canada*. Atomic and Molecular Physics (7g)

R. P. Hudson, Ph.D., *The National Bureau of Standards*. Electricity and Magnetism (5f)

Frederick V. Hunt, Ph.D., *Harvard University*. Acoustics (3c)

Hans Jaffe, Ph.D., *Gould, Inc.* Optics (6m)

T. L. Jobe, B.S., *The National Bureau of Standards*. Heat (4j)

Joseph Kaspar, Ph.D., *Aerospace Corporation*. Optics (6k)

R. Norris Keeler, Ph.D., *Lawrence Radiation Laboratory*. Heat (4d)

George C. Kennedy, Ph.D., *University of California*. Heat (4d)

Joseph Kestin, Ph.D., *Brown University*. Mechanics (2r)

Richard K. Kirby, B.S., *The National Bureau of Standards*. Heat (4f)

Max Klein, Ph.D., *The National Bureau of Standards*. Heat (4i)

C. C. Klick, Ph.D., *U.S. Naval Research Laboratories*. Solid State Physics (9h)

Karl R. Koch, Ph.D., *National Oceanic & Atmospheric Administration*. Mechanics (2h)

R. Bruce Lindsay, Ph.D., *Brown University*. Mechanics (2a, 2b, 2c, 2g)

Robert Lindsay, Ph.D., *Trinity College*. Mechanics (2l)

G. L. Link, Ph.D., *Bell Telephone Laboratories, Inc.* Electricity and Magnetism (5d)

Ernest Loewenstein, Ph.D., *Air Force Cambridge Research Laboratories.* Optics (6r)

Lewis G. Longsworth, Ph.D., *The Rockefeller University.* Mechanics (2p)

Walter Loveland, Ph.D., *Oregon State University.* Nuclear Physics (8g)

David MacAdam, Ph.D., *Eastman Kodak Company.* Optics (6a, 6j)

Nancy R. McClure, *Eastman Kodak Company.* Optics (6d)

T. R. McGuire, Ph.D., *IBM—Watson Research Center.* Electricity and Magnetism (5f)

John E. McKinney, Ph.D., *The National Bureau of Standards.* Mechanics (2l)

J. B. Marion, Ph.D., *University of Maryland.* Nuclear Physics (8a)

Robert S. Marvin, Ph.D., *The National Bureau of Standards.* Mechanics (2m)

W. P. Mason, Ph.D., *Columbia University.* Acoustics (3f, 3g); Solid State Physics (9a)

Frank Massa, M.S., *Dynamics Corporation of America.* Acoustics (3i)

W. J. Merz, Ph.D., *RCA Laboratories.* Solid State Physics (9f)

B. M. Miles, Ph.D., *The National Bureau of Standards.* Atomic and Molecular Physics (7i)

David Mintzner, Ph.D., *Northwestern University.* Mechanics (2a)

Karl Z. Morgan, Ph.D., *Oak Ridge National Laboratory.* Nuclear Physics (8i)

Edwin B. Newman, Ph.D., *Harvard University.* Acoustics (3k)

Wesley L. Nyborg, Ph.D., *University of Vermont.* Mechanics (2n, 2q)

Harry F. Olson, Ph.D., *RCA Laboratories.* Acoustics (3l, 3m)

Norman Pearlman, Ph.D., Purdue University. Heat (4e)

Karl B. Persson, Ph.D., *The National Bureau of Standards, Boulder.* Electricity and Magnetism (5e)

Harmon H. Plumb, Ph.D., *The National Bureau of Standards.* Heat (4a)

Robert L. Powell, Ph.D., *The National Bureau of Standards, Boulder.* Heat (4a, 4g)

Martin P. Reiser, Ph.D., *University of Maryland.* Nuclear Physics (8j)

B. W. Roberts, Ph.D., *General Electric Research and Development Center.* Solid State Physics (9g)

R. C. Roberts, Ph.D., *University of Maryland.* Mechanics (2s, 2t, 2u)

Arthur H. Rosenfeld, Ph.D., *University of California.* Nuclear Physics (8h)

Bruce D. Rothrock, B.S., *The National Bureau of Standards.* Heat (4f)

Hellmut H. Schmid, Ph.D., *National Oceanic & Atmospheric Administration.* Mechanics (2h)

R. H. Schumm, M.S., *The National Bureau of Standards.* Heat (4j)

Arthur F. Scott, Ph.D., *Reed College.* Mechanics (2n)

Philip A. Seeger, Ph.D., *Los Alamos Scientific Laboratory.* Nuclear Physics (8c)

J. M. H. Levelt Sengers, Ph.D., *The National Bureau of Standards.* Heat (4i)

J. C. Slater, Ph.D., *University of Florida.* Solid State Physics (9c)

Donald R. Smith, M.S., *Air Force Cambridge Research Laboratories.* Optics (6r)

W. R. Smythe, Ph.D., *California Institute of Technology.* Electricity and Magnetism (5a, 5b)

George A. Snow, Ph.D., *University of Maryland.* Nuclear Physics (8h)

Irene A. Stegun, M.S., *The National Bureau of Standards.* Mathematics Bibliography; SI Units (1a, 1b)

D. E. Stone, A.B., *Vertex Corporation.* Mechanics (2b, 2e)

Daniel R. Stull, Ph.D., *Dow Chemical Company.* Heat (4k)

Masahisa Sugiura, Ph.D., *NASA—Goddard Space Flight Center.* Electricity and Magnetism (5h, 5i)

James F. Swindells, A.B., *The National Bureau of Standards.* Heat (4a)

Paul Tamarkin, Ph.D., *RAND Corporation.* Mechanics (2a)

J. S. Thomsen, Ph.D., *The Johns Hopkins University.* Atomic and Molecular Physics (7f)

H. M. Trent, Ph.D. (Deceased), *U.S. Naval Research Laboratory.*   Mechanics (2b, 2e)

James E. Turner, Ph.D., *Oak Ridge National Laboratory.*   Nuclear Physics (8i)

Allyn C. Vine, Ph.D., *Woods Hole Oceanographic Institution.*   Mechanics (2j)

D. D. Wagman, Ph.D., *The National Bureau of Standards.*   Heat (4j)

David White, Ph.D., *University of Pennsylvania.*   Heat (4c)

J. E. White, Ph.D., *Globe Universal Sciences, Inc.*   Mechanics (2i)

W. L. Wiese, Ph.D., *The National Bureau of Standards.*   Atomic and Molecular Physics (7i)

Randolph C. Wilhoit, Ph.D., *Texas A and M University.*   Heat (4l)

Ferd E. Williams, Ph.D., *University of Delaware.*   Solid State Physics (9i)

E. A. Wood, Ph.D., *Bell Telephone Laboratories.*   Solid State Physics (9a)

Cavour Yeh, Ph.D., *University of California at Los Angeles.*   Electricity and Magnetism (5b)

Kenneth F. Young, B.S., *The National Bureau of Standards.*   Solid State Physics (9f)

Robert W. Young, Ph.D., *U.S. Naval Undersea Research and Development Center.*   Acoustics (3h)

Mark W. Zemansky, Ph.D., *The City College of the City University of New York.*   Heat (4b)

Fritz Zernike, M.S., *Perkin-Elmer Corporation.*   Optics (6n)

Bruno J. Zwolinski, Ph.D., *Texas A and M University.*   Heat (4l)

Preface

The *American Institute of Physics Handbook* has won wide acceptance among scientists and engineers. It is just such a degree of acceptance that has stimulated the issuance of this revised and updated third edition. This edition, like the previous two, continues the philosophy of supplying authoritative reference material—including tables of data, graphs, and bibliographies—selected and described with a minimum of narration by leaders in physical methods for research.

Among the entirely new sections in this edition are those on nonlinear optics, calibration energies for alpha particles and gamma rays, nonlinear acoustics, atomic mass formulas, particle accelerator principles, atomic transition probabilities, electric and magnetic fields in the earth's environment, and far infrared. Examples of topics in which especially extensive revisions have been made are: optical masers, various optical constants, virial coefficients, heats of combustion and formation, and superconductors. A number of sections were completely rewritten; these include radioastronomy, radiometry, various crystal properties, molecular constants and phase transitions. The mathematics section now consists of a special treatment of SI units and a bibliography that has been revised to include references to new methods, algorithms, and computer programs.

Publication of this Handbook was a mammoth undertaking that required the contributions and cooperation of many individuals and two organizations. Leading the individuals is Dr. Dwight E. Gray, who served as coordinating editor for this 1972 edition, as he also did for the 1957 and 1963 editions. Dr. Gray, who is a master of the pen and is well grounded in physics, was able to coordinate successfully the efforts of the eight section editors and the some 125 contributors. He did this work while also serving as the Washington Representative of the American Institute of Physics. Through his Washington office he was able to maintain contact with and coordinate the efforts of the many individuals concerned in the effort, as well as to handle the involvements of the sponsor—the American Institute of Physics—and the publisher—the McGraw-Hill Book Company.   Key McGraw-Hill individuals for this project included Mrs. Winifred C. Eisler, who copy-edited the manuscript, and Mr. Don A. Douglas, the Editing Manager.   To these individuals, editors and physicists alike, the scientific community is deeply indebted for their painstaking and conscientious contributions and acknowledges their efforts with thanks.

As with any user-oriented publication, comments, suggestions, and criticisms are solicited on this edition of the Handbook.   Only with such continuing contributions and cooperation can future Handbooks meet their responsibilities.

H. WILLIAM KOCH, *Director*
American Institute of Physics

AMERICAN INSTITUTE OF PHYSICS HANDBOOK

Section 1

MATHEMATICS
BIBLIOGRAPHY; SI UNITS

CONTENTS

The third edition of the AIP Handbook, like the second, presents a bibliography of mathematical references in lieu of an assortment of actual mathematical tables. Selection of such tables necessarily would have been arbitrary; they would have been bound to duplicate many tables already easily available to most physicists; and, most important, including them would have necessitated the omission of significant physics material. The basic pattern of the third-edition bibliography is described at the beginning of Sec. 1a. For reasons outlined in the first paragraph of Sec. 1b, it was believed neither practicable nor desirable to attempt exclusive use of the International System of Units in this edition of the Handbook. Section 1b outlines the background of SI Units, and presents a portion of a National Bureau of Standards bulletin on their interpretation.

1a. Mathematics Bibliography

The National Bureau of Standards

CONSTANCE CARTER

Library of Congress

In view of the appearance of large compendiums and the increasing use of computers with built-in functions or function subroutines in their compilers, many of the tables of elementary functions have been omitted from this bibliography. An effort has been made to include a dictionary; indexes of mathematical and statistical tables; compendiums of general tables, series, integrals, transforms, and differential equations; and references to numerical methods, new tables, and new disciplines.

For algorithms covering a wide variety of subjects such as the evaluation of systems of linear equations, estimations of definite integrals, sorting of data, etc., reference should be made to the "Collected Algorithms from CACM" (Communications of the Association for Computing Machinery, Inc.).

1. Abramowitz, Milton, and Irene A. Stegun, eds.: "Handbook of Mathematical Functions, with Formulas, Graphs, and Mathematical Tables," Dover Publications, Inc., New York, 1965, 1046 pages (Republication of National Bureau of Standards, Applied mathematics series, 55. Government Printing Office, Washington, D.C., 1964):
   A compendium containing most of the tables that have previously appeared in the United States, including the National Bureau of Standards Mathematical Tables, Applied Mathematics, and Columbia University Press Series. Contains mathematical properties, interrelations, and numerical methods, as well as an updated bibliography of textbooks and tables.
2. Arfken, George Brown: "Mathematical Methods for Physicists," Academic Press, Inc., New York, 1968, 704 pages:
   Includes bibliographies.
3. Bierens de Haan, David: "Nouvelles Tables d'Intégrales Définies" (New tables of definite integrals). Corrected 1867 edition, with an English translation of the introduction by J. F. Ritt. Hafner Publishing Company, Inc., New York, 1965, 716 pages:
   A special collection of some 8,400 integrals.
4. British Association for the Advancement of Science: "Mathematical Tables." Prepared under the auspices of the Royal Society, Cambridge University Press, London, 1931–1958:
   Vol. 1: "Circular and Hyperbolic Functions," 3d ed, 1951.
   Vol. 2: "Emden Functions," 1932. New edition in preparation.
   Vol. 3: "Minimum Decompositions into Fifth Powers," 1933.
   Vol. 4: "Cycles of Reduced Ideals in Quadratic Fields," 1934.
   Vol. 5: "Factor Table." 1935.
   Vol. 6: "Bessel Functions," pt. 1, 1958.
   Vol. 7: "The Probability Integral," 1939.
   Vol. 8: "Number-divisor Tables," 1940.
   Vol. 9: "Table of Powers Giving Integral Powers of Integers," 1950.
   Vol. 10: "Bessel Functions," pt. 2, 1952.
   "Auxiliary Tables," nos. 1–2, 1946.
   Continued by the Royal Society mathematical tables.

5. Burington, Richard S.: "Handbook of Mathematical Tables and Formulas," 4th ed., McGraw-Hill Book Company, New York, 1965, 448 pages:
   A companion to the "Handbook of Probability and Statistics with Tables," by Richard S. Burington and Donald C. May, the 4th edition includes new sections on sets, relations, and functions; algebraic structures; Boolean algebra; number systems; matrices; and statistics. A table of derivatives and a comprehensive table of integrals have been included.

6. Burington, Richard S., and Donald C. May: "Handbook of Probability and Statistics with Tables," 2d ed., McGraw-Hill Book Company, New York, 1970, 450 pages.

7. Byerly, William E.: "An Elementary Treatise on Fourier's Series, and Spherical, Cylindrical, and Ellipsoidal Harmonics, with Applications to Problems in Mathematical Physics," Dover Publications, Inc., New York, 1959, 287 pages.

8. Byrd, Paul F., and Morris D. Friedman: "Handbook of Elliptic Integrals for Engineers and Physicists," (Die Grundlehren der mathematischen Wissenschaften, Band 67), Springer Verlag, Berlin, 1954, 355 pages:
   A collection of over 3,000 integrals and formulas using Legendre's and Jacobi's notations.

9. Campbell, George A., and Ronald M. Foster: "Fourier Integrals for Practical Applications," D. Van Nostrand Company, Inc., Princeton, N.J., 1948, 177 pages: A large number of the known closed-form evaluations of Fourier integrals are compiled and tabulated in compact form for convenient use. Tables give coefficient pairs, admittances, and transient solutions.

10. "C. R. C. Standard Mathematical Tables," 16th ed, edited by Samuel Selby, Chemical Rubber Co., Cleveland, 1968, 692 pages:
   An expanded, revised edition of a standard work. The sections involving mensuration, trigonometry, analytic geometry, curves and graphs, and the algebra of sets have been completely rewritten, and sections to cover determinants and matrices have been added. An extension to the octal decimal conversion table to include hexadecimal and decimal conversion increases the usefulness of the volume.

11. David, F. N., M. G. Kendall, and D. E. Barton: "Symmetric Function and Allied Tables," Cambridge University Press, London, 1966, 278 pages:
   An elaborate set of 49 major tables accompanied by a detailed introduction of 63 pages, constituting a self-contained treatment of symmetric functions and their applications in statistics. A definitive compilation.

12. Davis, Harold T.: "The Summation of Series," Principia Press of Trinity University, San Antonio, Tex., 1962, 140 pages:
   Special emphasis placed upon the case of finite limits.

13. Davis, Harold T., comp.: "Tables of the Higher Mathematical Functions," Principia Press, Bloomington, Ind., 1933–1935, 2 vols.:
   Vol. I: Various tables of the gamma and psi functions as well as sections on classification and history of tables, interpolation and its uses, and interpolation tables. Vol. II: Tables of the polygamma functions (trigamma-hexagamma), the Bernoulli and Euler polynomials and numbers, gram polynomials, and polynomial approximation.

14. Davis, Philip J., and Philip Rabinowitz: "Numerical Integration," Blaisdell Publishing Company, Waltham, Mass., 1967, 230 pages:
   Includes bibliographies.

15. Doetsch, Gustav: "Handbuch der Laplace-Transformation" (Handbook of Laplace transforms). Verlag Birkhäuser, Basel, 1950–1956, 3 vols. ("Lehrbücher und Monographien aus dem Gebiete der exakten Wissenschaften, Mathematische Reihe," vols. 14, 15, and 19):
   Contents: Vol. 1, "Theory of Laplace Transforms"; vols. 2–3, "Applications of Laplace Transforms," including asymptotic expansions, convergent expansions, ordinary and partial differential equations, integral equations, and whole exponential functions.

16. Dwight, H. B.: "Tables of Integrals and Other Mathematical Data," 4th ed., The Macmillan Company, New York, 1961, 336 pages:
   Contains derivatives and integrals, classified as algebraic, trigonometric, inverse trigonometric, and exponential functions; probability integrals; logarithmic, hyperbolic, inverse hyperbolic, elliptic, and Bessel functions; surface zonal harmonics; definite integrals; and differential equations. Appendixes: A, Tables of Numerical Values; B, Bibliography.

17. Erdélyi, Arthur, and others: "Higher Transcendental Functions" (Based, in part, on notes left by Harry Bateman and compiled by the staff of the Bateman Manuscript Project, California Institute of Technology), McGraw-Hill Book Company, New York, 1953–1955, 3 vols.:

An account of the principal properties of such functions as gamma, hypergeometric, Legendre, Bessel, elliptic, automorphic, and generating functions, with extensive lists of references at the end of each chapter.

18. Erdélyi, Arthur, and others: "Tables of Integral Transforms" (Based, in part, on notes left by Harry Bateman and compiled by the staff of the Bateman Manuscript Project, California Institute of Technology) McGraw-Hill Book Company, New York, 1954, 2 vols:

Intended as a companion and sequel to "Higher Transcendental Functions." Contains Fourier, Laplace, and Mellin transforms and their inversions, as well as Hankel transforms. Also included are gamma, Legendre, Bessel, and hypergeometric functions. The entries are arranged in tabular form.

19. Fettis, Henry E., and James C. Caslin: "Elliptic Functions for Complex Arguments," Aerospace Research Laboratories, Office of Aerospace Research, Wright-Patterson Air Force Base, Ohio, 1967, 404 pages, ARL 67-0001 (Available from Clearinghouse for Federal Scientific and Technical Information, Springfield, Va. 22151):

These unique tables consist of 5D values of the Jacobian elliptic functions $sn(w,k)$, $cn(w,k)$, and $dn(w,k)$, where $w = u + iv$, as functions of Jacobi's nome q, which equals $exp\,(-K'/K)$, where K and K' are the quarter-periods (the complete elliptic integrals of the first kind of modulus k and of complementary modulus k', respectively). The range of parameters in the table is: $q = 0.005(0.005)0.480$, $u/K = 0(0.1)1$, and $v/K' = 0(0.1)1$.

20. Fettis, Henry E., and James C. Caslin: "Ten-place Tables of the Jacobian Elliptic Functions," pt. 1, Aerospace Research Laboratories, Office of Aerospace Research, Wright-Patterson Air Force Base, Ohio, 1965, 562 pages, ARL 65-180 (Available from Clearinghouse for Federal Scientific and Technical Information, Springfield, Va. 22151):

This report contains 10D tables of the Jacobi elliptic functions $am(u,k)$, $sn(u,k)$, $cn(u,k)$, and $dn(u,k)$, as well as the elliptic integral $E(am(u),k)$, $k^2 = 0(0.01)0.99$, $u = 0(0.01)K(k)$, and for $k^2 = 1$, $u = 0(0.01)3.69$.

21. Fettis, Henry E., and James C. Caslin: "An Extended Table of Zeros of Cross Products of Bessel Functions," Aerospace Research Laboratories, Office of Aerospace Research, Wright-Patterson Air Force Base, Ohio, 1966, 126 pages, ARL 66-0023 (Available from Clearinghouse for Federal Scientific and Technical Information, Springfield, Va. 22151):

This report presents 10D tables of the first five roots of the equations: (a) $J_0(\alpha)Y_0(k\alpha) - Y_0(\alpha)J_0(k\alpha) = 0$, (b) $J_1(\alpha)Y_1(k\alpha) - Y_1(\alpha)J_1(k\alpha) = 0$, (c) $J_0(\alpha)Y_1(k\alpha) - Y_0(\alpha)J_1(k\alpha) = 0$.

22. Fletcher, Alan, and others, eds.: "An Index of Mathematical Tables," 2d ed., Addison-Wesley Publishing Company, Inc., Reading, Mass., for Scientific Computing Service, Ltd., 1962, 2 vols.:

The second edition is more than double the size of the 1946 edition, and includes, as a new feature, a list of errors found in published tables. Contains an index according to function, giving for each table the range, tabular interval, number of significant figures in the values, whether or not tables of proportional parts are given, what orders of differences are shown, etc. Also includes an alphabetical list of references by author and publication year. Considered an important index to well-known tables of functions and to other less-known tables appearing in books and periodicals.

23. Forsythe, G. E., and P. C. Rosenbloom: "Numerical Analysis and Partial Differential Equations," John Wiley & Sons, Inc., New York, 1958, 204 pages.

24. Frazer, Robert A., W. J. Duncan, and A. R. Collar: "Elementary Matrices and Some Applications to Dynamics and Differential Equations," The Macmillan Company, New York, 1946, 416 pages.

25. Great Britain, National Physical Laboratory: "Mathematical Tables," H. M. Stationery Office, London, 1956——

Vol. 1: "The Use and Construction of Mathematical Tables," by L. Fox, 1956.
Vol. 2: "Tables of Everett Interpolation Coefficients," by L. Fox, 1956.
Vol. 3: "Tables of Generalized Exponential Integrals," by G. F. Miller, 1960.
Vol. 4: "Tables of Weber Parabolic Cylinder Functions and Other Functions for Large Arguments," by L. Fox, 1961.
Vol. 5: "Chebyshev Series for Mathematical Functions," 1962.
Vol. 6: "Tables for Bessel Functions of Moderate or Large Order," 1962.
Vol 7: "Tables of Jacobian Elliptic Functions Whose Arguments Are Rational Fractions of the Quarter Period," 1964.
This series contains tables of mathematical functions which may not come within the range of the more fundamental tables.

26. Greenwood, Joseph A., and H. O. Hartley: 'Guide to Tables in Mathematical Statistics." Princeton University Press, Princeton, N.J., 1962, 1014 pages.

27. Gröbner, Wolfgang, and N. Hofreiter: "Integraltafel (Integral table), pt. I, "Indefinite Integrals"; pt. II, "Definite Integrals," 2d improved ed., Springer Verlag, Vienna. 1965–1966, 2 vols., 166 and 204 pages:
   An extensive collection of integrals including a brief survey of methods of evaluation and transformation of integrals.
28. Hart, John F., and others: "Computer Approximations," John Wiley & Sons, Inc., New York, 1968, 343 pages:
   Extensive in its range in accuracy from a few digits up to 25D in the approximations; its wide selection of functions includes square root and cube root, exponential and hyperbolic, logarithm, trigonometric and inverse trigonometric functions, gamma function and its logarithm, error function, Bessel functions, and complete elliptic integrals; and in the range of methods, the book describes and compares them from the general methods of subroutine design to procedures for the design of maximum efficiency programs for commonly needed functions.
29. Hartree, Douglas R.: "Numerical Analysis," Oxford University Press, London, 1952, 287 pages:
   Includes interpolation and numerical integration formulas, finite differences, harmonic analysis, smoothing.
30. Harvard University Computation Laboratory: "Tables of the Bessel Functions of the First Kind," Harvard University Press, Cambridge, Mass., 1947–1951, 12 vols.: $J_n(x)$, $0 \leq x \leq 100$, 18D for $n = 1$; 10D for $n = 2$ through 135.
31. Harvard University Computation Laboratory: "Tables of the Cumulative Binomial Probability Distribution," Harvard University Press, Cambridge, Mass., 1955, 503 pages ("Annals of the Computation Laboratory of Harvard University," vol. 35).
32. Harvard University Computation Laboratory: "Tables of the Function arc sin z," Harvard University Press, Cambridge, Mass., 1956, 586 pages ("Annals of the Computation Laboratory of Harvard University," vol. 40).
33. Householder, Alston S.: "Principles of Numerical Analysis," McGraw-Hill Book Company, New York, 1953, 274 pages.
34. Jahnke, Eugene, Fritz Emde, and Friedrich Lösch: "Tables of Higher Functions," 7th ed, B. G. Teubner, Stuttgart, 1966, 322 pages:
   Text in German and English. Essentially a corrected version of the sixth edition, containing Bessel functions, circular and hyperbolic functions of a complex variable, cubic equations, miscellaneous conversion tables, Planck's radiation function, powers (2d to 15th), probability integral and related functions, reciprocals and square roots of complex numbers, Riemann zeta function, theta functions, transcendental equations, vector addition, and sine, cosine, and logarithmic integral.
35. James, Glenn, and Robert C. James: "Mathematics Dictionary," 3d ed., D. Van Nostrand Company, Inc., Princeton, N.J., 1968, 448 pages:
   Correlated condensation of mathematical concepts designed for time saving reference work.
36. Jolley, Leonard B. W.: "Summation of Series," Dover Publications, Inc., New York, 1961, 251 pages.
37. Kamke, Erich: "Differentialgleichungen, Lösungsmethoden und Lösungen (Differential equations, methods of solution, and solutions), 6th improved ed., Akademische Verlagsgesellschaft Geest & Portig, Leipzig, 1959, 666 pages ("Mathematik und ihre Anwendungen in Physik und Technik," Ser. A, vol. 18):
   A reference work containing general methods of solution and properties of solution, boundary-, and characteristic-value problems, and a dictionary of some 1,600 equations in lexicographical order with solutions, techniques for solving, and references.
38. Knuth, Donald E.: "The Art of Computer Programming," Addison-Wesley Publishing Company, Inc., Reading, Mass., 1968–1969, 2 vols.
39. Korn, Granino A., and Theresa M. Korn.: "Mathematical Handbook for Scientists and Engineers: Definitions, Theorems, and Formulas for Reference and Review," 2d, enlarged and rev. ed., McGraw-Hill Book Company, New York, 1968, 1129 pages.
40. Lehmer, Derrick H.: "Guides to Tables in the Theory of Numbers," National Academy of Sciences–National Research Council, Washington, D.C., 1941, 177 pages (National Research Council Bulletin 105).
41. Lehmer, Derrick Norman: "Factor Table for the First Ten Millions Containing the Smallest Factor of Every Number not Divisible by 2, 3, 5, or 7 between the Limits 0 and 10,017,000." Hafner Publishing Company, Inc., New York, 1956, 476 pages (Carnegie Institution of Washington Publ. 105):
   Introduction includes a list of errors in former tables by other authors.
42. Lehmer, Derrick Norman: "List of Prime Numbers from 1 to 10,006,721," Hafner Publishing Company, Inc., New York, 1956, 133 pages (Carnegie Institution of Washington Publ. 165):

The standard list of primes. Arranged in such a way that it is easy to find the nth prime for a given n.

43. Lieberman, Gerald J., and D. B. Owen: "Tables of the Hypergeometric Probability Distribution," Stanford University Press, Stanford, Calif., 1961, 726 pages (Stanford studies in mathematics and statistics no. 3):
    In addition to the following tables of the hypergeometric probability distribution: $N = 2$, $n = 1$ through $N = 100$, $n = 50$; $N = 1000$, $n = 500$; $k = n - 1$, n; $n = N/2$: $N = 100$, $n = 50$ through $N = 2000$, $n = 1000$, the theory, rationale, and specific applications of the hypergeometric probability are discussed.

44. Luke, Yudell: "Integrals of Bessel Functions," McGraw-Hill Book Company, New York, 1962, 424 pages:
    Designed to provide the research worker with basic information dealing with definite and indefinite integrals involving Bessel functions.

45. Madelung, Erwin: "Die Mathematischen Hilfsmittel des Physikers" (Mathematical tools for the physicist), 6th rev. ed., Springer Verlag, Berlin, 1957, 535 pages ("Die Grundlehren der Mathematischen Wissenschaften in Einzeldarstellungen," vol. IV):
    Comprehensive collection of formulas used in mathematical physics. Included are numbers, functions and operators, series, algebra, transformations, and statistics.

46. Magnus, Wilhelm, Fritz Oberhettinger, and Raj Pal Soni: "Formulas and Theorems for the Special Functions of Mathematical Physics," 3d enlarged ed., Springer Verlag, Berlin, 1966, 508 pages ("Die Grundlehren der Mathematischen Wissenschaften in Einzeldarstellungen," Band 52):
    Survey of the properties of a number of special functions including the following: gamma, hypergeometric, Bessel, Legendre, theta, and elliptic, as well as spherical harmonics, orthogonal polynomials, integral transforms and inversions, and coordinate transforms.

47. Mangulis, V.: "Handbook of Series for Scientists and Engineers," Academic Press, Inc., New York, 1965, 134 pages.

48. Margenau, Henry, and George M. Murphy: "The Mathematics of Physics and Chemistry," 2d ed., D. Van Nostrand Company, Inc., Princeton, N.J., 1956, 2 vols.

49. "Mathematics of Computation" (formerly: "Mathematical Tables and other Aids to Computation"), National Academy of Sciences National Research Council, quarterly, Washington, D.C.:
    A journal devoted to advances in numerical analysis, the application of computational methods, mathematical tables, high-speed calculators, and other aids to computation.

50. Meyer zur Capellen, Walther: "Integraltafeln, Sammlung unbestimmter Integrale elementarer Funktionen" (Tables of integrals; Collection of indefinite integrals of elemetary functions), Springer Verlag, Berlin, 1950, 292 pages:
    Lists some 3,000 integrals of algebraic and transcendental functions, as well as products of algebraic and transcendental functions. Tabulation permits use for differentiation purposes.

51. Morse, Philip M., and Herman Feshbach: "Methods of Theoretical Physics," McGraw-Hill Book Company, New York, 1953, 2 vols.

52. Oberhettinger, Fritz, "Tabellen zur Fourier Transformation" (Tables of Fourier transforms), Springer Verlag, Berlin, 1957, 213 pages.

53. Parke, Nathan Grier: "Guide to the Literature of Mathematics and Physics including Related Works on Engineering Science," 2d rev. ed., Dover Publications, Inc., New York, 1958, 436 pages:
    A useful handbook comprising chapters on principles of reading and study, searching the literature, types of materials, library usage, etc.; includes an annotated bibliography of some 5,000 titles arranged by subject with author and subject indexes.

54. Pearson, Karl: "Tables of the Incomplete Beta-function," 2d ed., with a new introduction by E. S. Pearson and N. L. Johnson, published for the Biometrika Trustees by the Cambridge University Press, Cambridge, Mass., 1968, 505 pages:
    Gives $I(u,p)$ with the argument u proceeding by increments of 0.1 from 0 up to that value of u which gives $I(u,p) = 1.0000000$ to the seventh decimal place. The argument p advances from -1.0 by increments of 0.05, from 1.0 to 5.0 by increments of 0.1, and from 5.0 to 50.00 by intervals of 0.2. Two new tables give some additional values to the integral computed a number of years ago but not hitherto published, and a list of references has been added.

55. Peirce, Benjamin O.: "A Short Table of Integrals," 4th ed., rev. by Ronald M. Foster, Ginn and Company, Boston, 1956, 189 pages:
    Fourth revision of Peirce's tables consisting of indefinite integrals, definite integrals, auxiliary formulas, and numerical tables, including common algebraic expressions; functions of angles in radians; differential equations; exponential functions; hyperbolic-function formulas; elliptic integrals; natural logs; tables of logs of numbers, logs of sines, cosines, etc.; probability integral and trigonometric formulas.

56. Riordan, John: "An Introduction to Combinatorial Analysis," John Wiley & Sons, Inc., New York, 1958, 244 pages.
57. Roberts, G. E., and H. Kaufman: "Table of Laplace Transforms," W. B. Saunders Company, Philadelphia, 1966, 367 pages:
    A comprehensive reference of Laplace transforms and their inverses which should prove useful to pure and applied mathematicians. The volume is in two parts—the first devoted to direct transforms and the second to inverse transforms.
58. Royal Society of London: "Royal Society Mathematical Tables," Cambridge University Press, London, 1950——.
    Vol. 1: "Farey Series of Order 1025," 1950.
    Vol. 2: "Rectangular-polar Conversion Tables," 1956.
    Vol. 3: "Tables of Binomial Coefficients," 1954.
    Vol. 4: "Tables of Partitions," 1958.
    Vol. 5: "Representations of Primes by Quadratic Forms," 1960.
    Vol. 6: "Tables of the Riemann Zeta Function," 1960.
    Vol. 7: "Bessel Functions," pt. 3, "Zeros and Associated Values," 1960.
    Vol. 8: "Tables of Natural and Common Logarithms to 110 Decimals," 1964.
    Vol. 9: "Tables of Indices and Primitive Roots," 1968.
    Vol. 10: "Bessel Functions," pt. 4, "Kelvin Functions," 1964.
    Vol. 11: "Coulomb Wave Functions," 1964.
59. Ryzhik, Iosif M., and I. S. Gradshteyn: "Table of Integrals, Series, and Products," translated from the 4th Russian ed., Academic Press Inc., New York, 1965, 1086 pages:
    An inclusive compilation, the work is advertised as the most comprehensive table of integrals ever published. New material on Mathieu, Struve, Lommel, as well as other special functions, has been added.
60. Slater, L. J.: "Confluent Hypergeometric Functions," Cambridge University Press, New York, 1960, 247 pages.
61. Smithsonian Institution: "Smithsonian Mathematical Formulae and Tables of Elliptic Functions," 3d reprinting, Washington, D.C., 1957, 314 pages.
62. Stroud, A. H., and D. Secrest: "Gaussian Quadrature Formulas," Prentice Hall, Inc., Englewood Cliffs, N.J., 1966, 374 pages:
    Valuable reference book for use and application of Gaussian quadrature formulas. Text is divided into five parts. Fortran programs to compute the abscissas and weights for quadrature formulas based on classical Jacobi, Laguerre, and Hermite polynomials are presented. Chapter 5 summarizes the tables of quadrature formulas found in the literature.
63. Todd, John, ed.: "Survey of Numerical Analysis," McGraw-Hill Book Company, New York, 1962, 608 pages.
64. U.S. National Bureau of Standards: "Basic Theorems in Matrix Theory," Marvin Marcus, Government Printing Office, Washington, D.C., 1960, 27 pages (Applied mathematics series, 57).
65. ——: "Experimental Statistics," Mary Gibbons Natrella, Government Printing Office, Washington, D.C., 1963, 1 vol. (various pagings) (Handbook 91):
    A collection of statistical procedures useful in the design, development, and testing of materials; the evaluation of equipment performance; and the conduct and interpretation of scientific experiments.
66. ——: "Guide to Tables of the Normal Probability Integral," Government Printing Office, Washington, D.C., 1952, 16 pages (Applied mathematics series, 21):
    A ready desk reference to the normal probability integral tabulated in standard statistical textbooks and other important sources. Provides a list of available tables as well as the form of the function tabulated.
67. ——: "Matrix Representations of Groups," Morris Newman, Government Printing Office, Washington, D.C., 1968, 79 pages (Applied mathematics series, 60).
68. ——: "Probability Tables for the Analysis of Extreme-value Data," Government Printing Office, Washington, D.C., 1953, 32 pages (Applied mathematics series, 22):
    Introduction outlines the theory and application of extreme values and describes nature, use, accuracy, and method of computation of tables. There are six tables for the asymptotic (cumulative) distribution of the largest value $\Phi = \exp(-e^{-y})$, its inverse, the corresponding density function, probability points for the asymptotic distribution of the mth largest values up to $m = 50$, and the asymptotic cumulative and density functions of the range.
69. ——: "Table for Conversion of X-ray Diffraction Angles to Interplanar Spacing," Government Printing Office, Washington, D.C., 1950, 159 pages (Applied mathematics series, no. 10):
    Tables of spacing values, $\theta = 0(0.01)90°$, $5S$, calculated by using the $K\alpha_1$ wavelengths for X-ray targets of molybdenum, copper, nickel, cobalt, iron, and chro-

mium. The wavelengths are those adopted at the International Conference sponsored by the British Institute of Physics in 1946.

70. ——: "Tables for the Analysis of Beta Spectra," Government Printing Office, Washington, D.C., 1952, 61 pages (Applied mathematics series, 13):

Tables of the values of the so-called Fermi function

$$F(Z,\eta) = \eta^{2+2s}e^{\pm\pi\delta} \cdot |\Gamma(i + S + i\delta)|^2$$

where the upper and lower signs of $\exp(\pm\pi\delta)$ apply to the spectra of negative and positive electrons, respectively; η = momentum of the electron after its ejection from the atom, in units of megaHertz; $S = \sqrt{1 - Z^2/137^2} - 1$; Z = atomic number: $\delta = Z\sqrt{1 + \eta^2/137\eta}$.

71. Wheelon, Albert D.: "Tables of Summable Series and Integrals Involving Bessel Functions," Holden-Day, Inc., Publisher, San Francisco, 1968, 125 pages.

1b. SI Units

With the continuing expansion of international cooperation and communication in science, the need for uniform usage of units and symbols has become increasingly critical. The culmination of a number of efforts in this direction is the International System of Units (abbreviated to SI for Système International) which was defined and given official status by the 11th General Conference on Weights and Measures, held in Paris in 1960. But achievement of widespread, let alone universal, acceptance and implementation of any international agreement takes time and requires the solution of a host of problems—some physical, some psychological. The SI units system has proved no exception. Many organizations have adopted SI officially; many have not. As one might expect, viewpoints of individual physicists within both groups vary from enthusiastic support through indifference to strong opposition. One British scientist has written, for example: "The reaction of a physicist asked to adopt the SI units is liable to be that which one could imagine from an abstemious early Christian anchorite asked by a Salvation Army tract bearer to sign the pledge . . . . Physicists have thought themselves particularly intelligent and virtuous about units, and they count the guardians of the units among their own number" [1]. Thus, even though the Paris conference occurred almost a decade ago, science in the United States is still very much in a period of transition with regard to the use of SI units. For this reason, the editors of the AIP Handbook did not think it practical to expect the more than 100 contributors to this edition to use SI units exclusively. On the other hand, it was clear that this new system could not be ignored. Inclusion of this descriptive subsection represents the compromise upon which the editors agreed.

In 1964, the National Bureau of Standards adopted SI units for use by its staff. Then, recognizing that there would have to be a transitional period, the Bureau appointed a Units and Usage Committee, and directed it to recommend interim practices for the NBS staff to follow. The rest of this subsection is quoted from a Bureau Technical News Bulletin titled NBS Interprets Policy on SI Units [2].

1b-1. Statement of Policy. Numerical data are used in NBS publications in two distinct ways: as descriptive data and as essential data. NBS policy for the transition period accepts different treatment of these two classes of data, although they may be presented in the same textbook, each with appropriate units. For example, it is acceptable to write "the interferometer mirror mounted on 1-in. rod, was advanced in 10-nanometer increments," or "a 200-in. telescope of 0.497 m effective aperture."

Descriptive data describe arrangements, environments, noncritical dimensions and shapes of apparatus, and similar measurements not entering into calculations or expression of results.

Essential data express, lead up to, or help to interpret the quantitative results of the activity that is being reported.

NBS policy also recognizes that communication of *scientific* results, via scientific papers, calls for more rigorous standards of units usage than does communication of *technological* results in technological papers.

Descriptive Data. Descriptive data should be expressed in the most useful and convenient manner. Forced translation into SI is not required. The units best understood by the expected audience are the most appropriate. Where non-SI units are used, the author may add SI equivalents in parentheses at his own discretion, but usage within a paper should be consistent on this point. Commercial gage designations or other standard nomenclatures, e.g., drill sizes, are acceptable. As SI units become more commonly used for commercial products, use of SI units in descriptive data should conform.

Essential Data. In technological papers the essential data may be expressed in the units customarily used in the relevant field of technology. SI equivalents should be added in parentheses, or in parallel columns in tables. If graphs are used as the primary or sole means of presenting essential data, the coordinates may be divided according to customary usage, but a secondary set of coordinate markings in SI units should be included. The top and right-hand sides of the graph are often appropriate for this purpose. If graphs are used only to indicate trends, or as supplements to tables, units customary in the field are adequate without SI translation. NBS authors should, however, use the SI as soon as the level of SI usage in the related field of technology renders it an efficient communication device. Familiarity with SI units (see Arts. 1.1 and 1.2 of the Appendix, Sec. 1b-2) is recommended to all NBS authors and all NBS staff.

In purely scientific papers, the essential data *shall* be expressed in SI units, or in units approved for use with the SI. The General Conference on Weights and Measures (CGPM) has designated names and symbols for many of the SI units. These and other names and units approved for NBS use but not yet acted upon by the General Conference are given in Arts. 1 to 6 of the Appendix. Values in other units may be added, in parentheses, where it is felt that this will improve the communication between authors and readers.

Reference data used generally in both science and technology should be expressed in SI units with appropriate indication of conversion factors into technological units or with parallel columns of converted values. Standard reference data applicable primarily to scientific interests (e.g., tables of X-ray atomic energy levels) should be expressed in SI units. Non-SI units may be included as parallel entries.

Where the general usage in any field does not recognize SI units, NBS authors should employ units comprehensible to their readers, but should try to increase reader familiarity with SI units as rapidly as possible. Again, the device of parallel columns (familiar plus SI units) is recommended.

In the lists in the Appendix, the short names for compound units, such as "coulomb" for "ampere-second," exist for convenience, and their use is not compulsory. For example, communication sometimes benefits if the author expresses magnetic flux in volt-seconds, instead of using the synonym webers, because of the descriptive value implicit in the former unit phrase.

The analysis, interpretation, or application of essential data may involve angles and the related values of natural functions (sine, tangent, log sin, etc.). In these cases, the angles may be expressed in degrees rather than in radians.

1b-2. Appendix. Articles 1 to 6 present those units that should be used by authors of *scientific* papers for *essential* data. Article 7 presents units that should *not* be used by authors of scientific papers for essential data.   Authors of *technological* papers are urged to become familiar with these guidelines, and to follow them as soon as their intended readers are ready to accept and understand them.

1.1.   *Official SI Units Names and Symbols*

Unit	Symbol	Unit	Symbol
meter	m	watt	W
kilogram	kg	coulomb	C
second	s	volt	V
ampere	A	ohm	Ω
kelvin[1]	K	farad	F
candela	cd	weber	Wb
radian	rad	henry	H
steradian	sr	tesla	T
hertz	Hz	lumen	lm
newton	N	lux	lx
joule	J		

1.2.   *Additional Names and Symbols Approved for NBS Use*

curie[2]	Ci
degree Celsius[3]	°C
gram	g
mho	mho
mole	mol
siemens[4]	S

1.3.   *Official Prefixes Indicating Decimal Multiples and Submultiples*

Multiples and submultiples	Prefix	Symbol
10^{12}	tera	T
10^{9}	giga	G
10^{6}	mega	M
10^{3}	kilo	k
10^{2}	hecto	h
10	deka	da
10^{-1}	deci	d
10^{-2}	centi	c
10^{-3}	milli	m
10^{-6}	micro	μ
10^{-9}	nano	n
10^{-12}	pico	p
10^{-15}	femto	f
10^{-18}	atto	a

Note.   Compound prefixes (e.g., millimicro) are not to be used.

[1] The same name and symbol are used for thermodynamic temperature and temperature interval.   (Adopted by the 13th General Conference on Weights and Measures, 1967.)
[2] Accepted by the General Conference on Weights and Measures for use with the SI.
[3] For expressing "Celsius temperature"; may also be used for a temperature interval.
[4] Adopted by IEC and ISO.

2. *Decimal multiples of SI units*, bearing coined names, are acceptable 'n their special fields only.[1] Use of SI units, however, is recommended. Examples include:

Unit	Symbol	SI equivalent
angstrom	Å	$= 10^{-10}$ m
bar	bar	$= 10^5$ N/m²
barn	b	$= 10^{-28}$ m²
kayser[2]	K	$=$ cm^{-1} $= 100$ m^{-1}
liter[3]	l	$= 10^{-3}$ m³
poise	P	$= 10^{-1}$ N · s/m²
rad	rd	$= 10^{-2}$ J/kg
stokes	St	$= 10^{-4}$ m²/s

3. *"Natural Units."* Natural units are acceptable. These are units tied directly to the fundamental Lorentz invariant constants of nature as well as to the properties of the microscopic constituents of matter. Although it is recognized that in casual conversation, dimensionally incorrect units are used, dimensionally correct units shall be used in published work. (Examples of conversational "shorthand" include: temperature in eV, mass in eV, momentum in fm^{-1}. These usages arise in expressing quantities by the value of *related* quantities: e.g., energy of a particle, wave number associated with momentum.) Acceptable natural units include:

electronic charge e
electron mass m_e
proton mass m_p
speed of light c
electron-volt   eV
Planck's constant h or $\hbar$
Bohr radius a_0
Bohr magneton μ_B
nuclear magneton μ_N
electron radius r_e
Compton wavelength of electron λ_C
atomic mass unit u
Faraday F

3a. *The term "X-unit"* is ambiguous, differing in American and European usage. It is acceptable, *provided* that its use is accompanied by its definition in terms of the $K \propto$ line of molybdenum or of tungsten. When accurate conversions to the SI become available, the usage of X-unit should be discontinued.

4. *Special cgs-esu multiple*, acceptable pending CGPM naming of a suitable replacement:

debye (10^{-18} statcoulomb-centimeter) $= 3.33564 \times 10^{-30}$ coulomb-meter

5. *Acceptable logarithmic measures:*

pH
decibel   (dB)
neper   (Np)

[1] The 13th General Conference canceled the name *micron* and its old symbol, μ; use *micrometer*, μm.

[2] Note the conflict with K for kelvin, which is an official symbol. It is felt that context will preclude confusion.

[3] To be used only for expressing volumes of gases and liquids; otherwise use cubic meter, etc.

6. *Acceptable units for essential data in expression of angles* in relation to their natural functions, and for naturally occurring geometrical relationships:

degree   °
minute   ′
second   ″

7. *Units NOT acceptable for expressing essential data* in scientific papers (see *Comments* below):

7a. *Unnecessary coined names:*

gamma   (for nanotesla)
gamma   (for microgram)
fermi   (for femtometer, fm)

7b. *Coined names for cgs units* (and therefore not compatible with SI):

gauss   gal
erg   stilb
dyne

7c. *Units not compatible with the SI,* nor any other metric system:

calorie	millimeter of mercury
British thermal unit	hour (time)[1]
entropy unit	minute (time and angle)[1]
roentgen	degree (angle)[2]
atmosphere	second (angle)[2]
torr	

Comments. The CGPM has not yet adopted a unit for "quantity of matter." Other organizations (ISO and IUPAC) have adopted a mole based on 0.012 kg of carbon-12; this is equivalent to the familiar gram-mole (symbol mol).

USASI and international usage require that letter symbols deriving from proper names be capitalized, although the unit names themselves are lowercase.

"Mho" and "siemens" are widely used coined names of derived SI units to which no name has yet been assigned by CGPM. They are, therefore, acceptable until CGPM acts to assign names.

The name "nit" (symbol nt) has been recommended by the International Commission on Illumination for the SI unit of luminance, "candela per square meter" (symbol cd/m^2). Although the name nit has not been assigned by CGPM and has not received extensive use, it may be used where the official name is felt to be awkward.

"Poise" and "stokes" are coined cgs names; there are no short names in use for the SI units of "viscosity" and "kinematic viscosity." (The compound names "newton-second per square meter" and "kilogram per meter-second" are long and awkward; a coined short name is needed. "Square meter per second" is also sometimes considered awkward.)

The lists of approved units given in Arts. 1 to 6 are not closed-ended. Other units *compatible with the SI* can and will be added on the basis of:

1. Actions taken by subsequent General Conferences
2. Clear indication of need in specialized fields, following approval of the NBS Editorial Review Boards and the NBS Units and Usage Committee

Artificial creations such as kiloangstrom and cubic angstrom are, for the most part, unnecessary and do more harm than good. They should be replaced by SI units as rapidly as possible.

Certain units based on natural constants, e.g., electron-volt, are both meaningful and convenient. These natural units are orders of magnitude outside the range of SI units with prefix.

Most tables of natural functions of angles give the angle in degrees, minutes, and

---

[1] Allowable when necessary for expression of extended time intervals.
[2] See Art. 6 for acceptable use.

seconds, or in degrees and decimals.   Wherever the mathematical or expository treatment or application of essential data requires that an angle and its natural functions be related, degrees may be used for expressing the angle.   Likewise, wherever naturally occurring geometrical phenomena (crystal data, bond angles, declination of the sun, etc.) are expressed quantitatively, degrees may be used.

The units in Art. 7c are convenient and acceptable for descriptive use, but *not* for essential data.   For example, the torr is widely used for describing an environment that does not enter into the calculation.

The transition from calories and kilocalories to joules will of course involve some distress.   NBS authors are urged to recognize their responsibilities for taking the lead in the acceptance of the joule, while still providing convenient communication to users, by parenthetical equivalents or parallel columns.

The use of "cycles per second" in place of hertz is deprecated.   It tends to perpetuate the common misuse of cycle by itself for the frequency unit.   When cycle is correctly used by itself, in the same paper as cycles per second, journal editors are prone to "correct" it!

Self-explanatory combinations of prefixed units are acceptable, e.g., milliampere per square centimeter, cubic decimeter, milligram.

The inclusion of the prefix "kilo" in the name of the base unit of mass creates an awkward situation.   Logically, the "gram" should be called the "millikilogram"!   On recommendation of the International Committee on Weights and Measures, the names of multiples of the kilogram are formed by adding prefixes to the word "gram."

Expression of extended time intervals (essential data) may sometimes be more readily comprehended if years, hours, and minutes are given in parentheses, as well as, e.g., kiloseconds.   However, as the author will recognize, this usage borders on descriptive data.   In any case, computations will always require use of seconds only.

Editorial Notes.   Words and symbols should *not* be mixed; if mathematical operations are indicated, only symbols should be used.   For example, one may write joules per mole, J/mol, $J \cdot mol^{-1}$, but *not* joules/mole, joules mol^{-1}, etc.

Note that exponents operate also on prefixes, as in cm^2, mm^3 which are *not* $10^{-2}m^2$, $10^{-3}m^3$.

In combinations such as meterkelvin, use of the product dot $(m \cdot K)$ avoids confusion with millikelvin.   It is good practice to indicate *all* unit products with multiplication dots, since some unit symbols consist of more than one letter, e.g., Wb for weber, *versus* $W \cdot b$ for watt-barn.

When a compound unit is formed by division of one unit by another, its symbol consists of the symbols for the separate units either separated by a solidus or multiplied by using negative powers (for example, m/s or $m \cdot s^{-1}$ for meter per second).   In simple cases use of the solidus is preferred, but in no case should more than one solidus be included in a combination unless parentheses are inserted to avoid ambiguity.   In complicated cases, negative powers should be used.

References

1. Frank, F. C.: A Flexible Policy, *Phys. Bull.* **20** (February, 1969), published by the British Institute of Physics and Physical Society.
2. National Bureau of Standards: NBS Interprets Policy on SI Units, *Tech. News Bull.* **52,** 6 (June, 1968).
3. National Bureau of Standards: NBS Adopts International System, *Tech. News Bull.* **48,** 61 (April, 1964).
4. Wolfe, Hugh C., and Paul J. Kliauga: Symbols, Units and Nomenclature in Physics, p. 694 in "The Encyclopedia of Physics," R. M. Besancon, ed., Reinhold Book Corporation, New York, 1966.
5. National Bureau of Standards: The English and Metric Systems of Measurement, *NBS Spec. Publ.* 304A, 1968.
6. International Union of Pure and Applied Physics: Symbols, Units and Nomenclature in Physics, *UIP* **11**; *SUN* **65-3** (1965).
7. ISO Recommendation R 1000 (1969): "Rules for the Use of . . . SI Units," (Available USASI, New York).

Section 2

MECHANICS

R. BRUCE LINDSAY, Editor

Brown University

CONTENTS

2a. Fundamental Concepts of Mechanics.   Units and Conversion Factors

DAVID MINTZER

Northwestern University

PAUL TAMARKIN

RAND Corporation

R. BRUCE LINDSAY

Brown University

Symbols

$\mathbf{a}$	vector acceleration
$a_x,\ a_y,\ a_z$	components of acceleration in rectangular coordinates
$a_r,\ a_\theta,\ a_z$	components of acceleration in cylindrical coordinates
$a_r,\ a_\theta,\ a_\phi$	components of acceleration in spherical coordinates
$\mathbf{C}$	elastic coefficient matrix
C_{ij}	elastic coefficient
$\mathbf{D}$	strain tensor
$\mathbf{e}_1,\ \mathbf{e}_2,\ \mathbf{e}_3$	base vectors of coordinate system
$e_{xx},\ e_{xy}$, etc.	components of strain tensor
E	Young's modulus
$\mathbf{f}_c$	inertial force (centrifugal or Coriolis)
$\mathbf{F}$	force vector
g	acceleration of gravity
G	shear modulus
$\mathfrak{g}$	impulse
$I_{xx},\ I_{yy},\ I_{zz}$	moment of inertia
$I_{xy},\ I_{yz},\ I_{xz}$	products of inertia
k	bulk modulus
K	kinetic energy
l	depth of liquid
$\mathbf{L}$	moment of momentum
$m,\ M$	mass
N	number of particles
p	pressure
r	radius vector
$\mathbf{r}$	position vector
$\mathbf{R}$	position vector of center of mass
$\mathfrak{R}$	radius of gyration

s	vector displacement
S	stress tensor
S_{ij}	elastic constant
t	time
T	torque
U	total mechanical energy
v	velocity vector
v_x, v_y, v_z	components of velocity in rectangular coordinates
v_r, v_θ, v_z	components of velocity in cylindrical coordinates
v_r, v_θ, v_ϕ	components of velocity in spherical coordinates
V	volume, potential energy, wave velocity
W	work
x, y, z	rectangular coordinates
X_x, X_y, etc.	components of stress tensor
α	angular acceleration
γ	surface tension
Δ	deformation displacement of a deformable medium
η	viscosity
θ	colatitude in spherical coordinates, azimuth in cylindrical coordinates
Θ	total dilatation
λ	Lamé elastic constant, wavelength in harmonic wave
ξ, η, ζ	components of deformation, displacement of a deformable medium
ρ	density
σ	Poisson's ratio
ϕ	longitude in spherical coordinates, velocity potential
ω	angular velocity
$\omega_x, \omega_y, \omega_z$	components of angular velocity in rectangular coordinates

2a-1. Newtonian Concepts of Mechanics. The science of mechanics deals with the motion of material bodies, which ideally can be considered as made up of point particles. In order to describe the motion of a particle three concepts are needed: *a frame of reference, distance,* and *time interval.* These concepts are left undefined as intuitive concepts with sufficiently universal meanings. Distance and time intervals are measured in terms of standards which have a wide range of acceptance, such as the *standard meter* and the *sidereal day.* (The important systems of units are tabulated in Secs. 2a-8 and 2a-9.) The frame of reference consists of a *reference point* and a *coordinate system* (whose origin may be at the reference point); a *reference* event is necessary as well as a frame of reference.

The position of a particle may be specified with respect to the reference point by considering a rectangular coordinate system whose origin is at the reference point. The position of any particle is then given in terms of the distances along the coordinate axes from the origin to the projection on these axes of the point representing the position of the particle.

The location of an event in time, or the time of an event, similarly is expressed in terms of the time interval with respect to the reference event. The terms "time interval" and "time" are usually used interchangeably.

The above concepts are usually referred to as Newtonian; they suffice for classical mechanics.

2a-2. Kinematics—The Space-Time Relationships in the Motions of Point Particles. *Velocity.* Velocity is the rate of change of position with respect to time. Two types of velocity are commonly used, instantaneous and average. Instantaneous velocity is the time rate of change of position calculated pointwise, thus being a derivative. Average velocity is the time rate of change of position calculated as the quotient of a finite distance and the corresponding finite time interval.

Velocity is a vector with components which depend in general on the coordinate

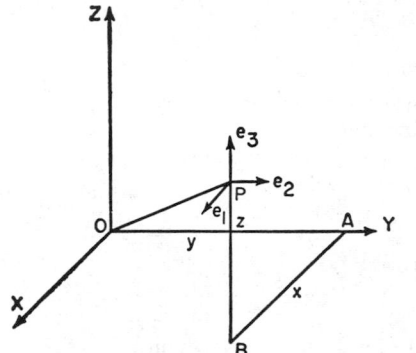

FIG. 2a-1. Base vectors in rectangular coordinates.

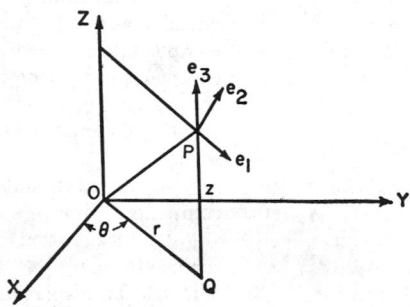

FIG. 2a-2. Base vectors in cylindrical coordinates.

system used. If e_1, e_2, e_3 are base vectors of the coordinate system under consideration, then, for three commonly used systems:

1. Rectangular coordinates (cf. Fig. 2a-1):

$$\mathbf{v} = e_1 v_x + e_2 v_y + e_3 v_z = e_1 \frac{dx}{dt} + e_2 \frac{dy}{dt} + e_3 \frac{dz}{dt} \qquad (2a\text{-}1)$$

2. Cylindrical coordinates (cf. Fig. 2a-2):

$$\mathbf{v} = e_1 v_r + e_2 v_\theta + e_3 v_z = e_1 \frac{dr}{dt} + e_2 r \frac{d\theta}{dt} + e_3 \frac{dz}{dt} \qquad (2a\text{-}2)$$

3. Spherical coordinates (cf. Fig. 2a-3):

$$\mathbf{v} = e_1 v_r + e_2 v_\theta + e_3 v_\phi = e_1 \frac{dr}{dt} + e_2 r \frac{d\theta}{dt} + e_3 r \sin\theta \frac{d\phi}{dt} \qquad (2a\text{-}3)$$

Acceleration. Acceleration is the rate of change of velocity with respect to time. Instantaneous and average acceleration may be defined analogously to instantaneous and average velocities; however, instantaneous acceleration, or the time derivative of velocity (or equivalently the second time derivative of position), is the more commonly used quantity. Acceleration is a vector with components which depend in general on the coordinate system used. If e_1, e_2, e_3 are the unit base vectors of the coordinate system under consideration, then for the commonly used systems:

1. Rectangular coordinates:

$$\mathbf{a} = e_1 a_x + e_2 a_y + e_3 a_z = e_1 \frac{d^2x}{dt^2} + e_2 \frac{d^2y}{dt^2} + e_3 \frac{d^2z}{dt^2} \qquad (2a\text{-}4)$$

2. Cylindrical coordinates:

$$\mathbf{a} = e_1 a_r + e_2 a_\theta + e_3 a_z = e_1 \left[\frac{d^2r}{dt^2} - r \left(\frac{d\theta}{dt} \right)^2 \right] + e_2 \left[r \frac{d^2\theta}{dt^2} + 2 \frac{dr}{dt} \frac{d\theta}{dt} \right] + e_3 \frac{d^2z}{dt^2} \qquad (2a\text{-}5)$$

3. Spherical coordinates:

$$\mathbf{a} = e_1 a_r + e_2 a_\theta + e_3 a_\phi = e_1 \left[\frac{d^2r}{dt^2} - r \left(\frac{d\theta}{dt} \right)^2 - r \sin^2\theta \left(\frac{d\phi}{dt} \right)^2 \right]$$

$$+ e_2 \left[r \frac{d^2\theta}{dt^2} + 2 \frac{dr}{dt} \frac{d\theta}{dt} - r \sin\theta \cos\theta \left(\frac{d\phi}{dt} \right)^2 \right]$$

$$+ e_3 \left[r \sin\theta \frac{d^2\phi}{dt^2} + 2r \cos\theta \frac{d\theta}{dt} \frac{d\phi}{dt} + 2 \sin\theta \frac{dr}{dt} \frac{d\phi}{dt} \right] \qquad (2a\text{-}6)$$

2a-3. Newtonian Dynamics of Particles—Relationship of the Motion of Particles to the Forces Acting upon Them. *Inertial Frames of Reference.* Not all frames of reference are equally useful in describing the motion of a body; of all possible frames there is a set, called "inertial frames of reference," in which particularly simple laws describe the motion of a particle. An intuitive definition of an inertial frame of reference regards such a frame as being one which is "embedded in space" with respect to an observer; more exactly, an inertial frame of reference is one in which an isolated body moves with constant velocity. It may be easily seen from Newton's second law of motion (below) that any inertial frame is transformed to any other by uniform motion in a straight line.

Definitions of Useful Concepts. MASS. The Newtonian mass of a particle may be defined by considering the acceleration associated with the mutual interaction of this particle with a second, a test particle, when the two form an isolated system. The mass of the first particle is defined as a constant times the ratio of the magnitude of the accelerations of the second and first particles, respectively. The constant depends only on the choice of the second particle, and by mutual consent the constant may

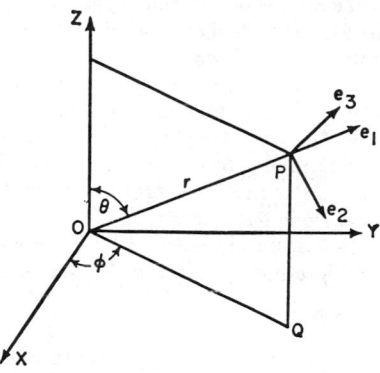

FIG. 2a-3. Base vectors in spherical coordinates.

arbitrarily be set equal to unity. The second particle then represents the standard unit of mass, and the mass of the first is thus determined by the above-mentioned ratio of accelerations. This method, although having the advantage of yielding an unequivocal definition of mass, is not usually a practicable one and is replaced by other methods (e.g., the balance) in actual determinations. Implicit in this definition is the assumption of additivity of masses, thus enabling the mass of a finite body, as an aggregate of particles, to be determined uniquely.

DENSITY. The density of a substance is defined as the mass per unit volume of the substance, and is calculated from the formula

$$\rho = \frac{m}{V} \tag{2a-7}$$

where ρ is the density, m is the mass, V is the volume occupied by mass m. Density is thus a measure of the volume concentration of mass.

MOMENTUM. The momentum of a particle is defined as the product of its mass and velocity and is therefore a vector quantity.

KINETIC ENERGY. The kinetic energy of a particle is defined as one-half the product of its mass and the square of its velocity, and is a scalar.

FORCE. The force acting upon a particle is assumed as the cause of the acceleration of the particle. It may be defined as that vector function which, in magnitude and direction, equals the time rate of change of momentum of the particle. Thus

$$\mathbf{F} = \frac{d}{dt}(m\mathbf{v}) \tag{2a-8}$$

where $\mathbf{F}$ represents the force, and m and $\mathbf{v}$ are the particle mass and velocity, respectively.

This force depends in general not only on the particle in question but also on the nature of other particles in, and properties of, the system of which the original particle is a part, the mutual separations and velocities of the particles and possibly of the time. Although force has been defined so far only for a particle, the definition may be extended to finite distributions of matter by considering infinitesimal portions as particles and integrating.

Newton's Laws. The dynamics of particles situated in an inertial frame of reference is governed by Newton's three laws of motion. The extension of these laws to a noninertial frame is, in principle, immediately forthcoming by considerations of the accelerations of the noninertial frame with respect to an inertial one; thus Newton's laws govern the dynamics of particles when Newtonian concepts are valid. Newton's laws are as follows:

1. A particle, not under the action of a force, will maintain its velocity unchanged in magnitude and direction.

2. A force acting on a particle causes a change of momentum of the particle, the rate of change of momentum being vectorially equal to the force.

3. If one particle exerts a force on a second, then the second exerts a force, equal in magnitude but opposite in direction, on the first.

Statics. The branch of dynamics which deals with particles undergoing no acceleration is termed "statics." We see from Newton's second law that, in this case,

$$\mathbf{F} = 0 \tag{2a-9}$$

where $\mathbf{F}$ refers to the vectorial sum of all the forces acting on the particle.

Noninertial Dynamics. At times it is convenient to consider the dynamics of a particle in a noninertial frame, e.g., motion relative to rotating or other moving axes. There will then be an apparent force acting on the particle which is the difference between the Newtonian force (that acting in the inertial system) and the inertial force $m\mathbf{a}_0$, where $\mathbf{a}_0$ is the acceleration of the noninertial system with respect to the inertial frame. Symbolically, $\mathbf{F}_d = \mathbf{F} - m\mathbf{a}_0$, where $\mathbf{F}_d$ is the apparent force, and $\mathbf{F}$ is the Newtonian force. We can set $\mathbf{F}_d = m\mathbf{a}_d$, where $\mathbf{a}_d$ is the acceleration of the particle with respect to the noninertial frame.

D'ALEMBERT'S PRINCIPLE. Often it is advantageous to choose a noninertial system such that $\mathbf{F}_d = 0$; the dynamical problem in the noninertial system then reduces to a statical one. That such a noninertial system can be chosen is one statement of D'Alembert's principle.

INERTIAL FORCES—CENTRIPETAL AND CORIOLIS FORCES. The difference between the Newtonian force and the apparent noninertial force can be termed the "inertial force." Centripetal and Coriolis forces are two commonly occurring examples of such inertial forces. The centripetal force is given for a particle of unit mass by

$$\mathbf{f}_c = \boldsymbol{\omega} \times (\boldsymbol{\omega} \times \mathbf{r}) \tag{2a-10}$$

where $\boldsymbol{\omega}$ is the instantaneous angular velocity of the moving axes about the axis of rotation and $\mathbf{r}$ is the position vector of the particle with respect to the moving axes. The Coriolis force is given for a particle of unit mass by

$$\mathbf{f}_c = 2\boldsymbol{\omega} \times \mathbf{v} \tag{2a-11}$$

where $\boldsymbol{\omega}$ has the same meaning as above and $\mathbf{v}$ is the apparent velocity of the particle with respect to the moving axes.

Conservation of Momentum. IMPULSE-MOMENTUM THEOREM. The impulse of a force acting between times t_0 and t_1 is defined by

$$\mathcal{J} = \int_{t_0}^{t_1} \mathbf{F} \, dt \tag{2a-12}$$

From Newton's second law, the impulse of the total force acting on a particle during some time interval is equal to the change in the momentum of the particle during the time interval, i.e.,

$$\mathcal{I} = m\mathbf{v}_1 - m\mathbf{v}_0 \qquad (2a\text{-}13)$$

CONSERVATION OF MOMENTUM. When the total force acting upon a particle is zero, the momentum of the particle is a constant; this follows directly from the impulse-momentum theorem.

Conservation of Energy. WORK-ENERGY THEOREM. The work done on a particle by a force acting during the displacement of a particle from position P_0 to position P_1 is defined as

$$W = \int_{P_0}^{P_1} \mathbf{F} \cdot d\mathbf{s} \qquad (2a\text{-}14)$$

where $d\mathbf{s}$ is an infinitesimal displacement along the path of the particle. From Newton's second law the work done by the total force acting on a particle during some displacement of the particle is equal to the change in kinetic energy of the particle:

$$W = \tfrac{1}{2}mv_1{}^2 - \tfrac{1}{2}mv_0{}^2 \qquad (2a\text{-}15)$$

POTENTIAL ENERGY. If the work done by a force acting on a particle does not depend upon the path of the particle, but only on the initial and end points of its motion, we call the force a "conservative force." The condition for a force to be conservative is that its curl shall vanish; i.e.,

$$\nabla \times \mathbf{F} = 0 \qquad (2a\text{-}16)$$

If the force is conservative, we may define a potential-energy function of position V such that

$$\mathbf{F} = -\nabla V \qquad (2a\text{-}17)$$

CONSERVATION OF ENERGY. If the total force acting upon a particle is conservative, the sum of the kinetic and potential energies is a constant; this follows from the work-energy theorem and the definition of the potential energy:

$$\tfrac{1}{2}mv^2 + V(x,y,z) = U \qquad (2a\text{-}18)$$

where U, the total mechanical energy, is a constant.

2a-4. Dynamics of Systems of Particles. In examining the dynamics of a system of point masses, consider N point particles, each of mass m_i, where $i = 1, 2, \ldots, N$. The total force acting on m_i due to m_j is $\mathbf{F}_{ij}$; in addition, a total external force $\mathbf{F}_i$ acts on m_i. At any time t, m_i has a position $\mathbf{r}_i$, a velocity $\dot{\mathbf{r}}_i$, and an acceleration $\ddot{\mathbf{r}}_i$, all relative to some inertial frame. (The dots denote differentiation with respect to time.)

Definition of Useful Concepts. CENTER OF MASS. The position of the center of mass of the above system is given by

$$\mathbf{R} = \frac{\displaystyle\sum_{i=1}^{N} m_i \mathbf{r}_i}{\displaystyle\sum_{i=1}^{N} m_i} \qquad (2a\text{-}19)$$

MOMENT OF MOMENTUM. The moment of momentum of the ith particle in the above system is defined as

$$\mathbf{L}_i = \mathbf{r}_i \times m_i \dot{\mathbf{r}}_i \qquad (2a\text{-}20)$$

The total moment of momentum of the system is

$$L = \sum_{i=1}^{N} L_i = \sum_{i=1}^{N} m_i(r_i \times \dot{r}_i) \qquad (2a\text{-}21)$$

If the collection of particles is a rigid body, the moment of momentum is called the "angular momentum" (cf. Sec. 2a-5).

TORQUE (MOMENT OF FORCE). The torque due to a force F_i acting on the ith particle in the above system is defined as

$$T_i = r_i \times F_i \qquad (2a\text{-}22)$$

The total torque acting on the system is $T = \sum_{i=1}^{N} T_i$. (The force F_i includes forces externally applied to the particle, as well as internal forces of interaction among the particles of the system.)

Application of Newton's Laws. We may apply Newton's second law to each particle of the system, and obtain

$$m_i\ddot{r}_i = F_i{}^i + F_i{}^e \qquad (2a\text{-}23)$$

where $F_i{}^i = \Sigma_{j \neq i} F_{ij}$ is the total internal force acting on m_i (due to all other particles), and $F_i{}^e$ is the external force on the ith particle.

If we sum over all particles of the system, we obtain, by use of Newton's third law,

$$\sum_{i=1}^{N} F_i{}^i = 0 \qquad (2a\text{-}24)$$

MOTION OF THE CENTER OF MASS. The analog of Newton's second law for the entire system is therefore

$$M\ddot{R} = \sum_{i=1}^{N} F_i{}^e \qquad (2a\text{-}25)$$

where $M = \sum_{i=1}^{N} m_i$ is the total mass of the system, $\ddot{R}$ is the acceleration of the center of mass of the system, and $\Sigma_i F_i{}^e$ is the total external force.

MOMENT OF MOMENTUM AND TORQUE. By forming the cross product of both sides of Eq. (2a-23) with r_i and summing over all particles we can show that

$$\frac{d}{dt} \sum [r_i \times (m_i\dot{r}_i)] = \sum T_i{}^e = T^e \qquad (2a\text{-}26)$$

provided that the internal force F_{ij} acts along the straight line connecting the particles i and j in each case.

In particular, if r_{ic} is the position of the ith particle with respect to the center of mass, so that

$$r_{ic} = r_i - R$$

it follows from Eq. (2a-26) that

$$\frac{d}{dt} \sum_{i=1}^{N} r_{ic} \times (m_i\dot{r}_{ic}) = \sum_{i=1}^{N} r_{ic} \times F^e \qquad (2a\text{-}27)$$

That is, the time rate of change of the moment of momentum is equal to the total external torque when both are taken with respect to the center of mass. The above equation is also true if the center of mass is replaced by any point moving with the velocity of the center of mass, which may, of course, also be at rest.

Conservation of Momentum. It follows from Eqs. (2a-25) and (2a-26) that:

1. If the total external force is zero, the linear momentum of the center of mass is constant.

2. If the total external torque about a fixed point, or one moving with velocity of the center of mass, is zero, the moment of momentum about that point is constant.

Conservation of Energy. WORK-ENERGY THEOREM. The total work done by the external and internal forces acting on the system is equal to the change in the total kinetic energy of the system (the sum of the kinetic energies of all particles)

$$\frac{1}{2} \sum_{i=1}^{N} m_i(\mathbf{v}_i^{\prime\prime 2} - \mathbf{v}_i^{\prime 2}) = \sum_{1=1}^{N} \int_{r_i'}^{r_i''} (\mathbf{F}_i^e + \mathbf{F}_i^i) \cdot d\mathbf{r}_i \qquad (2a\text{-}28)$$

where $\mathbf{v}_i'$, $\mathbf{v}_i''$ are the velocities of the ith particle at position $\mathbf{r}_i'$ and $\mathbf{r}_i''$, respectively, and $\mathbf{F}_i^i = \Sigma_{j \neq i}\mathbf{F}_{ij}$ is the total internal force acting on the ith particle.

CONSERVATION OF ENERGY. If the internal and external forces are conservative, so that they can be derived from potentials,

$$\mathbf{F}_i^i = -\nabla V_i^i \qquad \text{and} \qquad \mathbf{F}_i^e = -\nabla V_i^e \qquad (2a\text{-}29)$$

then the sum of the kinetic and potential energies of all the particles is a constant

$$\sum_{i=1}^{N} (\tfrac{1}{2}m_i v_i^2 + V_i^i + V_i^e) = U \qquad (2a\text{-}30)$$

where U is the total energy of the system.

2a-5. Dynamics of Rigid Bodies. *Definitions of Kinematical Concepts.* A rigid body is an aggregate of particles the distance between any two of which remains constant. The position of a rigid body in any frame of reference is completely determined by fixing the position of three noncollinear points. This means that the number of degrees of freedom of the rigid body is six. There are two principal types of motion of a rigid body: (1) *translation*, in which all particles move with the same velocity and acceleration in parallel paths, and (2) *rotation*, in which some point or line of points (axis) remains fixed in space. Every motion of a rigid body can be considered as a combination of translations and rotations.

The instantaneous angular velocity ω is the primary quantity descriptive of the kinematics of a rigid body. This is a vector lying along the instantaneous axis of rotation and having the magnitude such that its cross product with the position vector $\mathbf{r}_P$ of any point P of the rigid body relative to an origin on the axis yields the velocity of the point P. Symbolically

$$\mathbf{v}_P = \dot{\mathbf{r}}_P = \omega \times \mathbf{r}_P \qquad (2a\text{-}31)$$

The angular velocity can always be resolved into rectangular components ω_x, ω_y, ω_z, i.e.,

$$\omega = \mathbf{i}\omega_x + \mathbf{j}\omega_y + \mathbf{k}\omega_z \qquad (2a\text{-}32)$$

Angular acceleration is the time rate of change of angular velocity, i.e. (to use the dot notation),

$$\alpha = \dot{\omega} \qquad (2a\text{-}33)$$

Dynamical Concepts and Equations of Motion. The total moment of momentum **L** of the rigid body with respect to some fixed origin of coordinates either inside or outside the body [cf. Eq. (2a-20)] is called the angular momentum of the rigid body about the origin. By expansion of the summand in Eq. (2a-21) after employing Eq. (2a-31) there results

$$\begin{aligned} \mathbf{L} = &\ \mathbf{i}(\omega_x I_{xx} - \omega_y I_{xy} - \omega_z I_{xz}) \\ &+ \mathbf{j}(-\omega_x I_{yx} + \omega_y I_{yy} - \omega_z I_{yz}) \\ &+ \mathbf{k}(-\omega_x I_{zx} - \omega_y I_{zy} + \omega_z I_{zz}) \end{aligned} \tag{2a-34}$$

where I_{xx}, I_{yy}, I_{zz} are called the "moments of inertia" of the rigid body about the x, y, z axes, respectively, and I_{xy}, I_{yz}, I_{zx}, etc., are called "products of inertia." We have

$$\begin{aligned} I_{xx} &= \Sigma m_i(y_i{}^2 + z_i{}^2) \qquad \text{etc.} \\ I_{xy} &= \Sigma m_i x_i y_i \qquad \text{etc.} \end{aligned} \tag{2a-35}$$

By proper choice of axes (called "principal" axes) the products of inertia can be made to vanish. If we write

$$I_{xx} = MR^2 \tag{2a-36}$$

where

$$R^2 = \frac{\Sigma m_i(y_i{}^2 + z_i{}^2)}{M} \tag{2a-37}$$

and M is the total mass of the rigid body, R is termed the "radius of gyration" about the x axis.

The fundamental equation of motion (Newton's second law) of the rigid body about a *fixed* origin is

$$\dot{\mathbf{L}} = \mathbf{T} \tag{2a-38}$$

where **T** is the total torque about the instantaneous axis through the fixed origin. If the fixed origin is chosen as the center of mass, the total motion is obtained by superposing the translational motion of the center of mass on the rotational motion about the center of mass.

Static Equilibrium. A rigid body is in translational equilibrium if its center of mass moves with constant velocity in an inertial frame. It is in rotational equilibrium about any point if the resultant torque about the point vanishes. This means $\dot{\mathbf{L}} = 0$ and corresponds to conservation of angular momentum. The behavior of a rigid body under these conditions is the subject matter of rigid statics.

Moving Axes. Euler's Equation. For axes fixed in space, ω and the moments and products of inertia in general change with time as the rigid body moves. Simplification often results by using axes *fixed in the body*, since then I_{xx}, I_{xy}, etc., remain constant. Then, for motion about a fixed point the axes rotate, and we have

$$\dot{\mathbf{L}} = \mathbf{i}\dot{L}_x + \mathbf{j}\dot{L}_y + \mathbf{k}\dot{L}_z + \boldsymbol{\omega} \times \mathbf{L} \tag{2a-39}$$

where L_x, L_y, L_z are the components of angular momentum about the moving axes and ω is the instantaneous angular velocity of the body about the instantaneous axis of rotation. If we choose principal axes the equation of motion (2a-38) becomes

$$\begin{aligned} \mathbf{T} = &\ \mathbf{i}[I_{xx}\dot{\omega}_x + (I_{zz} - I_{yy})\omega_y\omega_z] \\ &+ \mathbf{j}[I_{yy}\dot{\omega}_y + (I_{xx} - I_{zz})\omega_z\omega_x] \\ &+ \mathbf{k}[I_{zz}\dot{\omega}_z + (I_{yy} - I_{xx})\omega_x\omega_y] \end{aligned} \tag{2a-40}$$

This is *Euler's equation.* The three component equations to which it reduces are usually called Euler's equations.

Kinetic Energy. Work-Energy Theorem. If a rigid body has one point fixed in space and the angular momentum about this point is **L**, while the angular velocity about an instantaneous axis through the point is ω, the kinetic energy of rotational motion is

$$K = \tfrac{1}{2}\omega \cdot \mathbf{L} \tag{2a 41}$$

The work done by the resultant torque **T** about the fixed point in time dt is

$$dW = \mathbf{T} \cdot \omega \, dt \tag{2a-42}$$

measured with respect to axes fixed in the body. Since

$$\omega \cdot d\mathbf{L} = dK \tag{2a-43}$$

it follows that the work done by the resultant torque in any time interval is equal to the change in kinetic energy of rotation during this same interval.

Total Energy. The total kinetic energy of a rigid body is the sum of the kinetic energy of translation of the center of mass (assuming all the mass to be concentrated there) and the kinetic energy of rotation about the center of mass. The total potential energy is the sum of the potential energy of the center of mass (with all the mass concentrated there) due to the external forces acting on the body and the potential energy of all the particles of the body due to the internal forces of cohesion that hold the body together. If the body remains really rigid throughout its motion, the last-named potential energy remains constant. With this understanding, the law of conservation of energy of a rigid body is phrased as precisely as that in the case of a particle.

2a-6. Dynamics of Deformable Media. *General Concepts of Strain and Stress.* Whenever an extended medium moves in such a way that the distance between any two particles constituting the medium changes, the medium is said to be *deformed*. Deformations are of two general types: (1) dilatational or extensional, in which a change in the density of the medium takes place (change in the size, if the medium is finite) and (2) shear, in which a change in the shape alone takes place. The corresponding *fractional* deformations (nondimensional quantities) are termed *strains*. Thus the *dilatational strain* is the negative of the change in density divided by the mean density. The *extensional strain* (in the case of a rod, string, or other linear medium) is the change in length divided by the mean length. The *shear strain* is the difference in displacement of two parallel planes in the medium divided by the perpendicular distance between them.

When a medium is deformed by the application of external forces, the dynamics of the deformation is best described in terms of internal *stresses* which are assumed to change with the deformation. A stress is a force per unit area with which the part of the medium on one side of an imaginary surface acts on the part on the other side. If the force is normal to the surface, the stress is dilatational; if the force is parallel to the surface, the stress is a shear. The stresses associated with deformations are strictly *excess* stresses (i.e., the change in stress produced by the application of the external force). The adjective is normally omitted.

Elastic Media. Hooke's Law. If when the deforming forces are removed a medium reverts to its original condition, it is said to be *elastic*. In such media the ratio of stress to strain is approximately a constant for a certain range of stress variation. This is Hooke's law. For all solid media the imposition of a sufficiently large deforming force leads to a breakdown of this linear relation; i.e., they possess an elastic limit (cf. Sec. 2e). Indeed even larger deforming forces may cause the solid to flow (strain dependent on time) and it becomes *plastic*. Even elastic substances do not always return *immediately* to their original condition after the removal of the deforming force (elastic lag or relaxation). Fluids can experience change of state under sufficiently high stresses.

For an elastic medium for which Hooke's law holds it is possible to define elastic moduli, i.e., ratios of stress to strain. Thus,

$$\frac{\text{Compressional stress}}{\text{Volume strain}} = k = \text{bulk modulus or modulus of volume elasticity}$$

$$\frac{\text{Tensile stress}}{\text{Linear strain}} = E = \text{Young's modulus}$$

$$\frac{\text{Shearing stress}}{\text{Shear strain}} = G = \text{shear modulus or rigidity}$$

The deformation of a homogeneous isotropic elastic medium can be completely described in terms of these three moduli. A fourth, Poisson's ratio σ, is usually added. This is the reciprocal of the ratio of linear extensional strain in a wire or rod to the concomitant lateral contractional strain. The following relations hold among the moduli:

$$E = 3k(1 - 2\sigma) = 2G(1 + \sigma)$$
$$E = \frac{9kG}{G + 3k} \tag{2a-44}$$

Evidently for such media

$$-1 < \sigma < \tfrac{1}{2} \tag{2a-45}$$

General Stress and Strain Expressions for an Arbitrary Medium. If the displacement from its equilibrium position of any particle of a deformable medium is denoted by the vector

$$\mathbf{\Delta} = \mathbf{i}\xi + \mathbf{j}\eta + \mathbf{k}\zeta \tag{2a-46}$$

where the displacement components ξ, η, ζ are in general functions of both space and time, the effective *strain* is denoted by the covariant tensor of the second order written in matrix form as follows:

$$\mathbf{D} = \left\| \begin{array}{l} \partial\xi/\partial x, \frac{1}{2}(\partial\eta/\partial x + \partial\xi/\partial y), \frac{1}{2}(\partial\zeta/\partial x + \partial\xi/\partial z) \\ \frac{1}{2}(\partial\eta/\partial x + \partial\xi/\partial y), \partial\eta/\partial y, \frac{1}{2}(\partial\eta/\partial z + \partial\zeta/\partial y) \\ \frac{1}{2}(\partial\zeta/\partial x + \partial\xi/\partial z), \frac{1}{2}(\partial\zeta/\partial y + \partial\eta/\partial z), \partial\zeta/\partial z \end{array} \right\| \tag{2a-47}$$

This is often written in the abbreviated symbolic form

$$D = \left\| \begin{array}{l} e_{xx}, \frac{1}{2}e_{xy}, \frac{1}{2}e_{xz} \\ \frac{1}{2}e_{xy}, e_{yy}, \frac{1}{2}e_{yz} \\ \frac{1}{2}e_{xz}, \frac{1}{2}e_{yz}, e_{zz} \end{array} \right\| \tag{2a-48}$$

The diagonal elements in this matrix are dilatational strain components, whereas the nondiagonal elements are shear strain components.

The total stress in a deformable medium is most adequately expressed in terms of the stress tensor $\mathbf{S}$ which is represented by the following matrix:

$$\mathbf{S} = \left\| \begin{array}{l} X_x, X_y, X_z \\ Y_x, Y_y, Y_z \\ Z_x, Z_y, Z_z \end{array} \right\| \tag{2a-49}$$

Here X_x is the tensile stress in the x direction on the surface normal to the x axis, X_y is the shear stress in the y direction on the surface normal to the x axis; X_z is the shear stress in the z direction on the surface normal to the x axis, etc. It should be noted that the stress tensor is symmetrical, i.e., $X_y = Y_x$, etc. The same is true of the strain tensor ($e_{xy} = e_{yx}$, etc.).

Hooke's Law in Tensor Form for a Homogeneous, Isotropic Elastic Medium. For this case Hooke's law takes the form

$$S = 2GD + \lambda D' \tag{2a-50}$$

where G is still the shear modulus, and $\lambda = k - 2G/3$. D' is the diagonal tensor

$$D' = \begin{Vmatrix} \Theta & 0 & 0 \\ 0 & \Theta & 0 \\ 0 & 0 & \Theta \end{Vmatrix} \tag{2a-51}$$

with

$$\Theta = e_{xx} + e_{yy} + e_{zz} \tag{2a-52}$$

Hooke's Law for an Arbitrary Crystalline Medium. If the medium is a crystal with different properties in different directions, Hooke's law takes the form of the following linear equations expressing the strain components in terms of the stress components.

$$
\begin{aligned}
e_{xx} &= S_{11}X_x + S_{12}Y_y + S_{13}Z_z + S_{14}Y_z + S_{15}Z_x + S_{16}X_y \\
e_{yy} &= S_{21}X_x + S_{22}Y_y + S_{23}Z_z + S_{24}Y_z + S_{25}Z_x + S_{26}X_y \\
e_{zz} &= S_{31}X_x + S_{32}Y_y + S_{33}Z_z + S_{34}Y_z + S_{35}Z_x + S_{36}X_y \\
e_{zz} &= S_{41}X_x + S_{42}Y_y + S_{43}Z_z + S_{44}Y_z + S_{45}Z_x + S_{46}X_y \\
e_{zy} &= S_{51}X_x + S_{52}Y_y + S_{53}Z_z + S_{54}Y_z + S_{55}Z_x + S_{56}X_y \\
e_{xy} &= S_{61}X_x + S_{62}Y_y + S_{63}Z_z + S_{64}Y_z + S_{65}Z_x + S_{66}X_y
\end{aligned}
\tag{2a-53}
$$

The 36 coefficients $S_{11}, S_{12}, \ldots, S_{ij}, \ldots, S_{66}$ are called the "elastic constants." If the above linear equations are solved for the stress components in terms of the strain components, the corresponding coefficients C_{ij} are called "elastic coefficients." It can be shown that, for any i_{ij}, $C_{ij} = C_{ji}$ and $S_{ij} = S_{ji}$.

For a cubic crystal the elastic coefficient matrix reduces to

$$C = \begin{Vmatrix} C_{11} & C_{12} & C_{12} & 0 & 0 & 0 \\ C_{12} & C_{11} & C_{12} & 0 & 0 & 0 \\ C_{12} & C_{12} & C_{11} & 0 & 0 & 0 \\ 0 & 0 & 0 & C_{44} & 0 & 0 \\ 0 & 0 & 0 & 0 & C_{44} & 0 \\ 0 & 0 & 0 & 0 & 0 & C_{44} \end{Vmatrix} \tag{2a-54}$$

Moreover for a cubic crystal $C_{44} = 1/S_{44}$. The bulk modulus in this case is given by

$$k = \frac{C_{11} + 2C_{12}}{3} \tag{2a-55}$$

Equation of Motion of a Deformed Homogeneous Isotropic Elastic Medium. The equation of motion of such a medium of density ρ, in which the displacement from equilibrium is the vector Δ, takes the form

$$\rho \ddot{\Delta} = \left(k + \frac{4G}{3} \right) \nabla \nabla \cdot \Delta - G \nabla \times \nabla \Delta \tag{2a-56}$$

If $\nabla \times \Delta = 0$ this is the equation of *irrotational* waves traveling with velocity

$$V_i = \sqrt{\frac{k + 4G/3}{\rho}} \tag{2a-57}$$

If $\nabla \cdot \Delta = 0$, this is the equation of solenoidal waves traveling with velocity

$$V_s = \sqrt{\frac{G}{\rho}} \tag{2a-58}$$

2a-7. Fluid Dynamics. *General Concepts. Fluids in Equilibrium.* A perfect fluid is a deformable medium in which deforming forces give rise only to dilatations and never to shears. This is an ideal concept and is realized only approximately for actual fluids. Gases manifest the property more nearly than liquids, though both are normally considered to be fluids. Liquids can present under many circumstances the phenomenon of a free surface.

The dilatational stress in the case of a fluid is termed the *pressure*, which is the force per unit area directed *against* any surface imagined to exist in the fluid. A perfect fluid in equilibrium under the influence of an external force F acting on unit mass is subject to the relation

$$\rho\mathbf{F} = \nabla p \tag{2a-59}$$

where p is the pressure (here treated for simplicity as a scalar since it acts *normally* to every surface when the fluid is in equilibrium) and ρ the density, all quantities being considered as functions of space alone. The solution of this equation for given $\mathbf{F}$ gives p as a function of position in space and yields Pascal's law of the transmissibility of pressure in a fluid in equilibrium. From this also follows at once the principle of Archimedes that any fluid in equilibrium exerts on a body immersed in it a buoyant force equal in magnitude to the weight of the fluid displaced by the body and directed upward through the center of gravity.

Flow Concepts. Equation of Continuity. In the Eulerian system to which this review is confined the flow velocity of a fluid is the vector $\mathbf{v}$ whose magnitude at any point and at any time is the volume flow per unit time per unit area placed normal to the direction of flow, the latter being the direction of $\mathbf{v}$. This quantity is a function of both space and time. In any continuous indestructible fluid of density ρ containing no sources or sinks $\mathbf{v}$ obeys the so-called equation of continuity

$$\nabla \cdot (\rho\mathbf{v}) = -\dot{\rho} \tag{2a-60}$$

where it is to be noted that ρ also is a function of space and time. For a homogeneous incompressible fluid this equation reduces to

$$\nabla \cdot \mathbf{v} = 0 \tag{2a-61}$$

i.e., $\mathbf{v}$ is a solenoidal vector. If further $\mathbf{v}$ is irrotational, so that $\nabla \times \mathbf{v} = 0$, it follows that

$$\mathbf{v} = \nabla \phi \tag{2a-62}$$

where ϕ is a scalar potential, called the "velocity potential," and the equation of continuity reduces to Laplace's equation

$$\nabla^2\phi = 0 \tag{2a-63}$$

Equation of Motion. Bernoulli's Principle. The vector equation of motion of a compressible fluid of density ρ subject to an external force $\mathbf{F}$ is

$$\dot{\mathbf{v}} + \mathbf{v} \cdot \nabla\mathbf{v} = \mathbf{F} - \frac{\nabla p}{\rho} \tag{2a-64}$$

where p is the pressure.

For irrotational flow in a conservative force field ($\mathbf{F} = -\nabla V$) it follows from the equation of motion that

$$\tfrac{1}{2}\rho v^2 + \rho V + p = \text{const} \tag{2a-65}$$

which is the principle of Bernoulli. It can also be shown that, even if the flow is not irrotational, as long as it is *steady* and in streamlines, so that $\mathbf{v}$ does not depend on the time, the above equation of Bernoulli will still hold as one proceeds along any given streamline, though the constant will in general be different for different streamlines.

Viscous Fluids. In contrast to a perfect fluid in which no shearing strains can exist, a viscous fluid is one in which the part of the medium flowing in one layer exerts a tangential or shearing stress on that flowing in the same direction in an adjacent layer. In the simplest type of viscous flow the tangential force is proportional to the velocity gradient normal to the layer and the coefficient of proportionality is called the viscosity η. Specifically

$$\eta = \frac{\text{shearing stress}}{\text{velocity gradient normal to flow}} \tag{2a-66}$$

The analogy between this relation and that defining the shear modulus for an elastic medium is obvious, the difference being that here the denominator is the rate of *change* of shear strain instead of the strain itself. The suggestion is immediate that the discussion of viscous flow can develop along the lines of the analysis of the behavior of deformable media in general (cf. Sec. 2a-6). This is indeed the case; it makes pressure appear as a tensor (analogous to the stress tensor). See also Secs. 2m, 2r, 2u.

A solid moving through a viscous fluid encounters increased resistance because of the viscosity. The simplest case is that in which a sphere of radius a moves through a fluid of vicosity η with *constant* velocity v. The resisting force is then given by Stokes' law

$$F = 6\pi\eta av \tag{2a-67}$$

Surface Tension in Liquids. This is the force per unit length γ in the surface separating a liquid from the material surrounding it. Details concerning this as well as numerical values will be found in Sec. 2n.

Surface Waves in Liquids. When the free surface of a liquid is deformed, the forces acting on the deformed elements are primarily surface tension and gravity. The velocity of the resulting surface wave, if it is harmonic and has wavelength λ, is

$$V = \sqrt{\left(\frac{g\lambda}{2\pi} + \frac{2\pi\gamma}{\rho\lambda}\right) \tanh \frac{2\pi l}{\lambda}} \tag{2a-68}$$

where g is the acceleration of gravity, ρ the density, γ the surface tension, and l the depth of the liquid. For a relatively shallow liquid, for which $l \ll \lambda$, and the surface tension not very large, we have

$$V \doteq \sqrt{gl} \tag{2a-69}$$

If the liquid is relatively deep, or $l \gg \lambda$,

$$V \doteq \sqrt{\frac{g\lambda}{2\pi} + \frac{2\pi\gamma}{\rho\lambda}} \tag{2a-70}$$

For long waves

$$V \doteq \sqrt{\frac{g\lambda}{2\pi}}$$

while for ripples (small λ), surface tension predominates and

$$V \doteq \sqrt{\frac{2\pi\gamma}{\rho\lambda}}$$

Compressional Waves in Fluids. The combination of the equation of motion (2a-64), the equation of continuity (2a-60), and the equation of state of the fluid, i.e., the relation connecting change in density with change in pressure, leads to the wave equation for compressional waves traveling with velocity

$$V = \sqrt{\frac{dp}{d\rho}} \qquad (2a\text{-}71)$$

The values of V for gases and liquids will be found in Sec. 3.

2a-8. Fundamental Units (mks System). These units are defined as follows:

Meter. Unit of length. By international agreement (Oct. 14, 1960) defined to be 1,650,763.73 wavelengths of the orange-red line of krypton 86. This replaces the definition in terms of the platinum-iridium meter bar in Paris.

Kilogram. Unit of mass. Defined to be the mass of a certain solid cylinder of platinum-iridium alloy preserved at the International Bureau of Weights and Measures in Paris.

Second. Unit of time. By earlier international agreement (October 14, 1960) defined to be 1/31,556,925.9747 of the tropical year 1900. (The *tropical year* is defined as the interval of time between two successive passages of the sun through the vernal equinox.) A more recent international conference (1964) adopted provisionally a new definition of the second as the time corresponding to 9.192631770 × 10^9 oscillations of the cesium atom in the so-called atomic clock.

2a-9. Supplementary Fundamental Units (cgs and English Systems). Definitions of these units follow:

Centimeter. Defined to be $\frac{1}{100}$ meter.

Gram. Defined to be 1/1,000 kilogram.

Second. Same as the second defined in Sec. 2a-8.

International Yard. Defined by agreement between the United States and the British Commonwealth (1959) to be 0.9144 meter.

International Pound. Defined by agreement between the United States and the British Commonwealth (1959) to be 0.45359237 kilogram.

2a-10. Angular Units. These units are defined as follows:

Degree. Angle subtended at the center by a circular arc which is $\frac{1}{360}$ of the circumference.

Minute of Arc. $\frac{1}{60}$ of a degree.

Second of Arc. $\frac{1}{60}$ of a minute of arc.

Radian. Angle subtended at the center by a circular arc which is equal in length to the radius of the circle.

Steradian. Solid angle subtended at the center by $1/4\pi$ of the surface area of a sphere of unit radius.

2a-11. Derived Units. These units are defined as follows:

Atmosphere. Pressure exerted by air at mean sea level under standard conditions = 1.013250 × 10^6 dynes/cm².

British Thermal Unit (Mean). Energy required to raise temperature of 1 lb mass of water 1°F (averaged from 32 to 212°F).

Calorie (Mean). Energy required to raise 1 g mass of water 1°C (averaged from 0 to 100°C).

Centimeters of Hg at 0°C. Pressure exerted by column of Hg of stated height at 0°C.

Dyne. Force necessary to give 1 g mass acceleration of 1 cm/sec².

Erg. Work done by force of 1 dyne moving a particle a distance of 1 cm.

Feet of Water at 4°C. Pressure exerted by column of water of stated height at 4°C.

Kilowatthour. Work done in 1 hr at power level or rate of 10^3 watts.

Newton. Force necessary to give 1 kg mass acceleration of 1 m/sec².

Poundal. Force necessary to give 1 lb mass acceleration of 1 ft/sec².

Watt. Rate of doing work, or power expended, in the amount of 10^7 ergs/sec.

TABLE 2a-1. UNITS AND CONVERSION FACTORS, LENGTH

	Angstrom	Centimeter	Fathom	Foot	Inch (U.S.)	Kilometer	Light-year
Angstrom..........	1	10^{-8}	......	3.281×10^{-10}	3.937×10^{-9}	10^{-13}	
Centimeter........	10^8	1	......	3.281×10^{-2}	0.3937	10^{-5}	
Fathom............			1	6	72		
Foot...............		30.48	0.1667	1	12		
Inch (U.S.)........	2.540×10^8	2.540		8.333×10^{-2}	1		
Kilometer.........		10^5		3.281×10^3		1	1.057×10^{-13}
Light-year........						9.46×10^{12}	1
Meter.............	10^{10}	10^2	0.5468	3.281	39.37	10^{-3}	
Micron............	10^4	10^{-4}			3.937×10^{-5}		
Mil................		2.540×10^{-3}			10^{-3}		
Mile (statute).....				5.280×10^3	6.336×10^4	1.609	1.69×10^{-13}
Millimeter........		10^{-1}			3.937×10^{-2}		
Millimicron........	10	10^{-7}					
Yard (U.S.)........		91.44		3	36		

	Meter	Micron	Mil	Mile (statute)	Millimeter	Milli-micron	Yard (U.S.)
Angstrom..............	10^{-10}	10^{-4}	3.937×10^{-6}		10^{-7}	10^{-1}	1.094×10^{-10}
Centimeter.............	10^{-2}	10^4	3.937×10^2		10	10^7	1.094×10^{-2}
Fathom................	1.829					...	2
Foot....................	0.3048			1.894×10^{-4}		...	0.3333
Inch (U.S.).............			10^3	1.578×10^{-5}	25.40	...	2.778×10^{-2}
Kilometer.............	10^3			0.6214		...	1.094×10^3
Light-year.............				5.9×10^{12}			
Meter..................	1			6.214×10^{-4}		10^9	1.094
Micron.................	10^{-6}	1	3.937×10^{-2}		10^{-3}	10^3	
Mil....................		25.40	1		2.450×10^2		
Mile (statute)...........	1.609×10^3			1		...	1.760×10^3
Millimeter..............	10^{-3}	10^3	39.37		1		
Millimicron..............	10^{-9}	10^{-3}				1	
Yard (U.S.)..............	0.9144			5.682×10^{-4}		...	1

TABLE 2a-2. UNITS AND CONVERSION FACTORS, AREA

	Circular mil	Square centimeter	Square foot (U.S.)	Square inch (U.S.)	Square kilometer	Square meter	Square mile	Square millimeter	Square yard
Circular mil	1	5.067×10^{-6}		7.854×10^{-7}				5.067×10^{-4}	
Square centimeter	1.974×10^{5}	1	1.076×10^{-3}	0.1550		10^{-4}		10^{2}	1.196×10^{-4}
Square foot (U.S.)		9.290×10^{2}	1	1.44×10^{2}		9.290×10^{-2}	3.587×10^{-8}		0.1111
Square inch (U.S.)	1.273×10^{6}	6.452	6.944×10^{-3}	1		6.452×10^{-4}		6.452×10^{2}	7.716×10^{-4}
Square kilometer			1.076×10^{7}		1	10^{6}	0.3861		1.196×10^{6}
Square meter		10^{4}	10.76	1.550×10^{3}	10^{-6}	1	3.861×10^{-7}	10^{6}	1.196
Square mile			2.788×10^{7}		2.590	2.590×10^{6}	1		3.098×10^{6}
Square millimeter	1.974×10^{3}	10^{-2}		1.550×10^{-3}		10^{-6}		1	
Square yard		8.361×10^{3}	9	1.296×10^{3}		0.8361	3.228×10^{-7}		1

TABLE 2a-3. UNITS AND CONVERSION FACTORS, VOLUME

	Cubic centimeter	Cubic foot	Cubic inch	Cubic meter	Cubic yard	Fluid ounce (U.S.)	Gallon (U.S.)	Liter*	Pint, dry (U.S.)	Pint, liquid (U.S.)	Quart, dry (U.S.)	Quart, liquid (U.S.)
Cubic centimeter	1	3.531×10^{-5}	6.102×10^{-2}	10^{-6}	1.308×10^{-6}	3.381×10^{-2}	2.642×10^{-4}	9.9997×10^{-4}	1.816×10^{-3}	2.113×10^{-3}	9.081×10^{-4}	1.057×10^{-3}
Cubic foot	2.832×10^{4}	1	1.728×10^{3}	2.832×10^{-2}	3.704×10^{-2}		7.481	28.32		59.84	25.71	29.92
Cubic inch	16.39	5.787×10^{-4}	1	1.639×10^{-5}	2.143×10^{-5}	0.5541	4.329×10^{-3}	1.639×10^{-2}	2.976×10^{-2}		1.488×10^{-2}	1.732×10^{-2}
Cubic meter	10^{6}	35.31	6.102×10^{4}	1	1.308		2.642×10^{2}	9.9997×10^{2}		2.113×10^{3}		1.057×10^{3}
Cubic yard	7.646×10^{5}	27	4.666×10^{4}	0.7646	1		2.020×10^{2}	7.645×10^{2}		1.616×10^{3}		8.079×10^{2}
Fluid ounce (U.S.)	29.57		1.805			1	7.813×10^{-3}	2.957×10^{-2}		6.250×10^{-2}		3.125×10^{-2}
Gallon (U.S.)	3.785×10^{3}	0.1337	2.310×10^{2}	3.785×10^{-3}	4.951×10^{-3}	1.280×10^{2}	1	3.785		8		4
Liter*	1.000×10^{3}	3.532×10^{-2}	61.03	1.000×10^{-3}	1.308×10^{-3}	33.81	0.2642	1	1.816	2.113	0.9081	1.057
Pint, dry (U.S.)	5.506×10^{2}		33.60					0.5506	1		0.5000	
Pint, liquid (U.S.)	4.732×10^{2}	1.671×10^{-2}	28.88		6.188×10^{-4}	16	0.1250	0.4732		1		0.5000
Quart, dry (U.S.)	1.101×10^{3}	3.889×10^{-2}	67.20					1.101	2		1	
Quart, liquid (U.S.)	9.464×10^{2}	3.342×10^{-2}	57.75			32	0.2500	0.9463		2		1

* 1 milliliter = 1.000027 cubic centimeters.

2b. Density of Solids

H. M. TRENT[1]

U.S. Naval Research Laboratory

D. E. STONE

Vertex Corporation[2]

R. BRUCE LINDSAY

Brown University

For the definition of density ρ consult Sec. 2a-3. The cgs unit of density is the gram per cubic centimeter and this is used throughout the tables in this subsection.

Densities of the elements in solid form are given in Table 2b-1. All data are taken from "Smithsonian Physical Tables" (9th revised edition, 1954) unless otherwise stated. The values marked * are calculated densities from X-ray crystallographic data at room temperature and are taken from International Critical Tables (1926) All others are measured values for polycrystalline condition, save when otherwise stated. Standard room temperature is understood, unless otherwise stated.

TABLE 2b-1. DENSITY OF THE ELEMENTS IN SOLID FORM

Element	Physical state	Density, g/cm³	Temp., °C
Aluminum	Commercial hard-drawn solid	2.70	20
Aluminum	Single crystal	2.692*	
Antimony	Vacuo-distilled solid	6.62	20
Antimony	Single crystal	6.73*	
Argon	Solid	1.65	−233
Argon	Single crystal	1.645*	−253
Arsenic	Crystallized solid	5.73	14
Arsenic	Single crystal	5.75*	
Barium	Solid	3.5	20
Beryllium	Solid	1.85	20
Beryllium	Single crystal	1.83*	
Bismuth	Vacuo-distilled solid	9.78	20
Bismuth	Single crystal	9.86*	
Boron	Crystallized solid	2.535	
Bromine	Solid	4.2	−273
Cadmium	Vacuo-distilled solid	8.65	20
Cadmium	Single crystal	8.56*	
Calcium	Solid	1.55	20
Calcium	Single crystal	1.54*	
Carbon	Diamond	3.52	20

[1] Deceased.
[2] H. M. Childers of the Vertex Corporation provided valuable consultant service.

TABLE 2b-1. DENSITY OF THE ELEMENTS IN SOLID FORM (*Continued*)

Element	Physical state	Density, g/cm³	Temp., °C
Carbon.............	Graphite	2.25	20
Cerium.............	Solid	6.90	20
Cerium.............	Cubic crystal	6.90*	
Cerium.............	Hexagonal crystal	6.73*	
Cesium.............	Solid	1.873	20
Chlorine............	Solid	2.2	−273
Chromium...........	Solid	7.14	20
Chromium...........	Crystal	7.22*	
Cobalt.............	Solid	8.71	21
Cobalt.............	Cubic crystal	8.67*	
Columbium..........	Solid	8.4	20
Copper.............	Vacuo-distilled solid	8.933	20
Copper.............	Single crystal	8.95*	
Erbium.............	Solid	4.77	
Fluorine............	Solid	1.5	−273
Gallium............	Solid	5.93	23
Germanium.........	Solid	5.46	
Germanium.........	Single crystal	5.38*	
Gold...............	Vacuo-distilled solid	18.88	20
Gold...............	Cast	19.3	20
Gold...............	Single crystal	19.4*	
Hafnium...........	Solid	13.3	20
Hafnium...........	Single crystal	11.3*	
Helium.............	Solid	0.19	−273
Hydrogen...........	Solid	0.0763	−260
Indium.............	Solid	7.28	
Indium.............	Single crystal	7.43*	
Iodine.............	Solid	4.94	20
Iridium............	Solid	22.42	17
Iridium............	Single crystal	22.8*	
Iron...............	Pure solid	7.86	
Iron...............	Single crystal Fe-α	7.92*	
Krypton............	Solid	3.4	−273
Lanthanum.........	Solid	6.15	
Lead...............	Vacuo-distilled	11.342	20
Lead...............	Single crystal	11.48*	
Lithium............	Solid	0.534	20
Lithium............	Single crystal	0.534*	
Magnesium.........	Solid	1.74	20
Magnesium.........	Single crystal	1.71*	
Manganese.........	Solid	7.3	
Manganese.........	Single crystal Mn-α	7.21*	
Mercury............	Solid	14.193	−38.8
Molybdenum........	Solid	9.01	
Molybdenum........	Single crystal	10.20*	
Neodymium.........	Solid	7.00	

TABLE 2b-1. DENSITY OF THE ELEMENTS IN SOLID FORM (*Continued*)

Element	Physical state	Density, g/cm³	Temp., °C
Neon.................	Solid	1.204	−245
Nickel...............	Solid	8.8	
Nickel...............	Single crystal	9.04*	
Nitrogen.............	Solid	1.14	−273
Osmium..............	Solid	22.5	
Osmium..............	Single crystal	22.8*	
Oxygen..............	Solid	1.568	−273
Palladium............	Solid	12.16	
Palladium............	Single crystal	12.25*	
Phosphorus...........	Solid, white	1.83	
Phosphorus...........	Solid, red	2.20	
Phosphorus...........	Solid, black	2.69	
Platinum.............	Solid	21.37	
Platinum.............	Single crystal	21.5*	
Potassium............	Solid	0.87	20
Praseodymium........	Solid	6.48	20
Radium..............	Solid	5(?)	
Rhenium.............	Solid	20.53	
Rhodium.............	Solid	12.44	
Rubidium............	Solid	1.53	20
Ruthenium...........	Solid	12.1	19
Samarium............	Solid	7.7–7.8	
Scandium............	Solid	3.02(?)	
Selenium.............	Solid	4.82	
Selenium.............	Single crystal	4.86*	
Silicon...............	Solid crystal	2.42	20
Silicon...............	Single crystal	2.32*	
Silver...............	Vacuo-distilled	10.492	20
Silver...............	Single crystal	10.49*	
Sodium..............	Solid	0.9712	20
Sodium..............	Single crystal	0.954*	
Strontium............	Solid	2.60	
Sulfur...............	Solid, rhombic	2.07	
Sulfur...............	Solid, monoclinic	1.96	
Sulfur...............	Single crystal	2.02*	
Tantalum.............	Solid	16.6	
Tantalum.............	Single crystal	17.1*	
Tellurium............	Solid, crystal	6.25	
Tellurium............	Single crystal	6.26*	
Thallium.............	Solid	11.86	
Thallium.............	Single crystal	11.7*	
Thorium.............	Solid	11.00	17
Thorium.............	Single crystal	12.0*	
Tin.................	Solid, white tetragonal	7.29	20
Tin.................	Solid, white rhombic	6.55	
Tin.................	Solid, gray	5.75	20

TABLE 2b-1. DENSITY OF THE ELEMENTS IN SOLID FORM (*Continued*)

Element	Physical state	Density, g/cm^3	Temp.. °C
Tin.................	White single crystal	7.30*	
Titanium............	Solid	4.5	18
Titanium............	Single crystal	4.58*	
Tungsten............	Solid	19.3	
Tungsten............	Single crystal	19.3*	
Uranium.............	Solid	18.7	13
Vanadium...........	Solid	5.87	15
Vanadium...........	Single crystal	5.98*	
Yttrium.............	Solid	3.8	
Zinc.................	Solid, vacuo-distilled	6.92	20
Zinc.................	Solid	4.32	−273
Zinc.................	Single crystal	7.04*	
Zirconium...........	Solid	6.44	
Zirconium...........	Single crystal	6.47*	

TABLE 2b-2. DENSITY OF COMMON SOLIDS AT 20°C*

Substance	Density, g/cm³	Substance	Density, g/cm³
Agate...................	2.5–2.7	Gypsum.................	2.31–2.33
Amber..................	1.06–1.11	Hematite...............	4.9–5.3
Anthracite..............	1.4–1.8	Hornblende.............	3.0
Aragonite...............	2.93	Ice....................	0.917
Asbestos................	2.0–2.8	Ivory..................	1.83–1.92
Basalt..................	2.4–3.1	Lava, basaltic...........	2.8–3.0
Beeswax................	0.96–0.97	Lava, trachytic..........	2.0–2.7
Beryl...................	2.69–2.7	Leather, dry...........	0.86
Bone...................	1.7–2.0	Leather, greased.........	1.02
Brick...................	1.4–2.2	Lime, mortar............	1.65–1.78
Butter..................	0.86–0.87	Lime, slaked............	1.3–1.4
Calcite.................	2.71	Limestone..............	2.68–2.76
Camphor................	0.99	Magnetite..............	4.9–5.2
Caoutchouc.............	0.92–0.99	Malachite..............	3.7 4.1
Celluloid...............	1.4	Marble.................	2.6–2.84
Cement (set)............	2.7–3.0	Mica...................	2.6–3.2
Chalk..................	1.9–2.8	Olivine.................	3.27–3.37
Charcoal, oak...........	0.57	Opal...................	2.2
Charcoal, pine..........	0.28–0.44	Paper..................	0.7–1.15
Cinnabar...............	8.12	Paraffin................	0.87–0.91
Clay...................	1.8–2.6	Pitch..................	1.07
Coal, soft..............	1.2–1.5	Porcelain...............	2.3–2.5
Coke...................	1.0–1.7	Pyrite.................	4.95–5.1
Cork...................	0.22–0.26	Quartz.................	2.65
Cork linoleum..........	0.55	Resin..................	1.07
Corundum..............	3.9–4.0	Rock salt..............	2.18
Dolomite...............	2.84	Rubber, hard...........	1.19
Ebonite................	1.15	Rubber, soft...........	1.1
Emery..................	4.0	Rutile.................	4.2
Feldspar................	2.55–2.75	Sandstone..............	2.19–2.36
Flint...................	2.63	Slate..................	2.6–3.3
Fluorite................	3.18	Soapstone..............	2.6–2.8
Garnet.................	3.15–4.3	Starch.................	1.53
Gelatin.................	1.27	Sugar..................	1.61
Glass, common..........	2.4–2.8	Talc...................	2.7–2.8
Glass, flint.............	2.9–5.9	Tallow.................	0.91–0.97
Glue...................	1.27	Tar....................	1.02
Granite................	2.64–2.76	Topaz.................	3.5–3.6
Graphite...............	2.30–2.72	Tourmaline.............	3.0–3.2
Gum arabic.............	1.3–1.4	Wax, sealing...........	1.8

* The density varies with the state and previous treatment of the solids. The figures quoted may be considered reasonable limits (taken largely from "Smithsonian Physical Tables," 9th ed.).

TABLE 2b-3. DENSITY OF STEELS*
(At room temperature)

Type of steel	ρ, g/cm³	Composition				Condition
		% C	% Si	% Mn	% Cr	
Carbon steel...........	7.871	0.06	0.01	0.38	....	Annealed at 1700°F
Carbon steel...........	7.859	0.23	0.11	0.635	....	Annealed at 1700°F
Carbon steel...........	7.844	0.435	0.20	0.69	....	Annealed at 1580°F
Carbon steel...........	7.830	1.22	0.16	0.35	....	Annealed at 1470°F
Low-Cr steel...........	7.84	0.31	....	0.74	1.00	Oil-quenched at 1650°F, tempered at 1350°F
Low-Cr steel...........	7.84	0.315	....	0.69	1.09	Annealed at 1580°F
Low-Cr steel...........	7.83	0.35	....	0.24	1.56	Annealed at 1580°F
Low-Cr steel...........	7.80	1.73	....	0.30	1.65	Annealed at 1580°F
Low-Cr steel...........	7.82	0.80	....	0.28	1.67	Annealed at 1580°F
Low-Cr steel...........	7.82	0.62	....	0.22	1.67	Annealed at 1580°F
Low-Cr steel...........	7.81	0.98	....	0.28	1.68	Annealed at 1580°F
Low-Cr steel...........	7.84	0.20	....	0.14	1.85	Oil-quenched at 1650°F, tempered at 1380°F
Low-Cr steel...........	7.82	0.22	....	0.10	2.80	Oil-quenched at 1650°F, tempered at 1380°F
Low-Cr steel...........	7.81	0.21	....	0.19	3.88	Oil-quenched at 1650°F, tempered at 1380°F
Low-Cr steel...........	7.79	0.30	....	0.08	5.54	Oil-quenched at 1650°F, tempered at 1380°F
Low-Cr steel...........	7.845	0.35	....	0.59	0.88 + 0.20 Mo	Annealed at 1580°F, tempered at 1185°F
					% Ni	
Low-alloy Ni-Cr steel...	7.85	0.33	....	0.53	0.80	3.38	Annealed at 1580°F, tempered at 1185°F
Low-alloy Ni-Cr steel...	7.85	0.325	....	0.55	0.71	3.41	Annealed at 1580°F, tempered at 1185°F
Low-alloy Ni-Cr steel...	7.92	1.28	....	0.24	1.80	3.46	Brine quenched at 2190°F
Low-alloy Ni-Cr steel...	7.82	1.28	....	0.24	1.80	3.46	Annealed at 1435°F
Low-alloy Ni-Cr steel...	7.855	0.325	....	0.55	0.17	3.47	Annealed at 1580°F
Low-alloy Ni-Cr steel...	7.835	0.51	....	0.22	1.72	3.52	Annealed at 1435°F
Low-alloy Ni-Cr steel...	7.86	0.34	....	0.55	0.78	3.53 + 0.39 Mo	Annealed at 1580°F, tempered at 1185°F

	ρ, g/cm³	% C	% Cr	% Ni	% Mo	% Zr	% Ti	% Cu	% Mn	Condition
Wrought stainless and heat-resisting steels...	7.93	0.10	18	9						
Wrought stainless and heat-resisting steels...	7.93	....	18	9	0.5					
Wrought stainless and heat-resisting steels...	7.98	....	23	13						
Wrought stainless and heat-resisting steels...	7.98	....	25	20.5						
Wrought stainless and heat-resisting steels...	7.98	....	17	12	2.25					
Wrought stainless and heat-resisting steels...	8.02	....	18	10.5						
Wrought stainless and heat-resisting steels...	7.75	....	12.5							
Wrought stainless and heat-resisting steels...	7.73	....	13	......	0.5					

* "Metals Handbook," 48th ed., American Society for Metals.

TABLE 2b-3. DENSITY OF STEELS (*Continued*)

Type of steel	ρ, g/cm³	Composition								Condition
		% C	% Cr	% Ni	% Mo	% Zr	% Ti	% Cu	% Mn	
Wrought stainless and heat-resisting steels...	7.70	....	13							
Wrought stainless and heat-resisting steels...	7.70	....	16							
Wrought stainless and heat-resisting steels...	7.68	....	17	......	0.6					
Wrought stainless and heat-resisting steels...	7.60	....	25							
Wrought stainless and heat-resisting steels...	7.77	....	17.88	......	....	..	..	....	8.26	
Wrought stainless and heat-resisting steels...	7.76	....	17.55	......	....	..	..	....	10.48	
Wrought stainless and heat-resisting steels...	7.91	....	18.40	4.07	....	..	..	0.78	5.33	
Wrought stainless and heat-resisting steels...	7.90	....	18.50	4.06	....	..	..	....	6.79	
Wrought stainless and heat-resisting steels...	7.78	....	18.04	2.06	....	..	..	....	7.90	
Wrought stainless and heat-resisting steels...	7.77	....	17.70	......	....	..	..	0.68	9.40	

	ρ, g/cm³	% W	% Cr	% V	% Mo	% Co	% C	Condition
Tool steel....................	8.67	18	4	1				
Tool steel....................	8.67	18	4	2				
Tool steel....................	7.925	1.64	3.68	1.00	8.24	....	0.80	Quenched at 2200°F
Tool steel....................	7.93	5.20	4.60	4.00	4.11	....	1.32	Hardened
Tool steel....................	7.76	....	4.39	4.10	7.75	....	1.20	Hardened
Tool steel..	8.89	20	4	2	....	12	....	Annealed
Tool steel....................	8.68	18	4	1	....	5	....	Annealed
Tool steel....................	8.16	6	......	2	5	....	....	Annealed
Tool steel....................	7.88	1.5	......	1	8			

	ρ, g/cm³	% Ni	% Al	% Co	% Cu			Condition
Permanent-magnet alloys.......	6.892	20	12	5	....	....	....	Alnico
Permanent-magnet alloys.......	7.086	17	10	12.5	6	....	....	Cast Alnico
Permanent-magnet alloys.......	6.892	25	12					
Permanent-magnet alloys.......	7.003	28	12	5				
Permanent-magnet alloys.......	7.307	14	8	24	3			
Permanent-magnet alloys.......	7.197	18	6	35	....	....	....	8% Ti

	ρ, g/cm³	% Ni	% C	% Mn				Condition
Miscellaneous ferrous alloys.....	8.16	28.37	......	......	....	....	....	Quenched at 1740°F
Miscellaneous ferrous alloys.....	8.00	36	......	......	....	....	....	Invar
Miscellaneous ferrous alloys.....	8.3	45	......	......	....	....	....	Radio metal
Miscellaneous ferrous alloys.....	8.25	50	......	......	....	....	....	Hipernik
Miscellaneous ferrous alloys.....	7.87	.....	1.2	13	....	....	....	Austenitic manganese steel. Air-cooled at 1920°F

TABLE 2b-4. DENSITY OF ALUMINUM ALLOYS*
(At 20°C)

Material	ρ, g/cm³	% Al	% Mn	% Cu	% Pb	% Bi	% Mg	% Si	% Ni	% Cr	% Zn
Wrought alloys:											
Pure aluminum..	2.6989	99.996									
(Commercially pure Al) 2S...	2.71	99.0+									
3S	2.73	98.8	1.2								
11S	2.82	93.5	....	5.5	0.5	0.5					
R-317	2.81	93.8	0.6	4.0	0.5	0.5	0.6				
14S	2.80	93.6	0.8	4.4	...	...	0.4	0.8			
R-30I (clad)	2.78	93.3	0.8	4.5	...	...	0.4	1.0			
17S	2.79	95.0	0.5	4.0	...	...	0.5				
18S	2.80	93.5	....	4.0	...	...	0.5				
24S	2.77	93.4	0.6	4.5	...	...	1.5				
25S	2.79	93.9	0.8	4.5	...	...	...	0.8			
32S	2.69	84.7	....	0.9	...	...	1.0	12.5	0.9		
A51S	2.69	98.15	....	....	...	...	0.6	1.0	...	0.25	
52S	2.68	97.25	....	....	...	...	2.5	....	...	0.25	
53S	2.69	97.75	....	....	...	...	1.3	0.7	...	0.25	
56S	2.64	94.6	0.1	....	...	...	5.2	....	...	0.10	
61S	2.70	97.9	....	0.25	...	...	1.0	0.6	...	0.25	
75S	2.80	90.0	0.20	1.5	...	...	2.5	....	...	0.30	5.5
R-303	2.82	89.9	....	1.2	...	...	2.5	....	...	....	6.4

Material	ρ, g/cm³	% Al	% Mn	% Mg	% Cu	% Zn	% Cr	% Si	% Ni	% Bi	% Sn	% Ti
Casting alloys:												
13 alloy	2.66	88	...	....	....	...	...	12				
43 alloy	2.69	95	...	....	....	...	...	5				
85 alloy	2.78	91	...	....	4	...	...	5				
108 alloy	2.79	93	...	....	4	...	...	3				
Allcast	2.76	92	...	...	3	...	...	5				
A108 alloy	2.79	90	...	....	4.5	...	...	5.5				
113 alloy	2.91	89.3	...	....	7	1.7	...	2				
C113 alloy	2.91	89.5	...	....	7	...	...	3.5				
122 alloy	2.95	89.8	...	0.2	10							
A132 alloy	2.68	83.5	...	1.2	0.8	...	...	12	2.5			
Red X-13	2.7	85.1	0.7	0.7	1.5	...	...	12				
142 alloy	2.81	92.5	...	1.5	4	...	...	....	2			
195 alloy	2.81	95.5	...	....	4.5							
B195 alloy	2.78	93.0	...	....	4.5	...	...	2.5				
214 alloy	2.65	96.2	...	3.8								
A214 alloy	2.65	94.4	...	3.8	....	1.8						
218 alloy	2.53	92.0	...	8								
220 alloy	2.58	90.0	...	10								
319 alloy	2.77	90.5	...	....	3.5	...	...	6				
355 alloy	2.70	93.2	...	0.5	1.3	...	...	5				
356 alloy	2.68	92.7	...	0.3	....	...	...	7				
Red X-8	2.73	89.9	0.3	0.3	1.5	...	...	8				
360 alloy	2.68	90.0	...	0.5	....	...	...	9.5				
380 alloy	2.76	88.0	...	...	3.5	...	...	8.5				
750 alloy	2.89	91.5	...	...	1.0	...	...	....	1.0	...	6.5	
40E alloy	2.81	93.2	...	0.6	....	5.5	0.5	....	...	...	...	0.2

* "Metals Handbook," 48th ed., American Society for Metals.

TABLE 2b-5. DENSITY OF COBALT ALLOYS*

Material	ρ, g/cm^3	% Co	% W	% Ni	% Cr	% Mo	% Cb	% Fe
Pure cobalt..........	8.9	100						
61 alloy (cast)........	8.54	70.0	5.0	2.0	23.0			
Vitallium.............	8.30	65.0	...	2.0	27.0	6.0		
X-40 alloy...........	8.61	60.0	7.0	10.0	23.0			
422-19 alloy..........	8.31	55.0	...	16.0	23.0	6.0		
S-816 alloy..........	8.59	50.0	4.0	20.0	19.0	...	4.0	3.0
6059................	8.21	39.0	...	32.0	23.0	6.0		

* "Metals Handbook," 48th ed., American Society for Metals.

MECHANICS

TABLE 2b-6. DENSITY OF COPPER ALLOYS*

Material	ρ, g/cm³	% Cu	% O	% P	% Zn	% Pb	% Sn	% Fe	% Mn	% Al	% Ni	% Si	% Be
Wrought alloys:													
Pure copper	8.96	100											
Electrolytic tough-pitch copper	8.89–8.94	99.92	0.04										
Deoxidized copper	8.94	99.94		0.02									
Gilding metal	8.86	95.0			5.0								
Commercial bronze	8.80	90.0			10.0								
Red brass	8.75	85.0			15.0								
Low brass	8.67	80.0			20.0								
Cartridge brass	8.53	70.0			30.0								
Yellow brass	8.47	65.0			35.0								
Muntz metal	8.39	60.0			40.0								
Leaded commercial bronze	8.83	89.0			9.25	1.75							
Low-leaded brass	8.47	64.5			35.0	0.5							
Low-leaded brass (tube)	8.50	67.0			32.5	0.5							
Medium-leaded brass	8.47	64.5			34.5	1.0							
High-leaded brass	8.47	62.5			35.75	1.75							
Extra-high-leaded brass	8.50	62.5			35.0	2.5							
Free-cutting brass	8.50	61.5			35.5	3.0							
Leaded muntz metal	8.41	60.0			39.5	0.5							
Free-cutting muntz metal	8.41	60.5			38.4	1.1							
Forging brass	8.44	60.0			38.0	2.0							
Architectural bronze	8.47	57.0			40.0	3.0							
Admiralty metal	8.53	71.0			28.0		1.00						
Naval brass	8.41	60.0			39.25	0.75	0.75						
Leaded naval brass	8.44	60.0			37.5	1.75	0.75						
Manganese bronze	8.53	58.5			39.0		1.00	1.4	0.1				

Material	Density									
Aluminum brass	8.33	76.0	22.0				2			
Aluminum brass	8.33									
Phosphor bronze	8.86	95.0			5.0					
Phosphor bronze 8% grade C	8.80	92.0			8.0					
Phosphor bronze 10% grade D	8.78	90.0			10.0					
Phosphor bronze 1.25% grade E	8.89	98.75			1.25					
Cupronickel, 30%	8.94	70.0						30.0		
Nickel silver, 18% alloy A	8.73	65.0	17.0					18.0		
Ni-Ag, 18%, alloy B	8.70	55.0	27.0					18.0		
Silicon bronze, type A	8.53	97.0							3.0	
Silicon bronze, type B	8.75	98.5							1.5	
5% aluminum bronze	8.17	95.0					5.0			
8% aluminum bronze	?	92.0					8.0			
10% aluminum bronze	7.58	90.0					10.0			
Aluminum bronze	7.58	82.5				2.50	10.0	5.0		
Constantan	8.9	55.0						45.0		
Beryllium copper	8.23 ± 0.02	97.65						0.35		2.0
Casting alloys (room temp.):										
Leaded tin bronze	8.7	88.0	4.5	1.5	6.0					
Leaded tin bearing bronze	8.80	87.0	4.0	1.0	8.0					
High-leaded tin bronze	8.87	85.0	1.0	9.0	5.0					
High-leaded tin bronze	8.93	83.0	3.0	7.0	7.0					
High-leaded tin bronze	8.80	80.0		10.0	10.0					
High-leaded tin bronze	9.25	78.0		15.0	7.0					
High-leaded tin bronze	9.30	70.0		25.0	5.0					
85-5-5-5	8.80	85.0	5.0	5.0	5.0					
Leaded red brass	8.6	83.0	7.0	6.0	4.0					
Leaded semired brass	8.70	81.0	9.0	7.0	3.0					
Leaded semired brass	8.6	76.0	15.0	6.0	3.0					
Leaded yellow brass	8.50	71.0	25.0	3.0	1.0					
Leaded yellow brass	8.4	66.0	30.0	3.0	1.0					

* "Metals Handbook," 1948 ed., American Society for Metals.

TABLE 2b-6. DENSITY OF COPPER ALLOYS* (Continued)

Material	ρ, g/cm³	% Cu	% O	% P	% Zn	% Pb	% Sn	% Fe	% Mn	% Al	% Ni	% Si	% Be
Leaded yellow brass	8.40	60.0	....	....	38.0	1.0	1.0						
High-strength yellow brass	7.9	62.0	....	....	26.0	....	....	3.0	3.5	5.5			
High-strength yellow brass	8.2	58.0	....	....	39.25	....	....	1.25	0.25	1.25			
Leaded manganese brass	8.2	59.0	....	....	37.0	....	0.75	1.25	0.50	0.75			
Nickel silver	8.8–8.9	66.0	....	....	2.0	1.5	5.0	....	....	....	25.0		
Nickel silver	8.85	64.0	....	....	8.0	4.0	4.0	....	....	....	20.0		
Nickel silver	8.95	57.0	....	....	20.0	9.0	2.0	....	....	....	12.0		
Leaded nickel brass	8.95	60.0	....	....	16.0	5.0	3.0	....	....	....	16.0		
Aluminum bronze	?	89.0	....	....	....	....	....	1.0	....	10.0			
Aluminum bronze	7.4	87.5	....	....	....	....	....	3.5	....	9.0			
Aluminum bronze	7.5	86.0	....	....	....	....	....	4.0	....	10.0			
Aluminum bronze	?	79.0	....	....	....	....	....	5.0	....	11.0	5.0		

* "Metals Handbook," 48th ed., American Society for Metals.

TABLE 2b-7. DENSITY OF LEAD ALLOYS*

Material	ρ, g/cm³	% Pb	% Ca	% Sb	% Sn	% As	% Co
Pure lead.............	11.34	99.73					
Chemically pure lead....	11.34						
Cable-sheath alloy......	11.34	99.8	0.028				
1% antimonial lead.....	11.27	99.0	.....	1.0			
Hard lead..............	11.04	96.0	.....	4.0			
Hard lead..............	10.88	94.0	.....	6.0			
8% antimonial lead.....	10.74	92.0	.....	8.0			
Grid metal.............	10.66	91.0	.....	9.0			
ASTM-12 bearing metal.	10.67	90.0	.....	10.0			
ASTM-11 bearing metal.	10.28	85.0	.....	15.0			
Lead-base babbitt.......	10.24	85.0	.....	10.0	5.0		
G lead-base babbitt.....	10.1	83.0	.....	12.75	0.75	3.0	
S lead-base babbitt.....	10.1	83.0	.....	15.0	1.0	1.0	
ASTM-10 bearing metal.	10.07	83.0	.....	15.0	2.0		
Lead-base babbitt......	10.04	80.0	.....	15.0	5.0		
Lead-base babbitt......	9.73	75.0	.....	15.0	10.0		
ASTM-6 bearing metal..	9.33	63.5	.....	15.0	20.0	...	1.5
Tin-lead solder.........	11.0	95.0	.....	.....	5.0		
Tin-lead solder.........	10.2	80.0	.....	.....	20.0		
50-50 half and half......	8.89	50.0	.....	.....	50.0		

* "Metals Handbook," 48th ed., American Society for Metals.

TABLE 2b-8. DENSITY OF MAGNESIUM ALLOYS*

Material	ρ, g/cm³	% Mg	% Al	% Mn	% Zn	% Sn	Remarks
Magnesium...	1.74	99.8					
A10 alloy.....	1.81	89.9	10.0	0.1	...	...	Wrought, sand cast, and permanent-mold cast
AZ91 alloy....	1.81	....	9.0	0.2	0.7	...	Die cast
AZ92 alloy....	1.82	....	9.0	0.1	2.0	...	Sand cast and permanent-mold cast
A8 alloy......	1.80	....	8.0	0.2	...	...	Sand cast
AZ61X alloy..	1.80	....	6.0	0.2	1.0	...	Wrought
AM244 alloy..	1.76	....	4.0	0.2	...	...	Sand cast
AM11 alloy...	1.70	....	1.25	1	...	...	Die cast
AZ80X alloy..	1.80	....	8.5	0.15	0.5	...	Wrought
AZ63 alloy....	1.84	....	6.0	0.2	3.0	...	Sand cast
AZ51X alloy..	1.79	....	5.0	0.25	1.0	...	Wrought
AZ31X alloy..	1.78	....	3.0	0.3	1.0	...	Wrought
M1..........	1.76	....	.....	1.5	...	...	Wrought
TA54........	1.84	....	3.0	0.5	...	5.0	Wrought
Mg-Al alloy...	1.75	98.0	2.0				
Mg-Al alloy...	1.77	96.0	4.0				
Mg-Al alloy...	1.78	94.0	6.0				
Mg-Al alloy...	1.80	92.0	8.0				
Mg-Al alloy...	1.81	90.0	10.0				
Mg-Al alloy...	1.82	88.0	12.0				

* "Metals Handbook," 48th ed., American Society for Metals.

TABLE 2b-9. DENSITY OF NICKEL ALLOYS*

Material	$\rho,$ g/cm³	% Ni	% Co	% Si	% Mn	% C	% Al	% Cu	% Fe	% Mo	% Cr	% W
Nickel	8.902	99.95										
A nickel	8.885	99.4										
Cast nickel	8.34	97.0	..	1.5	0.5	0.5						
D nickel	8.78	95.2	..	...	4.5							
Z nickel	8.75	94	..	...	...	....	4.5					
Monel	8.84	67	..	...	1.0	0.15	...	30	1.4			
Cast monel	8.63	63	..	1.6	...	0.2	...	32				
K monel	8.47	66	..	...	...	....	3	29				
S monel	8.36	63	..	4	...	....	...	30	2			
Hastelloy A	8.80	60	..	...	...	....	...	..	20	20		
Hastelloy B	9.24	65	..	...	...	....	...	..	5	30		
Hastelloy C	8.94	58	..	...	...	....	...	..	5	17	15	5
Hastelloy D	7.8	85	..	8–11	...	....	...	3				
Illium G	8.58	58	..	...	...	0.2	...	6	6	6	22	
Inconel	8.51	80	..	...	...	....	...	..	6	..	14	
Cast Inconel	8.3	77.5	..	2	...	....	...	..	6	..	13.5	
Chromel A	8.4	80	..	...	...	....	...	..	....	..	20	
Nichrome	8.25	60	..	...	...	....	...	..	24	..	16	
Chromax	7.95	35	..	...	...	....	...	..	50	..	15	
Constantan (wrought)	8.9	45	..	...	...	....	...	55				
Ni-Fe alloys	8.8	90	..	...	...	....	...	..	10			
Ni-Fe alloys	8.6	80	..	...	...	....	...	..	20			
Ni-Fe alloys	8.5	70	..	...	...	....	...	..	30			
Ni-Fe alloys	8.35	60	..	...	...	....	...	..	40			
Permalloy	8.6	78	..	...	...	....	...	..	22			
Numetal	8.6	76	..	...	...	....	...	6	16		2	

* "Metals Handbook," 48th ed., American Society for Metals.

TABLE 2b-10. DENSITY OF ZINC ALLOYS*

Material	$\rho,$ g/cm³	% Zn	% Al	% Cu	% Mg	% Pb	% Cd
Zinc	7.133	100					
Zamak (2)	6.7	92	4	3	0.03		
Zamak (3)	6.6	95	4	..	0.04		
Zamak (5)	6.7	94	4	1	0.04		
SAE 63, T-11 (cast)	6.9	86	4	10			
Commercial rolled zinc	7.14	99	..	..	....	0.08	
Commercial rolled zinc	7.14	99	..	..	....	0.06	0.06
Commercial rolled zinc	7.14	99	..	..	....	0.3	0.3
Zilloy 40 (rolled)	7.18	98	..	1	....	0.08	
Zilloy 15 (rolled)	7.18	98	..	1	0.01	0.1	

* "Metals Handbook," 48th ed., American Society for Metals.

TABLE 2b-11. DENSITY OF WOODS (OVEN-DRY)*

Common name	Botanical name	ρ, g/cm³
Applewood or wild apple	Pyrus malus	0.745
Ash, black	Fraxinus nigra	0.526
Ash, blue	Fraxinus quadrangulata	0.603
Ash, green	Fraxinus pennsylvanica lanceolata	0.610
Ash, white	Fraxinus americana	0.638
Aspen	Populus tremuloides	0.401
Aspen, large-toothed	Populus grandidentata	0.412
Balsa, tropical American	Ochroma	0.12–0.20†
Basswood	Tilia glabra or Tilia americanus	0.398
Beech	Fagus grandifolia or Fagus americana	0.655
Beech, blue	Carpinus caroliniana	0.717
Birch, gray	Betula populifolia	0.552
Birch, paper	Betula papyrifera	0.600
Birch, sweet	Betula lenta	0.714
Birch, yellow	Betula lutea	0.668
Buckeye, yellow	Aesculus octandra	0.383
Butternut	Juglans cinera	0.404
Cedar, eastern red	Juniperus virginiana	0.492
Cedar, northern white	Thuja occidentalis	0.315
Cedar, southern white	Chamaecyparis thyoides	0.352
Cedar, tropical American	Cedrela odorata	0.37–0.70†
Cedar, western red	Thuja plicata	0.344
Cherry, black	Prunus serotine	0.534
Cherry, wild red	Prunus pennsylvanica	0.425
Chestnut	Castanea dentata	0.454
Corkwood	Leitneria floridana	0.207
Cottonwood, eastern	Populus deltoides	0.433
Cypress, southern	Taxodium distichum	0.482
Dogwood (flowering)	Cornus florida	0.796
Douglas fir (coast type)	Pseudotsuga taxifolia	0.512
Douglas fir (mountain type)	Pseudotsuga taxifolia	0.446
Ebony, Andaman marblewood (India)	Diospyros Kurzii	0.978†
Ebony, Ebene marbre (Mauritius, East Africa)	Diospyros melanida	0.768†
Elm, American	Ulmus americana	0.554
Elm, rock	Ulmus racemosa or Ulmus thomasi	0.658
Elm, slippery	Ulmus fulva or Ulmus pubescens	0.568
Eucalyptus, Karri (west Australia)	Eucalyptus diversicolor	0.829†
Eucalyptus, mahogany (New South Wales)	Eucalyptus hemilampra	1.058†
Eucalyptus, west Australian mahogany	Eucalyptus marginata	0.787†
Fir, balsam	Abies balsamea	0.414
Fir, silver	Abies amabilis	0.415
Greenheart (British Guiana)	Nectandra rodioci	1.06–1.23†

See page 2-35 for footnotes.

TABLE 2b-11. DENSITY OF WOODS (OVEN-DRY)* (Continued)

Common name	Botanical name	ρ, g/cm³
Gum, black....................	*Nyssa sylvatica*	0.552
Gum, blue.....................	*Eucalyptus globulus*	0.796
Gum, red......................	*Liquidambar styraciflua*	0.530
Gum, tupelo...................	*Nussa aquatica*	0.524
Hemlock, eastern..............	*Tsuga canadensis*	0.431
Hemlock, mountain.............	*Tsuga martensiana*	0.480
Hemlock, western..............	*Tsuga heterophylla*	0.432
Hickory, bigleaf shagbark........	*Hicoria laciniosa*	0.809
Hickory, mockernut.............	*Hicoria alba*	0.820
Hickory, pignut................	*Hicoria glabra*	0.820
Hickory, shagbark..............	*Hicoria ovata*	0.836
Hornbeam.....................	*Ostryra virginiana*	0.762
Ironwood, black................	*Rhamnidium ferreum*	1.077
Jacaranda, Brazilian rosewood.....	*Dalbergia nigra*	0.85†
Larch, western.................	*Larix occidentalis*	0.587
Locust, black or yellow...........	*Robinia pseudacacia*	0.708
Locust, honey..................	*Gleditsia triacanthos*	0.666
Magnolia, cucumber.............	*Magnolia acuminata*	0.516
Mahogany (West Africa).........	*Khaya ivorensis*	0.668†
Mahogany (East India)..........	*Swietenia macrophylla*	0.54†
Mahogany (East India)..........	*Swietenia mahogani*	0.54†
Maple, black..................	*Acer nigrum*	0.620
Maple, red....................	*Acer rubrum*	0.546
Maple, silver..................	*Acer saccharinum*	0.506
Maple, sugar..................	*Acer saccharum*	0.676
Oak, black....................	*Quercus velutina*	0.669
Oak, bur.....................	*Quercus macrocarpa*	0.671
Oak, canyon live...............	*Quercus chrysolepsis*	0.838
Oak, chestnut..................	*Quercus montana*	0.674
Oak, laurel...................	*Quercus laurifolia*	0.703
Oak, live.....................	*Quercus virginiana*	0.977
Oak, pin.....................	*Quercus palustris*	0.677
Oak, post....................	*Quercus stellata* or *Quercus minor*	0.738
Oak, red.....................	*Quercus borealis*	0.657
Oak, scarlet..................	*Quercus coccinea*	0.709
Oak, swamp chestnut............	*Quercus prinus*	0.756
Oak, swamp white..............	*Quercus bicolor* or *Quercus platanoides*	0.792
Oak, white....................	*Quercus alba*	0.710
Persimmon....................	*Diospyros virginiana*	0.776
Pine, eastern white.............	*Pinus strobus*	0.373
Pine, jack....................	*Pinus banksiana* or *Pinus divaricata*	0.461
Pine, loblolly..................	*Pinus taeda*	0.593
Pine, longleaf.................	*Pinus palustris*	0.638
Pine, pitch...................	*Pinus rigida*	0.542
Pine, red.....................	*Pinus resinosa*	0.507

See page 2–35 for footnotes.

TABLE 2b-11. DENSITY OF WOODS (OVEN-DRY)* (*Continued*)

Common name	Botanical name	ρ, g/cm³
Pine, shortleaf...................	*Pinus echinata*	0.584
Poplar, balsam...................	*Populus balsamifera* or *Populus candicans*	0.331
Poplar, yellow...................	*Liriodendron tulipifera*	0.427
Redwood........................	*Sequoia sempervivens*	0.436
Sassafras.......................	*Sassafras variafolium*	0.473
Satinwood (Ceylon)..............	*Chloroxylon swietenia*	1.031†
Sourwood.......................	*Oxydendrum arboreum*	0.593
Spruce, black....................	*Picea mariana*	0.428
Spruce, red.....................	*Picea rubra* or *Picea rubens*	0.413
Spruce, white...................	*Picea glauca*	0.431
Sycamore.......................	*Platanus occidentalis*	0.539
Tamarack.......................	*Larix laricina* or *Larix americana*	0.558
Teak (India)....................	*Tectona grandis*	0.582†
Walnut, black...................	*Juglans nigra*	0.562
Willow, black...................	*Salix nigra*	0.408

* "Handbook of Chemistry and Physics," 30th ed.
† Air-dry.

TABLE 2b-12. DENSITY OF PLASTICS*

Resin group and subgroup	Trade names	ρ, g/cm³ Lower limit	ρ, g/cm³ Upper limit
Acrylate and methacrylate..........	Lucite, Crystalite, Plexiglas	1.16	1.20
Casein............................	Ameroid	1.34	1.35
Cellulose acetate (sheet)............	Bakelite, Lumarith, Plastecele, Protectoid	1.27	1.60
Cellulose acetate (molded)..........	Fibestos, Hercules, Nixonite, Tenite	1.27	1.60
Cellulose acetobutyrate.............	Tenite II	1.14	1.23
Cellulose nitrate..................	Celluloid, Nitron, Nixonoid, Pyralin	1.35	1.60
Ethyl cellulose....................	Ditzler, Ethocel, Ethofoil, Lumarith, Nixon, Hercules	1.05	1.25
Phenol-formaldehyde compounds:			
Wood-flour-filled (molded)........	Bakelite, Durez, Durite, Micarta, Catalin, Haveg, Indur, Makalot, Resinox, Textolite, Formica	1.25	1.52
Mineral-filled (molded)...........	Bakelite, Durez, Durite, Micarta, Catalin, Haveg, Indur, Makalot, Resinox, Textolite, Formica	1.59	2.09
Macerated-fabric-filled (molded)...	Bakelite, Durez, Durite, Micarta, Catalin, Haveg, Indur, Makalot, Resinox, Textolite, Formica	1.36	1.47
Paper-base (laminated)...........	Bakelite, Durez, Durite, Micarta, Catalin, Haveg, Indur, Makalot, Resinox, Textolite, Formica	1.30	1.40
Fabric base (laminated)..........	Bakelite, Durez, Durite, Micarta, Catalin, Haveg, Indur, Makalot, Resinox, Textolite, Formica	1.30	1.40
Cast (unfilled)..................	Bakelite, Catalin, Gemstone, Marblette, Opalon, Prystal	1.20	1.10
Phenolic furfural (filled)...........	Durite	1.3	2.0
Polyvinyl acetals (unfilled).........	Alvar, Formvar, Saflex, Butacite, Vinylite X, etc.	1.05	1.23
Polyvinyl acetate.................	Gelva, Vinylite A, etc.	1.19	(?)
Copolyvinyl chloride acetate........	Vinylite V, etc.	1.34	1.37
Polyvinyl chloride (and copolymer) plasticized.....................	Koroseal, Vinylite	1.2	1.7
Polystyrene......................	Bakelite, Loalin, Lustron, Styron	1.054	1.070

* "Handbook of Chemistry and Physics," 30th ed., p. 1282.

TABLE 2b-12. DENSITY OF PLASTICS (*Continued*)

Resin group and subgroup	Trade names	ρ, g/cm³	
		Lower limit	Upper limit
Modified isomerized rubber........	Plioform, Pliolite	1.06	(?)
Chlorinated rubber...............	Torneseit, Parlon	1.64	(?)
Urea formaldehyde...............	Bakelite, Beetle, Plascon	1.45	1.55
Melamine formaldehyde filled.......	Catalin, Melmac, Plaskon	1.49	1.86
Vinylidene chloride...............	Saran, Velon	1.68	1.75

TABLE 2b-13. DENSITY OF RUBBERS*

Rubber; raw polymer	Trade Name	At 25°C
Natural rubber......................	Hevea	0.92
Butadienestyrene copolymer..........................	....	0.94
Butadieneacrylonitrile copolymer........	....	1.00
Polychloroprene (neoprene)...............	....	1.25
Isobutylenediolefin copolymer (butyl).................................	....	0.91
Alkylene polysulfide...................	....	1.35

* "Handbook of Chemistry and Physics," 30th ed., p. 1282.

2c. Centers of Mass and Moments of Inertia

R. BRUCE LINDSAY

Brown University

TABLE 2c-1. CENTERS OF MASS*

Body	Center of Mass
1. Uniform circular wire of radius R, subtending angle 2θ at center	On axis of symmetry distant $(R \sin \theta)/\theta$ from center
2. Uniform triangular sheet	At intersection of the medians
3. Uniform rectangular sheet	At intersection of the diagonals
4. Uniform quadrilateral sheet	From each vertex lay off segments equal to $\frac{1}{3}$ the length of the corresponding sides meeting at this vertex. Draw extended lines through the ends of the segments associated with each vertex, respectively. These intersect to form a parallelogram. The intersection of the diagonals of this parallelogram is the center of mass of the quadrilateral
5. Uniform circular sector sheet of radius R subtending angle 2θ at center of circular arc	On axis of symmetry distant $(2R \sin \theta)/3\theta$ from center
6. Uniform circular segment sheet of radius R, subtending angle 2θ at center of circular arc and length of chord equal to $l = 2R \sin \theta$	On axis of symmetry distant $l^3/12A$ from center, where A = area of segment $= \dfrac{R^2(2\theta - \sin 2\theta)}{2}$
7. Uniform semielliptical sheet, major and minor axes of equivalent ellipse equal to $2a$ and $2b$, respectively	On axis of symmetry distant $4a/3\pi$ from center of equivalent ellipse if the semiellipse is bounded by minor axis. The distance is $4b/3\pi$ if the semiellipse is bounded by the major axis
8. Uniform quarter-elliptical sheet, major and minor axes of equivalent ellipse equal to $2a$ and $2b$, respectively	At point $4b/3\pi$ above major axis and $4a/3\pi$ above minor axis
9. Uniform parabolic sheet segment. Chord = $2l$ perpendicular to axis of symmetry distant h from vertex	On axis of symmetry distant $3h/5$ from vertex
10. Right rectangular pyramid (rectangular base with sides a and b and with height h)	On axis of symmetry distant $h/4$ from base
11. Pyramid (general)	On line joining apex with center of symmetry of base at distance three-quarters of its length from apex

* For definition see Sec. 2a-4. All bodies cited are homogeneous rigid bodies.

TABLE 2c-1. CENTERS OF MASS (*Continued*)

Body	Center of Mass
12. Frustum of pyramid with area of larger base S and smaller base s, and altitude h	On line joining apex of corresponding pyramid with center of symmetry of larger base and distant $$\frac{h(S + 2\sqrt{Ss} + 3s)}{4(S + \sqrt{Ss} + s)}$$ from the larger base
13. Right circular cone (height h)	On axis of symmetry distant $h/4$ from base
14. Frustum of right circular cone (altitude h, radii of larger and smaller bases R and r, respectively)	On axis of symmetry distant $$\frac{h[(R + r)^2 + 2r^2]}{4[(R + r)^2 - Rr]}$$ from the base
15. Cone (general)	On line joining apex with centroid of base at distance three-quarters of its length from apex
16. Frustum of cone with altitude h and radii of larger and smaller bases R and r, respectively	On line joining apex of corresponding cone with centroid of larger base and distant $$\frac{h[(R + r)^2 + 2r^2]}{4[(R + r)^2 - Rr]}$$ from the larger base
17. Spherical sector of radius R, with plane vertex angle equal to 2θ	On axis of symmetry distant $$\frac{3R}{8}(1 + \cos\theta)$$ from the vertex
18. Solid hemisphere of radius R	On axis of symmetry distant $3R/8$ from center of corresponding sphere
19. Spherical segment of radius R and maximum height from base equal to h	On axis of symmetry distant $\dfrac{h(4R - h)}{4(3R - h)}$ above the base of the segment
20. Octant of ellipsoid with semiaxes a, b, c, respectively, and center of corresponding ellipsoid at origin of system of rectangular coordinates	Point with coordinates $$\bar{x} = \frac{3a}{8} \qquad \bar{y} = \frac{3b}{8} \qquad \bar{z} = \frac{3c}{8}$$
21. Paraboloid of revolution with altitude h and radius of circular base equal to R	On axis of symmetry distant $h/3$ from the base
22. Uniform hemispherical shell of radius R (excluding base)	On axis of symmetry distant $R/2$ from center of corresponding sphere
23. Conical shell (excluding base)	On line joining the apex with the center of symmetry of the base at distance two-thirds its length from the apex

TABLE 2c-2. MOMENTS OF INERTIA*

Body	Axis	Moment of inertia
Uniform rectangular sheet of sides a and b	Through the center parallel to b	$m\dfrac{a^2}{12}$
Uniform rectangular sheet of sides a and b	Through the center perpendicular to the sheet	$m\dfrac{a^2+b^2}{12}$
Uniform circular sheet of radius r	Normal to the plate through the center	$m\dfrac{r^2}{2}$
Uniform circular sheet of radius r	Along any diameter	$m\dfrac{r^2}{4}$
Uniform circular ring, radii r_1 and r_2	Through center normal to plane of ring	$m\dfrac{r_1{}^2+r_2{}^2}{2}$
Uniform circular ring, radii r_1 and r_2	A diameter	$m\dfrac{r_1{}^2+r_2{}^2}{4}$
Uniform thin spherical shell, mean radius r	A diameter	$m\dfrac{2r^2}{3}$
Uniform cylindrical shell, radius r, length l	Longitudinal axis	mr^2
Right circular cylinder of radius r, length l	Longitudinal axis	$m\dfrac{r^2}{2}$
Right circular cone, altitude h, radius of base r	Axis of the figure	$m\dfrac{3}{10}r^2$
Spheroid of revolution, equatorial radius r	Polar axis	$m\dfrac{2r^2}{5}$
Ellipsoid, axes $2a$, $2b$, $2c$....	Axis $2a$	$m\dfrac{(b^2+c^2)}{5}$
Uniform thin rod..........	Normal to the length, at one end	$m\dfrac{l^2}{3}$
Uniform thin rod	Normal to the length, at the center	$m\dfrac{l^2}{12}$
Rectangular prism, dimensions $2a$, $2b$, $2c$	Axis $2a$	$m\dfrac{(b^2+c^2)}{3}$
Sphere, radius r	A diameter	$m\dfrac{2}{5}r^2$
Rectangular parallelepiped, edges a, b, and c	Through center perpendicular to face ab (parallel to edge c)	$m\dfrac{a^2+b^2}{12}$
Right circular cylinder of radius r, length l	Through center perpendicular to the axis of the figure	$m\left(\dfrac{r^2}{4}+\dfrac{l^2}{12}\right)$
Spherical shell, external radius r_1, internal radius r_2	A diameter	$m\dfrac{2}{5}\dfrac{(r_1{}^5-r_2{}^5)}{(r_1{}^3-r_2{}^3)}$
Hollow circular cylinder, length l, external radius r_1, internal radius r_2	Longitudinal axis	$m\dfrac{(r_1{}^2+r_2{}^2)}{2}$
Hollow circular cylinder, length l, radii r_1 and r_2	Transverse diameter	$m\left(\dfrac{r_1{}^2+r_2{}^2}{4}+\dfrac{l^2}{12}\right)$

* For definitions see Sec. 2a-5; m = mass of body. All bodies are homogeneous.

TABLE 2c-2. MOMENTS OF INERTIA (*Continued*)

Body	Axis	Moment of inertia
Hollow circular cylinder, length l, very thin, mean radius r	Transverse diameter	$m\left(\dfrac{r^2}{2} + \dfrac{l^2}{12}\right)$
Right elliptical cylinder, length $2a$, transverse axes $2b$, $2c$	Longitudinal axis $2a$ through center of mass	$m\dfrac{(b^2 + c^2)}{4}$
Right elliptical cylinder, length $2a$, transverse axes $2b$, $2c$	Transverse axis $2b$ through center of mass	$m\left(\dfrac{c^2}{4} + \dfrac{a^2}{3}\right)$
Frustum of right circular cone with radii of larger and smaller bases, equal to R and r, respectively	Axis of symmetry	$\dfrac{3m(R^5 - r^5)}{10(R^3 - r^3)}$
Right circular cone, radius of base r, altitude h	Perpendicular to axis of symmetry, through center of mass	$\dfrac{3m}{20}\left(r^2 + \dfrac{h^2}{4}\right)$
Solid hemisphere of radius r	Axis of symmetry	$\dfrac{2mr^2}{5}$
Spherical sector of radius r, with plane angle at vertex $= 2\theta$	Axis of symmetry through vertex	$\dfrac{mr^2(1 - \cos\theta)(2 + \cos\theta)}{5}$
Spherical segment of radius r and maximum height h	Axis of symmetry perpendicular to base	$m\left(r^2 - \dfrac{3rh}{4} + \dfrac{3h^2}{20}\right)\dfrac{2h}{3r - h}$
Torus or anchor ring mean radius R, radius of circular cross section r	Axis of symmetry perpendicular to plane of ring	$\dfrac{m(4R^2 + 3r^2)}{4}$
Torus mean radius R, radius of circular cross section r	Axis of symmetry in plane of ring	$\dfrac{m(4R^2 + 5r^2)}{8}$

2d. Coefficients of Friction

DUDLEY D. FULLER

Columbia University

Symbols

f_K coefficient of kinetic or sliding friction
f_R coefficient of rolling friction
f_s coefficient of static friction
P frictional resistance to rolling
r radius of roller
W load

2d-1. Static and Sliding Friction. All surfaces encountered in experience are more or less rough in the sense that as bodies move on them they exert forces parallel to the surface and in such direction as to resist motion. Such forces are termed "frictional." Frictional force is proportional to the normal thrust between body and surface; however, the coefficient of proportionality, known as the coefficient of friction, can for the same body and surface vary a great deal depending on the nature of the contact and the motion. It is customary to define

$$f_s = \frac{\text{magnitude of maximum frictional force}}{\text{magnitude of normal thrust}} \qquad (2d\text{-}1)$$

as the *coefficient of static friction* if motion is just on the point of starting. On the other hand, f_K, called the coefficient of *kinetic or sliding friction*, is the value of the ratio in Eq. (2d-1), when motion has once been established. In general $f_K < f_s$ for the same body and surface or the same two surfaces.

The friction between surfaces is dependent upon many variables. These include the nature of the materials themselves, surface finish and surface condition, atmospheric dust, humidity, oxide and other surface films, velocity of sliding, temperature, vibration, and extent of contamination.

In many instances the degree of contamination is perhaps the most important single variable. For example, Table 2d-1 lists values for the static coefficient of friction f_s for steel on steel under various test conditions.

TABLE 2d-1. COEFFICIENTS OF STATIC FRICTION FOR STEEL ON STEEL

Test condition	f_s	Ref.*
Degassed at elevated temp. in high vacuum.............	Weld on contact	20
Grease-free in vacuum................................	0.78	1
Grease-free in air...................................	0.39	8
Clean and coated with oleic acid......................	0.11	1
Clean and coated with solution of stearic acid...........	0.013	21

* References follow Table 2d-4.

The most effective lubricants for nonfluid lubrication are generally those which react chemically with the solid surface and form an adhering film that is attached to the surface with a chemical bond. This action depends upon the nature of the lubricant and upon the reactivity of the solid surface. Table 2d-2 indicates that a fatty acid such as those found in animal, vegetable, and marine oils reduces the coefficient

TABLE 2d-2. COEFFICIENTS OF STATIC FRICTION AT ROOM TEMPERATURE

Surfaces	Clean	Paraffin oil	Paraffin oil + 1% lauric acid	Degree of reactivity of solid
Nickel	0.7	0.3	0.28	Low
Chromium	0.4	0.3	0.3	Low
Platinum	1.2	0.28	0.25	Low
Silver	1.4	0.8	0.7	Low
Glass	0.9	....	0.4	Low
Copper	1.4	0.3	0.08	High
Cadmium	0.5	0.45	0.05	High
Zinc	0.6	0.2	0.04	High
Magnesium	0.6	0.5	0.08	High
Iron	1.0	0.3	0.2	Mild
Aluminum	1.4	0.7	0.3	Mild

of friction markedly only if it can react effectively with the solid surface. Paraffin oil is almost completely nonreactive. The data are taken from ref. 22.

It is generally recognized that coefficients of friction reduce on dry surfaces as sliding velocity increases. Dokos (ref. 4) has measured this for steel on steel. It is difficult to screen out the effect of temperature, however, which also increases with sliding velocity so that frequently, under these conditions, both variables are present. Table 2d-3 gives values which are the average of four tests at high contact pressures.

TABLE 2d-3. COEFFICIENTS OF FRICTION, STEEL ON STEEL, UNLUBRICATED

Velocity, in./sec	0.0001	0.001	0.01	0.1	1	10	100
Coefficient of friction f_K	0.53	0.48	0.39	0.31	0.23	0.19	0.18

Table 2d-4 presents typical values of the coefficients of static and sliding friction for various materials under a variety of conditions.

TABLE 2d-4. COEFFICIENTS OF STATIC AND SLIDING FRICTION*

Materials	Static friction		Sliding friction	
	Dry	Greasy	Dry	Greasy
Hard steel on hard steel..........	0.78(1)	0.11(1,a)	0.42(2)	0.029(5,h)
		0.23(1,b)	......	0.081(5,c)
		0.15(1,c)	......	0.080(5,i)
		0.11(1,d)	......	0.058(5,j)
		0.0075(18,p)	......	0.084(5,d)
		0.0052(18,h)	......	0.105(5,k)
				0.096(5,l)
				0.108(5,m)
				0.12(5,a)
Mild steel on mild steel...........	0.74(19)	....	0.57(3)	0.09(3,a)
				0.19(3,u)
Hard steel on graphite............	0.21(1)	0.09(1,a)		
Hard steel on babbitt (ASTM 1)...	0.70(11)	0.23(1,b)	0.33(6)	0.16(1,b)
		0.15(1,c)	......	0.06(1,c)
		0.08(1,d)	......	0.11(1,d)
		0.085(1,e)		
Hard steel on babbitt (ASTM 8)...	0.42(11)	0.17(1,b)	0.35(11)	0.14(1,b)
		0.11(1,c)	......	0.065(1,c)
		0.09(1,d)	......	0.07(1,d)
		0.08(1,e)	......	0.08(11,h)
Hard steel on babbitt (ASTM 10)..	......	0.25(1,b)	......	0.13(1,b)
		0.12(1,c)	......	0.06(1,c)
		0.10(1,d)	......	0.055(1,d)
Mild steel on cadmium silver......	......	....	......	0.097(2,f)
Mild steel on phosphor bronze.....	......	....	0.34(3)	0.173(2,f)
Mild steel on copper lead..........	......	....		0.145(2,f)
Mild steel on cast iron............	......	0.183(15,c)	0.23(6)	0.133(2,f)
Mild steel on lead................	0.95(11)	0.5(1,f)	0.95(11)	0.3(11,f)
Nickel on mild steel..............		....	0.64(3)	0.178(3,x)
Aluminum on mild steel...........	0.61(8)	....	0.47(3)	
Magnesium on mild steel..........	......	....	0.42(3)	
Magnesium on magnesium.........	0.6(22)	0.08(22,y)		

* Numbers in parentheses indicate references to data sources; letters identify lubricant in following list.

TABLE 2d-4. COEFFICIENTS OF STATIC AND SLIDING FRICTION (*Continued*)

Materials	Static friction		Sliding friction	
	Dry	Greasy	Dry	Greasy
Cadmium on mild steel............	....	....	0.46(3)	
Copper on mild steel.............	0.53(8)	....	0.36(3)	0.18(17,a)
Nickel on nickel.................	1.10(16)	0.28(22,y)	0.53(3)	0.12(3,w)
Brass on mild steel..............	0.51(8)	0.11(22,c)	0.44(6)	
Brass on cast iron...............	....	....	0.30(6)	
Zinc on cast iron................	0.85(16)	....	0.21(7)	
Magnesium on cast iron..........	....	....	0.25(7)	
Copper on cast iron..............	1.05(16)	....	0.29(7)	
Tin on cast iron.................	....	....	0.32(7)	
Lead on cast iron................	....	....	0.43(7)	
Aluminum on aluminum..........	1.05(16)	0.30(22,y)	1.4(3)	
Glass on glass...................	0.94(8)	0.35(22,y) 0.1(22,q)	0.4(3)	0.09(3,a)
Carbon on glass..................	....	....	0.18(3)	
Garnet on mild steel.............	....	....	0.39(3)	
Glass on nickel..................	0.78(8)	....	0.56(3)	
Copper on glass..................	0.68(8)	....	0.53(3)	
Cast iron on cast iron............	1.10(16)	0.2(22,y)	0.15(9)	0.070(9,d)
Bronze on cast iron..............	....	....	0.22(9)	0.077(9,n)
Oak on oak (parallel to grain)....	0.62(9)	....	0.48(9)	0.164(9,r) 0.067(9,s)
Oak on oak (perpendicular).......	0.54(9)	....	0.32(9)	0.072(9,s)
Leather on oak (parallel).........	0.61(9)	....	0.52(9)	
Cast iron on oak.................	....	....	0.49(9)	0.075(9,n)
Leather on cast iron.............	....	....	0.56(9)	0.36(9,t)
Teflon on Teflon.................	0.04(22)	....	0.04(22,f)	
Teflon on steel..................	0.04(22)	....	0.04(22,f)	
Fluted rubber bearing on steel....	....	....	....	0.05(13,t)
Laminated plastic on steel........	....	....	0.35(12)	0.05(12,t)
Tungsten carbide on tungsten carbide......................	0.2(22)	0.12(22,a)		
Tungsten carbide on steel........	0.5(22)	0.08(22,a)		

Materials	Sliding friction, dry
Nylon 6.6 on mild steel (no fibers).........................	0.40(23)
Nylon 6.6 on mild steel (30% by wt. carbon fibers)..........	0.35(23)
Copper-graphite (high copper) on hard steel................	0.40(23)
Copper-graphite (low copper) on hard steel.................	0.25(23)
Carbon-graphite(low graphite) on hard steel................	0.50(23)
Carbon-graphite (high graphite) on hard steel..............	0.25(23)
Carbon-Teflon on hard steel...............................	0.30(23)
Carbon-copper-Teflon on hard steel........................	0.29(23)

Lubricant References for Table 2d-4

a. Oleic acid
b. Atlantic spindle oil (light mineral)
c. Castor oil
d. Lard oil
e. Atlantic spindle oil plus 2 per cent oleic acid
f. Medium mineral oil
g. Medium mineral oil plus $\frac{1}{2}$ per cent oleic acid
h. Stearic acid
i. Grease (zinc oxide base)
j. Graphite
k. Turbine oil plus 1 per cent graphite
l. Turbine oil plus 1 per cent stearic acid

m. Turbine oil (medium mineral)
n. Olive oil
p. Palmitic acid
q. Ricinoleic acid
r. Dry soap
s. Lard
t. Water
u. Rape oil
v. 3-in-1 oil
w. Octyl alcohol
x. Triolein
y. 1 per cent lauric acid in paraffin oil

References for Table 2d-4

1. Campbell, W. E.: Studies in Boundary Lubrication, *Trans. ASME* **61** (7), 633–641 (1939).
2. Clark, G. L., B. H. Lincoln, and R. R. Sterrett: Fundamental Physical and Chemical Forces in Lubrication, *Proc. API* **16**, 68–80 (1935).
3. Beare, W. G., and F. P. Bowden: Physical Properties of Surfaces. 1, Kinetic Friction, *Trans. Roy. Soc.* (*London*), ser. A, **234**, 329–354 (June 6, 1935).
4. Dokos, S. J.: Sliding Friction under Extreme Pressures—1, *J. Appl. Mech.* **13**, A-148–156 (1946).
5. Boyd, J., and B. P. Robertson: The Friction Properties of Various Lubricants at High Pressures, *Trans. ASME* **67** (1), 51–56 (January, 1945).
6. Sachs, G.: Versuche über die Reibung fester Korper (Experiments about the Friction of Solid Bodies), *Z. angew. Math. Mech.* **4**, 1–32 (February, 1924).
7. Honda, K., and R. Yamada: Some Experiments on the Abrasion of Metals, *J. Inst. Metals* **33** (1), 49–69 (1925).
8. Tomlinson, G. A.: A Molecular Theory of Friction, *Phil. Mag.*, ser. 7, **7** (46), 905–939 (suppl., June, 1929).
9. Morin, A.: Nouvelles expériences sur le frottement (New Experiments on Friction) *Acad. roy. sci.*, Paris (a) **57**, 128 (1832); (b) **59**, 104 (1834); (c) **60**, 143 (1835); (d) **63**, 99 (1838).
10. Claypoole, W.: Static Friction, *Trans. ASME* **65**, 317–324 (May, 1943).
11. Tabor, D.: The Frictional Properties of Some White-metal Bearing Alloys: The Role of the Matrix and Hard Particles, *J. Appl. Phys.* **16** (6), 325–337 (June, 1945).
12. Eyssen, G. R.: Properties and Performance of Bearing Materials Bonded with Synthetic Resin, General Discussion on Lubrication and Lubricants, *Inst. Mech. Engrs.*, *J.* **1**, 84–92 (1937).
13. Brazier, S. A., and W. Holland-Bowyer: Rubber as a Material for Bearings, General Discussion on Lubrication and Lubricants, *Inst. Mech. Engrs.*, *J.* **1**, 30–37 (1937); *India-Rubber J.* **94** (22), 636–638 (Nov. 27, 1937).
14. Burwell, J. T.: The Role of Surface Chemistry and Profile in Boundary Lubrication, *J. SAE* **50** (10), 450–457 (1942).
15. Stanton, T. E.: "Friction," Longmans, Green & Co., Ltd., London, 1923.
16. Ernst, H., and M. E. Merchant: Surface Friction of Clean Metals—A Basic Factor in Metal Cutting Process, *Proc. Conf. Friction and Surface Finish* (MIT), June, 1940, pp. 76–101.
17. Gongwer, C. A.: *Proc. Conf. Friction and Surface Finish* (MIT), June, 1940, pp. 239–244.
18. Hardy, W., and I. Bircumshaw: Boundary Lubrication—Plane Surfaces and the Limitations of Amontons' Law, *Proc. Roy. Soc.* (*London*), ser. A, **108** (A 745), 1–27 (May, 1925).
19. Hardy, W. R., and J. K. Hardy: Note on Static Friction and on the Lubricating Properties of Certain Chemical Substances, *Phil. Mag.*, ser. 6, **38** (233), 32–48 (1919).

20. Bowden, F. P., and J. E. Young: Friction of Clean Metals and Influence of Adsorbed Films, *Proc. Roy. Soc. (London)*, ser. A, **208** (A 1094), 311–325 (September, 1951).
21. Hardy, W. B., and I. Doubleday: Boundary Lubrication—The Latent Period and Mixtures of Two Lubricants, *Proc. Roy. Soc. (London)*, ser. A, **104** (A 724), 25–38 (August, 1923).
22. Bowden, F. P., and D. Tabor: "The Friction and Lubrication of Solids," Oxford University Press, New York, 1950.
23. Lancaster, J. K.: Composite Self-lubricating Bearing Materials, *Proc. Inst. Mech. Engrs.* (London) **182**, 33–54 (1967–1968).

2d-2. Rolling Friction. Rolling is frequently substituted for sliding friction. The resistance to motion is substantially smaller than for sliding under nonfluid film conditions. The frictional resistance to rolling under the action of load W may be designated as P in Fig. 2d-1. The coefficient of rolling friction is then defined as

$$f_R = \frac{P}{W} \qquad (2\text{d-}2)$$

The frictional resistance P to the rolling of a cylinder under load is applied at the center of the roller and is inversely proportional to the radius r of the roller and proportional to a factor k, a function of the material and its surface condition. Thus

$$P = \frac{k}{r} W \qquad (2\text{d-}3)$$

If r is in inches, values of k may be taken as follows: hardwood on hardwood, 0.02; iron on iron, steel on steel, 0.002; hard polished steel on hard polished steel, 0.0002 to 0.0004. Noonan and Strange suggest, for steel rollers on steel plates: surfaces well

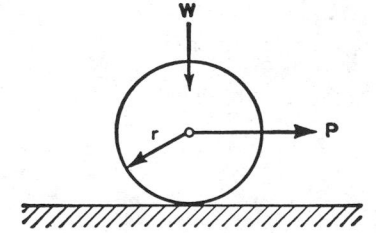

Fig. 2d-1. Rolling friction.

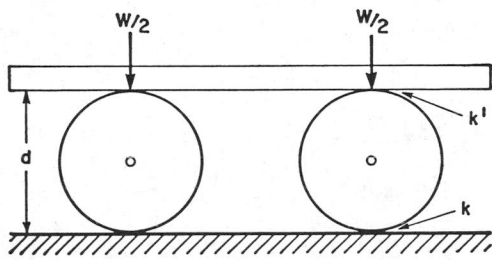

Fig. 2d-2. Load carried on rollers.

finished and clean, 0.005 to 0.001; surfaces well oiled, 0.001 to 0.002; surfaces covered with silt, 0.003 to 0.005; surfaces rusty, 0.005 to 0.01.

If the load is carried on rollers as in Fig. 2d-2, and k and k' are the respective factors for lower and upper surfaces, the force P is

$$P = \frac{(k + k')W}{d} \qquad (2\text{d-}4)$$

A comprehensive survey of rolling friction may be found in the following references presented at the annual meeting of the American Society of Mechanical Engineers, December 1 to 5, 1968.

Hersey, M. D.: Rolling Friction: I, Historical Introduction, Paper 68-LUB-B.

Hersey, M. D., and M. S. Downes: Rolling Friction: II, Cast Iron Car Wheels, Paper 68-LUB-C.

Hersey, M. D.: Rolling Friction: III, Review of Later Investigations. Paper 68-LUB-D.

2e. Elastic Constants, Hardness, Strength, Elastic Limits, and Diffusion Coefficients of Solids

H. M. TRENT[1]

U.S. Naval Research Laboratory

D. E. STONE

Vertex Corporation[2]

L. A. BEAUBIEN

U.S. Naval Research Laboratory

2e-1. Introduction. For the fundamental ideas connected with elasticity and for the definition of the elastic constants see Sec. 2a-6. For other definitions see Sec. 2e-3. The symbols and abbreviations used in this section are presented below.

E	Young's modulus
G	modulus of rigidity
σ	Poisson's ratio
ρ	density
C_{ij}	elastic constant (cf. Sec. 2a-6)
S_{ij}	elastic coefficient (cf. Sec. 2a-6)
T.S.	tensile strength
Y.S.	yield strength
Y.P.	yield point
S.S.	shear strength
El.	elongation
R.A.	reduction in area
Bhn	Brinell hardness number
R	Rockwell hardness number (often used with subscripts)
Vdh, Vhn	Vickers hardness number
D	diffusion coefficient
v	specific volume
p	pressure

2e-2. Elastic Constants and Coefficients of Crystals. Tables 2e-1 through 2e-6 contain tabulations of the elastic constants C_{ij} and coefficients S_{ij} of cubic, hexagonal, tetragonal, trigonal, orthorhombic, and monoclinic crystals (cf. Sec. 9a for X-ray crystallographic data). All temperatures are room temperatures unless otherwise specified. However, the original sources often contain values for a wide range of temperatures.

The two electrical boundary conditions for piezoelectric crystals are as follows: $D = 0$ denotes an electric field, generated piezoelectrically, parallel to the direction of wave propagation; $E = 0$ denotes a field perpendicular to this direction. Boundary

[1] Deceased.
[2] H. M. Childers of the Vertex Corporation provided valuable consultant service.

conditions are given only for those materials for which a change in boundary conditions produces a substantial change in one or more measured values.

References for these tables will be found immediately following Table 2e-6. References 1, 2, and 3 are published compilations from which the original sources can be obtained as well as references for values differing slightly from those given in these tables. In those cases in which two references are given, the first is for C_{ij} and the second for S_{ij}.

2e-3. Elastic Constants, Hardness, Strength, and Elastic Limits of Polycrystalline Solids. Tables 2e-7 through 2e-16 contain data on the Young's modulus, modulus of rigidity, hardness, etc., of various solids, metals, and alloys. The elastic constants, tensile strength, yield strength, shear strength, and all other quantities having the dimensions of stress are expressed in dynes per square centimeter. The definitions of these and other tabulated quantities are given in the following list.

1. *Tensile Strength.*[1] "The maximum tensile stress which a material is capable of developing."

Note: In practice, it is considered to be the maximum stress developed by a specimen representing the material in a tension test carried to rupture, under definite prescribed conditions. Tensile strength is calculated from the maximum load P carried during a tension test and the original cross-sectional area of the specimen A_0 from the formula

$$\text{Tensile strength} = \frac{P}{A_0}$$

2. *Yield Strength.*[1] "The stress at which a material exhibits a specified permanent set."

The yield strength is conventionally determined in either of two ways. In the first method, a specimen of the material is repeatedly loaded and unloaded with the load being increased at each cycle, the process being continued until a specified permanent set is obtained after one of the unloadings. The stress which produces this specified permanent set is called the yield strength.

In the second method, known as the offset method, a load-elongation curve is determined experimentally, the elongation being measured in units of extension per unit length of the undeformed specimen. A straight line is then drawn having a slope equal to the initial slope of the load-elongation curve and an intercept on the elongation axis equal to the specified offset, which is usually given in units of per cent elongation. The yield strength is taken to be that load defined by the interaction of the added straight line with the load-elongation curve.

Further discussion of yield strength can be found in ASTM E6-36.

3. *Yield Point.*[2] The stress at which a marked increase in deformation takes place without increase in the load.

4. *Shear Strength.*[3] "The stress, usually expressed in pounds per square inch, required to produce fracture when impressed perpendicularly upon the cross-section of a material."

5. *Elongation.*[4] "In tensile testing the elongation of a specimen is the increase in gage length, after rupture, referred to the original gage length. It is reported as percentage elongation."

6. *Reduction in Area.*[4] "In tensile testing the reduction in area of a specimen is the ratio of the difference between the original cross-sectional area of the specimen and the cross-sectional area after rupture, to the original cross-sectional area. It is reported as the percentage reduction of area."

[1] Standard Definitions of Terms Relating to Methods of Testing, ASTM E6-36.
[2] "Metals Handbook," 1948 ed., American Society for Metals.
[3] J. G. Henderson, "Metallurgical Dictionary."
[4] *Natl. Bur. Standards (U.S.) Circ.* C447.

TABLE 2e-1. ELASTIC CONSTANTS AND COEFFICIENTS OF CUBIC CRYSTALS
(C_{ij} in units of 10^{11} dynes/cm^2; S_{ij} in units of 10^{-13} cm^2/dyne)

Material	C_{11}	C_{12}	C_{44}	S_{11}	S_{12}	S_{44}	Ref.
Ag (silver)..............	12.40	9.34	4.61	22.9	−9.83	21.7	2
Ag, 25% Au..............				20.7	−8.91	20.5	4
Ag, 50% Au..............				19.7	−8.52	19.7	4
Ag, 75% Au..............				20.5	−9.09	20.6	4
AgBr...................	5.63	3.3	0.720	31.3	−11.7	139	2
AgCl...................	6.01	3.62	0.625	30.4	−11.4	160	2
Ag, 1.34% Cd..........	12.28	9.25	4.61	23.07	−9.91	21.69	5
Ag, 1.92% Cd..........	12.16	9.13	4.59	23.10	−9.91	21.77	5
Ag, 8.36% In..........	11.66	8.90	4.50	25.30	−10.95	22.20	5
Ag, 3.07% Mg..........	11.98	8.98	4.60	23.37	−10.01	21.74	5
Ag, 7.33% Mg..........	11.59	8.66	4.52	23.94	−10.24	22.10	5
Ag, 6.22% Pd..........	12.77	9.58	4.81	21.93	−9.40	20.79	5
Ag, 3.17% Sn..........	12.10	9.22	4.58	24.29	−10.51	21.83	5
Ag, 2.40% Zn..........	12.09	9.16	4.58	23.89	−10.30	21.85	5
Ag, 3.53% Zn..........	12.30	9.33	4.61	23.54	−10.16	21.68	5
Alum...................	2.56	1.07	0.85	52	−15	118	1
Aluminum..............	11.2	6.6	2.79	15.7	−5.8	35.9	1
Al, 5% Cu..............				15	−6.9	37	6
Ammonium alum........	2.50	1.06	0.80	53.5	−15.9	125	7, 3
Ammonium bromide.....	2.96	0.59	0.53	36.2	−6.0	189	1
Ammonium chloride......	3.90	0.72	0.68	27.2	−4.2	147	1
Au (gold)..............	18.6	15.7	4.20	23.3	−10.65	23.8	2
Barium nitrate..........	6.04	1.86	1.22	19.4	−4.6	82.0	1
CaF$_2$ (fluorspar)........	16.44	5.02	3.47	7.10	−1.66	28.8	2
Chromite...............	32.3	14.4	11.7	4.27	−1.31	8.56	2
Chromium alum........				54.2	−15.3	130	3
Cobalt zinc ferrite.......	26.6	15.3	7.8	6.49	−2.37	12.8	1
Copper.................	16.8	12.1	7.54	15.0	−6.3	13.26	1
Cu$_3$Au.................	19.07	13.83	6.63	13.4	−5.65	15.1	2
Cu, 4.1% Zn (α-brass)...	16.33	11.77	7.44				8
Cu, 9.1% Zn (α-brass)...	15.71	11.37	7.23				8
Cu, 17.1% Zn (α-brass)..	14.99	10.97	7.15				8
Cu, 22.7% Zn (α-brass)..	14.47	10.71	7.13				8
Cu, 47% Zn (α-brass)....	15.22	11.62	7.19				9
Cu, 44.9% Zn (β-brass)...	11.9	10.2	7.44	41.05	−19.0	13.4	2
Cu, 48.3% Zn (β-brass)...	12.91	10.97	8.24	35.3	−16.2	12.2	2
Cu, 48.9% Zn (β-brass)...	12.79	10.91	8.22	36.4	−16.8	12.2	1
Cu, 4.81% Al..........	16.58	12.16	7.49	15.9	−6.73	13.35	10
Cu, 9.98% Al..........	15.95	11.76	7.66	16.75	−7.11	13.05	10
Cu, 1.58% Ga..........	16.50	11.92	7.43	15.38	−6.45	13.46	10
Cu, 4.15% Ga..........	16.52	12.10	7.41	15.91	−6.73	13.50	10
Cu, 1.03% Ge..........	16.66	12.10	7.50	15.44	−6.50	13.33	10
Cu, 1.71% Ge..........	16.31	11.82	7.50	15.72	−6.60	13.33	10
Cu, 4.17% Si..........	16.78	12.42	7.48	16.10	−6.85	13.37	10
Cu, 5.16% Si..........	16.08	11.88	7.49	16.71	−7.10	13.35	10
Cu, 7.69% Si..........	16.58	12.64	7.41	17.72	−7.66	13.50	10
Cu, 4.59% Zn..........	16.34	11.92	7.42	15.91	−6.71	13.48	10

TABLE 2e-1. ELASTIC CONSTANTS AND COEFFICIENTS OF CUBIC CRYSTALS (*Continued*)

Material	C_{11}	C_{12}	C_{44}	S_{11}	S_{12}	S_{44}	Ref.
Cu, 28% Zn...............	.....	.....	.....	19.4	−8.4	13.9	11
Diamond................	107.6	12.50	57.58	0.953	−0.099	1.74	12
Diamond................	95	39	43	1.38	−0.40	2.3	13
Fe......................	23.7	14.1	11.6	7.72	−2.85	9.02	9, 14
Garnet 21.8% FeO.......	19.7	9.0	5.7	7.11	−2.2	17.5	15
Garnet 22.7% FeO.......	19.2	9.9	5.9	8.02	−2.7	16.9	15
Garnet 23.0% FeO.......	22.2	10.4	7.0	6.42	−2.1	14.3	15
Garnet 23.6% FeO.......	21.0	10.3	6.7	7.03	−2.3	14.9	15
Garnet 26.2% FeO......	22.6	12.6	6.2	7.36	−2.6	16.1	15
Garnet 28.7% FeO......	27.3	15.7	6.8	6.32	−2.3	14.7	15
Garnet 33.5% FeO.......	32.7	12.4	8.9	3.87	−1.1	11.2	15
Fe_3O_4 (magnetite).......	27.3	10.6	9.7	4.7	−1.31	10.3	2
FeS_2 (pyrite)...........	36.2	−4.4	10.4	2.85	0.39	9.6	2
GaAs..................	1.192	0.599	0.538	126.4	−42.34	186	2
GaSb..................	8.85	4.04	4.33	15.8	−4.96	23.1	2
Germanium.............	12.89	4.83	6.71	9.78	−2.66	14.90	2
Hexamethylene tetramine	1.5	0.3	0.7	70	−12	140	1
Lead nitrate............	4.56	3.09	1.37	48.5	−19.6	73.0	1
Indium antimonide.......	6.72	3.67	3.02	24.2	−8.55	33.1	2
Potassium alum.........	2.54	1.07	0.84	52.5	−15.6	119	2
K (potassium)..........	0.459	0.372	0.263	833	−370	380	9, 16
KBr..................	3.46	0.58	0.505	30.4	−4.35	198	2
KCl..................	3.98	0.62	0.625	26.2	−3.5	160	2
KF...................	6.58	1.49	1.28	.....	.....	.....	17
KI...................	2.67	0.43	0.421	39.2	−5.4	238	2
Li (195°K).............	1.320	1.102	0.960	316.4	−144	104	18
LiBr..................	3.94	1.88	1.91	.....	.....	.....	17
LiCl..................	4.94	2.26	2.49	.....	.....	.....	17
LiF...................	11.12	4.20	6.28	11.35	−3.1	15.9	2
LiI...................	2.85	1.40	1.35	.....	.....	.....	17
MgO..................	28.6	8.7	14.8	4.08	−0.95	6.76	2
Magnetite.............	27.5	10.4	9.55	4.59	−1.26	10.47	1
Molybdenum...........	46	17.6	11.0	2.8	−0.78	9.1	2
Na (sodium)...........	0.945	0.779	0.618	420	−190	162	2
NaBr..................	3.87	0.97	0.97	28.7	−5.8	103	2
$NaBrO_3$...............	5.73	1.76	1.52	20.4	−4.8	65.7	2
NaCl..................	4.87	1.24	1.26	22.9	−4.65	79.4	2
$NaClO_3$...............	4.99	1.41	1.17	22.9	−5.05	85.4	2
NaF..................	9.71	2.43	2.80	.....	.....	.....	17
NaI...................	3.035	0.90	0.72	.....	.....	.....	17
Ammonium alum........	2.50	1.06	0.8	53.5	−15.9	125	2
NH_4Br...............	2.96	0.59	0.53	36.2	−6.0	189	2
NH_4Cl...............	3.90	0.72	0.68	27.2	−4.2	147	2
Nickel................	24.65	14.73	12.47	7.34	−2.74	8.02	2
Palladium.............	22.71	17.60	7.173	.....	.....	.....	19
Pb (lead).............	5.03	3.93	1.40	63.2	−27.7	71.4	20
Pb...................	4.66	3.92	1.44	92.8	−42.4	69.4	2
PbS (galena)...........	10.2	3.8	2.5	12	−3	40	1

TABLE 2e-1. ELASTIC CONSTANTS AND COEFFICIENTS OF CUBIC CRYSTALS (*Continued*)

Material	C_{11}	C_{12}	C_{44}	S_{11}	S_{12}	S_{44}	Ref.
PbS	12.70	2.98	2.48	8.7	−1.64	40.3	1
RbBr	3.185	0.48	0.385	......	....	......	17
RbCl	3.645	0.61	0.475	......	....	......	17
RbF	5.7	1.25	0.91	......	....	......	17
RbI	2.585	0.375	0.281	......	....	......	17
Silicon	16.57	6.39	7.96	7.68	−2.14	12.56	2
Strontium nitrate	4.73	2.18	1.46	29.8	−9.4	68.5	1
Thallium bromide	3.78	1.48	0.756	33.9	−9.5	132	2
Thallium chloride	4.01	1.53	0.760	31.6	−8.7	132	2
Thallium alum	......	......	......	49.0	−15.5	115	3
Thallium bromide chloride	3.85	1.49	0.737	33.1	−9.2	136	1
Thallium bromide iodide	3.6	1.5	0.555	37	−11	180	1
Thorium	7.53	4.89	4.78	27.2	−10.7	20.9	2
W (tungsten)	50.1	19.8	15.14	2.57	−0.729	6.60	2
Zinc blende	10.0	6.5	3.4	20.5	−8.1	29.4	1
Zinc sulfide	10.79	7.22	4.12	20	−8.02	24.3	2

TABLE 2e-2. ELASTIC CONSTANTS AND COEFFICIENTS OF HEXAGONAL CRYSTALS
(C_{ij} in units of 10^{11} dynes/cm^2; S_{ij} in units of 10^{-13} cm^2/dyne)

Material	C_{11}	C_{33}	C_{44}	C_{12}	C_{13}	S_{11}	S_{33}	S_{44}	S_{12}	S_{13}	Ref.
Apatite	16.67	13.96	6.63	1.31	6.55	7.49	10.9	15.1	0.97	−4.0	23
BaTiO₃ ($D = 0$)	16.8	18.9	5.46	7.82	7.10	8.18	6.76	18.3	−2.98	−1.95	2
BaTiO₃ ($E = 0$)	16.6	16.2	4.29	7.66	7.75	8.55	8.93	23.3	−2.61	−2.85	2
BaTiO₃ 5% CaTiO₃ by wt. ($E = 0$)	17.41	16.88	4.74	7.93	8.00	8.05	8.42	21.1	−2.45	−2.65	2
Beryllium	30.8	35.7	11.0	−5.8	8.7	3.77	3.37	9.09	1.04	−1.17	1
Beryl I	27.81	24.8	6.61	10.01	6.77	4.27	4.47	15.1	−1.35	−0.80	1
Beryl II	29.71	26.5	7.54	10.26	7.39	3.97	4.21	13.3	−1.17	−0.78	1
Cadmium	11.0	4.69	1.56	4.04	3.83	12.9	36.9	64.0	−1.5	−9.3	2
CdS	8.1	8.0	1.43	4.9	4.8	22.2	21.9	70	−8.7	−8.0	21
Cobalt	30.7	35.81	7.53	16.5	10.3	4.72	3.19	13.24	−2.31	−0.69	2
Ice (−16°C)	1.33	1.42	0.306	0.63	0.46	101.3	82.8	326.5	−41.6	−19.3	22
Magnesium	5.97	6.17	1.64	2.62	2.17	22.0	19.7	61	−7.85	−5.0	2
SiO₂ (600°C) (β-quartz)	11.66	11.04	3.606	1.67	3.28	9.41	10.62	27.73	−0.60	−2.62	2
Yttrium	7.79	7.69	2.431	2.85	2.1	......	......	......	......	......	24
Zinc	16.1	6.10	3.83	3.42	5.01	8.38	28.38	26.1	0.53	−7.31	2

TABLE 2e-3. ELASTIC CONSTANTS AND COEFFICIENTS OF TETRAGONAL CRYSTALS
(C_{ij} in units of 10^{11} dynes/cm²)

Material	C_{11}	C_{33}	C_{44}	C_{66}	C_{12}	C_{13}	Ref.
Ammonium dihydrogen phosphate...	6.17	3.28	0.85	0.59	0.72	1.94	2
Ammonium dihydrogen phosphate...	7.58	2.96	0.87	0.614	−2.43	1.30	2
Ammonium dihydrogen phosphate ($D = 0$).....................	6.76	3.38	0.867	0.687	0.59	2.0	2
Ammonium dihydrogen phosphate ($E = 0$).....................	6.76	3.38	0.867	0.608	0.59	2.0	2
Ammonium dihydrogen phosphate (deuterated).................	6.2	3.0	0.91	0.61	−0.5	1.4	2
Barium titanate ($D = 0$)...........	28.3	17.8	8.05	11.3	18.7	14.2	2
Barium titanate ($E = 0$)..........	27.5	16.5	5.43	11.3	17.9	15.1	2
Indium........................	4.45	4.44	0.655	1.22	3.95	4.05	2
Nickel sulfate.................	3.21	2.93	1.16	1.78	2.31	0.21	1
Potassium dihydrogen arsenate......	5.3	3.7	1.2	0.7	−0.6	−0.2	1
Potassium dihydrogen phosphate....	7.14	5.62	1.27	0.628	−0.49	1.29	2
Potassium dihydrogen phosphate (0°C)......................	8.14	7.85	1.29	0.63	3.49	4.07	1
Sn (tin)......................	8.6	13.3	4.9	5.3	3.5	3.0	1
Sn...........................	7.35	8.7	2.2	2.265	2.34	2.8	2
Sn...........................	8.39	9.67	1.75	0.741	4.87	2.81	2
Zircon.......................	7.35	4.60	1.38	1.60	0.90	−0.54	2

TABLE 2e-3A. ELASTIC CONSTANTS AND COEFFICIENTS
OF TETRAGONAL CRYSTALS (*Continued*)
(S_{ij} in units of 10^{-13} cm²/dyne)

Material	S_{11}	S_{33}	S_{44}	S_{66}	S_{12}	S_{13}	Ref.
Ammonium dihydrogen arsenate..	16.9	44.5	152.9	124.0	−17.3	−11.1	3
Ammonium dihydrogen phosphate	20	45.7	117	169	1.7	−12.9	2
Ammonium dihydrogen phosphate	17.5	43.5	114	163	7.5	−11	2
Ammonium dihydrogen phosphate ($D = 0$)...................	18.1	43.5	115.3	145.5	1.9	−11.8	2
Ammonium dihydrogen phosphate ($E = 0$)...................	18.1	43.5	115.3	164.6	1.9	−11.8	2
Ammonium dihydrogen phosphate (deuterated).................	19	44	110	164	2	−11	2
Barium titanate ($D = 0$)........	7.25	10.8	12.4	8.84	−3.15	−3.26	2
Barium titanate ($E = 0$).........	8.05	15.7	18.4	8.84	−2.35	−5.24	2
Indium........................	149.4	187	152.7	82	−50.6	−90.2	2
Nickel sulfate.................	65	34.3	86.5	56.2	−46.8	−1.3	1
Potassium dihydrogen arsenate...	19	27	86.0	152	2	1	1
Potassium dihydrogen phosphate.	14.8	19.5	78.7	159.2	1.7	−3.79	2
Potassium dihydrogen phosphate (0°C)......................	17.5	20	77.5	159	−4	−7	1
Sn (tin)......................	14.6	8.5	20.6	19.0	−5.3	−2.07	1
Sn...........................	16.3	14.1	45.4	44.2	−3.6	−4.1	2
Sn...........................	18.5	11.8	57.0	135	−9.9	−2.5	2
Zircon.......................	13.9	22.1	72	62	−1.6	−1.4	2

TABLE 2e-4. ELASTIC CONSTANTS AND COEFFICIENTS OF TRIGONAL CRYSTALS
(C_{ij} in units of 10^{11} dynes/cm^2)

Material	C_{11}	C_{33}	C_{44}	C_{12}	C_{13}	C_{14}	Ref.
Alumina (corundum)	46.5	56.3	23.3	12.4	11.7	10.1	2
Antimony	7.92	4.27	2.85	2.48	2.61	1.05	2
Bismuth	6.28	4.40	1.08	3.50	2.11	−0.42	2
Calespar (calcite)	13.74	8.01	3.42	4.40	4.50	−2.03	2
Dextrose sodium bromide	2.06	2.40	0.634	0.53	0.79	0.03	1
Dextrose sodium chloride	2.20	1.77	0.771	1.09	0.75	−0.03	1
Dextrose sodium iodide	2.58	2.06	0.771	1.52	0.49	−0.03	1
Haematite	24.2	22.8	8.5	5.5	1.6	−1.3	2
Mercury (−190 °C)	3.60	5.05	1.29	2.89	3.03	0.5	2
α-Quartz	8.674	10.72	5.79	0.699	1.191	−1.791	2
α-Quartz	8.75	10.77	5.73	0.762	1.51	1.72	2
Sapphire	49.6	50.2	20.6	10.9	4.8	3.8	25
Sodium nitrate	8.67	3.74	2.13	1.63	1.60	0.82	2
Tellurium	......	7.00	.....	.....	2.31	......	2
Tourmaline	27.2	16.5	6.5	4.0	3.5	−0.68	1
Tourmaline I	26.3	15.1	5.95	6.1	4.9	−0.9	1
Tourmaline II	30.4	17.6	6.5	8.8	3.5	−0.4	1

TABLE 2e-4A. ELASTIC CONSTANTS AND COEFFICIENTS
OF TRIGONAL CRYSTALS (*Continued*)
(S_{ij} in units of 10^{-13} cm^2/dyne)

Material	S_{11}	S_{33}	S_{44}	S_{12}	S_{13}	S_{14}	Ref.
Alumina (corundum)	2.90	1.94	5.78	−1.05	−0.38	−1.71	2
Aluminum phosphate	16.1	16.1	53.0	−0.1	−8.3	8.9	3
Antimony	17.7	33.8	41.0	−3.8	−8.5	−8.0	2
Bismuth	26.9	28.7	104.8	−14.0	−6.2	16.0	2
Calespar (calcite)	11.0	17.3	39.4	−3.4	−4.3	8.6	2
Dextrose sodium bromide	56.9	52.3	158	−8.6	−16.0	−3.4	1
Dextrose sodium chloride	63.8	70.2	130	−26.1	−16	3.6	1
Dextrose sodium iodide	60.2	51.6	130	−34.3	−6.2	3.8	1
Haematite	4.42	4.44	11.92	−1.02	−0.23	0.80	2
Lithium trisodium chromate	78.7	35.0	.....	....	....	....	3
Lithium trisodium molybdate	29.5	27.1	.....	....	....	....	3
Mercury (−190 °C)	154	45	151	−119	−21	−100	2
α-Quartz	12.77	9.6	20.04	−1.79	−1.22	4.50	2
α-Quartz	12.69	9.71	20.05	−1.69	−1.54	−4.31	2
Sapphire	2.18	2.02	5.04	−0.50	−0.16	−0.49	25
Sodium nitrate	13.4	30.8	51.5	−2.2	−4.8	−6.0	2
Tellurium	48.7	23.4	58.1	−6.9	−13.8	....	2
Tourmaline	3.85	6.36	15.4	−0.48	−0.71	0.45	1
Tourmaline I	4.22	7.34	17.1	−0.80	−1.11	0.76	1
Tourmaline II	3.64	5.89	15.4	−1.00	−0.53	0.29	1

TABLE 2e-5. ELASTIC CONSTANTS AND COEFFICIENTS OF ORTHORHOMBIC CRYSTALS
(C_{ij} in units of 10^{11} dynes/cm^2)

Material	C_{11}	C_{22}	C_{33}	C_{44}	C_{55}	C_{66}	C_{12}	C_{13}	C_{23}	Ref.
Aragonite	16.0	8.7	8.5	4.12	2.56	4.27	3.73	0.17	1.57	2
Baryte	8.62	9.17	10.84	1.20	2.87	2.74	5.23	3.41	3.56	1
Celestite	10.44	10.61	12.86	1.35	2.79	2.66	7.73	6.05	6.19	2
Iodic acid	3.03	5.45	4.36	1.84	2.19	1.74	1.19	1.17	0.55	1
Lithium ammonium tartrate	3.86	5.39	3.63	1.19	0.67	2.33	1.65	0.87	2.01	1
Magnesium sulfate	6.98	5.29	8.22	1.07	2.33	2.22	3.90	2.82	2.83	1
Potassium pentaborate	5.82	3.59	2.55	1.64	0.463	0.57	2.29	1.74	2.31	2
Rochelle salt ($D = 0$)	2.55	3.81	3.71	1.34	0.321	0.979	1.41	1.16	1.46	2
Rochelle salt ($E = 0$)	2.55	3.81	3.71	.....	0.286	0.960	1.41	1.16	1.46	2
Rochelle salt ($D = 0$)	4.25	5.15	6.29	1.25	0.304	0.996	2.96	3.57	3.42	2
Rochelle salt ($E = 0$)	4.25	5.15	6.29	0.58	0.278	0.974	2.96	3.57	3.42	2
Sodium ammonium tartrate	3.68	5.09	5.54	1.06	0.303	0.87	2.72	3.08	3.47	1
Sodium tartrate	4.61	5.47	6.65	1.24	0.31	0.98	2.86	3.20	3.52	1
Staurolite	34.3	18.5	14.7	4.6	7.0	9.2	6.7	6.1	12.8	26
Strontium formate	4.39	3.48	3.74	1.54	1.07	1.72	1.04	−1.49	−0.14	1
Sulfur	2.40	2.05	4.83	0.43	0.87	0.76	1.33	1.71	1.59	2
Topaz	28.2	34.9	29.5	10.8	13.3	13.1	12.6	8.5	8.8	2
α-Uranium	21.47	19.86	26.71	12.44	7.342	7.433	4.65	2.18	10.76	27
Zinc sulfate	4.00	3.22	5.45	0.50	1.70	1.81	1.32	1.80	1.19	1

TABLE 2e-5A. ELASTIC CONSTANTS AND COEFFICIENTS
OF ORTHORHOMBIC CRYSTALS (*Continued*)
(S_{ij} in units of 10^{-13} cm^2/dyne)

Material	S_{11}	S_{22}	S_{33}	S_{44}	S_{55}	S_{66}	S_{12}	S_{13}	S_{23}	Ref.
Aragonite	6.95	13.2	12.2	24.3	39.0	23.4	−3.0	0.4	−2.4	2
Baryte	18.4	17.36	10.96	83.33	34.84	36.50	−9.45	−2.68	−2.73	1
Barium formate	.....	.....	.....	78.5	60.0	82.5	....	....	....	3
Celestite	22.0	21.9	11.4	74.1	35.8	37.6	−13.9	−3.7	−4.0	2
Iodic acid	39.8	20.1	25.6	54.5	45.6	57.6	−7.75	−9.7	−0.45	1
Lithium ammonium tartrate	30	25.6	35	84	150	43	−8.2	−2.7	−12.2	1
Magnesium sulfate	24.5	34.1	15.0	93.5	42.9	45.0	−16.6	−2.68	−6.05	1
Potassium pentaborate	23.2	73.6	98.3	61	215	175	−10.6	−6.1	−60	2
Rochelle salt ($D = 0$)	52.4	35.4	33.7	74.7	311	102	−15.4	−10.3	−9.1	2
Rochelle salt ($E = 0$)	52.4	35.4	33.7	......	350	104	−15.4	−10.3	−9.1	2
Rochelle salt ($D = 0$)	51.8	34.9	33.4	79.8	328	101	−15.3	−21.1	−10.3	2
Rochelle salt ($E = 0$)	51.8	34.9	33.4	174	360	103	−15.3	−21.1	−10.3	2
Sodium ammonium tartrate	57.0	38.5	40	94.5	330	115	−15.5	−22	−15.5	1
Sodium tartrate	37.1	31.6	26.4	80.6	323	102	−12.0	−11.5	−10.9	1
Strontium formate	28.4	31	31	65	93	58	−8	11	−2	1
Sulfur	71	83	30	232	115	132	−36	−13	−15	2
Topaz	4.43	3.53	3.84	9.23	7.53	7.63	−1.38	−0.86	−0.06	2
α-Uranium	4.91	6.73	4.79	8.04	13.62	13.45	−1.19	0.08	−2.61	27
Zinc sulfate	29.5	37.7	20.4	200	58.8	55.3	−10.8	−3.49	−6.10	1

TABLE 2e-6. ELASTIC CONSTANTS AND COEFFICIENTS OF MONOCLINIC CRYSTALS

(C_{ij} in units of 10^{11} dynes/cm²; S_{ij} in units of 10^{-13} cm²/dyne)

Material	C_{11}	C_{22}	C_{33}	C_{44}	C_{55}	C_{66}	C_{12}	C_{13}	C_{23}	C_{15}	C_{25}	C_{35}	C_{46}	Ref.
Dipotassium tartrate*	6.9	3.5	4.4	0.84	1.3	0.96	1.2	3.2	1.4	0.27	0.18	−0.13	−0.05	1
Ethylene diamine tartrate*	13.4	3.5	6.04	0.53	0.83	0.57	2.7	8.1	2.2	1.7	0.4	1.2	0.009	1
Lithium sulfate*	5.7	7.1	4.9	2.7	2.9	1.4	2.7	1.6	1.6	−0.22	1.6	0.14	0.17	1
Sodium thiosulfate*	3.31	3.02	4.57	0.57	1.11	0.60	1.83	1.84	1.68	0.25	1.04	−0.69	−0.27	1
Tartaric acid*	9.30	1.93	4.65	0.81	0.82	1.1	2.0	3.7	1.4	−1.2	−0.40	−0.036	0.14	1

Material	S_{11}	S_{22}	S_{33}	S_{44}	S_{55}	S_{66}	S_{12}	S_{13}	S_{23}	S_{15}	S_{25}	S_{35}	S_{46}	Ref.
Dipotassium tartrate*	22.4	33.7	38.6	119	81.5	104.1	−0.8	−16.4	−10.5	−6.4	−5.7	9.0	5.7	1
Ethylene diamine tartrate*	38.8	37	98	188	172	174	4.0	−52	−19	−7.0	1.0	−25	−3.0	1
Lithium sulfate*	23.9	21.3	23.1	36.9	41	74	−9.5	−5	−3.6	7.1	−12.0	0.5	−4.6	1
Sodium thiosulfate*	50.2	156	67.4	223	327	212	−32.3	−6.21	−71.9	15.2	−182	110	100	1
Tartaric acid*	21.6	77	38.5	130	180	96	−6.1	−15	−18	28	28	−29	−16	1
Dipotassium tartrate**	47.5	35.3	24.0	113.5	102	122.5	−17.4	−8	−6.2	−7.5	8.0	−14.0	−6.8	1
Ethylene diamine tartrate**	33.4	36.5	100	192	117	191	−3	−30	−18	−17	15	−26.5	3.8	1
Lithium sulfate**	22.9	22.5	22.8	71.3	64.0	36.1	−5.4	−7.5	−4.6	−2.1	−8.3	6.3	1.4	1

* The single-starred values of the S_{ij} correspond to the single-starred values of the C_{ij}; that is, $(C*)^{-1} = (S*)$.

** The double-starred values are referred to a differently oriented set of axes.

References for Tables 2e-1 through 2e-6

1. Hearmon: *Advances in Phys.* **5**, 323 (1956).
2. Huntington: "Solid State Physics," vol. 7, p. 213, Academic Press, Inc., New York, 1958.
3. Sundara Rao, Vedan, and Krishnan: "Progress in Crystal Physics," vol. 1, p. 73, S. Viswanathan, Madras, India, 1958.
4. Rohl: *Ann. Phys.* **16**, 887 (1933).
5. Bacon: *Dept. Phys., Case Inst. Technol., Tech. Rept.* 15, 1955.
6. Karnop and Sachs: *Z. Physik* **53**, 605 (1929).
7. Sundara Rao: *Current Sci. (India)* **17**, 50 (1948).
8. Rayne: *Phys. Rev.* **112**, 1125 (1958).
9. Jones: *Physica* **15**, 13 (1949).
10. Neighbors and Smith: *Acta Met.* **2**, 591 (1954).
11. Sundara Rao and Balakrishnan: *Proc. Indian Acad. Sci.* **28A**, 475 (1948).
12. McSkimmin and Bond: *Phys. Rev.* **105**, 116 (1957).
13. Bhagavantam and Bhimasenachar: *Proc. Roy. Soc.*, ser. A, **187**, 381 (1946).
14. Kimura: *Proc. Phys.-Math. Soc. Japan* **21**, 686, 786 (1939); **22**, 45, 219 (1940).
15. Ramachandra Rao: *Proc. Indian Acad. Sci.* **22A**, 194 (1945).
16. Seitz: *J. Appl. Phys.* **12**, 100 (1941).
17. Spangenberg and Haussuhl: *Z. Krist.* **109**, 422 (1957).
18. Nash and Smith: *Phys. Chem. Solids* **9**, 113 (1959).
19. Rayne: *Phys. Rev.* **118**, 1545 (1960).
20. Prasad and Wooster: *Acta Cryst.* **9**, 38 (1956).
21. Gutsche: *Naturwissenschaften* **45**, 566 (1958).
22. Bass, Rossberg, and Ziegler: *Z. Physik* **149**, 199 (1957).
23. Bhimasenachar: *Proc. Indian Acad. Sci.* **22** (sec. A), 209 (1945).
24. Smith and Gjevre: *J. Appl. Phys.* **31**, 647 (1960).
25. Mayer and Heidemann: *J. Acoust. Soc. Am.* **30**, 756 (1958).
26. Bhimasenachar and Venkata Rao: *J. Acoust. Soc. Am.* **29**, 343 (1957).
27. Fisher and McSkimmin: *J. Appl. Phys.* **29**, 1473 (1958).

Abbreviations in Tables 2e-7 *through* 2e-16

Abbreviation	Definition
H.R.	Hot rolled
C.R.	Cold rolled
W.Q.	Water quenched
O.Q.	Oil quenched
A.Q.	Air quenched
A.C.	Air cooled
F.C.	Furnace cooled
h-t	Heat-treated
wr	Wrought
ann	Annealed
art. aged	Artificially aged
nat. aged	Naturally aged
spec.	Specimen
G.S.	Grain size

TABLE 2e-7. ELASTIC AND STRENGTH CONSTANTS FOR SILVER, GOLD, PLATINUM, PALLADIUM ALLOYS

Material	Condition	E	σ	Tensile strength	Yield strength at 0.2% offset	Elongation	Reduction in area	Bhn	Ref.*
Ag	Strained 5%, heated 5 hr at 350°C	$7.1\text{-}7.8 \times 10^{11}$	...	...	...	...	...	...	1
Ag	Ann.	...	0.37	...	...	...	...	...	1
Ag + 80 Mo	...	...	...	55×10^{8}	...	...	...	190	1
Ag + 40 Mo	...	...	...	41×10^{8}	...	...	...	160	1
Ag + 20 Mo	...	...	...	24×10^{8}	...	...	...	40	1
Ag + 20 W	...	...	...	34×10^{8}	...	...	...	40	1
Ag + 40 W	...	...	...	41×10^{8}	...	...	...	150	1
Ag + 80 W	...	...	...	55×10^{8}	...	...	...	240	1
Ag + 40 Ni	Ann.	...	...	26×10^{8}	...	...	...	Vhn 70	1
Ag + 20 Ni	Ann.	...	...	21×10^{8}	...	...	...	Vhn 45	1
Ag + 1 graphite	...	...	...	...	...	...	...	$R_{15T}68$	1
Ag + 5 graphite	...	...	...	...	...	...	...	$R_{15T}55$	1
Ag + 10 graphite	...	...	...	...	...	...	...	$R_{15T}40$	1
Ag + 5 Cd	...	...	...	16×10^{8}	...	...	...	R_F30	1
Ag + 10 Cd	...	...	...	19×10^{8}	...	...	...	R_F44	1
Ag + 20 Cd	...	...	...	20×10^{8}	...	...	...	R_F55	1
33 Ag, 52 Hg, 12.5 Sn, 2 Cu, 0.5 Zn	Cast	1.0×10^{11}	...	$2.8\text{-}5.9 \times 10^{8}$	...	...	...	...	1
Au 99.99%	...	7.44×10^{11}	0.42	12.4×10^{8}	Nil	30	...	33	1
Au 99.99%	Wrought, ann.	8.00×10^{14}	0.42	13.1×10^{8}	...	45	...	25	1
58.3 Au, 4.9 Ag, 31.6 Cu, 5.2 Ni	Air cooled	...	...	56.9×10^{8}	33.1×10^{8} at 0.1% offset	41.0	36.0	R_B87	1

TABLE 2e-7. ELASTIC AND STRENGTH CONSTANTS FOR SILVER, GOLD, PLATINUM, PALLADIUM ALLOYS (Continued)

Material	Condition	E	σ	Tensile strength	Yield strength at 0.2% offset	Elongation	Reduction in area	Bhn	Ref.*
41.6 Au, 4.6 Ag, 43.4 Cu, 5.0 Ni, 5.4 Zn	Air cooled			46.8×10^8	26.7×10^8 at 0.1% offset	41.5	36.0	R_B68	1
69 Au, 25 Ag, 6 Pt	Ann.			37.6×10^8				Vhn 112	1
Pt 99.99%	Ann.	14.7×10^{11}	0.39	$12\text{–}13 \times 10^8$		25–40		Vhn 38–40	1
Pt + 5 Ir	Ann.			27×10^8				90	1
Pt + 10 Ir	Ann.			38×10^8				130	1
Pt + 25 Ir	Ann.			86×10^8				240	1
Pt + 3.5 Rh	Ann.			17×10^8				60	1
Pt + 5.0 Rh	Ann.			21×10^8				70	1
Pt + 10.0 Rh	Ann.			31×10^8		35		90	1
Pt + 20.0 Rh	Ann.			48×10^8		40		120	1
Pt + 5 Ru	Ann.			41×10^8				130	1
Pt + 10 Ru	Ann.			59×10^8				190	1
Pt + 1 Ni	Ann.			21×10^8				Vhn 60–65	1
Pt + 2 Ni	Ann.			28×10^8				Vhn 80–90	1
Pt + 5 Ni	Ann.			45×10^6				Vhn 130–140	1
84 Pt, 10 Pd, 6 Ru	Ann.			55×10^8		18–25		Vhn 150–170	1
96 Pt, 4 W	Ann.			$48\text{–}52 \times 10^8$		25		Vhn 140–150	1
Pd (pure)	Ann. and rolled	12.1×10^{11}		$\geq 15 \times 10^8$		24		Vhn 37–39	1
60 Pd, 40 Ag	Ann.			35×10^8		47		Vhn 100	1
60 Pd, 40 Cu	Ann.			52×10^8					1
95 Pd, 4 Ru, 1 Rh	Ann.			$38\text{–}41 \times 10^8$		25		Vhn 100–110	1

* References are on p. 2-76.

TABLE 2e-8. ELASTIC AND STRENGTH CONSTANTS FOR ALUMINUM ALLOYS

Alloys	Condition	E	G	σ	Tensile strength	Yield strength	Elongation	Bhn	Shear strength	Ref.*
Cast alloys:										
Al, 12 Si	Die cast	7.10×10^{11}	2.65×10^{11}	0.33	25.5×10^8	12.4×10^8	1.8†			1
Al, 5 Si	Die cast	7.10×10^{11}	2.65×10^{11}	0.33	20.7×10^8	9.65×10^8	7.0†	40†	9.65×10^8	1
Al, 5 Si	Sand cast	7.10×10^{11}	2.65×10^{11}	0.33	13.1×10^8	6.20×10^8	6.0‡			1
Al, 5 Si, 4 Cu	Die cast	7.10×10^{11}	2.65×10^{11}	0.33	27.6×10^8	15.2×10^8	3.5†	55†	13.8×10^8	1
Al, 4 Cu, 3 Si	Sand cast	7.10×10^{11}	2.65×10^{11}	0.33	14.5×10^8	9.65×10^8	2.5‡	Re65		1
Al, 5 Si, 3 Cu	Sand cast	7.10×10^{11}	2.65×10^{11}	0.33	18.6×10^8	9.65×10^8	2.5‡	Re80		1
Al, 5 Si, 3 Cu	Sand cast, h-t, aged	7.10×10^{11}	2.65×10^{11}	0.33	24.1×10^8	13.8×10^8	4.0‡	Re85		1
Al, 5 Si, 3 Cu	Perm. mold cast, h-t, aged	7.10×10^{11}	2.65×10^{11}	0.33	28.9×10^8	15.2×10^8	5.0‡			1
Al, 5.5 Si, 4.5 Cu	Perm. mold cast, h-t, aged	7.10×10^{11}	2.65×10^{11}	0.33	19.3×10^8	11.0×10^8	2.0‡	70†	17.2×10^8	1
Al, 7 Cu, 2 Si, 1.7 Zn	Sand cast	7.10×10^{11}	2.65×10^{11}	0.33	16.5×10^8	10.3×10^8	1.5‡	70†	13.8×10^8	1
Al, 7 Cu, 3.5 Si	Perm. mold cast	7.10×10^{11}	2.65×10^{11}	0.33	20.7×10^8	16.5×10^8	1.0‡	80†	15.2×10^8	1
Al, 10 Cu, 0.2 Mg	Sand cast (ann.)	7.10×10^{11}	2.65×10^{11}	0.33	18.6×10^8	13.8×10^8	1.0‡	80†	14.5×10^8	1
Al, 10 Cu, 0.2 Mg	H-t, artificially aged	7.10×10^{11}	2.65×10^{11}	0.33	27.6×10^8	20.7×10^8	0.5‡	115†	20.0×10^8	1
Al, 12 Si, 2.5 Ni, 1.2 Mg, 0.8 Cu	Perm. mold cast, art. aged	7.10×10^{11}	2.65×10^{11}	0.33	24.8×10^8	19.3×10^8	0.5‡	105†	16.5×10^8	1
Al, 12 Si, 1.5 Cu, 0.7 Mn, 0.7 Mg	Perm. mold cast (stress relieved)	7.10×10^{11}	2.65×10^{11}	0.33	24.8×10^8		0.5‡	100†		1
Al, 4 Cu, 2 Ni, 1.5 Mg	Ann. (sand cast)	7.10×10^{11}	2.65×10^{11}	0.33	18.6×10^8	12.4×10^8	1.0‡	70¶	14.5×10^8	1
Al, 4.5 Cu	H-t, nat. aged	7.10×10^{11}	2.65×10^{11}	0.33	22.0×10^8	11.0×10^8	8.5‡	60¶	16.5×10^8	1
Al, 4.5 Cu, 2.5 Si	H-t, nat. aged	7.10×10^{11}	2.65×10^{11}	0.33	27.6×10^8	15.2×10^8	10.0‡	75¶	20.7×10^8	1
Al, 3.8 Mg	Perm. mold cast	7.10×10^{11}	2.65×10^{11}	0.33	18.6×10^8	11.0×10^8	7.0‡	60¶	15.2×10^8	1
Al, 8 Mg	Die cast	7.10×10^{11}	2.65×10^{11}	0.33	28.9×10^8	15.8×10^8	7.0†			1
Al, 10 Mg	Sand cast, h-t, nat. aged	7.10×10^{11}	2.65×10^{11}	0.33	31.7×10^8	17.2×10^8	14.0‡	75¶	22.7×10^8	1
Al, 6 Si, 3.5 Cu	H-t, art. aged	7.10×10^{11}	2.65×10^{11}	0.33	24.8×10^8	16.5×10^8	2.0‡	80¶		1
Al, 6 Si, 3.5 Cu	As cast	7.10×10^{11}	2.65×10^{11}	0.33	18.6×10^8	12.4×10^8	2.0‡	70¶	16.5×10^8	1
Al, 5 Si, 1.3 Cu, 0.5 Mg	H-t, art. aged (sand cast)	7.10×10^{11}	2.65×10^{11}	0.33	24.1×10^8	17.2×10^8	2.5‡	80¶	20.8×10^8	1
Al, 5 Si, 1.3 Cu, 0.5 Mg	H-t, art. aged (perm. mold cast)	7.10×10^{11}	2.65×10^{11}	0.33	29.6×10^8	18.6×10^8	4.0‡	90¶	20.8×10^8	1
Al, 7 Si, 0.3 Mg	H-t, art. aged (sand cast)	7.10×10^{11}	2.65×10^{11}	0.33	22.7×10^8	16.5×10^8	4.0‡	70¶	18.6×10^8	1
Al, 7 Si, 0.3 Mg	H-t, art. aged (perm. mold cast)	7.10×10^{11}	2.65×10^{11}	0.33	27.6×10^8	18.6×10^8	5.0‡	90¶		1
Al, 8 Si, 1.5 Cu, 0.3 Mg, 0.3 Mn	Sand cast (stress relieved)	7.10×10^{11}	2.65×10^{11}	0.33	20.7×10^8	14.5×10^8	1.5‡	Re76		1
Al, 8 Si, 1.5 Cu, 0.3 Mg, 0.3 Mn	Perm. mold (stress relieved)	7.10×10^{11}	2.65×10^{11}	0.33	24.8×10^8		1.0‡	Re88		1
Al, 9.5 Si, 0.5 Mg	Die cast	7.10×10^{11}	2.65×10^{11}	0.33	28.9×10^8	15.8×10^8	1.8†			1
Al, 8.5 Si, 3.5 Cu	Die cast	7.10×10^{11}	2.65×10^{11}	0.33	31.0×10^8	17.2×10^8	2.0†			1
Al, 6.5 Sn, 1 Cu, 1 Ni	(Perm. mold cast) art. aged	7.10×10^{11}	2.65×10^{11}	0.33	15.2×10^8	6.89×10^8	12.0‡	45¶	9.65×10^8	1
Al, 5.5 Zn, 0.6 Mg, 0.5 Cr, 0.2 Ti	Sand cast	7.10×10^{11}	2.65×10^{11}	0.33	24.1×10^8	17.2×10^8	5.0‡	80¶	19.2×10^8	1

Wrought alloys:

Material	Condition	E (dyn/cm²)	G (dyn/cm²)	ν	$\times 10^8$	$\times 10^8$	Elong.	Hardness	Endurance	Ref.*
Aluminum 99.996 Al	Ann.	6.89×10^{11}	2.65×10^{11}	0.33	4.74×10^8	1.22×10^8	48.8§	17¶		1
Aluminum 99.996 Al	Cold rolled 75 %	6.89×10^{11}	2.65×10^{11}	0.33	11.2×10^8	10.6×10^8	5.5§	27¶		1
Aluminum 99.0⁺ Al	Ann.	6.89×10^{11}	2.65×10^{11}	0.33	8.96×10^8	3.45×10^8	35§	23¶	6.55×10^8	1
Aluminum 99.0⁺ Al	Hard H‖	6.89×10^{11}	2.65×10^{11}	0.33	16.6×10^8	14.5×10^8	5§	44¶	8.96×10^8	1
Al, 1.2 Mn	Ann.	6.89×10^{11}	2.65×10^{11}	0.33	11.0×10^8	4.14×10^8	30§	28¶	7.58×10^8	1
Al, 1.2 Mn	Hard H‖	6.89×10^{11}	2.65×10^{11}	0.33	20.0×10^8	17.2×10^8	4§	55¶	11.0×10^8	1
Al, 5.5 Cu, 0.5 Pb, 0.5 Bi	H-t, then cold-worked	7.10×10^{11}	2.65×10^{11}	0.33	36.5×10^8	32.4×10^8	15‡	95¶	20.7×10^8	1
Al, 5.5 Cu, 0.5 Pb, 0.5 Bi	H-t, then cold-worked, then art. aged	7.10×10^{11}	2.65×10^{11}	0.33	39.3×10^8	30.3×10^8	14‡	100¶	22.8×10^8	1
Al, 4 Cu, 0.6 Mn, 0.6 Mg, 0.5 Pb, 0.5 Bi	Quenched (h-t)	7.10×10^{11}	2.65×10^{11}	0.33	42.1×10^8	24.1×10^8	22§	100¶	12.4×10^8	1
Al, 4.4 Cu, 0.8 Si, 0.8 Mn, 0.4 Mg	Ann.	7.31×10^{11}	2.65×10^{11}	0.33	18.6×10^8	9.65×10^8	18‡	45¶	29.0×10^8	1
Al, 4.4 Cu, 0.8 Si, 0.8 Mn, 0.4 Mg	H-t, art. aged	7.31×10^{11}	2.65×10^{11}	0.33	48.3×10^8	41.4×10^8	13‡	135¶	12.4×10^8	1
Al, 4 Cu, 0.5 Mg, 0.5 Mn	Ann.	7.17×10^{11}	2.65×10^{11}	0.33	17.9×10^8	6.89×10^8	22§	45¶	26.2×10^8	1
Al, 4 Cu, 0.5 Mg, 0.5 Mn	H-t, nat. aged	7.17×10^{11}	2.65×10^{11}	0.33	42.7×10^8	27.6×10^8	17‡	105¶	16.6×10^8	1
Al, 4 Cu, 2 Ni, 0.5 Mg	Forged, h-t aged	7.10×10^{11}	2.65×10^{11}	0.33	43.4×10^8	32.4×10^8	1‡	115¶	12.4×10^8	2
Al, 4 Cu, 2 Ni, 1.5 Mg	Sand cast	7.31×10^{11}	2.65×10^{11}	0.33	19.3×10^8	16.6×10^8	19§	80¶	28.3×10^8	2
Al, 4.5 Cu, 1.5 Mg, 0.6 Mn	Ann.	7.31×10^{11}	2.65×10^{11}	0.33	18.6×10^8	7.58×10^8	11§	42¶	24.1×10^8	1
Al, 4.5 Cu, 1.5 Mg, 0.6 Mn	H-t, nat. aged	7.17×10^{11}	2.65×10^{11}	0.33	46.9×10^8	31.7×10^8	18‡	120¶	26.2×10^8	1
Al, 4.5 Cu, 0.8 Mn, 0.8 Si	H-t, art. aged	7.10×10^{11}	2.65×10^{11}	0.33	24.1×10^8	24.1×10^8	8‡	110¶	22.1×10^8	1
Al, 12.5 Si, 1.0 Mg, 0.9 Cu, 0.9 Ni	H-t, art. aged	7.03×10^{11}	2.65×10^{11}	0.33	38.6×10^8	31.7×10^8	20‡	125¶	12.4×10^8	1
Al, 1.0 Si, 0.6 Mg, 0.25 Cr	H-t, art. aged	7.03×10^{11}	2.65×10^{11}	0.33	32.4×10^8	27.6×10^8	25§	100¶	16.6×10^8	1
Al, 2.5 Mg, 0.25 Cr	Ann.	7.03×10^{11}	2.65×10^{11}	0.33	20.0×10^8	9.65×10^8	7§	45¶	7.58×10^8	1
Al, 2.5 Mg, 0.25 Cr	Strain hardened (H)	6.89×10^{11}	2.65×10^{11}	0.33	28.3×10^8	24.8×10^8	35‡	85¶	13.8×10^8	1
Al, 1.3 Mg, 0.7 Si, 0.25 Cr	Ann.	6.89×10^{11}	2.65×10^{11}	0.33	11.0×10^8	4.83×10^8	30‡	26¶		1
Al, 1.3 Mg, 0.7 Si, 0.25 Cr	H-t, nat. aged	6.89×10^{11}	2.65×10^{11}	0.33	22.8×10^8	13.8×10^8	35‡	65¶		1
Al, 5.2 Mg, 0.1 Mn, 0.1 Cr	Ann.	7.10×10^{11}	2.65×10^{11}	0.33	29.0×10^8	13.8×10^8				1
Al, 5.2 Mg, 0.1 Mn, 0.1 Cr	Hard H‖	7.10×10^{11}	2.65×10^{11}	0.33	40.0×10^8	33.1×10^8	7‡			1
Al, 1.0 Mg, 0.6 Si, 0.25 Cu, 0.25 Cr	Ann.	6.89×10^{11}	2.65×10^{11}	0.33	12.4×10^8	5.52×10^8	22§	30¶		1
Al, 1.0 Mg, 0.6 Si, 0.25 Cu, 0.25 Cr	H-t, nat. aged	6.89×10^{11}	2.65×10^{11}	0.33	24.1×10^8	14.5×10^8	22§	65¶		1
Al, 5.5 Zn, 2.5 Mg, 1.5 Cu, 0.3 Cr, 0.2 Mn	Ann.	7.17×10^{11}	2.65×10^{11}	0.33	22.8×10^8	10.3×10^8	17§			1
Al, 5.5 Zn, 2.5 Mg, 1.5 Cu, 0.3 Cr, 0.2 Mn	H-t, art. aged	7.17×10^{11}	2.65×10^{11}	0.33	56.5×10^8	49.6×10^8	11§	150¶	8.62×10^8	1
Al, 6.4 Zn, 2.5 Mg, 1.2 Cu	Ann. (0.064 sheet)	7.17×10^{11}	2.69×10^{11}	0.33	20.7×10^8	10.3×10^8	18	Re57–Re62	16.5×10^8	1

* References are on p. 2-76.
† ½-in. round specimen.
‡ ¾-in. round specimen.
¶ 10-mm ball, 500-kg load.
§ $\frac{1}{16}$-in. sheet specimen.
‖ H-strain hardened to a prescribed hardness.

Table 2e-9. Elastic and Strength Constants for Copper Alloys

Alloy	Condition	E	G	σ	Tensile strength	Yield strength	Elongation	Reduction in area	Bhn	Shear strength	Ref.
99.997 Cu, 0.0016 S	½-in. rod, cold drawn	12.77×10^{11}	4.68×10^{11}	0.364	35.1×10^8	34.0×10^{8}†	14	88	R_B37	....	2
99.996 Cu, 0.002 S, 0.002 Fe	Ann., ¾-in. rod	11.2×10^{11}	....		21.3×10^8	3.44×10^{8}†	60	92	......	....	2
99.950 Cu, 0.043 O$_2$, 0.002 Fe, 0.002 S	Ann., ¾-in. rod	10.9×10^{11}	....		21.7×10^8	3.79×10^{8}†	53	71	......	....	2
99.92 Cu, 0.04 O$_2$	H.R. (0.040-in. flat)	11.7×10^{11}	....	0.33 ± 0.01	23.4×10^8	6.89×10^{8}†	45	....	R_F45	15.8×10^8	1
99.94 Cu, 0.02 P	0.040 in. flat spec. (G.S. 0.050 mm)	11.7×10^{11}	....		22.0×10^8	6.89×10^{8}†	45	....	R_F40	15.2×10^8	1
95 Cu, 5 Zn	Rolled strip 0.040 in. (G.S. 0.050 mm)	11.7×10^{11}	....		23.4×10^8	6.89×10^{8}†	45	....	R_F46	....	1
95 Cu, 5 Zn	Rolled strip 0.040 in. (spring)	11.7×10^{11}	....		44.1×10^8	40.0×10^{8}†	4	....	R_B73	27.6×10^8	1
90 Cu, 10 Zn	Flat, 0.040 in. (spring)	11.7×10^{11}	....		49.6×10^8	42.7×10^{8}†	3	....	R_B78	28.9×10^8	1
90 Cu, 10 Zn	Flat, 0.040 in. as H.R.	11.7×10^{11}	....		26.9×10^8	9.65×10^{8}†	44	....	R_F60	21.4×10^8	1
85 Cu, 15 Zn	Flat, 0.040 in. (G.S. 0.050 mm)	11.7×10^{11}	....		27.6×10^8	8.27×10^{8}†	47	....	R_F59	21.4×10^8	1
85 Cu, 15 Zn	Flat, 0.040 in. (spring temper)	11.7×10^{11}	....		57.9×10^8	43.4×10^{8}†	3	....	R_B86	31.7×10^8	1
80 Cu, 20 Zn	Flat, 0.040 in. (G.S. 0.050 mm)	11.7×10^{11}	....		30.3×10^8	9.65×10^{8}†	50	....	R_F61	22.0×10^8	1
80 Cu, 20 Zn	Flat, 0.040 in. (spring temper)	11.0×10^{11}	....		62.7×10^8	44.8×10^{8}†	3	....	R_B91	33.1×10^8	1
70 Cu, 30 Zn	Flat, 0.040 in. (G.S. 0.070 mm)	11.0×10^{11}	....		31.7×10^8	9.65×10^{8}†	65	....	R_F58	22.0×10^8	1
70 Cu, 30 Zn	Flat, 0.040 in. (spring temper)	11.0×10^{11}	....		64.8×10^8	44.8×10^{8}†	3	....	R_B91	33.1×10^8	1
70 Cu, 30 Zn	Flat, 0.040 in. (extra spring temper)	11.0×10^{11}	....		68.2×10^8	44.8×10^{8}†	3	....	R_B93	....	1
65 Cu, 35 Zn	Flat, 0.040 in., ann.	10.3×10^{11}	....		33.8×10^8	11.7×10^{8}†	57	....	R_F68	23.4×10^8	1
65 Cu, 35 Zn	Flat, 0.040 in. (spring temper)	10.3×10^{11}	....		62.7×10^8	42.7×10^{8}†	3	....	R_B90	32.4×10^8	1
60 Cu, 40 Zn	Flat, 0.040 in., ann.	10.3×10^{11}	....		37.2×10^8	14.4×10^{8}†	45	....	R_F80	27.6×10^8	1
89 Cu, 9.25 Zn, 1.75 Pb	Rod, ann.	11.7×10^{11}	....		25.5×10^8	8.27×10^{8}†	45	70	R_F55	16.5×10^8	1
64.5 Cu, 35 Zn, 0.5 Pb	Flat specimen, ann.	10.3×10^{11}	....		33.8×10^8	11.7×10^{8}†	57	....	R_F68	23.4×10^8	1
67 Cu, 32.5 Zn, 0.5 Pb	Tubular specimen, ann.	10.3×10^{11}	....		32.4×10^8	10.3×10^{8}†	60	....	R_F64	....	1
64.5 Cu, 34.5 Zn, 1.0 Pb	Rolled, flat spec., ann.	10.3×10^{11}	....		33.8×10^8	11.7×10^{8}†	54	....	R_F68	23.4×10^8	1
62.5 Cu, 35.75 Zn, 1.75 Pb	Rolled, flat spec., ann.	10.3×10^{11}	....		33.8×10^8	11.7×10^{8}†	52	....	R_F68	22.0×10^8	1
62.5 Cu, 35 Zn, 2.5 Pb	Rolled, flat spec., ann.	9.65×10^{11}	....		33.8×10^8	11.7×10^{8}†	50	....	R_F68	21.4×10^8	1
61.5 Cu, 35.5 Zn, 3 Pb	Rod, ann.	9.65×10^{11}	....		33.8×10^8	12.4×10^{8}†	53	....	R_F68	20.7×10^8	1
60 Cu, 39.5 Zn, 0.5 Pb	H.R. 1-in. plate	10.3×10^{11}	....		37.2×10^8	13.8×10^8	45	....	R_F80	27.6×10^8	1
60.5 Cu, 38.4 Zn, 1.1 Pb	Light ann. 1.5-in. OD tubing	10.3×10^{11}	....		37.2×10^8	13.8×10^8	40	....	R_F80	....	1
60 Cu, 38 Zn, 2 Pb	Extruded 1-in. rod	10.3×10^{11}	....		35.8×10^8	13.8×10^8	45	....	R_F78	....	1
57 Cu, 40 Zn, 3 Pb	Extruded 1-in. section	9.65×10^{11}	....		41.3×10^8	13.8×10^8	30	....	R_B65	....	1
71 Cu, 28 Zn, 1 Sn	As H.R. (1-in. plate)	10.3×10^{11}	....		33.1×10^8	12.4×10^{8}†	65	....	R_F70	....	1
60 Cu, 39.25 Zn, 0.75 Sn	As H.R. (1-in. plate)	10.3×10^{11}	....		37.9×10^8	17.2×10^{8}†	50	....	R_B55	27.6×10^8	1

Composition, %	Form and treatment	$E \times 10^{11}$	Tensile $\times 10^3$	Yield $\times 10^3$	Elong., %	Red. of area, %	Hardness	Endurance $\times 10^8$	Ref.
60 Cu, 37.5 Zn, 1.75 Pb, 0.75 Sn	1-in. rod, soft ann.	10.3×10^{11}	39.3×10^3	20.7×10^3†	40	…	R_B55	24.8×10^8	1
58.5 Cu, 39 Zn, 1.4 Fe, 1 Sn, 0.1 Mn	1-in. rod, soft ann.	10.3×10^{11}	44.8×10^3	20.7×10^3†	33	…	R_B65	28.9×10^8	1
95 Cu, 5 Sn	Ann., flat spec.	11.0×10^{11}	32.4×10^3	13.1×10^3	64	…	R_F26	…	1
92 Cu, 8 Sn	Ann., flat plate (0.040 in.)	11.0×10^{11}	37.9×10^3	…	70	…	R_F75	…	1
92 Cu, 8 Sn	Spring temper plate (0.040 in.)	11.0×10^{11}	77.2×10^3	…	3	…	R_B93	…	1
90 Cu, 10 Sn	Ann. flat plate (0.040 in.)	11.0×10^{11}	45.5×10^3	…	68	…	R_B55	…	1
90 Cu, 10 Sn	Spring, flat plate (0.040 in.)	11.0×10^{11}	84.1×10^3	…	4	…	R_B101	…	1
98.75 Cu, 1.25 Sn	Ann., flat plate (0.040 in.)	11.7×10^{11}	27.6×10^3	9.65×10^3	48	…	R_F60	…	1
98.75 Cu, 1.25 Sn	Spring, flat plate (0.040 in.)	11.7×10^{11}	51.7×10^3	…	4	…	R_B79	…	1
70 Cu, 30 Ni	H.R. 1-in. plate.	15.2×10^{11}	37.9×10^3	13.8×10^3	45	…	R_B35	…	1
65 Cu, 18 Ni, 17 Zn	Ann., flat plate (0.040 in.)	12.4×10^{11}	40.0×10^3	17.2×10^3	40	…	R_B40	…	1
55 Cu, 27 Zn, 18 Ni	Ann., flat plate (0.040 in.)	12.4×10^{11}	41.3×10^3	18.6×10^3	40	…	R_B55	…	1
55 Cu, 27 Zn, 18 Ni	Spring, flat plate (0.040 in.)	12.4×10^{11}	79.2×10^3	64.1×10^3	2.5	…	R_B99	…	1
Cu, 3 Si	Flat plate (0.040 in.) (G.S. 0.070 mm)	…	38.6×10^3	14.5×10^3†	63	…	R_B40	28.9×10^8	1
Cu, 3 Si	Flat plate (0.040 in.) spring	…	75.8×10^3	42.7×10^3†	4	…	R_B97	43.4×10^8	1
Cu, 1.5 Si	1-in. rod (G.S. 0.035 mm)	10.3×10^{11}	27.6×10^3	17.6×10^3†	50	…	R_F55	…	1
94.88 Cu, 5.02 Al, 0.04 Fe, 0.06 Zn	0.041-in. sheet, ann. at 500°C	…	41.5×10^3	17.6×10^3†	65.8	…	$R_B48.5$	…	1
94.88 Cu, 5.02 Al, 0.04 Fe, 0.06 Zn	0.041-in. sheet, C.R., 44% reduction	…	68.9×10^3	44.0×10^3†	8.0	…	$R_B93.5$	…	1
91.74 Cu, 8.10 Al, 0.04 Fe, 0.02 Ni, 0.10 Zn	0.020-in. sheet, ann. at 400°C	…	53.9×10^3	29.1×10^3†	41.8	…	…	…	1
91.74 Cu, 8.10 Al, 0.04 Fe, 0.02 Ni, 0.10 Zn	0.020-in. sheet, C.R., 37% reduction	…	62.7×10^3	45.3×10^3†	12.8	…	…	…	1
92.65 Cu, 7.35 Al	H.R.	…	43.4×10^3	…	73	…	134‡	…	1
92 Cu, 7 Al, 1 Ni	C.R., ann.	…	85.4×10^3	…	4.5	…	…	…	1
89.25 Cu, 9.25 Al, 0.6 Fe, 0.5 Ni	Ann., rod	…	55.1×10^3	27.6×10^3†	22	…	…	…	1
87.46 Cu, 5.62 Al, 6.93 Ni	H.R.	…	75.7×10^3	71.2×10^3	20	…	241‡	…	1
85.75 Cu, 10.75 Al, 3.50 Fe	Sand cast	11.2×10^{11}	62.0×10^3†	25.5×10^3†	14	…	175§	…	1
81.3 Cu, 10.7 Al, 4.0 Fe, 4.0 Ni	Forged, ann. at 845°C	11.7×10^{11}	65.1×10^3	41.4×10^3	28.0	…	R_B90	…	1
Cu, 2 Be, 0.25 Co (or 0.35 Ni)	Solution treated, quenched	…	49.6×10^3	17.2×10^8¶	50	…	R_B90	…	1
88 Cu, 6 Sn, 1.5 Pb, 4.5 Zn	Sand cast 0.505-in. section	8.96×10^{11}	26.2×10^3	11.0×10^3†	35	…	66‡	…	1
87 Cu, 8 Sn, 1 Pb, 4 Zn	Sand cast 0.505-in. section	9.65×10^{11}	24.8×10^3	12.4×10^3†	30	35	68‡	…	1
85 Cu, 5 Sn, 9 Pb, 1 Zn	Sand cast 0.505-in. section	9.99×10^{11}	20.7×10^3	10.3×10^3†	15	16	60‡	…	1
83 Cu, 7 Sn, 7 Pb, 3 Zn	Sand cast 0.505-in. section	7.58×10^{11}	23.4×10^3	10.3×10^3†	20	18	60‡	…	1
80 Cu, 10 Sn, 10 Pb	Sand cast 0.505-in. section	7.23×10^{11}	22.0×10^3	11.7×10^3†	12	10.0	65‡	…	1
78 Cu, 7 Sn, 15 Pb	Sand cast	6.89×10^{11}	20.7×10^3	11.7×10^3†	15	15	55‡	…	1
70 Cu, 5 Sn, 25 Pb	Sand cast	9.30×10^{11}	14.5×10^3	11.0×10^3†	10	8	48‡	…	1
85 Cu, 5 Sn, 5 Pb, 5 Zn	Sand cast	…	23.4×10^3	11.7×10^3†	8	25	60‡	…	1
83 Cu, 4 Sn, 6 Pb, 7 Zn	Sand cast	…	22.0×10^3	10.3×10^3†	25	20	55‡	…	1
81 Cu, 3 Sn, 7 Pb, 9 Zn	Sand cast	8.96×10^{11}	22.0×10^3	10.3×10^3†	24	20	55‡	…	1

TABLE 2e-9. ELASTIC AND STRENGTH CONSTANTS FOR COPPER ALLOYS (*Continued*)

Alloy	Condition	B	G	σ	Tensile strength	Yield strength	Elongation	Reduction in area	Bhn	Shear strength	Ref.*
76 Cu, 3 Sn, 6 Pb, 15 Zn	Sand cast	8.27×10^{11}	.....	.....	22.0×10^8	10.3×10^8†	30	30	55‡	.....	1
71 Cu, 1 Sn, 3 Pb, 25 Zn	Sand cast 0.505-in. section	8.96×10^{11}	.....	.....	24.1×10^8	8.27×10^8†	35	30	48‡	.....	1
66 Cu, 1 Sn, 3 Pb, 30 Zn	Sand cast 0.505-in. section	8.96×10^{11}	.....	.....	23.4×10^8	8.96×10^8†	35	30	50‡	.....	1
60 Cu, 1 Sn, 1 Pb, 38 Zn	Sand cast	9.65×10^{11}	.....	.....	27.6×10^8	9.65×10^8†	25	25	65‡	.....	1
63 Cu, 26 Zn, 3 Fe, 5.5 Al, 3.5 Mn	Sand cast	10.7×10^{11}	.....	.....	79.2×10^8	48.2×10^8†	15	15	210§	.....	1
58 Cu, 39.25 Zn, 1.25 Fe, 1.25 Al, 0.25 Mn	Sand cast	10.3×10^{11}	.....	.....	48.2×10^8	19.3×10^8†	30	30	125‡	.....	1
59 Cu, 0.75 Sn, 0.75 Pb, 37 Zn, 1.25 Fe, 0.75 Al, 0.5 Mn	Sand cast	10.3×10^{11}	.....	.....	44.8×10^8	20.7×10^8†	18	20	85‡	.....	1
66 Cu, 5 Sn, 1.5 Pb, 2 Zn, 25 Ni	Sand cast	.....	.....	.....	34.4×10^8	16.5×10^8†	15	15	130‡	.....	1
64 Cu, 4 Sn, 4 Pb, 8 Zn, 20 Ni	Sand cast	.....	.....	.....	27.6×10^8	17.2×10^8†	15	14	105‡	.....	1
57 Cu, 2 Sn, 9 Pb, 20 Zn, 12 Ni	Sand cast	.....	.....	.....	23.4×10^8	10.3×10^8†	25	20	60‡	.....	1
60 Cu, 3 Sn, 5 Pb, 16 Zn, 16 Ni	Sand cast	.....	.....	.....	26.2×10^8	11.7×10^8†	15	25	75‡	.....	1
89 Cu, 1 Fe, 10 Al	Sand cast, cooled in sand	11.7×10^{11}	.....	.....	46.2×10^8	22.0×10^8†	15	15	140§	.....	1
87.5 Cu, 3.5 Fe, 9 Al	Sand cast	12.4×10^{11}	.....	.....	51.7×10^8	18.6×10^8†	35	32	120§	.....	1
86 Cu, 4 Fe, 10 Al	Sand cast, cooled in sand	11.7×10^{11}	.....	.....	51.7×10^8	24.1×10^8†	18	15	155§	.....	1
79 Cu, 5 Fe, 11 Al, 5 Ni	Sand cast	.....	.....	.....	65.5×10^8	31.0×10^8†	7	7	195§	.....	1

* References are on p. 2-76.
† At 0.5% extension.
‡ 10-mm ball, 500-kg load.
¶ At 0.01% offset.
§ 10-mm ball, 3,000-kg load.

TABLE 2e-10, ELASTIC AND STRENGTH CONSTANTS FOR VARIOUS SOLIDS

Material	Condition	E	G	Tensile strength	Yield strength at 0.2% offset	Elongation	Bhn	Ref.*
Iridium	Ann.	52×10^{11}					Vhn 170	1
Osmium	Ann.	56×10^{11}					Vhn 400	1
Rhodium	Ann.							1
Ruthenium	As cast	41×10^{11}		50×10^{8}			Vhn 390	1
Antimony		7.78×10^{11}		1.1×10^{8}			30–58	1
Beryllium	Vacuum cast	29×10^{11}		$12\text{–}15 \times 10^{8}$				1
Cadmium	Chill cast 1-in. section	5.5×10^{11}†		7.1×10^{8}		50	21–23	1
Calcium	Cast slab	$2\text{–}3 \times 10^{11}$		5.5×10^{8}	3.8×10^{8}	53–60	17	1
Chromium	As cast						110–170	1
Cobalt	Cast	21×10^{11}		23.7×10^{8}			125	1
Columbium	Sheet, ann. 0.01-in. section			34×10^{8}		30		1
Columbium	Sheet, worked 0.01-in. section			69×10^{8}		1		1
Lithium							Softer than pure lead	1
Manganese	Quenched	34×10^{11}		50×10^{8}	24×10^{8}	40	$R_C 35$	1
Molybdenum	Pressed + sintered (sheet)			69×10^{8}			156	1
Silicon	Chill cast 3.55 × 0.97 × 0.97 in.	11.26×10^{11}						1
Sodium							0.07‡	1
Tantalum	Ann. 0.010-in. sheet			34×10^{8}		40	$R_E 60$	1
Tantalum	Worked 0.010-in. sheet			76×10^{8}		1	$R_E 95$	1
Titanium	Ann.	11.6×10^{11}		54×10^{8}	43×10^{8}	25.2	$R_G 76$	1
Titanium	Hard, 60% reduction			76.82×10^{8}		1.5	$R_G 72$	1
Tungsten		34×10^{11}	13.5×10^{11}					1
Zirconium	Hard drawn	9.99×10^{11}		84×10^{8}	48×10^{8}	18¶	$R_B 87.4$	1

* References are on p. 2–76.
† Sand cast.
‡ 3.2-kg load, 10-mm ball.
¶ Per cent in 4 in.

TABLE 2e-11. ELASTIC AND STRENGTH CONSTANTS FOR IRON AND STEEL ALLOYS

% C	Alloy	Condition	E	G	σ	Tensile strength	Yield strength	Elongation	Reduction in area	Bhn	Shear strength	Ref.
	Iron:											
	2.50 C, 0.79 Si, 0.09 S, 0.04 P	Cast	13.8×10^{11}			32.8×10^8				266	30.7×10^8	2
	3.52 C, 2.55 Si, 1.01 Mn, 0.215 P, 0.086 S	⅞-in. cast, ann. bar	12.1×10^{11}			23.5×10^8				163	30.2×10^8	2
	3.52 C, 2.55 Si, 1.01 Mn, 0.215 P, 0.086 S	2-in. bar	8.27×10^{11}	5.10×10^{11}		15.5×10^8				164	25.1×10^8	2
	1.15–2.30 C, 0.85–1.20 Si, 0.40 Mn, 0.020 P, 0.012 S	Malleable, cast, ann.	17.2×10^{11}	8.61×10^{11}	0.17	39.3×10^8	25.8×10^8†	22		111–145	33.1×10^8	2
	Steel:											
	2.25–2.70 C, 0.80–1.10 Si		17.2×10^{11}	8.61×10^{11}	0.17	34.4×10^8	22.4×10^8†	14			33.1×10^8	12
0.03	0.12 Mn, 0.005 Si, 0.45 Cu, 0.07 Mo	H.R. at 540°C				34.3×10^8	24.0×10^8†	35.8	65			1
0.02	0.5 Cu	As normalized				34×10^8	27×10^8†	46	78			1
0.02	1.0 Cu	As normalized				39×10^8	31×10^8†	41	73			1
0.02	1.5 Cu	As normalized				48×10^8	43×10^8†	36	70			1
0.02	2.0 Cu	As normalized				55×10^8	50×10^8†	31	67			1
0.02	2.5 Cu	As normalized				56×10^8	52×10^8†	27	66			1
0.02	3.0 Cu	As normalized				56×10^8	52×10^8†	26	65			1
0.05	0.39 Si, 0.25 Mn, 0.014 P, 0.049 S	As rolled				40.0×10^8	27.9×10^8†	29.5§	72	117		1
0.07	1.17 Si, 0.32 Mn, 0.013 P, 0.034 S	As rolled				46.5×10^8	32.7×10^8†	29.5§	71.5	130		1
0.05	1.73 Si, 0.35 Mn, 0.014 P, 0.030 S	As rolled				50.0×10^8	37.6×10^8†	29.5§	64	140		1
0.06	2.39 Si, 0.16 Mn, 0.010 P, 0.016 S	As rolled				52.7×10^8	36.9×10^8†	24.5§	53.5	181		1
0.054	0.42 Mn, 0.025 Si, 0.031 Al, 0.265 Ti	H.R. 5% strained, aged				48.9×10^8	46.9×10^8†	11.9§	67.0			13
0.025	0.30 Mn, 0.010 P, 0.023 S, 0.09 Ni, 0.09 Cu, 0.26 V	Annealed				29.2×10^8	15.8×10^8†	28.1				1
0.08	1.01 Cr, 0.41 Cu, 0.80 Si, 27 Mn, 0.145 P, 0.020 S	H.R. ¾-in. bar	20.7×10^{11}	8.20×10^{11}		54.0×10^8	41.3×10^8‖	40	72	156		2
0.07	18.95 Cr, 7.69 Ni	C.R. ⅜-in. bar	17.2×10^{11}			98.5×10^8		21††		302		2
0.03	13.47 Cr, 0.27 V, 0.04 P, 0.01 S	H.R. 33⅜-in. bar	18.2×10^{11}	8.54×10^{11}		56.8×10^8		16§	26	175		2
0.08	1.07 Cu, 0.54 Ni, 0.43 Mn, 0.16 Si, 0.104 P, 0.022 S	H.R. ¾-in. bar	20.5×10^{11}	7.92×10^{11}		48.8×10^8	38.7×10^8**	38	69	145	47.2×10^8	2
0.08	1.46 Si, 0.102 Mn	W.Q. from 1830°F	20.9×10^{11}			63.7×10^8	47.1×10^8**	16		138		4
0.10	0.45 Mn, 3.71 Ni, 0.10 S	A.C. from 1550°F				60.4×10^8	34.4×10^8†	37	72			5
0.10	0.5 Cr, 0.3 Mo, 2.5 Ni	O.Q. from 820°C (carburized)				93.1×10^8	78.4×10^8†	13‡‡				5
0.10	0.6 Cr, 0.3 Mo, 3.3 Ni	O.Q. from 820°C (carburized)				122×10^8	108×10^8†	10‡‡				6
0.10	0.07 Si, 0.69 Mn, 0.092 P, 0.027 S, 0.16 Al	H.R. 4 hr at 540°C				52.8×10^8	39.2×10^8†	45.8				6
0.11	1.09 Cu, 0.15 Mo, 0.63 Ni	W.Q. from 900°C (carburized) 1⅛-in. diam C.R. bar	20.9×10^{11}			85.8×10^8	60.8×10^8†	12.8¶‖	52	Vhn 205		5
0.12	0.6 Mn, 1.4 Cr, 0.17 Mo, 1.0 Ni	Cast				57.4×10^8	52.4×10^8†	18	35.0	200		2
0.15	0.84 Mn, 0.12 Si, 0.099 P, 0.01 Si	Cast				68.9×10^8	44.0×10^8†	20.0	35.0	200		7
0.15	0.75 Mn, 0.30 Si, 1.75 Ni, 0.25 Mo	O.Q. from 780 to 180°				135×10^8	120×10^8†	15.8	50.0			8
0.16	0.75 Mn, 0.30 Si, 3.50 Ni	O.Q. from 1740°F, T at 1110°F	21.6×10^{11}			90.9×10^8	75.8×10^8†	21	63	85		9
0.15	0.4 Mn, 1.2 Cr, 0.25 Mo, 4.1 Ni	P.(O.Q.)(carburized)				83.8×10^8	66.8×10^8†	21.2	52.1			7
0.17	13.50 Cr, 0.11 Si					45×10^8	25×10^8†	32.0	53.0	130		9
0.18	0.5 Mn, 0.25 Mo, 1.8 Ni	Cast	20.5×10^{11}			96.5×10^8	82.7×10^8†	15.0	50.0			7
0.18	0.55 Mn, 0.25 Si	C.R.	22.6×10^{11}			67.6×10^8	45.5×10^8†	15	46			2
0.18	2.50 Cr, 0.55 Mn, 0.40 Si, 0.40 Mo, 0.20 V											2
0.18	0.92 Mn, 0.115 P, 0.12 S, 0.02 Si											2
0.20	16.17 Cr, 1.06 Mn, 0.30 Si	O.Q. from 1740°F, T at 840°F				130×10^8	61.2×10^8†	10		357		2

0.19	1.35 Mn, 0.10 S	W.Q. from 1550°F	20.9×10^{11}	82.0×10^{11}		89.6×10^8†	60.8×10^8†	16	44	251	58.0×10^8	4
0.20	0.45 Cr, 1.19 Mn, 0.67 Si, 0.033 P, 0.019 S	Rolled	20.4×10^{11}		0.273	59.0×10^8	37.4×10^8	30		156		2
0.25	0.45 Mn, 0.40 S, 0.03 Si, 0.012 P	Rolled, ¾-in. plate	20.3×10^{11}		0.306	43.5×10^8	22.3×10^8	40	70	122		2
0.25	to 0.35		20.53×10^{11}		0.297							2
0.15	to 0.25; 0.3–0.6 Mn, 0.045 P, 0.05 S	H.R.	20.12×10^{11}		0.313							10
0.15	to 0.25; 0.3–0.6 Mn, 0.015 P, 0.05 S	C.R.	18.9×10^{11}		0.286							10
0.27	0.72 Mn, 0.21 Si, 0.024 S, 0.014 P	Wr., ann. at 1450°F, F.C.	20.4×10^{11}		0.316	46.4×10^8	25.8×10^8	46	64	153	52.3×10^8	2
0.27	0.72 Mn, 0.21 Si, 0.024 S, 0.014 P	Wr., W.Q. from 1600°F, T at 1100°F			0.310	62.8×10^8	37.9×10^8	42	70	191		2
0.19	0.85 Mn, 0.05 (max) S, 0.045 (max) P	H.R. (trans. prop.)				42.6×10^8	22.5×10^8	36.0	53.7			1
0.19	0.85 Mn, 0.05 (max) S, 0.045 (max) P	H.R. (long. prop.)				44.1×10^8	25.0×10^8	43.5§§	66.5			1
0.10	0.75 Mn, 0.20 S, 0.10 P	H.R. (trans. prop.)				43.1×10^8	24.7×10^8	22.6§§	24.5			1
0.10	0.75 Mn, 0.20 S, 0.10 P	H.R. (long. prop.)				46.1×10^8	27.5×10^8	37.5§§	60.5			1
0.30	0.70 Mn, 3.5 Ni	Ann.				54.7×10^8	39.3×10^8	31.5	57.2			1
0.34	0.88 Mn, 0.35 Si, 0.035 S, 0.019 P	Rolled ¾-in. plate	20.5×10^{11}	8.06×10^{11}	0.291	59.5×10^8	24.1×10^8|||	33	58	168	55.0×10^8	2
0.38	0.65 Mn, 0.22 Si	Wr., ann. at 1450°F, F.C.	19.8×10^{11}	7.44×10^{11}	0.287	52.2×10^8	28.6×10^8|||	44	56	146		2
0.91	0.38 Mn, 0.16 Si, 0.036 P	O.Q. from 1575°F, T at 940°F	20.8×10^{11}	7.44×10^{11}		155×10^8	99.2×10^8|||	7		444		2
1.04	0.36 Mn, 0.16 Si, 0.018 S, 0.015 P	O.Q. from 1550°F, O.Q. from 120°F, T ½ hr at 800°F	20.5×10^{11}			163×10^8	99.2×10^8|||	5				2
0.37	0.50 Cr, 1.14 Mn, 0.84 Si, 0.033 S, 0.021 P	H.R. ¾-in. bar	21×10^{11}	8.27×10^{11}		86.1×10^8	55.6×10^8|||	23	58	255	60.2×10^8	2
0.60	0.56 Cr, 0.62 Mn, 0.26 Si	O.Q. from 1470°F, T at 750°F	21.1×10^{11}			164×10^8		2.5§	2.0	469		2
0.45	1.14 Cr, 0.69 Mn, 0.12 Si	N at 1525°F	21×10^{11}			83.4×10^8	61.7×10^8	12§	50	250		2
0.33	0.78 Cr, 0.24 Mo, 0.54 Mn, 0.21 Si, 0.025 P, 0.029 S	W., F.C. from 1450°F	19.7×10^{11}	8.27×10^{11}	0.288	52.8×10^8	29.3×10^8	48	66	170		2
0.33	0.78 Cr, 0.24 Mo, 0.54 Mn, 0.21 Si, 0.025 P, 0.029 S	Wr., O.Q. from 1600°F, T at 1100°F	19.8×10^{11}	8.13×10^{11}	0.272	86.8×10^8	62.4×10^8	28	60	229	78.5×10^8	2
0.34	0.46 Mn, 21.39 Cr, 10.95 Ni, 3.16 W, 1.39 Si	A.C. from 1740°F	20.1×10^{11}			88.2×10^8	30.9×10^8***	25	35	269		2
0.37	1.18 Cr, 0.16 V, 0.71 Mn, 0.33 Si, 0.037 S, 0.024 P	Wr., F.C. from 1450°F	20.3×10^{11}	8.13×10^{11}	0.289	61.1×10^8	33.9×10^8	42	62	179	61.7×10^8	2
0.31	1.66 Mn, 0.25 Si, 0.024 S, 0.015 P	Wr. F.C. from 1450°F	19.2×10^{11}	8.27×10^{11}	0.295	58.5×10^8	29.9×10^8	42	54	169	58.1×10^8	2
0.43	3.47 Ni, 0.64 Mn, 0.20 Si, 0.023 S, 0.015 P	Wr. F.C. from 1450°F	21×10^{11}	8.34×10^{11}	0.308	65.0×10^8	38.5×10^8	33	45	187	60×10^8	2
0.40	1.65 Ni, 0.99 Cr, 0.51 Mn, 0.20 Si, 0.028 S, 0.019 P	Wr., F.C. from 1450°F	19.8×10^{11}	7.78×10^{11}	0.299	61.9×10^8	30.2×10^8	40	54	170	62.4×10^8	2
0.32	1.52 Ni, 0.86 Cr, 0.30 Mo, 0.60 Mn, 0.16 Si, 0.019 S, 0.014 P	Wr., F.C. from 1450°F	19.8×10^{11}	7.92×10^{11}	0.288	66.2×10^8	34.2×10^8	37	58	202	66.0×10^8	2
0.32	2.42 Ni, 0.49 Cr, 0.38 Mo, 0.88 Mn, 0.23 Si, 0.13 Cu, 0.04 S, 0.03 P	Cast ann. at 1575°F, 6-in. bar, T at 1200°F	20.2×10^{11}	7.92×10^{11}		81.3×10^8	67.5×10^8	10†††	16	260	72.3×10^8	2
1.27	12.69 Mn, 0.12 Si	W.Q. from 1830°F				102×10^8	53.2×10^8	44	49			1
0.78	0.10 Mn	Ann. at 1472°F				68.2×10^8	65.4×10^8	12	35			1

* References are on p. 2-76.
† At yield point.
‡ At 0.2% offset.
§ % in 70 mm.
¶ % in 8 in.

|| At 0.005% permanent set.
** At 0.05% permanent set.
†† % in 1.5 in.
‡‡ % in 3.94 in.
¶¶ % in 1.97 in.

§§ % in 0.75 in.
||| At 0.001% permanent set.
*** At 0.1% offset.
††† % in 4/√ area.

TABLE 2e-12. ELASTIC AND STRENGTH CONSTANTS FOR LEAD AND LEAD ALLOYS

Alloy	Condition	E	σ	Tensile strength	Yield strength at 0.5% offset	Elongation, % in 2 in.	Reduction in area	Bhn	Shear strength	Ref.*
99.90 Pb	Rolled, aged	....	....	1.77×10^8	0.95×10^8	22	....	Rʙ75	....	1
99.73 Pb	Sand cast	1.38×10^{11}	....	$1.1–1.3 \times 10^8$	0.55×10^8	30	100	3.2–4.5	1.2×10^8	1
99.73 Pb	Chill cast	....	0.40–0.45	1.4×10^8	....	47	100	4.2	....	1
0.023–0.033 Ca, 0.02–0.1 Cu, 0.002–0.02 Ag		....	....	....	....	....	....	.....	....	
1 Sb	Extruded	1.38×10^{11}	....	2.1×10^8	....	40	....	7	....	1
4 Sb	Extruded and aged	....	....	2.1×10^8	....	50	....	8.1	....	1
6 Sb	Rolled, 95% reduction	....	....	2.77×10^8	....	48.3	....	13.0	....	1
6 Sb	Chill cast	....	....	4.71×10^8	....	24	....	10.7	....	1
6 Sb	Extruded	....	....	2.27×10^8	....	65	....		....	1
6 Sb	Cold rolled, 95% reduction	....	....	2.82×10^8	....	47	....	9.5†	....	1
8 Sb	Rolled, 95% reduction	....	....	3.20×10^8	....	31.3	....	15.4	....	1
9 Sb	Chill cast	....	....	5.2×10^8	....	17	....	8	....	1
4.5–5.5 Sn	....	....	....	2.3×10^8	1.0×10^8	50	80	11.3	....	1
20 Sn	....	....	....	4.0×10^8	2.51×10^8	16	50	14.5	4.04×10^8	1
50 Sn	....	....	....	4.2×10^8	3.3×10^8	60	70	19	....	1
4.50–5.50 Sn, 9.25–10.75 Sb	Chill cast	2.89×10^{11}	....	6.9×10^8	....	5	....	20	....	1
4.50 + 5.50 Sn, 14–16 Sb	Chill cast	2.89×10^{11}	....	6.9×10^8	....	5	....	22	....	1
9.3–10.7 Sn, 14–16 Sb	Cast	2.89×10^{11}	....	7.2×10^8	....	4	....	20	....	1
0.75–1.25 Sn, 0.8–1.4 As, 14.5–17.5 Sb	Chill cast	2.89×10^{11}	....	7.1×10^8	....	2	....	22	....	1
0.6–1.0 Sn, 1.5–3.0 As, 12.0–13.5 Sb	Chill cast	2.89×10^{11}	....	6.8×10^8	....	1.5	....	22	....	1

* References are on p. 2-76.
† 1/16-in. ball, 9.85-kg load for 30 sec.

TABLE 2e-13. ELASTIC AND STRENGTH CONSTANTS FOR MAGNESIUM ALLOYS

Alloy	Condition	B	G	σ	Tensile strength	Yield strength at 0.2% offset	Elongation, % in 2 in.	Bhn	Shear strength	Ref.*
99.9 + Mg..........	....	4.48×10^{11}	1.67×10^{11}	0.35						
8.3–9.7 Al, 0.10 Mn, 1.7–2.3 Zn, ≤0.3 Si, ≤0.05 Cu, ≤0.01 Ni, 0.3 other	Sand and permanent cast molds, as fabricated	4.48×10^{11}	1.67×10^{11}	0.35	16.5×10^{8}	9.65×10^{8}	2	65	13.1×10^{8}	14
8.3–9.7 Al, 0.10 Mn, 1.7–2.3 Zn, ≤0.3 Si, ≤0.05 Cu, ≤0.01 Ni, 0.3 other	Sand and permanent cast molds, cast and stabilized	4.48×10^{11}	1.67×10^{11}	0.35	16.5×10^{8}	9.65×10^{8}	2	..	13.1×10^{8}	14
8.3–9.7 Al, 0.10 Mn, 1.7–2.3 Zn, ≤0.3 Si, ≤0.05 Cu, ≤0.01 Ni, 0.3 other	Sand and permanent cast, solution h-t	4.48×10^{11}	1.67×10^{11}	0.35	27.6×10^{8}	9.65×10^{8}	10	63	13.8×10^{6}	14
5.3–6.7 Al, ≥0.15 Mn, 2.5–3.5 Zn, ≤0.3 Si, ≤0.05 Cu, ≤0.01 Ni, 0.3 other	Sand and permanent cast molds, as fabricated	4.48×10^{11}	1.67×10^{11}	0.35	20.0×10^{8}	9.65×10^{8}	6	50	12.4×10^{8}	14
5.3–6.7 Al, ≥0.15 Mn, 2.5–3.5 Zn, ≤0.3 Si, ≤0.05 Cu, ≤0.01 Ni, 0.3 other	Sand and permanent cast molds, cast and stabilized	4.48×10^{11}	1.67×10^{11}	0.35	20.0×10^{8}	9.65×10^{8}	5	..	13.1×10^{8}	14
5.3–6.7 Al, ≥0.15 Mn, 2.5–3.5 Zn, ≤0.3 Si, ≤0.05 Cu, ≤0.01 Ni, 0.3 other	Sand and permanent cast molds, solution h-t	4.48×10^{11}	1.67×10^{11}	0.35	27.6×10^{8}	9.65×10^{8}	12	55	13.1×10^{8}	14
8.3–9.7 Al, ≥0.13 Mn, 0.4–1.0 Zn, ≤0.5 Si, ≤0.10 Cu, ≤0.01 Ni, 0.3 other	Sand and permanent cast molds, as fabricated	4.48×10^{11}	1.67×10^{11}	0.35	16.5×10^{8}	9.65×10^{8}	2	52	....	14
8.3–9.7 Al, ≥0.13 Mn, 0.4–1.0 Zn, ≤0.5 Si, ≤0.10 Cu, ≤0.01 Ni, 0.3 other	Sand and permanent cast molds, solution h-t	4.48×10^{11}	1.67×10^{11}	0.35	27.6×10^{8}	9.65×10^{8}	11	53	....	14
8.3–9.7 Al, ≥0.13 Mn, 0.4–1.0 Zn, ≤0.5 Si, ≤0.10 Cu, ≤0.01 Ni, 0.3 other	Sand and permanent cast, solution h-t, aged	4.48×10^{11}	1.67×10^{11}	0.35	27.6×10^{8}	13.1×10^{8}	4	66	....	14
8.3–9.7 Al, ≥0.13 Mn, 0.4–1.0 Zn, ≤0.5 Si, 0.10 Cu, ≤0.01 Ni, 0.3 other	Die cast, as fabricated	4.48×10^{11}	1.67×10^{11}	0.35	22.7×10^{8}	15.2×10^{8}	3	60	13.8×10^{8}	14
8.3–9.7 Al, ≥0.10 Mn, 0.4–1.0 Zn, ≤0.5 Si, ≤0.3 Cu, ≤0.01 Ni, 0.3 other	Die cast, as fabricated	4.48×10^{11}	1.67×10^{11}	0.35	22.7×10^{8}	15.2×10^{8}	3	60	13.8×10^{8}	14
2.5–3.5 Al, 0.20 Mn, 0.6–1.4 Zn, 0.08–0.30 Ca, ≤0.3 Si, ≤0.05 Cu, ≤0.005 Fe, ≤0.005 Ni, 0.3 other.	Sheet, ann.	4.48×10^{11}	1.67×10^{11}	0.35	25.5×10^{8}	15.2×10^{8}	21	56	14.5×10^{8}	14
2.5–3.5 Al, ≥0.20 Mn, 0.6–1.4 Zn, 0.08–0.30 Ca, ≤0.3 Si, ≤0.05 Cu, ≤0.005 Fe, ≤0.005 Ni, 0.3 other.	Sheet, hard rolled	4.48×10^{11}	1.67×10^{11}	0.35	28.9×10^{8}	22.0×10^{8}	16	73	15.8×10^{8}	14

TABLE 2e-13. ELASTIC AND STRENGTH CONSTANTS FOR MAGNESIUM ALLOYS (*Continued*)

Alloy	Condition	E	G	σ	Tensile strength	Yield strength at 0.2% offset	Elongation, % in 2 in.	Bhn	Shear strength	Ref.*
2.5-3.5 Al, ≥0.20 Mn, 0.6-1.4 Zn, 0.08-0.30 Ca, ≤0.3 Si, ≤0.05 Cu, ≤0.005 Fe, ≤0.005 Ni, 0.3 other..........	Sheet, as fabricated	4.48×10^{11}	1.67×10^{11}	0.35	25.5×10^{8}	15.2×10^{8}	21	..	14.5×10^{8}	14
≥1.20 Mn, 0.08-0.14 Ca, ≤0.3 Si, ≤0.05 Cu, ≤0.01 Ni, 0.3 other..........	Sheet, ann.	4.48×10^{11}	1.67×10^{11}	0.35	22.7×10^{8}	12.4×10^{8}	16	48	12.4×10^{8}	14
≥1.20 Mn, 0.08-0.14 Ca, ≤0.3 Si, ≤0.05 Cu, ≤0.01 Ni, 0.3 other..........	Sheet, hard rolled	4.48×10^{11}	1.67×10^{11}	0.35	25.5×10^{8}	19.3×10^{8}	7	56	11.7×10^{8}	14
≥1.20 Mn, 0.08-0.14 Ca, ≤0.3 Si, ≤0.05 Cu, ≤0.01 Ni, ≤0.3 other..........	Sheet, as fabricated	4.48×10^{11}	1.67×10^{11}	0.35	22.7×10^{8}	....	..	..	....	14
5.8-7.2 Al, 0.15 Mn, 0.4-1.5 Zn, ≤0.3 Si, ≤0.05 Cu, ≤0.005 Ni, ≤0.005 Fe, +0.3 other	Extruded bars, rods, or shapes, as fabricated	4.48×10^{11}	1.67×10^{11}	0.35	30.3×10^{8}	20.7×10^{8}	14	60	13.1×10^{8}	14
7.8-9.2 Al, ≥0.15 Mn, 0.2-0.8 Zn, ≤0.3 Si, ≤0.05 Cu, ≤0.005 Ni, ≤0.005 Fe, 0.3 other	Extruded bars, rods, or shapes, as fabricated	4.48×10^{11}	1.67×10^{11}	0.35	33.1×10^{8}	22.0×10^{8}	12	60	15.2×10^{8}	14
7.8-9.2 Al, ≥0.15 Mn, 0.2-0.8 Zn, ≤0.3 Si, ≤0.05 Cu, ≤0.005 Ni, ≤0.005 Fe, 0.3 other	Extruded bars, rods, or shapes, aged	4.48×10^{11}	1.67×10^{11}	0.35	35.8×10^{8}	24.8×10^{8}	5	82	16.5×10^{8}	14
≥0.06 Mn, 4.3-6.2 Zn, ≥0.45 Zr, 0.3 other	Extruded bars, rods, or shapes, as fabricated	4.48×10^{11}	1.67×10^{11}	0.35	33.8×10^{8}	26.2×10^{8}	12	75	16.5×10^{8}	14
≥0.06 Mn, 4.3-6.2 Zn, ≥0.45 Zr, 0.3 other	Extruded bars, rods, or shapes, aged	4.48×10^{11}	1.67×10^{11}	0.35	35.1×10^{8}	28.9×10^{8}	10	82	17.2×10^{8}	14

* References are on p. 2-76.

Table 2e-14. Elastic and Strength Constants for Nickel and Nickel Alloys

Alloy	Condition	E	σ	Tensile strength	Yield strength at 0.2% offset	Elongation	Reduction in area	Bhn	Shear strength	Ref.*
63–70 Ni, ≤2.5 Fe, ≤2.0 Mn, remainder Cu	Wr., ann.			51.7×10^8	24.1×10^8	40	......	125	34–44×10^8	15
63–70 Ni, ≤2.5 Fe, ≤2.0 Mn, remainder Cu	Wr., H.R.	17.9×10^{11}	......	62.0×10^8	34.4×10^8	35	......	150	34–44×10^8	15
63–70 Ni, ≤2.5 Fe, ≤2.0 Mn, remainder Cu	Wr., cold drawn			68.9×10^8	55.1×10^8	25		190	34–44×10^8	15
63–70 Ni, ≤2.5 Fe, ≤2.0 Mn, remainder Cu	Wr., C.R. (hard temper)			75.8×10^8	68.9×10^8	5	......	240	34–44×10^8	15
63–70 Ni, 2.0–4.0 Al, 0.25–1.0 Ti, remainder Cu	H.R.	18×10^{11}	0.32	68.9×10^8	31.0×10^8	40		160	......	15
63–70 Ni, 2.0–4.0 Al, 0.25–1.0 Ti, remainder Cu	H.R., age hardened	18×10^{11}		103×10^8	75.8×10^8	25		280	......	15
63–70 Ni, 2.0–4.0 Al, 0.25–1.0 Ti, remainder Cu	Cold drawn	18×10^{11}	......	79.2×10^8	58.6×10^8	25	......	210	......	15
63–70 Ni, 2.0–4.0 Al, 0.25–1.0 Ti, remainder Cu	Cold drawn, age hardened	18×10^{11}		107×10^8	79.2×10^8	20		290	......	15
≥99.0 Ni, ≤0.15 C, ≤0.35 Mn, ≤0.40 Fe	Wr., ann.	21×10^{11}	0.31	48.2×10^8	13.8×10^8	40	......	100	36×10^8	15
≥99.0 Ni, ≤0.15 C, ≤0.35 Mn, ≤0.40 Fe	Wr., H.R.			51.7×10^8	17.2×10^8	40	......	110		15
≥99.0 Ni, ≤0.15 C, ≤0.35 Mn, ≤0.40 Fe	Wr., cold drawn			65.4×10^8	48.2×10^8	25	......	170		15
≥99.0 Ni, ≤0.15 C, ≤0.35 Mn, ≤0.40 Fe	Wr., cold rolled (hard temper)			72.3×10^8	65.4×10^8	5	......	210		15
≥99.0 Ni, ≤0.02 C	Ann.	21×10^{11}	0.31	41.3×10^8	10.3×10^8	50	......	90	......	15
≥93.0 Ni, 4.00–4.75 Al, 0.25–1.0 Ti, ≤0.30 C	H.R.	21×10^{11}	0.31	72.3×10^8	34.4×10^8	35		180		15
≥93.0 Ni, 4.00–4.75 Al, 0.25–1.0 Ti, ≤0.30 C	H.R., age hardened			117×10^8	89.6×10^8	15		320		15
≥93.0 Ni, 4.00–4.75 Al, 0.25–1.0 Ti, ≤0.30 C	Cold drawn			82.7×10^8	62.0×10^8	25		220	......	15
≥93.0 Ni, 4.00–4.75 Al, 0.25–1.0 Ti, ≤0.30 C	Cold drawn, age hardened			121×10^8	93.0×10^8	15		340	......	15

* References are on page 2-76.

TABLE 2e-14. ELASTIC AND STRENGTH CONSTANTS FOR NICKEL AND NICKEL ALLOYS (Continued)

Alloy	Condition	E	σ	Tensile strength	Yield strength at 0.2% offset	Elongation	Reduction in area	Bhn	Shear strength	Ref.*
≥72.0 Ni, 14.0–17.0 Cr, 6.0–10.0 Fe, ≤0.15 C	Wr., ann.	....	....	58.6×10^8	24.1×10^8	45	......	150	....	15
≥72.0 Ni, 14.0–17.0 Cr, 6.0–10.0 Fe, ≤0.15 C	Wr., H.R.	....	....	68.9×10^8	41.3×10^8	35	......	180	....	15
≥72.0 Ni, 14.0–17.0 Cr, 6.0–10.0 Fe, ≤0.15 C	Wr., cold drawn	....	....	79.2×10^8	62.0×10^8	20	......	200	....	15
≥72.0 Ni, 14.0–17.0 Cr, 6.0–10.0 Fe, ≤0.15 C	Wr., C.R. (hard temper)	....	....	93.0×10^8	75.8×10^8	5	......	260	....	15
≤70.0 Ni, 14.0–16.0 Cr, 5.0–9.0 Fe, 2.25–2.75 Ti, 0.4–1.0 Al, 0.7–1.2 Cb (+Ta)	Ann.	21×10^{11}	....	79.2×10^8	34.4×10^8	50	......	200	....	15
≥70.0 Ni, 14.0–16.0 Cr, 5.0–9.0 Fe, 2.25–2.75 Ti, 0.4–1.0 Al, 0.7–1.2 Cb (+Ta)	H.R., age hardened	21×10^{11}	....	124×10^8	82.7×10^8	25	......	360	....	15
63 Ni, 30 Cu, 4 Si, 2 Fe +	Sand cast	14.5×10^{11}	....	$76–100 \times 10^8$	$55–79 \times 10^8$	1–4	40–54	275–350	....	1
57 Ni, 20 Mo, 20 Fe +	Ann.	18.6×10^{11}	....	$76–83 \times 10^8$	$32.4–36 \times 10^8$	40–48	40–45	200–215	....	1
62 Ni, 30 Mo, 5 Fe +	Rolled, ann.	21.19×10^{11}	....	$90–96 \times 10^8$	$41–45 \times 10^8$	40–45	40–45	210–235	....	1
58 Ni, 17 Mo, 15 Cr, 5 W, 5 Fe +	Ann. plate	19.6×10^{11}	....	$79–88 \times 10^8$	$38–45 \times 10^8$	25–50	......	160–210	....	1
85 Ni, 10 Si, 3 Cu +	Sand cast	19.88×10^{11}	....	$25–27.9 \times 10^8$	....	....	......	....	....	1
80 Ni, 14 Cr, 6 Fe +	Ann.	21×10^{11}	....	....	19.3×10^8	....	......	....	....	1
58 Ni, 22 Cr, 6 Cu, 6 Mo, 6 Fe.	Sand cast	18.38×10^{11}	....	$41–50 \times 10^8$	....	4–9.5	8–11	160–210	41.9×10^8	1
80 Ni, 20 Cr +	Ann.	21×10^{11}	....	65.4×10^8	....	25–35	55	$R_B85–90$	....	1

* References are on p. 2-76.

TABLE 2e-15. ELASTIC AND STRENGTH CONSTANTS FOR TIN AND TIN ALLOYS

Alloy	Condition	E	σ	Tensile strength	Yield strength at 0.2% offset	Elonga-tion	Bhn	Shear strength	Ref.*
Pure tin	Cast	$4.1\text{--}4.5 \times 10^{11}$	....	2.14×10^8	....	55	5.3	2.00×10^8	1
Pure tin	Chill cast	....	....	1.45×10^8	....	69	....	....	1
Pure tin	0.1-in. sheet, ann.	....	....	1.65×10^8	....	96	....	....	1
99.8 Sn†	Cast	4.13×10^{11}	....	1.45×10^8	....	54†	Vhn 7.2	....	1
99.8 Sn†	Ann., 0.040-in. sheet	4.13×10^{11}	0.33	1.52×10^8	....	45		0.896×10^8	1
95 Sn, 5 Sb	Cast	....	....	4.06×10^8	....	38†	....	4.13×10^8	1
95 Sn, 5 Ag	0.040-in. sheet, aged at room temp.	....	....	3.17×10^8	2.48×10^8	49	....	....	1
70 Sn, 30 Pb	Cast	....	....	4.68×10^8	....	....	12	....	1
63 Sn, 37 Pb	Cast	....	....	5.17×10^8	....	32†	14	4.27×10^8	1
91 Sn, 4.5 Sb, 4.5 Cu	Chill cast	5.03×10^{11}	....	6.41×10^8	4.34×10^8‡	....	17	....	1
83.4 Sn, 8.3 Sb, 8.3 Cu	Chill cast	....	....	....	5.51×10^8‡	....	27	....	1

* References are on p. 2-76.
† % in 4 in.
‡ At 0.3% offset.

TABLE 2e-16. ELASTIC AND STRENGTH CONSTANTS FOR ZINC AND ZINC ALLOYS

Alloy	Condition	Tensile strength	Elongation, % in 2 in.	Bhn	Shear strength	Ref.*
3.5–4.3 Al, 0.03–0.08 Mg......	Die cast, ¼-in. section	28×10^8	10	82	21×10^8	1
3.5–4.3 Al, 0.75–1.25 Cu, 0.03–0.08 Mg	Die cast, ¼-in. section	33×10^8	7	91	26×10^8	1
3.5–4.5 Al, 2.5–3.5 Cu, 0.02–0.10 Mg	Die cast, ¼-in. section	35.9×10^8	8	100	32×10^8	1
4.5–5.0 Al, 0.2–0.3 Cu	Chill cast, ½-in. section	19×10^8	.....	...	....	1
5.25–5.75 Al................	Chill cast, ½-in. section	17×10^9	1	...	....	1
≤0.10 Pb..................	H.R. strip	$13.4–16 \times 10^8$	50–65	38	....	1
0.05–0.10 Pb, 0.05–0.08 Cd.....	H.R. strip	$14–17 \times 10^8$	30–52	43	....	1
0.25–0.50 Pb, 0.25–0.45 Cd.....	H.R. strip	$16–20 \times 10^8$	32–50	47	....	1
0.85–1.25 Cu................	H.R. strip	$16–22 \times 10^8$	15–20	52	....	1
0.85–1.25 Cu, 0.006–0.016 Mg...	H.R. strip	$19–25 \times 10^8$	10–20	61	....	1

* References are below.

References for Tables 2e-7 through 2e-16

1. "Metals Handbook," 1948 ed., American Society for Metals.
2. Natl. Bur. Standards (U.S.) Circ. C447, 1943.
3. Bain, E. C.: "Functions of the Alloying Elements in Steel," American Society for Metals, 1939.
4. Hoyt, S. L.: "Metals and Alloys Data Book" Reinhold Book Corporation, New York, 1943.
5. "Selection of Special Steels, Data Sheet," D.T.A. 72, Société de Commentry, Paris, France, 1946.
6. Halley, J. W.: Pat. 2,402,135, 1946.
7. "Nickel Alloy Steel," 2d ed., The International Nickel Co., Inc., New York, 1949.
8. "Fox Alloy Steels," Samuel Fox and Co. Ltd., Sheffield, England, 1942.
9. "Case Hardening of Nickel Alloy Steels," International Nickel Co., New York, 1941.
10. Everett, F. L., and J. Miklowitz: J. Appl. Phys. 15 (1944).
11. Climax Molybdenum Company Laboratory Records.
12. "Sheet Iron, a Primer," Republic Steel Corp., 1934.
13. Comstock, G. F.: J. Am. Ceram. Soc. 29 (1946).
14. "Magnesium Alloys and Products," Dow Chemical Co., 1950.
15. "Nickel," The International Nickel Co., Inc., rev. 1951.

TABLE 2e-17. DIFFUSION COEFFICIENTS FOR METALS

Metal	Test temp.	$D\left(\dfrac{cm^2}{sec}\right)$	Ref.
Ag into Ag	Room	0.895	1
Ag into Ag	460°C	8.0×10^{-14}	11
Ag into Ag	600°C	5.9×10^{-12}	11
Ag into Ag	666°C	2.45×10^{-11}	2
Ag into Ag	794°C	3.64×10^{-10}	2
Ag into Ag	936°C	4.61×10^{-9}	2
Al into Cu	Room	1.75×10^{-2}	1
Au into Au	Room	0.160	3
Au into Cu	Room	0.1 ± 0.06	4
Be into Cu	Room	2.32×10^{-4}	1
Bi into Pb	Room	0.018	3
Cd into Cu	Room	1.97×10^{-9}	3
Cd into Ag	Room	7.3×10^{-5}	3
Cd into Pb	Room	1.8×10^{-3}	3
Cl⁻ into NaCl single crystals	650°C	7.25×10^{-11}	5
Cl⁻ into NaCl single crystals	681°C	2.84×10^{-10}	5
Cl⁻ into NaCl single crystals	703°C	6.76×10^{-10}	5
Cl⁻ into NaCl single crystals	735°C	1.67×10^{-9}	5
Cl⁻ into NaCl single crystals	762°C	2.52×10^{-9}	5
Cu into Cu	Room	0.1–47	1
Cu into Cu	700°C	4.06×10^{-12}	7
Cu into Cu	900°C	3.58×10^{-10}	7
Cu into Cu	1000°C	1.95×10^{-9}	7
Cu into CuO	800°C	0.19×10^{-8}	6
Cu into CuO	900°C	0.77×10^{-8}	6
Cu into CuO	1000°C	3.2×10^{-8}	6
Cu into Ag	Room	5.95×10^{-5}	1
In into In	49.95°C	7–8.5×10^{-13}	9
In into In	87.25°C	1.4–1.5×10^{-11}	9
In into In	155.50°C	1.14×10^{-9}	9
In into In	155.81°C	1.70×10^{-7}	9
In into In	156.60°C	6.52×10^{-6}	9
In into In	157.30°C	1.23×10^{-5}	9
In into Ag	Room	4.85×10^{-5}	1
Liq. Hg into liq. Hg	2.5°C	1.52×10^{-5}	8
Liq. Hg into liq. Hg	16.4°C	1.68×10^{-5}	8
Lig. Hg into liq. Hg	23.0°C	1.79×10^{-5}	8
Liq. Hg into liq. Hg	31.9°C	1.88×10^{-5}	8
Liq. Hg into liq. Hg	41.5°C	1.98×10^{-5}	8
Liq. Hg into liq. Hg	66.1°C	2.24×10^{-5}	8
Liq. Hg into liq. Hg	91.2°C	2.57×10^{-5}	8
Mn into Cu	Room	0.72×10^{-5}	1
Ni into Cu	Room	6.5×10^{-5}	1
Ni into Pb	Room	0.66	1
Pd into Cu	Room	0.16×10^{-5}	1

TABLE 2e-17. DIFFUSION COEFFICIENTS FOR METALS (*Continued*)

Metal	Test temp.	$D\left(\dfrac{cm^2}{sec}\right)$	Ref.
Pt into Cu............................	Room	1.02×10^{-4}	1
Pb into Pb...........................	Room	6.6	1
Sb into Ag...........................	Room	5.31×10^{-5}	1
Si into ferrite.......................	$1435 \pm 5°C$	1.1×10^{-7}	10
Si into Cu...........................	Room	3.7×10^{-2}	1
Sn into Ag...........................	Room	7.82×10^{-5}	1
Sn into Cu...........................	Room	1.13	1
Sn into Pb...........................	Room	3.96	1
Ti into In...........................	49.27°C	1.4×10^{-12}	9
Ti into In...........................	74.19°C	9.2×10^{-12}	9
Ti into In...........................	101.55°C	$4.6-4.8 \times 10^{-11}$	9
Ti into In...........................	139.16°C	$2.8-3.2 \times 10^{-10}$	9
Ti into In...........................	155.60°C	2.17×10^{-9}	9
Ti into In...........................	155.91°C	1.87×10^{-7}	9
Ti into In...........................	157.80°C	2.27×10^{-5}	9
Ti into Pb...........................	Room	0.025	1

N. B.   The values quoted from ref. 1 are for D_0 in the equation $D = D_0 e^{-H/RT}$.   Cf. ref. 1 for values of H.

References for Table 2e-17

1. Nowick, A. S.: *J. Appl. Phys.* **22,** 1182 (1951).
2. Slifkin, L., D. Lazarus, and T. Tomizuka: *J. Appl. Phys.* **23,** 1032 (1952).
3. Smithells, C. J.: "Metals Reference Book."
4. Martin, A. B., and F. Asaro: *Phys. Rev.* **80,** 123A (1950).
5. Chemla, Marius: *Compt. rend.* **234,** 2601 (1952).
6. Moore, W. J., and Bernard Selikson: *J. Chem. Phys.* **19,** 1539 (1951).
7. Cohen, G., and G. C. Kuczynski: *J. Appl. Phys.* **21,** 1339L (1950).
8. Hoffman, R. E.: *J. Chem. Phys.* **20,** 1567 (1951).
9. Eckert, R. E., and H. G. Drickamer: *J. Chem. Phys.* **20,** 13 (1951).
10. Bradshaw, F. J., G. Hoyle, and K. Speight: *Nature* **171,** 488 (1953).
11. Kuczynski, G. C.: *J. Appl. Phys.* **21,** 632 (1950).

7. *Rockwell Hardness Number.*[1] "A hardness value indicated on a direct-reading dial when a designated load is imposed on a metallic material in the Rockwell hardness testing machine using a steel ball or a diamond penetrator. The value must be qualified by reference to the load and penetrator used. Several scales are in common use: Rockwell A hardness is determined with a minor load of 10 kg and a major load of 60 kg using the diamond cone (brale); Rockwell B hardness is determined with a minor load of 10 kg and a major load of 100 kg using a $\frac{1}{16}$-in. steel ball; Rockwell C hardness is determined with a minor load of 10 kg and a major load of 150 kg using the diamond cone"; Rockwell D hardness is determined with a minor load of 10 kg and a major load of 100 kg using a diamond cone indenter; Rockwell E hardness is determined with a minor load of 10 kg and a major load of 100 kg using a $\frac{1}{8}$-in. steel ball indenter; Rockwell F hardness is determined with a minor load of 10 kg and a major load of 60 kg using a $\frac{1}{16}$-in. steel ball; Rockwell G hardness is determined with a minor load of 10 kg and a major load of 150 kg, using a $\frac{1}{16}$-in. steel ball indenter.

A second set of Rockwell hardness numbers are the Rockwell superficial hardness numbers. One of these is the Rockwell 15T hardness, which is determined with a minor load of 3 kg and a major load of 15 kg, using a $\frac{1}{16}$-in. steel ball.

Note: The methods of determining the hardness values can be found in Standard Methods of Test for Rockwell Hardness and Rockwell Superficial Hardness of Metallic Materials, ASTM E18–42.

8. *Brinell Hardness Number.*[2] "A hard spherical indenter of diameter D mm is pressed into the metal surface under a load W kg and the mean chordal diameter of the resultant indentation measured (d mm). The Brinell hardness number (Bhn) is defined as

$$Bhn = \frac{W}{\text{curved area of indentation}}$$
$$= \frac{2W}{\pi D(D - \sqrt{D^2 - d^2})}$$

and is expressed in kg/mm²."

9. *Vickers Hardness Number.*[2] "A pyramidal diamond indenter is pressed into the surface of a metal under a load of W kg and the mean diagonal of the resultant indentation measured (d mm). The Vickers hardness number (Vhn), or Vickers diamond hardness (Vdh), is defined as

$$Vdh \text{ (or Vhn)} = \frac{W}{\text{pyramidal area of indentation}}$$

The indenter has an angle of 136° between opposite faces and 146° between opposite edges. From simple geometry, this means that the pyramidal area of the indentation is greater than the projected area of the indentation by the ratio 1:0.9272. Hence

$$Vdh = \frac{0.9272W}{\text{projected area of indentation}}$$
$$= 1.8544W/d^2$$

The value is expressed in kg/mm²."

10. *Diffusion Coefficient.* If the concentration (mass of solid per unit volume of solution) at one surface of a layer of liquid is d_1, and at the other surface d_2, the thickness of the layer is h, the area under consideration is A, and the mass of a given substance which diffuses through the cross section A in time t is m, then the diffusion coefficient is defined as

$$D = \frac{mh}{A(d_2 - d_1)t}$$

[1] J. G. Henderson, "Metallurgical Dictionary."
[2] D. Tabor, "The Hardness of Metals."

2e-4. Effect of High Pressure on the Specific Volume of Solids. Tables 2e-18 to 2e-22 present data on the change of specific volume of certain solids as a result of the imposition of very high pressure. The general reference in this field is P. W. Bridgman, "The Physics of High Pressure," G. Bell & Sons, Ltd, London, 1949.

Specific references are attached to each table.

TABLE 2e-18. VOLUME OF SOLID HELIUM AT 0°K*

Pressure, kg/cm²	Volume, ml/mole	Compressibility $(1/v)(\partial v/\partial p)_\tau$
52	19.0	184×10^{-5}
91	18.0	135
141	17.0	100
207	16.0	73
305	15.0	52
475	14.0	37
718	13.0	25
1,105	12.0	16
1,715	11.0	12
2,240	10.5	10

* J. S. Dugdale and F. E. Simon, *Proc. Roy. Soc. (London)* **218**, 291 (1953).

TABLE 2e-19. FRACTIONAL CHANGE OF VOLUME AT 25°C OF RELATIVELY INCOMPRESSIBLE METALS*

Pressure, kg/cm²	$\Delta V/V_0$						
	W	Pt	Fe	Cu	Ag	Au	Al
5,000	0.00155	0.00176	0.00289	0.00353	0.00473	0.00281	0.00668
10,000	0.00309	0.00351	0.00575	0.00696	0.00938	0.00558	0.01312
15,000	0.00475	0.00526	0.00856	0.01039	0.01385	0.00831	0.01932
20,000	0.00634	0.00701	0.01133	0.01370	0.01820	0.01101	0.02520
25,000	0.00797	0.00877	0.01407	0.01695	0.02236	0.01367	0.03090
30,000	0.00959	0.01048	0.01676	0.02010	0.02619	0.01626	0.03642

* P. W. Bridgman, *Proc. Am. Acad. Arts Sci.* **77**, 187 (1949).

TABLE 2e-20. RELATIVE VOLUMES OF VARIOUS SOLIDS AT 25°C*

Pressure, kg/cm²	Lucite	Cellulose acetate	Bakelite	Hard rubber	Nylon 6–10	Teflon	Orthoclase
1	1.0000	1.0000	1.0000	1.0000	1.0000	1.0000	1.0000
2,500	0.9633	0.9532	0.9760	0.9684	0.9615	0.9473	
5,000	0.9329	0.9216	0.9562	0.9390	0.9345	0.9153	
10,000	0.8903	0.8811	0.9240	0.8955	0.8940	0.8547	0.9829
15,000	0.8613	0.8514	0.8978	0.8655	0.8652	0.8306	
20,000	0.8329	0.8283	0.8765	0.8427	0.8430	0.8125	0.9667
30,000	0.8051	0.7935	0.8436	0.8083	0.8100	0.7857	0.9512
40,000	0.7816	0.7682	0.8188	0.7834	0.7861	0.7661	0.9366

Pressure, kg/cm²	Calcite	Garnet	Iodoform	Urea nitrate	Potassium phosphate	Potassium alum
1	1.0000	1.0000	1.0000	1.0000	1.0000	1.0000
5.000	......	......	0.9451	0.9628	0.9821	0.9718
10,000	0.9866	0.9929	0.9079	0.9358	0.9665	0.9486
15,000	tr.	......	0.8806	0.9145	0.9526	0.9296
20,000	0.9275	0.9862	0.8586	0.8966	0.9401	0.9131
30,000	0.9113	0.9800	0.8241	0.8669	0.9183	0.8843
40,000	0.8981	0.9743	0.7966	0.8431	0.9004	0.8607

* P. W. Bridgman, *Proc. Am. Acad. Arts Sci.* **76,** 71 (1948).

TABLE 2e-21. RELATIVE VOLUMES OF SOME OF THE MORE COMPRESSIBLE ELEMENTS, SALTS, AND OTHER SOLIDS AT 25°C*

In Tables 2e-21 and 2e-22 the symbol tr denotes a phase transition

Pressure, kg/cm²	Li	Na	K	Rb	Cs	Ca	Sr	Ba	C
1	1.000	1.000	1.000	1.000	1.000	1.000	1.000	1.000	1.000
10,000	0.928	0.889	0.814	0.802	0.761	0.942	0.925	0.914 tr	
20,000	0.874	0.816	0.723	0.708	0.656 tr	0.897	0.878	0.841 tr	
30,000	0.833	0.770	0.668	0.652	0.571 tr	0.861	0.828 tr	0.789 tr	0.940
40,000	0.801	0.737	0.628	0.612	0.521 tr	0.832	0.791	0.747	0.929
50,000	0.773	0.708	0.595	0.578	0.431 tr	0.805	0.761	0.712 tr	0.919
60,000	0.748	0.683	0.568	0.551	0.409	0.780 tr	0.734 tr	0.682 tr	0.911
70,000	0.727	0.661	0.546	0.528	0.392	0.748 tr	0.702 tr	0.639	0.903
80,000	0.707	0.641	0.528	0.507	0.381	0.732	0.683	0.618	0.896
90,000	0.689	0.623	0.513	0.489	0.375	0.716	0.665	0.598	0.890
100,000	0.672	0.606	0.500	0.473	0.368	0.702	0.648	0.580	0.885

* P. W. Bridgman, *Proc. Am. Acad. Arts Sci.* **76**, 55, 71 (1948); **74**, 425 (1942).

TABLE 2e-22. RELATIVE VOLUMES OF SOLIDS AT 25°C*

Pressure, kg/cm²	Mg	Sn	Pb	Bi	S	NaCl	NaI	CsCl	CsI
1	1.000	1.000	1.000	1.000	1.000	1.000	1.000	1.000	1.000
10,000	.....	0.982	0.978	0.972	0.917	0.962	0.944	0.952	0.935
20,000	.....	0.966	0.959	0.948 tr	0.869	0.932	0.902	0.914	0.887
30,000	0.935	0.951	0.941	0.842 tr	0.837	0.907	0.868	0.882	0.849
40,000	0.919	0.936	0.925	0.826 tr	0.812	0.885	0.840	0.856	0.818
50,000	0.904	0.923	0.901	0.808 tr	0.792	0.865	0.816	0.834	0.792
60,000	0.890	0.909	0.898	0.795 tr	0.775	0.848	0.795	0.816	0.770
70,000	0.878	0.897	0.885	0.778 tr	0.760	0.832	0.777	0.801	0.751
80,000	0.866	0.886	0.874	0.768	0.747	0.817	0.761	0.788	0.734
90,000	0.856	0.875	0.864	0.760 tr	0.736	0.803	0.747	0.777	0.719
100,000	0.847	0.864	0.855	0.739 tr	0.726	0.790	0.734	0.767	0.706

Pressure, kg/cm²	NaNO₃	PbS	PbTe	Quartz crystal	Quartz glass	Pyrex glass
1	1.000	1.000	1.000	1.000	1.000	1.000
10,000	0.966	0.980	0.978	0.976	0.970	0.969
20,000	0.938	0.962 tr	0.961	0.955	0.939	0.938
30,000	0.914	0.928 tr	0.939	0.939	0.909	0.907
40,000	0.893	0.918	0.930 tr	0.926	0.885	0.885
50,000	0.873	0.909	0.884 tr	0.914	0.864	0.867
60,000	0.846 tr	0.900	0.869	0.902	0.847	0.851
70,000	0.833 tr	0.892	0.855	0.892	0.832	0.838
80,000	0.820	0.886	0.842	0.883	0.819	0.827
90,000	0.809	0.881	0.831	0.875	0.808	0.817
100,000	0.799	0.876	0.820	0.868	0.798	0.809

*P. W. Bridgman, *Proc. Am. Acad. Arts Sci.* **76**, 55, 71 (1948); **74**, 425 (1942).

2f. Viscosity of Solids

EVAN A. DAVIS

Westinghouse Research Laboratory

Symbols

E elastic modulus
t time
T absolute temperature
u' elastic strain rate
u'' plastic strain rate
δ logarithmic decrement
ϵ' elastic strain
ϵ'' plastic strain
η viscosity
σ stress

2f-1. Anelasticity. A perfectly elastic solid is truly an ideal material. Actual materials contain structural imperfections which prohibit them from behaving in a perfectly elastic manner. Even when the stresses are low enough to ensure that no perceptible permanent deformation takes place the total strain is made up of a purely elastic part that is directly proportional to the load and a time-dependent but fully recoverable part that will vary with the rate of loading and the duration of the load. The behavior associated with the time-dependent part of the strain has been called "anelasticity" by Zener,[1] who has endeavored to explain this behavior in terms of the atomic arrangement and the microstructure of the material.

Anelastic behavior is observed in many ways, depending upon the manner in which the material is loaded. Its effect may be referred to as elastic hysteresis, internal friction, elastic aftereffect, specific damping capacity, or dynamic and static moduli of elasticity. The fact that the term anelasticity has been limited to the region of no permanent deformation does not exclude the existence of such behavior at higher stresses. When a material deforms permanently, however, the anelastic effects are overshadowed by and engulfed in the plastic behavior.

In the realm of small deformations a metal or a plastic can be represented qualitatively by the mechanical model of springs and dashpots shown in Fig. 2f-1. For the anelastic

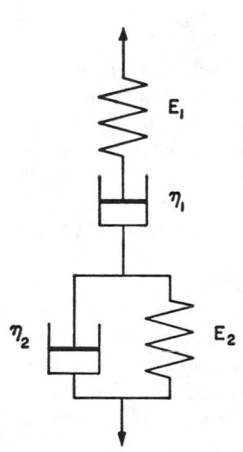

Fig. 2f-1. Mechanical model for demonstrating anelastic and creep behavior of solids.

[1] C. Zener, "Elasticity and Anelasticity of Metals," University of Chicago Press, Chicago, 1948.

2-83

behavior at low stresses the viscosity η_1 of the upper dashpot can be considered as infinite. The spring with the elastic modulus E_1 contributes the purely elastic strain. The time-dependent part of the strain comes from the parallel arrangement of spring E_2 and dashpot η_2. This model will exhibit, though not in a quantitative manner, the various anelastic effects of solids.

If the unit is elongated at a slow rate, dashpot η_2 will have little effect in resisting the deformation of spring E_2. The static or isothermal modulus of elasticity will be that of springs E_1 and E_2 connected in series. If the unit is elongated rapidly dashpot η_2 will tend to act as a rigid mechanism. The dynamic or adiabatic modulus of elasticity will be that of spring E_1 acting alone.

If the unit is put through a constant-rate loading and unloading cycle a hysteresis loop will be traced out in the stress-strain diagram. The area of the loop will be proportional to the amount of energy dissipated in dashpot η_2.

If the unit is loaded slowly and then unloaded rapidly the strain will not immediately return to zero. What appears to be a permanent strain or elastic aftereffect will be observed. The strain will return to zero when the stress trapped in the spring E_2 by dashpot η_2 has been relaxed.

If the mass is attached to the lower end of the unit and the entire mechanism is allowed to vibrate freely the amplitude of vibration will decrease with each cycle. The decrease in amplitude of vibration is due to the dissipation of energy in dashpot η_2. If the springs are linear and elastic and the dashpot behaves in a perfectly viscous manner the ratio of the decrease in amplitude for any given cycle to the amplitude at the beginning of the cycle will be a constant. This constant is called the logarithmic decrement δ, and it is probably the most-used measure of the anelastic behavior of materials.

The logarithmic decrement of actual materials is relatively high for dielectric materials and low for metals. Since this quantity depends upon imperfections in the atomic structure it will vary with such factors as heat-treatment, grain size, or the amount of cold working, and it will be impossible to assign a value to a specific material such as steel. The values listed by Kimball[1] and shown in Table 2f-1 and those listed by Gemant[2] and shown in Table 2f-2 are to be considered as representative values which give the order of magnitude of the decrement or internal friction.

The factors which affect the logarithmic decrement are discussed in detail by Zener and by Gemant. The decrement is influenced by such factors as frequency, temperature, amplitude, elastic modulus, grain size, annealing temperature, and aging time.

In general there is not much change in decrement with frequency. Gemant and Jackson[3] found slight increases in the decrement of ebonite and glass over rather narrow frequency ranges (Fig. 2f-2). Gemant shows a slight increase in the decrement for paraffin wax and a slight decrease in the decrement for steel (Fig. 2f-3). An exception to this rule was found by Rinehart,[4] who reported an appreciable increase in the decrement of Lucite at room temperature (Fig. 2f-4).

Certain materials show steep peaks in the log decrement vs. log frequency curve. These peaks are associated with frequencies that correspond to the reciprocal of some characteristic time for the material. Such a curve, taken from Gemant and based on the work of Zener and Bennewitz and Rötger,[5] is shown in Fig. 2f-5. In this case the peak in the internal-friction curve is due to the diffusion of heat from parts heated by compression to parts cooled by tensile stresses.

[1] A. L. Kimball, "Vibration Prevention in Engineering," John Wiley & Sons, Inc., New York, 1932.
[2] A. Gemant, "Frictional Phenomena," Chemical Publishing Company, Inc., New York, 1950.
[3] A. Gemant and W. Jackson, *Phil. Mag.* **23**, 960 (1937).
[4] J. S. Rinehart, *J. Appl. Phys.* **12**, 811 (1941).
[5] K. Bennewitz and H. Rötger, *Z. tech. Phys.* **19**, 521 (1938).

TABLE 2f-1. LOGARITHMIC DECREMENTS FOR VARIOUS MATERIALS*

Material	Logarithmic Decrement δ
Phosphor bronze, cold rolled	0.37×10^{-3}
Monel, cold rolled	1.43
Nickel steel, $3\frac{1}{2}\%$ swaged	2.3
Nickel, cold rolled	3.2
Phosphor bronze, annealed	3.2
Aluminum, cold rolled	3.4
Brass, cold rolled	4.8
Mild steel, cold rolled	4.9
Copper, cold rolled	5.0
Glass	6.4
Molybdenum, swaged	6.9
Swedish iron, annealed	7.9
Tungsten, swaged	16.5
Zinc, swaged	20
Maple wood	22
Celluloid	45
Tin, swaged	129
Rubber, 90% pure	260

* A. L. Kimball, "Vibration Prevention in Engineering," John Wiley & Sons, Inc., New York, 1932.

TABLE 2f-2. LOGARITHMIC DECREMENT OF VARIOUS MATERIALS*

Material	Logarithmic Decrement δ
Steel	0.6×10^{-3}
Quartz	2.6
Copper	3.2
Lead glass	4.2
Wood	27
Polystyrene	48
Ebonite	85
Paraffin wax	150

* A. Gemant, "Frictional Phenomena," Chemical Publishing Company, Inc., New York, 1950.

The logarithmic decrement usually increases with increasing temperature. The viscous behavior changes more rapidly than the elastic properties with temperature, with the result that at higher temperatures more energy is dissipated in the dashpot.

The decrement does not vary greatly with amplitude when the amplitudes are small. The decrement increases at higher amplitudes. This is evidence that the viscosity of materials is not of a pure viscous nature. The rate of strain increases more rapidly at the higher stresses than the linear viscous law would predict.

Materials with high elastic moduli have lower decrements than those with low moduli. There is some evidence to show that the product of the elastic modulus and the decrement is nearly a constant value.

The damping capacity of a structure depends upon the stress distribution in the structural members and the energy absorption characteristics of the material from which the members are made. This energy absorption may be brought about by plastic flow, thermoelastic effect, magnetoelastic effect, and atomic diffusion. The relative importance of these effects will depend upon the magnitude of the vibratory stresses.[1]

2f-2. Creep. When a material is subjected to the proper combination of high stress and temperature, it will deform permanently. A representative behavior will be produced by the model shown in Fig. 2f-1 if the viscosity of both dashpots η_1 and η_2 is finite. The continuing deformation of a material under a constant load is called

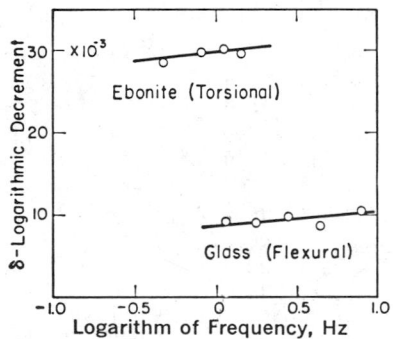

FIG. 2f-2. Logarithmic decrement vs. logarithm of frequency for ebonite and glass. (*Gemant and Jackson.*)

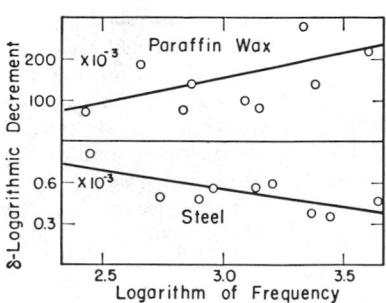

FIG. 2f-3. Logarithmic decrement vs. frequency at room temperature for steel and paraffin wax. (*Gemant.*)

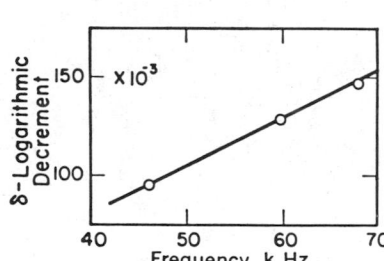

FIG. 2f-4. Logarithmic decrement vs. frequency for Lucite at 26°C. (*Rinehart.*)

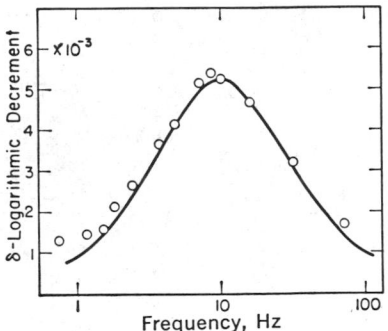

FIG. 2f-5. Logarithmic decrement vs. frequency for German silver. (*Measured points after Bennemitz and Rötger; theoretical curve after Zener.*)

"creep." If the model is loaded with a given load at $t = 0$, there will be an instantaneous elastic deflection ϵ' of spring E_1, dashpot η_1 will deform at some constant rate u_0'', and dashpot η_2 will deform at a decreasing rate.[2] The rate of strain in dashpot η_2

---

[1] This problem was discussed in detail during the early 1950s. See, for example, the following papers and their reference lists: B. J. Lazan, *J. Appl. Mech., Trans. ASME* **75** (1953); A. W. Cochardt, *J. Appl. Mech., Trans. ASME* **76** (1954).

[2] A prime (′) on a strain or strain rate indicates elastic deformation; a double prime (″) indicates plastic or permanent strain. The total strain, or strain rate, is the sum of the elastic and the plastic parts; i.e.,

$$\epsilon = \epsilon' + \epsilon'' \qquad \text{or} \qquad u = u' + u''$$

decreases because the load is gradually transferred to spring E_2 as the deformation takes place, and this part of the deformation stops at a strain ϵ_0'' when the spring E_2 carries the complete load. The creep curve for the model and for materials which are not stressed high enough to cause fracture will have the form shown in Fig. 2f-6 (the elastic strain ϵ' is not shown). The plastic strain starts at a rapid rate but approaches the asymptotic value given by

$$\epsilon'' = \epsilon_0'' + u_0''t \tag{2f-1}$$

The shape of the initial part of the creep curve or the manner in which the curve approaches the asymptote has been studied by Andrade[1] and by McVetty.[2] Andrade

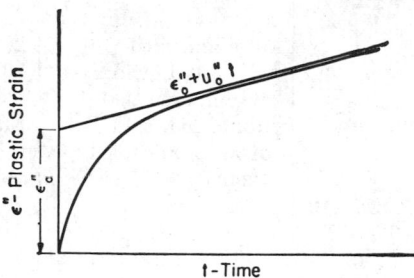

FIG. 2f-6. Typical creep curve.

found that the increase of strain during the first part of the test was proportional to the cube root of the time.

$$\epsilon'' = \beta t^{\frac{1}{3}} \tag{2f-2}$$

McVetty used an exponential relationship to describe the initial deformation.

$$\epsilon'' = \epsilon_0''(1 - e^{-\alpha t}) + u_0''t \tag{2f-3}$$

When creep tests are made to obtain design data for equipment having long service life, and most of the early creep tests were made under these conditions, the major part of the strain is accounted for by the $u_0''t$ term in Eq. (2f-1). The important relationship to be established, then, is that between the minimum creep rate u_0'' and the stress σ, and this is the only information reported by many investigations.

If shorter service times are considered, the initial part of the creep curve becomes more important, and it becomes desirable to know the relationship between the plastic intercept ϵ_0'' and the stress σ. McVetty shows a plot of this relationship for the lower stress range where a power function or hyperbolic sine relationship would be suitable.

$$\epsilon_0'' = A\sigma^n \qquad \text{or} \qquad \epsilon_0'' = B \sinh \frac{\sigma}{\sigma_0} \tag{2f-4}$$

Such relationships indicate that, if the model of Fig. 2f-1 is to represent actual materials, spring E_2 must be nonlinear. At higher stresses these relationships do not hold.

[1] E. N. da C. Andrade, *Proc. Roy. Soc. (London)*, ser. A, **84**, 1 (1911); **90**, 329 (1914).
[2] P. G. McVetty, *Mech. Eng.* **56**, 149 (March, 1934).

TABLE 2f-3. CREEP RATES FOR VARIOUS MATERIALS*

Material and composition	Condition	Temp		Stress for 0.001 strain in 1,000 hr, psi
		°C	°F	
Aluminum copper alloy, Cu 4.25, Mn 0.63, Mg 0.44, Fe 0.52, Si 0.25	$\frac{5}{8}$ diam rod, wrought, aged	150	302	22,000
		250	482	5,700
		350	662	1,500
Aluminum silicon alloy, Si 13.18, Ni 3.08, Cu 2.96, Mg 1.04, Fe 0.53	Wrought	205	400	8,800
		315	600	950
Electrocopper	Fully annealed	205	400	6,700
Deoxidized copper	$\frac{3}{4}$ diam rod, cold drawn, annealed	205	400	20,500
Copper nickel alloy, Ni 20.0, Zn 5.08, Mn 0.69	$\frac{3}{4}$ diam rod, cold drawn, annealed at 1200°F	315	600	27,800
Copper tin alloy, Sn 5.99, Zn 5.10, Pb 2.33, Ni 0.23, Fe 0.06	Cast	260	500	10,000
		315	600	3,000
Copper zinc alloy, Cu 96.43, Pb 0.05, Fe 0.01, Zn remainder	$\frac{1}{8}$ diam wire, drawn, fine-grained	149	300	50,000
		205	400	3,500
		260	500	700
Carbon steel, C 0.15, Mn 0.46, Si 0.28 (basic open hearth)	1 in. diam bar, wrought, annealed at 1500°F, grain size 5–6 ASTM	427	800	17,200
		538	1000	3,300
		648	1200	540
Carbon steel, C 0.15, Mn 0.50, Si 0.23 (basic electric furnace)	1 in. diam bar, wrought, annealed at 1550°F, grain size 4–5 ASTM	427	800	26,800
		482	900	16,900
		538	1000	5,750
		593	1100	1,800
		648	1200	620
Chromium steel, C 0.10, Cr 5.09, Mo 0.55, Mn 0.45, Si 0.18	1 in. diam bar, wrought, annealed at 1550°F, grain size 4–5 ASTM	482	900	15,200
		538	1000	10,100
		593	1100	5,850
		648	1200	2,800
Molybdenum steel, C 0.22, Mo 1.06, Mn 0.50, Si 0.13 (induction furnace)	Bar $1\frac{1}{4}$ sq. cast, annealed at 1650°F, grain size 7	427	800	28,000
		482	900	20,800
		538	1000	11,200
Nickel steel, C 0.36, Ni 1.19, Mn 0.58, Cr 0.51, Mo 0.51, Si 0.22 (induction furnace)	1 in. diam bar, hot rolled, normalized at 1600°F, tempered 3 hr at 1250°F	454	850	40,000
		538	1000	12,300
		593	1100	3,600
		648	1200	1,600
Lead	Grade 2	43	110	320
Magnesium alloy, Al 3, Zn 1	Sand cast, $\frac{1}{2}$ diam rods	150	302	4,900†
Nickel alloy, Cu 28.46, Fe 1.24, Mn 0.94, C 0.18, Si 0.10	Wrought	427	800	30,000
		482	900	23,000
		538	1000	3,700
		593	1100	1,300
		648	1200	450

* Mechanical Properties of Metals and Alloys, *Natl. Bur. Standards* (*U.S.*) *Circ.* C447, 1943.
† Stress for 0.005 strain in 1,000 hr.

TABLE 2f-3. CREEP RATES FOR VARIOUS MATERIALS (*Continued*)

Material and composition	Condition	Temp		Stress for 0.001 strain in 1,000 hr, psi
		°C	°F	
Zinc alloy, Cd 0.3, Pb 0.3	Rolled, soft, tested parallel to rolling direction	20	68	10,100
		40	104	8,000
		60	140	6,300
Zinc alloy, Cd 0.3, Pb 0.3	Rolled, soft, tested perpendicular to rolling direction	20	68	15,400
		40	104	12,100
		60	140	8,000

As the stress is increased, a maximum value is reached above which the value of ϵ_0'' decreases with increasing stress.

In the range of strain rates that can be tolerated in reasonable testing times the minimum creep rate u_0'' vs. stress σ curve can be approximated by a straight line on either a double-log or a semilog plot.

$$u_0'' = D\sigma^m \qquad \text{or} \qquad u_0'' = u_1'' \sinh \frac{\sigma}{\sigma_0} \tag{2f-5}$$

The hyperbolic sine relationship has been shown by Kauzmann[1] to have some theoretical foundation in terms of the "chemical rate theory." The power-function relationship has the advantage of being more workable from a mathematical point of view, but it suffers somewhat from the illogical conclusion that the viscosity of dashpot η_1 should approach infinity as the stress approaches zero. Creep properties, like anelastic properties, vary with many factors, and compilation of creep data means very little unless heat-treatment, grain size, and amount of cold working are also specified. A few representative values of the stress required for a creep rate of 10^{-6} per hour, taken from the 1943 compilation of the National Bureau of Standards,[2] are given in Table 2f-3.[3]

Materials held under constant load during long-time creep tests recover part of their plastic strain when the load is removed. According to the model of Fig. 2f-1 the recoverable strain should be equal to ϵ_0''. In actual practice, however, the recovery is usually much less than ϵ_0'' and is generally less than the elastic strain of unloading. If after the first unloading and subsequent recovery the specimen is loaded and unloaded the new plastic intercept ϵ_0'' and the recoverable strain are approximately equal.

Both constants in either of the expressions of Eqs. (2f-5) vary with temperature. According to the chemical rate theory of Kauzmann and the various theories based on

[1] W. Kauzmann, *Trans. AIME* **143**, 57–83 (1941).
[2] Mechanical Properties of Metals and Alloys, *Natl. Bur. Standards (U.S.) Circ.* C447, 1943.
[3] Recent compilations of creep test data are published by the American Society for Testing and Materials in their Data Publication Series.

diffusion phenomena the constants D and u_1 should decrease with increasing temperature according to an exponential expression

$$u_0'' = C_1 e^{-(C_2/T)} \qquad (2f\text{-}6)$$

This has been checked experimentally over reasonably wide temperature ranges. The constant σ_0, in the lower stress range, usually decreases slightly with increasing temperature. If the constant m changes with temperature caution must be observed in extrapolating toward regions where the curves for two different temperatures would cross.

2g. Astronomical Data

R. BRUCE LINDSAY

Brown University

TABLE 2g-1. PLANETARY ORBITS*

Planet	Mean distance to sun in terms of earth's distance†	Sidereal period, tropical years‡	Inclination to the ecliptic in degrees	Eccentricity
Mercury......	0.387,099	0.24085	7.00399	0.205,627
Venus.........	0.723,332	0.61521	3.39423	0.006,793
Earth.........	1.000,000	1.00004	—	0.016,726
Mars.........	1.523,691	1.88089	1.84991	0.093,368
Jupiter........	5.202,803	11.86223	1.30536	0.048,435
Saturn........	9.538,843	29.45772	2.48991	0.055,682
Uranus........	19.181,951	84.01331	0.77306	0.047,209
Neptune......	30.057,779	164.79345	1.77375	0.008,575
Pluto.........	39.438,71	247.686	17.1699	0.250,236

* Taken from "Explanatory Supplement to the Astronomical Ephemeris and the American Ephemeris and Nautical Almanac" H. M. Stationery Office, London, 1961.
† The mean distance from the earth to the sun is given as 1.00000003 astronomical units and is equal to 1.495×10^8 km.
‡ See Sec. 2a-8 for definition of tropical year.

TABLE 2g-2. PHYSICAL DATA FOR THE PLANETS AND THE MOON*

Planet	Mass (Earth = 1)	Mean diameter (Earth = 1)	Mean density, g/cm^3	Surface gravity (Earth = 1)	Velocity of escape, km/sec	Rotation period, days
Mercury	0.056	0.39	5.13	0.36	4.3	88
Venus	0.817	0.97	4.97	0.87	10.4	?
Earth	1.000	1.00	5.52	1.00	11.3	1
Mars	0.108	0.53	3.94	0.38	5.1	1.03
Jupiter	318.0	11.19	1.33	2.64	61.0	0.41
Saturn	95.2	9.47	0.69	1.13	36.7	0.43
Uranus	14.6	3.69	1.46	1.07	22.4	0.45
Neptune	17.3	3.50	2.27	1.41	25.6	0.58?
Pluto	0.9?	1.1?	4?	?	<5.3?	6.39?
Moon	0.012	0.27	3.34	0.16	2.4	27.3

* Taken from "Explanatory Supplement to the Astronomical Ephemeris and The American Ephemeris and Nautical Almanac," H. M. Stationery Office, London, 1961, except for the values of the velocity of escape, which are taken from "Smithsonian Physical Tables," 9th ed., 1954.

TABLE 2g-3. MISCELLANEOUS ASTRONOMICAL CONSTANTS*

Period of rotation of the earth with respect to the fixed stars:

24 hr 00 min 00100839 sec of mean sidereal time

23 hr 56 min 4.09895 sec of mean solar time

Mass of the earth.................... 5.98×10^{24} kg

Mass of the sun..................... 1.99×10^{30} kg

Mass of the moon.................... 7.35×10^{22} kg

Moon's mean distance from the earth.... 384,400 km

Moon's sidereal period................ 27.3 days

Earth's mean orbital speed............ 29.8 km/sec

Gravitational constant G.............. $(6.670 \pm 0.005) \times 10^{-8}$ dnye-cm^2/ge^2

Acceleration of gravity g.............. $980.64 - 2.59 \cos 2\phi$ cm/sec^2

where ϕ = latitude

Precession of the equinoxes............ $50.2564 + 0.000{,}222(t - 1{,}900)$ seconds of arc per year

where t = year in question

Sun's radius........................ 6.96×10^5 km

Solar parallax....................... 8.80 seconds of arc

Sun's mean density................... 1.41 g/cm^3

Obliquity of the ecliptic.............. $23°27'8.26'' - 0.4684(t - 1{,}900)$ seconds

where t is the year in question

* Taken from "Explanatory Supplement to the Astronomical Ephemeris and the American Ephemeris and Nautical Almanac, H. M. Stationery Office, London, 1961.

2h. Geodetic Data

HELLMUT H. SCHMID AND KARL R. KOCH

National Oceanic & Atmospheric Administration

2h-1. Introduction. The fundamental task of geodesy is the formulation of a three-dimensional mathematical model to which can be related uniquely:

1. The geometry of the physical surface of the earth which is truly the "shape" of the earth,

2. The mathematical description of the gravitational field associated with the earth's mass, where the detailed description of the equipotential surface representing mean sea level—the geoid—is of special interest, and

3. the Universal Time and the astronomical Right Ascension-Declination System.

Establishing the shape of the earth (1) requires the determination of three-dimensional coordinates for (ideally speaking) all points of the physical surface of the earth. Because of (2) it is convenient to establish the corresponding coordinate system in relation to the mass center of the earth. An expedient coordinate system is a geocentric equatorial cartesian (x,y,z) system, the origin of which coincides with the center of mass. In order to relate this system to both Universal Time and the astronomical reference system (3) it is necessary that the z axis coincide with the axis of rotation of the earth for a certain epoch, thus pointing toward a corresponding reference pole. The mean pole of the epoch 1900 to 1905, designated the Conventional International Origin (CIO), was adopted for this purpose by the International Association of Geodesy in 1968. The x axis points toward the meridian of Greenwich which is designated the null meridian for both the measurement of geographic longitude and Universal Time.

2h-2. Reference Ellipsoids. Mainly because of the uncertainty in the amounts of terrestial refraction (cf. Sec. 2h-3), geodetic surveys are generally based on horizontal angle measurements and projected onto reference surfaces. Ellipsoids of revolution, also called reference ellipsoids (in the United States sometimes reference spheroids) are used for the reduction of surveys covering extended continental areas. Portions of spheres or planes are introduced for more restricted surveys.

The method of triangulation for the purpose of surveying was introduced at the beginning of the seventeenth century. The horizontal angles in a triangle are measured with theodolites, and the size of the triangle—the scale of the triangulation—is determined by distance measurements. By connecting triangle to triangle, continents can be covered with triangulation nets. Chains of triangles along meridians and parallels were measured for determining the dimensions of the reference ellipsoids. Dimensions of reference ellipsoids are given in Table 2h-1.

The Clarke 1866 ellipsoid was adopted by the United States for the North American Datum 1927, while Hayford's ellipsoid of 1910 was accepted as International Ellipsoid by the International Association of Geodesy in 1924.

2h-3. Different Geodetic Systems. The conceptual approach to the establishment of a triangulation system begins with the selection of a datum point—ideally located near the center of the area under consideration. At this datum point the astronomical latitude, longitude, and azimuth of one side of a triangle, for which the datum point is a vertex, are determined. By setting these observed quantities equal to the corresponding ellipsoidal values, and with the additional assumption that the height above sea level is equal to the height above the reference ellipsoid, the surface of the ellipsoid provisionally becomes, neglecting the curvature of the plumb line, tangent to the geoid at this datum point. With the geodetic coordinates of one point thus fixed, the coordinates of other points in the triangulation net are then computed from the azimuths and lengths of the sides of the triangles.

Only horizontal angles are used in a triangulation, i.e., the angles measured in the plane perpendicular to the direction of local gravity, because they can be determined much more accurately than the typically small vertical angles which are distorted by

TABLE 2h-1. DIMENSIONS OF THE REFERENCE ELLIPSOID*
(a = semimajor axis, $f = (a - b)/a$ = flattening, b = semiminor axis)

Author	Year	a, meters	$1/f$
Bouguer, Maupertuis..........	1738	6,397,300	216.8
Delambre...................	1800	6,375,653	334.0
Walbeck....................	1819	6,376,896	302.8
Airy.......................	1830	6,376,542	299.3
Bessel.....................	1841	6,377,397	299.15
Clarke.....................	1866	6,378,206	295.0
Hayford....................	1910	6,378,388	297.0
Krassowski.................	1938	6,378,245	298.3
IAU adopted†	1964	6,378,160	298.25
Anderle‡...................	1967	6,378,144	298.23

* W. A. Heiskanen and F. A. Vening Meinesz, "The Earth and Its Gravity Field," p. 230, McGraw-Hill Book Company, New York, 1958.
† Cf. Sec. 2h-7.
‡ Cf. Sec. 2h-5.

refraction of the light path. As a consequence the triangulation computations must be based on a two-dimensional solution on the surface of a suitable reference ellipsoid. Once the observations have been made on points of the physical surface of the earth, the necessity arises to reduce these observations to the chosen reference ellipsoid. This reduction requires the deflection of the vertical (the small angle between the ellipsoid normal and the direction of gravity) and the height of the triangulation station above the ellipsoid. Some of these reduction corrections, being of small magnitude, were neglected in older triangulations but are at present considered significant in meeting modern accuracy requirements. The deflection at a triangulation station is obtained by observing astronomical latitude and longitude and comparing these data with the corresponding geodetic coordinates computed on the ellipsoid. Integrating the deflections along the path between two points and adding the difference in mean sea level elevations gives the difference in height above the reference ellipsoid between these two points. This approach is known as the *method of astrogeodetic deflections*.[1]

Since the deflections of the vertical are needed in the reduction of the observations to the ellipsoid, it is necessary that the geodetic coordinates—latitude and longitude—

[1] W. A. Heiskanen and H. Moritz: "Physical Geodesy," W. H. Freeman and Co., San Francisco, 1967.

be available, which can be computed only after the necessary reductions on the observations are made. Therefore an iterative procedure becomes necessary, an approach which is typical for the solution of many classical geodetic problems. In the course of such iterative steps, certain a priori assumptions are progressively modified. For example the condition of tangency of geoid and ellipsoid at the datum point may be relaxed and, at least in principle, the parameters of the originally chosen reference ellipsoid can be improved. However, despite the application of complex theoretical reduction methods, classical geodetic triangulation systems cannot establish ties between continents. Consequently triangulation systems on different continents have only partially related coordinate systems on, usually, different reference ellipsoids. Approximately a hundred different datums have been established in various parts of the earth, approximately eight of them being designated as major datums. One of these is the North American Datum (NAD 1927).

The quantities needed for reducing observations in triangulation nets can also be computed as functions of gravity anomalies. Furthermore, the gravimetric method (cf. Sec. 2h-6) also provides, at least in principle, a means for establishing a worldwide geodetic system by determining absolute geoidal undulations and deflections of the vertical with respect to a mass-centered reference ellipsoid.[1] Because of lack of sufficient observations over the oceans the usefulness of this method is impaired, particularly when considering modern accuracy requirements.

With the use of man-made satellites in geodesy the limitations of classical geodetic methods can be surmounted. In a strictly geometric method satellites serve as highly elevated target points for a three-dimensional triangulation of ground-based observation stations (cf. Sec. 2h-4). A dynamic interpretation of the observed satellite orbits leads to a simultaneous solution for the mass-center-referenced station coordinates, parameters of the orbital model, and certain gravitational parameters. Theoretical limitations arise from the necessary assumption that the effect of higher-order terms in the gravitational field is negligibly small. Practical difficulties result from the large number of unknowns solved for simultaneously, including, in addition to the geodetic parameters, nongravitational parameters for instance, for the air drag. The ultimate geodetic solution can therefore be expected when, in the foreseeable future, both the geometrically and dynamically obtained solutions are combined in a statistically significant result.

2h-4. Satellite Triangulation. The main objective of geometric satellite geodesy is the establishment of three-dimensional positions of a selected number of points on the physical surface of the earth. The significance of geometric satellite geodesy rests on the fact that, for the first time, such a spatial triangulation can be established on a worldwide basis with a minimum of a priori hypothesis; specifically without reference to either the direction or magnitude of the force of gravity. By simultaneously interpolating the satellite position, as seen from at least two observing stations, into the star background, the spatial directions are not only determined directly in terms of the astronomical system (cf. Sec. 2h-1) but are also interpolated in a physical sense into the astronomical refraction effect, thus providing a method essentially free of bias errors. This method—sometimes referred to as *stellar triangulation*[2,3]—is presently being applied in establishing a worldwide reference frame including some 40 stations, and, among other applications, in providing a precise spatial triangulation framework in the area of the North American Datum. Positional accuracy of one part per million

---

[1] *Ibid.*

[2] Y. Väisälä, An Astronomical Method of Triangulation, Helskinki, *Sitzber. Finn. Akad. Wiss.* **1947**.

[3] H. Schmid, Precision and Accuracy Considerations for the Execution of Geometric Satellite Triangulation, *Proc. 2d Intern. Symp. on Use of Artificial Satellites for Geodesy* **II**, Athens, Greece (1965).

is obtained for the worldwide triangulation net, and accuracies of ± 2 m are obtained for continental densification nets, where distances between stations are typically on the order of 1,000 to 1,500 km.

In common with all strictly geometric methods, satellite triangulation—executed as stellar triangulation or as a kind of three-dimensional trilateration, based on optical-electronic ranging—can only provide positions relative to an arbitrarily chosen origin. To obtain positions relative to the center of mass requires recourse either to potential theory, in the case of the gravimetric method, or to celestial mechanics in the dynamic method of satellite geodesy (cf. Sec. 2h-5).

2h-5. Dynamical Methods in Satellite Geodesy. The equations of motion of an artificial earth satellite are given by

$$\ddot{\mathbf{r}} = \mathbf{F}$$

where $\mathbf{r}$ is the position vector of the satellite in the geocentric equatorial coordinate system and $\mathbf{F}$ the force vector. $\mathbf{F}$ is a combination of individual terms

$$\mathbf{F} = \mathbf{F}_E + \mathbf{F}_{SM} + \mathbf{F}_D + \mathbf{F}_R$$

where $\mathbf{F}_E$ is the earth's gravitational effect, $\mathbf{F}_{SM}$ is the sun's and the moon's gravitational effects, $\mathbf{F}_D$ is the atmospheric drag effect, and $\mathbf{F}_R$ is the effect, due to solar radiation pressure.

Generally for earth satellites the terms $\mathbf{F}_{SM}$, $\mathbf{F}_D$, and $\mathbf{F}_R$ are small in comparison to $\mathbf{F}_E$ and can be computed from solar and lunar ephemerides and from models for the air drag and the radiation pressure.

The force term $\mathbf{F}_E$ is obtained as

$$\mathbf{F}_E = \text{grad } V$$

where V is the potential of gravitation of the earth, usually given by an expansion into spherical harmonics

$$V = \frac{GM}{r} \left[1 + \sum_{l=2}^{\infty} \sum_{m=0}^{l} \left(\frac{a}{r} \right)^l \bar{P}_{lm}(\sin \varphi)(\bar{C}_{lm} \cos m\lambda + \bar{S}_{lm} \sin m\lambda) \right]$$

G is the gravitational constant; M is the mass of the earth, r, φ, λ are polar coordinates in the geocentric equatorial coordinate system; a is the equatorial radius of the earth; $\bar{C}_{lM}$ and $\bar{S}_{lm}$ are normalized harmonic coefficients of degree l and order m; $\bar{P}_{lm}(\sin \varphi)$ is the associated Legendre functions, usually normalized in such a manner that

$$\frac{1}{4\pi} \int_0^{2\pi} \int_{-\pi/2}^{\pi/2} \left[\bar{P}_{lm}(\sin \varphi) \begin{Bmatrix} \cos m\lambda \\ \sin m\lambda \end{Bmatrix} \right]^2 \cos \varphi \, d\varphi \, d\lambda = 1$$

If the harmonic coefficients in the expression for V are known, orbits of earth satellites can be computed by numerical integration of the equations of motion or by perturbation theories, provided the initial position and velocity of the satellite are given. If, on the other hand, satellite orbits are observed by means of photographic cameras or electronic tracking devices, the initial positions and velocities of satellites and the harmonic coefficients in the expression for V can be determined.[1,2] Because of the restricted number and accuracy of the observations, only the harmonic coefficients of

[1] I. I. Mueller, "Introduction to Satellite Geodesy," Frederick Ungar Publishing Co., New York, 1964.
[2] W. M. Kaula, "Theory of Satellite Geodesy," Blaisdell Publishing Co., a division of Ginn and Company, Waltham, Mass., 1966.

low degrees are computed. To diminish the correlation between the coefficients, satellite orbits with different orbital parameters are used in the solution. At the present time the most complete set of harmonic coefficients is published by the Smithsonian Institution and given in Table 2h-2. More complete sets exist, but are unpublished.

TABLE 2h-2. HARMONIC COEFFICIENTS IN THE EARTH'S GRAVITATIONAL POTENTIAL*

l	m	$\bar{C} \times 10^6$	$\bar{S} \times 10^6$	l	m	$\bar{C} \times 10^6$	$\bar{S} \times 10^6$
2	0	−484.1735	0	8	4	−0.212	−0.012
2	2	2.379	−1.351	8	5	−0.053	0.118
				8	6	−0.017	0.318
3	0	0.9623	0	8	7	−0.0087	0.031
3	1	1.936	0.266	8	8	−0.248	0.102
3	2	0.734	−0.538				
3	3	0.561	1.620	9	0	0.0122	0
				9	1	0.117	0.012
4	0	0.5497	0	9	2	−0.0040	0.035
4	1	−0.572	−0.469				
4	2	0.330	0.661	10	00	0.0118	0
4	3	0.851	−0.190	10	01	0.105	−0.126
4	4	−0.053	0.230	10	02	−0.105	−0.042
				10	03	−0.065	0.030
5	0	0.0633	0	10	04	−0.074	−0.111
5	1	−0.079	−0.103				
5	2	0.631	−0.232	11	00	−0.0630	0
5	3	−0.520	0.007	11	01	−0.053	0.015
5	4	−0.265	0.064				
5	5	0.156	−0.592	12	00	0.0714	0
				12	01	−0.163	−0.071
6	0	−0.1792	0	12	02	−0.103	−0.0051
6	1	−0.047	−0.027	12	12	−0.031	0.0008
6	2	0.069	−0.366				
6	3	−0.054	0.031	13	00	0.0219	0
6	4	−0.044	−0.518	13	12	−0.059	0.050
6	5	−0.313	−0.458	13	13	−0.059	0.077
6	6	−0.040	−0.155				
				14	00	−0.0332	0
7	0	0.0860	0	14	01	−0.015	0.0053
7	1	0.197	0.156	14	11	0.0002	−0.0001
7	2	0.364	0.163	14	12	0.094	−0.028
7	3	0.250	0.018	14	14	−0.014	−0.003
7	4	−0.152	−0.102				
7	5	0.076	0.054	15	09	−0.0009	−0.0018
7	6	−0.209	0.063	15	12	−0.0619	0.0578
7	7	0.055	0.096	15	13	−0.058	−0.046
				15	14	0.0043	−0.0211
8	0	0.0655	0				
8	1	−0.075	0.065				
8	2	0.026	0.039				
8	3	−0.037	0.004				

* *Smithsonian Astrophys. Observatory Spec. Rept.* 200, p. 2, Cambridge, Mass., 1966.

Satellite observations are not only a function of the initial position and velocity of the satellite, and the harmonic coefficients in the expansion of the earth's potential, but also a function of the coordinates of the tracking stations. Hence, together with the harmonic coefficients, these coordinates can be determined in the geocentric equatorial coordinate system used to formulate the equation of motion of the satellite.

With the coordinates of the tracking stations the dimensions of the reference ellipsoid and datum shifts are obtained. These shifts are needed to transform the coordinates of the various datums to the geocentric equatorial system. With the value

$$GM = 398,601 \text{ km}^3/\text{sec}^2$$

as determined by lunar probes, it was found[1] that

$$a = 6,378,144 \text{ m} \quad \text{and} \quad 1/f = 298.23$$

and the datum shifts are given in Table 2h-3.

2h-6. Physical Geodesy. Another method for determining the earth's gravity potential is given by the solution of the geodetic boundary-value problem. The earth's potential W, consisting of the potential V of gravitation and the potential of

TABLE 2h-3. DATUM SHIFTS*

Datum	Rectangular coordinate shifts		
	Δx meters	Δy meters	Δz meters
North American 1927.......	-23	159	185
European..................	-81	-99	-118
Tokyo.....................	-147	530	676
Old Hawaiian..............	52	-262	-183

* Anderle et al., *op. cit.*, p. 13.

the centrifugal force, is separated into the potential U of a given reference ellipsoid, whose surface is an equipotential surface, and the disturbing potential T:

$$W = U + T$$

T can be regarded as a harmonic function if the reference ellipsoid closely approximates the geoid and if the rotational axes of the ellipsoid and the earth and their angular velocities are identical.

The unknown potential T is connected with the gravity anomalies Δg, measured at the surface of the earth, by the boundary condition, which is given here with the relative error of the flattening of the earth:

$$\Delta g = -\frac{\partial T}{\partial H} - \frac{2T}{R} - \frac{2}{R}(U_0 - W_0)$$

H is the height (i.e., the normal height), R the mean radius of the earth, U_0 the potential at the surface of the ellipsoid, and W_0 the potential of the earth at mean sea level. If the mass of the reference ellipsoid equals the earth's mass,

$$U_0 - W_0 = -\frac{R}{8\pi}\int\int \Delta g \cos \phi \, d\phi \, d\lambda$$

[1] R. J. Anderle, and S. J. Smith, "NWL-8 Geodetic Parameters Based on Doppler Satellite Observations," p. 7, U.S. Naval Weapons Laboratory, Dahlgren, Va., 1967.

The gravity anomalies Δg are computed by subtracting the gravity of the reference ellipsoid at height H above the ellipsoid from the gravity measured at the earth's surface at height H above sea level.

Usually gravity anomalies are referred to the International Ellipsoid whose gravity at its surface is defined by the International Gravity Formula adopted in 1930 by the International Association of Geodesy:

$$\gamma = 978.0490(1 + 0.005{,}2884 \sin^2 B - 0.000{,}0059 \sin^2 2B) \qquad \text{cm/sec}^2$$

γ is called the normal gravity, and B is the ellipsoidal latitude. Table 2h-4 shows the normal gravity from the equator to the pole. The value γ_H at height H above the ellipsoid is computed approximately by

$$\gamma_H = \gamma - 0.3086H \text{ cm/sec}^2 \qquad \text{with } H \text{ in km}$$

The units of gravity anomalies are usually milligals, abbreviated mgal $= 10^{-3}$ cm/sec^2.

The easiest way to obtain the disturbing potential T is by expressing T as the potential of a simple layer distributed over the surface of the earth. If T in the boundary condition is replaced by this expression, an integral equation is obtained, with the density of the surface layer as sought function and the gravity anomalies as absolute values. This integral equation has been derived by Molodenskii[1] who solved it by successive approximations. The first approximation T_0 of T is the well-known formula of Stokes,[2]

$$T_0 = \frac{R}{4\pi} \iint \Delta g \, S(\psi) \, \cos \phi \, d\phi \, d\lambda$$

$$\text{with } S(\psi) = \frac{1}{\sin(\psi/2)} - 6 \sin \frac{\psi}{2} + 1 - 5 \cos \psi - 3 \cos \psi \ln \left(\sin \frac{\psi}{2} + \sin^2 \frac{\psi}{2} \right)$$

This formula holds if the mass of the earth equals the mass of the reference ellipsoid and if both centers of mass coincide. ψ is the spherical distance between the fixed point where T_0 is computed and the variable point at the surface of the sphere with radius R on which the anomalies Δg are assumed to be given. Hence, by means of gravity anomalies the earth's gravitational potential can be computed.

The value of T/γ_H, where T is the value of the disturbing potential at the earth's surface, approximately equals the geoid undulation, i.e., the distance between the surface of the reference ellipsoid and the equipotential surface $W_0 = \text{const}$ at mean sea level, the geoid (cf. Sec. 2h-1). The deflection of the vertical is found by differentiating the disturbing potential T in the horizontal direction. The horizontal derivative of Stokes' formula is known as Vening Meinesz' formula. Thus, by means of gravity anomalies we are able to compute geoid undulations and deflections of the vertical with respect to an ellipsoid whose mass is identical with the mass of the earth and whose center coincides with the mass center of the earth. By knowing the undulations and the deflections of the vertical for the different datums of the world, all datums can be shifted into one common system.

To determine the earth's potential from the solution of the geodetic boundary-value problem requires that the earth's surface be covered with gravity measurements. At present, huge parts of the earth, especially the oceans, are without gravity anomalies. Hence, the gravity measurements have to be combined with the results of satellite observations to improve the knowledge about the earth's gravity field. Either given gravity anomalies are expanded into spherical harmonics and compared

[1] M. S. Molodenskii, V. F. Eremeev, and M. I. Yurkina, "Methods for Study of the External Gravitational Field and Figure of the Earth," Israel Program for Scientific Translations, Jerusalem, 1962.

[2] M. Hotine, "Mathematical Geodesy," *ESSA Monograph* 2, Government Printing Office, October, 1969.

TABLE 2h-4. NORMAL GRAVITY FROM THE EQUATOR TO THE POLE:
COMPUTED FROM THE INTERNATIONAL GRAVITY FORMULA
$[\gamma = 978.0490(1 + 0.0052884 \sin^2 B - 0.0000059 \sin^2 2B)$
cm/sec^2. Unit 1 milligal]

B, deg	Gravity	Difference	B, deg	Gravity	Difference	B, deg	Gravity	Difference
0	978,049.00		31	979,416.53	78.78	61	982,001.46	77.55
1	978,050.57	1.57	32	979,496.80	80.27	62	982,077.35	75.89
2	978,055.27	4.70	33	979,578.46	81.66	63	982,151.49	74.14
3	978,063.10	7.83	34	979,661.40	82.94	64	982,223.77	72.28
4	978,074.06	10.96	35	979,745.54	84.14	65	982,294.12	70.35
5	978,088.12	14.06	36	979,830.77	85.23	66	982,362.45	68.33
6	978,105.26	17.14	37	979,916.98	86.21	67	982,428.67	66.22
7	978,125.48	20.22	38	980,004.08	87.10	68	982,492.70	64.03
8	978,148.74	23.26	39	980,091.94	87.86	69	982,554.46	61.76
9	978,175.02	26.28	40	980,180.48	88.54	70	982,613.88	59.42
10	978,204.29	29.27	41	980,269.47	88.99	71	982,670.89	57.01
11	978,236.50	32.21	42	980,359.12	89.65	72	982,725.41	54.52
12	978,271.63	35.13	43	980,449.01	89.89	73	982,777.37	51.96
13	978,309.63	38.00	44	980,539.14	90.13	74	982,826.72	49.35
14	978,350.44	40.81	45	980,629.39	90.25	75	982,873.39	46.67
15	978,394.04	43.60	46	980,719.65	90.26	76	982,917.33	43.94
16	978,440.35	46.31	47	980,809.82	90.17	77	982,958.47	41.14
17	978,489.33	48.98	48	980,899.78	89.96	78	982,996.77	38.30
18	978,540.92	51.59	49	980,989.42	89.64	79	983,032.19	35.42
19	978,595.05	54.13	50	981,078.64	89.22	80	983,064.67	32.48
20	978,651.66	56.61	51	981,167.33	88.69	81	983,094.19	29.52
21	978,710.68	59.02	52	981,255.37	88.04	82	983,120.69	26.50
22	978,772.05	61.37	53	981,342.67	87.30	83	983,144.16	23.47
23	978,835.68	63.63	54	981,429.10	86.43	84	983,164.55	20.39
24	978,901.49	65.81	55	981,514.58	85.48	85	983,181.85	17.30
25	978,969.42	67.93	56	981,598.99	84.41	86	983,196.03	14.18
26	979,039.38	69.96	57	981,682.23	83.24	87	983,207.08	11.05
27	979,111.28	71.90	58	981,764.19	81.96	88	983,214.99	7.91
28	979,185.03	73.75	59	981,844.79	80.60	89	983,219.73	4.74
29	979,260.55	75.52	60	981,923.91	79.12	90	983,221.31	1.58
30	979,337.75	77.20						

with the harmonic coefficients found by satellite observations, or gravity anomalies are computed, using the harmonic coefficients obtained from satellites, and compared with given gravity anomalies, in order to compute corrected harmonic coefficients. The gecid map of Fig. 2h-1 and the gravity anomalies for 5° by 5° surface elements of Tables 2h-5 and 2h-6 were obtained by such a combination. Combination methods, using instead of the expansion into spherical harmonics the solution of the geodetic boundary-value problem to express the earth's potential, are under investigation.[1,2]

[1] K. Arnold, An Attempt to Determine the Unknown Parts of the Earth's Gravity Field by Successive Satellite Passages, *Bull. Géod.* no. 87, p. 97, Paris, 1968.
[2] Koch, K. R.: Alternate Representation of the Earth's Gravitational Field for Satellite Geodesy, *Boll. Geofisica teorica ed applicata* **10** (40) (1968).

TABLE 2h-5. 5° BY 5° MEAN GRAVITY ANOMALIES FROM A COMBINATION OF SATELLITE AND GRAVIMETRIC DATA REFERRED TO THE INTERNATIONAL GRAVITY FORMULA: EASTERN HEMISPHERE*

(Units milligals)

	0°						30°						60°						90°						120°						150°					180°
90°	6	7	2	3	3	3	3	3	2	3	3	3	−2	−3	2	2	3	3	4	5	5	5	5	5	5	5	5	5	5	5	5	5	5	5	5	5
	19	14	2	11	22	11	13	15	8	16	16	15	14	13	5	4	2	8	7	5	3	0	−2	−4	−5	−6	−5	−4	−2	0	3	6	8	11	14	14
	21	25	20	7	20	19	39	36	32	28	11	21	18	16	15	15	16	16	15	14	11	8	5	3	4	6	9	−8	18	22	24	25	25	24	−2	6
	10	17	27	18	12	9	20	20	17	17	4	7	12	8	4	2	−0	2	−12	−4	−14	−4	0	−2	6	1	7	8	9	10	10	10	9	10	12	16
	55	67	32	19	8	7	12	22	19	7	3	6	13	1	3	−4	−10	−6	−2	−2	−2	−1	−5	5	8	11	12	14	15	13	10	8	11	18	23	24
	24	25	8	−2	−15	0	9	4	10	14	5	21	33	10	3	−5	−5	−10	−11	−9	−7	−6	−7	−9	−6	3	17	29	33	31	25	21	22	25	26	22
60°	5	18	10	−4	−6	3	1	12	11	18	17	20	13	3	4	2	−1	−1	−30	−35	−21	−20	−8	16	13	12	17	15	9	13	25	30	25	14	5	2
	−3	11	16	8	8	22	8	14	13	8	11	3	8	10	5	−13	−21	−2	−3	−27	−19	−15	−0	−0	−3	4	16	22	6	13	32	35	18	11	15	18
	9	23	26	31	18	16	16	5	10	−6	−6	−5	−9	−9	−15	−22	−18	−1	0	9	10	1	−5	1	8	7	5	8	6	16	6	10	3	4	10	6
	20	22	17	38	57	46	8	1	43	−11	4	1	−8	−28	−24	−11	−8	−0	−25	−26	−23	−10	−8	−3	4	3	7	1	30	−1	−5	2	9	11	6	−3
	23	23	22	−5	−9	21	15	32	31	25	−5	−14	−25	−20	14	14	−10	−16	−1	18	−37	−15	−6	−11	2	13	17	35	19	17	−3	5	−16	−8	4	−6
	20	−14	18	2	−1	−35	−15	38	9	11	20	16	21	22	−45	25	5	62	39	39	16	−17	−26	−3	6	20	27	45	−25	3	−7	−9	−19	−13	−16	0
30°	−7	−24	−26	8	−7	−1	2	4	7	−2	−24	5	14	−3	1	−31	−49	−6	−58	−6	−14	−5	−7	1	1	10	−15	19	11	15	−7	−5	−8	−14	−10	8
	20	13	−2	−5	−12	−3	−3	13	19	−19	−24	17	−18	−3	4	6	−4	−8	−21	−17	−14	−11	1	3	13	5	7	1	3	−4	−3	1	−8	−11	−10	−10
	12	10	−15	−10	−3	−4	4	5	29	−11	−9	3	−12	−2	−20	−22	1	−30	−11	−8	−16	−10	7	13	41	9	−2	−0	3	−0	−0	−4	−5	−11	−2	8
	1	4	31	−6	4	7	9	10	6	−5	−6	−11	−15	−29	−31	−21	−8	−32	−16	−4	−5	−1	16	15	32	2	4	−6	−2	−22	−4	−11	−11	−2	−20	21
	16	7	11	−6	−14	3	3	13	−0	3	−51	−7	−14	−28	−45	−20	−8	−23	−30	3	3	7	−1	16	31	2	15	48	8	12	3	2	−6	0	−5	8
	−20	1	4	−14	−20	3	−3	−1	−34	−26	−28	−19	−21	−34	−42	−47	−50	−36	−16	9	14	18	19	27	28	19	52	3	15	10	3	9	−1	4	2	3
0°	−11	−1	−17	−27	−38	7	−4	22	−29	−22	−13	−16	−23	−30	−42	−49	−50	−51	−14	−1	22	32	29	28	4	11	25	16	17	33	13	22	9	10	9	−7
	−10	−28	−8	−12	−6	−3	−2	0	−31	−22	1	−6	−16	−10	−9	−30	−28	−11	−13	2	−3	12	18	11	−2	−18	−12	15	26	2	−20	21	20	−25	−1	7
	−14	−25	−7	9	−12	−10	1	−22	−21	2	−10	−14	−15	−4	−8	−25	−9	2	−11	5	8	2	−9	−14	−7	9	28	20	17	15	11	23	37	9	12	10
	−1	3	−2	−0	−4	−0	4	16	−14	7	−8	12	9	−4	−13	−12	−34	−25	−25	−34	−10	−14	−4	11	9	21	6	5	27	11	53	8	4	21	17	19
	7	8	8	17	−7	5	−10	−5	4	13	−3	11	25	35	19	−4	−6	−5	−3	−34	−28	−17	2	12	−6	−16	−14	0	12	24	10	5	36	23	22	21
	14	−1	3	23	8	26	17	1	3	−4	−16	1	12	13	31	3	−1	−4	−9	−11	−18	−32	4	10	4	−12	−24	−9	−0	2	20	6	11	21	18	11
−30°	8	13	14	12	30	18	13	2	5	9	11	17	2	15	32	7	−2	−3	−1	−25	−27	−32	−31	−3	−45	−25	−15	4	1	22	16	9	25	14	17	22
	5	7	10	13	12	13	21	7	2	3	9	14	14	20	13	3	−7	5	−6	−12	−24	−42	−15	−12	−22	−25	−14	1	8	22	21	0	2	9	27	14
	4	10	11	9	15	17	19	19	15	20	18	8	16	23	10	19	17	15	11	−0	−4	−6	−10	−12	−12	−9	−10	−14	9	27	8	−8	−6	−11	−3	3
	8	7	8	11	11	9	16	23	21	21	18	16	29	26	35	30	21	18	17	12	4	−4	−9	−11	−11	−3	4	−0	−1	6	2	−8	−4	7	−2	10
	25	16	14	22	24	20	21	25	24	23	23	20	19	28	32	30	20	11	11	13	9	4	3	3	−2	−1	5	9	9	7	−0	3	0	−2	13	2
	11	8	7	9	13	15	17	19	20	21	21	19	18	24	33	33	23	13	12	13	11	7	6	5	4	4	5	4	1	1	2	1	−4	−12	−16	−16
−60°	3	5	6	8	12	16	19	17	14	13	11	8	2	−1	1	5	6	6	5	5	6	6	4	−1	−4	−4	−1	−3	−6	−6	0	6	4	−2	−5	−3
	−6	−8	−10	−12	−12	−10	−5	36	15	29	38	37	28	15	10	14	23	21	13	28	26	25	21	23	20	19	19	9	−2	8	2	−3	−7	−12	−16	−18
	50	58	62	60	54	45	37	33	32	34	34	32	25	16	11	9	11	12	23	14	11	10	−37	7	6	5	4	−52	−11	−17	−23	−24	3	3	−16	−7
	15	14	11	7	3	−1	−4	−5	−4	−2	2	6	9	12	14	15	37	21	17	17	−45	1	11	7	4	0	−4	−35	−37	−35	−29	21	−36	−9	−13	−13
	−4	−6	−6	−5	−3	0	5	10	16	22	28	32	27	22	36	34	31	27	23	18	14	11	8	7	−25	−27	−19	−16	8	8	7	6	3	−0	−21	−16
	−8	5	5	5	5	5	6	6	6	6	6	6	6	6	5	5	4	4	−8	4	4	−10	5	6	7	9	10	−9	−11	−8	6	−7	−4	0	11	20
−90°																																				

* R. H. Rapp, Comparison of Two Methods for the Combination of Satellite and Gravimetric Data, *Ohio State Univ. Dept. Geod. Sci. Rept.* 113, p. 26, 1968.

Table 2h-6. 5° by 5° Mean Gravity Anomalies from a Combination of Satellite and Gravimetric Data Referred to the International Gravity Formula: Western Hemisphere*

(Units milligals)

Longitude 180° to 270° (divisions at 210°, 240°)

Latitude	180°						210°						240°					
90° to 60°	3	−0	−3	−3	6	0	6	4	8	10	13	15	14	12	10	9	8	13
	14	14	6	7	10	7	14	12	12	11	9	11	34	35	18	21	24	26
	16	−6	11	9	6	3	2	5	9	16	22	13	27	33	33	33	29	34
	20	24	6	9	11	−3	7	2	9	15	5	6	25	9	15	10	8	17
	22	19	3	−1	18	23	24	26	10	−1	−6	1	−9	2	−10	3	2	1
	19	8	7	19	24	16	42	59	17	5	−10	−13	−13	−11	4	4	0	−3
60° to 30°	7	14	18	42	29	29	12	10	4	9	1	−0	2	3	−9	0	−11	−4
	−18	−20	3	−16	−3	16	8	6	−1	−7	−36	3	27	6	11	19	7	3
	8	6	13	15	9	12	0	−5	−9	−5	−8	15	3	22	18	22	11	11
	−7	−11	−8	−2	3	1	−3	−14	−15	−5	−6	22	14	24	27	13	4	−3
	−6	−9	−13	−12	−6	−6	−5	−13	−18	−25	−21	−8	6	5	22	−1	−6	−3
	−6	−7	−37	−11	7	2	0	−8	−5	−12	−15	−20	−12	−10	7	−0	−1	2
30° to 0°	−13	−1	8	9	2	0	10	10	−7	−21	−25	−26	−15	−19	2	26	−4	2
	−5	2	11	36	26	3	−7	−8	−12	−16	−18	−19	−21	−11	−12	17	−4	7
	−7	−4	−8	−3	2	−7	−2	−14	−14	−29	−19	−23	−15	−10	−5	1	23	5
	17	−12	−4	1	9	−16	−3	−2	−5	−8	−18	10	−22	2	−3	−4	0	4
	−23	−8	−2	22	21	−12	−14	−10	−5	−4	−11	−12	−14	−1	−12	−1	0	1
	5	−7	−7	21	1	−8	−8	−6	−2	−2	−3	−7	−7	−3	−6	−2	−3	6
0° to −30°	−1	−7	−5	−8	−0	−2	−6	−8	−6	−5	−1	−4	−6	−8	−5	2	−5	2
	−2	−6	−7	7	12	10	−1	−1	−7	−5	−7	−4	−5	−4	−3	−4	1	−5
	11	2	−6	6	3	14	−2	−3	−13	−3	−3	2	−4	2	1	0	2	−2
	7	4	−1	5	9	13	6	8	−3	−5	−8	−1	−3	2	−7	2	1	−0
	−10	−16	0	5	5	4	−2	−8	−6	−1	3	0	−4	1	−5	4	−2	8
	10	9	4	3	−3	−2	−0	−6	−4	−6	−2	−4	−1	5	1	8	−4	6
−30° to −60°	−20	4	8	11	4	−2	−3	−6	−3	−3	−1	−5	−3	7	3	8	6	4
	−2	4	9	9	7	7	7	−1	−2	−2	−1	−2	−1	7	5	5	8	10
	9	9	7	8	10	8	6	3	4	0	−5	−1	3	3	5	8	9	12
	3	−0	0	7	8	4	3	2	1	1	−1	1	6	8	8	6	5	9
	−7	−8	−6	−4	−6	−3	3	4	2	3	5	6	8	8	5	5	10	13
	−12	−10	−10	−10	−7	−2	1	−1	−1	3	8	10	11	9	6	5	9	11
−60° to −90°	−1	−2	−6	−7	−6	−4	−6	−7	−6	−2	0	1	2	4	6	8	9	11
	−18	−18	−17	−15	−14	−15	−16	−16	−13	−10	−8	−7	−6	−3	1	3	2	−1
	−13	−19	−24	−28	−31	−32	−29	−24	−16	−8	−1	4	7	8	10	11	−6	27
	−18	−18	−22	−28	−12	−15	−8	−22	−16	−5	13	5	1	−9	−18	−11	−12	13
	−19	−18	−15	−19	−14	−12	−10	−7	−5	−19	−2	−22	−2	−23	−1	17	17	13
	−9	20	20	20	−11	20	21	20	20	20	20	−6	20	20	20	20	20	−23

Longitude 270° to 0° (divisions at 300°, 330°)

Latitude	270°						300°						330°					0°
90° to 60°	16	12	4	13	2	2	2	3	2	2	2	1	0	−1	1	1	0	1
	28	16	16	27	34	6	37	38	39	38	37	34	30	24	16	12	16	13
	34	32	29	13	10	39	33	15	−14	−11	−15	22	19	48	40	17	31	15
	3	12	−5	−2	0	−3	1	4	22	20	24	27	23	22	24	23	24	8
	2	5	9	4	1	2	4	−17	11	46	23	36	34	31	37	47	28	19
	−7	−10	−5	−7	−7	−2	−4	−17	−27	39	26	28	20	49	31	49	40	42
60° to 30°	1	3	−33	−22	−5	11	−2	−2	11	21	24	25	30	27	10	14	23	21
	−10	−15	−9	−6	17	0	−22	20	11	12	22	30	16	20	23	33	24	17
	−8	−5	−10	−11	−6	3	11	17	33	18	39	13	15	11	5	−5	−5	13
	−5	−2	−7	10	10	−15	−20	0	17	19	22	23	25	14	13	3	14	24
	1	−0	6	−16	−15	−10	−8	−7	−3	12	23	35	36	16	−2	11	30	19
	−0	11	−6	−31	−18	−2	−9	−7	−2	17	19	23	16	−5	14	−4	43	52
30° to 0°	12	23	−27	−28	−24	−18	−28	−19	11	24	−1	−10	−10	−7	8	24	10	−2
	22	−2	−5	−58	−27	−23	−28	−13	3	−4	−11	−13	−15	−8	−1	4	−7	−11
	15	−7	−20	4	−49	−16	−27	−26	−16	−17	−20	−9	−4	14	2	14	−3	11
	18	22	−11	−28	−36	−18	−50	−29	−34	−13	−14	−17	−12	−10	15	21	5	6
	12	17	34	12	18	−8	−17	−22	−37	−29	−8	4	1	3	5	16	29	18
	9	5	31	24	23	11	−1	−4	−11	−19	−7	4	6	2	−4	0	7	19
0° to −30°	9	3	11	24	19	4	−9	−6	−5	−15	−7	−1	−6	−8	−4	5	10	1
	−9	−6	13	3	−7	2	−7	−10	−36	−34	1	−19	−7	−10	−4	2	0	−2
	2	0	−8	26	18	13	5	−11	4	−9	−11	−10	−13	−12	−5	−4	−4	−4
	−3	−6	−20	−5	44	21	7	−2	−5	10	−11	−13	−9	−12	−4	−0	4	−7
	−12	21	−12	−2	48	−8	−7	−18	−7	−11	−16	−11	−10	−6	−4	5	0	−9
	−10	13	−4	21	27	16	10	8	−2	−21	−23	−8	−13	−5	−1	6	7	−2
−30° to −60°	−2	9	5	2	10	15	11	−0	16	−9	−4	11	0	0	6	4	0	16
	9	4	5	13	10	9	16	3	1	4	−5	−3	−5	−3	−4	5	3	9
	14	0	7	11	7	3	6	−4	−3	8	3	−1	−2	−3	4	8	8	5
	12	3	0	28	5	14	−1	−5	−4	0	0	−2	−2	2	9	12	11	9
	7	1	29	6	2	6	−1	15	15	16	16	12	9	10	14	18	23	28
	8	5	6	7	9	2	0	1	8	10	18	20	16	13	15	16	15	13
−60° to −90°	12	9	6	7	14	25	30	25	16	12	15	18	17	14	12	10	6	3
	8	−2	5	17	31	21	30	17	4	−3	−1	1	1	−3	−6	−8	−7	−6
	20	12	15	40	40	32	21	9	−1	−6	−6	−4	−1	2	6	14	25	38
	16	11	15	22	13	−10	8	3	0	−14	−13	2	4	7	9	11	13	14
	7	−9	−6	−4	18	−10	−22	−9	−14	7	7	7	6	5	4	2	0	−2
	−9	19	18	18	13	7	6	−14	−19	13	−27	11	10	9	8	−22	7	−26

* Rapp, *op. cit.*, p. 27.

2h-7. Geodetic Reference System: 1967. In 1967 the General Assembly of the International Union of Geodesy and Geophysics recommended replacing the International Ellipsoid and the International Gravity Formula with the Geodetic Reference System 1967 defined by[1]

$$a = 6\ 378\ 160\ \text{m}$$
$$GM = 398\ 603\ \text{km}^3/\text{sec}^2$$
$$J_2 = 10\ 827 \times 10^{-7}$$

with $J_2 = -\sqrt{5}\ \bar{C}_{20}$. This set of parameters is identical with the parameters adopted by the International Astronomical Union in 1964 as part of a system of new

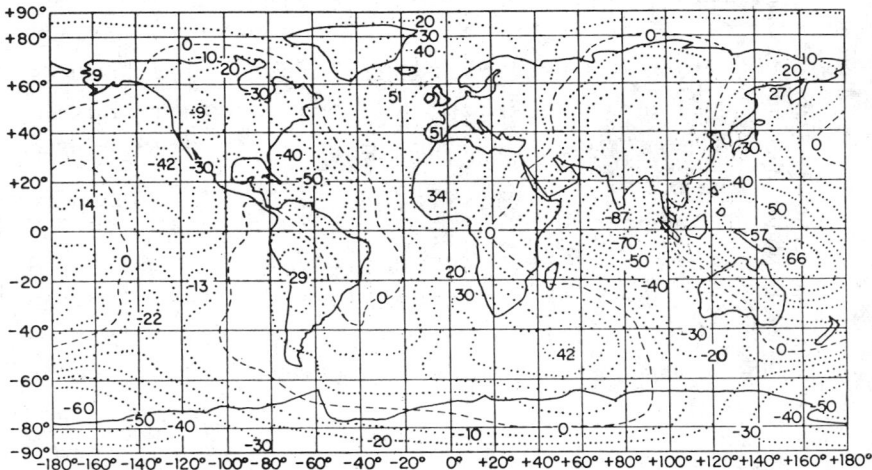

FIG. 2h-1. Geoid obtained by combining satellite and gravimetric data.   Units: meters. (*W. Kohnlein, Smithsonian Astrophysical Observatory Special Report* 264, *p.* 57, 1967.)

astronomical constants. The values for a, GM, and J_2, together with the value for the earth's rotational velocity, define an equipotential ellipsoid of revolution completely, so that the shape of the ellipsoid and its external gravity field are determined by the four constants. Only preliminary numerical values for the shape of the ellipsoid and the gravity formula of the Geodetic Reference System 1967 have been published until now.[2,3]

[1] *Bull. Geod.* no. 86, p. 367, Paris, 1967.

[2] A. H. Cook, The Polar Flattening and Gravity Formula in the Geodetic Reference System 1967, *Geophys. J.* **15,** p. 431, Oxford, 1968.

[3] H. Moritz, "The Geodetic Reference System 1967," *Allgem. Vermess.*, p. 2, Karlsruhe, 1968.

2i. Seismological and Related Data

B. GUTENBERG[1]

California Institute of Technology

J. E. WHITE

Globe Universal Sciences, Inc.

2i-1. List of Symbols

V velocity of longitudinal wave P
v velocity of transverse wave S
P symbol denoting longitudinal wave
S symbol denoting transverse wave
k bulk modulus or volume elasticity
μ rigidity or shear modulus
ρ density
τ Poisson's ratio
A ratio V/v
t temperature in degrees centigrade, time
p pressure in bars
h depth in the earth
T period of seismic disturbance
G symbol denoting surface shear waves
R_a symbol denoting Rayleigh waves
Δ epicentral distance
SH symbol denoting component of S wave in horizontal plane
SV symbol denoting component of S wave in vertical plane
i actual angle of incidence at a discontinuity
$\bar{\imath}$ apparent angle of incidence at a discontinuity
u ratio of horizontal ground displacement to incident amplitude

2i-2. Fundamental Equations for Elastic Constants and Wave Velocities. In purely elastic, isotropic, homogeneous media the velocity V of longitudinal waves P, v of transverse waves S, the bulk modulus k, the rigidity μ, the density ρ, and Poisson's ratio σ are connected by the following equations:

$$V^2 = \frac{k + \frac{4}{3}\mu}{\rho} \qquad v^2 = \frac{\mu}{\rho} \qquad (2i\text{-}1)$$

$$\sigma = \frac{\frac{1}{2}A^2 - 1}{A^2 - 1} \qquad A = \frac{V}{v} \qquad (2i\text{-}2)$$

$$k = \rho(V^2 - \tfrac{4}{3}v^2) \qquad \mu = v^2\rho \qquad (2i\text{-}3)$$

[1] Deceased.

2i-3. Elastic Constants and Wave Velocities in Rocks (Laboratory Experiments).
In rocks the elastic constants and the wave velocities usually increase with increasing
pressure p (Tables 2i-2 and 2i-3) and decrease with increasing temperature t and with
porosity. Phase changes affect all elastic quantities. Many sedimentary rocks show
significant anisotropy, with an axis of symmetry. Table 2i-4 gives an example of
velocity differences for vertical and horizontal travel[1] and for shear polarization.

TABLE 2i-1. CORRESPONDING VALUES OF POISSON'S RATIO σ AND V/v

σ	0.00	0.10	0.20	0.22	0.24	0.25	0.26	0.28	0.30	0.40	0.50
V/v	1.414	1.500	1.633	1.670	1.710	1.732	1.756	1.809	1.871	2.449	∞

TABLE 2i-2. ELASTIC CONSTANTS AND WAVE VELOCITIES IN ROCKS
AT ROOM TEMPERATURE[†]

	μ, 10^{11} dynes/cm^2		k, 10^{11} dynes/cm^2		σ	V, km/sec	v, km/sec
	1 atm	4,000 atm	1 atm	4,000 atm			
Dunite..........	$4\frac{3}{4}$–6	$6\frac{1}{2}$–7	?	$12\pm$	0.25–0.30	$7\frac{1}{2}$–$8\frac{1}{2}$	$4\frac{1}{4}$–$4\frac{3}{4}$
Gabbro..........	3–4	4–5	$6\pm$	$8\frac{3}{4}\pm$	0.2–0.3	5–7	$3\frac{1}{4}$–4
Granite..........	$1\frac{1}{2}$–$2\frac{1}{2}$	$3\frac{1}{4}$–$3\frac{1}{2}$	$2\frac{3}{4}$–$3\frac{1}{2}$	$5\frac{1}{4}\pm$	0.20–0.26	5–$6\frac{1}{4}$	2–$3\frac{1}{2}$
Obsidian glass......	$2\frac{3}{4}$–3	?	$3\frac{1}{4}\pm$	$3\frac{3}{4}$–4	0.1–0.2?	$5\pm$	$3\frac{1}{4}\pm$
Ice..............	$\frac{1}{4}$–$\frac{1}{2}$	?	$\frac{3}{4}$–1	?	0.3–0.4	$3\frac{1}{4}$–$3\frac{3}{4}$	$1\frac{1}{2}$–2

[†] F. Birch, ed., Handbook of Physical Constants, *Geol. Soc. Am., Spec. Paper* 36 (1942); L. H. Adams,
Elastic Properties of Materials of the Earth's Crust, in "Internal Constitution of the Earth," 2d ed.,
pp. 50–80, 1951. See also S. P. Clark, Jr., ed., Handbook of Physical Constants, rev. ed., *Geol. Soc.
Am., Mem.* 97 (1966).

TABLE 2i-3. LONGITUDINAL VELOCITIES, KM/SEC, AT PRESSURES p AND
TEMPERATURES t CORRESPONDING TO THE DEPTH h IN THE EARTH
AFTER LABORATORY MEASUREMENT[†]

p, bars	t, °C	h, km	Dunite	San Marcos gabbro	Texas gray granite	Woodbury granite
260	45	1	7.55	6.70	5.90	5.90
1,300	135	5	7.50	6.90	6.02	6.15
2,600	225	10	7.22	6.96	6.02	6.14
3,900	290	15	....	6.95	6.01	6.04
6,700	400	25	....	6.80		

[†] D. S. Hughes and C. Maurette, Variation of Elastic Wave Velocities in Basic Igneous Rocks with
Pressure and Temperature, *Geophysics* **22,** 23–31 (1957).

2i-4. Periods and Amplitudes of Seismic Waves. Seismological instrumentation
has made great advances in fidelity of observation, geographic distribution of stations,
and machine data reduction. Strain seismometers have uniform sensitivity from
periods of many hours down to a few seconds. Tilt meters and gravimeters also

[1] J. E. White and R. L. Sengbush, Velocity Measurements in Near-surface Formations,
Geophysics **18,** 54 (1963).

indicate earth motion down to "dc," i.e., periods much greater than the tidal period. A worldwide net of 125 stations has been established, recording three-component motion in 0.1-to-1-sec range and 10-to-100-sec range. A few array stations exist at which signals from dozens of seismometers in an array can be combined. This improved instrumentation gives an improved portrayal of earthquakes and more accurate knowledge of the structure of the earth.

Earthquakes create permanent displacements, which may be observed at great distances.[1] Great earthquakes excite the free oscillations of the earth to measurable amplitudes,[2] at periods of 3 to 54 min. Love waves and Rayleigh waves in the period range 10 to 100 sec are governed by velocity contrasts in the crust and mantle. Body waves display periods of 0.1 to 10 sec, depending on range, with shear waves tending to longer periods than compressional waves.

TABLE 2i-4. VELOCITIES IN SHALLOW SEDIMENTS, KM/SEC

	V vert.	V horiz.	v_{SV} vert.	v_{SH} horiz.
Chalk..........	2.6	3.0	1.1	1.2
Shale..........	1.8	2.4	0.4	0.6

Periods of natural microseisms (continuous motion from meteorological sources and ocean waves) range from a fraction of a second to a minute or more. The largest amplitudes of the most frequent types of microseisms (periods 4 to 10 sec) are a few microns at inland stations on rock and between 10 and 100 microns at stations near oceans during heavy storms.

After great earthquakes, waves through the earth's interior may reach the surface at great distances with amplitudes of over 10 microns and periods of the order of 5 sec, while the largest surface waves may have ground amplitudes of 10 mm with periods of 20 sec. Much greater amplitudes occur near the source. In motion from not too close artificial explosions, longitudinal waves usually carry the largest amplitudes; even waves through the earth's core have been identified on such records.[3]

2i-5. Travel Times of Earthquake Waves. Examples of travel times are given in Table 2i-5. Surface waves traveling a few times around the earth have travel times of several hours. No dispersion has been established for waves through the earth's body except for waves through the transition zone from the liquid outer core to the probably solid inner core.[4] However, the prevailing increase in the velocity of longitudinal and transverse waves with depth results in a prevailing increase in wave velocity of surface waves as their length (depth of energy penetration) increases. Surface waves of first, second, and third modes have been observed. The group velocity of surface waves of first mode has a minimum[5] for periods of several seconds, depending on the crustal structure.

[1] C. J. Wideman and M. W. Major, Strain Steps Associated with Earthquakes, *Bull. Seis. Soc. Am.* **57,** 1429 (1967).

[2] L. E. Alsop, Spheroidal Free Periods of the Earth Observed at Eight Stations around the World, *Bull. Seis. Soc. Am.* **54,** 755 (1964).

[3] B. Gutenberg, Travel Times of Longitudinal Waves from Surface Foci, *Proc. Natl. Acad. Sci. U.S.* **39,** 849 (1953).

[4] B. Gutenberg, Wave Velocities in the Earth's Core, *Bull. Seis. Soc. Am.* **48,** 301–314 (1958).

[5] M. Ewing and F. Press, Crustal Structure and Surface-wave Dispersion, *Bull. Seis. Soc. Am.* **40,** 271–280 (1950); **42,** 315–325 (1952); **43,** 137–144 (1953). Surface Waves and Guided Waves, "Encyclopedia of Physics," vol. 47, pp. 119–139, Springer-Verlag, Berlin, 1956.

2i-6. Reflection and Refraction of Waves. If a longitudinal wave P or a transverse wave S arrives at a discontinuity, one P and one S wave are reflected and one of each type is refracted if the velocity ratio V_r/V_i of the reflected or refracted (r) and incident (i) wave permits.[1]

$$\sin i_r = \frac{V_r}{V_i} \sin i_i \qquad (2i\text{-}4)$$

where i_i is the angle of incidence. Examples are given in Table 2i-6. Amplitudes of transverse waves (vibrations perpendicular to the ray) are frequently resolved into two components, SH in the horizontal plane, and SV (with a vertical component) perpendicular to SH. If an SH wave is incident, the reflected wave and the refracted wave (if it exists) are always of the SH type.

TABLE 2i-5. TRAVEL TIMES† (MIN:SEC) OF DIRECT LONGITUDINAL WAVES P
AND TRANSVERSE WAVES S THROUGH THE EARTH STARTING AT DEPTH h,
AND OF SURFACE SHEAR WAVES G AND RAYLEIGH WAVES R_a WITH
PERIODS OF ABOUT 1 MIN (INDEPENDENT OF FOCAL DEPTH)
(Δ = epicentral distance, deg; P waves arriving at $\Delta > 100$ deg enter the earth's core)

Δ	$h = 25$ km		G, min	R_a, min	$h = 300$ km		$h = 700$ km	
	P	S			P	S	P	S
0	0:04	0:07	....	....	0:39	1:08	1:20	2:24
2	0:32	0:55	....	....	0:46	1:24	1:24	2:30
4	0:59	1:56	....	....	1:07	1:51	1:32	2:48
10	2:28	.....	4.1	4.5	2:17	4:03	2:20	4:12
20	4:34	8:16	8.3	9.0	4:15	7:39	3:55	7:02
40	7:36	13:42	16.5	17.9	7:11	12:52	6:44	12:01
70	11:12	20:20	28.9	31.4	10:44	19:21	10:11	18:20
100	13:46	25:14	41.3	44.8	13:15	24:23	12:37	23:14
120	18:54	28:00	49.5	53.8	18:19	27:09	17:38	26:01
150	19:46	.....	61.9	67.2	19:11	.....	18:31	
180	20:10	.....	74.2	80.6	19:35	.....	18:54	

† B. Gutenberg, Travel Times of Longitudinal Waves from Surface Foci, *Proc. Natl. Acad. Sci, U.S.* **39**, 849 (1953); H. Jeffreys and K. E. Bullen, "Seismological Tables," British Association for the Advancement of Science, 1940; B. Gutenberg, and C. F. Richter, On Seismic Waves, *Gerlands Beitr. Geophys.* **43**, 56–133 (1934); **54**, 94–136 (1939).

If a wave arrives at the earth's surface (actual angle of incidence i) a wave of the same type is reflected (angle i), and one of the other type may be reflected [Eq. (2i-4)] (see Table 2i-7). As a consequence of these three waves, the apparent angle of incidence $\bar{\imath}$ calculated from records of horizontal H and vertical V instruments (tan $\bar{\imath} = H/V$) differs from i. In case of incident transverse waves the particles move in ellipses,[2] if $(V \sin i)/v > 1$. If an SH wave is incident, the reflected wave has the same amplitude as the incident wave, the ground displacement is twice the incident amplitude, and $\bar{\imath} = i$. For energy ratios of waves reflected and refracted at the boundary of the earth's core, see Table 2i-8. An SH wave incident upon the core is totally reflected.

[1] M. Ewing, W. S. Jardetzky, and F. Press, "Elastic Waves in Layered Media," pp. 74–93, McGraw-Hill Book Company, New York, 1957; B. Gutenberg, Energy Ratio of Reflected and Refracted Seismic Waves, *Bull. Seis. Soc. Am.* **34**, 85–102 (1944).
[2] B. Gutenberg, *SV* and *SH*, *Trans. Am. Geophys. Union* **33**, 573–584 (1952).

TABLE 2i-6. SQUARE ROOT OF ENERGY REFLECTED OR TRANSMITTED AT A
DISCONTINUITY WITH DENSITY RATIO (UPPER LAYER TO LOWER) 1.103,
CORRESPONDING VELOCITY RATIO 1.286 FOR P AND FOR S, POISSON'S
RATIO 0.25 IN BOTH LAYERS

(Incident energy taken as unity. Based on Slichter-Gabriel.[†] 1– indicates values
between 0.95 and 1.0. i = angle of incidence. P = longitudinal,
SV = component of transverse wave in plane of ray)

$i°$	Refracted waves								Reflected waves							
	P from				SV from				P from				SV from			
	Above		Below		Above		Below		Above		Below		Above		Below	
	P	SV	P	SV	P	SV	P	SV	P	SV	P	SV	P	SV	P	SV
0	1–	0.0	1–	0.0	0.0	0.2	0.0	1–	0.2	0.0	0.2	0.0	1.0	0.0	0.0	0.2
15	1–	0.1	1–	0.1	0.1	0.1	0.1	1–	0.2	0.1	0.2	0.1	1–	0.1	0.1	0.1
30	1–	0.1	1–	0.1	...	0.2	0.2	1–	0.1	0.1	0.1	0.1	0.9	0.2	0.1	0.0
45	0.5	0.2	0.9	0.1	...	0.4	0.3	1–	0.2	0.0	0.1	0.1	0.9	0.3	...	0.2
60	...	0.3	0.9	0.2	...	...	...	1–	0.9	0.1	0.2	0.1	...	...	...	0.3
75	...	0.4	0.8	0.3	...	...	...	0.8	0.9	0.1	0.4	0.1	...	...	...	0.5
90	...	0.0	0.0	0.0	...	...	...	0.0	1.0	0.0	1.0	0.0	...	...	...	1.0

† B. Gutenberg, *Bull. Seis. Soc. Am.* **34**, 85 (1944).

TABLE 2i-7. SQUARE ROOTS OF RATIO OF REFLECTED TO INCIDENT ENERGY a
AT EARTH'S SURFACE AS FUNCTION OF ANGLE OF INCIDENCE i AND RATIO
OF HORIZONTAL u AND VERTICAL w GROUND DISPLACEMENTS TO
INCIDENT AMPLITUDE FOR CONTINUOUS SINUSOIDAL WAVES
IF POISSON'S RATIO IS 0.25; i = APPARENT ANGLE OF
INCIDENCE CALCULATED FROM OBSERVED HORIZONTAL
AND VERTICAL COMPONENTS

(Elliptic motion of ground is indicated by *, and corresponding values for $\bar{\imath}$ are
calculated on the assumption that the vertical and horizontal component
reach their maximum simultaneously.[†] SV = component of transverse
wave in plane of ray)

i	Longitudinal wave P incident					SV incident				
	a of P	a of SV	u	w	$\bar{\imath}$, deg	a of P	a of SV	u	w	$\bar{\imath}$, deg
0°	1.0	0.0	0.0	2.0	0	0.0	1.0	2.0	0.0	0
20	0.8	0.6	0.8	1.9	23	0.9	0.4	1.8	0.8	23
30	0.6	0.8	1.2	1.7	34	1.0	0.0	1.7	1.0	30
35.3	0.5	0.9	1.3	1.5	39	0.0	1.0	4.9	0.0	±0
40	0.4	0.9	1.4	1.4	44	...	1.0	0.7*	1.6*	−64*
45	0.3	0.9	1.5	1.3	48	...	1.0	0.0	1.4	±90
60	0.0	1.0	1.7	1.0	60	...	1.0	0.5*	1.1*	66*
80	0.1	1.0	1.3	0.5	69	...	1.0	0.3*	0.5*	59*
90	1.0	0.0	0.0	0.0	71	...	1.0	0.0*	0.0*	60*

† B. Gutenberg, *SV and SH, Trans. Am. Geophys. Union* **33**, 573–584 (1952).

2i-7. Wave Types and Their Symbols. The main discontinuities of the earth (Fig. 2i-1) are its surface, the "Mohorovičić discontinuity" (depth 10 ± km below the surface in the deeper parts of the major oceans, 30 ± km under the lower parts of continents, up to about 70 km under high mountain ranges, e.g. North Pamir[1]), and the boundary of the earth's core at a depth of 2,900 ± 10 km (radius $r = 3,470$ km). The transition from the outer to the inner core is probably gradual. At a distance of about 1,500 km from the earth's center, the velocity of longitudinal waves begins to increase more rapidly with depth than in the outer core but becomes approximately constant about 300 km deeper. This transition zone between the outer and the inner core may correspond to a transition from the liquid to the solid state.

TABLE 2i-8. SQUARE ROOTS OF ENERGY RATIOS FOR WAVES REFRACTED
(REFR.) AND REFLECTED (REFL.) AT THE BOUNDARY OF THE
EARTH'S CORE[†]
[Assumed at the core boundary: densities 5.4 (mantle), 10.1 (core); longitudinal velocities 13.7 and 8.0 km/sec, respectively; transverse velocity in the mantle 7.25 km/sec, 0 in core. i = angle of incidence of the arriving wave]

P incident in mantle				P incident in core				SV incident in mantle			
i	Refr. P	Refl. P	Refl. S	i	Refr. P	Refr. S	Refl. P	i	Refr. P	Refl. P	Refl. S
0	0.999	0.04	0.00	0	0.999	0.00	0.04	0	0.00	0.00	1.00
20	0.96	0.12	0.24	20	0.90	0.44	0.08	20	0.50	0.39	0.78
40	0.87	0.29	0.39	$33\frac{1}{2}$	0.79	0.62	0.00	30	0.61	0.47	0.64
60	0.79	0.42	0.44	35	0.83	0.55	0.10	31	0.58	0.49	0.65
80	0.84	0.20	0.51	35.7	0.00	0.00	1.00	32.0	0.00	0.00	1.00
83.8	0.85	0.00	0.52	37	.....	0.85	0.53	33	0.84	....	0.54
85	0.85	0.10	0.52	50	.....	0.92	0.40	40	0.92	....	0.40
89	0.60	0.71	0.36	80	.....	0.62	0.78	64	0.55	....	0.84
90	0.00	1.00	0.00	90	.....	0.00	1.00	65.0	....	....	1.00

† After S. Dana, The Partition of Energy among Seismic Waves Reflected and Refracted at the Earth's Core, *Bull. Seis. Soc. Am.* **34**, 189–197 (1944).

By international agreement longitudinal waves in the mantle are indicated by P (starting downward at the source) or p (starting upward), transverse waves by S or s, longitudinal waves through the outer core by K, through the inner core by I, and (hypothetical) transverse waves through the inner core by J (Fig. 2i-2). Some authors use $P' \equiv PKP$, $P'' \equiv PKIKP$. For a source below the surface, there is one reflection at the surface near the epicenter, another about halfway between source and station. The symbols for these waves are, respectively, pP and PP, sP and SP, pS and PS, sS and SS. Similarly, for twice-reflected waves pPP, PPP, etc., are used. Time differences $pP - P$, $sP - P$, $sS - S$, etc., give a good indication for the focal depth (Table 2i-9).[2] Among observed waves through the core reflected at the

[1] I. P. Kominskaya, G. G. Mikhota, and Yu. V. Tulina, Crustal Structure of the Pamir-Alai Zone from Seismic Depth-sounding Data, *Izvest.*, *Geophys. Ser.*, trans. by Am. Geophys. Un., 1959, p. 673.
[2] B. Gutenberg and C. F. Richter, Materials for the Study of Deep-focus Earthquakes, *Bull. Seis. Sos. Am.* **26**, 341–390 (1936); see also H. Jeffreys and K. E. Bullen, "Seismological Tables," p. 24, British Association for the Advancement of Science, 1940.

surface of the earth are $pPKP$, $sPKP$, $P'P' \equiv PKPPKP$, $P'P'P'$, $P'P'P'P'$ (with a travel time of about $1\frac{1}{4}$ hr).

Waves in the mantle with a reflection at the core surface permit accurate determination of the radius of the core. They are indicated by c, e.g., PcP, PcS, ScS; $pPcP$, $ScSScS$, etc., are in addition, reflected at the surface. All these waves usually have

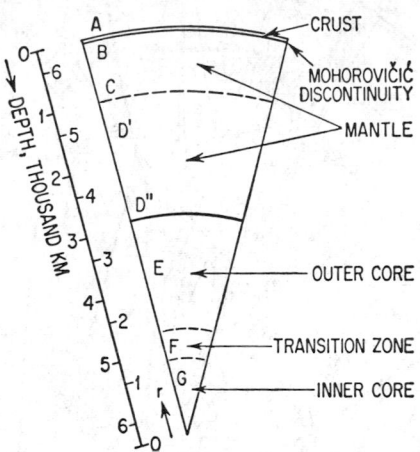

Fig. 2i-1. Main discontinuities of the earth. The letters A to G referring to the various regions in the earth have been suggested by Bullen and are used internationally. (*K. E. Bullen, Seismic Waves Transmission, vol. 47, p. 104, "Encyclopedia of Physics," Springer-Verlag, Berlin, 1956.*)

TABLE 2i-9. FOCAL DEPTH, KM, OF EARTHQUAKES FOR GIVEN TIME
DIFFERENCES $pP - P$, $sP - P$, AND $sS - S$ FOR EPICENTRAL
DISTANCES Δ OF 30, 80, AND 145 DEG
(* indicates that pP, sP, or sS, respectively, does not exist under given conditions)

Time diff., min:sec	$\Delta = 30$ deg			$\Delta = 80$ deg			$\Delta = 145$ deg	
	$pP - P$	$sP - P$	$sS - S$	$pP - P$	$sP - P$	$sS - S$	$pP' - P'$	$sP' - P'$
0:20	100	60	50	75	55	40	70	55
0:40	205	120	100	160	105	85	150	105
1:00	310	195	165	250	165	140	235	160
1:30	*	295	270	395	255	220	375	250
2:00	*	415	425	565	350	300	525	345
2:30	*	535	*	755	460	390	690	440
3:00	*	*	*	?	575	485	?	540

periods of 1 to 4 sec. Waves reflected inside the core are indicated by $PKKP$, $SKKS$, etc. Their periods, too, are small ($PKKP$ waves with wavelengths $L < 10$ km have been observed), indicating a sharp boundary of the core. Waves refracted through the core (in addition to PKP) are PKS, SKP, SKS, etc. All observed travel times agree within a few seconds with those following from the velocities for P, K, and S (see Table 2i-11).

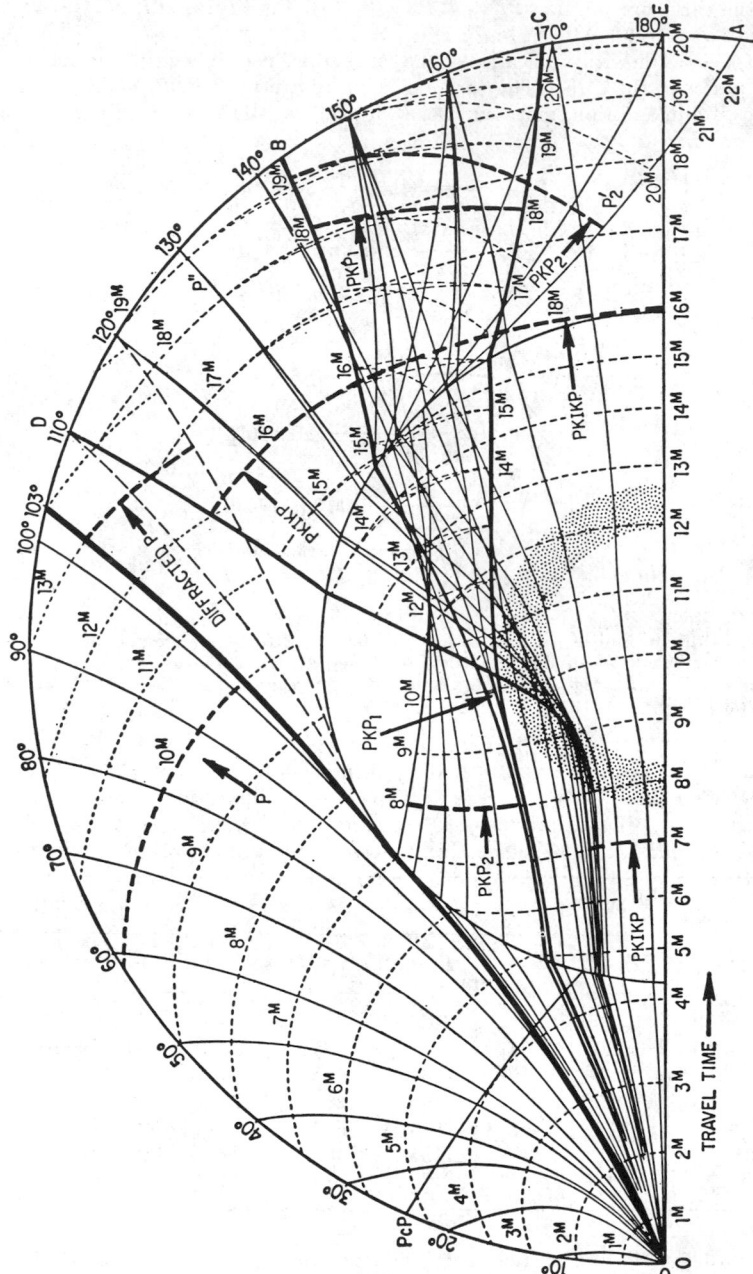

Fig. 2i-2. Paths and wavefronts (travel times) of longitudinal waves in the earth.

2i-8. Equations Used in Calculating Travel Times and Velocities. If i = angle of incidence (between ray and vertical), r = radius vector measured from center of earth, V = velocity, and if quantities at the surface of the earth are indicated by the index 0, the ray equation in a sphere in which the velocity depends on r only is

$$\frac{r \sin i}{V} = \frac{r_0 \sin i_0}{V_0} = \text{const} \tag{2i-5}$$

The radius R of curvature of the ray is given by

$$R = \frac{V}{(dV/dr) \sin i} \tag{2i-6}$$

If $dV/dr = V/r$, and $i = 90$ deg, $R = r$. If V decreases with depth at a greater rate, no ray can have its deepest point in the respective layer, and the travel-time curve is interrupted. The angle of incidence i_0 at the surface at a given epicentral distance Δ in kilometers is found from

$$\sin i_0 = \frac{V_0}{\bar{V}_0} \tag{2i-7}$$

where $\bar{V} = d\Delta/dt$. The angular distance Θ of a ray section (or the whole ray) and the corresponding travel time t are given by

$$\Theta = \int_{r_1}^{r_2} \frac{\tan i}{r} \, dr \qquad t = \int_{r_1}^{r_2} \frac{dr}{V \cos i} \tag{2i-8}$$

The radius r_S to the deepest point of a ray arriving at the distance Δ in degrees and the corresponding velocity V_S are found from

$$\log r_S = \log r_0 - 0.0024127 \int_0^{\Delta} q \, d\Delta \tag{2i-9}$$

where $\cosh q = \bar{V}_\Delta / \bar{V}(\Delta)$.

$$V_S = \bar{V} \frac{r_S}{r_0} \tag{2i-10}$$

$\bar{V}_\Delta = \bar{V}$ at the distance Δ; $\bar{V}(\Delta)$ is variable as a function of Δ.

2i-9. Wave Velocity, Elastic Constants, and Pressure in the Earth. Equations (2i-9) and (2i-10) or other methods are used to calculate V and v as a function of r. Poisson's ratio follows from Eqs. (2i-2). If the density ρ is known as a function of depth, Eqs. (2i-3) give the bulk modulus k and the rigidity μ. The pressure p and gravity g are given by

$$g = \frac{4\pi K}{r^2} \int_0^r \rho r^2 \, dr = \frac{3g_0}{\rho_m r_0 r^2} \int_0^r \rho r^2 \, dr \qquad p = \int_r^{r_0} g \, dr \tag{2i-11}$$

K is the gravitational constant (6.673×10^{-8} cgs), ρ_m is the mean density of the earth (5.517 g/cm^3), r_0 is the radius of the earth ($6{,}371$ km), and g_0 is the gravity at the surface (981 gals).

In sediments V ranges from $1 \pm$ km/sec for sand to $7 \pm$ in well-cemented rocks. In the continents frequently "granitic rocks," $V = 6 \pm$ km/sec, are below the sediments. At a depth of $15 \pm$ km V and v seem to have minima (compare Table 2i-3), thus permitting the propagation of guided channel waves. Under the continents, there is at least one such channel. Waves propagated in these crustal channels are designated, for example, by Lg, Li, and Rg. Their periods are usually between 1 and 10 sec; their velocities between $3\frac{1}{4}$ and $3\frac{3}{4}$ km/sec.[1]

---

[1] B. Gutenberg, Low Velocity Layers in the Earth's Mantle, *Bull. Geol. Soc. Am.* **65**, 337–347 (1954); Channel Waves in the Earth's Crust, *Geophysics* **20**, 283–294 (1955). Summary in B. Gutenberg, "Physics of the Earth's Interior," pp. 39–41, Academic Press, Inc., New York, 1959.

In the next deeper layer in the continents V is usually $6\frac{1}{2}$ to 7 km/sec, which is, e.g., characteristic of gabbro and olivine-gabbro (selected data in Table 2i-10; details differ appreciably). In some regions indications of velocities of 7 to $7\frac{1}{2}$ km/sec have been found immediately above the Mohorovičić discontinuity. The velocity possibly decreases slightly with depth in this layer owing to the increase in temperature (compare Table 2i-3).

Below the Mohorovičić discontinuity (depths M_0 in Table 2i-10) the velocity of longitudinal as well as transverse waves decreases with depth to minima near depths of 80 and 140 km, respectively (Table 2i-11). This low-velocity layer makes determinations of velocities below the Mohorovičić discontinuity rather difficult. It is

TABLE 2i-10. VELOCITY V, KM/SEC, OF LONGITUDINAL WAVES AT SELECTED DEPTH INTERVALS h, KM, OBSERVED IN VARIOUS REGIONS, 1950–1958[†]
(SE = source of energy, AE = artificial explosions, EQ = earthquake, RB = rock burst. M_0 is the depth of the Mohorovičić discontinuity below sea level in km; V_M, v_M are reported longitudinal and transverse velocities, respectively, just below M_0. Corresponding values of Poisson's ratio are 0.23 to 0.27)

Region	SE	h	V	h	V	M_0	V_M	v_M
N.W. Germany	AE	6–15	5.9±	15–28	6.5±	28±	8.2	?
Black Forest	AE	1–21	6.0	21–30	6.55	31	8.2	4.8
Southern Alps	EQ	0–35	5.7±	35–45	6.6±	45±	8.0	4.4
Northern Italy	EQ	0–15	5.3±	15–30	6.5±	40±	8.2	4.5±
South Africa	RB	4–36?	6.2	?	6.8	34	8.2	4.7
New York	AE	0–35	6.3	?	?	35	8.1	4.7
Eastern U.S.	AE	0–5	6.0	5–15	6.5±	40±	8.1	?
Wisconsin	AE	$\frac{1}{2}$–3	4.5	3–40±	6.0–6.9	42±	8.2	?
So. California	AE	1±	5.8	4–12	6.1–6.7	32±	8.2	?
So. California	EQ	1–25	6.4	25–35	7.1	35±	8.1	4.55
Canadian Shield	RB	0–30	6.2	30–35	7.1	37	8.2	4.85
Japan	AE	1–23	6.1±	23–32	7.4	32±	8.2	4.7±
N.E. India	EQ	1–25	5.6	25–46	6.6	46	7.9	4.5
Central Asia	AE	1–20	5.7	20–50	6.2	50	8.0	?
W. Atlantic	AE	Water	.....	5–10±	6.7	10±	8.0	?
Pacific Basin	AE	Water	.....	5–11±	6.8±	11±	8.2	?

[†] B. Gutenberg, "Physics of the Earth's Interior," pp. 32–35, Academic Press, Inc., New York, 1959.

another locus for channel waves (Pa and Sa), especially if the energy source is in or near these layers.[1]

New data on body-wave travel times, surface-wave dispersion, periods of free oscillations, and high-pressure laboratory experiments are producing more precise values for the parameters which are listed in Table 2i-11.[2] Changes of only a few percent in velocity or density may govern the conclusions as to chemical composition versus depth. As precision improves, zonal variations in the mantle may emerge, for which Table 2i-11 could only represent averages for a given radius.

[1] Summary in B. Gutenberg, "Physics of the Earth's Interior," pp. 86, 87, Academic Press, Inc., New York, 1959.
[2] Francis Birch, Density and Composition of Mantle and Core, *J. Geophys. Research* **69**, 4377–4388 (1964); S. P. Clark, Jr., and A. E. Ringwood, Density Distribution and Constitution of the Mantle, *Revs. Geophys.* **2**, 35–88 (1964); Frank Press, Density Distribution in Earth, *Science* **160**, 1218–1221 (1968).

2i-10. Intensity, Magnitude, and Energy of Earthquakes and Related Quantities.
The "intensity" of an earthquake refers to the effects of shaking at a given point. In the United States the modified Mercalli scale[1] (I to XII) is used; a few greatly condensed examples follow.

II. Felt by few persons at rest.

IV. Felt outdoors by few; some sleepers awakened; dishes, windows disturbed.

V. Some dishes, windows broken; unstable objects overturned; pendulum clocks may stop.

VI. Felt by all; some fallen plaster or damaged chimneys.

VII. Considerable damage in poorly built structures.

IX. Buildings shifted off foundations; ground cracked.

XI. Few structures remain standing; rails bent.

TABLE 2i-11. WAVE VELOCITIES V (LONGITUDINAL) AND v (TRANSVERSE), KM/SEC
[Poisson's ratio σ, Eqs. (2i-2); density ρ, g/cm^3; bulk modulus k and rigidity μ, both in 10^{12} dynes/cm^2, Eqs. (2i-3); gravitational acceleration g, cm/sec^2; and pressure p, million atm, in the earth as function of depth, km†]

Depth	V	v	σ	ρ	k	μ	g	p
Mantle:								
50	8.0	4.55	0.26	3.3	1.3	0.65	985	0.014
100	7.8	4.4	0.27	3.5	1.3	0.65	987	0.03
150	7.9	4.4	0.28	3.6	1.3	0.64	989	0.05
200	8.1	4.4	0.29	3.7	1.3	0.68	990	0.06
250	8.3	4.5	0.29	3.8	1.4	0.71	990	0.08
300	8.5	4.7	0.29	3.9	1.6	0.8	990	0.10
500	9.5	5.2	0.28	4.0	2	1.1	990	0.18
1,000	11.5	6.4	0.28	4.6	$3\frac{1}{2}$	1.9	990	0.39
1,500	12.2	6.7	0.28	5	$4\frac{1}{2}$	2.3	980	0.6
2,000	12.8	6.9	0.29	$5\frac{1}{4}$	5	2.6	980	0.9
2,900	13.7	7.3	0.30	$5\frac{3}{4}$	$6\frac{1}{2}$	3	1000	1.3
Outer core:								
2,900	8.0	≪1	0.5	$9\frac{1}{2}\pm$	6±	0?	1000	1.3
4,000	9.4	≪1	0.5	11±	10±	0?	800±	$2\frac{1}{4}\pm$
5,000	10.0	≪1	0.5	12±	12±	0?	600±	$3\frac{1}{4}\pm$
Inner core:								
5,400	11.1	?	0.4?	13?	15±	2?	500±	$3\frac{1}{2}\pm$
6,370	11.2	?	0.4?	13?	16±	2?	0	$3\frac{3}{4}\pm$

† Based on B. Gutenberg, "Physics of the Earth's Interior," Academic Press, Inc., New York, 1959. The probable errors of most quantities increase with depth.

The observed intensity depends on the depth of focus, the ground, the type of building, the density of population, etc. The intensity is useful for engineers but not for studies of seismicity, for which the earthquake magnitude is used. There are various scales. Magnitude M originally was defined[2] for southern California as the common logarithm of the maximum trace amplitude a or b in microns with which a seismograph of period 0.8 sec, magnification 2,800, damping 65:1 would record the shock at a distance of 100 km. Tables[3] permit the determination of M. In addition, for

[1] H. O. Wood and F. Neumann, Modified Mercalli Intensity Scale of 1931, *Bull. Seis. Soc. Am.* **21**, 277–283 (1931).
[2] C. F. Richter, An Instrumental Earthquake Magnitude Scale, *Bull. Seis. Soc. Am.* **25**, 1–32 (1935).
[3] M. Båth, Earthquake Seismology, *Earth-Sci. Revs.* **1**, 69–86 (1966).

$\Delta > 15°$, M_S is found from ground amplitudes b (in microns) of surface waves with periods of 20 sec in shallow earthquakes. The magnitude M is based on amplitudes a of P, PP, and S waves in shocks (focal depth h) recorded at the epicentral distance Δ:

$$M_S = \log b + F(\Delta) \qquad M = \log a - \log T + f(\Delta,h) \qquad (2\text{i-}12)$$
$$M = M_S - 0.37(M_S - 6.74) \qquad (\text{approximately})$$

For $F(\Delta)$ and $f(\Delta,h)$, see Table 2i-12; small station corrections are to be added. The amplitudes b of surface waves of length L decrease with increasing focal depth h

TABLE 2i-12. VALUES OF $f(\Delta,h)$ IN EQ. (2i-12) FOR VERTICAL COMPONENTS Z OF P AND PP, HORIZONTAL COMPONENT SH OF S, AND $F(\Delta)$ FOR HORIZONTAL COMPONENT OF MAXIMUM (MAX)

(h = focal depth; Δ = epicentral distance, deg*)

Δ	$h = 25$ km				$h = 300$ km			$h = 600$ km		
	PZ	PPZ	SH	Max	PZ	PPZ	SH	PZ	PPZ	SH
20	6.0	...	5.8	4.0	6.1	...	5.8	6.4	...	5.9
30	6.6	6.7	6.3	4.3	6.3	6.4	6.1	6.4	6.3	6.0
50	6.7	6.7	6.6	4.6	6.1	6.6	6.7	6.3	6.5	6.4
80	6.7	6.9	6.7	5.0	6.6	6.9	6.4	6.2	6.8	6.5
100	7.4	7.2	7.4	5.1	7.2	6.8	6.7	7.2	7.0	6.7
160	...	6.9	...	5.4	...	6.6	...	...	6.7	

* B. Gutenberg, Amplitudes of Surface Waves and Magnitudes of Shallow Earthquakes, *Bull. Seis. Soc. Am.* **35**, 3–12 (1945); Magnitude Determination for Deep-focus Earthquakes, *Bull. Seis. Soc. Am.* **35**, 117–130 (1945). B. Gutenberg and C. F. Richter, Magnitude and Energy of Earthquakes. *Ann. Geofis. Rome*, **9**, 1–15 (1956).

TABLE 2i-13. INTENSITY I AT THE EPICENTER, CORRESPONDING MAXIMUM ACCELERATION α, CM/SEC², MEAN RADIUS r_p OF AREA OF PERCEPTIBILITY, KM, FOR A GIVEN MAGNITUDE M IN AVERAGE SHOCKS IN SOUTHERN CALIFORNIA ($h = 16\pm$ KM)

(Values for I, α, r are based on empirical equations†)

M	2.2	3	4	5	6	7	8	$8\frac{1}{2}$
I	1.5	2.8	4.5	6.2	7.8	9.5	11.2	12.0
α	1	3	10	36	130	460	1,670	3,160
r_p	0	25	55	110	200	390	740	1,000

† B. Gutenberg and C. F. Richter, Earthquake Magnitude, Intensity, Energy, and Acceleration, *Bull. Seis. Soc. Am.* **32**, 163–191 (1942).

corresponding to a factor $e^{-qh/L}$, where q (about 2) depends on crustal structure. The average relationship of intensity to magnitude in California earthquakes is given in Table 2i-13.

The energy E corresponding to the magnitude M found from body waves is given to a first approximation[1] by

$$\log E = 12.24 + 1.44M \qquad (2\text{i-}13)$$

[1] M. Bâth, Earthquake Seismology, *Earth-Sci. Revs.* **1**, 69–86 (1966).

2i-11. Seismicity of the Earth. Earthquakes are divided into shallow shocks ($h \leq 60$ km), intermediate ($60 < h \leq 300$), and deep ($h > 300$, maximum $720 \pm$ km). Most shocks occur in narrow belts (Table 2i-14).[1] Deep and intermediate shocks are limited to the circumpacific belt and the trans-Asiatic (Alpide) belt.

For the magnitude of the largest observed shock and the relative frequency of earthquakes in various depth intervals, see Table 2i-15, which also shows examples of regional differences.

2i-12. Energy E of Earthquakes. Most calculations of E depend on Eq. (2i-13). This empirical formula is based on many observations, but is subject to adjustment.

TABLE 2i-14. Number of Shallow, Intermediate, and Deep-focus
Earthquakes, % of All Earthquakes in the Given Depth
Range, and Corresponding Energy Release (a) in the
Major Units of the Earth and (b) in
Selected Areas
(Averages 1904–1957)

Region	Number, %			Energy, %		
	Shallow	Inter-med.	Deep	Shallow	Inter-med.	Deep
(a) Circumpacific belt............	82	91	100	75	89	100
Trans-Asiatic belt.............	10	9	<1	23	11	0.3
Atlantic and Indian Oceans....	5	0	0	1	0	0
All others..................	3	0	0	1	0	0
Total.....................	100	100	100	100	100	100
(b) Pacific region, Alaska to U.S...	2	0	0	2	0	0
North and Central America, West Coast...............	12	10	0	12	8	0
South America, western part...	10	19	6	15	9	19
Kermadec-Tonga Is...........	3	3	41	4	5	25
New Hebrides and Solomon Is..	12	20	4	7	18	3
Marianas Is.................	2	6	6	1	8	3
Japan-Kamchatka............	15	16	35	19	22	44
Philippine Is................	5	3	4	6	2	3
Celebes-Sunda Is.............	8	11	4	6	15	3
Hindu Kush.................	0	5	0	0	6	0
Asia Minor to Italy..........	2	2	0	1	4	0
Total.....................	71	95	100	73	97	100

Shocks of magnitudes over 7 account for most of the total energy release (Table 2i-16). For annual extreme and average energy release, 1904 to 1957, see Table 2i-17. The annual energy release in shallow earthquakes decreased appreciably about 1907. While the average for 1897 to 1906 was about 20×10^{24} ergs, it was only 6×10^{24} ergs for 1907 to 1956 with no appreciable fluctuations in 10-year periods. As a consequence of the relatively short period of about 60 years for which data are available, averages for the annual energy release change noticeably with the period used for the calculation. The annual energy loss by heat flow through the earth's surface is of the order of 10^{28} ergs.

[1] B. Gutenberg and C. F. Richter, "Seismicity of the Earth," 2d ed., Princeton University Press, Princeton, N.J., 1954.

TABLE 2i-15. (a) MAGNITUDE m OF GREATEST KNOWN SHOCK (1905–1957) IN
DEPTH INTERVALS d, CENTERING AT h; (b) PERCENTAGE OF SHOCKS FOR
THE WHOLE EARTH; (c) CORRESPONDING FREQUENCY FOR SELECTED
PARTS OF THE CIRCUMPACIFIC BELT

d, km	60	60	100	100	100	100	100	50	50
h, km	30	90	175	275	375	475	575	650	700
(a) Largest observed m	8.2	8.0	8.0	7.8	7.8	7.5	7.5	7.5	6.9
(b) Number of shocks, %	72	12	7	2	2	2	2	1	$\frac{1}{3}$
(c) Mexico, Central America, %	73	20	6	1	0	0	0	0	0
Andes, %	36	30	20	5	0	0	4	4	0
New Zealand–Samoa, %	30	10	10	6	7	6	25	5	$\frac{1}{2}$
New Hebrides–New Guinea, %	43	30	20	4	3	1	0	0	0
Japan–Manchuria, %	36	16	11	6	15	9	6	$\frac{1}{4}$	0
Sunda Arc, %	30	26	20	1	4	1	10	2	5

TABLE 2i-16. AVERAGE ANNUAL ENERGY RELEASE IN ALL EARTHQUAKES
WITH $M_S \leq M^*$
(Units 10^{23} ergs. Ratios of figures are good approximations; absolute values may be
incorrect by factor 100)

M^*	6	7	8
Shallow shocks	0.2	1	5
Intermediate shocks	?	0.2	0.6
Deep shocks	?	0.05	0.1

TABLE 2i-17. MAXIMUM, MINIMUM, AND AVERAGE ANNUAL ENERGY RELEASE
IN EARTHQUAKES 1904–1957
(Units 10^{23} ergs. Accuracy as in Table 2i-16)

	Max	Year	Min	Year	Average
Shallow shocks	340†	1906	9	1954	70
Intermediate shocks	100	1911	1	1935	16
Deep shocks	35	1906	0.2±	Several	3
All shocks	390	1906	12	1930	90

† Possibly 500 in 1897.

2i-13. Aftershocks and Earthquake Sequences. Investigations by Benioff[1] show
that elastic strain-rebound increments in series of earthquake aftershocks follow two
types of functions:

$$(1) \quad S_1 = A + B \log t \qquad (2) \quad S_2 = C - De^{-\sqrt{t}} \tag{2i-14}$$

[1] H. Benioff, Earthquakes and Rock Creep, *Bull. Seis. Soc. Am.* **41**, 31–62 (1951).

where t is time from a selected zero point and A, B, C, D are constants of the process. (1) was given previously by Griggs for compressional recoverable creep strain, (2) by Michelson for shearing creep recovery. For series of earthquakes in certain areas and for all earthquakes in certain depth ranges Benioff[1] has found strain-rebound characteristics of forms similar to Eqs. (2i-14). Yearly strain rebound in all deep shocks shows a decrease between at least 1905 and 1950 following Eq. (1), whereas most great shallow shocks have occurred in five active periods. The units of the Pacific belt have different patterns of activity.[2]

2i-14. Nonelastic Properties of the Earth's Interior. Earthquake-generated surface waves, body waves, and free oscillations yield data on energy losses in the earth for periods of one to a few thousand seconds. Field and laboratory measurements on crustal rocks cover periods of a few seconds to fractions of a microsecond. For a given rock, losses can be expressed in terms of a specific dissipation constant $1/Q$ which is found to be independent of frequency.[3] This result strongly suggests a nonlinear stress-strain relation for the rock, even at infinitesimal values of strain. A hysteresis loop, in which energy expended per cycle is independent of the rate of traversal, would yield $1/Q$ independent of frequency. Since the nonlinear terms are small, one may assume that superposition applies and hence that stresses and strains are expressible as Fourier transforms. A modified Hooke's law can then be applied, in which stresses are proportional to strains through complex constants which have a one-to-one relationship with any elastic constants one would use for an elastic solid of a given geometry and wave type.[4] For example, shear stress $s(t)$ in a plane shear wave is related to strain $\epsilon(t)$ thus:

$$s(t) = (1/2\pi) \int_{-\infty}^{\infty} S(\omega) \exp{(i\omega t)} \, d\omega$$

$$\epsilon(t) = (1/2\pi) \int_{-\infty}^{\infty} E(\omega) \exp{(i\omega t)} \, d\omega \qquad (2\text{i-}15)$$

$$S(\omega) = (\mu + i \operatorname{sgn}\omega \, \mu^*) E(\omega)$$

This leads to a wave equation with a complex propagation constant $(a_S + i\omega/v)$. A plane wave along the x axis at angular frequency ω_0 is

$$\epsilon(t) = E_0 \exp{(-a_S x)} \cos\left(\omega_0 t - \frac{\omega_0 x}{v}\right) \qquad (2\text{i-}16)$$

With the loss parameter μ^* much smaller than the shear modulus μ,

$$\frac{1}{Q_S} = \frac{\mu^*}{\mu} = \frac{2 a_S v}{\omega} \qquad (2\text{i-}17)$$

With the additional loss parameter k^* introduced through the complex incompressibility $(k + i \operatorname{sgn}\omega \, k^*)$, the corresponding quantities for a plane compressional wave are

$$\frac{1}{Q_P} = \frac{k^* + 4\mu^*/3}{k + 4\mu/3} = \frac{V_P 2a}{\omega} \qquad (2\text{i-}18)$$

The two loss parameters Q_S and Q_P, applied to an isotropic solid, are sufficient to specify attenuation of Rayleigh waves, sharpness of resonance for free vibrations, etc.

[1] H. Benioff, Global Strain Accumulation and Release as Revealed by Great Earthquakes, *Bull. Geol. Soc. Am.* **62**, 331–338 (1951).
[2] H. Benioff, Orogenesis and Deep Crustal Structure—Additional Evidence from Seismology, *Bull. Geol. Soc. Am.* **65**, 385–400 (1954).
[3] S. P. Clark, Jr., "Handbook of Physical Constants," *Geol. Soc. Am. Mem.* **97**, 178 pp. (1966); L. Knopoff, Q, *Revs. Geophys.* **2**, 625–660 (1964).
[4] J. E. White, "Seismic Waves: Radiation, Transmission, and Attenuation," p. 94, McGraw-Hill Book Company, New York, 1965.

Since most data are not sufficiently complete and accurate to determine both, it makes some sense to attribute one parameter $(1/Q)$ to a given solid. Some ranges of $1/Q$ for common rock types appear in Table 2i-18.

On the assumption that Q is independent of frequency and varies with depth, attenuation versus period for surface waves and sharpness of resonance of free oscillations can yield the dependence of Q on depth.[1] Shear waves reflected from the core also give Q versus depth in the mantle.[2] Average Q for the mantle above 600 km is about 200; for the rest of the mantle, about 2,000. There is an indication of a minimum Q near the top of the mantle, probably correlating with the minimum in velocities (see Table 2i-11). Anderson et al.[3] suggest Q values of 200 to 600 for the crust, 50 to 110 for the minimum, increasing to 70 to 190 at 120 km, and continuing to increase with depth.

Slow adjustments of the crust indicate that the stress-strain relation for the crust and mantle must permit creep under steady load. The core is best described as a

TABLE 2i-18. RANGES OF SPECIFIC DISSIPATION CONSTANT $1/Q$
FOR SEVERAL COMMON ROCKS

Rock type	$1/Q$
Granite	0.002–0.02
Limestone	0.002–0.03
Sandstone	0.004–0.05
Shale	0.015–0.1

viscous fluid. A relation between shear stress s and shear strain ϵ which incorporates both possibilities is

$$\frac{s}{\mu} + \frac{1}{\eta} \int s \, dt = \epsilon + \lambda \frac{d\epsilon}{dt} \tag{2i-19}$$

In terms of transformed stress and strain, the complex shear modulus can be included to yield an equation which reduces to Eq. 2i-15 for intermediate frequencies:

$$\frac{S}{\mu + i \operatorname{sgn}\omega \, \mu^*} + \frac{S}{i\omega\eta} = E + i\omega\lambda E \tag{2i-20}$$

The rate of rise of Fennoscandia toward equilibrium following the melting of the Pleistocene ice masses (maximum thickness about $2\frac{1}{2}$ km) has been used by many authors to estimate the viscosity of the mantle. McConnell[4] has analyzed the detailed information on the shape of the upwarping of the area covered by the Fennoscandian ice sheet to deduce variation of viscosity with depth. His best model consists of an elastic layer overlying a layered elastoviscous mantle. His values of η are: 2×10^{21} poises from 120 to 400 km, 10^{22} poises from 400 to 800 km, increasing to almost 10^{23} poises at 1,500 km.

Where the product $\mu\lambda = \nu$ defines the fluid viscosity ν, data on torsional free oscillations and shear reflections ScS indicate that the viscosity of the core lies between 0.35×10^{11} and 4.7×10^{11} poises.[5]

[1] Don L. Anderson and C. B. Archambeau, The Anelasticity of the Earth, *J. Geophys. Research* **69**, 2071–2084 (1964).

[2] Robert L. Kovach and Don L. Anderson, Attenuation of Shear Waves in the Upper and Lower Mantle, *Bull. Seis. Soc. Am.* **54**, 1855–1864 (1964).

[3] Don L. Anderson, Ari Ben-Menahem, and C. B. Archambeau, Attenuation of Seismic Energy in the Upper Mantle, *J. Geophys. Research* **70**, 1441–1448 (1965).

[4] Robert K. McConnell, Jr., Isostatic Adjustment in a Layered Earth, *J. Geophys. Research* **70**, 5171–5188 (1965).

[5] R. Sato and A. F. Espinosa, Dissipation Factor of the Torsional Mode $_0T_2$, *J. Geophys. Research* **72**, 1761–1767 (1967).

2j. Oceanographic Data

A. C. VINE

Woods Hole Oceanographic Institution

Symbols

c	speed of sound, vapor pressure, speed of gravity wave
c_p	specific heat at constant pressure
g	acceleration of gravity
h	water depth, mean celestial longitude of sun
k	thermal conductivity
L	wavelength of gravity wave
n	refractive index
p	pressure, mean celestial longitude of lunar perigee
s	mean celestial longitude of moon
T	absolute temperature, hour angle of mean sun
v	specific volume
η	dynamic viscosity
κ	thermal diffusivity
ν	kinematic viscosity
ρ	density

2j-1. General. Oceanography is a composite of the marine aspects of science and engineering plus descriptive elements to incorporate spatial and temporal changes. This unusual grouping of disciplines has evolved special marine philosophies and experimental techniques that constitute much of the oceanographic profession.[1-3] This summary is primarily concerned with physical oceanography.

The great interconnected bodies of water constitute the world ocean. Although the salinity at the mouth of large rivers may approach zero, the salinity and density of the vast majority of ocean water is close to 35 parts per thousand of salt by weight and 1.030 g/cc density, or about 2.5 per cent greater density than that of fresh water at the same pressure.

Ocean currents are driven partly by the wind and partly as the working fluid of a great heat engine warmed near the equator and cooled near the poles. In the winter, and particularly in cold winters, high-latitude water cools and sinks to the bottom

[1] R. Fairbridge, "Encyclopedia of Oceanography," Reinhold Book Corporation, New York, 1966.

[2] G. Dietrich, "General Oceanography," Interscience Publishers, a division of John Wiley & Sons, Inc., New York, 1963.

[3] M. N. Hill, ed., "The Sea: Ideas and Observations on Progress in the Study of the Seas": vol. 1, "Physical Oceanography"; vol. 2, "Composition of Sea Water"; vol. 3, "The Earth beneath the Sea: History"; vol. 4 (to be published in 1971); Interscience Publishers, a division of John Wiley & Sons, Inc., New York and London, 1962, 1963.

where it slowly moves to warmer latitudes, mixing and rising as it goes. Water currents result from a combination of wind- and gravity-driven tidal oscillation and different-sized mixing eddies as modified by the earth's rotation.[1]

About half the ocean surface is warmer than 20°C—the upper limit of temperature being about 32°C in tropical regions, and the lower limit for liquid sea water the melting point −2°C. The temperature of sea ice ranges from 0°C down to something like −50°C.

Warm water is confined to the upper layers. For example, water deeper than 1 km is less than 10°C, and water deeper than 2 km is less than 4°C. (This excludes an interesting phenomenon recently found in the Red Sea where a few small bottom depressions are filled with very saline dense water reaching 56°C and 1.10 g/cc.) Typical temperature and salinity distributions appear in references, particularly those summing up major surveys.[2] Nontropical enclosed seas with shallow entrances to the open sea, such as the Mediterranean, have annual thermal patterns and circulatory systems intermediate between oceans and lakes. In drawing analogies to more familiar phenomena and professions, it seems reasonable to say:

1. Transoceanic depth profiles are remarkably similar to transcontinental elevation profiles.

2. The low compressibility, low thermal expansion, and high heat capacity of sea water result in a relatively more homogeneous and sluggish environment in the ocean than in the atmosphere.

3. Water masses in the ocean can be likened to air masses in the air and produce underwater weather that is remarkably similar to atmospheric weather, except that velocities are only 0.1 to 0.01 times as fast, with correspondingly longer time constants. However, except near the poles, the north-south-oriented continents block water currents from going around the world. This blocking effect plus Coriolis effects produces several major current systems such as the Gulf Stream.[3]

2j-2. Depth. Most of the surface of the earth's crust occurs at two prevailing levels, as shown by Table 2j-1. The higher of these levels is the continental platform, embracing the lower land and the submerged continental shelf out to depths of about 200 m. The lower oceanic platform lies about 5 km below the continental platform. The mean level of the crust's surface is 2.43 km below sea level, and the mean depth of the sea is 4.75 km.

The ocean floor appears to be more rugged than the dry land in both large-scale features[4,5] and small-scale features. The increased ruggedness results from less underwater weathering plus the buoyant effect on sediments and rock. Sedimentation is important at sea, and large flat areas may have slopes less than one in a thousand. Islands, seamounts, and submarine ridges rise from the oceanic platform, and about a dozen trenches extend below. The continental slopes between are frequently steep and cut by submarine canyons extending out from shore.

A more recent paper[6] has a recent and diversified breakdown of depths of the ocean in terms of both ocean areas and geologic provinces. Table 2j-2 gives one such breakdown.

[1] H. Stommel, "The Gulf Stream: a Physical and Dynamical Description," University of California Press, Berkeley, 1965.

[2] F. C. Fuglister, "Atlantic Ocean Atlas: Temperature and Salinity Profiles and Data," from the International Geophysical Year of 1957–1958, Woods Hole Oceanographic Intitution, Woods Hole, Mass., 1960.

[3] Stommel, *op. cit.*

[4] Hill, *op. cit.*

[5] P. H. Kuenen, "Marine Geology," John Wiley & Sons, Inc., New York, 1950; J. T. Wilson, The Development and Structure of the Crust, pp. 138–214 in "The Earth as a Planet," J. P. Kuiper, ed., University of Chicago Press, Chicago, 1954.

[6] H. W. Menard and S. M. Smith, Hypsometry of Ocean Basin Provinces, *J. Geophys. Research* **71**, 18, 4305 Sept. 15, 1966).

TABLE 2j-1. AREAS OF EARTH'S CRUST CLASSED ACCORDING TO HEIGHT OR DEPTH FROM SEA LEVEL. AFTER MEINARDUS*

Proportion in interval		Proportion above	
[Highest land†		8.85 km (314 mb‡)]	
Land above 5 km	0.1%	5 km (540 mb)	0.1%
Land 4–5 km	0.4%	4 km (616 mb)	0.5%
Land 3–4 km	1.1%	3 km (701 mb)	1.6%
Land 2–3 km	2.2%	2 km (795 mb)	3.8%
Land 1–2 km	4.5%	1 km (899 mb)	8.3%
Land 0–1 km	20.8%	0 km (1,013 mb)	29.1%
Ocean 0–1 km	8.5%	1 km (1,010 decibars¶)	37.6%
Ocean 1–2 km	2.9%	2 km (2,024 decibars)	40.5%
Ocean 2–3 km	4.7%	3 km (3,045 decibars)	45.2%
Ocean 3–4 km	14.1%	4 km (4,069 decibars)	59.3%
Ocean 4–5 km	23.9%	5 km (5,098 decibars)	83.2%
Ocean 5–6 km	16.0%	6 km (6,132 decibars)	99.2%
Ocean 6–7 km	0.7%	7 km (7,169 decibars)	99.9%
Ocean below 7 km	0.1%		
[Greatest depth§		10.86 km (11,216 decibars)]	

* Wilhelm Meinardus, Die bathygraphische Kurve des Tiefseebodens und die hypsographische Kurve der Erdkruste (Tabelle 6), *Ann. Hydrogr. mar. Meteor.* **70**, 225–244 (1942).
† Mt. Everest.
‡ Pressure according to NACA standard atmosphere. See, e.g., Smithsonian Meteorological Tables, 6th ed., *Smithsonian Misc. Collections* **114** (1951).
¶ Sea pressure in water at 0°C, salinity 35 per mille, gravity at sea level being 9.8 m/sec². From Vilhelm Bjerknes, Hydrographic Tables, *Carnegie Inst. Wash. Publ.* **88**, 1A–36A (1910).
§ Marianas Trench, adjacent to Mariana Islands, Pacific Ocean. Soundings from H.M. Survey Ship *Challenger* in 1951. T. F. Gaskell, J. C. Swallow, and G. S. Ritchie, Further Notes on the Greatest Oceanic Sounding and the Topography of the Marianas Trench, *Deep Sea Research* **1** 60–63 (1953).

TABLE 2j-2. DEPTH DISTRIBUTION OF WORLD OCEAN

Depth, km	Area, %	Cumulative area, %
0–0.2	7.49	7.49
0.2–1	4.42	11.91
1–2	4.38	16.29
2–3	8.50	24.79
3–4	20.94	45.73
4–5	31.69	77.42
5–6	21.20	98.62
6–7	1.23	99.85
7–8	0.10	99.96
8–9	0.03	99.99
9–10	0.01	100.00
10–11	0.00	100.00

2j-3. Properties of Sea Water. Because ocean currents are generated and maintained by very small differences in density of different water masses, much effort has been devoted to learning the relationships between temperature, pressure, salinity, and electrical conductivity in sea water. For many years salinity was generally determined by chemically titrating small samples of water. In the open ocean where horizontal and vertical gradients are low, water temperatures were hopefully measured to 0.02°C and salinities to 0.02 part per thousand, and densities were computed to about 2×10^{-5}. Since about 1960, electrical conductivity has gradually been replacing chemical titration, being much faster and more accurate in the field. This has resulted in oceanographers emphasizing conductivity measurements, although they still express total salt content in terms of salinity through the use of tables.

Recent studies and reports[1-3] on the physical properties of sea water have helped quantify and clarify many of the relations involved.

Pressure. Pressures in a standard atmosphere and ocean are included in parentheses in Table 2j-1. For the ocean, the quantity tabulated is sea pressure; the total pressure is sea pressure plus atmospheric pressure (10 decibars). The units are defined by

$$10^6 \text{ dynes/cm}^2 = \text{bar} = 10 \text{ decibars} = 10^3 \text{ mb}$$

Composition. Sea water, not including the suspended particles (inorganic matter, living organisms, and organic detritus), is a solution of a large number of constituents, which may be divided into four groups: water, major solids, minor solids (and liquids), and gases. The major solids are those which have appreciable influence on density. The minor solids compose only some 0.025 per cent of the total solids in typical sea water.

The major solids are composed of salts that are almost completely ionized, the proportions by mass being as follows:[4]

Na^+	30.61%	Cl^-	55.04%	H_3BO_3	0.07%
Mg^{++}	3.69	SO_4^-	7.68		
Ca^{++}	1.16	HCO_3^-	0.41		
K^+	1.10	Br^-	0.19		
Sr^{++}	0.04				

These proportions have been found to be highly constant throughout the ocean except where the water is nearly fresh (the salt in river water is very different from sea salt). Hence to a remarkable degree the measurement of the concentration of any of the major solids in sea water permits the calculation of the concentration of total solids. The abundance of the elements in sea water is presented in detail in Table 2j-3.

The constituent commonly measured (by chemical titration) is the sum of the halide ions (Cl^-, Br^-, I^-). The quantity *chlorinity* is approximately the ratio, by mass, of halides to total sample of sea water, but for the precise technical definition

[1] Physical and Chemical Properties of Sea Water, report of 1958 Easton, Md., Conference, *Natl. Acad. Sci.–Natl. Research Council Publ.* 600.

[2] N. P. Fofonoff, Physical Properties of Sea Water, in M. N. Hill, *op. cit.*, vol. 1, 1962. (Summary of physical relationships.)

[3] Handbook of Oceanographic Tables, *U.S. Naval Oceanographic Office Spec. Publ.* 68, 1966.

[4] H. U. Sverdrup, M. W. Johnson, and R. H. Fleming, "The Oceans, Their Physics, Chemistry, and General Biology," Table 33, Prentice-Hall, Inc., Englewood Cliffs, N.J., 1942.

TABLE 2j-3. ABUNDANCE OF THE ELEMENTS IN SEA WATER*

Element	μg/liter	Principal dissolved species	Element	μg/liter	Principal dissolved species
H	1.1×10^8	H_2O	Ru		
He	7×10^{-3}	He gas	Rh		
Li	1.7×10^2	Li^+	Pd		
Be	6×10^{-4}		Ag	0.3	Ag^+
B	4.5×10^3	$B(OH)_3$, $B(OH)_4{}^-$	Cd	0.1	Cd^{++}
C	2.8×10^4	$HCO_3{}^-$, $CO_3{}^{--}$	In	<20	
C(org)	1×10^2		Sn	0.8	
N	1.5×10^4	N_2 gas	Sb	0.3	
N	6.7×10^2	NO_3	Te		
O	8.8×10^8	H_2O	I	60	$IO_3{}^-$, I^-
O	6×10^3	O_2	Xe	5×10^{-2}	Xe gas
O	1.8×10^6	$SO_4{}^{--}$	Cs	0.3	Cs^+
F	1.3×10^3	F^-	Ba	20	Ba^{++}
Ne	0.12	Ne gas	La	3×10^{-3}	$La(OH)_3{}^0$
Na	1.1×10^7	Na^+	Ce	1×10^{-3}	$Ce(OH)_3{}^0$
Mg	1.3×10^6	Mg^{++}	Pr	0.6×10^{-3}	$Pr(OH)_3{}^0$
Al	1		Nd	3×10^{-3}	$Nd(OH)_3{}^0$
Si	3×10^3	$Si(OH)_4$, $SiO(OH)_3{}^-$	Sm	0.5×10^{-3}	$Sm(OH)_3{}^0$
P	90	$HPO_4{}^{--}$, $H_2PO_4{}^-$, $PO_4{}^{3-}$	Eu	0.1×10^{-3}	$Eu(OH)_3{}^0$
			Gd	0.7×10^{-3}	$Gd(OH)_3{}^0$
S	9.0×10^5	$SO_4{}^{--}$	Tb	1.4×10^{-3}	$Tb(OH)_3{}^0$
Cl	1.9×10^7	Cl^-	Dy	0.9×10^{-3}	$Dy(OH)_3{}^0$
Ar	4.5×10^2	Ar gas	Ho	0.2×10^{-3}	$Ho(OH)_3{}^0$
K	3.9×10^5	K^+	Er	0.9×10^{-3}	$Er(OH)_3{}^0$
Ca	4.1×10^5	Ca^{++}	Tm	0.2×10^{-3}	$Tm(OH)_3{}^0$
Sc	$<4 \times 10^{-3}$	$Sc(OH)_3$	Yb	0.8×10^{-3}	$Yb(OH)_3{}^0$
Ti	1	$Ti(OH)_4$	Lu	0.1×10^{-3}	$Lu(OH)_3{}^0$
V	2	$VO_2(OH)_3{}^{--}$	Hf	$<8 \times 10^{-3}$	
Cr	0.5	$CrO_4{}^{--}$, Cr^{3+}	Ta	$<3 \times 10^{-3}$	
Mn	2	Mn^{++}	W	0.1	$WO_4{}^{--}$
Fe	3		Re		
Co	0.4	Co^{++}	Os		
Ni	7	Ni^{++}	Ir		
Cu	3	Cu^{++}	Pt		
Zn	10	Zn^{++}	Au	1×10^{-2}	$AuCl_2{}^-$
Ga	3×10^{-2}		Hg	0.2	$HgCl_4{}^{--}$, $HgCl_2{}^0$
Ge	7×10^{-2}	$Ge(OH)_4$	Tl	<0.1	Tl^+
As	2.6	$HAsO_4{}^{--}$ $H_2AsO_4{}^-$	Pb	0.03	$PbCl_3{}^-$, $PbCl^+$, Pb^{++}
Se	9×10^{-2}	$SeO_4{}^{--}$	Bi	0.02	
Br	6.7×10^4	Br^-	Po		
Kr	0.2	Kr gas	At		
Rb	1.2×10^2	Rb^+	Rn	6×10^{-13}	
Sr	8×10^3	Sr^{++}	Ra	1×10^{-7}	Ra^{++}
Y	1×10^{-3}	$Y(OH)_3{}^0$	Ac		
Zr	3×10^{-2}		Th	$<5 \times 10^{-4}$	$Th(OH)_4$
Nb	0.01		Pa	Check	
Mo	10	$MoO_4{}^{--}$	U	3	$UO_2(CO_3)_3{}^{4-}$

* A. H. Seymour, ed., "Radioactivity in the Marine Environment," National Academy of Sciences–National Research Council, Washington, D.C. (to be published in 1971).

the reader is referred elsewhere.[1] Similarly, *salinity* is approximately the ratio of total solids to total sample of sea water, but the definition[2] used in practice is the one given by the empirical formula

$$\text{Salinity} = 0.00003 + 1.805 \times \text{chlorinity} \tag{2j-1}$$

Both chlorinity and salinity are customarily expressed in per mille, meaning 10^{-3}.

The salinity of most water in the open ocean lies between 33 and 37 per mille, and 35 per mille is often chosen as standard.

Because the major solids are uniform in composition, the density and some other physical properties of sea water depend on only three variables: temperature, salinity, and pressure. Some of these properties at a pressure of 1 atm are shown in Fig. 2j-1.

Density and Melting Point. Density at 1 atm is shown in Fig. 2j-1a.[3] The effect of pressure on the density of sea water of salinity 35 per mille at temperature 0°C is as follows:

Sea pressure, decibars.....	0	2,000	4,000	6,000	8,000	10,000
Density, g/ml............	1.02813	1.03748	1.04640	1.05495	1.06315	1.07104

Water that is more saline or warmer is less compressible.

The temperature of maximum density, shown for 1 atm on the graph, decreases as pressure increases. For pure water at 1 atm the decrease is 2.22°C per thousand decibars.[4]

The melting point decreases with increasing salinity or pressure. The melting-point depression at 1 atm equals 56.90°C times the salinity according to Miyake[5] and is shown as the dotted line on the graphs. The decrease with pressure for pure water at 1 atm is 0.742°C per thousand decibars.[6]

Many tables and other aids have been prepared for the routine calculation of density and specific volume of sea water. A selection follows.

References on Calculation of Density and Specific Volume of Sea Water

Density at a Pressure of 1 Atm

Knudsen, Martin: "Hydrographical Tables," Copenhagen, 63 pp., 1901. (Range −2 to 33°C, salinity 2 to 41 per mille.) Part of Knudsen's can be replaced with the following more detailed tables: Matthews, D. J.: "Tables for the Determination of the Density of Seawater under Normal Pressure, σ_t," Andr. Fred. Høst & Fils, Copenhagen, 56 pp., 1932.

Kalle, Kurt, und Hermann Thorade: Tabellen und Tafeln für die Dichte des Seewassers (σ_t), *Arch. deut. Seewarte Marineobs.* **60** (2), 49 pp. (1940). (Range −2 to 30°C, salinity 0 to 41.5 per mille.)

Ennis, C. C.: Note on Computation of Density of Sea Water and on Corrections for Deep-sea Reversing Thermometers, *Carnegie Inst. Wash. Publ.* **545A**, 23–45 (1944). (Range −2 to 30°C, salinity 34 to 36 per mille.)

[1] E.g., Sverdrup et al., *op. cit.*, p. 52.

[2] Bjørn Helland-Hansen, J. P. Jacobsen, and T. G. Thompson, Chemical Methods and Units, *Publ. sci. Ass. Océanogr. phys.* **9**, 28 (1948).

[3] Martin Knudsen, "Hydrographical Tables," Copenhagen, 1901; N. E. Dorsey, "Properties of Ordinary Water Substance," Reinhold Book Corporation, New York, 1940. For more recent work see M. S. Newton and G. C. Kennedy, An Experimental Study of the P-V-T-S Relations of Sea Water, *J. Marine Research* **23**, 89–103 (1965).

[4] Dorsey, *op. cit.*, p. 275.

[5] Yasuo Miyake, Chemical Studies of the Western Pacific Ocean, III, Freezing Point, Osmotic Pressure, Boiling Point and Vapour Pressure of Sea Water, *Bull. Chem. Soc. Japan* **14**, 58–62 (1939).

[6] Dorsey, *op. cit.*, Table 267.

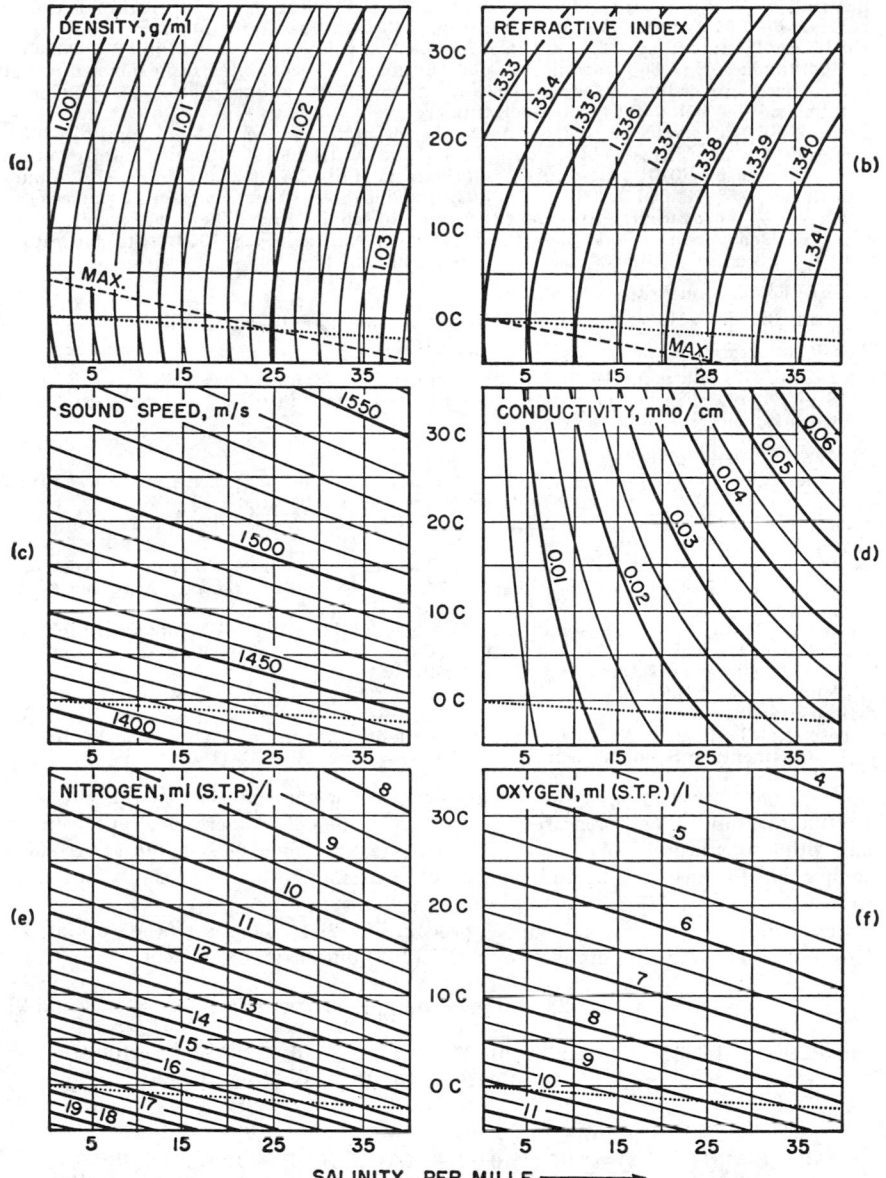

FIG. 2j-1. Temperature-salinity diagrams for sea water at 1 atm pressure: (a) density, (b) refractive index for sodium light (0.5893 micron) relative to air, (c) sound speed, (d) conductivity, (e) concentration of nitrogen in equilibrium with 1 atm (1,013.25 mb) of air saturated with aqueous vapor, (f) concentration of oxygen under same equilibrium conditions. Freezing point is shown by dotted line; values below it pertain to undercooled water.

LaFond, E. C.: Processing Oceanographic Data (Table X), U.S. Navy Hydrographic Office, H. O. Pub. 614, 1951. (Range -2 to 30°C, salinity 30 to 48 per mille.)

U.S. Navy Hydrographic Office: "Tables for Sea Water Density," H. O. Pub. 615, 265 pp., 1952 (range -2 to 30°C, salinity 0 to 40 per mille). Recomputed from same empirical formulas as preceding tables but expressed to one more decimal place (10^{-6} g/ml) and tabulated for each 0.01°C of temperature argument.

Bein, Willy: Physikalische und chemische Konstanten des Meerwassers (pp. 102–103), Veröffentl. Inst. Meeresk. Univ. Berlin, neue Folge, A, **28**, 36–190 (1935). The arguments are temperature (0, 1, . . . , 40°C) and the ratio of density at 17.5°C to density of pure water at 17.5°C (1.000, 1.002, . . . , 1.032). This table is based on Bein's own measurements, the most recent. Argument converted to salinity by G. Dietrich, Landolt-Börnstein Zahlenwerte und Funktionen, Ozeanographie, 6 Auflage, **3**, 428 (1952). (Temperature 0, 1, . . . , 32°C, salinity 0, 5, . . . , 40 per mille.)

Specific Volume at a Pressure of 1 Atm

LaFond, above, Table V. (Range -2 to 30°C, salinity 21 to 38 per mille.)

Density at Greater Pressures

Ekman, V. W.: Tables for Sea Water under Pressure, Publ. Circ. Cons. int. Explor. Mer **49**, 48 pp., 1910. [The arguments are density at 1 atm and 0°C, pressure (0 to 10,000 decibars), and temperature.]

Specific Volume at Greater Pressures

Bjerknes, Vilhelm: Hydrographic Tables, Carnegie Inst. Wash. Publ. **88**, 1A–36A (1910). (Range -2 to 30°C, salinity 0 to 40 per mille, 0 to 10,000 decibars.)

Subow, N. N., S. W. Brujewicz, and W. W. Shoulejkin: "Oceanographical Tables," Moscow, 208 pp., 1931.

Matthews, D. J.: "Tables for Calculating the Specific Volume of Seawater under Pressure," Andr. Fred. Høst & Fils, Copenhagen, 67 pp., 1938. [The arguments are density at 1 atm, pressure (0 to 12,000 decibars), and temperature.]

Sverdrup, H. U., M. W. Johnson, and R. H. Fleming: "The Oceans," Appendix, Prentice-Hall, Inc., Englewood Cliffs, N.J., 1942. [The arguments are density at 1 atm, pressure (0 to 10,000 decibars), temperature, and salinity.]

LaFond, above, Tables IV–VII. (Range -2 to 30°C, salinity 21 to 38 per mille, 0 to 10,000 decibars.)

Crease, J.: The Specific Volume of Sea Water under Pressure as Determined by Recent Measurements of Sound Velocity, Deep Sea Research **9**, 209–213 (1962).

Viscosity. Viscosity of sea water varies only a few per cent over the range of temperatures, salinities, and pressures encountered in the ocean, increasing with salinity and with temperature. With pressure it decreases at shallow depths; but at pressures comparable to ocean depths and depending upon temperature and salinity, it begins to increase.[1]

Sound Speed. The graph of sound speed in Fig. 2j-1c is based on the work of Del Grosso.[2] More work is summarized in Eq. (2j-2) due to W. D. Wilson.[3]

$$V = 1449.22 + \Delta V_T + \Delta V_P + \Delta V_S + \Delta V_{STP} \qquad (2j\text{-}2)$$

where $\Delta V_T = 4.6233T - 5.4585 \times 10^{-2}T^2 + 2.822 \times 10^{-4}T^3 - 5.07 \times 10^{-7}T^4$

$\Delta V_P = 1.60518 \times 10^{-1}P + 1.0279 \times 10^{-5}P^2 + 3.451 \times 10^{-9}P^3$
$$- 3.503 \times 10^{-12}P^4$$

$\Delta V_S = 1.391(S - 35) - 7.8 \times 10^{-2}(S - 35)^2$

$\Delta V_{STP} = (S - 35)(-1.197 \times 10^{-2}T + 2.61 \times 10^{-4}P - 1.96 \times 10^{-7}P^2$
$$- 2.09 \times 10^{-6}PT) + P(-2.796 \times 10^{-4}T + 1.3302 \times 10^{-5}T^2$$
$$- 6.644 \times 10^{-8}T^3) + P^2(-2.391 \times 10^{-7}T + 9.286 \times 10^{-10}T^2)$$
$$- 1.745 \times 10^{-10}P^3T$$

[1] R. A. Horne and D. S. Johnson, The Viscosity of Compressed Sea Water, J. Geophys. Research **71**, 22, (November, 1966).

[2] V. A. Del Grosso, The Velocity of Sound in Sea Water at Zero Depth, Naval Research Lab. Rept. 4002, 39 pp., 1952.

[3] W. D. Wilson, Speed of Sound in Sea Water as a Function of Temperature, Pressure, and Salinity, J. Acoust. Soc. Am., **32**, 6 (June, 1960).

TABLE 2j-4. CORRECTION TO BE ADDED TO $V_0 = 1449.22$ M/SEC TO ACCOUNT
FOR CHANGES IN TEMPERATURE
(T in °C, ΔV_T in m/sec)

T	ΔV_T	T	ΔV_T
−3	−14.37	13	51.48
−2	−9.47	14	54.78
−1	−4.68	15	58.00
0	0.00	16	61.12
0	0.00	17	64.17
1	4.57	18	67.13
2	9.03	19	70.01
3	13.39	20	72.81
4	17.64	21	75.53
5	21.79	22	78.18
6	25.84	23	80.75
7	29.78	24	83.25
8	33.64	25	85.68
9	37.39	26	88.04
10	41.05	27	90.32
11	44.62	28	92.54
12	48.10	29	94.69
		30	96.78

TABLE 2j-5. CORRECTION TO $V_0 = 1449.22$ M/SEC TO ACCOUNT FOR CHANGES
IN PRESSURE
(P in kg/cm², ΔV_P in m/sec)

P	ΔV_P	P	ΔV_P
0	....*	500	83.04
100	16.16	600	100.30
200	32.54	700	117.74
300	49.15	800	135.33
400	65.98	900	153.01

* At atmospheric pressure, $P = 1.0332$ kg/cm¹, $\Delta V_P = 0.16$ m/sec.

For greater pressures, Matthews[1] and Kuwahara[2] have computed the sound speed c from $c^2 = (dp/d\rho)_s$, which is the adiabatic change of pressure with density and can be expressed in terms of known properties (isothermal compressibility, thermal expansion, specific heat, temperature, and specific volume).

[1] D. J. Matthews, "Tables of the Velocity of Sound in Pure Water and Sea Water for Use in Echo-sounding and Sound-ranging," 2d ed., H. D. 282, Hydrographic Department, Admiralty, London, 52 pp., 1939. (Range −2 to 30°C, salinity 0 to 41 per mille, depth 0 to 10,900 m.)
[2] Susumu Kuwahara, The Velocity of Sound in Sea Water and Calculation of the Velocity for Use in Sonic Sounding, *Japan. J. Astron. Geophys.* **16**, 1–17 (1938). (Tables for range −2 to 30°C, salinity 30 to 40 per mille, pressure 0 to 10,000 decibars.) Kuwahara's tables are reproduced and extended in salinity down to 21 per mille by E. C. LaFond, "Processing Oceanographic Data," Table XIV, U.S. Navy Hydrographic Office, H.O. Pub. 614, 1951.

TABLE 2j-6. CORRECTION TO $V_0 = 1449.22$ M/SEC TO ACCOUNT FOR CHANGES
IN SALINITY

(S in parts per thousand, ΔV_S in m/sec)

S	ΔV_S
33	-3.09
34	-1.47
35	0.00
36	1.31
37	2.47

Additional ΔV_{TPS} tables correct for simultaneous changes in T, P, and S.

The transmission of sound in the sea can be very complex, and a large literature exists[1] on this subject.

Refractive Index. For given wavelength, temperature, and pressure the relation between refractive index and salinity is very nearly linear. The graph of refractive index for sodium light in Fig. 2j-1b is based on the formulas of Utterback et al.[2] The formulas have been adjusted slightly to agree at zero chlorinity with the measurements of Tilton and Taylor.[3] Measurements by Bein[4] give values higher than those of Utterback et al. by as much as 0.00012. The effect of pressure is roughly such that, if n is refractive index and ρ is density, $(n - 1)/\rho$ is constant for given temperature, salinity, and wavelength.[5]

Dissolved Nitrogen and Oxygen. The two principal atmospheric gases are differently distributed in the ocean. Because nitrogen is highly inert, its concentration is determined entirely by contact with the atmosphere. Oxygen is both released and consumed by biological processes, so that its concentration is much more variable. Right at the sea surface, there is equilibrium between the two phases; the nitrogen and oxygen in the liquid phase depend on their partial pressures in the gaseous phase, while the partial pressure of aqueous vapor depends on the salinity of the liquid phase (see below).

Figures 2j-1e and f show the concentrations of nitrogen and oxygen in equilibrium with saturated air at a pressure of 1 atm. For a given temperature, solubility decreases linearly with increasing salinity. The nitrogen graph has been calculated from Fox's[6] table for pure water and from Rakestraw and Emmel's[7] data for sea

[1] C. F. Officer, "Introduction to the Theory of Sound Transmission," McGraw-Hill Book Company, New York, 1958; R. J. Urick, "Principles of Underwater Sound for Engineers," McGraw-Hill Book Company, New York, 1967; L. M. Brekhovskikh, "Waves in Layered Media," Academic Press, Inc., New York, 1960.

[2] C. L. Utterback, T. G. Thompson, and B. D. Thomas, Refractivity-Chlorinity-Temperature Relationships of Ocean Waters, *J. Cons. int. Explor. Mer* **9**, 35–38 (1934). (Table for 0, 5, . . . , 25°C, chlorinity 1, 2, . . . , 22 per mille.)

[3] L. W. Tilton and J. K. Taylor, Refractive Index and Dispersion of Distilled Water for Visible Radiation, at Temperatures 0 to 60°C, *Natl. Bur. Standards J. Research* **20**, 419–477 (1938).

[4] Willy Bein, Physikalische und chemische Konstanten des Meerwassers, *Veröffentl. Inst. Meeresk. Univ. Berlin*, neue Folge, A, **28**, 162 (1935).

[5] Dorsey, *op. cit.*, Table 144.

[6] C. J. J. Fox, On the Coefficients of Absorption of Nitrogen and Oxygen in Distilled Water and Sea-water, and of Atmospheric Carbonic Acid in Sea-water, *Trans. Faraday Soc.* **5**, 68–87 (1909). (Table 1 gives nitrogen dissolved in pure water from 1 atm of pure nitrogen for 0, 1, . . . , 50°C.)

[7] N. W. Rakestraw and V. M. Emmel, The Solubility of Nitrogen and Argon in Sea Water, *J. Phys. Chem.* **42**, 1211–1215 (1938). (Table 2 gives nitrogen dissolved in sea water from 1 atm of saturated air for 0, 1, . . . , 28°C, chlorinity 15, 16, . . . , 21 per mille.) A personal communication from Dr. Rakestraw states that all values pertain to air saturated with aqueous vapor. Data in Table 1 used in present work.

water.   The oxygen graph is based on Fox's[1] table.   For temperatures below 0°C and for part of the high-temperature areas, the graphs depend on extrapolation.

The equilibrium concentration of dissolved nitrogen or oxygen is proportional to its partial pressure in the gaseous phase (Henry's law) up to several atmospheres.   As the partial pressure increases to 1,000 atm, however, the concentration attains only about half the value given by simple proportionality.[2]   Sea water has not been studied at pressures greater than 1 atm.

Carbon dioxide in the ocean is in dynamic balance with that in the atmosphere and represents the overwhelming majority in the air-water system.   The CO_2 residence time is longer than for most gases.[3]

Vapor Pressure.   The vapor-pressure lowering of an aqueous solution is related to the melting-point depression.   At the melting point of the solution the vapor pressure is the same as the vapor pressure of ice.

Let e be the vapor pressure of sea water of given salinity and temperature, and let e_0 be the vapor pressure of pure water at the same temperature.   Then

$$\frac{e_0 - e}{e_0} = 0.537 \times \text{salinity} \tag{2j-3}$$

so the vapor pressure for salinity 35 per mille is 98.12 per cent of that for pure water. This formula by Witting[4] is satisfactory for the range of conditions occurring at the natural ocean surface, but for greater salinity or higher temperature the results of recent measurements should be consulted.[5]

Latent Heats and Specific Heat.   The latent heats of fusion and vaporization are practically the same for sea water as for pure water.

The specific heat at constant pressure depends on salinity as follows at 17.5°C and 1 atm:[6]

Salinity, per mille...	0	5	10	15	20	25	30	35	40
c_p, cal g^{-1} °C^{-1}.....	1.000	0.982	0.968	0.958	0.951	0.945	0.939	0.932	0.926

The changes with temperature and pressure have not been measured.   The effect of pressure can be computed by use of the thermodynamic formula

$$\left(\frac{dc_p}{dp}\right)_T = -T\left(\frac{d^2v}{dT^2}\right)_p \tag{2j-4}$$

[1] C. J. J. Fox, On the Coefficients of Absorption of the Atmospheric Gases in Distilled Water and Sea Water, pt. I, Nitrogen and Oxygen, *Publ. Circ. Cons. int. Explor. Mer* **41**, 23 pp. (1907).   (Table 11 gives oxygen dissolved in sea water from 1 atm of dry air at −2, −1, . . . , 30°C, chlorinity 0, 1, . . . , 20 per mille.)   Table reproduced by K. Kalle, "Landolt-Börnstein Zahlenwerte und Funktionen," 6th ed., vol. 3, p. 478, 1952.

[2] Dorsey, *op. cit.*, Table 233.

[3] G. Skirrow, The Dissolved Gases—Carbon Dioxide, chap. 7 in "The Sea: Ideas and Observations on Progress in the Study of the Seas," vol. 2, "Composition of Sea Water," M. N. Hill, ed., Interscience Publishers, a division of John Wiley & Sons, Inc., New York and London, 1962.

[4] Rolf Witting, Untersuchungen zur Kenntnis der Wasserbewegungen und der Wasserumsetzung in den Finland umgebenden Meeren, I, *Finnländische Hydrographisch-biologische Untersuchungen* **2**, 173 (1908).

[5] A. B. Arons and C. F. Kientzler, Vapor Pressure of Sea-salt Solutions, *Trans. Am. Geophys. Union* **35**, 722–728 (1954).

[6] Otto Krummel, "Handbuch der Ozeanographie," vol. 1, p. 279, Stuttgart, 1907.

where v is specific volume and T is absolute temperature. The following values of the decrease in specific heat at 0°C and salinity 35 per mille are from Ekman's[1] table:

Sea pressure, decibars	0	2,000	4,000	6,000	8,000	10,000
$(c_p)_{0\ \text{decibars}} - c_p$, cal g^{-1} °C^{-1}	0	0.0159	0.0291	0.0401	0.0492	0.0566

Adiabatic Temperature Change. This quantity is computed from the thermo-dynamic formula

$$\left(\frac{dT}{dp}\right)_s = \frac{T}{c_p}\left(\frac{dv}{dT}\right)_p \tag{2j-5}$$

The following values for 0°C and salinity 35 per mille are converted from Ekman's paper:

Sea pressure, decibars	0	2,000	4,000	6,000	8,000	10,000
$(dT/dp)_s$, °C/1,000 decibars	0.035	0.072	0.104	0.133	0.159	0.181

Transport Phenomena. The values for a pressure of 1 atm are assembled in Table 2j-7. Measurements with sea water are restricted to viscosity; the other properties tabulated under sea water are from measurements with sodium chloride solutions. The diffusivities of nitrogen and oxygen are especially uncertain and may be incorrect by as much as 15 per cent.

Dynamic and kinematic viscosities and thermal conductivity change linearly with salinity. In contrast, both thermal diffusivity (associated with specific heat) and diffusivity of sodium chloride go through minima at salinities less than 35 per mille.

For pure water, pressure increasing to 10,000 decibars has a nonlinear effect on the dynamic viscosity, which decreases at 0°C by 8 per cent and increases at 30°C by 5 per cent.[2] The thermal conductivity at 30°C increases linearly with pressure and becomes 6 per cent greater at 10,000 decibars.[3]

2j-4. Gravity Waves. *Wave Speed.* Most of the ocean is stabilized by a downward increase of density, so that *internal waves* as well as *surface waves* are common. Only surface waves are discussed here.

Let L be wavelength, T be period, and c be wave speed. Then $L = Tc$. Let h be the depth of water (undisturbed surface to bottom) and g be gravity.

For a uniform train of long-crested sinusoidal waves of small amplitude in an ideal liquid of uniform depth, in general[4]

$$c^2 = \frac{gL}{2\pi} \tanh 2\pi \frac{h}{L} \tag{2j-6}$$

For $h/L \gg 1$, *deep-water waves*, the general formula reduces to $c^2 = gL/2\pi$, and the group speed equals half the wave speed. For $h/L \ll 1$, *shallow-water waves*, $c^2 = gh$, and group speed equals wave speed (there is no dispersion, and any waveform of small amplitude is propagated unchanged at this speed). Within 5 per cent, sufficient accuracy for many problems, the deep-water formula holds if $h/L \sim h/L_0 > \frac{1}{3}$ and the shallow-water formula holds if $h/L < \frac{1}{11}$ or $h/L_0 < \frac{1}{22}$; L_0 is defined below.

Change in depth along wave rays changes the speed and length of sufficiently long waves. Near shore, therefore, waves often experience refraction and accompanying

[1] V. W. Ekman, Der adiabatische Temperaturgradient im Meere, *Ann. Hydrogr. mar. Meteor.* **42**, 340–344 (1914).
[2] Dorsey, *op. cit.*, Table 86.
[3] *Ibid.*, Table 131.
[4] See, e.g., C. A. Coulson, "Waves," 6th ed., Oliver & Boyd, Ltd., Edinburgh, 1952.

TABLE 2j-7. TRANSPORT PHENOMENA IN WATER AT A PRESSURE OF 1 ATM

Name, symbol, units	Pure water		Sea water, salinity 35 per mille	
	0°C	20°C	0°C	20°C
Dynamic viscosity, η, g cm^{-1} sec^{-1} = poise..	0.01787^a	0.01002^b	0.01877^a	0.01075^a
Thermal conductivity, k, watt cm^{-1} °C^{-1}...	0.00566^c	0.00599^c	0.00563^c	0.00596^c
Kinematic viscosity, $\nu = \eta/\rho$, cm^2 sec^{-1}.....	0.01787	0.01004	0.01826	0.01049
Thermal diffusivity,^{d} $\kappa = k/c_p\rho$, cm^2 sec^{-1}...	0.00134	0.00143	0.00139	0.00149
Diffusivity, D, cm^2 sec^{-1}:				
NaCl.................................	0.0000074^e	0.0000141^f	0.0000068^e	0.0000129^f
N$_2$.................................	0.0000106^g	0.0000169^g		
O$_2$.................................		0.000021^h		
Prandtl number, $N_P = \nu/\kappa$...............	13.3	7.0	13.1	7.0

[a] Yasuo Miyake and Masami Koizumi, The Measurement of the Viscosity Coefficient of Sea Water, *J. Marine Research* **7**, 63–66 (1948). Values taken from their Table I and reduced by 0.00007 poise to agree with Swindells et al. (Table III presents smoothed values for 0, 1, . . . , 30°C, chlorinity 0, 1, . . . , 20 per mille.)

[b] J. F. Swindells, J. R. Coe, Jr., and T. B. Godfrey, Absolute Viscosity of Water at 20°C, *Natl. Bur. Standards J. Research*, **48**, 1–31 (1952).

[c] L. Riedel, Die Wärmeleitfähigkeit von wässrigen Lösungen starker Elektrolyte, *Chem.-Ing.-Technik* **23**, 59–64 (1951).

[d] Thermal diffusivity is also called thermometric conductivity.

[e] Values for 0°C calculated from those at 20°C by use of temperature coefficient of L. W. Öholm, Über die Hydrodiffusion der Elektrolyte, *Z. physik. Chem.* **50**, 309–349 (1904).

[f] A. R. Gordon, The Diffusion Constant of an Electrolyte, and Its Relation to Concentration, *J. Chem. Phys.* **5**, 522–526 (1937). Gordon used measurements by B. W. Clack, On the Study of Diffusion in Liquids by an Optical Method, *Proc. Phys. Soc. (London)* **36**, 313–335 (1924). R. H. Stokes, The Diffusion Coefficients of Eight Uni-univalent Electrolytes in Aqueous Solution at 25°, *J. Am. Chem. Soc.* **72**, 2243–2247 (1950).

[g] Gustav Tammann und Vitus Jessen, Über die Diffusionskoeffizienten von Gasen in Wasser und ihre Temperaturabhängigkeit, *Z. anorg. Chem.* **179**, 125–144 (1929).

[h] Tor Carlson, The Diffusion of Oxygen in Water, *J. Am. Chem. Soc.* **33**, 1027–1032 (1911); I. M. Kolthoff and C. S. Miller, The Reduction of Oxygen at the Dropping Mercury Electrode, *J. Am. Chem. Soc.* **63**, 1013–1017 (1941); H. A. Laitinen and I. M. Kolthoff, Voltammetry with Stationary Microelectrodes of Platinum Wire, *J. Phys. Chem.* **45**, 1061–1079 (1941).

convergence and divergence. Such phenomena are conveniently treated by relating the speed and length of waves of any given period to the speed and length for the same period in deep water, c_0 and L_0. As $T = L/c = L_0/c_0$ and $c_0{}^2 = gL_0/2\pi$, (2j-6) may be written[1,2]

$$\frac{c}{c_0} = \frac{L}{L_0} = \tanh 2\pi \frac{h}{L_0} \frac{L_0}{L} \tag{2j-7}$$

Functions of h/L_0 have been presented in an extensive table,[3] from which the following values are extracted:

h/L_0........	0	0.001	0.01	0.02	0.05	0.1	0.2	0.3	0.4	0.5	1
$c/c_0 = L/L_0$.	0	0.0792	0.2480	0.3470	0.5310	0.7093	0.8884	0.9611	0.9877	0.9964	1.0000

These elementary results are not suitable for direct application to the irregular aperiodic waves in areas of generation by wind.

[1] G. Neuman and W. Pierson, "Principles of Physical Oceanography," Prentice-Hall, Inc., Englewood Cliffs, N.J., 1966.

[2] J. J. Stoker, "Water Waves—Mathematical Theory and Applications," Interscience Publishers, Inc., New York, 1957.

[3] U.S. Department of the Army, Corps of Engineers, *Bulletin of the Beach Erosion Board*, Special Issue 2, Appendix D 1953.

TABLE 2j-8. SELECTED TIDAL CONSTITUENTS

Symbol	Name	Argument	Period	Speed, degrees per hour	Relative coefficient of equilibrium tide
Sa	Solar annual	h	1.0 year	0.0411	0.012
Ssa	Solar semiannual	$2h$	0.5 year	0.0821	0.073
Mm	Lunar monthly	$s - p$	27.55 day	0.5444	0.083
Mf	Lunar fortnightly	$2s$	13.66 day	1.0980	0.156
K_1	Lunisolar declinational diurnal	$T + h - 90°$	23.93 hr	15.0411	0.531
O_1	Lunar declinational diurnal	$T + h - 2s + 90°$	25.82 hr	13.9430	0.377
P_1	Solar declinational diurnal	$T - h + 90°$	24.07 hr	14.9589	0.176
Q_1	Lunar diurnal	$T + h - 3s + p + 90°$	26.87 hr	13.3987	0.072
M_2	Principal lunar semidiurnal	$2T + 2h - 2s$	12.42 hr	28.9841	0.908
S_2	Principal solar semidiurnal	$2T$	12.00 hr	30.0000	0.423
N_2	Larger lunar elliptic semidiurnal	$2T + 2h - 3s + p$	12.66 hr	28.4397	0.174
K_2	Lunisolar declinational semidiurnal	$2T + 2h$	11.97 hr	30.0821	0.115

Tidal Constituents.[1] The gravitational fluctuations that produce tides can be resolved into harmonic constituents. Some are listed in Table 2j-8, their periods being determined by the constant rates of change of four angles:

T = hour angle of mean sun (increasing by $15°$/hr)
h = mean celestial longitude of sun (increasing by $0.0411°$/hr)
s = mean celestial longitude of moon (increasing by $0.5490°$/hr)
p = mean celestial longitude of lunar perigee (increasing by $0.0046°$/hr)

Ocean Engineering. Increased activity at sea has brought a wide variety of engineers into the marine field. Recent books on ocean engineering provide useful information, tables, and references. The following are examples:

Brahtz, John F., ed. "Ocean Engineering, Systems Planning and Design," John Wiley & Sons, Inc., New York, 1968.
Handbook of Oceanographic Tables, *U.S. Naval Oceanographic Office Spec. Publ.* 68, 1966.
Myers, John J., ed.-in-chief: "Handbook of Ocean and Underwater Engineering," McGraw-Hill Book Company, New York, 1969.
Urick, Robert J.: "Principles of Underwater Sound for Engineers," McGraw-Hill Book Company, New York, 1967.

[1] Paul Schureman, "Manual of Harmonic Analysis and Prediction of Tides," U.S. Coast and Geodetic Survey, Special Publication 98, rev. ed., 1940; A. T. Doodson and H. D. Warburg, "Admiralty Manual of Tides," H.M. Stationery Office, London, 1941.

2k. Meteorological Information[1]

M. F. HARRIS

National Oceanic & Atmospheric Administration

2k-1. List of Symbols

c_p	specific heat of dry air at constant pressure
c_v	specific heat of dry air at constant volume
d	coefficient of molecular diffusion
D	coefficient of eddy diffusion
f	Coriolis parameter
g	acceleration of gravity
n	distance; spacing
p	pressure
Q	source strength; i.e., total amount of material released from a source
r	radius of curvature
R	gas constant for dry air
T	temperature
T_v	virtual temperature
T_{mv}	mean virtual temperature
U	west wind speed
v	speed
V	gradient wind speed
V_g	geostrophic wind speed
w	water-vapor mixing ratio
Z, z	height
β	northward variation of Coriolis parameter
γ	lapse rate of temperature
ζ	mean molecular speed; mixing velocity
λ	free path; mixing length; wavelength
ρ	density of air
σ	standard deviation
ϕ	latitude
Φ	geopotential
χ	concentration
ω	angular velocity of the earth

[1] All material not otherwise credited is abstracted either from List [23] or from Letestu [22]. These publications should be consulted for more complete explanations and additional references. For an encyclopedic summary of the status of knowledge (1950) in the principal fields of meteorology and atmospheric physics, including extensive references, see Malone [24]. More recent general references include [20], [1], [4], and [3]. For special annotated bibliographies see Rigby [30].

2k-2. Physical Constants

Pressure at mean sea level, 1 atm
$$= 1,013.250 \text{ millibars (mb)} = 1.013250 \times 10^6 \text{ dynes cm}^{-2}$$
$$= 76 \text{ cm Hg (at standard gravity of } 980.665 \text{ cm sec}^{-2} \text{ and temperature of } 0°C)$$
Mass of the atmosphere $= 5.14 \times 10^{21}$ g
Apparent molecular weight of dry air $M = 28.9644$[1]

Gas constant for 1 kg of dry air, $R = 287.05 \, J \text{ kg}^{-1} \text{ K}^{-1} = 6.8607 \times 10^{-2}$ cal*$\text{ g}^{-1} \text{ K}^{-1}$

Specific heat of dry air at constant pressure $c_p = \dfrac{7R}{2} = 0.2401$ cal $\text{g}^{-1} \text{ K}^{-1}$

Specific heat of dry air at constant volume $c_v = \dfrac{5R}{2} = 0.1715$ cal $\text{g}^{-1} \text{ K}^{-1}$

2k-3. Composition of the Atmosphere

TABLE 2k-1. NORMAL COMPOSITION OF CLEAN, DRY ATMOSPHERE AIR
NEAR SEA LEVEL*

Constituent gas and formula	Content, % by volume	Content variable relative to its normal	Molecular weight
Nitrogen (N_2)...............	78.084	–	28.0134
Oxygen (O_2).................	20.9476	–	31.9988
Argon (Ar).................	0.934	–	39.948
Carbon dioxide (CO)........	0.0314	†	44.00995
Neon (Ne).................	0.001818	–	20.183
Helium (He)................	0.000524	–	4.0026
Krypton (Kr)...............	0.000114	–	83.80
Xenon (Xe)................	0.0000087	–	131.30
Hydrogen (H_2)...............	0.00005	?	2.01594
Methane (CH_4).............	0.0002	†	16.04303
Nitrous oxide (N_2O)..........	0.00005	–	44.0128
Ozone (O_3).................	Summer: 0 to 0.000007	†	47.9982
	Winter: 0 to 0.000002	†	47.9982
Sulfur dioxide (SO_2)..........	0 to 0.0001	†	64.0628
Nitrogen dioxide (NO_2).......	0 to 0.000002	†	46.0055
Ammonia (NH_3).............	0 to trace	†	17.03061
Carbon monoxide (CO).......	0 to trace	†	28.01055
Iodine (I_2).................	0 to 0.000001	†	253.8088

* From "U.S. Standard Atmosphere, 1962" [34]; see also Glueckauf [16] and Keeling [19].
† The content of the gases marked with a dagger may undergo significant variations from time to time or from place to place *relative* to the normal indicated for those gases; for example, O_3 has a maximum concentration in the stratosphere.

2k-4. Geopotential. The geopotential Φ of a point at a height z above mean sea level is the work which must be done against gravity in raising a unit mass from sea level to height z.

$$\Phi = \int_0^z g \, dz \qquad (2k\text{-}1)$$

where g is the local acceleration of gravity at height z. For most meteorological work geopotential is measured in terms of the *geopotential meter* (gpm). By definition, 1 gpm $= 9.8 \times 10^4$ cm^{2}sec^{-2}. In the lower atmosphere, 1 gpm $\simeq$ 1 geometric meter.

[1] Carbon-12 isotope scale for which $C^{12} = 12$.
* Thermochemical calorie.

Table 2k-2 shows the relationship between geopotential and geometric height as a function of latitude.

2k-5. Hypsometry. The differential form of the hydrostatic equation, the equation expressing the relationship of pressure p, density ρ, and height z in the atmosphere, is

$$dp = -\rho g \, dz \qquad (2k-2)$$

Introducing the definition of geopotential, the hydrostatic equation becomes

$$dp = -\rho \, d\Phi \qquad (2k-3)$$

Substituting the equation of state for dry air, introducing the concept of virtual temperature,[1] and integrating, Eq. (2k-3) becomes

$$\Delta\Phi = RT_{mv} \log_e \frac{p_1}{p_2} \qquad (2k-4)$$

where $\Delta\Phi$ is the geopotential difference between levels having pressures p_1 and p_2, T_{mv} is the mean virtual temperature of the layer of air between p_1 and p_2, and R is

TABLE 2k-2. RELATION OF GEOPOTENTIAL TO GEOMETRIC HEIGHT

Latitude	Geopotential meters (gpm)										
	10,000	20,000	30,000	40,000	50,000	100,000	200,000	300,000	400,000	500,000	600,000
	m	m	m	m	m	m	m	m	m	m	m
0°	10,036	20,104	30,204	40,336	50,500	101,811	206,948	315,577	427,874	544,029	664,243
30°	10,023	20,077	30,163	40,282	50,432	101,672	206,656	315,115	427,225	543,174	663,161
45°	10,009	20,050	30,123	40,228	50,365	101,534	206,363	314,653	426,576	542,318	662,080
60°	9,996	20,024	30,083	40,174	50,297	101,395	206,071	314,191	425,927	541,465	661,000
90°	9,983	19,997	30,043	40,120	50,229	101,256	205,779	313,730	425,280	540,613	659,923

the gas constant for dry air. For temperatures in K and geopotential in gpm (i.e., geometric meters for most practical purposes) Eq. (2k-4) becomes

$$\Delta\Phi = 67.445 T_{mv} \log_{10} \frac{p_1}{p_2} \qquad (2k-5)$$

2k-6. Lapse Rates. The lapse rate γ in the atmosphere is defined as the rate of decrease of temperature with increasing height (or geopotential), $\gamma = -dT/dz$. γ is ordinarily expressed in °C per 100 m (or 100 gpm).

Dry-adiabatic Lapse Rate. Dry air, or moist air in which the water vapor enters into no change of state, which ascends (or descends) adiabatically in the atmosphere will decrease (or increase) in temperature at the rate of g/c_p. The dry adiabatic lapse rate is therefore 0.98°C/100 m.

---

[1] The virtual temperature T_v is defined to be the temperature that dry air would have at given pressure in order to have the same density as moist air at the same pressure but at temperature T and with specified moisture content. Approximately, $T_v = T(1 + 0.61w)$, where w is the water-vapor mixing ratio, or the ratio of water vapor to the dry air. The logarithmic mean virtual temperature is required in Eq. (2k-4).

Pseudoadiabatic Lapse Rate. Saturated air ascending adiabatically in the atmosphere, so that all condensation of water vapor is into liquid water which falls out immediately and all latent heat of condensation is realized in warming the air, decreases in temperature at the pseudoadiabatic (or moist-adiabatic) lapse rate for the water stage. Table 2k-3 gives the pseudoadiabatic lapse rate for the water stage as a function of temperature and pressure. The pseudoadiabat c lapse rate for the ice (snow) stage is different from that for the water stage.

2k-7. U.S. Standard Atmosphere. A revised standard atmosphere for levels up to 700 km (Table 2k-4) was adopted by the United States Committee on Extension to the Standard Atmosphere on March 15, 1962 [34]. The lower 32 km of this representation of the atmosphere was approved by the International Civil Aviation Organization on November 12, 1963, for international standardization. The values represent idealized, middle-latitude, year-round conditions for the range of solar activity between sunspot minimum and sunspot maximum. For representative values at the extremes of the solar cycle and at various latitudes and seasons, see [33] and [34].

TABLE 2k-3. PSEUDOADIABATIC LAPSE RATE FOR THE WATER STAGE, °C/100 GPM

Temp., °C	Pressure, mb									
	1000	900	800	700	600	500	400	300	200	100
−50	0.966	0.965	0.963	0.961	0.959	0.955	0.951	0.943	0.928	0.886
−40	0.950	0.947	0.944	0.939	0.934	0.925	0.913	0.896	0.863	0.775
−30	0.917	0.910	0.903	0.893	0.882	0.866	0.842	0.807	0.746	0.615
−20	0.855	0.844	0.830	0.814	0.794	0.767	0.730	0.677	0.596	0.454
−10	0.763	0.745	0.725	0.701	0.672	0.637	0.592	0.532	0.452	0.335
0	0.645	0.624	0.601	0.573	0.542	0.505	0.462	0.409	0.345	0.262
10	0.527	0.506	0.483	0.457	0.429	0.398	0.362	0.323	0.276	
20	0.426	0.408	0.389	0.368	0.346	0.322	0.296			
30	0.352	0.338	0.323	0.307	0.291	0.273				
40	0.301	0.290	0.279	0.267						
50	0.267	0.259								

The supplemental atmospheres in [33] conform closely to the Cospar International Reference Atmosphere [6].

Basic Assumptions. It is assumed that the air is dry, obeys the perfect-gas law, and is in hydrostatic equilibrium. (The other assumptions and necessary physical constants used are given in [34].)

Latitudinal Temperature Distribution. The average temperature in January and July in the Northern Hemisphere in the lowest 80 km of the atmosphere as a function of latitude and height is shown in Fig. 2k-1 taken from [33]. The heavy lines show the *mean* height of the tropopause. There may be more than one tropopause over a given point during certain meteorological conditions; during others it may be indistinct or missing altogether. The bold vertical lines indicate isothermal conditions.

2k-8. Other Properties and Phenomena of the Atmosphere. Space precludes the presentation of numerous meteorological data on the upper atmosphere recently obtained by satellite exploration [36,15,28,18]. However, schematic representations of the structure of the upper atmosphere, indicating the typical heights at which various phenomena have been observed, are shown in Fig. 2k-2. Figure 2k-3, after Hanson in [18], shows the dependence of electron concentration on night and day and on the 11-year sunspot cycle.

Table 2k-4. U.S. Standard Atmosphere*

Altitude, km	Temperature, K	Pressure, dynes/cm^2	Density, g/cm^3	Viscosity, poises	Speed of sound, m/sec	Molecular weight
−5	320.68	1.7776 + 6	1.9311 − 3	1.9422 − 4	358.986	28.964
−4	314.17	1.5960	1.7697	1.9123	355.324	28.964
−3	307.66	1.4297	1.6189	1.8820	351.625	28.964
−2	301.15	1.2778	1.4782	1.8515	347.888	28.964
−1	294.65	1.1393	1.3470	1.8206	344.111	28.964
0	288.15	1.0132	1.2250	1.7894	340.294	28.964
1	281.65	8.9876 + 5	1.1117	1.7579	336.435	28.964
2	275.15	7.9501	1.0066	1.7260	332.532	28.964
3	268.66	7.0121	9.0925 − 4	1.6938	328.583	28.964
4	262.17	6.1660	8.1935	1.6612	324.589	28.964
5	255.68	5.4048	7.3643	1.6282	320.545	28.964
6	249.19	4.7218	6.6011	1.5949	316.452	28.964
7	242.70	4.1105	5.9002	1.5612	312.306	28.964
8	236.22	3.5652	5.2579	1.5271	308.105	28.964
9	229.73	3.0801	4.6706	1.4926	303.848	28.964
10	223.25	2.6500	4.1351	1.4577	299.532	28.964
11	216.77	2.2700	3.6480	1.4223	295.154	28.964
12	216.65	1.9399	3.1194	1.4216	295.069	28.964
13	216.65	1.6580	2.6660	1.4216	295.069	28.964
14	216.65	1.4170	2.2786	1.4216	295.069	28.964
15	216.65	1.2112	1.9475	1.4216	295.069	28.964
16	216.65	1.0353	1.6647	1.4216	295.069	28.964
17	216.65	8.8497 + 4	1.4230	1.4216	295.069	28.964
18	216.65	7.5652	1.2165	1.4216	295.069	28.964
19	216.65	6.4675	1.0400	1.4216	295.069	28.964
20	216.65	5.5293	8.8910 − 5	1.4216	295.069	28.964
25	221.55	2.5492	4.0084	1.4484	298.389	28.964
30	226.51	1.1970	1.8410	1.4753	301.709	28.964
40	250.35	2.8714 + 3	3.9957 − 6	1.6009	317.189	28.964
50	270.65	7.9779 + 2	1.0269	1.7037	329.799	28.964
60	255.77	2.2461	3.0592 − 7	1.6287	320.606	28.964
70	219.70	5.5205 + 1	8.7535 − 8	1.4383	297.139	28.964
80	180.65	1.0366	1.9990	1.216	269.44	28.964
100	210.02	3.0075 − 1	4.974 − 10	†	†	28.88
150	892.79	5.0617 − 3	1.836 − 12			26.92
200	1235.95	1.3339	3.318 − 13			25.56
250	1357.28	4.6706 − 4	9.978 − 14			24.11
300	1432.11	1.8838	3.585			22.66
400	1487.38	4.0304 − 5	6.498 − 15			19.94
500	1499.22	1.0957	1.577			17.94
600	1506.13	3.4502 − 6	4.640 − 16			16.84
700	1507.61	1.1918	1.537			16.17

* From "U.S. Standard Atmosphere, 1962," NASA, USAF, USWB, Washington, D.C., December, 1962.

† Equations used to compute viscosity and speed of sound not applicable above 90 km.

Note. A one- or two-digit number (preceded by a plus or minus sign) following the initial entry indicates the power of 10 by which that entry and each succeeding entry of that column should be multiplied.

2k-9. Dynamical Relationships. *Coriolis Parameter.* The *apparent* force per unit mass acting upon a particle whose motion is described in a coordinate system fixed to the surface of the earth, due to the rotation of the earth in space, is proportional to the Coriolis parameter $f = 2\omega \sin \phi$, where ω is the angular velocity of the earth ($\omega = 7.292 \times 10^{-5}$ radian $\sec^{-1}$) and ϕ is latitude. If v is the speed of a particle of unit mass, the apparent force is equal to fv. This force is directed to the right of the direction of motion in the Northern Hemisphere and to the left in the Southern Hemisphere.

Geostrophic Wind. Steady, straight, frictionless air motion in an unchanging pressure field, with gravity as the only external force acting, so that the horizontal pressure-gradient force is balanced by the apparent force due to the earth's rotation (the

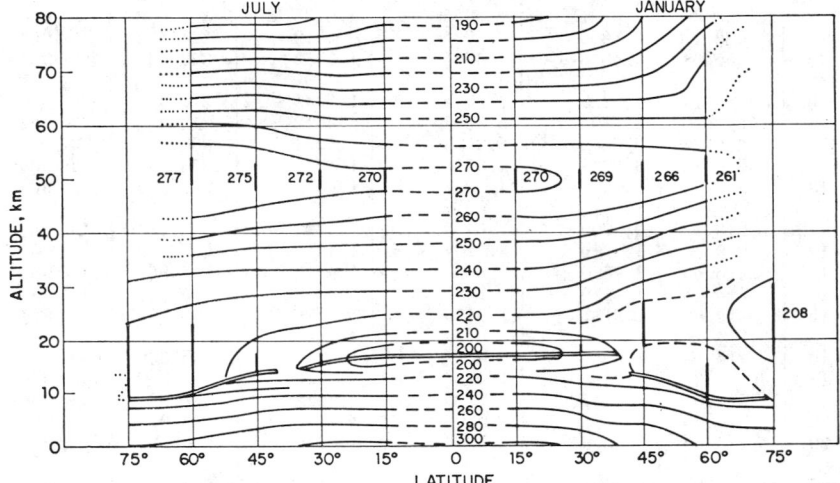

FIG. 2k-1. Temperature-altitude cross section for January and July. ("*U.S. Standard Atmosphere Supplements, 1966*," ref. 33.)

Coriolis force), is called the *geostrophic* wind. The geostrophic wind blows perpendicularly to the direction of the pressure gradient with low pressure to the left in the Northern Hemisphere, to the right in the Southern Hemisphere.

TABLE 2k-5. VALUE OF THE CORIOLIS PARAMETER f

Latitude	f, $\sec^{-1}$	Latitude	f, $\sec^{-1}$
0°	0	50°	1.1172×10^{-4}
10°	0.2533×10^{-4}	60°	1.2630×10^{-4}
20°	0.4988×10^{-4}	70°	1.3705×10^{-4}
30°	0.7292×10^{-4}	80°	1.4363×10^{-4}
40°	0.9375×10^{-4}	90°	1.4584×10^{-4}
45°	1.0313×10^{-4}		

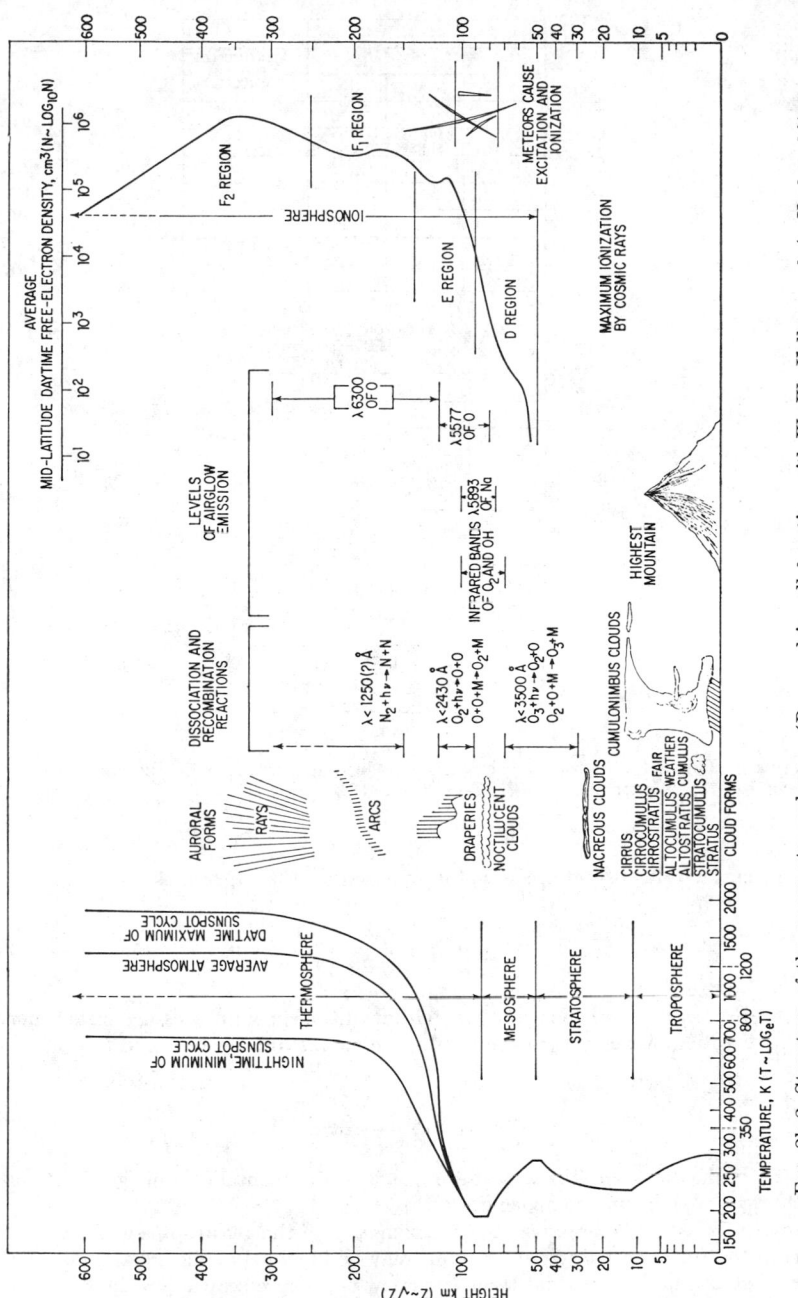

Fig. 2k-2. Structure of the upper atmosphere. (Prepared in collaboration with W. W. Kellogg and A. Kochanski.)

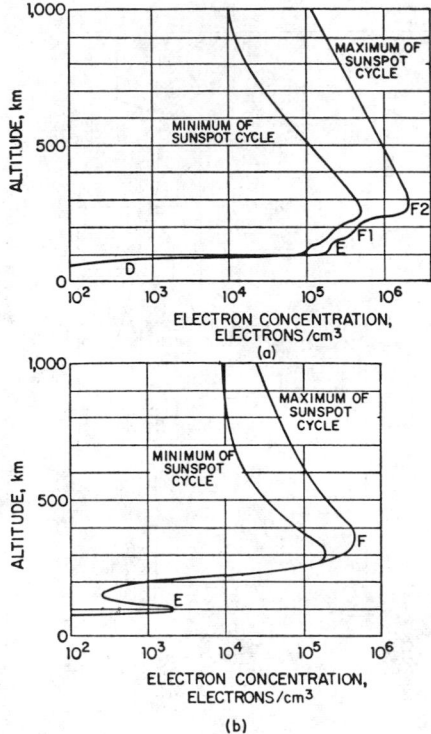

ELECTRON CONCENTRATION,
ELECTRONS/cm³
(a)

ELECTRON CONCENTRATION,
ELECTRONS/cm³
(b)

Fig. 2k-3. Diurnal and solar-cycle variations in the structure of the ionosphere. (a) Daytime. (b) Nighttime. (*After Henson*, ref. 18.)

On a surface of constant pressure, the equation for the speed of the geostrophic wind V_g is given by

$$V_g = \frac{1}{f} \frac{\partial \Phi}{\partial n} \tag{2k-6}$$

where $-\partial \Phi/\partial n$ is the gradient of geopotential on the constant-pressure surface normal to the direction of the geostrophic wind. On a constant-level surface,

$$V_g = \frac{1}{f\rho} \frac{\partial p}{\partial n} \tag{2k-7}$$

where ρ is the density of the air and $-\partial p/\partial n$ is the horizontal pressure gradient normal to the geostrophic wind component.

Gradient Wind. To improve the approximation of the geostrophic wind to the true wind in the free atmosphere, other terms may be included in the equation of motion. The most common additional term is that which expresses the acceleration arising from the curvature of the path of the moving air parcel. The addition of this term to the expression for the geostrophic wind speed gives the *gradient* wind speed V.

$$V = \frac{rf}{2} \left[-1 + \left(1 + \frac{4V_g}{rf} \right)^{1/2} \right] \tag{2k-8}$$

where r is the radius of curvature of the trajectory of the air parcel and the following sign convention is used: for cyclonic curvature $rf > 0$, for anticyclonic curvature $rf < 0$.

Zonal Motion. The average motion of the atmosphere is predominantly geostrophic and zonal. The zonal motion between sea level and 50 mb, for summer and winter, Northern and Southern Hemispheres, is shown in Fig. 2k-4, from Mintz [25].

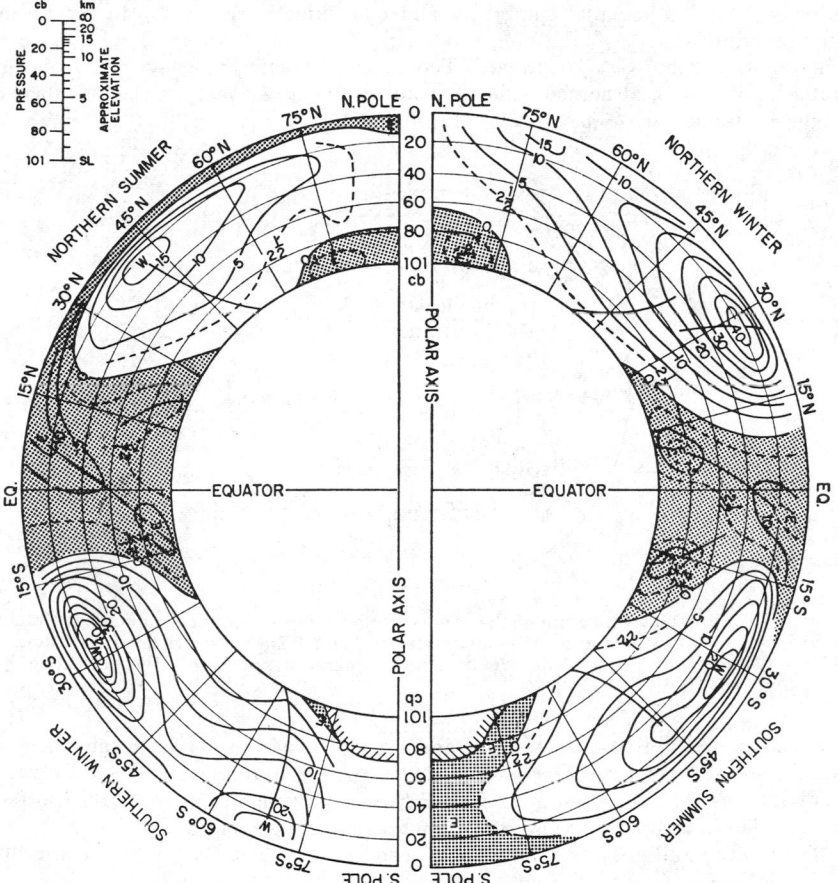

FIG. 2k-4. Zonal circulation of the atmosphere, m sec⁻¹, averaged over all longitudes. W represents motion from the west, E is motion from the east. (*From Mintz, ref. 25.*)

For levels above 35 km, see Murgatroyd in ref. [2]. A pronounced 26-month oscillation of the zonal wind in the equatorial stratosphere has been observed (see, e.g., Reed [29]).

Eddy Motion. Superimposed on the average zonal motion of the atmosphere are eddy circulations covering a wide spectrum, including cyclones and anticyclones in the lower troposphere and planetary or Rossby waves in the middle troposphere. Under *barotropic* conditions frequently observed in the middle troposphere, the speed c of planetary waves is given by

$$c = U - \frac{\beta\lambda^2}{4\pi^2} \qquad (2k\text{-}9)$$

where U is the west wind speed, β the northward change of the Coriolis parameter, and λ the wavelength. For an introduction to current numerical techniques of modeling and predicting atmospheric processes, especially atmospheric motions, see Thompson [31].

Energy Conversions. Figure 2k-5, from Oort [26], shows an estimate of the generation G, dissipation D, and conversion C rates for energy processes in the atmosphere. In the average, the energy cycle proceeds from mean available potential energy P_M via eddy available potential energy P_E and eddy kinetic energy K_E to the mean kinetic energy (K_M).

2k-10. Radiation. *Solar Constant.* The solar constant, the mean value of the total solar radiation, at normal incidence, outside the atmosphere at the mean solar distance $= 0.140$ w cm^{-2} (p.e. $= 2\%$) [18].

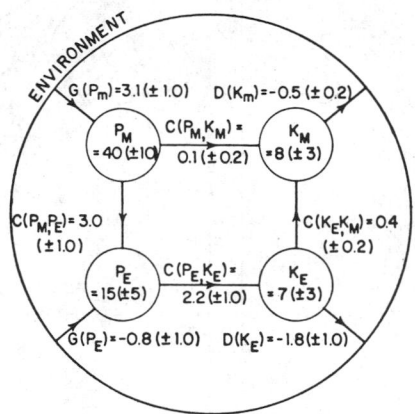

FIG. 2k-5. Tentative flow diagram of the atmospheric energy in the space domain. Values are averages over a year for the Northern Hemisphere. Energy units are in 10^5 joules m^{-2} ($= 10^8$ ergs cm^{-2}); energy transformation units are in watts m^{-2} ($= 10^3$ ergs cm^{-2} sec^{-1}). (*From Oort*, ref. 26.)

Insolation. Figure 2k-6 shows the average daily solar radiation received on a square centimeter of horizontal surface at the ground during January and July on cloudless days [11] (solid lines) and on days with average cloudiness [13] (dotted lines). The units are gram-calories per square centimeter per day.

Albedo. Table 2k-6 gives a range of albedo measurements[1] observed for various type of surface.

Heat Balance of the Atmosphere. Taking the incident solar radiation as 100 units, Byers [5] has computed the heat budget of the atmosphere as shown in Table 2k-8.

2k-11. Clouds.[2] The drop-size spectra of typical cloud types are given in Fig. 2k-7.

2k-12. Climatology. Space limitations preclude the presentation of climatological data. In addition to standard climatological texts, see [17], [32], [7], [8], [9], and [35]; the reports of World Data Center A, especially the subcenters on Meteorology, Upper Atmosphere Geophysics, and Rockets and Satellites; and various numbers in the key to Meteorological Records Documentation series, especially No. 4.11 [10].

---

[1] For a more complete list, including sources, see List, *op. cit.*, pp. 442–444.
[2] Data furnished by Dr. H. J. aufm Kampe, Signal Corps Engineering Laboratories, Ft. Monmouth, N.J.

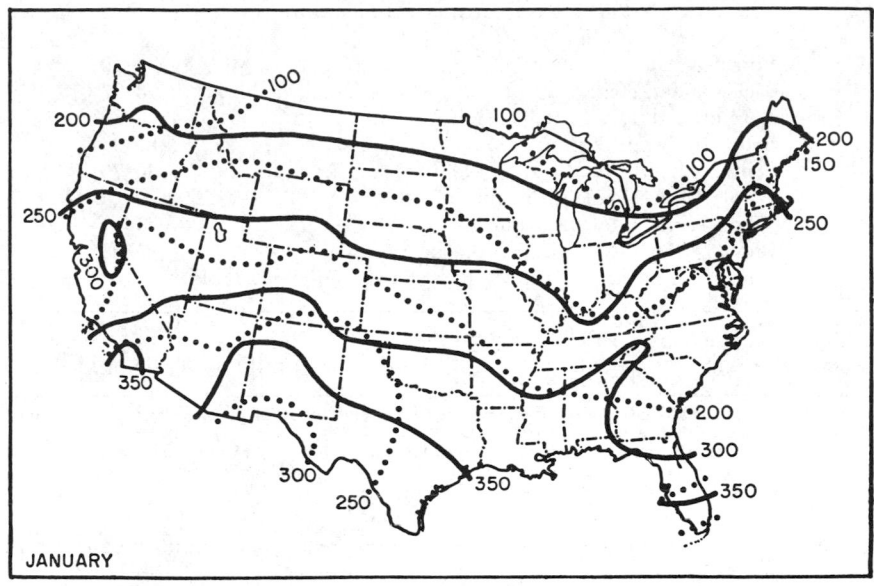

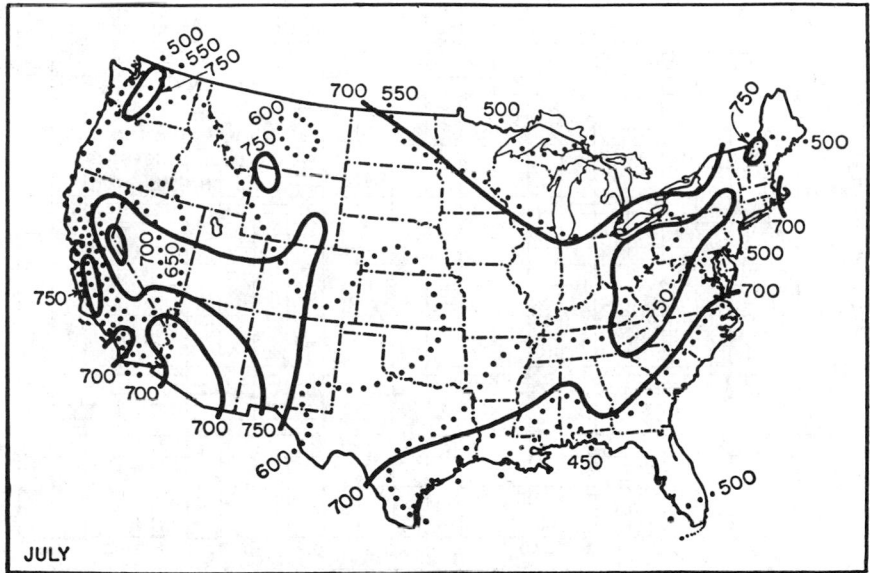

FIG. 2k-6. Average daily solar insolation (g-cal cm^{-2} day^{-1}) at the ground on cloudless days (solid lines) and on days of average cloudiness (dotted lines). (*After Fritz and MacDonald* [11, 13].)

TABLE 2k-6. ALBEDO MEASUREMENTS

	%
Forest...	3–10
Fields, grass, etc....................................	3–37
Bare ground...	3–30
Snow, fresh..	80–90
Snow, old...	45–70
Whole earth, visible spectrum.........................	39
Whole earth, total spectrum..........................	35
Clouds*...	5–85

Water (reflectivity values are given in the following table)†

Elevation of sun......	90°	70°	50°	40°	30°	20°	10°	5°	0°
Reflectivity, %.......	2.0	2.1	2.5	3.4	6.0	13.4	34.8	58.4	100.0

* For clouds in the absence of absorption the albedo is a function of the drop-size distribution, liquid water content, and cloud thickness. See S. Fritz [12].

† The reflectivity of a water surface for solar radiation is a function of the sun's elevation angle. The values given have been computed for a plane surface; however, the observed reflection from disturbed surfaces shows only small deviation from these values.

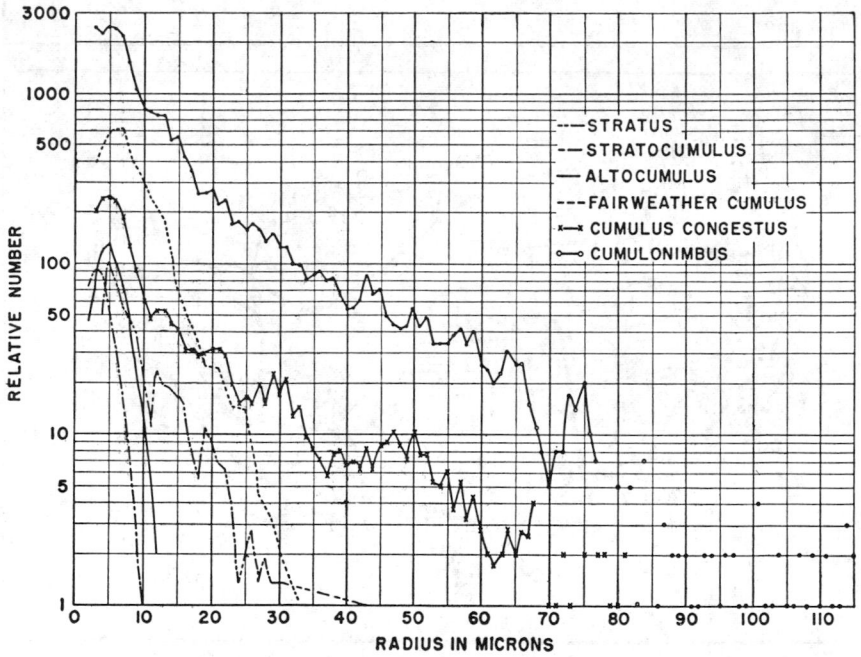

FIG. 2k-7. Cloud drop-size spectra. (*Prepared by H. J. Aufm Kampe.*)

2k-13. Atmospheric Diffusion.[1] In most meteorological problems, the effects of molecular diffusion are far outweighed by the turbulent eddies present in the atmosphere. One approach to this problem is to treat the phenomenon in a manner analogous to that of molecular diffusion. The coefficient of diffusion in such applications is a function of the size of the turbulent eddies and is therefore dependent on the

[1] For the definition of diffusion coefficient cf. p. **2-221.**

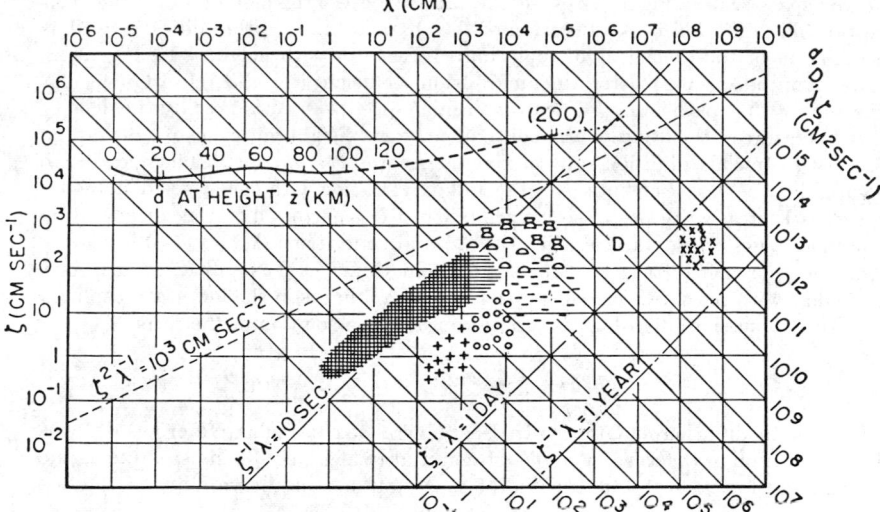

D AT HEIGHT z IS DENOTED BY CHARACTERISTIC AREAS ON THE DIFFUSION DIAGRAM:

≡≡≡ I–IO KM FOR ORDINARY TURBULENCE   ▦ O–I KM

ᗡᗡᗡ I–IO KM FOR CUMULUS CONVECTION   +₊+ 25–35 KM

ᗺ I–IO KM FOR CUMULONIMBUS CONVECTION   -⁻- 45–80 KM

°₀° IO–25 ,35–45 AND 80–IOO KM

ˣˣₓ HORIZONTAL GROSS–AUSTAUSCH OF THE GENERAL CIRCULATION

FIG. 2k-8. Diffusion diagram. (*From Lettau.*)

TABLE 2k-7. AVERAGE WATER CONTENT OF TYPICAL CLOUDS

Cloud type	Cirrus*	Cumulus congestus	Fair-weather cumulus	Stratus	Strato-cumulus
Water content, g m^{-3}	0.01–0.05	4	1	0.3	0.2

* Cirrus clouds consist mainly of column-shaped ice crystals. In cirrostratus, single, more or less completely built columns (twin crystals) of about 100 microns in length and 25 microns in diameter predominate. In dense cirrus and cirrocumulus clouds the columns are incompletely built and occur in clusters. The length of the individual crystals in such clusters is approximately 100 to 300 microns and the diameter 30 to 100 microns.

TABLE 2k-8. HEAT BUDGET OF THE ATMOSPHERE*

Absorbed	Units	Lost	Units
Solar radiation	13	Infrared radiation to earth's surface	106
Latent heat	23	Infrared radiation to space	47
Infrared radiation from earth's surface	120	Eddy transfer to ground	3

* After Byers [5].

time and space scale. Figure 2k-8 (see Lettau [21]) gives the magnitude of the coefficient of eddy diffusion D as a function of the characteristics of the eddies, as well as the variation of the coefficient of molecular diffusion d with height. In Fig. 2k-8, each point of the λ, ζ plane determines a diffusion coefficient (cm^2 sec^{-1}). In molecular diffusion, $\lambda \approx$ free path and $\zeta \approx$ mean molecular speed; $d = \lambda\zeta$ is fixed by the density and temperature of the atmosphere; consequently, the height variation of d is marked by a curve. In eddy diffusion, $\lambda \approx$ mixing length and $\zeta \approx$ mixing velocity; owing to the variability of these elements, $D = \lambda\zeta$ and its variation with height are denoted by characteristic areas when the possible variability of D is narrowed by the consideration of limiting values of eddy accelerations (ζ^2/λ) and time terms (λ/ζ).

Another approach to the problem of turbulent diffusion [Pasquill (27)] has been used to deal with the small-scale dispersion of contaminants in the lower atmosphere. The normal (gaussian) distribution function provides a solution of the form

$$\chi(x,y,z,t) = Q(2\pi\sigma_y)^{-\frac{3}{2}} \exp \frac{-r^2}{2\sigma_y{}^2}$$

to the Fickian diffusion equation. In this generalized formula (Gifford [14]), χ is the concentration in the *cloud* or *plume* of material (which may be invisible), Q is the source strength at an instantaneous point source, and $\sigma_y{}^2$ is the variance of the dis-

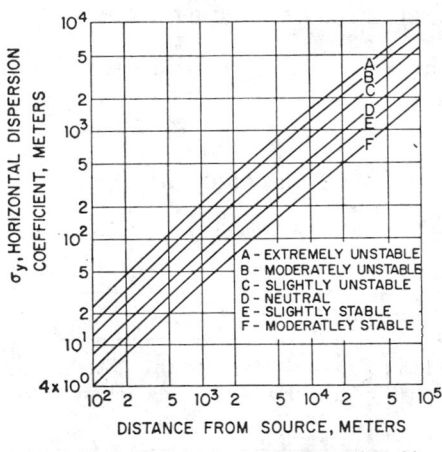

Fig. 2k-9. Graph of σ_y. (*From Gifford.*)

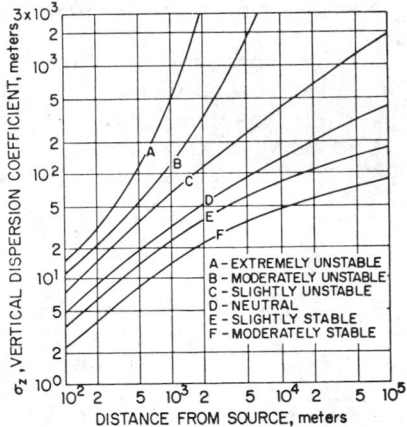

Fig. 2k-10. Graph of σ_z. (*From Gifford.*)

tribution of material in the plume. Since χ is assumed equal to $\bar{v}_x t$, where $\bar{v}_x$ is the unfluctuating wind velocity component and t is the travel time of the cloud, $r^2 = [(x - \bar{v}_x t)^2 + y^2 + z^2]$. (Initially it is assumed that the diffusion is isotropic, i.e., $\sigma_x = \sigma_y = \sigma_z$.) In practice, the assumption is made that the diffusion takes place independently in the three coordinate directions, so that with the graphs of σ_y and σ_z shown in Figs. 2k-9 and 2k-10 it is possible [14] to compute the concentration under different conditions of atmospheric stability represented by the curves A, B, etc.

References

1. Ackerman W. C.: *Trans. Am. Geophys. Union*, **48**, (2) 427–563 (June, 1967).
2. Bates, D. R., P. A. Sheppard, and R. C. Sutcliffe: *Proc. Roy. Soc. (London)*, ser. A, **288**, 478–588 (1965).
3. Blackadar, A. K., ed.: *Meteorol. Monographs.* **3**, (12–20) 283 pp. + iv, American Meteorological Society, Boston (July, 1957).
4. Bolin, Bert, ed.: "The Atmosphere and the Sea in Motion, The Rossby Memorial Volume," 509 pp., Rockefeller Institute Press in association with Oxford University Press, New York, 1959.

5. Byers, H. R.: The Atmosphere up to 30 Kilometers, pp. 299–370 in "The Earth as a Planet," G. P. Kuiper, ed., Chicago University Press, Chicago, 1954.
6. Committee on Space Research, International Council of Scientific Unions, *CIRA* **1965** (Cospar International Reference Atmosphere, 1965), North-Holland Publishing Company, Amsterdam, 1965.
7. Environmental Data Service, ESSA: *Climatological Data, National Summary* (issued, monthly, with an annual summary), Government Printing Office, Washington, D.C., 1950–.
8. Environmental Data Service, ESSA: "Climatic Atlas of the United States," Government Printing Office, Washington, D.C., 1968.
9. Environmental Data Service, ESSA: "World Weather Records" (monthly values and decadal means of pressure, temperature, and precipitation), Government Printing Office, Washington, D.C., 1968.
10. Environmental Data Service, ESSA: Selective Guide to Published Climatic Data Sources, *Key to Meteorological Records Documentation No.* 4.11, Government Printing Office, Washington, D.C., (1969).
11. Fritz, S.: *Heating and Ventilating* **46,** 69–74 (1940).
12. Fritz, S.: *J. Meteorol.* **11**(4), 291–300 (August, 1954).
13. Fritz, S., and T. H. MacDonald: *Heating and Ventilating* **46,** 61–64 (1949).
14. Gifford, F. A., Jr.: An Outline of Theories of Diffusion in the Lower Layers of the Atmosphere, pp. 65–116, in "Meteorology and Atomic Energy, 1968," D. H. Slade, ed., J. S. Atomic Energy Commission, Oak Ridge, Tenn., July 1968.
15. Glasstone, S.: "Source book on the Space Sciences," 937 pp. + xviii, D. Van Nostrand Co., Inc., Princeton, N.J., 1965.
16. Glueckauf, E.: "The Composition of Atmospheric Air," pp. 3–10, T. F. Malone, ed., American Meteorological Society, Boston, 1951.
17. Greathouse, G. A., and C. J. Wessel, eds.: Deterioration of Materials: Causes and Preventative Techniques, Chap. I in "Climate and Deterioration," Reinhold Book Corporation, New York, 1954.
18. Johnson, F. S., ed.: "Satellite Environment Handbook," 2d ed., 193 pp. + xiv, Stanford University Press, Stanford, Calif., 1965.
19. Keeling, C. D.: *Tellus* **XII**(2), 200–203 (1960).
20. Landsberg, H. E., and J. van Miegham, eds.: "Advances in Geophysics," vols. 1–12, Academic Press, Inc., New York, 1954–1967.
21. Lettau, H.: Diffusion in the Upper Atmosphere, pp. 320–333 in "Compendium of Meteorology," T. F. Malone, ed., American Meteorological Society, Boston, 1951.
22. Letestu, S., ed.: "International Meteorological Tables," WMO—No. 188 TP. 94, Secretariat of the World Meteorological Organization, Geneva, Switzerland, 1966.
23. List, R. J., ed.: "Smithsonian Meteorological Tables," 6th ed., 527 pp. + xi, Smithsonian Institution, Washington, D.C., 1951.
24. Malone, T. F., ed.: "Compendium of Meteorology," 1334 pp. + viii, American Meteorological Society, Boston, 1951.
25. Mintz, Y.: *Bull. Am. Meteorol. Soc.* **35**(5), 208–214 (May, 1954).
26. Oort, A. H.: *Monthly Weather Rev.* **92**(11), 1964, 483–493 (November, 1964).
27. Pasquill, F.: "Atmospheric Diffusion," D. Van Nostrand Company, Ltd., London, 1962.
28. Ratcliffe, J. A., ed.: "Physics of the Upper Atmosphere," Academic Press, Inc., New York, 1960.
29. Reed, R. J.: *Quart. J. Roy. Meteorol. Soc.* **90,** 441–466 (1964).
30. Rigby, M., ed.: "Meteorological and Geoastrophysical Abstracts," American Meteorological Society, Boston, 1950–.
31. Thompson, P. D.: "Numerical Weather Analysis and Prediction," The Macmillan Company, New York, 170 pp. + xiv, 1961.
32. U.S. Air Force, Geophysics Research Directorate: "Handbook of Geophysics," rev. ed., The Macmillan Company, New York, 1960.
33. U.S. Environmental Science Services Administration, U.S. National Aeronautics and Space Administration, U.S. Air Force: "*U.S. Standard Atmosphere Supplements*, 1966," 289 pp. + xx, Government Printing Office, Washington, D.C., 1967.
34. U.S. National Aeronautics and Space Administration, U.S. Air Force, U.S. Weather Bureau: "U.S. Standard Atmosphere, 1962," Government Printing Office, Washington, D.C., 1962.
35. U.S. Navy: "Marine Climatic Atlas of the World," vol. VIII, Government Printing Office, Washington, D.C., 1968.
36. van Zandt, T. E., and R. W. Knecht: The Structure and Physics of the Upper Atmosphere, pp. 166–225 in "Space Physics," D. P. LeGalley and A. Rosen, eds., John Wiley & Sons, Inc., New York, 1964.
37. World Data Center A, Reports, National Academy of Sciences, Washington, D.C., 1964–.

21. Density and Compressibility of Liquids[1]

JOHN E. McKINNEY

National Bureau of Standards

ROBERT LINDSAY

Trinity College

Symbols

B^*	complex compressibility
B'	Re B^*
B''	Im B^*
c_S	adiabatic velocity of sound
c	velocity of sound
C_P	specific heat at constant pressure
$d_{t_1}^{t_2}$	specific gravity (t_1 is temperature of liquid; t_2 is temperature of standard)
K	instantaneous bulk modulus
K^*	complex bulk modulus
K'	Re K^*
K''	Im K^*
P	pressure
S	entropy
T	absolute temperature
t	Celsius temperature
v	specific volume
v_0	specific volume in reference state
α_P	isobaric coefficient of volume expansivity
β	instantaneous compressibility
β_S	adiabatic compressibility
β_T	isothermal compressibility
ρ	density
ρ_0	density at reference state

21-1. Density of Liquids. Introduction. The density of a homogeneous liquid is defined as the mass per unit volume. The specific volume is the reciprocal of the density. Density can be expressed in either an absolute or a relative scale. The SI (Système International) absolute units are kilograms per cubic meter, and the cgs absolute units are grams per cubic centimeter. Before 1964 the liter was defined as the volume required to contain one kilogram of water at 3.98°C and 760 mm Hg pressure, equal to 1.000028 cubic decimeters. In 1964[2] the liter was redefined to be exactly equal to the cubic decimeter. This difference of 28 parts per million may be

[1] Contribution of the National Bureau of Standards, not subject to copyright.
[2] *Proc. 12th General Conf. Weights and Measures.* Paris, p. 21, Oct. 6–13, 1964.

significant in measurements of high accuracy. Accordingly, in order to avoid mis-understanding, it has been recommended that for volume measurements of high accuracy the units be expressed in cubic meters or their submultiples, in lieu of liters or their submultiples. The following tables which express densities in grams per milliliter have not been corrected. In cases where accurate data are required, the original source should be consulted to determine the correct units. Table 21-1 gives the conversion factors for the density units most commonly used.

TABLE 21-1. CONVERSION FACTORS FOR DENSITY UNITS*

Units	kg/m³ (SI)	g/cm³ (cgs)	g/ml (old)	lb/ft³	lb/in³
1 kg/m³ (SI).	1	10^{-3}	1.000028×10^{-3}	6.24280×10^{-2}	3.61273×10^{-5}
1 g/cm³ (cgs)	10^3	1	1.000028	62.4280	3.61273×10^{-2}
1 g/ml (old).	9.99972×10^2	0.999972	1	62.4262	3.61263×10^{-2}
1 lb/ft³......	16.0185	1.60185×10^{-2}	1.60189×10^{-2}	1	5.78704×10^{-4}
1 lb/in³.....	2.76799×10^4	27.6799	27.6807	1.72800×10^3	1

* Conversions to SI units taken from ASTM Metric Practice Guide, U.S. Department of Commerce, *Natl. Bur. Standards Handbook* **102**, 38 (Mar. 10, 1967).

For expressing densities on a relative scale the specific gravity is used. The specific gravity gives the ratio of the density of a liquid at a particular temperature to the density of a standard liquid (usually pure water) at a standard temperature. The conventional symbol for absolute density is ρ or d. The former will be used in this set of tables. The conventional symbol for specific gravity is $d_{t_1}^{t_2}$ where t_1 is the temperature of the liquid and t_2 is the temperature of the standard.

21-2. Methods of Measurement. The pycnometer method is most commonly used when precise density measurements on a particular liquid are desired at fixed temperature.[1] A pycnometer is a vessel made of glass with a low coefficient of expansion whose volume can be determined very precisely in terms of its capacity for a standard liquid. Most pycnometers have a capacity of about 30 ml. The general procedure consists in filling the pycnometer with the unknown liquid, thermostatting the system at the desired temperature, determining the volume of the pycnometer occupied by the liquid, and then weighing the pycnometer. For accurate work air buoyancy corrections should be applied. For determining densities of the same sample over a range of temperatures, the dilatometer method is sometimes used. In one variation of this method a secondary standard liquid such as mercury is placed in contact with the liquid sample. As the temperature is raised, the secondary liquid is displaced out of the dilatometer. The weight of the displaced secondary liquid is a measure of the change in volume of the unknown liquid. Another variation of this method involves the observation of the change in level of the unknown liquid in a narrow calibrated capillary attached to the main flask. (The measurements of densities of liquefied gases at or near their boiling points are more complicated, since a closed system may have to be used and significant corrections must be made for the density of the vapor in equilibrium with the liquid.[2]) Where less accuracy is required, hydrostatic weighing methods[3] and hydrometers are expedient. Hydrostatic weigh-

[1] A. Johnson, *J. Research Natl. Bur. Standards* **69C**, 1 (1965).

[2] W. H. Keeson, "Helium," pp. 206ff., American Elsevier Publishing Company, Inc., New York, 1942; E. R. Grilly, E. F. Hammel, and S. G. Sydoriak, *Phys. Rev.* **75**, 1103 (1949); E. R. Grilly, *J. Am. Chem. Soc.* **73**, 5307 (1951).

[3] H. A. Bowman and R. M. Schoonover, *J. Research Natl. Bur. Standards* **71C**, 179 (1967). (Although this reference is strictly applicable to solids, many of the procedures described here are also applicable to liquids.)

ing involves obtaining the apparent weight of solids (weights) of known mass and density submerged in the liquid from an analytical (or Westphal) balance on a thin wire or thread. A hydrometer is simply a calibrated float which reads the density directly. The performance of the last two methods is impaired by surface tension. Modifications which avoid this are the flotation and the elastic helix methods. The flotation method involves the adjustment of a submerged weight to the same average density as the density of the unknown liquid. At this point the weight will neither sink nor float. Alternatively, a balance is sometimes obtained by appropriately adjusting the temperature of the liquid. The method is tedious, but high accuracy can be obtained. In one version[1] a known electric current producing a magnetic field is adjusted until an iron weight suspended in the field and submerged in the liquid is stationary. With the elastic helix a weight is suspended in the unknown liquid from a completely submerged coil often made of quartz. The density of the liquid is related to the length of the helix.

The falling-drop method has been used recently on molten metals[2] and for the isotopic analysis of water.[3] This method involves measuring the transit time, usually at terminal velocity, of a drop of liquid sample falling within an immiscible liquid or gas. In another version[4] the volume of a falling drop of molten metal is determined by measuring the dimensions of its profile from a photograph.

The radiation method,[5] for which the gamma radiation from an irradiated isotope is claimed to be proportional to its density, has been used successfully on liquids at high temperatures. References[6] are recommended for more detailed and general discussion on most of the methods mentioned above.

21-3. Reliability. The reliability of the density measurements tabulated is variable. This compilation does not pretend to evaluate for extreme accuracy. Such factors as uncertainty in the temperature scale, possible impurities of the samples, and in some cases even changes in atomic weights must be taken into consideration when applying a critical analysis. The data are given as reported in the original literature or in other standard works and are to be interpreted in the spirit of being representative values. Reference to the original literature is recommended in cases of doubt.

21-4. Selected Reference Works with Density Data

"International Critical Tables," McGraw-Hill Book Company, New York, 1928.
Landolt-Börnstein: "Zahlenwerte und Funktionen aus Physik, Chemie, Astronomie, Geophysik, Technik," 6th ed., Springer Verlag, Berlin, 1950————.
Timmermans, J.: "Physico-chemical Constants of Pure Organic Compounds," American Elsevier Publishing Company, Inc., New York, vol. 1, 1950; vol. 2, supplement, 1965.
Timmermans, J.: "The Physico-chemical Constants of Binary Systems in Concentrated Solutions," 4 vols., Interscience Publishers, Inc., New York, 1959–1960.
Mellor, J. W.: "Comprehensive Treatise of Inorganic and Theoretical Chemistry," Longmans, Green & Co., Ltd., London, 16 vols., 1921–1937; supplements, 1956–1967.
Simons, J. H., ed.: "Fluorine Chemistry," Academic Press, Inc., New York, 1950.
Lyon, R. N., ed.: "Liquid Metals Handbook," U.S. Atomic Energy Commission and Bureau of Ships, Chap. 2, 1952; Sodium-NaK Supplement, 1955.

---

[1] F. J. Millero, Jr., *Rev. Sci. Instr.* **38**, 1441 (1967).
[2] Y. V. Naidich and V. N. Eremenko, *Fiz. Metal. i Metalloved.* (USSR) **11**, 883 (1961); English translation in *Phys. Metals Metallog.* **11**(6), 62 (1961).
[3] M. Pascalau, L. Blaga, and L. Blaga, *J. Sci. Instr.* **43**, 310 (1966).
[4] A. E. El-Mehairy and R. G. Ward, *Trans. Met. Soc. AIME* **227**, 1226 (1963).
[5] I. G. Dillon, P. A. Nelson, and B. S. Swanson, *J. Chem. Phys.* **44**, 4229 (1966); *Rev. Sci. Instr.* **37**, 614 (1966).
[6] P. Hidnert and E. L. Peffer, *Natl. Bur. Standards Circ.* 487 (Mar. 15, 1950); N. Bauer, Chap. 6 in "Physical Methods of Organic Chemistry," vol. I, A. Weissberger, ed., Interscience Publishers, Inc., New York, 1949; J. Reilly and W. N. Rae, "Physico-chemical Methods," vol. I, pp. 609–628, D. Van Nostrand Company, Inc., Princeton, N.J., 1953.

Stewart, R. B., and V. J. Johnson, eds.: "Compendium of the Properties of Materials at Low Temperature" (Phase I), Wright Air Force Development Division Technical Report, 1961.
Morey, G. W.: "The Properties of Glass," Reinhold Book Corporation, New York, 1954.
Janz, G. J.: "Molten Salts Handbook," Academic Press, Inc., New York, 1967.
Janz, G. J., F. W. Dampier, G. R. Lakshminarayanan, P. K. Lorenz, and R. P. T. Tomkins: Molten Salts: vol 1, Electrical Conductance, Density, and Viscosity Data, *Natl. Bur. Standards Ref. Data Ser.* 15, October, 1968.
The following contain tabulated sheets published periodically:
Zwolinski, B. J.: "Selected Values of Properties of Hydrocarbons and Related Compounds,"[1] Thermodynamics Research Center, Texas A&M University, College Station, Texas, 1966——.
Zwolinski, B. J.: "Selected Values of Properties of Chemical Compounds," Thermodynamics Research Center, Texas A&M University, College Station, Texas, 1966——.
Zwolinski, B. J.: "Selected Data on Thermodynamics and Spectroscopy," Thermodynamics Research Center, Texas A&M University College Station, Texas, 1969——.

21-5. Density of Water. A rather complete analysis of all the investigations of the physical properties of water is given by N. Ernest Dorsey.[2] He points out that the data from which the density tables are made up do not take into consideration the isotope effect. Because of this there may be uncertainties of the order of 8 parts in 10^7 introduced when the densities of samples from various sources are considered. The removal of deuterium from an average sample of distilled water increases the density by about 17 parts per million. There is also some reason to believe that the polymerization is a factor in the variability of the physical properties of water. Values of the density of water as a function of temperature are presented in Table 21-2.

[1] American Petroleum Institute Project 44.
[2] N. Ernest Dorsey, "Properties of Ordinary Water Substance," Reinhold Book Corporation, New York, 1948.

TABLE 2l-2. DENSITY OF PURE AIR-FREE H_2O AT ATMOSPHERIC PRESSURE
(ρ = g/ml*; t = °C)

Range 0–40°C†

t	ρ	t	ρ	t	ρ
0	0.9998676	5	0.9999919	10	0.9997281
1	0.9999265	6	0.9999683	11	0.9996336
2	0.9999678	7	0.9999297	12	0.9995261
3	0.9999922	8	0.9998765	13	0.9994059
4	1.0000	9	0.9998092	14	0.9992732

t	0.0	0.1	0.2	0.3	0.4	0.5	0.6	0.7	0.8	0.9
15	0.9991286	1134	0982	0828	0674	0518	0360	0202	0043	9882
16	0.9989721	9558	9394	9229	9062	8895	8726	8557	8386	8214
17	0.9988041	7867	7691	7515	7337	7158	6979	6798	6616	6433
18	0.9986248	6063	5877	5689	5501	5311	5120	4928	4735	4541
19	0.9984346	4150	3953	3754	3555	3355	3153	2950	2747	2542
20	0.9982336	2130	1922	1713	1503	1292	1080	0867	0653	0438
21	0.9980221	0004	9786	9567	9346	9125	8903	8679	8455	8230
22	0.9978003	7776	7547	7318	7088	6856	6624	6390	6156	5921
23	0.9975684	5447	5208	4969	4729	4487	4245	4002	3758	3512
24	0.9973266	3019	2771	2522	2272	2021	1769	1516	1262	1007
25	0.9970751	0494	0237	9978	9718	9458	9196	8934	8671	8406

t	ρ	t	ρ	t	ρ
26	0.9968141	31	0.9953722	36	0.9937159
27	0.9965437	32	0.9950575	37	0.9933604
28	0.9962642	33	0.9947344	38	0.9929970
29	0.9959757	34	0.9944030	39	0.9926260
30	0.9956783	35	0.9940635	40	0.9922473

Range 40–100°C‡

t	ρ	t	ρ	t	ρ
40	0.99224	65	0.98059	90	0.96534
45	0.99024	70	0.97781	95	0.96192
50	0.98807	75	0.97489	100	0.95838
55	0.98573	80	0.97183		
60	0.98324	85	0.96865		

Range 100–370°C§

t	ρ	t	ρ	t	ρ
100	0.95841	190	0.87639	280	0.75063
110	0.95099	200	0.86492	290	0.73237
120	0.94317	210	0.85290	300	0.71266
130	0.93494	220	0.84031	310	0.69118
140	0.92629	230	0.82712	320	0.66747
150	0.91721	240	0.81330	330	0.64095
160	0.90771	250	0.79881	340	0.61071
170	0.89776	260	0.78368	350	0.57497
180	0.88733	270	0.76769	360	0.52872

See page 2-153 for footnotes.

TABLE 2l-2. DENSITY OF PURE AIR-FREE H_2O AT
ATMOSPHERIC PRESSURE (*Continued*)
Range 0 to $-13°C$¶

t	ρ	t	ρ	t	ρ
0	0.999868	−5	0.999176	−10	0.997935
−1	0.999773	−6	0.998950	−11	0.997636
−2	0.999673	−7	0.998720	−12	0.997292
−3	0.999553	−8	0.998501	−13	0.997292
−4	0.999380	−9	0.998249		

For a comparative study and density values over this range expressed in SI units, see P. H. Bigg, *Brit. J. Appl. Phys.* **18**, 521 (1967).

* These data have not been corrected in terms of the new definition of the liter (Sec. 2l-1). For the conversion factor, see Table 2l-1.

† L. W. Tilton and J. K. Taylor, *J. Research Natl. Bur. Standards* **18**, 205 (1937).

‡ V. Stott and P. H. Bigg, "International Critical Tables," vol. 3, p. 24, McGraw-Hill Book Company, New York, 1928.

§ F. G. Keyes and L. B. Smith, *Mech. Eng.* **53**, 132 (1931).

¶ J. F. Mohler, *Phys. Rev.* **35**, 236 (1912).

TABLE 2l-3. DENSITY OF D_2O (100% D_2O WITH NORMAL OXYGEN ISOTOPE COMPOSITION)
(ρ = g/ml*; t = °C)
Range 3.8–20°C†

t	ρ
3.8	1.10533
5	1.10549
10	1.10588
15	1.10577
20	1.10527

Range 20–100°C‡

t	ρ	t	ρ	t	ρ
20	1.10530	50	1.09562	80	1.07815
25	1.10437	55	1.09316	85	1.07467
30	1.10315	60	1.09051	90	1.07104
35	1.10167	65	1.08766	95	1.06729
40	1.09989	70	1.08466	100	1.06339
45	1.09786	75	1.08148		

Range 90–250°C§

t	ρ	t	ρ	t	ρ
90	1.0708	150	1.0167	210	0.943
100	1.0630	160	1.0058	220	0.928
110	1.0547	170	0.9950	230	0.913
120	1.0459	180	0.9826	240	0.897
130	1.0366	190	0.970	250	0.881
140	1.0268	200	0.957		

For additional data on H_2O^{18}, D_2O, and D_2O^{18}, see: F. Steckel and S. Szapiro, *Trans. Faraday Soc.* **59**, 331 (1963).

* These data have not been corrected in terms of the new definition of the liter (Sec. 2l-1). For the conversion factor, see Table 2l-1.

† T. L.-Chang and J. Y. Chien, *J. Am. Chem. Soc.* **63**, 1709 (1941).

‡ R. Schrader and K. Wirtz, *Z. Naturforsch.* **6a**, 220 (1951).

§ J. R. Heiks, M. K. Barnett, L. V. Jones, and E. Orban, *J. Phys. Chem.* **58**, 488 (1954).

The maximum density of D_2O has been determined to be 1.10596 g/ml at 11.23°C. K. Stokland, E. Ronaess, and L. Tronstad, *Trans. Faraday Soc.* **35**, 312 (1938). This is based on a value for d_{25}^{35} of 1.10764. L. Tronstad and Brun, *Trans. Faraday Soc.* **34**, 766 (1938). See H. L. Johnston, *J. Am. Chem. Soc.* **61**, 878 (1939), for a discussion of these values.

The density of H_2O and D_2O at 370°C are approximately the same. E. H. Riesenfield and T. L.-Chang, *Z. physik. Chem.* **B30**, 61 (1935); **B28**, 408 (1935).

TABLE 2l-4. DENSITY OF MERCURY (Hg)

$(\rho = \text{g/ml}^*; t = °\text{C})$

Range -38.87 to $100°C$†

t	ρ	t	ρ
-38.87	13.691_9	24	13.536_4
-30	13.669_8	25	13.534_0
-20	13.645_0	30	13.521_8
-10	13.620_2	35	13.509_6
0	13.595_5	40	13.497_3
5	13.583_2	45	13.485_1
10	13.570_9	50	13.472_9
15	13.558_6	55	13.460_8
16	13.556_2	60	13.448_6
17	13.553_7	65	13.436_5
18	13.551_3	70	13.424_3
19	13.548_8	75	13.412_2
20	13.546_3	80	13.400_1
21	13.543_9	85	13.388_0
22	13.541_3	90	13.375_9
23	13.538_9	95	13.363_9
		100	13.351_8

Range 100-360°C‡

t	ρ	t	ρ	t	ρ
100	13.3518	200	13.113	300	12.875
120	13.304	220	13.065	320	12.827
140	13.256	240	13.018	340	12.779
160	13.208	260	12.970	357.1	12.737
180	13.160	280	12.922		

* These data have not been corrected in terms of the new definition of the liter (Sec. 2l-1). For the conversion factor, see Table 2l-1.

† Stott and Bigg, "International Critical Tables," vol. 2, p. 457, McGraw-Hill Book Company, New York, 1928; Sears, *Proc. Phys. Soc. (London)* **26**, 95 (1913).

‡ G. W. C. Kaye and T. H. Laby, "Tables of Physical and Chemical Constants," 10th ed., Longmans, Green & Co., Ltd., London, 1948; Chappuis, "Traveaux et mémoires du bureau international des poids et mesures," vol. 16, 1917.

TABLE 21-5. DENSITY OF METHYL ALCOHOL (CH_3OH)

(ρ = g/ml*; t = °C)

Density at Fixed Points

t	ρ
0	0.809985†
5	0.80535†
25	0.78654†
25	0.78655‡
30	0.78181†

Density as a Function of Temperature

Range 0–60°C§

t	ρ
0	0.80999
5	0.80536
10	0.80070
15	0.79602
20	0.79132
25	0.78660
30	0.78186
35	0.77710
40	0.77232
45	0.76753
50	0.76270
60	0.75300

These data fit a formula

$$\rho = 0.80999 - 0.0009253t - 0.00000041t^2 \tag{21-1}$$

Range −94.5 to 15°C¶

$$\rho = 0.81015 - 0.0010041t - 0.000001802t^2 - 0.00000001657t^3 \tag{21-2}$$

* These data have not been corrected in terms of the new definition of the liter (Sec. 21-1). For the conversion factor, see Table 21-1.

† A. Rakowski and A. B. Frost, *Trans. Inst. Pure Chem. Reagents U.S.S.R.* **9**(334), 95 (1930).

‡ R. E. Gibson, *J. Am. Chem. Soc.* **57**, 1551 (1935).

§ Brunel and Van Bibber, "International Critical Tables," vol. 3, p. 27, McGraw-Hill Book Company, New York, 1928.

¶ J. Timmermans, *Sci. Proc. Roy. Dublin Soc.* **13**, 310 (1912).

TABLE 21-6. DENSITY OF ETHYL ALCOHOL (C_2H_5OH)
(ρ = g/ml^a; t = °C)

Density at Fixed Points

t	ρ
0	0.806306^b
25	0.785063^b
25	0.78506^c
50	0.763137^b

Density as a Function of Temperature

Range 10–40°C^d

t	ρ
10	0.79784
15	0.79360
20	0.78934
25	0.78506
30	0.78075
35	0.77641
40	0.77203

These data fit a formula

$$\rho = 0.78506 - 0.0008591(t - 25) - 0.00000056(t - 25)^2 - 0.000000005(t - 25)^3 \tag{21-3}$$

Range 45–78°C^e

t	ρ
45	0.76773
50	0.76329
60	0.75423
70	0.74491
78	0.73720

These data fit a formula

$$\rho = 0.80625 - 0.0008461t + 0.000000160t^2 - 0.0000000085t^3 \tag{21-4}$$

Range below 0°C^f

t	ρ
−59	0.856
−78	0.872

^{a} These data have not been corrected in terms of the new definition of the liter (Sec. 21-1). For the conversion factor, see Table 21-1.
^{b} Kretschmer, Nowakowska, and Wieba, *J. Am. Chem. Soc.* **70**, 1785 (1948).
^{c} N. S. Osborne, E. C. McKelvey, and H. W. Bearce, *Natl. Bur. Standards (U.S.) Bull.* **9**, 327 (1913).
^{d} N. S. Osborne, E. C. McKelvey, and H. W. Bearce, *Natl. Bur. Standards (U.S.) Bull.* **9**, 327 (1913).
^{e} Brunel and Van Bibber, "International Critical Tables," vol. 3, p. 27, McGraw-Hill Book Company, New York, 1928.
^{f} Beilstein, "Organische Chemie," vol. 1, p. 148, 1928.

TABLE 21-7. DENSITIES OF SELECTED INORGANIC LIQUIDS
(Range 0–50°C; ρ = g/ml; t = °C; pressure atmospheric)

Substance	Formula	t	ρ	Year	Ref.
Antimony pentachloride.........	$SbCl_5$	20	2.336	1915	1
Antimony pentafluoride.........	SbF_5	23	2.99	1904	2
Arsenic tribromide..............	$AsBr_3$	25	3.540	1893	3
Arsenic trichloride..............	$AsCl_3$	20	2.161	1880	3
Arsenic trifluoride..............	AsF_3	20	2.590	1880	3
Boron tribromide...............	BBr_3	0	2.650	1893	4
Boron trichloride...............	BCl_3	11	1.3493	1927	5
Boron triiodide.................	BI_3	50	3.3	1891	6
Bromine pentafluoride..........	BrF_5	25	2.4604	1954	7
Bromine trifluoride.............	BrF_3	25	2.8030	1954	7
Carbonyl chloride..............	$COCl_2$	0	1.4187	1946	8
Dicarbon tetrachloride..........	C_2Cl_4	20	1.6226	1880	9
Carbon disulfide................	CS_2	20	1.2632	1926	10
Carbon oxysulfide..............	COS	0	1.073	1932	11
Carbon selenide sulfide..........	$CSeS$	20	1.9874	1929	12
Carbon suboxide................	C_3O_2	0	1.114	1908	13
Carbon tetrachloride...........	CCl_4	20	1.5940	1938	14
Chlorine dioxide................	ClO_2	0	1.642	1930	15
Perchloric acid.................	$HClO_4$	20	1.7676	1906	16
Chlorosulfonic acid.............	$HClSO_3$	20	1.753	1912	17
Chlorine trifluoride.............	ClF_3	0	1.891	1950	18
Chromium oxychloride..........	CrO_2Cl_2	20	1.923	1880	3
Germanium oxychloride.........	$GeOCl_2$	20	2.057	1931	19
Germanium tetrachloride........	$GeCl_4$	20	1.879	1926	20
Bromogermane.................	GeH_3Br	29.5	2.34	1929	21
Dibromogermane...............	GeH_2Br_2	0	2.80	1929	21
Trichlorogermane...............	$GeHCl_3$	0	1.93	1926	20
Hydrazine.....................	NH_2NH_2	0	0.9816	1950	22
Hydrogen disulfide.............	H_2S_2	25	1.3270	1930	23
Hydrogen fluoride.............	HF	0	1.0015	1933	24
Hydrogen pentasulfide..........	H_2S_5	16	1.67	1928	25
Hydrogen peroxide.............	H_2O_2	19.9	1.4419	1920	26
Hydrogen trisulfide.............	H_2S_3	15	1.496	1908	27
Iodine heptafluoride............	IF_7	6	2.8	1930	28
Iodine pentafluoride............	IF_5	0	3.29	1933	29
Iron pentacarbonyl.............	$Fe(CO)_5$	18	1.4644	1891	30
Lead tetrachloride..............	$PbCl_4$	0	3.18	1893	31
Molybdenum fluoride...........	MoF_6	27	2.503	1931	32
Nickel carbonyl................	$Ni(CO)_4$	20	1.310	1891	3
Nitric acid (100%).............	HNO_3	20	1.502	1919	3
Nitrogen dioxide...............	NO_2	20	1.348	1919	3
Dinitrogen oxide...............	N_2O_3	0	1.450	1888	3
Osmium tetraoxide.............	OsO_4	43	4.322	1931	33
Phosphorus tribromide..........	PBr_3	20	2.877	1845	3
Phosphorus trichloride..........	PCl_3	20	1.575	1880	3
Phosphorus oxychloride........	$POCl_3$	20	1.675	1880	3

TABLE 21-7. DENSITIES OF SELECTED INORGANIC LIQUIDS (*Continued*)

Substance	Formula	t	ρ	Year	Ref.
Rhenium hexafluoride............	ReF_6	18.8	3.616	1934	34
Rhenium oxytetrafluoride........	$ReOF_4$	39.7	3.717	1934	34
Selenic acid....................	H_2SeO_4	15	2.602	1889	35
Selenium monochloride..........	Se_2Cl_2	17.5	2.906	1884	36
Selenium oxychloride...........	$SeOCl_2$	20	2.434	1931	37
Selenium oxyfluoride...........	$SeOF_2$	20	2.67	1928	38
Selenium tetrafluoride..........	SeF_4	20	2.77	1928	39
Tribromosilane.................	$SiHBr_3$	17	2.7	1880	40
Tribromochlorosilane...........	$SiBr_3Cl$	20	2.434	1887	41
Trichlorosilane................	$SiHCl_3$	20	1.34	1905	42
Triiodosilane..................	$SiHI_3$	23	3.286	1908	43
Trisilane......................	Si_3H_8	0	0.725	1916	44
Tetrasilane....................	Si_4H_{10}	0	0.79	1916	44
Silicon tetrabromide...........	$SiBr_4$	18.6	2.7889	1931	45
Silicon tetrachloride...........	$SiCl_4$	20	1.4812	1926	46
Stannic chloride...............	$SnCl_4$	20	2.229	1932	47
Sulfuric acid..................	H_2SO_4	20	1.834	1923	3
Sulfur dichloride..............	SCl_2	19	1.606	1908	48
Sulfur monobromide...........	S_2Br_2	20	2.6355	1903	49
Sulfur monochloride...........	S_2Cl_2	20	1.678	1880	3
Sulfur trioxide................	SO_3	20.46	1.9207	1941	50
Disulfur decafluoride...........	S_2F_{10}	0	2.08	1934	51
Sulfuryl chloride..............	SO_2Cl_2	20	1.673	1897	3
Sulfuryl chloride fluoride.......	SO_2FCl	0	1.623	1936	52
Thionyl bromide...............	$SOBr_2$	18	2.68	1893	53
Thionyl chloride...............	$SOCl_2$	20	1.638	1880	3
Thiocarbonyl tetrabromide......	$CSBr_4$	20	3.0240	1929	54
Thiocarbonyl tetrachloride......	$CSCl_4$	20	1.6996	1929	54
Trithiocarbonic acid...........	H_2CS_3	17	1.47	1928	25
Pyrosulfurylchloride...........	$S_2O_5Cl_2$	20	1.837	1912	17
Thallium-mercury amalgam......	Tl_2Hg_5	25	12.94	1928	55
Titanium tetrachloride..........	$TiCl_4$	20	1.730	1932	47
Tungsten hexafluoride..........	WF_6	19	3.419	1931	32
Vanadium oxytrichloride........	$VOCl_3$	20	1.828	1910	3

For general references, see Sec. 21-4.
For data on molten optical glasses, see: L. Sharsis, and S. Spinner, *J. Research Natl. Bur. Standards* **46**, 176 (1951).

References for Table 21-7

1. Moles, C. O. E.: *Z. physik. Chem.* **90,** 87 (1915).
2. Ruff and Plato: *Ber. deut. chem. Ges.* **37,** 673 (1904).
3. Baxter, G. P.: "International Critical Tables," vol. 3, p. 22, McGraw-Hill Book Company, New York, 1928.
4. Ghira: *Gazz. chim.* **23** (II), 8 (1893).
5. Briscoe, Robinson, and Smith: *J. Chem. Soc. (London)* **1927,** 282.
6. Moissan, H.: *Compt. rend.* **112,** 717 (1891).
7. Stein, Vogel, and Ludewig: *J. Am. Chem. Soc.* **76,** 4287 (1954).
8. Davies, C. N.: *J. Chem. Phys.* **14,** 48 (1946).
9. Bruhl: *Liebigs Ann. Chem.* **200,** 173 (1880).
10. Mathews, J. H.: *J. Am. Chem. Soc.* **48,** 562 (1926).
11. Pearson, Robinson, and Trotter: *J. Chem. Soc. (London)* **1932,** 660.
12. Briscoe, Peel, and Robinson: *J. Chem. Soc. (London)* **1929,** 56.
13. Diels and Blumberg: *Ber. deut. chem. Ges.* **41,** 86 (1908).
14. Michielewicz, C.: *Roczniki Chem.* **18,** 718 (1938).
15. Cheesman: *J. Chem. Soc.* **1930,** 36.
16. vanWyk: *Z. anorg. allgem. Chem.* **48,** 1 (1906).
17. Sanger and Riegel: *Proc. Am. Acad. Arts Sci.* **47,** 699 (1912).
18. Simons, J. H., ed.: "Fluorine Chemistry," Academic Press, Inc., New York, 1950 (private communication to editor from C. F. Swinehart and F. J. Burton, Jr.).
19. Schwartz et al.: *Ber. deut. chem. Ges.* **64,** 365 (1931).
20. Laubengayer and Tabern: *J. Phys. Chem.* **30,** 1047 (1926).
21. Dennis and Judy: *J. Am. Chem. Soc.* **51,** 2321 (1929).
22. Hough, E. W., D. M. Mason, and B. H. Sage: *J. Am. Chem. Soc.* **72,** 5774 (1950).
23. Butler, K. H., and O. Maass: *J. Am. Chem. Soc.* **52,** 2184 (1930).
24. Simons, J. H., and J. W. Bouknight: *J. Am. Chem. Soc.* **54,** 129 (1932); **55,** 1458 (1933).
25. Mills and Robinson: *J. Chem. Soc. (London)* **1928,** 2326.
26. Maass, O., and W. H. Hatcher: *J. Am. Chem. Soc.* **42,** 2548 (1920).
27. Bloch and Höhn: *Ber. deut. chem. Ges.* **41,** 1971 (1908).
28. Ruff and Keim: *Z. anorg. allgem. Chem.* **193,** 183 (1930).
29. Ruff and Braida: *Z. anorg. allgem. Chem.* **206,** 63 (1932); **214,** 91 (1933).
30. Mond and Langer: *J. Chem. Soc. (London)* **59,** 1090 (1891).
31. Friedrich: *Ber. deut. chem. Ges.* **26,** (1893).
32. Ruff and Ascher: *Z. anorg. allgem. Chem.* **196,** 419 (1931).
33. Ogawa: *Bull. Chem. Soc. (Japan)* **6,** 302 (1931).
34. Ruff and Kwasnik: *Z. anorg. allgem. Chem.* **219,** 65 (1934).
35. Cameron and Macallan: *Chem. News* **59,** 219 (1889).
36. Divers and Shimose: *Ber. deut. chem. Ges.* **17,** 858 (1884).
37. Parker and Robinson: *J. Chem. Soc. (London)* **1931,** 1314.
38. Prideaux and Fox: *J. Chem. Soc. (London)* **1928,** 739.
39. Prideaux and Fox: *J. Chem. Soc. (London)* **1928,** 1603.
40. Gattermann: *Ber. deut. chem. Ges.* **22,** 193 (1880).
41. Reynolds: *J. Chem. Soc. (London)* **51,** 590 (1887).
42. Ruff and Albert: *Ber. deut. chem. Ges.* **38,** 53 (1905).
43. Ruff: *Ber. deut. chem. Ges.* **41,** 3738 (1908).
44. Stock and Somieski: *Ber. deut. chem. Ges.* **49,** 111 (1916).
    Stock, Somieski, Wintgen, and Ebenda: *Ber. deut. chem. Ges.* **50,** 1739 (1917).
45. Pohland: *Z. anorg. allgem. Chem.* **201,** 272 (1931).
46. Robinson and Smith: *J. Chem. Soc. (London)* **1926,** 1262, 3153.
47. Ulrich, Hertel, and Nespital: *Z. physik. chem.* (B) **17,** 372 (1932).
48. Beckmann: *Z. physik. Chem.* **65,** 289 (1908).
49. Ruff and Winterfeld: *Ber. deut. chem. Ges.* **36,** 2437 (1903).
50. Westrink, R.: Acad. Proefschrift Univ. Amsterdam, N.V. Drukkerj en Uitgeversgaak de Mercuur, Hilversum, 1941.
51. Denbigh and Whytlaw-Gray: *J. Chem. Soc. (London)* **1934,** 1346.
52. Booth and Herrmann: *J. Am. Chem. Soc.* **58,** 63 (1936).
53. Hartog and Sims: *Chem. News* **67,** 82 (1893).
54. Briscoe, Peel, and Robinson: *J. Chem. Soc. (London)* **1929,** 1048.
55. Biltz and Meyer: *Z. anorg. allgem. Chem.* **176,** 40 (1928).

TABLE 2l-8. LIQUID DENSITIES OF ELEMENTARY AND INORGANIC SUBSTANCES WHICH ARE NORMALLY GASEOUS UNDER STANDARD CONDITIONS
(Range below 0°C; ρ = g/ml; t = °C; pressure atmospheric*)

Substance	Formula	t	ρ	Year	Ref.
Air.................	20.9% oxygen	−194	0.92	....	1
	53.6% oxygen	−194	1.015	....	1
	72.15% oxygen	−194	1.068	....	1
	94.4% oxygen	−194	1.133	....	1
Ammonia...........	NH_3	−40	0.6900	1923	24
Argon..............	Ar	−189.38 (T.P.)	1.4195	1940	2
		−183.15	1.3740	1912	14
Boron trifluoride.....	BF_3	−101.0 (N.B.P.)	1.595	1932	3
Diborane...........	B_2H_6	−108.2	0.4542	1941	4
Carbon dioxide.......	CO_2	−56.6 (T.P.)	1.179	1928	5
Carbon monoxide.....	CO	−195.08 (ortho-baric)	0.80640	1936	6
Chlorine............	Cl_2	−33.7 (N.B.P.)	1.568	1909	7
		−40	1.574	1926	8
Fluorine............	F_2	−195.94	1.562_2	1954	9
Hydrogen bromide....	HBr	−68.7	2.157	1906	10
Hydrogen chloride....	HCl	−85.8	1.1937	1906	10
Hydrogen iodide......	HI	−35.7 (N.B.P)	2.799	1906	10
Hydrogen selenide....	H_2Se	−42	2.12	1902	13
		−27 (orthobaric)	1.961	1932	12
Hydrogen sulfide.....	H_2S	−60.1 (N.B.P.)	0.964	1906	10
		−63	0.9539	1932	11
Hydrogen telluride....	H_2Te	−17.7	2.701	1932	12
Krypton............	Kr	−157.21 (T.P.)	2.4525	1940	2
Neon...............	Ne	−245.9 (N.B.P.)	1.204	1915	14
Nitric oxide.........	NO	−150.2 (N.B.P.)	1.269	1910	16
		−153.6	1.227	1932	17
Nitrogen............	N_2	−195.84 (N.B.P.)	0.808_4	1915	14
		−198.3	0.8297	1902	15
Nitrous oxide........	N_2O	−89.4 (N.B.P.)	1.2257	1904	18
Dinitrogen trioxide...	N_2O_3	−8	1.464	1888	19
Nitrogen tetroxide....	N_2O_4	−5	1.5035	1888	19
Oxygen.............	O_2	−182.97 (N.B.P.)	1.144_7	1911	14
		−182.5	1.1181	1904	20
		−195.0	1.1953	1930	21
Ozone..............	O_3	−112.4 (N.B.P.)	1.63	1924	14
Radon..............	Rn	−62 (N.B.P.)	4.40	1912	14
Silicane............	SiH_4	−185	0.68	1916	22
Disilicane...........	Si_2H_6	−25	0.69	1916	22
Sulfur dioxide........	SO_2	−10	1.4601	1899	22
Uranium hexafluoride.	UF_6	−209.11 (T.P.)	3.630	1949	23
Xenon..............	Xe	−111.80 (T.P.)	3.0506	1932	2
		−106.9 (N.B.P.)	3.063	1912	14

N.B.P. = normal boiling point; T.P. = triple point.
See Table 2l-9 for liquid hydrogen and liquid helium.
* Unless specified as orthobaric (i.e., corresponding to thermodynamic equilibrium of coexistent liquid and vapor phases) or T.P.

References for Table 2l-8

1. Landolt-Börnstein: "Physikalisch-Chemische Tabellen," 5th ed., Springer-Verlag, Berlin, 1923 (Edwards Bros., Inc., Ann Arbor, Mich., 1943).
2. Clusius, K., and K. Wiegand: *Z. physik. Chem.* **B46**, 1 (1940).
3. Ruff, O., A. Braida, O. Breitschneider, W. Menzel, and H. Plaut: *Z. anorg. allgem. Chem.* **206**, 59 (1932).
4. Laubengayer, A. W., R. P. Ferguson, and A. W. Newkirk: *J. Am. Chem. Soc.* **63**, 559 (1941).
5. Keyes, F. G.: "International Critical Tables," vol. 3, p. 235, McGraw-Hill Book Company, New York, 1928.
6. Timmermans, J.: "Physico-chemical Constants of Pure Organic Compounds," p. 366, American Elsevier Publishing Company, Inc., New York, 1950; Mathias, E., and C. A. Crommelin: Comm. Inst. Intern. Froid. (le comm. 12th rap.) 1936.
7. Johnson, F. M. G., and D. McIntosh: *J. Am. Chem. Soc.* **31**, 1138 (1909).
8. Van Aubel: *Bull. acad. belges* **12** (5), 374 (1926).
9. White, D., J. H. Hu, and H. L. Johnston: *J. Am. Chem. Soc.* **76**, 2584 (1954).
10. McIntosh, S., B. D. Steele, and E. H. Archibald: *Z. physik. Chem.* **55**, 129 (1906).
11. Klemenc, A., and O. Bankowski: *Z. anorg. allgem. Chem.* **208**, 348 (1932); *Z. Elektrochem.* **38**, 592 (1932).
12. Robinson, P. L., and W. E. Scott: *J. Chem. Soc. (London)* **1932**, 972.
13. Fonzes-Diacon, H.: *Compt. rend.* **134**, 171 (1902).
14. Porter, A. W.: "International Critical Tables," vol. 3, p. 20, McGraw-Hill Book Company, New York, 1928.
15. Inglis and Coates: *J. Chem. Soc. (London)* **89**, 886 (1902).
16. Adwentowski: *Chem. Zentr.* **1910i**, 1107.
17. Cheesman, G. H.: *J. Chem. Soc. (London)* **1932**, 889.
18. Grunmach: *Berlin Sitzber.* **1904**, 1198; *Ann. Physik* **15**, 401 (1904).
19. Geuther: *Liebigs Ann. Chem.* **245**, 96 (1888).
20. Dewar: *Proc. Roy. Soc. (London)* **73**, 251 (1904).
21. Biltz, Fischer, and Wunnenberg: *Z. anorg. u. allgem. Chem.* **193**, 358 (1930).
22. Baxter, G. P.: "International Critical Tables," vol. 3, p. 22, McGraw-Hill Book Company, New York, 1928.
23. Hoge, J., and M. T. Wechsler: *J. Chem. Phys.*, **17**, 617 (1949).
24. Timmermans, J.: *Bull. soc. belges* **32**, 299 (1923).

TABLE 2l-9. DENSITIES OF CRYOGENIC LIQUIDS

Helium (isotope 4)[a]

N.B.P. = 4.216 K

T = 4.20 K; pressure = 1 atm; ρ = 0.1251 g/ml

T_λ = 2.186 K; pressure = 38.3 mm Hg; ρ = 0.1462 g/ml

T_λ = 2.178 K; pressure = 1 atm; ρ = 0.1473 g/ml

Helium (isotope 3)[b]

T = 3.20 K (N.B.P.); ρ = 0.0570 g/ml

Normal hydrogen (isotope 1)[c]

T = 20.39 K (N.B.P.); ρ = 0.07098 g/ml

Parahydrogen (isotope 1)[d]

T = 20.27 K (N.B.P.); ρ = 0.07076 g/ml

Hydrogen deuteride (HD)[e]

T = 16.604 K (T.P.); pressure = 92.8 mm Hg; ρ = 0.1234 g/ml

Deuterium (isotope 2)[f]

T = 18.72 K (T.P.); pressure = 128.5 mm Hg; ρ = 0.1739 g/ml

Tritium (isotope 3)[g]

T = 25.04 K (N.B.P.); ρ = 0.2571 g/ml

General reference: R. B. Stewart and V. J. Johnson, eds., "A Compendium of the Properties of Materials at Low Temperature" (Phase I), Wright Air Force Development Division Technical Report, 1961.
N.B.P. = normal boiling point; T.P. = triple point.
For a discussion of the provisional temperature scale in the liquid-hydrogen region see ref. e and H. J. Hoge and F. G. Brickwedde, *J. Research Natl. Bur. Standards* **22**, 351 (1939).
[a] W. H. Keesom, "Helium," pp. 207, 226, 240, American Elsevier Publishing Company, Inc., New York, 1940.   The temperature scale in the liquid-helium range must be considered when evaluating the reported results of density measurements.   A discussion of the problems involved and the most recent conventions adopted is given in C. F. Squire, "Low Temperature Physics,": p. 25, McGraw-Hill Book Company, 1953.
[b] E. R. Grilly, E. F. Hammel, and S. G. Sydoriak, *Phys. Rev.* **75**, 1103 (1949).   Interpolated value by private communication.
[c] R. B. Scott and F. G. Brickwedde, *J. Research Natl. Bur. Standards* **19**, 237 (1937).
[d] R. B. Scott and F. G. Brickwedde, *J. Research Natl. Bur. Standards* **19**, 237 (1937).
[e] H. W. Woolley, R. B. Scott, and F. G. Brickwedde, *J. Research Natl. Bur. Standards* **41**, 379 (1948).
[f] K. Clusius and E. Bartholome, Z. *phys. Chem.* **B30**, 1237 (1935).
[g] E. R. Grilly, *J. Am. Chem. Soc.* **73**, 5307 (1951).   Interpolated value by private communication.

TABLE 2l-10. DENSITIES OF SELECTED ORGANIC LIQUIDS
(Range 0 to 25°C; ρ = g/ml*; t = °C; pressure atmospheric)

Substance	Formula	t	ρ	Year	Ref.
Acetic acid.............	$CH_3 \cdot CO_2H$	20	1.04926	1930	1
Acetone...............	$CH_3 \cdot CO \cdot CH_3$	20	0.79053	1930	2
Alcohol, amyl..........	$C_5H_{11}OH$	15	0.81837	1932	3
Alcohol, n-butyl........	C_4H_9OH	25	0.80567	1943	4
Alcohol, ethyl†					
Alcohol, methyl‡					
Alcohol, n-propyl.......	C_3H_7OH	20	0.8035	1949	5
Alcohol, isopropyl......	$(CH_3)_2 \cdot CHOH$	25	0.78087	1935	6
Aniline................	$C_6H_5NH_2$	20	1.02173	1949	7
Benzene...............	C_6H_6	20	0.87903	1946	8
Bromobenzene.........	C_6H_5Br	20	1.49519	1930	9
Bromoform............	$CHBr_3$	20	2.8905	1935	10
Carbon disulfide........	CS_2	20	1.2632	1926	11
Carbon tetrachloride....	CCl_4	20	1.5940	1938	12
Chloroform............	$CHCl_3$	20	1.48913	1930	9
Chlorobenzene.........	C_6H_5Cl	20	1.10617	1930	9
Cyclohexane...........	C_6H_{12}	20	0.77853	1946	8
Cyclopentane..........	C_5H_{10}	20	0.74538	1946	8
Diethyl ether..........	$(C_2H_5)_2O$	15	0.71925	1928	13
Ethyl acetate.........	$CH_3 \cdot CO_2C_2H_5$	25	0.89468	1937	14
Ethyl formate.........	$H \cdot CO_2C_2H_5$	15	0.92892	1932	15
Formic acid...........	$H \cdot CO_2H$	15	1.22647	1930	1
Glycerol (glycerin)......	$CH_2OH \cdot CHOH \cdot CH_2OH$	20	1.2613	1937	16
Glycol, ethylene........	$(CH_2OH)_2$	15	1.11710	1935	17
n-Heptane.............	$n\text{-}C_7H_{16}$	20	0.68367	1946	8
Heptene-1.............	$CH_2{=}CH{-}(CH_2)_4{-}CH_3$	20	0.6972	1946	19
n-Hexane..............	$n\text{-}C_6H_{14}$	20	0.6595	1946	18
Hexene-1..............	$CH_2{=}CH{-}(CH_2)_3{-}CH_3$	20	0.6736	1946	19
Hydrogen cyanide......	HCN	20	0.6876	1932	20
Iodobenzene...........	C_6H_5I	30	1.81548	1932	21
Isoprene..............	$CH_2{=}C(CH_3){-}CH{=}CH_3$	20	0.6805	1936	22
Methyl formate........	$H \cdot CO_2CH_3$	20	0.97421	1930	1
Methyl iodide..........	CH_3I	15	2.29300	1934	23
Nicotine...............	$C_5H_4N \cdot C_4H_7N(CH_3)$	20	1.0093	1925	24
Nitrobenzene..........	$C_6H_5NO_2$	25	1.1983	1944	25
Nitroglycerin..........	$NH_2NO_3 \cdot CHNO_3 \cdot CH_2NO_3$	15	1.5964	1930	26
n-Nonane.............	$n\text{-}C_9H_{20}$	20	0.7174	1946	18
n-Octane..............	$n\text{-}C_8H_{18}$	20	0.70252	1946	8
n-Pentane.............	$n\text{-}C_5H_{12}$	20	0.62619	1947	27
Isopentane............	$iso\text{-}C_5H_{12}$	20	0.61963	1947	27
Pentene-1.............	$CH_2{=}CH{-}CH_2{-}CH_2{-}CH_3$	20	0.6406	1946	19
n-Propylbenzene........	$C_6H_5\text{-}n\text{-}C_3H_7$	20	0.8618	1946	28
Toluene...............	$C_6H_5{-}CH_3$	20	0.86683	1946	8

For extensive data on hydrocarbons see: A. F. Forziati and F. D. Rossini, *J. Research Natl. Bur. Standards* **43**, 473 (1949); A. F. Forziati, D. L. Camin, and F. D. Rossini, *J. Research Natl. Bur. Standards* **45**, 406 (1950).

For general references, see Sec. 2l-4.

* These data have not been corrected in terms of the new definition of the liter (Sec. 2l-1). For the conversion factor, see Table 2l-1.

† See Table 2l-6.

‡ See Table 2l-5.

References for Table 21-10

1. Bureau d'Étalons (International Bureau of Physico-chemical Standards), Brussels, 1930.
2. Zmaczynski, M. A.: *J. chim. phys.* **27**, 503 (1930).
3. Bureau d'Étalons, 1932.
4. Brunjes, A. S., and M. J. P. Bogart: *Ind. Eng. Chem.* **35**, 256 (1943).
5. Hatem, S.: *Compt. rend.* **1949**, 601.
6. Olsen, A. L., and E. R. Washburn: *J. Am. Chem. Soc.* **57**, 303 (1935).
7. Dreisbach, R. R., and R. A. Martin: *Ind. Eng. Chem.* **41**, 2875 (1949).
8. Forziati, A. F., A. R. Glasgow, Jr., C. B. Willingham, and F. D. Rossini: *Natl. Bur. Standards J. Research* **36**, 129 (1946).
9. Zmaczynski, M. A.: *J. chim. phys.* **27**, 503 (1930).
10. Desreux, V.: *Bull. soc. chim. Belges* **44**, 249 (1935).
11. Bureau d'Étalons, 1928.
12. Michielewicz, C.: *Roczniki Chem.* **18**, 718 (1938).
13. Bureau d'Étalons, 1928.
14. Wojciechowski, M., and E. Smith: *Roczniki Chem.* **17**, 118 (1937).
15. Bureau d'Étalons, 1932.
16. Albright, P. S.: *J. Am. Chem. Soc.* **59**, 2098 (1937).
17. Bureau d'Étalons, 1937.
18. Vogel, A. I.: *J. Chem. Soc. (London)* **1946**, 133.
19. Wibaut, J. P., and H. Geldof: *Rec. trav. chim.* **65**, 125 (1946).
20. Lowry, T. M., and S. T. Henderson: *Proc. Roy. Soc. (London)*, ser. A, **136**, 474 (1932).
21. Bureau d'Étalons, 1932.
22. Bekkadahl, N., L. A. Wood, and M. Wojciechowski: *Natl. Bur. Standards J. Research* **17**, 883 (1936).
23. Bureau d'Étalons, 1934.
24. Lowry, T. M., and B. K. Singh: *Compt. rend.* **181**, 909 (1925).
25. Coates, G. E., and J. E. Coates: *J. Chem. Soc. (London)* **1944**, 77.
26. Peterson, J. M.: *J. Am. Chem. Soc.* **52**, 3669 (1930).
27. Howard, F. L., T. W. Mears, A. Fookson, P. Pomerantz, and D. B. Brooks: *Natl. Bur. Standards J. Research* **38**, 365 (1947).
28. Gibbons, L. C., J. F. Thomson, T. W. Reynolds, J. I. Wright, H. H. Chanau, J. M. Lamberti, H. F. Hipsher, and J. V. Karabinas: *J. Am. Chem. Soc.* **68**, 1130 (1946).

TABLE 21-11. DENSITIES OF SELECTED FLUOROCARBON
AND CHLOROFLUORO LIQUIDS
(Range all temperatures; ρ = g/ml; t = °C)

Substance	Index	t	ρ	Year	Ref.
n-Butforane............	C_4F_{10}	20.8	1.47 (orthobaric)	1939	1
Cyclopentforane........	C_5F_{10}	20	1.648	1947	2
Ethforane..............	C_2F_6	−78.2	1.61	1933	3
Ethforene..............	C_2F_4	−76.3	1.519	1933	3
Fluoroform............	CF_3H	−84.4	1.465	1936	4
Freon-11..............	CCl_3F	15	1.4995	1940	5
Freon-12..............	CCl_2F_2	20	1.326 (orthobaric)	1942	6
Freon-13..............	$CClF_3$	−130	1.726	1931	7
Freon-21..............	$CHCl_2F$	15	1.3906 (orthobaric)	1940	5
Freon-22..............	$CHClF_2$	20	1.2130	1940	5
Freon-112.............	$C_2Cl_4F_2$	25	1.6447	1934	8
Hexforanes (mixture)...	C_6F_{14}	20	1.697	1947	2
Methforane............	CF_4	−130	1.62	1933	3
Octforanes (mixture)....	C_8F_{18}	20	1.802	1947	2
n-Pentforane..........	C_5F_{12}	20	1.634	1947	2
Propforane............	C_3F_8	0.2	1.45 (orthobaric)	1939	1
Benzo trifluoride........	$C_6H_5CF_3$	30	1.1762	1953	9
p-Fluorotoluene........	$CH_3C_6H_4F$	30	0.9869	1953	9
p-Fluorobromobenzene..	BrC_6H_4F	30	1.5859	1953	9

General reference: J. H. Simons, ed., "Fluorine Chemistry," vol. I, Academic Press, Inc., New York, 1950.
Pressure is atmospheric unless indicated as orthobaric conditions.

References for Table 21-11

1. Simons, J. H., and L. P. Block: J. Am. Chem. Soc. 61, 2962 (1939); 59, 1407 (1937).
2. Simons, J. H., ed.: "Fluorine Chemistry," vol. I, p. 412, Academic Press, Inc., New York, 1950.
3. Ruff, O., and O. Breitschneider: Z. anorg. allgem. Chem. 210, 173 (1933).
4. Ruff, O., O. Breitschneider, W. Luchsinger, and G. Millschitzky: Ber. 69, 299 (1936).
5. Simons, J. H., ed.: "Fluorine Chemistry," vol. I, Academic Press, Inc., New York, 1950; Benning, A. F., and R. C. McHarness: Ind. Eng. Chem. 31, 912 (1939); 32, 814 (1940).
6. Benning, A. F., and W. H. Markwood, Jr.: "Thermodynamic Properties of Freon 12," Kinetic Chemicals, Inc., Wilmington, Del., 1942.
7. Ruff, O., and R. Keim: Z. anorg. allgem. Chem. 201, 255 (1931).
8. Locke, E. G., W. R. Brode, and A. L. Henne: J. Am. Chem. Soc. 56, 1726 (1934).
9. Rutledge, G. P., and W. T. Smith, Jr.: J. Am. Chem. Soc. 75, 5762 (1953).

21-6. Volume of Liquids as a Function of Pressure and Temperature.   Introduction.
The density of a liquid in equilibrium must be a single-valued function of temperature
and pressure, though this is not necessarily true for solids.   Thermal equilibrium can
be attained in all liquids within a reasonable length of time, but for polymers and
other glass-forming liquids the viscosities may be so high that the reported "equilibrium" values of density often depend on the temperature and pressure history of
the sample.   Based on the extrapolation of observations at higher temperatures, it
is assumed that such systems are approaching a true equilibrium state, but approaching it so slowly that changes cannot be detected.   An example is optical glass near
room temperature.

In the following tables equilibrium[1] specific volumes of selected liquids are listed at
various pressures and temperatures.   These data were all determined by experimental measurement.   The range of pressures is from 1 to 50,000 atm (or to the freezing point).   The range of temperatures is from 0 to 200°C (in a few cases there are
points outside this range).   In addition a compilation of many other liquids for which
high-pressure data are available is given with references.

In the tables the volume as a function of pressure and temperature is expressed in
one of three ways:

1. Specific volume at a pressure and temperature.

2. Relative volume v/v_0, where v is the specific volume at the particular pressure
and temperature and v_0 is the specific volume in the reference state (usually 0°C and
760 mm).   Reference values are taken from the previous tables of this section,
where appropriate data are available.   The others are taken from: J. Timmermans,
"Physico-chemical Constants of Pure Organic Compounds," American Elsevier
Publishing Company, Inc., New York, 1950.

3. The change in volume of a given mass of liquid from a reference pressure of
5,000 kg/cm² along each experimental isotherm.   All the data in the very high
pressure range (5,000 to 50,000 kg/cm²) are expressed in this way.

From these data, isothermal compressibilities may be found (Sec. 21-12) and also
isobaric thermal expansions if the volumes are known for more than one temperature.
Other thermodynamic parameters, such as specific heats, have been directly determined as a function of pressure in very few cases and have to be inferred from the
volume relations by indirect methods.   No attempt is made here to give any of these
values.   References to the general subject of high pressures are cited below.[2]

21-7. Equations of State for Liquids.   From PVT (pressure-volume-temperature)
data equilibrium equations of state may be obtained.   An explicit, meaningful equation of state in closed form applicable over wide ranges of both temperature and
pressure is usually difficult to obtain from experimental data.   An empirical relation
can always be obtained from a statistical fit to a polynomial of the form

$$v = \sum_{i=0}^{\infty} \sum_{j=0}^{\infty} A_{ij} t^i P^j \qquad (21\text{-}5)$$

truncated to delete insignificant terms.   Over more limited ranges, volume isobars
often approximate linear dependence on temperature.   The Tait equation,

$$\frac{v_0 - v}{v_0} = c \log_{10} \left(1 + \frac{P}{b} \right) \qquad (21\text{-}6)$$

---

[1] Except, possibly, for glycerin at low temperatures, or high pressures (Table 21-24).

[2] P. W. Bridgman, "The Physics of High Pressure," G. Bell & Sons, Ltd., London, 1952;
Revs. Modern Phys. **18**, 1 (1946); J. Timmermans, "Les Constantes physiques des composés
organiques cristallisés," Masson et Cie., Paris, 1953.

is probably the most familiar semiempirical relation[1] used to approximate the volume isotherms. A statistical fit is facilitated because of the implication of linearity between bulk modulus and pressure,[2] which can be seen when Eq. (2l-6) is converted to the following form:[3]

$$-v_0 \left(\frac{\partial P}{\partial v}\right)_T = 2.302 \frac{b + P}{c} \tag{2l-7}$$

The constants b and c have been interpreted[4] in terms of the internal pressure and cohesive energy density. At high pressures the applicability of the Tait equation would appear to be limited in view of the fact that it predicts negative volumes at pressures exceeding a certain value. In practice this limitation is usually not serious because this value is in the order of several hundred million atmospheres.

Theoretical relations based on molecular structure are not very reliable in estimating PVT behavior. Considerable discussion on PVT estimation methods and their reliability is given by Bondi.[5]

21-8. Experimental Methods for Studying Compressibility of Liquids. A thorough description of the techniques employed in the experimental determination of the volume of liquids as a function of pressure and temperature is given by P. W. Bridgman in his text.[6] Even more extensive details are given in his original publications. The sylphon method, which was used in the pressure range 1 to 12,000 atm, is described in *Proc. Am. Acad. Arts Sci.* **66**, 185 (1931). The differential method, which was used in the pressure range 5,000 to 50,000 atm, is described in *Proc. Am. Acad. Arts Sci.* **74**, 21 (1940), and **74**, 399 (1942).

For more recent surveys with respect to liquids, see W. A. Steel and W. Webb, chap. 4i in "High Pressure Physics and Chemistry," vol. 1, R. S. Bradley, ed., Academic Press, Inc., New York, 1963. Recently the specific volumes of water[7] and mercury[8] have been evaluated from velocity of sound data. This involves the evaluation of the adiabatic compressibility as usual (see Sec. 2l-12). After the adiabatic compressibility is converted to the corresponding isothermal, the specific volume is evaluated by numerical integration. The data obtained from this technique are claimed to be even more accurate than those obtained directly from equilibrium measurements. Shock waves[9] are often used to obtain data at high pressures. For a review of nonequilibrium methods commonly used on polymer and glass-forming liquids, see R. S. Marvin and J. E. McKinney, chap. 9 in "Physical Acoustics," vol. 2B, W. P. Mason, ed., Academic Press, Inc., New York, 1965.

21-9. General Features of the Behavior of Liquids under Pressure[10]

1. Mercury is the least compressible of all liquids (in the range -30 to 200°C). In the nonmetallic group glycerin is the least compressible liquid.

2. At pressures above 10,000 atm, the relative volume change for all liquids is about the same.

[1] For critiques and comparative studies, see: J. R. MacDonald, *Revs. Modern Phys.* **38**, 669 (1966); A. T. J. Hayward, *Brit. J. Appl. Phys.* **18**, 965 (1967).
[2] L. A. Wood, *J. Polymer Sci. B* (Polymer Letters) **2**, 703 (1964).
[3] This definition of the bulk modulus is not fully equivalent to that given in Sec. 2l-12 (Eq. 2l-9) at isothermal conditions.
[4] M. A. Cook and L. A. Rogers, *J. Appl. Phys.* **34**, 2330 (1963).
[5] A. Bondi, "Physical Properties of Molecular Crystals, Liquids, and Glasses," Chap. 8, John Wiley & Sons, Inc., New York, 1968.
[6] P. W. Bridgman, "The Physics of High Pressure," *op. cit.*
[7] R. Vedam and G. Holton, *J. Acoust. Soc. Am.* **43**, 108 (1968).
[8] L. A. Davis and R. B. Gordon, *J. Chem. Phys.* **46**, 2650 (1967).
[9] G. E. Duvall and G. R. Fowles, Chap. 9 in "High Pressure Physics and Chemistry," vol. 2, R. S. Bradley, ed., Academic Press, Inc., New York, 1963.
[10] P. W. Bridgman, "The Physics of High Pressure," *op. cit.*; W. A. Steel and W. Webb, chap. 4i in "High Pressure Physics and Chemistry," vol. 1, *op. cit.*

3. The melting curve appears to exist up to the highest pressures experimentally obtainable with no indication of either a critical point or a maximum. However, at very high pressures, the viscosity of some liquids becomes so large that the approach to true equilibrium takes place infinitely slowly and a subcooling phenomenon appears.

4. The difference in volume between the solid and the liquid phase tends to decrease with increasing pressure but does not seem to approach zero at any finite pressure.

5. Differences in specific volumes among isomers (i.e., compounds having the same chemical formula but different structural formulas) tend to disappear at around 12,000 atm.

6. The compressibility decreases with increasing pressure, the rate of decrease becoming smaller with increasing pressure.

7. The compressibility increases with increasing temperature; this increase is less at higher pressures.

8. The volume isobars are more nearly linear than the isotherms; however, the bulk modulus often approximates a linear function of pressure. (This dependence is predicted by the Tait equation. See previous section.)

9. The sign of $(\partial^2 v/\partial T^2)_p$ changes from plus to minus with increasing temperature at constant pressure at pressures above about 3,000 to 4,000 atm.

10. The quantity $(\partial p/\partial T)_v$ decreases with increasing volume, but is not a function of volume alone.

11. The isothermal compressibility is unbounded at the critical point.[1]

21-10. Units and Conversion Factors.   The SI unit of pressure is the newton per square meter, and the cgs unit is the dyne per square centimeter. The bulk modulus is expressed in pressure units, and the compressibility, in reciprocal pressure units. Table 21-12 gives the conversion factors for the pressure units.

[1] See for example: P. A. Egelstaff, "An Introduction to the Liquid State," chap. 15, Academic Press, Inc., New York and London, 1967; P. A. Egglestaff and J. W. Ring, in "Physics of Simple Liquids," H. V. N. Temperley, J. S. Rowlinson, and G. S. Ruchbrooke, eds., North-Holland Publishing Company, Amsterdam, 1968.

TABLE 2l-12. CONVERSION FACTORS FOR PRESSURE OR BULK MODULUS UNITS*

Units	N/m² (SI)	bar	atm	cm Hg	dyn/cm² (cgs)	kg/cm²	psi	torr
1 N/m² (SI) ...	1	10^{-5}	9.86923×10^{-6}	7.50064×10^{-4}	10	1.01972×10^{-5}	1.45038×10^{-4}	7.50064×10^{-3}
1 bar.........	10^5	1	0.986923	75.0064	10^6	1.01972	14.5038	7.50064×10^2
1 atm.........	1.01325×10^5	1.01325	1	76.0002	1.01325×10^6	1.03323	14.6960	7.60002×10^2
1 cm Hg (0°C).	1.33322×10^3	1.33322×10^{-2}	1.31579×10^{-2}	1	1.33322×10^4	1.35951×10^{-2}	0.193367	10
1 dyn/cm² (cgs)	10^{-1}	10^{-6}	9.86923×10^{-7}	7.50064×10^{-5}	1	1.01972×10^{-6}	1.45038×10^{-5}	7.50064×10^{-4}
1 kg/cm²......	9.80665×10^4	0.980665	0.967841	73.5561	9.80665×10^5	1	14.2233	7.35561×10^2
1 psi.........	6.89476×10^3	6.89476×10^{-2}	6.80460×10^{-2}	5.17151	6.89476×10^4	7.03070×10^{-2}	1	51.7151
1 torr (mm Hg at °C)......	1.33322×10^2	1.33322×10^{-3}	1.31579×10^{-3}	10^{-1}	1.33322×10	1.35951×10^{-3}	1.93367×10^{-2}	1

* Conversion to SI units taken from ASTM Metric Practice Guide, U.S. Department of Commerce, *Natl. Bur. Standards Handbook* **102**, 39 (Mar. 10, 1967).

TABLE 2l-13. SPECIFIC VOLUME OF PURE AIR-FREE H_2O AS A FUNCTION OF
PRESSURE AND TEMPERATURE

Temp. range −20 to 100°C;* pressure range 1–12,000 kg/cm²; specific volume
in ml/g

$p,$ kg/cm²	−20°C	−15°C	−10°C	−5°C	0°C	20°C	40°C	60°C	80°C	100°C
1	......	......	......	......	1.0001	1.0018	1.0079	1.0171	1.0284	1.0435
500	......	......	......	......	0.9770	0.9819	0.9880	0.9959	1.0063	1.0183
1,000	......	......	......	0.9566	0.9576	0.9632	0.9706	0.9786	0.9883	0.9993
1,500	......	0.9370	0.9380	0.9394	0.9409	0.9476	0.9550	0.9632	0.9724	0.9826
2,000	0.9203	0.9214	0.9228	0.9246	0.9261	0.9328	0.9408	0.9492	0.9582	0.9679
2,500	0.9061	0.9080	0.9097	0.9116	0.9132	0.9199	0.9282	0.9365	0.9453	0.9545
3,000	......	0.8959	0.8977	0.9000	0.9015	0.9084	0.9167	0.9248	0.9334	0.9424
3,500	......	0.8851	0.8871	0.8892	0.8909	0.8984	0.9062	0.9142	0.9225	0.9312
4,000	......	......	0.8771	0.8794	0.8812	0.8888	0.8966	0.9044	0.9126	0.9208
5,000	......	......	0.8596	0.8622	0.8639	0.8709	0.8796	0.8874	0.8949	0.9028
6,000	......	......	......	......	0.8489	0.8565	0.8645	0.8721	0.8794	0.8871
7,000	......	......	......	......	......	......	0.8515	0.8586	0.8659	0.8731
8,000	......	......	......	......	......	......	0.8396	0.8564	0.8534	0.8604
9,000	......	......	......	......	......	......	0.8287	0.8354	0.8422	0.8490
10,000	......	......	......	......	......	......	0.8186	0.8252	0.8318	0.8385
11,000	......	......	......	......	......	......	0.8090	0.8157	0.8222	0.8385
12,000	......	......	......	......	......	......	0.8006	0.8070	0.8134	0.8199

TABLE 2l-13. RELATIVE OR DIFFERENTIAL VOLUME OF PURE AIR-FREE H_2O AS A FUNCTION OF PRESSURE AND TEMPERATURE (*Continued*)

Temp. range 25–175°C;† pressure range 5,000–36,560 kg/cm²; Δv in cm²/1.000 g from 5,000 kg/cm²

p, kg/cm²	25°C	75°C	125°C	175°C
5,000	0.000	0.000	0.000	0.000
9,800	0.057			
10,000		0.063	0.066	0.070
15,000	Ice VI	0.105	0.112	0.120
20,000		0.136	0.146	0.157
21,430		0.144		
25,000			0.173	0.185
28,140		Ice VII	0.186	
30,000				0.207
35,000			Ice VII	0.226
36,560				0.231
				Ice VII

Temp. range 0–360°C;‡ pressure range 1–350 atm; specific volume in ml/g

p, atm	0°C	20°C	40°C	60°C	80°C	100°C	120°C	140°C
1	1.0002	1.0020	1.0079	1.0170	1.0289	1.0434		
25	0.9991	1.0009	1.0068	1.0159	1.0277	1.0421	1.0590	1.0785
50	0.9980	0.9998	1.0057	1.0147	1.0265	1.0408	1.0576	1.0769
75	0.9968	0.9987	1.0046	1.0136	1.0253	1.0396	1.0562	1.0754
100	0.9957	0.9976	1.0034	1.0124	1.0241	1.0383	1.0548	1.0738
125	0.9946	0.9965	1.0024	1.0113	1.0230	1.0370	1.0535	1.0723
150	0.9935	0.9955	1.0013	1.0102	1.0218	1.0358	1.0521	1.0708
175	0.9935	0.9944	1.0002	1.0091	1.0207	1.0346	1.0508	1.0694
200	0.9914	0.9934	0.9992	1.0080	1.0195	1.0334	1.0495	1.0679
250	0.9893	0.9913	0.9971	1.0059	1.0173	1.0310	1.0469	1.0650
300	0.9873	0.9893	0.9950	1.0038	1.0151	1.0286	1.0444	1.0622
350	0.9853	0.9873	0.9930	1.0017	1.0129	1.0264	1.0419	1.0595

p, atm	160°C	180°C	200°C	220°C	240°C	260°C	280°C	300°C	320°C
1									
25	1.1007	1.1262	1.1555	1.1897					
50	1.0989	1.1241	1.1530	1.1866	1.2264	1.2747	1.3285		
75	1.0972	1.1221	1.1506	1.1836	1.2225	1.2694	1.3213		
100	1.0954	1.1200	1.1482	1.1806	1.2187	1.2644	1.3146	1.3965	
125	1.0937	1.1181	1.1458	1.1778	1.2150	1.2596	1.3082	1.3860	1.4882
150	1.0920	1.1161	1.1435	1.1749	1.2115	1.2549	1.3020	1.3764	1.4712
175	1.0904	1.1142	1.1412	1.1722	1.2080	1.2505	1.2962	1.3675	1.4563
200	1.0887	1.1123	1.1390	1.1694	1.2047	1.2461	1.2962	1.3591	1.4428
250	1.0855	1.1086	1.1346	1.1642	1.1982	1.2379	1.2854	1.3438	1.4192
300	1.0824	1.1050	1.1304	1.1592	1.1921	1.2302	1.2754	1.3303	1.3992
350	1.0793	1.1015	1.1263	1.1544	1.1862	1.2230	1.2662	1.3181	1.3816

TABLE 2l-13. SPECIFIC VOLUME OF PURE AIR-FREE H_2O AS A FUNCTION OF
PRESSURE AND TEMPERATURE (*Continued*)

Temp. range 0–360°C;‡ pressure range 1–350 atm; specific volume in ml/g

p, atm	340°C	360°C
1		
150	1.6287	
175	1.5943	
200	1.5671	1.8140
250	1.5243	1.6905
300	1.4908	1.6232
350	1.4631	1.5758

For more extensive collected data, see G. C. Kennedy, and W. T. Holser, sec. 16 in "Handbook of Physical Constants," S. P. Clark, Jr., ed., The Geological Society of America, 1966.

For data measured by acoustic techniques (Section 2l-8), see R. Vedam, and G. Holton, *loc. cit.*

* N. F. Dorsey, "Properties of Ordinary Water Substance," Reinhold Book Corporation, New York, 1948. Based on data of P. W. Bridgman, *J. Chem. Phys.* **3**, 597 (1936). See Dorsey for a further discussion of the factors involved in the interpretation of these data.

† P. W. Bridgman, *Proc. Am. Acad. Arts. Sci.* **74**, 419 (1942). These data were taken directly from the original publication.

‡ L. B. Smith and F. G. Keyes, *Proc. Am. Acad. Arts Sci.* **69**, 285 (1934). See Dorsey for a comment on these data.

Temp. range 200–1000°C;* pressure range 100–2,500 bars; specific volume in cm^3/g

Temp., °C	Pressure, bars						
	100	200	500	1,000	1,500	2,000	2,500
200	1.14830	1.13899	1.1145	1.0811	1.0533	1.0258	1.0027
300	1.39704	1.35992	1.2869	1.2131	1.1639	1.1257	1.0946
400	26.31	9.96	1.745	1.4443	1.3284	1.2591	1.2092
500	32.35	14.77	3.890	1.8794	1.5653	1.4402	1.3566
600	37.78	18.11	6.114	2.6802	1.9496	1.6630	1.5252
700	42.517	20.973	7.7651	3.5829	2.449	1.980	1.7346
800	46.082	23.391	9.0925	4.4338	2.994	2.350	2.000
900	49.54	25.74	10.28	5.208	3.531	2.738	2.296
1000	52.90	27.84	11.30	5.900	4.035	3.123	2.589

* G. C. Kennedy, *Am. J. Sci.* **248**, 540 (1950).

TABLE 2l-14. SPECIFIC VOLUME OF 99.9% D_2O AS A FUNCTION OF PRESSURE AND TEMPERATURE*

Temp. range −20 to 100°C; pressure range 1–12,000 kg/cm²; specific volume in ml/g

p, kg/cm²	−20°C	−15°C	−10°C	−5°C	0°C	20°C	40°C	60°C	80°C	100°C
1	......	......	......	......	0.9048	0.9049	0.9087	0.9169	0.9272	
500	......	......	......	......	0.8833	0.8857	0.8905	0.8979	0.9074	0.9187
1,000	......	......	......	0.8642	0.8652	0.8690	0.8744	0.8820	0.8912	0.9011
1,500	......	......	0.8475	0.8485	0.8495	0.8543	0.8605	0.8680	0.8769	0.8864
2,000	......	0.8318	0.8331	0.8344	0.8359	0.8415	0.8479	0.8553	0.8639	0.8731
2,500	0.8178	0.8193	0.8208	0.8222	0.8239	0.8298	0.8365	0.8440	0.8521	0.8613
3,000	0.8066	0.8082	0.8099	0.8116	0.8132	0.8194	0.8260	0.8335	0.8413	0.8502
3,500	......	0.7982	0.8001	0.8019	0.8036	0.8096	0.8165	0.8240	0.8317	0.8400
4,000	......	0.7892	0.7910	0.7928	0.7946	0.8009	0.8078	0.8153	0.8227	0.8305
5,000	......	......	......	0.7772	0.7789	0.7854	0.7924	0.7996	0.8064	0.8143
6,000	......	......	......	......	0.7665	0.7722	0.7787	0.7860	0.7926	0.8000
7,000	......	......	......	......	......	0.7597	0.7668	0.7736	0.7801	0.7870
8,000	......	......	......	......	......	0.7490	0.7559	0.7625	0.7690	0.7755
9,000	......	......	......	......	......	0.7391	0.7461	0.7526	0.7588	0.7653
10,000	......	......	......	......	......	......	0.7373	0.7432	0.7493	0.7558
11,000	......	......	......	......	......	......	0.7293	0.7348	0.7407	0.7470
12,000	......	......	......	......	......	......	0.7216	0.7271	0.7328	0.7393

* P. W. Bridgman, *J. Chem. Phys.* **3**, 597 (1936). These values were calculated from the original data assuming a molecular weight of 20.028 (chemical scale) for D_2O.

TABLE 2l-15. SPECIFIC VOLUME OF MERCURY AS A FUNCTION OF PRESSURE AND TEMPERATURE

Temp. range −30 to 20°C;* pressure range 1–12,000 atm; specific volume in ml/g†

p, atm	−30°C	−20°C	−10°C	0°C	10°C	20°C
1	0.073155	0.073288	0.073421	0.073554	0.073687	0.073820
1,000	0.072888	0.073016	0.073143	0.073270	0.073397	0.073524
2,000	0.072626	0.072748	0.072871	0.072993	0.073115	0.073237
3,000		0.072487	0.072605	0.072724	0.072842	0.072961
4,000		0.072233	0.072348	0.072463	0.072579	0.072696
5,000			0.072101	0.072213	0.072372	0.072440
6,000			0.071863	0.071973	0.072085	0.072196
7,000				0.071744	0.071853	0.071962
8,000					0.071632	0.071740
9,000					0.071422	0.071528
10,000					0.071223	0.071328
11,000						0.071140
12,000						0.070962

For critique, see K. E. Bett, K. E. Weale, and D. M. Newitt, *Brit. J. Appl. Phys.* **5**, 243 (1954).
For data measured by acoustic techniques (Sec. 2l-8), see L. A. Davis and R. B. Gordon, *J. Chem. Phys.* **46**, 2650 (1967).
* P. W. Bridgman, *Proc. Am. Acad. Arts Sci.* **47**, 345 (1911). These values were calculated from the original data, assuming the density of mercury at 0°C and 760 mm to be 13.5955 g/ml.
† These data have not been corrected in terms of the new definition of the liter (Sec. 2l-1). For the conversion factor, see Table 2l-1.

TABLE 21-16. RELATIVE OR DIFFERENTIAL VOLUME OF METHYL ALCOHOL (CH₃OH)
AS A FUNCTION OF PRESSURE AND TEMPERATURE
The Relative Volumes in Terms of the Volume at 0°C and 760 mm*

Temp. range 20–80°C; pressure range 1–12,000 atm; v/v_0; v = volume at (p,t);
$v_0 = 1.23459$ ml/g at 0°C and 760 mm

p, atm	20°C	40°C	60°C	80°C
1	1.0238	1.0483	1.0737	1.1005
500	0.9811	0.9987	1.0182	1.0400
1,000	0.9494	0.9651	0.9808	0.9993
1,500	0.9256	0.9393	0.9526	0.9672
2,000	0.9064	0.9189	0.9306	0.9429
2,500	0.8906	0.9019	0.9124	0.9231
3,000	0.8763	0.8870	0.8966	0.9065
3,500	0.8636	0.8733	0.8824	0.8915
4,000	0.8523	0.8613	0.8700	0.8782
4,500	0.8420	0.8505	0.8587	0.8663
5,000	0.8325	0.8407	0.8487	0.8559
6,000	0.8163	0.8240	0.8314	0.8381
7,000	0.8023	0.8099	0.8163	0.8231
8,000	0.7907	0.7973	0.8039	0.8102
9,000	0.7797	0.7859	0.7920	0.7981
10,000	0.7696	0.7756	0.7816	0.7875
11,000	0.7605	0.7664	0.7728	0.7785
12,000	0.7527	0.7587	0.7652	0.7709

The Change in Volume in cm³ per 0.792 g from a Reference Pressure of 5,000 kg/cm²
along Each Isotherm†

Temp. range 25–175°C; pressure range 5,000–50,000 kg/cm²; Δv in cm³/0.792 g

p, kg/cm²	25°C	75°C	125°C	175°C
5,000	0.000	0.000	0.000	0.000
10,000	0.062	0.066	0.073	0.082
15,000	0.099	0.106	0.117	0.128
20,000	0.125	0.135	0.139	0.161
25,000	0.145	0.157	0.174	0.187
30,000	0.161‡	0.173	0.194	0.208
35,000	0.173	0.187	0.210	0.226
40,000	0.183	0.198	0.223	0.240
45,000	0.191	0.208	0.234	0.253
50,000	0.199	0.218		

* P. W. Bridgman, "International Critical Tables," vol. 3, p. 41, McGraw-Hill Book Company, New York, 1928.
† P. W. Bridgman, *Proc. Am. Acad. Arts Sci.* **74**, 403 (1942).
‡ Displays subcooling at higher pressures.

TABLE 2l-17. RELATIVE OR DIFFERENTIAL VOLUME OF ETHYL ALCOHOL (C_2H_5OH) AS A FUNCTION OF PRESSURE AND TEMPERATURE
The Relative Volumes in Terms of the Volume at 0°C and 760 mm*

Temp. range 20–80°C; pressure range 1–12,000 atm; v/v_0; v = volume at (p,t); v_0 = 1.24022 ml/g at 0°C and 760 mm

p, atm	20°C	40°C	60°C	80°C
1	1.0212	1.0438	1.0679	1.0934
500	0.9782	0.9943	1.0121	1.0319
1,000	0.9479	0.9608	0.9760	0.9922
1,500	0.9247	0.9358	0.9482	0.9615
2,000	0.9059	0.9159	0.9266	0.9280
2,500	0.8899	0.8991	0.9088	0.9187
3,000	0.8760	0.8848	0.8935	0.9025
3,500	0.8634	0.8718	0.8800	0.8884
4,000	0.8517	0.8599	0.8678	0.8756
4,500	0.8410	0.8491	0.8567	0.8640
5,000	0.8314	0.8394	0.8467	0.8536
6,000	0.8149	0.8225	0.8291	0.8354
7,000	0.8009	0.8080	0.8139	0.8196
8,000	0.7888	0.7953	0.8005	0.8060
9,000	0.7776	0.7836	0.7884	0.7940
10,000	0.7671	0.7726	0.7776	0.7830
11,000	0.7574	0.7626	0.7682	0.7734
12,000	0.7485	0.7535	0.7600	0.7648

The Change in Volume in cm³ per 0.789 g from a Reference Pressure of 5,000 kg/cm²
along Each Isotherm†

Temp. range 25–175°C; pressure range 5,000 kg/cm² to 45,000 kg/cm²; Δv in cm³
per 0.789 g

p, kg/cm²	25°C	75°C	125°C	175°C
5,000	0.000	0.000	0.000	0.000
10,000	0.063	0.069	0.071	0.076
15,000	0.100	0.109	0.113	0.119
20,000	0.128‡	0.137	0.144	0.151
25,000		0.159	0.168	0.175
28,700		0.174‡		
30,000			0.187	0.195
35,000			0.203	0.211
40,000			0.217	0.225
45,000			0.230	0.238

* P. W. Bridgman, "International Critical Tables," vol. 3, p. 41, McGraw-Hill Book Company, New York, 1928.
† P. W. Bridgman, *Proc. Am. Acad. Arts. Sci.* **74**, 399 (1942).
‡ Solid below this.

TABLE 2l-18.   RELATIVE VOLUME OF ACETONE ($CH_2 \cdot CO \cdot CH_3$) AS A FUNCTION OF
PRESSURE AND TEMPERATURE*
The Relative Volumes in Terms of the Volume at 0°C and 760 mm

Temp. range 20–80°C; pressure range 1–12,000 atm; v/v_0; v = volume at (p,t);
v_0 = 1.23077 ml/g at 0°C and 760 mm

p, atm	20°C	40°C	60°C	80°C
1	1.0279	1.0585	1.0925	
500	0.9819	1.0032	1.0282	
1,000	0.9526	0.9706	0.9894	1.0082
1,500	0.9286	0.9441	0.9594	0.9736
2,000	0.9076	0.9217	0.9347	0.9467
2,500	0.8900	0.9028	0.9141	0.9253
3,000	0.8748	0.8868	0.8968	0.9073
3,500	0.8619	0.8729	0.8821	0.8920
4,000	0.8504	0.8607	0.8694	0.8786
4,500	0.8402	0.8498	0.8583	0.8666
5,000	0.8309	0.8398	0.8482	0.8558
6,000	0.8143	0.8225	0.8306	0.8370
7,000	0.7997	0.8072	0.8148	0.8209
8,000	0.7866	0.7935	0.8003	0.8066
9,000	......	0.7815	0.7876	0.7939
10,000	......	0.7707	0.7764	0.7821
11,000	Freezes	0.7607	0.7665	0.7715
12,000	......	0.7515	0.7577	0.7617

* P. W. Bridgman, "International Critical Tables," vol. 3, p. 42, McGraw-Hill Book Company,
New York, 1928.

TABLE 2l-19.   RELATIVE VOLUME OF BENZENE (C_6H_6) AS A FUNCTION OF PRESSURE
AND TEMPERATURE*

Pressure range 1–3,500 kg/cm²; v/v_0; v = volume at (p,t); v_0 = 1.11104 ml/g at
0°C and 760 mm

p, kg/cm²	50°C	95°C
0	1.0630	1.1295
500	1.0160	
1,000	0.9841	1.0201
1,500	0.9591	0.9916
2,000		0.9684
2,500		0.9494
3,000		0.9325
3,500		0.9177

* P. W. Bridgman, *Proc. Am. Acad. Arts. Sci.* **66,** 210 (1931).   Phase diagram of benzene given in
P. W. Bridgman, *Phys. Rev.* **3,** 171 (1914).

TABLE 2l-20. RELATIVE OR DIFFERENTIAL VOLUME OF CARBON DISULFIDE
(CS_2) AS A FUNCTION OF PRESSURE AND TEMPERATURE
The Relative Volumes in Terms of the Volume at 0°C and 760 mm*

Temp. range 20–80°C; pressure range 1–12 000 atm; v/v_0; v = volume at (p,t);
v_0 = 0.77357 ml/g at 0°C and 760 mm

p, atm	20°C	40°C	60°C	80°C
1	1.0235	1.0490	1.0774	1.1092
500	0.9854	1.0051	1.0243	1.0458
1,000	0.9567	0.9734	0.9887	1.0061
1,500	0.9338	0.9483	0.9615	0.9762
2,000	0.9151	0.9277	0.9397	0.9592
2,500	0.8994	0.9105	0.9215	0.9327
3,000	0.8852	0.8953	0.9055	0.9154
3,500	0.8730	0.8820	0.8916	0.9003
4,000	0.8620	0.8702	0.8790	0.8870
4,500	0.8521	0.8596	0.8679	0.8754
5,000	0.8429	0.8501	0.8578	0.8649
6,000	0.8265	0.8337	0.8405	0.8468
7,000	0.8119	0.8196	0.8258	0.8316
8,000	0.7990	0.8070	0.8130	0.8188
9,000	0.7875	0.7954	0.8014	0.8071
10,000	0.7774	0.7844	0.7906	0.7962
11,000	0.7686	0.7741	0.7802	0.7857
12,000	0.7609	0.7646	0.7706	0.7758

The Change in Volume in cm^3 per 1.261 g from a Reference Pressure of 5,000 kg/cm^2
along Each Isotherm †

Temp. range 25–175°C; pressure range 5,000–30,000 kg/cm^2; Δv in cm^3 per 1.261 g

p, kg/cm^2	25°C	75°C	125°C	175°C
5,000	0.000	0.000	0.000	0.000
10,000	0.063	0.068	0.073	0.078
12,600	0.086‡			
15,000		0.110	0.118	0.126
18,300		0.131‡		
20,000			0.148	0.159
24,400			0.170‡	
25,000				0.184
30,000				0.204
30,700				0.206‡

* P. W. Bridgman, "International Critical Tables," vol. 3, p. 41, McGraw-Hill Book Company
New York, 1928.
† P. W. Bridgman, *Proc. Am. Acad. Arts. Sci.* **74**, 415 (1941).
‡ Solid below this.

TABLE 2l-21.   RELATIVE VOLUME OF CARBON TETRACHLORIDE (CCl_4) AS A
FUNCTION OF PRESSURE AND TEMPERATURE*
The Relative Volumes in Terms of the Volume at 50°C and 760 mm Pressure

Pressure range 1–3,500 kg/cm^2; v/v_0; v = volume at (p,t); v_0 = 0.650995 ml/g, 50°C
and 760 mm

p, kg/cm^2	50°C	95°C
0	1.000	
500	0.9519	0.9928
1,000	0.9192	0.9540
1,500	0.8962	0.9362
2,000	......	0.9049
2,500	......	0.8872
3,000	......	0.8762
3,500	......	0.8603

* P. W. Bridgman, *Proc. Am. Acad. Arts Sci.* **66**, 212 (1931).

TABLE 2l-22.   DIFFERENTIAL VOLUME OF CHLOROFORM ($CHCl_3$) AS A FUNCTION
OF PRESSURE AND TEMPERATURE*
The Change in Volume in cm^3 per 1.489 g from a Reference Pressure of 5,000
kg/cm^2 along Each Isotherm

Temp. range 25–175°C; pressure range 5,000–18,400 kg/cm^2; Δv in cm^3 per
1,489 g from a reference pressure of 5,000 kg/cm^2

p, kg/cm^2	25°C	75°C	125°C	175°C
5,000	0.000	0.000	0.000	0.000
6,200	0.016†			
10,000	......	0.067†	0.073	0.079
14,000	......	......	0.109†	
15,000	......	......	......	0.124
18,400	......	......	......	0.148†
				... .

* P. W. Bridgman, *Proc. Am. Acad. Arts Sci.* **74**, 413 (1941).
† Solid below this.

TABLE 2l-23. RELATIVE VOLUME OF ETHER $((C_2H_5)_2O)$ AS A FUNCTION OF
PRESSURE AND TEMPERATURE[*]
The Relative Volumes in Terms of the Volume at 0°C and 760 mm

Temp. range 20–80°C, pressure range 1–12,000 atm; v/v_0; v = volume at (p,t);
$v_0 = 1.3583$ ml/g at 0°C and 760 mm

p, atm	20°C	40°C	60°C	80°C
1	1.0315	1.0669		
500	0.9668	0.9884	1.0123	1.0369
1,000	0.9337	0.9498	0.9683	0.9874
1,500	0.9070	0.9195	0.9336	0.9484
2,000	0.8850	0.8952	0.9069	0.9189
2,500	0.8663	0.8756	0.8860	0.8962
3,000	0.8503	0.8594	0.8688	0.8776
4,000	0.8246	0.8329	0.8407	0.8481
5,000	0.8044	0.8121	0.8189	0.8252
6,000	0.7883	0.7953	0.8017	0.8070
7,000	0.7743	0.7806	0.7865	0.7917
8,000	0.7613	0.7670	0.7725	0.7779
9,000	0.7492	9.7545	0.7597	0.7652
10,000	0.7380	0.7431	0.7482	0.7535
11,000	0.7275	0.7325	0.7377	0.7427
12,000	0.7178	0.7225	0.8280	0.7326

[*] P. W. Bridgman, "International Critical Tables," vol. 3, p. 41, McGraw-Hill Book Company,
New York, 1928. Additional data on ether are reported in the same temperature and pressure range
by Bridgman, *Proc. Am. Acad. Arts Sci.* **66**, 218 (1931). These data were obtained by a method different from that above.

TABLE 2l-24. RELATIVE VOLUME OF GLYCERIN[*]$(CH_2OHCHOHCH_2OH)$ AS A
FUNCTION OF PRESSURE AND TEMPERATURE[†]
The Relative Volumes in Terms of the Volume at 0°C and 760 mm

Temp. range 0–95°C; pressure range 1–12,000 kg/cm^2; v/v_0; v = volume at (p,t);
v_0 = volume at 0°C and 760 mm, $v(1, 50) = 0.80496$ ml/g

p, kg/cm^2	0°C	50°C	95°C
1	1.000	1.0266	
500	0.9900	1.0136	
1,000	0.9806	1.0025	1.0240
1,500	0.9721	0.9930	1.0125
2,000	0.9641	0.9843	1.0024
3,000	0.9501	0.9688	0.9853
4,000	0.9373	0.9548	0.9700
5,000	0.9264	0.9423	0.9565
6,000	0.9157	0.9310	0.9447
7,000	0.9057	0.9211	0.9342
8,000	0.8958	0.9121	0.9244
9,000	0.8867	0.9036	0.9152
10,000	0.8783	0.8955	0.9070
11,000	0.8712	0.8879	0.8994
12,000	0.8648	0.8800	0.8925

[*] The influence of large viscosities at low temperatures and high pressures (see Sec. 2l-6) introduced
some complications in obtaining these data.
[†] P. W. Bridgman, *Proc. Am. Acad. Arts Sci.* **67**, 11 (1931).

TABLE 21-25. RELATIVE VOLUMES OF *n*-OCTANE AND *n*-PENTANE AS
FUNCTIONS OF PRESSURE AND TEMPERATURE*
The Relative Volumes in Terms of the Volume at 0°C and 760 mm

Temp. range 0–95°C; pressure range 1–12,000 kg/cm²; v/v_0;
v = volume at (p,t); v_0 = volume at 0°C and 760 mm

p, kg/cm²	*n*-Octane v_0 = 1.39183 ml/g			*n*-Pentane v_0 = 1.54950 ml/g		
	0°C	50°C	95°C	0°C	50°C	95°C
1	1.0000	1.0595	1.1230	1.0000	(1.0837)	(1.1869)
500	0.9572	1.0005				
1,000	0.9311	0.9654	0.9943	0.9021	0.9395	0.9768
1,500						
2,000	0.8924	0.9200	0.9422	0.8546	0.8820	0.9078
3,000	0.8640	0.8882	0.9068	0.8229	0.8454	0.8671
4,000	......	0.8639	0.8802	0.7997	0.8193	0.8371
5,000	......	0.8428	0.8592	0.7811	0.7985	0.8125
6,000	......	0.8251	0.8416	0.7647	0.7807	0.7933
7,000	......	0.8103	0.8267	0.7506	0.7657	0.7775
8,000	......	......	0.8134	0.7381	0.7520	0.7641
9,000	......	......	0.8014	0.7281	0.7409	0.7527
10,000	......	......	0.7915	0.7192	0.7316	0.7433

* P. W. Bridgman, *Proc. Am. Acad. Arts Sci.* **66**, 185 (1931).

21-11. References to Compressibility Data for Other Substances

Reference: P. W. Bridgman, *Proc. Am. Acad. Arts Sci.* **67**, 6 (1932)
Pressure range: 0 to 12,000 kg/cm²
Substances:

Ethylene glycol
Trimethylene glycol
Propylene glycol
Diethylene glycol
Tri-*o*-cresyl phosphate
Tri-acetin
Ethyl dibenzyl malonate

Methyl oleate
Tri-caproin
n-Butyl phthalate
Eugenol
Isooctane (2,2,4-tri-methyl pentane)
Isoprene

Reference: P. W. Bridgman, *Proc. Am. Acad. Arts Sci.* **66**, 198 (1931)
Pressure range: 0 to 12,000 kg/cm²
Substances:

Isopentane
2-Methyl pentane
3-Methyl pentane
2-2-Dimethyl butane
2-3-Dimethyl butane
Normal heptane
Normal decane

Chlorobenzene
Bromobenzene
Bromoform
Isopropyl alcohol
Normal-butyl alcohol
Normal-hexyl alcohol

Reference: P. W. Bridgman, *Proc. Am. Acad. Arts Sci.* **68,** 1 (1933)
Pressure range: 0 to 12,000 kg/cm^2
Substances:

Triethanolamine	Normal-amyl bromide
Normal-propyl chloride	Normal-amyl iodide
Normal-propyl bromide	Octanol-3
Normal-propyl iodide	2-methyl heptanol-3
Normal-butyl chloride	2-methyl heptanol-5
Normal-butyl bromide	2-methyl heptanol-1
Normal-butyl iodide	3-methyl heptanol-4
Normal-amyl chloride	

Reference: P. W. Bridgman, *Proc. Am. Acad. Arts Sci.* **74,** 403 (1942)
Pressure range: 5,000 kg/cm^2 to 50,000 kg/cm^2
Substances:

Normal-propyl alcohol	Chloroform
Isopropyl alcohol	Chlorobenzene
Normal-butyl alcohol	Methylene chloride
Normal-amyl alcohol	Ethylene bromide
Ethyl bromide	Cyclohexane
Normal-propyl bromide	Methyl cyclohexane
Normal-butyl bromide	*p*-Xylene
Ethyl acetate	Benzene
Normal-amyl ether	

Reference: P. W. Bridgman, "International Critical Tables," vol. 3, p. 40, McGraw-Hill
Book Company, New York, 1928.
Pressure range: 1 to 12,000 atm
Substances:

Phosphorus trichloride
Ethyl iodide
Ethyl chloride
Isobutyl alcohol

Reference: R. S. Jessup, *Bur. Standards J. Research* **5,** 985 (1930), RP 244
Pressure range: 1 to 50 kg/cm^2
Temperature range: 0° to 300°C
Substances: 14 petroleum oils

Reference: F. R. Russell and H. C. Hottel, *Ind. Eng. Chem.* **30,** 372 (1938).
Pressure range: 1 to 400 kg/cm^2.   Max. temperature: 425°C
Substance: Liquid naphthalene

Reference: R. B. Owens, *J. Chem. Phys.* **44,** 3918 (1966).
Pressure range: 450 to 9,000 atm
Temperature range: 300 to 500°C
Substances: (molten nitrates[1])

Lithium nitrate
Sodium nitrate
Potassium nitrate
Rubidium nitrate
Silver nitrate

[1] For general reference on molten salts, see: G. J. Janz, "Molten Salts Handbook," Academic Press,
Inc., New York, 1967.

Reference: J. W. M. Boelhouer, *Physica* **34**, 484 (1967).
Pressure range: 1 to 1,400 atm
Temperature range: $-20°$ to $200°$C
Substances: (alkanes[1])

Heptane
Hexane
Octane
Nonane
Hexadecane

21-12. Compressibility of Liquids.   The instantaneous compressibility is defined by

$$\beta = -\frac{1}{v}\frac{dv}{dp} \tag{21-8}$$

and its reciprocal, the instantaneous bulk modulus, is accordingly defined by

$$K = -v\frac{dp}{dv} \tag{21-9}$$

In some definitions v preceding the derivatives in the above is replaced by v_0, the value of v at one atmosphere.   Bridgman[2] uses this form which he calls the *compressibility proper*.

As with the specific volume (Sec. 21-6) the compressibility depends upon the temperature and pressure history of the sample.   The simplest and most familiar thermodynamic paths are the isothermal and adiabatic (which are reversible when not influenced by viscosity) for which the instantaneous isothermal compressibility is

$$\beta_T = -\frac{1}{v}\left(\frac{\partial v}{\partial p}\right)_T \tag{21-10}$$

and the instantaneous adiabatic compressibility is

$$\beta_S = -\frac{1}{v}\left(\frac{\partial v}{\partial p}\right)_S \tag{21-11}$$

where S is the entropy.[3]   From reversible thermodynamics[4] the above may be shown to be related by

$$\beta_T - \beta_S = \frac{T\alpha_P^2}{\rho C_P} \tag{21-12}$$

where T is the absolute temperature, α_P is the isobaric thermal expansivity, and C_P is the isobaric heat capacity.   Isothermal compressibilities may be derived from equilibrium PVT isobars.   Adiabatic compressibilities are usually obtained directly from acoustic propagation using the relation

$$\beta_S = \frac{1}{\rho_0 c_S^2} \tag{21-13}$$

where c_S is the adiabatic phase velocity, and ρ_0 is the density in the absence of the acoustic field.   In order to apply Eq. (21-13), measurements must be made at frequency well removed from the dispersion region of the phase velocity.

The values of β_T in Tables 21-26 through 21-30 were obtained by first determining mean values of β_T from equilibrium PVT data using 1,000 kg/cm^2 intervals and then interpolating from the smoothed curve on a β_T versus pressure curve.

[1] Adiabatic compressibilities taken from sound velocity data.
[2] P. W. Bridgman, "The Physics of High Pressure," p. 169, G. Bell & Sons, Ltd., London, 1949.
[3] In reversible systems there is no distinction between the adiabatic and isentropic path.
[4] See, for example, M. W. Zemansky, "Heat and Thermodynamics," p. 260, McGraw-Hill Book Company, New York, 1957.

TABLE 2l-26. ISOTHERMAL COMPRESSIBILITY OF WATER
(Calculated from PVT data of Table 2l-13; all values in units of
10^{-12} cm^2/dyne; reliability $\pm5\%$)

Pressure, kg/cm^2	0°C	20°C	60°C	100°C
1	46	45	47	51
250	43	41	42	46
1,000	37	36	35	36
2,000	30	30	29	30
3,000	25	25	25	25
4,000	22	21	21	22
5,000	19	18	19	19
6,000	...	...	16	17
7,000	...	...	15	15
8,000	...	...	14	14
9,000	...	...	13	13
10,000	...	...	12	12
11,000	...	...	11	12

For adiabatic and isothermal values obtained from velocity of sound data, see R. Vedam and G. Holton, *J. Acoust. Soc. Am.* **43**, 108 (1968).

TABLE 2l-27. ISOTHERMAL COMPRESSIBILITY OF 99.9% D$_2$O
(Calculated from PVT data of Table 2l-14; all values in units of
10^{-12} cm^2/dyne; reliability $\pm5\%$)

Pressure, kg/cm^2	0°C	20°C	100°C
1	54	47	48
1,000	41	37	36
2,000	32	30	30
3,000	26	25	26
4,000	21	21	22
5,000	18	18	19
6,000	...	16	17
7,000	...	15	15
8,000	...	14	14
9,000	...	...	13
10,000	...	...	12
11,000	...	...	11

TABLE 21-28. ISOTHERMAL COMPRESSIBILITIES OF CERTAIN ORGANIC LIQUIDS
(β_T in units of 10^{-12} cm^2/dyne; reliability $\pm 5\%$)

Substance	Pressure, kg/cm^2	t(°C)	β_T	Ref.
Acetic acid.................	1	20	91	1
Acetone...................	1	20	126	2
	5,000	20	21	3
Aniline...................	1	20	45	1
Benzene..................	1	20	95	2
Carbon disulfide...........	1	20	93	2
	5,000	20	21	3
Carbon tetrachloride.......	1	20	106	2
Chlorobenzene.............	1	20	74	1
Chloroform...............	1	20	101	2
Cyclohexane..............	1	25	110	1
Ether....................	1	20	187	2
	5,000	20	22	3
Ethyl acetate.............	1	20	113	1
Ethyl alcohol.............	1	20	111	2
	5,000	20	22	3
Ethylene chloride..........	1	20	80	1
Glycerin.................	1	20	21	3
	5,000	20	12	3
Heptane..................	1	20	144	2
Methyl alcohol............	1	20	123	2
	5,000	20	21	3
Nitrobenzene..............	1	20	49	1
Toluene..................	1	20	91	2

References for Table 21-28

1. Data from "International Critical Tables," McGraw-Hill Book Company, New York.
2. Data from "Tables annuelles de constantes et données numériques," vol. 9, Gauthier-Villars, Paris, and McGraw-Hill Book Company, New York, 1929.
3. Calculated from PVT data in this section.

TABLE 21-29. ISOTHERMAL COMPRESSIBILITY OF SULFURIC AND NITRIC ACIDS*
Mean compressibility cofficient
$$\bar{\beta}_T = \frac{10^6}{v_1} \left(\frac{v_1 - v_2}{p_2 - p_1} \right) \text{atm}^{-1}$$

Substance	t, °C	p_1, p_2, atm	$\bar{\beta}_T$
Sulfuric acid............	12.6	1, 161	~33
Nitric acid..............	0	1, 32	~35

* L. Decombe and J. Decombe, p. 35 in "International Critical Tables," vol. 3, McGraw-Hill Book Company, New York, 1928.

21-13. Complex Compressibility. If a liquid is subjected to a sinusoidally time-dependent excitation, the oscillating pressure may not be in phase with the corresponding oscillating volume. If the amplitude of the pressure is sufficiently small to obtain linear response, the amplitude and phase relationships between pressure and dilatation may be given by a complex (frequency-dependent) compressibility,

$$B^* = B' - iB'' \qquad (21\text{-}14)$$

where B' is the ratio of the amplitude of the component of negative dilatation in phase with the pressure to pressure, and B'' is the corresponding ratio of negative

TABLE 2l-30. ISOTHERMAL COMPRESSIBILITY β_T AND ADIABATIC COMPRESSIBILITY β_S OF MERCURY*

(Units of bars^{-1})

Pressure, kbars	$t = 21.9°C$		$t = 40.5°C$		$t = 52.9°C$	
	$\beta_T \times 10^6$	$\beta_S \times 10^6$	$\beta_T \times 10^6$	$\beta_S \times 10^6$	$\beta_T \times 10^6$	$\beta_S \times 10^6$
1	3.881	3.395	3.963	3.444	4.018	3.477
2	3.751	3.289	3.827	3.334	3.878	3.365
3	3.632	3.192	3.702	3.234	3.749	3.262
4	3.522	3.102	3.587	3.141	3.632	3.167
5	3.419	3.018	3.481	3.055	3.523	3.080
6	3.324	2.941	3.383	2.975	3.422	2.998
7	3.235	2.868	3.290	2.900	3.327	2.921
8	3.15	2.799	3.20	2.829	3.24	2.850
9	3.07	2.735	3.12	2.763	3.16	2.782
10	3.00	2.674	3.05	2.701	3.08	2.719
11	2.93	2.616	2.98	2.641	3.01	2.659
12	2.87	2.562	2.91	2.585	2.94	2.602
13	2.80	2.510	2.84	2.532	2.87	2.548

Adiabatic values were obtained from acoustic propagation, using Eq. (2l-13). These were converted to isothermal values, using Eq. (2l-12).

For isothermal values over a wider temperature range, see: L. B. Smith and F. G. Keyes, *Proc. Am. Acad. Arts Sci.* **69**, 313 (1934).

* L. A. Davis and R. B. Gordon, *J. Chem. Phys.* **46**, 2650 (1967).

dilatation lagging the pressure by $\pi/2$. The reciprocal of the complex compressibility is the complex bulk modulus,

$$K^* = K' + iK'' \qquad (21\text{-}15)$$

All the components must be nonnegative quantities in both the above definitions.

The complex functions are usually used to study viscoelastic response.[1] Accordingly, it is necessary to separate the contributions from viscosity and heat of compression. The most convenient manner to obtain this resolution is by a selection of frequencies or boundary conditions to achieve (essentially) either adiabatic or isothermal conditions over a cycle. However, in obtaining extensive data over a wide

[1] See, for example: J. D. Ferry, "Viscoelastic Properties of Polymers," John Wiley & Sons, Inc., New York, 1961.

range of frequencies, achieving the above conditions is not always possible, and resolution may be extremely difficult.

Complex compressibility data may be obtained by acoustic propagation (using both bulk and shear waves) by means of the relations given by Herzfeld and Litovitz.[1] These relations reduce to Eq. (2l-13) at the low-frequency limit. Except for the molten metals, the conditions are approximately adiabatic for most liquids below 10 MHz. Marvin and McKinney[2] review other, more novel methods used in polymer and glass-forming liquids to obtain complex compressibilities and time-dependent response functions which are equivalent.

Table 2l-30 gives the compressibilities for mercury obtained from velocity of sound data at frequencies well below the dispersion region. The adiabatic compressibilities were obtained using Eq. (2l-13). These were converted to isothermal values by means of Eq. (2l-12).

2m. Viscosity of Liquids[3]

ROBERT S. MARVIN

National Bureau of Standards

DAVID L. HOGENBOOM

Lafayette College

2m-1. Units and Symbols

η viscosity
   Dimensions: $ML^{-1}T^{-1}$
   cgs unit: poise = dyne-s cm^{-2} = g cm^{-1} s^{-1}
   SI unit (no coined name) = kg m^{-1} s^{-1} = 10 poise
   (viscosity of water $\approx 10^{-2}$ poise = 10^{-3} SI unit)

ν kinematic viscosity = viscosity/density
   Dimensions: L^2T^{-1}
   cgs unit: stokes = cm^2 s^{-1}
   SI unit: (no coined name) = m^2 s^{-1} = 10^4 stokes
   (kinematic viscosity of water $\approx 10^{-2}$ stokes = 10^{-6} SI unit)

T Kelvin temperature
t Celsius temperature

2m-2. Introduction. Definition of Viscosity.

Viscosity is a material property involved in the relationships between the internal forces in a moving fluid and the kinematical quantities describing its motion. A more specific definition must be based on certain assumptions about the material which are normally expressed in terms of a constitutive equation—in effect, a nonequilibrium equation of state relating

---

[1] K. F. Herzfeld and T. A. Litovitz, "Absorption and Dispersion of Ultrasonic Waves," p. 450, Academic Press, Inc., New York, 1959.

[2] R. S. Marvin and J. E. McKinney, chap. 9 in "Physical Acoustics," vol. 2B, W. P. Mason, ed., Academic Press, Inc., New York, 1965.

[3] Contribution of the National Bureau of Standards, not subject to copyright.

stress (which reduces to the static pressure when the fluid is in static equilibrium), temperature, density, the rate of deformation, and perhaps other state properties which vanish at equilibrium.[1]

The Navier-Stokes equation includes the familiar Newtonian hypothesis about the resistance to flow in fluids. This can be formulated in terms of a constitutive equation involving two material parameters, a bulk and a shear viscosity, analogous to the bulk and shear moduli of classical elasticity theory. These viscosities are assumed to be functions of density and temperature, but not of kinematical quantities. The unmodified term viscosity implies shear viscosity.

We cannot, strictly speaking, be sure that any real liquid obeys this linear constitutive equation under all conditions, since in some types of flow the influence of various complicating factors such as inertial and viscous heating effects precludes an unambiguous evaluation of material properties. But, at least in the types of flow normally studied, the behavior of most low-molecular-weight homogeneous liquids is described by the Navier-Stokes equation. Such liquids are commonly termed *Newtonian*. Many other liquids show marked deviations from such behavior, and their properties must be described by a more complicated constitutive equation. In such liquids the stress associated with a steady flow is not proportional to the velocity gradient and, indeed, will not even be oriented in the same direction. Nonetheless, it is frequently possible to define a viscosity for such liquids as the ratio of stress to rate of strain measured in simple steady shear flow which has persisted for a sufficiently long period of time. The viscosity so defined is, of course, a function of the rate of shear.

Even in a non-Newtonian fluid the stress associated with a flow whose amplitude is sufficiently small or velocity sufficiently slow will be (at least within the precision of ordinary measurements) proportional to the magnitude of the velocity gradient, but time effects (or frequency effects) may be observed. We term the idealization which represents this type of behavior a linear viscoelastic material, and consider this one special case of a non-Newtonian material. One method of expressing the properties of a linear viscoelastic material is in terms of a (complex) dynamic viscosity (and dynamic bulk viscosity) defined as a ratio of stress to rate of strain for a strain varying sinusoidally in time.[2]

Values of dynamic viscosity (or other equivalent linear viscoelastic properties) are reviewed by Ferry[3] for a number of polymeric systems. Acoustic measurements of both shear and volume viscosity, and results for a number of different types of liquids, are discussed by Litovitz and Davis.[4] Many measurements of viscosity as a function of rate of shear, and attempts to relate this function to other rheological and/or structural properties, have been reported. The general, though not universal, pattern shows it decreasing monotonically with rate of shear, with (established or at least suggested) limiting values at both high and low rates. Changes in temperature, concentration, and molecular weight cause pronounced and often complex changes. Though some generalizations are suggested by recent work, nothing that permits a simple and concise tabulation of such behavior is yet established.

2m-3. Accuracy and Precision of Measurement. The remainder of this section will deal only with the shear viscosity (or simply viscosity) of Newtonian liquids.

---

[1] W. E. Langlois, "Slow Viscous Flow," The Macmillan Company, New York, 1964.

[2] At one time the term *dynamic viscosity* was used for the shear viscosity of a Newtonian liquid (to distinguish it from kinematic viscosity, the viscosity divided by density). Though still occasionally used in this sense, dynamic viscosity now generally denotes the frequency-dependent quantity defined above.

[3] J. D. Ferry, "Viscoelastic Properties of Polymers," 2d ed. John Wiley & Sons, Inc., N.Y., 1970.

[4] T. A. Litovitz and C. M. Davis, in "Physical Acoustics," W. P. Mason, ed., vol. 2A, Academic Press, Inc., N.Y., 1965.

Most measurements have been made in relative instruments which require calibration with a liquid of known viscosity, generally (either directly or indirectly) water at 20°C and one atmosphere. Normal variations in atmospheric pressure and in the exact composition of distilled water do not influence its viscosity appreciably.

Significantly different values have been used as the basis for such calibrations during the past 40 years. Dorsey, in compiling viscosities for the International Critical Tables, adopted the value of 0.01009 poise, based on his evaluation of measurements available at that time. Bingham and Jackson,[1] evaluating the same measurements, selected a value of 0.01005 poise which was rather generally accepted for a number of years. The value currently used throughout most of the world is 0.01002 poise (corresponding to $\eta/\rho = 0.01004$ stokes) due to Swindells, Coe, and Godfrey.[2] All these values are based on measurements in capillary viscometers with a precision of 0.1 per cent or better. The differences appear due to various systematic errors, some associated with end effects which cannot be calculated exactly. Experiments just completed at the National Bureau of Standards suggest that we should assign an uncertainty of ±0.2 per cent to the value 0.01002, this representing an unexplained disagreement between two absolute measurements, each with a precision of 0.1 per cent. One involved flow through a pipe; the other oscillations of a liquid contained within a sphere.

Ordinary low-molecular-weight liquids have viscosities no greater than a few centipoises at room temperature and one atmosphere. Measurements relative to water can be reproduced to ±0.1 per cent in this low-viscosity range. Comparative measurements of viscosities up to several hundred poises can be made with nearly this precision, and are often used in specifications for liquids whose exact composition is unknown. A determination of the viscosity of such liquids relative to water, however, requires additional measurements which increase the uncertainty to about ±0.5 per cent. Above about 10^3 poises, problems of measurement and temperature control normally limit the reproducibility of viscosity measurements to a figure greater than one per cent.

2m-4. Relationship to Structure, Temperature, and Pressure. Our understanding of the liquid state has advanced considerably during the past decade, but we cannot yet calculate transport properties from equilibrium properties, except for a few nearly spherical molecules, and even then with an accuracy significantly less than the accuracy of measurement.[3-5] Viscosities of the inert gases in both liquid and gaseous states can be represented rather successfully in terms of an empirical correlation involving reduced viscosities, temperature, and pressure (actual values divided by values at the critical point).[6] For more complex liquids, other correlations, whose form is suggested by theory but involving some empirical quantities, have been developed to express the influence of temperature, pressure, the variations between similar compounds, and the viscosities of mixtures.[7-9]

[1] E. C. Bingham and R. F. Jackson, Standard Substances for the Calibration of Viscometers, *Bull. Bur. Standards* **14**, 59 (1918–1919); Scientific Paper 298.

[2] J. F. Swindells, J. R. Coe, Jr., and T. B. Godfrey, *J. Research Natl. Bur. Standards* **48**, 1 (1952); RP 2279.

[3] S. G. Brush, *Chem. Rev.* **63**, 513 (1963).

[4] S. A. Rice, J. P. Boone, and H. T. Davis, in "Simple Dense Fluids," H. L. Frisch and Z. W. Salsburg, eds., Academic Press, Inc., New York, 1968.

[5] P. Gray, in "Physics of Simple Liquids," H. N. V. Temperley, J. S. Rowlinson, and G. S. Rushbrooke, eds., North-Holland Publishing Company, Amsterdam, 1968.

[6] H. Shimotake and G. Thodos, *A.I.Ch.E. Journal* **4**, 3, 257 (September, 1958).

[7] A. Bondi, in "Rheology: Theory and Applications," vol. 4, F. R. Eirich, ed., Academic Press, Inc., New York, 1967.

[8] A. Bondi, "Physical Properties of Molecular Crystals, Liquids, and Glasses," John Wiley & Sons, Inc., New York, 1968.

[9] R. C. Reid and T. K. Sherwood, "The Properties of Gases and Liquids," 2d ed., McGraw-Hill Book Company, New York, 1966.

In the higher-temperature portion of the normal liquid range, the prediction of the reaction-rate theory that viscosity is proportional to $\exp(1/T)$ is reasonably accurate. In the lower-temperature range, particularly for glass-forming liquids, it is more nearly proportional to $\exp[1/(T + A)]$. This latter proportionality can be predicted from the free-volume theory of Cohen and Turnbull,[1] the significant structure theory of Eyring and coworkers,[2] or the configurational entropy theory of Adam and Gibbs.[3] For most liquids no single expression of either type holds over the whole liquid range.[4]

Attempts to represent the influence of pressure by anything other than a strictly empirical expression have been relatively unsuccessful. The reaction-rate and free-volume theories mentioned above can represent such data, but only by adjusting the parameters involved so that their original meaning becomes questionable.

2m-5. Values of Viscosity. The various general tabulations or handbooks ("International Critical Tables," Landöldt-Börnstein, "Smithsonian Physical Tables," "Handbook of Chemistry and Physics," etc.) generally contain extensive data on the viscosity of liquids. Most of these tabulations were prepared by individuals well qualified to select the best available values, or were reproduced from earlier tabulations prepared by such individuals. The measurements on which most of them are based are equivalent to those commonly used today, but they may not be consistent with present measurements simply because of the different values for the viscosity of water to which relative measurements have been referred from time to time. Some extensive tabulations may not even be internally consistent because of this factor, and in some cases the basis used is not stated.

For most liquids, however, particularly those with viscosities of less than one poise, impurities in the samples measured are likely to introduce greater uncertainties than systematic errors in measurement. For many organic liquids viscosity is at least as sensitive a measure of purity as most of the methods normally employed.

We may reasonably assign an uncertainty of ± 0.2 per cent to the values given here for water, this representing a limitation due to possible systematic errors in measurement. Critical evaluations of the measurements on most other liquids have not been made. Where such evaluations have been attempted, the uncertainty estimates are rarely less than one percent. This point is stressed because values of viscosity are often reported and tabulated to 0.1 per cent—a reasonable precision or reproducibility of measurement, but seldom a realistic uncertainty in the values.

There is no general tabulation of viscosity values based on a current critical assessment of the purity of the material and the adequacy and reliability of the measurements. Nor is there likely to be such a tabulation produced by any single group. Both the magnitude of the problem and the special background required for a critical evaluation of the data for any given class of compounds suggest that such tabulations will probably be made by different individuals or groups for each class of compounds or materials.

2m-6. Selected Values at One Atmosphere. We emphasize here a few cases in which at least a start has been made toward a critical evaluation of the available data. Because of limited space we reproduce only a few values, and give references to more complete tabulations.

Water. The viscosity of water has been studied extensively, and careful relative measurements by various investigators agree to within about 0.1 per cent. No particular problem arises because of impurities, and any variation due to the quantity

[1] M. H. Cohen and D. Turnbull, *J. Chem. Phys.* **31**, 5, 1164 (1959).
[2] T. S. Ree, T. Ree, and H. Eyring, *J. Phys. Chem.* **68**(11), 3262 (1964).
[3] G. Adam and J. H. Gibbs, *J. Chem. Phys.* **43**(1), 139 (1965).
[4] A. Bondi, in "Rheology: Theory and Applications," *op. cit.*

TABLE 2m-1. VISCOSITY OF LIQUID WATER AT ATMOSPHERIC PRESSURE

Temperature, °C	Viscosity, cp	Temperature, °C	Viscosity, cp
5	1.518*	75	0.3784†
20	1.002†	100	0.2820†
40	0.6527†	125	0.2219†
60	0.4665†	150	0.1815†

* R. C. Hardy and R. L. Cottington, *J. Research Natl. Bur. Standards* **42**, 573 (1949) (adjusted to 1.002 cp at 20°)
† A. Korosi and B. M. Fabuss, *Anal. Chem.* **40**, 157 (1968).

of dissolved gas appears to be less than the ± 0.2 per cent uncertainty that we must assign owing to present limitations in absolute measurements.[1]

Both sets of measurements were carried out in relative capillary instruments. They agree to within 0.1 per cent where the temperature ranges overlap (20 to 125°C). Pressures up to those required to maintain the liquid state above 100° did not have any significant effect on measurements below 100°. The measured values are given with an error of less than 0.1 per cent up to 75° and a maximum of 0.16 per cent to 150° by

$$\log_{10} \frac{\eta_{20}}{\eta_t} = \frac{A(t - 20) + B(t - 20)^2}{C + t}$$

with $A = 1.37023$ η_{20} is viscosity at 20°C
      $B = 0.000836$ η_t is viscosity at t°C
      $C = 109$ t is temperature in °C

Simple Liquids. "Simple" liquids refers here to those whose molecules are essentially spherical, which current statistical mechanical theories attempt to describe. They include the noble gases, the homonuclear diatomic molecules, and some polyatomic but nearly spherically symmetrical molecules.

References

1. Johnson, V. J. ed.: "A Compendium of the Properties of Materials at Low Temperatures" (Phase I), part I, "Properties of Fluids," Wright Air Force Development Division Technical Report, 60–56, July, 1960 (available from Clearinghouse for Federal Scientific and Technical Information, Springfield, Va.). Includes tables and graphs of selected values of the viscosity of liquid (and gaseous):
Helium, hydrogen, neon, argon
Nitrogen, oxygen, carbon monoxide, fluorine, methane
2. Rice, S. A., J. P. Boone, and H. T. Davis: "Simple Dense Fluids," H. L. Frisch and Z. W. Salsburg, eds., Academic Press, Inc., New York, 1968. Includes a literature survey to November, 1966, of theoretical and experimental results on transport properties of simple fluids.
3. Gray, P.: "Physics of Simple Liquids," H. N. V. Temperley, J. S. Rowlinson, and G. S. Rushbrooke, eds., North-Holland Publishing Company, Amsterdam, 1968. Discusses measurements on liquid argon.
4. Shimotake, H. and G. Thodos, *A.I.Ch.E. Journal* **4**(3), 257 (September, 1958). Includes an often-reproduced Reduced State Viscosity Chart for the Inert Gases.

Organic Compounds. An evaluation of measurements on organic compounds requires that careful attention be given to the adequacy of the methods of purification, particularly for the higher members of a homologous series (from about C_9 or C_{10} on in most cases). Most existing tabulations make no statement about the criteria applied to the measurements selected. Two which do are the American Petroleum Institute Research Project 44 "Tables of Selected Values of Properties of Hydrocar-

[1] J. Kestin and J. H. Whitelaw, *Phys. of Fluids* **9**(5), 1032 (May, 1966).

bons and Related Compounds,"[1] and Timmermans' "Physico-chemical Constants of Pure Organic Compounds."[2] In both cases the compilers considered both the purity of the compounds measured and the techniques of measurement in selecting values they considered reliable. Timmermans presents the original data considered reliable; the API compilers present "average" values calculated from an empirical viscosity-temperature equation fitted to the selected data for each compound. Timmermans includes all organic compounds for which he found measurements which met his criteria in the literature to the end of 1964 (vol. 2; vol. 1 covers literature to 1950). The API tabulations aim at a more limited class of materials. The coverage of their present tables is shown in our Table 2m-2; additional issues on viscosity are planned for about 1971.

The estimated uncertainty in the API Tables is indicated in a general way as varying between ± 3 in the last figure listed to 10 times this amount (except where values are tabulated to 10^{-4} cp, in which case they estimate the uncertainty as between 1 and 3×10^{-3} cp). Though not stated specifically, these estimates presumably represent a lack of agreement between various measurements deemed equally reliable. These estimates are consistent with Timmermans' statement that few measurements of viscosity are reliable to better than one per cent. Timmermans' listing of separate measurements considered reliable permits some evaluation of uncertainty on the part of the reader, though generally the temperatures of the original measurements are not exactly the same, and so no direct and simple comparison of the values that he lists is possible.

A tabulation of the viscosities of over 300 organic compounds up to C_{31}, including many ring and many sulfur-containing compounds, is given in a report[3] of the work carried out at The Pennsylvania State University over a period of several years. The estimated accuracy of these measurements is ± 0.5 per cent.[4] The primary emphasis of this project was on the preparation and purification of the compounds used.[5] Measurements of the viscosity of normal paraffins from C_5 to C_{64} have been reported by Doolittle and Peterson.[6]

Glycerol-Water Solutions. Glycerol has been the subject of a number of studies. Unfortunately, it is difficult to obtain (and maintain) this compound free of water, and most measurements have been made on samples which apparently were not pure. Table 2m-3 shows the unusual sensitivity of the viscosity of this liquid to small amounts of water as a contaminant. Note that even the relative values (viscosities at various temperatures divided by the viscosity at a reference temperature) change significantly with composition.

Molten Salts. The first output of the National Standard Reference Data System to include critical evaluations of viscosity data is a recent publication by Janz,

[1] American Petroleum Institute Research Project 44, "Selected Values of Physical and Thermodynamic Properties of Hydrocarbons and Related Compounds," Carnegie Institute of Technology, Pittsburgh, Pa. (Present address: Thermodynamics Research Center, Texas A&M University, College Station, Tex.)

[2] J. Timmermans, "Physico-chemical Constants of Pure Organic Compounds," American Elsevier Publishing Company, Inc., New York, vol. 1, 1950; vol. 2, supplement, 1965; J. Timmermans, "The Physico-chemical Constants of Binary Systems in Concentrated Solutions," vols. 1 and 2, "Two Organic Compounds," 1959; vol. 3, "Systems with Metallic Compounds," 1960; vol. 4, "All Other Systems," 1960; Interscience Publishers, Inc., New York.

[3] "Properties of Hydrocarbons of High Molecular Weight Synthesized by Research Project 42 of the American Petroleum Institute," American Petroleum Institute, New York, 1967.

[4] Private communication, Professor J. A. Dixon, Director, API Research Project 42, 1955–1967.

[5] R. W. Schiessler and F. C. Whitmore, *Ind. Eng. Chem.* 47(8), 1660 (August, 1955).

[6] A. K. Doolittle and R. H. Peterson, *J. Am. Chem Soc.* 73, 2145 (1951).

TABLE 2m-2. VISCOSITIES OF HYDROCARBON LIQUIDS*

Compound	Formula	t_{min}, °C	Viscosity in centipoise† (multiply by 10^{-1} for SI Units)							t_{max}, °C
			t_{min}	0	20	25	30	100	t_{max}	
n-Paraffins (C_nH_{2n+2}), C_1 through C_{20}. Table 20c (Part 1), 1955										
Methane	CH_4	−185	0.225‡	·····	·····	·····	·····	·····	0.115§	−160
n-Pentane	C_5H_{12}	−130	3.62‡	0.278	0.234	0.224	0.215	·····	0.206	35
n-Octane	C_8H_{18}	−55	2.11	0.7104	0.5450	0.5136	0.4850	0.2547	0.2109	125
n-Dodecane	$C_{12}H_{26}$	−10	2.901‡	2.271	1.503	1.374	1.261	0.5168	0.2070	215
n-Hexadecane	$C_{16}H_{34}$	20	3.474‡	·····	3.474¶	3.086	2.758	0.8992	0.202	285
n-Eicosane	$C_{20}H_{42}$	35	4.685‡	·····	·····	·····	·····	1.406	0.20	340
Normal Alkyl Cyclopentanes (C_nH_{2n}), C_5 through C_{21}. Table 22c (Part 1), 1955										
Cyclopentane	C_5H_{10}	−25	0.78	0.553	0.438	0.415	0.393	·····	0.322§	50
Methylcyclopentane	C_6H_{12}	−25	0.93	0.648	0.505	0.477	0.451	·····	0.286§	75
n-Decylcyclopentane	$C_{15}H_{30}$	−20	8.60	5.52	3.55	3.19	2.86	0.96	0.85	110
n-Hexadecylcyclopentane	$C_{21}H_{42}$	20	9.60‡	·····	9.60‡	8.41	7.38	1.83	1.58	110
Normal Alkyl Cyclohexanes (C_nH_{2n}), C_6 through C_{22}. Table 23c (Part 1), 1955										
Cyclohexane	C_6H_{12}	5	1.296‡	·····	0.977	0.895	0.824	·····	0.410	80
Methylcyclohexane	C_7H_{14}	−25	1.55	0.990	0.732	0.683	0.639	0.30	0.30	100
n-Decylcyclohexane	$C_{16}H_{32}$	−5	10.7‡	9.23	5.24	4.59	4.04	1.12	0.97	110
n-Hexadecylcyclohexane	$C_{22}H_{44}$	35	8.73	·····	·····	·····	·····	2.03	1.69	110
Normal Monoölefins (1-Alkenes), (C_nH_{2n}), C_2 through C_{20}. Table 24c (Part 1), 1955										
Ethene	C_2H_4	−170	0.70‡	0.33	0.26	0.25	0.24	·····	0.15§	−100
1-Hexene	C_6H_{12}	−55	0.69	·····	·····	·····	·····	·····	0.19§	65
1-Dodecene	$C_{12}H_{24}$	0	1.95	1.95	1.30	1.203	1.114	0.475	0.41	115
1-Eicosene	$C_{20}H_{40}$	30	4.76‡	·····	·····	·····	4.76‡	1.35	1.10	115

TABLE 2m-2. VISCOSITIES OF HYDROCARBON LIQUIDS* (Continued)

Compound	Formula	t_{min}, °C	Viscosity in centipoise† (multiply by 10^{-1} for SI Units)							t_{max}, °C
			t_{min}	0	20	25	30	100	t_{max}	
Normal Alkyl Benzenes ($C_{6+n}H_{6+2n}$), C$_6$ through C$_{22}$, Table 21c (Part 1), 1955										
Benzene........	C_6H_6	5	0.8235‡	......	0.6468	0.6010	0.5604	......	0.301§	85
Methylbenzene (Toluene)........	C_7H_8	−25	1.17	0.771	0.5848	0.5500	0.5187	0.268	0.248	110
n-Decylbenzene........	$C_{16}H_{26}$	−20	16.0‡	6.70	3.79	3.36	3.01	0.974	0.56	150
n-Hexadecylbenzene........	$C_{22}H_{38}$	−25	8.96‡	......	......	8.96‡	7.74	1.817	0.92	150
Alkyl Benzenes ($C_{6+n}H_{6+2n}$), C$_6$ through C$_9$, Table 5c, 1955										
Ethylbenzene........	C_8H_{10}	−25	1.35	0.892	0.6763	0.6354	0.5985	0.307	0.230§	140
1,2-Dimethylbenzene (o-Xylene)........	C_8H_{10}	−5	1.211	1.105	0.807	0.754	0.706	0.344	0.244§	145
n-Propylbenzene........	C_9H_{12}	−25	1.90	1.178	0.8545	0.7962	0.7444	0.359	0.25	150

* A sampling of values given in American Petroleum Institute Research Project 44. "Tables of Selected Values of Properties of Hydrocarbons and Related Compounds op. cit. The original tables contain values of viscosity tabulated every 5° from t_{min} through t_{max}. In many but not all cases these limits approximate the normal liquid range. Values were calculated from empirical expressions with constants based on all data considered reliable on the basis of purity of material and technique of measurement. The estimated uncertainty is stated to range from 3 in the last figure tabulated to 10 times this. Additional tabulations of viscosities are planned for about 1971.

† Based on 1.002 for water at 20°C.

‡ Undercooled below normal freezing point.

§ At saturation pressure.

¶ Hardy [J. Research Natl. Bur. Standards 61 (1958)] found a value of 3.454 at 20° for n-hexadecane of 99.94 mole per cent purity, a value within the range of the estimated uncertainties of this tabulation. Doolittle and Peterson (loc. cit.) estimate that impurities (isomers or paraffins of slightly differing molecular weight) in even carefully purified samples can cause errors in viscosity of as much as 1 per cent in the range of C$_{24}$ to C$_{60}$.

TABLE 2m-3. VISCOSITIES OF GLYCEROL-WATER SOLUTIONS*

Glycerol, wt. %	Temperature °C					
	0	20	40	60	80	100
	Viscosity, centipoise					
10	2.43	1.31	0.824	0.573		
50	14.6	5.98	3.09	1.85	1.25	0.907
90	1,310	218	59.8	22.43	11.0	5.98
98	7,350	936	194	59.6	24.7	12.2
99	9,390	1,150	234	68.9	27.7	13.2
100	12,000	1,410	283	81.1	31.8	14.8

* From J. B. Segur and H. E. Oberstar, *Ind. Eng. Chem.* **43**(9), 2117 (September, 1951). Values from original reduced by 0.3 per cent (to adjust basis used to 1.002 for water at 20°C) and rounded to three significant figures. Original tabulation gives values every 10°C for 24 compositions.

Dampler, Lakshminarayanan, Lorenz, and Tomkins.[1] The authors have examined the data up to December, 1966, for 174 single-salt melts, selected the best data in each case, and presented these both as tabulations and empirical viscosity-temperature relations. For each compound a concise statement is given, citing the measurements on which the tabulated values are based, and comparing these with other measurements available. A measure of the precision with which each empirical equation represents the data is given, as is an estimated accuracy of the measurements themselves. This latter estimate is based on an evaluation of the measurement technique, purity of material, and agreement with other values.

2m-7. Viscosity at Elevated Pressure. Absolute measurements of viscosity at elevated pressure depend on factors like variations in dimensions of the instrument which are often not known with certainty. Most measurements have been made with falling-weight or rolling-ball viscometers, calibrated at atmospheric pressure. Such instruments can measure a wide range of viscosities with a precision of about one per cent, but uncertainties of several per cent may arise in introducing corrections to the calibration constant owing to increased pressure.

With the exception of water between 0 and 33°C, the viscosity of liquids increases monotonically with increasing pressure. Typically, the logarithm of viscosity versus pressure at constant temperature is concave toward the pressure axis at low pressures, becomes almost linear over an appreciable pressure range, and finally, if the sample does not freeze first, reverses curvature and shows an increasing slope at high pressure. The viscosity, especially at elevated pressure, is remarkably sensitive to molecular structure, in contrast to equilibrium properties such as density or compressibility which tend to follow a rather uniform pattern at high pressure.

Water. The lower-temperature isotherms for the viscosity of water versus pressure show minima. At 2°C the minimum value occurs at about 1,000 bars, where the

[1] G. J. Janz, F. W. Dampier, G. R. Lakshminarayanan, P. K. Lorenz, and R. P. T. Tomkins, Molten Salts, vol. 1, Electrical Conductance, Density and Viscosity Data, *Natl. Bur. Standards Ref. Data Ser.* **15**, October, 1968. See also G. J. Janz, "Molten Salts Handbook," Academic Press, Inc., New York, 1967. (Less detailed, but includes some mixtures.)

TABLE 2m-4. VISCOSITIES OF MOLTEN SALTS*

Potassium Chloride

T, K	1060	1070	1080	1090	1100	1110	1120	1130	1140	1150	1160	1170	1180	1190	1200
η, cp	1.14_9	1.10_8	1.07_1	1.03_6	1.00_4	0.97_5	0.94_8	0.92_4	0.90_1	0.88_1	0.86_3	0.84_7	0.83_2	0.81_9	0.80_7

Best equation: $\eta = 55.5632 - 0.127847T + 9.99580 \times 10^{-5}T^2 - 2.62035 \times 10^{-8}T^3$

s, centipoise: 0.0132   Uncertainty estimate, %: 1.5

Temperature range, K: 1056.5–1202.0

Lead (II) Chloride

T, K	780	790	800	810	820	830	840	850	860	870	880	890	900	910	920	930	940	950	960
η, cp	4.41	4.17	3.95	3.75	3.56	3.39	3.23	3.08	2.94	2.81	2.69	2.57	2.47	2.36	2.27	2.18	2.10	2.02	1.95

Best equation: $\eta = 5.619 \times 10^{-2} \exp (6{,}762/RT)$ s, centipoise: 0.0091   Uncertainty estimate, %: 1.5

Temperature range, K: 773.2–973.2

Sodium Bromide

T, K	1060	1070	1080	1090	1100	1110	1120	1130	1140	1150	1160	1170	1180	1190	1200	1210
η, cp	1.28_3	1.24_5	1.21_0	1.17_8	1.14_9	1.12_3	1.09_8	1.07_6	1.05_6	1.03_8	1.02_2	1.00_6	0.99_2	0.97_9	0.96_7	0.95_5

Best equation: $\eta = 64.3240 - 0.152525T + 1.23215 \times 10^{-4}T^2 - 3.34241 \times 10^{-8}T^3$ s, centipoise: 0.0040   Uncertainty estimate, %: 1.0

Temperature range, K: 1053.7–1212.7

Mercury (II) Iodide

T, K	550	560	570	580	590	600	610	620	630
η, cp	2.53	2.35	2.19	2.04	1.91	1.79	1.68	1.58	1.49

Best equation: $\eta = 4.00 \times 10^{-2} \exp (4531/RT)$ s, centipoise: 0.0367   Uncertainty estimate, %: 3.0

Temperature range, K: 541.2–631.2

TABLE 2m-4. VISCOSITIES OF MOLTEN SALTS (*Continued*)

Silicon Dioxide†

T, K	2210	2250	2300	2350	2400	2450	2500	2550
η, cp	717,000	515,000	325,000	195,000	114,000	69,000	50,200	46,400

Best equation: $\eta = 2.52255 \times 10^8 - 294897T + 114.9357T^2 - 1.49316 \times 10^{-2}T^3$

Temperature range, K: 2208–2595 s, centipoises: 55.9   Uncertainty estimate, %: 6.0

Tetrahexylammonium Tetrafluoroborate

T, K	380	390	400	410	420	430	440	450	460	470	480	490	500
η, cp	82.61	59.14	43.06	31.83	23.88	18.15	13.97	10.88	8.56	6.81	5.47	4.43	3.62

Best equation: $\eta = 1.806 \times 10^{-4} \exp (9841/RT)$

Temperature range, K: 376.0–502.8 s, centipoises: 2.90   Uncertainty estimate, %: —‡

$$s = \sqrt{\frac{\Sigma(X_e - X_t)^2}{n - p}}$$

where X_e and X_t are experimental and calculated values, n is the number of experimental values used, and p is the number of coefficients in "Best equation." Standard deviation s measures adequacy of this equation in reproducing experimental data. Uncertainty includes compilers' estimate of accuracy of measurements themselves. $R = 1.98717$ cal mol⁻¹ deg⁻¹.

* From Janz et al., Molten Salts, *op. cit.* Temperature range is for applicability of "Best equation."
† Original tabulation gives values every 10 K.
‡ In some cases, as here, the information available was considered insufficient to warrant a quantitative estimate of uncertainty. It is stated that ". . . the data can be considered reliable."

viscosity is about 93 or 94 per cent of its value at one atmosphere. This minimum ratio rises and shifts to lower pressures as the temperature is raised. Between 30 and 40°C the minimum disappears; at higher temperatures the isotherms show the normal monotonic increase with pressure. Between about 2 and 20°C, various measurements disagree by as much as 3 per cent, an amount significantly greater than their precision, though within their possible systematic error.

References

1. Horne, R. A. and D. S. Johnson: *J. Phys. Chem.* **70**(7), 2182 (1966).
2. Bett, K. E., and J. B. Cappi: *Nature* **207**, 620 (Aug. 7, 1965).
3. Wonham, J.: *Nature* **215**, 1053 (Sept. 2, 1967).
4. Bruges, E. A., B. Latto, and A. K. Ray: *Int. J. Heat Mass Transfer* **9**, 465(1966).

Other Liquids. All other liquids for which measurements are available show a monotonic increase in viscosity with pressure. In Table 2m-5 we list values for a few liquids selected from the measurements by Bridgman[1] (falling-weight viscometer) and by several investigators at The Pennsylvania State University[2] (rolling-ball viscometers). These two collections of data are the most extensive that are available on pure compounds. Bridgman obtained his liquids from various sources. Some were the purest available commercial liquids; some specially purified by various other workers. The Penn State measurements utilized the API-42 compounds mentioned earlier.[3] There are no objective grounds for assigning uncertainties to most of these

TABLE 2m-5A. VISCOSITY OF LIQUIDS UNDER ELEVATED PRESSURE[a]

Pressure { kg/cm² ... (atm)	500	1,000	2,000	4,000	6,000	8,000	10,000	12,000
bars* ..... (atm)	490	980	1,960	3,920	5,880	7,850	9,810	11,770
Temperature, °C	Viscosity in centipoises							
	n-Pentane:[b] η at 30°C, 1 atm = 0.215 cp, from API-44 tables							
30	0.215	0.326	0.444	0.719	1.51	2.78	4.94	8.86	15.1
75	0.139	0.222	0.313	0.516	1.02	1.74	2.83	4.42	6.69
	Toluene: η at 30°C, 1 atm = 0.5187 cp, from API-44 tables							
30	0.519	0.724	0.975	1.63	4.09	10.0	25.9	78.0	
75	0.324	0.451	0.602	0.959	2.05	4.08	7.96	16.6	35.2
	Ethyl alcohol: η at 30°C, 1 atm = 0.991 cp[c]							
30	0.991	1.27	1.57	2.29	4.10	6.68	10.4	16.1	24.3
75	0.450	0.59	0.74	1.10	1.93	2.94	4.27	5.94	8.22

* 1 bar = 0.9807 kg/cm². Values rounded to closest 10 bars.

[1] P. W. Bridgman, in *Proc. Am. Acad. Arts Sci.* **61**, 57 (1926); "The Physics of High Pressure," G. Bell & Sons, Ltd., London, 1952.
[2] D. L. Hogenboom, W. Webb, and J. A. Dixon, *J. Chem. Phys.* **46**(7), 2586 (1967); D. A. Lowitz, J. W. Spencer, W. Webb, and R. W. Schiessler, *J. Chem. Phys.* **30**(1), 73 (1959); E. M. Griest, W. Webb, and R. W. Schiessler, *J. Chem. Phys.* **30**(1), 73 (1958); results summarized in "Properties of Hydrocarbons of High Molecular Weight Synthesized by Research Project 42 of the American Petroleum Institute," *op. cit.*
[3] R. W. Schiessler and F. C. Whitmore, *Ind. Eng. Chem.* **47**(8), 1660 (August, 1955).

TABLE 2m-5B. VISCOSITY OF LIQUIDS UNDER ELEVATED PRESSURE[d]

Pressure, bars		(atm)	200	400	600	1,000	1,400	1,800	2,200	2,600	3,000	3,400
Compound	Temperature, °C	Viscosity in centipoises										
9-*n*-Octylheptadecane[e]	37.78	7.06	9.40	12.5	16.2	26.3	41.3	63.0	92.3	131	187	
$C(-C_8)_3$	60.00	3.91	5.13	6.65	8.45	13.1	19.5	28.1	39.7	55.0	75.5	
	79.44	2.60	3.35	4.26	5.34	8.07	11.6	16.2	22.6	30.7	41.2	
	98.89	1.87	2.37	2.96	3.65	5.38	7.58	10.3	13.7	18.2	23.7	
	115.00	1.48	1.85	2.31	2.83	4.17	5.80	7.80	10.2	13.2	16.7	
Perhydrochrysene	37.78	25.6	46.8	87.5	177	951	9,080					
	60.00	10.4	15.6	24.5	40.1	121	464	2,510				
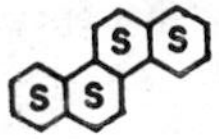	79.44	5.86	8.36	12.1	17.9	41.6	110	345	1,450	8,650		
	98.89	3.80	5.15	7.05	9.73	19.1	41.2	95.9	255	832	3,670	
	115.00	2.77	3.78	5.07	6.73	12.2	23.0	46.1	102	252	778	2,720
	135.00	2.09	2.74	3.56	4.60	7.67	13.1	23.5	43.6	87.5	194	488
1-α-Decalylpentadecane	60.00	8.56	11.8	16.1	21.4	36.6						
	79.44	5.24	7.18	9.46	12.2	19.8	31.0	47.6	71.8			
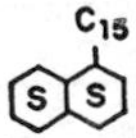	98.89	3.55	4.76	6.19	7.89	12.2	18.3	27.0	39.4	56.0	79.2	111
	115.00	2.64	3.57	4.63	5.85	8.75	12.8	18.3	25.9	36.3	49.9	67.9
	135.00	1.99	2.64	3.37	4.17	6.15	8.75	12.1	16.7	22.6	30.3	40.3
1-α-Naphthyl-pentadecane	60.00	8.41	10.9	14.2	18.2	29.6						
	79.44	5.05	6.49	8.18	10.3	15.8	23.9	35.3				
	98.89	3.41	4.24	5.22	6.43	9.48	13.7	19.5	27.4	38.1		
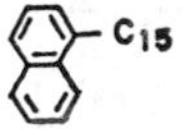	115.00	2.52	3.15	3.90	4.75	6.84	9.59	13.2	18.1	24.7	33.0	43.8
	135.00	1.90	2.36	2.89	3.47	4.83	6.64	8.87	11.7	15.3	20.0	25.9

TABLE 2m-5C. VISCOSITY OF LIQUIDS UNDER ELEVATED PRESSURE[f]

Compound	Temperature, °C	Pressure, bars (atm)	400	800	1,200	1,600	2,000	2,400	2,800	3,200	3,600
						Viscosity in centipoises					
n-Dodecane, n-C$_{12}$	37.78	1.102[a]	1.70	2.50	3.52	4.78	6.35				
	60.00	0.8026[a]	1.23	1.75	2.39	3.19	4.16	5.29	6.67	8.41	
	79.44	0.63	0.98	1.36	1.82	2.37	3.05	3.86	4.47	5.83	7.13
	98.89	0.5156[a]	0.80	1.11	1.46	1.87	2.36	2.95	3.62	4.38	5.23
	115.00	0.41	0.69	0.97	1.26	1.59	1.99	2.45	2.98	3.59	4.26
	135.00	0.34	0.58	0.82	1.07	1.34	1.64	1.98	2.39	2.86	3.36
n-Pentadecane, n-C$_{15}$	37.78	1.953[a]	3.20	4.85	7.00						
	60.00	1.335	2.10	3.11	4.37	5.93	7.89	10.43			
	79.44	1.01	1.56	2.27	3.14	4.19	5.47	7.02	8.90	11.16	
	98.89	0.7960[a]	1.24	1.75	2.37	3.13	4.02	5.09	6.33	7.78	
	115.00	0.67	1.06	1.48	1.98	2.58	3.27	4.09	5.05	6.20	
	135.00	0.54	0.87	1.22	1.60	2.05	2.59	3.22	3.93	4.73	
cis-Decahydronaphthalene	15.56	3.71	6.59	10.7	17.3	27.9	45.3	73.0			
	37.78	2.310[a]	3.76	5.81	8.81	13.3	20.2	30.3	45.5	68.7	106
	60.00	1.569[a]	2.57	3.95	5.89	8.60	12.5	18.2	26.6	38.7	56.5
	79.44	1.17	1.93	2.84	4.05	5.69	7.93	10.9	15.1	20.9	28.8
	98.89	0.9162[a]	1.50	2.19	3.06	4.17	5.67	7.59	10.2	13.6	18.1
trans-Decahydronaphthalene	37.78	1.546[a]	2.45	3.70	5.38	7.73	11.0	15.5			
	60.00	1.114[a]	1.76	2.58	3.63	5.06	6.89	9.38	12.7	17.3	
	79.44	0.86	1.35	1.94	2.70	3.70	4.98	6.58	8.64	11.4	15.0
	98.89	0.6960[a]	1.08	1.53	2.12	2.84	3.75	4.90	6.31	8.11	10.4
	115.00	0.59	0.90	1.30	1.78	2.36	3.09	3.99	5.07	6.44	8.15

Footnotes to Tables 2m-5A, 2m-5B, 2m-5C

[a] P. W. Bridgman, *Proc. Am. Acad. Arts Sci.* **61**, 57 (1926); "The Physics of High Pressure," G. Bell & Sons, Ltd., London, 1952.

[b] G. E. Babb and G. J. Scott [*J. Chem. Phys.* **40**, 3666 (1964)] report results which deviate by 4 per cent or less up to 8,000 bars.

[c] T. Titani, as quoted in J. Timmermans, "Physico-chemical Constants of Pure Organic Compounds," vol. 1 American Elsevier Publishing Company, Inc., New York, 1950.

[d] D. A. Lowitz, J. W. Spencer, W. Webb, and R. W. Schiessler, *J. Chem. Phys.* **30**(1), 73 (1959). Original also includes data for 7-*n*-hexyltridecane, 11-*n*-decylheneicosane, 13-*n*-dodecylhexacosane, 1,1-diphenylethane, 1,1-diphenylheptane, 1,1-diphenyltetradecane, 9(2-cyclohexylethyl)-heptadecane, 9(2-phenylethyl)heptadecane, 1,2,3,4,5,6,7,8,13,14,15,16-dodecahydrochrysene, 1,1-di(α-decalyl)hendecane.

[e] Confirmed measurements of E. M. Griest, W. Webb, and R. W. Schiessler, *J. Chem. Phys.* **29**(4), 711 (1958). Original also includes data for 1-phenyl-3(2-phenylethyl)hendecane; 1-cyclohexyl-3(2-cyclohexylethyl)hendecane, 9(3-cyclopentylpropyl)heptadecane, 1-cyclopentylpropyl)-heptadecane, 1-cyclopentyl-4(3-cyclopentylpropyl)dodecane, 1,7-dicyclopentyl-4(3-cyclopentylpropyl)heptane, 9-*n*-octyl(1,2,3,4-tetrahydro)-naphthacene.

[f] D. L. Hogenboom, W. Webb, and J. A. Dixon, *J. Chem. Phys.* **46**(7), 2586 (1967). Original also includes data for spiro(4,5)decane, spiro(5,5)-undecane, *cis*-octahydroindene, and *trans*-octahydroindene.

[g] Obtained with Cannon-Fenske capillary viscometers by American Petroleum Institute Research Project 42. "Properties of Hydrocarbons of High Molecular Weight Synthesized by Research Project 42 of the American Petroleum Institute," *op. cit.*, includes smoothed data from references d, e, and f for pressures in psi and temperatures in °F.

values. Based on the limited comparisons available for such measurements and the (at least partly subjective) estimates of the experimentalists involved, we estimate the uncertainty as ±3 per cent for values greater than 5 cp and up to ±5 per cent for lower values.

The viscosity of glycerol as a function of pressure has been measured by Bridgman,[1] by McDuffie and Kelly,[2] and by Tauke.[3] Unfortunately, all these measurements were on impure glycerol. Bridgman's glycerol apparently contained about 3 per cent water; that used by both McDuffie and Kelly and by Tauke, about $\frac{1}{2}$ per cent. As would be expected from the atmospheric-pressure results given in Table 2m-3, the values differ significantly and any predictions of the values for pure glycerol would be quite uncertain.

2n. Tensile Strength and Surface Tension of Liquids

WESLEY L. NYBORG

University of Vermont

ARTHUR F. SCOTT

Reed College

FRED D. AYRES

Reed College

Symbols

g	acceleration of gravity
h	rise of liquid in capillary tube
p	pressure
r	bore radius of capillary tube

[1] P. W. Bridgman, *Proc. Am. Acad. Arts Sci.* **61**, 57 (1926); "The Physics of High Pressure," G. Bell & Sons, Ltd., London, 1952.

[2] G. E. McDuffie, Jr., and M. V. Kelly, *J. Chem. Phys.* **41**(9), 2666 (1964).

[3] J. D. Tauke, Viscosity of Glycerol at Constant Density, M.S. thesis, Catholic University, October, 1964.

R_1, R_2 principal radii of curvature
V volume of liquid drop
W weight of liquid drop
γ surface tension
θ contact angle
λ wavelength of ripple wave
ρ density

2n-1. Tensile Strength. *Historical and General.* The maximum negative pressure (tensile strength) that a liquid can withstand has been the object of numerous investigations. Experimental values are quite discordant among themselves and are generally much lower than the theoretical estimates. The tensile strength of a liquid, measured in a device known as a tonometer, is taken as that stress (negative pressure) under which the liquid ruptures. A point of concern has been the possibility that rupture occurs at the wall of the container rather than in the body of the liquid and that therefore the observed negative pressure is a measure of adhesive force rather than of the assumed cohesive force.

Methods of Measuring Tensile Strength. Brief descriptions of these methods are given below, arranged according to the means used to produce the stress in the liquid. Each method is given a code designation for identification in Table 2n-1.

A. STRESS PRODUCED BY COOLING AND THUS CONTRACTING THE LIQUID. In Berthelot's method (A-1) the liquid, sealed in a thick-walled capillary tube, is first warmed until it just fills the tube and is then cooled until the liquid "breaks." The maximum negative pressure is calculated from the known mechanical properties of the liquid, assuming its extensibility to be the same as its compressibility. In Meyer's method (A-2) a spiral glass capillary is part of the tonometer and indicates the pressure exerted by the liquid, which completely fills the vessel. Meyer calibrated his spiral manometers under both positive and reduced pressure. Worthington, in a single experiment (A-3), measured the tension by means of a mercury-in-glass dilatometer, the bulb of which was enclosed within the tonometer. The calibration curve of the dilatometer, obtained previously by applying positive pressure, was extrapolated into the negative region. Vincent used a viscosity tonometer (A-4) in which the liquid completely filled a glass bulb and a fine capillary tube attached to it. By controlled cooling of the bulb, a gradually increasing tension is exerted on the liquid, measured at any time by the rate of flow through the capillary. The maximum tension can be calculated from the observed rates of flow before and after the liquid ruptures.

B. STRESS PRODUCED BY EXPANDING THE VOLUME OF TONOMETER. Vincent has described a new method (B-1) which employs a metal bellows completely filled with the liquid. Extension of the bellows exerts a pull on the contained liquid. An early method (B-2) involved the use of a long (2-m) tube closed at one end with a semipermeable membrane. After being filled with air-free water the tube is inverted and the open end is placed in a mercury trough. Evaporation of the water through the membrane causes the mercury to rise in the tube. The tension is estimated from the length of the column in excess of normal barometric height. Hulett (1903), in connection with an experiment of this type, observed a marked decrease in rate of evaporation as the mercury column rose and called attention to the analogy between negative pressure and osmotic pressure. This relationship forms the basis of a method for measuring the osmotic pressure of a solution. Budgett (B-3) measured the force required to pull apart flat steel surfaces wetted by a thin film of the liquid.

C. STRESS PRODUCED BY CENTRIFUGAL FORCE. Several experiments have been reported in which tension is developed by rotation of the tube containing the liquid. Reynolds (C-1) used U tubes sealed at both ends, with one arm longer than the other. One arm is filled completely with liquid; the other arm is only partially filled with liquid under its own vapor pressure. The tube is rotated about an axis positioned somewhat above the open part of the U. Temperly used a similar method (C-2), except that the short arm was open to the atmosphere. Recently Briggs (C-3) employed a Z-shaped capillary tube, open at both ends, rotating in the Z plane about

TABLE 2n-1. TENSILE STRENGTH OF LIQUIDS BY VARIOUS METHODS

Liquid and method	Max negative pressure, atm	Ref.	Liquid and method	Max negative pressure, atm	Ref.
Water:			Ether:		
A-1*	50–150	1	A-2	72	1
A-1†	157	2	B-1	2.2	1
A-1‡	17–56	3	Mineral oil:		
A-1	68	4	A-1	119	2
A-2	34	1	A-1¶	24	2
A-3	17	1	A-4	7.8	7
B-1	1.5	4	B-1	2.9	1
B-2	0.2–0.5	1	Acetic acid, C-3	288	8
B-3	4	1, 8	Benzene, C-3	150	8
C-1	4.8	1	Aniline, C-3	300	8
C-1	6.0	5	Carbon tetrachloride, C-3	276	8
C-2	5.6	5	Chloroform, C-3	317	8
C-3	277	6	Mercury, C-3	425	9
Alcohol:					
A-2	40	1			
A-3	17	1			
B-1	2.4	1			
C-1	7.9	1			

* Values reported prior to 1941.
† Tubes boiled for 8 hr to expel air.
‡ Tubes filled by vacuum technique to eliminate air.
¶ Tube sealed by liquid frozen in capillary side arm.

References for Table 2n-1

1. Vincent, R. S.: *Proc. Phys. Soc.* **53,** 141 (1941).
2. Vincent, R. S., and G. H. Simmonds: *Proc. Phys. Soc.* **55,** 376 (1943).
3. Scott, A. F., D. P. Shoemaker, K. N. Tanner, and J. G. Wendel: *J. Chem. Phys.* **16,** 495 (1948).
4. Scott, A. F., and G. M. Pound: *J. Chem. Phys.* **9,** 726 (1941).
5. Temperly, H. N. V., and L. G. Chambers: *Proc. Phys. Soc.* **58,** 420 (1946).
6. Briggs, Lyman J.: *J. Appl. Phys.* **21,** 721 (1950).
7. Vincent, R. S.: *Proc. Phys. Soc.* **55,** 41 (1943).
8. Briggs, Lyman J.: *J. Chem. Phys.* **19,** 970 (1951).
9. Briggs, Lyman J.: *J. Appl. Phys.* **24,** 488 (1953).

an axis passing through the center of the Z and perpendicular to the plane. The liquid menisci are located in the bent-back short arms of the Z. The speed of rotation is increased gradually until the liquid in the capillary "breaks."

A fairly complete summary of the experimental measurements of the tensile strength of pure liquids is tabulated above. Information and references pertaining to work prior to 1941 are to be found in ref. 1 (Table 2n-1), a paper which describes method B-1.

Another experimental method which gives insight into liquid tension involves measuring the pressure-amplitude threshold for sonically generated cavitation. This threshold varies with conditions; values quoted in the literature vary over the same range as for the maximum negative pressures recorded in Table 2n-1. It is found the preexisting gaseous bubbles or cavitation *nuclei* are important; recently it has been shown that such nuclei may be provided by cosmic rays or neutrons passing through liquid.[1,2]

2n-2. Surface Tension and Surface Energy of Liquids.[3] *Definitions.* Owing to molecular attraction two fluids in contact adjust themselves so that the area of their interface is a minimum consistent with other requirements. The work required to extend the surface by unit area is called the "free surface energy." In solving problems it is convenient to replace the concept of free surface energy by that of a hypothetical tension, acting parallel to the surface. Named the surface tension and its value denoted by γ, this is defined as the normal tensile force per unit of length across any line traced on the surface. The free surface energy and the surface tension have the same dimensions (MT^{-2}) and are numerically equal; the units of γ may be given either as dynes/cm or as ergs/cm^2.

Formulas Involving Surface Tension. When the interfacial surface between two fluids is curved the pressure p_1 on the concave side exceeds that, p_2, on the convex side by the amount

$$p_1 - p_2 = \gamma(R_1^{-1} + R_2^{-1}) \tag{2n-1}$$

where R_1, R_2 are the principal radii of curvature. The pressure p due to surface tension within a liquid drop or gas bubble of radius R surrounded by liquid is

$$p = \frac{2\gamma}{R} \tag{2n-2}$$

The velocity v of sinusoidal ripples on the surface of a liquid of great depth is given by[4]

$$v^2 = \frac{g\lambda}{2\pi} + \frac{2\pi\gamma}{\rho\lambda} \tag{2n-3}$$

where λ is the wavelength of the ripples, g is the acceleration due to gravity, and ρ is the density of the liquid (cf. Sec. 2a).

Methods of Measuring the Surface Tension of a Liquid Relative to a Gas Phase.
1. Capillary-height method. If a vertical capillary tube whose bore radius r is sufficiently small rests with its lower end below a liquid surface the liquid in it will rise to a height h given approximately by

$$h = \frac{2\gamma \cos \theta}{gr(\rho - \rho_v)} \tag{2n-4}$$

where ρ_v is the density of the gas above the liquid, and θ is the contact angle of the meniscus with the tube wall (θ is often zero). If the tube is not sufficiently small, corrections must be applied to the above formula.[5]

2. Maximum-bubble-pressure method. If a bubble is blown at the lower end of a tube of small bore dipping into a liquid the pressure in the bubble reaches a maximum value given by

$$p = \frac{2\gamma}{r} \tag{2n-5}$$

[1] H. G. Flynn, Physics of Acoustic Cavitation in Liquids, in "Physical Acoustics," vol. 1B, W. P. Mason, ed., Academic Press, Inc., New York, 1964.
[2] D. Sette and F. Wanderlingh, *Phys. Rev.* **125**, 409–417 (1962).
[3] General references: Neil K. Adam, "The Physics and Chemistry of Surfaces," chap. IX, Oxford University Press, New York, 1941; H. S. Taylor and S. Glasstone, "A Treatise on Physical Chemistry," D. Van Nostrand Company, Inc., Princeton, N.J., 1952.
[4] Rayleigh, *Phil. Mag.* **30**, 386 (1890).
[5] Adam, *loc. cit.*

where r, as before, is the bore radius. If r is not sufficiently small, corrections must be applied to the above formula.[1]

3. Drop-weight method. The weight W of a drop falling from the tip of a vertical tube is given by

$$W = \frac{r\gamma}{F} \qquad (2\text{n-6})$$

where F is an empirical function[1] of (V/r^3), V being the drop volume. When (V/r^3) is 5,000, F is 0.172; as (V/r^3) decreases to 1.55, F increases steadily to 0.26; further decrease of (V/r^3) causes F to oscillate slightly around 0.25.

TABLE 2n-2. SURFACE TENSION OF WATER AGAINST AIR*

Temp., °C	Surface tension, dynes/cm	Temp., °C	Surface tension, dynes/cm	Temp., °C	Surface tension, dynes/cm
−8	77.0	15	73.49	40	69.56
−5	76.4	18	73.05	50	67.91
0	75.6	20	72.75	60	66.18
5	74.9	25	71.97	70	64.4
10	74.22	30	71.18	80	62.6
				100	58.9

* General reference: "Handbook of Chemistry and Physics," Chemical Rubber Publishing Company, Cleveland.

TABLE 2n-3. SURFACE TENSION OF VARIOUS LIQUIDS

Name	Formula	In contact with	Temp., °C	Surface tension, dynes/cm	Ref.*
Acetic acid	$C_2H_4O_2$	Vapor	10	28.8	AC(22,23,25);
Acetic acid	$C_2H_4O_2$	Vapor	50	24.8	GC(1); JS(14); tPRS(1); ZC(1,6)
Acetone	C_3H_6O	Air or vapor	0	26.21	AC(20,24,25); AdC(1); BF(1);
Acetone	C_3H_6O	Air or vapor	40	21.16	JP(5); JS(4,14); ZC(6)
Ammonia	NH_3	Vapor	11.1	23.4	JP(7)
Ammonia	NH_3	Vapor	34.1	18.1	JP(7)
Argon	A	Vapor	−188	13.2	JS(15)
Benzene	C_6H_6	Air	10	30.22	AC(3,5,31,32,34);
Benzene	C_6H_6	Air	30	27.56	BF(2); JP(5); JS(4,9,10,11,14); PRS(2);tRIA(1); tPRS(1); ZC(4,5)
Benzophenone	$C_{13}H_{10}O$	Air or vapor	20	45.1	AC(27); AS(1); ZC(5)
Bromine	Br_2	Air or vapor	20	41.5	AC(17); AdP(3); GC(1)
n-Butyric acid	$C_4H_8O_2$	Air	20	26.8	AC(27); GC(1); JS(4)

[1] Adam, loc. cit.

TABLE 2n-3. SURFACE TENSION OF VARIOUS LIQUIDS (*Continued*)

Name	Formula	In contact with	Temp., °C	Surface tension, dynes/cm	Ref.*
Carbon bisulfide........	CS_2	Vapor	20	32.33	AC(17,28); GC(1); BF(2); JS(14); PRS(2), ZC(6)
Carbon dioxide.........	CO_2	Vapor	20	1.16	VK(1,2)
Carbon dioxide.........	CO_2	Vapor	−25	9.13	VK(1,2)
Carbon monoxide.......	CO	Vapor	−193	9.8	JS(15)
Carbon tetrachloride....	CCl_4	Vapor	20	26.95	AC(3,5,6,28,31);
Carbon tetrachloride....	CCl_4	Vapor	200	6.53	PRS(1,2); ZC(5)
Chlorine..............	Cl_2	Vapor	20	18.4	AC(11); JP(3)
Chlorine..............	Cl_2	Vapor	−60	31.2	AC(11); JP(3)
Chlorobenzene.........	C_4H_5Cl	Vapor	20	33.56	AC(6,20,28); JP(5); JS(11); PRS(2);tRIA(1); tPRS(1); ZC(5)
Chloroform...........	$CHCl_3$	Air	20	27.14	AC(6,28,31); AdC(1); PRS(2); tRIA(1); ZC(6)
Cyclohexane..........	C_6H_{12}	Air	20	25.5	PRS(1); ZA(1)
Ethyl acetate.........	$C_4H_8O_2$	Air	0	26.5	AC(26,33);
Ethyl acetate.........	$C_4H_8O_2$	Air	50	20.2	AdC(1); AS(2); JP(5); tPRS(1); ZC(6)
Ethyl alcohol.........	C_2H_6O	Air	0	24.05	AC(22,23,25,32);
Ethyl alcohol.........	C_2H_6O	Vapor	30	21.89	BF(2); JP(5); tRIA(1); tPRS(1)
Ethyl ether...........	$C_4H_{10}O$	Vapor	20	17.01	AC(4,15,28,31);
Ethyl ether...........	$C_4H_{10}O$	Vapor	50	13.47	AdC(1); tPRS(1)
Glycerol..............	$C_3H_8O_3$	Air	20	63.4	JR(1); MB(1);
Glycerol..............	$C_3H_8O_3$	Air	150	51.9	ZA(1); ZC(3)
Helium...............	He	Vapor	−269	0.12	cUL(2); PRA(2)
Helium...............	He	Vapor	−271.5	0.353	cUL(2); PRA(2)
n-Hexane.............	C_6H_{14}	Air	20	18.43	AC(5,6,16); AdC(1); AS(1)
Hydrogen.............	H_2	Vapor	−255	2.31	cUL(1); PRA(1)
Hydrogen peroxide......	H_2O_2	Vapor	18.2	76.1	AC(13)
Methyl alcohol........	CH_4O	Air	0	24.49	AC(22,23,25,32);
Methyl alcohol........	CH_4O	Vapor	50	20.14	tPRS(1)
Neon.................	Ne	Vapor	−248	5.50	cUL(3); PRA(3)
Nitric acid (98.8%)	HNO_3	Air	11.6	42.7	JS(2)
Nitrogen..............	N_2	Vapor	−183	6.6	JS(15)
Nitrogen..............	N_2	Vapor	−203	10.53	JS(15)
Nitrogen tetra oxide....	N_2O_4	Vapor	19.8	27.5	JS(4)
n-Octane..............	C_8H_{18}	Vapor	20	21.80	AC(4,5,34); JS(4)
n-Octyl alcohol........	$C_8H_{18}O$	Air	20	27.53	AC(4,5)
Oxygen...............	O_2	Vapor	−183	13.2	JS(15)

MECHANICS

TABLE 2n-3. SURFACE TENSION OF VARIOUS LIQUIDS (*Continued*)

Name	Formula	In contact with	Temp., °C	Surface tension, dynes/cm	Ref.*
Oxygen	O_2	Vapor	−203	18.3	JS(15)
Phenol	C_6H_6O	Air or vapor	20	40.9	AC(18,19,25); JS(2,6,13); JP(4)
Phosphorus trichloride	PCl_3	Vapor	20	29.1	AC(17); GC(1); JP(2); JS(4)
n-Propylamine	C_3H_9N	Air	20	22.4	GC(1); JS(3)
Sulfuric acid (98.5%)	H_2SO_4	Air or vapor	20	55.1	AC(17a); AdP(7); JS(2)
Toluene	C_7H_8	Vapor	10	27.7	AC(4,17,20,31)
Toluene	C_7H_8	Vapor	30	27.4	JP(5); PRS(2); ZC(5,6)

* General reference: "Handbook of Chemistry and Physics," Chemical Rubber Publishing Company, Cleveland. A reference key is at end of article.

TABLE 2n-4. SURFACE TENSION OF METALS

Name	Symbol	Gas	Temp., °C	Surface tension, dynes/cm	Ref.*
Aluminum	Al	Air	700	840	CR(1)
Antimony	Sb	H_2	750	368	ZA(4)
Antimony	Sb	H_2	640	350	PM(1)
Bismuth	Bi	H_2	300	388	PM(1)
Bismuth	Bi	H_2	583	354	ZA(4)
Bismuth	Bi	CO	700–800	346	AdP(2)
Cadmium	Cd	H_2	320	630	AC(10)
Copper	Cu	H_2	1131	1,103	ZA(4)
Gallium	Ga	CO_2	30	358	AC(30)
Gold	Au	H_2	1070	580–1,000	AdP(1); AdP(2); JI(1)
Lead	Pb	H_2	350	453	PM(1)
Lead	Pb	H_2	750	423	ZA(4)
Mercury	Hg	Vacuum	0	480.3	AC(7)
Mercury	Hg	Air	15	487	AC(9); AdP(5); AdP(6); CR(2)
Mercury	Hg	H_2	19	470	PM(1)
Mercury	Hg	Vacuum	60	467.1	AC(7)
Platinum	Pt	Air	2000	1,819	AdP(2)
Potassium	K	CO_2	62	411	AdP(3)
Silver	Ag	Air	970	800	AdP(2); AdP(4); JI(1)
Sodium	Na	CO_2	90	294	AdP(3)
Sodium	Na	Vacuum	100	206.4	PR(1)
Sodium	Na	Vacuum	250	199.5	PR(1)
Tin	Sn	H_2	253	526	PM(1)
Tin	Sn	H_2	878	508	ZA(4)
Zinc	Zn	H_2	477	753	AC(10)
Zinc	Zn	Air	590	708	JI(1)

* General reference: "Handbook of Chemistry and Physics," Chemical Rubber Publishing Company, Cleveland.

TABLE 2n-5. SURFACE TENSIONS OF AQUEOUS SOLUTIONS AGAINST AIR—ORGANIC*

Substance	°C		γ = surface tension for concentrations indicated							
Acetic acid.........	30	%	1.000	2.475	5.001	10.01	30.09	49.96	69.91	100.00
		γ	68.0	64.4	60.1	54.6	43.6	38.4	34.3	26.6
Acetone...........	25	%	5.00	10.0	20.00	25.00	50.00	75.0	95.0	100.00
		γ	55.5	48.9	41.1	38.3	30.4	26.8	24.2	23.0
Ethyl alcohol......	30	%	0.979	2.143	4.994	10.39	25.00	50.00	75.06	100.00
		γ	66.1	61.6	54.2	45.9	34.1	27.5	24.7	21.5
Sucrose...........	25	%	10.0	20.0	30.0	40.0	55.0			
		γ	72.5	73.0	73.4	74.1	75.7			

* General reference: "Handbook of Chemistry and Physics," Chemical Rubber Publishing Company, Cleveland.

TABLE 2n-6. SURFACE TENSION OF AQUEOUS SOLUTIONS AGAINST AIR— INORGANIC*

(f = gram formula weights per 1,000 g of solvent)

For these aqueous solutions the values of $\Delta\gamma$ are given. $\Delta\gamma$ is the difference between the surface tension of the solution and that of the solvent at the same temperature. Positive values of $\Delta\gamma$ mean that the surface tension of the solution is greater than that of the solvent; negative values the reverse. For convenience in computing the surface tension, the current accepted value for the surface tension of water at the stated temperature is given in the second column.

Formula	°C (γH_2O)		$\Delta\gamma$ for concentrations indicated							
CaCl₂	25 (71.97)	f	0.1	0.5	1.0	2.0	3.0	5.0	11.2	
		$\Delta\gamma$	0.35	1.5	3.2	6.9	11.0	18.4	35	
HCl	20 (72.75)	f	....	0.5	1.0	2.0	4.0	6.0	9.0	17.7
		$\Delta\gamma$	....	−0.2	−0.3	−0.5	−0.9	−1.3	−2.2	−7
NH₄OH	18 (73.05)	f	....	0.5	1.0	1.5	3.0	6.0	15.0	34.0
		$\Delta\gamma$	....	−1.4	−2.4	−3.1	−5.2	−7.8	−12.0	−16.0
HNO₃	20 (72.75)	f	....	0.7	1.5	2.8	....	....	8.5	
		$\Delta\gamma$	....	−0.6	−1.1	−1.8	....	....	−4	
KCl	20 (72.75)	f	0.1	0.5	1.0	2.0	3.0	4.0	4.4	
		$\Delta\gamma$	0.16	0.70	1.4	2.8	4.2	5.5	6.0	
KOH	18 (73.05)	f	....	0.5	1.0	2.0	3.8			
		$\Delta\gamma$	....	0.9	1.8	3.5	6.7			
MgCl₂	20 (72.75)	f	0.1	0.5	1.0	2.0	3.0	3.65		
		$\Delta\gamma$	0.32	1.52	3.0	6.4	10.2	13.0		
MgSO₄	20 (72.75)	f	0.1	0.5	1.0	2.0	2.7			
		$\Delta\gamma$	0.26	1.03	2.1	4.6	6.5			
NaBr	20 (72.75)	f	....	0.5	1.0	1.5	2.9			
		$\Delta\gamma$	....	0.7	1.3	2.0	3.8			
NaCl	20 (72.75)	f	0.1	0.5	1.0	2.0	3.0	5.0	6.0	
		$\Delta\gamma$	0.17	0.82	1.64	3.3	4.9	8.2	9.8	
Na₂CO₃	20 (72.75)	f	0.25	0.5	1.0	1.5				
		$\Delta\gamma$	0.7	1.3	2.7	4.0				
NaNO₃	20 (72.75)	f	0.1	0.5	1.0	2.0	3.0	5.0	7.0	12.2
		$\Delta\gamma$	0.12	0.60	1.2	2.4	3.5	5.6	7.5	11.3
NaOH	18 (73.05)	f	....	0.7	1.5	....	....	5.0	11.0	14.0
		$\Delta\gamma$	....	1.3	2.8	....	....	10.0	23	28
Na₂SO₄	20 (72.75)	f	0.2	0.5	1.0					
		$\Delta\gamma$	0.5	1.4	2.7					

* General reference: "Handbook of Chemistry and Physics," Chemical Rubber Publishing Company, Cleveland.

Reference Key to Surface-tension Data

AC.   *Journal of the American Chemical Society.*   (1) Baker and Gilbert, **62**, 2479–2480 1940.   (2) H. Brown, **56**, 2564–2568 (1938).   (3) Harkins and Brown, **41**, 449 (1919). (4) Harkins, Brown, and Davies, **39**, 354 (1917).   (5) Harkins and Cheng, **43**, 35 (1921). (6) Harkins, Clark, and Roberts, **42**, 700 (1920).   (7) Harkins and Ewing, **42**, 2539 (1920).   (8) Harkins and Feldman, **44**, 2665 (1922).   (9) Harkins and Grafton, **42**, 2534 (1920).   (10) Hogness, **43**, 1621 (1921).   (11) Johnson and McIntosh, **31**, 1139 (1909).   (12) Maass and Boomer, **44**, 1709 (1922).   (13) Maass and Hatcher, **42**, 2548 (1920).   (14) Maass and McIntosh, **31**, 1139 (1909).   (15) Maass and Wright, **43**, 1098 (1921).   (16) Morgan and Chazel, **35**, 1821 (1913).   (17) Morgan and Daghlian, **33**, 672 (1911).   (17a) Morgan and Davis, **38**, 555 (1916).   (18) Morgan and Egloff, **38**, 844 (1916).   (19) Morgan and Evans, **39**, 2151 (1917).   (20) Morgan and Griggs, **39**, 2261 (1917).   (21) Morgan and Kramer, **35**, 1834 (1913).   (22) Morgan and McAfee, **33**, 1275 (1911).   (23) Morgan and Neidle, **35**, 1856 (1913).   (24) Morgan and Owen, **33**, 1713 (1911).   (25) Morgan and Scarlett, **39**, 2275 (1917).   (26) Morgan and Schwartz, **33**, 1041 (1911).   (27) Morgan and Stone, **35**, 1505 (1913).   (28) Morgan and Thomssen, **33**, 657 (1911).   (29) Morgan and Woodward, **35**, 1249 (1913).   (30) Richards and Boyer, **43**, 274 (1921).   (31) Richards and Carver, **43**, 827 (1921).   (32) Richards and Coombs, **37**, 1656 (1915).   (33) Richards and Matthews, **30**, 8 (1908).   (34) Richards, Speyers, and Carver, **46**, 1196 (1924).

AD.   *Atti della reale accademia nazionale dei Lincei.*   (1) Magini, **1911**, 184, 10.

AdC.   *Annalen der Chemie, Justus Liebigs.*   (1) Schiff, **223**, (47) 84.

AdP.   *Annalen der Physik.*   (1) Heydweiller, **62**, 694 (1897).   (2) Quincke, **134**, 356 (1868).   (3) Quincke, **135**, 621 (1868).   (4) Gradenwitz, **67**, 467 (1899).   (5) Meyer, **66**, 523 (1898).   (6) Stöckle, **66**, 499 (1898).   (7) Röntgen and Schneider, **29**, 165 (1886).*

AS. *Archives des sciences physiques et naturelles.* (1) Dutoit and Friederich, **9**, 105 (1900). (2) Guye and Baud, **11**, 449 (1901). (3) Herzen, **14**, 232 (1902).
BD. *Berichte der deutschen chemischen Gesellschaft.* (1) Lorenz and Kauffler, **41**, 3727 (1908). (2) Schenck and Ellenberger, **37**, 3443 (1904).
BF. *Bulletin de la société chimique de France.* (1) Dutoit and Friederich, **19**, 321 (1898). (2) Santis Ann. Univ. Grenoble, **27**, 593 (1904).
CR. *Comptes rendus.* (1) A. Portevin and P. Bastien, **202**, 1072–1074 (1936). (2) Popesco, **172**, 1474 (1921).
cUL. *Communications from the Physical Laboratory at the University of Leiden.* (1) No. 142. (2) No. 179a. (3) No. 182b.
GC. *Gazzetta chimica italiana.* (1) Schiff, **14**, 368 (1884). (2) A. Giacolone and D. DiMaggio, **3**, 198–206 (1939).
JI. *Journal of the Institute of Metals.* (1) Smith, **12**, 168 (1914).
JP. *Journal de chimie physique.* (1) Dutoit and Fath, **1**, 358 (1903). (2) Dutoit and Moioiu, **7**, 169 (1909). (3) Marchand, **11**, 573 (1913). (4) Bolle and Guye, **3**, 38 (1905). (5) Renard and Guye. **5**, 81 (1907). (6) Berthoud, **21**, 143 (1924). (7) Berthoud, **16**, 429 (1918). (8) Homfray and Guye, **1**, 505 (1903). (9) Przyluska, **7**, 511 (1909).
JR. *Journal of the Russian Physical Chemical Society.* (1) Elisseev and Kurbatov, **41**, 1426 (1909).
JS. *Journal of the Chemical Society (London).* (1) Kellas, **113**, 903 (1918). (2) Aston and Ramsay, **65**, 167 (1894). (3) Turner and Merry, **97**, 2069 (1910). (4) Ramsay and Shields, **63**, 1089 (1893). (5) Sugden, Reed, and Wilkins, **127**, 1525 (1925). (6) Hewitt and Winmill, **91**, 441 (1907). (7) Smith, 105, 1703 (1914). (8) Atkins, **99**, 10 (1911). (9) Sugden, **119**, 1483 (1921). (10) Sugden, **121**, 858 (1922). (11) Sugden, **125**, 32 (1924). (12) Sugden, **125**, 1167 (1924). (13) Worley, **105**, 260 (1914). (14) Worley, **105**, 273 (1914). (15) Baly and Donnan, **81**, 907 (1902).
JSG. *Jahresberichte Schles. Ges. Vaterl. Kultur.* (1) Wilborn, **1912**, 56.
MB. *Metron. Beit.* (1) Weinstein, no. 6, 89.
MfC. *Monatshefte für Chemie und verwandte Teile anderer Wissenschaften.* (1) Kremann and Meingast, **35**, 1323 (1914).
PM. *The London, Edinburgh & Dublin Philosophical Magazine and Journal of Science.* (1) Bircumshaw, **2**, 341 (1926); **3**, 1286 (1927). (2) R. C. Brown, **13**, 578–584 (1932). (3) A. E. Bate, **28**, 252–255 (1939).
PR. *The Physical Review.* (1) Poindexter, **27**, 820 (1926).
PRA. *Proceedings of the Royal Academy of Sciences of Amsterdam.* (1) Kamerlingh Onnes, and Kuypers, **17**, 528 (1914). (2) Van Urk, Keesom, and Onnes, **28**, 958, (1925). (3) Van Urk, Keesom, and Nijhoff, **29**, 914 (1926). (4) Jaeger and Kahn, **18**, 75 (1915).
PRS. *Proceedings of the Royal Society (London).* (1) Hardy, **88**, 303 (1913). (2) Ramsay and Aston, **56**, 162, 182 (1894).
tPRS. *Philosophical Transactions of the Royal Society of London, Series A.* (1) Ramsay and Shields, **184**, 647 (1893).
tRIA. *Royal Irish Academy Transactions.* (1) Ramsay and Aston, **32A**, 93 (1902).
VK. *Verslag koninklijke Akademie van Wetenschappen te Amsterdam.* (1) Verschaffelt, no. 18, 74 (1895). (2) Verschaffelt, no. 28, 94 (1896).
WN. *Wissenschaftliche Natuurk., Tydschr.* (1) Verschaffelt, **2**, 231 (1925).
ZA. *Zeitschrift für anorganische und allgemeine Chemie.* (1) Jaeger, **101**, 1 (1917). (2) Motylewski, **38**, 410 (1904). (3) Lorenz, Liebmann, and Hochberg, **94**, 301 (1916). (4) Sauerwald and Drath, **154**, 79 (1926).
ZC. *Zeitschrift für physikalische Chemie.* (1) Bennett and Mitchell, **84**, 475 (1913). (2) Walden and Swinne, **82**, 271 (1913). (3) Drucker, **52**, 641 (1905). (4) Walden, **75**, 555 (1910). (5) Walden and Swinne, **79**, 700 (1912). (6) Whatmough, **39**, 129 (1902).
ZE. *Zeitschrift für Elektrochemie und angewandte physikalische Chemie.* (1) Bredig and Tiechmann, **31**, 449 (1925).
ZK. *Kolloid-Zeitschrift.* (1) N. Jermolanko, **48**, 14–146 (1929).

2o. Cavitation in Flowing Liquids

PHILLIP EISENBERG

Hydronautics, Incorporated

2o-1. Introduction—Status of Available Data. Although the possibility of occurrence of cavitation in hydrodynamic systems was recognized as long ago as 1754 by Euler,[1] significant researches on the physical phenomena have been developed only during the first half of the present century. This has resulted from the growing importance of the effects of cavitation (both useful and detrimental) in such diverse fields as underwater propulsion and hydraulic machinery (loss of efficiency, damage to materials, noise), underwater signaling (background noise, absorption of acoustical power), hydroballistics (increased drag and instability of missiles), medicine (divers' bends, bullet wounds), and chemical processing (acceleration of reactions and mixing processes, industrial cleaning). Because of the complexities of the phenomena—hydrodynamical and physicochemical—in cavitated regions, research activity continues to emphasize understanding and description of events. Consequently, this section is restricted to brief descriptions of the various factors involved in the cavitation process and to the presentation of data which, while consistent within themselves, are intended primarily to illustrate the text. In all cases, reference should be made to the original source for guidance in judging the limits of accuracy and applicability of these data.

The discussion given here is concerned particularly with phenomena associated with flowing liquids and excludes cavitation produced by heat addition (boiling) and acoustical pressure waves as well as problems of pure liquids (e.g., ultimate tensile strength). Rather complete discussions of cavitation in flowing liquids (and about forms moving through stationary liquids) have been given by Ackeret[2] and Eisenberg,[3] and extensive bibliographies will be found in the papers of these authors and in a compilation by Raven et al.[4]

2o-2. Definitions and Nomenclature

$$\sigma = \frac{P - p_v}{\frac{1}{2}\rho U^2} \quad \text{cavitation number}$$

P ambient pressure

---

[1] Leonhard Euler, Théorie plus complète des machines, qui sont mises en mouvement par la réaction de l'eau, *Historie de l'Academie Royale des Sciences et Belles Lettres*, Classe de Philosophie Experimentale, Mém. 10, pp. 227–295, 1754, Berlin, 1756.

[2] J. Ackeret, Kavitation (Hohlraumbildung), *Handbuch der Experimentalphysik* **IV** (1), 461–486 (Leipzig, 1932).

[3] Phillip Eisenberg, Kavitation, *Forschungshefte für Schiffstechnik* **3**, 111–124, 1953; **4**, 155–168 (1953); **5**, 201–212 (1954); On the Mechanism and Prevention of Cavitation, *David Taylor Model Basin, U.S. Navy Dept. Rept.* 712, July, 1950; A Brief Survey of Progress on the Mechanics of Cavitation, *David Taylor Model Basin, U.S. Navy Dept. Rept.* 842, June, 1953.

[4] F. A. Raven, A. M. Feiler, and Anna Jesperson, An Annotated Bibliography of Cavitation, *David Taylor Model Basin, U.S. Navy Dept. Rept.* R-81, December, 1947.

p_v	vapor pressure or actual pressure within a cavity
ρ	mass density of liquid
U	stream velocity
σ_i or K	cavitation number for inception of cavitation ("critical" cavitation number)
$Re = \dfrac{Ud}{\nu}$	Reynolds number
ν	kinematic viscosity
d	diameter of a body of revolution
d_m	maximum diameter of steady-state cavity
l	length of a steady-state cavity
R	radius of a transient cavity
h	altitude of a cone
$C_D = \dfrac{D}{\frac{1}{2}\rho U^2 A}$	drag coefficient
D	drag
A	area of body in plane normal to stream or cross-sectional area of circular cylinder
$C_D(\sigma)$	drag coefficient at cavitation number σ
α	total absolute air content
α_s	total saturation air content

2o-3. Inception of Cavitation. It is now generally agreed that cavitation originates with the growth of undissolved vapor or gas nuclei existing in the liquid or trapped on microscopic foreign particles. It is well known that the rupture forces of very clean and carefully degassed liquids are of the order of those predicted by kinetic theoretical formulations. Experimental evidence has also been obtained that water saturated with air, but denucleated by application of very high pressures, exhibits large tensile strength (of the order of several hundred atmospheres).[1] Thus the presence of nuclei is evidently necessary for the inception of cavitation at pressures of the order of vapor pressure. In supersaturated liquids, it is easy to account for the presence and stability of such nuclei, but in saturated and undersaturated liquids, the situation is not clear, and the presence of nuclei is usually accounted for on the basis that they are stabilized on suspended particles.[2] As a consequence, depending upon the size and number of these nuclei, cavitation may be expected to begin above as well as below the vapor pressure. The effect of total air content was shown in experiments of Crump[3] using a venturi nozzle having a diffuser of 5 deg included angle. Figure 2o-1 shows that in the undersaturated liquid it was possible to obtain tensions as the air content was reduced. Results in a nozzle with an abrupt expansion, however, show opposite trends in the pressures required for inception,[3] although here too tensions were obtained. Comparable results for sea water[4] are shown in Fig. 2o-2; since the water is supersaturated, thus presumably containing a large number of undissolved nuclei, bursts of cavitation are observed at pressures well above vapor pressure.

[1] Newton E. Harvey, W. D. McElroy, and A. H. Whiteley, On Cavity Formation in Water, *J. Appl. Phys.* **18**, 162–172 (February, 1947).

[2] Eisenberg, *loc. cit.;* P. S. Epstein and M. S. Plesset, On the Stability of Gas Bubbles in Liquid-Gas Solutions, *J. Chem. Phys.* **18** (11), 1505–1509 (November, 1950).

[3] S. F. Crump, Critical Pressures for the Inception of Cavitation in a Large-scale Numachi Nozzle as Influenced by the Air Content of the Water, *David Taylor Model Basin, U.S. Navy Dept. Rept.* 770, July, 1951.

[4] S. F. Crump, Determination of Critical Pressures for the Inception of Cavitation in Fresh and Sea Water as Influenced by Air Content of the Water, *David Taylor Model Basin, U.S. Navy Dept. Rept.* 575, October, 1949.

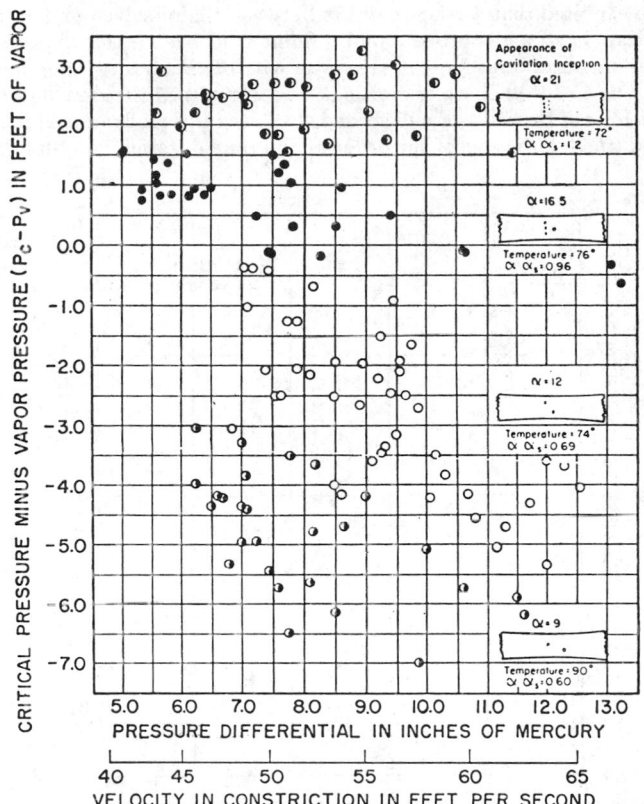

Fig. 2o-1. Cavitation inception in fresh water of varying air content. (*After Crump.*)

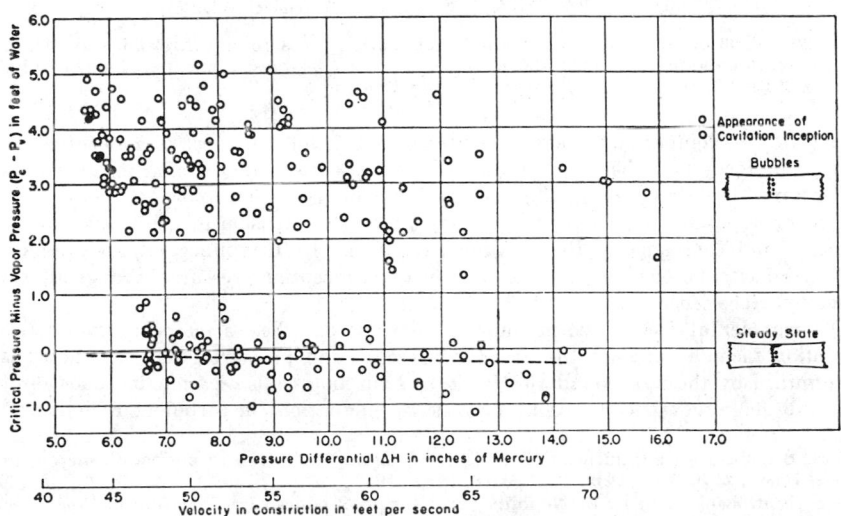

Fig. 2o-2. Critical pressure for inception of cavitation in sea water. (*After Crump.*)

It may be expected that a relation exists between the dissolved and entrained gas content, at least in an undisturbed liquid.   Some evidence for this assumption exists in the measurements of Strasberg[1] on tap water with ultrasonically induced cavitation. Since, according to the analysis of Noltingk and Neppiras,[2] the time duration of the pressure for times of the order of milliseconds has very little influence on the inception pressure, and since this is also of the order of the time duration in hydraulic applica-

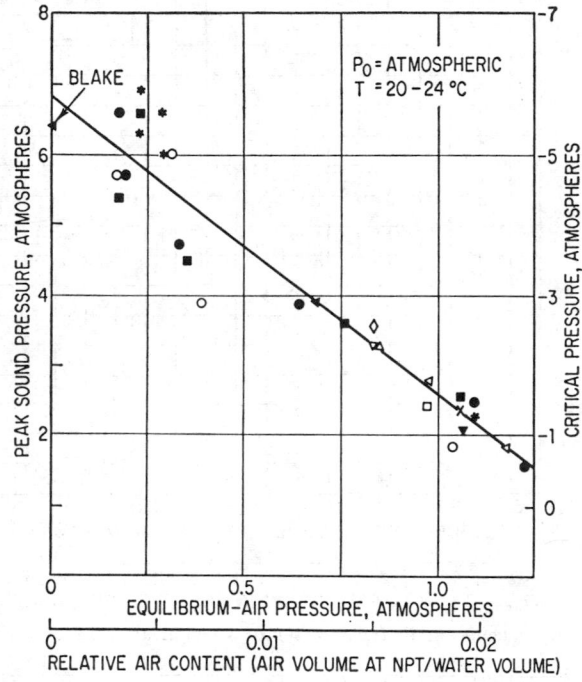

Fig. 2o-3. The effect of air content on the inception pressure for ultrasonic cavitation. (Each plotted point represents the average of 10 to 20 measurements.   Each symbol represents a different sample of water.)   (*After Strasberg.*)

tions, we can compare Strasberg's results as a basis for the effect of air content on cavitation inception.   These are shown in Fig. 2o-3.   Whether similar results would be obtained in flowing water using ultrasonic techniques is not known.

Properties of the liquid such as viscosity and surface tension influence the growth of nuclei and, consequently, the inception pressures.   In this connection, the presence of surface-active materials (detergents, etc.) affect inception pressures through alteration of surface tension.

Environmental factors which must be considered when attempting to predict inception include not only the average pressure and pressure-gradient conditions determined by the flow boundaries (such as bounding walls or a moving body) but also the magnitude and duration of pressure fluctuations in turbulent regions and

[1] M. Strasberg, The Influence of Air-filled Nuclei on Cavitation Inception, *David Taylor Model Basin, U.S. Navy Dept. Rept.* 1078, May, 1957.

[2] B. E. Noltingk and E. A. Neppiras, Cavitation Produced by Ultrasonics, *Proc. Roy. Soc (London)*, ser. B, **63**, 674–685 (1950); 1032–1038 (1951).

boundary-layer effects including flow in zones of separation. An example of the effects of the boundary layer and, in particular, local separation is shown in Fig. 2o-4 from the work of Rouse and McNown.[1] In this figure are compared the minimum pressure coefficients with the cavitation numbers at which the pressure distribution first showed a change. This change is attributed to microscale cavitation in locally separated flows and served to define the critical cavitation number. Effect of model size on inception has been studied by Kermeen[2] and others.[3] While the mechanisms

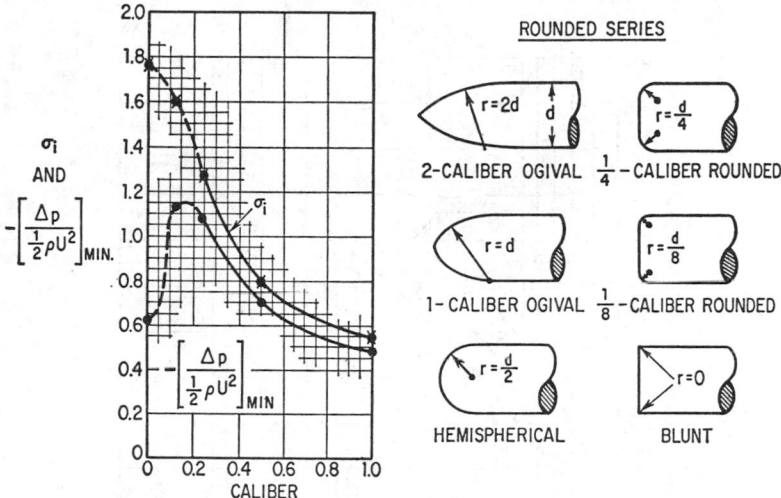

FIG. 2o-4. Critical cavitation number for first change in minimum pressure coefficient of bodies of revolution and minimum pressure coefficient vs. caliber of rounding. (*After Rouse and McNown.*)

are still only incompletely understood, trends are fairly well established and are consistent with the concept of nuclei and the role of the boundary layer.[4] An example of Kermeen's results is shown in Fig. 2o-5 wherein the average values of a large number of data are plotted for models of various diameters.

2o-4. Transient (Bubble) Cavities. These are small individual bubbles which grow, sometimes oscillate, and eventually collapse and disappear. Of particular interest here are the pressures produced in the vicinity of such cavities when they collapse. From studies of damage and acoustic radiation produced by such cavities, it is known that pressures of the order of thousands of atmospheres are developed. However, since the maximum pressure rise is confined to durations of the order of a microsecond, definitive measurements have not yet been achieved. The motion of such cavities depends not only upon the ambient pressure conditions but also upon the amount of permanent gas in the bubble and the condensation rates of the vapor as well as the properties of the liquid—compressibility, viscosity, surface tension.

---

[1] Hunter Rouse and John S. McNown, Cavitation and Pressure Distribution; Head Forms at Zero Angle of Yaw, *State Univ. Iowa Studies Eng. Bull.* 32, 1948.

[2] R. W. Kermeen, Some Observations of Cavitation on Hemispherical Head Models, *Calif. Inst. Technol. Hydrodynamics Lab. Rept.* E-35.1, June, 1952.

[3] Blaine R. Parkin, Scale Effects in Cavitating Flow, *Calif. Inst. Technol. Hydrodynamics Lab. Rept.* 21-8, July 31, 1952.

[4] Eisenberg, *loc. cit.;* Parkin, *loc. cit.*

Except for surface tension, all these factors tend to decrease the rate of collapse; in addition, distortion from spherical shape caused by pressure gradients or bubble-wall instability tends to result in reduced collapse rates and thus reduced pressures.

Plesset,[1] employing Rayleigh's[2] theoretical formulation for collapse of a spherical cavity in incompressible inviscid fluid but including effect of surface tension and comparing with the experimental results of Knapp and Hollander,[3] has shown that, in the region from maximum radius down to about one-quarter the maximum radius, the motion can be predicted with fair accuracy as long as the bubble is approximately

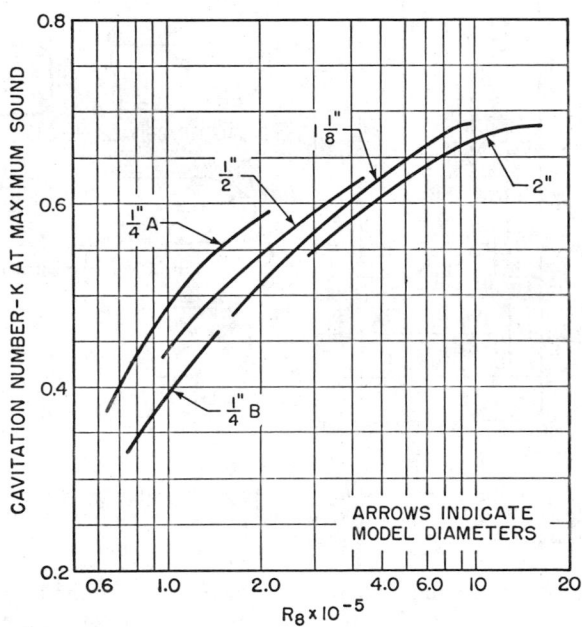

FIG. 2o-5. Cavitation number K for incipient cavitation (as defined by value at which noise disappears) as a function of Reynolds number for bodies with hemispherical heads and cylindrical middle bodies. (The $\frac{1}{4}$-in. A model was more accurately constructed than the $\frac{1}{4}$-in. B model.) (*After Kermeen.*)

spherical. This idealized theory, which predicts that the bubble-wall velocity is of the order of $R^{-\frac{3}{2}}$ as $R \to 0$ (and that the maximum pressure is infinite), is, of course, inadequate for the final stages of collapse where the effects mentioned above become important. For example, a further approximation carried out by Gilmore[4] shows that the effect of compressibility of the liquid is to reduce the wall velocity to the order of $R^{-\frac{1}{2}}$.

[1] M. S. Plesset, The Dynamics of Cavitation Bubbles, *J. Appl. Mech.* **16**, 277–282 (September, 1949).

[2] Lord Rayleigh, On the Pressure Developed in a Liquid during the Collapse of a Spherical Cavity, *Phil. Mag.* **34**, 94–98 (1917).

[3] R. T. Knapp and A. Hollander, Laboratory Investigations of the Mechanism of Cavitation, *Trans. Am. Soc. Mech. Engr.* **70** (5), 419–435 (July, 1948).

[4] Forrest R. Gilmore, The Growth or Collapse of a Spherical Bubble in a Viscous Compressible Liquid, *Calif. Inst. Technol. Hydrodynamics Lab. Rept.* 26–4, Apr. 1, 1952.

2o-5. Steady-state Cavities. Such cavities (also referred to as "fixed" and "sheet") are large stationary cavities observed behind blunt obstacles and on hydrofoil profiles with relatively sharp leading edges. While such cavities are, especially at low cavitation numbers, usually filled only with vapor phase and other gas, they are often observed to contain a mixture of individual bubbles and liquid phase. The surface usually oscillates, and often parts or the entire cavity is observed to grow and collapse; the average envelope, however, behaves essentially as the boundary of a time-independent flow.[1]

Reliable measurements of cavity shape have been made up to now only for axisymmetric cavities. Data for the principal dimensions of cavities formed behind truncated forms with the apex upstream (disks, cones, hemispheres, semiellipsoids, ogives) have been reported by Reichardt[2] and Eisenberg and Pond.[3] Such measurements for cavities about bodies of revolution composed of cylindrical middle bodies and various head shapes have been reported by Rouse and McNown.[4] Reichardt's data are particularly of interest, since they extend to the lowest cavitation numbers yet attained (as low as 0.013).

For the truncated forms for which the leading edge of the cavity is essentially fixed at the trailing edge of the form (cones, disks), measurements of the principal dimensions can be represented within the experimental error by formulas given by Reichardt.[2] The ratio of maximum cavity diameter to diameter of disk or base of cone is

$$\frac{d_m}{d} = \sqrt{C_D(0) \frac{1 + \sigma}{\sigma f}} \qquad (2\text{o-}1)$$

where

$$f = 1 - 0.132\sigma^{\frac{1}{2}} \qquad (2\text{o-}2)$$

and values of $C_D(0)$ are given in Table 2o-1. The ratio of maximum cavity diameter to cavity length is

$$\frac{d_m}{l} = \frac{0.066 + 1.70\sigma}{\sigma + 0.008} \qquad (2\text{o-}3)$$

2o-6. Drag in Cavitating Flow. Available data indicate that, for the truncated bodies discussed above, the drag coefficient is a linear function of the cavitation number. Available data may be represented by[5]

$$C_D(\sigma) = C_D(0)(1 + \beta\sigma) \qquad (2\text{o-}4)$$

where the value of β is given in Table 2o-1. This formula can also be used to represent available data for a circular cylinder with its axis normal to the flow. The value of $C_D(0)$ for the disk is the average of the extrapolated values of Reichardt[2] and Eisenberg and Pond.[3] The results for the cones are from Reichardt; the results for the hemisphere, semiellipsoid, and ogive are from Eisenberg and Pond. In each of these cases, the value of $C_D(0)$ is extrapolated from the experimental data from which the values of β were also obtained. The value of $C_D(0)$ for the circular cylinder is from a computation of Brodetsky.[6] The value of $\beta = 0.73$ for the circular cylinder

---

[1] Eisenberg, *loc. cit.*

[2] H. Reichardt, The Laws of Cavitation Bubbles at Axially Symmetrical Bodies in a Flow, *Ministry Aircraft Prod., Rept. Translations* 766, Aug. 15, 1946 (distributed in the United States by the Office of Naval Research, Washington, D.C.).

[3] Eisenberg, *loc. cit.*; Phillip Eisenberg and Hartley L. Pond, Water Tunnel Investigations of Steady State Cavities, *David Taylor Model Basin, U.S. Navy Dept. Rept.* 668, October. 1948.

[4] *Loc. cit.*

[5] Eisenberg, *loc. cit.*

[6] S. Brodetsky, Discontinuous Fluid Motion Past Circular and Elliptic Cylinders, *Proc. Roy. Soc. (London)*, ser. A, **102** (A718), 542–553 (February, 1923).

is given by Birkhoff[1] based on experiments of Martyrer. The other values of β for the circular cylinder are based on Kanstantinov's[2] experiments, which show differences depending on Reynolds number (based on cylinder diameter). It should be noted that Kanstantinov's results are for constant Reynolds number, whereas in Martyrer's tests the Reynolds number varied as the cavitation number was varied. There may be a question, however, as to the accuracy of Kanstantinov's results, since the forces were found by integrating pressure distributions rather than by direct measurement.

TABLE 2o-1. DATA FOR DRAG COEFFICIENT IN EQ. (2o-4)

Model	$C_D(0)$	Range of σ	β	Reynolds No.
Disk, $h/d = 0$	0.80	0.038–0.56	1.0	2.6–7.9×10^5*
Cones:				
$h/d = \frac{1}{4}$	0.63	0.033–0.125	1.0	
$h/d = \frac{1}{2}$	0.5	0.032–0.118	1.0	
$h/d = 1$	0.32	0.026–0.069	1.0	
$h/d = 2$	0.15	0.013–0.086	1.0	
Hemisphere	0.241	0.168–0.38	2.024	3–8.3×10^5
2:1 semiellipsoid and 2				
caliber ogive	0.114	0.133–0.394	3.65	≈ 3–9×10^5
Circular cylinder	≈ 0.55	....	0.81	2.72×10^5
			0.68	1.75×10^5
			0.73	2–6×10^5

* Phillip Eisenberg and Hartley L. Pond, Water Tunnel Investigations of Steady State Cavities, *David Taylor Model Basin, U.S. Navy Dept. Rept.* 668, October, 1948.

2o-7. Nonstationary Cavities and Other Topics. A third type of flow which may be defined as part of a general classification of cavitating flows is the "nonstationary" (or "unsteady") cavity. This is a cavity resembling steady-state cavities but varying in time as in the air-water entry of an air-dropped missile or as in the motion of an initially submerged but accelerating body. Although all three are free-boundary flows, in the transient cavity, the pressure at the boundary varies with time; in the steady-state cavity, the boundaries are free streamlines; and, in the third, the boundaries are such that the material lines are not necessarily free streamlines. The nomenclature used here was chosen to provide a consistent representation for both the physical phenomena and the corresponding mathematical descriptions. Further discussions of nonstationary cavities and references will be found in Eisenberg[3] and Birkhoff.[3]

For problems of lift in cavitating flows, especially supercavitating hydrofoils, see Tulin.[4] Information on such subjects as cavitation damage and measurement of air content will be found in Eisenberg.[5]

[1] Garrett Birkhoff, "Hydrodynamics," chap. 2, Princeton University Press, Princeton, N.J., 1950.
[2] W. A. Kanstantinov, Influence of the Reynolds Number on the Separation (Cavitation) Flow, *David Taylor Model Basin, U.S. Navy Dept. Translation* 233, November, 1950.
[3] *Loc. cit.*
[4] Marshall P. Tulin, Supercavitating Flows, part 2, sec. 12, "Handbook of Fluid Dynamics," V. L. Streeter, ed., McGraw-Hill Book Company, New York, 1961.
[5] Phillip Eisenberg, Mechanics of Cavitation, part 1, sec. 12, "Handbook of Fluid Dynamics," V. L. Streeter, ed., McGraw-Hill Book Company, New York, 1961.

2p. Diffusion in Liquids

L. G. LONGSWORTH

Rockefeller University

Symbols

c concentration of solution in moles per liter
D diffusion coefficient
F Faraday
k Boltzmann's gas constant
p concentration in grams per 100 milliliters
r_s Stokes radius
R gas constant
T Kelvin temperature
Z valence
η viscosity
λ ionic conductance

The diffusion coefficient in liquid solutions is defined as the coefficient D in Fick's equation

$$\frac{\partial c}{\partial t} = D \frac{\partial^2 c}{\partial x^2} \tag{2p-1}$$

in which c is the concentration of the solution and D is a function of the concentration. This coefficient is sometimes called the differential value of the diffusion. In the tables of this section it is always these values which are tabulated. The units of D throughout are cm²/sec multiplied by 10^5, and all values refer to a pressure of 1 atm. Moreover all values are *mutual* coefficients; i.e., in the absence of a volume change on mixing, the one value describes the diffusion of either solute or solvent. A recent compilation of *self-* and *tracer*-diffusion coefficients and the effect of pressure on diffusion in liquids has been made by P. A. Johnson and A. L. Babb, *Chem. Rev.* **56**, 387 (1956). In the tables that follow, the methods employed and the average deviations of the reported data from smooth interpolation curves are indicated by the following abbrevation scheme:

C   conductance ($\pm 0.2\%$)
D   diaphragm cell ($\pm 0.2\%$)
G   Gouy interference ($\pm 0.1\%$)
L   layer analysis ($\pm 0.2\%$)
M   Mach-Zehnder interference
R   Rayleigh interference ($\pm 0.1\%$)

In binary liquid systems, Table 2p-7, the reported uncertainty is about 1 per cent. Values for which no reference is given are previously unpublished results of the compiler.

TABLE 2p-1. DIFFUSION COEFFICIENTS OF DILUTE AQUEOUS SOLUTIONS
OF ELECTROLYTES AT 25°C
(Concentration, moles/liter)

Electrolyte	0.000	0.0006	0.001	0.002	0.003	0.005	0.007	0.010	Ref.	Method
LiCl	1.366	1.349	1.345	1.337	1.331	1.323	1.318	1.312	10	C
NaCl	1.612	......	1.586	1.576	1.570	1.561	1.554	1.545	10	C
KCl	1.994	......	1.964	1.952	1.944	1.933	1.924	1.915	10	C
RbCl	2.057	......	2.024	2.012	2.003	1.991	1.983	1.972	10	C
CsCl	2.046	......	2.013	2.001	1.992	1.978	1.969	1.958	10	C
KNO_3	1.931	......	1.899	1.887	1.879	1.866	1.856	1.844	10	C
$AgNO_3$	1.767	... ..	......	......	1.720	1.708	1.699	.....	10	C
$MgCl_2$	1.251	......	1.189	1.172	1.161	.....	.....	.....	9	C
$CaCl_2$	1.335	......	1.249	1.224	1.206	1.180	.....	.....	1	C
$SrCl_2$	1.336	......	1.269	1.249	1.236	1.219	1.210	.....	8	C
$BaCl_2$	1.387	1.332	1.320	1.299	1.285	1.264	.....	.....	9	C
Li_2SO_4	1.041	1.000	.990	0.975	0.965	0.950	.....	.....	2	C
Na_2SO_4	1.230	......	1.175	1.159	1.145	1.124	.....	.....	2	C
Cs_2SO_4	1.569	......	1.487	1.460	1.442	1.418	.....	.....	7	C
$MgSO_4$	0.849	0.784	0.767	0.741	0.726	0.708	0.700	.....	6	C
$ZnSO_4$	0.849	......	0.741	0.734	0.723	0.706	.....	.....	3	C
$LaCl_3$	1.294	......	1.173	1.144	1.125	1.102	1.087	.....	4	C
$K_4Fe(CN)_6$	1.473	......	.....	.....	1.211	1.183	.....	.....	5	C

References

1. Harned, H. S., and A. L. Levy: *J. Am. Chem. Soc.* **71**, 2781 (1949).
2. Harned, H. S., and C. A. Blake, Jr.: *J. Am. Chem. Soc.* **73**, 2448 (1951).
3. Harned, H. S., and R. M. Hudson: *J. Am. Chem. Soc.* **73**, 3781 (1951).
4. Harned, H. S., and C. A. Blake, Jr.: *J. Am. Chem. Soc.* **73**, 4255 (1951).
5. Harned, H. S., and R. M. Hudson: *J. Am. Chem. Soc.* **73**, 5083 (1951).
6. Harned, H. S., and R. M. Hudson: *J. Am. Chem. Soc.* **73**, 5880 (1951).
7. Harned, H. S., and C. A. Blake, Jr.: *J. Am. Chem. Soc.* **73**, 5882 (1951).
8. Harned, H. S., and F. M. Polestra: *J. Am. Chem. Soc.* **75**, 4168 (1953).
9. Harned, H. S., and F. M. Polestra: *J. Am. Chem. Soc.* **76**, 2064 (1954).
10. Harned, H. S.: *Proc. Natl. Acad. Sci. U.S.* **40**, 551 (1954).

TABLE 2p-2. DIFFUSION COEFFICIENTS OF CONCENTRATED AQUEOUS SOLUTIONS OF ELECTROLYTES AT 25°C

(Concentration, moles/liter)

Electrolyte	0.00	0.05	0.1	0.2	0.3	0.5	0.7	1.0	1.5	2.0	2.5	3.0	3.5	4.0	5.0	6.0	8.0	Ref.	Method
HCl	3.337	3.073	3.050	3.064	3.093	3.184	3.286	3.436	3.743	4.046	4.337	4.658	4.920	5.17				4	D
LiCl	1.366	1.280	1.269	1.267	1.269	1.278	1.288	1.302	1.331	1.363	1.397	1.430	1.464					4	D
NaCl	1.612	1.506	1.484	1.478	1.477	1.474	1.475	1.483	1.495	1.514	1.529	1.544	1.559	1.584				4	D
KCl	1.994	1.863	1.848	1.835	1.826	1.835	1.846	1.876	1.951	2.011	2.064	2.110	2.152					4	D
KCl		1.864	1.847	1.839	1.839	1.850	1.865	1.892	1.943	1.999	2.057	2.112	2.160	2.204				2	G
NH_4Cl			1.838	1.836	1.841	1.861	1.883	1.921	1.986	2.051	2.113	2.164	2.203	2.235	2.264			3	G
HBr	3.402	3.156	3.146	3.190	3.249	3.388	3.552	3.869										4	D
LiBr	1.377	1.300	1.279	1.285	1.296	1.328	1.360	1.404	1.473	1.542	1.597	1.650	1.693					4	D
NaBr	1.627	1.533	1.517	1.507	1.515	1.542	1.569	1.596	1.668	1.702								4	D
KBr	2.017	1.892	1.874	1.870	1.872	1.885	1.917	1.975	2.062	2.132	2.199	2.280	2.354	2.434				4	D
NaI	1.616	1.527	1.520	1.532	1.547	1.580	1.612	1.662	1.751	1.846	1.925	1.992						1	D
KI	2.000	1.891	1.865	1.859	1.884	1.955	2.001	2.065	2.166	2.254	2.347	2.440	2.533					1	D
$LiNO_3$	1.337		1.240	1.243		1.260		1.293	1.317	1.332	1.336	1.332		1.292	1.238	1.157		5	G
NH_4NO_3	1.928		1.769	1.749		1.724		1.690	1.661	1.633	1.605	1.578		1.524	1.472	1.421	1.320	5	G
$(NH_4)_2SO_4$	1.527	0.802	0.825	0.867		0.938		1.011	1.047	1.069	1.088	1.106						5	G
$CaCl_2$	1.335		1.110	1.111	1.118	1.140	1.166	1.203	1.263	1.307	1.306	1.265	1.195					3	G

References

1. Dunlop, P. J., and R. H. Stokes: *J. Am. Chem. Soc.* **73**, 5456 (1951).
2. Gosting, L. J.: *J. Am. Chem. Soc.* **72**, 4418 (1950).
3. Hall, J. R., B. F. Wishaw, and R. H. Stokes: *J. Am. Chem. Soc.* **75**, 1556 (1953).
4. Stokes, R. H.: *J. Am. Chem. Soc.* **72**, 2243 (1950).
5. Wishaw, B. F., and R. H. Stokes: *J. Am. Chem. Soc.* **76**, 2065 (1954).

TABLE 2p-3. DIFFUSION COEFFICIENTS OF AQUEOUS SOLUTIONS
OF NONELECTROLYTES
(Gouy interference method)

Concn. p, g/100 ml	Nonelectrolyte								
	Urea 25°C	Glycol-amide 25°C	Glycine		n-Butyl alcohol		α-Alanine 25°C	Sucrose	
			1°C	25°C	1°C	25°C		1°C	25°C
0.00	1.3817	1.1423	0.5200	1.0635	0.4523	0.9720	0.9145	0.2424	0.5233
0.25	......	......	0.5158	1.0571	0.4395	0.9610	0.9105	......	......
0.50	......	1.1359	0.5120	1.0507	0.4313	0.9500	0.9065	0.2403	0.5194
0.75	1.3720	1.1328	0.5083	1.0443	0.4242	0.9390	0.9026	0.2393	0.5175
1.00	1.3688	1.1296	0.5048	1.0379	0.4182	0.9282	0.8987	0.2383	0.5155
2	1.3561	1.1171	0.4914	1.0122	0.3968	0.8854	0.8834	0.2342	0.5078
3	1.3437	1.1047	0.4793	0.9866	0.3780	9.8436	0.8686	0.2302	0.5001
5	1.3197	1.0804	0.4590	0.9353	0.3462	0.7629	0.8405	0.2221	0.4846
10	1.2642	1.0222	......	......	......	......	0.7787		
15	1.2151	0.9676	......	......	......	......	0.7292		
20	1.1725	0.9167							
25	1.1363	0.8694							
30	......	0.8257							
Ref.	3	1	5	5	6	6	4	2	2

At 25°C the data from which this table was prepared may be represented analytically as follows:

Urea: $D \times 10^5 \pm 0.05\% = 1.3817 - 0.0130p + 0.0001288p^2$
for $p \leq 25$

Glycolamide: $D \times 10^5 \pm 0.08\% = 1.1423 - 0.01274p + 0.0000729p^2$
for $p \leq 30$

Glycine: $D \times 10^5 \pm 0.08\% = 1.0635 - 0.02563p$ for $p \leq 5$

n-Butyl alcohol: $D \times 10^5 \pm 0.03\% = 0.9720 - 0.0443p + 0.000496p^2$
for $p \leq 5$

α-Alanine: $D \times 10^5 \pm 0.09\% = 0.9145 - 0.01603p + 0.0002449p^2$
for $p \leq 15$

Sucrose: $D \times 10^5 \pm 0.04\% = 0.5233 - 0.007745p$ for $p \leq 5$

References

1. Dunlop, P. J., and L. J. Gosting: *J. Am. Chem. Soc.* **75**, 5073 (1953).
2. Gosting, L. J., and M. S. Morris: *J. Am. Chem. Soc.* **71**, 1998 (1949).
3. Gosting, L. J., and D. F. Akeley: *J. Am. Chem. Soc.* **74**, 2058 (1952).
4. Gutter, F. J., and G. Kegeles: *J. Am. Chem. Soc.* **75**, 3893 (1953).
5. Lyons, M. S., and J. V. Thomas: *J. Am. Chem. Soc.* **72**, 4506 (1950).
6. Lyons, P. A., and C. L. Sandquist: *J. Am. Chem. Soc.* **75**, 3896 (1953).

TABLE 2p-4. DIFFUSION OF ORGANIC COMPOUNDS IN DILUTE AQUEOUS SOLUTION AT 1° AND 25°C*

Compound	Wt. %	$D_1 \times 10^5$	$D_{25} \times 10^5$	Compound	Wt. %	$D_1 \times 10^5$	$D_{25} \times 10^5$
Methyl alcohol	0.00	0.76$_6$	1.58$_7$†	Glycylglycylglycine...	0.29	0.3175	0.6652
Ethyl alcohol	0.00	......	1.24$_8$†	Leucylglycylglycine...	0.30	......	0.5507
n-Propyl alcohol	0.59	......	1.02$_2$†	o-Aminobenzoic acid..	0.24	......	0.840
Isopropyl alcohol	0.59	......	1.02$_0$†	m-Aminobenzoic acid.	0.24	......	0.774
n-Butyl alcohol	0.49	......	0.95$_2$†	p-Aminobenzoic acid..	0.23	......	0.842
Isobutyl alcohol	0.49	......	0.93$_8$†				
sec-Butyl alcohol	0.49	......	0.92$_2$†	Proline	0.32	0.4187	0.8789
tert-Butyl alcohol	0.47	......	0.87$_9$†				
Glycine	0.30	0.5151	1.0554	Hydroxyproline	0.32	0.3930	0.8255
Glycolamide	0.30	......	1.1385	Histidine	0.28	0.3452	0.7328
α-Alanine	0.32	0.4317	0.9097				
β-Alanine	0.31	0.4500	0.9327	Phenylalanine	0.25	0.3244	0.7047
Sarcosine	0.32	......	0.9674				
				Tryptophane	0.23	0.3042	0.6592
Serine	0.31	0.4195	0.8802				
				$d(-)$Ribose	0.41	......	0.7769
α-Aminobutyric acid..	0.31	0.3891	0.8288	$1(+)$Arabinose	0.39	......	0.7599
β-Aminobutyric acid..	0.32	......	0.8367	$d(-)$Lyxose	0.40	......	0.7591
γ-Aminobutyric acid..	0.32	......	0.8259	$d(+)$Xylose	0.40	......	0.7462
α-Amino isobutyric acid	0.32	......	0.8130	$d(-)$Levulose	0.39	0.3230	0.6944
				$d(+)$Mannose	0.39	......	0.6875
Threonine	0.32	......	0.7984	$1(-)$Sorbose	0.39	......	0.6791
				$d(+)$Dextrose	0.39	0.3137	0.6728
Valine	0.30	0.3566	0.7725	$d(+)$Galactose	0.38	0.3131	0.6655
Norvaline	0.32	......	0.7682				
				$d(+)$Sucrose	0.39	0.2414	0.5209
Leucine	0.32	0.3333	0.7255	$d(+)$Lactose·H_2O....	0.40	......	0.5076
Norleucine	0.32	0.3328	0.7249	$d(+)$Cellobiose......	0.38	......	0.5039
				$d(+)$Melibiose·$2H_2O$.	0.41	......	0.5022
Asparagine	0.29	......	0.8300	$d(+)$Maltose·H_2O....	0.40	......	0.4929
Glycylglycine	0.29	0.3790	0.7909				
Glutamine	0.34	......	0.7623	$d(+)$Melezitose·$2H_2O$	0.40	......	0.4478
Glycylalanine	0.30	......	0.7221	$d(+)$Raffinose·$5H_2O$.	0.45	0.2011	0.4339
Alanylglycine	0.30	......	0.7207	Cycloheptaamylose...	0.39	......	0.3224
Glycylleucine	0.29	0.2869	0.6231	Bovine plasma			
Leucylglycine	0.31	0.2831	0.6129	albumin	0.25	......	0.0670‡

Isomers are in groups. Rayleigh interference method.
* L. G. Longsworth. *J. Am. Chem. Soc.* **74**, 4155 (1952); **75**, 5075 (1953).
† D strongly concentration-dependent.
‡ Extrapolated to salt-free solution.

TABLE 2p-5. DIFFUSION COEFFICIENTS IN AQUEOUS SOLUTION AT DIFFERENT TEMPERATURES

Solute	Wt. %	5 °C	15 °C	25 °C	35 °C	45 °C	55 °C	Ref.	Method
H$^+$	0.00	6.208	7.737	9.313	10.919	12.538	14.150	2	*
Li$^+$	0.00	0.5654	0.7769	1.0286	1.3197	1.6483	2.0142	2	
Na$^+$	0.00	0.7524	1.0218	1.3349	1.6928	2.0959	2.5439	2	
K$^+$	0.00	1.1604	1.5335	1.9565	2.4265	2.9403	3.4943	2	
Cl$^+$	0.00	1.1796	1.5801	2.0324	2.5368	3.0935	3.7031	2	
Br$^+$	0.00	1.2233	1.6259	2.0808	2.5869	3.1426	3.7465	2	
I$^-$	0.00	1.2066	1.6007	2.0457	2.5409	3.0850	3.6762	2	
Ca^{++}	0.00	......	0.6043	0.7919	1.0078	1.2528	....	1	
H^{1}H^{2}O^{16}	0.00	1.294	1.743	2.261	......	......	....	3	R
Urea	0.38	0.790	1.063	1.377	1.731	......	....	3	R
Glycolamide	0.46	0.637	0.869	1.140	1.451	1.794	....	...	R
Glycine	0.30	0.593	0.806	1.054	1.337	......	....	3	R
Alanine	0.32	0.500	0.688	0.909	1.164	......	....	3	R
Dextrose	0.38	0.3640	0.5038	0.6713	0.867	......	....	3	R
Cyclohepta-amylose	0.38	0.1738	0.2418	0.3225	0.4160	......	....	3	R
Bovine plasma albumin	0.25	0.0356	0.0493	0.0657	......	......	....	3	R

* D for ions computed from ionic conductances λ with the aid of the relation $D = RT\lambda/zF^2$, where $R = 8.3144$ joules/(deg)(mole), $T = °$Kelvin $= 273.13 + t$, $z = $ valence, and $F = 96,500$ coulombs/equivalent.

$$D_{\text{salt}} = \frac{(z_+ + z_-)D_+D_-}{z_+D_+ + z_-D_-}$$

Since the Stokes radius $r_s = kT/6\pi\eta D$ varies but little with temperature, a plot of r_s vs. t affords precise interpolation. Here $k = 1.3803 \times 10^{-16}$ erg/(deg)(mole), and η is the viscosity of the solvent in poises.

References

1. Benson, G. C., and A. R. Gordon: *J. Chem. Phys.* **13**, 470 (1945).
2. Harned, H. S., and B. B. Owen: "Physical Chemistry of Electrolytic Solutions," 2d ed., p. 590, Reinhold Book Corporation, New York, 1950.
3. Longsworth, L. G.: *J. Phys. Chem.* **58**, 770 (1954).

TABLE 2p-6. DIFFUSION COEFFICIENTS IN NONAQUEOUS SOLVENTS

Solvent Solute	t, °C	$D \times 10^5$	Ref.	Solvent Solute	t, °C	$D \times 10^5$	Ref.
Hexane				Carbon tetrachloride			
Iodine	25.0	4.05	7	Argon	0.0	2.44	5
n-Heptane				Argon	25.0	3.63	5
Iodine	25.0	3.42	7	Methane	0.0	2.05	5
Carbon tetrachloride	25.0	3.17	2	Methane	25.0	2.89	5
iso-Octane				Carbon tetrafluoride	25.0	2.04	4
Carbon tetrachloride	25.0	2.57	2	Silicon hexafluoride	25.0	1.71	3
Benzene				Heptane	25.0	1.349	3
Iodine	25.0	2.13	7	Dodecane	25.0	0.954	3
Diphenyl	25.0	1.558	6	Hexadecane	25.0	0.780	3
Diphenyl	35.0	1.847	6	Eicosane	25.0	0.664	3
Toluene				Docosane	25.0	0.620	3
Iodine	25.0	2.13	7	Octacosane	25.0	0.528	3
m-Xylene				Dotriacontane	25.0	0.479	3
Iodine	25.0	1.89	7	Tetrahydrofuran	25.0	1.468	3
Mesitylene				Naphthalene	25.0	1.200	3
Iodine	25.0	1.49	7	Anthracene	25.0	1.026	3
Dioxane				Diphenyl	25.0	1.074	3
Iodine	25.0	1.07	7	Diphenylmethane	25.0	0.985	3
Carbon tetrachloride	25.0	1.02	2	Triphenylmethane	25.0	0.694	3
sym-Tetrachlorethane*				Hexadecanol	25.0	0.741	3
sym-Tetrabromethane	0.4	0.351	1	Phenol	25.0	1.370	3
sym-Tetrabromethane	7.7	0.419	1	Benzohydrol	25.0	0.918	3
sym-Tetrabromethane	15.0	0.497	1	Triphenylmethanol	25.0	0.687	3
sym-Tetrabromethane	35.6	0.611	1	Benzoic acid†	25.0	0.882	3
sym-Tetrabromethane	51.1	0.741	1	Palmitic acid†	25.0	0.448	3
sym-Tetrabromethane		0.954	1	Hexachloroethane	25.0	1.007	3
Carbon tetrachloride				Hexachlorobenzene	25.0	0.922	3
Hydrogen	0.0	6.28	5	Hexachlorocyclohexane	25.0	0.843	3
Hydrogen	25.0	9.75	5	N,N-dimethylacetamide	25.0	1.228	3
Deuterium	25.0	7.71	5	N,N-dimethylpropionamide	25.0	1.135	3
Nitrogen	0.0	2.44	5				
Nitrogen	25.0	3.42	5				
Oxygen	25.0	3.82	4				

* Values in this solvent are at a concentration of 0.03 mole/liter. All other values in the table are limiting diffusion coefficients at zero concentration of solute.
† Dimer.

References for Table 2p-6

1. Cohen, E., and H. R. Bruin: Z. physik. Chem. 103, 404 (1923).   2. Hammond, B. R., and R. H. Stokes: Trans. Faraday Soc. 51, 1641 (1955).
3. Longsworth, L. G.: J. Colloid and Interface Sci. 22, 3 (1966).
4. Nakanishi, K., E. M. Voigt, and J. H. Hildebrand: J. Chem. Phys. 42, 1860 (1965).
5. Ross, M., and J. H. Hildebrand: J. Chem. Phys. 40, 2397 (1964).   6. Sandquist, C. L., and P. A. Lyons: J. Am. Chem. Soc. 76, 4641 (1954).
7. Stokes, R. H., P. J. Dunlop, and J. R. Hall: Trans. Faraday Soc. 49, 886 (1953).

Table 2p-7. Diffusion Coefficients of Binary Liquid Systems

System	t, °C	$D \times 10^5$ Mole % A					
		0.0	20.0	40.0	60.0	80.0	100.0
Hexane(A)-dodecane[6]*	25.1	1.43	1.61	1.80	2.02	2.33	2.73
Hexane(A)-hexadecane[5]	25.1	0.852	1.050	1.256	1.48	1.74	2.20
Heptane(A)-hexadecane[6]	25.1	0.745	0.888	1.05	1.23	1.46	1.78
Acetone(A)-benzene[1]	25.2	2.75	2.55	2.70	2.97	3.40	4.17
Cyclohexane(A)-benzene[14]	25.0	2.104	1.903	1.813	1.796	1.834	1.880
Hexane(A)-CCl4†[6]	25.1	1.47	1.85	2.26	2.73	3.26	3.87
Cyclohexane(A)-CCl4[12]	25.0	1.283	1.311	1.350	1.393	1.443	1.484
Toluene(A)-CCl4[13]	25.0	1.404	1.538	1.682	1.838	2.003	2.180
Mesitylene(A)-CCl4[13]	25.0	1.193	1.353	1.493	1.615	1.718	1.802
Acetone(A)-CCl4[1]	25.2	1.71	1.45	1.66	2.08	2.65	3.60
Methyl ethyl ketone(A)-CCl4[3]	25.0	1.57	1.44	1.68	2.04	2.49	3.02
Methyl alcohol(A)-CCl4[4,13]	25.0	2.61	0.556	0.489	0.799	1.278	2.258
Ethyl alcohol(A)-CCl4[4,10,13]	25.0	1.95	0.67	0.60	0.84	1.17	1.50
Benzyl alcohol(A)-CCl4[13]	25.0	1.280	0.386	0.349	0.344	0.342	0.324
Acetic acid(A)-CCl4[3,13]	25.0	1.42‡	1.19	0.92	0.82	0.88	1.28
Acetone(A)-chloroform[1]	25.2	2.34	2.95	3.30	3.45	3.55	3.63
Acetone(A)-chloroform[1]	40.0	2.88	3.58	4.05	4.23	4.26	4.31
Diethyl ether(A)-chloroform[2]	25.0	2.14	2.90	3.67	4.21	4.44	4.51
Water(A)-ethyl alcohol[9]	25.0	1.132	0.930	0.625	0.415	0.409	1.240

		Mole % A					
Water(A)-acetone[1]	25.0	3.04	19.64	51.07	76.08	91.69	99.78
		4.56	2.39	0.819	0.635	0.854	1.28

		Mole % A					
Methyl alcohol(A)-benzene[11]	27.1	0.48	11.43	25.48	50.32	75.17	99.82
		3.839	1.327	0.949	0.901	1.533	2.762
Methyl alcohol(A)-benzene[1]	40.0	4.67	1.85	1.30	1.28	1.98	3.16

		Mole % A					
Ethyl alcohol(A)-benzene[1]	25.2	0.37	9.57	20.34	34.15	70.22	99.61
		2.93	1.30	0.993	0.901	1.35	1.81
Ethyl alcohol(A)-benzene[1]	40.0	3.74	1.89	1.46	1.30	1.76	2.37

TABLE 2p-7. DIFFUSION COEFFICIENTS OF BINARY LIQUID SYSTEMS (*Continued*)

System	Mole % A	t, °C	$D \times 10^5$				
Benzene(A)–CCl$_4$[8]			2.15	25.02	50.51	74.98	98.15
		10.0	1.085	1.093	1.230	1.344	1.466
		25.3	1.419	1.519	1.651	1.759	1.912
		40.0	1.775	1.970	2.077	2.284	2.432
Toluene(A)–chlorobenzene[8]			1.34	25.01	49.92	74.92	98.62
		10.0	1.346	1.404	1.556	1.652	1.759
		27.0	1.756	1.852	1.985	2.128	2.264
		40.0	2.113	2.277	2.435	2.586	2.714
Chlorobenzene(A)–bromobenzene[8]			3.32	26.42	51.22	76.17	96.52
		10.0	1.007	1.069	1.146	1.226	1.291
		26.8	1.342	1.380	1.506	1.596	1.708
		40.0	1.584	1.691	1.806	1.902	1.996

* Reference no.
† Carbon tetrachloride.
‡ Acetic acid dimer.

References for Table 2p-7

1. Anderson, D. K., J. R. Hall, and A. L. Babb: *J. Phys. Chem.* **62**, 404 (1958).
2. Anderson, D. K., and A. L. Babb: *J. Phys. Chem.* **65**, 1281 (1961).
3. Anderson, D. K., and A. L. Babb: *J. Phys. Chem.* **66**, 899 (1962).
4. Anderson, D. K., and A. L. Babb: *J. Phys. Chem.* **67**, 1362 (1963).
5. Bidlack, D. L., and D. K. Anderson: *J. Phys. Chem.* **68**, 206 (1964).
6. Bidlack, D. L., and D. K. Anderson: *J. Phys. Chem.* **68**, 3790 (1964).
7. Caldwell, C. S., and A. L. Babb: *J. Phys. Chem.* **59**, 1113 (1955).
8. Caldwell, C. S., and A. L. Babb: *J. Phys. Chem.* **60**, 51 (1956).
9. Hammond, B. R., and R. H. Stokes: *Trans. Faraday Soc.* **49**, 890 (1953).
10. Hammond, B. R., and R. H. Stokes: *Trans. Faraday Soc.* **52**, 781 (1956).
11. Johnson, P. A., and A. L. Babb: *Chem. Reviews* **56**, 387 (1956).
12. Kulkarni, M. V., G. F. Allen, and P. A. Lyons: *J. Phys. Chem.* **69**, 2491 (1965).
13. Longsworth, L. G.: *J. Colloid and Interface Sci.* **22**, 3 (1966).
14. Rodwin, L., J. A. Harpst, and P. A. Lyons: *J. Phys. Chem.* **69**, 2783 (1965).

2q. Liquid Jets[1]

W. L. NYBORG

University of Vermont

2q-1. Circular Jet. We first deal with the laminar flow due to a circular jet of viscous fluid issuing from a point orifice into a space filled with the same fluid.

Symbols

J momentum crossing a plane normal to the axis of the jet per second
u, v x, y components, respectively, of fluid velocity in the jet
x distance parallel to the axis of the jet
y distance perpendicular to the axis of the jet
ρ fluid density
ν kinematic viscosity of the fluid

The flow-velocity components in the jet are given by the following formulas due to Schlichting:[2]

$$u = \frac{3}{8\pi} \frac{K}{\nu x} \frac{1}{(1 + \epsilon^2/4)^2}$$

$$v = \frac{1}{4} \sqrt{\frac{3K}{\pi}} \frac{1}{x} \frac{\epsilon(1 - \epsilon^2/4)}{(1 + \epsilon^2/4)^2} \tag{2q-1}$$

where

$$\epsilon = \frac{1}{4} \sqrt{\frac{3K}{\pi}} \frac{1}{\nu} \frac{y}{x}$$

$$K = \frac{J}{\rho} \tag{2q-2}$$

The formulas (2q-1) have been checked experimentally by Andrade and Tsien,[3] who found good agreement between the theory and experimental results for a jet of finite radius a at a distance of 8 jet diameters or more from the orifice, provided the x in (2q-1) is given by

$$x = x_o + 0.16u_o \frac{a^2}{\nu} \tag{2q-3}$$

where x_0 is the actual distance to the real orifice, and x may be interpreted as the distance to an effective point orifice upstream from the real one.

Figure 2q-1 shows a family of streamlines for a circular jet from a point orifice plotted from Eqs. (2q-1). (For reasons of clarity the figure is expanded in the y direction.) Typical velocity profiles (plots of u vs. y) are also given for two distances x from the orifice.

[1] For a general reference see H. Schlichting, "Boundary Layer Theory," 6th ed., translated by J. Kestin, McGraw-Hill Book Company, New York, 1968.
[2] H. Schlichting, *Z. angew. Math. Mech.* **13**, 260 (1933).
[3] E. N. da C. Andrade and L. C. Tsien, *Proc. Phys. Soc. (London)* **49**, 381 (1937).

2q-2. Plane Jet. Laminar flow due to a *plane* jet of viscous fluid issuing from a line orifice into a space filled with the same fluid is described by the following formulas:

$$u = 0.4543 \left(\frac{K^2}{\nu x}\right)^{\frac{1}{3}} \text{sech}^2 \epsilon$$

$$v = 0.5503 \left(\frac{K\nu}{x^2}\right)^{\frac{1}{3}} (2\epsilon \, \text{sech}^2 \epsilon - \tanh \epsilon)$$

(2q-4)

where

$$\epsilon = 0.2751 \left(\frac{K}{\nu^2}\right)^{\frac{1}{3}} y x^{-\frac{2}{3}}$$

$$K = \frac{J}{\rho}$$

Here x is distance from the line source, measured parallel to the plane of symmetry

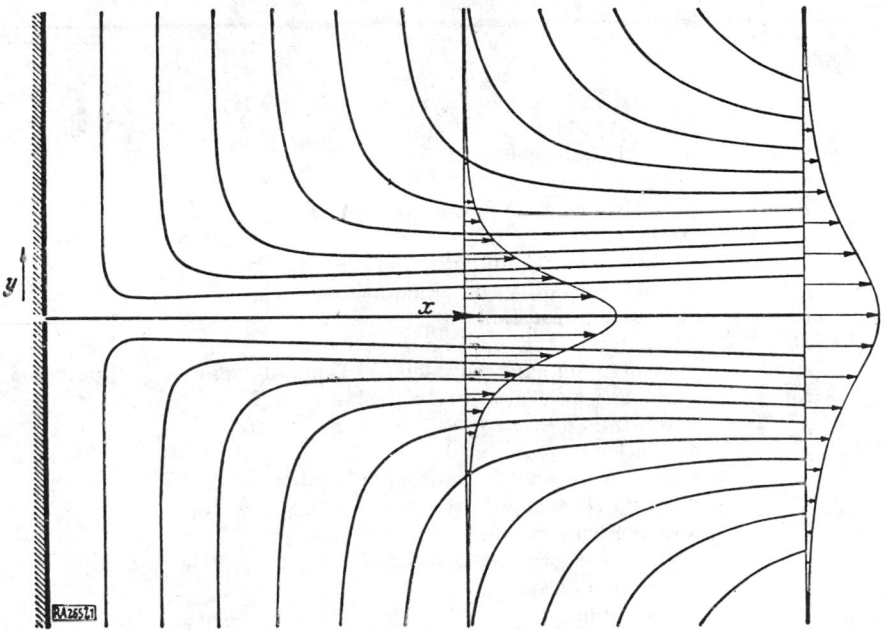

FIG. 2q-1. Streamlines for a circular jet from a point orifice.

of the jet, and y is measured normal to this plane; all other symbols have meanings analogous to those used in Eqs. (2q-1). This theoretical result due to Bickley[1] has been checked experimentally by Andrade[2] and found to be valid for jets from slits of finite width w, provided that x in Bickley's formula is given by

$$x = x_9 + \frac{0.65 K w}{\nu \nu}$$

(2q-5)

[1] W. G. Bickley, *Phil. Mag.* **23**, 727 (1937).
[2] E. N. da C. Andrade, *Proc. Phys. Soc. (London)* **51**, 784 (1939).

2r. Viscosity of Gases

J. KESTIN

Brown University

R. DI PIPPO

Southeastern Massachusetts University

Symbols

a	parameter in viscosity correlation, Eq. (2r-12)
b	virial coefficient, Eq. (2r-13)
b^*	reduced virial coefficient
c	virial coefficient, Eq. (2r-13)
E	potential energy of molecular binding
e_{ij}	strain tensor
F_μ	reduced viscosity function, Eq. (2r-12)
$f_\mu^{(n)}$	nth-order viscosity correction factor
k	Boltzmann's constant
M	molecular weight
m	mass of a molecule; parameter in Lennard-Jones $(m - 6)$ potential
n	number of molecules per unit volume
p	hydrostatic pressure
p_s	saturation pressure
r_m	molecular distance for maximum binding energy
s	viscosity stretching factor, Eq. (2r-12a)
T	absolute temperature
T^*	$(= kT/\epsilon)$ reduced temperature
t	Celsius temperature
t_{ij}	stress tensor
u	velocity of fluid flow
u_1, u_2, u_3	fluid-flow velocity components
δ_{ij}	Kronecker symbol
ϵ	maximum binding energy between molecules
θ	reduced temperature, Eq. (2r-12b)
λ	constant in constitutive equation
μ	viscosity of fluid; constant in constitutive equation
μ_0	viscosity at reference temperature T_0 or at zero density
μ_r	$(= \mu/\mu_0)$ viscosity ratio
ν	kinematic viscosity
ρ	density
ρ^*	reference density
σ	molecular distance for vanishing potential
τ	$(= 1/T)$ inverse temperature
τ_0	shear stress
$\Omega^{(2,2)*}$	reduced collision integral

TABLE 2r-1. ABSOLUTE VISCOSITY μ: UNITS AND CONVERSION FACTORS

Units	centipoises	kg/m-sec	kp sec/m²	lbf-sec/ft²	lbm/ft-hr	lbm/ft-sec	micropoises	poises	slugs/ft-sec
centipoises	1.0	1.0×10^{-3}	1.01972×10^{-4}	2.08854×10^{-6}	2.41909	6.71969×10^{-4}	1.0×10^{4}	1.0×10^{-2}	2.08854×10^{-6}
kg/m-sec	1.0×10^{3}	1.0	1.01972×10^{-1}	2.08854×10^{-2}	2.41909×10^{3}	6.71969×10^{-1}	1.0×10^{7}	1.0×10^{1}	2.08854×10^{-2}
kp sec/m²	9.80665×10^{3}	9.80665	1.0	2.04816×10^{-1}	2.37232×10^{3}	6.58976×10^{-1}	9.80665×10^{7}	9.80665×10^{1}	2.04816×10^{-1}
lbf-sec/ft²	4.78803×10^{4}	4.78803×10^{1}	4.88243	1.0	1.15826×10^{5}	3.21740×10^{1}	4.78803×10^{8}	4.78803×10^{2}	1.0
lbm/ft-hr	4.13379×10^{-1}	4.13379×10^{-4}	4.21529×10^{-5}	8.63361×10^{-6}	1.0	2.77778×10^{-4}	4.13379×10^{3}	4.13379×10^{-3}	8.63361×10^{-6}
lbm/ft-sec	1.48816×10^{3}	1.48816	1.51750×10^{-1}	3.10810×10^{-2}	3.60000×10^{3}	1.0	1.48816×10^{7}	1.48816×10^{1}	3.10810×10^{-2}
micropoises	1.0×10^{-4}	1.0×10^{-7}	1.01972×10^{-7}	2.08854×10^{-9}	2.41909×10^{-4}	6.71969×10^{-8}	1.0	1.0×10^{-6}	2.08854×10^{-9}
poises (g/cm-sec)	1.0×10^{2}	1.0×10^{-1}	1.01972×10^{-2}	2.08854×10^{-3}	2.41909×10^{2}	6.71969×10^{-2}	1.0×10^{6}	1.0	2.08854×10^{-3}
slugs/ft-sec	4.78803×10^{4}	4.78803×10^{1}	4.88243	1.0	1.15826×10^{5}	3.21740×10^{1}	4.78803×10^{8}	4.78803×10^{2}	1.0

Note. When the last significant digit is shown in boldface type, the conversion factor represents a conventional factor which is accurate by definition and involves no approximation.

2r-1. Definitions. The *viscosity* of a fluid is defined in relation to a *macroscopic* system which is assumed to possess the properties of a *continuum*. To obtain an *elementary definition* of viscosity (Fig. 2r-1) consider two infinite flat plates, a at rest and b moving at a constant velocity u, the space between

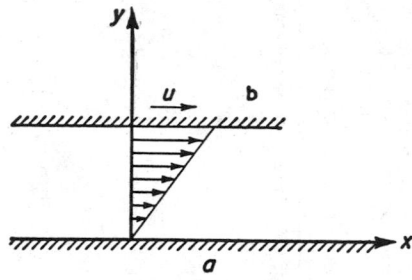

them being filled with the fluid under consideration. In the resulting *shear flow* the velocity distribution is linear with a constant transverse gradient du/dy. It is assumed (*Newton's law of fluid friction*) that the shearing stress τ_0 at either wall is proportional to the velocity gradient

$$\tau_0 = \mu \frac{du}{dy} \qquad (2r\text{-}1)$$

Fig. 2r-1. Illustration of Newton's law of fluid friction.

The coefficient of proportionality μ is known as the viscosity, or more precisely, as the *dynamic* or *absolute viscosity* of the fluid. The various units of viscosity and their conversion factors are given in Table 2r-1.

The ratio

$$\nu = \frac{\mu}{\rho} \qquad (2r\text{-}2)$$

is known as the *kinematic viscosity;* the respective units and conversion factors are given in Table 2r-2.

TABLE 2r-2. KINEMATIC VISCOSITY ν; UNITS AND CONVERSION FACTORS

Units	m²/sec	m²/hr	cm²/sec (stokes)	ft²/sec	ft²/hr
m²/sec............	1	3,600	1×10^4	10.7639	3.875×10^4
m²/hr.............	277.8×10^{-6}	1	2.778	299.0×10^{-5}	10.7639
cm²/sec (stokes)....	1×10^{-4}	0.36	1	10.7639×10^{-4}	3.875
ft²/sec............	0.092903	334.45	929.03	1	3,600
ft²/hr.............	25.806×10^{-6}	0.092903	0.25806	277.8×10^{-6}	1

From British Standard Code B.S. 1042: 1943 amended March, 1946. See Note to Table 2r-1.

In a general field of flow, u_1, u_2, u_3 of a homogeneous Newtonian incompressible fluid, the *shearing stresses* are proportional to the respective *rates of change of strain* (Stokes' law). The symmetric *stress tensor* t_{ij} is assumed to be a linear function of the *rate of strain tensor* e_{ij}. Taking into account that in a fluid at rest the stress is an isotropic tensor, we put

$$t_{ij} = -p\delta_{ij} + \lambda\delta_{ij}e_{kk} + 2\mu e_{ij}$$

where δ_{ij} is the Kronecker symbol ($\delta = 1$ for $i = j$ and $\delta = 0$ for $i \neq j$) and p is arbitrary. Since $t_{ij} = 0$ for $e_{ij} = 0$, we have $t_{ii} = -3p$ and $3\lambda + 2\mu = 0$ (Stokes' hypothesis). Consequently

$$t_{ij} = -p\delta_{ij} - \tfrac{2}{3}\mu\delta_{ij}e_{kk} + 2\mu e_{ij} \qquad (2r\text{-}3)$$

where now p denotes the hydrostatic pressure. The scalar μ is defined as *the absolute viscosity of the fluid.*

The viscosity is assumed to be a function of the thermodynamic state of the fluid and independent of the velocity field. For a homogeneous fluid μ is a function of *two properties.* It is customary to use either of the following two alternative representations:

$$\mu = \mu(p,T) \quad \text{or} \quad \mu = \mu(\rho,T) \tag{2r-4}$$

where T is the absolute temperature, p is the pressure, and ρ is the density of the fluid.

Numerical values of viscosity cannot be calculated with the aid of the equations of thermodynamics. They must be measured directly, the measurement being usually very difficult, particularly at higher pressures and temperatures. In principle, values of viscosity can be calculated by the methods of the kinetic theory of gases and statistical mechanics with quantum corrections where necessary.

In relation to a *microscopically* defined system the viscosity of a gas is assumed to be due to a transfer of momentum effected by molecules, their velocity being composed of the molecular (random) velocity and the macroscopic (ordered) velocity. In shear flow (Fig. 2r-2), the shearing stress acting on a small element of area aa is equal to the integral of the change in momentum effected by the particles moving across, both from above and from below it, the integral extending over all particles crossing.

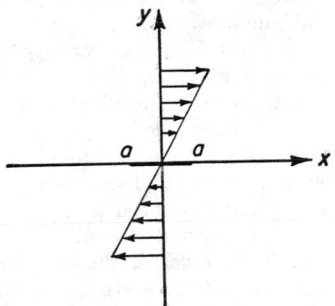

2r-2. Variation of Viscosity with Temperature and Pressure. The calculation of the viscosity of gases has so far met with only limited success, extensive experimental determinations still forming the basis for practical applications. The calculation of the viscosity of gases must make use of a *molecular model* for the gas, increasing refinements being possible.

Fig. 2r-2. Kinetic interpretation of viscosity.

On the simplest assumption of infinitely small, perfectly elastic molecules with zero fields of force (Maxwell) it is found that the absolute viscosity of a gas is independent of pressure and that it increases in proportion to $T^{\frac{1}{2}}$:

$$\mu = K_1 T^{\frac{1}{2}} \quad \left(\frac{\partial u}{\partial p}\right)_T = 0$$
$$\nu = K_2 T^{\frac{1}{2}} \quad p = \text{const} \tag{2r-5}$$

where K_1 and K_2 are empirical constants.

On the assumption of hard elastic spheres with a weak attraction force (Sutherland), it is found that

$$\mu = \frac{KT^{\frac{1}{2}}}{C + \tau} \quad \tau = \frac{1}{T} \tag{2r-6}$$

where K and C are empirical constants. Sutherland's equation (2r-6), as well as experimental results, show the increase with temperature to be *faster* than that in Maxwell's equation (2r-5).

The fact that the viscosity of a gas increases with temperature can be understood if it is realized that in gases the effects of molecular motion dominate over those due to intermolecular forces. In liquids cohesion forces are more important, and since the molecular bonds in a liquid are loosened as the temperature is increased, the absolute viscosity of a *liquid* decreases with temperature; that for a *gas* increases with temperature. Sutherland's equation (2r-6) is inadequate for the correlation of experimental data over large temperature intervals.

In problems of compressible fluid flow it is customary to use the empirical relation

$$\frac{\mu}{\mu_0} = \left(\frac{T}{T_0}\right)^\omega \tag{2r-7}$$

where μ_0 is the value of μ at a reference temperature T_0 and ω is an empirical constant ranging from 0.6 to 1.5. This correlation is less precise than those given later.

All preceding formulas relate to gases at low pressures (say atmospheric). Experimental results (which are still very scarce) show that the viscosity of gases at constant temperature *increases* with pressure, the increase being of the order of 20 to 40 per cent per 1,000 atm. For moderate pressure ranges it is possible to use a linear interpolation formula

$$\frac{\mu}{\mu_0} = 1 + k \tag{2r-8}$$

where μ_0 is the viscosity at temperature T, but at zero density, and k is an empirical constant.

More precisely, the viscosity of a gas increases as its density is increased. Since the viscosity of a gas consisting of molecules which exert no forces upon one another (Maxwell) is independent of density, this behavior is taken as evidence of the existence of intermolecular fields of forces. However, exceptions exist to this rule, notably steam and hydrocarbons, whose viscosity at constant temperature *decreases* with pressure, and therefore density, in certain ranges of states. In turn this is taken as evidence of the existence of some form of molecular association whose precise nature is not understood.

2r-3. Variation of Viscosity with Temperature and Pressure According to Kinetic Theory. There exists a rigorous kinetic theory of the equilibrium and transport properties of gases which is based on Boltzmann's equation. Thus, in particular, and in principle, the viscosity, thermal conductivity (see Sec. 4g) and virial coefficients of gases (see Sec. 4i) are calculated in a consistent and unified way. This theory is due to Chapman and Enskog (S. Chapman and T. G. Cowling, "Mathematical Theory of Non-uniform Gases," Cambridge University Press, New York, 1970; J. O. Hirschfelder, C. F. Curtiss, and R. B. Bird, "Molecular Theory of Gases and Liquids," John Wiley & Sons, Inc., New York, 1964.) The calculations are made on the basis of assumed semiempirical force potentials. For nonpolar gases the most widely used potentials have been the Lennard-Jones twelve-six potential and the modified Buckingham exp-six potential; that used for polar gases is the Stockmayer potential.

The viscosity at zero density is then calculated from the equation

$$\mu_0 = \frac{5\sqrt{m\mathbf{k}T}\, f_\mu^{(n)}(T^*)}{16\pi^{\frac{1}{2}}\sigma^2\Omega^{(2,2)*}(T^*)} \tag{2r-9}$$

or, with the values of the universal constants substituted

$$\frac{\mu_0}{\text{micropoise}} = 26.694\, \frac{\left(\dfrac{M}{\text{g/g-mole}} \cdot \dfrac{T}{K}\right)^{\frac{1}{2}} f_\mu^{(n)}(T^*)}{(\sigma^2/\text{Å}^2)\,\Omega^{(2,2)*}(T^*)} \tag{2r-10}$$

Here σ is the molecular distance at which the potential vanishes, M is the molecular weight, $\mathbf{k}$ is Boltzmann's constant, and $T^* = \mathbf{k}T/\epsilon$ is a dimensionless temperature with ϵ denoting the depth of the potential well. The collision integral $\Omega^{(2,2)*}$ and the factor $f_\mu^{(n)}$, both of which are unique functions of the dimensionless temperature T^*, are given in terms of the intermolecular force potential and must be tabulated for each one of them separately. Such tabulations for the more general $m - 6$ potential can be found in "Tables of Collision Integrals for the $(m - 6)$ Potential for Ten Values of m" by M. Klein and F. J. Smith (*Arnold Engineering Development Center Rept.*) AEDC-TR-68-92, May, 1968, Arnold Air Force Station, Tenn.), with m taking the values $m = 9, 12, 15, 18, 21, 24, 30, 40, 50$, and 75. Tables for the exp-six potential can be found in "Transport Properties of Gases Obeying a Modified Buckingham (Exp-Six) Potential" by E. A. Mason [*J. Chem. Phys.* **22**, 169 (1954)]. The factor $f_\mu^{(n)}$ with $n = 1, 2, \ldots$ represents successive approximations and it is usual to confine it to the third approximation, $f_\mu^{(3)}$, at most.

In principle, the form of and the constants in a potential can be determined by quantum mechanics from a knowledge of the structure of the molecule. However, the attendant mathematical difficulties preclude us from doing so, and potentials must be determined by fitting experimental data on a variety of properties to expressions like the one in Eq. (2r-9). The efforts to associate definite potentials and physically meaningful constants with even the simplest molecules have not yet met with complete success. One of the difficulties is connected with the fact that often several alternative potentials give equally good fits to a set of experimental data of a definite property of a gas, but none seems to reproduce all properties to within the experimental error. Thus, there exists no preferred or universal form of the potential, but, as a matter of experience, it can be stated that the viscosity of the simpler gases, except that of helium, is reproduced reasonably well by the potential family

$$E(r) = \frac{m\epsilon}{m - 6} \left[\frac{m}{6}\right]^{6/(m-6)} \left[\left(\frac{\sigma}{r}\right)^m - \left(\frac{\sigma}{r}\right)^6\right] \qquad (2r\text{-}11)$$

in which σ, ϵ, and m are treated as adjustable constants. The viscosity of helium is best reproduced by the exp-six potential with $r_m = 3.135$ Å, $\epsilon/k = 9.16$ K, and $\alpha = 12.4$ [E. A. Mason and W. E. Rice, *J. Chem. Phys.* **22**, 522, 843 (1954)]. Average, and to a certain extent preliminary, values of σ and ϵ for the Lennard-Jones potential are listed in Table 2r-3. A better representation is obtained with the aid of the semiempirical formula

$$F_\mu = 1.0 - \frac{1.0}{\theta} + \frac{0.5}{\theta^2} \qquad (2r\text{-}12)$$

where

$$F_\mu = \frac{s^2 \mu}{26.694 \sqrt{MT}} - a \qquad (2r\text{-}12a)$$

and

$$\theta = gT \qquad (2r\text{-}12b)$$

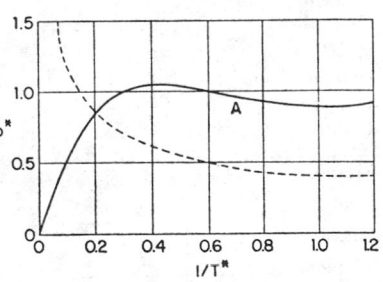

FIG. 2r-3. Dimensionless second virial coefficient for viscosity b^* as a function of reduced inverse temperature, according to Kim and Ross. [*J. Chem. Phys.* **42**, 263 (1965)].

The optimum values of the constants a, g, and s are listed in Table 2r-4 for several gases.

Except for the neighborhood of the critical point, the effect of density (i.e., pressure) on the viscosity of gases, even up to pressures of the order of several hundred atmospheres, can be accounted for with the aid of the virial expansion

$$\mu(\rho, T) = \mu_0(T) + b(T)\rho + c(T)\rho^2 + \cdots \qquad (2r\text{-}13)$$

containing three or four terms. Kim and Ross [*J. Chem. Phys.* **42**, 263 (1965)] provided a theory for the virial $b(T)$. The diagram in Fig. 2r-3 represents the universally valid relation between

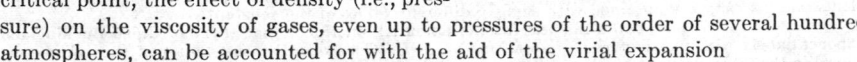
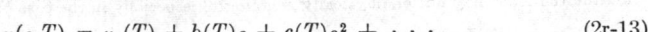

$$b^* = (T^*)^{-1} Q_{rei}(T^*) B_{\theta\delta}(T^*) / \Omega^{(2,2)*}(T^*) \qquad (2r\text{-}14)$$

and $1/T^*$. For the Lennard-Jones model, the expression reduces to

$$b^* = \frac{1}{15.20} \left(\frac{b}{g/cm^3}\right) \left(\frac{\mathring{A}}{\sigma}\right)^3 \left(\frac{K}{\epsilon/k}\right)^{\frac{1}{2}} \left(\frac{M}{g/g\text{-mole}}\right)^{\frac{1}{2}} \quad (2r\text{-}15)$$

In the range where $1/T^*$ exceeds 0.2 ($T^* < 5$ approximately), the virial coefficient b is nearly a constant with $b^* \approx 1$. Consequently, Eq. (2r-14) can be simplified to

$$\mu(\rho,T) - \mu_0(0,T) = \frac{15.20\mu P \text{ cm}^3}{g} \left(\frac{\sigma}{\mathring{A}}\right)^3 \left(\frac{\epsilon/k}{K}\right)^{\frac{1}{2}} \left(\frac{g/g\text{-mole}}{M}\right)^{\frac{1}{2}} + O(\rho^2) \quad (2r\text{-}16)$$

This form leads to an approximate equation for the excess viscosity $\mu(\rho,T) - \mu_0(0,T)$ which has often been used for correlations. This form is

$$\mu(\rho,T) - \mu_0(0,T) \approx f(\rho) \quad (2r\text{-}17)$$

in which $f(\rho)$ is a unique (empirical) function for each gas.

TABLE 2r-3. MOLECULAR-FORCE CONSTANTS FOR THE
LENNARD-JONES (12-6) POTENTIAL

$$E(r) = 4\epsilon \left[\left(\frac{\sigma}{r}\right)^{12} - \left(\frac{\sigma}{r}\right)^{6} \right]$$

Gas	Symbol	ϵ/k, K	σ, $\mathring{A}$	Ref.
Acetylene	C_2H_2	185	4.221	1
Air	—	$\left\{\begin{matrix} 84.0 \\ 117.5 \end{matrix}\right\}$	$\left\{\begin{matrix} 3.689 \\ 3.512 \end{matrix}\right\}$	1 2
Argon	Ar	$\left\{\begin{matrix} 124.0 \\ 152.8 \end{matrix}\right\}$	$\left\{\begin{matrix} 3.418 \\ 3.292 \end{matrix}\right\}$	1 2
Bromine	Br_2	520	4.268	1
Carbon dioxide	CO_2	261.1	3.705	2
Carbon monoxide	CO	110.3	3.590	3
Chlorine	Cl_2	257	4.40	1
Deuterium	D_2	39.3	2.948	1
Ethylene	C_2H_4	205	4.232	1
Helium	He	$\left\{\begin{matrix} 10.22 \\ 86.20 \end{matrix}\right\}$	$\left\{\begin{matrix} 2.576 \\ 2.158 \end{matrix}\right\}$	1 2
Hydrogen	H_2	38.0	2.915	1
Iodine	I_2	550	4.982	1
Krypton	Kr	206.4	3.522	2
Methane	CH_4	144	3.796	1
Neon	Ne	$\left\{\begin{matrix} 35.7 \\ 60.9 \end{matrix}\right\}$	$\left\{\begin{matrix} 2.789 \\ 2.648 \end{matrix}\right\}$	1 2
Nitric oxide	NO	119	3.470	1
Nitrogen	N_2	$\left\{\begin{matrix} 91.5 \\ 113.5 \end{matrix}\right\}$	$\left\{\begin{matrix} 3.681 \\ 3.566 \end{matrix}\right\}$	1 2
Oxygen	O_2	113	3.433	1
Propane	C_3H_8	254	5.061	1
Xenon	Xe	229	4.055	1

Note 1. Differences in the values in this table and the table in Sec. 4i are a measure of the uncertainties which still exist, as well as of the fact that the best fits to experimental values of virial coefficients and viscosity are obtained with slightly different values of the constants.

Note 2. In the case of helium the best form of potential function is that of the modified Buckingham exponential-six with parameters as quoted in the text. Consequently, the values of the parameters shown in the table may not be physically meaningful, especially in the case of those quoted from ref. 2.

References for Table 2r-3

1. Hirschfelder, J. O., C. F. Curtiss, and R. B. Bird: "Molecular Theory of Gases and Liquids," Table 1-A, p. 1110, John Wiley & Sons, Inc., New York, corrected edition, 1964.
2. DiPippo, R., and J. Kestin: Viscosity of Seven Gases up to 500°C and Its Statistical Interpretation, Proc. 4th Symp. on Thermophys. Properties, ASME, New York, 1968.
3. Natl. Bur. Standards Circ. 564, 1955.

TABLE 2r-4. PARAMETERS IN VISCOSITY CORRELATION, EQ. (2r-12)

Gas	Symbol	a	$g \times 10^3$, $(K)^{-1}$	s, Å	Temp. range, K
Air.....................	.....	1.3034	6.0906	3.484	298–773
Argon..................	Ar	1.0300	7.5793	2.970	298–573
Butane.................	C_4H_{10}	0.91040	5.5145	4.730	311–511
Carbon dioxide..........	CO_2	0.94147	5.3316	3.230	298–773
Ethane.................	C_2H_6	0.92669	6.2093	3.820	294–511
Ethylene...............	C_2H_4	0.71342	3.3598	2.235	303–368
Helium.................	He	1.5779	4.0302	2.250	298–673
Krypton................	Kr	0.83447	8.4746	2.935	298–473
Methane...............	CH_4	1.0532	5.2434	3.208	283–411
Neon...................	Ne	1.6602	6.6667	2.895	298–453
Nitrogen...............	N_2	1.3127	6.2232	3.548	298–773

Unpublished correlation prepared by authors of this article.

2r-4. Viscosity in the Neighborhood of the Critical Point. Contrary to earlier views, it has now become accepted that the viscosity of a gas does not increase anomalously in the neighborhood of the critical point, even though the representation in the form of Eq. (2r-13) breaks down there. The viscosity in the neighborhood of the critical point has been measured (rather sketchily) for a very small number of substances only. A qualitative idea of the resulting behavior can be obtained from the diagram for CO_2, given as Fig. 2r-4 [J. Kestin, J. H. Whitelaw and T. F. Zien, *Physica* **30**, 161 (1964)].

2r-5. Law of Corresponding States. Attempts have also been made to correlate the viscosity of gases with the aid of the law of corresponding states. The most promising correlation [J. M. J. Coremans and J. J. M. Beenakker, *Physica* **26**, 653 (1960)] makes use of molecular constants for the formation of reduced variables. The reference temperature is chosen as $T^* = kT/\epsilon$, the reference density being chosen as the fraction of volume occupied by the molecular core $\rho^* = \frac{1}{3}\pi n(\frac{1}{2}\sigma)^3$ where n is the number density. The viscosity μ is referred to μ_0 measured at zero density, so that $\mu_r = \mu/\mu_0$ and

$$\mu_r = f(T^*,\rho^*) \qquad (2r\text{-}18)$$

where f is an approximately universal function. It can be represented by the power series

$$\mu_r = 1 + \frac{0.55\rho^* + 0.96\rho^{*2} + 0.61\rho^{*3}}{T^{*0.59}} \qquad (2r\text{-}19)$$

from which it is seen that the relative excess viscosity $\mu_r - 1$ is a unique function of relative density ρ^* at constant relative temperature T^* according to Eq. (2r-17). Equation (2r-18) reproduces the experimental values for nonpolar or only slightly polar gases, with an error of the order of ± 3 per cent over a fairly large range of temperatures and densities. The error is negligible up to densities of approximately 200 amagat units, and the equation can be used up to about 500 amagat units.

2r-6. Mixtures of Gases. The viscosity of a gaseous mixture cannot be deduced from the knowledge of its composition and of the viscosities of its components by macroscopic methods, and methods of statistical mechanics must be used. In any case it should be noted that the viscosity of a mixture is not equal to the weighted mean of the viscosity of its components, it being possible for the viscosity of a mixture to be higher than that of its components. For example, a mixture of argon ($\mu_{Ar} = 222 \times 10^{-6}$ poise) and helium ($\mu_{He} = 195 \times 10^{-6}$ poise) containing 40 per cent He

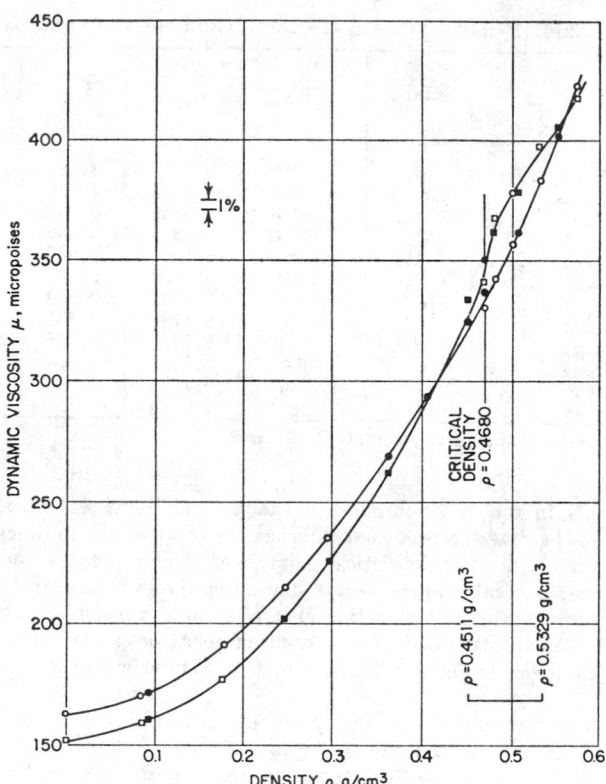

Fig. 2r-4. Viscosity of carbon dioxide as a function of density in the near-critical region according to Kestin, Whitelaw, and Zein [*Physica* **30**, 161 (1964)].

and 60 per cent Ar has a viscosity of $\mu = 230 \times 10^{-6}$ poise. Thus for a given pressure and temperature, the viscosity of a mixture can pass through a maximum when plotted as a function of composition. Maxima are also exhibited by the binary mixtures H_2-Xe, He-Xe, H_2-SO_2, H_2-C_3H_8, H_2-CO_2, H_2-C_3H_6, H_2-C_2H_6, H_2-NO, H_2-C_2H_4, He-Ar, H_2-NH_3, H_2-CH_4, NH_3-C_2H_4, HCl-CO_2, CH_4-NH_3, and possibly many others.

Even in the case of binary mixtures, the relation among the viscosity of the mixture, the viscosities of the pure components, and the composition is quite complex. At present the quality of the statistical approximation obtained by the methods of statistical mechanics is somewhat uncertain, and it is necessary to refer the reader to the treatise by J. O. Hirschfelder et al. (see footnote to Table 2r-3) for further details. Table 2r-5 gives sources of data on the viscosity of gaseous mixtures.

TABLE 2r-5. REFERENCES TO DATA ON BINARY GASEOUS MIXTURES

Mixture	Pressure range, atm	Temperature range, °C	Reference
Air-H_2O	1	25–75	1
He-Ar	1–50	20–30	2
He-Ne	1–35	20–30	3
He-Kr	1–25	20–30	4
He-H_2	1–25	20–30	5
He-N_2	1–25	20–30	4
He-O_2	1–25	20–30	5
He-CO_2	{ 1–25	20–30	6
	1–70	20	9
Ne-Ar	1–35	20–30	3
Ne-N_2	1–25	20–30	6
Ne-CO_2	1–25	20–30	7
Ar-NH_3	1–25	20–30	8
Ar-N_2	1–25	20–30	6
Ar-CO_2	1–25	20–30	4
Kr-CO_2	1–25	20–30	5
N_2-H_2	1–25	20–30	5
N_2-CO_2	1–25	20–30	4
CO_2-CH_4	1–25	20–30	5
CH_4-C_4H_{10}	1–25	20–30	5

References for Table 2r-5

1. Kestin, J., and J. H. Whitelaw: Measurement of the Viscosity of Dry and Humid Air, p. 301 in "Humidity and Moisture," vol. III, p. 301, Reinhold Book Corporation, New York, 1965.
2. Iwasaki, H., and J. Kestin: *Physica* **29**, 1345 (1963).
3. Kestin, J., and A. Nagashima: *J. Chem. Phys.* **40**, 3648 (1964).
4. Kestin, J., Y. Kobayashi, and R. T. Wood: *Physica* **32**, 1065 (1966).
5. Kestin, J., and J. Yata: *J. Chem. Phys.* **49**, 4780 (1968).
6. DiPippo, R., J. Kestin, and K. Oguchi: *J. Chem. Phys.* **46**, 4758 (1967).
7. Breetveld, J. D., R. DiPippo, and J. Kestin: *J. Chem. Phys.* **45**, 124 (1966).
8. Iwasaki, H., J. Kestin, and A. Nagashima: *J. Chem. Phys.* **40**, 2988 (1964).
9. Richardson, H. P., D. Cummins, and R. A. Guereca: Absolute Viscosity Determinations by Means of a Coiled-capillary Viscosimeter: Data for Helium, Carbon Dioxide Mixtures, *Proc. 4th Symp. Thermophys. Properties*, ASME, New York, 1968.

2r-7. Tables of Viscosity. The variation of the viscosity of several gases, all extrapolated to zero density (but accurate enough at atmospheric pressure), can be obtained from the correlation in Eq. (2r-12) and the data in Table 2r-4.

Table 2r-6 contains the best available data on the absolute viscosity μ of gases at 20°C *and atmospheric pressure* together with the temperature increment $(\Delta\mu)_T$ and the pressure increment $(\Delta\mu)_p$ at that point. Table 2r-7 lists the same values for the kinematic viscosity ν. The values have been carefully selected in each case, either mean values or preferred values having been chosen depending on the merits of the available experimental material. The estimated uncertainties are also based on a critical assessment of available data and are, to a certain extent, arbitrary. Experimental results for both high pressures and temperatures are, for all intents and purposes, nonexistent.

TABLE 2r-6. ABSOLUTE VISCOSITY μ OF GASES IN MICROPOISES

(10^{-6} g/cm sec $= 10^{-6}$ dyne sec/cm²; at 20°C and 1 atm)

Gas	Symbol	μ, μpoises	Estimated uncertainty $\pm \Delta\mu$, μpoises	Temp. increment $(\Delta\mu)_T$, μpoises/°C	Pressure increment $(\Delta\mu)_p$, μpoises/atm	Source
Acetylene	C_2H_2	93.5 (at 0°C)	......	......	......	"International Critical Tables"
Air	......	181.92	0.006	0.536	0.1224	Bearden, *Phys. Rev.* **56**, 1023 (1939)
Ammonia	NH_3	97.4	3	0.425	......	Wtd. mean of 2 values
Argon	Ar	222.86	0.1	0.704	0.1753	Ref. 1
Bromine	Br_2	149.5	......	0.500	......	Ref. 2
iso-Butane	C_4H_{10}	74.8	......	0.237	......	Ishida, *Phys. Rev.* **21** (1923)
n-Butane	C_4H_{10}	84.8	......	0.300	......	Kuenen and Visser, *Amsterdam Acad. Sci.* **22**, 336 (1913)
Carbon dioxide	CO_2	146.63	0.07	0.450	0.0046	Ref. 1
Carbon monoxide	CO	175.3	0.1	0.474	......	Wtd. mean of 4 values
Chlorine	Cl_2	133.0	......	0.451	......	Rankine, *Proc. Roy. Soc. (London)*, ser. A, **86**, 162 (1912)
Chloroform	$CHCl_3$	100.0	......	0.340	......	Ref. 2
Cyanogen	C_2N_2	100.2	......	0.360	......	Ref. 2
Deuterium	D_2	124.68	0.07	0.284	0.0082	Ref. 1
Deuteromethane	CD_4	129.0	......	0.580	......	Ref. 2
Ethane	C_2H_6	91.0	0.8	0.277	......	Wtd. mean of 2 values
Ethylene	C_2H_4	100.0	......	0.320	......	Van Cleave and Maass, *Can. J. Research* **13B**, 140 (1935)
Helium	He	196.14	0.1	0.464	-0.0093	Ref. 1
Hydrogen	H_2	88.73	0.05	0.200	0.0118	Ref. 1
Hydrogen bromide	HBr	184.3	......	0.680	......	Ref. 2
Hydrogen chloride	HCl	142.5	......	0.500	......	Ref. 2
Hydrogen deuteride	HD	111.8	0.3	......	......	Kestin and Nagashima, *Phys. Fluids* **7**, 730 (1964)

TABLE 2r-6. ABSOLUTE VISCOSITY μ OF GASES IN MICROPOISES (*Continued*)

Gas	Symbol	μ, μpoises	Estimated uncertainty $\pm\Delta\mu$, μpoises	Temp. increment $(\Delta\mu)_T$, μpoises/°C	Pressure increment $(\Delta\mu)_p$, μpoises/atm	Source
Hydrogen iodide......	HI	183.0	......	0.640	......	Ref. 2
Krypton...............	Kr	249.55	0.15	0.735	0.2816	Ref. 1
Mercury...............	Hg	450.0 (200°C)	......	......	......	Ref. 2
Methane...............	CH₄	109.8	0.1	0.330	0.016	Kestin and Leidenfrost, "Thermodynamic Properties of Gases, Liquids, Solids," p. 321, ASME 1958
Methyl bromide.......	CH₃Br	132.7	......	0.460	......	Ref. 2
Methyl chloride.......	CH₃Cl	107.0	......	0.425	......	Breitenbach, *Ann. Phys.* **5**, 166 (1901)
Neon.................	Ne	313.81	0.15	0.697	0.0354	Ref. 1
Nitric oxide..........	NO	189.8	0.1	0.538	......	Wtd. mean of 3 values
Nitrogen.............	N₂	175.69	0.09	0.454	0.1234	Ref. 1
Nitrous oxide........	N₂O	145.6	......	0.475	......	Johnston and McCloskey, *J. Phys. Chem.* **44**, 1038 (1940)
Oxygen..............	O₂	203.31	0.1	0.616	0.1205	Ref. 1
Propane.............	C₃H₈	80.0	......	0.22	......	Ref. 2
Sulfur dioxide.......	SO₂	125.0	......	0.400	......	Ref. 2
Xenon...............	Xe	227.40	0.14	0.725	0.2624	Ref. 1

References

1. Kestin, J., and W. Leidenfrost: *Physica* **25**, 1033 (1959).
2. Golubev, I. F.: "Viaz'kost' gazov i gazovykh smesei," Moscow, 1959. This reference contains extensive data whose accuracy, however, it is difficult to assess.

TABLE 2r-7. KINEMATIC VISCOSITY ν OF GASES
(10^{-3} cm²/sec; at 20°C and 1 atm)

Gas	Symbol	ν, 10^{-3} cm²/sec	Estimated uncertainty $\pm\Delta\nu$, 10^{-3} cm²/sec	Temp. increment $(\Delta\nu)_T$, 10^{-3} cm²/(sec)(°C)	Pressure increment $(\Delta\nu)_P$, 10^{-3} cm²/(sec)(atm)
Acetylene	C_2H_2	80.6 (at 0°C)	0.08	0.960	
Air	.....	151.1	4	1.07	−150.9
Ammonia	NH_3	138	0.06	0.882	
Argon	Ar	134.3			−134.1
Bromine	Br_2	22.50	.....	0.152	
iso-Butane	C_4H_{10}	31.0	.....	0.204	
n-Butane	C_4H_{10}	35.1	.....	0.244	
Carbon dioxide	CO_2	80.09	0.04	0.516	−80
Carbon monoxide	CO	150.6	0.09	0.921	
Chlorine	Cl_2	45.11	.....	0.307	
Chloroform	$CHCl_3$	20.16	.....	0.137	
Cyanogen	C_2N_2	46.35	.....	0.325	
Deuterium	D_2	744.2	0.4	4.24	−740
Deuteromethane	CD_4	154.7	.....	1.22	
Ethane	C_2H_6	72.9	0.6	0.471	
Ethylene	C_2H_4	85.84	.....	0.997	
Helium	He	1,179	0.6	6.81	−1,200
Hydrogen bromide	HBr	54.79	.....	0.389	
Hydrogen chloride	HCl	93.99	.....	0.651	
Hydrogen deuteride	HD	889.4	2.4		
Hydrogen	H_2	1,059	0.6	6.01	−1,060
Hydrogen iodide	HI	34.42	.....	0.237	
Krypton	Kr	72.44	0.044	0.460	−72.20
Mercury	Hg	87.12 (at 200°C)	.....		
Methane	CH_4	164.8	0.2	1.06	−160

TABLE 2r-7. KINEMATIC VISCOSITY ν OF GASES (*Continued*)

Gas	Symbol	ν, 10^{-3} cm²/sec	Estimated uncertainty $\pm \Delta\nu$, 10^{-3} cm²/sec	Temp. increment $(\Delta\nu)_T$, 10^{-3} cm²/(sec)(°C)	Pressure increment $(\Delta\nu)_P$, 10^{-3} cm²/(sec)(atm)
Methyl bromide	CH_3Br	33.64	.....	0.232	
Methyl chloride	CH_3Cl	50.97	.....	0.376	
Neon	Ne	374.1	0.18	2.11	−374
Nitric oxide	NO	152.1	0.08	0.950	
Nitrogen	N_2	150.9	0.08	0.905	−150.8
Nitrous oxide	N_2O	79.57	.....	0.531	
Oxygen	O_2	152.8	0.08	0.984	−152.6
Propane	C_3H_8	43.7	.....	0.269	
Sulfur dioxide	SO_2	46.94	.....	0.310	
Xenon	Xe	42.02	0.026	0.278	−42.38

TABLE 2-8. VISCOSITY OF COMPRESSED WATER AND SUPERHEATED STEAM (MICROPOISES)
Of each pair of figures the upper represents the adopted value and the lower the tolerance (±)

Pressure, bars	\ Temperature, °C: 0	50	100	150	200	250	300	350	375	400	425	450	475	500	550	600	650	700
1	17,500 / 400	5,440 / 140	121.1 / 1.2	141.5 / 1.4	161.8 / 1.6	182.2 / 1.8	202.5 / 2.0	223 / 7	233 / 7	243 / 7	253 / 8	264 / 8	274 / 8	284 / 8	304 / 9	325 / 10	345 / 10	365 / 11
5	17,500 / 400	5,440 / 140	2,790 / 70	1,810 / 50	160.2 / 1.6	181.4 / 1.8	202.3 / 2.0		234 / 9	244 / 10	254 / 10	264 / 11	274 / 11	284 / 11	305 / 12	325 / 13	345 / 14	366 / 15
10	17,500 / 400	5,440 / 140	2,790 / 70	1,810 / 50	158.5 / 1.6	180.6 / 1.8	202.2 / 2.0		234 / 9	244 / 10	255 / 10	265 / 11	275 / 11	285 / 11	305 / 12	326 / 13	346 / 14	366 / 15
25	17,500 / 400	5,440 / 140	2,800 / 70	1,820 / 50	1,340 / 30	177.8 / 1.8	201.6 / 2.0		236 / 9	246 / 10	256 / 10	266 / 11	276 / 11	287 / 12	307 / 12	327 / 13	347 / 14	367 / 15
50	17,500 / 400	5,450 / 140	2,800 / 70	1,820 / 50	1,350 / 30	1,070 / 30	200.6 / 2.0		240 / 10	250 / 10	259 / 10	269 / 11	279 / 11	289 / 12	309 / 12	329 / 13	349 / 14	369 / 15
75	17,500 / 400	5,450 / 140	2,800 / 70	1,830 / 50	1,350 / 30	1,080 / 30	199.2 / 2.0		244 / 10	253 / 10	263 / 10	273 / 11	282 / 11	292 / 12	312 / 12	332 / 13	352 / 14	372 / 15
100	17,500 / 400	5,450 / 140	2,810 / 70	1,830 / 50	1,360 / 30	1,080 / 30	905 / 23		249 / 10	258 / 10	267 / 11	276 / 11	286 / 11	295 / 12	315 / 13	334 / 13	354 / 14	374 / 15
125	17,500 / 400	5,460 / 140	2,810 / 70	1,840 / 50	1,360 / 30	1,090 / 30	911 / 23		254 / 10	263 / 10	271 / 11	280 / 11	289 / 12	299 / 12	318 / 13	337 / 14	357 / 14	376 / 15
150	17,400 / 400	5,460 / 140	2,820 / 70	1,840 / 50	1,370 / 30	1,100 / 30	917 / 23		262 / 11	269 / 11	276 / 11	285 / 11	294 / 12	302 / 12	321 / 13	340 / 14	359 / 14	379 / 15
175	17,400 / 400	5,460 / 140	2,820 / 70	1,850 / 50	1,380 / 30	1,100 / 30	924 / 23		273 / 11	276 / 11	282 / 11	290 / 12	298 / 12	307 / 12	324 / 13	343 / 14	362 / 14	381 / 15
200	17,400 / 400	5,460 / 140	2,830 / 70	1,860 / 50	1,380 / 40	1,110 / 30	930 / 23	735 / 29	291 / 12	286 / 11	289 / 12	296 / 12	303 / 12	311 / 12	328 / 13	346 / 14	365 / 15	384 / 15

Temp	1	2	3	4	5	6	7	8	9	10	11	12	13	14	15	16	17	18
225	386 / 15	368 / 15	350 / 14	332 / 13	316 / 13	309 / 12	302 / 12	298 / 12	299 / 12	491 / 20	747 / 30	936 / 23	1,120 / 30	1,390 / 40	1,860 / 50	2,830 / 70	5,460 / 140	17,400 / 400
250	389 / 16	371 / 15	353 / 14	336 / 13	321 / 13	315 / 13	310 / 12	309 / 12	321 / 13	597 / 24	760 / 30	943 / 24	1,120 / 30	1,390 / 40	1,870 / 50	2,840 / 70	5,470 / 140	17,400 / 400
275	392 / 16	374 / 15	357 / 14	341 / 14	327 / 13	322 / 13	320 / 13	324 / 13	367 / 15	633 / 25	772 / 31	949 / 24	1,130 / 30	1,400 / 40	1,870 / 50	2,840 / 70	5,470 / 140	17,400 / 400
300	395 / 16	377 / 15	361 / 14	346 / 14	334 / 13	330 / 13	331 / 13	345 / 14	458 / 18	657 / 26	785 / 31	955 / 24	1,130 / 30	1,400 / 40	1,880 / 50	2,850 / 70	5,470 / 140	17,400 / 400
350	401 / 16	385 / 15	369 / 15	357 / 14	349 / 14	351 / 14	363 / 14	416 / 17	573 / 23	693 / 28	805 / 32	968 / 24	1,150 / 30	1,420 / 40	1,890 / 50	2,860 / 70	5,480 / 140	17,400 / 400
400	408 / 16	392 / 16	379 / 15	369 / 15	369 / 15	379 / 15	411 / 16	503 / 20	628 / 25	721 / 29	825 / 33	981 / 39	1,160 / 50	1,430 / 60	1,900 / 80	2,870 / 120	5,480 / 200	17,300 / 400
450	415 / 17	401 / 16	389 / 16	383 / 15	393 / 16	415 / 17	468 / 19	565 / 23	664 / 27	743 / 30	837 / 33	993 / 40	1,170 / 50	1,440 / 60	1,910 / 80	2,880 / 120	5,490 / 220	17,300 / 700
500	423 / 17	410 / 16	401 / 16	400 / 16	421 / 17	456 / 18	521 / 21	609 / 24	693 / 28	762 / 30	850 / 34	1,010 / 40	1,180 / 50	1,450 / 60	1,920 / 80	2,890 / 120	5,490 / 220	17,200 / 700
550	431 / 17	420 / 17	414 / 16	418 / 17	453 / 18	497 / 20	564 / 23	643 / 26	716 / 29	780 / 31	860 / 34	1,020 / 40	1,200 / 50	1,460 / 60	1,930 / 80	2,900 / 120	5,500 / 220	17,200 / 700
600	439 / 18	430 / 17	428 / 17	439 / 18	485 / 19	534 / 21	600 / 24	670 / 27	736 / 29	795 / 32	870 / 35	1,030 / 40	1,210 / 50	1,480 / 60	1,940 / 80	2,910 / 120	5,500 / 220	17,200 / 700
650	448 / 18	441 / 18	442 / 18	460 / 18	516 / 21	567 / 23	629 / 25	698 / 28	754 / 30	809 / 32	882 / 35	1,040 / 40	1,220 / 50	1,490 / 60	1,960 / 80	2,920 / 120	5,510 / 220	17,200 / 700
700	458 / 18	453 / 18	458 / 18	482 / 19	545 / 22	596 / 24	654 / 26	713 / 28	770 / 31	822 / 33	895 / 36	1,060 / 40	1,230 / 50	1,500 / 60	1,970 / 80	2,930 / 120	5,510 / 220	17,100 / 700
750	468 / 19	466 / 19	474 / 19	504 / 20	572 / 23	621 / 25	676 / 27	732 / 29	784 / 31	835 / 33	905 / 36	1,070 / 40	1,240 / 50	1,510 / 60	1,980 / 80	2,940 / 120	5,520 / 220	17,100 / 700
800	478 / 19	478 / 19	491 / 20	526 / 21	596 / 24	644 / 26	695 / 28	748 / 30	798 / 32	846 / 34	915 / 37	1,080 / 40	1,260 / 50	1,520 / 60	1,990 / 80	2,950 / 120	5,520 / 220	17,100 / 700

Note 1. The entry shown for 0°C and 1 bar relates to a metastable liquid state. The stable state is here solid.

Note 2. The values and the tolerances in the region of the critical point do not take into account the possibility of an anomalous behavior of the viscosity in the immediate neighborhood of the critical point.

2r-8. Steam. The dynamic and kinematic viscosity of steam has been settled (subject to future amendment) by international agreement ["Supplementary Release on Transport Properties of the Sixth International Conference on the Properties of Steam," New York, 1963; obtainable from the Secretariat of the International Conference on the Properties of Steam, ASME, United Engineering Center, New York. See also E. Schmidt, "VDI-Wasserdampftafeln" (VDI-Steam Tables), 7th ed., Springer Verlag, 1968]. According to this internationally recognized correlation, the viscosity of steam and water can be represented empirically by the following equations, depending on the range of states under consideration:

Superheated steam at 1 bar pressure in temperature range $100°C < t < 700°C$:

$$\frac{\mu_1}{\text{micropoise}} = 80.4 + 0.407\,\frac{t}{°C} \tag{2r-20}$$

$$\text{Tolerance for }\mu_1: \quad \begin{array}{ll} \pm 1\%, & 100 < \dfrac{t}{°C} < 300 \\[2mm] \pm 3\%, & 300 < \dfrac{t}{°C} < 700 \end{array}$$

Superheated steam from 1 bar pressure to saturation in temperature range $100°C < t < 300°C$. (range of anomalous behavior where the viscosity decreases with density along an isotherm):

$$\frac{\mu - \mu_1}{\text{micropoise}} = -\frac{\rho}{\text{g/cm}^3}\left[1858 - 5.90\,\frac{t}{°C}\right] \tag{2r-21}$$

$$\text{Tolerance for }\mu: \pm 1\%$$

Supercritical steam from 1 to 800 bars pressure in temperature range $375°C < t < 700°C$:

$$\frac{\mu - \mu_1}{\text{micropoise}} = 353.0\,\frac{\rho}{\text{g/cm}^3} + 676.5\left(\frac{\rho}{\text{g/cm}^3}\right)^2 + 102.1\left(\frac{\rho}{\text{g/cm}^3}\right)^3 \tag{2r-22}$$

$$\text{Tolerance for }\mu: \pm 4\%$$

Liquid water along saturation line in temperature range $0°C < t < 300°C$:

$$\frac{\mu}{\text{micropoise}} = 241.4 \times /10^{247.8/(T/K-140)} \tag{2r-23}$$

$$\text{Tolerance for }\mu: \pm 2.5\%$$

Liquid water from saturation pressure to 800 bars in temperature range $0°C < t < 300°C$:

$$\frac{\mu}{\text{micropoise}} = \left(1 + \frac{p - p_s}{10^6\,\text{bars}}\,\phi\right) \times 241.4 \times 10^{247.8/(T/K-140)} \tag{2r-24}$$

where

$$\phi = 1.0467\left(\frac{T}{K} - 305\right)$$

$$\text{Tolerance for }\mu: \quad \begin{array}{ll} \pm 2.5\%, & 1 < \dfrac{p}{\text{bar}} < 350 \\[2mm] \pm 4.0\%, & 350 < \dfrac{p}{\text{bar}} < 800 \end{array}$$

The International Skeleton Table, reproduced as Table 2r-8, gives values of the viscosity of steam and water at agreed grid points, together with their tolerances (uncertainties). The dynamic viscosity μ of steam exhibits anomalous behavior below about 270°C in that an increase in density along an isotherm causes the viscosity to decrease.

2s. Molecular Diffusion of Gases

R. C. ROBERTS

University of Maryland—Baltimore County

Symbols

C	concentration of gas
D	diffusion coefficient
k	Boltzmann's constant
m	molecular mass
p	gas pressure
T	absolute temperature
t	time
$\bar{V}$	average velocity of molecules
x, y, z	rectangular position coordinates
λ	mean free path
σ_r	effective collision diameter

In the simple diffusion of one gas into another, the concentration of either component obeys the equation

$$\frac{\partial C}{\partial t} = \frac{\partial}{\partial x}\left(D\frac{\partial C}{\partial x}\right) + \frac{\partial}{\partial y}\left(D\frac{\partial C}{\partial y}\right) + \frac{\partial}{\partial z}\left(D\frac{\partial C}{\partial z}\right) \tag{2s-1}$$

where C = concentration of gas component
   t = time
x, y, z = position coordinates
   D = diffusion coefficient

Although the diffusion coefficient is, in general, a function of temperature, pressure, and concentration, it can often be considered as constant provided the variations of temperature, pressure, etc., are small. The usual cgs units for the diffusion coefficient are cm²/sec.

The elementary kinetic theory of gases shows that, for a two-component mixture,

$$D_{12} = \frac{1}{3}\frac{n_1\lambda_2\bar{V}_2 + n_2\lambda_1\bar{V}_1}{n_1 + n_2} \tag{2s-2}$$

where n = molecular density
   λ = mean free path
   $\bar{V}$ = average velocity of molecules
   1, 2 = subscripts denoting different gas components

The more exact theories show a quite complicated behavior for D. For example, in a model consisting of rigid elastic spheres

$$D_{12} = \frac{3}{8(n_1 + n_2)\sigma_r^2}\left[\frac{kT(m_1 + m_2)}{2\pi m_1 m_2}\right]^{\frac{1}{2}} \tag{2s-3}[1]$$

[1] Continued on p. 2-252.

TABLE 2s-1. DIFFUSION COEFFICIENTS D_0 AT STANDARD TEMPERATURES AND PRESSURE

$(p = 760 \text{ mm Hg}; T = 273°\text{K})$

Gas pair	D_0, cm²/sec	α
H_2O-CO_2	0.202 (34.4°C)	2
H_2O-air	0.219	1.75
H_2O-H_2	1.02 (34.4°C)	1.75
Ethyl alcohol–CO_2	0.0686	2
Ethyl alcohol–air	0.099	2
Ethyl alcohol–H_2	0.377	2
Ethyl ether–CO_2	0.0541	2
Ethyl ether–air	0.0786	2
Ethyl ether–H_2	0.299	2
Benzene-O_2	0.0797	1.75
Benzene-H_2	0.318	1.75
CCl_4-O_2	0.0636	
CCl_4-H_2	0.293	
Acetone-H_2	0.361	
Mercury-N_2	0.1190	2
Iodine-N_2	0.070	2
Iodine-air	0.0692	2
He-Ar	0.641	1.75
H_2-D_2	1.20	
H_2-O_2	0.697	1.75
H_2-N_2	0.674	1.75
H_2-CO	0.651	1.75
H_2-CO_2	0.550	1.75
H_2-CH_4	0.625	1.75
H_2-SO_2	0.480	1.75
H_2-N_2O	0.535	1.75
H_2-C_2H_4	0.602 (25°C)	1.75
H_2-Ar	0.77 (20°C)	
O_2-N_2	0.181	1.75
O_2-CO	0.185	1.75
O_2-CO_2	0.139	2
CO-N_2	0.192	
CO-CO_2	0.137	1.75
CO-C_2H_4	0.116	1.75
CO_2-N_2	0.144	
CO_2-CH_4	0.153	1.75
CO_2-N_2O	0.096	
H_2-air	0.611	1.75
O_2-air	0.178	1.75
CO_2-air	0.138	2
CH_4-air	0.196	
Ar-N_2	0.20 (20°C)	
Ar-O_2	0.20 (20°C)	
Ar-CO_2	0.14 (20°C)	
H_2-SF_6	0.420 (25°C)	
H_2-C_2H_6	0.537 (25°C)	
N_2-C_2H_6	0.148 (25°C)	
N_2-C_2H_4	0.163 (25°C)	
N_2-n-C_4H_{10}	0.0960 (25°C)	
N_2-iso-C_4H_{10}	0.0908 (25°C)	
H_2-cis-butene-2	0.378 (25°C)	
N_2-cis-butene-2	0.095 (25°C)	
H_2O-He	0.90 (34.4°C)	
H_2O-N_2	0.256 (34.4°C)	

Note. The values for the last ten pairs are taken from refs. 9 and 12.

TABLE 2s-2. DEPENDENCE OF DIFFUSION COEFFICIENTS ON CONCENTRATION

Pair of gases	n_1/n_2	D_{12}
First gas H_2; second gas CO_2	3	0.594
	1	0.605
	$\frac{1}{3}$	0.633
First gas He; second gas A	2.65	0.678
	2.26	0.693
	1.66	0.696
	1	0.706
	0.477	0.712
	0.311	0.731

TABLE 2s-3. DEPENDENCE OF DIFFUSION COEFFICIENT ON PRESSURE

Gas pair	D, cm²/sec	t, °C	p, mm Hg	$\dfrac{Dp}{760}$
CO_2-air.........	0.1653	17.6	751	0.163
CO_2-air.........	0.3376	15.2	364	0.162
CO_2-air.........	0.4139	15.7	309	0.164
CO_2-H_2.........	0.6142	12.8	757	0.612
CO_2-H_2.........	0.9184	15.4	510	0.616
H_2-O_2...........	0.8012	11.4	748	0.790
H_2-O_2...........	1.1718	15.8	512	0.791

TABLE 2s-4. COEFFICIENTS OF SELF-DIFFUSION*

Gas	Temp, °K	D, cm²/sec, experimental
Hydrogen (para-hydrogen into ortho-hydrogen)	273	1.285 ± 0.0025
	85	0.172 ± 0.008
	20.4	0.00816 ± 0.0002
Deuterium into hydrogen	288	1.24
Neon	293	0.473 ± 0.002
Argon	326.7	0.212 ± 0.002
	295.2	0.180 ± 0.001
	273.2	0.158 ± 0.002
	194.7	0.0833 ± 0.0009
	90.2	0.028 ± 0.0010
Krypton	294.0	0.09 ± 0.004
Xenon	292.1	0.0443 ± 0.002
Nitrogen	293	0.200 ± 0.008
Methane ($p = 60$ mm Hg)	292	26.32 ± 0.73
Hydrogen chloride	295.0	0.1246
Hydrogen bromide	295.3	0.0792
Uranium hexafluoride ($p = 10$ mm Hg)	303	$D \times \rho = (234 \pm 9) \times 10^{-6}$ g/cm $\times$ sec

* $p = 760$ mm Hg except where noted.

where m = mass of molecule

n = number of molecules per cm³

T = absolute temperature

σ_r = effective molecular collision diameter

k = Boltzmann's constant

1, 2 = subscripts denoting different gas components

It should be noted that the diffusion coefficient is symmetric with respect to its component variables; i.e., $D_{12} = D_{21}$. In the special case when both gas components are identical, we have the condition of self-diffusion where the coefficient is denoted by D_{11}. The diffusion of two different isotopes of the same gas is an example of self-diffusion, e.g., deuterium into hydrogen.

For most gases a convenient reduction formula may be given to reduce the diffusion coefficient to standard temperature T and pressure p. It is

$$D = D_9 \left(\frac{T}{T_0}\right)^\alpha \frac{p_0}{p} \qquad (2s\text{-}4)$$

where α varies between 1.75 and 2. This is reasonably valid over a range of normal temperature and pressure.

The preceding tables contain data on the diffusion coefficients for a number of gases and vapors. In Table 2s-1 values of α are given (if known) so that Eq. (2s-4) may be used to convert the coefficients to other than standard temperature and pressure. Tables 2s-2 and 2s-3 give certain data on the variation of D with pressure and concentration. Table 2s-4 gives some of the latest data on self-diffusion.

Chapman and Cowling (ref. 1) should be consulted for the advanced theory. A good bibliography may be found in Jost (ref. 2) and in ref. 10.

References

1. Chapman, S., and T. G. Cowling: "The Mathematical Theory of Non-uniform Gases," 2d ed., Cambridge University Press, London and New York, 1952.
2. Jost, W.: "Diffusion in Solids, Liquids, Gases," rev. ed., Academic Press, Inc., New York, 1960.
3. Jeans, J. H.: "An Introduction to the Kinetic Theory of Gases," Cambridge University Press, New York, 1940.
4. Kennard, E. H.: "Kinetic Theory of Gases," McGraw-Hill Book Company, New York, 1938.
5. Loeb, L. B.: "Kinetic Theory of Gases," McGraw-Hill Book Company, New York, 1934.
6. "International Critical Tables," vol. 5, pp. 62–63, McGraw-Hill Book Company, New York, 1928.
7. Hirschfelder, J. O., R. B. Bird, and E. L. Spotz: *Chem. Revs.* **44,** 205 (1949).
8. Hirschfelder, J. O., C. F. Curtiss, and R. B. Bird: "The Molecular Theory of Gases and Liquids," John Wiley & Sons, Inc., New York, 1954.
9. Rossini, F. D., ed.: "Thermodynamics and Physics of Matter," vol. 1 of "High Speed Aerodynamics and Jet Propulsion," Princeton, N.J., 1955.
10. American Society of Mechanical Engineers: "Thermodynamic and Transport Properties of Gases, Liquids, and Solids," McGraw-Hill Book Company, New York, 1959.
11. Boyd, C. A., N. Stein, V. Steingrimsson, and W. F. Rumpel: *J. Chem. Phys.* **19,** 548 (1951).
12. Schwartz, F. A., and J. E. Brow: *J. Chem. Phys.* **19,** 640 (1951).

2t. Compressible Flow of Gases

R. C. ROBERTS

University of Maryland—Baltimore County

Symbols

A	cross-sectional area
a	local sound velocity
C_p	pressure coefficient $(p - p_\infty)/\frac{1}{2}\rho_\infty U^2$
c_v	specific heat at constant volume
c_p	specific heat at constant pressure
E	internal energy per unit mass
M	Mach number
M_∞	free-stream Mach number
m	mass flow
p	pressure
p_∞	free-stream pressure
Q	external heat production rate per unit mass
q	resultant velocity of flow
R	gas constant
t	time
U	free-stream velocity
u, v, w	velocity components of fluid flow
X, Y, Z	rectangular components of external body force
x, y, z	rectangular coordinates
γ	ratio of specific heat at constant pressure to that at constant volume
ρ	density
ρ_∞	free-stream density
ϕ	velocity potential
ϕ_x	$\partial\phi/\partial x$
ϕ_{xx}	$\partial^2\phi/\partial x^2$
ϕ_{xy}	$\partial^2\phi/\partial x\,\partial y$
χ	similarity parameter
ψ	stream function

2t-1. Basic Equations in Rectangular Coordinates. The basic equations of motion for a compressible inviscid gas may be written as follows.

Momentum Equation. By applying Newton's laws of motion the Euler momentum equation may be derived in the form

$$\frac{\partial u}{\partial t} + u\frac{\partial u}{\partial x} + v\frac{\partial u}{\partial y} + w\frac{\partial u}{\partial z} = \frac{-1}{\rho}\frac{\partial p}{\partial x} + X$$

$$\frac{\partial v}{\partial t} + u\frac{\partial v}{\partial x} + v\frac{\partial v}{\partial y} + w\frac{\partial v}{\partial z} = \frac{-1}{\rho}\frac{\partial p}{\partial y} + Y \qquad (2t\text{-}1)$$

$$\frac{\partial w}{\partial t} + u\frac{\partial w}{\partial x} + v\frac{\partial w}{\partial y} + w\frac{\partial w}{\partial z} = \frac{-1}{\rho}\frac{\partial p}{\partial z} + Z$$

where x, y, z = rectangular coordinates

t = time

u, v, w = velocity components in direction of x, y, and z axes, respectively

p = pressure

ρ = density

X, Y, Z = rectangular components of external body force

Continuity Equation. The assumption that the gas is a continuous medium is expressed by the equation

$$\frac{\partial \rho}{\partial t} + \frac{\partial}{\partial x}(\rho u) + \frac{\partial}{\partial y}(\rho v) + \frac{\partial}{\partial z}(\rho w) = 0 \qquad (2t\text{-}2)$$

Energy Equation. The relationship between the kinetic and internal energy and the work done on the fluid by pressure and external forces is expressed by the equation

$$\rho \frac{DE}{Dt} + \rho \frac{D}{Dt}\left(\frac{1}{2} q^2\right) = \rho Q + \rho(uX + vY + wZ) - \frac{\partial}{\partial x}(pu) - \frac{\partial}{\partial y}(pv) - \frac{\partial}{\partial z}(pw)$$
$$(2t\text{-}3)$$

where $\dfrac{D}{Dt} \equiv \dfrac{\partial}{\partial t} + u\dfrac{\partial}{\partial x} + v\dfrac{\partial}{\partial y} + w\dfrac{\partial}{\partial z}$

E = internal energy per unit mass = $\int c_v \, dT$

$q^2 = u^2 + v^2 + w^2$

Q = external-heat-production rate per unit mass

c_v = specific heat at constant volume

Equation of State. For a complete specification of a flow it is necessary to give an equation of state. This commonly takes the form

$$p = f(\rho, T)$$

Many gases obey the equation of state of a perfect gas

$$p = \rho R T$$

under a great variety of conditions. In this equation R is a constant which depends on the particular gas. If the specific heat can be assumed constant, the gas is said to be calorically perfect and

$$E = c_v T$$

where T is the temperature on the absolute scale.

A specific case of great importance is that of isentropic flow. If the entropy is constant throughout the flow, the equation of state can be written as

$$p = K\rho^\gamma$$

where K is a constant and γ is the ratio of the specific heat at constant pressure c_p to that at constant volume c_v. Now the flow is completely determined by the momentum equations, the continuity equation, and the equation of state. Many practical flow problems are essentially cases of isentropic flow.

2t-2. Dynamic Similarity and Definition of Basic Flow Parameters. In the testing of models it is necessary to maintain a proper scaling of certain dynamic parameters in addition to the geometric scaling. For compressible inviscid flow with no heat sources and in which body forces are neglected, the only dynamic dimensionless parameter is the Mach number.

Definition of Mach Number. The local Mach number is defined as the ratio of the local flow velocity q to the local sound velocity a; i.e.,

$$M = \frac{q}{a} \qquad (2t\text{-}4)$$

Thus in a nonuniform flow the Mach number will vary from point to point.   The size
of the Mach number indicates whether the flow is subsonic, $M < 1$; transonic, $M \simeq 1$;
or supersonic, $M > 1$.   The term hypersonic is often used to describe flows where
$M > 5$.

Dynamic Similarity.   If the same gas flows around two geometrically similar bodies,
it might be expected that under the right conditions the streamline pattern would be
similar.   This is true if the Mach numbers of the two flows are equal.   It then follows
that all other dimensionless coefficients such as drag coefficient, lift coefficient, pressure
coefficient, etc., are also equal.

In determining the Mach number in a flow it is necessary to know not only the flow
velocity but the sound velocity as well.   For a perfect gas the sound velocity is proportional to the square root of the temperature; i.e.,

$$a = \sqrt{\gamma R T}$$

Table 2t-1 is based on this relationship.

2t-3. Basic Idea of One-dimensional Flow.   In many cases, as in a pipe of slowly
varying cross section, it is possible to make the assumption of constant flow properties
across any cross section perpendicular to the pipe axis.   Although strictly speaking
there are no one-dimensional flows, because of viscous effects on the boundaries,
it is still possible to get much valuable information of a practical nature from the
assumptions.

TABLE 2t-1. VARIATION OF VELOCITY OF SOUND WITH TEMPERATURE

T, °K	a, fps	a, m/sec
150	805	246
160	832	254
170	857	261
180	882	269
190	907	276
200	930	283
210	953	290
220	975	297
230	997	304
240	1,019	311
250	1,040	317
260	1,060	323
270	1,081	329
280	1,100	335
290	1,120	341
300	1,139	347
310	1,158	353
320	1,176	359
330	1,195	364
340	1,213	370
350	1,230	375

Basic Equations.   On the assumption of isentropic flow the equations of motion are

$$\frac{\partial u}{\partial t} + u \frac{\partial u}{\partial x} = -\frac{1}{\rho}\frac{\partial p}{\partial x} \qquad \text{(momentum)} \qquad (2t\text{-}5)$$

$$\frac{\partial \rho}{\partial t} + \frac{1}{A}\frac{\partial}{\partial x}(\rho u A) = 0 \qquad \text{(continuity)} \qquad (2t\text{-}6)$$

where A is the cross-sectional area. For unsteady one-dimensional flow in general and in particular for an excellent treatment of flow in pipes of constant area see ref. 3. The above equations also cover the case of cylindrical and spherically symmetric flow; i.e.,

$$\frac{1}{A}\frac{\partial A}{\partial x} = \frac{1}{x} \qquad \text{(for cylindrical flow)}$$

$$\frac{1}{A}\frac{\partial A}{\partial x} = \frac{2}{x} \qquad \text{(for spherically symmetric flow)}$$

In the important case of steady flow the equations can be integrated to give

$$\frac{\gamma}{\gamma - 1}\frac{p}{\rho} + \frac{1}{2}u^2 = \text{const} \tag{2t-7}$$

$$\rho u A = m = \text{const} \tag{2t-8}$$

where m is the mass flow. By taking logarithmic derivatives and remembering the definition of the Mach number M, the continuity equation may be written

$$\frac{du}{u}(1 - M^2) + \frac{dA}{A} = 0 \tag{2t-9}$$

Thus, if $du \neq 0$ and $M = 1$, we see that $dA = 0$. In other words, the Mach number becomes equal to unity only in a section of the pipe where the area is a minimum. This fact is of prime importance in the design of supersonic wind tunnels.

The dependence of the various flow variables on the Mach number for steady one-dimensional isentropic flow is given in Table 2t-2.

2t-4. Two-dimensional and Axially Symmetric Flow. Many important types of flow belong to the class of two-dimensional or axially symmetric flows. These include flows past wedges, cones, bodies of revolution, etc. The important distinctions to be made are those between subsonic and supersonic flow. Purely subsonic flow is qualitatively quite similar to incompressible flow, while supersonic flow exhibits many startlingly different properties. Among these are the appearance of shock waves (see Sec. 2v) and the existence of wavefronts. A general discussion of the above topics can be found in refs. 2, 3, and 6.

The greater bulk of the literature on two-dimensional and axially symmetric flow is concerned with steady flow. The unsteady cases are usually extremely difficult to solve.

Velocity Potential and Stream Function. In cases of irrotational or steady flow it is convenient to introduce the velocity potential or the stream function. This reduces the number of equations to one. The velocity potential exists whenever there is a state of steady or unsteady irrotational flow; i.e., the velocity components satisfy the equations

$$\frac{\partial w}{\partial y} - \frac{\partial v}{\partial z} = 0 \qquad \frac{\partial w}{\partial x} - \frac{\partial u}{\partial z} = 0 \qquad \frac{\partial v}{\partial x} - \frac{\partial u}{\partial y} = 0$$

Then the velocity components u, v, w can be expressed as the components of the gradient of the velocity potential ϕ. Thus

$$u = \frac{\partial \phi}{\partial x} \qquad v = \frac{\partial \phi}{\partial y} \qquad w = \frac{\partial \phi}{\partial z} \tag{2t-10}$$

For steady isentropic flow the equations of motion reduce to the single equation for ϕ,

$$\phi_{xx}\left(1 - \frac{\phi_x^2}{a^2}\right) + \phi_{yy}\left(1 - \frac{\phi_y^2}{a^2}\right) + \phi_{zz}\left(1 - \frac{\phi_z^2}{a^2}\right) - 2\phi_{yz}\frac{\phi_{yz}\phi}{a^2} - 2\phi_{zx}\frac{\phi_z\phi_x}{a^2}$$
$$- 2\phi_{xy}\frac{\phi_x\phi_y}{a^2} = 0 \tag{2t-11}$$

where
$$a^2 = \frac{\gamma - 1}{2}(q_{max}^2 - \phi_x^2 - \phi_y^2 - \phi_z^2)$$

TABLE 2t-2. DEPENDENCE OF FLOW VARIABLES ON MACH NUMBER FOR
ONE-DIMENSIONAL ISENTROPIC FLOW*

M	p/p_0	u/a_0	A/A^*	$\rho u^2/2p_0$	$\rho u/\rho_0 a_0$	ρ/ρ_0	T/T_0	a/a_0
0.0	1.00000	0.00000	∞	0.00000	0.00000	1.00000	1.00000	1.00000
0.1	0.99303	0.09990	5.822	0.00695	0.09940	0.99502	0.99800	0.99900
0.2	0.97250	0.19920	2.9635	0.02723	0.19528	0.98028	0.99206	0.99602
0.3	0.93947	0.29734	2.0351	0.05919	0.28437	0.95638	0.98232	0.99112
0.4	0.89561	0.39375	1.5901	0.10031	0.36393	0.92427	0.96899	0.98437
0.5	0.84302	0.48795	1.3398	0.14753	0.43192	0.88517	0.95238	0.97590
0.6	0.78400	0.57950	1.1882	0.19757	0.48704	0.84045	0.93284	0.96583
0.7	0.72093	0.66803	1.0944	0.24728	0.52880	0.79161	0.91075	0.95433
0.8	0.65602	0.75324	1.0382	0.29390	0.55739	0.73999	0.88652	0.94155
0.9	0.59126	0.83491	1.0089	0.33524	0.57362	0.68704	0.86059	0.92768
1.0	0.52828	0.91287	1.00000	0.36980	0.57870	0.63394	0.83333	0.91287
1.1	0.46835	0.98703	1.0079	0.39670	0.57415	0.58170	0.80515	0.89730
1.2	0.41238	1.0574	1.0304	0.41568	0.56161	0.53114	0.77640	0.88113
1.3	0.36091	1.1239	1.0663	0.42696	0.54272	0.48290	0.74738	0.86451
1.4	0.31424	1.1866	1.1149	0.43114	0.51905	0.43742	0.71839	0.84758
1.5	0.27240	1.2457	1.1762	0.42903	0.49203	0.39484	0.68966	0.83045
1.6	0.23527	1.3012	1.2502	0.42161	0.46288	0.35573	0.66138	0.81325
1.7	0.20259	1.3533	1.3376	0.40985	0.43264	0.31969	0.63371	0.79606
1.8	0.17404	1.4023	1.4390	0.39476	0.40216	0.28684	0.60680	0.77904
1.9	0.14924	1.4479	1.5553	0.37713	0.37210	0.25699	0.58072	0.76205
2.0	0.12780	1.4907	1.6875	0.35785	0.34294	0.23005	0.55556	0.74535
2.1	0.10935	1.5308	1.8369	0.33757	0.31504	0.20580	0.53135	0.72894
2.2	0.09352	1.5682	2.0050	0.31685	0.28863	0.18405	0.50813	0.71283
2.3	0.07997	1.6033	2.1931	0.29614	0.26387	0.16458	0.48591	0.69707
2.4	0.06840	1.6360	2.4031	0.27579	0.24082	0.14719	0.46468	0.68168
2.5	0.05853	1.6667	2.6367	0.25606	0.21948	0.13169	0.44444	0.66667
2.6	0.05012	1.6953	2.8960	0.23715	0.19983	0.11788	0.42517	0.65205
2.7	0.04295	1.7222	3.1830	0.21917	0.18181	0.10557	0.40683	0.63784
2.8	0.03685	1.7473	3.5001	0.20222	0.16534	0.09463	0.38941	0.62403
2.9	0.03165	1.7708	3.8498	0.18633	0.15032	0.08489	0.37286	0.61062
3.0	0.02722	1.7928	4.2346	0.17151	0.13666	0.07623	0.35714	0.59761
3.1	0.02345	1.8135	4.6573	0.15774	0.12426	0.06852	0.34223	0.58501
3.2	0.02023	1.8329	5.1210	0.14499	0.11301	0.06165	0.32808	0.57279
3.3	0.01748	1.8511	5.6287	0.13322	0.10281	0.05554	0.31466	0.56095
3.4	0.01512	1.8682	6.184	0.12239	0.09359	0.05009	0.30193	0.54948
3.5	0.01311	1.8843	6.790	0.11243	0.08523	0.04523	0.28986	0.53838
3.6	0.01138	1.8995	7.450	0.10328	0.07768	0.04089	0.27840	0.52763
3.7	0.00990	1.9137	8.169	0.09490	0.07084	0.03702	0.26752	0.51723
3.8	0.00863	1.9272	8.951	0.08722	0.06466	0.03355	0.25720	0.50715
3.9	0.00753	1.9398	9.799	0.08019	0.05906	0.03044	0.24740	0.49740
4.0	0.00659	1.9518	10.72	0.07379	0.05399	0.02766	0.23810	0.48795
4.1	0.00577	1.9631	11.71	0.06788	0.04940	0.02516	0.22925	0.47880
4.2	0.00506	1.9738	12.79	0.06250	0.04524	0.02292	0.22084	0.46994
4.3	0.00445	1.9839	13.95	0.05759	0.04147	0.02090	0.21286	0.46136
4.4	0.00392	1.9934	15.21	0.05309	0.03805	0.01909	0.20525	0.45305
4.5	0.00346	2.0025	16.56	0.04898	0.03494	0.01745	0.19802	0.44499
4.6	0.00305	2.0111	18.02	0.04521	0.03212	0.01597	0.19113	0.43719
4.7	0.00270	2.0192	19.58	0.04177	0.02955	0.01464	0.18457	0.42962
4.8	0.00239	2.0269	21.26	0.03862	0.02722	0.01343	0.17832	0.42228
4.9	0.00213	2.0343	23.07	0.03572	0.02509	0.01233	0.17235	0.41516
5.0	0.00189	2.0412	25.00	0.03308	0.02315	0.01134	0.16667	0.40825

* A more complete table may be found in refs. 4, 5, and 7.

and q_{max} is the velocity with which the gas flows into a vacuum. Other forms of this equation in different numbers of dimensions and for unsteady flow can be found in ref. 2.

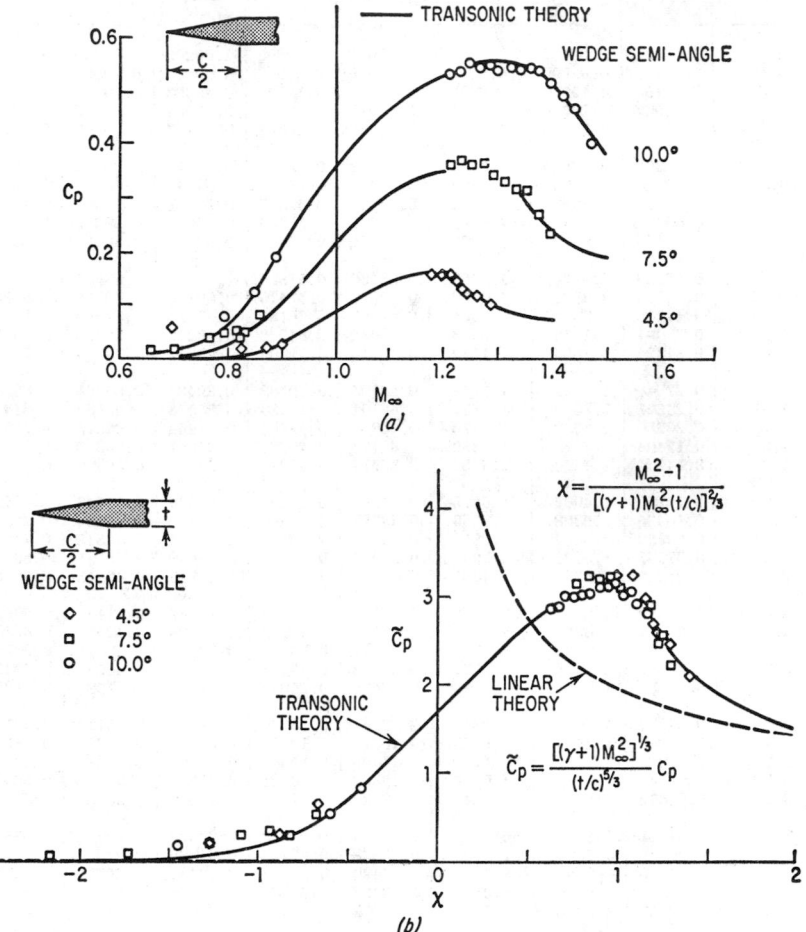

FIG. 2t-1. Comparison of the extended transonic similarity law with experiment. (a) Plotted in conventional coordinates. (b) Plotted in transonic similarity coordinates. (*After J. R. Spreiter, NACA; taken from ref. 6.*)

In compressible flow a stream function ψ exists only for steady two-dimensional or axially symmetric flow. The introduction of the function ψ causes the continuity equation to be satisfied identically. In two dimensions

$$u = \frac{1}{\rho}\psi_y \qquad v = -\frac{1}{\rho}\psi_x \qquad (2t\text{-}12)$$

If cylindrical coordinates (x, r, θ) are used and the flow is independent of θ, then the function ψ may be defined by

$$u = \frac{1}{\rho r}\psi_r \qquad v = -\frac{1}{\rho r}\psi_x \qquad (2t\text{-}13)$$

Note that u and v are now the velocity components in the x and r directions and $r = \sqrt{y^2 + z^2}$. Further details are given in ref. 2.

Equations of Small-perturbation Theory. For many slender or flat two- and three-dimensional bodies it may be assumed that the flow is disturbed very little from uniform flow. Thus if the free-stream velocity U is parallel to the x coordinate and M_∞ is the free-stream Mach number, the velocity components can be written in the form

$$u = U + \phi_x \qquad v = \phi_y \qquad w = \phi_z \tag{2t-14}$$

Here ϕ is called the disturbance potential. When Eqs. (2t-14) are put into Eq. (2t-11) and all nonlinear terms are neglected, the equation

$$(1 - M_\infty{}^2)\phi_{xx} + \phi_{yy} + \phi_{zz} = 0 \tag{2t-15}$$

is obtained. This equation holds for subsonic and moderate supersonic flows.

If M_∞ is very close to 1, Eq. (2t-15) is no longer valid and must be replaced by the equation

$$(1 - M_\infty{}^2)\phi_{xx} + \phi_{yy} + \phi_{zz} = \frac{M_\infty{}^2(\gamma + 1)}{U}\,\phi_x\phi_{xx} \tag{2t-16}$$

Similarity Rules. In many flows where the velocity perturbations are small, it is possible to show that the pressure, lift, drag, etc., depend on the various flow parameters in a simple manner. For example, in two-dimensional flow the pressure coefficient

$$C_p = \frac{p - p_\infty}{\frac{1}{2}\rho_\infty U^2}$$

is related to M_∞ and the thickness ratio τ by the formula

$$\frac{C_p[(\gamma + 1)M_\infty{}^2]^{\frac{1}{3}}}{\tau^{\frac{2}{3}}} = F\left(\frac{1 - M_\infty{}^2}{[\tau(\gamma + 1)M_\infty{}^2]^{\frac{2}{3}}}\right) = F(\chi)$$

This holds for subsonic, transonic, and supersonic flow. For hypersonic flow the similarity parameter is $K = M_\infty\tau$. Van Dyke in ref. 8 showed that the parameter $K' = \sqrt{M_\infty{}^2 - 1}\,\tau$ could be used as a unified similarity parameter. More information may be found in ref. 9.

The well-known Prandtl-Glauert rule can be found as a special case of the above formula. Further details can be found in ref. 6. The power of similarity rules is shown in Fig. 2t-1, where data and theory for flow past different wedges can be directly compared if plotted in terms of the similarity parameter χ.

References

1. Liepmann, H. W., and A. E. Puckett: "Introduction to Aerodynamics of a Compressible Fluid," John Wiley & Sons, Inc., New York, 1947.
2. Ferri, A.: "Elements of Aerodynamics of Supersonic Flows," The Macmillan Company, New York, 1949.
3. Courant, R., and K. O. Friedrichs: "Supersonic Flow and Shock Waves," Interscience Publishers, Inc., New York, 1948.
4. Emmons, H. W.: "Gas Dynamics Tables for Air," Dover Publications, Inc., New York, 1947.
5. Aeronautical Research Council: "Compressible Airflow Tables," Oxford University Press, New York, 1952.
6. Liepmann, H. W., and A. Roshko: "Elements of Gasdynamics," John Wiley & Sons, Inc., New York, 1957.
7. Ames Research Staff: Equations, Tables and Charts for Compressible Flow, *NACA Rept.* 1135, 1953.
8. Van Dyke, M. D.: A Study of Small Disturbance Theory, *NACA Rept.* 1194, 1954.
9. Hayes, W. D., and R. F. Probstein, "Hypersonic Flow Theory," 2d ed. Academic Press, Inc., New York, 1966.

2u. Laminar and Turbulent Flow of Gases

R. C. ROBERTS

University of Maryland—Baltimore County

Symbols

C_D	drag coefficient
c_f	skin-friction coefficient
c_p	specific heat at constant pressure
d	pipe diameter
E	internal energy per unit mass
G	mass rate of flow per unit cross-sectional area of pipe
G_r	Grashof number
g	acceleration of gravity
K_N	Nusselt number
k	coefficient of heat conductivity, surface roughness
L	reference length (for Reynolds number)
P_r	Prandtl number
p	pressure
Q	external-heat-production rate per unit mass
R	gas constant, Reynolds number
r	pipe radius
r/k	surface-roughness factor
S_t	Stanton number
T	absolute temperature
T_e	adiabatic wall temperature
T_w	wall temperature
T_∞	free-stream temperature
t	time
u, v, w	velocity components of fluid flow
u	free-stream velocity
X, Y, Z	rectangular components of external body force
x, y, z	rectangular coordinates
δ	boundary-layer thickness
θ	momentum thickness
μ	coefficient of viscosity
ν	kinematic viscosity
ρ	density
τ_w	wall shear stress per unit area

2u-1. Equations of Motion. The study of the motion of any real gas or fluid must of necessity take into consideration the effects of viscosity. The transfer of momentum due to viscosity and the transformation of kinetic energy into heat must be considered in formulating the equations of motion. The following equations govern the motion of a viscous, compressible, heat-conducting gas. The viscosity and heat conductivity are assumed to be functions of the temperature only.

Momentum Equations. In rectangular coordinates, the momentum equations can be written as

$$\rho\left(\frac{\partial u}{\partial t} + u\frac{\partial u}{\partial x} + v\frac{\partial u}{\partial y} + w\frac{\partial u}{\partial z}\right) = \rho X + \frac{\partial}{\partial x}\left[\frac{4}{3}\mu\frac{\partial u}{\partial x} - \frac{2}{3}\mu\left(\frac{\partial v}{\partial y} + \frac{\partial w}{\partial z}\right)\right]$$

$$+ \frac{\partial}{\partial y}\left[\mu\left(\frac{\partial u}{\partial y} + \frac{\partial v}{\partial x}\right)\right] + \frac{\partial}{\partial z}\left[\mu\left(\frac{\partial w}{\partial x} + \frac{\partial u}{\partial z}\right)\right] - \frac{\partial p}{\partial x}$$

$$\rho\left(\frac{\partial v}{\partial t} + u\frac{\partial v}{\partial x} + v\frac{\partial v}{\partial y} + w\frac{\partial v}{\partial z}\right) = \rho Y + \frac{\partial}{\partial x}\left[\mu\left(\frac{\partial v}{\partial x} + \frac{\partial u}{\partial y}\right)\right]$$

$$+ \frac{\partial}{\partial y}\left[\frac{4}{3}\mu\frac{\partial v}{\partial y} - \frac{2}{3}\mu\left(\frac{\partial u}{\partial x} + \frac{\partial w}{\partial z}\right)\right] + \frac{\partial}{\partial z}\left[\mu\left(\frac{\partial v}{\partial z} + \frac{\partial w}{\partial y}\right)\right] - \frac{\partial p}{\partial y}$$

$$\rho\left(\frac{\partial w}{\partial t} + u\frac{\partial w}{\partial x} + v\frac{\partial w}{\partial y} + w\frac{\partial w}{\partial z}\right) = \rho Z + \frac{\partial}{\partial x}\left[\mu\left(\frac{\partial w}{\partial x} + \frac{\partial u}{\partial z}\right)\right]$$

$$+ \frac{\partial}{\partial y}\left[\mu\left(\frac{\partial w}{\partial y} + \frac{\partial v}{\partial z}\right)\right] + \frac{\partial}{\partial z}\left[\frac{4}{3}\mu\frac{\partial w}{\partial z} - \frac{2}{3}\mu\left(\frac{\partial u}{\partial x} + \frac{\partial v}{\partial y}\right)\right] - \frac{\partial p}{\partial z} \quad (2\text{u}-1)$$

where μ is the coefficient of viscosity and the other terms are as defined in Sec. 2t.

Continuity Equation. The equation of continuity is

$$\frac{\partial \rho}{\partial t} + \frac{\partial}{\partial x}(\rho u) + \frac{\partial}{\partial y}(\rho v) + \frac{\partial}{\partial z}(\rho w) = 0 \quad (2\text{u}-2)$$

Energy Equation. By using the first law of thermodynamics and by considering that heat conduction may take place in the gas, the following energy equation may be written

$$\rho\left(\frac{\partial E}{\partial t} + u\frac{\partial E}{\partial x} + v\frac{\partial E}{\partial y} + w\frac{\partial E}{\partial z}\right) + p\left(\frac{\partial u}{\partial x} + \frac{\partial v}{\partial y} + \frac{\partial w}{\partial z}\right)$$

$$= \rho Q + \frac{\partial}{\partial x}\left(k\frac{\partial T}{\partial x}\right) + \frac{\partial}{\partial y}\left(k\frac{\partial T}{\partial y}\right) + \frac{\partial}{\partial z}\left(k\frac{\partial T}{\partial z}\right) + 2\mu\left[\left(\frac{\partial u}{\partial x}\right)^2 + \left(\frac{\partial v}{\partial y}\right)^2 + \left(\frac{\partial w}{\partial z}\right)^2\right]$$

$$- \frac{2}{3}\mu\left(\frac{\partial u}{\partial x} + \frac{\partial v}{\partial y} + \frac{\partial w}{\partial z}\right)^2 + \mu\left(\frac{\partial u}{\partial y} + \frac{\partial v}{\partial x}\right)^2 + \mu\left(\frac{\partial u}{\partial z} + \frac{\partial w}{\partial x}\right)^2 + \mu\left(\frac{\partial v}{\partial z} + \frac{\partial w}{\partial y}\right)^2 \quad (2\text{u}-3)$$

where k = heat-conductivity coefficient
E = internal energy per unit mass
Q = external-heat-production rate per unit mass
T = absolute temperature

The coefficients μ and k may be functions of the temperature T.

Equation of State. For a perfect gas the equation of state is

$$p = \rho RT \quad (2\text{u}-4)$$

Stream Function. For a steady flow in two dimensions or for axially symmetric flow a stream function may be defined as in Sec. 2t. It has great utility in boundary-layer work (see ref. 3).

2u-2. Definitions of Basic Parameters. The basic dimensionless parameters of a viscous, compressible, heat-conducting gas are usually considered to be the Mach number, the Reynolds number, the Prandtl number, and the Grashof number (see ref. 2). The Mach number has been defined in Sec. 2t. The other three parameters may be defined as follows:

Reynolds Number. In a flow with reference velocity u and reference length L, the Reynolds number R is defined as

$$R = \frac{uL}{\nu} \tag{2u-5}$$

where $\nu = \mu/\rho$ is the kinematic viscosity. Two viscous flows may not be dynamically similar unless their respective Reynolds numbers are the same.

Prandtl Number. The Prandtl number is defined as

$$P_r = \frac{\mu c_p}{k} \tag{2u-6}$$

where c_p is the specific heat at constant pressure. The Prandtl number depends only on the material properties of the gas.

The Prandtl number is primarily a function of the temperature only. For small temperature changes it is often assumed to be constant (see ref. 2). The variation of P_r with temperature is shown in Tables 2u-1 and 2u-2 for air and for molecular hydrogen H_2.

Grashof Number. The Grashof number may be defined as

$$G_r = \frac{L^3 g (T_1 - T_0)}{\nu^2 T_0} \tag{2u-7}$$

where g is the acceleration of gravity and T_1 and T_2 are two reference temperatures. The Grashof number is important in the study of flows with free convection, e.g., the flow of gas above a heated plate.

2u-3. Exact Solutions. Because of the extreme complexity of the equations of motion, few exact solutions have been found. Nearly all of these are limited to the incompressible steady flow case, with zero heat transfer through the walls bounding the flow. Since gases often behave as if they were nearly incompressible, these solutions may have practical importance.

Pipe Flow. The exact incompressible solution for two-dimensional or axially symmetric steady flow through a pipe of constant cross section is characterized by a parabolic velocity distribution. In the two-dimensional case the complete solution is given by

$$
\begin{aligned}
u &= -\frac{1}{2\mu} z(h - z) \frac{\partial p}{\partial x} \\
v &= w = 0 \\
\frac{\partial p}{\partial x} &= \text{const} \qquad \frac{\partial p}{\partial y} = \frac{\partial p}{\partial z} = 0
\end{aligned}
\tag{2u-8}
$$

where the boundaries are at $z = 0$ and $z = h$. In the case of flow through a circular pipe, the theoretical solution has been shown to coincide almost exactly with experiment for laminar flow.

Other Exact Solutions. There are a number of other exact solutions for the incompressible case such as steady flow between concentric cylinders and flow through tubes of noncircular cross section. These can be found by consulting refs. 1 and 3. Hamel (ref. 5) has found a number of nontrivial exact solutions.

2u-4. Boundary Layers. When the Reynolds number of the flow is large, most of the viscous effects take place in the immediate vicinity of the boundaries. The outer flow may then be considered determined by the inviscid flow equations while in the boundary layer certain simplifications of the equation of motion may be made. For the case of two-dimensional flow past flat or slowly curving surfaces the pressure may be assumed to be completely determined by the outer flow.

If the viscous effects are confined to a thin region next to a boundary, it then turns out that most of the viscous terms in Eqs. (2u-1) and (2u-3) can be neglected. The simplified equations are much easier to treat than the full equations.

TABLE 2u-1. PRANDTL NUMBER P_r FOR AIR

T, °K	P_r	T, °K	P_r
100	0.770	560	0.680
120	0.766	580	0.680
140	0.761	600	0.680
160	0.754	620	0.681
180	0.746	640	0.682
200	0.739	660	0.682
220	0.732	680	0.683
240	0.725	700	0.684
260	0.719	720	0.685
280	0.713	740	0.686
300	0.708	760	0.687
320	0.703	780	0.688
340	0.699	800	0.689
360	0.695	820	0.690
380	0.691	840	0.692
400	0.689	860	0.693
420	0.686	880	0.695
440	0.684	900	0.696
460	0.683	920	0.697
480	0.681	940	0.698
500	0.680	960	0.700
520	0.680	980	0.701
540	0.680	1000	0.702

Basic Equations. For two-dimensional steady flow as outlined above, the momentum, continuity, and energy equations are, respectively,

$$\rho \left(u \frac{\partial u}{\partial x} + v \frac{\partial u}{\partial y} \right) = \frac{\partial}{\partial y} \left(\mu \frac{\partial u}{\partial y} \right) - \frac{\partial p}{\partial x}$$

$$0 = \frac{\partial p}{\partial y}$$

$$\frac{\partial}{\partial x}(\rho u) + \frac{\partial}{\partial y}(\rho v) = 0 \tag{2u-9}$$

$$\rho \left(u \frac{\partial E}{\partial x} + v \frac{\partial E}{\partial y} \right) + p \left(\frac{\partial u}{\partial x} + \frac{\partial v}{\partial y} \right) = \frac{\partial}{\partial y} \left(k \frac{\partial T}{\partial y} \right) + \mu \left(\frac{\partial u}{\partial y} \right)^2$$

For a perfect gas the equation of state is $p = \rho R T$. In the above equations x may be considered as the distance along the boundary while y is the distance perpendicular to the boundary. The velocity components u and v are interpreted in like manner. The equations then hold also for a slowly curving boundary.

Blasius Flow. For incompressible steady flow past a flat plate with no pressure gradient, the equations of motion are

$$u \frac{\partial u}{\partial x} + v \frac{\partial u}{\partial y} = \nu \frac{\partial^2 u}{\partial y^2}$$

$$\frac{\partial u}{\partial x} + \frac{\partial v}{\partial y} = 0 \tag{2u-10}$$

with the boundary conditions $u = v = 0$ at $y = 0$ and $u = u_1 = $ const at $y = \infty$

TABLE 2u-2. PRANDTL NUMBER FOR MOLECULAR HYDROGEN H_2*

T, °K	P_r	T, °K	P_r
60	0.713	440	0.684
80	0.711	460	0.681
100	0.712	480	0.678
120	0.715	500	0.675
140	0.718	520	0.671
160	0.719	540	0.669
180	0.720	560	0.667
200	0.719	580	0.665
220	0.717	600	0.664
240	0.715	620	0.663
260	0.712	640	0.663
280	0.709	660	0.662
300	0.706	680	0.661
320	0.703	700	0.661
340	0.699	720	0.661
360	0.696	740	0.660
380	0.693	760	0.660
400	0.690	780	0.660
420	0.687	800	0.660

* The values in Tables 2u-1 and 2u-2 are taken from the National Bureau of Standards, "NACA Tables of Thermal Properties of Gases" (cf. ref. 6).

and at $x = 0$. u_1 is the free-stream velocity. Blasius solved this problem by means of the change of variable

$$\eta = \frac{1}{2}\left(\frac{u_1}{\nu x}\right)^{\frac{1}{2}} y \qquad u = \frac{1}{2} u_1 f' \qquad v = \frac{1}{2}\left(\frac{u_1 \nu}{x}\right)^{\frac{1}{2}} (\eta f' - f) \tag{2u-11}$$

This reduces the problem to the ordinary differential equation and boundary conditions

$$\frac{d^3 f}{d\eta^3} + f \frac{d^2 f}{d\eta^2} = 0$$

$$f = f' = 0 \text{ at } \eta = 0 \qquad \text{and} \qquad f' = 2 \text{ at } \eta = \infty \tag{2u-12}$$

2u-5. Turbulent Flow. For small values of the Reynolds number most flows are characterized by a certain uniformity of velocity distribution and smoothness of the streamline pattern. This type of flow is called laminar. As the Reynolds number is increased, the flow will remain laminar until a certain critical value of R is reached. At this time swirling or eddying motions begin to appear in the flow. These small-scale eddying motions move with the main flow but also possess an apparent random nature in the way they appear and decay. Such flows are called turbulent.

Turbulent flows also exhibit other striking features. The velocity distribution has a different behavior from that of laminar flow. The viscous drag and heat transfer

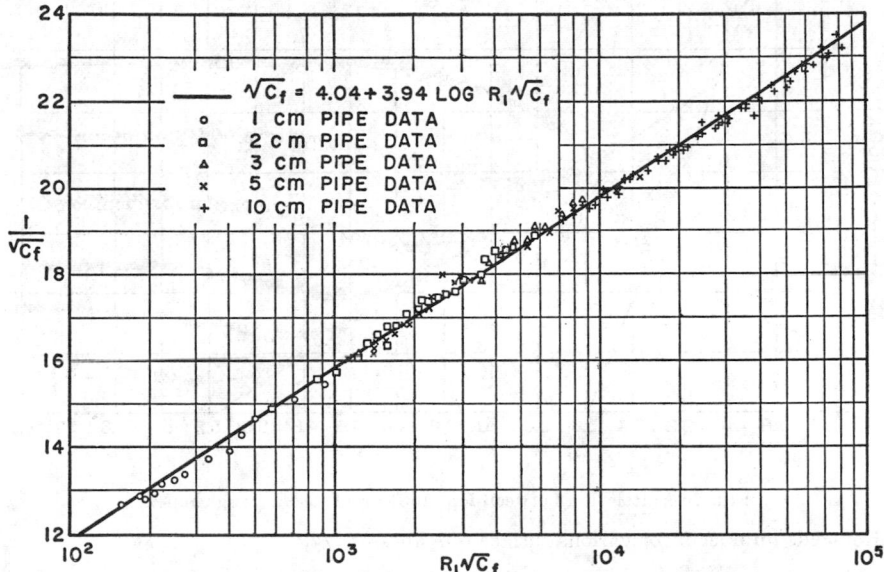

FIG. 2u-1. Universal wall-friction functional relation.

also undergo abrupt changes when turbulent flow begins. The sharp drop in the drag coefficient for the sphere shown in Fig. 2u-4 indicates the onset of turbulent flow.

2u-6. Data on Turbulent Flow through Pipes. The following data show the behavior of the skin friction for incompressible turbulent flow through smooth and rough pipes. These data come from Nikuradse (see refs. 7 and 8).

Smooth Pipes. The skin-friction coefficient c_f is a function of the Reynolds number R_1 for smooth pipes,

$$c_f = \frac{\tau_w}{\frac{1}{2}\rho u_1^2}$$

$$R_1 = \frac{r u_1}{\nu}$$

where τ_w = wall shear stress per unit area
  ρ = density
  ν = kinematic viscosity
  u_1 = velocity in center of pipe
  r = pipe radius

The behavior of c_f with R_1 is shown in Fig. 2u-1. An empirical curve which fits the data is also shown.

Rough Pipes. For rough pipes with average projection of the roughness k, the skin-friction data are shown in Fig. 2u-2. The friction factor λ is plotted against

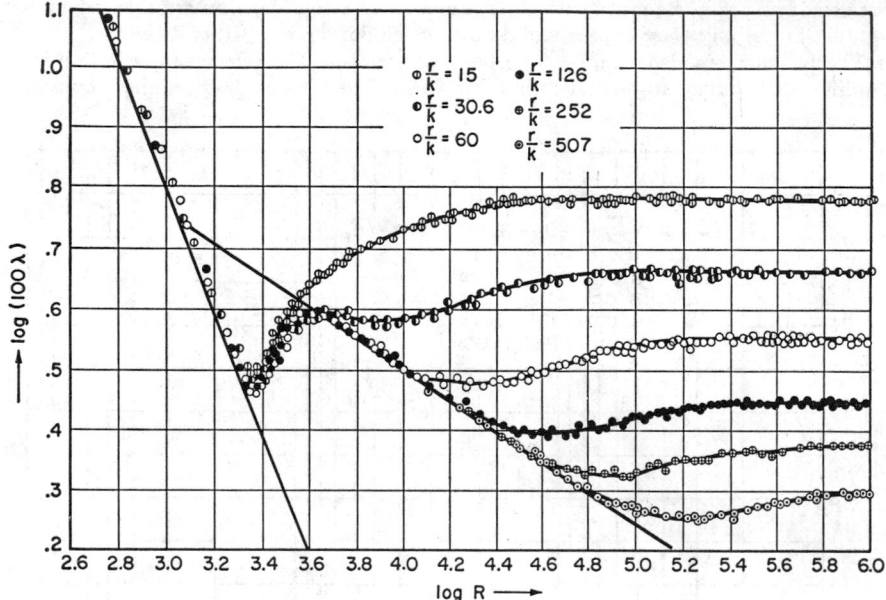

Fig. 2u-2. Relation between $\log (100\lambda)$ and $\log R$ (rough pipe).

Reynolds number R for various surface roughnesses r/k,

$$\lambda = 4c_f \left(\frac{u_1}{u}\right)^2$$

$\bar{u}$ = average velocity across pipe
d = pipe diameter
$\dfrac{r}{k}$ = roughness factor
r = pipe radius

2u-7. Drag Data for Spheres and Cylinders. For incompressible viscous steady flow the drag coefficient is a function of the Reynolds number only. The graphs of Figs. 2u-3 and 2u-4 give curves of the experimental data for C_D, the drag coefficient, for a cylinder in cross flow and for a sphere, respectively.

$$\text{Drag of cylinder } C_D = \frac{\text{drag force}}{\frac{1}{2}\rho u^2 d}$$

where d = diameter of cylinder
u = free-stream velocity
$R = ud/\nu$

$$\text{Drag of sphere } C_D = \frac{\text{drag force}}{\frac{1}{2}\rho u^2 (\pi d^2/4)}$$

where d = diameter of sphere
$R = ud/\nu$

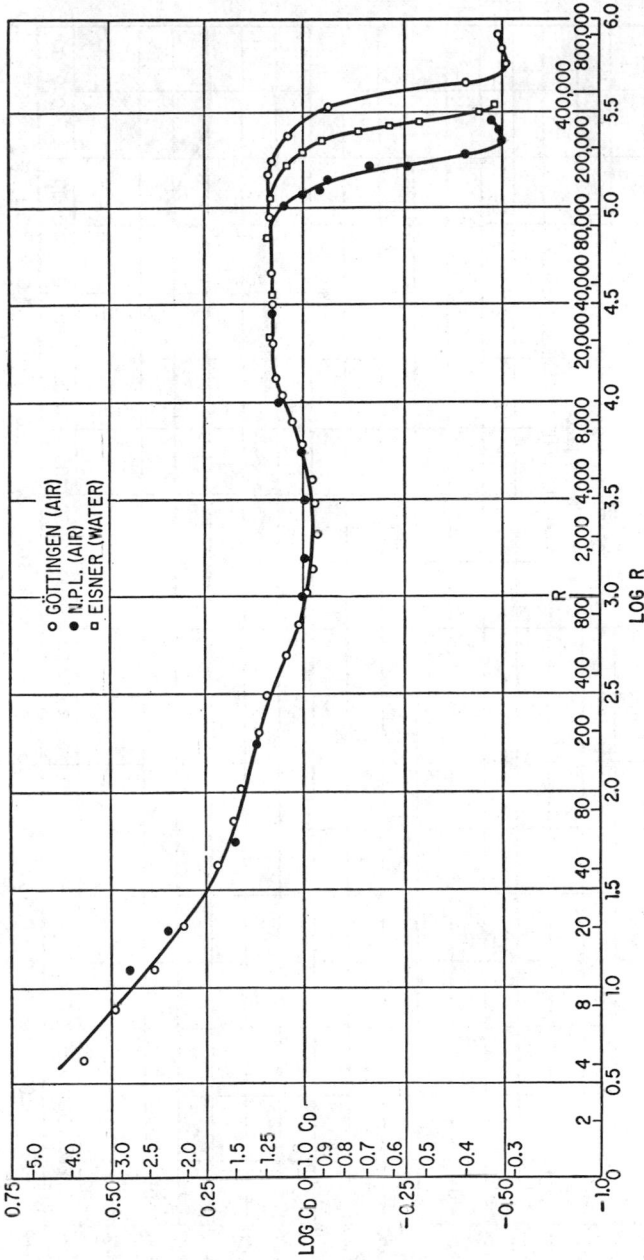

Fig. 2u-3. Relation between log C_D and log R (cylinder).

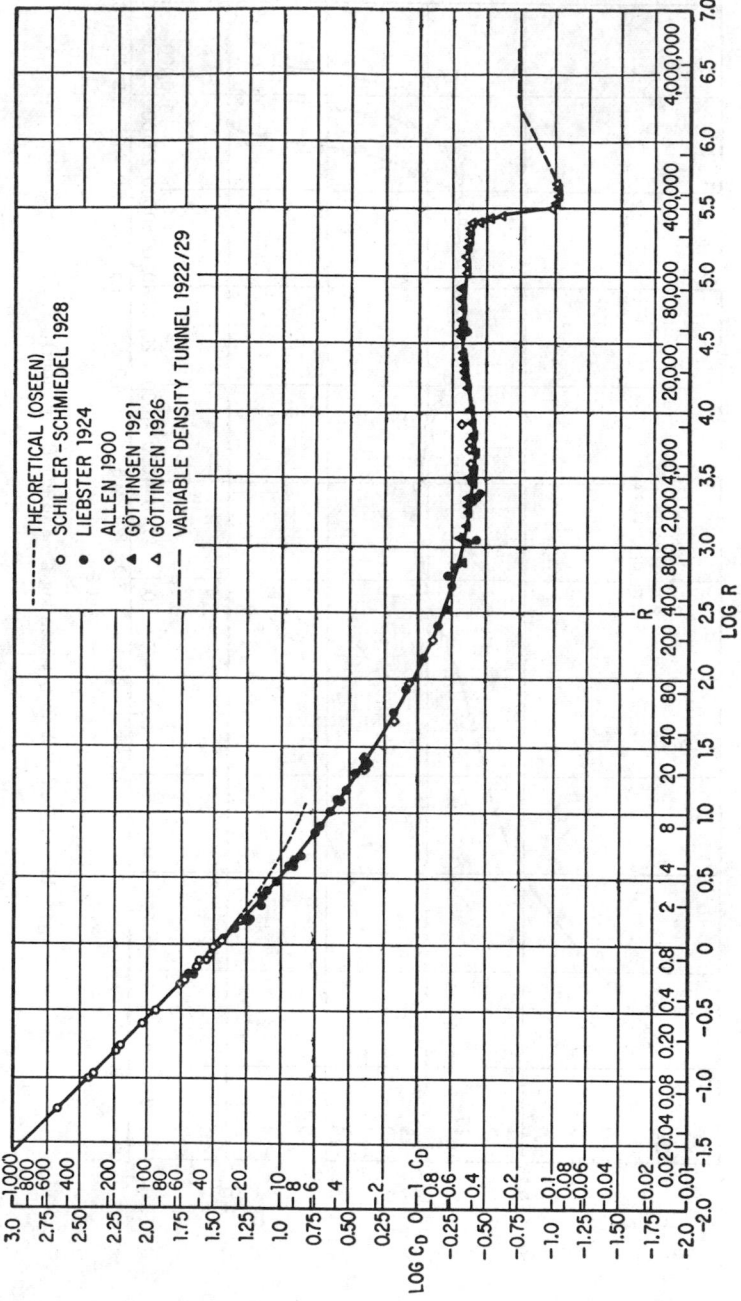

FIG. 2u-4. Relation between log C_D and log R (sphere).

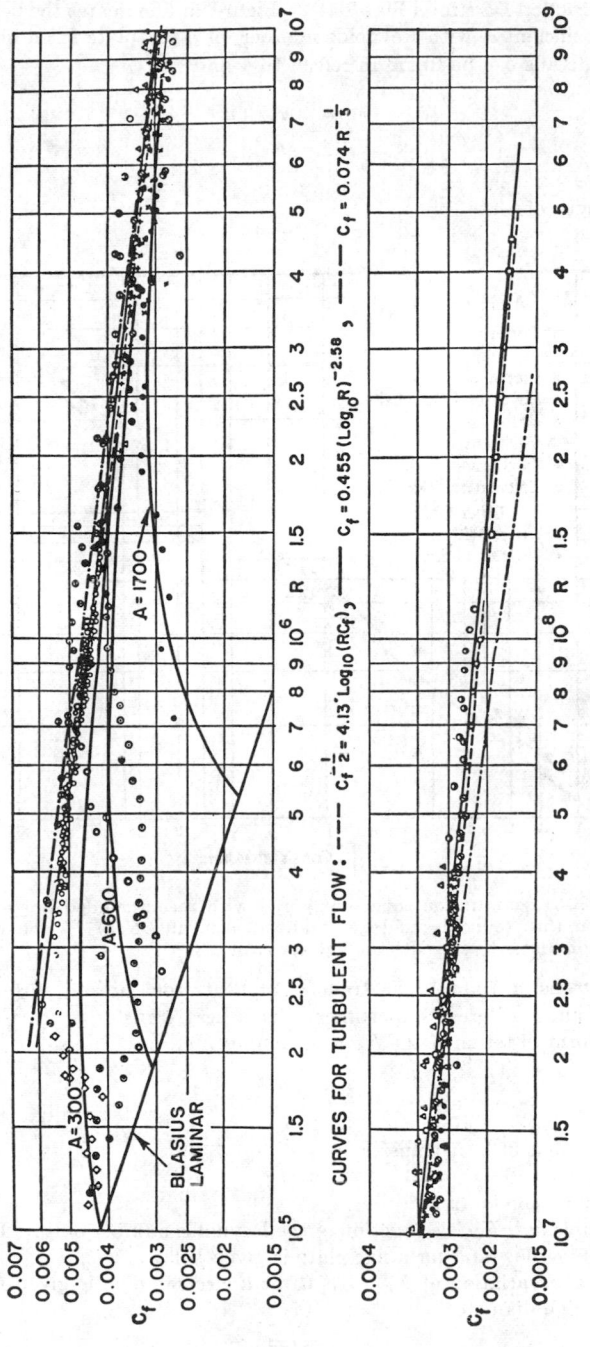

CURVES FOR TURBULENT FLOW: $---$ $c_f^{-\frac{1}{2}} = 4.13 \cdot \log_{10}(R c_f)$, $---$ $c_f = 0.455 (\log_{10} R)^{-2.58}$, $\longrightarrow$ $c_f = 0.074 R^{-\frac{1}{5}}$

Fig. 2u–5. Curves for turbulent flow along flat plate.

2u-8. Skin-friction Data for a Flat Plate. Figure 2u-5 indicates the behavior of the skin-friction coefficient c_f with Reynolds number for a flat plate in an incompressible fluid. (More details can be found in refs. 1 to 4 and 14.)

$$c_f = \frac{\tau_w}{\frac{1}{2}\rho u^2}$$

where $R = ul/\nu$
l = length of plate

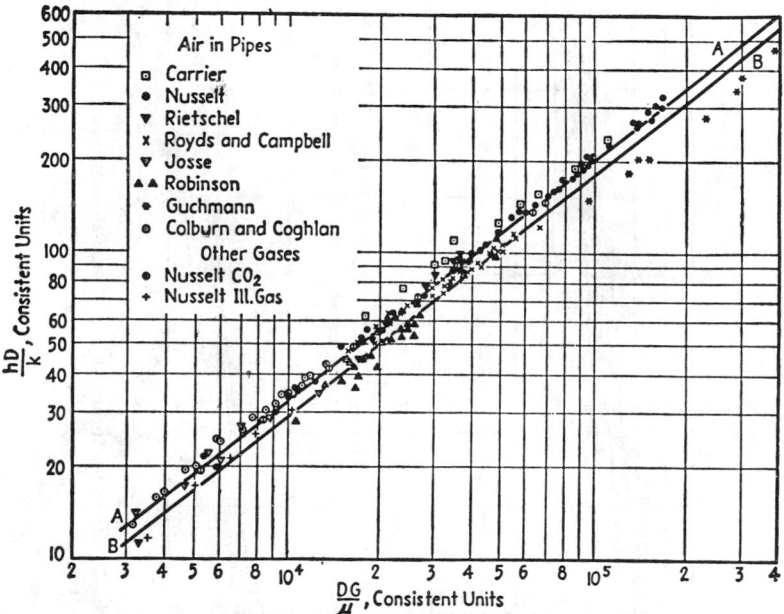

FIG. 2u-6. Data for gases inside tubes compared with recommended line AA. Line BB is obtained from the Reynolds analogy, taking $f = 0.049(DG/\mu)^{-0.2}$ and $c_p\mu/k = 0.74$. Line BB also represents the Prandtl analogy for r_v of 0.3.

2u-9. Heat-transfer Data. The transfer of heat from heated surfaces to gases moving past them is of great importance. This heat transfer is often expressed in dimensionless form in terms of the *Nusselt* number K_N,

$$K_N = \frac{hD}{k}$$

where h = coefficient of heat transfer
D = length
k = thermal conductivity
For incompressible flow K_N is a function of the Reynolds number only. The behavior K_N with R for pipe flow and for a flat plate is given below.

Pipe Flow. The variation of K_N with R for a circular pipe is given in Fig. 2u-6, where D is the pipe diameter.

$$R = \frac{DG}{\mu}$$

where $G = w/s$
  w = mass rate of flow
  s = cross-sectional area of pipe

Flat Plate. For a flat plate the variation of K_N with R for small R is shown in Fig. 2u-7, where D is the length of the flat plate. For higher values of R recourse must be made to empirical formulas converting the pipe-flow into equivalent flat-plate data (see page 117 of ref. 10) or to *Reynolds analogy*.

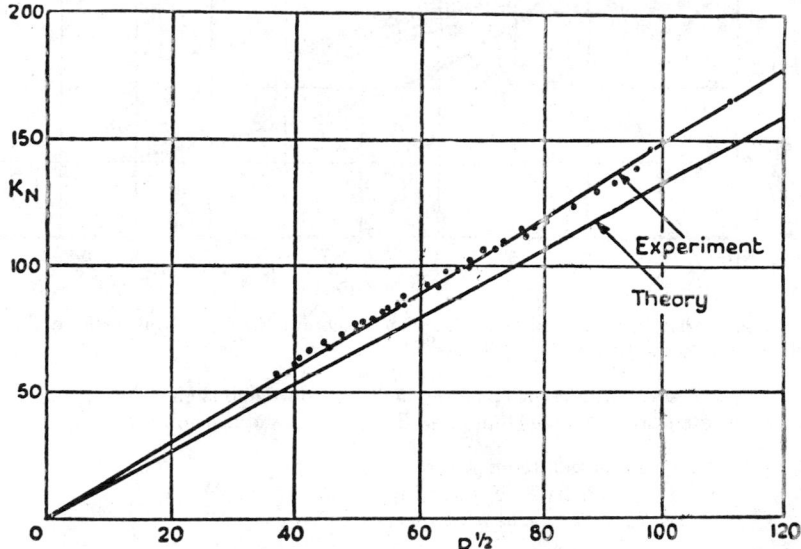

Fig. 2u-7. Comparison between theory and experiment for heat transfer from plate.

Reynolds analogy (see ref. 14) says that heat transfer and skin friction are related in the following way:

$$S_t = \tfrac{1}{2} c_f$$

where S_t is the Stanton number

$$S_t = \frac{q_w}{\rho_w c_p u_1 (T_w - T_1)}$$

and c_f is the skin-friction coefficient

$$c_f = \frac{\tau_w}{\tfrac{1}{2} \rho_w u_1{}^2}$$

The subscripts w and 1 refer to variables at the wall and at the outer edge of the boundary layer, respectively. This analogy holds only approximately and must be modified for compressible flow and high Mach numbers. The extensions of the analogy are given in ref. 14.

2u-10. Effect of Compressibility and Heat Transfer on Skin Friction. For a fixed Reynolds number the ratio of the local skin-friction coefficient c_f to the corresponding incompressible value c_{f_i} is a function of the Mach number and the heat transfer. The graph shown in Fig. 2u-8, taken from ref. 12, represents an excellent theoretical fit to data from refs. 11 and 13. The curves are plotted for zero heat transfer where

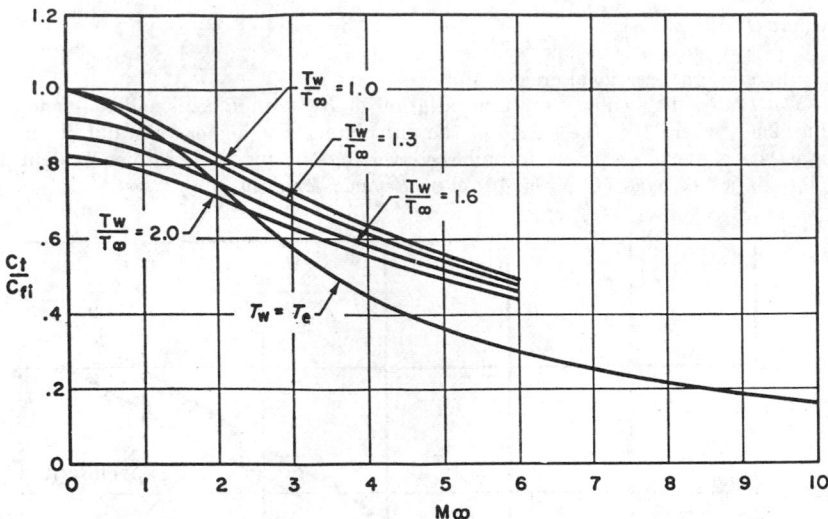

Fig. 2u-8. Variation of skin-friction ratio with Mach number for several constant values of wall-temperature ratio and $Re\theta = 13,500$.

$T_e = T_w$ and several different constant heat-transfer conditions. The graph is for a single representative Reynolds number R_θ based on momentum thickness.

$$T_w = \text{wall temperature}$$
$$T_e = \text{adiabatic wall temperature}$$
$$T_\infty = \text{free-stream temperature}$$
$$R_\theta = \frac{u_\infty \theta}{\nu}$$
$$\theta = \text{momentum thickness} = \int_0^\delta \frac{\rho u}{\rho_1 u_1}\left(1 - \frac{u}{u_1}\right) dy$$
$$\delta = \text{boundary-layer thickness}$$
$$\rho_1 = \text{density outside boundary layer}$$
$$u_1 = \text{velocity outside boundary layer}$$

References

1. Goldstein, S.: "Modern Developments in Fluid Dynamics," vol. I, Oxford University Press, New York, 1938.
2. Goldstein, S.: "Modern Developments in Fluid Dynamics," vol. II, Oxford University Press, New York, 1938.
3. Howarth, L.: "Modern Developments in Fluid Dynamics, High Speed Flow," vol. I, Oxford University Press, New York, 1953.
4. Howarth, L.: "Modern Developments in Fluid Dynamics, High Speed Flow," vol. II, Oxford University Press, New York, 1953.
5. Schlichting, H.: "Boundary Layer Theory," 6th ed., translated by J. Kestin, McGraw-Hill Book Company, New York, 1968.
6. "NACA Tables of Thermal Properties of Gases," Heat and Power Division, National Bureau of Standards.
7. Ross, D.: "Turbulent Flow in Smooth Pipes, A Reanalysis of Nikuradse's Experiments," Ordnance Research Laboratory, Pennsylvania State College, Serial No. NOrd 7958-246.
8. Nikuradse, J.: Laws of Flow in Rough Pipes, *NACA Tech. Mem.* 1292.
9. McAdams, W. H.: "Heat Transmission," 3d ed., McGraw-Hill Book Company, New York, 1954.
10. Eckert, E. R. G.: "Introduction to the Transfer of Heat and Mass," McGraw-Hill Book Company, New York, 1950.

11. Coles, D.: Measurements of Turbulent Friction on a Smooth Flat Plate in Supersonic Flow, *J. Aeronaut. Sci.* **21**, 7 (July, 1954).
12. Persh, Jerome: A Theoretical Investigation of Turbulent Boundary Layer Flow with Heat Transfer at Supersonic and Hypersonic Speeds, *NavOrd Rept.* 3854, U.S. Naval Ordnance Laboratory.
13. Lobb, R. K., E. M. Winkler, and J. Persh: Experimental Investigation of Turbulent Boundary Layers in Hypersonic Flow, *J. Aeronaut. Sci.* **22**, 1 (January, 1955). See also *NavOrd Rept.* 3880, U.S. Naval Ordnance Laboratory, by the same authors.
14. Lin, C. C., ed.: Turbulent Flows and Heat Transfer, vol. 5 in "High Speed Aerodynamics and Jet Propulsion," Princeton University Press, Princeton, N.J., 1959.

2v. Shock Waves

ELI FREEDMAN

Ballistic Research Laboratories, Aberdeen Proving Ground

EDWARD F. GREENE

Brown University

2v-1. List of Symbols

u flow velocity, measured in a coordinate system moving with the shock front

p pressure

ρ density

γ ratio of heat capacities $= C_P/C_V$

H enthalpy

E internal energy

T absolute temperature

S entropy

R^* gas constant per gram

c local sound velocity

M_1 Mach number of incident shock $= u_1/c_1$

n empirical constant in the Tait equation for liquids

$B(S)$ constant in the Tait equation for water

$\mathbf{n}$ unit vector normal to surface

$\mathbf{u}$ velocity vector

M_R Mach number of reflected shock $= u_{2R}/c_2$

Subscripts 1, 2, and 3 on any quantity (e.g., u_1, p_2, ρ_3) mean that the quantity is measured in front of an incident shock, behind the incident shock, or behind a reflected shock, respectively.

Primed and double-primed quantities (e.g., p', u'') are measured, respectively, on the two sides of a boundary between two media.

Subscript R on any quantity means that that quantity is measured in a coordinate system moving with a reflected shock.

2v-2. Introduction. Sound waves of infinitesimal amplitude in fluids always propagate without change of form (neglecting the effects of viscosity, thermal conductivity, and relaxation). For waves of finite amplitude this is no longer true. The denser regions move faster than the less dense, and hence the denser regions are always catching up with less dense ones in front of them, but since the velocity increases with density, the effect becomes more and more pronounced, the front of the wave becoming steeper and steeper until the density, temperature, and pressure changes across it are virtually discontinuous—a shock wave is formed. Mathematically, a shock wave is an actual discontinuity propagating with a velocity greater than the local sound velocity. Physically, although a shock transition is extremely abrupt (of the order of 10 mean free paths for a typical shock in a gas), it nevertheless is continuous, because of the action of dissipative forces. In what follows, attention will be focused exclusively on the regions behind or in front of the shock front. The relations that will be given are of general validity (except as noted) and are in any case independent of the actual course of events within the front itself.

It might be imagined that there could be a flow in which a shock moves from a dense region to a rarefied one. However, it can be shown from the energy-conservation law that steady-state flows of this type cannot exist in any fluid having an adiabat that is concave upward, the almost universally prevailing situation.

Another type of discontinuity occurring in gas flows is called a "contact discontinuity." It differs from a shock in that there is no mass flow across it, as there is in the case of a shock. Contact discontinuities cannot occur in steady-state flows and will not be further considered.

2v-3. Steady-state One-dimensional Flow. *General Relations.* Consider a shock propagating steadily in a fluid. Relative to a coordinate system moving with the shock, the equations of steady compressible flow are

$$u \frac{\partial \rho}{\partial x} + \rho \frac{\partial u}{\partial x} = 0 \tag{2v-1a}$$

$$u \frac{\partial u}{\partial x} + \frac{1}{\rho} \frac{\partial p}{\partial x} = 0 \tag{2v-1b}$$

Equation (2v-1a) leads to

$$\rho_2 u_2 = \rho_1 u_1 \tag{2v-2}$$

From Eqs. (2v-1) and (2v-2) we have

$$\rho_2 u_2{}^2 + p_2 = \rho_1 u_1{}^2 + p_1 \tag{2v-3}$$

Also, from (2v-1b),

$$\frac{1}{2} u^2 + \int \frac{dp}{\rho} = \text{const} \tag{2v-4a}$$

From the energy-conservation equation, it can be shown that

$$\tfrac{1}{2}u_2{}^2 + H_2 = \tfrac{1}{2}u_1{}^2 + H_1 \tag{2v-4b}$$

These equations lead at once to the *Rankine-Hugoniot relations:*

$$E_2 - E_1 = \Delta E = \frac{1}{2}(p_2 + p_1)\left(\frac{1}{\rho_1} - \frac{1}{\rho_2}\right) \tag{2v-5a}$$

$$H_2 - H_1 = \Delta H = \frac{1}{2}(p_2 - p_1)\left(\frac{1}{\rho_1} + \frac{1}{\rho_2}\right) \tag{2v-5b}$$

and

$$u_1 = \frac{1}{\rho_1}\left(\frac{p_2 - p_1}{1/\rho_1 - 1/\rho_2}\right)^{\frac{1}{2}} \tag{2v-5c}$$

Equations (2v-5a), (2v-5b), and (2v-5c) are based solely upon hydrodynamics and thermodynamics and are valid for all fluids. Further progress can now be made only when they are supplemented by an equation of state for the fluid.

Special Cases. THE IDEAL GAS

$$p = \rho R^* T$$

From Eqs. (2v-5a), (2v-5b), and (2v-5c) and the equation of state it can be shown that

$$\frac{p_2}{p_1} = \frac{\rho_2(\gamma + 1) - \rho_1(\gamma - 1)}{\rho_1(\gamma + 1) - \rho_2(\gamma - 1)} \tag{2v-6a}$$

$$\frac{\rho_2}{\rho_1} = \frac{p_2(\gamma + 1) + p_1(\gamma - 1)}{p_1(\gamma + 1) + p_2(\gamma - 1)} \tag{2v-6b}$$

and

$$\frac{T_2}{T_1} = \frac{p_2 \rho_1}{p_1 \rho_2} \tag{2v-6c}$$

In terms of the Mach number of the incident shock M_1,

$$\frac{p_2}{p_1} = \frac{2M_1^2 \gamma - \gamma + 1}{\gamma + 1} \tag{2v-7a}$$

and

$$\frac{\rho_2}{\rho_1} = \frac{M_1^2(\gamma + 1)}{M_1^2(\gamma - 1) + 2} \tag{2v-7b}$$

LIQUIDS. An often-used equation of state for liquids, especially water, is the Tait equation. A convenient form of it is

$$p = B(S) \left[\left(\frac{\rho(T,p)}{\rho(T,0)} \right)^n - 1 \right] \tag{2v-8a}$$

Approximately

$$B = \frac{\rho_1 c_1^2}{n} \tag{2v-8b}$$

It is a good approximation in liquids to assume that the initial and final states are connected by an adiabatic compression. With this assumption,

$$u_1 = c_1 \left(1 + \frac{n + 1}{4c} \sigma \right) \tag{2v-9a}$$

where

$$\sigma = \frac{2c_1}{n - 1} \left[\left(\frac{\rho_2}{\rho_1} \right)^{(n-1)/2} - 1 \right] \tag{2v-9b}$$

Systems Subject to Chemical Reaction. The Rankine-Hugoniot relation, Eq. (2v-5a), is plotted in the $(p, 1/\rho)$ plane in Fig. 2v-1 with an adiabat for comparison. This relation is of course valid when the system reacts chemically, if the chemical energy is included in ΔE. In this case the point (p_1, ρ_1) does not lie on the Rankine-Hugoniot curve, but either above or below it, depending on whether the chemical reaction is endothermic or exothermic. An especially interesting case, detonation, occurs when there is enough chemical energy alone to sustain the shock wave. Since the wave velocity is measured by the slope of the line through (p_1, ρ_1) which intersects the Rankine-Hugoniot curve [see Eqs. (2v-5)], there are usually an infinite number of possible velocities. However, in a steady-state detonation the lowest possible velocity, which corresponds to a line through (p_1, ρ_1) just tangent to the Rankine-Hugoniot curve, is the one that occurs. This is the *Chapman-Jouguet condition:*

$$u_{1_{\text{detonation}}} = \frac{1}{\rho_1} \left(\frac{p_2 - p_1}{1/\rho_1 - 1/\rho_2} \right)^{\frac{1}{2}}_{\text{min}} \tag{2v-10}$$

which provides the extra relation needed so that the detonation velocity can be calculated from Eqs. (2v-5).

When mechanical as well as chemical energy is available, the velocity increases from the Chapman-Jouguet value as (p_2,ρ_2) moves upward along the Rankine-Hugoniot curve. There is no common physical process corresponding to the value of (p_2,ρ_2) below the Chapman-Jouguet value. The other branch of the Rankine-Hugoniot curve for which $p_2 < p_1$ and $\rho_2 < \rho_1$ corresponds to a deflagration and is a subsonic process.

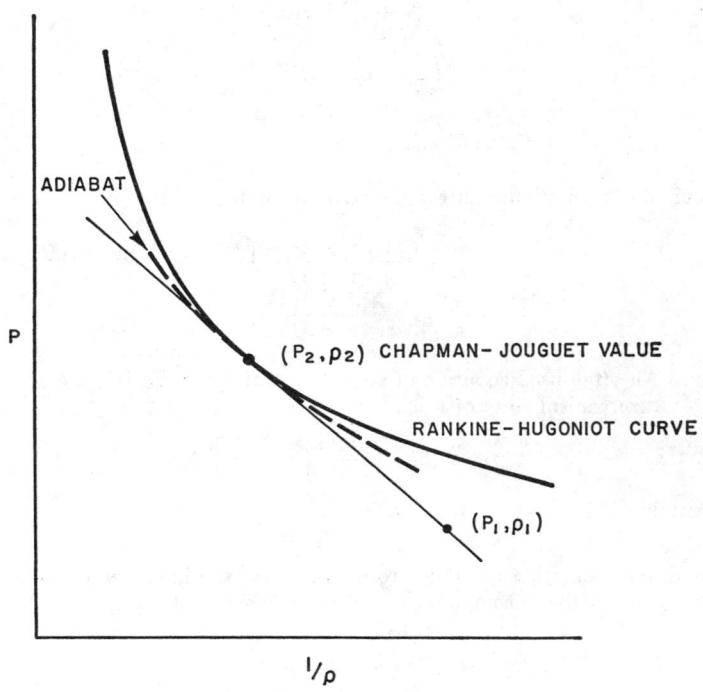

FIG. 2v-1. Plot of Rankine-Hugoniot relation.

2v-4. Reflection and Refraction at a Rigid Wall. At a rigid boundary, in addition to the previous Eqs. (2v-5) there must be added the condition

$$\mathbf{u} \cdot \mathbf{n} = 0 \qquad\qquad (2v\text{-}11)$$

Normal Incidence. The use of (2v-11) along with (2v-5) for a perfect gas leads to

$$\frac{p_3}{p_2} = \frac{(3\gamma - 1)(p_2/p_1) - \gamma + 1}{(\gamma - 1)(p_2/p_1) + \gamma + 1} \qquad\qquad (2v\text{-}12a)$$

and

$$-u_{3R} = c_1 \frac{2(p_2/p_1)(\gamma - 1) + 2}{\{2\gamma[(\gamma + 1)(p_2/p_1) + \gamma - 1]\}^{\frac{1}{2}}} \qquad\qquad (2v\text{-}12b)$$

which is the velocity of the reflected shock relative to the reflecting surface.

Oblique Incidence. In this case a second condition may be imposed: The incident and reflected waves should intersect at the surface. This condition cannot always be satisfied; when it is one speaks of *regular reflection*. Regular reflection always occurs at a sufficiently small angle of incidence (i.e., the angle between the normal to the surface and the normal to the shock front). The two boundary conditions then completely determine the direction and strength of the reflected shock.

There exists a critical angle of incidence above which regular reflection cannot occur. The point of intersection of the incident and reflected shocks rises above the surface and is joined to it by a third shock, called the Mach stem. This case is called "Mach reflection." Experimentally it is found that Mach reflection sets in at angles smaller than those predicted by theory.

2v-5. Reflection and Refraction at a Nonrigid Wall. There are now two boundary conditions that must be satisfied:

$$\mathbf{u}' \cdot \mathbf{n}' = \mathbf{u}'' \cdot \mathbf{n}'' \tag{2v-13a}$$

and
$$p' = p'' \tag{2v-13b}$$

Normal Incidence. In order to satisfy both (2v-13a) and (2v-13b) it is necessary that there be a transmitted and a reflected wave. The transmitted wave is always a shock, but the reflected wave may be either a shock or a rarefaction wave, depending on the properties of the two media and, in some cases, on the strength of the incident shock.

TABLE 2v-1. SOME PROPERTIES OF SHOCKS IN IDEAL GASES

M_1	Monatomic				Diatomic			
	p_2/p_1	ρ_2/ρ_1	T_2/T_1	M_R	p_2/p_1	ρ_2/ρ_1	T_2/T_1	M_R
1	1.000	1.000	1.000	1.000	1.000	1.000	1.000	1.000
1.5	2.562	1.714	1.495	1.397	2.458	1.862	1.320	1.426
2	4.750	2.286	2.078	1.648	4.500	2.667	1.688	1.732
2.5	7.562	2.703	2.798	1.808	7.125	3.333	2.138	1.949
3	11.00	3.000	3.667	1.915	10.33	3.857	2.679	2.104
4	19.75	3.368	5.863	2.041	18.50	4.571	4.047	2.297
5	31.00	3.571	8.680	2.104	29.00	5.000	5.800	2.408
6	44.75	3.692	12.12	2.142				
8	79.75	3.821	20.87	2.182				
10	124.8	3.884	32.12	2.201				
15	281.0	3.947	71.19	2.220				
20	499.8	3.970	125.9	2.227				

Oblique Incidence. There can occur either regular reflection or Mach reflection, of which the first case has been well investigated. It is shown that there is always a transmitted wave (i.e., total reflection of a shock wave cannot occur). If the second medium has a high acoustic impedance, the observed phenomena are similar to those found at a rigid surface; if the second medium has a low acoustic impedance, the observed phenomena are similar to those found at a free surface.

Free Surface (for Liquids Only). The condition (2v-13b) here becomes $p' = 0$. For a sufficiently small angle of incidence there is always a reflected rarefaction wave intersecting the incident shock at the surface. At some critical angle of incidence, determined by the strength of the incident shock as well as the properties of the liquid, this picture no longer applies. The phenomena in this case have not yet been intensively investigated.

Table 2v-1 lists some important properties of shock waves in ideal monatomic and diatomic gases. The following values have been used for γ, the ratio of heat capacities: For the monatomic gas, $\gamma = \frac{5}{3}$; for the diatomic gas, $\gamma = \frac{7}{5}$. For both gases, the possibility of electronic excitation has been neglected. In addition, for the diatomic gas, the possibilities of dissociation and the activation of the vibrational heat capacity have been neglected. Since the latter assumption becomes increasingly unrealistic at high temperatures, this part of the table has not been extended beyond $M_1 = 5$.

References

General

Courant, R., and K. O. Friedrichs: "Supersonic Flow and Shock Waves," Interscience Publishers, Inc., New York, 1948. Comprehensive, thorough treatment of entire subject, emphasizing mathematical aspects; 198 references.

Greene, E. F., and J. P. Toennies: "Chemische Reaktionen in Stosswellen," Steinkopf Verlag, Darmstadt, 1959. Over-all survey, emphasizing applications to chemistry.

Penney, W. G., and H. H. M. Pike: *Repts. Progr. in Phys.* **13**, 46–82 (1950). Over-all survey of problems and results, emphasizing physical aspects; 40 references.

Special

Bleakney, W., and A. H. Taub: *Revs. Modern Phys.* **21**, 584–605 (1949).

Cole, R. H.: "Underwater Explosions," Princeton University Press, Princeton, N.J., 1946. Thorough treatment of propagation of shocks in water.

Fletcher, C. H., A. H. Taub, and W. Bleakney: *Revs. Modern Phys.* **23**, 271–286 (1951). Mach reflection considered theoretically and experimentally.

Hirschfelder, J. O., C. F. Curtiss, and R. B. Bird: "Molecular Theory of Gases and Liquids," John Wiley & Sons, Inc., New York, 1954. Chapter 11 applies the rigorous kinetic theory of gases to detonations and shocks.

Lewis, Bernard, and Guenther von Elbe: "Combustion, Flames and Explosions of Gases," Academic Press, Inc., New York, 1951. Chapter XI treats detonation waves in gases.

Polachek, H., and R. J. Seeger: *Phys. Rev.* **84**, 922–929 (1951). Refraction at a gaseous interface.

Taub, A. H.: *Phys. Rev.* **72**, 51–60 (1947). Reflection and refraction of plane shocks.

See also "Waves of Finite Amplitude," Sec. 3c-6 of this book.

Section 3

ACOUSTICS

RICHARD K. COOK, Editor

National Bureau of Standards

CONTENTS

3a. Acoustical Definitions[1]

LEO L. BERANEK

Bolt Beranek and Newman Inc.

3a-1. General

Acceleration. Acceleration is a vector that specifies the time rate of change of velocity.

Acoustic, acoustical. The qualifying adjectives "acoustic" and "acoustical" mean containing, producing, arising from, actuated by, related to, or associated with sound. *Acoustic* is used when the term being qualified designates something that has the properties, dimensions, or physical characteristics associated with sound waves; *acoustical* is used when the term being qualified does not designate explicitly something that has such properties, dimensions, or physical characteristics. Usually the generic term is modified by *acoustical*, whereas the specific technical implication calls for *acoustic*.

Acoustics. (1) Acoustics is the science of sound, including its production, transmission, and effects. (2) The acoustics of a room are those qualities that together determine its character with respect to distinct hearing.

Ambient Noise. Ambient noise is the all-encompassing noise associated with a given environment, being usually a composite of sounds from many sources near and far.

Anechoic Space or Room. An anechoic space or room is one whose boundaries absorb effectively all the sound incident thereon, thereby affording essentially free-field conditions.

Audio Frequency. An audio frequency is any frequency corresponding to a normally audible sound wave.

Note 1: Audio frequencies range roughly from 15 to 20,000 Hz.
Note 2: The word "audio" can be used as a modifier to indicate a device or system intended to operate at audio frequencies, e.g., "audio amplifier."

Background Noise. Background noise is the total of all sources of interference in a system used for the production, detection, measurement, or recording of a signal, independent of the presence of the signal.

Note 1: Ambient noise detected, measured, or recorded with the signal becomes part of the background noise.
Note 2: Included in this definition is the interferenc resulting from primary power supplies, that separately is commonly described as *hum*.

Beats. Beats are periodic variations that result from the superposition of two simple harmonic quantities of different frequencies f_1 and f_2. They involve the periodic increase and decrease of amplitude at the *beat frequency* $(f_1 - f_2)$.

---

[1] From "American National Standard Acoustical Terminology" (including Mechanical Shock and Vibration), S1.1—American National Standards Institute. 1430 Broadway, New York, N.Y., 1960.

Continuous Spectrum. A continuous spectrum is the spectrum of a wave the components of which are continuously distributed over a frequency region.

Damping. Damping is the dissipation of energy with time or distance.

Decay, Rate of. The rate of decay is the time rate at which the sound pressure level (or other stated characteristic) decreases at a given point and at a given time. A commonly used unit is the decibel per second.

Echo. An echo is a wave that has been reflected or otherwise returned with sufficient magnitude and delay to be detected as a wave distinct from that directly transmitted.

Efficiency. The efficiency of a device with respect to a physical quantity which may be stored, transferred, or transformed by the device is the ratio of the useful output of the quantity to its total input.

Note: Unless specifically stated otherwise, the term "efficiency" means efficiency with respect to power.

Harmonic. A harmonic is a sinusoidal quantity having a frequency that is an integral multiple of the frequency of a periodic quantity to which it is related.

Infrasonic Frequency (Subsonic Frequency). An infrasonic frequency is a frequency lying below the audio-frequency range.

Note 1: The word "infrasonic" can be used as a modifier to indicate a device or system intended to operate at an infrasonic frequency.

Note 2: The term "subsonic" was once used in acoustics synonymously with infrasonic; such usage is now deprecated.

Jerk. Jerk is a vector that specifies the time rate of change of the acceleration; jerk is the third derivative of the displacement with respect to time.

Line Spectrum. A line spectrum is a spectrum whose components occur at a number of discrete frequencies.

Microbar, Dyne per Square Centimeter. A microbar is a unit of pressure commonly used in acoustics. One microbar is equal to one dyne per square centimeter.

Note: The term "bar" properly denotes a pressure of 10^6 dynes/cm².

Molecular Relaxation. Molecular relaxation is the equalization of energy among the degrees of freedom of a molecule following a disturbance that produces deviations from the equilibrium distribution law.

Noise. (1) Noise is any undesired sound. By extension, noise is any unwanted disturbance within a useful frequency band, such as undesired electric waves in a transmission channel or device. (2) Noise is an erratic, intermittent, or statistically random oscillation.

Note 1: If ambiguity exists as to the nature of the noise, a phrase such as "acoustic noise" or "electric noise" should be used.

Note 2: Since the above definitions are not mutually exclusive, it is usually necessary to depend upon context for the distinction.

Particle Velocity, Effective (Root-mean-square Particle Velocity). The effective particle velocity at a point is the root-mean-square value of the instantaneous particle velocities over a time interval at the point under consideration. For periodic particle velocities, the interval must be an integral number of periods or an interval that is long compared with a period. For nonperiodic particle velocities, the interval should be long enough to make the value obtained essentially independent of small changes in the length of the interval.

Power Spectral Density. Power spectral density of a function $X(t)$ is defined by the transform

$$S(\omega) = \frac{1}{2\pi} \int_{-\infty}^{\infty} R(\tau)e^{-i\omega\tau}\, d\tau$$

where $R(\tau)$ is the autocorrelation function of $X(t)$. $S(\omega)$ is proportional to the mean-square spectrum density as follows:

$$W(f) = 4\pi S(\omega) \qquad \omega = 2\pi f$$

where $W(f)$ is the *spectrum density.*

Note 1: The factor 4π arises from the difference in frequency units (2π) and the fact that both positive and negative frequencies are allowed for $S(\omega)$, which is an even function.

Note 2: Power spectral density is used in the field of random vibrations for every class of physical quantity, including displacement, velocity, pressure, and acceleration.

Radiation Pressure, Acoustic. The acoustic radiation pressure is a unidirectional, steady-state pressure exerted upon a surface exposed to an acoustic wave.

Random Noise. Random noise is an oscillation whose instantaneous magnitude is not specified for any given instant of time. The instantaneous magnitudes of a random noise are specified only by probability distribution functions giving the fraction of the total time that the magnitude, or some sequence of magnitudes, lies within a specified range.

Note: A random noise whose instantaneous magnitudes occur according to a "normal" or "gaussian" distribution is called "gaussian random noise."

Response. The response of a device or system is the motion (or other output) resulting from an excitation (stimulus) under specified conditions.

Note 1: Modifying phrases must be prefixed to the term response to indicate what kinds of input and output are being utilized.

Note 2: The response characteristic, often presented graphically, gives the response as a function of some independent variable such as frequency or direction. For such purposes it is customary to assume that other characteristics of the input (for example, voltage) are held constant.

Reverberation. (1) Reverberation is the persistence of sound in an enclosed space as a result of multiple reflections after the sound source has stopped. (2) Reverberation is the sound that persists in an enclosed space as a result of repeated reflection or scattering after the source of the sound has stopped.

Note: The repeated reflections of residual sound in an enclosure can alternatively be described in terms of the transient behavior of the modes of vibration of the medium bounded by the enclosure.

Sound Absorption. Sound absorption is the change of sound energy into some other form, usually heat, in passing through a medium or on striking a surface.

Sound Energy. The sound energy of a given part of a medium is the total energy in this part of the medium minus the energy which would exist in the same part of the medium with no sound waves present.

Sound-energy Density. The sound-energy density at a point in a sound field is the sound energy contained in a given infinitesimal part of the medium divided by the volume of that part of the medium.

Note 1: The terms "instantaneous energy density," "maximum energy density," and "peak energy density" have meanings analogous to the related terms used for sound pressure.

Note 2: In speaking of average energy density in general it is necessary to distinguish between the space average (at a given instant) and the time average (at a given point).

Note 3: In a plane wave at a point and in a standing-wave field, averaged in space over a distance in excess of the longest wavelength, the sound-energy density is

$$D = \frac{p^2}{\rho_0 c^2} = \frac{p^2}{\gamma P_0}$$

where γ is the ratio of specific heats for the gas and is equal to 1.4 for air and other diatomic gases and P_0 is the barometric pressure.

Sound Intensity (Sound-energy Flux Density, Sound-power Density). The sound intensity in a specified direction at a point is the average rate of sound energy transmitted in the specified direction through a unit area normal to this direction at the point considered.

Note 1: The sound intensity in any specified direction a of a sound field is the sound-energy flux through a unit area normal to that direction. This is given by the expression

$$I_a = \frac{1}{T} \int_0^T pv_a \, dt$$

where T = integral number of periods or time long compared with period
p = instantaneous sound pressure
v_a = component of instantaneous particle velocity in direction a
t = time

Note 2: In the case of a free plane or spherical wave having the effective sound pressure p, the speed of propagation c, in a medium of density ρ, the intensity in the direction of propagation is given by

$$I = \frac{p^2}{\rho c}$$

Sound Power of a Source. The sound power of a source is the total sound energy radiated by the source per unit of time.

Sound Pressure, Effective (Root-mean-square Sound Pressure). The effective sound pressure at a point is the root-mean-square value of the instantaneous sound pressures over a time interval at the point under consideration. For periodic sound pressures, the interval must be an integral number of periods or an interval that is long compared with a period. For nonperiodic sound pressures, the interval should be long enough to make the value obtained essentially independent of small changes in the length of the interval.

Note: The term "effective sound pressure" is frequently shortened to "sound pressure."

Spectrum. (1) The spectrum of a function of time is a description of its resolution into components, each of different frequency and (usually) different amplitude and phase. (2) "Spectrum" is also used to signify a continuous range of components, usually wide in extent, within which waves have some specified common characteristic, e.g., "audio-frequency spectrum."

Note: The term "spectrum" is also applied to functions of variables other than time, such as distance.

Spectrum (Spectral) Density. The spectrum density of a field quantity is the mean-square output of an ideal filter per unit bandwidth, in the limit as the bandwidth approaches zero.

Note: Examples of quantities often so used are sound pressure, voltage, and acceleration.

Speed of Sound. The speed of sound is the rate at which a sound wave travels through a medium.

Static Pressure. The static pressure P_0 at a point in the medium is the pressure that would exist at that point with no sound waves present. At normal barometric pressure, P_0 equals approximately 10^5 newtons/m² (10^6 dynes/cm²). This corresponds to a barometer reading of 0.751 m (29.6 in.) Hg (mercury) when the temperature of the mercury is 0°C. Standard atmospheric pressure is usually taken to be 0.760 m Hg at 0°C.

Strength of a Simple Sound Source. (1) The strength of a simple sound source is the maximum instantaneous rate of volume displacement produced by the source when emitting a wave with sinusoidal time variation. The unit is volume per second. (2) The strength of a simple sound source is the rms magnitude of the rate of volume displacement produced by the source when emitting a wave with sinusoidal time variation. The unit is volume per second.

Note 1: The user should specify (1) or (2) above.
Note 2: A simple sound source is one whose dimensions are small with respect to the wavelength.

Subharmonic. A subharmonic is a sinusoidal quantity having a frequency that is an integral submultiple of the fundamental frequency of a periodic quantity to which it is related.

Ultrasonic Frequency. An ultrasonic frequency is a frequency lying above the audio-frequency range. The term is commonly applied to elastic waves propagated in gases, liquids, or solids.

Note 1: The term "ultrasonic" can be used as a modifier to indicate a device or system intended to operate at an ultrasonic frequency.
Note 2: "Supersonic" was a term once used in acoustics synonymously with ultrasonic; such usage is now deprecated.

Velocity. Velocity is a vector that specifies the time rate of change of displacement with respect to a reference frame.

Note: If the reference frame is not inertial, the velocity is often designated "relative velocity."

Vibration. Vibration is an oscillation wherein the quantity is a parameter that defines the motion of a mechanical system.

Volume Velocity. Volume velocity is the rate of alternating flow of the medium through a specified surface due to a sound wave.

Note: Expressed mathematically the volume velocity V is

$$V = \int_S v \, d\sigma$$

where v is the component of particle velocity normal to the element of surface $d\sigma$. The integration is performed over surface S through which the medium is oscillating.

Wave. A wave is a disturbance which is propagated in a medium in such a manner that at any point in the medium the quantity serving as measure of disturbance is a function of the time while at any instant the displacement at a point is a function of the position of the point.

Any physical quantity that has the same relationship to some independent variable (usually time) that a propagated disturbance has, at a particular instant, with respect to space may be called a wave.

Wavelength. The wavelength of a periodic wave in an isotropic medium is the perpendicular distance between two wavefronts in which the displacements have a difference in phase of one complete period.

Wave Velocity (Velocity of Propagation). The velocity of propagation is a vector quantity that specifies the speed and direction with which a sound wave travels through a medium.

White Noise. White noise is a noise whose spectrum density (or spectrum level) is substantially independent of frequency over a specified range.

Note: White noise need not be random.

3a-2. Levels

Band Power Level. The band power level of a sound for a specified frequency band is the sound power level for the sound contained within the restricted band.

Band Pressure Level. The band pressure level of a sound for a specified frequency band is the sound pressure level for the sound contained within the restricted band. The reference pressure must be specified.

Note: The band may be specified by its lower and upper cutoff frequencies, or by its geometric center frequency and bandwidth. The width of the band may be indicated by a prefatory modifier; e.g., octave band (sound pressure) level, half-octave band level, third-octave band level, 50-Hz band level.

Decibel. The decibel is one tenth of a bel. Thus, the decibel is a unit of level when the base of the logarithm is the tenth root of 10, and the quantities concerned are proportional to power.

Note 1: Examples of quantities that qualify are power (any form), sound pressure squared, particle velocity squared, sound intensity, sound-energy density, voltage squared. Thus the decibel is a unit of sound-pressure-squared level; it is common practice, however, to shorten this to sound pressure level because ordinarily no ambiguity results from so doing.

Note 2: The logarithm to the base the tenth root of 10 is the same as ten times the logarithm to the base 10: e.g., for a number X^2, $\log_{10^{1/10}} X^2 = 10 \log_{10} X^2 = 20 \log_{10} X$. This last relationship is the one ordinarily used to simplify the language in definition of sound pressure level, etc.

Intensity Level (Sound-energy Flux Density Level). The intensity level, in decibels, of a sound is ten times the logarithm to the base 10 of the ratio of the intensity of this sound to the reference intensity. The reference intensity shall be stated explicitly.

Note 1: A common reference sound intensity is 10^{-16} watt/cm^2 in a specified direction.

Note 2: In a free progressive plane or spherical wave there is a known relation between sound intensity and sound pressure, so that sound intensity level can be deduced from a measurement of sound pressure level. In general, however, there is no simple relation between the two, and a measurement of sound pressure level should not be reported as one of intensity level.

Level. In acoustics, the level of a quantity is the logarithm of the ratio of that quantity to a reference quantity of the same kind. The base of the logarithm, the reference quantity, and the *kind* of level must be specified.

Note 1: Examples of kinds of levels in common use are electric power level, sound-pressure-squared level, voltage-squared level.

Note 2: The level as here defined is measured in units of the logarithm of a reference ratio. Also, a factor like 10 or 20 often precedes the logarithm to appropriately expand the scale; e.g., see Sound Pressure Level and Sound Power Level.

Note 3: In symbols,

$$L = \log_r \frac{q}{q_0}$$

where L = level of kind determined by the kind of quantity under consideration, measured in units of $\log_r r$

r = base of logarithms and the reference ratio

q = the quantity under consideration

q_0 = reference quantity of the same kind

Note 4: Differences in the levels of two like quantities q_1 and q_2 are described by the same formula because, by the rules of logarithms, the reference quantity is automatically divided out:

$$\log_r \frac{q_1}{q_0} - \log_r \frac{q_2}{q_0} = \log_r \frac{q_1}{q_2}$$

Neper. The neper is a unit of level when the logarithm is on the napierian base e. Use of the neper is restricted to levels of quantities analogous to electric current.

Note 1: Examples of quantities that qualify are voltage, current, particle velocity, sound pressure.
Note 2: One neper is equal to 8.686 dB.

Noise Level. (1) Noise level is the level of noise, the type of which must be indicated by further modifier or context.

Note: The physical quantity measured (e.g., voltage), the reference quantity, the instrument used, and the bandwidth or other weighting characteristic must be indicated.

(2) For airborne sound, unless specified to the contrary, noise level is the weighted sound pressure level called sound level; the weighting must be indicated.
Peak Level. The peak level is the maximum instantaneous level that occurs during a specified time interval. In acoustics, peak sound pressure level is to be understood, unless some other kind of level is specified.
Sound Level. Sound level is a weighted sound pressure level obtained by the use of metering characteristics and the weightings *A*, *B*, and *C* specified in Standard Specifications for Sound Level Meters, ANSI S1.4—1971 (revision of S1.4—1961). The weighting employed must always be stated. The reference pressure is 0.0002 microbar.

Note: A suitable method of stating the weighting is, for example, "The *A*-sound level was 43 dB."

Sound Power Level. The sound power level in decibels is ten times the logarithm to the base 10 of the ratio of a given power to a reference power.

Note 1: The internationally standardized reference power is 10^{-12} watt.
Note 2: In engineering acoustics in the United States, the reference power formerly was 10^{-13} watt, because sound pressure level approximately equals the sound power level (reference 10^{-13} watt) minus ten times the logarithm to the base 10 of the area through which the total sound power passes, when the area is expressed in square feet.

Sound Pressure Level. The sound pressure level, in decibels, of a sound is 20 times the logarithm to the base 10 of the ratio of the pressure of this sound to the reference pressure. The reference pressure shall be explicitly stated.

Note 1: The following reference pressures are in common use:
a. 2×10^{-4} microbar
b. 1 microbar
Reference pressure *a* is in general use for measurements concerned with hearing and with sound in air and liquids, while *b* has gained widespread acceptance for calibration of transducers and various kinds of sound measurements in liquids.
Note 2: Unless otherwise explicitly stated, it is to be understood that the sound pressure is the effective (rms) sound pressure.
Note 3: It is to be noted that in many sound fields the sound pressure ratios are not the square roots of the corresponding power ratios.

Spectrum Density Level (Spectrum Level). The spectrum density level of a specified signal at a particular frequency is equal to ten times the logarithm to the base 10 of the *spectrum density*. Ordinarily, this has significance only for a signal having a continuous distribution of components within the frequency range under consideration. The words "spectrum density level" cannot be used alone but must appear in combination with a prefatory modifier; e.g., pressure, velocity, voltage.

Note: For illustration, if L_{ps} be a desired pressure spectrum density level, p the effective pressure measured through the filter system, p_0 reference sound pressure, Δf the effective bandwidth of the filter system and Δf the reference bandwidth (1 Hz), then

$$L_{ps} = 10 \log_{10} \frac{p^2/\Delta f}{p_0^2/\Delta_0 f}$$

For computational purposes, if L_p is the band pressure level observed through the lter, the above relation re uces to

$$L_{ps} = L_p - 10 \log_{10} \frac{\Delta f}{\Delta_0 f}$$

Velocity Level. The velocity level, in decibels, of a sound is 20 times the logarithm to the base 10 of the ratio of the particle velocity of the sound to the reference particle velocity. The reference particle velocity shall be stated explicitly.

3a-3. Oscillation, Vibration, and Shock

Antiresonance. For a system in forced oscillation, antiresonance exists at a point when any change, however small, in the frequency of excitation causes an increase in the response at this point.

Coulomb Damping (Dry Friction Damping). Coulomb damping is the dissipation of energy that occurs when a particle in a vibrating system is resisted by a force whose magnitude is a constant independent of displacement and velocity, and whose direction is opposite to the direction of the velocity of the particle.

Critical Damping. Critical damping is the minimum viscous damping that will allow a displaced system to return to its initial position without oscillation.

Degrees of Freedom. The number of degrees of freedom of a mechanical system is equal to the minimum number of independent generalized coordinates required to define completely the positions of all parts of the system at any instant of time. In general, it is equal to the number of independent generalized displacements that are possible.

Dynamic Vibration Absorber (Tuned Damper). A tuned damper is a device for reducing vibration of a primary system by the transfer of energy to an auxiliary resonant system which is tuned to the frequency of the vibration. The force exerted by the auxiliary system is opposite in phase to the force acting on the primary system.

Forced Oscillation (Forced Vibration). The oscillation of a system is forced if the response is imposed by the excitation. If the excitation is periodic and continuing, the oscillation is steady state.

Free Vibration. Free oscillation of a system is oscillation that occurs in the absence of forced oscillation.

Fundamental Frequency. (1) The fundamental frequency of a periodic quantity is the frequency of a sinusoidal quantity which has the same period as the periodic quantity. (2) The fundamental frequency of an oscillating system is the lowest natural frequency. The normal mode of vibration associated with this frequency is known as the fundamental mode.

Impact. An impact is a single collision of one mass in motion with a second mass which may be either in motion or at rest.

Impulse. Impulse is the product of a force and the time during which the force is applied; more specifically, the impulse is $\int_{t_1}^{t_2} F \, dt$ where the force F is time-dependent and equal to zero before time t_1 and after time t_2.

Isolation. Isolation is a reduction in the capacity of a system to respond to an excitation attained by the use of a resilient support. In steady-state forced vibration, isolation is expressed quantitatively as the complement of transmissibility.

Mechanical Shock. Mechanical shock occurs when the position of a system is significantly changed in a relatively short time in a nonperiodic manner. It is characterized by suddenness and large displacements, and develops significant internal forces in the system.

Natural Frequency. Natural frequency is the frequency of free oscillation of a system. For a multiple-degree-of-freedom system, the natural frequencies are the frequencies of the normal modes of vibration.

Normal Mode of Vibration. A normal mode of vibration is a mode of free vibration of an undamped system. In general, any composite motion of the system is analyzable into a summation of its normal modes.

Note 1: The characteristic pattern of motion typically consists of a space distribution one part of which is negative in relation to the other part. Thus, at the same time that the particles in one part are moving outward in the positive direction from their positions of equilibrium, the particles in the other part are moving outward in the negative direction, and conversely.

Note 2: Vibration in a no mal mode occurs at a natural frequency of the undamped system.

Note 3: The terms "natural mode," "characteristic mode," and "eigen mode" are synonymous with normal mode.

Pulse Rise Time. The pulse rise time is the interval of time required for the leading edge of a pulse to rise from some specified small fraction to some specified larger fraction of the maximum value.

Q (Quality Factor). The quantity Q is a measure of the sharpness of resonance or frequency selectivity of a resonant vibratory system having a single degree of freedom, either mechanical or electrical.

Note 1: In a mechanical system, this quantity is equal to one-ha f the reciprocal of the damping ratio. It is commonly used only with reference to a lightly damped system, and is then approximately equal to the foll wing:

(1) Transmissibility at resonance
(2) π/δ, where δ is the logarithmic decrement
(3) $2\pi W/\Delta W$, where W is the stored energy, and ΔW the energy dissipation per cycle
(4) $f_r/\Delta f$, where f_r is the resonance freq ency, and Δf is the bandwidth between the half-power points

Note 2: Historically the letter Q was an arbitrarily chosen symbol to designate the ratio of reactance to resistance of a circuit element. The name "quality fa tor" was introduced later.

Resonance. Resonance of a system in forced oscillation exists when any change, however small, in the frequency of excitation causes a decrease in the response of the system.

Note: Velocity resonance, for example, may occur at a frequenc different from that of displacement resonance; see Tab e 3a-1.

Resonance Frequency (Resonant Frequency). A resonance frequency is a frequency at which resonance exists.

Note 1: In case of possible confusion the type of resonance must be indicated, e.g., velocity resonance frequency.

Note 2: See Table 3a-1.

Self-induced (Self-excited) Vibration. The vibration of a mechanical system is self-induced if it results from conversion, within the system, of nonoscillatory excitation to oscillatory excitation.

Shock Spectrum. A shock spectrum is a plot of the maximum acceleration experienced by a single-degree-of-freedom system, as a function of the natural frequency of that system, in response to an applied shock.

Transmissibility. Transmissibility is the nondimensional ratio of the response amplitude of a system in steady-state forced vibration to the excitation amplitude. The ratio may be one of forces, displacements, velocities, or accelerations.

Velocity Shock. Velocity shock is a mechanical shock resulting from a nonoscillatory change in velocity of an entire system.

Vibration Isolator. A vibration isolator is a resilient support that tends to isolate a system from steady-state excitation.

TABLE 3a-1. RESONANCE RELATIONS

In the case of a system whose motion can be described by the equation

$$M \frac{d^2x}{dt^2} + R \frac{dx}{dt} + Kx = A \cos \omega t$$

the characteristics of the different kinds of resonance in terms of the constants of the above equation are as follows:

	At velocity resonance	At displacement resonance	Damped natural frequency
Frequency.............	$\dfrac{1}{2\pi}\sqrt{\dfrac{K}{M}}$	$\dfrac{1}{2\pi}\sqrt{\dfrac{K}{M} - \dfrac{R^2}{2M^2}}$	$\dfrac{1}{2\pi}\sqrt{\dfrac{K}{M} - \dfrac{R^2}{4M^2}}$
Amplitude of displacement........	$\dfrac{A}{R\sqrt{\dfrac{K}{M}}}$	$\dfrac{A}{R\sqrt{\dfrac{K}{M} - \dfrac{R^2}{4M^2}}}$	$\dfrac{A}{R\sqrt{\dfrac{K}{M} - \dfrac{3R^2}{16M^2}}}$
Amplitude of velocity..	$\dfrac{A}{R}$	$\dfrac{A}{R\sqrt{1 + \dfrac{R^2}{4MK - 2R^2}}}$	$\dfrac{A}{R\sqrt{1 + \dfrac{R^2}{16MK - 4R^2}}}$
Phase of displacement with reference to applied force.........	$\dfrac{\pi}{2}$	$\tan^{-1}\sqrt{\dfrac{4MK}{R^2} - 2}$	$\tan^{-1}\sqrt{\dfrac{16MK}{R^2} - 4}$

For values of R, small compared with $\sqrt{KM}$, there is little difference between the three cases discussed above. The frequency at velocity resonance is equal to the undamped natural frequency of the system. Other symbols are also employed depending upon the quantity represented by x; for example, c is frequently used instead of R.

Viscous Damping. Viscous damping is the dissipation of energy that occurs when a particle in a vibrating system is resisted by a force that has a magnitude proportional to the magnitude of the velocity of the particle and direction opposite to the direction of the particle velocity.

3a-4. Sound Transmission and Propagation

Absorption Loss. Absorption loss is that part of the transmission loss due to the dissipation or conversion of sound energy into other forms of energy (e.g., heat), either within the medium or attendant upon a reflection.

Acoustic Attenuation Constant. The acoustic attenuation constant is the real part of the acoustic propagation constant. The commonly used unit is the neper per section or per unit distance.

Acoustic Dispersion. Acoustic dispersion is the change of speed of sound with frequency.

Acoustic Phase Constant. The acoustic phase constant is the imaginary part of the acoustic propagation constant. The commonly used unit is the radian per section or per unit distance.

Acoustic Propagation Constant. The acoustic propagation constant of a uniform system or of a section of a system of recurrent structures is the natural logarithm of the complex ratio of the steady-state particle velocity, volume velocities, or pressures at two points separated by unit distance in the uniform system (assumed to be of infinite length), or at two successive corresponding points in the system of recurrent structures (assumed to be of infinite length). The ratio is determined by dividing the value at the point nearer the transmitting end by the corresponding value at the more remote point.

Acoustic Refraction. Acoustic refraction is the process by which the direction of sound propagation is changed due to spatial variation in the speed of sound in the medium.

Acoustic Scattering. Acoustic scattering is the irregular reflection, refraction, or diffraction of a sound in many directions.

Antinode (Loop). An antinode is a point, line, or surface in a standing wave where some characteristic of the wave field has maximum amplitude.

Note: The appropriate modifier should be used before the word "antinode" to signify the type that s intende ', e.g., disp ɛ cement antinode, velocity antinode, pressure antinode.

Backscattering Cross Section. The acoustic backscattering cross section of an object is an area equal to 4π times the product of the square of a unit distance and the square of the sound pressure scattered by the object, back in the direction from which the sound has come as observed at unit distance from the acoustic center of the object, divided by the square of the sound pressure of the plane wave incident on the object. The unit of the cross section is the square of the unit distance.

Note 1: In symbols, if σ_b is the backscattering cross section, p_{sb}^2 is the square of the backscattered sound pressure, r_0 the un't distance and p_i^2 the square of th. incident sound pressu e,

$$\sigma_b = \frac{4\pi p_{sb}^2 r_0^2}{p_i^2}$$

Note 2: The scattering cross section or any other direction is milarly defined; the direction must be sp cified.

Compressional Wave. A compressional wave is a wave in an elastic medium which causes an element of the medium to change its volume without undergoing rotation.

Note 1: Mathematically, the velocity field of a compressional wave has zero curl.
Note 2: A compressional plane wave is a longitudinal wave.

Diffraction. Diffraction is that process that produces a diffracted wave.

Diffracted Wave. A diffracted wave is one whose front has been changed in direction by an obstacle or other inhomogeneity in a medium, otherwise than by reflection or refraction.

Divergence Loss. Divergence loss is that part of the transmission loss due to the divergence on spreading of the sound rays in accordance with the geometry of the system (e.g., spherical waves emitted by a point source).

Doppler Effect. The Doppler effect is the phenomenon evidenced by the change in the observed frequency of a wave in a transmission system caused by a time rate of change in the effective length of the path of travel between the source and the point of observation.

Note: The effect is described quantitatively by

$$f_r = \frac{1 + v_r/c}{1 - v_s/c} f_s$$

where f_r = observed frequency
f_s = frequency at source
v_r = component of velocity (relative to the medium) of observation point toward source
v_s = component of velocity (relative to the medium) of source toward observation point
c = speed of sound in a stationary medium

Insertion Loss.   The insertion loss, in decibels, resulting from insertion of a transducer in a transmission system is ten times the logarithm to the base 10 of the ratio of the power delivered to that part of the system that will follow the transducer, before the insertion of the transducer, to the power delivered to that same part of the system after insertion of the transducer.

Note: If the input power or the output power or both consist of more than one component, the particular component must be specified

Longitudinal Wave.   A longitudinal wave is a wave in which the direction of displacement at each point of the medium is normal to the wavefront.

Node.   A node is a point, line, or surface in a standing wave where some characteristic of the wave field has essentially zero amplitude.

Note: The appropriate modifier should be used before the wo d "node" to signify the type that is intended, e.g., displacement node, velocity node, pressure node.

Partial Node.   A partial node is the point, line, or surface in a standing wave system where some characteristic of the wave field has a minimum amplitude differing from zero.

Note: The appropriate modifier should be used wi h the words "partial node" to signify the type that is intended; e.g., displacement partial node, velocity partial node, pressure partial node.

Plane Wave.   A plane wave is a wave in which the wavefronts are everywhere parallel planes normal to the direction of propagation.

Rayleigh Wave.   A Rayleigh wave is a surface wave associated with the free boundary of a solid, such that a surface particle describes an ellipse whose major axis is normal to the surface, and whose center is at the undisturbed surface.   At maximum particle displacement away from the solid surface the motion of the particle is opposite to that of the wave.

Note: The propagation velocity of a Rayleigh wave is slightly less than that of a shear wave in the solid; the wave amplitude o. the Rayleigh wave diminishes exponentially with depth.

Refraction.   Acoustic refraction is the process by which the direction of sound propagation is changed because of spatial variation in the speed of sound in the medium.

Refraction Loss.   Refraction loss is that part of the transmission loss due to refraction resulting from nonuniformity of the medium.

Refracted Wave.   A refracted wave is one whose front has been changed in direction owing to refraction.

Scattering Cross Section.   The acoustic scattering cross section of an object is an area equal to 4π times the product of the mean-square sound pressure scattered by the object, averaged over a sphere of unit radius surrounding the object, and the square of the unit radius, divided by the square of the sound pressure of the plane wave incident upon the object.   The unit of the cross section is the square of the unit radius.

Note 1: In symbols, if σ is the scattering cross section, p_s^2 the average mean-square scattered sound pressure, r_0 the unit radius, and p_i^2 the square of the incident sound pressure,

$$\sigma = \frac{4\pi p_s^2 r_0^2}{p_i^2}$$

Note 2: Actual measurements must be made at a distance sufficiently great that the sound appears to be scattering from a single point called the acoustic center.

Scattering Differential. The scattering differential is the amount by which the level of the scattered mean-square sound pressure averaged over all directions at a specified unit distance from the effective acoustic center of the object exceeds the plane-wave free-field pressure level of the sound incident upon the object. The scattering differential of an object is ten times the logarithm to the base 10 of the ratio of the scattering cross section to the area of the sphere of unit radius surrounding the object.

Note 1: In symbols, if Δ is the scattering differential, and the other symbols are those of *scattering cross section*:

$$\Delta = 10 \log \frac{\sigma}{4\pi r_0{}^2} = 10 \log \frac{p_s{}^{'}}{p_i{}^2}$$

Scattering Loss. Scattering loss is that part of the transmission loss due to scattering within the medium or due to roughness of the reflecting surface.

Shear Wave (Rotational Wave). A shear wave is a wave in an elastic medium which causes an element of the medium to change its shape without a change of volume.

Note 1: Mathematically, the velocity field of a shear wave has zero divergence.
Note : A plane shear wave in an isotropic medium is a transverse wave.

Sound Field, Free (Free Field). A free sound field is a field in a homogeneous, isotropic medium free from boundaries. In practice it is a field in which the effects of the boundaries are negligible over the region of interest.

Note: The actual pressure impinging on an object (e.g., electroacoustic transducer) placed in an otherwise free sound field will differ from the pressure which would exist at that point with the object removed unless the acoustic impedance of the object matches the acoustic impedance of the medium.

Standing Wave. A standing wave is a periodic wave having a fixed distribution in space which is the result of interference of progressive waves of the same frequency and kind. Such waves are characterized by the existence of nodes or partial nodes and antinodes that are fixed in space.

Stationary Wave. A stationary wave is a standing wave in which the net energy flux is zero at all points.

Note: Stationary waves can be only approximated in practice.

Streaming (Acoustic). Acoustic streaming is the name given to unidirectional flow currents in a fluid that are due to the presence of acoustic waves.

Transmission Loss. Transmission loss is the reduction in the magnitude of some characteristic of a signal between two stated points in a transmission system.

Note 1: The characteristic is often some kind of level, such as power level or voltage level. In acoustics the characteristic that is commonly measured is sound pressure level. Thus, if the levels are expressed in decibels, the transmission-level loss is likewise in decibels.
Note 2: It is imperative that the characteristic concerned (such as the sound pressure level) be clearly identified because in all transmission systems more than one characteristic is propagated.

Wavefront. (1) The wavefront of a progressive wave in space is a continuous surface which is a locus of points having the same phase at a given instant. (2) The wavefront of a progressive surface wave is a continuous line which is a locus of points having the same phase at a given instant.

Wave Interference. Wave interference is the phenomenon which results when waves of the same or nearly the same frequency are superposed and is characterized by a spatial or temporal distribution of amplitude of some specified characteristic differing from that of the individual superposed waves.

3a-5. Complex Parameters of Linear Systems

Acoustic Compliance. (1) The acoustic compliance of an enclosed volume of gas is equal to the magnitude of the ratio of the volume displacement of a piston forming one side of the volume to the pressure caused by the displacement (units cm^5/dyne or m^5/newton). (2) Acoustic compliance is the reciprocal of acoustic stiffness.

Acoustic Impedance. The acoustic impedance of a fluid medium on a given surface lying in a wavefront is the complex ratio of the sound pressure (force per unit area) on that surface to the flux (volume velocity or particle velocity multiplied by the area) through the surface. When concentrated rather than distributed impedances are considered, the impedance of a portion of the medium is based on the pressure difference effective in driving that portion and the flux (volume velocity). The acoustic impedance may be expressed in terms of mechanical impedance divided by the square of the area of the surface considered (units $dyne-sec/cm^5$ or $newton-sec/m^5$).

Note 1: Velocities in the direction along which the impedance is to be specified are considered positive.
Note 2: The real part of an acoustic impedance is *acoustic resistance*, and the imaginary part is *acoustic reactance*.

Acoustic Mass (Acoustic Inertance). Acoustic mass is the quantity which, when multiplied by 2π times the frequency, gives the acoustic reactance.

Acoustic Mobility. The acoustic mobility of a fluid medium on a given surface lying in a wavefront is the complex ratio of the flux (volume velocity, or particle velocity multiplied by the area) to the acoustic stress acting normal to the surface. When concentrated rather than distributed mobilities are considered, the mobility of a portion of the medium is based on the flux and the acoustic stress acting through that portion (units cm^5/dyne-sec or m^5/newton-sec).

Note: The acoustic mobility may be expressed in terms of the mechanical mobility multiplied by the square of the area of the surface considered.

Acoustic Ohm (cgs). An acoustic resistance, reactance, or impedance has a magnitude of one acoustic (cgs) ohm when a sound pressure of one microbar produces a volume velocity of one cubic centimeter per second.

Note: In the mks system the expression mks acoustic ohm is used.

Acoustic Resistance. Acoustic resistance is the real component of the acoustic impedance.

Acoustic Reactance. Acoustic reactance is the imaginary component of the acoustic impedance.

Acoustic Stiffness. Acoustic stiffness is the quantity which, when divided by 2π times the frequency, gives the acoustic reactance.

Analogous Impedance. Analogous impedances have classically been defined as the quotients of the following complex amplitudes:

Mechanical impedance Z_M = force across/velocity through
Rotation impedance Z_R = torque across/angular velocity through
Acoustic impedance Z_A = sound pressure across/volume velocity through
Electric impedance Z_E = voltage across/current through

The real part of each impedance above is called the *resistance*, and the imaginary part the *reactance*. The reciprocal of any impedance is an *admittance* whose real part is a *conductance* and imaginary part a *susceptance*.

Analogy. An analogy is a recognized relationship of consistent mutual similarity between the equations and structures appearing within two or more fields of knowledge, and an identification and association of the quantities and structural elements that play mutually similar roles in these equations and structures, for the purpose of facilitating transfer of knowledge of mathematical procedures of analysis and behavior of the structures between these fields.

Characteristic Impedance (Intrinsic Impedance). The characteristic impedance of a medium is the ratio of the effective sound pressure at a given point to the effective particle velocity at that point in a free plane progressive sound wave.

Note 1: The characteristic impedance is equal to the product of the density and the speed of sound in the medium.

Note 2: The characteristic impedance of an acoustic medium is analogous to the characteristic impedance of an infinitely long electrical transmission line.

Impedance. An impedance is the ratio of two complex quantities whose arguments increase linearly with time and whose real (or imaginary) parts represent a forcelike and velocitylike quantity respectively.

Note 1: Examples of forcelike quantities are force, sound pressure, voltage, temperature, electric field strength. Examples of velocitylike quantities are velocity, volume velocity, current, heat flow, magnetic flux.

Note 2: The terms and definitions related to impedance pertain to single-frequency quantities in the steady state and to systems whose properties are independent of the magnitudes of these quantities. These quantities can be represented mathematically by complex exponential functions of time. Under these conditions the factors involving time cancel out in the ratios called for, leaving complex numbers independent of time. Solutions based on complex exponential functions under these conditions give the solution for real sinusoidal oscillations.

Because of the similarity of electrical, mechanical, and acoustical transmission theory, the same terminology is used in the three cases. Where confusion is likely to occur, the proper term should be prefixed to the general term, e.g., "acoustic transfer impedance," but unless otherwise specified, the definitions apply not only in acoustics but in mechanics as well. While acoustics is a branch of mechanics, it is found convenient to distinguish an acoustic system from a mechanical one whenever elastic wave motion is an essential feature.

Intrinsic Impedance. See Characteristic Impedance.

Mechanical Compliance. The mechanical compliance of a springlike device is equal to the magnitude of the ratio of the displacement of the device to the force that produced the displacement (units cm/dyne or m/newton).

Mechanical Impedance. The mechanical impedance at a point or an interface of a mechanical system is the complex ratio of a sine-wave force acting on the system at that point or interface to the sine-wave velocity resulting from that force (units dyne-sec/cm or newton-sec/m).

Note 1: The ratio of force to displacement is sometimes also called mechanical impedance; this usage is deprecated.

Note 2: If the force and velocity are measured at the same point, the ratio is designated driving-point impedance; if they are measured at different points, the ratio is designated *transfer impedance.*

Mechanical Ohm (cgs). A mechanical resistance, reactance, or impedance has a magnitude of a mechanical (cgs) ohm when a force of 1 dyne/cm^2 produces a velocity of 1 cm/sec.

Note: In the mks system, the expression mks mechanical ohm is used.

Mechanical Reactance. Mechanical reactance is the imaginary part of the mechanical impedance.

Mechanical Resistance. Mechanical resistance is the real part of the mechanical impedance.

Mobility. Analogous mobilities are defined as the quotients of the following complex amplitudes:

Mechanical mobility Z_M = velocity across/force through
Rotational mobility Z_R = angular velocity across/torque through
Acoustic mobility Z_A = volume velocity across/sound pressure
Through electric impedance Z_E = voltage across/current through

The real part of each mobility is called the *responsiveness* and the imaginary part the *excitability.* The reciprocal of any mobility is an *immobility* whose real part is an *unresponsiveness* and imaginary part an *unexcitability.*

Rayl. The rayl is the magnitude of a specific acoustic resistance, reactance, or impedance for which a sound pressure of one microbar produces a linear velocity of one centimeter per second (dyne-sec/cm³). When expressed in newton-sec/m³ it is called the mks rayl.

Rotational Impedance. A rotational impedance is the complex ratio of a torque to angular velocity (or a relative angular velocity).

Specific Acoustic Compliance. The specific acoustic compliance of a springlike device or an enclosed volume of gas is equal to the magnitude of the ratio of the displacement of the device or of a piston forming one side of the volume to the pressure that produced the displacement (units cm³/dyne or m³/newton).

Specific Acoustic Impedance (Unit Area Acoustic Impedance). The specific acoustic impedance at a point in the medium is the complex ratio of sound pressure to particle velocity.

Specific Acoustic Mass. The specific acoustic mass is the quantity which when multiplied by 2π times the frequency gives the specific acoustic reactance associated with the kinetic energy of the medium (units g/cm² or kg/m²).

Specific Acoustic Reactance. Specific acoustic reactance is the imaginary component of the specific acoustic impedance (units dyne-sec/cm³ or newton-sec/m³).

Specific Acoustic Stiffness. The specific acoustic stiffness is the reciprocal of the specific acoustic compliance.

Specific Acoustic Resistance. Specific acoustic resistance is the real component of the specific acoustic impedance.

3a-6. Transducer Parameters

Acoustical Reciprocity Theorem. In an acoustic system comprising a fluid medium having bounding surfaces $S_1, S_2, S_3, \ldots$ and subject to no impressed body forces, if two distributions of normal velocities v_n' and v_n'' of the bounding surfaces produce pressure fields p' and p'', respectively, throughout the region, then the surface integral of $(p''v_n' - p'v_n'')$ over all the bounding surfaces $S_1, S_2, S_3, \ldots$ vanishes.

Note: If the region contains only one simple source, the theorem reduces to the form ascribed to Helmholtz; viz., in a region as described, a simple source at A produces the same sound pressure at another point B as would have been produced at A had the source been located at B.

Available Power. (1) The available power of a linear source of electric energy is the quotient of the mean square of the open-circuit terminal voltage of the source divided by four times the resistive component of the internal impedance of the source.

Note: The available power would be delivered to a load impedance that is the conjugate of the internal impedance of the source and is the maximum power that can be delivered by that source.

(2) The available power of a sound-field, with respect to a given object placed in it, is the power which would be abstracted from the acoustic medium by an ideal trans-

ducer having the same dimensions and the same orientation as the given object. The dimensions and their orientation with respect to the sound field must be specified.

Note: The acoustic power available to an electroacoustic transducer, in a plane-wave sound field of a given frequency, is the product of the free-field sound intensity by the effective area of the transducer.

For this purpose the effective area of an electroacoustic transducer, for which the surface velocity distribution is independent of the manner of excitation of the transducer, is $1/4\pi$ times the product of the receiving directivity factor by the square of the wavelength of a free progressive wave in the medium.

If the physical dimensions of the transducer are small in comparison with the wavelength, the directivity factor is near unity and the effective area varies inversely as the square of the frequency. If the physical dimensions are large in comparison with the wavelength, the directivity factor is nearly proportional to the square of the frequency and the effective area approaches the actual area of the active face of the transducer.

Available Power Efficiency.   The available power efficiency of an electroacoustic transducer used for sound reception is the ratio of the electric power available at the electric terminals of the transducer to the acoustic power available to the transducer.

Note 1: For an electroacoustic transducer which obeys the reciprocity principle, the available power efficiency in sound reception is equal to the transmitting efficiency.

Note 2: In a given narrow-frequency band, the available power efficiency is numerically equal to the fraction of the open-circuit mean-square thermal noise voltage present at the electric terminals which is contributed by thermal noise in the acoustic medium.

Available Power Response.   The available power response of an electroacoustic transducer used for sound emission is the ratio of the mean-square sound pressure apparent at a distance of 1 meter in a specified direction from the effective acoustic center of the transducer to the available electric power from the source.

Note 1: The sound pressure apparent at a distance of 1 meter can be found by multiplying the sound pressure observed at a remote point where the sound field is spherically divergent by the number of meters from the effective acoustic center of the transducer to that point.

Note 2: The available power response is a function not only of the transducer but also of some source impedance, either actual or hypothetical, the value of which must be specified.

Beam Width.   The beam width of a directional transducer, at a given frequency in a given plane including the beam axis, is the angle included between the two directions, one to the left and the other to the right of the axis, at which the angular deviation loss has a specified value.

Note: Beam widths are commonly specified for an angular deviation loss of 3, 6, or 10 dB, the choice depending upon the directivity of the transducer or upon its intended application. The particular angular deviation loss can be indicated conveniently by use of a term such as "3-dB beam width."

Directional Gain (Directivity Index).   The directional gain of a transducer, in decibels, is ten times the logarithm to the base 10 of the directivity factor.

Directional Response Pattern (Beam Pattern).   The directional response pattern of a transducer used for sound emission or reception is a description, often presented graphically, of the response of the transducer as a function of the direction of the transmitted or incident sound waves in a specified plane and at a specified frequency.

Note 1: A complete description of the directional response pattern of a transducer would require three-dimensional presentation.

Note 2: The directional response pattern is often shown as the response relative to the maximum response.

Directivity Factor.   (1) The directivity factor of a transducer used for sound emission is the ratio of the sound pressure squared, at some fixed distance and specified direction, to the mean-square sound pressure at the same distance averaged over all directions from the transducer. The distance must be great enough so that the sound appears to diverge spherically from the effective acoustic center of the source. Unless otherwise specified, the reference direction is understood to be that of maximum response. (2) The directivity factor of a transducer used for sound reception is the ratio of the square of the open-circuit voltage produced in response to sound waves arriving in a specified direction to the mean-square voltage that would be produced in a perfectly diffused sound field of the same frequency and mean-square sound pressure.

Note 1: This definition can be extended to cover the case of finite frequency bands whose spectrum can be specified.
Note 2: The average free-field response can be obtained in various ways, such as
a. By the use of a spherical integrator
b. By numerical integration of a sufficient number of directivity patterns corresponding to different planes
c. By integration of one or two directional patterns whenever the pattern of the transducer is known to possess adequate symmetry

Dynamic Range. The dynamic range of an electroacoustic transducer used for sound reception is the difference between the overload pressure level and the equivalent pressure level of the noise.

Note 1: The useful dynamic range is limited at the low-level end by noise in the medium (acoustic noise) or by electrical circuit noise. The nature of the noise limit must be stated explicitly (e.g., ambient noise, equipment noise, thermal noise, etc).
Note 2: The method of overload determination and the type of overload must be specified (i.e., signal distortion, overheating, damage, etc).

Effective Acoustic Center. The effective acoustic center of an acoustic generator is the point from which the spherically divergent sound waves, observable at remote points, appear to diverge.

Effective Bandwidth. The effective bandwidth of a specified transmission system is the bandwidth of an ideal system which (1) has uniform transmission in its passband equal to the maximum transmission of the specified system and (2) transmits the same power as the specified system when the two systems are receiving equal input signals having a uniform distribution of energy at all frequencies.

Note: This can be expressed mathematically as follows:

$$\text{Effective bandwidth} = \int_0^\infty G \, df$$

where f is the frequency in hertz, and G is the ratio of the power transmission at the frequency f to the transmission at the frequency of maximum transmission.

Electroacoustical Reciprocity Theorem. For an electroacoustic transducer satisfying the reciprocity principle, the quotient of the magnitude of the ratio of the open-circuit voltage at the output terminals (or the short-circuit output current) of the transducer, when used as a sound receiver, to the free-field sound pressure referred to an arbitrarily selected reference point on or near the transducer, divided by the magnitude of the ratio of the sound pressure apparent at a distance d from the reference point to the current flowing at the transducer input terminals (or the voltage applied at the input terminals), when used as a sound emitter, is a constant, called the "reciprocity constant," independent of the type or constructional details of the transducer.

Note: The reciprocity constant is given by

$$\left| \frac{M_0}{S_0} \right| = \left| \frac{M_s}{S_s} \right| = \frac{2d}{\rho f} \cdot 10^{-7}$$

where M_0 = free-field voltage sensitivity as a sound receiver referred to arbitrary refer-
          ence point on or near transducer, open-circuit volts/microbar
     M_s = free-field current sensitivity referred to arbitrary reference point on or near
          transducer, short-circuit amp/microbar
     S_0 = sound pressure produced at distance d cm from arbitrary reference point,
          microbars/amp of input current
     S_s = sound pressure produced at distance d cm from arbitrary reference point,
          microbars/volt applied at the input terminals
     f = frequency, Hz
     ρ = density of medium, in g/cm³
     d = distance from arbitrary reference point on or near transducer to point at
          which sound pressure established by transducer when emitting is evaluated,
          cm

Equivalent Noise Pressure (Inherent Noise Pressure).   The equivalent noise pressure
of an electroacoustic transducer or system used for sound reception is the root-mean-
square sound pressure of a sinusoidal plane progressive wave, which, if propagated
parallel to the principal axis of the transducer, would produce an open-circuit signal
voltage equal to the root mean square of the inherent open-circuit noise voltage of
the transducer in a transmission band having a bandwidth of 1 Hz and centered on
the frequency of the plane sound wave.

Note: If the equivalent noise pressure of the transducer is a function of secondary
variables, such as ambient temperature or pressure, the applicable value of these quantities
should be stated explicitly.

Insertion Loss.   The insertion loss, in decibels, resulting from insertion of a transducer in a transmission system is ten times the logarithm to the base 10 of the ratio
of the power delivered to that part of the system that will follow the transducer
before the insertion of the transducer to the power delivered to that same part of the
system after insertion of the transducer.

Note: If the input power or the output power or both consist of more than one component,
the particular component must be specified.

Principal Axis.   The principal axis of a transducer used for sound emission or
reception is a reference direction for angular coordinates used in describing the directional characteristics of the transducer.   It is usually an axis of structural symmetry
or the direction of maximum response, but if these do not coincide, the reference
direction must be described explicitly.

Relative Response.   The relative response of a transducer, in decibels, is the amount
by which the response under some particular condition exceeds the response under a
reference condition that should be stated explicitly.

Sensitivity Level; Response Level (Sensitivity)(Response).   The sensitivity (or
response) level of a transducer, in decibels, is 20 times the logarithm to the base 10
of the ratio of the amplitude sensitivity to the reference sensitivity, where the amplitude is a quantity proportional to the square root of power.   The kind of sensitivity
and the reference sensitivity must be indicated.

Note: For a microphone, the free-field voltage-pressure sensitivity is the kind often used,
and a common reference sensitivity is $s_0 = 1$ volt per microbar.   The square of the sensitivity is proportional to a power ratio.   The free-field voltage sensitivity-squared level,
in decibels, is therefore $S = 10 \log (s^2/s_0{}^2) = 20 \log (s/s_0)$.   Often, sensitivity-squared
level in decibels can be shortened, without ambiguity, to sensitivity level in decibels, or
simply sensitivity in decibels.

3a-7. Underwater Sound

Active Sonar (Echo-ranging Sonar).   In an active sonar system, a pulse of acoustic
energy is generated by an underwater source and radiated outward; such pulse being
reflected in part by an object and transmitted back to an underwater receiver; said

reflected pulse yielding information as to distance, bearing angle, and character of object.

Passive Sonar (Listening Sonar). Passive sonar is the method or equipment by which information concerning a distant object is obtained by evaluation of sound generated by the object.

Sea Noise. Sea noise is that portion of the ambient noise in the sea that remains if the noise components contributed by marine life, terrestrial noise, and precipitation, and by ships, traffic, and other identifiable man-made sources are excluded.

Sonar. Sonar is the method or equipment for determining, by underwater sound, the presence, location, or nature of objects in the sea.

Note: The word "sonar" is an acronym derived from the expression "SOund NAvigation and Ranging."

Sonar Background Noise. Sonar background noise is the total noise that interferes with the reception of the desired signal. The noise is that presented to the final receiving element, such as a recorder or the ear of a listener.

Target Strength. In underwater sound, the target strength of an object is the backscattering differential of an object for sound scattered back along the path of the incident sound. Unless otherwise specified, the reference distance is 1 yard. (See *Scattering Differential* under Sec. 3a-4.)

3a-8. Sonics

Cavitation. Sonically induced cavitation in a liquid is the formation, growth, and collapse of gaseous and vapor bubbles due to the action of intense sound waves.

Cavitation Noise. Cavitation noise is the noise produced in a liquid by gaseous or vaporous cavitation.

Electrostriction. Electrostriction is the phenomenon wherein some dielectric materials experience an elastic strain when subjected to an electric field, this strain being independent of the polarity of the field.

Hydrodynamic Oscillator. A hydrodynamic oscillator is a transducer for generating sound waves in fluids, in which a continuous flow through an orifice is modulated by a reciprocating valve system controlled by acoustic feedback.

Jet-edge Generator. A jet-edge sonic generator is a fluid dynamic transducer, involving vortex formation, in which stabilization is achieved by hydrodynamic feedback between a jet and an edge.

Macrosonics. Macrosonics is the technology of sound at signal amplitudes so large that linear approximations are not valid.

Note: Processing techniques usually involve macrosonics.

Magnetostriction. Magnetostriction is the phenomenon wherein ferromagnetic materials experience an elastic strain when subjected to an external magnetic field. Also, magnetostriction is the converse phenomenon in which mechanical stresses cause a change in the magnetic induction of a ferromagnetic material.

Oseen Force. An Oseen force is a steady force exerted on a suspended particle by second-order velocity effects resulting from second harmonic content in a distorted wave.

Piezoelectricity. Piezoelectricity is the property exhibited by some asymmetrical crystalline materials which when subjected to strain in suitable directions develop electric polarization proportional to the strain. Inverse piezoelectricity is the effect in which mechanical strain is produced in certain asymmetrical crystalline materials when subjected to an external electric field; the strain is proportional to the electric field.

Sonics. Sonics is the technology of sound in processing and analysis. Sonics includes the use of sound in any noncommunication process.

Ultrasonics. Ultrasonics is the technology of sound at frequencies above the audio range.

Note: Supersonics is the general subject covering phenomena associated with speed higher than the speed of sound (as in the case of aircraft and projectiles traveling faster than sound). This term was once used in acoustics synonymously with "ultrasonics"; such usage is now deprecated.

3a-9. Architectural Acoustics

Anechoic Room (Free-field Room). An anechoic room is one whose boundaries absorb effectively all the sound incident thereon, thereby affording essentially free-field conditions.

Dead Room. A dead room is a room that is characterized by an unusually large amount of sound absorption.

Decay Constant. The decay constant is the exponential power by which sound decays after the source is stopped (units sec^{-1}).

Note: If p_0 is the effective sound pressure at $t = 0$, $p(t)$ is the effective sound pressure at time t, and the two are related by

$$p(t) = p_0 e^{-kt}$$

then k is the decay constant.

Diffuse Sound Field. A diffuse sound field is one in which the time average of the mean-square sound pressure is everywhere the same and the flow of energy in all directions is equally probable.

Direct Sound Wave. A direct sound wave in an enclosure is a wave emitted from a source prior to the time it has undergone its first reflection from a boundary of the enclosure.

Note: Frequently, a sound wave is said to be direct if it contains reflections that have occurred from surfaces within about 0.05 sec after the sound was first emitted.

Echo. An echo is a wave that has been reflected or otherwise returned with sufficient magnitude and delay to be detected as a wave distinct from that directly transmitted. In architectural acoustics the word "echo" is generally restricted to mean "unwanted echo." The word "reflection" is generally used to mean "desired echo."

Equivalent Absorption Area of an Object or of a Surface. Area of a surface having a sound power absorption coefficient of unity that would absorb sound energy at the same rate as the object or the surface. In the case of a surface, the equivalent absorption area is the product of the area of the surface and its sound power absorption coefficient. In the case of an object in a given situation in a room, the equivalent absorption area is the increase of the equivalent absorption area produced in the room by the introduction of the object.

Eyring Coefficient. Equivalent sound absorption area attributed to a surface by the Eyring reverberation time formula, divided by the area of the surface.

Flutter Echo. A flutter echo is a rapid succession of reflected pulses resulting from a single initial pulse.

Live Room. A live room is a room that is characterized by an unusually small amount of sound absorption.

Mean Free Path. The mean free path for sound waves in an enclosure is the average distance sound travels between successive reflections in the enclosure.

Noise Reduction. In architectural acoustics, noise reduction generally is the difference between the effective sound pressure levels (in decibels) between the noise fields on opposite sides of a noise-reducing panel, with all sources of sound being on one side of the panel.

Random Incidence. If an object is in a diffuse sound field, the sound is said to strike the object at random incidence.

Rate of Decay. The rate of decay is the time rate at which the sound pressure level (or other stated characteristic) decreases at a given point and at a given time. A commonly used unit is the decibel per second.

Reflection. A reflection is an echo that occurs in combination with the direct sound or with other reflections or both to produce desired acoustical effects in a room, such as enhancement of the direct sound, reverberation, etc.

Reverberant Sound. Reverberant sound is that part of the sound in an enclosure that has undergone one or more reflections from the boundaries of the enclosure.

Reverberation Room. A reverberation room is a room having a long reverberation time, especially designed to make the sound field therein as diffuse as possible.

Reverberation Time. The reverberation time of an enclosure, for a sound of given frequency or frequency band, is the time after the source has been stopped that would be required for the sound pressure level in the enclosure to decrease by 60 dB.

Room Absorption. Room absorption is the sum of Sabine absorptions due to objects and surfaces in a room, and of dissipation in the medium within the room.

Room Constant. The room constant is equal to the product of the average absorption coefficient of the room and the total internal area of the room divided by the quantity one minus the average absorption coefficient.

Sabin. The sabin is a unit of absorption having the dimensions of square feet.

Note: The *metric sabin* has the dimension of square meters.

Sabine Absorption. Sabine absorption is that absorption defined by the Sabine reverberation time equation. Sabine absorption is equal to 24 times the volume of a room divided by the product of the reverberation time therein, the speed of sound, and the common logarithm of the Naperian base.

Note 1: The unit of absorption is the sabin when the unit of area is the square foot, or the metric sabin when the unit of area the square meter.

Sabine Coefficient. The Sabine coefficient of a surface, is the increase in Sabine absorption, due to introduction of the surface into a room, divided by the area of the surface.

Sound Absorption. Sound absorption is the property possessed by materials and objects of absorbing sound energy, due to either the propagation in a medium or the dissipation when sound strikes a surface.

Sound Power Absorption Coefficient. The SPAC at a given frequency and for specified conditions, of a surface, is the fraction of incident sound power not reflected from the surface. Unless otherwise specified, a diffuse sound field at the surface is to be understood.

Sound Power Reflection Coefficient. This coefficient of a surface, at a given frequency and for specified conditions, is the fraction of incident sound power reflected by the surface.

Sound Pressure Reflection Coefficient. This coefficient of a surface, at a given frequency and for specified conditions, is the fraction of incident sound pressure reflected by the surface.

Sound Reduction between Rooms. The sound reduction, in decibels, between two rooms is the amount by which the mean-square sound pressure level averaged throughout the source room exceeds the same level averaged throughout the receiving room.

Sound-transmission Coefficient. The sound-transmission coefficient of a partition is the fraction of incident sound transmitted through it. Unless otherwise specified, transmission of sound energy between two diffuse sound fields is assumed.

Sound-transmission Loss of a Partition. The sound-transmission loss of a partition is a measure of sound insulation. Expressed in decibels, it is ten times the logarithm to the base 10 of the reciprocal of the sound-transmission coefficient of the partition. Unless otherwise specified, the sound fields on both sides of the partition are assumed to be diffuse.

Statistical Absorption Coefficient. The SAC is the absorption coefficient measured or calculated with plane waves at randomly distributed angles of incidence.

3a-10. Hearing and Speech

Air Conduction. Air conduction is the process by which sound is conducted to the inner ear through the air in the outer ear canal as part of the pathway.

Articulation (Percent Articulation) and Intelligibility (Percent Intelligibility). Percent articulation or percent intelligibility of a communication system is the percentage of the speech units spoken by a talker or talkers that is correctly repeated, written down, or checked by a listener or listeners.

The word "articulation" is used when the units of speech material are meaningless syllables or fragments; the word "intelligibility" is used when the units of speech material are complete, meaningful words, phrases, or sentences.

Note 1: It is important to specify the type of speech material and the units into which it is analyzed for the purpose of computing the percentage. The units may be fundamental speech sounds, syllables, words, sentences, etc.

Note 2: The percent articulation or percent intelligibility is a property of the entire communication system: talker, transmission equipment or medium, and listener. Even when attention is focused upon one component of the system (e.g., a talker, a radio receiver), the other components of the system should be specified.

Note 3: The kind of speech material used is identified by an appropriate adjective in phrases such as "syllable articulation," "individual sound articulation," "vowel (or consonant) articulation," "monosyllabic word intelligibility," "discrete word intelligibility," "discrete sentence intelligibility."

Audiogram (Threshold Audiogram). An audiogram is a graph showing hearing loss as a function of frequency.

Auditory Sensation Area. The auditory sensation area is the region enclosed by the curves defining the threshold of pain and the threshold of audibility as functions of frequency.

Aural Critical Band. The aural critical band is that frequency band of sound, being a portion of a continuous-spectrum noise covering a wide band, that contains sound power equal to that of a simple (pure) tone centered in the critical band and just audible in the presence of the wide-band noise.

Note 1: By "just audible" is meant audible in a specified fraction of the trials.

Note 2: The use of the aural critical band to estimate masking should be limited to masking by noises having continuous spectra without excessive slopes or irregularities and to cases where masking exceeds 15 dB.

Note 3: In order to be just audible in a wide-band continuous noise, the level of a simple tone in decibels must exceed the spectrum level of the continuous noise (at the same frequency) by ten times the logarithm to the base 10 of the ratio of the critical bandwidth to unit bandwidth.

Aural Harmonic. An aural harmonic is a harmonic generated in the auditory mechanism.

Average Speech Power. The average speech power for a stated time interval is the average value of the instantaneous speech power over that interval.

Bone Conduction. Bone conduction is the process by which sound is conducted to the inner ear through the cranial bones.

Composite Noise-exposure Index. The sum of partial noise-exposure indices for all relevant sound levels over a working week (40 hours).

Difference Limen (Differential Threshold, Just Noticeable Difference). A difference limen is the increment in a stimulus that is just noticed in a specified fraction of the trials. The relative difference limen is the ratio of the difference limen to the absolute magnitude of the stimulus to which it is related.

Electrophonic Effect. Electrophonic effect is the sensation of hearing produced when an alternating current of suitable frequency and magnitude from an external source is passed through an animal.

Equivalent Continuous Sound Level. That sound level in dB(A) which, if present for 40 hours in one week, produces the same composite noise-exposure index as the various measured sound levels over one week.

Formant. A formant of a complex sound is a frequency range of the spectrum of the sound within which the partials have relatively large amplitudes.

Note: The central frequency within the formant range is called the formant frequency.

Hearing Level (Hearing Loss, Hearing Threshold Level). The hearing level of an ear at a specified frequency is the amount, in decibels, by which the threshold of audibility for that ear exceeds the standard audiometric threshold.

Note: See International Standards Organization (ISO) Recommendation R-389 (E) for standard reference zero for pure-tone audiometers.

Impairment of Hearing for Conversational Speech. The hearing of a subject for conversational speech is considered to be impaired if his hearing level is shifted by 25 dB or more, averaged over the test frequencies 500, 1,000 and 2,000 Hz compared with the threshold given in ISO Recommendation R-389, Standard Reference Zero for the Calibration of Pure-tone Audiometers. Percent of total impairment is generally taken to equal 1.5 times the number of decibels of hearing impairment for speech. Thus total impairment occurs at 92 dB average hearing level at the three test frequencies.

Instantaneous Speech Power. The instantaneous speech power is the rate at which sound energy is being radiated by a speech source at any given instant.

Level above Threshold (Sensation Level). The level above threshold of a sound is the pressure level of the sound in decibels above its threshold of audibility for the individual observer or for a specified group of individuals.

Loudness. Loudness is the intensive attribute of an auditory sensation in terms of which sounds may be ordered on a scale extending from soft to loud.

Note: Loudness depends primarily upon the sound pressure of the stimulus, but it also depends upon the frequency and waveform of the stimulus.

Loudness Contour. A loudness contour is a curve that shows the related values of sound pressure level and frequency required to produce a given loudness sensation for the typical listener.

Loudness Level. The loudness level of a sound, in phons, is numerically equal to the median sound pressure level, in decibels, relative to 0.0002 microbar, of a free progressive wave of frequency 1,000 Hz presented to listeners facing the source, which in a number of trials is judged by the listeners to be equally loud.

Note: The manner of listening to the unknown sound, which must be stated, may be considered one of the characteristics of that sound.

Loudness-level Contour. A loudness-level contour is a curve that shows the related values of sound pressure level and frequency required to produce a given loudness level for the typical listener.

Masking. (1) Masking is the process by which the threshold of audibility for one sound is raised by the presence of another (masking) sound. (2) Masking is the amount by which the threshold of audibility of a sound is raised by the presence of another (masking) sound.   The unit customarily used is the decibel.

Masking Audiogram.   A masking audiogram is a graphical presentation of the masking due to a stated noise.   This is plotted, in decibels, as a function of the frequency of the masked tone.

Mel.   The mel is a unit of pitch.   By definition, a simple tone of frequency 1,000 Hz, 40 dB above a listener's threshold, produces a pitch of 1,000 mels.   The pitch of any sound that is judged by the listener to be n times that of 1-mel tone is n mels.

Partial Noise-exposure Index.   An index determined by a sound level and its duration within a working week (40 hours).

Peak Speech Power.   The peak speech power is the maximum value of the instantaneous speech power within the time interval considered.

Phon.   The phon is the unit of loudness level, as specified in definition of loudness level.

Pitch.   Pitch is that attribute of auditory sensation in terms of which sounds may be ordered on a scale extending from low to high.   Pitch depends primarily upon the frequency of the sound stimulus, but it also depends upon the sound pressure and waveform of the stimulus.

Note: The pitch of a sound can be described by the frequency or frequency level of that simple tone having a specified sound pressure level which is judged by listeners to produce the same pitch.

Recognition Differential.   The recognition differential for a specified aural detection system is that amount by which the signal level exceeds the noise level presented to the ear when there is a 50 percent probability of detection of the signal.   The bandwidth of the system, within which signal and noise are presented and measured, must be specified.

Note 1: The signal and noise need not be measured actually at the ear but may be measured at any convenient point in the system, provided it is established that the difference between the signal level and noise level at that point is the same as the difference at the ear.
Note 2: The psychophysical method chosen for testing probability of detection must adequately control errors of commission as well as errors of omission.

Risk.   Risk of hearing impairment is the difference between the percentage of people with impaired hearing in a noise-exposed group and the percentage of people with impaired hearing in a non-noise-exposed (but otherwise equivalent) group.

Risk of Hearing Impairment for Conversational Speech.   The particular value of risk when the impairment of hearing in question is as described in Impairment of Hearing for Conversational Speech.

Sone.   The sone is a unit of loudness.   By definition, a simple tone of frequency 1,000 Hz, 40 dB above a listener's threshold, produces a loudness of 1 sone.   The loudness of any sound that is judged by the listener to be n times that of the 1-sone tone is n sones.

Note 1: A millisone is equal to 0.001 sone.
Note 2: The loudness scale is a relation between loudness and level above threshold (sensation level) for a particular listener.   In presenting data relating loudness in sones to sound pressure level or in averaging the loudness scales of several listeners, the thresholds (measured or assumed) should be specified.

Threshold of Audibility (Threshold of Detectability).   The threshold of audibility for a specified signal is the minimum effective sound pressure level of the signal that is capable of evoking an auditory sensation in a specified fraction of the trials.   The characteristics of the signal, the manner in which it is presented to the listener, and the point at which the sound pressure level is measured must be specified.

Note 1: Unless otherwise indicated, the ambient noise reaching the ears is assumed to be negligible.

Note 2: The threshold is usually given as a sound pressure level in decibels relative to 0.0002 microbar.

Note 3: Instead of the method of constant stimuli, which is implied by the phrase "a specified fraction of the trials," another psychophysical method (which should be specified) may be employed.

Threshold of Discomfort.   The threshold of discomfort for a specified signal is the minimum effective sound pressure level of that signal which, in a specified fraction of the trials, will stimulate the ear to a point at which the sensation of feeling becomes uncomfortable.

Threshold of Feeling (or Tickle).   The threshold of feeling (or tickle) for a specified signal is the minimum sound pressure level at the entrance to the external auditory canal which, in a specified fraction of the trials, will stimulate the ear to a point at which there is a sensation of feeling that is different from the sensation of hearing.

Threshold of Pain.   The threshold of pain for a specified signal is the minimum effective sound pressure level of that signal which, in a specified fraction of the trials, will stimulate the ear to a point at which the discomfort gives way to definite pain that is distinct from mere nonnoxious feeling of discomfort.

Timbre.   Timbre is that attribute of auditory sensation in terms of which a listener can judge that two sounds similarly presented and having the same loudness and pitch are dissimilar.

Note: Timbre depends primarily upon the spectrum of the stimulus, but it also depends upon the waveform, the sound pressure, the frequency location of the spectrum, and the temporal characteristics of the stimulus.

3a-11. Music

Cent.   The cent is the interval between two sounds having as a basic frequency ratio the twelve-hundredth root of 2.

Note: The interval in cents between any two frequencies is 1,200 times the logarithm to the base 2 of the frequency ratio.   Thus, 1,200 cents equal 12 equally tempered semitones equal 1 octave.

Complex Tone.   (1) A complex tone is a sound wave containing simple sinusoidal components of different frequencies.   (2) A complex tone is a sound sensation characterized by more than one pitch.

Equally Tempered Scale.   An equally tempered scale is a musical scale formed by a division of the octave into a number (usually 12) of equal intervals (see Table 3a-2).

TABLE 3a-2. EQUALLY TEMPERED INTERVALS

Name of interval	Frequency ratio	Cents
Unison................................	1:1	0
Minor second or semitone................	1.059463:1	100
Major second or whole tone..............	1.122462:1	200
Minor third............................	1.189207:1	300
Major third............................	1.259921:1	400
Perfect fourth.........................	1.334840:1	500
Augmented fourth; diminished fifth........	1.414214:1	600
Perfect fifth..........................	1.498307:1	700
Minor sixth............................	1.587401:1	800
Major sixth............................	1.681793:1	900
Minor seventh.........................	1.781797:1	1,000
Major seventh.........................	1.887749:1	1,100
Octave................................	2:1	1,200

Fundamental.   The fundamental is the component in a periodic wave corresponding to the fundamental frequency.

Harmonic.   A harmonic is a partial whose frequency is an integral multiple of the fundamental frequency.

Note: The term "overtone" has frequently been used in place of "harmonic," the nth harmonic being called the $(n - 1)$ overtone.   The term overtone is now deprecated in order to reduce ambiguity in the numbering of the components of a complex tone.

Harmonic Series of Sounds.   A harmonic series of sounds is one in which each basic frequency in the series is an integral multiple of a fundamental frequency.

Interval.   The interval between two sounds is their spac'ng in pitch or frequency, whichever is indicated by the context.   The frequency interval is expressed by the ratio of the frequencies or by a logarithm of this ratio.

Octave.   (1) An octave is the interval between two sounds having a basic frequency ratio of 2.   (2) An octave is the pitch interval between two tones such that one tone may be regarded as duplicating the basic musical import of the other tone at the nearest possible higher pitch.

Note 1: The interval in octaves between any two frequencies is the logarithm to the base 2 (for 3.322 times the logarithm to the base 10) of the frequency ratio.
Note 2: The frequency ratio corresponding to an octave pitch interval is approximately, but not always exactly, 2:1.

Overtone.   (1) An overtone is a physical component of a complex sound having a frequency higher than that of the basic frequency (see Partial below).   (2) An overtone is a component of a complex tone having a pitch higher than that of the fundamental pitch.

Note: The term overtone is now deprecated.   See Harmonic.

Partial.   (1) A partial is a physical component of a complex tone.   (2) A partial is a component of a sound sensation which can be distinguished as a simple tone that cannot be further analyzed by the ear and which contributes to the timbre of the complex sound.

Note 1: The frequency of a partial may be either higher or lower than the basic frequency and may or may not be an integral multiple or submultiple of the basic frequency.   If the frequency is not a multiple or submultiple, the partial is inharmonic.
Note 2: When a system is maintained in steady forced vibration at a basic frequency equal to one of the frequencies of the normal modes of vibration of the system, the partials in the resulting complex tone are not necessarily identical in frequency with those of the other normal modes of vibration.

Scale.   A musical scale is a series of notes (symbols, sensations, or stimuli) arranged from low to high by a specified scheme of intervals, suitable for musical purposes.

Semitone (Semit, Half Step).   The semitone is the interval between two sounds having a basic frequency ratio approximately the twelfth root of 2.

Note: The interval in equally tempered semits between any two frequencies is 12 times the logarithm to the base 2 (or 39.86 times the logarithm to the base 10) of the frequency ratio.

Simple Tone (Pure Tone).   (1) A simple tone is a sound wave the instantaneous sound pressure of which is a simple sinusoidal function of the time.   (2) A simple tone is a sound sensation characterized by its singleness of pitch.

Note: Whether a listener hears a tone as simple or complex (see Complex Tone) is dependent upon ability, experience, and listening attitude.

Standard Tuning Frequency (Standard Musical Pitch).   The standard tuning frequency is the frequency for the note A_4, namely, 440 Hz.

TABLE 3a-3. RELATIONS AMONG VARIOUS ACOUSTICAL UNITS

Quantity	Dimension	cgs unit	mks unit	Conversion factor*	British unit	Conversion factor†
Mass	M	gram	kilogram	10^{-3}	slug	6.854×10^{-5}
Velocity (linear)	LT^{-1}	cm per sec	meter per sec	10^{-2}	ft per sec	3.281×10^{-2}
Force	MLT^{-2}	dyne	newton	10^{-5}	lb weight	2.248×10^{-6}
Sound pressure	$ML^{-1}T^{-2}$	dyne per sq cm [microbar]	newton per sq meter	10^{-1}	lb per sq ft	2.089×10^{-3}
Volume velocity	$L^{3}T^{-1}$	cu cm per sec	cu meter per sec	10^{-6}	cubic foot per sec	3.531×10^{-5}
Sound energy	$ML^{2}T^{-2}$	erg	joule	10^{-7}	ft-lb	7.376×10^{-8}
Sound-energy density	$ML^{-1}T^{-2}$	erg per cu cm	joule per cu meter	10^{-1}	ft-lb per cu ft	2.089×10^{-3}
Sound-energy flux [sound power of source]	$ML^{2}T^{-3}$	erg per sec	watt	10^{-7}	ft-lb per sec	7.376×10^{-8}
Sound-energy-flux density [sound intensity]	MT^{-3}	erg per sec per sq cm	watt per sq meter	10^{-3}	(ft-lb per sec) per sq ft	6.847×10^{-5}
Mechanical impedance	MT^{-1}	mechanical ohm [dyne-sec per cm]	mks mechanical ohm [newton-sec per meter]	10^{-3}	lb-sec per ft	6.854×10^{-2}
Acoustic impedance [resistance, reactance]	$ML^{-4}T^{-1}$	acoustical ohm [dyne-sec per cm⁵] [dyne-sec per cm^5]	mks acoustical ohm [newton-sec per m^5]	10^{5}	(lb per sq ft) per (cu ft per sec)	59.61
Specific acoustic impedance	$ML^{-2}T^{-1}$	rayl [acoustical ohm × sq cm]	mks rayl [mks acoustical ohm × sq meter]	10	(lb per sq ft) per (ft per sec)	6.366×10^{-5}
Acoustic inertance	ML^{-4}	gram per cm to the fourth power	kilogram per meter to the fourth power	10^{5}	slug per (ft to the fourth power)	59.16
Acoustic stiffness	$ML^{-4}T^{-2}$	(gram per cm to the fourth power) per sq sec	(kilogram per meter to the fourth power) per sq sec	10^{5}	(slug per ft to the fourth power) per sq sec	59.16
Acoustic compliance	$M^{-1}L^{4}T^{2}$	(cm to the fifth power) per dyne	(meter to the fifth power) per newton	10^{5}	(ft to the fifth power) per lb	1.690×10^{-2}

* Multiply a magnitude expressed in cgs units by the tabulated conversion factor to obtain magnitude in mks units.

† Multiply a magnitude expressed in cgs units by the tabulated conversion factor to obtain magnitude in British units. These conversion factors were calculated on the basis of standard acceleration due to gravity.

Note: M, L, T represent mass, length, and time, respectively, in the sense of the theory of dimensions. Mks mechanical ohm, rayl, and mks rayl are proposed terms. Alternate terms and units are in square brackets.

Note: It is recommended that tuning and retuning of musical instruments be within an accuracy of plus or minus 0.5 Hz at the standard tuning frequency when the instruments are played where the ambient temperature is 22°C (71.6°F).

Tone.   (1) A tone is a sound wave capable of exciting an auditory sensation having pitch.   (2) A tone is a sound sensation having pitch.

Vibrato.   The vibrato is a family of tonal effects in music that depend upon periodic variations of one or more characteristics of the sound wave.

Note: When the particular characteristics are known, the term "vibrato" should be modified accordingly: e.g., frequency vibrato, amplitude vibrato, phase vibrato, and so forth.

3a-12. Acoustical Units

Acoustical Units.   In different sections of acoustics at least three systems of units are in common use: the centimeter-gram-second (cgs), the meter-kilogram-second (mks), and the Brit'sh. Table 3a-3 is provided to facilitate conversion from one system of units to another.

3b. Standard Letter Symbols and Conversion Factors for Acoustical Quantities

Bolt Beranek and Newman Inc.

Symbols

T	absolute temperature, degrees Kelvin
a	absorption, energy, acoustic, total in a room
α	absorption coefficient, energy
$\bar{\alpha}$	absorption coefficient, energy, average
Y_A	acoustic admittance (complex)
C_A	acoustic compliance
G_A	acoustic conductance
Z_A	acoustic impedance (complex)
M_A	acoustic mass (inertance)
P_A	acoustic power
X_A	acoustic reactance
R_A	acoustic resistance
B_A	acoustic susceptance
Y_A	admittance, acoustic (complex)
Y_E	admittance, electric (complex)
Y_M	admittance, mechanical (complex)
Y_R	admittance, rotational (complex)
Y_S	admittance, specific acoustic (complex)
A	amplitude of velocity potential
Ω	angle, solid
ϕ	angular displacement
ω	angular frequency $(2\pi f)$
k	angular wave number
f_A	antiresonance frequency
S	area (diaphragm, tube, room, or radiator)
p_s	atmospheric (static) pressure
α	attenuation constant (coefficient)
$\bar{\alpha}$	average absorption coefficient, energy
k	Boltzmann constant
C_E	capacitance, electrical
$\rho_0 c$	characteristic impedance
Q	charge, electrical
k	circular wave number
α	coefficient of absorption
C_A	compliance, acoustic
C_S	compliance, specific acoustic

C_M	compliance, mechanical
C_R	compliance, rotational
$\xi, \eta, \zeta; \xi_x, \xi_y, \xi_z$	components of the particle displacement in the x, y, z directions
$u, v, w; u_x, u_y, u_z$	components of the particle velocity in x, y, z directions
s	condensation
G_A	conductance, acoustic
G_E	conductance, electric
G_M	conductance, mechanical
G_R	conductance, rotational
G_S	conductance, specific acoustic
κ	conductivity, thermal
i	current, electric
q, U	current, volume (volume per second) (volume velocity)
δ	decay constant (damping coefficient)
dB	decibel
E, w	density, energy
ρ	density of the medium (instantaneous)
ρ_0	density of the medium (static)
ϵ	dielectric coefficient
Δ	dilatation
D_i	directivity index
R_θ	directivity ratio
ϕ	displacement, angular
ξ_x, x	displacement, linear
ξ	displacement, particle
X	displacement, volume
δ	dissipation (damping) coefficient (energy)
r	distance from source
s	distance, linear
μ	elasticity, shear
Y_E	electric admittance (complex)
C_E, C	electric capacitance
Q	electric charge
G_E	electric conductance
i	electric current
Z_E	electric impedance (complex)
P_E	electric power
X_E	electric reactance
R_E	electric resistance
ρ	electric resistivity
B_E	electric susceptance
e	electromotive force, voltage
J	energy
E, w	energy density
T, E_K	energy, kinetic
V, E_P	energy, potential
H	field strength, magnetic
m	flare coefficient in a horn
B	flux density, magnetic
f_M, F	force

f	frequency
ω	frequency, angular $(2\pi f)$
f_R	frequency, resonance
Z_A	impedance, acoustic (complex)
$\rho_0 c$	impedance, characteristic acoustic
Z_E	impedance, electric (complex)
Z_M	impedance, mechanical (complex)
Z_R	impedance, rotational (complex)
Z_S	impedance, specific acoustic (complex)
n	index of refraction
L	inductance
M_A	inertance, (acoustic mass)
I	inertia, moment of
I, J	intensity, sound
L_I	intensity level, decibels
ν	kinematic viscosity
T, E_K	kinetic energy (inductive energy)
σ	leakage coefficient, magnetic
l	length of a vibrating string, pipe, or rod
L	level in decibels, general
x, ξ	linear displacement
s	linear distance
Λ	logarithmic decrement
N	loudness, sones
L_N	loudness level, decibels or phons
H	magnetic field strength
Φ	magnetic flux
B	magnetic flux density
σ	magnetic leakage coefficient
$\mathfrak{F}$	magnetomotive force
K	magnetostriction constant
m, M_M	mass
M_A	mass, acoustic
M_S	mass, specific acoustic
Y_M	mechanical admittance
C_M	mechanical compliance
G_M	mechanical conductance
Z_M	mechanical impedance (complex)
P_M	mechanical power
X_M	mechanical reactance
R_M	mechanical resistance
B_M	mechanical susceptance
Y, E	modulus of elasticity
I	moment of inertia
L_{NR}	noise reduction, decibels
N	number of turns
ξ	particle displacement
$\xi, \eta, \zeta; \xi_x, \xi_y, \xi_z$	particle-displacement components in the x, y, z directions
u_a	particle velocity (average)

$u, v, w; u_x, u_y, u_z$	particle-velocity components in the x, y, z directions
u_i	particle velocity (instantaneous)
u_m	particle velocity (maximum)
u_p	particle velocity (peak)
u	particle velocity (rms)
P	perimeter
T	period $T = 1/f$
θ	phase angle
β	phase constant (coefficient)
f_{ij}, g_{ij}, d_{ij}	piezoelectric constants
σ	Poisson's ratio
Y, P	porosity (of an acoustical material)
V, E_P	potential energy (capacitive energy)
ϕ	potential velocity
P	power
P_A, W_A	power, acoustic
P_E	power, electric
P_M	power, mechanical
P_R	power, rotational
p_s	pressure, atmospheric (static)
p_a	pressure, sound (average)
p_i	pressure, sound (instantaneous)
p_m	pressure, sound (maximum)
p_p	pressure, sound (peak)
p	pressure, sound (rms)
$\gamma = \alpha + j\beta$	propagation constant (coefficient)
Q	quality factor
a	radius of a diaphragm, tube, or radiator
Q	ratio of reactance to resistance
γ	ratio of specific heats
X_A	reactance, acoustic
X_E	reactance, electric
X_M	reactance, mechanical
X_R	reactance, rotational
X_S	reactance, specific acoustic
r	reflection coefficient, energy
n	refraction, index of
τ	relaxation time
$\mathfrak{R}$	reluctance
R_A	resistance, acoustic
R_E	resistance, electric
R_M	resistance, mechanical
R_R	resistance, rotational
R_S	resistance, specific acoustic
ρ	resistivity, electrical
f_R	resonance frequency
T	reverberation time
R	room constant $\bar{a}S/(1 - \bar{a})$
Y_R	rotational admittance
C_R	rotational compliance
G_R	rotational conductance

Z_R	rotational impedance (complex)
P_R	rotational power
X_R	rotational reactance
R_R	rotational resistance
B_R	rotational susceptance
L_S	sensation level, decibels
μ	shear elasticity, shear modulus (modulus of rigidity)
A	simple source strength
Ω	solid angle
I, J	sound intensity
L_P, L_W	sound power level, decibels
p_a	sound pressure (average)
p_i	sound pressure (instantaneous)
p_M	sound pressure (maximum)
p_p	sound pressure (peak)
p	sound pressure (rms)
L_p	sound pressure level, decibels
A	source, simple, strength of
r	source, distance from
Y_S	specific acoustic admittance
C_S	specific acoustic compliance
G_S	specific acoustic conductance
Z_S	specific acoustic impedance (complex)
M_S	specific acoustic mass
X_S	specific acoustic reactance
R_S	specific acoustic resistance
γ	specific heats, ratio of
c	speed of sound
s	stiffness
A, U_0	strength of a simple source
B_A	susceptance, acoustic
B_E	susceptance, electric
B_M	susceptance, mechanical
B_R	susceptance, rotational
B_S	susceptance, specific acoustic
T	temperature, absolute, kelvins
F	tension (force) in a membrane or string
κ	thermal conductivity
t	thickness
t	time
τ	time, relaxation
T	time, reverberation
T	torque
a	total acoustical (energy) absorption in a room
τ	transmission coefficient, energy, barriers
L_{TL}	transmission loss
R	transmission loss of building structures, decibels
N	turns, number of
u	velocity
c	velocity of sound

ω	velocity, angular
u_a	velocity, particle (average)
u_i	velocity, particle (instantaneous)
u_m	velocity, particle (maximum)
u_p	velocity, particle (peak)
u	velocity, particle (rms)
ϕ	velocity potential
A	velocity potential amplitude
$q,\ U$	velocity, volume
η	viscosity, dissipative or frictional
ν	viscosity, kinematic
e	voltage, electromotive force
V	volume
$q,\ U$	volume current; volume velocity
X	volume displacement
$q,\ U$	volume velocity; volume current
λ	wavelength
k	wave number (phase constant),

$$\frac{\omega}{c} = \frac{2\pi f}{c} = \frac{2\pi}{\lambda} = k$$

w	width
J	work
$Y,\ E$	Young's modulus

TABLE 3b-1. CONVERSION FACTORS FOR ACOUSTICAL QUANTITIES

Multiply the number of	By	To obtain the number of	Conversely multiply by
Acoustic ohms	10^5	Mks acoustic ohms	10^{-5}
Atmospheres	406.80	Inches of water at 4°C	2.458×10^{-3}
Centimeters	10^{-2}	Meters	10^2
Cubic centimeters	10^{-6}	Cubic meters	10^6
Dynes	10^{-5}	Newtons	10^5
Dynes/cm²	10^{-1}	Newtons per square meter	10
Ergs	10^{-7}	Joules	10^7
Ergs per second	10^{-7}	Watts	10^7
Ergs per second/cm²	10^{-3}	Watts per square meter	10^3
Gauss	10^{-4}	Webers per square meter	10^4
Kilograms	10^3	Grams	10^{-3}
Mechanical ohms	10^{-3}	Mks mechanical ohms	10^3
Meters	10^2	Centimeters	10^{-2}
Microbars	10^{-1}	Newtons per square meter	10
Newtons	10^5	Dynes	10^{-5}
Newtons per square meter	10	Dynes per square centimeter	10^{-1}
Pounds per square foot	0.4882	Grams per square centimeter	2.0482
Rayls	10	Mks rayls	10^{-1}
Watts per square meter	10^{-4}	Watts per square centimeter	10^4
Webers per square centimeter	10^4	Gauss	10^{-4}

3c. Propagation of Sound in Fluids

FREDERICK V. HUNT

Harvard University

3c-1. Glossary of Symbols[1]

$a, a_1; a_i$	material coordinate (31); surface element (12)
$A; \mathbf{A}_1$	surface (12), attenuation per wavelength (76), Avogadro's number (95); first order vector potential
B	coefficient relating $\nabla\rho$ and ∇p (58)
$c, c_0; c^0, c^\infty$	speed of sound, reference speed (25); low- and high-frequency limit speeds (84)
c'	speed of thermal wave (78b)
C_p, C_v	specific heats at constant pressure, constant volume (14)
d_{ij}	rate of deformation tensor (9)
D	material differential operator (2)
E, F, G, H	algebraic abbreviations (74)
$E, E_k, E_I; E_{\text{diss}}$	energy densities per unit mass (60), (12); degraded component of internal energy (66)
$f, \mathbf{f}_v, f(\ \ ), f(h)$	frequency, sum of viscosity terms (62), "function of" (45), special tabulated function (75)
Δf_c	critical bandwidth (98)
$F_i, \mathbf{F}$	vector body force per unit mass (6)
$g(h)$	tabulated function (75)
h	material mass coordinate (37), argument of tabulated function (75), Planck's constant (89)
i, j, k	coordinate indexes (1)
I	average sound-energy-flux density = sound intensity (64)
j	designation of imaginary axis, $[e^{+j\omega t}]$ (69)
$\mathbf{J}$	sound-energy flux vector (54)
k, k_0	phase constant $= \omega/c = 2\pi/\lambda$, Boltzmann's constant (89), $k_0 = \omega/c_0 = 2\pi/\lambda_0$ (17)
$K; K_s, K_0, K_T$	elastic modulus $= -V(DP/DV)$ (25), material constant $= c^0/c^\infty$ (84); isentropic modulus, reference modulus, isothermal modulus
L	mean free path (86), a sum of linear dimensions (90)
M	peak particle-velocity Mach number $= \omega\xi_0/c_0$ (49), molecular weight (95)
n_V	total number of molecules per unit volume (95)
N	number of modes of vibration (90)
$O(\ \ )$	additive terms of indicated order of magnitude (76)

[1] Numbers indicate equation number in or near which quantity is defined.

p; p_1, p_2 incremental, or sound, pressure; first- and second-order sound pressures (25)

P, P_0; P_m, P_{th} total pressure (7), equilibrium or reference pressure (25); mean pressure (7), thermodynamic pressure (14)

P_1, P_2; $\mathcal{P}$ rms fundamental and second-harmonic pressure (49a); Prandtl number (72)

q, q_i; q heat flux vector (12); Stokes radiation coefficient (21b)

q; q^E, q^L exemplar of state or condition variable (39); superscript indicates function of spatial (E) variables, or material (L) variables (32b)

$\mathbf{R}$, R; R_1, R_2 vorticity $= \frac{1}{2}\nabla \times \mathbf{u}$ (11d), real part of complex impedance; first- and second-order components of vorticity (57)

s, s_1 specific entropy per unit mass (14), first-order condensation $= \rho_1/\rho_0$ (59)

S; S'; S_{irr} Stokes number $= \omega\eta/\rho_0c_0{}^2$ (72), total interior surface (90); frequency number for radiation $= \omega/\mathsf{q}$ (72); entropy generated irreversibly (15a)

t; t_{ij} time (2); stress tensor (6)

T absolute temperature (14)

$\mathbf{u}$, u_1; u_1, u_2, u_3 particle velocity (1); velocity components

$\mathbf{u}_1$, $\mathbf{u}_2$ first- and second-order components of particle velocity (25)

v; $\bar{v}$ specific volume $= \rho^{-1}$ (1); mean molecular velocity (86)

$\mathcal{U}$ viscosity number $= 2 + \eta'/\eta$ (10)

V; V_{ij} volume (1); residual stress tensor (7)

x_1, x_2, x_3 cartesian coordinates (1)

X; X' frequency number $= \omega\eta\mathcal{U}/\rho_0c_0{}^2$ (72), specific acoustic reactance (69); frequency number for relaxation (84)

Y thermoviscous number $= \kappa/\eta\mathcal{U}C_p$ (72)

z, Z specific acoustic impedance ratio (87), and impedance (69)

α; α_K, α_C attenuation constant (69); "Kirchhoff" and "classical" attenuation (79a,b)

β; β_{noise} coefficient of thermal expansion $= \rho(\partial v/\partial T)_P$ (22); spectrum level $= 10\log_{10}[d(p^2/p_0{}^2)/df]$ (98)

γ ratio of specific heats $= C_p/C_v$ (14)

δ; δ_{ij}; Δ finite increment (32); Kronecker delta (7); dilatation rate $= \nabla \cdot \mathbf{u}$ (4)

ϵ specific internal energy per unit mass (13)

η, η', η_B coefficient of shear viscosity (10), "second" or dilatational viscosity (10), bulk viscosity (10)

θ_1, θ_2 first- and second-order variational components of temperature (25)

κ thermal conductivity (21a)

λ; λ_0 wavelength $= c/f$ (47); $\lambda_0 = c_0/f$

ν, ν', ν_B kinematic viscosity coefficients (10) $= \eta/\rho$, etc.

ξ; ξ_t displacement of particle from equilibrium (31); partial derivative with respect to subscript variable (41b)

ρ, ρ_0; ρ_1, ρ_2 densities: total, equilibrium; first- and second-order variational components

τ_r, τ_v, τ_k relaxation times (83, 85)

φ_2; ϕ_η, ϕ_k scalar velocity potential (55); viscous and thermal dissipation functions (16, 18)

χ complex propagation constant $= \alpha + jk$ (69)

ψ functional relation (71)

ω; ω_r, ω_v, ω_k angular frequency $= 2\pi f$; relaxation angular frequencies (84)

∇, $\nabla\cdot$, $\nabla\times$ gradient, divergence, and curl operators

$\langle\ \rangle$ time average

3c-2. The Motion of Viscous Fluids. The motions of a fluid medium that comprise sound waves are governed by equations that include (1) a continuity equation expressing the conservation of mass, (2) a force equation expressing the conservation of momentum, (3) a heat-exchange equation expressing the conservation of energy, and (4) one or more defining equations expressing the constitutive relations that characterize the medium and its response to thermal or mechanical stress. These equations will first be presented in their complete exact form in order to provide a rigorous point of departure for the approximations that must ultimately be made in formulating the linearized, or small-signal, acoustic equations.

The transformation properties of these equations can be indicated by writing them in either vectorial or tensorial form, and both forms will be exhibited in order to facilitate contacts with the rich literature dealing with the motion of fluids.[1]

Cartesian spatial coordinates will be designated x_1, x_2, x_3, and the vector velocity of a material particle will be identified as $\mathbf{u}$ with components u_1, u_2, u_3. These will also be written as x_i and u_i, where it is implied that the subscript i, j, or k takes on successively the values 1, 2, 3. The term "material particle" denotes a finite mass element of the medium small enough for the values assumed by the state variables at every interior point of the particle not to differ significantly from the values they have at the interior reference point whose coordinates "locate" the particle.

Equation of Continuity. The conservation of mass requires that $\rho V = \rho_0 V_0$, where ρ_0 and V_0 are initial and ρ and V are subsequent values assumed by the density and volume of a particular material element of the medium. It follows that

$$\rho\, DV + V\, D\rho = 0 \qquad \frac{DV}{V} = -\frac{D\rho}{\rho} \tag{3c-1}$$

If $\rho_0 V_0$ is set equal to 1, V_0 becomes the *specific volume*, $v \equiv 1/\rho$, whence the relation between the total logarithmic time derivatives of v and ρ is

$$\frac{1}{v}\frac{Dv}{Dt} = -\frac{1}{\rho}\frac{D\rho}{Dt} = \frac{D\log v}{Dt} = -\frac{D\log\rho}{Dt} \tag{3c-2}$$

where $D(\ )/Dt$ denotes the "material" derivative, i.e., one that follows the motion of a material "particle" of the medium relative to a fixed spatial coordinate system, and is defined by

$$\frac{D(\ )}{Dt} \equiv \frac{\partial(\ )}{\partial t} + \mathbf{u}\cdot\mathbf{grad}\ (\ ) \equiv \frac{\partial(\ )}{\partial t} + u_i\frac{\partial(\ )}{\partial x_i} \tag{3c-3}$$

Analysis of the rate of deformation of a volume element yields the kinematical relation

$$\frac{1}{v}\frac{Dv}{Dt} = \operatorname{div}\mathbf{u} \equiv \Delta = \frac{\partial u_i}{\partial x_i} \tag{3c-4}$$

where Δ is the *dilatation rate*. Note that in the last terms of (3c-3) and (3c-4) summation is implied over all the allowable values of the subscript index. Equations (3c-2), (3c-3), and (3c-4) can be combined to yield the following equivalent forms of Euler's *continuity equation:*

[1] A definitive restatement of the classical-continuum point of view, with critical comments on more than 800 bibliographical references, has been given by C. Truesdell, The Mechanical Foundations of Elasticity and Fluid Dynamics, *J. Rational Mechanics and Analysis* **1**, 125–300 (January and April, 1952), and Corrections and Additions . . . , *J. Rational Mechanics and Analysis* **2**, 593–616 (July, 1953). See also Lamb, "Hydrodynamics," 6th ed., Dover Publications, New York, 1945; Rayleigh, "Theory of Sound," 2d ed., rev., Dover Publications, New York, 1945; and L. Howarth, ed., "Modern Developments in Fluid Dynamics," vol. I, chap. III, Oxford University Press, New York, 1953.

$$\frac{D\rho}{Dt} + \rho \frac{\partial u_i}{\partial x_i} = \frac{\partial \rho}{\partial t} + u_i \frac{\partial \rho}{\partial x_i} + \rho \frac{\partial u_i}{\partial x_i} = \frac{D\rho}{Dt} + \rho \operatorname{div} \mathbf{u} = 0$$

$$= \frac{1}{\rho}\frac{D\rho}{Dt} + \Delta = \frac{\partial \rho}{\partial t} + \mathbf{u} \cdot \operatorname{grad} \rho + \rho \operatorname{div} \mathbf{u}$$

$$= \frac{\partial \rho}{\partial t} + \mathbf{u} \cdot \nabla \rho + \rho \nabla \cdot \mathbf{u} = \frac{\partial \rho}{\partial t} + \nabla \cdot (\rho \mathbf{u}) \tag{3c-5}$$

In the last line of (3c-5), the Gibbs-Hamilton notation has been used for the differential vector operators, $\nabla \equiv \operatorname{grad}$; $\nabla \cdot \equiv \operatorname{div}$; $\nabla \times \equiv \operatorname{curl}$.

Force Equation. The linear-momentum principle can be stated in terms of Cauchy's first law of motion,

$$\rho \frac{Du_i}{Dt} = \rho F_i + \frac{\partial t_{ij}}{\partial x_j} \tag{3c-6}$$

where the vector F_i is an extraneous body force per unit mass, and where t_{ij} is a second-rank *stress tensor* that represents the net mechanical action of contiguous material on a volume element of the medium due to the actual forces of material continuity. For an isotropic medium in which the stress is a linear function of the rate of deformation, as here assumed, the stress tensor can be resolved arbitrarily as the sum of a scalar, or hydrostatic, pressure function P and a residual stress tensor V_{ij} defined by

$$t_{ij} = -P\delta_{ij} + V_{ij} \qquad t_{ij} = t_{ji} \tag{3c-7}$$

where δ_{ij} is the Kronecker delta which equals unity if $i = j$, but is zero otherwise. Unless V_{ii} vanishes, P is *not* identical with the mean pressure, $P_m = -\frac{1}{3}t_{ii}$. The resolution given by (3c-7) is both unique and useful, however, if P is made equal to the thermodynamic pressure P_{th} defined below. Then the residual stress tensor is given, to a first approximation, by the linear terms of an expansion in powers of the viscosity coefficients,

$$V_{ij} = \eta' d_{kk}\delta_{ij} + 2\eta d_{ij} \qquad V_{ij} = V_{ji} \tag{3c-8}$$

in which d_{ij} is the *rate of deformation* tensor defined by

$$d_{ij} = \frac{1}{2}\left(\frac{\partial u_i}{\partial x_j} + \frac{\partial u_j}{\partial x_i}\right) \tag{3c-9}$$

and where η is the "first," or conventional shear, viscosity coefficient. In accordance with current proposals for standardization, η' replaces λ, the symbol used by Stokes, Rayleigh, Lamb, et al., to designate the "second," or dilatational, viscosity coefficient. The term "bulk" viscosity is reserved for $(\lambda + \frac{2}{3}\mu) \to (\eta' + \frac{2}{3}\eta)$, the linear combination of coefficients that vanishes when the *Stokes relation* holds. Thus, $\eta \equiv$ first, or shear, viscosity; $\eta' \equiv$ second, or dilatational, viscosity; $\eta_B \equiv \eta' + \frac{2}{3}\eta =$ bulk viscosity; $\nu \equiv \eta/\rho$; $\nu' \equiv \eta'/\rho$; $\nu_B \equiv \eta_B/\rho$ (kinematic viscosities);

$$(\lambda + 2\mu) \to \eta' + 2\eta = \eta_B + \frac{4}{3}\eta = \eta\left(\frac{4}{3} + \frac{\eta_B}{\eta}\right) = \eta\mathcal{U} \tag{3c-10}$$

$$\mathcal{U} \equiv \frac{4}{3} + \frac{\eta_B}{\eta} = 2 + \frac{\eta'}{\eta} \equiv \text{viscosity number}$$

Putting (3c-7), (3c-8), (3c-9) into (3c-6) yields the vector *force equation* in the following equivalent forms:

$$\rho \frac{\partial u_i}{\partial t} + \rho u_j \frac{\partial u_i}{\partial x_j} = \rho F_i - \frac{\partial P}{\partial x_i} + \frac{\partial}{\partial x_j}\left(\eta' d_{kk}\delta_{ij} + 2\eta d_{ij}\right)$$

$$= \rho F_i - \frac{\partial P}{\partial x_i} + \eta' \frac{\partial^2 u_k}{\partial x_i \partial x_k} + \eta \frac{\partial}{\partial x_j}\left(\frac{\partial u_i}{\partial x_j} + \frac{\partial u_j}{\partial x_i}\right)$$

$$+ \frac{\partial u_k}{\partial x_k}\frac{\partial \eta'}{\partial x_i} + \frac{\partial u_i}{\partial x_j}\frac{\partial \eta}{\partial x_j} + \frac{\partial u_j}{\partial x_i}\frac{\partial \eta}{\partial x_j} \tag{3c-11a}$$

$$\rho \frac{D\mathbf{u}}{Dt} = \rho\mathbf{F} - \text{grad } P + (\eta' + \eta) \text{ grad (div } \mathbf{u}) + \eta\nabla^2(\mathbf{u})$$
$$+ (\text{div } \mathbf{u}) \text{ grad } \eta' + 2 (\text{grad } \eta \cdot \text{grad}) \mathbf{u} + \text{grad } \eta \times \text{curl } \mathbf{u} \quad (3c\text{-}11b)$$

$$\rho \frac{\partial\mathbf{u}}{\partial t} = \rho\mathbf{F} - \rho(\mathbf{u} \cdot \nabla)\mathbf{u} - \nabla P + (\eta' + 2\eta)\nabla(\nabla \cdot \mathbf{u}) - \eta\nabla \times (\nabla \times \mathbf{u})$$
$$+ (\nabla \cdot \mathbf{u})\nabla\eta' + 2(\nabla\eta \cdot \nabla)\mathbf{u} + \nabla\eta \times (\nabla \times \mathbf{u}) \quad (3c\text{-}11c)$$

The vorticity, defined by $\mathbf{R} = \frac{1}{2} \text{ curl } \mathbf{u} = \frac{1}{2}(\nabla \times \mathbf{u})$, and the dilatation rate, $\Delta \equiv \nabla \cdot \mathbf{u}$, can be introduced as useful abbreviations. A somewhat more symmetrical expression in terms of the mass transport velocity $\rho\mathbf{u}$ is obtained if the last form of the continuity equation (3c-5) is multiplied by $\mathbf{u}$ and added to (3c-11c), giving

$$\frac{\partial(\rho\mathbf{u})}{\partial t} + \mathbf{u}(\nabla \cdot \rho\mathbf{u}) + (\rho\mathbf{u} \cdot \nabla)\mathbf{u} = \rho\mathbf{F} - \nabla P + \eta \mathcal{U}\nabla\Delta - 2\eta\nabla \times \mathbf{R} + \Delta\nabla\eta'$$
$$+ 2(\nabla\eta \cdot \nabla)\mathbf{u} + 2\nabla\eta \times \mathbf{R} \quad (3c\text{-}11d)$$

These equations reduce to the so-called *Navier-Stokes equations* when it is assumed that η and η' are constant ($\nabla\eta = \nabla\eta' = 0$) and that the Stokes relation holds ($\eta_B = 0$, $\mathcal{U} = \frac{4}{3}$); and still further simplification follows if the motion is assumed irrotational so that $\mathbf{R} = 0$. If the viscosity coefficients are to be regarded as functions of one or more of the state variables, however, the gradients of the η's must be retained so that the implicit functional dependence can be introduced by writing, for example, $\nabla\eta = (\partial\eta/\partial T)\nabla T + \cdots$.

Energy Relations and Equations of State. The conservation of energy requires that the following power equation be satisfied:

$$\frac{D(E_k + E_I)}{Dt} = \int_V \rho F_i u_i \, dV + \int_A t_{ij} u_j \, da_i - \int_A q_i \, da_i \quad (3c\text{-}12)$$

where E_k is the kinetic energy associated with the material velocity, E_I is the total internal energy, V is a volume bounded by the surface A, da_i is the projection of a surface element of A on the plane normal to the $+x_i$ axis, F_i is the extraneous body force (per unit mass), and q_i is the total heat flux vector (mechanical units). After the surface integrals are converted to volume integrals by using the divergence theorem, and with the help of (3c-6), this equation reduces to the Fourier–Kirchhoff–C. Neumann[1] energy equation,

$$\rho \frac{D\epsilon}{Dt} = t_{ij}d_{ij} - \frac{\partial q_i}{\partial x_i} \quad (3c\text{-}13)$$

where ϵ is the local value of the specific internal energy (per unit mass) defined through $E_I = \int_V \rho\epsilon \, dV$. It is now postulated that the state of the fluid is completely specified by ϵ and two other local state variables, which can be taken as the specific entropy s (per unit mass) and the specific volume $v = \rho^{-1}$, in terms of which the thermodynamic pressure and temperature and the specific heats can be defined by

$$\epsilon = \epsilon(s,v) \qquad P_{\text{th}} \equiv -\left(\frac{\partial\epsilon}{\partial v}\right)_s \qquad T \equiv \left(\frac{\partial\epsilon}{\partial s}\right)_v$$
$$C_p \equiv T\left(\frac{\partial s}{\partial T}\right)_p \qquad C_v \equiv T\left(\frac{\partial s}{\partial T}\right)_v \qquad \gamma \equiv \frac{C_p}{C_v} \quad (3c\text{-}14)$$

The second law of thermodynamics can be introduced in the form of an equality, which replaces the classical Clausius-Duhem inequality, through the expedient of accounting explicitly for the creation of entropy S_{irr} (per unit volume) by irreversible

[1] See footnote, p. 3-39.

dissipative processes;[1] thus

$$\frac{D}{Dt} \int_V \rho s \, dV = - \int_A \frac{q_i}{T} \, da_i + \int_V \frac{DS_{irr}}{Dt} \, dV \tag{3c-15a}$$

This relation states that the increase of entropy in a material element is accounted for by the influx of heat and by the irreversible production of entropy within the element. The left-hand side of (3c-15a) can also be written, with the help of the continuity relation, as $\int_V \rho(Ds/Dt) \, dV$. Then, after converting the surface integral to a volume integral, the second law can be given in differential form as

$$\begin{aligned}
\rho \frac{Ds}{Dt} &= - \frac{\partial}{\partial x_i} \frac{q_i}{T} + \frac{DS_{irr}}{Dt} \\
&= - \frac{1}{T} \frac{\partial q_i}{\partial x_i} + \frac{q_i}{T^2} \frac{\partial T}{\partial x_i} + \frac{DS_{irr}}{Dt}
\end{aligned} \tag{3c-15b}$$

A thermal-dissipation function ϕ_k can be defined by

$$\phi_\kappa = - \frac{q_i}{T} \frac{\partial T}{\partial x_i} \tag{3c-16}$$

whereupon multiplying (3c-15b) by T yields the second-law equality in the form

$$\rho T \frac{Ds}{Dt} = - \frac{\partial q_i}{\partial x_i} - \phi_\kappa + T \frac{DS_{irr}}{Dt} \tag{3c-15c}$$

Taking the material derivative of the basic equation of state $(3c-14_1)$ (where the subscript added to an equation number indicates the serial number of the equality sign to which reference is made when several relations are grouped under one marginal identification number), introducing the definitions for P_{th} and T, multiplying by ρ, and using (3c-4), gives

$$\rho T \frac{Ds}{Dt} = \rho \frac{D\epsilon}{Dt} + P_{th}\Delta \tag{3c-17}$$

The energy equation (3c-13) can be recast, using (3c-7) and (3c-9), in the form

$$\rho \frac{D\epsilon}{Dt} + P\Delta + \frac{\partial q_i}{\partial x_i} = V_{ij}d_{ij} = \phi_\eta \tag{3c-18}$$

in which $V_{ij}d_{ij}$, the dissipative component of the stress power $t_{ij}d_{ij}$, is defined as the viscous dissipation function ϕ_η. The usefulness of specifying the arbitrary scalar in (3c-7) as the thermodynamic pressure, so that $P \cdot = P_{th}$, becomes apparent when $\rho \cdot D\epsilon/Dt$ is eliminated between (3c-18) and (3c-17), giving

$$\begin{aligned}
\rho T \frac{Ds}{Dt} &= (P_{th} - P)\Delta + \phi_\eta - \frac{\partial q_i}{\partial x_i} \\
&= \phi_\eta - \frac{\partial q_i}{\partial x_i}
\end{aligned} \tag{3c-19}$$

The viscous dissipation function (dissipated energy per unit volume) is thus seen to account for either an efflux of heat or an increase of entropy. Subtracting (3c-19) from (3c-15c) then allows the rate of irreversible production of entropy to be evaluated directly in terms of the two dissipation functions,

$$T \frac{DS_{irr}}{Dt} = \phi_\eta + \phi_\kappa \tag{3c-20}$$

The total heat flux vector q_i, whose divergence is the energy transferred *away* from the volume element, must account for energy transport by either conduction or radi-

[1] Tolman and Fine, *Revs. Modern Phys.* **20**, 51–77 (1948).

ation. The part due to conduction is given by the Fourier relation, which serves also to define the heat conductivity κ,

$$(q_i)_{\text{cond}} = -\kappa \frac{\partial T}{\partial x_i}$$

$$\frac{\partial (q_i)_{\text{cond}}}{\partial x_i} = -\frac{\partial (\kappa \partial T/\partial x_i)}{\partial x_i} = -\kappa \frac{\partial^2 T}{\partial x_i{}^2} - \frac{\partial T}{\partial x_i} \frac{\partial \kappa}{\partial x_i} \tag{3c-21a}$$

The last term, containing the gradient of κ, must be retained if implicit dependence of κ on the state variables is to be represented. On the other hand, if κ is assumed to be constant, (3c-21a) reduces to the more familiar form

$$\boldsymbol{\nabla} \cdot \mathbf{q}_{\text{cond}} = -\kappa \boldsymbol{\nabla}^2 T$$

The component of heat flux due to radiation can be approximated, for small temperature differences, by Newton's law of cooling,

$$\frac{\partial (q_i)_{\text{rad}}}{\partial x_i} = \rho C_v \mathfrak{q}(T - T_0) = \boldsymbol{\nabla} \cdot \mathbf{q}_{\text{rad}} \tag{3c-21b}$$

where $(T - T_0)$ is the local temperature excess and $\mathfrak{q}$ is a radiation coefficient introduced by Stokes.[1] The foregoing thermal relations can be combined with the equations of continuity and momentum more readily if the term $T(Ds/Dt)$ appearing in (3c-19) is expressed in terms of the variables $\mathbf{u}$, v, and T. The defining equations (3c-14) establish that $P = P(v,s)$ and $T = T(v,s)$, from which it follows that one may also write $s = s(T,v)$ or $s = s(T,P)$. Using both of the latter leads, after some manipulation,[2] to the identity

$$\rho T \frac{Ds}{Dt} = \rho C_v \left[(\gamma - 1) \frac{\Delta}{\beta} + \frac{DT}{Dt} \right] \tag{3c-22}$$

in which β is the coefficient of thermal expansion, $\beta \equiv \rho(\partial v/\partial T)_P$. After (3c-22) and (3c-21) are combined with (3c-19), the energy equation can be written in the alternate forms

$$\frac{\rho C_v DT}{Dt} + \rho C_v \frac{\gamma - 1}{\beta} \frac{\partial u_i}{\partial x_i} + \frac{\partial q_i}{\partial x_i} - \phi_\eta = 0$$

$$\rho C_v \left(\frac{\partial T}{\partial t} + \mathbf{u} \cdot \boldsymbol{\nabla} T \right) + \frac{\rho(C_p - C_v)}{\beta} \Delta - \boldsymbol{\nabla} \cdot (\kappa \boldsymbol{\nabla} T) + \rho C_v \mathfrak{q}(T - T_0) - \phi_\eta = 0 \tag{3c-23}$$

$$\frac{\partial T}{\partial t} + \mathbf{u} \cdot \boldsymbol{\nabla} T + \frac{(\gamma - 1)}{\beta} \Delta - \frac{\kappa}{\rho C_v} \boldsymbol{\nabla}^2 T - \frac{\boldsymbol{\nabla} T \cdot \boldsymbol{\nabla} \kappa}{\rho C_v} + \mathfrak{q}(T - T_0) - \frac{\phi_\eta}{\rho C_v} = 0$$

The viscous dissipation function ϕ_η can be evaluated, with the aid of (3c-8) and (3c-9) in the explicit form

$$\phi_\eta = V_{ij} d_{ji} = \eta' d_{kk} d_{ii} + 2\eta d_{ij} d_{ji}$$

$$= \eta_B \Delta^2 + \frac{4}{3} \eta \left[\left(\frac{\partial u_1}{\partial x_1} \right)^2 + \left(\frac{\partial u_2}{\partial x_2} \right)^2 + \left(\frac{\partial u_3}{\partial x_3} \right)^2 - \frac{\partial u_1}{\partial x_1} \frac{\partial u_2}{\partial x_2} - \frac{\partial u_2}{\partial x_2} \frac{\partial u_3}{\partial x_3} - \frac{\partial u_3}{\partial x_3} \frac{\partial u_1}{\partial x_1} \right]$$

$$+ \eta \left[\left(\frac{\partial u_1}{\partial x_2} + \frac{\partial u_2}{\partial x_1} \right)^2 + \left(\frac{\partial u_2}{\partial x_3} + \frac{\partial u_3}{\partial x_2} \right)^2 + \left(\frac{\partial u_3}{\partial x_1} + \frac{\partial u_1}{\partial x_3} \right)^2 \right] \tag{3c-24a}$$

The thermal dissipation function ϕ_κ due to heat conduction can be evaluated, with the aid of (3c-16) and (3c-21a), in the form

$$\phi_\kappa = -\frac{q_i}{T} \frac{\partial T}{\partial x_i} = +\frac{\kappa}{T} \left(\frac{\partial T}{\partial x_i} \right)^2 = \frac{\kappa}{T} (\boldsymbol{\nabla} T)^2 \tag{3c-24b}$$

It does not appear explicitly in (3c-23), but it is there implicitly as a consequence of the heat-transfer processes described by (3c-23).

[1] *Phil. Mag.* (4) **1**, 305–317 (1851).

[2] See, for example, Zemansky, "Heat and Thermodynamics," 3d ed., pp. 246–255, McGraw-Hill Book Company, New York, 1951.

Summary of Assumptions. The fluid considered is assumed to be continuous except at boundaries or interfaces, locally homogeneous and isotropic when at rest, viscous, thermally conducting, and chemically inert, and its local thermodynamic condition is assumed to be completely determined by specifying three "state" variables, any two of which determine the third uniquely through an equation of state. No structural or thermal "relaxation" mechanism has been presumed up to this point in the analysis, except to the extent that ordinary heat conduction and viscous losses may be described in such terms. Local thermodynamic reversibility has been assumed in using conventional thermodynamic identities based on the second law, but the irreversible production of entropy by dissipative processes has been accounted for explicitly. It is also assumed that the stress tensor is a linear function of the rate of deformation, and that the tractions due to viscosity can be represented by the linear terms of an expansion in powers of the viscosity coefficients. The viscosity and heat-exchange parameters of the fluid η, η', κ, and q may depend in any continuous way on the state variables and hence may be implicit functions of time and the spatial coordinates. Within the scope thus defined the equations given are exact.

The functional dependence on time and the spatial coordinates of the condition and motion variables P, T, ρ, and u can be evaluated, in a formal sense at least, by solving the set of four simultaneous equations connecting these variables [Eqs. (3c-5), (3c-11), (3c-23), and (3c-15) or one of its alternates]. No general solution of these complete equations has been given, however, and one or another of the least important terms is usually omitted in order to render the equations tractable for dealing with specific problems.

3c-3. The Small-signal Acoustic Equations. The physical theory of sound waves deals with systematic motions of a material medium relative to an equilibrium state and thus comprises the variational aspects of elasticity and fluid dynamics. Such perturbations of state can be described by incremental, or acoustic, variables and approximate equations governing them can be obtained by arbitrarily "linearizing" the general equations of motion. These results, as well as higher-order approximations, can be derived in an orderly way by invoking a modified perturbation analysis.[1] This consists of replacing the dependent variables appearing in (3c-5), (3c-11), and (3c-23) by the sum of their equilibrium or zero-order values and their first- and second-order variational components, and then forming the separate equations that must be satisfied by the variables of each order. Two of the composite state variables, for example ρ and T, can be defined arbitrarily, whereupon the third, P, is determined by the functional equation of state. These definitions, some self-evident manipulations, and the subscript notation identifying the orders can be exhibited as follows:

$$\rho \equiv \rho_0 + \rho_1 + \rho_2 \qquad T \equiv T_0 + \theta_1 + \theta_2$$
$$\nabla\rho = \nabla\rho_1 + \nabla\rho_2 \qquad \nabla T = \nabla\theta_1 + \nabla\theta_2$$
$$P(\rho,T) \equiv P_0(\rho_0,T_0) + p_1 + p_2$$

$$p_1 + p_2 = \left[\left(\frac{\partial P}{\partial \rho}\right)_T\right]_0 (\rho - \rho_0) + \left[\left(\frac{\partial P}{\partial T}\right)_\rho\right]_0 (T - T_0) + \cdots \qquad (3c\text{-}25)$$

$$K = K_T \equiv \rho\left(\frac{\partial P}{\partial \rho}\right)_T \qquad \beta \equiv -\frac{1}{\rho}\left(\frac{\partial \rho}{\partial T}\right)_P \qquad c_0{}^2 \equiv \left[\left(\frac{\partial P}{\partial \rho}\right)_s\right]_0 \equiv \frac{(K_s)_0}{\rho_0}$$

$$\gamma = \frac{K_s}{K_T} = \frac{C_p}{C_v}$$

$$p_1 = \frac{c_0{}^2}{\gamma}(\rho_1 + \beta_0\rho_0\theta_1) \qquad p_2 = \frac{c_0{}^2}{\gamma}(\rho_2 + \beta_0\rho_0\theta_2)$$

$$\mathbf{u} \equiv 0 + \mathbf{u}_1 + \mathbf{u}_2 \qquad \nabla\cdot\mathbf{u} \equiv \Delta \equiv \Delta_1 + \Delta_2 = \nabla\cdot\mathbf{u}_1 + \nabla\cdot\mathbf{u}_2$$
$$\rho\mathbf{u} = [\rho_0\mathbf{u}_1]_1 + [\rho_1\mathbf{u}_1 + \rho_0\mathbf{u}_2]_2 + \cdots$$
$$\nabla\cdot(\rho\mathbf{u}) = [\rho_0\nabla\cdot\mathbf{u}_1]_1 + [\rho_1\nabla\cdot\mathbf{u}_1 + \mathbf{u}_1\cdot\nabla\rho_1 + \rho_0\nabla\cdot\mathbf{u}_2]_2 + \cdots$$

[1] Eckart, *Phys. Rev.* **73**, 68–76 (1948).

Terms containing $\nabla \rho_0$ have been omitted in writing out $\nabla \cdot (\rho \mathbf{u})$, on the assumption that ρ_0, T_0, and P_0 are constant and $\mathbf{u}_0 = 0$. The reference state need not be so restricted to one of static equilibrium provided its time and space rates of change are presumed small in comparison with the corresponding change rates of the acoustic variables. The extraneous body force $\mathbf{F}$ will also be omitted hereafter; it would become important in cases involving electromagnetic interaction, but it usually derives from a gravitation potential and affects primarily the equilibrium configuration.[1] Little generality is sacrificed by omitting $\mathbf{F}$ and assuming a static reference, moreover, since the basic equations characterize directly the equilibrium condition and since the "cross-modulation" effects brought in by nonlinearity are dealt with adequately through second- or higher-order approximations.

Notice that the foregoing represents a mathematical-approximation procedure that is concerned only with the *precision* achieved in interpreting the content of the basic equations. The *accuracy* with which the basic equations themselves delineate the behavior of a real fluid is an entirely different question that must be considered independently on its own merits. It follows that, while good judgment may restrain the effort, there is no impropriety involved in pursuing higher-order solutions of the acoustic equations, even though the equations themselves may embody first-order approximations to reality such as that represented by assuming linear dependence on the viscosity coefficients and the deformation rate.

When the appropriate relations from (3c-25) are substituted in (3c-5), (3c-11), and (3c-23), the *first-order acoustic equations* can be separated out in the form

$$\frac{\partial \rho_1}{\partial t} + \rho_0 (\nabla \cdot \mathbf{u}_1) = 0 \tag{3c-26a}$$

$$\rho_0 \frac{\partial \mathbf{u}_1}{\partial t} + \frac{c_0^2}{\gamma} \left(1 + \beta_0 \rho_0 \frac{\nabla \theta_1}{\nabla \rho_1} \right) \nabla \rho_1 - (\eta_0 \mathcal{U}) \nabla (\nabla \cdot \mathbf{u}_1) + \eta_0 \nabla \times (\nabla \times \mathbf{u}_1) = 0 \tag{3c-26b}$$

$$\rho_0 C_v \frac{\partial \theta_1}{\partial t} + \frac{\rho_0 C_v (\gamma - 1)}{\beta_0} (\nabla \cdot \mathbf{u}_1) - \kappa_0 \nabla^2 \theta_1 + \rho_0 C_v \mathfrak{q} \theta_1 = 0 \tag{3c-26c}$$

Inasmuch as the first-order effects of both shear and dilatational viscosity and of heat conduction and radiation have been included, these equations comprehend a *viscothermal theory* of small-signal sound waves. The sound absorption and velocity dispersion predicted by this theory are discussed below. Note especially that taking heat exchange into account explicitly by including (3c-26c) has precluded the conventional adiabatic assumption and denied the simplifying assumption that $P = P(\rho)$.

Adiabatic behavior would be assured, on the other hand, if it were assumed at the outset that $\kappa = \mathfrak{q} = 0$, but the behavior would *not* at the same time be strictly isentropic so long as irreversible viscous losses are still present and accounted for. The difference between adiabatic and isentropic behavior in this case is of second order, however, as indicated by the fact that the second-order dissipation functions ϕ do not appear in the first-order energy equation (3c-26c), which is thereby reduced to yielding just the isentropic relation between dilatation and excess temperature. It is allowable, therefore, in this first-order approximation, to replace the quotient $(\nabla \theta_1 / \nabla \rho_1)$ appearing in (3c-26b) with the isentropic derivative $(\partial T / \partial \rho)_s = (\gamma - 1)/\rho \beta$, whereupon the first-order equation of motion for an *adiabatic viscous* fluid can be written as

$$\rho_0 \frac{\partial \mathbf{u}_1}{\partial t} + c_0^2 \nabla \rho_1 - \eta_0 \mathcal{U} \nabla (\nabla \cdot \mathbf{u}_1) + 2\eta_0 (\nabla \times \mathbf{R}_1) = 0 \tag{3c-27}$$

If the effects of viscosity, as well as of heat exchange, are to be neglected, the divergence of what is left of (3c-27) can be subtracted from the time derivative of (3c-26a)

[1] But, for a case in which $\mathbf{F}$ and $\nabla \rho_0$ cannot be neglected, see Haskell, *J. Appl. Phys.* **22**, 157–168 (February, 1951).

to yield the typical *small-signal scalar wave equation* of classical acoustics,

$$\frac{\partial^2 \rho_1}{\partial t^2} = \left(\frac{\partial P}{\partial \rho}\right)_s \nabla^2 \rho_1 \qquad (3\text{c-}28a)$$

and, with the help of the first-order isentropic relation $p_1 = c_0{}^2(\rho_1)_s$, this wave equation becomes, in terms of the sound pressure,

$$\frac{\partial^2 p_1}{\partial t^2} = c_0{}^2 \nabla^2 p_1 \qquad (3\text{c-}28b)$$

3c-4. The Second-order Acoustic Equations. The same substitution of composite variables that delivered (3c-26a), (3c-26b), and (3c-26c) will also yield directly the second-order equations of acoustics, which can now be marshaled as follows:

$$\frac{\partial \rho_2}{\partial t} + \rho_0 (\nabla \cdot \mathbf{u}_2) + \nabla \cdot (\rho_1 \mathbf{u}_1) = 0 \qquad (3\text{c-}29a)$$

$$\rho_0 \frac{\partial \mathbf{u}_2}{\partial t} + \frac{\partial (\rho_1 \mathbf{u}_1)}{\partial t} + \rho_0 \mathbf{u}_1 (\nabla \cdot \mathbf{u}_1) + \rho_0 (\mathbf{u}_1 \cdot \nabla) \mathbf{u}_1$$

$$+ \frac{c_0{}^2}{\gamma} \left(1 + \beta_0 \rho_0 \frac{\nabla \theta_2}{\nabla \rho_2}\right) \nabla \rho_2 - \eta_0 \mho \nabla (\nabla \cdot \mathbf{u}_2) + 2\eta_0 (\nabla \times \mathbf{R}_2)$$

$$- (\nabla \eta_1')(\nabla \cdot \mathbf{u}_1) - 2(\nabla \eta_1 \cdot \nabla)\mathbf{u}_1 - 2(\nabla \eta_1) \times \mathbf{R}_1 = 0 \qquad (3\text{c-}29b)$$

$$\frac{\partial \theta_2}{\partial t} + \mathbf{u}_1 \cdot (\nabla \theta_1) + \frac{\gamma - 1}{\beta_0} (\nabla \cdot \mathbf{u}_2) - \frac{\kappa_0}{\rho_0 C_v} \nabla^2 \theta_2$$

$$+ \frac{\kappa_0}{\rho_0{}^2 C_v} \rho_1 \nabla^2 \theta_1 - \frac{\nabla \theta_1 \cdot \nabla \kappa_1}{\rho_0 C_v} + \mathsf{q}\theta_2 - \frac{\phi_\eta}{\rho_0 C_v} = 0 \qquad (3\text{c-}29c)$$

The subscripts appended to κ and the η's imply that each may be expressed in the generic form

$$\eta(T, \rho, \cdots) = \eta_0(T_0, \rho_0, \cdots) + \eta_1 \qquad \eta_1 = \frac{\partial \eta}{\partial T} \theta_1 + \frac{\partial \eta}{\partial \rho} \rho_1 + \cdots \qquad (3\text{c-}30)$$

No general solution of these complete second-order equations has been given, but they provide a useful point of departure for making approximations and for investigating some second-order phenomena that cannot be predicted by the first-order equations alone.

3c-5. Spatial and Material Coordinates. Equations (3c-26) and (3c-29) are couched in terms of the local values assumed by the dependent variables ρ, P, T, and $\mathbf{u}$ at *places* identified by their coordinates x_i in a fixed *spatial* reference frame, commonly called *Eulerian* coordinates (in spite of their first use by d'Alembert). As an alternate method of representation, the behavior of the medium can be described in terms of the sequence of values assumed by the dependent condition and state variables pertaining to identified *material* particles of the medium no matter how these particles may move with respect to the spatial coordinate system. The independent variables in this case are the identification coordinates a_i, rather than the position coordinates; the latter then become dependent variables that describe, as time progresses, the travel history of each particle of the medium. Such a representation in terms of *material* coordinates is commonly called *Lagrangian* (in spite of its first introduction and use by Euler).

The Wave Equation in Material Coordinates. The use of material coordinates can be demonstrated by deriving the exact equations governing one-dimensional (plane-wave) propagation in a *nonviscous adiabatic* fluid. Consider a cylindrical segment of the medium of unit cross section with its axis along $+x$, the direction of propagation, and let x and $x + \delta x$ define the boundaries of a thin laminar "particle" whose undisturbed equilibrium position is given by a and $a + \delta a$. The difference $x - a = \xi$ defines the displacement of the a particle from its equilibrium position and provides a convenient incremental, or acoustic, dependent variable in terms of which to describe the position, velocity, and acceleration of the particle; thus

$$x(a,t) = a + \xi(a,t) \qquad \frac{\partial x}{\partial t} = u^L(a,t) = \frac{\partial \xi}{\partial t} \qquad \frac{\partial u^L}{\partial t} = \frac{\partial^2 \xi}{\partial t^2} \qquad (3c\text{-}31)$$

Continuity requires that the mass of the particle remain constant during any displacement, which means that

$$\rho_0 \delta a = \rho^L \delta x = \rho^L \left(\delta a + \frac{\partial \xi}{\partial a} \delta a \right) \qquad \frac{\rho_0}{\rho^L} = \frac{\partial x}{\partial a} = 1 + \frac{\partial \xi}{\partial a} \qquad (3c\text{-}32a)$$

or, for three-dimensional disturbances and in general,

$$\frac{\rho_0}{\rho^L} = \frac{\partial(x_1,x_2,x_3)}{\partial(a_1,a_2,a_3)} \qquad (3c\text{-}32b)$$

in which the symbolic derivative stands for the Jacobian functional determinant. The superscript L is used here and below as a reminder that the dependent variable so tagged adheres to, or "follows" in the Lagrangian sense, a specific particle, and that it is a function of the independent identification coordinates. When not so tagged, or with superscript E added for emphasis, the state variables ρ, P, T and the condition variable u are each assumed to be functions of time and the spatial coordinate x.

The net force per unit mass acting on the particle at time t is $-(\rho^L)^{-1}\partial P^L/\partial x$, where ρ^L and P^L are the density and pressure at x, the "now" position of the moving particle. However, inasmuch as x is not an independent variable in this case, the pressure gradient must be rewritten as $(\partial P^L/\partial a)(\partial a/\partial x)$, from which the second factor can be eliminated by recourse to (3c-32a). The momentum equation then becomes just

$$\frac{\rho_0 \partial^2 \xi}{\partial t^2} = \frac{-\partial P^L}{\partial a} \qquad (3c\text{-}33)$$

The adiabatic assumption makes available the simplified equation of state, $P = P(\rho)$, and this relation, in turn, allows the material gradient, $\partial P^L/\partial a$, to be written as

$$-\frac{\partial P^L}{\partial a} = -\left(\frac{\partial P^L}{\partial \rho^L}\right)_s \frac{\partial \rho^L}{\partial a} = -c^2 \frac{\partial \rho^L}{\partial a} \qquad (3c\text{-}34)$$

from which the last factor can be eliminated by using (3c-32a) again. This leads at once to the exact wave equation[1]

$$\frac{\partial^2 \xi}{\partial t^2} = \left(\frac{c\rho^L}{\rho_0}\right)^2 \frac{\partial^2 \xi}{\partial a^2} = c^2 \left(1 + \frac{\partial \xi}{\partial a}\right)^{-2} \frac{\partial^2 \xi}{\partial a^2} \qquad (3c\text{-}35)$$

The pressure-density relation for a perfect adiabatic gas is $P = P_0(\rho/\rho_0)^\gamma$, from which it can be deduced that

$$c^2 = \left(\frac{\partial P}{\partial \rho}\right)_s = \frac{\gamma P_0}{\rho_0}\left(\frac{\rho}{\rho_0}\right)^{\gamma-1} = c_0^2 \left(\frac{\rho}{\rho_0}\right)^{\gamma-1} \qquad (3c\text{-}36)$$

No generalization of comparable simplicity is available for liquids.[2] When (3c-36) is introduced in (3c-35), the exact "Lagrangian" wave equation for an adiabatic perfect gas becomes

$$\frac{\partial^2 \xi}{\partial t^2} = c_0^2 \left(\frac{\rho^L}{\rho_0}\right)^{\gamma+1} \frac{\partial^2 \xi}{\partial a^2} = c_0^2 \left(1 + \frac{\partial \xi}{\partial a}\right)^{-(\gamma+1)} \frac{\partial^2 \xi}{\partial a^2} \qquad (3c\text{-}37)$$

In the Lagrangian formulation illustrated above, the choice of a, the initial-position coordinate, as the independent variable is useful but any other coordinate that

[1] Rayleigh, "Theory of Sound," vol. II, §249; Lamb, "Hydrodynamics," §§13–15, 279–284.
[2] But see Courant and Friedrichs, "Supersonic Flow and Shock Waves," p. 8, Interscience Publishers, Inc., New York, 1948.

identifies the particles would serve the same purpose. For example, the particle located momentarily at x can be uniquely identified by the material coordinate $h \equiv \int_0^x \rho \, dx$, where h represents the mass of fluid contained between the origin and the particle. Inasmuch as this included mass will not change as the particle moves, the use of h as an independent "mass" variable automatically satisfies the requirements of continuity, with some attendant simplification in the analysis of transient disturbances. In the undisturbed condition, $\rho = \rho_0$ and $x = a$, whence the relation $a = h/\rho_0$ allows the independent variables to be interchanged by direct substitution in (3c-37).

Material and Spatial Coordinate Transforms. It is useful to have available a systematic procedure for converting a functional expression for one of the state variables from the form involving material coordinates to the corresponding form in spatial coordinates, or the inverse. One should avoid, however, the trap of referring to the state variables themselves as Lagrangian or Eulerian *quantities;* density and pressure, for example, are scalar point functions that can have only one value at a given place and time. On the other hand, it is of prime importance to distinguish carefully (and to specify!) the independent variables when computing the derivatives of these quantities.

The E and L functions are tied together by the displacement variable ξ, which provides a single-valued connection between the a particle and its instantaneous position coordinate x and which may therefore be regarded as a function of either of its terminal coordinates a or x. This can be indicated [cf. (3c-31)] by writing $x(a,t) = a + \xi(a,t)$, or the inverse relation $a(x,t) = x - \xi(x,t)$, from which follow the alternate expressions

$$a = x - \xi(a,t) \qquad x = a + \xi(x,t) \tag{3c-38}$$

The desired coordinate transforms can then be established by means of Taylor series expansions, the two forms following according to whether the expansion is centered on the instantaneous particle position or spatial coordinate x, or on the particle's equilibrium position or material coordinate a. Thus, if q is used to represent any one of the variables ρ, P, T, or u, one of the expansions can be based on the obvious identity

$$q^L(a,t) = q^E(x,t)_{x=a+\xi(x,t)}$$
$$= q^E(x,t)_{x=a} + \left[\xi(x,t) \, \frac{\partial q^E(x,t)}{\partial x} \right]_{x=a} + \frac{1}{2} \left[\xi^2(x,t) \, \frac{\partial^2 q^E(x,t)}{\partial x^2} \right]_{x=a} + \cdots \tag{3c-39}$$

Note that all terms on the right of (3c-39) are functions of the spatial coordinates and that each is to be evaluated at the equilibrium position coordinate a. This transform yields, therefore, the instantaneous value in material coordinates of the variable represented by q, in terms of the local value of q modified by correction terms (comprising the succeeding terms of the series) based on the spatial rate of change of q and the instantaneous displacement.

The inverse transform is derived in a similar way from the identity

$$q^E(x,t) = [q^L(a,t)]_{a=x-\xi(a,t)}$$
$$q^E(x,t) = [q^L(a,t)]_{a=x} - \left[\xi(a,t) \, \frac{\partial q^L(a,t)}{\partial a} \right]_{a=x} + \frac{1}{2} \left[\xi^2(a,t) \, \frac{\partial^2 q^L(a,t)}{\partial a^2} \right]_{a=x} - \cdots \tag{3c-40}$$

In symmetrical contrast with (3c-39), all terms on the right in (3c-40) are functions of the material coordinates and are to be evaluated for $a = x$. This transform, therefore, yields the instantaneous local value of the variable q at the place x, in terms of the instantaneous value of q for the now-displaced particle whose equilibrium position or material coordinate is $a = x$, modified by the succeeding terms of the series in accordance with the material-coordinate rate of change of q and the instantaneous displacement.

The transforms (3c-39) and (3c-40) indicate that the differences between q^L and q^E are of second order, which explains why the troublesome distinction between spatial and material coordinates does not intrude when only first-order effects are being considered. It also follows that the first two terms of these transforms are sufficient to deliver all terms of q^L or q^E through the second order. The use of these transforms can be illustrated by writing them out explicitly for u and ρ, including all second-order terms,

$$u^L \equiv \xi_t \qquad u^E = u^L - \xi u_a{}^L = \xi_t - \xi\xi_{ta} \tag{3c-41a}$$
$$\rho^L = \rho_0(1 + \xi_a)^{-1} = \rho_0(1 - \xi_a + \xi_a{}^2 - \cdots)$$
$$\rho^E = \rho_0(1 - \xi_a + \xi_a{}^2 + \xi\xi_{aa}) = \rho_0[1 - \xi_a + (\xi\xi_a)_a] \tag{3c-41b}$$

in which the subscripts indicate partial differentiation with respect to a or t. The product of (3c-41a) and (3c-41b) gives at once the relation between the material and spatial coordinate expressions for the mass transport ρu; thus, through second order,

$$\rho^E u^E = \rho^L u - \xi(\rho^L u^L)_a + \xi^2(\rho_a{}^L u_a{}^L) = \rho_0[\xi_t - (\xi\xi_t)_a] = \rho_0[\xi - \xi\xi_a]_t \tag{3c-42}$$

It is then straightforward to show that, if the particle velocity w is simple harmonic, the time average of the local mass transport $\rho^E u^E$ will vanish through the second order, even though the average value of u^E is not zero. Note, however, that the displacement velocity ξ_t is measured from an equilibrium position that is here assumed to be static; the average mass transport may indeed take on nonvanishing values if the wave motion as a whole leads to gross streaming (see Sec. 3c-7).

3c-6. Waves of Finite Amplitude.[1] A distinguished tradition adheres to the study of the propagation of unrestricted compressional waves. That the particle velocity is forwarded more rapidly in the condensed portion of the wave was known early (Poisson, 1808; Earnshaw, 1858; Riemann, 1859); and that this should lead eventually to the formation of a discontinuity or shock wave was recognized by Stokes (1848), interpreted by Rayleigh,[2] discussed more recently by Fubini,[3] and has been reviewed still more recently with heightened interest by modern students of blast-wave transmission.[4]

By virtue of the adiabatic assumption underlying $P = P(\rho)$, the speed of sound is also a function of density alone and may be approximated by the leading terms of its expansion about the equilibrium density:

$$c^2 \doteq c_0{}^2 \left[1 - 2\xi_a \frac{\rho_0}{c_0} \left(\frac{Dc}{D\rho} \right)_0 + \cdots \right] \tag{3c-43}$$

When (3c-43) is introduced in the exact wave equation in material coordinates, (3c-35), the latter can be recast in the following form, using the subscript convention for partial differentiation and retaining only, but all, terms through second order:

$$\xi_{tt} - c_0{}^2\xi_{aa} = -c_0{}^2 \left[1 + \frac{\rho_0}{c_0} \left(\frac{Dc}{D\rho} \right)_0 \right] (\xi_a{}^2)_a \tag{3c-44}$$

If it is then assumed that an arbitrary plane displacement $\xi(0,t) = f(t)$ is impressed at the origin, it can be verified by direct substitution that a solution of (3c-44) is

$$\xi(a,t) = f\left(t - \frac{a}{c_0} \right) + \frac{a}{2c_0{}^2} \left[1 + \frac{\rho_0}{c_0} \left(\frac{Dc}{D\rho} \right)_0 \right] \left[f'\left(t - \frac{a}{c_0} \right) \right]^2 \tag{3c-45}$$

The density variations associated with these displacements are to be found by entering (3c-45) in (3c-32), and the variational pressure can then be evaluated in terms of the adiabatic compressibility of the medium.

Relatively more attention has been devoted to the analysis of solutions of (3c-37) for the case of an adiabatic perfect gas. For an arbitrary initial displacement, as

[1] For more recent developments see Sec. 3n, Nonlinear Acoustics (Theoretical), pp. 3-183 to 3-205.
[2] "Theory of Sound," vol. II, §§249–253. *Proc. Roy. Soc. (London)* **84**, 247–284 (1910).
[3] *Alta Frequenza* **4**, 530–581 (1935).
[4] See also Sec. 2y of this book, Shock Waves, pp. 2-273 to 2-278.

above, the solution of the corresponding wave equation (3c-37), again including all terms through second order, is

$$\xi(a,t) = f\left(t - \frac{a}{c_0}\right) + \frac{a}{2c_0{}^2}\frac{\gamma + 1}{2}\left[f'\left(t - \frac{a}{c_0}\right)\right]^2 \tag{3c-46}$$

Technological interest in this problem centers on the generation of spurious harmonics, which can be studied by assuming the initial displacement to be simple harmonic, viz., $f(t) = \xi_0(1 - \cos \omega t)$ at the origin. The solution then takes the explicit form

$$\xi(a,t) = \xi_0[1 - \cos (\omega t - k_0 a)] + \frac{\gamma + 1}{8} k_0{}^2\xi_0{}^2 a[1 - \cos 2(\omega t - k_0 a)] \tag{3c-47}$$

in which k_0 is written for the phase constant, $k_0 = \omega/c_0 = 2\pi/\lambda_0$.

The most striking feature of the solutions (3c-45) and (3c-47) is the appearance of the material coordinate a in the coefficient of the second-harmonic term. As a consequence, the condensation wave front becomes progressively steeper as the wave propagates, the energy supplied at fundamental frequency being gradually diverted toward the higher harmonic components. The compensating diminution of the fundamental-frequency component would be exhibited explicitly if third-order terms had been retained in (3c-46) and (3c-47) inasmuch as all odd-order terms include a "contribution" to the fundamental. When such higher terms are retained it is predicted that propagation will always culminate in the formation of a shock wave at a distance from the source given approximately by $a \doteq 2\xi_0/(\gamma + 1)M^2$, where M is the peak value of the particle-velocity Mach number.[1] On the other hand, when dissipative mechanisms are taken into account, the fact that attenuation increases with frequency for either liquids or gases leads to the result that, except for very large in tial disturbances, the wavefront will achieve a maximum steepness when the propagation distance is such that the rate of energy conversion to higher frequencies by nonlinearity is just compensated by the increase of absorption at higher frequencies. Note, however, that this steepest wave front does not qualify as a "disturbance propagated without change of form." When attention is centered on the fundamental component, the diversion of energy to higher frequencies appears as an attenuation and accounts for the relatively more rapid absorption sometimes observed near a sound source.[2]

The variational or acoustic pressure, in material coordinates, can be expressed generally as a function of the displacement gradients by using the adiabatic pressure-density relation $P^L = P_0(\rho^L/\rho_0)^\gamma$ in conjuction with the continuity relation (3c-32); thus,

$$P^L - P_0 = p^L = \gamma P_0[-\xi_a + \tfrac{1}{2}(\gamma + 1)\xi_a{}^2] = \langle p^L \rangle + p_1{}^L + p_2{}^L \tag{3c-48}$$

in which the last member identifies the steady-state alteration of the average pressure and the fundamental and second-harmonic components of sound pressure. When the harmonic solution (3c-47) is introduced in (3c-48), the two alternating components of pressure for $a^2 \gg (\lambda/4\pi)^2$ can be shown, after some algebraic manipulation, to be

$$p_1{}^L = +\gamma P_0 M \sin (\omega t - k_0 a) = +\sqrt{2}\,P_1 \sin (\omega t - k_0 a) \tag{3c-49a}$$

$$p_2{}^L = \gamma P_0 M^2 k_0 a \tfrac{1}{4}(\gamma + 1) \sin 2(\omega t - k_0 a) = \sqrt{2}\,P_2 \sin 2(\omega t - k_0 a) \tag{3c-49b}$$

in which P_1 and P_2 are the rms values of the fundamental and second-harmonic sound pressures, and $M = k_0\xi_0 = \omega\xi_0/c_0$ is again the peak value of the particle-velocity Mach number at the origin. The relative magnitude of P_2 ncreases linearly with distance from the origin and is directly proportional to the peak Mach number, as may be deduced from (3c-49a) and (3c-49b); thus

$$\frac{P_2}{P_1} = \frac{1}{4}(\gamma + 1)Mk_0a \qquad P_2 = \frac{P_1{}^2 k_0 a(\gamma + 1)}{2\sqrt{2}\,\gamma P_0} \tag{3c-50}$$

[1] Fubini, *Alta Frequenza* **4**, 530–581 (1935).
[2] Fox and Wallace, *J. Acoust. Soc. Am.* **26**, 994–1006 (1954). Blackstock, *J. Acoust. Soc. Am.* **36**, 534–542 (1964).

Various experimental studies of second-harmonic generation have given results in reasonably good agreement with the predictions of (3c-50).[1]

The sound-induced alteration of mean total pressure, or "average" acoustic pressure, is given by the time-independent terms yielded by the substitution of (3c-47) in (3c-48), viz.,

$$\langle p^L \rangle = + \frac{\gamma P_0 M^2 (\gamma + 1)}{8} \qquad (3c\text{-}51)$$

Note that this pressure increment is given as a function of the material coordinates, which means that it pertains to a *moving* element of the fluid. The *local* value of the pressure change can be found by means of the transform (3c-40), which gives, through second-order terms, the following replacement for (3c-48),

$$p^E = p^L - \xi \frac{\partial p^L}{\partial a} = \gamma P_0 \left[-\xi_a + \frac{1}{2} (\gamma + 1)\xi_a{}^2 + \xi \xi_{aa} \right] \qquad (3c\text{-}52)$$

When (3c-47) is introduced in (3c-52), the time-independent terms give the local change in mean pressure as

$$\langle p^E \rangle = + \frac{\gamma P_0 M^2 (\gamma - 3)}{8} \qquad (3c\text{-}53)$$

and since γ is usually less than 2, it follows that the local value of mean pressure will be *reduced* by the presence of the sound wave, in striking contrast to the *increase* of mean pressure that would be observed when following the motion of a particle of the medium. Negative pressure increments as large as 10 newtons m^{-2} (100 dynes cm^{-2}) have been reported experimentally, in reasonably good agreement with (3c-53).

The mean value of the material particle velocity, $u^L \equiv \xi_t$, vanishes, as may be seen by differentiating (3c-47). The local particle velocity that would be observed at a fixed spatial position does not similarly vanish, however, and may be shown, by using the transform (3c-40) again, to be

$$u^E = \xi_t - \xi \xi_{ta} \qquad \langle u^E \rangle = -\frac{1}{2} c_0 M^2 = -\frac{\rho_0 c_0 \omega^2 \xi_0{}^2}{2 \rho_0 c_0{}^2} = -(\rho_0 c_0{}^2)^{-1} \langle J \rangle \qquad (3c\text{-}54)$$

where $\langle J \rangle$ is the average sound energy flux, or sound intensity.[2]

3c-7. Vorticity and Streaming. As suggested above, and with scant respect for the traditional symmetry of simple-harmonic motion, sound waves are found experimentally to exert net time-independent forces on the surfaces on which they impinge, and there is often aroused in the medium a pattern of steady-state flow that includes the formation of streams and eddies. The exact wave equation considered in the preceding section has been solved only for one-parameter waves (i.e., plane or spherical), and these solutions do not embrace some of the gross rotational flow patterns that are observed to occur. It is necessary, therefore, to revert for the study of these phenomena to the perturbation procedures introduced by the first- and second-order equations (3c-26) and (3c-29).

It is plausible that vortices and eddies should arise, if there is any net transport at all, inasmuch as material continuity would require that any net flow in the direction of sound propagation must be made good in the steady state by recirculation toward the source. Streaming effects can be studied most usefully, therefore, in terms of the generation and diffusion of circulation, or vorticity. More specifically, the time average of the second-order velocity $\mathbf{u}_2$ will be a first-order measure of the streaming

[1] Thuras, Jenkins, and O'Neil, *J. Acoust. Soc. Am.* **6**, 173–180 (1935); Fay, *J. Acoust. Soc. Am.* **3**, 222–241 (October, 1931); O. N. Geertsen, unpublished (ONR) Tech. Report no. III, May, 1951, U.C.L.A.; D. T. Blackstock, Report of the Fourth International Congress on Acoustics, Part I, 1962.

[2] Westervelt, *J. Acoust. Soc. Am.* **22**, 319–327 (1950).

velocity. The vector function describing $\mathbf{u}_2$ can always be resolved into solenoidal and lamellar components defined by

$$\mathbf{u}_2 \equiv -\nabla\varphi_2 + \nabla \times \mathbf{A}_2 \qquad \nabla^2\varphi_2 \equiv -\nabla \cdot \mathbf{u}_2 \qquad \nabla^2\mathbf{A}_2 = -(\nabla \times \mathbf{u}_2) \qquad (3\text{c-}55)$$

The irrotational component that represents the compressible, or acoustic, part of the fluid motion is derived from the scalar potential φ_2. The vector potential $\mathbf{A}_2$ is associated with the rotational component comprising the incompressible circulatory flow that is of primary interest in streaming phenomena.

The failure of the first-order equations to predict streaming can be demonstrated by writing directly the curl of the first-order force equation (3c-26b). The gradient terms are eliminated by this operation, since $\nabla \times \nabla(\ \ ) \equiv 0$, leaving just

$$\frac{\partial \mathbf{R}_1}{\partial t} - \nu_0 \nabla^2 \mathbf{R}_1 = 0 \qquad (3\text{c-}56)$$

Thus the first-order vorticity, $\mathbf{R}_1 \equiv \frac{1}{2}(\nabla \times \mathbf{u}_1)$, if it has any value other than zero, obeys a typical homogeneous diffusion equation. On the other hand, it would appear to follow that, if $\mathbf{R}_1$ were ever zero everywhere, its time derivative would also vanish everywhere and $\mathbf{R}_1$ would be constrained always thereafter to remain zero. This is *not* a valid proof of the famous Lagrange-Cauchy proposition on the permanence of the irrotational state, but the absence of any source terms on the right-hand side of (3c-56) does indicate correctly[1] that first-order vorticity cannot be generated in the interior of a fluid even when viscosity and heat conduction are taken into account. Instead, first-order vorticity, if it exists at all, must diffuse inward from the boundaries under control of (3c-56).

A notably different result is obtained when the second-order equations are dealt with in the same way. It is useful, before taking the curl of (3c-29b), to eliminate the second and third terms of this equation by subtracting from it the product of (ρ_1/ρ_0) and (3c-26b), and the product of $\mathbf{u}_1$ and (3c-26a). In effect this raises the first-order equations to second order and then combines the information in both sets. The augmented second-order force equation can then be arranged in the form

$$\rho_0 \frac{\partial \mathbf{u}_2}{\partial t} + 2\eta_0(\nabla \times \mathbf{R}_2) + \nu_0 \mho \rho_1 \nabla(\nabla \cdot \mathbf{u}_1) - 2\nu_0 \rho_1(\nabla \times \mathbf{R}_1) - 2\rho_0(\mathbf{u}_1 \times \mathbf{R}_1)$$

$$-2[(\nabla\eta_1 \cdot \nabla)\mathbf{u}_1 + \nabla\eta_1 \times (\nabla \times \mathbf{u}_1)] + 2(\nabla\eta_1 \times \mathbf{R}_1) + \rho_0\nabla\left(\frac{1}{2}\mathbf{u}_1 \cdot \mathbf{u}_1\right) + B_2\nabla\rho_2$$

$$- B_1\nabla\left(\frac{1}{2}\rho_1{}^2\right) - \eta_0\mho\nabla(\nabla \cdot \mathbf{u}_2) - \nabla\eta_1'(\nabla \cdot \mathbf{u}_1) = 0 \qquad (3\text{c-}57)$$

The following abbreviations have been used for the coefficients of $\nabla\rho_1$ in (3c-26b) and of $\nabla\rho_2$ in (3c-29b):

$$B_1 \equiv \frac{c_0{}^2}{\gamma}\left[1 + \beta_0\rho_0\left(\frac{D\theta_1}{D\rho_1}\right)_0\right] \qquad B_2 \equiv \frac{c_0{}^2}{\gamma}\left[1 + \beta_0\rho_0\left(\frac{D\theta_2}{D\rho_2}\right)_0\right] \qquad (3\text{c-}58)$$

in which the quotients $(\nabla\theta_1/\nabla\rho_1)$ and $(\nabla\theta_2/\nabla\rho_2)$ have been replaced by the corresponding material derivatives $D\theta/D\rho$, which must be evaluated, of course, for the particular conditions of heat exchange satisfying the energy equations (3c-26c) and (3c-29c). This evaluation can be evaded temporarily (at the cost of neglecting ∇B_1 and ∇B_2) by observing that each of the last five terms of (3c-57) contains a gradient. These disappear on taking the curl of (3c-57), whereupon the vorticity equation emerges as

$$\frac{\partial \mathbf{R}_2}{\partial t} - \nu_0 \nabla^2 \mathbf{R}_2 = \frac{1}{2}\nu_0\mho\left(\nabla s_1 \times \nabla \frac{\partial s_1}{\partial t}\right) + \rho_0{}^{-1}\nabla \times (\mathbf{u}_1 \cdot \nabla)\nabla\eta_1 + \nu_0 s_1 \nabla^2 \mathbf{R}_1$$

$$- \nu_0\nabla s_1 \times (\nabla \times \mathbf{R}_1) - \nabla \times (\mathbf{u}_1 \times \mathbf{R}_1) + \rho_0{}^{-1}\nabla \times (\nabla\eta_1 \times \mathbf{R}_1) \qquad (3\text{c-}59)$$

[1] St. Venant, *Compt. rend.* **68**, 221–237 (1869).

in which s_1 has been introduced as an abbreviation for the first-order condensation, $s_1 = \rho_1/\rho_0$. This inhomogeneous diffusion equation puts in evidence various second-order sources of vorticity: four vanish if the first-order motion is irrotational ($\mathbf{R}_1 = 0$), and two drop out when the shear viscosity is constant ($\nabla \eta_1 = 0$). It is notable that the dilatational viscosity η' does not appear in any of these source terms except through the ratio η'/η that forms part of the dimensionless viscosity number $\mathcal{U} \equiv 2 + (\eta'/\eta)$.

Except for the third source term, which (3c-56) shows to be one order smaller than the change rate of $\mathbf{R}_1$, *all* the vorticity sources would vanish—and the streaming would "stall"—if the wave front were strictly plane with $\mathbf{u}_1$, s_1, and η functions of only one space coordinate. Wave fronts cannot remain strictly plane at grazing incidence, however,[1] and rapid changes in the direction and magnitude of $\mathbf{u}_1$ will occur near reflecting surfaces, in the neighborhood of sound-scattering obstacles, and in thin viscous boundary layers. As a consequence, the "surface" source terms containing $\mathbf{R}_1$ become relatively more important in these cases.[2] In other circumstances, when the sound field is spatially restricted by source directionality, the first source term in (3c-59) dominates and leads to a steady-state streaming velocity proportional to the *ratio* of the dilatational and shear viscosity coefficients—and hence to a unique independent method of measuring this moot ratio.[3] Both the force that drives the fluid circulation and the viscous drag that opposes it are proportional to the kinematic viscosity, which does not therefore control the final value of streaming velocity but only the time constant of the motion, i.e., the time required to establish the steady state.[4]

Evaluating the second-order vorticity source terms in any specific case requires that the first-order velocity field be known, and this calls in the usual way for solutions that satisfy the experimental boundary conditions and the wave equation. Unusual requirements of exactness are imposed on such solutions, moreover, by the fact that even the second-order acoustic equations yield only a first approximation to the mean particle velocity.

The analysis of vorticity can be recast, by skillful abbreviation and judicious regrouping of the elements of (3c-57), in such a way as to yield a general law of rotational motion, according to which the average rate of increase of the moment of momentum of a fluid element responds to the difference between the sound-induced torque and a viscous torque arising from the induced flow.[5] A close relation has also been shown to exist in some cases between the streaming potential and the attenuation of sound by the medium without regard for whether the attenuation is caused by viscosity, heat conduction, or by some relaxation process; in effect the average momentum of the stream "conserves" the momentum diverted from the sound wave by absorption.[6] This principle has so far been established rigorously only for the adiabatic assumption under which $P = P(\rho)$, and under restrictive assumptions on the variability of η and $\mathcal{U}$, but its prospective importance would appear to justify efforts to extend the generalization.

3c-8. Acoustical Energetics and Radiation Pressure. If the kinetic energy density that appeared briefly in (3c-12) is restored to (3c-18), the change rate of the specific

[1] Morse, "Vibration and Sound," 2d ed., pp. 368–371, McGraw-Hill Book Company, New York, 1948.

[2] Medwin and Rudnick, *J. Acoust. Soc. Am.* **25**, 538–540 (1953).

[3] Liebermann, *Phys. Rev.* **75**, 1415–1422 (1949); Medwin, *J. Acoust. Soc. Am.* **26**, 332–341 (1954).

[4] Eckart, *Phys. Rev.* **73**, 68–76 (1948).

[5] Nyborg, *J. Acoust. Soc. Am.* **25**, 938–944 (1953); Westervelt. *J. Acoust. Soc. Am.* **25**, 60–67 and *errata*, 799 (1953).

[6] Nyborg, *J. Acoust. Soc. Am.* **25**, 68–75 (1953); Doak, *Proc. Roy. Soc. (London),* ser. A, **226**, 7–16 (1954); Piercy and Lamb, *Proc. Roy. Soc. (London),* ser. A, **226**, 43–50 (1954).

total energy density (per unit mass), E/ρ, can be formulated in terms of

$$\rho\frac{D(E/\rho)}{Dt} = \rho\frac{D(\frac{1}{2}\mathbf{u}\cdot\mathbf{u})}{Dt} + \rho\frac{D\epsilon}{Dt}$$

$$= \rho\frac{D(\frac{1}{2}\mathbf{u}\cdot\mathbf{u})}{Dt} - \rho P\frac{Dv}{Dt} - \boldsymbol{\nabla}\cdot\mathbf{q} + \phi_\eta \qquad (3\text{c-}60)$$

Material derivatives are used here so that the energy balance reckoned for a particular volume element will continue to hold as the derivatives "follow" the motion of the material particles. The mechanical work term on the right in (3c-60) can be resolved into two components by writing $P = P_0 + p$, where the excess, or sound, pressure p now represents the sum of the variational components of all orders

$$(p = p_1 + p_2 + \cdots)$$

Thus

$$\rho\frac{D(E/\rho)}{Dt} = \rho\frac{D(\frac{1}{2}\mathbf{u}\cdot\mathbf{u})}{Dt} - \rho p\frac{Dv}{Dt} + \rho P_0\frac{Dv}{Dt} - \boldsymbol{\nabla}\cdot\mathbf{q} + \phi_\eta \qquad (3\text{c-}61)$$

A second equation involving the first two terms on the right of (3c-61) can be formed by multiplying the continuity equation (3c-5) by p and adding it to the scalar product of the vector $\mathbf{u}$ and the vector force equation (3c-11b); thus

$$\rho\mathbf{u}\cdot\frac{D\mathbf{u}}{Dt} + \mathbf{u}\cdot\boldsymbol{\nabla}p + p\left(\frac{1}{\rho}\frac{D\rho}{Dt} + \boldsymbol{\nabla}\cdot\mathbf{u}\right) = \mathbf{u}\cdot\mathbf{f}_v(\eta, \eta', \mathbf{u})$$

$$= \rho\mathbf{u}\cdot\frac{D\mathbf{u}}{Dt} - p\rho\frac{Dv}{Dt} + \mathbf{u}\cdot\boldsymbol{\nabla}p + p\boldsymbol{\nabla}\cdot\mathbf{u} \qquad (3\text{c-}62)$$

where $\mathbf{f}_v$ stands for the sum of the five viscosity terms that appear on the right-hand side of (3c-11b). Combining this result with (3c-61) gives

$$\rho\frac{D(\frac{1}{2}\mathbf{u}\cdot\mathbf{u})}{Dt} - \rho p\frac{Dv}{Dt} + \boldsymbol{\nabla}\cdot(p\mathbf{u}) = +\mathbf{u}\cdot\mathbf{f}_v$$

$$\rho\frac{D(E/\rho)}{Dt} + \boldsymbol{\nabla}\cdot(p\mathbf{u}) = -\rho P_0\frac{Dv}{Dt} - \boldsymbol{\nabla}\cdot\mathbf{q} + \phi_\eta + \mathbf{u}\cdot\mathbf{f}_v$$

$$(3\text{c-}63)$$

The significance of this result can be made more apparent by using the continuity equation again, this time in the form $(E/\rho)[\partial\rho/\partial t + \boldsymbol{\nabla}\cdot(\rho\mathbf{u})] = 0$. Adding this "zero" to the left-hand side of (3c-63), after first using (3c-3) to express the material derivative in terms of fixed spatial coordinates, allows the continuity of acoustic energy to be expressed by

$$\rho\frac{D(E/\rho)}{Dt} + \boldsymbol{\nabla}\cdot(\rho\mathbf{u}) = \rho\frac{\partial(E/\rho)}{\partial t} + \rho\mathbf{u}\cdot\boldsymbol{\nabla}\frac{E}{\rho} + \boldsymbol{\nabla}\cdot(p\mathbf{u})$$

$$+ \left[\frac{E}{\rho}\frac{\partial\rho}{\partial t} + \frac{E}{\rho}\boldsymbol{\nabla}\cdot(\rho\mathbf{u})\right]$$

$$\frac{\partial E}{\partial t} = -\boldsymbol{\nabla}\cdot(p\mathbf{u} + E\mathbf{u}) - P_0\Delta - \boldsymbol{\nabla}\cdot\mathbf{q} + \mathbf{u}\cdot\mathbf{f}_v + \phi_\eta \qquad (3\text{c-}64)$$

The acoustic energy-flux vector can be identified as $p\mathbf{u} = \mathbf{J}$, inasmuch as this term represents the instantaneous rate at which one portion of the medium does mechanical work on a contiguous portion in the process of forwarding the sound energy. The time average of the sound-energy flux through unit area normal to $\mathbf{u}$ is defined as the *sound intensity*, $\langle\mathbf{J}\rangle \equiv \mathbf{I}$. Ordinarily it is only the time average of each term of (3c-64) that is of interest, but the equation itself holds at every instant and asserts that growth of the total energy density of a volume element is accounted for by the influx of acoustic and thermal energy across the boundaries of the element, by the energy dissipated in viscous losses, and by the work done by the equilibrium pressure on the volume element during condensation. The latter component is represented by $(-P_0\Delta)$ and by a corresponding linear term contained implicitly in E [cf. (3c-19)]. It is omitted in most textbook descriptions of acoustic energy density, the neglect being justified if at all on the grounds that the stored energy varies linearly with the dilatation and hence will have a vanishing net value when averaged over an integral number of periods or wavelengths, or over the entire region occupied by the sound field. Care must be taken to ensure that it does indeed vanish rigorously on the average inasmuch as the peak values of this component of energy storage are larger than the acoustic energy in the ratio P_0/p.

Acoustic Radiation Pressure. The appearance of the product $E\mathbf{u}$ as an additive term in the first right-hand member of (3c-64) is notable and represents the net energy density carried across the boundary of a volume element by convection, the net flow being measured by the divergence of the particle velocity.[1] No approximations have been made in deducing (3c-64), which holds, therefore, within the scope of validity of the basic assumptions.

It is significant to remark the fact that E is directly additive to p when the divergence term is written as $\nabla \cdot (p + E)\mathbf{u}$, thereby identifying the additive term as a *radiation pressure* whose magnitude at every instant is just equal to the total energy density, $E = \frac{1}{2}\rho\mathbf{u} \cdot \mathbf{u} + \rho\epsilon$. This interpretation can be fortified by revising (3c-64) by expanding $\nabla \cdot (E\mathbf{u}) = E(\nabla \cdot \mathbf{u}) + \mathbf{u} \cdot \nabla E$. The last term can be used to restore the material time derivative of E and the other can be merged with the linear term in P_0, yielding a revised power equation in the form

$$\frac{DE}{Dt} = -\nabla \cdot (p\mathbf{u}) - (P_0 + E)\Delta - \nabla \cdot \mathbf{q} + \phi_\eta + \mathbf{u} \cdot \mathbf{f}_v \qquad (3c\text{-}65)$$

The role of E as an additive or radiation pressure is thus retained in (3c-65) where its time-independent part is now exhibited appropriately as a slight change in the equilibrium pressure.

When seeking to evaluate the net mechanical force due to radiation pressure on a material obstacle or screen exposed to a sound field, care must be taken to specify the boundary conditions and to account for *all* the reaction forces involved, including the steady-state interaction of the obstacle with the medium as well as the dynamic interaction of the obstacle with the sound field itself. Thus, for example, if a long tube is "filled" with a progressive plane wave, the walls of the tube, which interact only with the medium, would experience only the mean increment of the equilibrium pressure [cf. (3c-53)], and this would disappear if the walls were permeable to the medium, but not to the sound wave (e.g., with capillary holes). On the other hand, if a sound-absorbing screen were freely suspended athwart the wavefronts, it would experience just the pressure E shown by (3c-64) to be additive to p; but if the screen were to form an impermeable termination of the tube it would experience both components of pressure, including changes due to the enhancement of $\langle E \rangle$ by the reflected wave.[2]

3c-9. Sound Absorption and Dispersion. The basic manifestation of the absorption or attenuation of sound is the conversion of organized systematic motions of the particles of the medium into the uncoordinated random motions of thermal agitation.

[1] Schock, *Acustica* **3**, 181–184 (1953).
[2] Suggested references: On fundamentals, see L. Brillouin, "Les Tenseurs en mécanique et en élasticité," Dover Publications, New York, 1946. On influence of oblique incidence and of obstacle's reflection coefficient, see F. E. Borgnis, On the Forces upon Plane Obstacles Produced by Acoustic Radiation, *J. Madras Inst. Technol.* **1** (2), 171–210 (November, 1953), and (3), 1–33 (September, 1954); also condensed in *Revs. Modern Phys.*, **25**, 653–664 (1953). For review, critical bibliography, and sophisticated analysis of general topic, see E. J. Post, *J. Acoust. Soc. Am.* **25**, 55–60 (1953); *Phys. Rev.* **118**, 1113–1118 (1960).

Various agencies of conversion can be identified as viscosity, heat conduction, or as some other mechanism that gives rise to a delay in the establishment of thermodynamic equilibrium; but all are mechanisms of interaction that lead to the same result, viz. that the energy of mass motion imparted intermittently to the medium by the sound source becomes increasingly disordered and "unavailable." Describing this in terms of the irreversible production of entropy leads to the definition of dissipation functions and paves the way for formulating an acoustic energy balance.

Equation of Continuity for Acoustic Energy. This may take the form of a statement that the mean net influx of sound energy across the boundaries of a volume element situated in a sound field must just balance the average time rate at which this energy is degraded, or made unavailable, throughout the volume element by irreversible increase of entropy; thus, by extension of (3c-20),

$$- \int_A J_i \, da_i = \int_V \frac{DE_{\text{diss}}}{Dt} \, dV = \int_V T \frac{DS_{\text{irr}}}{Dt} \, dV = \int_V (\phi_\kappa + \phi_\eta) \, dV \quad (3\text{c-}66)$$

where the sound energy flux vector is $J_i = pu_i$, and E_{diss} is the degraded component of internal energy associated with the irreversible entropy S_{irr}.

The differential form of (3c-66) can be obtained in the usual way by using the divergence theorem to convert the surface integral to a volume integral. Then, after introducing the explicit forms of the dissipation functions, (3c-24a) and (3c-24b), the acoustic energy continuity relation becomes

$$-\nabla \cdot \mathbf{J} = - \frac{\partial(pu_i)}{\partial x_i} = \phi_\kappa + \phi_\eta = \frac{\kappa}{T} \left(\frac{\partial T}{\partial x_i} \right)^2 + \eta' \frac{\partial u_k}{\partial x_k} \frac{\partial u_i}{\partial x_i}$$
$$+ \frac{1}{2} \eta \left[\left(\frac{\partial u_i}{\partial x_j} \right)^2 + \left(\frac{\partial u_j}{\partial x_i} \right)^2 + 2 \frac{\partial u_i}{\partial x_j} \frac{\partial u_j}{\partial x_i} \right] \quad (3\text{c-}67a)$$

where it is understood that only the time-independent parts of each side of (3c-67a) are to be retained. The algebraic complexity of dealing with (3c-67a) is considerably abated by considering only plane waves, for which case the running subscripts each reduce to unity and can be dropped. The plane-wave form of the acoustic-energy relation then becomes, after introducing P as an implicit variable in ∇T,

$$- \frac{\partial(pu)}{\partial x} = \frac{\kappa}{T} \left(\frac{DT}{DP} \right)^2 \left(\frac{\partial p}{\partial x} \right)^2 + \eta \mathcal{U} \left(\frac{\partial u}{\partial x} \right)^2 \quad (3\text{c-}67b)$$

in which $\eta \mathcal{U}$ has been written for $\eta' + 2\eta$ [cf. (3c-10)]. The thermal dissipation term can then be maneuvered into more suggestive form by further manipulation involving the equation of state $T = T(P,\rho)$ and various thermodynamic identities including the useful relation that holds for all fluids, $T\beta^2 c^2 = C_p(\gamma - 1)$. This leads, still without approximation, and with the time average explicitly indicated, to

$$\left\langle - \frac{\partial(pu)}{\partial x} \right\rangle = \left\langle \eta \mathcal{U} \left(\frac{\partial u}{\partial x} \right)^2 \right\rangle + \left\langle \frac{\kappa}{\rho C_p} \frac{[(\rho c^2/K_T) - 1]^2}{(\gamma - 1)\rho c^2} \left(\frac{\partial p}{\partial x} \right)^2 \right\rangle \quad (3\text{c-}68)$$

It can now be observed that p, u, and their derivatives must be known throughout the sound field in order to evaluate the sound energy flux and the dissipation functions that make up (3c-67a) or its reduced form (3c-68). On the other hand, if these field variables are known explicitly, the effects of dissipation will already be in evidence without recourse to (3c-68). Such a continuity equation for *acoustic* energy is therefore redundant, as might have been expected inasmuch as the conservation of energy has already been incorporated in the basic equations (3c-5), (3c-15), and (3c-23). Nevertheless, (3c-68) retains some logical utility as an auxiliary relation, even though it no longer needs to be relied on for the pursuit of absorption measures, at least for plane waves.

Exact Solution of the First-order Equations. An exact solution of the complete first-order equations (3c-26a), (3c-26b), (3c-26c) for the plane-wave case and a definitive discussion of its implications have been given recently by Truesdell.[1] The specific problem considered is that of forced plane damped waves in a viscous, conducting fluid medium. It is assumed that each of the first-order incremental state and field variables can be described by the real parts of

$$u_1 = u_{10}e^{j\omega t}e^{-(\alpha+jk)x} \tag{3c-69}$$

and of similar equations for ρ_1, p_1, θ_1. It is assumed that $(u_1)_{x=0} = u_{10}e^{j\omega t}$ is the simple-harmonic velocity imparted to the medium by the vibrating surface of a source located at $x = 0$, but the other amplitude coefficients may be complex in order to embody the phase angles by which these variables lead or lag u_1. The exponent expressing time dependence is written $+j\omega t$, as required in order to preserve both the conventional form $R + jX$ for complex impedances *and* the positive sign for inductive or mass reactance. The attenuation constant α and the phase constant $k \equiv \omega/c$, or $k_0 \equiv \omega/c_0$, are the real and imaginary parts of the complex propagation constant $\chi \equiv \alpha + jk$; and $c_0 \equiv (\partial P/\partial \rho)_s^{\frac{1}{2}}$ is the *reference* value of sound speed.

When the assumed solutions (3c-69) are systematically introduced in (3c-26a), (3c-26b), and (3c-26c), three algebraic equations in ρ_1, u_1, θ_1 are obtained, as follows:

$$\begin{aligned}
\rho_0(\alpha + jk)u_1 \qquad\qquad -j\omega\rho_1 \qquad\qquad &= 0 \\
[j\omega\rho_0 - \eta \mathcal{U}(\alpha + jk)^2]u_1 \quad -(\alpha + jk)\left[\frac{c_0^2}{\gamma}(\rho_1 + \beta_0\rho_0\theta_1)\right] &= 0 \\
-\frac{\gamma - 1}{\beta_0}(\alpha + jk)u_1 \quad +\left[j\omega - \frac{\kappa}{\rho_0 C_v}(\alpha + jk)^2 + \mathfrak{q}\right]\theta_1 &= 0
\end{aligned} \tag{3c-70}$$

If these equations are indeed to admit solutions of the assumed form (3c-69), the determinant of the coefficients of u_1, ρ_1, and θ_1 must vanish. The characteristic or *secular equation* formed in this way (Kirchhoff, for perfect gases, 1868; extended to any fluid with arbitrary equation of state by P. Langevin[2]) turns out to be a biquadratic in the dimensionless complex propagation variable $(\alpha + jk)/k_0$. Writing this out in full, however, will be facilitated by first considering the question of how best to specify the properties of the medium.

Dimensional Analysis and Absorption Measure. Examination of (3c-70) reveals that, in addition to $(\alpha + jk)/k_0$ and the three independent variables, there are 10 parameters that pertain to the behavior of the medium at the angular frequency ω. One of these could be eliminated, in principle at least, by using the relation $T\beta^2c^2 = (\gamma - 1)C_p$, leaving 9 that are independent: C_p, C_v, η, η', κ, ρ_0, c_0, $\mathfrak{q}$, and ω. Then, since each of these can be expressed in terms of 4 basic dimensional units (e.g., mass, length, time, and temperature), it follows from the pi theorem of dimensional analysis[3] that just 5 independent dimensionless ratios can be formed out of combinations of these 9 parameters. This leads to a functional expression of the absorption measure in the symbolic form

$$\frac{\alpha + jk}{k_0} = \psi\left(\frac{C_p}{C_v}, \frac{\eta'}{\eta}, \frac{\eta C_p}{\kappa}, \frac{\omega\eta}{\rho_0 c_0^2}, \frac{\mathfrak{q}}{\omega}\right) \tag{3c-71}$$

The first two ratios have already been incorporated in γ and the viscosity number $\mathcal{U} \equiv 2 + \eta'/\eta$; the third is the Prandtl number $\mathcal{P} \equiv \eta C_p/\kappa$, and the fourth and fifth can be identified as Stokes numbers $S \equiv \omega\eta/\rho_0 c_0^2$ and $S' \equiv \omega/\mathfrak{q}$. The present purpose

[1] C. A. Truesdell, Precise Theory of the Absorption and Dispersion of Forced Plane Infinitesimal Waves According to the Navier-Stokes Equations, *J. Rational Mechanics and Analysis* **2**, 643–741 (October, 1953).
[2] Reported by Biquard, *Ann. phys.* (11) **6**, 195–304 (1936).
[3] E. Buckingham, *Phys. Rev.* **4**, 345 (1914); *Phil. Mag.* (6) **42**, 696 (1921).

is served somewhat better by substituting for the third and fourth ratios their products with the dimensionless viscosity number, thus defining a frequency number X and thermoviscous number Y through

$$X \equiv \mathcal{U}S = \frac{\omega \eta \mathcal{U}}{\rho_0 c_0^2} \qquad Y \equiv (\mathcal{P}\mathcal{U})^{-1} = \frac{\kappa}{\eta \mathcal{U} C_p} \qquad XY = \frac{\omega \kappa}{\rho_0 c_0^2 C_p} \qquad (3c\text{-}72)$$

The frequency parameter X also provides a natural criterion for designating frequencies as "low," "medium," or "high" according to whether X is much less than, comparable with, or much greater than unity. It may also be noted that, for nearly perfect gases, $\rho_0 c_0^2 \doteq \gamma P_0$, from which it follows that $X_{\text{gas}} \doteq (\omega/P_0)(\eta \mathcal{U}/\gamma)$. Hence variation of pressure may be used to extend in effect the accessible range of frequency in measurements on gases, and the ratio ω/P_0 is a proper parameter in terms of which to report such results.

Solutions of the Characteristic Equation. If the dimensionless ratios discussed above are now introduced in the expanded determinant of the coefficients of (3c-70), the resulting Kirchhoff-Langevin secular equation can be written as

$$\left(1 - \frac{j}{S'}\right) + \left(\frac{\alpha + jk}{k_0}\right)^2 \left[1 + jX(1 + \gamma Y) + \frac{\gamma X - j}{\gamma S'}\right]$$
$$+ \left(\frac{\alpha + jk}{k_0}\right)^4 XY(j - \gamma X) = 0 \qquad (3c\text{-}73)$$

The standard "quadratic formula" can be used at once to solve (3c-73) for the reciprocal square of the propagation constant,

$$-2\left(1 - \frac{j}{S'}\right)\left(\frac{k_0}{\alpha + jk}\right)^2 = 1 + \frac{X}{S'} + j\left[X(1 + \gamma Y) - \frac{1}{\gamma S'}\right]$$
$$\pm \left[\left(1 + \frac{X}{S'}\right)^2 - \left[X(1 - \gamma Y) - \frac{1}{\gamma S'}\right]^2\right.$$
$$\left. + 2j\left\{X[1 - (2 - \gamma)Y] + X^2 \frac{1 - \gamma Y}{S'} - \frac{[1 + (X/S')]}{\gamma S'}\right\}\right]^{\frac{1}{2}} \qquad (3c\text{-}74a)$$

Skillful abbreviation might allow this complete solution to be carried somewhat further but no algebraic magic can lighten very much the burden of depicting the behavior of α and k as a function of *four* independent parameters—and it might have been five but for the welcome fact that $\mathcal{U}$ does not appear except as embodied in X and Y. Moreover, each parameter that does appear in (3c-74a) occurs in one or more product combinations, and hence it can *not* be assumed in general that the effects of viscosity and heat exchange will be linearly additive. The common practice of assessing these one at a time and then superimposing the results must therefore be considered unreliable unless justified explicitly and quantitatively. Nevertheless, something must give, and it is customary to abandon first the radiant-heat exchange, at least temporarily, by letting S' become infinite in (3c-74a). With this simplification, and with some new abbreviations, (3c-74a) becomes

$$-2\left(\frac{k_0}{\alpha + jk}\right)^2 = 1 + jX(1 + \gamma Y) \pm \{1 - X^2(1 - \gamma Y)^2 + j2X[1 - (2 - \gamma)Y]\}^{\frac{1}{2}}$$
$$\equiv G + jH = 1 + jX(1 + \gamma Y) \pm (E + jF)^{\frac{1}{2}} \qquad (3c\text{-}74b)$$
$$E \equiv 1 - X^2(1 - \gamma Y)^2 \qquad F \equiv 2X[1 - (2 - \gamma)Y]$$

This equation has two pairs of noncoincident complex roots, but only the one of each pair that has a nonnegative real part corresponding to real attenuation is to be retained. These two physical solutions comprise the two branches of a complex square root; one branch pertains to typical compressional sound waves identified as type I, the other to so-called thermal waves identified as type II. It is an unwarranted oversimplification, however, to describe these simply as "pressure" waves and "thermal" waves inasmuch as *all* the state and condition variables—pressure, density, velocity, temperature, heat flux, etc.—are simultaneously entrained and propagated by *each* wave type, and waves of *both* types are always excited simultaneously by any source. On the other hand, the absorption and dispersion measures for waves of type I and type II will, in general, be quite different and will vary differently with the frequency parameter X and with the thermoviscous parameters γ and Y that characterize the fluid. For example, type II waves are so rapidly attenuated in ordinary fluids at accessible frequencies that they cannot be observed, whereas in strongly conducting liquids such as mercury (and perhaps in liquid helium II) the absorption for type II waves becomes less than for type I waves when the frequency is high enough for X to exceed $\frac{1}{3}$.

It should be noticed, parenthetically, that if the basic first-order equations (3c-70) had not been restricted to plane waves, the last term of (3c-26b) would not have dropped out. Instead, there would have turned up eventually in (3c-70) a pair of terms in the first-order vector velocity potential $\mathbf{A}_1$ [see (3c-55)] on the basis of which it would have been predicted that still another type of allowed wave motion can exist in viscous fluids—a transverse *viscous wave* that is propagated by virtue of the transverse shear reactions due to viscosity.[1]

Viscothermal Absorption and Dispersion Measures. The problem of branch determination arising in the solution of (3c-74b) has been discussed thoroughly by Truesdell.[2] One view of it can be expressed by writing the formal solution in the explicit form

$$\frac{\alpha}{k} \equiv \frac{A}{2\pi} = \frac{H}{+(G^2 + H^2)^{\frac{1}{2}} + G} \qquad \left(\frac{c}{c_0}\right)^2 = \frac{2(G^2 + H^2)}{+ (G^2 + H^2)^{\frac{1}{2}} + G}$$

$$2G = 1 \pm f(h)(+E^{\frac{1}{2}}) \qquad 2H = X(1 + \gamma Y) \pm (\mathrm{sgn}\, F)g(h)(+E^{\frac{1}{2}}) \qquad (3c\text{-}75a)$$

(upper signs yield type I waves, lower signs type II waves)

$$h \equiv \frac{F}{-E} \qquad f(h) \equiv + \sqrt{2}\,[+(1 + h^2)^{\frac{1}{2}} + 1] = + \cosh \tfrac{1}{2}(\sinh^{-1} h)$$

$$g(h) \equiv + \sqrt{2}\,[+(1 + h^2)^{\frac{1}{2}} - 1] = + \sinh \tfrac{1}{2}(\sinh^{-1} h) \qquad (3c\text{-}75b)$$

where the plus signs associated with roots denoted by fractional exponents indicate that the principal or positive root is to be used. The solution (3c-75a) can now be attacked frontally, either by means of power-series expansions for large or small values of X or by resorting to brute-force numerical computation for intermediate frequencies. The several square-root operations on complex quantities required by the latter procedure are often facilitated by using the f and g functions defined by (3c-75b), for which the principal values have been tabulated.[3]

The clue to a basis for classifying fluids according to their viscothermal behavior is afforded by noting that the algebraic sign of F appears in (3c-75a) in such a way as to interchange the wave types when F changes sign, and that this occurs when $(2 - \gamma)Y$ passes through unity. On this basis, one may categorize fluids as *strong* conductors if Y is greater than $(2 - \gamma)^{-1}$. The contrary alternative can be further subdivided usefully[2] into *weak* conductors for which Y is less than γ^{-1}, and *moderate* conductors for which Y has intermediate values. Most liquids (including the liquefied noble gases) qualify as weak conductors, most gases as moderate conductors. On the other hand, the fact that mercury, the molten metals, and liquid helium II rank as strong

[1] Rayleigh, "Theory of Sound," vol. II, §§347; Mason, *Trans. ASME* **69**, 359–367 (1947); Epstein and Carhart, *J. Acoust. Soc. Am.* **25**, 553–565 [557] (1953).

[2] C. A. Truesdell, *Precise Theory of the Absorption and Dispersion of Forced Plane Infinitesimal Waves According to the Navier-Stokes Equations*, *J. Rational Mechanics and Analysis* **2**, 643–741 (October, 1953).

[3] G. W. Pierce, *Proc. Am. Acad. Arts Sci.* **57**, 175–191 (1922).

conductors emphasizes the value of including a wide range of parameter values in any general survey of thermoviscous behavior.

For weak or moderate conductors, the absorption and dispersion measures for type I waves at moderately low frequencies can be expressed with any desired precision by means of power-series expansions in the frequency number X:

$$\left(\frac{c}{c_0}\right)^2 = 1 + \frac{1}{4}X^2[3 + 10(\gamma - 1)Y - (\gamma - 1)(7 - 3\gamma)Y^2] + 0(X^4)$$

$$\frac{\alpha}{k_0} \equiv \frac{A_0}{2\pi} = \frac{1}{2}X\left\{1 + (\gamma - 1)Y - \frac{1}{8}X^2[5 + 35(\gamma - 1)Y + (\gamma - 1)(35\gamma - 63)Y^2\right.$$
$$\left. + (\gamma - 1)(5\gamma^2 - 30\gamma + 33)Y^3]\right\} + O(X^5) \quad (3c\text{-}76)$$

$$\frac{\alpha}{k} \equiv \frac{A}{2\pi} = \frac{1}{2}X\left\{1 + (\gamma - 1)Y - \frac{1}{4}X^2[1 + 11(\gamma - 1)Y - (\gamma - 1)(23 - 11\gamma)Y^2\right.$$
$$\left. + (\gamma - 1)(\gamma^2 - 10\gamma + 13)Y^3]\right\} + O(X^5)$$

Note that $\alpha/k \equiv \alpha\lambda/2\pi \equiv A/2\pi$, where A is the amplitude attenuation per wavelength, and that α/k_0 is similarly related to the attenuation per reference wavelength λ_0. The series (3c-76) can be used with confidence for almost any values of γ and Y so long as the frequency is low enough to keep $X < 0.1$, and for a somewhat wider range of X when certain restrictions on γ and Y are satisfied.[1]

On the other hand, for frequencies high enough to make $X^{-2} \ll 1$, the absorption and dispersion are given, within $O(X^{-2})$, by

$$\frac{(c/c_0)^2}{2X} = \frac{\alpha}{k} \equiv \frac{A}{2\pi} = \frac{A_0^2X}{2\pi^2} = \left(\frac{\alpha}{k_0}\right)^2 2X$$
$$= 1 - \frac{1 - Y}{(1 - \gamma Y)X} \quad (3c\text{-}77)$$

It can be inferred at once from (3c-77) that, for sufficiently high frequencies, dispersion is always anomalous (i.e., speed *increases* with frequency) regardless of γ and Y; that $\alpha/k = A/2\pi$ approaches the limit 1, and that α/k_0 and A_0 recede to zero as the actual wavelength decreases with respect to the reference wavelength λ_0. It also follows, from comparison of this result with (3c-76₃), that as frequency increases, $\alpha = A/\lambda = A_0/\lambda_0$ *will always have at least one maximum that is characteristic of viscothermal resonance.* The frequency at which this resonance occurs lies in the range $X = 1$ to 1.7, but the peak is relatively broad and flat and often cannot be located experimentally with high precision.

It can also be deduced from (3c-77) that the asymptotic speed of sound at very high frequencies will always be determined by viscosity alone, without regard for the form of the equation of state; thus,

$$(c^2)_{X \to \infty} = \frac{2\omega\eta\mathcal{V}}{\rho} \quad (3c\text{-}78a)$$

Under the same limiting conditions, the asymptotic speed of type II, or "thermal," waves is similarly determined by thermal conductivity alone, according to

$$(c'^2)_{X \to \infty} = \frac{2\omega\kappa}{\rho_0 C_p} \quad (3c\text{-}78b)$$

The steady increase of c' with $\omega^{\frac{1}{2}}$ predicted by (3c-78b) has sometimes been cited as a basis for denying that second sound in helium II, which displays small dispersion and low attenuation,[2] can be a type II thermal wave of the sort predicted by viscothermal

---

[1] Truesdell, *J. Rational Mechanics and Analysis* **2**, 643–741 (October, 1953).

[2] Peshkof, *J. Phys. (U.S.S.R.)* **8**, 381 (1944); **10**, 389–398 (1946); Lane, Fairbank, and Fairbank, *Phys. Rev.* **71**, 600–605 (1947).

theory. This conclusion is probably correct but the argument is faulty inasmuch as the vanishing viscosity of the superfluid would make it more appropriate to use as a type criterion the behavior predicted for the limiting condition $X \rightarrow 0$. Thus, if the Kirchhoff-Langevin secular equation (3c-73) is reduced by letting $X \rightarrow 0$ while XY is held fixed, and if XY is then allowed to increase indefinitely as required by the super-conductivity of helium II, what is left of (3c-73) *does* have a pair of roots for which the attenuation vanishes and the speed is nondispersive, viz., $\alpha = A_0 = 0$ and $c = c_0/\gamma^{\frac{1}{2}}$. This result looks, at first sight, like just an isothermal velocity for type I waves, as might be expected to prevail if uniform temperature were enforced by infinite conductivity. On the other hand, the wave types would be expected to interchange, according to (3c-75a), as Y becomes very large; and one has also to deal with the standing conclusion that any viscosity however small will eventually take over control of dispersion when X departs sufficiently from zero. These remarks are intended to emphasize primarily the fact that the problem of branch determination, or type identification, under such extreme circumstances needs probably to be attacked by considering the relative *rates* at which the various limiting conditions are approached. Other considerations need also to be taken into account, of course, in dealing with the two-fluid-mixture theory of liquid helium; but it seems clear that further inquiry is warranted concerning the relevance of classical viscothermal concepts now that a more exact theory of these effects is available.

The Kirchhoff approximation for weak or moderate conductors at low frequencies can be obtained directly from (3c-76) by neglecting terms in X^2 or higher. The dispersion is thereby predicted to be negligible, so that $c \doteq c_0$; and the "Kirchhoff" attenuation α_K is given by

$$\alpha_K = \frac{1}{2} k_0 [X + (\gamma - 1)XY] = \frac{1}{2} k_0 S \left(\mathcal{V} + \frac{\gamma - 1}{\mathcal{P}} \right)$$

$$= \frac{\omega^2}{2\rho_0 c_0^3} \left[\eta \mathcal{V} + \frac{(\gamma - 1)\kappa}{C_p} \right] \tag{3c-79a}$$

If the Stokes relation is then presumed, by setting $\mathcal{V} = \frac{4}{3}$ (which neither Kirchhoff nor Stokes himself did in this connection), (3c-79a) becomes

$$\alpha_C = \frac{1}{2} k_0 S \left(\frac{4}{3} + \frac{\gamma - 1}{\mathcal{P}} \right) = \frac{\omega^2}{2\rho_0 c_0^3} \left[\frac{4}{3} \eta + \frac{(\gamma - 1)\kappa}{C_p} \right] \tag{3c-79b}$$

The absorption predicted by (3c-79b) is commonly, but not very appropriately, referred to as "classical"; but such an emasculated theoretical prediction neither accounts adequately for the attenuation observed experimentally, except in the case of a few monatomic gases, nor does justice to the essential content of the classical theory of viscous conducting fluids.

Even when terms through X^2 are included, no change occurs in the odd function α/k_0, but dispersion is then predicted according to (3c-76₁) which accounts for the second-order effects of both compressional and shear viscosity, heat conduction, and their interaction. This dispersion is anomalous for weak or moderate conductors (small Y) but becomes normal if the speed-reducing influence of thermal conductivity becomes large enough to make $(7 - 3\gamma)Y > 10$. On the other hand, if heat exchange were to be ignored altogether, the first two terms of (3c-76₁) would give, for the dispersion due to viscosity alone,

$$\left(\frac{c}{c_0} \right)^2 \doteq 1 + \frac{3}{4} X^2 = 1 + \frac{3}{4} \left(\frac{\omega \eta \mathcal{V}}{\rho_0 c_0^2} \right)^2$$

$$c \doteq c_0 \left[1 + \frac{3}{8} \left(\frac{\omega \eta \mathcal{V}}{\rho_0 c_0^2} \right)^2 \right] \tag{3c-80}$$

Absorption and Dispersion Due to Heat Radiation. The effects of heat exchange by radiation, which were abandoned above in order to make (3c-74) more manageable, can now be assessed by reverting to (3c-73). The nonlinear interaction between radiation and viscosity will be neglected, for the sake of expediency, even though (3c-74) suggests that it may be as large as second order. The primary effects of viscosity and heat conduction can be eliminated from (3c-73) by letting both X and XY go to zero while holding the frequency variable $S' = \omega/q$ finite. This reduces the characteristic secular equation to the simple quadratic form

$$\gamma(S' - j) + \left(\frac{\alpha + jk}{k_0}\right)^2 (\gamma S' - j) = 0 \qquad (3c\text{-}81)$$

which can be solved directly to yield the following exact expressions for the attenuation and dispersion due to radiation alone:

$$\frac{A}{2\pi}\left(\frac{c_0}{c}\right)^2 = \frac{\alpha}{k}\left(\frac{c_0}{c}\right)^2 = \gamma S' \frac{\gamma - 1}{2[1 + (\gamma S')^2]}$$

$$\left(\frac{A_0}{2\pi}\right)^2 = \left(\frac{\alpha}{k_0}\right)^2 = \frac{1}{2}\gamma \frac{(1 + S')^{\frac{1}{2}}(1 + \gamma^2 S'^2)^{\frac{1}{2}} - (1 + \gamma S'^2)}{1 + (\gamma S')^2} \qquad (3c\text{-}82)$$

$$\left(\frac{c}{c_0}\right)^2 = \frac{2}{\gamma} \frac{1 + (\gamma S')^2}{(1 + \gamma S'^2) + (1 + S'^2)^{\frac{1}{2}}(1 + \gamma^2 S'^2)^{\frac{1}{2}}}$$

These equations indicate that both attenuation and dispersion become vanishingly small for either very large or very small values of S', and that a maximum of attenuation occurs in mid-range, near the single point of inflection of the dispersion curve. This absorption peak is characterized by

$$\left(\frac{\alpha}{k}\right)_{\max} = \frac{\gamma^{\frac{1}{4}} - 1}{\gamma^{\frac{1}{4}} + 1} \qquad S'_{\max A} = \gamma^{-\frac{1}{2}} \qquad \tau_{\text{rad}} = \frac{2\pi\gamma^{\frac{1}{2}}}{q}$$

$$\left(\frac{\alpha}{k_0}\right)_{\max} = \left(\frac{A_0}{2\pi}\right)_{\max} = \frac{\gamma - 1}{[8(\gamma + 1)]^{\frac{1}{2}}} \qquad S'_{\max A_0} = \gamma^{-1}\frac{(3\gamma + 1)^{\frac{1}{2}}}{(\gamma + 3)^{\frac{1}{2}}} \qquad (3c\text{-}83)$$

There is a curious dearth of quantitative information concerning the radiation coefficient q, and little is added to this by noticing the low attenuation and negligible dispersion observed for a wide range of audible sounds in air since these might correspond to values of S' either far above or far below the resonance peak described by (3c-83). The choice $S' \gg 1$ is unambiguously dictated, however, by the fact that the observed speed of sound is very close to the isentropic value c_0, whereas (3c-82₅) indicates that the isothermal speed $c_0/\gamma^{\frac{1}{2}}$ would prevail if q were large enough to make S' small for all audio frequencies. Truesdell[1] has pointed out that these conclusions leave still in effect a prediction that at some lower subaudible frequency a peak of attenuation should appear with a magnitude $A_0 = 0.185\pi$ (≈ 5 dB per reference wavelength). This absorption peak has not been observed yet, at least deliberately, although its possible bearing on the acoustical character of thunder might be worth investigating.

Relaxation Processes and Sound Absorption. The foregoing analysis of heat exchange by radiation puts in evidence the first example of what would now be called a typical relaxation process. The characteristic feature of such a process, in so far as the gross hydrodynamical response of the medium is concerned, is the existence of two relations among the state variables, one of which prevails asymptotically for slow variations, the other for rapid changes. Such bivalent behavior is typical of fluid mixtures containing two interacting components, such as a partly dissociated gas[2] or an ionic solution.[3] In these cases the relative concentrations of the two components either follow faithfully, in quasi-static equilibrium, the dictates of slowly changing external variables, or else, at the other asymptotic limit, they do not change at all

[1] C. A. Truesdell, *J. Rational Mechanics and Analysis* **2**, 643–741 [666] (October, 1953).
[2] Einstein, *Sitzber. deut. Akad. Wiss. Berlin Math.-Phys. Kl.* **1920**, 380–385.
[3] Liebermann, *Phys Rev.* **76**, 1520–1524 (1949).

when the finite reaction rate is such that the external variables can complete cyclic changes too rapidly for the concentrations to "follow." A different but comparable kind of mixture is exemplified by an ensemble of atoms or molecules capable of being excited to different energy levels, of which the most common example is a diatomic gas in which the rotational degrees of freedom may or may not share the cyclic work of compression depending on whether an appropriately normalized frequency variable is "low" or "high."

The *physical* problem of characterizing the rate-dependent properties of mixtures can be studied without regard for its acoustical consequences, and various approaches to this problem have turned on the assignment of two or more different internal or "partial" temperatures, different compressibilities, specific heats, etc. All the physical theories of pure relaxation appear to converge, however, in predicting the same *acoustical* behavior; viz., at low frequencies an asymptotic speed of sound c^0, a transition region of anomalous dispersion ($dc/d\omega > 0$) within which a maximum of attenuation occurs, and at high frequencies an asymptotic sound speed c^∞ which can be related to c^0 by writing $K \equiv c^0/c^\infty \leq 1$, where K is a material constant of the two-component medium. It follows then that, when the constant K and a dimensionless frequency variable X' can be properly identified and interpreted in terms of the *physical* mechanism involved, the *acoustical* behavior for any pure relaxation process will be described exactly by the following expressions derived from (3c-82) and (3c-83) by substitution:

$$\left(\frac{c}{c^0}\right)^2 = \frac{2(1 + X'^2)}{1 + K^2X'^2 + [(1 + K^4X'^2)(1 + X'^2)]^{\frac{1}{2}}}$$

$$\doteq \frac{1 + X'^2}{1 + K^2X'^2}$$

$$\frac{\alpha}{k}\left(\frac{c^0}{c}\right)^2 = \frac{1}{2}\frac{(1 - K^2)X'}{1 + X'^2} \tag{3c-84}$$

$$\left(\frac{\alpha}{k}\right)_{\text{max}} = \frac{1 - K}{1 + K} \qquad \left(\frac{\alpha}{k_0}\right)_{\text{max}} = \frac{1 - K^2}{[8(1 + K^2)]^{\frac{1}{2}}}$$

$$X'_{\text{max A}} = K^{-1} = \frac{c^\infty}{c^0} \qquad X'_{\text{max A0}} = \left(\frac{3 + K^2}{1 + 3K^2}\right)^{\frac{1}{2}}$$

These equations revert exactly to (3c-82) and (3c-83) when the substitutions $K^2 = \gamma^{-1}$ and $X' = \gamma S'$ are made, and when a factor γ^{-1} is introduced to convert the low-frequency reference speed c^0 to the usual isentropic reference c_0.

The "resonance" frequency characterizing a relaxation process is usually defined as the angular frequency at which the maximum attenuation per wavelength, $A = \alpha\lambda$, occurs; thus, $\omega_r \equiv 2\pi/\tau_r = (\omega/X')X'_{\text{max A}}$, where τ_r is the related "relaxation period." It has been pointed out that *any* mechanism of sound absorption can be interpreted as a relaxation phenomenon by suitably defining its relaxation time. For example, viscosity and heat-conduction "relaxation times" and their associated "resonance frequencies" can be defined by writing

$$\tau_v = \frac{2\pi}{\omega_v} = \frac{X}{\omega}\frac{4}{3\mho} = \frac{\frac{4}{3}\eta}{\rho_0c_0^2} \qquad \tau_\kappa = \frac{2\pi}{\omega_\kappa} = \frac{XY}{\omega} = \frac{\kappa}{\rho_0c_0^2C_p} \tag{3c-85}$$

Note that ω_v is specified in such a way that it reduces to ω/X when $\mho$ has the Stokes-relation value $\frac{4}{3}$. When these relaxation frequencies are introduced in (3c-79) and (3c-80), the second-order dispersion and the Kirchhoff linear approximation for attenuation become

$$c \doteq c_0\left[1 + \frac{3}{8}(2\pi)^2\left(\frac{3\mho}{4}\right)^2\frac{\omega^2}{\omega_v^2}\right]$$

$$\alpha_K = \pi k_0\left[\frac{3\mho}{4}\frac{\omega}{\omega_v} + (\gamma - 1)\frac{\omega}{\omega_\kappa}\right] \tag{3c-86}$$

When the fluid medium consists of an ideal monatomic gas, the physical significance of the relaxation times τ_v and τ_κ can readily be interpreted as the time required for subsidence of a momentary departure from the equilibrium distribution of energy among the translational degrees of freedom. In the classical kinetic theory of gases, this recovery time is shown to be approximately $L/\bar{v}$, the mean free path divided by the mean molecular velocity.[1] The conformity of the definitions (3c-85) with this concept can then be verified by recalling the kinetic-theory evaluations of viscosity $[\eta \doteq \frac{1}{2}\rho\bar{v}L]$, thermal conductivity $[(\kappa/C_p) \doteq (5/4\gamma)\rho\bar{v}L]$, and the speed of sound $[c \doteq 0.74\bar{v}]$. These considerations show, incidentally, that for such a gas the attenuation per reference wavelength is contributed almost equally by viscosity and heat conduction, and is proportional to the ratio of mean free path to wavelength.

The precise physical significance of τ_v and τ_κ is less obvious for polyatomic gases and liquids; but if this is glossed over, the frequency ratios $\frac{3}{2}\pi\mho\omega/\omega_v$, $\frac{4}{3}\omega_v/\omega_\kappa\mho$, and $2\pi\omega/\omega_\kappa$ can be substituted directly for X, Y, and XY in any of the viscothermal relations deduced above. Merely introducing these "relaxation" frequencies, however, does not invest heat conduction or viscosity with any new or different relaxation-like properties, and the exact viscothermal theory, in whatever symbols expressed, continues to predict that sound speed will increase indefinitely with frequency, that A_0 will display a typical broad maximum for some X in the range 1 to 1.7 (depending on the thermoviscous parameters γ and Y), that $(A_0)_{max}$ will always have about the same magnitude $(\alpha/k_0 \approx \frac{1}{3})$, and that the peak in A_0 can be made to occur at any chosen actual frequency by suitable assignment of the viscosity number $\mho$ [cf. (3c-72), (3c-85)]. In contrast with this behavior, a pure relaxation phenomenon would call for the sound speed to level off at the high-frequency limit given by K^{-1}, and would display a maximum in A_0 that increases in height and retreats toward higher frequencies as the speed increment $c^\infty - c^0$ increases and K varies from 1 toward zero.

Allusion has already been made to the established fact that measured values of attenuation usually exceed the "classical" prediction (3c-79b) and often exhibit one or more maxima at finite frequencies. As a matter of fact, even when the complete consequences of the classical theory are taken into account, and when the viscosity number is adjusted to make the predicted attenuation at low frequencies correspond with experiment, the classical viscothermal theory still fails to account for all the experimental facts, but *for a reason that is just the opposite of that usually advanced*, namely, because it then predicts *too much* attenuation at the resonance peak and at higher frequencies! In spite of this latent contradiction, the alleged failure of "classical" theory as represented by (3c-79b) (which is, after all, only *part* of an *approximate* solution of the *linearized first-order* equations) has stimulated widespread efforts to repair its deficiency by invoking a wide variety of relaxation and other theories,[2] many of which have been marred by an *ad hoc* flavor that renders them little more than examples of ingenuity in curve fitting.

Measurements of absorption and dispersion in rarefied helium gas over a wide range of the frequency variable S have confirmed in all essential details the pattern of behavior predicted by the *exact* viscothermal theory.[3] Unless the classical concepts of viscosity and heat conduction are to be abandoned altogether, therefore, logic demands that the *exact* viscothermal theory be accepted as the foundation on which to erect any more complete analysis of sound absorption in media less idealized than rarefied

---

[1] Jeans, "Dynamical Theory of Gases," 2d ed., pp. 260–262, Cambridge University Press, Cambridge, England, 1916.

[2] For reviews of what has been called the "exuberant literature" dealing with relaxation and other theories of sound absorption, see Kneser, *Ergeb. exakt. Naturwiss.* **22**, 121–185 (1949); Markham, Beyer, and Lindsay, *Revs. Modern Phys.* **23**, 353–411 (1951); Kittel, *Phys. Soc. (London), Repts. Progr. in Phys.* **11**, 205–247 (1948); see also, for background, W. T. Richards, *Revs. Modern Phys.* **11**, 36–64 (1939).

[3] Greenspan, *Phys. Rev.* **75**, 197–198 (1949); *J. Acoust. Soc. Am.* **22**, 568–571 (1950).

helium. A good many "honest" relaxation mechanisms do exist and must be accounted for, but in the accounting these effects should presumably be regarded as factors perturbing the fundamental thermoviscous behavior rather than the converse. The two-fluid-mixture theory of relaxation effects seems best adapted for inclusion in such a compound analysis, and a start in this direction has already been made.[1] Much remains to be done, however, before this basic acoustical problem can be said to be understood.

3c-10. Characteristic Acoustic Impedance of a Thermoviscous Medium. When the first-order sound pressure p_1 is put back into (3c-70$_2$) [by tracing its last term back through (3c-25$_{11}$)], this equation of motion can be rewritten at once in terms of the specific acoustic impedance, as follows:

$$[j\omega\rho_0 - (\alpha + jk)^2\eta\mathcal{U}]u_1 - (\alpha + jk)p_1 = 0$$

$$\frac{p_1}{u_1} \equiv Z = jk\rho_0 c(\alpha + jk)^{-1} - \eta\mathcal{U}(\alpha + jk)$$

$$= \rho_0 c\left(1 - j\frac{\alpha}{k}\right)^{-1} - j\rho_0 c\,\frac{\omega\eta\mathcal{U}}{\rho_0 c_0^2}\left(\frac{c_0}{c}\right)^2\left(1 - j\frac{\alpha}{k}\right) \qquad (3\text{c-}87)$$

$$\frac{p_1}{\rho_0 c u_1} \equiv z = \left(1 - j\frac{\alpha}{k}\right)^{-1} - jX\left(\frac{c_0}{c}\right)^2\left(1 - j\frac{\alpha}{k}\right)$$

The normalized specific impedance, or *specific impedance ratio*, $(p_1/\rho_0 c u_1) \equiv z$, which would be unity in the nondissipative case, is now in a form to be evaluated by direct substitution of the series expansions (3c-76). After some manipulation, and retaining only terms through X^2 and Y^2, the impedance ratio can be put in the form

$$\frac{p_1}{\rho_0 c u_1} = 1 - \frac{\alpha}{k}\left[\frac{\alpha}{k} + X\left(\frac{c_0}{c}\right)^2\right] + j\left[\frac{\alpha}{k} - X\left(\frac{c_0}{c}\right)^2\right]$$

$$= 1 - \frac{1}{4}X^2[3 + 4(\gamma - 1)Y + (\gamma - 1)^2Y^2] + O(X^4)$$

$$- j\left\{\frac{1}{2}X[1 - (\gamma - 1)Y] + O(X^3)\right\} \qquad (3\text{c-}88)$$

It follows that sound pressure *lags* the particle velocity when $(\gamma - 1)\kappa/\eta\mathcal{U}C_p$ is less than unity, as it is for the common fluids under ordinary conditions; but pressure *leads* the particle velocity when the ratio of heat conductivity to viscosity is high enough to make $(\gamma - 1)\kappa > \eta\mathcal{U}C_p$.

3c-11. Thermal Noise in the Acoustic Medium. The mode of motion that is heat furnishes a restless background of noise that underlies all acoustical phenomena. The magnitude and nature of this thermal noise can be assessed by appealing to concepts drawn from such apparently unrelated sources as architectural acoustics, elementary quantum theory, and the classical kinetic theory of gases.

The scheme of analysis can be described simply: the thermoacoustic noise energy density, as measured by the mean-square sound pressure, is set equal to the density of the internal energy of thermal agitation associated with the translational degrees of freedom of the molecules composing the medium. It is then postulated that these molecular motions of thermal agitation can be regarded as a vector summation of the motions associated with a three-dimensional manifold of compressional standing waves, each behaving as it would in an ideal continuous medium having the same gross mechanical and elastic properties that characterize the actual medium. Each of these standing-wave systems thus constitutes an allowed, thermally excited, normal mode of vibration, or degree of freedom, to which can be assigned, in accordance with elementary quantum theory, the average energy

[1] Z. Sakadi, *Proc. Phys.-Math. Soc. Japan* (3) **23**, 208–213 (1941); Meixner, *Acustica* **2**, 101–109 (1952).

$$\frac{\text{Energy}}{\text{Mode}} = \frac{hf}{\exp{(hf/kT)} - 1} \tag{3c-89}$$

where h is Planck's constant, k is Boltzmann's constant, T is the absolute temperature, and f is the frequency in hertz.

The incremental number of such energy-bearing modes of vibration is given by the count of normal frequencies lying between f and $f + df$; and this is given, as in the theory of room acoustics,[1] by

$$dN = \left(\frac{4\pi V f^2}{c^3} + \frac{\pi S f}{2c^2} + \frac{L}{2c} \right) df \tag{3c-90}$$

where V is the volume, S the total surface, and L the sum of the three dimensions of the region under consideration, and where the three terms represent, respectively, the normal-frequency "points" distributed throughout the volume, over the coordinate planes, and along the coordinate axes of an octant of frequency space. If the three dimensions of the region are not too disparate, S can be approximated by $6V^{\frac{2}{3}}$, and L by $3V^{\frac{1}{3}}$, giving

$$dN = \frac{4\pi V f^2 df}{c^3} \left[1 + \frac{3\lambda}{4V^{\frac{1}{3}}} + \frac{3\lambda^2}{8\pi V^{\frac{2}{3}}} \right] \tag{3c-91}$$

For sufficiently high frequencies, this reduces to the classical expression (Rayleigh, 1900; Jeans, 1905) for the distribution of normal frequencies,

$$dN = \frac{4\pi V f^2 df}{c^3} \tag{3c-92}$$

an aymptotic form that can be shown (Weyl, 1911) to be independent of the shape of V and rigorously valid in the limit when $\lambda = c/f$ becomes small in comparison with $V^{\frac{1}{3}}$.

If attention is confined for the moment to finite frequency bands that do not include the lower frequencies, the incremental translational energy density of thermal agitation will be given by the product of (3c-89) and (3c-92). Then, by hypothesis, this can be set equal to the incremental energy density of the diffuse sound field, which is given by $d(\langle p^2 \rangle / \rho c^2)$, where p is the rms sound pressure; thus

$$\begin{aligned} d \frac{\langle p^2 \rangle}{\rho c^2} &= \frac{(4\pi f^2 df/c^3)hf}{\exp{(hf/kT)} - 1} \\ &= \frac{(4\pi kT/c^3)f^2 df(hf/kT)}{\exp{(hf/kT)} - 1} \\ &= \frac{4\pi kT}{c^3} f^2 df \left[1 - \frac{1}{2} \frac{hf}{kT} + \frac{1}{12} \left(\frac{hf}{kT} \right)^2 - \cdots \right], \quad \left(\frac{hf}{kT} \right)^2 < 4\pi^2 \end{aligned} \tag{3c-93} \tag{3c-94}$$

The total energy density associated with all the allowed modes of vibration is then to be found by extending the integral of (3c-94) over all frequencies less than the upper limiting frequency for which the mode count [by (3c-92)] is just equal to three times n_V, the total number of molecules in unit volume. This upper frequency limit, f_{lim}, is given, for either liquids or gases, by the integral of (3c-92),

$$\frac{N_{\text{lim}}}{V} = \frac{4\pi f_{\text{lim}}^3}{3c^3} = 3n_V = 3A \frac{\rho}{M} \qquad f_{\text{lim}}^3 = \frac{9c^3 A \rho}{4\pi M} \tag{3c-95}$$

where A is Avogadro's number (6.025×10^{26} molecules/kg mole), ρ is in kg/m³, and M is the molecular weight (numeric, $O_2 = 32$). At ordinary room temperature, $f_{\text{lim}} \approx 2 \times 10^{10}$ Hz for air, $\approx 4 \times 10^{12}$ Hz for water. These frequencies are well outside the range so far accessible for acoustical experimentation and need not be

---

[1] Maa, *J. Acoust. Soc. Am.* **10**, 235–238 (1939); Bolt, *J. Acoust. Soc. Am.* **10**, 228–234 (1939).

considered further except when the foregoing notions are used as the basis for a theory of specific heats, in which case it is necessary also to take into account vibrational and rotational degrees of freedom, and to reexamine the equilibrium statistics that underlie (3c-89). Note in passing that the *phonon* of specific-heat theory merely identifies the burden of internal energy carried by each of the normal modes of vibration postulated above.

Within the ranges of frequency and temperature ordinarily of interest in the assessment of thermal noise, the exponent hf/kT is so small that even the linear term in the series expansion of (3c-94) can be omitted. This amounts to a reversion to the classical analysis of energy partition in continuous media[1] and to the assignment of an energy kT to each allowed mode of vibration. With this simplification, (3c-94) can be integrated at once to yield the mean-square sound pressure, in the frequency band $f_2 - f_1$, as

$$\langle p^2 \rangle = \frac{4}{3} \pi k T \frac{\rho}{c} (f_2{}^3 - f_1{}^3) \qquad \text{(newtons/m}^2)^2 \qquad \text{(3c-96)}$$

in which Boltzmann's constant $k = 1.380 \times 10^{-23}$ joule/K, T is in kelvins, ρ in kg/m³, and c in m/sec. To facilitate computation, it is useful to rearrange (3c-96) in the following forms:

$$p_{rms} = 1.3 \times 10^{-12} \left(\frac{\rho}{c}\right)^{\frac{1}{2}} \left[\frac{T}{293} (f_2{}^3 - f_1{}^3) \right]^{\frac{1}{2}} \qquad \text{newtons/m}^2 \qquad \text{(3c-97a)}$$

$$(p_{rms})_{air} = 0.76 \times 10^{-10} \left[\frac{T}{293} (f_2{}^3 - f_1{}^3) \right]^{\frac{1}{2}} \qquad \text{dynes/cm}^2 = \mu b \qquad \text{(3c-97b)}$$

$$(p_{rms})_{sea\ water} = 10.6 \times 10^{-10} \left[\frac{T}{293} (f_2{}^3 - f_1{}^3) \right]^{\frac{1}{2}} \qquad \mu b \qquad \text{(3c-97c)}$$

in which the constants have been adjusted to make the temperature factor reduce to unity at 20°C, and where ρ/c has been taken as 0.00345 for air and 0.67 for sea water. It follows, for example, that the rms thermal noise pressure, for the wide-range audio-frequency band extending to 19 kHz in air, is just equal to the reference sound pressure, $p_0 = 0.0002\ \mu b$.

The power spectrum of thermal noise can be deduced from either (3c-94) or (3c-97b) and may be expressed as a *sound spectrum level* by writing

$$\beta_{noise} = 10 \log_{10} \left(\frac{d(\langle p^2 \rangle / p_0{}^2)}{df} \right) = 10 \log_{10} \frac{4\pi k T f^2 \rho}{c p_0{}^2}$$

$$= 10 \log_{10} \left[4.33 \times 10^{-7} (f_{kc/s})^2 \frac{T}{293} \right]$$

$$= -63.6 + 20 \log_{10} f_{kc/s} + 10 \log_{10} \frac{T}{293} \qquad \text{db} \qquad \text{(3c-98)}$$

Note that this noise spectrum is *not* "white" but has instead a uniform positive slope of 6 dB/octave, corresponding to an rms thermal-noise sound pressure that is directly proportional to frequency. On the other hand, for frequencies low enough to make the additive "correction" terms of (3c-91) significant, the noise spectrum level tends increasingly to lie *above* the +6 dB/octave line as the frequency approaches the low-frequency cutoff at which only the gravest mode of vibration can be excited. The noise spectrum level can also be expected to vary erratically as the low-frequency limit is approached and the population of normal frequencies becomes sparse, in much the same way that the steady-state pressure response of small rooms varies irregularly with frequency when only a few normal modes of vibration are available for excitation. It does not follow, however, that thermal noise in such a small enclosure could be

[1] Jeans, "Dynamical Theory of Gases," 2d ed., pp. 381-391.

"quieted" by the application of sound absorbents. The boundary surfaces, without regard for their acoustical character, will always reach the same radiative equilibrium with the interior medium if both are at the same temperature; otherwise there would be a net flow of thermal "noise" energy across the boundaries in the guise of ordinary heat transfer.

The possibility that thermal noise might be the factor that limits human hearing acuity can be assessed with the help of (3c-98). If the critical-band theory of masking by wide-band noise continues to hold for subliminal stimuli, the effective masking level of thermal noise can be found by adding, at any frequency, the critical bandwidth (expressed as $10 \log_{10} \Delta f_c$) and the spectrum level given by (3c-98). Comparing this result with the binaural threshold for random incidence then leads to the conclusion that thermal noise remains about 11 to 13 dB below threshold at the frequency of greatest vulnerability (ca. 3 to 5 kHz), even for young people with exceptionally acute hearing. On this basis human hearing might be assigned a "noise figure" of approximately 12 dB. It is probable that some at least of this failure to achieve ideal function can be ascribed to internal noise of physiological origin. The near miss on thermal noise limiting gives comforting reassurance, however, that not more than a few decibels of additional hearing acuity could be utilized effectively by humans even if biological adaptation were to make it available.

3d. Acoustic Properties of Gases

LEO L. BERANEK

Bolt Beranek and Newman Inc.

A number of the physical properties of a gas are important in determining its acoustic characteristics. These include density, pressure, temperature, specific heats, and coefficients of viscosity. These properties, and others, are presented and discussed below.

3d-1. Density. The density ρ_0 of a number of common gases at standard temperature and pressure is given in Table 3d-1. The density at any temperature and pressure can be obtained from the expression

$$\rho = \rho_0 \frac{P}{760} \frac{273.16}{T}$$

where P is the barometric pressure in millimeters of mercury, and T is the absolute temperature in kelvins.

3d-2. Atmospheric Pressure and Temperature. The atmospheric pressure and air temperatures, and consequently the air density, vary with elevation above the surface of the earth. Table 3d-2 gives the air pressure, temperature, density, and mean molecular weight as a function of altitude. This is the U.S. standard atmos-

TABLE 3d-1. DENSITY OF GASES ρ_0 AT 0°C, 1 ATM*

Gas	Formula	ρ_0, kg/m^3	ρ_0, lb/ft^3
Acetylene.................	C_2H_2	1.173	0.0732
Air........................	.....	1.2929	0.08072
Ammonia..................	NH_3	0.7710	0.0481
Argon.....................	A	1.7837	0.1114
Carbon dioxide.............	CO_2	1.977	0.1234
Carbon monoxide...........	CO	1.250	0.0780
Chlorine...................	Cl_2	3.214	0.2006
Ethane (10°C)..............	C_2H_6	1.356	0.0846
Ethylene...................	C_2H_4	1.260	0.0786
Helium....................	He	0.1785	0.01114
Hydrogen..................	H_2	0.0899	0.00561
Hydrogen sulfide............	H_2S	1.539	0.0961
Methane...................	CH_4	0.7168	0.0447
Neon......................	Ne	0.9003	0.0562
Nitric oxide (10°C)..........	NO	1.340	0.0836
Nitrogen...................	N_2	1.2506	0.0781
Nitrous oxide..............	N_2O	1.977	0.1234
Oxygen....................	O_2	1.429	0.0892
Propane...................	C_3H_8	2.009	0.1254
Sulfur dioxide..............	SO_2	2.927	0.1827
Steam (100°C).............	H_2O	0.598	0.0373

* "Handbook of Chemistry and Physics," 48th ed.

phere used for the calibration of aeronautical instruments. The actual atmosphere differs from summer to winter.[1]

At 288.16 K (15.0°C) and at standard gravity of 9.80665 m/sec^2 a 0.760-m column of mercury exerts a pressure of 1.01325×10^5 newtons/m^2. This is the standard ICAO atmosphere. Other ICAO values are: density 1.225014 kg/m^3, kinematic viscosity 1.4607413×10^{-5} m^2/sec, mean free path 6.6317223×10^{-8} m, molecular weight 28.966 (dimensionless), sound speed 340.29205 m/sec, specific weight 12.013284 kg/(m^2-sec^2), coefficient of viscosity 1.7894285×10^{-5} kg/(m-sec).

3d-3. Specific Heat. For several common gases the values of C_p, the specific heat at constant pressure, and γ, the ratio of C_p to C_v, are given in Table 3d-3. C_v is the specific heat at constant volume. C_p is expressed in calories per g-°C.

3d-4. Viscosity. The coefficient of viscosity η of a number of gases is given in Table 3d-4. The units of η are dyne-seconds per square centimeter, or poises. For example, the coefficient of viscosity for air at 0°C is 1.708×10^{-4} poises (dyne-sec/cm^2).

The ratio η/ρ of viscosity to density occurs frequently and is known as the *kinematic viscosity coefficient*. It is usually designated by the letter ν and has the dimensions square centimeters per second in the cgs system.

For a plane acoustic wave propagating in an unbounded gas a small attenuation will occur because of viscosity. The attenuation factor is $e^{-\alpha_\eta x}$ for the pressure (or particle velocity), where

$$\alpha_\eta = \frac{2}{3} \frac{\eta}{\rho} \frac{\omega^2}{c^3} = \frac{2}{3} \nu \frac{\omega^2}{c^3} \quad \text{nepers}$$

where c is the speed of sound, and ω the angular frequency of the wave.

[1] L. L. Beranek, "Acoustic Measurements," p. 42, John Wiley & Sons, Inc., New York, 1040.

TABLE 3d-2. U.S. STANDARD ATMOSPHERE, 1962*

Altitude, km	Temp. T, K	Pressure P, newtons/m²	Density ρ, kg/m³	Molecular weight M
−5	320.676	1.77762 +5	1.9311 +0	28.964
−4	314.166	1.59598	1.7697	28.964
−3	307.659	1.42973	1.6189	28.964
−2	301.154	1.27783	1.4782	28.964
−1	294.651	1.13931	1.3470	28.964
0	288.150	1.01325	1.2250	28.964
0.5	284.900	9.54612 +4	1.1673	28.964
1.0	281.651	8.98762	1.1117	28.964
1.5	278.402	8.45596	1.0581	28.964
2	275.154	7.95014	1.0066	28.964
2.5	271.906	7.46917	9.5695 −1	28.964
3	268.659	7.01211	9.0925	28.964
4	262.166	6.16604	8.1935	28.964
5	255.676	5.40482	7.3643	28.964
6	249.187	4.72176	6.6011	28.964
7	242.700	4.11052	5.9002	28.964
8	236.215	3.56516	5.2579	28.964
9	229.733	3.08007	4.6706	28.964
10	223.252	2.64999	4.1351	28.964
15	216.650	1.21118	1.9475	28.964
20	216.650	5.52930 +3	8.8910 −2	28.964
25	221.552	2.54922	4.0084	28.964
30	226.509	1.19703	1.8410	28.964
40	250.350	2.87143 +2	3.9957 −3	28.964
50	270.650	7.97790 +1	1.0269	28.964
60	255.772	2.24606	3.0592 −4	28.964
70	219.700	5.52047 +0	8.7535 −5	28.964
80	180.65	1.0366	1.999	28.964
100	210.02	3.0075 −2	4.974 −7	28.964
150	892.79	5.0617 −4	1.836 −9	28.964
200	1235.95	1.3339	3.318 −10	28.964
250	1357.28	4.6706 −5	9.978 −11	28.964
300	1432.11	1.8838	3.585	28.961
400	1487.38	4.0304 −6	6.498 −12	27.97
500	1499.22	1.0957	1.577	26.86
600	1506.13	3.4502 −7	4.640 −13	26.06
700	1507.61	1.1918	1.537 −13	25.17

* See also Sec. 2k, and Tables 2k-4, and 3d-9. Data taken from "U.S. Standard Atmosphere, 1962," published by the U.S. Committee on Extension to the Standard Atmosphere (COESA), Washington, D.C., 1962.

Note. A one- or two-digit number (preceded by a plus or minus sign) following the initial entry indicates the power of 10 by which that entry and each succeeding entry of that column should be multiplied.

3d-5. Thermal Conductivity. The thermal conductivity κ of a number of gases is given in Table 3d-5. The units of κ are calories per centimeter-second-degree.

The quantity $\kappa/\rho C_v$ frequently appears in heat-conduction equations. It is often designated by the symbol α and is called the *thermal diffusivity*. In the cgs system the units of α are square centimeters per second. For air $\alpha = 0.27$ cm²/sec at 18°C and 760 mm of mercury.

TABLE 3d-3. SPECIFIC HEAT AT CONSTANT PRESSURE C_p AND THE RATIO γ
OF C_p TO THE SPECIFIC HEAT AT CONSTANT VOLUME C_v*

$[C_p \text{ (cal/g-deg)}; \gamma = C_p/C_v]$

Gas	Temp., °C (atm)	C_p	Temp., °C (atm)	γ
Air.....................	−120 (10)	0.2719	−118 (1)	1.415
	(20)	0.3221		
	(40)	0.4791		
	(70)	0.7771	− 78 (1)	1.408
	− 50 (10)	0.2440		
	(20)	0.2521		
	(40)	0.2741		
	(70)	0.3121		
	0 (1)	0.2398	0 (1)	1.403
	(20)	0.2484		
	(60)	0.2652	17 (1)	1.403
	50 (20)	0.2480		
	(100)	0.2719		
	(220)	0.2961		
	100 (1)	0.2404	100 (1)	1.401
	(20)	0.2471		
	(100)	0.2600	200 (1)	1.398
	(220)	0.2841		
	400 (1)	0.2430	400 (1)	1.393
	1000 (1)	0.2570	1000 (1)	1.365
	1400 (1)	0.2699	1400 (1)	1.341
	1800 (1)	0.2850	1800 (1)	1.316
Ammonia, NH_3.........	15 (1)	0.5232	15 (1)	1.310
Argon, Ar..............	15 (1)	0.1253	15 (1)	1.668
Carbon dioxide, CO_2.....	15 (1)	0.1989	15 (1)	1.304
Carbon monoxide, CO...	15 (1)	0.2478	15 (1)	1.404
Chlorine, Cl_2...........	15 (1)	0.1149	15 (1)	1.355
Ethane, C_2H_6..........	15 (1)	0.3861	15 (1)	1.22
Ethylene, C_2H_4.........	15 (1)	0.3592	15 (1)	1.255
Helium, He............	−180 (1)	1.25	−180 (1)	1.660
Hydrogen, H_2..........	15 (1)	3.389	15 (1)	1.410
Hydrogen sulfide, H_2S...	15 (1)	0.2533	15 (1)	1.32
Methane, CH_4..........	15 (1)	0.5284	15 (1)	1.31
Neon, Ne..............	25 (1)	0.246	19 (1)	1.64
Nitric oxide, NO........	15 (1)	0.2329	15 (1)	1.400
Nitrogen, N_2...........	15 (1)	0.2477	15 (1)	1.404
Nitrous oxide, N_2O......	15 (1)	0.2004	15 (1)	1.303
Oxygen, O_2............	15 (1)	0.2178	15 (1)	1.401
Propane, C_3H_8.........	....	.....	16 (0.5)	1.13
Steam, H_2O............	100 (1)	0.4820	100 (1)	1.324
Sulfur dioxide, SO_2.......	15 (1)	0.1516	15 (1)	1.29

* "Handbook of Chemistry and Physics," 41st ed.

A plane acoustic wave propagating in an unbounded gas will be attenuated slightly because of thermal conduction effects. The attenuation constant α_T is

$$\alpha_T = \frac{\kappa(\gamma - 1)\omega^2}{2\gamma\rho C_v c^3} \quad \text{nepers}$$

where $\kappa/\rho C_v$ is the thermal diffusivity, γ the ratio of specific heats, c the propagation speed, and ω the angular frequency of the wave.

TABLE 3d-4. COEFFICIENT OF VISCOSITY η FOR DIFFERENT GASES AS A FUNCTION OF TEMPERATURE*

Gas	Formula	Temp., °C	Viscosity, micropoises (cgs)
Air.....................	....	−104.0	113.0
		0	170.8
		18	182.7
		40	190.4
		54	195.8
		74	210.2
		229	263.8
		357	317.5
		409	341.3
		620	391.6
		810	441.9
		1034	490.6
Argon...................	Ar	0	209.6
		20	221.7
		100	269.5
		401	411.5
Carbon dioxide..........	CO_2	−60.0	106.1
		0	139.0
		20	148.0
		40	157.0
		104	188.9
		302	268.2
Carbon monoxide........	CO	−191.5	56.1
		0	166
		15.	172
		126.7	218.3
		276.9	271.4
Helium.................	He	−191.6	87.1
		0	186.0
		20	194.1
		100	228.1
		407	343.6
Hydrogen...............	H	−198.4	33.6
		0	83.5
		20.7	87.6
		129.4	108.6
		412	155.4
Neon...................	Ne	0	297.3
		20	311.1
		100	364.6
		429	545.4
Nitric oxide..............	NO	0	178
		20	187.6
		200	268.2
Nitrogen................	N	−21.5	156.3
		10.9	170.7
		27.4	178.1
		490	337.4
Nitrous oxide............	N_2O	0	135
		26.9	148.8
		126.9	194.3
Oxygen.................	O_2	0	189
		19.1	201.8
		127.7	256.8
		227.0	301.7
		402	369.3

* "Handbook of Chemistry and Physics," 48th ed.

TABLE 3d-5. THERMAL CONDUCTIVITY κ OF GASES AT 0°C*

Gas	Formula	Thermal conductivity κ at 0°C, cal/cm-sec-deg
Air............................	....	0.0576×10^{-3}
Argon......................	A	0.039×10^{-3}
Carbon dioxide.............	CO_2	0.034×10^{-3}
Carbon monoxide..........	CO	0.053×10^{-3}
Helium....................	He	0.343×10^{-3}
Hydrogen.................	H_2	0.419×10^{-3}
Neon......................	Ne	0.110×10^{-3}
Nitric oxide...............	NO	0.046×10^{-3}
Nitrogen..................	N	0.057×10^{-3}
Nitrous oxide.............	N_2O	0.036×10^{-3}
Oxygen...................	O_2	0.058×10^{-3}
Steam (100°C)............	H_2O	0.055×10^{-3}

* Condon and Odishaw, "Handbook of Physics," p. 5–66, McGraw-Hill Book Co., New York, 1958.

3d-6. Speed (Velocity) of Propagation. The speed of sound for small sound amplitudes can be written exactly as[1]

$$c = \left[\frac{RT}{M}\left(f + \frac{gR}{hC_v^\infty}\right)\right]^{\frac{1}{2}}$$

where

$$f = -\frac{V^2}{RT}\left(\frac{\partial p}{\partial V}\right)_T$$

$$g = \left(\frac{V}{R}\frac{\partial p}{\partial T}\right)_v^2$$

$$h = \frac{C_v}{C_v^\infty} = 1 + \frac{T}{C_v^\infty}\int_V^\infty \left(\frac{\partial^2 p}{\partial T^2}\right)_v dV$$

C_v^∞ is the specific heat for constant volume as the volume approaches infinity; M, the molecular weight of the gas, has been substituted for ρV; and R, the gas constant, puts the equation in a useful form. The quantities f, g, h are dimensionless and differ only slightly from unity as determined by the imperfection of the gas.

Thus, if the molecular weight, the specific heat, and the equation of state are known, the speed of sound under any conditions can be calculated.

For an ideal gas, where $PV = RT$ one can write

$$c = \left(\frac{RT\gamma}{M}\right)^{\frac{1}{2}} = \left(\frac{\gamma p}{\rho}\right)^{\frac{1}{2}}$$

where $\gamma = C_p/C_v$, and p is the ambient pressure.

The accepted value of c_0, the speed at standard conditions of temperature and pressure, for a number of gases is given in Table 3d-6.

The accepted value of the speed of sound in air, c, as calculated and checked on the average by several reported determinations, is[1]

$$c_0 = 331.45 \pm 0.05 \text{ m/sec}$$
$$c_0 = 1{,}087.42 \pm 0.16 \text{ fps}$$

[1] See Hardy, Telfair, and Pielemeier, *J. Acoust. Soc. Am.* **13**, 226 (1942).

TABLE 3d-6. SPEED (VELOCITY) OF SOUND IN GASES*

Gas	Formula	Speed, m/sec at 0°C	Speed, fps at 0°C
Air (dry)...............	.....	331.45	1,087.42
Ammonia..............	NH_3	415	1,362
Argon.................	A	319	1,047
Carbon dioxide.........	CO_2	259	850
Carbon monoxide........	CO	338	1,189
Carbon disulfide.........	CS_2	189	606
Chlorine...............	Cl_2	206	676
Ethylene...............	C_2H_4	317	1,040
Helium.................	He	965	3,166
Hydrogen..............	H_2	1,284	4,213
Illuminating gas (coal)...	.....	453	1,486
Methane...............	CH_4	430	1,411
Neon..................	Ne	435	1,427
Nitric oxide (10°C).......	NO	324	1,063
Nitrogen...............	N_2	334	1,096
Nitrous oxide...........	N_2O	263	863
Oxygen................	O_2	316	1,037
Steam (134°C)...........	H_2O	494	1,621
Sulfur dioxide...........	SO_2	213	699

* "Handbook of Chemistry and Physics," 48th ed.

under the conditions (1) audible frequency range, (2) temperature at 0°C, (3) 1 atm pressure, (4) 0.03 percent mole content of CO_2, (5) 0 percent water content. To calculate the speed of sound at various temperatures one can write

$$c = \frac{R\gamma}{M}\left(273.16\right)^{\frac{1}{2}}\sqrt{\frac{T}{273.16}}$$

$$= 331.45\sqrt{\frac{T}{273.16}} \quad \text{m/sec}$$

$$= 331.45\sqrt{1 + \frac{T°C}{273.16}} \doteq 331.45 + 0.6\,(T°C)\ \text{m/sec}\left(\frac{T°C}{273} \ll 1\right)$$

where T is the absolute temperature, and $T°C$ is the temperature in degrees centigrade. If the gas is made up of a mixture of gases or if water vapor is present, the expression

$$c = \left[\frac{RT}{M}\left(1 + \frac{R}{C_v}\right)\right]^{\frac{1}{2}}$$

can be used to calculate the velocity. The molecular weight M of the mixture can be calculated, or realizing that $RT/M = p/\rho$, the density of the mixture can be used.

In addition to correcting M (or ρ), it is necessary to correct C_v also. It is incorrect to take the weighted average of the ratios of the specific heats, γ. The weighted average of the specific heats themselves must be used.

For rough calculations of the variation with humidity or composition, a first approximation can be obtained by correcting for the density of the mixture.

Recent studies in molecular acoustics have yielded new values for the speed of sound at a wide range of temperatures, pressures, and frequencies. Some results are shown in Tables 3d-7 and 3d-8. Speeds of sound in helium and argon for a wide range of temperatures and pressures are shown in Figs. 3d-1 through 3d-5.

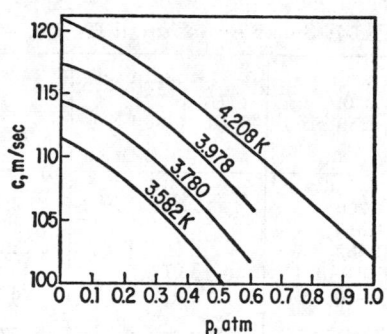

FIG. 3d-1. Speed of sound vs. static pressure in helium (gas) at 510 kHz.   The parameter is absolute temperature.   (*After van Itterbeek and Forrez.*)

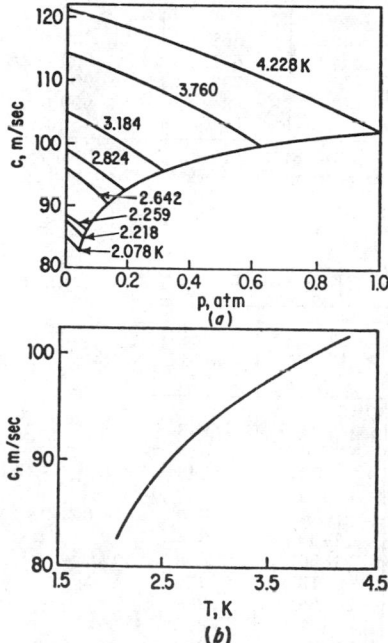

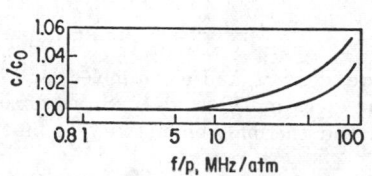

FIG. 3d-2. Speed of sound in helium (gas) at audio frequencies: (*a*) vs. static pressure with absolute temperature as parameter, and (*b*) vs. absolute temperature at vapor pressure.   (*After van Itterbeek and de Laet.*)

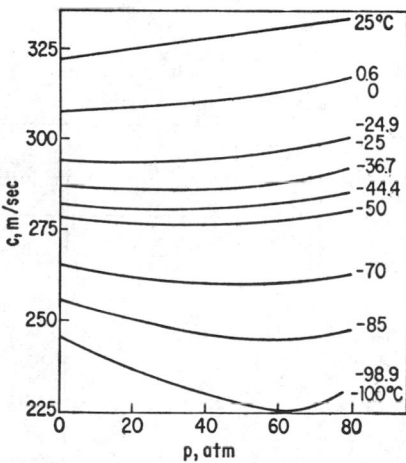

FIG. 3d-3. Speed of sound vs. static pressure in argon (gas).   The parameter is temperature.   (*After van Itterbeek.*)

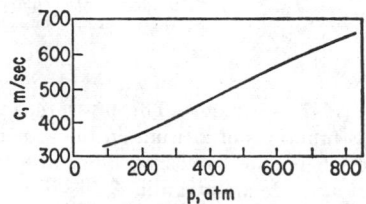

FIG. 3d-4. Relative speed of sound vs. ratio of frequency to static pressure in argon (gas) at 803.3 kHz.   (*After van Itterbeek and Boyer.*)

FIG. 3d-5. Speed of sound in argon (gas) vs. static pressure at 24°C and 600 to 900 kHz. (*After Lacam and Noury.*)

TABLE 3d-7. SPEED OF SOUND IN H_2, He, AND N_2*

p, atm	H_2 at 0°C, m/sec (286 kHz)	He at 27°C, m/sec (286 kHz)	N_2 at 27°C	
			m/sec	kHz
1	1,200	946	353	286; 486
20	1,216	954	356	286; 486
40	1,232	963	361	286 ;486
60	1,249	972	366	286; 486
80	1,265	982	372	286; 486
100	1,281	...	379	286; 486
140	.....	...	390	900
200	.....	...	420	900
340	.....	...	496	900
480	.....	...	568	900
580	.....	...	621	900
780	.....	...	737	900
1,080	.....	...	884	900

* W. Schaafs, Landolt-Börnstein New Series, Group II, "Atomic and Molecular Physics," vol. 5, "Molecular Acoustics," Springer-Verlag New York Inc., New York, 1967.

TABLE 3d-8. SPEED OF SOUND IN CO_2, CO, N_2O, AND SO_2 AT 1 ATMOSPHERE*

Gas	Frequency, kHz	T, °C	Speed, m/sec
CO_2..........	53–147	25	283
		100	313
		200	348
		500	430
CO...........	8–27	1000	710
		1400	814
		1800	898
N_2O..........	94	19	273
		56	288
		128	316
SO_2...........	111	20	222
		79	244
		124	260

* W. Schaafs, Landolt-Börnstein New Series, Group II, "Atomic and Molecular Physics," vol. 5, "Molecular Acoustics," Springer-Verlag New York Inc., New York, 1967.

3d-7. Altitude. The pressure, temperature, density, and mean molecular weight as functions of altitude in the atmosphere are given in Table 3d-2. Speed of sound, coefficient of viscosity, kinematic viscosity, and thermal conductivity as functions of altitude are given in Table 3d-9.

3d-8. Characteristic Impedance. The characteristic impedance is equal to the ratio of the sound pressure to the particle velocity in a plane wave traveling in an unbounded medium. It is equal to the density times the velocity of propagation,

TABLE 3d-9. U.S. STANDARD ATMOSPHERE, 1962*

Altitude, km	Sound speed c, m/sec	Coefficient of viscosity η, dyne-sec/cm^2	Kinematic viscosity ν, cm^2/sec	Thermal conductivity κ, cal/cm-sec-deg
−5	358.986	1.9422 −4	1.0058 −4	6.6545 −5
−4	355.324	1.9123	1.0806	6.5356
−3	351.625	1.8820	1.1625	6.4161
−2	347.888	1.8515	1.2525	6.2958
−1	344.111	1.8206	1.3516	6.1748
0	340.294	1.7894	1.4607	6.0530
0.5	338.370	1.7737	1.5195	5.9919
1.0	336.435	1.7579	1.5813	5.9305
1.5	334.489	1.7420	1.6463	5.8690
2	332.532	1.7260	1.7147	5.8073
2.5	330.563	1.7099	1.7868	5.7451
3	328.583	1.6938	1.8628	5.6833
4	324.589	1.6612	2.0275	5.5586
5	320.545	1.6282	2.2110	5.4331
6	316.452	1.5949	2.4162	5.3068
7	312.306	1.5612	2.6461	5.1798
8	308.105	1.5271	2.9044	5.0520
9	303.848	1.4926	3.1957	4.9235
10	299.532	1.4577	3.5251	4.7942
15	295.069	1.4216	7.2995	4.6617
20	295.069	1.4216	1.5989	4.6617
25	298.389	1.4484	3.6135	4.7602
30	301.709	1.4753	8.0134	4.8593
40	317.189	1.6009	4.0067	5.3295
50	329.799	1.7037	1.6591	5.7214
60	320.606	1.6287	5.3241	5.4349
70	297.139	1.4389	1.6431	4.7230
80	269.44	1.216	6.085	3.925

* Data taken from: "U.S. Standard Atmosphere, 1962," published by the U.S. Committee on Extension to the Standard Atmosphere (COESA), Washington, D.C., 1962.   For T, P, and ρ see Table 3d-2.
Note.   A single-digit number (preceded by a plus or minus sign) following the initial entry indicates the power of 10 by which that entry and each succeeding entry of that column should be multiplied.

that is, ρc.   The variation of ρc with temperature and pressure can be calculated from the expression

$$\rho c = \rho_0 c_0 \left(\frac{273.16}{T}\right)^{\frac{1}{2}} \frac{P}{760} \quad \text{mks rayls}$$

where $\rho_0 c_0$ is the value at 0°C and 1 atm pressure.   For air $\rho_0 c_0 = 428.5$ newton-sec/m^3.   Table 3d-10 contains values of $\rho_0 c_0$ for several common gases.

3d-9. Attenuation.   In addition to the dispersion of sound due to wind, turbulence in the atmosphere, and temperature gradients, two properties of the medium combine to attenuate a wave which is propagated in free space.   The first of these attenuations is caused by molecular absorption and dispersion in polyatomic gases involving an exchange of translational and vibrational energy between colliding molecules.   The second is due to viscosity and heat conduction in the medium, discussed earlier in this section.

TABLE 3d-10. CHARACTERISTIC IMPEDANCE $\rho_0 c_0$ OF COMMON GASES AT 0°C (273.16 K) TEMPERATURE AND 0.760 M HG BAROMETRIC PRESSURE

Gas	Formula	$\rho_0 c_0$, newton-sec/m³ at 0°C, 0.76 m Hg
Air.....................	....	428.5
Argon...................	A	569
Carbon dioxide............	CO_2	512
Carbon monoxide..........	CO	421
Helium....................	He	173.1
Hydrogen.................	H_2	114.1
Neon.....................	Ne	385
Nitric oxide...............	NO	435
Nitrogen.................	N_2	421
Nitrous oxide.............	N_2O	518
Oxygen...................	O_2	453

Knudsen[1] says that "the attenuation of sound is greatly dependent upon location and weather conditions, that is, upon the humidity and temperature of the air. For the hot and relatively dry summer air of the desert, such as at Greenland Ranch, Inyo County, California, where the relative humidity may drop as low as 2.4 percent, the attenuation at 3,000 Hz is 0.14 dB/m, and at 10,000 Hz it is 0.48 dB/m."

Data on the absorption of audible sound in air are valuable because they are needed to calculate the reverberation time for high-frequency sound in rooms, for determining the amplification characteristics of public-address systems for use outdoors, and for predicting the range of effectiveness of apparatus for sound signaling and sound ranging in the atmosphere.

Kneser[2] treated analytically the problem of absorption and dispersion of sound by molecular collision and summarized his results in the form of a nomogram which has been reprinted along with comments by Pielemeier.[3] Recent data by Harris[4] show larger values for molecular absorption at most relative humidities than are yielded rom Kneser's nomogram.

The attenuation caused by heat conduction and viscosity of the air α_e is not known so accurately. The classical absorption due to these causes,[5] as discussed earlier, is given by

$$\alpha_c = \alpha_\eta + \alpha_T = \frac{\omega^2}{2\rho c^3}\left[\frac{4\eta}{3} + (\gamma - 1)\frac{\kappa}{C_p}\right] \quad \text{nepers/m}$$

where $\omega/2\pi$ is the frequency in hertz, ρ is the density in kilograms per meter cubed, c is the speed of sound in meters per second, η is the coefficient of viscosity in mks

[1] V. O. Knudsen, The Propagation of Sound in the Atmosphere: Attenuation and Fluctuations, J. Acoust. Soc. Am. 18, 90–96 (1946).

[2] H. O. Kneser, The Interpretation of the Anomalous Sound-absorption in Air and Oxygen in Terms of Molecular Collisions, J. Acoust. Soc. Am. 5, 122–126 (1933); A Nomogram for Determination of the Sound Absorption Coefficient in Air, Akust. Z. 5, 256–257 (1940) (in German).

[3] W. H. Pielemeier, Kneser's Sound Absorption Nomogram and Other Charts, J. Acoust. Soc. Am. 16, 273–274 (1945).

[4] C. M. Harris, Absorption of Sound in Air versus Humidity and Temperature, J. Acoust. Soc. Am. 40, 148–159 (1966).

[5] Lord Rayleigh, "Theory of Sound," The Macmillan Company, New York, 1929, and Dover Publication, Inc., New York, 1945.

units, γ is the ratio of specific heats, κ is the coefficient of thermal conductivity in mks units, and C_p is the specific heat at constant pressure in mks units. Measured values of α_c are as much as 40 to 50 percent higher than those calculated from the equation above.[1,2]

The total attenuation α_A due to both types of absorption is therefore

$$\alpha_A = \alpha_m + \alpha_c \quad \text{nepers/m}$$

where α_m is the absorption in nepers/m arising from molecular resonance. To convert from nepers per meter to decibels per meter, multiply by 8.686.

Harris[3] has measured the total attenuation for a sound wave traveling through air having a carbon dioxide content of 300 parts per million (0.03 percent). The temperature, relative humidity, and frequency were varied over a wide range. The results are shown in Figs. 3d-6 and 3d-7a to f. In Fig. 3d-6, m is defined as the

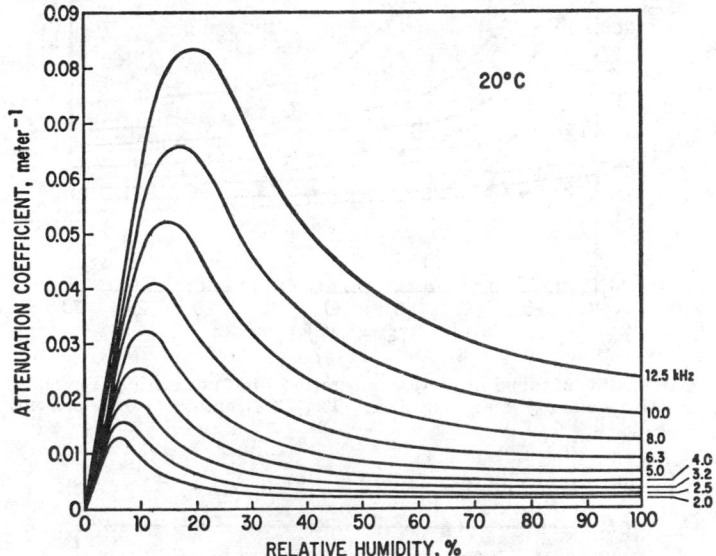

Fig. 3d-6. Values of the total attenuation coefficient m (in meters^{-1}) versus percent relative humidity for air at 20°C and normal atmospheric pressure for frequencies between 2,000 and 12,500 Hz at one-third-octave intervals. To convert to decibels per meter, multiply ordinate by 4.343. (*After Harris.*)

attenuation coefficient per meter as expressed in the equation $I = I_0 e^{-mx}$, where I_0 is the sound intensity (in watts/m^2) at $x = 0$, and I is that at x. To convert from m to decibels per meter, multiply by 4.343.

Harris[4] has also presented data on the absorption of sound in air at pressures in the range from 0.2 to 0.9 atm at 20°C. The results show that, at a given frequency,

[1] L. J. Sivian, High Frequency Absorption in Air and in Other Gases, *J. Acoust. Soc. Am.* **19**, 914–916 (1947).
[2] P. E. Krasnooshkin, On Supersonic Waves in Cylindrical Tubes and the Theory of the Acoustical Interferometer, *Phys. Rev.* **65**, 190 (1944). See also W. H. Pielemeier, Observed Classical Sound Absorption in Air, *J. Acoust. Soc. Am.* **17**, 24–28 (1945).
[3] C. M. Harris, Absorption of Sound in Air versus Humidity and Temperature, *Acoust. J. Soc. Am.* **40**, 148–159 (1966).
[4] C. M. Harris, On the Absorption of Sound in Humid Air at Reduced Pressures, *J. Acoust. Soc. Am.* **43**, 530–532 (1968).

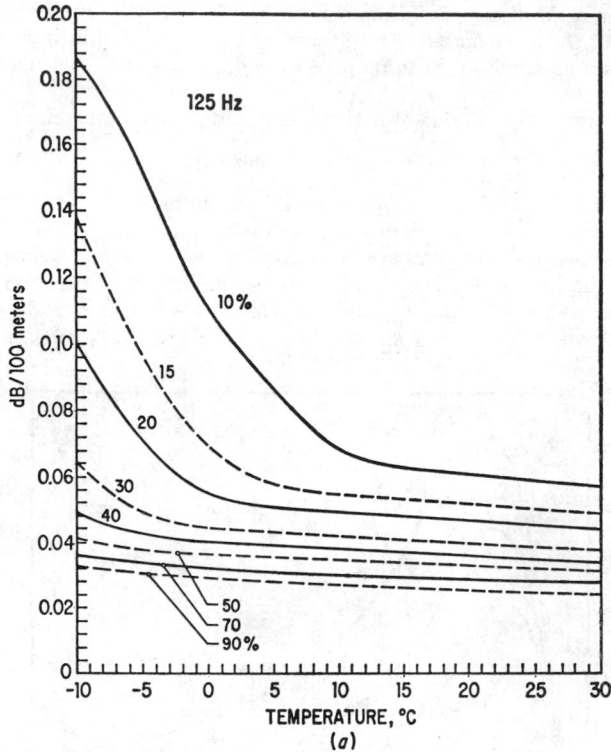

FIG. 3d-7. Attenuation of sound in air vs. temperature, at atmospheric pressure, for various values of relative humidity and frequency. The CO_2 content is 0.03 percent. (*After Harris.*)

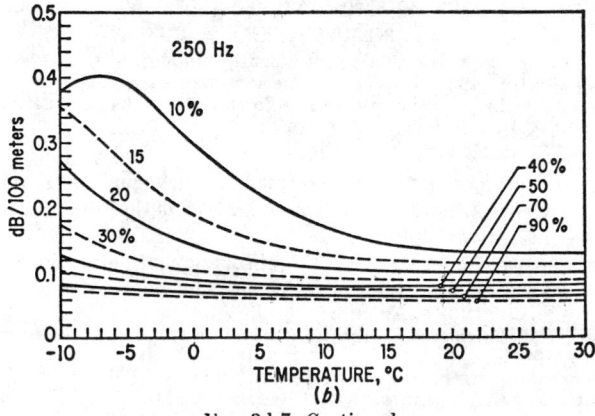

FIG. 3d-7. *Continued.*

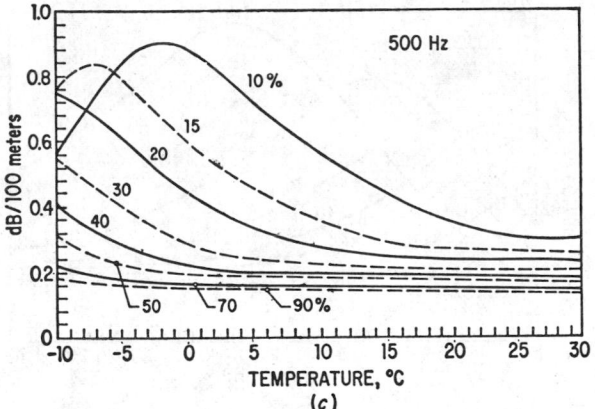

FIG. 3d-7. *Continued.*

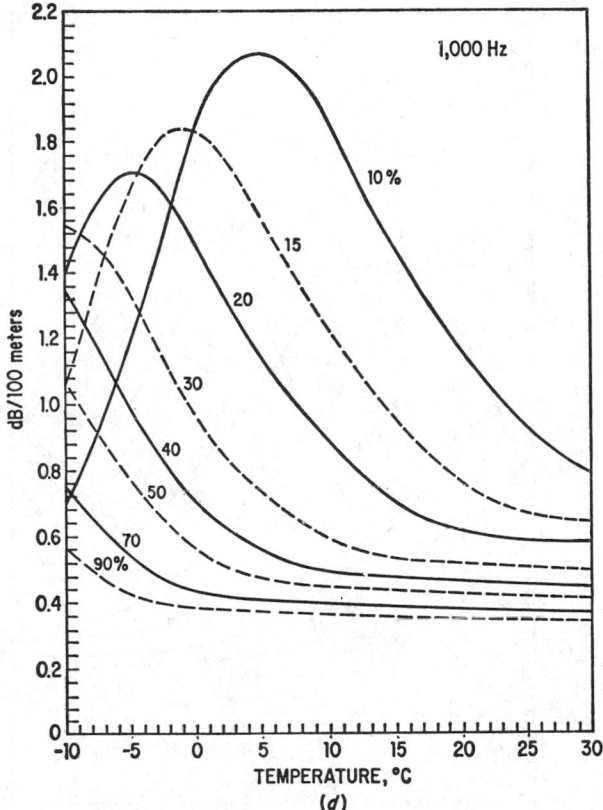

FIG. 3d-7. *Continued.*

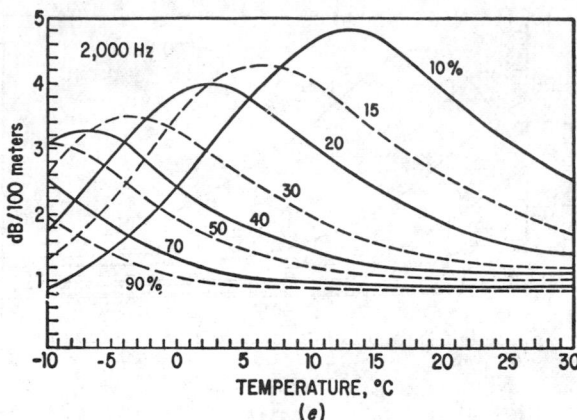

FIG. 3d-7 (*Continued*)

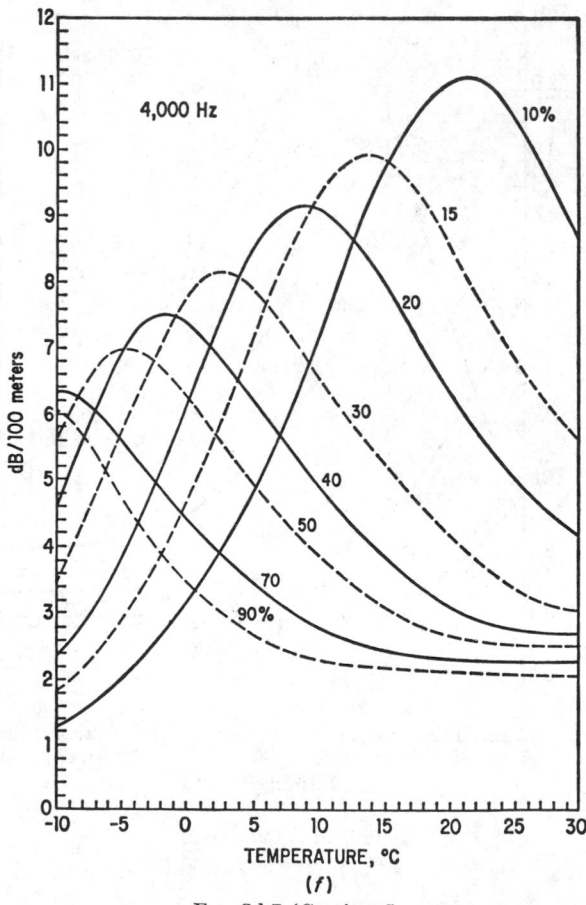

FIG. 3d-7 (*Continued*)

a plot of molecular absorption versus humidity has a maximum value that is independent of pressure. Lowering the pressure shifts the peaks in the curves of absorption versus humidity to lower values of relative humidity. The relations among frequency of maximum absorption, relative humidity, and frequency are given in Fig. 3d-8.

Other studies of the molecular absorption process are reported by Monk,[1] Shields and Faughn,[2] Henderson and Herzfeld,[3] and Connelly.[4]

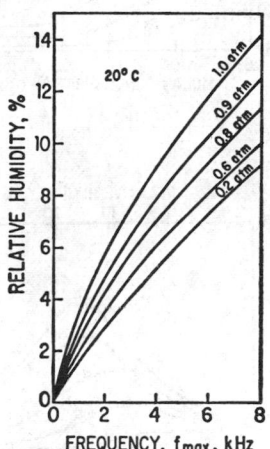

FIG. 3d-8. Frequency of maximum total absorption, f_{max}, as a function of relative humidity. The parameter is atmospheric pressure. The temperature is 20°C.

Below 1,000 Hz the attenuation of sound in air is much less than above 1,500 Hz. Harris and Tempest[5] have measured attenuation coefficients for air in this frequency range. Their data for a range of moisture contents, temperatures, and barometric pressures are given in Fig. 3d-9a through e.

[1] R. G. Monk, Thermal Relaxation in Humid Air, J. Acoust. Soc. Am. 46, 580–586 (1969).
[2] F. D. Shields and J. Faughn, Sound Velocity and Absorptions in Low-pressure Gases Confined to Tubes of Circular Cross Sections, J. Acoust. Soc. Am. 46, 158–163 (1968).
[3] M. C. Henderson and K. P. Herzfeld, Effect of Water Vapor on the Napier Frequency of Oxygen and Air, J. Acoust. Soc. Am. 37, 986–988 (1965).
[4] J. H. Connolly, Combined Effect of Shear Viscosity, Thermal Conduction, and Thermal Relaxation on Acoustic Propagation in Linear-molecular Ideal Gases, J. Acoust. Soc. Am. 36, 2374–2381 (1964).
[5] C. M. Harris and W. Tempest, Absorption of Sound below 1000 Hz, J. Acoust. Soc. Am. 36, 2390–2394 (1964).

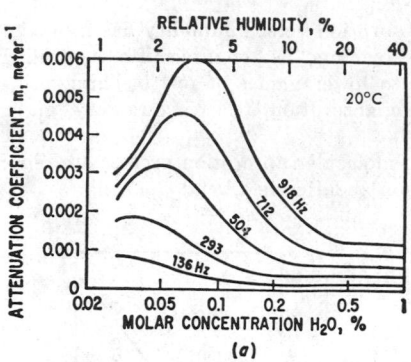

(a)

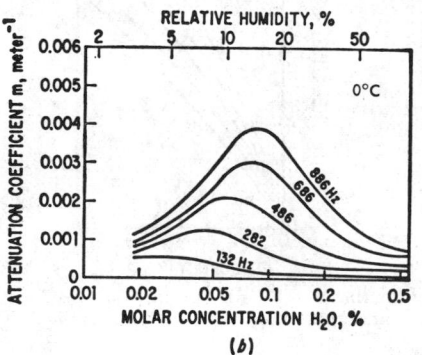

(b)

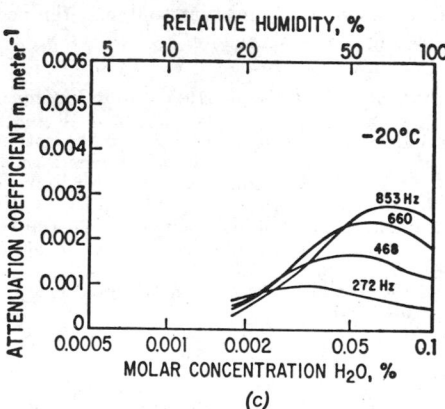

(c)

FIG. 3d-9. Attenuation coefficient m vs. percent relative humidity for air at various frequencies and temperatures. Pressure is atmospheric for (a), (b), and (c), and as shown on the graphs for (d) and (e). To convert to decibels per 100 meters, multiply ordinate by 434.

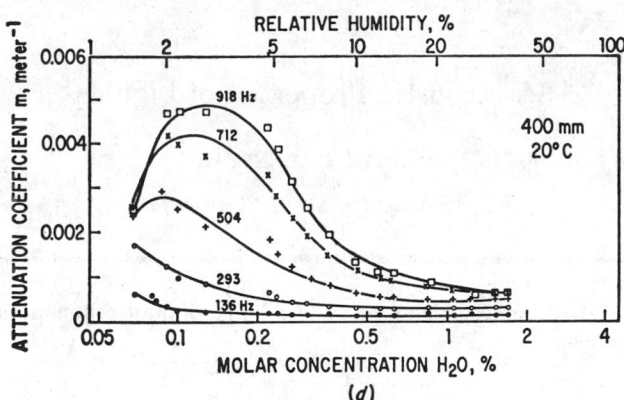

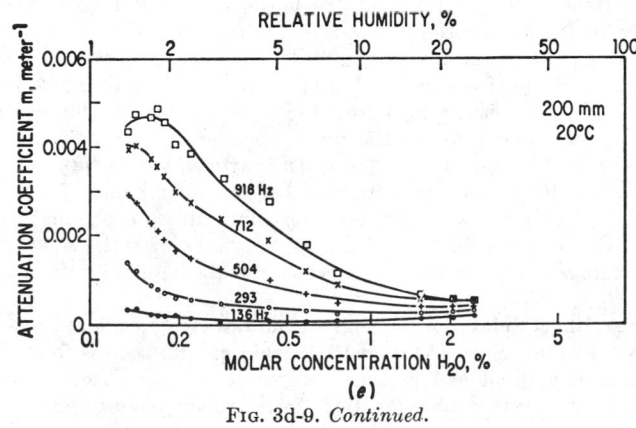

Fig. 3d-9. *Continued.*

3e. Acoustic Properties of Liquids[1]

MARTIN GREENSPAN

National Bureau of Standards

3e-1. General. The acoustic property of a liquid of most common interest is the propagation constant

$$k = \alpha + i\beta = \alpha + \frac{i\omega}{c}$$

in which α is the attenuation and c the phase speed in a plane progressive wave, $\omega = 2\pi f$ being the angular frequency. The liquid is thus characterized by α and c, both of which are in general functions of temperature, pressure, and frequency.

The total attenuation α consists of two parts, a "classical" part, α_{class}, and an "excess" part, α_{exc}. The classical attenuation is that due to the effects of viscosity and heat conduction (the latter is negligible except for liquid metals), and it varies with the square of the frequency, although presumably this relationship would be "relaxed out" at sufficiently high frequencies, as yet inaccessible experimentally. In the accessible frequency region the phase speed c is independent of frequency; i.e., the dispersion is inappreciable. The excess attenuation is supposed to be connected with a relaxation mechanism induced by slow interchange of energy between various modes. In most cases the relaxation times are so short that at accessible frequencies there is no dispersion and α_{exc} again varies with the square of the frequency; therefore α/f^2 and c completely characterize the liquid at a given temperature and pressure.

3e-2. Mechanism of Relaxation. Herzfeld and Litovitz[1] recognize four mechanisms of relaxation. In the "Kneser" liquids, the main contribution is from slow energy exchange between internal and external degrees of freedom (thermal relaxation), as in a gas. In a few cases the relaxation times are long enough so that dispersion can be observed;[2] these are not considered here. In many associated liquids, a slow change in structure occurs, and in some there are a slow formation and dissociation of chemical complexes. Not considered here are the effects of isomerism in organic liquids and of rigidity in polymeric liquids or of solutions of polymers in ordinary liquids or of associated liquids cooled to the glassy state.

We consider only the normal, Kneser, and associated liquids and only in the nondispersive regime (c and α/f^2 independent of f). The available body of data is very extensive, and much of it represents isolated observations. We present here only such results as illustrate the variation of c and α/f^2 with an important variable. References to other data are given in Sec. 3e-7. For the most part the available data are not critical, and little can be said about the accuracy. Various methods of measurement yield different results, and in many cases the purity of the tested liquids is questionable.

The units used are for the most part those of the original authors. For interconversion of pressure units see Table 2l-12.

---

[1] Contribution of the National Bureau of Standards, not subject to copyright.
[2] K. F. Herzfeld and T. A. Litovitz, "Absorption and Dispersion of Ultrasonic Waves," Academic Press, Inc., New York, 1959.

3e-3. Normal Liquids. These are the liquids for which α/α_{class} is approximately unity. Monatomic liquids (Hg, He I well above the λ-point, A, etc.) and certain liquefied diatomic gases (N_2, O_2, H_2) are the most conspicuous examples. See Tables 3e-1 and 3e-2 for temperature and pressure effects in He I; Table 3e-3 for A, N_2, H_2, and O_2; and Table 3e-4 for Hg.

TABLE 3e-1. SPEED AND ATTENUATION OF SOUND
IN LIQUID HE I AT THE VAPOR PRESSURE

T, K	f, MHz	c, m/sec	$10^{15}\alpha/f^2$, sec^2/m	Ref.
5.1	15	.....	790	a
4.47	15	181	300	a
4.22	15	183	260	a
4.22	1.3	179.8	...	b
4.0	15	211	230	a
4.0	1.3	198.2	...	b
3.6	1.3	206.5	...	b
3.52	15	207	170	a

(a) J. R. Pellam and C. F. Squire, *Phys. Rev.* **72,** 1245 (1947).
(b) J. C. Findlay, A. Pitt, H. Grayson Smith, and J. O. Wilhelm, *Phys. Rev.* **54,** 506 (1938).

TABLE 3e-2. INFLUENCE OF PRESSURE ON THE SPEED OF SOUND
IN LIQUID `HE I AT 3.2 K*
($f = 1.3$ MHz)

p, atm............	Vapor	1	2.47	5.55
c, m/sec............	212	221	238	266

* J. C. Findlay, A. Pitt, H. Grayson Smith, and J. O. Wilhelm, *Phys. Rev.* **56,** 122 (1939).

TABLE 3e-3. SPEED AND ATTENUATION OF SOUND IN SOME
LIQUEFIED GASES AT THE VAPOR PRESSURE*
($f = 44.4$ MHz)

Liquid	T, K	c, m/sec	$10^{15}\alpha/f^2$, sec^2/m
A	85.2	853	10.1
N_2	73.9	962	10.6
H_2	17	1,187	5.6
O_2	87	952	8.6
O_2	70	1,094	8.6
O_2	60	1,119	8.6

* J. K. Galt, *J. Chem. Phys.* **16,** 505 (1948); R. T. Beyer, *J. Chem. Phys.* **19,** 788 (1951).

3e-4. Kneser Liquids. The Kneser liquids are characterized by a value of α/α_{class} greater than about 3 (in some cases as high as several thousand) and a positive temperature coefficient of attenuation. It follows from the approximately additive and constitutive nature of the molar sound speed[1] that in a homologous series of

[1] B. B. Kudriavtsev, *Soviet Physics-Acoustics* **2,** 354 (1956).

TABLE 3e-4. SPEED AND ATTENUATION OF SOUND IN HG AT 1 ATM

T, °C	f, MHz	c, m/sec	$10^{15}\alpha/f^2$, sec²/m	Ref.
20	0.5	1,451	...	a
24.3	22	.....	6.3	b
24.3	54	.....	6.4	b
23.8	152	1,449	5.8	c
24.0	291	1,451	5.5	c
28.2	390	1,450	5.7	c
27.2	774	1,470	4.7	c
26.9	996	1,440	6.0	c

(a) J. C. Hubbard and A. L. Loomis, *Phil. Mag.* **5**, 1177 (1928).
(b) P. Rieckmann, *Physik. Z.* **40**, 582 (1939).
(c) G. R. Ringo, J. W. Fitzgerald, and B. G. Hurdle, *Phys. Rev.* **72**, 87 (1947).

TABLE 3e-5. SPEED OF SOUND AT 20°C AND ATMOSPHERIC PRESSURE
IN SOME STRAIGHT-CHAIN HYDROCARBONS

Alkanes,* C_nH_{2n+2}		Alkenes,† C_nH_{2n}	
Compound	c, m/sec	Compound	c, m/sec
Pentane	1,008	1-Heptene	1,128
Hexane	1,083	1-Octene	1,184
Heptane	1,162	1-Nonene	1,218
Octane	1,197	1-Decene	1,250
Nonane	1,248	1-Undecene	1,275
		1-Tridecene	1,313
		1-Pentadecene	1,351

* W. Schaafs, *Z. Physik. Chem.* **194**, 28 (1944).
† R. T. Lagemann, D. R. McMillan, and M. Woolsey, *J. Chem. Phys.* **16**, 247 (1948).

TABLE 3e-6. SPEED AND ATTENUATION OF SOUND IN
CS_2, C_6H_6, AND CCl_4 AT 1 ATM*

f, MHz	2.0		5.0		4.85 and 10.0	
Compound	CS_2		C_6H_6		CCl_4	
T, °C	c, m/sec	$10^{15}\alpha/f^2$, sec²/m	c, m/sec	$10^{15}\alpha/f^2$, sec²/m	c, m/sec	$10^{15}\alpha/f^2$, sec²/m
0	1,220	5,510				
25	1,140	5,680	1,310	873	930	538
40	1,090	5,930				
50	.....	.....	1,190	964	852	590
70	.....	.....	1,100	1,050		

* J. F. Mifsud and A. W. Nolle, *J. Acoust. Soc. Am.* **28**, 469 (1956).

compounds the speed of sound will increase smoothly with molecular weight. Examples are given in Table 3e-5.

The liquids carbon disulfide, dibromomethane, dichloromethane, benzene, and carbon tetrachloride are of special interest because for them the values of α/α_{class} are among the highest known; at room temperature and atmospheric pressure the values are about 1,150, 354, 183, 100, and 26.6, respectively.[1] See Table 3e-6 for

TABLE 3e-7. SPEED AND ATTENUATION OF SOUND IN CS_2, C_6H_6, AND CCl_4
AT 25°C*
(f as in Table 3e-6)

Compound	CS_2		C_6H_6		CCl_4	
p, psi	c, m/sec	$10^{15}\alpha/f^2$, sec²/m	c, m/sec	$10^{15}\alpha/f^2$, sec²/m	c, m/sec	$10^{15}\alpha/f^2$, sec²/m
14.7	1,140	5,700	1,310	870	930	540
5,000	1,270	4,200	1,450	650	1,050	380
10,000	1,350	3,300	1,590	520	1,140	290
15,000	1,440	2,700	.....	...	1,230	220
20,000	1,500	2,400	.....	...	1,300	180

* J. F. Mifsud and A. W. Nolle, *J. Acoust. Soc. Am.* **28**, 469 (1956).

TABLE 3e-8. SPEED OF SOUND IN ETHER* AND ACETONE† AT $p = 1$ ATM

T, °C	c, m/sec	
	Ether	Acetone
16	1,023	
20	.....	1,203
25	976	
30	945	
30.5	.....	1,158
41	.....	1,097
44	862	

* E. G. Richardson and R. I. Tait, *Phil. Mag.* **2**, 441 (1957).
† H. F. Eden and E. G. Richardson, *Acustica* **10**, 309 (1960).

the temperature dependence of c and α/f^2 at 1 atm for CS_2, C_6H_6, and CCl_4 and Table 3e-7 for the pressure dependence at 25°C.

Tables 3e-8 and 3e-9 give some results for ether and acetone, Tables 3e-10 and 3e-11 for *n*-pentane and isopentane, Table 3e-12 for *n*-hexane and cyclohexane, and Tables 3e-13 to 3e-15 for the monohalogenated benzenes.

3e-5. Associated Liquids. The associated liquids are characterized by a value of α/α_{class} less than about 4 and a negative temperature coefficient of attenuation.

[1] K. F. Herzfeld and T. A. Litovitz, "Absorption and Dispersion of Ultrasonic Waves," Academic Press, Inc., New York, 1959.

TABLE 3e-9. SPEED AND ATTENUATION OF SOUND IN ETHER* AND
ACETONE† AS A FUNCTION OF PRESSURE

T, °C	Ether		Acetone	
	16	16	30.5	30
p, psi	c, m/sec	$10^{15}\alpha/f^2$, sec^2/m	c, m/sec	$10^{15}\alpha/f^2$, sec^2/m
0	1,023	60	1,158	54.2
2,000	1,117	47.5	1,228	42.8
4,000	1,198	39.8	1,295	33.2
6,000	1,268	33.4	1,353	27.9
8,000	1,328	28.8	1,402	22.7
10,000	.....	25.5	1,450	21.5

* E. G. Richardson and R. I. Tait, *Phil. Mag.* **2,** 441 (1957).
† H. F. Eden and E. G. Richardson, *Acustica* **10,** 309 (1960).

TABLE 3e-10. SPEED OF SOUND IN n-PENTANE* AND ISOPENTANE† AT
p = 1 ATM

T, °C	c, m/sec	
	n-Pentane	Isopentane
0	.....	951
3	.....	932
6.5	.....	917
8	.....	900
10.5	.....	883
15	1,027	
25	986	
35	944	
44	908	

* E. G. Richardson and R. I. Tait, *Phil. Mag.* **2,** 441 (1957).
† H. F. Eden and E. G. Richardson, *Acustica* **10,** 309 (1960).

Some Alcohols and the Waters. The speed of sound c (m/sec) is given, for temperature T (°C) and pressure p (psi), by the equation

$$\sum_{i,j} c_{ij} T^i p^j \qquad (3e-1)$$

The nonzero coefficients c_{ij} are given in Table 3e-16. The limits of applicability of Eq. (3e-1) are

For the alcohols,　　$0 < T < 58°C$　　$14.7 < p < 14,000$ psi
For heavy water,　　$0 < T < 95°C$　　$14.7 < p < 14,000$ psi
For light water,　　$16 < T < 94°C$　　$14.7 < p < 14,000$ psi

TABLE 3e-11. SPEED OF SOUND IN n-PENTANE* AND
ISOPENTANE† AND ATTENUATION IN n-PENTANE
AS A FUNCTION OF PRESSURE

p, psi	n-Pentane $T = 15°C$		Isopentane $T = 0°C$
	c, m/sec	$10^{15}\alpha/f^2$, sec^2/m	c, m/sec
0	1,027	100	951
2,000	1,122	93	1,029
4,000	1,208	86	1,096
6,000	1,282	81	1,153
8,000	1,350	77	1,200
10,000	.....	74	1,242

* E. G. Richardson and R. I. Tait, *Phil. Mag.* **2**, 441 (1957).
† H. F. Eden and E. G. Richardson, *Acustica* **10**, 309 (1960).

TABLE 3e-12. SPEED AND ATTENUATION OF SOUND IN n-HEXANE AND
CYCLOHEXANE AS A FUNCTION OF PRESSURE*

p, psi	n-Hexane			Cyclohexane	
	$T = 20°C$	$T = 30°C$		$T = 19°C$	$T = 19°C$
	c, m/sec	$10^{15}\alpha/f^2$, sec^2/m		c, m/sec	$10^{15}\alpha/f^2$, sec^2/m
0	1,103	87		1,280	330
2,000	1,183	79		1,329	299
4,000	1,260	72		1,377	270
6,000	1,327	67			
8,000	1,391	65			
10,000	1,442	63			

* H. F. Eden and E. G. Richardson, *Acustica* **10**, 309 (1960).

TABLE 3e-13. SPEED OF SOUND IN THE MONOHALOGENATED
BENZENES AT $p = 1$ ATM*

T, °C	c, m/sec			
	Fluoro	Chloro	Bromo	Iodo
20	1,183	1,311	1,169	1,114
30	1,144	1,282	1,136	1,085
40	1,105	1,254	1,105	1,058
50	1,066	1,226	1,074	1,030
60	1,028	1,197	1,042	1,003

* H. F. Eden and E. G. Richardson, *Acustica* **10**, 309 (1960).

TABLE 3e-14. SPEED OF SOUND IN THE MONOHALOGENATED
BENZENES AT 22°C*

p, psi	c, m/sec			
	Fluoro	Chloro	Bromo	Iodo
0	1,177	1,304	1,167	1,104
2,000	1,228	1,346	1,204	1,135
4,000	1,279	1,382	1,239	1,164
6,000	1,321	1,418	1,270	1,190
8,000	1,357	1,449	1,300	1,210
10,000	1,388	1,479	1,326	

* H. F. Eden and E. G. Richardson, *Acustica* **10**, 309 (1960).

TABLE 3e-15. ATTENUATION OF SOUND IN THE MONOHALOGENATED
BENZENES AT 22°C*

p, psi	$10^{15}\alpha/f^2$, sec^2/m			
	Fluoro	Chloro	Bromo	Iodo
0	317	167	163	242
2,000	282	148	136	210
4,000	256	134	119	188
6,000	235	123	104	172
8,000	219	116	96	160

* H. F. Eden and E. G. Richardson, *Acustica* **10**, 309 (1960).

3e. Acoustic Properties of Liquids

TABLE 3e-16. COEFFICIENTS OF EQ. (3e-1) FOR THE SPEED OF SOUND IN METHYL, ETHYL, n-PROPYL, AND n-BUTYL ALCOHOL* AND HEAVY† AND LIGHT‡ WATER§

Coefficient	CH_3OH	C_2H_5OH	n-C_3H_7OH	n-C_4H_9OH	99.82 mole % D_2O	H_2O
C_{00}	1,188.68	1,231.54	1,294.40	1,326.42	1,300.96	1,401.97
C_{10}	−3.45382	−3.57641	−3.67732	−3.38784	+5.16714	+5.05172
C_{20}	+1.26504(−3)	+2.38718(−3)	+4.00023(−3)	−9.86829(−3)	−5.50243(−2)	−5.84853(−2)
C_{30}	+7.49187(−5)	+2.65362(−5)	+3.53481(−5)	+3.62362(−4)	+2.23939(−4)	+3.38108(−4)
C_{40}	−8.33143(−7)	−2.44228(−7)	−7.40770(−7)	−3.37830(−6)	−3.95228(−7)	−1.48486(−6)
C_{50}						+3.09107(−9)
C_{01}	+3.80159(−2)	+4.01142(−2)	+3.78515(−2)	+3.68124(−2)	+9.05259(−3)	+8.37365(−3)
C_{11}	+1.58829(−4)	+1.68911(−4)	+1.59730(−4)	+1.32570(−4)	+4.38713(−5)	+4.32593(−4)
C_{21}	+3.95282(−7)	+2.19721(−7)	−6.87332(−8)	−2.42387(−7)	+9.95828(−8)	−2.02913(−5)
C_{31}	−1.14480(−9)	−3.43214(−11)	+3.29403(−9)	+9.83530(−9)	−6.17920(−10)	+4.49332(−7)
C_{41}						−4.49886(−9)
C_{51}						+1.67496(−11)
C_{02}	−1.36252(−6)	−1.57386(−6)	−1.26114(−6)	−1.17082(−6)	+1.90159(−7)	+4.96904(−7)
C_{12}	−1.06463(−8)	−1.03789(−8)	−9.74285(−9)	−6.42786(−9)	−4.94379(−9)	−1.00058(−7)
C_{22}	−1.18491(−11)	−1.21958(−11)	−8.57457(−12)	−2.93630(−11)	+1.99021(−11)	+5.47803(−9)
C_{32}						−1.29087(−10)
C_{42}						+1.35726(−12)
C_{52}						−5.22954(−15)
C_{03}	+6.31651(−11)	+7.87874(−11)	+5.62621(−11)	+4.80566(−11)	−4.93662(−12)	+2.05170(−11)
C_{13}	+3.17911(−13)	+3.01401(−13)	+2.85090(−13)	+2.00600(−13)	+5.21311(−14)	+3.78760(−12)
C_{23}						−3.37361(−13)
C_{33}						+9.46064(−15)
C_{43}						−1.09054(−16)
C_{53}						+4.44147(−19)
C_{04}	−1.53080(−15)	−1.97697(−15)	−1.34003(−15)	−1.04146(−15)	+5.81145(−17)	−3.38357(−15)
C_{14}						+1.45263(−16)
C_{24}						+7.83063(−19)
C_{34}						−1.24589(−19)
C_{44}						−1.98453(−21)
C_{54}						−9.31747(−24)

* W. Wilson and D. Bradley, *J. Acoust. Soc. Am.* **36**, 333 (1964).
† W. D. Wilson, *J. Acoust. Soc. Am.* **33**, 314 (1961).
‡ A. J. Barlow and E. Yazgan, *Brit. J. Appl. Phys.* **18**, 645 (1967).
§ All coefficients to be multiplied by the power of 10 in the parenthesis.

Table 3e-17 illustrates the temperature dependence of c in the alcohols at $p = 1$ atm, and Table 3e-18 the pressure dependence at $T = 20°C$.

The pressure dependence of α/f^2 in four alcohols is given in Table 3e-19; the temperature dependence of α/f^2 in n-butyl alcohol is given in Table 3e-20. For other alcohols see references in Sec. 3e-7.

Typical values for ordinary and heavy water are given in Tables 3e-21 to 3e-24. The speeds of sound for H_2O at 1 atm given in Table 3e-21 are more accurate (within 0.1 m/sec) than other such values given in this section. For instance, the values calculated from Eq. (3e-1) and Table 3e-16 may be in error by 0.3 m/sec or more. Lovett[1] has reviewed recent measurements. He recommends that the equation of Greenspan and Tschiegg[2] be modified to read

$$c = 1402.336 + 5.03358T - 5.79506 \times 10^{-2}T^2$$
$$+ 3.31636 \times 10^{-4}T^3 - 1.45262 \times 10^{-6}T^4$$
$$+ 3.0449 \times 10^{-9}T^5 \text{ m/sec} \qquad (3e-2)$$

[1] J. R. Lovett, *J. Acoust. Soc. Am.* **45**, 1051 (1969).
[2] M. Greenspan and C. E. Tschiegg, *J. Research NBS* **59C**, 249 (1957).

TABLE 3e-17. SPEED OF SOUND IN FOUR ALCOHOLS AT $p = 1$ ATM[*]

T, °C	c, m/sec			
	Methyl	Ethyl	n-Propyl	n-Butyl
0	1,189.2	1,232.1	1,295.0	1,327.0
10	1,154.9	1,196.7	1,258.6	1,292.5
20	1,121.2	1,161.8	1,223.2	1,257.7
30	1,088.2	1,127.6	1,188.7	1,223.6
40	1,055.9	1,094.1	1,154.7	1,190.3
50	1,024.0	1,061.2	1,121.0	1,157.2

[*] W. D. Wilson, *J. Acoust. Soc. Am.* **36**, 333 (1964).

TABLE 3e-18. SPEED OF SOUND IN FOUR ALCOHOLS AT $T = 20°C$[*]

p, psi	c, m/sec			
	Methyl	Ethyl	n-Propyl	n-Butyl
14.7	1,121.2	1,161.8	1,223.2	1,257.7
2,000	1,197.5	1,241.8	1,299.3	1,331.1
4,000	1,264.7	1,311.8	1,367.1	1,397.0
6,000	1,324.8	1,374.1	1,428.0	1,456.5
8,000	1,379.5	1,430.8	1,483.8	1,511.1
10,000	1,430.2	1,483.3	1,535.7	1,562.1
12,000	1,477.5	1,532.4	1,584.3	1,610.0
14,000	1,521.6	1,577.9	1,629.7	1,655.2

[*] W. D. Wilson, *J. Acoust. Soc. Am.* **36**, 333 (1964).

TABLE 3e-19. ATTENUATION OF SOUND IN FOUR ALCOHOLS AT $T = 30°C$[*]

p, kg/cm²	$10^{15}\alpha/f^2$, sec²/m			
	Methyl f = 45 MHz	Ethyl f = 45 MHz	n-Propyl f = 25 MHz	n-Butyl f = 25 MHz
1	30.2	48.5	64.5	74.3
500	18.2	31.2	48.5	60.5
1,000	13.5	24.5	41.5	55.8
1,500	11.2	21.4	39.2	54.0
2,000	9.9	19.9	39.0	53.5

[*] E. H. Carnevale and T. A. Litovitz, *J. Acoust. Soc. Am.* **27**, 547 (1955).

for T in °C. The values for H_2O in Table 3e-21 were calculated from Eq. (3e-2). Some results at very high pressures, up to 10,000 kg/cm², with reduced accuracy, are also available.[1,2]

[1] G. Holton et al., *J. Acousi. Soc. Am.* **43**, 102 (1968).
[2] W. H. Johnson, Jr., and G. Holton, *Rev. Sci. Instr.* **39**, 1247 (1968).

TABLE 3e-20. ATTENUATION OF SOUND IN n-BUTYL
ALCOHOL AT $p = 2,000$ KG/CM^2*
($f = 25$ MHz)

T, °C	0	15	30	45
$10^{15}\alpha/f^2$, sec^2/m	124.0	79.6	53.5	39.0

* E. H. Carnevale and T. A. Litovitz, *J. Acoust. Soc. Am.*, **27**, 547 (1955).

TABLE 3e-21. SPEED OF SOUND IN H_2O* AND
99.82 MOLE % D_2O† AT $p = 1$ ATM

T, °C	c, m/sec	
	H_2O	D_2O
0	1,402.3	
4	1,421.6	1,320.9
10	1,447.2	1,347.5
20	1,482.3	1,384.2
30	1,509.0	1,412.3
40	1,528.8	1,433.1
50	1,542.5	1,447.4
60	1,550.9	1,456.3
70	1,554.7	1,460.5
74	1,555.1	
80	1,554.4	1,460.8
90	1,550.4	1,457.8
100	1,543.0	1,452.0

* J. R. Lovett, *J. Acoust. Soc. Am.* **45**, 1052 (1969).   M. Greenspan and C. E. Tschiegg, *J. Research NBS* **59C**, 249 (1957).
† W. D. Wilson, *J. Acoust. Soc. Am.* **33**, 374 (1961).

TABLE 3e-22. SPEED OF SOUND IN H_2O* AND
99.82 MOLE % D_2O† NEAR ROOM TEMPERATURE

p, psi	c, m/sec	
	H_2O at 30.68° C	D_2O at 30°C
14.7	1,510.6	1,412.3
1,450	1,527.8	
2,000		1,433.3
2,901	1,544.5	
4,000		1,454.7
4,351	1,561.6	
5,802	1,578.4	
6,000		1,476.3
7,252	1,595.3	
8,000		1,498.0
8,702	1,611.8	
10,000		1,519.8
10,150	1,628.8	
11,600	1,645.4	
12,000		1,541.5
14,000		1,563.0

* A. J. Barlow and E. Yazgan, *Brit. J. App. Phys.* **18**, 645 (1967).
† W. D. Wilson, *J. Acoust. Soc. Am.* **33**, 374 (1961).

Sea Water. Wilson[1] has measured the speed of sound in sea water from the Bermuda–Key West area of the Atlantic over the following range: temperature $-3 < T < 30°C$, pressure $1.033 < p < 1,000$ kg/cm^2, and salinity $3.3 < S < 3.7$ percent. Typical results are given in Tables 3e-25 and 3e-26. There is some reason to believe that these values are high. Lovett[2] recommends that they be reduced by 0.65 m/sec.

TABLE 3e-23. ATTENUATION OF SOUND IN H_2O AT $p = 1$ ATM*
(f varied from 8 to 67 MHz)

T, °C	$10^{15}\alpha/f^2$, sec^2/m
0	56.9
5	44.1
10	36.1
15	29.6
20	25.3
30	19.1
40	14.6
50	12.0
60	10.2
70	8.7
80	7.9
90	7.2

* J. M. M. Pinkerton, *Nature* **160**, 128 (1947).

TABLE 3e-24. ATTENUATION OF SOUND IN H_2O AT $T = 30°C$*

p, atm	0	500	1,000	1,500	2,000
$10^{15}\alpha/f^2$, sec^2/m	18.5	15.4	12.7	11.1	9.9

* T. A. Litovitz and E. H. Carnevale, *J. Appl. Phys.* **26**, 816 (1955).

TABLE 3e-25. SPEED OF SOUND IN SEA WATER AT $p = 1$ ATM*

T, °C	c, m/sec		
	$S = 3.3\%$	$S = 3.5\%$	$S = 3.7\%$
-3	1,431.9	1,435.0	1,437.6
0	1,446.3	1,449.4	1,451.9
5	1,468.2	1,471.2	1,473.5
10	1,487.6	1,490.4	1,492.7
15	1,504.6	1,507.4	1,509.5
20	1,519.6	1,522.2	1,524.2
25	1,531.9	1,535.1	1,536.9
30	1,443.8	1,546.2	1,547.9

* W. D. Wilson, *J. Acoust. Soc. Am.* **32**, 641 (1960).

Electrolytes. Monovalent ions in most cases affect the attenuation only slightly and increase the speed of sound to an extent depending on the concentration. Polyvalent ions introduce dispersion. References are given in Sec. 3e-7.

[1] W. D. Wilson, *J. Acoust. Soc. Am.* **32**, 641 (1960).
[2] J. R. Lovett, *J. Acoust. Soc. Am.* **45**, 1051 (1969).

3e-6. Mixtures. The behavior of liquid mixtures is varied. If the two constituents are both Kneser liquids, then c and α are in most cases intermediate to those of the constituents themselves. A mixture of two associated liquids will generally have a maximum in α at some composition and also, especially if one component is water, a maximum in c at some other composition. A mixture of one Kneser and one associated liquid may behave like a mixture of two Kneser liquids or in even a more complex fashion than a mixture of two associated liquids. The literature is extensive; references are given in Sec. 3e-7.

3e-7. Sources of Other Data. By far the best single source of data is Schaafs' book [1]. The coverage is through 1963. Data are given on inorganic, organic, and silico-organic liquids; on supercooled liquids; crystalline liquids; fatty acids; and molten metals and salts. Also treated are binary (and some ternary) mixtures, and aqueous and some nonaqueous solutions of electrolytes. Bergmann's book [2] is a good source for miscellaneous substances, as is a recent report by Turk and Hunter

TABLE 3e-26. SPEED OF SOUND IN SEA WATER AT $T = 20°C$*

p, kg/cm²	c, m/sec		
	$S = 3.3\%$	$S = 3.5\%$	$S = 3.7\%$
1.033	1,519.6	1,522.2	1,524.1
100	1,535.4	1,537.1	1,540.1
200	1,551.5	1,554.2	1,556.3
300	1,567.7	1,571.5	1,572.5
400	1,584.0	1,586.8	1,588.9
500	1,602.0	1,603.1	1,605.2
600	1,616.8	1,619.5	1,621.6
700	1,633.1	1,635.8	1,637.9
800	1,649.4	1,652.1	1,654.2
900	1,665.5	1,668.2	1,670.3

* W. D. Wilson, *J. Acoust. Soc. Am.* **32**, 641 (1960).

[16]. Sette has published four compilations, treating absorption [3, 6] and velocity [5, 7] in pure liquids [3, 7] and in mixtures [5, 6]. Velocity as related to molecular constitution is treated by Markham et al. [4], Herzfeld and Litovitz [15], Schaafs [17], and Nozdrev [18]. Del Grosso and Smura [8] give the speed of sound in and impedances of liquids suitable for certain applications; liquids simulating sea water and liquids having unusually low or high speeds of sound are included. Weissler and coworkers present considerable data in connection with their work on molecular structure, especially for alcohols [9], linear polymethyl siloxanes [10], cyclic compounds [11], inorganic halides [12], acetylene derivatives [13], and polyethylene glycols [14]. The acoustic and some other properties of many alcohols are given by Marks [19].

1. Schaafs, W.: Landolt-Börnstein New Series, Group II, "Atomic and Molecular Physics," vol. 5, "Molecular Acoustics," Springer-Verlag New York Inc., New York, 1967.
2. Bergmann, L.: "Der Ultraschall und seine Anwendung in Wissenschaft und Technik," 6th ed., S. Hirzel Verlag KG, Leipzig, 1954.
3. Sette, D.: *Nuovo Cimento (Suppl.)* **6**, 1 (1949).
4. Markham, J. J., R. T. Beyer, and R. B. Lindsay: *Rev. Mod. Phys.* **23**, 353 (1951).
5. Sette, D.: *Ricerca Sci.* **19**, 1338 (1949).
6. Sette, D.: *Nuovo Cimento (Suppl.)* 2) **7**, 318 (1950).
7. Sette, D.: *Ricerca Sci.* **20**, 102 (1950).
8. Del Grosso, V. A., and E. J. Smura: *NRL Rept.* 4193, 1953.

9. Weissler, A.: *J. Am. Chem. Soc.* **70**, 1634 (1948).
10. Weissler, A.: *J. Am. Chem. Soc.* **71**, 93 (1949).
11. Weissler, A.: *J. Am. Chem. Soc.* **71**, 419 (1949).
12. Weissler, A.: *J. Am. Chem. Soc.* **71**, 1272 (1949).
13. Weissler, A., and V. A. Del Grosso: *J. Am. Chem. Soc.* **72**, 4209 (1950).
14. Weissler, A., J. W. Fitzgerald, and I. Resnick: *J. Appl. Phys.* **18**, 434 (1947).
15. Herzfeld, K. F., and T. A. Litovitz: "Absorption and Dispersion of Ultrasonic Waves," Academic Press, Inc., New York, 1959.
16. Turk, R. A., and J. L. Hunter: The Velocity and Absorption of Sound in Various Liquids, *ONR Tech. Rept.* 8, Contract NONR 2577(01) (AD) 651 978), Department A, Clearinghouse, Springfield, Va.
17. Schaafs, W.: "Molekularakustik," Springer-Verlag OHG, Berlin, 1963.
18. Nozdrev, V. F.: "Application of Ultrasonics in Molecular Physics," Gordon and Breach, Science Publishers, Inc., New York, 1963.
19. Marks, G. W., *J. Acoust. Soc. Am.* **41**, 104 (1967).

3f. Acoustic Properties of Solids

W. P. MASON

Columbia University

3f-1. Elastic Constants, Densities, Velocities, and Impedances. Solids are used for conducting acoustic waves in such devices as delay lines useful for storing information and as resonating devices for controlling and selecting frequencies. Acoustic-wave propagation in solids has been used to determine the elastic constants of single crystals and polycrystalline materials. Changes in velocity with frequency and changes in attenuation with frequency have been used to analyze various intergrain, interdomain, and imperfection motions as discussed in Sec. 3f-2.

In an infinite isotropic solid and also in a finite solid for which the wavefront is a large number of wavelengths, plane and nearly plane longitudinal and shear waves can exist which have the velocities

$$v_{long} = \sqrt{\frac{\lambda + 2\mu}{\rho}} \qquad v_{shear} = \sqrt{\frac{\mu}{\rho}} \qquad (3f\text{-}1)$$

where μ and λ are the two Lamé elastic moduli, μ is the shearing modulus, and $\lambda + 2\mu$ has been called the plate modulus. For a rod whose diameter is a small fraction of a wavelength, extensional and torsional waves can be propagated with velocities

$$v_{ext} = \sqrt{\frac{Y_0}{\rho}} \qquad v_{tor} = \sqrt{\frac{\mu}{\rho}}$$

where
$$Y_0 = \mu \left(\frac{3\lambda + 2\mu}{\lambda + \mu} \right) \qquad (3f\text{-}2)$$

For anisotropic media, three waves will, in general, be propagated, but it is only in special cases that the particle motions will be normal and perpendicular to the direction of propagation.   The three velocities satisfy an equation[1]

$$\begin{vmatrix} \lambda_{11} - \rho v^2 & \lambda_{12} & \lambda_{13} \\ \lambda_{12} & \lambda_{22} - \rho v^2 & \lambda_{23} \\ \lambda_{13} & \lambda_{23} & \lambda_{33} - \rho v^2 \end{vmatrix} = 0 \qquad (3f\text{-}3)$$

where ρ is the density, v the velocity, and the λ's are related to the elastic constants of the crystal by the formulas

$$\begin{aligned}
\lambda_{11} &= l^2 c_{11} + m^2 c_{66} + n^2 c_{55} + 2mn c_{56} + 2nl c_{15} + 2lm c_{16} \\
\lambda_{12} &= l^2 c_{16} + m^2 c_{26} + n^2 c_{45} + mn(c_{46} + c_{25}) + nl(c_{14} + c_{56}) + lm(c_{12} + c_{66}) \\
\lambda_{13} &= l^2 c_{15} + m^2 c_{46} + n^2 c_{35} + mn(c_{45} + c_{36}) + nl(c_{13} + c_{55}) + lm(c_{14} + c_{56}) \\
\lambda_{23} &= l^2 c_{56} + m^2 c_{24} + n^2 c_{34} + mn(c_{44} + c_{23}) + nl(c_{36} + c_{45}) + lm(c_{25} + c_{46}) \\
\lambda_{22} &= l^2 c_{66} + m^2 c_{22} + n^2 c_{44} + 2mn c_{24} + 2nl c_{46} + 2lm c_{26} \\
\lambda_{33} &= l^2 c_{55} + m^2 c_{44} + n^2 c_{33} + 2mn c_{34} + 2nl c_{35} + 2lm c_{45}
\end{aligned} \qquad (3f\text{-}4)$$

In these formulas c_{11} to c_{66} are the 21 elastic constants and l, m, n the direction cosines of the direction of propagation with respect to the crystallographic x, y, and z axes which are related to the a, b, c crystallographic axes as discussed in an IRE publication.[2]

In Eq. (3f-3), we solve for the quantity ρv^2.   It was shown by Christoffel[1] that the direction cosines for the particle motion ξ, i.e., α, β, γ, are related to the λ constants and a solution of ρv^2 by the equations

$$\alpha \lambda_{11} + \beta \lambda_{12} + \gamma \lambda_{13} = \alpha \rho v_i{}^2; \ \alpha \lambda_{12} + \beta \lambda_{22} + \gamma \lambda_{23} = \beta \rho v_i{}^2; \ \alpha \lambda_{13} + \beta \lambda_{23} + \gamma \lambda_{33} = \gamma \rho v_i{}^2$$
$$(3f\text{-}5)$$

where $i = 1, 2, 3$.   Hence, solutions of Eq. (3f-3) are related to particle motions by the equations of (3f-5).

Most metals crystallize in the cubic and hexagonal systems.   Furthermore, when a metal is produced by rolling, an alignment of grains occurs such that the rolling direction is a unique axis.   This type of symmetry, known as transverse isotropy, results in the same set of constants as that for hexagonal symmetry.   For cubic crystals, the resulting elastic constants are

$$c_{11} = c_{22} = c_{33} \qquad c_{12} = c_{13} = c_{23} \qquad c_{44} = c_{55} = c_{66} \qquad (3f\text{-}6)$$

while for hexagonal symmetry or transverse isotropy, the resulting elastic constants are

$$c_{11} = c_{22}, \qquad c_{12}, \qquad c_{13} = c_{23}, c_{33}, \qquad c_{44} = c_{55}, \qquad c_{66} = \frac{c_{11} - c_{12}}{2} \qquad (3f\text{-}7)$$

For cubic symmetry, the waves transmitted along the [100] direction and the [110] direction have purely longitudinal and shear components with the elastic-constant values and particle direction ξ given by

[100] direction

$$v_{\text{long}} = \sqrt{\frac{c_{11}}{\rho}} \quad \xi \text{ along } [100] \qquad v_{\text{shear}} = \sqrt{\frac{c_{44}}{\rho}}$$
$$\xi \text{ along any direction in the } [100] \text{ plane}$$

[1] Love, "Theory of Elasticity," 4th ed., p. 298, Cambridge University Press, New York, 1934.
[2] Standards on Piezoelectric Crystals, *Proc. IRE* **37** (12), 1378–1395 (December, 1949).

[110] direction

$$v_{\text{long}} = \sqrt{\frac{c_{11} + c_{12} + 2c_{44}}{2\rho}} \qquad \xi \text{ along } [110]$$

$$v_{1\,\text{shear}} = \sqrt{\frac{c_{44}}{\rho}} \quad \xi \text{ along } [001] \qquad v_{2\,\text{shear}} = \sqrt{\frac{c_{11} - c_{12}}{2\rho}}$$

$$\xi \text{ along } [1\bar{1}0]$$

[111] direction

$$v_1 = v_{\text{long}} = \sqrt{\frac{c_{11} + 2c_{12} + 4c_{44}}{3\rho}} \qquad \xi \text{ along } [111]$$

$$v_2 = v_3 = v_{\text{shear}} = \sqrt{\frac{c_{11} - c_{12} + c_{44}}{3\rho}}$$

ξ can be in any direction in the (111) plane.

For hexagonal or transverse isotropy, waves transmitted along the unique axis and any axis perpendicular to this will have the values

[001] direction

$$v_{\text{long}} = \sqrt{\frac{c_{33}}{\rho}} \quad \xi \text{ along } [001] \qquad v_{\text{shear}} = \sqrt{\frac{c_{44}}{\rho}}$$

$$\xi \text{ along any direction in the } [001] \text{ plane}$$

[100] direction

$$v_{\text{long}} = \sqrt{\frac{c_{11}}{\rho}} \quad \xi \text{ along } [100] \qquad v_{1\,\text{shear}} = \sqrt{\frac{c_{44}}{\rho}}$$

$$\xi \text{ along } [001] \qquad v_{2.\text{shear}} = \sqrt{\frac{c_{11} - c_{12}}{2\rho}} \qquad \xi \text{ along } [010]$$

The fifth constant is measured by transmitting a wave 45 deg between the [100] and [001] directions; i.e., $l = n = 1/\sqrt{2}$; $m = 0$. For this case

$$\lambda_{11} = \frac{c_{11} + c_{44}}{2} \qquad \lambda_{12} = \lambda_{23} = 0 \qquad \lambda_{13} = \frac{c_{13} + c_{44}}{2} \qquad \lambda_{22} = \frac{c_{11} - c_{12} + 2c_{44}}{4}$$

$$\lambda_{33} = \frac{c_{44} + c_{33}}{2} \tag{3f-8}$$

The three solutions of Eq. (3f-3) are

$$\rho v_1{}^2 = \frac{c_{11} - c_{12} + 2c_{44}}{4}$$

$$\rho v_{2,3}{}^2 = \frac{[(c_{11} + c_{33} + 2c_{44})/2] \pm \sqrt{[(c_{11} - c_{33})/2]^2 + (c_{13} + c_{44})^2}}{2} \tag{3f-9}$$

For these three velocities, the particle velocities have the direction cosines

For v_1, $\beta = 1$

For v_2, $\alpha = \gamma \left\{ \dfrac{c_{11} - c_{33}}{2(c_{13} + c_{44})} + \sqrt{1 + \left[\dfrac{c_{11} - c_{33}}{2(c_{13} + c_{44})}\right]} \right\}$ (3f-10)

For v_3, $\alpha = -\gamma \left\{ \dfrac{c_{33} - c_{11}}{2(c_{13} + c_{44})} + \sqrt{1 + \left[\dfrac{(c_{11} - c_{33})^2}{2(c_{13} + c_{44})}\right]} \right\}$

Hence, unless c_{11} is nearly equal to c_{33}, a longitudinal or shear crystal will generate both types of waves. Experimentally, however, it is found that a good discrimination can be obtained against the type of wave that is not primarily generated and a single velocity can be measured. A resonance technique can also be used to evaluate all the elastic constants of a crystalline material.

TABLE 3f-1. DENSITIES OF GLASSES, PLASTICS, AND METALS IN
POLYCRYSTALLINE AND CRYSTALLINE FORM (X-RAY DENSITIES
FOR CRYSTALS)*

Materials	Composition	Temp., °C	Density, 10^3 kg/m^3 or g/cm^3
Aluminum			
Hard-drawn		20	2.695
Crystal		25	2.697
Aluminum and copper	10 Al, 90 Cu	..	7.69
	5 Al, 95 Cu	..	8.37
	3 Al, 97 Cu	..	8.69
Beryllium		20	1.87
Crystal		18	1.871
Brass:			
Yellow	70 Cu, 30 Zn	..	8.5–8.7
Red	90 Cu, 10 Zn	..	8.6
White	50 Cu, 50 Zn	..	8.2
Bronze	90 Cu, 10 Sn	..	8.78
	85 Cu, 15 Sn	..	8.89
	80 Cu, 20 Sn	..	8.74
	75 Cu, 25 Sn	..	8.83
Chromium		20	6.92–7.1
Crystal		18	7.193
Cobalt		21	8.71
Crystal		..	8.788
Constantan	60 Cu, 40 Ni	..	8.88
Copper		..	8.3–8.93
Crystal		18	8.936
Duralumin	17ST = 4 Cu, 0.5 Mg, 0.5 Mn	..	2.79
Germanium		..	5.3
Crystal		20	5.322
German silver	26.3 Cu, 36.6 Zn, 36.8 Ni	..	8.30
	52 Cu, 26 Zn, 22 Ni	..	8.45
	59 Cu, 30 Zn, 11 Ni	..	8.34
	63 Cu, 30 Zn, 6 Ni	..	8.30
Gold		..	18.9–19.3
Crystal		20	19.32
Indium		..	7.28
Crystal		..	7.31
Invar	63.8 Fe, 36 Ni, 0.20 C	..	8.0
Iron		20	7.6–7.85
Crystal		20	7.87
Lead		20	11.36
Crystal		18	11.34
Lead and tin	87.5 Pb, 12.5 Sn	..	10.6
	84 Pb, 16 Sn	..	10.33
	72.8 Pb, 22.2 Sn	..	10.05
	63.7 Pb, 36.3 Sn	..	9.43
	46.7 Pb, 53.3 Sn	..	8.73
	30.5 Pb, 69.5 Sn	..	8.24

TABLE 3f-1. DENSITIES OF GLASSES, PLASTICS, AND METALS IN
POLYCRYSTALLINE AND CRYSTALLINE FORM (X-RAY DENSITIES
FOR CRYSTALS)* (*Continued*)

Materials	Composition	Temp., °C	Density, 10^3 kg/m^3 or g/cm^3
Magnesium.................	....	..	1.74
Crystal....................	....	25	1.748
Manganese.................	....	..	7.42
Crystal....................	....	..	7.517
Mercury.....................	....	20	13.546
Monel metal..............	71 Ni, 27 Cu, 2 Fe	..	8.90
Molybdenum...............	....	..	10.1
Crystal....................	....	25	10.19
Nickel......................	....	..	8.6–8.9
Crystal....................	....	25	8.905
Nickel silver.............	....	..	8.4
Phosphor bronze..........	79.7 Cu, 10 Sn, 9.5 Sb, 0.8 P	..	8.8
Platinum...................	....	20	21.37
Crystal....................	....	18	21.62
Silicon......................	....	15	2.33
Crystal....................	....	25	2.332
Silver.......................	....	..	10.4
Crystal....................	....	25	10.49
Steel K9....................	....	..	7.84
347 stainless steel..........	....	..	7.91
Tin..........................	....	..	7–7.3
Crystal....................	....	..	7.3
Titanium...................	....	..	4.50
Tungsten...................	....	..	18.6–19.1
Crystal....................	....	25	19.2
Tungsten carbide..........	....	..	13.8
Zinc........................	....	..	7.04–7.18
Crystal....................	....	25	7.18
Fused silica...............	....	..	2.2
Pyrex glass (702)..........	....	..	2.32
Heavy silicate flint........	....	..	3.879
Light borate crown........	....	..	2.243
Lucite......................	....	..	1.182
Nylon 6-6..................	....	..	1.11
Polyethylene...............	....	..	0.90
Polystyrene................	....	..	1.056

* See also Tables 2b-1 through 2b-13.

When a longitudinal or shear wave is reflected at an angle from a plane surface, both a longitudinal and a shear wave will in general be reflected from the surface, the angles of reflection and refraction satisfying Snell's law

$$\frac{\sin \beta}{v_S} = \frac{\sin \alpha}{v_l}$$

(3f-11)

where α and β are the angles of incidence and refraction with respect to a normal to the reflecting surface. Exceptions to this rule occur if a shear wave has its direction of particle displacement parallel to the reflecting surface, in which case only a pure shear wave is reflected, with the angle of reflection being equal to the angle of incidence. Use is made of this result in constructing delay lines which can be contained in a small volume. When the direction of transmission is normal to the surface, the incident wave is reflected without change of mode. If the transmitting medium is connected to another medium with different properties, the transmission and reflection factors are determined by the relative impedances of the two media. The impedance is given by the formula

$$Z = \rho v = \sqrt{E\rho} \qquad (3f\text{-}12)$$

where E is the appropriate elastic stiffness and ρ the density. The reflection and transmission coefficients between medium 1 and medium 2 are given by the equations

$$R = \frac{Z_1 - Z_2}{Z_1 + Z_2} \qquad T = 1 - R = \frac{2Z_2}{Z_1 + Z_2} \qquad (3f\text{-}13)$$

Tables 3f-1 to 3f-4 list the densities, elastic constants, velocities, and impedances for a number of materials used in acoustic-wave propagation.

3f-2. Attenuation Due to Thermal Effects, Relaxations, and Scattering. When sound is propagated through a solid, it suffers a conversion of mechanical energy into heat. While all the causes of conversion are not known, a number of them are, and tables for these effects are given in this section.

3f-3. Loss Due to Heat Flow. When a sound wave is sent through a body, a compression or rarefaction occurs which heats or cools the body. This heat causes thermal expansions which alter slightly the elastic constants of the material. Since the compressions and rarefactions occur very rapidly, there is not time for much heat to flow and the elastic constants measured by sound propagation are the adiabatic constants. For an isotropic material, the adiabatic constants are related to the isothermal constants by the formulas[1]

$$\lambda^\sigma = \lambda^\theta + \frac{9\alpha^2 B^2 \Theta}{\rho C_v} \qquad \mu^\sigma = \mu^\theta \qquad Y_0{}^\sigma = Y_0{}^\theta + \left(\frac{\mu}{\lambda + \mu}\right)^2 \frac{9\alpha^2 B^2 \Theta}{\rho C_v} \qquad (3f\text{-}14)$$

where the superscripts σ and θ indicate adiabatic and isothermal constants, α is the linear temperature coefficient of expansion, B the bulk modulus ($B = \lambda + \frac{2}{3}\mu$), Θ the absolute temperature in kelvins, ρ the density, and C_v the specific heat at constant volume. Table 3f-5 shows these quantities for a number of materials.

The difference between λ^σ and λ^θ should be taken account of when one compares the elastic constants measured by ultrasonic means with those measured by static means. From the data given in Table 3f-5, it is evident that this effect can produce errors as high as 10 percent in the case of zinc. Adiabatic elastic constants are measured from frequencies somewhat less than those for which thermal equilibrium is established during the cycle up to a frequency[1] $f \doteq (\sigma C_v v^2 / 2\pi K)$ for which wave propagation takes place isothermally. This latter frequency is approximately 10^{12} Hz for most metals.

When account is taken of the energy lost by heat flow between the hot and cool parts, this adds an attenuation for longitudinal waves equal to

$$A = \frac{2\pi f^2}{\rho v^3} \left[\frac{K}{C_v} \left(\frac{E^\sigma - E^\theta}{E^\theta} \right) \right] \qquad \text{nepers/m} \qquad (3f\text{-}15)$$

[1] W. P Mason, "Piezoelectric Crystals and Their Application to Ultrasonics," pp. 480–481, D. Van Nostrand Company, Inc., Princeton, N.J. 1950.

TABLE 3f-2. ELASTIC CONSTANTS, WAVE VELOCITIES, AND CHARACTERISTIC IMPEDANCES OF METALS, GLASSES, AND PLASTICS

Materials	$Y_0 \times 10^{-10}$ newton/m²	$\mu \times 10^{-10}$ newton/m²	$\lambda \times 10^{-10}$ newton/m²	Poisson's ratio, σ	$V_l = \sqrt{(\lambda + 2\mu)/\rho}$, m/sec	$V_s = \sqrt{\mu/\rho}$, m/sec	$V_{ext} = \sqrt{Y_0/\rho}$, m/sec	$Z_l = \sqrt{\rho(\lambda + 2\mu)}$, 10^6 kg/sec m²	$Z_s = \sqrt{\rho\mu}$, 10^6 kg/sec m²
Aluminum, rolled	6.8–7.1	2.4–2.6	6.1	0.355	6,420	3,040	5,000	17.3	8.2
Beryllium	30.8	14.7	1.6	0.05	12,890	8,880	12,870	24.1	16.6
Brass, yellow, 70 Cu, 30 Zn	10.4	3.8	11.3	0.374	4,700	2,110	3,480	40.6	18.3
Constantan	16.1	6.09	11.4	0.327	5,177	2,625	4,270	45.7	23.2
Copper, rolled	12.1–12.8	4.6	13.1	0.37	5,010	2,270	3,750	44.6	20.2
Duralumin 17S	7.15	2.67	5.44	0.335	6,320	3,130	5,150	17.1	8.5
Gold, hard-drawn	8.12	2.85	15.0	0.42	3,240	1,200	2,030	62.5	23.2
Iron, cast	15.2	5.99	6.92	0.27	4,994	2,809	4,480	37.8	21.35
Iron electrolytic	20.6	8.2	11.3	0.29	5,950	3,240	5,120	46.4	25.3
Armco	21.2	8.24	11.35	0.29	5,960	3,240	5,200	46.5	25.3
Lead, rolled	1.5–1.7	0.54	3.3	0.43	1,960	690	1,210	22.4	7.85
Magnesium, drawn, annealed	4.24	1.62	2.56	0.306	5,770	3,050	4,940	10.0	5.3
Monel metal	16.5–18	6.18–6.86	12.4	0.327	5,350	2,720	4,400	47.5	24.2
Nickel	21.4	8.0	16.4	0.336	6,040	3,000	4,900	53.5	26.6
Nickel silver	10.7	3.92	11.2	0.37	4,760	2,160	3,575	40.0	18.1
Platinum	16.7	6.4	9.9	0.303	3,260	1,730	2,800	69.7	37.0
Silver	7.5	2.7	8.55	0.38	3,650	1,610	2,680	38.0	16.7
Steel, K9	21.6	8.29	10.02	0.276	5,941	3,251	5,250	46.5	25.4
347 stainless steel	19.6	7.57	11.3	0.30	5,790	3,100	5,000	45.7	24.5
Tin, rolled	5.5	2.08	4.04	0.34	3,320	1,670	2,730	24.6	11.8
Titanium	11.6	4.40	7.79	0.32	6,070	3,125	5,090	27.3	14.1
Tungsten, drawn	36.2	13.4	31.3	0.35	5,410	2,640	4,320	103	50.5
Tungsten carbide	53.4	21.95	17.1	0.22	6,655	3,984	6,240	91.8	55.0
Zinc, rolled	10.5	4.2	4.2	0.25	4,210	2,440	3,850	30	17.3
Fused silica	7.29	3.12	1.61	0.17	5,968	3,764	5,760	13.1	8.29
Pyrex glass	6.2	2.5	2.3	0.24	5,640	3,280	5,170	13.1	7.6
Heavy silicate flint	5.35	2.18	1.77	0.224	3,980	2,380	3,720	15.4	9.22
Light borate crown	4.61	1.81	2.2	0.274	5,100	2,840	4,540	11.4	6.35
Lucite	0.40	0.143	0.562	0.4	2,680	1,100	1,840	3.16	1.3
Nylon 6–6	0.355	0.122	0.511	0.4	2,620	1,070	1,800	2.86	1.18
Polyethylene	0.076	0.026	0.288	0.458	1,950	540	920	1.75	0.48
Polystyrene	0.360	0.133	0.319	0.353	2,350	1,120	1,840	2.49	1.19

TABLE 3f-3. ELASTIC CONSTANTS OF CUBIC SINGLE CRYSTALS*

(s = compliance modulus, m²/newton; c = stiffness modulus, newtons/m²; for cgs units of dynes/cm², multiply the c tabular entries by 10; divide the s tabular entries by 10 to obtain cm²/dyne)

Crystal	$s_{11} \times 10^{11}$	$s_{12} \times 10^{11}$	$s_{44} \times 10^{11}$	$c_{11} \times 10^{-10}$	$c_{12} \times 10^{-10}$	$c_{44} \times 10^{-10}$	$B = [(c_{11} + 2c_{12})/3] \times 10^{-10}$	Anisotropy $2c_{44}/(c_{11} - c_{12})$
Ag	2.32	−0.993	2.29	11.9	8.94	4.37	9.93	2.95
Al	1.59	−0.58	3.52	10.82	6.13	2.85	7.69	1.24
Au	2.33	−1.07	2.38	19.6	16.45	4.20	17.5	2.67
Cu	1.49	−0.625	1.33	17.02	12.3	7.51	13.9	3.18
Fe	0.757	−0.282	0.862	23.7	14.1	11.6	17.3	2.37
Ge	0.964	−0.260	1.49	12.92	4.79	6.70	7.50	1.65
K	83.3	−37.0	38.0	0.416	0.333	0.263	0.361	6.34
Na	48.3	−20.9	16.85	0.615	0.469	0.592	0.518	8.11
Ni (sat.)	0.80	−0.312	0.844	25.0	16.0	11.85	19.0	2.63
Pb	9.30	−4.26	6.94	4.85	4.09	1.44	4.34	3.79
Si	0.768	−0.214	1.26	16.57	6.39	7.956	9.783	1.56
W	0.257	−0.073	0.66	50.2	19.9	15.15	30.0	1.0
Diamond†	0.0958	−0.01	0.174	107.6	12 5	57.6	44.2	1.21
NaCl	2.4	−0.50	7.8	4.9	1.24	1.26	2.5	0.688
KBr	4.0	−1.2	7.5	3.5	0.58	0.50	1.6	0.342
KCl	2.7	−0.3	15.6	4.0	0.62	0.62	1.7	0.361

Elastic Constants of Copper Alloys‡

Alloy	Atom % of second component	$s_{11} \times 10^{11}$	$s_{12} \times 10^{11}$	$s_{44} \times 10^{11}$	$c_{11} \times 10^{-10}$	$c_{12} \times 10^{-10}$	$c_{44} \times 10^{-10}$	$B = [(c_{11} + 2c_{12})/3] \times 10^{-10}$	Anisotropy $2c_{44}/(c_{11} - c_{12})$
CuZn	4.53	1.59	−0.671	1.348	16.34	11.92	7.42	13.39	3.36
CuAl	4.81	1.59	−0.674	1.335	16.58	12.16	7.49	13.63	3.39
	9.98	1.67	−0.711	1.305	15.95	11.77	7.66	13.16	3.66
CuGa	1.58	1.55	−0.65	1.346	16.49	11.93	7.43	13.45	3.25
	4.15	1.59	−0.672	1.349	16.51	12.10	7.41	13.57	3.36
CuSi	4.17	1.61	−0.685	1.336	16.78	12.42	7.48	13.87	3.43
	5.16	1.67	−0.709	1.335	16.09	11.88	7.49	13.28	3.56
	7.69	1.73	−0.745	1.350	16.64	12.60	7.41	13.95	3.67
CuGe	1.03	1.52	−0.637	1.333	16.66	12.00	7.50	13.62	3.29
	1.71	1.57	−0.663	1.333	16.30	11.83	7.50	13.32	3.35

* See also Tables 2e-1 through 2e-6.
† Recent data by W. L. Bond and H. J. McSkimin.
‡ Data from C. S. Smith.

TABLE 3f-4. ELASTIC CONSTANTS OF HEXAGONAL CRYSTALS

s = compliance moduli, m²/newton; c = stiffness moduli, newtons/m²; for cgs units of dynes/cm² multiply the c tabular entries by 10; divide the s tabular entries by 10 to obtain cm²/dyne)

Crystal	$s_{11} \times 10^{11}$	$s_{12} \times 10^{11}$	$s_{13} \times 10^{11}$	$s_{33} \times 10^{11}$	$s_{44} \times 10^{11}$
Cd	1.23	−0.15	−0.93	3.55	5.40
Mg	2.21	−0.77	−0.49	1.97	6.03
Zn	0.84	+0.11	−0.78	2.87	2.64
Co	0.473	−0.231	−0.07	0.319	1.325

	$c_{11} \times 10^{-10}$	$c_{12} \times 10^{-10}$	$c_{13} \times 10^{-10}$	$c_{33} \times 10^{-10}$	$c_{44} \times 10^{-10}$	$B = \dfrac{1}{2(s_{11} + s_{12}) + s_{33} + 4s_{13}} \times 10^{-10}$
Cd	12.12	4.81	4.42	4.45	1.85	5.03
Mg	5.86	2.49	2.08	6.60	1.65	3.46
Zn	16.35	2.64	5.17	5.31	3.78	8.26
Co	30.71	16.5	10.27	35.81	7.55	19.01

TABLE 3f-5. ADIABATIC AND ISOTHERMAL ELASTIC CONSTANTS AND ATTENUATION DUE TO HEAT FLOW

Material	$10^{-3} \times$ density, kg/m^3	C_v, joules/kg/°C $\times 10^{-3}$	$\alpha \times 10^6$ 1/°C	$K \times 10^{-3}$ watts/m^2/m/°C	$\lambda^\theta \times 10^{-10}$ newtons/m^2	$\mu \times 10^{-10}$ newtons/m^2	$(\lambda^\sigma - \lambda^\theta) \times 10^{-9}$ newtons/m^2	$(Y_0^\sigma - Y_0^\theta) \times 10^{-8}$ newtons/m^2	A/f^2, nepers/m
Aluminum	2.699	0.9	23.9	2.22	6.1	2.5	3.8	3.2	2.3×10^{-16}
Beryllium	1.82	2.17	12.4	1.58	1.6	14.7	1.4	11.4	2.1×10^{-18}
Copper	8.96	0.384	16.5	3.93	13.1	4.6	5.5	3.7	4.45×10^{-16}
Gold	19.32	0.13	14.2	2.97	15.0	2.85	6.1	1.5	1.95×10^{-15}
Iron	7.87	0.46	11.7	0.75	11.3	8.2	2.7	4.8	1.88×10^{-17}
Lead	11.4	0.128	29.4	0.344	3.3	0.54	2.12	0.36	2.95×10^{-15}
Magnesium	1.74	1.04	26	1.59	2.56	1.62	1.3	2.1	2.0×10^{-16}
Nickel	8.90	0.44	13.3	0.92	16.4	8.0	5.7	6.1	3.8×10^{-17}
Silver	10.49	0.234	19.7	4.18	8.55	2.7	4.5	2.6	1.95×10^{-15}
Tin	7.3	0.225	23	0.67	4.04	2.08	3.5	4.0	9.7×10^{-16}
Tungsten	19.3	0.134	4.3	2.0	31.3	13.4	3.1	2.8	5.0×10^{-17}
Zinc	7.1	0.382	29.7	1.12	4.2	4.2	4.3	10.7	3.8×10^{-16}
Fused silica	2.2	0.92	0.5	0.01	1.61	3.12	0.00045	0.002	2.6×10^{-22}

TABLE 3f-6. FACTORS GOVERNING INTERGRAIN HEAT FLOW IN METALS

Metal	Pb	Ag	Cu	Au	Fe	Al	W
R	0.035	0.031	0.031	0.014	0.022	0.0009	10^{-1}
$(C_p - C_v)/C_v$	0.067	0.040	0.028	0.038	0.016	0.046	0.006
Product	4.4×10^{-3}	1.2×10^{-3}	8.7×10^{-4}	5.3×10^{-4}	3.5×10^{-4}	4×10^{-5}	6×10^{-9}

where f is the frequency, v the velocity, K the heat conductivity, and E the appropriate elastic constant for the mode of propagation considered. Since $Q = B/2A$, it becomes

$$Q = \frac{\rho C_v v^2}{2fK[(E^\sigma - E^\theta)/E^\theta]} \tag{3f-16}$$

where Q is the ratio of 2π times the energy stored to energy dissipated per cycle and B is the phase shift per unit length. Table 3f-5 shows the attenuation for a number of solids due to thermoelastic loss.

The thermoelastic effect produces about half the thermal attenuation for metals but only about 4 percent for dielectric crystals. The largest source of loss for these crystals is the Akhieser effect which results from an instantaneous separation of the phonon modes, followed by an equilibration of these temperatures which occurs with a relaxation time τ. This effect produces a loss of about 40 times the thermoelastic loss for insulators. According to a recent theory[1] this loss is

$$\alpha_{(\text{nep.}/\text{cm})} = \frac{\omega^2 D(E_0 K/C_v \bar{v}^2)}{2\rho v^3(1 + \omega^2 \tau^2)} \qquad \tau = \frac{3K}{C_v \bar{v}^2} \tag{3f-17}$$

where the ratio of the total thermal energy E_0 to the specific heat C_v is proportional to a factor F times the absolute temperature T. F varies from 0.25 at very low temperatures to unity above the Debye temperature. D is a nonlinear constant which can be calculated when the third-order moduli of the material are known, K is the thermal conductivity, ρ the crystal density, v is the sound velocity, and $\bar{v}$ the Debye average velocity. A number of third-order moduli have been measured for at least six crystals, and the agreement with Eq. (3f-17) is good.

Figure 3f-1 shows typical measurements of the attenuation of the two shear waves and the longitudinal wave in a single crystal of aluminum oxide Al_2O_3. Below 20 to 30 K the attenuation is independent of the temperature. This region is assumed to be controlled by scattering losses due to imperfections in the crystal and transducers. This loss is a good measure of the imperfections in the crystal. Above this region the attenuation for the slow shear wave increases as the fourth power of the temperature from 20 to 80 K. This is in agreement with the theory of Landau and Rumer (1937),[2] which considers the direct interactions of the acoustic waves with the thermal phonons. This formula can be put into the form

$$\alpha = 60\gamma^2 \frac{kT}{Mv^2} \left(\frac{T}{\theta}\right)^3 \frac{2\pi}{\lambda} \tag{3f-18}$$

where α is the attenuation in nepers per cm, γ the Grueneisen constant, k the Boltzmann constant, M the average atomic mass, $\bar{v}$ an average sound velocity, T the absolute temperature, θ the Debye temperature, and λ the acoustic wavelength. The agreement with the formula is quite good. The fast shear wave and the longitudinal wave behave in a different manner with slopes proportional to T^7 and T^9, respectively. Explanations for these values have not yet been obtained.

For higher temperatures when the product of the angular frequency ω times the thermal relaxation time τ is much less than unity, individual interactions between sound waves and phonons can no longer be followed. In this region the two effects causing the thermal attenuation are the thermoelastic effect and the Akhieser effect, discussed above.

3f-4. Loss Due to Intergrain Heat Flow. A related thermal loss that occurs in polycrystalline material is the thermoelastic relaxation loss which arises from heat flow

[1] See W. P. Mason, "Physical Acoustics," vol. IIIB, chap. 6, Academic Press, Inc., New York, 1965.
[2] L. Landau and G. Rumer, *Physik. Z. Sowjetunion* **11**, 18 (1937).

from grains that have received more compression or extension in the course of the wave motion than do adjacent grains. The Q from this source has been shown to be[1]

$$\frac{1}{Q} = \frac{C_p - C_v}{C_v} R \frac{f_0 f}{f_0{}^2 + f^2} \tag{3f-19}$$

where R is that fraction of the total strain energy which is associated with the fluctuations of dilations, and f_0, the relaxation frequency, is approximately

$$f_0 = \frac{D}{L_c{}^2} = \frac{K}{\rho C_p L_c{}^2} \tag{3f-20}$$

where L_c is the mean diameter of the crystallites and D the diffusion constant.

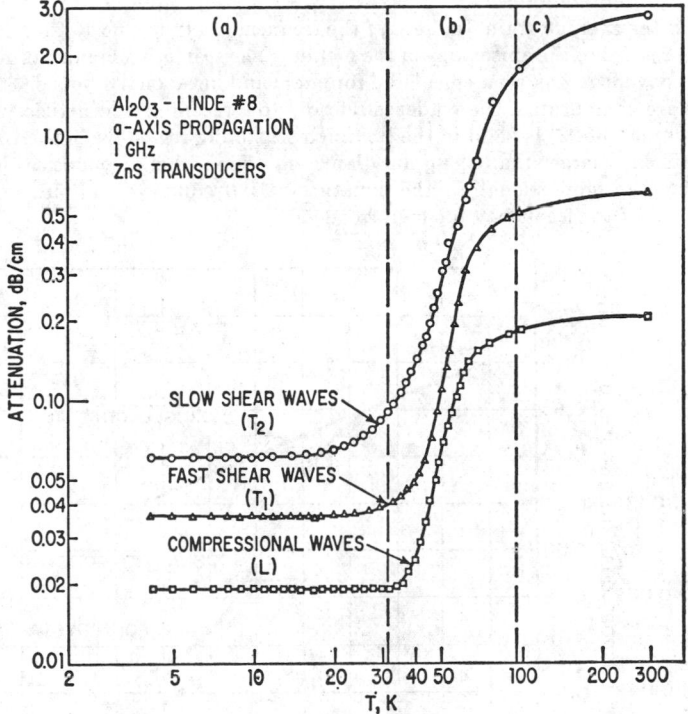

Fig. 3f-1. Attenuation of plane waves in aluminum oxide.

For most materials, the relaxation frequencies are under 100 kHz. Table 3f-6 gives the product $[(C_p - C_v)/C_v]R$ for a number of metals.

3f-5. Loss Due to Grain Rotation. Another source of loss due to grain structure in metals is the loss due to the viscosity of the boundary layer between grains. This allows a relative rotation of grains provided the relaxation time is comparable to the time of the applied force. Figure 3f-1 shows the elastic modulus and the associated Q of a polycrystalline aluminum rod in torsional vibration at a frequency of 0.8 Hz as compared with similar measurements for a single crystal. The relaxation time for grain-boundary rotation is a function of temperature according to the equation

$$\tau = \tau_0 e^{H/kT} \tag{3f-21}$$

[1] C. Zener, "Elasticity and An elasticity of Metals," p. 84, University of Chicago Press Chicago, 1948.

where H, the activation energy, is of the same order as that found for creep and self-diffusion.

3f-6. Loss Due to Grain Scattering of Sound. Another effect of grain structure in solids is a loss of energy from the main wave due to the scattering of sound when the sound wavelength is of the same order as the grain size. This scattering occurs because adjacent grains have different orientations, and a reflection of sound occurs because of the resulting impedance difference between grains. An approximate formula[1] holding when the wavelength is larger than three times the grain size, and multiple scattering is neglected, is

$$\alpha_s \doteq \frac{8\pi^4 L_c{}^3 f^4}{9v^4} S \qquad \text{nepers/m} \qquad (3f\text{-}22)$$

where L_c is the average grain diameter, f the frequency, v the velocity, and S a scattering factor related to the anisotropy of the metal. The scattering factor taking account of mode conversion has been calculated for cubic and hexagonal crystals.[2] Since the formulas are complicated, the reader is referred to a recent review article.[3]

The formula (3f-22) is valid in the Rayleigh scattering region when the wavelength is three times or larger than the grain diameter. For higher frequencies the attenuation increases proportional to the square of the frequency and finally becomes independent of the frequency for high frequencies.

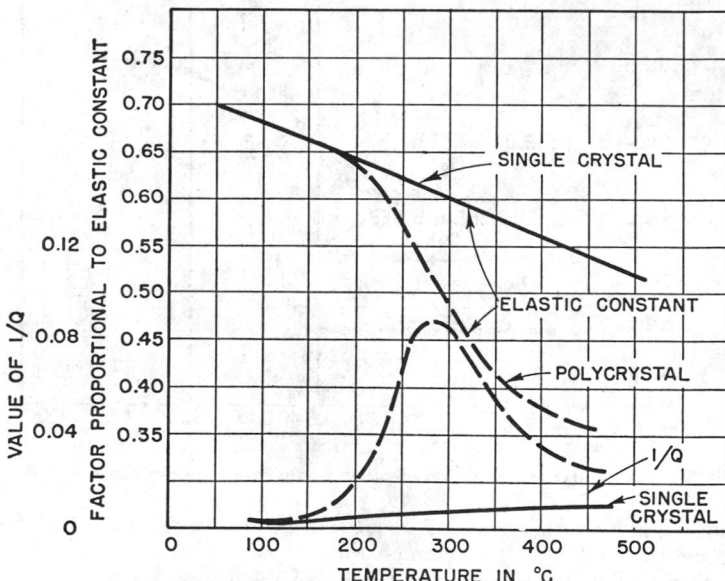

FIG. 3f-2. Elastic constants and Q for single-crystal and polycrystal aluminum. *(After Kê.)*

3f-7. Acoustic Losses in Ferromagnetic and Ferroelectric Materials. Stresses in ferromagnetic and ferroelectric materials can cause motion of domain walls or rota tion of domain directions. These occur in such a manner that domains are strengthened in directions parallel, antiparallel, or perpendicular to the direction of the stress. The

[1] Mason, *op. cit.*, p. 422.
[2] L. G. Merkulov, *Soviet Phys.—Tech. Phys.* (English Translation) **1**, 59–69 (1950).
[3] See Emmanuel P. Papadakis, "Physical Acoustics," vol. IVB, chap. 15, Academic Press, Inc., New York, 1968.

increased polarization in the direction of the stress produces increased strains which are the same sign in both parallel and antiparallel domains since magnetostriction and electrostriction are square-law effects and hence the elastic stiffnesses of demagnetized materials are less than those of completely magnetized materials. For polarizations directed along cube axes, the difference in elastic constants for the saturated and depolarized states, i.e., the ΔE effect, is[1]

$$\frac{\Delta E}{E_D} = \frac{9\mu\lambda_s{}^2 E_s}{20\pi P_s{}^2} \tag{3f-23}$$

where μ is the initial permeability or dielectric constant, λ_s the saturated change in length along a polycrystalline rod, E_s and E_D the saturated and demagnetized elastic-stiffness constant, and P_s the saturated magnetic or electric polarization. When the polarization lies along a cube diagonal—as in nickel—λ_s is replaced by $\frac{2}{3}\lambda_{111}[5c_{44}/(c_{11} - c_{12} + 3c_{44})]$ where λ_{111} is the saturated increase in length along the [111] direction and $5c_{44}/(c_{11} - c_{12} + 3c_{44})$ is a ratio of elastic constants.

The motion of walls or the rotation of domains in metallic ferromagnetic materials generates eddy currents and hence causes an acoustic loss. It has been shown that the permeability follows a relaxation equation

$$\mu = \mu_0 \frac{1 - jf/f_0}{1 + f^2/f_0{}^2} \tag{3f-24}$$

where $f_0 \doteq 4R/25\mu_0 L_c{}^2$, R is the resistivity, L_c is the domain diameter, and $j^2 = -1$. For a distribution of domain sizes

$$\mu = \mu_0 \sum_{i=1}^{m} \frac{V_i}{V} \frac{1 - jf/f_i}{1 + f^2/f_i{}^2} \tag{3f-25}$$

where V_i is the volume occupied by domains of size L_i and V is the total volume. Inserting in Eq. (3f-23) the $\Delta E/E_D$ and Q are given by

$$\frac{\Delta E}{E_D} = \frac{9\lambda_s{}^2 E_s}{20\pi P_s{}^2} \left(\sum_{i=1}^{m} \frac{V_i/V}{1 + f^2/f_i{}^2} \right); \qquad \frac{1}{Q} = \frac{9\lambda_s{}^2 E_s}{20\pi P_s{}^2} \left[\sum \frac{(V_i/V)(f/f_i)}{1 + (f/f_i)^2} \right] \tag{3f-26}$$

Figure 3f-3 shows measurements of the ΔE effect and the decrement $\delta = \pi/Q$ plotted over a frequency range, for a polycrystalline nickel rod.

Another effect causing losses in ferromagnetic and ferroelectric materials is the microhysteresis effect. In this effect the domain walls or domain rotations lag behind the applied stress and produce a hysteresis loop. Hence the initial susceptibility has a hysteresis component which is a function of the amount of stress. Average values of the parameters can be written in the form

$$\mu = \mu_0[1 - jf(A)] \tag{3f-27}$$

where $f(A)$ is a function of the amplitude. Inserting this value of μ in Eq. (3f-23), the value of the microhysteresis loss is given. This type of loss is present in ferroelectric materials and is the principal cause of the low mechanical Q.

3f-8. Other Types of Losses. In addition to these recognized types of losses, other types exist which appear to be associated with the motion of dislocations. Figure 3f-4 shows the Q of a number of materials measured[2] in a frequency range from 10^3 to

[1] R. M. Bozorth, "Ferromagnetism," p. 691, D. Van Nostrand Company, Inc., Princeton, N.J., 1951.
[2] R. L. Wegel and H. Walter, *Physics* **6**, 141 (1953).

10^5 Hz for small strains. Except for nickel and iron rods, whose decrease in Q with increase in frequency is accounted for by microeddy-current effects, the materials have a Q nearly independent of frequency. When a single or polycrystal sample is strained, an internal friction peak develops, as shown by Fig. 3f-5, whose peak temperature depends on the frequency. This peak, known as the Bordoni peak after its discoverer,[1] is believed to be due to the motion of dislocation segments from one minimum energy position in the crystal to adjacent ones under the combined action of thermal and mechanical applied stresses. This action takes place over the Peierl barrier which determines the forces returning the dislocations to their minimum energy positions. In fact, the Bordoni peak measurements provide the most reliable estimates of the Peierl barrier values. At high frequencies and for pure materials, the internal friction is determined by the damping of dislocation loops[2] by loss of energy to phonons and electrons.

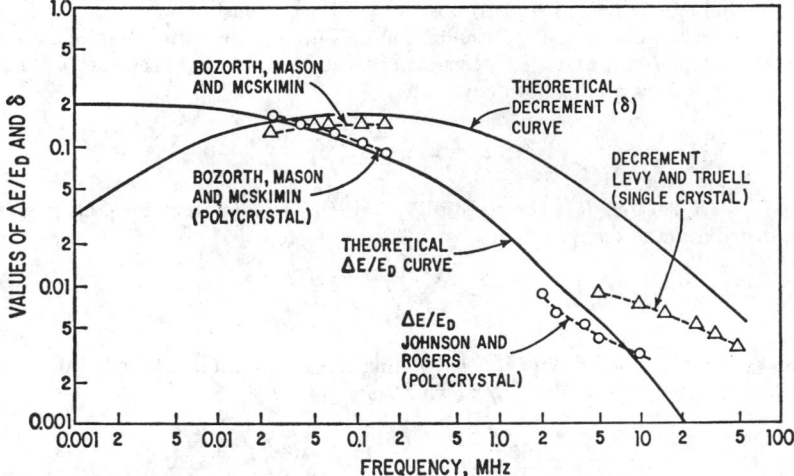

Fig. 3f-3. Decrement and ΔE effect as a function of frequency for polycrystal rod. (*After Bozorth, Mason, and McSkimin; Johnson and Rogers; and Levy and Truell.*)

Other internal friction effects and modulus changes occur for higher strain levels and elevated temperatures as shown by Fig. 3f-6. These are usually ascribed[3] to the pulling of dislocations away from pinning impurity atoms by stresses and thermal agitation. For still higher stresses, Frank-Read sources can be actuated, and these result in a region of very rapidly rising internal friction accompanied by fatigue of the material which occurs for a sufficiently large number of cycles.

Attenuation at Low Temperatures. At very low temperatures, the ultrasonic attenuation of pure normally conducting metals becomes high. Figure 3f-7 shows measurements of pure tin for two directions in the crystal and for two frequencies. Above 10 K, the ultrasonic attenuation is relatively small and increases as the square of the frequency. At 4 K, at which tin is still in the normal state, the attenuation is high and increases in proportion to the frequency. It has been shown that the

[1] P. G. Bordoni, *J. Acoust. Soc. Am.* **26**(4), 495 (July, 1954).
[2] See A. V. Granato and K. Lücke, "Physical Acoustics," vol. IVA, chap. 6, Academic Press, Inc., New York, 1968.
[3] See Warren P. Mason, "Physical Acoustics and the Properties of Solids," chap. 9, D. Van Nostrand Company, Inc., Princeton, N.J., 1958.

added attenuation in the normal state is due to the transfer of momentum and energy from the acoustic wave to the free electrons in the metal. If the acoustic wavelength is greater than the electronic mean free path, this transfer determines an effective viscosity, and the attenuation increases in proportion to the square of the frequency. When the mean free path becomes longer than the acoustic wavelength, as it does at low temperatures, the energy communicated to the electron is not returned to the

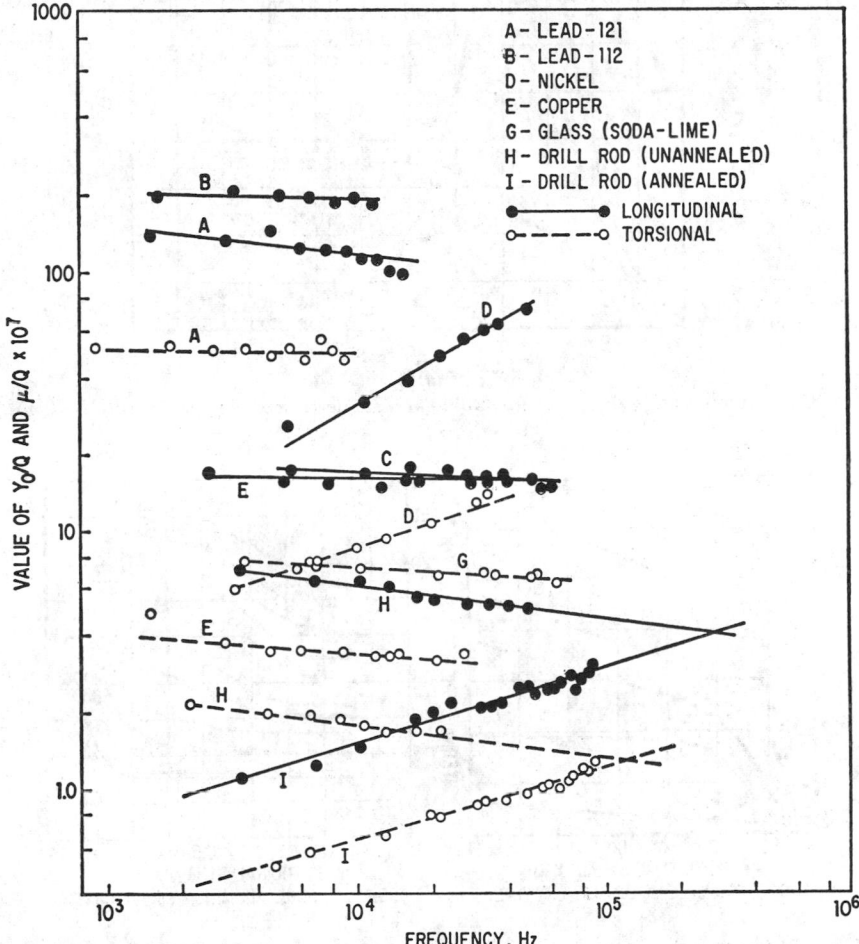

FIG. 3f-4. Values of Y_0/Q and μ/Q as function of frequency for a number of polycrystalline materials. (*After Wegel and Walther.*)

acoustic wave and a high attenuation results. The attenuation is proportional to the number of times the crystal lattice vibrates and hence to the frequency.

As the temperature drops below the temperature at which tin becomes superconductive (3.71 K), this source of attenuation drops rapidly to zero. The form of the curve has been used to confirm the Bardeen-Cooper-Schrieffer energy-gap theory of superconductivity. However, at lower frequencies—i.e., from 10 to 100 MHz— losses due to dislocations can occur. These are different for the normal and super-

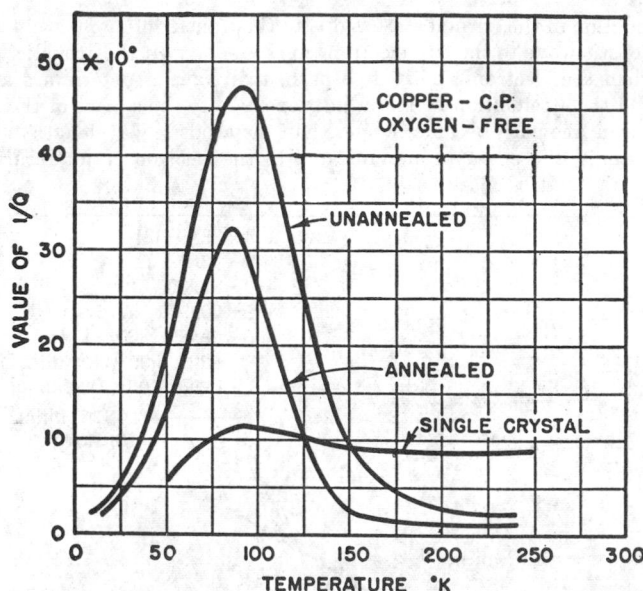

FIG. 3f-5. Attenuation peak in polycrystalline and single-crystal copper.   (*After Bordoni.*

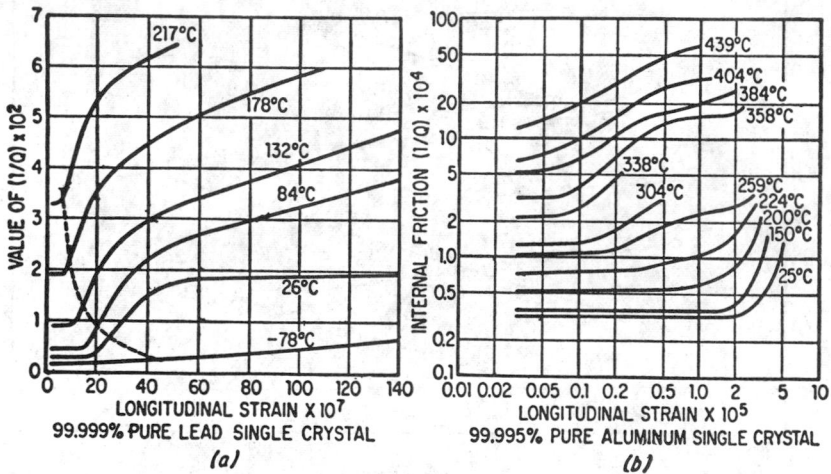

99.999% PURE LEAD SINGLE CRYSTAL   99.995% PURE ALUMINUM SINGLE CRYSTAL

(a)   (b)

FIG. 3f-6. Decrement as a function of strain and temperature for single crystals of lead and aluminum.

conducting states, and this difference has to be taken account of in order to determine the form of the energy-gap relation.   For frequencies above 100 MHz, the attenuation due to dislocations is small compared to the electron-phonon loss, and direct measurements give the shape of the energy-gap curves.

Acoustic measurements are also useful for type II or high-field superconductors (HFS).[1]   For these types of superconductors—which are of use for superconducting

[1] See V. Shapira, "Physical Acoustics," vol. V, chap. 1, Academic Press, Inc., New York, 1968.

magnets—there are two critical fields—as shown by Fig. 3f-8—rather than the single field of type I superconductors. Figure 3f-8 compares the magnetization curves of type I and type II superconductors when the magnetic field H is directed along the axis of the cylinder. In type I the magnet flux is completely excluded from the interior of the material below H_c. For type II superconductors the magnetic flux

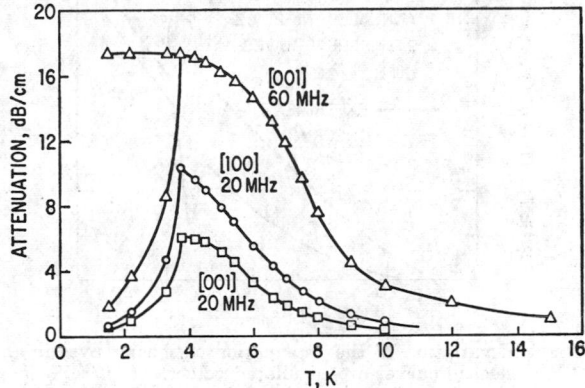

Fig. 3f-7. Longitudinal sound-wave attenuation measurements for a single crystal of tin along the [001] axis and along the [100] axis.

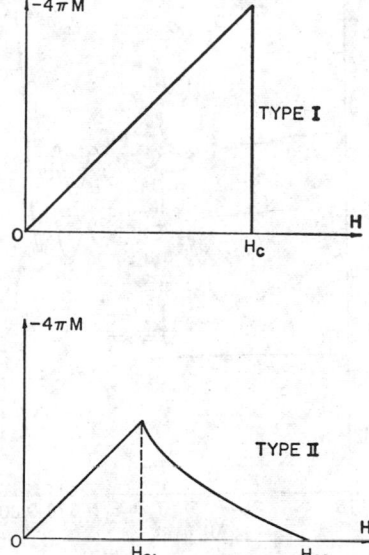

Fig. 3f-8. Magnetization curves of long cylinders of type I and type II superconductors. The applied field H is directed along the axis of the cylinder.

is completely excluded from the interior only below H_{c1}. Between H_{c1} and H_{c2} the magnetic flux consists of flux vortices in the form of filaments directed along H, embedded in a superconducting material. When a d-c electric current flows in a direction normal to H, each vortex experiences a force normal to its length which causes it to move. The vortices are pinned by defects, and it requires a finite current density before the vortices move. An alternating current or alternating stress causes motions of the pinned vortices which lag behind the applied forces. The result is an acoustic attenuation. Figure 3f-9 shows the change in attenuation of a 9.1 MHz shear wave plotted as a function of the magnetic field. A sharp dip occurs near the

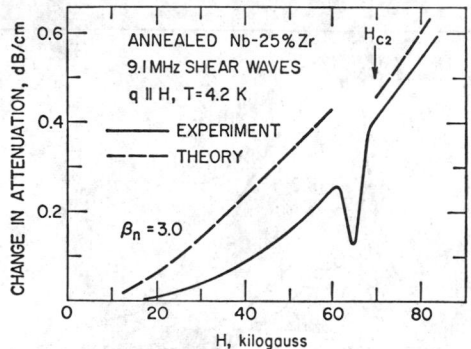

FIG. 3f-9. Magnetic-field variation of the attenuation of shear waves in annealed Nb–25 atom. % Zr at 4.2 K. Dashed curves are calculated values. (*After Y. Shapira.*)

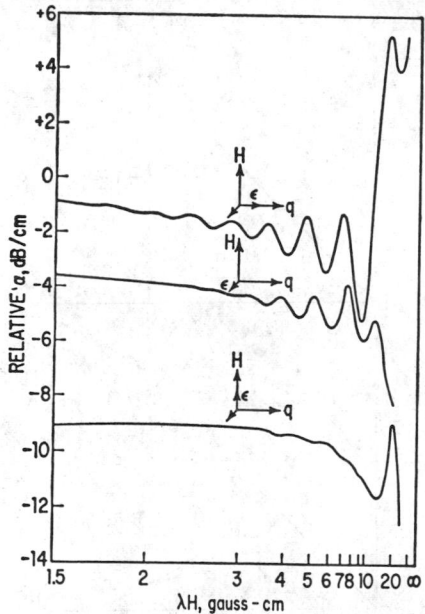

FIG. 3f-10. Relative attenuation in pure single-crystal copper as a function of the product of the wavelength times the magnetic field for several orientations of magnetic field and wave direction. (*After Morse.*)

superconducting field H_{c2}. Above H_{c2} the material is in the normal state, and the attenuation rises rapidly with the field.

Magnetoacoustics and Fermi Surface Determinations. In the presence of a magnetic field, the attenuation in metals in the normal state shows variations which are cyclic when plotted as a function of λH, where H is the magnetic field. Figure 3f-10 shows measurements in a very pure copper single crystal at 4.2 K.   These cyclic variations can be related to the shape of the Fermi surface, which is a constant-energy surface that bounds the occupied states of electrons in momentum space.   The electrical effects in a metal are primarily determined by the electrons whose energy is near the Fermi surface, since these are the only ones free to move.   For free electrons, such surfaces are spherical with a radius determined by the Fermi energy.

The effect of the periodic crystal potential in the band-theory approximation is to distort the Fermi surface from a spherical surface.   Electrons of the same energy (which all lie on the Fermi surface) will then have different momenta.   Figure 3f-11 shows the probable Fermi surfaces for monovalent copper, gold, and silver, and their relation to the Brillouin zone.   If an electron's orbit in momentum space carries it

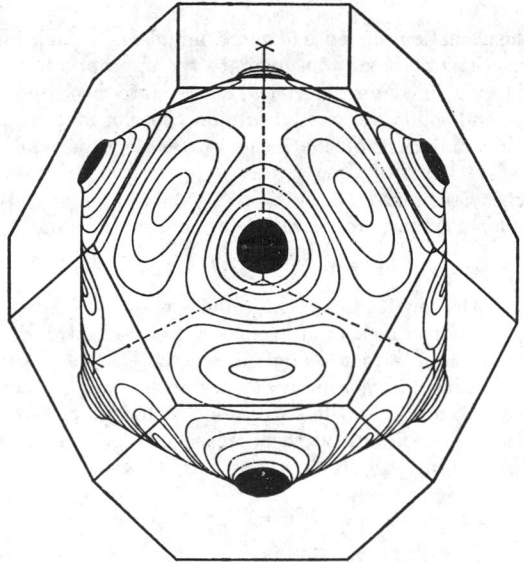

FIG. 3f-11. Fermi surfaces for copper, gold, and silver, and their relation to the Brillouin zone.   (*After Pippard.*)

to the Brillouin zone face, the electron will be refracted to the opposite Brillouin zone face.   In momentum space, this has the effect of repeating the zone over and over in an extended zone scheme.   The effect of a magnetic field is to localize the electrons that can move onto a plane perpendicular to the magnetic field in momentum space.   It can be shown that the periodicity of the attenuation-λH curves can be related to the linear dimension of the Fermi surface perpendicular to the magnetic field and perpendicular in momentum space to the direction of wave propagation in real space.   The various measurements of Fig. 3f-10 give details of the Fermi surface for different directions in momentum space.

Several other types of oscillations in the attenuation occur.[1]   These are the de Haas–van Alphen oscillations of the attenuation, the giant quantum oscillations, acoustic cyclotron resonance, and open orbit resonances.

[1] See B. W. Roberts, "Physical Acoustics," vol. IVB, chap. 10, Academic Press, Inc. New York, 1968.

3g. Properties of Transducer Materials

W. P. MASON

Columbia University

To determine the acoustic properties of gases, liquids, and solids and to utilize them in acoustic systems, it is necessary to generate the appropriate waves by means of transducer materials which convert electrical energy into mechanical energy and vice versa. For liquids and solids, the most common types of materials are piezoelectric crystals, ferroelectric materials of the barium titanate type, and magnetostrictive materials.

3g-1. Piezoelectric Crystals. The static relations for a piezoelectric quartz crystal producing a single longitudinal mode are for rationalized mks units

$$S_2 = s_{22}{}^E T_2 + d_{21} E_x \qquad D_x = d_{21} T_2 + \epsilon_1{}^T E_x \tag{3g-1}$$

where S_2 and T_2 are the longitudinal strain and stress, respectively, $s_{22}{}^E$ the elastic compliance along the length measured at constant electric field, d_{21} the piezoelectric constant relating the strain with the applied field E_x, D_x the electric displacement, and $\epsilon_1{}^T$ the dielectric constant measured at constant stress. Equations of this type suffice to determine the static and low-frequency behavior of piezoelectric crystals. Using the first equation, one finds that the increase in length for no external stress and the external force for no increase in length are, respectively,

$$\Delta l = d_{21} \frac{Vl}{t} \qquad F = T_2 tw = -d_{21} \frac{Vw}{s_{22}{}^E} \tag{3g-2}$$

where V is the applied potential, l, w, and t are the length, width, and thickness of the crystal, and F is the force which is considered positive for an extensional stress. From the second equation one finds that the open-circuit voltage and the short-circuited charge for a given applied force are, respectively,

$$V = -\left(\frac{d_{21}}{\epsilon^T}\right)\frac{lF}{tw} \qquad Q = \int_0^l \int_0^w D_x \, dl \, dw = d_{21}\frac{Fl}{t} \tag{3g-3}$$

Another important criterion for transducer use is the electromechanical-coupling factor k whose square is defined as the ratio of the energy stored in mechanical form to the total input electrical energy. Using Eqs. (3g-1), this can be shown to be

$$k^2 = \frac{d_{21}{}^2}{s_{22}{}^E \epsilon^T} \tag{3g-4}$$

It is readily shown that the clamped dielectric constant ϵ^S, obtained by setting $S_2 = 0$, and the constant-displacement elastic compliance s^D, obtained by setting $D_x = 0$, are related to the constant-stress dielectric constant ϵ^T and the constant-field elastic compliance $s_{22}{}^E$ by the equations

$$\frac{\epsilon_1{}^S}{\epsilon_1{}^T} = \frac{s_{22}{}^D}{s_{22}{}^E} = 1 - k^2 \tag{3g-5}$$

Equivalent circuits in which the properties of the crystal are expressed in terms of equivalent electrical elements are often useful (see Sec. 3l). An equivalent circuit for a piezoelectric crystal for static conditions is shown by Fig. 3g-1A. In this network the compliance $C_1 = s_{22}{}^E l/wt$ represents the compliance of the crystal with the electrodes short-circuited, the capacitance C_0 is the capacitance of the clamped crystal, i.e., $C_0 = lw\epsilon_1{}^S/t$, while the transformer shown is a perfect transformer, i.e., a transformer having no loss between zero frequency and the highest frequency for which the piezoelectric effect is operative, having a turns ratio of φ to 1 where

$$\varphi = -d_{21}\frac{w}{s_{22}{}^E} \qquad (3g\text{-}6)$$

The fact that this equivalent circuit presents the same information as Eq. (3g-1) is readily verified by substitution and integration over the area of the crystal.

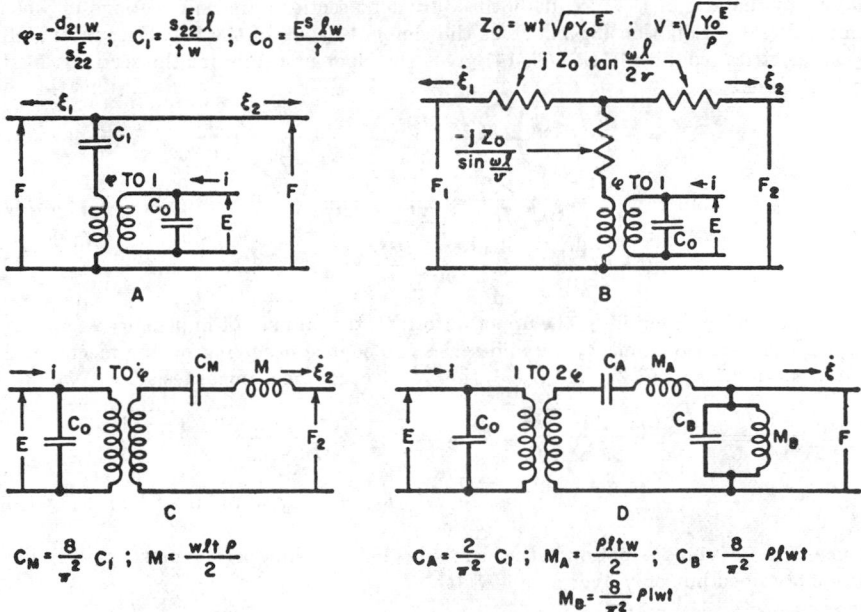

FIG. 3g-1. Equivalent circuit for a piezoelectric crystal for clamped and free conditions.

As an example of the use of such a network, one can calculate from it the efficiency of transformation of mechanical to electrical energy, or vice versa, under various conditions. Suppose that we clamp one end of the crystal and apply a force through the sending-end mechanical resistance R_M and receive the power generated into an electrical resistance R_E. Solving the network equations and obtaining the conditions for maximum power output, it is readily shown that the maximum power is obtained if

$$R_M = \frac{1}{\omega C_1 \sqrt{1 - k^2}} \qquad R_E = \frac{\sqrt{1 - k^2}}{\omega C_0} \qquad (3g\text{-}7)$$

where $\omega = 2\pi$ times the frequency f. With these values the power in the termination is

$$P_0 = \frac{F^2 k^4}{4\varphi^2 R_E} \qquad (3g\text{-}8)$$

The available power that can be obtained from a source having an open-circuit force F with an internal impedance R_M is maximum when $\varphi^2 R_E = R_M$. This power is then

$$i_2{}^2 R_E = \frac{F^2}{4\varphi^2 R_E} \tag{3g-9}$$

and hence the power-conversion efficiency is

$$P_E = k^4 \tag{3g-10}$$

Hence, unless the coupling is high, the efficiency of conversion by static means is low.

This efficiency can be improved by resonating the capacitance C_0 by an electric coil L_0 at the frequency of operation and can be further improved by mechanically resonating the static compliance of the crystal. The simplest way to analyze these circuits for their optimum conditions is to observe that, if the perfect transformer is moved to the end of the circuit, both equivalent sections are half sections of well-known filters. Equation (3g-11) gives the element values of the first filter resonated by an electrical coil, while Eq. (3g-14) gives the element values for the section tuned on both ends.

$$C_1' = \frac{s_{22}{}^E l}{wt}\left(\frac{d_{21}w}{s_{22}{}^E}\right)^2 = \frac{lw}{t}\frac{d_{21}{}^2}{s_{22}{}^E} = \frac{f_1 + f_2}{2\pi f_1 f_2 Z_0'} = \frac{1}{2\pi f_1 Z_0}$$

$$C_0 = \frac{\epsilon^S lw}{t} = \frac{f_1}{2\pi f_2(f_2 - f_1)Z_0'} = \frac{f_1}{2\pi(f_2{}^2 - f_1{}^2)Z_0} \tag{3g-11}$$

$$L_0 = \frac{(f_2 - f_1)Z_0'}{2\pi f_1 f_2} = \frac{(f_2{}^2 - f_1{}^2)Z_0}{2\pi f_1 f_2{}^2}$$

where f_1 is the lower cutoff, f_2 the upper cutoff, Z_0 the mid-shunt impedance occurring on the electrical side, and Z_0 the mid-series impedance occurring on the mechanical side. Solving for f_1, f_2, Z_0, and $Z_0'(\varphi^2)$, i.e., the actual mechanical resistance, we find

$$f_2 = \frac{1}{2\pi\sqrt{L_0 C_0}} \qquad f_1 = \frac{\sqrt{1-k^2}}{2\pi\sqrt{L_0 C_0}} \qquad Z_0 = R_E = \frac{1-k^2}{2\pi f_1 C_0}\cdot\frac{1+\sqrt{1-k^2}}{k^2}$$

$$R_M = \varphi^2 Z_0' = 2\pi f_1(ls_{22}{}^E/tw) \tag{3g-12}$$

Hence, if there is no dissipation in the elements of the crystal, perfect power conversion can be obtained but only over a bandwidth of

$$\frac{f_2 - f_1}{f_2} = 1 - \sqrt{1-k^2} \tag{3g-13}$$

We consider next a wider bandpass filter having the element values

$$C_1' = \frac{lw}{t}\frac{d_{21}{}^2}{s_{22}{}^E} = \frac{f_2 - f_1}{2\pi f_1 f_2 Z_0} \qquad L_1' = \frac{\rho lt}{w}\left(\frac{s_{22}{}^E}{d_{21}}\right)^2 = \frac{Z_0}{2\pi(f_2 - f_1)}$$

$$C_0 = \frac{\epsilon^S lw}{t} = \frac{1}{2\pi(f_2 - f_1)Z_0} \qquad L_0 = \frac{(f_2 - f_1)Z_0}{2\pi f_1 f_2} \tag{3g-14}$$

Solving for the bandwidth and the impedances

$$\frac{f_2 - f_1}{f_m} = \frac{k}{\sqrt{1-k^2}} \qquad f_m = \sqrt{f_1 f_2} = \frac{1}{2\pi\sqrt{L_0 C_0}} = \frac{1}{2\pi\sqrt{L_1 C_1}}$$

$$Z_0 = R_E = \frac{\sqrt{1-k^2}}{2\pi f_m C_0 k} \qquad R_M = \varphi^2 Z_0 = \frac{k}{\sqrt{1-k^2}}\frac{1}{2\pi f_m s_{22}{}^E}\frac{wt}{l} \tag{3g-15}$$

This filter section can efficiently transform mechanical into electrical energy and vice versa with a loss determined only by the dissipation in the elements of the crystal.

The simplest method for mechanically resonating the crystal is to use it near its natural mechanical resonance. An exact equivalent circuit for a vibrating crystal is shown by Fig. 3g-1B. Near the first resonant frequency, the equivalent circuit for a clamped quarter-wave crystal is shown by Fig. 3g-1C while the equivalent circuit for a half-wave crystal is shown by Fig. 3g-1D. When the half-wave crystal resonated by a shunt coil is applied to converting electrical into mechanical energy, the same formulas given in Eqs. (3g-14) and (3g-15) are applicable except that $k^2/(1 - k^2)$ is replaced by $(8/\pi^2)[k^2/(1 - k^2)]$. By using the complete representation of Fig. 3g-1B the effect can be calculated by using various backing plates on the radiation from the front surface.

The general form of Eq. (3g-1) holds for any single mode whether it is longitudinal or transverse as long as the appropriate constants are used. For longitudinal thickness modes when the radiating surface is a number of wavelengths in diameter, s_{22}^E is replaced by $1/c_{11}^E$ and d_{21} by e_{21}/c_{11}^E, the appropriate thickness piezoelectric constant. For a thickness shear mode, the appropriate shear stiffness (c_{44}, c_{55}, or c_{66})

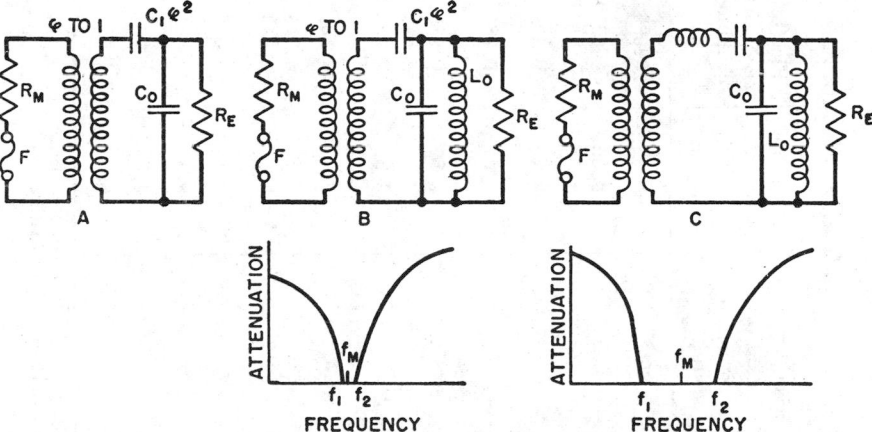

FIG. 3g-2. Use of equivalent circuit in determining the optimum conditions for energy transmission.

replaces $1/s_{22}$ and the appropriate shear piezoelectric constant replaces d_{21}. Table 3g-1 lists the constants in mks units for a number of standard crystal cuts.

3g-2. Electrostrictive and Magnetostrictive Materials. Other types of materials that have been used in transducers are ferroelectric crystals and ceramics of the barium titanate type and ferromagnetic crystals, polycrystals, and sintered materials of the ferrite type. All these materials have changes in lengths proportional to squares and even powers of the polarization and to obtain a linear response they have to be polarized. These polarized materials have relations between stresses, strains, electric and magnetic fields, and electric displacement and magnetic flux similar to those for a piezoelectric crystal shown by Eq. (3g-1) and hence these materials can be said to have "equivalent" constants which depend not only on the material but also on the degree of poling and in some cases on aging effects. The dielectric and permeability constants are those associated with the polarized medium as are also the elastic constants.

To obtain these equivalent piezoelectric and piezomagnetic constants, one can start with the more fundamental potential equations which have the same form for either electrostrictive or magnetostrictive materials. For polycrystalline or sintered materials, these potential equations can be written in the form

TABLE 3g-1. PROPERTIES OF PIEZOELECTRIC CRYSTALS IN MKS UNITS

Crystal and cut	Mode	Elastic constant, 10^{-11} m²/newton	Piezoelectric constant d, 10^{-12} coulomb/newton	Dielectric capacitivity ϵ, 10^{-11} farad/m	Electromechanical coupling k	Open-circuit voltage $g = d/\epsilon$, volt-meters/newton	Force factor d/s, newtons/volt-meter	Density, 10^3 kg/m³
Quartz X cut, length Y	L.L.	$s_{22}^E = 1.27$	$d_{21} = 2.25$	4.06	0.099	0.055	0.177	2.65
X cut	T.L.	$\dfrac{1}{c_{11}^E} = 1.16$	$\dfrac{e_{11}}{c_{11}^E} = -2.04$	4.06	0.093	0.050	0.175	2.65
Y cut	T.S.	$\dfrac{1}{c_{66}^E} = 2.57$	$\dfrac{e_{26}}{c_{66}^E} = +4.4$	4.06	0.137	0.108	0.171	2.65
Rochelle salt, 45-deg X cut	L.L.	$s_{22}^{\prime E} = 6.7$	$\dfrac{d_{14}}{2} = 435$	444.0	0.78	0.098	6.5	1.77
45-deg Y cut	L.L.	$s_{11}^{\prime E} = 9.89$	$\dfrac{d_{25}}{2} = -28.4$	9.85	0.288	0.29	0.287	1.77
ADP, 45-deg Z cut	L.L.	$s_{11}^{E\prime} = 5.3$	$\dfrac{d_{36}}{2} = 24.6$	13.8	0.29	0.178	0.465	1.804
KDP, 45-deg Z cut	L.L.	$s_{11}^{E\prime} = 4.85$	$\dfrac{d_{36}}{2} = 10.7$	19.6	0.12	0.058	0.22	2.31
EDT, Y cut, length X	L.L.	$s_{11}^{E} = 3.88$	$d_{21} = 11.3$	7.4	0.215	0.152	0.29	1.538
DKT, 45-deg Z cut	L.L.	$s_{11}^{E\prime} = 4.25$	$d_{31}' = -12.2$	5.8	0.245	0.21	0.287	1.988
L.H., Y cut	T.L.	$\dfrac{1}{c_{22}^E} = 2$	$\dfrac{e_{22}}{c_{22}^E} = 15$	9.15	0.35	0.165	0.75	2.06
L.H., hydrostatic	H.		$d_{21} + d_{22} + d_{23} = 13$	9.15		0.143		
Tourmaline, Z cut	T.L.	$\dfrac{1}{c_{33}^E} = 0.61$	$\dfrac{e_{33}}{c_{33}^E} = -1.84$	6.65	0.092	0.0275	0.3	3.1
Tourmaline, hydrostatic	H.		$d_{31} + d_{33} = -2.16$	6.65		0.0325		

Abbreviations: L.L. = length longitudinal; T.L. = thickness longitudinal; T.S. = thickness shear; ADP = ammonium dihydrogen phosphate; KDP = potassium dihydrogen phosphate; EDT = ethylene diamine tartrate; L.H. = lithium sulfate monohydrate.

$$G = -\tfrac{1}{2}[s_{11}{}^D(T_1{}^2 + T_2{}^2 + T_3{}^2) + 2s_{12}{}^D(T_1T_2 + T_1T_3 + T_2T_3)$$
$$+ 2(s_{11}{}^D - s_{12}{}^D)(T_4{}^2 + T_5{}^2 + T_6{}^2)] - \{Q_{11}(D_1{}^2T_1 + D_2{}^2T_2 + D_3{}^2T_3)$$
$$+ Q_{12}[T_1(D_2{}^2 + D_3{}^2) + T_2(D_1{}^2 + D_3{}^2) + T_3(D_1{}^2 + D_2{}^2)]$$
$$+ 2(Q_{11} - Q_{12})(T_4D_2D_3 + T_5D_1D_3 + T_6D_1D_2)\} + \tfrac{1}{2}\beta_{11}{}^T(D_1{}^2 + D_2{}^2 + D_3{}^2)$$
$$+ K_{11}{}^T(D_1{}^4 + D_2{}^4 + D_3{}^4) + K_{12}{}^T(D_1{}^2D_2{}^2 + D_1{}^2D_3{}^2 + D_2{}^2D_3{}^2)$$
$$+ K_{111}{}^T(D_1{}^6 + D_2{}^6 + D_3{}^6) + K_{112}{}^T[D_1{}^4(D_2{}^2 + D_3{}^2) + D_2{}^4(D_1{}^2 + D_3{}^2)$$
$$+ D_3{}^4(D_1{}^2 + D_2{}^2)] + K_{123}{}^TD_1{}^2D_2{}^2D_3{}^2 \qquad \text{(3g-16)}\dagger$$

where T_1, T_2, T_3 are the three extensional stresses, T_4, T_5, T_6 the three shearing stresses, D_1, D_2, D_3 the three components of the electrical displacement for ferro-electric materials or the three components of the magnetic flux B for ferromagnetic materials, the s constants are the compliance constants for an isotropic material measured at constant electric or magnetic displacement, the Q's are the electrostrictive or magnetostrictive constants, $\beta_{11}{}^T$ the inverse of the initial dielectric constant or permeability measured at constant stress, and the K^T's are constants determining the total energy stored for higher polarizations. The static equations can be obtained by differentiation of G according to the relations

$$S_i = -\frac{\partial G}{\partial T_i} \qquad E_m = \frac{\partial G}{\partial D_m} \qquad \text{(3g-17)}$$

Since linear equations are obtained only if a permanent polarization P_0 is introduced, we assume that

$$D_3 = P_0 + D_3{}^* \qquad \text{(3g-18)}$$

where $D_3{}^*$ is a small variable component superposed on P_0. Also, D_1 and D_2 are small so that their squares and higher powers can be neglected compared with P_0. Introducing these into (3g-16) and differentiating, we have

$$S_1 = s_{11}{}^DT_1 + s_{12}{}^D(T_2 + T_1) + Q_{12}(P_0{}^2 + 2P_0D_3{}^*)$$
$$S_2 = s_{11}{}^DT_2 + s_{12}{}^D(T_1 + T_3) + Q_{12}(P_0{}^2 + 2P_0D_3{}^*)$$
$$S_3 = s_{11}{}^DT_3 + s_{12}{}^D(T_1 + T_2) + Q_{11}(P_0{}^2 + 2P_0D_3{}^*)$$
$$S_4 = 2(s_{11}{}^D - s_{12}{}^D)T_4 + 2(Q_{11} - Q_{12})P_0D_2$$
$$S_5 = 2(s_{11}{}^D - s_{12}{}^D)T_5 + 2(Q_{11} - Q_{12})P_0D_1 \qquad \text{(3g-19)}$$
$$S_6 = 2(s_{11}{}^D - s_{12}{}^D)T_6$$
$$E_1 = -2(Q_{11} - Q_{12})P_0T_5 + D_1(\beta_{11}{}^T + 2K_{12}{}^TP_0{}^2 + 2K_{112}{}^TP_0{}^4)$$
$$E_2 = -2(Q_{11} - Q_{12})P_0T_4 + D_2(\beta_{11}{}^T + 2K_{12}{}^TP_0{}^2 + 2K_{112}{}^TP_0{}^4)$$
$$E_3 = -2Q_{11}P_0T_3 - 2Q_{12}P_0(T_1 + T_2) + D_3{}^*(\beta_{11}{}^T + 12K_{11}{}^TP_0{}^2 + 30K_{111}{}^TP_0{}^4)$$

It is obvious that the variable components of Eq. (3g-19) follow the same rule as for a piezoelectric crystal. There are three longitudinal modes and a shearing mode. The length longitudinal mode has the following constants:

L.L. mode $\quad s_{11}{}^E = s_{11}{}^D\left[1 + \dfrac{4Q_{12}{}^2P_0{}^2}{\beta_{33}{}^T(P_0)s_{11}{}^D}\right] \qquad d_{31} = \dfrac{2Q_{12}P_0}{\beta_{33}{}^T(P_0)}$

$$\epsilon_{33}{}^T(P_0) = \frac{1}{\beta_{33}(P_0)} \qquad \text{(3g-20)}$$

where $\beta_{33}{}^T(P_0) = (\beta_{11}{}^T + 12K_{11}{}^TP_0{}^2 + 30K_{111}{}^TP_0{}^4)$ is the dielectric impermeability of the ceramic when it has a permanent polarization P_0

L.T. bar $\quad s_{11}{}^E = s_{11}{}^D\left[1 + \dfrac{4Q_{11}{}^2P_0{}^2}{\beta_{33}{}^T(P_0)s_{11}{}^D}\right] \qquad d_{33} = \dfrac{2Q_{11}P_0}{\beta_{33}{}^T(P_0)}$

$$\epsilon_{33}{}^T(P_0) = \frac{1}{\beta_{33}{}^T(P_0)} \qquad \text{(3g-21)}$$

† If higher-order terms than those considered here are used, second-order electrostrictive and magnetostrictive terms and the change in elastic constants with polarization can be taken care of. For example, see W. P. Mason, *Phys. Rev.* **82** (5), 715–723 (June 1, 1951).

These formulas hold for a bar which is long in the direction of vibration compared with the cross-sectional dimensions. When a plate is used which is a number of wavelengths across, the sidewise motions S_1 and S_2 are zero and the constants are

L.T. plate $\qquad\qquad \dfrac{1}{c_{11}{}^E} \qquad d_{33}' \qquad \epsilon_{33}'(P_0)$ $\qquad\qquad$ (3g-22)

where

$$\frac{1}{c_{11}{}^E} = \frac{1}{c_{11}{}^P} + d_{33}'^2 \epsilon_{33}'(P_0) \qquad d_{33}' = 2P_0 \left(Q_{11} - \frac{2s_{12}{}^D}{s_{11}{}^D + s_{12}{}^D} Q_{12} \right) \epsilon_{33}'(P_0)$$

$$\epsilon_{33}^{T'}(P_0) = \frac{1}{\beta_{33}{}^T(P_0) + [4Q_{12}{}^2 P_0{}^2/(s_{11}{}^D + s_{12}{}^D)]} \quad \text{and} \quad c_{11}{}^D$$

$$= \frac{s_{11}{}^D + s_{12}{}^D}{(s_{11}{}^D - s_{12}{}^D)(s_{11}{}^D + 2s_{12}{}^D)}$$

The thickness shear mode has the fundamental constants $2(s_{11}{}^E - s_{12}{}^E)$; d_{14}; $\epsilon_{11}{}^T(P_0)$,

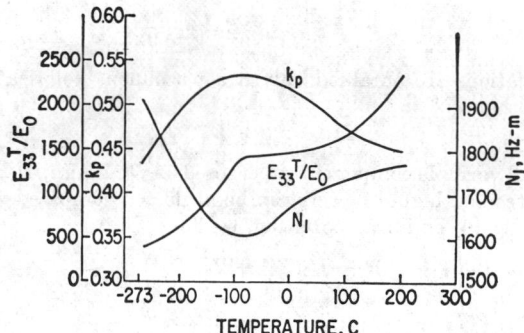

Fig. 3g-3. Temperature variation of k_p, permittivity, and bar frequency constant for prestabilized PZT-4.

i.e., the dielectric constant perpendicular to the poling direction, where

$$d_{14} = \frac{2(Q_{11} - Q_{12})P_0}{\epsilon_{11}{}^T(P_0)} \qquad 2(s_{11}{}^E - s_{12}{}^E) = 2(s_{11}{}^D - s_{12}{}^D) + \frac{4(Q_{11} - Q_{12})^2 P_0{}^2}{\beta_{11}{}^T(P_0)}$$

$$\epsilon_{11}{}^T(P_0) = \frac{1}{(\beta_{11}{}^T + 2K_{12}{}^T P_0{}^2 + 2K_{112}{}^T P_0{}^4)} \qquad (3g\text{-}23)$$

Two other modes have been used in electrostrictive and magnetostrictive materials,

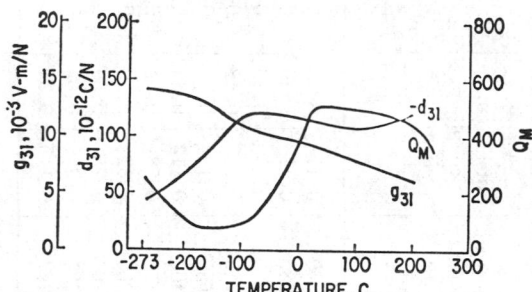

FIG. 3g-4. Temperature variation of g_{31}, d_{31}, and mechanical Q for prestabilized PZT-4.

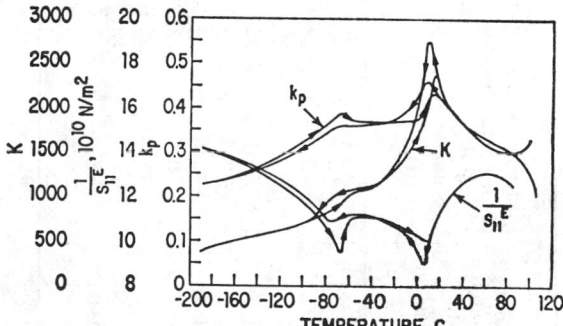

FIG. 3g-5. Temperature variation of k_p, permittivity K, and Young's modulus ($1/s_{11}{}^E$) for BaTiO$_3$ ceramic.

TABLE 3g-2. PROPERTIES OF FERROELECTRIC CERAMICS AT 25°C

Material	Young's modulus, 10^{11} newtons/m²	Piezoelectric constant d_{31}, 10^{-11} coulomb/newton	Piezoelectric constant d_{33}, 10^{-11} coulomb/newton	Dielectric capacitivity $\epsilon_{33}{}^T$, 10^{-11} farad/m	Electromechanical coupling factors		Open-circuit voltage $g = d_{33}/\epsilon_{33}{}^T$, volt-meters/newton	Force factor $d_{33}Y_0{}^E$, newtons/volt-meter
					k_{31}	k_{33}		
Commercial BaTiO₃ ceramics	1.18	−5.6	16	1,250	0.17	0.45	0.0106	13.5
97% BaTiO₃, 3% CaTiO₃	1.22	−5.3	13.5	1,230	0.17	0.43	0.0111	11.0
96% BaTiO₃, 4% PbTiO₃	1.14	−3.8	10.5	880	0.14	0.39	0.012	9.2
90% BaTiO₃, 4% PbTiO₃, 6% CaTiO₃	1.24	−4.0	11.5	710	0.167	0.48	0.016	9.3
84% BaTiO₃, 8% PbTiO₃, 8% CaTiO₃	1.31	−2.7	8.0	530	0.124	0.4	0.015	6.1
80% BaTiO₃, 12% PbTiO₃, 8% CaTiO₃	1.28	−2.0	6.0	400	0.113	0.34	0.015	4.7
PZT #4*	0.815	−9.7	23.5	875	0.28	0.63	0.0268	19.2
PZT #5*	0.675	−14.0	32.0	1,200	0.32	0.70	0.0266	21.6
PZT #6*	0.865	−7.8	19.1	860	0.25	0.60	0.022	16.5
NbO₃ (K 50%; Na 50%)†	1.02	−3.2	8.0	235	0.226	0.52	0.034	8.15
Pb(NbO₃)₂‡	0.29	−3.3	9.0	240	0.115	0.31	0.037	2.6

* Data from Clevite Brush Company.
† L. Egerton, *J. Am. Ceram. Soc.* **42**, 438 (1959).
‡ G. Goodman, *J. Am. Ceram. Soc.* **36**, 368 (1953).

the radial mode and the torsional mode. The first is driven by polarizing the disk perpendicular to the major surface and involves the same fundamental constants as the length longitudinal mode of Eq. (3g-20). It has been shown[1] that the effective coupling and the resonant frequency of such disks are given by the equations

$$k^2 = \frac{2}{1 - \sigma} \frac{4Q_{11}{}^2 P_0{}^2 \epsilon^T(P_0)}{s_{11}{}^E} \qquad f_R = \frac{2.03}{2\pi a} \sqrt{\frac{1}{s_{11}{}^E \rho(1 - \sigma^2)}} \qquad (3g\text{-}24)$$

where σ is Poisson's ratio, which is approximately 0.3 for barium titanate ceramics. The torsional mode is generated in electrostrictive and magnetostrictive materials when the alternating displacement is at right angles to the polarization. This is easily accomplished for a magnetostrictive material by polarizing a cylinder radially by one set of windings and driving the cylinder by a set of windings coaxial with the cylinder. In an electrostrictive material, a torsional vibration can be obtained by inducing a permanent polarization in different directions on two sides of the cylinder and driving the cylinder by a set of two electrodes with the two gaps between them coming in the region of greatest permanent polarization. The fundamental elastic constant is the shear constant ($s_{44}{}^E = s_{55}{}^E$) while the fundamental piezoelectric constant is the shear piezoelectric constant d_{15} or the similar magnetostrictive constants.

Table 3g-2 gives some typical constants for a number of barium titanate compositions with lead and calcium titanate additions. A number of new ceramics, particularly lead zirconate titanate (trade name PZT), sodium potassium niobate, and lead metaniobate, have recently appeared. These have higher Curie temperatures than barium titanate combinations but lower values of electrical and mechanical Q's. The stored electrical polarization in lead zirconate titanate (nearly 30 microcoulombs/cm^2) is higher than in any other ceramic and such materials are especially useful for producing a high current when depolarized by a mechanical shock (E.E.T. transducers). Figures 3g-3, 4, and 5 show how the fundamental constants vary with temperature over a wide temperature range for the most used ceramic PZT-4, and for the original BaTiO$_3$ ceramic. Table 3g-3 gives some typical constants for a number of magnetostrictive materials.

3g-3. Equivalent Circuits for Magnetostrictive Transducers. The energy equation (3g-16) is the same for magnetostrictive and electrostrictive materials, provided the electric field and displacement are replaced by the magnetic field H and the magnetic flux density B. Hence the equivalent circuit of Fig. 3g-1 also applies to a magnetostrictive material, provided we replace E and i by $\int_0^l H_i\,dl = U$, the magnetomotive force, and $\dot{B}S = \dot{\Phi}$, where S is the cross-sectional area, Φ the total flux through the magnetostrictive transducer, and $\dot{\Phi}$ the time rate of change of this flux. Hence all the fundamental quantities and coupling factors can be expressed in terms of the analogous quantities as shown by Table 3g-3. These hold for materials having a closed magnetic circuit such as a ring or a rod with closing magnetic circuit having a reluctance small compared with that for the rod. If this is not true, demagnetizing factors and additional reluctance values have to be taken account of and the value of Φ is the average value determined by all these factors.

In a transducer, however, it is not U and $\dot{\Phi}$ that we deal with, but rather the input voltage and current. These quantities are related by equations of the type

$$E = N \frac{d\Phi}{dt} \qquad U = Ni \qquad (3g\text{-}25)$$

where N is the number of turns and the voltage, current, flux, and magnetomotive forces are directed as shown by Fig. 3g-6. These are the equations of a gyrator, shown

[1] W. P. Mason, "Piezoelectric Crystals and Their Application to Ultrasonics," chap. XII, D. Van Nostrand Company, Inc., Princeton, N.J., 1950.

TABLE 3g-3. MAGNETOSTRICTIVE PROPERTIES OF METALS AND FERRITES

Data from C. M. Van der Burgt, *Phillips Research Repts.* **8**, 91–132, 1953

Material	$d_{33} \times 10^9$ webers/newton	$d_{14} \times 10^9$ webers/newton	Rev. per. long. $\mu^T(P_0) \times 10^4$ henrys/m	$\frac{1}{s^H} \times 10^{-11}$ newtons/m²	$Y_0{}^H = \times 10^{-11}$ newtons/m²	k_{33}	Rev. per shear $\mu^T(P_0) \times 10^4$ henrys/m	Shear stiffness $G^E \times 10^{-11}$ newton/m²	Torsional coupling k_T	Energy stored $\frac{1}{2}(d_{33}{}^2/\mu^T) \times 10^{12}$ joules-m/newton²	Density kg/m³ $\times 10^{-3}$
99.9 nickel	−5.3	......	2.84	2.0	2.1	0.14	......	......	.....	0.05	8.9
50 Co; 0.5 Cr; 49.5 Fe	12.3	......	8.3	2.2	...	0.20	......	......	.....	0.09	8.2
35 Co; 0.5 Cr; 64.5 Fe	13.4	......	19.2	2.1	...	0.14	......	......	.....	0.047	8.1
NiO (15%); ZnO (35%); Fe_2O_3 (50%)	−11.1	−28.5	190	1.8	1.6	0.034	139	0.68	0.063	0.003	5.06
NiO (18%); ZnO (32%); Fe_2O_3 (50%)	−16.0	−39.5	77.5	1.62	...	0.073	74	0.62	0.115	0.0165	4.9
NiO (25%); ZnO (25%); Fe_2O_3 (50%)	−9.8	−20.3	22.0	1.53	...	0.082	20	0.59	0.110	0.022	4.85
NiO (32%); ZnO (18%); Fe_2O_3 (50%)	−8.7	−15.8	13.4	1.5	2.3	0.093	13.2	0.58	0.105	0.0282	4.85
NiO (40%); ZnO (10%); Fe_2O_3 (50%)	−5.9	−13.0	5.5	1.37	...	0.112	5.35	0.54	0.13	0.0315	4.76
NiO (50%); Fe_2O_3 (50%)	−4.4	......	2.8	0.93	...	0.08	2.4	0.36	0.09	0.0344	4.20

Data from R. M. Bozorth, E. A. Nesbit, and H. J. Williams

Material	Flux density B, webers/m²	Long. rev. per $\mu^T(P_0) \times 10^4$ henrys/m	Young's modulus $Y_0{}^A \times 10^{-11}$ newtons/m²	$d_{33} \times 10^9$ webers/newton	Longitudinal coupling k_{33}	Energy stored $\frac{1}{2}(d_{33}{}^2/\mu^T) \times 10^{12}$ joules-m/newton²	Density kg/m³ $\times 10^{-3}$
99.9 % nickel	0.4	0.98	2.1	−5.0	0.232	0.127	8.9
	0.5	0.515	...	−3.26	0.208	0.103	
	0.55	0.317	...	−2.18	0.177	0.075	
45 % Ni, 55 % Fe, i.e., 45 % Permalloy	0.722	8.94	1.6	11.5	0.154	0.074	8.17
	0.965	7.36	...	12.2	0.179	0.101	
	1.2	4.45	...	9.4	0.178	0.099	
	1.4	1.97	...	5.3	0.15	0.071	
2V Permindur, 2 % V, 50 % Co, 48 % Fe	1.5	3.54	2.3	9.35	0.238	0.123	8.3
	1.6	2.61	...	7.5	0.222	0.108	
	1.8	2.23	...	6.3	0.202	0.089	
	2.0	1.14	...	4.0	0.18	0.07	

by the symbol of Fig. 3g-6, which does not satisfy the reciprocity relationship. If we call Z_M the magnetic impedance defined by

$$Z_M = \frac{U}{d\Phi/dt} \tag{3g-26}$$

it is evident that the electrical impedance at the terminals of the transducer is equal to

$$Z_E = \frac{E}{i} = \frac{N^2}{Z_M} \tag{3g-27}$$

Hence the effect of the gyrator coupling is to invert all the elements of the equivalent

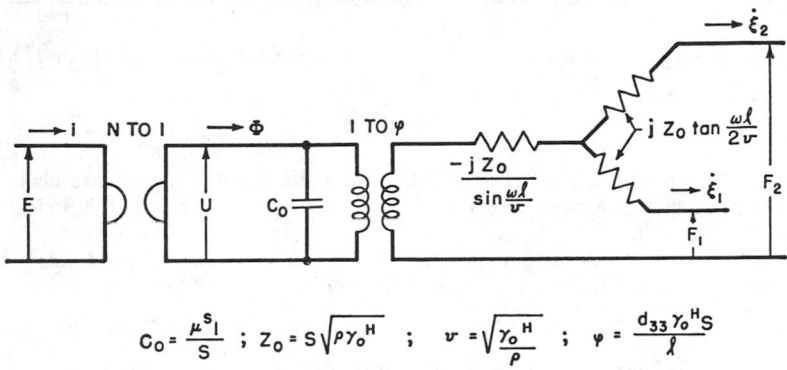

$$C_0 = \frac{\mu^S l}{S} \quad ; \quad z_0 = S\sqrt{\rho\gamma_0{}^H} \quad ; \quad v = \sqrt{\frac{\gamma_0{}^H}{\rho}} \quad ; \quad \varphi = \frac{d_{33}\gamma_0{}^H S}{l}$$

FIG. 3g-6. Equivalent circuit of a magnetostrictive rod.

circuit. Hence one should determine the element values of Fig. 3g-6 for the appropriate terminating conditions and then invert the values in accordance with Eq. (3g-27) to determine the elements of a magnetostrictive transducer. The values given in Fig. 3g-6 are for a longitudinally vibrating rod where S is the cross-sectional area and l the length. μ^S is the average value of the permeability in the equations for the reluctance R

$$R = \frac{l}{\mu^S S} \tag{3g-28}$$

where μ^S is for the constant stress condition.

3h. Frequencies of Simple Vibrators. Musical Scales

ROBERT W. YOUNG

U.S. Naval Undersea Research and Development Center

3h-1. Strings. The fundamental frequency of vibration of an ideal string is

$$f_0 = \frac{1}{2l} \sqrt{\frac{F}{m}} \tag{3h-1}$$

where f_0 is the frequency, l is the free length, F is the force (tension) stretching the string, and m is the mass per unit length. Values of m for steel and gut strings are given in Table 3h-1.

In addition to the vibration in a single loop which gives rise to the fundamental frequency, the ideal string may vibrate in harmonics whose frequencies are

$$f_n = n f_0 \tag{3h-2}$$

where n is the integer denoting the particular mode of vibration. The length of each vibration loop is l/n. These successive lengths and the corresponding periods of vibration (i.e., reciprocals of the frequencies) constitute a harmonic series according to the strict mathematical definition; nowadays, however, the frequencies themselves are usually said to make up a harmonic series.

The frequencies of actual strings depart somewhat from the frequencies computed from the simple formula because actual strings are stiff, they may be partially clamped at the ends, they are not infinitely thin, the tension increases with amplitude of vibration, the mass per unit length is not exactly uniform, there is internal damping and damping due to the surrounding air and supports, and the supports are not infinitely rigid. In the formulas which follow damping has been neglected.

For an actual string set

$$f = n f_0 (1 + G) \tag{3h-3}$$

where the factor $(1 + G)$ is a measure of the departure (i.e., the inharmonicity) from the ideal harmonic values. Table 3h-2 lists values of G for various small perturbations. The approximations are valid only when G is small.

For musical purposes it is often convenient to give the inharmonicity in cents (hundredths of an equally tempered semitone) by setting

$$1 + G = 2^{\delta/1,200} = e^{\delta/1,731} \tag{3h-4}$$

where δ is the inharmonicity. To a usually acceptable approximation, $\delta = 1,731 G$.

If the stiff string listed in Table 3h-2 is of steel music wire, $Y/\rho = 25.5 \times 10^6$ m²/sec², Y being Young's modulus and ρ the density. The tension is very nearly $F = l^2 \rho f_0^2 \pi d^2$. Thus for steel wire, and by virtue of the stiffness formula, the inharmonicity in cents is $\delta = 3.4 \times 10^{13} d^2 n^2 / f_0^2 l^4$, provided that the diameter and length are in centimeters.

3–130

TABLE 3h-1. MASS PER UNIT LENGTH OF STEEL AND GUT STRINGS*

Diam		Steel, g/m	Gut, g/m	Diam		Steel, g/m	Gut, g/m	Diam		Steel, g/m	Gut, g/m
mm	in.			mm	in.			mm	in.		
0.20	0.0079	0.25	0.04	1.00	0.0394	6.15	1.10	1.80	0.0709	19.9	3.56
0.22	0.0087	0.30	0.05	1.02	0.0402	6.40	1.14	1.82	0.0717	20.4	3.64
0.24	0.0094	0.35	0.06	1.04	0.0409	6.65	1.19	1.84	0.0724	20.8	3.72
0.26	0.0102	0.42	0.07	1.06	0.0417	6.91	1.24	1.86	0.0732	21.3	3.80
0.28	0.0110	0.48	0.09	1.08	0.0425	7.17	1.28	1.88	0.0740	21.7	3.88
0.30	0.0118	0.55	0.10	1.10	0.0433	7.44	1.33	1.90	0.0748	22.2	3.97
0.32	0.0126	0.63	0.11	1.12	0.0441	7.71	1.38	1.92	0.0756	22.7	4.05
0.34	0.0134	0.71	0.13	1.14	0.0449	7.99	1.43	1.94	0.0764	23.1	4.14
0.36	0.0142	0.80	0.14	1.16	0.0457	8.27	1.48	1.96	0.0772	23.6	4.22
0.38	0.0150	0.89	0.16	1.18	0.0465	8.56	1.53	1.98	0.0780	24.1	4.31
0.40	0.0157	0.98	0.18	1.20	0.0472	8.86	1.58	2.00	0.0787	24.6	4.40
0.42	0.0165	1.08	0.19	1.22	0.0480	9.15	1.64	2.02	0.0795	25.1	4.49
0.44	0.0173	1.19	0.21	1.24	0.0488	9.46	1.69	2.04	0.0803	25.6	4.58
0.46	0.0181	1.30	0.23	1.26	0.0496	9.76	1.75	2.06	0.0811	26.1	4.67
0.48	0.0189	1.42	0.25	1.28	0.0504	10.1	1.80	2.08	0.0819	26.6	4.76
0.50	0.0197	1.54	0.27	1.30	0.0512	10.4	1.86	2.10	0.0827	27.1	4.85
0.52	0.0205	1.66	0.30	1.32	0.0520	10.7	1.92	2.12	0.0835	27.6	4.94
0.54	0.0213	1.79	0.32	1.34	0.0528	11.1	1.97	2.14	0.0843	28.2	5.04
0.56	0.0220	1.93	0.34	1.36	0.0535	11.4	2.03	2.16	0.0850	28.7	5.13
0.58	0.0228	2.07	0.37	1.38	0.0543	11.7	2.09	2.18	0.0858	29.2	5.23
0.60	0.0236	2.21	0.40	1.40	0.0551	12.1	2.16	2.20	0.0866	29.8	5.32
0.62	0.0244	2.36	0.42	1.42	0.0559	12.4	2.22	2.22	0.0874	30.3	5.42
0.64	0.0252	2.52	0.45	1.44	0.0567	12.8	2.28	2.24	0.0882	30.9	5.52
0.66	0.0260	2.68	0.48	1.46	0.0575	13.1	2.34	2.26	0.0890	31.4	5.62
0.68	0.0268	2.84	0.51	1.48	0.0583	13.5	2.41	2.28	0.0898	32.0	5.72
0.70	0.0276	3.01	0.54	1.50	0.0591	13.8	2.47	2.30	0.0906	32.5	5.82
0.72	0.0283	3.19	0.57	1.52	0.0598	14.2	2.54	2.32	0.0913	33.1	5.92
0.74	0.0291	3.37	0.60	1.54	0.0606	14.6	2.61	2.34	0.0921	33.7	6.02
0.76	0.0299	3.55	0.64	1.56	0.0614	15.0	2.68	2.36	0.0929	34.3	6.12
0.78	0.0307	3.74	0.67	1.58	0.0622	15.4	2.74	2.38	0.0937	34.8	6.23
0.80	0.0315	3.94	0.70	1.60	0.0630	15.7	2.81	2.40	0.0945	35.4	6.33
0.82	0.0323	4.14	0.74	1.62	0.0638	16.1	2.89	2.42	0.0953	36.0	6.44
0.84	0.0331	4.34	0.78	1.64	0.0646	16.5	2.96	2.44	0.0961	36.6	6.55
0.86	0.0339	4.55	0.81	1.66	0.0654	16.9	3.03	2.46	0.0968	37.2	6.65
0.88	0.0346	4.76	0.85	1.68	0.0661	17.4	3.10	2.48	0.0976	37.8	6.76
0.90	0.0354	4.98	0.89	1.70	0.0669	17.8	3.18	2.50	0.0984	38.4	6.87
0.92	0.0362	5.20	0.93	1.72	0.0677	18.2	3.25	2.52	0.0992	39.1	6.98
0.94	0.0370	5.43	0.97	1.74	0.0685	18.6	3.33	2.54	0.1000	39.7	7.09
0.96	0.0378	5.67	1.01	1.76	0.0693	19.0	3.41	2.56	0.1008	40.3	7.21
0.98	0.0386	5.91	1.06	1.78	0.0701	19.5	3.48	2.58	0.1016	40.9	7.32

* This table is based on a density of steel of 7.83 g/cm³. Density of gut is assumed to be 1.4 g/cm³, about one-sixth that of steel. This is only approximate, since the density of gut varies from sample to sample, and increases markedly with humidity. Brass wire has a density of 8.7 g/cm³, about 1.1 times that of steel.

3h-2. Air Columns and Rods. The air within a simple tube of constant cross section, open at both ends or closed at both ends, vibrates freely at a frequency near

$$f = \frac{nc}{2l} \tag{3h-5}$$

where n is an integer (mode of vibration number), c is the speed of sound in the contained air, and l is the length of the tube. (See Sec. 3d for speed of sound in air and its dependence on temperature.) The diameter of the tube must be relatively small;

TABLE 3h-2. PERTURBATION IN FREQUENCY OF A STRING

Cause	G	Explanation
Stiffness	$\dfrac{n^2\pi^3 d^4 Y}{128 l^2 F}$	Y is Young's modulus, d is the diameter of the string
Yielding support	$\dfrac{4ml}{4\pi^2 n^2 M - K/f_0^2}$	The support consists of a mass M on a spring of transverse force constant K. Multiply by 2 if there are two such supports
Variable density	$-\dfrac{1}{l}\displaystyle\int_0^l g(x)\,\sin^2\frac{\pi nx}{l}\,dx$	The mass per unit length is $m = m_0[1 + g(x)]$ where m_0 is the mean value over the string and x is the distance from one end of the string; the function $g(x)$ must be small in comparison with unity

plane sound waves propagated longitudinally are assumed. The same formula applies to thin rods vibrating longitudinally and suitably supported (say, at distances $l/2n$ from the ends) so that the vibration is not inhibited. (See Sec. 3f for speed of sound in solids.)

An open organ pipe is an example of a doubly open tube of constant cross section. To calculate its frequency adequately it must be recognized, however, that the air beyond the physical ends of the tube partakes of the vibration and adds inertia to the vibrating system. (This does *not* mean, however, that there is a velocity antinode beyond the end of the tube.) The necessary corrections to the simple formula are usually introduced as empirical "end corrections" to be added to the geometrical length; thus

$$f = \frac{nc}{2(l + x_1 + x_2)} \tag{3h-6}$$

where $x_1 = 0.3d$ is the correction for the unimpeded end (d being the inside diameter of the pipe) and $x_2 = 1.4d$ is the correction for the mouth of the pipe. These are rough approximations; the literature on the end correction is extensive.[1]

The air inside a cylindrical tube that is closed at one end and open at the other vibrates at frequency

$$f = \frac{nc}{4(l + x)} \tag{3h-7}$$

where $x = 0.3d$ if the open end is unimpeded. In the case of the "closed" organ pipe (meaning closed at one end only), for the mouth $x = 1.4d$.

[1] E. G. Richardson ed , "The Technical Aspects of Sound," vol. I, pp. 493–496, 578, Elsevier Publishing Company, Amsterdam, 1953; Harold Levine, *J. Acoust. Soc. Am.* **26**, 200–211 (1954).

The speed of sound c (and thus the frequency of vibration) in a gas contained within a tube is reduced somewhat from its value c_0 in free space, as a consequence of friction and loss of heat to the wall of the tube. If the frequency of vibration f and the tube diameter d are such that $df^{\frac{1}{2}} > 2\nu^{\frac{1}{2}}$, ν being the kinematic viscosity of the gas, the speed of sound (longitudinal phase velocity) within the tube is[1]

$$c = \frac{c_0}{[1 + 2(\nu/\pi f)^{\frac{1}{2}}/d]^{\frac{1}{2}}[1 + 2(\gamma - 1)(\nu/\pi f P_r)^{\frac{1}{2}}/d]^{\frac{1}{2}}}$$

where γ is the ratio of specific heats, and P_r the Prandtl number for the gas. For air at 20°C, and when $df^{\frac{1}{2}} > 0.8$ with d in cm and f in hertz, with slight approximation the Helmholtz-Kirchhoff correction for the speed of sound is

$$c = c_0 \left(1 - \frac{0.33}{df^{\frac{1}{2}}}\right)$$

Correspondingly the interval by which the frequency of vibration is lowered owing to friction and heat conduction is $572/df^{\frac{1}{2}}$ cents. As $df^{\frac{1}{2}}$ becomes less than $2\nu^{\frac{1}{2}}$ a transition[1] occurs to an even more marked reduction in the speed of sound propagation in the tube.

The air in a conical tube is resonant in some cases at the same frequencies as a doubly open cylindrical tube of the same length, but there is the important difference that the contained sound waves are spherical rather than plane. Table 3h-3 gives equations[2] to be solved for each combination of end conditions; $k = 2\pi f/c$. "Closed-open," for example, means that the smaller end of the truncated cone is closed while the larger end is open; r_1 is the slant distance from the extrapolated apex of the cone to the smaller end and r_2 is the slant distance to the larger end. The slant length of the resonator is thus $r_2 - r_1$. When $r_1 = 0$, the length is r_2 and the cone is complete to the apex. Formulas for computing frequency when the cone is complete are shown at the right of Table 3h-3. As in the case of cylindrical tubes, the length should be

TABLE 3h-3. FREQUENCIES OF CONICAL RESONATORS

Ends	Equation	For $r_1 = 0$
Closed-closed	$kr_2 - \tan^{-1} kr_2 = kr_1 - \tan^{-1} kr_1$	$\tan kr_2 = kr_2$
Closed-open	$\tan k(r_2 - r_1) = -kr_1$	$f_1 = \dfrac{nc}{2r_2}$
Open-closed	$\tan k(r_2 - r_1) = kr_2$	$\tan kr_2 = kr_2$
Open-open	$f = \dfrac{nc}{2(r_2 - r_1)}$	$f = \dfrac{nc}{2r_2}$

slightly modified by end corrections. As the angle of the cone increases the correction decreases and may even become negative.[3]

3h-3. Volume Resonators. The Helmholtz resonator consists of a nearly closed cavity of volume V with an opening of acoustical conductance C. If the opening is

[1] A. H. Benade, *J. Acoust. Soc. Am.* **44**, 616–623 (1968). Multiplication by the correction term is erroneously shown there in eq. (13c), instead of division.

[2] Eric J. Irons, *Phil. Mag.* **9**, 346–360 (1930).

[3] A. E. Bate and E. T. Wilson, *Phil. Mag.* **26**, 752–757 (1938).

in a thin wall the conductance is simply d, the diameter of the hole.   If the opening is through a short neck of length l, approximately

$$C = \frac{\pi d^2}{4(l + 0.8d)} \tag{3h-8}$$

The natural frequency of the resonator is

$$f = \frac{c}{2\pi} \sqrt{\frac{C}{V}} \tag{3h-9}$$

the speed of sound in the opening being c.   The equation is valid for wavelengths large in comparison with the dimensions of the resonator.

The ocarina may be recognized as an instrument of the resonator type because the *position* of an open hole of given size is immaterial; when the holes are all equal, they can be opened in any order to give the same scale.   The total conductance for use in the formula given above is the sum of the conductance of individual holes, provided that they are separated far enough that there is no interaction.

TABLE 3h-4. FREQUENCIES OF TRANSVERSE VIBRATION OF BARS

Ends	Frequency	Ratio			Cents		
	Mode → 1	2	3	4	2	3	4
Clamped-free	$f_1 = \dfrac{0.5597\kappa}{l^2} \sqrt{\dfrac{Y}{\rho}}$	6.267	17.548	34.387	3,177	4,960	6,124
Free-free, or clamped-clamped	$f_1 = \dfrac{3.561\kappa}{l^2} \sqrt{\dfrac{Y}{\rho}}$	2.756	5.404	8.933	1,755	2,921	3,791

3h-4. Bars.   A long thin bar clamped and/or free at the end(s) can vibrate transversely at the fundamental frequencies listed in Table 3h-4 under mode 1.   The length of the bar is l, Y is Young's modulus, ρ is the density, and κ is the radius of gyration about the neutral axis of the cross section.   For a round bar $\kappa = d/4$, where d is the diameter.   For a flat bar of thickness t (in the plane of vibration) $\kappa = t/\sqrt{12}$; the width is immaterial.   The frequency of a bar clamped at both ends is the same as that of a bar free at both ends.   The frequency of a higher mode of vibration can be found by multiplying the fundamental frequency by the ratio indicated in Table 3h-4; the intervals in cents corresponding to these ratios are given at the extreme right of the table.   These are the classic[1] values for thin bars; the frequencies of actual bars are lowered slightly as a consequence of rotatory inertia, lateral inertia, and shear.[2]   For example, for a steel bar whose length is 40 times the thickness, the frequencies of the first four modes of vibration are expected to be 0.997, 0.992, 0.984, and 0.974 times the corresponding "thin" values (i.e., lowered 5, 14, 28, and 46 cents, respectively).

[1] Lord Rayleigh, "Theory of Sound," vol. I, p. 280, Macmillan & Co., Ltd., London, 1894. The interval erroneously given as 2.4359 octaves has been corrected here to 2.4340 octaves = 2,921 cents.
[2] William T. Thomson, *J. Acoust. Soc. Am.* **11,** 199–204 (1939).   There is an error: $m = \beta/[1 + \beta^2(k/L)^2]^{\frac{1}{4}}$, not $m = \beta/[1 + \beta^2(k/L)^2]^{\frac{1}{2}}$.

TABLE 3h-5. FREQUENCIES OF THE EQUALLY TEMPERED SCALE, BASED ON THE
INTERNATIONAL STANDARD A = 440 HERTZ

Note	S	f	$2\pi f$	Note	S	f	$2\pi f$	Note	S	f	$2\pi f$
C_0	0	16.352	102.74	C_3	36	130.81	821.92	C_6	72	1,046.5	6,575.4
	1	17.324	102.74		37	138.59	870.79		73	1,108.7	6,966.4
D_0	2	18.354	115.32	D_3	38	146.83	922.58	D_6	74	1,174.7	7,380.6
	3	19.445	122.18		39	155.56	977.43		75	1,244.5	7,819.5
E_0	4	20.602	129.44	E_3	40	164.81	1,035.6	E_6	76	1,318.5	8,284.4
F_0	5	21.827	137.14	F_3	41	174.61	1,097.1	F_6	77	1,396.9	8,777.1
	6	23.125	145.30		42	185.00	1,162.4		78	1,480.0	9,299.0
G_0	7	24.500	153.93	G_3	43	196.00	1,231.5	G_6	79	1,568.0	9,851.9
	8	25.957	163.09		44	207.65	1,304.7		80	1,661.2	10,438
A_0	9	27.500	172.59	A_3	45	220.00	1,382.3	A_6	81	1,760.0	11,058
	10	29.135	183.06		46	233.08	1,464.5		82	1,864.7	11,716
B_0	11	30.868	193.95	B_3	47	246.94	1,551.6	B_6	83	1,975.5	12,413
C_1	12	32.703	205.48	C_4	48	261.63	1,643.8	C_7	84	2,093.0	13,151
	13	34.648	217.70		49	277.18	1,741.6		85	2,217.5	13,933
D_1	14	36.708	230.64	D_4	50	293.66	1,845.2	D_7	86	2,349.3	14,761
	15	38.891	244.36		51	311.13	1,954.9		87	2,489.0	15,639
E_1	16	41.203	258.89	E_4	52	329.63	2,071.1	E_7	88	2,637.0	16,569
F_1	17	43.654	274.28	F_4	53	349.23	2,194.3	F_7	89	2,793.8	17,554
	18	46.249	290.59		54	369.99	2,324.7		90	2,960.0	18,598
G_1	19	48.999	307.87	G_4	55	392.00	2,463.0	G_7	91	3,136.0	19,704
	20	51.913	326.18		56	415.30	2,609.4		92	3,322.4	20.875
A_1	21	55.000	345.58	A_4	57	440.00	2,764.6	A_7	93	3,520.0	22,117
	22	58.270	366.12		58	466.16	2,929.0		94	3,729.3	23,432
B_1	23	61.735	387.90	B_4	59	493.88	3,103.2	B_7	95	3,951.1	24,825
C_2	24	65.406	410.96	C_5	60	523.25	3,287.7	C_8	96	4,186.0	26,301
	25	69.296	435.40		61	554.37	3,483.2		97	4,434.9	27,865
D_2	26	73.416	461.29	D_5	62	587.33	3,690.3	D_8	98	4,698.6	29,522
	27	77.782	488.72		63	622.25	3,909.7		99	4,978.0	31,278
E_2	28	82.407	517.78	E_5	64	659.26	4,142.2	E_8	100	5,274.0	33,138
F_2	29	87.307	548.57	F_5	65	698.46	4,388.5	F_8	101	5,587.7	35,108
	30	92.499	581.19		66	739.99	4,649.5		102	5,919.9	37,196
G_2	31	97.999	615.74	G_5	67	783.99	4,926.0	G_8	103	6,271.9	39,408
	32	103.83	652.36		68	830.61	5,218.9		104	6,644.9	41,751
A_2	33	110.00	691.15	A_5	69	880.00	5,529.2	A_8	105	7,040.0	44,234
	34	116.54	732.25		70	932.33	5,858.0		106	7,458.6	46,864
B_2	35	123.47	775.79	B_5	71	987.77	6,206.3	B_8	107	7,902.1	49,651

Numerous subscript notations have been employed to distinguish the notes of one octave from those of another. The particular scheme used here assigns to C_0 a frequency which corresponds roughly to the lowest audible pitch. S is the number of semitones counted from this C_0.

The simple tuning fork may be recognized as an example of dual clamped-free bars. The frequency of a tuning fork made of ordinary steel can be computed approximately from

$$f = \frac{80,000t}{l^2} \text{ Hz} \tag{3h-10}$$

provided that the thickness t and length l of the prongs are given in centimeters.

It is evident from Table 3h-4 that the different modes of vibration of a uniform bar are inharmonic. However, the cross section of the bar in the modern xylophone or marimba is often given an empirical lengthwise "undulation" such that the second

TABLE 3h-6. INTERVALS IN CENTS CORRESPONDING TO CERTAIN
FREQUENCY RATIOS

Name of interval	Frequency ratio	Cents
Unison..............................	1:1	0
Minor second or semitone............	1.059463:1	100
Semitone............................	16:15	111.731
Minor tone or lesser whole tone.......	10:9	182.404
Major second or whole tone..........	1.122462:1	200
Major tone or greater whole tone......	9:8	203.910
Minor third........................	1.189207:1	300
Minor third........................	6:5	315.641
Major third........................	5:4	386.314
Major third........................	1.259921:1	‹ 00
Perfect fourth......................	4:3	498.045
Perfect fourth......................	1.334840:1	500
Augmented fourth...................	45:32	590.224
Augmented fourth...................	1.414214:1	600
Diminished fifth....................	1.414214:1	600
Diminished fifth....................	64:45	609.777
Perfect fifth.......................	1.498307:1	700
Perfect fifth.......................	3:2	701.955
Minor sixth........................	1.587401:1	800
Minor sixth........................	8:5	813.687
Major sixth........................	5:3	884.359
Major sixth........................	1.681793:1	900
Harmonic minor seventh.............	7:4	968.826
Grave minor seventh................	16:9	996.091
Minor seventh......................	1.781797:1	1,000
Minor seventh......................	9:5	1,017.597
Major seventh......................	15:8	1,088.269
Major seventh......................	1.887749:1	1,100
Octave.............................	2:1	1,200.000

mode of vibration of the free-free bar is changed in frequency to 3 or 4 times the fundamental frequency.[1] The frequencies of the higher modes of vibration are also modified by variation in cross section for special purposes such as the simulation of the sound of a bell.[2]

3h-5. Membranes. The membrane often assumed for vibration calculations is flexible, thin, and of uniform mass per unit area σ. The membrane is stretched by a tension T, this being the force per unit length anywhere in the membrane. The

---

[1] See U.S. Pats. 1,838,502 (1931) and 1,632,751 (1927).
[2] See U.S. Pats. 2,273,333 (1942), 2,516,725 (1950), 2,536,800 (1951), and 2,606,474 (1952).

characteristic frequencies of transverse vibration for such a rectangular membrane clamped at its edges are given by

$$f = \frac{c}{2}\left[\left(\frac{m}{a}\right)^2 + \left(\frac{n}{b}\right)^2\right]^{\frac{1}{2}} \tag{3h-11}$$

where

$$c = \sqrt{\frac{T}{\sigma}} \tag{3h-12}$$

is the speed of propagation of transverse wave motion, a and b are the lengths of the sides, and m and n are integers. Note the similarity of Eq. (3h-11) to Eq. (3h-1).

TABLE 3h-7. RATIOS FOR INTERVALS TO 100 CENTS

Cents	Ratio	Cents	Ratio	Cents	Ratio	Cents	Ratio
0	1.000000	25	1.014545	50	1.029302	75	1.044274
1	1.000578	26	1.015132	51	1.029896	76	1.044877
2	1.001158	27	1.015718	52	1.030492	77	1.045481
3	1.001734	28	1.016305	53	1.031087	78	1.046085
4	1.002313	29	1.016892	54	1.031683	79	1.046689
5	1.002892	30	1.017480	55	1.032079	80	1.047294
6	1.003472	31	1.018068	56	1.032876	81	1.047899
7	1.004052	32	1.018656	57	1.033473	82	1.048505
8	1.004632	33	1.019244	58	1.034070	83	1.049111
9	1.005212	34	1.019833	59	1.034667	84	1.049717
10	1.005793	35	1.020423	60	1.035265	85	1.050323
11	1.006374	36	1.021012	61	1.035863	86	1.050930
12	1.006956	37	1.021602	62	1.036462	87	1.051537
13	1.007537	38	1.022192	63	1.037060	88	1.052145
14	1.008120	39	1.022783	64	1.037660	89	1.052753
15	1.008702	40	1.023374	65	1.038259	90	1.053361
16	1.009285	41	1.023965	66	1.038859	91	1.053970
17	1.009868	42	1.024557	67	1.039459	92	1.054579
18	1.010451	43	1.025149	68	1.040060	93	1.055188
19	1.011035	44	1.025741	69	1.040661	94	1.055798
20	1.011619	45	1.026334	70	1.041262	95	1.056408
21	1.012204	46	1.026927	71	1.041864	96	1.057018
22	1.012789	47	1.027520	72	1.042466	97	1.057629
23	1.013374	48	1.028114	73	1.043068	98	1.058240
24	1.013959	49	1.028708	74	1.043671	99	1.058851

The characteristic frequencies of a circular membrane clamped at its boundary are given by

$$f = \frac{c}{2a}\,\beta_{mn}$$

where a is the radius of the membrane. For $n = 1$, 2, and 3, $\beta_{0n} = 0.766$, 1.757, and 2.755, these numbers being the first three roots divided by π of the Bessel function of zero order set equal to zero. Similarly, $\beta_{1m} = 1.220$, 2.233, and 3.238 are from the Bessel function of first order and $\beta_{2n} = 1.635$, 2.679, and 3.699 are from the Bessel function of second order. The number of diametral nodes is m; the number of circular nodes is n, including the node at the boundary. The modes of vibration are not in general harmonics; the lowest characteristic frequencies are in the proportions 1.000:1.593:2.135. For a circular membrane constrained to certain radial (not diametral) nodes, harmonics are, however, possible.

The tambourine is a musical instrument that consists of a free membrane nearly of the kind discussed above. In most drums, however, the membrane closes a cavity; in the case of the kettledrum (and some kinds of capacitor microphones) this cavity is relatively rigid and airtight. If the speed of transverse waves in the membrane is significantly less than the speed of sound in the contained air, the cavity has little effect on those modes of vibration with diametral nodes. The frequencies of other modes of vibration are increased[1] by the stiffness of the contained air.

3h-6. Musical Scales. By international agreement the standard tuning frequency for musical performance is the A of 440 Hz. The frequencies of the equally tempered scale based on this frequency appear in Table 3h-5. Middle C thus has a frequency of 261.6 Hz. The C of 256 Hz, frequently used in the past for demonstrations in physics, has never been adopted for practical musical performance.

For many calculations with musical intervals it is convenient to deal with logarithmic units that can be added instead of the ratios which must be multiplied. The octave is equal to 1,200 logarithmic cents, and the equally tempered semitone is 100 cents. The interval in cents corresponding to any two frequencies f_1 and f_2 is $1,200 \log_2(f_2/f_1) = 3,986 \log_{10}(f_2/f_1)$. Table 3h-6 lists certain common intervals in cents and the corresponding ratios; the frequency ratios for intervals up to 100 cents are given in Table 3h-7.

---

[1] Philip M. Morse, "Vibration and Sound," 2d ed., p. 193, McGraw-Hill Book Company, New York, 1948.

3i. Radiation of Sound

FRANK MASSA

Massa Division, Dynamics Corporation of America

3i-1. Introduction. Radiation of sound may take place in a number of ways, but basically, all sound generators cause an alternating pressure to be set up in the fluid medium within which the sound energy is established. The sound energy that is set up in a medium depends not only on the physical characteristics of the medium and the oscillatory volume displacement of the fluid set up by the vibrating source but also upon the size and shape of the generator. The acoustic power generated by any vibrating source can be expressed by

$$P = U^2 R_A \times 10^{-7} \quad \text{watts} \tag{3i-1}$$

where U = rate of volume displacement of fluid, cc/sec

R_A = acoustic radiation resistance seen by source, acoustic ohms

If the rate of volume displacement is taken in peak cc/sec, Eq. (3i-1) will yield peak watts of power. If the volume displacement is taken in rms cc/sec, the power will be given in rms watts.

Of the many possible methods for generating sound, two types of generators will effectively serve to typify most of them. These basic generators are (1) pulsating sphere and (2) vibrating piston.

Each type of generator has its own acoustic impedance characteristic which depends on the dimensions of the source and on the frequency of vibration.

3i-2. Acoustic Impedance. *Pulsating Sphere.* The specific acoustic impedance of a pulsating sphere is given by

$$z = \frac{\rho c}{1 + [1/(\pi D/\lambda)]^2} + j \frac{\rho c/(\pi D/\lambda)}{1 + [1/(\pi D/\lambda)]^2} \quad \text{acoustic ohms/cm}^2 \tag{3i-2}$$

where ρ = density of the medium, g/cc

c = velocity of sound in the medium, cm/sec

D = diameter of the sphere, cm

$\lambda = c/f$

f = frequency, Hz

It can be seen from inspection that at high frequencies, where D/λ becomes very large, the specific acoustic impedance becomes a pure resistance equal to ρc and the reactance term vanishes. At low frequencies, where D/λ is small, the specific acoustic impedance becomes

$$z = \rho c \left(\frac{\pi D}{\lambda}\right)^2 + j \rho c \frac{\pi D}{\lambda} \quad \text{acoustic ohms/cm}^2 \tag{3i-3}$$

A plot of the specific acoustic resistance and reactance of a pulsating sphere as a function of D/λ is shown in Fig. 3i-1. To obtain the total acoustic radiation resistance

R_A of the sphere, it is necessary to divide the specific acoustic resistance by the total surface area of the sphere in cm². The value of R_A thus determined, when substituted in Eq. (3i-1), will give the actual acoustic watts being generated by the spherical source.

Vibrating Piston. The specific acoustic impedance of a circular piston set in an infinite rigid baffle and radiating sound from one of its surfaces is given by

$$z = \rho c \left[1 - \frac{J_1(2\pi D/\lambda)}{\pi D/\lambda} \right] + j\rho c \frac{K_1(2\pi D/\lambda)}{2(\pi D/\lambda)^2} \qquad \text{acoustic ohms/cm}^2 \qquad (3\text{i-}4)$$

where D is the diameter of the piston in centimeters, J_1 and K_1 are Bessel functions, and the remaining symbols are defined under Eq. (3i-2).

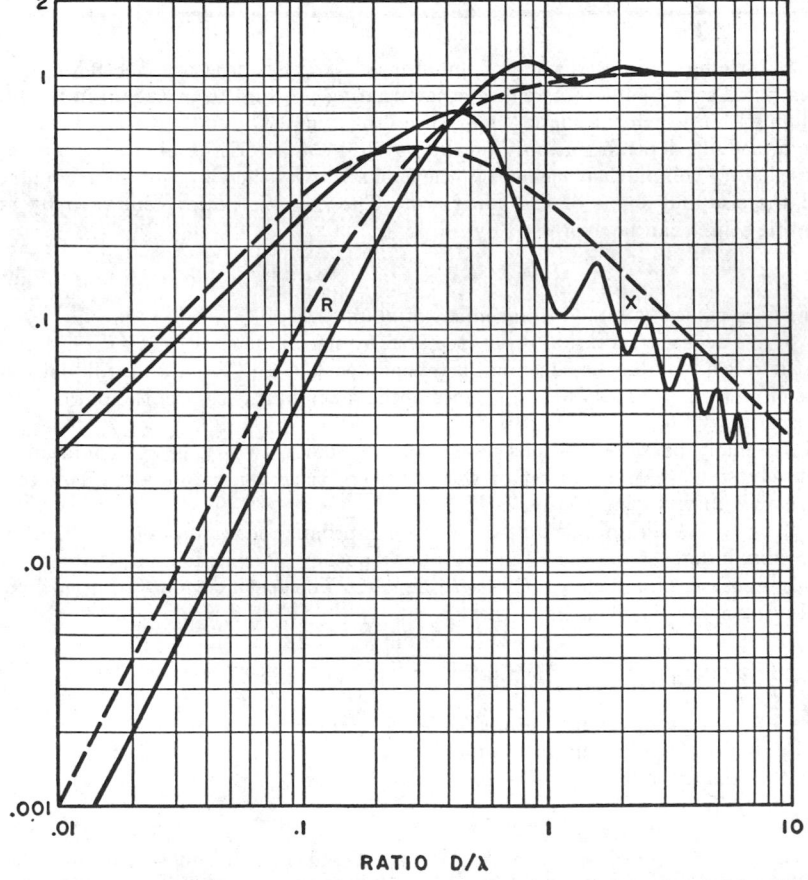

RATIO D/λ

Fɪɢ. 3i-1. Specific acoustic resistance R and reactance X of a pulsating sphere (dashed curves) and a vibrating piston set in an infinite baffle (solid curves). To obtain magnitude of R or X multiply ordinates by ρc of the medium.

At high frequencies, where D/λ is large, Eq. (3i-4) reduces to a pure resistance equal to ρc. At low frequencies, where D/λ is small, the specific acoustic impedance for a piston set in an infinite baffle with one side radiating becomes

$$z = \frac{\rho c (\pi D/\lambda)^2}{2} + j\rho c \frac{8D}{3\lambda} \qquad \text{acoustic ohms/cm}^2 \qquad (3\text{i-}5)$$

A plot of the specific acoustic resistance and reactance for a vibrating piston mounted in an infinite baffle is shown in Fig. 3i-1. To obtain the total acoustic radiation resistance of the piston, it is necessary to divide the specific resistance by the piston area in cm^2. The value of R_A so determined, when substituted in Eq. (3i-1), will give the actual acoustic watts being generated by a piston.

Summary of Radiation Impedance Characteristics. In Table 3i-1 are shown the magnitudes of the acoustic radiation resistance and reactance for a sphere and piston for both low-frequency (D/λ small) and high-frequency (D/λ large) operation.

TABLE 3i-1. TABULATED VALUES OF THE TOTAL ACOUSTIC RADIATION
RESISTANCE AND REACTANCE OF A SPHERE AND PISTON
IN ACOUSTIC OHMS

	$D/\lambda \ll 1$		$D/\lambda \gg 1$	
	R_A	X_A	R_A	X_A
Pulsating sphere	$\rho c \dfrac{\pi}{4\lambda^2}$	$\dfrac{\rho c}{\pi D\lambda}$	$\dfrac{\rho c}{A}$	0
Vibrating piston (in infinite baffle)	$\rho c \dfrac{\pi}{2\lambda^2}$	$\rho c \dfrac{8}{3\pi D\lambda}$	$\dfrac{\rho c}{A}$	0

ρ = density of the medium, g/cm^3
c = velocity of sound in the medium, cm/sec
λ = wavelength of sound in the medium, cm
$\lambda = c/f$

f = frequency of the sound vibration, Hz
D = diameter of sphere or piston, cm
A = surface area of sphere or piston, cm^2

3i-3. Directional Radiation of Sound. Whenever sound energy is generated from a source whose dimensions are small compared with the wavelength of the vibration in the medium, the intensity will be uniform in all angular directions and the generator is generally defined as a point source. When the dimensions of the vibrating surface are large compared with the wavelength, phase interferences will be experienced at different points in space due to the differences in time arrival of the vibrations originating from different portions of the surface, which results in a nonuniform directional radiation pattern. Practical use is made of this phenomenon when it is desired to produce special directional patterns by arranging the geometry and size of the vibrating surfaces of a sound generator to create the desired characteristic.

In many instances, a transmitter is designed so that the sound is radiated in a relatively sharp beam so that the energy is concentrated only within a specific desired angular region. When such a directional structure is employed as a receiver, the transducer will be more capable of picking up weak signals from a specified direction than would be the case from a nondirectional transducer. The reason for this improvement is the reduced sensitivity of the directional receiver to random background noises that will be present in all directions from the source. The number of decibels by which the signal-to-noise ratio is improved by a directional receiver over a nondirectional receiver is known as the *directivity index* (directional gain) of the transducer. It will be defined more fully later. The following will show the directional radiation characteristics of several common structures.

Uniform Line Source. If a uniform long line is vibrating at uniform amplitude, the radiated sound intensity will be a maximum in a plane which is the perpendicular bisector of the line. At angles removed from the perpendicular bisector of the line, the intensity will fall off to a series of nulls and secondary maxima of diminishing amplitudes as the angle of incidence to the axis of the line deviates from the normal bisector of the line. For a line of length L vibrating uniformly over its entire length at a frequency corresponding to a wavelength of sound λ in the medium, the ratio of the sound pressure p_θ produced at an angle θ removed from the normal axis of maximum response to the sound pressure p_0 on the normal axis is given by

$$\frac{p_\theta}{p_0} = \frac{\sin \left[(\pi L/\lambda) \sin \theta\right]}{(\pi L/\lambda) \sin \theta} \tag{3i-6}$$

If L is large compared with λ, the response as a function of θ will go through a series of nulls and secondary maxima of successively diminishing amplitudes.

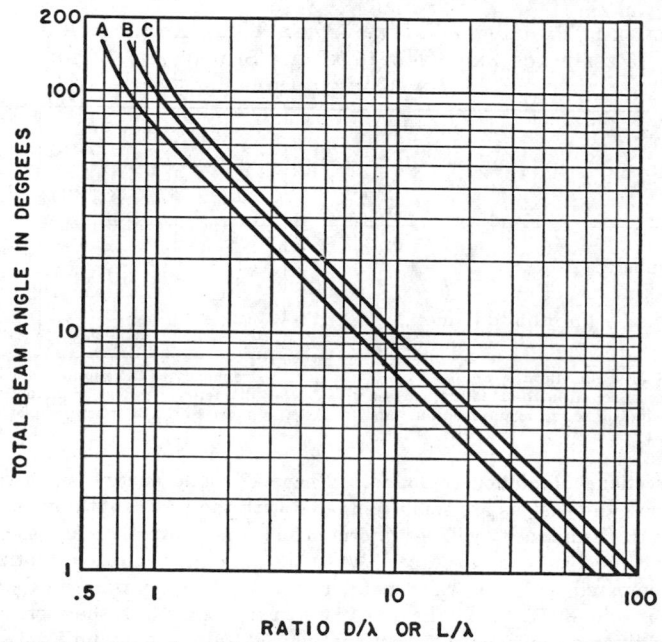

FIG. 3i-2. Total beam angle for a piston, ring, and line source as a function of size of source to wavelength of sound being radiated. *A*, thin ring of diameter *D*. *B*, uniform line of length *L*. *C*, piston of diameter *D*. (*Curves A and C from Massa, "Acoustic Design Charts," The Blakiston Division, McGraw-Hill Book Company, Inc., New York, 1942.*)

Circular Piston in Infinite Baffle. The directional radiation pattern from a large circular piston vibrating at constant amplitude and phase and set into an infinite rigid baffle may be obtained from the expression

$$\frac{p_\theta}{p_0} = \frac{2J_1[(\pi D/\lambda) \sin \theta]}{(\pi D/\lambda) \sin \theta} \tag{3i-7}$$

where p_θ = sound pressure at an angle θ from the normal axis of the piston
p_0 = sound pressure on normal axis of piston
D = diameter of piston
λ = wavelength of sound
J_1 = Bessel function of order 1

From this equation, it can be seen that, as D/λ increases, the beam width becomes smaller and the sound pressure goes through a series of nulls and secondary maxima as θ progressively departs from the normal axis to the piston.

Thin Circular Ring. The directional radiation pattern from a large narrow circular ring of diameter D vibrating at constant amplitude and fitted into an infinite plane baffle may be obtained from the expression

$$\frac{p_\theta}{p_0} = J_0 \left(\frac{\pi D}{\lambda} \sin \theta \right) \tag{3i-8}$$

where J_0 = Bessel function of order zero and all other symbols are defined under Eq. (3i-7).

Beam Width for Line, Piston, and Ring. From Eqs. (3i-6), (3i-7), and (3i-8), the total beam width has been computed for the radiation from each of the three types of sound generators. The total beam width is here defined as the angle 2θ at which the pressure p_θ is reduced 10 dB in magnitude from the maximum on axis reponse p_0. By setting p_θ/p_0 equal to -10 dB or 0.316 in magnitude in these equations, the three curves plotted in Fig. 3i-2 were computed.

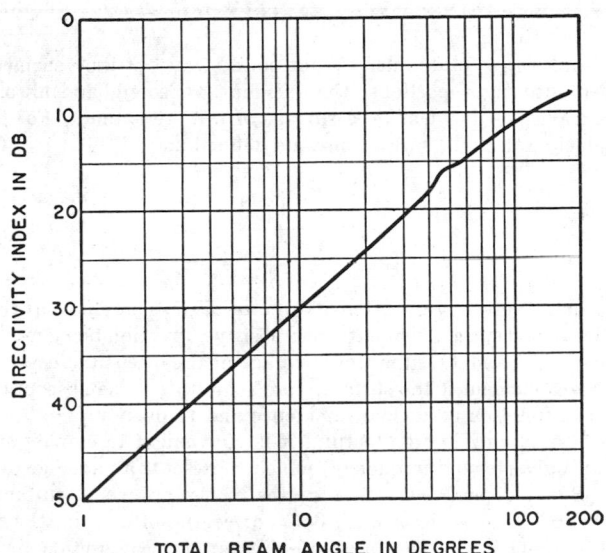

FIG. 3i-3. Directivity index of a piston or ring as a function of total beam angle where beam angle is defined as the included angle of the main beam between the 10-decibel-down points in the directional response. (*Computed from Massa, "Acoustic Design Charts," The Blakiston Division, McGraw-Hill Book Company, Inc., New York, 1942.*)

3i-4. Directivity Index. It has already been mentioned that a directional transducer has an advantage over a nondirectional structure whenever it is desired to send or receive signals from a particular localized direction only. The fact that the directional transducer is less sensitive to sounds coming from random undesired directions makes it possible for it to detect weaker signals than would be possible with a nondirectional unit. The measure of this improvement in decibels corresponds to the *directivity index* of the transducer, which is 10 times the logarithm (to the base 10) of the ratio of intensity of the response along the axis of maximum sensitivity to the average intensity of the response over the entire spherical region surrounding the transducer. See Sec. 3a for a more detailed definition.

The directivity index of a transducer is expressed in decibels, and a plot of the directivity index as a function of beam width for a piston or ring is shown in Fig. 3i-3.

3j. Architectural Acoustics

CYRIL M. HARRIS

Columbia University

3j-1. Sound-absorptive Materials. When sound waves strike a surface, the energy may be divided into three portions: the incident, reflected, and absorbed energy. Suppose plane waves are incident on a surface of infinite extent. For this case, the absorption coefficient α of the surface may be defined as

$$\alpha = \frac{\int_s I_p \cdot ds}{\int_s I_A \cdot ds} \tag{3j-1}$$

where I_p is the time average of the intensity vector of the sound field at the absorptive surface, ds is the vector surface element—the positive direction being into the material from the incident side, and I_A is the time average of the intensity vector which would exist at the surface element if the surface were removed. The absorption coefficient defined above is a function of angle of incidence and frequency.

For acoustical designing in architecture, it is convenient to employ an absorption coefficient α (at a given sound frequency) which represents an average over all angles of incidence. But α depends also on the area of the absorbent surface; the larger the area of a sound absorber on a wall, floor, or ceiling of a room, the smaller is its sound absorption coefficient. The data for α presented in this section are for measurements made on areas of about 72 sq ft, but we assume these are valid for all areas. A surface of S ft^2 is said to have an absorption of αS sabins. Thus the *sabin* (sometimes called a square-foot unit of absorption) is the absorption equivalent of 1 ft^2 of material having an absorption coefficient of unity.

A quantity which describes the acoustical properties of a material that is more fundamental than absorption coefficient is its *acoustic impedance*, defined as the complex ratio of sound pressure to the corresponding particle velocity at the surface of the material. Because of the complexities involved in the solutions to problems of room acoustics by boundary-value theory in terms of boundary impedances, the simpler concept of absorption coefficient is usually employed in calculating the acoustical properties of rooms, as indicated in the following section.

Most manufactured acoustical materials depend largely on their porosity for their acoustic absorption, the sound waves being converted into heat as they are propagated into the interstices of the material and also by vibration of the small fibers of the material. Another important mechanism of absorption is panel vibration; when sound waves force a panel into motion, the resulting flexural vibration converts a fraction of the incident sound energy into heat.

The average value of absorption coefficient of a material varies with frequency. Tables usually list the values of α at 125, 250, 500, 1,000, 2,000, and 4,000 Hz, or at

128, 256, 512, 1,024, 2,048, and 4,096 Hz, which for practical purposes are identical. In comparing materials which are used for noise-reduction purposes in offices, banks, corridors, etc., it is sometimes useful to employ a single figure called the noise-reduction coefficient (abbreviated NRC) of the material which is the average of the absorption coefficients at 250, 500, 1,000, and 2,000 Hz, to the nearest multiple of 0.05.

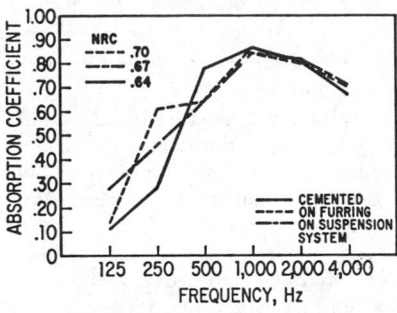

Figures 3j-1 through 3j-3 give the absorption coefficient vs. frequency for several types of acoustical material.[1] The absorption-frequency characteristics of regularly perforated cellulose fiber tile $\frac{3}{4}$ in. thick is shown in Fig. 3j-1. These curves represent average coefficients for materials of the same type, thickness, and method of mounting but of different manufacture. Similar data are shown in Fig. 3j-2 for fissured mineral tile $\frac{13}{16}$ in. thick. Values of noise-reduction coefficient are shown to the right of the graph. Values of absorption coefficient for various types of building materials are given in Table 3j-1.[1] The

FIG. 3j-1. The absorption vs. frequency characteristic for regularly perforated cellulose fiber acoustical tile. These data represent average values for $\frac{3}{4}$-in.-thick tile, mounted in the same way but of different manufacture. (*After H. J. Sabine, chap. 18 in "Handbook of Noise Control," C. M. Harris, ed., McGraw-Hill Book Company, New York,* 1957.)

equivalent absorption of individuals and seats, expressed in sabins, is given in Table 3j-2. More complete data and data for other types of material are given in the literature.[1,2]

Sound-absorptive materials and structures may be classified in the following way: (1) prefabricated units, including acoustical tile, tile boards, and certain mechanically

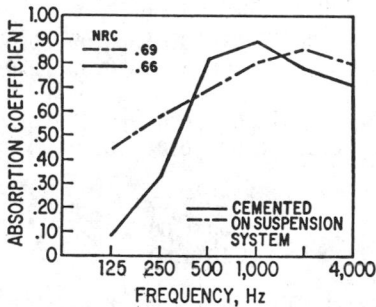

FIG. 3j-2. The absorption vs. frequency characteristic for fissured mineral tile. These data represent average values for $\frac{13}{16}$-in.-thick tile, mounted in the same way but of different manufacture. (*After H. J. Sabine, chap. 18 in "Handbook of Noise Control," C. M. Harris, ed., McGraw-Hill Book Company, New York,* 1957.)

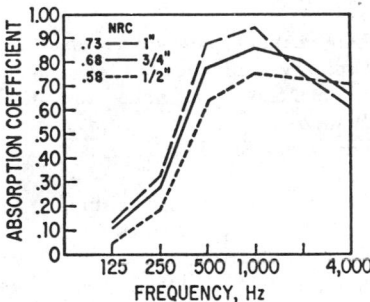

FIG. 3j-3. The absorption vs. frequency characteristic for regularly perforated cellulose fiber acoustical tile which has been spot-cemented to a rigid surface. These data represent the average value for tiles of different manufacture, mounted in the same way and having different thicknesses. (*After H. J. Sabine, chap. 18 in "Handbook of Noise Control," C. M. Harris, ed., McGraw-Hill Book Company, New York,* 1957.)

perforated units backed with absorptive material; (2) acoustical plasters; (3) acoustical blankets, consisting of mineral wool, glass fibers, hair felt, or wood fibers held together in blanket form by a suitable binder; (4) panel absorbers, including panels of plywood, paperboard, and pressed-wood fiber; (5) membrane absorbers consisting of a membrane of negligible stiffness backed by an enclosed air space; (6) resonator absorbers of the Helmholtz type; and (7) special types.

[1] *Acoust. Materials Assoc., Bull.* XXIX, New York, 1969.

[2] For example, see V. O. Knudsen and C. M. Harris, "Acoustical Designing in Architecture," John Wiley & Sons, Inc., New York, 1950.

TABLE 3j-1. ABSORPTION COEFFICIENTS FOR BUILDING MATERIALS*

Materials	Frequency, Hz					
	125	250	500	1,000	2,000	4,000
Brick, unglazed.....................	0.03	0.03	0.03	0.04	0.05	0.07
Brick, unglazed, painted.............	0.01	0.01	0.02	0.02	0.02	0.03
Carpet, heavy, on concrete...........	0.02	0.06	0.14	0.37	0.60	0.65
Same, on 40-oz hairfelt or foam rubber.	0.08	0.24	0.57	0.69	0.71	0.73
Same, with impermeable latex backing on 40-oz hairfelt or foam rubber...	0.08	0.27	0.39	0.34	0.48	0.63
Concrete block, coarse...............	0.36	0.44	0.31	0.29	0.39	0.25
Concrete block, painted..............	0.10	0.05	0.06	0.07	0.09	0.08
Fabrics:						
Light velour, 10 oz/yd² hung straight, in contact with wall.............	0.03	0.04	0.11	0.17	0.24	0.35
Medium velour, 14 oz/yd², draped to half area.......................	0.07	0.31	0.49	0.75	0.70	0.60
Heavy velour, 18 oz/yd², draped to half area.......................	0.14	0.35	0.55	0.72	0.70	0.65
Floors:						
Concrete or terrazzo...............	0.01	0.01	0.015	0.02	0.02	0.02
Linoleum, asphalt, rubber, or cork tile on concrete.....................	0.02	0.03	0.03	0.03	0.03	0.02
Wood...........................	0.15	0.11	0.10	0.07	0.06	0.07
Wood parquet in asphalt on concrete.	0.04	0.04	0.07	0.06	0.06	0.07
Glass:						
Large panes of heavy plate glass.....	0.18	0.06	0.04	0.03	0.02	0.02
Ordinary window glass..............	0.35	0.25	0.18	0.12	0.07	0.04
Gypsum board, $\frac{1}{2}$ in. nailed to 2 × 4's 16 in. o.c.........................	0.29	0.10	0.05	0.04	0.07	0.09
Marble or glazed tile................	0.01	0.01	0.01	0.01	0.02	0.02
Openings:						
Stage, depending on furnishings......			0.25–0.75			
Deep balcony, upholstered seats.....			0.50–1.00			
Grills, ventilating.................			0.15–0.50			
Plaster, gypsum or lime, smooth finish on tile or brick....................	0.013	0.015	0.02	0.03	0.04	0.05
Plaster, gypsum, or lime, rough finish on lath...........................	0.02	0.03	0.04	0.05	0.04	0.03
Same, with smooth finish...........	0.02	0.02	0.03	0.04	0.04	0.03
Plywood paneling, $\frac{3}{8}$ in. thick..........	0.28	0.22	0.17	0.09	0.10	0.11
Water surface, as in a swimming pool...	0.008	0.008	0.013	0.015	0.020	0.025

* From *Acoust. Materials Assoc. Bull.* XXIX, New York, 1969.

Some tables list the "ceiling attenuation factor" of acoustical materials designed for use in suspended ceilings. This factor is a measure of the reduction of sound level between two contiguous rooms when the transmission path of the sound is through the two suspended ceilings and the plenum common to both.

3j-2. Reverberation-time Calculations. After sound has been produced in or enters an enclosed space, it will be reflected by the boundaries of the enclosure.

Although some energy is lost at each reflection, several seconds may elapse before the sound decays to inaudibility. This prolongation of sound after the original source has stopped is called *reverberation*, a certain amount of which is found to add a pleasing characteristic to the acoustical qualities of a room. On the other hand, excessive reverberation can ruin the acoustical properties of an otherwise well-designed room.

TABLE 3j-2. ABSORPTION OF SEATS AND AUDIENCE*
(In sabins per person or unit of seating)

	125 Hz	250 Hz	500 Hz	1,000 Hz	2,000 Hz	4,000 Hz
Audience, seated in upholstered seats...	3.3	4.1	4.8	5.3	5.1	4.7
Unoccupied seats, cloth-covered, upholstered...	2.7	3.6	4.4	4.8	4.5	3.9
Unoccupied seats, leather-covered, upholstered...	2.4	3.0	3.3	3.4	3.2	2.6
Wooden pews, occupied...	3.1	3.4	4.1	4.7	5.0	4.7
Chairs, metal or wood seats...	0.15	0.19	0.22	0.39	0.38	0.30

* Based on values given in *Acoust. Materials Assoc. Bull.* XXIX, New York, 1969, modified by author. Materials and methods of fabrication can greatly influence the above values.

Because of the importance of the proper control of reverberation in rooms, a standard of measure called *reverberation time* (abbreviated t_{60}) has been established. It is one of the important parameters in architectural acoustics. This is the time required for a specified sound to die away to one-thousandth of its initial pressure, a drop in sound pressure level of 60 dB. It is given by the following equation:

$$t_{60} = \frac{0.049V}{S\bar{\alpha} + 4mV} \quad \text{sec} \tag{3j-2}$$

where V = volume of the room, ft³
  S = total surface area, ft²
  $\bar{\alpha}$ = average absorption coefficient given by

$$\bar{\alpha} = \frac{\alpha_1 S_1 + \alpha_2 S_2 + \alpha_3 S_3 + \cdots}{S_1 + S_2 + S_3 + \cdots} = \frac{a}{S} \tag{3j-3}$$

  α_1 = absorption coefficient of area S_1, etc.
  a = total absorption in the room, sabins

The quantity m is the attenuation coefficient for air given by Fig. 3d-6. For relatively small auditoriums and frequencies below 2,000 Hz, the mV term can usually be neglected so that Eq. (3j-2) reduces to

$$t_{60} = \frac{0.049V}{S\bar{\alpha}} \quad \text{sec} \tag{3j-4}$$

3j-3. Optimum Reverberation Time. A certain amount of reverberation in a room adds a pleasing quality to music. Since the reverberation time one would consider to be optimum is a matter of personal preference, it is not a quantity that can be calculated from a formula. On the other hand, useful engineering-design data can be obtained from a critical evaluation of empirical data based upon the preference evaluations of large groups of individuals. The results of such information from all available sources considered reliable, in this country and abroad, have been carefully evaluated by Knudsen and Harris,[1] who have published the curves for optimum

[1] *Ibid.*

reverberation time shown in Figs. 3j-4 and 3j-5. The data in Fig. 3j-4 give the optimum reverberation times at 500 Hz as a function of volume for rooms and auditoriums that are used for different purposes. Since the optimum reverberation time for music depends on the type of music, it is represented by a broad band. The optimum reverberation time for a room used primarily for speech is considerably shorter; a reverberation time longer than those shown results in a decrease in speech intelligibility.

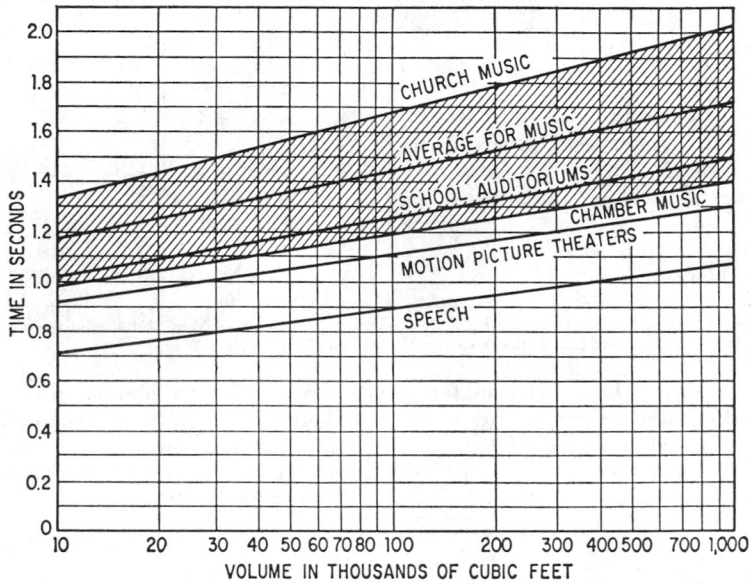

FIG. 3j-4. Optimum reverberation time at 500 Hz for different types of rooms as a function of room volume. This figure should be used in conjunction with Fig. 3j-5 to obtain optimum reverberation time as a function of frequency. (*After V. O. Knudsen and C. M. Harris, "Acoustical Designing in Architecture," John Wiley & Sons, Inc., New York, 1950.*)

The optimum reverberation times at frequencies other than 500 Hz are obtained by multiplying the values given in Fig. 3j-4 by the ratio R from Fig. 3j-5 for the desired frequency. These data indicate that below 500 Hz the optimum reverberation time may fall anywhere in a wide range shown by the crosshatched band; smaller rooms usually have preferred ratios that are in the lower part of the band.

3j-4. Structure-borne Sound Transmission. Noise in a building may originate from sources in air, it may be generated by impacts against the building structure, or it may result from mechanical vibration imparted to the building structure. The transmission path from one location in the building to another may be by either one, or a combination, of the following mechanisms: (1) sound may be transmitted along a direct air path, for example through a ventilation duct from one room to the next; (2) mechanical energy may be imparted to the structure; such energy then travels through the structure (usually with relatively little attenuation) to surfaces elsewhere in the building which it sets into vibration, thereby radiating noise; or (3) sound may force a partition into vibration, thereby transmitting acoustic energy into an adjacent room.

Sound also may be transmitted from one room to an adjacent room by a path other than through the common intervening partition, for example, along other walls or along the floor or ceiling. Such indirect transmission of sound is called *flanking transmission*.

Structures that provide good isolation against airborne noise do not necessarily provide good isolation against structure-borne noise. In general, structure-borne sound insulation techniques are designed to prevent vibratory energy from entering the building structure, for example, by the use of resilient flooring, carpeting, or "floating floor" constructions.

Data giving the structure-borne noise insulation values of many types of floor and ceiling constructions are given by Berendt[1] et al.

3j-5. Air-borne Sound Transmission through Partitions. The fraction of incident sound energy transmitted through a partition is called its transmission coefficient τ.

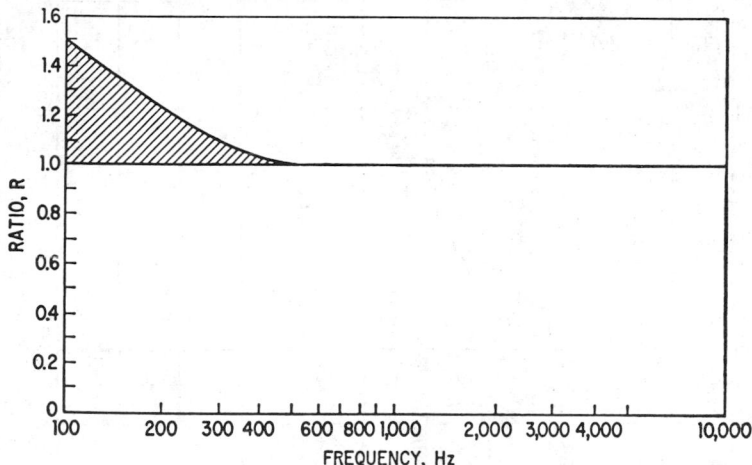

FIG. 3j-5. Chart for computing optimum reverberation time as a function of frequency. The time at any frequency is given in terms of a ratio R which should be multiplied by the optimum time at 500 Hz (from Fig. 3j-4) to obtain the optimum time at that frequency. (*After V. O. Knudsen and C. M. Harris, "Acoustical Designing in Architecture." John Wiley & Sons, Inc., New York, 1950.*)

In rating the noise-insulating value of partitions, windows, and doors, it is generally convenient to employ a logarithmic quantity, transmission loss T.L., which is equal to the number of decibels by which sound energy that is incident on a partition is reduced in transmission through it. The two quantities are related by the equation

$$\text{T.L.} = 10 \log \frac{1}{\tau} \quad \text{dB} \qquad (3j\text{-}5)$$

Air-borne sound is transmitted through a so-called "rigid" partition, such as a wall of concrete or brick, by forcing it into vibration; then the vibrating partition becomes a secondary source, radiating sound to the side opposite the original source. Over a large portion of the audible range, such a partition, on the average, approximates a mass-controlled system so that its transmission loss should increase 6 dB each time the weight of the partition is doubled. In most actual partitions the increase is usually less, say 4 to 5 dB for the average frequency range between 125 and 2,000 Hz. This is illustrated by Fig. 3j-6, which gives the transmission loss (averaged over

[1] R. D. Berendt, G. E. Winzer, and C. B. Burroughs, "A Guide to Airborne, Impact, and Structure Borne Noise-Control in Multifamily Dwellings," U.S. Department of Housing and Urban Development, Washington, D.C., September, 1967.

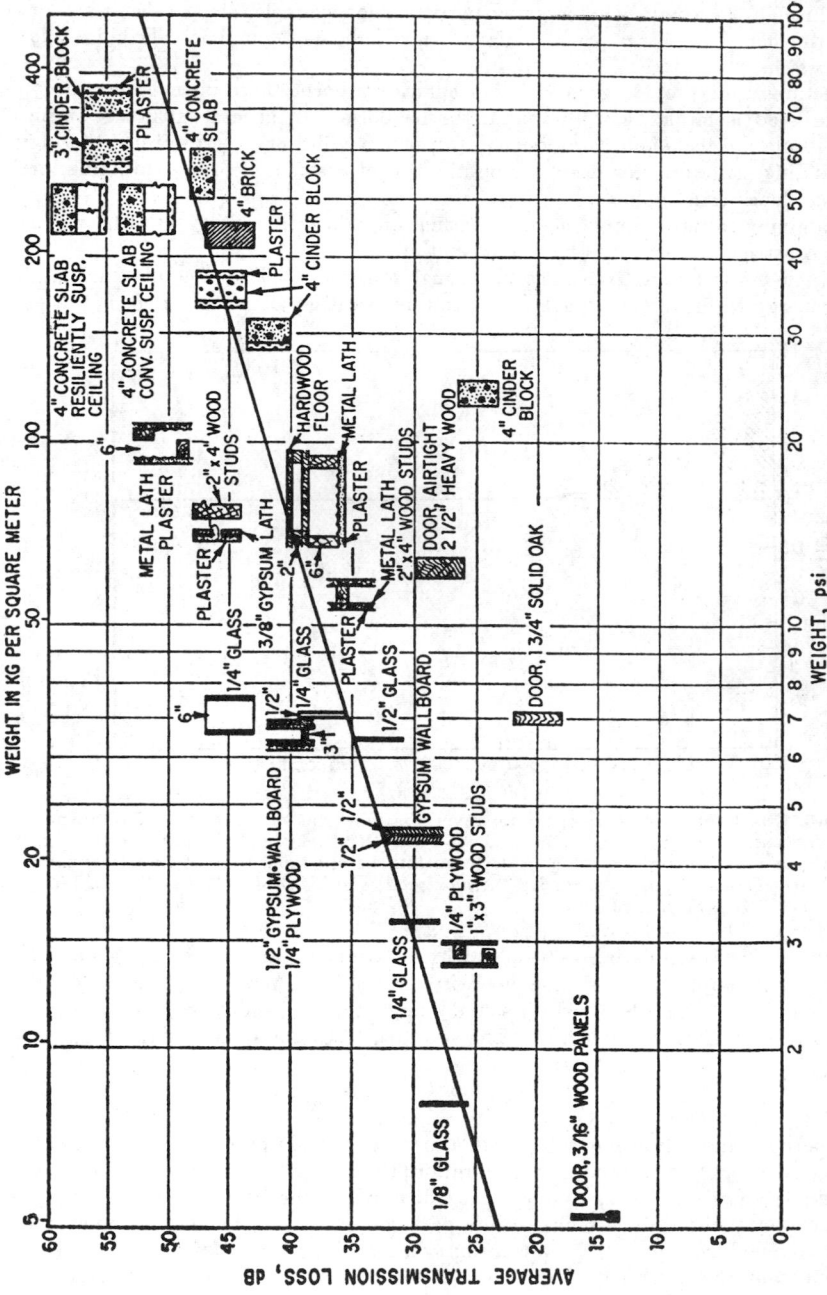

FIG. 3j-6. The mass-law relation between average sound transmission loss and mass per unit area of a homogeneous partition. The transmission loss, expressed in decibels, is averaged over the frequency range from 125 to 4,000 Hz. (R. K. Cook and P. Chrzanowski, from C. M. Harris, ed., "Handbook of Noise Control," chap. 20, McGraw-Hill Book Company, New York, 1957.)

frequency in the range from 125 to 4,000 Hz) as a function of weight of the partition in pounds per square foot of surface area. The straight line represents an average of the experimental data showing that the average transmission loss increases approximately 4.4 dB for each doubling of mass per unit area of a homogeneous partition. The transmission loss for a partition is not constant with frequency, increasing usually 3 to 6 dB/octave.

A single number which represents the sound transmission loss of a partition averaged over frequency may correlate rather poorly with the subjective assessment of the insulation value of the partition. Therefore another rating is frequently employed to represent the sound insulation value of a partition by a single number; "sound transmission class" (STC). The STC value of a partition is determined by comparing the curve of transmission loss vs. frequency for the partition with a set of standardized transmission loss vs. frequency contours.[1]

Sound insulation values for various types of walls and floors employed in ordinary building construction are given in Table 3j-3. Note that a compound-wall construction can yield relatively high sound insulation with relatively low mass per unit wall area. The double-wall construction is one such example. It is important that the separation between the walls be as complete as possible—structural ties will greatly reduce the effectiveness of such a structure.

3j-6. Noise Level within a Room. The sound level of noise which is transmitted into a room from the outside depends on (1) the noise-insulating properties of its bounding surfaces, (2) the total absorption in the room, and (3) the characteristics of the noise source. The following formula gives a rating of the overall noise reduction provided by the enclosure. It represents, approximately, the difference between the noise level outside a room and the noise level inside a room.

$$\text{Level difference} = 10 \log \frac{a}{T} \quad \text{dB} \qquad (3\text{j-}6)$$

where a represents the total absorption in the room in sabins defined by Eq. (3j-4), and T represents the total transmittance of the enclosure given by

$$T = \tau_1 S_1 + \tau_2 S_2 + \tau_3 S_3 + \cdots \qquad (3\text{j-}7)$$

where τ_1 is equal to the transmission coefficient of area S_1, etc.

If a source of noise is within a room, then at distances near to the source the sound pressure decreases inversely with increasing distance from the source; there is a decrease in sound pressure level of 6 dB for each doubling of the distance from the source, just as if the source were in the open air. However, at every point in the room there will be an additional contribution to the total pressure as a result of reflections from the walls. As one recedes from the source, the reflected contributions become more and more important until direct sound from the source becomes negligible by comparison.

Then if the sound field is diffuse (perfect diffusion is said to exist if the sound pressure everywhere in the room is the same, and it is equally probable that the waves are traveling in every direction), the sound pressure level in the room will be given approximately by

$$L_p = 10 \log \frac{W}{a} + 136.4 \quad \text{dB} \qquad (3\text{j-}8)$$

[1] R. D. Berendt, G. E. Winzer, and C. B. Burroughs, "A Guide to Airborne, Impact, and Structure Borne Noise-Control in Multifamily Dwellings," U.S. Department of Housing and Urban Development, Washington, D.C., September, 1967. See also *ASTM Rept.* E90-66T, Tentative Recommended Practice for Laboratory Measurement of Airborne Sound Transmission Loss of Building Partitions.

TABLE 3j-3. INSULATION VALUES FOR VARIOUS TYPES OF WALL AND FLOOR CONSTRUCTION*

Type of construction	Weight, lb/ft²	STC rating, dB	Transmission Loss, dB								
			125 Hz	175 Hz	250 Hz	350 Hz	500 Hz	700 Hz	1,000 Hz	2,000 Hz	4,000 Hz
Solid concrete, 3 in. thick	39	47	35	37	40	41	44	49	52	59	64
Solid concrete, 6 in. thick, both sides plastered	80	53	39	42	42	47	50	55	58	64	
Hollow concrete block, 6 in. thick	34	43	32	33	33	37	40	43	47	51	48
Same as above except painted	34	45	37	35	36	39	42	47	49	55	58
Hollow gypsum block, 3 in. thick, one side plastered, other side plaster on resilient clips	27	45	48	43	41	43	47	48	44	55	62
Double brick wall, 6 in. cavity. Overall thickness 18 in.	120	62	48	54	54	56	56	60	64	69	
Wood stud, gypsum wallboard	6	38	20	21	27	33	37	38	43	48	43
Wood stud, gypsum lath and plaster	15	46	32	34	37	40	42	46	48	48	63
Same as above but with perforated lath	14	44	42	34	32	38	42	47	49	50	62
Staggered wood stud, gypsum board and insulation	14	46	39	38	40	41	42	44	48	56	51
Wood stud, plastered gypsum lath on resilient clips	13	52	46	44	46	53	54	56	57	50	62
Metal channel stud, gypsum board	5	39	20	24	30	33	37	43	47	48	44
Metal channel stud, 2 layers of gypsum board	9	47	31	35	38	41	45	51	53	54	54

* Values based on data taken from R. D. Berendt, G. E. Winzer, and C. B. Burroughs, "A Guide to Airborne, Impact, and Structure Borne Noise-Control in Multi-family Dwellings," U.S. Department of Housing and Urban Development, Washington, D.C., September, 1967. For average values for other types of construction including doors and windowpane materials, see Fig. 3j-6. For the definition of STC used to obtain the ratings (above), see ASTM Rept. E90-66T, Tentative Recommended Practice for Laboratory Measurement of Airborne Sound Transmission Loss of Building Walls and Floors.

if a value of ρc = 40.8 rayls is assumed for air; W = power of the sound source in watts, and a = total absorption of the room in sabins. A consideration of the above formula shows that, if the acoustic-power output of the noise source remains constant, and if the total absorption in the room is increased from a_1 to a_2, the reduction in noise level is given by

$$\text{Noise reduction} = 10 \log \frac{a_2}{a_1} \quad \text{dB} \qquad (3\text{j-}9)$$

According to this equation, which should be regarded as an engineering approximation to actual conditions, if the absorption in a room is increased by a factor of 4, the noise reduction will be 6 dB. It shows that the addition of absorption in a room will provide substantial noise reduction in average level in a room that is relatively bare but little decrease in level in a highly damped room. The reduction will be different at different frequencies since the total absorption is a function of frequency. However, it is sometimes convenient to employ the noise-reduction coefficient of a material to obtain a single noise-reduction figure. Besides reducing the steady-state noise level, the addition of absorptive treatment in a room also provides beneficial effects by reducing the reverberation time in the room and by localizing the source of noise to the area in which it originates—thereby minimizing unexpected noises.

TABLE 3j-4. RECOMMENDED ACCEPTABLE AVERAGE NOISE LEVELS IN
UNOCCUPIED ROOMS*

Decibels†

Radio, recording, and television studios	25–30
Music rooms	30–35
Legitimate theaters	30–35
Hospitals	35–40
Motion-picture theaters, auditoriums	35–40
Churches	35–40
Apartments, hotels, homes	35–45
Classrooms, lecture rooms	35–40
Conference rooms, small offices	40–45
Courtrooms	40–45
Private offices	40–45
Libraries	40–45
Large public offices, banks, stores, etc	45–55
Restaurants	50–55

* V. O. Knudsen and C. M. Harris, "Acoustical Designing in Architecture," John Wiley & Sons, Inc., New York, 1950.
† The levels given in this table are "weighted"; i.e., they are the levels measured with a standard sound-level meter employing the "A" (40-dB) frequency-weighting network.

3j-7. Acceptable Noise Levels for Various Types of Room. Table 3j-4 gives values of recommended acceptable average noise levels for unoccupied rooms with the ventilation system in operation. These values often are used for design purposes, for example, in computing the amount of overall noise insulation that should be provided for a room. Although even lower noise levels than those which are listed may provide some advantage under certain circumstances and may be desirable if cost is not a factor, this table gives values which represent a combination of acceptability and economic practicality. For certain types of room the values which are recommended are lower than those which are commonly found.

The single numbers given in Table 3j-4 represent "weighted" sound pressure levels. In another system for rating noise conditions in a room a single number is obtained by comparing the octave-band spectrum of the room noise with a set of octave-band level curves called "noise criterion" or "NC" curves. In many cases, description by a single number will not suffice. Then the noise spectra must be expressed in terms of octave-band levels.

3k. Speech and Hearing

EDWIN B. NEWMAN[1]

Harvard University

The data concerning hearing are, without exception, empirical in derivation. Consequently, the values reported always represent some parameter of a population, most often a mean, and the reader is warned to bear constantly in mind the many sources of variability that attach to any particular measurement.

3k-1. Physical Dimensions of the Ear

TABLE 3k-1. PHYSICAL DIMENSIONS OF THE EAR*

Pinna:
  Mean length, young men, 65.0 mm
  Range, 52–79 mm
Auditory meatus:
  Cross section, 0.3–0.5 cm²
  Diameter, 0.7 cm
  Length, 2.7 cm
  Volume, 1.0 cc
Tympanic membrane:
  Area, 0.5–0.9 cm² (roughly circular)
  Thickness, about 0.1 mm
  Volume elasticity for 10 Hz, equivalent to about 8 cc air
  Displacement amplitude for 1,000-Hz tone (at threshold), 10^{-9} cm
  Displacement amplitude for low-frequency tones (threshold of feeling), about 10^{-2} cm

Middle ear:
  Total volume, about 2 cc
  Malleus:
    Weight, 23 mg
    Length, 5.5–6.0 mm
  Incus: weight, 27 mg
  Stapes:
    Weight, 215 mg
    Length of footplate, 3.2 mm
    Width of footplate, 1.4 mm
    Area of footplate, 3.2 mm²
    Width of elastic ligament, 0.015–0.1 mm
Cochlea:
  Length of cochlear channels, 35 mm
  Height of scala vestibuli or scala tympani, about 1 mm (great variability)
  Round window: area, 2 mm²
  Basilar membrane:
    Width at stapes, 0.04 mm
    Width at helicotrema, 0.5 mm
  Helicotrema: area of opening, 0.25–0.4 mm²

* S. S. Stevens, ed., "Handbook of Experimental Psychology," John Wiley & Sons, Inc., New York, 1951.

3k-2. Acoustic Impedance of the Ear.

Reasonable agreement on measurements below 1,000 Hz has been obtained. The reference point for measurements is just

[1] This section benefited from the advice and assistance of Dr. S. S. Stevens and Mrs. Nancy C. Waugh.

within the external meatus. The values in Table 3k-2 are representative but are subject to wide variations among individuals.[1]

3k-3. Minimum Audible Sound. Table 3k-3 lists the minimum audible (threshoïd, sound pressures of pure tones measured at the entrance to the external meatus. The pressure measurements were made when the subject heard the tone one-half the time it was presented via an earphone applied to his ear with a standard static force. Observations were made on young persons, eighteen to twenty-five years of age, with no record of hearing impairment. Sound pressures were determined with a probe-tube microphone and are given in decibels relative to 2×10^{-4} dyne/cm^2.

The results of such measurements made in various laboratories show a considerable amount of variation. The pressures in Table 3k-3 are based on measurements made in two independent laboratories; see the first footnote for details.

The variance in the threshold sound pressures measured at the entrance to the meatus has been so great that such pressures cannot serve usefully as standards for audiometry. Experience has shown that the most accurate method for storing audiometric standard threshold information is as follows. Measurements of threshold voltages on an earphone applied to a number of young persons at the various audiometric frequencies are the primary data. The sound pressures which are produced by these voltages when the earphone is applied to an artificial ear (coupler) then serve as the standard thresholds for that particular earphone-coupler combination. This method of measuring and storing standard threshold sound pressures is now in use in several countries. A comparison of the standard thresholds was completed under the auspices of Technical Committee 43 on Acoustics of the International Organization for Standardization (ISO). An internationally agreed-upon standard threshold has been issued by ISO in its Recommendation R389, Standard Reference Zero for the Calibration of Pure Tone Audiometers. The standard data in it are sound pressures corresponding to the threshold of hearing for five earphone-coupler combinations now in use in several countries.[2]

3k-4. Threshold of Feeling or Discomfort. The upper limit for a tolerable intensity of sound rises substantially with increasing habituation. Moreover, a variety of subjective effects are reported, such as discomfort, tickle, pressure, and pain, each at a slightly different level. As a simple engineering estimate it can be said that naïve listeners reach a limit at about 125 dB SPL and experienced listeners at 135 to 140 dB. These are overall measures of sound falling within the audible range and are roughly independent of frequency.

3k-5. Differential Thresholds for Pure Tones and Noise. A differential threshold represents a careful determination by laboratory methods of the ability of a subject to just detect, and report, a difference in any specific property of a sound, all other factors presumably being held constant.

The method for determining the differential threshold for intensity of pure tones employed one tone beating with a second tone at 3 beats per second.[3] Much evidence is available to support what should be kept always in mind, that thresholds determined by other methods are a function of numerous psychological parameters and will differ systematically from the values in Table 3k-4. A more conventional method was used to determine the thresholds for white noise, with the results given in the last column.[4]

[1] E. Waetzmann and L. Keibs, Hörschwellenbestimmungen mit dem Thermophon und Messungen am Trommelfell, *Ann. Physik* **26**, 141–144 (1936); O. Metz, The Acoustic Impedance Measured on Normal and Pathological Ears, *Acta Oto-Laryngol.*, *Suppl.* 63, 1–254 (1946); A. H. Inglis, C. H. G. Gray, and R. T. Jenkins, A Voice and Ear for Telephone Measurements, *Bell System Tech. J.* **11**, 293–317 (1932).

[2] P. G. Weissler, International Standard Reference Zero for Audiometers, *J. Acoust. Soc. Am.* **44**, 264–275 (1968).

[3] R. R. Reisz, Differential Intensity Sensitivity of the Ear for Pure Tones, *Phys. Rev.* **31**, 867–875 (1928).

[4] G. A. Miller, Sensitivity to Changes in the Intensity of White Noise and Its Relation to Masking and Loudness, *J. Acoust. Soc. Am.* **19**, 609–619 (1947).

ACOUSTICS

The ability to distinguish pitch is subject to a greater range of individual variability than other functions reported here. The data given are for three trained listeners and have been smoothed in both directions. Untrained listeners usually require a greater

TABLE 3k-2. ACOUSTIC IMPEDANCE OF THE EAR IN ACOUSTIC OHMS, MEASURED JUST WITHIN THE MEATUS

Frequency	Total impedance	Resistive component	Reactive component
250	200	50	-190
350	150	40	-145
500	125	35	-115
700	70	25	-65
1,000	55	25	-50

Above 1,000 Hz measurements depend increasingly on the method of measurement.

TABLE 3k-3. MINIMUM AUDIBLE (THRESHOLD) PRESSURE AT ENTRANCE TO EXTERNAL EAR CANAL (MAC)*
(In dB re 2×10^{-4} dyne/cm²)

	Frequency, Hz										
	125	250	500	1,000	1,500	2,000	3,000	4,000	6,000	8,000	10,000
MAC	35	22	14	8	9	9	10	9	14	17	16

The following quantities are to be added in order to obtain threshold pressures for other conditions:

a. MAC to Threshold Pressure at Tympanic Membrane†

	Frequency, Hz							
	125	250	500	1,000	2,000	4,000	6,000	8,000
Add..........	0.0	0.0	-0.5	-1.0	-4.5	-10.5	-4.0	-2.5

b. MAC to Free Field (MAF) (plane wave, 0° azimuth in absence of head)‡

	Frequency, Hz								
	125	250	500	1,000	2,000	4,000	6,000	8,000	10,000
Add....	$+1.0$	$+0.5$	-2.0	-4.0	-11.0	-12.5	-7.0	-3.0	-3.0

* J. P. Albrite, R. E. Shutts, M. B. Whitlock, R. K. Cook, E. L. R. Corliss, and M. D. Burkhard, Research in Normal Threshold of Hearing, *AMA Arch. Otolaryngol.* **68,** 194–198 (1958).
† F. M. Wiener and D. A. Ross, The Pressure Distribution in the Auditory Canal in a Progressive Sound Field, *J. Acoust. Soc. Am.* **18,** 401–408 (1946).
‡ L. J. Sivian and S. D. White, On Minimum Audible Sound Fields, *J. Acoust. Soc. Am.* **4,** 288–321 (1933).

TABLE 3k-3. MINIMUM AUDIBLE (THRESHOLD) PRESSURE AT ENTRANCE TO
EXTERNAL EAR CANAL (MAC) (*Continued*)

c. Mean Monaural to Mean Binaural Listening§

	Frequency, Hz				
	125–2,000	4,000	6,000	8,000	10,000
Add...................	−2.0	−3.0	−4.0	−5.0	−6.0

d. Reference Age Group (18–25) to Older Age Groups ¶

	Frequency, Hz					
	125–1,000	2,000	4,000	6,000	8,000	10,000
Add for:						
Men 30–39.........	+1.0	+2.0	+5.0	+6.0	+6.0	+7.0
Men 40–49.........	+2.0	+5.0	+13.0	+13.0	+11.0	+13.0
Men 50–59.........	+5.0	+13.0	+27.0	+32.0	+35.0	+35.0
Women 30–39......	+1.0	+2.0	+3.0	+4.0	+4.0	+4.0
Women 40–49......	+3.0	+5.0	+6.0	+8.0	+9.0	+9.0
Women 50–59......	+5.0	+9.0	+13.0	+18.0	+20.0	+22.0

§ H. Fletcher, "Speech and Hearing in Communication," p. 131, D. Van Nostrand Company, Inc.,
Princeton, N.J., 1953.
¶ J. C. Steinberg, H. C. Montgomery, and M. B. Gardner, Results of the World's Fair Hearing Tests,
J. Acoust. Soc. Am. **12,** 291–301 (1940); J. C. Webster, H. W. Himes, and M. Lichtenstein, San Diego
County Fair Hearing Survey, *J. Acoust. Soc. Am.* **22,** 473–483 (1950).

TABLE 3k-4. DIFFERENTIAL THRESHOLD FOR INTENSITY, IN DECIBELS

Sensation level, dB above absolute threshold	Pure tones, frequency in Hz							
	35	70	200	1,000	4,000	7,000	10,000	White noise
5	....	....	4.75	3.03	2.48	4.05	4.72	1.80
10	7.24	4.22	3.44	2.35	1.70	2.83	3.34	1.20
20	4.31	2.38	1.93	1.46	0.97	1.49	1.70	0.47
30	2.72	1.52	1.24	1.00	0.68	0.90	1.10	0.44
40	1.76	1.04	0.86	0.72	0.49	0.68	0.86	0.42
50	....	0.75	0.68	0.53	0.41	0.61	0.75	0.41
60	....	0.61	0.53	0.41	0.29	0.53	0.68	0.41
70	....	0.57	0.45	0.33	0.25	0.49	0.61	
80	....	....	0.41	0.29	0.25	0.45	0.57	
90	....	....	0.41	0.29	0.21	0.41		
100	....	....	....	0.25	0.21			
110	....	....	....	0.25				

frequency difference than that reported here. Note also that individual listeners commonly show idiosyncrasies at particular frequencies.

TABLE 3k-5. DIFFERENTIAL THRESHOLD FOR FREQUENCY, IN $\Delta F/F$*

Sensation level, dB above absolute threshold	Pure tones, frequency in Hz						
	60	125	250	500	1,000	2,000	4,000
5	0.0252	0.0110	0.0097	0.0065	0.0049	0.0040	0.0077
10	0.0140	0.0060	0.0053	0.0035	0.0027	0.0022	0.0042
15	0.0092	0.0040	0.0035	0.0024	0.0018	0.0014	0.0028
20	0.0073	0.0032	0.0028	0.0019	0.0014	0.0012	0.0022
30		0.0032	0.0028	0.0019	0.0014	0.0011	0.0022

* J. D. Harris, Pitch Discrimination, *J. Acoust. Soc. Am.* **24**, 750–755 (1952).

3k-6. Masking. Masking refers to our inability to hear a weak sound in the presence of a louder sound. It is usually measured by the amount of change in the threshold of the weaker sound, i.e., how much more intense the weak sound must be made in order to be heard over the masking sound than it needed to be when the masking sound was not present. The masking of one pure tone by another is a complex function of the particular frequencies and of the absolute level of the respective tones. See any standard text on hearing for the curves describing this relationship.

The masking of a pure tone by a noise with a reasonably flat and continuous spectrum is a linear function (except at levels below 10 dB) of the total intensity within a "critical band" centered on the masked tone. The width of the critical band of frequencies whose total energy is just equal to the energy of the masked tone is given by Table 3k-6.

TABLE 3k-6. WIDTH OF "CRITICAL BAND" ΔF AS A FUNCTION OF CENTER
FREQUENCY F (10 log ΔF)*

	Frequency, Hz							
	100	250	500	1,000	2,000	4,000	8,000	10,000
ΔF, dB	19.4	17.1	17.1	18.0	19.9	23.1	27.7	29.2

* N. R. French and J. C. Steinberg, Factors Governing the Intelligibility of Speech Sounds, *J. Acoust. Soc. Am.* **19**, 90–119 (1947).

The masking of a narrow-band noise by two tones, one higher and one lower than the noise, shows a similar relationship. The masking produced by the two tones overlaps unless the tones are separated by more than a "critical band," at which point the masking begins to fall off sharply. The critical band measured in this way is 3 to 4 dB wider than the values given in Table 3k-6.[1]

The masking of one continuous noise by another can be thought of as a case of differential sensitivity to change in the intensity of a noise (see last column of Table 3k-4). Thus, above 40 dB SPL, if a weak noise is more than 10 dB less intense than a very similar masking noise, the weak noise will not be heard; its presence or absence

[1] E. Zwicker, G. Flottorp, and S. S. Stevens, Critical Band Width in Loudness Summation, *J. Acoust. Soc. Am.* **29**, 548–557 (1957). See especially the summary of the concept of critical bands, pp. 554–557.

does not produce a discriminable difference in intensity. If the spectral compositions of the two noises, masking and masked, are quite different, then the critical-band concept must be employed.

3k-7. Sounds of Short Duration. Acoustic disturbances of very short duration, i.e., less than 0.0001 sec, are heard only to the extent that they transmit energy to the ear. Short pulses at ultrasonic frequencies are generally not heard unless they are rectified. Impulse or step functions excite the ear, but not efficiently.

At the opposite extreme, tones, or continuous noise, of duration greater than from 0.2 to 0.5 sec are generally heard independently of duration. Between these limits relatively complex relations are found.[1]

As a first approximation for both tones and noise, the effective intensity of short sounds is a function of total energy integrated over the duration of the sound. More accurately, the threshold is defined by[2]

$$I_t = kIt^{0.8} \tag{3k-1}$$

For some short tones and for many types of impulse noise, account must be taken of the frequency distribution of energy. Inasmuch as the ear varies in sensitivity as a function of frequency, any change in the shape or duration of a short acoustic pulse will also change its effectiveness because of the altered spectral composition.

3k-8. Loudness. Loudness and pitch are ways in which a listener reacts to sounds. Furthermore, within limits, a listener can use numbers to describe how much of a response he makes to the sound. These numbers usefully describe how loud or how high in pitch a sound seems to be. It is then necessary to relate how loud it is (subjective response) to how intense it is in physical terms. The loudness of a pure tone of 1,000 Hz is described by the following relationship:

$$\log L = 0.0301N - 1.204 \tag{3k-2}$$

in which L is the loudness measured in sones and N is the loudness level in phons (equal to the sound pressure level of the tone in decibels above 0.0002 dyne/cm²).[3] Another way of putting this is to say that loudness doubles for each 10-dB change in sound pressure level.

There is some evidence that the loudness of a noise grows more rapidly than that of a tone with an increase in sound pressure level, especially at low levels. The exact relations are less well known than those for a tone.

The loudness of tones at other frequencies than 1,000 Hz is given by determining the loudness level in the manner described below and converting to sones by Eq. (3k-2).

The loudness of noises can be measured by direct subjective comparison with a standard, such as a tone of 1,000 Hz, but such comparisons are difficult and need to be repeated by a number of judges. An approximation to the loudness of a noise can be calculated from measurements of the sound pressure level in a series of bands, usually a third-octave, a half-octave, or an octave in width, covering the audible spectrum.

The total loudness of the noise is given by the formula

$$L_t = S_m + F \left(\sum^i S_i - S_m \right) \tag{3k-3}$$

[1] S. S. Stevens, ed., "Handbook of Experimental Psychology," pp. 1020–1021, John Wiley & Sons, Inc., New York, 1951.
[2] D. B. Yntema, "The Probability of Hearing a Short Tone Near Threshold," Ph.D. Dissertation, Harvard University, 1954, 43 pp.
[3] S. S. Stevens, The Measurement of Loudness, *J. Acoust. Soc. Am.* **27**, 815–829 (1955).

The calculated loudness L_t should be qualified by the width of the bands used for its calculation. The terms S are empirical values of a loudness index shown as the parameter of the curves in Fig. 3k-1. The figure is entered with the geometric mean frequency of each band and the band pressure level as arguments. The loudness index S_i is estimated for each of the i bands. The band having the greatest index S_m is determined by inspection.

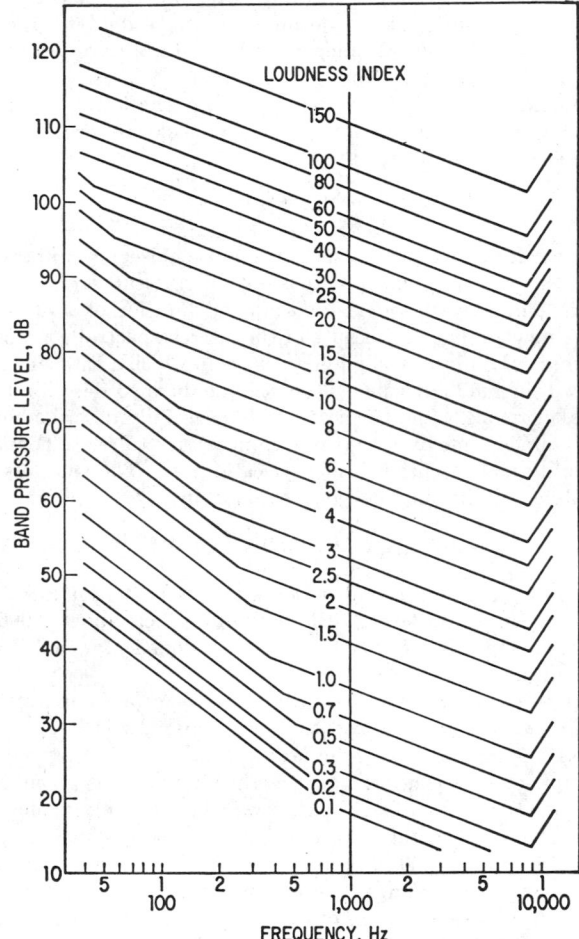

FIG. 3k-1. Loudness index S, as a function of geometric mean frequency of band measured and band pressure level (sound pressure level in third-octave, half-octave, or octave band under measurement). (*Taken from S. S. Stevens, "Procedure for Calculating Loudness," Mark VI, Psycho-Acoustic Laboratory Report PNR-253, Mar. 1, 1961.*)

As the formula indicates, total loudness is linearly additive except for a constant factor F that represents the reduction due to mutual masking of all bands except the loudest. The value of F depends on the width of the bands used. It has the value of 0.15 for third-octave, 0.2 for half-octave, and 0.3 for octave bands.

The loudness L_t can be converted to loudness level by Eq. (3k-2).

3k-9. Loudness Level. The loudness level of a tone of 1,000 Hz, expressed in phons, is defined as the sound pressure level in decibels above the reference level of 0.0002 dyne/cm².

The loudness level of tones of other frequencies is given by the empirical relations in Table 3k-7.

TABLE 3k-7. LOUDNESS LEVEL AS A FUNCTION OF SOUND PRESSURE LEVEL AND FREQUENCY*

Sound pressure level	Frequency, Hz							
	125	250	500	1,000	2,000	4,000	8,000	10,000
10	.....	.....	.....	10.0	18.0	18.0		
20	.....	6.3	16.0	20.0	28.0	28.0	11.0	
30	4.0	18.0	26.5	30.0	37.0	36.5	20.5	17.0
40	17.0	31.0	38.5	40.0	45.5	45.0	29.5	26.0
50	34.0	45.5	52.0	50.0	55.0	54.0	38.0	35.0
60	52.0	59.5	64.5	60.0	64.0	63.5	47.0	43.5
70	70.0	72.5	76.0	70.0	73.5	72.5	56.0	53.5
80	86.0	84.5	86.0	80.0	84.5	83.0	66.0	63.5
90	98.0	95.5	96.0	90.0	95.0	94.5	77.0	73.5
100	108.0	105.5	105.0	100.0	106.0	106.0	88.0	85.5
110	118.0	115.5	113.0	110.0	117.0	117.5	101.5	98.0

* American Standard for Noise Measurement, ASA Z24.2—1942.

Note that this table is based on the ASA standard and presumes the "free-field" measurement of sound pressure. This requires a measurement of a plane progressive wave at the listener's position before the listener is placed in the field. More meaningful measurements would doubtless be obtained from pressure measurements at the ear. For this purpose, apply the corrections contained in Table 3k-3b to the ear canal pressures before entering Table 3k-7.

To enter the table with sound pressure levels measured under other conditions, first add the corrections in Table 3k-3b, then subtract rather than add corrections in Tables 3k-3a and 3k-3c. Note, however, that corrections given for presbycusis in Table 3k-3d may give quite misleading results because of recruitment at high frequencies in some elderly people.

3k-10. Pitch. The relation between frequency and the subjective magnitude of perceived pitch is shown by Table 3k-8. By definition, the pitch of a tone of 1,000 Hz at 40 dB SPL is 1,000 mels.[1]

3k-11. Localization of Sound. The localization of complex sounds is primarily a function of time differences of arrival at the two ears, and to a first approximation, such differences can be calculated by assuming the ears on either end of the diameter of a sphere of 7.5 cm radius.

The localization of tones of low frequency (below 1,500 Hz) is possible on the basis of phase differences, which may be interpreted in terms of time differences.

The localization of tones of high frequency is possible on the basis of intensity differences resulting from the sound shadow of the head. Exact measurements here are difficult at best.

[1] S. S. Stevens and J. Volkmann, The Relation of Pitch to Frequency: A Revised Scale, *Am. J. Psychol.* **53**, 329–353 (1940).

TABLE 3k-8. PITCH OF A PURE TONE, IN MELS, AS A FUNCTION OF FREQUENCY

Frequency	Mels	Frequency	Mels	Frequency	Mels
20	0	350	460	1,750	1,428
30	24	400	508	2,000	1,545
40	46	500	602	2,500	1,771
60	87	600	690	3,000	1,962
80	126	700	775	3,500	2,116
100	161	800	854	4,000	2,250
150	237	900	929	5,000	2,478
200	301	1,000	1,000	6,000	2,657
250	358	1,250	1,154	7,000	2,800
300	409	1,500	1,296	10,000	3,075

Sound localization is greatly aided when the head or body can be rotated or moved about in the sound field while the observer hears the appropriate sequence of sounds.[1]

Sound localization in reverberant rooms or with so-called "stereophonic-sound sources" depends critically upon a "precedence effect," by which the localization determined by the primary sound or sound from the nearer of two sound sources is overriding in its effect.[2]

In experiments where time differences are used to balance out intensity differences in the opposite direction, 1.0×10^{-5} sec priority offsets a 6-dB difference in intensity; 2.3×10^{-5} sec offsets a 14-dB difference in intensity between the two ears.[3]

3k-12. Speech Power. The total radiated speech power, averaged over a 15-sec interval for a sample including both men and women at conversational levels used for telephone talking, has been estimated as 32 microwatts.

When measured at the face of a telephone transmitter, this power produces the sound pressure levels given in Table 3k-9 for different distances from the mouth of the speaker.[4]

TABLE 3k-9. AVERAGE SOUND PRESSURE LEVEL PRODUCED BY CONVERSATIONAL SPEECH AS A FUNCTION OF DISTANCE FROM LIPS TO MICROPHONE

	Distance, cm								
	Touching	0.5	1.0	2.5	5.0	10.0	25.0	50.0	100.0
Sound pressure level.....	104	102	99	95	90	85	78	72	66

A second source of variability lies in the essentially statistical distribution of speech power in time. If speech power is measured in successive $\frac{1}{8}$-sec intervals (a time slightly shorter than a syllable and slightly longer than a phoneme), a distribution is obtained with the mean values given in Table 3k-9 and variability that can be

---

[1] H. Wallach, Ueber die Wahrnehmung der Schallrichtung, *Psychol. Forsch.* **22**, 238–266 (1938).

[2] H. Wallach, E. B. Newman, and M. R. Rosenzweig, The Precedence Effect in Sound Localization, *Am. J. Psychol.* **62**, 313–336 (1949).

[3] J. H. Shaxby and F. H. Gage, Studies in the Localization of Sound. A. The Localization of Sounds in the Median Plane: An Experimental Investigation of the Physical Processes Concerned, *Med. Research Council (Brit.) Spec. Rept. Ser.* no. 166 (1932), 32 pp.

[4] M. H. Abrams, S. J. Goffard, J. Miller, F. H. Sanford, and S. S. Stevens, The Effect of Microphone Position on the Intelligibility of Speech in Noise, *OSRD Rept.* 4023 (1944), 16 pp.

attributed to time sampling equal to a standard deviation of 7.0 dB.[1] The distribution is badly skewed, so that the value 7.0 dB indicates only a rough order of magnitude. The variability is also greater when particular frequency bands are measured.

A third source of variability is the variation in effort expended by the person who is talking. As a rough approximation, a raised voice level is 6 dB above conversational level, the loudest level that can be maintained is 12 dB above conversational level, and the loudest shout is 18 dB above conversational level. In the other direction, a whisper may be 20 dB below conversational level.

3k-13. Speech Sounds

TABLE 3k-10. CHARACTERISTICS OF SOUNDS IN GENERAL AMERICAN SPEECH

Symbol	Example	Power,* dB re long time average†	Relative frequency of sound, %‡	Formant frequencies for men and women¶					
				First		Second		Third	
				M	W	M	W	M	W
u	cool	+0.6	1.60	300	370	870	950	2,240	2,670
ʊ	cook	+2.3	0.69	440	470	1,020	1,160	2,240	2,680
o	conc	+2.5	0.33	500	. . .	820			
ɔ	talk	+4.1	1.26	570	590	840	920	2,410	2,710
ɒ	cloth }	+3.7	{2.81	730	850	1,090	1,220	2,440	2,810
ɑ	calm }		0.49}						
a	ask }	+2.5	3.95	660	860	1,720	2,050	2,410	2,850
æ	bat }								
ɛ	bet	+1.6	3.44	530	610	1,840	2,330	2,480	2,990
e	tape	+1.4	1.84						
ɪ	bit	0.0	8.53	390	430	1,990	2,480	2,550	3,070
i	beet	0.0	2.12	270	310	2,290	2,790	3,010	3,310
ɚ	bird	−0.5	0.53	490	500	1,350	1,640	1,690	1,960
ə	sofa	. . . . .	4.63						
ʌ	bun	+2.9	2.33	640	760	1,190	1,400	2,390	2,780
eɪ	laid	+1.4	see e						
aɪ	bite	+2.5	1.59						
ju	you	+0.6	0.31						
oʊ	soap	+2.5	1.30						
aʊ	about	+2.3	0.59						
ɔɪ	boil	+3.0	0.09						

* The power measurements do not represent the peak instantaneous power but the average over the sustained portion of the phoneme where such a period can be defined. In this case, as with the formant frequencies, the absolute values are highly variable, but intercomparisons among the various sounds are generally more reliable.

† H. Fletcher, "Speech and Hearing in Communication," p. 86, D. Van Nostrand Company, Inc., Princeton, N.J., 1953.

‡ G. Dewey, "Relative Frequency of English Speech Sounds," Harvard University Press, Cambridge, Mass., 1923.

¶ E. G. Richardson, ed., "Technical Aspects of Sound," pp. 215–217, Elsevier Press, Inc., New York 1953.

[1] H. K. Dunn and S. D. White, Statistical Measurements on Conversational Speech, *J. Acoust. Soc. Am.* **11**, 278–288 (1940).

TABLE 3k-10. CHARACTERISTICS OF SOUNDS IN GENERAL
AMERICAN SPEECH (*Continued*)

Symbol	Example	Power,* dB re long time average†	Relative frequency of sound, %‡	Formant frequencies for men and women ¶			
				First	Second	Third	Fourth
l	lip	−3.0	3.74	450	1,000	2,550	2,950
m	me	−5.8	2.78	140	1,250	2,250	2,750
n	nip	−7.4	7.24	140	1,450	2,300	2,750
ŋ	sing	−4.4	0.96	140	2,350	2,750	
w	we	0.0	2.08				
r	rip	−1.0	6.35	500	1,350	1,850	3,500
j	yes	0.0	0.60	270	2,040		
p	pie	−15.2	2.04	...	800	1,350	
t	tie	−11.2	7.13	...	1,700	2,450	
k	key	−11.9	2.71	...	Variable		
b	by	−14.6	1.81	140	800	1,350	
d	die	−14.6	4.31	140	1,700	2,450	
g	guy	−11.2	0.74	140	Variable		
v	vie	−12.2	2.28	140	1,150	2,500	3,650
f	foe	−16.0	1.84	...	1,150	2,500	3,650
θ	thin	−23.0	0.37	...	1,450	2,550	
ð	then	−12.6	3.43	140	1,450	2,550	
s	sip	−11.0	4.55	...	2,000	2,700	
z	is	−11.0	2.97	140	2,000	2,700	
ʃ	shy	−4.0	0.82	...	2,150	2,650	
ʒ	measure	−10.0	0.05	140	2,150	2,650	
h	hit	−13.0	1.81				
tʃ	chop	−6.8	0.52				
dʒ	Joe	−9.4	0.44				

* The power measurements do not represent the peak instantaneous power but the average over the sustained portion of the phoneme where such a period can be defined. In this case, as with the formant frequencies, the absolute values are highly variable, but intercomparisons among the various sounds are generally more reliable.

† H. Fletcher, "Speech and Hearing in Communication," p. 86, D. Van Nostrand Company, Inc., Princeton, N.J., 1953.

‡ G. Dewey, "Relative Frequency of English Speech Sounds," Harvard University Press, Cambridge, Mass., 1923.

¶ E. G. Richardson, ed., "Technical Aspects of Sound," pp. 215–217, Elsevier Press, Inc., New York, 1953.

3k-14. Articulation Index. The articulation index is a set of numbers that makes possible the prediction of the efficiency of some types of voice-communication systems by the addition of suitably chosen values. The operations involve (1) dividing the speech spectrum into a series of bands having an equal possible contribution ΔA to the total efficiency, and (2) determining what proportion of the ΔA each band will contribute under the particular noise and speech conditions being tested.

Under (1) it is customary to use no more than 20 such bands. The frequency limits of 20 such bands are given in Table 3k-11.

TABLE 3k-11. TWENTY FREQUENCY BANDS CONTRIBUTING EQUALLY TO
EFFICIENCY OF SPEECH COMMUNICATION*

Band No.	Frequency range	Band No.	Frequency range	Band No.	Frequency range
1	395	8	1,250–1,425	15	2,930–3,285
2	395–540	9	1,425–1,620	16	3,285–3,700
3	540–675	10	1,620–1,735	17	3,700–4,200
4	675–810	11	1,735–2,075	18	4,200–4,845
5	810–950	12	2,075–2,335	19	4,845–5,790
6	950–1,095	13	2,335–2,620	20	5,790
7	1,095–1,250	14	2,620–2,930		

* H. Fletcher, "Speech and Hearing in Communication," D. Van Nostrand Company, Inc., Princeton, N.J., 1953.

For conditions where substantial wide-band noise is present, the second requirement may be approximated by the formula

$$w_i = \tfrac{1}{30}(S_i - N_i + 6) \tag{3k-4}$$

in which w_i is a weight having a maximum value of 1.0, S_i is the signal level in band i in decibels, N_i is the noise level in the same band i in decibels referred to the same base as S_i.[1]

TABLE 3k-12. ARTICULATION SCORES AS A FUNCTION OF ARTICULATION INDEX*

Articulation index	CVC syllables, %	Monosyllabic words (PB lists), %
0.10	7	7
0.20	22	22
0.30	38	40
0.40	55	61
0.50	68	77
0.60	79	87
0.70	87	93
0.80	93	96
0.90	96	98
1.00	98	99

* E. G. Richardson, ed., "Technical Aspects of Sound," Elsevier Press, Inc., New York, 1953.   "CVC syllables" are estimated from sets of words that vary the initial consonants, the vowel, and the final consonant separately.   PB words are lists of monosyllables phonetically balanced so that the proportion of phonemes roughly equals that in general speech.

The articulation index A is then described by the summation

$$A = \frac{1}{n} \sum_{i=1}^{i=n} w_i \tag{3k-5}$$

Articulation scores are related to the articulation index according to the Table 3k-12.

[1] N. R. French and J. C. Steinberg, Factors Governing the Intelligibility of Speech Sounds, *J. Acoust. Soc. Am.* **19,** 90–119 (1947).

31. Classical Dynamical Analogies

HARRY F. OLSON

RCA Laboratories

Analogies are useful when it is desired to compare an unfamiliar system with one that is better known. The relations and actions are more easily visualized, the mathematics more readily applied, and the analytical solutions more readily obtained in the familiar system. Analogies make it possible to extend the line of reasoning into unexplored fields. In view of the tremendous amount of study which has been directed toward the solution of circuits, particularly electric circuits, and the engineer's familiarity with electric circuits, it is logical to apply this knowledge to the solutions of vibration problems in other fields by the same theory as that used in the solution of electric circuits. The objective in this section is the establishment of analogies between electrical, mechanical, and acoustical systems.

31-1. Resistance. *Electric Resistance.* Electric energy is changed into heat by the passage of an electric current through an electric resistance. Electric resistance R_E, in abohms, is defined as

$$R_E = \frac{e}{i} \tag{31-1}$$

where e = voltage across the electric resistance, abvolts
i = current through the electric resistance, abamp

Mechanical Rectilineal Resistance. Mechanical rectilineal energy is changed into heat by a rectilinear motion which is opposed by mechanical rectilineal resistance (friction). Mechanical rectilineal resistance (termed mechanical resistance when there is no ambiguity) R_M, in mechanical ohms, is defined as

$$R_M = \frac{f_M}{u} \tag{31-2}$$

where f_M = applied mechanical force, dynes
u = velocity at the point of application of the force, cm/sec

Mechanical Rotational Resistance. Mechanical rotational energy is changed into heat by a rotational motion which is opposed by a rotational resistance (rotational friction). Mechanical rotational resistance (termed rotational resistance when there is no ambiguity) R_R, in rotational ohms, is defined as

$$R_R = \frac{f_R}{\Omega} \tag{31-3}$$

where f_R = applied torque, dyne-cm
Ω = angular velocity about the axis at the point of the torque, radians/sec

Acoustic Resistance. Acoustic energy is changed into heat either by a motion in a fluid which is opposed by acoustic resistance due to a fluid resistance incurred by viscosity or by the radiation of sound. Acoustic resistance R_A, in acoustical ohms, is defined as

$$R_A = \frac{p}{U} \tag{31-4}$$

where p = pressure, dynes/cm^2
U = volume velocity, cm^3/sec

31-2. Inductance, Mass, Moment of Inertia, Inertance. *Inductance.* Electromagnetic energy is associated with inductance. Inductance is the electric-circuit element that opposes a change in current. Inductance L, in abhenrys, is defined as

$$e = L \frac{di}{dt} \tag{31-5}$$

where e = voltage, emf, or driving force, abvolts
$\frac{di}{dt}$ = rate of change of current, abamp/sec

Mass. Mechanical rectilineal inertial energy is associated with mass in the mechanical rectilineal system. Mass is the mechanical element which opposes a change in velocity. Mass m, in grams, is defined as

$$f_M = m \frac{du}{dt} \tag{31-6}$$

where $\frac{du}{dt}$ = acceleration, cm/sec^2
f_M = driving force, dynes

Moment of Inertia. Mechanical rotational energy is associated with moment of inertia in the mechanical rotational system. Moment of inertia is the rotational element which opposes a change in angular velocity. Moment of inertia I, in gram (centimeter)^2, is defined as

$$f_R = I \frac{d\Omega}{dt} \tag{31-7}$$

where $\frac{d\Omega}{dt}$ = angular acceleration, radians/sec^2
f_R = torque, dyne-cm

Inertance. Acoustic inertial energy is associated with inertance in the acoustic system. Inertance is the acoustic element which opposes a change in volume velocity. Inertance M, in grams per (centimeter)^4, is defined as

$$p = M \frac{dU}{dt} \tag{31-8}$$

where $\frac{dU}{dt}$ = rate of change of volume velocity, cm^3/sec^2
p = driving pressure, dynes/cm^2

31-3. Electric Capacitance, Rectilineal Compliance, Rotational Compliance, Acoustic Capacitance. *Electric Capacitance.* Electric capacitance is associated with capacitance. Electric capacitance is the electric-circuit element which opposes a change in voltage. Electric capacitance C_E, in abfarads, is defined as

$$i = C_E \frac{de}{dt} \tag{31-9}$$

$$e = \frac{1}{C_E} \int i\, dt = \frac{Q}{C_E} \tag{31-10}$$

where Q = charge on the electrical capacitance, abcoulombs
e = emf, abvolts

Rectilineal Compliance. Mechanical rectilineal potential energy is associated with the compression of a spring or compliant element. Rectilineal compliance is the mechanical element which opposes a change in the applied force. Rectilineal compliance (termed compliance when there is no ambiguity) C_M, in centimeters per dyne, is defined as

$$f_M = \frac{x}{C_M} \tag{31-11}$$

where x = displacement, cm
f_M = applied force, dynes

Rotational Compliance. Mechanical rotational potential energy is associated with the twisting of a spring or compliant element. Rotational compliance is the mechanical element that opposes a change in the applied torque. Rotational compliance C_R, in radians per centimeter per dyne, is defined as

$$f_R = \frac{\phi}{C_R} \tag{31-12}$$

where ϕ = angular displacement, radians
f_R = applied torque, dyne-cm

Acoustic Capacitance. Acoustic potential energy is associated with the compression of a fluid or a gas. Acoustic capacitance is the acoustic element which opposes a change in the applied pressure. The acoustic capacitance C_A, in (centimeters)^5 per dyne, is defined as

$$p = \frac{X}{C_A} \tag{31-13}$$

where X = volume displacement, cm^3
p = pressure, dynes/cm^2

31-4. Representation of Electrical, Mechanical Rectilineal, Mechanical Rotational, and Acoustical Elements. Electrical, mechanical rectilineal, mechanical rotational,

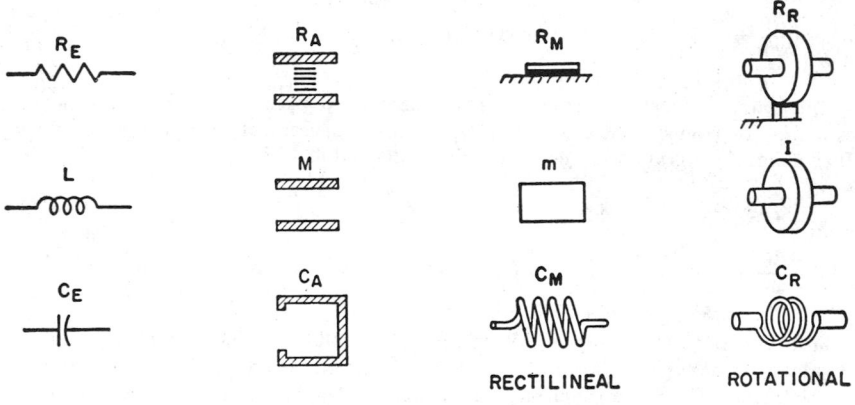

ELECTRICAL   ACOUSTICAL   MECHANICAL

FIG. 31-1. Graphical representation of the three basic elements in electrical, mechanical rectilineal, mechanical rotational, and acoustical systems.

and acoustical elements have been defined in the preceding sections. Figure 31-1 illustrates schematically the three elements in each of the four systems.

The electrical elements, electric resistance, inductance, and electric capacitance, are represented by the conventional symbols.

Mechanical rectilineal resistance is represented by sliding friction which causes dissipation. Mechanical rotational resistance is represented by a wheel with a sliding-friction brake which causes dissipation. Acoustic resistance is represented by narrow slits which cause dissipation due to viscosity when fluid is forced through the slits. These elements are analogous to electric resistance in the electrical system.

Inertia in the mechanical rectilineal system is represented by a mass. Moment of inertia in the mechanical rotational system is represented by a flywheel. Inertance in the acoustical system is represented as the fluid contained in a tube in which all the particles move with the same phase when actuated by a force due to pressure. These elements are analogous to inductance in the electrical system.

Compliance in the mechanical rectilineal system is represented as a spring. Rotational compliance in the mechanical rotational system is represented as a spring. Acoustic capacitance in the acoustical system is represented as a volume which acts as a stiffness or spring element. These elements are analogous to electric capacitance in the electrical system.

Table 31-1 shows the quantities, units, and symbols in the four systems.

31-5. Description of Systems of One Degree of Freedom. Electrical, mechanical rectilineal, mechanical rotational, and acoustical systems of one degree of freedom are shown in Fig. 31-2. In one degree of freedom the activity in every element of the

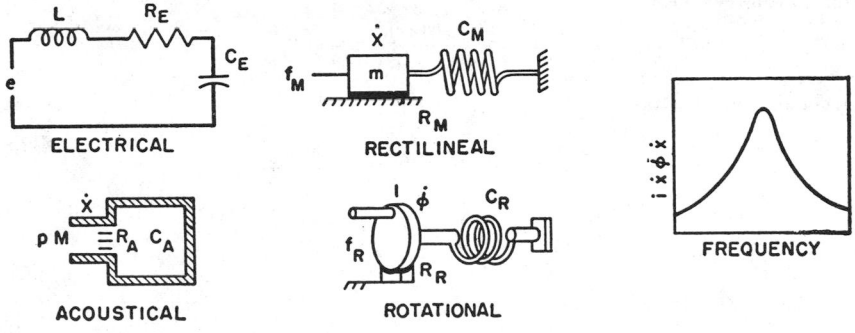

FIG. 31-2. Electrical, mechanical rectilineal, mechanical rotational, and acoustical systems of one degree of freedom and the current, velocity, angular velocity and volume velocity response characteristics.

system can be expressed in terms of one variable. In the electrical system an electromotive force e acts upon an inductance L, an electric resistance R_E, and an electric capacitance C_E connected in series. In the mechanical rectilineal system a driving force f_M acts upon a particle of mass m fastened to a spring of compliance C_M and sliding upon a plate with a frictional force which is proportional to the velocity and designated as the mechanical rectilineal resistance R_M. In the mechanical rotational system a driving torque f_R acts upon a flywheel of moment of inertia I connected to a spring or rotational compliance C_R and the periphery of the wheel sliding against a brake with a frictional force which is proportional to the velocity and designated as the mechanical rotational resistance R_R. In the acoustical system, an impinging sound wave of pressure p acts upon an inertance M and an acoustic resistance R_A comprising the air in the tubular opening which is connected to the volume or acoustical capacitance C_A. The acoustic resistance R_A is due to viscosity.

The differential equations describing the four systems of Fig. 31-2 are as follows:
Electrical

$$L\ddot{q} + R_E\dot{q} + \frac{Q}{C_E} = E\epsilon^{j\omega t} \qquad (31\text{-}14)$$

Mechanical rectilineal

$$m\ddot{x} + R_M\dot{x} + \frac{x}{C_M} = F_M\epsilon^{j\omega t} \tag{31-15}$$

Mechanical rotational

$$I\ddot{\phi} + R_R\dot{\phi} + \frac{\phi}{C_R} = F_R\epsilon^{j\omega t} \tag{31-16}$$

Acoustical

$$M\ddot{X} + R_A\dot{X} + \frac{X}{C_A} = P\epsilon^{j\omega t} \tag{31-17}$$

E, F_M, F_R, and P are the amplitudes of the driving forces in the four systems. $E\epsilon^{j\omega t} = e$, $F_M\epsilon^{j\omega t} = f_M$, $F_R\epsilon^{j\omega t} = f_R$ and $P\epsilon^{j\omega t} = p$.

The steady-state solutions of Eqs. (31-14) to (31-17) are:

Electrical

$$\dot{q} = i = \frac{E\epsilon^{j\omega t}}{R_E + j\omega L - (j/\omega C_E)} = \frac{e}{Z_E} \tag{31-18}$$

Mechanical rectilineal

$$\dot{x} = \frac{F\epsilon^{j\omega t}}{R_M + j\omega m - (j/\omega C_M)} = \frac{f_M}{Z_M} \tag{31-19}$$

Mechanical rotational

$$\dot{\phi} = \frac{F\epsilon^{j\omega t}}{R_R + j\omega I - (j/\omega C_R)} = \frac{f_R}{Z_R} \tag{31-20}$$

Acoustical

$$\dot{X} = \frac{P\epsilon^{j\omega t}}{R_A + j\omega M - (j/\omega C_A)} = \frac{p}{Z_A} \tag{31-21}$$

The vector electric impedance is

$$Z_E = R_E + j\omega L - \frac{j}{\omega C_E} \tag{31-22}$$

The vector mechanical rectilineal impedance is

$$Z_M = R_M + j\omega m - \frac{j}{\omega C_M} \tag{31-23}$$

The vector mechanical rotational impedance is

$$Z_R = R_R + j\omega I - \frac{j}{\omega C_R} \tag{31-24}$$

The vector acoustic impedance is

$$Z_A = R_A + j\omega M - \frac{j}{\omega C_A} \tag{31-25}$$

31-6. Applications of Classical Electrodynamical Analogies. The fundamental principles relating to electrical, mechanical rectilineal, mechanical rotational, and acoustical analogies have been established in the preceding sections. Employing these fundamental principles, the vibrations produced in mechanical and acoustical systems owing to impressed forces can be solved as follows: Draw the electrical network which is analogous to the problem to be solved; solve the electrical network by conventional electrical circuit theory; convert the electrical answer into the original system. In this procedure any problem involving vibrating systems is reduced to the solution of an electrical network. In the illustrations in the preceding sections, the elements in the electrical network have been labeled $r_{E,L}$ and C_E. However, when analogies are used in actual practice, the conventional procedure is to label the elements in the analogous electrical network with r_M, M, and C_M for a

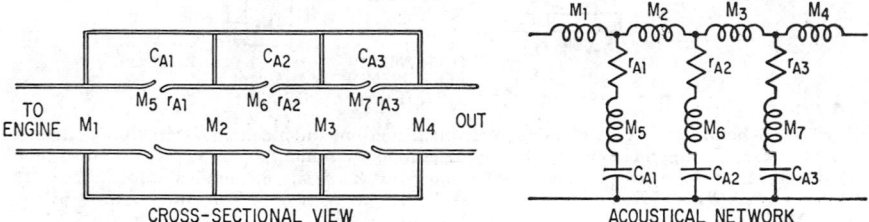

FIG. 31-3. Cross-sectional view and acoustical network of an automobile muffler. In the acoustical network: M_1, M_2, M_3, and M_4, the inertances of the series elements; r_{A1}, r_{A2}, and r_{A3}, the acoustical resistances of the shunt elements; M_5, M_6, and M_7, the inertances of the shunt elements; C_{A1}, C_{A2}, and C_{A3}, the acoustical capacitances of the shunt elements. (*After Olson, "Dynamical Analogies," D. Van Nostrand Company, Inc., Princeton, N.J., 1959.*)

mechanical rectilineal system; with r_R, I, and C_R for a mechanical rotational system; and with r_A, M, and C_A for an acoustical system. This procedure will be followed in this section in labeling the elements in the analogous electrical network. The customary procedure is to label the network with the caption mechanical network or rotational network or acoustical network as the case may be. When there is only one path, the term circuit will be used instead of network. A complete treatment of the examples of the use of analogies in the solution of problems in mechanical and acoustical systems is beyond the scope of this section. However, a few typical examples will serve to illustrate the principles and method.

Acoustical—Automobile Muffler. The sound output from the exhaust of an automobile engine contains all audible frequencies in addition to frequencies below and above the audible range. The purpose of a muffler is to reduce the sound output in the audible frequency range without increasing the exhaust back pressure.

By the application of acoustical principles employing analogies improved mufflers have been developed in which the following advantages have been obtained: smaller size, higher attenuation in the audible frequency range, and reduction of back pressure at the engine. A cross-sectional view of the improved muffler is shown in Fig. 31-3. The acoustical network shows that the system is essentially a low-pass acoustical filter. The main channel is of the same diameter as the exhaust pipe. Therefore, there is no increase in the direct flow of exhaust gases as compared with a plain pipe. In order not to impair the efficiency of the engine, the muffler should not increase the acoustical impedance to subaudible frequencies. The system of Fig. 31-3 can be designed so that the subaudible frequencies are not attenuated and at the same time high attenuation is introduced in the audible frequency range.

The terminations at the two ends of the network are not ideal. Therefore, it is necessary to use shunt arms tuned to different frequencies in the low-frequency range. Acoustical resistance is obtained by employing slit-type openings into the side chambers.

In a development of this kind, the frequency spectrum of the sound which issues from the exhaust is usually determined. From these data the amount of suppression required for each part of the audible frequency range can be ascertained. The acoustical network can be determined from these data and the terminating acoustical

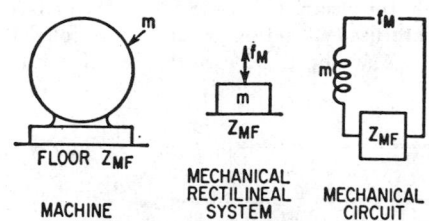

Fig. 31-4. Schematic view, mechanical rectilineal system, and mechanical circuit of a machine mounted directly upon the floor. In the mechanical circuit: f_M, the vibrating force developed by the machine; m, the mass of the machine; Z_{MF}, the mechanical impedance of the floor. (*After Olson, "Dynamical Analogies," D. Van Nostrand Company, Inc., Princeton, N.J., 1959.*)

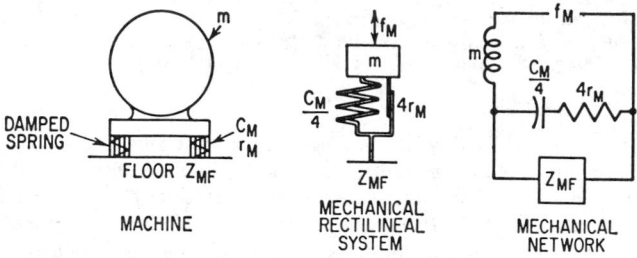

Fig. 31-5. Schematic view, mechanical rectilineal system, and mechanical network of a machine mounted upon a vibration isolating system. In the mechanical network: f_M, the vibrating force developed by the machine; m, the mass of the machine; C_M, the compliance of one of the four spring mounts; r_M, the mechanical rectilineal resistance of one of the spring mounts. (*After Olson, "Dynamical Analogies," D. Van Nostrand Company, Inc., Princeton, N.J., 1959.*)

networks. In general, changes are required to compensate for the approximations. In this empirical work the acoustical network serves as the guide in directing the appropriate changes.

Mechanical Rectilineal—Machine Vibration Isolator. The vibration of a machine is transmitted from its supports to all parts of the surrounding building structure. In many cases, the vibrations are so intense as to be intolerable. The reduction of the transmission of machinery vibrations is one of the most common problems in noise control. For these conditions, the solution of the problem is to provide suitable vibrational isolation between the machine and the floor upon which it is placed.

A machine mounted directly on the floor is shown in Fig. 31-4. The mechanical rectilineal system and the mechanical circuit for vertical vibrations are shown in Fig. 31-4. The driving force f_M is due to the vibrations of the machine. The mechanical circuit shows that the only isolation in the system of Fig. 3l-4 is due to the mass of the machine.

In the simple isolating system of Fig. 3l-5 the machine is mounted on springs with mechanical resistance added to serve as damping. The compliance and mechanical resistance of each support are C_M and r_M. Since there are four supports, these values become $C_M/4$ and $4r_M$ in the mechanical rectilineal system and the mechanical network for vertical vibrations. The mechanical network depicts the action of the shunt circuit $C_M r_M$ in reducing the force of the vibration transmitted to the floor z_{MF}.

Mechanical Rotational—Vibration Damper. In reciprocating engines and other rotating machinery, rotational vibrations of large amplitudes occur at certain speeds. These rotational vibrations are sometimes of such high amplitude that the shafts

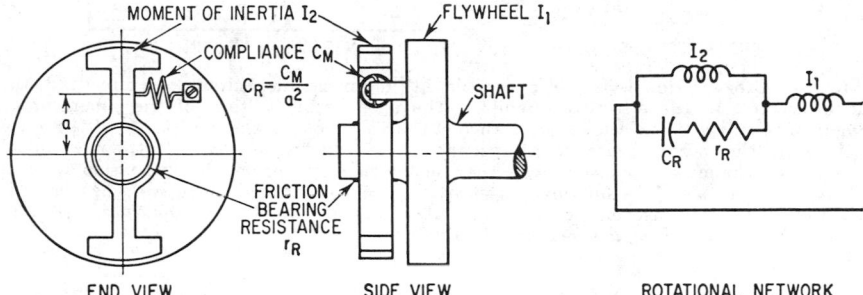

END VIEW   SIDE VIEW   ROTATIONAL NETWORK

FIG. 3l-6. End and side views and the rotational network of a vibration damper. In the rotational network: I_1, the moment of inertia of the flywheel; I_2, the moment of inertia of the damper; C_R, the rotational compliance of the damper; r_R, the mechanical rotational resistance between the damper and the shaft. (*After Olson, "Dynamical Analogies,"* D. Van Nostrand Company, Inc., Princeton, N.J., 1959.)

will fail after a few hours of operating. A number of various rotational dampers have been developed for reducing these rotational vibrations. A typical example of a vibration damper used to control the vibrations of the flywheel is shown in Fig. 3l-6. The damper consists of a rotational element having a moment of inertia I_2 rotating on a shaft with a mechanical rotational resistance r_R between the inertia element and shaft. The inertial element is coupled to the flywheel by means of a spring of compliance C_M. The rotational compliance is $C_R = C_M/a^2$, where a is the radius at the point of attachment of the spring with respect to the center line of the shaft. Referring to the rotational network it will be seen that the rotational damper forms a shunt mechanical rotational system. The shunt rotational circuit $C_R r_R I_2$ is tuned to the frequency of the vibration. Since the mechanical rotational impedance of the shunt resonant rotational circuit is very high at the resonant frequency, the angular velocity (or amplitude) of vibration of the flywheel will be reduced. A consideration of the rotational network illustrates the principle of the device.

Electrical Mechanical—Direct Radiator Dynamic Loudspeaker. The direct radiator dynamic loudspeaker shown in Fig. 3l-7 is almost universally used for radio, phonograph, television, and other small-scale sound reproductions.

The mechanical circuit of the loudspeaker is shown in Fig. 3l-7. The mechanical rectilineal impedance at the voice coil, where a force f_M is applied, can be determined from the constants of the elements of the mechanical circuit. The mass m_2 and the mechanical resistance r_{M2} of the air load can be obtained from Sec. 3i-2 on the acoustic impedance of vibrating pistons.

The electrical circuit of the loudspeaker is also shown in Fig. 3l-7. The motional

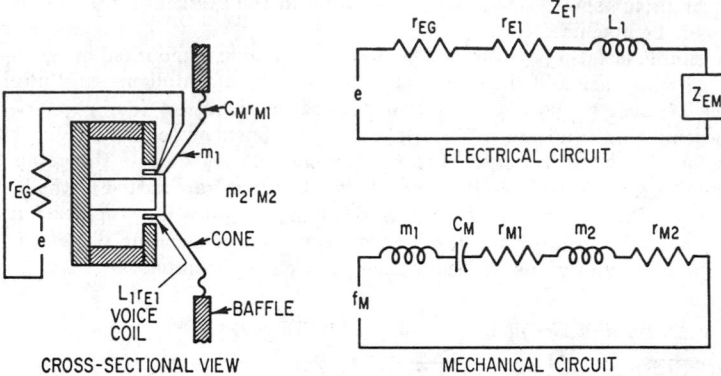

CROSS-SECTIONAL VIEW   MECHANICAL CIRCUIT

Fig. 31-7. Cross-sectional view, electrical circuit, and mechanical circuit of a direct radiator loudspeaker. In the electrical circuit: e, the open-circuit voltage of the generator or vacuum tube; r_{EG}, the electrical resistance of the voice coil; L, the inductance of the voice coil; Z_{EM}, the motional electrical impedance of the driving system. In the mechanical circuit: m_1, the mass of the cone; r_{M1}, the mechanical resistance of the suspension system; C_M, the compliance of the suspension system; m_2, the mass of the air load; r_{M2}, the mechanical rectilineal resistance of the air load. (*After Olson, "Dynamical Analogies," D. Van Nostrand Company, Inc., Princeton, N.J., 1959.*)

electrical impedance in the electrical circuit is given by

$$z_{EM} = \frac{(Bl)^2}{z_{MT}} \tag{31-26}$$

where z_{EM} = motional electrical impedance, abohms
   B = flux density in air, gauss
   l = length of conductor in voice coil
   z_{MT} = mechanical impedance at location f_M in mechanical circuit, mechanical ohms

The mechanical driving force is given by

$$f_M = Bli \tag{31-27}$$

where f_M = driving force, dynes
   i = current in voice coil, abamp

The velocity can be determined from the mechanical circuit of Fig. 31-7 and the following equation:

$$\dot{x} = \frac{f_M}{z_{MT}} \tag{31-28}$$

where $\dot{x}$ is the velocity in centimeters per second.   The sound output is given by

$$P = r_M \dot{x}^2 \tag{31-29}$$

where P = sound power output, ergs/sec
   r_M = mechanical ohms
   $\dot{x}$ = velocity of cone from Eq. (31-28)

The object is to select the constants so that the power output as given by Eq. (31-29) is practically independent of the frequency over the desired frequency range.

TABLE 31-1. QUANTITIES, UNITS, AND SYMBOLS FOR ELECTRICAL, MECHANICAL
RECTILINEAL, MECHANICAL ROTATIONAL, AND ACOUSTICAL ELEMENTS

Electrical			Mechanical rectilineal		
Quantiy	Unit	Symbol	Quantity	Unit	Symbol
Electromotive force........	Volts $\times 10^8$	e	Force	Dynes	f_M
Charge or quantity	Coulombs $\times 10^{-1}$	Q	Linear displacement	Centimeters	x
Current........	Amperes $\times 10^{-1}$	i	Linear velocity	Centimeters per second	$\dot{x}$ or u
Electric impedance	Ohms $\times 10^9$	Z_E	Mechanical impedance	Mechanical ohms	Z_M
Electric resistance	Ohms $\times 10^9$	R_E	Mechanical resistance	Mechanical ohms	R_M
Electric reactance	Ohms $\times 10^9$	X_E	Mechanical reactance	Mechanical ohms	X_M
Inductance....	Henrys $\times 10^9$	L	Mass	Grams	m
Electric capacitance	Farads $\times 10^{-9}$	C_E	Compliance	Centimeters per dyne	C_M
Power.........	Ergs per second	P_E	Power	Ergs per second	P_M

Mechanical rotational			Acoustical		
Quantity	Unit	Symbol	Quantity	Unit	Symbol
Torque........	Dyne-centimeters	f_R	Pressure	Dynes per square centimeter	p
Angular displacement	Radians	ϕ	Volume displacement	Cubic centimeters	X
Angular velocity	Radians per second	ϕ or Ω	Volume velocity	Cubic centimeters per second	$\dot{X}$ or U
Rotational impedance	Rotational ohms	Z_R	Acoustic impedance	Acoustic ohms	Z_A
Rotational resistance	Rotational ohms	R_R	Acoustic resistance	Acoustic ohms	R_A
Rotational reactance	Rotational ohms	X_R	Acoustic reactance	Acoustic ohms	X_A
Moment of inertia	(Gram) (centimeter)^2	I	Inertance	Grams per (centimeter)^4	M
Rotational compliance	Radians per dyne per centimeter	C_R	Acoustic capacitance	(Centimeter)^5 per dyne	C_A
Power.........	Ergs per second	P_R	Power	Ergs per second	P_A

3m. Mobility Analogy

HARRY F. OLSON

RCA Laboratories

The analogies that have been presented and considered in Sec. 3l have been formal ones owing to the similarity of the differential equations of electrical, mechanical and acoustical vibrating systems. For this reason these analogies have been termed the classical impedance analogies; they are, however, not the only ones possible of development for useful applications. For example, mechanical impedance has been defined by some authors—in addition to the ratio of force to velocity as developed in Sec. 3l— as the ratio of pressure to velocity, the ratio of force to displacement, and the ratio of pressure to displacement. During the past three decades the developments in the field of analogies have been reported in publications[1] by many investigators. In this connection a useful analogy, developed by Firestone and designated by him as the "mobility analogy," has been employed on a wide scale to solve problems in mechanical vibrating systems. In the mobility analogy mechanical mobility is defined as the complex ratio of velocity to force. Although the mobility analogy can be applied and used with all types of vibrating systems, its most direct and useful application is in the field of mechanical vibrating systems. Therefore, in order to make the subject of analogies complete in this handbook, it seems logical to include the mobility analogy. Accordingly, it is the purpose of this chapter to develop the mobility analogy, particularly as applied to mechanical rectilineal systems.[2]

3m-1. Mechanical Rectilineal Mobility. Mechanical rectilineal mobility is the inverse of mechanical rectilineal impedance. Mechanical rectilineal mobility z_I, in mechanical mhos, is defined as the complex ratio of linear velocity to linear force as follows:

$$z_I = \frac{v}{f_M} \tag{3m-1}$$

where v = velocity, cm/sec
f_M = force, dynes

It will be evident that a mechanical element in the mechanical mobility sense is analogous to the electric element if velocity difference across the mechanical element is analogous to the voltage difference across the electric element and if the force through the mechanical element is analogous to the electric current through the electric element.

---

[1] See the end of Section 3 for a list of references.

[2] The considerations in this section will be confined to mechanical rectilineal systems. The mobility analogy is equally applicable to mechanical rotational systems. In this connection mechanical rectilineal and mechanical rotational systems are not sufficiently different to warrant a separate treatment for the mechanical rotational system, particularly in view of the fact that fundamental aspects of the two systems have been considered from the classical impedance analogy viewpoint in this book.

Mechanical rectilineal mobility z_I, in mechanical mhos, is a complex quantity and may be written as follows:

$$z_I = r_I + jx_I \qquad (3m\text{-}2)$$

where r_I = responsivity, mechanical mhos

x_I = excitability, mechanical mhos

3m-2. Responsivity (Mobility Resistance). In the mechanical rectilineal mobility system mechanical rectilineal responsivity (mobility resistance) r_I, in mechanical mhos, is defined as

$$r_I = \frac{v}{f_M} = \frac{1}{r_M} \qquad (3m\text{-}3)$$

where v = velocity, cm/sec

f_M = force, dynes

r_M = mechanical impedance, mechanical ohms

3m-3. Mass (Mobility Capacitance). In the mechanical rectilineal mobility system the mass (mobility capacitance) m_I, in grams, is analogous to electric capacitance C_E.

The mechanical rectilineal excitability x_I of a mass (mobility capacitance), in mechanical mhos, is defined as

$$x_I = \frac{1}{\omega m_I} \qquad (3m\text{-}4)$$

where $\omega = 2\pi f$

f = frequency, hertz

Equation (3m-4) shows that the mass (mobility capacitance) m_I in the mechanical rectilineal mobility system is analogous to electric capacitance C_E in the electric system.

Mass (mobility capacitance) m_I in the mechanical rectilineal mobility system may also be defined as follows:

$$f_M = m_I \frac{dv}{dt} \qquad (3m\text{-}5)$$

$$v = \frac{1}{m_I} \int f_M \, dt \qquad (3m\text{-}6)$$

In the electric system electric capacitance C_E may be defined as follows:

$$i = C_E \frac{de}{dt} \qquad (3m\text{-}7)$$

where i = electric current, abamp

C_E = electric capacitance, abfarads

e = electromotive force, abvolts

t = time, sec

$$e = \frac{1}{C_E} \int i \, dt \qquad (3m\text{-}8)$$

where i = current in abamperes.

It will be seen that Eqs. (3m-5) and (3m-6) in the mechanical rectilineal mobility system are analogous to Eqs. (3m-7) and (3m-8) in the electric system.

3m-4. Compliance (Mobility Inertia). In the mechanical rectilineal mobility system the compliance (mobility inertia) C_I, in centimeters per dyne, is analogous to electric inductance L.

The mechanical rectilineal excitability x_I of a compliance (mobility inertia), in mechanical mhos, is defined as

$$x_I = \omega C_I \qquad (3m\text{-}9)$$

where $\omega = 2\pi f$

f = frequency, Hz

Equation (3m-9) shows that compliance (mobility inertia) C_I, in centimeters per dyne, is analogous to inductance.

Compliance (mobility inertia) C_I in the mechanical rectilineal mobility system may also be defined as

$$v = C_I \frac{df_M}{dt} \tag{3m-10}$$

In the electric system inductance may be defined as

$$e = L \frac{di}{dt} \tag{3m-11}$$

where L = inductance in abhenrys.

It will be seen that Eq. (3m-10) in the mechanical rectilineal mobility system is analogous to Eq. (3m-11) in the electric system.

3m-5. Representation of Electrical and Mechanical Rectilineal Mobility Elements. Electric elements have been defined in Sec. 3l. Elements in the mechanical rectilineal mobility system have been described in this section.

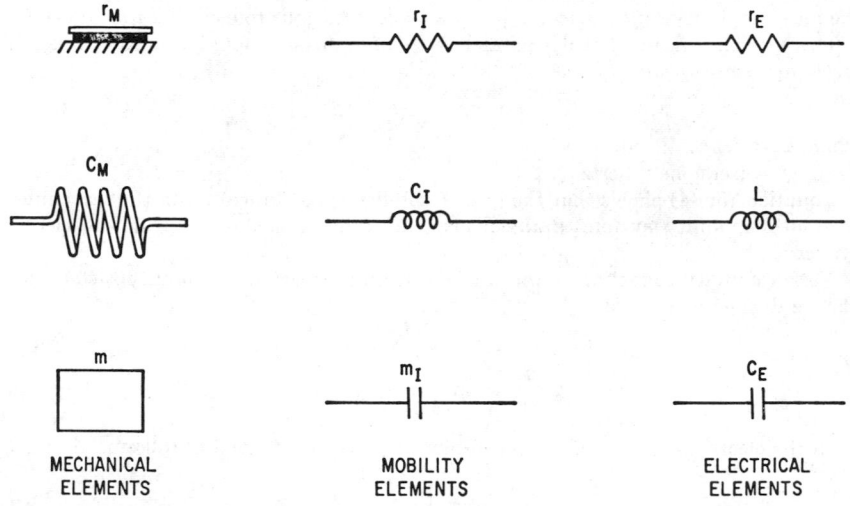

Fig. 3m-1. Graphical representation of the three basic elements in mechanical rectilineal, mobility, and electric systems.

r_M = mechanical rectilineal resistance	r_I = responsivity	r_E = electrical resistance
C_M = compliance	C_I = mobility inertia	L = inductance
m = mass	m_I = mobility capacitance	C_E = electric capacitance

(*After Olson, "Solutions of Engineering Problems by Dynamical Analogies," D. Van Nostrand Co., Princeton, N.J., 1966.*)

Figure 3m-1 illustrates schematically the mechanical elements and the analogous elements in the electric and mechanical rectilineal mobility systems.

Mechanical rectilineal resistance r_M in the mechanical rectilineal system is represented as sliding or viscous friction. Mechanical rectilineal responsivity (mobility resistance) r_I in the mechanical rectilineal mobility system is the reciprocal of mechanical rectilineal resistance r_M and is analogous to electrical resistance r_E.

Compliance C_M in the mechanical rectilineal system is represented as a spring. Compliance (mobility inertia) C_I in the mechanical rectilineal mobility system is analogous to inductance L in the electric system.

Mass m in the mechanical rectilineal system is represented as a mass or weight. Mass (mobility capacitance) m_I in the mechanical rectilineal mobility system is analogous to electric capacitance C_E in the electric system.

The electrical and the mechanical rectilineal quantities in the mobility system are shown in Table 3m-1. The units and the analogous elements and symbols also are shown in Table 3m-1.

3m-6. Mechanical Vibrating System Consisting of a Mass, Compliance, and Mechanical Resistance. The vibrating system[1] of one degree of freedom consisting of a mass, compliance, and mechanical resistance has been considered from the standpoint of the classical mechanical impedance analogy in Sec. 3l. It is the purpose of this section to consider the same mechanical vibrating system from the standpoint of the mechanical mobility analogy.[2]

TABLE 3m-1. CORRESPONDENCE BETWEEN ELECTRICAL AND MECHANICAL QUANTITIES IN THE MOBILITY SYSTEM

Electrical			Mechanical rectilineal mobility		
Quantity	Unit	Symbol	Quantity	Unit	Symbol
Electromotive force	Volts $\times 10^{-8}$	e	Velocity	Centimeters per second	$\dot{x}$ or v
Charge or quantity	Coulombs $\times 10^{-1}$	q	Impulse or momentum	Gram-centimeter per second	Q
Current	Amperes $\times 10^{-1}$	i	Force	Dynes	f_M
Electrical impedance	Ohms $\times 10^9$	z_E	Mechanical mobility	Mechanical mhos	z_I
Electrical resistance	Ohms $\times 10^9$	r_E	Responsivity	Mechanical mhos	r_I
Electrical reactance	Ohms $\times 10^9$	x_E	Excitability	Mechanical mhos	x_I
Inductance	Henrys $\times 10^9$	L	Compliance or mobility inertia	Centimeters per dyne	C_I
Electrical capacitance	Farads $\times 10^9$	C_E	Mass or mobility capacitance	Grams	m_I
Power	Joules per second	P_E	Power	Ergs per second	P_I

The mechanical system consisting of a mass, compliance, and mechanical resistance is shown in Fig. 3m-2A. The mechanical vibrating system may be rearranged to form the equivalent as shown in Fig. 3m-2B. From the mechanical vibrating system of Fig. 3m-2B it is a relatively simple matter to develop the mobility analogy of Fig. 3m-2C.

[1] The preceding paragraphs have been concerned with fundamental considerations. Therefore, the modifier rectilineal has been employed for the sake of accuracy. Since the remainder of this section will be concerned with applications of the mechanical rectilineal mobility, the modifier rectilineal will be dropped.

[2] In view of the fact that this section is concerned with mechanical systems, the modifier mechanical in relation to the mechanical mobility analogy is also superfluous and need not be used.

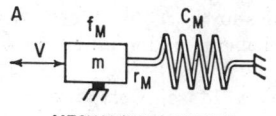

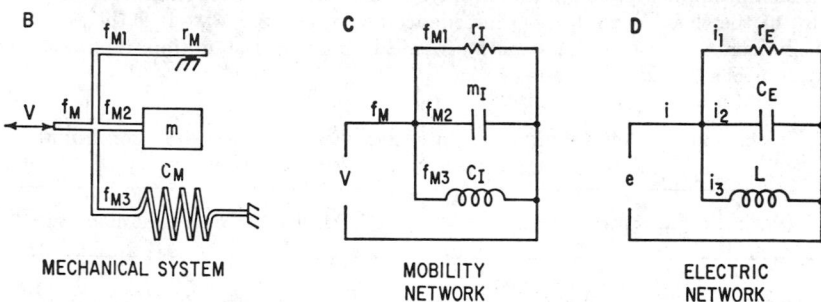

FIG. 3m-2. A mechanical vibrating system consisting of a mass, compliance, and mechanical resistance. *A*. Mechanical system. *B*. Mechanical system equivalent to the mechanical system of *A*. *C*. Mobility network of the mechanical system. *D*. Electric network analog of the mobility system. (*After Olson, "Solution of Engineering Problems by Dynamical Analogies," D. Van Nostrand Company, Princeton, N.J., 1966.*)

The sum of the forces through the three branches of the mobility network[1] of Fig. 3m-2*C* is

$$f_M = f_{M1} + f_{M2} + f_{M3} \tag{3m-12}$$

where

$$f_{M1} = \frac{v}{r_I} \tag{3m-13}$$

$$f_{M2} = m_I \frac{dv}{dt} \tag{3m-14}$$

$$f_{M3} = \frac{1}{C_I} \int v \, dt \tag{3m-15}$$

From the sum of Eqs. (3m-13) to (3m-15) the differential equation of the mobility network of Fig. 3m-2*C* is

$$f_M = m_I \frac{dv}{dt} + \frac{v}{r_I} + \frac{1}{C_I} \int v \, dt \tag{3m-16}$$

The sum of the electric currents of the electric network of Fig. 3m-2*D* is

$$i = i_1 + i_2 + i_3 \tag{3m-17}$$

where

$$i_1 = \frac{e}{r_E} \tag{3m-18}$$

$$i_2 = C_E \frac{de}{dt} \tag{3m-19}$$

$$i_3 = \frac{1}{L} \int e \, dt \tag{3m-20}$$

[1] In establishing analogies between electric and mechanical systems the elements in the electric network have been labeled r_E, L, and C_E. However, in using analogies in actual practice, the conventional procedure is to label the elements in the analogous electric network as r_M, m, and C_M for the classical mechanical rectilineal system and as r_I, C_I, and m_I for the mobility mechanical rectilineal system. This procedure will be followed in this section in labeling the elements of the analogous electric network. It is literally accurate to label the network with the caption "Analogous electric network of the mechanical rectilineal system" (or, of the mobility mechanical rectilineal system). For the sake of brevity, these networks will be labeled "mechanical network" and "mobility network." Where there is only one path, "circuit" will be used instead of "network."

From the sum of Eqs. (3m-18) to (3m-20) the differential equation of the electric network of Fig. 3m-2D is

$$i = C_E \frac{de}{dt} + \frac{e}{r_E} + \frac{1}{L} \int e \, dt \qquad (3m\text{-}21)$$

Comparing the variables and coefficients of the mobility and electric networks in the differential equations (3m-16) and (3m-21) establishes the analogous variables and quantities in the two systems as given in Table 3m-1.

The classical mechanical impedance analogy of the mechanical system of Fig. 3m-2 has been considered in Sec. 3l and will not be repeated here.

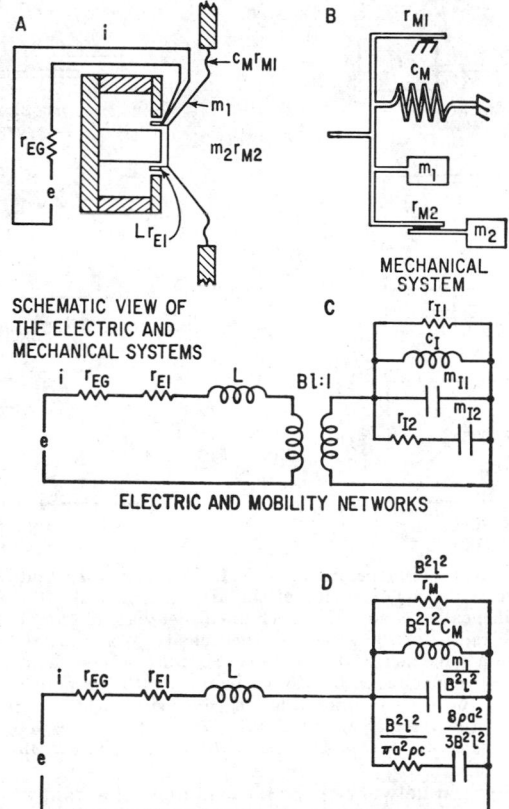

FIG. 3m-3. Cross-sectional view, the mechanical system, the electric and mobility networks, and the electric network of a direct radiator dynamic loudspeaker. In the electric and mechanical networks: e, the electromotive force of the electric generator. r_{EG}, the electrical resistance of the electric generator. L, the inductance of the voice coil. R_{E1}, the electrical resistance of the voice coil. m_1, the mass of the cone. C_M, and r_{M1}, the compliance and mechanical resistance of the suspension. m_2 and r_{M2}, the mass and mechanical resistance of the air load. m_I, the mobility capacitance of the cone. C_I and r_{I1}, the mobility inertia and responsivity of the suspension. m_{I2} and r_{I2}, the mobility capacitance and responsivity of the air load. B, the flux density in the air gap. l, the length of the voice coil conductor. a, the radius of the cone. ρ, the density of air. (*After Olson, "Solutions of Engineering Problems by Dynamical Analogies," D. Van Nostrand Company, Princeton, N.J., 1966.*)

3m-7. Direct Radiator Loudspeaker. The direct radiator dynamic loudspeaker shown in Fig. 3m-3 is almost universally used for radio, phonograph, television, and other small-scale sound reproduction.

The electric and mechanical systems of the complete loudspeaker are shown in Fig. 3m-3A. The mechanical vibrating system consisting of the voice coil, cone, suspension, and air load is presented in Fig. 3m-3B.

The mass m_1 of the cone and voice coil, and the compliance C_M and mechanical resistance of the suspension system, can be obtained from measurements of the vibrating system.

The mechanical system of the air load—namely, the mechanical resistance r_{M2} and mass m_2 of the air load upon the front of the cone—is depicted in Fig. 3m-4A and

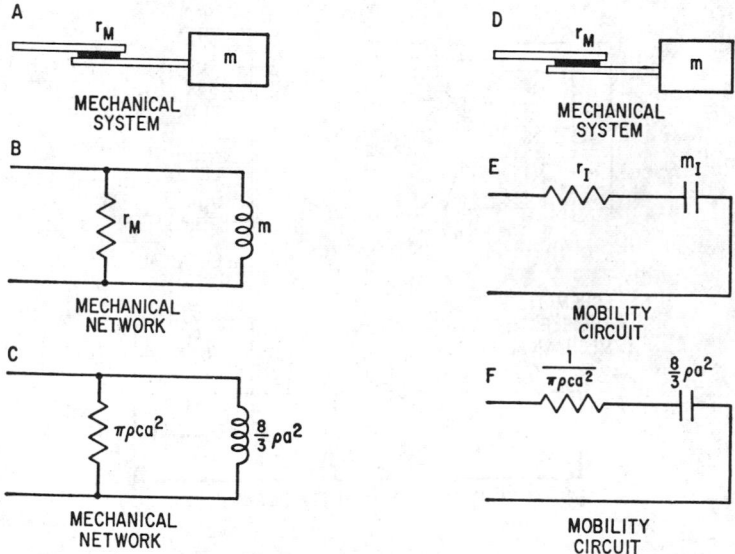

Fig. 3m-4. Air load upon a loudspeaker cone. *A.* Mechanical system: m, the mass of the air load. r_M, the mechanical resistance of the air load. *B.* Mechanical network of the air load upon a loudspeaker cone. *C.* Mechanical network of the air load upon a loudspeaker cone: a, the radius of the cone. ρ, the density of air. c, the velocity of sound. *D.* Mechanical system same as *A.* *E.* Mobility circuit of the air load upon a loudspeaker cone: m_I, the mobility capacitance of the air load. r_I, the responsivity of the air load. *F.* Mobility circuit of the air load upon a loudspeaker cone: a, the radius of the cone. ρ, the density of air. c, the velocity of sound. *(After Olson, "Solution of Engineering Problems by Dynamical Analogies." D. Van Nostrand Company, Princeton, N.J., 1966.)*

3m-4D. The mechanical network of the air load upon the front of the cone is shown in Fig. 3m-4B. The constants of the mechanical resistance and mass of the air load upon the front of the cone are shown in the mechanical network of Fig. 3m-4C. The mobility circuit of the air load upon the front of the cone appears in Fig. 3m-4E. The constants of the responsivity and compliance are given in the mobility circuit of Fig. 3m-4F.

The electric and mobility networks with the ideal transformer connecting the electric and mobility sections are shown in Fig. 3m-3.

In Fig. 3m-3D the ideal transformer has been eliminated, and the entire vibrating system reduced to an electric network. The electrical impedance due to the mechanical system is given by Eq. (3l-26) as follows:

$$z_{EM} = \frac{(Bl)^2}{z_M} \tag{3m-22}$$

where z_{EM} = electrical impedance due to the mechanical system, abohms

z_M = mechanical impedance of the mechanical system, mechanical ohms

B = flux density in the air gap, gauss

l = length of the voice coil conductor, cm

Since $1/z_M = z_I$, Eq. (3m-22) may be written as

$$z_{EM} = (Bl)^2 z_I \tag{3m-23}$$

where z_I = mobility in mechanical mhos.

By means of Eq. (3m-23) it is possible to convert the combined electric and mobility networks to the electric network, as shown in Fig. 3m-3.

The process employing the mobility analysis of this section may be compared with the classical impedance analysis of Sec. 3l.

References

1. Olson, H. F.: "Dynamical Analogies" 2d ed., D. Van Nostrand Company, Inc., Princeton, N.J., 1958.
2. Olson, H. F.: "Solution of Engineering Problems by Dynamical Analogies," Van Nostrand Reinhold Co., New York, N.Y., 1968.

3n. Nonlinear Acoustics (Theoretical)

DAVID T. BLACKSTOCK

University of Texas

Until the early 1950s most of what was known about sound waves of finite amplitude was confined to propagation, and to a lesser extent reflection, of plane waves in lossless gases. Since that time a great deal has been learned about propagation in other media, about nonplanar propagation (still chiefly in one dimension), about the effect of losses, and about standing waves. Inroads have been made on problems of refraction. Diffraction is still relatively untouched.

In this section the exact equations of motion for thermoviscous fluids will first be stated. Various retreats from the full generality of these equations will then be discussed. No attempt will be made to cover streaming and radiation pressure. See Secs. 3c-7 and 3c-8 for a discussion of those topics.

GENERAL EQUATIONS FOR FLUIDS

The basic conservation equations will be stated briefly for viscous fluids with heat flow. Other compressible media, such as solids and relaxing fluids, are discussed later in the section.

3n-1. Conservation of Mass, Momentum, and Energy. In Eulerian (spatial) coordinates the continuity and momentum equations are respectively

$$\frac{D\rho}{Dt} + \rho \frac{\partial u_i}{\partial x_i} = 0 \tag{3n-1}$$

$$\rho \frac{Du_i}{Dt} + \frac{\partial p}{\partial x_i} = \frac{\partial}{\partial x_j}(\eta' d_{kk}\delta_{ij} + 2\eta d_{ij}) \tag{3n-2}$$

An entropy equation is stated here in place of the usual energy equation:

$$\rho T \frac{DS}{Dt} = C_v \left[\rho \frac{D\Im}{Dt} - \frac{\gamma - 1}{\beta_e} \frac{D\rho}{Dt} \right] = \psi^{(\eta)} - \frac{\partial Q_i}{\partial x_i} \tag{3n-3}$$

Here ρ is the density, u_i is the ith (cartesian) component of particle velocity, p is pressure, δ_{ij} is the Kronecker delta, $d_{ij} = \frac{1}{2}(\partial u_i/\partial x_j + \partial u_j/\partial x_i)$ is the rate-of-deformation tensor, η and η' are the shear and dilatational coefficients of viscosity, C_v and C_p are the specific heats at constant volume and pressure, $\Im$ is absolute temperature, S is entropy per unit mass, $\gamma = C_p/C_v$ is the ratio of specific heats, $\beta_e = -\rho^{-1}(\partial \rho/\partial \Im)_p$ is the coefficient of thermal expansion, $\psi^{(\eta)} = 2\eta d_{ij}d_{ji} + \eta' d_{kk}d_{ii}$ is the viscous energy dissipation function, and Q_i is the ith component of the total heat flux. The material derivative $D(\ )/Dt$ stands for $\partial(\ )/\partial t + u_i \partial(\ )/\partial x_i$. If the flow of heat is due to conduction,

$$Q_i = -\kappa \frac{\partial \Im}{\partial x_i} \tag{3n-4}$$

where κ is the coefficient of thermal conduction. For heat radiation the relation between q and $\Im$ is generally quite complicated; see, for example, Vincenti and Baldwin (ref. 1). The model used by Stokes (ref. 2) amounts to Newton's law of cooling and may be expressed by

$$\frac{\partial Q_i}{\partial x_i} = \rho C_v q(\Im - \Im_0) \tag{3n-5}$$

where $\Im_0$ is the ambient temperature, and q is the radiation coefficient. Although too simple to describe radiant heat transfer in a fluid adequately, this equation is worth considering because of (1) its analytical simplicity and (2) its application as a convenient model for relaxation processes.

3n-2. Equation of State. To the conservation equations must be added an equation of state.

Perfect Gas. The gas law for a perfect gas is

$$p = R\rho\Im \tag{3n-6}$$

where R is the gas constant. An approximate form of this equation will now be derived. For a perfect gas the small-signal sound speed c_0 is given by $c_0{}^2 = \gamma R\Im_0 = \gamma p_0/\rho_0$, where p_0 and ρ_0 are the ambient values of p and ρ. Let $\Im = \beta_{e0}(1 + \theta)$, $p = p_0 + \rho_0 c_0{}^2 P$, and $\rho = \rho_0(1 + s)$, where β_{e0} is the ambient value of β_e (for perfect gases $\beta_{e0}\Im_0 = 1$). Assume that θ, P, and s are small quantities of first order. Expansion of Eq. (3n-6) to second order yields

$$\theta = \gamma P - s + s^2 - \gamma P s \tag{3n-7}$$

First-order relations are now defined to be those that hold in linear, lossless acoustic theory; examples are $\rho_t = -\rho_0 \nabla \cdot u$ and $p - p_0 = c_0{}^2(\rho - \rho_0)$. At this point we assert that any factor in a second-order term in Eq. (3n-7) may be replaced by its first-order equivalent. The justification is that any more precise substitution would result in the appearance of third- or higher-order terms, and such terms have already been excluded from Eq. (3n-7). Thus in the last second-order term in Eq. (3n-7) P may be replaced by s to give

$$\theta = \gamma P - s - (\gamma - 1)s^2 \tag{3n-8}$$

correct to second order. This is a useful approximate form of the perfect gas law.

One of the most fruitful special cases to consider is the isentropic perfect gas. When a perfect gas is inviscid and there is no heat flow, Eq. (3n-3) can be used to reduce the gas law, Eq. (3n-6), to

$$\frac{p}{p_0} = \left(\frac{\rho}{\rho_0}\right)^{\gamma} \tag{3n-9}$$

The square of the sound speed, which by definition is,

$$c^2 \equiv \left(\frac{\partial p}{\partial \rho}\right)_s \tag{3n-10}$$

becomes

$$c^2 = \frac{\gamma p}{\rho} = c_0{}^2 \left(\frac{p}{p_0}\right)^{(\gamma-1)/\gamma} \tag{3n-11}$$

An expanded form of Eq. (3n-9) is as follows:

$$P = s + \tfrac{1}{2}(\gamma - 1)s^2 + \cdots \tag{3n-12}$$

Other Fluids. For liquids and for gases that are not perfect, one can start with a general equation of state $\mathfrak{I} = \mathfrak{I}(p,\rho)$. Recognizing that $(\partial \mathfrak{I}/\partial p)_\rho = \gamma(\rho c^2 \beta_e)^{-1}$, one obtains the exact expression

$$\theta_t = \frac{\beta_{e0}}{\beta_e}(1 + s)^{-1}\left[\gamma\left(\frac{c_0}{c}\right)^2 P_t - s_t\right] \tag{3n-13}$$

In order to obtain an approximation analogous to Eq. (3n-8), it is first necessary to set down a general isentropic equation of state,

$$p - p_0 = \rho_0 c_0{}^2\left(s + \frac{B}{2A}s^2 + \frac{C}{3A}s^3 + \cdots\right) \tag{3n-14}$$

where the coefficients B/A, C/A, etc., are to be determined experimentally (see Sec. 3o). With the help of this expression and some elementary thermodynamic relations, one invokes the approximation procedure described following Eq. (3n-7) and reduces Eq. (3n-13) to (ref. 3)

$$\theta = \gamma P - s - (h - 1)s^2 \tag{3n-15}$$

correct to second order, where

$$h = 1 + \frac{\gamma B}{2A} + \tfrac{1}{2}(\gamma - 1)\left(1 - \frac{B}{2A}\right) - (\gamma - 1)^2(4\beta_{e0}\mathfrak{I})^{-1} \tag{3n-16}$$

If Eqs. (3n-14) and (3n-12) are compared, it will be seen that B/A replaces the quantity $\gamma - 1$ in describing second-order nonlinearity of the $p - \rho$ relation. For a perfect gas, therefore, replace B/A by $\gamma - 1$ and β_{e0} by $\mathfrak{I}_0^{-1}$ in Eq. (3n-16). The quantity h then reduces to γ, and Eq. (3n-7) is recovered.

PROPAGATION IN LOSSLESS FLUIDS

For isentropic flow (taken here to mean that the entropy of every particle is the same and remains so) Eqs. (3n-1) and (3n-2) reduce to

$$\frac{D\rho}{Dt} + \frac{\rho \partial u_i}{\partial x_i} = 0 \tag{3n-17a}$$

$$\frac{\rho D u_i}{Dt} + \frac{\partial p}{\partial x_i} = 0 \tag{3n-17b}$$

and the equation of state may be expressed simply by $p = p(\rho)$. If the new thermodynamic quantity

$$\lambda \equiv \int_{\rho_0}^{\rho} \frac{c}{\rho'} d\rho' \tag{3n-18}$$

is introduced, Eqs. (3n-17) take the following symmetric form:

$$\frac{D\lambda}{Dt} + \frac{c\partial u_i}{\partial x_i} = 0 \tag{3n-19a}$$

$$\frac{Du_i}{Dt} + \frac{c\partial\lambda}{\partial x_i} = 0 \tag{3n-19b}$$

Very little has been done in the way of solving these general equations.

3n-3. Plane Waves in Lossless Fluids. For one-dimensional flow in the x direction Eqs. (3n-19) become

$$\lambda_t + u\lambda_x + cu_x = 0 \tag{3n-20a}$$
$$u_t + uu_x + c\lambda_x = 0 \tag{3n-20b}$$

where subscripts x and t now denote partial differentiation, and u represents the particle velocity in the x direction. Hyperbolic equations of this form have been studied in great detail (ref. 4). Their solutions are of two general types: (1) those representing simple waves (waves propagating in one direction only), and (2) those representing compound waves (waves propagating in both directions).

Simple Waves. Simple-wave flow is characterized by the existence of a unique relationship between the particle velocity and the local thermodynamic state of the fluid. For simple waves traveling into a medium at rest, this relationship is (ref. 5)

$$\lambda = \pm u \tag{3n-21}$$

where the $(+)$ sign holds for outgoing waves (waves traveling in the direction of increasing x), and the $(-)$ sign for incoming waves (waves traveling in the direction of decreasing x). Hereinafter when multiple signs are used, the upper sign pertains to outgoing waves. Equations (3n-20) now reduce to the single equation

$$u_t + (u \pm c)u_x = 0 \tag{3n-22}$$

which becomes autonomous once the equation of state is specified, since Eqs. (3n-18) and (3n-21) imply a relationship $c = c(u)$. Note that the linearized version of Eq. (3n-22), $u_t \pm c_0 u_x = 0$, possesses the familiar traveling-wave solution $u = f(x \mp c_0 t)$ of linear acoustics.

The most important nonlinear effect in simple-wave flow can be readily identified directly from Eq. (3n-22). Combine that equation with the differential expression $du = u_x\,dx + u_t\,dt$ to obtain

$$\left(\frac{dx}{dt}\right)_{u=\text{const}} = -\frac{u_t}{u_x} = u \pm c \tag{3n-23}$$

This relation states that the propagation speed of a given point on the waveform (the point being identified by the value of u there) is $u \pm c$. In linear theory the propagation speed of all points is the same, namely, $\pm c_0$. The ramifications of the variable propagation speed are discussed in Sec. 3n-4.

Compound Waves. When waves traveling in both directions are present, there is no fixed relationship between u and λ. A propagation speed can still be defined, however. New dependent variables $\mathfrak{r}$ and $\mathfrak{s}$, called "Riemann invariants," may be defined by

$$2\mathfrak{r} = \lambda + u \qquad 2\mathfrak{s} = \lambda - u \tag{3n-24}$$

If Eqs. (3n-20) are first added and then subtracted, the results are respectively

$$\mathfrak{r}_t + (u + c)\mathfrak{r}_x = 0 \tag{3n-25a}$$
$$\mathfrak{s}_t + (u - c)\mathfrak{s}_x = 0 \tag{3n-25b}$$

Thus, as first found by Riemann (ref. 6),

$$\left(\frac{dx}{dt}\right)_{r=\text{const}} = u + c \qquad (3\text{n-}26a)$$

$$\left(\frac{dx}{dt}\right)_{s=\text{const}} = u - c \qquad (3\text{n-}26b)$$

Despite its apparent simplicity, this result is much more complicated to apply than Eq. (3n-23).

3n-4. Plane, Simple Waves in Lossless Gases. For perfect gases the isentropic equation of state is given by Eq. (3n-9). For this case $\lambda = 2(c - c_0)/(\gamma - 1)$, and the simple-wave relation Eq. (3n-21) becomes

$$c = c_0 \pm (\beta - 1)u \qquad (3\text{n-}27)$$

where $\beta = \frac{1}{2}(\gamma + 1)$. Combination of this equation with Eq. (3n-11) leads to

$$p - p_0 = p_0 \left\{ \left[1 \pm (\beta - 1)\frac{u}{c_0} \right]^{2\gamma/(\gamma-1)} - 1 \right\} \qquad (3\text{n-}28)$$

which can be used to obtain the characteristic impedance for finite-amplitude waves. For weak waves, i.e., $u/c_0 \ll 1$, this expression reduces to the traditional one,

$$p - p_0 = \pm \rho_0 c_0 u \qquad (3\text{n-}29)$$

The nonlinear differential equation for simple waves, Eq. (3n-22), becomes

$$u_t + (\beta u \pm c_0)u_x = 0 \qquad (3\text{n-}30)$$

If we restrict ourselves momentarily to outgoing waves, the propagation speed is

$$\left(\frac{dx}{dt}\right)_{u=\text{const}} = \beta u + c_0 \qquad (3\text{n-}31a)$$

which shows quite clearly that the peaks of the wave travel fastest, the troughs slowest. Equivalently, as the wave travels from one point to another, the peaks suffer the least delay, the troughs the most. This latter view is illustrated in Fig. 3n-1,

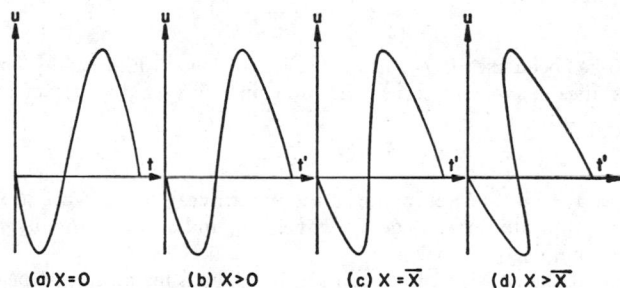

(a) X = 0   (b) X > 0   (c) X = $\overline{X}$ (d) X > $\overline{X}$

FIG. 3n-1. Progressive distortion of a finite-amplitude wave. Symbols are: u = particle velocity, x = spatial coordinate, t = time, $t' = t - x/c_0$ (delay time), $\bar{x}$ = point at which a shock begins to form.

which shows the time waveform of an outgoing disturbance at various distances from the source. The progressive distortion is quite striking, leading eventually to the curious waveform shown in Fig. 3n-1d. The interpretation of Fig. 3n-1d will be discussed presently.

Why physically does the exact propagation speed differ from c_0, the accepted value in linear theory? Two effects are at work: one kinematic, the other thermodynamic.

3-188 ACOUSTICS

The sound wave travels with speed c with respect to the fluid particles. But these particles are themselves in motion, moving with velocity u. To a fixed observer, therefore, the net speed is $u + c$. This is the kinematic effect and is frequently referred to as *convection* (the fluid particles convect the wave along as a result of their own motion). The thermodynamic effect is the deviation from constancy of the sound speed c. Where the acoustic pressure is positive, the gas is a little hotter. Consequently c is greater. Conversely, in the wave troughs, where the gas is expanded and therefore colder, c is less. The variation of c from point to point along the wave can be traced to nonlinearity of the pressure-density relation. As Eq. (3n-10) shows, c would be constant if p were linearly related to ρ. This would be true, for example, for an isothermal gas.

For an incoming wave the propagation speed is

$$\left(\frac{dx}{dt}\right)_{u=\text{const}} = \beta u - c_0 \tag{3n-31b}$$

Similar arguments apply in this case. A difference is that the troughs of the particle velocity wave travel fastest (in a backward direction), the peaks slowest. Because pressure and particle velocity are out of phase in an incoming wave, however, it is still true that the peaks of the pressure wave proceed most rapidly and the troughs least so.

General Solutions. Three forms of the general solution of Eq. (3n-30) are now given. First is what might be called the "Poisson solution" (ref. 7)

$$u = f[x - (\beta u \pm c_0)t] \tag{3n-32}$$

which is implied by Eq. (3n-31); f is an arbitrary function. This result is most easily interpreted as the solution of an initial-value problem for which the spatial dependence of the particle velocity is prescribed everywhere at $t = 0$, i.e., $u(x,0) = f(x)$. The problem is somewhat artificial, however, because the progressive wave motion must already exist at $t = 0$. Of more practical interest are boundary-value problems involving a source; then simple waves arise quite naturally. If the time history of the particle velocity is known at a particular place, say $u(0,t) = g(t)$, the solution is

$$u = g\left(t - \frac{x}{\beta u \pm c_0}\right) \tag{3n-33}$$

This equation has been used to construct the waveforms in Fig. 3n-1. To make such constructions, it is convenient to use the following "inverted" form of the solution:

$$t' = g^{-1}(u) - \frac{\beta u}{c_0 \pm \beta u}\frac{x}{c_0} \tag{3n-34}$$

where $t' = t \mp x/c_0$ is the delay (for outgoing waves) or advance (for incoming waves) time appropriate for zeros of the waveform, and $g^{-1}(u)$ is the inverse function corresponding to g, i.e., $g^{-1}[g(u)] = u$.

The solution of the classic piston problem, in which a piston at rest begins at time $t = 0$ to move smoothly with a given displacement $X(t)$ in a lossless tube, is more complicated because of the moving boundary condition

$$u[X(t),t] = X'(t)H(t) \tag{3n-35}$$

where $H(t)$ is the unit step function. The solution of this problem may be given in parametric form as follows (refs. 5, 8):

$$u = X(\phi)H\left(\frac{t \mp x}{c_0}\right) \tag{3n-36a}$$

where $$\phi = t - \frac{x - X(\phi)}{\beta X'(\phi) \pm c_0} \tag{3n-36b}$$

The parameter ϕ represents the time at which a given signal (i.e., given value of u) left the piston.

It is generally quite difficult to convert any of the three general solutions into an explicit analytical expression $u(x,t)$. One can, however, always obtain a sketch of the waveform through use of the inversion procedure indicated by Eq. (3n-34).

Shock Formation. A more far-reaching limitation, both mathematically and physically, is that these solutions contain the seeds of their own destruction. Except for a wave of pure expansion, the dependence of the propagation speed on u will cause steepening of the waveform. Steepening eventually leads to multivalued shapes like that shown in Fig. 3n-1d. But these must be rejected because pressure disturbances in nature cannot be multivalued, either in time or in space. In fact, once any section of the waveform attains a vertical tangent, as in Fig. 3n-1c, results cannot in general be continued further (ref. 9). Physically, what happens is that a shock wave begins to form. For reasons discussed in detail in Sec. 3n-8, this formally marks the end of validity of lossless, simple-wave theory. For mathematical analyses of shock formation see, for example, refs. 4 and 8.

Fubini Solution. A problem of special interest in acoustics is the propagation of a finite-amplitude wave that is sinusoidal at its point of origin. Suppose that the wave is produced by sinusoidal vibration of a piston in a lossless tube. Let the piston displacement be given by $X(t) = (u_0/\omega)(1 - \cos \omega t)$ where u_0 is the velocity amplitude of the piston, and ω is the angular frequency. The solution is given by applying Eqs. (3n-36). For the outgoing wave we have

$$\frac{u}{u_0} = \sin \omega\phi H\left(t - \frac{x}{c_0}\right) \tag{3n-37a}$$

where

$$\omega\phi = \omega t - \frac{kx - \epsilon(1 - \cos \omega\phi)}{1 + \beta\epsilon \sin \omega\phi} \tag{3n-37b}$$

Here $k = \omega/c_0$ is the wave number, and $\epsilon = u_0/c_0$ is the velocity amplitude expressed as a Mach number.

An explicit solution is now sought by writing u as a Fourier series,

$$\frac{u}{u_0} = \Sigma A_n \cos n(\omega t - kx) + \Sigma B_n \sin n(\omega t - kx) \tag{3n-38}$$

Although the exact expressions for all the coefficients A_n and B_n have not been obtained, an approximate computation is available. First expand Eq. (3n-37b), writing σ for $\beta\epsilon kx$, and t' for $t - x/c_0$, and rearrange as follows:

$$\omega\phi - \omega t' = \sigma \sin \omega\phi + \epsilon(1 - \cos \omega\phi - \beta\sigma \sin^2 \phi) + O(\epsilon^2)$$

If $\sigma \gg \epsilon$ (i.e., $\beta kx \gg 1$), and $\epsilon \ll 1$, this equation reduces to

$$\omega\phi = \omega t' + \sigma \sin \omega\phi \tag{3n-39}$$

Under this approximation the Fourier coefficients A_n vanish, and the B_n can be evaluated in terms of Bessel functions. The final result is (ref. 8)

$$\frac{u}{u_0} = \sum_{n=1}^{\infty} \frac{2}{n\sigma} J_n(n\sigma) \sin n(\omega t - kx) \tag{3n-40}$$

which is generally referred to as the Fubini solution (ref. 10).

The acoustic pressure signal is found by substituting the value of u given by Eq. (3n-40) in the linear impedance elation, Eq. (3n 20). Use of a more accurate expansion of Eq. (3n-28) for this purpose would not be consistent with the approximations that led to Eq. (3n-39).

The shock formation distance for this problem can be deduced by inspection of Eqs. (3n-39) [or, alternatively, the exact expression Eqs. (3n-37b)] and (3n-37a). The relationship of u to t' is one-to-one only if $\sigma < 1$. For $\sigma \geq 1$ the waveform curve $u(t')$ is multivalued. Hence a shock starts to form at $\sigma = 1$, i.e., at

$$\bar{x} = (\beta e k)^{-1} \tag{3n-41}$$

where the overbar signifies shock formation. The physical interpretation of σ is therefore that it is a spatial variable scaled in terms of the shock formation distance. The Fubini solution is not valid beyond the point $\sigma = 1$.

3n-5. An Approximate Theory of Lossless Simple Waves. The approximations leading to the Fubini solution can be used to obtain a general approximate theory of traveling waves of finite amplitude. The mathematical restrictions required are

$$\sigma \gg \epsilon \tag{3n-42a}$$
$$\epsilon \ll 1 \tag{3n-42b}$$

where the definitions of σ and ϵ are generalized to

$$\sigma = \frac{\beta \epsilon x}{x_c} \qquad \epsilon = \frac{u_0}{c_0} \tag{3n-43}$$

Here x_c is a characteristic distance defined so that significant distortion (for example, shock formation) takes place over the range $0 < \sigma < 1$, and u_0 is the maximum particle velocity that occurs in the flow. The physical implications of these restrictions are as follows:

1. The finite displacement of the source can be neglected. In other words, the exact boundary condition given by Eq. (3n-35) can be replaced by

$$u(0,t) = X'(t)H(t) \tag{3n-44}$$

Any error thus committed is made small by inequality (3n-42a).

2. The linear impedance relation, Eq. (3n-29), can be used to obtain the acoustic pressure, once the particle velocity waveform is known.

3. The nonlinear effect that *must* be taken into account is the nonconstancy of the propagation speed. But this effect is approximated by writing Eqs. (3n-31) as follows:

$$\left(\frac{dx}{dt}\right)_{u=\text{const}} \doteq \frac{\pm c_0}{1 \mp \beta u/c_0} \tag{3n-45}$$

Retention of nonconstancy of the propagation speed as the only important nonlinear effect gives recognition to the fact that this effect is the only *cumulative* one. It is the cause of the progressive distortion that engulfs the wave. We neglect the other nonlinear effects because they are *noncumulative*, or local. The distortion they cause does not grow with distance.

The formal theory based on these ideas will now be developed. An approximate differential equation may be derived by applying the method used earlier to convert Eq. (3n-7) to (3n-8). For simple waves the appropriate first-order relation is $u_x = \mp c_0^{-1}u_t$. When this is substituted in the nonlinear term in Eq. (3n-30), the result is

$$c_0 u_x \pm u_t - \beta c_0^{-1} u u_t = 0 \tag{3n-46}$$

This differential equation could also have been deduced from Eq. (3n-45).

Next let x and $t' = t \mp x/c_0$ be new independent variables. Equation (3n-46) reduces to

$$c_0^2 u_x - \beta u u_{t'} = 0 \tag{3n-47}$$

For the boundary condition

$$u\Big|_{x=0} = g(t)H(t) = g(t') \tag{3n-48}$$

where it is assumed that $g(t) = 0$ for $t < 0$, the solution is

$$u = g(\phi) \tag{3n-49a}$$
$$\phi = t' + \beta c_0^{-2} x g(\phi) \tag{3n-49b}$$

When the excitation is sinusoidal, i.e. $g(t) = u_0 \sin \omega t$, the Fubini solution follows exactly. It is also worth noting that within the limits of the approximate theory the difference between Lagrangian and Eulerian coordinates is negligible. As a general rule, the approximate theory is useful when $\epsilon < 0.1$ (ref. 8).

3n-6. Plane, Simple Waves in Liquids and Solids. *Liquids.* For lossless fluids whose isentropic equation of state is not given by Eq. (3n-9), we may proceed by using Eq. (3n-14). The propagation speed is (ref. 8)

$$\left(\frac{dx}{dt}\right)_{u=\text{const}} = u \pm c_0(1 + c_1 U + c_2 U^2 + \cdots) \tag{3n-50}$$

where $U = u/c_0$ and $c_1 = B/2A$, $c_2 = C/2A + B/4A - (B/2A)^2$, etc. Thus, in the exact solution of the piston problem [Eqs. (3n-36)], the parameter ϕ is given by

$$\phi = t - \frac{x - X(\phi)}{u \pm c_0(1 + c_1 U + c_2 U^2 \cdots)} \tag{3n-51}$$

where U is to be interpreted as $c_0^{-1} x_t(\phi)$.

Solids. The mathematical formalism for plane, longitudinal elastic waves in solids, either crystalline or isotropic, is very similar to that for liquids and gases (refs. 11–13). The wave equation is given in Lagrangian coordinates as

$$\xi_{tt} = c_0^2 G(\xi_a) \xi_{aa} \tag{3n-52}$$

where
$$G(\xi_a) = 1 + \left(\frac{M_3}{M_2}\right)\xi_a + \left(\frac{M_4}{M_2}\right)\xi_{aa} \cdots \tag{3n-53}$$

Here a represents the rest position of a particle; ξ is partical displacement; and M_2, M_3, M_4, etc., are quantities involving the second-, third-, fourth-, and higher-order elastic coefficients (ref. 12). The quantity $c_0^2 G$ plays the same role that $(\rho c/\rho_0)^2$ does for fluids (ref. 14). By the Lagrangian equation of continuity, $\rho_0/\rho = 1 + \xi_a$; thus replace Eq. (3n-18) by

$$\lambda = -c_0 \int_0^{\xi_a} [G(\xi_a')]^{\frac{1}{2}} \, d\xi_a' \tag{3n-54}$$
$$= -c_0[\xi_a - \tfrac{1}{4}m_3\xi_a^2 + (\tfrac{1}{8} - \tfrac{1}{6}m_4)m_3^2\xi_a^3 \cdots] \tag{3n-55}$$

where $m_3 = -M_3/M_2$, $m_4 = 1 - M_4/M_2 m_3^2$, etc. Riemann invariants are defined as before by Eq. (3n-24). Note that $u = \xi_t$ in Lagrangian coordinates.

Simple-wave fields are again specified by Eq. (3n-21), which when combined with Eq. (3n-5) leads to

$$\xi_a = \mp U + \tfrac{1}{4}m_3 U^2 \mp \tfrac{1}{6}m_4 m_3^2 U^3 \cdots \tag{3n-56}$$

The propagation speed for simple waves is

$$\left(\frac{da}{dt}\right)_{u=\text{const}} = \pm c_0 G^{\frac{1}{2}} \tag{3n-57}$$

The factor u, which appears in Eq. (3n-23), is absent here because the coordinate system is Lagrangian. Equation (3n-57) expanded in series form is

$$\left(\frac{da}{dt}\right)_{u=\text{const}} = \pm c_0[1 \pm \tfrac{1}{2}m_3 U + \tfrac{1}{4}m_3^2(1 - 2m_4)U^2 \cdots] \tag{3n-58}$$

Therefore, the solution of the piston problem, given $u(0,t) = X_t(t)$, is

$$\phi = t \mp \frac{a/c_0}{1 \pm \frac{1}{2}m_3 U + \frac{1}{4}m_3{}^2(1 - 2m_4)U^2 \cdots} \qquad (3n\text{-}59)$$

where U is to be interpreted, as in Eq. (3n-51), as $c_0{}^{-1}X_t(\phi)$. More complete versions of some of the series expansions given above can be found in ref. 12.

Approximate Theory. The approximate theory of simple waves described in Sec. 3n-5 is very easily generalized to apply to liquids and solids. For liquids $\gamma - 1$ is replaced by B/A, as mentioned after Eq. (3n-16). For solids $\gamma + 1$ is replaced by $-M_3/M_2$ (see ref. 12 for other useful associations). Therefore, let

$$\beta = \frac{1}{2}(\gamma + 1) \qquad \text{for gases} \qquad (3n\text{-}60a)$$

$$\beta = 1 + \frac{B}{2A} \qquad \text{for liquids} \qquad (3n\text{-}60b)$$

$$\beta = \frac{-M_3}{2M_2} \qquad \text{for solids} \qquad (3n\text{-}60c)$$

and all results stated in Sec. 3n-5 become applicable for a very wide range of continuous media. For many liquids and solids the first "nonlinearity coefficient" (B/A for liquids, M_3/M_2 for solids) is known, but higher-order ones are not. In such cases it is difficult to justify using anything more precise than the approximate theory. But see ref. 12 for a discussion related to this point.

3n-7. Nonplanar Simple Waves. In this section one-dimensional nonplanar waves are considered, namely, spherical and cylindrical waves, and waves in horns. The general theory is not very highly developed. One fundamental difficulty is that simple waves of arbitrary waveform do not generally exist for nonplanar waves (ref. 15). Consider, for example, the wave motion generated by a pulsating sphere in an infinite medium. Most of the wave field consists of outgoing radiation, but there is also some backscatter (ref. 15). In the far field, however, simple waves do occur as an approximation. This is the case treated here. The results represent an extension of the approximate theory developed in Secs. 3n-5 and 3n-6.

Spherical and Cylindrical Waves. For large values of the radial coordinate r (actually large kr, where k is an appropriate wave number of the disturbance), the following approximate equation for simple waves in a fluid can be obtained (ref. 16):

$$c_0{}^2 w_z - \beta w w_{t'} = 0 \qquad (3n\text{-}61)$$

where $t' = t \mp (r - r_0)/c_0$, r_0 is a reference distance, and β is given by Eq. (3n-60a) or (3n-60b). This equation may also apply to longitudinal waves in an isotropic solid, but so far no derivation has been given. The dependent variable w equals $(r/r_0)^{\frac{1}{2}}u$ and $(r/r_0)u$ for cylindrical and spherical waves, respectively. The independent variable z is given for the two cases by

Cylindrical: $\qquad z = 2(\sqrt{r} - \sqrt{r_0})\sqrt{r_0} \qquad (3n\text{-}62a)$

Spherical: $\qquad z = r_0 \ln \dfrac{r}{r_0} \qquad (3n\text{-}62b)$

Note that $z > 0$ for diverging waves ($r > r_0$), but $z < 0$ for converging waves ($r < r_0$).

Equation (3n-61) is solved by recognizing that it has the same form as the plane-wave equation (3n-47). For the boundary condition take $u(r_0,t) = g(t)$, which may represent either the motion of a source at r_0 or the measured time signal of a wave as it passes by the point r_0. Since $z = 0$ and $t' = t$ when $r = r_0$, the condition on w is

$$w(0,t') = g(t') \qquad (3n\text{-}63)$$

Therefore, for the two kinds of waves the solution is

Cylindrical:
$$u = \left(\frac{r_0}{r}\right)^{\frac{1}{2}} g(\phi) \tag{3n-64a}$$

$$\phi = t' + 2\beta c_0^{-2} \sqrt{r_0}(\sqrt{r} - \sqrt{r_0})g(\phi) \tag{3n-64b}$$

Spherical:
$$u = \frac{r_0}{r} g(\phi) \tag{3n-65a}$$

$$\phi = t' + \beta c_0^{-2} r_0 \ln \frac{r}{r_0} g(\phi) \tag{3n-65b}$$

Some applications of these results are given in refs. 16 to 18. It has been shown (ref. 19) that Eq. (3n-65b) corresponds to a second-order approximation of results obtained using the Kirkwood-Bethe hypothesis (ref. 20).

Many special solutions for spherical and cylindrical waves have also been found. Most are of the similarity type. The most famous is Taylor's solution for the compression wave generated by a sphere that expands at a constant rate (refs. 21, 22).

Waves in Horns. For waves traveling in ducts whose cross-sectional area $A = A(x)$ does not vary rapidly, the waves may be assumed to be quasi-plane. It is assumed that the effect of variations in the cross section can be accounted for simply by correcting the continuity equation as follows:

$$\frac{D(A\rho)}{Dt} + \rho A u_x = 0 \tag{3n-66}$$

The one-dimensional formalism is thereby retained.

By the same methods used for spherical and cylindrical waves it is possible to derive an equation exactly like Eq. (3n-61). However, w and z are now defined as

$$w = \left(\frac{A}{A_0}\right)^{\frac{1}{2}} u \tag{3n-67a}$$

$$z = \int_{x_0}^{x} \left(\frac{A_0}{A}\right)^{\frac{1}{2}} dx' \tag{3n-67b}$$

where x_0 is a reference distance, $A_0 = A(x_0)$, and $t' = t \pm (x - x_0)/c_0$. The sign of z identifies the wave as outgoing ($x > x_0$) or incoming ($x < x_0$). Note that a conical horn ($A \propto x^2$) gives results identical with those for spherical waves, and a parabolic horn ($A \propto x$) gives results identical with those for cylindrical waves.

The general solution for a boundary condition of the form given by Eq. (3n-63) is (ref. 23)

$$w = \left(\frac{A}{A_0}\right)^{\frac{1}{2}} u = g(\phi) \tag{3n-68a}$$

$$\phi = t' + \beta c_0^{-2} z g(\phi) \tag{3n-68b}$$

For reference the value of the stretched coordinate z for an exponential horn ($A \propto e^{2lx}$) is

$$z = l^{-1}(1 - e^{-l(x-x_0)}) \tag{3n-69a}$$

and for a catenoidal horn ($A \propto \cosh^2 lx$) is

$$z = 2l^{-1}(\tan^{-1} e^{lx} - \tan^{-1} e^{lx_0}) \cosh lx_0 \tag{3n-69b}$$

All the results previously obtained for plane waves (approximate theory) may now be applied to nonplanar one-dimensional waves simply by replacing u and x by w and z, as given by Eqs. (3n-67). For example, for sinusoidal excitation at $x = x_0$ the shock formation distance is found by putting $\bar{z} = \pm (\beta \epsilon k)^{-1}$ and then making use of Eq. (3n-67b).

Parametric Array. An application of particular interest is the so-called parametric, end-fired array, conceived by Westervelt (ref. 53). A source such as a baffled piston emits radiation consisting of two high-frequency carrier waves into an open medium. The carriers, whose frequencies are ω_1 and ω_2, interact nonlinearly to produce a difference-frequency wave (frequency $\omega_d = \omega_2 - \omega_1$). Also produced, of course, but not of interest here, are the harmonics of the two carriers as well as the sum-frequency and other intermodulation components (ref. 54). In Westervelt's original treatment the two carrier waves were assumed to be collinear beams of collimated plane waves. More recently, Muir (ref. 55) has taken the directivity and spherical spreading of the carriers into account. In any case, however, the interaction to produce the difference-frequency wave amounts to setting into operation a line of virtual sources of frequency ω_d, all phased so as to constitute an end-fired array. The result is that the difference-frequency wave has a very high directivity. In other words, a low-frequency beam is produced that is much more highly directive than would have been the case had the source emitted the difference-frequency signal directly. Typically, too, there are no minor lobes. Absorption by the medium may be relied upon to filter out the two carrier waves and the sum-frequency component, eventually leaving the difference-frequency wave as the most prominent signal. Experiments have confirmed the remarkable properties of the parametric array (refs. 55, 56), and many further studies of it have been done (ref. 57).

WEAK-SHOCK THEORY

3n-8. General Discussion. The appearance of shocks in a flow poses a serious challenge to the theory of simple waves as developed thus far. In the first place, the waveform gradient at a shock is so high that the dissipation terms in Eqs. (3n-2) and (3n-3), heretofore deemed negligible, are in fact very large. A second problem is that since the shock is (at least approximately) a discontinuity in the medium, it can cause partial reflection of signals that catch up with it. The presence of reflected waves invalidates the simple-wave assumption. Strictly speaking, therefore, the flow cannot be simple wave, once shocks form (ref. 9).

The situation is not quite so bad as it seems, however, provided we restrict ourselves to relatively weak waves, i.e., $u_0/c_0 < 0.1$, approximately. Under this condition the signals that are reflected from a shock in the waveform are so feeble as to be negligible. The simple-wave model may therefore be retained as a good approximation. Next, triple-valued waveforms of the kind shown in Fig. 3n-1 must be avoided. This requires that provision be made for dissipation. There are two approaches. First, one can take explicit account of the dissipation terms. This leads to Burgers' equation, or variations thereof; the method is described in Sec. 3n-12. Alternatively, one can postulate mathematical discontinuities—shocks—at places where the waveform would otherwise be triple valued. The Rankine-Hugoniot relations are invoked to relate conditions on either side of each shock. In this way dissipation is accounted for indirectly. A tacit assumption, it will be noted, is that all the dissipation takes place at the shocks.

The mathematical method is more fully appreciated if the physical aspects of the process are first understood. The history of a typical waveform is depicted in Fig. 3n-2 (taken from ref. 27). Figure 3n-2a shows the initial waveform. Numbered dots indicate initial phase points (values of ϕ) on the wave. In the beginning, distortion takes place as described in Sec. 3n-4 (Fig. 3n-2b and c). After the shock is born (Fig. 3n-2c), it travels supersonically. In consequence of Eq. (3n-72), however, phase points just behind, such as number 5, travel faster. As they catch up with the shock, it grows because the top of the discontinuity is always determined by the amplitude of the phase point that just caught up with it. (Conversely, the bottom of the discontinuity always coincides with the phase point just overtaken by the shock.)   The top reaches a maximum when phase point 5 catches up.   After that, the top decays (Fig. 3n-2e).   In Fig. 3n-2f the decay has progressed to the extent that all phase points of the original waveform between 4 and 6 have disappeared.   Eventually all that remains (Fig. 3n-2g) is the shock and a linear section connecting it with the zero, phase point 7.   This is the asymptotic shape toward which many waveforms or waveform sections tend (ref. 26).

3n-9. Mathematical Formulation of Weak-shock Theory.   For the continuous sections of the waveform the most general solution from the approximate theory of simple waves is adopted, namely, Eqs. (3n-68), where w and z are given by Eqs. (3n-67).   Plane, cylindrical, and spherical waves, which are not really "quasi-plane," are nevertheless included formally within the framework of this solution by taking $A = 1$, x, and x^2, respectively.

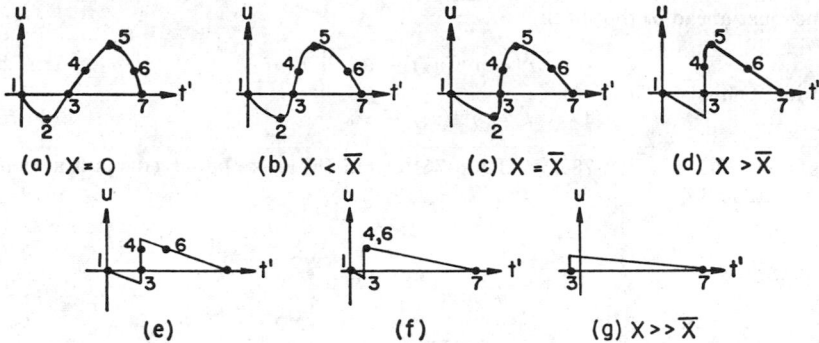

FIG. 3n-2. Development and decay of a finite-amplitude wave.   Numbered points refer to initial phases (values of ϕ) of the wave.   (*From ref. 27.*)

Suppose now that a shock begins to form at time $\bar{t}$ and distance $\bar{x}$.   It will arrive at a subsequent point x at time t_s given by

$$t_s = \bar{t} + \int_{\bar{x}}^{x} v^{-1}\, d\mu \tag{3n-71}$$

where v is the shock's propagation speed.   The Rankine-Hugoniot relations can be combined to give v in terms of u_a and u_b, the particle velocities just ahead of and just behind the shock, respectively.   An approximation of the required relation is

$$v = \pm c_0 + \tfrac{1}{2}\beta(u_a + u_b) \tag{3n-72}$$

or, to the same order,

$$v^{-1} = \pm c_0^{-1} - \tfrac{1}{2}\beta c_0^{-2}(u_a + u_b) \tag{3n-73}$$

Substitution of this value in Eq. (3n-71) leads to

$$t'_s = \bar{t}' - \tfrac{1}{2}\beta c_0^{-2}\int_{\bar{x}}^{x} (u_a + u_b)\, d\mu \tag{3n-74}$$

where overbars continue to indicate values at the instant of shock formation, and primes denote retarded (or advanced) time.   In terms of the generalized dependent and independent variables w and z, Eq. (3n-74) becomes

$$t'_s = \bar{t}' - \tfrac{1}{2}\beta c_0^{-2}\int_{\bar{z}}^{z} (w_a + w_b)\, d\mu \tag{3n-75}$$

An equivalent relation is

$$\frac{dt'_s}{dz} = -\tfrac{1}{2}\beta c_0^{-2}(w_a + w_b) \tag{3n-76}$$

Once the particle velocity u has been determined, the linear impedance relation, Eq. (3n-29), is used to find the pressure signal (ref. 23).

This completes the formal solution, except for some interpretation. The waveform in the continuous sections between shocks is prescribed by Eqs. (3n-68). For each shock the path and amplitude are determined by Eq. (3n-75) or Eq. (3n-76) together with Eqs. (3n-68), which are to be evaluated just ahead of the shock ($u = u_a$, $\phi = \phi_a$, $t' = t'_s$) and just behind it ($u = u_b$, $\phi = \phi_b$, $t' = t'_s$). In principle, Eqs. (3n-68) can be combined to eliminate the parameter ϕ as follows:

$$t' = g^{-1}(w) - \beta c_0^{-2}zw \tag{3n-77}$$

Hence just ahead of the shock

$$t'_s = g^{-1}(w_a) - \beta c_0^{-2}zw_a \tag{3n-78a}$$

and just behind

$$t'_s = g^{-1}(w_b) - \beta c_0^{-2}zw_b \tag{3n-78b}$$

Equations (3n-78a), (3n-78b), and (3n-75) or (3n-76) are to be solved simultaneously for w_a, w_b, and t'_s.

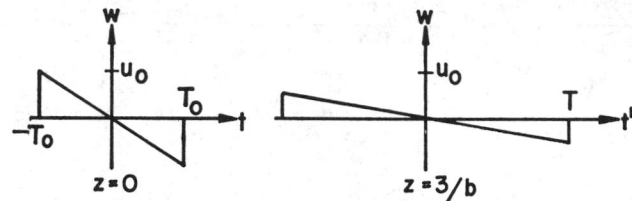

FIG. 3n-3. N wave.

3n-10. Applications of Weak-shock Theory. N *Wave.* Perhaps the most famous application is to the wave shaped like the letter N. The sonic boom is a cylindrical N wave in the far field. For the present consider outgoing waves only. Refer to Fig. 3n-3 for notation. At $t = 0$, $u = -u_0t/T_0$ for $-T_0 < t < T_0$. Thus $g(\phi) = -u_0\phi/T_0$, and Eq. (3n-68b) yields $\phi = t'/(1 + bz)$, where $b = \beta u_0/c_0^2 T_0$. The solution is given by Eq. (3n-68a) as

$$w = -\frac{t'}{T_0}\frac{u_0}{1 + bz} \qquad -T < t' < T$$

To determine T, make use of Eq. (3n-76) for the head shock: that is,

$$\frac{dt'_s}{dz} = -\tfrac{1}{2}\beta c_0^{-2}w_b = \frac{\tfrac{1}{2}bt'_s}{1 + bz}$$

Integration gives

$$-t'_s = T = T_0(1 + bz)^{\frac{1}{2}}$$

The amplitude of the wave is therefore given by

$$u_b = \left(\frac{A_0}{A}\right)^{\frac{1}{2}}\frac{u_0}{(1 + bz)^{\frac{1}{2}}}$$

Next consider incoming waves. The major difference in the results is that z is replaced by $-z$. But z itself also changes sign [see the discussion following Eqs. (3n-67)]. The following formulas cover both incoming and outgoing waves:

$$w = \mp \frac{u_0}{1 + b|z|} \frac{t'}{T_0} \qquad -T < t' < T \tag{3n-79}$$

$$T = T_0(1 + b|z|)^{\frac{1}{2}} \tag{3n-80}$$

$$|u_b| = \left(\frac{A_0}{A}\right)^{\frac{1}{2}} \frac{u_0}{(1 + b|z|)^{\frac{1}{2}}} \tag{3n-81}$$

The growth of a converging wave ($A < A_0$) and the diminution of a diverging wave ($A > A_0$) are not comparable because the factor $(1 + b|z|)^{-\frac{1}{2}}$ acts to diminish both types of waves. Both waves spread at the same rate, however. From Eq. (3n-81) one obtains the classical results that outgoing plane, cylindrical, and spherical waves decay at great distances as $x^{-\frac{1}{2}}$, $r^{-\frac{3}{4}}$, and $r^{-1}(\ln r)^{-\frac{1}{2}}$, respectively.

Sawtooth Wave. Assume that the wave shown in Fig. 3n-3a is repetitive. The magnitude of the jump at the shock is now $2u_0$ to begin with. Because of the symmetry, we have $u_a = -u_b$, which means that, by Eq. (3n-72), the shocks all travel at sonic speed. Unlike the N wave, therefore, the sawtooth does not stretch out as it travels. The decay is more rapid however. Proceeding as before, we find the wave amplitude to be given by

$$|w_b| = \frac{\pi u_0}{\pi + \beta \epsilon k|z|} \tag{3n-82}$$

where k is the fundamental wave number of the wave. See ref. 28 for a discussion of power loss and related topics for sawtooth waves in an exponential horn.

Originally Sinusoidal Wave. It will be recalled that a sinusoidally vibrating piston gives rise to periodic waves whose mathematical description, for outgoing waves, is given by Eq. (3n-40), the Fubini solution. Weak-shock theory makes it possible to obtain a solution of this problem for distances beyond the point of shock formation. It turns out that after forming at $x = \bar{x} = (\beta \epsilon k)^{-1}$, the shocks reach a maximum amplitude at $x = \pi \bar{x}/2$ and thereafter decay. For distance greater than $3\bar{x}$ the wave is effectively a sawtooth of amplitude

$$u_b = \frac{\pi u_0}{1 + \sigma} \tag{3n-83}$$

where (see Sec. 3n-4) $\sigma = \beta \epsilon k x = x/\bar{x}$. This problem is treated in full in ref. 27, as is the similar one of an isolated sine-wave cycle. To generalize Eq. (3n-83) to other one-dimensional outgoing waves it is merely necessary to replace u_b by w_b and σ by $\beta \epsilon k z$.

3n-11. Limitations of Weak-shock Theory. The primary advantage of weak-shock theory over the method based on Burgers' equation (see below) is that results are obtained quickly and easily. Details of the actual profile of the wave in the neighborhood of each shock are suppressed simply by approximating the shock as a mathematical discontinuity. The method's strength is also its weakness, however. At great distances the shocks may become so weak that they become dispersed and are no longer approximate discontinuities.

As a test we may compare the shock rise time (ref. 29) τ with a characteristic period or time duration T of the wave. Thus consider the ratio

$$\frac{\tau}{T} = \frac{12\delta}{c_0|u_b|T} = \frac{12\delta}{c_0|w_b|T} \left(\frac{A}{A_0}\right)^{\frac{1}{2}} \tag{3n-84}$$

where δ is proportional to the viscosity and heat conduction coefficients of the fluid [see Eq. (3n-86)]. For an N wave $|w_b|T$ is a constant ($= u_0 T_0$) so that τ/T is simply proportional to $(A/A_0)^{\frac{1}{2}}$. Therefore, if the N wave is plane, τ/T is constant, which means that the validity of the weak-shock computation does not change with distance. The wave simply spreads out as rapidly as the shock. For all other outgoing N waves, however, the shock disperses more rapidly, and eventually $\tau \sim T$, beyond which point weak-shock theory should not be trusted. Let r_{max} designate the distance at which $\tau/T = 1$. For spherical N waves we obtain,

$$\frac{r_{max}}{r_0} = \frac{\beta u_0 c_0 T_0}{12\delta} \qquad (3n\text{-}85a)$$

The comparable result for cylindrical N waves is

$$\frac{r_{max}}{r_0} = \left(\frac{\beta u_0 c_0 T_0}{12\delta}\right)^2 \qquad (3n\text{-}85b)$$

For an outgoing sawtooth wave τ/T is proportional to $(1 + \beta\epsilon k|z|)(A/A_0)^{\frac{1}{2}}$, which means that weak-shock theory is always limited, even when the wave is plane. Even for converging waves τ may approach T in certain instances (refs. 17, 18). Care must therefore be exercized in using asymptotic formulas based on Eq. (3n-82). Calculations of r_{max} for sawtooth waves based on taking $\tau = T$ are in agreement with estimates obtained by other methods (ref. 27).

The importance of the limitation on weak-shock theory varies a great deal in practice. For sonic booms the limitation is apparently not significant. Typically at ground level τ is of the order of milliseconds, whereas T is measured in tenths of a second. For long-range propagation of pulses from underwater explosions (ref. 30), however, the limitation can be crucial.

In conclusion we remark that "weak-shock theory" is in some respects a misnomer. The theory is valid for weak shocks but not, in general, for very weak ones.

BURGERS' EQUATION AND OTHER MODELS

We now consider explicitly the effects that viscosity, heat conduction, and relaxation have on the propagation of finite-amplitude waves. The full-fledged equations—(3n-1), (3n-2), (3n-3), and (3n-6) or other equation of state—must be dealt with. Successful attacks on these equations have been mainly directed at specific problems, such as the profile of a steady shock wave (ref. 29). General exact results analogous to those for lossless waves are not known. The only general approach presently available, that based on Burgers' equation, is limited to relatively weak waves. For our purposes, however, this method is a fitting companion for weak-shock theory and its predecessor, the approximate theory of lossless simple waves.

3n-12. Thermoviscous Fluids. Burgers' Equation. *Plane Waves.* By employing an approximation procedure similar to that used to change Eq. (3n-7) into (3n-8), Lighthill (ref. 29) reduced the equations of motion for outgoing plane waves in a thermoviscous perfect gas to Burgers' equation,

$$u_t + \beta u u_{x'} = \delta u_{x'x'} \qquad (3n\text{-}86a)$$

Here $x' = x - c_0 t$, $\delta = \frac{1}{2}\nu[\mathcal{V} + (\gamma - 1)/\text{Pr}]$, $\nu = \eta/\rho_0$ is the kinematic viscosity, $\mathcal{V} = (\eta' + 2\eta)/\eta$ is the viscosity number, and $\text{Pr} = \eta C_p/\kappa$ is the Prandtl number. The equation applies as well to fluids of the arbitrary equation of state (refs. 31, 32); simply let β be given by Eq. (3n-60b). In certain cases it applies also to solids (ref. 33).

Equation (3n-86a) is convenient for initial-value problems because the moving coordinate x' reduces to $x' = x$ at $t = 0$. For boundary-value problems a more convenient, yet equally valid, form is (refs. 31, 3, 34)

$$c_0^2 u_x - \beta u u_{t'} = \pm \delta c_0^{-1} u_{t't'} \qquad (3n\text{-}86b)$$

where $t' = t \mp x/c_0$. [To make Eq. (3n-86a) apply to incoming as well as outgoing waves, redefine x' as $x \mp c_0 t$.]

Burgers' equation has a known exact solution. The introduction of the logarithmic potential ζ by

$$u = \pm \frac{2\delta}{\beta c_0} (\ln \zeta)_t = \pm \frac{2\delta}{\beta c_0} \frac{\zeta_{t'}}{\zeta} \tag{3n-87}$$

causes Eq. (3n-86b) to be reduced to

$$\pm c_0{}^3 \zeta_x - \delta \zeta_{t't'} = 0 \tag{3n-88}$$

which is a diffusion equation with the usual roles of space and "time" reversed. To avoid confusion we drop the multiple-sign notation at this point and focus attention on outgoing waves. It is clear that an incoming wave can be considered simply by replacing δ with $-\delta$. The solution of Eq. (3n-88) [with the (+) sign] is

$$\zeta = \sqrt{\frac{K}{\pi}} \int_{-\infty}^{\infty} \zeta_0(\lambda) \exp\left[-K(\lambda - t')^2\right] d\lambda \tag{3n-89}$$

where $K = c_0{}^3/4\delta x$. The quantity $\zeta_0(t') = \zeta(0,t')$ represents the transformed boundary condition. If the original boundary condition is given by Eq. (3n-48), then, by Eq. (3n-87),

$$\zeta_0(t') = \exp\left[\int_{-\infty}^{t'} \frac{\beta c_0}{2\delta} g(\mu) \, d\mu\right] \tag{3n-90}$$

Normally one takes $g(t) = 0$ for $t < 0$, in which case $\zeta_0 = 1$ for $t' < 0$, and the integral's lower limit is zero. The solution of Burgers' equation has been applied to a number of specific problems (refs. 29, 32).

The only solution reviewed here is the one for which the piston motion is sinusoidal (refs. 31, 34, 35): $u(0,t) = u_0 \sin \omega t H(t)$. Equation (3n-90) gives $\zeta_0 = \exp\left[\frac{1}{2}\Gamma(1 - \cos \omega t')\right]$ for $t' > 0$ ($\zeta_0 = 1$ otherwise), where

$$\Gamma = \frac{\beta c_0 u_0}{\delta \omega} = \frac{\beta \epsilon}{\alpha \lambda} \tag{3n-91}$$

and $\alpha\lambda = \alpha/k$ is the dimensionless small-signal attenuation coefficient ($\alpha\lambda = \omega\delta/c_0{}^2$). The dimensionless parameter Γ characterizes the importance of nonlinear effects relative to dissipation. The value $\Gamma = 1$ roughly marks the dividing line between the importance and unimportance of nonlinearity in a periodic wave (ref. 36). When the value of ζ_0 is substituted in Eq. (3n-89), the potential ζ can be separated into transient and steady-state parts. The steady-state part, to which we restrict ourselves, may be expressed as an infinite series,

$$\zeta = I_0(\tfrac{1}{2}\Gamma) + 2 \sum_{n=1}^{\infty} (-1)^n I_n(\tfrac{1}{2}\Gamma) e^{-n^2\alpha x} \cos n\omega t' \tag{3n-92}$$

where I_n is the Bessel function of imaginary argument.

The most interesting case is that of strong waves, i.e., $\Gamma \gg 1$. In this circumstance ζ reduces to a theta function, and the logarithmic differentiation required by Eq. (3n-87) is easy to carry out. The result is (ref. 35)

$$\frac{u}{u_0} = \frac{2}{\Gamma} \sum \frac{\sin n\omega t'}{\sinh n(1 + \sigma)/\Gamma} \tag{3n-93}$$

which is Fay's solution (ref. 37) with Fay's constant α_0 taken to be Γ^{-1}. If σ is not large, the hyperbolic sine function may be approximated by its argument, giving

$$u = \frac{2u_0}{1 + \sigma} \Sigma n^{-1} \sin n\omega t' \tag{3n-94}$$

which represents a sawtooth wave of amplitude

$$u_b = \frac{\pi u_0}{1 + \sigma}$$

This is exactly the same result found by means of weak-shock theory; see Eq. (3n-83).

For strong waves at great distances, i.e., $\sigma \gg \Gamma \gg 1$, the waveform is found, either by the Fay solution or directly by Eqs. (3n-92) and (3n-87), to be

$$u \cong 4\alpha\lambda c_0\beta^{-1}e^{-\alpha x} \sin \omega t' \qquad (3n-95)$$

The simple exponential decay is expected because the wave has now become quite weak. What is remarkable is the absence of the original amplitude factor u_0. The wave amplitude at great distances is independent of the source strength. In other words saturation is reached. This result is obviously of great importance. Saturation has been observed experimentally (refs. 15, 55, 58). Note from Eq. (3n-83) that the asymptotic amplitude given by weak-shock theory is (ref. 26)

$$u_b \cong \frac{\pi c_0^2}{\beta \omega x} \qquad (3n-96)$$

but this result is accurate only in the sawtooth region, which is defined roughly by $3\bar{x} < x < \alpha^{-1}$ (ref. 35).

Nonplanar Waves. For other one-dimensional waves the analog of Eq. (3n-86b) is

$$c_0^3 (u_x + uA_x/2A) - \beta c_0 u u_{t'} = \delta u_{t't'} \qquad (3n-97)$$

(again, for incoming waves replace δ by $-\delta$). It is necessary to make the far-field assumption in deriving this equation. The transformations that have proved so helpful in previous cases, namely, Eqs. (3n-67), lead to

$$c_0^3 w_z - \beta c_0 w w_{t'} = \delta \left(\frac{A}{A_0}\right)^{\frac{1}{2}} w_{t't'} \qquad (3n-98)$$

which is similar to Burgers' equation, but has one variable coefficient. No exact solutions are known.

For periodic spherical and cylindrical waves, solutions of Eq. (3n-98) have been obtained that are valid in the shock-free region ($z < \bar{z}$) and in the sawtooth region (refs. 17, 18). These solutions correspond, respectively, to the Fubini solution for spherical and cylindrical waves and to the related weak-shock solutions (ref. 27). The latter are improved upon, however, because the detailed configuration of the waveform in the vicinity of the shocks is obtained. The behavior of the shock thickness is strongly dependent upon whether the wave is a diverging or a converging one. This can be seen from the form of Eq. (3n-98). A diverging wave ($A > A_0$) is equivalent to a plane wave in a medium in which the dissipation increases with distance. Conversely, for a converging wave ($A < A_0$) the dissipation seems to decrease with distance (refs. 17, 18).

3n-13. Equations for Other Forms of Dissipation. If dissipation is due to an agency other than the thermoviscous effects discussed in the last section, it may still be possible to derive an approximate unidirectional-wave equation similar to Burgers'.

Relaxing Fluids. An elementary example of a relaxing fluid is one that radiates heat in accordance with Eq. (3n-5)(ref. 38). For simplicity take the fluid to be a perfect gas, and let it be inviscid and thermally nonconducting. At very low frequencies infinitesimal waves travel at the isothermal speed of sound, given by $b_0^2 = p_0/\rho_0$. At very high frequencies the speed is the adiabatic value, given by $b_\infty^2 =$

$\gamma p_0/\rho_0$ (the notation b_∞ is used here in place of c_0 to emphasize the role played by frequency). The dispersion m, defined by

$$m \equiv \frac{b_\infty^2 - b_0^2}{b_0^2} \qquad (3n\text{-}99)$$

is equal to $\gamma - 1$ for the radiating gas. If the dispersion is very small, i.e., $m \ll 1$ (which in this case implies $\gamma \doteq 1$), the following approximate equation for plane waves can be derived:

$$\left(q + \frac{\partial}{\partial t'}\right) u_x - b_0^{-2}\left(\beta_i q + \beta_a \frac{\partial}{\partial t'}\right) u u_t' = \pm \frac{m}{2b_0} u_{t't'} \qquad (3n\text{-}100)$$

where $t' = t \mp x/b_0$. It is seen that the radiation coefficient q [see Eq. (3n-5)] is the reciprocal of a relaxation time. Subscripts a and i used with β indicate adiabatic and isothermal values, respectively; that is, $\beta_a = (\gamma + 1)/2$ and $\beta_i = (1 + 1)/2 = 1$. The two values are essentially the same, since it has been assumed that $\gamma \doteq 1$. At either very low frequencies ($\omega q^{-1} \ll 1$) or very high frequencies ($\omega q^{-1} \gg 1$) the left-hand side of the equation takes on the same form as Eq. (3n-47). If the equation is linearized, a dispersion relation can be found that gives the expected behavior for a relaxation process (the actual formulas for the attenuation and phase velocity agree with the exact ones for a radiating gas only for $m \ll 1$).

Polyakova, Soluyan, and Khokhlov considered a relaxation process directly and obtained a pair of equations that can be merged to form a single equation exactly like Eq. (3n-100) except that β_i and β_a are equal (ref. 39). Some solutions (refs. 39, 40) have been found. One represents a steady shock wave. The shock profile is single-valued for very weak shocks. But when the shock is strong enough that its propagation speed [see Eq. (3n-72)] exceeds b_∞, the solution breaks down (a triple-valued waveform is predicted). This illustrates an important fact about the role of relaxation in nonlinear propagation: Relaxation absorption can stand off weak nonlinear effects, but not strong ones. In frequency terms, relaxation offers high attenuation to a broad mid-range of frequencies. If the wave is quite weak, the distortion components are easily absorbed because their frequencies fall in the range of high attenuation. But if the wave is strong, many more very high frequency components are produced, and these are not attenuated efficiently by the relaxation process. To keep the waveform from becoming triple valued, it is necessary to include a viscosity term in the approximate wave equation. In ref. 40 the problem of an originally sinusoidal wave is treated. Quantitative approximate solutions are obtained for cases in which the source frequency is either very low or very high, and a qualitative discussion is given for source frequencies in between.

Marsh, Mellen, and Konrad (ref. 30) postulated a "Burgers-like" equation for spherical waves. It is similar to Eq. (3n-100) but is corrected to take account of spherical divergence. A viscosity term is added, and β_i and β_a are the same. At either very low or very high frequencies the equation takes on the form of Eq. (3n-98) [for spherical waves $(A/A_0)^{\frac{1}{2}} = r/r_0 = e^{z/r_0}$], and some initial attempts at solving this equation were described.

Boundary-layer Effects. Consider the propagation of a plane wave in a thermoviscous fluid contained in a tube. The wave can never be truly plane because the phase fronts curve a great deal as they pass through the viscous and thermal boundary layers at the wall of the tube. If the boundary-layer thicknesses are small compared with the tube radius, however, the curvature of the phase fronts is restricted to very narrow regions, and the wave may be considered quasi-plane. The boundary layers still affect the wave, causing an attenuation that is proportional to $\sqrt{\omega}$ and a comparable dispersion. If the frequency is low, the attenuation from this source is much more important than that due to thermoviscous effects in the mainstream (central core of the fluid), and so it makes sense to find a Burgers-like equation for this case.

A one-dimensional model of time-harmonic wave propagation in ducts with boundary-layer effects treated as a body force has been given by Lamb (ref. 41). Chester (ref. 42) has generalized this model and applied it to compound flow in a closed tube. His method can be used to obtain the following equation for simple-wave flow:

$$u_x - \frac{\beta}{c_0^2} u u_{t'} = \mp \frac{1 + (\gamma - 1)/\sqrt{\text{Pr}}}{c_0 D/2} \left(\frac{\nu}{\pi}\right)^{\frac{1}{2}} \int_0^\infty u_{t'}(x, t' - \mu) \frac{d\mu}{\sqrt{\mu}} \quad (3\text{n-}101)$$

where D is the hydraulic diameter of the duct (four times the cross-sectional area divided by the circumference). No solutions are presently available. But the equation does have proper limiting forms. If the effect of the boundary layers (right-hand side) is neglected, the result is Eq. (3n-47). If the nonlinear term is dropped, the time-harmonic solution can be found, and this solution yields the correct attenuation and dispersion. Because of the relative weakness of boundary-layer attenuation (the dimensionless attenuation $\alpha\lambda$ varies as $1/\sqrt{\omega}$), the higher spectral components generated as a manifestation of steepening of the waveform are not efficiently absorbed. Thus discontinuous solutions, modified somewhat by the attenuation and dispersion, are to be expected.

REFLECTION, STANDING WAVES, AND REFRACTION

3n-14. Reflection and Standing Waves. For plane interacting waves in lossless fluids we return to Eqs. (3n-24) to (3n-26). For perfect gases the Riemann invariants are given by

$$\mathfrak{r} = \frac{c}{\gamma - 1} + \frac{u}{2} \quad (3\text{n-}102a)$$

$$\mathfrak{s} = \frac{c}{\gamma - 1} + \frac{u}{2} \quad (3\text{n-}102b)$$

Equations (3n-26) tell us that the quantity $\mathfrak{r}$ is forwarded unchanged with speed $u + c = \frac{1}{2}(\gamma + 1)\mathfrak{r} - \frac{1}{3}(3 - \gamma)\mathfrak{s}$. Similarly, the speed for the invariant $\mathfrak{s}$ is $u - c = \frac{1}{2}(3 - \gamma)\mathfrak{r} - \frac{1}{2}(\gamma + 1)\mathfrak{s}$. The roles of independent and dependent variables can be reversed to give the following differential equation for the flow:

$$t_{\mathfrak{r}\mathfrak{s}} + N(\mathfrak{r} + \mathfrak{s})^{-1}(t_{\mathfrak{r}} + t_{\mathfrak{s}}) = 0 \quad (3\text{n-}103)$$

where $N = \frac{1}{2}(\gamma + 1)/(\gamma - 1)$. For monatomic and diatomic gases $N = 2$ and $N = 3$, respectively. An exact solution of this equation in terms of arbitrary functions $f(\mathfrak{r})$ and $g(\mathfrak{s})$ is known, but it is usually difficult to determine f and g from the initial conditions (ref. 4).

Reflection. Certain valuable information about reflection can be obtained without solving for the entire flow field. Consider the problem of reflection from a rigid wall. For the moment we need not be specific about the equation of state. Let the incident wave be an outgoing simple wave. The Riemann invariant $\mathfrak{r}$ for a particular signal in this wave is, by Eqs. (3n-21) and (3n-24),

$$2\mathfrak{r} = \lambda_i + u_i = 2\lambda_i$$

But $\mathfrak{r}$ can also be evaluated at the wall during the interaction of the incident and reflected waves: i.e.,

$$2\mathfrak{r} = \lambda_{\text{wall}} + u_{\text{wall}} = \lambda_{\text{wall}}$$

Elimination of $\mathfrak{r}$ between these two expressions gives

$$\lambda_{\text{wall}} = 2\lambda_i$$

This is an exact statement of the law of reflection for continuous finite-amplitude waves at a rigid wall: The quantity λ doubles, not the acoustic pressure.

To see what happens to the pressure, we must specify an equation of state. Take the case of a perfect gas, for which $\lambda = 2(c - c_0)/(\gamma - 1)$ (thus $c - c_0$ doubles at a rigid wall). Using Eq. (3n-11), we obtain

$$\left(\frac{p}{p_0}\right)_{\text{wall}} = \left[2\left(\frac{p_i}{p_0}\right)^{1/\mu} - 1\right]^\mu \tag{3n-105}$$

where $\mu = 2\gamma/(\gamma - 1)$. Now define a wall amplification factor α by

$$\alpha = \frac{p_{\text{wall}} - p_0}{p_i - p_0}$$

Substitution from Eq. (3n-105) gives

$$\alpha = \frac{[2(p_i/p_0)^{1/\mu} - 1]^\mu - 1}{p_i/p_0 - 1} \tag{3n-106}$$

An analogous result in terms of the source that generated the incident simple wave is given in ref. 43; Eq. (3n-106) was first obtained by Pfriem (ref. 44). For weak waves $(p_i - p_0 \ll p_0)$ $\alpha = 2$, in agreement with linear theory. The limiting value for very strong waves is $\alpha = 2^\mu$ ($= 2^7$ for air), a quite startling result. It is only of passing interest, however, because a wave this strong would already have deformed into a shock by the time it reached the wall [for shocks the expression for α is entirely different; the limiting value for strong shocks is $\alpha = 2 + (\gamma + 1)/(\gamma - 1) = 8$ for air (ref. 4)]. In fact, the deviation from pressure doubling is small even for fairly strong waves. For an originally sinusoidal wave of sound pressure level 174 dB, the maximum deviation is about 6 percent (ref. 43).

For a pressure release surface the law of reflection for finite-amplitude waves is the same as for infinitesimal waves. To see this, evaluate r as before, first in the incident wave $(2r = \lambda_i + u_i = 2u_i)$ and then at the pressure-release surface $(2r = \lambda_{\text{surface}} + u_{\text{surface}} = u_{\text{surface}}$, since $\lambda = 0$ when $p = p_0$, $\rho = \rho_0$). The result is

$$u_{\text{surface}} = 2u_i$$

that is, the particle velocity doubles at the surface. The reflection has an interesting effect on the wave, however. Consider a finite wave train so that after interaction the reflected signal is a simple wave. To a good approximation, the acoustic pressure wave suffers phase inversion as a result of the reflection. A wave that distorts as it travels toward the surface therefore tends to "undistort" after reflection. This effect has been observed experimentally (ref. 45).

Reflection from and transmission through other types of surfaces, such as gaseous interfaces, are considered in ref. 43.

Oblique reflection of continuous waves from a plane surface has not been solved in any general way; see ref. 46 for a perturbation treatment.

Standing Waves. First consider finite-amplitude wave motion in a tube closed at one end and containing a vibrating piston in the other end. This problem is one of the few in which much experimental evidence is available (refs. 47, 48, 50). At resonance, if the piston amplitude is sufficiently high, shocks occur traveling to and fro between the piston and the closed end. Slightly off resonance, again for high enough amplitude, the waveform exhibits cusps. Below resonance the cusps occur at the troughs of the waveform, above resonance at the peaks. It would seem that such rich phenomena would have stimulated intensive theoretical treatments of the problem.

In fact, the theoretical problem has proved a difficult nut to crack. The Riemann solution [of Eq. (3n-103)] is of no avail because of the presence of shocks. There is no well-developed weak-shock theory for compound waves as there is for simple waves. For weak waves perturbation treatments have been used (ref. 48). For strong waves one approach has been to assume the existence of shocks at the outset. The Rankine-Hugoniot relations are used to provide boundary conditions for the continuous-wave flow in between shocks (refs. 47, 49).

A more fundamental approach has been taken by Chester (ref. 42). His treatment is of general interest because of the way the effect of the boundary layer is assimilated in the one-dimensional model [see Eq. (3n-101) for an adaptation to simple waves]. An "inviscid solution" is first obtained; it contains discontinuities at and near resonance, and cusps at one point on either side of resonance. General agreement with experimental observation is thus good (ref. 50). Improved solutions are then considered in which thermoviscous effects, first in the mainstream and then in the boundary layers, are taken into account.

3n-15. Refraction. Treatments of oblique reflection and refraction at interfaces have mainly been confined to shock waves in which the flow behind the shock is basically steady. Slow, continuous refraction, such as that caused by gradual changes in the medium or by gradual variations along the phase fronts of the wave, has been treated, however (refs. 26, 51, 52). The basis of the method is ordinary ray acoustics. The propagation speed along each ray tube and the cross-sectional area of the tube are modified to take account of nonlinear effects. The approach is similar to that given in Sec. 3n-7 except that the cross-sectional area of the horn varies in a manner that depends on the wave motion.

Acknowledgment. Support for the preparation of this review came from the Aeromechanics Division, Air Force Office of Scientific Research.

References

1. Vincenti, W. G., and B. S. Baldwin, Jr.: *J. Fluid Mech.* **12**, 449–477 (1962).
2. Stokes, G. G.: *Phil. Mag.*, ser. 4, **1**, 305–317 (1851).
3. Blackstock, D. T.: Approximate Equations Governing Finite-amplitude Sound in Thermoviscous Fluids, *Suppl. Tech. Rept.* AFOSR-5223 (AD 415 442), May, 1963.
4. Courant, R., and K. O. Friedrichs: "Supersonic Flow and Shock Waves" Interscience Publishers, Inc., New York, 1948.
5. Earnshaw, S.: *Trans. Roy. Soc. (London)* **150**, 133–148 (1860).
6. Riemann, B.: *Abhandl. Ges. Wiss. Göttingen, Math.-Physik. Kl.* **8**, 43 (1860), or "Gesammelte Mathematische Werke," 2d ed., pp. 156–175, H. Weber, ed., Dover Publications, Inc., New York, 1953.
7. Poisson, S. D.: *J. École Polytech. (Paris)* **7**, 364–370 (1808). However, Poisson's solution is for the special case of a constant-temperature gas, which in our notation corresponds to $\beta = 1$.
8. Blackstock, D. T.: *J. Acoust. Soc. Am.* **34**, 9–30 (1962).
9. Stokes, G. G.: *Phil. Mag.*, ser. 3, **33**, 349–356 (1848).
10. Fubini, E.: *Alta Frequenza* **4**, 530–581 (1935). Fubini was the first to render the Fourier coefficients in terms of Bessel functions. He used Lagrangian coordinates, not Eulerian as in the derivation here, and attempted to calculate some of the higher-order terms. The mathematical similarity of this problem to Kepler's problem in astronomy is discussed in ref. 8.
11. Gol'dberg, Z. A.: *Akust. Zh.* **6**, 307–310 (1960); English translation: *Soviet Phys.—Acoust.* **6**, 306–310 (1961).
12. Thurston, R. N., and M. J. Shapiro: *J. Acoust. Soc. Am.* **41**, 1112–1125 (1967).
13. Breazeale, M. A., and Joseph Ford: *J. Appl. Phys.* **36**, 3486–3490 (1965).
14. Compare Eq. (3n-51) with Eq. (1), p. 481 in H. Lamb, "Hydrodynamics" 6th ed., Dover Publications, Inc., New York, 1945.
15. Laird, D. T., E. Ackerman, J. B. Randels, and H. L. Oestreicher: Spherical Waves of Finite Amplitude, *WADC Tech. Rept.* 57–463 (AD 130 949), July, 1957.
16. Blackstock, D. T.: *J. Acoust. Soc. Am.* **36**, 217–219 (1964).
17. Naugol'nykh, K. A., S. I. Soluyan, and R. V. Khokhlov: *Vestn. Mosk. Univ. Fiz. Astron.* **4**, 65–71 (1962)(in Russian).
18. Naugol'nykh, K. A., S. I. Soluyan, and R. V. Khokhlov: *Akust. Zh.* **9**, 54–60 (1963); English translation: *Soviet Phys.—Acoust.* **9**, 42–46 (1963).
19. Akulichev, V. A., Yu. Ya. Boguslavskii, A. I. Ioffe, and K. A. Naugol'nykh: *Akust. Zh.* **13**, 321–328 (1967); English translation: *Soviet Phys.—Acoust.* **13**, 281–285 (1968)

20. Cole, R. H.: "Underwater Explosions," Dover Publications, Inc., New York, 1965.
21. Taylor, G. I.: *Proc. Roy. Soc. (London)*, ser. A, **186**, 273–292 (1946).
22. Naugol'nykh, K. A.: *Akust. Zh.* **11**, 351–358 (1965) English translation: *Soviet Phys.—Acoust.* **11**, 296–301 (1966).
23. This solution has been derived by G. B. Whitham, *J. Fluid Mech.* **1**, 290–318, (1956), on a somewhat different basis.
24. Landau, L. D.: *J. Phys. U.S.S.R.* **9**, 496–500 (1945).
25. Friedrichs, K. O.: *Commun. Pure Appl. Math.* **1**, 211–245 (1948).
26. Whitham, G. B.: *Commun. Pure Appl. Math.* **5**, 301–348 (1952).
27. Blackstock, D. T.: *J. Acoust. Soc. Am.* **39**, 1019–1026 (1966).
28. Rudnick, I.: *J. Acoust. Soc. Am.* **30**, 339–342 (1958).
29. Lighthill, M. J.: In "Surveys in Mechanics," pp. 250–351, edited by G. K. Batchelor and R. M. Davies, eds., Cambridge University Press, Cambridge, England, 1956.
30. See, for example, H. W. Marsh, R. H. Mellen, and W. L. Konrad, *J. Acoust. Soc. Am.* **38**, 326–338 (1965).
31. Mendousse, J. S.: *J. Acoust. Soc. Am.* **25**, 51–54 (1953).
32. Hayes, W. D.: "Fundamentals of Gas Dynamics," chap. D, H. W. Emmons, ed., Princeton University Press, Princeton, N.J., 1958.
33. Pospelov, L. A.: *Akust. Zh.* **11**, 359–362 (1965); English translation: *Soviet Phys.—Acoust.* **11**, 302–304 (1966).
34. Soluyan, S. I., and R. V. Khokhlov: *Vestn. Mosk. Univ. Fiz. Astron.* **3**, 52–61 (1961) (in Russian).
35. Blackstock, D. T.: *J. Acoust. Soc. Am.* **36**, 534–542 (1964).
36. Gol'dberg, Z. A.: *Akust. Zh.* **2**, 325–328 (1956); **3**, 322–328 (1957); English translation: *Soviet Phys.—Acoust.* **2**, 346–350 (1956); **3**, 340–347 (1957).
37. Fay, R. D.: *J. Acoust. Soc. Am.* **3**, 222–241 (1931). Fay was concerned with a viscous gas.
38. Truesdell, C. A.: *J. Math. Mech.* **2**, 643–741 (1953).
39. Polykova, A. L., S. I. Soluyan, and R. V. Khokhlov: *Akust. Zh.* **8**, 107–112 (1962); English translation: *Soviet Phys.—Acoust.* **8**, 78–82 (1962).
40. Soluyan, S. I., and R. V. Khokhlov: *Akust. Zh.* **8**, 220–227 (1962); English translation *Soviet Phys.—Acoust.* **8**, 170–175 (1962).
41. Ref. 14, art. 360b.
42. Chester, W.: *J. Fluid Mech.* **18**, 44–64 (1964).
43. Blackstock, D. T.: Propagation and Reflection of Plane Sound Waves of Finite Amplitude in Gases, *Harvard Univ. Acoust. Res. Lab. Tech. Mem.* 43 (AD 242 729), June, 1960.
44. Pfriem, H.: *Forsch. Gebeite Ingenieurw.* **B12**, 244–256 (1941).
45. See, for example, R. H. Mellen and D. G. Browning: *J. Acoust. Soc. Am.* **44**, 646–647 (1968).
46. Shao-sung, F.: *Akust. Zh.* **6**, 491–493 (1960): English translation: *Soviet Phys.—Acoust.* **6**, 488–490 (1961).
47. Saenger, R. A., and G. E. Hudson: *J. Acoust. Soc. Am.* **32**, 961–970 (1960).
48. Coppens, A. B., and J. V. Sanders: *J. Acoust. Soc. Am.* **43**, 516–529 (1968).
49. Betchov, R.: *Phys. Fluids* **1**, 205–212 (1958).
50. Cruikshank, D. B.: An Experimental Investigation of Finite-amplitude Oscillations in a Closed Tube at Resonance, *Univ. Rochester Acoust. Phys. Lab. Tech. Rept.* AFOSR 69-1869 (AD 693 635), July 31, 1969.
51. Whitham, G. B.: *J. Fluid Mech.* **2**, 145–171 (1957).
52. Friedman, M. P., E. J. Kane, and A. Sigalla: *AIAA Journal* **1**, 1327–1335 (1963).
53. Westervelt, P. J.: *J. Acoust. Soc. Am.* **35**, 535–537 (1963).
54. Thuras, A. L., R. T. Jenkins, and H. T. O'Neil: *J. Acoust. Soc. Am.* **6**, 173–180 (1935).
55. Muir, T. G.: "An analysis of the parametric acoustic array for spherical wave fields," Ph.D. dissertation, University of Texas at Austin, Texas (1971).
56. Bellin, J. L. S. and R. T. Beyer: *J. Acoust. Soc. Am.* **34**, 1051–1054 (1962).
57. See, for example, Berktay, H. O.: *J. Sound Vib.* **5**, 155–163 (1967).
58. Lester, W. W.: *J. Acoust. Soc. Am.* **40**, 847–851 (1966).

3o. Nonlinear Acoustics (Experimental)

ROBERT T. BEYER

Brown University

3o-1. Fluids. In the experimental study of nonlinear acoustics, three types of quantities have been measured. These are the effective sound absorption for waves of finite amplitude, the growth of harmonic content, and the nonlinear variation terms in the isentropic expansion of the pressure in the medium in terms of the density changes.

Since the comparison of the first two of these properties with theory depends on the third, it is most effective to consider first the nonlinearity of the equation of state.

This isentropic equation of state can be expanded in a Taylor series in the condensation $s = (\rho - \rho_0)/\rho_0$:

$$p - p_0 = As + \frac{B}{2!} s^2 + \frac{C}{3!} s^3 + \cdots \tag{3o-1}$$

Here p_0 and ρ_0 are the equilibrium values of the pressure and the density. Also, $A = \rho_0 c_0^2$. By application of thermodynamics (ref. 1) the ratio B/A can be written

$$\frac{B}{A} = 2\rho_0 c_0 \left(\frac{\partial c}{\partial p}\right)_T + \frac{2c_0 T\beta}{c_p} \left(\frac{\partial c}{\partial T}\right)_P \tag{3o-2}$$

In this equation, β is the coefficient of thermal expansion, c_p the specific heat at constant pressure, and the derivatives are evaluated under condition of sound waves of infinitesimal amplitude. B/A is sometimes known as the parameter of nonlinearity.

Evaluation of C/A is more involved. It can be shown that (ref. 2)

$$\frac{C}{A} = \frac{3}{2} \left(\frac{B}{A}\right)^2 + 2\rho_0^2 c_0^3 \left(\frac{\partial^2 c}{\partial p^2}\right)_s \tag{3o-3}$$

At a hydrostatic pressure of one atmosphere, the second term on the right is generally quite small compared with the first, although it is likely to become appreciable at higher hydrostatic pressures (ref. 3).

For an ideal gas, we can expand the adiabatic equation of state

$$p = p_0 \left(\frac{\rho}{\rho_0}\right)^\gamma = p_0 \left[1 + \gamma s + \frac{\gamma(\gamma - 1)}{2!} s^2 + \cdots\right] \tag{3o-4}$$

where γ is the ratio of specific heats.

By comparing coefficients in Eqs. (3o-1) and (3o-4) we find

$$A = \gamma p_0 \qquad B = \gamma(\gamma - 1)p_0$$

whence

$$\frac{B}{A} = \gamma - 1 \qquad \text{for an ideal gas} \tag{3o-5}$$

The ratio B/A has now been measured for a considerable number of liquids at atmospheric pressure and, in some instances, over a modest temperature range. A number of these experimental values are given in Table 3o-1. The error in these measurements is generally of the order of 2 to 3 percent, except for the liquid metals, where the larger uncertainties are listed in the table.

The few samples of temperature dependence of B/A shown indicate that B/A can increase or decrease with temperature, depending on the material, but that the temperature variation is usually quite slight.

TABLE 3o-1. VALUES OF B/A FOR VARIOUS LIQUIDS

Liquid	T, °C	B/A	Reference	Liquid	T, °C	B/A	Reference	Estimated error, %
Acetone.......	20	9.2	2	Sulfur.................	121	9.5	6	
Alcohol								
Methyl......	20	9.6	2	Water (distilled).........	0	4.1	3	
					10	4.6	1	
Ethyl........	20	10.5	2		20	5.0	1	
					30	5.2	3	
n-Propyl.....	20	10.7	2		40	5.5	3	
					50	5.55	3	
n-Butyl......	20	10.7	2		60	5.6	3	
					80	5.7	3	
Benzene.......	30	9.0	2					
	40	9.2	2					
	50	9.3	2	Water (sea, 33 % salinity).	0	4.9	2	
	60	9.45	2		10	5.1	2	
	70	9.5	2		20	5.2	2	
					30	5.4	2	
Benzyl alcohol..	30	10.2	2					
Chlorobenzene..	30	9.3	2	Liquid Metals				
				Bismuth...............	318	7.1	8	15
Cyclohexane....	30	10.1	2	66 Bi (wt %), 34 In	125	6.1	8	5
	40	10.1	2					
	50	10.1	2	48 Bi 52 In	125	5.1	8	5
	60	9.85	2	34 Bi 76 In	125	5.1	8	5
	70	9.75	2	17 Bi 83 In	125	4.9	8	5
Diethylamine...	30	10.3	2	Indium................	160	4.55	8	5
Ethylene glycol.	30	9.7	5	Mercury...............	30	2.9	8	3
Ethyl formate..	30	9.8	5	Potassium.............	100	2.9	7	15
Heptane......	30	10.0	4	Sodium................	110	2.7	8	2
Hexane........	30	9.9	4	Tin....................	240	4.4	8	11
Methyl acetate.	30	9.7	2					
Methyl iodide..	30	8.2	5					

The dependence of B/A on hydrostatic pressure is shown in Table 3o-2 for several liquids. Table 3o-3 gives the few known values of the third-order ratio, C/A, all under the approximation

$$\left(\frac{\partial^2 c}{\partial p^2}\right)_S \approx \left(\frac{\partial^2 c}{\partial p^2}\right)_T$$

The general form of the acoustic wave equation for a fluid satisfying Eq. (3o-1) (with neglect of the s^3 and higher terms) is, in Lagrangian coordinates,

$$\frac{\partial^2 \xi}{\partial t^2} = \frac{c_0{}^2}{(1 + \partial \xi/\partial x)^{2+B/A}} \frac{\partial^2 \xi}{\partial x^2} \tag{3o-6}$$

where ξ is the particle displacement, and c_0 is the speed of sound for infinitesimal ξ.

In approximate solutions of this equation [such as Eqs. (3n-40) and (3n-92)], the ratio B/A always appears in the form

$$\beta = 1 + \frac{B}{2A} \tag{3o-7}$$

Hence distortions of the wave form of an initial sinusoid can be used to determine the ratio B/A.

Finally, the effective absorption coefficient for a finite-amplitude wave can be written for a nonrelaxing medium as

$$\frac{\alpha_{\text{eff}}}{\alpha} = 1 + \frac{3\omega^2 \xi_0}{4\alpha c^2} \left(1 + \frac{B}{2A} \right) e^{-2\alpha x}(1 - e^{-2\alpha x})^2 + \text{higher-order terms} \quad (3\text{o-}8)$$

where α is the absorption coefficient for infinitesimal displacement amplitude ξ_0. The ratio B/A could therefore be obtained from this equation, although with reduced accuracy.

TABLE 3o-2. VALUES OF B/A AT VARIOUS PRESSURES

Temperature T, °C	Pressure p, kg/cm²						
	1	250	500	1,000	2,000	4,000	8,000
Water [3]							
0	4.08	4.90	5.58	6.35	6.78	6.60	
40	5.49	5.59	5.69	5.84	6.00	6.06	5.79
80	5.74	5.79	5.84	5.86	5.82	5.64	5.50
1-Propyl alcohol [9]							
30	10.4	....	8.9	8.0	7.3	6.4	5.7
Mercury [9]							
40.5	8.33	....	....	....	7.84	7.37	7.01

TABLE 3o-3. VALUES OF C/A

Pressure at 30°C	$\frac{3}{2}(B/A)^2$	$2\rho_0^2 c_0^3 (\partial^2 c / \partial p^2)_T$ at 30°C	C/A
Water [3]			
1 atm	40.7	−8.7	32.0
2,000 kg/cm²	55.5	−16.9	38.6
4,000 kg/cm²	57.5	−25.0	32.5
8,000 kg/cm²	52.7	−26.7	26.0
1-Propyl alcohol [9]			
1 kg/cm²	162	−87	75
8,000 kg/cm²	49	−24	25

3o-2. Solids. Equation (3n-60c) indicates that the coefficient $\beta = 1 + B/2A$ for liquids must be replaced by $\beta = -M_3/2M_2$ for solids, where M_2 and M_3 are elastic-constant combinations that appear in the partial differential equation for purely longitudinal waves in solids (ref. 10),

$$\frac{\partial^2 u}{\partial t^2} = \frac{1}{\rho_0} \frac{\partial^2 u}{\partial x^2} \left(M_2 + M_3 \frac{\partial u}{\partial x} + \text{higher-order terms} \right) \quad (3\text{o-}9)$$

where u is the displacement velocity. The constants M_2 and M_3 are often written in terms of other so-called second- and third-order elastic coefficients K_2 and K_3:

$$M_2 \equiv K_2 \qquad M_3 = K_3 + 2K_2$$

The coefficients K_2 and K_3 are in turn related to the more familiar second- and third-order elastic constants C_{ij} and C_{ijk}. The connections for the [100], [110], and [111] directions are shown in Table 3o-4.   More detailed relations of this sort are given in ref. 12.

By measurement of the distortion of an initially sinusoidal longitudinal wave through a solid, it is therefore possible to determine the third-order elastic constants. A number of these constants have been determined.   Their values are given in Table 3o-5 (ref. 13).

TABLE 3o-4. K_2 AND K_3 FOR THE [100], [110], AND [111] DIRECTIONS [11]

Direction	K_2	K_3
[100]	C_{11}	C_{111}
[110]	$\dfrac{C_{11} + C_{12} + 2C_{44}}{2}$	$\dfrac{C_{111} + 3C_{112} + 12C_{166}}{4}$
[111]	$\dfrac{C_{11} + 2C_{12} + 4C_{44}}{3}$	$\dfrac{C_{111} + 6C_{112} + 12C_{144} + 24C_{166} + 2C_{123} + 16C_{456}}{9}$

TABLE 3o-5. MEASURED THIRD-ORDER ELASTIC CONSTANTS OF SOME CUBIC
CRYSTALS AT ROOM TEMPERATURE [13]
(x 10^{12} dynes/cm^2)

Crystal	C_{111}	C_{112}	C_{123}	C_{144}	C_{166}	C_{456}	Ref.
Ge	−7.10	−3.89	−0.18	−0.23	−2.92	−0.53	14
Si	−8.25	−4.51	−0.64	+0.12	−3.10	−0.64	14
GaAs	−6.22	−3.87	+0.57	+0.02	−2.69	−0.39	15
GaAs	−6.72	−4.02	−0.04	−0.70	−3.20	−0.69	16
InSb	−3.14	−2.10	−0.48	+0.09	−1.18	+0.002	17
Cu	−15.0	−8.5	−2.5	−1.35	−6.45	−0.16	18
Cu	−12.71	−8.14	−0.50	−0.03	−7.80	−0.95	19
Ge	−7.32	−2.90	−2.2	−0.08	−3.03	−0.41	20
Ge	−7.16	−4.03	−0.18	−0.53	−3.15	−0.47	21
MgO	−48.9	−0.95	−0.69	+1.13	−6.6	+1.47	21
NaCl	−8.3						22
KCl	−7.1						22
NaCl	−8.80	−0.57	0.284	0.257	−0.611	0.271	23
KCl	−7.01	−0.224	0.133	0.127	−0.245	0.118	23
BaF$_2$	−5.84	−2.99	−2.06	−1.21	−0.889	0.271	24
Approx. accuracy, %	±5	±10	±50	±50	±3	±15	

References

1. Beyer, R. T.: *J. Acoust. Soc. Am.* **32**, 719–721 (1960).
2. Coppens, A. B., R. T. Beyer, M. B. Seiden, J. Donohue, F. Guepin, R. D. Hodson and C. Townsend: *J. Acoust. Soc. Am.* **38**, 797–804 (1963).
3. Hagelberg, M. P., G. Holton, and S. Kao: *J. Acoust. Soc. Am.* **41**, 564–567 (1967).
4. Maki, W. C.: M.A.T. thesis, Brown University, Providence, R.I., June, 1966.
5. Freeman, R. A.: M.A.T. thesis, Brown University, Providence, R.I., June, 1966.
6. Dunn, F. W.: M.A.T. thesis, Brown University, Providence, R.I., June, 1967.
7. Sander, C. F.: M.A.T. thesis, Brown University, Providence, R.I., June, 1969.
8. Coppens, A. B., R. T. Beyer, and J. Ballou: *J. Acoust. Soc. Am.* **41**, 1443–1448 (1967).
9. Hagelberg, M. P.: *J. Acoust. Soc. Am.* **47**, 158–162 (1970).
10. Thurston, R. N., and M. J. Shapiro: *J. Acoust. Soc. Am.* **41**, 1112–1125 (1967).
11. Breazeale, M. A., and Joseph Ford: *J. Appl. Phys.* **36**, 3486, 3490 (1965).

12. Thurston, R. N., and K. Brugger: *Phys. Rev.* **133A**, 1604–1610 (1964); erratum, *ibid.* **135**(AB7), 3 (1964).
13. Beyer, R. T., and S. V. Letcher: "Physical Ultrasonics," p. 255, Academic Press, Inc., New York, 1969.
14. McSkimin, H. J., and P. Andreatch, Jr.: *J. Appl. Phys.* **35**, 3312 (1964).
15. McSkimin, H. J., and P. Andreatch, Jr.: *J. Appl. Phys.* **38**, 2610 (1967).
16. Drabble, J. R., and A. J. Brammer: *Solid State Commun.* **4**, 467 (1966).
17. Drabble, J. R., and A. J. Brammer: *Proc. Phys. Soc. (London)* **91**, 959 (1967).
18. Salama, K., and G. A. Alers: *Phys. Rev.* **161**, 673 (1967).
19. Hiki, Y., and A. V. Granato: *Phys. Rev.* **144**, 411 (1966).
20. Bateman, T., W. P. Mason, and H. J. McSkimin: *J. Appl. Phys.* **32**, 928 (1961).
21. Bogardus, E. H.: *J. Appl. Phys.* **36**, 2504 (1965).
22. Stanford, A. L., Jr., and S. P. Zehner: *Phys. Rev.* **153**, 1025 (1967).
23. Chang, Z. P.: *Phys. Rev.* **140A**, 1788 (1965).
24. Gerlich, D.: *Phys. Rev.* **168**, 947 (1968).

3p. Selected References on Acoustics

LEO L. BERANEK

Bolt Beranek and Newman Inc.

Acoustical Materials Association: Sound Absorption Coefficients of Architectural Acoustical Materials, *Acoust. Materials Assoc. Bull.* XXIX, New York, 1969.

Adam, N.: "Akustik," Verlag Paul Haupt, Bern, 1958.

Albers, V. M.: "Underwater Acoustics Handbook," 2d ed. Pennsylvania State University Press, University Park, Pa., 1965.

Albers, V. M.: "Underwater Acoustics," vols. 1 and 2, Plenum Publishing Corporation, New York, 1963, 1967.

ASHRAE: "Guide and Data Book: Systems and Equipment," chap. 31, Sound and Vibration Control, 1967.

Babikov, O. I.: "Ultrasonics and Its Industrial Applications," translated from Russian, Consultants Bureau, Plenum Publishing Corporation, New York, 1960.

Bartholomew, W. T.: "Acoustics of Music," Prentice-Hall, Inc., Englewood Cliffs, N.J., 1946.

Beranek, L. L.: "Acoustics," McGraw-Hill Book Company, New York, 1954.

Beranek, L. L.: "Acoustic Measurements," John Wiley & Sons, Inc., New York, 1960.

Beranek, L. L.: "Noise Reduction," McGraw-Hill Book Company, New York, 1960.

Beranek, L. L.: "Music, Acoustics and Architecture," John Wiley & Sons, Inc., New York, 1962.

Beranek, L. L.: "Noise and Vibration Control," McGraw-Hill Book Company, New York, 1971.

Bergmann, L.: "Der Ultraschall und seine Anwendung in Wissenschaft und Technik," 6th ed., S. Hirzel Verlag KG, Stuttgart, 1954.

Brekhovskikh, L. M.: "Waves in Layered Media," Academic Press, Inc., New York, 1960.

Burris-Meyer, H., and L. S. Goodfriend: "Acoustics for the Architect," Reinhold Publishing Corporation, New York, 1957.

Canac, F., ed.: "Acoustique musicale," Editions du Centre National de la Recherche Scientifique, Paris, 1959.

Carlin, B.: "Ultrasonics," 2d ed. McGraw-Hill Book Company, New York, 1960.

Chalupnik, J. D.: "Transportation Noises," University of Washington Press, Seattle, Washington, 1970.

Crede, C. E.: "Shock and Vibration Concepts in Engineering Design," Prentice-Hall, Inc., Englewood Cliffs, N.J., 1965.

Crede, C. E.: "Vibration and Shock Isolation," John Wiley & Sons, Inc., New York, 1951.
Cremer, L.: "Die wissenschaftlichen Grundlagen der Raumakustik," vol. I, S. Hirzel Verlag KG, Stuttgart, 1949.
Cremer, L.: "Die wissenschaftlichen Grundlagen der Raumakustik," vol. II, S. Hirzel Verlag KG, Stuttgart, 1961.
Cremer, L.: "Die wissenschaftlichen Grundlagen der Raumakustik," vol. III, S. Hirzel Verlag KG, Stuttgart, 1950.
Cremer, L., and M. Heckl: "Körperschall," Springer-Verlag OHG, Berlin, 1967.
Culver, C. A.: "Musical Acoustics," 4th ed., McGraw-Hill Book Company, New York, 1956.
Davis, H. and S. R. Silverman, eds.: "Hearing and Deafness," rev. ed., Holt, Rinehart and Winston, Inc., New York, 1960.
Eckart, C., ed.: "Principles and Applications of Underwater Acoustics." U.S. Government Printing Office, Washington, D.C., 1968.
Fant, G.: "On the Acoustics of Speech," 3 vols, Mouton & Co., The Hague, 1960.
Fletcher, H.: "Speech and Hearing in Communication," D. Van Nostrand Company, Inc., Princeton, N.J., 1953.
Flügge, S., ed.: "Handbuch der Physik," vol. XI/1, Akustik I; vol. XI/2, Akustik II, Springer-Verlag OHG, Berlin, 1961, 1962.
Frayne, J. G., and H. Wolfe: "Sound Recording," John Wiley & Sons, Inc., New York, 1949.
Frederick, J. R.: "Ultrasonic Engineering," John Wiley & Sons, Inc., New York, 1965.
Furrer, W.: "Room and Building Acoustics and Noise Abatement," (Butterworths) Plenum Publishing Corporation, New York, 1964.
Hansen, H. M., and P. F. Chenea: "Mechanics of Vibrations," John Wiley & Sons, Inc., New York, 1952.
Harris, C. M., ed.: "Handbook of Noise Control," McGraw-Hill Book Company, New York, 1957.
Harris, C. M., and E. Crede: "Shock and Vibration Handbook," 3 vols, McGraw-Hill Book Company, New York, 1961.
Helmholtz, H. L. F.: "On the Sensations of Tone as a Physiological Basis for the Theory of Music," translated from 3d German ed. by A. J. Ellis, Longmans, Green & Co., Ltd., London, 1875; 5th rev. ed., 1930.
Herzfeld, K. F., and T. A. Litovitz: "Absorption and Dispersion of Ultrasonic Waves," Academic Press, Inc., New York, 1959.
Hirsch, I. J.: "The Measurement of Hearing," McGraw-Hill Book Company, New York, 1952.
Hueter, T. F., and R. H. Bolt: "Sonics," John Wiley & Sons, Inc., New York, 1955.
Hunt, F. V.: "Electroacoustics," Harvard University Press, Cambridge, Mass., and John Wiley & Sons, Inc., New York, 1954.
Hunter, J. L.: "Acoustics," Prentice-Hall, Inc., Englewood Cliffs, N.J., 1957.
Kacherovich, A. N., and E. E. Khomootov: "Acoustics and Architecture of Cinema Theaters," State Publishing House "Art," Moscow, 1961.
Keast, D. N.: "Measurements in Mechanical Dynamics," McGraw-Hill Book Company, New York, 1967.
Kikuchi, Y.: "Ultrasonic Transducers," Corona Publishing Company, Tokyo, 1969.
Kinsler, L. E., and A. R. Frey: "Fundamentals of Acoustics," 2d ed., John Wiley & Sons, Inc., New York, 1962; 5th printing, 1967.
Knudsen, V., and C. Harris: "Acoustical Designing in Architecture," John Wiley & Sons, Inc., New York, 1950.
Krasil'nikov, V. A.: "Sound and Ultrasound Waves." 3d rev. ed., translated from the Russian by N. Kaner and M. Segal, Israel Program for Scientific Translations Ltd., Jerusalem, 1963.
Kryter, K. D.: The Effects of Noise on Man, *J. Speech Hearing Disorders, Monograph Suppl.* 1, September, 1950.
Kryter, K. D.: "The Effects of Noise on Man," Academic Press, New York, 1970.
Kurtze, G.: "Physics and Techniques of Noise Control," (in German) Verlag G. Braun, Karlsruhe, Germany, 1964.
Lamb, H.: "The Dynamical Theory of Sound," 2d ed., N.Y., Dover Publications, Inc., New York, 1960.
Lamb, H.: "Hydrodynamics," 6th ed., Dover Publications, Inc., New York, 1945.
Lindsay, R. B.: "Mechanical Radiation," McGraw-Hill Book Company, New York, 1960.
Lyon, R. H.: "Random Noise and Vibration in Space Vehicles," U.S. Government Printing Office, Washington, D.C., 1967.
Malecki, I.: "Physical Foundations of Technical Acoustics," Pergamon Press, New York, 1968.

Mason, W. P.: "Electro-mechanical Transducers and Wave Filters," 2d ed., D. Van Nostrand Company, Inc., Princeton, N.J., 1948.

Mason, W. P.: "Piezoelectric Crystals and Their Application to Ultrasonics," D. Van Nostrand Company, Inc., Princeton, N.J., 1950.

Mason, W. P.: "Physical Acoustics," 7 vols., Academic Press, Inc., New York, 1964–1971.

Mason, W. P.: "Physical Acoustics and the Properties of Solids," D. Van Nostrand Company, Inc., Princeton, N.J., 1958.

Miller, G. A.: "Language and Communication," McGraw-Hill Book Company, New York, 1951.

Morse, P. M.: "Vibration and Sound," 2d ed., McGraw-Hill Book Company, New York, 1948.

Morse, P. M., and K. U. Ingard: "Theoretical Acoustics," McGraw-Hill Book Company, New York, 1968.

Ol'shevskii, V. V.: "Characteristics of Sea Reverberations," translated from Russian, Consultants Bureau, Plenum Publishing Corporation, New York, 1967.

Olson, H. F.: "Acoustical Engineering," 3d ed., D. Van Nostrand Company, Inc., Princeton, N.J., 1957.

Olson, H. F.: "Solution of Energy Problems by Dynamical Analogies," D. Van Nostrand Company, Inc., Princeton, N.J., 1958.

Olson, H. F.: "Musical Engineering," McGraw-Hill Book Company, New York, 1952.

Parkin, P. H., and H. R. Humphreys: "Acoustics, Noise and Buildings," Faber & Faber, Ltd., London, 1958.

Parkin, P. H., H. J. Purkis, and W. E. Scholes: "Field Measurements of Sound Insulation Between Dwellings," Her Majesty's Stationery Office, London, 1960.

Peterson, A. P. G., and E. E. Gross, Jr.: "Handbook of Noise Measurement," 6th ed., General Radio Company, West Concord, Mass., 1967.

Pierce, J. R., and E. E. David Jr.: "Man's World of Sound," Doubleday & Company Inc., Garden City, N.Y., 1958.

Purkis, H. J.: "Building Physics: Acoustics," Pergamon Press, New York, 1966.

Lord Rayleigh: "The Theory of Sound," 2d ed., vols. 1 and 2, Dover Publications, Inc., New York, 1945.

Rettinger, M.: "Acoustics: Room Design and Noise Control," Chemical Publishing Company, Inc., New York, 1968.

Reichardt, W.: "Foundations of Technical Acoustics" (in German), Portig K. G., Leipzig, 1968.

Richardson, E. G.: "Technical Aspects of Sound," 3 vols., American Elsevier Publishing Company, Inc., New York, 1962.

Rschevkin, S. N.: "A Course of Lectures on the Theory of Sound," translated from the Russian by O. M. Blunn, edited by P. E. Doak, Pergamon Press, New York, 1963.

Schaafs, W.: "Landolt-Börnstein New Series, Group II," "Atomic and Molecular Physics," vol. 5, "Molecular Acoustics," Springer-Verlag New York Inc., New York, 1967.

Skudrzyk, E.: "Die Grundlagen der Akustik," Springer-Verlag HG, Vienna, 1954.

Skudrzyk, E.: "Simple and Complex Vibrating Systems," Pennsylvania State University Press, University Park, Pa., 1968.

Stephens, R. W. B., and A. E. Bate: "Acoustics and Vibrational Physics," St. Martin's Press, Inc., New York, 1966.

Stevens, S. S., J. G. C. Loring, and D. Cohen: "Bibliography on Hearing," Harvard University Press, Cambridge, Mass., 1955.

Stevens, S. S., ed.: "Handbook of Experimental Psychology," John Wiley & Sons, Inc. New York, 1951.

Swenson, G. W., Jr.: "Principles of Modern Acoustics," D. Van Nostrand Company, Inc., Princeton, N.J., 1953.

Tolstoy, I. and P. S. Clay: "Ocean Acoustics," McGraw-Hill Book Company, New York, 1966.

Trapp, W. J., and D. M. Forney, Jr., eds.: "Acoustical Fatigue in Aerospace Structures," Syracuse University Press, Syracuse, New York, 1965.

Tucker, D. G., and B. Z. Gazey: "Applied Underwater Acoustics," Pergamon Press, New York, 1966.

Urick, R. J.: "Principles of Underwater Sound for Engineers," McGraw-Hill Book Company, New York, 1967.

Wever, E. G., and M. Lawrence: "Physiological Acoustics," Princeton University Press, Princeton, N.J., 1954.

Wiethaup, H.: "Noise Abatement in Western Germany" (in German), Carl Heymanns Verlag KG, Cologne, 1961.

Wood, A.: "Acoustics," Dover Publications, Inc., New York, 1966.

Zwikker, C., and C. W. Kosten: "Sound Absorbing Materials," American Elsevier Publishing Company, Inc., New York, 1949.

Section 4

HEAT

MARK W. ZEMANSKY, Editor

The City College of the City University of New York

CONTENTS

4a. Temperature Scales, Thermocouples, and Resistance Thermometers[1]

H. H. PLUMB, R. L. POWELL, W. J. HALL, AND J. F. SWINDELLS

The National Bureau of Standards

The Comité International des Poids et Mesures (CIPM) in October, 1968, agreed to adopt the International Practical Temperature Scale of 1968[2] (IPTS-68) in accordance with the decision of the 13th General Conference of Weights and Measures, Resolution 8, of October, 1967. IPTS-68 has replaced IPTS-48 (amended edition of 1960). It was formulated in such a way that temperature measured on it closely approximates the thermodynamic temperature, and extends the range of definition down to 13.81 kelvins. (The previous scale, IPTS-48, terminated at $-183°C$.)

The basic temperature is the thermodynamic temperature, symbol T, the unit of which is the kelvin, symbol K. The kelvin is the fraction $1/273.16$ of the thermodynamic temperature of the triple point of water.[3] The Celsius temperature, symbol t, is defined by

$$t = T - T_0$$

where $T_0 = 273.15$ K (the ice point). The unit employed to express a Celsius temperature is the degree Celsius, symbol °C, which is equal to the kelvin. A difference of temperature is expressed in kelvins; it may also be expressed in degrees Celsius.

The International Practical Temperature Scale of 1968 distinguishes between the International Practical Kelvin Temperature with the symbol T_{68} and the International Practical Celsius Temperature with the symbol t_{68}. The relation between T_{68} and t_{68} is

$$t_{68} = T_{68} - 273.15 \text{ K}$$

The units of T_{68} and t_{68} are the kelvin, symbol K, and degree Celsius, symbol °C, as in the case of the thermodynamic temperature T and the Celsius temperature t.

The IPTS-68 is based on the assigned values of the temperatures of a number of reproducible equilibrium states (defining fixed points) and on standard instruments calibrated at those temperatures. Interpolation between the fixed-point temperatures is provided by formulas used to establish the relation between indications of the standard instruments and values of the International Practical Temperature. The defining fixed points are given in Table 4a-1.

---

[1] Acknowledgment is made of the previous contributions to this section in the second edition of the Handbook by H. F. Stimson, J. F. Swindells, and R. E. Wilson. Data on optical pyrometry and thermal radiation are given in Sec. 6.

[2] The text in French of this scale is published in *Compt. rend. 13ème conf. gén. poids mesures*, 1967–1968, Annexe 2, and Comité Consultatif de Thermométrie, 8ᵉ session, 1967, Annexe 18. The English text is published in *Metrologia* **5**(2), 35 (1969).

[3] 13th General Conference of Weights and Measures, 1967, Resolutions 3 and 4.

TABLE 4a-1. DEFINING FIXED POINTS OF THE IPTS-68*

Equilibrium state	Assigned value of International Practical Temperature	
	T_{68} K	t_6 °C
Equilibrium between the solid, liquid, and vapor phases of equilibrium hydrogen (triple point of equilibrium hydrogen)	13.81	−259.34
Equilibrium between the liquid and vapor phases of equilibrium hydrogen at a pressure of 33 330.6 N/m² (25/76 standard atmosphere)	17.042	−256.108
Equilibrium between the liquid and vapor phases of equilibrium hydrogen (boiling point of equilibrium hydrogen)	20.28	−252.87
Equilibrium between the liquid and vapor phases of neon (boiling point of neon)	27.402	−246.048
Equilibrium between the solid, liquid, and vapor phases of oxygen (triple point of oxygen)	54.361	−218.789
Equilibrium between the liquid and vapor phases of oxygen (boiling point of oxygen)	90.188	−182.962
Equilibrium between the solid, liquid, and vapor phases of water (triple point of water) †	273.16	0.01
Equilibrium between the liquid and vapor phases of water (boiling point of water) †‡	373.15	100
Equilibrium between the solid and liquid phases of zinc (freezing point of zinc)	692.73	419.58
Equilibrium between the solid and liquid phases of silver (freezing point of silver)	1235.08	961.93
Equilibrium between the solid and liquid phases of gold (freezing point of gold)	1337.58	1064.43

* Except for the triple points and one equilibrium hydrogen point (17.042 K) the assigned values of temperature are for equilibrium states at a pressure $p_0 = 1$ standard atmosphere (101 325 N/m²). In the realization of the fixed points small departures from the assigned temperatures will occur as a result of the differing immersion depths of thermometers or the failure to realize the required pressure exactly. If due allowance is made for these small temperature differences, they will not affect the accuracy of realization of the Scale.

† The water used should have the isotopic composition of ocean water.

‡ The equilibrium state between the solid and liquid phases of tin (freezing point of tin) has the assigned value of $t_{68} = 231.9681$ °C and may be used as an alternative to the boiling point of water.

In the range 13.81 to 273.15 K, the interpolating instrument is a platinum thermometer and T_{68} is defined by the relation

$$W(T_{68}) = W_{\text{CCT-68}}(T_{68}) + \Delta W(T_{68}) \qquad (4a\text{-}1)$$

where $W(T_{68})$ is the resistance ratio of the platinum thermometer as defined by

$$W(T_{68}) = \frac{R(T_{68})}{R(273.15 \text{ K})}$$

and $W_{\text{CCT-68}}(T_{68})$ is the resistance ratio as given by the reference function in Table 4a-2.

The deviations $\Delta W(T_{68})$ at the temperatures of the defining fixed points are the differences between the measured values of $W(T_{68})$ and the corresponding values of $W_{\text{CCT-68}}(T_{68})$.

TABLE 4a-2. THE REFERENCE FUNCTION $W_{CCT\text{-}68}(T_{68})$ FOR PLATINUM RESISTANCE THERMOMETERS FOR THE RANGE FROM 13.81 TO 273.15 K*

$$T_{68} = \left\{ A_0 + \sum_{i=1}^{20} A_i[\ln W_{CCT\text{-}68}(T_{68})]^i \right\} \text{ K}$$

Coefficients A_i:

i	A_i	i	A_i
0	$0.273\ 15 \times 10^3$	11	$0.767\ 976\ 358\ 170\ 845\ 8 \times 10$
1	$0.250\ 846\ 209\ 678\ 803\ 3 \times 10^3$	12	$0.213\ 689\ 459\ 382\ 850\ 0 \times 10$
2	$0.135\ 099\ 869\ 964\ 999\ 7 \times 10^3$	13	$0.459\ 843\ 348\ 928\ 069\ 3$
3	$0.527\ 856\ 759\ 008\ 517\ 2 \times 10^2$	14	$0.763\ 614\ 629\ 231\ 648\ 0 \times 10^{-1}$
4	$0.276\ 768\ 548\ 854\ 105\ 2 \times 10^2$	15	$0.969\ 328\ 620\ 373\ 121\ 3 \times 10^{-2}$
5	$0.391\ 053\ 205\ 376\ 683\ 7 \times 10^2$	16	$0.923\ 069\ 154\ 007\ 007\ 5 \times 10^{-3}$
6	$0.655\ 613\ 230\ 578\ 069\ 3 \times 10^2$	17	$0.638\ 116\ 590\ 952\ 653\ 8 \times 10^{-4}$
7	$0.808\ 035\ 868\ 559\ 866\ 7 \times 10^2$	18	$0.302\ 293\ 237\ 874\ 619\ 2 \times 10^{-5}$
8	$0.705\ 242\ 118\ 234\ 052\ 0 \times 10^2$	19	$0.877\ 551\ 391\ 303\ 760\ 2 \times 10^{-7}$
9	$0.447\ 847\ 589\ 638\ 965\ 7 \times 10^2$	20	$0.117\ 702\ 613\ 125\ 477\ 4 \times 10^{-8}$
10	$0.212\ 525\ 653\ 556\ 057\ 8 \times 10^2$		

* The reference function $W_{CCT\text{-}68}(T_{68})$ is continuous at $T_{68} = 273.15$ K in its first and second derivatives with the function $W(t_{68})$ given by Eqs. (4a-6) and (4a-7) for $\alpha = 3.9259668 \times 10^{-3}(°C)^{-1}$ and $\delta = 1.496334°C$. A tabulation of this reference function, sufficiently detailed to allow interpolation to an accuracy of 0.0001 K, is available from the Bureau International des Poids et Mesures, 92-Sèvres, France.

The following interpolation formulas are used to determine $\Delta W(T_{68})$ at intermediate temperatures:

13.81 to 20.28 K: $\quad \Delta W(T_{68}) = A_1 + B_1 T_{68} + C_1 T_{68}^2 + D_1 T_{68}^3 \quad$ (4a-2)[1]
20.28 to 54.361 K: $\quad \Delta W(T_{68}) = A_2 + B_2 T_{68} + C_2 T_{68}^2 + D_2 T_{68}^3 \quad$ (4a-3)[2]
54.361 to 90.188 K: $\quad \Delta W(T_{68}) = A_3 + B_3 T_{68} + C_3 T_{68}^3 \quad$ (4a-4)[3]
90.188 to 273.15 K: $\quad \Delta W(T_{68}) = A_4 t_{68} + C_4 t_{68}^3(t_{68} - 100°C) \quad$ (4a-5)[4]

In the range 0°C (273.15 K) to 630.74°C, t_{68} is defined by

$$t_{68} = t' + 0.045 \frac{t'}{100°C}\left(\frac{t'}{100°C} - 1\right)$$
$$\left(\frac{t'}{419.58°C} - 1\right)\left(\frac{t'}{630.74°C} - 1\right)°C \qquad (4a\text{-}6)$$

where t' is defined as

$$t' = \frac{1}{\alpha}[W(t') - 1] + \delta\left(\frac{t'}{100°C}\right)\left(\frac{t'}{100°C} - 1\right) \qquad (4a\text{-}7)$$

[1] Constants for Eq. (4a-2) are determined by the three measured deviations—at the triple point of equilibrium hydrogen, the temperature of 17.042 K, and the boiling point of equilibrium hydrogen—and by the derivative of the deviation function at the boiling point of equilibrium hydrogen as derived from Eq. (4a-3).
[2] Constants for Eq. (4a-3) are determined by the three measured deviations—at the boiling point of equilibrium hydrogen, the boiling point of neon, and the triple point of oxygen—and by the derivative of the deviation function at the triple point of oxygen as derived from Eq. (4a-4).
[3] Constants for Eq. (4a-4) are determined by the two measured deviations—at the triple point and the boiling point of oxygen and by the derivative of the deviation function at the boiling point of oxygen as derived from Eq. (4a-5).
[4] Constants for Eq. (4a-5) are determined by the two measured deviations at the boiling point of oxygen and the boiling point of water.

The resistance ratio $W(t') = R(t')/R(0°C)$ and the constants $R(0°C)$, α and δ are determined by measurement of three resistances—at the triple point of water, the boiling point of water (or the freezing point of tin), and the freezing point of zinc. Equation (4a-7) is equivalent to

$$W(t') = 1 + At' + Bt'^2 \tag{4a-8}$$

when $\quad A = \alpha \left(1 + \dfrac{\delta}{100°C}\right) \quad$ and $\quad B = -10^{-4}\,\alpha\delta(°C)^{-2}$

From 630.74 to 1064.43°C, t_{68} is defined by

$$E(t_{68}) = a + bt_{68} + ct_{68}^2 \tag{4a-9}$$

where $E(t_{68})$ is the electromotive force of a standard thermocouple of rhodium-platinum alloy and platinum, when one junction is at the temperature $t_{68} = 0°C$ and the other junction is at temperature t_{68}. The constants a, b, and c are calculated from the values of E at $630.74 \pm 0.2°C$, as determined by a platinum resistance thermometer, and at the freezing points of silver and gold.

Above 1337.58 K (1064.43°C) the temperature T_{68} is defined by

$$\frac{L_\lambda(T_{68})}{L_\lambda(T_{68}(\text{Au}))} = \frac{\exp\left[c_2/\lambda T_{68}(\text{Au})\right] - 1}{\exp\left[c_2/\lambda T_{68}\right] - 1} \tag{4a-10}$$

where $L_\lambda(T_{68})$ and $L_\lambda(T_{68}(\text{Au}))$ are the spectral concentrations at temperature T_{68} and at the freezing point of gold, $T_{68}(\text{Au})$ of the radiance of a black body at the wavelength[1] λ; $c_2 = 0.014388$ meter kelvin.

Table 4a-3 (p. 4-6) lists the approximate differences between the IPTS-68 and IPTS-48 and should prove to be a utility for many references to this section.

To avoid creating conflicting statements, the preceding description of IPTS-68 has for the most part been taken from the English language version of the International Practical Temperature Scale of 1958 as it appeared in *Metrologia*. For a more complete description of the IPTS-68 and pertinent supplementary information the reader should refer to the defining text.[2]

In Tables 4a-4, 4a-5, and 4a-6 which follow, values have been adjusted to agree with IPTS-68.

[1] Since $T_{68}(\text{Au})$ is close to the thermodynamic temperature of the freezing point of gold, and c_2 is close to the second radiation constant of the Planck equation, it is not necessary to specify the value of the wavelength to be employed in the measurements [see *Metrologia* **3**, 28 (1967)].

[2] *Metrologia* **5** (2), 35 (1969).

TABLE 4a-3. APPROXIMATE DIFFERENCES ($t_{68} - t_{48}$), IN KELVINS, BETWEEN THE VALUES OF TEMPERATURE GIVEN BY THE IPTS OF 1968 AND THE IPTS OF 1948

t_{68}°C	0	−10	−20	−30	−40	−50	−60	−70	−80	−90	−100
−100	0.022	0.013	0.003	−0.006	−0.013	−0.013	−0.005	0.007	0.012		
0	0.000	0.006	0.012	0.018	0.024	0.029	0.032	0.034	0.033	0.029	0.022

t_{68}°C	0	10	20	30	40	50	60	70	80	90	100
0	0.000	−0.004	−0.007	−0.009	−0.010	−0.010	−0.010	−0.008	−0.006	−0.003	0.000
100	0.000	0.004	0.007	0.012	0.016	0.020	0.025	0.029	0.034	0.038	0.043
200	0.043	0.047	0.051	0.054	0.058	0.061	0.064	0.067	0.069	0.071	0.073
300	0.073	0.074	0.075	0.076	0.077	0.077	0.077	0.077	0.077	0.076	0.076
400	0.076	0.075	0.075	0.075	0.074	0.074	0.074	0.075	0.076	0.077	0.079
500	0.079	0.082	0.085	0.089	0.094	0.100	0.108	0.116	0.126	0.137	0.150
600	0.150	0.165	0.182	0.200	0.23	0.25	0.28	0.31	0.34	0.36	0.39
700	0.39	0.42	0.45	0.47	0.50	0.53	0.56	0.58	0.61	0.64	0.67
800	0.67	0.70	0.72	0.75	0.78	0.81	0.84	0.87	0.89	0.92	0.95
900	0.95	0.98	1.01	1.04	1.07	1.10	1.12	1.15	1.18	1.21	1.24
1000	1.24	1.27	1.30	1.33	1.36	1.39	1.42	1.44			

t_{68}°C	0	100	200	300	400	500	600	700	800	900	1000
1000	3.2	1.5	1.7	1.8	2.0	2.2	2.4	2.6	2.8	3.0	3.2
2000	5.9	3.5	3.7	4.0	4.2	4.5	4.8	5.0	5.3	5.6	5.9
3000		6.2	6.5	6.9	7.2	7.5	7.9	8.2	8.6	9.0	9.3

TABLE 4a-4. THERMAL EMF OF CHEMICAL ELEMENTS RELATIVE TO PLATINUM*

Temp., °C	Lithium, mV	Sodium, mV	Potassium, mV	Rubidium, mV	Cesium, mV	Calcium, mV	Cerium, mV
−200	−1.12	+1.00	+1.61	+1.09	+0.22		
−100	−1.00	+0.29	+0.78	+0.46	−0.13		
0	0	0	0	0	0	0	0
+100	+1.82	....	....	......	......	−0.51	+1.14
200	......	....	....	......	......	−1.13	2.46
300	......	....	....	......	......	−1.85	

Temp., °C	Magnesium, mV	Zinc, mV	Cadmium, mV	Mercury, mV	Indium, mV	Thallium, mV	Aluminum, mV
−200	+0.37	−0.07	−0.04	......	......	......	+0.45
−100	−0.09	−0.33	−0.31	......	......	......	+0.06
0	0	0	0	0	0	0	0
+100	+0.44	+0.76	+0.90	−0.60	+0.69	+0.58	+0.42
200	+1.10	1.89	2.35	−1.33	......	1.30	1.06
300	......	3.42	4.24	......	......	2.16	1.88
400	......	5.29	....	......	......	......	2.84
500	......	....	....	......	......	......	3.93
600	......	....	....	......	......	......	5.15

Temp. °C	Carbon, mV	Silicon, mV	Germanium, mV	Tin, mV	Lead, mV	Antimony, mV	Bismuth, mV
−200	......	+63.13	−46.00	+0.26	+0.24	......	+12.39
−100	......	+37.17	−26.62	−0.12	−0.13	......	+7.54
0	0	0	0	0	0	0	0
+100	+0.70	−41.56	+33.9	+0.42	+0.44	+4.89	−7.34
200	1.54	−80.57	72.4	1.07	1.09	10.14	−13.57
300	2.55	−110.07	91.8	......	1.91	15.44	
400	3.72	....	82.3	......	......	20.53	
500	5.15	....	63.5	......	......	25.10	
600	6.79	....	43.9	......	......	28.87	
700	8.82	....	27.9				
800	10.98						
900	13.55						
1000	16.46						
1100	19.46						

TABLE 4a-4. THERMAL EMF OF CHEMICAL ELEMENTS RELATIVE TO PLATINUM*
(*Continued*)

Temp., °C	Copper, mV	Silver, mV	Gold, mV	Cobalt, mV	Nickel, mV
−200	−0.19	−0.21	−0.21	......	+2.28
−100	−0.37	−0.39	−0.39	......	+1.22
0	0	0	0	0	0
+100	+0.76	+0.74	+0.78	−1.33	−1.48
200	1.83	1.77	1.84	−3.08	−3.10
300	3.15	3.05	3.14	−5.10	−4.59
400	4.68	4.57	4.63	−7.24	−5.45
500	6.41	6.36	6.29	−9.35	−6.16
600	8.34	8.41	8.12	−11.28	−7.04
700	10.47	10.73	10.11	−12.87	−8.10
800	12.81	13.33	12.26	−13.99	−9.33
900	15.37	16.16	14.58	−14.49	−10.67
1000	18.16	......	17.05	−14.21	−12.11
1100	......	......	......	−13.01	−13.60
1200	......	......	......	−10.70	

Temp., °C	Iridium, mV	Rhodium, mV	Palladium, mV	Molybdenum, mV	Tungsten, mV	Tantalum, mV	Thorium, mV
−200	−0.25	−0.20	+0.81	......	+0.43	+0.21	
−100	−0.35	−0.34	+0.48	......	−0.15	−0.10	
0	0	0	0	0	0	0	0
+100	+0.65	+0.70	−0.57	+1.45	+1.12	+0.33	−0.13
200	1.49	1.61	−1.23	3.19	2.62	0.93	−0.26
300	2.47	2.68	−1.99	5.23	4.48	1.79	−0.40
400	3.55	3.91	−2.82	7.57	6.70	2.91	−0.50
500	4.78	5.28	−3.84	10.20	9.30	4.30	−0.53
600	6.10	6.77	−5.03	13.13	12.26	5.95	−0.45
700	7.55	8.39	−6.40	16.33	15.58	7.86	−0.21
800	9.10	10.14	−7.96	19.83	19.25	10.02	+0.22
900	10.77	12.01	−9.69	23.63	23.30	12.45	+0.86
1000	12.57	14.02	−11.61	27.74	27.73	15.15	+1.72
1100	14.45	16.15	−13.67	32.15	32.53	18.13	+2.78
1200	16.45	18.39	−15.86	36.86	37.72	21.37	+4.03
1300	18.45	20.69	−18.11	......	......	......	+5.41
1400	20.47	22.99	−20.40				
1500	22.51	25.36	−22.75				

* A positive sign means that in a simple thermoelectric circuit the resultant emf given is in such a direction as to produce a current from the element to the platinum at the reference junction (0°C).

The values below 0°C, in most cases, have not been determined on the same samples as the values above 0°C.

Based upon the original table in American Institute of Physics, "Temperature, Its Measurement and Control in Science and Industry," pp. 1309–1310, Reinhold Book Corporation, New York, 1941. Values of the emf have been adjusted to correspond to temperatures expressed on the International Practical Temperature Scale of 1968.

TABLE 4a-5. THERMAL EMF OF IMPORTANT THERMOCOUPLE MATERIALS RELATIVE TO PLATINUM*

Temp., °C	Chromel P, mV	Alumel, mV	Copper, mV	Iron, mV	Constantan, mV
−200	−3.36	+2.39	−0.19	−2.92	+5.35
−100	−2.20	+1.29	−0.37	−1.84	+2.98
0	0	0	0	0	0
+100	+2.81	−1.29	+0.76	+1.89	−3.51
200	5.96	−2.17	1.83	3.54	−7.45
300	9.32	−2.89	3.15	4.85	−11.71
400	12.75	−3.64	4.68	5.88	−16.19
500	16.21	−4.43	6.41	6.79	−20.79
600	19.61	−5.28	8.34	7.80	−25.46
700	22.94	−6.18	10.47	9.11	−30.15
800	26.20	−7.07	12.81	10.84	−34.81
900	29.37	−7.94	15.37	12.82	−39.39
1000	32.47	−8.78	18.16	14.28	−43.85
1100	35.52	−9.57			
1200	38.48	−10.33			
1300	41.38	−11.06			
1400	44.04	−11.77			

* American Institute of Physics, "Temperature, Its Measurement and Control in Science and Industry," p. 1308, Reinhold Book Corporation, New York, 1941. Values of the emf have been adjusted to correspond to temperatures expressed on the International Practical Temperature Scale of 1968.

TABLE 4a-6. THERMAL EMF OF SOME ALLOYS RELATIVE TO PLATINUM*

Temp., °C	Manganin, mV	Gold-chromium, mV	Copper-beryllium, mV	Yellow brass, mV	Phosphor bronze, mV	Solder 50 Sn–50 Pb, mV	Solder 96.5 Sn–3.5 Ag, mV
0	0	0	0	0	0	0	0
+100	+0.61	−0.17	+0.67	+0.60	+0.55	+0.46	+0.45
200	1.55	−0.32	1.62	1.49	1.34		
300	2.77	−0.44	2.81	2.58	2.34		
400	4.25	−0.55	4.19	3.85	3.50		
500	5.95	−0.63	......	5.30	4.81		
600	7.84	−0.66	......	6.96	6.30		

TABLE 4a-6. THERMAL EMF OF SOME ALLOYS RELATIVE TO PLATINUM*
(*Continued*)

Temp., °C	18-8 stainless steel, mV	Spring steel, mV	80 Ni– 20 Cr, mV	60 Ni– 24 Fe– 16 Cr, mV	Copper coin (95 Cu– 4 Sn– 1 Zn), mV	Nickel coin (75 Cu– 25 Ni), mV	Silver coin (90 Ag– 10 Cu), mV
0	0	0	0	0	0	0	0
+100	+0.44	+1.32	+1.14	+0.85	+0.60	−2.76	+0.80
200	1.04	2.63	2.62	2.01	1.48	−6.01	1.90
300	1.76	3.81	4.34	3.41	2.60	−9.71	3.25
400	2.60	4.84	6.25	5.00	3.91	−13.78	4.81
500	3.56	5.80	8.31	6.76	5.44	−18.10	6.59
600	4.67	6.86	10.53	8.68	7.14	−22.59	8.64
700	5.92	......	12.89	10.76			
800	7.35	......	15.41	13.03			
900	8.96	......	18.07	15.47			
1000	......	......	20.87	18.06			

* American Institute of Physics, "Temperature, Its Measurement and Control in Science and Industry," p. 1310, Reinhold Book Corporation, New York, 1941. Values of the emf have been adjusted to correspond to temperatures expressed on the International Practical Temperature Scale of 1968.

Thermocouple Reference Tables. Tables 4a-7 through 4a-12 contain abbreviated data on the thermoelectric voltages of six thermocouple combinations, two noble-metal types S and R and four base-metal types E, J, K, and T. The full tables, functional representations, approximations, and material descriptions appear in NBS Monograph 125, "Thermocouple Reference Tables Based on the IPTS-68" by R. L. Powell, W. J. Hall, C. H. Hyink, L. L. Sparks, G. W. Burns, and H. H. Plumb, U.S. Government Printing Office, Washington, D.C., 1972.

TABLE 4a-7. TYPE S. PLATINUM–10% RHODIUM VS. PLATINUM THERMOCOUPLES
[Emf, absolute millivolts; temp., °C (IPTS-68); reference junctions at 0°C]

°C	0	10	20	30	40	50	60	70	80	90	100
0	0.000	0.055	0.113	0.173	0.235	0.299	0.365	0.432	0.502	0.573	0.645
100	0.645	0.719	0.795	0.872	0.950	1.029	1.109	1.190	1.273	1.356	1.440
200	1.440	1.525	1.611	1.698	1.785	1.873	1.962	2.051	2.141	2.232	2.323
300	2.323	2.414	2.506	2.599	2.692	2.786	2.880	2.974	3.069	3.164	3.260
400	3.260	3.356	3.452	3.549	3.645	3.743	3.840	3.938	4.036	4.135	4.234
500	4.234	4.333	4.432	4.532	4.632	4.732	4.832	4.933	5.034	5.136	5.237
600	5.237	5.339	5.442	5.544	5.648	5.751	5.855	5.960	6.064	6.169	6.274
700	6.274	6.380	6.486	6.592	6.699	6.805	6.913	7.020	7.128	7.236	7.345
800	7.345	7.454	7.563	7.672	7.782	7.892	8.003	8.114	8.225	8.336	8.448
900	8.448	8.560	8.673	8.786	8.899	9.012	9.126	9.240	9.355	9.470	9.585
1000	9.585	9.700	9.816	9.932	10.048	10.165	10.282	10.400	10.517	10.635	10.754
1100	10.754	10.872	10.991	11.110	11.229	11.348	11.467	11.587	11.707	11.827	11.947
1200	11.947	12.067	12.188	12.308	12.429	12.550	12.671	12.792	12.913	13.034	13.155
1300	13.155	13.276	13.397	13.519	13.640	13.761	13.883	14.004	14.125	14.247	14.368
1400	14.368	14.489	14.610	14.731	14.852	14.973	15.094	15.215	15.336	15.456	15.576
1500	15.576	15.697	15.817	15.937	16.057	16.176	16.296	16.415	16.534	16.653	16.771
1600	16.771	16.890	17.008	17.125	17.243	17.360	17.477	17.594	17.711	17.826	17.942
1700	17.942	18.056	18.170	18.282	18.394	18.504	18.612				

TABLE 4a-8. TYPE R. PLATINUM–13% RHODIUM VS. PLATINUM THERMOCOUPLES
[Emf, absolute millivolts; temp., °C (IPTS-68); reference junctions at 0°C]

°C	0	10	20	30	40	50	60	70	80	90	100
0	0.00	0.054	0.111	0.171	0.232	0.296	0.363	0.431	0.501	0.573	0.647
100	0.647	0.723	0.800	0.879	0.959	1.041	1.124	1.208	1.294	1.380	1.468
200	1.468	1.557	1.647	1.738	1.830	1.923	2.017	2.111	2.207	2.303	2.400
300	2.400	2.498	2.596	2.695	2.795	2.896	2.997	3.099	3.201	3.304	3.407
400	3.407	3.511	3.616	3.721	3.826	3.933	4.039	4.146	4.254	4.362	4.471
500	4.471	4.580	4.689	4.799	4.910	5.021	5.132	5.244	5.356	5.469	5.582
600	5.582	5.696	5.810	5.925	6.040	6.155	6.272	6.388	6.505	6.623	6.741
700	6.741	6.860	6.979	7.098	7.218	7.339	7.460	7.582	7.703	7.826	7.949
800	7.949	8.072	8.196	8.320	8.445	8.570	8.696	8.822	8.949	9.076	9.203
900	9.203	9.331	9.460	9.589	9.718	9.848	9.978	10.109	10.240	10.371	10.503
1000	10.503	10.636	10.768	10.902	11.035	11.170	11.304	11.439	11.574	11.710	11.846
1100	11.846	11.983	12.119	12.257	12.394	12.532	12.669	12.808	12.946	13.085	13.224
1200	13.224	13.363	13.502	13.642	13.782	13.922	14.062	14.202	14.343	14.483	14.624
1300	14.624	14.765	14.906	15.047	15.188	15.329	15.470	15.611	15.752	15.893	16.035
1400	16.035	16.176	16.317	16.458	16.599	16.741	16.882	17.022	17.163	17.304	17.445
1500	17.445	17.585	17.726	17.866	18.006	18.146	18.286	18.425	18.564	18.703	18.842
1600	18.842	18.981	19.119	19.257	19.395	19.533	19.670	19.807	19.944	20.080	20.215
1700	20.215	20.350	20.483	20.616	20.748	20.878	21.006				

TABLE 4a-9. TYPE E. CHROMEL VS. CONSTANTAN THERMOCOUPLES
[Emf, absolute millivolts; temp., °C (IPTS-68); reference junctions at 0°C]

°C	0	10	20	30	40	50	60	70	80	90	100
−200	−8.824	−9.063	−9.274	−9.455	−9.604	−9.719	−9.797	−9.835			
−100	−5.237	−5.680	−6.167	−6.516	−6.907	−7.279	−7.631	−7.963	−8.273	−8.561	−8.824
(−)0	0.00	−0.581	−1.151	−1.709	−2.254	−2.787	−3.306	−3.811	−4.301	−4.777	−5.237
(+)0	0.00	0.591	1.192	1.801	2.419	3.047	3.683	4.329	4.983	5.646	6.317
100	6.317	6.996	7.683	8.377	9.078	9.787	10.501	11.222	11.949	12.681	13.419
200	13.419	14.161	14.909	15.661	16.417	17.178	17.942	18.710	19.481	20.256	21.033
300	21.033	21.814	22.597	23.383	24.171	24.961	25.754	26.549	27.345	28.143	28.943
400	28.943	29.744	30.546	31.350	32.155	32.960	33.767	34.574	35.382	36.190	36.999
500	36.999	37.808	38.617	39.426	40.236	41.045	41.853	42.662	43.470	44.278	45.085
600	45.085	45.891	46.697	47.502	48.306	49.109	49.911	50.713	51.513	52.312	53.110
700	53.110	53.907	54.703	55.498	56.291	57.083	57.873	58.663	59.451	60.237	61.022
800	61.022	61.806	62.588	63.368	64.147	64.924	65.700	66.473	67.245	68.015	68.783
900	68.783	69.549	70.313	71.075	71.835	72.593	73.350	74.104	74.857	75.608	76.358
1000	76.358										

TABLE 4a-10. TYPE J. IRON VS. CONSTANTAN THERMOCOUPLES
[Emf, absolute millivolts; temp., °C (IPTS-68); reference functions at 0°C]

°C	0	10	20	30	40	50	60	70	80	90	100
−200	−7.890	−8.096									
−100	−4.632	−5.036	−5.426	−5.801	−6.159	−6.499	−6.821	−7.122	−7.402	−7.659	−7.890
(−)0	0.00	−0.501	−0.995	−1.481	−1.960	−2.431	−2.892	−3.344	−3.785	−4.215	−4.632
(+)0	0.00	0.507	1.019	1.536	2.058	2.585	3.115	3.649	4.186	4.725	5.268
100	5.268	5.812	6.359	6.907	7.457	8.008	8.560	9.113	9.667	10.222	10.777
200	10.777	11.332	11.887	12.442	12.998	13.553	14.108	14.663	15.217	15.771	16.325
300	16.325	16.879	17.432	17.984	18.537	19.089	19.640	20.192	20.743	21.295	21.846
400	21.846	22.397	22.949	23.501	24.054	24.607	25.161	25.716	26.272	26.829	27.388
500	27.388	27.949	28.511	29.075	29.642	30.210	30.782	31.356	31.933	32.513	33.096
600	33.096	33.683	34.273	34.867	35.464	36.066	36.671	37.280	37.893	38.510	39.130
700	39.130	39.754	40.382	41.013	41.647	42.283	42.922				

TABLE 4a-11. TYPE K. CHROMEL VS. ALUMEL THERMOCOUPLES
[Emf, absolute millivolts; temp., °C (IPTS-68); reference junctions at 0°C]

°C	0	10	20	30	40	50	60	70	80	90	100
−200	−5.891	−6.035	−6.158	−6.262	−6.344	−6.404	−6.441	−6.458			
−100	−3.553	−3.852	−4.138	−4.410	−4.669	−4.912	−5.141	−5.354	−5.550	−5.730	−5.891
(−)0	0.00	−0.392	−0.777	−1.156	−1.527	−1.889	−2.243	−2.586	−2.920	−3.242	−3.553
(+)0	0.00	0.397	0.798	1.203	1.611	2.022	2.436	2.850	3.266	3.681	4.095
100	4.095	4.508	4.919	5.327	5.733	6.137	6.539	6.939	7.338	7.737	8.137
200	8.137	8.537	8.938	9.341	9.745	10.151	10.560	10.969	11.381	11.793	12.207
300	12.207	12.623	13.039	13.456	13.874	14.292	14.712	15.132	15.552	15.974	16.395
400	16.395	16.818	17.241	17.664	18.088	18.513	18.938	19.363	19.788	20.214	20.640
500	20.640	21.066	21.493	21.919	22.346	22.772	23.198	23.624	24.050	24.476	24.902
600	24.902	25.327	25.751	26.176	26.599	27.022	27.445	27.867	28.288	28.709	29.128
700	29.128	29.547	29.965	30.383	30.799	31.214	31.629	32.042	32.455	32.866	33.277
800	33.277	33.686	34.095	34.502	34.909	35.314	35.718	36.121	36.524	36.925	37.325
900	37.325	37.724	38.122	38.519	38.915	39.310	39.703	40.096	40.488	40.879	41.269
1000	41.269	41.657	42.045	42.432	42.817	43.202	43.585	43.968	44.349	44.729	45.108
1100	45.108	45.486	45.863	46.238	46.612	46.985	47.356	47.726	48.095	48.462	48.828
1200	48.828	49.192	49.555	49.916	50.276	50.633	50.990	51.344	51.697	52.049	52.398
1300	52.398	52.747	53.093	53.439	53.782	54.125	54.466	54.807			

TABLE 4a-12. TYPE T. COPPER VS. CONSTANTAN THERMOCOUPLES

[Emf, absolute millivolts; temp., °C (IPTS-68); reference junctions at 0°C]

°C	0	10	20	30	40	50	60	70	80	90	100
−200	−5.603	−5.753	−5.889	−6.007	−6.105	−6.181	−6.232	−6.258			
−100	−3.378	−3.656	−3.923	−4.177	−4.419	−4.648	−4.865	−5.069	−5.261	−5.439	−5.603
(−)0	0.00	−0.383	−0.757	−1.121	−1.475	−1.819	−2.152	−2.475	−2.788	−3.089	−3.378
(+)0	0.00	0.391	0.789	1.196	1.611	2.035	2.467	2.908	3.357	3.813	4.277
100	4.277	4.749	5.227	5.712	6.204	6.702	7.207	7.718	8.235	8.757	9.286
200	9.286	9.820	10.360	10.905	11.456	12.011	12.572	13.137	13.707	14.281	14.860
300	14.860	15.443	16.030	16.621	17.217	17.816	18.420	19.027	19.638	20.252	20.869
400	20.869										

TABLE 4a-13. ELECTRICAL RESISTIVITY OF SOME ELEMENTS AND ALLOYS AS A FUNCTION OF TEMPERATURE*

[At 0°C both the relative R_t/R_0 and actual resistivity (microhm cm) are given]

Temp., °C	Platinum (R_t/R_0)	Copper (R_t/R_0)	Nickel (R_t/R_0)	Iron (R_t/R_0)	Silver (R_t/R_0)	90 Pt–10 Rh (R_t/R_0)	87 Pt–13 Rh (R_t/R_0)
−200	0.177	0.117	. . . . .	. . . . . .	0.176		
−100	0.599	0.557	. . . . .	. . . . . .	0.596		
0	1.000	1.000	1.000	1.000	1.000	1.000	1.000
	(9.83)	(1.56)	(6.38)	(8.57)	(1.50)	(18.4)	(19.0)
+100	1.392	1.431	1.663	1.650	1.408	1.166	1.156
200	1.773	1.862	2.501	2.464	1.827	1.330	1.308
300	2.142	2.299	3.611	3.485	2.256	1.490	1.456
400	2.499	2.747	4.847	4.716	2.698	1.646	1.601
500	2.844	3.210	5.398	6.162	3.150	1.798	1.744
600	3.178	3.695	5.882	7.839	3.616	1.947	1.885
700	3.499	4.207	6.326	9.785	4.093	2.093	2.023
800	3.809	4.750	6.749	12.003	4.584	2.233	2.156
900	4.108	5.332	7.154	12.788	5.089	2.369	2.286
1000	4.395	5.959	7.541	13.070	. . . . .	2.503	2.414
1100	4.672	. . . . .	. . . . .	. . . . . .	. . . . .	2.633	2.538
1200	4.937	. . . . .	. . . . .	. . . . . .	. . . . .	2.762	2.661
1300	5.190	. . . . .	. . . . .	. . . . . .	. . . . .	2.888	2.781
1400	5.431	. . . . .	. . . . .	. . . . . .	. . . . .	3.013	2.900
1500	5.660	. . . . .	. . . . .	. . . . . .	. . . . .	3.136	3.017

* American Institute of Physics, "Temperature, Its Measurement and Control in Science and Industry," p. 1312, Reinhold Book Corporation, New York, 1941. The values below 0°C, in most cases, were not determined on the same samples as the values above 0°C.

TABLE 4a-14. ELECTRICAL RESISTIVITY OF SOME ALLOYS AS A
FUNCTION OF TEMPERATURE*

[At 0°C both the relative (R_t/R_0) and actual resistivity (microhm cm) are given]

Temp., °C	80 Ni–20 Cr (R_t/R_0)	60 Ni–24 Fe–16 Cr (R_t/R_0)	50 Fe–30 Ni–20 Cr (R_t/R_0)	Chromel P (90 Ni–10 Cr) (R_t/R_0)	Alumel (95 Ni–bal. Al, Si, and Mn) (R_t/R_0)	Constantan (55 Cu–45 Ni) (R_t/R_0)	Manganin (R_t/R_0)
0	1.000	1.000	1.000	1.000	1.000	1.000	1.000
	(107.6)	(111.6)	(99.0)	(70.0)	(28.1)	(48.9)	(48.2)
100	1.021	1.025	1.037	1.041	1.239	0.999	1.002
200	1.041	1.048	1.073	1.086	1.428	0.996	0.996
300	1.056	1.071	1.107	1.134	1.537	0.994	0.991
400	1.068	1.092	1.137	1.187	1.637	0.994	0.983
500	1.073	1.108	1.163	1.222	1.726	1.007	
600	1.071	1.115	1.185	1.248	1.814	1.024	
700	1.067	1.119	1.204	1.275	1.899	1.040	
800	1.066	1.127	1.221	1.304	1.982	1.056	
900	1.071	1.138	1.237	1.334	2.066	1.074	
1000	1.077	1.149	1.251	1.365	2.150	1.092	
1100	1.083	.....	.....	1.397	2.234	1.110	
1200	.....	.....	.....	1.430	2.318		

* American Institute of Physics, "Temperature, Its Measurement and Control in Science and Industry," p. 1312, Reinhold Book Corporation, New York, 1941. The values below 0°C, in most cases, were not determined on the same samples as the values above 0°C.

Tables 4a-15, 4a-16, and 4a-17 give the thermoelectric voltage E and Seebeck coefficient dE/dT, for the three thermocouple combinations that are most useful between liquid helium temperatures and the ice point: ANSI types E, T, and EP vs. Au–0.07 at.% Fe (often referred to as Chromel vs. constantan, copper vs. constantan, and Chromel vs. Au–0.07 at.% Fe, respectively). Type E thermocouples are recommended for general use above the normal boiling point of hydrogen (20 K). Both components of this thermocouple combination have low thermal conductivity and reasonably good homogeneity. For operation below 20 K, the combination EP vs. Au–0.07 at.% Fe is the most sensitive combination commonly available for which there is a standard table. The values given for the last combination are interim: the *gold*-iron material is not yet standardized. Additional details and discussion for these three types of thermocouples have been published.[1,2] Seebeck coefficients for each type are shown in Fig. 4a-1.

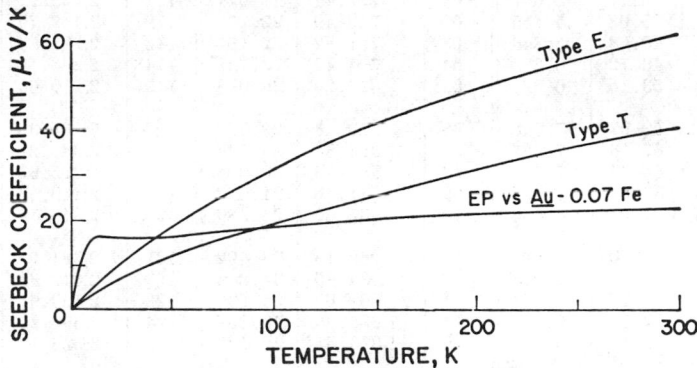

FIG. 4a-1. Seebeck coefficients for USASI thermocouple types E, T, and EP vs. Au–0.07 at.% Fe.

[1] For standardized thermocouples: NBS Monograph 124, "Reference Tables for Low Temperature Thermocouples," by L. L. Sparks, R. L. Powell, and W. J. Hall, U.S. Government Printing Office, Washington, D.C., 1972.

[2] For nonstandardized combinations: L. L. Sparks and R. L. Powell, *J. Res. Nat. Bur. Stand. (U.S.)* **76 A**, No. 3, 1972.

TABLE 4a-15. TYPE E THERMOCOUPLES (CHROMEL VS. CONSTANTAN)

T, K	E, μV	dE/dT, μV/K	T, K	E, μV	dE/dT, μV/K	T, K	E, μV	dE/dT, μV/K
1	0.41	0.660	51	522.68	19.001	101	1,806.73	31.648
2	1.31	1.149	52	541.83	19.303	102	1,838.49	31.865
3	2.70	1.623	53	561.28	19.603	103	1,870.46	32.081
4	4.56	2.085	54	581.04	19.900	104	1,902.65	32.295
5	6.87	2.535	55	601.08	20.194	105	1,935.05	32.509
6	9.62	2.975	56	621.42	20.486	106	1,967.67	32.722
7	12.81	3.406	57	642.05	20.776	107	2,000.50	32.934
8	16.43	3.829	58	662.97	21.063	108	2,033.54	33.145
9	20.47	4.244	59	684.18	21.348	109	2,066.79	33.355
10	24.92	4.653	60	705.67	21.630	110	2,100.25	33.565
11	29.77	5.057	61	727.44	21.911	111	2,133.92	33.773
12	35.03	5.455	62	749.49	22.188	112	2,167.79	33.981
13	40.68	5.848	63	771.82	22.464	113	2,201.88	34.187
14	46.72	6.238	64	794.42	22.737	114	2,236.17	34.393
15	53.15	6.623	65	817.29	23.008	115	2,270.66	34.598
16	59.97	7.005	66	840.43	23.277	116	2,305.36	34.802
17	67.16	7.385	67	863.84	23.544	117	2,340.27	35.005
18	84.74	7.761	68	887.52	23.809	118	2,375.37	35.207
19	82.69	8.135	69	911.46	24.072	119	2,410.68	35.408
20	91.01	8.506	70	935.66	24.333	120	2,446.19	35.609
21	99.70	8.876	71	960.13	24.592	121	2,481.90	35.809
22	108.76	9.243	72	984.85	24.849	122	2,517.81	36.007
23	118.18	9.608	73	1,009.82	25.104	123	2,553.91	36.205
24	127.97	9.971	74	1,035.05	25.357	124	2,590.22	36.402
25	138.12	10.332	75	1,060.54	25.608	125	2,626.72	36.599
26	148.64	10.691	76	1,086.27	25.858	126	2,663.41	36.794
27	159.51	11.049	77	1,112.25	26.106	127	2,700.30	36.989
28	170.73	11.404	78	1,138.48	26.353	128	2,737.39	37.182
29	182.31	11.758	79	1,164.96	26.598	129	2,774.67	37.375
30	194.25	12.110	80	1,191.68	26.841	130	2,812.14	37.567
31	206.53	12.460	81	1,218.64	27.083	131	2,859.80	37.759
32	219.17	12.808	82	1,245.84	27.323	132	2,887.66	37.949
33	232.15	13.154	83	1,273.28	27.562	133	2,925.70	38.139
34	245.47	13.498	84	1,300.96	27.799	134	2,963.94	38.328
35	259.14	13.840	85	1,328.88	28.035	135	3,002.36	38.516
36	273.15	14.179	86	1,357.03	28.270	136	3,040.97	38.703
37	287.50	14.517	87	1,385.42	28.503	137	3,079.76	38.890
38	302.19	14.853	88	1,414.04	28.735	138	3,118.75	39.075
39	317.20	15.186	89	1,442.89	28.966	139	3,157.91	39.260
40	332.56	15.517	90	1,471.97	29.196	140	3,197.27	39.445
41	348.24	15.846	91	1,501.28	29.424	141	3,236.80	39.628
42	364.25	16.172	92	1,530.82	29.652	142	3,276.52	39.811
43	380.58	16.496	93	1,560.59	29.878	143	3,316.42	39.993
44	397.24	16.818	94	1,590.58	30.103	144	3,356.51	40.174
45	414.22	17.137	95	1,620.79	30.327	145	3,396.77	40.355
46	431.51	17.454	96	1,651.23	30.549	146	3,437.22	40.534
47	449.13	17.769	97	1,681.89	30.771	147	3,477.84	40.713
48	467.05	18.081	98	1,712.77	30.992	148	3,518.64	40.892
49	485.29	18.390	99	1,743.87	31.212	149	3,559.62	41.070
50	503.83	18.697	100	1,775.19	31.430	150	3,600.78	41.247

TABLE 4a-15. TYPE E THERMOCOUPLES (CHROMEL vs. CONSTANTAN) (*Continued*)

T, K	E, μV	dE/dT, μV/K	T, K	E, μV	dE/dT, μV/K	T, K	E, μV	dE/dT, μV/K
151	3,642.12	41.423	196	5,675.23	48.743	241	8,010.92	54.888
152	3,683.63	41.599	197	5,724.05	48.893	242	8,065.87	55.014
153	3,725.31	41.774	198	5,773.02	49.041	243	8,120.95	55.140
154	3,767.18	41.948	199	5,822.13	49.189	244	8,176.15	55.265
155	3,809.21	42.122	200	5,871.40	49.336	245	8,231.48	55.390
156	3,851.42	42.295	201	5,920.81	49.485	246	8,286.93	55.515
157	3,893.80	42.467	202	5,970.36	49.629	247	8,342.51	55.639
158	3,936.35	42.639	203	6,020.06	49.775	248	8,398.21	55.763
159	3,979.08	42.811	204	6,059.91	49.919	249	8,454.04	55.887
160	4,021.97	42.981	205	6,119.90	50.064	250	8,509.99	56.010
161	4,065.04	43.151	206	6,170.04	50.207	251	8,566.06	56.133
162	4,108.28	43.321	207	6,220.32	50.350	252	8,622.25	56.255
163	4,151.68	43.490	208	6,270.74	50.492	253	8,678.57	56.377
164	4,195.26	43.658	209	6,321.30	50.634	254	8,735.01	56.498
165	4,239.00	43.825	210	6,372.01	50.775	255	8,791.56	56.619
166	4,282.91	43.993	211	6,422.85	50.915	256	8,848.24	56.739
167	4,326.98	44.159	212	6,473.84	51.055	257	8,905.04	56.858
168	4,371.22	44.325	213	6,524.96	51.194	258	8,961.96	56.977
169	4,415.63	44.591	214	6,576.22	51.333	259	9,019.00	57.095
170	4,460.21	44.655	215	6,627.63	51.471	260	9,076.15	57.212
171	4,504.94	44.820	216	6,679.17	51.608	261	9,133.42	57.329
172	4,549.84	44.984	217	6,730.84	51.745	262	9,190.81	57.445
173	4,594.91	45.147	218	6,782.66	51.882	263	9,248.31	57.559
174	4,640.14	45.309	219	6,834.61	52.017	264	9,305.93	57.673
175	4,685.53	45.471	220	6,886.69	52.152	265	9,363.66	57.786
176	4,731.08	45.633	221	6,938.91	52.287	266	9,421.50	57.898
177	4,776.79	45.794	222	6,991.27	52.421	267	9,479.45	58.009
178	4,822.67	45.954	223	7,043.75	52.555	268	9,537.51	58.119
179	4,868.70	46.114	224	7,096.38	52.688	269	9,595.69	58.227
180	4,914.90	46.274	225	7,149.13	52.821	270	9,653.97	58.335
181	4,961.25	46.432	226	7,202.02	52.953	271	9,712.36	58.442
182	5,007.76	46.590	227	7,255.04	53.085	272	9,770.85	58.547
183	5,054.43	46.748	228	7,308.19	53.216	273	9,829.45	58.651
184	5,101.26	46.905	229	7,361.47	53.347	274	9,888.15	58.755
185	5,148.24	47.061	230	7,414.88	53.477	275	9,946.96	58.857
186	5,195.38	47.217	231	7,468.42	53.607	276	10,005.87	58.958
187	5,242.67	47.373	232	7,522.09	53.737	277	10,064.88	59.059
188	5,290.12	47.527	233	7,575.89	53.866	278	10,123.98	59.159
189	5,337.73	47.681	234	7,629.83	53.995	279	10,183.19	59.258
190	5,385.49	47.835	235	7,683.88	54.125	280	10,242.50	59.356
191	5,433.40	47.988	236	7,738.07	54.252			
192	5,481.46	48.140	237	7,792.39	54.380			
193	5,529.68	48.292	238	7,846.83	54.507			
194	5,578.05	48.443	239	7,901.40	54.634			
195	5,626.56	48.593	240	7,956.10	54.761			

TABLE 4a-16. TYPE T THERMOCOUPLES (COPPER VS. CONSTANTAN)

T, K	E, μV	dE/dT, μV/K	T, K	E, μV	dE/dT, μV/K	T, K	E, μV	dE/dT, μV/K
1	−0.09	0.147	51	343.30	12.345	101	1,147.25	19.498
2	0.28	0.586	52	355.73	12.519	102	1,166.81	19.629
3	1.07	0.985	53	368.33	12.690	103	1,186.51	19.758
4	2.24	1.351	54	381.11	12.859	104	1,206.33	19.888
5	3.76	1.690	55	394.05	13.025	105	1,226.28	20.017
6	5.61	2.006	56	407.16	13.189	106	1,246.37	20.147
7	7.77	2.304	57	420.43	13.350	107	1,266.58	20.275
8	10.21	2.587	58	433.86	13.510	108	1,286.92	20.404
9	12.94	2.859	59	447.44	13.668	109	1,307.38	20.532
10	15.93	3.121	60	461.19	13.824	110	1,327.98	20.660
11	19.18	3.377	61	475.09	13.978	111	1,348.71	20.788
12	22.68	3.628	62	489.15	14.130	112	1,369.56	20.916
13	26.43	3.876	63	503.35	14.281	113	1,390.54	21.043
14	30.43	4.120	64	517.71	14.431	114	1,411.64	21.170
15	34.67	4.364	65	532.21	14.579	115	1,432.88	21.297
16	39.16	4.606	66	546.86	14.726	116	1,454.24	21.424
17	43.89	4.848	67	561.66	14.872	117	1,475.73	21.551
18	48.85	5.091	68	576.61	15.017	118	1,497.34	21.677
19	54.07	5.333	69	591.70	15.160	119	1,519.08	21.803
20	59.52	5.576	70	606.93	15.303	120	1,540.95	21.929
21	65.22	5.818	71	622.30	15.445	121	1,562.94	22.055
22	71.16	6.062	72	637.82	15.587	122	1,585.06	22.181
23	77.34	6.305	73	653.48	15.727	123	1,607.30	22.306
24	83.77	6.548	74	669.27	15.868	124	1,629.67	22.431
25	90.44	6.791	75	685.21	16.007	125	1,652.16	22.557
26	97.35	7.033	76	701.29	16.146	126	1,674.78	22.682
27	104.50	7.274	77	717.50	16.284	127	1,697.53	22.807
28	111.90	7.515	78	733.86	16.422	128	1,720.39	22.932
29	119.53	7.754	79	750.35	16.559	129	1,743.39	23.056
30	127.40	7.991	80	766.97	16.696	130	1,766.51	23.181
31	135.51	8.227	81	783.74	16.833	131	1,789.75	23.305
32	143.86	8.461	82	800.64	16.969	132	1,813.12	23.430
33	152.43	8.692	83	817.68	17.105	133	1,836.61	23.554
34	161.24	8.921	84	834.85	17.241	134	1,860.23	23.678
35	170.27	9.147	85	852.16	17.376	135	1,883.97	23.802
36	179.53	9.371	86	869.60	17.511	136	1,907.83	23.926
37	189.01	9.591	87	887.18	17.645	137	1,931.82	24.050
38	198.72	9.809	88	904.89	17.779	138	1,955.93	24.174
39	208.63	10.023	89	922.74	17.913	139	1,980.17	24.297
40	218.76	10.234	90	940.72	18.047	140	2,004.52	24.420
41	229.10	10.442	91	958.83	18.180	141	2,029.01	24.544
42	239.64	10.647	92	977.08	18.314	142	2,053.61	24.667
43	250.39	10.848	93	995.46	18.446	143	2,078.34	24.790
44	261.34	11.046	94	1,013.97	18.579	144	2,103.19	24.913
45	272.48	11.241	95	1,032.62	18.711	145	2,128.17	25.035
46	283.82	11.433	96	1,051.39	18.843	146	2,153.26	25.158
47	295.35	11.621	97	1,070.30	18.975	147	2,178.48	25.280
48	307.06	11.807	98	1,089.34	19.106	148	2,203.82	25.402
49	318.96	11.989	99	1,108.51	19.237	149	2,229.28	25.524
50	331.04	12.168	100	1,127.82	19.368	150	2,254.87	25.646

TABLE 4a-16. TYPE T THERMOCOUPLES (COPPER VS. CONSTANTAN) (*Continued*)

T, K	E, μV	dE/dT, μV/K	T, K	E, μV	dE/dT, μV/K	T, K	E, μV	dE/dT, μV/K
151	2,280.58	25.767	196	3,559.13	30.975	241	5,061.31	35.688
152	2,306.40	25.889	197	3,590.16	31.086	242	5,097.05	35.787
153	2,332.35	26.010	198	3,621.30	31.197	243	5,132.89	35.886
154	2,358.42	26.130	199	3,652.56	31.307	244	5,168.82	35.986
155	2,384.61	26.251	200	3,683.92	31.417	245	5,204.86	36.084
156	2,410.93	26.371	201	3,715.39	31.527	246	5,240.99	36.183
157	2,437.36	26.491	202	3,746.97	31.637	247	5,277.22	36.282
158	2,463.91	26.611	203	3,778.66	31.746	248	5,313.55	36.381
159	2,490.58	26.731	204	3,810.46	31.856	249	5,349.99	36.480
160	2,517.37	26.850	205	3,842.38	31.965	250	5,386.51	36.579
161	2,544.28	26.969	206	3,874.39	32.074	251	5,423.15	36.677
162	2,571.31	27.088	207	3,906.52	32.182	252	5,459.87	36.776
163	2,598.46	27.206	208	3,938.76	32.291	253	5,496.69	36.874
164	2,625.72	27.325	209	3,971.10	32.399	254	5,533.62	36.972
165	2,653.10	27.442	210	4,003.56	32.506	255	5,570.64	37.070
166	2,680.61	27.560	211	4,036.12	32.614	256	5,607.75	37.167
167	2,708.22	27.677	212	4,068.78	32.721	257	5,644.97	37.264
168	2,735.96	27.795	213	4,101.56	32.828	258	5,682.28	37.360
169	2,763.81	27.911	214	4,134.44	32.934	259	5,719.69	37.456
170	2,791.78	28.028	215	4,167.43	33.040	260	5,757.20	37.551
171	2,819.87	28.144	216	4,200.52	33.146	261	5,794.79	37.645
172	2,848.07	28.260	217	4,233.72	33.252	262	5,832.49	37.739
173	2,876.39	28.376	218	4,267.02	33.357	263	5,870.27	37.831
174	2,904.82	28.491	219	4,300.43	33.462	264	5,908.15	37.923
175	2,933.37	28.606	220	4,333.95	33.566	265	5,946.12	38.014
176	2,962.04	28.721	221	4,367.56	33.670	266	5,984.17	38.103
177	2,990.81	28.836	222	4,401.29	33.774	267	6,022.32	38.193
178	3,019.71	28.950	223	4,435.11	33.877	268	6,060.56	38.281
179	3,048.71	29.065	224	4,469.04	33.980	269	6,098.88	38.368
180	3,077.84	29.179	225	4,503.07	34.083	270	6,137.30	38.455
181	3,107.07	29.292	226	4,537.21	34.185	271	6,175.79	38.541
182	3,136.42	29.406	227	4,571.44	34.287	272	6,214.37	38.627
183	3,165.88	29.519	228	4,605.78	34.388	273	6,253.05	38.714
184	3,195.46	29.632	229	4,640.22	34.490	274	6,291.80	38.802
185	3,225.15	29.745	230	4,674.76	34.501	275	6,330.65	38.891
186	3,254.95	29.858	231	4,709.40	34.691	276	6,369.59	38.983
187	3,284.86	29.971	232	4,744.14	34.792	277	6,408.62	39.078
188	3,314.89	30.083	233	4,778.98	34.892	278	6,447.74	39.178
189	3,345.03	30.195	234	4,813.93	34.992	279	6,486.97	39.283
190	3,375.28	30.307	235	4,848.97	35.092	280	6,526.31	39.397
191	3,405.64	30.419	236	4,884.11	35.192			
192	3,436.12	30.531	237	4,919.35	35.291			
193	3,466.71	30.642	238	4,954.69	35.391			
194	3,497.40	30.753	239	4,990.13	35.490			
195	3,528.21	30.864	240	5,025.67	35.589			

TABLE 4a-17. TYPE EP VS. AU–0.07 FE (CHROMEL VS. AU–0.07 AT.% FE)
(INTERIM VALUES)

T, K	E, μV	dE/dT, μV/K	T, K	E, μV	dE/dT, μV/K	T, K	E, μV	dE/dT, μV/K
1	7.86	8.645	51	785.00	16.402	101	1,665.72	18.731
2	17.21	10.035	52	801.42	16.450	102	1,684.47	18.770
3	27.86	11.220	53	817.89	16.498	103	1,703.26	18.809
4	39.59	12.226	54	834.42	16.548	104	1,722.09	18.848
5	52.26	13.073	55	850.99	16.598	105	1,740.95	18.886
6	65.69	13.782	56	867.61	16.648	106	1,759.86	18.924
7	79.78	14.369	57	884.29	16.699	107	1,778.80	18.961
8	94.40	14.852	58	901.01	16.750	108	1,797.78	18.998
9	109.45	15.243	59	917.79	16.801	109	1,816.80	19.035
10	124.86	15.555	60	934.61	16.852	110	1,835.85	19.071
11	140.54	15.801	61	951.49	16.903	111	1,854.94	19.107
12	156.44	15.989	62	968.42	16.953	112	1,874.06	19.143
13	172.50	16.128	63	985.40	17.004	113	1,893.22	19.178
14	188.68	16.228	64	1,002.43	17.055	114	1,912.42	19.213
15	204.94	16.293	65	1,019.51	17.105	115	1,931.65	19.247
16	221.26	16.331	66	1,036.64	17.155	116	1,950.91	19.281
17	237.60	16.347	67	1,053.82	17.205	117	1,970.21	19.315
18	253.95	16.346	68	1,071.05	17.255	118	1,989.54	19.349
19	270.29	16.330	69	1,088.33	17.304	119	2,008.91	19.382
20	286.60	16.305	70	1,105.66	17.354	120	2,028.31	19.415
21	302.89	16.272	71	1,123.03	17.402	121	2,047.74	19.447
22	319.15	16.235	72	1,140.46	17.451	122	2,067.20	19.480
23	335.36	16.195	73	1,157.94	17.499	123	2,086.70	19.512
24	351.54	16.154	74	1,175.46	17.547	124	2,106.23	19.543
25	367.67	16.114	75	1,193.03	17.595	125	2,125.78	19.575
26	383.77	16.076	76	1,210.65	17.642	126	2,145.37	19.606
27	399.82	16.040	77	1,228.31	17.689	127	2,165.00	19.637
28	415.85	16.008	78	1,246.03	17.736	128	2,184.65	19.668
29	431.84	15.980	79	1,263.79	17.783	129	2,204.33	19.698
30	447.81	15.957	80	1,281.59	17.829	130	2,224.04	19.728
31	463.76	15.938	81	1,299.44	17.875	131	2,243.79	19.758
32	479.69	15.924	82	1,317.34	17.921	132	2,263.56	19.788
33	495.61	15.915	83	1,335.29	17.966	133	2,283.36	19.818
34	511.52	15.911	84	1,353.28	18.011	134	2,303.20	19.847
35	527.43	15.912	85	1,371.31	18.056	135	2,323.06	19.876
36	543.35	15.917	86	1,389.39	18.101	136	2,342.95	19.905
37	559.27	15.927	87	1,407.51	18.145	137	2,362.87	19.834
38	575.20	15.941	88	1,425.68	18.189	138	2,382.82	19.963
39	591.15	15.960	89	1,443.89	18.233	139	2,402.80	19.991
40	607.12	15.982	90	1,462.14	18.276	140	2,422.80	20.020
41	623.12	16.007	91	1,480.44	18.319	141	2,442.83	20.048
42	639.14	16.036	92	1,498.78	18.362	142	2,462.90	20.076
43	655.19	16.068	93	1,517.16	18.404	143	2,482.99	20.104
44	671.28	16.103	94	1,535.59	18.446	144	2,503.10	20.131
45	687.40	16.140	95	1,554.06	18.488	145	2,523.25	20.159
46	703.56	16.180	96	1,572.56	18.529	146	2,543.42	20.186
47	719.76	16.221	97	1,591.11	18.570	137	2,563.62	20.213
48	736.00	16.264	98	1,609.70	18.611	148	2,583.85	20.240
49	752.29	16.309	99	1,628.34	18.651	149	2,604.10	20.267
50	768.62	16.355	100	1,647.01	18.691	150	2,624.38	20.293

TABLE 4a-17. TYPE EP VS. AU–0.07 FE (CHROMEL VS. AU–0.07 AT.% FE)
(INTERIM VALUES)
(*Continued*)

T, K	E, μV	dE/dT, μV/K	T, K	E, μV	dE/dT, μV/K	T, K	E, μV	dE/dT, μV/K
151	2,644.69	20.320	196	3,582.42	21.288	241	4,556.11	21.933
152	2,665.02	20.346	197	3,603.72	21.305	242	4,578.05	21.944
153	2,685.38	20.372	198	3,625.03	21.322	243	4,600.00	21.955
154	2,705.76	20.398	199	3,646.36	21.340	244	4,621.96	21.966
155	2,726.18	20.423	200	3,667.71	21.357	245	4,643.94	21.977
156	2,746.61	20.449	201	3,689.08	21.374	246	4,665.92	21.987
157	2,767.07	20.474	202	3,710.46	21.391	247	4,687.91	21.998
158	2,787.56	20.499	203	3,731.86	21.408	248	4,709.91	22.009
159	2,808.07	20.524	204	3,753.27	21.424	249	4,731.93	22.019
160	2,828.61	20.548	205	3,774.71	21.441	250	4,753.95	22.030
161	2,849.17	20.573	206	3,796.16	21.457	251	4,775.99	22.040
162	2,869.75	20.597	207	3,817.62	21.474	252	4,798.03	22.050
163	2,890.36	20.621	208	3,839.10	21.490	253	4,820.09	22.060
164	2,910.99	20.644	209	3,860.60	21.506	254	4,842.15	22.071
165	2,931.65	20.668	210	3,882.12	21.522	255	4,864.23	22.081
166	2,952.33	20.691	211	3,903.65	21.538	256	4,886.31	22.090
167	2,973.03	20.714	212	3,925.19	21.554	257	4,908.41	22.100
168	2,993.76	20.736	213	3,946.75	21.569	258	4,930.51	22.109
169	3,014.50	20.759	214	3,968.33	21.585	259	4,952.63	22.118
170	3,035.27	20.781	215	3,989.92	21.600	260	4,974.75	22.127
171	3,056.06	20.803	216	4,011.53	21.615	261	4,996.88	22.135
172	3,076.88	20.825	217	4,033.15	21.630	262	5,019.03	22.143
173	3,097.71	20.846	218	4,054.79	21.645	263	5,041.17	22.150
174	3,118.57	20.867	219	4,076.44	21.659	264	5,063.32	22.157
175	3,139.45	20.888	220	4,098.11	21.674	265	5,085.48	22.164
176	3,160.35	20.909	221	4,119.79	21.688	266	5,107.64	22.170
177	3,181.27	20.930	222	4,141.49	21.702	267	5,129.82	22.175
178	3,202.21	20.950	233	4,163.19	21.716	268	5,152.00	22.180
179	3,223.17	20.970	224	4,184.92	21.729	269	5,174.18	22.185
180	3,244.15	20.990	225	4,206.65	21.743	270	5,196.37	22.190
181	3,265.15	21.010	226	4,228.40	21.756	271	5,218.56	22.194
182	3,286.17	21.030	227	4,250.17	21.769	272	5,240.75	22.108
183	3,307.21	21.049	228	4,271.94	21.782	273	5,262.96	22.203
184	3,328.27	21.068	229	4,293.73	21.794	274	5,285.16	22.209
185	3,349.34	21.088	230	4,315.53	21.807	275	5,307.37	22.216
186	3,370.44	21.106	231	4,337.34	21.819	276	5,329.59	22.224
187	3,391.56	21.125	232	4,359.17	21.831	277	5,351.83	22.235
188	3,412.69	21.144	233	4,381.00	21.843	278	5,374.06	22.248
189	3,433.85	21.162	232	4,402.85	21.855	279	5,396.32	22.266
190	3,455.02	21.180	235	4,424.71	21.866	280	5,418.60	22.290
191	3,476.21	21.199	236	4,446.59	21.878			
192	3,497.41	21.217	237	4,468.47	21.889			
193	3,518.64	21.235	238	4,490.36	21.900			
194	3,539.88	21.252	239	4,512.27	21.911			
195	3,561.14	21.270	240	4,534.19	21.922			

4b. Thermodynamic Symbols, Definitions, and Equations

MARK W. ZEMANSKY

The City College of the City University of New York

4b-1. Simple Systems. A simple system is defined as one of constant mass whose equilibrium states are described with the aid of only three thermodynamic coordinates, one of which is the kelvin temperature. The simple systems most often used are listed in Tables 4b-1 and 4b-2, and the rules for converting any equation holding for a hydrostatic system into the analogous equation for another simple system are given in Table 4b-7. Tables 4b-3 to 4b-6 contain the most useful thermodynamic equations involving first derivatives only. Table 4b-8 refers to phase transitions.

TABLE 4b-1. THERMODYNAMIC SYSTEMS AND COORDINATES

System	Intensive coordinate		Extensive coordinate	
Hydrostatic system........	Pressure	P	Volume	V
Stretched wire............	Tension	$\mathscr{F}$	Length	L
Surface film..............	Surface tension	$\mathscr{S}$	Area	A
Electric cell..............	Emf	$\mathscr{E}$	Charge	Z
Capacitor.................	Electric intensity	E	Polarization	P'
Magnetic substance........	Magnetic intensity	$\mathscr{H}$	Magnetization	M

TABLE 4b-2. WORK DONE BY THERMODYNAMIC SYSTEMS

System	Intensive quantity (generalized force)	Extensive quantity (generalized displacement)	Work
Hydrostatic system.........	P in N/m²	V in m³	$P\,dV$ in J
Stretched wire..............	$\mathscr{F}$ in N	L in m	$-\mathscr{F}\,dL$ in J
Surface film................	$\mathscr{S}$ in dynes/cm	A in cm²	$-\mathscr{S}\,dA$ in ergs
Electric cell................	$\mathscr{E}$ in V	Z in C	$-\mathscr{E}\,dZ$ in J
Capacitor...................	E in N/C	P' in C · m	$-E\,dP'$ in J
Magnetic substance..........	$\mathscr{H}$ in A/m	M in Wb · m	$-\mathscr{H}\,dM$ in J

TABLE 4b-3. DEFINITIONS AND SYMBOLS FOR THERMAL QUANTITIES

Thermal quantity	Symbol	Definition
Heat	Q	
Internal energy	U	
Entropy	S	
Enthalpy (also called heat content, heat function, total heat)	H	$U + PV$
Helmholtz function (also called free energy and work function, with symbol A used)	F	$U - TS$
Gibbs function (also called free energy, free enthalpy, thermodynamic potential, with symbol F used)	G	$H - TS$
Volume expansivity (coefficient of volume expansion)	β	$\dfrac{1}{V}\left(\dfrac{\partial V}{\partial T}\right)_P$
Isothermal bulk modulus	B	$-V\left(\dfrac{\partial P}{\partial V}\right)_T$
Adiabatic bulk modulus	B_S	$-V\left(\dfrac{\partial P}{\partial V}\right)_S$
Isothermal compressibility	k	$-\dfrac{1}{V}\left(\dfrac{\partial V}{\partial P}\right)_T$
Adiabatic compressibility	k_S	$-\dfrac{1}{V}\left(\dfrac{\partial V}{\partial P}\right)_S$
Heat capacity at constant volume	C_V	$\left(\dfrac{dQ}{dT}\right)_V = T\left(\dfrac{\partial S}{\partial T}\right)_V$
Heat capacity at constant pressure	C_P	$\left(\dfrac{dQ}{dT}\right)_P = T\left(\dfrac{\partial S}{\partial T}\right)_P$
Ratio of heat capacities	γ	$\dfrac{C_P}{C_V}$
Joule coefficient	η	$\left(\dfrac{\partial T}{\partial V}\right)_U$
Joule-Thomson (Kelvin) coefficient	μ	$\left(\dfrac{\partial T}{\partial P}\right)_H$
Linear expansivity	α	$\dfrac{1}{L}\left(\dfrac{\partial L}{\partial T}\right)_{\mathscr{F}}$
Isothermal Young's modulus	Y	$\dfrac{L}{A}\left(\dfrac{\partial \mathscr{F}}{\partial L}\right)_T$
Adiabatic Young's modulus	Y_S	$\dfrac{L}{A}\left(\dfrac{\partial \mathscr{F}}{\partial L}\right)_S$

TABLE 4b-4. THERMODYNAMIC EQUATIONS FOR A HYDROSTATIC
SYSTEM OF CONSTANT MASS

First law of thermodynamics:

$$Q = U_2 - U_1 + W$$
$$dQ = dU + dW \tag{4b-1}$$

Second law of thermodynamics:

$$dQ = T\,dS \tag{4b-2}$$

Third law of thermodynamics:

$$\lim_{T \to 0} \Delta S_T = 0 \tag{4b-3}$$

$$dU = T\,dS - P\,dV \tag{4b-4}$$
$$dH = T\,dS + V\,dP \tag{4b-5}$$
$$dF = -S\,dT - P\,dV \tag{4b-6}$$
$$dG = -S\,dT + V\,dP \tag{4b-7}$$

Maxwell's equations:

$$\left(\frac{\partial T}{\partial V}\right)_S = -\left(\frac{\partial P}{\partial S}\right)_V \tag{4b-8}$$

$$\left(\frac{\partial T}{\partial P}\right)_S = \left(\frac{\partial V}{\partial S}\right)_P \tag{4b-9}$$

$$\left(\frac{\partial S}{\partial V}\right)_T = \left(\frac{\partial P}{\partial T}\right)_V \tag{4b-10}$$

$$\left(\frac{\partial S}{\partial P}\right)_T = -\left(\frac{\partial V}{\partial T}\right)_P \tag{4b-11}$$

Basic thermodynamic equations:

$$C_V = \left(\frac{\partial U}{\partial T}\right)_V = T\left(\frac{\partial S}{\partial T}\right)_V = -T\left(\frac{\partial P}{\partial T}\right)_V \left(\frac{\partial V}{\partial T}\right)_S \tag{4b-12}$$

$$C_P = \left(\frac{\partial H}{\partial T}\right)_P = T\left(\frac{\partial S}{\partial T}\right)_P = T\left(\frac{\partial V}{\partial T}\right)_P \left(\frac{\partial P}{\partial T}\right)_S \tag{4b-13}$$

$$T\,dS = C_V\,dT + T\left(\frac{\partial P}{\partial T}\right)_V dV = C_V\,dT + \frac{\beta T}{k}\,dV \tag{4b-14}$$

$$T\,dS = C_P\,dT - T\left(\frac{\partial V}{\partial T}\right)_P dP = C_P\,dT - V\beta T\,dP \tag{4b-15}$$

$$T\,dS = C_V\left(\frac{\partial T}{\partial P}\right)_V dP + C_P\left(\frac{\partial T}{\partial V}\right)_P dV = \frac{C_V k}{\beta}\,dP + \frac{C_P}{\beta V}\,dV \tag{4b-16}$$

$$\left(\frac{\partial C_V}{\partial V}\right) = T\left(\frac{\partial^2 P}{\partial T^2}\right)_V \tag{4b-17}$$

$$\left(\frac{\partial C_P}{\partial P}\right)_T = -T\left(\frac{\partial^2 V}{\partial T^2}\right)_P \tag{4b-18}$$

$$C_P - C_V = T\left(\frac{\partial P}{\partial T}\right)_V \left(\frac{\partial V}{\partial T}\right)_P = -T\left(\frac{\partial V}{\partial T}\right)_P^2 \left(\frac{\partial P}{\partial V}\right)_T = \frac{TV\beta^2}{k} \tag{4b-19}$$

$$\gamma = \frac{C_P}{C_V} = \frac{(\partial P/\partial V)_S}{(\partial P/\partial V)_T} = \frac{k}{k_S} \tag{4b-20}$$

$$\mu = \left(\frac{\partial T}{\partial P}\right)_H = \frac{1}{C_P}\left[T\left(\frac{\partial V}{\partial T}\right)_P - V\right] = \frac{V}{C_P}(\beta T - 1) \tag{4b-21}$$

$$\eta = \left(\frac{\partial T}{\partial V}\right)_U = -\frac{1}{C_V}\left[T\left(\frac{\partial P}{\partial T}\right)_V - P\right] = -\frac{1}{C_V}\left(\frac{\beta T}{k} - P\right) \tag{4b-22}$$

TABLE 4b-5. FIRST DERIVATIVES OF T, P, V, AND S

$$\left(\frac{\partial T}{\partial P}\right)_V = \frac{k}{\beta}$$

$$\left(\frac{\partial T}{\partial V}\right)_P = \frac{1}{V\beta}$$

$$\left(\frac{\partial T}{\partial S}\right)_P = \frac{T}{C_P}$$

$$\left(\frac{\partial P}{\partial T}\right)_V = \frac{\beta}{k}$$

$$\left(\frac{\partial P}{\partial V}\right)_T = -\frac{1}{Vk}$$

$$\left(\frac{\partial P}{\partial S}\right)_T = -\frac{1}{V\beta}$$

$$\left(\frac{\partial V}{\partial T}\right)_P = V\beta$$

$$\left(\frac{\partial V}{\partial P}\right)_T = -Vk$$

$$\left(\frac{\partial V}{\partial S}\right)_T = \frac{k}{\beta}$$

$$\left(\frac{\partial S}{\partial T}\right)_V = \frac{C_V}{T}$$

$$\left(\frac{\partial S}{\partial P}\right)_T = -V\beta$$

$$\left(\frac{\partial S}{\partial V}\right)_P = \frac{C_P}{V\beta T} = \frac{\gamma\beta}{(\gamma-1)k}$$

$$\left(\frac{\partial T}{\partial P}\right)_S = \frac{V\beta T}{C_P} = \frac{(\gamma-1)k}{\gamma\beta}$$

$$\left(\frac{\partial T}{\partial V}\right)_S = -\frac{\beta T}{C_V k} = -\frac{\gamma-1}{V\beta}$$

$$\left(\frac{\partial T}{\partial S}\right)_V = \frac{T}{C_V}$$

$$\left(\frac{\partial P}{\partial T}\right)_S = \frac{C_P}{V\beta T} = \frac{\gamma\beta}{(\gamma-1)k}$$

$$\left(\frac{\partial P}{\partial V}\right)_S = -\frac{\gamma}{Vk}$$

$$\left(\frac{\partial P}{\partial S}\right)_V = \frac{\beta T}{C_V k} = \frac{\gamma-1}{V\beta}$$

$$\left(\frac{\partial V}{\partial T}\right)_S = -\frac{C_V k}{\beta T} = -\frac{V\beta}{\gamma-1}$$

$$\left(\frac{\partial V}{\partial P}\right)_S = -\frac{Vk}{\gamma}$$

$$\left(\frac{\partial V}{\partial S}\right)_P = \frac{V\beta T}{C_P} = \frac{(\gamma-1)k}{\gamma\beta}$$

$$\left(\frac{\partial S}{\partial T}\right)_P = \frac{C_P}{T}$$

$$\left(\frac{\partial S}{\partial P}\right)_V = \frac{C_V k}{\beta T} = \frac{V\beta}{\gamma-1}$$

$$\left(\frac{\partial S}{\partial V}\right)_T = \frac{\beta}{k}$$

TABLE 4b-6. FIRST DERIVATIVES OF U, H, F, AND G WITH RESPECT TO T, P, V, AND S

Internal energy U	Enthalpy H	Helmholtz function F	Gibbs function G
$\left(\dfrac{\partial U}{\partial T}\right)_P = C_P - PV\beta$	$\left(\dfrac{\partial H}{\partial T}\right)_P = C_P$	$\left(\dfrac{\partial F}{\partial T}\right)_P = -S - PV\beta$	$\left(\dfrac{\partial G}{\partial T}\right)_P = -S$
$\left(\dfrac{\partial U}{\partial T}\right)_V = C_V$	$\left(\dfrac{\partial H}{\partial T}\right)_V = C_V + \dfrac{V\beta}{k}$	$\left(\dfrac{\partial F}{\partial T}\right)_V = -S$	$\left(\dfrac{\partial G}{\partial T}\right)_V = -S + \dfrac{V\beta}{k}$
$\left(\dfrac{\partial U}{\partial T}\right)_S = \dfrac{PC_Vk}{\beta T} = \dfrac{PV\beta}{\gamma - 1}$	$\left(\dfrac{\partial H}{\partial T}\right)_S = \dfrac{C_P}{\beta T}$	$\left(\dfrac{\partial F}{\partial T}\right)_S = \begin{cases} -S + \dfrac{PV\beta}{\gamma - 1} \\[2mm] -S + \dfrac{PC_Vk}{\beta T} \end{cases}$	$\left(\dfrac{\partial G}{\partial T}\right)_S = \begin{cases} -S + \dfrac{C_P}{\beta T} \\[2mm] -S + \dfrac{V\beta}{(\gamma - 1)k} \end{cases}$
$\left(\dfrac{\partial U}{\partial P}\right)_T = -V\beta T + VkP$	$\left(\dfrac{\partial H}{\partial P}\right)_T = V(1 - \beta T) = -C_P\mu$	$\left(\dfrac{\partial F}{\partial P}\right)_T = PVk$	$\left(\dfrac{\partial G}{\partial P}\right)_T = V$
$\left(\dfrac{\partial U}{\partial P}\right)_V = \dfrac{C_Vk}{\beta} = \dfrac{V\beta T}{\gamma - 1}$	$\left(\dfrac{\partial H}{\partial P}\right)_V = V + \dfrac{C_Vk}{\beta}$	$\left(\dfrac{\partial F}{\partial P}\right)_V = -\dfrac{Sk}{\beta}$	$\left(\dfrac{\partial G}{\partial P}\right)_V = -\dfrac{Sk}{\beta} + V$
$\left(\dfrac{\partial U}{\partial P}\right)_S = \dfrac{PVk}{\gamma}$	$\left(\dfrac{\partial H}{\partial P}\right)_S = V$	$\left(\dfrac{\partial F}{\partial P}\right)_S = \begin{cases} -\dfrac{SV\beta T}{C_P} + \dfrac{PVk}{\gamma} \\[2mm] -\dfrac{Sk(\gamma - 1)}{\gamma\beta} + \dfrac{PVk}{\gamma} \end{cases}$	$\left(\dfrac{\partial G}{\partial P}\right)_S = \begin{cases} -\dfrac{SV\beta T}{C_P} + V \\[2mm] -\dfrac{S(\gamma - 1)k}{\gamma\beta} + V \end{cases}$
$\left(\dfrac{\partial U}{\partial V}\right)_T = \dfrac{\beta T}{k} - P$	$\left(\dfrac{\partial H}{\partial V}\right)_T = \dfrac{\beta T}{k} - \dfrac{1}{k}$	$\left(\dfrac{\partial F}{\partial V}\right)_T = -P$	$\left(\dfrac{\partial G}{\partial V}\right)_T = -\dfrac{1}{k}$

TABLE 4b-6. FIRST DERIVATIVES OF U, H, F, AND G WITH RESPECT TO T, P, V, AND S (Continued)

Internal energy U	Enthalpy H	Helmholtz function F	Gibbs function G
$\left(\dfrac{\partial U}{\partial V}\right)_P = \dfrac{C_P}{V\beta} - P = \dfrac{\gamma\beta T}{(\gamma-1)k} - P$	$\left(\dfrac{\partial H}{\partial V}\right)_P = \dfrac{C_P}{V\beta}$	$\left(\dfrac{\partial F}{\partial V}\right)_P = -\dfrac{S}{V\beta} - P$	$\left(\dfrac{\partial G}{\partial V}\right)_P = -\dfrac{S}{V\beta}$
$\left(\dfrac{\partial U}{\partial V}\right)_S = -P$	$\left(\dfrac{\partial H}{\partial V}\right)_S = -\dfrac{\gamma}{k}$	$\left(\dfrac{\partial F}{\partial V}\right)_S = \begin{cases} \dfrac{S\beta T}{C_V k} - P \\[2mm] \dfrac{S(\gamma-1)}{V\beta} - P \end{cases}$	$\left(\dfrac{\partial G}{\partial V}\right)_S = \begin{cases} \dfrac{S\beta T}{C_V k} - \dfrac{\gamma}{k} \\[2mm] \dfrac{S(\gamma-1)}{V\beta} - \dfrac{\gamma}{k} \end{cases}$
$\left(\dfrac{\partial U}{\partial S}\right)_T = T - \dfrac{Pk}{\beta}$	$\left(\dfrac{\partial H}{\partial S}\right)_T = T - \dfrac{1}{\beta}$	$\left(\dfrac{\partial F}{\partial S}\right)_T = -\dfrac{Pk}{\beta}$	$\left(\dfrac{\partial G}{\partial S}\right)_T = \dfrac{1}{\beta}$
$\left(\dfrac{\partial U}{\partial S}\right)_P = \begin{cases} T - \dfrac{PV\beta T}{C_P} \\[2mm] T - \dfrac{(\gamma-1)Pk}{\gamma\beta} \end{cases}$	$\left(\dfrac{\partial H}{\partial S}\right)_P = T$	$\left(\dfrac{\partial F}{\partial S}\right)_P = \begin{cases} -\dfrac{ST}{C_P} - \dfrac{PV\beta T}{C_P} \\[2mm] -\dfrac{ST}{C_P} - \dfrac{Pk(\gamma-1)}{\gamma\beta} \end{cases}$	$\left(\dfrac{\partial G}{\partial S}\right)_P = -\dfrac{ST}{C_P}$
$\left(\dfrac{\partial U}{\partial S}\right)_V = T$	$\left(\dfrac{\partial H}{\partial S}\right)_V = T + \dfrac{\gamma-1}{\beta}$	$\left(\dfrac{\partial F}{\partial S}\right)_V = -\dfrac{ST}{C_V}$	$\left(\dfrac{\partial G}{\partial S}\right)_V = \begin{cases} -\dfrac{ST}{C_V} + \dfrac{V\beta T}{C_V k} \\[2mm] -\dfrac{ST}{C_V} + \dfrac{V(\gamma-1)}{V\beta} \end{cases}$

System	Replacement for P	Replacement for V
Stretched wire............	$-\mathscr{F}$	L
Surface film.............	$-\mathscr{S}$	A
Electric cell.............	$-\mathscr{E}$	Z
Capacitor...............	$-E$	P'
Magnetic substance........	$-\mathscr{H}$	M

TABLE 4b-8. PHASE TRANSITIONS

First-order transition from 1 to 2

$$L_{1\rightarrow2} = \text{latent heat} = H_2 - H_1 = T(S_2 - S_1)$$

$$\frac{dP}{dT} = \frac{L_{1\rightarrow2}}{T(V_2 - V_1)} \qquad \text{(Clapeyron's equation)}$$

$$\frac{dL_{1\rightarrow2}}{dT} = C_{P2} - C_{P1} + [V_2(1 - \beta_2 T) - V_1(1 - \beta_1 T)]\frac{dP}{dT}$$

Second-order transition from 1 to 2

$$\frac{dP}{dT} = \frac{C_{P2} - C_{P1}}{TV(\beta_2 - \beta_1)}$$

$$\frac{dP}{dT} = \frac{\beta_2 - \beta_1}{k_2 - k_1} \qquad \text{(Ehrenfest's equations)}$$

Lambda transition

Let P'_λ represent the slope of the PT curve at the λ point, and let S'_λ represent the slope of the ST curve at the λ point. At a temperature T, a *small amount* above or below T_λ,

$$\frac{C_P}{T} = S'_\lambda + V\beta P'_\lambda$$

If V'_λ is the slope of the VT curve at T_λ,

$$\frac{C_V}{T} = S'_\lambda - \frac{\beta}{k} V'_\lambda$$

$$\beta = \frac{V'_\lambda}{V} - kP'_\lambda$$

TABLE 4b-9. EQUATIONS FOR SYSTEMS OF VARIABLE
COMPOSITION AND VARIABLE MASS

Consider a system which is a homogeneous mixture of any number of constituents $1, 2, \ldots, k, \ldots$ with n_1 moles of substance 1, n_2 moles of substance 2, etc., at a uniform temperature T, a uniform pressure P, and a volume V. The chemical potential of the kth constituent is

$$\mu_k = \left(\frac{\partial U}{\partial n_k}\right)_{S,V,\text{ other } n\text{'s}} \tag{4b-23}$$

$$\mu_k = \left(\frac{\partial H}{\partial n_k}\right)_{S,P,\text{ other } n\text{'s}} \tag{4b-24}$$

$$\mu_k = \left(\frac{\partial A}{\partial n_k}\right)_{T,V,\text{ other } n\text{'s}} \tag{4b-25}$$

$$\mu_k = \left(\frac{\partial G}{\partial n_k}\right)_{T,P,\text{ other } n\text{'s}} \tag{4b-26}$$

$$dU = T\,dS - P\,dV + \Sigma \mu_k\,dn_k \tag{4b-27}$$

$$dH = T\,dS + V\,dP + \Sigma \mu_k\,dn_k \tag{4b-28}$$

$$dA = -S\,dT - P\,dV + \Sigma \mu_k\,dn_k \tag{4b-29}$$

$$dG = -S\,dT + V\,dP + \Sigma \mu_k\,dn_k \tag{4b-30}$$

$$G = U + PV - TS = \Sigma n_k \mu_k \tag{4b-31}$$

$$-S\,dT + V\,dP = \Sigma n_k\,d\mu_k \tag{4b-32}$$

4b-2. Equations for Ideal Gas Reactions. Assume a mixture of four reacting substances whose chemical symbols are A, B, C, and D, where A and B are the initial constituents (reactants) and C and D the final constituents (products), the reaction being represented by

$$aA + bB \rightleftharpoons cC + dD$$

Four substances are chosen only for convenience; the equations to be developed can easily be applied to reactions in which any number of substances participate. The lower-case letters a, b, c, and d are the *stoichiometric coefficients*, which are always positive integers or fractions.

If we start with arbitrary amounts of *both* initial and final constituents and imagine the reaction to proceed completely to the right, at least one of the initial constituents, say A, will completely disappear. Then it is possible to find a positive number n_0 such that the original number of moles of each of the initial constituents is expressed in the form

$$n_A \text{ (original)} = n_0 a$$
$$n_B \text{ (original)} = n_0 b + i_B$$

where i_B is a constant representing the number of moles of B that cannot combine. If we imagine the reaction to proceed completely to the left, at least one of the final constituents, say C, will completely disappear. In this event, another positive number n_0' can be found such that the original number of moles of each final constituent is expressed in the form

$$n_C \text{ (original)} = n_0' c$$
$$n_D \text{ (original)} = n_0' d + i_D$$

If the reaction is imagined to proceed completely to the left, there is the maximum amount possible of each initial constituent and the minimum amount of each final constituent. Thus,

$$n_A \text{ (max)} = (n_0 + n_0')a \qquad n_C \text{ (min)} = 0$$
$$n_B \text{ (max)} = (n_0 + n_0')b + i_B \qquad n_D \text{ (min)} = i_D$$

If the reaction is imagined to proceed completely to the right, there is the minimum amount possible of each initial constituent and the maximum amount of each final constituent. Thus,

$$n_A \text{ (min)} = 0 \qquad n_C \text{ (max)} = (n_0 + n_0')c$$
$$n_B \text{ (min)} = i_B \qquad n_D \text{ (max)} = (n_0 + n_0')d + i_D$$

Suppose the reaction proceeds partially either to the right or to the left to such an extent that there are n_A moles of A, n_B moles of B, n_C moles of C, and n_D moles of D present at a given moment. We define the *degree of reaction* ϵ in terms of any one of the initial constituents, say A, as the fraction

$$\epsilon = \frac{n_A \text{ (max)} - n_A}{n_A \text{ (max)} - n_A \text{ (min)}}$$

It follows that $\epsilon = 0$ when the reaction is completely to the left and $\epsilon = 1$ when it is completely to the right. Expressing n_A (max) and n_A (min) in terms of the constants that determine the original amounts, we get

$$\epsilon = \frac{(n_0 + n_0')a - n_A}{(n_0 + n_0')a}$$

or

$$n_A = (n_0 + n_0')a(1 - \epsilon)$$

The number of moles of each of the constituents is given, therefore, by the expressions

$$n_A = (n_0 + n_0')a(1 - \epsilon) \qquad n_C = (n_0 + n_0')c\epsilon$$
$$n_B = (n_0 + n_0')b(1 - \epsilon) + i_B \qquad n_D = (n_0 + n_0')d\epsilon + i_D$$

Since all the n's are functions of ϵ only, it follows that in a homogeneous mixture, all the mole fractions are functions of ϵ only. Consider, for example, a vessel containing n_0 moles of water vapor only, with no hydrogen or oxygen present. If dissociation occurs until the degree of dissociation is ϵ, then the n's and the x's (mole fractions) are functions of ϵ as shown in Table 4b-10.

TABLE 4b-10. $H_2O \rightleftharpoons H_2 + \frac{1}{2}O_2$

Constituent	Stoichio- metric coefficient	Number of moles n	Mole fraction x
$A = H_2O$	$a = 1$	$n_A = n_0(1 - \epsilon)$	$x_A = \dfrac{1 - \epsilon}{1 + \epsilon/2}$
$C = H_2$	$c = 1$	$n_C = n_0\epsilon$	$x_C = \dfrac{\epsilon}{1 + \epsilon/2}$
$D = O_2$	$d = \frac{1}{2}$	$n_D = n_0\epsilon/2$	$x_D = \dfrac{\epsilon/2}{1 + \epsilon/2}$
		$\Sigma n = n_0(1 + \epsilon/2)$	

In the case of a mixture of four reacting *ideal* gases in equilibrium, the *equilibrium constant* K is given by the *law of mass action*,

$$\frac{(x_C)^c \cdot (x_D)^d}{(x_A)^a \cdot (x_B)^b} P^{c+d-a-b} = K$$

Since each x is a function of ϵ, K is a function of P and the equilibrium value of ϵ, the function being different for each reaction. The values of K as functions of P and ϵ for all possible reactions among two, three, or four constituents, with stoichiometric coefficients taking on the value 1, 2, or 3, are given in Table 4b-11.

TABLE 4b-11. EQUILIBRIUM CONSTANTS EXPRESSED AS FUNCTIONS
OF P AND ϵ

Reaction	K	Reaction	K
$A \rightleftharpoons C$	$\dfrac{\epsilon}{1 - \epsilon}$	$3A \rightleftharpoons 2C + 2D$	$\dfrac{16\epsilon^4 P}{27(1 - \epsilon)^3(3 + \epsilon)}$
$A \rightleftharpoons 2C$	$\dfrac{4\epsilon^2 P}{1 - \epsilon^2}$	$3A \rightleftharpoons C + 3D$	$\dfrac{\epsilon^4 P}{(1 - \epsilon)^3(3 + \epsilon)}$
$A \rightleftharpoons 3C$	$\dfrac{27\epsilon^3 P^2}{(1 - \epsilon)(1 + 2\epsilon)^2}$	$3A \rightleftharpoons 2C + 3D$	$\dfrac{4\epsilon^5 P^2}{(1 - \epsilon)^3(3 + 2\epsilon)^2}$
$2A \rightleftharpoons C$	$\dfrac{\epsilon(2 - \epsilon)}{4(1 - \epsilon)^2 P}$	$A + B \rightleftharpoons C$	$\dfrac{\epsilon(2 - \epsilon)}{(1 - \epsilon)^2 P}$
$2A \rightleftharpoons 3C$	$\dfrac{27\epsilon^3 P}{4(2 + \epsilon)(1 - \epsilon)^2}$	$A + B \rightleftharpoons 2C$	$\dfrac{4\epsilon^2}{(1 - \epsilon)^2}$
$3A \rightleftharpoons C$	$\dfrac{\epsilon(3 - 2\epsilon)^2}{27(1 - \epsilon)^3 P^2}$	$A + B \rightleftharpoons 3C$	$\dfrac{27\epsilon^3 P}{(1 - \epsilon)^2(2 + \epsilon)}$
$3A \rightleftharpoons 2C$	$\dfrac{4\epsilon^2(3 - \epsilon)}{27(1 - \epsilon)^3 P}$	$A + 2B \rightleftharpoons C$	$\dfrac{\epsilon(3 - 2\epsilon)^2}{4(1 - \epsilon)^3 P^2}$
$A \rightleftharpoons C + D$	$\dfrac{\epsilon^2 P}{1 - \epsilon^2}$	$A + 2B \rightleftharpoons 2C$	$\dfrac{\epsilon^2(3 - \epsilon)}{(1 - \epsilon)^3 P}$
$A \rightleftharpoons C + 2D$	$\dfrac{4\epsilon^3 P^2}{(1 - \epsilon)(1 + 2\epsilon)^2}$	$A + 2B \rightleftharpoons 3C$	$\dfrac{27\epsilon^3}{4(1 - \epsilon)^3}$
$A \rightleftharpoons 2C + 2D$	$\dfrac{16\epsilon^4 P^3}{(1 - \epsilon)(1 + 3\epsilon)^3}$	$2A + 2B \rightleftharpoons C$	$\dfrac{\epsilon(4 - 3\epsilon)^3}{16(1 - \epsilon)^4 P^3}$
$A \rightleftharpoons C + 3D$	$\dfrac{27\epsilon^4 P^3}{(1 - \epsilon)(1 + 3\epsilon)^3}$	$2A + 2B \rightleftharpoons 3C$	$\dfrac{27\epsilon^3(4 - \epsilon)}{16(1 - \epsilon)^4 P}$
$A \rightleftharpoons 2C + 3D$	$\dfrac{108\epsilon^5 P^4}{(1 - \epsilon)(1 + 4\epsilon)^4}$	$A + 3B \rightleftharpoons C$	$\dfrac{\epsilon(4 - 3\epsilon)^3}{27(1 - \epsilon)^4 P^3}$
$A \rightleftharpoons 3C + 3D$	$\dfrac{729\epsilon^6 P^5}{(1 - \epsilon)(1 + 5\epsilon)^5}$	$A + 3B \rightleftharpoons 2C$	$\dfrac{4\epsilon^2(4 - 2\epsilon)^2}{27(1 - \epsilon)^4 P^2}$
$2A \rightleftharpoons C + D$	$\dfrac{\epsilon^2}{4(1 - \epsilon^2)}$	$A + 3B \rightleftharpoons 3C$	$\dfrac{\epsilon^3(4 - \epsilon)}{(1 - \epsilon)^4 P}$
$2A \rightleftharpoons C + 2D$	$\dfrac{\epsilon^3 P}{(1 - \epsilon)^2(2 + \epsilon)}$	$2A + 3B \rightleftharpoons C$	$\dfrac{\epsilon(5 - 4\epsilon)^4}{108(1 - \epsilon)^5 P^4}$
$2A \rightleftharpoons C + 3D$	$\dfrac{27\epsilon^4 P^2}{16(1 - \epsilon^2)^2}$	$2A + 3B \rightleftharpoons 2C$	$\dfrac{\epsilon^2(5 - 3\epsilon)^3}{27(1 - \epsilon)^5 P^3}$
$2A \rightleftharpoons 2C + 3D$	$\dfrac{27\epsilon^5 P^3}{(1 - \epsilon)^2(2 + 3\epsilon)^3}$	$2A + 3B \rightleftharpoons 3C$	$\dfrac{\epsilon^3(5 - 2\epsilon)^2}{4(1 - \epsilon)^5 P^2}$
$2A \rightleftharpoons 3C + 3D$	$\dfrac{729\epsilon^6 P^4}{4(1 - \epsilon)^2(2 + 4\epsilon)^4}$	$3A + 3B \rightleftharpoons C$	$\dfrac{\epsilon(6 - 5\epsilon)^5}{729(1 - \epsilon)^6 P^5}$
$3A \rightleftharpoons C + D$	$\dfrac{\epsilon^2(3 - \epsilon)}{27(1 - \epsilon)^3 P}$	$3A + 3B \rightleftharpoons 2C$	$\dfrac{4\epsilon^2(6 - 4\epsilon)^4}{729(1 - \epsilon)^6 P^4}$
$3A \rightleftharpoons C + 2D$	$\dfrac{4\epsilon^3}{27(1 - \epsilon)^3}$	$A + B \rightleftharpoons C + D$	$\dfrac{\epsilon^2}{(1 - \epsilon)^2}$

TABLE 4b-11. EQUILIBRIUM CONSTANTS EXPRESSED AS FUNCTIONS
OF P AND ϵ (*Continued*)

Reaction	K	Reaction	K
$A + B \rightleftharpoons C + 2D$	$\dfrac{4\epsilon^3 P}{(1 - \epsilon)^2(2 + \epsilon)}$	$A + 3B \rightleftharpoons C + 2D$	$\dfrac{4\epsilon^3(4 - \epsilon)}{27(1 - \epsilon)^4 P}$
$A + B \rightleftharpoons 2C + 2D$	$\dfrac{4\epsilon^4 P^2}{(1 - \epsilon)^2(1 + \epsilon)^2}$	$A + 3B \rightleftharpoons 2C + 2D$	$\dfrac{16\epsilon^4}{27(1 - \epsilon)^4}$
$A + B \rightleftharpoons C + 3D$	$\dfrac{27\epsilon^4 P^2}{4(1 - \epsilon)^2(1 + \epsilon)^2}$	$A + 3B \rightleftharpoons C + 3D$	$\dfrac{\epsilon^4}{(1 - \epsilon)^4}$
$A + B \rightleftharpoons 2C + 3D$	$\dfrac{108\epsilon^5 P^3}{(1 - \epsilon)^2(2 + 3\epsilon)^3}$	$A + 3B \rightleftharpoons 2C + 3D$	$\dfrac{4\epsilon^5 P}{(1 - \epsilon)^4(4 + \epsilon)}$
$A + B \rightleftharpoons 3C + 3D$	$\dfrac{729\epsilon^6 P^4}{16(1 - \epsilon)^2(1 + 2\epsilon)^4}$	$A + 3B \rightleftharpoons 3C + 3D$	$\dfrac{27\epsilon^6 P^2}{4(1 - \epsilon)^4(2 + \epsilon)^2}$
$A + 2B \rightleftharpoons C + D$	$\dfrac{\epsilon^2(3 - \epsilon)}{4(1 - \epsilon)^3 P}$	$2A + 3B \rightleftharpoons C + D$	$\dfrac{\epsilon^2(5 - 3\epsilon)^3}{108(1 - \epsilon)^5 P^3}$
$A + 2B \rightleftharpoons C + 2D$	$\dfrac{\epsilon^3}{(1 - \epsilon)^3}$	$2A + 3B \rightleftharpoons C + 2D$	$\dfrac{\epsilon^3(5 - 2\epsilon)^2}{27(1 - \epsilon)^5 P^2}$
$A + 2B \rightleftharpoons 2C + 2D$	$\dfrac{4\epsilon^4 P}{(1 - \epsilon)^3(3 + \epsilon)}$	$2A + 3B \rightleftharpoons 2C + 2D$	$\dfrac{4\epsilon^4(5 - \epsilon)}{27(1 - \epsilon)^5 P}$
$A + 2B \rightleftharpoons C + 3D$	$\dfrac{27\epsilon^4 P}{4(1 - \epsilon)^3(3 + \epsilon)}$	$2A + 3B \rightleftharpoons C + 3D$	$\dfrac{\epsilon^4(5 - \epsilon)}{4(1 - \epsilon)^5 P}$
$A + 2B \rightleftharpoons 2C + 3D$	$\dfrac{27\epsilon^5 P^2}{(1 - \epsilon)^3(3 + 2\epsilon)^2}$	$2A + 3B \rightleftharpoons 2C + 3D$	$\dfrac{\epsilon^5}{(1 - \epsilon)^5}$
$A + 2B \rightleftharpoons 3C + 3D$	$\dfrac{27\epsilon^6 P^3}{4(1 - \epsilon)^3(1 + \epsilon)^3}$	$2A + 3B \rightleftharpoons 3C + 3D$	$\dfrac{27\epsilon^6 P}{4(1 - \epsilon)^5(5 + \epsilon)}$
$2A + 2B \rightleftharpoons C + D$	$\dfrac{\epsilon^2(2 - \epsilon)^2}{4(1 - \epsilon)^4 P^2}$	$3A + 3B \rightleftharpoons C + D$	$\dfrac{16\epsilon^2(3 - 2\epsilon)^4}{729(1 - \epsilon)^6 P^4}$
$2A + 2B \rightleftharpoons C + 2D$	$\dfrac{\epsilon^3(4 - \epsilon)}{4(1 - \epsilon)^4 P}$	$3A + 3B \rightleftharpoons C + 2D$	$\dfrac{4\epsilon^3(2 - \epsilon)^3}{27(1 - \epsilon)^6 P^3}$
$2A + 2B \rightleftharpoons C + 3D$	$\dfrac{27\epsilon^4}{16(1 - \epsilon)^4}$	$3A + 3B \rightleftharpoons 2C + 2D$	$\dfrac{64\epsilon^4(3 - \epsilon)^2}{729(1 - \epsilon)^6 P^2}$
$2A + 2B \rightleftharpoons 2C + 3D$	$\dfrac{27\epsilon^5 P}{4(1 - \epsilon)^4(4 + \epsilon)}$	$3A + 3B \rightleftharpoons C + 3D$	$\dfrac{4\epsilon^4(3 - \epsilon)^2}{27(1 - \epsilon)^6 P^2}$
$2A + 2B \rightleftharpoons 3C + 3D$	$\dfrac{729\epsilon^6 P^2}{64(1 - \epsilon)^4(2 + \epsilon)^2}$	$3A + 3B \rightleftharpoons 2C + 3D$	$\dfrac{4\epsilon^5(6 - \epsilon)}{27(1 - \epsilon)^6 P}$
$A + 3B \rightleftharpoons C + D$	$\dfrac{4\epsilon^2(2 - \epsilon)^2}{27(1 - \epsilon)^4 P^2}$		

4c. Critical Constants

DAVID WHITE

University of Pennsylvania

The critical temperature, critical pressure, and critical density of elements and inorganic compounds are given in Table 4c-1; those for organic compounds are listed in Table 4c-2. The numbers in the reference column refer to authors and journals listed after Table 4c-2.

TABLE 4c-1. CRITICAL TEMPERATURE, PRESSURE, AND DENSITY OF ELEMENTS AND INORGANIC COMPOUNDS

Element or compound	T_c, K	P_c, atm	ρ_c, g/cm³	V_c, cm³/mole	Ref.
Ammonia	405.51	111.3	0.235	72.5	1
Antimony trichloride	794.1	....	0.842	271	2
Argon	150.72	48.00	0.5308	75.25	1
Arsenic trichloride	311.29	....	0.720	252	2
Bismuth tribromide	1220	....	1.487	301.8	3
Bismuth trichloride	1178	....	1.210	261	4
Boron tribromide	573	....	0.90	278	1
Boron trichloride	452.0	38.2	....	.....	1
Boron trifluoride	260.9	49.2	0.59	115	5
Bromine	584	102	1.18	135	1
Carbon dioxide	304.20	72.85	0.468	94.0	1
Carbon diselenide	612	69	0.850	200	6
Carbon disulfide	552	78	0.441	173	1
Carbon monoxide	133.0	34.5	0.3010	93.06	1
Carbonyl sulfide	378	61	....	.....	1
Cesium	2056	130.8	0.451	295	7
Chlorine	417.2	76.1	0.573	124	1
Chlorotrifluorosilane	307.64	34.20	....	.....	1
Cyanogen	400	59	....	.....	1
Deuterium (equilibrium)	38.26	16.28	0.0668	60.33	1
Deuterium (normal)	38.35	16.43	....	.....	1
Dichlorodifluorosilane	368.93	35.54	....	.....	1
Dichlorosilane	470	44.7	0.515	196	8
Fluorine	144	55	....	.....	9
Gallium	5410	250	1.58	44.1	10
Germanium tetrachloride	550.1	38	....	.....	1
Helium-3	3.38	1.22	0.041	97.6	11
Helium-4	5.21	2.26	0.0693	57.76	1
Hydrazone	653	145	....	.....	1
Hydrogen (equilibrium)	32.94	12.77	0.0308	65.45	1
Hydrogen (normal)	33.24	12.80	0.03102	64.99	1

TABLE 4c-1. CRITICAL TEMPERATURE, PRESSURE, AND DENSITY OF ELEMENTS
AND INORGANIC COMPOUNDS (*Continued*)

Element or compound	T_c, K	P_c, atm	ρ_c, g/cm³	V_c, cm³/mole	Ref.
Hydrogen bromide...............	362.96	84.00	....	......	1
Hydrogen chloride................	324.7	81.5	0.45	81.0	12
Hydrogen cyanide................	456.7	53.2	0.195	139	1
Hydrogen deuteride..............	35.91	14.65	0.0481	62.99	1
Hydrogen fluoride...............	461	64.1	0.29	96.0	13
Hydrogen iodide.................	423.2	80.8	....	......	1
Hydrogen selenide...............	411	88	....	......	1
Hydrogen sulfide................	373.6	88.9	0.3488	97.71	1
Iodine.........................	785	116	....	......	9
Krypton........................	209.39	54.27	0.9085	92.24	1
Lead...........................	5400	850	2.2	94.2	14
Mercuric chloride...............	972	....	1.555	174.6	15
Monochlorosilane................	409	47.5	0.444	150	8
Neon...........................	44.44	26.86	0.4835	41.74	1
Niobium pentabromide...........	1009	....	1.05	469	16
Niobium pentachloride..........	807	46	0.68	397	17
Nitric oxide....................	180.3	64.6	0.52	58	1
Nitrogen.......................	126.3	33.54	0.3110	90.10	1
Nitrogen dioxide................	431	100	0.56	82.2	1
Nitrogen trifluoride.............	233.90	44.72	....	......	18
Nitrous oxide...................	309.59	71.596	0.4525	97.28	19
Nitryl fluoride..................	349.5	....	....	......	20
Oxygen.........................	154.78	50.14	0.41	78.0	1
Oxygen fluoride.................	215.2	48.9	0.553	97.6	21
Ozone..........................	285.3	54.6	....	......	22
Perchloryl fluoride..............	368.4	53.0	0.637	161	23
Phosgene.......................	455	56	0.52	190	1
Phosphine......................	324.5	64.5	....	......	1
Phosphonium chloride...........	322.3	72.7	....	......	1
Phosphorous....................	993.8	120.8	....	......	9
Phosphorous trichloride.........	793	....	0.520	264	2
Radon..........................	377.16	62.0	....	......	9
Rubidium.......................	2111	....	0.334	256	7
Silane..........................	270	42.2	0.309	104	8
Silicon tetrachloride............	506.8	37.1	0.584	291	8, 24
Silicon tetrafluoride.............	259.01	36.66	....	......	1
Silver..........................	7500	....	1.85	58.3	14
Stannic chloride......	591.9	36.95	0.7419	351.2	1
Sulfur..........................	1313	116	....	......	1
Sulfur dioxide..................	430.7	77.808	0.525	122	25
Sulfur hexafluoride.............	318.71	37.11	0.7517	194.3	1, 26
Sulfur tetrafluoride.............	364.1	....	....	......	27
Sulfur trioxide.................	491.4	83.8	0.633	126	1
Tantalum pentabromide.........	973	....	1.26	461	16
Tantalum pentachloride.........	767	....	0.89	402	17
Titanium tetrachloride..........	....	45.7	....	......	28
Trichlorofluorosilane............	438.42	35.33	....	......	1
Trichlorosilane.................	495	41.2	0.533	254	8
Tritium.........................	40.0	....	0.109	27.7	29
Uranium hexafluoride...........	503.4	45.5	....	......	1
Water..........................	647.4	218.3	0.326	55.3	1
Water (heavy)..................	644.1	216	....	......	30
Xenon..........................	289.75	58.0	1.105	118.8	1

TABLE 4c-2. CRITICAL TEMPERATURE, PRESSURE, AND DENSITY
OF ORGANIC COMPOUNDS

Compound	T_c, K	P_c, atm	ρ_c, g/cm³	V_c, cm³/mole	Ref.
Acetic acid...................	594.8	57.1	0.351	171	1
Acetic anhydride.................	569	46.2	.....	...	1
Acetone......................	508.7	46.6	0.273	213	1
Acetonitrile....................	547.9	47.7	0.237	173	1
Acetylene.....................	309.5	61.6	0.231	113	1
Aniline.......................	698.8	52.3	0.340	274	1
Benzene......................	562.7	48.6	0.300	260	1
Bromobenzene..................	670.9	44.6	0.458	343	1
n-Butane.....................	425.17	37.47	0.228	255	1
Butanol.......................	560.11	48.60	0.270	275	31
1-Butene.....................	419.6	39.7	0.234	240	1
2-Butene (cis).................	428.2	40.5	0.236	238	1
2-Butene (trans)...............	433.2	41.5	0.240	234	1
Carbon tetrachloride............	556.4	44.97	0.558	276	1
Carbon tetrafluoride............	227.9	.....	0.60	147	32
Chlorobenzene.................	632.4	44.6	0.365	308	1
Chlorodifluoromethane..........	369.6	48.48	0.525	165	1
Chloroform....................	536.6	54	0.496	241	1
Chlorotrifluoroethylene.........	379	40	0.55	212	1
Chlorotrifluoromethane.........	302.02	38.2	0.578	181	1
1-Chloro-1,1-difluoroethane.......	410.3	40.7	0.435	231	33
2-Chloro-1,1-difluoroethylene.......	400.6	44.0	0.499	197	33
Cyclohexane..................	554.2	40.57	0.273	308	1
Cyclopentane.................	511.8	44.55	0.27	260	1
Cyclopropane.................	397.81	54.23	.....	...	34
Dibromomethane..............	583.0	70.6	.....	...	1
1,1-Dichloroethane.............	523	50	.....	...	1
1,2-Dichloroethane.............	561	53	0.44	225	1
1,1-Dichloro-1,2,2,2-tetrafluoroethane	418.7	32.6	0.582	294	33
Dichlorodifluoromethane...........	384.7	39.6	0.555	218	1
Dichlorofluoromethane..........	451.7	51.0	0.522	197	1
Diethyl ether..................	467.8	35.6	0.265	280	1
Diethyl ketone................	561.0	36.9	0.256	336	35
1,1-Difluoroethane.............	386.7	44.4	0.365	181	33
1,1-Difluoroethylene............	303.3	43.8	0.417	154	33
Dimethylamine................	437.7	52.4	.....	...	1
2,2-Dimethylbutane............	489.4	30.67	0.240	359	36
2,3-Dimethylbutane............	550.3	30.99	0.241	358	36
Dimethyl ether................	400.1	52.6	0.246	187	1
Dimethyl oxalate..............	628	39.3	.....	...	37
Dioxane......................	585	50.7	0.36	245	1
Ethane.......................	305.43	48.20	0.203	148	1
Ethyl acetate..................	523.3	37.8	0.308	286	1
Ethyl alcohol..................	516	63.0	0.276	167	1
Ethylamine...................	456.4	55.54	.....	...	1
Ethyl bromide.................	503.9	61.5	0.507	215	1
Ethyl chloride.................	460.4	51.72	.....	...	1
Ethyl cyclopentane.............	569.5	33.53	0.262	268	18
Ethyl fluoride.................	375.32	46.62	.....	...	1
Ethyl formate.................	508.5	46.8	0.323	229	1
Ethyl mercaptan...............	499	54.2	0.300	207	1
Ethyl methyl ether.............	437.9	43.4	0.272	221	1
Ethyl methyl ketone............	533.7	39.46	0.252	286	1
Ethyl propyl ether.............	500.6	32.1	0.260	339	1
Ethyl sulfide..................	498.7	54.2	0.300	301	1
Ethylene.....................	283.06	50.50	0.227	124	1

TABLE 4c-2. CRITICAL TEMPERATURE, PRESSURE, AND DENSITY
OF ORGANIC COMPOUNDS *(Continued)*

Compound	T_c, K	P_c, atm	ρ_c, g/cm^3	V_c, cm^3/mole	Ref.
Ethylene oxide...................	469.0	70.97	0.32	137	1
Fluorobenzene...................	560.08	44.91	0.269	357	39
Hexafluorobenzene...............	516.91	.....	.....	...	40
n-Hexane.......................	507.9	29.94	0.234	368	1
Iodobenzene....................	721	44.6	0.581	351	1
Isobutane......................	408.14	36.00	0.221	263	41
Isopentane.....................	461.0	32.9	0.234	308	1
Methane.......................	191.1	45.80	0.162	99	1
Methyl acetate..................	506.9	46.3	0.325	228	1
Methyl alcohol..................	513.2	78.47	0.272	118	1
Methylamine....................	430.1	73.6	.....	...	1
Methyl borate...................	501.7	35.4	.....	...	42
Methyl bromide..................	464	.....	.....	...	1
Methyl butyrate.................	554.5	34.3	0.300	340	1
Methyl chloride.................	416.28	65.93	0.353	143	1
Methylcyclopentane..............	532.77	37.36	0.264	212	38
Methyl fluoride.................	317.71	58.0	0.300	113	1
Methyl formate.................	487.2	59.2	0.349	172	1
Methyl iodide...................	528	.....	.....	...	1
Methyl isobutyl ketone...........	571.5	32.3	.....	...	35
Methyl isopropyl ketone..........	553.4	38.0	0.278	310	35
2-Methylpentane.................	497.9	29.95	0.235	367	36
3-Methylpentane.................	504.4	30.83	0.235	367	36
Methyl n-propyl ketone...........	564.0	38.4	0.286	301	35
Methyl sulfide	503.1	54.6	0.309	201	1
Methylene chloride...............	510.2	59.97	.....	...	1
Neopentane.....................	433.76	31.57	0.238	303	43
Nitromethane...................	588	62.3	0.352	173	1
n-Octane.......................	569.4	24.64	0.235	486	1
n-Pentane......................	569.78	33.31	0.232	311	1
Perfluorobutane.................	386.4	22.93	0.600	397	44
Perfluorocyclohexane.............	457.2	24	.....	...	45
Perfluoro-n-heptane.............	474.8	16.0	0.584	664	1
Perfluorohexane.................	447.7	.....	.....	...	46
Perfluoromethylcyclohexane........	486.8	23	.....	...	45
Phenol.........................	692.4	60.5	.....	...	1
Propane........................	370.0	42.01	0.220	200	1
Propene........................	365.0	45.6	0.233	181	1
Propionic acid..................	612	53	0.32	232	1
Propionitrile...................	564.4	41.3	0.240	230	1
n-Propyl acetate................	549.4	32.9	0.296	345	1
n-Propyl alcohol................	537.3	50.2	0.273	220	1
Propyl formate..................	538.1	40.1	0.309	285	1
Propyne........................	401	52.8	.....	...	1
Pyridine........................	617.4	60.0	.....	...	1
Toluene........................	594.0	41.6	0.29	318	1
Trichlorotrifluoromethane..........	471.2	43.2	0.554	189	1
Trichlorotrifluoroethane...........	487.3	33.7	0.576	325	1
1,1,1-Trifluoroethane..............	346.3	37.1	0.434	194	33
Trimethylamine..................	433.3	40.2	0.233	254	1

References for Table 4c-1 and 4c-2

1. Kobe, K. A., and R. E. Lynn, Jr.: *Chem. Revs.* **52**, 117 (1953).
2. Nisel'son, L. A., U. V. Maguchiva, and T. D. Sokolova: *Zhur. Neorg. Khim.* **10**, 592 (1965).
3. Johnson, J. W., D. Cubicciotti, and W. J. Silva: *J. Phys. Chem.* **69**, 1989 (1965).
4. Johnson, J. W., and D. Cubicciotti: *J. Phys. Chem.* **68**, 2235 (1964).
5. Smith, C. R. F.: *U.S. AEC Rept.* NAA-SR-5286, 1960.
6. Gattow, G., and M. Draeger: *Z. anorg. allgem. Chem.* **343**, 11 (1966).
7. Hochman, J. M., and C. F. Bonila: *Symp. Thermophys. Prop. Am. Assoc. Mech. Engrs. (Purdue)*, 122 (1965).
8. Lapidus, I. I., A. L. Seifer, and L. A. Nisel'son: *Izvest. Uysshikh Tcheben Zavedenii Tsvetn Met* **9**, 92 (1966).
9. Gates, D. S., and G. Thodos: *A. I. Ch. E. Journal* **6**, 50 (1960).
10. Nizhenko, V. I., L. I. Sklyarenko, and U. N. Eremenko: *Ukrain. Khim. Zhur.* **31**, 559 (1965).
11. Peshkov, V. P.: *Zhur. Ekspl. i Teoret. Fiz.* **33**, 833 (1957).
12. Frank, E. U., M. Brose, and K. Mangold: *Progr. Intern. Research Thermodyn. Transport Properties Symp.*, *Thermophys. Properties 2d*, p. 159, 1962.
13. Frank, E. U., and W. Spalthoff: *Z. Electrochem.* **61**, 348 (1957).
14. Grosse, A. V., and A. D. Kirshenbaum: *J. Inorg. & Nuclear Chem.* **24**, 739 (1963).
15. Johnson, J. W., W. J. Silva, and D. Cubicciotti: *J. Phys. Chem.* **70**, 1169 (1966).
16. Nisel'son, L. A., and T. D. Sokolova: *Zhur. Neorg. Khim.* **9**, 2066 (1964).
17. Nisel'son, L. A., A. I. Pustil'nik, and T. D. Sokolova: *Zhur. Neorg. Khim.* **9**, 1049 (1964).
18. Jarry, R. L., and H. C. Miller: *J. Phys. Chem.* **60**, 1412 (1956).
19. Couch, E. J., and K. A. Kobe: *J. Chem. Eng. Data* **6**, 229 (1961).
20. Hetherington, G., and P. L. Robinson: *J. Chem. Soc.*, 2230 (1955).
21. Anderson, R., J. G. Schnizlein, R. C. Toole, and T. D. O'Brien: *J. Phys. Chem.* **56**, 473–474 (1952).
22. Jenkins, A. C., and C. N. Birdsall: *J. Chem. Phys.* **20**, 1158 (1952).
23. Engelbrecht, A., and H. Atzwanger: *J. Inorg. & Nuclear Chem.* **2**, 348 (1956), and R. L. Jarry: *J. Phys. Chem.* **61**, 498 (1957).
24. Menzer, W.: *Naturwissenschaften* **45**, 126 (1958).
25. Kang, T. L., L. J. Hirth, K. A. Kobe, and J. J. Mcketta: *J. Chem. Eng. Data* **6**, 220 (1960).
26. Otto, J. and W. Thomas: *Z. physik. Chem. (Frankfurt)* **23**, 84 (1960).
27. Tullock, C. W., F. S. Fawcett, W. C. Smith, and D. D. Coffman: *J. Am. Chem. Soc.* **82**, 539 (1960).
28. Menzer, W.: *Naturwissenchaften* **45**, 126 (1958).
29. Rogers, J. D., and F. G. Brickwedde: *J. Chem. Phys.* **42**, 2822 (1965).
30. Oliver, G. D., and J. W. Gisard: *J. Am. Chem. Soc.* **78**, 561 (1956).
31. Singh, R., and L. W. Shemilt: *J. Chem. Phys.* **23**, 1370 (1955).
32. McCormack, K. E., and W. G. Schmeider: *J. Chem. Phys.* **19**, 849 (1951).
33. Mears, W. H., R. F. Stahl, S. R. Orfes, R. C. Shair, L. F. Kells, W. Thompson, and H. McCann: *Ind. Eng. Chem.* **47**, 1449 (1955).
34. Booth, H. S., and W. C. Monic: *J. Phys. Chem.* **62**, 875 (1958).
35. Kobe, K. A., H. R. Crawford, and R. W. Stephenson: *Ind. Eng. Chem.* **47**, 1767 (1955).
36. Kay, W. B.: *J. Am. Chem. Soc.* **68**, 1136 (1946).
37. Stern, S. A., and W. B. Kay: *J. Phys. Chem.* **61**, 374 (1955).
38. Kay, W. B.: *J. Am. Chem. Soc.* **69**, 1273 (1947).
39. Doulsen, D. R., R. T. Moore, J. P. Dawson, and G. Waddington: *J. Am. Chem. Soc.* **80**, 2031 (1958).
40. Counsell, J. F., J. H. S. Green, J. L. Hales, and J. F. Martin: *Trans. Faraday Soc.* **61**, 212 (1965).
41. Beattie, J. A., D. G. Edwards, and S. Marple: *J. Chem. Phys.* **17**, 576 (1949).
42. Griskev, R. G., W. E. Gorgas, and L. N. Canjar: *A. I. Ch. E. Journal* **6**, 128 (1960).
43. Beattie, J. A., D. R. Doulson, and S. W. Levine: *J. Chem. Phys.* **19**, 948 (1951).
44. Brown, J. A., and W. H. Mears: *J. Phys. Chem.* **62**, 960 (1958).
45. Rowlenson, J. S., and R. Thacker: *Trans. Faraday Soc.*, **53**, 1 (1957).
46. Dunlap, R. D., C. J. Murphy, Jr., and R. G. Bedford: *J. Am. Chem. Soc.* **80**, 83 (1958).

4d. Compressibility

GEORGE C. KENNEDY[1]

Institute of Geophysics and Planetary Physics, University of California, Los Angeles

R. NORRIS KEELER[2]

Lawrence Radiation Laboratory, University of California, Livermore

4d-1. Compressibilities below 250 Kilobars.[3] The data on the compressibility of solids are widely scattered through the scientific literature. Further, these data are given at various pressure intervals and for various pressure units. Bridgman normally published work in units of kilograms per square centimeter, whereas most modern high-pressure data are published in units of bars or kilobars. Bridgman further examined the compressibility of some substances a number of different times with substantially differing results. Data at the upper end of one pressure range determined with one kind of apparatus do not overlap well with data in another pressure range determined with another kind of apparatus. A large fraction, if not most, of the available data on compressibility of solids, liquids, and gases, where data extends to 10 kb and beyond, has been extracted from the technical literature. All this has been plotted. Where data are in conflict, we have plotted the various results and attempted to fit the best smooth curves through them. From these curves we have read off points and tabulated the results: pressure P in kilobars, and relative volume, the ratio of the volume V to the volume V_0 at standard conditions. Many of the results in the following tables are interpolations and smoothed values, so that the tabulated results are not identical in many cases to those found in the source material. A substantial amount of judgment in selection of data has had to be used.

In addition, a large amount of data has recently become available from the extensive program of shock-wave research carried out at Los Alamos Laboratory, Lawrence Radiation Laboratory, and various foreign laboratories. Dr. R. N. Keeler has reduced the shock-wave data for a number of selected substances and presented them as a separate set of tables in a following section.

The reduction of data from shock-wave experiments is crucially dependent on assumptions of an equation of state. Consequently, these assumptions are set out by R. N. Keeler. It should be emphasized that the assumptions used in the reduction of these data differ from those used by a number of other laboratories. For a number of substances, specifically for such substances as indium and calcium, the shock-wave results are quite different from the static compression results. Where conflicts occur, the shock-wave results are to be preferred.

References are given by number after the table titles or underneath the column heads. Temperatures are 25°C, unless otherwise marked.

---

[1] Compressibilities below 250 kilobars.
[2] High-pressure compressibilities.
[3] *Unv. Calif. (Los Angeles) Inst. Geophys. and Planetary Phys. Publi.* 732.

TABLE 4d-1. V/V_0 OF ELEMENTS*

P, kilobars	H_2 at 30°C [1]	H_2 at 65°C [1]	He at 65°C [1]	N_2 at 23.5°C [7]	N_2 at 65°C [1]	Ar at 25°C [11]	Ar at 55°C [7]
1	.....	.....	....	....	....	1.06	
2	13.89	.....	....	....	....	0.85	0.88
3	11.55	12.17	5.54	1.24	1.29	0.77	0.80
4	10.52	11.04	4.76	1.16	1.20	0.73	0.75
5	9.81	10.29	4.31	1.11	1.14	0.69	0.71
6	9.29	9.72	4.00	1.06	1.09	....	0.68
7	8.87	9.29	3.74	....	1.06	....	0.66
8	8.55	8.97	3.58	....	1.03	....	0.64
9	8.25	9.71	3.44	....	1.00	....	0.63
10	8.01	8.49	3.31	....	0.98	....	0.62
11	7.78	8.29	3.21	....	0.96	....	0.61
12	7.54	8.13	3.12	....	0.95	....	0.60
13	7.32	7.95	3.04	....	0.93	....	0.59
14	.....	.....	2.98	....	0.92	....	0.58
15	.....	.....	2.92	....	0.91	....	0.57

* For references see p. 4-96.

TABLE 4d-1. V/V_0 OF ELEMENTS (*Continued*)

P, kilobars	Ag [13]	Al [2]	As [2]	Au [13]	Ba [12]	Be [5]	Bi at 25°C [12]	Bi at −78.5°C [10]	C (graphite) [6, 12]	Ca [12]	Cd [12]	Ce [2, 5]
0	1.000	1.000	1.000	1.000	1.000	1.000	1.000	1.000	1.000	1.000	1.000	1.000
5	0.995	0.993	0.988	0.997	0.955	0.996	0.985	0.985	0.984	0.968	0.987	0.976
10	0.990	0.986	0.977	0.994	0.908	0.991	0.971	0.972	0.972	0.942	0.977	0.953
												f
15	0.986	0.980	0.967	0.990	0.872	0.987	0.959	0.960	0.962	0.917	0.966	0.835
20	0.981	0.974	0.960	0.988	0.865	0.982	0.948 *b*	0.948	0.954	0.896	0.957	0.813
25	0.977	0.969	0.952	0.985	0.813	0.978	0.848	0.937 *d*	0.946	0.877	0.947	0.798
30	0.972	0.964	0.945	0.983	0.788	0.975	0.840	0.843	0.939	0.860	0.938	0.787
35	.....	0.960	0.938	.....	0.765	0.971	0.833	0.839	0.933	0.844	0.930	0.777
40	.....	0.955	0.933	.....	0.744	0.967	0.825	0.835	0.927	0.829	0.922	0.769
45	.....	0.951	0.926	.....	0.725	0.965	0.814	0.826	0.923	0.815	0.915	0.762
50	.....	0.947	0.920	.....	0.707	0.963	0.807	0.823	0.917	0.801	0.907	0.755
55	.....	0.943	0.915	.....	0.691	0.960	0.800	.....	0.913	0.788	0.900	0.699
60	.....	0.938	0.910	.....	0.658 *a*	0.958	0.794	.....	0.908	0.778	0.895	0.693
65	.....	0.935	0.906	.....	0.647	0.956	0.781	.....	0.905	0.745 *e*	0.887	0.687
70	.....	0.932	0.902	.....	0.636	0.953	0.776	.....	0.901	0.737	0.883	0.682
75	.....	0.929	0.897	.....	0.625	0.951	0.771	.....	0.897	0.728	0.876	0.676
80	.....	0.927	0.895	.....	0.615	0.949	0.766	.....	0.895	0.722	0.871	0.671
85	.....	0.923	0.892	.....	0.605	0.947	0.762 *c*	.....	0.891		0.866	0.667
90	.....	0.920	0.888	.....	0.595	0.945	0.746	.....	0.889	0.713	0.860	0.663
95	.....	0.917	0.886	.....	0.586	0.944	0.742	.....	0.887	0.706	0.856	0.660
100	.....	0.914	0.883	.....	0.576	0.943	0.737	.....	0.884	0.699	0.851	0.657

* For references see p. 4-96.

a Transition at 55.5 kb; volumes 0.682 and 0.633.

b Two transitions at 25.4 and 27.0: extreme volumes 0.936 and 0.850.

c Transition at 77.5; volumes 0.760 and 0.748.

d At 27.7 the volumes are 0.931 and 0.846.

e Transition at 62.7; volumes 0.771 and 0.758.

f Transition at 12.2; volumes 0.926 and 0.850.

TABLE 4d-1. V/V_0 OF ELEMENTS (Continued)

P, kilobars	Co [13]	Cs at 25°C [2, 5]	Cs at 50°C (solid) [15]	Cs at 75°C (liquid) [15]	Cu [13]	Dy [17]	Er [17]	Fm [13]	Ge [2, 13]	Gd [17]	Hg [19]	Ho [17]
0	1.000	1.000	1.000	1.000	1.000	1.000	1.000	1.000	1.000	1.000	1.000	1.000
5	0.997	0.840	0.813	0.810	0.996	0.986	0.987	0.997	0.992	0.987	0.981	0.987
10	0.994	0.756	0.727	....	0.993	0.974	0.976	0.994	0.985	0.974	0.966	0.975
15	0.991	0.700	0.672	....	0.989	0.963	0.965	0.991	0.980	0.963	....	0.965
20	0.989	0.650	....	....	0.985	0.953	0.955	0.989	0.975	0.953	....	0.955
25	0.987	*g* 0.606	....	....	0.982	0.943	0.946	0.987	0.970	0.943		0.945
30	0.984	0.570	....	....	0.979	0.934	0.937	0.986	0.965	0.935		0.936
35	....	0.542	....	....	....	0.925	0.928	....	0.960	0.927		0.928
40	....	0.519	....	....	....	0.917	0.921	....	0.956	0.920		0.919
45	....	*h* 0.445	....	....	....	....	....	....	0.951			
50	....	0.428	....	....	....	....	....	....	0.947			
55	....	0.415	....	....	....	....	....	....	0.943			
60	....	0.405	....	....	....	....	....	....	0.940			
65	....	0.397	....	....	....	....	....	....	0.937			
70	....	0.390	....	....	....	....	....	....	0.934			
75	....	0.385	....	....	....	....	....	....	0.930			
80	....	0.380	....	....	....	....	....	....	0.927			
85	....	0.375	....	....	....	....	....	....	0.924			
90	....	0.372	....	....	....	....	....	....	0.921			
95	....	0.370	....	....	....	....	....	....	0.919			
100	....	0.368	....	....	....	....	....	....	0.917			

* For references see p. 4-96.
g Transition at 22.6; volumes 0.628 and 0.622.
h Discontinuity of volume at 44.7.

TABLE 4d-1. V/V_0 OF ELEMENTS (Continued)

P, kilobars	In at 25°C [9]	In at −78.5°C [10]	Ir [13]	K [2]	La [2, 5]	Li [2]	Lu [17]	Mg [2]	Mn [2]	Mo [13]	Na [2]	Nb [13]
0	1.000	1.000	1.000	1.000	1.000	1.000	1.000	1.000	1.000	1.000	1.000	1.000
5	0.987	0.987	0.998	0.875	0.980	0.962	0.988	0.987	0.990	0.997	0.931	0.996
10	0.975	0.975	0.997	0.810	0.963	0.938	0.976	0.975	0.982	0.995	0.883	0.993
15	0.965	0.965	0.995	0.759	0.947	0.899	0.965	0.963	0.974	0.993	0.846	0.990
20	0.955	0.955	0.994	0.720	0.931	0.873	0.955	0.952	0.967	0.991	0.825	0.988
					i							
25	0.946	0.946	0.992	0.689	0.915	0.850	0.946	0.942	0.961	0.990	0.788	0.985
30	0.937	0.937	0.991	0.663	0.903	0.831	0.938	0.933	0.955	0.989	0.767	0.983
35	0.928	0.930	......	0.641	0.890	0.815	0.930	0.926	0.950	......	0.749	
40	0.920	0.923	......	0.622	0.880	0.798	0.922	0.917	0.946	......	0.734	
45	0.912	0.917	......	0.605	0.870	0.783	......	0.910	0.942	......	0.718	
50	0.904	0.910	......	0.591	0.861	0.769	......	0.902	0.938	......	0.705	
55	0.896	......	......	0.578	0.852	0.757	......	0.895	0.935	......	0.692	
60	0.890	......	......	0.565	0.844	0.745	......	0.887	0.932	......	0.680	
65	0.882	......	......	0.554	0.836	0.732	......	0.881	0.929	......	0.670	
70	0.876	......	......	0.543	0.828	0.722	......	0.875	0.926	......	0.658	
75	0.870	......	......	0.534	0.821	0.711	......	0.869	0.923	......	0.648	
80	0.865	......	......	0.525	0.815	0.700	......	0.864	0.921	......	0.637	
85	0.859	......	......	0.518	0.810	0.692	......	0.858	0.919	......	0.627	
90	0.854	......	......	0.511	0.806	0.684	......	0.854	0.917	......	0.620	
95	0.845	......	......	0.505	0.802	0.676	......	0.850	0.916	......	0.610	
100	0.840	......	......	0.499	0.798	0.669	......	0.845	0.915	......	0.602	

* For references see p. 4-96.
i Transition at 22.9; volumes 0.924 and 0.922.

TABLE 4d-1. V/V_0 of Elements (*Continued*)

P, kilobars	Nd [2]	Ni [13]	P (red) [8]	P (black) [2]	P (violet) [2]	Pb 25°C [9]	Pb at 75°C [16]	Pd [13]	Pr [2]	Pt [13]	Pu [20]	Rb at 25°C [5]	Rb at 50°C [15]
0	1.000	1.000	1.000	1.000	1.000	1.000	1.000	1.000	1.000	1.000	1.000	1.000	1.000
5	0.984	0.997	0.977	0.984	0.977	0.988	0.988	0.997	0.982	0.998	0.991	0.877	0.840
10	0.970	0.994	0.958	0.970	0.955	0.977	0.976	0.994	0.965	0.995	0.983	0.805	0.765
15	0.958	0.991	.....	0.957	0.935	0.966	0.965	0.992	0.950	0.993	0.975	0.753	0.718
20	0.946	0.988	.....	0.946	0.917	0.957	0.955	0.988	0.937	0.992	0.968	0.710	
25	0.936	0.986	.....	0.935	0.899	0.948	0.945	0.986	0.925	0.990	0.961	0.675	
30	0.925	0.984	.....	0.926	0.883	0.940	0.935	0.984	0.915	0.899	0.955	0.648	
35	0.915	.....	.....	0.917	0.867	0.932	.....	.....	0.905	.....	0.949	0.626	
40	0.906	.....	.....	0.911	0.852	0.924	.....	.....	0.895	.....	0.943	0.607	
45	0.896	.....	.....	0.904	0.837	0.915	.....	.....	0.885	.....	0.938	0.591	
50	0.887	.....	.....	*j* 0.848	0.826	0.908	.....	.....	0.876	.....	0.933	0.575	
55	0.879	.....	.....	0.841	0.815	0.901	.....	.....	0.864	.....	0.929	0.561	
60	0.871	.....	.....	0.835	0.807	0.899	.....	.....	0.861	.....	0.924	0.548	
65	0.863	.....	.....	0.829	0.799	0.889	.....	.....	0.853	.....	0.920	0.536	
70	0.856	.....	.....	0.824	0.791	0.883	.....	.....	0.846	.....	0.915	0.525	
75	0.848	.....	.....	0.818	0.784	0.877	.....	.....	0.840	.....	0.911	0.515	
80	0.842	.....	.....	0.814	0.777	0.872	.....	.....	0.834	.....	0.907	0.505	
85	0.836	.....	.....	0.810	*k* 0.668	0.867	.....	.....	0.826	.....	0.904	0.496	
90	0.830	.....	.....	0.806	0.665	0.862	.....	.....	0.820	.....	0.901	0.487	
95	0.825	.....	.....	0.802	0.662	0.857	.....	.....	0.815	.....	0.989	0.477	
100	0.820	.....	.....	0.797	0.659	0.852	.....	.....	0.808	.....	0.987	0.470	

* For references see p. 4-96.
j Reversible transition in this region.
k Irreversible transition at 83.3 from violet to black; volumes 0.773 and 0.670.

TABLE 4d-1. V/V_0 OF ELEMENTS (Continued)

P, kilobars	Rh [13]	Ru [13]	S at 25°C [9]	S at -78.5°C [10]	Sb at 25°C [12]	Sb at -78.5°C [10]	Se at 25°C [12]	Se at -78.5°C [10]	Si [2]	Sm [17]	Sm [12]	Ta [13]
0	1.000	1.000	1.000	1.000	1.000	1.000	1.000	1.000	1.000	1.000	1.000	1.000
5	0.997	0.998	0.950	0.958	0.986	0.987	0.945	0.952	0.995	0.982	0.991	0.996
10	0.995	0.996	0.915	0.925	0.975	0.976	0.907	0.915	0.990	0.966	0.982	0.994
15	0.993	0.994	0.888	0.900	0.963	0.965	0.876	0.885	0.985	0.953	0.973	0.992
20	0.991	0.992	0.869	0.881	0.953	0.956	0.850	0.860	0.981	0.940	0.965	0.990
25	0.990	0.991	0.851	0.864	0.943	0.946	0.830	0.839	0.977	0.928	0.958	0.988
30	0.989	0.989	0.837	0.850	0.934	0.937	0.813	0.825	0.975	0.915	0.950	0.981
35	.....	.....	0.822	0.839	0.925	0.928	0.798	0.811	0.971	0.910	0.942	
40	.....	.....	0.810	0.831	0.916	0.920	0.786	0.800	0.967	0.894	0.935	
45	.....	.....	0.800	0.825	0.908	0.913	0.776	0.791	0.964	.....	0.928	
50	.....	.....	0.791	0.821	0.900	0.905	0.767	.....	0.960	.....	0.921	
55	.....	.....	0.782	.....	0.894	.....	0.760	.....	0.957	.....	0.915	
60	.....	.....	0.774	.....	0.887	.....	0.751	.....	0.955	.....	0.908	
65	.....	.....	0.766	.....	0.880	.....	0.745	.....	0.953	.....	0.902	
70	.....	.....	0.758	.....	0.875	.....	0.738	.....	0.950	.....	0.895	
75	.....	.....	0.752	.....	0.869	.....	0.731	.....	0.948	.....	0.890	
80	.....	.....	0.745	.....	0.865	.....	0.725	.....	0.946	.....	0.884	
					l							
85	.....	.....	0.740	.....	0.816	.....	0.719	.....	0.945	.....	0.878	
90	.....	.....	0.735	.....	0.815	.....	0.714	.....	0.943	.....	0.822	
95	.....	.....	0.730	.....	0.814	.....	0.708	.....	0.942	.....	0.816	
100	.....	.....	0.725	.....	0.813	.....	0.702	.....	0.941	.....	0.810	

* For references see p. 4-96.
l Transition at 83.3; volumes 0.858 and 0.821.

Table 4d-1. V/V_0 of Elements (Continued)

P, kilobars	Te at 25°C [12]	Te at −78.5°C [10, 13]	Th [2]	Ti [2, 13]	Tl [12]	Tm [17]	U [2]	U [13]	Y [16]	Yb [17]	Zn [12]	Zr [2]
0	1.000	1.000	1.000	1.000	1.000	1.000	1.000	1.000	1.000	1.000	1.000	1.000
5	0.975	0.976	0.990	0.994	0.987	0.987	0.995	0.998	0.986	0.962	0.992	0.994
10	0.955	0.958	0.981	0.989	0.975	0.975	0.990	0.996	0.973	0.928	0.988	0.987
15	0.930	0.942	0.972	0.985	0.965	0.965	0.985	0.994	0.961	0.889	0.975	0.982
20	0.918	0.928	0.963	0.980	0.955	0.955	0.981	0.993	0.950	0.874	0.967	0.975
25	0.902	0.915	0.955	0.977	0.946	0.946	0.976	0.992	0.940	0.852	0.960	0.970
30	0.888	0.903	0.947	0.973	0.937	0.937	0.973	0.991	0.930	0.832	0.952	0.965
35	0.876	0.892	0.940	0.968	0.929	0.928	0.969	.....	0.921	0.814	0.944	0.959
40	*m* 0.865	*o* 0.882	0.932	0.965	*p* 0.911	0.921	0.966	.....	0.913	0.797	0.937	0.954
45	0.791	0.873	0.926	0.962	0.903	.....	0.963	.....	.....	.....	0.930	0.950
50	0.785	0.869	0.920	0.958	0.895	.....	0.960	.....	.....	.....	0.923	0.945
55	0.779	.....	0.916	0.955	0.887	.....	0.957	.....	.....	.....	0.917	0.940
60	0.774	.....	0.911	0.953	0.880	.....	0.955	.....	.....	.....	0.910	0.935
65	0.770	.....	0.907	0.950	0.872	.....	0.952	.....	.....	.....	0.904	0.931
70	*n* 0.760	.....	0.903	0.947	0.865	.....	0.950	.....	.....	.....	0.897	0.927
75	0.754	.....	0.900	0.945	0.859	.....	0.948	.....	.....	.....	0.891	0.924
80	0.748	.....	0.896	0.944	0.852	.....	0.945	.....	.....	.....	0.886	0.920
85	0.744	.....	0.894	0.942	0.846	.....	0.944	.....	.....	.....	0.881	0.917
90	0.740	.....	0.890	0.940	0.840	.....	0.943	.....	.....	.....	0.876	0.915
95	0.735	.....	0.888	0.938	0.834	.....	0.942	.....	.....	.....	0.872	0.912
100	0.730	.....	0.885	0.936	0.829	.....	0.941	.....	.....	.....	0.867	0.909

* For references see p. 4-96.

m Transition at 40.1; volumes 0.848 and 0.893.

n Transition at 68.6; volumes 0.766 and 0.759.

o Transition at 40.3; volumes 0.881 to 0.837.

p Transition at 36.7; volumes 0.921 and 0.914.

TABLE 4d-2. V/V_0 of Inorganic Compounds*

P, kilobars	AgBr at 25°C [9]	AgBr at -78.5°C [10]	AgBrO₃ [9]	AgCl at 25°C [9]	AgCl at -78.5°C [10]	AgCN [5]	AgI at 25°C [9]	AgI at -78.5°C [10]	AgNO₃ [9]	BaS at 25°C [10]	BaS at -78.5°C [10]
0	1.000	1.000	1.000	1.000	1.000	1.000	1.000	1.000	1.000	1.000	1.000
5	0.989	0.989	0.985	0.990	0.990	0.955	c 0.820	d 0.822	0.983	0.985	0.985
10	0.987	0.978	0.972	0.979	0.979	0.922	0.810	0.812	e 0.955	0.974	0.974
15	0.968	0.968	0.959	0.969	0.970	0.895	0.800	0.802	0.937	0.963	0.963
20	0.959	0.960	0.947	0.959	0.961	0.873	0.790	0.793	0.921	0.953	0.954
25	0.950	0.951	0.937	0.950	0.953	0.854	0.779	0.785	0.907	0.945	0.945
30	0.941	0.943	0.928	0.941	0.945	0.787	0.770	0.776	0.895	0.937	0.937
35	0.932	0.935	0.919	0.933	0.937	0.772	0.761	0.769	0.883	0.930	0.931
40	0.923	0.927	0.910	0.924	0.930	0.760	0.752	0.762	0.872	0.924	0.925
45	0.915	0.920	0.902	0.915	0.923	....	0.743	0.756	0.862	0.918	0.920
50	0.906	0.913	0.895	0.907	0.917	....	0.735	0.750	0.852	0.912	0.914
55	0.899	....	0.888	0.900	....	....	0.727	....	0.843		
60	0.892	....	0.881	0.893	....	....	0.720	....	0.835		
65	0.885	....	0.875	0.887	....	....	0.713	....	0.828		
70	0.880	....	0.870	0.880	....	....	0.706	....	0.822		
75	0.875	....	0.865	0.874	....	....	0.700	....	0.816		
80	0.870	....	0.860	0.868	....	....	0.693	....	0.810		
85	a 0.847	....	0.856	b 0.863	....	....	0.687	....	0.805		
90	0.841	....	0.852	0.743	....	....	0.681	....	0.800		
95	0.835	....	0.848	0.737	....	....	0.675	....	0.796		
100	0.829	....	0.845	0.732	....	....	0.670	....	0.792		

* For references see p. 4-96.
a Transition at 84.3; volumes 0.859 and 0.848.
b Transition at 88.2; volumes 0.860 and 0.744.
c Transition at 2.9 kb; volumes 0.989 ± 2.9 and 0.826.
d Transition in this region.
e Transition at 9.3; volumes 0.970 and 0.957.

TABLE 4d-2. V/V_0 OF INORGANIC COMPOUNDS (Continued)

P, kilobars	BaSe at 25°C [10]	BaSe at −78.5°C [10]	BaTe at 25°C [10]	BaTe at −78.5°C [10]	CaS at 25°C [10]	CaS at −78.5°C [10]	CaSe at 25°C [10]	CaSe at −78.5°C [10]	CaTe at 25°C [10]	CaTe at −78.5°C [10]	CsBr at 25°C [9]	CsBr at −78.5°C [10]
0	1.000	1.000	1.000	1.000	1.000	1.000	1.000	1.000	1.000	1.000	1.000	1.000
5	0.985	0.987	0.983	0.985	0.987	0.988	0.990	0.991	0.988	0.989	0.972	0.973
10	0.974	0.976	0.969	0.973	0.975	0.978	0.980	0.982	0.978	0.980	0.947	0.948
15	0.964	0.967	0.957	0.962	0.966	0.969	0.972	0.974	0.969	0.971	0.924	0.926
20	0.954	0.958	0.945	0.951	0.957	0.961	0.964	0.967	0.961	0.963	0.904	0.907
25	0.945	0.950	0.935	0.943	0.949	0.955	0.956	0.960	0.953	0.956	0.885	0.890
30	0.937	0.943	0.927	0.935	0.943	0.948	0.950	0.954	0.947	0.950	0.868	0.875
35	0.930	0.936	0.918	0.929	0.937	0.943	0.943	0.949	0.940	0.943	0.851	0.862
40	0.923	0.930	0.909	0.923	0.932	0.938	0.938	0.945	0.934	0.937	0.837	0.850
45	0.916	0.925	0.901	0.916	0.929	0.934	0.932	0.939	0.928	0.932	0.823	0.839
50	0.910	0.920	f	f	0.925	0.930	0.927	0.935	0.922	0.926	0.810	0.830
55	......	......	......	......	......	......	......	......	......	......	0.799	
60	......	......	......	......	......	......	......	......	......	......	0.789	
65	......	......	......	......	......	......	......	......	......	......	0.780	
70	......	......	......	......	......	......	......	......	......	......	0.770	
75	......	......	......	......	......	......	......	......	......	......	0.762	
80	......	......	......	......	......	......	......	......	......	......	0.753	
85	......	......	......	......	......	......	......	......	......	......	0.746	
90	......	......	......	......	......	......	......	......	......	......	0.738	
95	......	......	......	......	......	......	......	......	......	......	0.731	
100	......	......	......	......	......	......	......	......	......	......	0.724	

* For references, see p. 4-96.
f Transition at 49.

TABLE 4d-2. V/V_0 OF INORGANIC COMPOUNDS* (Continued)

P, kilobars	CsCl at 25°C [9]	CsCl at −78.5°C [10]	CsClO₃ [6]	CsClO₄ [6]	CsI at 25°C [9]	CsI at −78.5°C [10]	CsMnO₄ [5]	CsNO₃ [9]	CuBr [5]	CuI [5]	HgS at 25°C [2]	HgS at −78.5°C [10]
0	1.000	1.000	1.000	1.000	1.000	1.000	1.000	1.000	1.000	1.000	1.000	1.000
5	0.975	0.975	0.974	0.976	0.964	0.966	0.973	0.975	0.986	0.985	0.977	0.977
10	0.952	0.953	0.950	0.956	0.934	0.938	0.950	0.953	0.975	0.975	0.958	0.960
15	0.930	0.935	0.931	0.940	0.907	0.914	0.905	0.933	...	0.928 g	0.944	0.946
20	0.913	0.918	0.915	0.924	0.884	0.892	0.889	0.914	...	0.915	0.934	0.935
25	0.895	0.902	0.903	0.910	0.865	0.872	0.874	0.897	...	0.903	0.926	0.928
30	0.880	0.888	...	...	0.848	0.855	0.860	0.883	...	0.892	0.920	0.922
35	0.867	0.876	...	...	0.832	0.839	0.846	0.869	...	0.882	0.916	0.918
40	0.855	0.865	...	...	0.816	0.826	0.834	0.856	...	0.872	0.912	0.914
45	0.844	0.855	...	...	0.802	0.815	...	0.845	...	...	0.910	0.911
50	0.833	0.847	...	...	0.790	0.805	...	0.833	...	...	0.907	0.907
55	0.823	...	...	...	0.778	...	...	0.824	...	...	0.904	0.903
60	0.814	...	...	...	0.767	...	...	0.815	...	...	0.901	0.900
65	0.805	...	...	...	0.757	...	...	0.808	...	...	0.898	
70	0.798	...	...	...	0.747	...	...	0.801	...	...	0.896	
75	0.791	...	...	...	0.738	...	...	0.795	...	...	0.894	
80	0.785	...	...	...	0.730	...	...	0.789	...	...	0.891	
85	0.780	...	...	...	0.728	...	...	0.784	...	...	0.889	
90	0.774	...	...	...	0.717	...	...	0.779	...	...	0.887	
95	0.769	...	...	...	0.710	...	...	0.775	...	...	0.885	
100	0.764	...	...	...	0.704	...	...	0.771	...	...	0.883	

* For references, see p. 4-96.
g Transition at 14.1; volumes 0.996 and 0.930.

TABLE 4d-2. V/V_0 of Inorganic Compounds* (Continued)

P, kilobars	HgSe at 25°C [2,10]	HgSe at -78.5°C [10]	HgTe [2]	HIO₃ [21]	KAl(SO₄)₂·12H₂O [5]	KB₅O₈·4H₂O [5]	KBr at 25°C [9]	KBr at -78.5°C [10]	KCl at 25°C [9]	KCl at -78.5°C [10]
0	1.000	1.000	1.000	1.000	1.000	1.000	1.000	1.000	1.000	1.000
5	0.991[h]	0.991	0.948	0.978	0.970	0.923[k]	0.970	0.973	0.974	0.976
10	0.888[i]	0.890[j]	0.917	0.959	0.947	0.900	0.944	0.948	0.951	0.955
15	0.872	0.872[i]	0.896	0.943	0.929	0.880[l]	0.923	0.926	0.932	0.933
20	0.862	0.861	0.881	0.927	0.912	0.856	0.800[m]	0.804[n]	0.915	0.918
25	0.852	0.851	0.868	0.915	0.897	0.844	0.785	0.790	0.787[o]	0.793[p]
30	0.843	0.842	0.808	0.903	0.883	0.832	0.770	0.778	0.775	0.782
35	0.834	0.834	0.800		0.870	0.821	0.757	0.767	0.764	0.773
40	0.827	0.828	0.792		0.858	0.811	0.745	0.757	0.754	0.765
45	0.820	0.822	0.785				0.739	0.750	0.745	0.758
50	0.813	0.817	0.778				0.724	0.742	0.737	0.750[p]
55	0.808		0.771				0.715		0.729	
60	0.803		0.761				0.707		0.722	
65	0.798		0.758				0.699		0.715	
70	0.794		0.751				0.693		0.708	
75	0.790		0.745				0.686		0.702	
80	0.786		0.739				0.681		0.697	
85	0.784		0.732				0.675		0.691	
90	0.782		0.726				0.671		0.687	
95	0.779		0.720				0.667		0.682	
100	0.777		0.714				0.663		0.678	

* For references, see p. 4-96.
h Volume at 6.9 = 0.987 and at 7.1 = 0.891.
i Volume at 10.7 = 0.8.
j Volume at 7.1 = 0.9.
k Transition at 3.5; volumes 0.979 and 0.934.
l Transition at 19.9; volumes 0.865 and 0.859.
m Transition at 18.0; volumes 0.912 ad 0.807.
n Transition in this region.
o Transition at 19.7; volumes 0.915 and 0.803.
p Transition in this region.

TABLE 4d-2. V/V_0 of Inorganic Compounds* (Continued)

P, kilobars	$KClO_3$ [6]	$KClO_4$ [6]	KCN [5]	KI at 25°C [9]	KI at −78.5°C [10]	KIO_3 [6]	KIO_4 [6]	$KNaC_4H_4O_6$ [21]	KNO_3 [9]	KPO_3 [21]	$LiClO_4$ [6]
0	1.000	1.000	1.000	1.000	1.000	1.000	1.000	1.000	1.000	1.000	1.000
5	0.973	0.978	0.970	0.965	0.967	0.984	0.990	0.975	0.877 [u]	0.983	0.980
10	0.944 [q]	0.959	0.945	0.935	0.937	0.970	0.981	0.953	0.861	0.967	0.964
15	0.927	0.942	0.921	0.907	0.912	0.958	0.972	0.935	0.845	0.954	0.948
20	0.912	0.926	0.807 [r]	0.802 [s]	0.805 [t]	0.947	0.964	0.919	0.830	0.941	0.918 [v]
25	0.898	.....	0.795	0.785	0.789	0.938	0.955	0.905	0.817	0.930	0.907
30	.....	.....	0.783	0.770	0.774	.....	0.948	0.894	0.803	0.920	.....
35	.....	.....	0.775	0.756	0.760	.....	0.940	.....	0.791	.....	.....
40	.....	.....	0.768	0.743	0.746	.....	.....	.....	0.780	.....	.....
45	.....	.....	.....	0.730	0.733	.....	.....	.....	0.770	.....	.....
50	.....	.....	.....	0.718	0.722	.....	.....	.....	0.761	.....	.....
55	.....	.....	.....	0.708	.....	.....	.....	.....	0.752	.....	.....
60	.....	.....	.....	0.698	.....	.....	.....	.....	0.745	.....	.....
65	.....	.....	.....	0.690	.....	.....	.....	.....	0.737	.....	.....
70	.....	.....	.....	0.682	.....	.....	.....	.....	0.730	.....	.....
75	.....	.....	.....	0.676	.....	.....	.....	.....	0.724	.....	.....
80	.....	.....	.....	0.670	.....	.....	.....	.....	0.719	.....	.....
85	.....	.....	.....	0.665	.....	.....	.....	.....	0.714	.....	.....
90	.....	.....	.....	0.660	.....	.....	.....	.....	0.710	.....	.....
95	.....	.....	.....	0.655	.....	.....	.....	.....	0.705	.....	.....
100	.....	.....	.....	0.650	.....	.....	.....	.....	0.700	.....	.....

* For references, see p. 4-96.

q Transition at 7.5; volumes 0.961 and 0.906.

r Transition at 19.9; volumes 0.899 and 0.811.

s Transition at 17.8; volumes 0.895 and 0.810.

t Transition in this region.

u Transition at 3.6; volumes 0.977 and 0.887.

v Transition at 16.1; volumes 0.944 and 0.928.

TABLE 4d-2. V/V_0 OF INORGANIC COMPOUNDS* (Continued)

P, kilobars	LiNaCr₂O₇ [21]	MgSO₄ [5]	NaBr at 25°C [9]	NaBr at −78.5°C [10]	NaBrO₃ [6]	NaCl at 25°C [9]	NaCl at −78.5°C [10]	NaClO₃ [5]	NaClO₄ [6]	NaI at 25°C [9]	NaI at −78.5°C [10]	NaIO₃ [6]
0	1.000	1.000	1.000	1.000	1.000	1.000	1.000	1.000	1.000	1.000	1.000	1.000
5	0.982	0.975	0.978	0.980	0.983	0.980	0.982	0.981	0.979	0.970	0.970	0.983
		w										
10	0.965	0.954	0.953	0.959	0.967	0.962	0.966	0.964	0.961	0.944	0.944	0.968
15	0.951	0.915	0.937	0.940	0.954	0.947	0.950	0.949	0.946	0.922	0.922	0.955
20	0.938	0.900	0.969	0.923	0.942	0.932	0.935	0.936	0.933	0.902	0.903	0.943
		x										
25	0.925	0.856	0.954	0.957	0.931	0.919	0.922	0.923	0.921	0.883	0.885	0.932
30	0.917	0.845	0.940	0.944		0.907	0.911	0.913		0.866	0.870	
35		0.834	0.927	0.932		0.895	0.900	0.902		0.851	0.857	
40		0.824	0.916	0.922		0.884	0.890	0.894		0.837	0.844	
45			0.905	0.912		0.874	0.881			0.825	0.834	
50			0.895	0.905		0.864	0.873			0.813	0.825	
55			0.886			0.855				0.802		
60			0.877			0.846				0.792		
65			0.870			0.838				0.783		
70			0.862			0.830				0.774		
75			0.855			0.822				0.766		
80			0.848			0.815				0.758		
85			0.842			0.808				0.751		
90			0.836			0.801				0.745		
95			0.830			0.794				0.739		
100			0.825			0.788				0.733		

* For references, see p. 4-96.
w Very sluggish transition between 9.8 and 14.7.
x Probably two sluggish transitions in neighborhood of 24.5. Volume discontinuity of one about 4 times that of other.

TABLE 4d-2. V/V_0 OF INORGANIC COMPOUNDS* (Continued)

P, kilobars	NaIO₄ [6]	NaNH₄C₄H₄O₆ [21]	NaNO₃ [9]	NH₄B₅O₈·4H₂O [5]	NH₄Br at 25°C [9]	NH₄Br at -78.5°C [10]	NH₄CHO₂ [5]	NH₄Cl at 25°C [9]	NH₄Cl at -78.5°C [10]	NH₄IO₄ [6]	NH₄NO₃ [9]
0	1.000	1.000	1.000	1.000	1.000	1.000	1.000	1.000	1.000	1.000	1.000
5	0.981	0.974	0.982	0.964	0.973	0.973	0.965	0.973	0.978	0.980	0.972
10	0.966	0.952	0.965	0.938	0.950	0.951	0.932	0.952	0.960	0.963	0.948
							aa				
15	0.953	0.933	0.950	0.917	0.929	0.932	0.805	0.933	0.945	0.946	0.928
20	0.942	0.917	0.937	0.900	0.910	0.915	0.794	0.917	0.931	0.931	0.912
				z							
25	.....	.....	0.912	0.845	0.878	0.885	0.773	0.888	0.906	.....	0.897
30	.....	0.889	0.912	0.845	0.878	0.885	0.773	0.888	0.906	.....	0.882
35	.....	.....	0.901	0.834	0.863	0.873	0.764	0.875	0.895	.....	0.870
40	.....	.....	0.890	0.824	0.850	0.862	0.754	0.864	0.885	.....	0.857
45	.....	.....	0.881	.....	0.838	0.851	.....	0.853	0.875	.....	0.846
50	.....	.....	0.871	.....	0.827	0.842	.....	0.843	0.867	.....	0.835
			y								
55	.....	.....	0.852	.....	0.817	.....	.....	0.835	.....	.....	0.826
60	.....	.....	0.843	.....	0.808	.....	.....	0.826	.....	.....	0.817
65	.....	.....	0.836	.....	0.800	.....	.....	0.818	.....	.....	0.810
70	.....	.....	0.830	.....	0.892	.....	.....	0.810	.....	.....	0.804
75	.....	.....	0.823	.....	0.885	.....	.....	0.803	.....	.....	0.797
80	.....	.....	0.817	.....	0.878	.....	.....	0.796	.....	.....	0.792
85	.....	.....	0.812	.....	0.872	.....	.....	0.790	.....	.....	0.787
90	.....	.....	0.807	.....	0.865	.....	.....	0.783	.....	.....	0.784
95	.....	.....	0.802	.....	0.859	.....	.....	0.776	.....	.....	0.780
100	.....	.....	0.797	.....	0.852	.....	.....	0.769	.....	.....	0.777

* For references, see p. 4-96.
y Transition at 53.9; volumes 0.864 and 0.853.
z Transition at 22.8; volumes 0.892 and 0.868.
aa Transition at 11.2; volumes 0.926 and 0.815.

TABLE 4d-2. V/V_0 of Inorganic Compounds* (Continued)

P, kilobars	NH$_4$PO$_4$ [5]	NiSO$_4$ [21]	PbI$_2$ [5]	NH$_4$SO$_3$H [21]	NH$_4$ClO$_4$ [6]	NH$_4$H$_2$PO$_4$ [21]	NH$_4$I at 25°C [9]	NH$_4$I at −78.5°C [10]	NH$_4$IO$_3$ [6]	PbS at 25°C [2]	PbS at −78.5°C [10]
0	1.000	1.000	1.000 *bb*	1.000	1.000	1.000	1.000 *cc*	1.000	1.000	1.000	1.000
5	0.981	0.983	0.897	0.979	0.971	0.982	0.832	0.967	0.986	0.983	0.989
10	0.965	0.967	0.878	0.963	0.948	0.966	0.807	0.941	0.973	0.969	0.980
15	0.951	0.953	0.863	0.948	0.927	0.952	0.781	0.920	0.961	0.956	0.971
20	0.939	0.940	0.850	0.935	0.910	0.940	0.767	0.901	0.950	0.945	0.962 *dd*
25	0.877	0.886	0.838	0.923	0.895	0.929	0.754	0.885	0.940	0.935	0.933
30	0.867	0.861	0.827	0.913	.....	0.919	0.740	0.870	.....	0.928	0.925
35	0.857	.....	0.818	.....	.....	.....	0.728	0.858	.....	0.921	0.918
40	0.848	.....	.....	.....	.....	.....	0.716	0.846	.....	0.915	0.913
45	.....	.....	.....	.....	.....	.....	0.705	0.837	.....	0.909	0.909
50	.....	.....	.....	.....	.....	.....	0.695	0.828	.....	0.903	0.905
55	.....	.....	.....	.....	.....	.....	0.686	.....	.....	0.899	.....
60	.....	.....	.....	.....	.....	.....	0.678	.....	.....	0.896	.....
65	.....	.....	.....	.....	.....	.....	0.670	.....	.....	0.892	.....
70	.....	.....	.....	.....	.....	.....	0.662	.....	.....	0.890	.....
75	.....	.....	.....	.....	.....	.....	0.655	.....	.....	0.887	.....
80	.....	.....	.....	.....	.....	.....	0.648	.....	.....	0.885	.....
85	.....	.....	.....	.....	.....	.....	0.642	.....	.....	0.882	.....
90	.....	.....	.....	.....	.....	.....	0.635	.....	.....	0.880	.....
95	.....	.....	.....	.....	.....	.....	0.628	.....	.....	0.878	.....
100	.....	.....	.....	.....	.....	.....	0.622	.....	.....	0.876	.....

* For references, see p. 4-96.
bb Transition at 5.0; volumes 0.963 and 0.924.
cc Transition at 0.5; volumes 0.997 ± and 0.856.
dd Volume at 24.2 = 0.958 and at 22.3 = 0.937.

TABLE 4d-2. V/V_0 of Inorganic Compounds* (Continued)

P, kilobars	PbSe at 25°C [2]	PbSe at -78.5°C [10]	PbTe at 25°C [2]	PbTe at -78.5°C [10]	PCl₃ [23]	RbBr at 25°C [9]	RbBr at -78.5°C [10]	Rb₂C₄H₄O₆ [21]	RbCl at 25°C [9]	RbCl at -78.5°C [10]	RbClO₄ [6]
0	1.000	1.000	1.000	1.000	1.000	1.000	1.000	1.000	1.000	1.000	1.000
	ee	*ff*	*gg*			*hh*	*ii*		*jj*	*kk*	
5	0.983	0.986	0.984	0.985	0.833	0.830	0.830	0.979	0.830	0.831	0.975
10	0.967	0.974	0.970	0.973	0.774	0.811	0.812	0.960	0.811	0.812	0.954
15	0.955	0.962	0.960	0.962	.....	0.794	0.796	0.944	0.795	0.796	0.934
20	0.945	0.951	0.950	0.952	.....	0.777	0.782	0.935	0.780	0.782	0.917
25	0.937	0.941	0.943	0.944	.....	0.762	0.768	0.916	0.765	0.770	0.901
30	0.930	0.931	0.937	0.936	.....	0.748	0.756	0.904	0.752	0.758	
35	0.925	0.924	0.933	0.929	.....	0.735	0.745	.....	0.740	0.748	
40	0.922	0.916	0.930	0.923	.....	0.722	0.736	.....	0.728	0.740	
45	0.892	0.900	0.943	0.917	.....	0.711	0.728	.....	0.717	0.733	
50	0.886	0.892	0.933	0.913	.....	0.701	0.720	.....	0.706	0.726	
55	0.881	.....	0.925	.....	.....	0.692	.....	.....	0.696		
60	0.875	.....	0.916	.....	.....	0.683	.....	.....	0.687		
65	0.870	.....	0.909	.....	.....	0.675	.....	.....	0.678		
70	0.865	.....	0.902	.....	.....	0.668	.....	.....	0.670		
75	0.861	.....	0.895	.....	.....	0.661	.....	.....	0.663		
80	0.856	.....	0.890	.....	.....	0.655	.....	.....	0.657		
85	0.852	.....	0.884	.....	.....	0.650	.....	.....	0.650		
90	0.848	.....	0.879	.....	.....	0.645	.....	.....	0.645		
95	0.843	.....	0.873	.....	.....	0.639	.....	.....	0.640		
100	0.840	.....	0.868	.....	.....	0.634	.....	.....	0.635		

* For references, see p. 4-96.
ee Transition at 44.1; volumes 0.917 and 0.893.
ff Volume at 42.45 = 0.916 and at 42.49 = 0.906.
gg Transition at 44.1; volumes 0.925 and 0.892.
hh Transition at 4.5; volumes 0.967 and 0.834.
ii Transition in this region.
jj Transition at 4.9; volumes 0.970 and 0.830.
kk Transition in this region.

TABLE 4d-2. V/V_0 OF INORGANIC COMPOUNDS* (Continued)

P, kilobars	RbI at 25°C [9]	RbI at -78.5°C [10]	RbIO₄ [6]	RbNO₃ [9]	Sr(CHO₂)₂ [21]	SrS at 25°C [10]	SrS at -78.5°C [10]	SrSe at 25°C [10]	SrSe at -78.5°C [10]	SrTe at 25°C [10]	SrTe at -78.5°C [10]
0	1.000 *ll*	1.000 *mm*	1.000	1.000	1.000	1.000	1.000	1.000	1.000	1.000	1.000
5	0.832	0.834	0.978	0.978	0.983	0.984	0.985	0.988	0.988	0.985	0.987
10	0.807	0.811	0.959	0.956	0.966	0.969	0.973	0.978	0.979	0.972	0.976
15	0.783	0.790	0.943	0.937	0.953	0.957	0.963	0.969	0.970	0.960	0.965
20	0.762	0.774	0.930	0.920	0.941	0.948	0.955	0.961	0.962	0.949	0.955
25	0.743	0.759	0.918	0.904	0.931	0.940	0.949	0.953	0.954	0.940	0.946
30	0.725	0.745	0.907	0.889	0.921	0.934	0.944	0.946	0.947	0.931	0.937
35	0.710	0.735	.....	0.875	.....	0.928	0.940	0.939	0.941	0.923	0.929
40	0.695	0.724	.....	0.863	.....	0.924	0.937	0.933	0.935	0.916	0.922
45	0.683	0.715	.....	0.851	.....	0.920	0.934	0.928	0.930	0.908	0.915
50	0.672	0.706	.....	0.841	.....	0.917	0.932	0.923	0.925	0.902	0.909
55	0.661	.....	.....	0.832	.....						
60	0.651	.....	.....	0.823	.....						
65	0.643	.....	.....	0.815	.....						
70	0.635	.....	.....	0.808	.....						
75	0.628	.....	.....	0.802	.....						
80	0.621	.....	.....	0.796	.....						
85	0.615	.....	.....	0.792	.....						
90	0.609	.....	.....	0.787	.....						
95	0.605	.....	.....	0.783	.....						
100	0.600	.....	.....	0.780	.....						

* For references, see p. 4-96.
ll Transition at 4.0; volumes 0.965 and 0.839.
mm Transition in this region.

TABLE 4d-2. V/V_0 OF INORGANIC COMPOUNDS* (Continued)

P, kilobars	TlBr at 25°C [9]	TlBr at -78.5°C [10]	TlCl at 25°C [9]	TlCl at -78.5°C [10]	TlI at 25°C [9]	TlI at -78.5°C [10]	TlNO₃ [9]	ZnS at 25°C [2]	ZnS at -78.5°C [10]	ZnSe at 25°C [2]	ZnSe at -78.5°C [10]	ZnTe at 25°C [2]	ZnTe at -78.5°C [10]
0	1.000	1.000	1.000	1.000	1.000	1.000	1.000	1.000	1.000	1.000	1.000	1.000	1.000
5	0.978	0.980	0.978	0.980	0.973	0.973	0.980	0.993	0.993	0.988	0.990	0.987	0.988
10	0.957	0.960	0.959	0.962	0.950	0.950	0.962	0.987	0.987	0.977	0.979	0.975	0.978
15	0.939	0.942	0.942	0.946	0.928	0.930	0.945	0.982	0.980	0.968	0.970	0.965	0.968
20	0.922	0.925	0.927	0.933	0.910	0.912	0.930	0.977	0.975	0.960	0.963	0.957	0.959
25	0.906	0.911	0.912	0.920	0.892	0.897	0.917	0.972	0.970	0.953	0.957	0.950	0.950
30	0.892	0.899	0.899	0.909	0.875	0.884	0.903	0.967	0.964	0.949	0.951	0.944	0.943
35	0.878	0.888	0.881	0.899	0.860	0.871	0.891	0.963	0.959	0.947	0.948	0.940	0.935
40	0.867	0.880	0.875	0.890	0.847	0.860	0.880	0.958	0.954	*nn* 0.938	*oo* 0.940	*pp* 0.932	*qq* 0.922
45	0.857	0.872	0.864	0.882	0.833	0.850	0.870	0.955	0.949	0.933	0.937	0.925	0.916
50	0.847	0.866	0.854	0.875	0.822	0.841	0.861	0.951	0.945	0.928	0.935	0.920	0.910
55	0.837	.....	0.844	.....	0.810	.....	0.853	0.948	.....	0.924	.....	0.914	
60	0.829	.....	0.835	.....	0.800	.....	0.846	0.945	.....	0.920	.....	0.908	
65	0.821	.....	0.827	.....	0.790	.....	0.840	0.942	.....	0.915	.....	0.902	
70	0.813	.....	0.820	.....	0.781	.....	0.834	0.938	.....	0.910	.....	0.897	
75	0.805	.....	0.813	.....	0.773	.....	0.828	0.936	.....	0.905	.....	0.892	
80	0.798	.....	0.807	.....	0.766	.....	0.823	0.933	.....	0.901	.....	0.886	
85	0.792	.....	0.801	.....	0.760	.....	0.818	0.930	.....	0.897	.....	0.881	
90	0.786	.....	0.795	.....	0.753	.....	0.814	0.928	.....	0.893	.....	0.875	
95	0.780	.....	0.790	.....	0.746	.....	0.809	0.926	.....	0.890	.....	0.870	
100	0.774	.....	0.785	.....	0.740	.....	0.805	0.924	.....	0.886	.....	0.866	

* For references, see p. 4-96.
nn Small transition at 37.2.
oo Volume at 37.7 = 0.946 and at 35.6 = 0.944.
pp Small transition here.
qq Transition at 35.3: volumes 0.933 and 0.928.

TABLE 4d-3. V OF H_2O, IN CM^3/G [29]

Temperature, °C

P, kilobars	0	10	20	30	40	50	60	70	80	90	100	200	300	400	500	600	700	800	900	1000
0	1.00013	1.00027	1.00177	1.00434	1.00781	1.01208	1.01706	1.02271	1.02900	1.0359	1.0434	2171.0	2638.6	3102.6	3565.7	4028.	4491.	4951.	5413.	5875.
1	0.9564	0.9589	0.9616	0.9649	0.9687	0.9729	0.9774	0.9824	0.9878	0.9937	0.9999	1.0811	1.2131	1.442	1.892	2.670	3.547	4.406	5.163	5.863
2	0.925	0.928	0.932	0.936	0.940	0.944	0.948	0.953	0.958	0.963	0.968	1.032	1.127	1.260	1.435	1.666	1.980	2.348	2.722	3.097
3	0.900	0.904	0.907	0.911	0.915	0.919	0.924	0.928	0.933	0.938	0.943									
4	0.879	0.883	0.887	0.890	0.894	0.898	0.902	0.906	0.911	0.916	0.921	0.944	0.995	1.065	1.139					1.616
5	0.862	0.865	0.869	0.872	0.876	0.880	0.884	0.888	0.893	0.897	0.902									
6	0.847	0.849	0.853	0.857	0.861	0.865	0.869	0.873	0.877	0.881	0.885									
7		0.835	0.839	0.843	0.847	0.851	0.855	0.859	0.862	0.866	0.870									
8			0.826	0.830	0.834	0.838	0.842	0.846	0.849	0.853	0.857									
9				0.819	0.823	0.827	0.831	0.835	0.838	0.842	0.846									
10				0.809	0.813	0.817	0.821	0.825	0.829	0.833	0.837	0.872	0.907	0.937	0.978					1.189
15							0.780	0.784	0.787	0.791	0.794	0.821	0.846	0.873	0.900					1.040
20									0.757	0.760	0.762	0.783	0.806	0.826	0.848					0.959
25												0.754	0.774	0.794	0.814					0.914
30												0.732	0.751	0.770	0.789					0.886
40												0.700	0.713	0.731	0.749					0.839
50													0.681*	0.698	0.715					0.799
100													0.575*	0.587	0.599					0.658
150													0.514*	0.523*	0.531*					0.573
200													0.478*	0.483*	0.489*					0.519
250													0.454*	0.459*	0.463*					0.484

* Probably supercooled with respect to ice.

TABLE 4d-4. V/V_0 OF PLASTICS [5]

P, kilobars	Aero-glass 4121[a]	Bakelite	α-Beetle filler[b]	Cellulose acetate	Fluorine plastic[c]	Laminac 4201[d]	Lucite	Melmac 404[e]	Melmac 1079[f]	Melmac S-6004[g]	Nylon 6-10[h]	Nylon 6-6 un-oriented[i]
0	1.000	1.000	1.000	1.000	1.000	1.000	1.000	1.000	1.000	1.000	1.000	1.000
2	0.960	0.979	0.979	0.964	0.967	0.967	0.968	0.980	0.980	0.986	0.967	0.971
4	0.934	0.962	0.961	0.936	0.943	0.944	0.941	0.963	0.962	0.973	0.944	0.948
6	0.914	0.957	0.957	0.914	0.924	0.926	0.920	0.948	0.947	0.962	0.924	0.930
8	0.898	0.933	0.934	0.896	0.909	0.911	0.903	0.935	0.934	0.952	0.907	0.915
10	0.884	0.922	0.922	0.881	0.896	0.899	0.889	0.929	0.929	0.943	0.893	0.903
12	0.872	0.911	0.912	0.868	0.885	0.887	0.877	0.914	0.912	0.935	0.880	0.891
14	0.861	0.901	0.902	0.856	0.876	0.877	0.866	0.904	0.903	0.927	0.868	0.881
16	0.851	0.892	0.894	0.846	0.867	0.868	0.856	0.895	0.894	0.920	0.858	0.872
18	0.842	0.883	0.886	0.836	0.858	0.859	0.847	0.887	0.886	0.914	0.848	0.863
20	0.833	0.875	0.878	0.827	0.851	0.851	0.839	0.879	0.878	0.907	0.840	0.855
22	0.825	0.867	0.871	0.819	0.845	0.843	0.831	0.872	0.871	0.902	0.832	0.847
24	0.817	0.860	0.864	0.811	0.838	0.836	0.823	0.865	0.864	0.897	0.826	0.840
26	0.810	0.853	0.858	0.804	0.832	0.830	0.817	0.858	0.858	0.892	0.819	0.834
28	0.803	0.847	0.852	0.797	0.827	0.824	0.810	0.852	0.852	0.887	0.813	0.827
30	0.797	0.841	0.847	0.792	0.822	0.817	0.804	0.847	0.847	0.883	0.807	0.822
32	0.792	0.835	0.842	0.786	0.817	0.812	0.798	0.842	0.842	0.878	0.802	0.816
34	0.787	0.830	0.837	0.780	0.813	0.807	0.793	0.837	0.837	0.875	0.798	0.811
36	0.737	0.826	0.832	0.775	0.809	0.802	0.788	0.833	0.833	0.872	0.793	0.806
38	0.728	0.821	0.828	0.771	0.806	0.797	0.784	0.828	0.828	0.868	0.789	0.802
40	0.724	0.817	0.824	0.767	0.803	0.793	0.780	0.824	0.824	0.865	0.785	0.797

[a] Clear and copolymer type, no filler.
[b] Urea-formaldehyde, α-cellulose filler.
[c] Experimental, developed during World War II.
[d] Clear cast copolymer-type resin similar to that used in laminating applications, no filler.
[e] Melamine-formaldehyde, no filler, in 1948 an experimental product.
[f] Melamine-formaldehyde, α-cellulose filler.
[g] Melamine-formaldehyde, asbestos filler.
[h] Cut from sheet about 0.25 in. thick.
[i] Provided in rod 0.5 in. in diameter (unoriented). Oriented specimen produced by upsetting cylinder originally 1 in. long by one-sided compression between platens of a hydraulic press until thickness reduced to 0.085 in.

TABLE 4d-4. V/V_0 OF PLASTICS [5] (Continued)

P, kilobars	Nylon 6-6 oriented[i]	Poly-styrene	Silicone 160[j]	Silicone 181[k]	Silicone 180[l]	Silicone 167[m]	Silicone 120[n]	Silicone 125[n]	Silicone 150	Silicone, glass-filled	Teflon[o]	48,000[q]
0	1.000	1.000	1.000	1.000	1.000	1.000	1.000	1.000	1.000	1.000	1.000	1.000
2	0.968	0.958	0.921	0.932	0.936	0.934	0.915	0.910	0.930	0.971	0.953	0.956
4	0.946	0.931	0.892	0.900	0.910	0.905	0.883	0.880	0.900	0.951	0.925	0.928
6	0.927	0.910	0.872	0.879	0.891	0.886	0.862	0.859	0.880	0.935	0.904	0.906
8	0.911	0.894	0.857	0.862	0.875	0.872	0.847	0.842	0.863	0.925	[p] 0.865	0.888
10	0.897	0.879	0.844	0.848	0.864	0.860	0.834	0.827	0.850	0.908	0.853	0.873
12	0.885	0.866	0.834	0.836	0.853	0.851	0.822	0.815	0.838	0.897	0.842	0.860
14	0.873	0.854	0.825	0.825	0.843	0.842	0.813	0.805	0.828	0.887	0.833	0.848
16	0.863	0.843	0.816	0.815	0.835	0.835	0.804	0.796	0.819	0.878	0.825	0.838
18	0.854	0.833	0.810	0.806	0.826	0.829	0.796	0.788	0.810	0.819	0.818	0.829
20	0.844	0.824	0.803	0.798	0.820	0.823	0.789	0.781	0.803	0.811	0.811	0.820
22	0.836	0.816	0.797	0.790	0.813	0.818	0.783	0.775	0.796	0.803	0.805	0.812
24	0.828	0.807	0.792	0.783	0.807	0.814	0.777	0.769	0.790	0.796	0.799	0.805
26	0.821	0.800	0.787	0.776	0.801	0.809	0.772	0.764	0.785	0.789	0.794	0.798
28	0.814	0.794	0.782	0.770	0.796	0.805	0.767	0.760	0.780	0.782	0.789	0.791
30	0.808	0.787	0.778	0.764	0.790	0.801	0.763	0.756	0.775	0.776	0.784	0.785
32	0.802	0.782	0.774	0.758	0.786	0.798	0.758	0.751	0.770	0.770	0.780	0.780
34	0.797	0.776	0.770	0.753	0.781	0.794	0.755	0.748	0.766	0.765	0.776	0.775
36	0.792	0.772	0.766	0.749	0.777	0.791	0.751	0.744	0.763	0.760	0.772	0.770
38	0.788	0.767	0.762	0.744	0.773	0.788	0.748	0.741	0.759	0.756	0.768	0.766
40	0.784	0.763	0.759	0.740	0.765	0.780	0.745	0.737	0.756	0.752	0.765	0.762

[i] 30 % zinc oxide, 30 % titanium oxide, 40 % high-polymeric methyl silicone.
[k] 60 % silica, 40 % high-polymeric methyl silicone.
[l] 30 % titanium oxide, 30 % silica, 40 % high-polymeric methyl silicone.
[m] 60 % titanium oxide, 40 % high-polymeric methyl silicone.
[n] 50 % titanium oxide and 50 % high-polymeric methyl silicone.
[o] Polytetrafluorethylene.
[p] Transition at 6.4; volumes 0.8996 and 0.8770.
[q] Phenolic formaldehyde types.

TABLE 4d-4. V/V_0 OF PLASTICS [5] (Continued)

P, kilobars	41,000[q]	46,000[r]	2060-39[s]	M-1805[t]	M-2348[u]	M-1364[v]	M-2347[w]	M-2343[x]	M-2345[y]	M-2346[z]	M-2344[aa]	M-2342[bb]
0	1.000	1.000	1.000	1.000	1.000	1.000	1.000	1.000	1.000	1.000	1.000	1.000
2	0.977	0.970	0.965	0.939	0.951	0.946	0.954	0.963	0.970	0.961	0.952	0.961
4	0.958	0.945	0.940	0.910	0.921	0.920	0.928	0.937	0.933	0.936	0.926	0.934
6	0.941	0.926	0.920	0.888	0.899	0.900	0.909	0.916	0.911	0.916	0.906	0.912
8	0.927	0.910	0.905	0.872	0.881	0.884	0.893	0.900	0.893	0.899	0.889	0.894
10	0.914	0.896	0.890	0.857	0.865	0.869	0.879	0.885	0.878	0.884	0.874	0.879
12	0.902	0.883	0.877	0.845	0.852	0.857	0.866	0.873	0.864	0.871	0.862	0.864
14	0.891	0.872	0.865	0.835	0.840	0.845	0.855	0.862	0.852	0.859	0.859	0.852
16	0.881	0.861	0.853	0.825	0.829	0.834	0.844	0.850	0.841	0.848	0.839	0.841
18	0.872	0.852	0.843	0.817	0.820	0.824	0.835	0.840	0.832	0.838	0.829	0.831
20	0.863	0.843	0.834	0.809	0.811	0.815	0.825	0.831	0.822	0.829	0.820	0.822
22	0.856	0.835	0.825	0.802	0.803	0.807	0.817	0.822	0.814	0.821	0.812	0.813
24	0.849	0.827	0.817	0.796	0.796	0.799	0.810	0.814	0.807	0.814	0.804	0.806
26	0.842	0.820	0.810	0.790	0.789	0.792	0.803	0.806	0.799	0.807	0.797	0.799
28	0.836	0.814	0.803	0.785	0.782	0.786	0.797	0.799	0.793	0.800	0.791	0.792
30	0.830	0.808	0.797	0.779	0.777	0.780	0.791	0.793	0.787	0.794	0.785	0.787
32	0.824	0.802	0.792	0.774	0.771	0.775	0.785	0.787	0.782	0.789	0.779	0.781
34	0.820	0.797	0.786	0.770	0.766	0.770	0.780	0.782	0.777	0.784	0.774	0.776
36	0.815	0.792	0.781	0.765	0.761	0.765	0.775	0.777	0.772	0.779	0.770	0.771
38	0.810	0.788	0.777	0.761	0.756	0.761	0.771	0.773	0.768	0.775	0.765	0.767
40	0.806	0.784	0.773	0.756	0.752	0.757	0.766	0.769	0.764	0.771	0.761	0.763

[q] Glyptal, made by reaction of a dibasic acid, such as phthalic, with glycerin or a like compound.
[r] Polyester resin, made by reaction of a dibasic acid containing maleic acid with a glycol, hardened with a vinyl material such as styrene.
[s] Polyethylene, experimental-low-molecular-weight, 100 % base resin.
[t] Polyethylene, standard DYNH 100 % base resin.
[u] Polyethylene, experimental high-molecular-weight, 100 % base resin.
[v] Vinylite VYNW-5 resin, 98 % base resin, 2 % organotin heat stabilizer by weight.
[w] Vinylite VYNW-5 resin, 73.6 % base resin, 1.7 % organotin heat stabilizer, 24.7 % tricresyl phosphate by weight.
[x] Vinylite VYNS resin, 98 % base resin, 2 % organotin heat stabilizer by weight.
[y] Vinylite VYNW-5 resin, 76.0 % base resin, 1.7 % organotin heat stabilizer, 22.3 % resinous products G-25 (resinous-type plasticizer reported to be a condensation product of sebacic acid, glycerol, and rincinoleic acid).
[aa] Vinylite VYHH resin, 98 % base resin, 2 % organotin heat stabilizer by weight.
[bb] Vinylite VYNW-5 resin, 76.6 % base resin, 1.7 % organotin heat stabilizer, 21.7 % dioctyl phthalate by weight.

TABLE 4d-5. V/V_0 OF RUBBERS [6]

P, kilobars	Americapol 0-7700	Buna S 8774	Butyl gum	Butyl tread	Du-prene	Goodrich D-402	Goodrich D-420	Goodrich D-453	Goodrich D-453	Hard rubber (new, 2–3 yrs)	Hard rubber (old, 25 yrs)	Hevea gum	Hevea tread	Hood 844 A	Koroseal 89023	Neoprene 832	Rubber A	Rubber B	Rubber C [30]
0	1.000	1.000	1.000	1.000	1.000	1.000	1.000	1.000	1.000	1.000	1.000	1.000	1.000	1.000	1.000	1.000	1.000	1.000	1.000
2	0.965	0.952	0.937	0.956	0.973	0.956	0.956	0.872	0.857	0.968	0.966	0.945	0.945	0.953	0.948	0.950	0.983	0.977	0.974
4	0.941	0.925	0.912	0.929	0.952	0.930	0.932	0.950	0.931	0.944	0.943	0.914	0.921	0.929	0.913	0.923	0.955	0.941	0.937
6	0.922	0.906	0.895	0.910	0.936	0.911	0.914	0.933	0.910	0.925	0.924	0.892	0.903	0.910	0.891	0.903	0.931	0.918	0.910
8	0.908	0.890	0.880	0.895	0.922	0.894	0.899	0.918	0.893	0.909	0.907	0.873	0.888	0.894	0.873	0.886	0.912	0.900	0.892
10	0.894	0.876	0.868	0.882	0.910	0.885	0.887	0.906	0.879	0.895	0.893	0.857	0.875	0.881	0.859	0.872	0.898	0.887	0.877
12	0.884	0.865	0.857	0.872	0.899	0.868	0.875	0.895	0.867	0.883	0.880	0.845	0.863	0.870	0.847	0.859	0.874	0.874	0.855
14	0.873	0.854	0.847	0.863	0.890	0.857	0.865	0.885	0.855	0.871	0.868	0.833	0.853	0.860	0.836	0.848			
16	0.864	0.844	0.838	0.856	0.882	0.848	0.856	0.877	0.846	0.860	0.858	0.822	0.843	0.850	0.826	0.837			
18	0.855	0.840	0.830	0.849	0.874	0.839	0.847	0.869	0.837	0.851	0.849	0.813	0.835	0.841	0.817	0.827			
20	0.847	0.827	0.822	0.843	0.867	0.831	0.840	0.862	0.829	0.842	0.840	0.805	0.826	0.833	0.810	0.818			
22	0.840	0.818	0.815	0.839	0.861	0.824	0.832	0.856	0.822	0.834	0.832	0.797	0.819	0.825	0.803	0.810			
24	0.833	0.811	0.809	0.834	0.855	0.817	0.825	0.850	0.815	0.826	0.824	0.790	0.812	0.818	0.796	0.802			
26	.....	.....	.....	.....	.....	.....	.....	.....	.....	0.819	0.816								
28	.....	.....	.....	.....	.....	.....	.....	.....	.....	0.812	0.810								
30	.....	.....	.....	.....	.....	.....	.....	.....	.....	0.806	0.803								
32	.....	.....	.....	.....	.....	.....	.....	.....	.....	0.800	0.797								
34	.....	.....	.....	.....	.....	.....	.....	.....	.....	0.795	0.792								
36	.....	.....	.....	.....	.....	.....	.....	.....	.....	0.790	0.787								
38	.....	.....	.....	.....	.....	.....	.....	.....	.....	0.786	0.782								
40	.....	.....	.....	.....	.....	.....	.....	.....	.....	0.782	0.778								

Notes:

A. Hard rubber from panel made by the Goodrich Company. It is a rubber-sulfur compound containing no inorganic fillers. The total sulfur amounts to 27.4%, of which 0.21% is free sulfur. Density equals 1.149 at 27°C.

B. A rubber-sulfur compound containing 90% smoked rubber and 10% sulfur, and vulcanized 105 min at 300°F. Density = 0.990 at 25°C.

C. 90.75% pale crepe rubber, 5% zinc oxide, 4% sulfur, 0.25% tetramethylthiuram disulfide. It was vulcanized 30 min at 260°F. Density = 0.990 at 27°C.

TABLE 4d-6. V/V_0 OF GLASSES [2]

P, kilobars	Basalt glass [31]	Borax glass	Quartz glass	Pyrex glass	Glass A	Glass C	Glass D
0	1.000	1.000	1.000	1.000	1.000	1.000	1.000
5	0.988	0.965	0.982	0.981	0.984	0.988	0.983
10	0.978	0.936	0.963	0.963	0.969	0.975	0.967
15	0.968	0.913	0.946	0.947	0.956	0.964	0.953
20	0.959	0.894	0.932	0.931	0.944	0.954	0.940
25	0.951	0.877	0.918	0.917	0.933	0.945	0.930
30	0.942	0.864	0.905	0.905	0.923	0.935	0.921
35	0.934	0.851	0.893	0.894	0.913	0.927	0.912
40	0.927	0.840	0.882	0.883	0.903	0.918	0.904
45	0.921	0.830	0.872	0.874	0.895	0.910	0.897
50	0.915	0.821	0.862	0.865	0.887	0.902	0.890
55	.....	0.812	0.853	0.857	0.879	0.894	0.883
60	.....	0.803	0.844	0.849	0.872	0.887	0.876
65	.....	0.791	0.836	0.842	0.865	0.880	0.870
70	.....	0.788	0.829	0.835	0.859	0.873	0.865
75	.....	0.781	0.822	0.829	0.853	0.867	0.859
80	.....	0.775	0.816	0.824	0.847	0.862	0.853
85	.....	0.769	0.810	0.819	0.841	0.856	0.847
90	.....	0.863	0.805	0.815	0.836	0.861	0.842
95	.....	0.757	0.800	0.811	0.830	0.846	0.837
100	.....	0.752	0.797	0.807	0.825	0.841	0.833

Notes:
A. A potash-lead silicate of very high lead content.
C. A soda-potash-lime silicate.
D. A soda-zinc borosilicate.

TABLE 4d-7. V/V_0 OF DIMETHYLSILOXANE POLYMERS [24]

P, kilobars	Dimer	Trimer	Tetra-mer	Penta-mer	Hexa-mer	Hepta-mer	Octa-mer
0	1.000	1.000	1.000	1.000	1.000	1.000	1.000
2	0.857	0.863	0.869	0.871	0.877	0.876	0.880
4	0.760 *a*	0.809	0.816	0.818	0.824	0.823	0.828
6	0.738	0.789	0.797	0.800	0.806	0.805	0.810
8	0.720	0.740	0.758	0.763	0.765	0.762	0.764
10	0.707	0.722	0.740	0.745	0.748	0.747	0.750
12	0.695	0.708	0.725	0.730	0.734	0.734	0.736
14	0.685	0.695	0.712	0.717	0.722	0.723	0.725
16	0.676	0.685	0.701	0.706	0.712	0.712	0.715
18	0.668	0.675	0.692	0.697	0.702	0.704	0.706
20	0.660	0.666	0.683	0.688	0.693	0.695	0.698
22	0.653	0.658	0.676	0.680	0.686	0.689	0.691
24	0.647	0.651	0.669	0.673	0.679	0.682	0.685
26	0.641	0.644	0.663	0.667	0.673	0.675	0.678
28	0.635	0.638	0.657	0.661	0.667	0.670	0.673
30	0.630	0.633	0.652	0.656	0.662	0.664	0.668
32	0.625	0.628	0.647	0.651	0.657	0.660	0.663
34	0.620	0.623	0.643	0.647	0.652	0.655	0.659
36	0.616	0.618	0.639	0.642	0.648	0.651	0.655
38	0.612	0.614	0.635	0.638	0.645	0.647	0.652
40	0.607	0.610	0.632	0.635	0.641	0.643	0.648

a Freezes at 3.7; volumes 0.805 and 0.768.

TABLE 4d-8. V/V_0 OF "DOW-CORNING FLUIDS"* [24]

P, kilobars	500-0.65	500-1.00	500-2.00	500-12.8	200-100	200-350	200-1,000	200 12,500	550-112
0	1.000	1.000	1.000	1.000	1.000	1.000	1.000	1.000	1.000
2	0.854	0.866	0.873	0.887	0.880	0.880	0.888	0.885	0.918
	a								
4	0.756	0.811	0.813	0.831	0.833	0.837	0.836	0.834	0.875
6	0.734	0.777	0.780	0.800	0.803	0.809	0.806	0.807	0.852
8	0.717	0.754	0.757	0.780	0.781	0.787	0.785	0.786	0.835
10	0.703	0.736	0.739	0.764	0.764	0.769	0.768	0.771	0.82)
12	0.691	0.722	0.725	0.750	0.750	0.754	0.754	0.757	0.807
14	0.681	0.710	0.712	0.738	0.738	0.740	0.743	0.745	0.795
16	0.671	0.699	0.701	0.727	0.728	0.728	0.733	0.740	0.785
18	0.663	0.690	0.691	0.717	0.719	0.717	0.724	0.726	0.776
20	0.655	0.682	0.682	0.709	0.712	0.709	0.716	0.717	0.767
22	0.649	0.675	0.675	0.701	0.705	0.698	0.709	0.710	0.759
24	0.643	0.668	0.668	0.694	0.698	0.690	0.702	0.703	0.752
26	0.637	0.663	0.663	0.687	0.692	0.682	0.696	0.697	0.745
28	0.632	0.657	0.657	0.682	0.686	0.676	0.690	0.690	0.739
30	0.628	0.652	0.652	0.676	0.681	0.669	0.685	0.685	0.733
32	0.624	0.647	0.646	0.672	0.676	0.664	0.680	0.680	0.727
34	0.619	0.643	0.642	0.667	0.671	0.658	0.675	0.676	0.722
36	0.615	0.639	0.637	0.663	0.667	0.654	0.671	0.671	0.718
38	0.611	0.636	0.633	0.659	0.663	0.650	0.667	0.667	0.714
40	0.607	0.632	0.629	0.656	0.659	0.646	0.664	0.663	0.710

* 500 and 200 series are primarily dimethylsiloxane polymers of varying viscosities. 550 series has a portion of the methyl groups replaced by phenyl groups.
a Freezes at 3.9; volumes 0.796 and 0.760.

TABLE 4d-9. V/V_0 OF OIL AND KEROSENE

P, kilobars	Fluorocarbons [24]		P, kilobars	Kerosene [32]
	Light	Oil		
0	1.000	1.000	0	1.000
2	0.848	0.918	5	0.898
4	0.792	0.879	10	0.856
	a			
6	0.743		15	0.825
8	0.725		20	0.800
10	0.711		25	0.776
12	0.700		30	0.763
14	0.690		35	
16	0.682		40	
18	0.675		45	
20	0.669		50	
22	0.664		55	
24	0.658		60	
26	0.655		65	
28	0.650		70	
30	0.647		75	
32	0.643		80	
34	0.641		85	
36	0.638		90	
38	0.636		95	
40	0.634		100	

a Freezes at 4.3; volumes 0.786 and 0.761.

TABLE 4d-10. V/V_0 OF ALLOYS AND INTERMETALLIC COMPOUNDS*

Pressure, kilobars	Ag$_2$Al [13]	Ag-Au alloys† [14]			Ag 98.70 Cd 1.30 [13]	Ag 91.40 In 8.60 [37]	Ag 96.92 Mg 3.08 [37]	Ag-Mn system [37]		
		75 Ag 25 Au	50 Ag 50 Au	25 Ag 75 Au				100 Ag	96.15 Ag 3.85 Mn	85.41 Ag 14.59 Mn
0	1.000	1.000	1.000	1.000	1.000	1.000	1.000	1.000	1.000	1.000
2	0.997	0.998	0.998	0.998	0.997	0.997	0.998	0.998	0.998	0.998
4	0.995	0.997	0.997	0.997	0.995	0.995	0.996	0.996	0.996	0.996
6	0.993	0.995	0.996	0.996	0.993	0.993	0.994	0.994	0.994	0.993
8	0.991	0.994	0.995	0.994	0.992	0.991	0.992	0.992	0.992	0.991
10	0.989	0.993	0.994	0.993	0.990	0.990	0.990	0.990	0.990	0.989
12	0.987	0.992	0.993	0.992	0.988	0.988	0.988	0.988	0.988	0.987
14	0.986	. . . . .	. . . . .	. . . . .	0.987	0.986	0.987	0.987	0.986	0.985
16	0.984	. . . . .	. . . . .	. . . . .	0.985	0.985	0.985	0.985	0.984	0.983
18	0.982	. . . . .	. . . . .	. . . . .	0.983	0.983	0.983	0.983	0.982	0.981
20	0.980	. . . . .	. . . . .	. . . . .	0.981	0.981	0.982	0.982	0.980	0.979
22	0.979	. . . . .	. . . . .	. . . . .	0.980	0.980	0.980	0.980	0.978	0.977
24	0.977	. . . . .	. . . . .	. . . . .	0.978	0.978	0.978	0.978	0.977	0.975
26	0.976	. . . . .	. . . . .	. . . . .	0.977	0.976	0.977	0.976	0.975	0.973
28	0.974	. . . . .	. . . . .	. . . . .	0.975	0.975	0.975	0.975	0.973	0.972
30	0.972	. . . . .	. . . . .	. . . . .	0.974	0.973	0.973	0.973	0.971	0.970

* For references see p. 4-96.
† At 30°.

TABLE 4d-10. V/V_0 OF ALLOYS AND INTERMETALLIC COMPOUNDS* (Continued)

Pressure, kilobars	Ag-Pd system [33]					Ag-Pt system [37]		AgZn [13]	Ag₃Zn₈ [13]	Ag 96.44 Zn 3.56 [13]	Al-Mg system [33]	
	100 Ag	79.0 Ag 21.0 Pd	48.9 Ag 51.1 Pd	29.5 Ag 70.5 Pd	100 Pd	100 Ag	92.75 Ag 7.25 Pt				100 Al	85.7 Al 14.3 Mg
0	1.000	1.000	1.000	1.000	1.000	1.000	1.000	1.000	1.000	1.000	1.000	1.000
2	0.998	0.998	0.999	0.999	0.998	0.998	0.998	0.998	0.998	0.998	0.997	0.997
4	0.996	0.996	0.997	0.997	0.997	0.996	0.996	0.996	0.996	0.996	0.994	0.993
6	0.994	0.995	0.996	0.996	0.996	0.994	0.994	0.993	0.993	0.994	0.991	0.990
8	0.992	0.993	0.994	0.994	0.995	0.992	0.992	0.991	0.992	0.992	0.989	0.987
10	0.990	0.991	0.993	0.993	0.994	0.990	0.990	0.989	0.989	0.991	0.986	0.984
12	0.988	0.989	0.991	0.992	0.993	0.988	0.988	0.987	0.987	0.989	0.984	0.981
14	0.987	0.988	0.990	0.991	0.992	0.987	0.987	0.985	0.985	0.987	0.982	0.978
16	0.985	9.986	0.988	0.990	0.991[a]	0.985	0.985	0.983	0.983	0.985	0.979	0.975
18	0.983	0.985	0.987	0.988	0.990	0.983	0.983	0.981	0.981	0.983	0.977	0.973
20	0.982	0.983	0.986	0.987	0.989	0.982	0.982	0.980	0.979	0.982	0.974	0.970
22	0.980	0.982	0.984	0.986	0.988	0.980	0.980	0.978	0.977	0.980	0.972	0.967
24	0.978	0.981	0.983	0.985	0.987	0.978	0.978	0.976	0.975	0.979[b]	0.970	0.965
26	0.976	0.979	0.982	0.984	0.986	0.976	0.977	0.975	0.973	0.976	0.967	0.963
28	0.975	0.978	0.981	0.983	0.985	0.975	0.975	0.973	0.972	0.975	0.965	0.961
30	0.973	0.977	0.980	0.982	0.985	0.973	0.974	0.972	0.970	0.973	0.963	0.958

* For references see p. 4-96.
a Cusp at 14.7.
b Slight discontinuity here.

TABLE 4d-10. V/V_0 OF ALLOYS AND INTERMETALLIC COMPOUNDS* (Continued)

Pressure, kilobars	Al-Zn system [33]		Au-Ag system [37]		Au-Mn system [37]			Au-Pd system [37]		Au-Pt system [37]		AuZn [13]
	100 Al	90 Al 10 Zn	100 Au	94.65 Au 5.35 Ag	100 Au	93.17 Au 6.83 Mn	81.36 Au 18.64 Mn	100 Au	96.01 Au 3.99 Pd	100 Au	95.96 Au 4.04 Pt	
0	1.000	1.000	1.000	1.000	1.000	1.000	1.000	1.000	1.000	1.000	1.000	1.000
2	0.998	0.998	0.999	0.999	0.999	0.998	0.998	0.999	0.999	0.999	0.998	0.997
4	0.995	0.995	0.997	0.997	0.997	0.997	0.997	0.997	0.997	0.997	0.997	0.996
6	0.992	0.992	0.996	0.996	0.996	0.996	0.996	0.996	0.996	0.996	0.996	0.994
8	0.989	0.989	0.995	0.995	0.995	0.995	0.994	0.995	0.995	0.995	0.994	0.993
10	0.987	0.987	0.994	0.993	0.994	0.994	0.993	0.994	0.994	0.994	0.993	0.992
12	0.984	0.984	0.993	0.992	0.993	0.993	0.992	0.993	0.993	0.993	0.992	0.990
14	0.982	0.982	0.992	0.991	0.992	0.992	0.991	0.992	0.992	0.992	0.991	0.988
16	0.979	0.979	0.991	0.990	0.991	0.990	0.989	0.991	0.991	0.991	0.990	0.987
18	0.977	0.977	0.990	0.989	0.990	0.989	0.988	0.990	0.990	0.990	0.989	0.986
20	0.975	0.974	0.988	0.987	0.988	0.988	b 0.986	0.988	0.989	0.988	0.988	0.984
22	0.972	0.971	0.987	0.986	0.987	0.987	0.985	0.987	0.988	0.987	0.987	0.983
24	0.970	0.969	0.986	0.985	0.986	0.986	0.984	0.986	0.987	0.986	0.986	0.982
26	0.968	0.967	0.985	0.984	0.985	0.985	0.983	0.985	b 0.985	0.985	0.985	0.980
28	0.966	0.965	0.984	0.983	0.984	0.984	0.982	0.984	0.984	0.984	0.984	0.799
30	0.964	0.963	0.983	0.982	0.983	0.983	0.981	0.983	0.983	0.983	0.983	0.798

* For references see p. 4-96.
b Slight discontinuity here.

TABLE 4d-10. V/V_0 OF ALLOYS AND INTERMETALLIC COMPOUNDS* (Continued)

Pressure, kilobars	Bi-Ag system [34]			Bi-In system [34]					Bi-Pb system [34]			
	100 Bi	50 Bi 50 Ag	100 Ag	100 Bi	80 Bi 20 In	50 Bi 50 In	25 Bi 75 In	100 In	100 Bi	75 Bi 25 Pb Inc.	75 Bi 25 Pb Dec.	50 Bi 50 Pb
0	1.000	1.000	1.000	1.000	1.000	1.000	1.000	1.000	1.000	1.000	1.000	1.000
2	0.993	0.995	0.997	0.993	0.995	0.994	0.995	0.995	0.993	0.994	0.993	0.994
4	0.987	0.990	0.995	0.987	0.989	0.988	0.990	0.990	0.987	0.989	0.987	0.990
6	0.981	0.985	0.993	0.981	0.983	0.983	0.985	0.985	0.981	0.984	0.975 [i]	0.985
8	0.976	0.980	0.991	0.976	0.977	0.977	0.980	0.981	0.976	0.979	0.968	0.980
10	0.970	0.976	0.990	0.970	0.972	0.972	0.976	0.976	0.970	0.974	0.961	0.975
12	0.965	0.971	0.988	0.965	0.967	0.967	0.971	0.972	0.965	0.969	0.955	0.971
14	0.961	0.967	0.986	0.961	0.962	0.963	0.967	0.968	0.961	0.964	0.950	0.966
16	0.956	0.963	0.984	0.956	0.957	0.958	0.962	0.963	0.956	0.959	0.945	0.962
18	0.952	0.959	0.982	0.952	0.952	0.954	0.958	0.959	0.952	0.955	0.940	0.957
20	0.948	0.955	0.981	0.948	0.948 [e]	0.949	0.953	0.955	0.948	0.950	0.935	0.953, 0.938
22	0.944	0.951	0.979	0.944	0.944	0.945	0.950	0.951	0.944	0.945	0.931	0.948, 0.935
24	0.940 [c]	0.948 [d]	0.978	0.940 [c]	0.940 [f]	0.941	0.945	0.948	0.940 [c]	0.941 [h]	0.928	0.944, 0.931
26	0.902	0.880	0.976	0.902	0.883	0.937	0.941	0.944	0.902	0.885	0.887 [j]	0.928
28	0.899	0.876	0.974	0.899	0.880	0.933	0.937	0.941	0.899	0.882	0.884	0.924
30	0.896	0.874	0.973	0.896	0.876 [g]	0.929	0.933	0.937	0.896	0.878	0.880	0.921
32	0.893	0.871	0.971	0.893	0.873	0.925	0.930	0.933	0.893	0.875	0.877	0.918
34	0.890	0.868	0.970	0.890	0.870	0.921	0.926	0.930	0.890	0.872	0.873	0.915
36	0.887	0.866	0.968	0.887	0.867	0.918	0.922	0.927	0.887	0.869	0.870	0.912
38	0.885	0.863	0.967	0.885	0.865	0.914	0.918	0.924	0.885	0.866	0.867	0.909
40	0.882	0.861	0.966	0.882	0.862	0.911	0.915	0.920	0.882	0.863	0.864	0.906

* For references see p. 4-96.
c Transition in this region.
d Transition; volumes at 24.5 = 0.948 and 0.833.
e Volumes at 19.6 = 0.950 and 0.938.
f Volumes at 24.5 = 0.939 and 0.887.
g Volumes at 29.4 = 0.895 and 0.878.
h Volumes at 24.5 = 0.940, 0.905, 0.888.
i Volumes at 4.9 = 0.985, 0.978.
j Volumes at 24.5 = 0.926, 0.904, 0.890.

TABLE 4d-10. V/V_0 OF ALLOYS AND INTERMETALLIC COMPOUNDS* (Continued)

	Bi-Pb system [34]		Bi-Sb system [34]					Bi-Te system [34]		Ca-Cd system [36]		
Pressure, kilobars	25 Bi / 75 Pb	100 Pb	100 Bi	80 Bi / 20 Sb	50 Bi / 50 Sb	20 Bi / 80 Sb	100 Sb	100 Bi	99.00 Bi / 1.00 Te	100 Ca	95 Ca / 5 Cd	100 Cd
0	1.000	1.000	1.000	1.000	1.000	1.000	1.000	1.000	1.000	1.000	1.000	1.000
2	0.995	0.995	0.993	0.995	0.995	0.995	0.995	0.993	0.994	0.987	0.987	0.994
4	0.990	0.990	0.987	0.990	0.991	0.990	0.989	0.987	0.988	0.974	0.977	0.990
6	0.985	0.985	0.981	0.985	0.987	0.985	0.984	0.981	0.983	0.962	0.967	0.986
8	0.980	0.981	0.976	0.980	0.983	0.981	0.980	0.976	0.978	0.952	0.958	0.982
10	0.976	0.977	0.970	0.976	0.979	0.976	0.975	0.970	0.973	0.941	0.949	0.977
12	0.971	0.972	0.965	0.971	0.975	0.972	0.971	0.965	0.968	0.931	0.941	0.973
14	0.967	0.968	0.961	0.967	0.971	0.968	0.966	0.961	0.963	0.922	0.932	0.969
16	0.963	0.965	0.956	0.962	0.967	0.964	0.962	0.956	0.958	0.913	0.924	0.966
18	0.959	0.961	0.952	0.958	0.964	0.960	0.958	0.952	0.953	0.904	0.916	0.962
20	0.955	0.957	0.948	0.954	0.960	0.957	0.954	0.948	0.949	0.896	0.909	0.958
22	0.951	0.954	0.944	0.949	0.956	0.953	0.951	0.944	0.947	0.888	0.902	0.955
24	0.947	0.951	0.940	0.945	0.953	0.949	0.947	0.940	0.942	0.881	0.895	0.951
			c					c	l			
26	0.944	0.948	0.902	0.941	0.949	0.945	0.943	0.902	.....	0.873	0.889	0.948
28	0.940	0.945	0.899	0.937	0.946	0.942	0.940	0.899	0.849	0.867	0.882	0.944
30	0.937	0.942	0.896	c	0.942	0.939	0.937	0.896	0.846	0.860	0.876	0.941
32	0.934	0.939	0.893	.....	0.938	0.935	.....	0.893	0.843	0.853	0.870	0.938
34	0.931	0.936	0.890	.....	0.935	0.932	.....	0.890	0.840	0.848	0.864	0.935
36	0.928	0.933	0.887	0.887	c, k	0.929	.....	0.887	0.838	0.842	0.859	0.932
38	0.925	0.930	0.885	0.877	.....	0.926		0.885	0.835	0.836	0.853	0.929
40	0.922	0.928	0.882	0.867	.....	0.923		0.882	0.833	0.830	0.848	0.926

* For references see p. 4-96.
c Transition in this region.
k Volume at 39.2 = 0.913.
l At 26.0 V_{I-II} = 0.048, and V_{II-III} = 0.034.

TABLE 4d-10. V/V_0 of Alloys and Intermetallic Compounds* (Continued)

Pressure, kilobars	Ca-Mg system [36]				Carboloy 999† [13]	0.05 C 0.09 Mn 0.01 Si 36.0 Ni 63.88 Fe [37]	Carbon steel [37]		100.00 Cd	Cd-Bi system [35]			
	100 Ca	61.9 Ca 38.1 Mg	28.6 Ca 71.4 Mg	100 Mg			100 Fe	95.69 Fe 4.31 C		75.10 Cd 24.90 Bi	50.05 Cd 49.95 Bi	24.40 Cd 75.58 Bi	100.00 Bi
0	1.000	1.000	1.000	1.000	1.000	1.000	1.000	1.000	1.000	1.000	1.000	1.000	1.000
2	0.987	0.989	0.990	0.994	0.999	0.997	0.998	0.998	0.995	0.994	0.994	0.994	0.993
4	0.974	0.980	0.982	0.988	0.998	0.995	0.997	0.997	0.991	0.990	0.989	0.988	0.987
6	0.962	0.971	0.974	0.983	0.998	0.993	0.996	0.996	0.987	0.985	0.983	0.983	0.981
8	0.952	0.963	0.968	0.977	0.997	0.992	0.995	0.995	0.983	0.980	0.978	0.978	0.976
10	0.941	0.955	0.961	0.972	0.997	0.990	0.993	0.993	0.979	0.976	0.974	0.973	0.970
12	0.931	0.947	0.955	0.968	0.996	0.988	0.992	0.992	0.975	0.972	0.970	0.968	0.965
14	0.922	0.940	0.948	0.963	0.996	0.986	0.991	0.991	0.972	0.968	0.965	0.964	0.961
16	0.913	0.933	0.942	0.958	0.995	0.984	0.990	0.990	0.968	0.964	0.961	0.959	0.956
18	0.904	0.926	0.937	0.953	0.995	0.983	0.989	0.989	0.965	0.960	0.957	0.955	0.952
20	0.896	0.920	0.932	0.949	0.995	0.981	0.988	0.988	0.962	0.956	0.953	0.950	0.948
22	0.888	0.914	0.926	0.945	0.994	0.980	0.987	0.987	0.959	0.952	0.949	0.946	0.944
24	0.881	0.908	0.921	0.941	0.994	0.978	0.986	0.986	0.956	0.949	0.946	0.942	0.940
26	0.873	0.902	0.917	0.936	0.993	0.976	0.985	0.985	.....	0.928 [m]	0.910 [o]	0.899 [q]	0.902 [c]
28	0.867	0.897	0.912	0.932	0.993	0.975	0.984	0.984	.....	0.925	0.902	0.897	0.899
30	0.860	0.892	0.907	0.928	0.992	0.974	0.983	0.983	.....	0.905 [n]	0.884 [p]	0.863 [r]	0.896
32	0.853	0.887	0.903	.....	.....	.....	.....	.....	.....	0.902	0.880	0.860	0.893
34	0.848	0.882	0.899	.....	.....	.....	.....	.....	.....	0.899	0.877	0.857	0.890
36	0.842	0.877	0.895	.....	.....	.....	.....	.....	.....	0.897	0.874	0.854	0.887
38	0.836	0.873	0.891	.....	.....	.....	.....	.....	.....	0.895	0.872	0.852	0.885
40	0.830	0.869	0.888	.....	.....	.....	.....	.....	.....	0.892	0.870	0.850	0.882

* For references see p. 4-96.

† WC with 3 % Co binder.

c Transition in this region.

p Volumes at 28.4 = 0.901 and 0.888.

m Volumes at 24.5 = 0.948 and 0.931.   o Volumes at 24.5 = 0.925 and 0.910.   q Volumes at 24.5 = 0.945 and 0.915.

n Volumes at 24.5 = 0.941 and 0.901.   r Volumes at 28.4 = 0.901 and 0.868.   Volumes at 28.4 = 0.897 and 0.901.

TABLE 4d-10. V/V_0 of Alloys and Intermetallic Compounds* (Continued)

Pressure, kilobars	Cd-Pb system [36]			Cd-Sn system [36]					Cd-Zn system [36]		
	100 Cd	50 Cd 50 Pd	100 Pd	100 Cd	75 Cd 25 Sn	50 Cd 50 Sn	25 Cd 75 Sn	100 Sn	100 Cd	50 Cd 50 Zn	100 Zn
0	1.000	1.000	1.000	1.000	1.000	1.000	1.000	1.000	1.000	1.000	1.000
2	0.995	0.994	0.994	0.995	0.995	0.996	0.996	0.996	0.995	0.995	0.996
4	0.990	0.991	0.990	0.990	0.991	0.992	0.992	0.922	0.990	0.992	0.993
6	0.987	0.987	0.986	0.987	0.987	0.988	0.988	0.988	0.987	0.988	0.990
8	0.983	0.983	0.982	0.983	0.983	0.984	0.984	0.984	0.983	0.985	0.987
10	0.979	0.979	0.977	0.979	0.979	0.980	0.981	0.981	0.979	0.981	0.984
12	0.976	0.975	0.973	0.976	0.976	0.977	0.977	0.977	0.976	0.978	0.981
14	0.973	0.971	0.969	0.973	0.972	0.973	0.974	0.974	0.973	0.974	0.977
16	0.969	0.967	0.966	0.969	C.969	0.970	0.971	0.971	0.969	0.971	0.975
18	0.965	0.963	0.962	0.965	0.965	0.966	0.968	0.967	0.965	0.968	0.972
20	0.962	0.960	0.958	0.962	0.962	0.963	0.965	0.964	0.962	0.965	0.969
22	0.959	0.957	0.955	0.959	0.958	0.960	0.961	0.961	0.959	0.962	0.966
24	0.956	0.953	0.951	0.956	9.955	0.957	0.958	0.958	0.956	0.959	0.964
26	0.952	0.949	0.948	0.952	0.952	0.953	0.956	0.956	0.952	0.956	0.961
28	0.950	0.946	0.944	0.950	0.949	0.951	0.953	0.953	0.950	0.953	0.958
30	0.947	0.943	0.941	0.947	0.946	0.948	0.950	0.950	0.947	0.950	0.956
32	0.944	0.940	0.938	0.944	0.943	0.945	0.948	0.948	0.944	0.947	0.954
34	0.942	0.937	0.935	0.942	0.941	0.942	0.945	0.945	0.942	0.945	0.951
36	0.939	0.934	0.932	0.939	0.939	0.940	0.942	0.943	0.939	0.942	0.949
38	0.937	0.931	0.929	0.937	0.937	0.937	0.940	0.940	0.937	0.940	0.947
40	0.935	0.929	0.926	0.935	0.934	0.934	0.938	0.938	0.935	0.938	0.945

* For references see p. 4-96.

TABLE 4d-10. V/V_0 OF ALLOYS AND INTERMETALLIC COMPOUNDS* (Continued)

Pressure, kilobars	Co-Fe system [33]			Cu-Ag system [33]		Cu-Al system [33]		Cu-Au system [33]			
	100 Co	59.06 Co 40.94 Fe	100 Fe	100 Cu	96.0 Cu 4.0 Ag	100 Cu	90.02 Cu 9.90 Al	100 Cu	93 Cu 7 Au	85 Cu 15 Au	75 Cu 25 Au
0	1.000	1.000	1.000	1.000	1.000	1.000	1.000	1.000	1.000	1.000	1.000
2	0.998	0.998	0.998	0.998	0.998	0.998	0.998	0.998	0.998	0.998	0.998
4	0.997	0.997	0.997	0.997	0.996	0.997	0.997	0.997	0.996	0.997	0.997
6	0.996	0.996	0.996	0.995	0.995	0.995	0.995	0.995	0.995	0.995	0.995
8	0.995	0.995	0.995	0.994	0.993	0.994	0.994	0.994	0.993	0.994 *b*	0.994
10	0.994	0.994	0.993	0.992	0.992	0.992	0.993	0.992	0.992	0.993	0.993
12	0.993	0.993	0.992	0.990	0.990	0.990	0.991	0.990	0.991	0.992	0.991
14	0.992	0.992	0.991	0.989	0.988	0.989	0.990	0.989	0.990	0.990	0.990
16	0.991	0.991	0.990	0.987	0.987	0.987	0.988	0.987	0.988	0.989	0.989
18	0.990	0.990	0.989	0.986	0.986	0.986	0.987	0.986	0.987	0.988	0.987
20	0.989	0.989	0.988	0.984	0.984	0.984	0.985	0.984	0.985 *b*	0.987	0.986
22	0.988	0.987	0.987	0.983	0.983	0.983	0.984	0.983	0.984	0.985	0.985
24	0.987	0.986	0.986	0.982	0.982	0.982	0.983	0.982	0.982	0.984	0.984
26	0.986	0.985	0.985	0.980	0.980	0.980	0.980 *c*	0.980	0.981	0.983	0.983
28	0.985	0.984	0.984	0.979	0.979	0.979	0.979	0.979	0.980	0.982	0.981
30	0.984	0.983	0.983	0.978	0.978	0.978	0.978	0.978	0.979	0.980	0.980

* For references see p. 4-96.
b Slight discontinuity here.
c Transition in this region.

TABLE 4d-10. V/V_0 of Alloys and Intermetallic Compounds* (Continued)

Pressure, kilobars	Cu$_3$Au [13]	Cu$_5$Cd$_8$ R.T. [13]	Cu-Cr system [37]		Cu-Ga system [33]		Cu-Ge system [33]		Cu-Mn system [37]		
			100 Cu	99.818 Cu 0.182 Cr	100 Cu	95.85 Cu 4.15 Ga	100 Cu	98.29 Cu 1.71 Ge	100 Cu	95.40 Cu 4.60 Mn	90.86 Cu 9.14 Mn
0	1.000	1.000	1.000	1.000	1.000	1.000	1.000	1.000	1.000	1.000	1.000
2	0.999	0.997	0.998	0.998	0.998	0.998	0.998	0.998	0.998	0.998	0.998
4	0.997	0.995	0.997	0.997	0.997	0.997	0.997	0.997	0.997	0.997	0.996
6	0.996	0.992	0.995	0.995	0.995	0.995	0.995	0.995	0.995	0.995	0.995
8	0.995	0.990	0.994	0.994	0.994	0.994	0.994	0.993	0.994	0.994	0.993
10	0.993	0.988	0.992	0.992	0.992	0.992	0.992	0.992	0.992	0.992	0.992
12	0.992	0.985	0.990	0.990	0.990	0.991	0.990	0.990	0.990	0.991	0.990
14	0.990	0.983	0.989	0.989	0.989	0.989	0.989	0.989	0.989	0.989	0.989
16	0.989	0.981	0.987	0.987	0.987	0.988	0.987	0.988	0.987	0.988	0.987
18	0.988	0.978	0.986	0.986	0.986	0.987	0.986	0.986	0.986	0.987	0.986
20	0.986	0.975	0.984	0.984	0.984	0.985	0.984	0.985	0.984	0.985	0.984
22	0.985	0.973	0.983	0.983	0.983	0.984	0.983	0.984	0.983	0.984	0.983
24	0.984	0.971	0.982	0.982	0.982	0.982	0.982	0.982	0.982	0.982	0.981
26	0.983	0.969	0.980	0.980	0.980	0.981	0.980	0.981	0.980	0.981	0.980
28	0.982	0.966	0.979	0.979	0.979	0.980	0.979	0.980	0.979	0.980	0.978
30	0.980	0.964	0.978	0.978	0.978	0.978	0.978	0.979	0.978	0.978	0.977

* For references see p. 4-96.

TABLE 4d-10. V/V_0 of Alloys and Intermetallic Compounds* (Continued)

Pressure, kilobars	Cu-Ni system [33]					Cu-Pd system [37]		Cu-Pt system [37]		Cu-Si system [33]		Cu₃₁Sn₈ [13]
	100 Cu	60 Cu 40 Ni	50 Cu 50 Ni	40 Cu 60 Ni	100 Ni	100 Cu	95.91 Cu 4.09 Pd	100 Cu	98.662 Cu 1.338 Pt	100 Cu	89.86 Cu 10.14 Si	
0	1.000	1.000	1.000	1.000	1.000	1.000	1.000	1.000	1.000	1.000	1.000	1.000
2	0.998	0.998	0.998	0.999	0.999	0.998	0.998	0.998	0.998	0.998	0.998	0.998
4	0.997	0.996	0.997	0.998	0.998	0.997	0.996	0.997	0.996	0.997	0.996	0.996
6	0.995	0.995	0.996	0.996	0.997	0.995	0.995	0.995	0.995	0.995	0.995	0.994
8	0.994	0.993	0.995	0.995	0.996	0.994	0.994	0.994	0.993	0.994	0.994	0.992
10	0.992	0.992	0.993	0.994	0.995	0.992	0.993	0.992	0.992	0.992	0.992	0.991
12	0.990	0.991	0.992	0.993	0.993 s	0.990	0.991	0.990	0.991	0.990	0.991	0.989
14	0.989	0.989	0.991	0.991	0.992	0.989	0.990	0.989	0.990	0.989	0.990	0.987
16	0.987	0.988	0.990	0.990	0.991	0.987	0.988	0.987	0.989	0.987	0.988	0.985
18	0.986	0.987	0.988	0.989	0.990	0.986	0.987	0.986	0.987	0.986	0.987	0.984
20	0.984	0.986	0.987	0.988	0.988	0.984	0.986	0.984	0.986	0.984	0.986	0.983
22	0.983	0.985	0.986	0.987	0.987	0.983	0.984	0.983	0.985	0.983	0.985	0.981
24	0.982	0.984	0.985	0.986	0.986	0.982	0.983	0.982	0.984	0.982	0.983	0.980
26	0.980	0.983	0.984	0.985	0.985	0.980	0.982	0.980	0.982 b	0.980	0.982	0.978
28	0.979	0.982	0.982	0.983	0.985	0.979	0.980	0.979	0.981	0.979	0.981	0.976
30	0.978	0.981	0.981	0.982	0.984	0.978	0.979	0.978	0.980	0.978	0.980	0.974

* For references see p. 4-96.
b Slight discontinuity here.
s Cusp at 10.1.

TABLE 4d-10. V/V_0 OF ALLOYS AND INTERMETALLIC COMPOUNDS* (Continued)

Pressure, kilobars	Cu-Zn system [33]						Fe-Ni alloys [37]			Fe-Si system [33]	
	100 Cu	90 Cu 10 Zn	80 Cu 20 Zn	52.7 Cu 47.3 Zn	CuZn [13]	Cu$_5$Zn$_8$ [13]	85.58 Fe 14.42 Ni	76.16 Fe 23.84 Ni	63.0 Fe 37.0 Ni	100 Fe	94.25 Fe 5.75 Si
0	1.000	1.000	1.000	1.000	1.000	1.000	1.000	1.000	1.000	1.000	1.000
2	0.998	0.998	0.998	0.997	0.998	0.998	0.998	0.999	0.998	0.998	0.999
4	0.997	0.996	0.996	0.995	0.996	0.996	0.997	0.997	0.996	0.997	0.997
6	0.995	0.995	0.994	0.993	0.994	0.994	0.996	0.996	0.994	0.996	0.996
8	0.994	0.993	0.992	0.992	0.992	0.991	t 0.994	0.994	0.993	0.995	0.995
10	0.992	0.992	0.991	0.990	0.990	0.990	0.993	0.993	0.991	0.994	0.994
12	0.990	0.990	0.989	0.988	0.989	0.988	0.992	0.992	u 0.990	0.993	0.993
14	0.989	0.989	0.988	0.986	0.987	0.986	0.990	0.991	0.988	0.992	0.992
16	0.987	0.988	0.986	0.985	0.985	0.984	0.989	0.990	0.986	0.990	0.991
18	0.986	0.986	0.985	0.983	0.984	0.982	0.988	0.988	0.985	0.989	0.990
20	0.984	0.985	0.983	0.982	0.982	0.980	0.987	0.987	0.983	0.988	0.989
22	0.983	0.983	0.982	0.980	0.981	0.978	0.986	0.986	0.982	0.987	0.988
24	0.982	0.982	0.980	0.978	0.979	0.976	0.985	0.985	0.980	0.986	0.987
26	0.980	0.981	0.979	0.977	0.977	0.975	0.984	0.984	0.978	0.985	0.986
28	0.979	0.980	0.978	0.976	0.976	0.973	0.983	0.983	0.977	0.984	0.985
30	0.978	0.978	0.976	0.975	0.974	0.972	0.982	0.982	0.975	0.983	0.984

* For references see p. 4-96.
t Cusp at 6.8; volume 0.996.
u Cusp at 11.0; volume 0.991.

TABLE 4d-10. V/V_0 OF ALLOYS AND INTERMETALLIC COMPOUNDS* (Continued)

Pressure, kilobars	In-Pb system [36]					100 Li	Li-Mg system [36]				100 Mg	Martensite [13]
	100 In	75 In 25 Pb	50 In 50 Pb	25 In 75 Pb	100 Pb		80 Li 20 Mg	60 Li 40 Mg	40 Li 60 Mg	20 Li 80 Mg		
0	1.000	1.000	1.000	1.000	1.000	1.000	1.000	1.000	1.000	1.000	1.000	1.000
2	0.995	0.995	0.995	0.995	0.995	0.982	0.987	0.990	0.992	0.993	0.994	0.998
4	0.990	0.990	0.990	0.990	0.990	0.967	0.975	0.980	0.984	0.985	0.988	0.997
6	0.985	0.985	0.985	0.985	0.986	0.954	0.963	0.970	0.977	0.978	0.983	0.996
8	0.981	0.980	0.980	0.981	0.982	0.940	0.952	0.961	0.969	0.971	0.977	0.995
10	0.976	0.975	0.975	0.977	0.978	0.927	0.941	0.953	0.962	0.965	0.973	0.994
12	0.972	0.971	0.971	0.972	0.974	0.915	0.931	0.944	0.955	0.959	0.968	0.992
14	0.968	0.966	0.967	0.968	0.970	0.904	0.921	0.936	0.948	0.953	0.963	0.991
16	0.963	0.962	0.963	0.965	0.966	0.892	0.911	0.928	0.942	0.947	0.958	0.990
18	0.959	0.958	0.959	0.961	0.962	0.882	0.902	0.920	0.935	0.941	0.953	0.989
20	0.955	0.954	0.955	0.957	0.958	0.872	0.893	0.914	0.929	0.935	0.949	0.987
22	0.952	0.950	0.952	0.954	0.954	0.862	0.885	0.906	0.924	0.930	0.944	0.986
24	0.948	0.947	0.948	0.950	0.951	0.853	0.877	0.900	0.918	0.925	0.940	0.985
26	0.944	0.943	0.945	0.947	0.947	0.845	0.869	0.394	0.913	0.920	0.936	0.984
28	0.940	0.939	0.941	0.943	0.944	0.835	0.862	0.888	0.908	0.915	0.932	0.983
30	0.937	0.936	0.938	0.940	0.941	0.828	0.855	0.882	0.903	0.910	0.929	0.982
32	0.933	0.933	0.934	0.937	0.937	0.820	0.848	0.877	0.899	0.906		
34	0.930	0.930	0.931	0.934	0.935	0.813	0.841	0.872	0.895	0.902		
36	0.927	0.927	0.928	0.931	0.932	0.807	0.835	0.866	0.890	0.898		
38	0.923	0.923	0.925	0.928	0.929	0.800	0.830	0.861	0.887	0.894		
40	0.921	0.921	0.922	0.925	0.927	0.794	0.825	0.856	0.884	0.890		

* For references see p. 4-96.

TABLE 4d-10. V/V_0 of Alloys and Intermetallic Compounds* (Continued)

Pressure, kilobars	35% Ni 65% Fe [13]	Ni-Mn system [33]		Ni-Si system [33]		Nirex [13]	Pb-Sb system [36]					
		100 Ni	71.0 Ni 29.0 Mn	100 Ni	94.2 Ni 5.8 Si		100 Pb	80 Pb 20 Sb	60 Pg 40 Sb	40 Pb 60 Sb	20 Pb 80 Sb	100 Sb
0	1.000	1.000	1.000	1.000	1.000	1.000	1.000	1.000	1.000	1.000	1.000	1.000
2	0.998	0.999	0.998	0.999	0.998	0.998	0.995	0.995	0.995	0.995	0.995	0.995
4	0.996	0.998	0.997	0.998	0.997	0.997	0.990	0.990	0.990	0.990	0.991	0.990
6	0.994	0.997	0.995	0.997	0.996	0.996	0.986	0.985	0.986	0.985	0.986	0.985
8	0.992	0.996	0.994	0.996	0.995	0.994	0.982	0.980	0.982	0.980	0.982	0.980
10	0.990	0.995	0.993	0.995	0.994	0.993	0.978	0.976	0.977	0.976	0.978	0.976
12	0.988	0.994	0.991	0.994	0.993	0.992	0.974	0.972	0.973	0.972	0.973	0.972
14	0.986	0.992	0.990	0.992	0.992	0.991	0.970	0.968	0.969	0.968	0.969	0.967
16	0.985	0.991	0.989	0.991	0.991	0.990	0.966	0.964	0.965	0.964	0.965	0.963
18	0.983	0.990	0.987	0.990	0.990	0.989	0.962	0.960	0.961	0.960	0.962	0.959
20	0.982	0.989	0.986	0.989	0.989	0.988	0.958	0.956	0.957	0.957	0.958	0.955
22	0.981	0.988	0.985	0.988	0.988	0.987	0.955	0.953	0.953	0.953	0.955	0.951
24	0.979	0.987	0.983	0.987	0.987	0.986	0.951	0.949	0.950	0.950	0.951	0.947
26	0.978	0.986	0.982	0.986	0.986	0.985	0.948	0.946	0.946	0.947	0.948	0.944
28	0.977	0.985	0.981	0.985	0.985	0.984	0.944	0.943	0.943	0.943	0.945	0.940
30	0.975	0.984	0.979	0.984	....	....	0.941	0.940	0.940	0.941	0.941	0.937
32	....	....	....	....	....	....	0.938	0.937	0.936	0.938	0.938	0.933
34	....	....	....	....	....	....	0.935	0.934	0.933	0.935	0.936	0.930
36	....	....	....	....	....	....	0.932	0.931	0.930	0.932	0.933	0.927
38	....	....	....	....	....	....	0.929	0.928	0.927	0.929	0.930	0.925
40	....	....	....	....	....	....	0.927	0.925	0.925	0.927	0.928	0.922

* For references see p. 4-96.

TABLE 4d-10. V/V_0 OF ALLOYS AND INTERMETALLIC COMPOUNDS* (Continued)

Pressure, kilobars	Pb-Sn system [36]					Pb-Zn system [36]			SbSn [13]	Sb$_2$Tl$_7$ [13]
	100 Pb	75 Pb 25 Sn	50 Pb 50 Sn	25 Pb 75 Sn	100 Sn	100 Pb	50 Pb 50 Zn	100 Zn		
0	1.000	1.000	1.000	1.000	1.000	1.000	1.000	1.000	1.000	1.000
2	0.995	0.995	0.996	0.996	0.996	0.995	0.996	0.997	0.995	0.994
4	0.990	0.990	0.991	0.992	0.992	0.990	0.992	0.993	0.991	0.988
6	0.986	0.986	0.987	0.988	0.988	0.986	0.988	0.990	0.987	0.983
8	0.982	0.982	0.983	0.984	0.985	0.982	0.984	0.987	0.983	0.978
10	0.978	0.978	0.979	0.980	0.981	0.978	0.980	0.985	0.979	0.973
12	0.974	0.975	0.975	0.977	0.978	0.974	0.976	0.982	0.976	0.968
14	0.970	0.971	0.972	0.973	0.975	0.970	0.972	0.978	0.972	0.964
16	0.966	0.967	0.968	0.970	0.972	0.966	0.969	0.976	0.969	0.959
18	0.962	0.964	0.965	0.966	0.968	0.962	0.965	0.973	0.966	0.955
20	0.958	0.960	0.962	0.963	0.965	0.958	0.962	0.970	0.962	0.951
22	0.955	0.957	0.958	0.960	0.962	0.955	0.959	0.967	0.959	0.947
24	0.951	0.953	0.956	0.957	0.959	0.951	0.955	0.965	0.956	0.943
26	0.948	0.950	0.953	0.953	0.956	0.948	0.952	0.962	0.953	0.939
28	0.944	0.947	0.950	0.950	0.953	0.944	0.949	0.959	0.950	0.935
30	0.941	0.943	0.947	0.948	0.951	0.941	0.946	0.957	0.947	0.932
32	0.938	0.940	0.944	0.945	0.948	0.938	0.943	0.954		
34	0.935	0.937	0.942	0.942	0.945	0.935	0.941	0.952		
36	0.932	0.934	0.939	0.940	0.943	0.932	0.938	0.950		
38	0.929	0.932	0.937	0.937	0.941	0.929	0.935	0.947		
40	0.927	0.929	0.934	0.935	0.939	0.927	0.933	0.945		

* For references see p. 4-96.

Table 4d-10. V/V_0 of Alloys and Intermetallic Compounds* (Continued)

Pressure, kilobars	Sn-Zn system [36]					Stainless steel [13]		Tl-Bi system [38]				
	100 Sn	80 Sn 20 Zn	50 Sn 50 Zn	20 Sn 80 Zn	100 Zn	H26‡	H29§	100 Tl	80 Tl 20 Bi	50 Tl 50 Bi	20 Tl 80 Bi	100 Bi
0	1.000	1.000	1.000	1.000	1.000	1.000	1.000	1.000	1.000	1.000	1.000	1.000
2	0.996	0.996	0.996	0.996	0.996	0.999	0.999	0.994	0.994	0.994	0.994	0.993
4	0.992	0.992	0.993	0.993	0.993	0.997	0.997	0.989	0.989	0.989	0.989	0.987
6	0.988	0.989	0.990	0.989	0.990	0.996	0.996	0.984	0.984	0.984	0.983	0.981
8	0.984	0.985	0.986	0.986	0.987	0.995	0.995	0.979	0.979	0.979	0.978	0.976
10	0.981	0.981	0.983	0.982	0.983	0.994	0.994	0.974	0.974	0.974	0.974	0.970
12	0.977	0.978	0.979	0.979	0.980	0.993	0.992	0.969	0.969	0.969	0.968	0.965
14	0.974	0.974	0.976	0.976	0.977	0.992	0.991	0.964	0.965	0.965	0.964	0.961
16	0.970	0.971	0.973	0.973	0.975	0.990	0.990	0.960	0.960	0.960	0.960	0.956
18	0.967	0.968	0.970	0.970	0.972	0.989	0.989	0.955	0.956	0.956	0.955	0.952
20	0.964	0.965	0.967	0.967	0.969	0.988	0.988	0.951	0.951	0.952	0.951	0.948
22	0.961	0.962	0.963	0.963	0.966	0.987	0.986	0.947	0.947	0.948	0.947	0.944
24	0.958	0.958	0.960	0.961	0.963	0.986	0.985	0.943	0.943	0.944	0.942	0.940
											v	c
26	0.955	0.956	0.957	0.958	0.961	0.985	0.984	0.938	0.939	0.940	0.902	0.902
28	0.952	0.953	0.954	0.955	0.958	0.984	0.983	0.934	0.936	0.936	0.898	0.899
30	0.950	0.950	0.952	0.952	0.956	0.983	0.982	0.930	0.932	0.933	0.895	0.896
32	0.947	0.947	0.949	0.950	0.953	....	....	0.926	0.928	0.930	0.892	0.893
34	0.945	0.945	0.947	0.947	0.951	....	....	0.923	0.925	0.926	0.889	0.890
36	0.942	0.942	0.944	0.945	0.949	....	....	0.919	0.922	0.923	0.886	0.887
38	0.940	0.940	0.942	0.942	0.947	....	....	0.915	0.918	0.920	0.883	0.885
40	0.938	0.938	0.939	0.940	0.945	....	....	0.912	0.915	0.917	0.881	0.882

* For references see p. 4-96.

For references see p. 4-96.

‡ Stainless steel H26: 0.094 C, 0.36 Mn, 0.023 P, 0.022 S, 0.35 Si, 12.26 Cr, 0.46 Ni, 0.50 Mo, N.D. Cu.
§ Stainless steel H29: 0.058 C, 0.70 Mn, 0.030 P, 0.013 S, 0.85 Si, 18.51 Cr, 8.95 Ni, N.D. Mo, 0.20 Cu.
ᵛ Volumes at 24.5: phase I = 0.942, phase II = 0.921, phase III = 0.905.
ᶜ Transition in this region.

TABLE 4d-10. V/V_0 OF ALLOYS AND INTERMETALLIC COMPOUNDS* (Continued)

Pressure, kilobars	Tl-Cd system [38]						Tl-In system [38]				
	100 Tl	80 Tl 20 Cd	60 Tl 40 Cd	40 Tl 60 Cd	20 Tl 80 Cd	100 Cd	100 Tl	77 Tl 23 In	50 Tl 50 In	20 Tl 80 In	100 In
0	1.000	1.000	1.000	1.000	1.000	1.000	1.000	1.000	1.000	1.000	1.000
2	0.994	0.995	0.995	0.995	0.996	0.995	0.994	0.994	0.995	0.995	0.995
4	0.989	0.990	0.990	0.990	0.991	0.991	0.989	0.988	0.990	0.990	0.990
6	0.984	0.985	0.986	0.986	0.987	0.987	0.984	0.983	0.985	0.985	0.985
8	0.979	0.980	0.981	0.981	0.983	0.983	0.979	0.977	0.980	0.980	0.981
10	0.974	0.975	0.977	0.977	0.979	0.979	0.974	0.973	0.975	0.975	0.976
								w			
12	0.969	0.970	0.972	0.972	0.975	0.976	0.969	0.962	0.971	0.970	0.972
14	0.964	0.965	0.968	0.968	0.971	0.972	0.964	0.958	0.966	0.966	0.967
16	0.960	0.960	0.963	0.963	0.967	0.969	0.960	0.954	0.962	0.961	0.963
18	0.955	0.956	0.959	0.960	0.963	0.966	0.955	0.950	0.958	0.957	0.958
20	0.951	0.952	0.955	0.956	0.959	0.962	0.951	0.945	0.954	0.953	0.954
22	0.947	0.948	0.951	0.952	0.956	0.959	0.947	0.941	0.950	0.949	0.950
24	0.943	0.944	0.947	0.948	0.952	0.956	0.943	0.937	0.946	0.945	0.946
26	0.938	0.940	0.943	0.945	0.949	0.953	0.938	0.933	0.942	0.942	0.943
28	0.934	0.936	0.940	0.941	0.946	0.950	0.934	0.930	0.938	0.938	0.939
30	0.930	0.933	0.936	0.938	0.942	0.948	0.930	0.926	0.934	0.935	0.936
32	0.926	0.930	0.932	0.934	0.939	0.945	0.926	0.923	0.931	0.932	0.932
34	0.923	0.926	0.929	0.932	0.936	0.942	0.923	0.920	0.927	0.928	0.929
36	0.919	0.923	0.925	0.928	0.933	0.940	0.918	0.916	0.924	0.925	0.926
38	0.915	0.921	0.922	0.926	0.931	0.937	0.915	0.913	0.921	0.922	0.922
40	0.912	0.918	0.919	0.923	0.928	0.935	0.912	0.910	0.918	0.919	0.920

* For references see p. 4-96.
w Transition at 11.5: volumes 0.969 and 0.963.

TABLE 4d-10. V/V_0 OF ALLOYS AND INTERMETALLIC COMPOUNDS* *(Continued)*

Pressure, kilobars	Tl-Pb system [38]						Tl-Sn system [38]					
	100 Tl	80 Tl 20 Pb	60 Tl 40 Pb	40 Tl 60 Pb	20 Tl 80 Pb	100 Pb	100 Tl	80 Tl 20 Sn	60 Tl 40 Sn	40 Tl 60 Sn	20 Tl 80 Sn	100 Sn
0	1.000	1.000	1.000	1.000	1.000	1.000	1.000	1.000	1.000	1.000	1.000	1.000
2	0.994	0.994	0.995	0.995	0.995	0.995	0.994	0.994	0.995	0.995	0.996	0.996
4	0.989	0.989	0.990	0.990	0.990	0.990	0.989	0.989	0.991	0.991	0.992	0.992
6	0.984	0.984	0.985	0.986	0.986	0.986	0.984	0.984	0.987	0.986	0.987	0.988
8	0.979	0.979	0.981	0.981	0.981	0.982	0.979	0.979	0.982	0.982	0.984	0.985
10	0.974	0.974	0.976	0.977	0.977	0.977	0.974	0.975	0.978	0.978	0.980	0.981
12	0.969	0.970	0.972	0.972	0.973	0.973	0.969	0.970	0.973	0.973	0.976	0.977
14	0.964	0.965	0.967	0.968	0.968	0.969	0.964	0.966	0.969	0.970	0.972	0.974
16	0.960	0.961	0.963	0.963	0.964	0.965	0.960	0.961	0.965	0.966	0.969	0.971
18	0.955	0.956	0.958	0.959	0.961	0.962	0.955	0.957	0.961	0.962	0.966	0.968
20	0.951	0.952	0.954	0.955	0.957	0.958	0.951	0.953	0.957	0.959	0.962	0.964
22	0.947	0.948	0.950	0.951	0.953	0.954	0.947	0.949	0.953	0.955	0.959	0.961
24	0.943	0.945	0.947	0.947	0.949	0.951	0.943	0.945	0.950	0.952	0.956	0.958
26	0.938	0.940	0.942	0.943	0.945	0.947	0.938	0.942	0.946	0.949	0.953	0.956
28	0.934	0.937	0.938	0.939	0.942	0.944	0.934	0.938	0.942	0.946	0.950	0.953
30	0.930	0.933	0.935	0.935	0.938	0.941	0.930	0.935	0.939	0.942	0.947	0.950
32	0.926	0.929	0.931	0.932	0.935	0.938	0.926	0.931	0.936	0.939	0.944	0.947
34	0.923	0.926	0.928	0.928	0.931	0.935	0.923	0.928	0.932	0.936	0.941	0.945
36	0.919	0.923	0.924	0.925	0.928	0.933	0.919	0.925	0.929	0.933	0.938	0.943
38	0.915	0.920	0.921	0.922	0.925	0.930	0.915	0.922	0.926	0.931	0.936	0.941
40	0.912	0.917	0.918	0.919	0.922	0.927	0.912	0.919	0.923	0.928	0.933	0.938

* For references see p. 4-96.

TABLE 4d-11. V/V_0 OF ORGANIC COMPOUNDS*

Pressure, kilobars	Ethyl acetate [3]	Ace-napathy-lene [24]	Acetone [23]	Ethyl alcohol [23]	Methyl alcohol [23]	Propyl alcohol [23]	c-Propyl alcohol [26]	n-Amyl iodide [25]	Amyl alcohol [23]	n-Amyl bromide [25]	n-Amyl chloride [25]	n-Amyl ether [24]
0	1.000	1.000	1.000	1.000	1.000	1.000	1.000	1.000	1.000	1.000	1.000	1.000
2	0.887	0.975	0.885	0.893	0.892	0.895	0.906	0.915	0.888	0.907	0.892	0.888
4	0.830	0.955	0.831	0.831	0.836	0.854	0.853	0.860	0.842	0.860	0.844	0.843
6	0.795	0.937	0.795	0.795	0.797	0.825	0.820	0.830	0.816	0.828	0.815	0.810
8	0.770	0.923	0.768	0.769	0.771	0.802	0.798	0.806	0.796	0.802	0.792	0.786
10	0.750	0.910	.....	0.749	0.752	0.785	0.780	0.787	0.777	0.782	0.773	0.766
												a
12	0.736	0.899	.....	0.735	0.737	0.769	.....	.....	0.761	.....	.....	0.728
14	0.724	0.888	.....	.....	.....	.....	.....	.....	.....	.....	.....	0.719
16	0.713	0.879	.....	.....	.....	.....	.....	.....	.....	.....	.....	0.710
18	0.704	0.871	.....	.....	.....	.....	.....	.....	.....	.....	.....	0.702
20	0.695	0.863	.....	.....	.....	.....	.....	.....	.....	.....	.....	0.695
22	0.687	0.856	.....	.....	.....	.....	.....	.....	.....	.....	.....	0.688
24	0.680	0.849	.....	.....	.....	.....	.....	.....	.....	.....	.....	0.680
26	0.672	0.843	.....	.....	.....	.....	.....	.....	.....	.....	.....	0.674
28	0.666	0.838	.....	.....	.....	.....	.....	.....	.....	.....	.....	0.668
30	0.660	0.832	.....	.....	.....	.....	.....	.....	.....	.....	.....	0.662
32	0.654	0.827	.....	.....	.....	.....	.....	.....	.....	.....	.....	0.657
34	0.649	0.822	.....	.....	.....	.....	.....	.....	.....	.....	.....	0.652
36	0.643	0.818	.....	.....	.....	.....	.....	.....	.....	.....	.....	0.647
38	0.639	0.813	.....	.....	.....	.....	.....	.....	.....	.....	.....	0.643
40	0.635	0.808	.....	.....	.....	.....	.....	.....	.....	.....	.....	0.639

* For references see p. 4-96.
a Freezes at 10.8; volumes 0.776 and 0.755.

TABLE 4d-11. V/V_0 OF ORGANIC COMPOUNDS* (Continued)

Pressure, kilobars	Nitroaniline [5]			Anthracene [24]	Anthraquinone [6]	Benzene [24]	Bromobenzene [26]	Chlorobenzene [24]	Diphenylbenzene [5]			Hexaethylbenzene [24]
	ortho-	meta-	para-						ortho-	meta-	para-	
0	1.000	1.000	1.000	1.000	1.000	1.000	1.000	1.000	1.000	1.000	1.000	1.000
2	0.983	0.980	0.977	0.972	0.978	0.857	0.930	0.895	0.974	0.974	0.974	0.980
4	0.967	0.964	0.960	0.951	0.961	0.805	.....	0.860	0.950	0.950	0.950	0.960
6	0.952	0.948	0.944	0.935	0.947	0.770	.....	0.834	0.930	0.931	0.931	0.940
								b				
8	0.939	0.935	0.931	0.920	0.935	0.745	.....	0.784	0.912	0.917	0.917	0.923
10	0.928	0.924	0.920	0.910	0.924	0.725	.....	0.772	0.896	0.905	0.905	0.908
12	0.917	0.913	0.910	0.898	0.915	0.712	.....	0.759	0.884	0.893	0.894	0.895
14	0.907	0.903	0.900	0.888	0.906	0.700	.....	0.748	0.872	0.883	0.884	0.883
16	0.898	0.895	0.892	0.880	0.898	0.699	.....	0.738	0.862	0.874	0.875	0.872
18	0.890	0.887	0.884	0.872	0.890	0.682	.....	0.730	0.852	0.865	0.867	0.862
20	0.883	0.880	0.877	0.865	0.883	0.675	.....	0.722	0.845	0.857	0.860	0.853
22	0.876	0.872	0.869	0.857	0.877	0.667	.....	0.715	0.837	0.849	0.853	0.845
24	0.870	0.866	0.863	0.851	0.871	0.660	.....	0.708	0.830	0.842	0.846	0.837
26	0.865	0.860	0.857	0.845	.....	0.654	.....	0.703	0.823	0.835	0.840	0.830
28	0.859	0.854	0.851	0.838	.....	0.648	.....	0.697	0.817	0.829	0.834	0.823
												c
30	0.853	0.848	0.845	0.832	.....	0.644	.....	0.692	0.812	0.824	0.830	0.807
32	0.849	0.843	0.840	0.827	.....	0.638	.....	0.687	0.805	0.818	0.823	0.802
34	0.845	0.838	0.835	0.822	.....	0.634	.....	0.683	0.800	0.813	0.818	0.798
36	0.840	0.833	0.830	0.817	.....	0.630	.....	0.679	0.796	0.808	0.814	
38	0.837	0.829	0.826	0.812	.....	0.625	.....	0.675	0.792	0.805	0.809	
40	0.832	0.825	0.821	0.808	.....	0.622	.....	0.672	0.789	0.801	0.805	

* For references see p. 4-96.
b Freezes at 7.4; volumes 0.817 and 0.788.
c Sluggish transition here, not complete.

TABLE 4d-11. V/V_0 OF ORGANIC COMPOUNDS* (Continued)

Pressure, kilobars	Nitrobromobenzene [5] ortho-	meta-	para-	Nitrochlorobenzene [5] ortho-	para-	Nitroiodobenzene [5] ortho-	meta-	para-	Aminobenzenesulfonic acid [5] ortho-	meta-	para-	Benzil [5]
0	1.000	1.000	1.000	1.000	1.000	1.000	1.000	1.000	1.000	1.000	1.000	1.000
2	0.970	0.980	0.975	0.970[e]	0.970	0.973	0.975	0.975	0.983	0.987	0.985[f]	0.953
4	0.950	0.962	0.957	0.944	0.949	0.953	0.957	0.956	0.968	0.976	0.949	0.925
6	0.935[d]	0.946	0.942	0.920	0.930	0.937	0.942	0.940	0.955	0.965	0.935	0.904
8	0.915	0.933	0.928	0.902	0.915	0.923	0.928	0.926	0.943	0.956	0.916	0.888
10	0.898	0.923	0.915	0.888	0.903	0.910	0.916	0.914	0.932	0.947	0.902[g]	0.874
12	0.885	0.911	0.905	0.875	0.890	0.899	0.905	0.903	0.922	0.940	0.885	0.865
14	0.873	0.901	0.895	0.865	0.880	0.888	0.895	0.893	0.913	0.934	0.873	0.855
16	0.863	0.892	0.886	0.855	0.871	0.879	0.887	0.885	0.905	0.928	0.862	0.846
18	0.855	0.885	0.877	0.846	0.862	0.870	0.879	0.877	0.897	0.922	0.852	0.838
20	0.847	0.877	0.868	0.839	0.853	0.863	0.871	0.869	0.890	0.918	0.843	0.831
22	0.840	0.870	0.862	0.832	0.846	0.855	0.864	0.862	0.883	0.913	0.835	0.824
24	0.832	0.864	0.855	0.825	0.839	0.848	0.857	0.855	0.876	0.908	0.828	0.818
26	0.827	0.858	0.848	0.819	0.831	0.842	0.851	0.849	0.871	0.904	0.821	0.812
28	0.821	0.952	0.842	0.813	0.825	0.836	0.845	0.843	0.865	0.900	0.815	0.808
30	0.816	0.846	0.836	0.810	0.820	0.831	0.840	0.838	0.861	0.897	0.810	0.813
32	0.810	0.841	0.831	0.804	0.814	0.825	0.835	0.833	0.856	0.893	0.805	0.800
34	0.806	0.837	0.825	0.799	0.809	0.820	0.830	0.827	0.852	0.889	0.801	0.796
36	0.802	0.832	0.820	0.795	0.804	0.815	0.826	0.823	0.848	0.886	0.797	0.793
38	0.798	0.828	0.814	0.792	0.800	0.811	0.822	0.817	0.845	0.883	0.793	0.790
40	0.793	0.825	0.810	0.787	0.795	0.807	0.818	0.814	0.841	0.880	0.790	0.787

* For references see p. 4-96.
d Transition at 7.3; volumes 0.926 and 0.920.
e Transition at 3.9; volumes 0.952 and 0.945.
f Transition at 3.9; volumes 0.973 and 0.962.
g Transition at 11.2; volumes 0.894 and 0.892.

TABLE 4d-11. V/V_0 OF ORGANIC COMPOUNDS* (Continued)

Pressure kilobars	Aminobenzoic acid [5]			Bromobenzoic acid [5]			Chlorobenzoic acid [5]			Iodobenzoic acid [5]		
	ortho-	meta-	para-	ortho-	meta-	para-	ortho-	meta-	para-	ortho-	meta-	para-
0	1.000	1.000	1.000	1.000	1.000	1.000	1.000	1.000	1.000	1.000	1.000	1.000
2	0.981	0.986	0.981	0.980	0.975	0.980	0.977	0.977	0.977	0.977	0.979	0.978
4	0.964	0.975	0.964	0.961	0.953	0.961	0.958	0.958	0.958	0.957	0.960	0.960
6	0.947	0.965	0.947	0.945	0.935	0.945	0.942	0.942	0.942	0.940	0.941	0.944
8	0.932	0.957	0.932	0.930	0.929	0.930	0.928	0.927	0.927	0.927	0.927	0.931
10	0.918	0.949	0.918	0.917	0.907	0.917	0.915	0.915	0.915	0.916	0.914	0.919
12	0.907	0.942	0.907	0.907	0.894	0.907	0.905	0.903	0.903	0.905	0.903	0.908
14	0.896	0.935	0.896	0.897	0.883	0.897	0.895	0.893	0.893	0.895	0.893	0.898
16	0.887	0.929	0.887	0.888	0.874	0.888	0.886	0.883	0.884	0.887	0.884	0.889
18	0.878	0.923	0.878	0.880	0.865	0.880	0.878	0.875	0.876	0.880	0.875	0.880
20	0.871	0.918	0.872	0.872	0.857	0.872	0.870	0.867	0.869	0.872	0.868	0.872
22	0.863	0.912	0.863	0.860	0.849	0.860	0.862	0.860	0.862	0.865	0.860	0.865
24	0.856	0.907	0.856	0.858	0.842	0.858	0.855	0.853	0.855	0.859	0.854	0.857
26	0.850	0.903	0.850	0.852	0.836	0.852	0.849	0.846	0.849	0.853	0.847	0.850
28	0.845	0.898	0.845	0.846	0.830	0.846	0.843	0.840	0.843	0.847	0.841	0.844
30	0.840	0.895	0.840	0.841	0.824	0.841	0.838	0.833	0.837	0.842	0.837	0.838
32	0.835	0.890	0.834	0.835	0.818	0.835	0.832	0.827	0.831	0.837	0.830	0.831
34	0.831	0.886	0.829	0.830	0.813	0.830	0.827	0.822	0.825	0.832	0.825	0.826
36	0.827	0.883	0.825	0.825	0.808	0.825	0.823	0.817	0.820	0.827	0.821	0.821
38	0.823	0.880	0.821	0.820	0.804	0.820	0.818	0.812	0.816	0.823	0.816	0.816
40	0.820	0.877	0.817	0.816	0.800	0.816	0.814	0.807	0.812	0.819	0.812	0.811

* For references see p. 4-96.

TABLE 4d-11. V/V_0 OF ORGANIC COMPOUNDS* (Continued)

Pressure, kilobars	Benzophenone [6]	Bibenzyl [24]	Bromoform [26]	2,2-Dimethylbutane [26]	2,3-Dimethylbutane [26]	Isobutyl alcohol [23]	n-Butyl alcohol [26]	n-Butyl bromide [25]	n-Butyl chloride [25]	n-Butyl iodide [25]	d-Camphor [5]	Carbon disulfide [23]
0	1.000	1.000	1.000	1.000	1.000	1.000	1.000	1.000	1.000	1.000	1.000	1.000
2	0.972	0.957	0.902	0.904	0.869	0.878	0.913	0.900	0.893	0.912	0.950	0.892
											ʰ	
4	0.950	0.930	....	0.846	0.816	0.833	0.867	0.853	0.841	0.862	0.873	0.842
6	0.932	0.909	....	0.807	0.782	0.802	....	0.819	0.807	0.830	0.855	0.808
8	0.917	0.892	....	....	....	0.777	....	0.794	0.783	0.805	0.841	0.781
10	0.902	0.877	....	....	....	0.758	....	0.775	0.763	0.785	0.828	0.772
12	0.890	0.865	....	....	....	0.743	....	0.760	....	0.766	0.817	0.762
14	0.879	0.853	....	....	....	....	....	....	....	....	0.808	0.745
16	0.869	0.842	....	....	....	....	....	....	....	....	0.800	
18	0.860	0.833	....	....	....	....	....	....	....	....	0.791	
20	0.850	0.824	....	....	....	....	....	....	....	....	0.783	
22	0.842	0.815	....	....	....	....	....	....	....	....	0.777	
24	0.835	0.808	....	....	....	....	....	....	....	....	0.770	
26	....	0.801	....	....	....	....	....	....	....	....	0.764	
28	....	0.795	....	....	....	....	....	....	....	....	0.758	
30	....	0.789	....	....	....	....	....	....	....	....	0.752	
32	....	0.783	....	....	....	....	....	....	....	....	0.747	
34	....	0.728	....	....	....	....	....	....	....	....	0.743	
36	....	0.723	....	....	....	....	....	....	....	....	0.738	
38	....	0.718	....	....	....	....	....	....	....	....	0.733	
40	....	0.714	....	....	....	....	....	....	....	....	0.729	

* For references see p. 4-96.
ʰ Transition at 3.5; volumes 0.925 and 0.877.

TABLE 4d-11. V/V_0 OF ORGANIC COMPOUNDS* (Continued)

Pressure, kilobars	Carbon tetrachloride [26]	Dicresyl carbonate [5]			Chloroacetanilide [5]			Chloroform [24]	Methylnitrocinnamate [5]			Cumene [24]
		ortho-	meta-	para-	ortho-	meta-	para-		ortho-	meta-	para-	
0	1.000	1.000	1.000	1.000	1.000	1.000	1.000	1.000	1.000	1.000	1.000	1.000
2	0.885	0.975	0.975	0.975	0.976	0.976	0.978	0.892	0.974	0.974	0.977	0.901
4	......	0.952	0.952	0.953	0.955	0.955	0.961	0.841 ^{i}	0.953	0.956	0.958	0.850
6	......	0.932	0.932	0.937	0.935	0.936	0.945	0.743	0.937	0.940	0.943	0.822
8	......	0.915	0.915	0.922	0.916	0.919	0.931	0.727	0.922	0.926	0.928	0.800
10	......	0.900	0.901	0.909	0.898	0.902	0.918	0.715	0.910	0.914	0.916	0.783
12	......	0.887	0.889	0.892	0.885	0.890	0.907	0.704	0.899	0.903	0.905	0.768
14	......	0.876	0.877	0.887	0.873	0.878	0.897	0.694	0.889	0.893	0.895	0.755
16	......	0.866	0.868	0.877	0.861	0.867	0.887	0.685	0.880	0.883	0.885	0.744
18	......	0.857	0.858	0.868	0.852	0.858	0.878	0.677	0.871	0.874	0.876	0.734
20	......	0.849	0.850	0.860	0.844	0.849	0.870	0.670	0.863	0.866	0.869	0.725
22	......	0.842	0.842	0.853	0.835	0.842	0.863	0.663	0.856	0.858	0.861	0.717
24	......	0.835	0.835	0.846	0.827	0.834	0.857	0.657	0.848	0.851	0.854	0.710
26	......	0.829	0.829	0.840	0.820	0.828	0.850	0.651	0.842	0.845	0.847	0.702
28	......	0.823	0.823	0.834	0.814	0.821	0.845	0.646	0.835	0.838	0.842	0.696
30	......	0.817	0.816	0.828	0.808	0.815	0.838	0.641	0.830	0.833	0.836	0.690
32	......	0.812	0.810	0.823	0.803	0.810	0.834	0.636	0.824	0.827	0.830	0.685
34	......	0.807	0.805	0.818	0.798	0.805	0.829	0.632	0.818	0.822	0.825	0.680
36	......	0.802	0.799	0.814	0.793	0.800	0.825	0.627	0.813	0.817	0.821	0.676
38	......	0.797	0.794	0.809	0.789	0.795	0.821	0.623	0.808	0.812	0.816	0.672
40	......	0.793	0.789	0.805	0.785	0.791	0.817	0.619	0.804	0.808	0.812	0.669

* For references see p. 4-96.

^{i} Freezes at 5.4; volumes, 0.815 and 0.748.

TABLE 4d-11. V/V_0 OF ORGANIC COMPOUNDS* *(Continued)*

Pressure, kilobars	Cyanamide [5]	n-Decane [24]	Dextrin [6]	Dextrose [6]	Di-ethylene glycol [27]	Diphenyl [24]	Diphenyl amine [28]	n-Do-decane [24]	Ethyl ether [23]	Ethyl bromide [23]	Ethyl chloride [23]	Ethyl iodide [23]
0	1.000	1.000	1.000	1.000	1.000	1.000	1.000	1.000	1.000	1.000	1.000	1.000
2	0.985	0.892 [j]	0.980	0.990	0.949	0.964	0.964	0.800 [k]	0.859	0.910	0.849	0.877
4	0.971	0.766	0.964	0.980	0.915	0.939	0.938	0.783	0.790	0.830	0.793	0.835
6	0.959	0.756	0.948	0.972	.....	0.920	0.920	0.770	0.761	0.789	0.761	0.805
8	0.947	0.741	0.935	0.964	.....	0.905	0.914	0.758	0.734	0.766	0.735	0.779
10	0.938	0.728	0.923	0.956	.....	0.891	0.890	0.747	0.714	0.746	0.715	0.758
12	0.928	0.717	0.912	0.949	.....	0.880	0.878	0.737	0.695	0.731	0.695	0.740
14	0.920	0.707	0.903	0.942	.....	0.869	.....	0.727				
16	0.911	0.699	0.895	0.935	.....	0.860	.....	0.719				
18	0.904	0.690	0.886	0.928	.....	0.850	.....	0.711				
20	0.897	0.684	0.879	0.922	.....	0.842	.....	0.703				
22	0.890	0.677	0.872	0.916	.....	0.834	.....	0.696				
24	0.882	0.672	0.866	0.911	.....	0.827	.....	0.690				
26	0.876	0.666	0.860	.....	.....	0.820	.....	0.684				
28	0.870	0.662	.....	.....	.....	0.813	.....	0.628				
30	0.865	0.657	.....	.....	.....	0.808	.....	0.623				
32	0.859	0.652	.....	.....	.....	0.802	.....	0.618				
34	0.853	0.649	.....	.....	.....	0.797	.....	0.615				
36	0.848	0.645	.....	.....	.....	0.792	.....	0.611				
38	0.843	0.642	.....	.....	.....	0.788	.....	0.607				
40	0.838	0.638	.....	.....	.....	0.784	.....	0.604				

* For references see p. 4-96.
j Freezes at 3.0; volumes 0.863 and 0.789.
k Freezes at 1.65; volumes 0.916 and 0.813.

TABLE 4d-11. V/V_0 OF ORGANIC COMPOUNDS* *(Continued)*

Pressure, kilobars	Ethylene bromide [5]	Ethylene glycol [27]	Eugenol [27]	Fluor-anthene [24]	Fluorene [16]	Glycerin [27]	Guanidine sulfate [5]	n-Heptane [24]	3-Methyl-heptanol-1 [25]	2-Methyl-heptanol-3 [25]	3-Methyl-heptanol [25]	2-Methyl-heptanol-5 [25]
0	1.000 *l*	1.000	1.000	1.000	1.000	1.000	1.000	1.000	1.000	1.000	1.000	1.000
2	0.829	0.948	0.989	0.971	0.975	0.964	.....	0.865	0.918	0.914	0.916	0.912
4	0.804	0.915	0.978	0.950	0.952	0.935	.....	0.816	0.876	0.865	0.873	0.867
6	0.782	.....	0.967	0.931	0.933	0.912	*m* 0.818	0.799	0.846	.....	0.842	
8	0.765	.....	6.957	0.915	0.917	0.893	0.810	0.755				
10	0.751	.....	0.947	0.901	0.902	0.877	0.803	0.734				
12	0.740	.....	0.938	0.890	0.890	0.863	*n* 0.792	*o* 0.675				
14	0.727	.....	0.930	0.879	0.879	.....	0.786	0.665				
16	0.717	.....	0.923	0.869	0.868	.....	0.780	0.655				
18	0.708	.....	0.916	0.860	0.859	.....	0.775	0.646				
20	0.700	.....	0.910	0.852	0.850	.....	0.770	0.638				
22	0.693	.....	0.904	0.844	0.842	.....	0.766	0.630				
24	0.687	.....	0.898	0.837	0.835	.....	0.762	0.623				
26	0.682	.....	0.893	0.830	0.829	.....	0.757	0.616				
28	0.677	.....	0.887	0.824	0.824	.....	0.754	0.610				
30	0.672	.....	0.883	0.818	0.819	.....	0.750	0.604				
32	0.668	.....	0.878	0.813	0.815	.....	0.747	0.591				
34	0.666	.....	0.873	0.808	0.812	.....	0.744	0.594				
36	0.663	.....	0.870	0.805	0.810	.....	0.741	0.590				
38	0.661	.....	0.866	0.801	0.809	.....	0.738	0.586				
40	0.660	.....	0.862	0.798	0.808	.....	0.735	0.583				

* For references see p. 4-96.
l Freezes at 0.5; volumes 0.967 and 0.870.
m Three transitions below 4.9.
n Transition at 10.3; volumes 0.803 and 0.797.
o Freezes at 11.2; volumes 0.722 and 0.680.

TABLE 4d-11. V/V_0 of Organic Compounds* (Continued)

Pressure, kilobars	n-Hexane [26]	Cyclohexane [24]	Methylcyclohexane [24]	n-Hexadecane [24]	Hexamethylenetetramine [5]	n-Hexyl alcohol [26]	Iodoform [5]	Isoprene [27]	Levulose [6]	dl-Limonene [24]	Ethyl dibenzyl malonate [27]	Melamine [5]
0	1.000	1.000	1.000	1.000	1.000	1.000	1.000	1.000	1.000	1.000	1.000	1.000
2	0.876	0.862 p	0.886	0.828 r	0.980	0.918	0.977	0.860	0.990	0.896	0.929	0.983
4	0.823	0.825	0.840	0.803	0.962	......	0.955	0.819	0.981	0.853	......	0.969
6	0.790	0.799	0.810	0.783	0.947	......	0.937	0.789	0.972	0.826	......	0.953
8	......	0.747 q	0.786	0.768	0.935	......	0.922	0.764	0.963	0.806	......	0.947
10	......	0.729	0.767	0.755	0.925	......	0.908	0.743	0.955	0.790	......	0.938
12	......	0.715	0.751	0.744	0.915	......	0.896	0.725	0.947	0.775	......	0.929
14	......	0.704	0.737	0.735	0.906	......	0.885	......	0.940	0.763	......	0.921
16	......	0.694	0.725	0.725	0.898	......	0.875	......	0.934	0.752	......	0.914
18	......	0.685	0.715	0.717	0.890	......	0.865	......	0.927	0.742	......	0.907
20	......	0.678	0.706	0.710	0.883	......	0.857	......	0.920	0.733	......	0.900
22	......	0.672	0.698	0.703	0.876	......	0.848	......	0.915	0.725	......	0.894
24	......	0.665	0.690	0.697	0.870	......	0.841	......	0.909	0.717	......	0.888
26	......	0.660	0.684	0.692	0.863	......	0.833	......	......	0.711	......	0.882
28	......	0.655	0.677	0.698	0.857	......	0.827	......	......	0.705	......	0.877
30	......	0.650	0.672	0.682	0.851	......	0.820	......	......	0.700	......	0.872
32	......	0.645	0.666	0.678	0.845	......	0.815	......	......	0.695	......	0.867
34	......	0.641	0.660	0.674	0.840	......	0.809	......	......	0.691	......	0.864
36	......	0.637	0.655	0.670	0.836	......	0.804	......	......	0.687	......	0.860
38	......	0.633	0.650	0.666	0.831	......	0.799	......	......	0.683	......	0.856
40	......	0.630	0.646	0.663	0.827	......	0.794	......	......	0.679	......	0.852

* For references see p. 4-96.

p Freezes at 0.3; volumes 0.967 and 0.926. (Volume liquid = 0.977 at 0.2.)

q Transition at 7.4; volumes 0.784 and 0.757.

r Freezes at 0.4; volumes 0.970 and 0.861.

TABLE 4d-11. V/V_0 OF ORGANIC COMPOUNDS* (Continued)

Pressure, kilobars	Menthol [6]	Mesitylene [24]	Triphenylmethane [24]	Methylamine hydrochloride [5]	Methylene chloride [24]	Morpholine hydrogen tartrate [5]	Naphthalene [24]	β-Methylnaphthalene [24]	Tetrahydronaphthalene [24]	n-Octane [24]	Iso-octane [27]	n-Octacosane [24]
0	1.000	1.000	1.000	1.000	1.000	1.000	1.000	1.000	1.000	1.000	1.000	1.000
2	0.966	0.909	0.974	0.982	0.910	0.985	0.970	0.965	0.926	0.883	0.893	0.955
4	0.941	0.825	0.952	0.967	0.860	0.971	0.946	0.937	0.829	0.828	0.835	0.930
6	0.921	0.802	0.935	0.900	0.819	0.959	0.928	0.915	0.813	0.750	0.803	0.912
8	0.905	0.784	0.919	0.890	0.787	0.949	0.912	0.896	0.800	0.725	...	0.896
10	0.888	0.764	0.906	0.880	0.762	0.939	0.899	0.881	0.788	0.707	...	0.883
12	0.875	0.756	0.893	0.871	0.744	0.930	0.887	0.868	0.777	0.695	...	0.872
14	0.861	0.745	0.883	0.862	0.690	0.922	0.877	0.856	0.768	0.683	...	0.861
16	0.849	0.735	0.872	0.853	0.678	0.914	0.867	0.845	0.760	0.674	...	0.851
18	0.837	0.725	0.862	0.845	0.669	0.907	0.858	0.835	0.751	0.665	...	0.842
20	0.826	0.717	0.854	0.838	0.660	0.900	0.849	0.826	0.744	0.658	...	0.834
22	0.816	0.710	0.845	0.830	0.652	0.893	0.841	0.818	0.737	0.650	...	0.826
24	0.806	0.702	0.838	0.823	0.645	0.886	0.833	0.810	0.731	0.645	...	0.818
26	...	0.697	0.832	0.801	0.639	0.880	0.826	0.803	0.725	0.639	...	0.813
28	...	0.691	0.825	0.795	0.633	0.874	0.820	0.797	0.720	0.634	...	0.806
30	...	0.685	0.819	0.788	0.628	0.869	0.813	0.790	0.715	0.630	...	0.800
32	...	0.680	0.814	0.782	0.623	0.864	0.807	0.785	0.710	0.625	...	0.795
34	...	0.676	0.809	0.775	0.619	0.858	0.802	0.779	0.706	0.621	...	0.789
36	...	0.672	0.804	0.769	0.615	0.854	0.797	0.774	0.702	0.618	...	0.784
38	...	0.668	0.800	0.763	0.612	0.849	0.792	0.770	0.698	0.615	...	0.779
40	...	0.665	0.796	0.758	0.609	0.845	0.787	0.765	0.695	0.611	...	0.775

* For references see p. 4-96.
s Freezes at 3.4; volumes 0.871 and 0.837.
u Transition at 24.8; volumes 0.821 and 0.805.
w Freezes at 2.99; volumes 0.902 and 0.840.

t Transition at 5.4; volumes 0.956 and 0.904.
v Freezes at 12.2; volumes 0.741 and 0.701.
x Freezes at 5.4; volumes 0.741 and 0.701.

TABLE 4d-11. V/V_0 OF ORGANIC COMPOUNDS* (Continued)

Pressure, kilobars	n-Octa-decane [24]	Octanol-3 [25]	Octylene [24]	Methyl oleate [27]	Oxalic acid anhydrous [5]	n-Pentane [26]	Iso-pentane [26]	2-Methyl-pentane [26]	3-Methyl-pentane [26]	Phenylenediamine [5] ortho-	meta-	para-
0	1.000	1.000	1.000	1.000	1.000	1.000	1.000	1.000	1.000	1.000	1.000	1.000
2	0.966	0.916	0.905	0.932	0.985	0.852	0.857	0.863	0.867	0.977	0.978	0.977
4	0.935	0.871	0.845	.....	0.971	0.802	0.802	0.816	0.813	0.959	0.960	0.959
6	0.913	.....	0.805	.....	0.958	0.765	0.765	0.784	0.780	0.944	0.945	0.944
8	0.895	.....	0.778	.....	0.947	0.738	.....	.....	0.755	0.930	0.932	0.930
10	0.880	.....	0.757	.....	0.937	0.717	.....	.....	0.736	0.918	0.920	0.918
12	0.868	.....	0.742	.....	0.928	.....	.....	.....	.....	0.907	0.910	0.907
14	0.857	.....	0.728	.....	0.920	.....	.....	.....	.....	0.897	0.900	0.897
16	0.846	.....	0.715	.....	0.912	.....	.....	.....	.....	0.888	0.890	0.888
18	0.837	.....	0.705	.....	0.905	.....	.....	.....	.....	0.879	0.882	0.880
20	0.828	.....	0.695	.....	0.898	.....	.....	.....	.....	0.871	0.874	0.873
22	0.820	.....	0.686	.....	0.892	.....	.....	.....	.....	0.864	0.867	0.865
24	0.813	.....	0.678	.....	0.885	.....	.....	.....	.....	0.856	0.860	0.858
26	0.806	.....	0.670	.....	0.880	.....	.....	.....	.....	0.850	0.854	0.852
28	0.800	.....	0.663	.....	0.874	.....	.....	.....	.....	0.844	0.848	0.847
30	0.794	.....	0.657	.....	0.868	.....	.....	.....	.....	0.838	0.843	0.841
32	0.787	.....	0.650	.....	0.863	.....	.....	.....	.....	0.833	0.838	0.836
34	0.782	.....	0.644	.....	0.858	.....	.....	.....	.....	0.828	0.834	0.832
36	0.776	.....	0.638	.....	0.853	.....	.....	.....	.....	0.823	0.830	0.828
38	0.772	.....	0.633	.....	0.849	.....	.....	.....	.....	0.819	0.826	0.824
40	0.767	.....	0.628	.....	0.845	.....	.....	.....	.....	0.816	0.822	0.821

* For references see p. 4-96.

TABLE 4d-11. V/V_0 of Organic Compounds* (Continued)

Pressure, kilobars	Phenylenediamine hydrochloride [5]			Aminophenol [5]			2,4-Dichloro-phenol [5]	Nitrophenol [5]			Tri-o-cresyl phosphate [27]	Normal butyl phthalate [27]	n-Propyl bromide [25]
	ortho-	meta-	para-	ortho-	meta-	para-		ortho-	meta-	para-			
0	1.000	1.000	1.000	1.000	1.000	1.000	1.000	1.000	1.000	1.000	1.000	1.000	1.000
2	0.979	0.979	0.979	0.979	0.980	0.982	0.972	0.973	0.980	0.979	0.9478	0.931	0.902
4	0.962	0.962	0.962	0.961	0.963	0.966	0.953	0.950	0.961	0.960	.....	0.892	0.850
6	0.946	0.946	0.946	0.945	0.950	0.952	0.936	0.934	0.946	0.943	.....	0.864	0.812
8	0.932	0.932	0.932	0.932	0.937	0.940	0.923	0.918	0.932	0.929	.....	0.841	0.786
10	0.919	0.919	0.919	0.920	0.926	0.928	0.910	0.905	0.920	0.916	.....	.....	0.766
12	0.908	0.908	0.908	0.911	0.916	0.917	0.899	0.893	0.910	0.905	.....	.....	0.750
14	0.898	0.898	0.898	0.902	0.908	0.908	0.889	0.883	0.900	0.895			
16	0.888	0.890	0.890	0.894	0.900	0.900	0.880	0.873	0.892	0.885			
18	0.880	0.882	0.881	0.886	0.892	0.891	0.871	0.865	0.884	0.877			
20	0.871	0.874	0.873	0.879	0.885	0.884	0.862	0.856	0.876	0.868			
22	0.863	0.867	0.866	0.872	0.879	0.876	0.855	0.849	0.870	0.861			
24	0.856	0.861	0.860	0.866	0.873	0.869	0.847	0.842	0.863	0.854			
26	0.850	0.855	0.854	0.860	0.867	0.862	0.840	0.836	0.857	0.847			
28	0.844	0.849	0.848	0.855	0.862	0.857	0.834	0.830	0.851	0.841			
30	0.838	0.844	0.843	0.849	0.857	0.851	0.828	0.825	0.845	0.835			
32	0.833	0.839	0.838	0.845	0.853	0.846	0.822	0.820	0.840	0.830			
34	0.829	0.835	0.834	0.843	0.849	0.843	0.817	0.816	0.835	0.825			
36	0.824	0.831	0.830	0.836	0.845	0.836	0.812	0.812	0.831	0.821			
38	0.820	0.827	0.825	0.832	0.841	0.831	0.808	0.808	0.827	0.817			
40	0.816	0.823	0.821	0.827	0.837	0.827	0.804	0.804	0.823	0.813			

* For references see p. 4-96.

TABLE 4d-11. V/V_0 OF ORGANIC COMPOUNDS* (Continued)

Pressure, kilobars	n-Propyl chloride [25]	n-Propyl iodide [25]	Propylene glycol [27]	Quinone [3]	Semi-carbazide hydro-chloride [5]	Styrene [24]	Succinic acid [6]	Sucrose [6]	Thymol [6]	Toluic acid [5] ortho-	meta-	para-
0	1.000	1.000	1.000	1.000	1.000	1.000	1.000	1.000	1.000	1.000	1.000	1.000
2	0.878	0.902	0.943	0.978	0.987	0.907	0.985	0.985	0.966	0.972	0.974	0.972
4	0.831	0.854	0.906	0.957	0.977	0.821 aa	0.973	0.972	0.942	0.950	0.953	0.950
6	0.798	0.824	0.877	0.931 y	0.967	0.797	0.960	0.961	0.922	0.932	0.936	0.932
8	0.771	0.800	0.855	0.915	0.959	0.779	0.948	0.950	0.905	0.918	0.922	0.916
10	0.750	0.780	0.836	0.900	0.946 z	0.764	0.936	0.940	0.890	0.905	0.910	0.902
12	0.734	0.762	0.820	0.886	0.939	0.752	0.925	0.932	0.877	0.894	0.899	0.890
14	.....	.....	.....	0.874	0.932	0.740	0.915	0.924	0.865	0.884	0.888	0.878
16	.....	.....	.....	0.862	0.925	0.731	0.905	0.916	0.855	0.874	0.878	0.868
18	.....	.....	.....	0.852	0.918	0.722	0.896	0.910	0.846	0.865	0.869	0.859
20	.....	.....	.....	0.842	0.912	0.714	0.888	0.903	0.838	0.856	0.860	0.851
22	.....	.....	.....	0.833	0.907	0.707	0.881	0.897	0.830	0.849	0.852	0.843
24	.....	.....	.....	0.825	0.901	0.700	0.874	0.893	0.823	0.841	0.845	0.835
26	.....	.....	.....	0.817	0.896	0.695	.....	0.883	0.817	0.834	0.838	0.829
28	.....	.....	.....	0.810	0.891	0.689	.....	0.854	0.811	0.828	0.832	0.822
30	.....	.....	.....	0.803	0.886	0.684	.....	0.850	0.806	0.822	0.826	0.817
32	.....	.....	.....	0.797	0.882	0.679	.....	.....	.....	0.816	0.820	0.810
34	.....	.....	.....	0.791	0.877	0.674	.....	.....	.....	0.810	0.815	0.805
36	.....	.....	.....	0.786	0.874	0.670	.....	.....	.....	0.806	0.810	0.800
38	.....	.....	.....	0.781	0.869	0.666	.....	.....	.....	0.802	0.806	0.795
40	.....	.....	.....	0.777	0.866	0.662	.....	.....	.....	0.797	0.801	0.790

* For references see p. 4-96.
y Transition at 4.4; volumes 0.952 and 0.943.
z Transition at 9.3; volumes 0.953 and 0.950.
aa Freezes at 3.1; volumes 0.884 and 0.835.

TABLE 4d-11. V/V_0 OF ORGANIC COMPOUNDS* (Continued)

Pressure, kilobars	Acetyl toluidine [5] ortho-	meta-	para-	Toluidine hydrochloride ortho-	meta-	para-	Nitro-urea [5]	Thiourea [5]	Urea nitrate [5]	n-Xylene [24]	o-Xylene [24]	p-Xylene [24]
0	1.000	1.000	1.000	1.000	1.000	1.000	1.000	1.000	1.000	1.000	1.000	1.000
2	0.974	0.972	0.974	0.982	0.972	0.969	0.984	0.982 cc	0.985	0.903	0.910 ee	ff 0.782
4	0.955	0.949	0.955	0.967	0.948	0.945	0.970 bb	0.945	0.971	0.856	0.812	0.760
6	0.937	0.930	0.937	0.954	0.930	0.927	0.906	0.931	0.958	0.824, 0.779 dd	0.792	0.754
8	0.925	0.913	0.925	0.943	0.913	0.913	0.895	0.918	0.947	0.798, 0.765	0.775	0.730
10	0.912	0.898	0.914	0.932	0.900	0.900	0.884	0.907	0.937	0.752	0.761	0.719
12	0.901	0.885	0.903	0.923	0.888	0.889	0.874	0.897	0.927	0.741	0.749	0.708
14	0.891	0.874	0.894	0.914	0.877	0.879	0.865	0.887	0.918	0.732	0.738	0.699
16	0.882	0.863	0.885	0.906	0.867	0.870	0.857	0.878	0.910	0.723	0.730	0.690
18	0.873	0.853	0.877	0.898	0.858	0.861	0.849	0.870	0.902	0.715	0.726	0.683
20	0.865	0.844	0.869	0.891	0.850	0.854	0.842	0.863	0.895	0.707	0.714	0.676
22	0.858	0.836	0.862	0.885	0.842	0.847	0.835	0.857	0.888	0.700	0.707	0.669
24	0.851	0.828	0.855	0.878	0.835	0.840	0.829	0.850	0.882	0.695	0.702	0.663
26	0.845	0.820	0.849	0.873	0.828	0.834	0.822	0.844	0.876	0.688	0.697	0.658
28	0.838	0.814	0.843	0.867	0.822	0.828	0.817	0.838	0.870	0.683	0.692	0.653
30	0.833	0.807	0.838	0.862	0.816	0.823	0.811	0.832	0.865	0.678	0.687	0.648
32	0.827	0.802	0.833	0.857	0.811	0.818	0.806	0.827	0.860	0.673	0.683	0.644
34	0.822	0.796	0.828	0.852	0.806	0.814	0.802	0.822	0.855	0.668	0.679	0.641
36	0.818	0.791	0.823	0.848	0.801	0.810	0.797	0.817	0.850	0.664	0.676	0.637
38	0.814	0.787	0.820	0.844	0.796	0.806	0.793	0.813	0.846	0.660	0.673	0.634
40	0.810	0.783	0.816	0.840	0.792	0.802	0.790	0.809	0.842	0.656	0.670	0.631

* For references see p. 4-96.
bb Transition at 5.4; volumes 0.961 and 0.909.
dd Freezes in this region; first figure for solid, second for liquid.
ff Freezes at 0.3; volumes 0.975 and 0.811.
cc Transition at 3.5; volumes 0.969 and 0.949.
ee Freezes at 2.3; volumes 0.902 and 0.831.

TABLE 4d-11. V/V_0 OF ORGANIC COMPOUNDS* (Continued)

Pressure, kilobars	Triacetin [27]	Tricaproin [26]	Triethanolamine [25]	Trimethylene glycol [27]	Tricresyl thiophosphate [5]		Urea [5]	Sodium xylene sulfonate [5]		
					ortho-	para-		ortho-	meta-	para-
0	1.000	1.000	1.000	1.000	1.000	1.000	1.000	1.000	1.000	1.000
2	0.937	0.929	0.952	0.943	0.973	0.970	0.982	0.982	0.979	0.986
4	0.900	0.885	.....	0.905	0.951	0.947	0.968	0.966	0.961	0.972
6	.....	0.855	.....	.....	0.933	0.927	ɑɑ 0.890	0.951	0.946	0.960
8	.....	0.832	.....	.....	0.919	0.911	0.878	0.938	0.933	0.959
10	.....	.....	.....	.....	0.906	0.897	0.867	0.927	0.921	0.949
12	.....	.....	.....	.....	0.895	0.885	0.856	0.916	0.910	0.930
14	.....	.....	.....	.....	0.885	0.873	0.847	0.906	0.900	0.922
16	.....	.....	.....	.....	0.875	0.862	0.838	0.897	0.892	0.913
18	.....	.....	.....	.....	0.866	0.852	0.830	0.889	0.883	0.907
20	.....	.....	.....	.....	0.858	0.843	0.823	0.881	0.871	0.900
22	.....	.....	.....	.....	0.850	0.835	0.815	0.874	0.869	0.894
24	.....	.....	.....	.....	0.843	0.827	0.809	0.867	0.862	0.888
26	.....	.....	.....	.....	0.837	0.320	0.803	0.862	0.857	0.882
28	.....	.....	.....	.....	0.830	0.813	0.797	0.855	0.851	0.877
30	.....	.....	.....	.....	0.824	0.307	0.792	0.850	0.845	0.872
32	.....	.....	.....	.....	0.818	0.302	0.787	0.845	0.841	0.868
34	.....	.....	.....	.....	0.813	0.797	0.782	0.839	0.837	0.864
36	.....	.....	.....	.....	0.808	0.792	0.777	0.835	0.832	0.860
38	.....	.....	.....	.....	0.803	0.788	0.773	0.831	0.829	0.856
40	.....	.....	.....	.....	0.799	0.784	0.769	0.827	0.825	0.852

* For references see p. 4-96.

ɑɑ Transition at 5.4; volumes 0.959 and 0.895.

References for Tables 4d-1 to 4d-11

1. Bridgman, P. W.: *Proc. Am. Acad. Arts Sci.* **59**, 173–211 (1924).
2. Bridgman, P. W.: *Proc. Am. Acad. Arts Sci.* **76**, 55–70 (1948).
3. Bridgman, P. W.: *Proc. Natl. Acad. Sci. U.S.* **21**, 109–113 (1935).
4. Swenson, C. A.: *Phys. Rev.* **99**, 423–430 (1955).
5. Bridgman, P. W.: *Proc. Am. Acad. Arts Sci.* **76**, 71–87 (1948).
6. Bridgman, P. W.: *Proc. Am. Acad. Arts Sci.* **76**, 9–24 (1945).
7. Bridgman, P. W.: *Proc. Am. Acad. Arts Sci.* **70**, 1–32 (1935).
8. Bridgman, P. W.: *Proc. Am. Acad. Arts. Sci.* **62**, 207–226 (1927).
9. Bridgman, P. W.: *Proc. Am. Acad. Arts Sci.* **76**, 1–7 (1945).
10. Bridgman, P. W.: *Proc. Am. Acad. Arts Sci.* **74**, 21–51 (1940).
11. Bridgman, P. W.: *Proc. Am. Acad. Arts Sci.* **58**, 165–242 (1923).
12. Bridgman, P. W.: *Phys. Rev.* **60**, 351–354 (1941).
13. Bridgman, P. W.: *Proc. Am. Acad. Arts Sci.* **77**, 189–234 (1949).
14. Bridgman, P. W.: *Proc. Am. Acad. Arts Sci.* **68**, 95–123 (1933).
15. Bridgman, P. W.: *Proc. Am. Acad. Arts Sci.* **60**, 385–421 (1925).
16. Bridgman, P. W.: *Proc. Am. Acad. Arts Sci.* **84**, 112–129 (1955).
17. Bridgman, P. W.: *Proc. Am. Acad. Arts Sci.* **83**, 3–21 (1954).
18. Bridgman, P. W.: *Proc. Am. Acad. Arts Sci.* **44**, 255–279 (1909).
19. Bridgman, P. W.: *Proc. Am. Acad. Arts Sci.* **47**, 347–438 (1911).
20. Bridgman, P. W.: *J. Appl. Phys.* **30**, 214–217 (1959).
21. Bridgman, P. W.: *Proc. Am. Acad. Arts Sci.* **76**, 89–99 (1948).
22. Bridgman, P. W.: *Proc. Am. Acad. Arts Sci.* **67**, 345–375 (1932).
23. Bridgman, P. W.: *Proc. Am. Acad. Arts Sci.* **49**, 3–114 (1913).
24. Bridgman, P. W.: *Proc. Am. Acad. Arts Sci.* **77**, 129–146 (1949).
25. Bridgman, P. W.: *Proc. Am. Acad. Arts Sci.* **68**, 1–25 (1933).
26. Bridgman, P. W.: *Proc. Am. Acad. Arts Sci.* **66**, 185–233 (1931).
27. Bridgman, P. W.: *Proc. Am. Acad. Arts Sci.* **67**, 1–27 (1932).
28. Bridgman, P. W.: *Proc. Am. Acad. Arts Sci.* **64**, 51–73 (1929).
29. Kennedy, George C., and William T. Holser: "Handbook of Physical Constants," *G.S.A. Mem.* **97**, 373–383, 1966.
30. Adams, L. H., and R. E. Gibson: *J. Wash. Acad. Sci.* **20**, 213 (1930).
31. Bridgman, P. W.: *Am. J. Sci.* **237**, 7–19 (1939).
32. Bridgman, P. W.: *Proc. Am. Acad. Arts Sci.* **48**, 309–362 (1912).
33. Bridgman, P. W.: *Proc. Am. Acad. Arts Sci.* **84**, 131–177 (1957).
34 Bridgman, P. W.: *Proc. Am. Acad. Arts Sci.* **84**, 43–109 (1955).
35. Bridgman, P. W.: *Proc. Am. Acad. Arts Sci.* **82**, 101–156 (1953).
36. Bridgman, P. W.: *Proc. Am. Acad. Arts Sci.* **83**, 151–190 (1954).
37. Bridgman, P. W.: *Proc. Am. Acad. Arts Sci.* **84**, 179–216 (1957).
38. Bridgman, P. W.: *Proc. Am. Acad. Arts Sci.* **84**, 1–42 (1955).

4d-2. High-pressure Compressibilities.[1] The high-pressure 25°C isotherms presented here were calculated from dynamic equation-of-state measurements made at Lawrence Radiation Laboratory, Livermore; Los Alamos Scientific Laboratory; Ballistics Research Laboratory; and in the Soviet Union. The shock-wave data were chosen on the basis of completeness, accuracy, and absence of effects which would tend to introduce large errors into the calculations. The isotherms were all calculated from experimental data compiled in the Compendium of Shock Wave Data [1].

In dynamic high-pressure experiments, a high-pressure shock wave is passed through the material under investigation. The one-dimensional mass and momentum conservation relationships [7]

$$V = V_0 \left(1 - \frac{U_p}{U_s}\right) \tag{4d-1}$$

and

$$P = P_0 + \rho_0 U_p U_s \tag{4d-2}$$

were used to calculate the pressure and specific volume behind the shock front from the experientally determined shock-wave velocity U_s and the bulk material velocity (particle velocity) behind the shock front U_p. Hugoniot curves were obtained from

[1] Work performed under the auspices of the U. S. Atomic Energy Commission.

a large number of experiments in which the shock strengths were varied. A *Hugoniot* is defined as the locus of all points that can be reached by shocking a material from a given initial state.

The conversion of the Hugoniot to the 25°C isotherm is done by means of the Grüneisen equation of state $P = P(V,E)$ in the form

$$P_H(V) - P = \gamma(V) \frac{E_H(V) - E}{V} \qquad (4d\text{-}3)$$

where the subscript H refers to conditions on the Hugoniot, and E_H is calculated from the conservation relationship

$$E_H(V) = E_0 + [P_H(V) + P_0] \frac{V_0 - V}{2} \qquad (4d\text{-}4)$$

$\gamma(V)$ is the Grüneisen gamma and is assumed to be a function of V only.

Since the 0 K isotherm and isentrope coincide,

$$E_{0\,\mathrm{K}} = - \int_{V_0}^{V} P \, dV \qquad (4d\text{-}5)$$

and $P_{0\,\mathrm{K}}(V)$ can be calculated as soon as $\gamma(V)$ is known. Several models have been proposed for relating $\gamma(V)$ to the curvature of the 0 K isotherm. Some of them are contained in the formula

$$\gamma(V) = \frac{t - 2}{3} - \frac{V}{2} \frac{(d^2/dV^2)(PV^{2t/3})_{0\,\mathrm{K}}}{(d/dV)(PV^{2t/3})} \qquad (4d\text{-}6)$$

When $t = 0$, a formula derived by Slater [2] is given; $t = 1$ yields a formula proposed by Dugdale and MacDonald [3] and rederived by Rice [4] et al.; and $t = 2$ gives a relationship derived by Zubarev and Vashchenko [5]. Rice et al. have shown that the Dugdale-MacDonald form gives results that are in agreement with thermodynamic data on metals. The Dugdale-MacDonald form was used to calculate the isotherms given here.

Once the 0 K curve and $\gamma(V)$ are calculated, the 25°C isotherm is obtained by adding to the 0 K isotherm the correction obtained from the Grüneisen equation:

$$\begin{aligned} \Delta P(V) &= \gamma(V) \frac{E_{25°\mathrm{C}}(V) - E_{0\,\mathrm{K}}(V)}{V} \\ &= \gamma(V) \int_{0\,\mathrm{K}}^{25°\mathrm{C}} C_V(V) \frac{dT}{V} \\ &\approx \gamma(V) \frac{E_0}{V} \end{aligned} \qquad (4d\text{-}7)$$

Hugoniot measurements cannot be used indiscriminately for generating hydrostatic isotherms for comparison with static high-pressure work. Careful evaluations of the assumptions and possible sources of error should be made. First, it is important to note that the Grüneisen equation as it is normally derived is based on a model of the crystalline solid state, and its application to a Hugoniot is consistent with this derivation only if the material remains in the same solid phase all along the Hugoniot; i.e., if no phase transitions are encountered on the Hugoniot. The Grüneisen gamma is assumed to be a function of volume only, and this is a good approximation for temperatures at and above the Debye temperature. In addition, the presence of effects of finite yield strength can cause the measured Hugoniot to be offset above the hydrostatic Hugoniot; i.e., the longitudinal stress measured in a shock-wave experiment is not identical with the hydrostatic pressure.

The materials for which isotherms were calculated have a relatively low yield strength so that corrections for finite yield strength are low and can be neglected, with the exception of Al_2O_3. For this material, corrections were applied, and so the Hugoniot data used as inputs were essentially hydrostatic. The temperatures involved are of the order of magnitude or well above the Debye temperatures of the materials. Therefore, the assumption that the Grüneisen gamma is a function of volume only is valid, at least in the solid phase as presented in Table 4d-12. The only part of these calculations that merits serious scrutiny is the use of the model in the situation where melting occurs along the Hugoniot, as it does with the alkali metals at shock pressures less than 100 kilobars. For most materials, the Hugoniot is characterized by a linear relationship between shock and particle velocity. This behavior is characteristic of materials that do not experience a phase transition along the Hugoniot (except for liquids at very low pressures) and is true of almost all metals. Extrapolation of shock velocity versus particle velocity to zero pressure ($U_p = 0$), yields a value of U_s within 5 per cent of that calculated from the ordinary elastic constants for the solid metals at 1 atm. This indicates: either (1) that the change in volume and enthalpy on melting is negligible at high pressures, or (2) that the effect of the change in volume cancels the effect of the change in enthalpy in determining the Hugoniot. At high pressures, experimental evidence indicates that both alternatives are true to some extent. A careful comparison of the Bridgman [6] data and the Hugoniot data reveals no systematic differences that can be attributed to melting on the Hugoniot. Thus, the error in assuming that the isotherm derived from the experimental Hugoniot represents the solid is certainly no worse than ±5 per cent and probably far less.

Although the values of Grüneisen gamma calculated by the three models mentioned previously differ successively from one another by one-third at zero pressure, the correction to the Hugoniot at low pressures is small, and the uncertainty in gamma does not affect the calculated 0 K curve. At high pressure, the correction is important, but here gamma is reasonably well known. In fact, the values of the three gammas differ by less than 10 per cent in the high-pressure range. Thus, the 0 K curve is probably calculated to ±5 per cent in pressure in the range of interest.

The experimental measurements are probably precise to ±2 per cent. Consideration of the errors mentioned previously leads to the conclusion that the calculated 25°C isotherms are probably accurate to ±5 per cent in pressure, and certainly better than ±10 per cent.

References for Sec. 4d-2

1. van Thiel, M.: Compendium of Shock Wave Data, *Univ. Calif., Lawrence Radiation Lab.* (Livermore) *Rept.* UCRL-50108, 1966.
2. Slater, J. C.: "Introduction to Chemical Physics," chaps. 13 and 14, McGraw-Hill Book Company, New York, 1939.
3. Dugdale, J. S., and D. MacDonald: *Phys. Rev.* **89**, 832 (1953).
4. Rice, M. H., *et al.*: *Solid State Phys.* **6**, 1 (1958).
5. Zubarev, V. N., and V. Ya. Vashchenko: *Fiz. Tverd. Tela* **5**, 886 (1963); *Soviet Phys.— Solid State* **5**, 653 (1963).
6. Bridgman, P. W.: *Phys. Rev.* **46**, 930 (1934).
Other documents include:
7. Duvall, G. E. and G. R. Fowles: "High Pressure Chemistry and Physics," R. S. Bradley, ed., vol. 2, p. 209, Academic Press Inc., London, 1963.
8. Al'tshuler, L. V.: *Uspekhi Fiz. Nauk* **85**, 197 (1965) [English transl.: *Soviet Phys.— Usp.* **8**, 52 (1965)].
9. Skidmore, I. C.: *Appl. Mater. Research* **4**, 131 (1965).
10. Hamann, S. D.: "Advances in High Pressure Research," R. S. Bradley, ed., vol 1, p. 85, Academic Press, Inc., London, 1966.

TABLE 4d-12. RELATIVE VOLUMES OF SOLIDS AT 25°C

P, kilobars	Li	Be	Na	Mg	Al	K	Ca	Ti	V
5	0.958	....	0.931	....	....	0.878	0.975		
10	0.922	0.992	0.878	0.973	0.987	0.802	0.953	0.990	0.994
15	0.891	0.988	0.837	0.961	0.981	0.747	0.932	0.985	0.991
20	0.864	0.984	0.803	0.949	0.976	0.704	0.912	0.981	0.988
25	0.841	0.980	0.775	0.938	0.970	0.669	0.894	0.976	0.985
30	0.819	0.976	0.750	0.928	0.964	0.640	0.877	0.971	0.982
35	0.800	0.973	0.728	0.918	0.959	0.615	0.861	0.967	0.979
40	0.783	0.969	0.708	0.909	0.954	0.593	0.846	0.963	0.976
45	0.766	0.965	0.691	0.900	0.949	0.574	0.832	0.958	0.973
50	0.751	0.962	0.675	0.891	0.944	0.557	0.819	0.954	0.971
60	0.725	0.955	0.647	0.875	0.935	0.527	0.794	0.946	0.965
70	0.701	0.948	0.622	0.860	0.926	0.502	0.771	0.938	0.960
80	0.680	0.942	0.601	0.847	0.918	0.480	0.750	0.931	0.955
90	0.661	0.935	0.583	0.834	0.910	0.461	0.731	0.924	0.950
100	0.644	0.929	0.566	0.822	0.902	0.445	0.713	0.917	0.945
120	0.614	0.918	0.537	0.800	0.888	0.417	0.680	0.903	0.935
140	0.588	0.906	0.513	0.780	0.875	0.393	0.652	0.891	0.926
160	0.566	0.896	0.492	0.762	0.862	....	0.626	0.879	0.917
180	0.546	0.886	0.474	0.746	0.851	....	0.604	0.867	0.909
200	0.528	0.876	0.458	0.731	0.840	....	0.583	0.857	0.901
220	....	0.867	....	0.717	0.830	....	0.564	0.846	0.893
240	....	0.858	....	0.705	0.820	....	0.547	0.837	0.886
260	....	0.850	....	0.693	0.811	....	0.530	0.827	0.879
280	....	0.841	....	0.681	0.802	....	0.516	0.818	0.872
300	....	0.834	....	0.671	0.794	....	0.502	0.810	0.865
320	....	0.826	....	0.661	0.786	....	0.489	0.802	0.859
340	....	0.819	....	0.652	0.778	....	0.476	0.794	0.853
360	....	0.812	....	0.643	0.771	....	0.465	0.786	0.846
380	....	0.805	....	0.634	0.764	....	....	0.779	0.841
400	....	0.798	....	0.626	0.757	....	....	0.772	0.835
420	....	0.792	....	0.619	0.751	....	....	0.765	0.829
440	....	0.786	....	0.612	0.744	....	....	0.758	0.824
460	....	0.780	....	0.605	0.738	....	....	0.752	0.819
480	....	0.774	....	0.598	0.732	....	....	0.746	0.813
500	....	0.768	....	0.591	0.726	....	....	0.740	0.808
550	....	0.755	....	0.576	0.713	....	....	0.725	0.796
600	....	0.742	....	....	0.700	....	....	0.712	0.785
650	....	0.730	....	....	0.688	....	....	0.700	0.774
700	....	0.719	....	....	0.677	....	....	0.688	0.764
750	....	0.708	....	....	0.667	....	....	0.677	0.755
800	....	0.698	....	....	0.657	....	....	0.666	0.746
850	....	....	....	....	....	....	....	0.657	0.737
900	....	....	....	....	....	....	....	0.647	0.729
950	....	....	....	....	....	....	....	0.638	0.721
1,000	....	....	....	....	....	....	....	0.630	0.713
1,200	....	....	....	....	....	....	....	0.599	
1,400	....	....	....	....	....	....	....	0.573	
1,600	....	....	....	....	....	....	....	0.550	
1,800	....	....	....	....	....	....	....	0.530	
2,000	....	....	....	....	....	....	....	0.512	

P, kilobars	Cr	Co	Ni	Cu	Zn	Rb	Zr	Nb	Mo
5	.....	.....	.....	.....	.....	0.838	0.995		
10	.....	.....	.....	.....	.....	0.750	0.990	0.994	0.996
15	0.992	0.993	0.992	09900	.....	0.691	0.985	0.991	0.994
20	0.990	0.990	0.990	0.986	0.970	9.646	0.980	0.989	0.993
25	0.987	0.988	0.987	0.983	0.964	0.611	0.975	0.986	0.991
30	0.985	0.985	0.985	0.980	0.957	0.582	0.970	0.983	0.989
35	0.983	0.983	0.982	0.977	0.951	0.557	0.965	0.980	0.987
40	0.980	0.981	0.980	0.974	0.945	0.536	0.961	0.978	0.986
45	0.978	0.978	0.978	0.971	0.939	0.517	0.956	0.975	0.984
50	0.976	0.976	0.975	0.968	0.934	0.501	0.952	0.972	0.982
60	0.967	0.972	0.971	0.962	0.924	0.472	0.943	0.957	0.979
70	0.967	0.967	0.966	0.956	0.914	0.449	0.935	0.962	0.975
80	0.963	0.963	0.962	0.951	0.905	0.429	0.927	0.958	0.972
90	0.958	0.959	0.958	0.945	0.896	0.411	0.919	0.953	0.969
100	0.954	0.955	0.954	0.940	0.888	0.396	0.911	0.948	0.966
120	0.947	0.947	0.946	0.930	0.873	0.370	0.897	0.939	0.960
140	0.939	0.940	0.938	0.921	0.859	0.349	0.883	0.931	0.954
160	0.932	0.932	0.931	0.912	0.847	.....	0.870	0.922	0.948
180	0.925	0.925	0.924	0.904	0.835	.....	0.857	0.915	0.942
200	0.918	0.919	0.917	0.896	0.825	.....	0.846	0.907	0.937
220	0.912	0.912	0.911	0.889	0.815	.....	0.834	0.900	0.931
240	0.906	0.906	0.904	0.881	0.805	.....	0.824	0.893	0.926
260	0.900	0.900	0.898	0.874	0.797	.....	0.813	0.886	0.921
280	0.894	0.894	0.893	0.868	0.788	.....	0.803	0.879	0.916
300	0.889	0.888	0.887	0.861	0.781	.....	0.794	0.873	0.911
320	0.883	0.883	0.881	0.855	0.773	.....	0.785	0.866	0.906
340	0.878	0.877	0.876	0.849	0.766	.....	0.776	0.866	0.902
360	0.873	0.872	0.871	0.843	0.760	.....	0.767	0.855	0.897
380	0.868	0.867	0.866	0.838	0.753	.....	0.759	0.849	0.893
400	0.864	0.862	0.861	0.832	0.747	.....	0.751	0.845	0.889
420	0.859	0.857	0.857	0.827	0.741	.....	0.744	0.838	0.884
440	0.854	0.853	0.852	0.822	0.736	.....	0.736	0.833	0.880
460	0.850	0.848	0.848	0.817	0.730	.....	0.729	0.823	0.876
480	0.846	0.844	0.843	0.812	0.725	.....	0.722	0.823	0.872
500	0.842	0.839	0.839	0.808	0.720	.....	0.715	0.818	0.868
550	0.832	0.829	0.829	0.797	0.708	.....	0.699	0.806	0.859
600	0.822	0.819	0.820	0.786	0.698	.....	0.684	0.795	0.850
650	0.813	0.810	0.811	0.777	0.688	.....	0.670	0.785	0.841
700	0.805	0.801	0.802	0.768	0.679	.....	0.657	0.775	0.833
750	0.797	0.792	0.794	0.759	0.670	.....	0.644	0.766	0.825
800	0.789	0.784	0.786	0.751	0.662	.....	0.632	0.757	0.818
850	0.782	0.777	0.779	0.743	0.654	.....	0.621	0.748	0.811
900	0.775	0.769	0.772	0.736	0.647	.....	0.611	0.740	0.804
950	0.769	0.762	0.765	0.729	0.640	.....	0.600	0.732	0.797
1,000	0.762	0.755	0.758	0.722	0.634	.....	0.591	0.725	0.790
1,200	0.739	0.730	0.735	0.697	0.611	.....	0.556	.....	0.766
1,400	.....	.....	.....	0.677	0.592	.....	0.527	.....	0.745
1,600	.....	.....	.....	0.658	0.576	.....	.....	.....	0.726
1,800	.....	.....	.....	0.642	0.561	.....	.....	.....	0.709
2,000	.....	.....	.....	0.627	0.548	.....	.....	.....	0.693
2,500	.....	.....	.....	0.596	0.521	.....	.....	.....	0.659
3,000	.....	.....	.....	0.571	.....	.....	.....	.....	0.631
3,500	.....	.....	.....	0.550	.....	.....	.....	.....	0.606
4,000	.....	.....	.....	0.532					
4,500	.....	.....	.....	0.516					

P, kilobars	Pd	Ag	Cd	In	Sn	Ta	Pt	Au
10	.....	.....	.....	0.977	0.979	0.995		
15	0.993	0.987	0.973	0.966	0.969	0.993	0.995	0.992
20	0.990	0.982	0.965	0.957	0.960	0.990	0.993	0.990
25	0.988	0.978	0.957	0.947	0.951	0.988	0.991	0.987
30	0.985	0.974	0.950	0.938	0.943	0.985	0.990	0.985
35	0.983	0.970	0.943	0.930	0.935	0.983	0.988	0.982
40	0.981	0.966	0.936	0.922	0.928	0.981	0.986	0.980
45	0.978	0.963	0.930	0.915	0.920	0.978	0.985	0.977
50	0.976	0.959	0.924	0.907	0.914	0.976	0.983	0.975
60	0.972	0.952	0.912	0.894	0.901	0.972	0.980	0.970
70	0.968	0.945	0.901	0.882	0.889	0.967	0.977	0.966
80	0.963	0.939	0.891	0.870	0.878	0.963	0.974	0.962
90	0.959	0.932	0.882	0.859	0.868	0.959	0.971	0.957
100	0.955	0.926	0.873	0.840	0.858	0.955	0.968	0.953
120	0.948	0.915	0.857	0.831	0.841	0.947	0.962	0.945
140	0.941	0.905	0.843	0.814	0.825	0.939	0.956	0.938
160	0.934	0.895	0.829	0.800	0.811	0.932	0.951	0.930
180	0.927	0.886	0.817	0.786	0.798	0.924	0.946	0.923
200	0.921	0.877	0.806	0.774	0.786	0.917	0.941	0.917
220	0.914	0.869	0.796	0.762	0.775	0.911	0.936	0.910
240	0.908	0.861	0.786	0.752	0.764	0.904	0.931	0.904
260	0.903	0.854	0.777	0.742	0.755	0.898	0.927	0.898
280	0.897	0.847	0.768	0.732	0.746	0.892	0.922	0.893
300	0.892	0.840	0.760	0.724	0.737	0.886	0.918	0.887
320	0.887	0.834	0.753	0.715	0.729	0.880	0.913	0.882
340	0.882	0.828	0.746	0.708	0.722	0.875	0.909	0.877
360	0.877	0.822	0.739	0.700	0.715	0.869	0.905	0.871
380	0.872	0.816	0.732	0.693	0.708	0.864	0.901	0.867
400	0.867	0.811	0.726	0.687	0.701	0.859	0.898	0.862
420	0.863	0.806	0.720	0.680	0.695	0.854	0.894	0.857
440	0.859	0.800	0.715	0.674	0.689	0.849	0.890	0.853
460	0.854	0.796	0.709	0.668	0.684	0.844	0.886	0.846
480	0.850	0.791	0.704	0.663	0.678	0.839	0.883	0.844
500	0.846	0.786	0.699	0.657	0.673	0.835	0.880	0.840
550	0.837	0.775	0.687	0.645	0.661	0.823	0.871	0.830
600	0.828	0.765	0.676	0.633	0.649	0.813	0.863	0.821
650	0.819	0.756	0.666	0.623	.....	0.803	0.856	0.812
700	0.811	0.747	0.657	0.613	.....	0.794	0.849	0.804
750	0.803	0.739	0.648	0.604	.....	0.785	0.842	0.796
800	0.796	0.731	0.640	0.595	.....	0.776	0.835	0.789
850	0.789	0.724	0.633	0.587	.....	0.768	0.829	0.781
900	0.782	0.717	0.626	0.580	.....	0.760	0.823	0.775
950	0.776	0.710	0.619	.....	.....	0.752	0.817	0.768
1,000	0.770	0.704	0.613	.....	.....	0.745	0.811	0.762
1,200	0.747	0.681	.....	.....	.....	0.718	0.791	0.739
1,400	0.728	0.662	.....	.....	.....	0.694	0.773	0.719
1,600	0.711	0.645	.....	.....	.....	0.674	0.756	0.702
1,800	.....	0.630	.....	.....	.....	0.655	0.742	0.686
2,000	.....	0.617	.....	.....	.....	.....	0.728	

TABLE 4d-12. RELATIVE VOLUMES OF SOLIDS AT 25°C (*Continued*)

P, kilobars	Tl	Pb	Th	LiF	LiCl	LiBr	LiI	NaF
5	.....	.....	.....	.....	.....	.....	0.985	
10	0.974	0.979	0.982	.....	0.972	0.960	0.971	0.980
15	0.963	0.969	0.973	0.978	0.959	0.942	0.958	0.970
20	0.952	0.960	0.965	0.971	0.947	0.926	0.945	0.961
25	0.942	0.952	0.957	0.964	0.935	0.912	0.933	0.953
30	0.933	0.943	0.950	0.958	0.925	0.898	0.921	0.944
35	0.924	0.936	0.942	0.952	0.914	0.886	0.910	0.936
40	0.915	0.928	0.935	0.946	0.905	0.874	0.899	0.929
45	0.908	0.921	0.929	0.940	0.896	0.863	0.888	0.921
50	0.900	0.914	0.922	0.934	0.887	0.853	0.878	0.914
60	0.886	0.901	0.910	0.924	0.870	0.834	0.859	0.900
70	0.873	0.890	0.898	0.914	0.855	0.817	0.842	0.888
80	0.861	0.878	0.887	0.904	0.841	0.801	0.825	0.876
90	0.850	0.868	0.877	0.895	0.828	0.787	0.809	0.865
100	0.840	0.858	0.867	0.887	0.816	0.773	0.795	0.854
120	0.822	0.841	0.848	0.871	0.794	0.750	0.767	0.834
140	0.805	0.825	0.832	0.857	0.774	0.729	0.743	0.816
160	0.790	0.810	0.816	0.843	0.756	0.711	0.720	
180	0.777	0.797	0.802	0.831	0.740	0.694	0.699	
200	0.765	0.785	0.789	0.820	0.725	0.679	0.680	
220	0.753	0.773	0.777	0.809	0.712	0.666	0.662	
240	0.743	0.763	0.765	0.799	.....	0.653	0.646	
260	0.733	0.753	0.754	0.789	.....	.....	0.631	
280	0.724	0.744	0.744	0.780	.....	.....	0.616	
300	0.715	0.735	0.734	0.772				
320	0.707	0.727	0.725	0.764				
340	0.700	0.719	0.716	0.756				
360	.....	0.712	0.708	0.749				
380	.....	0.705	0.700	0.742				
400	.....	0.698	0.693	0.735				
420	.....	0.692	0.685	0.729				
440	.....	0.686	0.678	0.722				
460	.....	0.680	0.672	0.716				
480	.....	0.674	0.665	0.711				
500	.....	0.669	0.659	0.705				
550	.....	0.656	0.644	0.692				
600	.....	0.645	0.631	0.680				
650	.....	0.634	0.619	0.669				
700	.....	0.624	0.607	0.659				
750	.....	0.615	0.596	0.649				
800	.....	.....	0.586	0.640				
850	.....	.....	0.577					
900	.....	.....	0.568					
950	.....	.....	0.560					
1,000	.....	.....	0.552					

TABLE 4d-12. RELATIVE VOLUMES OF SOLIDS AT 25°C (*Continued*)

P, kilobars	NaCl	NaBr	NaI	KF	KI	RbF	RbCl	RbBr
5	.....	.....	0.976	.....	0.954	.....	.....	0.946
10	0.963	0.958	0.955	0.936	0.916	0.944	0.886	0.903
15	0.946	0.939	0.936	0.910	0.886	0.921	0.849	0.870
20	0.932	0.923	0.918	0.888	0.859	0.901	0.820	0.842
25	0.918	0.907	0.902	0.869	0.837	0.883	0.796	0.818
30	0.905	0.893	0.887	0.852	0.816	0.867	0.775	0.797
35	0.894	0.880	0.873	0.837	0.798	0.852	0.757	0.779
40	0.883	0.868	0.860	0.823	0.782	0.838	0.741	0.762
45	0.873	0.856	0.848	0.811	0.767	0.826	0.727	0.748
50	0.863	0.846	0.836	0.800	0.753	0.814	0.713	0.734
60	0.845	0.826	0.815	0.779	0.729	0.793	0.690	0.710
70	0.829	0.808	0.796	0.761	0.707	0.774	0.671	0.689
80	0.815	0.792	0.778	0.746	0.689	0.757	0.654	0.671
90	0.802	0.777	0.762	0.731	0.672	0.741	0.639	0.655
100	0.789	0.763	0.748	0.719	0.657	0.728	0.625	0.640
120	0.767	0.738	0.721	0.696	0.630	0.703	0.602	0.615
140	0.748	0.717	0.698	0.677	0.608	0.681	.....	0.594
160	0.731	0.698	0.678	0.660	0.588	0.663	.....	0.575
180	0.715	0.680	0.659	0.646	0.571	0.646		
200	0.701	0.665	0.643	0.633	.....	0.631		
220	.....	0.651	0.627	0.621	.....	0.618		
240	.....	0.638	0.613	0.610	.....	0.605		

TABLE 4d-12. RELATIVE VOLUMES OF SOLIDS AT 25°C (*Continued*)

P, kilobars	RbI	CsCl	CsBr	CSI	H_2O	MgO	Al_2O_3	Brass
5	0.954	.....	0.979	0.966				
10	0.917	0.952	0.959	0.937	.....	.....	0.996	
15	0.886	0.932	0.941	0.913	.....	0.990	0.994	
20	0.860	0.914	0.925	0.892	0.737	0.987	0.992	0.984
25	0.838	0.898	0.910	0.873	0.707	0.984	0.990	0.980
30	0.817	0.884	0.896	0.855	0.684	0.981	0.988	0.976
35	0.799	0.870	0.883	0.840	0.664	0.978	0.987	0.973
40	0.783	0.858	0.871	0.826	0.647	0.975	0.985	0.969
45	0.768	0.846	0.859	0.813	0.632	0.973	0.983	0.965
50	0.754	0.836	0.848	0.801	0.618	0.970	0.981	0.962
60	0.730	0.816	0.828	0.779	0.595	0.964	0.977	0.955
70	0.709	0.799	0.810	0.760	.....	0.959	0.974	0.949
80	0.690	0.783	0.794	0.743	.....	0.954	0.970	0.942
90	0.673	0.768	0.778	0.727	.....	0.949	0.967	0.936
100	0.658	0.755	0.764	0.713	.....	0.944	0.964	0.931
120	0.631	0.732	0.739	.....	.....	0.935	0.957	0.919
140	0.609	0.712	0.717	.....	.....	0.927	0.951	0.909
160	0.589	0.694	0.697	.....	.....	0.919	0.945	0.899
180	0.572	0.678	0.680	.....	.....	0.911	0.939	0.890
200	.....	0.664	0.663	.....	.....	0.904	0.933	0.881
220	.....	0.651	0.649	.....	.....	0.897	0.927	0.873
240	.....	.....	0.635	.....	.....	0.890	0.922	0.865
260	.....	.....	0.623	.....	.....	0.884	0.916	0.857
280	.....	.....	0.611	.....	.....	0.878	0.911	0.850
300	.....	.....	0.600	.....	.....	0.872	0.906	0.843
320	.....	.....	0.590	.....	.....	0.867	0.901	0.836
340	.....	.....	0.581	.....	.....	0.861	0.896	0.830
360	.....	.....	0.572	.....	.....	0.856	0.892	0.823
380	.....	.....	0.563	.....	.....	0.851	0.887	0.817
400	.....	.....	0.555	.....	.....	0.846	0.883	0.812
420	.....	.....	0.547	.....	.....	0.841	0.878	0.806
440	.....	.....	0.540	.....	.....	0.837	0.874	0.800
460	.....	.....	0.533	.....	.....	0.832	0.870	0.795
480	.....	.....	0.526	.....	.....	0.828	0.866	0.790
500	.....	.....	0.520	.....	.....	0.824	0.862	0.785
550	.....	.....	0.505	.....	.....	0.814	0.852	0.774
600	.....	.....	.....	.....	.....	0.805	0.843	0.763
650	.....	.....	.....	.....	.....	0.796	0.834	0.752
700	.....	.....	.....	.....	.....	0.788	0.825	0.743
750	.....	.....	.....	.....	.....	0.780	0.817	0.734
800	.....	.....	.....	.....	.....	0.772	0.809	0.725
850	.....	.....	.....	.....	.....	0.765	0.802	0.717
900	.....	.....	.....	.....	.....	0.759	0.795	
950	.....	.....	.....	.....	.....	.....	0.788	
1,000	.....	.....	.....	.....	.....	.....	0.781	
1,200	.....	.....	.....	.....	.....	.....	0.757	

4e. Heat Capacities

GEORGE T. FURUKAWA AND THOMAS B. DOUGLAS

The National Bureau of Standards

NORMAN PEARLMAN

Purdue University

In both Tables 4e-1 and 4e-2, temperatures are given in kelvins (K).[1] The formula weights accompanying the symbols of the elements are based on the International Atomic Weights of 1961 (C^{12} = 12.0000). The heat capacity is given in calories per kelvin per gram-formula-weight (1 cal = 4.1840 joules). Except for separate listings for allotropic modifications of a few elements, the heat capacity is given only for the physical state in which the element is stable at one atmosphere pressure and at the temperature in question. (For condensed gases of Table 4e-1, the values are given for the saturation pressures.) This state of the element (crystalline, liquid, or gaseous) is indicated in parentheses by the appropriate letters, except that when two or more crystalline forms of an element are known, these are distinguished as α, β, γ, etc. In Table 4e-1 the values given are for the crystalline phase unless indicated otherwise; changes in phase are identified. The asterisks (*) indicate that the values given are in the region of sharp transitions or "heat effects," and the parentheses enclose values that were obtained by extrapolation or interpolation over a broad temperature range. These values may contain large errors. No attempt was made to resolve the discrepancies in the existing data for the few elements where Tables 4e-1 and 4e-2 disagree at the temperature 298.15 K.

In the preparation of Table 4e-1, the results of an unpublished critical analysis of elemental substances being conducted as a part of the National Standard Reference Data System have largely been used. Wherever data more recent than the above analysis were known to have been published, they were examined for any major changes. When the authors tabulated smoothed values at even temperatures, their values were freely used with minor adjustments whenever needed. The following compilations were also used in the preparation of the table: K. K. Kelley, Contributions to the Data on Theoretical Metallurgy: XIV, Entropies of the Elements and Inorganic Compounds, *U.S. Bur. Mines Bull.* 592, 1961; and R. Hultgren, R. L. Orr, P. D. Anderson, and K. K. Kelley, "Selected Values of Thermodynamic Properties of Metals and Alloys," John Wiley & Sons, Inc., New York, 1963.

In Table 4e-2, the values cover, in general, only the temperature range of experimental measurements (or of statistical-thermodynamic calculation in the case of the

---

[1] The name of the unit of thermodynamic temperature was changed from degree Kelvin (symbol: °K) to kelvin (symbol: K); and kelvin is now defined as the fraction 1/273.16 of the thermodynamic temperature of the triple point of water [*NBS Tech. News Bull.* **52**, 10 (1968)]. The temperatures and temperature intervals used in the experimental measurements of heat capacity are taken to be consistent with the above definition within the accuracy of the values of heat capacity given in Table 4e-1 and 4e-2.

TABLE 4e-1. MOLAR HEAT CAPACITY AT CONSTANT PRESSURE OF ELEMENTAL SUBSTANCES AT LOW TEMPERATURES, CAL/MOLE·K

As revised December, 1967

Element	10	15	20	25	30	50	70	100	150	200	250	298.15
Aluminum Al 26.9815	0.0098	0.026	0.054	0.11	0.20	0.91	1.85	3.12	4.43	5.16	5.56	5.82
Antimony Sb 121.75	0.10	0.36	0.74	1.17	1.59	3.09	4.10	4.92	5.55	5.82	5.95	6.03
Argon A 39.948	0.90	1.89	2.82	3.70	4.39	5.93	6.96(c)	4.97(g)	4.97	4.97	4.97	4.97
Arsenic As 74.9216	(0.03)	0.12	0.27	0.50	0.78	1.90	2.88	3.98	4.94	5.43	5.75	5.89
Barium Ba 137.34	0.44	1.11	1.90									(6.30)
Beryllium Be 9.0122	0.0006	0.002	0.003	0.006	0.009	0.04	0.12	0.43	1.36	2.41	3.30	3.93
Bismuth Bi 208.980	0.51	1.21	1.80	2.32	2.86	4.29	4.99	5.52	5.85	5.99	6.11	6.19
Boron (crystalline) B 10.811			0.006	0.009	0.01	0.02	0.08	0.26	0.77	1.45	2.11	2.65
Boron (amorphous) B 10.811			0.02	0.026	0.071	0.16	0.41	0.33	0.86	1.55	2.22	2.86
Bromine Br₂ 159.818	(0.55)		3.04	4.30	5.36	7.97	9.23	10.42	11.75	12.85	14.16(c)	18.08(l)
Cadmium Cd 112.40	0.22	0.69	1.23	1.79	2.30	3.80	4.64	5.28	5.74	5.94	6.08	6.22
Calcium Ca 40.08	0.05	0.15	0.35	0.62	0.97	2.60	3.64	4.66	5.49	5.94	6.08	6.30
Carbon (graphite) C 12.01115	0.004	0.01	0.02	0.03	0.04	0.12	0.23	0.40	0.77	1.18	1.63	2.04
Carbon (diamond) C 12.01115	0.0001	0.0002	0.0003	0.0007	0.001	0.005	0.016	0.059	0.24	0.56	0.99	1.46
Cerium Ce 140.12	1.05*	1.14*	1.76	2.46	3.08	3.10	5.83	6.46	6.71*	(6.90)		(6.44)
Cesium Cs 132.905	2.64	3.91	4.67	5.08	5.36	5.80	5.97	6.16	6.41*	6.59*	6.94	7.67*
Chlorine Cl₂ 70.906		0.89	1.85	2.89	3.99	6.99	8.68	10.10	12.20(c)	15.95(l)	15.95(l)	8.11(g)
Chromium Cr 51.996	0.006	0.014	0.026	0.054	0.095	0.45	1.20	2.39	3.94	4.81	5.30	5.56
Cobalt Co 58.9332	0.013	0.037	0.071	0.14	0.24	0.98	2.05	3.34	4.65	5.33	5.75	5.95
Copper Cu 63.54	0.018	0.044	0.110	0.230	0.405	1.47	2.60	3.82*	4.90	5.41	5.68	5.84
Dysprosium Dy 162.50	(0.18)	0.60	1.34	2.20	3.04	5.52	6.97*	8.32*	10.88*	6.96	6.73	6.73
Erbium Er 167.26	(0.47)	1.60	2.38	3.73	4.60	6.78*	7.35*	5.87	6.59	6.46	6.59	6.72
Europium Eu 151.96		(0.47)										(6.48)
Fluorine F₂ 37.9968	0.93	1.75	3.10	4.61	6.03(cz)	11.79(cz)	13.56(l)	6.96(g)	6.99	7.10	7.28	7.49
Gadolinium Gd 157.25	0.18	0.46	1.06	1.77	2.44	4.64	5.40	6.91	7.81	8.64	9.95*	8.86*
Gallium Ga 69.72	0.058	0.25	0.54	0.84	1.19	2.45	3.47	4.43	5.26	5.69	5.95	6.24
Germanium Ge 72.59	0.014	0.077	0.22	0.41	0.63	1.48	2.27	3.30	4.44	5.03	5.37	5.59
Gold Au 196.967	0.103	0.352	0.762	1.25	1.76	3.41	4.38	5.12	5.64	5.84	5.97	6.08
Hafnium Hf 178.49	0.04	0.16	0.41	0.77	1.20	2.90	4.00	4.92	5.55	5.83	6.01	6.15
Helium He 4.0026	4.97(g)	4.97	4.97	4.97	4.97	4.97	4.97	4.97	4.97	4.97	4.97	4.97
Holmium Ho 164.930	0.64	1.57*	2.29*	2.98	3.67	5.86	7.41	9.39	6.34	6.33	6.42	6.49
n-Hydrogen H₂ 2.01594	0.47(c)	3.32(l)	4.55	6.22	10.10(l)	4.98(g)	5.06	5.53	6.07	6.52	6.77	6.89

(between n-Hydrogen and n-Deuterium: ← 99.8% para →   ← 97.8% ortho →)

Element	10	15	20	25	30	50	70	100	150	200	250	298.15
n-Deuterium D₂ 4.02820	0.54(c)	1.64(c)	5.49(l)		5.06(g)	5.95	6.89	7.19	7.03	6.98	6.98	6.98
Indium In 114.82	0.43	1.01	1.67	2.36	2.94	4.41	5.08	5.58	5.99	6.17	6.31	6.38
Iodine I₂ 253.8088	0.96	2.45	3.87	5.14	6.16	8.57	9.96	10.96	11.86	12.32	12.73	13.01
Iridium Ir 192.2	(0.014)	0.038	0.094	0.22	0.43	1.75	2.98	4.15	5.17	5.58	5.87	(6.12)
Iron Fe 55.847	0.017	0.034	0.061	0.11	0.18	0.73	1.61	2.88	4.33	5.13	5.63	5.99
Krypton Kr 83.80	1.46	2.90	3.67	4.43	5.01	6.01	6.57	7.55(c)	4.97(g)	4.97	4.97	4.97
Lanthanum La 138.91	0.26	0.80	1.48	2.20	2.81	4.41	5.12	5.64	6.04	6.17		6.36
Lead Pb 207.19	0.66	1.68	2.58	3.37	3.94	5.12	5.57	5.85	6.04	6.16	6.27	6.42
Lithium Li 6.939	0.015	0.043	0.095	0.17	0.28	0.97	1.86	3.19*	4.48*	5.15	5.60	5.91
Lutetium Lu 174.97	(0.12)	0.40	0.88	1.46	2.02	3.79	4.72	5.40	5.91		6.31	5.95
Magnesium Mg 24.312	0.010	0.034	0.086	0.18	0.33	1.36	2.51	3.77	4.90	5.43	5.75	5.95
Manganese (α) Mn 54.9380	0.04	0.068	0.12	0.19	0.33	1.16	2.24	3.52*	4.79	5.51	5.97	6.29

Element	1	2	3	4	5	6	7	8	9	10	11	12
Mercury Hg 200.59	6.69	6.78(l)	6.52(c)	6.19	5.80	5.36	4.76	3.53	3.04	2.52	1.85	1.12
Molybdenum Mo 95.94	5.68	5.50	5.15	4.52	3.21	1.92	0.90	0.18	0.099	0.050	0.024	0.010
Neodymium Nd 144.24	(6.57)				(6.36)	(5.99)	5.16	3.39	2.85	2.45*	1.79	1.25
Neon Ne 20.183	4.97	4.97	4.97	4.97	4.97(g)			8.44(l)	4.37(c)			
Neptunium Np (237)	(7.0)											1.25
Nickel Ni 58.71	5.90	5.83	5.37	4.62	3.28	1.97	0.96	0.23	0.14	0.077	0.043	0.024
Niobium (columbium) Nb(Cb) 92.906	5.96	5.77	5.52	5.10	4.17	3.02	1.88	0.65	0.42	0.24	0.12	0.050
Nitrogen N₂ 28.0134	(5.90)	6.96	6.96	6.96	6.96(g)	13.4(l)	9.92(c₂)	8.26(c₁)	6.50	4.50	2.87	(1.06)
Osmium Os 190.2	7.02	6.98	6.96	6.96	6.96	6.96(g)	12.7(l)	11.0(c₃)	5.30(c₂)	3.27(c₁)	1.59	(0.60)
Oxygen O₂ 31.9988	6.21	6.06	5.79	5.30	4.26	3.10	1.96	0.67	0.41	0.24	0.12	0.050
Palladium Pd 106.4	4.98											
Phosphorus (red) P 30.9738	5.63			5.47*	4.65	3.68	2.56	1.03	0.65	0.36	0.17	0.03
Phosphorus (white) P 30.9738	6.19	6.05	5.83	5.50*	4.56*							
Platinum Pt 195.09	7.65*	7.06*	6.34*									
Plutonium Pu (239)	(6.30)	(6.56)	(6.51)	6.22	5.89	5.52	5.01	3.70	3.08	2.34	1.51	0.66
Polonium Po (210)	7.06	6.70	5.80	6.42	6.22	6.30*	6.18*	5.07	4.36	3.18	2.02	0.99
Potassium K 39.102	(6.59)	6.08	5.41	5.33	4.31	3.08	1.89	0.54	0.32	0.15	0.033	0.016
Praseodymium Pr 140.907	6.14	5.75	5.56	4.82	3.61	2.32	1.20	0.26	0.14	0.067	0.022	1.73
Rhenium Re 186.2	5.97	5.85	5.17	6.34	6.10	5.88	5.60	4.90	4.49	3.83	2.95	0.63*
Rhodium Rh 102.905	7.39	5.54	6.47	6.28	3.23*	1.92	0.89	0.17	0.084	0.042	1.35*	(0.010)
Rubidium Rb 85.47	7.06	5.74	5.59	5.05	9.08*	7.01	5.39	3.11	2.40	1.77	0.022	0.63*
Ruthenium Ru 101.07	6.12	5.89	5.58	5.18	3.92	2.70	1.54	(0.47)	(0.29)	(0.17)	(0.087)	(0.038)
Samarium Sm 150.35	6.07	5.80	3.74	2.86	4.34	3.42	2.62	1.51	1.17	0.83	(0.45)	(0.18)
Scandium Sc 44.956	4.78	4.36	3.77	5.47	1.74	1.02	0.53	0.12	0.057	0.0023	0.0073	0.0018
Selenium Se 78.96	6.06	5.95	5.77	5.91	4.80	3.90	2.79	1.14	0.733	0.394	0.160	0.044
Silicon Si 28.086	6.74	6.45	6.20		5.36	4.65	3.70	2.00	1.44	0.88	0.42	0.14
Silver Ag 107.870	(6.30)			5.91							0.57	0.17
Sodium Na 22.9898	5.64	5.29	4.80	4.06	3.10	2.36	1.77	(1.08)	(0.86)	1.14	(0.31)	(0.10)
Strontium Sr 87.62	5.42	5.08	4.64	3.96	3.06	2.35	1.77	1.08	0.86	(0.61)	(0.31)	(0.10)
Sulfur (monoclinic) S 32.064	6.05	5.94	5.76	5.45	4.74	3.75	2.60	1.00	0.62	0.34	0.61	0.050
Sulfur (rhombic) S 32.064	(5.80)										0.15	
Tantalum Ta 180.948	6.15	6.07	5.91	5.69	5.09	4.47	3.55	2.00	1.57	1.09	0.62	(0.21)
Technetium Tc (99)	6.91	7.55	11.20	8.90	7.56	6.46	5.10	2.64	1.85	1.07	0.47	(0.12)
Tellurium Te 127.60	6.29	6.23	6.15	6.06	5.48	5.60	5.01	3.80	3.18	2.40	1.59	(0.81)
Terbium Tb 158.924	6.53	6.39	6.22	5.97	5.48	4.88	4.05	2.40*	1.80	1.11	0.52	0.16
Thallium Tl 204.37	6.46	6.39	6.34*	6.20*	6.04*	5.84*	9.12*	5.27*	3.93*	2.54	1.32	0.47
Thorium Th 232.038	6.16	6.02	5.81	5.44	4.67	3.70	3.70	1.58	1.24	0.89	0.50	(0.16)
Thulium Tm 168.934	6.30	6.20	6.08	5.85	5.35	4.53	3.68	2.20	1.65	1.11	0.64	0.22
Tin (gray) Sn 118.69	5.97	5.72	5.32	4.68	3.43	2.17	1.13	0.27	0.15	0.079	0.036	0.015
Tin (white) 118.69	5.81	5.66	5.37	4.90	3.83	2.57	1.39	0.32	0.17	0.077	0.032	0.011
Titanium Ti 47.90	6.61	6.40	6.17	5.87	5.32	4.59	3.75	1.93	1.31	0.73	0.31	0.084
Tungsten W 183.80	4.97	4.97	5.16(g)	4.41	3.07	1.83	0.89	0.23	0.13	0.087	0.052	0.028
Uranium U 238.03	6.39	6.25	4.97(g)	8.04(c)	6.75	6.32	5.99	5.19	4.73	4.00	3.24	1.94
Vanadium V 50.942	6.34	6.20	6.00	5.64	5.74	5.36	4.78	3.34	2.70	1.89	1.05	0.35
Xenon Xe 131.30	6.07	5.94	5.74	5.43	4.95	4.02	2.91	1.26	0.70	0.45	0.19	(0.055)
Ytterbium Yb 173.04	6.34	5.94	6.00	5.64	4.61	3.67	2.65	1.16	0.77	0.42	0.17	0.039
Yttrium Y 88.905	6.08	5.97	5.74	5.30	4.49	3.37	2.18	0.75	0.45	0.25	0.10	0.03

TABLE 4e-2. MOLAR HEAT CAPACITY AT CONSTANT PRESSURE OF THE CHEMICAL ELEMENTS AT HIGHER THAN ROOM TEMPERATURE, CAL/MOLE·K

As revised December, 1967

Element	298.15	400	500	600	700	800	1000	1200	1500	2000	2500	3000
Aluminum Al 26.9815	5.81(c)	6.16	6.45	6.72	7.00	7.37(c)	7.59(l)	7.59(l)				4.971(g)
Antimony Sb 121.75	6.03(c)	6.22	6.38	6.56	6.88	7.15(c)	7.50(c)	7.50(l)				
Argon Ar 39.948	4.968(g)	4.968	4.968	4.968	4.968	4.968	4.968	4.968	4.968	4.968	4.968	4.968(g)
Arsenic As 74.9216	5.89(α)	6.15	6.32	6.50	6.74	7.02(α)	6.54(c)				4.968	5.020(g)
Beryllium Be 9.0122	3.93(c)	4.73	5.20	5.54	5.82	6.06		6.26	6.67			
Bismuth Bi 208.980	6.20(c)	6.45	6.69(c)	7.6(l)	7.6	7.6(l)		6.39	6.75			
Boron (crystalline) B 10.811	2.65(c)	3.72	4.49	4.99	5.32	5.56	5.95			7.12(c)		
Boron (amorphous) B 10.811	2.86(c)	3.78	4.40	4.88	5.26	5.57	6.05			7.07(c)		
Bromine Br₂ 159.818	18.09(l)	8.78(g)	8.86	8.91	8.94	8.97	9.01(g)					4.969(g)
Cadmium Cd 112.40	6.20(c)	6.49	6.78(c)	7.10(l)	7.10	7.10	7.10(l)	4.968(g)	4.968	4.968	4.968	
Calcium Ca 40.08	6.26(α)	6.62	7.03	7.45	7.87(α)	7.86(β)	9.32(β)	7.4(l)		5.008(g)	5.219	5.796(g)
Carbon (graphite) C 12.01115	2.04(c)	2.85	3.50	4.04	4.44	4.74	5.15	5.43	5.67	5.86	5.97	6.06(c)
Carbon (diamond) C 12.0115	1.46(c)	2.45	3.24	3.85	4.31	4.66	5.16	5.00(c)		5.10	5.61	
Cerium Ce 140.12	6.44(β)	6.76	7.10	7.46	7.84	8.25	9.14(β)	9.35(l)				
Chlorine Cl₂ 70.906	8.11(c)	8.44	8.62	8.74	8.82	8.88	8.96	9.01(g)				7.358(g)
Chromium Cr 51.996	5.58(c)	6.02	6.41	6.73	7.00	7.22	7.66	8.48	9.68(c)			
Cobalt Co 58.9332	5.93(α)	6.34	6.74	7.09	7.42(α)	7.75(β)	8.84	10.33	9.50(β)			
Copper Cu 63.54	5.84(c)	6.08	6.25	6.39	6.52	6.62	6.82	7.00(c)	7.5(l)			
Erbium Er 167.26	6.71(c)	6.79	6.87	6.97	7.11	7.27	7.67	8.18	9.14(c)			
Europium Eu 151.96	6.48(c)	6.68	*	7.24	7.52	7.88	9.09(c)	9.11(l)				
Fluorine F₂ 37.9968	7.49(g)	7.90	8.21	8.43	8.59	8.71(g)						6.010(g)
Germanium Ge 72.59	5.58(c)	5.85	5.95	6.03	6.10	6.19	6.50	6.86(c)	6.60(l)			6.74(g)
Gold Au 196.967	6.06(c)	6.17	6.29	6.40	6.51	6.65	6.89	7.13(c)	7.0(l)			
Hafnium Hf 178.49	6.15(α)	6.34	6.52	6.70	6.88	7.06	7.43	7.79(α)				
Helium He 4.0026	4.968(g)	4.968	4.968	4.968	4.968	4.968	4.968	4.968	4.968	4.968	4.968	4.968(g)
Holmium Ho 164.930	6.49(α)	6.65	6.74	6.76	6.80	6.95	7.61	8.58	10.69(α)			
n-Hydrogen H₂ 2.01594	6.892(g)	6.975	6.993	7.008	7.035	7.078	7.215	7.401	7.706	8.162(g)		
n-Deuterium D₂ 4.02820	6.978(g)	6.989	7.018	7.078	7.171	7.288	7.557	7.824	8.164	8.568(g)		
Indium In 114.82	6.39(c)	6.93(c)	7.03(l)	6.99	6.96	6.93(l)	6.96	7.24				
Iodine I₂ 253.8088	13.01(c)	19.28(l)	8.95(g)	8.98	9.00	9.02(g)					5.709(g)	5.509(g)
Iridium Ir 192.2	6.00(c)	6.14	6.27	6.41	6.55	6.69			7.65(c)			
Iron Fe 55.847	5.97(α)	6.54	7.10	7.66	8.27	9.07	13.01(α)	8.13(γ)	8.73(γ)			
Krypton Kr 83.80	4.968(g)	4.968	4.968	4.968	4.968	4.968	4.968	4.968	4.968	4.968	4.968	4.968(g)
Lanthanum La 138.91	6.65(β)	6.81	6.97	7.13	7.29	7.45(β)	7.03	6.88(l)				
Lead Pb 207.19	6.32(c)	6.56	6.79	7.02(c)	7.25(l)	7.17	6.89	6.87(l)				
Lithium Li 6.939	5.78(c)	6.62(c)	7.20(l)	7.06	6.93	6.92	7.24	7.85			6.951(g)	7.925(g)
Lutetium Lu 174.97	6.40(c)	6.42	6.46	6.50	6.61	6.79			9.09(c)			
Magnesium Mg 24.312	5.095(c)	6.24	6.52	6.80	7.08	7.36(c)	7.8(l)	7.8(l)	4.968(g)	4.969	4.977	5.022(g)

Manganese Mn 54.9380	6.28(c)	6.76	7.21	7.63	8.01	8.35(α)	9.01(β)	9.21(β)	10.99(δ)	.....	5.039(g)	5.252(g)
Mercury Hg 200.59	6.69(l)	6.54	6.48	6.48(l)	4.968(g)	4.968	4.968	4.968	4.968	4.968	4.968	4.968(g)
Molybdenum Mo 95.94	5.73(c)	6.05	6.25	6.38	6.48	6.55	6.70	6.93	7.47	8.63	10.46(c)	
Neodymium Nd 144.24	6.55(α)	6.88	7.24	7.66	8.14	8.71	10.03(α)	10.65(γ)				
Neon Ne 20.183	4.968(g)	4.968	4.968	4.968	4.968	4.968	4.968	4.968	4.968	4.968	4.968	4.968(g)
Nickel Ni 58.71	6.23(c)	6.80	7.37	8.31	7.37	7.44	7.88	8.34	8.65(c)	10.30(l)	4.968	
Niobium (columbium) Nb(Cb) 92.906	5.88(c)	6.09	6.18	6.28	6.38	6.48	6.68	6.88(c)				
Nitrogen N₂ 28.0134	6.96(g)	6.99	7.07	7.20	7.35	7.51	7.82	8.06	8.33	8.60	8.76	8.86(g)
Osmium Os 190.2	5.90(c)	5.99	6.09	6.18	6.27	6.63	6.54	6.72	7.00(c)			
Oxygen O₂ 31.9988	7.02(g)	7.20	7.43	7.67	7.88	8.06	8.34	8.53	8.74	9.03(g)		
Ozone O₃ 47.9982	9.38(g)	10.46	11.30	11.92	12.37	12.70	13.15	13.43	13.68(g)			
Palladium Pd 106.4	6.21(c)	6.35	6.49	6.62	6.76	6.90	7.17	7.44	7.86(c)			
Phosphorus (red, triclinic) P 30.9738	5.07(c)	5.54	5.85	6.16	6.50(c)							
Phosphorus (white) P 30.9738	5.70(β)	6.29(l)	6.29(l)									
Platinum Pt 195.09	6.18(c)	6.31	6.44	6.57	6.70	6.82	7.08	7.34	7.72	8.37(c)		
Plutonium Pu[239]	7.64(α)	8.03(β)	8.53(γ)	9.00(δ)	9.00(δ)	8.40(ε)	7.26(l)					
Potassium K 39.102	0.70(c)	7.53(l)	7.34	7.20	7.13	7.11	6.95		10.03(α)			
Radon Rn [222]	4.968(g)	4.968	4.968	4.968	4.968	4.968	4.968	4.968	4.968	4.968	4.968	4.968(g)
Rhenium Re 186.2	6.16(c)	6.22	6.32	6.43	6.56	6.69	6.95	7.22(c)				
Rhodium Rh 102.905	5.97(c)	6.21	6.45	6.69	6.93	7.17	7.65(c)					
Ruthenium Ru 101.07	5.75(c)	5.82	5.91	6.04	6.23	6.42	6.75	7.11	7.24(c)		6.06(g)	6.08(g)
Samarium Sm 150.35	7.06(α)	7.93	8.94	9.75	10.19	10.52	10.82(α)	11.22(β)	9.14(α)			
Scandium Sc 44.956	6.10(α)	6.29	6.41	6.57	6.75	6.96	7.46	8.06				
Selenium (metallic) Se 78.96	6.06(α)	6.65(c)	8.40(l)	8.40(l)								
Silicon Si 28.086	4.78(c)	5.30	5.61	5.82	5.99	6.13	6.35	6.49	6.66(c)			
Silver Ag 107.870	6.07(c)	6.18	6.30	6.42	6.56	6.72	7.15	7.62(c)				
Sodium Na 22.9898	6.72(c)	7.53(l)	7.30	7.12	7.00	6.92	6.92(l)					
Sulfur S 32.064	5.40(rh)	7.73(l)	9.08	8.20	7.80(l)	6.45	6.57	6.69	6.87	7.17	7.46(c)	6.173(g)
Tantalum Ta 180.948	6.06(c)	6.22	6.30	6.33	6.39	6.45	6.57					
Tellurium Te 127.60	6.14(c)	6.68	7.21	7.73	8.26(c)	9.00(l)	9.00(l)					
Thallium Tl 204.37	6.29(α)	7.03(α)	7.2(l)	7.2(l)		8.06	8.67	9.28(α)		5.420(g)	5.830	5.32(g)
Thorium Th 232.038	6.53(α)	7.15	7.45	7.76	7.76	7.08	7.52	7.89	8.43(c)		5.20(g)	6.266(g)
Thulium Tm 168.934	6.46(c)	6.51	6.59	6.87(l)	6.85	6.85(l)	6.75(c)					
Tin (white) Sn 118.69	6.45(c)	7.32(c)	6.53	6.77	7.01	7.25	7.73(α)	7.10(β)	7.85(β)			9.80(c)
Titanium Ti 47.90	5.98(α)	6.31	6.06	6.16	6.27	6.37	6.59	6.80	7.14	7.71	8.30	
Tungsten (wolfram) W 183.85	5.81(c)	5.96	7.65	8.31	9.08	9.99(α)	10.26(β)	9.15(γ)	11.45(l)			
Uranium U 238.03	6.61(α)	7.10	6.06	6.57	6.70	6.85	7.27	7.85	8.69(c)			
Vanadium V 50.942	5.95(c)	6.27	6.44	6.57	6.70	6.85			6.96(c)			
Xenon Xe 130.30	4.968(g)	4.968	4.968	4.968	4.968	4.968	4.968	4.968	4.968	4.968	4.968	4.968(g)
Ytterbium Yb 173.04	6.39(α)	6.60	6.41	7.13	7.25	7.37	7.64(α)	8.79(l)	4.97(g)	4.97	5.01	5.16(g)
Yttrium Y 8.905	6.34(α)	6.49	6.65	6.82	7.00	7.18	7.53	7.90	8.43(α)			
Zinc Zn 65.37	6.07(c)	6.31	6.55	6.79(c)	7.5(l)	7.5(l)	7.5(l)	4.968(g)	4.968	4.968	4.968	4.968(g)
Zirconium Zr 91.22	6.06(α)	6.54	6.78	7.01	7.23	7.45	7.90(α)	7.50(β)	7.50(β)			4.969(g)

gases). With the exception of B (amorphous), C (diamond), Se, Te, and the gases H_2, D_2, Eu, Sm, Tm, and Yb, the tabulated values are based on (1) R. Hultgren, R. L. Orr, P. D. Anderson, and K. K. Kelley, "Selected Values of Thermodynamic Properties of Metals and Alloys," John Wiley & Sons, Inc., New York, 1963 (and later looseleaf supplements); (2) JANAF Thermochemical Tables, Clearinghouse, U.S. Department of Commerce, Springfield, Va. (*PB Rept.* 168370, 1965; *PB Rept.* 168-370-1, 1966) (and later looseleaf supplements); and (3) J. Hilsenrath, C. G. Messina, and W. H. Evans, Ideal Gas Thermodynamic Functions for 73 Atoms and Their First and Second Ions to 10,000 K, *Air Force Weapons Lab. Rept.* TDR-64-44, Kirtland Air Force Base, N.Mex., 1964.

TABLE 4e-3. HEAT CAPACITY OF WATER
(Osborne, Stimson, and Ginnings, National Bureau of Standards)

Temp, °C	$\dfrac{J}{g \cdot K}$	Temp., °C	$\dfrac{J}{g \cdot K}$
0	4.2177	50	4.1807
5	4.2022	55	4.1824
10	4.1922	60	4.1844
15	4.1858	65	4.1868
20	4.1819	70	4.1896
25	4.1796	75	4.1928
30	4.1785	80	4.1964
35	4.1782	85	4.2005
40	4.1786	90	4.2051
45	4.1795	95	4.2103
50	4.1807	100	4.2160

As a first approximation in explaining the temperature dependence of the heat capacity of solids, Einstein made the assumption that all oscillators in the lattice vibrated with the same frequency ν_0. If h is Planck's constant and k is Boltzmann's constant, let

$$\Theta_E = \frac{h\nu_0}{k} \qquad x = \frac{\Theta_E}{T}$$

and denote the zero-point energy per mole by U_0. Then the Einstein theory of specific heat yields for the molar energy U at the temperature T

$$\text{(Einstein)} \quad \frac{U - U_0}{3RT} = \frac{x}{e^x - 1} \tag{4e-1}$$

where R is the universal gas constant. The Einstein molar heat capacity at constant volume is given by $C_V = dU/dT$, or

$$\text{(Einstein)} \quad \frac{C_V}{3R} = \frac{x^2 e^x}{(e^x - 1)^2} \tag{4e-2}$$

The molar entropy S is equal to $\int (C_V/T)\, dT$, whence

$$\text{(Einstein)} \quad \frac{S}{3R} = \frac{x}{e^x - 1} - \ln(1 - e^{-x}) \tag{4e-3}$$

Numerical values of the quantities in Eqs. (4e-1), (4e-2), and (4e-3) are given in Tables 4e-4, 4e-5 and 4e-6, taken from "Contributions to the Thermodynamic Functions by a Planck-Einstein Oscillator in One Degree of Freedom," prepared by Herrick L. Johnston, Lydia Savedoff, and Jack Belzer, of the Cryogenic Laboratory of the

TABLE 4e-4. $\dfrac{U - U_0}{3RT}$ (EINSTEIN)

$\dfrac{\Theta_E}{T}$	0.0	0.1	0.2	0.3	0.4	0.5	0.6	0.7	0.8	0.9
0	1.00000	0.95083	0.90333	0.85749	0.81330	0.77075	0.72982	0.69050	0.65277	0.61661
1	0.58198	0.54886	0.51722	0.48702	0.45824	0.43083	0.40475	0.37998	0.35646	0.33416
2	0.31304	0.29304	0.27414	0.25629	0.23945	0.22356	0.20861	0.19453	0.18129	0.16886
3	0.15719	0.14624	0.13598	0.12638	0.11739	0.10898	0.10113	0.09380	0.08695	0.08057
4	0.07463	0.06909	0.06394	0.05915	0.05469	0.05055	0.04671	0.04314	0.03983	0.03676
5	0.03392	0.03128	0.02885	0.02658	0.02450	0.02257	0.02079	0.01914	0.01761	0.01621
6	0.01491	0.01371	0.01261	0.01159	0.01065	0.00979	0.00899	0.00826	0.00758	0.00696
7	0.00639	0.00586	0.00538	0.00494	0.00453	0.00415	0.00381	0.00349	0.00320	0.00293
8	0.00269	0.00246	0.00225	0.00206	0.00189	0.00173	0.00158	0.00145	0.00133	0.00121
9	0.00111	0.00102	0.00093	0.00085	0.00078	0.00071	0.00065	0.00059	0.00054	0.00050
10	0.00045	0.00042	0.00038	0.00035	0.00032	0.00029	0.00026	0.00024	0.00022	0.00020
11	0.00018	0.00017	0.00015	0.00014	0.00013	0.00012	0.00011	0.00010	0.00009	0.00008
12	0.00007	0.00007	0.00006	0.00006	0.00005	0.00005	0.00004	0.00004	0.00004	0.00003
13	0.00003	0.00003	0.00002	0.00002	0.00002	0.00002	0.00002	0.00002	0.00001	0.00001
14	0.00001	0.00001	0.00001	0.00001	0.00001	0.00001	0.00001	0.00001	0.00001	0.00001

TABLE 4e-5. $\dfrac{C_V}{3R}$ (EINSTEIN)

$\dfrac{\Theta_E}{T}$	0.0	0.1	0.2	0.3	0.4	0.5	0.6	0.7	0.8	0.9
0	1.00000	0.99917	0.99667	0.99253	0.98677	0.97942	0.97053	0.96015	0.94833	0.93515
1	0.92067	0.90499	0.88817	0.87031	0.85151	0.83185	0.81143	0.79035	0.76869	0.74657
2	0.72406	0.70127	0.67827	0.65515	0.63200	0.60889	0.58589	0.56307	0.54049	0.51820
3	0.49627	0.47473	0.45363	0.43301	0.41289	0.39331	0.37429	0.35584	0.33799	0.32073
4	0.30409	0.28806	0.27264	0.25783	0.24363	0.23004	0.21704	0.20462	0.19277	0.18149
5	0.17074	0.16053	0.15083	0.14162	0.13290	0.12464	0.11683	0.10944	0.10247	0.09588
6	0.08968	0.08383	0.07833	0.07315	0.06828	0.06371	0.05942	0.05539	0.05162	0.04808
7	0.04476	0.04166	0.03876	0.03605	0.03351	0.03115	0.02894	0.02687	0.02495	0.02316
8	0.02148	0.01993	0.01848	0.01713	0.01587	0.01471	0.01362	0.01261	0.01168	0.01081
9	0.01000	0.00925	0 00855	0.00791	0.00731	0.00676	0.00624	0.00577	0.00533	0.00492
10	0.00454	0 00419	0 00387	0.00357	0.00329	0.00304	0.00280	0.00258	0.00238	0.00219
11	0.00202	0.00186	0.00172	0.00158	0.00145	0.00134	0.00123	0.00114	0.00104	0.00096
12	0.00088	0.00081	0.00075	0.00069	0.00063	0.00058	0.00054	0.00049	0.00045	0.00042
13	0.00038	0.00035	0.00032	0.00030	0.00027	0.00025	0.00023	0.00021	0.00019	0.00018
14	0.00016	0.00015	0.00014	0.00013	0.00012	0.00011	0.00010	0.00009	0.00008	0.00008

Ohio State University, under contract between the Office of Naval Research and the Ohio State University Research Foundation, 1949.

Debye assumed that the oscillators occupying the lattice points in a crystalline solid vibrated with a continuous spectrum of frequencies from zero to a maximum value ν_m. Defining the "Debye temperature" Θ_D and y by the equations

$$\Theta_D = \frac{h\nu_m}{k} \qquad y = \frac{\Theta_D}{T}$$

TABLE 4e-6. $\dfrac{S}{3R}$ (Einstein)

$\dfrac{\Theta_E}{T}$	0.0	0.1	0.2	0.3	0.4	0.5	0.6	0.7	0.8	0.9
0	∞	3.30300	2.61110	2.20772	1.92293	1.70350	1.52569	1.37684	1.24939	1.13845
1	1.04066	0.95363	0.87560	0.80521	0.74139	0.68331	0.63027	0.58171	0.53714	0.49617
2	0.45845	0.42367	0.39158	0.36194	0.33455	0.30921	0.28579	0.26410	0.24403	0.22546
3	0.20826	0.19234	0.17760	0.16396	0.15133	0.13964	0.12884	0.11883	0.10958	0.10102
4	0.09312	0.08580	0.07905	0.07281	0.06704	0.06172	0.05681	0.05228	0.04809	0.04423
5	0.04068	0.03740	0.03438	0.03159	0.02903	0.02666	0.02450	0.02249	0.02064	0.01896
6	0.01739	0.01596	0.01464	0.01343	0.01232	0.01130	0.01035	0.00949	0.00869	0.00797
7	0.00730	0.00669	0.00613	0.00562	0.00514	0.00470	0.00431	0.00394	0.00361	0.00330
8	0.00303	0.00276	0.00252	0.00231	0.00211	0.00193	0.00176	0.00162	0.00148	0.00135
9	0.00123	0.00113	0.00103	0.00094	0.00086	0.00078	0.00072	0.00065	0.00060	0.00055
10	0.00050	0.00046	0.00042	0.00038	0.00035	0.00032	0.00028	0.00026	0.00024	0.00022
11	0.00020	0.00019	0.00016	0.00015	0.00014	0.00013	0.00012	0.00011	0.00010	0.00009
12	0.00008	0.00008	0.00007	0.00006	0.00005	0.00005	0.00004	0.00004	0.00004	0.00003
13	0.00003	0.00003	0.00002	0.00002	0.00002	0.00002	0.00002	0.00002	0.00001	0.00001
14	0.00001	0.00001	0.00001	0.00001	0.00001	0.00001	0.00001	0.00001	0.00001	0.00001

TABLE 4e-7. $\dfrac{U - U_0}{3RT}$ (Debye)

$\dfrac{\Theta_D}{T}$	0.0	0.1	0.2	0.3	0.4	0.5	0.6	0.7	0.8	0.9
0	1.000000	.963000	.926999	.891995	.857985	.824963	.792923	.761858	.781759	.702615
1	.674416	.647148	.620798	.595351	.570793	.547107	.524275	.502280	.481103	.460726
2	.441128	.422291	.404194	.386816	.370137	.354136	.338793	.324086	.309995	.296500
3	.283580	.271215	.259385	.248070	.237252	.226911	.217029	.207589	.198571	.189959
4	.181737	.173888	.166396	.159246	.152424	.145914	.139704	.133780	.128129	.122739
5	.117597	.112694	.108016	.103555	.099300	.095241	.091369	.087675	.084152	.080789
6	.077581	.074520	.071598	.068809	.066146	.063604	.061177	.058858	.056644	.054528
7	.052506	.050573	.048726	.046960	.045271	.043655	.042109	.040630	.039214	.037858
8	.036560	.035317	.034126	.032984	.031890	.030840	.029834	.028869	.027942	.027053
9	.026200	.025380	.024593	.023837	.023110	.022411	.021739	.021092	.020470	.019872
10	.019296	.018741	.018207	.017692	.017196	.016718	.016257	.015812	.015384	.014970
11	.014570	.014185	.013813	.013453	.013106	.012770	.012445	.012131	.011828	.011534
12	.011250	.010975	.010709	.010452	.010202	.009960	.009726	.009499	.009279	.009066
13	.008859	.008658	.008463	.008275	.008091	.007913	.007740	.007572	.007409	.007251
14	.007097	.006947	.006801	.006660	.006522	.006388	.006258	.006132	.006008	.005888

$\dfrac{\Theta_D}{T}$	0	1	2	3	4	5	6	7	8	9
10	.019296	.014570	.011250	.008859	.007097	.005771	.004756	.003965	.003340	.002840
20	.002435	.002104	.001830	.001601	.001409	.001247	.001108	.000990	.000887	.000799
30	.000722	.000654	.000595	000542	000496	.000454	.000418	.000385	.000355	.000328
40	.000304	.000283	.000263	.000245	.000229	.000214	.000200	.000188	.000176	.000166

TABLE 4e-8. $\dfrac{C_V}{3R}$ (DEBYE)

$\dfrac{\Theta_D}{T}$	0.0	0.1	0.2	0.3	0.4	0.5	0.6	0.7	0.8	0.9
0	1.000000	.999500	.998003	.915514	.992046	.987611	.982229	.975922	.968717	.960643
1	.951732	.942020	.931545	.920346	.908467	.895950	.882842	.869186	.855031	.840422
2	.825408	.810034	.794347	.778392	.762213	.745853	.729355	.712759	.696103	.679424
3	.662758	.646137	.629593	.613154	.596848	.580700	.564732	.548966	.533421	.518113
4	.503059	.488272	.473763	.459543	.445620	.432002	.418693	.405700	.393024	.380669
5	.368635	.356922	.345529	.334456	.323698	.313255	.303121	.293293	.283767	.274536
6	.265597	.256943	.248568	.240466	.232631	.225056	.217735	.210662	.203828	.197229
7	.190856	.184704	.178766	.173035	.167505	.162169	.157021	.152055	.147264	.142644
8	.138187	.133889	.129744	.125746	.121890	.118172	.114585	.111126	.107790	.104572
9	.101467	.098472	.095583	.092795	.090105	.087509	.085004	082585	.080251	.077997
10	.075821	.073719	.071690	.069729	.067835	.066005	.064236	.062526	.060874	.059276
11	.057731	.056237	.054791	.053393	.052039	.050730	.049462	.048235	.047046	.045895
12	.044780	.043700	.042653	.041639	.040655	.039702	.038777	.037880	.037010	.036166
13	.035347	.034552	.033781	.033031	.032304	.031597	.030910	.030243	.029595	.028964
14	.028352	.027756	.027177	.026613	.026065	.025532	.025013	.024508	.024016	.023537

$\dfrac{\Theta_D}{T}$	0	1	2	3	4	5	6	7	8	9
10	.075821	.057731	.044780	.035347	.028352	.023071	.019018	.015859	.013361	.011361
20	.009741	.008414	.007318	.006405	.005637	.004987	.004434	.003959	.003550	.003195
30	.002886	.002616	.002378	.002168	.001983	.001818	.001670	.001538	.001420	.001314
40	.001218	.001131	.001052	.000980	.000915	.000855	.000801	.000751	.000705	.000662

the values of molar energy, molar heat capacity at constant volume and molar entropy were found to be

$$\text{(Debye)} \quad \frac{U - U_0}{3RT} = D(y) = \frac{3}{y^2} \int_0^y \frac{z^3\, dz}{e^z - 1} \tag{4e-4}$$

$$\text{(Debye)} \quad \frac{C_V}{3R} = 4D(y) - \frac{3y}{e^y - 1} \tag{4e-5}$$

$$\text{(Debye)} \quad \frac{S}{3R} = \frac{4}{3} D(y) - \ln (1 - e^{-y}) \tag{4e-6}$$

Values of the quantities in Eqs. (4e-4), (4e-5) and (4e-6) are given in Tables 4e-7, 4e-8 and 4e-9, prepared by John E. Kilpatrick and Robert H. Sherman, of the Los Alamos Scientific Laboratory, under contract with the U.S. Atomic Energy Commission, 1964.

To calculate $U - U_0$ in joules/mole, and C_V and S in joules/mole kelvin, take as the value of R

$$R = 8.3143 \text{ joules/mole kelvin}$$

To convert to calories, it must be kept in mind that there are three different calories:

The 15-degree calorie = 4.1858 joules
The International steam table calorie = 4.1868 joules
The thermochemical calorie = 4.1840 joules

TABLE 4e-9. $\dfrac{S}{3R}$ (DEBYE)

$\dfrac{\Theta_D}{T}$	0.0	0.1	0.2	0.3	0.4	0.5	0.6	0.7	0.8	0.9
0	∞	3.636168	2.943771	2.539553	2.253613	2.032703	1.853102	1.702152	1.572296	1.458656
1	1.357896	1.267635	1.186113	1.111987	1.044212	0.981958	0.924550	0.871435	0.822152	0.776313
2	0.733585	0.693684	0.656362	0.621403	0.588616	0.557832	0.528900	0.501685	0.476064	0.451928
3	0.429176	0.407715	0.387463	0.368341	0.350279	0.333211	0.317077	0.301819	0.287386	0.273728
4	0.260801	0.248562	0.236970	0.225990	0.215585	0.205724	0.196375	0.187510	0.179103	0.171126
5	0.163557	0.156373	0.149554	0.143077	0.136926	0.131083	0.125530	0.120252	0.115234	0.110462
6	0.105924	0.101605	0.097495	0.093583	0.089858	0.086310	0.082930	0.079709	0.076639	0.073712
7	0.070920	0.068257	0.065715	0.063289	0.060972	0.058760	0.056646	0.054626	0.052695	0.050849
8	0.049083	0.047393	0.045775	0.044227	0.042744	0.041324	0.039963	0.038658	0.037407	0.036208
9	0.035057	0.033952	0.032892	0.031873	0.030895	0.029956	0.029053	0.028184	0.027349	0.026546
10	0.025773	0.025029	0.024313	0.023623	0.022959	0.022318	0.021701	0.021106	0.020532	0.019978
11	0.019444	0.018928	0.018431	0.017950	0.017485	0.017037	0.016603	0.016184	0.015778	0.015386
12	0.015007	0.014639	0.014284	0.013940	0.013607	0.013284	0.012972	0.012669	0.012375	0.012090
13	0.011814	0.011546	0.011286	0.011034	0.010790	0.010552	0.010321	0.010097	0.009880	0.009669
14	0.009463	0.009263	0.009069	0.008881	0.008697	0.008519	0.008345	0.008176	0.008011	0.007851

$\dfrac{\Theta_D}{T}$	0	1	2	3	4	5	6	7	8	9
10	0.025773	0.019444	0.015007	0.011814	0.009463	0.007695	0.006341	0.005287	0.004454	0.003787
20	0.003247	0.002805	0.002439	0.002135	0.001879	0.001662	0.001478	0.001320	0.001183	0.001065
30	0.000962	0.000872	0.000793	0.000723	0.000661	0.000606	0.000557	0.000513	0.000473	0.000438
40	0.000406	0.000377	0.000351	0.000327	0.000305	0.000285	0.000267	0.000250	0.000235	0.000221

At values of the temperature less than $\Theta_D/100$, the Debye theory can be relied upon to give correctly the contribution to the heat capacity attributable to lattice vibrations. In some cases it holds well at temperatures up to $\Theta_D/50$. At these low temperatures the Debye expression for C_V reduces to

$$C_V = \frac{12\pi^4 R}{5\Theta_D{}^3} \cdot T^3 \qquad (4e\text{-}7)$$

or

$$C_V = 124.8 \ \frac{\text{mJ}}{\text{mole} \cdot \text{K}} \left(\frac{T}{\Theta_D}\right)^3 \qquad (4e\text{-}8)$$

For metals there is a contribution to the heat capacity due to the free electrons equal to γT, where γ is known as the electronic constant. The total heat capacity of a metal is therefore

$$C_V = \frac{12\pi^4 R}{5\Theta_D{}^3} \cdot T^3 + \gamma T \qquad (4e\text{-}9)$$

The most reliable values of Θ_D and γ are obtained from heat-capacity measurements in the liquid helium region. It is customary in such work to plot C_V/T against T^2. On such a plot a nonmetal gives a straight line through the origin, whereas a metal gives a straight line with a positive intercept.

Values of Θ_D and γ are given in Table 4e-10. They were obtained in almost all cases by calorimetric measurements in the liquid helium range or lower. Values of Θ_D are given in kelvins, and those of γ in millijoules per mole kelvin squared.

TABLE 4e-10. VALUES OF Θ_D AND γ, IN THE HEAT-CAPACITY EQUATION (4e-9).

Substance	Symbol	Θ_D, K	$\gamma, \dfrac{mJ}{mole \cdot K^2}$	Refs.
Aluminum	Al	428	1.35	1, 2, 89
Antimony	Sb	211	0.112	3
Argon	A	93		
Arsenic	As	282	0.19	3, 4
Barium	Ba	110	2.7	5
Beryllium	Be	1440	0.17	6, 7
Bismuth	Bi	119	0.021	8, 9
Bismuth telluride	Bi_2Te_3	155	.....	31
Cadmium	Cd	209	0.69	10
Calcium	Ca	230	2.9	11, 5
Calcium fluoride	CaF_2	510	.....	32
Carbon (graphite)	C	420	.....	12
Carbon (diamond)	C	2230	.....	13, 14, 90
Cesium	Cs	38	3.2	15
Chlorine	Cl	115		
Chromium	Cr	630	1.40	16
Cobalt	Co	445	4.7	17, 18, 19
Copper	Cu	343	0.688	20, 21, 22, 54
Dysprosium	Dy	210	.....	23
Gadolinium	Gd	195	.....	93
Gallium	Ga	320	0.60	10, 24
Germanium	Ge	370	.....	25, 26
Germanium telluride	GeTe	166	1.32	85
Gold	Au	165	0.69	20, 27, 91
Hafnium	Hf	252	2.16	28
Helium⁴ (hcp)	He⁴ (hcp)	26.4	.....	84
Helium³ (bcc)	He³ (bcc)	16		
Hydrogen	H	105		
Hydrogen²	H²	97		
Ice	H_2O	192		
Indium	In	108	1.6	29, 30
Indium antimonide	InSb	200	.....	44
Iodine	I	106		
Iridium	Ir	420	3.1	59
Iron	Fe	467	5.0	60, 61, 92
Iron oxide	Fe_2O_3	660	.....	33
Iron selenide	$FeSe_2$	366	.....	66
Iron sulfide	FeS_2	637	.....	66
Krypton	Kr	72		
Lanthanum	La	142	10	62, 95
Lead	Pb	105	3.0	56
Lead selenide	PbSe	135–160		
Lead sulfide	PbS	194		
Lead telluride	PbTe	124–135		
Lithium	Li	344	1.63	15
Lithium chloride	LiCl	422	.....	86
Lithium fluoride	LiF	732	.....	34, 35, 36
Magnesium	Mg	400	1.3	63, 64
Magnesium cadmide	Mg_3Cd	290	0.8	37
Magnesium oxide	MgO	946	.....	38
Manganese	Mn	410	14	48, 49, 50
Mercury	Hg	71.9	1.79	65, 75
Molybdenum	Mo	450	2.0	59, 67, 68, 94
Neon	Ne	75	.....	45
Nickel	Ni	450	7.1	16, 69
Nickel selenide	$NiSe_2$	297	.....	66
Niobium	Nb	275	7.79	52, 58, 94

TABLE 4e-10. VALUES OF Θ AND γ, IN THE HEAT-CAPACITY EQUATION (4e-9)
(*Continued*)

Substance	Symbol	Θ_D, K	$\gamma, \dfrac{mJ}{mole \cdot K^2}$	Refs.
Niobium-tin	Nb₃Sn	228	13.1	87
Nitrogen	N	68		
Osmium	Os	500	2.4	59
Oxygen	O	91		
Palladium	Pd	274	9.42	53
Platinum	Pt	240	6.8	27
Potassium	K	91	2.1	15
Potassium bromide	KBr	174	.....	39
Potassium chloride	KCl	235	.....	39, 40, 41
Potassium fluoride	KF	336	.....	86
Potassium iodide	KI	132	.....	39
Rhenium	Re	430	2.3	59, 70, 94
Rhodium	Rh	480	4.9	59
Rubidium	Rb	56	2.4	15
Rubidium bromide	RbBr	131	.....	86
Rubidium chloride	RbCl	165	.....	86
Rubidium iodide	RbI	103	.....	86
Ruthenium	Ru	600	3.3	59
Scandium	Sc	360	10.7	46, 96
Selenium	Se	90	.....	71
Silicon	Si	640	.....	25, 26
Silicon dioxide	SiO₂	470	.....	42
Silver	Ag	225	0.650	54
Silver bromide	AgBr	144		
Silver chloride	AgCl	183		
Sodium	Na	158	1.4	5, 72, 73
Sodium bromide	NaBr	225	.....	86
Sodium chloride	NaCl	321	.....	39
Sodium fluoride	NaF	492	.....	86
Sodium iodide	NaI	164	.....	39
Strontium	Sr	147	3.6	5
Tantalum	Ta	240	5.9	59, 68, 74, 94
Tellurium	Te	153	.....	71
Thallium	Tl	78.5	1.47	75
Thorium	Th	163	4.3	76, 97
Tin (white)	Sn	199	1.78	67, 77, 78
Tin (gray)	Sn	210	.....	41
Titanium	Ti	420	3.5	28, 79, 80
Titanium dioxide	TiO₂	760	.....	43
Tungsten	W	400	1.3	59, 68, 81
Uranium	U	207	10.0	57, 97
Uranium dioxide	UO₂	160		
Vanadium	V	380	9.8	47
Xenon	Xe	64	.....	45
Yttrium	Y	280	10.2	51, 94
Yttrium iron garnet	YIG	510		
Zinc	Zn	327	0.65	24, 82, 83
Zinc sulfide	ZnS	315	.....	35
Zirconium	Zr	291	2.80	28

References for Table 4e-10

1. Phillips, N. E.: *Phys. Rev.* **114**, 676 (1959).
2. Zavaritskii, N. V.: *Zhur. Eksp. i. Teoret. Fiz.* **34**, 1116 (1958) [transl. *Soviet Phys. JETP* **7**, 773 (1958)].
3. Culbert, H. V.: *Phys. Rev.* **157**, 560 (1967).
4. Taylor, W. A., D. C. McCollum, B. C. Passenheim, and H. W. White: *Phys. Rev.* **161**, 652 (1967).
5. Roberts, L. M.: *Proc. Phys. Soc. (London)*, ser. B, **70**, 738 (1957).
6. Ahlers, G.: *Phys. Rev.* **145**, 419 (1966).
7. Gmelin, M. E.: *Compt. rend.* **259**, 3459 (1964).
8. Kalinkina, I. N., and P. G. Strelkov: *Zhur. Eksp. i. Teoret. Fiz.* **34**, 616 (1958) [transl. *Soviet Phys. JETP* **6**, 426 (1958)].
9. Phillips, N. E., *Phys. Rev.* **118**, 644 (1960).
10. Phillips, N. E.: *Phys. Rev.* **134**, A385 (1964).
11. Griffel, M., R. W. Vest, and J. F. Smith: *J. Chem. Phys.* **27**, 1267 (1957).
12. Flubacher, P., A. J. Leadbetter, and J. A. Morrison: *Phys. Chem. Solids* **13**, 160 (1960).
13. Burk, D. L., and S. A. Friedberg: *Phys. Rev.* **111**, 1275 (1958).
14. Desnoyers, J. E., and J. A. Morrison: *Phil. Mag.* **3**(8), 42 (1958).
15. Martin, D. L.: *Phys. Rev.* **139**, A150 (1965).
16. Rayne, J. A., and W. R. G. Kemp: *Phil. Mag.* **1**(8), 918 (1956).
17. Arp, V., N. Kurti, and R. Petersen: *Bull. Am. Phys. Soc.* **2**(II), 388 (1957).
18. Duyckaerts, G.: *Physica* **6**, 817 (1939).
19. Heer, C. V., and R. A. Erickson: *Phys. Rev.* **108**, 896 (1957).
20. Corak, W. S., M. P. Garfunkel, C. B. Satterthwaite, and A. Wexler: *Phys. Rev.* **98**, 1699 (1955).
21. Phillips, N. E.: *Proc. 5th Intern. Conf. Low Temp. Phys. Chem.*, pp. 414–416, University of Wisconsin Press, Madison, Wis., 1958.
22. Rayne, J. A.: *Australian J. Phys.* **9**, 189 (1956).
23. Dreyfus, B., B. B. Goodman, G. Troillet, and L. Weil: *Compt. rend.* **253**, 1085 (1961).
24. Seidel, G., and P. H. Keesom: *Phys. Rev.* **112**, 1083 (1958).
25. Flubacher, P., A. J. Leadbetter, and J. A. Morrison: *Phil. Mag.* **4**(8), 273 (1959).
26. Keesom, P. H., and G. Seidel: *Phys. Rev.* **113**, 33 (1959).
27. Ramanathan, K. G., and T. M. Srinivasan: *Proc. Indian Acad. Sci.* **49**, 55 (1959).
28. Kneip, G. D., Jr., J. O. Betterton, Jr., and J. O. Scarbrough, *Phys. Rev.* **130**, 1687 (1963).
29. Bryant, C. A., and P. H. Keesom: *Phys. Rev. Letters* **4**, 460 (1960).
30. Clement, J. R., and E. H. Quinnell: *Phys. Rev.* **92**, 258 (1953).
31. Itskevich, E. S.: *Zhur. Eksp. i. Teoret. Fiz.* **38**, 351 (1960).
32. Huffman, D. R., and M. H. Norwood: *Phys. Rev.* **117**, 709 (1960).
33. Kouvel, J. S.: *Phys. Rev.* **102**, 1489 (1956).
34. Jones, G. O., and D. L. Martin: *Phil. Mag.* **45**(7), 649 (1954).
35. Martin, D. L.: *Phil. Mag.* **46**(7), 751 (1955).
36. Scales, W. W.: *Phys. Rev.* **112**, 59 (1958).
37. Bergenlid, U. M., R. S. Craig, and W. E. Wallace: *J. Am. Chem. Soc.* **79**, 2019 (1957).
38. Barron, T. H. K., W. T. Berg, and J. A. Morrison: *Proc. Roy. Soc. (London)*, ser. A, **250**, 70 (1959).
39. Berg, W. T., and J. A. Morrison: *Proc. Roy. Soc. (London)*, ser. A, **242**, 467, 478 (1957).
40. Keesom, P. H., and N. Pearlman: *Phys. Rev.* **91**, 1354 (1953).
41. Webb, F. J., and J. Wilks: *Proc. Roy. Soc. (London)*, ser. A, **230**, 549 (1955).
42. Jones, G. H. S., and A. C. Hollis-Hallett: *Can. J. Phys.* **38**, 696 (1960).
43. Keesom, P. H., and N. Pearlman: *Phys. Rev.* **112**, 800 (1958).
44. Keesom, P. H., and N. Pearlman: Unpublished data.
45. Fenichel, H., and B. Serin: *Phys. Rev.* **142**, 490 (1966).
46. Wohlleben, D. (quoted by M. A. Jensen and J. P. Maita: *Phys. Rev.* **149**, 410 (1966).
47. Radebaugh, R., and P. H. Keesom: *Phys. Rev.* **149**, 209 (1966).
48. Guthrie, G. L., S. A. Friedberg, and J. E. Goldman: *Phys. Rev.* **139**, A1200 (1965).
49. Stetsenko, P. N., and Y. I. Avsebt'ev: *Soviet Phys. JETP* **20**, 539 (1965).
50. Scurlock, R. G., and W. N. R. Stevens: *Proc. Phys. Soc. (London)* **86**, 331 (1965).
51. Heiniger, F., E. Bucher, and J. Muller: *Phys. Kond. Materie,* **5**, 243 (1966).
52. van der Hoeven, B. J. C., and P. H. Keesom: *Phys. Rev.* **134**, A1320 (1964).
53. Veal, B. W., and J. A. Rayne: *Phys. Rev.* **135**, A442 (1964).
54. Martin, D. L.: *Phys. Rev.* **141**, 576 (1966).
55. Zimmerman, J. E., A. Arrott, and S. Shinozaki: in *Proc. Low Temp. Calorimetry Conf. (Helsinki)*, O. V. Lounasmaa, ed., p. 147, 1966.
56. van der Hoeven, B. J. C., and P. H. Keesom, *Phys. Rev.* **137**, A103 (1965).

57. Ho, J. C., and N. E. Phillips: *Phys. Rev. Letters* **17**, 694 (1966).
58. Leupold, H. A., and H. A. Boorse: *Phys. Rev.* **134**, A1322 (1964).
59. Wolcott, N. M.: *Bull. inst. intern. du froid*, Annexe 1955–3, pp. 286–289.
60. Duyckaerts, G.: *Physica* **6**, 401 (1939).
61. Keesom, W. H., and B. Kurrelmeyer: *Physica* **6**, 633 (1939).
62. Berman, A., M. W. Zemansky, and H. A. Boorse: *Phys. Rev.* **109**, 70 (1958).
63. Logan, J. K., J. R. Clement, and H. R. Jeffers: *Phys. Rev.* **105**, 1435 (1957).
64. Smith, P. L.: *Phil. Mag.* **46**(7), 744 (1955).
65. Douglass, R. L., W. H. Lien, R. G. Peterson, and N. E. Phillips: *Proc. 7th Intern. Conf. Low Temp. Phys. Chem.*, pp. 242–243, University of Toronto Press, 1960.
66. Grønvold, F. C., and E. F. Westrum, Department of Chemistry, University of Michigan.
67. Bryant, C. A.: Thesis, Purdue University, 1960 (unpublished).
68. P. H. Keesom and N. Pearlman: in "Encyclopedia of Physics," S. Flugge, ed., vol. XIV, "Low Temperature Heat Capacity of Solids," Tables 1–13, Springer-Verlag OHG, Berlin, 1956.
69. Keesom, W. H., and C. W. Clark: *Physica* **6**, 513 (1939).
70. Keesom, P. H., and C. A. Bryant: *Phys. Rev. Letters* **2**, 260 (1959).
71. Smith, P. L.: *Bull. inst. intern. du froid*, Annexe 1955–3, pp. 281–283.
72. Gaumer, R. E., and C. V. Heer: *Phys. Rev.* **118**, 955 (1960).
73. Lien, W. H., and N. E. Phillips: *Phys. Rev.* **118**, 958 (1960).
74. Chou, C., D. White, and H. L. Johnston: *Phys. Rev.* **109**, 788 (1958).
75. van der Hoeven, B. J. C., and P. H. Keesom: *Phys. Rev.* **135**, A631 (1964).
76. Smith, P. L., and N. M. Wolcott: *Bull. inst. intern. du froid*, Annexe 1955–3, pp. 283–286.
77. Corak, W. S., and C. B. Satterthwaite: *Phys. Rev.* **102**, 662 (1956).
78. Zavaritskii, N. V.: *Zhur. Eksp. i. Teoret. Fiz.* **33**, 1085 (1957) [transl. *Soviet Phys. JETP* **6**, 837 (1958)].
79. Aven, M. H., R. S. Craig, T. R. Waite, and W. E. Wallace: *Phys. Rev.* **102**, 1263 (1956).
80. Wolcott, N. M.: *Phil. Mag.* **2**(8), 1246 (1957).
81. Waite, T. R., R. S. Craig, and W. E. Wallace: *Phys. Rev.* **104**, 1240 (1956).
82. Garland, C. W., and J. Silverman: *J. Chem. Phys.* **34**, 781 (1961).
83. Martin, D. L.: *Phys. Rev.* **167**, 640 (1968).
84. Edwards, D. O., and R. C. Pandorf: *Phys. Rev.* **140**, A816 (1965).
85. Finegold, L.: *Phys. Rev. Letters* **13**, 233 (1964).
86. Lewis, J. T., A. Lehoczky, and C. V. Briscoe, *Phys. Rev.* **161**, 877 (1967).
87. Vieland, L. J., and A. W. Wicklund: *Phys. Rev.* **166**, 424 (1968).
88. Rayne, J. A.: *Phys. Rev.* **95**, 1428 (1954).
89. Berg, W. T.: *Phys. Rev.* **167**, 583 (1968).
90. van der Hoeven, B. J. C., and P. H. Keesom: *Phys. Rev.* **130** 1318 (1963).
91. Martin, D. L.: *Phys. Rev.* **170**, 650 (1968).
92. Shinozaki, S. S., and A. Arrott: *Phys. Rev.* **152**, 611 (1966).
93. Donald, D. K., L. T. Crane, and J. E. Zimmerman, unpublished [quoted by O. V. Lounasmaa and L. J. Sundstrom, *Phys. Rev.* **150**, 399 (1966)].
94. Morin, F. J., and J. P. Maita: *Phys. Rev.* **129**, 1115 (1963).
95. Ohtsuka, T., and T. Satoh: in *Proc. Low Temp. Calorimetry Conf.* (*Helsinki*), O. V. Lounasmaa, ed., p. 92, 1966.
96. Flotow, H. E., and D. W. Osborne: *Phys. Rev.* **160**, 467 (1967).
97. Gordon, J. E., H. Montgomery, R. J. Noer, G. R. Pickett, and R. Tobon: *Phys. Rev.* **152**, 432 (1966).

4f. Thermal Expansion

RICHARD K. KIRBY, THOMAS A. HAHN, AND BRUCE D. ROTHROCK

The National Bureau of Standards

In Table 4f-1, the coefficients of linear thermal expansion, $\alpha = (1/L_{293})dL/dT$, are given in units of 10^{-6} K^{-1}; and the expansion, $\epsilon = (L_T - L_{293})/L_{293}$, is given in units of 10^{-6}. When data are given for two or more crystalline forms of the element, the forms are designated (α), (β), (γ), etc. The coefficient of cubical expansion may be computed from the following equations:

$$\beta = 3\alpha$$
$$\beta = 2\alpha_a + \alpha_c$$
$$\beta = \alpha_a + \alpha_b + \alpha_c$$

where α_a, α_b, and α_c are the coefficients of linear expansion in the a, b, and c directions.

In Table 4f-2, the coefficients of linear thermal expansion, $\alpha = (1/L_0)dL/dT$, are given in units of 10^{-8} K^{-1}. The data designated by the symbols $\parallel$ or $\perp$ are in directions parallel or perpendicular to the c axis of the crystal. An (S) denotes data for the material in the superconducting state. The data for the material in the normal state in the superconducting region were measured in a magnetic field high enough to destroy superconductivity. An asterisk (*) denotes a region where large errors may result because of the coefficient changing rapidly.

In Table 4f-3, the coefficient of linear thermal expansion, $\alpha = (1/L_{20})dL/dt$, is given in units of 10^{-6}/K^{-1}; and the expansion, $\epsilon = (L_t - L_{20})/L_{20}$ is given in units of 10^{-5} where t stands for the Celsius temperature and L_{20} is the length at 20°C.

References used in the compilation of these tables and data on or references to publications on the thermal expansion of other materials can be obtained from the National Bureau of Standards.

An approximate relation between the coefficient of volume expansion

$$\beta = \frac{1}{V}\left(\frac{\partial V}{\partial T}\right)_P$$

and the temperature is given by Grüneisen's equation

$$\beta = \frac{C_V}{Q_0[1 - k(U/Q_0)]^2}$$

where C_V is the molar heat capacity at constant volume, U is the energy of the lattice vibrations, and Q_0 and k are constants. If the Debye temperature Θ_D is known, both C_V and U may be calculated at any temperature T from the equations*

$$C_V = 3R\left[12\left(\frac{T}{\Theta_D}\right)^3 \int_0^{\Theta_D/T} \frac{y^3\,dy}{e^y - 1} - 3\frac{\Theta_D/T}{e^{\Theta_D/T} - 1}\right]$$
$$U = \int_0^T C_V\,dT$$

* This material is concluded on page 4-142.

TABLE 4f-1. COEFFICIENTS OF LINEAR THERMAL EXPANSION, α $(10^{-6})(K^{-1})$,

Temperature, K	Aluminum		Antimony		Antimony a axis		Antimony c axis	
	α	ϵ	α	ϵ	α	ϵ	α	ϵ
25	0.5	−4160	......	......	......	......	......	......
50	3.5	−4120	......	......				
75	8.1	−3970	8.2	−2220	4.7	−1580	15.1	−3500
100	12.0	−3720	9.3	−2000	6.0	−1440	15.8	−3110
150	17.1	−2980	10.1	−1520	7.1	−1120	16.1	−2310
200	20.2	−2040	10.5	−1000	7.7	−750	16.1	−1500
250	22.0	−980	10.8	−470	8.1	−360	16.2	−700
293	23.0	0	11.0	0	8.4	0	16.2	0
350	24.1	1340	11.2	630	......	......	......	......
400	24.9	2560	11.3	1200	......	......	......	......
500	26.5	5150	11.6	2350	......	......	......	......
600	28.2	7890	11.8	3520	......	......	......	......
700	30.4	10890	11.9	4700	......	......	......	......
800	33.5	14110	12.0	5890	......	......	......	......
1000								

TABLE 4f-1. COEFFICIENTS OF LINEAR THERMAL EXPANSION, α $(10^{-6})(K^{-1})$,

Temperature, K	Beryllium c axis		Bismuth		Bismuth a axis		Bismuth c axis	
	α	ϵ	α	ϵ	α	ϵ	α	ϵ
25	....	.....	4.8	−3200	....	......	....	......
50	....	.....	9.2	−3010	....	......	....	......
75	0.1	−970	11.1	−2760	9.0	−2390	15.3	−3500
100	0.7	−960	11.9	−2470	9.9	−2150	15.7	−3110
150	2.9	−870	12.6	−1860	10.9	−1630	16.1	−2320
200	5.2	−670	12.9	−1220	11.2	−1070	16.2	−1510
250	7.4	−350	13.1	−570	11.5	−500	16.2	−700
293	8.9	0	13.2	0	11.7	0	16.2	0
350	10.2	550	13.4	760	11.9	680	16.3	930
400	11.1	1080	13.4	1430	11.9	1270	16.3	1740
500	12.3	2260	13.5	2780	12.1	2470	16.4	3380
600	13.4	3540	....	......	....	......	....	......
700	14.4	4940	....	......	....	......	....	......
800	15.7	6440	....	......	....	......	....	......
1000	....	.....	....	......	....	......	....	......
1200	....	.....	....	......	....	......	....	......
1400	....	.....	....	......	....	......	....	......

AND THE EXPANSION, ϵ (10^{-6}), OF ELEMENTS

Argon		Arsenic		Barium		Beryllium		Beryllium a axis	
α	ϵ	α	ϵ	α	ϵ	α	ϵ	α	ϵ
220	1950								
460	10870								
590	24140	...	...	...	....	0.5	−1300	0.7	−1470
...	.....	...	...	...	....	1.3	−1280	1.6	−1440
...	.....	...	...	...	....	4.1	−1150	4.7	−1290
...	.....	...	...	...	....	7.1	−870	8.0	−970
...	.....	...	...	...	....	9.6	−450	10.7	−500
...	.....	5.6	0	13	0	11.2	0	12.3	0
...	.....	...	320	18	880	12.7	690	13.9	750
...	.....	...	...	21	1850	13.7	1350	15.0	1480
...	.....	...	...	24	4110	15.2	2790	16.6	3060
...	.....	...	...	...	....	16.4	4380	18.0	4790
...	.....	...	...	...	....	17.7	6080	19.3	6650
...	.....	...	...	...	....	19.0	7920	20.6	8650
...	.....	...	...	...	....	21.6	11970		

AND THE EXPANSION, ϵ (10^{-6}), OF ELEMENTS (*Continued*)

Cadmium		Cadmium a axis		Cadmium c axis		Calcium		Carbon (diamond)	
α	ϵ	α	ϵ	α	ϵ	α	ϵ	α	ϵ
12.0	−7370	−3.8	−3460	43.5	−15450				
21.4	−7010	+2.7	−3480	58.8	−14090				
25.3	−6420	7.8	−3340	60.4	−12580	14.0	−4260		
27.1	−5760	10.7	−3110	59.9	−11080	16.7	−3870	0.05	−84
29.2	−4340	14.4	−2460	58.7	−8110	18.9	−2970	0.20	−78
30.2	−2860	16.5	−1690	57.5	−5200	20.4	−1990	0.41	−63
30.7	−1330	18.2	−820	55.8	−2370	21.4	−940	0.70	−36
31.3	0	19.8	0	54.3	0	22.1	0	1.00	0
32.0	1800	22.1	1190	51.7	3020	22.7	1280	1.5	71
33.0	3420	24.9	2360	49.1	5550	23.0	2420	1.8	153
38.4	6940	36.7	5350	41.9	10120	23.5	4750	2.5	369
....	......	.....	......	....	....	23.8	7120	3.0	640
....	......	.....	......	....	....	24.0	9510	3.4	960
....	......	.....	......	....	....	....	......	3.7	1320
....	......	.....	......	....	....	....	......	4.3	2120
....	......	.....	......	....	....	....	......	4.7	3030
....	......	.....	......	....	....	....	......	5.1	4020

TABLE 4f-1. COEFFICIENTS OF LINEAR THERMAL EXPANSION, α $(10^{-6})(K^{-1})$,

Temperature K	Carbon (graphite)		Carbon (graphite) a axis		Carbon (graphite) c axis		Cerium	
	α	ϵ	α	ϵ	α	ϵ	α	ϵ
25	....	......	.....	.....	....	......	...	....
50	....	......	.....	.....	....	......	...	....
75	....	......	.....	.....	....	......	...	....
100	4.9	−1300	−0.4	152	15.4	−4220	...	....
150	6.1	−1030	−0.6	127	19.5	−3340	...	....
200	7.0	−700	−0.8	92	22.6	−2280	...	....
250	7.6	−330	−1.0	47	24.8	−1090	...	....
293	7.8	0	−1.2	0	25.9	0	5.2	0
350	8.1	460	−1.2	−68	26.8	1500	5.6	310
400	8.4	870	−1.1	−128	27.4	2860	5.8	590
500	8.9	1730	−0.7	−226	28.2	5640	6.4	1210
600	9.4	2640	−0.2	−276	28.6	8480	7.0	1870
700	9.8	3600	+0.2	−278	28.9	11360	7.6	2600
800	10.0	4580	0.5	−246	29.1	14260	8.2	3390
1000	10.4	6630	0.8	−118	29.6	20140	9.4	5140
1200	10.6	8730	0.9	+52	30.1	26100	...	....
1400	10.9	10890	1.0	250	30.6	32180	...	....
1600	11.1	13090	1.1	460	31.1	38340		
1800	11.3	15330	1.2	690	31.6	44620		
2000	11.5	17610	1.2	930	32.1	50980		

TABLE 4f-1. COEFFICIENTS OF LINEAR THERMAL EXPANSION, α $(10^{-6})(K^{-1})$,

Temperature, K	Dysprosium a axis		Dysprosium c axis		Erbium		Erbium a axis	
	α	ϵ	α	ϵ	α	ϵ	α	ϵ
25								
50								
75								
100	...	....	....	....	....	....	...	....
150	...	....	....	....	8.4	−1300	...	....
200	...	....	....	....	9.0	−860	...	....
250	...	....	....	....	9.3	−410	...	....
293	5.5	0	14.8	0	9.5	0	6.0	0
350	5.6	320	15.8	870	9.7	550	6.1	340
400	5.7	600	16.5	1680	9.8	1040	6.2	650
500	5.9	1170	18.2	3410	10.1	2040	6.4	1280
600	6.1	1770	19.6	5310	10.5	3070	6.7	1930
700	6.3	2390	21.6	7370	10.9	4140	7.1	2620
800	6.7	3040	23.5	9620	11.6	5260	...	....
1000	...	....	....	....	14.8	7830	...	....

AND THE EXPANSION, ϵ (10^{-6}), OF ELEMENTS (*Continued*)

Cesium	Chromium		Cobalt		Copper		Dysprosium	
α	α	ϵ	α	ϵ	α	ϵ	α	ϵ
. . .	0.1	−980	. . . . . .	. . . . .	0.6	−3252		
. . .	0.6	−980	. . . . . .	. . . . .	3.8	−3214		
. . .	1.5	−950	. . . . . .	. . . . .	7.6	−3067		
. . .	2.5	−900	. . . . . .	. . . . .	10.5	−2836	4.9(α)	−1120
. . .	4.0	−740	. . . . . .	. . . . .	13.6	−2218	1.9	−940
. . .	5.1	−510	. . . . . .	. . . . .	15.2	−1492	7.7(β)	−760
. . .	5.6	−240	. . . . . .	. . . . .	16.1	−707	8.3	−360
100	5.0	0	13.7	0	16.7	0	8.6	0
. . .	7.1	290	13.8	780	17.3	970	9.0	500
. . .	8.0	670	13.9	1480	17.6	1840	9.3	960
. . .	9.0	1530	14.2	2880	18.3	3640	10.0	1920
. . .	9.7	2470	14.9	4330	18.9	5500	10.6	2950
. . .	10.4	3470	18(α)	5900	19.6	7420	11.4	4050
. . .	10.9	4530	14.3(β)	8380	20.4	9420	12.3	5230
. . .	12.0	6830	14.5	11260	22.4	13700	14.2	7870
. . .	. . . .	. . . . .	14.7	14170	24.8	18410		

AND THE EXPANSION, ϵ (10^{-6}), OF ELEMENTS (*Continued*)

Erbium c axis		Euro- pium	Gadolinium		Gadolinium a axis		Gadolinium c axis	
α	ϵ	α	α	ϵ	α	ϵ	α	ϵ
. . . .	. . . .	. . .	5.2	−460				
. . . .	. . . .	. . .	6.2	−160				
. . . .	. . . .	. . .	6.0(α)	+150				
. . . .	. . . .	. . .	−1	350				
16.6	0	25	−2	0	. . .	0	. . . .	0
16.9	960	. . .	+5.5(β)	220	5.5	310	5.4	40
17.1	1800	. . .	6.7	530	5.6	590	9.0	410
17.7	3540	. . .	7.8	1270	6.0	1170	11.5	1490
18.2	5340	. . .	8.4	2080	6.4	1790	12.4	2690
18.7	7180	. . .	8.9	2940	6.9	2450	12.8	3950
. . . .	. . . .	. . .	9.4	3850	7.7	3180	13.0	5240
. . . .	. . . .	. . .	11.7	5890				

TABLE 4f-1. COEFFICIENTS OF LINEAR THERMAL EXPANSION, α $(10^{-6})(K^{-1})$,

Temperature, K	Gallium		Gallium a axis	Gallium b axis	Gallium c axis	Germanium		Gold	
	α	ϵ	α	α	α	α	ϵ	α	ϵ
25	....	.....	....	....	....	−0.1	−950	3.2	−3250
50	....	.....	....	....	....	+0.2	−950	7.8	−3120
75	....	.....	....	....	....	1.1	−930	10.5	−2880
100	....	.....	....	....	....	2.4	−890	11.9	−2600
150	....	.....	....	....	....	4.1	−720	13.1	−1970
200	....	.....	....	....	....	4.9	−490	13.6	−1300
250	19.2	−840	....	....	....	5.6	−240	14.0	−610
293	19.7	0	16.6	11.5	31.0	5.7	0	14.2	0
350	....	.....	....	....	....	6.0	330	14.5	820
400	....	.....	....	....	....	6.2	640	14.7	1550
500	....	.....	....	....	....	6.5	1280	15.2	3040
600	....	.....	....	....	....	6.7	1940	15.8	4590
700	....	.....	....	....	....	6.9	2620	16.4	6200
800	....	.....	....	....	....	7.2	3320	17.1	7870
1000	....	.....	....	....	....	.....	.....	18.8	11440
1200	....	.....	....	....	....	.....	.....	21.1	15400

TABLE 4f-1. COEFFICIENTS OF LINEAR THERMAL EXPANSION, α $(10^{-6})(K^{-1})$,

Temperature, K	Iodine		Iodine a axis		Iodine b axis		Iodine c axis	
	α	ϵ	α	ϵ	α	ϵ	α	ϵ
25	11	−16600	...	....	...	....	...	....
50	34	−16100	...	....	...	....	...	....
75	47	−15000	...	....	...	....	...	....
100	55	−13800	...	....	...	....	...	....
150	65	−10700	...	....	...	....	...	....
200	71	−7300	...	....	...	....	...	....
250	79	−3600	...	....	...	....	...	....
293	87	0	133	0	94	0	34	0
350	...	5360	...	7580	...	5870	...	2560
400	...	....	...	....	...	....	...	....
500	...	....	...	....	...	....	...	....
600	...	....	...	....	...	....	...	....
700	...	....	...	....	...	....	...	....
800	...	....	...	....	...	....	...	....
1000	...	....	...	....	...	....	...	....
1200	...	....	...	....	...	....	...	....
1400	...	....	...	....	...	....	...	....
1600	...	....	...	....	...	....	...	....
1800	...	....	...	....	...	....	...	....

AND THE EXPANSION, ϵ (10^{-6}), OF ELEMENTS (*Continued*)

Hafnium		Hol-mium	Indium		Indium a axis		Indium c axis	
α	ϵ	α	α	ϵ	α	ϵ	α	ϵ
...	......	...	10.1	−6970	4.4	−9130	21.4	−2660
...	......	...	19.3	−6580	17.7	−8840	22.6	−2090
...	......	...	23.0	−6050	24.2	−8310	20.5	−1550
4.8	−1090	...	24.9	−5450	28.4	−7650	18.0	−1060
5.4	−830	...	26.7	−4150	34.0	−6080	12.2	−300
5.7	−550	...	28.2	−2780	39.5	−4250	5.6	+150
5.9	−260	...	30.0	−1330	46.0	−2120	−2.0	+240
6.0	0	10	32.1	0	52.9	0	−9.6	0
6.1	350	...	35.9	1930	64.6	3320	−21.5	−870
6.2	660	...	39.9	3830	77.3	6860	−34.8	−2240

AND THE EXPANSION, ϵ (10^{-6}), OF ELEMENTS (*Continued*)

Iridium		Iron		Krypton		Lanthanum		Lead	
α	ϵ	α	ϵ	α_0	ϵ_0	α	ϵ	α	ϵ
....	......	0.2	−2040	192	2170	......	.....	14.2	−6940
....	......	1.3	−2020	275	8230	......	.....	21.7	−6470
....	......	3.5	−1970	343	16040	......	.....	24.4	−5890
4.4	−1110	5.7	−1850	430	25780	......	.....	25.4	−5270
5.3	−860	8.4	−1490	...	.....	3.8	−620	26.6	−3970
5.9	−580	10.1	−1020	...	.....	4.2	−420	27.5	−2610
6.2	−270	11.1	−490	...	.....	4.6	−200	28.2	−1220
6.5	0	11.8	0	...	.....	4.9	0	28.7	0
6.7	380	12.6	700	...	.....	5.4	290	29.3	1650
6.8	710	13.2	1340	...	.....	6.0	580	29.8	3130
7.2	1410	14.3	2720	...	.....	6.8(β)	1230	32.1	6200
7.4	2140	15.2	4190	...	.....	3	950		
7.7	2900	16.1	5760	...	.....	8.8(γ)	1720		
7.9	3680	16.5	7390	...	.....	10.2	2670		
8.4	5320	15.5(α)	10660	...	.....	13.0	4970		
8.8	7040	24(γ)	11800						
9.2	8840								
9.6	10730								
10.1	12700								

TABLE 4f-1. COEFFICIENTS OF LINEAR THERMAL EXPANSION, α $(10^{-6})(\mathrm{K}^{-1})$,

Temperature, K	Lithium		Lutetium	Magnesium		Magnesium a axis		Magnesium c axis	
	α	ϵ	α	α	ϵ	α	ϵ	α	ϵ
25	1.3	−7930	...	0.8	−4970	0.8	−4860	0.8	−5180
50	5.0	−7870	...	5.6	−4900	5.5	−4790	5.8	−5110
75	13.0	−7660	...	11.5	−4680	11.3	−4580	12.0	−4890
100	21.2	−7230	...	15.6	−4340	15.3	−4240	16.3	−4530
150	32.5	−5850	...	20.6	−3410	20.1	−3340	21.5	−3560
200	39.3	−4040	...	23.3	−2310	22.8	−2260	24.3	−2410
250	43.8	−1950	...	25.0	−1090	24.4	−1070	26.1	−1140
293	46.6	0	10	25.9	0	25.3	0	27.0	0
350	....	......	...	26.9	1500	26.3	1470	28.1	1570
400	....	......	...	27.7	2870	27.1	2800	28.9	3000
500	....	......	...	29.2	5710	....	......	....	......
600	....	......	...	31.1	8730	....	......	....	......
700	....	......	...	33.2	11940	....	......	....	......
800	....	......	...	35.6	15380	....	......	....	......
1000	....	......	...	....	......	....	......	....	......
1200	....	......	...	....	......	....	......	....	......
1400	....	......	...	....	......	....	......	....	......
1600	....	......	...	....	......	....	......	....	......
1800	....	......	...	....	......	....	......	....	......
2000	....	......	...	....	......	....	......	....	......

TABLE 4f-1. COEFFICIENTS OF LINEAR THERMAL EXPANSION, α $(10^{-6})(\mathrm{K}^{-1})$,

Temperature, K	Neodymium		Nickel		Niobium		Osmium	
	α	ϵ	α	ϵ	α	ϵ	α	ϵ
25	....	.....	0.2	−2290				
50	....	.....	1.7	−2270				
75	....	.....	4.3	−2200	3.8	−1310		
100	....	.....	6.6	−2060	4.7	−1200	...	....
150	6.7	−980	9.6	−1650	5.8	−940	...	....
200	6.8	−640	11.2	−1120	6.4	−630	...	....
250	6.9	−300	12.2	−540	6.8	−300	...	....
293	6.9	0	12.8	0	7.1	0	4.7	0
350	7.1	400	13.5	750	7.3	410	4.8	270
400	7.2	750	14.1	1440	7.5	780	4.9	520
500	7.6	1490	15.1	2910	7.7	1540	5.2	1020
600	8.1	2270	17.4	4520	7.9	2320	5.5	1560
700	8.8	3110	16.3	6170	8.1	3120	6.0	2130
800	9.5	4030	16.6	7820	8.3	3940	6.5	2760
1000	11.3	6110	17.5	11220	8.5	5620	...	....
1200	....	.....	....	......	8.9	7350	...	....
1400	....	.....	....	......	9.2	9160	...	....
1600	....	.....	....	......	9.4	11010	...	....
1800	....	.....	....	......	9.7	12920	...	....
2000	....	.....	....	......	10.0	14890		

AND THE EXPANSION, ϵ (10^{-6}), OF ELEMENTS (*Continued*)

Manganese		Mercury		Mercury a axis		Mercury c axis		Molybdenum	
α	ϵ	α_{100}	ϵ_{100}	α_{100}	ϵ_{100}	α_{100}	ϵ_{100}	α	ϵ
-0.7	-3500								
-1.2	-3520								
$+0.7$	-3540								
6.2	-3460	36.9	0	33.8	0	43.2	0	2.8	-830
16.0	-2830	41.2	1960	37.3	1780	49.0	2300	4.0	-660
19.0	-1950	49.0	4160	....	....	....	....	4.6	-450
21.1	-940	....	....	....	....	....	....	4.9	-210
22.6	0	....	....	....	....	....	....	5.0	0
24.1	1330	....	....	....	....	....	....	5.1	290
25.3	2570	....	....	....	....	....	....	5.2	550
27.2	5200	....	....	....	....	....	....	5.2	1070
28.9	8010	....	....	....	....	....	....	5.4	1600
30.6	10990	....	....	....	....	....	....	5.5	2150
32.2(α)	14130	....	....	....	....	....	....	5.7	2710
......	20890								
	27010	....	....	....	....	....	....	6.1	3890
......	......							6.6	5170
43.2(β)	31230	....	....	....	....	....	....	7.2	6550
45.2(γ)	42780	....	....	....	....	....	....	7.8	8050
......	......	....	....	....	....	....	....	8.5	9680
......	......	....	....	....	....	....	....	9.3	11450

AND THE EXPANSION, ϵ (10^{-6}), OF ELEMENTS (*Continued*)

Osmium a axis		Osmium c axis		Palladium		Phosphorous		Platinum	
α	ϵ	α	ϵ	α	ϵ	α	ϵ	α	ϵ
...	....	...	....	8.0	-2010	...	......	6.7	-1590
...	....	...	....	9.9	-1560	...	......	7.9	-1220
...	....	...	....	10.8	-1050	...	......	8.4	-810
...	....	...	....	11.3	-490	...	-5460	8.7	-380
4.0	0	5.9	0	11.6	0	127	0	8.9	0
4.2	230	6.1	340	11.9	670	...	......	9.1	513
4.3	450	6.2	650	12.1	1270	...	......	9.2	972
4.6	890	6.6	1290	12.6	2510	...	......	9.5	1909
4.9	1360	7.0	1970	13.1	3790	...	......	9.7	2871
5.2	1860	7.5	2690	13.5	5120	...	......	10.0	3856
5.7	2410	8.1	3480	14.0	6500	...	......	10.2	4866
...	....	...	....	14.8	9380	...	......	10.8	6971
...	....	...	....	15.7	12430	...	......	11.6	9210
...	....	...	....	....	......	...	......	12.4	11600
...	....	...	....	....	......	...	......	13.4	14170
...	....	...	....	....	......	...	......	14.4	16950

TABLE 4f-1. COEFFICIENTS OF LINEAR THERMAL EXPANSION, α $(10^{-6})(K^{-1})$,

Temperature, K	Plutonium		Polonium		Potassium		Praseodymium	
	α	ϵ	α	ϵ	α	ϵ	α	ϵ
25	5	-8870						
50	13	-8640						
75	20	-8220	...	-5000	...	......	...	....
100	26	-7650	...	......	...	......	...	....
150	34	-6130	...	......	...	......	...	....
200	41	-4250	...	......	...	......	...	....
250	46	-2080	23	......	77	-3400	...	....
293	50	0	...	0	82	0	4.4	0
350	56(α)	3030	...	......	...	......	4.6	260
400	36(β)	34600	...	......	...	......	4.8	490
500	35(γ)	45600	...	......	...	......	5.3	990
600	$-10(\delta)$	69500	...	......	...	......	5.8	1540
700	$-10(\delta)$	68500	...	......	...	......	6.5	2160
800	$+26(\epsilon)$	52000	...	......	...	......	7.4	2850
1000	......	......	...	......	...	......	...	....
1200	......	......	...	......	...	......	...	....
1400	......	......	...	......	...	......	...	....
1600	......	......	...	......	...	......	...	....
1800	......	......	...	......	...	......	...	....
2000	......	......	...	......	...	......	...	....

TABLE 4f-1. COEFFICIENTS OF LINEAR THERMAL EXPANSION, α $(10^{-6})(K^{-1})$,

Temperature, K	Ruthenium c axis		Scandium		Scandium a axis		Scandium c axis	
	α	ϵ	α	ϵ	α	ϵ	α	ϵ
25	....	....	....	.....	....	....	....	.....
50	....	....	....	.....	....	....	....	.....
75	....	....	....	.....	....	....	....	.....
100	....	....	....	.....	....	....	....	.....
150	....	....	....	.....	....	....	....	.....
200	....	....	....	.....	....	....	....	.....
250	....	....	....	.....	....	....	....	.....
293	8.6	0	10.1	0	7.6	0	15.2	0
350	8.9	500	10.3	580	7.8	440	15.2	870
400	9.2	950	10.4	1100	8.1	840	15.1	1620
500	9.8	1900	10.7	2150	8.6	1670	15.1	3130
600	10.4	2910	11.0	3240	9.0	2550	15.2	4650
700	10.9	3970	11.4	4360	9.5	3480	15.3	6170
800	11.5	5090	11.8	5520	10.0	4460	15.5	7710
1000	....	....	12.7	7970	10.9	6560	16.3	10880
1200	....	....	13.7	10610	11.9	8840	17.4	14240

AND THE EXPANSION, ϵ (10^{-6}), OF ELEMENTS (*Continued*)

Rhenium		Rhodium		Rubidium	Ruthenium		Ruthenium a axis	
α	ϵ	α	ϵ	α	α	ϵ	α	ϵ
...	.....	3.6	−1470					
...	.....	5.0	−1360					
...	.....	6.5	−1070					
...	.....	7.3	−730					
...	.....	7.9	−350					
5.9	0	8.2	0	91	6.7	0	5.7	0
6.0	340	8.6	480	...	6.9	390	5.9	330
6.1	640	8.8	910	...	7.1	740	6.1	630
6.3	1260	9.3	1820	...	7.5	1470	6.4	1260
6.4	1900	9.8	2780	...	8.0	2240	6.8	1920
6.6	2550	10.3	3780	...	8.4	3060	7.1	2610
6.8	3220	10.8	4840	...	8.8	3920	7.5	3340
7.1	4620	11.9	7120					
7.5	6080	13.0	9600					
7.8	7610	14.2	12320					
8.2	9210	15.4	15270					
8.5	10870							
8.8	12610							

AND THE EXPANSION, ϵ (10^{-6}), OF ELEMENTS (*Continued*)

Selenium		Selenium a axis		Selenium c axis		Selenium (amorphous)		Silicon	
α	ϵ	α	ϵ	α	ϵ	α	ϵ	α	ϵ
...	....	...	....	......	......	...	......	0.0	−220
...	....	...	....	......	......	...	......	−0.2	−223
...	....	...	....	......	......	...	......	−0.5	−233
...	....	...	....	......	......	34	−7740	−0.3	−244
...	....	...	....	......	......	38	−5950	+0.5	−242
...	....	...	....	......	......	40	−4030	1.4	−192
...	....	...	....	......	......	43	−1950	2.1	−101
44	0	68	0	−3.8	0	51	0	2.5	0
48	2600	77	4100	−10.9	−400	...	......	2.9	156
50	5100	85	8200	−19.7	−1150	...	......	3.2	310
...	....	...	....	......	......	...	......	3.5	640
...	....	...	....	......	......	...	......	3.8	1010
...	....	...	....	......	......	...	......	4.0	1400
...	....	...	....	......	......	...	......	4.1	1800
...	....	...	....	......	......	...	......	4.3	2650
...	....	...	....	......	......	...	......	4.5	3530

TABLE 4f-1. COEFFICIENTS OF LINEAR THERMAL EXPANSION, α $(10^{-6})(K^{-1})$,

Tempera-ture, K	Silver		Sodium		Strontium		Sulfur	
	α	ϵ	α	ϵ	α	ϵ	α	ϵ
25	2.4	−4090	...	......	...	......	19	−13300
50	7.9	−3970	...	......	...	......	28	−12700
75	12.3	−3710	35	−13060	...	......	36	−11900
100	14.4	−3370	46	−12040	...	−4400	43	−10900
150	16.7	−2590	58	−9400	...	......	52	−8400
200	17.9	−1720	65	−6300	23	......	56	−5700
250	18.6	−810	68	−2950	...	......	61	−2800
293	19.0	0	69	0	...	0	70	0
350	19.5	1100	...	......	...	......	88	4450
400	19.9	2080	...	......	...	......	...	......
500	20.7	4110	...	......	...	......	...	......
600	21.5	6220	...	......	...	......	...	......
700	22.4	8420	...	......	...	......	...	......
800	23.4	10720	...	......	...	......	...	......
1000	26.0	15640	...	......	...	......	...	......
1200	29.6	21180	...	......	...	......	...	......
1400	....	......	...	......	...	......	...	......
1600	....	......	...	......	...	......	...	......
1800	....	......	...	......	...	......	...	......
2000	....	......	...	......	...	......	...	......

TABLE 4f-1. COEFFICIENTS OF LINEAR THERMAL EXPANSION, α $(10^{-6})(K^{-1})$,

Tempera-ture, K	Thallium		Thorium		Thu-lium	Tin (white)		Tin (white) a axis	
	α	ϵ	α	ϵ	α	α	ϵ	α	ϵ
25	14.6	−6680	....	......	...	4.2	−4640	0.4	−3290
50	20.2	−6230	....	......	...	10.8	−4440	5.1	−3230
75	22.3	−5690	7.9	−2140	...	14.3	−4120	9.2	−3050
100	23.7	−5120	8.7	−1930	...	16.2	−3740	11.4	−2790
150	25.4	−3880	9.6	−1470	...	18.3	−2870	13.6	−2160
200	26.7	−2580	10.1	−980	...	19.6	−1920	14.7	−1450
250	27.8	−1210	10.6	−460	...	20.8	−920	15.6	−690
293	28.7	0	11.0	0	12	21.9	0	16.5	0
350	29.7	1660	11.5	640	...	23.5	1290	17.8	980
400	30.4	3160	11.8	1220	...	25.0	2500	19.0	1900
500	31.8	6270	12.5	2440	...	....	......	....	......
600	....	......	13.1	3730	...	....	......	....	......
700	....	......	13.7	5070	...	....	......	....	......
800	....	......	14.2	6460	...	....	......	....	......
1000	....	......	....	......	...	....	......	....	......
1200	....	......	....	......	...	....	......	....	......
1400	....	......	....	......	...	....	......	....	......
1600	....	......	....	......	...	....	......	....	......
1800	....	......	....	......	...	....	......	....	......
2000	....	......	....	......	...	....	......	....	......

AND THE EXPANSION, ϵ (10^{-6}), OF ELEMENTS (*Continued*)

Tantalum		Tellurium		Tellurium a axis		Tellurium c axis		Terbium	
α	ϵ	α	ϵ	α	ϵ	α	ϵ	α	ϵ
5.2	−1180								
5.9	−900	....	....	....	....	....	....	4.6(α)	+420
6.3	−600	....	....	....	....	....	....	−2	+530
6.4	−280	....	....	....	....	....	....	+3.0(β)	−240
6.5	0	18.2	0	28.1	0	−1.7	0	6.6	0
6.6	370	18.5	1040	28.7	1620	−1.7	−100	7.7	410
6.6	700	....	....	....	....	....	....	8.4	810
6.8	1380	....	....	....	....	....	....	9.3	1700
6.9	2060	....	....	....	....	....	....	9.9	2660
7.0	2750	....	....	....	....	....	....	10.4	3680
7.1	3450	....	....	....	....	....	....	11.0	4750
7.3	4890	....	....	....	....	....	....	14.7	7190
7.5	6370								
7.7	7890								
8.0	9460								
8.2	11070								
8.4	12730								

AND THE EXPANSION, ϵ (10^{-6}), OF ELEMENTS (*Continued*)

Tin (white) c axis		Titanium		Tungsten		Uranium		Uranium a axis	
α	ϵ	α	ϵ	α	ϵ	α	ϵ	α	ϵ
11.7	−7320	0.2	−1530						
22.1	−6860	1.2	−1520						
24.6	−6270	3.0	−1470						
25.8	−5640	4.4	−1370	2.6	−770				
27.7	−4300	6.4	−1100	3.6	−610	13.1	−1930	22	−3200
29.4	−2880	7.4	−750	4.1	−400	13.4	−1270	22	−2100
31.1	−1370	8.1	−360	4.4	−190	13.7	−600	22	−970
32.6	0	8.6	0	4.5	0	14.1	0	23	0
34.8	1920	9.1	500	4.6	260	14.6	820	24	1320
37.0	3720	9.4	970	4.6	490	15.2	1560	24	2500
....	......	9.9	1940	4.6	950	16.8	3160	28	5100
....	......	10.4	2950	4.7	1420	18.8	4940	32	8100
....	......	10.7	4000	4.8	1890	21.4	6950	38	11600
....	......	11.0	5090	4.8	2370	24.5(α)	9240	47	15800
....	......	11.6(α)	7360	5.0	3350	17.3(β)	17400		
....	......	......	9170	5.1	4370	21.3(γ)	24100		
....	......	12.5(β)	11550	5.3	5410	23.7	28600		
....	......	13.0	14110	5.6	6500				
....	......	......	......	6.0	7660				
....	......	......	......	6.4	8890				

TABLE 4f-1. COEFFICIENTS OF LINEAR THERMAL EXPANSION, α $(10^{-6})(K^{-1})$,

Temperature, K	Uranium b axis		Uranium c axis		Vanadium		Xenon	
	α	ϵ	α	ϵ	α	ϵ	α_0	ϵ_0
25	......	......	...	......	0.2	−1500	123	1340
50	......	......	...	......	1.1	−1490	188	5400
75	......	......	...	......	2.8	−1440	220	10500
100	......	......	...	......	4.6	−1350	250	16400
150	1.3	−130	16	−2500	6.5	−1060	340	30800
200	1.1	−70	17	−1700	7.4	−710	...	.....
250	0.8	−26	18	−800	7.7	−330	...	.....
293	0.5	0	19	0	7.8	0	...	.....
350	0.0	17	20	1120	...	......	...	.....
400	−0.7	2	22	2200	...	......	...	.....
500	−3.0	−180	26	4600	...	......	...	.....
600	−6.5	−630	31	7400	...	......	...	.....
700	−11	−1500	37	10800	...	......	...	.....
800	−17	−2900	44	14800	...	......	...	.....
1000	......	......	...	......	...	......	...	.....

AND THE EXPANSION, ϵ (10^{-6}), OF ELEMENTS (*Continued*)

Ytterbium		Yttrium		Yttrium a axis		Yttrium c axis		Zinc	
α	ϵ	α	ϵ	α	ϵ	α	ϵ	α	ϵ
....	.....	0.6	−1940	0.2	−880	1.5	−4060	5.6	−6790
....	.....	2.5	−1910	0.4	−880	6.6	−3970	15.9	−6520
....	.....	4.8	−1810	1.8	−850	10.9	−3750	21.7	−6040
....	.....	6.6	−1670	3.0	−780	13.7	−3440	24.5	−5460
....	.....	8.2	−1290	3.8	−600	16.9	−2660	27.4	−4150
....	.....	8.9	−860	4.1	−400	18.5	−1770	28.8	−2740
....	.....	9.3	−400	4.4	−190	19.1	−830	29.6	−1280
24.6	0	9.5	0	4.6	0	19.3	0	30.1	0
24.9	1410	9.7	550	4.8	270	19.4	1100	30.6	1740
25.3	2670	9.8	1040	5.0	520	19.4	2070	31.0	3280
26.5	5250	10.1	2030	5.4	1030	19.5	4020	31.9	6410
28.7	8000	10.4	3050	5.8	1580	19.6	5980	34.2	9690
31.3	11010	10.8	4100	6.3	2180	19.8	7950		
32.9	14240	11.4	5210	7.0	2840	20.2	9940		
36.5	21140	13.5	7670	9.4	4450	21.8	14100		

TABLE 4f-1. COEFFICIENTS OF LINEAR THERMAL EXPANSION, α $(10^{-6})(K^{-1})$, AND THE EXPANSION, ϵ (10^{-6}), OF ELEMENTS (*Continued*)

Tempera-ture, K	Zinc a axis		Zinc c axis		Zirconium	
	α	ϵ	α	ϵ	α	ϵ
25	−3.4	−1790	23.6	−16800	0.2	−1160
50	−3.3	−1890	54.0	−15760	1.6	−1140
75	+0.4	−1930	64.3	−14250	3.3	−1080
100	3.9	−1880	65.8	−12620	4.0	−980
150	8.1	−1560	65.9	−9330	4.8	−760
200	10.4	−1090	65.5	−6040	5.2	−510
250	11.9	−540	64.9	−2780	5.5	−240
293	13.0	0	64.3	0	5.7	0
350	14.2	780	63.5	3640	6.0	340
400	15.2	1520	62.6	6800	6.2	640
500	17.8	3150	60.0	12950	6.5	1280
600	23.4	5160	55.8	18760	6.8	1950
700	.....	......	....	......	7.0	2640
800	.....	......	....	......	7.3	3360
1000	.....	......	....	......	7.7(α)	4870
1200	.....	......	....	......	9.5(β)	4530
1400	.....	......	....	......	10.2	6520
1600	.....	......	....	......	10.7	8600

TABLE 4f-2. COEFFICIENTS OF LINEAR EXPANSION, $(10^{-8})(K^{-1})$, OF MATERIALS AT LOW TEMPERATURES

Temperature, K...	2	4	6	8	10	12	14	16	18	20
Aluminum........	0.21	0.54	1.1	2.1	3.6	5.7	8.5	12.3	17.0	22.9
Argon............	......	......	......	......	4,000	6,000	8,000	10,000	12,000	15,000
Beryllium........	0.05	0.09	0.14	0.18	0.23	0.28	0.32	0.37	0.41	0.46
Bismuth ‖........	......	3	7	20	45	72	100	140	170	210
⊥........	......	8	23	66	125	210	310	410	500	590
Cadmium ‖......	......	9	80	310	720	1,200	1,800	2,370	2,920	3,400
⊥......	......	+0.4	−7	−40	−110	−180	−260	−320	−360	−385
Cerium...........	−140	−300	−480	−680	−850	−1,060	−90*	+620		
Chromium........	0.7	1.4	2.1	2.8	3.5	4.2	4.9	5.6	6.3	7.0
Cobalt............		1.1	1.7	2.5	3.5	4.5	5.8	7.4	9.2	11.8
Copper...........	0.06	0.25	0.74	1.6	3.0	5.2	8.3	12.8	18.6	26
Gadolinium.......	2.2	+1.4*	−3.3	−3.9	−2.6	+1.5				
Germanium.......	0.01	0.05	0.17	0.38	0.45	0.50	0.19	−0.42	−1.30	−2.35
Gold.............	......	......	......	......	20	48	78	115	160	210
Indium ‖..........	−20	−105	−99	+77	401	730	1,047	1,341	1,600	1,809
⊥.........	13	70	115	112	94	77	82	109	163	236
Iron.............	......	1.4	2.1	3.0	4.0	5.1	6.3	7.8	9.5	11.6
Krypton..........	......	500	1,700	3,400	5,600	7,800	10,000	12,100	13,900	15,700
Lanthanum.......	+1.7	−3.6	−12	−28	−45	−59	−75	−87	−95	−89
(S)........	2.6	16	23							
Lead.............	1.5	10	49	144	290	420	572	750	950	1,140
(S).............	1.0	7.9	44	(88 at 7 K)						
Magnesium ‖.....	......	......	1.0	1.8	3.2	5.2	8.4	13	20	31
⊥.....			1.6	2.8	5.0	8.3	13	18	26	36
α Manganese.....	......	−11	−17	−23	−28	−34	−40	−46	−52	−58
Molybdenum......	0.09	0.20	0.34	0.54	0.81	1.2				
Neodynium.......	−40	−80	+400*	380*	320	300	300			
Neon.............	......	2,000	8,200	19,100	33,900	51,300	70,300	90,000	110,400	137,000
Nickel...........	......	1.7	2.5	3.6	5.0	6.5	8.2	10.2	12.5	15.5
Niobium..........	......	......	......	......	4.4	6.5	9.0	12.4	17	22
Palladium........	0.77	1.8	3.2	5.2	8.0	12	17			
Platinum..........	0.5	1.3	2.6	4.8	8.1	12.8				
Rhenium..........	0.2	0.4	0.7	1.1	1.6	2.3				
Silicon............	0.001	0.006	0.02	0.04	0.09	0.15	0.24	0.32	+0.18	−0.25
Silver.............	......	......	......	......	10	26	47	74	107	140
Tantalum.........	0.2	0.6	1.4	2.8	4.8	7.5	11.6	17	24	32
Thallium ‖........	......	240	390	580	820	1,100	1,320	1,470	1,590	1,700
⊥........	......	60	100	140	180	230	280	360	450	570
Tin ‖.............	......	7.5	28	84	165	280	410	550	690	840
⊥.............		−1	−4	−7	−10	−12	−14	−14	−10	−4
Titanium.........	......	......	......	......	2.5	3.4	4.4	5.8	7.5	9.8
Tungsten.........	0.01	0.04	0.16	0.25	0.48	0.81				
Vanadium.........	......	......	2.5	3.4	4.5	5.8	7.2	8.9	11	13
Xenon............	......	1,000	1,700	2,400	3,300	4,300	5,500	6,700	8,000	9,400
Ytterbium........	13	10	27	65	130	230				
Yttrium..........	......	......	......	......			9	13	19	27	36
Zinc ‖............	......	3.4	7.5	23	76	190	380	640	940	1,290
⊥............		−0.3	+0.2	−0.2	−6.3	−23	−52	−97	−150	−205
CsBr.............	......	6.2	23	70	155	275	430	610	810	1,000
GaAs.............	0.007	0.06	0.19	0.44	0.45	0.1	−1.0	−2.8	−4.9	−7.5
GaSb.............	0.008	0.07	+0.23	−0.1	−2.6	−7.0	−12.5	−18.3	−24.2	−29.0
InAs.............	0.001	+0.025	−0.25	−1.6	−5.8	−12.5	−22	−33	−45	−56
InSb.............	+0.013	0	−1.5	−9.4	−24.0	−43.2	−65	−87	−108	−123
KBr.............	......	......	2.5	7	16	36	64	103	154	218
KCl.............	......	......	0.9	2.2	5.4	10	18	30	47	72
KI.............	......	1.8	6.6	21	53	100	180	260	360	460
LiF.............	......	......	......	......		1.5	2.3	3.4	4.6	6.3
NaCl.............	......	......	1.2	3.1	6	11	18	28	42	61
NaI.............	......	3.2	12	31	70	133	210	295	390	490
RbI.............	......	−1.4	−4.0	+1.5	31	90	180	300	440	600
Fused silica......	......	−1.9	−6.6	−14.4	−24	−33	−41	−48	−54	−58
Quartz ‖..........	......	......	−0.25	−0.5	−0.6	0	+1.8	4.4	8.0	13.3
⊥..........	......	......	0.4	1.0	2.7	6.9	13	22	35	53
35.1 at. % Ni, 64.2 at. % Fe....	......	−37	−56	−75	−93	−110	−124	−135	−145	−150
41.9 at. % Ni, 57.5 at. % Fe...	......	−12	−17	−23	−29	−35	−37	−39	−41	−42
49.9 at. % Ni, 49 at. % Fe....	......	−0.6	−0.8	−0.7	−0.4	0	+1	3	5	8

TABLE 4f-3. EXPANSION, ϵ (10^{-5}), AND COEFFICIENTS OF LINEAR EXPANSION,

Temperature, °C	Al_2O_3 $\|c$ axis		Al_2O_3 $\perp c$ axis		BeO		CsBr	
	ϵ	α	ϵ	α	ϵ	α	ϵ	α
−250	−64	0.0	−54	0.0	....	....	−1089	13.5
−225	−64	0.0	−54	0.0	....	....	−1031	29.6
−200	−63	0.2	−54	0.1	....	....	−947	36.3
−150	−60	1.2	−52	0.9	....	....	−753	40.6
−100	−50	2.6	−44	2.2	....	....	−544	43.1
−50	−34	4.0	−30	3.5	....	....	−323	45.0
0	−11	5.2	−10	4.6	....	....	−94	46.3
20	0	5.6	0	5.0	0	6.5	0	47.4
100	50	6.8	45	6.2	54	6.9	392	50.6
200	124	7.8	112	7.1	126	7.5	918	54.8
300	206	8.5	186	7.7	204	8.0	1486	58.8
400	293	8.9	266	8.2	287	8.6	......	....
500	384	9.2	349	8.4	375	9.1	......	....
600	478	9.4	434	8.6	469	9.6	......	....
700	573	9.6	522	8.8	567	10.1	......	....
800	670	9.8	611	9.1	670	10.5	......	....
1000	870	10.2	796	9.5	888	11.3	......	....
1200	1079	10.6	989	9.9	1123	12.1	......	....
1400	1295	11.0	1190	10.3	1373	12.9	......	....
1600	1518	11.4	1400	10.7	1640	13.7	......	....
1800	....	....	....	....	1922	14.5	......	....
2000	....	....	....	....	2220	15.3	......	....

TABLE 4f-3. EXPANSION, ϵ (10^{-5}), AND COEFFICIENTS OF LINEAR EXPANSION,

Temperature °C	$Li_2O \cdot Al_2O_3 \cdot 2SiO_2$ $\perp$ principal axis		Inconel		InSb		Invar	
	ϵ	α	ϵ	α	ϵ	α	ϵ	α
−250	...	...	−228	0.0	....	.....	−37	−1.5
−225	...	...	−227	1.7	−85	−0.4	−40	−0.7
−200	...	...	−220	4.1	−85	+1.1	−40	+0.8
−150	...	...	−188	8.1	−73	3.4	−31	2.2
−100	...	...	−142	10.4	−54	4.0	−19	2.2
−50	...	...	−86	11.7	−33	4.4	−10	1.7
0	...	...	−25	12.5	−10	4.8	−2	1.3
20	0	8.2	0	12.8	0	4.9	0	1.2
100	66	8.2	106	13.8	41	5.2	12	2.0
200	148	8.2	248	14.6	95	5.5	45	5.5
300	230	8.2	398	15.4	....	.....	149	14.4
400	312	8.2	554	16.1	....	.....	305	16.4
500	394	8.2	719	16.9	....	.....	472	16.9
600	476	8.2	892	17.7	....	.....	644	17.2
700	558	8.2	1073	18.5	....	.....	817	17.5
800	640	8.2	1262	19.3	....	.....	992	17.6

α $(10^{-6})(K^{-1})$, OF SOME COMPOUNDS AND ALLOYS

CsCl		CsI		CO_2	H_2O	HfC		$Li_2O \cdot Al_2O_3 \cdot 2SiO_2$ ∥ principal axis	
ϵ	α	ϵ	α	α	α	ϵ	α	ϵ	α
−1018	8.7	−1135	16.8	37					
−976	23.7	−1068	33.2	80					
−906	31.4	−978	37.1	124	6.1				
−729	38.2	−777	42.2	221					
−530	41.6	−560	44.4	355	30.9				
−316	43.8	−333	46.3	...					
−92	45.7	−97	48.2	...	55.8				
0	46.3	0	49.0	...	....	0	6.2	0	−17.6
......	....	......	....	...	....	50	6.3	−141	−17.6
......	....	......	....	...	....	114	6.5	−317	−17.6
......	....	......	....	...	....	180	6.7	−493	−17.6
......	....	......	....	...	....	247	6.7	−669	−17.6
......	....	......	....	...	....	315	6.8	−845	−17.6
......	....	......	....	...	....	383	6.8	−1021	−17.6
......	....	......	....	...	....	452	6.9	−1197	−17.6
......	....	......	....	...	....	521	7.0	−1373	−17.6
......	....	......	....	...	....	661	7.1		
......	....	......	....	...	....	804	7.2		
......	....	......	....	...	....	949	7.3		
......	....	......	....	...	....	1096	7.4		
......	....	......	....	...	....	1246	7.6		
......	....	......	....	...	....	1399	7.7		

α $(10^{-6})(K^{-1})$, OF SOME COMPOUNDS AND ALLOYS (*Continued*)

KBr		KCl		KI		LiBr		LiCl	
ϵ	α	ϵ	α	ϵ	α	ϵ	α	ϵ	α
−833	3.7	−745	1.3	−893	7.4				
−809	16.2	−731	10.5	−858	19.9				
−757	24.5	−693	20.0	−798	27.7				
−612	31.8	−568	28.2	−640	33.8	−756	36.8	−654	30.6
−444	35.0	−418	31.8	−464	36.4	−558	42.2	−486	36.4
−265	36.8	−252	34.5	−277	38.4	−336	46.0	−294	40.1
−77	38.2	−74	36.5	−81	40.1	−99	48.9	−86	42.8
0	38.7	0	37.1	0	40.8	0	49.8	0	43.8
316	40.4	306	39.3	336	43.3				
732	42.8	712	41.9	811	46.7				
1174	45.9	1145	44.6	1297	50.7				
1657	50.9	1606	47.7	1829	56.0				
2196	57.3	2099	52.0	2421	62.8				
2806	64.8	2627	54.7	3091	71.5				
3212	73.4	3192	58.4						
.....	....	3794	62.3						

TABLE 4f-3. EXPANSION, ϵ (10^{-5}), AND COEFFICIENTS OF LINEAR EXPANSION,

Temperature, °C	LiF		LiI		MgO		NaBr	
	ϵ	α	ϵ	α	ϵ	α	ϵ	α
−250	−505	0.2	....	....	−141	0.0	....	....
−225	−503	1.2	....	....	−141	0.2	....	....
−200	−497	4.4	....	....	−139	0.8	....	....
−150	−449	14.8	−902	43.8	−129	3.6	−656	33.6
−100	−352	23.4	−665	50.4	−104	6.3	−479	37.2
−50	−220	29.0	−401	54.9	−67	8.5	−287	39.6
0	−66	32.3	−118	58.2	−20	9.9	−84	41.5
20	0	33.2	0	59.4	0	10.4	0	42.3
100	279	36.4	....	....	89	11.7	....	....
200	662	40.4	....	....	211	12.6	....	....
300	1087	44.6	....	....	341	13.3	....	....
400	1555	49.2	....	....	476	13.7	....	....
500	2072	54.3	....	....	615	14.1	....	....
600	2643	60.0	....	....	757	14.4	....	....
700	3274	66.4	....	....	903	14.7	....	....
800	3973	73.5	....	....	1051	15.0	....	....
1000	....	....	....	....	1357	15.6	....	....
1200	....	....	....	....	1675	16.2	....	....
1400	....	....	....	....	2006	16.9	....	....
1600	....	....	....	....	....	....	....	....
1800	....	....	....	....	....	....	....	....

TABLE 4f-3. EXPANSION, ϵ (10^{-5}), AND COEFFICIENTS OF LINEAR EXPANSION,

Temperature, °C	SiO₂, crystalline ⊥ to axis		SiO₂, vitreous*		Steel, AISI 304		Steel, AISI 410	
	ϵ	α	ϵ	α	ϵ	α	ϵ	α
−250	−244	0.9	7	−0.70	−294	−0.7	−176	0.1
−225	−239	3.5	5	−0.86	−293	+1.7	−175	0.8
−200	−227	5.6	3	−0.79	−283	6.0	−171	2.7
−150	−191	8.5	−1	−0.52	−239	10.9	−149	6.1
−100	−144	10.3	−2	−0.13	−178	13.4	−112	8.1
−50	−89	11.8	−2	+0.15	−108	14.7	−68	9.3
0	−27	13.1	−1	0.35	−32	15.6	−20	10.0
20	0	13.6	0	0.41	0	15.9	0	10.3
100	117	15.6	4	0.55	131	16.7	85	11.0
200	284	17.9	10	0.60	302	17.5	199	11.6
300	477	20.8	16	0.59	480	18.1	318	12.1
400	707	26.0	22	0.56	664	18.8	441	12.6
500	1023	39.9	27	0.53	855	19.5	570	13.2
600	1752	4.0	32	0.51	....	....	705	13.7
700	1747	−0.5	37	0.48	....	....	845	14.3
800	1742	−0.6	42	0.47	....	....	....	....
1000	....	....	...	......	....	....	....	....
1200	....	....	...	......	....	....	....	....
1400	....	....	...	......	....	....	....	....
1600	....	....	...	......	....	....	....	....
1800	....	....	...	......	....	....	....	....
2000	....	....	...	......	....	....	....	....

* These values are an average of values reported in the literature and may differ from a given specimen of vitreous SiO₂, depending on method of fabrication and heat treatment.

α $(10^{-6})(K^{-1})$, OF SOME COMPOUNDS AND ALLOYS (*Continued*)

NaCl		NaF		NaI		SiC		SiO$_2$, crystalline $\parallel$ to axis	
ϵ	α	ϵ	α	ϵ	α	ϵ	α	ϵ	α
−766	0.3	−547	0.3	−967	6.5	...	...	−117	0.2
−758	7.3	−544	2.5	−933	20.4	...	...	−115	1.2
−726	18.2	−532	7.9	−870	29.2	...	...	−111	2.2
−604	28.0	−462	19.0	−703	36.1	...	...	−96	3.8
−445	33.8	−350	26.0	−513	39.7	...	...	−74	4.9
−268	36.8	−213	28.9	−308	42.4	...	...	−47	6.0
−79	38.9	−63	31.1	−90	44.6	...	...	−14	7.0
0	39.7	0	31.7	0	45.5	0	3.5	0	7.4
329	42.3	262	33.8	.....	....	29	3.7	64	8.8
766	45.2	612	36.2	.....	....	68	4.1	160	10.4
1233	48.3	984	38.3	.....	....	110	4.4	272	12.3
1733	51.9	1378	40.5	.....	....	155	4.6	409	15.4
2272	56.1	1794	42.7	.....	....	203	4.9	593	23.3
2853	60.9	2232	45.0	.....	....	254	5.1	997	2
3492	67.3	.....	....	.....	....	306	5.3	982	−1.5
4207	76.1	.....	....	.....	....	361	5.6	967	−1.5
.....	....	.....	....	.....	....	474	5.8		
.....	....	.....	....	.....	....	593	6.1		
.....	....	.....	....	.....	....	716	6.2		
.....	....	.....	....	.....	....	942	6.4		
.....	....	.....	....	.....	....	971	6.5		

α $(10^{-6})(K^{-1})$, OF SOME COMPOUNDS AND ALLOYS (*Continued*)

Steel, SAE 1020		TaC		ThO$_2$		TiC		TiO$_2$ $\parallel$ c axis	
ϵ	α	ϵ	α	ϵ	α	ϵ	α	ϵ	α
−202	0.2							−152	3.2
−200	1.3							−130	5.4
−194	3.4	....	...	....	....	....	....	−98	7.0
−168	7.0	....	...	....	....	....	....	−61	8.1
−127	9.2	....	...	....	....	....	....	−18	8.9
−78	10.4	....	..	....	....	....	....		
−23	11.4	....	..	....	....	....	....		
0	11.7	0	5.1	0	7.6	0	5.6	0	9.1
98	12.7	44	5.7	64	8.2	47	6.2	76	9.8
230	13.8	103	6.1	149	8.7	112	6.8	177	10.3
373	14.7	165	6.3	238	9.1	182	7.3	282	10.6
524	15.6	228	6.5	330	9.4	257	7.7	389	10.8
684	16.4	294	6.6	425	9.6	336	8.0	498	10.9
851	17.0	361	6.7	522	9.8	417	8.2	607	11.0
1023	17.2	429	6.8	620	10.0	501	8.5	718	11.1
.....	....	498	6.9	721	10.2	587	8.7	830	11.2
.....	....	639	7.1	928	10.5	764	9.0	1057	11.4
.....	....	784	7.3	1142	10.9	947	9.3		
.....	....	933	7.5	1366	11.4	1136	9.5		
.....	....	1086	7.8	1597	11.8	1330	9.8		
.....	....	1246	8.2	1838	12.3	1529	10.1		
.....	....	1414	8.5	2089	12.8	1734	10.3		

TABLE 4f-3. EXPANSION, ϵ (10^{-5}), AND COEFFICIENTS OF LINEAR EXPANSION, α $(10^{-6})(K^{-1})$, OF SOME COMPOUNDS AND ALLOYS (*Continued*)

Temperature, °C	TiO₂ ⊥ c axis		VC		WC		Yellow Brass		ZrC	
	ϵ	α	ϵ	α	ϵ	α	ϵ	α	ϵ	α
−250	.....	...	...	...	...	...	−383	0.8		
−225	.....	...	...	...	...	...	−376	5.4		
−200	−124	2.9	...	...	...	...	−357	9.6		
−150	−105	4.6	...	...	...	...	−294	14.5		
−100	−78	5.7	...	...	...	...	−216	16.7		
−50	−48	6.5	...	...	...	...	−129	17.8		
0	−14	7.0	...	...	...	...	−38	18.7		
20	0	7.1	0	5.6	0	4.4	0	19.0	0	6.2
100	59	7.6	47	6.0	35	4.4	156	19.9	50	6.2
200	137	7.9	110	6.6	80	4.5	360	20.9	113	6.3
300	217	8.2	179	7.1	126	4.6	574	22.0	177	6.4
400	300	8.3	253	7.7	172	4.6	799	23.0	242	6.5
500	383	8.4	333	8.2	219	4.7	1034	24.0	308	6.6
600	468	8.5	428	8.8	268	4.8	.....	....	375	6.7
700	554	8.6	...	...	317	4.9	.....	....	443	6.8
800	641	8.7	...	...	368	5.1	.....	....	512	6.9
1000	816	8.8	...	...	473	5.4	.....	....	653	7.1
1200	.....	...	...	...	582	5.6	.....	....	798	7.3
1400	.....	...	...	...	695	5.8	.....	....	947	7.6
1600	.....	...	...	...	...	...	.....	....	1100	7.8
1800	.....	...	...	...	...	...	.....	....	1257	8.0
2000	.....	...	...	...	...	...	.....	....	1418	8.2

TABLE 4f-4. THERMAL EXPANSION OF LIQUIDS*

(If V_0 is the volume at 0° then at $t°$ the expansion formula is $V_t = V_0(1 + at + bt^2 + ct^3)$. The table gives values of a, b, and c and β, the true coefficient of cubical expansion at 20° for some liquids and solutions. Δt is the temperature range of the observation.)

Liquid	Δt, °C	a, $10^{-3}(°C)^{-1}$	b, $10^{-6}(°C)^{-2}$	c, $10^{-8}(°C)^{-3}$	β at 20°C, $10^{-3}(°C)^{-1}$
Acetic acid.............	16 to 107	1.0630	0.12636	1.0876	1.071
Acetone...............	0 to 54	1.3240	3.8090	−0.87983	1.487
Alcohol:					
Amyl................	−15 to 80	0.9001	0.6573	1.18458	0.902
Ethyl, 30% by vol....	18 to 39	0.2928	10.790	−11.87	
Ethyl, 50% by vol....	0 to 39	0.7450	1.85	0.730	
Ethyl, 99.3% vol.....	27 to 46	1.012	2.20	....	1.12
Ethyl, 500 atm					
pressure..........	0 to 40	0.866			
Ethyl, 3,000 atm					
pressure..........	0 to 40	0.524			
Methyl.............	0 to 61	1.1342	1.3635	0.8741	1.199
Benzene...............	11 to 81	1.17626	1.27776	0.80648	1.237
Bromine...............	0 to 59	1.06218	1.87714	−0.30854	1.132
Calcium chloride:					
5.8% solution........	18 to 25	0.07878	4.2742	....	0.250
40.9% solution.......	17 to 24	0.42383	0.8571	....	0.458
Carbon disulfide........	−34 to 60	1.13980	1.37065	1.91225	1.218
500 atm pressure.....	0 to 50	0.940			
3,000 atm pressure....	0 to 50	0.581			
Carbon tetrachloride....	0 to 76	1.18384	0.89881	1.35135	1.236
Chloroform............	0 to 63	1.10715	4.66473	−1.74328	1.273
Ether.................	−15 to 38	1.51324	2.35918	4.00512	1.656
Glycerin...............	....	0.4853	0.4895	....	0.505
Hydrochloric acid,					
33.2% solution.......	0 to 33	0.4460	0.215	....	0.455
Mercury...............	0 to 300	0.18152	0.008325	0.01338	0.18175
Olive oil...............	....	0.6821	1.1405	−0.539	0.721
Pentane...............	0 to 33	1.4646	3.09319	1.6084	1.608
Petroleum, density					
0.8467..............	24 to 120	0.8994	1.396	....	0.955
Potassium chloride,					
24.3% solution.......	16 to 25	0.2695	2.080	....	0.353
Phenol................	36 to 157	0.8340	0.10732	0.4446	1.090
Sodium chloride, 20.6%					
solution.............	0 to 29	0.3640	1.237	....	0.414
Sodium sulfate, 24%					
solution.............	11 to 40	0.3599	1.258	....	0.410
Sulfuric acid:					
10.9% solution.......	0 to 30	0.2835	2.580	....	0.387
100.0%..............	0 to 30	0.5758	−0.432	....	0.558
Turpentine............	−9 to 106	0.9003	1.9595	−0.44998	0.973
Water.................	0 to 33	−0.06427	8.5053	−6.7900	0.207

* Reprinted by permission from "Smithsonian Physical Tables," 9th ed.

TABLE 4f-5. CONSTANTS IN GRÜNEISEN'S EQUATION FOR THERMAL EXPANSION

Element	$\Theta_D{}^E$, K	Q_0, $\dfrac{\text{kcal}}{\text{mole}}$	k	T_m, K
Ag	210	111.9	3.5	1234
Al	373	89.1	4.1	933
Be	899	122.4	4.2	1623
C (diamond)	1976	389.7	2.4	3800
Cu	325	118.8	2.8	1357
Li	442	42.1	2.1	459
Mg	325	81.1	3.4	924
Mo	350	404.4	9.5	2893
Na	298	28.7	0.7	371
Nb	292	275.2	3.9	2773
Pb	89	79.2	3.3	601
Pd	282	172.9	3.9	1825
Pt	227	229.8	4.5	2042
Rh	333	242.9	7.7	2233
Sb	182	184.7	2.3	904
Ta	218	305.5	3.2	3278
Th	178	198.7	7.8	2118
Ti	410	225.4	6.2	2085
W	319	446.2	5.8	3653

Values of C_V for many values of Θ_D/T are given in Table 4e-8, and values of U in Table 4e-7. Thus, if Θ_D, Q_0 and k are known, β can be calculated. Table 4f-5 lists values for Θ_D, Q_0, and k for 19 elements. Values of $\Theta_D{}^E$ from expansion measurements are in reasonable agreement with room-temperature values of Θ_D from heat-capacity measurements.

A consequence of Grüneisen's theory of the solid state is the approximate proportionality of Q_0 with the melting temperature T_m. Values of T_m are listed in the last column of Table 4f-5.

4g. Thermal Conductivity

ROBERT L. POWELL AND GREGG E. CHILDS

Cryogenics Division, NBS Institute for Basic Standards

Symbols and Units

A cross-sectional area, meters^2
c_p specific heat at constant pressure, joules/kilogram · kelvin
c_v specific heat at constant volume, joules/kilogram · kelvin
J heat current density, watts/meter^2
L Lorenz ratio $\equiv \lambda/\sigma\,T$, volts^2/kelvin^2
l mean free path, meters

M molecular weight, kilograms/mole
$\dot{Q}$ heat current, watts
R gas constant per mole, 8.3143 joule/kelvin · mole
T temperature, kelvins
t time, seconds
v velocity, meters/second
x space coordinate, meters
α thermal diffusivity, meters²/second
λ thermal conductivity, watts/meter · kelvin
μ dynamic viscosity, Newton · second/meter², (= 10 poise)
ρ density, kilograms/meter³
σ electrical conductivity, 1/ohm · meter

4g-1. General Definitions and Units. The thermal conductivity is a nonequilibrium property usually determined in a steady-state experiment utilizing the Fourier law for linear heat flow in a homogeneous, isotropic substance:

$$\dot{Q} = -\lambda A \frac{dT}{dx}$$

where $\dot{Q}$ is the thermal energy current, A is the cross-sectional area, dT/dx is the temperature gradient, and λ is the thermal conductivity coefficient. Commonly used units and their conversion factors are given in Table 4g-1. For nonisotropic bodies such as some dielectric crystals the basic differential equation is modified to

$$J = -\lambda \operatorname{grad} T$$

where J is the vector thermal current density, and λ is a symmetric tensor of second order. Heat-conduction equations and their solutions for nonhomogeneous and nonlinear systems are discussed at length by McAdams [38], Schneider [62], and Carslaw and Jaeger [10].

In general, the total heat *transport* is affected by radiation, convection, and conduction mechanisms and may depend on temperature, pressure, density, material, and temperature gradient, etc. However, the coefficient λ, as defined by the above equations, refers only to heat transport by conduction mechanisms. It is usually assumed that the thermal conductivity is not a function of the temperature gradient, but is a function of the temperature, composition, purity, perfection, and other similar intensive parameters of the system. It is also assumed that the conductivity is not size- or shape-dependent, though this is not always true. Size and shape effects become significant whenever the size of the conductor is comparable to the mean free path for motion of the particles (or quasi-particles) that transport the thermal energy. These effects have been observed for conduction by molecules in rarefied gases and for conduction by phonons (quantized normal modes of lattice vibration) in small, high-purity dielectric crystals at low temperatures.

Representative values for the temperature dependence of several substances are given in Fig. 4g-1. They are typical curves for a high-purity metal (copper), high-purity crystalline dielectric (sapphire), nonferrous alloy (aluminum alloy), ferrous alloy (stainless steel), disordered dielectric (glass), fluid (helium), and water.

The Thermophysical Properties Research Center at Purdue University has published a large compilation of thermal conductivity data and graphs over large temperature ranges for many solids and fluids [72]. A general survey of the experimental and theoretical aspects of thermal conductivity is given in the book edited by Tye [73]. Proceedings of the annual conferences on thermal conductivity [71] are usually available from the sponsoring agency; sometimes they are formally published.

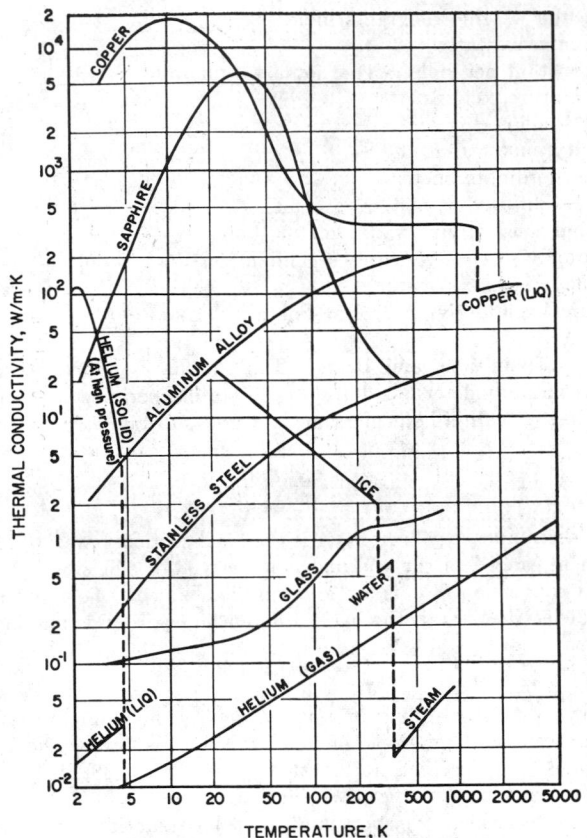

FIG. 4g-1. Typical curves showing temperature dependence of thermal conductivity.

A similar coefficient useful in *transient* heat-flow problems is the thermal diffusivity α defined by

$$\alpha \equiv \frac{\lambda}{\rho c_p}$$

where ρ is the density, and c_p is the specific heat at constant pressure. For an isotropic, homogeneous body without local heat sources or sinks, the basic partial differential equation for transient heat conduction is

$$\frac{\partial T}{\partial t} = \alpha \, \nabla^2 T$$

The more complicated equations and their solutions are discussed by Jakob [29], Schneider [62], and Carslaw and Jaeger [10]. Values of the diffusivity α are not tabulated, but may be calculated, using conductivity values from this section, densities from Sec. 2b, and specific heats from Sec. 4e.

4g-2. Heat Transport in Fluids. In fluids three types of heat transport can occur: radiation, convection, and conduction. Thermal radiation becomes more important as the temperature of a system is increased. For most systems above about 1000 K it becomes a significant contribution to the total heat transfer. Radiative heat transfer is also significant, however, in low-temperature apparatus if any of the critical components are exposed to room-temperature radiation. Radiative constants are given in Sec. 6g of this Handbook. Convection is particularly important in systems with fluid density inversions, heated vertical surfaces, or forced fluid flow. Convection is not actually a separate mode of heat transfer, but rather a complex combination of fluid conduction, solid-to-fluid boundary conduction, and fluid flow. Therefore it is not surprising that solutions for realistic convective heat-transfer problems are complicated but inexact. The physical parameters entering the equations are complex, for they depend not only on the temperature and pressure but also on the shape, position, material, and roughness of the surfaces; the composition and density of the fluids; and the velocity and the type of fluid flow, whether laminar or turbulent, forced, or free. Convective heat transfer is discussed in detail by McAdams [38]; Bird, Stewart, and Lightfoot [6]; and Rohsenow and Choi [58].

For a dilute gas, the thermal conductivity increases slowly with temperature ($\sim T^{0.6}$), and is in principle independent of density or pressure. A definition of "diluteness" is given by Childs and Hanley [12]. A gas is essentially dilute up to about 10 atm at room temperatures and about 40 atm at 1000 K.

A convenient equation for estimating the thermal conductivity of a dilute monatomic gas is

$$\lambda = \frac{15}{4} \frac{R}{M} \mu$$

where R is the gas constant, M the molecular weight, and μ the viscosity. For dilute polyatomic gases a correction factor is needed. The simplest is the Eucken formula,

$$\lambda = \frac{15}{4} \frac{R}{M} \mu \left(\frac{3}{5} + \frac{4}{15} \frac{c_v}{R}\right)$$

where c_v is the specific heat at constant volume. A survey of more sophisticated corrections and of the theories and equations for thermal conductivities of gases is given by Hirschfelder, Curtiss, and Bird [25]. More recent work on dense gas theories is discussed by Sengers [64]. The thermal conductivities for dilute inorganic gases are given in Table 4g-2; for dilute organic gases in Table 4g-3.

Near the critical or condensation region, the thermal conductivity of a fluid is very density- and pressure-dependent, as is shown by the typical set of curves in Fig. 4g-2. Most classical fluids show a similar behavior. Water, however, is an exception. Its conductivity along the liquidus curve above the dome has a broad maximum near 140°C. Green and Sengers [19] review recent work on the anomalous behavior of the thermal conductivity of fluids near the critical point.

The conductivity of a gas is also density- and pressure-dependent at high pressures. Data on the effect of pressure on the conductivity of four gases are given in Table 4g-4. Rough estimates for the conductivity of other gases at high pressures can be made, using the principle of corresponding states as explained by Hirschfelder, Curtiss, and Bird [25].

At low pressures where the effective mean free path of the molecule is limited by the dimensions of the container (below about 10^{-4} atm for many systems), heat conduction through a gas is directly proportional to the pressure. Conduction is then not a property of the gas alone, but also depends upon the gas-wall interactions as represented by the accommodation coefficients or temperature discontinuities at the walls. The conductivity values listed in Tables 4g-2 to 4g-4 do not apply under the above conditions. This transport phenomenon at low pressures is called *free-molecule* or *Knudsen gas conduction*. Formulas, discussions, and coefficients for this effect have been given by Kennard [32] and, more recently, by Corruccini [14], von Ubisch [74], and Devienne [15]. Corruccini also discusses the transition region between free-molecule and regular gas conduction.

The thermal conductivity of classical liquids decreases with increasing temperatures, although water is again an exception. Data for the conductivities of normal liquids near room temperature and at atmospheric pressure are given in Table 4g-5; of cryogenic liquids at saturation pressures in Table 4g-6; and of liquid metals in Table 4g-7. Liquid conductivities at very high pressures are given by Bridgman [7]. For most substances, the conductivity is about 10 times larger in the liquid phase than in the gaseous phase. Similarly, solid conductivity near the melting point is considerably larger than liquid conductivity, liquid bismuth and tellurium being exceptions.

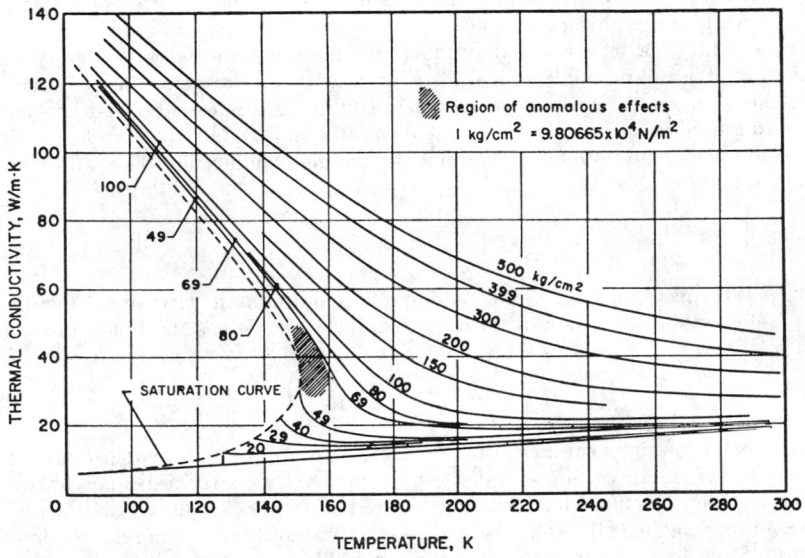

FIG. 4g-2. Effect of pressure on the thermal conductivity of argon. [B. J. Bailey and K. Kellner, Physica 39, 444 (1968).]

4g-3. Heat Transport in Solids. Two principal mechanisms are responsible for the transport of heat energy in a solid. The first is the drift motion of conduction electrons; the second is the directional cooperative vibration of interacting lattice ions, represented by the quasi-particle concept, phonons. Other mechanisms such as internal radiation or excitons may be important in some materials. Electron conduction is predominant in metals and alloys; phonon conduction in dielectrics and some highly disordered alloys. Both electron and phonon conductions are limited at low temperatures by impurity and imperfection scattering. Therefore the thermal conductivity at low temperatures is critically dependent on the exact amount and types of impurities and imperfections. At high temperatures electron conduction is limited by phonon or lattice scattering primarily and therefore is not critically dependent on the impurities. The conductivities of various metals or alloys of approximately the same composition tend to converge at high temperatures. Phonon conduction in crystalline dielectrics is limited at high temperatures by phonon-phonon scattering. Therefore the thermal conductivities of crystalline dielectrics also tend to converge to common values at high temperatures. Phonon conduction in disordered dielectrics is highly limited at most temperatures by imperfection scattering.

TABLE 4g-1. CONVERSION FACTORS FOR THERMAL CONDUCTIVITY*

	Watt cm / cm² °K	Watt m / m² °K	Watt in. / in.² °R	Cal cm / cm² sec °K	Kcal m / m² hr °R	Cal in. / in.² sec °R	Btu in. / in.² sec °R	Btu in. / in.² hr °R	Btu ft / ft² hr °R	Btu in. / ft² hr °R
Watt cm / cm²°K =	1.000	100.0	1.411	0.2390	86.04	0.3373	1.338×10^{-3}	4.818	57.82	693.8
Watt m / m²°K =	1.000×10^{-2}	1.000	1.411×10^{-2}	2.390×10^{-3}	0.8604	3.373×10^{-3}	1.338×10^{-5}	4.818×10^{-2}	0.5782	6.938
Watt in. / in.²°R =	0.7087	70.87	1.000	0.1694	60.97	0.2390	9.485×10^{-4}	3.414	40.97	491.7
Cal cm / cm² sec °K =	4.184	418.4	5.904	1.000	360.0	1.411	5.600×10^{-3}	20.16	241.9	2903
Kcal m / m² hr °R =	1.162×10^{-2}	1.162	1.640×10^{-2}	2.778×10^{-3}	1.000	3.920×10^{-3}	1.555×10^{-5}	5.600×10^{-2}	0.6720	8.064
Cal in. / in.² sec °R =	2.965	296.5	4.184	0.7087	255.1	1.000	3.968×10^{-3}	14.29	171.4	2057
Btu in. / in.² sec °R =	747.2	7.472×10^{4}	1054	178.6	6.429×10^{4}	252.0	1.000	3600	4.320×10^{4}	5.184×10^{5}
Btu in. / in.² hr °R =	0.2075	20.75	0.2929	4.961×10^{-2}	17.86	7.000×10^{-2}	2.778×10^{-4}	1.000	12.00	144.0
Btu ft / ft² hr °R =	1.730×10^{-2}	1.730	2.441×10^{-2}	4.134×10^{-3}	1.488	5.833×10^{-3}	2.315×10^{-5}	8.333×10^{-2}	1.000	12.00
Btu in. / ft² hr °R =	1.441×10^{-3}	0.1441	2.034×10^{-3}	3.445×10^{-4}	0.1240	4.861×10^{-4}	1.929×10^{-6}	6.944×10^{-3}	8.333×10^{-2}	1.000

* Units are given in terms of (1) the absolute joule per second or watt, (2) the defined thermochemical calorie = 4.184 joules, or (3) the defined British thermal unit (Btu) where 1.8 Btu/lb = 1 cal/g and therefore 1 Btu = 1,054.35 joules.

Therefore the thermal conductivity of disordered dielectrics is very low compared with that of metals or crystalline dielectrics. A review of the phenomena and various mechanisms for metals at low temperatures was given by Powell [44], and a more detailed study of the theories and concepts for solids by Rosenberg [60]. The book edited by Tye [73] also contains reviews of both the theoretical and experimental aspects for solids.

The thermal conductivity of a metal or alloy can be estimated by using the Wiedemann-Franz-Lorenz law,

$$\lambda = L\sigma T$$

where σ is the electrical conductivity, λ is the thermal conductivity, T is temperature, and L is the Lorenz ratio whose Sommerfeld classical value is

$$L \approx 2.45 \times 10^{-8} \text{ (volt/kelvin)}^2$$

For both pure metals and alloys the Lorenz ratio generally approaches the Sommerfeld value at high and very low temperatures. For pure metals L is lower than the above number at temperatures below the ice point; for alloys it is higher, as much as 10 times greater (near 20 K) for very disordered, multicomponent alloys.

Data for the conductivity of metals are given in Table 4g-8. It should be noted that the values quoted at 4.2 K, and often at 20 K, are for the most pure specimen that has been measured at the present. Future measurements on more pure metals may give substantially higher conductivites. Above 20 K the quoted values should not change substantially (more than 5 per cent) as more pure metals are measured with more refined techniques. Data for the conductivity of some commercial alloys are given in Table 4g-9; of semiconductors in Table 4g-10; of crystalline dielectrics and optical materials in Table 4g-11; and of disordered materials in Table 4g-12.

A few selected values for solids below 1 K are given in Table 4g-13. The conductivity of normal metals varies linearly with temperature at temperatures below their conductivity maximum. In superconductors the electrons do not contribute significantly to the transport of heat. Therefore their conductivity is governed by phonon processes and usually varies as T^3. The conductivity of crystalline solids is usually size-dependent below 1 K and therefore depends on the exact specimen configuration.

A recent review of the literature by Childs et al. [46] includes data, tables, and graphs on most solids for temperatures at and below 300 K.

TABLE 4g-2. THERMAL CONDUCTIVITY OF DILUTE INORGANIC GASES
(In milliwatts/meter · kelvin)

Gas	Ref.	20 K	60 K	80 K	100 K	200 K	300 K	400 K	600 K	800 K	1000 K
Ar	21	....	....	.....	6.44	12.5	17.7	22.3	29.7	36.1	41.8
Air	24	....	....	.....	9.22	18.2	26.1	33.0	45.6	56.9	67.2
Br$_2$	72	....	....	.....	.....	......	5.7*				
Cl$_2$	72	....	....	.....	.....	......	8.89	12.4	19.0		
CO	24, 72	....	....	.....	8.75	17.4	25.2	32.3	44.4	54.9	64.4
CO$_2$	24, 72	....	....	.....	.....	9.53	16.6	24.6	38.0	54.0	67.0
D$_2$	72	....	36.0	47.5	57.7	101	141	176			
F$_2$	72	....	....	.....	8.60	18.3	27.9	37.1	52.7	61.8	
Freon-12	72	....	....	.....	.....	......	9.70	15.1			
H$_2$ (normal)	72	15.5	42.6	55.2	67.6	128	182	228	291	360	428
H$_2$ (para)	30	15.5	43.0	57.8	75.0	146	187				
He	52	25.8	52.1	63.1	73.0	115	150	180	247	307	363
Hg	72	....	....	.....	.....	......	......	7.7†			
Kr	22	....	....	.....	.....	6.48	9.48	12.2	16.8	20.7	
N$_2$	12	....	....	7.62	9.76	18.7	26.1	32.4	43.9	55.2	66.0
Ne	22	....	14.8	18.6	21.7	37.0	48.9	59.1	76.7	92.0	105
NH$_3$	52	....	....	.....	.....	15.3	24.6	36.4	65.6	97.9	
O$_2$	12	....	....	.....	9.04	18.3	26.6	34.1	47.4	59.4	71.8
Steam	28	....	....	.....	.....	......	18.1	26.4	46.4	68.0	
Xe	22	....	....	.....	.....	3.82	5.52	7.17	10.2	12.8	15.1

* At 350 K.
† At 476 K.
Various authors differ from 2 to 10 % on the experimental results for the conductivities of gases.

TABLE 4g-3. THERMAL CONDUCTIVITY OF DILUTE ORGANIC GASES
(In milliwatts/meter · kelvin)

Gas	Formula	200 K	300 K	400K	500 K	1000 K
Acetone	(CH$_3$)$_2$CO	.....	11.5	20.1	31.0	
Benzene	C$_6$H$_6$	.....	10.4	19.5	33.5	
Carbon tetrachloride	CCl$_4$	.....	6.73	9.89	12.6	
Ethane	C$_2$H$_6$	10.2	21.8	36.0	51.6	164
Ethyl alcohol	C$_2$H$_5$OH	.....	.....	24.5	32.7	
Ethylene	C$_2$H$_4$	8.80	20.4	35.0		
Ethyl ether	C$_4$H$_{10}$O	.....	......	25.0	37.1	
Methane	CH$_4$	21.8	34.3	48.4	67.1	169
Methyl alcohol	CH$_3$OH	.....	.....	24.9	35.1	
Propane	C$_3$H$_8$	.....	18.3	29.5	41.7	

Various authors differ from 2 to 10 % on the experimental results for the conductivities of gases.
Values quoted are from ref. 72.

Table 4g-4. Pressure Effect on Thermal Conductivity of Gases
(In milliwatts/meter · kelvin)

Gas	Ref.	T, K	Pressure, atm[†]			
			1	10	100	300
Ar..........	3, 24, 79	90	5.19	119*	124	136
		100	6.44	106*	113	126
		120	7.70	.....	91.5	108
		140	8.95	.....	70.1	90.5
		160	10.2	.....	48.4	75.4
		180	11.4	11.7	30.0	62.9
		200	12.5	13.0	23.9	53.1
		300	17.7	18.1	22.4	34.0
N₂..........	24, 79	80	7.62	131*	139	154
		100	9.76	95*	110	129
		120	11.3	13.4	82	107
		140	13.0	14.2	58	88
		160	14.7	15.9	41	75
		180	16.7	17.6	34	64
		200	18.7	19.3	30	58
		300	26.1	.....	32	
		400	32.4	.....	37	
		600	43.9	.....	48	
		800	55.2	.....	59	
O₂..........	24, 79	80	163*	164*	170	
		90	8.4	151*	158	
		100	9.04	137*	146	
		120	11.3	13.4	121	
		140	13.0	15.1	95	
		160	15.1	16.7	66	
		180	16.7	18.0	42	
		200	18.3	19.7	33	
Steam.......	14, 28	380	24.5	680*	690*	700
		400	26.4	680*	690*	710
		420	28.2	680*	690*	710
		440	30.0	670*	690*	700
		460	31.7	33.3	680*	700
		480	33.7	34.4	670*	690
		500	35.0	35.6	660*	670
		550	41.1	39.1	590*	620
		600	46.4	42.9	61.6	530
		650	51.8	46.9	57.0	212
		700	57.2	51.0	57.2	113
		750	63.0	55.1	59.3	85.5
		800	68.0	59.3	62.2	76.7

* Indicates a liquid below its critical pressure.
† 1 atm = 1.01325×10^5 N/m².
Various authors differ from 2 to 10 % on the experimental results for the conductivities of gases.

TABLE 4g-5. THERMAL CONDUCTIVITY OF LIQUIDS NEAR ROOM TEMPERATURE
(In milliwatts/meter · kelvin)

Liquid	−20°C	0°C	20°C	40°C	60°C	Ref.
Acetone	177	169	162	155	.....	72
Benzene	...	...	146	141	136	72
Benzene (ortho-dichloro)	...	...	...	127	122	51
Carbon tetrachloride	113	109	105	101	97.2	72
Ethyl alcohol	179	174	168	162	156	72
Glycerol	...	...	287	290	292	57
Kerosene	...	...	...	147	142	61
Methyl alcohol	216	210	204	198	193	72
Oil, mineral	...	...	131	129	127	57
Oil, petroleum	...	...	150	...	.....	61
Oil, silicone (mol. wt. 162)	...	...	...	...	99.3	61
Oil, silicone (mol. wt. 1,200)	...	...	...	...	132	61
Oil, silicone (mol. wt. 15,800)	...	...	...	...	160	61
Oil, transformer	...	136	134	132	131	61
Toluene	146	141	136	131	126	72
Water	...	562	597	627	652	72
Water, heavy	...	554	579	600	620	51

Various authors differ from 2 to 15 % on the experimental results for the conductivities of liquids near room temperature.

For additional results on liquids see refs. 28, 51, and 61.

TABLE 4g-6. THERMAL CONDUCTIVITY OF CRYOGENIC LIQUIDS
AT SATURATION PRESSURE
(In milliwatts/meter · kelvin)

Liquid	Ref.	T, K	Conductivity	Liquid	Ref.	T, K	Conductivity
Ar..........	3	85	125	N$_2$........	79	70	150
		90	117			72	147
		95	111			74	144
		100	105			76	142
		105	100			78	139
		110	93.0			80	136
		115	88.5			82	134
		120	80.9			84	131
		125	74.5			86	128
		130	69.5			88	125
		135	62.9				
		140	56.8				
		145	50.0				
D$_2$..........	55	21	128	Ne........	37	25	117
		22	130			26	116
		23	132			27	114
						28	112
H$_2$..........	30	16	109			29	106
		18	113				
		20	118				
		22	123				
		24	127				
		26	132				
He^3........	36	1.2	10.5	O$_2$........	79	80	163
		1.4	11.2			85	157
		1.6	12.2			90	150
		1.8	13.0			95	144
		2.0	13.5			100	137
		2.1	13.8			105	129
						110	122
He^4(I)*.....	36	2.4	19.0			115	115
		2.8	19.5			120	109
		3.2	20.8			125	103
		3.6	23.3			130	96.7
		4.0	26.8			135	89.1
						140	79.1

* Heat conduction in liquid helium II is not governed by the usual heat-conduction mechanisms and equations.

Various authors differ from 2 to 8 % on the experimental results for the conductivities of liquids at saturation pressure.

TABLE 4g-7. THERMAL CONDUCTIVITY OF LIQUID METALS
(In watts/meter · kelvin)

Metal	Ref.	t, °C	Conductivity	Metal	Ref.	t, °C	Conductivity
Al	54	700	90.0	K-Na........	17	150	24
		750	91.6	(23 wt. % Na)		200	25
		800	93.2			300	26
		850	94.8			400	26
		900	96.4			500	26
		950	98.0			600	26
		1000	98.8			700	25
Bi	43	300	11.3	K-Na	17	150	24
		350	11.8	(43.5 wt. % Na)		200	25
		400	12.3			300	26
		450	12.8			400	27
		500	13.3			500	27
		550	13.9	Li	72	250	44
Bi-Pb eutectic....	53	150	9.3			300	43
(44.5 wt. % Pb)		200	10.1			400	40
		250	10.9			500	34
		300	11.7				
		350	12.4	Na	16	200	82
		400	13.1			300	76
		450	13.7			400	71
		500	14.2			500	67
Cu........	72	1100	160	Na-Hg.....	53	100	22
		1500	172	(6.3 wt. % Hg)		150	25
		1700	176	Na-Hg.....	53	100	9
		2000	177	(30 wt. % Hg)		150	11
Ga	8	50	33.1	Pb	53	350	16.0
		100	42.5			400	16.9
		150	53.8			450	17.6
		200	54.5			500	18.1
		250	57.3			550	18.4
Hg	17	0	8.4			600	18.7
		100	9.5				
		200	10.7	Sb	72	700	22
		300	11.8				
		400	12.6	Sn	43	300	34
		500	13.3			400	33
						500	33
K	16	200	45				
		300	42	Sn-Pb......	72	250	24
		400	40	(38 wt. % Pb)		300	26
		500	38			400	29
		600	35				
				Te	1	460	20
						500	13
				Zn	72	450	59
						500	59
						600	58
						700	57

Various authors differ from 5 to 75 % on the experimental results for the conductivities of liquid metals.
See the article by R. W. Powell (ref. 47) for a review.

TABLE 4g-8. THERMAL CONDUCTIVITY OF SOLID ELEMENTS
(In watts/meter · kelvin)

Element*	4.2 K†	20 K†	77 K	194 K	273 K	373 K	573 K	973 K‡
Ag.......	14,500	5,100	481	430	428	422	407	376
Al.......	17,000	11,500	440	238	235	234	233	
Au.......	2,190	1,570	354	328	318	313	306	279
B........	40	350	270	54	32	22	17	
Be.......	....	3,500	1,800	318	220	172	134	96
Bi (⊥c)...	1,590	100	27	13	11	9		
Cd.......	9,500	226	107	99	98	95	89	
Ce.......	0.5	1.9	5.0	......	11			
Co.......	95	450	205	130	100	85		
Cr.......	165	575	192	112	95	87	81	66
Cs.......	110	61						
Cu.......	11,800	10,500	610	419	401	393	384	359
Dy......	2.8	15	12	9	11			
Er......	5	12	11	14	14			
Fe.......	8,500	15,000	218	95	83.5	71.9	56.3	33.9
Ga (∥b)...	16,000	630	105	90	85			
Gd......	16	32	16	13	14			
Ge.......	900	1,500	330	98	67	47	29	18
Hf.......	3.6	18	26	......	22	22	21	
Hg.......	180	....	41.5	34				
Ho......	5.8	18	7.7	9	10.5			
In.......	850	183	....	100	87			
Ir.......	550	1,800	230	162	160	157	152	
K........	650	150	114	112	109			
La......	9							
Li......	256	720	152	90	82	80		
Lu (∥c)...	20	40	28	25	23			
Mg......	510	1,390	200	156	153	150	146	
Mn......	0.94	2.4	5.5	7.0	7.7			
Mo......	66	277	215	138	135	132	130	113
Na......	4,750	590	137	128	125			
Nb......	27	85	56	51	51	53	55	60
Nd......	....	....	....	......	16			
Ni.......	300	800	195	110	91	81	67	71
Os.......	125	540	135	91	88	85	85	
P........	0.6	27	35	18	13			
Pb.......	2,100	59	41	37	35	34	32	
Pd......	800	600	82	76	76	76		
Pr.......	....	....	6.8	11	12.5			
Pt.......	910	490	87	75	73	72	72	74
Pu.......	....	....	2.8	4.6	6.2	8.8		
Rb......	185	70	61	60	60			
Re.......	700	840	64	51	49	47	44	44
Rh......	1,110	3,800	250	155	151	147	137	
Ru......	660	2,300	190	120	117	117	117	
S........	11	2.4	0.67	0.37	0.29	0.15		
Sb......	200	220	58	31	26	21	19	
Sc.......	2.8	12	14	18	22			
Se (∥c)...	150	59	14					
Si.......	260	4,900	1,400	270	170	108	65	32
Sm......	4.8	6.9	7.1	11	13			
Sn......	7,500	230	89	73	67	63		
Ta.......	48	147	60	57	57	57	59	61
Tb......	6.0	20	12	10	13			
Tc.......	....	....	....	......	51	50	50	
Te (⊥c)..	850	90	15	6.2	4.4	3.2	2.4	
Th.......	17	54	50	39	37	37	37	38
Ti.......	5.8	28	33	25	22	21	19	20
Tl.......	1,800	80	58	49	47			
Tm......	9.0	18	12	16	17			
U........	4.6	15	21	25	27	28	30	40
V........	2.3	12	23	27	30	32	34	
W.......	4,000	5,400	264	180	170	166	141	122
Y........	2.2	10	....	......	15	15	15	15
Zn.......	970	690	138	124	119	113	103	
Zr.......	40	105	37	25	22	20	19	21

* Solid A, H_2, and He are in Table 4g-11.　Symbols in parentheses indicate heat flow parallel or perpendicular to b or c axes.　For other elements the results are for isotropic or polycrystalline elements.
　† The low-temperature conductivity depends critically on the exact amount and types of impurities and imperfections.　The values quoted are for the most pure specimen that has been measured at the present.
　‡ For temperatures greater than 973 K see reviews by refs. 46 and 52.
　Values quoted are from the reviews and compilations by R. L. Powell and Blanpied (ref. 46) and R. W. Powell et al. (ref. 52).　The values are generally for the highest-purity metals tested.　Disagreements among different modern authors are caused primarily by differences in sample purity and preparation.

TABLE 4g-9. THERMAL CONDUCTIVITY OF SELECTED COMMERCIAL ALLOYS
(In watts/meter · kelvin)

Alloy*†	Ref.	4.2 K	20 K	77 K	194 K	273 K	373 K	573 K	973 K
Aluminum:									
1100............	45	50	240	270	220	220			
2024............	45	3.2	17	56	95	130			
3003............	45	11	58	140	150	160			
5052............	45	4.8	25	77	120	140			
5083, 5086......	45	3.0	17	55	95	120			
Duralumin......	72	5.5	30	91	140	160	180		
Bismuth:									
Rose metal......	46	......	5.5	8.3	14	16			
Wood's metal....	72	4	17	23					
Copper:									
Electrolytic tough pitch...	45	330	1,300	550	400	390	380	370	350
Free cutting, leaded........	45	200	800	460	380	380			
Phosphorus deoxidized....	45	7.5	42	120	190	220			
Brass, leaded....	45	2.3	12	39	70	120			
Beryllium.......	72	2.0	17	36	70	90	113	172	
German silver...	46, 72	0.75	7.5	17	20	23	25	30	40
Silicon bronze, A.	45	......	3.4	11	23	30			
Manganin.......	46	0.48	3.2	14	17	22			
Constantan.....	46	0.9	8.6	17	19	22			
Ferrous:									
Commercial pure iron.....	46, 48	15	72	106	82	76	66	54	34
SAE 1020.......	46	13	20	58	65	65			
SAE 1095.......	46	......	8.5	31	41	45			
3 Ni, 0.7 Cr, 0.6 Mo.......	48	......	6	22	.....	33	35	36	30
4 Si............	48	......	.....	.....	.....	20	24	28	26
Stainless........	11	0.3	2	8	13	14	16	19	25
27 Ni, 15 Cr.....	48	......	1.7	55	.....	11	12	16	21
Gold:									
Gold-cobalt thermo-couple........	45	1.2	8.6	20					
Lead:									
60 Pb, 40 Sn....	72	......	28	44					
Nickel:									
80 Ni, 20 Cr.....	48	......	.....	.....	.....	12	14	17	23
Contracid.......	46	0.2	2	7.3	9.5	13			
Inconel.........	27, 48	0.5	4.2	12.5	13	15	16	19	26
Monel..........	46	0.9	7.1	15	20	21	24	30	43
Platinum:									
10 Ir...........	46	......	.....	.....	.....	31	31.4		
10 Rh..........	46	......	.....	.....	.....	30.1	30.5		
Silver:									
Silver solder.....	46	......	12	34	58				
Normal Ag ther-mocouple.....	46	48	230	310					
Tin (60 Sn, 40 Pb)..	72	16	55	51					
Titanium:									
5.5 Al, 2.5 Sn, 0.2 Fe........	27	......	1.8	4.3	6.4	7.8	8.4	10.8	
4.7 Mn, 3.99 Al, 0.14 C........	46	......	1.7	4.5	6.5	8.5			

* Commercial alloys of the same nominal composition may vary in conductivity from 5 to 25 % because of differences in heat treatment and uncontrolled impurities. Contracid, Inconel, and Monel are registered trade names for nickel alloys. See ref. 46 for additional data.
† When composition is given, it is by weight percent.

Table 4g-10. Thermal Conductivity of Semiconductors*
(In watts/meter · kelvin)

Substance	Ref.	Dopant	Electrical conductivity at room temp., ohm-m	Type	Carriers at room temp., no./m³	Temperature					
						4K	20K	77K	194K	273K	373K
Graphite	50		9.8×10^{-6}	⊥C		0.25	15	190	300	250	540
	50, 72		4.1×10^{-5}	∥C		0.2	7.4	68	84	80	28
	50, 72		2.4×10^{-5}	Ext.		0.02	0.7	~15			50
	70		5.7×10^{-5}	Molded			0.08	1.2	4	5	
	50, 72		1.2×10^{-6}	Natural						160	
	50, 72		2.4×10^{-4}	Deposit		6	150	610	270	580	
	70		2.1×10^{-3}	Coke					90	28	
Ge	9			p,n	10^{20}	1000	1500	300		70	
	9	Cu	3×10^{-2}	p	10^{21}	290	1300	310			
		In	3×10^{-5}	p	10^{25}	8.5	270	200			
		Ga		p							
Si	56	O_2	5.5×10^{-2}	n	7.8×10^{20}	310	1900	900	270	150	
	66	P	4.6×10^{-1}	p	4×10^{20}	200	4200	1500	270	150	
	9	Au	2.2×10^{-1}	p	4×10^{26}	87	1300	840	250	150	
	66	B	4.5×10^{-2}	p	4×10^{21}	140	3500	1500	270	150	
Bi_2Te_3	72			n	3×10^{23}			6	3	3	4.9
GaAs	26			n	7×10^{15}	1000	1900	300	80	50	5.0
InAs	72			n	10^{25}					6.7	5.2
InSb	26			n	3×10^{22}		1000	110		6.7	12
	26			p	10^{23}					17	14
PbS	20			n	2×10^{24}	17	46	8		18	
PbSe	20, 56			p	8×10^{23}	0.7	2.8	3.9		2.4	
				n	5×10^{24}	24	33	5.5			
PbTe	20			p	2×10^{24}	11	31	7		2.3	
				n	6×10^{24}			8	1.3		
SiC	66	N	10×10^{-2}	n	10^{23}	30	2000	3000	1000	500	1.7
	66	Al	6×10^{-3}	p	4×10^{25}	0.08	1.5	50	150	140	2.4

* Pure Ge and Si are in Table 4g-8.

Various authors differ from 5 to 10% on the experimental results for the conductivities of semiconductors.

See also the review by E. F. Steigmeier in vol. 2, pp. 203–251, of ref. 73.

TABLE 4g-11. THERMAL CONDUCTIVITY OF CRYSTALLINE DIELECTRICS
AND OPTICAL MATERIALS
(In watts/meter · kelvin)

Material	Ref.	T, K	Conductivity	Material	Ref.	T, K	Conductivity
Ar	52	8	6.0	Glass (plastic perspex)	46	4.2	0.058
		10	3.7			20	0.074
		20	1.4	Glass (Pyrex)	46	77	0.44
		77	0.31			194	0.88
AgCl	39	223	1.3			273	1.0
		273	1.2	H_2 (para + 0.5 % ortho)	46	2.5	100
		323	1.1			3.0	150
		373	1.1			4.0	200
Al_2O_3 (sapphire) 36 deg to c axis	46	4.2	110			6.0	30
		20	3,500			10	3
		35	6,000	H_2O (ice)	72	173	3.5
		77	1,100			223	2.8
Al_2O_3 (sapphire) ⊥ to c axis	48, 72	373	2.6			273	2.2
		523	3.9	He^3	52	0.6	25
		773	5.8			1.0	2
Al_2O_3 (sintered)	46, 48	4.2	0.5			1.5	0.57
		20	23			2.0	0.21
		77	150	He^4	52	0.5	42
		194	48			0.8	120
		273	35			1.0	24
		373	26			2.0	0.18
		973	8	I	52	300	0.45
As_2S_3 (glass)	72	283	0.16			325	0.42
		323	0.21			350	0.40
		373	0.27	KBr	39, 75	2	150
BaF_2	39	225	20			4.2	360
		260	13.4			100	12
		305	10.9			273	5.0
		370	10.5			323	4.8
BeO	46, 48	4.2	0.3			373	4.8
		20	16	KCl	39, 75	4.2	500
		77	270			25	140
		373	210			80	35
		573	120			194	10
		1,273	29			273	7.0
C (diamond)	46	4.2	75			323	6.5
		20	1,600			373	6.3
		77	3,400	KI	75	4.2	700
		194	870			80	13
		273	660			194	4.6
$CaCO_3$ ∥ to c axis	72	83	25			273	3.1
		273	5.5	Kr	52	4.2	0.48
$CaCO_3$ ⊥ to c axis	72	83	17			10	1.7
		194	6.5			20	1.2
		273	4.6			77	0.36
		373	3.6	LiF	5, 72	4.2	620
CaF_2	39	83	39			20	1,800
		223	18			77	150
		273	10			373	13
		323	9.2			773	8.5
		373	9.0	$MgO \cdot Al_2O_3$ (spinel)	48, 68		
CsBr	39	223	1.2	MnO	72	4.2	0.25
		273	0.94			40	55
		323	0.81			120	8.0
		373	0.77			573	3.5
CsI	39	223	1.4	NaCl	39, 77	4.2	440
		273	1.2			20	300
		323	1.0			77	30
		373	0.95			273	6.4
Glass (phoenix)	46	4.2	0.095			323	5.6
		20	0.13			373	5.4
		77	0.37	NaF	69, 75	5	1,100
						50	250
						100	90

TABLE 4g-11. THERMAL CONDUCTIVITY OF CRYSTALLINE DIELECTRICS
AND OPTICAL MATERIALS (*Continued*)
(In watts/meter · kelvin)

Material	Ref.	T, K	Conductivity	Material	Ref.	T, K	Conductivity
Ne....................	52	2	3.0	SiO$_2$ (quartz) ⊥ to c axis.	72	20	370
		3	4.6			194	10
		4.2	4.2			273	6.8
		10	0.8	SiO$_2$ (fused)...........	72	4.2	0.25
		20	0.3			20	0.7
NH$_4$Cl...............	46	77	17			77	0.8
		194	23			194	1.2
		230	38			273	1.4
		273	27			373	1.6
NH$_4$H$_2$PO$_4$ ∥ to optic axis	72	315	0.71			673	1.8
		339	0.71	TlBr................	72	316	0.59
NH$_4$H$_2$PO$_4$ ⊥ to optic axis................	72	313	1.26	TlCl.................	72	311	0.75
		342	1.34	TiO$_2$ (rutile) ∥ to optic axis................	72	4.2	200
NiO..................	72	4.2	5.9			20	1,000
		40	400			273	13
		194	82	TiO$_2$ (rutile) ⊥ to optic axis................	72	4.2	160
SiO$_2$ (quartz) ∥ to c axis.	72	20	720			20	690
		194	20			273	9
		273	12				

Various authors differ from 5 to 25 % on the experimental results for the conductivities of crystalline dielectrics and optical materials.
See ref. 46 for additional data.

TABLE 4g-12. THERMAL CONDUCTIVITY OF DISORDERED DIELECTRICS, CERAMICS, REFRACTORY OXIDES, AND INSULATING MATERIALS
(In watts/meter · kelvin)

Material	Density g/cc	t, °C	Conductivity	Material	Density g/cc	t, °C	Conductivity
Alumina	3.8	100	30	Magnesium oxide	....	100	36
		400	13			400	18
		1,300	6			1,200	5.8
		1,800	7.4			1,700	9.2
	3.5	100	17	Magnesium + SiO_2	....	100	5.3
		800	7.6			400	3.5
Alumina + MgO	....	100	15			1,500	2.3
		400	10	Mica, muscovite	....	100	0.72
		1,000	5.6			300	0.65
Asbestos	0.4	−100	0.07			600	0.69
		0	0.09	Phlogopite	....	100	0.66
		100	0.1	Canadian	....	300	0.19
Asbestos + 85 % MgO	0.3	30	0.08			600	0.20
Barium titanate	....	50	3	Micanite	....	30	0.3
		100	2.8	Mineral wool	0.15	30	0.04
		150	2.6	Paper, fiber glass and Al foil layers	0.1	−200 to 20	0.0001
		200	2.4				
Beryllia	2.8	100	210	Perlite, expanded	0.1	−200 to 20	0.002
		400	90				
		1,000	20	Plastic:			
		1,800	15	Celluloid	1.4	30	0.02
	1.85	50	64	Polystyrene foam	0.05	−200 to 20	0.033
		200	40				
		600	23	Aluminized Mylar foil	0.05	−200 to 20	0.0001
Brick, dry	1.54	0	0.04				
Brick refractory:				Porcelain	....	90	1
Aloxite	....	1,000	1.3	Rock, basalt	....	20	2
Aluminous	1.99	400	1.2	Chalk	....	20	0.92
		1,000	1.3	Granite	2.8	20	2.2
Diatomaceous	0.77	100	0.20	Limestone	2.0	20	1
		500	0.24	Sandstone	2.2	20	1.3
	0.40	100	0.08	Slate, ⊥	....	95	1.4
		500	0.10	Slate, ∥	....	95	2.5
Fireclay	2.0	400	1	Rubber, sponge	0.2	20	0.05
		1,000	1.2	Rubber, 92 %	....	25	0.16
Silicon carbide	2	200	2	Sawdust	0.2	30	0.06
		600	2.4	Shellac	....	20	0.23
Vermiculite	0.77	200	0.26	Silica aerogel	0.1	−200 to 20	0.003
		600	0.31				
Calcium oxide	....	100	16	Silica aerogel + 50 % Al	0.1	−200 to 20	0.003
		400	9				
		1,000	7.5	Snow	0.25	0	0.16
Cement mortar	2.0	90	0.55	Steel wool	0.1	55	0.09
Charcoal	0.2	20	0.055	Thoria	....	100	10
Concrete	1.6	0	0.8			400	5.8
Cork	0.05	0	0.03			1,500	2.4
		100	0.04	Titanium dioxide	....	100	6.5
	0.35	0	0.06			400	3.8
		100	0.08			1,200	3.3
Cotton wool	0.08	30	0.04	Uranium dioxide	....	100	9.8
Diatomite	0.2	0	0.05			400	5.5
		400	0.09			1,000	3.4
	0.5	0	0.09	Wood, balsa, ⊥	0.11	30	0.04
		400	0.16	Fir, ⊥	0.54	20	0.14
Ebonite	1.2	0	0.16	Fir, ∥	0.54	20	0.35
Felt, flax	0.2	30	0.05	Pine, ⊥	0.45	60	0.11
	0.3	30	0.04	Pine, ∥	0.45	60	0.26
	0.3	30	0.05	Walnut, ⊥	0.65	20	0.14
Fuller's earth	0.53	30	0.10	Wool	0.09	30	0.04
Glass wool	0.2	−200 to 20	0.005	Zinc oxide	....	200	17
		50	0.01			800	5.3
		100	0.05	Zirconia	....	100	2
		300	0.08			400	2
Graphite, 100 mesh	0.48	40	0.18			1,500	2.5
20–40 mesh	0.70	40	1.29	Zirconia + SiO_2	....	200	5.6
Linoleum, cork	0.54	20	0.08			600	4.6
						1,500	3.7

Values quoted are from W. D. Kingery et al. (ref. 33), R. W. Powell (ref. 48), and "International Critical Tables" (ref. 42).
Various authors differ from 10 to 50 % on the experimental results for the conductivities of disordered dielectrics, ceramics, refractory oxides, and insulating materials.

TABLE 4g-13. THERMAL CONDUCTIVITY OF SOME MATERIALS BELOW 1 K†
(In watts/meter · kelvin)

Material	Ref.	Temperature				
		0.2 K	0.4 K	0.6 K	0.8 K	1.0 K
Cr K Alum..............	72	0.05	0.3*	0.8*	2*	3.2
Epibond 104‡...........	2	0.00018	0.001	0.0025	0.0048	0.006*
In (superconducting).....	18, 72	0.1*	1.2	7	15	23
KCl...................	4, 65	0.2*	1.6	5	16	30
KI....................	65	1.5*	6	22	50	100
Kel-F‡................	2	0.00029	0.0012	0.0027	0.0045	0.006*
LiF...................	23	0.09	0.8	2	6.5	10
Nb (normal, superconducting).................	13, 72	4.5*	9*	12*	20*	25
	13	0.1*	0.5	1.5	3.5	6
Nylon‡................	2	0.00018	0.0006	0.0011	0.0015*	0.002*
Pyrex.................	2	0.001	0.003	0.0065	0.011	0.016*
Rubber (hard).........	2	0.0025	0.007	0.015	0.02*	0.03*
Sn (normal, superconducting).................	52, 78	100*	2000*	4500	6000	7000
	52, 35	0.4	2.5	15	80	150
Ta (normal, superconducting)	13, 72	3*	6*	8*	10*	11
	13	0.15*	0.9	3	9	10
Teflon‡................	2	0.00008	0.0004	0.0013	0.0022*	0.0035*
Tl (normal, superconducting)	52, 78	2000*	4000*	6000*	7000*	8500*
	52, 78	0.06	5	120	600	1200

The conductivities of materials below 1 K are strongly dependent upon their chemical purity and physical perfection and, for small specimens, depend on the actual dimensions and surfaces. The values quoted are typical for these materials. See ref. 46 for additional data.

* Extrapolated values.

† The thermal conductivity of normal metals in the impurity range varies as $\lambda = aT$.

‡ Epibond is a diglycidal ether of bisphenol which is a fluid mixed in a ratio of 4:1 with powdered Bentonite clay. Kel-F is a polychlorotrifluorethylene. Nylon is a polyamide. Teflon is a polytetrafluorethylene.

General References for Sec. 4g

1. Amirkhanov, Kh. I., G. B. Bagduev, and M. A. Kazhlaev: *Soviet Phys. Doklady* **2**, 556 (1957).
2. Anderson, A. C., W. Reese, and J. C. Wheatley: *Rev. Sci. Instr.* **34**, 1386 (1963).
3. Bailey, B. J., and K. Kellner: *Physica* **39**, 444 (1968).
4. Berman, A.: *Cryogenics* **5**, 297 (1965).
5. Berman, R., and J. C. F. Brock: *Proc. Roy. Soc. (London)*, ser. A, **289**, 46 (1965).
6. Bird, R. B., W. E. Stewart, and E. N. Lightfoot: "Transport Phenomena," John Wiley & Sons, Inc., New York, 1960.
7. Bridgman, P. W.: *Proc. Am. Acad. Arts Sci.* **59**, 158 (1932).
8. Briggs, L. J.: *J. Chem. Phys.* **26**, 784 (1957).
9. Carruthers, J. A., T. H. Geballe, H. M. Rosenberg, and J. M. Ziman: *Proc. Roy. Soc. (London)*, ser. A, **238**, 502 (1957).
10. Carslaw, H. S., and J. C. Jaeger: "Conduction of Heat in Solids," 2d ed., Oxford University Press, London, 1959.
11. Chari, M. S. R., and J. de Nobel: *Physica* **25**, 73 (1959).
12. Childs, G. E., and H. J. M. Hanley: *NBS Tech. Note* 350, October, 1966; *Cryogenics* **8**, 94 (1968).
13. Connolly, A., and K. Mendelssohn: *Proc. Roy. Soc. (London)*, ser. A, **266**, 429 (1962)
14. Corruccini, R. J.: *Vacuum* **7–8**, 19 (1959).
15. Devienne, F. M.: *Mem. sci. phys.* **56**, 1 (1953).
16. Ewing, C. T., J. A. Grand, and R. R. Miller: *J. Am. Chem. Soc.* **74**, 11 (1952).
17. Ewing, C. T., R. E. Seebold, J. A. Grand, and R. R. Miller: *J. Phys. Chem.* **59**, 524 (1955).
18. Graham, G. M.: *Proc. Roy. Soc. (London)* ser. A, **248**, 522 (1958).

THERMAL EXPANSION   4-161

19. Green, M. S., and J. V. Sengers, eds.: Critical Phenomena: Proceedings of a Conference, *NBS Misc. Publ.* 273, 1966.
20. Greig, D.: *Phys. Rev.* **120**, 358 (1960).
21. Hanley, H. J. M.: *NBS Tech. Note* 333, March, 1966.
22. Hanley, H. J. M., and G. E. Childs: *NBS Tech. Note* 352 March, 1967.
23. Harrison, J. P.: *Rev. Sci. Instr.* **39**, 145 (1968).
24. Hilsenrath, J., et al.: "Tables of Thermodynamic and Transport Properties," Pergamon Press, New York, 1960.
25. Hirschfelder, J. O., C. F. Curtiss, and R. B. Bird: "Molecular Theory of Gases and Liquids," 2d ed., John Wiley & Sons, Inc., New York, 1964.
26. Holland, M. G.: *Phys. Rev.* **A134**, 471 (1964).
27. Hust, J. G., and R. L. Powell: *8th Conf. on Thermal Conductivity*, Purdue University, 1968.
28. Institute of Mechanical Engineers, *Proc. Joint Conf. Thermodyn. and Transport Properties of Fluids*, London, 1958.
29. Jakob, Max: "Heat Transfer," vols. I and II, John Wiley & Sons, Inc., New York, 1949, 1957.
30. Johnson, V. J., ed.: "A Compendium of the Properties of Materials at Low Temperatures (Phase I)," Office of Technical Services, Department of Commerce, Washington, D.C., 1960.
31. Kannaluik, W. G., and E. H. Carman: *Proc. Roy. Soc. (London)*, ser B, **65**, 701 (1952).
32. Kennard, E. H.: "Kinetic Theory of Gases," McGraw-Hill Book Company, New York, 1938.
33. Kingery, W. D., J. Francl, R. L. Coble, and T. Vasilos: *J. Am. Ceram. Soc.* **37**, 107 (1954).
34. "Landolt-Börnstein Zahlenwerte und Funktionen aus Physik, Chemie, Astronomie, Geophysik und Technik," II Band, 5 Teil, Bandteil b, Wärmetechnik, Springer-Verlag, Berlin, 1968.
35. Laredo, S. J.: *Proc. Roy. Soc. (London)*, ser. A, **229**, 473 (1955).
36. Lee, D. M., and H. A. Fairbank: *Proc. Intern. Conf. Low Temp. Phys. 5th*, p. 90, Plenum Press, New York, 1957.
37. Lochtermann, E.: *Cryogenics* **3**, 44 (1963).
38. McAdams, W. H.: "Heat Transmission," 3d ed., McGraw-Hill Book Company, New York, 1954.
39. McCarthy, K. A., and S. S. Ballard: *J. Appl. Phys.* **31**, 1410 (1960).
40. Mendelssohn, K., and H. M. Rosenberg: The Thermal Conductivity of Metals at Low Temperatures in "Solid State Physics," vol. 12, F. Seitz and D. Turnbull, eds., Academic Press, Inc., New York, 1961.
41. Metals Handbook Committee: "Metals Handbook," 8th ed., vol. 1, American Society for Metals, Novelty, Ohio, 1961.
42. National Academy of Sciencies and National Research Council: "International Critical Tables," McGraw-Hill Book Company, New York, 1926–1930.
43. Pashaev, B. P.: *Soviet Phys.—Solid State* **3**, 303 (1961).
44. Powell, R. L.: *ASTM Spec. Tech. Publ.* 387, 1966.
45. Powell, R. L., et al.: *J. Appl. Phys.* **28**, 1282 (1957); **31**, 496, 504, 1221 (1960).
46. Powell, R. L., and W. A. Blanpied: Thermal Conductivity of Metals and Alloys at Low Temperatures, *NBS Circ.* 556, 1954. Revised and expanded by G. E. Childs, L. J. Ericks, and R. L. Powell: Thermal Conductivity of Solids at Low Temperatures, NBS Monograph, to be published in 1971.
47. Powell, R. W.: *J. Iron Steel Inst.* **162**, 315 (1949).
48. Powell, R. W.: *Research* **7**, 492 (1954).
49. Powell, R. W.: *Bull. inst. intern. du froid*, Annexe 1955–2 p. 115, 1955.
50. Powell, R. W.: "Industrial Carbon and Graphite," p. 46, Society of Chemical Industry, London, 1958.
51. Powell, R. W.: *Advan. Phys. (Phil. Mag. Suppl.)* **7**, 276 (1958); A. R. Challoner and R. W. Powell: *Proc. Roy. Soc. (London)*, ser. A, **238**, 90 (1956); A. R. Challoner, H. A. Gundry, and R. W. Powell; *Proc. Roy. Soc. (London)*, ser. A, **245**, 259 (1958).
52. Powell, R. W., C. Y. Ho, and P. E. Liley: "Thermal Conductivity of Selected Materials," National Standard Reference Data Series, National Bureau of Standards: no. 8, part 1, 1966; C. Y. Ho, et al.: no. 16, part 2, 1968; no. 18, part 3 (to be published).
53. Powell, R. W., and R. P. Tye: "Joint Conference on Thermodynamic and Transport Properties of Fluids," p. 182, Institute of Mechanical Engineers, London, 1958.
54. Powell, R. W., R. P. Tye, and S. C. Metcalf: *ASME 3d Symp. on Thermophys. Properties*, p. 289, 1965.
55. Powers, R. W., R. W. Mattox, and H. L. Johnston: *J. Am. Chem. Soc.* **76**, 5968, 5972, 5974 (1954).
56. *Prague Intern. Confer. on Semiconductors Proc.*, 1960.

57. Riedel, L.: *Chem. Ing. Tech.* **22**, 107 (1950), and **23**, 321, 465 (1951); Wärmeleitfähigkeitmessungen an Flüssigkeiten, and *Mitt. kältetech. Inst. tech. Hochschule Karlsruhe*, no. 2.
58. Rohsenow, W. M., and H. Y. Choi: "Heat, Mass, and Momentum Transfer," Prentice-Hall, Inc., Englewood Cliffs, N.J., 1961.
59. Rosenberg, H. M.: *Phil. Trans. Roy. Soc. (London)*, ser. A, **247**, 441 (1955).
60. Rosenberg, H. M.: "Low Temperature Solid State Physics," Oxford University Press, London, 1963.
61. Sakiadis, B. C., and J. Coates: *Louisiana State Univ. Expt. Sta. Bulls.* 34, 1952, and 48, 1954; *A. I. Ch. E. Journal* **1**, 275 (1955) (literature surveys).
62. Schneider, P. J.: "Conduction Heat Transfer," Addison-Wesley Press, Inc., Cambridge, Mass., 1955.
63. Schön, M., and H. Welker: "Semiconductors and Phosphors," Fredrig Vieweg und Sohn, Brunswick, Germany, 1958.
64. Sengers, J. V.: Transport Properties of Compressed Gases, in *Proc. 4th Tech. Meeting Soc. Eng. Sci.*, Gordon and Breach, Science Publishers, Inc., New York, 1968.
65. Seward, W. D.: *Proc. Intern. Conf. Low Temp. Phys.* 9th, p. 1130, Plenum Press, Plenum Publishing Corporation, New York, 1965.
66. Slack, G. A.: *J. Appl. Phys.* **35**, 3460 (1964).
67. Slack, G. A., and C. Glassbrenner: *Phys. Rev.* **120**, 782 (1960).
68. Smakula, A.: Physical Properties of Optical Crystals, *Office of Tech. Serv. Doc.* PB 111053, 1952.
69. Smirnov, I. A.: *Soviet Phys.—Solid State* **9**, 1454 (1967).
70. Smith, A. W., and N. S. Rasor: *Phys. Rev.* **104**, 885 (1956).
71. Thermal Conductivity Conferences, unpublished proceedings: e.g., 6th—Dayton, Ohio, 1966; 7th—National Bureau of Standards, Washington, D.C., 1967 (published as *NBS Spec. Publ.* 302); 8th—Lafayette, Inc., 1968.
72. Touloukian, Y. S., director: "Thermophysical Properties Research Center Data Book," vols. I, II, and III, Purdue University, Lafayette, Ind., 1964.
73. Tye, R. P., ed., "Thermal Conductivity," 2 vols., Academic Press, Inc., London, 1969.
74. von Ubisch, H.: *Appl. Sci. Research*, **A2**, 364 (1951).
75. Walker, C. T.: *Phys. Rev.* **132**, 1963 (1963).
76. White, G. K., and S. B. Woods: *Phil. Trans. Roy. Soc. (London)*, ser. A, **251**, 273 (1959).
77. Worlock, J. M.: *Phys. Rev.* **147**, 636 (1966).
78. Zavaritskii, N. V.: *Soviet Phys. JETP* **12**, 1093 (1961).
79. Ziebland, H., and J. T. A. Burton: *Brit. J. Appl. Phys.* **6**, 416 (1955); **9**, 52 (1958).
80. Ziman, J. M.: "Electrons and Phonons," Oxford University Press, London, 1960.

4h. Thermodynamic Properties of Gases

JOSEPH HILSENRATH

The National Bureau of Standards

4h-1. Thermodynamic Properties of Nonionized Gases. The thermodynamic properties of air, argon, hydrogen, nitrogen, oxygen at temperatures below 3000 K (Tables 4h-3 through 4h-27) are an abridged version of a collection of tables computed and published at the National Bureau of Standards [1]. The tables of compressibility and density were computed from equations of state which were fitted to the existing PVT data. In most instances the method of fitting permitted simultaneous consideration of other experimental data, such as Joule-Thomson coefficients, specific heat, and sound-velocity measurements. The tables for entropy, enthalpy, and specific heats were obtained by combining these properties of the ideal gas with corrections for the gas imperfection obtained, through the thermodynamic identities, from the equation of state. A fuller discussion and more extensive tabulations in the temperature argument are to be found in the above-cited circular of the National Bureau of Standards.

The tables are presented in dimensionless form. Conversion factors given in Tables 4h-1 and 4h-2 permit ready conversion to some of the more frequently used units. Values of the gas constant R are listed for frequently used units in order to facilitate the use of the tables of the compressibility factor in calculating, by means of the equation $Z = PV/RT$, the pressure P, the specific volume V (or density $1/V$), or the temperature T, when any two of these are known. The molecular weights given in Table 4h-2 permit extension of the tabulated values of R to still other units.

Pressure entries have been chosen to facilitate four-point Lagrangian interpolation, when linear interpolation is not valid. A convenient rule of thumb for determining the adequacy of linear interpolation is the following: "The error introduced by linear interpolation is approximately $\frac{1}{8}$ of the second difference." Where this error greatly exceeds the uncertainty of the table, nonlinear interpolation is recommended.

4h-2. Thermodynamic Properties of Ionized Gases. The thermodynamic properties of air, nitrogen, and argon are given in Tables 4h-28, 4h-29, and 4h-30. The properties are given for one mole of low-temperature gas whose molecules, atoms, and ions are in chemical equilibrium with the electrons. The tables include the effect of second virial forces and the limiting-law Debye-Hückel effect upon both the equilibrium compositions and the thermodynamic properties of the mixture.

The tables for air are given at 2000 K intervals from 4000 to 14,000 K at uniform intervals in log (ρ/ρ_0) from $-5.$ to 2. They contain the dimensionless quantities: compressibility factor, $Z = PV/RT$; internal energy, E/RT; enthalpy, H/RT; entropy, S/R; the logarithm of the pressure, $\log_{10} P_{atm}$; and Z^*, the number of moles of dissociated gas per mole of low-temperature (undissociated) air. The data are taken from more extensive tables by Hilsenrath and Klein [2] who present the equilibrium composition in addition to the above-enumerated properties, and the equations from which the tables were computed.

The tables for nitrogen and argon contain, in addition, the specific heat at constant pressure, c_p/R; the specific heat at constant volume, c_v/R; and the sound velocity ratio, a/a_0. These tables are from a more extensive set by Hilsenrath, Messina, Klein, and Thompson [3], to which the reader is referred for a detailed discussion of the computation of equilibrium properties of a gas mixture undergoing dissociation and ionization under the influence of both ionic and virial forces.

Table 4h-31 gives $Z^* = PV/RT$, E^*/RT, P (atm), and ρ/ρ_0 for highly ionized air, nitrogen, and oxygen as a function of the electron concentration C_e and temperature*

TABLE 4h-1. VALUES OF THE GAS CONSTANT R IN VARIOUS UNITS

P	V	T	R
N/m²................	m³/mole	K	8.3143 Nm/mole·K
atm.................	cm³/mole	K	82.0567 atm cm³/mole·K
kg/cm².............	cm³/mole	K	84.7832 (kg/cm²)cm³/mole·K
bars................	cm³/mole	K	83.1440 bars cm³/mole·K
mm Hg.............	cm³/mole	K	62,363.1 (mm Hg)cm³/mole·K
atm.................	liters/mole	K	0.0820544 atm liters/mole·K
kg/cm².............	liters/mole	K	0.0847809 (kg/cm²) liters/mole·K
mm Hg.............	liters/mole	K	62.3613 (mm Hg) liters/mole·K
atm.................	ft³/(lb)mole	°R	0.730228 atm ft³/mole °R
mm Hg.............	ft³/(lb)mole	°R	554.973 (mm Hg) ft³/mole °R

* Concluded on page 4-204.

TABLE 4h-2. CONVERSION FACTORS FOR TABLES 4h-4 THROUGH 4h-30

To convert tabulated value of	To	Having the dimensions indicated below	Air	Argon	CO₂	H₂	N₂	O₂	Steam
						Multiply by			
$(H - E_0°)/RT_0$	$(H - E_0°)$	cal mole⁻¹	542.821	542.821	542.821	542.821	542.821	542.821	542.821
		cal g⁻¹	18.7399	13.5996	12.3340	269.256	19.3754	16.9632	30.1299
		joules g⁻¹	78.4079	56.8589	51.6056	1126.57	81.0669	70.9742	126.064
		Btu (lb mole)⁻¹	976.437	976.437	976.437	976.437	976.437	976.437	976.437
		Btu lb⁻¹	33.7098	24.4451	22.1867	484.344	34.8528	30.5137	54.1893
C_p, S	$C_p/R, S/R$	cal mole⁻¹ °K⁻¹ (or °C⁻¹)	1.98719	1.98719	1.98719	1.98719	1.98719	1.98719	1.98719
		cal g⁻¹ °K⁻¹ (or °C⁻¹)	0.0686042	0.0497494	0.0451531	0.985709	0.0709305	0.0620997	0.110301
		joules g⁻¹ °K⁻¹ (or °C⁻¹)	0.287041	0.208152	0.188921	4.12422	0.296774	0.259826	0.461500
		Btu (lb mole)⁻¹ °R⁻¹ (or °F⁻¹)	1.98588	1.98588	1.98588	1.98588	1.98588	1.98588	1.98588
		Btu lb⁻¹ °R⁻¹ (or °F⁻¹)	0.0685590	0.0497166	0.0451234	0.985060	0.0708838	0.0620588	0.110229
ρ/ρ_0 and for steam of ρ in g cm⁻³	ρ	g cm⁻³	1.29304×10^{-3}	1.78377×10^{-3}	1.9770×10^{-3}	8.99854×10^{-5}	1.25046×10^{-3}	1.42900×10^{-3}	1
		mole cm⁻³	4.46400×10^{-5}	4.46568×10^{-5}	4.4922×10^{-5}	4.45860×10^{-5}	4.46338×10^{-5}	4.46562×10^{-5}	0.055506
		g liter⁻¹	1.29308	1.78382	1.9771	8.98879×10^{-2}	1.25050	1.42904	1.00003×10^{3}
		lb in⁻³	4.67143×10^{-6}	6.44432×10^{-6}	7.1424×10^{-6}	3.24734×10^{-6}	4.51760×10^{-5}	5.16262×10^{-6}	3.61275×10^{-2}
		lb ft⁻³	8.07223×10^{-2}	0.111358	0.12342	5.61140×10^{-3}	7.80641×10^{-2}	8.92101×10^{-2}	62.4283
Molecular weight			28.966	39.944	44.010	2.016	28.016	32.000	18.016

TABLE 4h-3. COMPRESSIBILITY FACTOR FOR AIR, $Z = \dfrac{PV}{RT}$

T, K	1 atm	4 atm	7 atm	10 atm	40 atm	70 atm	100 atm
100	0.98090						
200	0.99767	0.99067	0.98367	0.97666	0.9080	0.8481	0.8105
300	0.99970	0.99879	0.99797	0.99717	0.99135	0.9900	0.9933
400	1.00019	1.00079	1.00141	1.00205	1.00946	1.0188	1.0299
500	1.00034	1.00137	1.00242	1.00348	1.01454	1.0265	1.0393
600	1.00038	1.00152	1.00267	1.00385	1.01574	1.0281	1.0408
700	1.00038	1.00153	1.00268	1.00385	1.01558	1.0275	1.0397
800	1.00037	1.00148	1.00259	1.00371	1.01493	1.0263	1.0379
900	1.00035	1.00140	1.00246	1.00351	1.01411	1.0248	1.0356
1000	1.00033	1.00132	1.00231	1.00331	1.01325	1.0233	1.0333
1100	1.00031	1.00124	1.00218	1.00311	1.01245	1.0218	1.0312
1200	1.00029	1.00117	1.00205	1.00293	1.01170	1.0205	1.0292
1300	1.00028	1.00110	1.00193	1.00275	1.01100	1.0192	1.0275
1400	1.00026	1.00104	1.00182	1.00259	1.01037	1.0181	1.0259
1500	1.00024	1.00098	1.00171	1.00245	1.00978	1.0171	1.0244
1600	1.00023	1.00094	1.00163	1.00233	1.0093	1.0162	1.0232
1700	1.00023	1.00090	1.00157	1.00223	1.0088	1.0154	1.0220
1800	1.00024	1.00087	1.00152	1.00213	1.0083	1.0146	1.0208
1900	1.00027	1.00085	1.00146	1.00204	1.0079	1.0138	1.0198
2000	1.00035	1.00085	1.00140	1.00196	1.0076	1.0132	1.0188
2100	1.0006	1.0010	1.0014	1.0019	1.0073	1.0126	1.0180
2200	1.0008	1.0010	1.0014	1.0019	1.0070	1.0121	1.0172
2300	1.0014	1.0013	1.0016	1.0020	1.0067	1.0116	1.0165
2400	1.0023	1.0017	1.0019	1.0022	1.0067	1.0113	1.0160
2500	1.0036	1.0024	1.0024	1.0026	1.0066	1.0110	1.0155
2600	1.0056	1.0034	1.0031	1.0032	1.0067	1.0108	1.0151
2700	1.0086	1.0048	1.0042	1.0041	1.0068	1.0107	1.0148
2800	1.0124	1.0068	1.0057	1.0053	1.0071	1.0108	1.0145
2900	1.0178	1.0096	1.0079	1.0071	1.0079	1.0111	1.0147
3000	1.0252	1.0133	1.0107	1.0095	1.0092	1.0119	1.0151

TABLE 4h-4. RELATIVE DENSITY OF AIR, ρ/ρ_0

T, K	1 atm	4 atm	7 atm	10 atm	40 atm	70 atm	100 atm
100	2.7830						
200	1.3681	5.511	9.713	13.976	60.13	112.66	168.40
300	0.9102	3.644	6.383	9.125	36.72	64.34	91.61
400	0.6823	2.7277	4.771	6.811	27.043	46.89	66.27
500	0.5458	2.1809	3.813	5.441	21.526	37.23	52.53
600	0.4548	1.8171	3.176	4.532	17.917	30.977	43.71
700	0.3898	1.5575	2.7226	3.885	15.360	26.567	37.51
800	0.3411	1.3629	2.3825	3.400	13.449	23.274	32.879
900	0.3032	1.2115	2.1180	3.023	11.964	20.720	29.290
1000	0.2729	1.0905	1.9065	2.721	10.777	18.675	26.419
1100	0.24809	0.9914	1.7334	2.474	9.805	17.001	24.066
1200	0.22742	0.9089	1.5892	2.268	8.994	15.605	22.103
1300	0.20993	0.8390	1.4671	2.094	8.308	14.422	20.438
1400	0.19494	0.7791	1.3625	1.945	7.720	13.406	19.007
1500	0.18195	0.7272	1.2718	1.815	7.209	12.525	17.766
1600	0.17058	0.6818	1.1924	1.702	6.762	11.753	16.675
1700	0.16054	0.6417	1.1223	1.602	6.367	11.070	15.712
1800	0.15162	0.6061	1.0600	1.513	6.016	10.463	14.857
1900	0.14364	0.5742	1.0043	1.434	5.702	9.921	14.089
2000	0.13645	0.5455	0.9541	1.362	5.419	9.430	13.398
2100	0.12992	0.5194	0.9087	1.297	5.162	8.986	12.770
2200	0.12399	0.4958	0.8674	1.239	4.929	8.582	12.199
2300	0.11852	0.4741	0.8295	1.185	4.716	8.213	11.676
2400	0.11348	0.4542	0.7947	1.135	4.520	7.873	11.195
2500	0.10880	0.4357	0.7625	1.089	4.339	7.560	10.753
2600	0.10441	0.4185	0.7327	1.047	4.172	7.271	10.343
2700	0.10024	0.4024	0.7048	1.007	4.017	7.003	9.963
2800	0.09630	0.3873	0.6786	0.970	3.872	6.752	9.610
2900	0.09249	0.3729	0.6538	0.935	3.736	6.517	9.277
3000	0.08876	0.3592	0.6302	0.901	3.607	6.295	8.964

TABLE 4h-6. ENTHALPY OF AIR, $(H - E_0^\circ)/RT_0$

T, K	1 atm	4 atm	7 atm	10 atm	40 atm	70 atm	100 atm
100	1.2552						
200	2.5465	2.5281	2.5094	2.4908	2.2922	2.0794	1.8734
300	3.8292	3.8204	3.8118	3.8034	3.7194	3.6411	3.5699
400	5.1167	5.1125	5.1079	5.1039	5.0623	5.0252	4.9926
500	6.4195	6.4176	6.4154	6.4137	6.3951	6.3795	6.3670
600	7.7463	7.7459	7.7454	7.7449	7.7408	7.7388	7.7390
700	9.1023	9.1027	9.1035	9.1037	9.1096	9.1168	9.1253
800	10.489	10.490	10.491	10.492	10.505	10.519	10.534
900	11.904	11.906	11.908	11.909	11.928	11.947	11.968
1000	13.348	13.350	13.352	13.354	13.377	13.400	13.424
1100	14.817	14.819	14.822	14.824	14.851	14.877	14.904
1200	16.310	16.312	16.316	16.318	16.347	16.376	16.405
1300	17.826	17.828	17.832	17.834	17.866	17.897	17.928
1400	19.363	19.365	19.370	19.373	19.407	19.440	19.471
1500	20.922	20.924	20.929	20.932	20.968	21.003	21.036
1600	22.504	22.506	22.511	22.514	22.551	22.587	22.621
1700	24.110	24.112	24.116	24.118	24.156	24.193	24.228
1800	25.740	25.740	25.744	25.746	25.784	25.821	25.857
1900	27.397	27.392	27.394	27.396	27.434	27.472	27.509
2000	29.086	29.071	29.070	29.072	29.108	29.146	29.183
2100	30.813	30.781	30.774	30.775	30.806	30.844	30.881
2200	32.592	32.527	32.510	32.509	32.530	32.566	32.603
2300	34.443	34.318	34.286	34.279	34.282	34.315	34.349
2400	36.393	36.169	36.107	36.093	36.067	36.092	36.123
2500	38.470	38.093	37.994	37.961	37.888	37.901	37.927
2600	40.713	40.110	39.955	39.895	39.750	39.744	39.764
2700	43.172	42.246	42.008	41.911	41.661	41.627	41.638
2800	45.901	44.528	44.173	44.026	43.630	43.556	43.554
2900	48.960	46.985	46.474	46.262	45.666	45.539	45.518
3000	52.403	49.655	48.940	48.650	47.784	47.583	47.537

TABLE 4h-5. SPECIFIC HEAT OF AIR, C_p/R

T, K	1 atm	4 atm	7 atm	10 atm	40 atm	70 atm	100 atm
100	3.5824						
200	3.5062	3.5495	3.5950	3.6427	4.256	5.132	6.079
300	3.5059	3.5220	3.5383	3.5546	3.722	3.889	4.046
400	3.5333	3.5416	3.5500	3.5583	3.640	3.717	3.788
500	3.5882	3.5932	3.5983	3.6032	3.652	3.697	3.739
600	3.6626	3.6660	3.6693	3.6726	3.705	3.735	3.763
700	3.7455	3.7479	3.7502	3.7525	3.775	3.797	3.817
800	3.828	3.830	3.832	3.834	3.851	3.867	3.882
900	3.906	3.908	3.909	3.910	3.924	3.936	3.947
1000	3.979	3.980	3.982	3.983	3.993	4.003	4.012
1100	4.046	4.047	4.048	4.049	4.057	4.065	4.072
1200	4.109	4.110	4.111	4.111	4.118	4.125	4.130
1300	4.171	4.172	4.172	4.173	4.179	4.184	4.189
1400	4.230	4.231	4.231	4.232	4.236	4.241	4.245
1500	4.289	4.290	4.290	4.290	4.294	4.298	4.302
1600	4.352	4.351	4.351	4.351	4.354	4.357	4.361
1700	4.418	4.414	4.413	4.414	4.416	4.419	4.421
1800	4.487	4.480	4.479	4.478	4.477	4.479	4.481
1900	4.566	4.549	4.544	4.543	4.540	4.540	4.542
2000	4.662	4.626	4.617	4.613	4.603	4.604	4.605
2100	4.781	4.715	4.699	4.692	4.674	4.670	4.671
2200	4.947	4.823	4.791	4.780	4.745	4.738	4.734
2300	5.179	4.969	4.918	4.893	4.828	4.814	4.806
2400	5.484	5.149	5.067	5.026	4.922	4.897	4.886
2500	5.882	5.373	5.247	5.186	5.028	4.987	4.971
2600	6.40	5.661	5.474	5.389	5.152	5.088	5.062
2700	7.06	6.019	5.753	5.634	5.295	5.203	5.172
2800	7.87	6.455	6.088	5.930	5.467	5.341	5.297
2900	8.86	6.993	6.497	6.300	5.668	5.496	5.434
3000	9.96	7.605	6.991	6.724	5.906	5.678	5.602

TABLE 4h-8. COMPRESSIBILITY FACTOR FOR ARGON, $Z = PV/RT$

T, K	1 atm	4 atm	7 atm	10 atm	40 atm	70 atm	100 atm
100	0.9782						
200	0.99706	0.98818	0.97923	0.97023	0.8778	0.7838	0.6917
300	0.99937	0.99750	0.99565	0.99382	0.9773	0.9643	0.9553
400	0.99998	0.99991	0.99986	0.99982	1.0002	1.0022	1.0057
500	1.00018	1.00072	1.00127	1.00183	1.0079	1.0147	1.0224
600	1.00025	1.00101	1.00178	1.00255	1.0105	1.0190	1.0279
700	1.00027	1.00111	1.00194	1.00278	1.0113	1.0201	1.0292
800	1.00028	1.00111	1.00195	1.00279	1.0113	1.0199	1.0288
900	1.00027	1.00109	1.00191	1.00273	1.0110	1.0194	1.0279
1000	1.00026	1.00104	1.00183	1.00261	1.0105	1.0185	1.0265
1100	1.00025	1.00100	1.00174	1.00249	1.0100	1.0176	1.0252
1200	1.00024	1.00095	1.00166	1.00237	1.0095	1.0167	1.0239
1300	1.00023	1.00090	1.00158	1.00225	1.0090	1.0158	1.0226
1400	1.00021	1.00085	1.00149	1.00213	1.0085	1.0149	1.0213
1500	1.00020	1.00081	1.00142	1.00203	1.0081	1.0142	1.0203
1600	1.00019	1.00077	1.00135	1.00193	1.0077	1.0135	1.0193
1700	1.00018	1.00073	1.00128	1.00183	1.0073	1.0128	1.0183
1800	1.00018	1.00070	1.00123	1.00175	1.0070	1.0123	1.0175
1900	1.00017	1.00067	1.00117	1.00167	1.0067	1.0117	1.0167
2000	1.00016	1.00064	1.00111	1.00159	1.0064	1.0111	1.0159
2100	1.00015	1.00061	1.00107	1.00153	1.0061	1.0107	1.0153
2200	1.00015	1.00058	1.00102	1.00146	1.0058	1.0102	1.0146
2300	1.00014	1.00056	1.00098	1.00140	1.0056	1.0098	1.0140
2400	1.00014	1.00054	1.00095	1.00135	1.0054	1.0095	1.0135
2500	1.00013	1.00052	1.00091	1.00130	1.0052	1.0091	1.0130
2600	1.00013	1.00050	1.00088	1.00125	1.0050	1.0088	1.0125
2700	1.00012	1.00048	1.00084	1.00120	1.0048	1.0084	1.0120
2800	1.00012	1.00046	1.00081	1.00116	1.0046	1.0081	1.0116
2900	1.00011	1.00045	1.00078	1.00112	1.0045	1.0078	1.0112
3000	1.00011	1.00043	1.00076	1.00108	1.0043	1.0076	1.0108

TABLE 4h-7. ENTROPY OF AIR, S/R

T, K	1 atm	4 atm	7 atm	10 atm	40 atm	70 atm	100 atm
100	20.049						
200	22.497	21.091	20.513	20.139	18.551	17.767	17.184
300	23.917	22.524	21.958	21.594	20.138	19.513	19.095
400	24.929	23.539	22.976	22.616	21.194	20.602	20.214
500	25.723	24.335	23.773	23.414	22.006	21.428	21.056
600	26.383	24.995	24.434	24.077	22.677	22.104	21.736
700	26.954	25.567	25.006	24.649	23.253	22.685	22.320
800	27.460	26.073	25.512	25.155	23.762	23.196	22.833
900	27.915	26.528	25.968	25.610	24.219	23.655	23.293
1000	28.330	26.944	26.384	26.025	24.634	24.071	23.709
1100	28.713	27.327	26.767	26.408	25.018	24.454	24.093
1200	29.068	27.682	27.122	26.763	25.373	24.809	24.448
1300	29.399	28.013	27.463	27.093	25.702	25.138	24.777
1400	29.711	28.324	27.764	27.404	26.013	25.448	25.087
1500	30.005	28.618	28.058	27.698	26.306	25.741	25.380
1600	30.284	28.897	28.337	27.977	26.585	26.020	25.659
1700	30.549	29.162	28.602	28.242	26.850	26.287	25.926
1800	30.804	29.416	28.856	28.496	27.104	26.542	26.181
1900	31.048	29.660	29.100	28.740	27.348	26.785	26.424
2000	31.284	29.896	29.335	28.974	27.582	27.019	26.658
2100	31.514	30.124	29.563	29.201	27.808	27.245	26.884
2200	31.740	30.346	29.784	29.421	28.027	27.463	27.102
2300	31.964	30.563	29.999	29.636	28.240	27.676	27.314
2400	32.191	30.778	30.212	29.847	28.447	27.883	27.520
2500	32.423	30.992	30.422	30.055	28.650	28.084	27.721
2600	32.663	31.208	30.632	30.263	28.849	28.281	27.918
2700	32.917	31.428	30.844	30.471	29.046	28.476	28.111
2800	33.188	31.654	31.059	30.681	29.242	28.669	28.302
2900	33.481	31.889	31.279	30.895	29.438	28.861	28.491
3000	33.799	32.136	31.507	31.114	29.634	29.052	28.678

TABLE 4h-10. SPECIFIC HEAT OF ARGON, C_p/R

T, K	1 atm	4 atm	7 atm	10 atm	40 atm	70 atm	100 atm
100	2.6077						
200	2.5154	2.5626	2.612	2.663	3.31	4.2	5.2
300	2.5057	2.5230	2.5404	2.5581	2.74	2.93	3.12
400	2.5029	2.5118	2.5206	2.5294	2.61	2.70	2.79
500	2.5018	2.5071	2.5124	2.5176	2.570	2.621	2.670
600	2.5012	2.5047	2.5082	2.5117	2.546	2.579	2.611
700	2.5008	2.5033	2.5058	2.5082	2.532	2.555	2.578
800	2.5006	2.5025	2.5043	2.5062	2.524	2.541	2.558
900	2.5005	2.5020	2.5033	2.5047	2.519	2.531	2.544
1000	2.5004	2.5015	2.5026	2.5037	2.515	2.525	2.536
1100	2.5003	2.5012	2.5021	2.5030	2.512	2.520	2.528
1200	2.5002	2.5010	2.5017	2.5024	2.510	2.516	2.523
1300	2.5002	2.5008	2.5014	2.5020	2.508	2.514	2.519
1400	2.5002	2.5007	2.5012	2.5017	2.507	2.512	2.516
1500	2.5001	2.5006	2.5010	2.5014	2.506	2.510	2.513
1600	2.5001	2.5005	2.5009	2.5012	2.505	2.509	2.511
1700	2.5001	2.5004	2.5007	2.5011	2.504	2.507	2.511
1800	2.5001	2.5004	2.5006	2.5009	2.504	2.506	2.509
1900	2.5001	2.5003	2.5006	2.5008	2.503	2.506	2.508
2000	2.5001	2.5003	2.5005	2.5007	2.503	2.505	2.507
2100	2.5001	2.5002	2.5004	2.5006	2.502	2.504	2.506
2200	2.5001	2.5002	2.5004	2.5005	2.502	2.504	2.505
2300	2.5000	2.5002	2.5003	2.5005	2.502	2.503	2.505
2400	2.5000	2.5002	2.5003	2.5004	2.502	2.503	2.504
2500	2.5000	2.5002	2.5003	2.5004	2.502	2.503	2.504
2600	2.5000	2.5001	2.5002	2.5003	2.501	2.502	2.503
2700	2.5000	2.5001	2.5002	2.5003	2.501	2.502	2.503
2800	2.5000	2.5001	2.5002	2.5003	2.501	2.502	2.503
2900	2.5000	2.5001	2.5002	2.5003	2.501	2.502	2.503
3000	2.5000	2.5001	2.5002	2.5002	2.501	2.502	2.502

TABLE 4h-9. RELATIVE DENSITY OF ARGON, ρ/ρ_0

T, K	1 atm	4 atm	7 atm	10 atm	40 atm	70 atm	100 atm
100	2.79						
200	1.3685	5.5232	9.754	14.064	62.18	121.9	197.3
300	0.91023	3.6477	6.3954	9.1531	37.23	66.03	95.22
400	0.68226	2.7292	4.7764	6.8237	27.28	47.65	67.84
500	0.54570	2.1816	3.8157	5.4480	21.66	37.65	53.38
600	0.45471	1.8175	3.1781	4.5367	18.00	31.25	44.25
700	0.38975	1.5577	2.7237	3.8877	15.42	26.75	37.88
800	0.34103	1.3630	2.3832	3.4017	13.49	23.41	33.16
900	0.30314	1.2116	2.1185	3.0239	12.00	20.82	29.50
1000	0.27283	1.0905	1.9068	2.7219	10.80	18.76	26.59
1100	0.24803	0.99136	1.7336	2.4747	9.825	17.07	24.20
1200	0.22736	0.90879	1.5893	2.2688	9.011	15.66	22.21
1300	0.20987	0.83893	1.4671	2.0945	8.322	14.47	20.53
1400	0.19489	0.77904	1.3625	1.9451	7.731	13.44	19.09
1500	0.18189	0.72714	1.2717	1.8156	7.219	12.56	17.83
1600	0.17053	0.68172	1.1923	1.7023	6.770	11.78	16.73
1700	0.16050	0.64164	1.1223	1.6023	6.375	11.09	15.76
1800	0.15158	0.60601	1.0600	1.5134	6.022	10.48	14.90
1900	0.14361	0.57414	1.0042	1.4339	5.707	9.938	14.13
2000	0.13643	0.54544	0.95408	1.3623	5.423	9.447	13.43
2100	0.12993	0.51949	0.90868	1.2975	5.167	9.000	12.80
2200	0.12403	0.49589	0.86742	1.2386	4.933	8.595	12.23
2300	0.11863	0.47434	0.82974	1.1848	4.720	8.225	11.70
2400	0.11369	0.45458	0.79520	1.1355	4.524	7.885	11.22
2500	0.10914	0.43641	0.76342	1.0902	4.344	7.572	10.78
2600	0.10495	0.41963	0.73408	1.0483	4.178	7.283	10.37
2700	0.10106	0.40410	0.70692	1.0095	4.024	7.016	9.987
2800	0.097452	0.38967	0.68169	0.9735	3.881	6.768	9.635
2900	0.094092	0.37624	0.65820	0.93997	3.747	6.536	9.306
3000	0.090956	0.36371	0.63628	0.90868	3.623	6.320	8.999

Table 4h-12. Entropy of Argon, S/R

T, K	1 atm	4 atm	7 atm	10 atm	40 atm	70 atm	100 atm
100	15.8425						
200	17.6069	16.2012	15.6218	15.245	13.64	12.83	12.2
300	18.6245	17.2308	16.6637	16.2995	14.8389	14.2067	13.781
400	19.3449	17.9548	17.3913	17.0308	15.6067	15.0118	14.618
500	19.9032	18.5146	17.9527	17.5937	16.1850	15.6037	15.2261
600	20.3593	18.9715	18.4104	18.0522	16.6513	16.0776	15.7072
700	20.7449	19.3575	18.7969	18.4391	17.0426	16.4732	16.1070
800	21.0787	19.6917	19.1313	18.7739	17.3802	16.8134	16.4498
900	21.3733	19.9864	19.4263	19.0690	17.6772	17.1122	16.7503
1000	21.6368	20.2500	19.6900	19.3328	17.9423	17.3785	17.0179
1100	21.8751	20.4884	19.9285	19.5715	18.1819	17.6190	17.2592
1200	22.0923	20.7060	20.1462	19.7892	18.4003	17.8381	17.4789
1300	22.2927	20.9062	20.3464	19.9895	18.6010	18.0394	17.6807
1400	22.4780	21.0916	20.5318	20.1749	18.7869	18.2256	17.8673
1500	22.6505	21.2640	20.7043	20.3474	18.9597	18.3988	18.0408
1600	22.8119	21.4254	20.8657	20.5089	19.1214	18.5607	18.2029
1700	22.9635	21.5771	21.0174	20.6606	19.2733	18.7128	18.3552
1800	23.1064	21.7200	21.1603	20.8035	19.4165	18.8561	18.4987
1900	23.2415	21.8551	21.2955	20.9387	19.5518	18.9915	18.6343
2000	23.3698	21.9834	21.4238	21.0670	19.6802	19.1201	18.7630
2100	23.4917	22.1053	21.5457	21.1890	19.8022	19.2422	18.8851
2200	23.6080	22.2217	21.6620	21.3053	19.9187	19.3587	19.0017
2300	23.7192	22.3329	21.7732	21.4165	20.0299	19.4701	19.1131
2400	23.8256	22.4393	21.8797	21.5229	20.1364	19.5766	19.2197
2500	23.9276	22.5413	21.9817	21.6249	20.2385	19.6787	19.3218
2600	24.0257	22.6394	22.0798	21.7231	20.3366	19.7769	19.4201
2700	24.1200	22.7337	22.1741	21.8174	20.4310	19.8713	19.5145
2800	24.2109	22.8246	22.2650	21.9083	20.5219	19.9623	19.6055
2900	24.2987	22.9124	22.3528	21.9961	20.6098	20.0501	19.6934
3000	24.3834	22.9971	22.4375	22.0808	20.6945	20.1349	19.7782

Table 4h-11. Enthalpy of Argon, $(H - E_0^\circ)/RT_0$

T, K	1 atm	4 atm	7 atm	10 atm	40 atm	70 atm	100 atm
100	0.8935						
200	1.8236	1.8029	1.7819	1.7606	1.53	1.3	
300	2.7422	2.7319	2.7217	2.7114	2.610	2.512	2.42
400	3.6590	3.6532	3.6476	3.6418	3.586	3.533	3.48
500	4.5750	4.5718	4.5686	4.5654	4.535	4.506	4.48
600	5.4907	5.4891	5.4874	5.4859	5.471	5.457	5.445
700	6.4063	6.4057	6.4052	6.4047	6.400	6.397	6.395
800	7.3218	7.3220	7.3222	7.3226	7.326	7.330	7.335
900	8.2372	8.2380	8.2388	8.2396	8.249	8.258	8.268
1000	9.1525	9.1538	9.1551	9.1564	9.170	9.184	9.198
1100	10.0679	10.0696	10.0712	10.0729	10.090	10.107	10.125
1200	10.9832	10.9852	10.9871	10.9891	11.009	11.029	11.049
1300	11.8985	11.9007	11.9029	11.9051	11.927	11.950	11.972
1400	12.8138	12.8162	12.8186	12.8210	12.845	12.869	12.894
1500	13.7291	13.7316	13.7342	13.7367	13.763	13.788	13.815
1600	14.6443	14.6470	14.6497	14.6524	14.680	14.707	14.735
1700	15.5595	15.5624	15.5652	15.5680	15.597	15.625	15.654
1800	16.4749	16.4778	16.4808	16.4837	16.513	16.543	16.572
1900	17.3901	17.3931	17.3962	17.3992	17.430	17.460	17.491
2000	18.3053	18.3085	18.3116	18.3147	18.346	18.377	18.409
2100	19.2206	19.2238	19.2269	19.2301	19.262	19.294	19.326
2200	20.1358	20.1390	20.1423	20.1456	20.178	20.211	20.243
2300	21.0510	21.0543	21.0576	21.0609	21.094	21.127	21.160
2400	21.9662	21.9696	21.9729	21.9763	22.010	22.044	22.077
2500	22.8815	22.8849	22.8884	22.8918	22.926	22.960	22.994
2600	23.7967	23.8002	23.8036	23.8071	23.842	23.876	23.911
2700	24.7120	24.7154	24.7189	24.7224	24.757	24.792	24.827
2800	25.6272	25.6307	25.6342	25.6377	25.673	25.708	25.743
2900	26.5424	26.5459	26.5495	26.5530	26.589	26.624	26.659
3000	27.4576	27.4612	27.4647	27.4683	27.504	27.540	27.575

TABLE 4h-13. COMPRESSIBILITY FACTOR FOR HYDROGEN, $Z = PV/RT$

T, K	1 atm	4 atm	7 atm	10 atm	40 atm	70 atm	100 atm
40	0.9845	0.9362	0.8853	0.8317			
60	0.9955	0.9822	0.9691	0.9564	0.8757	0.8700	0.9395
80	0.9986	0.9946	0.9908	0.9872	0.9682	0.9782	1.0174
100	0.9998	0.9992	0.9987	0.9983	1.0029	1.0222	1.0560
120	1.0003	1.0012	1.0021	1.0030	1.0176	1.0405	1.0726
140	1.0005	1.0020	1.0036	1.0052	1.0243	1.0488	1.0786
160	1.0006	1.0024	1.0043	1.0062	1.0271	1.0516	1.0798
180	1.0007	1.0028	1.0048	1.0067	1.0283	1.0523	1.0785
200	1.0007	1.0028	1.0048	1.0068	1.0283	1.0513	1.0760
220	1.0007	1.0028	1.0048	1.0067	1.0276	1.0497	1.0730
240	1.0007	1.0027	1.0047	1.0066	1.0269	1.0480	1.0698
260	1.0006	1.0024	1.0044	1.0064	1.0259	1.0459	1.0667
280	1.0006	1.0024	1.0042	1.0061	1.0247	1.0439	1.0636
300	1.0006	1.0024	1.0042	1.0059	1.0238	1.0420	1.0607
320	1.0006	1.0024	1.0041	1.0057	1.0229	1.0402	1.0579
340	1.0005	1.0021	1.0037	1.0054	1.0217	1.0384	1.0553
360	1.0005	1.0020	1.0036	1.0052	1.0209	1.0367	1.0529
380	1.0005	1.0020	1.0035	1.0050	1.0201	1.0353	1.0507
400	1.0005	1.0020	1.0034	1.0048	1.0193	1.0339	1.0486
420	1.0005	1.0019	1.0033	1.0046	1.0185	1.0325	1.0466
440	1.0004	1.0017	1.0030	1.0045	1.0180	1.0314	1.0448
460	1.0004	1.0016	1.0029	1.0043	1.0172	1.0301	1.0431
480	1.0004	1.0016	1.0028	1.0041	1.0165	1.0289	1.0415
500	1.0004	1.0016	1.0028	1.0040	1.0160	1.0280	1.0400
520	1.0004	1.0016	1.0028	1.0039	1.0155	1.0271	1.0385
540	1.0004	1.0016	1.0026	1.0037	1.0148	1.0260	1.0372
560	1.0004	1.0015	1.0026	1.0036	1.0144	1.0252	1.0360
580	1.0003	1.0013	1.0024	1.0035	1.0140	1.0244	1.0348
600	1.0003	1.0012	1.0023	1.0034	1.0136	1.0237	1.0337

TABLE 4h-14. RELATIVE DENSITY OF HYDROGEN, ρ/ρ_0

T, K	1 atm	4 atm	7 atm	10 atm	40 atm	70 atm	100 atm
40	6.9408	29.195	54.029	82.160			
60	4.5761	18.552	32.905	47.632	208.08	366.53	484.88
80	3.4214	13.740	24.138	34.609	141.15	244.49	335.82
100	2.7338	10.942	19.158	27.379	109.01	187.17	258.83
120	2.2771	9.0999	15.910	22.709	89.532	153.23	212.36
140	1.9514	7.7937	13.617	19.422	76.240	130.30	181.01
160	1.7073	6.8167	11.907	16.978	66.528	113.71	158.21
180	1.5174	6.0569	10.578	15.084	59.067	101.01	140.80
200	1.3657	5.4512	9.5206	13.574	53.160	90.995	127.01
220	1.2415	4.9557	8.6551	12.341	48.361	82.849	115.79
240	1.1381	4.5431	7.9347	11.314	44.361	76.068	106.46
260	1.0506	4.1949	7.3265	10.446	40.988	70.358	98.553
280	0.97559	3.8953	6.8045	9.7026	38.105	65.457	91.780
300	0.91055	3.6356	6.3509	9.0575	35.596	61.204	85.896
320	0.85364	3.4084	5.9546	8.4931	33.401	57.479	80.740
340	0.80351	3.2088	5.6065	7.9959	31.473	54.192	76.178
360	0.75887	3.0309	5.2956	7.5532	29.748	51.265	72.110
380	0.71893	2.8714	5.0174	7.1571	28.204	48.632	68.458
400	0.68298	2.7278	4.7670	6.8006	26.815	46.286	65.165
420	0.65046	2.5982	4.5404	6.4780	25.558	44.120	62.181
440	0.62095	2.4806	4.3353	6.1842	24.408	42.160	59.457
460	0.59396	2.3729	4.1473	5.9165	23.365	40.377	56.964
480	0.56921	2.2741	3.9749	5.6711	22.407	38.740	54.675
500	0.54644	2.1831	3.8159	5.4448	21.522	37.223	52.563
520	0.52542	2.0991	3.6691	5.2359	20.704	35.823	50.615
540	0.50596	2.0214	3.5339	5.0430	19.951	34.533	48.801
560	0.48789	1.9494	3.4077	4.8634	19.246	33.326	47.113
580	0.47112	1.8826	3.2908	4.6961	18.590	32.202	45.541
600	0.45541	1.8200	3.1815	4.5400	17.977	31.150	44.070

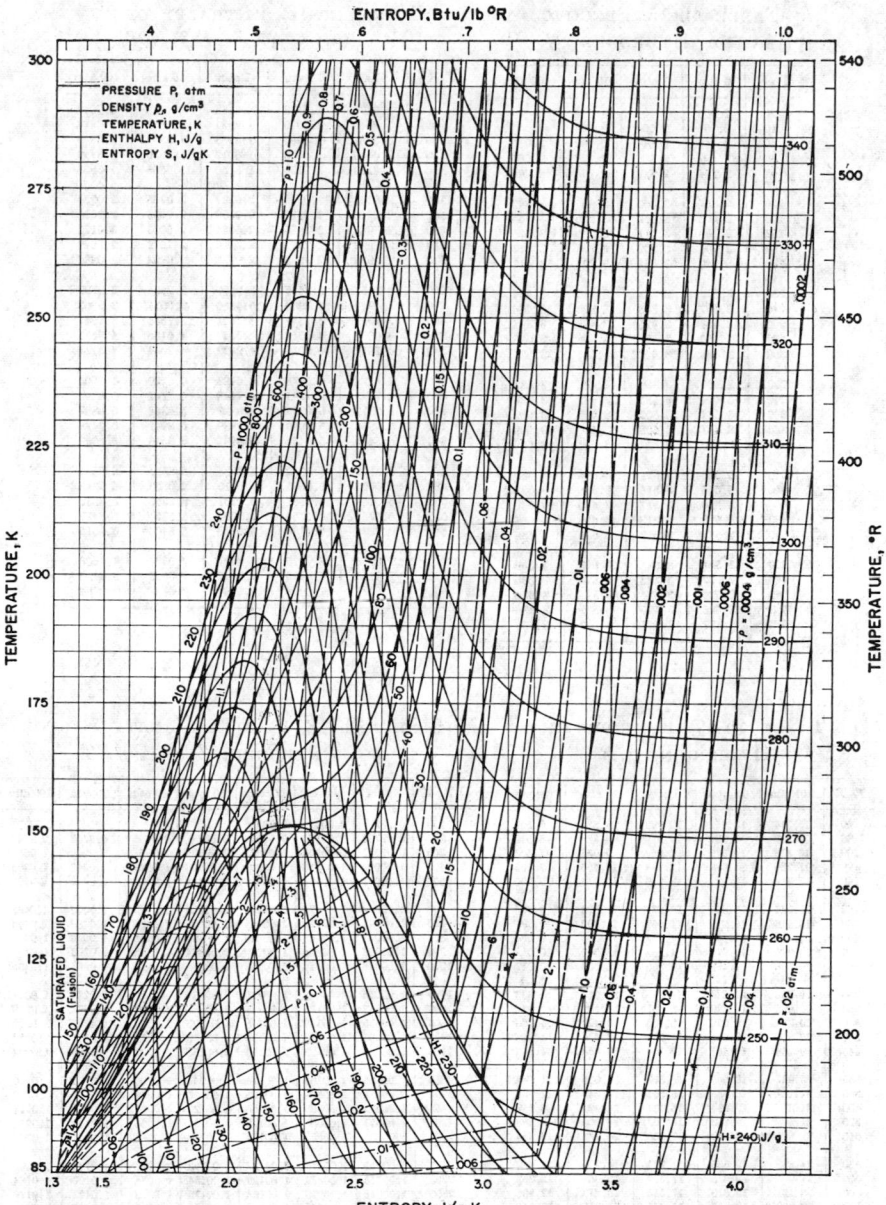

FIG. 4h-1. Temperature-entropy diagram for argon. (*From NSRDS—NBS 27 entitled Thermodynamic Properties of Argon from the Triple Point to 300 K at Pressures to 1,000 Atmospheres, authored by A. L. Gosman, R. D. McCarthy, and J. G. Hurt, U.S. Government Printing Office, Washington, D.C.*)

TABLE 4h-15. SPECIFIC HEAT OF HYDROGEN, C_p/R

T, K	0 atm	1 atm	10 atm	100 atm
20	2.500			
40	2.501	2.564	3.463	
60	2.519	2.544	2.780	3.957
80	2.591	2.605	2.723	3.564
100	2.714	2.722	2.790	3.295
120	2.857	2.862	2.905	3.242
140	2.993	2.996	3.026	3.264
160	3.108	3.111	3.135	3.326
180	3.204	3.206	3.226	3.377
200	3.280	3.282	3.296	3.413
220	3.340	3.341	3.355	3.454
240	3.387	3.388	3.399	3.486
260	3.424	3.425	3.433	3.504
280	3.450	3.451	3.458	3.516
300	3.469	3.470	3.476	3.526
320	3.483	3.484	3.489	3.532
340	3.494	3.495	3.499	3.536
360	3.501	3.502	3.506	3.539
380	3.507	3.508	3.510	3.539
400	3.510	3.511	3.514	3.539
420	3.513	3.514	3.516	3.539
440	3.515	3.516	3.518	3.538
460	3.516	3.517	3.519	3.538
480	3.518	3.518	3.520	3.537
500	3.519	3.519	3.521	3.536
520	3.521	3.521	3.523	3.536
540	3.522	3.522	3.524	3.536
560	3.524	3.524	3.526	3.536
580	3.525	3.525	3.527	3.536
600	3.527	3.527	3.529	3.536

TABLE 4h-16. ENTHALPY OF HYDROGEN, $(H - E_0°)/RT_0$

K	0.01 atm	0.1 atm	1 atm	10 atm	100 atm
60	1.0175	1.0172	1.0142	0.9833	0.7818
80	1.2042	1.2040	1.2021	1.1837	1.0577
100	1.3981	1.3980	1.3968	1.3852	1.3059
120	1.6020	1.6020	1.6012	1.5936	1.5449
140	1.8163	1.8163	1.8158	1.8108	1.7825
160	2.0398	2.0398	2.0394	2.0365	2.0234
180	2.2710	2.2710	2.2708	2.2695	2.2690
200	2.5085	2.5085	2.5084	2.5083	2.5178
220	2.7509	2.7509	2.7510	2.7519	2.7692
240	2.9973	2.9973	2.9975	2.9993	3.0236
260	3.2467	3.2467	3.2470	3.2495	3.2792
280	3.4983	3.4984	3.4986	3.5017	3.5363
300	3.7517	3.7517	3.7521	3.7556	3.7941
320	4.0063	4.0063	4.0067	4.0106	4.0525
340	4.2617	4.2617	4.2622	4.2664	4.3114
360	4.5178	4.5178	4.5183	4.5229	4.5705
380	4.7743	4.7744	4.7748	4.7797	4.8296
400	5.0312	5.0312	5.0317	5.0368	5.0887
420	5.2883	5.2883	5.2889	5.2941	5.3478
440	5.5455	5.5456	5.5461	5.5516	5.6067
460	5.8029	5.8030	5.8035	5.8091	5.8659
480	6.0604	6.0605	6.0610	6.0669	6.1249
500	6.3180	6.3181	6.3187	6.3246	6.3839
520	6.5757	6.5758	6.5764	6.5824	6.6427
540	6.8335	6.8336	6.8342	6.8404	6.9015
560	7.0915	7.0915	7.0921	7.0984	7.1606
580	7.3495	7.3496	7.3502	7.3565	7.4194
600	7.6077	7.6078	7.6084	7.6147	7.6784

TABLE 4h-17. ENTROPY OF HYDROGEN, S/R

T, K	0.01 atm	0.1 atm	1 atm	10 atm	100 atm
60	15.554	13.251	10.938	8.535	5.557
80	16.287	13.984	11.676	9.324	6.642
100	16.878	14.575	12.269	9.937	7.400
120	17.386	15.083	12.778	10.456	7.996
140	17.836	15.533	13.229	10.913	8.496
160	18.244	15.941	13.637	11.324	8.935
180	18.616	16.313	14.009	11.699	9.331
200	18.958	16.655	14.352	12.043	9.688
220	19.273	16.970	14.667	12.359	10.015
240	19.566	17.263	14.960	12.653	10.317
260	19.838	17.535	15.232	12.926	10.596
280	20.093	17.790	15.487	13.182	10.857
300	20.331	18.029	15.726	13.421	11.100
320	20.556	18.254	15.951	13.646	11.328
340	20.768	18.465	16.162	13.858	11.542
360	20.967	18.665	16.362	14.058	11.744
380	21.157	18.854	16.552	14.248	11.936
400	21.337	19.034	16.731	14.428	12.117
420	21.508	19.206	16.903	14.600	12.290
440	21.671	19.369	17.066	14.763	12.454
460	21.828	19.525	17.223	14.919	12.612
480	21.977	19.675	17.372	15.069	12.762
500	22.121	19.818	17.515	15.213	12.906
520	22.260	19.957	17.655	15.352	13.046
540	22.392	20.090	17.787	15.484	13.179
560	22.520	20.218	17.915	15.612	13.308
580	22.644	20.341	18.038	15.736	13.431
600	22.764	20.461	18.158	15.856	13.552

TABLE 4h-18. COMPRESSIBILITY FACTOR FOR NITROGEN, $Z = PV/RT$

T, K	1 atm	4 atm	7 atm	10 atm	40 atm	70 atm	100 atm
100	0.981	0.909	0.783				
200	0.99788	0.99150	0.98514	0.9788	0.9185	0.8705	0.844
300	0.99982	0.99930	0.99882	0.99838	0.9962	0.9984	1.0054
400	1.00028	1.00113	1.00201	1.00290	1.01292	1.0248	1.0383
500	1.00041	1.00164	1.00289	1.00414	1.01726	1.0313	1.0461
600	1.00044	1.00174	1.00306	1.00439	1.01795	1.0320	1.0465
700	1.00043	1.00171	1.00301	1.00430	1.01744	1.0309	1.0446
800	1.00041	1.00163	1.00286	1.00409	1.0165	1.0292	1.0420
900	1.00038	1.00154	1.00269	1.00384	1.0155	1.0273	1.0391
1000	1.00036	1.00144	1.00252	1.00360	1.0145	1.0255	1.0365
1100	1.00034	1.00135	1.00236	1.00337	1.0135	1.0238	1.0341
1200	1.00032	1.00126	1.00221	1.00316	1.0127	1.0223	1.0319
1300	1.00030	1.00119	1.00208	1.00297	1.0119	1.0209	1.0299
1400	1.00028	1.00112	1.00195	1.00279	1.0112	1.0196	1.0280
1500	1.00026	1.00105	1.00184	1.00263	1.0105	1.0185	1.0264
1600	1.00025	1.00100	1.00174	1.00249	1.0100	1.0175	1.0250
1700	1.00024	1.00094	1.00165	1.00235	1.0094	1.0165	1.0236
1800	1.00022	1.00089	1.00156	1.00223	1.0089	1.0156	1.0223
1900	1.00021	1.00085	1.00148	1.00212	1.0085	1.0148	1.0212
2000	1.00020	1.00081	1.00141	1.00202	1.0081	1.0141	1.0202
2100	1.00019	1.00077	1.00135	1.00193	1.0077	1.0135	1.0193
2200	1.00018	1.00074	1.00129	1.00184	1.0074	1.0129	1.0184
2300	1.00018	1.00070	1.00123	1.00176	1.0070	1.0123	1.0176
2400	1.00017	1.00068	1.00118	1.00169	1.0068	1.0118	1.0169
2500	1.00016	1.00065	1.00113	1.00162	1.0065	1.0113	1.0162
2600	1.00016	1.00062	1.00109	1.00156	1.0062	1.0109	1.0156
2700	1.00015	1.00060	1.00105	1.00150	1.0060	1.0105	1.0150
2800	1.00015	1.00058	1.00102	1.00145	1.0058	1.0102	1.0145
2900	1.00014	1.00056	1.00097	1.00139	1.0056	1.0097	1.0139
3000	1.00014	1.00054	1.00095	1.00135	1.0054	1.0095	1.0135

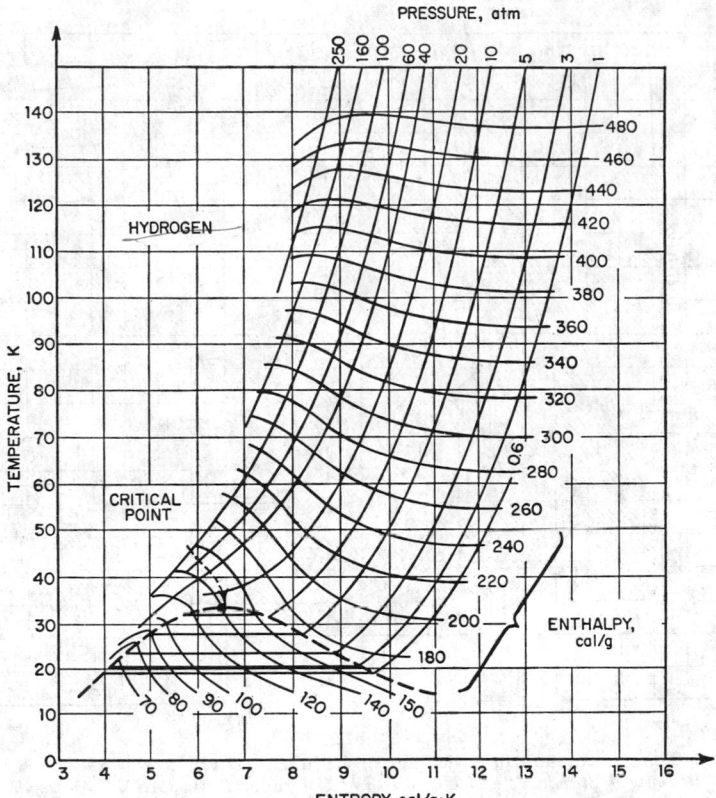

Fig. 4h-2. Temperature-entropy diagram for hydrogen. [*H. W. Woolley, R. B. Scott, and F. G. Brickwedde, NBS J. Research* **41** (1948).]

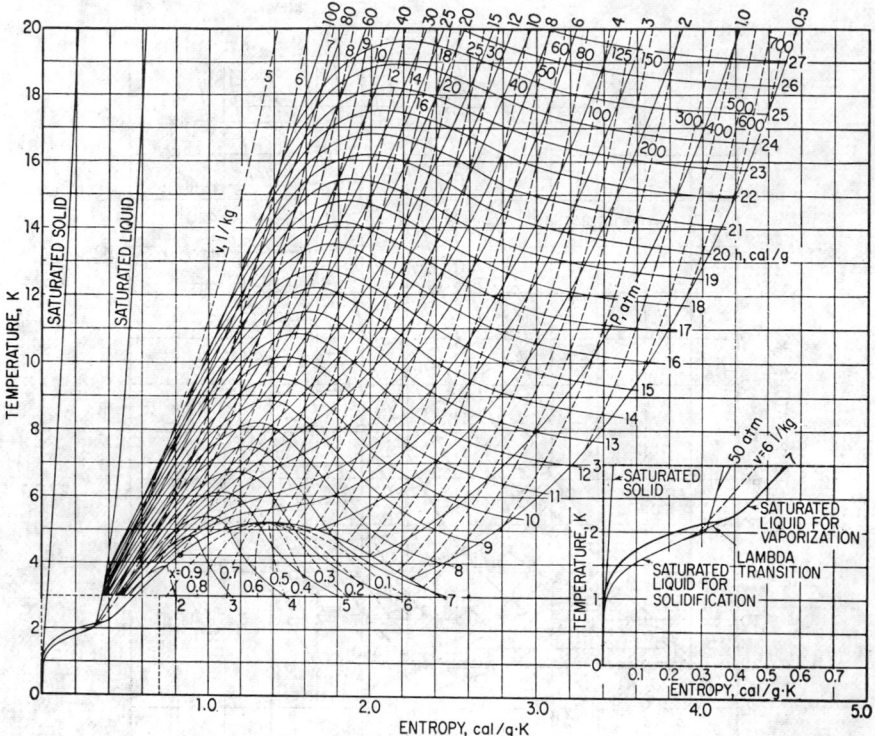

FIG. 4h-3. Temperature-entropy diagram of helium. (*D. B. Mann and R. B. Stewart, National Bureau of Standards and University of Colorado, Trans. ASME, November*, 1959.)

Table 4h-20. Specific Heat of Nitrogen, C_p/R

T, K	1 atm	4 atm	7 atm	10 atm	40 atm	70 atm	100 atm
100	3.613	3.5569	3.6009	3.6466	4.1865	4.860	5.64
200	3.5146	3.5243	3.5404	3.5565	3.7195	3.878	4.021
300	3.5083	3.5289	3.5372	3.5454	3.6260	3.7023	3.773
400	3.5207	3.5645	3.5694	3.5744	3.6225	3.6680	3.7104
500	3.5595						
600	3.6225	3.6258	3.6292	3.6324	3.6642	3.6944	3.7229
700	3.6998	3.7021	3.7045	3.7067	3.7293	3.7506	3.7709
800	3.7812	3.7829	3.7846	3.7863	3.8029	3.8188	3.8338
900	3.8600	3.8614	3.8627	3.8640	3.8766	3.8888	3.9004
1000	3.9329	3.9340	3.9350	3.9361	3.9460	3.9556	3.9647
1100	3.9985	3.9993	4.0001	4.0010	4.0089	4.0166	4.0239
1200	4.0564	4.0571	4.0578	4.0584	4.0649	4.0712	4.0772
1300	4.1074	4.1079	4.1085	4.1091	4.1144	4.1197	4.1247
1400	4.1520	4.1524	4.1529	4.1533	4.1578	4.1621	4.1663
1500	4.1910	4.1914	4.1918	4.1922	4.1960	4.1995	4.2031
1600	4.2253	4.2256	4.2260	4.2263	4.2295	4.2326	4.2356
1700	4.2555	4.2558	4.2561	4.2563	4.2591	4.2618	4.2644
1800	4.2822	4.2824	4.2827	4.2829	4.2852	4.2875	4.2896
1900	4.3058	4.3060	4.3062	4.3064	4.3084	4.3103	4.3122
2000	4.3269	4.3270	4.3272	4.3274	4.3292	4.3309	4.3325
2100	4.3458	4.3459	4.3461	4.3462	4.3478	4.3492	4.3507
2200	4.3627	4.3629	4.3630	4.3632	4.3645	4.3658	4.3671
2300	4.3780	4.3782	4.3783	4.3784	4.3796	4.3807	4.3818
2400	4.3920	4.3921	4.3922	4.3924	4.3934	4.3944	4.3953
2500	4.4047	4.4048	4.4049	4.4050	4.4059	4.4068	4.4076
2600	4.4163	4.4164	4.4165	4.4166	4.4174	4.4182	4.4189
2700	4.4270	4.4271	4.4272	4.4272	4.4280	4.4287	4.4293
2800	4.4369	4.4370	4.4370	4.4371	4.4377	4.4384	4.4389
2900	4.4460	4.4461	4.4461	4.4462	4.4467	4.4473	4.4478
3000	4.4545	4.4546	4.4546	4.4547	4.4551	4.4556	4.4561

Table 4h-19. Relative Density of Nitrogen, ρ/ρ_0

T, K	1 atm	4 atm	7 atm	10 atm	40 atm	70 atm	100 atm
100	2.783	12.010	24.40	13.947	59.45	109.77	161.7
200	1.36809	5.50755	9.7004	9.1160	36.543	63.810	90.523
300	0.91029	3.64304	6.3783	6.8061	26.955	46.625	65.741
400	0.68240	2.72729	4.76856	5.4382	21.472	37.065	52.200
500	0.54585	2.18072	3.81150				
600	0.45486	1.81708	3.17571	4.5307	17.881	30.866	43.484
700	0.38989	1.55755	2.72218	3.8838	15.334	26.485	37.339
800	0.34116	1.36296	2.38226	3.3990	13.429	23.212	32.754
900	0.30326	1.21163	2.11792	3.0223	11.949	20.673	29.195
1000	0.27294	1.09058	1.90645	2.7206	10.765	18.631	26.342
1100	0.24813	0.99152	1.73342	2.4737	9.796	16.971	24.003
1200	0.22746	0.90898	1.58020	2.2681	8.987	15.579	22.049
1300	0.20997	0.83912	1.46715	2.0940	8.302	14.401	20.393
1400	0.19497	0.77923	1.36253	1.9448	7.714	13.389	18.971
1500	0.18198	0.72733	1.27183	1.8155	7.205	12.510	17.734
1600	0.17061	0.68191	1.19246	1.7022	6.758	11.739	16.648
1700	0.16057	0.64184	1.12242	1.6023	6.364	11.060	15.690
1800	0.15165	0.60621	1.06016	1.5135	6.014	10.455	14.837
1900	0.14367	0.57433	1.00444	1.4340	5.699	9.912	14.072
2000	0.13649	0.54563	0.95428	1.3625	5.416	9.423	13.381
2100	0.12999	0.51967	0.90890	1.2977	5.161	8.980	12.755
2200	0.12409	0.49606	0.86763	1.2388	4.927	8.576	12.186
2300	0.11869	0.47451	0.82996	1.1850	4.715	8.208	11.665
2400	0.11375	0.45475	0.79542	1.1357	4.519	7.870	11.187
2500	0.10920	0.43658	0.76364	1.0904	4.340	7.559	10.747
2600	0.10500	0.41980	0.73430	1.0485	4.174	7.271	10.340
2700	0.10111	0.40426	0.70713	1.0097	4.020	7.005	9.963
2800	0.09750	0.38983	0.68190	0.9737	3.878	6.757	9.611
2900	0.09414	0.37639	0.65842	0.9402	3.745	6.527	9.286
3000	0.09100	0.36385	0.63648	0.9089	3.620	6.310	8.980

Table 4h-21. Enthalpy of Nitrogen, $(H - E_0^\circ)/RT_0$

T, K	1 atm	4 atm	7 atm	10 atm	40 atm	70 atm	100 atm
100	1.2589						
200	2.5535	2.5358	2.5179	2.4999	2.3140	2.125	1.94
300	3.8385	3.8302	3.8221	3.8140	3.7351	3.662	3.596
400	5.1244	5.1203	5.1164	5.1125	5.0756	5.0426	5.013
500	6.4194	6.4178	6.4162	6.4147	6.4005	6.3891	6.3802
600	7.7334	7.7333	7.7332	7.7331	7.7332	7.7354	7.7393
700	9.0735	9.0744	9.0752	9.0762	9.0861	9.0977	9.1103
800	10.4428	10.4444	10.4460	10.477	10.4647	10.4829	10.5020
900	11.8416	11.8438	11.8459	11.8482	11.8705	11.8937	11.9177
1000	13.2683	13.2708	13.2734	13.2760	13.3025	13.3296	13.3573
1100	14.7203	14.7232	14.7261	14.7290	14.7588	14.7897	14.8197
1200	16.1950	16.1982	16.2014	16.2046	16.2369	16.2697	16.3029
1300	17.6894	17.6929	17.6963	17.6997	17.7343	17.7691	17.8043
1400	19.2014	19.2050	19.2086	19.2122	19.2486	19.2851	19.3221
1500	20.7288	20.7325	20.7363	20.7400	20.7779	20.8159	20.8542
1600	22.2695	22.2734	22.2773	22.2812	22.3203	22.3597	22.3992
1700	23.8219	23.8259	23.8299	23.8340	23.8742	23.9146	23.9550
1800	25.3848	25.3889	25.3930	25.3971	25.4382	25.4795	25.5209
1900	26.9568	26.9610	26.9652	26.9693	27.0113	27.0533	27.0954
2000	28.5370	28.5413	28.5455	28.5498	28.5924	28.6352	28.6779
2100	30.1246	30.1290	30.1333	30.1376	30.1808	30.2241	30.2674
2200	31.7187	31.7230	31.7274	31.7318	31.7755	31.8193	31.8632
2300	33.3187	33.3231	33.3275	33.3319	33.3761	33.4203	33.4647
2400	34.9240	34.9284	34.9329	34.9374	34.9819	35.0266	35.0712
2500	36.5342	36.5387	36.5432	36.5477	36.5926	36.6377	36.6827
2600	38.1488	38.1533	38.1579	38.1624	38.2076	38.2530	38.2983
2700	39.7676	39.7722	39.7767	39.7813	39.8268	39.8723	39.9179
2800	41.3901	41.3947	41.3993	41.4039	41.4496	41.4954	41.5413
2900	43.0160	43.0206	43.0252	43.0298	43.0758	43.1218	43.1678
3000	44.6452	44.6499	44.6545	44.6591	44.7053	44.7514	44.7976

Table 4h-22. Entropy of Nitrogen, S/R

T, K	1 atm	4 atm	7 atm	10 atm	40 atm	70 atm	100 atm
100	19.1705	17.607	16.55				
200	21.6249	20.2208	19.6431	19.2682	17.6905	16.932	16.382
300	23.0482	21.6549	21.0884	20.7248	19.2706	18.6461	18.230
400	24.0586	22.6687	22.1055	21.7454	20.3246	19.7322	19.3448
500	24.8479	23.4595	22.8977	22.5390	21.1322	20.5532	20.1781
600	25.5020	24.1144	23.5534	23.1953	21.7958	21.2236	20.8548
700	26.0662	24.6790	24.1184	23.7607	22.3654	21.7970	21.4319
800	26.5656	25.1786	24.6183	24.2609	22.8682	22.3022	21.9396
900	27.0154	25.6286	25.0685	24.7113	23.3203	22.7561	22.3949
1000	27.4260	26.0393	25.4793	25.1223	23.7323	23.1693	22.8094
1100	27.8039	26.4173	25.8574	25.5004	24.1114	23.5491	23.1899
1200	28.1543	26.7678	26.2080	25.8511	24.4627	23.9010	23.5424
1300	28.4811	27.0947	26.5349	26.1780	24.7901	24.2289	23.8707
1400	28.7872	27.4007	26.8410	26.4842	25.0965	24.5357	24.1779
1500	29.0751	27.6887	27.1290	26.7721	25.3848	24.8242	24.4666
1600	29.3467	27.9603	27.4006	27.0438	25.6567	25.0964	24.7390
1700	29.6037	28.2173	27.6577	27.3009	25.9140	25.3537	24.9965
1800	29.8477	28.4613	27.9017	27.5449	26.1582	25.5981	25.2410
1900	30.0799	28.6936	28.1339	27.7772	26.3905	25.8306	25.4736
2000	30.3013	28.9150	28.3553	27.9986	26.6120	26.0522	25.6953
2100	30.5129	29.1266	28.5670	28.2102	26.8238	26.2640	25.9072
2200	30.7154	29.3291	28.7695	28.4128	27.0264	26.4667	26.1100
2300	30.9097	29.5234	28.9638	28.6071	27.2207	26.6611	26.3043
2400	31.0963	29.7100	29.1504	28.7937	27.4074	26.8478	26.4911
2500	31.2759	29.8896	29.3300	28.9733	27.5870	27.0275	26.6708
2600	31.4488	30.0625	29.5029	29.1462	27.7600	27.2004	26.8438
2700	31.6157	30.2294	29.6698	29.3131	27.9269	27.3674	27.0108
2800	31.7769	30.3906	29.8310	29.4743	28.0882	27.5287	27.1721
2900	31.9327	30.5464	29.9868	29.6301	28.2440	27.6846	27.3280
3000	32.0836	30.6973	30.1377	29.7810	28.3949	27.8355	27.4790

TABLE 4h-24. RELATIVE DENSITY OF OXYGEN, ρ/ρ_0

T, K	1 atm	4 atm	7 atm	10 atm	40 atm	70 atm	100 atm
100	2.79257						
200	1.36860	5.5245	9.7584	14.073	62.4	123	198.5
300	0.91023	3.6474	6.39455	9.151	37.231	66.082	95.34
400	0.68225	2.72885	4.77519	6.8212	27.246	47.557	67.69
500	0.54568	2.18129	3.81474	5.4459	21.628	37.556	53.217
600	0.45470	1.81723	3.17736	4.5351	17.9767	31.165	44.098
700	0.38974	1.55750	2.72307	3.8864	15.3980	26.684	37.747
800	0.34102	1.36282	2.38269	3.4006	13.4746	23.355	33.045
900	0.30313	1.21143	2.11809	3.0231	11.9823	20.776	29.404
1000	0.27282	1.09035	1.90646	2.7211	10.7901	18.717	26.505
1100	0.24802	0.99129	1.73331	2.4741	9.8150	17.034	24.131
1200	0.22736	0.90872	1.58903	2.26828	9.0024	15.631	22.154
1300	0.20987	0.83887	1.46694	2.09409	8.3145	14.443	20.480
1400	0.19488	0.77899	1.36228	1.94476	7.7247	13.423	19.041
1500	0.18189	0.72710	1.27157	1.81533	7.2131	12.539	17.794
1600	0.17053	0.68168	1.19219	1.70206	6.7653	11.764	16.700
1700	0.16050	0.64161	1.12214	1.60210	6.3700	11.080	15.735
1800	0.15158	0.60599	1.05987	1.51324	6.0184	10.472	14.874
1900	0.14361	0.57412	1.00415	1.43373	5.7037	9.927	14.105
2000	0.13643	0.54543	0.95400	1.36215	5.4202	9.436	13.410
2100	0.12993	0.51947	0.90862	1.29737	5.1637	8.991	12.781
2200	0.12403	0.49587	0.86736	1.23849	4.9303	8.587	12.208
2300	0.11863	0.47432	0.82968	1.18473	4.7173	8.217	11.686
2400	0.11369	0.45457	0.79515	1.13543	4.5218	7.878	11.206
2500	0.10915	0.43640	0.76337	1.09007	4.3419	7.566	10.763
2600	0.10495	0.41962	0.73404	1.04820	4.1758	7.278	10.355
2700	0.10106	0.40409	0.70888	1.00943	4.0219	7.010	9.976
2800	0.09745	0.38966	0.68165	0.97342	3.8790	6.762	9.624
2900	0.09409	0.37623	0.65817	0.93989	3.7458	6.531	9.296
3000	0.09096	0.36370	0.63625	0.90861	3.6216	6.315	8.990

TABLE 4h-23. COMPRESSIBILITY FACTOR FOR OXYGEN, $Z = PV/RT$

T, K	1 atm	4 atm	7 atm	10 atm	40 atm	70 atm	100 atm
100	0.97724						
200	0.99701	0.99796	0.97880	0.96956	0.8734	0.7764	0.6871
300	0.99939	0.99759	0.99580	0.99402	0.97731	0.9636	0.9541
400	1.00001	1.00006	1.00012	1.00019	1.00161	1.0042	1.0079
500	1.00022	1.00088	1.00154	1.00222	1.00942	1.0173	1.0256
600	1.00029	1.00116	1.00204	1.00292	1.01205	1.0216	1.0314
700	1.00031	1.00124	1.00218	1.00312	1.01275	1.0227	1.0328
800	1.00031	1.00124	1.00218	1.00311	1.01265	1.0224	1.0323
900	1.00030	1.00121	1.00211	1.00302	1.01223	1.0216	1.0312
1000	1.00029	1.00115	1.00202	1.00288	1.01167	1.0206	1.0296
1100	1.00027	1.00109	1.00192	1.00274	1.01107	1.0195	1.0281
1200	1.00026	1.00104	1.00182	1.00260	1.01047	1.0184	1.0265
1300	1.00025	1.00098	1.00172	1.00246	1.00991	1.0174	1.0250
1400	1.00023	1.00093	1.00163	1.00233	1.00938	1.0165	1.0237
1500	1.00022	1.00088	1.00155	1.00221	1.00890	1.0156	1.0224
1600	1.00021	1.00084	1.00147	1.00210	1.00845	1.0149	1.0213
1700	1.00020	1.00080	1.00140	1.00200	1.00803	1.0141	1.0202
1800	1.00019	1.00076	1.00133	1.00190	1.00765	1.0134	1.0193
1900	1.00018	1.00072	1.00127	1.00181	1.00728	1.0128	1.0183
2000	1.00017	1.00069	1.00121	1.00173	1.00696	1.0122	1.0175
2100	1.00017	1.00066	1.00116	1.00166	1.00666	1.0117	1.0167
2200	1.00016	1.00063	1.00111	1.00159	1.00638	1.0112	1.0161
2300	1.00015	1.00061	1.00107	1.00152	1.00610	1.0107	1.0153
2400	1.00015	1.00058	1.00102	1.00146	1.00586	1.0103	1.0147
2500	1.00014	1.00056	1.00098	1.00141	1.00564	1.0099	1.0142
2600	1.00014	1.00054	1.00095	1.00135	1.00543	1.0095	1.0136
2700	1.00013	1.00052	1.00091	1.00130	1.00523	1.0092	1.0131
2800	1.00013	1.00050	1.00088	1.00126	1.00505	1.0089	1.0127
2900	1.00012	1.00049	1.00085	1.00122	1.00488	1.0086	1.0122
3000	1.00012	1.00047	1.00082	1.00117	1.00471	1.0083	1.0118

TABLE 4h-26. ENTHALPY OF OXYGEN, $(H - E_0^\circ)/RT_0$

T, K	1 atm	4 atm	7 atm	10 atm	40 atm	70 atm	100 atm
100	1.254						
200	2.5523	2.5308	2.5091	2.4871	2.248	1.972	1.659
300	3.8424	3.8319	3.8213	3.8108	3.705	3.602	3.505
400	5.1523	5.1464	5.1406	5.1349	5.078	5.023	4.971
500	6.5000	6.4968	6.4936	6.4905	6.460	6.431	6.403
600	7.8919	7.8903	7.8888	7.8873	7.873	7.860	7.848
700	9.3254	9.3250	9.3245	9.3242	9.321	9.319	9.318
800	10.7951	10.7956	10.7960	10.7965	10.802	10.807	10.814
900	12.2949	12.2960	12.2970	12.2981	12.309	12.321	12.333
1000	13.8198	13.8213	13.8228	13.8243	13.840	13.857	13.874
1100	15.3653	15.3672	15.3691	15.3710	15.391	15.411	15.431
1200	16.9285	16.9307	16.9329	16.9351	16.958	16.981	17.004
1300	18.5067	18.5092	18.5116	18.5141	18.539	18.565	18.591
1400	20.0985	20.1012	20.1038	20.1065	20.134	20.161	20.189
1500	21.7025	21.7054	21.7082	21.7111	21.740	21.769	21.799
1600	23.3181	23.3211	23.3241	23.3271	23.358	23.388	23.419
1700	24.9447	24.9479	24.9510	24.9541	24.986	25.018	25.050
1800	26.5820	26.5852	26.5885	26.5917	26.625	26.658	26.691
1900	28.2299	28.2333	28.2366	28.2399	28.274	28.308	28.342
2000	29.8880	29.8915	29.8949	29.8983	29.933	29.968	30.003
2100	31.5566	31.5601	31.5636	31.5671	31.602	31.638	31.674
2200	33.2353	33.2389	33.2424	33.2460	33.282	33.318	33.355
2300	34.9239	34.9275	34.9312	34.9348	34.971	35.008	35.045
2400	36.6229	36.6266	36.6303	36.6340	36.671	36.708	36.745
2500	38.3314	38.3352	38.3389	38.3426	38.380	38.418	38.455
2600	40.0500	40.0537	40.0575	40.0613	40.099	40.137	40.175
2700	41.7778	41.7816	41.7854	41.7892	41.827	41.866	41.904
2800	43.5151	43.5189	43.5227	43.5266	43.565	43.604	43.643
2900	45.2614	45.2653	45.2691	45.2730	45.312	45.351	45.390
3000	47.0165	47.0204	47.0243	47.0282	47.067	47.107	47.146

TABLE 4h-25. SPECIFIC HEAT OF OXYGEN, C_p/R

T, K	1 atm	4 atm	7 atm	10 atm	40 atm	70 atm	100 atm
200	3.519	3.5681	3.6196	3.6739	4.415	5.66	7.6
300	3.5403	3.5584	3.5766	3.5951	3.7862	3.981	4.165
400	3.6243	3.6335	3.6427	3.6520	3.7453	3.836	3.921
500	3.7415	3.7470	3.7526	3.7582	3.8134	3.8677	3.920
600	3.8611	3.8648	3.8685	3.8722	3.9087	3.9445	3.980
700	3.9681	3.9707	3.9733	3.9759	4.0016	4.0266	4.052
800	4.0583	4.0603	4.0622	4.0641	4.0830	4.1017	4.120
900	4.1332	4.1347	4.1361	4.1376	4.1521	4.1664	4.180
1000	4.1952	4.1964	4.1975	4.1987	4.2101	4.2213	4.232
1100	4.2472	4.2481	4.2491	4.2500	4.2591	4.2681	4.277
1200	4.2915	4.2922	4.2930	4.2937	4.3012	4.3085	4.316
1300	4.3302	4.3308	4.3315	4.3321	4.3382	4.3442	4.350
1400	4.3653	4.3658	4.3663	4.3669	4.3721	4.3771	4.382
1500	4.3976	4.3981	4.3985	4.3990	4.4034	4.4076	4.412
1600	4.4283	4.4287	4.4291	4.4295	4.4332	4.4369	4.440
1700	4.4579	4.4582	4.4586	4.4589	4.4621	4.4652	4.468
1800	4.4869	4.4872	4.4875	4.4878	4.4905	4.4933	4.496
1900	4.5154	4.5156	4.5159	4.5161	4.5185	4.5209	4.523
2000	4.5437	4.5439	4.5441	4.5443	4.5464	4.5485	4.551
2100	4.5716	4.5717	4.5719	4.5721	4.5739	4.5758	4.578
2200	4.5993	4.5995	4.5997	4.5999	4.6016	4.6032	4.605
2300	4.6268	4.6269	4.6271	4.6272	4.6287	4.6301	4.631
2400	4.6540	4.6542	4.6543	4.6544	4.6558	4.6570	4.658
2500	4.6808	4.6810	4.6811	4.6812	4.6824	4.6835	4.685
2600	4.7071	4.7072	4.7073	4.7074	4.7085	4.7095	4.710
2700	4.7328	4.7329	4.7330	4.7331	4.7341	4.7349	4.736
2800	4.7579	4.7580	4.7581	4.7582	4.7590	4.7598	4.761
2900	4.7824	4.7825	4.7826	4.7826	4.7834	4.7841	4.785
3000	4.8062	4.8063	4.8064	4.8064	4.8072	4.8077	4.808

TABLE 4h-27. ENTROPY OF OXYGEN, S/R

T, K	1 atm	4 atm	7 atm	10 atm	40 atm	70 atm	100 atm
100	20.794						
200	23.2553	21.8488	21.2686	20.8908	19.2709	18.431	17.74
300	24.6839	23.2899	22.7224	22.3579	20.8928	20.2555	19.825
400	25.7127	24.3224	23.7587	23.3980	21.9719	21.3733	20.9789
500	26.5337	25.1450	24.5830	24.2239	22.8139	22.2311	21.8517
600	27.2266	25.8387	25.2775	24.9193	23.5176	22.9429	22.5712
700	27.8299	26.4425	25.8819	25.5241	24.1272	23.5571	23.1900
800	28.3659	26.9788	26.4185	26.0610	24.6670	24.0999	23.7357
900	28.8484	27.4615	26.9013	26.5440	25.1521	24.5869	24.2246
1000	29.2872	27.9005	27.3404	26.9833	25.5926	25.0287	24.6678
1100	29.6896	28.3029	27.7430	27.3859	25.9963	25.4334	25.0733
1200	30.0610	28.6744	28.1146	27.7576	26.3685	25.8064	25.4471
1300	30.4061	29.0196	28.4598	28.1029	26.7144	26.1527	25.7939
1400	30.7283	29.3419	28.7821	28.4252	27.0372	26.4760	26.1176
1500	31.0307	29.6442	29.0845	28.7276	27.3399	26.7790	26.4209
1600	31.3155	29.9290	29.3693	29.0125	27.6250	27.0644	26.7067
1700	31.5848	30.1984	29.6387	29.2819	27.8946	27.3342	26.9766
1800	31.8404	30.4540	29.8943	29.5375	28.1505	27.5902	27.2328
1900	32.0838	30.6974	30.1377	29.7810	28.3941	27.8339	27.4767
2000	32.3161	30.9297	30.3701	30.0133	28.6265	28.0664	27.7094
2100	32.5385	31.1521	30.5925	30.2358	28.8490	28.2890	27.9320
2200	32.7518	31.3655	30.8058	30.4491	29.0625	28.5025	28.1456
2300	32.9568	31.5705	31.0108	30.6541	29.2675	28.7077	28.3508
2400	33.1543	31.7680	31.2083	30.8516	29.4651	28.9053	28.5485
2500	33.3449	31.9586	31.3990	31.0422	29.6558	29.0960	28.7393
2600	33.5289	32.1426	31.5830	31.2263	29.8399	29.2802	28.9235
2700	33.7071	32.3208	31.7612	31.4045	30.0181	29.4585	29.1018
2800	33.8796	32.4933	31.9337	31.5770	30.1907	29.6310	29.2744
2900	34.0470	32.6607	32.1011	31.7444	30.3581	29.7985	29.4419
3000	34.2096	32.8233	32.2637	31.9070	30.5207	29.9612	29.6047

HEAT

TABLE 4h-28. THERMODYNAMIC PROPERTIES OF IONIZED AIR
$T = 4000$ K

$\log \rho/\rho_0$	Z	E/RT	H/RT	S/R	$\log_{10} P$, atm	Z^*
−5.00	1.27113	7.98674	9.25788	50.3307	−3.72990	1.27113
−4.80	1.25871	7.63908	8.89779	49.4006	−3.53416	1.25871
−4.60	1.24863	7.35804	8.60667	48.5423	−3.33765	1.24863
−4.40	1.24045	7.13130	8.37175	47.7425	−3.14051	1.24045
−4.20	1.23380	6.94846	8.18225	46.9900	−2.94284	1.23379
−4.00	1.22834	6.80080	8.02914	46.2755	−2.74477	1.22834
−3.80	1.22381	6.68107	7.90488	45.5912	−2.54637	1.22381
−3.60	1.21997	6.58320	7.80317	44.9306	−2.34774	1.21997
−3.40	1.21662	6.50204	7.71866	44.2884	−2.14893	1.21662
−3.20	1.21356	6.43313	7.64669	43.6600	−1.95003	1.21356
−3.00	1.21060	6.37242	7.58303	43.0411	−1.75108	1.21060
−2.80	1.20755	6.31608	7.52363	42.4279	−1.55218	1.20755
−2.60	1.20419	6.26016	7.46434	41.8167	−1.35339	1.20418
−2.40	1.20024	6.20044	7.40068	41.2033	−1.15482	1.20024
−2.20	1.19544	6.13223	7.32767	40.5834	−0.95656	1.19543
−2.00	1.18943	6.05042	7.23985	39.9524	−0.75875	1.18941
−1.80	1.18189	5.94985	7.13174	39.3058	−0.56151	1.18187
−1.60	1.17257	5.82615	6.99871	38.6399	−0.36495	1.17252
−1.40	1.16133	5.67706	6.83839	37.9533	−0.16913	1.16127
−1.20	1.14833	5.50363	6.65196	37.2480	0.02598	1.14822
−1.00	1.13396	5.31076	6.44471	36.5296	0.22051	1.13380
−0.80	1.11884	5.10642	6.22526	35.8065	0.41468	1.11859
−0.60	1.10369	4.90000	6.00369	35.0884	0.60876	1.10330
−0.40	1.08918	4.70034	5.78952	34.3839	0.80301	1.08856
−0.20	1.07585	4.51438	5.59022	33.6994	0.99766	1.07488
0.	1.06407	4.34655	5.41062	33.0390	1.19288	1.06256
0.20	1.05411	4.19894	5.25305	32.4037	1.38880	1.05174
0.40	1.04615	4.07180	5.11795	31.7931	1.58551	1.04242
0.60	1.04038	3.96410	5.00448	31.2050	1.78311	1.03453
0.80	1.03713	3.87406	4.91120	30.6367	1.98175	1.02791
1.00	1.03695	3.79961	4.83656	30.0848	2.18167	1.02242
1.20	1.04079	3.73861	4.77940	29.5456	2.38328	1.01787
1.40	1.05028	3.68908	4.73936	29.0148	2.58722	1.01412
1.60	1.06811	3.64930	4.71741	28.4877	2.79453	1.01100
1.80	1.09856	3.61790	4.71646	27.9580	3.00674	1.00837
2.00	1.14848	3.59393	4.74241	27.4175	3.22604	1.00606

Z^* is the number of moles of dissociated gas corresponding to one mole of low-temperature gas.

TABLE 4h-28. THERMODYNAMIC PROPERTIES OF IONIZED AIR (Continued)

$T = 6000$ K

$\log \rho/\rho_0$	Z	E/RT	H/RT	S/R	$\log_{10} P$, atm	Z^*
−5.00	1.98639	19.99067	21.97706	68.7753	−3.35993	1.98639
−4.80	1.98008	19.86188	21.84196	67.7331	−3.16131	1.98008
−4.60	1.97116	19.68615	21.65732	66.6475	−2.96327	1.97117
−4.40	1.95832	19.43841	21.39673	65.4948	−2.76611	1.95832
−4.20	1.93995	19.08795	21.02790	64.2465	−2.57021	1.93995
−4.00	1.91430	18.60197	20.51627	62.8727	−2.37599	1.91430
−3.80	1.87994	17.95327	19.83321	61.3500	−2.18385	1.87994
−3.60	1.83631	17.13119	18.96749	59.6719	−1.99405	1.83631
−3.40	1.78420	16.15055	17.93475	57.8573	−1.80655	1.78420
−3.20	1.72580	15.05251	16.77831	55.9509	−1.62100	1.72580
−3.00	1.66421	13.89543	15.55964	54.0131	−1.43679	1.66421
−2.80	1.60268	12.74058	14.34326	52.1061	−1.25315	1.60268
−2.60	1.54398	11.63995	13.18393	50.2811	−1.06935	1.54398
−2.40	1.49003	10.62989	12.11991	48.5727	−0.88480	1.49002
−2.20	1.44187	9.73045	11.17232	46.9984	−0.69907	1.44186
−2.00	1.39984	8.94817	10.34801	45.5620	−0.51192	1.39983
−1.80	1.36376	8.27996	9.64372	44.2577	−0.32326	1.36373
−1.60	1.33312	7.71677	9.04989	43.0737	−0.13313	1.33307
−1.40	1.30722	7.24649	8.55371	41.9956	0.05835	1.30715
−1.20	1.28534	6.85589	8.14122	41.0082	0.25102	1.28522
−1.00	1.26671	6.53186	7.79857	40.0967	0.44468	1.26653
−0.80	1.25062	6.26209	7.51271	39.2473	0.63913	1.25035
−0.60	1.23641	6.03531	7.27172	38.4480	0.83417	1.23598
−0.40	1.22347	5.84143	7.06490	37.6877	1.02960	1.22280
−0.20	1.21125	5.67145	6.88270	36.9572	1.22524	1.21020
0.	1.19932	5.51750	6.71682	36.2482	1.42094	1.19767
0.20	1.18737	5.37295	6.56032	35.5541	1.61659	1.18478
0.40	1.17526	5.23261	6.40787	34.8697	1.81214	1.17121
0.60	1.16316	5.09308	6.25624	34.1918	2.00765	1.15683
0.80	1.15159	4.95296	6.10456	33.5187	2.20330	1.14169
1.00	1.14152	4.81288	5.95440	32.8507	2.39949	1.12605
1.20	1.13447	4.67497	5.80944	32.1889	2.59680	1.11030
1.40	1.13262	4.54223	5.67485	31.5344	2.79609	1.09487
1.60	1.13913	4.41768	5.55681	30.8872	2.99858	1.08015
1.80	1.15861	4.30382	5.46243	30.2449	3.20594	1.06639
2.00	1.19794	4.20247	5.40041	29.6019	3.42044	1.05372

TABLE 4h-28. THERMODYNAMIC PROPERTIES OF IONIZED AIR (*Continued*)

$T = 8000$ K

log ρ/ρ_0	Z	E/RT	H/RT	S/R	$\log_{10} P$, atm	Z^*
-5.00	2.17208	20.12586	22.29794	74.4276	-3.19618	2.17228
-4.80	2.13622	19.32981	21.46602	72.6399	-3.00341	2.13640
-4.60	2.10717	18.68526	20.79243	71.0185	-2.80936	2.10733
-4.40	2.08370	18.16490	20.24859	69.5334	-2.61422	2.08385
-4.20	2.06474	17.74534	19.81007	68.1587	-2.41819	2.06487
-4.00	2.04939	17.40674	19.45612	66.8730	-2.22143	2.04950
-3.80	2.03685	17.13225	19.16910	65.6577	-2.02410	2.03696
-3.60	2.02645	16.90733	18.93378	64.4973	-1.82632	2.02654
-3.40	2.01753	16.71903	18.73656	63.3778	-1.62824	2.01761
-3.20	2.00945	16.55509	18.56455	62.2867	-1.42998	2.00953
-3.00	2.00151	16.40300	18.40452	61.2110	-1.23170	2.00158
-2.80	1.99288	16.24878	18.24166	60.1370	-1.03358	1.99293
-2.60	1.98248	16.07573	18.05820	59.0485	-0.83585	1.98252
-2.40	1.96897	15.86333	17.83229	57.9262	-0.63882	1.96900
-2.20	1.95069	15.58703	17.53772	56.7471	-0.44287	1.95072
-2.00	1.92584	15.22003	17.14587	55.4872	-0.24844	1.92585
-1.80	1.89274	14.73824	16.63098	54.1259	-0.05597	1.89273
-1.60	1.85048	14.12817	15.97865	52.6536	0.13423	1.85044
-1.40	1.79940	13.39465	15.19404	51.0793	0.32207	1.79932
-1.20	1.74129	12.56340	14.30470	49.4326	0.50782	1.74116
-1.00	1.67906	11.67587	13.35493	47.7575	0.69201	1.67884
-0.80	1.61594	10.77874	12.39468	46.1017	0.87537	1.61560
-0.60	1.55484	9.91370	11.46854	44.5068	1.05863	1.55432
-0.40	1.49788	9.11130	10.60918	43.0017	1.24242	1.49708
-0.20	1.44632	8.38943	9.83575	41.6021	1.42721	1.44509
0.	1.40063	7.75491	9.15555	40.3123	1.61327	1.39875
0.20	1.36078	7.20639	8.56717	39.1282	1.80073	1.35789
0.40	1.32640	6.73723	8.06362	38.0405	1.98962	1.32195
0.60	1.29703	6.33794	7.63497	37.0374	2.17989	1.29017
0.80	1.27233	5.99796	7.27029	36.1060	2.37154	1.26172
1.00	1.25222	5.70678	6.95900	35.2337	2.56462	1.23578
1.20	1.23714	5.45484	6.69197	34.4088	2.75936	1.21162
1.40	1.22829	5.23393	6.46222	33.6205	2.95625	1.18866
1.60	1.22807	5.03758	6.26565	32.8590	3.15617	1.16650
1.80	1.24054	4.86115	6.10169	32.1147	3.36056	1.14487
2.00	1.27229	4.70179	5.97408	31.3777	3.57153	1.12367

TABLE 4h-28. THERMODYNAMIC PROPERTIES OF IONIZED AIR (*Continued*)

$T = 10,000$ K

log ρ/ρ_0	Z	E/RT	H/RT	S/R	$\log_{10} P$, atm	Z*
−5.00	3.13590	34.42541	37.56131	94.5501	−2.93978	3.13820
−4.80	2.98152	31.65851	34.64003	90.3748	−2.76171	2.98385
−4.60	2.83477	29.02681	31.86158	86.4040	−2.58363	2.83708
−4.40	2.70036	26.61470	29.31506	82.7180	−2.40472	2.70260
−4.20	2.58086	24.46888	27.04974	79.3568	−2.22438	2.58300
−4.00	2.47709	22.60469	25.08178	76.3287	−2.04220	2.47911
−3.80	2.38866	21.01522	23.40388	73.6195	−1.85799	2.39054
−3.60	2.31438	19.67975	21.99413	71.2017	−1.67171	2.31612
−3.40	2.25270	18.57046	20.82315	69.0413	−1.48344	2.25430
−3.20	2.20192	17.65711	19.85903	67.1026	−1.29335	2.20338
−3.00	2.16037	16.90999	19.07036	65.3514	−1.10162	2.16169
−2.80	2.12649	16.30150	18.42800	63.7561	−0.90848	2.12769
−2.60	2.09889	15.80687	17.90577	62.2888	−0.71416	2.09997
−2.40	2.07632	15.40423	17.48055	60.9250	−0.51885	2.07729
−2.20	2.05764	15.07428	17.13192	59.6433	−0.32278	2.05850
−2.00	2.04181	14.79967	16.84149	58.4248	−0.12613	2.04258
−1.80	2.02780	14.56421	16.59201	57.2524	0.07088	2.02847
−1.60	2.01453	14.35175	16.36629	56.1091	0.26803	2.01510
−1.40	2.00076	14.14511	16.14587	54.9779	0.46505	2.00123
−1.20	1.98502	13.92488	15.90990	53.8398	0.66162	1.98536
−1.00	1.96556	13.66889	15.63445	52.6740	0.85734	1.96575
−0.80	1.94043	13.35298	15.29341	51.4585	1.05175	1.94041
−0.60	1.90775	12.95423	14.86198	50.1733	1.24437	1.90744
−0.40	1.86624	12.45686	14.32310	48.8066	1.43482	1.86553
−0.20	1.81580	11.85896	13.67475	47.3606	1.62292	1.81450
0.	1.75782	11.17605	12.93388	45.8546	1.80883	1.75567
0.20	1.69501	10.43853	12.13354	44.3219	1.99303	1.69159
0.40	1.63071	9.68368	11.31439	42.8013	2.17623	1.62540
0.60	1.56819	8.94681	10.51500	41.3280	2.35925	1.56007
0.80	1.51018	8.25507	9.76525	39.9277	2.54288	1.49781
1.00	1.45879	7.62528	9.08407	38.6146	2.72785	1.44000
1.20	1.41573	7.06479	8.48053	37.3925	2.91484	1.38714
1.40	1.38272	6.57376	7.95649	36.2576	3.10459	1.33914
1.60	1.36208	6.14771	7.50979	35.2001	3.29806	1.29550
1.80	1.35741	5.77984	7.13726	34.2067	3.49657	1.25553
2.00	1.37455	5.46289	6.83744	33.2617	3.70201	1.21845

TABLE 4h-28. THERMODYNAMIC PROPERTIES OF IONIZED AIR (*Continued*)

$T = 12,000$ K

log ρ/ρ_0	Z	E/RT	H/RT	S/R	$\log_{10} P$, atm	Z^*
−5.00	3.86391	40.69274	44.55665	107.9424	−2.76994	3.86758
−4.80	3.80741	39.84992	43.65733	105.3326	−2.57633	3.81182
−4.60	3.73023	38.69570	42.42592	102.4419	−2.38523	3.73543
−4.40	3.63012	37.19560	40.82571	99.2462	−2.19704	3.63612
−4.20	3.50778	35.35895	38.86673	95.7652	−2.01193	3.51451
−4.00	3.36735	33.24726	36.61461	92.0699	−1.82968	3.37468
−3.80	3.21572	30.96381	34.17953	88.2705	−1.64969	3.22348
−3.60	3.06089	28.62933	31.69022	84.4904	−1.47112	3.06888
−3.40	2.91034	26.35611	29.26645	80.8427	−1.29302	2.91837
−3.20	2.76976	24.23093	27.00069	77.4102	−1.11452	2.77766
−3.00	2.64278	22.30907	24.95185	74.2427	−0.93490	2.65040
−2.80	2.53107	20.61675	23.14782	71.3597	−0.75366	2.53833
−2.60	2.43486	19.15766	21.59252	68.7579	−0.57049	2.44169
−2.40	2.35334	17.92033	20.27367	66.4186	−0.38528	2.35970
−2.20	2.28514	16.88443	19.16957	64.3152	−0.19805	2.29103
−2.00	2.22862	16.02552	18.25414	62.4174	−0.00893	2.23402
−1.80	2.18207	15.31821	17.50028	60.6949	0.18190	2.18700
−1.60	2.14383	14.73796	16.88179	59.1189	0.37422	2.14830
−1.40	2.11234	14.26199	16.37433	57.6631	0.56780	2.11636
−1.20	2.08614	13.86943	15.95557	56.3040	0.76238	2.08973
−1.00	2.06387	13.54099	15.60486	55.0202	0.95772	2.06701
−0.80	2.04413	13.25829	15.30243	53.7916	1.15354	2.04681
−0.60	2.02548	13.00293	15.02840	52.5992	1.34956	2.02762
−0.40	2.00628	12.75541	14.76169	51.4233	1.54543	2.00777
−0.20	1.98469	12.49447	14.47916	50.2433	1.74073	1.98533
0.	1.95864	12.19723	14.15587	49.0379	1.93499	1.95812
0.20	1.92615	11.84127	13.76742	47.7871	2.12773	1.92397
0.40	1.88572	11.40910	13.29482	46.4769	2.31851	1.88113
0.60	1.83703	10.89385	12.73098	45.1043	2.50715	1.82890
0.80	1.78138	10.30382	12.08519	43.6808	2.69379	1.76804
1.00	1.72172	9.66068	11.38239	42.2310	2.87900	1.70072
1.20	1.66222	8.99461	10.65683	40.7858	3.06372	1.62987
1.40	1.60767	8.33607	9.94374	39.3747	3.24923	1.55847
1.60	1.56321	7.70993	9.27314	38.0189	3.43705	1.48888
1.80	1.53453	7.13311	8.66764	36.7296	3.62901	1.42261
2.00	1.52846	6.61560	8.14406	35.5078	3.82729	1.36034

TABLE 4h-28. THERMODYNAMIC PROPERTIES OF IONIZED AIR (*Continued*)

$T = 14{,}000$ K

log ρ/ρ_0	Z	E/RT	H/RT	S/R	$\log_{10} P$, atm	Z*
−5.00	3.96453	37.19404	41.15858	110.4755	−2.69183	3.96768
−4.80	3.95531	37.07956	41.03487	108.5372	−2.49284	3.95924
−4.60	3.94140	36.90620	40.84760	106.5454	−2.29437	3.94630
−4.40	3.92053	36.64408	40.56461	104.4727	−2.09667	3.92661
−4.20	3.88963	36.25307	40.14270	102.2829	−1.90011	3.89710
−4.00	3.84491	35.68344	39.52835	99.9317	−1.70513	3.85400
−3.80	3.78231	34.88151	38.66382	97.3728	−1.51226	3.79320
−3.60	3.69844	33.80203	37.50047	94.5700	−1.32200	3.71123
−3.40	3.59192	32.42548	36.01739	91.5140	−1.13469	3.60657
−3.20	3.46445	30.77278	34.23723	88.2358	−0.95038	3.48078
−3.00	3.32103	28.90780	32.22884	84.8080	−0.76874	3.33872
−2.80	3.16891	26.92442	30.09333	81.3301	−0.58911	3.18754
−2.60	3.01587	24.92472	27.94059	77.9061	−0.41061	3.03498
−2.40	2.86883	22.99891	25.86775	74.6258	−0.23231	2.88800
−2.20	2.73281	21.21359	23.94640	71.5512	−0.05341	2.75165
−2.00	2.61079	19.60888	22.21968	68.7168	0.12675	2.62901
−1.80	2.50400	18.20170	20.70570	66.1325	0.30862	2.52138
−1.60	2.41233	16.99150	19.40383	63.7909	0.49242	2.42872
−1.40	2.33481	15.96627	18.30108	61.6732	0.67823	2.35013
−1.20	2.26997	15.10755	17.37752	59.7547	0.86600	2.28418
−1.00	2.21612	14.39394	16.61006	58.0086	1.05558	2.22920
−0.80	2.17154	13.80342	15.97497	56.4081	1.24675	2.18348
−0.60	2.13453	13.31458	15.44911	54.9280	1.43928	2.14528
−0.40	2.10343	12.90702	15.01045	53.5449	1.63291	2.11293
−0.20	2.07663	12.56120	14.63783	52.2367	1.82734	2.08474
0.	2.05249	12.25780	14.31028	50.9826	2.02226	2.05892
0.20	2.02925	11.97699	14.00624	49.7620	2.21732	2.03354
0.40	2.00501	11.69771	13.70272	48.5537	2.41210	2.00640
0.60	1.97779	11.39771	13.37551	47.3365	2.60616	1.97505
0.80	1.94576	11.05519	13.00095	46.0904	2.79907	1.93701
1.00	1.90776	10.65244	12.56020	44.8001	2.99050	1.89018
1.20	1.86404	10.18093	12.04497	43.4599	3.18044	1.83349
1.40	1.81692	9.64525	11.46216	42.0767	3.36932	1.76732
1.60	1.77093	9.06324	10.83417	40.6687	3.55818	1.69358
1.80	1.73261	8.46230	10.19491	39.2615	3.74868	1.61516
2.00	1.70947	7.87630	9.58577	37.8837	3.94284	1.53540

TABLE 4h-29. THERMODYNAMIC PROPERTIES OF NITROGEN
$T = 4000$ K

$\log \rho/\rho_0$	Z	Moles	E/RT	H/RT	S/R	$\log_{10} P$	$N(E)$†	C_p/R	C_v/R	a/a_0
−5.0	1.07027	1.07027	5.1506	6.2209	44.3262	−3.8047	2.57 + 08	34.3362	30.3825	3.4994
−4.8	1.05623	1.05623	4.7552	5.8114	43.4412	−3.6104	3.00 + 08	28.4237	25.1844	3.4844
−4.6	1.04494	1.04493	4.4368	5.4817	42.6391	−3.4151	3.51 + 08	23.6420	20.9369	3.4752
−4.4	1.03586	1.03586	4.1811	5.2170	41.9044	−3.2189	4.13 + 08	19.7917	17.4869	3.4712
−4.2	1.02859	1.02859	3.9763	5.0049	41.2243	−3.0219	4.89 + 08	16.7015	14.6981	3.4716
−4.0	1.02278	1.02278	3.8124	4.8352	40.5882	−2.8244	5.82 + 08	14.2273	12.4520	3.4761
−3.8	1.01814	1.01814	3.6816	4.6998	39.9875	−2.6264	6.98 + 08	12.2500	10.6483	3.4839
−3.6	1.01444	1.01443	3.5773	4.5917	39.4151	−2.4279	8.41 + 08	10.6719	9.2031	3.4945
−3.4	1.01148	1.01148	3.4941	4.5056	38.8655	−2.2292	1.02 + 09	9.4138	8.0473	3.5073
−3.2	1.00913	1.00913	3.4278	4.4370	38.3340	−2.0302	1.25 + 09	8.4115	7.1241	3.5215
−3.0	1.00726	1.00726	3.3751	4.3824	37.8170	−1.8310	1.53 + 09	7.6136	6.3877	3.5365
−2.8	1.00577	1.00577	3.3331	4.3389	37.3115	−1.6317	1.88 + 09	6.9786	5.8008	3.5517
−2.6	1.00459	1.00459	3.2998	4.3044	36.8152	−1.4322	2.32 + 09	6.4736	5.3333	3.5665
−2.4	1.00365	1.00365	3.2732	4.2769	36.3263	−1.2326	2.88 + 09	6.0720	4.9612	3.5804
−2.2	1.00291	1.00290	3.2521	4.2550	35.8432	−1.0329	3.58 + 09	5.7527	4.6651	3.5932
−2.0	1.00232	1.00230	3.2354	4.2377	35.3647	−.8332	4.46 + 09	5.4989	4.4295	3.6048
−1.8	1.00185	1.90183	3.2220	4.2239	34.8899	−.6334	5.57 + 09	5.2972	4.2423	3.6149
−1.6	1.00149	1.00145	3.2114	4.2129	34.4180	−.4335	6.97 + 09	5.1369	4.0934	3.6236
−1.4	1.00121	1.00115	3.2030	4.2042	33.9485	−.2336	8.73 + 09	5.0095	3.9750	3.6311
−1.2	1.00101	1.00092	3.1963	4.1974	33.4808	−.0337	1.09 + 10	4.9083	3.8810	3.6375
−1.0	1.00087	1.00073	3.1910	4.1919	33.0145	.1662	1.37 + 10	4.8279	3.8062	3.6428
−.8	1.00081	1.00058	3.1868	4.1876	32.5494	.3662	1.72 + 10	4.7641	3.7469	3.6474
−.6	1.00082	1.00046	3.1835	4.1843	32.0852	.5662	2.17 + 10	4.7133	3.6997	3.6513
−.4	1.00094	1.00037	3.1808	4.1818	31.6216	.7662	2.72 + 10	4.6730	3.6622	3.6549
−.2	1.00121	1.00029	3.1787	4.1800	31.1585	.9664	3.42 + 10	4.6410	3.6325	3.6584
−.0	1.00168	1.00023	3.1771	4.1788	30.6957	1.1666	4.31 + 10	4.6155	3.6089	3.6622
.2	1.00249	1.00018	3.1758	4.1783	30.2330	1.3669	5.42 + 10	4.5953	3.5903	3.6667
.4	1.00379	1.00015	3.1749	4.1787	29.7701	1.5675	6.83 + 10	4.5793	3.5756	3.6727
.6	1.00590	1.00012	3.1742	4.1801	29.3067	1.7684	8.62 + 10	4.5666	3.5641	3.6812
.8	1.00926	1.00009	3.1738	4.1831	28.8423	1.9698	1.09 + 11	4.5566	3.5553	3.6940
1.0	1.01460	1.00007	3.1737	4.1883	28.3763	2.1721	1.38 + 11	4.5489	3.5488	3.7136
1.2	1.02308	1.00006	3.1740	4.1971	27.9076	2.3757	1.75 + 11	4.5430	3.5445	3.7440
1.4	1.03653	1.00005	3.1747	4.2113	27.4343	2.5814	2.23 + 11	4.5392	3.5424	3.7914
1.6	1.05786	1.00004	3.1762	4.2341	26.9539	2.7903	2.87 + 11	4.5379	3.5427	3.8653
1.8	1.09166	1.00003	3.1788	4.2704	26.4621	3.0039	3.73 + 11	4.5405	3.5463	3.9799
2.0	1.14526	1.00002	3.1829	4.3282	25.9522	3.2247	4.96 + 11	4.5505	3.5544	4.1560

† $N(E)$ is the number of electrons per cm³, expressed in the form $a + b$, meaning $a \times 10^b$.

TABLE 4h-29. THERMODYNAMIC PROPERTIES OF NITROGEN (*Continued*)

$T = 6000$ K

log ρ/ρ_0	Z	Moles	E/RT	H/RT	S/R	$\log_{10} P$	$N(E)$†	C_p/R	C_v/R	a/a_0
−5.0	1.98949	1.98949	21.9731	23.9626	67.4937	−3.3593	1.27 + 12	14.2250	11.4527	6.1982
−4.8	1.98000	1.98000	21.7840	23.7640	66.3905	−3.1614	1.59 + 12	16.9096	13.8083	6.1281
−4.6	1.96630	1.96630	21.5179	23.4842	65.2154	−2.9644	1.99 + 12	21.1530	17.5086	6.0481
−4.4	1.94649	1.94649	21.1387	23.0852	63.9350	−2.7688	2.48 + 12	27.3503	22.8446	5.9653
−4.2	1.91840	1.91840	20.6050	22.5234	62.5110	−2.5751	3.08 + 12	35.7322	29.9229	5.8809
−4.0	1.87992	1.87992	19.8771	21.7570	60.9081	−2.3839	3.81 + 12	46.0825	38.4336	5.7917
−3.8	1.82974	1.82974	18.9297	20.7595	59.1061	−2.1957	4.67 + 12	57.4674	47.4846	5.6933
−3.6	1.76801	1.76800	17.7661	19.5341	57.1136	−2.0106	5.68 + 12	68.2367	55.7094	5.5828
−3.4	1.69676	1.69676	16.4240	18.1207	54.9734	−1.8285	6.85 + 12	76.4862	61.7151	5.4604
−3.2	1.61954	1.61954	14.9701	16.5896	52.7557	−1.6487	8.20 + 12	80.7954	64.6166	5.3294
−3.0	1.54057	1.54056	13.4836	15.0242	50.5416	−1.4704	9.73 + 12	80.7463	64.2934	5.1948
−2.8	1.46376	1.46376	12.0382	13.5020	48.4046	−1.2926	1.15 + 13	76.9116	61.2648	5.0624
−2.6	1.39214	1.39213	10.6904	12.0825	46.3994	−1.1144	1.35 + 13	70.4385	56.3665	4.9371
−2.4	1.32757	1.32756	9.4755	10.8030	44.5585	−.9350	1.58 + 13	62.5714	50.4523	4.8226
−2.2	1.27091	1.27090	8.4095	9.6804	42.8945	−.7540	1.86 + 13	54.3400	44.2225	4.7211
−2.0	1.22225	1.22223	7.4938	8.7161	41.4051	−.5709	2.18 + 13	46.4481	38.1688	4.6336
−1.8	1.18114	1.18112	6.7204	7.9015	40.0785	−.3858	2.56 + 13	39.2959	32.5899	4.5599
−1.6	1.14688	1.14684	6.0755	7.2224	38.8979	−.1986	3.01 + 13	33.0555	27.6345	4.4996
−1.4	1.11861	1.11855	5.5434	6.6620	37.8443	−.0094	3.56 + 13	27.7517	23.3485	4.4516
−1.2	1.09549	1.09539	5.1078	6.2033	36.8990	.1815	4.24 + 13	23.3262	19.7134	4.4147
−1.0	1.07670	1.07655	4.7534	5.8301	36.0447	.3740	5.07 + 13	19.6816	16.6756	4.3877
−.8	1.06154	1.06130	4.4666	5.5281	35.2656	.5679	6.09 + 13	16.7080	14.1651	4.3693
−.6	1.04938	1.04901	4.2354	5.2847	34.5484	.7629	6.37 + 13	14.2982	12.1082	4.3586
−.4	1.03970	1.03912	4.0495	5.0892	33.8816	.9588	8.95 + 13	12.3549	10.4340	4.3545
−.2	1.03211	1.03120	3.9005	4.9326	33.2556	1.1556	1.09 + 14	10.7933	9.0784	4.3559
−.0	1.02629	1.02486	3.7812	4.8075	32.6624	1.3532	1.23 + 14	9.5418	7.9850	4.3622
.2	1.02204	1.01979	3.6859	4.7080	32.0956	1.5514	1.66 + 14	8.5406	7.1059	4.3728
.4	1.01930	1.01574	3.6100	4.6293	31.5496	1.7502	2.05 + 14	7.7407	6.4007	4.3876
.6	1.01813	1.01251	3.5494	4.5676	31.0200	1.9497	2.55 + 14	7.1022	5.8360	4.4067
.8	1.01881	1.00993	3.5014	4.5202	30.5030	2.1500	3.19 + 14	6.5927	5.3846	4.4314
1.0	1.02192	1.00787	3.4633	4.4853	29.9952	2.3513	4.00 + 14	6.1860	5.0240	4.4640
1.2	1.02845	1.00623	3.4334	4.4619	29.4934	2.5541	5.04 + 14	5.8613	4.7363	4.5088
1.4	1.04009	1.00492	3.4103	4.4504	28.9942	2.7590	6.39 + 14	5.6018	4.5071	4.5726
1.6	1.05954	1.00386	3.3930	4.4526	28.4938	2.9670	8.18 + 14	5.3946	4.3250	4.6663
1.8	1.09117	1.00301	3.3811	4.4723	27.9872	3.1798	1.06 + 15	5.2303	4.1815	4.8064
2.0	1.14194	1.00233	3.3745	4.5165	27.4674	3.3996	1.40 + 15	5.1030	4.0700	5.0712

† $N(E)$ is the number of electrons per cm³, expressed in the form $a + b$, meaning $a \times 10^b$.

TABLE 4h-29. THERMODYNAMIC PROPERTIES OF NITROGEN (*Continued*)

T = 8000 K

log ρ/ρ_0	Z	Moles	E/RT	H/RT	S/R	log$_{10}$ P	N(E)†	C_P/R	C_v/R	a/a_0
−5.0	2.18555	2.18576	21.9006	24.0862	73.4742	−3.1936	4.99 + 13	57.8164	49.5308	7.1542
−4.8	2.14878	2.14897	21.0770	23.2258	71.6529	−3.0009	6.35 + 13	47.9289	41.1283	7.1139
−4.6	2.11897	2.11914	20.4097	22.5287	70.0032	−2.8070	8.06 + 13	39.8826	34.2034	7.0883
−4.4	2.09484	2.09499	19.8705	21.9653	68.4939	−2.6120	1.02 + 14	33.3888	28.5537	7.0759
−4.2	2.07530	2.07544	19.4349	21.5102	67.0982	−2.4161	1.29 + 14	28.1914	23.9896	7.0749
−4.0	2.05938	2.05951	19.0820	21.1414	65.7936	−2.2194	1.63 + 14	24.0755	20.3459	7.0838
−3.8	2.04625	2.04636	18.7940	20.8402	64.5603	−2.0222	2.06 + 14	20.8709	17.4880	7.1001
−3.6	2.03514	2.03524	18.5549	20.5900	63.3814	−1.8245	2.61 + 14	18.4544	15.3165	7.1207
−3.4	2.02528	2.02537	18.3499	20.3751	62.2415	−1.6267	3.28 + 14	16.7528	13.7713	7.1408
−3.2	2.01589	2.01596	18.1643	20.1801	61.1254	−1.4287	4.14 + 14	15.7475	12.8372	7.1540
−3.0	2.00601	2.00607	17.9820	19.9880	60.0171	−1.2308	5.21 + 14	15.4825	12.5497	7.1526
−2.8	1.99447	1.99453	17.7843	19.7787	58.8981	−1.0333	6.55 + 14	16.0696	12.9991	7.1298
−2.6	1.97977	1.97981	17.5477	19.5274	57.7462	−.8365	8.22 + 14	17.6847	14.3226	7.0828
−2.4	1.96000	1.96003	17.2436	19.2036	56.5348	−.6409	1.03 + 15	20.5348	16.6689	7.0142
−2.2	1.93292	1.93295	16.8391	18.7720	55.2336	−.4469	1.28 + 15	24.7702	20.1165	6.9286
−2.0	1.89629	1.89630	16.3009	18.1972	53.8133	−.2552	1.59 + 15	30.3278	24.5437	6.8290
−1.8	1.84843	1.84841	15.6043	17.4527	52.2540	−.0663	1.96 + 15	36.7543	29.5100	6.7156
−1.6	1.78897	1.78893	14.7437	16.5327	50.5555	.1195	2.41 + 15	43.1458	34.2646	6.5872
−1.4	1.71942	1.71934	13.7401	15.4595	48.7437	.3022	2.93 + 15	48.3501	37.9607	6.4446
−1.2	1.64295	1.64282	12.6385	14.2815	46.8678	.4825	3.54 + 15	51.3902	39.973	6.2911
−1.0	1.56368	1.56347	11.4978	13.0614	44.9887	.6610	4.25 + 15	51.8389	40.1100	6.1325
−.8	1.48571	1.48538	10.3759	11.8616	43.1648	.8388	5.08 + 15	49.8989	38.5859	5.9753
−.6	1.41231	1.41182	9.3196	10.7320	41.4415	1.0168	6.06 + 15	46.1972	35.8563	5.8257
−.4	1.34571	1.34497	8.3601	9.7058	39.8472	1.1958	7.23 + 15	41.4889	32.4268	5.6885
−.2	1.28704	1.28592	7.5129	8.7999	38.3941	1.3765	8.62 + 15	36.4363	28.7324	5.5669
−.0	1.23664	1.23494	6.7815	8.0181	37.0820	1.5591	1.03 + 16	31.5133	25.0899	5.4626
.2	1.19427	1.19169	6.1611	7.3554	35.9022	1.7440	1.24 + 16	27.0023	21.6986	5.3764
.4	1.15944	1.15549	5.6422	6.8016	34.8416	1.9311	1.49 + 16	23.0350	18.6634	5.3084
.6	1.13161	1.12552	5.2127	6.3443	33.8849	2.1206	1.81 + 16	19.6429	16.0224	5.2586
.8	1.11031	1.10091	4.8602	5.9705	33.0164	2.3123	2.21 + 16	16.7982	13.7710	5.2277
1.0	1.09541	1.08081	4.5728	5.6683	32.2214	2.5064	2.71 + 16	14.4444	11.8803	5.2173
1.2	1.08722	1.06447	4.3397	5.4270	31.4860	2.7032	3.36 + 16	12.5140	10.3097	5.2308
1.4	1.08677	1.05122	4.1516	5.2383	30.7976	2.9030	4.20 + 16	10.9397	9.0151	5.2746
1.6	1.09618	1.04047	4.0004	5.0966	30.1442	3.1067	5.32 + 16	9.6596	7.9535	5.3591
1.8	1.11922	1.03174	3.8799	4.9991	29.5142	3.3158	6.84 + 16	8.6204	7.0855	5.5012
2.0	1.16221	1.02463	3.7852	4.9474	28.8951	3.5321	9.00 + 16	7.7781	6.3770	5.7266

† N(E) is the number of electrons per cm³, expressed in the form a + b, meaning a × 10^b.

TABLE 4h-29. THERMODYNAMIC PROPERTIES OF NITROGEN (*Continued*)

$T = 10,000$ K

log ρ/ρ_0	Z	Moles	E/RT	H/RT	S/R	$\log_{10} P$	$N(E)$†	C_p/R	C_v/R	a/a_0
−5.0	3.18098	3.18339	36.5970	39.7780	94.4282	−2.9337	3.18 + 14	149.0162	118.0254	9.6774
−4.8	3.02429	3.02674	33.7675	36.7918	90.1701	−2.7556	4.37 + 14	146.4103	115.6397	9.4318
−4.6	2.87425	2.87669	31.0574	33.9316	86.1022	−2.5777	5.92 + 14	137.4614	108.7631	9.1941
−4.4	2.73608	2.73845	28.5607	31.2968	82.3143	−2.3991	7.90 + 14	124.4353	98.9949	8.9726
−4.2	2.61275	2.61502	26.3316	28.9444	78.8543	−2.2191	1.04 + 15	109.5346	87.8303	8.7728
−4.0	2.50536	2.50750	24.3902	26.8956	75.7350	−2.0374	1.36 + 15	94.4607	76.4281	8.5980
−3.8	2.41365	2.41565	22.7319	25.1455	72.9447	−1.8536	1.77 + 15	80.3112	65.5650	8.4492
−3.6	2.33651	2.33836	21.3367	23.6732	70.4563	−1.6677	2.28 + 15	67.6674	55.6889	8.3260
−3.4	2.27238	2.27407	20.1766	22.4490	68.2355	−1.4797	2.93 + 15	56.7445	47.0036	8.2270
−3.2	2.21951	2.22106	19.2205	21.4400	66.2455	−1.2900	3.75 + 15	47.5314	39.5509	8.1501
−3.0	2.17619	2.17759	18.4376	20.6138	64.4508	−1.0985	4.79 + 15	39.8975	33.2769	8.0938
−2.8	2.14079	2.14206	17.7989	19.9397	62.8183	−.9057	6.10 + 15	33.6624	28.0792	8.0550
−2.6	2.11183	2.11297	17.2782	19.3900	61.3187	−.7116	7.75 + 15	28.6395	23.8389	8.0315
−2.4	2.08796	2.08899	16.8522	18.9401	59.9259	−.5165	9.82 + 15	24.6607	20.4418	8.0205
−2.2	2.06793	2.06885	16.4997	18.5676	58.6166	−.3207	1.24 + 16	21.5911	17.7925	8.0185
−2.0	2.05053	2.05134	16.2011	18.2516	57.3698	−.1244	1.58 + 16	19.3381	15.8243	8.0210
−1.8	2.03452	2.03522	15.9374	17.9719	56.1655	.0722	1.99 + 16	17.8586	14.5065	8.0217
−1.6	2.01849	2.01909	15.6885	17.7070	54.9834	.2688	2.51 + 16	17.1647	13.8494	8.0131
−1.4	2.00081	2.00130	15.4322	17.4330	53.8015	.4650	3.17 + 16	17.3223	13.9028	7.9870
−1.2	1.97948	1.97983	15.1426	17.1221	52.5952	.6603	3.99 + 16	18.4357	14.7390	7.9377
−1.0	1.95218	1.95237	14.7906	16.7428	51.3376	.8543	5.01 + 16	20.5976	16.4067	7.8634
−.8	1.91650	1.91648	14.3460	16.2625	50.0018	1.0463	6.26 + 16	23.7932	18.8518	7.7656
−.6	1.87042	1.87011	13.7838	15.6543	48.5673	1.2357	7.80 + 16	27.7743	21.8277	7.6464
−.4	1.81310	1.81237	13.0927	14.9058	47.0276	1.4222	9.66 + 16	31.9827	24.8683	7.5072
−.2	1.74544	1.74414	12.2821	14.0275	45.3972	1.6057	1.19 + 17	35.6334	27.3928	7.3503
−.0	1.67024	1.66811	11.3829	13.0531	43.7113	1.7866	1.46 + 17	37.9783	28.9133	7.1802
.2	1.59152	1.58820	10.4404	12.0319	42.0177	1.9656	1.78 + 17	38.5956	29.2077	7.0037
.4	1.51350	1.50853	9.5027	11.0163	40.3651	2.1438	2.17 + 17	37.5083	28.3484	6.8295
.6	1.44028	1.43263	8.6105	10.0508	38.7930	2.3222	2.64 + 17	35.0854	26.6047	6.6662
.8	1.37447	1.36297	7.7926	9.1670	37.3273	2.5019	3.22 + 17	31.8393	24.3122	6.5218
1.0	1.31821	1.30088	7.0644	8.3826	35.9795	2.6838	3.95 + 17	28.2515	21.7767	6.4037
1.2	1.27299	1.24680	6.4311	7.7041	34.7500	2.8686	4.87 + 17	24.6832	19.2292	6.3194
1.4	1.24027	1.20049	5.8899	7.1302	33.6307	3.0573	6.06 + 17	21.3594	16.8201	6.2779
1.6	1.22206	1.16131	5.4337	6.6558	32.6081	3.2509	7.64 + 17	18.3922	14.6318	6.2914
1.8	1.22165	1.12842	5.0532	6.2748	31.6657	3.4507	9.81 + 17	15.8165	12.6971	6.3782
2.0	1.24465	1.10089	4.7387	5.9834	30.7843	3.6588	1.29 + 18	13.6213	11.0172	6.5650

† $N(E)$ is the number of electrons per cm³, expressed in the form $a + b$, meaning $a \times 10^b$.

TABLE 4h-29. THERMODYNAMIC PROPERTIES OF NITROGEN (*Continued*)

$T = 12{,}000$ K

log ρ/ρ_0	Z	Moles	E/RT	H/RT	S/R	$\log_{10} P$	$N(E)$†	C_p/R	C_v/R	a/a_0
−5.0	3.89384	3.89756	42.3391	46.2329	107.6566	−2.7667	5.10 + 14	35.2358	27.8584	12.2802
−4.8	3.84199	3.84648	41.5624	45.4044	105.0980	−2.5725	7.86 + 14	46.0096	36.8252	12.0582
−4.6	3.77010	3.77544	40.4828	44.2529	102.2648	−2.3807	1.20 + 15	59.4348	47.7151	11.8445
−4.4	3.67524	3.68143	39.0550	42.7303	99.1219	−2.1918	1.80 + 15	74.5153	59.5572	11.6258
−4.2	3.55714	3.56413	37.2743	40.8314	95.6750	−2.0059	2.65 + 15	89.2729	70.7073	11.3921
−4.0	3.41909	3.42676	35.1897	38.6087	91.9834	−1.8231	3.83 + 15	101.2202	79.3388	11.1399
−3.8	3.26763	3.27579	32.8994	36.1671	88.1532	−1.6428	5.43 + 15	108.2871	84.1379	10.8726
−3.6	3.11094	3.11939	30.5275	33.6385	84.3126	−1.4642	7.56 + 15	109.6088	84.7330	10.5988
−3.4	2.95702	2.96554	28.1950	31.1520	80.5832	−1.2862	1.03 + 16	105.6704	81.6338	10.3291
−3.2	2.81223	2.82063	25.9985	28.8107	77.0588	−1.1080	1.39 + 16	97.8534	75.8521	10.0732
−3.0	2.68073	2.68887	24.0019	26.6827	73.7980	−.9288	1.85 + 16	87.8025	68.5000	9.8388
−2.8	2.56464	2.57240	22.2375	24.8021	70.8265	−.7480	2.44 + 16	76.9582	60.5325	9.6308
−2.6	2.46437	2.47168	20.7124	23.1767	68.1441	−.5653	3.19 + 16	66.3548	52.6473	9.4515
−2.4	2.37925	2.38606	19.4166	21.7959	65.7336	−.3806	4.13 + 16	56.6156	45.2877	9.3011
−2.2	2.30792	2.31422	18.3302	20.6381	63.5684	−.1938	5.34 + 16	48.0408	38.6928	9.1785
−2.0	2.24869	2.25448	17.4280	19.6767	61.6175	−.0051	6.86 + 16	40.7144	32.9578	9.0818
−1.8	2.19980	2.20508	16.6837	18.8835	59.8493	.1853	8.80 + 16	34.5958	28.0871	9.0082
−1.6	2.15947	2.16426	16.0714	18.2308	58.2335	.3773	1.13 + 17	29.5848	24.0357	8.9549
−1.4	2.12603	2.13035	15.5665	17.6925	56.7421	.5705	1.44 + 17	25.5652	20.7392	8.9182
−1.2	2.09787	2.10172	15.1463	17.2442	55.3496	.7647	1.83 + 17	22.4324	18.1344	8.8941
−1.0	2.07338	2.07676	14.7890	16.8624	54.0319	.9596	2.33 + 17	20.1095	16.1739	8.8773
−.8	2.05092	2.05379	14.4730	16.5240	52.7664	1.1549	2.97 + 17	18.5591	14.8355	8.8609
−.6	2.02865	2.03095	14.1759	16.2045	51.5298	1.3502	3.77 + 17	17.7873	14.1263	8.8367
−.4	2.00448	2.00608	13.8729	15.8774	50.2981	1.5450	4.80 + 17	17.8368	14.0758	8.7959
−.2	1.97601	1.97672	13.5370	15.5130	49.0454	1.7387	6.08 + 17	18.7592	14.7108	8.7316
−.0	1.94072	1.94022	13.1398	15.0805	47.7461	1.9309	7.70 + 17	20.5531	16.0044	8.6409
.2	1.89640	1.89419	12.6561	14.5525	46.3784	2.1209	9.73 + 17	23.0720	17.8091	8.5242
.4	1.84185	1.83722	12.0703	13.9122	44.9315	2.3082	1.23 + 18	25.9489	19.8184	8.3842
.6	1.77763	1.76950	11.3836	13.1612	43.4110	2.4928	1.54 + 18	28.6180	21.6135	8.2255
.8	1.70635	1.69318	10.6160	12.3223	41.8410	2.6750	1.94 + 18	30.4802	22.7989	8.0556
1.0	1.63231	1.61182	9.8022	11.4345	40.2584	2.8558	2.43 + 18	31.1283	23.1454	7.8856
1.2	1.56070	1.52949	8.9821	10.5428	38.7033	3.0363	3.06 + 18	30.4801	22.6414	7.7302
1.4	1.49687	1.44987	8.1916	9.6885	37.2091	3.2181	3.87 + 18	28.7446	21.4420	7.6069
1.6	1.44616	1.37563	7.4572	8.9034	35.7977	3.4032	4.96 + 18	26.2816	19.7759	7.5362
1.8	1.41417	1.30831	6.7946	8.2088	34.4773	3.5935	6.48 + 18	23.4606	17.8661	7.5435
2.0	1.40779	1.24849	6.2102	7.6180	33.2443	3.7915	8.70 + 18	20.5783	15.8904	7.6608

† $N(E)$ is the number of electrons per cm³, expressed in the form $a + b$, meaning $a \times 10^b$.

TABLE 4h-29. THERMODYNAMIC PROPERTIES OF NITROGEN (*Continued*)

T = 14,000 K

log ρ/ρ_0	Z	Moles	E/RT	H/RT	S/R	$\log_{10} P$	$N(E)$†	C_p/R	C_v/R	a/a_0
−5.0	3.98442	3.98759	38.4405	42.4250	110.0090	−2.6897	5.34 + 14	13.4990	9.1410	14.6494
−4.8	3.97617	3.98013	38.3386	42.3148	108.0739	−2.4906	8.43 + 14	14.7994	10.2533	14.4552
−4.6	3.96375	3.96869	38.1846	42.1483	106.0914	−2.2920	1.33 + 15	16.8835	12.0340	14.2098
−4.4	3.94507	3.95121	37.9511	41.8962	104.0366	−2.0940	2.09 + 15	20.0705	14.7420	13.9361
−4.2	3.91726	3.92484	37.6007	41.5180	101.8755	−1.8971	3.26 + 15	24.7569	18.6854	13.6577
−4.0	3.87667	3.88592	37.0854	40.9621	99.5650	−1.7016	5.07 + 15	31.3375	24.1420	13.3903
−3.8	3.81914	3.83028	36.3503	40.1695	97.0572	−1.5081	7.79 + 15	40.0270	31.2001	13.1364
−3.6	3.74084	3.75399	35.3444	39.0852	94.3097	−1.3171	1.18 + 16	50.5831	39.5436	12.8883
−3.4	3.63954	3.65471	34.0372	37.6767	91.3023	−1.1291	1.77 + 16	62.0583	48.3114	12.6342
−3.2	3.51598	3.53301	32.4367	35.9526	88.0533	−.9441	2.60 + 16	72.8262	56.2191	12.3656
−3.0	3.37439	3.39296	30.5967	33.9711	84.6263	−.7619	3.74 + 16	81.0447	61.9814	12.0803
−2.8	3.22182	3.24149	28.6084	31.8302	81.1189	−.5820	5.29 + 16	85.3575	64.7957	11.7834
−2.6	3.06640	3.08667	26.5777	29.6441	77.6405	−.4035	7.33 + 16	85.3746	64.5639	11.4842
−2.4	2.91560	2.93601	24.6030	27.5186	74.2887	−.2254	1.00 + 17	81.6518	61.7724	11.1936
−2.2	2.77512	2.79524	22.7593	25.5344	71.1352	−.0468	1.35 + 17	75.2978	57.1972	10.9213
−2.0	2.64848	2.66798	21.0937	23.7422	68.2214	.1329	1.80 + 17	67.5224	51.6344	10.6744
−1.8	2.53725	2.55588	19.6278	22.1651	65.5621	.3143	2.37 + 17	59.3377	45.7417	10.4573
−1.6	2.44152	2.45911	18.3638	20.8053	63.1523	.4976	3.10 + 17	51.4482	39.9878	10.2715
−1.4	2.36039	2.37685	17.2907	19.6511	60.9741	.6829	4.04 + 17	44.2666	34.6643	10.1168
−1.2	2.29238	2.30766	16.3900	18.6824	59.0026	.8702	5.24 + 17	37.9846	29.9255	9.9914
−1.0	2.23575	2.24982	15.6397	17.8754	57.2100	1.0593	6.78 + 17	32.6513	25.8312	9.8926
−.8	2.18865	2.20149	15.0165	17.2051	55.5684	1.2501	8.74 + 17	28.2368	22.3849	9.8170
−.6	2.14923	2.16082	14.4973	16.6466	54.0507	1.4422	1.13 + 18	24.6792	19.5624	9.7604
−.4	2.11566	2.12591	14.0598	16.1755	52.6315	1.6353	1.45 + 18	21.9149	17.3337	9.7177
−.2	2.08604	2.09480	13.6817	15.7677	51.2860	1.8292	1.87 + 18	19.8990	15.6769	9.6828
−.0	2.05840	2.06537	13.3400	15.3984	49.9901	2.0234	2.40 + 18	18.6156	14.5871	9.6481
.2	2.03056	2.03525	13.0108	15.0413	48.7193	2.2175	3.10 + 18	18.0776	14.0754	9.6049
.4	2.00012	2.00174	12.6679	14.6680	47.4481	2.4110	4.00 + 18	18.3090	14.1534	9.5451
.6	1.96466	1.96196	12.2842	14.2489	46.1514	2.6032	5.16 + 18	19.2999	14.7975	9.4634
.8	1.92215	1.91324	11.8352	13.7573	44.8070	2.7937	6.68 + 18	20.9371	15.9001	9.3593
1.0	1.87168	1.85378	11.3038	13.1755	43.4018	2.9821	8.66 + 18	22.9394	17.2342	9.2371
1.2	1.81434	1.78337	10.6872	12.5016	41.9363	3.1686	1.13 + 19	24.8634	18.4806	9.1067
1.4	1.75368	1.70374	9.9992	11.7529	40.4267	3.3539	1.48 + 19	26.2200	19.3242	8.9836
1.6	1.69564	1.61818	9.2670	10.9626	38.9005	3.5392	1.96 + 19	26.6528	19.5647	8.8907
1.8	1.64812	1.53063	8.5239	10.1720	37.3881	3.7269	2.67 + 19	26.0586	19.1666	8.8586
2.0	1.62076	1.44479	7.8019	9.4227	35.9144	3.9196	3.76 + 19	24.5826	18.2331	8.9257

† $N(E)$ is the number of electrons per cm³, expressed in the form $a + b$, meaning $a \times 10^b$.

TABLE 4h-29.　THERMODYNAMIC PROPERTIES OF NITROGEN (Continued)
$T = 16,000$ K

log ρ/ρ_0	Z	Moles	E/RT	H/RT	S/R	$\log_{10} P$	$N(E)$†	C_p/R	C_v/R	a/a_0
−5.0	4.00486	4.00751	34.7983	38.8031	111.2459	−2.6295	5.39 + 14	16.8199	12.2744	15.1260
−4.8	3.99908	4.00239	34.6959	38.6950	109.3006	−2.4301	8.53 + 14	14.9838	10.5851	15.3666
−4.6	3.99363	3.99778	34.6139	38.6075	107.3781	−2.2307	1.35 + 15	14.0217	9.6837	15.5310
−4.4	3.98750	3.99270	34.5354	38.5229	105.4619	−2.0314	2.13 + 15	13.7384	9.3885	15.5969
−4.2	3.97961	3.98612	34.4450	38.4246	103.5369	−1.8323	3.37 + 15	14.0680	9.6320	15.5578
−4.0	3.96862	3.97675	34.3252	38.2938	101.5868	−1.6335	5.31 + 15	15.0567	10.4430	15.4208
−3.8	3.95269	3.96282	34.1541	38.1068	99.5916	−1.4352	8.36 + 15	16.8539	11.9360	15.2059
−3.6	3.92940	3.94194	33.9031	37.8325	97.5253	−1.2378	1.31 + 16	19.6999	14.2950	14.9418
−3.4	3.89553	3.91094	33.5348	37.4303	95.3547	−1.0415	2.04 + 16	23.8884	17.7333	14.6574
−3.2	3.84725	3.86598	33.0044	36.8517	93.0409	−.8470	3.16 + 16	29.6715	22.4049	14.3724
−3.0	3.78057	3.80295	32.2649	36.0454	90.5443	−.6545	4.84 + 16	37.0841	28.2583	14.0925
−2.8	3.69234	3.71857	31.2783	34.9707	87.8362	−.4648	7.32 + 16	45.7253	34.8830	13.8123
−2.6	3.58163	3.61160	30.0314	33.6131	84.9136	−.2780	1.09 + 17	54.6326	41.4715	13.5231
−2.4	3.45067	3.48402	28.5476	31.9983	81.8099	−.0942	1.59 + 17	62.4355	47.0102	13.2193
−2.2	3.30494	3.34100	26.8877	30.1926	78.5940	.0871	2.27 + 17	67.8093	50.6395	12.9019
−2.0	3.15193	3.18987	25.1366	28.2885	75.3562	.2665	3.20 + 17	69.9751	51.9549	12.5780
−1.8	2.99944	3.03837	23.3840	26.3835	72.1874	.4449	4.42 + 17	68.9216	51.0660	12.2580
−1.6	2.85408	2.89317	21.7068	24.5608	69.1628	.6234	6.03 + 17	65.2614	48.4401	11.9526
−1.4	2.72049	2.75903	20.1596	22.8801	66.3326	.8025	8.12 + 17	59.8968	44.6797	11.6707
−1.2	2.60128	2.63871	18.7740	21.3753	63.7222	.9831	1.08 + 18	53.7136	40.3514	11.4183
−1.0	2.49733	2.53323	17.5615	20.0588	61.3364	1.1654	1.43 + 18	47.4093	35.9009	11.1985
−.8	2.40828	2.44237	16.5193	18.9276	59.1652	1.3496	1.89 + 18	41.4486	31.6346	11.0119
−.6	2.33296	2.36503	15.6350	17.9680	57.1898	1.5358	2.48 + 18	36.0932	27.7384	10.8573
−.4	2.26977	2.29967	14.8912	17.1610	55.3866	1.7239	3.24 + 18	31.4597	24.3085	10.7320
−.2	2.21689	2.24445	14.2677	16.4846	53.7303	1.9136	4.23 + 18	27.5755	21.3828	10.6326
−.0	2.17243	2.19740	13.7434	15.9158	52.1957	2.1048	5.53 + 18	24.4242	18.9677	10.5546
.2	2.13446	2.15643	13.2968	15.4313	50.7577	2.2972	7.23 + 18	21.9773	17.0581	10.4928
.4	2.10100	2.11926	12.9060	15.0070	49.3918	2.4903	9.48 + 18	20.2155	15.6511	10.4412
.6	2.07000	2.08338	12.5480	14.6180	48.0734	2.6839	1.25 + 19	19.1374	14.7523	10.3932
.8	2.03934	2.04597	12.1981	14.2375	46.7773	2.8774	1.65 + 19	18.7556	14.3712	10.3420
1.0	2.00698	2.00390	11.8307	13.8377	45.4781	3.0704	2.19 + 19	19.0712	14.5017	10.2831
1.2	1.97130	1.95408	11.4206	13.3919	44.1519	3.2626	2.95 + 19	20.0236	15.0864	10.2166
1.4	1.93189	1.89399	10.9479	12.8798	42.7802	3.4539	4.01 + 19	21.4329	15.9824	10.1495
1.6	1.89045	1.82254	10.4033	12.2937	41.3555	3.6445	5.58 + 19	22.9770	16.9630	10.0976
1.8	1.85158	1.74076	9.7937	11.6453	39.8845	3.8354	8.03 + 19	24.2625	17.7828	10.0857
2.0	1.82286	1.65216	9.1447	10.9675	38.3899	4.0286	1.23 + 20	25.0006	18.2986	10.1454

† $N(E)$ is the number of electrons per cm³, expressed in the form $a + b$, meaning $a \times 10^b$.

TABLE 4h-30. THERMODYNAMIC PROPERTIES OF ARGON

$T = 4000$ K

$\log \rho/\rho_0$	Z	Moles	E/RT	H/RT	S/R	$\log_{10} P$	$N(E)$†	C_p/R	C_v/R	a/a_0
−5.0	1.00000	1.00000	1.5000	2.5000	33.9295	−3.8339	1.54 + 08	2.5007	1.5006	3.8258
−4.8	1.00000	1.00000	1.5000	2.5000	33.4690	−3.6339	1.94 + 08	2.5005	1.5005	3.8259
−4.6	1.00000	1.00000	1.5000	2.5000	33.0085	−3.4339	2.44 + 08	2.5004	1.5004	3.8259
−4.4	1.00000	1.00000	1.5000	2.5000	32.5480	−3.2339	3.07 + 08	2.5003	1.5003	3.8260
−4.2	1.00000	1.00000	1.5000	2.5000	32.0875	−3.0339	3.87 + 08	2.5003	1.5003	3.8260
−4.0	1.00000	1.00000	1.5000	2.5000	31.6269	−2.8339	4.87 + 08	2.5002	1.5002	3.8260
−3.8	1.00000	1.00000	1.5000	2.5000	31.1664	−2.6339	6.13 + 08	2.5002	1.5002	3.8261
−3.6	1.00000	1.00000	1.5000	2.5000	30.7059	−2.4339	7.72 + 08	2.5001	1.5001	3.8261
−3.4	1.00000	1.00000	1.5000	2.5000	30.2454	−2.2339	9.72 + 08	2.5001	1.5001	3.8261
−3.2	1.00000	1.00000	1.5000	2.5000	29.7849	−2.0339	1.22 + 09	2.5001	1.5001	3.8261
−3.0	1.00000	1.00000	1.5000	2.5000	29.3244	−1.8339	1.54 + 09	2.5001	1.5001	3.8261
−2.8	1.00000	1.00000	1.5000	2.5000	28.8638	−1.6339	1.94 + 09	2.5001	1.5001	3.8261
−2.6	1.00000	1.00000	1.5000	2.5000	28.4033	−1.4339	2.44 + 09	2.5000	1.5000	3.8261
−2.4	1.00000	1.00000	1.5000	2.5000	27.9428	−1.2339	3.07 + 09	2.5000	1.5000	3.8261
−2.2	1.00001	1.00000	1.5000	2.5000	27.4823	−1.0339	3.87 + 09	2.5000	1.5000	3.8261
−2.0	1.00001	1.00000	1.5000	2.5000	27.0218	−.8339	4.87 + 09	2.5000	1.5000	3.8262
−1.8	1.00002	1.00000	1.5000	2.5000	26.5612	−.6339	6.13 + 09	2.5000	1.5000	3.8262
−1.6	1.00003	1.00000	1.5000	2.5000	26.1007	−.4339	7.72 + 09	2.5000	1.5000	3.8262
−1.4	1.00005	1.00000	1.5000	2.5000	25.6402	−.2339	9.72 + 09	2.5000	1.5000	3.8263
−1.2	1.00007	1.00000	1.5000	2.5001	25.1796	−.0339	1.22 + 10	2.5000	1.5000	3.8264
−1.0	1.00012	1.00000	1.5000	2.5001	24.7191	.1661	1.54 + 10	2.5000	1.5000	3.8266
−.8	1.00019	1.00000	1.5000	2.5002	24.2585	.3661	1.94 + 10	2.5000	1.5000	3.8268
−.6	1.00030	1.00000	1.5000	2.5003	23.7979	.5662	2.45 + 10	2.5000	1.5001	3.8272
−.4	1.00047	1.00000	1.5000	2.5005	23.3372	.7663	3.08 + 10	2.5001	1.5001	3.8278
−.2	1.00075	1.00000	1.5000	2.5008	22.8764	.9664	3.89 + 10	2.5001	1.5002	3.8288
−.0	1.00118	1.00000	1.5001	2.5012	22.4155	1.1666	4.90 + 10	2.5001	1.5002	3.8304
.2	1.00187	1.00000	1.5001	2.5020	21.9543	1.3669	6.20 + 10	2.5002	1.5004	3.8330
.4	1.00297	1.00000	1.5002	2.5031	21.4927	1.5673	7.85 + 10	2.5003	1.5006	3.8369
.6	1.00470	1.00000	1.5002	2.5049	21.0306	1.7681	9.98 + 10	2.5005	1.5009	3.8433
.8	1.00745	1.00000	1.5004	2.5078	20.5675	1.9693	1.28 + 11	2.5008	1.5015	3.8532
1.0	1.01181	1.00000	1.5006	2.5124	20.1028	2.1712	1.65 + 11	2.5013	1.5024	3.8690
1.2	1.01872	1.00000	1.5010	2.5197	19.6357	2.3741	2.15 + 11	2.5022	1.5038	3.8940
1.4	1.02968	1.00000	1.5015	2.5312	19.1648	2.5788	2.88 + 11	2.5039	1.5060	3.9333
1.6	1.04703	1.00000	1.5024	2.5494	18.6878	2.7860	4.00 + 11	2.5069	1.5095	3.9950
1.8	1.07454	1.00000	1.5038	2.5783	18.2012	2.9973	5.88 + 11	2.5127	1.5150	4.0915
2.0	1.11814	1.00000	1.5060	2.6242	17.6993	3.2146	9.46 + 11	2.5242	1.5237	4.2413

† $N(E)$ is the number of electrons per cm³, expressed in the form $a + b$, meaning $a \times 10^b$.

TABLE 4h-30. THERMODYNAMIC PROPERTIES OF ARGON (*Continued*)

$T = 6000$ K

log ρ/ρ_0	Z	Moles	E/RT	H/RT	S/R	$\log_{10} P$	$N(E)$†	C_p/R	C_v/R	a/a_0
−5.0	1.00162	1.00162	1.5519	2.5535	34.5928	−3.6572	4.35 + 11	3.3885	2.3335	4.3758
−4.8	1.00128	1.00128	1.5412	2.5425	34.1210	−3.4573	5.47 + 11	3.2059	2.1623	4.4211
−4.6	1.00102	1.00102	1.5327	2.5338	33.6515	−3.2574	6.89 + 11	3.0608	2.0262	4.4623
−4.4	1.00081	1.00081	1.5260	2.5268	33.1838	−3.0575	8.68 + 11	2.9455	1.9181	4.4990
−4.2	1.00064	1.00064	1.5207	2.5213	32.7176	−2.8576	1.09 + 12	2.8539	1.8322	4.5309
−4.0	1.00051	1.00051	1.5164	2.5169	32.2526	−2.6576	1.38 + 12	2.7812	1.7639	4.5584
−3.8	1.00041	1.00041	1.5130	2.5134	31.7885	−2.4577	1.73 + 12	2.7234	1.7097	4.5817
−3.6	1.00032	1.00032	1.5104	2.5107	31.3251	−2.2577	2.18 + 12	2.6774	1.6666	4.6011
−3.4	1.00026	1.00026	1.5082	2.5085	30.8623	−2.0577	2.74 + 12	2.6410	1.6323	4.6173
−3.2	1.00020	1.00020	1.5065	2.5067	30.4000	−1.8578	3.46 + 12	2.6120	1.6051	4.6306
−3.0	1.00016	1.00016	1.5052	2.5054	29.9381	−1.6578	4.35 + 12	2.5890	1.5835	4.6414
−2.8	1.00013	1.00013	1.5041	2.5043	29.4764	−1.4578	5.48 + 12	2.5707	1.5663	4.6502
−2.6	1.00010	1.00010	1.5033	2.5034	29.0150	−1.2578	6.90 + 12	2.5561	1.5527	4.6574
−2.4	1.00009	1.00008	1.5026	2.5027	28.5538	−1.0578	8.69 + 12	2.5446	1.5419	4.6631
−2.2	1.00007	1.00006	1.5021	2.5021	28.0927	−.8578	1.09 + 13	2.5354	1.5333	4.6678
−2.0	1.00006	1.00005	1.5016	2.5017	27.6317	−.6578	1.38 + 13	2.5282	1.5264	4.6715
−1.8	1.00006	1.00004	1.5013	2.5014	27.1708	−.4578	1.73 + 13	2.5224	1.5210	4.6745
−1.6	1.00006	1.00003	1.5010	2.5011	26.7100	−.2578	2.18 + 13	2.5178	1.5167	4.6769
−1.4	1.00007	1.00003	1.5008	2.5009	26.2492	−.0578	2.75 + 13	2.5141	1.5133	4.6789
−1.2	1.00009	1.00002	1.5007	2.5008	25.7885	.1422	3.46 + 13	2.5112	1.5105	4.6805
−1.0	1.00013	1.00002	1.5005	2.5007	25.3278	.3422	4.36 + 13	2.5089	1.5084	4.6819
−.8	1.00019	1.00001	1.5004	2.5006	24.8671	.5422	5.49 + 13	2.5071	1.5067	4.6831
−.6	1.00030	1.00001	1.5004	2.5007	24.4064	.7423	6.92 + 13	2.5056	1.5053	4.6843
−.4	1.00046	1.00001	1.5003	2.5008	23.9457	.9423	8.72 + 13	2.5045	1.5043	4.6857
−.2	1.00073	1.00001	1.5003	2.5010	23.4849	1.1425	1.10 + 14	2.5036	1.5035	4.6873
−.0	1.00115	1.00001	1.5003	2.5014	23.0239	1.3426	1.39 + 14	2.5028	1.5029	4.6895
.2	1.00182	1.00000	1.5003	2.5021	22.5628	1.5429	1.75 + 14	2.5022	1.5025	4.6928
.4	1.00288	1.00000	1.5004	2.5033	22.1013	1.7434	2.21 + 14	2.5018	1.5023	4.6976
.6	1.00456	1.00000	1.5005	2.5051	21.6392	1.9441	2.81 + 14	2.5014	1.5022	4.7052
.8	1.00722	1.00000	1.5008	2.5080	21.1763	2.1453	3.57 + 14	2.5012	1.5025	4.7169
1.0	1.01145	1.00000	1.5012	2.5127	20.7120	2.3471	4.57 + 14	2.5010	1.5031	4.7354
1.2	1.01814	1.00000	1.5019	2.5200	20.2454	2.5500	5.91 + 14	2.5010	1.5043	4.7645
1.4	1.02875	1.00000	1.5030	2.5317	19.7754	2.7545	7.75 + 14	2.5013	1.5062	4.8102
1.6	1.04557	1.00000	1.5047	2.5502	19.2998	2.9615	1.04 + 15	2.5024	1.5094	4.8820
1.8	1.07222	1.00000	1.5074	2.5796	18.8153	3.1724	1.45 + 15	2.5054	1.5145	4.9943
2.0	1.11446	1.00000	1.5117	2.6262	18.3169	3.3892	2.15 + 15	2.5124	1.5228	5.1685

† $N(E)$ is the number of electrons per cm³, expressed in the form $a + b$, meaning $a \times 10^b$.

TABLE 4h-30. THERMODYNAMIC PROPERTIES OF ARGON (*Continued*)

$T = 8000$ K

log ρ/ρ_0	Z	Moles	E/RT	H/RT	S/R	$\log_{10} P$	$N(E)$†	C_p/R	C_v/R	a/a_0
−5.0	1.08764	1.08771	3.6428	4.7304	37.2914	−3.4964	2.36 + 13	30.8686	26.6115	4.6163
−4.8	1.07028	1.07034	3.2185	4.2887	36.3703	−3.3034	3.00 + 13	25.3133	21.8364	4.5941
−4.6	1.05625	1.05630	2.8755	3.9317	35.5378	−3.1091	3.80 + 13	20.7930	17.9039	4.5817
−4.4	1.04495	1.04500	2.5993	3.6443	34.7779	−2.9138	4.81 + 13	17.1368	14.6903	4.5786
−4.2	1.03588	1.03592	2.3775	3.4134	34.0771	−2.7176	6.09 + 13	14.1926	12.0800	4.5843
−4.0	1.02861	1.02865	2.1998	3.2284	33.4241	−2.5207	7.70 + 13	11.8293	9.9698	4.5984
−3.8	1.02279	1.02283	2.0577	3.0805	32.8097	−2.3231	9.72 + 13	9.9369	8.2701	4.6207
−3.6	1.01815	1.01818	1.9442	2.9623	32.2263	−2.1251	1.23 + 14	8.4243	6.9050	4.6506
−3.4	1.01445	1.01447	1.8536	2.8680	31.6677	−1.9267	1.55 + 14	7.2169	5.8110	4.6878
−3.2	1.01149	1.01152	1.7814	2.7929	31.1290	−1.7280	1.95 + 14	6.2541	4.9360	4.7315
−3.0	1.00914	1.00916	1.7239	2.7330	30.6063	−1.5290	2.46 + 14	5.4870	4.2370	4.7806
−2.8	1.00727	1.00729	1.6781	2.6853	30.0962	−1.3298	3.10 + 14	4.8761	3.6792	4.8339
−2.6	1.00578	1.00580	1.6416	2.6474	29.5962	−1.1304	3.91 + 14	4.3899	3.2346	4.8898
−2.4	1.00460	1.00461	1.6126	2.6172	29.1044	−.9309	4.93 + 14	4.0031	2.8804	4.9468
−2.2	1.00366	1.00367	1.5896	2.5932	28.6189	−.7313	6.22 + 14	3.6953	2.5983	5.0030
−2.0	1.00291	1.00292	1.5712	2.5741	28.1385	−.5316	7.84 + 14	3.4506	2.3738	5.0571
−1.8	1.00232	1.00232	1.5566	2.5590	27.6622	−.3319	9.88 + 14	3.2560	2.1951	5.1077
−1.6	1.00186	1.00184	1.5450	2.5469	27.1892	−.1321	1.24 + 15	3.1013	2.0530	5.1539
−1.4	1.00150	1.00147	1.5358	2.5373	26.7187	.0677	1.57 + 15	2.9783	1.9400	5.1953
−1.2	1.00123	1.00117	1.5285	2.5297	26.2502	.2676	1.98 + 15	2.8805	1.8501	5.2317
−1.0	1.00103	1.00093	1.5227	2.5237	25.7833	.4675	2.49 + 15	2.8028	1.7786	5.2632
−.8	1.00091	1.00074	1.5181	2.5190	25.3177	.6675	3.15 + 15	2.7410	1.7218	5.2901
−.6	1.00086	1.00059	1.5144	2.5153	24.8532	.8675	3.97 + 15	2.6918	1.6766	5.3130
−.4	1.00090	1.00047	1.5115	2.5124	24.3893	1.0675	5.01 + 15	2.6528	1.6408	5.3324
−.2	1.00107	1.00037	1.5092	2.5103	23.9261	1.2675	6.32 + 15	2.6217	1.6123	5.3489
−.0	1.00140	1.00030	1.5074	2.5088	23.4632	1.4677	7.99 + 15	2.5970	1.5897	5.3635
.2	1.00199	1.00024	1.5060	2.5080	23.0005	1.6679	1.01 + 16	2.5774	1.5718	5.3769
.4	1.00296	1.00019	1.5050	2.5080	22.5379	1.8684	1.28 + 16	2.5618	1.5577	5.3902
.6	1.00455	1.00015	1.5043	2.5088	22.0749	2.0691	1.62 + 16	2.5494	1.5467	5.4049
.8	1.00710	1.00012	1.5039	2.5110	21.6114	2.2702	2.07 + 16	2.5395	1.5382	5.4230
1.0	1.01116	1.00010	1.5038	2.5150	21.1467	2.4719	2.65 + 16	2.5316	1.5319	5.4475
1.2	1.01762	1.00008	1.5042	2.5218	20.6800	2.6747	3.41 + 16	2.5254	1.5276	5.4827
1.4	1.02786	1.00007	1.5052	2.5330	20.2102	2.8790	4.45 + 16	2.5206	1.5253	5.5357
1.6	1.04411	1.00006	1.5070	2.5511	19.7352	3.0858	5.92 + 16	2.5173	1.5253	5.6168
1.8	1.06988	1.00005	1.5101	2.5800	19.2520	3.2964	8.11 + 16	2.5162	1.5279	5.7420
2.0	1.11073	1.00004	1.5153	2.6260	18.7558	3.5127	1.16 + 17	2.5191	1.5344	5.9353

† $N(E)$ is the number of electrons per cm³, expressed in the form $a + b$, meaning $a \times 10^b$.

TABLE 4h-30. THERMODYNAMIC PROPERTIES OF ARGON (*Continued*)

$T = 10,000$ K

log ρ/ρ_0	Z	Moles	E/RT	H/RT	S/R	$\log_{10} P$	$N(E)$†	C_P/R	C_v/R	a/a_0
−5.0	1.64223	1.64320	14.2637	15.9059	49.7398	−3.2206	1.73 + 14	86.0704	69.1039	6.3474
−4.8	1.56361	1.56461	12.7038	14.2674	47.4418	−3.0419	2.40 + 14	87.5084	69.8199	6.1903
−4.6	1.48602	1.48703	11.1639	12.6499	45.1999	−2.8640	3.29 + 14	84.4647	67.2871	6.0341
−4.4	1.41277	1.41377	9.7099	11.1227	43.0787	−2.6859	4.43 + 14	78.0709	62.3681	5.8848
−4.2	1.34609	1.34705	8.3859	9.7320	41.1199	−2.5069	5.88 + 14	69.7159	56.0343	5.7472
−4.0	1.28712	1.28804	7.2148	8.5019	39.3427	−2.3264	7.74 + 14	60.6334	49.1289	5.6246
−3.8	1.23616	1.23701	6.2022	7.4384	37.7495	−2.1439	1.01 + 15	51.7164	42.2718	5.5185
−3.6	1.19289	1.19368	5.3424	6.5353	36.3307	−1.9594	1.31 + 15	43.5056	35.8580	5.4293
−3.4	1.15667	1.15740	4.6225	5.7792	35.0701	−1.7728	1.68 + 15	36.2615	30.0997	5.3567
−3.2	1.12668	1.12736	4.0263	5.1530	33.9485	−1.5842	2.16 + 15	30.0553	25.0792	5.2998
−3.0	1.10208	1.10269	3.5369	4.6390	32.9461	−1.3938	2.76 + 15	24.8464	20.7949	5.2575
−2.8	1.08202	1.08258	3.1379	4.2199	32.0443	−1.2018	3.52 + 15	20.5376	17.1969	5.2287
−2.6	1.06575	1.06626	2.8142	3.8799	31.2262	−1.0083	4.47 + 15	17.0100	14.2119	5.2125
−2.4	1.05262	1.05308	2.5527	3.6053	30.4771	−.8137	5.68 + 15	14.1436	11.7583	5.2078
−2.2	1.04206	1.04247	2.3422	3.3843	29.7844	−.6181	7.20 + 15	11.8268	9.7558	5.2141
−2.0	1.03358	1.03394	2.1732	3.2068	29.1375	−.4217	9.12 + 15	9.9616	8.1305	5.2306
−1.8	1.02680	1.02711	2.0378	3.0645	28.5277	−.2245	1.15 + 16	8.4644	6.8169	5.2567
−1.6	1.02137	1.02165	1.9294	2.9507	27.9477	−.0268	1.46 + 16	7.2651	5.7588	5.2914
−1.4	1.01705	1.01728	1.8427	2.8598	27.3918	.1713	1.85 + 16	6.3058	4.9088	5.3338
−1.2	1.01361	1.01379	1.7736	2.7872	26.8551	.3699	2.34 + 16	5.5395	4.2272	5.3826
−1.0	1.01089	1.01101	1.7184	2.7293	26.3338	.5687	2.96 + 16	4.9278	3.6815	5.4364
−.8	1.00875	1.00879	1.6744	2.6832	25.8248	.7678	3.74 + 16	4.4399	3.2452	5.4935
−.6	1.00710	1.00702	1.6394	2.6465	25.3256	.9671	4.74 + 16	4.0508	2.8967	5.5522
−.4	1.00586	1.00560	1.6114	2.6173	24.8342	1.1665	6.00 + 16	3.7406	2.6185	5.6106
−.2	1.00500	1.00448	1.5892	2.5942	24.3489	1.3662	7.60 + 16	3.4934	2.3965	5.6675
−.0	1.00451	1.00359	1.5715	2.5760	23.8686	1.5660	9.64 + 16	3.2965	2.2196	5.7216
.2	1.00444	1.00287	1.5575	2.5619	23.3920	1.7659	1.22 + 17	3.1396	2.0786	5.7726
.4	1.00487	1.00231	1.5464	2.5513	22.9183	1.9661	1.56 + 17	3.0146	1.9665	5.8206
.6	1.00600	1.00185	1.5377	2.5437	22.4466	2.1666	1.98 + 17	2.9151	1.8774	5.8664
.8	1.00814	1.00150	1.5310	2.5391	21.9761	2.3675	2.54 + 17	2.8360	1.8070	5.9120
1.0	1.01181	1.00121	1.5260	2.5378	21.5061	2.5691	3.26 + 17	2.7732	1.7516	5.9605
1.2	1.01786	1.00099	1.5225	2.5403	21.0353	2.7717	4.23 + 17	2.7234	1.7087	6.0168
1.4	1.02762	1.00082	1.5205	2.5481	20.5625	2.9758	5.54 + 17	2.6844	1.6762	6.0885
1.6	1.04323	1.00069	1.5202	2.5634	20.0857	3.1824	7.39 + 17	2.6545	1.6531	6.1869
1.8	1.06809	1.00060	1.5219	2.5900	19.6017	3.3926	1.01 + 18	2.6330	1.6389	6.3291
2.0	1.10757	1.00054	1.5264	2.6340	19.1059	3.6084	1.45 + 18	2.6212	1.6342	6.5404

† $N(E)$ is the number of electrons per cm³, expressed in the form $a + b$, meaning $a \times 10^b$.

TABLE 4h-30. THERMODYNAMIC PROPERTIES OF ARGON (*Continued*)

$T = 12,000$ K

log ρ/ρ_0	Z	Moles	E/RT	H/RT	S/R	$\log_{10} P$	$N(E)$†	C_p/R	C_v/R	a/a_0
−5.0	1.96966	1.97102	17.8020	19.7717	56.3866	−3.0624	2.61 + 14	13.8208	10.7732	8.1027
−4.8	1.95381	1.95548	17.5399	19.4937	55.2208	−2.8659	4.07 + 14	18.1228	14.4797	7.9430
−4.6	1.93076	1.93279	17.1577	19.0884	53.9438	−2.6711	6.30 + 14	24.0504	19.5007	7.7997
−4.4	1.89835	1.90078	16.6191	18.5174	52.5232	−2.4784	9.64 + 14	31.6594	25.7961	7.6657
−4.2	1.85484	1.85768	15.8943	17.7491	50.9338	−2.2885	1.45 + 15	40.5410	32.9236	7.5311
−4.0	1.79956	1.80280	14.9718	16.7714	49.1695	−2.1017	2.16 + 15	49.6798	39.9900	7.3879
−3.8	1.73355	1.73714	13.8687	15.6022	47.2525	−1.9179	3.14 + 15	57.6164	45.8652	7.2322
−3.6	1.65961	1.66346	12.6312	14.2908	45.2335	−1.7368	4.48 + 15	62.9469	49.5971	7.0651
−3.4	1.58165	1.58568	11.3251	12.9067	43.1811	−1.5577	6.27 + 15	64.8736	50.7572	6.8913
−3.2	1.50381	1.50791	10.0195	11.5233	41.1652	−1.3796	8.61 + 15	63.4370	49.5009	6.7176
−3.0	1.42960	1.43366	8.7734	10.2030	39.2439	−1.2016	1.17 + 16	59.3379	46.3780	6.5507
−2.8	1.36147	1.36543	7.6285	8.9899	37.4566	−1.0228	1.56 + 16	53.5596	42.0742	6.3961
−2.6	1.30081	1.30461	6.6080	7.9088	35.8235	−.8426	2.06 + 16	47.0373	37.2186	6.2577
−2.4	1.24809	1.25167	5.7202	6.9683	34.3492	−.6606	2.69 + 16	40.4836	32.2930	6.1377
−2.2	1.20313	1.20648	4.9623	6.1654	33.0272	−.4765	3.50 + 16	34.3527	27.6174	6.0366
−2.0	1.16535	1.16846	4.3249	5.4902	31.8447	−.2904	4.53 + 16	28.8818	23.3735	5.9544
−1.8	1.13398	1.13684	3.7949	4.9289	30.7856	−.1022	5.83 + 16	24.1545	19.6418	5.8900
−1.6	1.10817	1.11079	3.3583	4.4665	29.8329	.0878	7.48 + 16	20.1604	16.4355	5.8425
−1.4	1.08710	1.08947	3.0012	4.0883	28.9705	.2794	9.57 + 16	16.8387	13.7273	5.8104
−1.2	1.06998	1.07212	2.7107	3.7807	28.1835	.4726	1.22 + 17	14.1068	11.4692	5.7927
−1.0	1.05616	1.05807	2.4754	3.5315	27.4587	.6669	1.56 + 17	11.8780	9.6048	5.7881
−.8	1.04504	1.04672	2.2854	3.3304	26.7850	.8623	1.99 + 17	10.0700	8.0769	5.7956
−.6	1.03615	1.03757	2.1324	3.1686	26.1529	1.0586	2.54 + 17	8.6094	6.8319	5.8140
−.4	1.02910	1.03022	2.0095	3.0386	25.5545	1.2556	3.23 + 17	7.4329	5.8221	5.8423
−.2	1.02356	1.02432	1.9109	2.9345	24.9833	1.4533	4.12 + 17	6.4874	5.0057	5.8793
−.0	1.01932	1.01960	1.8319	2.8513	24.4340	1.6515	5.27 + 17	5.7286	4.3476	5.9237
.2	1.01623	1.01581	1.7688	2.7850	23.9022	1.8502	6.73 + 17	5.1205	3.8183	5.9741
.4	1.01424	1.01278	1.7184	2.7326	23.3843	2.0493	8.63 + 17	4.6337	3.3936	6.0296
.6	1.01340	1.01036	1.6782	2.6916	22.8774	2.2490	1.11 + 18	4.2443	3.0535	6.0892
.8	1.01392	1.00844	1.6464	2.6604	22.3788	2.4492	1.43 + 18	3.9335	2.7820	6.1531
1.0	1.01624	1.00691	1.6215	2.6377	21.8865	2.6502	1.86 + 18	3.6858	2.5664	6.2225
1.2	1.02110	1.00570	1.6023	2.6234	21.3983	2.8522	2.43 + 18	3.4895	2.3965	6.3005
1.4	1.02975	1.00477	1.5881	2.6178	20.9120	3.0559	3.22 + 18	3.3352	2.2649	6.3931
1.6	1.04422	1.00406	1.5785	2.6227	20.4251	3.2620	4.34 + 18	3.2163	2.1663	6.5105
1.8	1.06775	1.00356	1.5735	2.6413	19.9344	3.4716	6.04 + 18	3.1290	2.0979	6.6683
2.0	1.10549	1.00327	1.5741	2.6795	19.4351	3.6867	8.79 + 18	3.0740	2.0608	6.8895

† $N(E)$ is the number of electrons per cm³, expressed in the form $a + b$, meaning $a \times 10^b$.

TABLE 4h-30. THERMODYNAMIC PROPERTIES OF ARGON (*Continued*)

$T = 14,000$ K

log ρ/ρ_0	Z	Moles	E/RT	H/RT	S/R	$\log_{10} P$	$N(E)$†	C_p/R	C_v/R	a/a_0
−5.0	2.00182	2.00297	16.2050	18.2069	57.4144	−2.9884	2.70 + 14	9.4008	7.0151	9.0622
−4.8	1.99786	1.99929	16.1296	18.1275	56.4181	−2.7893	4.26 + 14	8.4235	6.1023	9.1895
−4.6	1.99368	1.99546	16.0595	18.0532	55.4288	−2.5902	6.72 + 14	8.1468	5.8214	9.2404
−4.4	1.98849	1.99072	15.9807	17.9691	54.4330	−2.3914	1.06 + 15	8.5071	6.1086	9.1991
−4.2	1.98137	1.98415	15.8778	17.8592	53.4160	−2.1929	1.67 + 15	9.5486	6.9931	9.0803
−4.0	1.97111	1.97455	15.7326	17.7037	52.3606	−1.9952	2.62 + 15	11.4085	8.5810	8.9179
−3.8	1.95615	1.96039	15.5218	17.4779	51.2452	−1.7985	4.09 + 15	14.2902	11.0251	8.7442
−3.6	1.93458	1.93974	15.2168	17.1514	50.0441	−1.6033	6.34 + 15	18.3997	14.4624	8.5767
−3.4	1.90430	1.91049	14.7866	16.6909	48.7296	−1.4101	9.74 + 15	23.8214	18.9035	8.4182
−3.2	1.86344	1.87073	14.2035	16.0669	47.2786	−1.2196	1.48 + 16	30.3428	24.0973	8.2626
−3.0	1.81108	1.81946	13.4529	15.2640	45.6815	−1.0319	2.20 + 16	37.3149	29.4596	8.1018
−2.8	1.74784	1.75721	12.5426	14.2904	43.9514	−.8474	3.22 + 16	43.7039	34.1758	7.9307
−2.6	1.67606	1.68624	11.5060	13.1821	42.1262	−.6656	4.63 + 16	48.4041	37.4782	7.7485
−2.4	1.59939	1.61013	10.3954	11.9948	40.2614	−.4859	6.53 + 16	50.6621	38.9300	7.5591
−2.2	1.52189	1.53293	9.2696	10.7915	38.4170	−.3075	9.40 + 16	50.3277	38.5294	7.3689
−2.0	1.44720	1.45828	8.1817	9.6289	36.6457	−.1293	1.23 + 17	47.8011	36.6121	7.1850
−1.8	1.37798	1.38888	7.1711	8.5491	34.9849	.0494	1.66 + 17	43.7771	33.6659	7.0135
−1.6	1.31585	1.32641	6.2619	7.5777	33.4557	.2293	2.20 + 17	38.9803	30.1724	6.8589
−1.4	1.26150	1.27157	5.4643	6.7258	32.0650	.4110	2.91 + 17	33.9995	26.5204	6.7237
−1.2	1.21488	1.22440	4.7786	5.9935	30.8094	.5947	3.80 + 17	29.2345	22.9808	6.6091
−1.0	1.17554	1.18445	4.1982	5.3737	29.6790	.7804	4.96 + 17	24.9120	19.7177	6.5149
−.8	1.14275	1.15102	3.7130	4.8557	28.6602	.9681	6.43 + 17	21.1295	16.8129	6.4402
−.6	1.11571	1.12332	3.3111	4.4268	27.7385	1.1577	8.32 + 17	17.9011	14.2916	6.3837
−.4	1.09361	1.10053	2.9808	4.0744	26.8997	1.3490	1.08 + 18	15.1935	12.1435	6.3440
−.2	1.07571	1.08190	2.7109	3.7866	26.1305	1.5418	1.39 + 18	12.9506	10.3389	6.3194
−.0	1.06135	1.06673	2.4914	3.5528	25.4190	1.7360	1.79 + 18	11.1092	8.8390	6.3086
.2	1.05003	1.05443	2.3136	3.3636	24.7552	1.9313	2.32 + 18	9.6072	7.6027	6.3103
.4	1.04134	1.04449	2.1700	3.2114	24.1302	2.1277	3.00 + 18	8.3881	6.5908	6.3233
.6	1.03505	1.03649	2.0546	3.0896	23.5367	2.3251	3.90 + 18	7.4026	5.7674	6.3470
.8	1.03113	1.03007	1.9621	2.9933	22.9686	2.5234	5.10 + 18	6.6092	5.1015	6.3812
1.0	1.02977	1.02495	1.8886	2.9184	22.4207	2.7229	6.70 + 18	5.9735	4.5671	6.4267
1.2	1.03154	1.02091	1.8309	2.8625	21.8884	2.9236	8.90 + 18	5.4680	4.1429	6.4858
1.4	1.03749	1.01778	1.7867	2.8242	21.3680	3.1261	1.20 + 19	5.0716	3.8131	6.5631
1.6	1.04942	1.01546	1.7545	2.8039	20.8556	3.3311	1.65 + 19	4.7699	3.5671	6.6667
1.8	1.07031	1.01392	1.7341	2.8044	20.3475	3.5396	2.36 + 19	4.5569	3.4024	6.8091
2.0	1.10481	1.01328	1.7270	2.8318	19.8402	3.7534	3.57 + 19	4.4420	3.3310	7.0070

† $N(E)$ is the number of electrons per cm³, expressed in the form $a + b$, meaning $a \times 10^b$.

TABLE 4h-30. THERMODYNAMIC PROPERTIES OF ARGON (*Continued*)

T = 16,000 K

log ρ/ρ_0	Z	Moles	E/RT	H/RT	S/R	$\log_{10} P$	$N(E)$†	C_p/R	C_v/R	a/a_0
−5.0	2.10148	2.10270	16.7027	18.8042	60.0565	−2.9094	2.96 + 14	51.3510	43.5643	9.1366
−4.8	2.06759	2.06901	15.9723	18.0399	58.3666	−2.7164	4.55 + 14	37.9023	32.1211	9.1189
−4.6	2.04341	2.04508	15.4552	17.4986	56.9032	−2.5215	7.05 + 14	27.6685	23.2395	9.1467
−4.4	2.02639	2.02840	15.0976	17.1240	55.6088	−2.3252	1.10 + 15	20.3499	16.7869	9.2207
−4.2	2.01427	2.01672	14.8515	16.8657	54.4324	−2.1278	1.72 + 15	15.3947	12.3643	9.3367
−4.0	2.00512	2.00814	14.6776	16.6827	53.3331	−1.9297	2.71 + 15	12.2402	9.5191	9.4783
−3.8	1.99738	2.00111	14.5452	16.5426	52.2791	−1.7314	4.26 + 15	10.4361	7.8696	9.6114
−3.6	1.98967	1.99430	14.4293	16.4190	51.2452	−1.5331	6.71 + 15	9.6847	7.1537	9.6899
−3.4	1.98066	1.98640	14.3081	16.2888	50.2097	−1.3351	1.06 + 16	9.8374	7.2328	9.6807
−3.2	1.96888	1.97598	14.1598	16.1287	49.1519	−1.1377	1.65 + 16	10.8736	8.0740	9.5860
−3.0	1.95262	1.96136	13.9605	15.9132	48.0495	−.9413	2.58 + 16	12.8644	9.7163	9.4368
−2.8	1.92992	1.94056	13.6837	15.6136	46.8783	−.7463	4.01 + 16	15.9107	12.2114	9.2662
−2.6	1.89871	1.91148	13.3014	15.2001	45.6141	−.5534	6.15 + 16	20.0386	15.5326	9.0918
−2.4	1.85723	1.87227	12.7897	14.6469	44.2372	−.3630	9.33 + 16	25.0616	19.4696	8.9163
−2.2	1.80466	1.82198	12.1363	13.9410	42.7403	−.1755	1.39 + 17	30.4743	23.5737	8.7355
−2.0	1.74169	1.76111	11.3481	13.0898	41.1352	.0091	2.05 + 17	35.4890	27.2312	8.5452
−1.8	1.67963	1.69181	10.4531	12.1237	39.4543	.1910	2.95 + 17	39.2635	19.8632	8.3450
−1.6	1.59500	1.61749	9.4951	11.0901	37.7443	.3709	4.17 + 17	41.2109	31.1302	8.1386
−1.4	1.51870	1.54198	8.5235	10.0422	36.0559	.5496	5.80 + 17	41.1887	31.0114	7.9324
−1.2	1.44519	1.46877	7.5829	9.0281	34.4330	.7280	7.95 + 17	39.4686	29.7379	7.7334
−1.0	1.37705	1.40048	6.7069	8.0839	32.9074	.9071	1.08 + 18	36.5555	27.6609	7.5479
−.8	1.31584	1.33877	5.9162	7.2320	31.4971	1.0873	1.44 + 18	32.9919	25.1353	7.3801
−.6	1.26221	1.28437	5.2202	6.4824	30.2078	1.2692	1.92 + 18	29.2300	22.4534	7.2326
−.4	1.21617	1.23737	4.6196	5.8358	29.0369	1.4531	2.54 + 18	25.5861	19.8239	7.1065
−.2	1.17729	1.19738	4.1094	5.2866	27.9758	1.6390	3.35 + 18	22.2469	17.3771	7.0014
−.0	1.14493	1.16380	3.6812	4.8261	27.0132	1.8269	4.40 + 18	19.2997	15.1819	6.9164
.2	1.11837	1.13586	3.3255	4.4439	26.1366	2.0167	5.79 + 18	16.7662	13.2644	6.8500
.4	1.09691	1.11283	3.0325	4.1294	25.3337	2.2083	7.62 + 18	14.6294	11.6230	6.8008
.6	1.07999	1.09398	2.7929	3.8729	24.5930	2.4015	1.01 + 19	12.8530	10.2408	6.7676
.8	1.06722	1.07869	2.5985	3.6657	23.9044	2.5964	1.33 + 19	11.3939	9.0938	6.7497
1.0	1.05847	1.06642	2.4422	3.5007	23.2588	2.7928	1.78 + 19	10.2097	7.1565	6.7474
1.2	1.05395	1.05674	2.3186	3.3725	22.6490	2.9909	2.42 + 19	9.2635	7.4061	6.7626
1.4	1.05437	1.04935	2.2234	3.2778	22.0685	3.1911	3.33 + 19	8.5272	6.8261	6.7991
1.6	1.06109	1.04415	2.1545	3.2156	21.5128	3.3939	4.72 + 19	7.9869	6.4112	6.8633
1.8	1.07634	1.04136	2.1128	3.1892	20.9794	3.6001	7.01 + 19	7.6550	6.1803	6.9645
2.0	1.10300	1.04211	2.1070	3.2100	20.4722	3.8107	1.13 + 20	7.6148	6.2191	7.1095

† $N(E)$ is the number of electrons per cm³. expressed in the form $a + b$, meaning $a \times 10^b$.

TABLE 4h-31. THERMODYNAMIC PROPERTIES OF HIGHLY IONIZED NITROGEN, OXYGEN, AND AIR†

$\log T = 4.4 \quad T = 25{,}119$ K

$\log C_e$	Nitrogen				Oxygen				Air			
	ρ/ρ_0	Z^*	E^*/RT	P, atm	ρ/ρ_0	Z^*	E^*/RT	P, atm	ρ/ρ_0	Z^*	E^*/RT	P, atm
-5	.....	.....	.....	.....	.....	.....	.....	.....	2.54(−6)	5.934	50.209	1.386(−3)
-4	5.493(−5)	2.8203	22.4494	1.4247(−2)	6.658(−5)	2.5019	18.6759	1.5319(−2)	2.852(−5)	5.506	43.302	1.445(−2)
-3	7.621(−4)	2.3121	14.7619	1.5204(−1)	9.163(−4)	2.0914	11.5787	1.7622(−1)	3.951(−4)	4.531	28.177	1.646(−1)
-2	9.667(−3)	2.0344	10.6227	1.8086	9.914(−3)	2.0087	10.1592	1.8313	4.859(−3)	4.058	21.049	1.813
-1	0.1087	1.9197	9.4752	1.91894(1)	0.1012	1.9877	9.9100	1.8499(1)	5.353(−2)	3.868	19.134	1.9042(1)
0	1.9252	1.5194	6.4984	2.6899(2)	1.1668	1.8570	8.9608	1.9925(2)	0.8462	3.182	14.038	2.50(2)

$\log T = 4.6 \quad T = 39{,}811$ K

$\log C_e$	Nitrogen				Oxygen				Air			
	ρ/ρ_0	Z^*	E^*/RT	P, atm	ρ/ρ_0	Z^*	E^*/RT	P, atm	ρ/ρ_0	Z^*	E^*/RT	P, atm
-5	.....	.....	.....	.....	.....	.....	.....	.....	1.68(−6)	7.957	68.511	1.949(−3)
-4	3.608(−5)	3.7715	30.2600	1.9833(−2)	4.047(−5)	3.4711	27.3687	2.0475(−2)	1.846(−5)	7.416	59.297	1.995(−2)
-3	4.460(−4)	3.2434	21.8559	2.1084(−1)	4.837(−4)	3.0672	20.3529	2.1623(−1)	2.266(−4)	6.412	43.076	2.118(−1)
-2	5.087(−3)	2.9657	17.7766	2.1989	5.327(−3)	2.8772	17.7458	2.2339	2.568(−3)	5.894	35.540	2.207
-1	6.320(−2)	2.5823	13.7697	2.3786(1)	7.186(−2)	2.3917	12.2432	2.5050(1)	3.242(−2)	5.084	26.894	2.402(1)
0	0.9484	2.0544	8.8117	2.8398(2)	1.0159	1.9843	8.1263	2.9381(2)	0.4810	4.079	17.333	2.860(2)
$+1$	1.71167(1)	1.5842	6.3699	3.95214(3)	1.77836(1)	1.5623	5.9211	4.04935(3)	8.6281	3.159	12.550	3.973(3)

$\log T = 4.8 \quad T = 63{,}096$ K

$\log C_e$	Nitrogen				Oxygen				Air			
	ρ/ρ_0	Z^*	E^*/RT	P, atm	ρ/ρ_0	Z^*	E^*/RT	P, atm	ρ/ρ_0	Z^*	E^*/RT	P, atm
-5	.....	.....	.....	.....	.....	.....	.....	.....	1.11(−6)	11.006	98.885	2.822(−3)
-4	2.456(−5)	5.0711	40.7265	2.8770(−2)	2.589(−5)	4.8626	39.9829	2.9080(−2)	1.242(−5)	10.054	81.138	2.884(−2)
-3	2.755(−4)	4.6294	33.6748	0.2946	3.000(−4)	4.3333	31.2774	0.3003	1.402(−4)	9.134	66.335	0.2958
-2	3.252(−3)	4.0752	25.5689	3.0614	3.423(−3)	3.9217	25.0767	3.1009	1.643(−3)	8.085	50.930	3.070
-1	3.879(−2)	3.5778	19.8189	3.2057(1)	4.184(−2)	3.3899	18.7528	3.2762(1)	1.970(−2)	7.076	39.187	3.221(1)
0	0.5297	2.8880	13.4425	3.5337(2)	0.5383	2.8576	13.4707	3.5531(2)	0.26574	5.763	26.897	3.538(2)
$+1$	8.6176	2.1604	9.0014	4.30041(3)	8.8718	2.1272	8.8848	4.35934(3)	4.335	4.307	17.953	4.31288(3)

* Air is taken to be $0.78847 N_2 + 0.21153 O_2$.

† The symbols Z^* and E^* refer to the compressibility factor and energy, respectively, of the gas mixture in the ideal gas approximation, with dissociation and ionization effects included, but without intermolecular and ionic force corrections.

Table 4h-31. THERMODYNAMIC PROPERTIES OF HIGHLY IONIZED NITROGEN, OXYGEN, AND AIR (Continued)

$\log T = 5.0$ $T = 100{,}000$ K

$\log C_e$	Nitrogen				Oxygen				Air			
	ρ/ρ_0	Z^*	E^*/RT	P, atm	ρ/ρ_0	Z^*	E^*/RT	P, atm	ρ/ρ_0	Z^*	E^*/RT	P, atm
-5	......	......	......	......	......	......	......	......	9.62(-7)	12.400	88.347	4.367(-3)
-4	2.002(-5)	5.9941	39.9015	4.3932(-2)	1.775(-5)	6.6336	54.5680	4.3108(-2)	9.75(-6)	12.259	86.008	4.376(-2)
-3	2.023(-4)	5.9442	39.2902	0.4402	1.985(-4)	6.0370	44.7463	0.4387	1.007(-4)	11.928	80.888	0.4398
-2	2.166(-3)	5.6159	35.2938	4.4532	2.264(-3)	5.4160	35.7412	4.4891	1.093(-3)	11.147	70.777	4.461
-1	2.530(-2)	4.9529	27.6217	4.5876(1)	2.653(-2)	4.7695	27.5515	4.6323(1)	1.2774(-2)	9.828	55.214	4.596(1)
0	0.31486	4.1760	20.2730	4.8137(2)	0.3312	4.0197	19.9701	4.8739(2)	0.1591	8.286	40.418	4.826(2)
+1	4.5067	3.2189	13.8801	5.3108(3)	4.7320	3.1133	13.6880	5.3934(3)	2.2736	6.393	27.679	5.328(3)

$\log T = 5.2$ $T = 158{,}490$ K

$\log C_e$	Nitrogen				Oxygen				Air			
	ρ/ρ_0	Z^*	E^*/RT	P, atm	ρ/ρ_0	Z^*	E^*/RT	P, atm	ρ/ρ_0	Z^*	E^*/RT	P, atm
-5	......	......	......	......	......	......	......	......	......	......	......	......
-4	2.000(-5)	5.9999	28.5422	6.9627(-2)	1.667(-5)	6.9990	42.1981	6.769(-2)	9.595(-6)	12.422	62.862	6.92(-2)
-3	2.000(-4)	5.9994	28.5379	0.6962	1.669(-4)	6.9902	42.1008	0.6770	9.599(-5)	12.418	62.814	0.692
-2	2.003(-3)	5.9937	28.4949	6.9641	1.692(-3)	6.9085	41.2041	6.7823	9.6395(-4)	12.374	62.367	6.92
-1	2.025(-2)	5.9392	28.0852	6.9784(1)	1.835(-2)	6.4490	36.2326	6.8664(1)	9.9069(-3)	12.094	59.617	6.95(1)
0	0.2194	5.5589	25.2956	7.0764(2)	0.2175	5.5974	27.9500	7.0637(2)	0.1095	11.134	51.714	7.074(2)
+1	2.8280	4.5360	18.8699	7.4431(3)	2.8802	4.4719	19.7684	7.4734(3)	1.4194	9.045	38.120	7.449(3)

$\log T = 5.4$ $T = 251{,}190$ K

$\log C_e$	Nitrogen				Oxygen				Air			
	ρ/ρ_0	Z^*	E^*/RT	P, atm	ρ/ρ_0	Z^*	E^*/RT	P, atm	ρ/ρ_0	Z^*	E^*/RT	P, atm
-5	......	......	......	......	......	......	......	......	9.5(-7)	12.481	48.109	1.090(-2)
-4	1.998(-5)	6.0038	21.4332	1.1031(-1)	1.667(-5)	7.0000	30.5070	0.1073	9.59(-6)	12.429	46.705	0.110
-3	2.000(-4)	6.0003	21.3407	1.1036	1.667(-4)	6.9998	30.5059	1.0731	9.594(-5)	12.423	46.559	1.097
-2	2.000(-3)	5.9995	21.3298	1.1034(1)	1.667(-3)	6.9981	30.4954	1.0728(1)	9.596(-4)	12.421	46.537	1.097(1)
-1	2.002(-2)	5.9945	21.3127	1.1035(2)	1.672(-2)	6.9814	30.3921	1.0734(2)	9.609(-3)	12.407	46.467	1.097(2)
0	0.2022	5.9465	21.1549	1.1057(3)	0.1715	6.8305	29.4678	1.0772(3)	9.740(-2)	12.267	45.827	1.099(3)
+1	2.1844	5.5780	19.9145	1.12054(4)	1.9859	6.0354	24.9055	1.10221(4)	1.070	11.350	41.941	1.1170(4)

TABLE 4h-31. THERMODYNAMIC PROPERTIES OF HIGHLY IONIZED NITROGEN, OXYGEN, AND AIR (Continued)

$\log T = 5.6$, $T = 398{,}110$ K

$\log C_e$	Nitrogen				Oxygen				Air			
	ρ/ρ_0	Z^*	E^*/RT	P, atm	ρ/ρ_0	Z^*	E^*/RT	P, atm	ρ/ρ_0	Z^*	E^*/RT	P, atm
−5	......				......				......			
−4	1.553(−5)	7.4409	43.5716	0.1684	1.591(−5)	7.2854	29.2764	1.689(−1)	7.80(−6)	14.816	81.096	0.1685
−3	1.673(−4)	6.9782	34.2161	1.7016	1.656(−4)	7.0382	23.9473	1.6987	8.346(−5)	13.982	64.088	1.701
−2	1.822(−3)	6.4876	25.3735	1.7227(1)	1.666(−3)	7.0037	23.2082	1.7006(1)	8.934(−4)	13.194	49.831	1.718(1)
−1	1.967(−2)	6.0831	18.2776	1.7437(2)	1.667(−2)	6.9973	23.1256	1.7001(2)	9.476(−3)	12.553	38.606	1.734(2)
0	0.20046	5.9883	16.9174	1.7496(3)	0.1675	6.9700	23.0593	1.70157(3)	9.623(−2)	12.392	36.433	1.739(3)
+1	2.0749	5.8194	16.5657	1.7599(4)	1.7409	6.7442	22.5449	1.7112(4)	0.9970	12.300	35.661	1.7500(4)

$\log T = 5.8$, $T = 630{,}960$ K

$\log C_e$	Nitrogen				Oxygen				Air			
	ρ/ρ_0	Z^*	E^*/RT	P, atm	ρ/ρ_0	Z^*	E^*/RT	P, atm	ρ/ρ_0	Z^*	E^*/RT	P, atm
−5	......				......				6.9(−7)	16.422	83.609	2.62(−2)
−4	1.429(−5)	7.9995	39.3190	0.2640	1.253(−5)	8.9802	50.7353	0.2599	6.94(−6)	16.414	83.468	0.2631
−3	1.430(−4)	7.9953	39.2610	2.6410	1.277(−4)	8.8308	48.1188	2.6049	6.972(−5)	16.344	82.269	2.633
−2	1.438(−3)	7.9547	38.7050	2.6423(1)	1.371(−3)	8.2929	38.7624	2.6264	7.116(−4)	16.052	77.434	2.639(1)
−1	1.499(−2)	7.6706	34.8243	2.6560(2)	1.485(−2)	7.7337	29.6449	2.6530(2)	7.481(−3)	15.368	67.457	2.655(2)
0	0.1657	7.0362	26.4910	2.6932(3)	0.1621	7.1689	21.2538	2.6844(3)	0.08245	14.128	50.766	2.691(3)
+1	1.8873	6.2987	18.8607	2.7459(4)	1.7010	6.8787	18.6828	2.7028(4)	0.9225	12.843	37.646	2.7367(4)

$\log T = 6.0$, $T = 1{,}000{,}000$ K

$\log C_e$	Nitrogen				Oxygen				Air			
	ρ/ρ_0	Z^*	E^*/RT	P, atm	ρ/ρ_0	Z^*	E^*/RT	P, atm	ρ/ρ_0	Z^*	E^*/RT	P, atm
−5	......				......				6.93(−7)	16.423	61.855	4.17(−2)
−4	1.429(−5)	8.0000	29.2413	0.4184	1.250(−5)	9.0000	37.2123	0.41186	6.933(−6)	16.423	61.855	0.417
−3	1.429(−4)	8.0000	29.2410	4.184	1.250(−4)	8.9997	37.2095	4.1186	6.933(−5)	16.423	61.853	4.17
−2	1.429(−3)	7.9996	29.2388	4.1839(1)	1.250(−3)	8.9971	37.1809	4.1186(1)	6.934(−4)	16.421	61.838	4.17(1)
−1	1.4286(−2)	7.9961	29.2170	4.1820(2)	1.254(−2)	8.9719	36.9010	4.1186(2)	6.942(−3)	16.405	61.685	4.17(2)
0	0.1436	7.9618	29.0039	4.1856(3)	0.1288	8.7657	34.6534	4.133(3)	0.07011	16.264	60.398	4.175(3)
+1	1.498	7.6748	27.2651	4.2090(4)	1.4231	8.0268	27.3929	4.1819(4)	0.74082	15.498	54.584	4.2033(4)

TABLE 4h-31. THERMODYNAMIC PROPERTIES OF HIGHLY IONIZED NITROGEN, OXYGEN, AND AIR (*Continued*)

$\log T = 6.2 \quad T = 1{,}584{,}900$ K

log C_e	Nitrogen				Oxygen				Air			
	ρ/ρ_0	Z^*	E^*/RT	P, atm	ρ/ρ_0	Z^*	E^*/RT	P, atm	ρ/ρ_0	Z^*	E^*/RT	P, atm
−5	1.429(−5)	8.0000	22.8785	0.6631	1.250(−5)	9.0000	28.4617	0.65276	6.993(−6)	16.423	48.119	0.661
−4	1.429(−4)	8.0000	22.8785	6.631	1.250(−4)	9.0000	28.4616	6.5276	6.933(−5)	16.423	48.119	6.61
−3	1.429(−3)	7.9999	22.8784	6.631(1)	1.250(−3)	8.9999	28.4614	6.5276(1)	6.934(−4)	16.423	48.119	6.61(1)
−2	1.429(−2)	7.9988	22.8777	6.630(2)	1.250(−2)	8.9985	28.4587	6.5265(2)	6.935(−3)	16.420	48.116	6.61(2)
−1	0.1431	7.9881	22.8705	6.6326(3)	0.1252	8.9851	28.4318	6.527(3)	6.945(−2)	16.398	48.094	6.609(3)
0	1.4510	7.8916	22.8025	6.6440(4)	1.2717	8.8631	28.1745	6.5402(4)	0.7045	16.194	47.878	6.6200(4)
+1												

$\log T = 6.4 \quad T = 2{,}511{,}900$ K

log C_e	Nitrogen				Oxygen				Air			
	ρ/ρ_0	Z^*	E^*/RT	P, atm	ρ/ρ_0	Z^*	E^*/RT	P, atm	ρ/ρ_0	Z^*	E^*/RT	P, atm
−5	1.429(−5)	8.0000	18.8639	1.051	1.250(−5)	9.0000	22.9401	1.035				
−4	1.429(−4)	8.0000	18.8639	1.051(1)	1.250(−4)	9.0000	22.9402	1.0346(1)	6.933(−5)	16.423	39.452	1.047(1)
−3	1.429(−3)	7.9999	18.8638	1.051(2)	1.250(−3)	8.9999	22.9401	1.0346(2)	6.933(−4)	16.423	39.452	1.047(2)
−2	1.429(−2)	7.9990	18.8637	1.051(3)	1.250(−2)	8.9989	22.9399	1.0345(3)	6.934(−3)	16.421	39.452	1.047(3)
−1	0.14306	7.9899	18.8618	1.0511(4)	0.12516	8.9896	22.9372	1.0338(4)	6.943(−2)	16.403	39.448	1.0474(4)
0	1.4477	7.9074	18.8450	1.0527(5)	1.2651	8.9043	22.9124	1.0359(5)	0.7024	16.237	39.411	1.0490(5)
+1												

at uniform logarithmic intervals to 2.5 million kelvins. The tables taken from Hilsenrath, Green, and Beckett [4] represent the properties of atoms in equilibrium with their ions. The formulation in terms of electron concentration permits a solution of the equations for equilibrium properties in closed form and allows the computation of properties of a mixture directly from the equilibrium properties of the constituent gases. In these tables the asterisk refers to properties of the gas mixture in the ideal-gas approximation (with dissociation and ionization effects included but without intermolecular and ionic force corrections).

References

1. Hilsenrath, J., et al.: Tables of Thermal Properties of Gases, *NBS Circ.* 564, 1955. Reprinted in 1960 by Pergamon Press under the title "Tables of Thermodynamic and Transport Properties of Air, Argon, Carbon Dioxide, Carbon Monoxide, Hydrogen, Nitrogen, Oxygen, and Steam."
2. Hilsenrath, J. and M. Klein: "Tables of Thermodynamic Properties of Air in Chemical Equilibrium including Second Virial Corrections from 1,500°K to 15,000°K," *Arnold Eng. Develop. Center Rept*, AEDC-TR-65-58, March, 1965. Available under the designation AD 612301 from the Clearinghouse for Federal Scientific and Technical Information, U.S. Department of Commerce, Springfield, Virginia 22151 (price $3.00).
3. Hilsenrath, J., C. G. Messina, M. Klein, and R. C. Thompson: Thermodynamic and Shock Wave Properties of Argon and Nitrogen in Chemical Equilibrium with Virial and Ionic Corrections, vols. I, II, and III. *Air Force Weapons Lab. Rept.* AFWL-TR-68-60, Kirtland Air Force Base, New Mexico, May, 1969.
4. Hilsenrath, J., M. S. Green, and C. W. Beckett: Internal Energy of Highly Ionized Gases, *Proc. IXth Intern. Astronaut. Congr.* (Amsterdam), 1958, pp. 120–136, Springer-Verlag OHG Vienna, 1959.

4i. Pressure-Volume-Temperature Relationships of Gases; Virial Coefficients[1]

J. M. H. LEVELT SENGERS, MAX KLEIN, AND JOHN S. GALLAGHER

The National Bureau of Standards

4i-1. Definition. Virial coefficients are the coefficients in the expansion of the compressibility factor PV of a gas in powers of the density $1/V$,

$$PV = RT \left(1 + \frac{B_V}{V} + \frac{C_V}{V^2} + \cdots \right) \qquad (4i\text{-}1)$$

or in powers of the pressure P,

$$PV = RT(1 + B_P P + C_P P^2 + \cdots) \qquad (4i\text{-}2)$$

The density expansion is the more fundamental of the two. It can be proved that such an expansion exists for gases at moderate densities, and its consecutive coefficients can be related to interactions between pairs, triplets, etc., of molecules [1]. The pressure expansion is often more practical, the pressure being more readily mea-

[1] Supported in part by the Air Force Systems Command, Arnold Engineering Development Center, Tullahoma Tenn., on Delivery Order no. (40-600) 66-938.

sured than the volume, but it usually converges more slowly, and its coefficients are not as simply related to molecular interaction. In what follows, the emphasis will be on the expansion (4i-1).

4i-2. Units. The units of the virials depend on the units of volume (4i-1) or pressure (4i-2) chosen. We will express the volume in cm^3/mol and give the virials in the corresponding units. However, a practical unit of volume frequently used is the amagat unit; the volume in amagat units is the ratio of the actual volume of a gas over the normal volume, i.e., that which it would occupy at 0°C and 1 atm (1.013250 bars). The normal volume for a mole of a real gas differs slightly from the normal volume $V_0 = 22,413.6$ cm^3/mol of a perfect gas, owing to deviations from ideality at 0°C and 1 atm. The virial expansion used in conjunction with amagat units of volume is

$$PV_A = A_A + \frac{B_A}{V_A} + \frac{C_A}{V_A^2} + \cdots \qquad (4i-3)$$

In Table 4i-1, the virials B_P, C_P; A_A, B_A, C_A are expressed in terms of B_V, C_V.

TABLE 4i-1. RELATIONS BETWEEN VOLUME AND PRESSURE VIRIAL COEFFICIENTS

Gas constant	Ideal-gas normal volume per mole
$R = 8.3143$ J K^{-1} mol^{-1}	$V_0 = 22,413.6^{\bullet} cm^3$ mol^{-1}
($= 82.056$ cm^3 at K^{-1} mol^{-1})	

(Both on unified scale)

Pressure virials (4i-2)	Amagat virials (4i-3)
	$V_n = V_0/A_0$
	$A_0 = 1 - B_A$ (0°C) $- C_A$ (0°C) $\cdots$
	$A_A = A_0 T/273.15$
$B_P = B_V/RT$	$B_A = B_V A_A/V_n$
$C_P = (C_V - B_V^2)/(RT)^2$	$C_A = C_V A_A/V_n^2$

4i-3. Theoretical Interest. Of great interest is the fundamental relationship of B_V, C_V, . . . to the molecular interaction. If the molecular field is represented by a function $\phi(\mathbf{r})$ where $\mathbf{r}$ specifies the relative coordinates of two molecules, then

$$B_V(T) = \frac{N}{2} \int_0^\infty (1 - e^{-\phi(\mathbf{r})/kT})\, d\mathbf{r} \qquad (4i-4)$$

The virial $B_V(T)$ is uniquely determined through Eq. (4i-4) if the molecular interaction $\phi(\mathbf{r})$ is known but the reverse is not true. Higher virials can be likewise related to interactions between triplets, etc., of interacting molecules. These expressions for the higher virials are less useful in practice, not only because the higher virials are poorly known experimentally, but also because the influence of potential function nonadditivity [1] on these virials is poorly known theoretically. We have used the relationship (4i-4) between second virial and potential function for smoothing the experimental $B(T)$ values; for obtaining derivatives dB/dT, d^2B/dT^2; and where reasonable, for extrapolating the $B(T)$ tables beyond the temperature range where experimental data are available.

4i-4. Practical Importance. The virials B_V and C_V represent the initial deviations of the equation of state from ideality as a gas is compressed [Eqs. (4i-1 and 4i-2)]. Functions of these virials serve to estimate the initial density dependence of thermodynamic properties. Thus, the internal energy $U_i = U(V,T) - U(\infty,T)$ is given by

$$U_i = -RT\left(\frac{T}{V}\frac{dB_V}{dT} + \frac{T}{2V^2}\frac{dC_V}{dT} + \cdots\right)$$

Similar expressions are valid for other thermodynamic functions [2].

HEAT

TABLE 4i-2. THE SECOND VIRIAL COEFFICIENT OF HELIUM

T, K	B, cm³/mol	T, K	B, cm³/mol
9.00	−26.0	35.00	5.4
10.00	−21.7	40.00	6.6
11.00	−18.1	45.00	7.5
12.00	−15.2	50.00	8.2
13.00	−12.7	60.00	9.2
14.00	−10.5	80.00	10.6
15.00	−8.7	100.00	11.4
16.00	−7.1	120.00	11.8
17.00	−5.6	160.00	12.3
18.00	−4.3	200.00	12.3
19.00	−3.2	273.15	12.0
20.00	−2.2	373.15	11.3
22.00	−0.5	400.00	11.1
22.64	0.0	600.00	10.4
24.00	0.9	800.00	9.8
26.00	2.0	1000.00	9.3
28.00	3.0	1200.00	8.8
30.00	3.8	1400.00	8.4

TABLE 4i-3. THE SECOND VIRIAL COEFFICIENT OF NEON AND ITS TEMPERATURE DERIVATIVES

T, K	B, cm³/mol	$T\,dB/dT$, cm³/mol	$T^2\,d^2B/dT^2$, cm³/mol
80.00	−11.8	37	−87
90.00	−7.8	31	−73
100.00	−4.8	27	−63
110.00	−2.3	24	−55
120.00	−0.4	21	−49
122.11	0.0	21	−48
130.00	1.2	19	−44
140.00	2.6	18	−40
160.00	4.8	15	−34
200.00	7.6	11	−26
240.00	9.4	9	−20
273.15	10.4	7	−17
280.00	10.6	7	−17
320.00	11.5	6	−14
360.00	12.1	5	−12
373.15	12.3	5	−12
400.00	12.6	4	−11
500.00	13.3	3	−8
600.00	13.8	2	−6
700.00	14.0	1	−5
800.00	14.2	1	−4
900.00	14.3	1	−3
1000.00	14.3	0	−3

4i-5. Determination; Errors. Virial coefficients, in the majority of cases, are not directly measured but are obtained by analysis of PVT data of gases. The most common practice is a least-squares fit of the PV values along isotherms with either density or pressure as an independent variable. Using this procedure, the precision of the virials can then be obtained from linear least-squares estimates of their standard deviations. For a single experimental set in the very best cases, it may be better

TABLE 4i-4. THE SECOND VIRIAL COEFFICIENT OF ARGON
AND ITS TEMPERATURE DERIVATIVES

T, K	B, cm³/mol	$T\,dB/dT$, cm³/mol	$T^2\,d^2B/dT^2$, cm³/mol	T, K	B, cm³/mol	$T\,dB/dT$, cm³/mol	$T^2\,d^2B/dT^2$, cm³/mol
80.00	−288.0	577	−1,954	172.00	−66.9	135	−345
82.00	−274.2	544	−1,820	176.00	−63.8	130	−331
84.00	−261.4	514	−1,700	180.00	−60.9	126	−318
86.00	−249.7	488	−1,592	190.00	−54.4	116	−290
88.00	−238.7	463	−1,495	200.00	−48.7	107	−266
90.00	−228.6	441	−1,408	210.00	−43.7	100	−245
92.00	−219.1	420	−1,328	220.00	−39.2	93	−228
94.00	−210.3	401	−1,256	230.00	−35.2	88	−212
96.00	−202.0	384	−1,190	240.00	−31.5	83	−198
98.00	−194.3	367	−1,129	250.00	−28.2	78	−186
100.00	−187.0	352	−1,074	260.00	−25.3	74	−176
102.00	−180.2	338	−1,023	273.15	−21.7	69	−163
104.00	−173.8	325	−976	280.00	−20.1	67	−157
106.00	−167.7	313	−932	300.00	−15.7	61	−142
108.00	−161.9	302	−892	320.00	−11.9	56	−130
110.00	−156.5	292	−855	340.00	−8.7	51	−119
112.00	−151.3	282	−820	360.00	−5.8	48	−110
114.00	−146.4	272	−788	373.15	−4.2	46	−105
116.00	−141.7	264	−757	380.00	−3.4	44	−102
118.00	−137.3	255	−729	400.00	−1.1	42	−96
120.00	−133.1	247	−702	411.52	0.0	40	−92
124.00	−125.2	233	−654	450.00	3.4	36	−82
128.00	−118.0	220	−612	500.00	6.9	31	−71
132.00	−111.4	209	−573	550.00	9.7	28	−63
136.00	−105.4	198	−539	600.00	11.9	25	−57
140.00	−99.8	188	−509	700.00	15.4	20	−47
144.00	−94.6	180	−481	800.00	17.8	17	−40
148.00	−89.8	172	−456	900.00	19.7	14	−34
152.00	−85.3	164	−433	1000.00	21.1	12	−30
156.00	−81.1	157	−413	1100.00	22.2	11	−27
160.00	−77.2	151	−394	1300.00*	23.8	8	−21
164.00	−73.5	145	−376	1500.00	24.8	7	−18
168.00	−70.1	140	−360				

* Data below the dashed line in this and succeeding tables are extrapolations.

than 0.1 cm³/mol for B and 50 cm⁶/mol² for C. However, virial data from different experiments usually differ by much more than their combined precision because of the presence of systematic errors. The main sources of systematic errors are:

1. *Experimental:* (a) errors in the value of RT because of temperature errors or the use of scales other than the thermodynamic scale, (b) systematic errors in the volume because of calibration problems, and (c) difficulties with extrapolation to zero density, especially with data obtained by the Burnett method [3].

2. *Cutoff problems:* A finite polynomial has to be used rather than the theoretically correct infinite series [Eqs. (4i-1) and (4i-2)], but errors arise if the powers omitted would have contributed in the density range studied.

To minimize systematic errors, if there is evidence that any existed, we have refitted the experimental data when available. The data refitted are indicated by asterisks in the literature reference for the tables. Wherever feasible, we reduced temperatures

TABLE 4i-5. THE SECOND VIRIAL COEFFICIENT OF KRYPTON
AND ITS TEMPERATURE DERIVATIVES

T, K	B, cm³/mol	$T\,dB/dT$, cm³/mol	$T^2\,d^2B/dT^2$, cm³/mol	T, K	B, cm³/mol	$T\,dB/dT$, cm³/mol	$T^2\,d^2B/dT^2$, cm³/mol
106.00	−394.3	807	−2,813	215.00	−102.4	198	−521
108.00	−379.6	771	−2,659	220.00	−97.9	191	−499
110.00	−365.8	737	−2,518	230.00	−89.8	178	−459
112.00	−352.8	706	−2,389	240.00	−82.4	166	−425
114.00	−340.5	677	−2,270	250.00	−75.9	156	−396
116.00	−329.0	650	−2,160	260.00	−69.9	147	−369
118.00	−318.1	625	−2,059	270.00	−64.5	139	−346
120.00	−307.8	601	−1,965	273.15	−62.9	136	−340
122.00	−298.0	579	−1,878	280.00	−59.6	132	−326
124.00	−288.8	559	−1,797	290.00	−55.1	125	−308
126.00	−280.0	539	−1,722	300.00	−51.0	119	−291
128.00	−271.6	521	−1,651	310.00	−47.2	113	−276
130.00	−263.7	504	−1,586	320.00	−43.7	108	−263
132.00	−256.1	488	−1,524	340.00	−37.4	100	−239
134.00	−248.9	473	−1,467	360.00	−31.9	92	−219
136.00	−242.0	458	−1,413	373.15	−28.7	87	−207
138.00	−235.4	444	−1,362	380.00	−27.1	85	−202
140.00	−229.1	432	−1,314	400.00	−22.9	80	−187
144.00	−217.3	408	−1,226	420.00	−19.1	75	−175
148.00	−206.4	386	−1,148	440.00	−15.8	70	−163
152.00	−196.4	366	−1,078	460.00	−12.7	66	−153
156.00	−187.1	348	−1,015	500.00	−7.5	59	−137
160.00	−178.5	332	−959	550.00	−2.2	52	−120
164.00	−170.5	317	−907	575.00	0.0	49	−113
168.00	−163.1	303	−860	600.00	2.0	47	−107
172.00	−156.1	290	−817	650.00	5.6	42	−96
176.00	−149.5	279	−778	700.00	8.5	38	−88
180.00	−143.4	268	−742	800.00	13.2	32	−74
186.00	−134.9	253	−693	900.00	16.7	28	−64
192.00	−127.1	239	−650	1000.00	19.5	24	−56
198.00	−119.9	227	−611	1100.00	21.6	21	−50
205.00	−112.2	214	−570	1300.00	24.8	17	−40
210.00	−107.2	206	−545	1500.00	27.0	14	−34

to the thermodynamic scale, using the known relation between this scale and the IPTS [4]. If a laboratory maintained its own gas scale, this scale was used. In a few cases, notably He at low temperatures, where one of the purposes of the experiment was gas thermometry, we had to leave the intercept free. Regarding the cutoff criterion [5], we chose the maximum density range in which the $(k + 1)$th virial does not contribute beyond experimental error, and fitted this range with a polynomial of degree $k − 1$. Depending on the amount of low-density data available, we took $k = 3$ or 4. In order to do this, an estimate of the size of the $(k + 1)$th virial was necessary. In cases where it could not be obtained from the data, we used the theoretical value as calculated for the Lennard-Jones six-twelve potential. This procedure was justified since only order-of-magnitude estimates were needed.

After second and third virials had been obtained from each set of experimental data for a given substance, and after obviously wrong results had been eliminated, a smoothing or averaging procedure was established. Use was made of the fundamental relation (4i-4) between the second virial coefficient and the intermolecular potential.

4i-6. Potential Functions—Determination, Use. Equation (4i-4) applies to substances for which quantum effects are negligible. For such substances, Eq. (4i-4) is

TABLE 4i-6. THE SECOND VIRIAL COEFFICIENT OF XENON
AND ITS TEMPERATURE DERIVATIVES

T, K	B, cm^3/mol	$T\, dB/dT$, cm^3/mol	$T^2\, d^2B/dT^2$, cm^3/mol	T, K	B, cm^3/mol	$T\, dB/dT$, cm^3/mol	$T^2\, d^2B/dT^2$, cm^3/mol
220.00	−230.7	429	−1,225	380.00	−78.5	175	−433
225.00	−221.2	411	−1,166	390.00	−74.0	169	−415
230.00	−212.4	395	−1,111	400.00	−69.8	163	−398
235.00	−204.0	380	−1,061	420.00	−62.2	152	−368
240.00	−196.2	366	−1,015	440.00	−55.4	142	−342
245.00	−188.7	353	−972	460.00	−49.2	133	−319
250.00	−181.7	341	−933	480.00	−43.7	126	−299
255.00	−175.1	329	−896	500.00	−38.8	119	−282
260.00	−168.8	319	−861	525.00	−33.1	111	−262
265.00	−162.8	308	−829	550.00	−28.1	104	−245
270.00	−157.2	299	−799	575.00	−23.6	98	−230
273.15	−153.7	293	−781	600.00	−19.6	93	−216
280.00	−146.6	281	−745	650.00	−12.5	84	−193
290.00	−137.0	266	−697	700.00	−6.6	76	−175
300.00	−128.3	252	−654	768.03	0.0	67	−154
310.00	−120.2	239	−616	800.00	2.7	64	−146
320.00	−112.8	227	−582	900.00	9.6	55	−125
330.00	−106.0	217	−551	1000.00	15.0	48	−110
340.00	−99.7	207	−523	1100.00	19.3	42	−97
350.00	−93.8	198	−497	1200.00	22.8	38	−87
360.00	−88.4	190	−474				
370.00	−83.3	182	−452	1300.00	25.6	34	−79
373.15	−81.7	180	−446	1400.00	28.0	31	−72
				1500.00	30.1	28	−66

exact. Quantum effects become important for only the lightest gases [6], e.g., helium, hydrogen, etc. These latter are generally referred to as *quantum gases*. For such gases, Eq. (4i-4) can represent only a first approximation whose quality goes down with decreasing molecular weight. In either case, i.e., whether Eq. (4i-4) is exact or an approximation, the use of it requires a knowledge of the intermolecular potential function $\phi(r)$. In principle, such functions can be obtained by direct quantum-mechanical calculation. In practice, this procedure is not feasible even for the simplest system. This has required, in effect, the partial reversal of the process. Thus, instead of using Eq. (4i-4) with a known function $\phi(r)$ to predict $B(T)$, one uses (4i-4), in part, to produce information on $\phi(r)$ and, in part, to predict $B(T)$. This is done by assuming a form for $\phi(r)$ (often referred to as a *potential model*), based on whatever fundamental knowledge is available, inserting a number of parameters in this form [for example, ϵ and σ of (4i-5)], and varying the values of these parameters to obtain the best agreement between the $B(T)$ values calculated from Eq. (4i-4) and those determined from the analysis of PVT data described above. The predictive power of (4i-4) remains essentially intact, provided the number of experimental points used is far in excess of the number of parameters sought. Frequently used in this way to describe the intermolecular potential of simple nonpolar substances is the Lennard Jones twelve-six potential, a member of the more general class of spherically symmetric $m - 6$ potentials:

$$\phi(r) = \frac{m\epsilon}{m - 6} \left(\frac{m}{6}\right)^{6/(m-6)} \left[\left(\frac{\sigma}{r}\right)^m - \left(\frac{\sigma}{r}\right)^6 \right] \qquad (4i\text{-}5)$$

TABLE 4i-7. THE SECOND VIRIAL COEFFICIENT OF NITROGEN
AND ITS TEMPERATURE DERIVATIVES

T, K	B, cm³/mol	$T\, dB/dT$, cm³/mol	$T^2\, d^2B/dT^2$, cm³/mol	T, K	B, cm³/mol	$T\, dB/dT$, cm³/mol	$T^2\, d^2B/dT^2$, cm³/mol
100.00	−160.0	304	−874	210.00	−31.1	94	−224
102.00	−154.1	293	−837	220.00	−26.9	88	−209
104.00	−148.5	283	−802	230.00	−23.2	82	−195
106.00	−143.2	273	−769	240.00	−19.7	78	−183
108.00	−138.2	264	−739	260.00	−13.8	70	−163
110.00	−133.4	256	−711	273.15	−10.5	65	−152
112.00	−128.7	248	−684	280.00	−8.9	63	−147
116.00	−120.4	233	−637	300.00	−4.7	58	−134
120.00	−112.7	220	−594	320.00	−1.2	53	−122
124.00	−105.7	208	−557	327.22	0.0	52	−119
128.00	−99.3	197	−524	340.00	1.9	49	−113
132.00	−93.4	188	−494	360.00	4.6	46	−105
136.00	−87.9	179	−467	373.15	6.2	44	−100
140.00	−82.8	171	−442	380.00	7.0	43	−97
144.00	−78.1	163	−420	400.00	9.1	40	−91
148.00	−73.7	156	−400	450.00	13.5	34	−78
152.00	−69.7	150	−381	500.00	16.8	30	−68
156.00	−65.8	144	−364	550.00	19.5	26	−61
160.00	−62.3	139	−349	600.00	21.7	24	−54
166.00	−57.3	131	−327	700.00	25.0	19	−45
172.00	−52.7	125	−308	800.00	27.3	16	−38
178.00	−48.6	118	−291	900.00	29.1	14	−33
184.00	−44.7	113	−276	1000.00	30.4	12	−29
190.00	−41.2	108	−262	1200.00	32.3	9	−23
200.00	−35.9	100	−242	1400.00	33.5	7	−19

where ϵ and σ are parameters to be determined for each substance. This expression, with proper choice of m, adequately describes the second virial coefficient of simple nonpolar substances.

It should be noted that once a "best" set of parameters has been decided upon, a potential function exists which can serve as a representation for the "actual" potential function appropriate to the gas of interest. The use of such potential functions need not be restricted to Eq. (4i-4). They can also be employed in various statistical mechanical theories for calculating macroscopic thermodynamic quantities from molecular properties. In short, these potential functions have their own importance.

Various methods, of which the use of Eq. (4i-4) is only one example, by means of which potential parameters are determined from experimental data have recently been subjected to close scrutiny [7]. In that study it was determined that all reasonable three-parameter potential models should produce essentially the same set of second virial coefficients. Because of this it was reasonable to fix on one particular model, and we chose Eq. (4i-5) for that purpose. A second result of the study of methods for determining potential parameters was the discovery of a reduced temperature range, for each property, over which that property cannot be used to distinguish between

TABLE 4i-8. THE SECOND VIRIAL COEFFICIENT OF OXYGEN
AND ITS TEMPERATURE DERIVATIVES

T, K	B, cm³/mol	$T\,dB/dT$, cm³/mol	$T^2\,d^2B/dT^2$, cm³/mol	T, K	B, cm³/mol	$T\,dB/dT$, cm³/mol	$T^2\,d^2B/dT^2$, cm³/mol
100.00	−197.5	383	−1,201	210.00	−44.8	104	−259
102.00	−190.1	367	−1,141	220.00	−40.1	97	−240
104.00	−183.1	352	−1,087	230.00	−35.9	91	−223
106.00	−176.5	339	−1,036	240.00	−32.1	86	−208
108.00	−170.3	326	−989	250.00	−28.7	81	−195
110.00	−164.4	314	−946	260.00	−25.6	77	−184
112.00	−158.9	303	−906	273.15	−22.0	72	−171
114.00	−153.6	293	−868	280.00	−20.2	69	−164
116.00	−148.6	283	−833	300.00	−15.7	63	−148
120.00	−139.3	265	−770	320.00	−11.8	58	−135
124.00	−130.9	249	−715	340.00	−8.4	53	−124
128.00	−123.2	235	−667	360.00	−5.5	49	−114
132.00	−116.2	222	−624	373.15	−3.7	47	−109
136.00	−109.7	210	−585	380.00	−2.9	46	−106
140.00	−103.8	200	−551	400.00	−0.6	43	−99
144.00	−98.3	190	−520	405.88	0.0	42	−97
148.00	−93.2	182	−492	450.00	4.1	37	−85
154.00	−86.2	170	−454	500.00	7.7	32	−74
160.00	−79.9	159	−422	550.00	10.6	29	−65
166.00	−74.2	150	−393	600.00	12.9	25	−58
172.00	−69.1	142	−368	700.00	16.5	21	−48
178.00	−64.3	134	−346	800.00	19.1	18	−41
184.00	−60.0	127	−326	1000.00	22.4	13	−31
190.00	−56.0	121	−308	1200.00	24.5	10	−25
200.00	−50.0	112	−281	1400.00	25.9	8	−20

potential functions. For the second virial coefficient, this range is given approximately by $0.6 < T/T_{\text{Boyle}} < 3.0$. We have included a table (Table 4i-17) of experimental Boyle temperatures to facilitate the conversion of these numbers into experimental temperatures for the various gases studied. The second result mentioned states, in effect, that one should not use Eq. (4i-4) with a potential function determined by data entirely contained in the insensitive range to predict $B(T)$ outside that range; nor should one use the resulting potential function in other theories. On the other hand, potential functions determined with data entirely outside the insensitive range can be used in an extrapolation to predict $B(T)$ values within that range.

4i-7. Construction of the Tables. Using linear and nonlinear [8] least-squares techniques, calculated second virial coefficients based on the function (4i-5) were fitted to the experimental second virial coefficient data for eight substances. Each value of m was taken to define a separate potential, with ϵ and σ in (4i-5) the adjustable parameters for the fit. The value of m was varied until the standard deviation of the fit was a minimum. The "best" $m - 6$ potential was used to generate a table of B, $T\,dB/dT$ and $T^2\,d^2B/dT^2$ values at various temperatures. Furthermore, it was used for extrapolation beyond the range of experimental data. Such extrapolations are indicated in each case by a dashed line across the tables.

TABLE 4i-9. THE SECOND VIRIAL COEFFICIENT OF DRY CO_2-FREE AIR
AND ITS TEMPERATURE DERIVATIVES

T, K	B, cm³/mol	$T\,dB/dT$, cm³/mol	$T^2\,d^2B/dT^2$, cm³/mol	T, K	B, cm³/mol	$T\,dB/dT$, cm³/mol	$T^2\,d^2B/dT^2$, cm³/mol
100.00	-167.3	318	-935	210.00	-34.5	95	-230
102.00	-161.2	307	-893	220.00	-30.2	89	-214
104.00	-155.3	295	-854	230.00	-26.4	84	-200
106.00	-149.8	285	-818	240.00	-22.9	79	-187
108.00	-144.6	275	-785	250.00	-19.8	75	-176
110.00	-139.6	266	-754	260.00	-16.9	71	-166
112.00	-134.9	258	-725	273.15	-13.5	66	-155
114.00	-130.4	249	-698	280.00	-11.9	64	-150
116.00	-126.0	242	-673	300.00	-7.7	58	-136
118.00	-122.0	235	-649	320.00	-4.1	54	-124
120.00	-118.2	228	-627	340.00	-1.0	50	-114
124.00	-110.9	215	-586	346.81	0.0	48	-111
128.00	-104.3	204	-550	360.00	1.7	46	-106
132.00	-98.1	193	-517	373.15	3.4	44	-101
136.00	-92.5	184	-488	380.00	4.2	43	-98
140.00	-87.3	176	-462	400.00	6.3	40	-92
144.00	-82.5	168	-438	450.00	10.7	35	-79
148.00	-78.0	161	-416	500.00	14.1	30	-69
152.00	-73.8	154	-396	550.00	16.8	27	-61
156.00	-69.9	148	-378	600.00	19.0	24	-55
160.00	-66.2	142	-361	650.00	20.8	21	-50
166.00	-61.1	134	-339	700.00	22.3	19	-45
172.00	-56.5	127	-319	800.00	24.7	16	-38
178.00	-52.2	121	-301	900.00	26.4	14	-33
184.00	-48.3	115	-284	1000.00	27.8	12	-29
190.00	-44.7	110	-270	1200.00	29.7	9	-23
200.00	-39.3	102	-248	1400.00	30.9	7	-19

Equation (4i-4) and the procedures described above were used for the quantum gases He, H_2, and D_2 as well. In these cases, however, the methods were used only to facilitate smoothing and interpolation of virial data. For H_2O, D_2O, and CO_2 a potential of the form (4i-5) was found to be inadequate. These substances were therefore treated as were the quantum gases; that is, the methods outlined were used only for smoothing and interpolation. The tables prepared for these six substances consist only of smoothed experimental $B(T)$ values, with no extrapolations attempted. Tables of $T\,dB/dT$ and $T^2\,d^2B/dT^2$ are not given, nor are potential parameters used in the smoothing process reported since they are without clear meaning. Since the $B(T)$ tables for these six substances are so closely tied to the experimental values, minor departures from smoothness in the tables may be detected.

Third virials, in all cases, were obtained by graphical interpolation of the (refitted) experimental values for C. They are summarized in Table 4i-16.

Table 4i-17 contains values for the Boyle temperature and the inversion temperature. In those cases where the form (4i-5) for the intermolecular potential applies, the potential parameters and the value of m are summarized in Table 4i-18.

We note that the optimum value of m is much closer to 18 than to the popular value of 12.

4i-8. Accuracy of the Tables. From a computational point of view, in all cases, the temperature spacing is sufficiently fine to allow for an interpolation to be made

TABLE 4i-10. THE SECOND VIRIAL COEFFICIENT OF HYDROGEN

T, K	B, cm³/mol	T, K	B, cm³/mol
24.00	−112.8	74.00	−12.9
25.00	−106.2	78.00	−10.9
26.00	−100.3	82.00	−8.9
27.00	−94.8	86.00	−7.2
28.00	−89.6	90.00	−5.7
29.00	−85.0	100.00	−2.5
30.00	−80.7	110.00	−0.0
31.00	−76.7	110.04	0.0
32.00	−73.0	120.00	2.0
33.00	−69.5	130.00	3.7
34.00	−66.2	140.00	5.1
35.00	−63.2	150.00	6.4
36.00	−60.2	160.00	7.6
38.00	−55.0	170.00	8.6
40.00	−50.3	180.00	9.5
42.00	−46.2	190.00	10.2
44.00	−42.5	200.00	10.8
46.00	−39.2	250.00	13.0
48.00	−36.2	273.15	13.7
50.00	−33.4	300.00	14.4
54.00	−28.6	350.00	15.3
58.00	−24.5	373.15	15.6
62.00	−21.0	400.00	15.9
66.00	−17.9	420.00	16.1
70.00	−15.2		

using a quadratic formula without the introduction of errors. Furthermore, linear interpolation can be used without introducing an error of more than 0.3 cm³/mol in $B(T)$ owing to the neglect of quadratic terms. It should be noted that where B, $T\,dB/dT$, and $T^2\,d^2B/dT^2$ are available, a Taylor series expansion can be used for interpolation.

It is much harder to assess the absolute accuracy of the tables in any general way. Where data from many sources are available for one substance, as is the case for most of the noble gases and for nitrogen, one usually finds discrepancies up to 1.5 cm³/mol in B and up to 30 percent in C between data from different laboratories. Discrepancies in B may become much larger at temperatures below critical. The main source

TABLE 4i-11. THE SECOND VIRIAL COEFFICIENT OF DEUTERIUM

T, K	B, cm^3/mol	T, K	B, cm^3/mol
84.00	− 10.4	200.00	10.2
88.00	− 8.7	220.00	11.3
92.00	− 7.0	240.00	12.2
96.00	− 5.6	260.00	12.8
100.00	− 4.2	273.15	13.1
110.00	− 1.3	280.00	13.2
115.00	0.0	300.00	13.5
120.00	1.0	320.00	14.0
130.00	3.0	340.00	14.4
140.00	4.6	360.00	14.7
150.00	6.0	373.15	14.9
160.00	7.1	380.00	15.0
170.00	8.1	400.00	15.2
180.00	8.9	420.00	15.5
190.00	9.5		

TABLE 4i-12. THE SECOND VIRIAL COEFFICIENT OF WATER VAPOR (H_2O)

T, K	B, cm^3/mol	T, K	B, cm^3/mol
432.00	− 311.2	500.00	− 176.2
434.00	− 304.5	505.00	− 170.4
436.00	− 298.1	510.00	− 165.0
438.00	− 291.9	515.00	− 160.0
440.00	− 285.5	520.00	− 155.3
442.00	− 279.7	530.00	− 146.7
444.00	− 273.9	540.00	− 139.1
446.00	− 268.5	550.00	− 132.0
448.00	− 263.2	560.00	− 125.3
450.00	− 258.2	570.00	− 119.0
452.00	− 253.4	580.00	− 113.1
454.00	− 248.9	590.00	− 107.6
456.00	− 244.7	600.00	− 102.5
458.00	− 240.5	610.00	− 97.6
460.00	− 236.5	620.00	93.0
462.00	− 232.6	630.00	− 88.6
464.00	− 228.9	640.00	− 84.4
466.00	− 225.4	650.00	− 80.4
468.00	− 222.1	660.00	− 76.6
470.00	− 218.5	670.00	− 72.9
475.00	− 210.2	680.00	− 69.4
480.00	− 202.5	690.00	− 66.1
485.00	− 195.4	700.00	− 62.9
490.00	− 188.6	710.00	− 59.9
495.00	− 182.2	720.00	− 57.0

of oxygen data (L. A. Weber) is particularly precise, $\sim$0.1 cm^3/mol in B, and agrees with the others within combined precision. For hydrogen and deuterium, problems with the temperature scale between 100 and 274 K may cause errors in B as large as 0.5 cm^3/mol. For H_2O and D_2O, there is only one source for which the precision ranges from several cm^3/mol at the lower temperatures to 0.2 cm^3/mol at the higher ones. For CO_2, discrepancies of several cm^3/mol in B exist between data of different sources, and for CH_4, of 0.7 cm^3/mol in B and of 10 percent in C.

TABLE 4i-13. THE SECOND VIRIAL COEFFICIENT OF HEAVY WATER VAPOR (D_2O)

T, K	B, cm^3/mol	T, K	B, cm^3/mol
432.00	−314.5	500.00	−177.6
434.00	−307.8	505.00	−171.8
436.00	−301.3	510.00	−166.3
438.00	−295.0	515.00	−161.1
440.00	−288.6	520.00	−156.4
442.00	−282.6	530.00	−147.7
444.00	−276.9	540.00	−140.0
446.00	−271.3	550.00	−132.8
448.00	−265.9	560.00	−126.1
450.00	−260.8	570.00	−119.6
452.00	−255.9	580.00	−113.6
454.00	−251.4	590.00	−108.1
456.00	−247.0	600.00	−103.0
458.00	−242.8	610.00	−98.0
460.00	−238.8	620.00	−93.3
462.00	−234.8	630.00	−88.8
464.00	−231.1	640.00	−84.6
466.00	−227.6	650.00	−80.6
468.00	−224.2	660.00	−76.7
470.00	−220.5	670.00	−73.1
475.00	−212.2	680.00	−69.5
480.00	−204.4	690.00	−66.2
485.00	−197.2	700.00	−63.0
490.00	−190.3	710.00	−59.9
495.00	−183.8	720.00	−57.0

4i-9. Use of the Tables. The averaged virials presented here can be used for calculations of precise PV products at low pressures. However, in the process of separately averaging and rounding the second and third virials, correlations in their experimental errors have been obliterated; thus, they cannot be used to represent the PVT data from which they were derived within experimental precision over the entire density range. If precise PVT values are needed at higher densities, it is usually preferable to interpolate in the original data.

The tables of virials and their temperature derivatives can be used to calculate the initial density dependence of other thermodynamic properties [2].

TABLE 4i-14. THE SECOND VIRIAL COEFFICIENT OF CARBON DIOXIDE (CO_2)

T, K	B, cm³/mol	T, K	B, cm³/mol
250.00	−181.8	420.00	−52.6
255.00	−174.1	430.00	−49.1
260.00	−166.8	440.00	−45.9
265.00	−160.0	450.00	−42.8
270.00	−153.5	460.00	−40.0
273.15	−149.7	480.00	−34.7
275.00	−147.4	500.00	−30.0
280.00	−141.7	520.00	−25.8
285.00	−136.2	540.00	−21.9
290.00	−131.1	560.00	−18.4
295.00	−126.2	580.00	−15.3
300.00	−121.5	600.00	−12.4
310.00	−112.8	620.00	−9.8
320.00	−104.8	640.00	−7.4
330.00	−97.5	660.00	−5.1
340.00	−90.8	680.00	−3.1
350.00	−84.7	700.00	−1.3
360.00	−79.0	714.81	0.0
370.00	−73.8	750.00	2.7
373.15	−72.2	800.00	6.0
380.00	−68.9	850.00	8.8
390.00	−64.4	900.00	11.1
400.00	−60.2	950.00	13.0
410.00	−56.3	1000.00	14.6

TABLE 4i-15. THE SECOND VIRIAL COEFFICIENT OF METHANE
AND ITS TEMPERATURE DERIVATIVES

T, K	B, cm^3/mol	$T\,dB/dT$, cm^3/mol	$T^2\,d^2B/dT^2$, cm^3/mol	T, K	B, cm^3/mol	$T\,dB/dT$, cm^3/mol	$T^2\,d^2B/dT^2$, cm^3/mol
110.00	−334.0	671	−2,244	210.00	−95.3	193	−505
112.00	−322.2	643	−2,132	220.00	−86.6	179	−463
114.00	−311.0	618	−2,029	230.00	−78.9	167	−428
116.00	−300.5	594	−1,934	240.00	−72.0	157	−397
118.00	−290.5	571	−1,846	250.00	−65.8	147	−369
120.00	−281.1	550	−1,764	260.00	−60.2	139	−346
122.00	−272.2	531	−1,688	270.00	−55.2	131	−324
124.00	−263.7	512	−1,617	273.15	−53.6	129	−318
126.00	−255.6	495	−1,551	280.00	−50.5	124	−306
128.00	−248.0	479	−1,490	290.00	−46.3	118	−289
130.00	−240.7	464	−1,432	300.00	−42.3	113	−273
132.00	−233.7	449	−1,378	320.00	−35.4	103	−247
134.00	−277.0	435	−1,327	340.00	−29.4	94	−225
136.00	−220.7	422	−1,280	360.00	−24.2	87	−207
140.00	−208.8	398	−1,193	373.15	−21.2	83	−106
144.00	−197.9	377	−1,115	380.00	−19.7	81	−191
148.00	−187.8	357	−1,046	400.00	−15.7	76	−177
152.00	−178.5	340	−984	450.00	−7.4	65	−150
156.00	−169.9	323	−928	500.00	−1.1	56	−130
160.00	−161.9	309	−877	509.66	0.0	55	−126
164.00	−154.5	295	−831	550.00	4.0	50	−114
168.00	−147.5	282	−789	600.00	8.1	45	−102
172.00	−141.0	271	−751	650.00	11.5	40	−92
176.00	−134.9	260	−716	700.00	14.3	37	−83
180.00	−129.2	250	−683	800.00	18.8	31	−70
184.00	−123.8	241	−653	900.00	22.1	26	−61
188.00	−118.7	232	−626	1000.00	24.7	23	−53
192.00	−113.9	224	−600	1100.00	26.8	20	−47
196.00	−109.4	216	−576	1300.00	29.8	16	−39
200.00	−105.1	209	−554	1500.00	31.9	13	−32

TABLE 4i-16. THE THIRD VIRIAL COEFFICIENTS OF VARIOUS SUBSTANCES
(C in units of $10^2 cm^6/mol^2$)

T, K	He	Ne	Ar	Kr	Xe	N_2	O_2	Air	H_2	D_2	H_2O	D_2O	CO_2	CH_4
25	...	...	...	...	...	...	...	...	14.0					
30	...	...	...	...	...	...	...	...	16.0					
35	...	...	...	...	...	...	...	...	14.3					
40	...	...	...	...	...	...	...	...	12.1					
45	...	...	...	...	...	...	...	...	10.7					
50	...	...	...	...	...	...	...	...	9.6					
55	...	...	...	...	...	...	...	...	8.9					
60	2.7	4	...	...	...	...	...	...	8.4					
70	2.5	4	...	...	...	...	...	...	7.4					
80	2.4	4	7	...	...	...	...	...	6.9					
90	2.3	4	9	...	...	...	...	...	6.4					
100	2.2	4	12	...	...	...	...	...	6.1	6				
110	2.1	3	16	...	...	...	...	...	5.9	5				
120	2.0	3	20	...	...	...	...	...	5.7	5				
130	1.9	3	23	...	...	...	...	...	5.5	5				
140	1.8	3	25	...	...	...	...	28	5.4	5				
150	1.7	3	23	...	...	...	...	26	5.3	5				
160	1.6	3	22	...	...	26	23	24	5.2	5				
180	1.5	3	20	...	...	21	20	21	5.0	5				
200	1.3	3	18	...	...	19	17	19	4.8	5				
220	1.2	3	16	33	...	17	15	18	4.6	5				
240	1.1	3	15	30	...	16	13	17	4.5	5				
260	1.1	3	13	28	...	15	12	16	4.4	5				
273	1.1	3	12	27	62	15	11	15	4.2	5	.....	.....	57	29
280	1.0	3	12	26	59	15	11	15	4.1	5	.....	.....	56	28
300	1.0	2	11	24	54	14	10	15	3.9	5	.....	.....	52	26
320	1.0	2	11	23	50	14	...	14	3.6	5	.....	.....	49	24
340	0.9	2	10	21	46	14	...	14	3.4	5	.....	.....	45	22
360	0.8	2	9	20	41	13	...	...	3.2	5	.....	.....	42	21
380	0.8	2	9	19	36	13	...	...	3.0	4	.....	.....	38	19
400	0.7	2	9	18	34	13	...	...	2.9	4	.....	.....	36	18
420	0.7	...	9	18	32	12	...	...	...	3	.....	.....	32	17
440	...	...	8	17	30	12	...	...	...	...	.....	.....	...	16
460	...	...	8	16	28	12	...	...	...	...	.....	.....	...	16
480	...	...	8	16	26	12	...	...	...	...	.....	.....	...	15
500	...	...	7	15	24	12	...	...	...	...	− 100	− 150	...	15
525	...	...	7	15	22	...	...	...	...	...	− 53	− 64	...	14
550	...	...	7	14	20	...	...	...	...	...	− 17	− 20	...	14
575	...	...	7	14	18	...	...	...	...	...	+ 2	0	...	14
600	...	...	7	13	...	...	...	...	...	...	9	8	...	13
650	...	...	...	13	...	...	...	...	...	...	12	12		
700	...	...	...	12	...	...	...	...	...	...	10	12		

TABLE 4i-17. THE BOYLE TEMPERATURE AND THE INVERSION TEMPERATURE
OF VARIOUS SUBSTANCES

Substance	Boyle temperature, K	Inversion temperature, K
Helium	22.64	
Neon	122.11	231.42
Argon	411.52	779.91
Krypton	575.00	1089.72
Xenon	768.03	1455.79
Nitrogen	327.22	620.63
Oxygen	405.88	764.43
Air	346.81	658.79
Hydrogen	110.04	
Deuterium	115.30	
Carbon dioxide	714.81	
Methane	509.66	967.81

TABLE 4i-18. POTENTIAL PARAMETERS FOR THE $m - 6$ POTENTIAL
OF SELECTED SUBSTANCES

Substance	m	ϵ/k, K	$b_0 \left(= \dfrac{2\pi N}{3} \sigma^3 \right)$, cm^3/mol
Neon	18	47.74	22.83
Argon	18	160.87	43.74
Krypton	18	224.78	53.78
Xenon	18	300.29	73.32
Nitrogen	21	139.41	54.41
Oxygen	21	172.93	44.49
Air	21	147.76	50.95
Methane	21	217.14	57.96

References

1. Hirschfelder, J. O., C. F. Curtiss, and R. B. Bird: "Molecular Theory of Gases and Liquids," chap. 3, John Wiley & Sons, Inc., New York.
2. Appendix 3B of ref. 1.
3. Burnett, E. S.: J. Appl. Mech., Trans. ASME **58**, A136 (1936).
   Hoover, A. E., F. B. Canfield, R. Kobayashi, and Th. W. Leland, Jr.: J. Chem. Eng. Data **9**, 568 (1964).
4. Thomas, H. Preston, and C. G. M. Kirby: Metrologia **4**, 30 (1968).
5. Sengers, J. M. H. Levelt: ASME Proc. 4th Symp. Thermophys. Properties, p. 37, 1968.
6. de Boer, J.: Rept. Progr. Phys. **12**, 305 (1948).
7. Klein, Max, and H. J. M. Hanley: NBS Tech. Note 360; Trans. Faraday Soc. **64**, 2927 (1968).
8. Marquardt, D. W.: J. Soc. Ind. Appl. Math. **11**, 431 (1963).

References for Table 4i-2 to 4i-15

Experimental PVT data from sources marked by an asterisk were fitted.

Table 4i-2: Helium

1. Holborn, L., and H. Schultze: Ann. Physik **47**, 1089 (1915).
*2. Holborn, L., and J. Otto: Z. Physik **30**, 320 (1924).

3. Michels, A., and H. Wouters: *Physica* **8**, 923 (1941).
4. Kistemaker, J., and W. H. Keesom: *Physica* **12**, 227 (1946).
5. Schneider, W. G., and J. A. H. Duffie: *J. Chem. Phys.* **17**, 751 (1949).
6. Yntema, J. L., and W. G. Schneider: *J. Chem. Phys.* **18**, 641 (1950).
7. Keller, W. E.: *Phys. Rev.* **97**, 1 (1955).
8. Silberberg, J. J., K. A. Kobe, and J. J. McKetta: *J. Chem. Eng. Data* **4**, 314 (1959).
9. Stroud, L., J. E. Miller, and L. W. Brandt: *J. Chem. Eng. Data* **5**, 51 (1960).
*10. White, D., Th. Rubin, P. Camky, and H. L. Johnston: *J. Phys. Chem.* **64**, 1607 (1960).
11. Canfield, F. B., T. W. Leland, and R. Kobayashi: *Advances in Cryog. Eng.* **8**, 146 (1963).
12. Witonski, R. J., and J. G. Miller: *J. Am. Chem. Soc.* **85**, 282 (1963).
13. Hoover, A. E., F. B. Canfield, R. Kobayashi, and Th. W. Leland, Jr.: *J. Chem. Eng. Data* **9**, 568 (1964).
14. Boyd, M. E., S. Y. Larsen, and H. Plumb: *J. Research NBS* **72A**, 155 (1968).
15. Cataland, G., and H. Plumb: To be published.

Table 4i-3: Neon

*1. Holborn, L., and J. Otto: *Z. Physik* **33**, 1 (1925).
*2. Holborn, L., and J. Otto: *Z. Physik* **38**, 359 (1926).
3. Nicholson, G. A., and W. G. Schneider: *Can. J. Chem.* **33**, 589 (1955).
*4. Michels, A., T. Wassenaar, and P. Louwerse: *Physica* **26**, 539 (1960).

Table 4i-4: Argon

*1. Holborn, L., and H. Schultze: *Ann. Physik* **47**, 1089 (1915).
*2. Holborn, L., and J. Otto: *Z. Physik* **23**, 77 (1924).
*3. Holborn, L., and J. Otto: *Z. Physik* **30**, 320 (1924).
4. Tanner, C. C., and I. Masson: *Proc. Roy. Soc. (London)*, ser. A, **123**, 268 (1930).
*5. Michels, A., H. Wijker, and H. Wijker: *Physica* **15**, 627 (1949).
6. Whalley, E., Y. Lupien, and W. G. Schneider: *Can. J. Chem.* **31**, 722 (1953).
*7. Michels, A., J. M. H. Levelt and W. de Graaff: *Physica* **24**, 659 (1958).
8. Lecocq, A., and J. Rech: *CNRS* **50**, 55 (1960).
*9. Crain, R. W., and R. E. Sonntag: *Advances in Cryog. Eng.* **11**, 379 (1965).
10. Weir, R. D., I. Wynn Jones, J. S. Rowlinson, and G. Saville: *Trans. Faraday Soc.* **63**, 1320 (1967).
11. Byrne, M. A., M. R. Jones, and L. A. K. Staveley: *Trans. Faraday Soc.* **64**, 1747 (1968).

Table 4i-5: Krypton

*1. Beattie, J. A., J. S. Brierley, and R. J. Barriault: *J. Chem. Phys.* **20**, 1613 (1952).
2. Whalley, E., and W. G. Schneider: *Trans. ASME*, **76**, 1001 (1954).
3. Fender, B. E. F., and G. D. Halsey: *J. Chem. Phys.* **36**, 1881 (1962).
4. Thomaes, G., and R. Van Steenwinkel: *Nature* **13**, 160 (1962).
*5. Trappeniers, N. J., T. Wassenaar, and G. J. Wolkers: *Physica* **32**, 1503 (1966).
6. Weir, R. D., I. Wynn Jones, J. S. Rowlinson, and G. Saville: *Trans. Faraday Soc.* **63**, 1320 (1967).
7. Byrne, M. A., M. R. Jones, and L. A. K. Staveley: *Trans. Faraday Soc.* **64**, 1741 (1968).

Table 4i-6: Xenon

*1. Beattie, J. M., R. J. Barriault, and J. S. Brierley: *J. Chem. Phys.* **19**, 1219 (1951).
*2. Michels, A., T. Wassenaar, and P. Louwerse: *Physica* **20**, 99 (1954).
3. Whalley, E., Y. Lupien, and W. G. Schneider: *Can. J. Chem.* **33**, 633 (1955).

Table 4i-7: Nitrogen

*1. Holborn, L., and J. Otto: *Z. Physik* **10**, 367 (1922); **23**, 77 (1924); **30**, 320 (1924).
*2. Michels, A., H. Wouters, and J. de Boer: *Physica* **1**, 587 (1934).
3. Saurel, J., and J. Rech: *CNRS* **42**, 21 (1958).
4. Canfield, F. B., T. W. Leland, and R. Kobayashi: *Advances in Cryog. Eng.* **8**, 146 (1963).
5. Witonsky, R. J., and J. G. Miller: *J. Am. Chem. Soc.* **85**, 282 (1963).
*6. Crain, R. W., and R. E. Sonntag: *Advances in Cryog. Eng.* **11**, 379 (1965).

Table 4i-8: Oxygen

*1. Holborn, L., and J. Otto: *Z. Physik* **10**, 367 (1922).
*2. Michels, A., H. W. Schamp, and W. de Graaff: *Physica* **20**, 1141 (1954).

*3. Weber, L. A.: *NBS Rept.* 9710, 1968.   (We acknowledge the help of Dr. Weber in refitting his data for our purpose.)

Table 4i-9: *Air*

*1. Holborn, L., and H. Schultze: *Ann. Physik* **47**, 1089 (1915).
*2. Michels, A., T. Wassenaar and W. Van Seventer: *Appl. Sci. Research* **A4**, 52 (1953).
*3. Michels, A., T. Wassenaar, J. M. H. Levelt, and W. de Graaff: *Appl. Sci. Research* **A4**, 381 (1954).

Table 4i-10: *Hydrogen*

*1. Michels, A., W. de Graaff, T. Wassenaar, J. M. H. Levelt, and P. Louwerse: *Physica* **25**, 25 (1959).
2. Goodwin, R. D., D. E. Diller, H. M. Roder, and L. A. Weber: *J. Research NBS* **68A** 121 (1964).

Table 4i-11: *Deuterium*

*1. Michels, A., W. de Graaff, T. Wassenaar, J. M. H. Levelt, and P. Louwerse: *Physica* **25**, 25 (1959).
2. Knaap, H. F. P., M. Knoester, C. M. Knobler, and J. J. M. Beenakker: *Physica* **28**, 21 (1962).

Table 4i-12: *Water Vapor* (H_2O)

1. Kell, G. S., G. E. McLaurin, and E. Whalley: *J. Chem. Phys.* **48**, 3805 (1968).

Table 4i-13: *Heavy Water Vapor* (D_2O)

1. Kell, G. S., G. E. McLaurin, and E. Whalley: *J. Chem. Phys.* **49**, 2839 (1968).

Table 4i-14: *Carbon Dioxide* (CO_2)

1. Michels, A., and C. Michels: *Proc. Roy. Soc. (London),* ser. A, **153**, 201 (1935).
2. MacCormack, K. E., and W. G. Schneider: *J. Chem. Phys.* **18**, 1269 (1950).
3. Kendall, B. J., and B. H. Sage: *Petroleum* **14**, 184 (1951).
4. Pfefferle, W. C., J. A. Goff, and J. G. Miller: *J. Chem. Phys.* **23**, 509 (1955).
5. Dadson, R. S., E. J. Evans, and J. H. King: *Proc. Phys. Soc.* **92**, 1115 (1967).
6. Vukalovich, M. P., and V. V. Altunin: "Thermophysical Properties of Carbon Dioxide," United Kingdom Atomic Energy Authority Translation, Collet's Publishers, London and Wellingborough, England, 1968.

Table 4i-15: *Methane*

1. Michels, A., and G. W. Nederbragt: *Physica* **3**, 569 (1936).
2. Schamp, H. W., E. A. Mason, A. C. B. Richardson, and A. Altman: *Phys. Fluids* **1**, 329 (1958).
3. Douslin, D. R.: *Progr. Intern. Research Thermodyn. Transport Properties,* p. 135, ASME, Princeton University, 1962.
4. Byrne, M. A., M. R. Jones, and L. A. K. Staveley: *Trans. Faraday Soc.* **64**, 1747 (1968).

4j. Temperatures, Pressures, and Heats of Transition, Fusion and Vaporization

D. D. WAGMAN, T. L. JOBE, E. S. DOMALSKI, AND R. H. SCHUMM

The National Bureau of Standards

Table 4j-1 summarizes the data on the temperatures, pressures, and heats for the processes of transition, fusion, and vaporization for a selected list of substances. The table comprises data on stoichiometric inorganic compounds and a small number of organic compounds containing one carbon atom. We have included all the chemical elements for which data are available. We have also included data for halides, oxides, and some nitrates, sulfates, sulfides, and other miscellaneous salts. In some cases, thermodynamic data for vaporization have not been given because of vapor dissociation or decomposition. Noncongruent melting data have also not been included.

Symbols in Table 4j-1

c	crystal
liq	liquid
g	gas
tr	transition
fus	fusion
vap	vaporization
sub	sublimation
equil.	equilibrium mixture of molecular species
g, std.	gas in the standard state (ideal gas at 1 atm)
orthorh.	orthorhombic
monocl.	monoclinic

Units in Table 4j-1

The units of energy in Table 4j-1 are the kilojoule (kJ) and the kilocalorie (kcal), connected by the relation

$$1 \text{ kcal} = 4.1840 \text{ kJ}$$

The unit of mass is the mole (mol) based on the mass in grams corresponding to the formula as written in the column headed "Substance." The atomic weights are taken from A. E. Cameron and E. Wichers, *J. Am. Chem. Soc*, **84**, 4175 (1962).

The equilibrium saturation pressure is given in mm Hg (1 mm Hg $= 133.322$ N/m²). When needed, exponents of the base 10 are indicated in parentheses. For example, 2.66 (E-9) means 2.66×10^{-9}. The equilibrium temperature is given in kelvins (K) on the International Temperature Scale (1948).

Sources of the Data

The data on transition properties of inorganic substances were summarized in *NBS Circ.* 500 (see ref. 320). Selected references to data published since 1950 are indicated in Table 4j-2, in which the numbers following the chemical formulas refer to the bibliography. We have also made considerable use of such reviews as those by Hultgren et al. [164] and Glushko [129].

TABLE 4j-1. TEMPERATURES, PRESSURES, AND HEATS

Substance	Process	State		P	T	ΔH	
		Initial	Final	mm Hg	K	kcal/mol	kJ/mol
Ac.........	fus	c	liq	....	1323		
	vap	liq	g	760	3473		
Ag.........	fus	c	liq	....	1234	2.70	11.30
	vap	liq	g	760	2436	59.90	250.63
AgBr.......	fus	c	liq	....	697	2.32	9.707
	vap	liq	g	760	1778	44.0	184.1
AgCN.......	fus	c	liq	....	619		
AgCl.......	fus	c	liq	....	728.6	3.04	12.72
	vap	liq	g	760	1818	45.5	190.4
AgF........	fus	c	liq	....	708		
	vap	liq	g				
AgI........	tr	c, β	c, α	....	423	1.45	6.067
	fus	c, α	liq	....	831	2.25	9.41
	vap	liq	g	760	1777	34.4	143.9
AgNO₃......	tr	c	c	....	432.5	0.57	2.38
	fus	c	liq	....	483	2.89	12.09
Ag₂S........	tr	, β	c, α	....	450	1.0	4.18
Ag₂SO₄......	tr	c	c	....	703	1.9	7.95
	fus	c	liq	....	933	4	16.7
Ag₂Se.......	tr	c	c	....	406	1.68	7.029
	fus	c	liq	0.11	1163		
Al..........	fus	c	liq	2.66(E − 9)	933.2	2.58	10.79
	vap	liq	g	760	2793	70.13	293.43
Al₂Br₆.......	fus	c	liq	4.38	371.1	5.4	22.6
	sub	c	g	4.38	371.1	18.4	76.98
	vap	liq	g	760	528	11.6	48.53
Al₂Cl₆......	fus	c	liq	1,690	465.6		
	sub	c	g	1,690	465.6	27.1	113.4
AlF₃........	tr	c	c	....	728	0.135	0.5648
AlI₃........	fus	c	liq	....	461.4		
Al₂O₃.......	fus	c	liq	....	2323		
AlPO₄.......	tr	c	c	....	978	0.26	1.098
Am.........	fus	c	liq	1.4(E − 3)	1268	2.9	12.1
Ar..........	fus	c	liq	516.8	83.81	0.284	1.188
	vap	liq	g	760	87.29	1.555	6.506
As..........	sub	c	g, equil.	760	885		
AsCl₃.......	fus	c	liq	....	257	2.44	10.21
	vap	liq	g	760	404.5	8.20	34.31
AsF₃........	fus	c	liq	....	267.21	2.486	10.401
	vap	liq	g	142.6	292.50	8.566	35.840
	vap	liq	g	760	331	8.00	33.472

TABLE 4j-1. TEMPERATURES, PRESSURES, AND HEATS (*Continued*)

Substance	Process	State		P	T	ΔH	
		Initial	Final	mm Hg	K	kcal/mol	kJ/mol
AsF₅........	fus	c	liq	149	192.9	2.71	11.34
	vap	liq	g	149	192.9		
	vap	liq	g	760	220.6	4.96	20.75
AsF₃O.......	fus	c	liq	....	204.9		
	vap	liq	g	760	248	5.0	20.92
AsH₃........	tr	c, III	c, II	....	32	0.024	0.100
	tr	c, II	c, I	....	105.55	0.131	0.5481
	fus	c, I	liq	22.38	156.23	0.286	1.197
	vap	liq	g	22.38	156.23		
	vap	liq	g	760	210.68	3.998	16.728
AsI₃........	fus	c	liq	1.1	413.6	5.21	21.80
	vap	liq	g	760	643.7	13.45	56.28
As₄O₆........	fus	c, octahed.	liq	28	551	11.9	49.79
	sub	c, octahed.	g	28	551	26.1	109.2
	fus	c, monocl.	liq	67	587	8.8	36.8
	vap	liq	g	760	734	13.40	56.06
Au.........	fus	c	liq	2.15(E − 5)	1336	2.955	
	vap	liq	g	760	3081	80.88	335.03
B..........	fus	c	liq	....	2340	5	20.9
	vap	liq	g	760	4075		
BBr₃.......	fus	c	liq	0.686	227.3		
	vap	liq	g	760	363.1	7.72	32.30
B(CH₃)₃.....	fus	c	liq	46.5	199.92	0.777	3.250
	vap	liq	g	46.5	199.92	5.52	23.09
BCl₃........	fus	c	liq	....	165.16	1.627	6.807
	vap	liq	g	760	285.7	5.727	23.962
BF₃........	fus	c	liq	61	144.79	1.10	4.602
	vap	liq	g	61	144.79	4.48	18.74
	vap	liq	g	760	173.2	4.16	17.40
B₂H₆.......	fus	c	liq	....	108.30	1.069	4.473
	vap	liq	g	760	180.57	3.412	14.276
B₅H₉.......	tr	c	c	....	136.7	0.45	1.88
	fus	c	liq	....	226.34	1.466	6.134
	vap	liq	g	190.2	296	7.259	30.372
B₂O₃........	fus	c	liq	....	723	5.85	24.48
	vap	liq	g	0.020	1500	94	393
Ba.........	tr	c, α	c, β	58.(E − 6)	648		
	fus	c, β	liq	0.0107	1002		
	sub	c, β	g	1.1(E − 3)	900	40.5	169.5
BaBr₂.......	fus	c	liq	....	1130	7.63	31.92
	vap	liq	g	0.0037	1200	67.1	280.7
	vap	liq	g	760	2120		
BaCO₃.......	tr	c, orthorh.	c, hexag.	....	1079	4.5	18.8
	tr	c, hexag.	c, cubic	....	1241	0.7	2.9
BaCl₂........	tr	c	c	....	1193	4.10	17.2
	fus	c	liq	....	1233	3.90	16.3
	vap	liq	g	760	2450		

TABLE 4j-1. TEMPERATURES, PRESSURES, AND HEATS (*Continued*)

| Substance | Process | State | | P | T | ΔH | |
		Initial	Final	mm Hg	K	kcal/mol	kJ/mol
BaF₂........	fus	c	liq	1.58(E − 4)	1617	6.8	28.5
	sub	c	g	1.58(E − 4)	1617	86	360
BaI₂........	fus	c	liq	1.18(E − 4)	984	6.34	26.53
	vap	liq	g	0.016	1200	53.6	224.3
Ba(NO₃)₂....	fus	c	liq	....	865	9.9	41.4
BaO........	fus	c	liq	....	2190		
	sub	c	g	0.0030	1700	103	431
BaTiO₃......	tr	c	c	....	201.6	0.012	0.050
	tr	c	c	....	285	0.024	0.100
	tr	c	c	....	390	0.050	0.209
	tr	c, cubic	c, tetrag.	....	1548		
	fus	c	liq	....	1970		
Be..........	tr	c, α	c, β	....	1527	0.611	2.556
	fus	c, β	liq	0.037	1560	2.92	12.21
	vap	liq	g	760	2745	69.89	292.41
BeCl₂........	tr	c, β	c, α	....	676	1.32	5.523
	fus	c, α	liq	....	688	2.07	8.661
	sub	c, β	g	7.6(E − 4)	504	33.0	138.1
	vap	liq	g	760	754	28.9	120.9
BeF₂........	fus	c	liq	1.3(E − 3)	825	1.13	4.728
	sub	c	g	9.8(E − 3)	880	52.9	221.3
BeO........	tr	c	c	....	2323	1.25	5.23
	fus	c	liq	....	2820		
BeSO₄.......	tr	c, α	c, β	....	861	1.2	5.02
	tr	c, β	c, γ	....	912	0.5	2.1
Bi..........	fus	c	liq	....	544.52	2.70	11.30
	vap	liq	g, equil.	760	1837		
BiBr₃........	tr	c	c	....	431	0.74	3.10
	fus	c	liq	2.59	492.0	5.10	21.34
	vap	liq	g	987	741	17.26	72.22
BiCl₃........	fus	c	liq	....	506	5.64	23.60
	vap	liq	g	760	713	17.0	71.13
BiF₃........	fus	c	liq	....	1033		
Bi₂O₃........	tr	c, monocl.	c, cubic	....	1003	7.31	30.58
	fus	c, cubic	liq	....	1100	3.99	16.69
Bi₂S₃........	fus	c	liq	....	1036	19.0	79.50
Br₂..........	fus	c	liq	45.83	265.90	2.527	10.573
	vap	liq	g	760	332.35	7.06	29.45
BrF₃........	fus	c	liq	....	281.92	2.875	12.029
	vap	liq	g, equil.	760	398.90	9.65	40.376
BrF₅........	fus	c	liq	....	212.6		
	vap	liq	g	760	314.44	6.96	29.12
C..........	sub	c, graphite	g, std.	760	298.15	171.291	716.682
	sub	c, graphite	g, equil.	760	4100		

TABLE 4j-1. TEMPERATURES, PRESSURES, AND HEATS (*Continued*)

Substance	Process	State		P	T	ΔH	
		Initial	Final	mm Hg	K	kcal/mol	kJ/mol
CBr_4........	tr	c, II	c, I	....	320.1	1.41	5.90
	fus	c, I	liq	....	365.7	0.94	3.93
	vap	liq	g	760	460	10.4	43.5
CCl_4........	tr	c, II	c, I	....	225.5	1.09	4.56
	fus	c, I	liq	....	250.28	0.59	2.47
	vap	liq	g	....	349.9	7.17	30.00
CF_4........	tr	c, II	c, I	....	76.23	0.35	1.46
	fus	c, I	liq	....	89.57	0.167	0.699
	vap	liq	g	760	145.14	3.01	12.59
CH_4........	tr	c, II	c, I	....	20.44	0.0181	0.0757
	fus	c, I	liq	87.7	90.68	0.225	0.941
	vap	liq	g	760	111.66	1.955	8.18
CH_3Br.......	tr	c, II	c, I	....	173.80	0.113	0.473
	fus	c, I	liq	....	179.49	1.429	5.98
	vap	liq	g	760	276.71	5.715	23.911
CH_3Cl......	fus	c	liq	65.66	175.43	1.537	6.431
	vap	liq	g	760	248.93	5.14	21.50
CH_3F........	fus	c	liq	....	131.4		
	vap	liq	g	760	195.0	4.06	16.99
CH_3I........	fus	c	liq	....	206.70		
	vap	liq	g	760	315.65	6.73	28.16
CH_3OH......	tr	c, III	c, I	....	157.6	0.17	0.71
	fus	c, I	liq	....	175.4	0.755	3.159
	vap	liq	g	760	337.8	8.43	35.27
	vap	liq	g, std.	760	298.15	9.08	37.99
CH_2Cl_2......	fus	c	liq	....	176	1.1	4.60
	vap	liq	g	....	312.94	6.69	27.99
CH_2F_2.......	vap	liq	g	760	221.46	5.0	20.92
CH_2I_2.......	fus	c, II	liq	....	278.75	3.00	12.55
	fus	c, I	liq	....	279.25	2.87	12.01
	vap	liq	g	15	340.7	10.2	
CH_2O (for-maldehyde)	fus	c	liq	....	154.9		
	vap	liq	g	760	253.9	5.7	23.8
$CHBr_3$.......	fus	c	liq	....	281.2	2.65	11.09
	vap	liq	g	760	422.7	8.7	36.4
$CHCl_3$.......	fus	c	liq	....	209.7	2.27	9.50
	vap	liq	g	760	334.4	7.10	29.71
CHF_3........	fus	c	liq	0.456	117.97	0.970	4.058
	vap	liq	g	760	190.97	3.994	16.711
CO..........	tr	c, II	c, I	....	61.57	0.151	0.632
	fus	c, I	liq	115.3	68.10	0.200	0.837
	vap	liq	g	760	81.66	1.444	6.042
CO_2.........	sub	c	g	760	194.640	6.031	25.234
	fus	c	liq	....	217.0	1.99	8.33
$COBr_2$.......	vap	liq	g	760	333	7.2	30.1

TABLE 4j-1. TEMPERATURES, PRESSURES, AND HEATS (*Continued*)

Substance	Process	State		P	T	ΔH	
		Initial	Final	mm Hg	K	kcal/mol	kJ/mol
COCl₂.......	fus	c, III	liq	....	139.19	1.131	4.732
	fus	c, II	liq	....	142.09	1.336	5.590
	fus	c, I	liq	....	145.37	1.371	5.736
	vap	liq	g	760	280.66	5.832	24.401
COF₂........	fus	c	liq	....	161.89	1.603	6.707
	vap	liq	g	760	188.58	4.368	18.276
CS₂..........	fus	c	liq	....	161.2	1.05	4.39
	vap	liq	g	760	319.37	6.390	26.736
COS........	fus	c	liq	0.8	134.31	1.130	4.728
	vap	liq	g	760	222.87	4.423	18.506
Ca.........	tr	c, α	c, β	....	720	0.22	0.920
	fus	c, β	liq	6.0(E − 5)	1112	2.04	8.54
	vap	liq	g	760	1757	36.72	153.64
CaB₂O₄......	fus	c	liq	....	1435	17.67	73.93
Ca₂B₂O₅.....	tr	c, α	c, β	....	804	1.10	4.60
	fus	c, β	liq	....	1585	24.09	100.79
CaBr₂.......	fus	c	liq	....	1014	6.90	28.87
	vap	liq	g	0.079	1250	56.6	236.8
	vap	liq	g	760	2088		
CaC₂........	tr	c, tetrag.	c, cubic	....	720	1.33	5.565
	fus	c, cubic	liq	....	2430		
CaCO₃.......	tr	c, aragon.	c, calcite	....	753	0.05	0.21
CaCl₂........	fus	c	liq	7.3(E − 3)	1055	6.78	28.37
	vap	liq	g	....	1195	62.1	259.8
CaF₂........	tr	c, α	c, β	....	1424	1.14	4.77
	fus	c, β	liq	0.08	1691	7.1	29.7
	sub	c, β	g	0.029	1625	92.0	384.9
CaO........	fus	c	liq	....	2887	12	50
	sub	c	g	2.6(E − 7)	1675	125	523
CaSO₄.......	tr	c, α	c, β	....	1486	5.0	20.9
	fus	c, β	liq	....	1738	6.7	28.0
CaSiO₃.......	tr	c	c	....	1398		
	fus	c	liq	....	1817	13.4	56.1
Ca₂SiO₄......	tr	c, β	c, α′	....	970	0.44	1.84
	tr	c, γ	c, α′	....	1120	3.44	14.39
	tr	c, α′	c, α	....	1710	3.39	14.18
	fus	c, α	liq	....	2403		
CaTiO₃......	tr	c, II	c, I	....	1530	0.55	2.30
	fus	c, I	liq	....	2188		
Cd.........	fus	c	liq	0.109	594.18	1.48	6.19
	vap	liq	g	760	1040	23.79	99.54
CdBr₂.......	fus	c	liq	12.8	841.2	7.97	33.35
	vap	liq	g	53.2	921	27.5	115.1
CdCl₂........	tr	c	c	....	733		
	fus	c	liq	2.04	842	7.22	30.21
	vap	liq	g	17.5	950	31.7	132.6

TABLE 4j-1. TEMPERATURES, PRESSURES, AND HEATS (*Continued*)

Substance	Process	State		P	T	ΔH	
		Initial	Final	mm Hg	K	kcal/mol	kJ/mol
CdF$_2$........	fus	c	liq	....	1322		
	sub	c	g	0.024	1185	64.5	269.9
	vap	liq	g	760	2021	52.3	218.8
CdI$_2$........	fus	c	liq	0.52	661.2	4.95	20.71
	vap	liq	g	760	1013	26.4	110.46
Ce..........	tr	c, α	c, β	....	125		
	tr	c, β	c, γ	....	350		
	tr	c, γ	c, δ	....	999	0.715	2.992
	fus	c, δ	liq	....	1071	1.305	5.460
	vap	liq	g	760	3699	99	414
CeO$_2$........	fus	c	liq	1.2(E − 3)	2670		
	sub	c	g	1.2(E − 3)	2670	88	368
Ce$_2$O$_3$........	fus	c	liq	....	2415		
Cl$_2$..........	fus	c	liq	10.1	172.12	1.531	6.406
	vap	liq	g	760	239.05	4.878	20.410
ClF........	fus	c	liq	....	119		
	vap	liq	g	760	172.9	5.34	22.34
ClF$_3$........	tr	c	c	....	190.50	0.36	1.51
	fus	c	liq	....	196.84	1.819	7.611
	vap	liq	g	760	284.90	6.580	27.531
ClO$_2$........	fus	c	liq	20.8	214		
	vap	liq	g	760	282.8	6.2	25.9
Co..........	tr	c, α	c, β	....	700	0.108	0.452
	fus	c, β	liq	....	1768	3.87	16.19
	vap	liq	g	760	3201	90.0	376.6
CoCl$_2$........	fus	c	liq	....	1000	7.4	30.9
	vap	liq	g	760	1323	27.2	113.8
CoF$_2$........	fus	c	liq	3.01	1400	10.72	44.85
	sub	c	g	3.01	1400	68.1	284.9
CoO........	fus	c	liq	....	2078		
Cr..........	fus	c	liq	3.25	2130	4.047	16.93
	vap	liq	g	760	2945	82.3	344.3
CrBr$_3$........	sub	c	g	0.076	890	56.6	236.8
Cr(CO)$_6$.....	sub	c	g	724	420	15.7	65.69
CrF$_3$........	tr	c	c	....	45.6		
	tr	c	c	....	69.8		
	sub	c	g	2.7(E − 3)	1000	57.8	241.8
Cr$_2$O$_3$........	tr	c	c	....	305	0.10	0.418
	fus	c	liq	....	2548		
Cs..........	fus	c	liq	1.4(E − 6)	301.8	0.52	2.18
	vap	liq	g, equil.	760	955		
CsBr........	fus	c	liq	0.20	909	5.64	23.60
	sub	c	g	5.9(E − 3)	800	46.6	195.0
	vap	liq	g	760	1576	36.0	150.6

TABLE 4j-1. TEMPERATURES, PRESSURES, AND HEATS (*Continued*)

Substance	Process	State		P	T	ΔH	
		Initial	Final	mm Hg	K	kcal/mol	kJ/mol
CsCl.........	tr	c, II	c, I	....	743	0.90	3.76
	fus	c, I	liq	101.1	918	4.82	20.16
	vap	liq	g	760	1573	35.4	148.1
CsF.........	sub	c	g	0.008	800	46.4	194.1
	fus	c	liq	....	976	5.19	21.71
	vap	liq	g	760	1524	34.3	143.5
CsI.........	fus	c	liq	0.260	899	5.90	24.68
	sub	c	g	0.260	899	45.5	190.37
	vap	liq	g	760	1524	34.3	143.5
CsNO₃.......	tr	c, hexag.	c, cubic	....	424.7	0.89	3.72
	fus	c, cubic	liq	....	678	3.37	14.10
CsOH.......	tr	c, α	c, β	....	488	1.76	7.363
	fus	c, β	liq	....	619	1.6	6.69
Cs₂SO₄.......	tr	c	c	....	1005		
	fus	c	liq	....	1286	9.6	40.17
Cu.........	fus	c	liq	4.49(E − 4)	1356.5	3.14	13.14
	vap	liq	g	760	2839	71.77	300.29
(CuBr)₂......	tr	c, γ	c, β	....	658	4.2	17.57
	tr	c, β	c, α	....	743	2.1	8.79
	fus	c, α	liq	0.276	756		
	sub	c, α	g	0.276	756	29	121
(CuCl)₂......	fus	c	liq	....	703		
	sub	c	g	2.54(E − 4)	550	35.4	148.1
CuF₂.......	fus	c	liq	7.9(E − 6)	1058		
	sub	c	g	8.85(E − 3)	960	59.5	248.9
(CuI)₂.......	tr	c, γ	c, β	....	644	5.1	21.3
	tr	c, β	c, α	....	682	2.3	9.62
	fus	c, α	liq	....	871		
Cu₂O.......	tr	c	c	....	329		
	fus	c	liq	....	1515	15.35	64.224
Cu₂S.......	tr	c, III	c, II	....	376	0.92	3.85
	tr	c, II	c, I	....	....	0.20	0.837
Dy.........	tr	c, α	c, β	....	1657	0.955	3.996
	fus	c, β	liq	0.591	1682	2.64	11.06
	vap	liq	g	760	2835	55.0	230.1
Er.........	fus	c	liq	0.317	1795	4.76	19.92
	vap	liq	g	760	3136	62.47	261.37
ErCl₃.......	fus	c	liq	....	1049	7.8	32.6
	vap	liq	g	2.07	1250	53.6	224.3
ErF₃.......	tr	c	c	....	1369		
	fus	c	liq	....	1413		
Eu.........	fus	c	liq	0.72	1090	2.20	9.21
	vap	liq	g	760	1870	34.30	143.49
EuCl₃.......	fus	c	liq	....	891		
	vap	liq	g	1.0	1140	31	130

TABLE 4j-1. TEMPERATURES, PRESSURES, AND HEATS (*Continued*)

Substance	Process	State		P	T	ΔH	
		Initial	Final	mm Hg	K	kcal/mol	kJ/mol
Eu_2O_3........	tr	c	c	....	1373		
	fus	c	liq	....	2510		
F_2...........	tr	c	c	....	45.55	0.174	0.728
	fus	c	liq	1.66	53.54	0.122	0.5104
	vap	liq	g	760	85.02	1.562	6.535
F_2O........	fus	c	liq	....	49.4		
	vap	liq	g	760	128.1	2.41	10.08
Fe...........	tr	c, α	c, β	....	1033	0.0	0.0
	tr	c, β	c, γ	....	1184	0.215	0.8996
	tr	c, γ	c, δ	....	1665	0.200	0.837
	fus	c, δ	liq	0.026	1809	3.30	13.81
	vap	liq	g	760	3135	83.55	349.56
$FeBr_2$........	fus	c	liq	21	962		
	vap	liq	g	102	1073	31.6	132.2
$Fe(CO)_5$.....	fus	c	liq	....	252.88		
	vap	liq	g	760	378	8.7	36.4
$FeCl_2$........	fus	c	liq	8.84	950	10.28	43.011
	vap	liq	g	760	1299	30.2	126.4
$(FeCl_3)_2$......	sub	c	g	126	550	30.6	128.0
	fus	c	liq	547	577	18.3	76.57
	vap	liq	g	610	583	12.3	51.46
	vap	liq	g	760	592	12.1	50.63
FeF_2........	tr	c	c	....	78.35		
	fus	c	liq	6.0	1373		
	sub	c	g	2.1(E − 3)	1060	72.4	302.9
FeF_3........	sub	c	g	9.6(E − 3)	880	52.7	220.5
FeI_2........	fus	c	liq	5.8	867	13.3	55.65
	vap	liq	g	5.8	867	35.6	148.9
	vap	liq	g	760	1208		
$Fe_{0.947}O$......	tr	c	c	....	189	0.06	0.25
	fus	c	liq	....	1650	7.5	31.4
Fe_2O_3........	tr	c, III	c, II	....	960	0.16	0.669
	tr	c, II	c, I	....	1050	0.0	0.0
Fe_3O_4........	tr	c, II	c, I	....	880	0.0	0.0
	fus	c, I	liq	....	1867	33	138
FeS........	tr	c, III	c, II	....	411	0.57	2.38
	tr	c, II	c, I	....	598	0.12	0.502
	fus	c, I	liq	....	1468	7.73	32.34
Ga..........	fus	c	liq	....	302.9	1.335	5.585
	vap	liq	g	760	2520	61.46	257.16
$(GaCl_3)_2$.....	fus	c	liq	10.4	350.9	5.2	21.8
	sub	c	g	10.4	350.9	17.4	72.80
	vap	liq	g	760	474.4	10.5	43.93
GaI_3........	fus	c	liq	17.2	485	3.1	13.0
	vap	liq	g	760	619	16.5	69.04

TABLE 4j-1. TEMPERATURES, PRESSURES, AND HEATS (*Continued*)

Substance	Process	State		P	T	ΔH	
		Initial	Final	mm Hg	K	kcal/mol	kJ/mol
Gd.........	tr	c, α	c, β	....	1533	0.935	3.912
	fus	c, β	liq	....	1585	2.40	10.04
	vap	liq	g	760	3539	85.9	359.4
GdBr₃.......	fus	c	liq	....	1058	8.7	36.4
GdCl₃.......	fus	c	liq	....	875	9.6	40.2
	vap	liq	g	0.37	1183	44.0	184.1
GdF₃........	tr	c	c	....	1280		
	fus	c	liq	....	1301		
Gd₂O₃.......	tr	c, monocl.	c, cubic	....	1473		
	fus	c	liq	....	2595		
Ge.........	fus	c	liq	....	1210.4	8.83	36.94
	vap	liq	g	760	3107	79.1	330.9
GeBr₄.......	fus	c	liq	....	299.3		
	vap	liq	g	760	462	9.4	39.3
GeCl₄.......	fus	c	liq	0.59	221.6	1.8	7.53
	sub	c	g	0.59	221.6	10.9	45.61
	vap	liq	g	760	356.4	7.2	30.12
GeF₄........	sub	c	g	760	236.6	7.8	32.6
	fus	c	liq	3032	258.1		
GeH₄........	tr	c, III	c, II	....	73.2	0.050	0.209
	tr	c, II	c, I	....	76.6	0.086	0.360
	fus	c, I	liq	....	107.25	0.200	0.8367
	vap	liq	g	760	184.79	3.361	14.062
GeI₄........	sub	c	g	0.22	380	19.5	81.6
	fus	c	liq	....	417		
GeO₂.......	tr	c, II	c, I	....	1306	5.05	21.13
	fus	c, I	liq	....	1389	3.59	15.02
H₂.........	fus	c	liq	54.0	13.957	0.028	0.117
	vap	liq	g	54.0	13.957	0.219	0.9163
	vap	liq	g	760	20.38	0.219	0.9163
HBr........	tr	c, III	c, rhombic	....	89.8		
	tr	c, rhombic	c, cubic	....	116.9		
	fus	c, cubic	liq	285	186.24	0.575	2.406
	vap	liq	g	760	206.38	4.210	17.615
HCN........	tr	c, II	c, I	170.42	0.004		
	fus	c, I	liq	140.4	259.91	2.009	8.4057
	vap	liq	g	760	298.85	6.027	25.217
HCl........	tr	c, rhomb.	c, cubic	....	98.36	0.284	1.188
	fus	c, cubic	liq	103.4	158.91	0.476	1.992
	vap	liq	g	760	188.07	3.860	16.150
HF.........	fus	c	liq	4.03	189.79	0.939	3.929
	vap	liq	g, equil.	760	292.67	1.790	7.4894
HI.........	tr	c, III	c, II	....	70.1		
	tr	c, II	c, I	....	125.7		
	fus	c, I	liq	371	222.31	0.686	2.870
	vap	liq	g	760	237.75	4.724	19.705

TABLE 4j-1. TEMPERATURES, PRESSURES, AND HEATS (*Continued*)

Substance	Process	State		P	T	ΔH	
		Initial	Final	mm Hg	K	kcal/mol	kJ/mol
HNO₃.......	fus	c	liq	....	231.55	2.503	10.473
	vap	liq	g	48	293.1	9.42	39.41
H₂O........	fus	c	liq	4.58	273.16	1.436	6.0082
	vap	liq	g	4.58	273.16	10.767	45.0491
	vap	liq	g	23.75	298.15	10.514	43.9906
	vap	liq	g, std.	760	298.15	10.520	44.0157
	vap	liq	g	760	373.15	9.717	40.656
H₂S........	tr	c, II	c, I	....	103.50	0.365	1.527
	fus	c, I	liq	174	187.61	0.568	2.377
	vap	liq	g	174	187.61	4.67	19.54
	vap	liq	g	760	212.80	4.463	18.673
H₂SO₄.......	fus	c	liq	....	283.5	2.560	10.711
H₂Se........	tr	c, II	c, I	....	82.3	0.309	1.293
	fus	c, I	liq	205.4	207.46	0.601	2.514
	vap	liq	g	205.4	207.46	5.48	22.93
	vap	liq	g	760	231.8	4.76	19.91
H₂Te........	fus	c	liq	70	222	1.0	4.18
	vap	liq	g	760	270.9	5.6	23.4
H₃PO₄.......	fus	c	liq	....	315.5	3.07	12.84
¹H²H........	fus	c	liq	93	16.62	0.038	0.159
	vap	liq	g	760	22.14	0.257	1.075
¹H²HO.......	vap	liq	g	22.0	298.15	10.65	44.56
	vap	liq	g	760	374.0		
²H₂O........	fus	c	liq	5.01	276.96	1.508	6.309
	vap	liq	g	5.01	276.96	11.105	46.463
	vap	liq	g	760	374.58	9.933	41.559
He..........	fus	c	liq	22.5(E + 3)	1.764	0.002	0.0084
	tr	liq, II	liq, I	37.8	2.172		
	vap	liq, I	g	760	4.214	0.020	0.084
Hf..........	tr	c, α	c, β	....	2013	1.61	6.736
	fus	c, β	liq	1.1(E − 3)	2500	5.75	24.06
	vap	liq	g	760	4876	137	573.2
HfBr₄........	sub	c	g	83.3	531	23.5	98.40
	fus	c	liq	15,270	693		
HfCl₄........	fus	c	liq	2.2(E − 4)	705		
	sub	c	g	2.2(E − 4)	705	23.8	99.58
	vap	liq	g	2.2(E − 4)	705	14.1	58.99
HfF₄........	sub	c	g	54.1	1112	56.9	238.1
	sub	c	g	760	1240		
HfI₄.........	tr	c, α	c, β	....	697	14.4	60.25
	tr	c, β	c, γ		745	5.4	22.6
	sub	c, γ	g	760	667	28.2	118.0
HfO₂........	fus	c	liq	....	3026		
Hg..........	fus	c	liq	....	234.29	0.548	2.292
	vap	liq	g	760	629.73	14.172	59.296
HgBr₂.......	fus	c	liq	....	511.2	4.28	17.91
	vap	liq	g	760	592	14.08	58.91

TABLE 4j-1. TEMPERATURES, PRESSURES, AND HEATS (*Continued*)

Substance	Process	State		P	T	ΔH	
		Initial	Final	mm Hg	K	kcal/mol	kJ/mol
HgCl₂.......	tr	c	c	....	428	0.077	0.322
	fus	c	liq	....	553.2	4.55	19.04
	vap	liq	g	760	575.0	14.08	58.91
HgF₂........	sub	c	g	178	575	16	66.9
	fus	c	liq	....	918		
HgI₂........	tr	c, red	c, yellow	0.20	404.6	0.65	2.72
	sub	c, yellow	g	8.8	530	19.95	83.47
	fus	c, yellow	liq	8.8	530	4.53	18.95
	vap	liq	g	760	627	14.26	59.664
HgS........	tr	c, red	c, black	....	659	1.0	4.18
Ho..........	tr	c, α	c, β	....	1701	1.12	4.686
	fus	c, β	liq	....	1743	2.91	12.17
	vap	liq	g	760	2968	57.6	241.0
HoCl₃.......	fus	c	liq	....	993	7.0	29.3
	vap	liq	g	0.25	1143	62.7	262.3
HoF₃.......	fus	c	liq	1.57(E − 3)	1416		
	sub	c	g	1.57(E − 3)	1416	105.0	439.32
	vap	liq	g	1.57(E − 3)	1416	85.1	356.1
Ho₂O₃.......	fus	c	liq	....	2640		
I₂..........	sub	c	g	0.31	298.15	14.93	62.467
	fus	c	liq	92.0	386.75	3.71	15.52
	vap	liq	g	760	458.39	9.99	41.80
ICl.........	fus	c	liq	32.62	300.53	2.76	11.55
	sub	c	g	32.62	300.53	12.62	52.80
IF₅.........	fus	c	liq	10.45	282.58		
	vap	liq	g	760	374	9.04	37.82
IF₇.........	tr	c	c	....	153		
	sub	c	g	760	277	7.46	31.21
In..........	fus	c	liq	....	429.76	0.78	
	vap	liq	g	760	2343	55.4	231.8
InBr₃.......	fus	c	liq	392	709		
	sub	c	g	9.9(E − 4)	460	33.5	140.2
InCl........	tr	c, II	c, I	....	393		
	fus	c, I	liq	0.038	498		
	vap	liq	g	6.63	656	21.2	88.70
	vap	liq	g	760	926		
InCl₃.......	fus	c	liq	....	859		
	sub	c	g	6.3(E − 4)	510	37.0	154.8
InI₃........	fus	c	liq	0.26	480		
	vap	liq	g	0.26	480	19.2	80.33
In₂O₃.......	fus	c	liq	....	2183		
Ir..........	fus	c	liq	....	2716	6.3	2.64
	vap	liq	g	....	4662	146.3	612.3
IrF₆........	tr	c	c	61.7	273.5	1.70	7.11
	fus	c	liq	531.3	316.9	0.7	2.93
	vap	liq	g	531.3	316.9	7.65	32.01

TABLE 4j-1. TEMPERATURES, PRESSURES, AND HEATS (*Continued*)

Substance	Process	State Initial	State Final	P mm Hg	T K	ΔH kcal/mol	ΔH kJ/mol
K..........	fus	c	liq	....	336.4	0.562	2.351
	vap	liq	g	760	1031	19.18	80.23
KBr........	fus	c	liq	....	1007	6.1	25.5
	vap	liq	g, equil.	760	1657	30.8	128.9
KCN.......	tr	c, II	c, I	....	168.3	0.30	1.26
	fus	c, I	liq	....	908	3.5	14.6
KCl........	fus	c	liq	0.40	1044	6.282	26.284
	vap	liq	g, equil.	760	1700	28.7	120.1
KF.........	fus	c	liq	....	1130	6.75	28.24
	vap	liq	g	760	1775		
KI.........	fus	c	liq	0.36	954	5.7	23.8
	vap	liq	g, equil.	760	1617	26.9	112.5
KNO₃.......	tr	c, II	c, I	....	401.1	1.22	5.104
	fus	c, I	liq	....	610	2.413	10.096
KOH........	tr	c, II	c, I	....	522	1.52	6.360
	fus	c, I	liq	....	677	1.8	7.53
	vap	liq	g	760	1600	30.8	128.9
K₂SO₄.......	tr	c, II	c, I	....	856	1.94	8.12
	fus	c, I	liq	....	1342	8.76	36.65
Kr..........	fus	c	liq	549	115.78	0.392	1.640
	vap	liq	g	760	119.93	2.162	9.046
KrF₂........	sub	c	g	29	273	9.9	41.42
KrF₄.......	sub	c	g	760	341	8.3	34.73
La..........	tr	c, α	c, β	....	550	0.087	0.364
	tr	c, β	c, γ	....	1134	0.746	3.121
	fus	c, γ	liq	....	1193	1.481	6.196
	vap	liq	g	760	3730	98.9	413.7
LaBr₃.......	sub	c	g	0.0032	1026	70.7	295.8
	fus	c	liq	0.0102	1061	13.0	54.39
LaCl₃.......	sub	c	g	0.0010	1067	72.3	302.5
	fus	c	liq	0.0072	1131	13.0	54.39
LaF₃........	sub	c	g	8.9(E − 3)	1495	99.4	415.9
	fus	c	liq	1.46	1763		
LaI₃........	fus	c	liq	9.0(E − 3)	1034		
	sub	c	g	9.0(E − 3)	1034	69.9	292.5
La₂O₃.......	fus	c	liq	....	2490		
Li..........	tr	c, II	c, I	....	77		
	fus	c, I	liq	....	453.69	0.717	3.000
	vap	liq	g	760	1597	35.40	148.13
LiBr........	fus	c	liq	....	823	4.22	17.65
	vap	liq	g, equil.	760	1555	27.0	113.0
LiCl........	fus	c	liq	....	883	4.74	19.83
	vap	liq	g	760	1656		
LiF..........	fus	c	liq	....	1121	6.474	27.087
	vap	liq	g, equil.	760	1966		

TABLE 4j-1. TEMPERATURES, PRESSURES, AND HEATS (*Continued*)

Substance	Process	State		P	T	ΔH	
		Initial	Final	mm Hg	K	kcal/mol	kJ/mol
LiI.........	fus	c	liq	....	742	3.50	14.64
	vap	liq	g, equil.	760	1415	26.4	110.4
LiNO₃.......	fus	c	liq	....	525	6.1	25.5
LiOH........	fus	c	liq	....	744.3	5.01	20.96
Li₂SO₄.......	tr	c, II	c, I	....	859	6.5	27.2
	fus	c, I	liq	....	1132	1.8	7.53
Lu..........	fus	c	liq	0.011	1936	4.46	18.65
	vap	liq	g	760	3668	85.06	355.89
LuCl₃.......	fus	c	liq	....	1165		
	vap	liq	g	0.89	915	57.2	239.3
LuF₃........	tr	c	c	....	1200		
	sub	c	g	1.1(E − 3)	1368	96.1	402.1
	fus	c	liq	....	1455		
Lu₂O₃........	fus	c	liq	....	2740		
Mg..........	fus	c	liq	3.10	922	2.140	8.954
	vap	liq	g	760	1363	30.45	127.40
MgBr₂.......	sub	c	g	0.017	842	50.3	210.5
	fus	c	liq	....	984	8.3	34.7
MgCl₂.......	fus	c	liq	0.120	987	10.30	43.095
	sub	c	g	0.120	987	57.7	241.4
	vap	liq	g	30.7	1310	43.08	180.25
MgF₂........	fus	c	liq	0.077	1525	13.90	58.158
	vap	liq	g	0.077	1525	72.6	303.8
MgI₂........	sub	c	g	0.015	757	45.0	188.3
Mg₃N₂.......	tr	c, III	c, II	....	823	0.22	0.920
	tr	c, II	c, I	....	10.61	0.26	1.09
MgO........	fus	c	liq	....	3125	18.5	77.40
MgSO₄.......	tr	c, II	c, I	....	1283		
	fus	c, I	liq	....	1400	3.5	14.6
Mn..........	tr	c, α	c, β	....	980	0.532	2.226
	tr	c, β	c, γ	....	1360	0.507	2.121
	tr	c, γ	c, δ	....	1410	0.449	1.879
	fus	c, δ	liq	1.03	1517	2.88	12.05
	vap	liq	g	760	2335	54.0	225.9
MnBr₂.......	fus	c	liq	....	971		
MnCl₂.......	fus	c	liq	0.24	923	8.97	37.53
	vap	liq	g	0.24	923	40.0	167.4
	vap	liq	g	760	1511		
MnF₂........	fus	c	liq	0.031	1203		
	sub	c	g	0.031	1203	72.0	301.2
MnI₂........	fus	c	liq	....	911		
MnO........	fus	c	liq	....	2088		
Mn₃O₄.......	tr	c, II	c, I	....	1445	4.97	20.79
	fus	c, I	liq	....	1840		

TABLE 4j-1. TEMPERATURES, PRESSURES, AND HEATS (*Continued*)

| Substance | Process | State | | P | T | ΔH | |
		Initial	Final	mm Hg	K	kcal/mol	kJ/mol
Mo.........	fus	c	liq	0.031	2890	6.65	27.82
	vap	liq	g	760	4880	141.6	592.45
Mo(CO)₆.....	sub	c	g	48	375	16.3	68.20
MoF₃........	fus	c	liq	2.67	340.1		
	vap	liq	g	760	486.7	11.9	49.79
MoF₆........	tr	c, II	c, I	....	263.50	1.953	8.171
	fus	c, I	liq	408.5	290.76	1.034	4.326
	vap	liq	g	760	307.2	6.75	28.242
MoO₃........	fus	c	liq	1.76(E − 2)	1074	11.69	48.911
N₂..........	tr	c, II	c, I	....	35.61	0.055	0.230
	fus	c, I	liq	93.9	63.15	0.172	0.719
	vap	liq	g	93.9	63.15	1.446	6.050
	vap	liq	g	760	77.35	1.335	5.586
NH₃.........	fus	c	liq	45.37	195.40	1.351	5.652
	vap	liq	g	45.37	195.40	6.061	25.359
	vap	liq	g	760	239.73	5.581	23.351
N₂H₄........	fus	c	liq	....	274.69	3.025	12.656
	vap	liq	g	764	386.7	9.70	40.58
NH₄Br.......	tr	c, II	c, I	....	411.0	0.77	3.22
	fus	c, I	liq	....	815		
NH₄Cl.......	tr	c, III	c, II	....	243	0.27	1.13
	tr	c, II	c, I	....	457.7	1.0	4.18
	fus	c, I	liq	2.62(E + 4)	793		
NH₄F.......	tr	c, II	c, I	....	289.1	0.81	3.39
NH₄I........	tr	c, II	c, I	....	260	0.70	2.93
	fus	c, I	liq	....	824		
NH₄NO₃.....	tr	c, V	c, IV	....	256.2	0.111	0.464
	tr	c, IV	c, III	....	305.4	0.410	1.715
	tr	c, III	c, II	....	357.4	0.32	1.34
	tr	c, II	c, I	....	398.4	1.01	4.23
	fus	c, I	liq	....	442.8	1.3	5.44
NO..........	fus	c	liq	164.4	109.50	0.550	2.301
	vap	liq	g	164.4	109.50	3.43	14.35
	vap	liq	g	760	121.4	3.293	13.778
N₂O.........	fus	c	liq	659	182.1	1.56	6.527
	vap	liq	g	659	182.1	3.97	16.61
	vap	liq	g	760	184.6	3.958	16.560
Na..........	fus	c	liq	....	370.98	0.622	2.601
	vap	liq	g	760	1156	23.43	98.01
NaBr........	fus	c	liq	0.4	1020	6.25	26.15
	vap	liq	g, equil.	760	1665		
NaCN.......	tr	c, III	c, II	....	172.1	0.15	0.628
	tr	c, II	c, I	....	288.5	0.70	2.93
	fus	c, I	liq	....	836	4	17
	vap	liq	g	760	1770	37	155
NaCl........	fus	c	liq	....	1074	6.73	28.16
	vap	liq	g, equil.	760	1730		

TABLE 4j-1. TEMPERATURES, PRESSURES, AND HEATS (*Continued*)

Substance	Process	State		P	T	ΔH	
		Initial	Final	mm Hg	K	kcal/mol	kJ/mol
NaF.........	fus	c	liq	....	1269	7.92	33.14
	vap	liq	g, equil.	760	1977		
NaI.........	fus	c	liq	....	933	5.64	23.60
	vap	liq	g, equil.	760	1577		
Na₂MoO₄....	tr	c, II	c, I	....	713	14.6	61.09
	fus	c, I	liq	....	960	3.6	15.1
NaNO₃......	tr	c, II	c, I	....	549	0.94	3.93
	fus	c, I	liq	....	579.5	3.696	15.464
NaOH......	tr	c, II	c, I	....	566.0	1.520	6.3597
	fus	c, I	liq	....	592.3	1.52	6.360
Na₂SO₄......	tr	c, V	c, III	....	450	0.74	3.10
	tr	c, III	c, I	....	515	1.79	7.489
	fus	c, I	liq	....	1157	5.70	23.85
Na₂TiO₃.....	tr	c, II	c, I	....	560	0.4	1.7
	fus	c, I	liq	....	1303	16.8	70.29
Nb..........	fus	c	liq	....	2740	6.30	26.36
	vap	liq	g	760	5017	163	682.0
NbCl₅.......	fus	c	liq	260	478.9	8.09	33.85
	sub	c	g	260	478.9	21.3	89.12
	vap	liq	g	760	520.5	12.6	52.72
NbF₅........	fus	c	liq	2.44	350.7	2.92	12.217
	vap	liq	g	58.0	423	12.9	53.97
NbO₂........	tr	c, α	c, β	....	1090	0.72	3.01
	tr	c, β	c, γ	....	1200	0.0	0.0
	fus	c, γ	liq	5.0(E − 4)	1900	21	87.9
Nb₂O₅......	fus	c	liq	....	1780	24.69	103.30
Nd..........	tr	c, α	c, β	....	1128	0.72	3.01
	fus	c, β	liq	....	1289	1.71	7.15
	vap	liq	g	760	3341	65.2	272.8
NdBr₃.......	fus	c	liq	1.06(E − 4)	955	10.8	45.19
	sub	c	g	1.06(E − 4)	955	67.6	282.8
NdCl₃.......	fus	c	liq	2.2(E − 3)	1032	12.0	50.21
	sub	c	g	2.2(E − 3)	1032	69.1	289.1
NdF₃........	sub	c	g	0.012	1460	85.7	358.6
	fus	c	liq	0.35	1647		
NdI₃........	tr	c	c	....	847	3.4	14.2
	sub	c	g	4.5(E − 3)	978	66.3	277.4
	fus	c	liq	0.063	1060	9.7	40.6
Nd₂O₃.......	tr	c, α	c, β	....	1395	0.14	0.586
	fus	c, β	liq	....	2485		
Ne..........	fus	c	liq	324	24.544	0.08	0.33
	vap	liq	g	324	24.544	0.431	1.803
	vap	liq	g	760	27.15	0.429	1.795
Ni..........	fus	c	liq	3.1(E − 3)	1726	4.176	17.472
	vap	liq	g	760	3187	88.5	370.3

TABLE 4j-1. TEMPERATURES, PRESSURES, AND HEATS (*Continued*)

Substance	Process	State		P	T	ΔH	
		Initial	Final	mm Hg	K	kcal/mol	kJ/mol
NiBr₂........	sub	c	g	0.044	823	52.5	219.7
	sub	c	g	760	1193		
	fus	c	liq	....	1236		
Ni(CO)₄.....	fus	c	liq	46.6	253.86	3.306	13.832
	vap	liq	g	760	315.4	7.0	29.3
NiCl₂........	sub	c	g	0.045	850	53.0	221.7
	sub	c	g	760	1243		
	fus	c	liq	....	1303	18.47	77.28
NiF₂.........	sub	c	g	2.5(E − 3)	1080	77.3	323.4
NiI₂.........	sub	c	g	0.43	750	36.5	152.7
	fus	c	liq	....	1070		
NiO.........	tr	c, III	c, II	....	525	0.0	0.0
	tr	c, II	c, I	....	565	0.0	0.0
	fus	c, I	liq	87	2263		
Np..........	tr	c, III	c, II	....	533	2	4.4
	tr	c, II	c, I	....	850		
	fus	c, I	liq	....	910		
NpF₆........	fus	c	liq	748.6	327.92	4.189	17.527
	vap	liq	g	748.6	327.92	7.133	29.844
	vap	liq	g	760	328.33		
O₂..........	tr	c, III	c, II	....	23.85	0.022	0.0920
	tr	c, II	c, I	....	43.77	0.178	0.745
	fus	c, I	liq	1.14	54.363	0.106	0.4435
	vap	liq	g	1.14	54.363	1.828	7.648
	vap	liq	g	760	90.180	1.630	6.820
O₃..........	fus	c	liq	0.86	80.65	0.5	2.1
	vap	liq	g	760	161.3	3.58	14.98
Os..........	fus	c	liq	....	3323		
	sub	c	g	6.2(E − 6)	2550	187.4	784.1
OsF₅........	fus	c	liq	0.566	343.1		
	vap	liq	g	15.1	400	15.69	65.647
	vap	liq	g	760	499.0		
OsF₆........	tr	c	c	81.3	272.7	2.0	8.37
	fus	c	liq	463.6	306.5	1.6	6.69
	sub	c	g	463.6	306.5	8.40	35.15
	vap	liq	g	760	320.6	6.70	28.03
OsOF₅.......	tr	c, II	c, I	....	305.6		
	fus	c, I	liq	175.6	332.3	1.62	6.778
	vap	liq	g	394.6	354.0	8.74	36.57
P₄..........	tr	c, IV	c, III	....	195.35	0.500	2.092
	fus	c, III	liq	....	317.30	0.628	2.628
	vap	liq	g	....	317.30	13.32	55.731
	vap	liq	g	760	530	12.48	52.216
PBr₃........	fus	c	liq	....	232.7		
	vap	liq	g	760	446.4	9.33	39.04
PCl₃........	fus	c	liq	....	183		
	vap	liq	g	760	348.3	7.17	30.00

TABLE 4j-1. TEMPERATURES, PRESSURES, AND HEATS (*Continued*)

| Substance | Process | State | | P | T | ΔH | |
		Initial	Final	mm Hg	K	kcal/mol	kJ/mol
PCl$_5$........	fus	c	liq	....	437.7	6.1	25.5
	sub	c	g, equil.	760	432	18.1	75.73
PF$_3$........	tr	c, III	c, II	....	83.7	0.060	0.251
	tr	c, II	c, I	....	110.6	0.55	2.30
	fus	c, I	liq	9.80	121.8	0.224	0.9372
	vap	liq	g	760	171.8	3.48	14.56
PF$_5$........	fus	c	liq	427	179.4	2.7	11.3
	vap	liq	g	427	179.4	4.2	17.6
	vap	liq	g	760	188.7	4.1	17.2
PH$_3$........	tr	c, IV	c, III	....	30.31	0.0196	0.08200
	tr	c, III	c, II	....	49.46	0.186	0.7782
	tr	c, II	c, I	....	88.15	0.115	0.4812
	fus	c, I	liq	27.2	139.40	0.270	1.130
	vap	liq	g	760	185.43	3.486	14.585
P$_4$O$_6$........	fus	c	liq	1.7	297.1	3.36	14.06
	vap	liq	g	1.7	297.1	11.14	46.610
	vap	liq	g	760	448.5	10.38	43.430
P$_4$O$_{10}$.......	fus	c, hexag.	liq	3690	693	5.0	20.9
	sub	c, hexag.	g	3690	693	13.9	58.16
	fus	c, rhomb.	liq	570	844	16.1	67.36
	sub	c, rhomb.	g	570	844	36.4	152.3
Pb..........	fus	c	liq	....	600.45	1.147	4.7990
	vap	liq	g	760	2023	42.5	177.8
PbBr$_2$.......	tr	c, II	c, I	....	617		
	fus	c, I	liq	0.011	643.1	5.0	20.9
	vap	liq	g	760	1166	30.2	126.4
Pb(CH$_3$)$_4$....	fus	c	liq	....	242.92	2.58	10.79
	vap	liq	g	760	383.2	7.87	32.93
PbCl$_2$........	tr	c, α	c, β	....	695		
	fus	c, β	liq	....	773	5.25	21.97
	vap	liq	g	760	1227	30.4	127.2
PbF$_2$........	tr	c, rhomb.	c, cubic	....	723		
	fus	c, cubic	liq	....	1099	3.0	12.6
	vap	liq	g	760	1566	38.4	160.7
PbI$_2$........	tr	c, II	c, I	....	645		
	fus	c, I	liq	0.23	685	3.9	16.3
	sub	c, I	g	0.23	685	36.8	154.0
PbO........	tr	c, red	c, yellow	....	762	0.394	1.648
	fus	c	liq	0.35	1158	6.57	27.49
	vap	liq	g, equil.	760	1813		
PbS........	fus	c	liq	....	1382	4.2	17.6
PbSO$_4$.......	tr	c, II	c, I	....	1139	4.06	16.99
	fus	c, I	liq	....	1360	9.6	40.2
Pd..........	fus	c	liq	0.031	1825	4.20	17.56
	vap	liq	g	760	3237	85.4	357.3
PdCl$_2$.......	fus	c	liq	....	953	5	21

TABLE 4j-1. TEMPERATURES, PRESSURES, AND HEATS (*Continued*)

Substance	Process	State		P	T	ΔH	
		Initial	Final	mm Hg	K	kcal/mol	kJ/mol
Po.........	tr	c, II	c, I	....	327		
	fus	c, I	liq	....	527	3.0	12.5
	vap	liq	g	760	1235		
Pr.........	tr	c, α	c, β	....	1068	0.76	3.18
	fus	c, β	liq	....	1204	1.65	6.904
	vap	liq	g	760	3785	70.9	296.6
PrBr₃......	fus	c	liq	....	966	11.3	47.28
	sub	c	g	....	966	68.1	284.9
PrCl₃......	fus	c	liq	3.5(E − 3)	1059	12.1	50.62
	sub	c	g	3.5(E − 3)	1059	70.3	294.1
	vap	liq	g	23	1523	54.7	228.9
PrF₃.......	fus	c	liq	....	1668		
	sub	c	g	1.3(E − 3)	1400	82.3	344.3
PrI₃.......	fus	c	liq	....	1011	12.7	53.14
	sub	c	g	....	1011	66.5	278.2
Pt.........	fus	c	liq	....	2043	4.7	19.7
	vap	liq	g	760	4097	121.8	509.6
PtF₆.......	tr	c, orthorh.	c, cubic	....	276.15	2.14	8.954
	fus	c, cubic	liq	....	334.45	1.08	4.519
	vap	liq	g	760	342.29	7.06	29.54
Pu.........	tr	c, VI	c, V	....	395	0.80	3.35
	tr	c, V	c, IV	....	480	0.14	0.586
	tr	c, IV	c, III	....	588	0.13	0.544
	tr	c, III	c, II	....	730	0.02	0.084
	tr	c, II	c, I	....	753	0.44	1.84
	fus	c, I	liq	....	913	0.68	2.85
	vap	liq	g	760	3503	82.1	343.7
PuBr₃......	fus	c	liq	2.1(E − 3)	954	11.6	48.53
	vap	liq	g	2.1(E − 3)	954	57.3	239.7
PuCl₃......	fus	c	liq	1.9(E − 3)	1033	13.3	55.65
	vap	liq	g	1.9(E − 3)	1033	58.6	245.2
PuF₃.......	fus	c	liq	0.72	1698		
	sub	c	g	2.33(E − 3)	1400	93.0	389.1
PuF₄.......	sub	c	g	4.3(E − 4)	1123	45.9	192.0
	fus	c	liq	8.2(E − 3)	1310		
PuF₆.......	fus	c	liq	533.0	324.74	4.456	18.644
	vap	liq	g	760	335.31	7.03	29.41
Ra.........	fus	c	liq	....	973		
Rb.........	fus	c	liq	....	312	0.54	2.26
	vap	liq	g, equil.	760	967		
RbBr.......	fus	c	liq	....	965	5.57	23.30
	vap	liq	g	760	1625	37.1	155.2
RbCl.......	fus	c	liq	0.27	995	5.67	23.72
	vap	liq	g	760	1654	36.9	154.4
RbF........	fus	c	liq	0.6	1068	5.5	23.0
	sub	c	g	0.6	1068	52.3	218.8

TABLE 4j-1. TEMPERATURES, PRESSURES, AND HEATS (*Continued*)

Substance	Process	State		P	T	ΔH	
		Initial	Final	mm Hg	K	kcal/mol	kJ/mol
RbI.........	fus	c	liq	0.4	920	5.27	22.05
	sub	c	g	0.4	920	46.7	195.4
	vap	liq	g	760	1578	35.9	150.2
RbNO₃......	tr	c, IV	c, III	....	437	0.90	3.77
	tr	c, III	c, II	....	501		
	tr	c, II	c, I	....	564	0.88	3.68
	fus	c, I	liq	....	589	1.10	4.602
RbOH.......	tr	c, II	c, I	....	518	1.70	7.113
	fus	c, I	liq	....	656		
Re..........	fus	c	liq	0.024	3453	7.9	33.1
	vap	liq	g	760	5960	171	715.5
(ReBr₃)₃.....	sub	c	g	....	550	47.6	199.2
(ReCl₃)₃......	sub	c	g	....	550	49	205
ReF₅........	fus	c	liq	0.37	321.1		
	vap	liq	g	5.61	367	13.9	58.16
	vap	liq	g	760	494		
ReF₆........	tr	c, II	c, I	153.1	271.2	2.09	8.745
	fus	c, I	liq	426.5	291.8	1.10	4.602
	vap	liq	g	760	306.9	6.8	28.5
ReF₇........	tr	c, II	c, I	....	163		
	fus	c, I	liq	311.6	321.4	1.80	7.531
	vap	liq	g	311.6	321.4	7.35	30.75
Re₂O₇........	fus	c	liq	72	573.5	14.7	61.50
	sub	c	g	72	573.5	32	134
	vap	liq	g	760	634	16.8	70.29
Rh..........	fus	c	liq	....	2233	5.15	21.55
	vap	liq	g	760	4000	118	493.7
Rn..........	fus	c	liq	502	202	0.69	2.89
	vap	liq	g	760	211	4.0	16.7
Ru..........	fus	c	liq	....	2700	6.2	25.9
	vap	liq	g	760	4390	141	589.9
RuF₅........	fus	c	liq	5.71	379	10.4	43.51
	vap	liq	g	5.71	379	15.6	65.27
	vap	liq	g	760	500		
RuF₆........	tr	c, II	c, I	....	275.6		
	sub	c, I	g	40	281	9.1	38.1
	fus	c, I	liq	....	327		
RuO₄........	fus	c	liq	10.6	298.5	2.6	10.9
	vap	liq	g	10.6	298.5	10.6	44.35
S............	tr	c, rhomb.	c, monocl.	3.8(E − 3)	368.46	0.096	0.402
	tr	c, rhomb.	c, monocl.	....	374.15	0.0	0.0
	fus	c, monocl.	liq	....	388.33	0.411	1.711
	vap	liq	g, equil.	760	717.75	2.2	9.20
SF₄..........	fus	c	liq	0.54	152.1		
	vap	liq	g	41.7	192	6.3	26.4

TABLE 4j-1. TEMPERATURES, PRESSURES, AND HEATS (*Continued*)

Substance	Process	State Initial	State Final	P mm Hg	T K	ΔH kcal/mol	ΔH kJ/mol
SF$_6$.........	tr	c, II	c, I	....	94.26	0.384	1.607
	sub	c, I	g	760	209.5	5.70	23.85
	fus	c, I	liq	1,700	222.5	1.20	5.021
SO$_2$.........	fus	c	liq	12.56	197.69	1.769	7.4015
	vap	liq	g	760	263.13	5.955	24.916
Sb..........	fus	c	liq	....	904	4.75	.19.87
	vap	liq	g, equil.	760	1860		
SbBr$_3$.......	fus	c	liq	1.65	369.8	3.5	14.6
	vap	liq	g	760	562	12.6	52.72
SbCl$_3$.......	fus	c	liq	....	346.4	3.0	12.5
	vap	liq	g	760	494	10.80	45.187
SbCl$_5$.......	fus	c	liq	....	276.2	2.4	10.0
	vap	liq	g	30	358	11.7	48.95
SbF$_5$........	fus	c	liq	1.47	281.4		
	vap	liq	g	1.47	281.4	11.1	46.44
	vap	liq	g	760	416		
SbH$_3$........	fus	c	liq	....	179		
	vap	liq	g	760	255	5.1	21.3
SbI$_3$........	fus	c	liq	1.6	443.3		
	vap	liq	g	760	675	14.8	61.92
Sb$_4$O$_6$.......	tr	c, cubic	c, orthorh.	0.52	843	2.8	11.7
	fus	c, orthorh.	liq	2.5	928	27	113
	vap	liq	g	760	1729	17.8	74.48
Sc..........	tr	c, II	c, I	....	1608	0.96	4.02
	fus	c, I	liq	0.084	1812	3.37	14.10
	vap	liq	g	760	3104	75.1	314.2
ScBr$_3$.......	sub	c	g	162	1134	63.0	263.6
	fus	c	liq	1,530	1233		
ScCl$_3$.......	fus	c	liq	1,260	1240		
	sub	c	g	1,260	1240	63	264
ScF$_3$........	sub	c, II	g	1.8(E − 3)	1290	89	372
	tr	c, II	c, I	....	1620		
	fus	c, I	liq	....	1803		
ScI$_3$........	sub	c	g	112	1100	61	255
	fus	c	liq	....	1218		
Se..........	tr	c, II	c, I	....	398	0.18	0.753
	fus	c, I	liq	....	494	1.25	5.230
	vap	liq	g, equil.	760	958		
SeF$_4$........	fus	c	liq	1.65	263.6		
	vap	liq	g	760	380	10.0	41.84
SeF$_6$........	sub	c	g	760	226.6	6.27	26.23
	fus	c	liq	1,500	238.6	1.78	7.448
	vap	liq	g	1,500	238.6	4.30	17.99
SeO$_2$........	sub	c	g	760	629	21.1	88.28
Si..........	fus	c	liq	....	1685	21.1	50.62
	vap	liq	g, equil.	760	3540		

TABLE 4j-1. TEMPERATURES, PRESSURES, AND HEATS (Continued)

Substance	Process	State		P	T	ΔH	
		Initial	Final	mm Hg	K	kcal/mol	kJ/mol
SiBr₄........	fus	c	liq	1.83	278.0		
	vap	liq	g	760	426	9.1	38.1
Si(CH₃)₄.....	fus	c	liq	0.2	174.12	1.648	6.8952
	vap	liq	g	760	299.8	5.79	24.23
SiCl₄........	fus	c	liq	....	205	1.84	7.699
	vap	liq	g	760	330.4	6.81	28.49
SiF₄........	fus	c	liq	1340	183.0	2.27	9.498
	sub	c	g	1340	183.0	6.33	26.48
SiH₄........	tr	c, II	c, I	....	63.5	0.147	0.6150
	fus	c, I	liq	....	88.5	0.159	0.6653
	vap	liq	g	760	161.8	2.9	12.1
SiH₃F.......	vap	liq	g	760	185.1	4.3	18.0
SiO₂........	tr	quartz, III	quartz, II	....	91		
	tr	quartz, II	quartz, I	....	846	0.15	0.628
	tr	quartz, I	tridym., I	....	1140	0.12	0.502
	fus	quartz, I	liq	....	1883	2.04	8.535
	tr	tridym., IV	tridym., III	....	390	0.07	0.29
	tr	tridym., III	tridym., II	....	436	0.04	0.18
	tr	tridym., II	tridym., I	....	498	0.05	0.21
	tr	tridym., I	cristob., I	....	1743	0.05	0.21
	fus	tridym., I	liq	....	1953		
	tr	cristob., II	cristob., I	....	522	0.20	0.837
	fus	cristob., I	liq	....	2001	1.84	7.699
Sm..........	tr	c, II	c, I	....	1190	0.74	3.10
	fus	c, I	liq	3.18	1345	2.06	8.619
	vap	liq	g	760	2064	39.8	166.5
Sm₂O₃.......	tr	c, monocl.	c, cubic	....	1148		
	fus	c, cubic	liq	....	2535		
Sn..........	tr	c, white	c, grey	....	286.2	0.500	2.092
	fus	c, grey	liq	....	505.06	1.67	6.987
	vap	liq	g	760	2896	70.8	296.2
SnBr₂........	fus	c	liq	....	505	1.7	7.11
	vap	liq	g	....	911	2.2	92.1
SnBr₄........	tr	c, II	c, I	....	288.5	0.304	1.272
	fus	c, I	liq	0.66	302.5	2.80	11.71
	vap	liq	g	0.66	302.5	12.2	51.04
	vap	liq	g	760	477	10.7	44.77
SnCl₂........	fus	c	liq	....	521	3.0	12.5
	vap	liq	g	760	888	21.0	87.86
SnCl₄........	fus	c	liq	....	239.9	2.19	9.163
	vap	liq	g	760	386.8	8.5	35.5
SnF₂........	vap	liq	g, equil.	760	1126		
SnH₄........	fus	c	liq	....	123.3		
	vap	liq	g	760	220.8	4.4	18.4
SnI₂........	fus	c	liq	....	593		
	vap	liq	g	760	1000	22.4	93.72
SnI₄........	fus	c	liq	....	417	4.53	18.95
	vap	liq	g	760	621	12.4	51.88

TABLE 4j-1. TEMPERATURES, PRESSURES, AND HEATS (*Continued*)

| Substance | Process | State | | P | T | ΔH | |
		Initial	Final	mm Hg	K	kcal/mol	kJ/mol
SnS.........	tr	c, II	c, I	....	875	0.160	0.669
	fus	c, I	liq	....	1153	7.55	31.59
	vap	liq	g, equil.	760	1500		
Sr...........	tr	c, α	c, β	....	505		
	tr	c, β	c, γ	....	893		
	fus	c, γ	liq	1.8	1043		
	vap	liq	g	760	1648	33.2	138.9
SrBr₂........	tr	c, II	c, I	....	918	2.90	12.13
	fus	c, I	liq	....	930	2.50	10.46
	vap	liq	g	5.9(E − 3)	1200	58.2	243.5
SrCO₃.......	tr	c, III	c, II	....	1203	4.7	19.7
	tr	c, II	c, I	....	1689	0.8	3.3
	fus	c, I	liq	....	1770		
SrCl₂........	tr	c, II	c, I	1.07(E − 6)	1003	0.65	2.72
	sub	c, I	g	1.07(E − 6)	1003	71.5	299.2
	fus	c	liq	2.0(E − 4)	1146	3.80	15.90
	vap	liq	g	4.0(E − 3)	1245	66.0	276.1
SrF₂........	sub	c	g	3.02(E − 6)	1270	98.4	411.7
	fus	c	liq	....	1736	7.135	29.853
SrI₂........	fus	c	liq	....	811	4.70	19.66
	vap	liq	g	0.040	1200	56.8	237.7
Sr(NO₃)₂.....	fus	c	liq	....	891	12.7	5.3
SrO........	fus	c	liq	....	2688		
SrSO₄........	tr	c, II	c, I	....	1425		
	fus	c, I	liq	....	1878		
SrTiO₃.......	fus	c	liq	....	2313		
SrWO₄.......	fus	c	liq	....	1843		
Ta..........	fus	c	liq	....	3250	7.5	31.4
	vap	liq	g	760	5638	182.1	761.91
TaBr₅.......	fus	c	liq	....	528		
TaCl₅........	fus	c	liq	....	489.0	7.1	29.7
	vap	liq	g	760	506.0	12.8	53.56
Ta₂O₅........	tr	c	c, tetrag.	....	1633		
	fus	c, tetrag.	liq	....	2160		
Tb..........	tr	c, II	c, I	....	1560	1.20	5.021
	fus	c, I	liq	8.1(E − 4)	1630	2.58	10.79
	vap	liq	g	760	3496	79.1	331.0
TbCl₃........	fus	c	liq	....	855		
	vap	liq	g	0.275	1223	42.0	175.7
TbF₃........	fus	c	liq	....	1446		
Tb₂O₃.......	fus	c	liq	....	2565		
Tb₄O₇.......	fus	c	liq	....	2610		
Tc..........	sub	c	g	2.0(E − 6)	2150	164	686.2
	fus	c	liq	2.0(E − 4)	2443		

TABLE 4j-1. TEMPERATURES, PRESSURES, AND HEATS (*Continued*)

Substance	Process	State Initial	State Final	P mm Hg	T K	ΔH kcal/mol	ΔH kJ/mol
TcF$_6$........	tr	c, I	c, II	....	267.8	1.8	7.53
	fus	c, II	liq	400	310.5	1.1	4.60
	vap	liq	g	400	310.5	7.44	31.13
	vap	liq	g	760	328.4	7.22	30.21
Tc$_2$O$_7$........	sub	c	g	0.7	392.6	30.2	126.4
	vap	liq	g	0.7	392.6	18.8	78.66
Te..........	fus	c	liq	0.176	722.95	4.18	17.49
	vap	liq	g, equil.	760	1261		
TeF$_4$........	fus	c	liq	....	402.7	5.5	23.0
TeO$_2$.........	fus	c	liq	0.11	1006	6.95	29.08
	vap	liq	g	0.11	1006	53.1	222.2
Th..........	tr	c, α	c, β	....	1636	0.65	2.72
	fus	c, β	liq	....	2028	3.85	16.11
	vap	liq	g	760	5061	123.0	514.63
ThCl$_4$.......	tr	c, α	c, β	....	679	1.20	5.021
	fus	c, β	liq	....	1042	14.69	61.463
ThF$_4$........	fus	c	liq	0.52	1375	4	17
	vap	liq	g	0.52	1375	71.3	298.3
ThI$_4$.........	sub	c	g	7.2(E − 4)	623	36.1	151.0
ThO$_2$........	sub	c	g	1.8(E − 4)	2400	162	677.8
	fus	c	liq	....	3490		
Ti...........	tr	c, α	c, β	....	1167	0.99	4.15
	fus	c, β	liq	4.4(E − 3)	1943	3.7	15.5
	vap	liq	g	760	3562	100.6	420.91
TiBr$_4$........	fus	c	liq	0.411	311.4	3.08	12.89
	vap	liq	g	0.411	311.4	13.10	54.810
	vap	liq	g	760	506.6	10.60	44.350
TiCl$_4$........	fus	c	liq	....	249.9	2.23	9.330
	vap	liq	g	....	249.9	10.34	43.263
	vap	liq	g	760	410.6	8.15	34.10
TiF$_4$.........	sub	c	g	760	456.3	21.6	90.37
TiI$_4$.........	tr	c, α	c, β	....	379	2.37	9.916
	fus	c, β	liq	....	428	4.68	19.58
	vap	liq	g	760	650	13.98	58.492
TiO.........	tr	c, II	c, I	....	1264	0.82	3.43
TiO$_2$.........	fus	c	liq	....	2113	11	46.0
Tl...........	tr	c, α	c, β	....	507	0.09	0.38
	fus	c, β	liq	....	577	0.98	4.10
	vap	liq	g	760	1760	39.4	164.8
TlBr.........	sub	c	g	1.9	733	30.5	127.6
	fus	c	liq	1.9	733	3.92	16.40
	vap	liq	g, equil.	760	1092	23.9	100.0
TlCl.........	fus	c	liq	....	704	3.72	15.56
	vap	liq	g, equil.	760	1093	24	100

TABLE 4j-1. TEMPERATURES, PRESSURES, AND HEATS (*Continued*)

Substance	Process	State Initial	State Final	P mm Hg	T K	ΔH kcal/mol	ΔH kJ/mol
TlF.........	fus	c	liq	....	595.4	3.315	13.870
	vap	liq	g, equil.	760	1099		
TlI.........	tr	c, II	c, I	....	451	0.22	0.92
	fus	c, I	liq	1.0	715	3.52	14.73
	vap	liq	g	1.0	715	27.3	114.2
	vap	liq	g, equil.	760	1099		
TlNO₂.......	tr	c, II	c, I	....	416	0.91	3.81
	fus	c, I	liq	....	479.8	2.264	9.473
Tl₂O........	fus	c	liq	....	852	7.24	30.29
Tl₂O₃.......	fus	c	liq	....	998	3	12.5
Tm.........	fus	c	liq	....	1818	4.02	16.82
	vap	liq	g	760	2220	45.6	190.8
TmCl₃.......	fus	c	liq	....	1103		
	vap	liq	g	0.554	1173	77.5	324.3
	vap	liq	g	760	1763		
TmF₃.......	tr	c, α	c, β	....	1316		
	fus	c, β	liq	2.7(E − 3)	1431		
	sub	c, β	g	2.7(E − 3)	1431	88.9	372.0
Tm₂O₃.......	fus	c	liq	....	2665		
U...........	tr	c, α	c, β	....	941	0.667	2.791
	tr	c, β	c, γ	....	1048	1.137	4.7572
	fus	c, γ	liq	....	1405	2.036	8.5186
	vap	liq	g	760	4407	110.9	464.01
UBr₃........	tr	c	liq	0.013	1003	15	62.8
UBr₄........	fus	c	liq	5.7	792	16	66.9
	vap	liq	g	5.7	792	33.9	141.8
	vap	liq	g	760	1039	30.5	127.6
UCl₄........	tr	c, II	c, I	....	820		
	fus	c, I	liq	32.6	863	11	46.0
	vap	liq	g	760	1075	20.4	85.35
UCl₆........	sub	c	g	1.8	370	17.3	72.38
UF₄.........	tr	c	c	....	1110	3.4	14.2
	fus	c	liq	7.03	1330	10.24	42.844
	vap	liq	g	7.03	1330	57.1	238.9
UF₅.........	tr	c, II	c, I	....	408		
	fus	c, I	liq	13.4	621	11.1	46.44
	vap	liq	g	13.4	621	25.1	105.0
	vap	liq	g	760	776	23.2	97.07
UF₆.........	sub	c	g	760	329.7	11.5	48.12
	sub	c	g	1,138	337.2	11.4	47.70
	vap	liq	g	1,138	337.2	6.9	28.9
UI₄.........	fus	c	liq	4.5	779	19.3	80.75
UO₂........	fus	c	liq	....	3115	18.2	76.15
V..........	fus	c	liq	2.0(E − 3)	2175	5.00	20.92
	vap	liq	g	760	3682	108.0	451.87

TABLE 4j-1. TEMPERATURES, PRESSURES, AND HEATS (*Continued*)

Substance	Process	State		P	T	ΔH	
		Initial	Final	mm Hg	K	kcal/mol	kJ/mol
VCl_4.........	fus	c	liq	....	252.6	2.3	9.62
	vap	liq	g	760	426	9.5	39.7
$VOCl_3$.......	fus	c	liq	....	196	2.29	9.581
	vap	liq	g	760	400	8.45	35.35
V_2O_5.........	fus	c	liq	....	947	15.6	65.27
W...........	fus	c	liq	0.039	3653	8.46	35.40
	vap	liq	g	760	5828	197.0	824.25
WBr_5........	fus	c	liq	....	568	4	17
	vap	liq	g	760	665	13.9	58.16
$WOBr_4$......	fus	c	liq	640	595.5	14	58.6
	vap	liq	g	640	595.5	13.4	56.07
	vap	liq	g	760	604.5	13.2	55.23
WCl_6........	tr	c, III	c, II	....	458		
	tr	c, II	c, I	....	503.1	3.39	14.18
	fus	c, I	liq	213	554.6	1.6	6.69
	vap	liq	g	213	554.6	15.0	62.76
WF_6........	tr	c, II	c, I	240	264.9	1.0	4.18
	sub	c, II	g	240	264.9	8.8	36.8
	fus	c, I	liq	413	275.1	1.3	5.44
	vap	liq	g	413	275.1	7.70	32.22
	vap	liq	g	760	290.2	6.25	26.15
WOF_4.......	fus	c	liq	25.1	377.8	1.4	5.86
	vap	liq	g	25.1	377.8	14.8	61.92
	vap	liq	g	760	459.0	13.8	57.74
WO_3.........	tr	c, α	c, β	....	1050	0.410	1.715
	fus	c, β	liq	....	1745	17.55	73.43
Xe..........	fus	c	liq	611	161.36	0.548	2.293
	vap	liq	g	760	165.03	3.021	12.640
XeF_2........	fus	c	liq	1412	402.2		
	sub	c	g	1412	402.2	13.0	54.39
XeF_4........	fus	c	liq	811.3	390.25		
	sub	c	g	811.3	390.25	14.40	60.250
XeF_6........	fus	c	liq	159	319		
	sub	c	g	159	319	15.5	64.85
Y...........	tr	c, α	c, β	....	1752	1.193	4.9915
	fus	c, β	liq	2.2(E − 3)	1799	2.724	11.397
	vap	liq	g	760	3611	86.8	363.2
YCl_3........	fus	c	liq	....	973		
	vap	liq	g	....	1100	30.9	129.3
YF_3........	tr	c, II	c, I	....	1325		
	sub	c, I	g	....	1325	100	418.4
	fus	c, I	liq	....	1420		
YI_3..........	sub	c	g	....	890	53.6	224.3
	fus	c	liq	....	1237		
Y_2O_3........	fus	c, cubic	liq	....	2556		

TABLE 4j-1. TEMPERATURES, PRESSURES, AND HEATS (*Continued*)

Substance	Process	State Initial	State Final	P mm Hg	T K	ΔH kcal/mol	ΔH kJ/mol
Yb.........	tr	c, α	c, β	....	1033	0.418	1.749
	fus	c, β	liq	19.8	1097	1.83	7.657
	vap	liq	g	760	1467	30.8	128.9
YbCl$_2$.......	fus	c	liq	....	981		
	vap	liq	g	1.41	1573	59.8	250.2
YbF$_3$.......	sub	c	g	....	1362	85.5	357.7
Yb$_2$O$_3$.......	fus	c, cubic	liq	....	2645		
Zn.........	fus	c	liq	0.15	692.65	1.765	7.3848
	vap	liq	g	760	1184	27.62	115.56
ZnBr$_2$.......	fus	c	liq	....	675.2	3.74	15.65
	vap	liq	g, equil.	760	928.6		
ZnCl$_2$........	fus	c	liq	0.021	590	2.45	10.25
	vap	liq	g, equil.	760	989.4		
ZnO........	fus	c ·	liq	....	2248		
ZnSO$_4$.......	tr	c, α	c, β	....	1007	4.8	20.1
Zr..........	tr	c, α	c, β	1.8(E − 18)	1136	0.94	3.93
	fus	c, β	liq	1.2(E − 5)	2125	4.0	16.9
	vap	liq	g	760	4682	139	581.6
	sub	c, α	g	1.8(E − 18)	1136	144.7	605.42
ZrBr$_4$........	sub	c	g	40	550	27.2	113.8
ZrC........	fus	c	liq	....	3765		
ZrCl$_2$........	fus	c	liq	....	995		
ZrCl$_4$........	sub	c	g	760	605	24.4	102.1
	fus	c	liq	15,800	710	6.9	28.9
	vap	liq	g	15,800	710	16.8	70.29
ZrF$_4$.........	tr	c, α	c, β	....	678		
	sub	c, β	g	2.1(E − 3)	800	53.0	221.8
	sub	c, β	g	760	1181	50.4	210.9
	fus	c, β	liq	819	1185		
ZrI$_4$........	sub	c	g	1.3(E − 3)	425	26.2	109.6
ZrN........	fus	c	liq	....	3225		
ZrO$_2$........	tr	c, II	c, I	....	1473	1.42	5.941
	sub	c, I	g	3(E − 4)	2400	165	690.4
	fus	c, I	liq	....	2979	20.8	87.0

TABLE 4j-2. SELECTED REFERENCES*

Substance	Reference	Substance	Reference
Ac	164	BeF_2	130, 157, 343, 384
		BeO	8, 103, 131, 182, 363
Ag	164	$BeSO_4$	20
AgBr	31, 33, 411		
AgCN	294	Bi	164
AgCl	31, 33, 205, 209	$BiBr_3$	76, 305, 391, 410
AgF	417	$BiCl_3$	81, 83, 173, 246, 305, 390, 399
AgI	15, 24, 178, 223, 239, 271, 289	BiF_3	81
$AgNO_3$	5, 84, 90, 169, 170, 204, 308, 318	Bi_2O_3	76, 119, 222, 370
Ag_2S	321, 383	Bi_2S_3	71, 127
Ag_2SO_4	152		
Ag_2Se	10, 275, 321, 383, 412	Br_2	129
		BrF_3	276
Al	164	BrF_5	225, 315
Al_2Br_6	97, 175, 387		
Al_2Cl_6	112, 267	C	164
AlF_3	46, 93, 104, 215	CBr_4	320
AlI_3	387	CCl_4	320
Al_2O_3	48, 57, 125, 183, 270, 324, 336	CF_4	320
$AlPO_4$	329	CH_4	320
		CH_3Br	320
Am	164	CH_3Cl	320
		CH_3F	108
Ar	129	CH_3I	253
		CH_3OH	110, 230, 359, 398
As	164	CH_2Cl_2	320
$AsCl_3$	212, 266	CH_2F_2	240
AsF_3	384	CH_2I_2	320
AsF_5	320	CH_2O	320
AsF_3O	249	$CHBr_3$	320
AsH_3	352, 385	$CHCl_3$	219, 312
AsI_3	75, 120	CHF_3	162
As_4O_6	374	CO	320
		CO_2	320
Au	164	$COBr_2$	320
		$COCl_2$	124
B	164, 196	COF_2	284
BBr_3	14, 160	CS_2	320
$B(CH_3)_3$	117	COS	320
BCl_3	4, 133		
BF_3	214	Ca	164
B_2H_6	287, 407	CaB_2O_4	200
B_5H_9	143, 176, 408	$Ca_2B_2O_5$	200
B_2O_3	30, 262, 313, 333, 368	$CaBr_2$	100, 165, 171
		CaC_2	320
Ba	164	$CaCO_3$	306
$BaBr_2$	100, 165, 171	$CaCl_2$	58, 100, 156
$BaCO_3$	11, 220, 307	CaF_2	37, 86, 301, 340
$BaCl_2$	100, 217, 273, 171	CaO	9, 270, 335
BaF_2	19, 150, 292, 293, 301, 317	$CaSO_4$	135, 144, 394
BaI_2	100, 165	$CaSiO_3$	126
$Ba(NO_3)_2$	203	Ca_2SiO_4	41, 70
BaO	166, 260	$CaTiO_3$	66, 186
$BaTiO_3$	105, 354, 372, 389		
		Cd	164
Be	164	$CdBr_2$	33, 391
$BeCl_2$	111, 116, 132, 154, 208, 231	$CdCl_2$	35, 36, 391
		CdF_2	28
		CdI_2	33, 391

* Numbers in Reference column refer to items in the list that follows this table.

TABLE 4j-2. SELECTED REFERENCES (*Continued*)

Substance	Reference	Substance	Reference
Ce	164	$Fe_{0.947}O$	388
CeO_2	218, 254	Fe_2O_3	67
Ce_2O_3	254	Fe_3O_4	67
		FeS	68
Cl_2	129		
ClF	320	Ga	164
ClF_3	137, 232	$(GaCl_3)_2$	133
ClO_2	139	GaI_3	311, 362
Co	164	Gd	164
$CoCl_2$	325	$GdBr_3$	101
CoF_2	29. 181	$GdCl_3$	101
CoO	320	GdF_3	301, 369
		Gd_2O_3	254, 401
Cr	164		
$CrBr_3$	357	Ge	164
$Cr(CO)_6$	65, 309	$GeBr_4$	320
CrF_3	149, 416	$GeCl_4$	12
Cr_2O_3	335	GeF_4	320
		GeH_4	320
Cs	164	GeI_4	177
CsBr	99, 332, 365	GeO_2	236, 238, 261
CsCl	99, 339, 365		
CsF	99, 332	H_2	129
CsI	99, 332, 335	HBr	129
$CsNO_3$	257	HCN	62
CsOH	319	HCl	129
Cs_2SO_4	22, 291	HF	129
		HI	129
Cu	164	HNO_3	98
$(CuBr)_3$	151, 351	H_2O	320
$(CuCl)_3$	234, 351	H_2S	320
CuF_2	145, 190	H_2SO_4	129
$(CuI)_3$	250, 351	H_2Se	129
Cu_2O	237	H_2Te	320
Cu_2S	163, 310	H_3PO_4	129
		$^1H^2H$ ·	129
Dy	164	$^1H^2HO$	129
		^2H_2O	129
Er	164		
$ErCl_3$	101, 255, 274	He	129
ErF_3	369		
		Hf	164
Eu	164	$HfBr_4$	331
$EuCl_3$	255, 298	$HfCl_4$	268, 285
Eu_2O_3	254, 334, 401	HfF_4	111
		HfI_4	376
F_2	129	HfO_2	270
F_2O	337		
		Hg	164
Fe	164	$HgBr_2$	168
$FeBr_2$	233	$HgCl_2$	77, 174, 391
$Fe(CO)_5$	227	HgF_2	320
$FeCl_2$	328	HgI_2	120, 234
$(FeCl_3)_2$	243	HgS	320
FeF_2	56, 192		
FeF_3	416	Ho	164
FeI_2	326	$HoCl_3$	101, 255
		HoF_3	27
		Ho_2O_3	254

TABLE 4j-2. SELECTED REFERENCES (*Continued*)

Substance	Reference	Substance	Reference
I_2	129	Mn	164
ICl	52	$MnBr_2$	320
IF_5	316	$MnCl_2$	235, 325
IF_7	47	MnF_2	136, 191
In	164	MnI_2	320
$InBr_3$	362	MnO	360
InCl	106, 362	Mn_3O_4	146
$InCl_3$	362		
InI_3	109, 362	Mo	164
In_2O_3	334	$Mo(CO)_6$	320
		MoF_3	51
Ir	164	MoF_6	282
IrF_6	50	MoO_3	140, 201
K	164	N_2	320
KBr	34, 99	NH_3	320
KCN	320	N_2H_4	129
KCl	18, 35, 339, 392	NH_4Br	320
KF	292, 304	NH_4Cl	129
KI	35, 99	NH_4F	129
KNO_3	204, 367	NH_4I	320
KOH	320	NH_4NO_3	129
K_2SO_4	320	NO	320
		N_2O	320
Kr	129		
KrF_2	141, 142	Na	164
KrF_4	138	NaBr	99, 118
		NaCN	320
La	164	NaCl	85, 99
$LaBr_3$	101, 353	NaF	113, 301, 304
$LaCl_3$	100, 353	NaI	99, 118
LaF_3	244, 301	Na_2MoO_4	320
LaI_3	353	$NaNO_3$	114, 204, 258
La_2O_3	254	NaOH	92
		Na_2SO_4	69, 300
Li	164	Na_2TiO_3	320
LiBr	99		
LiCl	99, 314	Nb	164
LiF	91, 301	$NbCl_5$	3, 189, 267
LiI	99	NbF_5	40, 107
$LiNO_3$	114, 204	NbO_2	199, 347
LiOH	355	Nb_2O_5	121, 281
Li_2SO_4	290, 397		
		Nd	164
Lu	164	$NdBr_3$	353, 101
$LuCl_3$	255	$NdCl_3$	100, 272, 353
LuF_3	369, 415	NdF_3	369, 418
Lu_2O_3	254	NdI_3	353, 100
		Nd_2O_3	254, 286
Mg	164		
$MgBr_2$	26	Ne	129
$MgCl_2$	155, 339		
MgF_2	155, 317	Ni	164
MgI_2	26	$NiBr_2$	229, 327
Mg_3N_2	320	$Ni(CO)_4$	371
MgO	335	$NiCl_2$	49, 229, 325
$MgSO_4$	320	NiF_2	55, 102
		NiO	197

Table 4j-2. Selected References (*Continued*)

Substance	Reference	Substance	Reference
Np	164	RbI	42, 99
NpF_6	283, 405	$RbNO_3$	6, 115, 204
		RbOH	38
O_2	129		
O_3	129	Re	164
		$(ReBr_3)_3$	45
Os	54, 164, 395	$(ReCl_3)_3$	45
OsF_5	51	ReF_5	51
OsF_6	50	ReF_6	50, 241
$OsOF_5$	17	ReF_7	241
		Re_2O_7	128, 366
P_4	129		
PBr_3	288	Rh	164
PCl_3	288		
PCl_5	288	Rn	129
PF_3	288		
PF_5	129	Ru	164
PH_2	129	RuF_5	159
P_4O_6	129	RuF_6	61
P_4O_{10}	129	RuO_4	264
Pb	164		
$PbBr_2$	32, 33	S	129
$Pb(CH_3)_4$	373	SF_4	43
$PbCl_2$	13, 18, 35, 251	SF_6	259
PbF_2	13	SO_2	320
PbI_2	32, 96, 252		
PbO	207	Sb	164
PbS	358	$SbBr_3$	78, 356
$PbSO_4$	320	$SbCl_3$	266
		$SbCl_5$	279
Pd	164	SbF_5	158
$PdCl_2$	21, 280	SbH_3	25
		SbI_3	120, 356
Po	129	Sb_4O_6	288
Pr	164	Sc	164
$PrBr_3$	101, 353	$ScBr_3$	320
$PrCl_3$	100, 299, 353	$ScCl_3$	64
PrF_3	369, 379	ScF_3	193, 213
PrI_3	100, 353	ScI_3	320
Pt	164	Se	288
PtF_6	403	SeF_4	79, 129
		SeF_6	129
Pu	164	SeO_2	245
$PuBr_3$	296		
$PuCl_3$	296	Si	164
PuF_3	296	$SiBr_4$	44, 322
PuF_4	296	$Si(CH_3)_4$	382
PuF_6	296	$SiCl_4$	288
		SiF_4	288
Ra	320	SiH_4	320
		SiF_3H	378
Rb	164	SiO_2	320
RbBr	99		
RbCl	99, 392	Sm	164
RbF	99, 304, 344, 365	Sm_2O_3	254, 270

TABLE 4-2. SELECTED REFERENCES (Continued)

Substance	Reference	Substance	Reference
Sn	164	Tm	164
$SnBr_2$	320	$TmCl_3$	255
$SnBr_4$	185	TmF_3	369, 415
$SnCl_2$	122	Tm_2O_3	254
$SnCl_4$	265, 277		
SnF_2	111, 414	U	164
SnH_4	320	UBr_3	134
SnI_2	184	UBr_4	134
SnI_4	180		
SnS	63, 206	UCl_4	134, 194, 350
		UCl_6	172
Sr	164	UF_4	194, 198, 221
$SrBr_2$	100, 165	UF_5	2, 409
$SrCO_3$	11	UF_6	195, 393
$SrCl_2$	100, 171, 224, 273	UI_4	134
SrF_2	19, 293, 301	UO_2	153
SrI_2	100, 165		
$Sr(NO_3)_2$	203	V	164
SrO	320	VCl_4	277
$SrSO_4$	320	$VOCl_3$	277, 278
$SrTiO_3$	95	V_2O_5	161, 210
$SrWO_4$	361		
		W	164, 380
Ta	164	WBr_5	346
$TaBr_5$	23	$WOBr_4$	211
$TaCl_5$	3, 263, 345	WCl_6	349, 375, 406
Ta_2O_5	400	WF_6	50
		WF_4O	51
Tb	164	WO_3	201
$TbCl_3$	255		
TbF_3	369	Xe	248
Tb_2O_3	254	XeF_2	338
Tb_4O_7	254	XeF_4	338
		XeF_6	242, 404
Tc	216		
TcF_6	341	Y	164
Tc_2O_7	364	YCl_3	94, 255
		YF_3	193, 301, 386
Te	164	YI_3	89
TeF_4	179	Y_2O_3	270
TeO_2	247, 302, 303, 413		
		Yb	39, 164
Th	164	$YbCl_2$	297
$ThCl_4$	58	YbF_3	415
ThF_4	80, 301	Yb_2O_3	254
ThI_4	123		
ThO_2	1, 82	Zn	164
		$ZnBr_2$	74, 187
Ti	164, 396	$ZnCl_2$	74, 122, 187
$TiBr_4$	147, 185, 322	ZnO	320
$TiCl_4$	226, 256, 277, 402	$ZnSO_4$	167
TiF_4	148		
TiI_4	202	Zr	164
TiO	320	$ZrBr_4$	330
TiO_2	323	ZrC	320
		$ZrCl_2$	381
Tl	164	$ZrCl_4$	87, 88, 269, 285
TlBr	16, 205, 419	ZrF_4	53, 59, 111, 342
TlCl	16, 72, 205, 420	ZrI_4	123
TlF	188, 419	ZrN	320
TlI	16, 73, 419	ZrO_2	60, 228, 270
$TlNO_3$	7, 204		
Tl_2O	76		
Tl_2O_3	348		

References

1. Ackerman, R. J., R. J. Thorn, and P. W. Gilles: *J. Am. Chem. Soc.* **78**, 1767 (1956).
2. Agron, P. A.: *U.S. AEC Rept.* TID 5290, 1958.
3. Ainscough, J. B., R. J. W. Holt, and F. W. Trowse: *J. Chem. Soc.* **1957**, 1034.
4. Apple, E. F., and T. Wartik: *J. Am. Chem. Soc.* **80**, 6158 (1958).
5. Arell, A.: *Ann. Acad. Sci. Fennicae*, Ser. A, VI, **100** (1962).
6. Arell, A., and M. Varteva: *Ann. Acad. Sci. Fennicae*, Ser. A, VI, **88** (1961).
7. Arell, A., and M. Varteva: *Ann. Acad. Sci. Fennicae*, Ser. A, VI, **98** (1962).
8. Austerman, S. B.: *U.S. AEC Rept.* NAA-SR-7654, 1963.
9. Babeliowsky, T. P. J. H.: *J. Chem. Phys.* **38**, 2035 (1963).
10. Baer, Y., G. Busch, C. Frölich, and E. Steigmeier: *Z. Naturforsch.* **17a**, 886 (1962).
11. Baker, E. H.: *J. Chem Soc.* **1962**, 2525.
12. Balk, P., and D. Dong: *J. Phys. Chem.* **68**, 960 (1964).
13. Banashek, E. I., N. N. Patsakova, and I. S. Rassonskaya: *Izvest. Sektora Fiz.-Khim. Anal. Akad. Nauk S.S.S.R.* **27**, 223 (1956).
14. Barber, W. F., C. F. Boynton, and P. E. Gallagher: *PB Rept.* 148374, 1959.
15. Barrall, E. M., and L. B. Rogers: *Anal. Chem.* **36**, 1405 (1964).
16. Barrow, R. F., E. A. N. S. Jeffries, and M. Swinstead: *Trans. Faraday Soc.* **51**, 1650 (1955).
17. Bartlett, N., and N. K. Jha: *J. Chem. Soc.* **A1968**, 536.
18. Barton, J. L., and H. Bloom: *J. Phys. Chem.* **60**, 1413 (1956).
19. Bautista, R. G., and J. L. Margrave: *J. Phys. Chem.* **69**, 1770 (1965).
20. Bear, I. J., and A. G. Turnbull: *Australian J. Chem.* **19**, 751 (1966).
21. Bell, W. E., V. Merten, and M. Tagami: *J. Phys. Chem.* **65**, 510 (1961).
22. Belyaev, I. U., and N. N. Chikova: *Zhur. Neorg. Khim.* **8**, 1442 (1963).
23. Berdonosov, S. S., A. V. Lapitskii, and E. K. Bakov: *Zhur. Neorg. Khim.* **10**, 322 (1965).
24. Berger, C., M. Richard, and L. Eyrand: *Bull. soc. chim. France* **5**, 1491 (1965).
25. Berka, L., T. Briggs, M. Millard, and W. L. Jolly: *J. Inorg. Nuclear Chem.* **14**, 190 (1960).
26. Berkowitz, J., and J. R. Marquart: *J. Chem. Phys.* **37**, 1853 (1962).
27. Besenbruch, G., T. V. Charles, K. F. Zmbov, and J. L. Margrave: *J. Less-Common Metals* **12**, 335 (1967).
28. Besenbruch, G., A. S. Kana'an, and J. L. Margrave: *J. Phys. Chem.* **69**, 3174 (1965).
29. Binford, J. S., J. M. Strohmeyer, and T. H. Herbert: *J. Phys. Chem.* **71**, 2404 (1967).
30. Blackburn, P. E., and A. Büchler: *J. Phys. Chem.* **69**, 4250 (1965).
31. Blanc, M.: *Compt. rend.* **247**, 273 (1958).
32. Blanc, M., and G. Petet: *Compt. rend.* **248**, 1305 (1959).
33. Bloom, H., J. O'M. Bockris, N. E. Richards, and R. G. Taylor: *J. Am. Chem. Soc.* **80**, 2044 (1958).
34. Bloom, H., and J. W. Hastie: *Australian J. Chem.* **21**, 583 (1968).
35. Bloom, H., and S. B. Tricklebank: *Australian J. Chem.* **19**, 187 (1966).
36. Bloom, H., and B. J. Welsh: *J. Phys. Chem.* **62**, 1594 (1958).
37. Blue, G. D., J. W. Green, R. G. Bautista, and J. L. Margrave: *J. Phys. Chem.* **67**, 877 (1963).
38. Bogart, D.: *J. Phys. Chem.* **58**, 1168 (1954).
39. Bohdansky, J., and H. E. J. Schins: *J. Less-Common Metals* **12**, 248 (1967).
40. Brady, A. P., O. E. Myers, and J. K. Clauss: *J. Phys. Chem.* **64**, 588 (1960).
41. Bredig, M. A.: *J. Am. Ceram. Soc.* **33**, 188 (1950).
42. Bridgers, H. E.: Thesis, Ohio State University, 1953.
43. Brown, F., and P. L. Robinson: *J. Chem. Soc.* **1955**, 3147.
44. Brown, H. C., and W. J. Wallace: *J. Am. Chem. Soc.* **75**, 6279 (1953).
45. Büchler, A., P. E. Blackburn, and J. L. Stauffer: *J. Phys. Chem.* **70**, 685 (1966).
46. Büchler, A., E. P. Marram, and J. L. Stauffer; *J. Phys. Chem.* **71**, 4139 (1967).
47. Burbank, R. D., and F. R. Bensey: *J. Chem. Phys.* **27**, 981 (1957).
48. Burns, R. P.: *J. Chem. Phys.* **44**, 3307 (1966).
49. Busey, R. H., and W. F. Giauque: *J. Am. Chem. Soc.* **74**, 4443 (1952).
50. Cady, G. H., and G. B. Hargreaves: *J. Chem. Soc.* **1961**, 1563.
51. Cady, G. H., and G. B. Hargreaves: *J. Chem. Soc.* **1961**, 1568.
52. Calder, G. V., and W. F. Giauque: *J. Phys. Chem.* **69**, 2443 (1965).
53. Cantor, S., R. F. Newton, W. R. Grimes, and F. F. Blankenship: *J. Phys. Chem.* **62**, 96 (1958).
54. Carrera, N. J., R. F. Walker, and E. R. Plante: *J. Research NBS* **68A**, 325 (1964).
55. Catalano, E., and J. W. Stout: *J. Chem. Phys.* **23**, 1284 (1955).
56. Catalano, E., and J. W. Stout: *J. Chem. Phys.* **23**, 1803 (1955).

57. Chekovskoi, V. Ya., and V. A. Petrov: *Izmeritelnaya Tekh.* **1963**, 26.
58. Chiotti, P., G. J. Gartner, E. R. Stevens, and Y. Saito: *J. Chem. Eng. Data* **11**, 571 (1966).
59. Chretien, A., and B. Gaudreau: *Compt. rend.* **246**, 2266 (1958).
60. Chupka, W. A., J. Berkowitz, and M. G. Inghram: *J. Chem. Phys.* **26**, 1207 (1957).
61. Claasen, H. H., H. Selig, J. G. Malm, C. L. Chernick, and B. Weinstock: *J. Am. Chem. Soc.* **83**, 2390 (1961).
62. Coates, J. E., and R. H. Davies: *J. Chem. Soc.* **1950**, 1194.
63. Colin, R., and J. Drowart: *J. Chem. Phys.* **37**, 1120 (1962).
64. Corbett, J. D., and B. N. Ramsey: *Inorg. Chem.* **4**, 260 (1965).
65. Cordes, J. F., and S. Schreiner: *Z. anorg. allgem. Chem.* **299**, 87 (1959).
66. Coughanour, L. W., R. S. Roth, and V. A. DeProsse: *J. Research NBS* **52**, 37 (1954).
67. Coughlin, J. P.: *U.S. Bur. Mines Bull.* **542** (1954).
68. Coughlin, J. P.: *J. Am. Chem. Soc.* **72**, 5445 (1950).
69. Coughlin, J. P.: *J. Am. Chem. Soc.* **77**, 868 (1955).
70. Coughlin, J. P., and C. J. O'Brien: *J. Phys. Chem.* **61**, 767 (1957).
71. Cubicciotti, D.: *J. Phys. Chem.* **66**, 1205 (1962).
72. Cubicciotti, D.: *J. Phys. Chem.* **68**, 1528 (1964).
73. Cubicciotti, D.: *J. Phys. Chem.* **69**, 1410 (1965).
74. Cubicciotti, D., and H. Eding: *J. Chem. Phys.* **40**, 978 (1964).
75. Cubicciotti, D., and H. Eding: *J. Phys. Chem.* **69**, 2743 (1965).
76. Cubicciotti, D., and H. Eding: *J. Chem. Eng. Data* **12**, 548 (1967).
77. Cubicciotti, D., H. Eding, and J. W. Johnson: *J. Phys. Chem.* **70**, 2989 (1966).
78. Cushen, D. W., and R. Hulme: *J. Chem. Soc.* **1964**, 4162.
79. Dagron, C.: *Compt. rend.* **255**, 122 (1962).
80. Darnell, A. J., and F. J. Keneshea, Jr.: *J. Phys. Chem.* **62**, 1143 (1958).
81. Darnell, A. J., and W. A. McCollum: *J. Phys. Chem.* **72**, 1327 (1968).
82. Darnell, A. J., and W. A. McCollum: *U.S. AEC Rept.* NAA-SR-6498, 1961.
83. Darnell, A. J., and S. J. Yosim: *U.S. AEC Rept.* NAA-SR-3827, 1959.
84. Davis, W. J., S. E. Rogers, and A. R. Ubbelohde: *Proc. Roy. Soc. (London)*, ser. A, **220**, 14 (1953).
85. Dawson, R., E. B. Brackett, and T. E. Brackett: *J. Phys. Chem.* **67**, 1669 (1963).
86. Delbove, F.: *Compt. rend.* **252**, 2192 (1961).
87. Denisova, N. D., E. K. Safronov, and O. N. Bystrova: *Zhur. Neorg. Khim.* **11**, 2185 (1966).
88. Denisova, N. D., E. K. Safronov, A. I. Pustilnik, and O. N. Bystrova: *Zhur. Fiz. Khim.* **41**, 59 (1967).
89. Dennison, D. H., F. H. Spedding, and A. A. Daane: *U.S. AEC Rept.* IS-57, 1959.
90. Doucet, Y., J. A. LeDuc, and G. Pannetier: *Compt. rend.* **236**, 1018 (1953).
91. Douglas, T. B., and J. L. Dever: *J. Am. Chem. Soc.* **76**, 4826 (1954).
92. Douglas, T. B., and J. L. Dever: *J. Research NBS* **53**, 81 (1954).
93. Douglas, T. B., and D. A. Ditmars: *J. Research NBS* **71A**, 185 (1967).
94. Drobst, D. V., G. P. Ankina, L. V. Dusinina, and B. G. Korshunov: *Zhur. Neorg. Khim.* **10**, 562 (1965).
95. Drys, M., and W. Trzebiatowski: *Roczniki Chem.* **31**, 489 (1957).
96. Duncan, J. F., and F. G. Thomas: *J. Chem. Soc.* **1964**, 360.
97. Dunne, T. G., and N. W. Gregory: *J. Am. Chem. Soc.* **80**, 1526 (1958).
98. Dunning, W. J., and C. W. Nutt: *Trans. Faraday Soc.* **47**, 15 (1951).
99. Dworkin, A. S., and M. A. Bredig: *J. Phys. Chem.* **64**, 269 (1960).
100. Dworkin, A. S., and M. A. Bredig: *J. Phys. Chem.* **67**, 697 (1963).
101. Dworkin, A. S., and M. A. Bredig: *J. Phys. Chem.* **67**, 2499 (1963).
102. Ehlert, T. C., R. A. Kent, and J. L. Margrave: *J. Am. Chem. Soc.* **86**, 5093 (1964).
103. Engberg, C. J., and E. H. Zehms: *J. Am. Ceram. Soc.* **42**, 300 (1959).
104. Erokhin, E. V., N. A. Zhegulskaya, L. N. Sidorov, and N. A. Akishin: *Izvest. Akad. Nauk S.S.S.R., Neorg. Mater.* **3**, 873 (1967).
105. Eru, V.: *J. Am. Ceram. Soc.* **46**, 295 (1963).
106. Fadeev, V. N., and P. I. Fedorov: *Zhur. Neorg. Khim.* **9**, 381 (1964).
107. Fairbrother, F., and W. C. Firth: *J. Chem. Soc.* **1951**, 3051.
108. Farhat-Aziz and E. A. Moelwyn-Hughes: *J. Chem. Soc.* **1961**, 1523.
109. Fedorov, P. I., A. G. Dudareva, and N. G. Drobst: *Zhur. Neorg. Khim.* **8**, 1287 (1963).
110. Feher, F., and G. Hitzemann: *Z. anorg. allgem. Chem.* **294**, 50 (1958).
111. Fischer, W., T. Petzel, and S. Lauter: *Z. anorg. allgem. Chem.* **333**, 226 (1964).
112. Foster, L. M.: *J. Am. Chem. Soc.* **72**, 1902 (1950).
113. Frank, W. B.: *J. Phys. Chem.* **65**, 2081 (1961).
114. Franzosini, P., and C. Sinistri: *Ric. Sci. Rend.* **3A**, 411 (1963).
115. Freeman, E. S., and D. A. Anderson: *Nature* **199**, 63 (1963).

116. Furby, E., and K. L. Wilkinson: *J. Inorg. & Nuclear Chem.* **14**, 123 (1960).
117. Furukawa, G. T., and R. P. Park: *NBS Rept.* 3649, 1954.
118. Gardner, T. E., and A. R. Taylor: *U.S. Bur. Mines Rept. Invest.* 7040, 1967.
119. Gattow, G., and D. Schnetze: *Z. anorg. allgem. Chem.* **328**, 44 (1964).
120. Gäumann, A.: *Helv. Chim. Acta* **51**, 543 (1968).
121. Geld, P. V., and F. G. Kasenko: *Izvest. Akad. Nauk S.S.S.R., Otdel. Tekh. Metallurg.* **2**, 79 (1960).
122. George, L. C., J. W. Jensen, and R. M. Doerr: *U.S. Bur. Mines Rept. Invest.* 7022, 1967.
123. Gerlach, J., J. P. Krumme, F. Pawlek, and H. Probst: *Z. physik. Chem.* [NF] **53**, 135 (1967).
124. Giauque, W. F., and J. B. Ott: *J. Am. Chem. Soc.* **82**, 2689 (1960).
125. Gitlesen, G., and K. Motzfeldt: *Acta Chem. Scand.* **19**, 661 (1965).
126. Glasser, F. P.: *J. Am. Ceram. Soc.* **45**, 242 (1962).
127. Glatz, A. C., and K. E. Cordo: *J. Phys. Chem.* **70**, 3757 (1966).
128. Glemser, O., A. Müller, and U. Stöcker: *Z. anorg. allgem. Chem.* **333**, 25 (1964).
129. Glushko, V. P., ed.: "Thermal Constants of Substances," vols. 1, 2, and 3, VINITI, Moscow, U.S.S.R.
130. Greenbaum, M. A., J. N. Foster, M. L. Arim, and M. Farber: *J. Phys. Chem.* **67**, 36 (1963).
131. Greenbaum, M. A., J. Weiker, and M. Farber: *J. Phys. Chem.* **69**, 4035 (1965).
132. Greenbaum, M. A., E. E. Yates, and M. Farber: *J. Phys. Chem.* **67**, 1802 (1963).
133. Greenwood, N. N., and K. Wade: *J. Chem. Soc.* **1956**, 1527.
134. Gregory, N. W.: *U.S. AEC Rept.* TID 5290, 1958.
135. Grieveson, P., and E. T. Turkdogan: *Trans. AIME* **224**, 1086 (1962).
136. Griffel, M., and J. W. Stout: *J. Am. Chem. Soc.* **72**, 4351 (1950).
137. Grisard, J. W., H. A. Bernhardt, and G. D. Oliver: *J. Am. Chem. Soc.* **73**, 5725 (1951).
138. Grosse, A. V., A. D. Kirshenbaum, A. G. Streng, and L. V. Streng: *Science* **139**, 1047 (1963).
139. Grubitsch, H., and E. Suppan: *Monatsh. Chem.* **93**, 246 (1962).
140. Gulbransen, E. A., K. F. Andrew, and F. A. Brassart: *J. Electrochem. Soc.* **110**, 242 (1963).
141. Gunn, S. R.: *J. Phys. Chem.* **71**, 2934 (1967).
142. Gunn, S. R.: *J. Am. Chem. Soc.* **88**, 5924 (1966).
143. Gunn, S. R., and L. G. Green: *J. Phys. Chem.* **65**, 2173 (1961).
144. Gutt, W., and M. A. Smith: *Trans. Brit. Ceram. Soc.* **66**, 337 (1967).
145. Haendler, H. M.: *Rept. NYO* 6121 (1953).
146. Hahn, W. C., and A. Muan: *Am. J. Sci.* **258**, 66 (1960).
147. Hall, E. H., J. M. Blocher, and I. E. Campbell: *J. Electrochem. Soc.* **105**, 271 (1958).
148. Hall, E. H., J. M. Blocher, and I. E. Campbell: *J. Electrochem. Soc.* **105**, 275 (1958).
149. Hansen, W. N., and M. Griffel: *J. Chem. Phys.* **28**, 902 (1958).
150. Hart, P. E., and A. W. Searcy: *J. Phys. Chem.* **70**, 2763 (1966).
151. Hashino, S.: *J. Phys. Soc. Japan* **7**, 560 (1952).
152. Hedvall, J. A., R. Lindner, and N. Hartler: *Acta Chem. Scand.* **4**, 1099 (1950).
153. Hein, R. A., P. N. Flagella, and J. B. Conway: *J. Am. Ceram. Soc.* **51**, 291 (1968).
154. Hildenbrand, D. L.: Private communication, 1965.
155. Hildenbrand, D. L., W. F. Hall, F. Ju, and N. D. Potter: *J. Chem. Phys.* **40**, 2882 (1964).
156. Hildenbrand, D. L., and N. D. Potter: *J. Phys. Chem.* **67**, 2231 (1963).
157. Hildenbrand, D. L., and L. P. Theard: *J. Chem. Phys.* **42**, 3230 (1965).
158. Hoffman, C. J., and W. L. Jolly: *J. Phys. Chem.* **61**, 1574 (1957).
159. Holloway, J. H., and R. D. Peacock: *J. Chem. Soc.* **1963**, 527.
160. Holmes, R. R.: *Thesis*, Purdue University, 1958.
161. Holtzberg, F., A. Reisman, M. Berry, and M. Berkenblit: *J. Am. Chem. Soc.* **78**, 1536 (1956).
162. Hou, Y. C., and J. J. Martin: *Am. Inst. Chem. Eng.* **5**, 125 (1959).
163. Hu, J. H., and H. L. Johnston: Unpublished data, Ohio State University.
164. Hultgren, R. R., R. L. Orr, P. D. Anderson, and K. K. Kelley: "Selected Values of Thermodynamic Properties of Metals and Alloys," John Wiley & Sons, Inc., New York, and supplements.
165. Hutchinson, J. F.: *U.S. AEC Rept.* IS-T-50, 1965.
166. Inghram, M. G., W. A. Chupka, and R. F. Porter: *J. Chem. Phys.* **23**, 2159 (1955).
167. Ingraham, T. R., and H. H. Kellogg: *Trans. AIME* **227**, 1419 (1963).
168. Janz, G. J., and J. Goodkin: *J. Phys. Chem.* **63**, 1975 (1959).
169. Janz, G. J., D. W. James, and J. Goodkin: *J. Phys. Chem.* **64**, 937 (1960).
170. Janz, G. J., and F. J. Kelly: *J. Phys. Chem.* **67**, 2848 (1963).

171. Janz, G. J., F. J. Kelly, and J. L. Perano: *Trans. Faraday Soc.* **59**, 2718 (1963).
172. Johnson, O., T. Butler, and A. S. Newton: *U.S. AEC Rept.* TID 5290, 1958.
173. Johnson, J. W., W. J. Silva, and D. Cubicciotti: *J. Phys. Chem.* **69**, 3916 (1965).
174. Johnson, J. W., W. J. Silva, and D. Cubicciotti: *J. Phys. Chem.* **70**, 2985 (1966).
175. Johnson, J. W., W. J. Silva, and D. Cubicciotti: *J. Phys. Chem.* **72**, 1669 (1968).
176. Johnston, H. L., H. L. Kerr, E. C. Clarke, and J. T. Hallett: *ONR Rept.* RF-309, 1949.
177. Jolly, W. L., and W. M. Latimer: *J. Am. Chem. Soc.* **74**, 5754 (1952).
178. Jost, W., H. J. Oel, and G. Schneidermann: *Z. physik. Chem.* [NF] **17**, 175 (1958).
179. Junkins, J. H., H. A. Bernhardt, and E. J. Barber: *J. Am. Chem. Soc.* **74**, 5749 (1952).
180. Kabesh, A., and R. S. Nyholm: *J. Chem. Soc.* **1951**, 3245.
181. Kana'an, A. S., G. Besenbruch, and J. L. Margrave: *J. Inorg. & Nuclear Chem.* **28**, 1035 (1966).
182. Kandyba, V. V., P. B. Kantor, R. M. Krasovitskaya, and E. N. Fomichev: *Doklady Akad. Nauk S.S.S.R.* **131**, 566 (1960).
183. Kantor, P. B., E. N. Fomichev, and V. V. Kandyba: *Izmeritelnayr Tekh.* **5**, 27 (1966).
184. Karnenko, N. V.: *Zhur. Neorg. Khim.* **12**, 3248 (1967).
185. Keavney, J. J., and N. O. Smith: *J. Phys. Chem.* **64**, 737 (1960).
186. Kelley, K. K., and A. D. Mah: *U.S. Bur. Mines Rept. Invest.* 5490, 1959.
187. Keneshea, F. J., and D. Cubicciotti: *J. Chem. Phys.* **40**, 191 (1964).
188. Keneshea, F. J., and D. Cubicciotti: *J. Phys. Chem.* **71**, 1958 (1967).
189. Keneshea, F. J., D. Cubicciotti, G. Withers, and H. Eding: *J. Phys. Chem.* **72**, 1272 (1968).
190. Kent, R. A., J. D. McDonald, and J. L. Margrave: *J. Phys. Chem.* **70**, 874 (1966).
191. Kent, R. A., T. C. Ehlert, and J. L. Margrave: *J. Am. Chem. Soc.* **86**, 5090 (1964).
192. Kent, R. A., and J. L. Margrave: *J. Am. Chem. Soc.* **87**, 4754 (1965).
193. Kent, R. A., K. F. Zmbov, A. S. Kana'an, G. Besenbruch, J. D. McDonald, and J. L. Margrave: *J. Inorg. & Nuclear Chem.* **28**, 1419 (1966).
194. Khripin, L. A., Yu. V. Gazarinski, and L. A. Luk'yanova: *Izvest. Sibir. Otdel. Akad. Nauk S.S.S.R., Khim. Nauk* **11**, 14 (1965).
195. Kigorski, K.: *Bull. Chim. Soc. Japan* **23**, 67 (1950).
196. Kimpel, R. F., and R. G. Moss: *J. Chem. Eng. Data* **13**, 231 (1968).
197. King, E. G., and A. U. Christensen: *J. Am. Chem. Soc.* **80**, 1800 (1958).
198. King, E. G., and A. U. Christensen: *U.S. Bur. Mines Rept. Invest.* 5709, 1961.
199. King, E. G., and A. U. Christensen: *U.S. Bur. Mines Rept. Invest.* 5789, 1961.
200. King, E. G., D. R. Torgeson, and O. A. Cook: *J. Am. Chem. Soc.* **70**, 2160 (1948).
201. King, E. G., W. W. Weller, and A. U. Christensen: *U. S. Bur. Mines Rept. Invest.* 5664, 1960.
202. King, E. G., W. W. Weller, A. U. Christensen, and K. K. Kelley: *U.S. Bur. Mines Rept. Invest.* 5799, 1961.
203. Kleppa, O. J.: *Phys. Chem. Solids,* **23**, 819 (1962).
204. Kleppa, O. J., and F. G. McCarty: *J. Chem. Eng. Data* **8**, 331 (1963).
205. Kleppa, O. J., and S. V. Meschel: *J. Phys. Chem.* **67**, 668 (1963).
206. Klushin, D. N., and V. Ya. Chernykh: *Zhur. Neorg. Khim.* **5**, 1409 (1960).
207. Knache, O., and K. E. Prescher: *Z. Erzbergbau u. Metallhüttenw.* **17**, 28 (1964).
208. Ko, H. C., M. A. Greenbaum, M. Farber, and C. C. Selph: *J. Phys. Chem.* **71**, 254 (1967).
209. Kobayashi, K.: *Sci. Repts. Tohoku Imp. Univ.* [I] **35**, 112 (1950).
210. Kohlmuller, R., and J. Martin: *Bull. soc. chim. France* **1961**, 748.
211. Kokovin, G. A.: *Zhur. Neorg. Khim.* **12**, 15 (1967).
212. Kolditz, L.: *Z. anorg. allgem. Chem.* **289**, 118 (1957).
213. Komissarova, L. N., and B. I. Pokrovskii: *Doklady Akad. Nauk S.S.S.R.* **149**, 599 (1963).
214. Kostryukov, V. N., O. P. Samorukov, and P. G. Strelkov: *Zhur. Fiz. Khim.* **34**, 1354 (1958).
215. Krause, R. F., and T. B. Douglas: *J. Phys. Chem.* **72**, 475 (1968).
216. Krikorian, O. H., J. H. Carpenter, and R. S. Newbury: *U.S. AEC Rept.* UCRL 1226-T, 1964.
217. Krohn, C.: *Acta Chem. Scand.* **20**, 255 (1966).
218. Kul'varskaya, B. S., and K. S. Maslovskaya: *Radiotekh. i Elektron.* **5**, 1254 (1960).
219. Kusano, K.: *Nippon Kagaku Zasshi* **78**, 614 (1958).
220. Lander, J. J.: *J. Am. Chem. Soc.* **73**, 5794 (1951).
221. Langer, S., and F. F. Blankenship: *J. Inorg. Nuclear Chem.* **14**, 26 (1960).
222. Levin, E. M., and R. S. Roth: *J. Research NBS* **A68**, 189 (1964).
223. Lieser, K. H.: *Z. physik. Chem.* [NF] **2**, 238 (1954).
224. Loehman, R. E., R. A. Kent, and J. L. Margrave: *J. Chem. Eng. Data* **10**, 296 (1965).
225. Long, R. D., J. J. Martin, and R. C. Vogel: *J. Chem. Eng. Data* **3**, 28 (1958).

226. Luchinskii, G. P.: *Zhur. Fiz. Khim.* **40**, 593 (1966).
227. Ludbetter, A. J., and J. E. Spice: *Can. J. Chem.* **37**, 1923 (1959).
228. Lynch, C. T., F. W. Pahldiek, and L. B. Robinson: *J. Am. Ceram. Soc.* **44**, 147 (1961).
229. McCreary, J. R., and R. J. Thorn: *J. Chem. Phys.* **48**, 3290 (1968).
230. McCurdy, K. G., and K. J. Laidler: *Can. J. Chem.* **41**, 1867 (1963).
231. McDonald, R. A., and F. L. Oetting: *CPIA Proc.* **1**, 101 (1965).
232. McGill, R., W. S. Wendolkowski, and E. J. Barber: *J. Phys. Chem.* **61**, 1101 (1957).
233. MacLaren, R. O., and N. W. Gregory: *J. Phys. Chem.* **59**, 184 (1955).
234. Magee, D. W.: Thesis, Ohio State University, 1955.
235. Mah, A. D.: *U.S. Bur. Mines Rept. Invest.* 5600, 1960.
236. Mah, A. D., and L. H. Adami: *U.S. Bur. Mines Rept. Invest.* 6034, 1962.
237. Mah, A. D., L. B. Pankratz, W. W. Weller, and E. G. King: *U.S. Bur. Mines Rept. Invest.* 7026, 1967.
238. Majumdar, A. J., and R. Roy: *J. Inorg. & Nuclear Chem.* **27**, 1961 (1965).
239. Majumdar, A. J., and R. Roy: *J. Phys. Chem.* **63**, 1858 (1959).
240. Malbrunot, P. F., P. A. Mennier, G. M. Scatema, W. H. Mears, K. P. Murphy, and J. V. Sinka: *J. Chem. Eng. Data* **13**, 16 (1968).
241. Malm, J. G., and H. Selig: *J. Inorg. & Nuclear Chem.* **20**, 189 (1961).
242. Malm, J. G., I. Sheft, and C. L. Chernick: *J. Am. Chem. Soc.* **85**, 110 (1963).
243. Mapes, W. H., and N. W. Gregory: *J. Chem. Eng. Data* **13**, 249 (1968).
244. Mar, R. W., and A. W. Searcy: *J. Phys. Chem.* **71**, 888 (1967).
245. Margulis, E. V., L. S. Getskin, and N. S. Milskaya: *Zhur. Neorg. Khim.* **7**, 729 (1962).
246. Mayer, S. W., S. J. Yosim, and L. E. Topol: *J. Phys. Chem.* **64**, 238 (1960).
247. Mezaki, R., and J. L. Margrave: *J. Phys. Chem.* **66**, 1713 (1962).
248. Michels, A., and C. Prins: *Physica* **28**, 101 (1962).
249. Mitra, G.: *J. Am. Chem. Soc.* **80**, 5639 (1958).
250. Miyake, S., S. Hashino, and T. Takenaka: *J. Phys. Soc. Japan* **7**, 19 (1952).
251. Modestova, T. P.: *Zhur. Neorg. Khim.* **5**, 1655 (1960).
252. Modestova, T. P., and T. Sumarokova: *Zhur. Neorg. Khim.* **3**, 1655 (1958).
253. Moelwyn-Hughes, E. A., and R. W. Missen: *Trans. Faraday Soc.* **53**, 607 (1957).
254. Mordovin, O. A., N. I. Timofeeva, and L. N. Drozdova: *Izvest. Akad. Nauk S.S.S.R., Neorg. Mater.* **3**, 187 (1967).
255. Moriarty, J. L.: *J. Chem. Eng. Data* **8**, 422 (1963).
256. Morozov, I. S., and D. Ya. Toptygin: *Zhur. Neorg. Khim.* **1**, 2601 (1956).
257. Mustajoki, A.: *Ann. Acad. Sci. Fennicae,* ser. A, **6**(7), (1957).
258. Mustajoki, A.: *Ann. Acad. Sci. Fennicae,* ser. A, **6**(9), (1958).
259. Neudorffer, J.: *Compt. rend.* **232**, 2102 (1951).
260. Newbury, R. S., G. W. Bartow, and A. W. Searcy: *J. Chem. Phys.* **48**, 793 (1968).
261. Newns, G. R., and R. Hanks: *J. Chem. Soc.* **A1966**, 954.
262. Nikitin, D. T., and P. A. Akishin: *Doklady Akad. Nauk S.S.S.R.* **145**, 1294 (1962).
263. Nikolaev, R. K., Z. N. Orshanskaya, O. R. Gavrilov, and L. A. Nisel'son: *Zhur. Neorg. Khim.* **12**, 556 (1967).
264. Nikolskii, A. B.: *Zhur. Neorg. Khim.* **8**, 1045 (1963).
265. Nisel'son, L. A., I. I. Lapidus, Yu. V. Golubkov, and V. V. Mogucheva: *Zhur. Neorg. Khim.* **12**, 1952 (1967).
266. Nisel'son, L. A., and V. V. Mogucheva: *Zhur. Neorg. Khim.* **11**, 144 (1966).
267. Nisel'son, L. A., A. I. Pustilnik, O. R. Gavrilov, and V. A. Rodin: *Zhur. Neorg. Khim.* **10**, 2339 (1965).
268. Nisel'son, L. A., T. D. Sokolova, and V. I. Stolyarov: *Zhur. Fiz. Khim.* **41**, 1654 (1967).
269. Nisel'son, L. A., V. I. Stolyarov, and T. D. Sokolova: *Zhur. Fiz. Khim.* **39**, 3025 (1965).
270. Noguchi, T., M. Mizuno, and T. Kozuka: *Kogyo Kagaku Zasshi* **69**, 1705 (1966).
271. Nölting, J.: *Ber. Bunsenges. Phys. Chem.* **67**, 172 (1963).
272. Novikov, G. I., and A. K. Baev: *Zhur. Neorg. Khim.* **7**, 1349 (1962).
273. Novikov, G. I., and F. G. Gavryuchenkov: *Zhur. Neorg. Khim.* **9**, 475 (1964).
274. Novikov, G. I., and F. G. Gavryuchenkov: *Zhur. Neorg. Khim.* **10**, 1668 (1965).
275. Novoselova, A. V., Zh. G. Shleifman, V. P. Zhomanov, and R. K. Sloma: *Izvest. Akad. Nauk S.S.S.R., Neorg. Mater.* **3**, 1143 (1967).
276. Oliver, G. D., and J. W. Grisard: *J. Am. Chem. Soc.* **74**, 2705 (1952).
277. Oppermann, H.: *Z. physik. Chem.* **236**, 161 (1967).
278. Oppermann, H.: *Z. anorg. allgem. Chem.* **351**, 113 (1967).
279. Oppermann, H.: *Z. anorg. allgem. Chem.* **356**, 1 (1967).
280. Oranskaya, M. A., and N. A. Makhovlova: *Zhur. Neorg. Khim.* **5**, 12 (1960).
281. Orr, R. L.: *J. Am. Chem. Soc.* **75**, 2808 (1953).
282. Osborne, D. W., F. Schreiner, J. G. Malm, H. Selig, and L. Rochester: *J. Chem. Phys.* **44**, 2802 (1966).

283. Osborne, D. W., B. Weinstock, and J. H. Burns: Preliminary communication 1963.
284. Pace, E. L., and M. A. Reno: *J. Chem. Phys.* **48**, 1231 (1968).
285. Palko, A. A., A. D. Ryon, and D. W. Kuhn: *J. Phys. Chem.* **62**, 319 (1958).
286. Pankratz, L. B., E. G. King, and K. K. Kelley: *U.S. Bur. Mines Rept. Invest.* 6033 1962.
287. Paridon, L. J., G. E. MacWood, and J. H. Hu: *J. Phys. Chem.* **63**, 1998 (1959).
288. Parker, V.: Private communication, 1966.
289. Perrott, C. M., and N. H. Fletcher: *J. Chem. Phys.* **48**, 2143 (1968).
290. Petit, G., and C. Bourlange: *Compt. rend.* **245**, 1788 (1957).
291. Petit, G., and N. N. Chikova: *Zhur. Neorg. Khim.* **8**, 1442 (1963).
292. Petit, G., and A. Cremieu: *Compt. rend.* **243**, 360 (1956).
293. Petit, G., and F. Delbové: *Compt. rend.* **254**, 1388 (1962).
294. Pistorius, C. W. F. T.: *J. Inorg. & Nuclear Chem.* **19**, 367 (1961).
295. Platteeuw, J. C., and S. Meyer: *Trans. Faraday Soc.* **52**, 1066 (1956).
296. Plutonium Handbook, vol. I, chap. 12, C. E. Wick, ed., Gordon and Breach, Science Publishers, Inc., New York, 1967.
297. Polyachenok, O. G., and G. I. Novikov: *Zhur. Neorg. Khim.* **8**, 2631 (1963).
298. Polyachenok, O. G., and G. I. Novikov: *Zhur. Neorg. Khim.* **9**, 773 (1964).
299. Polyachenok, O. G., and G. I. Novikov: *Vestinik Leningrad Univ.* **18**, Ser. Fiz. Khim. **3**, 133 (1963).
300. Popov, M. M., and M. Ginzburg: *J. Gen. Chem. U.S.S.R.* **26**, 1107 (1956).
301. Porter, B., and E. A. Brown: *J. Am. Ceram. Soc.* **45**, 49 (1962).
302. Prescher, K. E., and W. Schrödter: *Z. Erzbergbau u. Metallhüttenw.* **15**, 299 (1962).
303. Prescher, K. E., and W. Schrödter: *Z. Erzbergbau u. Metallhüttenw.* **16**, 352 (1963).
304. Pugh, A. C. P., and R. F. Barrow: *Trans. Faraday Soc.* **54**, 671 (1958).
305. Pustilnik, A. I., N. D. Denisova, L. G. Nekhamkin, and L. A. Nisel'son: *Zhur. Neorg. Khim.* **12**, 103 (1967).
306. Rao, G. V. S., M. Natorjan, and C. N. R. Rao: *J. Am. Ceram. Soc.* **51**, 179 (1968).
307. Rapoport, E., and C. W. F. T. Pistorius: *J. Geophys. Res.* **72**, 6353 (1967).
308. Reinsborough, V. C., and F. E. W. Wetmore: *Australian J. Chem.* **20**, 1 (1967).
309. Rezukhina, T. N., and V. V. Shvyrev: *Vestnik Moskov. Univ.* **7**, Fiz. Mat. i Estestven. Nauk **4**, 41 (1952).
310. Richardson, F. D., and J. E. Anthill: *Trans. Faraday Soc.* **51**, 22 (1955).
311. Riebling, E. F., and C. E. Erickson: *J. Phys. Chem.* **67**, 509 (1963).
312 Röck, H., and W. Schröder: *Z. physik. Chem.* [NF] **11**, 41 (1957).
313. Rockett, T. S., and W. R. Foster: *J. Am. Ceram. Soc.* **48**, 75 (1965).
314. Rodigino, E. N., K. Z. Gomelskii, and V. F. Luginina: *Zhur. Neorg. Khim.* **4**, 975 (1959).
315. Rogers, M. T., and J. L. Spiers: *J. Phys. Chem.* **60**, 1462 (1956).
316. Rogers, M. T., J. L. Spiers, H. B. Thomson, and M. P. Pannish: *J. Am. Chem. Soc.* **76**, 4843 (1954).
317. Rolin, M., and M. Clausier: *Rev. Int. Hautes Temp.* **4**, 39 (1967).
318. Rolla, M., and P. Franzosini: *Ann. chim. (Rome)* **48**, 723 (1958).
319. Rollet, A. P., R. Cohn-Adad, and C. Ferlin: *Compt. rend.* **256**, 5580 (1963).
320. Rossini, F. D., D. D. Wagman, W. H. Evans, S. Levine, and I. Jaffe: Selected Values of Chemical Thermodynamic Properties, NBS Circ. 500, 1952.
321. Roy, R., A. J. Majumdar, and C. W. Hulber: *Econ. Geol.* **54**, 1278 (1959).
322. Sackman, H., D. Demus, and D. Pankow: *Z. anorg. allgem. Chem.* **318**, 257 (1962).
323. St. Pierre, P. D. S.: *J. Am. Ceram. Soc.* **35**, 188 (1952).
324. Sata, T.: *Rev. Int. Hautes Temp.* **3**, 337 (1966).
325. Schäfer, H., L. Bayer, G. Breil, K. Etzel, and K. Krehl: *Z. anorg. allgem. Chem.* **278**, 300 (1955).
326. Schäfer, H., and W. J. Hönes: *Z. anorg. allgem. Chem.* **288**, 62 (1956).
327. Schäfer, H., and H. Jacob: *Z. anorg. allgem. Chem.* **286**, 56 (1956).
328. Schäfer, H., and K. Krehl: *Z. anorg. allgem. Chem.* **268**, 35 (1952).
329. Schäfer, E. C., and R. Roy: *Z. physik. Chem.* [NF] **11**, 30 (1957).
330. Schäfer, H. L., and H. Skoludek: *Z. Elektrochem.* **66**, 367 (1962).
331. Schäfer, H. L., and H. W. Wills: *Z. anorg. allgem. Chem.* **351**, 279 (1967).
332. Scheer, M. D., and J. Fine: *J. Chem. Phys.* **36**, 1647 (1962).
333. Schmidt, N. E.: *Zhur. Neorg. Khim.* **11**, 441 (1966).
334. Schneider, S. J.: *J. Research NBS* **65A**, 429 (1961).
335. Schneider, S. J.: *J. Am. Ceram. Soc.* **46**, 354 (1963).
336. Schneider, S. J., and C. L. McDaniel: *J. Research NBS* **71A**, 317 (1967).
337. Schnizlein, J. G., J. L. Sheard, R. C. Toole, and T. D. O'Brien: *J. Phys. Chem.* **56**, 233 (1952).
338. Schreiner, F., G. N. McDonald, and C. L. Chernick: *J. Phys. Chem.* **72**, 1162 (1968).

339. Schrier, E. E., and H. M. Clark: *J. Phys. Chem.* **67**, 1259 (1963).
340. Schultz, D. A., and A. W. Searcy: *J. Phys. Chem.* **67**, 103 (1963).
341. Selig, H., and J. G. Malm: *J. Inorg. & Nuclear Chem.* **24**, 641 (1962).
342. Sense, K. A., M. J. Snyder, and R. B. Filbert, Jr.: *J. Phys. Chem.* **58**, 995 (1954).
343. Sense, K. A., and R. W. Stone: *J. Phys. Chem.* **62**, 453 (1958).
344. Sense, K. A., and R. W. Stone: *J. Phys. Chem.* **62**, 1411 (1958).
345. Shchukarev, S. A., and A. R. Kurbanov: *Vestnik Leningrad Univ.* **17**, *Fiz. Khim.* 144 (1962).
346. Shchukarev, S. A., G. I. Novikov, and G. A. Kokovin: *Zhur. Neorg. Khim.* **4**, 2185 (1959).
347. Shchukarev, S. A., G. A. Semenov, and K. E. Frantseva: *Zhur. Neorg. Khim.* **11**, 233 (1966).
348. Shchukarev, S. A., G. A. Semenov, and I. A. Ratkovskii: *Zhur. Neorg. Khim.* **6**, 2817 (1961).
349. Shchukarev, S. A., and A. V. Suvorov: *Vestnik Leningrad Univ.* **16**, 87 (1961).
350. Shchukarev, S. A., I. V. Vasil'kova, A. I. Efimov, and V. P. Kerdyashev: *Zhur. Neorg. Khim.* **1**, 2272 (1956).
351. Shelton, P. A. J.: *Trans. Faraday Soc.* **57**, 2113 (1961).
352. Sherman, R. H., and W. F. Giauque: *J. Am. Chem. Soc.* **77**, 2154 (1955).
353. Shimazaki, E., and K. Niwa: *Z. anorg. allgem. Chem.* **314**, 21 (1962).
354. Shirane, G., and A. Takeda: *J. Phys. Soc. Japan* **7**, 1 (1952).
355. Shomate, C. H., and A. J. Cohen: *J. Am. Chem. Soc.* **77**, 285 (1955).
356. Sime, R. J.: *J. Phys. Chem.* **67**, 501 (1963).
357. Sime, R. J., and N. W. Gregory: *J. Am. Chem. Soc.* **82**, 93 (1960).
358. Simpson, D. R.: *Econ. Geol.* **59**, 150 (1964).
359. Singh, J., and G. C. Benson: *Can. J. Chem.* **46**, 1249 (1968).
360. Singleton, E. J., L. Carpenter, and R. V. Lundquist: *U.S. Bur. Mines Rept. Invest.* 5938, 1962.
361. Smirnova, I. N., and I. P. Kislyakov: *Izvest. Akad. Nauk S.S.S.R. Neorg. Mater,* **1**, 1162 (1965).
362. Smith, F. J., and R. F. Barrow: *Trans. Faraday Soc.* **54**, 826 (1958).
363. Smith, D. K., C. F. Cline, and V. D. Frechetti: *J. Nuclear Mater.* **6**, 265 (1962).
364. Smith, W. T., Jr., J. W. Cobble, and G. E. Boyd: *J. Am. Chem. Soc.* **75**, 5773 (1953).
365. Smith, D. F., C. E. Kaylor, G. E. Walden, A. R. Taylor, and J. B. Gayle: *U.S. Bur. Mines Rept. Invest.* 5832, 1961.
366. Smith, W. T., L. E. Line, and W. A. Bell: *J. Am. Chem. Soc.* **74**, 4964 (1952).
367. Sokolov, V. A., and N. E. Schmidt: *Izvest. Sektora. Fiz.-Khim. Anal. Inst. Obshch. i. Neorg. Khim. Akad. Nauk S.S.S.R.* **27**, 217 (1956).
368. Sommer, A.: Thesis, Ohio State University, 1962.
369. Spedding, F. H., and A. H. Daane: *Iowa State Coll. Rept.* IS-902, 1957.
370. Speranskaya, E. I., and A. A. Arshakuni: *Zhur. Neorg. Khim.* **9**, 414 (1964).
371. Spice, J. E., L. A. K. Staveley, and G. A. Harrow: *J. Chem. Soc.* **1955**, 100.
372. Statton, W. O.: *J. Chem. Phys.* **19**, 33 (1951).
373. Staveley, L. A. K., J. B. Warren, H. P. Paget, and D. J. Dowrick: *J. Chem. Soc.* **1954**, 1992.
374. Stevenson, F. D., and C. E. Wicks: *U.S. Bur. Mines Rept. Invest.* 6212, 1963.
375. Stevenson, F. D., C. E. Wicks, and F. E. Block: *U.S. Bur. Mines Rept. Invest.* 6367, 1964.
376. Stevenson, F. D., C. E. Wicks, and F. E. Block: *J. Chem. Eng. Data* **10**, 33 (1965).
377. Stull, D. R., ed: JANAF Thermochemical Tables, *PB Rept.* 168370, 1965.
378. Sujishi, S., and S. Witz: *J. Am. Chem. Soc.* **79**, 2447 (1957).
379. Suvorov, A. V., E. V. Krzhizhanovskaya, and G. I. Novikov: *Zhur. Neorg. Khim.* **11**, 2685 (1966).
380. Swarc, R., E. R. Plante, and J. J. Diamond: *J. Research NBS* **69A**, 417 (1965).
381. Swaroop, B., and S. N. Flengas: *Can. J. Chem.* **44**, 199 (1966).
382. Tannenbaum, S., S. Kaye, and G. F. Lewenz: *J. Am. Chem. Soc.* **75**, 3753 (1953).
383. Tavernier, B. H., J. Vervechev, P. Messieu, and M. Baiwir: *Z. anorg. allgem. Chem.* **356**, 77 (1967).
384. Taylor, A. R., and T. E. Gardner: *U.S. Bur. Mines Rept. Invest.* 6664, 1965.
385. Thiloaud, E., and P. Flögel: *Z. anorg. allgem. Chem.* **329**, 244 (1964).
386. Thomas, R. E., C. F. Weaver, H. A. Friedman, H. Imsley, L. A. Harris, and H. A. Yokel, Jr.: *J. Phys. Chem.* **65**, 1096 (1961).
387. Thonstad, J.: *Can. J. Chem.* **42**, 2739 (1964).
388. Todd, S. S., and K. R. Bonnickson: *J. Am. Chem. Soc.* **73**, 3894 (1951).
389. Todd, S. S., and R. E. Lorenson: *J. Am. Chem. Soc.* **74**, 2043 (1952).
390. Topol, L. E., S. W. Mayer, and L. D. Ransom: *J. Phys. Chem.* **64**, 862 (1960).
391. Topol, L. E., and L. D. Ransom: *J. Phys. Chem.* **64**, 1339 (1960).

392. Treadwell, W. D., and W. Werner: *Helv. Chim. Acta* **36**, 1436 (1953).
393. Trevorrow, L. E., M. J. Steindler, D. V. Steidl, and J. T. Savage: *Inorg. Chem.* **6**, 1060 (1967).
394. Trzebiatowski, W., J. Damm, and T. Romotowski: *Roczniki Chem.* **30**, 431 (1956).
395. Tylkina, M. A., V. P. Polyakova, and O. Kh. Khamidov: *Zhur. Neorg. Khim.* **8**, 776 (1963).
396. Vollmer, O., M. Braun, and R. Kohlhaus: *Z. Naturforsch.* **22A**, 833 (1967).
397. Voskresenskaya, N. K., and E. I. Banashek: *Izvest. Sektora Fiz.-Khim. Anal. Inst. Obshch. Neorg. Khim. Akad. Nauk S.S.S.R.* **25**, 150 (1954).
398. Wadsö, I.: *Acta Chem. Scand.* **20**, 544 (1966).
399. Walden, G. E., and D. F. Smith: *U.S. Bur. Mines Rept. Invest.* 5859, 1961.
400. Waring, J. L., and R. S. Roth: *J. Research NBS* **72A**, 175 (1968).
401. Warshaw, I., and R. Roy: *J. Phys. Chem.* **65**, 2048 (1961).
402. Weed, H. C.: Thesis, Ohio State University, 1957.
403. Weinstock, B., J. G. Malm, and E. E. Weaver: *J. Am. Chem. Soc.* **83**, 4310 (1961).
404. Weinstock, B., E. E. Weaver, and C. P. Knop: *Inorg. Chem.* **5**, 2189 (1966).
405. Weinstock, B., E. E. Weaver, and J. G. Malm: *J. Inorg. & Nuclear Chem.* **11**, 104 (1959).
406. Welty, J. R.: Thesis, Oregon State University, 1962.
407. Wirth, H. E., and E. D. Palmer: *J. Phys. Chem.* **60**, 911 (1956).
408. Wirth, H. E., and E. D. Palmer: *J. Phys. Chem.* **60**, 914 (1956).
409. Wolf, A. S., J. C. Posey, and K. E. Rapp: *Inorg. Chem.* **4**, 751 (1965).
410. Wolten, G. M., and S. W. Mayer: *Acta Cryst.* **11**, 739 (1958).
411. Zakharchenko, G. A.: *Zhur. Obshchei Khim.* **21**, 453 (1951).
412. Zhitaneva, G. M., Y. V. Rumyantsev, and F. M. Bolondz: *Trudy Vostochno-Siber. Filiala Akad. Nauk S.S.S.R.* **41**, 121 (1962).
413. Zlomanov, V. P., A. V. Novoselova, A. S. Pashinkin, Yu. P. Simanov, and K. H. Semenenko: *Zhur. Neorg. Khim.* **3**, 1473 (1958).
414. Zmbov, K. F., J. W. Hastie, and J. L. Margrave: *Trans. Faraday Soc.* **64**, 861 (1968).
415. Zmbov, K. F., and J. L. Margrave: *J. Less-Common Metals* **12**, 494 (1967).
416. Zmbov, K. F., and J. L. Margrave: *J. Inorg. & Nuclear Chem.* **29**, 673 (1967).
417. Zmbov, K. F., and J. L. Margrave: *J. Phys. Chem.* **71**, 446 (1967).
418. Zmbov, K. F., and J. L. Margrave: *J. Chem. Phys.* **45**, 3167 (1966).
419. Cubicciotti, D., and H. Eding: *J. Chem. Eng. Data* **10**, 343 (1965).
420. Eding, H. and D. Cubicciotti: *J. Chem. Eng. Data* **9**, 524 (1964).

4k. Vapor Pressure

DANIEL R. STULL

The Dow Chemical Company

Tables 4k-1 to 4k-4, Vapor Pressures of Inorganic and Organic Compounds, were compiled by the author at The Dow Chemical Company and were published in *Ind. Eng. Chem.* **39**(4), 517 (April, 1947), and **39**(12), 1684 (December, 1947). A much more extensive list and references can be found in this journal. The numbers represent temperatures in degrees Celsius at which the vapor pressure is the value appearing at the top of the column.

Symbols

d	decomposes	M.P.	melting point
d	dextrorotatory	P_c	critical pressure
dl	inactive (50% d and 50% l)	p	polymerizes
e	explodes	s	solid
l	levorotatory	T_c	critical temperature

TABLE 4k-1. VAPOR PRESSURE OF INORGANIC COMPOUNDS—PRESSURES LESS THAN 1 ATMOSPHERE

Formula	Name	Temp., °C										M.P.
		1 mm	5 mm	10 mm	20 mm	40 mm	60 mm	100 mm	200 mm	400 mm	760 mm	
AlB₃H₁₂	Aluminum borohydride	s	-52.2	-42.9	-32.5	-20.9	-13.4	-3.9	+11.2	28.1	45.9	-64.5
AlBr₃	Aluminum bromide	81.3s	103.8	118.0	134.0	150.6	161.7	176.1	199.8	227.0	256.3	97.5
AlCl₃	Aluminum chloride	100.0s	116.4s	123.8s	131.8s	139.9s	145.4s	152.0s	161.8s	171.6s	180.2s	192.4
AlF₃	Aluminum fluoride	1238	1298	1324	1350	1378	1398	1422	1457	1496	1537	1040
AlI₃	Aluminum iodide	178.0s	207.7	225.8	244.2	265.0	277.8	294.5	322.0	354.0	385.5	
Al₂O₃	Aluminum oxide	2148	2306	2385	2465	2549	2599	2665	2766	2874	2977	2050
NH₃	Ammonia	-109.1s	-97.5s	-91.9s	-85.8s	-79.2s	-74.3	-68.4	-57.0	-45.4	-33.6	-77.7
ND₃	Deutero ammonia						-74.0	-67.4	-57.0	-45.4	-33.4	-74.0
NH₄N₃	Ammonium azide	29.2s	49.4s	59.2s	69.4s	80.1s	86.7s	95.2s	107.7s	120.4s	133.8s	
NH₄Br	Ammonium bromide	198.3s	234.5s	252.0s	270.6s	290.0s	303.8s	320.0s	345.3s	370.9s	396.0s	
NH₄CO₂NH₂	Ammonium carbamate	-26.1s	-10.4s	-2.9s	+5.3s	14.0s	19.6s	26.7s	37.2s	48.0s	58.3s	
NH₄Cl	Ammonium chloride	160.4s	193.8s	209.8s	226.1s	245.0s	256.2s	271.5s	293.3s	316.5s	337.8s	520
NH₄HS	Ammonium hydrogen sulfide	-51.1	-36.0	-28.7	-20.8	-12.3	-7.0	0.0	+10.5	21.8	33.3	
NH₄I	Ammonium iodide	210.9s	247.0s	263.5s	282.8s	302.8s	316.0s	331.8s	355.8s	381.0s	404.9s	
NH₄CN	Ammonium cyanide	-50.6s	-35.7s	-28.6s	-20.9s	-12.6s	-7.4s	-0.5s	+9.6s	20.5s	31.7s	36
SbBr₃	Antimony tribromide	93.9	126.0	142.7	158.3	177.4	188.1	203.5	225.7	250.2	275.0	96.6
SbCl₃	Antimony trichloride	49.2s	71.4s	85.2	100.6	117.8	128.3	143.3	165.9	192.2	219.0	73.4
SbCl₅	Antimony pentachloride	22.7	48.6	61.8	75.8	91.0	101.0	114.1	d			2.8
SbI₃	Antimony triiodide	163.6s	203.8	223.5	244.8	267.8	282.5	303.5	333.8	368.5	401.0	167
Sb₂O₃	Antimony trioxide	574s	626s	666	729	812	873	957	1085	1242	1425	656
AsBr₃	Arsenic tribromide	41.8	70.6	85.2	101.3	118.7	130.0	145.2	167.7	193.6	220.0	
AsCl₃	Arsenic trichloride	-11.4	+11.4	+23.5	36.0	50.0	58.7	70.9	89.2	109.7	130.4	-18
AsF₃	Arsenic trifluoride					2.5	+4.2	13.2	26.7	41.4	56.3	-5.9
AsF₅	Arsenic pentafluoride	-117.9s	-108.0s	-103.1s	-98.0s	-92.4s	-88.5s	-84.3s	-75.5	-64.0	-52.8	-79.8
AsH₃	Arsenic hydride (arsine)	-142.6s	-130.8s	-124.7s	-117.7s	-110.2s	-104.8s	-98.0s	-87.2	-75.2	-62.1	-116.3
As₂O₃	Arsenic trioxide	212.5s	242.6s	259.7s	279.2s	299.2s	310.3s	332.5	370.0	412.2	457.2	312.8
BeB₂H₈	Beryllium borohydride	+1.0s	19.8s	28.1s	36.6s	46.2s	51.7s	58.6s	69.0s	79.7s	90.0s	123
BeBr₂	Beryllium bromide	289s	325s	342s	361s	379s	390s	405s	427s	451s	474s	490
BeCl₂	Beryllium chloride	291s	328s	346s	365s	384s	395s	411	435s	461s	487s	405
BeI₂	Beryllium iodide	283s	322s	341s	361s	382s	394s	411s	435s	461s	487s	488
BiBr₃	Bismuth tribromide	s	261	282	305	327	340	360	392	425	461	218
BiCl₃	Bismuth trichloride		242	264	287	311	324	343	372	405	441	230
BH₃CO	Borine carbonyl	-139.2	-127.3	-121.1	-114.1	-106.6	-101.9	-95.3	-85.5	-74.8	-64.0	-137.0
BBr₃	Boron tribromide	-41.4	-20.4	-10.1	+1.5	14.0	22.1	33.5	50.3	70.0	91.7	-45
BCl₃	Boron trichloride	-91.5	-75.2	-66.9	-57.9	-47.8	-41.2	-32.4	-18.9	+3.6	+12.7	-107
BF₃	Boron trifluoride	-154.6s	-145.4s	-141.3s	-136.4s	-131.0s	-127.6s	-123.0	-115.9	-108.3	-100.7	-126.8

Table of vapor-pressure temperatures (°C). For each substance the eleven values are given in the order printed (top → bottom in the original column): melting point, then temperatures at 760, 400, 200, 100, 60, 40, 20, 10, 5, 1 mm Hg. ("s" = solid/sublimation, "d" = decomposes, "p" marker as printed; blank = no value listed.)

Formula	Name	Values (as printed, top → bottom)
B₂H₆	Dihydrodiborane	−169 · −86.5 · −99.6 · −111.2 · −120.9 · −127.2 · −131.6 · −138.5 · −144.3 · −149.5 · −159.7
B₂BrH₅	Diborane hydrobromide	−104.2 · +16.3 · 0.0 · −15.4 · −29.0 · −38.2 · −45.4 · −56.4 · −66.3 · −75.3 · −93.3
B₃H₆N₃	Triborine triamine	−58.2 · +50.6 · +34.3 · +18.5 · +4.0 · −5.8 · −13.2 · −25.0 · −35.3 · −45.0 · −63.0s
B₄H₁₀	Tetrahydrotetraborane	−119.9 · +16.1 · +0.8 · −14.0 · −28.1 · −37.4 · −44.3 · −54.8 · −64.3 · −73.1 · −90.9
B₅H₉	Dihydropentaborane	−47.0 · +58.1 · +40.8 · +24.6 · +9.6 · +0.4 · −8.0 · −20.0 · −30.7 · −40.4 · −50.2
B₅H₁₁	Tetrahydropentaborane	· +67.0 · +51.2 · +34.8 · +20.1 · +10.2 · +2.7 · −9.2 · −19.9 · −29.9 · −60.0s
B₁₀H₁₄	Dihydrodecaborane	99.6 · 40.0 · d · 163.8 · 142.3 · 127.8 · 117.4 · 100.0 · 90.2s · 80.8s · 60.3s
BrF₅	Bromine pentafluoride	−61.4 · 40.0 · 25.7 · +9.9 · −4.5 · −14.0 · −21.0 · −32.0 · −41.9 · −51.0 · −69.3s
CdCl₂	Cadmium chloride	568 · 967 · 908 · 847 · 797 · 762 · 736 · 695 · 656 · 618 · 1385ᵃ
CdF₂	Cadmium fluoride	520 · 2024 · 1924 · 1834 · 1759 · 1709 · 1673 · 1617 · 1559 · 1504 · 1385
CdI₂	Cadmium iodide	385 · 796 · 742 · 688 · 640 · 608 · 584 · 546 · 512 · 481 · 416
CdO	Cadmium oxide	· 1559s · 1484s · 1409s · 1341s · 1295s · 1257s · 1200s · 1149s · 1100s · 1000s
CBr₄	Carbon tetrabromide	90.1 · 189.5 · 163.5 · 139.7 · 119.7 · 106.3 · 96.3 · s · 19.6s · 30.0s · 50.0s
CCl₄	Carbon tetrachloride	−22.6 · 76.7 · 57.8 · 38.3 · 23.0 · 12.3 · 4.3 · −8.2 · −19.6 · −30.0s · −50.0s
CF₄	Carbon tetrafluoride	−183.7 · −127.7 · −135.5 · −143.6 · −150.7 · −155.4 · −158.8 · −164.3 · −169.3 · −174.1 · −184.6s
C₃O₂	Carbon suboxide	−107 · +6.3 · −23.3 · −23.3 · −36.9 · −45.5 · −52.0 · −62.2 · −71.0 · −79.0 · −94.8
CS₂	Carbon disulfide	−110.8 · 46.5 · 28.0 · 10.4 · −5.1 · −15.3 · −22.5 · −34.3 · −44.7 · −54.3 · −73.8
C₃S₂	Carbon subsulfide	0.4 · 85.6 · 65.2 · 45.7 · 28.3 · 17.0 · 8.6 · −4.4 · −16.0 · −26.5 · −47.3
CSSe	Carbon selenosulfide	75.2 · 85.6 · p · 130.8s · 109.9 · 96.0 · 85.6 · 69.3 · 54.9 · 41.2 · 14.0
CO	Carbon monoxide	−205.0 · −191.3 · −196.3 · −200.0 · −205.7s · −208.1s · −210.0s · −212.8s · −215.0s · −217.2s · −222.0s
COCl₂	Carbonyl chloride	−104 · 8.3 · −7.6 · −22.3 · −35.6 · −44.0 · −50.3 · −60.3 · −69.3 · −77.0 · −92.9
COSe	Carbonyl selenide	· −21.9 · −35.6 · −49.8 · −61.7 · −70.2 · −76.4 · −86.3 · −95.0 · −102.3 · −117.1
COS	Carbonyl sulfide	−138.8 · −49.9 · −62.7 · −75.0 · −85.9 · −93.0 · −98.3 · −106.0 · −113.3 · −119.8 · −132.4
CCl₃NO₂	Chloropicrin	−64 · 111.9 · 91.8 · 71.8 · 53.8 · 42.3 · 33.8 · 20.0 · 7.8 · +3.3 · −25.5
CClF₃	Chlorotrifluoromethane	−181 · −81.2 · −92.7 · −102.5 · −111.7 · −121.9 · −128.5 · −134.1 · −139.2 · −149.5 ·
C₂N₂	Cyanogen	−34.4 · −21.0 · −42.0s · −51.1 · −57.9s · −62.7s · −70.1s · −76.8s · −83.2s · −95.8s ·
CBrN	Cyanogen bromide	58 · 61.5 · 46.0s · 33.8s · 22.6s · 14.7s · 8.6s · 1.0s · −10.0s · −18.3s · −35.7s
CClN	Cyanogen chloride	6.5 · 13.1 · 2.3 · 14.1 · 24.9s · 32.1 · 37.5s · 46.1 · 53.8s · 61.4s · 76.7s
CFN	Cyanogen fluoride	· 72.6s · 80.5s · 89.2s · 97.0s · 106.4s · 112.8s · 118.5s · 123.8s ·	
CIN	Cyanogen iodide	· 141.1s · 126.1s · 111.5s · 97.6s · 88.0s · 80.3s · 68.6s · 57.7s · 47.2s · 25.2s
CDN	Deuterocyanic acid	12 · 26.2 · 5.4 · −17.5 · −24.7 · −30.1 · −38.8 · −46.7 · −57.7 · −68.9s ·
CCl₂F₂	Dichlorodifluoromethane	12 · 29.8 · 43.9 · 57.0 · 68.6 · 76.1 · 81.6 · 90.1 · 97.8 · 104.6 · 118.5
CHCl₂F	Dichlorofluoromethane	· 8.9 · 6.2 · 20.9 · 33.9 · 42.6 · 48.8 · 58.6 · 67.5 · 75.5 · 91.3
CHClF₂	Chlorodifluoromethane	−135 · 40.8 · 53.6 · 65.8 · 76.4 · 83.4 · 88.6 · 96.5 · 103.7 · 110.2 · 122.8
CCl₃F	Trichlorofluoromethane	−160 · 23.7 · 6.8 · 9.1 · 23.0 · 32.3 · 39.0 · 49.7 · 59.0 · 67.6 · 84.3
CsBr	Cesium bromide	636 · 1300 · 1221 · 1140 · 1072 · 1026 · 993 · 938 · 887 · 838 · 748
CsCl	Cesium chloride	646 · 1300 · 1217 · 1139 · 1069 · 1023 · 989 · 934 · 884 · 837 · 744
CsF	Cesium fluoride	683 · 1251 · 1170 · 1092 · 1025 · 980 · 947 · 893 · 844 · 798 · 712
CsI	Cesium iodide	621 · 1280 · 1200 · 1124 · 1055 · 1009 · 976 · 923 · 873 · 828 · 738
ClF	Chlorine fluoride	−145 · −100.5 · −107.0 · −114.4 · −120.8 · −125.3 · −128.5 · −134.3 · −139.0 · −143.4 · s
ClF₃	Chlorine trifluoride	83 · +11.5 · +4.9 · −20.7 · −34.7 · −44.1 · −51.3 · −62.3 · −71.8 · −80.4 · −98.5
ClO	Chlorine monoxide	−116 · +2.2 · −12.5 · −26.5 · −39.4 · −48.0 · −54.3 · −64.3 · −73.1 · −81.6 · s
ClO₂	Chlorine dioxide	59 · +11.1 · +17.8 · −17.8 · −29.4 · −37.2 · −42.8 · −51.2 · −59.0 · · +42.0
Cl₂O₆	Dichlorine hexoxide	3.5 · 142.0 · 123.8 · 104.7 · 87.7 · 76.3 · 68.0 · 54.3 · 42.0 · 30.5 · +7.5
Cl₂O₇	Chlorine heptoxide	91 · 78.8 · 62.2 · 44.6 · 29.1 · 18.2 · 10.3 · 2.1 · −13.2 · −23.8 · −45.3

TABLE 4k-1. VAPOR PRESSURE OF INORGANIC COMPOUNDS—PRESSURES LESS THAN 1 ATMOSPHERE (Continued)

Formula	Name	Temp., °C 1 mm	5 mm	10 mm	20 mm	40 mm	60 mm	100 mm	200 mm	400 mm	760 mm	M.P.
HSO_3Cl	Chlorosulfonic acid	32.0	53.5	64.0	75.3	87.6	95.2	105.3	120.0	136.1	151.0d	−80
$Cr(CO)_6$	Chromium carbonyl	36.0	58.0	68.3	79.5	91.2	98.3	108.0	121.8	137.2	151.0	
CrO_2Cl_2	Chromyl chloride	−18.4	+3.2	13.8	25.7	38.5	46.7	58.0	75.2	95.2	117.1	
$CoCl_2$	Cobaltous chloride	s	s	s	s	770	801	843	904	974	1050	735
$Co(CO)_3NO$	Cobalt nitrosyl tricarbonyl	s	s	s	−1.3s	+11.0	18.5	29.0	44.4	62.0	80.0	−11
CbF_5	Columbium pentafluoride	s	s	86.3	103.0	121.5	133.2	148.5	172.2	198.0	225.0	75.5
Cu_2Br_2	Cuprous bromide	572	666	718	777	844	887	951	1052	1189	1355	504
Cu_2Cl_2	Cuprous chloride	546	645	702	766	838	886	960	1077	1249	1490	422
Cu_2I_2	Cuprous iodide		610	656	716	786	836	907	1018	1158	1336	605
$FeCl_3$	Ferric chloride	194.0s	221.8s	235.5s	246.0s	256.8s	263.7s	272.5s	285.0s	298.0s	319.0	304
$FeCl_2$	Ferrous chloride			700	737	779	805	842	897	961	1026	
F_2O	Fluorine monoxide	−196.1	−186.6	−182.3	−177.8	−173.0	−170.0	−165.8	−159.0	−151.9	−144.6	−223.9
$GaCl_3$	Gallium trichloride	48.0s	67.8s	76.5s	91.3	107.5	118.0	132.0	152.8	176.3	200.0	77.0
GeH_4	Germanium hydride	−163.0	−151.0	−145.3	−139.2	−131.6	−126.7	−120.3	−111.2	−100.2	−88.9	−165
$GeBr_4$	Germanium bromide	s	43.3	56.8	71.8	88.1	98.8	113.2	135.4	161.6	189.0	26.1
$GeCl_4$	Germanium chloride	−45.0	−24.9	−15.0	−4.1	+8.0	16.2	27.5	44.4	63.8	84.0	−49.5
$GeHCl_3$	Trichlorogermane	−41.3	−22.3	−13.0	−3.0	+8.8	16.2	26.5	41.6	58.3d	75.0d	−71.1
$Ge(CH_3)_4$	Tetramethylgermanium	−73.2	−54.6	−45.2	−35.0	−23.4	−16.2	−6.3	+8.8	26.0	44.0	−88
Ge_2H_6	Digermane	−88.7	−69.8	−60.1	−49.9	−38.2	−30.7	−20.3	−4.7	+13.3	31.5	−109
Ge_3H_8	Trigermane	−36.9	−12.8	−0.9	+11.8	26.3	35.5	47.9	67.0	88.6	110.8	−105.6
HD	Hydrogen deuteride			−259.8	−259.1	−258.2	−257.6	−256.6	−255.0	−253.0	−251.0	
HBr	Hydrogen bromide	−138.8s	−127.4s	−121.8s	−115.4s	−108.3s	−103.8s	−97.7s	−88.1s	−78.0	−66.5	−87.0
HCl	Hydrogen chloride	−150.8s	−140.7s	−135.6s	−130.0s	−123.8s	−119.6s	−114.0	−105.2	−95.3	−84.8	−114.3
HCN	Hydrogen cyanide	s	−55.3s	−47.7s	−39.7s	−30.9s	−25.1s	−17.8s	−5.3	+10.2	25.9	−13.2
HF	Hydrogen fluoride	s	−74.7	−65.8	−56.0	−45.0	−37.9	−28.2	−13.2	+2.5	19.7	−83.7
HI	Hydrogen iodide	−123.3s	−109.6s	−102.3s	−94.5s	−85.6s	−79.8s	−72.1s	−60.3s	−48.3	−35.1	−50.9
H_2O_2	Hydrogen peroxide	15.3	38.8	50.4	63.3	77.0	85.8	97.9	116.5	137.4d	158.0d	−0.9
H_2Se	Hydrogen selenide	−115.3s	−103.4s	−97.9s	−91.8s	−84.7s	−80.2s	−74.2s	−65.2s	−53.6	−41.1	−64
H_2S	Hydrogen sulfide	−134.3s	−122.4s	−116.3s	−109.7s	−102.3s	−97.9s	−91.6s	−82.3	−71.8	−60.4	−85.5
H_2S_2	Hydrogen disulfide	−43.2	−24.4	−15.2	−5.1	+6.0	12.8	22.0	35.3	49.6	64.0	−89.7
H_2Te	Hydrogen telluride	−96.4s	−82.4s	−75.4s	−67.8s	−59.1s	−53.7s	−45.7s	−32.4	−17.2	−2.0	−49.0
NH_2OH	Hydroxylamine	s	39.0	47.2	55.8	64.6	70.0	77.5	87.9	99.2	110.0	34.0
IF_5	Iodine pentafluoride	−15.2s	+1.5s	8.5	20.0	32.2	40.0	50.0	65.4	81.2	97.0	8.0
IF_7	Iodine heptafluoride	−87.0s	−70.7s	−63.0s	−54.5s	−45.3s	−39.1s	−31.9s	−20.7s	−8.3s	+4.0s	5.5
$Fe(CO)_5$	Iron pentacarbonyl		−6.5	4.6	16.7	30.3	39.1	50.3	60.3	86.1	105.0	−21
$PbBr_2$	Lead bromide	513	578	610	646	686	711	745	796	856	914	373
$PbCl_2$	Lead chloride	547	615	648	684	725	750	784	833	893	954	501

Vapor pressure (temperatures in °C for the indicated pressures in mm Hg; final column = melting point). "s" = sublimation, "d" = decomposition.

Formula	Substance	1 mm	5 mm	10 mm	20 mm	40 mm	60 mm	100 mm	200 mm	400 mm	760 mm	M.P.
PbF_2	Lead fluoride	s	861	904	950	1003	1036	1080	1144	1219	1293	855
PbI_2	Lead iodide	479	540	571	605	644	668	701	750	807	872	402
PbO	Lead oxide	943	1039	1085	1134	1189	1222	1265	1330	1402	1472	890
PbS	Lead sulfide	852s	928s	975s	1005s	1048s	1074s	1108s	1160	1221	1281	1114
$LiBr$	Lithium bromide	748	840	888	939	994	1028	1076	1147	1226	1310	547
$LiCl$	Lithium chloride	783	880	932	987	1045	1081	1129	1203	1290	1382	614
LiF	Lithium fluoride	1047	1156	1211	1270	1333	1372	1425	1503	1591	1681	870
LiI	Lithium iodide	723	802	841	883	927	955	993	1049	1110	1171	446
$MgCl_2$	Magnesium chloride	778	877	930	988	1050	1088	1142	1223	1316	1418	712
$MnCl_2$	Manganous chloride	—	736	778	825	879	913	960	1028	1108	1190	650
$HgBr_2$	Mercuric bromide	136.5s	165.3s	179.8s	194.3s	211.5s	221.0s	237.8	262.7	290.0	319.0	237
$HgCl_2$	Mercuric chloride	136.2s	166.0s	180.2s	195.8s	212.5s	222.2s	237.0s	256.5s	275.5s	304.0	277
HgI_2	Mercuric iodide	157.5s	189.2s	204.5s	220.0	238.2s	249.0s	261.8	291.0	324.2	354.0	259
MoF_6	Molybdenum hexafluoride	-65.5s	-49.0s	-40.8s	-32.0s	-22.1s	-16.2s	-8.0s	+4.1s	17.2	36.0	17
MoO_3	Molybdenum trioxide	734s	785s	814	851	892	917	955	1014	1082	1151	795
$NiCl_2$	Nickel chloride	671s	731s	759s	789s	821s	840s	866s	904s	945s	987s	1001
$Ni(CO)_4$	Nickel carbonyl	—	—	—	—	-23.0	-15.9	6.0	8.8	25.8	42.5	-25
NF_3	Nitrogen trifluoride	-184.5	-175.5	-170.7	-165.7	-160.2	-156.5	-152.3	-145.2	-137.4	-129.0	-183.7
NO	Nitric oxide	—	-180.6s	-178.2s	-175.3s	-171.7s	-168.9s	-166.0s	-162.3s	-156.8s	-151.7	-161
N_2O	Nitrous oxide	-143.4s	-133.4s	-128.7s	-124.0s	-118.3s	-114.9s	-110.3s	-103.6s	-96.2s	-88.5	-90.9
N_2O_4	Nitrogen tetroxide	-55.6s	-42.7s	-36.7s	-30.4s	-23.9s	-19.9s	-14.7s	5.0	8.0	21.0	-9.3
N_2O_5	Nitrogen pentoxide	-36.8s	1.8s	2.9s	7.4	10.0s	15.6s	16.7s	23.0s	24.4s	32.4	30
$NOCl$	Nitrosyl chloride	—	—	—	—	—	—	—	—	-20.3	-6.4	-64.5
NOF	Nitrosyl fluoride	-132.0	-120.3	-114.3	-107.8	-102.8	-95.7	-88.8	-79.2	-68.2	-56.0	-134
NO_2F	Nitroxyl fluoride	-143.7s	-132.1	-126.2	-119.8	-110.2	-108.4	-102.3	-93.5	-83.2	-72.0	-139
OsO_4	Osmium tetroxide (white)	5.6s	15.6s	26.0s	37.4s	50.5	59.4	71.5	89.5	109.3	130.0	42
OsO_4	Osmium tetroxide (yellow)	3.2s	22.0s	31.3s	41.0s	51.7s	59.4	71.5	89.5	109.3	130.0	56
O_3	Ozone	-180.4	-168.6	-163.2	-157.2	-150.7	-146.7	-141.0	-132.6	-122.5	-111.1	-251
PBr_3	Phosphorous tribromide	7.8	34.4	47.8	62.4	79.0	89.8	103.6	125.2	149.7	175.3	-40
PCl_3	Phosphorous trichloride	—	—	—	—	—	—	—	—	56.9	74.2	-111.8
PCl_5	Phosphorous pentachloride	55.5s	74.0s	83.2s	92.5s	102.5s	108.3s	117.0s	131.3s	147.2s	162.0s	—
PH_3	Phosphorous hydride (phosphene)	—	—	—	—	—	—	—	—	-98.3	-87.5	-132.5
PH_4Br	Phosphonium bromide	-43.7s	-28.5s	-21.2s	-13.3s	-5.0s	0.3s	7.4s	17.6s	28.0s	38.3d	28.5
PH_4Cl	Phosphonium chloride	-91.0s	79.6s	74.0s	68.0s	61.5s	57.3s	52.0s	44.0s	35.4s	27.0s	—
PH_4I	Phosphonium iodide	25.2s	9.0s	1.1s	7.3s	16.1s	21.9s	29.3s	39.9s	51.6s	62.3s	—
P_2O_3	Phosphorous trioxide	s	39.7	53.0	67.8	84.0	94.2	108.3	129.0	150.3	173.1	22.5
$POCl_3$	Phosphorous oxychloride	—	—	2.0	13.6	27.3	35.8	47.4	65.0	84.3	105.1	2
P_2O_5	Phosphorous pentoxide (stable form)	384s	424s	442s	462s	481s	493s	510s	532s	556s	591	569
P_2O_5	Phosphorous pentoxide (metastable form)	189	220	236	253	270	280	294	314	336	358	—
$PSBr_3$	Phosphorous thiobromide	50.0	72.4	83.6	95.5	108.0	116.0	126.3	141.8	157.8	175.0d	38
$PSCl_3$	Phosphorous thiochloride	-18.3	4.6	16.1	29.0	42.7	51.8	63.8	82.0	102.3	124.0	-36.2
KBr	Potassium bromide	795	892	940	994	1050	1087	1137	1212	1297	1383	730

TABLE 4k-1. VAPOR PRESSURE OF INORGANIC COMPOUNDS—PRESSURES LESS THAN 1 ATMOSPHERE (*Continued*)

Formula	Name	Temp., °C 1 mm	5 mm	10 mm	20 mm	40 mm	60 mm	100 mm	200 mm	400 mm	760 mm	M.P.
KCl	Potassium chloride	821	919	968	1020	1078	1115	1164	1239	1322	1407	790
KF	Potassium fluoride	885	988	1039	1096	1156	1193	1245	1323	1411	1502	880
KOH	Potassium hydroxide	719	814	863	918	976	1013	1064	1142	1233	1324	380
KI	Potassium iodide	745	840	887	938	995	1030	1080	1152	1238	1324	723
Re$_2$O$_7$	Rhenium heptoxide	212.5s	237.5s	248.0s	261.0s	272.0s	280.0s	289.0s	307.0	336.0	362.4s	296
RbBr	Rubidium bromide	781	876	923	975	1031	1066	1114	1186	1267	1352	682
RbCl	Rubidium chloride	792	887	937	990	1047	1084	1133	1207	1294	1381	715
RbF	Rubidium fluoride	921	982	1016	1052	1096	1123	1168	1239	1322	1408	760
RbI	Rubidium iodide	748	839	884	935	991	1026	1072	1141	1223	1304	642
SeO$_2$	Selenium dioxide	157.0s	187.7s	202.5s	217.5s	234.1s	244.6s	258.0s	277.0s	297.7s	317.0s	340
SeF$_6$	Selenium hexafluoride	−118.6s	−105.2s	−98.9s	−92.3s	−84.7s	−80.0s	−73.9s	−64.8s	−55.2s	−45.8s	−34.7
SeOCl$_2$	Selenium oxychloride	34.8	59.8	71.9	84.2	98.0	106.5	118.0	134.6	151.7	168.0	8.5
SeCl$_4$	Selenium tetrachloride	74.0s	96.3s	107.4s	118.1s	130.1s	137.8s	147.5s	161.0s	176.4s	191.5d	
SiH$_4$	Silane	−179.3	−168.6	−163.0	−156.9	−150.3	−146.3	−140.5	−131.6	−122.0	−111.5	−185
SiO$_2$	Silicon dioxide	s	s	1732	1798	1867	1911	1969	2053	2141	2227	1710
SiCl$_4$	Silicon tetrachloride	−63.4	−44.1	−34.4	−24.0	−12.1	−4.8	+5.4	21.0	38.4	56.8	−68.8
SiF$_4$	Silicon tetrafluoride	−144.0s	−134.8s	−130.4s	−125.9s	−120.8s	−117.5s	−113.3s	−107.2s	−100.7s	−94.8s	−90
SiH$_3$Br	Bromosilane	s	−85.7	−77.3	−68.3	−57.8	−51.1	−42.3	−28.6	−13.3	−2.4	−93.9
SiH$_3$Cl	Chlorosilane	−117.8	−104.3	−97.7	−90.1	−81.8	−76.0	−68.5	−57.0	−44.5	−30.4	
SiH$_3$F	Fluorosilane	−153.0	−145.5	−141.2	−136.3	−130.8	−127.2	−122.4	−115.2	−106.8	−98.0	
SiH$_3$I	Iodosilane	−86.5	−53.0	−43.7	−33.4	−21.8	−14.3	−4.4	+10.7	27.9	45.4	−57.0
SiBrCl$_2$F	Bromodichlorofluorosilane	−144.0s	−133.0	−127.0	−120.5	−112.8	−108.2	−101.7	−91.7	−81.0	35.4	−112.3
SiBrF$_3$	Bromotrifluorosilane	65.2	−45.5	−35.6	−24.5	−12.0	−4.7	+6.3	23.0	43.0	−41.7	−70.5
SiClF$_3$	Chlorotrifluorosilane	s	s	−66.8	−57.7	−47.4	−41.0	−31.9	−18.2	−2.6	−70.0	142
SiBr$_2$ClF	Dibromochlorofluorosilane	60.9	40.0	29.4	18.0	5.2	+3.2	31.6	50.7	59.5	−99.3	
SiBr$_2$F$_2$	Dibromodifluorosilane	−124.7	−110.5	−102.9	−94.5	−85.0	−78.6	−70.3	−58.0	−45.0	13.7	−66.9
SiH$_2$Br$_2$	Dibromosilane	−146.7	−136.0	−130.4	−124.3	−117.6	−113.3	−107.3	−98.3	−87.6	70.5	−70.2
SiCl$_2$F$_2$	Dichlorodifluorosilane	s	+3.8	18.0	34.1	52.6	64.0	79.4	101.8	125.5	+2.6	−139.7
SiH$_2$F$_2$	Difluorosilane	−114.8	99.3	91.4	82.7	72.8	66.4	57.5	44.6	29.0	−70.2	
SiH$_2$I$_2$	Diiodosilane	−112.5	95.8	88.2	79.8	70.4	64.2	55.9	43.5	29.3	149.5	1.0
SiH$_6$	Disilane	92.6	76.4	68.3	59.0	48.8	42.2	33.2	19.3	4.0	−14.3	−132.6
(SiH$_3$)$_2$O	Disiloxane	+4.0	27.4	38.8	51.5	65.3	73.9	85.4	102.2	120.6	−15.4	−144.2
SiCl$_3$F	Fluorotrichlorosilane	5.0	+17.8	29.4	41.5	55.2	63.8	75.4	92.5	113.6	12.2	−120.8
Si$_2$Cl$_6$	Hexachlorodisilane	81.0s	68.8s	63.1s	57.0s	50.6s	46.7s	41.7s	34.2s	26.4s	139.0	1.2
(SiCl$_3$)$_2$O	Hexachlorodisiloxane	46.3	74.7	89.3	104.2	121.5	132.0	146.0	166.2	189.5	135.6	33.2
Si$_2$F$_6$	Hexafluorodisilane	−27.7	−6.2	+4.3	15.8	28.4	36.6	47.4	63.6	81.7	−18.9s	18.6
Si$_2$Cl$_3$	Octachlorotrisilane										211.4	
SiH$_{10}$	Tetrasilane										100.0	−93.6

The following is a vapor-pressure table. The first numeric column is the melting point (°C); the remaining ten columns give the temperature (°C) at successively higher pressures.

Formula	Name	m.p.										
SiBr₃F	Tribromofluorosilane	-82.5	-46.1	-25.4	-15.1	-3.7	9.2	17.4	28.6	45.7	64.6	83.8
SiHBr₃	Tribromosilane	-73.5	-30.5	-8.0	+3.4	16.0	30.0	39.2	51.6	70.2	90.2	111.8
SiHCl₃	Trichlorosilane	-126.6	-80.7	-62.6	-53.4	-43.8	-32.9	-25.8	-16.4	-1.8	+14.5	31.8
SiHF₃	Trifluorosilane	-131.4	-152.0s	-142.7s	-138.2s	-132.9s	-127.3	-123.7	-118.7	-111.3	-102.8	-95.0
SiH₄s	Trisilane	-117.2	-68.9	-49.7	-40.0	-29.0	-16.9	-9.0	+1.6	17.8	35.5	53.1
(SiH₃)₃N	Disilazane	-105.7	-68.7	-49.9	-40.4	-30.0	-18.5	-11.0	+1.1	+14.0	31.0	48.7
AgCl	Silver chloride	455	912	1019	1074	1134	1200	1242	1297	1379	1467	1564
AgI	Silver iodide	552	820	927	983	1045	1111	1152	1210	1297	1400	1506
NaBr	Sodium bromide	755	806	903	952	1005	1063	1099	1148	1220	1304	1392
NaCl	Sodium chloride	800	865	967	1017	1072	1131	1169	1220	1296	1379	1465
NaCN	Sodium cyanide	564	817	928	983	1046	1115	1156	1214	1302	1401	1497
NaF	Sodium fluoride	992	1077	1186	1240	1300	1363	1403	1455	1531	1617	1704
NaOH	Sodium hydroxide	318	739	843	897	953	1017	1057	1111	1192	1286	1378
NaI	Sodium iodide	651	767	857	903	952	1005	1039	1083	1150	1225	1304
SnBr₄	Stannic bromide	31.0	s	58.3	72.7	88.1	105.5	116.2	131.0	152.8	177.7	204.7
SnCl₄	Stannic chloride	-30.2	-22.7	-1.0	+10.0	22.0	35.2	43.5	54.7	72.0	92.1	113.0
SnH₄	Stannic hydride	-149.9	-140.0	-125.8	-118.5	-111.2	-102.3	-96.6	-89.2	-78.0	-65.2	-52.3
SnI₄	Stannic iodide	144.5	s	156.0	175.8	196.2	218.8	234.2	254.2	283.5	315.5	348.0
SnCl₂	Stannous chloride	246.8	316	366	391	420	450	467	493	533	577	623
SrO	Strontium oxide		2068s	2198s	2262s	2333_4	2410_6					2430
SF₆	Sulfur hexafluoride	-50.2	-132.7s	-120.6s	-114.7s	-108.4s	-101.5s	-96.8s	-90.9s	-82.3s	-72.6s	-63.5s
SO₂	Sulfur dioxide	-73.2	-95.5s	-83.0s	-76.8s	-69.7	-60.5	-54.6	-46.9	-35.4	-23.0	-10.0
S₂Cl₂	Sulfur monochloride		7.4	15.7	27.5	40.0	54.1	63.2	75.3	93.5	115.4	138.0
SO₂Cl₂	Sulfuryl chloride	-54.1	s	s	-24.8	-13.4	-1.0	+7.2	17.8	33.7	51.3	69.2
SO₃	Sulfur trioxide (α)	16.8	-39.0s	-23.7s	-16.5s	-9.1s	+1.0s	4.0s	10.5s	20.5	32.6	44.8
SO₃	Sulfur trioxide (β)	32.3	-34.0s	-19.2s	-12.3s	-4.9s	+3.2s	8.0s	14.3s	23.7	32.6	44.8
SO₃	Sulfur trioxide (γ)	62.1	-15.3s	-2.0s	+4.3s	11.1s	17.9s	21.4s	28.0s	35.8s	44.0s	51.6s
H₂SO₄	Sulfuric acid	10.5	145.8	178.0	194.2	211.5	229.7	241.5	257.0	279.8	305.0	330.0d
SOBr₂	Thionyl bromide	-52.2	6.7	18.4	31.0	44.1	58.8	68.3	80.6	99.0	119.2	139.5
SOCl₂	Thionyl chloride	-104.5	-52.9	-32.4	-21.9	-10.5	+2.2	10.3	21.4	37.9	56.5	75.4
TaF₅	Tantalum pentafluoride	96.8	s	s	s	130.0	159.9	194.0	230.0			
TeCl₄	Tellurium tetrachloride	224	s	s	253	273	287	304	330	360	392	
TeF₆	Tellurium hexafluoride	-37.8	-111.3s	-98.8s	-92.4	-86.0s	-78.4s	-73.8s	-67.9s	-57.3s	-48.2s	-38.6s
TlBr	Thallium bromide	460	490	522	559	598	621	653	703	759	807	819
TlCl	Thallium chloride	430	487	517	550	589	612	645	694	748	763	807
TlI	Thallium iodide	440	502	531	567	607	631	663	712	748	763	823
TiCl₄	Titanium tetrachloride	-30	-13.9	9.4	21.3	34.2	48.4	58.0	71.0	90.5	112.7	136.0
WF₆	Tungsten hexafluoride	0.5	-71.4s	-56.5s	-49.2s	-41.5s	-27.5s	-20.3s	-10.0	+1.2	10.0	17.3
UF₆	Uranium hexafluoride	69.2	-38.8s	-22.0s	-13.8s	-5.2s	+4.4s	10.4s	18.2s	30.0s	42.7s	55.7
VOCl₃	Vanadyl trichloride		-23.2	0.2	12.2	26.6	40.0	49.8	62.5	82.0	103.5	127.2
ZnCl₂	Zinc chloride	365	428	481	508	536	566	584	610	648	689	732
ZnF₂	Zinc fluoride	872	1243	1328	1359	1402	1448	1480	1527	1602	1690	1770
ZrBr₄	Zirconium tetrabromide	450	207s	237s	250s	266s	281s	289s	301s	318s	337s	357s
ZrCl₄	Zirconium tetrachloride	437	190s	217s	230s	243s	253s	263s	279s	295s	312s	331s
ZrI₄	Zirconium tetraiodide	499	264s	297s	311s	329s	344s	355s	369s	389s	409s	431s

TABLE 4k-2. VAPOR PRESSURE OF INORGANIC COMPOUNDS—PRESSURES GREATER THAN 1 ATMOSPHERE

Formula	Name	Temp., °C									T_c	P_c
		1 atm	2 atm	5 atm	10 atm	20 atm	30 atm	40 atm	50 atm	60 atm		
NH_3	Ammonia	-33.6	-18.7	+4.7	25.7	50.1	66.1	78.9	89.3	98.3	132.4	111.5
A	Argon	-185.6	-179.0	-166.7	-154.9	-141.3	-132.0	-124.9	.....	.....	-122.0	48.0
BCl_3	Boron trichloride	12.7	33.2	66.0	96.7	135.4	161.5	.....	.....	.....	178.8	38.2
BF_3	Boron trifluoride	-100.7	-89.4	-72.6	-57.7	-40.0	-28.4	-19.0	.....	.....	-12.2	49.2
Br_2	Bromine	58.2	78.8	110.3	139.8	174.0	197.0	216.0	230.0	243.5	302.2	121
CCl_4	Carbontetrachloride	76.7	102.0	141.7	178.0	222.0	251.2	276.0	.....	.....	283.1	45.0
CO_2	Carbon dioxide	-78.2s	-69.1s	-56.7	-39.5	-18.9	-5.3	+5.9	14.9	22.4	31.1	73.0
CS_2	Carbon disulfide	46.5	69.1	104.8	136.3	175.5	201.5	222.8	240.0	256.0	273.0	72.9
CO	Carbon monoxide	-191.3	-183.5	-170.7	-161.0	-149.7	-141.9	.....	.....	.....	-138.7	34.6
$COCl_2$	Carbonyl chloride	+8.3	27.3	57.2	85.0	119.0	141.8	159.8	174.0	.....	181.7	56.0
$CClF_3$	Chlorotrifluoromethane	-81.2	-66.7	-42.7	-18.5	+12.0	34.8	52.8	.....	.....	53	40.3
C_2N_2	Cyanogen	-21.0	-4.4	+21.4	44.6	72.6	91.6	106.5	118.2	.....	126.6	58.2
CCl_2F_2	Dichlorodifluoromethane	-29.8	-12.2	+16.1	42.4	74.0	95.6	.....	.....	.....	111.5	39.6
$CHCl_2F$	Dichlorofluoromethane	8.9	28.4	59.0	87.0	121.2	144.0	162.6	177.5	.....	178.5	51.0
$CHClF_2$	Chlorodifluoromethane	-40.8	-24.7	+0.3	24.0	52.0	70.3	85.3	.....	.....	96.0	48.7

CCl₃F	Trichlorofluoromethane	23.7	44.1	77.3	108.2	146.7	172.0	194.0			198.0	43.2
Cl₂	Chlorine	−33.8	−16.9	+10.3	35.6	65.0	84.8	101.6	115.2	127.1	144.0	76.1
HBr	Hydrogen bromide	−66.5	−51.5	−29.1	−8.4	16.8	33.9	48.1	60.0	70.6	90.0	84.4
HCl	Hydrogen chloride	−84.8	−71.4	−50.5	−31.7	−8.8	+5.9	17.8	27.9	36.2	51.4	81.6
HCN	Hydrogen cyanide	25.9	45.8	75.8	102.7	135.0	153.8	169.9	183.5		183.5	50.0
HI	Hydrogen iodide	−35.1	−18.9	+7.3	32.0	62.2	83.2	100.7	116.2	127.5	151.0	82.0
H₂S	Hydrogen sulfide	−60.4	−45.9	−22.3	+0.4	25.5	41.9	55.8	66.7	76.3	100.3	88.9
H₂Se	Hydrogen selenide	−41.1	−25.2	0.0	23.4	50.8	69.7	84.6	97.2	108.7	137	91.0
Kr	Krypton	−152.0	−143.5	−130.0	−118.0	−101.7	−88.8	−78.4	−66.5		−63	54
NO	Nitric oxide	−151.7	−145.1	−135.7	−127.3	−116.8	−109.0	−103.2	−99.0	−94.8	−92.9	64.6
N₂O	Nitrous oxide	−88.5	−76.8	−58.0	−40.7	−18.8	+4.3	8.0	18.0	27.4	36.5	71.7
N₂O₄	Nitrogen tetroxide	21.0	37.3	59.8	79.4	100.3	112.3	121.4	127.0	132.2	158	99
SiF₄	Silicon tetrafluoride	−94.8$_s$	−84.4	−67.9	−52.6	−33.4	−21.2				−14.2	36.7
SiClF₃	Chlorotrifluorosilane	−70.0	−57.3	−37.2	−18.6	−4.1	19.4				34.8	34.2
SiCl₂F₂	Dichlorodifluorosilane	−31.8	−15.1	+11.6	36.6	66.2	86.0				95.8	34.5
SiCl₃F	Fluorotrichlorosilane	12.2	32.4	64.6	94.2	131.8	156.0				165.3	35.3
SnCl₄	Stannic chloride	113.0	141.3	184.3	223.0	270.0	299.8				318.7	37.9
SO₂	Sulfur dioxide	−10.0	6.3	32.1	55.5	83.8	102.6	118.0	130.2	141.7	157.2	77.7
SO₃	Sulfur trioxide	44.8	60.0	82.5	104.0	138.0	157.8	175.0	187.8	198.0	218.3	83.6

TABLE 4k-3. VAPOR PRESSURE OF ORGANIC COMPOUNDS—PRESSURES LESS THAN 1 ATMOSPHERE

Formula	Name	Temp., °C										M.P.
		1 mm	5 mm	10 mm	20 mm	40 mm	60 mm	100 mm	200 mm	400 mm	760 mm	
$CClF_3$	Chlorotrifluoromethane	−149.5	−139.2	−134.1	−128.5	−121.9	−117.3	−111.7	−102.5	−92.7	−81.2	
CCl_2F_2	Dichlorodifluoromethane	−118.5	−104.6	−97.8	−90.1	−81.6	−76.1	−68.6	−57.0	−43.9	−29.8	−104
CCl_2O	Carbonyl chloride	−92.9	−77.0	−69.3	−60.3	−50.3	−44.0	−35.6	−22.3	−7.6	+8.3	
CCl_3F	Trichlorofluoromethane	−84.3	−67.6	−59.0	−49.7	−39.0	−32.3	−23.0	−9.1	+6.8	+23.7	
CCl_4	Carbontetrachloride	−50.0s	−30.0s	−19.6	−8.2	+4.3	12.3	23.0	38.3	57.8	76.7	−22.6
$CHClF_2$	Chlorodifluoromethane	−122.8	−110.2	−103.7	−96.5	−88.8	−83.4	−76.4	−65.8	−53.6	−40.8	−160
$CHCl_2F$	Dichlorofluoromethane	−91.3	−75.5	−67.5	−58.6	−48.8	−42.6	−33.9	−20.9	−6.2	+8.9	−135
$CHCl_3$	Trichloromethane	−58.0	−39.1	−29.7	−19.0	−7.1	+0.5	10.4	25.9	42.7	61.3	−63.5
CHN	Hydrocyanic acid	−70.8s	−55.6s	−48.2s	−40.3s	−31.3s	−25.8s	−18.8s	−5.9	+9.8	25.8	−14
CH_2O	Formaldehyde			−88.0	−79.6	−70.6	−65.0	−57.3	−46.0	−33.0	−19.5	−92
CH_2O_2	Formic acid	−20.0s	−5.0s	+2.1s	10.3	24.0	32.4	43.8	61.4	80.3	100.6	8.2
CH_3Br	Methyl bromide	−96.3s	−80.6s	−72.8	−64.0	−54.2	−48.0	−39.4	−26.5	−11.9	+3.6	−93
CH_3Cl	Methyl chloride		−99.5s	−92.4	−84.8	−76.0	−70.4	−63.0	−51.2	−38.0	−24.0	−97.7
CH_3F	Methyl fluoride	−147.3	−137.0	−131.6	−125.9	−119.1	−115.0	−109.0	−99.9	−89.5	−78.2	
CH_3I	Methyl iodide		−55.0	−45.8	−35.6	−24.2	−16.9	−7.0	8.0	25.3	42.4	−64.4
CH_3NO_2	Nitromethane	−29.0	−7.9	+2.8	14.1	27.5	35.5	46.6	63.5	82.0	101.2	−29
CH_4	Methane	−205.9s	−199.0s	−195.5s	−191.8s	−187.7s	−185.1s	−181.4	−175.5	−168.8	−161.5	−182.5
CH_4O	Methanol	−44.0	−25.3	−16.2	−6.0	+5.0	12.1	21.2	34.8	49.9	64.7	−97.8
CH_4S	Methanethiol	−90.7	−75.3	−67.5	−58.8	−49.2	−43.1	−34.8	−22.1	−7.9	+6.8	−121
CH_5N	Methylamine	−95.8s	−81.3	−73.8	−65.9	−56.9	−51.3	−43.7	−32.4	−19.7	−6.3	−93.5
CO	Carbon monoxide	−222.0s	−217.2s	−215.0s	−212.8s	−210.0s	−208.1s	−205.7s	−201.3	−196.3	−191.3	−205.0
CS_2	Carbon disulfide	−73.8	−54.3	−44.7	−34.3	−22.5	−15.3	−5.1	10.4	28.0	46.5	−110.8
C_2ClF_3	1-Chloro-1,2,2-trifluoro-ethylene	−116.0	−102.5	−95.9	−88.2	−79.7	−74.1	−66.7	−55.0	−41.7	−27.9	−157.5
$C_2Cl_2F_4$	1,2-Dichloro-1,1,2,2-tetra-fluoroethane	−95.4	−80.0	−72.3	−63.5	−53.7	−47.5	−39.1	−26.3	−12.0	+3.5	−94
$C_2Cl_3F_3$	1,1,2-Trichloro-1,2,2-tri-fluoroethane	−68.0s	−49.4s	−40.3s	−30.0	−18.5	−11.2	+1.7	13.5	30.2	47.6	−35
C_2H_2	Acetylene	−142.9s	−133.0s	−128.2s	−122.8s	−116.7s	−112.8s	−107.9s	−100.3s	−92.0s	−84.0s	−81.5
$C_2H_2Cl_2$	cis-1,2-Dichloroethylene	−58.4	−39.2	−29.9	−19.4	−7.9	−0.5	9.5	24.6	41.0	59.0	−80.5
$C_2H_2Cl_2$	trans-1,2-Dichloroethylene	−65.4s	−47.2	−38.0	−28.0	−17.0	−9.4	+0.2	14.3	30.8	47.8	−50.0
C_2H_4	Ethylene	−168.3	−158.3	−153.2	−147.6	−141.3	−137.3	−131.8	−123.4	−113.9	−103.7	−169
$C_2H_4Br_2$	1,2-Dibromoethane	−27.0s	+4.7s	18.6	32.7	48.0	57.9	70.4	89.8	110.1	131.5	10
$C_2H_4Cl_2$	1,1-Dichloroethane	−60.7	−41.9	−32.3	−21.9	−10.2	−2.9	+7.2	22.4	39.8	57.4	−96.7

Temperatures (°C) at which the indicated vapor pressure (mm Hg) is reached, with melting point (mp, °C).

Formula	Name	1	5	10	20	40	60	100	200	400	760	mp
$C_2H_4Cl_2$	1,2-Dichloroethane	-44.5_5	-24.0	-13.6	-2.4	$+10.0$	18.1	29.4	45.7	64.0	82.4	-35.3
$C_2H_4O_2$	Acetic acid	-17.2_2	$+6.3_5$	17.5	29.9	43.0	51.7	63.0	80.0	99.0	118.1	16.7
$C_2H_4O_2$	Methyl formate	-74.3	-57.0	-48.6	-39.2	-28.7	-21.9	-12.9	0.8	16.0	32.0	-99.8
C_2H_5Br	Ethyl bromide	-74.3	-56.4	-47.5	-37.8	-26.7	-19.5	-10.0	4.5	21.0	38.4	-117.8
C_2H_5Cl	Ethyl chloride	-89.8	-73.9	-65.8	-56.8	-47.0	-40.6	-32.0	-18.6	$+3.9$	$+12.3$	-139
C_2H_5F	Ethyl fluoride	-117.0_5	-103.8	-97.7	-90.0	-81.8	-76.4	-69.3	-58.0	-45.5	-32.0	—
C_2H_6	Ethane	-159.5	-148.5	-142.9	-136.7	-129.8	-125.4	-119.3	-110.2	-98.0	-88.6	-183.2
C_2H_6O	Ethanol	-31.3	-12.0	-2.3	$+8.0$	19.0	26.0	34.9	48.4	63.5	78.4	-112
C_2H_6O	Dimethyl ether	-115.7	-101.1	-93.3	-85.2	-76.2	-70.4	-62.7	-50.9	-37.8	-23.7	-138.5
C_2H_6S	Dimethyl sulfide	-75.6	-58.0	-49.2	-39.4	-28.4	-21.4	-12.0	2.6	18.7	36.0	-83.2
C_2H_6S	Ethanethiol	-76.7	-59.1	-50.2	-40.7	-29.8	-22.4	-13.0	1.5	17.7	35.0	-121
C_2H_7N	Ethylamine	-82.3_3	-66.4	-58.3	-48.6	-39.8	-33.4	-25.1	-12.3	2.0	16.6	-80.6
C_2H_7N	Dimethylamine	-87.7	-72.2	-64.6	-56.0	-46.7	-40.7	-32.6	-20.4	-7.1	7.4	-96
C_2N_2	Cyanogen	-95.8_8	-83.2_8	-76.8_8	-70.1_8	-62.7_8	-57.9_8	-51.8_5	-42.6_8	-33.0	-21.0	-34.4
C_3H_4	Propadiene	-120.6	-108.0	-101.0	-93.4	-85.2	-78.8	-72.5	-61.3	-48.3	-34.5	-136
C_3H_4	Propyne	-111.0_6	-97.5	-90.5	-82.9	-74.3	-68.8	-61.3	-49.8	-37.2	-23.3	-102.7
$C_3H_5N_3O_9$	Nitroglycerine	127	167	188	210	235	251 (d)	(e)	—	—	—	11
C_3H_6	Propylene	-131.9	-120.7	-112.1	-104.7	-96.5	-91.3	-79.6	-68.4	-55.6	-47.7	-185
C_3H_6O	Acetone	-59.4	-40.5	-31.1	-20.8	-9.4	-2.0	7.7	22.7	39.5	56.5	-94.6
$C_3H_6O_2$	Propionic acid	4.6	28.0	39.7	52.0	65.8	74.1	85.8	102.5	122.0	141.1	-22
$C_3H_6O_2$	Methyl acetate	-57.2	-38.6	-29.3	-19.1	-7.9	0.5	9.4	24.0	40.0	57.8	-98.7
$C_3H_6O_2$	Ethyl formate	-60.5	-42.2	-33.0	-22.7	-11.5	-4.3	5.4	20.0	37.1	54.3	-79
C_3H_8	Propane	-128.9	-115.4	-108.5	-100.9	-92.4	-87.0	-79.6	-68.4	-55.6	-42.1	-187.1
C_3H_8O	1-Propanol	-15.0	$+5.0$	14.7	25.3	36.4	43.5	52.8	66.8	82.0	97.8	-127
C_3H_8O	2-Propanol	-26.1	-7.0	2.4	12.7	23.8	30.5	39.5	53.0	67.8	82.5	-85.8
C_3H_8O	Ethyl methyl ether	-91.0	-75.6	-67.8	-59.1	-49.4	-43.3	-34.8	-22.0	-7.8	7.5	—
$C_3H_8O_3$	Glycerol	125.5	153.8	167.2	182.2	198.0	208.0	220.1	240.0	263.0	290.0	17.9
C_3H_9N	Propylamine	-64.4	-46.3	-37.2	-27.1	-16.0	-9.0	0.5	15.0	31.5	48.5	-83
C_3H_9N	Trimethylamine	-97.1	-81.7	-73.8	-65.0	-55.2	-48.8	-40.3	-27.0	-12.5	2.9	-117.1
C_4H_2	1,3-Butadiyne	-82.5_8	-68.0_8	-61.2_8	-53.8_8	-45.9_8	-41.0_8	-34.0	-20.9	6.1	9.7	-34.9
C_4H_6	1,2-Butadiene	-89.0	-72.7	-64.2	-54.9	-44.3	-37.5	-28.3	-14.2	1.8	18.5	—
C_4H_6	1,3-Butadiene	-102.8	-87.6	-79.7	-71.0	-61.3	-55.1	-46.8	-33.9	-19.3	-4.5	-108.9
C_4H_6	Cyclobutene	-99.1	-83.4	-75.4	-66.6	-56.4	-50.0	-41.2	-27.8	-12.2	2.4	-130
C_4H_6	1-Butyne	-92.5	-76.7	-68.7	-59.9	-50.0	-43.4	-34.9	-21.6	-6.9	8.7	—
C_4H_6	2-Butyne	-73.0_5	-57.9_8	-50.5_8	-42.5_8	-33.9_8	-27.8	-18.8	5.0	10.6	27.2	-32.5
$C_4H_6O_3$	Acetic anhydride	1.7	24.8	36.0	48.3	62.1	70.8	82.2	100.0	119.8	139.6	-73
$C_4H_6O_4$	Dimethyl oxalate	20.0	44.0	56.0	69.4	83.6	92.8	104.8	123.3	143.3	163.3	—
$C_4H_8O_2$	Butyric acid	25.5	49.8	61.5	74.0	88.0	96.5	108.0	125.5	144.5	163.5	-4.7
$C_4H_8O_2$	Isobutyric acid	14.7	39.3	51.2	64.0	77.8	86.3	98.0	115.8	134.5	154.5	-47
$C_4H_8O_2$	Ethyl acetate	-43.4	-23.5	-13.5	3.0	9.1	16.6	27.0	42.0	59.3	77.1	-82.4
$C_4H_8O_2$	Methyl propionate	-42.0	-21.5	-11.8	-1.0	11.0	18.7	29.0	44.2	61.8	79.8	-87.5

TABLE 4k-3. VAPOR PRESSURE OF ORGANIC COMPOUNDS—PRESSURES LESS THAN 1 ATMOSPHERE (Continued)

Formula	Name	Temp., °C										M.P.
		1 mm	5 mm	10 mm	20 mm	40 mm	60 mm	100 mm	200 mm	400 mm	760 mm	
$C_3H_6O_2$	Propyl formate	−43.0	−22.7	−12.6	−1.7	+10.8	18.8	29.5	45.3	62.6	81.3	−92.9
C_4H_{10}	Butane	−101.5	−85.7	−77.8	−68.9	−59.1	−52.8	−44.2	−31.2	−16.3	−0.5	−135
C_4H_{10}	2-Methylpropane	−109.2	−94.1	−86.4	−77.9	−68.4	−62.4	−54.1	−41.5	−27.1	−11.7	−145
$C_4H_{10}O$	Butyl alcohol	−1.2	20.0	30.2	41.5	53.4	60.3	70.1	84.3	100.8	117.5	−79.9
$C_4H_{10}O$	sec-Butyl alcohol	−12.2	+7.2	16.9	27.3	38.1	45.2	54.1	67.9	83.9	99.5	−114.7
$C_4H_{10}O$	Isobutyl alcohol	−9.0	11.0	21.7	32.4	44.1	51.7	61.5	75.9	91.4	108.0	−108
$C_4H_{10}O$	tert-Butyl alcohol	-20.4_s	-3.0_s	$+5.5_s$	14.3_s	24.5_s	31.0	39.8	52.7	68.0	82.9	+25.3
$C_4H_{10}O$	Diethyl ether	−74.3	−56.9	−48.1	−38.5	−27.7	−21.8	−11.5	+2.2	17.9	34.6	−116.3
$C_4H_{10}S$	Diethyl sulfide	−39.6	−18.6	−8.0	+3.5	16.1	24.2	35.0	51.3	69.7	88.0	−99.5
$C_4H_{11}N$	Diethylamine	s	s	−33.0	−22.6	−11.3	−4.0	+6.0	21.0	38.0	55.5	−38.9
$C_4H_{12}Si$	Tetramethylsilane	−83.8	−66.7	−58.0	−48.3	−37.4	−30.3	−20.9	−6.5	+10.0	27.0	−102.1
$C_5H_{10}O_2$	Ethyl propionate	−28.0	−7.2	+3.4	14.3	27.2	35.1	45.2	61.7	79.8	99.1	−72.6
$C_5H_{10}O_2$	Propyl acetate	−26.7	−5.4	5.0	16.0	28.8	37.0	47.8	64.0	82.0	101.8	−92.5
$C_5H_{10}O_2$	Methyl butyrate	−26.8	−5.5	5.0	16.7	29.6	37.4	48.0	64.3	83.1	102.3	−84.7
$C_5H_{10}O_2$	Methyl isobutyrate	−34.1	−13.0	2.9	8.4	21.0	28.9	39.6	55.7	73.6	92.6	
$C_5H_{10}O_2$	Isobutyl formate	−32.7	−11.4	0.8	11.0	24.1	32.4	43.4	60.0	79.0	98.2	−95.3
C_5H_{12}	Pentane	−76.6	−62.5	−50.1	−40.2	−29.2	−22.2	−12.6	+1.9	18.5	36.1	−129.7
C_5H_{12}	2-Methylbutane	−82.9	−65.8	−57.0	−47.3	−36.5	−29.6	−20.2	−5.9	10.5	27.8	−159.7
C_5H_{12}	2,2-Dimethylpropane	-102.0_s	-85.4_s	-76.7_s	-67.2_s	-56.1_s	-49.0_s	-39.1_s	-23.7_s	−7.1	+9.5	−16.6
$C_5H_{12}O$	Ethyl propyl ether	−64.3	−45.0	−35.0	−24.0	−12.0	−4.0	+6.8	23.3	41.6	61.7	
C_6H_5Br	Bromobenzene	+2.9	27.8	40.0	53.8	68.6	78.1	90.8	110.1	132.3	156.2	−30.7
C_6H_5Cl	Chlorobenzene	−13.0	10.6	22.2	35.3	49.7	58.3	70.7	89.4	110.0	132.2	−45.2
C_6H_5F	Fluorobenzene	-43.4_s	−22.8	−12.4	+1.2	11.5	19.6	30.4	47.2	65.7	84.7	−42.1
C_6H_5I	Iodobenzene	+24.1	50.6	64.0	78.3	94.4	105.0	118.3	139.8	163.9	188.6	−28.5
C_6H_6	Benzene	-36.7_s	-19.6_s	-11.5_s	-2.6_s	+7.6	15.4	26.1	42.2	60.6	80.1	+5.5
C_6H_6O	Phenol	$+40.1_s$	62.5	73.8	86.0	100.1	108.4	121.4	139.0	160.0	181.9	+40.6
C_6H_7N	Aniline	+34.8	57.9	69.4	82.0	96.7	106.0	119.9	140.1	161.9	184.4	−6.2
C_6H_{12}	Cyclohexane	-45.3_s	-25.4_s	-15.9_s	-5.0_s	+6.7	14.7	25.5	42.0	60.8	80.7	+6.6
C_6H_{14}	Hexane	−53.9	−34.5	−25.0	−14.1	−2.3	5.4	15.8	31.6	49.6	68.7	−95.3
C_6H_{14}	2,3-Dimethylbutane	−63.6	−44.5	−34.9	−24.1	−12.4	−4.9	+5.4	21.1	39.0	58.0	−128.2
C_7H_8	Toluene	−26.7	−4.4	6.4	18.4	31.8	40.3	51.9	69.5	89.5	110.6	−95.0
C_7H_{16}	Heptane	−34.0	−12.7	2.1	9.5	22.3	30.6	41.8	58.7	78.0	98.4	−90.6
C_8H_{10}	Ethylbenzene	−9.8	13.9	25.9	38.6	52.8	61.8	74.1	92.7	113.8	136.2	−94.9
C_8H_{18}	Octane	−14.0	8.3	19.2	31.5	45.1	53.8	65.7	83.6	104.0	125.6	−56.8
$C_{12}H_{26}$	Dodecane	+47.8	75.8	90.0	104.6	121.7	132.1	146.2	167.2	191.0	216.2	−9.6

Tables 4k-5 to 4k-14, Vapor Pressures of Special Gases, listing values of the vapor pressures of He_4, He_3, normal and equilibrium H_2, Ne, N_2, and O_2, were taken from Thermometry at Low Temperature, a master's essay at the University of Pittsburgh, 1965, by Edward R. Simco. This booklet is also entitled Research Report 4 and was supported in part by the National Science Foundation.

Table 4k-15, Vapor Pressures of the Chemical Elements, lists values of the vapor pressure, temperature, and heat associated with the phase transitions for the chemical elements. The numbers represent temperature in degrees Celsius at which the vapor pressure is the value appearing at the top of the column. A circled dot between columns indicates a change of phase. The six columns on the right side list the following information:

ΔH_{v298}	heat of vaporization at 25°C, or atmospheric boiling temperature if the value contains an asterisk (*), cal/mol
T_m	melting temperature
ΔH_m	heat of melting, cal/mol
T_t	transition temperature
ΔH_t	heat of transition, cal/mol
Trans	designates solid-state transition

Equilibrium vapor pressures are listed for substances with polymorphic vapor or condensed forms (As, Sb, Bi, P, Po, S, Se, Te). The basic sources should be consulted for vapor pressures of the various polymorphic forms. The sources for this table are: (1) Ralph Hultgren, Raymond L. Orr, Philip D. Anderson, and Kenneth K. Kelley, "Selected Values of Thermodynamic Properties of Metals and Alloys," John Wiley & Sons, Inc., New York, 1963 (updated by privately distributed supplements); (2) Daniel R. Stull and Gerard C. Sinke, "Thermodynamic Properties of the Elements," *Advances in Chem. Ser. No. 18*: (3) Richard E. Honig, "Vapor Pressure Data for the Solid and Liquid Elements," *RCA Rev.* **23**(4), 567–586 (1962); (4) Richard E. Honig and H. O. Hook, "Vapor Pressure Data for Some Common Gases," *RCA Rev.* **21**(3), 360–368 (1960).

Table 4k-16, Vapor Pressure of Ice, has been taken from the *NBS Circ.* 564, Tables of Thermal Properties of Gases, by J. Hilsenrath, C. W. Beckett, W. S. Benedict, L. Fano, H. J. Hoge, J. F. Masi, R. L. Nuttall, Y. S. Touloukian, and H. W. Woolley, U.S. Government Printing Office, Washington, D.C., 1955. The values were smoothed, and adjusted to agree with the ice-point value adopted in Table 4k-17.

Table 4k-17, Vapor Pressure of Liquid Water below 100°C, and Table 4k-18, Vapor Pressure of Liquid Water above 100°C, have been taken from the recent work of M. R. Gibson and E. A. Bruges, *J. Mech. Eng. Sci.*, **9**(1), 24–35 (February, 1967).

Table 4k-19, Vapor Pressure of Mercury, is taken from the compilation of J. Johnston, F. Fenwick, and H. G. Leopold, "International Critical Tables," vol. III, McGraw-Hill Book Company, New York, 1928.

Table 4k-20, Vapor Pressure of Carbon Dioxide, is from C. H. Meyers and M. S. Van Dusen, *J. Research NBS*, **10**, 409 (1933).

Table 4k-21, Vapor Pressure of Ethyl Alcohol, and Table 4k-22, Vapor Pressure of Methyl Alcohol, are reprinted by permission from the "Smithsonian Physical Tables," 9th ed. Smithsonian Institution, Washington, D.C., 1954.

Table 4k-23, Constants in the Equation for the Rate of Evaporation of Metals, is taken by permission from pages 752–754 of "Scientific Foundations of Vacuum Technique," by S. Dushman, John Wiley & Sons, Inc., New York, 1949.

TABLE 4k-4. VAPOR PRESSURE OF ORGANIC COMPOUNDS—PRESSURES GREATER THAN 1 ATMOSPHERE

Formula	Name	Temp., °C									T_c	P_c
		1 atm	2 atm	5 atm	10 atm	20 atm	30 atm	40 atm	50 atm	60 atm		
CClF₃	Chlorotrifluoromethane	− 81.2	− 66.7	− 42.7	− 18.5	+ 12.0	34.8	52.8	......	......	53	40.3
CCl₂F₂	Dichlorodifluoromethane	− 29.8	− 12.2	+ 16.1	42.4	74.0	95.6	......	......	......	111.5	39.6
CCl₂O	Carbonyl chloride	8.3	27.3	57.2	85.0	119.0	141.8	159.8	174.0	......	181.7	56.0
CCl₃F	Trichlorofluoromethane	23.7	44.1	77.3	108.2	146.7	172.0	194.0	......	......	198.0	43.2
CCl₄	Carbontetrachloride	76.7	102.0	141.7	178.0	222.0	251.2	276.0	......	......	283.1	45.0
CHClF₂	Chlorodifluoromethane	− 40.8	− 24.7	+ 0.3	24.0	52.0	70.3	85.3	......	......	96	48.7
CHCl₂F	Dichlorofluoromethane	8.9	28.4	59.0	87.0	121.2	144.0	162.6	177.5	......	178.5	51.0
CHCl₃	Trichloromethane	61.3	83.9	120.0	152.3	191.8	216.5	237.5	254.0	......	260	54.9
CHN	Hydrocyanic acid	25.8	45.5	75.5	103.5	134.2	154.0	170.2	183.5	......	183.5	50.6
CH₃Br	Methyl bromide	3.6	23.3	54.8	84.0	121.7	147.5	170.2	190.0	......	194	51.6
CH₃Cl	Methyl chloride	− 24.0	− 6.4	+ 22.0	47.3	77.3	97.5	113.8	126.0	137.5	143.8	65.8
CH₃F	Methyl fluoride	− 78.2	− 64.5	− 42.0	− 21.0	+ 2.6	15.5	26.5	36.0	43.5	44.9	62.0
CH₃I	Methyl iodide	42.4	65.5	101.8	138.0	176.5	206.0	228.5	248.0	......	255	54.6
CH₄	Methane	−161.5	−152.3	−138.3	−124.8	−108.5	− 96.3	− 86.3	......	......	− 82.1	45.8
CH₄O	Methanol	64.7	84.0	112.5	138.0	167.8	186.5	203.5	214.0	224.0	240.0	78.7
CH₄S	Methanethiol	6.8	26.1	55.9	83.4	117.5	140.0	157.7	172.0	185.0	196.8	71.4
CH₅N	Methylamine	− 6.3	+ 10.1	36.0	59.5	87.8	106.3	121.8	133.7	144.6	156.9	73.6
CO	Carbon monoxide	−191.3	−183.5	−170.7	−161.0	−149.7	−141.9	......	......	......	−138.7	34.6
CS₂	Carbon disulfide	46.5	69.1	104.8	136.3	175.5	201.5	222.8	240.0	256.0	273.0	72.9
C₂ClF₃	1-Chloro-1,2,2-trifluoroethylene	− 27.9	− 11.1	+ 15.5	40.0	71.1	91.9	......	......	......	107.0	39.0
C₂Cl₂F₄	1,2-Dichloro-1,1,2,2-tetrafluoroethane	3.5	22.8	54.0	82.3	117.5	140.9	......	......	......	145.7	32.3
C₂Cl₃F₃	1,1,2-Trichloro-1,2,2-trifluoroethane	47.6	70.0	105.5	138.0	177.7	205.0	......	......	......	214.1	33.7

C_2H_2	Acetylene	-84.0$_5$	-71.6	-50.2	-32.7	-10.0	+4.8	16.8	26.8	34.8	36.0	62.0
$C_2H_2Cl_2$	cis-1,2-Dichloroethylene	59.0	82.1	119.3	152.3	194.0	221.5	244.5	260.0		271.0	57.9
$C_2H_2Cl_2$	trans-1,2,Dichloroethylene	47.8	69.8	104.0	135.7	174.0	199.8	220.0	236.5		243.3	54.5
C_2H_4	Ethylene	-103.7	-90.8	-71.1	-52.8	-29.1	-14.2	-1.5	+8.9		9.6	50.7
$C_2H_4Br_2$	1,2-Dibromoethane	131.5	157.7	200.0	237.0	269.0	286.0	295.0	300.0	304.5	309.8	70.6
$C_2H_4Cl_2$	1,1-Dichloroethane	57.3	80.2	117.3	150.3	192.7	220.0	243.0	261.5		261.5	50.0
$C_2H_4Cl_2$	1,2-Dichloroethane	83.7	108.1	147.8	183.5	226.5	254.0	272.0	285.0		288.4	53.0
$C_2H_4O_2$	Acetic acid	118.1	143.5	180.3	214.0	252.0	276.5	297.0	312.5		321.6	57.2
$C_2H_4O_2$	Methyl formate	32.0	51.9	83.5	112.0	147.2	169.7	188.5	213.0		214.0	59.1
C_2H_5Br	Ethyl bromide	38.4	60.2	95.0	126.8	164.3	188.0	206.5	220.0	229.5	230.8	61.5
C_2H_5Cl	Ethyl chloride	12.3	32.5	64.0	92.6	127.3	149.5	167.0	180.5		187.2	52.0
C_2H_5F	Ethyl fluoride	-32.0	-16.7	+7.7	30.2	57.5	75.7	90.0			102.2	49.6
C_2H_6	Ethane	-88.6	-75.0	-52.8	-32.0	-6.4	+10.0	23.6			32.3	48.2
C_2H_6O	Ethanol	78.4	97.5	126.0	151.8	183.0	203.0	218.0	230.0	242.0	243.5	63.1
C_2H_6O	Dimethyl ether	-23.7	-6.4	+20.8	45.5	75.7	96.0	112.1	125.2		126.9	52.0
C_2H_6S	Ethanethiol	35.0	56.6	90.7	121.9	159.5	184.3	204.7	220.0		225.5	54.2
C_2H_6S	Dimethyl sulfide	36.0	57.8	92.3	124.5	163.8	188.5	209.0	224.5		229.9	54.6
C_2H_7N	Ethylamine	16.6	35.7	65.3	91.8	124.0	146.0	163.0	176.0		183.2	55.5
C_2H_7N	Dimethylamine	7.4	25.0	53.9	80.0	111.7	132.2	149.8	162.6		164.5	52.4
C_2N_2	Cyanogen	-21.0	-4.4	+21.4	44.6	72.6	91.6	106.5	118.2		126.6	58.2
C_3H_4	Propadiene	-35.0	-18.4	+8.0	33.2	64.5	85.5	108.5	118.0		120.7	51.8
C_3H_4	Propyne	-23.3	-7.1	+19.5	43.8	74.0	94.0	111.5	125.0		128	52.8
C_3H_6	Propylene	-47.7	-31.4	-4.8	+19.8	49.5	70.0	85.0			91.4	45.4
C_3H_6O	Acetone	56.5	78.6	113.0	144.5	181.0	205.0	214.5			235.0	47.0
$C_3H_6O_2$	Propionic acid	141.1	160.0	186.0	203.5	220.0	228.0	233.0	238.0		239.5	53.0

TABLE 4k-4. VAPOR PRESSURE OF ORGANIC COMPOUNDS—PRESSURES GREATER THAN 1 ATMOSPHERE (*Continued*)

Formula	Name	Temp., °C									T_c	P_c
		1 atm	2 atm	5 atm	10 atm	20 atm	30 atm	40 atm	50 atm	60 atm		
$C_3H_6O_2$	Methyl acetate	57.8	79.5	113.1	144.2	181.0	205.0	225.0	.....	.....	233.7	46.3
$C_3H_6O_2$	Ethyl formate	54.3	76.0	110.5	142.2	180.0	205.0	225.0	.....	.....	235.3	46.8
C_3H_8	Propane	−42.1	−25.6	+1.4	26.9	58.1	78.7	94.8	.....	.....	96.8	42.0
C_3H_8O	1-Propanol	97.8	117.0	149.0	177.0	210.8	232.3	250.0	.....	.....	263.7	49.9
C_3H_8O	2-Propanol	82.5	101.3	130.2	155.7	186.0	205.0	220.2	232.0	.....	235	53
C_3H_8O	Ethyl methyl ether	7.5	26.5	56.4	84.0	108.0	141.4	160.0	.....	.....	164.7	43.4
C_3H_9N	Propylamine	48.5	69.8	102.8	133.4	170.0	194.3	214.5	.....	.....	223.8	46.8
C_4H_6	1,3-Butadiene	−4.5	+15.3	47.0	76.0	114.0	139.8	158.0	.....	.....	161.8	42.6
$C_4H_6O_3$	Acetic anhydride	139.6	162.0	194.0	221.5	253.0	272.8	288.5	.....	.....	296	46
$C_4H_6O_4$	Dimethyl oxalate	163.3	189.6	228.7	.....	.....	.....	.....	.....	.....	260	9.5
$C_4H_8O_2$	Butyric acid	163.5	188.3	225.0	257.0	295.0	319.0	338.0	352.0	.....	355	52.0
$C_4H_8O_2$	Isobutyric acid	154.5	179.8	217.0	250.0	289.0	315.0	336.0	.....	.....	336	40.0
$C_4H_8O_2$	Ethyl acetate	77.1	100.6	136.6	169.7	209.5	235.0	.....	.....	.....	250.1	37.9
$C_4H_8O_2$	Methyl propionate	79.8	103.0	139.8	172.6	212.5	239.0	.....	.....	.....	257.4	39.3
$C_4H_8O_2$	Propyl formate	81.3	104.3	142.0	176.4	217.5	245.0	.....	.....	.....	264.8	39.5
C_4H_{10}	Butane	−0.5	+18.8	50.0	79.5	116.0	140.6	.....	.....	.....	152.8	36.0
C_4H_{10}	2-Methylpropane	−11.7	+7.5	39.0	66.8	99.5	120.5	.....	.....	.....	134.0	37.0
$C_4H_{10}O$	Butyl alcohol	117.5	139.8	172.5	203.0	237.0	259.0	277.0	.....	.....	287	48.4
$C_4H_{10}O$	sec-Butyl alcohol	99.5	118.2	147.5	172.0	204.0	230.0	251.0	.....	.....	265	48
$C_4H_{10}O$	Isobutyl alcohol	108.0	127.3	156.2	182.0	212.5	232.0	251.0	.....	.....	265	48
$C_4H_{10}O$	tert-Butyl alcohol	82.9	102.0	130.0	154.2	184.2	207.0	222.5	.....	.....	235	49
$C_4H_{10}O$	Diethyl ether	34.6	56.0	90.0	122.0	159.0	183.3	.....	.....	.....	193.8	35.5
$C_4H_{10}S$	Diethyl sulfide	88.0	112.0	153.8	190.2	234.0	263.0	.....	.....	.....	283.8	39.1
$C_4H_{11}N$	Diethylamine	55.5	77.8	113.0	145.3	184.5	210.0	.....	.....	.....	223.3	36.6
$C_4H_{12}Si$	Tetramethylsilane	27.0	48.0	82.0	113.0	152.0	178.0	.....	.....	.....	185	33

$C_5H_{10}O_2$	Ethyl propionate	99.1	123.8	162.7	197.8	240.0	264.5				272.8	33.2
$C_5H_{10}O_2$	Propyl acetate	101.8	126.8	165.7	200.5	242.8	269.0				276.2	33.2
$C_5H_{10}O_2$	Isobutyl formate	98.2	121.8	157.8	192.4	234.0	261.0				278.0	38.0
$C_5H_{10}O_2$	Methyl butyrate	102.3	127.5	166.7	203.0	244.5	272.0				281.2	34.2
$C_5H_{10}O_2$	Methyl isobutyrate	92.6	116.7	155.2	190.2	232.0	259.5				267.5	33.9
C_5H_{12}	Pentane	36.1	58.0	92.4	124.7	164.3	191.3				197.2	33.0
C_5H_{12}	2-Methylbutane	27.8	48.8	82.8	114.5	154.0	180.3				187.8	32.8
C_5H_{12}	2,2-Dimethylpropane	+9.5	29.5	61.1	90.7	127.6	152.5				159.0	33.0
$C_5H_{12}O$	Ethyl propyl ether	61.7	85.3	123.1	156.2	197.2	223.0				227.4	32.1
C_6H_5Br	Bromobenzene	156.2	186.2	232.5	274.5	327.0	359.8	387.5			397	44.6
C_6H_5Cl	Chlorobenzene	132.2	160.2	205.0	245.3	292.8	324.4	349.8			359.2	44.6
C_6H_5F	Fluorobenzene	84.7	109.9	148.5	184.4	227.6	257.0	279.3			286.5	44.7
C_6H_5I	Iodobenzene	188.6	220.0	270.0	315.7	371.5	406.0	437.2			448	44.7
C_6H_6	Benzene	80.1	103.8	142.5	178.8	221.5	249.5	272.3	290.3		290.5	50.1
C_6H_6O	Phenol	181.9	208.0	248.2	283.8	328.7	358.0	382.1	400.0	418.7	419	60.5
C_6H_7N	Aniline	184.4	212.8	254.8	292.7	342.0	375.5	400.0	422.4		426	52.4
C_6H_{12}	Cyclohexane	80.7	106.0	146.4	184.0	228.4	257.5				279.9	39.8
$C_6H_{12}O_2$	Ethyl isobutyrate	110.1	135.5	174.2	210.0	253.0	280.0				280.0	30.0
C_6H_{14}	Hexane	68.7	93.0	131.7	166.6	209.4					234.8	29.6
C_6H_{14}	2,3-Dimethylbutane	58.0	82.0	120.3	155.7	198.7	225.5				227.4	30.7
C_7H_8	Toluene	110.6	136.5	178.0	215.8	262.5	292.8	319.0			320.6	41.6
C_7H_{16}	Heptane	98.4	124.8	165.7	202.8	247.5					266.8	26.9
C_8H_{10}	Ethylbenzene	136.2	163.5	207.5	246.3	294.5	326.5				346.4	38.1
C_8H_{18}	Octane	125.6	152.7	196.2	235.8	181.4					296.2	24.7
$C_{12}H_{26}$	Dodecane	216.2	249.2	300.0	345.8						385	17.5

The 1958 He^4 temperature scale is defined by the equation[1]

$$\ln P = i_0 - \frac{L_0}{RT} + \frac{5}{2}\ln T - \frac{1}{RT}\int_0^T S_l \, dT + \frac{1}{RT}\int_0^P V_l \, dP + \epsilon$$

where

$$i_0 \equiv \ln (2\pi m)^{\frac{3}{2}} \frac{k^{\frac{5}{2}}}{h^3},$$

and

$$\epsilon \equiv \ln \frac{PV}{NRT} - \frac{2BN}{V} - \frac{3}{2}\frac{CN^2}{V^2}$$

L_0 is heat of vaporization of liquid He^4 at 0 K, S_l and V_l are the molar entropy and volume of liquid He^4, m is the mass of a He^4 atom, and B and C are virial coefficients of He^4.

The scale has been approved by the International Committee on Weights and Measures and is used for temperature measurements between the boiling point of helium (4.2150 K) and about 1.0 K and can be used up to the critical point of helium ($T = 5.1994$ K, $P = 1{,}718$ mm Hg). It is in agreement with the thermodynamic scale to within ± 2 millikelvins.

The vapor-pressure–temperature relation for He^3 is based on the equation

$$\ln P_3 = \frac{2.49174}{T} + 4.80386 - 0.286001T + 0.198608T^2 - 0.0502237T^3$$

$$+ 0.00505486T^4 + 2.24846 \ln T \qquad 0.2 \leq T \leq 3.324 \text{ K}$$

which defines the T_{62} He^3 temperature scale. The T_{62} He^3 temperature scale is the result of the work done by Sydoriak, Roberts, and Sherman[2] at the Los Alamos Scientific Laboratory.

Table 4k-7, which gives the temperature as a function of vapor pressure for He^3, is taken from the work of R. H. Sherman, S. G. Sydoriak, and T. R. Roberts.[3]

Temperature measurements using the vapor pressure of liquid hydrogen are complicated by the phenomenon of ortho-para conversion. For ortho-hydrogen the proton spins are parallel while for para-hydrogen the spins are antiparallel. Due to the different energies of the two states, the equilibrium composition varies from 75% ortho-H_2 and 25% para-H_2 at room temperature to 99.79% para-H_2 and 0.21% ortho-H_2 at 20.4 K.[4] However in the absence of a catalyst the rate of conversion from the ortho to the para form is very slow; thus it is possible to liquefy hydrogen and preserve for many hours the equilibrium composition at room temperature.[5] Hydrogen having the composition 75% ortho-H_2 and 25% para-H_2 is generally called *normal* hydrogen and hydrogen having the composition 99.79% para-H_2 and 0.21% ortho-H_2 *equilibrium* hydrogen.

The vapor-pressure–temperature relation for equilibrium hydrogen (99.79% para-H_2 and 0.21% ortho-H_2) is based on an equation proposed by Durieux,[6]

$$\log P \text{ (mm)} = 4.635384 - \frac{44.2674}{T} + 0.021669T - 0.000021T^2$$

[1] F. G. Brickwedde, H. Van Dijk, M. Durieux, J. R. Clement, and J. K. Logan, *J. Research NBS* **64A**, 1 (1960).

[2] S. G. Sydoriak, T. R. Roberts, and R. H. Sherman, *J. Research NBS* **68A**, 559 (1964).

[3] R. H. Sherman, S. G. Sydoriak, and T. R. Roberts, *Los Alamos Rept.* LAMS 2701, pp. 17–21, 1962; *J. Research NBS* **68A**, 579 (1964).

[4] G. K. White, "Experimental Techniques in Low Temperature Physics," p. 41, Oxford University Press, London, 1959.

[5] R. P. Hudson, in "Experimental Cryophysics," p. 224, F. E. Hoare, L. C. Jackson, and N. Kurti, eds., Butterworth & Company (Publishers), Ltd., London, 1961.

[6] M. Durieux, Thesis, p. 95, Leiden, 1960.

TABLE 4k-5. VAPOR PRESSURE OF HELIUM 4 (1958 SCALE)

Vapor pressure of ⁴He. Unit 10^{-3} mm Hg at 0°C, $g = 980.665$ cm/sec²

T	0.00	0.01	0.02	0.03	0.04	0.05	0.06	0.07	0.08	0.09
0.5	.016342	.022745	.031287	.042561	.057292	.076356	.10081	.13190	.17112	.22021
	.28121	.35649	.44877	.56118	.69729	.86116	1.0574	1.2911	1.5682	1.8949
	2.2787	2.7272	3.2494	3.8549	4.5543	5.3591	6.2820	7.3365	8.5376	9.9013
	11.445	13.187	15.147	17.348	19.811	22.561	25.624	29.027	32.800	36.974
	41.581	46.656	52.234	58.355	65.059	72.386	80.382	89.093	98.567	108.853
1.0	120.000	132.070	145.116	159.198	174.375	190.711	208.274	227.132	247.350	269.006
	292.169	316.923	343.341	371.512	401.514	433.437	467.365	503.396	541.617	582.129
	625.025	670.411	718.386	769.057	822.527	878.916	938.330	1000.87	1066.67	1135.85
	1208.51	1284.81	1364.83	1448.73	1536.61	1628.62	1724.91	1825.58	1930.79	2040.67
	2155.35	2274.99	2399.73	2529.72	2665.09	2805.99	2952.60	3105.04	3263.48	3428.07
1.5	3598.97	3776.32	3960.32	4151.07	4348.79	4553.58	4765.68	4985.18	5212.26	5447.11
	5689.88	5940.76	6199.90	6467.42	6743.57	7028.47	7322.31	7625.21	7937.40	8259.02
	8590.22	8931.18	9282.06	9643.02	10014.3	10395.9	10788.2	11191.2	11605.1	12030.1
	12466.1	12913.7	13372.8	13843.6	14326.1	14820.7	15327.3	15846.3	16377.7	16921.7
	17478.2	18047.7	18630.1	19225.5	19834.1	20455.9	21091.1	21739.7	22402.0	23077.9
2.0	23767.4	24470.9	25188.1	25919.2	26664.2	27423.3	28196.3	28983.2	29784.2	30599.1
	31428.1	32271.1	33128.0	33998.6	34882.8	35780.3	36690.9	37614.3	38550.2	39500.3
	40465.6	41446.6	42443.5	43456.5	44485.7	45531.3	46593.5	47672.5	48768.6	49881.8
	51012.3	52160.2	53325.8	54509.2	55710.5	56930.0	58167.8	59423.8	60698.8	61992.0
	63304.3	64635.2	65985.4	67354.8	68743.5	70152.0	71580.2	73028.1	74496.0	75984.2
2.5	77493.1	79022.2	80572.2	82142.9	83734.6	85347.2	86981.2	88636.7	90313.8	92012.6
	93733.4	95476.0	97240.8	99028.2	100838	102669	104525	106403	108304	110228
	112175	114145	116139	118156	120198	122263	124353	126465	128603	130765
	132952	135164	137401	139663	141949	144260	146597	148961	151349	153763
	156204	158671	161164	163684	166230	168802	171402	174028	176682	179364
3.0	182073	184810	187574	190366	193187	196037	198914	201820	204755	207719
	210711	213732	216783	219864	222975	226115	229285	232484	235714	238974
	242266	245587	248939	252322	255736	259182	262658	266166	269706	273278
	276880	280516	284183	287883	291615	295380	299178	303008	306871	310768
	314697	318659	322654	326684	330747	334845	338976	343141	347341	351575
3.5	355844	360147	364485	368860	373269	377714	382194	386710	391262	395849
	400471	405130	409825	414556	419324	424128	428968	433846	438760	443713
	448702	453729	458794	463897	469038	474218	479435	484691	489985	495317
	500688	506098	511547	517036	522564	528132	533739	539387	545075	550805
	556574	562383	568234	574126	580059	586034	592051	598110	604210	610352
4.0	616537	622764	629033	635345	641700	648099	654541	661026	667554	674125
	680740	687399	694103	700851	707643	714479	721360	728285	735255	742269
	749328	756431	763579	770772	778010	785294	792623	799999	807422	814893
	822411	829978	837592	845255	852966	860725	868533	876390	884296	892252
	900258	908313	916418	924573	932778	941033	949338	957693	966099	974556
4.5	983066	991628	1000239	1008905	1017621	1026390	1035213	1044087	1053014	1061995
	1071029	1080114	1089254	1098449	1107699	1117002	1126359	1135772	1145239	1154761
	1164339	1173972	1183662	1193407	1203209	1213066	1222981	1232955	1242983	1253069
	1263212	1273414	1283673	1293991	1304367	1314802	1325297	1335850	1346462	1357136
	1367870	1378662	1389516	1400429	1411404	1422438	1433533	1444690	1455911	1467191
5.0	1478535	1489940	1501409	1512940	1524535	1536192	1547912	1559698	1571546	1583458
	1595437	1607481	1619589	1631761	1644000	1656305	1668673	1681108	1693612	1706180
	1718817	1731521	1744290							

TABLE 4k-6. 1958 He⁴ Vapor-pressure–Temperature Scale, T in K as a Function of P in millimeters mercury at 0°C and Standard Gravity, 980.665 cm/sec²

P	0	d	1	d	2	d	3	d	4	d	5	d	6	d	7	d	8	d	9	d
0.01	0.7907	65	7972	61	8033	57	8090	53	8143	50	8193	47	8240	45	8285	43	8328	40	8368	39
0.02	0.8407	38	8445	35	8480	35	8515	33	8548	32	8580	32	8612	30	8642	29	8671	28	8699	28
0.03	0.8727	27	8754	26	8780	25	8805	25	8830	24	8854	24	8878	23	8901	22	8923	22	8945	22
0.04	0.8967	21	8988	21	9009	20	9029	20	9049	19	9068	20	9088	18	9106	19	9125	18	9143	18
0.05	0.9161	18	9179	17	9196	17	9213	17	9230	16	9246	17	9263	17	9279	15	9294	16	9310	15
0.06	0.9325	15	9340	15	9355	15	9370	15	9385	14	9399	14	9413	14	9427	14	9441	14	9455	13
0.07	0.9468	14	9482	13	9495	13	9508	13	9521	13	9534	12	9546	13	9559	12	9571	12	9583	12
0.08	0.9595	12	9607	12	9619	12	9631	12	9643	11	9654	11	9665	12	9677	11	9688	11	9699	11
0.09	0.9710	11	9721	11	9732	10	9742	11	9753	10	9763	11	9774	10	9784	10	9794	10	9804	10
0.10	0.9814	10	9824	10	9834	10	9844	10	9854	9	9863	10	9873	10	9883	9	9892	9	9901	10
0.11	0.9911	9	9920	9	9929	9	9938	9	9947	9	9956	9	9965	9	9974	9	9983	8	9991	9
0.12	1.0000	9	0009	8	0017	9	0026	8	0034	8	0042	9	0051	8	0059	8	0067	8	0075	8
0.13	1.0083	8	0091	8	0099	8	0107	8	0115	8	0123	8	0131	8	0139	7	0146	8	0154	8
0.14	1.0162	7	0169	8	0177	7	0184	8	0192	7	0199	7	0206	8	0214	7	0221	7	0228	8
0.15	1.0236	7	0243	7	0250	7	0257	7	0264	7	0271	7	0278	7	0285	7	0292	7	0299	6
0.16	1.0305	7	0312	7	0319	7	0326	6	0332	7	0339	7	0346	6	0352	7	0359	6	0365	7
0.17	1.0372	6	0378	7	0385	6	0391	7	0398	6	0404	6	0410	7	0417	6	0423	6	0429	6
0.18	1.0435	6	0441	7	0448	6	0454	6	0460	6	0466	6	0472	6	0478	6	0484	6	0490	6
0.19	1.0496	6	0502	6	0508	5	0513	6	0519	6	0525	6	0531	6	0537	5	0542	6	0548	6
0.2	1.0554	55	0609	54	0663	52	0715	49	0764	49	0813	46	0859	45	0904	44	0948	43	0991	41
0.3	1.1032	41	1073	39	1112	38	1150	38	1188	36	1224	36	1260	35	1295	34	1329	33	1362	33
0.4	1.1395	32	1427	32	1459	31	1490	30	1520	30	1550	29	1579	29	1608	28	1636	27	1663	28
0.5	1.1691	27	1718	26	1744	26	1770	26	1796	25	1821	25	1846	25	1871	24	1895	24	1919	23
0.6	1.1942	24	1966	23	1989	22	2011	23	2034	22	2056	22	2078	21	2099	21	2120	21	2141	21
0.7	1.2162	21	2183	20	2203	20	2223	20	2243	20	2263	20	2283	19	2302	19	2321	19	2340	19
0.8	1.2359	18	2377	18	2395	19	2414	18	2432	17	2449	18	2467	18	2485	17	2502	17	2519	17
0.9	1.2536	17	2553	17	2570	16	2586	17	2603	16	2619	16	2635	16	2651	16	2667	16	2683	16
1.0	1.2699	15	2714	16	2730	15	2745	15	2760	15	2775	15	2790	15	2805	15	2820	14	2834	15
1.1	1.2849	14	2863	15	2878	14	2892	14	2906	14	2920	14	2934	14	2948	13	2961	14	2975	14
1.2	1.2989	13	3002	13	3015	14	3029	13	3042	13	3055	13	3068	13	3081	13	3094	13	3107	12
1.3	1.3119	13	3132	13	3145	12	3157	12	3169	13	3182	12	3194	12	3206	12	3218	12	3230	12
1.4	1.3242	12	3254	12	3266	12	3278	12	3290	11	3301	12	3313	12	3325	11	3336	12	3348	11
1.5	1.3359	11	3370	11	3381	12	3393	11	3404	11	3415	11	3426	11	3437	11	3448	11	3459	10
1.6	1.3469	11	3480	11	3491	10	3501	11	3512	11	3523	10	3533	11	3544	10	3554	10	3564	11
1.7	1.3575	10	3585	10	3595	10	3605	10	3615	10	3625	10	3635	10	3645	10	3655	10	3665	10
1.8	1.3675	10	3685	10	3695	9	3704	10	3714	10	3724	9	3733	10	3743	9	3752	10	3762	9
1.9	1.3771	9	3780	10	3790	9	3799	10	3809	9	3818	9	3827	9	3836	9	3845	9	3854	9
2	1.3863	89	3952	86	4038	82	4120	80	4200	77	4877	75	4352	73	4425	71	4496	69	4565	67
3	1.4632	65	4697	63	4760	63	4823	60	4883	60	4943	58	5001	56	5057	56	5113	55	5108	53
4	1.5221	53	5274	51	5325	51	5376	49	5425	49	5474	48	5522	47	5569	47	5616	46	5662	45
5	1.5707	44	5751	44	5795	43	5838	42	5880	42	5922	41	5963	41	6004	40	6044	40	6084	39
6	1.6123	39	6162	38	6200	38	6238	37	6275	37	6312	36	6348	36	6384	36	6420	35	6455	35
7	1.6490	35	6525	34	6559	34	6593	33	6626	33	6659	33	6692	32	6724	32	6756	32	6788	32
8	1.6820	31	6851	31	6882	31	6913	30	6943	30	6973	30	7003	29	7033	29	7062	29	7091	29
9	1.7120	28	7148	29	7177	28	7205	28	7233	28	7261	27	7288	28	7316	27	7343	27	7370	26
10	1.7396	27	7423	26	7449	26	7475	26	7501	26	7527	25	7552	26	7578	25	7603	25	7628	25
11	1.7653	25	7678	24	7702	25	7727	24	7751	24	7775	24	7799	24	7823	23	7846	24	7870	23
12	1.7893	23	7916	23	7939	23	7962	23	7985	23	8008	22	8030	23	8053	22	8075	22	8097	22
13	1.8119	22	8141	22	8163	21	8184	22	8206	21	8227	22	8249	21	8270	21	8291	21	8312	21
14	1.8333	20	8353	21	8374	21	8395	20	8415	20	8435	21	8456	20	8476	20	8496	20	8516	20
15	1.8536	19	8555	20	8575	20	8595	19	8614	20	8634	19	8653	19	8672	19	8691	19	8710	19
16	1.8729	19	8748	19	8767	18	8785	19	8804	19	8823	18	8841	19	8860	18	8878	18	8896	18
17	1.8914	18	8932	18	8950	18	8968	18	8986	18	9004	18	9022	17	9039	18	9057	17	9074	18
18	1.9092	17	9109	17	9126	18	9144	17	9161	17	9178	17	9195	17	9212	17	9229	17	9246	16
19	1.9262	17	9279	17	9296	16	9312	17	9329	16	9345	17	9362	16	9378	16	9394	17	9411	16
20	1.9427	16	9443	16	9459	16	9475	16	9491	16	9507	16	9523	16	9539	15	9554	16	9570	16

TABLE 4k-6. 1958 He⁴ Vapor-pressure–Temperature Scale, T in K as a Function of P in millimeters mercury at 0°C and Standard Gravity, 980.665 cm/sec² (*Continued*)

P	0	1	2	3	4	5	6	7	8	9
21	1.9586 16	9602 15	9617 15	9632 16	9648 15	9663 16	9679 15	9694 15	9709 15	9724 16
22	1.9740 15	9755 15	9770 15	9785 15	9800 15	9815 15	9830 14	9844 15	9859 15	9874 15
23	1.9889 14	9903 15	9918 14	9932 15	9947 14	9961 15	9976 14	9990 15	0005 14	0019 14
24	2.0033 15	0048 14	0062 14	0076 14	0090 14	0104 14	0118 14	0132 14	0146 14	0160 14
25	2.0174 14	0188 14	0202 13	0215 14	0229 14	0243 14	0257 13	0270 14	0284 13	0297 14
26	2.0311 13	0324 14	0338 13	0351 14	0365 13	0378 13	0391 14	0405 13	0418 13	0431 13
27	2.0444 14	0458 13	0471 13	0484 13	0497 13	0510 13	0523 13	0536 13	0549 13	0562 13
28	2.0575 13	0588 12	0600 13	0613 13	0626 13	0639 13	0652 12	0664 13	0677 13	0690 12
29	2.0702 13	0715 12	0727 13	0740 12	0752 13	0765 12	0777 13	0790 12	0802 12	0814 13
30	2.0827 12	0839 12	0851 12	0863 13	0876 12	0888 12	0900 12	0912 12	0924 12	0936 13
31	2.0949 12	0961 12	0973 12	0985 12	0997 12	1009 12	1021 11	1032 12	1044 12	1056 12
32	2.1068 12	1080 12	1092 11	1103 12	1115 12	1127 12	1139 11	1150 12	1162 12	1174 11
33	2.1185 12	1197 11	1208 12	1220 11	1231 12	1243 11	1254 12	1266 11	1277 12	1289 11
34	2.1300 12	1312 11	1323 11	1334 12	1346 11	1357 11	1368 11	1379 12	1391 11	1402 11
35	2.1413 11	1424 12	1436 11	1447 11	1458 11	1469 11	1480 11	1491 11	1502 11	1513 11
36	2.1524 11	1535 11	1546 11	1557 11	1568 11	1579 11	1590 11	1601 11	1612 11	1623 11
37	2.1634 10	1644 11	1655 11	1666 11	1677 11	1688 10	1698 11	1709 11	1720 11	1731 10
38	2.1741 11	1752 11	1763 10	1773 11	1784 11	1795 10	1805 11	1816 10	1826 11	1837 11
39	2.1848 10	1858 11	1869 11	1880 10	1890 10	1900 10	1910 11	1921 10	1931 11	1942 10
40	2.1952 10	1962 11	1973 10	1983 10	1993 11	2004 10	2014 10	2024 10	2034 10	2044 11
41	2.2055 10	2065 10	2075 10	2085 10	2095 10	2105 10	2115 11	2126 10	2136 10	2146 10
42	2.2156 10	2166 10	2176 10	2186 10	2196 10	2206 10	2216 9	2225 10	2235 10	2245 10
43	2.2255 10	2265 10	2275 10	2285 9	2294 10	2304 10	2314 10	2324 10	2334 9	2343 10
44	2.2353 10	2363 9	2372 10	2382 10	2392 9	2401 10	2411 10	2421 9	2430 10	2440 10
45	2.2450 9	2459 9	2468 10	2478 10	2488 10	2498 9	2507 9	2516 9	2525 10	2535 9
46	2.2544 10	2554 9	2563 10	2573 9	2582 9	2591 10	2601 9	2610 9	2619 10	2629 9
47	2.2638 9	2647 9	2656 10	2666 9	2675 9	2684 9	2693 10	2703 9	2712 9	2721 9
48	2.2730 9	2739 9	2748 9	2757 10	2767 9	2776 9	2785 9	2794 9	2803 9	2812 9
49	2.2821 9	2830 9	2839 9	2848 9	2857 9	2866 9	2875 9	2884 9	2893 9	2902 9
50	2.2911 88	2999 87	3086 86	3172 85	3257 84	3341 83	3424 82	3506 81	3587 79	3666 79
60	2.3745 78	3823 78	3901 76	3977 75	4052 75	4127 74	4201 73	4274 73	4347 71	4418 71
70	2.4489 71	4560 69	4629 69	4698 68	4766 68	4834 67	4901 66	4967 66	5033 66	5099 64
80	2.5163 64	5227 64	5291 63	5354 63	5417 62	5479 61	5540 61	5601 61	5662 60	5722 59
90	2.5781 60	5841 58	5899 59	5958 57	6015 57	6073 57	6130 56	6186 57	6243 55	6298 56
100	2.6354 55	6409 55	6464 54	6518 54	6572 53	6625 54	6679 53	6732 52	6784 52	6836 52
110	2.6888 52	6940 51	6991 51	7042 51	7093 50	7143 50	7193 50	7243 49	7292 49	7341 49
120	2.7390 49	7439 48	7487 48	7535 48	7583 48	7631 47	7678 47	7725 47	7772 46	7818 47
130	2.7865 46	7911 46	7957 45	8002 46	8048 45	8093 45	8138 44	8182 45	8227 44	8271 44
140	2.8315 44	8359 43	8402 44	8446 43	8489 43	8532 43	8575 42	8617 42	8659 43	8702 42
150	2.8744 41	8785 42	8827 42	8869 41	8910 41	8951 41	8992 40	9032 41	9073 40	9113 40
160	2.9153 40	9193 40	9233 40	9273 39	9312 40	9352 39	9391 39	9430 39	9469 39	9508 38
170	2.9546 39	9585 38	9623 38	9661 38	9699 38	9737 37	9774 38	9812 38	9850 36	9886 38
180	2.9924 37	9961 36	9997 37	0034 37	0071 36	0107 36	0143 36	0179 36	0215 36	0251 36
190	3.0287 36	0323 35	0358 35	0393 36	0429 35	0464 35	0499 35	0534 34	0568 35	0603 34
200	3.0637 35	0672 34	0706 34	0740 34	0774 34	0808 34	0842 34	0876 33	0909 34	0943 33
210	3.0976 34	1010 33	1043 33	1076 33	1109 33	1142 32	1174 33	1207 33	1240 32	1272 32
220	3.1304 33	1337 32	1369 32	1401 32	1433 32	1465 31	1496 32	1528 32	1560 31	1591 31
230	3.1622 32	1654 31	1685 31	1716 31	1747 31	1778 31	1809 31	1840 30	1870 31	1901 30
240	3.1931 31	1962 30	1992 30	2022 30	2052 30	2082 30	2112 30	2142 30	2172 30	2202 29
250	3.2231 30	2261 30	2291 29	2320 29	2349 30	2379 29	2408 29	2437 29	2466 29	2495 29
260	3.2524 28	2552 29	2581 29	2610 28	2638 29	2667 28	2695 29	2724 28	2752 28	2780 28
270	3.2808 28	2836 28	2864 28	2892 28	2920 28	2948 28	2976 27	3003 28	3031 27	3058 28
280	3.3086 27	3113 28	3141 27	3168 27	3195 27	3222 27	3249 27	3276 27	3303 27	3330 27
290	3.3357 27	3384 26	3410 27	3437 26	3463 27	3490 26	3516 27	3543 26	3569 26	3595 27
300	3.3622 26	3648 26	3674 26	3700 26	3726 26	3752 26	3778 25	3803 26	3829 26	3855 25
310	3.3880 26	3906 25	3931 26	3957 25	3982 26	4008 25	4033 25	4058 25	4083 26	4109 25
320	3.4134 25	4159 25	4184 25	4209 24	4233 25	4258 25	4283 25	4308 24	4332 25	4357 25
330	3.4382 24	4406 25	4431 24	4455 24	4479 25	4504 24	4528 24	4552 24	4576 24	4601 24

TABLE 4k-6. 1958 He⁴ VAPOR-PRESSURE–TEMPERATURE SCALE, T IN K AS A
FUNCTION OF P IN MILLIMETERS MERCURY AT 0°C AND STANDARD
GRAVITY, 980.665 CM/SEC² (*Continued*)

P	0	1	2	3	4	5	6	7	8	9
340	3.4625 24	4649 24	4673 24	4697 24	4721 23	4744 24	4768 24	4792 24	4816 23	4839 24
350	3.4863 23	4886 24	4910 23	4933 24	4957 23	4980 24	5004 23	5027 23	5050 23	5073 24
360	3.5097 23	5120 23	5143 23	5166 23	5189 23	5212 23	5235 23	5258 22	5280 23	5303 23
370	3.5326 23	5349 22	5371 23	5394 23	5417 22	5439 23	5462 22	5484 22	5506 23	5529 22
380	3.5551 22	5573 23	5596 22	5618 22	5640 22	5662 22	5684 22	5706 22	5728 22	5750 22
390	3.5772 22	5794 22	5816 22	5838 22	5860 21	5881 22	5903 22	5925 22	5947 21	5968 22
400	3.5990 21	6011 22	6033 21	6054 22	6076 21	6097 22	6119 21	6140 21	6161 21	6182 22
410	3.6204 21	6225 21	6246 21	6267 21	6288 21	6309 21	6330 21	6351 21	6372 21	6393 21
420	3.6414 21	6435 21	6456 21	6477 20	6497 21	6518 21	6539 20	6559 21	6580 21	6601 20
430	3.6621 21	6642 20	6662 21	6683 20	6703 21	6724 20	6744 20	6764 21	6785 20	6805 20
440	3.6825 20	6845 21	6866 20	6886 20	6906 20	6926 20	6946 20	6966 20	6986 20	7006 20
450	3.7026 20	7046 20	7066 20	7086 19	7105 20	7125 20	7145 20	7165 19	7184 20	7204 20
460	3.7224 19	7243 20	7263 19	7282 20	7302 20	7322 19	7341 19	7360 20	7380 19	7399 20
470	3.7419 19	7438 19	7457 20	7477 19	7496 19	7515 19	7534 19	7553 20	7573 19	7592 19
480	3.7611 19	7630 19	7649 19	7668 19	7687 19	7706 19	7725 19	7744 19	7763 18	7781 19
490	3.7800 19	7819 19	7838 19	7857 18	7875 19	7894 19	7913 18	7931 19	7950 19	7969 18
500	3.7987 19	8006 18	8024 19	8043 18	8061 19	8080 18	8098 19	8117 18	8135 18	8153 19
510	3.8172 18	8190 18	8208 19	8227 18	8245 18	8263 18	8281 18	8299 18	8317 19	8336 18
520	3.8354 18	8372 18	8390 18	8408 18	8426 18	8444 18	8462 18	8480 18	8498 18	8516 17
530	3.8533 18	8551 18	8569 18	8587 18	8605 17	8622 18	8640 18	8658 18	8676 17	8693 18
540	3.8711 17	8728 18	8746 18	8764 17	8781 18	8799 17	8816 18	8834 17	8851 18	8869 17
550	3.8886 17	8903 18	8921 17	8938 17	8955 18	8973 17	8990 17	9007 18	9025 17	9042 17
560	3.9059 17	9076 17	9093 18	9111 17	9128 17	9145 17	9162 17	9179 17	9196 17	9213 17
570	3.9230 17	9247 17	9264 17	9281 17	9298 17	9315 17	9332 17	9349 16	9365 17	9382 17
580	3.9399 17	9416 17	9433 16	9449 17	9466 17	9483 16	9499 17	9516 17	9533 16	9549 17
590	3.9566 17	9583 16	9599 17	9616 16	9632 17	9649 16	9665 17	9682 16	9698 17	9715 16
600	3.9731 16	9747 17	9764 16	9780 17	9797 16	9813 16	9829 17	9846 16	9862 16	9878 16
610	3.9894 17	9911 16	9927 16	9943 16	9959 16	9975 16	9991 16	0007 17	0024 16	0040 16
620	4.0056 16	0072 16	0088 16	0104 16	0120 16	0136 16	0152 16	0168 16	0184 15	0199 16
630	4.0215 16	0231 16	0247 16	0263 16	0279 16	0295 15	0310 16	0326 16	0342 16	0358 15
640	4.0373 16	0389 16	0405 15	0420 16	0436 16	0452 15	0467 16	0483 15	0498 16	0514 16
650	4.0530 15	0545 16	0561 15	0576 16	0592 15	0607 16	0623 15	0638 15	0653 16	0669 15
660	4.0684 16	0700 15	0715 15	0730 16	0746 15	0761 15	0776 16	0792 15	0807 15	0822 15
670	4.0837 16	0853 15	0868 15	0883 15	0898 15	0913 15	0928 16	0944 15	0959 15	0974 15
680	4.0989 15	1004 15	1019 15	1034 15	1049 15	1064 15	1079 15	1094 15	1109 15	1124 15
690	4.1139 15	1154 15	1169 15	1184 14	1198 15	1213 15	1228 15	1243 15	1258 15	1273 14
700	4.1287 15	1302 15	1317 15	1332 14	1346 15	1361 15	1376 15	1391 14	1405 15	1420 15
710	4.1435 14	1449 15	1464 14	1478 15	1493 15	1508 14	1522 15	1537 14	1551 15	1566 14
720	4.1580 15	1595 14	1609 15	1624 14	1638 15	1653 14	1667 14	1681 15	1696 14	1710 15
730	4.1725 14	1739 14	1753 15	1768 14	1782 14	1796 15	1811 14	1825 14	1839 14	1853 15
740	4.1868 14	1882 14	1896 14	1910 15	1925 14	1939 14	1953 14	1967 14	1981 14	1995 14
750	4.2009 15	2024 14	2038 14	2052 14	2066 14	2080 14	2094 14	2108 14	2122 14	2136 14
760	4.2150 14	2164 14	2178 14	2192 14	2206 14	2220 14	2234 14	2248 14	2262 13	2275 14
770	4.2289 14	2303 14	2317 14	2331 14	2345 14	2358 14	2372 14	2386 14	2400 14	2414 13
780	4.2427 14	2441 14	2455 14	2469 13	2482 13	2496 14	2510 13	2523 14	2537 14	2551 13
790	4.2564 14	2578 14	2592 13	2605 14	2619 13	2632 14	2646 13	2659 14	2673 13	2686 14
800	4.2700 135	2835 133	2968 132	3100 131	3231 131	3362 129	3491 128	3619 127	3746 126	3872 125
900	4.3997 124	4121 123	4244 122	4366 122	4488 120	4608 120	4728 118	4846 118	4964 117	5081 116
1000	4.5197 116	5313 114	5427 114	5541 113	5654 112	5766 112	5878 111	5989 110	6099 109	6208 109
1100	4.6317 108	6425 107	6532 107	6639 106	6745 105	6850 105	6955 104	7059 103	7162 103	7265 102
1200	4.7367 102	7469 101	7570 100	7670 100	7770 100	7870 98	7968 99	8067 99	8164 97	8261 97
1300	4.8358 96	8454 96	8550 95	8645 94	8739 94	8833 94	8927 93	9020 92	9112 92	9204 92
1400	4.9296 91	9387 91	9478 90	9568 90	9658 89	9747 89	9836 89	9925 88	0013 88	0101 87
1500	5.0188 87	0275 86	0361 86	0447 86	0533 85	0618 85	0703 84	0787 84	0871 84	0955 83
1600	5.1038 83	1121 82	1203 83	1286 81	1367 82	1449 81	1530 81	1611 80	1691 80	1771 80
1700	5.1851 79	1930 79	2009 79	2088 78	2166 78					

TABLE 4k-7. $T_{62}\text{He}^3$ TEMPERATURES IN K AS A FUNCTION OF VAPOR PRESSURE
P AT 0°C AND STANDARD GRAVITY, 980.665 CM/SEC²

P in micrometers (10^{-3} mm) of Mercury

P	0	1	2	3	4	5	6	7	8	9
0.01	0.1974 13	1987 12	1999 11	2010 10	2020 10	2030 9	2039 9	2048 8	2056 7	2063 8
0.02	0.2071 7	2078 7	2085 6	2091 7	2098 6	2104 6	2110 5	2115 6	2121 5	2126 5
0.03	0.2131 6	2137 4	2141 5	2146 5	2151 4	2155 5	2160 4	2164 4	2168 4	2172 4
0.04	0.2176 4	2180 4	2184 4	2188 4	2192 3	2195 4	2199 3	2202 4	2206 3	2209 4
0.05	0.2213 3	2216 3	2219 3	2222 3	2225 3	2228 3	2231 3	2234 3	2237 3	2240 3
0.06	0.2243 3	2246 2	2248 3	2251 3	2254 3	2257 2	2259 3	2262 2	2264 3	2267 2
0.07	0.2269 3	2272 2	2274 2	2276 3	2279 2	2281 2	2283 3	2286 2	2288 2	2290 2
0.08	0.2292 3	2295 2	2297 2	2299 2	2301 2	2303 2	2305 2	2307 2	2309 2	2311 2
0.09	0.2313 2	2315 2	2317 2	2319 2	2321 2	2323 2	2325 2	2327 1	2328 2	2330 2
0.10	0.2332 18	2350 16	2366 15	2381 14	2395 13	2408 12	2420 12	2432 11	2443 11	2454 10
0.20	0.2464 10	2474 9	2483 9	2492 9	2501 8	2509 9	2518 8	2526 7	2533 8	2541 7
0.30	0.2548 7	2555 7	2562 6	2568 7	2575 6	2581 6	2587 6	2593 6	2599 6	2605 5
0.40	0.2610 6	2616 5	2621 5	2626 6	2632 5	2637 5	2642 5	2647 4	2651 5	2656 5
0.50	0.2661 4	2665 5	2670 4	2674 5	2679 4	2683 4	2687 4	2691 4	2695 4	2699 4
0.60	0.2703 4	2707 4	2711 4	2715 4	2719 3	2722 4	2726 4	2730 3	2733 4	2737 3
0.70	0.2740 4	2744 3	2747 3	2750 4	2754 3	2757 3	2760 4	2764 3	2767 3	2770 3
0.80	0.2773 3	2776 3	2779 3	2782 3	2785 3	2788 3	2791 3	2794 3	2797 3	2800 3
0.90	0.2803 2	2805 3	2808 3	2811 3	2814 2	2816 3	2819 3	2822 2	2824 3	2827 2
1	0.2829 25	2854 23	2877 21	2898 20	2918 19	2937 18	2955 17	2972 16	2988 16	3004 14
2	0.3018 14	3032 14	3046 13	3059 13	3072 12	3084 12	3096 11	3107 11	3118 11	3129 11
3	0.3140 10	3150 10	3160 9	3169 10	3179 9	3188 9	3197 9	3206 8	3214 9	3223 8
4	0.3231 8	3239 8	3247 8	3255 7	3262 8	3270 7	3277 7	3284 8	3292 7	3299 6
5	0.3305 7	3312 7	3319 6	3325 7	3332 6	3338 6	3344 6	3350 6	3356 6	3362 6
6	0.3368 6	3374 6	3380 6	3386 5	3391 6	3397 5	3402 6	3408 5	3413 5	3418 5
7	0.3423 5	3428 6	3434 5	3439 4	3443 5	3448 5	3453 5	3458 5	3463 5	3468 4
8	0.3472 5	3477 4	3481 5	3486 4	3490 5	3495 4	3499 5	3504 4	3508 4	3512 4
9	0.3516 5	3521 4	3525 4	3529 4	3533 4	3537 4	3541 4	3545 4	3549 4	3553 4
10	0.3557 37	3594 34	3628 33	3661 30	3691 29	3720 27	3747 26	3773 24	3797 24	3821 23
20	0.3844 21	3865 21	3886 21	3907 19	3926 19	3945 18	3963 18	3981 17	3998 17	4015 16
30	0.4031 16	4047 15	4062 15	4077 15	4092 15	4107 14	4121 13	4134 14	4148 13	4161 13
40	0.4174 13	4187 12	4199 12	4211 12	4223 12	4235 11	4246 12	4258 11	4269 11	4280 11
50	0.4291 10	4301 11	4312 10	4322 10	4332 10	4342 10	4352 10	4362 10	4372 9	4381 10
60	0.4391 9	4400 9	4409 9	4418 9	4427 9	4436 8	4444 9	4453 8	4461 9	4470 8
70	0.4478 8	4486 9	4495 8	4503 7	4510 8	4518 8	4526 8	4534 7	4541 8	4549 7
80	0.4556 8	4564 7	4571 7	4578 8	4586 7	4593 7	4600 7	4607 7	4614 7	4621 7
90	0.4628 6	4634 7	4641 7	4648 6	4654 7	4661 6	4667 7	4674 7	4680 6	4686 7

P in millimeters of Mercury

P	0	1	2	3	4	5	6	7	8	9
0.1	0.4693 60	4753 56	4809 53	4862 50	4912 47	4959 44	5003 43	5046 41	5087 39	5126 37
0.2	0.5163 37	5200 34	5234 34	5268 33	5301 31	5332 31	5363 30	5393 29	5422 28	5450 27
0.3	0.5477 27	5504 26	5530 26	5556 24	5580 25	5605 24	5629 23	5652 23	5675 23	5698 22
0.4	0.5720 21	5741 22	5763 20	5783 21	5804 20	5824 20	5844 19	5863 20	5883 19	5902 18
0.5	0.5920 19	5939 18	5957 18	5975 17	5992 18	6010 17	6027 17	6044 16	6060 17	6077 16
0.6	0.6093 16	6109 16	6125 16	6141 15	6156 16	6172 15	6187 15	6202 14	6216 15	6231 15
0.7	0.6246 14	6260 14	6274 14	6288 14	6302 14	6316 14	6330 13	6343 14	6357 13	6370 13
0.8	0.6383 13	6396 13	6409 13	6422 12	6434 13	6447 12	6459 13	6472 12	6484 12	6496 12
0.9	0.6508 12	6520 12	6532 12	6544 12	6556 11	6567 12	6579 11	6590 11	6601 12	6613 11
1.0	0.6624 11	6635 11	6646 11	6657 11	6668 10	6678 11	6689 11	6700 10	6710 11	6721 10
1.1	0.6731 11	6742 10	6752 10	6762 10	6772 10	6782 10	6792 10	6802 10	6812 10	6822 10
1.2	0.6832 10	6842 9	6851 10	6861 9	6870 10	6880 9	6889 10	6899 9	6908 9	6917 9
1.3	0.6926 10	6936 9	6945 9	6954 9	6963 9	6972 9	6981 9	6990 8	6998 9	7007 9
1.4	0.7016 9	7025 8	7033 9	7042 9	7051 8	7059 9	7068 8	7076 8	7084 9	7093 8
1.5	0.7101 8	7109 9	7118 8	7126 8	7134 8	7142 8	7150 8	7158 8	7166 8	7174 8
1.6	0.7182 8	7190 8	7198 8	7206 7	7213 8	7221 8	7229 8	7237 7	7244 8	7252 8
1.7	0.7260 7	7267 8	7275 7	7282 8	7290 7	7297 8	7305 7	7312 7	7319 8	7327 7
1.8	0.7334 7	7341 7	7348 8	7356 7	7363 7	7370 7	7377 7	7384 7	7391 7	7398 7
1.9	0.7405 7	7412 7	7419 7	7426 7	7433 7	7440 7	7447 7	7454 6	7460 7	7467 7

TABLE 4k-7. $T_{62}\mathrm{He}^3$ Temperatures in K as a Function of Vapor Pressure P at 0°C and Standard Gravity, 980.665 cm/sec² (Continued)
P in millimeters of Mercury

P	0	1	2	3	4	5	6	7	8	9
2.0	0.7474 7	7481 7	7488 6	7494 7	7501 7	7508 6	7514 7	7521 6	7527 7	7534 6
2.1	0.7540 7	7547 6	7553 7	7560 6	7566 7	7573 6	7579 7	7586 6	7592 6	7598 7
2.2	0.7605 6	7611 6	7617 6	7623 7	7630 6	7636 6	7642 6	7648 6	7654 7	7661 6
2.3	0.7667 6	7673 6	7679 6	7685 6	7691 6	7697 6	7703 6	7709 6	7715 6	7721 6
2.4	0.7727 6	7733 6	7739 6	7745 5	7750 6	7756 6	7762 6	7768 6	7774 6	7780 5
2.5	0.7785 6	7791 6	7797 6	7803 5	7808 6	7814 6	7820 5	7825 6	7831 6	7837 5
2.6	0.7842 6	7848 5	7853 6	7859 5	7864 6	7870 6	7876 5	7881 6	7887 5	7892 5
2.7	0.7897 6	7903 5	7908 6	7914 5	7919 6	7925 5	7930 5	7935 6	7941 5	7946 5
2.8	0.7951 6	7957 5	7962 5	7967 6	7973 5	7978 5	7983 5	7988 6	7994 5	7999 5
2.9	0.8004 5	8009 5	8014 5	8019 6	8025 5	8030 5	8035 5	8040 5	8045 5	8050 5
3	0.8055 50	8105 49	8154 48	8202 47	8249 46	8295 45	8340 44	8384 43	8427 43	8470 42
4	0.8512 41	8553 40	8593 40	8633 39	8672 38	8710 38	8748 38	8786 36	8822 37	8859 35
5	0.8894 36	8930 34	8964 35	8999 33	9032 34	9066 33	9099 32	9131 33	9164 31	9195 32
6	0.9227 31	9258 31	9289 30	9319 30	9349 30	9379 29	9408 29	9437 29	9466 29	9495 28
7	0.9523 28	9551 28	9579 27	9606 27	9633 27	9660 27	9687 26	9713 27	9740 26	9766 25
8	0.9791 26	9817 25	9842 25	9867 25	9892 25	9917 24	9941 25	9966 24	9990 24	0014 24
9	1.0038 23	0061 24	0085 23	0108 23	0131 23	0154 22	0176 23	0199 22	0221 23	0244 22
10	1.0266 22	0288 22	0310 21	0331 22	0353 21	0374 21	0395 21	0416 21	0437 21	0458 21
11	1.0479 21	0500 20	0520 20	0540 21	0561 20	0581 20	0601 19	0620 20	0640 20	0660 19
12	1.0679 20	0699 19	0718 19	0737 19	0756 19	0775 19	0794 19	0813 19	0832 18	0850 19
13	1.0869 18	0887 18	0905 19	0924 18	0942 18	0960 18	0978 18	0996 17	1013 18	1031 18
14	1.1049 17	1066 17	1083 18	1101 17	1118 17	1135 17	1152 17	1169 17	1186 17	1203 17
15	1.1220 17	1237 16	1253 17	1270 16	1286 17	1303 16	1319 16	1335 17	1352 16	1368 16
16	1.1384 16	1400 16	1416 16	1432 15	1447 16	1463 16	1479 15	1494 16	1510 15	1525 16
17	1.1541 15	1556 15	1571 16	1587 15	1602 15	1617 15	1632 15	1647 15	1662 15	1677 15
18	1.1692 15	1707 14	1721 15	1736 15	1751 14	1765 15	1780 14	1794 15	1809 14	1823 14
19	1.1837 14	1851 15	1866 14	1880 14	1894 14	1908 14	1922 14	1936 14	1950 14	1964 14
20	1.1978 13	1991 14	2005 14	2019 13	2032 14	2046 14	2060 13	2073 14	2087 13	2100 13
21	1.2113 14	2127 13	2140 13	2153 13	2166 14	2180 13	2193 13	2206 13	2219 13	2232 13
22	1.2245 13	2258 13	2271 13	2284 12	2296 13	2309 13	2322 13	2335 12	2347 13	2360 13
23	1.2373 12	2385 13	2398 12	2410 13	2423 12	2435 12	2447 13	2460 12	2472 12	2484 12
24	1.2496 13	2509 12	2521 12	2533 12	2545 12	2557 12	2569 12	2581 12	2593 12	2605 12
25	1.2617 12	2629 12	2641 12	2653 11	2664 12	2676 12	2688 12	2700 11	2711 12	2723 11
26	1.2734 12	2746 12	2758 11	2769 12	2781 11	2792 11	2803 12	2815 11	2826 12	2838 11
27	1.2849 11	2860 11	2871 12	2883 11	2894 11	2905 11	2916 11	2927 11	2938 11	2949 11
28	1.2960 11	2971 11	2982 11	2993 11	3004 11	3015 11	3026 11	3037 11	3048 11	3059 10
29	1.3069 11	3080 11	3091 11	3102 10	3112 11	3123 11	3134 10	3144 11	3155 10	3165 11
30	1.3176 11	3187 10	3197 10	3207 11	3218 10	3228 11	3239 10	3249 11	3260 10	3270 10
31	1.3280 10	3290 11	3301 10	3311 10	3321 10	3331 11	3342 10	3352 10	3362 10	3372 10
32	1.3382 10	3392 10	3402 10	3412 10	3422 10	3432 10	3442 10	3452 10	3462 10	3472 10
33	1.3482 10	3492 10	3502 10	3512 10	3522 9	3531 10	3541 10	3551 10	3561 9	3570 10
34	1.3580 10	3590 9	3599 10	3609 10	3619 9	3628 10	3638 9	3647 10	3657 10	3667 9
35	1.3676 10	3686 9	3695 10	3705 9	3714 9	3723 10	3733 9	3742 10	3752 9	3761 9
36	1.3770 10	3780 9	3789 9	3798 10	3808 9	3817 9	3826 9	3835 9	3844 10	3854 9
37	1.3863 9	3872 9	3881 9	3890 9	3899 10	3909 9	3918 9	3927 9	3936 9	3945 9
38	1.3954 9	3963 9	3972 9	3981 9	3990 9	3999 9	4008 8	4016 9	4025 9	4034 9
39	1.4043 9	4052 9	4061 9	4070 8	4078 9	4087 9	4096 9	4105 8	4113 9	4122 9
40	1.4131 9	4140 8	4148 9	4157 9	4166 8	4174 9	4183 9	4192 8	4200 9	4209 8
41	1.4217 9	4226 8	4234 9	4243 8	4251 9	4260 8	4268 9	4277 8	4285 9	4294 8
42	1.4302 9	4311 8	4319 9	4328 8	4336 8	4344 9	4353 8	4361 8	4369 9	4378 8
43	1.4386 8	4394 9	4403 8	4411 8	4419 8	4427 9	4436 8	4444 8	4452 8	4460 8
44	1.4468 9	4477 8	4485 8	4493 8	4501 8	4509 8	4517 8	4525 8	4533 8	4541 9

TABLE 4k-7. $T_{62}\mathrm{He}^3$ TEMPERATURES IN K AS A FUNCTION OF VAPOR PRESSURE P AT 0°C AND STANDARD GRAVITY, 980.665 CM/SEC^2 (Continued)

P in millimeters of Mercury

P	0	1	2	3	4	5	6	7	8	9
45	1.4550 8	4558 8	4566 8	4574 8	4582 8	4590 8	4598 8	4606 8	4614 8	4622 8
46	1.4630 7	4637 8	4645 8	4653 8	4661 8	4669 8	4677 8	4685 8	4693 8	4701 7
47	1.4708 8	4716 8	4724 8	4732 8	4740 7	4747 8	4755 8	4763 8	4771 7	4778 8
48	1.4786 8	4794 8	4802 7	4809 8	4817 8	4825 7	4832 8	4840 8	4848 7	4855 8
49	1.4863 7	4870 8	4878 8	4886 7	4893 8	4901 7	4908 8	4916 8	4924 7	4931 8
50	1.4939 74	5013 74	5087 73	5160 72	5232 71	5303 70	5373 69	5442 69	5511 68	5579 67
60	1.5645 66	5712 66	5778 65	5843 64	5907 64	5971 62	6033 63	6096 62	6158 61	6219 60
70	1.6279 60	6339 60	6399 59	6458 58	6516 58	6574 57	6631 57	6688 56	6744 56	6800 56
80	1.6856 55	6911 54	6965 55	7020 53	7073 54	7127 52	7179 53	7232 52	7284 52	7336 51
90	1.7387 51	7438 51	7489 50	7539 50	7589 49	7638 49	7687 49	7736 49	7785 48	7833 48
100	1.7881 47	7928 48	7976 47	8023 46	8069 47	8116 46	8162 46	8208 45	8253 45	8298 45
110	1.8343 45	8388 45	8433 44	8477 44	8521 43	8564 44	8608 43	8651 43	8694 43	8737 42
120	1.8779 43	8822 42	8864 41	8905 42	8947 41	8988 42	9030 41	9071 40	9111 41	9152 40
130	1.9192 40	9232 40	9272 40	9312 39	9351 40	9391 39	9430 39	9469 39	9508 38	9546 39
140	1.9585 38	9623 38	9661 38	9699 38	9737 37	9774 38	9812 37	9849 37	9886 37	9923 36
150	1.9950 37	9996 36	0032 37	0069 36	0105 36	0141 35	0176 36	0212 36	0248 35	0283 35
160	2.0318 35	0353 35	0388 35	0423 35	0458 34	0492 34	0526 35	0561 34	0595 34	0629 34
170	2.0663 33	0696 34	0730 33	0763 34	0797 33	0830 33	0863 33	0896 33	0929 32	0961 33
180	2.0994 33	1027 32	1059 32	1091 32	1123 32	1155 32	1187 32	1219 32	1251 31	1282 32
190	2.1314 31	1345 31	1376 32	1408 31	1439 30	1469 31	1500 31	1531 31	1562 30	1592 31
200	2.1623 30	1653 30	1683 30	1713 30	1743 30	1773 30	1803 30	1833 30	1863 29	1892 30
210	2.1922 29	1951 29	1980 30	2010 29	2039 29	2068 29	2097 29	2126 28	2154 29	2183 29
220	2.2212 28	2240 28	2268 29	2297 28	2325 28	2353 28	2381 28	2409 28	2437 28	2465 28
230	2.2493 28	2521 27	2548 28	2576 27	2603 28	2631 27	2658 27	2685 27	2712 28	2740 27
240	2.2767 26	2793 27	2820 27	2847 27	2874 27	2901 26	2927 27	2954 26	2980 26	3006 27
250	2.3033 26	3059 26	3085 26	3111 26	3137 26	3163 26	3189 26	3215 26	3241 25	3266 26
260	2.3292 26	3318 25	3343 26	3369 25	3394 25	3419 26	3445 25	3470 25	3495 25	3520 25
270	2.3545 25	3570 25	3595 25	3620 25	3645 24	3669 25	3694 24	3718 25	3743 25	3768 24
280	2.3792 24	3816 25	3841 24	3865 24	3889 24	3913 24	3937 24	3961 24	3985 24	4009 24
290	2.4033 24	4057 24	4081 23	4104 24	4128 24	4152 23	4175 24	4199 23	4222 24	4246 23
300	2.4269 23	4292 24	4316 23	4339 23	4362 23	4385 23	4408 23	4431 23	4454 23	4477 23
310	2.4500 23	4523 22	4545 23	4568 23	4591 22	4613 23	4636 22	4658 23	4681 22	4703 23
320	2.4726 22	4748 22	4770 23	4793 22	4815 22	4837 22	4859 22	4881 22	4903 22	4925 22
330	2.4947 22	4969 22	4991 22	5013 21	5034 22	5056 22	5078 21	5099 22	5121 22	5143 21
340	2.5164 21	5185 22	5207 21	5228 22	5250 21	5271 21	5292 21	5313 22	5335 21	5356 21
350	2.5377 21	5398 21	5419 21	5440 21	5461 21	5482 21	5503 21	5524 20	5544 21	5565 21
360	2.5586 20	5606 21	5627 21	5648 20	5668 21	5689 20	5709 21	5730 20	5750 21	5771 20
370	2.5791 20	5811 20	5831 21	5852 20	5872 20	5892 20	5912 20	5932 20	5952 20	5972 20
380	2.5992 20	6012 20	6032 20	6052 20	6072 20	6092 20	6112 19	6131 20	6151 20	6171 19
390	2.6190 20	6210 20	6230 19	6249 20	6269 19	6288 20	6308 19	6327 19	6346 20	6366 19
400	2.6385 19	6404 20	6424 19	6443 19	6462 19	6481 19	6500 20	6520 19	6539 19	6558 19
410	2.6577 19	6596 19	6615 19	6634 18	6652 19	6671 19	6690 19	6709 19	6728 18	6746 19
420	2.6765 19	6784 18	6802 19	6821 19	6840 18	6858 19	6877 18	6895 19	6914 18	6932 19
430	2.6951 18	6969 18	6987 19	7006 18	7024 18	7042 19	7061 18	7079 18	7097 18	7115 18
440	2.7133 18	7151 19	7170 18	7188 18	7206 18	7224 18	7242 18	7260 18	7278 17	7295 18
450	2.7313 18	7331 18	7349 18	7367 18	7385 17	7402 18	7420 18	7438 17	7455 18	7473 18
460	2.7491 17	7508 18	7526 17	7543 18	7561 17	7578 18	7596 17	7613 18	7631 17	7648 17
470	2.7665 18	7683 17	7700 17	7717 18	7735 17	7752 17	7769 17	7786 17	7803 18	7821 17
480	2.7838 17	7855 17	7872 17	7889 17	7906 17	7923 17	7940 17	7957 17	7974 17	7991 17
490	2.8008 17	8025 16	8041 17	8058 17	8075 17	8092 16	8108 17	8125 17	8142 17	8159 16
500	2.8175 17	8192 17	8209 16	8225 17	8242 16	8258 17	8275 16	8291 17	8308 16	8324 17
510	2.8341 16	8357 16	8373 17	8390 16	8406 17	8423 16	8439 16	8455 16	8471 17	8488 16
520	2.8504 16	8520 16	8536 16	8552 17	8569 16	8585 16	8601 16	8617 16	8633 16	8649 16
530	2.8665 16	8681 16	8697 16	8713 16	8729 16	8745 16	8761 16	8777 15	8792 16	8808 16
540	2.8824 16	8840 16	8856 15	8871 16	8887 16	8903 16	8919 15	8934 16	8950 16	8966 15

TABLE 4k-7. $T_{62}He^3$ TEMPERATURES IN K AS A FUNCTION OF VAPOR PRESSURE P AT 0°C AND STANDARD GRAVITY, 980.665 CM/SEC^2 (Continued)
P in millimeters of Mercury

P	0	1	2	3	4	5	6	7	8	9
550	2.8981 16	8997 15	9012 16	9028 16	9044 15	9059 16	9075 15	9090 16	9106 15	9121 15
560	2.9136 16	9152 15	9167 16	9183 15	9198 15	9213 16	9229 15	9244 15	9259 16	9275 15
570	2.9290 15	9305 15	9320 15	9335 16	9351 15	9366 15	9381 15	9396 15	9411 15	9426 15
580	2.9441 15	9456 15	9471 15	9486 15	9501 15	9516 15	9531 15	9546 15	9561 15	9576 15
590	2.9591 15	9606 15	9621 15	9636 14	9650 15	9665 15	9680 15	9695 15	9710 14	9724 15
600	2.9739 15	9754 15	9769 14	9783 15	9798 14	9812 15	9827 15	9842 14	9856 15	9871 14
610	2.9885 15	9900 15	9915 14	9929 15	9944 14	9958 14	9972 15	9987 14	0001 15	0016 14
620	2.0030 15	0045 14	0059 14	0073 15	0088 14	0102 14	0116 15	0131 14	0145 14	0159 14
630	3.0173 15	0188 14	0202 14	0216 14	0230 14	0244 14	0258 15	0273 14	0287 14	0301 14
640	3.0315 14	0329 14	0343 14	0357 14	0371 14	0385 14	0399 14	0413 14	0427 14	0441 14
650	3.0455 14	0469 14	0483 14	0497 14	0511 13	0524 14	0538 14	0552 14	0566 14	0580 14
660	3.0594 13	0607 14	0621 14	0635 14	0649 13	0662 14	0676 14	0690 13	0703 14	0717 14
670	3.0731 13	0744 14	0758 14	0772 13	0785 14	0799 13	0812 14	0826 13	0839 14	0853 13
680	3.0866 14	0880 13	0893 14	0907 13	0920 14	0934 13	0947 14	0961 13	0974 13	0987 14
690	3.1001 13	1014 13	1027 14	1041 13	1054 13	1067 14	1081 13	1094 13	1107 13	1120 14
700	3.1134 13	1147 13	1160 13	1173 14	1187 13	1200 13	1213 13	1226 13	1239 13	1252 13
710	3.1265 13	1278 14	1292 13	1305 13	1318 13	1331 13	1344 13	1357 13	1370 13	1383 13
720	3.1396 13	1409 13	1422 13	1435 13	1448 12	1460 13	1473 13	1486 13	1499 13	1512 13
730	3.1525 13	1538 12	1550 13	1563 13	1576 13	1589 13	1602 12	1614 13	1627 13	1640 13
740	3.1653 12	1665 13	1678 13	1691 12	1703 13	1716 13	1729 12	1741 13	1754 13	1767 12
750	3.1779 13	1792 12	1804 13	1817 13	1830 12	1842 13	1855 12	1867 13	1880 12	1892 13
760	3.1905 12	1917 13	1930 12	1942 13	1955 12	1967 12	1979 13	1992 12	2004 13	2017 12
770	3.2029 12	2041 13	2054 12	2066 12	2078 13	2091 12	2103 12	2115 13	2128 12	2140 12
780	3.2152 12	2164 13	2177 12	2189 12	2201 12	2213 13	2226 12	2238 12	2250 12	2262 12
790	3.2274 12	2286 12	2298 13	2311 12	2323 12	2335 12	2347 12	2359 12	2371 12	2383 12
800	3.2395 12	2407 12	2419 12	2431 12	2443 12	2455 12	2467 12	2479 12	2491 12	2503 12
810	3.2515 12	2527 12	2539 12	2551 12	2563 11	2574 12	2586 12	2598 12	2610 12	2622 12
820	3.2634 12	2646 11	2657 12	2669 12	2681 12	2693 11	2704 12	2716 12	2728 12	2740 11
830	3.2751 12	2763 12	2775 12	2787 11	2798 12	2810 12	2822 11	2833 12	2845 12	2857 11
840	3.2868 12	2880 11	2891 12	2903 12	2915 11	2926 12	2938 11	2949 12	2961 11	2972 12
850	3.2984 11	2995 12	3007 11	3018 12	3030 11	3041 12	3053 11	3064 12	3076 11	3087 12
860	3.3099 11	3110 11	3121 12	3133 11	3144 12	3156 11	3167 11	3178 12	3190 11	3201 11
870	3.3212 12	3224 11	3235 11	3246						

and for normal hydrogen (75% ortho-H$_2$ and 25% para-H$_2$),

$$\log P \text{ (mm)} = 4.658334 - \frac{44.8793}{T} + 0.021276T - 0.0000021T^2$$

The equation for equilibrium hydrogen is within experimental accuracy in agreement with the vapor-pressure data of Hoge and Arnold[1] over the whole temperature region (14 to 33 K). The equation for normal hydrogen is in agreement with the vapor-pressure data of Woolley, Scott, and Brickwedde.[2]

For solid equilibrium hydrogen the vapor-pressure–temperature relation is based on the equation,

$$\log P \text{ (mm)} = 4.62438 - \frac{47.0172}{T} + 0.03635T$$

[1] H. J. Hoge and R. D. Arnold, J. Research NBS, 47, 63 (1951).
[2] H. W. Woolley, R. B. Scott, and F. G. Brickwedde, J. Research NBS 41, 379 (1948).

and for solid normal hydrogen,

$$\log P \text{ (mm)} = 4.56488 - \frac{47.2059}{T} + 0.03939T$$

These equations were obtained by Woolley, Scott, and Brickwedde[1] from their measurements on the vapor pressure of hydrogen.

Tables 4k-8 to 4k-11 give the temperature in kelvins for integral values of vapor pressure in millimeters of mercury at 0°C and standard gravity, 980.665 cm/sec², as calculated from the above equations.

TABLE 4k-8. SOLID EQUILIBRIUM HYDROGEN TEMPERATURES IN K FOR INTEGRAL VALUES OF VAPOR PRESSURE P, IN MILLIMETERS OF MERCURY AT 0°C AND STANDARD GRAVITY, 980.665 CM/SEC²

P	0	1	2	3	4	5	6	7	8	9
0	......	9.463	10.029	10.391	10.662	10.881	11.067	11.228	11.371	11.501
10	11.619	11.727	11.828	11.922	12.010	12.093	12.172	12.247	12.318	12.386
20	12.452	12.514	12.575	12.633	12.689	12.743	12.795	12.846	12.896	12.944
30	12.990	13.035	13.080	13.123	13.165	13.206	13.246	13.285	13.323	13.361
40	13.398	13.434	13.469	13.504	13.538	13.571	13.604	13.636	13.668	13.699
50	13.730	13.760	13.789	13.819						

TABLE 4k-9. SOLID NORMAL HYDROGEN TEMPERATURES IN K FOR INTEGRAL VALUES OF VAPOR PRESSURE P, IN MILLIMETERS OF MERCURY AT 0°C AND STANDARD GRAVITY, 980,665 CM/SEC²

P	0	1	2	3	4	5	6	7	8	9
0	......	9.554	10.124	10.488	10.761	10.982	11.169	11.331	11.475	11.605
10	11.723	11.832	11.933	12.028	12.116	12.200	12.279	12.354	12.426	12.494
20	12.559	12.622	12.683	12.741	12.797	12.852	12.904	12.955	13.004	13.052
30	13.099	13.145	13.189	13.232	13.274	13.315	13.355	13.394	13.433	13.470
40	13.507	13.543	13.579	13.613	13.647	13.681	13.713	13.746	13.777	13.808
50	13.839	13.869	13.899	13.928	13.957	13.985				

The vapor-pressure–temperature relation for neon is based on the equation

$$\log P \text{ (mm)} = 8.746376 - \frac{126.780}{T} - 0.0436834T$$

This equation was obtained by Henning and Otto[2] from their experimental measurements on the vapor pressure of neon.

Table 4k-12 gives the temperature in K for integral values of vapor pressure in millimeters of mercury at 0°C and standard gravity, 980.665 cm/sec².

The vapor-pressure–temperature relation for nitrogen is based on the equation

$$\log P \text{ (mm)} = 6.49594 - \frac{255.821}{T - 6.600}$$

This equation was obtained by Armstrong[3] from his experimental measurements on the vapor pressure of nitrogen.

[1] H. W. Woolley, R. B. Scott, and F. G. Brickwedde, *J. Research NBS* **41**, 379 (1948).
[2] F. Henning and J. Otto, *Physik. Z.* **37**, 633 (1936).
[3] G. T. Armstrong, *J. Research NBS* **53**, 263 (1954).

TABLE 4k-10. LIQUID EQUILIBRIUM HYDROGEN TEMPERATURES IN K FOR INTEGRAL VALUES OF VAPOR PRESSURE P, IN MILLIMETERS OF MERCURY AT 0°C AND STANDARD GRAVITY, 980.665 CM/SEC^2

P	0	1	2	3	4	5	6	7	8	9
50	. . . . . .	. . . . . .	13.774	13.806	13.839	13.870	13.901	13.932	13.963	13.993
60	14.022	14.051	14.080	14.109	14.137	14.165	14.192	14.219	14.246	14.273
70	14.299	14.325	14.351	14.376	14.401	14.426	14.451	14.475	14.499	14.523
80	14.547	14.570	14.594	14.617	14.640	14.662	14.685	14.707	14.729	14.751
90	14.772	14.794	14.815	14.836	14.857	14.878	14.898	14.919	14.939	14.959
100	14.979	14.999	15.019	15.038	15.058	15.077	15.096	15.115	15.134	15.152
110	15.171	15.189	15.208	15.226	15.244	15.262	15.280	15.298	15.315	15.333
120	15.350	15.367	15.384	15.401	15.418	15.435	15.452	15.469	15.485	15.502
130	15.518	15.534	15.550	15.567	15.582	15.598	15.614	15.630	15.646	15.661
140	15.677	15.692	15.707	15.722	15.738	15.753	15.768	15.783	15.797	15.812
150	15.827	15.841	15.856	15.870	15.885	15.899	15.913	15.928	15.942	15.956
160	15.970	15.984	15.997	16.011	16.025	16.039	16.052	16.066	16.079	16.093
170	16.106	16.119	16.133	16.146	16.159	16.172	16.185	16.198	16.211	16.224
180	16.237	16.249	16.262	16.275	16.287	16.300	16.312	16.325	16.337	16.349
190	16.362	16.374	16.386	16.398	16.410	16.422	16.434	16.446	16.458	16.470
200	16.482	16.494	16.506	16.517	16.529	16.541	16.552	16.564	16.575	16.587
210	16.598	16.609	16.621	16.632	16.643	16.654	16.666	16.677	16.688	16.699
220	16.710	16.721	16.732	16.743	16.753	16.764	16.775	16.786	16.797	16.807
230	16.818	16.829	16.839	16.850	16.860	16.871	16.881	16.891	16.902	16.912
240	16.923	16.933	16.943	16.954	16.964	16.974	16.984	16.994	17.004	17.014
250	17.024	17.034	17.044	17.054	17.064	17.074	17.084	17.093	17.103	17.113
260	17.123	17.132	17.142	17.152	17.161	17.171	17.180	17.190	17.200	17.209
270	17.219	17.228	17.237	17.247	17.256	17.265	17.275	17.284	17.293	17.302
280	17.312	17.321	17.330	17.339	17.348	17.357	17.366	17.376	17.385	17.394
290	17.403	17.411	17.420	17.429	17.438	17.447	17.456	17.465	17.474	17.482
300	17.491	17.500	17.509	17.517	17.526	17.534	17.543	17.552	17.560	17.569
310	17.577	17.586	17.594	17.603	17.611	17.620	17.628	17.637	17.645	17.653
320	17.662	17.670	17.678	17.687	17.695	17.703	17.711	17.720	17.728	17.736
330	17.744	17.752	17.760	17.768	17.777	17.785	17.793	17.801	17.809	17.817
340	17.825	17.833	17.841	17.849	17.856	17.864	17.872	17.880	17.888	17.896
350	17.904	17.911	17.919	17.927	17.935	17.942	17.950	17.958	17.966	17.973
360	17.981	17.988	17.996	18.004	18.011	18.019	18.026	18.034	18.041	18.049
370	18.056	18.064	18.071	18.079	18.086	18.094	18.101	18.108	18.116	18.123
380	18.131	18.138	18.145	18.153	18.160	18.167	18.174	18.182	18.189	18.196
390	18.203	18.210	18.218	18.225	18.232	18.239	18.246	18.254	18.261	18.268
400	18.275	18.282	18.289	18.296	18.303	18.310	18.317	18.324	18.331	18.338
410	18.345	18.352	18.359	18.366	18.373	18.379	18.386	18.393	18.400	18.407
420	18.414	18.421	18.427	18.434	18.441	18.448	18.455	18.461	18.468	18.475
430	18.481	18.488	18.495	18.502	18.508	18.515	18.522	18.528	18.535	18.541
440	18.548	18.555	18.561	18.568	18.574	18.581	18.587	18.594	18.600	18.607
450	18.613	18.620	18.626	18.633	18.639	18.646	18.652	18.659	17.665	18.671
460	18.678	18.684	18.691	18.697	18.703	18.710	18.716	18.722	18.729	18.735
470	18.741	18.748	18.754	18.760	18.766	18.773	18.779	18.785	18.791	18.798
480	18.804	18.810	18.816	18.822	18.828	18.834	18.841	18.847	18.853	18.859
490	18.865	18.871	18.877	18.883	18.889	18.896	18.902	18.908	18.914	18.920
500	18.926	18.932	18.938	18.944	18.950	18.956	18.962	18.968	18.974	18.979

TABLE 4k-10. LIQUID EQUILIBRIUM HYDROGEN TEMPERATURES IN K FOR INTEGRAL
VALUES OF VAPOR PRESSURE P, IN MILLIMETERS OF MERCURY AT 0°C AND
STANDARD GRAVITY, 980.665 CM/SEC^2 (*Continued*)

P	0	1	2	3	4	5	6	7	8	9
510	18.985	18.991	18.997	19.003	19.009	19.015	19.021	19.027	19.033	19.038
520	19.044	19.050	19.056	19.062	19.068	19.073	19.079	19.085	19.091	19.096
530	19.102	19.108	19.114	19.120	19.125	19.131	19.137	19.142	19.148	19.154
540	19.160	19.165	19.171	19.177	19.182	19.188	19.193	19.199	19.205	19.210
550	19.216	19.222	19.227	19.233	19.238	19.244	19.249	19.255	19.261	19.266
560	19.272	19.277	19.283	19.288	19.294	19.299	19.305	19.310	19.316	19.321
570	19.327	19.332	19.338	19.343	19.349	19.354	19.359	19.365	19.370	19.376
580	19.381	19.386	19.392	19.397	19.403	19.408	19.413	19.419	19.424	19.429
590	19.435	19.440	19.445	19.451	19.456	19.461	19.467	19.472	19.477	19.482
600	19.488	19.493	19.498	19.503	19.509	19.514	19.519	19.524	19.530	19.535
610	19.540	19.545	19.550	19.556	19.561	19.566	19.571	19.576	19.581	19.587
620	19.592	19.597	19.602	19.607	19.612	19.617	19.623	19.628	19.633	19.638
630	19.643	19.648	19.653	19.658	19.663	19.668	19.673	19.678	19.683	19.688
640	19.693	19.699	19.704	19.709	19.714	19.719	19.724	19.729	19.734	19.738
650	19.743	19.748	19.753	19.758	19.763	19.768	19.773	19.778	19.783	19.788
660	19.793	19.798	19.803	19.808	19.813	19.817	19.822	19.827	19.832	19.837
670	19.842	19.847	19.852	19.856	19.861	19.866	19.871	19.876	19.881	19.885
680	19.890	19.895	19.900	19.905	19.909	19.914	19.919	19.924	19.928	19.933
690	19.938	19.943	19.948	19.952	19.957	19.962	19.967	19.971	19.976	19.981
700	19.985	19.990	19.995	20.000	20.004	20.009	20.014	20.018	20.023	20.028
710	20.032	20.037	20.042	20.046	20.051	20.056	20.060	20.065	20.069	20.074
720	20.079	20.083	20.088	20.093	20.097	20.102	20.106	20.111	20.116	20.120
730	20.125	20.129	20.134	20.138	20.143	20.147	20.152	20.157	20.161	20.166
740	20.170	20.175	20.179	20.184	20.188	20.193	20.197	20.202	20.206	20.211
750	20.215	20.220	20.224	20.229	20.233	20.238	20.242	20.246	20.251	20.255
760	20.260	20.264	20.269	20.273	20.278	20.282	20.286	20.291	20.295	20.300
770	20.304	20.308	20.313	20.317	20.322	20.326	20.330	20.335	20.339	20.343
780	20.348	20.352	20.356	20.361	20.365	20.370	20.374	20.378	20.383	20.387
790	20.391	20.396	20.400	20.404	20.409	20.413	20.417	20.421	20.426	20.430
800	20.434	20.438	20.443	20.447	20.451	20.456	20.460	20.464	20.468	20.473
810	20.477	20.481	20.485	20.490	20.494	20.498	20.502	20.506	20.511	20.515
820	20.519	20.523	20.527	20.532	20.536	20.540	20.544	20.548	20.553	20.557
830	20.561	20.565	20.569	20.573	20.578	20.582	20.586	20.590	20.594	20.598
840	20.602	20.607	20.611	20.615	20.619	20.623	20.627	20.631	20.635	20.640
850	20.644	20.648	20.652	20.656	20.660	20.664	20.668	20.672	20.676	20.680
860	20.684	20.688	20.693	20.696	20.701	20.705	20.709	20.713	20.717	20.721
870	20.725	20.729	20.733	20.737	20.741	20.745	20.749	20.753	20.757	20.761
880	20.765	20.769	20.773	20.777	20.781	20.785	20.789	20.793	20.797	20.801
890	20.805	20.809	20.813	20.817	20.821	20.825	20.829	20.832	20.836	20.840
900	20.844	20.848	20.852	20.856	20.860	20.864	20.868	20.872	20.876	20.880
910	20.884	20.887	20.891	20.895	20.899	20.903	20.907	20.911	20.915	20.918
920	20.922	20.926	20.930	20.934	20.938	20.942	20.945	20.949	20.953	20.957
930	20.961	20.965	20.969	20.972	20.976	20.980	20.984	20.988	20.992	20.995
940	20.999	21.003	21.007	21.011	21.014	21.018	21.022	21.026	21.030	21.033
950	21.037	21.041	21.045	21.049	21.052	21.056	21.060	21.064	21.067	21.071
900	21.075	21.079	21.082	21.086	21.090	21.094	21.097	21.101	21.105	21.109
970	21.112	21.116	21.120	21.124	21.127	21.131	21.135	21.138	21.142	21.146
980	21.149	21.153	21.157	21.160	21.164	21.168	21.172	21.175	21.179	21.183
990	21.186	21.190	21.194	21.197	21.201	21.205	21.208	21.212	21.216	21.219

TABLE 4k-11. LIQUID NORMAL HYDROGEN TEMPERATURES IN K FOR INTEGRAL VALUES OF VAPOR PRESSURE P, IN MILLIMETERS OF MERCURY AT 0°C AND STANDARD GRAVITY, 980.665 CM/SEC^2

P	0	1	2	3	4	5	6	7	8	9
50					13.944	13.976	14.007	14.038	14.068	14.099
60	14.128	14.157	14.186	14.215	14.243	14.271	14.299	14.326	14.353	14.379
70	14.406	14.432	14.458	14.483	14.508	14.533	14.558	14.582	14.607	14.631
80	14.654	14.678	14.701	14.724	14.747	14.770	14.792	14.815	14.837	14.859
90	14.880	14.902	14.923	14.944	14.965	14.986	15.007	15.027	15.048	15.068
100	15.088	15.108	15.127	15.147	15.166	15.186	15.205	15.224	15.243	15.261
110	15.280	15.298	15.317	15.335	15.353	15.371	15.389	15.407	15.424	15.442
120	15.459	15.477	15.494	15.511	15.528	15.545	15.562	15.578	15.595	15.611
130	15.628	15.644	15.660	15.676	15.692	15.708	15.724	15.740	15.756	15.771
140	15.787	15.802	15.817	15.833	15.848	15.863	15.878	15.893	15.908	15.923
150	15.937	15.952	15.966	15.981	15.995	16.010	16.024	16.038	16.052	16.066
160	16.081	16.094	16.108	16.122	16.136	16.150	16.163	16.177	16.190	16.204
170	16.217	16.230	16.244	16.257	16.270	16.283	16.296	16.309	16.322	16.335
180	16.348	16.361	16.373	16.386	16.399	16.411	16.424	16.436	16.449	16.461
190	16.473	16.486	16.498	16.510	16.522	16.534	16.546	16.558	16.570	16.582
200	16.594	16.606	16.617	16.629	16.641	16.652	16.664	16.676	16.687	16.698
210	16.710	16.721	16.733	16.744	16.755	16.766	16.778	16.789	16.800	16.811
220	16.822	16.833	16.844	16.855	16.866	16.877	16.887	16.898	16.909	16.920
230	16.930	16.941	16.952	16.962	16.973	16.983	16.994	17.004	17.015	17.025
240	17.035	17.046	17.056	17.066	17.076	17.087	17.097	17.107	17.117	17.127
250	17.137	17.147	17.157	17.167	17.177	17.187	17.197	17.206	17.216	17.226
260	17.236	17.245	17.255	17.265	17.275	17.284	17.294	17.303	17.313	17.322
270	17.332	17.341	17.351	17.360	17.369	17.379	17.388	17.397	17.407	17.416
280	17.425	17.434	17.443	17.453	17.462	17.471	17.480	17.489	17.498	17.507
290	17.516	17.525	17.534	17.543	17.552	17.561	17.570	17.578	17.587	17.596
300	17.605	17.613	17.622	17.631	17.639	17.648	17.657	17.666	17.674	17.683
310	17.691	17.700	17.708	17.717	17.725	17.734	17.742	17.751	17.759	17.767
320	17.776	17.784	17.792	17.801	17.809	17.817	17.825	17.834	17.842	17.850
330	17.858	17.866	17.875	17.883	17.891	17.899	17.907	17.915	17.923	17.931
340	17.939	17.947	17.955	17.963	17.971	17.979	17.987	17.994	18.002	18.010
350	18.018	18.026	18.034	18.041	18.049	18.057	18.064	18.072	18.080	18.088
360	18.095	18.103	18.111	18.118	18.126	18.133	18.141	18.148	18.156	18.163
370	18.171	18.179	18.186	18.193	18.201	18.208	18.216	18.223	18.231	18.238
380	18.245	18.253	18.260	18.267	18.275	18.282	18.289	18.297	18.304	18.311
390	18.318	18.326	18.333	18.340	18.347	18.354	18.361	18.368	18.376	18.383
400	18.390	18.397	18.404	18.411	18.418	18.425	18.432	18.439	18.446	18.453
410	18.460	18.467	18.474	18.481	18.488	18.495	18.502	18.508	18.515	18.522
420	18.529	18.536	18.543	18.549	18.556	18.563	18.570	18.577	18.583	18.590
430	18.597	18.604	18.610	18.617	18.624	18.630	18.637	18.644	18.650	18.657
440	18.663	18.670	18.677	18.683	18.690	18.696	18.703	18.709	18.716	18.722
450	18.729	18.735	18.742	18.748	18.755	18.761	18.768	18.774	18.781	18.787
460	18.794	18.800	18.806	18.813	18.819	18.825	18.832	18.838	18.844	18.851
470	18.857	18.863	18.870	18.876	18.882	18.888	18.895	18.901	18.907	18.913
480	18.919	18.926	18.932	18.938	18.944	18.950	18.957	18.963	18.969	18.975
490	18.981	18.987	18.993	18.999	19.005	19.011	19.018	19.024	19.030	19.036
500	19.042	19.048	19.054	19.060	19.066	19.072	19.078	19.084	19.090	19.095

TABLE 4k-11. LIQUID NORMAL HYDROGEN TEMPERATURES IN K FOR INTEGRAL
VALUES OF VAPOR PRESSURE P, IN MILLIMETERS OF MERCURY AT 0°C
AND STANDARD GRAVITY, 980.665 CM/SEC^2 (Continued)

P	0	1	2	3	4	5	6	7	8	9
510	19.102	19.107	19.113	19.119	19.125	19.131	19.137	19.143	19.149	19.154
520	19.160	19.166	19.172	19.178	19.184	19.190	19.195	19.201	19.207	19.213
530	19.219	19.224	19.230	19.236	19.242	19.247	19.253	19.259	19.264	19.270
540	19.276	19.282	19.287	19.293	19.299	19.304	19.310	19.316	19.321	19.327
550	19.332	19.338	19.344	19.349	19.355	19.360	19.366	19.371	19.377	19.383
560	19.388	19.394	19.399	19.405	19.410	19.416	19.421	19.427	19.432	19.438
570	19.443	19.449	19.454	19.460	19.465	19.471	19.476	19.482	19.487	19.492
580	19.498	19.503	19.509	19.514	19.519	19.525	19.530	19.535	19.541	19.546
590	19.551	19.557	19.562	19.567	19.573	19.578	19.583	19.589	19.594	19.599
600	19.604	19.610	19.615	19.620	19.626	19.631	19.636	19.641	19.646	19.652
610	19.657	19.662	19.667	19.673	19.678	19.683	19.688	19.693	19.698	19.704
620	19.709	19.714	19.719	19.724	19.729	19.734	19.740	19.745	19.750	19.755
630	19.760	19.765	19.770	19.775	19.780	19.785	19.790	19.795	19.801	19.806
640	19.811	19.816	19.821	19.826	19.831	19.836	19.841	19.846	19.851	19.856
650	19.861	19.866	19.871	19.876	19.880	19.885	19.890	19.895	19.900	19.905
660	19.910	19.915	19.920	19.925	19.930	19.935	19.940	19.944	19.949	19.954
670	19.959	19.964	19.969	19.974	19.979	19.983	19.988	19.993	19.998	20.003
680	20.008	20.012	20.017	20.022	20.027	20.032	20.036	20.041	20.046	20.051
690	20.055	20.060	20.065	20.070	20.075	20.079	20.084	20.089	20.093	20.098
700	20.103	20.108	20.112	20.117	20.122	20.126	20.131	20.136	20.140	20.145
710	20.150	20.155	20.159	20.164	20.168	20.173	20.178	20.182	20.187	20.192
720	20.196	20.201	20.206	20.210	20.215	20.219	20.224	20.229	20.233	20.238
730	20.242	20.247	20.252	20.256	20.261	20.265	20.270	20.274	20.279	20.283
740	20.288	20.292	20.297	20.301	20.306	20.310	20.315	20.320	20.324	20.329
750	20.333	20.337	20.342	20.346	20.351	20.355	20.360	20.364	20.369	20.373
760	20.378	20.382	20.387	20.391	20.395	20.400	20.404	20.409	20.413	20.418
770	20.422	20.426	20.431	20.435	20.440	20.444	20.448	20.453	20.457	20.461
780	20.466	20.470	20.474	20.479	20.483	20.488	20.492	20.496	20.501	20.505
790	20.509	20.514	20.518	20.522	20.527	20.531	20.535	20.539	20.544	20.548
800	20.552	20.557	20.561	20.565	20.569	20.574	20.578	20.582	20.586	20.591
810	20.595	20.599	20.603	20.608	20.612	20.616	20.620	20.625	20.629	20.633
820	20.637	20.642	20.646	20.650	20.654	20.658	20.663	20.667	20.671	20.675
830	20.679	20.683	20.688	20.692	20.696	20.700	20.704	20.708	20.712	20.717
840	20.721	20.725	20.729	20.733	20.737	20.741	20.745	20.750	20.754	20.758
850	20.762	20.766	20.770	20.774	20.778	20.782	20.787	20.791	20.795	20.799
860	20.803	20.807	20.811	20.815	20.819	20.823	20.827	20.831	20.835	20.839
870	20.843	20.847	20.851	20.855	20.859	20.863	20.867	20.871	20.875	20.880
880	20.884	20.888	20.891	20.895	20.900	20.904	20.908	20.911	20.915	20.919
890	20.923	20.927	20.931	20.935	20.939	20.943	20.947	20.951	20.955	20.959
900	20.963	20.967	20.971	20.975	20.979	20.983	20.986	20.990	20.994	20.998
910	21.002	21.006	21.010	21.014	21.018	21.022	21.026	21.029	21.033	21.037
920	21.041	21.045	21.049	21.053	21.057	21.060	21.064	21.068	21.072	21.076
930	21.080	21.084	21.087	21.091	21.095	21.099	21.103	21.107	21.110	21.114
940	21.118	21.122	21.126	21.129	21.133	21.137	21.141	21.145	21.149	21.152
950	21.156	21.160	21.164	21.167	21.171	21.175	21.179	21.182	21.186	21.190
960	21.194	21.198	21.201	21.205	21.209	21.213	21.216	21.220	21.224	21.227
970	21.231	21.235	21.239	21.242	21.246	21.250	21.254	21.257	21.261	21.265
980	21.268	21.272	21.276	21.280	21.283	21.287	21.291	21.294	21.298	21.302
990	21.305	21.309	21.313	21.316	21.320	21.324	21.327	21.331	21.335	21.338

TABLE 4k-12. LIQUID NEON TEMPERATURES IN K FOR INTEGRAL VALUES OF VAPOR PRESSURE P, IN MILLIMETERS OF MERCURY AT 0°C AND STANDARD GRAVITY, 980.665 CM/SEC^2

P	0	1	2	3	4	5	6	7	8	9
320					24.554	24.562	24.570	24.578	24.586	24.594
330	24.602	24.610	24.618	24.626	24.634	24.642	24.650	24.657	24.665	24.673
340	24.681	24.689	24.696	24.704	24.712	24.719	24.727	24.735	24.742	24.750
350	24.758	24.765	24.773	24.780	24.788	24.796	24.803	24.811	24.818	24.825
360	24.833	24.840	24.848	24.855	24.863	24.870	24.877	24.885	24.892	24.899
370	24.907	24.914	24.921	24.929	24.936	24.943	24.950	24.958	24.965	24.972
380	24.979	24.986	24.993	25.001	25.008	25.015	25.022	25.029	25.036	25.043
390	25.050	25.057	25.064	25.071	25.078	25.085	25.092	25.099	25.106	25.113
400	25.120	25.127	25.134	25.140	25.147	25.154	25.161	25.168	25.175	25.181
410	25.188	25.195	25.202	25.208	25.215	25.222	25.229	25.235	25.242	25.249
420	25.255	25.262	25.269	25.275	25.282	25.289	25.295	25.302	25.308	25.315
430	25.322	25.328	25.335	25.341	25.348	25.354	25.361	25.367	25.374	25.380
440	25.387	25.393	25.399	25.406	25.412	25.419	25.425	25.431	25.438	25.444
450	25.451	25.457	25.463	25.470	25.476	25.482	25.488	25.495	25.501	25.507
460	25.514	25.520	25.526	25.532	25.538	25.545	25.551	25.557	25.563	25.569
470	25.576	25.582	25.588	25.594	25.600	25.606	25.612	25.618	25.624	25.631
480	25.637	25.643	25.649	25.655	25.661	25.667	25.673	25.679	25.685	25.691
490	25.697	25.703	25.709	25.715	25.721	25.727	25.733	25.738	25.744	25.750
500	25.756	25.762	25.768	25.774	25.780	25.786	25.791	25.797	25.803	25.809
510	25.815	25.820	25.826	25.832	25.838	25.844	25.849	25.855	25.861	25.867
520	25.872	25.878	25.884	25.890	25.895	25.901	25.907	25.912	25.918	25.924
530	25.929	25.935	25.941	25.946	25.952	25.957	25.963	25.969	25.974	25.980
540	25.985	25.991	25.997	26.002	26.008	26.013	26.019	26.024	26.030	26.035
550	26.041	26.046	26.052	26.057	26.063	26.068	26.074	26.079	26.085	26.090
560	26.096	26.101	26.107	26.112	26.117	26.123	26.128	26.134	26.139	26.144
570	26.150	26.155	26.161	26.166	26.171	26.177	26.182	26.187	26.193	26.198
580	26.203	26.209	26.214	26.219	26.224	26.230	26.235	26.240	26.246	26.251
590	26.256	26.261	26.267	26.272	26.277	26.282	26.287	26.293	26.298	26.303
600	26.308	26.313	26.319	26.324	26.329	26.334	26.339	26.344	26.350	26.355
610	26.360	26.365	26.370	26.375	26.380	26.385	26.391	26.396	26.401	26.406
620	26.411	26.416	26.421	26.426	26.431	26.436	26.441	26.446	26.451	26.456
630	26.461	26.466	26.471	26.476	26.481	26.486	26.491	26.496	26.501	26.506
640	26.511	26.516	26.521	26.526	26.531	26.536	26.541	26.546	26.551	26.556
650	26.561	26.566	26.570	26.575	26.580	26.585	26.590	26.595	26.600	26.605
660	26.609	26.614	26.619	26.624	26.629	26.634	26.639	26.643	26.648	26.653
670	26.658	26.663	26.667	26.672	26.677	26.682	26.687	26.691	26.696	26.701
680	26.706	26.710	26.715	26.720	26.725	26.729	26.734	26.739	26.744	26.748
690	26.753	26.758	26.763	26.767	26.772	26.777	26.781	26.786	26.791	26.795
700	26.800	26.805	26.809	26.814	26.819	26.823	26.828	26.833	26.837	26.842
710	26.847	26.851	26.856	26.860	26.865	26.870	26.874	26.879	26.883	26.888
720	26.893	26.897	26.902	26.906	26.911	26.915	26.920	26.925	26.929	26.934
730	26.938	26.943	26.947	26.952	26.956	26.961	26.965	26.970	26.974	26.979
740	26.983	26.988	26.992	26.997	27.001	27.006	27.010	27.015	27.019	27.024
750	27.028	27.033	27.037	27.042	27.046	27.050	27.055	27.059	27.064	27.068
760	27.073	27.077	27.081	27.086	27.090	27.095	27.099	27.103	27.108	27.112
770	27.117	27.121	27.125	27.130	27.134	27.138	27.143	27.147	27.151	27.156
780	27.160	27.165	27.169	27.173	27.178	27.182	27.186	27.191	27.195	27.199
790	27.203	27.208	27.212	27.216	27.221	27.225	27.229	27.234	27.238	27.242
800	27.246	27.251	27.255	27.259	27.263	27.268	27.272	27.276	27.280	27.285

TABLE 4k-12. LIQUID NEON TEMPERATURES IN K FOR INTEGRAL VALUES OF
VAPOR PRESSURE P, IN MILLIMETERS OF MERCURY AT 0°C AND STANDARD
GRAVITY, 980.665 CM/SEC^2 (Continued)

P	0	1	2	3	4	5	6	7	8	9
810	27.289	27.293	27.297	27.302	27.306	27.310	27.314	27.318	27.323	27.327
820	27.331	27.335	27.339	27.344	27.348	27.352	27.356	27.360	27.365	27.369
830	27.373	27.377	27.381	27.385	27.390	27.394	27.398	27.402	27.406	27.410
840	27.414	27.419	27.423	27.427	27.431	27.435	27.439	27.443	27.447	27.452
850	27.456	27.460	27.464	27.468	27.472	27.476	27.480	27.484	27.488	27.492
860	27.497	27.501	27.505	27.509	27.513	27.517	27.521	27.525	27.529	27.533
870	27.537	27.541	27.545	27.549	27.553	27.557	27.561	27.565	27.569	27.573
880	27.577	27.581	27.585	27.589	27.593	27.597	27.601	27.605	27.609	27.613
890	27.617	27.621	27.625	27.629	27.633	27.637	27.641	27.645	27.649	27.653
900	27.657	27.661	27.665	27.669	27.673	27.677	27.681	27.685	27.689	27.692
910	27.696	27.700	27.704	27.708	27.712	27.716	27.720	27.724	27.728	27.732
920	27.735	27.739	27.743	27.747	27.751	27.755	27.759	27.763	27.767	27.770
930	27.774	27.778	27.782	27.786	27.790	27.794	27.798	27.801	27.805	27.809
940	27.813	27.817	27.821	27.824	27.828	27.832	27.836	27.840	27.844	27.847
950	27.851	27.855	27.859	27.863	27.866	27.870	27.874	27.878	27.882	27.885
960	27.889	27.893	27.897	27.901	27.904	27.908	27.912	27.916	27.919	27.923
970	27.927	27.931	27.935	27.938	27.942	27.946	27.950	27.953	27.957	27.961
980	27.965	27.968	27.972	27.976	27.980	27.983	27.987	27.991	27.994	27.998

Table 4k-13 gives the temperature in kelvins for integral values of vapor pressure in millimeters of mercury at 0°C and standard gravity, 980.665 cm/sec^2.

The following interpolation equation has been derived for oxygen by a least-squares fit of the vapor-pressure data of Hoge[1]

$$\log P \text{ (mm)} = 8.01602 - \frac{415.6909}{T} - 0.0058382T$$

the equation being valid for the region from the triple point (54.363 K) to 90.827 K. Table 4k-14 gives the temperature in kelvins for integral values of vapor pressure P, in millimeters of mercury at 0°C and standard gravity, 980.665 cm/sec^2.

[1] H. J. Hoge, *J. Research NBS* **44**, 326 (1950).

TABLE 4k-13. LIQUID NITROGEN TEMPERATURES IN K FOR INTEGRAL VALUES
OF VAPOR PRESSURE P, IN MILLIMETERS OF MERCURY AT 0°C AND
STANDARD GRAVITY, 980.665 CM/SEC²

P	0	1	2	3	4	5	6	7	8	9
90					63.162	63.220	63.277	63.333	63.389	63.445
100	63.500	63.555	63.609	63.663	63.717	63.770	63.822	63.875	63.926	63.978
110	64.029	64.080	64.130	64.180	64.230	64.279	64.328	64.376	64.425	64.473
120	64.520	64.568	64.615	64.661	64.708	64.754	64.799	64.845	64.890	64.935
130	64.980	65.024	65.068	65.112	65.156	65.199	65.242	65.285	65.327	65.370
140	65.412	65.453	65.495	65.536	65.578	65.618	65.659	65.700	65.740	65.780
150	65.820	65.859	65.899	65.938	65.977	66.015	66.054	66.092	66.131	66.169
160	66.206	66.244	66.281	66.319	66.356	66.393	66.429	66.466	66.502	66.538
170	66.574	66.610	66.646	66.681	66.717	66.752	66.787	66.822	66.856	66.891
180	66.925	66.960	66.994	67.028	67.061	67.095	67.129	67.162	67.195	67.228
190	67.261	67.294	67.327	67.359	67.392	67.424	67.456	67.488	67.520	67.552
200	67.583	67.615	67.646	67.677	67.709	67.740	67.771	67.801	67.832	67.862
210	67.893	67.923	67.953	67.984	68.013	68.043	68.073	68.103	68.132	68.162
220	68.191	68.220	68.249	68.278	68.307	68.336	68.365	68.393	68.422	68.450
230	68.479	68.507	68.535	68.563	68.591	68.619	68.647	68.674	68.702	68.729
240	68.757	68.784	68.811	68.838	68.865	68.892	68.919	68.946	68.972	68.999
250	69.025	69.052	69.078	69.104	69.131	69.157	69.183	69.209	69.235	69.260
260	69.286	69.312	69.337	69.363	69.388	69.413	69.439	69.464	69.489	69.514
270	69.539	69.564	69.588	69.613	69.638	69.662	69.687	69.711	69.736	69.760
280	69.784	69.808	69.833	69.857	69.881	69.904	69.928	69.952	69.976	69.999
290	70.023	70.047	70.070	70.093	70.117	70.140	70.163	70.186	70.209	70.232
300	70.255	70.278	70.301	70.324	70.347	70.369	70.392	70.414	70.437	70.459
310	70.482	70.504	70.526	70.549	70.571	70.593	70.615	70.637	70.659	70.681
320	70.702	70.724	70.746	70.768	70.789	70.811	70.832	70.854	70.875	70.897
330	70.918	70.939	70.960	70.981	71.003	71.024	71.045	71.066	71.086	71.107
340	71.128	71.149	71.170	71.190	71.211	71.232	71.252	71.273	71.293	71.313
350	71.334	71.354	71.374	71.394	71.415	71.435	71.455	71.475	71.495	71.515
360	71.535	71.555	71.574	71.594	71.614	71.634	71.653	71.673	71.692	71.712
370	71.731	71.751	71.770	71.790	71.809	71.828	71.847	71.867	71.886	71.905
380	71.924	71.943	71.962	71.981	72.000	72.019	72.038	72.057	72.075	72.094
390	72.113	72.131	72.150	72.169	72.187	72.206	72.224	72.243	72.261	72.279
400	72.298	72.316	72.334	72.353	72.371	72.389	72.407	72.425	72.443	72.461
410	72.479	72.497	72.515	72.533	72.551	72.569	72.586	72.604	72.622	72.640
420	72.657	72.675	72.692	72.710	72.728	72.745	72.762	72.780	72.797	72.815
430	72.832	72.849	72.867	72.884	72.901	72.918	72.935	72.952	72.970	72.987
440	73.004	73.021	73.038	73.055	73.071	73.088	73.105	73.122	73.139	73.156
450	73.172	73.189	73.206	73.222	73.239	73.256	73.272	73.289	73.305	73.322
460	73.338	73.354	73.371	73.387	73.404	73.420	73.436	73.452	73.469	73.485
470	73.501	73.517	73.533	73.549	73.566	73.582	73.598	73.614	73.630	73.646
480	73.661	73.677	73.693	73.709	73.725	73.741	73.756	73.772	73.788	73.804
490	73.819	73.835	73.850	73.866	73.882	73.897	73.913	73.928	73.944	73.959
500	73.975	73.990	74.005	74.021	74.036	74.051	74.067	74.082	74.097	74.112
510	74.127	74.143	74.158	74.173	74.188	74.203	74.218	74.233	74.248	74.263
520	74.278	74.293	74.308	74.323	74.338	74.353	74.367	74.382	74.397	74.412
530	74.427	74.441	74.456	74.471	74.485	74.500	74.515	74.529	74.544	74.558
540	74.573	74.587	74.602	74.616	74.631	74.645	74.660	74.674	74.688	74.703
550	74.717	74.731	74.746	74.760	74.774	74.788	74.803	74.817	74.831	74.845

TABLE 4k-13. LIQUID NITROGEN TEMPERATURES IN K FOR INTEGRAL VALUES OF VAPOR PRESSURE P, IN MILLIMETERS OF MERCURY AT 0°C AND STANDARD GRAVITY, 980.665 CM/SEC² (*Continued*)

P	0	1	2	3	4	5	6	7	8	9
560	74.859	74.873	74.888	74.902	74.916	74.930	74.944	74.958	74.972	74.986
570	75.000	75.014	75.027	75.041	75.055	75.069	75.083	75.097	75.111	75.124
580	75.138	75.152	75.165	75.179	75.193	75.207	75.220	75.234	75.247	75.261
590	75.275	75.288	75.302	75.315	75.329	75.342	75.356	75.369	75.383	75.396
600	75.409	75.423	75.436	75.450	75.463	75.476	75.490	75.503	75.516	75.529
610	75.543	75.556	75.569	75.582	75.595	75.609	75.622	75.635	75.648	75.661
620	75.674	75.687	75.700	75.713	75.726	75.739	75.752	75.765	75.778	75.791
630	75.804	75.817	75.830	75.843	75.855	75.868	75.881	75.894	75.907	75.919
640	75.932	75.945	75.958	75.970	75.983	75.996	76.008	76.021	76.034	76.046
650	76.059	76.071	76.084	76.097	76.109	76.122	76.134	76.147	76.159	76.172
660	76.184	76.197	76.209	76.221	76.234	76.246	76.259	76.271	76.283	76.296
670	76.308	76.320	76.333	76.345	76.357	76.369	76.382	76.394	76.406	76.418
680	76.430	76.443	76.455	76.467	76.479	76.491	76.503	76.515	76.527	76.539
690	76.551	76.564	76.576	76.588	76.600	76.612	76.623	76.635	76.647	76.659
700	76.671	76.683	76.695	76.707	76.719	76.731	76.742	76.754	76.766	76.778
710	76.790	76.801	76.813	76.825	76.837	76.848	76.860	76.872	76.883	76.895
720	76.907	76.918	76.930	76.942	76.953	76.965	76.977	76.988	77.000	77.011
730	77.023	77.034	77.046	77.057	77.069	77.080	77.092	77.103	77.115	77.126
740	77.137	77.149	77.160	77.172	77.183	77.194	77.206	77.217	77.228	77.240
750	77.251	77.262	77.274	77.285	77.296	77.307	77.319	77.330	77.341	77.352
760	77.363	77.375	77.386	77.397	77.408	77.419	77.430	77.441	77.453	77.464

TABLE 4k-14. LIQUID OXYGEN TEMPERATURES IN K FOR INTEGRAL VALUES
OF VAPOR PRESSURE P, IN MILLIMETERS OF MERCURY AT 0°C AND
STANDARD GRAVITY, 980.665 CM/SEC^2

P	0	1	2	3	4	5	6	7	8	9
0		53.980	56.278	57.719	58.790	59.650	60.373	60.998	61.551	62.048
10	62.499	62.913	63.297	73.654	63.989	64.303	64.601	64.883	65.151	65.407
20	65.652	65.887	66.112	66.329	66.539	66.741	66.936	67.125	67.308	67.486
30	67.659	67.827	67.991	68.150	68.305	68.457	68.605	68.749	68.891	69.029
40	69.164	69.297	69.427	69.554	69.679	69.802	69.922	70.040	70.156	70.270
50	70.383	70.493	70.601	70.708	70.813	70.917	71.019	71.120	71.219	71.316
60	71.413	71.508	71.601	71.694	71.785	71.875	71.964	72.052	72.139	72.225
70	72.309	72.393	72.476	72.558	72.639	72.719	72.798	72.876	72.953	73.030
80	73.106	73.181	73.255	73.329	73.402	73.474	73.545	73.616	73.686	73.756
90	73.824	73.893	73.960	74.027	74.094	74.159	74.225	74.289	74.354	74.417
100	74.480	74.543	74.605	74.667	74.728	74.788	74.849	74.908	74.968	75.026
110	75.085	75.143	75.200	75.257	75.314	75.370	75.426	75.482	75.537	75.591
120	75.646	75.700	75.753	75.807	75.860	75.912	75.964	76.016	76.068	76.119
130	76.170	76.221	76.271	76.321	76.371	76.420	76.469	76.518	76.566	76.615
140	76.663	76.710	76.758	76.805	76.852	76.898	76.945	76.991	77.036	77.082
150	77.127	77.172	77.217	77.262	77.306	77.350	77.394	77.438	77.481	77.525
160	77.568	77.611	77.653	77.695	77.738	77.780	77.821	77.863	77.904	77.945
170	77.986	78.027	78.068	78.108	78.148	78.188	78.228	78.268	78.307	78.347
180	78.386	78.425	78.463	78.502	78.540	78.579	78.617	78.655	78.693	78.730
190	78.768	78.805	78.842	78.879	78.916	78.953	78.989	79.025	79.062	79.098
200	79.134	79.169	79.205	79.241	79.276	79.311	79.346	79.381	79.416	79.451
210	79.485	79.520	79.554	79.588	79.622	79.656	79.690	79.724	79.757	79.791
220	79.824	79.857	79.890	79.923	79.956	79.989	80.021	80.054	80.086	80.118
230	80.150	80.182	80.214	80.246	80.278	80.309	80.341	80.372	80.403	80.435
240	80.466	80.497	80.527	80.558	80.589	80.619	80.650	80.680	80.710	80.741
250	80.771	80.801	80.831	80.860	80.890	80.920	80.949	80.979	81.008	81.037
260	81.066	81.095	81.124	81.153	81.182	81.211	81.239	81.268	81.296	81.324
270	81.353	81.381	81.409	81.437	81.465	81.493	81.521	81.548	81.576	81.604
280	81.631	81.658	81.686	81.713	81.740	81.767	81.794	81.821	81.848	81.875
290	81.902	81.928	81.955	81.981	82.008	82.034	82.060	82.087	82.113	82.139
300	82.165	82.191	82.217	82.242	82.268	82.294	82.319	82.345	82.370	82.396
310	82.421	82.446	82.472	82.497	82.522	82.547	82.572	82.597	82.622	82.646
320	82.671	82.696	82.720	82.745	82.769	82.794	82.818	82.842	82.867	82.891
330	82.915	82.939	82.963	82.987	83.011	83.035	83.058	83.082	83.106	83.129
340	83.153	83.177	83.200	83.223	83.247	83.270	83.293	83.316	83.340	83.363
350	83.386	83.409	83.432	83.454	83.477	83.500	83.523	83.545	83.568	83.591
360	83.613	83.636	83.658	83.680	83.703	83.725	83.747	83.769	83.792	83.814
370	83.836	83.858	83.880	83.902	83.923	83.945	83.967	83.989	84.010	84.032
380	84.054	84.075	84.097	84.118	84.139	84.161	84.182	84.203	84.225	84.246
390	84.267	84.288	84.309	84.330	84.351	84.372	84.393	84.414	84.435	84.455
400	84.476	84.497	84.518	84.538	84.559	84.579	84.600	84.620	84.641	84.661

TABLE 4k-14. LIQUID OXYGEN TEMPERATURES IN K FOR INTEGRAL VALUES
OF VAPOR PRESSURE P, IN MILLIMETERS OF MERCURY AT 0°C AND
STANDARD GRAVITY, 980.665 CM/SEC^2 (*Continued*)

P	0	1	2	3	4	5	6	7	8	9
410	84.681	84.702	84.722	84.742	84.762	84.783	84.803	84.823	84.843	84.863
420	84.883	84.903	84.922	84.942	84.962	84.982	85.002	85.021	85.041	85.061
430	85.080	85.100	85.119	85.139	85.158	85.178	85.197	85.216	85.236	85.255
440	85.274	85.293	85.313	85.332	85.351	85.370	85.389	85.408	85.427	85.446
450	85.465	85.484	85.503	85.521	85.540	85.559	85.578	85.596	85.615	85.634
460	85.652	85.671	85.689	85.708	85.726	85.745	85.763	85.782	85.800	85.818
470	85.836	85.855	85.873	85.891	85.909	85.927	85.946	85.964	85.982	86.000
480	86.018	86.036	86.054	86.071	86.089	86.107	86.125	86.143	86.161	86.178
490	86.196	86.214	86.231	86.249	86.266	86.284	86.302	86.319	86.337	86.354
500	86.371	86.389	86.406	86.424	86.441	86.458	86.475	86.493	86.510	86.527
510	86.544	86.561	86.578	86.596	86.613	86.630	86.647	86.664	86.681	86.697
520	86.714	86.731	86.748	86.765	86.782	86.799	86.815	86.832	86.849	86.865
530	86.882	86.899	86.915	86.932	86.948	86.965	86.981	86.998	87.014	87.031
540	87.047	87.064	87.080	87.096	87.113	87.129	87.145	87.162	87.178	87.194
550	87.210	87.226	87.242	87.259	87.275	87.291	87.307	87.323	87.339	87.355
560	87.371	87.387	87.403	87.419	87.434	87.450	87.466	87.482	87.498	87.513
570	87.529	87.545	87.561	87.576	87.592	87.608	87.623	87.639	87.654	87.670
580	87.686	87.701	87.717	87.732	87.747	87.763	87.778	87.794	87.809	87.824
590	87.840	87.855	87.870	87.886	87.901	87.916	87.931	87.947	87.962	87.977
600	87.992	88.007	88.022	88.037	88.052	88.067	88.082	88.097	88.112	88.127
610	88.142	88.157	88.172	88.187	88.202	88.217	88.232	88.246	88.261	88.276
620	88.291	88.306	88.320	88.335	88.350	88.364	88.379	88.394	88.408	88.423
630	88.437	88.452	88.466	88.481	88.496	88.510	88.524	88.539	88.553	88.568
640	88.582	88.597	88.611	88.625	88.640	88.654	88.668	88.683	88.697	88.711
650	88.725	88.740	88.754	88.768	88.782	88.796	88.810	88.824	88.839	88.853
660	88.867	88.881	88.895	88.909	88.923	88.937	88.951	88.965	88.979	88.993
670	89.007	89.020	89.034	89.048	89.062	89.076	89.090	89.103	89.117	89.131
680	89.145	89.158	89.172	89.186	89.200	89.213	89.227	89.241	89.254	89.268
690	89.281	89.295	89.309	89.322	89.336	89.349	89.363	89.376	89.390	89.403
700	89.417	89.430	89.443	89.457	89.470	89.484	89.497	89.510	89.524	89.537
710	89.550	89.564	89.577	89.590	89.603	89.617	89.630	89.643	89.656	89.669
720	89.683	89.696	89.709	89.722	89.735	89.748	89.761	89.774	89.787	89.800
730	89.813	89.826	89.839	89.852	89.865	89.878	89.891	89.904	89.917	89.930
740	89.943	89.956	89.969	89.982	89.994	90.007	90.020	90.033	90.046	90.058
750	90.071	90.084	90.097	90.109	90.122	90.135	90.147	90.160	90.173	90.185
760	90.198	90.211	90.223	90.236	90.248	90.261	90.274	90.286	90.299	90.311
770	90.324	90.336	90.349	90.361	90.374	90.386	90.398	90.411	90.423	90.436
780	90.448	90.460	90.473	90.485	90.497	90.510	90.522	90.534	90.547	90.559
790	90.571	90.584	90.596	90.608	90.620	90.632	90.645	90.657	90.669	90.681
800	90.693	90.705	90.718	90.730	90.742	90.754	90.766	90.778	90.790	90.802
810	90.814	90.826	90.838	90.850	90.862	90.874	90.886	90.898	90.910	90.922

TABLE 4k-15. VAPOR PRESSURES OF THE CHEMICAL ELEMENTS

Temperature, °C (at the stated Pressure, atm)

Element	10^{-10}	10^{-9}	10^{-8}	10^{-7}	10^{-6}	10^{-5}	10^{-4}	10^{-3}	10^{-2}	10^{-1}	1	ΔH_{v298}	T_m, °C	ΔH_m	T_t, °C	ΔH_t	Trans.
Actinium Ac	(1025)⊙	(1111)	(1211)	(1325)	(1459)	(1617)	(1806)	(2038)	(2328)	(2702)	3200	(95,000)*	1050	3400			
Aluminum Al	744	812	889	979	1084	1209	1360	1546	1782	2093	2520	78,700	660	2580			
Americium Am	625	689	767	855	958	1085	1245	1444	1710	2074	2614	66,000	995	2900	600		α-β
Antimony Sb	308	343	381	426	476	534	603	738	947	1220	1587	12,340	631	4750			
Argon Ar	-245	-243	-240	-238	-234	-230	-226	-220	-212	-201⊙	-186	1,558*	-189	281			
Arsenic As	123	145	170	198	231	269	313	366	429	508⊙	612	8,600	817	6620			
Astatine At	-10	+5	21	40	62	87	121	156	201	259⊙	335	(21,600)*	(300)	(5,700)			
Barium Ba	360	409	466	533	614	712⊙	843	1015	1250	1590	2125	40,490†	729	2300			
Beryllium Be	747	812	887	974	1074	1193⊙	1336	1520	1752	2056	2472	77,500	1287	(2919)	1245	(611)	α-β
Bismuth Bi	366	409	459	518	586	669	768	893	1053	1266	1564	50,100	271	2700			
Boron B	1450	1560	1620	1750	1900⊙	2076	2291	2550	2870	3275	3802	136,500	2027	5290			
Bromine Br	-139	-131	-122	-111	-100	-86	-71	-52	-28⊙	+4	58	7,170*	-7	2520			
Cadmium Cd	91	115	143	175	213	257	310	382	473	595	767	26,720	321	1480			
Calcium Ca	314	354	399	453⊙	516	591	683	799	955	1171	1484	42,600	839	2040	447	220	α-β
Carbon C	1750	1859	1983	2115	2267	2439	2636	2864	3130	3446	3827	171,290	(3850)				
Cerium Ce	1091	1183	1288	1410	1552	1721	1923	2172	2483	2886	3426	101,000	798	1305	726	715	γ-δ
Cesium Cs	0	22⊙	47	75	109	159	202	269	359	486	682	18,670	29	520			
Chlorine Cl	-184	-179	-172	-165	-157	-147	-135	-120	-103	-76	-34	4,878*	-101	1531			
Chromium Cr	902	974	1056	1149	1257⊙	1384	1535	1719	1952	2259	2672	95,000	1857	4047			
Cobalt Co	995	1072	1161	1262	1379⊙	1517	1686	1894	2152	2484	2928	102,400	1495	3870	427	108	α-β
Copper Cu	774	841	918	1006⊙	1109	1237	1389	1578	1818	2133	2566	80,500	1083	3140			
Dysprosium Dy	667	731	803	888	987	1150	1259	1432	1683	2033	2562	69,400	1409	2643	1384	995	α-β
Erbium Er	757	827	907	1001	1111	1242	1403	1610	1893	2283	2863	75,800	1522	4757			
Europium Eu	316	356	404	459	524	602	699	821	991	1230	1597	41,900	817	2202			
Fluorine F	-253	-251	-248	-245	-241	-236	-230	-223	-215	-205	-188	1,562*	-220	122			
Francium Fa	-15	+4	30⊙	57	90	131	182	246	334	467	(674)	(15,200)*	(27)	(500)			
Gadolinium Gd	987	1070	1165	1270	1406	1564	1757	1996	2302	2706	3266	95,000	1312	2403	1260	935	α-β
Gallium Ga	617	678	748	830	926	1038	1176	1348	1565	1855	2247	65,300	30	1335			
Germanium Ge	851	926	1013	1114	1232	1372	1542	1751	2016	2362	2834	89,500	937	8830			
Gold Au	863	938	1024	1122	1237	1374	1542	1750	2010	2348	2808	87,300	1063	2955			
Hafnium Hf	1608	1726	1862	2017	2197	2417	2682	3000	3406	3896	4603	148,000	2227	5750	1740	1610	α-β
Helium He	⊙	⊙	⊙	⊙	⊙	⊙	⊙	0.95	1.23	2.49	4.22	20*	(3.5)	5			
Holmium Ho	703	769	844	932	1035	1159	1311	1503	1769	2139	2695	71,900	1470	2911	1428	1121	α-β
Hydrogen H	-269	-269	-268	-268	-267	-266	-265	-264	-262	-259	-253	216	-259	28			
Indium In	533	589	654	729	819	924	1053	1213	1419	1691	2070	58,000	156	780			
Iodine I	-86	-75	-63	-49	-32	-14	+9	35	67	109⊙	183	9,970	114	3770			

The eleven left-hand numerical columns give temperatures (°C) at successive vapor pressures (pressure headers are carried on the facing page). The three following columns give ΔHv, the melting point (°C), and ΔHm. The final columns give solid-state transition data.

Element	1	2	3	4	5	6	7	8	9	10	11	ΔHv	m.p.	ΔHm	Transition temp, °C	ΔHt	Transition
Iridium Ir	1673	1789	1921	2070	2242	2441	2685	2980	3343	3800	4389	160,000	2454	6300			
Iron Fe	948	1024	1109	1208	1321	1455⊙	1617	1820	2075	2407	2862	99,300	1536	3300	911, 1392	215, 200	α-γ, γ-δ
Krypton Kr	-234	-231	-228	-224	-220	-214	-208	-199	-189	-174	-153	2,158	-157	391			
Lanthanum La	1102	1194	1299	1421	1564	1733	1937	2187	2501	2908	3457	103,000	920	1481	277, 861	87, 746	α-β, β-γ
Lead Pb	404	425	478	573	616	706	817	957	1140	1389	1750	46,620	327	1147			
Lithium Li	265	303	344	397	456	531	619	730	871	1068	1324	38,584	180	717			
Lutetium Lu	1056	1141	1239	1351	1481	1635	1829	2071	2383	2800	3395	102,200	1663	4457			
Magnesium Mg	204	236	273	315	365	424	496	584⊙	699	860	1090	35,000	649	2140			
Manganese Mn	586	640	700⊙	774	858	956	1074⊙	1222⊙	1419	1683	2062	67,700	1244	2882	707, 1087, 1137	532, 507, 449	α-β, β-γ, γ-δ
Mercury Hg	-60	-46⊙	-29	-9	+14	42	77⊙	120⊙	176	251	357	14,692	-39	549			
Molybdenum Mo	1670	1862	1925	2080	2259	2469	2722	3041	3436	3942	4610	157,300	2617	6650			
Neodymium Nd	796	869	954⊙	1054	1173	1318	1498	1725	2026	2443	3068	78,300	1016	1707	855	724	α-β
Neon Ne	-265	-265	-264	-263	-262	-261	-260	-258	-255	-252	-246	422	-249	80			
Nickel Ni	994	1070	1155	1259	1373	1511	1678	1883	2139	2469	2914	102,800	1453	4176			
Niobium Nb	1833	1959	2101	2262	2448⊙	2671	2935	3252	3640	4124	4744	172,400	2467	6302			
Nitrogen N	-248	-246	-244	-242	-239	-236	-232	-227	-220	-211	-196	1,335	-210	172	-238	55	α-β
Osmium Os	2002	2140	2288	2464	2662	2888	3151	3483	3874	4370	4987	187,400	3045	7000			
Oxygen O	-243	-241	-239	-237	-234	-230	-226	-220⊙	-212	-200	-183	1,630	-219	106	-249, -229	22, 178	α-β, β-γ
Palladium Pd	917	995	1085	1188	1308	1450	1625	1844	2124	2483	2964	90,000	1552	(4197)			
Phosphorus P	67	86	106	126	153	182	216	256	303	362	431	42,700	597	4500			
Platinum Pt	1385	1484	1600	1731⊙	1882	2058	2278	2544	2871	3286	3824	134,970	1770	4700			
Plutonium Pu	895	979	1076	1189	1323	1485	1681	1927	2240	2356	3230	84,100	640	680	122, 207, 315, 457, 480	800, 140, 130, 20, 440	α-β, β-γ, γ-δ, δ-δ', δ'-ε
Polonium Po	132	156	183	216	258	308	373	458	572	744	947	34,500	254	3000			
Potassium K	39	62⊙	88	120	157	203	260	333	430	563	758	21,415	63	562			
Praseodymium Pr	902⊙	985	1082	1196	1331	1495	1698	1958	2301	2782	3512	85,000	931	1646	795	757	α-β
Radium Ra	275	312	359	410	474	547	634⊙	769	928	1181	1527	32,700*	700	2000			
Rhenium Re	2052	2196	2362	2552	2771	3032	3344	3737	4232	4857	5687	186,100	3180	7900			
Rhodium Rh	1358	1456	1567	1694	1841	2017	2223	2476	2743	3196	3727	133,100	1966	5150			
Rubidium Rb	16	35⊙	59	88	123	165	218	286	376	503	694	19,600	39	540			

ΔHv Heat of vaporization, kcal/mol.
ΔHm Heat of melting, kcal/mol.
ΔHt Heat of transition, kcal/mol.
⊙ Change of phase.
* At the normal boiling point.
† ΔHt at 627°C.

TABLE 4k-15. VAPOR PRESSURES OF THE CHEMICAL ELEMENTS (*Continued*)

Pressure, atm: Element	Temperature, °C											ΔH_{v298}	T_m, °C	ΔH_m	T_t, °C	ΔH_t	Trans.
	10^{-10}	10^{-9}	10^{-8}	10^{-7}	10^{-6}	10^{-5}	10^{-4}	10^{-3}	10^{-2}	10^{-1}	1						
Ruthenium Ru	1592	1702	1826	1967	2128	2316 ⊙	2540 ⊙	2815 ⊙	3153	3575	4119	155,000	2427	6210			
Samarium Sm	407	453	507	568	641	728	834 ⊙	968 ⊙	1148	1405	1791	49,400	1072	2060	917	744	α-β
Scandium Sc	879	952	1035	1130	1241 ⊙	1372	1531 ⊙	1734	1996	2343	2831	90,320	1539	3369	1335	958	α-β
Selenium Se	80	104	130	160	194 ⊙	237	290	354	435	538	679	55,000	217	1300			
Silicon Si	1071	1154	1244	1348 ⊙	1476	1636	1830	2067	2365	2750	3267	108,900	1412	12,082			
Silver Ag	617	674	740	815	904	1010	1141	1304	1510	1783	2163	67,900	961	2700			
Sodium Na	95 ⊙	121	152	188	231	282	347	428	535	681	883	25,852	98	622	589	200	α-γ
Strontium Sr	270	307	350	400	458 ⊙	528 ⊙	614	721 ⊙	867	1074	1375	39,300	770	2400	96	90	α-β
Sulfur S	1	15	35	52	77 ⊙	105 ⊙	142	183	238	321	445	66,000	115	337			
Tantalum Ta	2079	2222	2384	2568	2779 ⊙	3025 ⊙	3327	3686	4124	4670	5365	186,900	2977	7475			
Technetium Tc	1664	1772	1910	2058 ⊙	2233	2458	2721	3053	3463	3958	4627	138,000	2127	5500			
Tellurium Te	196	223	253	289	327	373	428 ⊙	505	617	768	988	20,120	449	4180			
Terbium Tb	950	1030	1122	1228 ⊙	1352 ⊙	1507	1695	1931	2234	2642	3223	92,900	1357	2580	1287	1200	α-β
Thallium Tl	321	362 ⊙	410	465	530	609	706	825	980	1190	1487	43,550	304	975	234	90	α-β
Thorium Th	1527	1646 ⊙	1782	1944	2134	2361	2636	2978	3413	3989	4788	137,500	1755	3853	1363	654	α-β
Thulium Tm	487	539	599	667	748	845	963	1109	1298 ⊙	1559	1947	55,500	1545	4025			
Tin Sn	736	807	888	982	1093	1225	1385	1584	1836	2167	2623	72,200	13	500			
Titanium Ti	1124	1211	1310	1423	1554 ⊙	1709	1900	2132	2421	2793	3289	112,300	1670	3692	882	1017	α-β
Tungsten W	2235	2383	2550	2739	2955	3205	3502	3866	4309	4857	5555	203,000	3407	8460			
Uranium U	1320	1426	1548	1689	1855	2053	2292	2588 ⊙	2964	3457	4134	125,000	1132	2036	668 / 775	667 / 1137	α-β / β-γ
Vanadium V	1228	1318	1420	1537	1675	1827 ⊙	2017	2252	2543	2916	3409	122,900	1902	5002			
Xenon Xe	−219	−215	−211	−206	−200	−192	−183	−171 ⊙	−157	−137 ⊙	−108	3,021	−112	549			
Ytterbium Yb	228	262	301	346	400	463	541	638 ⊙	775	934	1194	36,350	824	1830	760	418	α-β
Yttrium Y	1040	1124	1220	1331	1458 ⊙	1616	1809	2049	2356	2765	3338	101,500	1526	2724	1479	1193	α-β
Zinc Zn	148	176	208	244	286	338	398 ⊙	479	581	718	911	31,245	420	1765			
Zirconium Zr	1574	1690	1822 ⊙	1976	2156	2367	2620	2926	3304	3783	4409	145,500	1852	4038			

TABLE 4k-16. VAPOR PRESSURE OF ICE
(Pressure of aqueous vapor over ice from −120 to 0°C)

Temp., °C	Bars	mm of Hg	Temp., °C	Bars	mm of Hg
−120	0.000 000 0001	0.000 000 09	−70	0.000 002 577	0.001 933
−119	0.000 000 0002	0.000 000 11	−69	0.000 002 992	0.002 245
−118	0.000 000 0002	0.000 000 15	−68	0.000 003 469	0.002 603
−117	0.000 000 0003	0.000 000 19	−67	0.000 004 017	0.003 013
−116	0.000 000 0003	0.000 000 25	−66	0.000 004 643	0.003 483
−115	0.000 000 0004	0.000 000 32	−65	0.000 005 360	0.004 021
−114	0.000 000 0005	0.000 000 41	−64	0.000 006 179	0.004 635
−113	0.000 000 0007	0.000 000 52	−63	0.000 007 113	0.005 336
−112	0.000 000 0009	0.000 000 66	−62	0.000 008 178	0.006 135
−111	0.000 000 0011	0.000 000 84	−61	0.000 009 389	0.007 043
−110	0.000 000 0014	0.000 001 07	−60	0.000 010 765	0.008 076
−109	0.000 000 0018	0.000 001 35	−59	0.000 012 328	0.009 248
−108	0.000 000 0023	0.000 001 69	−58	0.000 014 098	0.010 576
−107	0.000 000 0028	0.000 002 13	−57	0.000 016 103	0.012 080
−106	0.000 000 0035	0.000 002 66	−56	0.000 018 369	0.013 780
−105	0.000 000 0044	0.000 003 32	−55	0.000 020 93	0.015 70
−104	0.000 000 0055	0.000 004 13	−54	0.000 023 82	0.017 87
−103	0.000 000 0068	0.000 005 13	−53	0.000 027 07	0.020 31
−102	0.000 000 0085	0.000 006 36	−52	0.000 030 73	0.023 05
−101	0.000 000 0105	0.000 007 85	−51	0.000 034 85	0.026 14
−100	0.000 000 0129	0.000 009 68	−50	0.000 039 47	0.029 61
−99	0.000 000 0159	0.000 011 90	−49	0.000 044 66	0.033 50
−98	0.000 000 0194	0.000 014 59	−48	0.000 050 47	0.037 86
−97	0.000 000 0238	0.000 017 85	−47	0.000 056 97	0.042 74
−96	0.000 000 0290	0.000 021 78	−46	0.000 064 24	0.048 19
−95	0.000 000 0354	0.000 026 53	−45	0.000 072 36	0.054 28
−94	0.000 000 0430	0.000 032 24	−44	0.000 081 42	0.061 08
−93	0.000 000 0521	0.000 039 09	−43	0.000 091 52	0.068 66
−92	0.000 000 0630	0.000 047 29	−42	0.000 102 77	0.077 09
−91	0.000 000 0761	0.000 057 10	−41	0.000 115 28	0.086 48
−90	0.000 000 0917	0.000 068 79	−40	0.000 129 18	0.096 91
−89	0.000 000 1103	0.000 082 71	−39	0.000 144 62	0.108 49
−88	0.000 000 1323	0.000 099 24	−38	0.000 161 74	0.121 33
−87	0.000 000 1584	0.000 118 85	−37	0.000 180 72	0.135 57
−86	0.000 000 1894	0.000 142 05	−36	0.000 201 72	0.151 33
−85	0.000 000 2259	0.000 1694	−35	0.000 2250	0.1688
−84	0.000 000 2689	0.000 2018	−34	0.000 2506	0.1880
−83	0.000 000 3196	0.000 2398	−33	0.000 2790	0.2093
−82	0.000 000 3792	0.000 2844	−32	0.000 3103	0.2328
−81	0.000 000 4490	0.000 3368	−31	0.000 3447	0.2586
−80	0.000 000 5307	0.000 3981	−30	0.000 3827	0.2871
−79	0.000 000 6262	0.000 4697	−29	0.000 4245	0.3184
−78	0.000 000 7376	0.000 5533	−28	0.000 4704	0.3529
−77	0.000 000 8673	0.000 6506	−27	0.000 5209	0.3907
−76	0.000 001 0182	0.000 7638	−26	0.000 5762	0.4323
−75	0.000 001 1934	0.000 8952	−25	0.000 6370	0.4778
−74	0.000 001 3964	0.001 0476	−24	0.000 7035	0.5277
−73	0.000 001 6314	0.001 2238	−23	0.000 7764	0.5824
−72	0.000 001 9030	0.001 4275	−22	0.000 8561	0.6422
−71	0.000 002 2162	0.001 6625	−21	0.000 9433	0.7076

TABLE 4k-16. VAPOR PRESSURE OF ICE (*Continued*)

Temp., °C	Bars	mm of Hg	Temp., °C	Bars	mm of Hg
−20	0.001 0385	0.7790	−5	0.004 023	3.018
−19	0.001 1424	0.8570	−4	0.004 379	3.285
−18	0.001 2558	0.9421	−3	0.004 763	3.573
−17	0.001 3794	1.0348	−2	0.005 178	3.884
−16	0.001 5140	1.1358	−1	0.005 625	4.220
−15	0.001 661	1.246	0	0.006 107	4.581
−14	0.001 820	1.365			
−13	0.001 993	1.495			
−12	0.002 181	1.636			
−11	0.002 386	1.790			
−10	0.002 607	1.956			
−9	0.002 847	2.136			
−8	0.003 107	2.331			
−7	0.003 389	2.542			
−6	0.003 694	2.771			

TABLE 4k-17. VAPOR PRESSURE OF WATER BELOW 100°C
(Pressure of aqueous vapor over water from −15.0 to 100.0°C)

Temp., °C	Bars	mm of Hg	Temp., °C	Bars	mm of Hg
−15.0	0.001 914	1.436	−5.0	0.004 216	3.162
−14.8	0.001 946	1.459	−4.8	0.004 280	3.210
−14.6	0.001 978	1.484	−4.6	0.004 345	3.259
−14.4	0.002 011	1.508	−4.4	0.004 411	3.308
−14.2	0.002 044	1.533	−4.2	0.004 478	3.359
−14.0	0.002 078	1.558	−4.0	0.004 545	3.409
−13.8	0.002 112	1.584	−3.8	0.004 614	3.461
−13.6	0.002 147	1.610	−3.6	0.004 684	3.513
−13.4	0.002 182	1.637	−3.4	0.004 754	3.566
−13.2	0.002 218	1.663	−3.2	0.004 826	3.620
−13.0	0.002 254	1.691	−3.0	0.004 898	3.674
−12.8	0.002 291	1.718	−2.8	0.004 972	3.729
−12.6	0.002 328	1.746	−2.6	0.005 046	3.785
−12.4	0.002 366	1.775	−2.4	0.005 121	3.841
−12.2	0.002 404	1.803	−2.2	0.005 198	3.899
−12.0	0.002 443	1.833	−2.0	0.005 275	3.957
−11.8	0.002 483	1.862	−1.8	0.005 353	4.015
−11.6	0.002 523	1.893	−1.6	0.005 433	4.075
−11.4	0.002 564	1.923	−1.4	0.005 513	4.135
−11.2	0.002 605	1.954	−1.2	0.005 595	4.196
−11.0	0.002 647	1.985	−1.0	0.005 677	4.258
−10.8	0.002 689	2.017	−0.8	0.005 761	4.321
−10.6	0.002 732	2.049	−0.6	0.005 846	4.385
−10.4	0.002 776	2.082	−0.4	0.005 932	4.449
−10.2	0.002 820	2.115	−0.2	0.006 019	4.515
−10.0	0.002 865	2.149	0.0	0.006 107	4.581
−9.8	0.002 911	2.183	0.2	0.006 196	4.648
−9.6	0.002 957	2.218	0.4	0.006 287	4.716
−9.4	0.003 003	2.253	0.6	0.006 379	4.785
−9.2	0.003 051	2.288	0.8	0.006 472	4.854
−9.0	0.003 099	2.324	1.0	0.006 566	4.925
−8.8	0.003 148	2.361	1.2	0.006 661	4.996
−8.6	0.003 197	2.398	1.4	0.006 758	5.069
−8.4	0.003 248	2.436	1.6	0.006 856	5.142
−8.2	0.003 298	2.474	1.8	0.006 955	5.217
−8.0	0.003 350	2.513	2.0	0.007 055	5.292
−7.8	0.003 402	2.552	2.2	0.007 157	5.368
−7.6	0.003 455	2.592	2.4	0.007 260	5.445
−7.4	0.003 509	2.632	2.6	0.007 364	5.523
−7.2	0.003 564	2.673	2.8	0.007 469	5.602
−7.0	0.003 619	2.715	3.0	0.007 576	5.683
−6.8	0.003 675	2.757	3.2	0.007 684	5.764
−6.6	0.003 732	2.799	3.4	0.007 794	5.846
−6.4	0.003 790	2.842	3.6	0.007 905	5.929
−6.2	0.003 848	2.886	3.8	0.008 017	6.013
−6.0	0.003 907	2.931	4.0	0.008 131	6.099
−5.8	0.003 967	2.976	4.2	0.008 246	6.185
−5.6	0.004 028	3.021	4.4	0.008 363	6.273
−5.4	0.004 090	3.067	4.6	0.008 481	6.361
−5.2	0.004 152	3.114	4.8	0.008 600	6.451

TABLE 4k-17. VAPOR PRESSURE OF WATER BELOW 100°C (*Continued*)

Temp., °C	Bars	mm of Hg	Temp., °C	Bars	mm of Hg
5.0	0.008 721	6.542	15.0	0.017 049	12.788
5.2	0.008 844	6.633	15.2	0.017 270	12.954
5.4	0.008 968	6.726	15.4	0.017 493	13.121
5.6	0.009 093	6.821	15.6	0.017 719	13.290
5.8	0.009 220	6.916	15.8	0.017 947	13.462
6.0	0.009 349	7.012	16.0	0.018 178	13.635
6.2	0.009 479	7.110	16.2	0.018 412	13.810
6.4	0.009 611	7.209	16.4	0.018 648	13.987
6.6	0.009 745	7.309	16.6	0.018 887	14.166
6.8	0.009 880	7.410	16.8	0.019 128	14.347
7.0	0.010 016	7.513	17.0	0.019 373	14.531
7.2	0.010 155	7.617	17.2	0.019 620	14.716
7.4	0.010 295	7.722	17.4	0.019 869	14.903
7.6	0.010 437	7.828	17.6	0.020 122	15.093
7.8	0.010 580	7.936	17.8	0.020 377	15.284
8.0	0.010 725	8.045	18.0	0.020 635	15.478
8.2	0.010 872	8.155	18.2	0.020 896	15.673
8.4	0.011 021	8.267	18.4	0.021 160	15.871
8.6	0.011 172	8.379	18.6	0.021 427	16.071
8.8	0.011 324	8.494	18.8	0.021 696	16.274
9.0	0.011 478	8.609	19.0	0.021 969	16.478
9.2	0.011 634	8.726	19.2	0.022 245	16.685
9.4	0.011 792	8.845	19.4	0.022 523	16.894
9.6	0.011 952	8.965	19.6	0.022 805	17.105
9.8	0.012 113	9.086	19.8	0.023 090	17.319
10.0	0.012 277	9.209	20.0	0.023 378	17.535
10.2	0.012 442	9.333	20.2	0.023 669	17.753
10.4	0.012 610	9.458	20.4	0.023 963	17.974
10.6	0.012 779	9.585	20.6	0.024 261	18.197
10.8	0.012 951	9.714	20.8	0.024 562	18.423
11.0	0.013 124	9.844	21.0	0.024 866	18.651
11.2	0.013 300	9.976	21.2	0.025 173	18.881
11.4	0.013 477	10.109	21.4	0.025 483	19.114
11.6	0.013 657	10.243	21.6	0.025 797	19.350
11.8	0.013 838	10.380	21.8	0.026 115	19.588
12.0	0.014 022	10.518	22.0	0.026 435	19.828
12.2	0.014 208	10.657	22.2	0.026 759	20.071
12.4	0.014 396	10.798	22.4	0.027 087	20.317
12.6	0.014 587	10.941	22.6	0.027 418	20.565
12.8	0.014 779	11.085	22.8	0.027 753	20.816
13.0	0.014 974	11.231	23.0	0.028 091	21.070
13.2	0.015 171	11.379	23.2	0.028 433	21.326
13.4	0.015 370	11.529	23.4	0.028 778	21.585
13.6	0.015 572	11.680	23.6	0.029 127	21.847
13.8	0.015 776	11.833	23.8	0.029 480	22.112
14.0	0.015 982	11.988	24.0	0.029 836	22.379
14.2	0.016 191	12.144	24.2	0.030 197	22.649
14.4	0.016 402	12.302	24.4	0.030 561	22.922
14.6	0.016 615	12.462	24.6	0.030 928	23.198
14.8	0.016 831	12.624	24.8	0.031 300	23.477

TABLE 4k-17. VAPOR PRESSURE OF WATER BELOW 100°C (*Continued*)

Temp., °C	Bars	mm of Hg	Temp., °C	Bars	mm of Hg
25.0	0.031 676	23.759	35.0	0.056 237	42.181
25.2	0.032 055	24.043	35.2	0.056 862	42.650
25.4	0.032 439	24.331	35.4	0.057 493	43.123
25.6	0.032 826	24.621	35.6	0.058 130	43.601
25.8	0.033 217	24.915	35.8	0.058 773	44.083
26.0	0.033 613	25.212	36.0	0.059 422	44.570
26.2	0.034 013	25.512	36.2	0.060 077	45.062
26.4	0.034 416	25.814	36.4	0.060 739	45.558
26.6	0.034 824	26.120	36.6	0.061 407	46.059
26.8	0.035 236	26.429	36.8	0.062 081	46.565
27.0	0.035 653	26.742	37.0	0.062 762	47.075
27.2	0.036 073	27.057	37.2	0.063 449	47.591
27.4	0.036 498	27.376	37.4	0.064 143	48.111
27.6	0.036 928	27.698	37.6	0.064 843	48.636
27.8	0.037 361	28.023	37.8	0.065 549	49.166
28.0	0.037 800	28.352	38.0	0.066 263	49.701
28.2	0.038 242	28.684	38.2	0.066 983	50.241
28.4	0.038 689	29.019	38.4	0.067 710	50.786
28.6	0.039 141	29.358	38.6	0.068 443	51.337
28.8	0.039 597	29.700	38.8	0.069 184	51.892
29.0	0.040 058	30.046	39.0	0.069 931	52.453
29.2	0.040 524	30.395	39.2	0.070 686	53.019
29.4	0.040 994	30.748	39.4	0.071 447	53.590
29.6	0.041 469	31.104	39.6	0.072 216	54.166
29.8	0.041 948	31.464	39.8	0.072 991	54.748
30.0	0.042 433	31.827	40.0	0.073 774	55.335
30.2	0.042 922	32.195	40.2	0.074 564	55.928
30.4	0.043 417	32.565	40.4	0.075 362	56.526
30.6	0.043 916	32.940	40.6	0.076 166	57.130
30.8	0.044 421	33.318	40.8	0.076 979	57.739
31.0	0.044 930	33.700	41.0	0.077 798	58.354
31.2	0.045 444	34.086	41.2	0.078 626	58.974
31.4	0.045 964	34.476	41.4	0.079 460	59.600
31.6	0.046 488	34.869	41.6	0.080 303	60.232
31.8	0.047 018	35.267	41.8	0.081 153	60.870
32.0	0.047 553	35.668	42.0	0.082 011	61.513
32.2	0.048 094	36.073	42.2	0.082 876	62.162
32.4	0.048 639	36.483	42.4	0.083 750	62.818
32.6	0.049 190	36.896	42.6	0.084 631	63.479
32.8	0.049 747	37.313	42.8	0.085 521	64.146
33.0	0.050 309	37.735	43.0	0.086 418	64.819
33.2	0.050 876	38.160	43.2	0.087 324	65.498
33.4	0.051 449	38.590	43.4	0.088 237	66.184
33.6	0.052 028	39.024	43.6	0.089 159	66.875
33.8	0.052 612	39.462	43.8	0.090 090	67.573
34.0	0.053 201	39.904	44.0	0.091 028	68.277
34.2	0.053 797	40.351	44.2	0.091 975	68.987
34.4	0.054 398	40.802	44.4	0.092 931	69.704
34.6	0.055 005	41.257	44.6	0.093 894	70.427
34.8	0.055 618	41.717	44.8	0.094 867	71.156

TABLE 4k-17. Vapor Pressure of Water Below 100°C (*Continued*)

Temp., °C	Bars	mm of Hg	Temp., °C	Bars	mm of Hg
45.0	0.095 848	71.892	55.0	0.157 45	118.09
45.2	0.096 838	72.635	55.2	0.158 96	119.23
45.4	0.097 837	73.384	55.4	0.160 49	120.38
45.6	0.098 844	74.139	55.6	0.162 03	121.53
45.8	0.099 861	74.902	55.8	0.163 58	122.70
46.0	0.100 886	75.671	56.0	0.165 15	123.87
46.2	0.101 921	76.447	56.2	0.166 72	125.05
46.4	0.102 964	77.230	56.4	0.168 31	126.25
46.6	0.104 017	78.019	56.6	0.169 92	127.45
46.8	0.105 079	78.816	56.8	0.171 53	128.66
47.0	0.106 150	79.619	57.0	0.173 16	129.88
47.2	0.107 231	80.430	57.2	0.174 81	131.12
47.4	0.108 321	81.248	57.4	0.176 46	132.36
47.6	0.109 421	82.072	57.6	0.178 13	133.61
47.8	0.110 530	82.904	57.8	0.179 81	134.87
48.0	0.111 649	83.744	58.0	0.181 51	136.14
48.2	0.112 777	84.590	58.2	0.183 22	137.43
48.4	0.113 916	85.444	58.4	0.184 94	138.72
48.6	0.115 064	86.305	58.6	0.186 68	140.02
48.8	0.116 222	87.174	58.8	0.188 43	141.34
49.0	0.117 390	88.050	59.0	0.190 20	142.66
49.2	0.118 568	88.934	59.2	0.191 98	144.00
49.4	0.119 757	89.825	59.4	0.193 77	145.34
49.6	0.120 955	90.724	59.6	0.195 58	146.70
49.8	0.122 164	91.630	59.8	0.197 40	148.06
50.0	0.123 38	92.545	60.0	0.199 24	149.44
50.2	0.124 61	93.467	60.2	0.201 09	150.83
50.4	0.125 85	94.398	60.4	0.202 96	152.23
50.6	0.127 10	95.336	60.6	0.204 84	153.64
50.8	0.128 37	96.282	60.8	0.206 73	155.06
51.0	0.129 64	97.236	61.0	0.208 64	156.50
51.2	0.130 92	98.198	61.2	0.210 57	157.94
51.4	0.132 21	99.169	61.4	0.212 51	159.40
51.6	0.133 52	100.147	61.6	0.214 47	160.86
51.8	0.134 83	101.134	61.3	0.216 44	162.34
52.0	0.136 16	102.129	62.0	0.218 42	163.83
52.2	0.137 50	103.133	62.2	0.220 43	165.33
52.4	0.138 85	104.145	62.4	0.222 44	166.85
52.6	0.140 21	105.166	62.6	0.224 48	168.37
52.8	0.141 58	106.195	62.8	0.226 53	169.91
53.0	0.142 96	107.232	63.0	0.228 59	171.46
53.2	0.144 36	108.278	63.2	0.230 67	173.02
53.4	0.145 77	109.333	63.4	0.232 77	174.59
53.6	0.147 18	110.397	63.6	0.234 88	176.18
53.8	0.148 61	111.470	63.8	0.237 01	177.77
54.0	0.150 06	112.551	64.0	0.239 16	179.38
54.2	0.151 51	113.642	64.2	0.241 32	181.00
54.4	0.152 98	114.741	64.4	0.243 50	182.64
54.6	0.154 45	115.850	64.6	0.245 69	184.29
54.8	0.155 94	116.967	64.8	0.247 91	185.94

TABLE 4k-17. VAPOR PRESSURE OF WATER BELOW 100°C (*Continued*)

Temp., °C	Bars	mm of Hg	Temp., °C	Bars	mm of Hg
65.0	0.250 13	187.62	75.0	0.385 53	289.17
65.2	0.252 38	189.30	75.2	0.388 77	291.60
65.4	0.254 64	191.00	75.4	0.392 03	294.05
65.6	0.256 92	192.71	75.6	0.395 32	296.51
65.8	0.259 22	194.43	75.8	0.398 62	298.99
66.0	0.261 54	196.17	76.0	0.401 95	301.49
66.2	0.263 87	197.92	76.2	0.405 31	304.00
66.4	0.266 22	199.68	76.4	0.408 68	306.54
66.6	0.268 59	201.46	76.6	0.412 08	309.09
66.8	0.270 97	203.25	76.8	0.415 51	311.66
67.0	0.273 38	205.05	77.0	0.418 96	314.24
67.2	0.275 80	206.87	77.2	0.422 43	316.85
67.4	0.278 24	208.70	77.4	0.425 92	319.47
67.6	0.280 70	210.54	77.6	0.429 45	322.11
67.8	0.283 17	212.40	77.8	0.432 99	324.77
68.0	0.285 67	214.27	78.0	0.436 56	327.45
68.2	0.288 18	216.15	78.2	0.440 15	330.14
68.4	0.290 71	218.05	78.4	0.443 77	332.86
68.6	0.293 27	219.27	78.6	0.447 42	335.59
68.8	0.295 84	221.90	78.8	0.451 09	338.34
69.0	0.298 43	223.84	79.0	0.454 78	341.12
69.2	0.301 03	225.79	79.2	0.458 50	343.91
69.4	0.303 66	227.76	79.4	0.462 25	346.71
69.6	0.303 31	229.75	79.6	0.466 02	349.54
69.8	0.308 97	231.75	79.8	0.469 82	352.39
70.0	0.311 66	233.76	80.0	0.473 64	355.26
70.2	0.314 37	235.79	80.2	0.477 49	358.15
70.4	0.317 09	237.84	80.4	0.481 37	361.05
70.6	0.319 84	239.90	80.6	0.485 27	363.98
70.8	0.322 60	241.97	80.8	0.489 20	366.93
71.0	0.325 39	244.06	81.0	0.493 15	369.89
71.2	0.328 20	246.17	81.2	0.497 13	372.88
71.4	0.331 02	248.29	81.4	0.501 14	375.89
71.6	0.333 87	250.42	81.6	0.505 18	378.92
71.8	0.336 74	252.57	81.8	0.509 24	381.96
72.0	0.339 63	254.74	82.0	0.513 33	385.03
72.2	0.342 54	256.92	82.2	0.517 45	388.12
72.4	0.345 47	259.12	82.4	0.521 60	391.23
72.6	0.348 42	261.34	82.6	0.525 77	394.36
72.8	0.351 39	263.57	82.8	0.529 97	397.51
73.0	0.354 39	265.81	83.0	0.534 20	400.68
73.2	0.357 40	268.07	83.2	0.538 46	403.88
73.4	0.360 44	270.35	83.4	0.542 75	407.09
73.6	0.363 50	272.65	83.6	0.547 06	410.33
73.8	0.366 58	274.96	83.8	0.551 40	413.59
74.0	0.369 68	277.29	84.0	0.555 78	416.87
74.2	0.372 81	279.63	84.2	0.560 18	420.17
74.4	0.375 96	281.99	84.4	0.564 61	423.49
74.6	0.379 13	284.37	84.6	0.569 07	426.84
74.8	0.382 32	286.76	84.8	0.573 56	430.20

TABLE 4k-17. VAPOR PRESSURE OF WATER BELOW 100°C (*Continued*)

Temp., °C	Bars	mm of Hg	Temp., °C	Bars	mm of Hg
85.0	0.578 08	433.59	93.0	0.784 91	588.73
85.2	0.582 62	437.00	93.2	0.790 78	593.14
85.4	0.587 20	440.44	93.4	0.796 69	597.57
85.6	0.591 81	443.89	93.6	0.802 63	602.02
85.8	0.596 45	447.37	93.8	0.808 61	606.51
86.0	0.601 12	450.88	94.0	0.814 63	611.02
86.2	0.605 82	454.40	94.2	0.820 68	615.56
86.4	0.610 55	457.95	94.4	0.826 78	620.13
86.6	0.615 31	461.52	94.6	0.832 90	624.73
86.8	0.620 10	465.11	94.8	0.839 07	629.36
87.0	0.624 92	468.73	95.0	0.845 28	634.01
87.2	0.629 78	472.37	95.2	0.851 52	638.69
87.4	0.634 67	476.04	95.4	0.857 80	643.40
87.6	0.639 58	479.73	95.6	0.864 12	648.14
87.8	0.644 53	483.44	95.8	0.870 48	652.91
88.0	0.649 51	487.18	96.0	0.876 87	657.71
88.2	0.654 53	490.94	96.2	0.883 31	662.54
88.4	0.659 57	494.72	96.4	0.889 79	667.39
88.6	0.664 65	498.53	96.6	0.896 30	672.28
88.8	0.669 76	502.36	96.8	0.902 85	677.20
89.0	0.674 91	506.22	97.0	0.909 45	682.14
89.2	0.680 08	510.10	97.2	0.916 08	687.12
89.4	0.685 29	514.01	97.4	0.922 76	692.12
89.6	0.690 53	517.94	97.6	0.929 47	697.16
89.8	0.695 81	521.90	97.8	0.936 22	702.23
90.0	0.701 12	525.88	98.0	0.943 02	707.32
90.2	0.706 46	529.89	98.2	0.949 86	712.45
90.4	0.711 84	533.93	98.4	0.956 73	717.61
90.6	0.717 25	537.98	98.6	0.963 65	722.80
90.8	0.722 70	542.07	98.8	0.970 61	728.02
91.0	0.728 18	546.18	99.0	0.977 61	733.27
91.2	0.733 69	550.32	99.2	0.984 66	738.55
91.4	0.739 24	554.48	99.4	0.991 74	743.87
91.6	0.744 83	558.67	99.6	0.998 87	749.21
91.8	0.750 45	562.88	99.8	1.006 04	754.59
92.0	0.756 10	567.12	100.0	1.013 25	760.00
92.2	0.761 79	571.39			
92.4	0.767 52	575.69			
92.6	0.773 28	580.01			
92.8	0.779 08	584.36			

TABLE 4k-18. VAPOR PRESSURE OF WATER ABOVE 100°C
(Pressure of aqueous vapor over water from 100° to the critical temperature, 374.15°C)

Temp., °C	Bars	mm of Hg	Temp., °C	Bars	mm of Hg
100	1.0133	760.0	150	4.7597	3,570.1
101	1.0500	787.5	151	4.8887	3,666.8
102	1.0878	815.9	152	5.0205	3,765.7
103	1.1267	845.1	153	5.1551	3,866.7
104	1.1667	875.1	154	5.2926	3,969.8
105	1.2080	906.1	155	5.4331	4,075.1
106	1.2504	937.9	156	5.5765	4,182.7
107	1.2941	970.6	157	5.7228	4,292.5
108	1.3390	1,004.3	158	5.8723	4.404.6
109	1.3851	1,038.9	159	6.0248	4,519.0
110	1.4326	1,074.5	160	6.1805	4,635.8
111	1.4814	1,111.1	161	6.3393	4,754.9
112	1.5316	1,148.8	162	6.5014	4,876.5
113	1.5831	1,187.4	163	6.6668	5,000.5
114	1.6361	1,227.2	164	6.8355	5,127.1
115	1.6905	1,268.0	165	7.0076	5,256.1
116	1.7064	1,309.9	166	7.1831	5,387.8
117	1.8038	1,353.0	167	7.3621	5,522.0
118	1.8627	1,397.2	168	7.5446	5,658.9
119	1.9232	1,442.5	169	7.7306	5,798.4
120	1.9853	1,489.1	170	7.9203	5,940.7
121	2.0490	1,536.9	171	8.1136	6,085.7
122	2.1144	1,585.9	172	8.3107	6,233.5
123	2.1815	1,636.2	173	8.5115	6,384.2
124	2.2503	1,687.8	174	8.7161	6,537.6
125	2.3208	1,740.7	175	8.9247	6,694.0
126	2.3931	1,795.0	176	9.1371	6,853.4
127	2.4673	1,850.6	177	9.3535	7,015.7
128	2.5433	1,907.7	178	9.5739	7,181.1
129	2.6213	1,966.1	179	9.7985	7,349.5
130	2.7011	2,026.0	180	10.0271	7,520.9
131	2.7829	2,087.4	181	10.2599	7,695.6
132	2.8667	2,150.2	182	10.4969	7,873.4
133	2.9525	2,214.6	183	10.7383	8,054.4
134	3.0405	2,280.5	184	10.9839	8,238.6
135	3.1305	2,348.1	185	11.234	8,426
136	3.2226	2,417.2	186	11.489	8,617
137	3.3170	2,487.9	187	11.748	8,811
138	3.4136	2,560.4	188	12.011	9,009
139	3.5124	2,634.5	189	12.279	9,210
140	3.6135	2,710.3	190	12.552	9,415
141	3.7170	2,787.9	191	12.830	9,623
142	3.8228	2,867.3	192	13.112	9,835
143	3.9310	2,948.5	193	13.399	10,050
144	4.0417	3,031.5	194	13.692	10,270
145	4.1549	3,116.4	195	13.989	10,492
146	4.2706	3,203.2	196	14.291	10,719
147	4.3889	3,292.0	197	14.598	10,949
148	4.5098	3,382.7	198	14.910	11,184
149	4.6334	3,475.4	199	15.228	11,422

TABLE 4k-18. VAPOR PRESSURE OF WATER ABOVE 100°C (*Continued*)

Temp., °C	Bars	mm of Hg	Temp., °C	Bars	mm of Hg
200	15.550	11,664	250	39.776	29,834
201	15.879	11,910	251	40.452	30,341
202	16.212	12,160	252	41.137	30,855
203	16.551	12,414	253	41.830	31,375
204	16.895	12,672	254	42.533	31,902
205	17.245	12,935	255	43.244	32,436
206	17.601	13,202	256	43.965	32,976
207	17.962	13,472	257	44.695	33,524
208	18.329	13,748	258	45.434	34,078
209	18.701	14,027	259	46.182	34,640
210	19.080	14,311	260	46.940	35,208
211	19.464	14,599	261	47.707	35,783
212	19.855	14,892	262	48.484	36,366
213	20.251	15,190	263	49.270	36,955
214	20.654	15,492	264	50.066	37,553
215	21.062	15,798	265	50.872	38,157
216	21.477	16,109	266	51.687	38,769
217	21.899	16,425	267	52.513	39,388
218	22.326	16,746	268	53.349	40,015
219	22.760	17,072	269	54.195	40,650
220	23.201	17,402	270	55.051	41,292
221	23.648	17,738	271	55.917	41,941
222	24.102	18,078	272	56.794	42,599
223	24.562	18,423	273	57.681	43,264
224	25.030	18,774	274	58.579	43,938
225	25.504	19,129	275	59.487	44,619
226	25.985	19,490	276	60.406	45,308
227	26.473	19,856	277	61.336	46,006
228	26.968	20,227	278	62.277	46,712
229	27.470	20,604	279	63.229	47,426
230	27.979	20,986	280	64.192	48,148
231	28.495	21.373	281	65.166	48,878
232	29.019	21,766	282	66.151	49,617
233	29.550	22,164	283	67.147	50,365
234	30.088	22,568	284	68.155	51,121
235	30.634	22,978	285	69.175	51,885
236	31.188	23,393	286	70.206	52,659
237	31.749	23,814	287	71.249	53,441
238	32.318	24,241	288	72.304	54,232
239	32.895	24,674	289	73.370	55,032
240	33.480	25,112	290	74.449	55,841
241	34.073	25,557	291	75.539	56,659
242	34.673	26,007	292	76.642	57,486
243	35.282	26,464	293	77.757	58,322
244	35.899	26,926	294	78.884	59,168
245	36.524	27,395	295	80.024	60,023
246	37.157	27,870	296	81.177	60,888
247	37.799	28,352	297	82.342	61,762
248	38.450	28,840	298	83.521	62,646
249	39.109	29,334	299	84.712	63,539

TABLE 4k-18. VAPOR PRESSURE OF WATER ABOVE 100°C (*Continued*)

Temp., °C	Bars	mm of Hg	Temp., °C	Bars	mm of Hg
300	85.916	64,442	340	146.08	109,569
301	87.133	65,355	341	147.92	110,949
302	88.363	66,278	342	149.78	112,344
303	89.606	67,210	343	151.66	113,753
304	90.863	68,153	344	153.56	115,177
305	92.134	69,106	345	155.48	115,616
306	93.419	70,070	346	157.41	118,070
307	94.717	71,044	347	159.37	119,539
308	96.029	72,028	348	161.35	121,023
309	97.356	73,023	349	163.35	122,523
310	98.696	74,028	350	165.37	124,038
311	100.050	75,044	351	167.40	125,563
312	101.418	76,070	352	169.46	127,106
313	102.801	77,107	353	171.54	128,665
314	104.199	78,156	354	173.64	130,242
315	105.611	79,215	355	175.77	131,835
316	107.039	80,286	356	177.91	133,446
317	108.481	81,368	357	180.08	135,075
318	109.939	82,461	358	182.28	136,721
319	111.412	83,566	359	184.50	138,385
320	112.900	84,682	360	186.74	140,067
321	114.403	85,809	361	189.00	141,761
322	115.921	86,948	362	191.28	143,475
323	117.456	88,099	363	193.60	145,209
324	119.006	89,262	364	195.93	146,963
325	120.57	90,437	365	198.30	148,736
326	122.15	91,624	366	200.69	150,530
327	123.75	92,823	367	203.11	152,344
328	125.37	94,035	368	205.55	154,179
329	127.00	95,259	369	208.03	156,034
330	128.65	96,495	370	210.53	157,911
331	130.31	97,743	371	213.06	159,808
332	131.99	99,003	372	215.62	161,728
333	133.69	100,277	373	218.21	163,671
334	135.41	101,564	374	220.84	165,644
335	137.14	102,864	374.15	221.23	165,936
336	138.89	104,178			
337	140.66	105,505			
338	142.45	106,846			
339	144.26	108,201			

TABLE 4k-19. VAPOR PRESSURE OF MERCURY*

(Vapor pressure of mercury in mm of Hg for temperatures from −38 to 400°C.
Note that the values for the first four lines only are to be multiplied by 10^{-6})

Temp., °C	0	2	4	6	8
	10^{-6}	10^{-6}	10^{-6}	10^{-6}	10^{-6}
−30	4.78	3.59	2.66	1.97	1.45
−20	18.1	14.0	10.8	8.28	6.30
−10	60.6	48.1	38.0	29.8	23.2
− 0	185	149	119	95.4	76.2
+ 0	0.000185	0.000228	0.000276	0.000335	0.000406
+10	0.000490	0.000588	0.000706	0.000846	0.001009
20	0.001201	0.001426	0.001691	0.002000	0.002359
30	0.002777	0.003261	0.003823	0.004471	0.005219
40	0.006079	0.007067	0.008200	0.009497	0.01098
50	0.01267	0.01459	0.01677	0.01925	0.02206
60	0.02524	0.02883	0.03287	0.03740	0.04251
70	0.04825	0.05469	0.06189	0.06993	0.07889
80	0.08880	0.1000	0.1124	0.1261	0.1413
90	0.1582	0.1769	0.1976	0.2202	0.2453
100	0.2729	0.3032	0.3366	0.3731	0.4132
110	0.4572	0.5052	0.5576	0.6150	0.6776
120	0.7457	0.8198	0.9004	0.9882	1.084
130	1.186	1.298	1.419	1.551	1.692
140	1.845	2.010	2.188	2.379	2.585
150	2.807	3.046	3.303	3.578	3.873
160	4.189	4.528	4.890	5.277	5.689
170	6.128	6.596	7.095	7.626	8.193
180	8.796	9.436	10.116	10.839	11.607
190	12.423	13.287	14.203	15.173	16.200
200	17.287	18.437	19.652	20.936	22.292
210	23.723	25.233	26.826	28.504	30.271
220	32.133	34.092	36.153	38.318	40.595
230	42.989	45.503	48.141	50.909	53.812
240	56.855	60.044	63.384	66.882	70.543
250	74.375	78.381	82.568	86.944	91.518
260	96.296	101.28	106.48	111.91	117.57
270	123.47	129.62	136.02	142.69	149.64
280	156.87	164.39	172.21	180.34	188.79
290	197.57	206.70	216.17	226.00	236.21
300	246.80	257.78	269.17	280.98	293.21
310	305.89	319.02	332.62	346.70	361.26
320	376.33	391.92	408.04	424.71	441.94
330	459.74	478.13	497.12	516.74	537.00
340	557.90	579.45	601.69	624.64	648.30
350	672.69	697.83	723.73	750.43	777.92
360	806.23	835.38	865.36	896.23	928.02
370	960.66	994.34	1028.9	1064.4	1100.9
380	1138.4	1177.0	1216.6	1257.3	1299.1
390	1341.9	1386.1	1431.3	1477.7	1525.2
400	1574.1				

* From the compilation of J. Johnston, F. Fenwick, and H. G. Leopold. "International Critical Tables," Vol. III, p. 206, McGraw-Hill Book Company, New York, 1928.

TABLE 4k-20. VAPOR PRESSURE OF CARBON DIOXIDE*

SOLID
Pressure, Microns of Mercury

Temp., °C	0	1	2	3	4	5	6	7	8	9
−180	0.013	0.008	0.006	0.004	0.003	0.0017	0.0011	0.0007	0.0005	0.0003
−170	0.37	0.27	0.20	0.14	0.10	0.074	0.052	0.037	0.026	0.018
−160	5.9	4.6	3.6	2.7	2.1	1.58	1.19	0.90	0.67	0.50
−150	60.5	48.8	39.2	31.4	25.1	19.9	15.8	12.4	9.8	7.6
−140	431	359	298	247	204	168	138	113	92	75

Pressure, Mm of Mercury

Temp., °C	0	1	2	3	4	5	6	7	8	9
−130	2.31	1.97	1.68	1.43	1.22	1.03	0.87	0.73	0.61	0.51
−120	9.81	8.57	7.46	6.49	5.63	4.88	4.22	3.64	3.13	2.69
−110	34.63	30.76	27.27	24.14	21.34	18.83	16.58	14.58	12.80	11.22
−100	104.81	94.40	84.91	76.27	68.43	61.30	54.84	48.99	43.71	38.94
−90	279.5	254.7	231.8	210.8	191.4	173.6	157.3	142.4	128.7	116.2
−80	672.2	618.3	568.2	521.7	478.5	438.6	401.6	367.4	335.7	306.5
−70	1486.1	1377.3	1275.6	1180.5	1091.7	1008.9	931.7	859.7	792.7	730.3
−60	3073.1	2865.1	2669.7	2486.3	2314.2	2152.8	2001.5	1859.7	1726.9	1602.5
−50	......	......	......	......	......	......	......	3780.9	3530.2	3294.6

LIQUID

Temp., °C	0	1	2	3	4	5	6	7	8	9
−50	5127.8	4922.7	4723.9	4531.1	4344.3	4163.2	3987.9	3818.2†	3653.9†	3495.0†
−40	7545	7271	7005	6746	6494	6250	6012	5781	5557	5339
−30	10718	10363	10017	9679	9350	9029	8716	8412	8115	7826
−20	14781	14331	13891	13461	13040	12630	12229	11838	11455	11082
−10	19872	19312	18764	18228	17703	17189	16686	16194	15712	15241
−0	26142	25457	24786	24127	23482	22849	22229	21622	21026	20443
0	26142	26840	27552	28277	29017	29771	30539	31323	32121	32934
10	33763	34607	35467	36343	37236	38146	39073	40017	40980	41960
20	42959	43977	45014	46072	47150	48250	49370	50514	51680	52871
30	54086	55327								

* From C. H. Meyers and M. S. Van Dusen, *Natl. Bur. Standards J. Research* **10**, 409 (1933).
Mercury column density = 13.5951 g/cm³; g = 980.665 cm/sec².
† Undercooled liquid.
Critical temperature = 31.0°C.   Triple point, −56.602 ± 0.005°C; 3885.2 ± 0.4 mm.

Table 4k-21. Vapor Pressure of Ethyl Alcohol*

Temp., °C	0	1	2	3	4	5	6	7	8	9
	Vapor pressure, mm Hg at 0°C									
0	12.24	13.18	14.15	15.16	16.21	17.31	18.46	19.68	20.98	22.34
10	23.78	25.31	27.94	28.67	30.50	32.44	34.49	36.67	38.97	41.40
20	44.00	46.66	49.47	52.44	55.56	58.86	62.33	65.97	69.80	73.83
30	78.06	82.50	87.17	92.07	97.21	102.60	108.24	114.15	120.35	126.86
40	133.70	140.75	148.10	155.80	163.80	172.20	181.00	190.10	199.65	209.60
50	220.00	230.80	242.50	253.80	265.90	278.60	291.85	305.65	319.95	334.85
60	350.30	366.40	383.10	400.40	418.35	437.00	456.45	476.45	497.25	518.85
70	541.20	564.35	588.35	613.20	638.95	665.55	693.10	721.55	751.00	781.45

* Ramsay and Young, *Trans. Roy. Soc.* (*London*) **177**, part 1, 123 (1886).

Table 4k-22. Vapor Pressure of Methyl Alcohol*

Temp., °C	0	1	2	3	4	5	6	7	8	9
	Vapor pressure, mm Hg at 0°C									
0	29.97	31.6	33.6	35.6	37.8	40.2	42.6	45.2	47.9	50.8
10	53.8	57.0	60.3	63.8	67.5	71.4	75.5	79.8	84.3	89.0
20	94.0	99.2	104.7	110.4	116.5	122.7	129.3	136.2	143.4	151.0
30	158.9	167.1	175.7	184.7	194.1	203.9	214.1	224.7	235.8	247.4
40	259.4	271.9	285.0	298.5	312.6	327.3	342.5	358.3	374.7	391.7
50	409.4	427.7	446.6	466.3	486.6	507.7	529.5	552.0	575.3	599.4
60	624.3	650.0	676.5	703.8	732.0	761.1	791.1	822.0		

* Ramsay and Young, *Trans. Roy. Soc.* (*London*), **178**, 313 (1887); see also Young, *Sci. Proc. Roy. Dublin Soc.*, **12**, 374–443 (1910).

Table 4k-23 concerns the evaporation of metals. The rate of evaporation W of a metal is given by the equation

$$\log W = A - \frac{B}{T} - \frac{1}{2} \log T + c$$

where W is expressed in g/sec cm². The values of A, B, and c given in Table 4k-23 are chosen to yield the value of W in these units.

TABLE 4k-23. CONSTANTS IN THE EQUATION FOR THE RATE OF EVAPORATION OF METALS*

Metal	A	$10^{-3} \times B$	$c + 4$	Metal	A	$10^{-3} \times B$	$c + 4$
Li......	10.50(l)	7.480	0.1867	Si......	13.20(s)	19.72	0.4900
Na......	10.71(l)	5.480	0.4468		12.55(l)	18.55	0.4900
K......	10.36(l)	4.503	0.5621	Ti......	11.25(s)	18.64	0.6061
Rb......	10.42(l)	4.132	0.7319		11.98(l)	20.11	
	[10.53(l)	4.291]		Zr......	12.38(s)	25.87	0.7460
Cs......	9.86(l)	3.774	0.8278		13.04(l)	27.43	
	[10.02(l)	3.883]		Th......	12.52(l)	28.44	0.9488
				Ge......	10.94(l)	15.15	0.6965
Cu......	12.81(s)	18.06	0.6678				
	11.72(l)	16.58		Sn......	9.97(l)	13.11	0.8032
Ag......	12.28(s)	14.85	0.7825	Pb......	10.69(l)	9.60	0.9242
	11.66(l)	14.09		V......	13.32	26.62	0.6195
Au......	11.65(l)	18.52	0.9135	Nb......	14.37(s)	40.40	0.7500
Be......	12.99(s)	18.22	0.2436	Ta......	13.00(s)	40.21	0.8947
	11.95(l)	16.59					
Mg......	11.82(s)	7.741	0.4590	Sb₂......	11.42	9.913	0.9592
				Bi......	11.14(l)	9.824	0.9260
Ca......	11.30(s)	9.055	0.5675	Cr......	12.88(s)	17.56	0.6240
Sr......	11.13(s)	8.324	0.7373	Mo......	11.80(s)	30.31	0.7570
Ba......	10.88	8.908	0.8349	W......	12.24(s)	40.26	0.8983
Zn......	11.94(s)	6.744	0.6737				
Cd......	11.78(s)	5.798	0.7914	U......	12.88(l)	25.80	0.9544
				Mn......	12.25(s)	14.10	0.6359
B......	14.13(s)	21.37	0.2831	Fe......	12.63(s)	20.00	0.6395
Al......	11.99(l)	15.63	0.4814		13.41(l)	21.40	
Sc......	11.94	18.57	0.5931	Co......	12.43	21.96	0.6512
Y......	12.43	21.97	0.7405	Ni......	13.28(s)	21.84	0.6503
La......	11.88(l)	18.00	0.8374		12.55(l)	20.60	
Ce......	13.74(l)	20.10	0.8392	Ru......	13.50	33.80	0.7696
Ga......	10.79(l)	13.36	0.6877	Rh......	13.55	30.40	0.7722
In......	10.93(l)	12.15	0.7959	Pd......	11.46	19.23	0.7801
Tl......	11.15(l)	8.92	0.9212	Os......	13.59	37.00	0.9056
C......	14.06(s)	38.57	0.3056	Ir......	13.06	34.11	0.9089
				Pt......	12.633	27.50	0.9112

* From Saul Dushman, "Scientific Foundations of Vacuum Technique," pp. 752–754, John Wiley & Sons, Inc., New York, 1949.

41. Heats of Formation and Heats of Combustion

BRUNO J. ZWOLINSKI AND RANDOLPH C. WILHOIT

Thermodynamics Research Center,
Texas A&M University

Tables 41-1, 41-2, and 41-3 list values of the enthalpy of formation, $\Delta Hf°$, and enthalpy of combustion, $\Delta Hc°$, of pure elements and compounds in their standard states at one atmosphere pressure and 25°C in units of kilocalories per mole. Data on "key" substances, which play important roles in evaluating the data on other compounds, are collected in Table 41-1. Enthalpies of formation of elements and inorganic compounds are given in Table 41-2. They are arranged in a standard order, based on the order of elements in the periodic table. The organic compounds in Table 41-3 are arranged first by standard order of the elements of which they are composed and then by classes which have certain common molecular structural features or functional groups.

41-1. Sources of Data. All reported values were derived from published experimental measurements, and most of the data were selected from the following compilations: (1) Selected Values of Chemical Thermodynamic Properties: part 1, *NBS Tech. Note* 270-1, 1965; part 2, *NBS Tech. Note* 270-2, 1966; (2) Selected Values of Properties of Hydrocarbons and Related Compounds, *Am. Petroleum Inst. Research Proj.* 44, Thermodynamics Research Center, Texas A&M University, College Station, Texas (looseleaf data sheets, extant 1967); (3) Selected Values of Properties of Chemical Compounds, *Thermodyn. Research Center Data Proj.*, Texas A&M University, College Station, Texas (looseleaf data sheets, extant 1967).

These sources were supplemented by information in the files of the Thermodynamics Research Center at Texas A&M University. Data in all three tables are internally consistent, and, wherever necessary, original data have been converted to the units and conventions listed below.

41-2. Symbols and Units

calorie   the thermochemical calorie defined as equal to 4.184 joules (exactly)
mole   a unit of mass equal to the formula (molecular) weight in grams, calculated from the 1961 table of unified atomic weights based on carbon-12
standard state   for condensed phases, the specified crystal or liquid form at one atmosphere pressure; for gases, the hypothetical ideal gas at one atmosphere pressure

g   gas
l liquid
c   crystal
aq   aqueous (water) solution
H enthalpy, $H = U + PV$, for a change from an initial to a final state, $\Delta H = H(\text{final}) - H(\text{initial})$, which is equal to the heat absorbed by the system at constant pressure

$\Delta Hf°$ the heat of formation of one mole of compound or element in its standard state from the elements in their reference states. [For an organic oxygen compound this corresponds to the chemical reaction,

$$a\text{C(graphite)} + \tfrac{1}{2}b\text{H}_2\text{(gas)} + \tfrac{1}{2}c\text{O}_c\text{(gas)} \rightarrow \text{C}_a\text{H}_b\text{O}_c\text{(standard state)}.$$

Reference states for elements are identified by a zero enthalpy of formation in the tables.]

$\Delta Hc°$, gross   the heat of combustion of a compound with excess oxygen gas to produce pure, thermodynamically stable products at 25°C and one atmosphere, with all components in their standard states. [The products of combustion are: CO_2(gas), H_2O(liquid), HF(gas), Cl_2(gas), Br_2(liquid), I_2(crystal), H_2SO_4(liquid), and N_2(gas), as appropriate for the stoichiometry of the combustion reaction.]

$\Delta Hc°$, net   the heat of combustion of a compound with excess oxygen to produce the following products: CO_2(gas), H_2O(gas), HF(gas), Cl_2(gas), Br_2(gas), I_2(gas), SO_2(gas), and N_2(gas). (These are the principal products formed when a compound is burned in an open flame in the air.)

41-3. Uncertainties. The number of significant figures used in reporting a value of $\Delta Hf°$ or $\Delta Hc°$ is related to the estimated uncertainty according to the following scheme.

Estimated uncertainty in $\Delta Hf°$ or $\Delta Hc°$, kcal mole^{-1}	Value written to
0.005–0.05	0.001
0.05–0.5	0.01
0.5–2	0.1
2–10	1.

TABLE 4l-1. HEATS OF FORMATION AND HEATS OF COMBUSTION OF KEY COMPOUNDS

Substance name	Formula and state	Mol. weight	kcal mole^{-1} at 25°C		
				$-\Delta Hc°$	
			$\Delta Hf°$	Gross	Net
Water...................	H_2O,g	18.015	-57.796		
	H_2O,l	....	-68.315		
Hydrogen fluoride.........	HF,g	20.006	-64.8		
	HF,l	....	71.65		
in ∞ H_2O..............	HF,aq	....	-79.54		
Hydrogen chloride........	HCl,g	36.461	-22.062	12.096	6.836
in ∞ H_2O..............	HCl,aq	....	-39.952		
Hydrogen bromide........	HBr,g	80.917	-8.70	25.46	16.50
in ∞ H_2O..............	HBr,aq	....	-29.05		
Hydrogen Iodide..........	HI,g	127.912	6.33	40.49	27.77
Sulfur dioxide............	SO_2,g	64.063	-70.944		
	SO_2,l	....	-76.6		
Sulfuric acid..............	H_2SO_4,l	98.078	-194.548		
in ∞ H_2O..............	H_2SO_4,aq	....	-217.32		
in 115 H_2O..............	H_2SO_4,aq	....	-212.192		
Orthophosphoric acid......	H_3PO_4,c	97.995	-305.7		
	H_3PO_4,l	....	-302.8		
in ∞ H_2O..............	H_3PO_4,aq	....	-307.92		
Carbon dioxide...........	CO_2,g	44.010	-94.051		
Butanedioic acid (succinic acid)	$C_4H_6O_4$,g	118.090	-196.8	384.35	352.79
	$C_4H_6O_4$,c	....	-224.86	356.29	324.73
Benzoic acid..............	$C_7H_6O_2$,g	122.125	-70.19	793.11	761.55
	$C_7H_6O_2$,c	....	-92.04	771.26	739.70
Carbon tetrafluoride (tetra-fluoromethane)........	CF_4,g	88.005	-221		
p-Fluorobenzoic acid......	$C_8H_5O_2F$,c	140.115	-139.56	720.22	699.19
α,α,α-Trifluoro-m-toluic acid.................	$C_8H_5O_2F_3$,c	190.123	-253.68	761.44	750.92
Carbon disulfide..........	CS_2,g	76.139	21.44	....	263.99
	CS_2,l	....	28.05	....	257.38
Thianthrene (diphenylene disulfide).............	$C_{12}H_8S_2$,c	216.326	43.12	1,697.46	1,544.80
N-Benzoylaminoethanoic acid (hippuric acid)...	$C_9H_9O_3N$,c	179.177	-145.49	1,008.39	961.05
Boric oxide...............	B_2O_3,c	69.620	-304.20		
amorphous.............	B_2O_3,c	....	-299.84		
Boron trifluoride	BF_3,g	67.806	-271.03		
Silicon dioxide					
quartz.................	SiO_2,c	60.085	-217.72		
cristobalite.............	SiO_2,c	....	-217.37		
tridymite..............	SiO_2,c	....	-217.27		
amorphous.............	SiO_2,c	....	-215.95		
Silicon tetrafluoride........	SiF_4,g	104.080	-385.98		

TABLE 4l-2. HEATS OF FORMATION OF ELEMENTS AND INORGANIC COMPOUNDS

Substance name	Formula	Mol. weight	$\Delta Hf°$, kcal mole^{-1} at 25°C		
			Gas	Liquid	Solid
Oxygen and Hydrogen					
Oxygen..................	O_2	31.999	0.0		
Ozone..................	O_3	47.998	34.1		
Hydrogen...............	H_2	2.016	0.0		
Hydrogen peroxide........	H_2O_2	34.015	....	−44.88	
Halogens					
Fluorine.................	F_2	37.997	0.0		
Chlorine.................	Cl_2	70.906	0.0		
Chlorine monoxide........	ClO	51.452	24.36		
Chlorine dioxide..........	ClO_2	67.452	24.5		
Dichlorine monoxide......	Cl_2O	86.905	19.2		
Perchloric acid...........	$HClO_4$	100.459	....	−9.70	
Chlorine monofluoride.....	ClF	54.451	−11.92		
Chlorine trifluoride........	ClF_3	92.448	−38.0	−44.3	
Bromine.................	Br_2	159.818	7.39	0.0	
Bromine monoxide........	BrO	95.908	30.06		
Bromine dioxide..........	BrO_2	111.908	....	....	11.6
Bromine trifluoride.......	BrF_3	136.904	....	−71.9	
Bromine pentafluoride.....	BrF_5	174.901	−102.5	−109.6	
Bromine chloride........	$BrCl$	115.362	3.50		
Iodine...................	I_2	253.809	14.92	....	0.0
Iodic acid...............	HIO_3	175.911	....	....	−55.0
Iodine monofluoride.......	IF	145.903	−2.10		
Iodine pentafluoride.......	IF_5	221.896	−196.58	−206.7	
Iodine heptafluoride.......	IF_7	259.893	−225.6		
Iodine monochloride......	ICl	162.357	4.25	−5.71	−8.4
Iodine trichloride.........	ICl_3	233.263	....	....	−21.4
Iodine monobromide.....	IBr	206.813	9.76	....	−2.5
Sulfur					
Sulfur...................	S	32.064	66.64		
rhombic..............	....	......	....	....	0.0
monoclinic............	....	......	....	....	0.08
Sulfur...................	S_2	64.128	30.68		
Sulfur...................	S_3	96.192	31.7		
Sulfur trioxide...........	SO_3	80.062	−94.58	−105.41	−108.63
Hydrogen sulfide..........	H_2S	34.080	−4.93		
Sulfur tetrafluoride........	SF_4	108.058	−185.2		
Sulfur hexafluoride........	SF_6	146.054	−289.		
Disulfur dichloride........	S_2Cl_2	135.034	−4.4	−14.2	
Thionyl chloride..........	$SOCl_2$	118.969	−50.8	−58.7	
Sulfuryl chloride..........	SO_2Cl_2	134.969	−87.0	−94.2	
Thionyl bromide..........	$SOBr_2$	207.881	−17.7		
Nitrogen					
Nitrogen.................	N_2	28.013	0.0		
Nitric oxide..............	NO	30.006	21.57		
Nitrogen dioxide..........	NO_2	46.006	7.93		

TABLE 4l-2. HEATS OF FORMATION OF ELEMENTS AND
INORGANIC COMPOUNDS (*Continued*)

Substance name	Formula	Mol. weight	$\Delta Hf°$, kcal mole^{-1} at 25°C		
			Gas	Liquid	Solid
Nitrogen (Cont.)					
Nitrous oxide............	N₂O	44.013	19.61		
Nitrogen trioxide........	N₂O₃	76.012	20.01	12.02	
Nitrogen tetroxide.......	N₂O₄	29.011	2.19	−4.66	
Nitrogen pentoxide.......	N₂O₅	108.010	2.7	....	−10.3
Ammonia.................	NH₃	17.031	−11.02		
Hydrazine...............	N₂H₄	32.045	22.80	12.10	
Hydrogen azide..........	HN₃	43.028	70.3	63.1	
Nitrous acid.............	HNO₂	47.014	−19.0		
Nitric acid..............	HNO₃	63.013	−32.28	−41.61	
Hydroxylamine...........	NH₂OH	33.030	....	....	−27.3
Ammonium hydroxide.....	NH₄OH	35.046	....	−86.33	
Ammonium nitrate........	NH₄NO₃	80.044	....	....	−87.37
Nitrogen trifluoride.......	NH₃	71.002	−29.8		
Nitrosyl fluoride.........	NOF	49.005	−15.9		
Ammonium fluoride......	NH₄F	37.037	....	....	−110.89
Nitrogen trichloride......	NCl₃	120.366	....	55	
Nitrosyl chloride.........	NOCl	65.459	12.36		
Ammonium chloride......	NH₄Cl	53.492	....	....	−75.15
Hydrazine hydrochloride...	N₂H₅Cl	68.506	....	....	−47.0
Ammonium perchlorate....	NH₄ClO₄	117.489	....	....	−70.58
Nitrosyl bromide.........	NOBr	109.915	19.64		
Ammonium bromide......	NH₄Br	94.924	....	....	−64.73
Ammonium iodide........	NH₄I	144.943	....	....	−48.14
Ammonium hydrogen sulfide...............	NH₄HS	51.111	....	....	−37.5
Sulfamic acid.............	H₂NSO₃H	97.093	....	....	−161.3
Sulfamide...............	SO₂(NH₂)₂	96.108	....	....	−129.3
Ammonium hydrogen sulfate...............	NH₄HSO₄	115.108	....	....	−245.45
Ammonium sulfate........	(NH₄)₂SO₄	132.139	....	....	−282.23
Phosphorus					
Phosphorus α, white..............	P	30.974	....	....	0.0
triclinic, red...........	....	....	....	....	−4.2
black.................	....	....	....	....	−9.4
amorphous, red........	....	....	....	....	−1.8
Phosphorus.............	P₂	61.948	34.5		
Phosphorus.............	P₄	123.895	14.08		
Phosphorus trioxide.......	P₄O₆	219.892	....	....	−392.0
Phosphorus pentoxide.....	P₄O₁₀	283.889	....	....	−713.2
Phosphine...............	PH₃	33.998	1.3		
Metaphosphoric acid......	HPO₃	79.980	....	....	−226.7
Pyrophosphoric acid.......	H₄P₂O₇	177.975	....	....	−535.6
Phosphorus trifluoride.....	PF₃	87.969	−219.6		
Phosphorus pentafluoride..	PF₅	125.966	−381.4		
Phosphorus oxyfluoride....	POF₃	103.968	−289.5		
Phosphorus trichloride.....	PCl₃	137.333	....	−76.4	
Phosphorus pentachloride..	PCl₅	208.239	−89.6	....	−106.0
Phosphorus oxychloride...	POCl₃	153.332	....	−142.7	
Phosphorus tribromide....	PBr₃	270.701	−33.3	−44.1	
Phosphorus pentabromide..	PBr₅	430.494	....	....	−64.5
Phosphorus oxybromide...	POBr₃	286.700	....	....	−109.6

TABLE 4l-2. HEATS OF FORMATION OF ELEMENTS AND
INORGANIC COMPOUNDS (*Continued*)

Substance name	Formula	Mol. weight	$\Delta Hf°$, kcal mole^{-1} at 25°C		
			Gas	Liquid	Solid
Phosphorus (Cont.)					
Phosphorus triiodide	PI₃	411.687	....	....	−10.9
Ammonium dihydrogen phosphate	NH₄H₂PO₄	115.026	....	....	−345.94
Ammonium hydrogen phosphate	(NH₄)₂HPO₄	132.057	....	....	−347.50
Ammonium phosphate	(NH₄)₃PO₄	149.087	....	....	−399.6
Boron					
Boron	B	10.811	....	....	0.0
amorphous	....	....	....	....	0.9
Diborane	B₂H₆	27.670	8.5		
Boric acid	H₃BO₃	61.833	....	....	−261.55
Boron trichloride	BCl₃	117.170	−96.50	−102.1	
Silicon					
Silicon	Si	28.086	....	....	0.0
amorphous	....	....	....	....	1.0
Silicon	Si₂	56.172	142		
Silicon monoxide	SiO	44.085	−23.8		
Silane	SiH₄	32.118	8.2		
Disilane	Si₂H₆	62.220	19.2		
Metasilic acid	H₂SiO₃	78.100	....	....	−284.1
Orthosilic acid	H₄SiO₄	96.116	....	....	−354.0
Silicon tetrachloride	SiCl₄	169.898	−157.03	−164.2	
Silicon tetrabromide	SiBr₄	347.722	−99.3	−109.3	
Silicon tetraiodide	SiI₄	535.704	....	....	−45.3
Tetramethylsilane	Si(CH₃)₄	88.226	−57.15	−63	
Hexamethyldisiloxane	[(CH₃)₃Si]₂O	162.382	−185.88	−194.8	
Beryllium, Sodium, Potassium					
Beryllium	Be	9.012	78.0	....	0.0
Beryllium oxide	BeO	25.012	30.2	....	−145.0
Beryllium fluoride	BeF₂	47.009	−186.1	....	−245.3
Beryllium chloride	BeCl₂	79.918	−85.7	....	−117.2
Sodium	Na	22.990	25.9	....	0.0
Sodium oxide	Na₂O	61.979	....	....	−99.4
Sodium hydride	NaH	23.998	29.88	....	−13.7
Sodium hydroxide	NaOH	39.997	....	....	−101.72
Sodium fluoride	NaF	41.988	−70.1	....	−136.6
Sodium chloride	NaCl	58.443	−43.7	....	−98.5
Sodium carbonate	Na₂CO₃	105.989	....	....	−269.8
Sodium formate	NaCHO₂	68.008	....	....	−155.03
Sodium acetate	NaC₂H₃O₂	82.035	....	....	−169.8
Potassium	K	39.102	21.52	....	0.0
Potassium oxide	K₂O	94.203	....	....	−86.4
Potassium hydride	KH	40.110	30.0	....	−15.6
Potassium hydroxide	KOH	56.109	....	....	−101.52
Potassium fluoride	KF	58.100	−78.2	....	−134.4
Potassium chloride	KCl	74.555	−51.6	....	−104.1

TABLE 4l-3. HEATS OF FORMATION AND HEATS OF
COMBUSTION OF COMPOUNDS OF CARBON

Substance name	Formula and state	Mol. weight	Kcal mole^{-1} at 25°C		
			$\Delta Hf°$	$-\Delta Hc°$	
				Gross	Net
Carbon and Carbon-Oxygen					
Carbon...................	C,g	12.011	171.29	265.34	265.34
graphite.................	C,c	....	0.0	94.05	94.05
diamond.................	C,c	....	0.45	94.50	94.50
Carbon...................	C₂,g	24.021	199.03	387.13	387.13
Carbon monoxide...........	CO,g	28.011	−26.42	67.64	67.64
Carbon suboxide............	C₃O₂,g	68.032	−22.20	259.95	259.95
	C₃O₂,l	....	−28.03	254.12	254.12
Carbon-Hydrogen, Alkanes					
Methane...................	CH₄,g	16.043	−17.88	212.80	191.76
Ethane....................	C₂H₆,g	30.070	−20.23	372.82	341.26
Propane...................	C₃H₈,g	44.097	−24.81	530.60	488.52
	C₃H₈,l	....	−28.69	526.72	484.64
n-Butane..................	C₄H₁₀,g	58.124	−30.14	687.64	635.04
	C₄H₁₀,l	....	−35.31	682.47	629.87
2-Methylpropane (isobutane)..	C₄H₁₀,g	58.124	−32.14	685.64	633.04
	C₄H₁₀,l	....	−36.88	680.89	628.30
n-Pentane.................	C₅H₁₂,g	72.151	−34.98	845.16	782.05
	C₅H₁₂,l	....	−41.37	838.78	775.66
2-Methylbutane (isopentane)..	C₅H₁₂,g	72.151	−36.90	843.24	780.13
	C₅H₁₂,l	....	−42.92	837.22	774.11
2,2-Dimethylpropane (neopentane)	C₅H₁₂,g	72.151	−39.66	840.49	777.38
	C₅H₁₂,l	....	−45.00	835.14	772.03
n-Hexane..................	C₆H₁₄,g	86.178	−39.92	1,002.59	928.95
	C₆H₁₄,l	....	−47.50	995.01	921.38
2-Methylpentane............	C₆H₁₄,g	86.178	−41.62	1,000.89	927.26
	C₆H₁₄,l	....	−48.80	993.71	920.08
3-Methylpentane............	C₆H₁₄,g	86.178	−40.99	1,001.52	927.89
	C₆H₁₄,l	....	−48.26	994.25	920.62
2,2-Dimethylbutane.........	C₆H₁₄,g	86.178	−44.32	998.19	924.56
	C₆H₁₄,l	....	−50.99	991.52	917.89
2,3-Dimethylbutane.........	C₆H₁₄,g	86.178	−42.46	1,000.06	926.42
	C₆H₁₄,l	....	−49.46	993.05	919.42
n-Heptane.................	C₇H₁₆,g	100.206	−44.85	1,160.02	1,075.87
	C₇H₁₆,l	....	−53.61	1,151.27	1,067.12
2-Methylhexane.............	C₇H₁₆,g	100.206	−46.57	1,158.31	1,074.16
	C₇H₁₆,l	....	−54.91	1,149.97	1,065.82
3-Methylhexane.............	C₇H₁₆,g	100.206	−45.92	1,158.96	1,074.80
	C₇H₁₆,l	....	−54.32	1,150.55	1,066.40
3-Ethylpentane.............	C₇H₁₆,g	100.206	−45.29	1,159.59	1,075.44
	C₇H₁₆,l	....	−53.75	1,151.13	1,066.98
2,2-Dimethylpentane.........	C₇H₁₆,g	100.206	−49.25	1,155.63	1,071.48
	C₇H₁₆,l	....	−57.03	1,147.85	1,063.70
2,3-Dimethylpentane.........	C₇H₁₆,g	100.206	−46.78	1,158.10	1,073.95
	C₇H₁₆,l	....	−55.79	1,149.09	1,064.94
2,4-Dimethylpentane.........	C₇H₁₆,g	100.206	−48.26	1,156.62	1,072.47
	C₇H₁₆,l	....	−56.15	1,148.73	1,064.58

TABLE 4l-3. HEATS OF FORMATION AND HEATS OF
COMBUSTION OF COMPOUNDS OF CARBON (*Continued*)

Substance name	Formula and state	Mol. weight	$\Delta Hf°$	$-\Delta Hc°$ Gross	$-\Delta Hc°$ Net
			Kcal mole^{-1} at 25°C		
Carbon-Hydrogen, Alkanes (Cont.)					
3,3-Dimethylpentane.........	C_7H_{16},g	100.206	−48.12	1,156.75	1,072.60
	C_7H_{16},l	....	−56.05	1,148.83	1,064.68
2,2,3-Trimethylbutane........	C_7H_{16},g	100.206	−48.92	1,155.96	1,071.81
	C_7H_{16},l	....	−56.61	1,148.27	1,064.12
n-Octane...................	C_8H_{18},g	114.233	−49.79	1,317.45	1,222.78
	C_8H_{18},l	....	−59.71	1,307.53	1,212.86
2-Methylheptane............	C_8H_{18},g	114.233	−51.47	1,315.77	1,221.10
	C_8H_{18},l	....	−60.96	1,306.28	1,211.61
3-Methylheptane............	C_8H_{18},g	114.233	−50.80	1,316.45	1,221.78
	C_8H_{18},l	....	−60.32	1,306.92	1,212.25
4-Methylheptane............	C_8H_{18},g	114.233	−50.66	1,316.58	1,221.91
	C_8H_{18},l	....	−60.15	1,307.09	1,212.42
3-Ethylhexane..............	C_8H_{18},g	114.233	−50.37	1,316.87	1,222.20
	C_8H_{18},l	....	−59.85	1,307.39	1,212.72
2,2-Dimethylhexane.........	C_8H_{18},g	114.233	−53.68	1,313.56	1,218.89
	C_8H_{18},l	....	−62.60	1,304.64	1,209.97
2,3-Dimethylhexane.........	C_8H_{18},g	114.233	−51.10	1,316.14	1,221.47
	C_8H_{18},l	....	−60.38	1,306.86	1,212.19
2,4-Dimethylhexane.........	C_8H_{18},g	114.233	−52.41	1,314.83	1,220.16
	C_8H_{18},l	....	−61.44	1,305.80	1,211.13
2,5-Dimethylhexane.........	C_8H_{18},g	114.233	−53.19	1,314.06	1,219.38
	C_8H_{18},l	....	−62.24	1,305.00	1,210.33
3,3-Dimethylhexane.........	C_8H_{18},g	114.233	−52.58	1,314.66	1,219.99
	C_8H_{18},l	....	−61.56	1,305.68	1,211.01
3,4-Dimethylhexane.........	C_8H_{18},g	114.233	−50.88	1,316.36	1,221.69
	C_8H_{18},l	....	−60.20	1,307.04	1,212.37
2-Methyl-3-ethylpentane......	C_8H_{18},g	114.233	−50.45	1,316.79	1,222.12
	C_8H_{18},l	....	−59.66	1,307.58	1,212.91
3-Methyl-3-ethylpentane......	C_8H_{18},g	114.233	−51.36	1,315.88	1,221.21
	C_8H_{18},l	....	−60.44	1,306.80	1,212.13
2,2,3-Trimethylpentane.......	C_8H_{18},g	114.233	−52.58	1,314.66	1,219.99
	C_8H_{18},l	....	−61.41	1,305.83	1,211.16
2,2,4-Trimethylpentane.......	C_8H_{18},g	114.233	−53.55	1,313.69	1,219.02
	C_8H_{18},l	....	−61.95	1,305.29	1,210.62
2,3,3-Trimethylpentane.......	C_8H_{18},g	114.233	−51.70	1,315.54	1,220.87
	C_8H_{18},l	....	−60.60	1,306.64	1,211.97
2,3,4-Trimethylpentane......	C_8H_{18},g	114.233	−51.94	1,315.30	1,220.63
	C_8H_{18},l	....	−60.96	1,306.28	1,211.61
2,2,3,3-Tetramethylbutane....	C_8H_{18},g	114.233	−53.97	1,313.28	1,218.61
	C_8H_{18},l	....	−64.21	1,303.03	1,208.36
n-Nonane..................	C_9H_{20},g	128.260	−54.56	1,475.05	1,369.86
	C_9H_{20},l	....	−65.66	1,463.95	1,358.76
2,2-Dimethylheptane.........	C_9H_{20},g	128.260	−58.74	1,470.87	1,365.68
	C_9H_{20},l	....	−68.85	1,460.76	1,355.57
2,2,3-Trimethylhexane........	C_9H_{20},g	128.260	−57.59	1,472.02	1,366.83
	C_9H_{20},l	....	−67.56	1,462.05	1,356.86
2,2,4-Trimethylhexane........	C_9H_{20},g	128.260	−57.85	1,471.76	1,366.57
	C_9H_{20},l	....	−67.58	1,462.03	1,356.84

TABLE 4l-3. HEATS OF FORMATION AND HEATS OF
COMBUSTION OF COMPOUNDS OF CARBON (*Continued*)

Substance name	Formula and state	Mol. weight	Kcal mole^{-1} at 25°C		
			$\Delta Hf°$	$-\Delta Hc°$	
				Gross	Net
Carbon-Hydrogen, Alkanes (Cont.)					
2,2,5-Trimethylhexane........	C_9H_{20},g	128.260	−60.36	1,469.24	1,364.06
	C_9H_{20},l	....	−69.97	1,459.64	1,354.45
2,3,3-Trimethylhexane........	C_9H_{20},g	128.260	−57.13	1,472.48	1,367.29
	C_9H_{20},l	....	−67.18	1,462.43	1,357.24
2,3,5-Trimethylhexane........	C_9H_{20},g	128.260	−57.91	1,471.70	1,366.51
	C_9H_{20},l	....	−67.81	1,461.80	1,356.61
2,4,4-Trimethylhexane........	C_9H_{20},g	128.260	−57.06	1,472.55	1,367.36
	C_9H_{20},l	....	−66.87	1,462.74	1,357.55
3,3,4-Trimethylhexane........	C_9H_{20},g	128.260	−56.20	1,473.41	1,368.22
	C_9H_{20},l	....	−66.33	1,463.28	1,358.09
2,2-Dimethyl-3-ethylpentane..	C_9H_{20},g	128.260	−55.21	1,474.40	1,369.21
	C_9H_{70},l	....	−65.17	1,464.44	1,359.25
2,4-Dimethyl-3-ethylpentane..	C_9H_{20},g	128.260	−54.30	1,475.31	1,370.12
	C_9H_{20},l	....	−64.42	1,465.19	1,360.00
n-Decane..................	$C_{10}H_{22}$,g	142.287	−59.64	1,632.34	1,516.63
	$C_{10}H_{22}$,l	....	−71.92	1,620.06	1,504.35
Carbon-Hydrogen, Cycloalkanes					
Cyclopropane..............	C_3H_6,g	42.081	12.75	499.85	468.29
Cyclobutane..............	C_4H_8,g	56.108	6.32	655.78	613.70
	C_4H_8,l	....	0.76	650.22	608.14
Cyclopentane.............	C_5H_{10},g	70.135	−18.41	793.42	740.83
	C_5H_{10},l	....	−25.28	786.55	733.96
Methylcyclopentane.........	C_6H_{12},g	84.163	−25.34	948.86	885.75
	C_6H_{12},l	....	−32.92	941.28	878.17
Ethylcyclopentane..........	C_7H_{14},g	98.190	−30.33	1,106.23	1,032.60
	C_7H_{14},l	....	−39.06	1,097.50	1,023.87
1,1-Dimethylcyclopentane....	C_7H_{14},g	98.190	−33.02	1,103.54	1,029.91
	C_7H_{14},l	....	−41.12	1,095.44	1,021.81
1-cis-2-Dimethylcyclopentane..	C_7H_{14},g	98.190	−30.94	1,105.62	1,031.99
	C_7H_{14},l	....	−39.50	1,097.06	1,023.43
1-trans-2-Dimethylcyclopentane	C_7H_{14},g	98.190	−32.64	1,103.92	1,030.29
	C_7H_{14},l	....	−40.92	1,095.64	1,022.01
1-cis-3-Dimethylcyclopentane..	C_7H_{14},g	98.190	−32.44	1,104.12	1,030.49
	C_7H_{14},l	....	−40.66	1,095.90	1,022.27
1-trans-3-Dimethylcyclopentane	C_7H_{14},g	98.190	−31.90	1,104.66	1,031.03
	C_7H_{14},l	....	−40.17	1,096.39	1,022.76
n-Propylcyclopentane........	C_8H_{16},g	112.217	−35.37	1,263.56	1,179.41
	C_8H_{16},l	....	−45.19	1,253.74	1,169.59
n-Butylcyclopentane........	C_9H_{18},g	126.244	−40.19	1,421.10	1,326.43
	C_9H_{18},l	....	−51.19	1,410.10	1,315.43
n-Pentylcyclopentane........	$C_{10}H_{20}$,g	140.271	−45.12	1,578.54	1,473.35
	$C_{10}H_{20}$,l	....	−57.30	1,566.36	1,461.17
n-Hexylcyclopentane........	$C_{11}H_{22}$,g	154.298	−50.04	1,735.99	1,620.28
	$C_{11}H_{22}$,l	....	−63.40	1,722.63	1,606.92
n-Heptylcyclopentane........	$C_{12}H_{24}$,g	168.325	−54.96	1,893.43	1,767.20
	$C_{12}H_{24}$,l	....	−69.50	1,878.89	1,752.66

TABLE 4l-3. HEATS OF FORMATION AND HEATS OF
COMBUSTION OF COMPOUNDS OF CARBON (*Continued*)

Substance name	Formula and state	Mol. weight	Kcal mole^{-1} at 25°C		
			$\Delta Hf°$	$-\Delta Hc°$ Gross	Net
Carbon-Hydrogen, Cycloalkanes (Cont.)					
n-Octylcyclopentane	$C_{13}H_{26}$,g	182.352	−59.89	2,050.87	1,914.12
	$C_{13}H_{26}$,l	....	−75.61	2,035.15	1,898.40
n-Nonylcyclopentane	$C_{14}H_{28}$,g	196.379	−64.80	2,208.32	2,061.05
	$C_{14}H_{28}$,l	....	−81.70	2,191.42	2,044.15
n-Decylcyclopentane	$C_{15}H_{30}$,g	210.406	−69.73	2,365.76	2,207.98
	$C_{15}H_{30}$,l	....	−87.81	2,347.68	2,189.90
Cyclohexane	C_6H_{12},g	84.163	−29.42	944.78	881.67
	C_6H_{12},l	....	−37.33	936.87	873.76
Methylcyclohexane	C_7H_{14},g	98.190	−36.98	1,099.58	1,025.95
	C_7H_{14},l	....	−45.43	1,091.13	1,017.50
Ethylcyclohexane	C_8H_{16},g	112.217	−41.03	1,257.90	1,173.75
	C_8H_{16},l	....	−50.70	1,248.23	1,164.08
1,1-Dimethylcyclohexane	C_8H_{16},g	112.217	−43.24	1,255.69	1,171.54
	C_8H_{16},l	....	−52.28	1,246.65	1,255.69
1-cis-2-Dimethylcyclohexane	C_8H_{16},g	112.217	−41.12	1,257.81	1,173.66
	C_8H_{16},l	....	−50.62	1,248.31	1,164.16
1-trans-2-Dimethylcyclohexane	C_8H_{16},g	112.217	−42.98	1,255.95	1,171.80
	C_8H_{16},l	....	−52.16	1,246.77	1,162.62
1-cis-3-Dimethylcyclohexane	C_8H_{16},g	112.217	−44.12	1,254.81	1,170.66
	C_8H_{16},l	....	−53.28	1,245.65	1,161.50
1-trans-3-Dimethylcyclohexane	C_8H_{16},g	112.217	−42.17	1,256.76	1,172.61
	C_8H_{16},l	....	−51.55	1,247.38	1,163.23
1-cis-4-Dimethylcyclohexane	C_8H_{16},g	112.217	−42.19	1,256.74	1,172.59
	C_8H_{16},l	....	−51.53	1,247.40	1,163.25
1-trans-4-Dimethylcyclohexane	C_8H_{16},g	112.217	−44.08	1,254.85	1,170.70
	C_8H_{16},l	....	−53.15	1,245.78	1,161.63
Cycloheptane	C_7H_{14},g	98.190	−28.26	1,108.30	1,034.67
	C_7H_{14},l	....	−37.47	1,099.09	1,025.46
Cyclooctane	C_8H_{16},g	112.217	−30.22	1,268.71	1,184.56
	C_8H_{16},l	....	−40.58	1,258.35	1,174.20
Cyclotetradecane	$C_{14}H_{28}$,g	196.379	−54.36	2,218.76	2,071.49
	$C_{14}H_{28}$,l	....	−86.57	2,186.55	2,039.28
Carbon-Hydrogen, Alkenes					
Ethene (ethylene)	C_2H_4,g	28.054	12.50	337.23	316.19
Propene (propylene)	C_3H_6,g	42.081	4.89	491.99	460.43
1-Butene	C_4H_8,g	56.108	−0.01	649.45	607.37
	C_4H_8,l	....	−4.98	644.49	602.41
cis-2-Butene	C_4H_8,g	56.108	−1.65	647.81	605.73
	C_4H_8,l	....	−7.06	642.41	600.33
trans-2-Butene	C_4H_8,g	56.108	−2.65	646.81	604.73
	C_4H_8,l	....	−7.88	641.58	599.50
2-Methylpropene (isobutene)	C_4H_8,g	56.108	−4.03	645.43	603.35
	C_4H_8,l	....	−9.11	640.36	598.28
1-Pentene	C_5H_{10},g	70.135	−4.98	806.85	754.26
	C_5H_{10},l	....	−11.14	800.68	748.09

TABLE 4l-3. HEATS OF FORMATION AND HEATS OF
COMBUSTION OF COMPOUNDS OF CARBON (*Continued*)

Substance name	Formula and state	Mol. weight	Kcal mole^{-1} at 25°C		
			$\Delta Hf°$	$-\Delta Hc°$	
				Gross	Net
Carbon-Hydrogen, Alkenes (Cont.)					
cis-2-Pentene...............	C_5H_{10},g	70.135	-6.49	805.34	752.74
	C_5H_{10},l	......	-12.96	798.87	746.27
trans-2-Pentene..............	C_5H_{10},g	70.135	-7.57	804.26	751.66
	C_5H_{10},l	......	-14.02	797.81	745.22
2-Methyl-1-butene...........	C_5H_{10},g	70.135	-8.66	803.17	750.58
	C_5H_{10},l	......	-14.92	796.92	744.32
3-Methyl-1-butene..........	C_5H_{10},g	70.135	-6.90	804.93	752.34
	C_5H_{10},l	......	-12.76	799.07	746.47
2-Methyl-2-butene..........	C_5H_{10},g	70.135	-10.15	801.68	749.08
	C_5H_{10},l	......	-16.68	795.15	742.55
1-Hexene...................	C_6H_{12},g	84.163	-9.92	964.28	901.16
	C_6H_{12},l	......	-17.28	956.92	893.81
cis-2-Hexene................	C_6H_{12},g	84.163	-12.49	961.71	898.60
	C_6H_{12},l	......	-20.04	954.16	891.05
trans-2-Hexene..............	C_6H_{12},g	84.163	-12.86	961.34	898.23
	C_6H_{12},l	......	-20.43	953.77	890.66
cis-3-Hexene................	C_6H_{12},g	84.163	-11.35	962.85	899.74
	C_6H_{12},l	......	-18.85	955.35	892.24
trans-3-Hexene..............	C_6H_{12},g	84.163	-12.98	961.22	898.11
	C_6H_{12},l	......	-20.55	953.65	890.54
2-Methyl-1-pentene..........	C_6H_{12},g	84.163	-14.16	960.04	896.93
	C_6H_{12},l	......	-21.48	952.72	889.61
3-Methyl-1-pentene..........	C_6H_{12},g	84.163	-11.80	962.40	899.29
	C_6H_{12},l	......	-18.66	955.54	892.43
4-Methyl-1-pentene..........	C_6H_{12},g	84.163	-12.22	961.98	898.87
	C_6H_{12},l	......	-19.11	955.09	891.98
2-Methyl-2-pentene..........	C_6H_{12},g	84.163	-15.96	958.24	895.13
	C_6H_{12},l	......	-23.54	950.66	887.55
3-Methyl-*cis*-2-pentene.......	C_6H_{12},g	84.163	-15.03	959.17	896.06
	C_6H_{12},l	......	-22.55	951.65	888.54
3-Methyl-*trans*-2-pentene.....	C_6H_{12},g	84.163	-14.89	959.31	896.20
	C_6H_{12},l	......	-22.59	951.61	888.50
4-Methyl-*cis*-2-pentene.......	C_6H_{12},g	84.163	-13.70	960.50	897.39
	C_6H_{12},l	......	-20.78	953.42	890.31
4-Methyl-*trans*-2-pentene.....	C_6H_{12},g	84.163	-14.66	959.54	896.43
	C_6H_{12},l	......	-21.86	952.34	889.23
2-Ethyl-1-butene.............	C_6H_{12},g	84.163	-13.36	960.84	897.73
	C_6H_{12},l	......	-20.80	953.40	890.29
2,3-Dimethyl-1-butene........	C_6H_{12},g	84.163	-15.48	958.72	895.61
	C_6H_{12},l	......	-22.49	951.71	888.60
3,3-Dimethyl-1-butene........	C_6H_{12},g	84.163	-14.48	959.72	896.61
	C_6H_{12},l	......	-20.89	953.31	890.20
2,3-Dimethyl-2-butene........	C_6H_{12},g	84.163	-16.41	957.79	894.67
	C_6H_{12},l	......	-24.22	949.98	886.87
1-Heptene...................	C_7H_{14},g	98.190	-14.66	1,121.90	1,048.27
	C_7H_{14},l	......	-23.19	1,113.37	1,039.74
cis-2-Heptene...............	C_7H_{14},g	98.190	-16.47	1,120.09	1,046.46
	C_7H_{14},l	......	-25.09	1,111.47	1,037.84

TABLE 4l-3. HEATS OF FORMATION AND HEATS OF
COMBUSTION OF COMPOUNDS OF CARBON (*Continued*)

Substance name	Formula and state	Mol. weight	Kcal mole⁻¹ at 25°C		
			$\Delta Hf°$	$-\Delta Hc°$	
				Gross	Net
Carbon-Hydrogen, Alkenes (Cont.)					
trans-2-Heptene..............	C_7H_{14},g	98.190	−17.47	1,119.09	1,045.46
	C_7H_{14},l	....	−26.09	1,110.47	1,036.84
cis-3-Heptene................	C_7H_{14},g	98.190	−16.47	1,120.09	1,046.46
	C_7H_{14},l	....	−24.99	1,111.57	1,037.94
trans-3-Heptene..............	C_7H_{14},g	98.190	−17.47	1,119.09	1,045.46
	C_7H_{14},l	....	−25.99	1,110.57	1,036.94
2-Methyl-1-hexene............	C_7H_{14},g	98.190	−18.48	1,118.08	1,044.45
	C_7H_{14},l	....	−26.90	1,109.66	1,036.03
3-Methyl-1-hexene............	C_7H_{14},g	98.190	−15.94	1,120.62	1,046.99
	C_7H_{14},l	....	−24.16	1,112.40	1,038.77
4-Methyl-1-hexene............	C_7H_{14},g	98.190	−15.94	1,120.62	1,046.99
	C_7H_{14},l	....	−24.26	1,112.30	1,038.67
5-Methyl-1-hexene............	C_7H_{14},g	98.190	−16.58	1,119.98	1,046.35
	C_7H_{14},l	....	−24.80	1,111.76	1,038.13
2-Methyl-2-hexene............	C_7H_{14},g	98.190	−19.88	1,116.68	1,043.05
	C_7H_{14},l	....	−28.40	1,108.16	1,034.53
3-Methyl-*cis*-2-hexene........	C_7H_{14},g	98.190	−19.24	1,117.32	1,043.69
	C_7H_{14},l	....	−27.76	1,108.80	1,035.17
3-Methyl-*trans*-2-hexene......	C_7H_{14},g	98.190	−19.24	1,117.32	1,043.69
	C_7H_{14},l	....	−27.76	1,108.80	1,035.17
4-Methyl-*cis*-2-hexene........	C_7H_{14},g	98.190	−17.54	1,119.02	1,045.39
	C_7H_{14},l	....	−25.86	1,110.70	1,037.07
4-Methyl-*trans*-2-hexene......	C_7H_{14},g	98.190	−18.54	1,118.02	1,044.39
	C_7H_{14},l	....	−26.96	1,109.60	1,035.97
5-Methyl-*cis*-2-hexene........	C_7H_{14},g	98.190	−18.18	1,118.38	1,044.75
	C_7H_{14},l	....	−26.50	1,110.06	1,036.43
5-Methyl-*trans*-2-hexene......	C_7H_{14},g	98.190	−19.18	1,117.38	1,043.75
	C_7H_{14},l	....	−27.50	1,109.06	1,035.43
2-Methyl-*cis*-3-hexene........	C_7H_{14},g	98.190	−18.18	1,118.38	1,044.75
	C_7H_{14},l	....	−26.40	1,110.16	1,036.53
2-Methyl-*trans*-3-hexene......	C_7H_{14},g	98.190	−19.18	1,117.38	1,043.75
	C_7H_{14},l	....	−27.40	1,109.16	1,035.53
3-Methyl-*cic*-3-hexene........	C_7H_{14},g	98.190	−18.99	1,117.57	1,043.94
	C_7H_{14},l	....	−27.72	1,108.84	1,035.21
3-Methyl-*trans*-3-hexene......	C_7H_{14},g	98.190	−18.37	1,118.19	1,044.56
	C_7H_{14},l	....	−26.95	1,109.61	1,035.98
2-Ethyl-1-pentene............	C_7H_{14},g	98.190	−17.84	1,118.72	1,045.09
	C_7H_{14},l	....	−26.26	1,110.30	1,036.67
3-Ethyl-1-pentene............	C_7H_{14},g	98.190	−15.32	1,121.24	1,047.61
	C_7H_{14},l	....	−23.54	1,113.02	1,039.39
2,3-Dimethyl-1-pentene.......	C_7H_{14},g	98.190	−19.50	1,117.06	1,043.43
	C_7H_{14},l	....	−27.72	1,108.84	1,035.21
2,4-Dimethyl-1-pentene.......	C_7H_{14},g	98.190	−20.04	1,116.52	1,042.89
	C_7H_{14},l	....	−27.97	1,108.59	1,034.96
3,3-Dimethyl-1-pentene.......	C_7H_{14},g	98.190	−18.15	1,118.41	1,044.78
	C_7H_{14},l	....	−26.17	1,110.39	1,036.76
3,4-Dimethyl-1-pentene.......	C_7H_{14},g	98.190	−17.60	1,118.96	1,045.33
	C_7H_{14},l	....	−25.72	1,110.84	1,037.21
4,4-Dimethyl-1-pentene.......	C_7H_{14},g	98.190	−18.96	1,117.60	1,043.97
	C_7H_{14},l	....	−26.43	1,110.13	1,036.50

TABLE 4l-3. HEATS OF FORMATION AND HEATS OF COMBUSTION OF COMPOUNDS OF CARBON (*Continued*)

Substance name	Formula and state	Mol. weight	Kcal mole⁻¹ at 25°C		
			$\Delta Hf°$	$-\Delta Hc°$	
				Gross	Net
Carbon-Hydrogen, Alkenes (Cont.)					
3-Ethyl-2-pentene............	C_7H_{14},g	98.190	−18.62	1,117.94	1,044.31
	C_7H_{14},l	....	−27.14	1,109.42	1,035.79
2,3-Dimethyl-2-pentene.......	C_7H_{14},g	98.190	−21.10	1,115.46	1,041.83
	C_7H_{14},l	....	−29.62	1,106.94	1,033.31
2,4-Dimethyl-2-pentene.......	C_7H_{14},g	98.190	−21.20	1,115.36	1,041.73
	C_7H_{14},l	....	−29.42	1,107.14	1,033.51
3,4-Dimethyl-*cis*-2-pentene....	C_7H_{14},g	98.190	−20.90	1,115.66	1,042.03
	C_7H_{14},l	....	−29.22	1,107.34	1,033.71
3,4-Dimethyl-*trans*-2-pentene..	C_7H_{14},g	98.190	−20.90	1,115.66	1,042.03
	C_7H_{14},l	....	−29.32	1,107.24	1,033.61
4,4-Dimethyl-*cis*-2-pentene....	C_7H_{14},g	98.190	−17.36	1,119.20	1,045.57
	C_7H_{14},l	....	−25.17	1,111.39	1,037.76
4,4-Dimethyl-*trans*-2-pentene..	C_7H_{14},g	98.190	−21.24	1,115.32	1,041.69
	C_7H_{14},l	....	−29.11	1,107.45	1,033.82
3-Methyl-2-ethyl-1-butene....	C_7H_{14},g	98.190	−19.04	1,117.52	1,043.89
	C_7H_{14},l	....	−27.26	1,109.30	1,035.67
2,3,3-Trimethyl-1-butene	C_7H_{14},g	98.190	−20.43	1,116.13	1,042.50
	C_7H_{14},l	....	−28.13	1,108.43	1,034.80
1-Octene..................	C_8H_{16},g	112.217	−19.41	1,279.52	1,195.37
	C_8H_{16},l	....	−29.11	1,269.82	1,185.67
2,2-Dimethyl-*cis*-3-hexene.....	C_8H_{16},l	112.217	−30.21	1,268.72	1,184.56
2,2-Dimethyl-*trans*-3-hexene...	C_8H_{16},l	112.217	−34.65	1,264.28	1,180.13
2-Methyl-3-ethyl-1-pentene...	C_8H_{16},l	112.217	−32.92	1,266.01	1,181.86
2,2,4-Trimethyl-1-pentene....	C_8H_{16},l	112.217	−34.81	1,264.12	1,179.97
2,4,4-Trimethyl-2-pentene	C_8H_{16},l	112.217	−34.04	1,264.89	1,180.74
1-Nonene..................	C_9H_{18},g	126.244	−24.70	1,436.59	1,341.92
	C_9H_{18},l	....	−35.58	1,425.71	1,331.04
1-Decene	$C_{10}H_{20}$,g	140.271	−29.48	1,594.18	1,488.99
	$C_{10}H_{20}$,l	....	−41.54	1,582.12	1,476.93
2,2,5,5-Tetramethyl-*cis*-3-hexene	$C_{10}H_{20}$,l	140.271	−39.09	1,584.57	1,479.38
2,2,5,5-Tetramethyl-*trans*-3-hexene...............	$C_{10}H_{20}$,l	140.271	−49.62	1,574.04	1,468.83
Carbon-Hydrogen, Alkadienes					
Propadiene (allene)	C_3H_4,g	40.065	45.93	464.71	443.67
1,2-Butadiene	C_4H_6,g	54.092	38.78	619.93	588.37
	C_4H_6,l	....	32.98	614.13	582.58
1,3-Butadiene	C_4H_6,g	54.092	26.34	607.49	575.93
	C_4H_6,l	....	21.17	602.32	570.76
1,2-Pentadiene..............	C_5H_8,g	68.120	33.62	777.14	735.06
	C_5H_8,l	....	26.74	770.26	728.18
1-*cis*-3-Pentadiene...........	C_5H_8,g	68.120	19.78	763.30	721.22
	C_5H_8,l	....	12.97	756.48	714.41
1-*trans*-3-Pentadiene.........	C_5H_8,g	68.120	18.12	761.64	719.56
	C_5H_8,l	....	11.44	754.95	712.88

TABLE 4l-3. HEATS OF FORMATION AND HEATS OF
COMBUSTION OF COMPOUNDS OF CARBON (*Continued*)

Substance name	Formula and state	Mol. weight	Kcal mole^{-1} at 25°C		
			$\Delta Hf°$	$-\Delta Hc°$	
				Gross	Net
Carbon-Hydrogen, Alkadienes (Cont.)					
1,4-Pentadiene...............	C_5H_8,g	68.120	25.42	768.94	726.86
	C_5H_8,l	....	19.33	762.84	720.76
2,3-Pentadiene...............	C_5H_8,g	68.120	31.80	775.32	733.24
	C_5H_8,*l*	....	24.72	768.23	726.16
3-Methyl-1,2-butadiene.......	C_5H_8,g	68.120	31.00	774.51	732.43
	C_5H_8,l	....	23.26	766.78	724.70
2-Methyl-1,3-butadiene (isoprene)	C_5H_8,g	68.120	18.10	761.62	719.54
	C_5H_8,l	....	11.72	755.24	713.16
Carbon-Hydrogen, Cycloalkenes					
Cyclopropene................	C_3H_4,g	40.065	66.	485.	464.
Cyclopentene................	C_5H_8,g	68.120	7.75	751.27	709.19
	C_5H_8,l	....	1.04	744.55	702.47
1-Methylcyclopentene........	C_6H_{10},g	82.147	−0.96	904.92	852.32
	C_6H_{10},l	....	−8.71	897.17	844.58
3-Methylcyclopentene........	C_6H_{10},l	82.147	−5.66	900.22	847.62
4-Methylcyclopentene........	C_6H_{10},l	82.147	−4.20	901.68	849.08
1-Ethylcyclopentene.........	C_7H_{12},g	96.174	−4.89	1,063.36	1,000.25
	C_7H_{12},l	....	−13.93	1,054.32	991.21
3-Ethylcyclopentene.........	C_7H_{12},l	96.174	−11.77	1,056.48	993.37
4-Ethylcyclopentene.........	C_7H_{12},l	96.174	−10.31	1,057.94	994.83
1-n-Propylcyclopentene.......	C_8H_{14},g	110.201	−9.81	1,220.80	1,147.17
	C_8H_{14},l	....	−20.03	1,210.58	1,136.95
3-n-Propylcyclopentene.......	C_8H_{14},l	110.201	−17.87	1,112.74	1,039.11
3-n-Propylcyclopentene.......	C_8H_{14},l	110.201	−16.41	1,214.20	1,140.57
1-n-Butylcyclopentene........	C_9H_{16},g	124.228	−14.73	1,378.25	1,294.10
	C_9H_{16},l	....	−26.13	1,366.85	1,282.70
3-n-Butylcyclopentene........	C_9H_{16},l	124.228	−23.97	1,369.01	1,284.86
4-n-Butylcyclopentene........	C_9H_{16},l	124.228	−22.51	1,370.47	1,286.32
1-n-Pentylcyclopentene.......	$C_{10}H_{18}$,g	138.255	−19.66	1,535.69	1,441.02
	$C_{10}H_{18}$,l	....	−32.24	1,523.11	1,428.44
3-n-Pentylcyclopentene.......	$C_{10}H_{18}$,l	138.255	−30.08	1,525.27	1,430.60
4-n-Pentylcyclopentene.......	$C_{10}H_{18}$,l	138.255	−28.62	1,526.73	1,432.06
Cyclohexene.................	C_6H_{10},g	82.147	−1.08	904.80	852.21
	C_6H_{10},l	....	−9.13	896.75	844.16
1-Methylcyclohexene.........	C_7H_{12},g	96.174	−10.36	1,057.89	994.78
	C_7H_{12},l	....	−19.40	1,048.85	985.74
1-Ethylcyclohexene..........	C_8H_{14},g	110.201	−15.17	1,215.44	1,141.81
	C_8H_{14},l	....	−25.50	1,205.11	1,131.48
1-n-Propylcyclohexene.......	C_9H_{16},g	124.228	−20.10	1,372.88	1,288.73
	C_9H_{16},l	....	−31.61	1,361.37	1,277.22
1-n-Butylcyclohexene........	$C_{10}H_{18}$,g	138.255	−25.02	1,530.33	1,435.66
	$C_{10}H_{18}$,l	....	−37.71	1,517.64	1,422.97
1-n-Pentylcyclohexene.......	$C_{11}H_{20}$,g	152.282	−29.94	1,687.77	1,582.58
	$C_{11}H_{20}$,l	....	−43.71	1,674.00	1,568.81

TABLE 4l-3. HEATS OF FORMATION AND HEATS OF COMBUSTION OF COMPOUNDS OF CARBON (*Continued*)

Substance name	Formula and state	Mol. weight	$\Delta Hf°$	$-\Delta Hc°$ Gross	$-\Delta Hc°$ Net
			Kcal mole⁻¹ at 25°C		

Kcal mole⁻¹ at 25°C

Substance name	Formula and state	Mol. weight	$\Delta Hf°$	Gross	Net
Carbon-Hydrogen, Alkynes					
Ethyne (acetylene)...........	C_2H_2,g	26.038	54.20	310.62	300.10
Propyne (methylacetylene)....	C_3H_4,g	40.065	44.33	463.11	442.07
1-Butyne (ethylacetylene).....	C_4H_6,g	54.092	39.49	620.64	589.08
	C_4H_6,l	....	33.70	614.85	583.30
2-Butyne (dimethylacetylene).	C_4H_6,g	54.092	34.98	616.13	584.57
	C_4H_6,l	....	28.54	609.69	578.13
1-Pentyne..................	C_5H_8,g	68.120	34.52	778.03	735.95
	C_5H_8,l	....	27.67	771.18	729.10
2-Pentyne..................	C_5H_8,g	68.120	30.81	774.33	732.25
	C_5H_8,l	....	23.43	766.95	724.87
3-Methyl-1-butyne...........	C_5H_8,g	68.120	32.62	776.13	734.05
	C_5H_8,l	....	26.35	769.86	727.79
1-Hexyne..................	C_6H_{10},g	82.147	29.5	935.45	882.86
1-Heptyne.................	C_7H_{12},g	96.174	24.64	1,092.89	1,029.78
1-Octyne..................	C_8H_{14},g	110.201	19.73	1,250.34	1,176.71
1-Nonyne..................	C_9H_{16},g	124.228	14.80	1,407.78	1,323.63
1-Decyne..................	$C_{10}H_{18}$,g	138.255	9.88	1,565.22	1,470.55
1,8-Cyclotetradecadiyne......	$C_{14}H_{20}$,g	188.315	76.97	2,076.83	1,971.64
	$C_{14}H_{20}$,c	....	37.29	2,037.15	1,931.96
Carbon-Hydrogen-Miscellaneous Ring Compounds					
Spiropentane................	C_5H_8,g	68.120	44.26	787.77	745.69
	C_5H_8,l	....	37.66	781.17	739.09
cis-Decahydronaphthalene....	$C_{10}H_{18}$,g	138.255	−40.36	1,514.99	1,420.32
	$C_{10}H_{18}$,l	....	−52.42	1,502.92	1,408.25
trans-Decahydronaphthalene..	$C_{10}H_{18}$,g	138.255	−43.54	1,511.80	1,417.13
	$C_{10}H_{18}$,l	....	−55.12	1,500.23	1,405.56
9-Methyl-*trans*-decalin........	$C_{11}H_{20}$,l	152.282	−59.67	1,658.04	1,552.85
9-Methyl-*cis*-decalin	$C_{11}H_{20}$,l	152.282	−58.28	1,659.43	1,554.24
1,3,5,7-Cyclooctatetraene.....	C_8H_8,l	104.153	60.83	1,086.50	1,044.42
Carbon-Hydrogen, Benzenes and Naphthalenes					
Benzene....................	C_6H_6,g	78.115	19.80	789.05	757.49
	C_6H_6,l	....	11.71	780.96	749.40
Methylbenzene (toluene)......	C_7H_8,g	92.142	11.95	943.57	901.49
	C_7H_8,l	....	2.87	934.49	892.41
Ethylbenzene...............	C_8H_{10},g	106.169	7.14	1,101.13	1,048.53
	C_8H_{10},l	....	−2.95	1,091.03	1,038.44
1,2-Dimethylbenzene (*o*-xylene)	C_8H_{10},g	106.169	4.56	1,098.54	1,045.95
	C_8H_{10},l	....	−5.82	1,088.16	1,035.56
1,3-Dimethylbenzene (*m*-xylene)	C_8H_{10},g	106.169	4.13	1,098.12	1,045.52
	C_8H_{10},l	....	−6.06	1,087.92	1,035.33
1,4-Dimethylbenzene (*p*-xylene)	C_8H_{10},g	106.169	4.30	1,098.29	1,045.69
	C_8H_{10},l	....	−5.82	1,088.16	1,035.56

TABLE 4l-3. HEATS OF FORMATION AND HEATS OF
COMBUSTION OF COMPOUNDS OF CARBON (*Continued*)

Substance name	Formula and state	Mol. weight	Kcal mole⁻¹ at 25°C		
			$\Delta Hf°$	$-\Delta Hc°$ Gross	Net
Carbon-Hydrogen, Benzenes and Naphthalenes (Cont.)					
n-Popylbenzene	C_9H_{12},g	120.196	1.89	1,258.24	1,195.12
	C_9H_{12},l	......	−9.16	1,247.19	1,184.08
Isopropylbenzene (cumene)	C_9H_{12},g	120.196	0.96	1,257.31	1,194.20
	C_9H_{12},l	......	−9.83	1,246.52	1,183.41
1-Methyl-2-ethylbenzene	C_9H_{12},g	120.196	0.31	1,256.66	1,193.55
	C_9H_{12},l	......	−11.09	1,245.26	1,182.15
1-Methyl-3-ethylbenzene	C_9H_{12},g	120.196	−0.43	1,255.92	1,192.81
	C_9H_{12},l	......	−11.64	1,244.71	1,181.60
1-Methyl-4-ethylbenzene	C_9H_{12},g	120.196	−0.49	1,255.86	1,192.75
	C_9H_{12},l	......	−11.90	1,244.45	1,181.34
1,2,3-Trimethylbenzene (hemimellitene)	C_9H_{12},g	120.196	−2.26	1,254.08	1,190.97
	C_9H_{12},l	......	−13.99	1,242.36	1,179.25
1,2,4-Trimethylbenzene (pseudocumene)	C_9H_{12},g	120.196	−3.31	1,253.04	1,189.92
	C_9H_{12},l	......	−14.77	1,241.58	1,178.47
1,3,5-Trimethylbenzene (mesitylene)	C_9H_{12},g	120.196	−3.81	1,252.54	1,189.42
	C_9H_{12},l	......	−15.16	1,241.19	1,178.08
n-Butylbenzene	$C_{10}H_{14}$,g	134.223	−3.30	1,415.42	1,341.78
	$C_{10}H_{14}$,l	......	−15.28	1,403.44	1,329.80
n-Pentylbenzene	$C_{11}H_{16}$,g	148.250	−8.23	1,572.85	1,488.70
	$C_{11}H_{16}$.l	......	−21.39	1,559.69	1,475.54
n-Hexylbenzene	$C_{12}H_{18}$,g	162.277	−13.15	1,730.30	1,635.63
	$C_{12}H_{18}$,l	......	−27.49	1,715.96	1,621.29
n-Heptylbenzene	$C_{13}H_{20}$,g	176.304	−18.08	1,887.73	1,782.54
	$C_{13}H_{20}$,l	......	−33.60	1,872.21	1,767.02
n-Octylbenzene	$C_{14}H_{22}$,g	190.331	−23.00	2,045.18	1,929.47
	$C_{14}H_{22}$,l	......	−39.70	2,028.48	1,912.77
n-Nonylbenzene	$C_{15}H_{24}$,g	204.358	−27.93	2,202.92	2,076.69
	$C_{15}H_{24}$,l	......	−45.51	2,185.03	2,058.81
n-Decylbenzene	$C_{16}H_{26}$,g	218.385	−32.86	2,360.05	2,223.30
	$C_{16}H_{26}$,l	......	−51.92	2,340.99	2,204.24
Ethenylbenzene (styrene)	C_8H_8,g	104.153	35.23	1,060.90	1,018.82
	C_8H_8,l	......	24.84	1,050.51	1,008.43
Biphenyl	$C_{12}H_{10}$,g	154.214	43.52	1,513.71	1,461.11
	$C_{12}H_{10}$,c	......	24.02	1,494.21	1,441.61
Naphthalene	$C_{10}H_8$,g	128.175	36.14	1,249.91	1,207.83
	$C_{10}H_8$,c	......	18.77	1,232.54	1,190.46
1-Methylnaphthalene	$C_{11}H_{10}$,l	142.202	13.54	1,389.59	1,337.00
2-Methylnaphthalene	$C_{11}H_{10}$,l	142.202	10.74	1,386.88	1,334.29
Carbon-Oxygen-Hydrogen, Alkanols					
Methanol (methyl alcohol)	CH_4O,g	32.042	−48.06	182.62	161.58
	CH_4O,l	......	−57.13	173.55	152.51
Ethanol (ethyl alcohol)	C_2H_6O,g	46.070	−56.03	337.02	305.46
	C_2H_6O,l	......	−66.20	326.85	295.29
1-Propanol (n-propyl alcohol)	C_3H_8O,g	60.097	−61.28	494.13	452.06
	C_3H_8O,l	......	−72.66	484.75	440.68

TABLE 4l-3. HEATS OF FORMATION AND HEATS OF
COMBUSTION OF COMPOUNDS OF CARBON (Continued)

Substance name	Formula and state	Mol. weight	Kcal mole^{-1} at 25°C		
			$\Delta Hf°$	$-\Delta Hc°$	
				Gross	Net
Carbon-Oxygen-Hydrogen, Alkanols (Cont.)					
2-Propanol (isopropyl alcohol).	C_3H_8O,g	60.097	−65.11	490.30	448.23
	C_3H_8O,l	......	−75.97	479.44	437.37
1-Butanol.................	$C_4H_{10}O$,g	74.124	−65.65	652.13	599.53
	$C_4H_{10}O$,l	......	−78.18	639.60	587.00
2-Butanol	$C_4H_{10}O$,g	74.124	−69.94	647.84	595.24
	$C_4H_{10}O$,l	......	−81.88	635.90	583.30
2-Methyl-1-propanol	$C_4H_{10}O$,g	74.124	−67.69	650.09	597.49
	$C_4H_{10}O$,l	......	−79.85	637.93	585.33
2-Methyl-2-propanol........	$C_4H_{10}O$,g	74.124	−74.67	643.11	590.51
	$C_4H_{10}O$,l	......	−85.86	631.92	579.32
1-Pentanol.................	$C_5H_{12}O$,g	88.151	−71.94	808.20	745.09
	$C_5H_{12}O$,l	......	−85.55	794.60	731.48
2-Pentanol.................	$C_5H_{12}O$,g	88.151	−75.0	805.1	742.0
	$C_5H_{12}O$,l	......	−87.70	792.44	729.33
3-Pentanol.................	$C_5H_{12}O$,g	88.151	−75.8	804.3	741.2
	$C_5H_{12}O$,l	......	−88.46	791.68	728.57
2-Methyl-1-butanol..........	$C_5H_{12}O$,g	88.151	−72.2	807.9	744.8
	$C_5H_{12}O$,l	......	−85.19	794.96	731.84
3-Methyl-1-butanol..........	$C_5H_{12}O$,g	88.151	−71.8	808.3	745.2
	$C_5H_{12}O$,l	......	−85.13	795.01	731.90
2-Methyl-2-butanol..........	$C_5H_{12}O$,g	88.151	−78.7	801.3	738.2
	$C_5H_{12}O$,l	......	−90.66	789.48	726.37
3-Methyl-2-butanol..........	$C_5H_{12}O$,g	88.151	−75.18	804.96	741.85
	$C_5H_{12}O$,l	......	−87.58	792.57	729.45
1-Hexanol.................	$C_6H_{14}O$,g	102.178	−75.9	966.6	893.0
	$C_6H_{14}O$,l	......	−90.7	951.8	878.2
1-Heptanol.................	$C_7H_{16}O$,g	116.205	−79.4	1,125.4	1,041.3
	$C_7H_{16}O$,l	......	−95.3	1,109.6	1,025.4
1-Octanol.................	$C_8H_{18}O$,g	130.232	−84.4	1,282.8	1,188.2
	$C_8H_{18}O$,l	......	−101.6	1,265.6	1,171.0
1-Nonanol.................	$C_9H_{20}O$,g	144.259	−91.2	1,438.4	1,333.2
	$C_9H_{20}O$,l	......	−109.7	1,419.9	1,314.7
1-Decanol.................	$C_{10}H_{22}O$,g	158.286	−94.2	1,597.8	1,482.1
	$C_{10}H_{22}O$,l	......	−114.3	1,577.7	1,462.0
Carbon-Oxygen-Hydrogen, Polyhydroxy Alkanols					
1,2-Ethanediol (ethylene glycol.................	$C_2H_6O_2$,l	62.069	−108.72	284.33	252.77
1,2-Propanediol.............	$C_3H_8O_2$,l	76.096	−118.10	437.3	345.2
1,2-Butanediol.............	$C_4H_{10}O_2$,l	90.123	−123.17	594.61	542.01
1,2,3-Propanetriol (glycerol)...	$C_3H_8O_3$,g	92.095	−139.8	415.6	373.5
	$C_3H_8O_3$,l	......	−160.3	395.1	353.0
anti-1,2,3,4-Butanetetrol (erythritol)..............	$C_4H_{10}O_4$,c	122.122	−217.6	500.2	447.6
1,2,3,4,5,6-Hexanehexol (dulcitol)...............	$C_6H_{14}O_6$,c	182.175	−321.87	720.64	646.74
1,2,3,4,5,6-Hexanehexol (mannitol).............	$C_6H_{14}O_6$,c	182.175	−319.58	722.93	649.03

TABLE 4l-3. HEATS OF FORMATION AND HEATS OF
COMBUSTION OF COMPOUNDS OF CARBON (Continued)

Substance name	Formula and state	Mol. weight	Kcal mole⁻¹ at 25°C		
			$\Delta Hf°$	$-\Delta Hc°$	
				Gross	Net
Carbon-Oxygen-Hydrogen, Cycloalkanols					
Cyclopentanol................	$C_5H_{10}O$,g	86.135	−57.62	754.21	701.62
	$C_5H_{10}O$,l	....	−71.36	740.47	687.88
Cyclohexanol................	$C_6H_{12}O$,g	100.162	−68.39	905.81	842.70
	$C_6H_{12}O$,l	....	−83.21	890.99	827.88
cis-2-Methylcyclohexanol.....	$C_7H_{14}O$,l	114.189	−90.9	1,045.7	927.0
trans-2-Methylcyclohexanol...	$C_7H_{14}O$,l	114.189	−97.0	1,039.6	965.9
cis-3-Methylcyclohexanol.....	$C_7H_{14}O$,l	114.189	−91.9	1,044.7	971.0
trans-3-Methylcyclohexanol...	$C_7H_{14}O$,l	114.189	−97.1	1,039.5	965.8
Carbon-Oxygen-Hydrogen, Ethers					
Dimethyl ether..............	C_2H_6O,g	46.070	−43.99	349.06	317.50
	C_2H_6O,l	....	−48.6	344.4	312.9
Methyl ethyl ether..........	C_3H_8O,g	60.097	−51.73	503.68	461.61
	C_3H_8O,l	....	−57.4	498.0	455.9
Diethyl ether...............	$C_4H_{10}O$,g	74.124	−60.26	657.52	604.92
	$C_4H_{10}O$,l	....	−66.83	650.95	598.35
Methyl-n-propyl ether........	$C_4H_{10}O$,g	74.124	−56.82	660.96	608.36
	$C_4H_{10}O$,l	....	−63.61	654.17	601.57
Methyl isopropyl ether.......	$C_4H_{10}O$,g	74.124	−60.24	657.54	604.94
Methyl tert-butyl ether.......	$C_5H_{12}O$,g	88.151	−69.1	811.0	747.9
	$C_5H_{12}O$,l	....	−76.6	803.5	740.4
Di-n-Propyl ether...........	$C_6H_{14}O$,g	102.178	−70.07	972.44	898.81
	$C_6H_{14}O$,l	....	−78.58	963.93	890.30
Diisopropyl ether...........	$C_6H_{14}O$,g	102.178	−76.24	966.27	892.64
	$C_6H_{14}O$,l	....	−84.00	958.51	884.88
Ethyl tert-butyl ether........	$C_6H_{14}O$,l	102.178	−88.5	954.0	880.4
Di-n-butyl ether............	$C_8H_{18}O$,g	130.232	−79.8	1,287.4	1,192.8
	$C_8H_{18}O$,l	....	−90.2	1,277.0	1,182.4
Ethylene oxide..............	C_2H_4O,g	44.054	−12.58	312.15	291.11
Propylene oxide.............	C_3H_6O,g	58.081	−22.17	464.93	433.37
	C_3H_6O,l	....	−28.84	458.26	426.70
Furan......................	C_4H_4O,g	68.076	−8.29	504.54	483.50
	C_4H_4O,l	....	−14.77	498.07	477.03
1,3-Dioxane................	$C_4H_8O_2$,g	88.107	−81.5	568.0	525.9
	$C_4H_8O_2$,l	....	−89.99	559.47	517.39
1,4-Dioxane................	$C_4H_8O_2$,g	88.107	−75.30	574.16	532.09
	$C_4H_8O_2$,l	....	−84.50	564.96	522.89
Carbon-Oxygen-Hydrogen, Alkanals and Alkanones					
Methanal (formaldehyde).....	CH_2O,g	30.027	−27.70	134.67	124.15
Ethanal (acetaldehyde).......	C_2H_4O,g	44.054	−39.68	285.05	264.01
n-Propanal.................	C_3H_6O,g	58.081	−45.56	441.54	409.98
	C_3H_6O,l	....	−52.65	434.45	402.89
n-Butanal...	C_4H_8O,g	72.108	−48.80	600.66	558.59
	C_4H_8O,l	....	−56.85	592.61	550.54

TABLE 4l-3. HEATS OF FORMATION AND HEATS OF
COMBUSTION OF COMPOUNDS OF CARBON (Continued)

Substance name	Formula and state	Mol. weight	Kcal mole^{-1} at 25°C		
			$\Delta Hf°$	$-\Delta Hc°$	
				Gross	Net
Carbon-Oxygen-Hydrogen, Alkanals and Alkanones (Cont.)					
Ketene....................	C_2H_2O,g	42.038	-14.6	241.8	231.3
2-Propanone (acetone)........	C_3H_6O,g	58.081	-51.80	435.30	403.74
	C_3H_6O,l	....	-58.99	428.11	396.55
2-Butanone................	C_4H_8O,l	72.108	-66.85	582.61	540.54
Cyclopentanone.............	C_5H_8O,g	84.119	-46.40	697.12	655.04
	C_5H_8O,l	......	-56.61	686.91	644.83
Cyclohexanone.............	$C_6H_{10}O$,g	98.146	-54.07	851.81	799.22
	$C_6H_{10}O$,l	......	-64.84	841.04	788.44
Carbon-Oxygen-Hydrogen, Sugars					
D-Ribose....................	$C_5H_{10}O_5$,c	150.121	-251.16	560.67	508.08
D-Arabinose.................	$C_5H_{10}O_5$,c	150.132	-252.84	558.99	506.40
D-Xylose....................	$C_5H_{10}O_5$,c	150.132	-251.4	560.4	507.8
α-D-Galactose...............	$C_6H_{12}O_6$,c	180.159	-307.4	666.8	603.7
α-D-Galactose monohydrate...	$C_6H_{14}O_7$,c	198.174	-373.3	669.2	595.6
α-D-Glucose.................	$C_6H_{12}O_6$,c	180.159	-304.60	669.60	606.48
β-D-Glucose.................	$C_6H_{12}O_6$,c	180.159	-303.07	671.13	608.01
α-D-Glucose monohydrate.....	$C_6H_{14}O_7$,c	198.174	-375.50	667.01	593.38
β-D-Fructose.................	$C_6H_{12}O_6$,c	180.159	-302.5	671.7	608.6
L-Sorbose...................	$C_6H_{12}O_6$,c	180.159	-303.8	670.4	607.3
α-Lactose...................	$C_{12}H_{22}O_{11}$,c	342.303	-531.0	1,349.1	1,233.4
β-Lactose...................	$C_{12}H_{22}O_{11}$,c	342.303	-534.6	1,345.5	1,229.8
α-Lactose monohydrate.......	$C_{12}H_{24}O_{12}$,c	360.318	-606.2	1,342.2	1,216.0
Maltose....................	$C_{12}H_{22}O_{11}$,c	342.303	-592.2	1,350.9	1,235.2
β-Maltose monohydrate......	$C_{12}H_{24}O_{12}$,c	360.318	-606.9	1,341.5	1,215.3
Sucrose....................	$C_{12}H_{22}O_{11}$,c	342.303	-531.0	1,349.1	1,233.4
Carbon-Oxygen-Hydrogen, Alkanoic and Alkanedioic Acids, Anhydrides and Esters					
Methanoic acid (formic acid)..	CH_2O_2,g	46.026	-90.48	71.89	61.57
	CH_2O_2,l	....	-101.51	60.86	50.54
Ethanoic acid (acetic acid)....	$C_2H_4O_2$,g	60.053	-103.83	220.96	199.86
	$C_2H_4O_2$,l	......	-115.73	209.00	187.96
n-Propanoic acid (propionic acid)..................	$C_3H_6O_2$,l	74.080	-122.12	364.98	333.42
n-Butanoic acid (butyric acid).	$C_4H_8O_2$,l	88.107	-127.9	521.6	479.5
n-Pentanoic acid.............	$C_5H_{10}O_2$,l	102.134	-133.88	677.95	625.36
Hexadecanoic acid (palmitic acid)	$C_{16}H_{32}O_2$,g	240.433	-176.0	2,421.9	2,253.6
	$C_{16}H_{32}O_2$,c	....	-212.9	2,385.0	2,216.7
Octadecanoic acid (stearic acid)	$C_{18}H_{36}O_2$,g	268.487	-118.5	2,804.1	2,614.7
	$C_{18}H_{36}O_2$,c	......	-158.3	2,764.3	2,574.9
Ethanoic (acetic) anhydride...	$C_4H_6O_3$,g	102.091	-137.62	433.53	411.97
	$C_4H_6O_3$,l	......	-149.16	431.99	400.43
Methyl methanoate (methyl formate)	$C_2H_4O_2$,g	60.053	-83.6	241.1	220.1
	$C_2H_4O_2$,l	......	-90.4	234.4	213.3
Ethyl ethanoate (ethyl acetate)	$C_4H_8O_2$,g	88.107	-108.0	541.5	499.4
	$C_4H_8O_2$,l	......	-116.3	533.2	491.1
Ethanedioic acid.............	$C_2H_2O_4$,g	90.036	-174.5	81.9	71.4
α-Oxalic acid...............	$C_2H_2O_4$,c	......	-197.9	58.5	48.0

TABLE 4l-3. HEATS OF FORMATION AND HEATS OF
COMBUSTION OF COMPOUNDS OF CARBON (*Continued*)

Substance name	Formula and state	Mol. weight	Kcal mole⁻¹ at 25°C		
			ΔHf°	$-\Delta Hc^\circ$	
				Gross	Net
Carbon-Oxygen-Hydrogen, Alkanoic and Alkanedioic Acid, Anhydrides and Esters (Cont.)					
β-Oxalic acid................	$C_2H_2O_4$,c	....	− 196.7	59.7	49.2
Ethanedioic acid dihydrate....	$C_2H_6O_6$,c	126.067	− 339.7	53.3	21.8
Propanedioic acid (malonic acid)................	$C_3H_4O_4$,c	104.063	− 212.93	205.85	184.82
Pentanedioic acid (glutaric acid)................	$C_5H_8O_4$,c	132.117	− 229.3	514.2	472.1
Carbon-Oxygen-Hydrogen, Hydroxy Acids and Esters					
L(+)-2-Hydroxypropanoic acid (L(+)-lactic acid)	$C_3H_6O_3$,l $C_3H_6O_3$,c	90.079	− 161.9 − 165.88	325.2 321.21	293.6 289.66
L-2-Hydroxybutanedioic acid (L-malic acid)............	$C_4H_6O_5$,c	134.089	− 263.7	317.4	285.9
DL-2-Hydroxybutanedioic acid (DL-malic acid)..........	$C_4H_6O_5$,c	134.089	− 264.2	319.9	285.4
D-1,2-Dihydroxybutanedioic acid (D-tartaric acid).....	$C_4H_6O_6$,c	150.089	− 306.1	275.0	243.5
DL-1,2-Dihydroxybutanedioic acid (DL-tartaric acid)....	$C_4H_6O_6$,c	150.089	− 308.2	272.9	241.4
DL-1,2-Dihydroxybutanedioic acid monohydrate (DL-tartaric acid monohydrate)..	$C_4H_8O_7$,c	168.104	− 377.9	203.2	171.7
meso-1,2-Dihydroxybutanedioic acid (meso-tartaric acid)................	$C_4H_6O_6$,c	150.089	− 306.7	274.4	242.9
D-Dimethyltartrate..........	$C_6H_{10}O_6$,c	178.143	− 286.2	619.7	567.1
DL-Dimethyltartrate.........	$C_6H_{10}O_6$,g $C_6H_{10}O_6$,c	178.143	− 260.4 − 287.6	645.5 618.3	592.9 565.7
meso-Dimethyltartrate.......	$C_6H_{10}O_6$,g $C_6H_{10}O_6$,c	178.143	− 258 − 286.1	648 619.8	595 567.2
2-Hydroxy-1,2,3-propanetricarboxylic acid (citric acid)................	$C_6H_8O_7$,c	192.126	− 396.0	468.6	426.5
2-Hydroxy-1,2,3-propanetricarboxylic acid monohydrate (citric acid monohydrate)...............	$C_6H_{10}O_8$,c	210.142	− 439.4	466.5	413.9
Carbon-Oxygen-Hydrogen, Unsaturated Compounds					
2-Propen-1-ol (allyl alcohol)..	C_3H_6O,l	58.081	− 41.60	445.50	413.94
Propenoic acid (acrylic acid)..	$C_3H_4O_2$,g $C_3H_4O_2$,l	72.064	− 80.56 − 91.77	338.22 327.01	317.19 305.98
cis-Butenedioic acid (maleic acid)	$C_4H_4O_4$,g $C_4H_4O_4$,c	116.074	− 161.6 − 188.65	351.2 324.18	330.2 303.15
Maleic anhydride............	$C_4H_2O_3$,g $C_4H_2O_3$,c	74.036	− 99.6 − 112.43	344.9 332.09	334.4 321.57
trans-Butenedioic acid (fumaric acid)	$C_4H_4O_4$,g $C_4H_4O_4$,c	116.074	− 160.6 − 193.75	352.2 319.08	331.2 298.05

TABLE 4l-3. HEATS OF FORMATION AND HEATS OF
COMBUSTION OF COMPOUNDS OF CARBON (*Continued*)

Substance name	Formula and state	Mol. weight	Kcal mole^{-1} at 25°C		
			$\Delta Hf°$	$-\Delta Hc°$ Gross	Net

Carbon-Oxygen-Hydrogen, Hydroxy Benzenes

Substance name	Formula and state	Mol. weight	$\Delta Hf°$	Gross	Net
Phenol	C_6H_6O,g	94.114	−22.98	746.27	714.71
	C_6H_6O,c	....	−39.44	729.81	698.25
2-Methylphenol (*o*-cresol)	C_7H_8O,g	108.141	−30.70	900.92	858.84
	C_7H_8O,c	....	−48.87	882.75	840.67
3-Methylphenol (*m*-cresol)	C_7H_8O,g	108.141	−31.53	900.09	858.01
	C_7H_8O,l	....	−46.28	885.34	843.26
4-Methylphenol (*p*-cresol)	C_7H_8O,g	108.141	−29.9	901.7	859.64
	C_7H_8O,c	....	−47.61	884.01	841.93
2-Ethylphenol	$C_8H_{10}O$,g	122.168	−34.7	1,059.3	1,006.7
	$C_8H_{10}O$,l	....	−49.91	1,044.07	991.48
3-Ethyphenol	$C_8H_{10}O$,g	122.168	−35.9	1,058.1	1,005.5
	$C_8H_{10}O$,l	....	−51.21	1,042.77	990.18
4-Ethylphenol	$C_8H_{10}O$,g	122.168	−34.4	1,059.6	1,007.0
	$C_8H_{10}O$,c	....	−54.63	1,040.35	987.76
2,3-Dimethylphenol (2,3-xylenol)	$C_8H_{10}O$,g	122.168	−37.6	1,056.4	1,003.8
	$C_8H_{10}O$,c	....	−57.65	1,036.33	983.74
2,4-Dimethylphenol (2,4-xylenol)	$C_8H_{10}O$,g	122.168	−38.9	1,055.1	1,002.5
	$C_8H_{10}O$,l	....	−54.67	1,039.31	986.72
2,5-Dimethylphenol (2,5-xylenol)	$C_8H_{10}O$,g	122.168	−38.6	1,055.4	1,002.8
	$C_8H_{10}O$,c	....	−58.94	1,035.04	982.45
2,6-Dimethylphenol (2,6-xylenol)	$C_8H_{10}O$,g	122.168	−38.7	1,055.3	1,002.7
	$C_8H_{10}O$,c	....	−56.73	1,037.25	984.66
3,4-Dimethylphenol (3,4-xylenol)	$C_8H_{10}O$,g	122.168	−37.4	1,056.6	1,004.0
	$C_8H_{10}O$,c	....	−57.91	1,036.07	983.48
3,5-Dimethylphenol (3,5-xylenol)	$C_8H_{10}O$,g	122.168	−38.6	1,055.4	1,002.8
	$C_8H_{10}O$,c	....	−58.41	1,035.57	982.98
1,2-Dihydroxybenzene (pyrocatechol)	$C_6H_6O_2$,g	110.114	−65.0	704.3	672.7
	$C_6H_6O_2$,c	....	−84.4	684.9	653.3
1,3-Dihydroxybenzene (resorcinol)	$C_6H_6O_2$,g	110.114	−65.7	703.6	672.0
	$C_6H_6O_2$,c	....	−87.95	681.30	649.74
1,4-Dihydroxybenzene (hydroquinone)	$C_6H_6O_2$,g	110.114	−65.9	703.4	671.8
	$C_6H_6O_2$,c	....	−87.35	681.90	650.34
Benzyl alcohol	C_7H_8O,l	108.141	−38.41	893.21	851.13
α-Hydroxy-α-phenylacetophenone (benzoin)	$C_{14}H_{12}O_2$,c	212.251	−59.21	1,667.39	1,604.28

Carbon-Fluorine and Oxygen

Substance name	Formula and state	Mol. weight	$\Delta Hf°$	Gross	Net
Tetrafluoroethene	C_2F_4,g	100.016	−155.0		
Hexafluoroethane	C_2F_6,g	138.013	−316.8		
Carbonyl fluoride	COF_2,g	66.007	−151.7		

Carbon-Hydrogen-Fluorine

Substance name	Formula and state	Mol. weight	$\Delta Hf°$	Gross	Net
Difluoromethane	CH_2F_2,g	52.024	−106.8	86.2	80.9
Trifluoromethane	CHF_3,g	70.014	−164.5		
1-Fluoropropane	C_3H_7F,g	62.088	−66.8	585.1	453.5

TABLE 4l-3. HEATS OF FORMATION AND HEATS OF
COMBUSTION OF COMPOUNDS OF CARBON (*Continued*)

Substance name	Formula and state	Mol. weight	Kcal mole⁻¹ at 25°C		
			$\Delta Hf°$	$-\Delta Hc°$	
				Gross	Net
Carbon-Hydrogen-Fluorine (Cont.)					
2-Fluoropropane.............	C₃H₇F,g	62.088	−68.6	483.3	451.7
Perfluoro-*n*-heptane..........	C₇F₁₆,l	388.052	−808.2		
Perfluoroethylcyclohexane.....	C₈F₁₆,l	388.052	−789.5		
Perfluoromethylcyclohexane...	C₇F₁₄,l	350.056	−692.8		
Fluorobenzene...............	C₆H₅F,g	96.105	−27.09	738.64	717.61
	C₆H₅F,l	....	−35.36	730.37	709.34
1,2-Difluorobenzene..........	C₆H₄F₂,g	114.096	−68.95	693.27	682.75
	C₆H₄F₂,l	....	−77.60	684.62	674.10
1,3-Difluorobenzene..........	C₆H₄F₂,g	114.096	−72.00	690.22	679.70
	C₆H₄F₂,l	....	−80.94	681.93	671.41
1,4-Difluorobenzene..........	C₆H₄F₂,g	114.096	−72.03	690.19	679.67
	C₆H₄F₂,l	....	−80.54	681.68	671.16
4-Fluorotoluene..............	C₇H₇F,g	110.132	−34.65	893.45	861.89
	C₇H₇F,l	....	−44.07	884.03	852.47
m-Fluorobenzotrifluoride......	C₇H₄F₄,g	164.104	−186.77	730.78	730.78
	C₇H₄F₄,l	....	−195.84	721.71	721.71
Benzotrifluoride..............	C₇H₅F₃,g	146.113	−141.08	779.99	769.47
	C₇H₅F₃,l	....	−150.06	771.00	760.49
Carbon-Oxygen-Hydrogen-Fluorine					
m-Fluorobenzoic acid.........	C₇H₅O₂F,c	140.115	−138.46	721.32	700.29
o-Fluorobenzoic acid..........	C₇H₅O₂F,c	140.115	−135.00	724.78	703.75
Carbon-Chlorine and Oxygen					
Tetrachloromethane (carbon tetrachloride)	CCl₄,g	153.823	−24.6	69.5	69.5
	CCl₄,l	....	−32.37	61.68	61.68
Tetrachloroethene............	C₂Cl₄,g	165.834	−3.6	184.5	184.5
Hexachloroethane, triclinic....	C₂Cl₆,g	236.740	−35.3	152.8	152.8
	C₂Cl₆,c	....	−49.4	138.7	138.7
Carbonyl chloride (phosgene)..	COCl₂,g	98.917	−52.3	41.8	41.8
Carbon-Hydrogen-Chlorine					
Chloromethane...............	CH₃Cl,g	50.488	−19.32	177.20	161.42
Dichloromethane.............	CH₂Cl₂,g	84.933	−22.10	140.27	129.75
	CH₂Cl₂,l	....	−29.02	133.34	122.82
Trichloromethane............	CHCl₃,g	119.378	−24.65	103.56	98.30
	CHCl₃,l	....	−32.14	96.07	90.81
Chloroethane................	C₂H₅Cl,g	64.515	−25.5	333.4	307.1
1,2-Dichloroethane...........	C₂H₄Cl₂,g	98.960	−31.1	293.6	272.6
	C₂H₄Cl₂,l	....	−39.6	285.1	264.1
1-Chloropropane.............	C₃H₇Cl,g	78.542	−31.2	490.1	453.2
	C₃H₇Cl,l	....	−38.2	483.0	446.2

TABLE 4l-3. HEATS OF FORMATION AND HEATS OF
COMBUSTION OF COMPOUNDS OF CARBON (*Continued*)

Substance name	Formula and state	Mol. weight	Kcal mole^{-1} at 25°C		
			$\Delta Hf°$	$-\Delta Hc°$	
				Gross	Net
Carbon-Hydrogen-Chlorine (Cont.)					
2-Chloropropane.............	C₃H₇Cl,g	78.542	−32.9	488.4	451.5
	C₃H₇Cl,l	....	−39.2	482.0	445.2
2-Chloro-2-methylpropane....	C₄H₉Cl,g	92.569	−43.0	640.6	593.3
Chloroethene...............	C₂H₃Cl,g	62.499	8.1	298.7	282.9
1,1-Dichloroethene...........	C₂H₂Cl₂,g	96.944	0.3	256.7	246.2
	C₂H₂Cl₂,l	....	−6.0	250.4	239.9
Chlorobenzene...............	C₆H₅Cl,g	112.560	12.5	747.6	721.3
	C₆H₅Cl,l	....	2.5	737.6	711.3
o-Dichlorobenzene...........	C₆H₄Cl₂,l	147.005	−4.4	696.5	675.5
p-Dichlorobenzene	C₆H₄Cl₂,l	147.005	−10.1	690.8	669.8
m-Dichlorobenzene..........	C₆H₄Cl₂,l	147.005	−4.8	696.1	675.1
Carbon-Fluorine-Chlorine					
Fluorotrichloromethane.......	CFCl₃,g	137.369	−66.		
	CFCl₃,l	....	−72.02		
Difluorodichloromethane......	CF₂Cl₂,g	120.914	−114		
Trifluorochloromethane.......	CF₃Cl,g	104.459	−166		
Carbon-Bromine, and Oxygen, Hydrogen, Fluorine, and Chlorine					
Tetrabromomethane (carbon tetrabromide)............	CBr₄,g	331.647	19	113	98
Carbonyl bromide............	COBr₂,g	187.829	−23.0	71.1	63.7
	COBr₂,l	....	−30.4	63.7	56.3
Bromomethane..............	CH₃Br,g	94.944	−8.4	188.1	168.6
	CH₃Br,l	....	−14.1	182.5	163.0
Tribromomethane...........	CHBr₃,g	252.746	4	132	116
	CHBr₃,l	....	−6.8	121.4	105.1
Bromoethane...............	C₂H₅Br,g	108.971	−14.8	344.1	314.1
	C₂H₅Br,l	....	−21.5	337.3	307.4
1-Bromopropane.............	C₃H₇Br,g	122.998	−19.9	501.4	460.8
	C₃H₇Br,l	....	−27.5	493.7	453.2
2-Bromopropane.............	C₃H₇Br,g	122.998	−22.7	498.6	458.0
	C₃H₇Br,l	....	−29.9	491.3	450.8
Bromoethene...............	C₂H₃Br,g	106.955	18.7	309.3	289.8
	C₂H₃Br,l	....	13.3	303.9	284.4
Bromobenzene..............	C₆H₅Br,g	157.016	27.2	762.3	732.3
	C₆H₅Br,l	....	16.5	751.6	721.6
Trifluorobromomethane.......	CF₃Br,g	148.915	−153.6		
Trichlorobromomethane......	CCl₃Br,g	198.279	−11.0	83.0	79.4
Carbon-Hydrogen-Iodine					
Iodomethane................	CH₃I,g	191.940	3.1	199.6	176.4
	CH₃I,l	....	−3.7	192.8	169.6
Diiodomethane..............	CH₂I₂,g	267.836	27.0	189.4	163.9
	CH₂I₂,l	....	16.0	178.4	152.9

TABLE 4l-3. HEATS OF FORMATION AND HEATS OF
COMBUSTION OF COMPOUNDS OF CARBON (*Continued*)

Substance name	Formula and state	Mol. weight	Kcal mole⁻¹ at 25°C		
			$\Delta Hf°$	$-\Delta Hc°$	
				Gross	Net
Carbon-Hydrogen-Iodine (Cont.)					
Triiodomethane.............	CHI₃,c	393.732	33.7	161.9	134.3
2-Iodo-2-methylpropane......	C₄H₉I,g	184.021	−17.2	666.4	611.6
1,2-Diiodoethane............	C₂H₄I₂,g	281.863	15.9	340.6	304.7
	C₂H₄I₂,c	....	0.8	325.5	289.6
Iodobenzene................	C₆H₅I,g	204.011	40.5	775.6	741.8
	C₆H₅I,l	....	29.0	764.1	730.3
Carbon-Oxygen-Sulfur					
Carbon oxysulfide...........	COS,g	60.075	−33.96	....	131.04
Carbon-Hydrogen-Sulfur, Alkanethiols and Cycloalkanethiols					
Methanethiol (methyl	CH₄S,g	48.107	−5.34	351.57	275.25
mercaptan)	CH₄S,l	....	−11.08	345.83	269.51
Ethanethiol................	C₂H₆S,g	62.134	−11.01	508.27	421.42
	C₂H₆S,l	....	−17.59	501.69	414.84
1-Propanethiol..............	C₃H₈S,g	76.161	−16.21	665.44	568.07
	C₃H₈S,l	....	−23.86	657.79	560.43
2-Propanethiol..............	C₃H₈S,g	76.161	−18.21	663.44	566.07
	C₃H₈S,l	....	−25.29	656.36	558.99
1-Butanethiol...............	C₄H₁₀S,g	90.188	−21.05	822.96	715.08
	C₄H₁₀S,l	....	−29.78	814.23	706.35
2-Butanethiol...............	C₄H₁₀S,g	90.188	−23.15	820.86	712.98
	C₄H₁₀S,l	....	−31.29	812.72	704.84
2-Methyl-1-propanethiol......	C₄H₁₀S,g	90.188	−23.24	820.77	712.59
	C₄H₁₀S,l	....	−31.54	812.47	704.29
2-Methyl-2-propanethiol......	C₄H₁₀S,g	90.188	−26.17	817.84	709.96
	C₄H₁₀S,l	....	−33.56	810.45	702.57
1-Pentanethiol..............	C₅H₁₂S,g	104.215	−25.92	980.46	862.06
	C₅H₁₂S,l	....	−35.74	970.63	852.23
3-Methyl-1-butanethiol.......	C₅H₁₂S,g	104.215	−27.45	978.93	860.52
2-Methyl-2-butanethiol.......	C₅H₁₂S,g	104.215	−30.31	976.07	857.67
	C₅H₁₂S,l	....	−38.85	967.53	849.13
Cyclopentanethiol...........	C₅H₁₀S,g	102.199	−11.46	926.60	818.72
	C₅H₁₀S,l	....	−21.39	916.67	808.79
Cyclohexanethiol............	C₆H₁₂S,g	116.227	−22.95	1,077.48	959.08
Carbon-Hydrogen-Sulfur, Thiaalkanes, Dithiaalkanes, and Thiacycloalkanes					
2-Thiapropane (dimethyl	C₂H₆S,g	62.134	−8.96	510.32	423.47
sulfide)	C₂H₆S,l	....	−15.62	503.66	416.82
2-Thiabutane...............	C₃H₈S,g	76.161	−14.24	667.41	570.04
	C₃H₈S,l	....	−21.88	659.77	562.40
2-Thiapentane..............	C₄H₁₀S,g	90.188	−19.54	824.47	716.59
	C₄H₁₀S,l	....	−28.19	815.82	707.94
3-Thiapentane..............	C₄H₁₀S,g	90.188	−19.95	824.06	716.18
	C₄H₁₀S,l	....	−28.50	815.51	707.63

TABLE 4l-3. HEATS OF FORMATION AND HEATS OF
COMBUSTION OF COMPOUNDS OF CARBON (*Continued*)

Substance name	Formula and state	Mol. weight	Kcal mole⁻¹ at 25°C		
			$\Delta Hf°$	$-\Delta Hc°$ Gross	Net

Carbon-Hydrogen-Sulfur, Thiaalkanes, Dithiaalkanes, and Thiacycloalkanes (Cont.)

Substance name	Formula and state	Mol. weight	$\Delta Hf°$	Gross	Net
3-Methyl-2-thiabutane........	$C_4H_{10}S$,g	90.188	−21.66	822.35	643.52
	$C_4H_{10}S$,l	....	−29.81	814.20	635.37
2-Thiahexane...............	$C_5H_{12}S$,g	104.215	−24.43	981.95	863.54
	$C_5H_{12}S$,l	....	−34.16	972.22	853.81
3-Thiahexane...............	$C_5H_{12}S$,g	104.215	−25.01	981.37	862.96
	$C_5H_{12}S$,l	....	−34.59	971.79	853.38
3,3-Dimethyl-2-thiabutane....	$C_5H_{12}S$,g	104.215	−28.88	977.50	859.09
	$C_5H_{12}S$,l	....	−37.44	968.94	850.53
4-Thiaheptane..............	$C_6H_{14}S$,g	118.242	−29.98	1,138.86	1,009.94
	$C_6H_{14}S$,l	....	−40.76	1,127.98	999.06
2,4-Dimethyl-3-thiapentane...	$C_6H_{14}S$,g	118.242	−33.89	1,134.85	1,005.93
	$C_6H_{14}S$,l	....	−43.33	1,125.41	996.49
5-Thianonane...............	$C_8H_{18}S$,g	146.297	−40.02	1,453.46	1,303.50
2,3-Dithiabutane.............	$C_2H_6S_2$,g	94.198	−5.74	639.77	497.64
	$C_2H_6S_2$,l	....	−14.91	630.60	488.47
3,4-Dithiahexane.............	$C_4H_{10}S_2$,g	122 252	−17.82	952.42	789.25
	$C_4H_{10}S_2$,l	....	−28.71	941.54	778.36
4,5-Dithiaoctane.............	$C_6H_{14}S_2$,g	150.306	−28.01	1,266.97	1,082.76
	$C_6H_{14}S_2$,l	....	−40.56	1,254.42	1,070.21
Thiacyclobutane.............	C_3H_6S,g	74.145	14.59	627.92	541.08
	C_3H_6S,l	....	6.01	619.34	532.50
Thiacyclopentane............	C_4H_8S,g	88.172	−8.20	767.50	670.13
	C_4H_8S,l	....	−17.48	758.22	660.85
2-Methylthiacyclopentane.....	$C_5H_{10}S$,g	102.199	−15.16	922.90	815.02
Thiacyclohexane.............	$C_5H_{10}S$,g	102.199	−15.16	922.90	815.02
	$C_5H_{10}S$,l	....	−25.38	912.68	804.80

Carbon-Hydrogen-Sulfur, Miscellaneous

Substance name	Formula and state	Mol. weight	$\Delta Hf°$	Gross	Net
Cyclopentyl-1-thiaethane.....	$C_6H_{12}S$,g	116.227	−15.46	1,084.96	966.57
Benzenethiol (thiophenol).....	C_6H_6S,g	110.179	26.94	922.42	835.58
	C_6H_6S,l	....	15.30	910.78	823.94
Thiophene (thiofuran)........	C_4H_4S,g	84.140	27.78	666.85	590.52
	C_4H_4S,l	....	19.49	658.56	582.23

Carbon-Hydrogen-Nitrogen, Amines

Substance name	Formula and state	Mol. weight	$\Delta Hf°$	Gross	Net
Aminomethane (methylamine)	CH_5N,g	31.058	−5.49	259.35	233.05
	CH_5N,l	....	−11.3	253.5	227.2
Aminoethane (ethylamine)....	C_2H_7N,g	45.085	−11.0	416.20	379.39
	C_2H_7N,l	....	−17.71	409.49	372.68
1-Aminopropane (*n*-propyl-amine)	C_3H_9N,g	59.112	−17.3	572.3	524.9
	C_3H_9N,l	....	−24.8	564.8	517.4
1-Aminobutane (*n*-butyl-amine)	$C_4H_{11}N$,g	73.139	−22.10	729.84	671.98
	$C_4H_{11}N$,l	....	−30.53	721.41	663.55
2-Aminobutane (*sec*-butyl-amine)	$C_4H_{11}N$,g	73.139	−24.9	727.0	669.2
	$C_4H_{11}N$,l	....	−32.88	719.06	661.20

TABLE 4l-3. HEATS OF FORMATION AND HEATS OF
COMBUSTION OF COMPOUNDS OF CARBON (*Continued*)

Substance name	Formula and state	Mol. weight	Kcal mole⁻¹ at 25°C		
			$\Delta Hf°$	$-\Delta Hc°$ Gross	Net

Substance name	Formula and state	Mol. weight	$\Delta Hf°$	Gross	Net
Carbod-Oxygen-Hydrogen-Nitrogen (Cont.)					
2-Amino-2-methylpropane	$C_4H_{11}N$,g	73.139	−28.65	723.29	665.43
(*tert*-butylamine)	$C_4H_{11}N$,l	......	−35.95	715.99	658.13
Dimethylamine.............	C_2H_7N,g	45.085	−4.5	422.70	385.89
	C_2H_7N,l	......	−10.50	416.70	379.89
Diethylamine...............	$C_4H_{11}N$,g	73.139	−17.3	734.64	676.78
	$C_4H_{11}N$,l	......	−24.79	727.15	669.29
Trimethylamine............	C_3H_9N,g	59.112	−5.70	583.87	536.54
	C_3H_9N,l	......	−10.94	578.63	531.30
Triethylamine..............	$C_6H_{15}N$,g	101.193	−23.8	1,052.9	974.0
	$C_6H_{15}N$,l	......	−32.07	1,044.60	965.71
Aminobenzene (aniline).......	C_6H_7N,g	93.129	20.76	824.17	787.35
	C_6H_7N,l	......	7.43	810.84	774.02
Carbon-Hydrogen-Nitrogen, Cyanides					
Hydrogen cyanide...........	HCN,g	27.026	32.3	160.5	155.2
	HCN,l	......	26.02	154.23	148.97
Cyanamide.................	CH_2N_2,c	42.041	14.1	176.5	165.9
Ammonium cyanide..........	NH_4CN,c	44.056	0.10	230.78	209.74
Acetonitrile................	C_2H_3N,g	41.053	21.0	311.6	295.8
	C_2H_3N,l	......	12.3	302.9	287.1
Acrylonitrile................	C_3H_3N,g	53.064	44.2	428.8	404.7
	C_3H_3N,l	......	35.9	420.5	413.0
Carbon-Hydrogen-Nitrogen, Cyclic Compounds					
Pyrrolidine	C_4H_9N,g	71.123	−0.86	682.76	635.43
	C_4H_9N,l	......	−9.84	673.78	626.45
Pyridine...................	C_5H_5N,g	79.102	33.50	674.54	648.24
	C_5H_5N,l	......	23.89	664.93	638.64
6-Aminopurine (adenine)......	$C_5H_5N_5$,g	135.129	48.	689.	663.
	$C_5H_5N_5$,c	......	23.2	664.2	637.9
2-Picoline.................	C_6H_7N,g	93.129	23.65	827.06	790.24
	C_6H_7N,l	......	13.50	816.91	780.09
3-Picoline.................	C_6H_7N,g	93.129	25.37	828.78	791.96
	C_6H_7N,l	......	14.75	818.16	781.34
Carbon-Oxygen-Hydrogen-Nitrogen					
Ammonium formate..........	CH_5O_2N,c	63.057	−135.63	129.2	102.9
Ammonium bicarbonate......	CH_4O_3N,c	79.056	−203.0	61.8	35.5
Ammonium cyanate..........	CH_4ON_2,c	60.056	−72.75	157.9	136.9
Methyl nitrate..............	CH_3O_3N,g	77.040	−29.8	166.7	150.9
	CH_3O_3N,l	......	−38.0	158.5	142.7
Methyl nitrite..............	CH_3O_2N,g	61.041	−16.5	180.0	164.2

TABLE 4l-3. HEATS OF FORMATION AND HEATS OF
COMBUSTION OF COMPOUNDS OF CARBON (*Continued*)

Substance name	Formula and state	Mol. weight	Kcal mole^{-1} at 25°C		
			$\Delta Hf°$	$-\Delta Hc°$	
				Gross	Net
Carbon-Oxygen-Hydrogen-Nitrogen (Cont.)					
Nitromethane...............	CH_3O_2N,g	61.041	−17.86	178.66	162.89
	CH_3O_2N,l	......	−27.03	169.49	153.72
Nitroethane................	$C_2H_5O_2N$,g	75.068	−24.2	334.7	308.4
	$C_2H_5O_2N$,l	......	−33.9	325.0	298.7
1-Nitropropane.............	$C_3H_7O_2N$,g	89.095	−29.8	491.5	454.6
	$C_3H_7O_2N$,l	......	−40.15	481.11	444.29
2-Nitropropane.............	$C_3H_7O_2N$,g	89.095	−33.5	487.8	450.9
	$C_3H_7O_2N$,l	......	−43.3	478.0	441.1
1-Nitrobutane..............	$C_4H_9O_2N$,g	103.122	−34.4	649.2	601.9
	$C_4H_9O_2N$,l	......	−46.03	637.59	590.26
2-Nitrobutane..............	$C_4H_9O_2N$,g	103.122	−39.1	644.5	597.2
	$C_4H_9O_2N$,l	......	−49.61	634.01	586.68
Trinitromethane............	CHO_6N_3,l	151.036	−5.1	123.1	117.5
Formamide.................	CH_3ON,l	45.041	−60.7	135.8	120.0
Urea......................	CH_4ON_2,c	60.056	−79.58	151.10	130.06
Aminoethanoic acid (glycine)..	$C_2H_5O_2N$,g	75.068	−94.4	264.5	238.2
	$C_2H_5O_2N$,c	......	−128.4	230.5	204.2
DL-2-Aminopropanoic acid (DL-alanine).............	$C_3H_7O_2N$,c	89.095	−134.7	386.6	349.7
L-2-Aminopropanoic acid (L-alanine)	$C_3H_7O_2N$,g	89.095	−101.0	420.3	383.4
	$C_3H_7O_2N$,c	......	−134.5	386.8	349.9
Carbon-Nitrogen, and Hydrogen, Chlorine, Bromine, Iodine, and Sulfur					
Methylamine hydrochloride...	CH_6ClN,c	67.519	−71.20	227.8	196.2
Trinitrochloromethane........	CO_6ClN_3,l	185.481	−5.6	88.5	88.5
Cyanogen bromide...........	$CBrN$,g	105.927	44.5	138.6	134.9
	$CBrN$,c	......	33.58	127.63	123.93
Cyanogen iodide.............	CIN,g	152.922	53.9	148.0	140.5
	CIN,c	......	39.71	133.8	126.3
Thiourea...................	CH_4SN_2,c	76.120	−21.1	335.8	259.5
Ammonium thiocyanate......	CH_4SN_2,c	76.120	−18.8	338.1	261.8

Section 5

ELECTRICITY AND MAGNETISM

D. F. BLEIL, Editor

U.S. Naval Ordnance Laboratory, White Oak

CONTENTS

5a. Definitions, Units, Nomenclature, Symbols, Conversion Tables

W. R. SMYTHE

California Institute of Technology

5a-1. Fundamental Definitions Based on Mechanical Measurements

Capacitivity or Dielectric Constant. The capacitivity in farads per meter is the ratio of the force between two charged conductors measured in vacuum to that measured when the vacuum is replaced by a homogeneous fluid insulating medium, multiplied by 8.85434×10^{-12}. In a homogeneous solid it is the product of 8.85434×10^{-12} by the ratio of the force on a given small charge measured at the center of a thin disk-shaped evacuated cavity placed normal to a uniform electric field to that on the same charge measured at the center of a thin needle-shaped evacuated cavity aligned with the same field.

Charge. One coulomb is that charge which, when carried by each of two bodies whose distance apart r in meters is very large compared with their dimensions, produces in a vacuum a mutual repulsion of $8.98740r^{-2} \times 10^9$ newton. A charge of one coulomb is transported by a current of one ampere in one second. There are two kinds of charge. Electrons carry a negative charge and protons a positive charge.

Current. An ampere is that current which, flowing in the same direction in each of two identical coaxial circular loops of wire whose distance apart r in meters is very large compared with their radius a, produces in a vacuum a mutual attraction of $6\pi^2 a^4 r^{-4} \times 10^{-7}$ newton. A current of one ampere transports one coulomb of charge per second. Current direction is defined as that in which a positive charge moves.

Electric Intensity. The electric intensity in volts per meter is the vector force in newtons acting on a very small body carrying a very small positive charge placed at the field point, divided by the charge in coulombs. In a homogeneous solid the measurement is carried out at the center of a thin evacuated needle-shaped cavity aligned so that the force lies along the axis.

Electromotance or Electromotive Force. The electromotance in volts around a closed path is the work in joules required to carry a very small positive charge around that path, divided by the charge in coulombs.

Magnetic Induction or Magnetic Flux Density. The magnetic induction in webers per square meter is a vector whose direction is that in which the axis of a small circular current-carrying test loop that rests in stable equilibrium at the field point would advance if it were a right-hand screw rotated in the sense of the current circulation and whose magnitude equals the torque in newton meters on the loop when its axis is normal to the induction, divided by the product of loop current by loop area. In a homogeneous solid the measurement is carried out at the center of a thin evacuated disk-shaped cavity oriented so that the induction is normal to its faces.

Permeability. The permeability in henrys per meter is the ratio of the force between two linear circuits carrying fixed current measured in a homogeneous fluid insulating medium to that measured in a vacuum, multiplied by $4\pi \times 10^{-7}$. In a homogeneous solid it is the product of $4\pi \times 10^{-7}$ by the ratio of the magnetic induction at the center of a thin evacuated disk-shaped cavity oriented so that the induction is normal to its faces to that at the center of a thin evacuated needle-shaped cavity oriented so that the induction is directed along its axis.

Potential. The potential in volts at a point in an electrostatic field is the work in joules done in bringing a very small positive charge to the point from a point arbitrarily chosen at zero potential, divided by the charge in coulombs.

5a-2. Basic Laws

Ampère's Law. At any field point near a linear circuit, each circuit element contributes to the magnetic induction an amount inversely proportional to the square of the distance **r** from it to the point, directly proportional to its length, current, and the sine of the angle between **ds** and **r**, and in the direction of **ds** $\times$ **r**.

Coulomb's Law. The force in a homogeneous isotropic medium of infinite extent between two point charges is proportional to the product of their magnitudes divided by the square of the distance between them.

Faraday's Law of Induction. The electromotance induced in a circuit is proportional to the rate of change of the magnetic flux linking the circuit.

Joule's Law. The rate of production of heat in a constant-resistance electric circuit is proportional to the square of the current.

Kirchhoff's Laws. (1) The algebraic sum of the currents flowing into any point in a network is zero. (2) The algebraic sum of the products of current by resistance around any closed path in a network equals the algebraic sum of the electromotances in that path.

Lenz's Law. The current induced in a circuit due to a change in the magnetic flux through it or to its motion in a magnetic field is so directed as to oppose the change in flux or to exert a mechanical force opposing the motion.

Ohm's Law. The current in an electric circuit is directly proportional to the electromotance in it.

5a-3. Definitions of Some Descriptive Terms. For quantitative terms, see Table 5a-1.

Anode. The positive electrode in such devices as the arc, vacuum tube, and electrolytic cell.

Antiferroelectric Materials. Those in which spontaneous electric polarization occurs in lines of ions; adjacent lines are polarized in an antiparallel arrangement.

Antiferromagnetic Materials. Those in which spontaneous magnetic polarization occurs in equivalent sublattices; the polarization in one sublattice is aligned antiparallel to the other.

Cathode. The negative electrode in such devices as the arc, vacuum tube, and electrolytic cell.

Coercive Force. The value of the reverse magnetic intensity needed to destroy the magnetic moment of the specimen.

Conductors. Bodies in which differences of potential, if not maintained by some driving electromotance, disappear rapidly with a flow of current.

Curie Point. The point, as the temperature increases, at which the transition from ferromagnetic to paramagnetic properties of a substance is complete.

Diamagnetic Bodies. Those which, when placed in an inhomogeneous magnetic field, tend to move toward its weaker regions.

Dielectric Bodies. Those which can support an electric strain and in which differences of potential disappear very slowly or not at all because of current flow.

Eddy or *Foucault Currents.* Circulating currents set up in conducting masses or sheets by varying magnetic fields.

Edison or *Richardson Effect.* The thermionic emission of electrons from hot bodies at a rate which increases rapidly with temperature.

Electric Circuit. The path taken by an electric current. Elements of the circuit which possess the properties of capacitance, inductance, resistance, etc. (Table 5a-1) are known as capacitors, inductors, resistors, etc., respectively.

Electric Lines of Force. Curves in an electric field whose tangents at any point give the direction of the field at that point.

Electric Tubes of Flux. Charge-free regions in isotropic space whose sides are everywhere tangent to the electric intensity and whose ends terminate on charges or charged areas or may meet to form closed rings.

Electrodes. Terminals by which current may enter or leave a region.

Electrolysis. The process of passing current through a substance when so doing liberates one or more of its constituents at the electrodes.

Electrolyte. A substance capable of electrolysis.

Electrostriction. The change of dimensions of a dielectric body when placed in an electric field.

Ettinghausen Effect. The phenomenon observed when a conductor carries current in a transverse magnetic field and a temperature gradient appears in a direction normal to both.

Ferrimagnetic Materials. Those in which spontaneous magnetic polarization occurs in nonequivalent sublattices; the polarization in one sublattice is aligned antiparallel to the other.

Ferroelectric Materials. Those in which the electric polarization (see Table 5a-1) is produced by cooperative action between groups or domains of collectively oriented molecules.

Ferromagnetic Materials. Those in which the magnetization is produced by cooperative action between groups or domains of collectively oriented molecules.

Gyromagnetic Effects. The phenomena of magnetization by rotation (Barnett effect) and rotation by magnetization (Einstein–de Haas effect).

Hall Effect. The production of a transverse potential gradient in a material by a steady electric current which has a component normal to a magnetic field.

Hysteresis Curves. These show the steady-state relation between the magnetic induction in a material and the steady-state alternating magnetic intensity (see Table 5a-1) that produces it.

Image Force. The force on a charge due to that charge or polarization which it induces on neighboring conductors or dielectrics.

Magnetic Lines of Force. Curves in a magnetic field whose tangents at any point give the direction of the magnetic intensity there.

Magnetic Saturation. A condition in which further increases in the magnetizing field produce no increase in magnetization.

Magnetic Tubes of Flux. Regions in space whose sides are everywhere tangent to the magnetic induction and whose ends may meet to form closed rings.

Magnetostriction. The change in dimensions of a body when placed in a magnetic field.

Nernst Effect. The production of a transverse electric field by a heat current.

Parallel Connections. These are so arranged that current divides between elements, no portion passing through more than one element.

Paramagnetic Bodies. When placed in an inhomogeneous magnetic field, these bodies tend to move toward its stronger regions.

Peltier Effect. The phenomenon of absorption or generation of heat according to the direction of passage of current across a junction of two conductors.

Permanent Magnets. Strongly magnetized bodies whose magnetization is little affected by the action of internal or external magnetic fields or by moderate mechanical shocks.

Photoconductivity. The property of a material which causes its resistivity (see Table 5a-1) to change when light falls upon it.

Photoelectric Effect. The liberation of electrons from a surface when light falls upon it.

Piezoelectric Effects. The phenomena of separation of charge in a crystal by mechanical stresses and the converse.

Proximity Effect. The distortion of alternating-current flow in one conductor due to that in neighboring conductors.

Pyroelectric Effect. The phenomenon of separation of charge in a crystal by heating.

Rectifiers. Devices which offer higher resistance (see Table 5a-1) to current passing in one direction than the other.

Seebeck or *Thermoelectric Effect.* The flow of current in a circuit consisting of two or more conductors caused by temperature differences at the junctions.

Semiconductor. A rather poor conductor whose conductivity may be changed radically by small changes in its physical condition.

Series Connections. These are so arranged that current must pass through all the elements in succession.

Skin Effect. The concentration of high-frequency alternating current near the surface of a conductor.

Thomson Effects. Phenomena in which potential gradients are produced in a material by differences of temperature.

Triboelectricity. The electric charges separated by friction between bodies.

Volta or *Contact-potential Effect.* The appearance of opposite charges on two dissimilar uncharged metals when placed in contact and the existence of a difference of potential between them.

Work Function. The energy needed to carry a charge across a metal vacuum boundary.

Note on Tables 5a-2, 5a-3, *and* 5a-4. These tables are presented to facilitate transposition of formulas from one system of units into another. In such systems as the Gaussian, the formula to be transposed must be written for a medium in which μ and ϵ are not unity before using the tables. For example, the force on a moving charge in static fields is

$$\mathbf{F} = Q\mathbf{E} + c^{-1}Q(\mathbf{v} \times \mathbf{B}') \qquad \text{(Gaussian)}$$

where $\mathbf{F}$ is in dynes, Q and $\mathbf{E}$ in esu, $\mathbf{v}$ in cm/sec, $\mathbf{B}'$ in emu or gauss, and $c \approx 3 \times 10^{10}$ cm/sec. The equivalent formula in cgs emu is found from Table 5a-3, where, using primes for emu quantities, we write, according to directions, cQ' for Q, $c^{-1}\mathbf{E}'$ for $\mathbf{E}$ and obtain

$$\mathbf{F}' = Q'\mathbf{E}' + Q'(\mathbf{v} \times \mathbf{B}') \qquad \text{(cgs emu)}$$

For mks units, written with a double prime, we use the same table but write $10^{-5}\mathbf{F}''$ for $\mathbf{F}$, $10Q''$ for Q', $10^{-2}\mathbf{v}''$ or $10^{-2}\mathbf{l}''/t$ for $\mathbf{v}$ or $\mathbf{l}/t$, and $10^{-4}\mathbf{B}''$ for $\mathbf{B}'$, giving, after cancellation of 10^{-5} throughout,

$$\mathbf{F}'' = Q''\mathbf{E}'' + Q''(\mathbf{v}'' \times \mathbf{B}'') \qquad \text{(mks)}$$

In this formula $\mathbf{F}''$ is in newtons, Q'' in coulombs, $\mathbf{E}''$ in volts per meter, $\mathbf{v}''$ in meters per second, and $\mathbf{B}''$ in webers per square meter.

TABLE 5a-1. SYMBOLS. MKS^a UNIT NAMES. SYMBOLIC DEFINITIONS. DIMENSIONS^b

Quantity	Symbol^c		Mks unit	Equivalents	Dimensions
Admittance	$\breve{Y}$		mho	$\breve{Z}^{-1} = G + jB$	$m^{-1}l^{-2}t^3I^2$
Attenuation			decibels	$10 \log (A_1/A_2)$	0
Attenuation constant	α	a	parts/m	$(x_2 - x_1)^{-1} \ln (A_1/A_2)$	l^{-1}
Capacitance	C		farad	QV^{-1}	$m^{-1}l^{-2}t^4I^2$
Mutual	C_m, C_{rs}		farad	$Q_s V_r^{-1}$ if $V_t = 0,\ t \neq r$	$m^{-1}l^{-2}t^4I^2$
Self	C, C_{rr}		farad	$Q_r V_r^{-1}$ if $V_t = 0,\ t \neq r$	$m^{-1}l^{-2}t^4I^2$
Capacitivity	ϵ		farad/m	Defined in Sec. 5a-1	$m^{-1}l^{-3}t^4I^2$
Capacitivity of vacuum	ϵ_v	ϵ_0	farad/m	8.85434×10^{-12}	$m^{-1}l^{-3}t^4I^2$
Capacitivity, relative	K_e, K		ratio	$\epsilon \epsilon_v^{-1}$	0
Charge	$Q, \breve{Q}, q$		coulomb	Defined in Sec. 5a-1	tI
Charge density, line	λ		coulomb/m	dQ/ds	$l^{-1}tI$
Surface	$\sigma, (\rho_s)$		coulomb/m^2	dQ/dS	$l^{-2}tI$
Volume	ρ		coulomb/m^3	dQ/dv	$l^{-3}tI$
Conductance	G		mho	$R^{-1} = IV^{-1}$	$m^{-1}l^{-2}t^3I^2$
Conductivity	$\gamma, (\sigma)$		mho/m	iE^{-1}	$m^{-1}l^{-3}t^3I^2$
Surface	$\gamma', (\sigma')$		mho	$i_s E_s^{-1}$	$m^{-1}l^{-2}t^3I^2$
Current	$I, \breve{I}$		ampere	Fundamental	I
Current density	$i, \breve{i}, J$		ampere/m^2	$\gamma E, \gamma \breve{E}$	$l^{-2}I$
Surface	$i', \breve{i}', J'$		ampere/m	$\gamma' E_s, \gamma' \breve{E}_s$	$l^{-1}I$
Dielectric constant	ϵ		farad/m	Defined in Sec. 5a-1	$m^{-1}l^{-3}t^4I^2$
Displacement, electric	$\mathbf{D}, \breve{\mathbf{D}}$		coulomb/m^2	$\epsilon \mathbf{E}, \epsilon \breve{\mathbf{E}},\ \epsilon_v \mathbf{E} + \mathbf{P}$	$l^{-2}tI$
Elastance	S		daraf	C^{-1}, VQ^{-1}	$ml^2t^{-4}I^{-2}$
Mutual	S_m, S_{rs}		daraf	$V_s Q_r^{-1}$ if $Q_t = 0,\ t \neq r$	$ml^2t^{-4}I^{-2}$
Self	S, S_{rr}		daraf	$V_r Q_r^{-1}$ if $Q_t = 0,\ t \neq r$	$ml^2t^{-4}I^{-2}$
Elastivity	σ		daraf-m	ϵ^{-1}	$ml^3t^{-4}I^{-2}$
Electromotance (electromotive force)	$\mathcal{E}, \breve{\mathcal{E}}$	E	volt	Defined in Sec. 5a-1	$ml^2t^{-3}I^{-1}$
Electronic charge	e		coulomb	1.6020×10^{-19}	tI
Energy	W		joule	$I\Phi, QV, \frac{1}{2}\int \mathbf{E} \cdot \mathbf{D}\, dv,$ $\frac{1}{2}\int \mathbf{H} \cdot \mathbf{B}\, dv$	ml^2t^{-2}
Flux, electric	ψ	Ψ	coulomb	$\int \mathbf{n} \cdot \mathbf{D}\, dS$	tI
Flux, magnetic	Φ		weber	$\int \mathbf{n} \cdot \mathbf{B}\, dS$	$ml^2t^{-2}I^{-1}$
Force	$\mathbf{F}$		newton	$Q\mathbf{E}, \int i \times \mathbf{B}\, dv$	mlt^{-2}
Frequency	ν	f	hertz	$v\lambda^{-1}, \omega(2\pi)^{-1}$	t^{-1}
Frequency, angular	ω		radian/sec	$2\pi\nu, 2\pi v\lambda^{-1}$	t^{-1}
Impedance	$\breve{Z}$		ohm	$\mathcal{E}\breve{I}^{-1},\ R + jX$	$ml^2t^{-3}I^{-2}$
Intrinsic, vacuum	η		ohm	$\mu_v^{\frac{1}{2}}\epsilon_v^{-\frac{1}{2}} \approx 120\pi$	$ml^2t^{-3}I^{-2}$
Mutual	$\breve{Z}_m, \breve{Z}_{rs}$		ohm		$ml^2t^{-3}I^{-2}$
Self	$\breve{Z}, \breve{Z}_{rr}$		ohm		$ml^2t^{-3}I^{-2}$

^{a} The meter, kilogram, second, and ampere (mksa) were adapted as the fundamental units by the members of the Tenth General Conference on Weights and Measures, Paris, October, 1954. The vote for the ampere was not unanimous but a strong three-fourths majority. *Compt. rend.* **239**, 64 (1954).

^{b} Space vectors are printed in boldface. Phasors, which are complex numbers used in solving algebraically for the steady-state value of a sinusoidally time-dependent quantity, are designated by a flat v over the symbol. For conjugate phasors, an inverted flat v is used. The symbol j is used for $(-1)^{\frac{1}{2}}$.

^{c} The symbols listed in the left-hand column are those recommended in "Letter Symbols for Physics," Z10.6 1948, American Standards Association, American Society of Mechanical Engineers, New York. Those symbols recommended by the "International Electrotechnical Commission (IEC)," Publication 27, Geneva, Switzerland, 1953, which are different from those of the ASA, are shown on the right-hand side of this column. The ASA symbols have been used throughout Sec. 5 except for Sec. 5g, where the symbols are those recommended by the International Union of Pure and Applied Chemistry. Conversion to IEC symbols was not possible at this time because the Commission has not completed its list.

TABLE 5a-1. SYMBOLS. MKS UNIT NAMES. SYMBOLIC
DEFINITIONS. DIMENSIONS (*Continued*)

Quantity	Symbol	Mks unit	Equivalents	Dimensions
Inductance	L	henry	....	$ml^2t^{-2}I^{-2}$
Mutual	M, L_m, L_{rs}	henry	$(I_1 I_2)^{-1}\int \mathbf{B}_2 \cdot \mathbf{n}\, dS_1$	$ml^2t^{-2}I^{-2}$
Self	L, L_{rr}	henry	$(\mu I^2)^{-1}\int B^2\, dv$	$ml^2t^{-2}I^{-2}$
Induction, magnetic	$\mathbf{B}, \check{\mathbf{B}}$	weber/m²	Defined in Sec. 5a-1	$mt^{-2}I^{-1}$
Intensity, electric	$\mathbf{E}, \check{\mathbf{E}}$	volt/m	Defined in Sec. 5a-1	$mlt^{-3}I^{-1}$
Intensity, magnetic	$\mathbf{H}, \check{\mathbf{H}}$	amp-turn/m	$\mu^{-1}\mathbf{B}, \mu_v^{-1}\mathbf{B} - \mathbf{M}$	$l^{-1}I$
Length	l	meter	Fundamental	l
Magnetization (loop)	$\mathbf{M}$	amp-turn/m	$(K_m - 1)\mathbf{H}$	$l^{-1}I$
Magnetization (dipole).. (See Polarization, magnetic)	$\mathbf{M}$			
Magnetomotance (magnetomotive force)	$\mathcal{F}$ F	amp-turn	$\mu^{-1}\oint\mathbf{B}\cdot\mathbf{ds}, \oint\mathbf{H}\cdot\mathbf{ds}$	I
Mass	m	kilogram	Fundamental	m
Moment, electric	$\mathbf{p},\check{\mathbf{p}}$	coulomb-m	$Q\,\mathbf{ds}$	ltI
Moment, magnetic loop	$\mathbf{m},\check{\mathbf{m}}$	amp-m²	$\pi a^2 I\mathbf{n}$	$l^2 I$
Moment, magnetic (dipole)	$\mathbf{m}$	weber-m	$m\,\mathbf{ds}$	$ml^3t^{-2}I^{-1}$
Period	T	second	$\omega^{-1}2\pi, \nu^{-1}, \lambda v^{-1}$	t
Permeance	$\mathcal{P}$ Λ	henry	$\mathcal{R}^{-1}, \mathcal{F}^{-1}\Phi$	$ml^2t^{-2}I^{-1}$
Permeability	μ	henry/m	Defined in Sec. 5a-1	$mlt^{-2}I^{-2}$
Vacuum	μ_v μ_0	henry/m	$4\pi \times 10^{-7}$	$mlt^{-2}I^{-2}$
Relative	K_m	....	$\mu_v^{-1}\mu$	0
Phase angle	φ	radian	....	0
Phase constant (see Wave number)				
Polarization, electric	$\mathbf{P}$	coulomb/m²	$(K_e - 1)\epsilon_v\mathbf{E}$	$l^{-2}tI$
Polarization, magnetic.. (Magnetization dipole)	$\mathbf{M}$ J	weber/m²	$(K_m - 1)\mu_v\mathbf{H}$	$mt^{-2}I^{-1}$
Pole strength	m	weber	....	$ml^2t^{-2}I^{-1}$
Potential, electrostatic..	V U	volt	Defined in Sec. 5a-1	$ml^2t^{-3}I^{-1}$
Electrodynamic	$\Phi, \check{\Phi}$	volt	$\mathbf{E} = -\nabla\Phi - d\mathbf{A}/dt$	$ml^2t^{-3}I^{-1}$
Vector magnetic	$\mathbf{A}, \check{\mathbf{A}}$	weber/m	$\mathbf{B} = \nabla\times\mathbf{A}$	$mlt^{-2}I^{-1}$
Power	P	watt	dW/dt	ml^2t^{-3}
Poynting vector	$\mathbf{\Pi}$	watt/m²	$\mu^{-1}\mathbf{E}\times\mathbf{B}$	mt^{-3}
Propagation constant	$\dot{\Gamma}, (\check{\gamma})$	parts/m	$\alpha + j\beta$	l^{-1}
Quality factor	Q	a ratio	$\omega L R^{-1}$	0
Reactance	X	ohm	$\omega L - (\omega C)^{-1}$	$ml^2t^{-3}I^{-2}$
Reluctance	$\mathcal{R}$ R	amp-turn/weber	$\mathcal{F}\Phi^{-1}$	$m^{-1}l^{-2}t^2I^2$
Reluctivity	ν	m/henry	μ^{-1}	$m^{-1}l^{-1}t^2I^2$
Resistance	R	ohm	VI^{-1}	$ml^2t^{-3}I^{-2}$
Resistivity	ρ	ohm-m	$\mathbf{E}\mathbf{i}^{-1}$	$ml^3t^{-3}I^{-2}$
Susceptance	B	mho	$\check{Y} = G + jB$	$m^{-1}l^{-2}t^3I^2$
Susceptibility, electric..	χ_e κ_e	....	$K_e - 1$	0
Magnetic	χ_m κ_m	....	$K_m - 1$	0
Time	t	second	Fundamental	t
Time constant	τ	second	LR^{-1}, RC	t
Velocity of light	c	m/sec	2.99790×10^8	lt^{-1}
Wavelength	λ	meter	$2\pi\beta^{-1}, 2\pi v\omega^{-1}$	l
Wave number (phase constant)	β, k	radian/m	$2\pi\lambda^{-1}, \omega v^{-1}, \gamma = \alpha + j\beta$	l^{-1}
Work	W	joule	$\oint\mathbf{F}\cdot\mathbf{ds}$	ml^2t^{-2}

TABLE 5a-2. REDUCTION OF FORMULA TO CGS ESU[a]

Quantity	Esu	Emu	Practical cgs and rationalized mks
Capacitance	C	$c^{-2}C$	$9^{-1}10^{-11}C$ farad
Capacitivity	ϵ	$c^{-2}\epsilon$	$9^{-1}10^{-11}\epsilon(4\pi \text{ farad})/\text{cm}$ $\epsilon_v\epsilon$ farad/m
Charge, quantity	Q	$c^{-1}Q$	$3^{-1}10^{-9}Q$ coulomb
Conductance	G	$c^{-2}G$	$9^{-1}10^{-11}G$ mho
Conductivity, area	γ'	$c^{-2}\gamma'$	$9^{-1}10^{-11}\gamma'$ mho
Conductivity, volume	γ	$c^{-2}\gamma$	$9^{-1}10^{-11}\gamma$ mho/cm $9^{-1}10^{-9}\gamma$ mho/m
Current	I	$c^{-1}I$	$3^{-1}10^{-9}I$ amp
Current density, area	i'	$c^{-1}i'$	$3^{-1}10^{-9}i'$ amp/cm $3^{-1}10^{-7}i'$ amp/m
Current density, volume	i	$c^{-1}i$	$3^{-1}10^{-9}i$ amp/cm² $3^{-1}10^{-5}i$ amp/m²
Displacement	D	$c^{-1}D$	$3^{-1}10^{-9}D$ $3\epsilon_v 10^4D$ coulomb/m²
			(4π coulomb)/cm²
Elastance	S	c^2S	$9 \times 10^{11}S$ daraf
Electromotance	$\mathcal{E}$	$c\mathcal{E}$	$300\mathcal{E}$ volt
Energy	W	W	W erg $10^{-7}W$ joule
Force	F	F	F dyne $10^{-5}F$ newton
Impedance	$\check{Z}$	$c^2\check{Z}$	$0 \times 10^{11}\check{Z}$ ohm
Inductance	L	c^2L	$9 \times 10^{11}L$ henry
Intensity, electric	E	cE	$300E$ volt/cm $30{,}000E$ volt/m
Length	l	l	l centimeter $10^{-2}l$ meter
Mass	m	m	m gram $10^{-3}m$ kilogram
Polarization, electric	P	$c^{-1}P$	$3^{-1} \times 10^{-9}P$ $3^{-1} \times 10^{-5}P$
			coulomb/cm² coulomb/m²
Potential, electric	V	cV	$300V$ volt
Power	P	P	P erg/sec $10^{-7}P$ watt
Reactance	X	c^2X	$9 \times 10^{11}X$ ohm
Resistance	R	c^2R	$9 \times 10^{11}R$ ohm
Resistivity, area	σ	$c^2\sigma$	$9 \times 10^{11}\sigma$ ohm
Resistivity, volume	ρ	$c^2\rho$	$9 \times 10^{11}\rho$ ohm-cm $9 \times 10^9\rho$ ohm-m

[a] A formula given in emu, unrationalized practical cgs or rationalized mks units, in which the capacitivity or permeability, if relevant, appears explicitly, is expressed in cgs esu by replacing each symbol by the value given in the emu, practical cgs or rationalized mks column, respectively. Each line may be read as an equation relating the size of the units involved. Here c is 2.9979×10^{10}. For precise work the 3 and 9 factors should be replaced by 2.9979 and 8.9874. See inside cover for more precise values.

Table 5a-3. Reduction of Formula to cgs emu[a]

Quantity	Emu	Esu	Practical cgs and rationalized mks	
Capacitance	C	c^2C*	10^9C farad	
Charge, quantity	Q	$cQ*$	$10Q$ coulomb	
Conductance	G	c^2G*	10^9G mho	
Conductivity, area	γ'	$c^2\gamma'*$	$10^9\gamma'$ mho	
Conductivity, volume	γ	$c^2\gamma*$	$10^9\gamma$ mho/cm	$10^{11}\gamma$ mho/m
Current	I	$cI*$	$10I$ amp	
Current density, area	i'	$ci'*$	$10i'$ amp/cm	$10^3i'$ amp/m
Current density, volume	i	$ci*$	$10i$ amp/cm^2	10^5i amp/m^2
Elastance	S	$c^{-2}S*$	$10^{-9}S$ daraf	
Electromotance	ε	$c^{-1}\varepsilon*$	$10^{-8}\varepsilon$ volt	
Energy	W	W	W erg	$10^{-7}W$ joule
Flux, magnetic	$\Phi*$	$c^{-1}\Phi$	Φ maxwell	$10^{-8}\Phi$ weber
Force	F	F	F dyne	$10^{-5}F$ newton
Impedance	$\check{Z}$	$c^{-2}\check{Z}*$	$10^{-9}\check{Z}$ ohm	
Inductance	L	$c^{-2}L*$	$10^{-9}L$ henry	
Induction, magnetic	$B*$	$c^{-1}B$	B gauss	$10^{-4}B$ weber/m^2
Intensity, electric	E	$c^{-1}E*$	$10^{-8}E$ volt/cm	$10^{-6}E$ volt/m
Intensity, magnetic	$H*$	cH	H oersted	$(4\pi)^{-1}10^3H$ amp-turn/m
Length	l	l	l centimeter	$10^{-2}l$ meter
Magnetic moment (dipole)	$m'*$	$c^{-1}m'$	$4\pi m'$ maxwell-cm	$4\pi10^{-10}m'$ weber-m
Magnetic moment (loop)	$m*$	cm	$10m$ amp-cm^2	$10^{-3}m$ amp-m^2
Magnetization (dipole)	$M'*$	$c^{-1}M'$	$4\pi M'$ maxwell/cm^2	$4\pi10^{-4}M'$ weber/m^2
Magnetization (loop)	$M*$	cM	$10M$ amp/cm	$1000M$ amp/m
Magnetomotance	$\mathfrak{F}*$	$c\mathfrak{F}$	$\mathfrak{F}$ gilbert	$(4\pi)^{-1}10\mathfrak{F}$ amp-turn
Mass	m	m	m gram	$10^{-3}m$ kilogram
Permeability	$\mu*$	$c^{-2}\mu$	μ gauss/oersted	$4\pi10^{-7}\mu$ henry/m
Pole strength, magnetic	$m*$	$c^{-1}m$	$4\pi m$ maxwell	$4\pi10^{-8}m$ weber
Potential, electric	V	$c^{-1}V*$	$10^{-8}V$ volt	
Potential, vector	$A*$	$c^{-1}A$	A gauss-cm	$10^{-6}A$ weber/m
Power	P	P	P erg/sec	$10^{-7}P$ watt
Reactance	X	$c^{-2}X*$	$10^{-9}X$ ohm	
Reluctance	$\mathfrak{R}*$	$c^2\mathfrak{R}$	$\mathfrak{R}$ gilbert/max	$(4\pi)^{-1}10^9\mathfrak{R}$ amp-turn/weber
Resistance	R	$c^{-2}R*$	$10^{-9}R$ ohm	
Resistivity, area	σ	$c^{-2}\sigma*$	$10^{-9}\sigma$ ohm	
Resistivity, volume	ρ	$c^{-2}\rho*$	$10^{-9}\rho$ ohm-cm	$10^{-11}\rho$ ohm-m

[a] A formula given in esu, Gaussian (starred), unrationalized practical cgs, or rationalized mks units, in which the capacitivity or permeability, if relevant, appears explicitly, is expressed in cgs emu by replacing each symbol by the value given in the esu, starred, practical cgs, or rationalized mks columns, respectively. Each line may be read as an equation relating the size of the units involved. Here c is 2.9979×10^{10}. (See inside cover.)

TABLE 5a-4. REDUCTION OF FORMULA TO RATIONALIZED MKS UNITS[a]

Quantity	(a) Practical cgs (b) mks	Emu	Esu
Capacitance.............	C farad	$10^{-9}C$	$9 \times 10^{11}C*$
Capacitivity.............	(a) ϵ (4π farad)/cm	$10^{-9}\epsilon$	$9 \times 10^{11}\epsilon*$
	(b) ϵ farad/m	$4\pi 10^{-11}\epsilon$	$\epsilon_v{}^{-1}\epsilon*$
Charge, quantity........	Q coulomb	$10^{-1}Q$	$3 \times 10^9 Q*$
Conductance............	G mho	$10^{-9}G$	$9 \times 10^{11}G*$
Conductivity, area........	γ' mho	$10^{-9}\gamma'$	$9 \times 10^{11}\gamma'*$
Conductivity, volume.....	(a) γ mho/cm	$10^{-9}\gamma$	$9 \times 10^{11}\gamma*$
	(b) γ mho/m	$10^{-11}\gamma$	$9 \times 10^9\gamma*$
Current.................	I ampere	$10^{-1}I$	$3 \times 10^9 I*$
Current density, area.....	(a) i' amp/cm	$10^{-1}i'$	$3 \times 10^9 i'*$
	(b) i' amp/m	$10^{-3}i'$	$3 \times 10^7 i'*$
Current density, volume...	(a) i amp/cm²	$10^{-1}i$	$3 \times 10^9 i*$
	(b) i amp/m²	$10^{-5}i$	$3 \times 10^5 i*$
Displacement, electric.....	(a) D (4π coulomb)/cm²	$10^{-1}D$	$3 \times 10^9 D*$
	(b) D coulomb/m²	$4\pi 10^{-5}D$	$(3\epsilon_v)^{-1}10^{-4}D*$
Elastance...............	S daraf	$10^9 S$	$9^{-1}10^{-11}S*$
Electromotance..........	ε volt	$10^8\varepsilon$	$(300)^{-1}\varepsilon*$
Energy..................	(a) W erg	W	W
	(b) W joule	$10^7 W$	$10^7 W$
Flux, magnetic...........	(a) Φ maxwell	$\Phi*$	$3^{-1}10^{-10}\Phi$
	(b) Φ weber	$10^8\Phi*$	$(300)^{-1}\Phi$
Force...................	(a) F dyne	F	F
	(b) F newton	$10^5 F$	$10^5 F$
Impedance..............	$\check{Z}$ ohm	$10^9\check{Z}$	$9^{-1}10^{-11}\check{Z}*$
Inductance.............	L henry	$10^9 L$	$9^{-1}10^{-11}L*$
Induction, magnetic......	(a) B gauss	$B*$	$3^{-1}10^{-10}B$
	(b) B weber/m²	$10^4 B*$	$3^{-1}10^{-6}B$
Intensity, electric........	(a) E volt/cm	$10^8 E$	$(300)^{-1}E*$
	(b) E volt/m	$10^6 E$	$3^{-1}10^{-4}E*$
Intensity, magnetic.......	(a) H oersted	$H*$	$3 \times 10^{10}H$
	(b) H amp-turn/m	$4\pi 10^{-3}H*$	$12\pi 10^7 H$
Length.................	(a) l centimeter	l	l
	(b) l meter	$10^2 l$	$10^2 l$
Magnetic moment (dipole).	(a) m' maxwell-cm	$(4\pi)^{-1}m'*$	$(12\pi)^{-1}10^{-10}m'$
	(b) m' weber-m	$(4\pi)^{-1}10^{10}m'*$	$(12\pi)^{-1}m'$
Magnetic moment (loop)..	(a) m amp-cm²	$10^{-1}m*$	$3 \times 10^9 m$
	(b) m amp-m²	$10^3 m*$	$3 \times 10^{13}m$
Magnetization (dipole)....	(a) M' maxwell/cm²	$(4\pi)^{-1}M'*$	$(12\pi)^{-1}10^{-10}M'$
	(b) M' weber/m²	$(4\pi)^{-1}10^4 M'*$	$(12\pi)^{-1}10^{-6}M'$
Magnetization (loop)......	(a) M amp/cm	$10^{-1}M*$	$3 \times 10^9 M$
	(b) M amp/m	$10^{-3}M*$	$3 \times 10^7 M$

[a] A formula given in cgs emu, cgs esu, or Gaussian (starred) units, in which the capacitivity or permeability, if relevant, appears explicitly, may be expressed in (a) unrationalized cgs practical units or (b) rationalized mks units by replacing each symbol with its value in the emu, esu, or starred column, respectively. Each line may be read as an equation relating the size of the units involved. In precise work, replace 3 by 2.9979 and 9 by 8.9874.

TABLE 5a-4. REDUCTION OF FORMULA TO RATIONALIZED MKS UNITS (*Continued*)

Quantity	(a) Practical cgs (b) mks	Emu	Esu
Magnetomotance.........	(a) $\mathfrak{F}$ gilbert	$\mathfrak{F}^*$	$3 \times 10^{10}\mathfrak{F}$
	(b) $\mathfrak{F}$ amp-turn	$4\pi10^{-1}\mathfrak{F}^*$	$12\pi \times 10^9\mathfrak{F}$
Mass...................	(a) m gram	m	m
	(b) m kilogram	10^3m	10^3m
Permeability............	(a) μ gauss/oersted	μ^*	$9^{-1}10^{-20}\mu$
	(b) μ henry/m	$(4\pi)^{-1}10^7\mu^*$	$(36\pi)^{-1}10^{-13}\mu$
Polarization, electric......	(a) **P** coulomb/cm²	$10^{-1}\mathbf{P}$	$3 \times 10^9\mathbf{P}^*$
	(b) **P** coulomb/m²	$10^{-5}\mathbf{P}$	$3 \times 10^5\mathbf{P}^*$
Pole strength...........	(a) m maxwell	$(4\pi)^{-1}m^*$	$(12\pi)^{-1}10^{-10}m$
	(b) m weber	$(4\pi)^{-1}10^8m^*$	$(1200\pi)^{-1}m$
Potential, electric........	V volt	10^8V	$(300)^{-1}V^*$
Potential, vector.........	(a) **A** gauss-cm	$\mathbf{A}^*$	$3^{-1}10^{-10}\mathbf{A}$
	(b) **A** weber/m	$10^6\mathbf{A}^*$	$3^{-1}10^{-4}\mathbf{A}$
Power..................	(a) P erg/sec	P	P
	(b) P watt	10^7P	10^7P
Reactance..............	X ohm	10^9X	$9^{-1}10^{-11}X^*$
Reluctance..............	(a) $\mathfrak{R}$ gilbert/max	$\mathfrak{R}^*$	$9 \times 10^{20}\mathfrak{R}$
	(b) $\mathfrak{R}$ amp-turn/weber	$4\pi10^{-9}\mathfrak{R}^*$	$36\pi \times 10^{11}\mathfrak{R}$
Resistance..............	R ohm	10^9R	$9^{-1}10^{-11}R^*$
Resistivity, area..........	σ ohm	$10^9\sigma$	$9^{-1}10^{-11}\sigma^*$
Resistivity, volume.......	(a) ρ ohm-cm	$10^9\rho$	$9^{-1}10^{-11}\rho^*$
	(b) ρ ohm-m	$10^{11}\rho$	$9^{-1}10^{-9}\rho^*$

5b. Formulas

W. R. SMYTHE[1]

California Institute of Technology

C. YEH[2]

University of California, Los Angeles

STATIC-FIELD FORMULAS

Note. In the following formulas $\approx$ designates an approximate equality, $K(k)$ and $E(k)$ are complete elliptic integrals of modulus k, $F(\phi,k)$ and $E(\phi,k)$ are incomplete elliptic integrals, ln x is the natural logarithm of x, $\delta_n{}^m$ is the Kronecker delta which is zero unless m equals n when it is one, $J_n(x)$ is a Bessel function, $\Gamma(x)$ is a gamma function, $(2n - 1)!!$ means $1 \cdot 3 \cdot 5 \cdots (2n - 1)$, $(2n)!!$ means $2 \cdot 4 \cdot 6 \cdots (2n)$. Vectors are written boldface unless only the strength or magnitude is involved when the same symbol is used without boldface. The positive value of a difference $x - y$ is indicated by $|x - y|$.

5b-1. Capacitance Formulas in MKS Units

Single Body Remote from Earth

Sphere of radius a $C = 4\pi\epsilon a \approx 1.1128 \times 10^{-10}a$

Oblate spheroid of semiaxes a and c, $a > c$ $C = 4\pi\epsilon(a^2 - c^2)^{\frac{1}{2}}[\tan^{-1}(a^2c^{-2} - 1)^{\frac{1}{2}}]^{-1}$

Prolate spheroid of semiaxes a and b, $a > b$ $C = 4\pi\epsilon(a^2 - b^2)^{\frac{1}{2}}[\tanh^{-1}(1 - b^2a^{-2})^{\frac{1}{2}}]^{-1}$

Ellipsoid of semiaxes a, b, and c, $a > b > c$

$$C = 4\pi\epsilon(a^2 - c^2)^{\frac{1}{2}}[F(k,\phi)]^{-1}$$

where $\phi = \sin^{-1}(1 - c^2a^{-2})^{\frac{1}{2}}$ and $k = (a^2 - b^2)^{\frac{1}{2}}(a^2 - c^2)^{-\frac{1}{2}}$.

Circular disk of radius a $C = 8\epsilon a$

Elliptic disk of semiaxes a and b, $a > b$ $C = 4\pi\epsilon a\{K[(1 - b^2a^{-2})^{\frac{1}{2}}]\}^{-1}$

Two spheres of radius a in contact $C = 8\pi\epsilon a \ln 2$

Two spheres of radii a and b in contact

$$C = -4\pi\epsilon ab(a + b)^{-1}\{2\gamma + \psi[b(a + b)^{-1}] + \psi[a(a + b)^{-1}]\}$$

where $\psi(z) = \Gamma'(z)/\Gamma(z)$ and γ is Euler's constant 0.5772.

Circular solid cylinder of radius a and length $2b$

$$C = [8 + 6.95(b/a)^{0.76}]\epsilon a$$

This formula is accurate to 0.2 per cent when $0 \lesssim b/a \lesssim 8$.

---

[1] Static-field formulas.
[2] Dynamic-field formulas.

Two spheres of radius a, distance between centers c, connected by thin wire

$$C = 8\pi\epsilon a \sinh\beta \sum_{n=1}^{\infty} (-1)^{n+1} \operatorname{csch} m\beta$$

where $\cosh\beta = \frac{1}{2}ca^{-1}$.

Two spheres of radii a and b, distance between centers c, connected by thin wire

$$C = 8\pi\epsilon ab \sinh\alpha \sum_{n=1}^{\infty} \{(c \sinh n\alpha)^{-1} + [a \sinh n\alpha + b \sinh (n-1)\alpha]^{-1}\}$$

where $\cosh\alpha = \frac{1}{2}a^{-1}b^{-1}(c^2 - a^2 - b^2)$.

Two spherical caps with a common rim which meet at an external angle π/m where m is a positive integer

$$C = 4\pi\epsilon a\{1 + \sin\alpha \sum_{s=1}^{\infty} [\csc (m^{-1}s\pi + \alpha) - \csc (m^{-1}s\pi)]\}$$

The sphere of which the flatter cap is a portion has a radius a and the rim subtends an angle α at its center.

Same as above but with external angle $3\pi/2$.[1]

$$C = 4\pi\epsilon 3^{-\frac{1}{4}}a \sin\alpha\{3^{\frac{1}{2}} - 3^{-1}4 + [2 \sin\tfrac{1}{3}\alpha(\sin\tfrac{1}{3}\alpha + \sin\tfrac{1}{3}\pi)]^{-1}$$
$$+ [2 \cos\tfrac{1}{3}\alpha(\cos\tfrac{1}{3}\alpha + \cos\tfrac{1}{3}\pi)]^{-1}\}$$

Spherical bowl whose chord, drawn from center to rim, subtends an angle α at the center of the sphere of radius a on which it lies

$$C = 4\epsilon a(\alpha + \sin\alpha)$$

Torus formed by rotation of a circle of radius a about a coplanar line a distance b from its center

$$C = 8\pi\epsilon b(1 - a^2b^{-2})^{\frac{1}{2}} \sum_{n=1}^{\infty} (2 - \delta_n{}^0) \frac{Q_n}{P_n}$$

where $P_0 = 2k^{\frac{1}{2}}K(k')$, $Q_0 = 2k^{\frac{1}{2}}K(k)$, $P_1 = 2k^{-\frac{1}{2}}E(k')$, and $Q_1 = 2k^{-\frac{1}{2}}[K(k) - E(k)]$ and the moduli of the complete elliptic functions are given by

$$k = a[b + (b^2 - a^2)^{\frac{1}{2}}]^{-1} = (1 - k'^2)^{\frac{1}{2}}$$

When $n > 1$, the following recurrence formula may be used to find both P_n and Q_n

$$(2n + 1)P_{n+1} - 4na^{-1}bP_n + (2n - 1)P_{n-1} = 0$$

A capacitance table is given in *Australian J. Phys.* [**7**, 350 (1954)].

Torus formed by rotation of a circle of diameter d about a tangent line

$$C = 8\pi\epsilon d \sum_{n=1}^{\infty} [J_1(k_nd)]^{-1} S_{0,0}(k_nd) \approx 0.970 \times 10^{-10}d$$

where $S_{0,0}(k_nd)$ is a Lommel function and $J_0(k_nd) = 0$.

[1] For additional intersecting sphere-capacitance formulas, see Snow, *J. Research Natl. Bur. Standards* **43**, 377–407 (1949).

Aichi's formula for a nearly spherical surface

$$C \approx 3.139 \times 10^{-11} S^{\frac{1}{2}}$$

where S is surface area.

Cube of side a. Close lower limit

$$C \approx 0.7283 \times 10^{-10} a$$

Figure of rotation, $z = a(\cos u + k \cos 2u)$, $\rho = a(\sin u - k \sin 2u)$, $0 < k < \frac{1}{2}$

$$C \approx 1.11278 \times 10^{-10} a(1 - 0.06857k^2 - 0.00559k^4)$$

Flat circular annulus, with edges at $\rho = a$, $\rho = b$. $a < b$

$$C \approx 4.510 \times 10^{-11} b \left[\cos^{-1} \frac{a}{b} + \left(1 - \frac{a^2}{b^2} \right)^{\frac{1}{2}} \tanh^{-1} \frac{a}{b} \right] \left(1 + \frac{0.0143b}{a} \tan^3 \frac{1.28a}{b} \right)$$

Error varies from about $\pm 0.001C$ at $b = 1.1a$ to zero at $b = \infty$.

$$C \approx 17.48 \times 10^{-12} (a + b) \{ \ln [16(a + b)(b - a)^{-1}] \}^{-1}$$

Error varies from about $\pm 0.001C$ at $b = 1.1a$ to zero at $b = a$.

Thin torus generated by rotation of a circle of radius a about a coplanar line a distance b from its center

$$C \approx 3.49066 \times 10^{-10} b \left(\ln \frac{8b}{a} \right)^{-1}$$

Capacitance between Two Bodies Remote from All Others and Carrying Equal and Opposite Charges

Two spheres of radii a and b with distance r between centers

$$C = (c_{11}c_{22} - c_{12}{}^2)(c_{11} + c_{22} + 2c_{12})^{-1}$$

where c_{11} or $c_{22} = 4\pi\epsilon ab \sinh \alpha \sum_{n=1}^{\infty} [(b \text{ or } a) \sinh n\alpha + (a \text{ or } b) \sinh (n - 1)\alpha]$.

$$c_{12} = -4\pi\epsilon abr^{-1} \sinh \alpha \sum_{n=1}^{\infty} \operatorname{csch} n\alpha \quad \text{and} \quad \cosh \alpha = \tfrac{1}{2}(r^2 - a^2 - b^2)a^{-1}b^{-1}$$

Two equal spheres of radius a with distance r between centers

$$C = 2\pi\epsilon a \sinh \beta \sum_{n=1}^{\infty} [\operatorname{csch} (2n - 1)\beta + \operatorname{csch} 2n\beta]$$

where $\cosh \beta = \tfrac{1}{2}ra^{-1}$.

Kirchhoff's formula for two identical plane parallel coaxial circular disks of thickness t and radius r with square edges and a distance d between adjacent faces

$$C \approx 8.855 \times 10^{-12} (\pi r^2 d^{-1} + r\{ -1 + \ln [16\pi rd^{-1}(1 + td^{-1})] + 4\pi td^{-1} \ln (1 + t^{-1}d) \})$$

Two identical oppositely charged plane parallel coaxial infinitely thin circular disks at a distance d apart

$$C \approx 8.855 \times 10^{-12} \{ \pi r^2 d^{-1} + r[\ln (16\pi rd^{-1}) - 1] \}$$

Two thin oppositely charged coaxial rings generated by rotating two coplanar circles of radius a about a line parallel to and at a distance b from the line of length c that joins their centers

$$C \approx 1.7480 \times 10^{-10} \left\{ \frac{1}{2b} \ln \frac{8b}{a} + \frac{1}{(4a^2 + c^2)^{\frac{1}{2}}} K[(1 + 4c^2b^{-2})^{-\frac{1}{2}}] \right\}^{-1}$$

Capacitance between Two Bodies, One Enclosing the Other

Concentric spheres of radii a and b, $a < b$ $C = 4\pi\epsilon ab(b - a)^{-1}$
Spheres of radii a and b with distance c between centers

$$C = 4\pi\epsilon ab \sinh \alpha \sum_{s=0}^{\infty} [b \sinh n\alpha - a \sinh (n - 1)\alpha]^{-1}$$

where $\cosh \alpha = \frac{1}{2}(a^2 + b^2 - c^2)(ab)^{-1}$.
Confocal ellipsoids with semiaxes $a > b > c$, $a' > b' > c'$, and $a > a'$

$$C = 4\pi\epsilon a'(a - a')^{-1}(a^2 - c^2)^{\frac{1}{2}} \{F[(a^2 - b^2)^{\frac{1}{2}}(a^2 - c^2)^{-\frac{1}{2}}, \sin^{-1} (1 - c^2a^{-2})^{\frac{1}{2}}]\}^{-1}$$

Small sphere of radius a midway between planes a distance $2c$ apart

$$C \approx 1.1128 \times 10^{-10} \left(\frac{1}{a} - \frac{1}{c} \ln 2 \right)^{-1}$$

Sphere of radius b on axis of infinite cylinder of radius a

$$C = b[1.11285 - 0.9277r - 0.114r^2 - 0.1955r^3 + 1.8858r(1 - r)^{-0.5463}] \times 10^{-10}$$

where $r = b/a$.   The error is less than 1 part in 4,000 for $0 \lessgtr r \lessgtr 0.95$.

Two-dimensional Formulas for Capacitance per Meter Length

Let $U + jV = f(x + jy)$; then if V_1 and V_2 form two closed curves in the xy plane such that all U lines originate inside one and terminate inside the other and are continuous in the intermediate regions, V_1 and V_2 are sections of two cylindrical conductors and the capacitance per meter between them is

$$C_1 = \epsilon[U]|V_2 - V_1|^{-1}$$

where $[U]$ is the increment in U in passing once around V_1 or V_2 in the positive direction.
Two circular cylinders of radii a and b with a distance c between centers

$$C_1 = 2\pi\epsilon \left(\cosh^{-1} \frac{|c^2 - a^2 - b^2|}{2ab} \right)^{-1}$$

One cylinder may enclose the other or they may be mutually external.
Cylinder of radius a and plane at a distance c from its center

$$C_1 = 2\pi\epsilon[\cosh^{-1} (ca^{-1})]^{-1}$$

Coaxial circular cylinders of radii a and b, $b > a$, $C_1 = 2\pi\epsilon[\ln (a^{-1}b)]^{-1}$.
Confocal elliptic cylinders semiaxes a, b and a', b', $b > a$, $b' > a'$, $a > a'$

$$C_1 = 2\pi\epsilon[\tanh^{-1} (b^{-1}a) - \tanh^{-1} (b'^{-1}a')]^{-1}$$

Rectangular prism of n width, a sides, inside coaxial circular cylinder of radius b.   If $b \gg a$, $C \approx 2\pi\epsilon[\ln (a^{-1}bN)]^{-1}$, where $N = 2\pi n^{-1}\Gamma(1 + 2n^{-1})[\Gamma(1 + n^{-1})]^{-2}$.

The capacitance per unit length of conductor systems 1 to 12 given below is

$$C_1 = A \epsilon K(k) \{K[(1 - k^2)^{\frac{1}{2}}]\}^{-1}$$

where A is 2 or 4 as indicated, and $K(k)$ is a complete elliptic integral of modulus k. This is given below in terms of the arrangement of straight lines or circular arcs or both that are formed by taking a normal cross section of the two-dimensional conductor system. Any of these configurations, if used as a transmission line and perfectly conducting, has the characteristic high-frequency impedance $\eta \epsilon C_1^{-1}$ (see Table 5b-1).

1. Two collinear lines of lengths a and b with a gap c between. Also valid if $b = \infty$.

$$A = 2 \qquad k = (ab)^{\frac{1}{2}}(a + c)^{-\frac{1}{2}}(b + c)^{-\frac{1}{2}}$$

2. A circle of radius R whose center lies on an interior line of length a, or in a gap of width a between two external collinear semi-infinite lines at a distance c from the near one. Valid for an infinite line normal to a semi-infinite one if $R = \infty$.

$$A = 4 \qquad k = aR(aR - c^2 + c|2R - a|)^{-1}$$

3. A radial line of length a at a distance c from a circle of radius R inside $(-)$ with $a + c < R$, or outside $(+)$ with $a \lessgtr \infty$.

$$A = 4 \qquad k = aR[R(2c + a) \pm c(a + R)]^{-1}$$

4. A vertical line of width $2a$ bisected by a gap of width $2b$ in an infinite horizontal line whose near-end distance is c. Set $b = \infty$ to remove half of horizontal line.

$$A = 2 \qquad k^2 = \frac{2a\{(c^2 + a^2)^{\frac{1}{2}} + [(b - c)^2 + a^2]^{\frac{1}{2}}\}}{[(c^2 + a^2)^{\frac{1}{2}} + a]\{[(b - c)^2 + a^2]^{\frac{1}{2}} + a\}}$$

5. A vertical line of width $2a$ between two horizontal infinite lines a distance b apart with its center a distance c from the nearer horizontal line.

$$A = 2 \qquad k = 2\left(\sin\frac{\pi a}{b} \sin\frac{\pi c}{b}\right)^{\frac{1}{2}} \left(\sin\frac{\pi a}{b} + \sin\frac{\pi c}{b}\right)^{-1}$$

6. A line of length b on the x axis and a line of length $2a$ on the y axis centered at the origin. The gap between lines is c. Valid also for $b = \infty$.

$$A = 2 \qquad k^2 = \frac{2a\{[(c + b)^2 + a^2]^{\frac{1}{2}} - (c^2 + a^2)^{\frac{1}{2}}\}}{\{[(c + b)^2 + a^2]^{\frac{1}{2}} - a\}[(c^2 + a^2)^{\frac{1}{2}} + a]}$$

7. A line of length $2a$ which lies on a diameter of length d that bisects an opening of width s in a circle. From line center to circle is c. Use system 6 but with

$$\frac{d(c + a)}{d - c - a} \text{ for } c + b \qquad \frac{d(c - a)}{d - c + a} \text{ for } c \qquad \frac{d[d^2 + (d^2 - s^2)^{\frac{1}{2}}]}{s} \text{ for } a$$

8. A line of length a whose near end is a distance c from the point where the ends of two semi-infinite lines meet at an angle 2α and which lies on a bisector of this angle. Here α lies between 0 and π.

$$A = 2 \qquad k^2 = 1 - c^{\pi/\alpha}(c + a)^{-\pi/\alpha}$$

9. A line lying on the x axis with gaps b between its edges and the points $x = a$ and $x = -a$ at both of which the ends of two semi-infinite lines, only one to a quadrant, meet the x axis at an obtuse angle α. Here A is 4, and k is found from

$$bB\left(\frac{\alpha}{\pi}, \frac{3}{2}\right) = aB_{(1-k^2)^{\frac{1}{2}}}\left(\frac{\alpha}{\pi}, \frac{3}{2}\right)$$

where $B(m,n)$ and $B_x(m,n)$ are complete and incomplete beta functions.

10. Two lines of length $2a$, normal to the x axis, which bisects them, and a distance c apart. Here A is 4, and k is given implicitly by the equations[1]

$$E(k)F(\phi,k) - K(k)E(\phi,k) = \frac{\pi a}{c} \qquad \sin^2 \phi = \frac{K(k) - E(k)}{k^2 K(k)}$$

11. A line of length $2a$ midway between and parallel to two infinite lines $2b$ apart

$$A = 2 \qquad k = (1 - e^{2\pi a/b})^{\frac{1}{2}}$$

12. A $2c$ by $2b$ rectangle midway between two infinite parallel lines at a distance $2a$ apart to which the $2b$ sides are parallel. First determine the modulus h of the complete elliptic integral $K(h)$ so that it satisfies

$$(a - c)K(h) = bK[(1 - h^2)^{\frac{1}{2}}]$$

Now write N for $\pi b[2ah^2 K(h)]^{-1}$ to obtain

$$A = 4 \qquad k^2 = \frac{1}{2}\{1 + h^2 N^2 - [1 + 2N^2(h^2 - 2) + N^4]^{\frac{1}{2}}\}$$

Approximate formula for system 10 above is

$$C_1 \approx \epsilon b^{-1} a\{1 + b(\pi a)^{-1}[1 + \ln (2\pi b^{-1} a)]\}$$

Circular cylinder of radius a midway between earthed parallel plates at a distance $2b$ apart, $C_1 \approx 4\epsilon K(\sin \theta)[K(\cos \theta)]^{-1}$, where $\sin \theta = \tanh [\pi a\theta(2b\theta - \pi a)^{-1}]$. This is an upper limit which is about 0.1 percent above the true value when $a = \frac{1}{2}b$, and approaches the true value as a/b diminishes.

Square coaxial line with faces of inner square section of width $2a$ parallel to faces of outer square section of width $2b$.

$$C_1 = 2\epsilon \frac{K[(k_1{}^2 - k_2{}^2)^{\frac{1}{2}}k_1{}^{-1}(1 - k_2{}^2)^{-\frac{1}{2}}]}{K[k_2(1 - k_1{}^2)^{\frac{1}{2}}k_1{}^{-1}(1 - k_2{}^2)^{-\frac{1}{2}}]}$$

where k_1 and k_2 are found from

$$\frac{K(k_1)}{K[(1 - k_1{}^2)^{\frac{1}{2}}]} = \frac{K[(1 - k_2{}^2)^{\frac{1}{2}}]}{K(k_2)} = \frac{b + a}{b - a}$$

Small wire of radius a parallel to and at a distance c from the nearer of two parallel earthed plates at a distance b apart. $a \ll c$.

$$C_1 \approx 2\pi\epsilon \left[\ln \left(\frac{2b}{\pi a} \sin \frac{\pi c}{b} \right) \right]^{-1}$$

Capacitance Edge Corrections. Consider a thin, charged semi-infinite plate with straight edge parallel to and halfway between two infinite conducting plates at potential zero spaced a distance b apart. Increased capacitance per unit length of edge due to bulging of field is equivalent to adding strip of width $\pi^{-1} \ln 2$ to the edge and assuming no bulging. Same as above but infinite plates a distance $2B$ apart and central plates of thickness $2A$ with square edge. Increased capacitance per unit length due to bulging of field is equivalent to adding to central plate a strip of thickness $2A$ and width

$$\frac{2}{\pi}\left\{ B \ln \frac{2B - A}{B - A} - A \ln \frac{[A(2B - A)]^{\frac{1}{2}}}{B - A} \right\}$$

and assuming no bulging or charge on edge.

[1] For other two-strip configurations see A. E. H. Love, *Proc. London Math. Soc.* **22**, 339–369 (1923).

Parallel-plate capacitor with rectangular step in one plate, spacing on one side of step a and on other b. $b > a$. Additional capacitance per unit length of step above that from assumption of uniform field on each side of step is

$$2\pi\epsilon \left(\frac{a^2 + b^2}{ab} \ln \frac{b + a}{b - a} + 2 \ln \frac{b^2 - a^2}{4ab} \right)^{-1}$$

Two infinite sheets, each of which has one half bent at right angles to the other, are placed with the edges of the bends parallel so that the distance between sheets on one side of the bend is a and on the other b. The additional capacitance per unit length of bend over that given by the assumption of a uniform field over each a half of the inner sheet and no field in the corner rectangle is

$$\frac{2\epsilon}{\pi} \left(\ln \frac{a^2 + b^2}{4ab} + \frac{a}{b} \tan^{-1} \frac{b}{a} + \frac{b}{a} \tan^{-1} \frac{a}{b} \right)$$

Capacitance and Elastance Coefficients

In a system of n conductors the charge on conductor m is

$$Q_m = c_{1m}V_1 + c_{2m}V_2 + \cdots + c_{mm}V_m + \cdots + c_{nm}V_n$$

In a system of n conductors the potential of conductor m is

$$V_m = s_{1m}Q_1 + s_{2m}Q_2 + \cdots + s_{mm}Q_m + \cdots + s_{nm}Q_n$$

The force or torque tending to increase distance or angle x is

$$-\frac{1}{2} \sum_{p=1}^{n} \sum_{q=1}^{n} \frac{\partial c_{pq}}{\partial x} Q_p Q_q = +\frac{1}{2} \sum_{p=1}^{n} \sum_{q=1}^{n} \frac{\partial s_{pq}}{\partial x} V_p V_q$$

The energy of a system of n conductors is

$$W = \frac{1}{2} \sum_{p=1}^{n} \sum_{q=1}^{n} c_{pq}V_p V_q = \frac{1}{2} \sum_{p=1}^{n} \sum_{q=1}^{n} s_{pq}Q_p Q_q$$

For two distant conductors, $s_{pq} = s_{qp} \approx (4\pi\epsilon r)^{-1}$. If conductor 2 encloses conductor 1 only, then, $c_{11} = -c_{12}$ and $s_{1r} = s_{2r}$, where $1 < r$. For two spheres of radii a_1 and a_2 with centers a distance c apart, far from all other bodies

$$c_{11} = 4\pi\epsilon a_1 a_2 \sinh \alpha \sum_{n=1}^{\infty} [a_2 \sinh n\alpha \pm a_1 \sinh (n - 1)\alpha]^{-1}$$

where $\cosh \alpha = \frac{1}{2}|c^2 - a^2 - b^2|a^{-1}b^{-1}$ and the upper sign is used unless a_2 encloses a_1. If spheres are mutually external

$$c_{12} = -4\pi\epsilon a_1 a_2 c^{-1} \sinh \alpha \sum_{n=1}^{\infty} \operatorname{csch} n\alpha$$

If the capacitances to earth of two distant bodies when alone are C_1 and C_2, the capacitance coefficients are approximately

$$c_{11} \approx \frac{16\pi^2\epsilon^2 r^2 C_1}{16\pi^2\epsilon^2 r^2 - C_1 C_2} \qquad c_{12} = c_{21} \approx -\frac{C_1 C_2}{4\pi\epsilon r} \qquad c_{22} \approx \frac{16\pi^2\epsilon^2 r^2 C_2}{16\pi^2\epsilon^2 r^2 - C_1 C_2}$$

5b-2. Electrostatic-force Formulas. The force in the direction of the unit vector **m** on a conductor with surface charge density σ in a dielectric of capacitivity ϵ is

$$F_m = \tfrac{1}{2}\epsilon^{-1} \int_S \sigma^2 \mathbf{m} \cdot \mathbf{n} \, dS$$

where **n** is a unit vector normal to the surface.

When a uniform isotropic dielectric body of capacitivity ϵ occupies the volume v, where, before its advent, the field due to a fixed distribution of charge was **E** and after its advent **E'**, its energy is

$$W = \tfrac{1}{2} \int_v (\epsilon_v - \epsilon) \mathbf{E} \cdot \mathbf{E'} \, dv$$

The force or torque tending to increase the distance or angle x of the above body is

$$F_x = -\frac{\partial W}{\partial x}$$

The torque tending to increase the angle α which the normal to a disk of radius a makes with a field that would be uniform and of strength E except for the disk is

$$T = \frac{8}{3} \epsilon a^3 E^2 \sin 2\alpha$$

The torque tending to increase the angle α between the field and the major axis of an oblate dielectric spheroid of capacitivity ϵ with semiaxes a and b, where $b > a$, placed in a field that would be uniform and of strength E except for the spheroid is

$$T = \frac{2\pi\epsilon_v(K - 1)^2 b^2 a E^2 (3P - 2) \sin 2\alpha}{3[(K - 1)^2 P^2 + (K - 1)(2 - K)P - 2K]}$$

where $P = A[(1 + A^2) \cot^{-1} A - A]$, $A = a(b^2 - a^2)^{-\frac{1}{2}}$, and $K = \epsilon\epsilon_v^{-1}$.
If the above oblate spheroid is conducting, the torque is

$$T = \frac{2\pi\epsilon_v b^2 a E^2 (3P - 2) \sin 2\alpha}{3P(P - 1)}$$

The torque tending to increase the angle α between the field and the major axis of a prolate dielectric spheroid of capacitivity ϵ with semiaxes a and b where $b < a$ placed in a field that would be uniform and of strength E except for the spheroid is

$$T = \frac{2\pi\epsilon_v(K - 1)^2 b^2 a E^2 (2 - 3Q) \sin 2\alpha}{3[(K - 1)^2 Q^2 + (K - 1)(2 - K)Q - 2K]}$$

where $Q = C[(1 - C^2) \coth^{-1} C + C]$. $C = a(a^2 - b^2)^{-\frac{1}{2}}$ and $K = \epsilon\epsilon_v^{-1}$.

If the above prolate spheroid is conducting, the torque becomes[1]

$$T = \frac{2\pi\epsilon_v b^2 a E^2 (2 - 3Q) \sin 2\alpha}{3Q(Q - 1)}$$

The axis of rotational symmetry of a right circular solid conducting cylinder of radius a and length $2b$ makes an angle θ with a field that would be uniform and of strength E except for the cylinder. The torque tending to align the axis with the field is

$$T = \pi\epsilon a^2 b E^2 \sin 2\theta(\alpha_1 - \alpha_t)$$

---

[1] For torque on general ellipsoid, see Stratton, "Electromagnetic Theory," p. 215, McGraw-Hill Book Company, New York, 1941.

where
$$\alpha_1 = 1 + 2.1444 \left(\frac{b}{a}\right)^{0.828} + 0.7171 \left(\frac{b}{a}\right)^{1.6752}$$

$$\alpha_t = 2 + 0.84883 \left(\frac{a}{b}\right) + 0.369 \left(\frac{a}{b}\right)^{0.548} \tanh^{0.5}\left(\frac{a}{b}\right)^{0.712}$$

The torque vanishes at $(a/b) = 1.1958$. The errors in these formulas are less than 1 part in 4,000 for $0.25 \lessgtr (a/b) \lessgtr \infty$.

Two parallel cylinders of radii a and b carry charges $+Q$ and $-Q$, and their axes are a distance c apart. The force per unit length tending to increase c is

$$F_1 = \frac{\pm Q^2 c}{2\pi\epsilon[(c^2 - a^2 - b^2)^2 - 4a^2 b^2]^{\frac{1}{2}}}$$

The plus sign is used if one cylinder encloses the other and the minus sign if they are mutually external.

A complete elliptic integral formula given earlier (page 5-16) for capacitance per unit length used a different modulus in each of the cases 1 to 12 involving a distance c. The force tending to decrease c when the system charges are $+Q$ and $-Q$ is

$$F = \frac{\pi A Q^2}{4k(1 - k^2)K[(1 - k^2)^{\frac{1}{2}}]} \left|\frac{\partial k}{\partial c}\right|$$

Two identical infinite coplanar parallel conducting strips carry equal positive charges Q, the distance between their near edges being $2a$ and between their far edges $2b$. The repulsive force per unit length between them is

$$F_1 = \frac{Q^2}{2\pi\epsilon(a + b)}$$

The force on a point charge at a distance b from the center of a sphere of radius a at zero potential is

$$F = \frac{abQ^2}{4\pi\epsilon(a^2 - b^2)^2}$$

When $b > a$, the force is toward the center; and when $b < a$, it is away from the center. The repulsive force between a point charge q at a distance b from the center of a sphere of radius a carrying a total charge Q is, when $b > a$,

$$F = \frac{q}{4\pi\epsilon b^2}\left[Q + \frac{a^3(a^2 - 2b^2)q}{b(b^2 - a^2)^2}\right]$$

At the point x_0, y_0, z_0 inside a rectangular conducting box bounded by the planes $x = 0$, a, $y = 0$, b, $z = 0$, c, the image force on a charge Q is

$$F_z = -\frac{2Q^2}{\epsilon ab}\sum_{n=1}^{\infty}\sum_{m=1}^{\infty}\frac{\sinh A_{mn}(c - 2z_0)}{\sinh A_{mn}c}\sin^2\frac{n\pi x_0}{a}\sin^2\frac{m\pi y_0}{b}$$

in the z direction, where $A_{mn} = \pi(ab)^{-1}(m^2 a^2 + n^2 b^2)^{\frac{1}{2}}$. The other force components are given by cyclic permutation of the symbols x, y, z; a, b, c; and x_0, y_0, z_0. At a distance c from one of two parallel uncharged plates at a distance b apart, the image force on a charge Q is

$$F = \frac{Q^2}{16\pi\epsilon a^2}\left[\varsigma\left(2, \frac{1}{2} - \frac{c}{b}\right) - \varsigma\left(2, \frac{c}{b}\right)\right]$$

where $\varsigma(z,a)$ is a Riemann zeta function.

On the axis and at a distance b from the center of a conducting disk of radius a carrying a charge Q, the repulsive force on a point charge q is

$$F = \frac{q}{4\pi\epsilon(a^2 + b^2)}\left[Q - \frac{a(3b^2 + a^2)q}{2\pi b(a^2 + b^2)} + \frac{3b^2 - a^2}{2\pi b^2}q\tan^{-1}\frac{a}{b}\right]$$

At a distance c from the center of an uncharged dielectric sphere of radius a and relative capacitivity K, the attractive force on a charge Q is

$$F = \frac{(K-1)Q^2}{4\pi\epsilon_v c^2}\sum_{n=1}^{\infty}\frac{n(n+1)}{Kn+n+1}\left(\frac{a}{c}\right)^{2n+1}$$

At a distance c from the plane face of an infinite block of dielectric of relative capacitivity K, the attractive force on a point charge Q is

$$F = \frac{Q^2}{16\pi\epsilon_v c^2}\frac{K-1}{K+1}$$

The attractive force on a point charge Q at a distance a from the plane face of a dielectric slab of thickness c and relative capacitivity K is

$$F = \frac{\beta Q^2}{16\pi\epsilon_v}\left[\frac{1}{a^2} - (1 - \beta^2)\sum_{n=1}^{\infty}\frac{\beta^{2(n-1)}}{(a+nc)^2}\right]$$

where $\beta = (K-1)(K+1)^{-1}$.

The attractive force per unit length on a line charge of strength λ per unit length parallel to and at a distance c from the axis of an uncharged circular cylinder of radius a and relative capacitivity K is

$$F_1 = \frac{K-1}{K+1}\frac{\lambda^2 a^2}{2\pi\epsilon_v c(c^2 - a^2)}$$

For a conductor, $K = \infty$; so the first factor is unity.

The force toward the wall per unit length on a line charge of strength λ per unit length parallel to and at a distance c from the axis of a circular cylindrical hole of radius a in an infinite block of dielectric of relative capacitivity K is

$$F_1 = \frac{K-1}{K+1}\frac{c\lambda^2}{2\pi\epsilon_v(a^2 - c^2)}$$

For a conductor, $K = \infty$; so the first factor is unity.

The attractive force per unit length on a line charge of strength λ per unit length parallel to and at a distance a from the nearer face of a dielectric slab of thickness c and relative capacitivity K is

$$F_1 = \frac{\beta\lambda^2}{4\pi\epsilon_v}\left[\frac{1}{a} - (1 - \beta^2)\sum_{n=1}^{\infty}\frac{\beta^{2(n-1)}}{a+nc}\right]$$

where $\beta = (K-1)(K+1)^{-1}$.

In the foregoing case, if $a = mc$ where m is an integer, the force per unit length is expressible in finite terms; thus

$$F_1 = \frac{\beta\lambda^2}{4\pi\epsilon_v c}\left\{\frac{1}{m} - \frac{1-\beta^2}{\beta^{2(m+1)}}\left[\ln(1-\beta^2) + \sum_{n=1}^{m}\frac{\beta^{2n}}{n}\right]\right\}$$

The attractive force per unit length on a line charge of strength λ per unit length parallel to and at a distance a from an uncharged conducting plane is

$$F_1 = \frac{\lambda^2}{4\pi\epsilon a}$$

The attractive force between a line charge of strength λ per unit length and an uncharged conducting sphere of radius a whose center is at a distance b from it is

$$F = \frac{\lambda^2 a^2}{\pi\epsilon_v b(b^2 - a^2)^{\frac{1}{2}}} \sin^{-1}\frac{a}{b}$$

The attractive force between a line charge of strength λ per unit length and an uncharged dielectric sphere of relative capacitivity K and radius a is

$$F = \frac{(K-1)\lambda^2}{\pi\epsilon_v} \sum_{n=1}^{\infty} \frac{n(2n-2)!!}{(2n-1)!!(Kn+n+1)} \left(\frac{a}{b}\right)^{2n+1}$$

5b-3. Multipole Formulas. The potential of a point charge Q is

$$V = \frac{Q}{4\pi\epsilon r}$$

where r is the distance from the charge to the field point.
The force on a point charge in a field of electric intensity $\mathbf{E}$ is

$$\mathbf{F} = Q\mathbf{E}$$

The potential of a dipole of moment $\mathbf{p}$ is

$$V = \frac{p\cos\theta}{4\pi\epsilon r^2} = \frac{\mathbf{p}\cdot\mathbf{r}}{4\pi\epsilon r^3}$$

where $\mathbf{r}$ is measured from the dipole to the field point.
The force on a dipole in a field $\mathbf{E}$ is $\mathbf{F} = (\mathbf{p}\cdot\nabla)\mathbf{E}$.
The torque on a dipole in a field $\mathbf{E}$ is $\mathbf{T} = \mathbf{p}\times\mathbf{E}$.
The mutual energy of two dipoles of moment $\mathbf{p}_1$, $\mathbf{p}_2$ which make angles θ_1 and θ_2 with the vector $\mathbf{r}$ that joins them and whose planes intersect along $\mathbf{r}$ at an angle ψ is

$$W = \frac{p_1 p_2}{4\pi\epsilon r^3}(\sin\theta_1\sin\theta_2\cos\psi - 2\cos\theta_1\cos\theta_2)$$

The components of force and torque between two dipoles are

$$F_r = -\frac{\partial W}{\partial r} \qquad T_\alpha = -\frac{\partial W}{\partial\alpha}$$

The potential of a multipole of the nth order and moment strength $p^{(n)}$ is

$$V_n = \frac{(-1)^n p^{(n)}}{4\pi\epsilon n!}\frac{\partial^n}{\partial l_1 \cdots \partial l_n}\left(\frac{1}{r}\right)$$

$$= \sum_{m=0}^{n}(a_{nm}\cos m\varphi + b_{nm}\sin m\varphi)r^{-n-1}P_n^m(\cos\theta)$$

5b-4. Dielectric-boundary Formulas. If V' and V'' are the electrostatic potentials in the dielectrics ϵ' and ϵ'', then at their uncharged interface

$$V' = V'' \qquad \text{and} \qquad \epsilon' \frac{\partial V'}{\partial n} = \epsilon'' \frac{\partial V''}{\partial n}$$

where n is a coordinate normal to the interface.
The normal stress, directed from ϵ'' to ϵ', on the above interface is

$$F_n = \frac{\epsilon'' - \epsilon'}{2\epsilon'} \left(\frac{D_t'^2}{\epsilon'} + \frac{D_n'^2}{\epsilon''} \right)$$

where D_t' and D_n' are the tangential and normal components of the displacement in ϵ'.

5b-5. Dielectric Bodies in Electrostatic Fields. A sphere of radius a and capacitivity ϵ is placed in a uniform field of intensity **E**. The uniform field intensity inside and the potential outside due to its polarization are, respectively,

$$\mathbf{E}_i = \frac{3\epsilon_v \mathbf{E}}{\epsilon + 2\epsilon_v} \qquad V_p = E \frac{\epsilon_v - \epsilon}{\epsilon + 2\epsilon_v} \frac{a^3}{r^2} \cos \theta$$

where r is measured from the center of the sphere and **E** is directed along $\theta = 0$.
An oblate dielectric spheroid of capacitivity ϵ whose minor (rotational) axis on $\theta = 0$ is $2a$ and whose focal circle is of radius c is placed in a uniform electric field **E** parallel to $\theta = 0$. The uniform field inside and the potential outside due to its polarization are, respectively,

$$\mathbf{E}_i = \mathbf{E} \, \epsilon_v c^3 M \qquad V_p = M(\epsilon - \epsilon_v)a(a^2 + c^2)E(\cot^{-1} \zeta - \zeta^{-1})r \cos \theta$$

where
$$M = \{a(\epsilon_v - \epsilon)[(a^2 + c^2) \cot^{-1} (c^{-1}a) - ac] + \epsilon c^3\}^{-1}$$
and
$$\zeta^2 = \tfrac{1}{2}c^{-2}\{r^2 - c^2 + [(r^2 - c^2)^2 + 4r^2c^2 \cos^2 \theta]^{\frac{1}{2}}\}$$

The above spheroid is placed in a field **E'** in the $\varphi = 0$ direction, normal to the $\theta = 0$ axis. The uniform field intensity inside and the potential outside due to its polarization are, respectively,

$$\mathbf{E}_i' = 2\mathbf{E}'\epsilon_v c^3 M' \qquad V_p' = M'(\epsilon - \epsilon_v)a(c^2 + a^2)E'[\cot^{-1} \zeta - \zeta(1 + \zeta^2)^{-1}]r \sin \theta \cos \varphi$$

where
$$M' = \{a(\epsilon - \epsilon_v)[(a^2 + c^2) \cot^{-1} (c^{-1}a) - ac] + 2\epsilon_v c^3\}^{-1}$$

The above spheroid is placed in a uniform field $\mathbf{E}_0$ which makes an angle α with its rotational $\theta = 0$ axis. The uniform field inside and the potential outside due to its polarization are, respectively,

$$\mathbf{E}_{0i} = \mathbf{E}_0 \epsilon_v c^3 [M \cos \alpha + M' \sin \alpha] \qquad V_{0p} = E_0[V_p E^{-1} \cos \alpha + V_p' E'^{-1} \sin \alpha]$$

where V_p', V_p, **E**, and **E'** are given in the preceding formulas.
A prolate spheroid of capacitivity ϵ whose major (rotational) axis on $\theta = 0$ is $2b$ and whose focal distance is $2c$ is placed in a uniform electric field **E** parallel to $\theta = 0$. The uniform field intensity inside and the potential outside due to its polarization are, respectively,

$$\mathbf{E}_i = \mathbf{E}\epsilon_v c^3 N \qquad V_p = N(\epsilon - \epsilon_v)b(c^2 - b^2)E(\coth^{-1} \eta - \eta^{-1})r \cos \theta$$

where
$$N = \{b(\epsilon_v - \epsilon)[(c^2 - b^2) \coth^{-1} (c^{-1}b) + bc] + \epsilon c^3\}^{-1}$$
and
$$\eta^2 = \tfrac{1}{2}c^{-2}\{r^2 + c^2 + [(r^2 + c^2)^2 - 4c^2r^2 \cos^2 \theta]^{\frac{1}{2}}\}$$

The above spheroid is placed in a field **E'** in the $\varphi = 0$ direction normal to the $\theta = 0$ axis. The uniform field inside and the potential outside due to its polarization are, respectively,

$$\mathbf{E}_i = \mathbf{E}'\epsilon_v c^3 N' \qquad V_p' = N'(\epsilon - \epsilon_v)b(b^2 - c^2)E'[\coth^{-1} \eta - \eta(1 - \eta^2)^{-1}]r \sin \theta \cos \varphi$$

where $N' = \{b(\epsilon_v - \epsilon)[(b^2 - c^2) \coth^{-1} c^{-1}b - bc] + 2\epsilon_v c^3\}^{-1}$.

The above prolate spheroid is placed in a uniform field $\mathbf{E}_0$ which makes an angle α with its rotational $\theta = 0$ axis. The uniform field inside and the potential outside due to its polarization are, respectively,

$$\mathbf{E}_{0i} = \mathbf{E}_0 \epsilon_v c^3 [N \cos \alpha + N' \sin \alpha] \qquad V_{0p} = E_0 [V_p E^{-1} \cos \alpha + V_p' E'^{-1} \sin \alpha]$$

where V_p, V_p', $\mathbf{E}$, and $\mathbf{E}'$ are given in the foregoing formulas.

5b-6. Static-current-flow Formulas. *Linear-circuit Formulas.* See steady-state alternating-current formulas.

Currents in Extended Media (Three Dimensions). The following formulas assume the medium to be uniform, homogeneous, and isotropic and to have a resistivity ρ which obeys Ohm's law.

The resistance between a single perfectly conducting electrode immersed in an infinite medium and the concentric infinite sphere is related to the capacitance of the same electrode by the formula

$$R = \rho \epsilon_v C^{-1}$$

where the capacitance C for a sphere, prolate or oblate spheroid, ellipsoid, circular disk, elliptic disk, two spheres in contact, two spheres connected by a wire, two spheres intersecting at an angle π/m, a spherical bowl, torus, cube, and circular plane annulus is given in the electrostatic section. The resistance between widely separated source and sink electrodes immersed in an infinite medium is

$$R_{12} \approx R_1 + R_2 - \rho(2\pi r)^{-1}$$

where R_1 and R_2 are the resistances to infinity of each alone and r, the distance between them, is large compared with their dimensions. The resistance to infinity of a single electrode, sunk into the plane surface of a semi-infinite medium such as the earth in such a way that the submerged part, if combined with its mirror image in the surface, would form one of the above electrodes, is

$$R = 2\rho \epsilon_v C^{-1}$$

When both source and sink electrodes are half submerged in the plane face just described, the resistance between them is

$$R = R_{12} \approx 2[R_1 + R_2 - \rho(2\pi r)^{-1}]$$

where R_{12}, R_1, and R_2 have the same significance as before and r, the distance between them, is much larger than the electrode dimensions. In the preceding case, if the medium has a resistivity ρ_1 to a depth a and ρ_2 below this depth, then the resistance between electrodes is

$$R \approx 2 \left\{ R_1 + R_2 - \frac{\rho_1}{2\pi r} + \frac{\rho_1}{\pi} \sum_{n=1}^{\infty} \left[\frac{(-\beta)^n}{2na} - \frac{(-\beta)^n}{(4n^2a^2 + r^2)^{\frac{1}{2}}} \right] \right\}$$

where $\beta = (\rho_1 - \rho_2)(\rho_1 + \rho_2)^{-1}$ and both a and r are large compared with the electrode dimensions.

Two perfectly conducting disk electrodes of radii a and b are applied to the plane horizontal face of a semi-infinite homogeneous medium whose horizontal and vertical resistivities are ρ_1 and ρ_2. If the electrode spacing r is much greater than a and b, the resistance between them is

$$R \approx (\rho_1\rho_2)^{\frac{1}{2}}[(4a)^{-1} + (4b)^{-1} - (\pi r)^{-1}]$$

Two conical perfectly conducting electrodes of half angle β with an angle α between their axes pass normally through a spherical shell of thickness b and resistivity ρ.

The resistance between them is rigorously

$$R = \rho(\pi b)^{-1} \cosh^{-1} (\csc \beta \sin \tfrac{1}{2}\alpha)$$

A cylindrical column of length l and radius a of material of resistivity ρ connects normally the plane faces of two semi-infinite masses of the same resistivity. The resistance R between the infinite hemispherical perfectly conducting electrodes bounding the masses lies within the limits

$$\frac{\rho l}{\pi a^2} + \frac{\rho}{2a} < R < \frac{\pi\rho}{2[\pi a - l \ln (1 + \pi a/l)]}$$

This formula is most accurate for small values of l/a and is exact at $l = 0$. For large values of l

$$R \approx \rho a^{-1}(0.31831 l a^{-1} + 0.522)$$

Perfectly conducting disk electrodes of radius b are applied concentrically to the ends of a solid right circular cylinder of radius a, length $2c$, and resistivity ρ. The resistance between them is

$$R \approx 2\rho(\pi a^2)^{-1}[c + f(b)]$$

where $f(0.25a) = 2.05164a$, $f(0.50a) = 0.5336a$, and $f(0.75a) = 0.1060a$. The errors are less than 0.05 per cent if c is greater than $4a$.

Currents in Extended Media (Two Dimensions). The resistance between perfectly conducting plane electrodes covering the ends and orthogonal to the sides of a bar of rectangular section, resistivity ρ, and thickness b bent in a circular arc with inner radius a and outer radius c, which subtends an angle α at the center, is

$$R = \rho\alpha b^{-1}[\ln (a^{-1}c)]^{-1}$$

The resistance between two small cylindrical electrodes of radius r passing normally through a strip of width a, thickness b, and resistivity ρ at a distance $2c$ apart on a line midway between its edges is, if $r \ll a$ and $r \ll c$,

$$R \approx \frac{\rho}{\pi b} \ln \frac{a \sinh 2\pi a^{-1}c}{\pi r}$$

The resistance between the electrodes in the above strip when they are equidistant from its center on a line normal to its edges is

$$R \approx \frac{\rho}{\pi b} \ln \frac{2a \tan \pi a^{-1}c}{\pi r}$$

In the following six configurations the bars of resistivity ρ have rectangular cross sections and are of uniform thickness b. Perfectly conducting electrodes cover the ends which are at right angles to the sides. For 1 percent accuracy the interval between each end and the beginning of the boundary perturbation should exceed about twice the width of the intervening straight bar.

A bar of width a has an infinitely narrow cut of depth c normal to one side. The additional resistance due to the cut is

$$\Delta R = -4\rho(\pi b)^{-1} \ln \cos \tfrac{1}{2}\pi a^{-1}c$$

One side of a bar is straight and the other has a rectangular step in it. The width on one side of the step is a and on the other c where $a > c$. The additional resistance due to the distortion of the flow near the step over the sum of the resistances of the two straight portions alone is

$$\Delta R = \frac{\rho}{\pi b} \left(\frac{a^2 + c^2}{ac} \ln \frac{a + c}{a - c} + 2 \ln \frac{a^2 - c^2}{4ac} \right)$$

In the preceding case the corner of the step is cut off at 45 deg so that the width increases linearly from c to a. The additional resistance due to the tapered section over that of the two straight portions alone is

$$\Delta R = \frac{2\rho}{\pi b}\left(\frac{a^2 + c^2}{ac}\tanh^{-1}\frac{c}{a} + \frac{a^2 - c^2}{ac}\tan^{-1}\frac{c}{a} + \ln\frac{a^4 - c^4}{8a^2c^2}\right)$$

A straight rectangular bar has a right-angle bend, the width on one side of the bend being a and on the other c. The increase of resistance over the sum of the resistances of the two straight portions alone, the corner rectangle common to both being excluded, is

$$\Delta R = \frac{2\rho}{\pi b}\ln\left(\frac{a^2 + c^2}{4ac} + \frac{a}{c}\tan^{-1}\frac{c}{a} + \frac{c}{a}\tan^{-1}\frac{a}{c}\right)$$

A straight rectangular bar of width a has a hole drilled through it equidistant from its edges. The increase in resistance due to the hole is less than

$$\Delta R \approx -2\rho c(ab\theta)^{-1}\ln\cos\theta$$

where θ is a parameter chosen so that $\sin\theta = \tanh[\pi c\theta(a\theta - \pi c)^{-1}]$.
These formulas are practically exact for small holes far from the ends. When the diameter of the hole is half the strip width R is about 0.1 per cent too large. For small values of c/a the parameter is given by

$$\theta \approx \frac{2\pi c}{a}\left(1 - \frac{\pi^2 c^2}{3a^2} + \frac{\pi^4 c^4}{3a^4}\right)$$

The value of ΔR given above is unchanged if the hole is replaced by two semicircular notches of the same radius in opposite edges of the strip.

Perfectly conducting electrodes are applied to a block of thickness b, width a, length c, and resistivity ρ in such a way as to cover the full thickness over a band of width w at the center of opposite ends. The resistance between the electrodes lies between the limits

$$\frac{2\rho}{\pi b}\cosh^{-1}\frac{\cosh\frac{1}{2}\pi a^{-1}c}{\sin\frac{1}{2}\pi a^{-1}w} > R > \frac{2}{\pi b}\sinh^{-1}\frac{\sinh\frac{1}{2}\pi a^{-1}c}{\sin\frac{1}{2}\pi a^{-1}w}$$

5b-7. Static-magnetic-field Formulas. *Magnetic Field of Various Circuit Configurations.* The magnetic induction due to a current density $\mathbf{i}$ flowing in a volume v is

$$\mathbf{B} = \frac{\mu}{4\pi}\nabla\times\int_v\frac{\mathbf{i}\,dv}{r}$$

The magnetic induction of a thin linear circuit with total current I is

$$B = \frac{\mu I}{4\pi}\oint\frac{\sin\theta\,ds}{r^2}$$

where θ is the angle between $\mathbf{ds}$ and $\mathbf{r}$ and B is normal to the plane of $\mathbf{ds}$ and $\mathbf{r}$.
The magnetic induction due to a long straight cylinder carrying current parallel to its axis, when both current density and permeability are independent of the azimuth angle θ, is $B_\theta = \mu_a I_a(2\pi a)^{-1}$ where a is distance of field point from axis, I_a is current inside radius a, and μ is the permeability at the field point.
The edges of a flat strip lie at $x = a$ and $x = -a$ and it carries a uniformly distributed current I in the z direction. The distances of a field point in the positive quadrant from the near and far edges are, respectively, r_1 and r_2 and the angle between r_1 and r_2

is α. The magnetic induction components are

$$B_y = \frac{\mu I}{4\pi a} \ln \frac{r_2}{r_1} \qquad B_x = -\frac{\mu I}{4\pi a}\, \alpha$$

A conductor of rectangular section of area A is bounded by the planes $x = a$, $x = -a$, $y = b$, and $y = -b$ and carries a uniformly distributed current I in the z direction. The distances from a field point in the positive quadrant to the corners, starting with the nearest and proceeding clockwise about the z axis, are r_1, r_2, r_3, and r_4. The angles between successive r's are α_1, α_2, α_3, and α_4, and the x and y components of r_1 and r_3 are x_1, y_1 and x_3, y_3. If all the above quantities are taken positive, the magnetic-induction components are

$$B_x = -\frac{1}{2}\,\mu I (\pi A)^{-1} \left(y_3\alpha_4 - y_1\alpha_1 + x_3 \ln \frac{r_3}{r_2} - x_1 \ln \frac{r_4}{r_1} \right)$$

$$B_y = \frac{1}{2}\,\mu I (\pi A)^{-1} \left(x_3\alpha_2 - x_1\alpha_3 + y_3 \ln \frac{r_3}{r_4} - y_1 \ln \frac{r_2}{r_1} \right)$$

The space inside and outside the conductor has the same permeability μ.

The magnetic induction outside the conductors of a long bifilar line that consists of a cylinder whose axis is $y = a$ which carries a uniformly distributed x-directed current I and another cylinder whose axis is $y = -a$ that carries the same current in the opposite direction is

$$B_y = \tfrac{1}{2}\pi^{-1}\mu Iz(r_2^{-2} - r_1^{-2}) \qquad B_z = -\tfrac{1}{2}\pi^{-1}\mu I[r_2^{-2}(y + a) - r_1^{-2}(y - a)]$$

where r_1 and r_2 are the distances from positive and negative wire axes, respectively, and μ is the permeability of the conductors and surrounding space.

The magnetic induction of bifilar lines composed of flat strips or rectangular bars can be found by taking the vector sum of the inductions already given for each conductor alone.

A long circular conducting cylinder of radius b has a longitudinal hole of radius a whose axis is displaced a distance c from the cylinder axis. If a longitudinal current I is uniformly distributed over the conducting area, the induction **B** in the hole is uniform and normal to c and its magnitude is

$$B = \mu c I[2\pi(b^2 - a^2)]^{-1}$$

A circular loop of wire lies at $z = 0$, $\rho = a$ and carries a current clockwise about the z axis. The magnetic-induction components are

$$B_z = A(I_1 - a^{-1}\rho I_2) \qquad B_\rho = Aa^{-1}zI_2$$

where[1] $I_1 = \pi^{-1} \int_0^\pi (1 - b \cos \theta)^{-\frac{3}{2}}\, d\theta$, $I_2 = \pi^{-1} \int_0^\pi (1 - b \cos \theta)^{\frac{3}{2}} \cos \theta\, d\theta$, $A = \frac{1}{2}\mu Ia^2(a^2 + z^2 + \rho^2)^{-\frac{3}{2}}$, and $b = 2a\rho(a^2 + z^2 + \rho^2)^{-1}$.

Two coaxial wire loops of radius a at a distance a apart carry currents I in the same direction and constitute a Helmholtz coil which gives a nearly uniform field on the axis midway between them. For a small distance r around this point the field varies as $(r/a)^4$. The induction there is

$$B = 8\mu I 5^{-\frac{3}{2}}a^{-1}$$

Accurate values of B may be found by a superposition of the fields calculated separately by the preceding formula for a single loop.

[1] Six-place tables of I_1 and I_2 suitable for linear interpolation are given by C. L. Bartberger, *J. Appl. Phys.* **21**, 1108 (1950).

The magnetic-induction components at a great distance from a small loop of wire at $\theta = \frac{1}{2}\pi$, $r = a$ which carries a current I are

$$B_r = \tfrac{1}{2}\mu I r^{-3}a^2 \cos\theta \qquad B_\theta = \tfrac{1}{4}\mu I r^{-3}a^2 \sin\theta$$

A rectangular loop of wire lies at $x = \pm a$, $y = \pm b$ and carries a current I clockwise about the z axis. The distances of the field point at x, y, z in the positive octant from successive corners, starting with the nearest, are r_1, r_2, r_3, and r_4 and the components of r_1 and r_3 are x_1, y_1, z and x_3, y_3, z. The components of the magnetic induction are

$$B_x = \tfrac{1}{4}\pi^{-1}\mu I z \{[r_1(r_1 - y_1)]^{-1} + [r_3(r_3 + y_3)]^{-1} - [r_4(r_4 + y_3)]^{-1} - [r_2(r_2 - y_1)]^{-1}\}$$
$$B_y = \tfrac{1}{4}\pi^{-1}\mu I z \{[r_3(r_3 + x_3)]^{-1} + [r_1(r_1 - x_1)]^{-1} - [r_4(r_4 - x_1)]^{-1} - [r_2(r_2 + x_3)]^{-1}\}$$
$$B_z = \tfrac{1}{4}\pi^{-1}\mu I \{x_1[r_1(r_1 - y_1)]^{-1} - x_1[r_4(r_4 + y_3)]^{-1} + x_3[r_2(r_2 - y_1)]^{-1} - x_3[r_3(r_3 + y_3)]^{-1}$$
$$+ y_1[r_1(r_1 - x_1)]^{-1} - y_1[r_2(r_2 + x_3)]^{-1} + y_3[r_4(r_4 - x_1)]^{-1} - y_3[r_3(r_3 + x_3)]^{-1}\}$$

All lengths are to be taken positive. If the single wire of the preceding formulas is replaced by N wires, the fields may be found rigorously by superimposing N solutions of the type given, one for each wire, or by integration over the section. In case the area of this section is small compared with other coil dimensions, a sufficiently accurate result is often given by substitution of NI for I in these formulas and the use of the dimensions of the center turn for that of the loop.

A helix of pitch α is wound on a cylinder of radius a. The angles between the positive axis and vectors drawn from the field point to the ends of the helix wire are β_1 and β_2. The axial component of the induction is then given rigorously by

$$B_a = \tfrac{1}{4}\mu I \cot\alpha(\pi a)^{-1}(\cos\beta_2 - \cos\beta_1)$$

There is also a component normal to the axis which becomes negligible when α is small. The axial component of the induction on the axis of a solenoid with n turns per unit length is, using the notation of the preceding formula,

$$B_a = \tfrac{1}{2}\mu n I(\cos\beta_2 - \cos\beta_1)$$

The induction approaches uniformity everywhere inside an infinitely long solenoid as the pitch decreases and its limiting value is $B_2 = n\mu I$.

When any figure, such as a torus, generated by the rotation of a closed curve about a coplanar external line, is closely and uniformly wound with N turns of wire so that each turn nearly coincides with one position of the generating curve, then, when carrying a current I, the exterior induction is zero and the interior induction is

$$B_\varphi = \tfrac{1}{2}\mu N I(\pi r)^{-1}$$

A coil of N circular turns wound closely over the entire surface of an oblate spheroid whose major and minor semiaxes are a and b will give a uniform induction B inside, provided that the projections of these turns on the b axis are uniformly spaced. The total number of ampere-turns needed is

$$NI = -\frac{2B}{\mu}\left[\frac{a^2 - b^2}{b - a^2(a^2 - b^2)^{-\frac{1}{2}}\cos^{-1}(b/a)}\right]$$

When $b = a$, this becomes $NI = 3bB/\mu$.

A coil of N circular turns wound closely over the entire surface of a prolate spheroid whose major and minor semiaxes are b and a will give a uniform induction B inside, provided that the projections of these turns on the b axis are uniformly spaced. The total number of ampere-turns needed is

$$NI = \frac{2B}{\mu}\left[\frac{b^2 - a^2}{b - a^2(b^2 - a^2)^{-\frac{1}{2}}\cosh^{-1}(b/a)}\right]$$

Self- and Mutual Inductance for Static Fields. The mutual inductance between two circuits is given by the formulas

$$M = L_{12} = 10^{-7} \oint_1 \oint_2 r^{-1}\, \mathbf{ds}_1 \cdot \mathbf{ds}_2 = \tfrac{1}{4}\pi^{-1}10^7 \int_v \mathbf{B}_1 \cdot \mathbf{B}_2\, dv$$

where $\mathbf{ds}_1$ and $\mathbf{ds}_2$ are elements of circuit 1 and circuit 2 and $\mathbf{B}_1$ and $\mathbf{B}_2$ are their separate magnetic inductions for unit current. One line integral covers each circuit and the volume integral covers the whole field region.

The self-inductance of a circuit is a special case of the above formula

$$L = \tfrac{1}{4}\pi^{-1}10^7 \int_v B^2\, dv$$

where B is the magnetic induction per unit current and v includes the entire field region.

The energy in the field of n circuits carrying currents $I_1, I_2, \ldots, I_n$ is

$$W = \tfrac{1}{2} \sum_{p=0}^{n} \sum_{q=0}^{n} L_{pq} I_p I_q$$

Note. In the following material there are many references to Grover. These refer to F. W. Grover, "Inductance Calculations," Dover Publications, Inc., New York, 1962. In this book most inductances are given in microhenrys and lengths in centimeters. In the following formulas mks units are used; so the inductances are in henrys and the lengths in meters. Unless otherwise stated, the permeability throughout is that of a vacuum.

The self-inductance of a round wire of relative permeability K_m and length l in a vacuum is

$$L \approx 2l[\ln\,(2a^{-1}l) - 1 + \tfrac{1}{4}K_m] \times 10^{-7}$$

The self-inductance of a rectangular bar of perimeter p is

$$L \approx 2l[\ln\,(4p^{-1}l) + \tfrac{1}{2} + 0.1118l^{-1}p] \times 10^{-7}$$

The self-inductance of a bar of elliptical section, semiaxes a and b, is

$$L \approx 2l\{\ln\,[2l(a+b)^{-1}] - 0.05685\} \times 10^{-7}$$

The self-inductance of a tube of external and internal radii a and b is

$$L \approx 2l\left[\ln\frac{2l}{a} + \frac{b^4}{(a^2-b^2)^2}\ln\frac{a}{b} + \frac{7b^2-5a^2}{4(a^2-b^2)}\right] \times 10^{-7}$$

Note. In the following formulas for bifilar lines the inductance per unit length is found by setting $l = 1$. In all cases l is supposed to be much greater than the pair spacing. The current densities are taken uniform. The current goes out on one element and returns on the other.

The self-inductance of two parallel cylinders of radii a and b and length l with a distance d between axes is

$$L = l\{1 + 2\ln\,[(ab)^{-1}d^2]\} \times 10^{-7}$$

The self-inductance of two similar parallel wires of radius a and relative permeability K_m with a distance d between axes is

$$L \approx l[4\ln\,(a^{-1}d) + K_m - 4d] \times 10^{-7}$$

The self-inductance of two similar parallel rectangular wires of perimeter p with a distance d between centers is

$$L \approx [4l \ln (2p^{-1}d) + 6l + 0.447p - 4d] \times 10^{-7}$$

The self-inductance of two similar parallel tubes, external radius a, internal radius b, with a distance d between centers is

$$L \approx l \left[4 \ln \frac{d}{a} + \frac{4b^4}{(a^2 - b^2)^2} \ln \frac{a}{b} + \frac{3b^2 - a^2}{a^2 - b^2} \right] \times 10^{-7}$$

The self-inductance of a coaxial line when the external radii of the inside conductor, insulation space, and outside conductor are c, b, and a, respectively, and the relative permeabilities K_m, K_m', and K_m'' and when the length l is great compared with a is

$$L = 2l \left\{ \frac{1}{4} K_m + K_m' \ln \frac{b}{c} + \frac{1}{4} K_m'' \left[\frac{4a^4}{(a^2 - b^2)^2} \ln \frac{a}{b} - \frac{3a^2 - b^2}{a^2 - b^2} \right] \right\} \times 10^{-7}$$

If $K_m = K_m' = K_m''$, this formula also holds for a noncoaxial line provided the axes are parallel.

The self-inductance of a wire of radius r and relative permeability K_m which is bent into a circular loop of mean radius a, neglecting small terms in r^4/a^4, is

$$L \approx 4\pi a \left[\left(1 + \frac{r^2}{8a^2} \right) \ln \frac{8a}{r} + \frac{r^2}{24a^2} - 2 + \frac{1}{4} K_m \right] \times 10^{-7}$$

The self-inductance of a wire of radius r and relative permeability K_m which is bent into a rectangular loop with sides a and b and diagonal $d = (a^2 + b^2)^{\frac{1}{2}}$ is[1]

$$L \approx 4 \left[a \ln \frac{2ab}{r(a + d)} + b \ln \frac{2ab}{r(b + d)} + 2d - \left(2 - \frac{1}{4} K_m \right) (a + b) \right] \times 10^{-7}$$

The self-inductance of a wire with rectangular section of perimeter p which is bent into a rectangular loop with sides a and b and diagonal d is

$$L \approx 4 \left[a \ln \frac{4ab}{p(a + d)} + b \ln \frac{4ab}{p(b + d)} + 2d + \frac{1}{2} (a + b) + 0.223p \right] \times 10^{-7}$$

The self-inductance of a thin band of radius a and width b is

$$L \approx 4\pi a [\ln (8b^{-1}a) - \tfrac{1}{2}] \times 10^{-7}$$

The mutual inductance of two thin coaxial circular loops of radii a and b, when r_1 and r_2 are the farthest and nearest distances between the loops, is given in terms of complete elliptic integrals by[2]

$$M = 8\pi k^{-1} a^{\frac{1}{2}} b^{\frac{1}{2}} [(1 - \tfrac{1}{2}k^2)K(k) - E(k)] \times 10^{-7}$$
$$= 8\pi k_1^{-1} a^{\frac{1}{2}} b^{\frac{1}{2}} [K(k_1) - E(k_1)] \times 10^{-7}$$

where $k^2 = r_1^{-2}(r_1^2 - r_2^2)$ and $k_1^2 = (r_1 - r_2)(r_1 + r_2)^{-1}$.

The mutual inductance between a long straight wire and a loop of radius a whose diameter it intersects at right angles at a distance c from the loop center is

$$M = 4\pi [c \sec \alpha - (c^2 \sec^2 \alpha - a^2)^{\frac{1}{2}}] \times 10^{-7} \qquad c > a$$
$$M = 4\pi c \tan (\tfrac{1}{4}\pi - \tfrac{1}{2}\alpha) \times 10^{-7} \qquad c < a$$

[1] Tables are given by Grover, pp. 59–65.
[2] Grover gives tables on pp. 77–87.

where α is the acute angle between the plane of the loop and the plane defined by its center and the straight wire.

The mutual inductance of two parallel coaxial identical rectangular loops whose sides are a and b and which are spaced so that the distance from any corner of one loop to the most distant corner of the other is d is[1]

$$M = 4\left[a \ln \frac{(a+A)B}{(a+D)d} + b \ln \frac{(b+B)A}{(b+D)d} + 8(D - A - B + d)\right] \times 10^{-7}$$

where $A^2 = a^2 + d^2$, $B^2 = b^2 + d^2$ and $D^2 = a^2 + b^2 + d^2$.

The mutual inductance between two circular loops of wire whose axes intersect at an angle γ at a point where the radius a of one loop subtends an angle α and the radius b of the other an angle β is

$$M = 4\pi^2 a \sum_{n=1}^{\infty} \frac{a^n \sin \alpha \sin \beta}{n(n+1)b^n} P_n{}^1(\cos \alpha) P_n{}^1(\cos \beta) P_n(\cos \gamma) \times 10^{-7}$$

where the last terms include two associated Legendre functions and one polynomial.[2] The mutual inductance of two circular loops with parallel axes can be calculated from tables in Grover, pages 177 to 192.

Note. The self- or mutual inductance of thin coils whose cross section is small compared with other dimensions is given approximately by insertion of the factor N^2 or $N_1 N_2$, respectively, where N is the total number of turns, in the corresponding loop formula and the use of the mean coil dimensions for the corresponding loop dimensions.

A circular ring encircles or is encircled by a coaxial helix, the larger radius being A and the smaller a. The distances from the plane of the ring to the farther and nearer ends of the helix are b_1 and b_2 and n is the number of turns per meter on the helix. The mutual inductance is

$$M = 2\pi n(A + a)\{c[k_1{}^{-1}(K_1 - E_1) \pm k_2{}^{-1}(K_2 - E_2)] + (A - a)(b_1{}^{-1}\psi_1 \pm b_2{}^{-1}\psi_2)\}$$
$$\times 10^{-7}$$

where the subscript 1 or 2 indicates the use of b_1 or b_2 for b in the following formulas:

$$k^2 = 4Aa[(A+a)^2 + b^2]^{-1} \quad k' = (1 - k^2) \quad c^2 = 4Aa(A+a)^{-2} \quad k' \sin \beta = (1 - c^2)^{\frac{1}{2}}$$
$$\psi = K(k)E(k',\beta) - [K(k) - E(k)]F(k',\beta) - \tfrac{1}{2}\pi$$

The upper sign in the $\pm$ is taken when the plane of the ring cuts the helix; otherwise the lower sign is used. Complete elliptic integrals of modulus k are indicated by K or $K(k)$ and E or $E(k)$ and $E(k',\beta)$ and $F(k',\beta)$ are incomplete elliptic integrals of modulus k' and amplitude β.[3]

Note. The following current-sheet formulas assume that the current density on the shell is uniform and flows around the cylinder normal to the axis in an infinitely thin sheet. A correction may be added to take account of the fact that the current is actually concentrated in wires of definite radius and spacing as in Grover, pages 148 to 150, but is often not needed for close windings. By a process equivalent to integration of the preceding formula, an exact formula for the mutual inductance between a cylindrical current sheet or helix and a coaxial concentric current sheet can be derived.[4]

---

[1] For tables, see Grover, pp. 66–69.
[2] For tables, see Grover, pp. 193–208.
[3] For tables, see Grover, pp. 114–118.
[4] Louis Cohen, *Bull. Natl. Bur. Standards* **3**, 298 (1907). **For practical purposes, tables given in Grover, pp. 122–141, are better.**

The self-inductance of a current sheet of radius a, length b, and diagonal $d = (4a^2 + b^2)^{\frac{1}{2}}$ having a total number of turns N is[1]

$$L = \tfrac{4}{3}\pi b^{-2}N^2[d(4a^2 - b^2)E(k) - b^2dK(k) - 8a^3] \times 10^{-7}$$

where $k = 2d^{-1}a$.

A current sheet is wound on the surface of the toroid formed by the rotation in the φ direction of a plane area S about an external line. If there are N turns and if the current density is independent of φ and has no φ component, then the self-inductance is

$$L = 2K_mN^2 \int_s r^{-1}\, dS \times 10^{-7}$$

where K_m is the relative permeability inside the current sheet and r is the distance of the area element dS from the rotational axis. The self-inductance in the above case, if S is a circle of radius a whose center is at a distance b from the rotational axis, is

$$L = 4\pi K_mN^2[b - (b^2 - a^2)^{\frac{1}{2}}] \times 10^{-7}$$

The self-inductance, if S is a rectangular section with sides parallel to the axis of length a and sides normal to it of length b and with the inside surface a distance R from the axis, is

$$L = 2N^2aK_m \ln (1 + R^{-1}b) \times 10^{-7}$$

The self-inductance of a circular coil of N turns and circular section is

$$L \approx 4\pi N^2a[(1 + \tfrac{1}{8}r^2a^{-2}) \ln (8r^{-1}a) + r^2(24a^2)^{-1} - 1.75] \times 10^{-7}$$

where r is the radius of the section, a the radius of the axis of the section, and $(r/a)^n$ is neglected when $n > 2$. The self-inductance of the above coil if it has a square section of side c is, if $c \ll a$,

$$L \approx 4\pi aN^2\{\tfrac{1}{2}[1 + c^2(24a^2)^{-1}] \ln (32c^{-2}a^2) - 0.84834 + 0.051a^{-2}c^2\} \times 10^{-7}$$

The self-inductance of coils of rectangular section can be calculated from tables given in Grover, pages 94 to 113.

The mutual inductance of coils of rectangular section and parallel axes can be calculated from tables given in Grover, pages 225 to 235. The mutual inductance of coils of rectangular section with inclined axes can be found from tables given by Grover on pages 209 to 214.

The increase in self-inductance of a circuit due to the placement of a sphere of radius a and relative permeability K_m in a position near it where the induction B per unit current is nearly uniform is

$$\Delta L \approx a^3B^2(K_m - 1)(K_m + 2)^{-1} \times 10^7$$

The increase of self-inductance of a loop of radius a due to the insertion concentrically of a sphere of radius b and infinite permeability is

$$\Delta L = 8\pi a^{-2}b^3K(a^{-2}b^2) \times 10^{-7}$$

The mutual inductance between two coaxial loops of radii a and b when the distance between centers is c and there is an infinite slab of thickness t and relative permeability K_m between and parallel to them is

$$M = 8\pi (ab)^{\frac{1}{2}}(1 - \beta^2) \sum_{n=0}^{\infty} k_n^{-1}\beta^{2n}[(1 - k_n^2)K(k_n) - E(k_n)]$$

$$k_n^2 = 4ab[(a + b)^2 + (c + 2nt)^2]^{-1} \qquad \beta = (K_m - 1)(K_m + 2)^{-1}$$

[1] For most purposes the tables given in Grover, pp. 142–162, are more practical than the formula.

Magnetic Forces on Circuits. The component of force in newtons tending to displace one of a pair of circuits in the x direction, the other being fixed, is

$$F_x = I_1 I_2 \frac{\partial M}{\partial x}$$

where I_1 and I_2 are the currents and M is the mutual inductance. The torque in newton meters tending to rotate one of a pair of circuits through an angle α, the other being fixed, is

$$T_\alpha = I_1 I_2 \frac{\partial M}{\partial \alpha}$$

Thus any desired forces or torques may be computed from the mutual-inductance formulas of the last few pages by differentiation, provided that it is possible to express M explicitly in terms of x or α. When this is not possible the difference in the mutual-inductance values calculated for the position x or α and the position $x + dx$ or $\alpha + d\alpha$ using the Grover tables may be multiplied by $I_1 I_2$ and divided by dx or $d\alpha$. In many cases the tabular intervals are small enough so this will give adequate accuracy; in other cases careful interpolation will be needed. Notice that in Grover's tables distances are in centimeters.

The force per unit length between two long parallel circular cylinders or tubes carrying uniformly distributed currents I_1 and I_2 is

$$F_1 = 2 I_1 I_2 a^{-1} \times 10^{-7}$$

The force is attractive when I_1 and I_2 have the same direction; otherwise it is repulsive.

The force per unit length between two parallel strips[1] of width a symmetrically placed with their faces a uniform distance b apart and carrying currents I_1 and I_2 is

$$4 I_1 I_2 a^{-1} [\tan^{-1} (b^{-1}a) - \tfrac{1}{2} a^{-1} b \ln (1 + b^{-2} a^2)] \times 10^{-7}$$

The force is attractive when I_1 and I_2 have the same direction; otherwise it is repulsive.

The force between two coaxial loops of radii a and b with centers at a distance c apart that carry currents I_1 and I_2 is

$$F = I_1 I_2 \pi c k [a^{\frac{1}{2}} b^{\frac{1}{2}} (1 - k^2)]^{-1} [(2 - k^2) E(k) - 2(1 - k^2) K(k)] \times 10^{-7}$$

where $k^2 = 4ab[(a + b)^2 + c^2]^{-1}$. The force is attractive when I_1 and I_2 encircle the axis in the same direction.

The axial force between a circular loop of radius a and a coaxial helix of radius b (a may be greater or less than b) and n turns per meter is

$$F = I_1 I_2 n (M - M') \times 10^{-7}$$

The loop center may lie inside or outside the helix. Here M and M' are the mutual inductances between a loop of radius a and coaxial loops of radius b whose planes pass through the extreme near end and extreme far end of the helix, respectively. The force is toward the center of the helix if the currents circle the axis in the same direction.

The force between a helix and a coaxial circular coil of mean radius a, square section of side c, and N turns is given approximately by the foregoing formula if $N I_1$ is used for I_1 and $a[1 + c^2(24a^2)^{-1}]$ for a. The force between two coaxial single-layer coils may be calculated by a formula in Grover on page 258 and a table on page 115.

The torque on a circular coil of rectangular section with internal and external radii a and b and any length which carries a current I, has N turns, and whose axis makes

---

[1] The force between two parallel rectangular bus bars is given by B. Hague, "Electromagnetic Problems in Electrical Engineering," p. 338, Oxford University Press, New York, 1929.

an angle α with a uniform field of induction **B** is

$$T = \tfrac{1}{3}\pi BNI(a^2 + ab + b^2) \sin \alpha$$

The torque on the above coil if it has a circular section of radius b whose center is at a distance a from the axis is

$$T = \tfrac{1}{4}\pi BNI(4a^2 - b^2) \sin \alpha$$

The torque on one of two concentric circular loops of wire of radii a and b which carry currents I_1 and I_2 is

$$T = 4\pi^2 aI_1I_2 \times 10^{-7} \sum_{n=0}^{\infty} \frac{2n + 2}{2n + 1} \left[\frac{(2n + 1)!!}{(2n + 2)!!} \right]^2 \left(\frac{a}{b} \right)^{2n+1} P_{2n+1}{}^1 (\cos \alpha)$$

where α is the angle between their axes and $P_{2n+1}{}^1 (\cos \alpha)$ is a Legendre function. It is directed so as to set one current parallel to the other.

The force on any circuit near the plane face of a semi-infinite block of material having a uniform relative permeability K_m which is independent of field strength equals the force between the circuit carrying a current I and its mirror-image circuit in the plane face carrying a current $I' = (K_m - 1)(K_m + 1)^{-1}I$. The direction of I', if K_m is greater than one, is such that the projections of I and I' on the interface coincide in position and direction. It is evident that if $K_m \gg 1$ then $I \approx I'$ and the exact value of K_m need not be known.

The force per unit length on an infinite wire carrying a current I parallel to the walls of an infinite evacuated rectangular conduit of infinite permeability is

$$F_x = 4\pi b^{-1}I^2 \times 10^{-7} \sum_{m=1}^{\infty} \operatorname{csch} (m\pi ab^{-1}) \sinh [m\pi b^{-1}(2c - a)] \cos^2 (m\pi db^{-1})$$

where the walls of the conduit are at $x = 0$, $x = a$ and $y = 0$, $y = b$. The wire lies at $x = c$, $y = d$. To get F_y, interchange a with b and c with d. The series converges very rapidly unless the wire is near the wall. The force per unit length toward the nearest wall on an infinite wire parallel to and at a distance c from the axis of an evacuated cylindrical hole of radius a in a block of material of relative permeability K_m is

$$F_1 = 2(a^2 - c^2)^{-1}cI^2(K_m - 1)(K_m + 1)^{-1} \times 10^{-7}$$

Permeable Bodies in Magnetic Fields. The energy of an unmagnetized body of volume v when placed in a field of induction **B** produced by fixed sources in a region of constant permeability μ is

$$W = \tfrac{1}{2} \int_v (\mu^{-1} - \mu_i^{-1})\mathbf{B} \cdot \mathbf{B}_i \, dv$$

where $\mathbf{B}_i$ and μ_i are the final values of the magnetic induction and permeability in the volume element dv inside the body and the integration is over the volume of the body. The torque tending to decrease the angle α between **B** and the major axis of an oblate permeable spheroid of relative permeability K_m with semiaxes a and b, where $b > a$, placed in a uniform field of induction **B** produced by fixed sources in a vacuum is

$$T = \frac{(K_m - 1)^2 b^2 aB^2(3P - 2) \sin 2\alpha}{6[(K_m - 1)^2 P^2 + (K_m - 1)(2 - K_m)P - 2K_m]} \times 10^7$$

where $P = A[(1 + A^2) \cot^{-1} A - A]$ and $A = a(b^2 - a^2)^{-\frac{1}{2}}$.

The torque tending to decrease the angle α between **B** and the major axis of a prolate permeable spheroid of relative permeability K_m with semiaxis a and b where $b < a$ placed in a uniform field of induction **B** produced by fixed sources in a vacuum is

$$T = \frac{(K_m - 1)^2 b^2 a B^2 (2 - 3Q) \sin 2\alpha}{6[(K_m - 1)^2 Q^2 + (K_m - 1)(2 - K_m)Q - 2K_m]} \times 10^7$$

where $Q = c[(1 - c^2) \coth^{-1} c + c]$ and $c = a(a^2 - b^2)^{-\frac{1}{2}}$.

The attractive force between a long cylinder carrying a uniformly distributed current I and an external sphere of relative permeability K_m and radius a whose center is at a distance b from the cylinder axis is

$$F = 4I^2 \times 10^{-7} \sum_{n=1}^{\infty} \frac{(2n - 2)!!n(K_m - 1)}{(2n - 1)!!(nK_m + n + 1)} \left(\frac{a}{b}\right)^{2n+1}$$

If the permeability is very large in the above case, the force is

$$F = 4I^2 a^2 b^{-1} (b^2 - a^2)^{-\frac{1}{2}} \sin^{-1} (b^{-1}a) \times 10^7$$

Magnetic Shielding. Two long wires of a bifilar lead at $\rho = c$, $\varphi = 0$ and $\rho = c$, $\varphi = \pi$ carry currents I and $-I$ and are shielded by a cylinder of relative permeability K_m of internal and external radius a and b. The components of the induction outside the shield are

$$B_\rho = -16I \times 10^{-7} \sum_{n=0}^{\infty} \frac{b^{4n+2} c^{2n+1} \rho^{-2n-2} \sin (2n + 1)\theta}{(K_m + 1)^2 b^{4n+2} - (K_m - 1)^2 a^{4n+2}}$$

$$B_\varphi = 16I \times 10^{-7} \sum_{n=0}^{\infty} \frac{b^{4n+2} c^{2n+1} \rho^{-2n-2} \cos (2n + 1)\theta}{(K_m + 1)^2 b^{4n+2} - (K_m - 1)^2 a^{4n+2}}$$

A long cylindrical shield of internal and external radius a and b and relative permeability K_m is placed across a uniform field of induction **B**. The induction B_i inside is uniform and of magnitude

$$B_i = \frac{4K_m b^2 B}{4K_m b^2 + (K_m - 1)^2 (b^2 - a^2)}$$

A spherical shield of internal and external radius a and b and relative permeability K_m is placed in a uniform field of induction **B**. The induction B_i inside is uniform and its magnitude is

$$B_i = \frac{9K_m b^3 B}{9K_m b^3 + 2(K_m - 1)^2 (b^3 - a^3)}$$

The Magnetic Circuit. The reluctance $\mathfrak{R}$ of a magnetic circuit is well defined only when all the magnetic flux Φ links all N turns of the magnetizing coils which when carrying a current I generate the magnetomotance $\mathfrak{F}$. Then

$$\mathfrak{F} = \mathfrak{R}\Phi = NI$$

The reluctance of a toroid of such high and uniform relative permeability K_m that there is no flux leakage can be calculated regardless of the position of the magnetizing coil from the current-sheet self-inductance formulas for N turns already given for toroids of various sections. Thus

$$\mathfrak{R} = N^2 L^{-1}$$

The change in reluctance of a closed magnetic plane circuit of thickness b, rectangular section and uniform relative permeability K_m so high that leakage is negligible due to the presence of corners, steps, tapered sections, and circular holes can be calculated from the formulas already given for resistance change ΔR for two-dimensional current flow in media of resistivity ρ. Thus

$$\Delta \mathcal{R} = 4\pi \times 10^7 K_m \rho^{-1} \Delta R$$

If a gap of uniform width a is cut out of a magnetic circuit of high relative permeability K_m, normal to the induction $\mathbf{B}$, and if a is small compared with all dimensions of the section of area A cut, then the increase in reluctance is

$$\Delta \mathcal{R} \approx 4\pi a A^{-1}(K_m - 1) \times 10^{-7}$$

where the surrounding space is empty and the fringing field at the edge of the gap is neglected.

The fringing field may be calculated when the region of negative x is filled with an infinitely permeable medium except for a gap bounded by $y = \frac{1}{2}a$ and $y = -\frac{1}{2}a$ which extends to $x = -\infty$. A magnetomotance is applied across the gap so that far from the edge the induction is B_0. The induction B_y anywhere on the x axis is then given implicitly by

$$x = \pi^{-1}a[B_0 B_y^{-1} - \tanh^{-1}(B_y B_0^{-1})]$$

where $0 \lessgtr B_y \lessgtr B_0$.

If the magnetomotance across a gap with faces at $z = \frac{1}{2}b$ and $z = -\frac{1}{2}b$ in an infinitely permeable cylinder bounded by $\rho = a$ is $\mathfrak{F}_0$, then the magnetomotance in the gap when $\rho < a$ is

$$\mathfrak{F} \approx \mathfrak{F}_0 \left[\frac{z}{b} + \sum_{n=1}^{\infty} C_n \frac{I_0(\frac{1}{2}n\pi\rho/b)}{I_0(\frac{1}{2}n\pi a/b)} \sin \frac{n\pi z}{2b} \right]$$

where $C_1 = -0.17232$ and when $n > 1$.

$$C_n = \frac{(-1)^n}{n} \left[0.5836 \frac{0.1775 \cdot 1.1775 \cdots (n - 0.8225)}{0.8225 \cdot 1.8225 \cdots (n - 0.1775)} - 0.0201n^{-2} \right]$$

The induction is $\mathbf{B} = -4\pi \times 10^{-7}\nabla\mathfrak{F}$. This formula assumes that the field across the edge of the gap is two-dimensional. If this is the only gap in an infinitely permeable circuit, then $\mathfrak{F}_0 = NI$ where N is the number of turns of the magnet coil and I is its current.[1]

Permanent Magnets. In the following formulas it is assumed that the magnetization M of a permanent magnet is absolutely rigid and that any magnetization induced in it by external fields is negligible compared with M. The energy of such a magnet when placed in an external field of induction $\mathbf{B}$ in a vacuum is $W = -\int \mathbf{M} \cdot \mathbf{B}\, dv$, where the integration is over the volume of the magnet and the "loop" definition of $\mathbf{M}$ is used rather than the "pole" definition. The forces and torques acting on the magnet are

$$F_x = \frac{\partial W}{\partial x} \qquad T = \frac{\partial W}{\partial \theta}$$

The moment of a magnet is $\mathbf{m} = \int \mathbf{M}\, dv$ where the integration is over the magnet volume.

The mutual (apparently potential) energy of two thin needles magnetized lengthwise at a distance a apart large compared with their length and having loop moments of magnitude m_1 and m_2, when immersed in a medium of relative permeability K_m, is

$$W = m_1 m_2 K_m^{-1} r^{-3}(\sin \theta_1 \sin \theta_2 \cos \psi - 2\cos \theta_1 \cos \theta_2) \times 10^{-7}$$

[1] Tables of $\frac{1}{2}C_n$ are given by W. R. Smythe, *Revs. Modern Phys.* **20**, 176 (1948).

where θ_1 and θ_2 are the angles between m_1 and m_2, respectively, and r. The angle between the planes that contain m_1 and m_2 and intersect in r is ψ. The repulsive force between two needles is $-\partial W/\partial r$ and if α is the azimuth angle about any line the torque on either magnet about that line is $-\partial W/\partial \alpha$, the other magnet being fixed. In a vacuum where K_m is unity this formula applies to magnets of moments m_1 and m_2 of any shape provided their dimensions are small compared with r. In other media the mutual energy depends on the shape.

Uniformly magnetized bodies may be replaced by their equivalent current sheets for the purpose of calculating fields and mutual torques in a vacuum. The current sheet coincides with the surface of the body and the current density encircles the body in a path normal to the direction x of magnetization and is uniform in terms of x and numerically equal to M. Thus the fields of thin disks magnetized normal to their faces and the torques and forces between them are identical with those between circular loops already given, if I_1 and I_2 are replaced by M_1 and M_2. Similarly, in a vacuum the fields and forces involving uniformly magnetized bars may be calculated from the formulas already given for solenoids provided nI, where n is the number of turns per meter, is replaced by M. The mutual-inductance tables given by Grover and already referred to may be used.

A right circular cylinder of length b and radius a uniformly magnetized lengthwise with an intensity M, when placed with its flat end against an infinitely permeable flat surface, adheres with a force

$$F = 8\pi abM^2\{k^{-1}[K(k) - E(k)] - k_1^{-1}[K(k_1) - E(k_1)]\} \times 10^{-7}$$

where the moduli of the complete elliptic integrals are $k = 2a(4a^2 + b^2)^{-\frac{1}{2}}$ and $k_1 = a(a^2 + b^2)^{-\frac{1}{2}}$. If M is very large, this gives approximately

$$F \approx 2\pi^2 a^2 M^2 \times 10^{-7}$$

The same force is experienced by two identical cylindrical magnets placed N to S. The same force, but repulsive, appears if they are placed N to N or S to S.

A long straight bar of uniform cross-sectional area S has a uniform lengthwise magnetization M. The flat end, when placed in contact with an infinitely permeable flat block, adheres with a force

$$F \approx 2\pi SM^2 \times 10^{-7}$$

The above bar bent in the shape of a horseshoe with coplanar ends will, if the magnetization remains uniform, adhere with twice this force.

The torque on a sphere with uniform magnetization M immersed in a medium of relative permeability K_m in a field of induction B such that the angle between B and M is α is

$$T = \frac{4\pi a^3 M B \sin \alpha}{2K_m + 1}$$

The torque on any body of volume v with a uniform magnetization M when placed in a uniform field of induction B in a vacuum so that the angle between B and M is α is

$$T = BMv \sin \alpha$$

DYNAMIC-FIELD FORMULAS

5b-8. The Electromagnetic Field Equations. In this section some basic relations and concepts of the classic electromagnetic field are given. The mks or Giorgi system of units will be used throughout.

Maxwell's Equations. The basic equations governing the field vectors are Maxwell's equations.

DIFFERENTIAL FORMS

$$\nabla \times \mathbf{E}(\mathbf{r},t) = -\frac{\partial}{\partial t}\mathbf{B}(\mathbf{r},t) \tag{5b-1}$$

$$\nabla \times \mathbf{H}(\mathbf{r},t) = \mathbf{J}(\mathbf{r},t) + \frac{\partial}{\partial t}\mathbf{D}(\mathbf{r},t) \tag{5b-2}$$

$$\nabla \cdot \mathbf{B}(\mathbf{r},t) = 0 \tag{5b-3}$$

$$\nabla \cdot \mathbf{D}(\mathbf{r},t) = \rho(\mathbf{r},t) \tag{5b-4}$$

where $\mathbf{E}$ is the electric field intensity vector in volts/meter, $\mathbf{H}$ is the magnetic field intensity vector in amperes/meter, $\mathbf{B}$ is the magnetic-induction vector in webers/meter², $\mathbf{D}$ is the electric displacement vector in coulombs/meter², $\mathbf{J}$ is the current density vector in amperes/meter², ρ is the volume density of charge in coulombs/meter³, $\mathbf{r}$ is the position vector in meters, and t is the time in seconds. The vector $\mathbf{J}$ and the volume density of charge ρ are source quantities, and the vectors $\mathbf{E}, \mathbf{H}, \mathbf{B}, \mathbf{D}$ are field quantities. The conservation of charge is expressed by the equation of continuity

$$\nabla \cdot \mathbf{J}(\mathbf{r},t) = -\frac{\partial}{\partial t}\rho(\mathbf{r},t) \tag{5b-5}$$

which is a corollary of Eq. (5b-4) and the divergence of Eq. (5b-2).

INTEGRAL FORMS. Integral forms of Maxwell's equations follow readily from Eqs. (5b-1) to (5b-4) with the aid of Stokes' theorem and the divergence theorem. They are:

$$\oint_{\Gamma} \mathbf{E} \cdot dl = -\int_{S} \frac{\partial \mathbf{B}}{\partial t} \cdot \mathbf{n}\, dS \qquad \text{(Faraday's emf law)} \tag{5b-6}$$

$$\oint_{\Gamma} \mathbf{H} \cdot dl = \int_{S} \left(\mathbf{J} + \frac{\partial \mathbf{D}}{\partial t}\right) \cdot \mathbf{n}\, dS \qquad \text{(Generalized Ampères' law)} \tag{5b-7}$$

$$\oint_{S} \mathbf{D} \cdot \mathbf{n}\, dS = \int_{V} \rho\, dV \qquad \text{(Gauss' law)} \tag{5b-8}$$

$$\oint_{S} \mathbf{B} \cdot \mathbf{n}\, dS = 0 \qquad \text{(magnetic flux conservation law)} \tag{5b-9}$$

where Γ is a closed curve spanned by an arbitrary surface S, both stationary in the observer's frame of reference; $\mathbf{n}$ is a unit vector normal to S; and V is the volume enclosed by a closed surface S.

DUALITY, MAGNETIC SOURCES. For a "simple" medium in which $\mathbf{D}(\mathbf{r},t) = \epsilon \mathbf{E}(\mathbf{r},t)$ and $\mathbf{B}(\mathbf{r},t) = \mu\mathbf{H}(\mathbf{r},t)$, where ϵ and μ are respectively the dielectric constant and the permeability of the medium, Maxwell's equations possess a certain duality in $\mathbf{E}$ and $\mathbf{H}$ provided that the mathematical artifice of magnetic charge and magnetic current are introduced. Hence, the generalized Maxwell's equations are:

$$\nabla \times \mathbf{E} = -\mathbf{J}_m - \mu\frac{\partial \mathbf{H}}{\partial t} \tag{5b-10}$$

$$\nabla \times \mathbf{H} = \mathbf{J} + \epsilon\frac{\partial \mathbf{E}}{\partial t} \tag{5b-11}$$

$$\nabla \cdot \mathbf{E} = \frac{1}{\epsilon}\rho \tag{5b-12}$$

$$\nabla \cdot \mathbf{H} = \frac{1}{\mu}\rho_m \tag{5b-13}$$

where J_m and ρ_m are respectively the fictitious magnetic current source and magnetic charge source. Substituting the duality transformation,

$$\mathbf{E}^d = \pm \left(\frac{\mu}{\epsilon}\right)^{\frac{1}{2}} \mathbf{H} \qquad \mathbf{H}^d = \mp \left(\frac{\epsilon}{\mu}\right)^{\frac{1}{2}} \mathbf{E} \qquad \mathbf{J}^d = \pm \left(\frac{\epsilon}{\mu}\right)^{\frac{1}{2}} \mathbf{J}_m$$

$$\mathbf{J}_m{}^d = \mp \left(\frac{\mu}{\epsilon}\right)^{\frac{1}{2}} \mathbf{J} \qquad \rho^d = \pm \left(\frac{\epsilon}{\mu}\right)^{\frac{1}{2}} \rho_m \qquad \rho_m{}^d = \mp \left(\frac{\mu}{\epsilon}\right)^{\frac{1}{2}} \rho$$

into Eqs. (5b-10) to (5b-13) gives

$$\nabla \times \mathbf{E}^d = -\mathbf{J}_m{}^d - \mu \frac{\partial \mathbf{H}^d}{\partial t} \qquad \nabla \times \mathbf{H}^d = \mathbf{J}^d + \epsilon \frac{\partial \mathbf{E}^d}{\partial t}$$

$$\nabla \cdot \mathbf{E}^d = \frac{1}{\epsilon} \rho^d \qquad\qquad\qquad \nabla \cdot \mathbf{H}^d = \frac{1}{\mu} \rho_m{}^d$$

Thus to every electromagnetic field $(\mathbf{E},\mathbf{H})$ produced by electric current $\mathbf{J}$, there is a dual field $(\mathbf{H}^d,\mathbf{E}^d)$ produced by a fictive magnetic current[1] $\mathbf{J}_m{}^d$.

TIME-PERIODIC FIELD. If all quantities have time dependence $e^{-i\omega t}$, the time factor can be suppressed and Maxwell's equations in simple, linear, time-independent media become relations between complex amplitudes. The differential forms of Maxwell's equation are:

$$\nabla \times \mathbf{E}(\mathbf{r}) = i\omega \mathbf{B}(\mathbf{r}) \qquad\qquad (5b\text{-}14)$$

$$\nabla \times \mathbf{H}(\mathbf{r}) = \mathbf{J}(\mathbf{r}) - i\omega \mathbf{D}(\mathbf{r}) \qquad (5b\text{-}15)$$

$$\nabla \cdot \mathbf{B}(\mathbf{r}) = 0 \qquad\qquad\qquad (5b\text{-}16)$$

$$\nabla \cdot \mathbf{D}(\mathbf{r}) = \rho(\mathbf{r}) \qquad\qquad\quad (5b\text{-}17)$$

It is understood that $\mathbf{E}(\mathbf{r},t) = \text{Re}\ [\mathbf{E}(\mathbf{r})e^{-i\omega t}]$, $\mathbf{H}(\mathbf{r},t) = \text{Re}[H(\mathbf{r})e^{-i\omega t}]$, . . . , etc. Re is shorthand for "the real part of."

Covariance of Maxwell's Equations. According to the theory of relativity, the Maxwell's equations are covariant under the Lorentz transformation. In other words, Maxwell's equations have the same form in all inertial frames of reference.

LORENTZ TRANSFORMATIONS. The Lorentz transformations between an inertial frame $S(\mathbf{r},t)$ and another inertial frame $S'(\mathbf{r}',t')$ which is moving at a uniform velocity $\mathbf{v}$ with respect to S can be written in the general form

$$\mathbf{r}' = \mathbf{r} - \gamma \mathbf{v}t + (\gamma - 1) \frac{\mathbf{r} \cdot \mathbf{v}}{v^2} \mathbf{v} \qquad (5b\text{-}18)$$

$$t' = \gamma \left(t - \frac{\mathbf{r} \cdot \mathbf{v}}{c^2}\right) \qquad\qquad (5b\text{-}19)$$

where $\gamma = (1 - \beta^2)^{-\frac{1}{2}}$, $\beta = v/c$, $\mathbf{r} = x\mathbf{e}_x + y\mathbf{e}_y + z\mathbf{e}_z$, and c is the velocity of light in vacuum.

FIELD AND SOURCE TRANSFORMATIONS. To assure the covariance of Maxwell's equations between S and S' systems, the following transformations for the field vectors, the current density vector, and the charge density must be used:

$$\mathbf{E}' = \gamma(\mathbf{E} + \mathbf{v} \times \mathbf{B}) + (1 - \gamma) \frac{\mathbf{E} \cdot \mathbf{v}}{v^2} \mathbf{v} \qquad (5b\text{-}20)$$

$$\mathbf{B}' = \gamma \left(\mathbf{B} - \frac{1}{c^2} \mathbf{v} \times \mathbf{E}\right) + (1 - \gamma) \frac{\mathbf{B} \cdot \mathbf{v}}{v^2} \mathbf{v} \qquad (5b\text{-}21)$$

$$\mathbf{D}' = \gamma \left(\mathbf{D} + \frac{1}{c^2} \mathbf{v} \times \mathbf{H}\right) + (1 - \gamma) \frac{\mathbf{D} \cdot \mathbf{v}}{v^2} \mathbf{v} \qquad (5b\text{-}22)$$

$$\mathbf{H}' = \gamma(\mathbf{H} - \mathbf{v} \times \mathbf{D}) + (1 - \gamma) \frac{\mathbf{H} \cdot \mathbf{v}}{v^2} \mathbf{v} \qquad (5b\text{-}23)$$

$$\mathbf{J}' = \mathbf{J} - \gamma \mathbf{v}\rho + (\gamma - 1) \frac{\mathbf{J} \cdot \mathbf{v}}{v^2} \mathbf{v} \qquad (5b\text{-}24)$$

$$\rho' = \gamma \left(\rho - \frac{1}{c^2} \mathbf{J} \cdot \mathbf{v}\right) \qquad\qquad (5b\text{-}25)$$

[1] This duality property is intimately related to Babinet's principle discussed in Sec. 5b-12.

Constitutive Relations. Only two of the four Maxwell's equations (5b-1) to (5b-4) are independent, since the two divergence equations (5b-3) and (5b-4) can be obtained from the two curl equations (5b-1) and (5b-2) and the continuity equation (5b-5). Therefore, the number of field vectors required to describe an electromagnetic field must be reduced to two from the original four. This reduction is accomplished by the introduction of constitutive parameters which provide a mathematical description of the macroscopic electromagnetic properties of matter.

ELECTRIC AND MAGNETIC POLARIZATION VECTORS. The behavior of a material medium in an electromagnetic field can be described in terms of distributions of electric and magnetic dipoles. The medium can be characterized by two polarization density functions: $\mathbf{P}$, electric dipole moment per unit volume, and $\mathbf{M}$, magnetic dipole moment per unit volume. The polarization may be induced under action of the field from other sources, or it may be virtually permanent and independent of external fields. The permanent polarizations will be designated by $\mathbf{P}_0$ and $\mathbf{M}_0$. The relationships between the field vectors and the polarization vectors are

$$\mathbf{D} = \epsilon_0\mathbf{E} + \mathbf{P} + \mathbf{P}_0 \tag{5b-26}$$
$$\mathbf{B} = \mu_0(\mathbf{H} + \mathbf{M} + \mathbf{M}_0) \tag{5b-27}$$

where ϵ_0 and μ_0 are respectively the permittivity and permeability of free space.

ISOTROPIC MEDIA. In simple isotropic media, the polarization vectors are proportional to the field (i.e., $\mathbf{P} = \epsilon_0\chi\mathbf{E}$ and $\mathbf{M} = \chi_m\mathbf{H}$), and the constitutive parameters are scalar quantities:

$$\mathbf{D} = \epsilon_0(1 + \chi)\mathbf{E} = K\epsilon_0\mathbf{E} = \epsilon\mathbf{E} \tag{5b-28}$$
$$\mathbf{B} = \mu_0(1 + \chi_m)\mathbf{H} = K_m\mu_0\mathbf{H} = \mu\mathbf{H} \tag{5b-29}$$

where χ is the electric susceptibility, K is the relative permittivity of the medium (or the dielectric constant), ϵ is its absolute permittivity, χ_m is the magnetic susceptibility, K_m is the relative permeability of the medium, and μ is its absolute permeability. For isotropic inhomogeneous media, ϵ and μ may be functions of positions. Strictly speaking, the relations (5b-28) and (5b-29) are definably only for time-periodic phenomena, since in general ϵ and μ are functions of the frequency. (The frequency dependence of the constitutive parameters is known as the *dispersive property* of the medium.) Hence, these relations are applicable to other than time-periodic, time-varying fields only when over the significant part of the frequency spectrum covered by the Fourier components of the time dependence the constitutive parameters ϵ and μ are sensibly independent of frequency.

ANISOTROPIC MEDIA. The constitutive relations for an anisotropic medium have the form

$$\mathbf{D} = \boldsymbol{\varepsilon} \cdot \mathbf{E} \tag{5b-30}$$
$$\mathbf{B} = \boldsymbol{\mu} \cdot \mathbf{H} \tag{5b-31}$$

where $\boldsymbol{\varepsilon}$ and $\boldsymbol{\mu}$ are second-rank tensors having ϵ_{ij} and μ_{ij} as their components. For inhomogeneous and anisotropic medium, ϵ_{ij} and μ_{ij} are functions of positions. For anisotropic and dispersive medium, ϵ_{ij} and μ_{ij} are functions of the frequency; the relationships (5b-30) and (5b-31) then become relationships between complex amplitudes.

CONDUCTING MEDIA. A conducting medium is characterized by a linear relation between current density and the electric vector: For isotropic conducting medium

$$\mathbf{J} = \sigma\mathbf{E} \tag{5b-32}$$

where σ is a scalar. For anisotropic conducting medium

$$\mathbf{J} = \boldsymbol{\sigma} \cdot \mathbf{E} \tag{5b-33}$$

where σ is a second-rank tensor having components σ_{ij}. Again σ may be position-dependent or frequency-dependent.

UNIFORMLY MOVING MEDIA. Assume that an inertial frame $S'(\mathbf{r}',t')$ is moving at a uniform velocity $\mathbf{v}$ with respect to an observer's inertial frame $S(\mathbf{r},t)$. If the constitutive relations in S' frame are $\mathbf{D}' = \epsilon'\mathbf{E}'$ and $\mathbf{B}' = \mu'\mathbf{H}'$, then with the aid of Eqs. (5b-20) to (5b-23) we may find the constitutive relations in the observer's S frame from the following equations:

$$\mathbf{D} - \epsilon'(\mathbf{v} \times \mathbf{B}) = \epsilon'\mathbf{E} - \frac{1}{c^2}\mathbf{v} \times \mathbf{H} \tag{5b-34}$$

$$\mathbf{v} \times \mathbf{D} + \frac{1}{\mu'}\mathbf{B} = \frac{1}{\mu'}\left(\frac{1}{c^2}\mathbf{v} \times \mathbf{E}\right) + \mathbf{H} \tag{5b-35}$$

Note that in uniformly moving medium $\mathbf{D}$ is linearly related to $\mathbf{E}$ as well as $\mathbf{H}$, and $\mathbf{B}$ is also linearly related to $\mathbf{H}$ as well as $\mathbf{E}$.[1]

NONLINEAR MEDIA. The constitutive relations for a nonlinear medium have the form

$$\mathbf{D} = \epsilon(\mathbf{E})\mathbf{E} \tag{5b-36}$$
$$\mathbf{B} = \mu(\mathbf{H})\mathbf{H} \tag{5b-37}$$

where $\epsilon(\mathbf{E})$ and $\mu(\mathbf{H})$ are functions of the field strengths. Substituting these constitutive relations into Eqs. (5b-1) and (5b-2) gives a set of equations that are nonlinear. Because of the field-dependent characteristics of the permittivity and the permeability of the medium, there is energy exchange between a number of electromagnetic fields of different frequencies.[2]

Boundary Conditions. BOUNDARIES BETWEEN STATIONARY MEDIA. Let Γ be a smooth surface separating two media, 1 and 2; let the unit vector normal to the boundary be $\mathbf{n}$, pointing from medium 1 into medium 2.

1. Media 1 and 2 are dielectrics having constitutive parameters ϵ_1, μ_1, σ_1, and ϵ_2, μ_2, σ_2, respectively. The boundary conditions are:
Tangential components of the electric field vector $\mathbf{E}$ are continuous:

$$\mathbf{n} \times (\mathbf{E}_2 - \mathbf{E}_1) = 0 \tag{5b-38}$$

Tangential components of the magnetic field vector $\mathbf{H}$ are continuous:

$$\mathbf{n} \times (\mathbf{H}_2 - \mathbf{H}_1) = 0 \tag{5b-39}$$

Normal component of $\mathbf{D}$ is discontinuous by an amount equal to the electric surface-charge density:

$$\mathbf{n} \cdot (\mathbf{D}_2 - \mathbf{D}_1) = \rho_s \tag{5b-40}$$

Normal component of $\mathbf{B}$ is continuous:

$$\mathbf{n} \cdot (\mathbf{B}_2 - \mathbf{B}_1) = 0 \tag{5b-41}$$

Note that Eqs. (5b-40) and (5b-41) can be derived from Eqs. (5b-38) and (5b-39) with the aid of Maxwell's equations and the continuity equation. Hence, either Eqs. (5b-38) and (5b-39) or Eqs. (5b-40) and (5b-41) are sufficient to specify the boundary conditions.

2. Medium 1 is a perfect conductor, and medium 2 is a dielectric. The boundary conditions are

$$\mathbf{n} \times \mathbf{E}_2 = 0 \qquad \mathbf{n} \times \mathbf{H}_2 = K \qquad \mathbf{n} \cdot \mathbf{B}_2 = 0 \qquad \mathbf{n} \cdot \mathbf{D}_2 = \rho_s \tag{5b-42}$$

[1] C. Moller, "The Theory of Relativity," Oxford University Press, London, 1952.
[2] N. Bloembergen, "Nonlinear Optics," W. A. Benjamin, Inc., New York, 1965.

where K is the electric surface current density. A surface having these boundary conditions is said to be an "electric wall." By duality a surface displaying the boundary conditions

$$\mathbf{n} \times \mathbf{E}_2 = -K_m \qquad \mathbf{n} \times \mathbf{H}_2 = 0 \qquad \mathbf{n} \cdot \mathbf{D}_2 = 0 \qquad \mathbf{n} \cdot \mathbf{B}_2 = \rho_{sm} \qquad (5b\text{-}43)$$

is said to be a "magnetic wall." K_m is the magnetic surface current density, and ρ_{sm} is the magnetic charge density.

3. Medium 1 has a surface impedance Z_s which is defined as the ratio of the tangential electric field to the tangential magnetic field at the surface, and electromagnetic fields are impenetrable into medium 1. Medium 2 is a dielectric. The boundary condition is

$$\mathbf{n} \times \mathbf{E}_2 = Z_s \mathbf{H}_2 \qquad (5b\text{-}44)$$

If medium 1 is a good conductor, in which σ_1 is larger but finite,[1] then $Z_s = (1 - i)(\omega\mu_1/2\sigma_1)^{\frac{1}{2}}$ with $i = \sqrt{-1}$. The surface-impedance boundary condition is valid only for time-harmonic fields.

BOUNDARIES BETWEEN MOVING MEDIA. Let medium 1 be moving with respect to medium 2 with a velocity $\mathbf{v}$; let $S'(\mathbf{r}',t')$ and $S(\mathbf{r},t)$ be inertial frames for medium 1 and medium 2, respectively. The boundary conditions are

$$\mathbf{n} \times (\mathbf{E}_1 + \mathbf{v} \times \mathbf{B}_1) = \mathbf{n} \times (\mathbf{E}_2 + \mathbf{v} \times \mathbf{B}_2) \qquad (5b\text{-}45)$$
$$\mathbf{n} \times (\mathbf{H}_1 - \mathbf{v} \times \mathbf{D}_1) = \mathbf{n} \times (\mathbf{H}_2 - \mathbf{v} \times \mathbf{D}_2) \qquad (5b\text{-}46)$$

where $\mathbf{n}$ is a unit vector, in inertial frame S, normal to the boundary and pointing from medium 1 into medium 2. $\mathbf{E}_1$, $\mathbf{B}_1$, $\mathbf{H}_1$, and $\mathbf{D}_1$, and $\mathbf{E}_2$, $\mathbf{B}_2$, $\mathbf{H}_2$, and $\mathbf{D}_2$ are respectively field vectors in medium 1 and medium 2 as observed from the inertial frame[2] $S(\mathbf{r},t)$.

RADIATION CONDITIONS. The field associated with a finite distribution of sources or the field scattered from obstacles must satisfy conditions at infinity which pertain to the finiteness of the energy radiated by the sources or scattered by obstacles as well as the assurance that the field at infinity represents an outgoing wave. For time-periodic field in a homogeneous medium the condition at infinity take the form

$$\lim_{r \to \infty} r \left[\mathbf{H} - \left(\frac{\epsilon}{\mu} \right)^{\frac{1}{2}} (\mathbf{e}_r \times \mathbf{E}) \right] = 0 \qquad (5b\text{-}47)$$

$$\lim_{r \to \infty} r\mathbf{E} \text{ is finite} \qquad (5b\text{-}48)$$

where r is the radial distance from an arbitrary origin in the neighborhood of the sources or the scattering bodies, and $\mathbf{e}_r$ is a unit vector directed from the origin in the radial direction.

EDGE CONDITION. At sharp edges the field vectors may become infinite. But the order of this singularity is restricted by the Bonwkamp-Meixner edge condition: The energy density must be integrable over any finite domain even if this domain happens to include field singularities: i.e., the energy in any finite region of space must be finite. For example, when applied to a perfectly conducting sharp edge, this condition states that the singular components of the electric and magnetic vectors are of order $\xi^{-\frac{1}{2}}$, where ξ is the distance from the edge, whereas the parallel components are always finite.

UNIQUENESS THEOREM. A field in a lossy region is uniquely specified by the sources within the region plus the tangential components of $\mathbf{E}$ over the boundary, or the tangential components of $\mathbf{H}$ over the boundary, or the former over part of the bound-

---

[1] M. A. Leontovich, "Investigation of Propagation of Radiowaves," part II, Moscow, 1948; T. B. A. Senior, *Appl. Sci. Research* B-8, 418 (1960).
[2] A recent example of the interaction of electromagnetic waves with moving media was given by C. Yeh, *J. Appl. Phys.*, **36**, 3513 (1965); **38**, 5194 (1967).

ary and the latter over the rest of the boundary. The uniqueness theorem for the lossy case can be carried over to the lossless case if we consider the field in a lossless medium to be the limit of the corresponding field in a lossy medium as the loss goes to zero.

Energy Relations. POYNTING'S VECTOR THEOREM. Taking the scalar product of Maxwell's equations (5b-1) and (5b-2) with **H** and **E**, respectively, and subtracting the resultant equations gives the following energy relation:

$$\mathbf{\nabla} \cdot \mathbf{S}(t) + \mathbf{E}(t) \cdot \mathbf{J}(t) = - \frac{\partial}{\partial t}\left(w_e(t) + w_m(t)\right) \tag{5b-49}$$

where $\mathbf{S}(t) = \mathbf{E}(t) \times \mathbf{H}(t)$, defined as the instantaneous Poynting's vector representing the flow of energy associated with an electromagnetic field; $w_e(t) = \frac{1}{2}\mathbf{E}(t) \cdot \mathbf{D}(t)$, defined as the electric energy per unit volume; and $w_m(t) = \frac{1}{2}\mathbf{H}(t) \cdot \mathbf{B}(t)$, defined as the magnetic energy per unit volume. Equation (5b-49) is the differential form of Poynting's vector theorem. Taking the volume integral of Eq. (5b-49) gives the integral form of Poynting's vector theorem:

$$\int \mathbf{S}(t) \cdot \mathbf{n}\, dA + \int (\mathbf{E} \cdot \mathbf{J})\, dv = - \frac{\partial}{\partial t}\int (w_e + w_m)\, dv \tag{5b-50}$$

The first integral represents the electromagnetic energy flowing out or in per second from a volume v bounded by a surface A. The second integral represents power generated within the volume v; or, if $\mathbf{J} = \sigma\mathbf{E}$, it represents power dissipated as Joules heat in the volume v. The third integral represents the time rate of change of electric and magnetic energy in the volume v.

For time-harmonic fields, we have the following relations in terms of complex quantities:

$$\tfrac{1}{2}\mathbf{\nabla} \cdot \mathbf{S} + \tfrac{1}{2}\mathbf{J} \cdot \mathbf{E}^* = 2i\omega(\bar{w}_m - \bar{w}_e) \tag{5b-51}$$

and

$$\tfrac{1}{2}\int \mathbf{S} \cdot \mathbf{n}\, dA + \tfrac{1}{2}\int(\mathbf{J} \cdot \mathbf{E}^*)\, dv = 2i\omega\int(\bar{w}_m - \bar{w}_e)\, dv \tag{5b-52}$$

where $\mathbf{S} = \mathbf{E} \times \mathbf{H}^*$ is the complex Poynting's vector, $\bar{w}_e$ which is $\frac{1}{4}\mathbf{E} \cdot \mathbf{D}^*$ and $\bar{w}_m$ which is $\frac{1}{4}\mathbf{H} \cdot \mathbf{B}^*$ are time-average energy densities. The asterisks denote complex-conjugate values. Real part of $\frac{1}{2}\int\mathbf{S} \cdot \mathbf{n}\, dA$ gives the time-averaged power generated within the volume v, or, if $\mathbf{J} = \sigma\mathbf{E}$, it represents the time-averaged power dissipated as Joules heat within the volume v.

MAXWELL'S STRESS TENSOR. Neglecting the contribution due to electrostriction and magnetostriction, components of the second-rank Maxwell's stress tensor in a material medium are

$$T_{\alpha\beta} = E_\alpha D_\beta - \tfrac{1}{2}\delta_{\alpha\beta}E_\gamma D_\gamma + H_\alpha B_\beta - \tfrac{1}{2}\delta_{\alpha\beta}H_\gamma B_\gamma \tag{5b-53}$$

where $\delta_{\alpha\beta}$ is the Kronecker delta; $\alpha = x, y, z$; $\beta = x, y, z$; $\gamma = x, y, z$. The volume density of force in the medium is

$$f_\alpha = \frac{\partial T_{\alpha\beta}}{\partial x_\beta} - \frac{\partial}{\partial t}(\mathbf{D} \times \mathbf{B}) \tag{5b-54}$$

where $\partial T_{\alpha\beta}/\partial x_\beta$ is the tensor divergence of $T_{\alpha\beta}$. The total force on a volume element V is given by

$$\mathbf{F}_\alpha = \int_V f_\alpha\, dv \tag{5b-55}$$

The above expression is particularly useful for computation of forces acting on dielectric or magnetic materials by electromagnetic waves.

The Wave Equations for the Field Vectors. By combining Maxwell's equations with the constitutive relations, equations for the field vectors $\mathbf{E}$ and $\mathbf{H}$ can be derived:

IN HOMOGENEOUS ISOTROPIC MEDIA

$$\nabla \times \nabla \times \mathbf{E} + \mu\epsilon \frac{\partial^2 \mathbf{E}}{\partial t^2} = -\mu \frac{\partial \mathbf{J}}{\partial t} - \nabla \times \mathbf{J}_m \tag{5b-56}$$

$$\nabla \times \nabla \times \mathbf{H} + \mu\epsilon \frac{\partial^2 \mathbf{H}}{\partial t^2} = -\epsilon \frac{\partial \mathbf{J}_m}{\partial t} + \nabla \times \mathbf{J} \tag{5b-57}$$

IN INHOMOGENEOUS ISOTROPIC MEDIA

$$\nabla \times \nabla \times \mathbf{E} - \frac{\nabla\mu(\mathbf{r})}{\mu(\mathbf{r})} \times \nabla \times \mathbf{E} + \epsilon(\mathbf{r})\mu(\mathbf{r}) \frac{\partial^2 \mathbf{E}}{\partial t^2} = -\mu(\mathbf{r}) \frac{\partial \mathbf{J}}{\partial t} - \nabla \times \mathbf{J}_m \tag{5b-58}$$

$$\nabla \times \nabla \times \mathbf{H} - \frac{\nabla\epsilon(\mathbf{r})}{\epsilon(\mathbf{r})} \times \nabla \times \mathbf{H} + \epsilon(\mathbf{r})\mu(\mathbf{r}) \frac{\partial^2 \mathbf{E}}{\partial t^2} = -\epsilon(\mathbf{r}) \frac{\partial \mathbf{J}_m}{\partial t} + \nabla \times \mathbf{J} \tag{5b-59}$$

$\epsilon(\mathbf{r})$ and $\mu(\mathbf{r})$ are respectively the inhomogeneous dielectric permittivity and the inhomogeneous magnetic permeability of the medium.[1]

IN HOMOGENEOUS ANISOTROPIC MEDIA. In a general anisotropic medium with $\mathbf{B} = \mathbf{\mu} \cdot \mathbf{H}$ and $\mathbf{D} = \mathbf{\varepsilon} \cdot \mathbf{E}$, the wave equations are expressible only as two coupled second-order differential equations. These equations are usually very involved; only in the special case of a gyromagnetic ferrite medium[2] or a gyroelectric plasma medium[3] have the solutions for these wave equations been found.

IN HOMOGENEOUS ISOTROPIC MOVING MEDIA. In the observer's S system in which a simple homogeneous isotropic medium is moving at velocity $\mathbf{v}$, the wave equations are

$$\left[\nabla^2 - \frac{1}{c^2}\frac{\partial^2}{\partial t^2} - K\gamma^2 \left(\frac{\partial}{\partial t} + \mathbf{v}\cdot\nabla \right)^2 \right] \mathbf{A}(\mathbf{r},t) = -\mu'\mathbf{J} - \frac{\mu'K}{n'^2}\gamma\mathbf{v}(\gamma\mathbf{J}\cdot\mathbf{v} - \gamma c^2\rho) \tag{5b-60}$$

$$\left[\nabla^2 - \frac{1}{c^2}\frac{\partial^2}{\partial t^2} - K\gamma^2 \left(\frac{\partial}{\partial t} + \mathbf{v}\cdot\nabla \right)^2 \right] \phi(\mathbf{r},t) = -\mu'c^2\rho - \frac{\mu'K}{n'^2}\gamma c^2(\gamma\mathbf{J}\cdot\mathbf{v} - \gamma c^2\rho) \tag{5b-61}$$

$$\mathbf{E} = -\nabla\phi - \frac{\partial}{\partial t}\mathbf{A}$$

$$\mathbf{B} = \nabla \times \mathbf{A}$$

where $K = (c^2\epsilon'\mu' - 1)/c^2$, $n' = c\sqrt{\epsilon'\mu'}$, $\gamma = (1 - \beta^2)^{-\frac{1}{2}}$, and $\beta = v/c$. ϵ' and μ' are respectively the dielectric constant and the permeability of the medium in the S' system which is stationary with respect to the medium.

The Vector and Scalar Potentials. THE $\mathbf{A}$ AND ϕ POTENTIALS. The electromagnetic field can, in general, be divided into two parts, one associated with electric-type sources $\mathbf{J}$ and ρ, the other associated with magnetic-type sources $\mathbf{J}_m$ and ρ_m. Each part can be developed by means of vector and scalar potentials as follows:

$$\mathbf{E}_e = -\nabla\phi_e - \frac{\partial \mathbf{A}_e}{\partial t} \tag{5b-62}$$

$$\mathbf{B}_e = \nabla \times \mathbf{A}_e \tag{5b-63}$$

$$\mathbf{D}_m = -\nabla \times \mathbf{A}_m \tag{5b-64}$$

$$\mathbf{H}_m = -\nabla\phi_m - \frac{\partial \mathbf{A}_m}{\partial t} \tag{5b-65}$$

---

[1] J. R. Wait, "Electromagnetic Waves in Stratified Media," Pergamon Press, New York, 1964.

[2] B. Lax, and K. J. Button, "Microwave Ferrites and Ferrimagnetics," McGraw-Hill Book Company, New York, 1962.

[3] M. A. Heald, and C. B. Wharton, "Plasma Diagnostics with Microwaves," John Wiley & Sons, Inc., New York, 1965.

The general representation of the field in terms of potentials is accordingly

$$\mathbf{E} = -\nabla \phi_e - \frac{\partial \mathbf{A}_e}{\partial t} - \frac{1}{\epsilon} \nabla \times \mathbf{A}_m \qquad (5b\text{-}66)$$

$$\mathbf{H} = \frac{1}{\mu} \nabla \times \mathbf{A}_e - \nabla \phi_m - \frac{\partial \mathbf{A}_m}{\partial t} \qquad (5b\text{-}67)$$

For homogeneous isotropic media the differential equations relating the potentials to the source functions are:

$$\nabla^2 \mathbf{A}_e - \mu\epsilon \frac{\partial^2 \mathbf{A}_e}{\partial t^2} = -\mu \mathbf{J} \qquad (5b\text{-}68)$$

$$\nabla^2 \phi_e - \mu\epsilon \frac{\partial^2 \phi_e}{\partial t^2} = -\frac{\rho}{\epsilon} \qquad (5b\text{-}69)$$

$$\nabla^2 \mathbf{A}_m - \mu\epsilon \frac{\partial^2 \mathbf{A}_m}{\partial t^2} = -\epsilon \mathbf{J}_m \qquad (5b\text{-}70)$$

$$\nabla^2 \phi_m - \mu\epsilon \frac{\partial^2 \phi_m}{\partial t^2} = -\frac{\rho_m}{\mu} \qquad (5b\text{-}71)$$

with the auxiliary conditions

$$\nabla \cdot \mathbf{A}_e + \mu\epsilon \frac{\partial \phi_e}{\partial t} = 0 \qquad (5b\text{-}72)$$

$$\nabla \cdot \mathbf{A}_m + \mu\epsilon \frac{\partial \phi_m}{\partial t} = 0 \qquad (5b\text{-}73)$$

THE DEBYE POTENTIALS. For source-free region, it is sometimes more convenient to derive the time-harmonic electromagnetic fields from two scalar potentials as follows:[1]

$$\mathbf{E} = \nabla \times (\mathbf{a}\Psi) + \frac{i}{\omega\epsilon} \nabla \times \nabla \times (\mathbf{a}\Phi) \qquad (5b\text{-}74)$$

$$\mathbf{H} = \nabla \times (\mathbf{a}\Phi) + \frac{1}{i\omega\mu} \nabla \times \nabla \times (\mathbf{a}\Psi) \qquad (5b\text{-}75)$$

where $\mathbf{a}$ is a unit vector or the position vector $\mathbf{r}$. For example, in spherical coordinates $\mathbf{a} = \mathbf{r}$, the radial position vector; in cylindrical coordinates $\mathbf{a} = \mathbf{e}_z$, the axial vector; in rectangular coordinates $\mathbf{a} = \mathbf{e}_x$ or $\mathbf{e}_y$ or $\mathbf{e}_z$, the unit vector in x or y or z direction, respectively. The two scalar functions Ψ and Φ are the Debye potentials which satisfy a pair of second-order differential equations. These differential equations are obtained by substituting Eqs. (5b-74) and (5b-75) into the wave equations. In homogeneous isotropic medium, Ψ and Φ satisfy the scalar Helmholtz equation,

$$(\nabla^2 + k^2) \left\{ \begin{matrix} \Psi \\ \Phi \end{matrix} \right\} = 0 \qquad (5b\text{-}76)$$

with $k^2 = \omega^2 \mu\epsilon$. By choosing $\mathbf{a}$ appropriately, one may also apply Eqs. (5b-74) and (5b-75) to the case of an inhomogeneous medium.[2]

Basic Wave Types

1. Transverse electromagnetic waves (*TEM* waves)—containing neither an electric nor a magnetic field component in the direction of propagation.
2. Transverse magnetic waves (*TM* or *E* waves)—containing an electric field component but not a magnetic field component in the direction of propagation.

---

[1] The vector $\mathbf{a}\Phi$ may be identified as the electric Hertz vector and the vector $\mathbf{a}\Psi$ may be identified as the magnetic Hertz vector. $\mathbf{a}$ in this case is a constant vector.
[2] C. Yeh, *Phys. Rev.* **131**, 2350 (1963).

3. Transverse electric waves (*TE* or *H* waves)—containing a magnetic field component but not an electric field component in the direction of propagation.

4. Hybrid waves (*HE* waves)—containing all components of electric and magnetic fields. These hybrid waves are obtainable by linear superposition of *TE* and *TM* waves.

Formal Solutions for the Time-harmonic Vector Wave Equation. INTEGRAL REPRESENTATIONS. Upon direct integration of the wave equation in homogeneous isotropic medium, integral solutions in terms of the sources can be obtained. A harmonic time dependence of $e^{-i\omega t}$ is assumed and suppressed in this section.

Direct integration of Eqs. (5b-68) to (5b-71) gives

$$\mathbf{A}_e = \frac{\mu}{4\pi} \int_V \mathbf{J}(\mathbf{r}') \frac{e^{ik|\mathbf{r}-\mathbf{r}'|}}{|\mathbf{r}-\mathbf{r}'|} \, dv' \tag{5b-77}$$

$$\phi_e = \frac{1}{4\pi\epsilon} \int_V \rho(\mathbf{r}') \frac{e^{ik|\mathbf{r}-\mathbf{r}'|}}{|\mathbf{r}-\mathbf{r}'|} \, dv' \tag{5b-78}$$

$$\mathbf{A}_m = \frac{\epsilon}{4\pi} \int_V \mathbf{J}_m(\mathbf{r}') \frac{e^{ik|\mathbf{r}-\mathbf{r}'|}}{|\mathbf{r}-\mathbf{r}'|} \, dv' \tag{5b-79}$$

$$\phi_m = \frac{1}{4\pi\mu} \int_V \rho_m(\mathbf{r}') \frac{e^{ik|\mathbf{r}-\mathbf{r}'|}}{|\mathbf{r}-\mathbf{r}'|} \, dv' \tag{5b-80}$$

where $\mathbf{r}$ is the coordinate of the observation point and $\mathbf{r}'$ is the coordinate of the source point. The integration with respect to the primed coordinates extends throughout the volume V occupied by the source.

Direct integration of Eqs. (5b-56) and (5b-57) gives

$$\mathbf{E} = i\omega\mu \int_V \mathbf{\Gamma}(\mathbf{r},\mathbf{r}') \cdot \mathbf{J}(\mathbf{r}') \, dv' - \int_V \nabla G(\mathbf{r},\mathbf{r}') \times \mathbf{J}_m(\mathbf{r}') \, dv' \tag{5b-81}$$

$$\mathbf{H} = i\omega\epsilon \int_V \mathbf{\Gamma}(\mathbf{r},\mathbf{r}') \cdot \mathbf{J}_m(\mathbf{r}') \, dv' + \int_V \nabla G(\mathbf{r},\mathbf{r}') \times \mathbf{J}(\mathbf{r}') \, dv' \tag{5b-82}$$

with

$$\mathbf{\Gamma}(\mathbf{r},\mathbf{r}') = \left(u + \frac{1}{k^2} \nabla\nabla \right) G(\mathbf{r},\mathbf{r}') \tag{5b-83}$$

$$G(\mathbf{r},\mathbf{r}') = \frac{e^{ik|\mathbf{r}-\mathbf{r}'|}}{4\pi|\mathbf{r}-\mathbf{r}'|} \tag{5b-84}$$

The properties of u and $\nabla\nabla$ are $u \cdot \mathbf{c} = \mathbf{c}$ and $(\nabla\nabla) \cdot \mathbf{c} = \nabla(\nabla \cdot \mathbf{c})$ where $\mathbf{c}$ is any vector function. The gradient operator ∇ is defined as

$$\nabla = \mathbf{e}_x \frac{\partial}{\partial x} + \mathbf{e}_y \frac{\partial}{\partial y} + \mathbf{e}_z \frac{\partial}{\partial z}$$

where $\mathbf{e}_x$, $\mathbf{e}_y$, $\mathbf{e}_z$ are unit vectors for the unprimed coordinates. It is noted that the radiation condition has been met by Eqs. (5b-77) to (5b-82). Hence, these equations represent the integral expressions for the fields in unbounded homogeneous region.

More general integral expressions for the field vectors in terms of the current sources and the field values over the bounding surfaces S_i are also available:

$$\mathbf{E}(\mathbf{r}) = i\omega\mu \int_V \mathbf{\Gamma}(\mathbf{r},\mathbf{r}') \cdot \mathbf{J}(\mathbf{r}') \, dv' + \int_V \nabla' G(\mathbf{r},\mathbf{r}') \times \mathbf{J}_m(\mathbf{r}') \, dv'$$
$$+ \int_{S_1\cdots S_n} [(\mathbf{n}' \cdot \mathbf{E}(\mathbf{r}'))\nabla' G(\mathbf{r},\mathbf{r}') + (\mathbf{n}' \times \mathbf{E}(\mathbf{r}')) \times \nabla' G(\mathbf{r},\mathbf{r}')$$
$$+ i\omega\mu G(\mathbf{r},\mathbf{r}')(\mathbf{n}' \times \mathbf{H}(\mathbf{r}'))] \, dS' \tag{5b-85}$$

$$\mathbf{H}(\mathbf{r}) = i\omega\epsilon \int_V \mathbf{\Gamma}(\mathbf{r},\mathbf{r}') \cdot \mathbf{J}_m(\mathbf{r}') \, dv' - \int_V \nabla' G(\mathbf{r},\mathbf{r}') \times \mathbf{J}(\mathbf{r}') \, dv'$$
$$+ \int_{S_1\cdots S_n} \{[\mathbf{n}' \times \mathbf{H}(\mathbf{r}')] \times \nabla' G(\mathbf{r},\mathbf{r}') + [\mathbf{n}' \cdot \mathbf{H}(\mathbf{r}')]\nabla' G(\mathbf{r},\mathbf{r}')$$
$$- i\omega\epsilon G(\mathbf{r},\mathbf{r}')[\mathbf{n}' \times \mathbf{E}(\mathbf{r}')]\} \, dS \tag{5b-86}$$

where region V is assumed to be bounded by the surfaces $S_1 \cdots S_n$, S_n is a surface enclosing V, and $\mathbf{n}'$ is the inward-drawn normal from any boundary surface S_i into the volume V. The gradient operator ∇' is with respect to the primed coordinates. $G(\mathbf{r},\mathbf{r}')$ is the scalar Green's function. For an unbounded region, S_n recedes to infinity, $G(\mathbf{r},\mathbf{r}')$ is given by Eq. (5b-84), and Eqs. (5b-85) and (5b-86) represent fields in unbounded region. For a bounded region, the scalar Green's function is

$$G(\mathbf{r},\mathbf{r}') = \frac{e^{ik|\mathbf{r}-\mathbf{r}'|}}{4\pi|\mathbf{r}-\mathbf{r}'|} + g(\mathbf{r},\mathbf{r}') \tag{5b-87}$$

where $g(\mathbf{r},\mathbf{r}')$ satisfies

$$\nabla'^2 g + k^2 g = 0 \tag{5b-88}$$

everywhere in the volume V and over the boundary surfaces $S_1 \cdots S_n$.

SEPARATION OF VARIABLES. Only in five coordinate systems is the method of separation of variables applicable to the source-free vector-wave equation in homogeneous medium [i.e., Eq. (5b-56) or (5b-57) with $\mathbf{J} = \mathbf{J}_m = 0$]. For some specific variation of $\epsilon(\mathbf{r})$ and $\mu(\mathbf{r})$, the method of separation of variables may also be used to solve the source-free vector-wave equation in inhomogeneous medium.[1]

1. **Rectangular Coordinates.** In the rectangular coordinates x, y, z the three distinct types of basic rectangular wave functions, characterized by the relationship between the field vectors and the z axis (the direction of propagation), are:

TEM Waves

$$\mathbf{E} = \mathbf{E}_0 e^{ikz} \tag{5b-89}$$

$$\mathbf{H} = \left(\frac{\epsilon}{\mu}\right)^{\frac{1}{2}} (\mathbf{e}_z \times \mathbf{E}_0) e^{ikz}$$

TM Waves

$$\mathbf{H} = \nabla \times (\mathbf{e}_z \Phi) \qquad \mathbf{E} = \frac{i}{\omega\epsilon} \nabla \times \nabla \times (\mathbf{e}_z \Phi) \tag{5b-90}$$

$$\Phi = e^{\pm ik_x x} e^{\pm ik_y y} e^{\pm i\gamma z}$$
$$\gamma = (k^2 - k_x^2 - k_y^2)^{\frac{1}{2}}$$

TE Waves

$$\mathbf{E} = \nabla \times (\mathbf{e}_z \Psi) \qquad \mathbf{H} = \frac{1}{i\omega\mu} \nabla \times \nabla \times (\mathbf{e}_z \Psi) \tag{5b-91}$$

$$\Psi = e^{\pm ik_x x} e^{\pm ik_y y} e^{\pm i\gamma z}$$
$$\gamma = (k^2 - k_x^2 - k_y^2)^{\frac{1}{2}}$$

2. **Circular Cylindrical Coordinates.** In the circular cylindrical coordinates ρ, ϕ, z the three distinct types of basic circular cylindrical wave functions, characterized by the relationship between the field vectors and the z axis (the direction of propagation), are:

TEM Waves

$$E_z = H_z = 0$$

$$E_\rho = \frac{\partial U_n}{\partial \rho} e^{ikz} \qquad H_\phi = \frac{1}{\rho} \frac{\partial U_n}{\partial \phi} e^{ikz} \tag{5b-92}$$

or

$$H_\rho = \frac{\partial V_n}{\partial \rho} e^{ikz} \qquad E_\phi = \frac{1}{\rho} \frac{\partial V_n}{\partial \phi} e^{ikz} \tag{5b-93}$$

where U_n or V_n are solutions of Laplace's equation in two dimensions; explicitly

$$U_n \text{ or } V_n = \begin{Bmatrix} \rho^n \\ \rho^{-n} \end{Bmatrix} e^{in\phi} \tag{5b-94}$$

$$U_0 \text{ or } V_0 = \begin{Bmatrix} \ln \rho \\ \text{constant} \end{Bmatrix} \begin{Bmatrix} \theta \\ \text{constant} \end{Bmatrix} \tag{5b-95}$$

[1] C. Yeh, and K. F. Casey, *IEEE Trans. Microwave Theory and Tech.* **MTT-13**, 297 (1965).

TM Waves

$$\mathbf{H} = \nabla \times (\mathbf{e}_z \Phi) \qquad \mathbf{E} = \frac{i}{\omega \epsilon} \nabla \times \nabla \times (\mathbf{e}_z \Phi) \tag{5b-96}$$

$$\Phi = \{Z_\nu(\Lambda \rho)\} \{e^{\pm i\nu\phi}\} \{e^{\pm i\gamma z}\} \tag{5b-97}$$

$$\gamma = (k^2 - \Lambda^2)^{\frac{1}{2}} \tag{5b-98}$$

where $Z_\nu(\Lambda \rho)$ are two linearly independent solutions to the Bessel differential equation of order ν.

TE Waves

$$\mathbf{E} = \nabla \times (\mathbf{e}_z \Psi) \qquad \mathbf{H} = \frac{1}{i\omega\mu} \nabla \times \nabla \times (\mathbf{e}_z \Psi) \tag{5b-99}$$

$$\Psi = \{Z_\nu(\Lambda \rho)\} \{e^{\pm i\nu\phi}\} \{e^{\pm i\gamma z}\} \tag{5b-100}$$

$$\gamma = (k^2 - \Lambda^2)^{\frac{1}{2}} \tag{5b-101}$$

where $Z_\nu(\Lambda \rho)$ have been defined earlier.

3. **Spherical Coordinates.** In the spherical coordinates r, θ, ϕ the three distinct types of basic spherical wave functions, characterized by the relationship between the field vectors and the radial r direction (the direction of propagation), are:

TEM Waves

$$E_r = H_r = 0$$

$$E_\theta = \frac{e^{\pm ikr}}{r \sin \theta} \qquad H_\phi = \pm \sqrt{\frac{\epsilon}{\mu}} E_\theta \tag{5b-102}$$

TM Waves

$$\mathbf{H} = \nabla \times (\mathbf{e}_r r\Phi) \qquad \mathbf{E} = \frac{i}{\omega\epsilon} \nabla \times \nabla \times (\mathbf{e}_r r\Phi) \tag{5b-103}$$

$$\Phi = \{z_n(kr)\} \begin{Bmatrix} P_n{}^m \\ Q_n{}^m \end{Bmatrix} (\cos \theta) \} \{e^{\pm im\phi}\} \tag{5b-104}$$

where $z_n(kr)$ are two linearly independent solutions to the spherical Bessel differential equation and are related to the cylinder function by the expression

$$z_n(kr) = \left(\frac{\pi}{2kr}\right)^{\frac{1}{2}} Z_{n+\frac{1}{2}}(kr)$$

$P_n{}^m(\cos \theta)$ and $Q_n{}^m(\cos \theta)$ are two linearly independent solutions to the associated Legendre differential equation.[1]

TE Waves

$$\mathbf{E} = \nabla \times (\mathbf{e}_r r\Psi) \qquad \mathbf{H} = \frac{1}{i\omega\mu} \nabla \times \nabla \times (\mathbf{e}_r r\Psi) \tag{5b-105}$$

$$\Psi = \{z_n(kr)\} \begin{Bmatrix} P_n{}^m \\ Q_n{}^m \end{Bmatrix} (\cos \theta) \} \{e^{\pm im\phi}\} \tag{5b-106}$$

4. **Elliptical Cylinder Coordinates.** In the elliptical coordinates[2] ξ, η, z the two distinct types of basic elliptical cylindrical wave functions, characterized by the relationship between the field vectors and the z axis (the direction of propagation), are

TM Waves

$$\mathbf{H}^{(e,0)} = \nabla \times (\mathbf{e}_z \Phi^{(e,0)}) \qquad \mathbf{E}^{(e,0)} = \frac{i}{\omega\epsilon} \nabla \times \nabla \times (\mathbf{e}_z \Phi^{(e,0)}) \tag{5b-107}$$

$$\Phi^{(e)} = \{Mc_n{}^{(1),(2)}(\xi,q)\} \{ce_n(\eta,q)\} \{e^{\pm i\gamma z}\} \tag{5b-108}$$

$$\Phi^{(0)} = \{Ms_n{}^{(1),(2)}(\xi,q)\} \{se_n(\eta,q)\} \{e^{\pm i\gamma z}\} \tag{5b-109}$$

$$q = \frac{c}{2}(k^2 - \gamma^2)^{\frac{1}{2}}$$

[1] W. Magnus and F. Oberhettinger, "Formulas and Theorems for the Functions of Mathematical Physics," Chelsea Publishing Company, New York, 1954.

[2] In terms of the rectangular coordinates x, y, the elliptical coordinates ξ, η are defined by the following relations: $x = c \cosh \xi \cos \eta$, $y = c \sinh \xi \sin \eta$ $(0 \leq \xi < \infty, 0 \leq \eta \leq 2\pi)$, where c is the semifocal length.

where $ce_n(\eta,q)$ and $se_n(\eta,q)$ are respectively even and odd periodic angular Mathieu functions. The radial Mathieu functions corresponding to the even function $ce_n(\eta,q)$ having the same characteristic values are $Mc_n^{(1),(2)}(\xi,q)$, and those corresponding to the odd function[1] $se_n(\eta,q)$ are $Ms_n^{(1),(2)}(\xi,q)$.

TE Waves

$$\mathbf{E}^{(e,0)} = \nabla \times (\mathbf{e}_z \Psi^{(e,0)}) \qquad \mathbf{H}^{(e,0)} = \frac{1}{i\omega\mu} \nabla \times \nabla \times (\mathbf{e}_z \Psi^{(e,0)}) \qquad (5b\text{-}110)$$

$$\Psi^{(e)} = \{Mc_n^{(1),(2)}(\xi,q)\}\{ce_n(\eta,q)\}\{e^{\pm i\gamma z}\}$$
$$\Psi^{(0)} = \{Ms_n^{(1),(2)}(\xi,q)\}\{se_n(\eta,q)\}\{e^{\pm i\gamma z}\}$$

$$q = \frac{c}{2}(k^2 - \gamma^2)^{\frac{1}{2}}$$

5. Parabolic Cylinder Coordinates. In the parabolic coordinates[2] ξ, η, z the two distinct types of basic parabolic cylindrical wave functions, characterized by the relationship between the field vectors and the z axis (the direction of propagation), are:

TM Waves

$$\mathbf{H} = \nabla \times (\mathbf{e}_z\Phi) \qquad \mathbf{E} = \frac{i}{\omega\epsilon}\nabla \times \nabla \times (\mathbf{e}_z\Phi) \qquad (5b\text{-}111)$$

$$\Phi = \{U_m^{(1),(2)}(\xi)\}\{V_m^{(1),(2)}(\eta)\}\{e^{\pm i\gamma z}\} \qquad (5b\text{-}112)$$

where U and V satisfy Weber's equation of the confluent hypergeometric type,[3]

$$\left[\frac{\partial^2}{\partial\xi^2} + (q^2\xi^2 + m)\right]U(\xi) = 0 \qquad (5b\text{-}113)$$

$$\left[\frac{\partial^2}{\partial\eta^2} + (q^2\eta^2 - m)\right]V(\eta) = 0 \qquad (5b\text{-}114)$$

$$q^2 = k^2 - \gamma^2$$

TE Wave

$$\mathbf{E} = \nabla \times (\mathbf{e}_z\Psi) \qquad \mathbf{H} = \frac{1}{i\omega\mu}\nabla \times \nabla \times (\mathbf{e}_z\Psi) \qquad (5b\text{-}115)$$

$$\Psi = \{U_m^{(1),(2)}(\xi)\}\{V_m^{(1),(2)}(\eta)\}\{e^{\pm i\gamma z}\} \qquad (5b\text{-}116)$$

Polarization of Waves. Consider a plane wave in free space propagating in the z direction and having the following components:

$$\mathbf{E} = \mathbf{e}_x E_1 e^{+ikz-i\omega t} + \mathbf{e}_y E_2 e^{ikz-i\omega t} \qquad (5b\text{-}117)$$

$$\mathbf{B} = -\mathbf{e}_x E_2 \sqrt{\mu_0\epsilon_0}\, e^{ikz-i\omega t} + \mathbf{e}_y E_1 \sqrt{\mu_0\epsilon_0}\, e^{ikz-i\omega t} \qquad (5b\text{-}118)$$

with $k = \omega\sqrt{\mu_0\epsilon_0}$. Note that (E_x, B_y) and (E_y, B_x) are linearly independent fields, and E_1 and E_2 are complex constants.

LINEARLY POLARIZED WAVE. E_1 and E_2 have the same phase. In this case $\mathbf{E}$ at any point in space oscillates along a directional line which makes a constant angle ϕ with the x axis, this angle being given by $\phi = \tan^{-1}(E_2/E_1)$.

CIRCULARLY POLARIZED WAVE. E_1 and E_2 have the same magnitude but their phases differ by 90°. Hence

$$\mathbf{E} = \text{Re }(\mathbf{e}_x \pm i\mathbf{e}_y)E_1 e^{ikz-i\omega t}$$
$$= E_1[\mathbf{e}_x \cos(\omega t - kz) \pm \mathbf{e}_y \sin(\omega t - kz)]$$

Hence $\mathbf{E}$ at any point in space does not oscillate. Its magnitude is constant, but its direction rotates at the angular velocity ω. When $E_2 = -iE_1$, the wave is said to be

[1] J. Meixner and F. W. Schäfke, "Mathieu-funktionen und Spharoid-funktionen," Springer-Verlag OHG, Berlin, 1954.
[2] In terms of the rectangular coordinates x, y, the parabolic coordinates ξ, η are defined by the following relations: $x = \frac{1}{2}(\xi^2 - \eta^2)$, $y = \xi\eta$ $(-\infty < \xi < \infty, 0 \leq \eta < \infty)$.
[3] S. O. Rice, *Bell System Tech. J.* **33**, 417 (1954).

right-handed circularly polarized. When $E_2 = iE_1$, the wave is said to be left-handed circularly polarized.

ELLIPTICALLY POLARIZED WAVE. E_1 and E_2 have arbitrary relative amplitudes and phases. At any point in space the tip of **E** describes a locus which is an ellipse.

References

1. Stratton, J. A.: "Electromagnetic Theory," McGraw-Hill Book Company, New York, 1941.
2. Smythe, W. R.: "Static and Dynamic Electricity," 3d ed., McGraw-Hill Book Company, New York, 1968.
3. Elliott, R. S.: "Electromagnetics," McGraw-Hill Book Company, New York, 1966.
4. Papas, C. H.: "Theory of Electromagnetic Wave Propagation," McGraw-Hill Book Company, New York, 1965.
5. Van Bladel, J.: "Electromagnetic Fields," McGraw-Hill Book Company, New York, 1964.
6. Panofsky, W. K. H., and M. Phillips: "Classical Electricity and Magnetism," 2d ed., Addison-Wesley Publishing Company, Inc., Reading, Mass., 1962.
7. Kraichman, M. B., "Handbook of Electromagnetic Propagation in Conducting Media," NAVMAT P-2302, 1970, U.S. Government Printing Office, Washington, D.C. 20402.

5b-9. Guided Waves. In this section some basic properties of guided waves are given. These properties are found from the solutions that satisfy the source-free Maxwell's equations and the appropriate boundary conditions. When the guided waves propagate along a straight-line path, one may assume that every component of the electromagnetic wave may be represented in the form

$$f(u,v)e^{i\gamma z}e^{-i\omega t} \tag{5b-119}$$

in which z is chosen as the propagation direction and u, v are generalized orthogonal coordinates in a transverse plane.[1] γ is the propagation constant. Under this assumption, the transverse field components in homogeneous isotropic medium (ϵ,μ) are

$$E_u = \frac{1}{\omega^2\mu\epsilon - \gamma^2}\left(\frac{i\gamma}{h_1}\frac{\partial E_z}{\partial u} + \frac{i\omega\mu}{h_2}\frac{\partial H_z}{\partial v}\right) \tag{5b-120}$$

$$E_v = \frac{1}{\omega^2\mu\epsilon - \gamma^2}\left(\frac{i\gamma}{h_2}\frac{\partial E_z}{\partial v} - \frac{i\omega\mu}{h_1}\frac{\partial H_z}{\partial u}\right) \tag{5b-121}$$

$$H_u = \frac{1}{\omega^2\mu\epsilon - \gamma^2}\left(\frac{-i\omega\epsilon}{h_2}\frac{\partial E_z}{\partial v} + \frac{i\gamma}{h_1}\frac{\partial H_z}{\partial u}\right) \tag{5b-122}$$

$$H_v = \frac{1}{\omega^2\mu\epsilon - \gamma^2}\left(\frac{i\omega\epsilon}{h_1}\frac{\partial E_z}{\partial u} + \frac{i\gamma}{h_2}\frac{\partial H_z}{\partial v}\right) \tag{5b-123}$$

and the longitudinal field components satisfy the following equation:

$$\left[\frac{1}{h_1h_2}\left(\frac{\partial}{\partial u}\frac{h_2}{h_1}\frac{\partial}{\partial u} + \frac{\partial}{\partial v}\frac{h_1}{h_2}\frac{\partial}{\partial v}\right) + \Gamma^2\right]\begin{Bmatrix}E_z\\H_z\end{Bmatrix} = 0 \qquad \Gamma^2 = \omega^2\mu\epsilon - \gamma^2 \tag{5b-124}$$

Only discrete values of Γ^2 will satisfy the boundary conditions. These allowed Γ^2 values are called *eigenvalues;* and corresponding to these eigenvalues are the *eigenfunctions*. The orthogonality properties of the field components can therefore be found according to the well-known orthogonality properties of the eigenfunction.

It will be recalled from Sec. 5b-8 (Basic Wave Types) that *TM* modes refer to waves having $H_z = 0$, *TE* modes having $E_z = 0$, HE modes having all field components $\neq 0$, and *TEM* modes having $E_z = 0$ and $H_z = 0$.

Propagation Characteristics. Propagation characteristics of guided waves refer to the behavior of the propagation constant γ as a function of frequency. In general,

[1] J. A. Stratton, "Electromagnetic Theory," chap. 1, McGraw-Hill Book Company, New York, 1941.

γ may be complex: $\gamma = i\alpha + \beta$, where α is the attenuation constant and β is a phase constant. Several commonly used terms to describe guided waves are defined as follows:

Cutoff frequency,
$$f_c = \frac{\Gamma}{2\pi\sqrt{\mu\epsilon}}$$

Cutoff wavelength,
$$\lambda_c = \frac{1}{f_c\sqrt{\mu\epsilon}}$$

Guide wavelength,
$$\lambda_g = \frac{2\pi}{\gamma} = \frac{2\pi}{(k^2-\Gamma^2)^{\frac{1}{2}}} \qquad k = \omega\sqrt{\mu\epsilon}$$

Phase velocity,
$$v_p = \frac{\omega}{\gamma} \qquad (5b\text{-}125)$$

Group velocity,
$$v_g = \frac{d\omega}{d\gamma} = \frac{\gamma}{\omega\epsilon\mu}\cdot\frac{1}{1-(\Gamma/k)(d\Gamma/dk)}$$

The above considerations are applicable for TE, TM, or HE waves only. For TEM modes, we have $\gamma = k$, $f_c = 0$, $v_p = \omega/k$, and $\lambda_g = 2\pi/k$ with $k = \omega\sqrt{\mu\epsilon}$.

Bounded Waveguides. Only TM waves and TE waves are physically possible in a cylindrical region bounded by a simply connected conducting region. However, in a coaxial region with perfectly conducting walls, a TEM as well as TM and TE waves can be present.

The propagation parameters for cylindrical waveguides bounded by good (but not perfectly) conducting walls are summarized as follows:

FOR PROPAGATING MODES, $f > f_c$

$$\alpha = 0$$
$$\gamma = \beta = k\left[1-\left(\frac{f_c}{f}\right)^2\right]^{\frac{1}{2}}$$
$$v_p = \frac{\omega}{\beta} = \frac{1}{\sqrt{\mu\epsilon(1-(f_c/f)^2)}} \qquad (5b\text{-}126)$$
$$v_g = \frac{1}{\mu\epsilon v_p}$$

FOR NONPROPAGATING MODES (THE EVANESCENT WAVES), $f < f_c$

$$\beta = 0$$
$$\gamma = i\alpha = ik\left[\left(\frac{f_c}{f}\right)^2-1\right]^{\frac{1}{2}} \qquad (5b\text{-}127)$$

ATTENUATION DUE TO IMPERFECTLY CONDUCTING WALLS

$$\alpha_w = \frac{\text{power loss}}{2\text{ power transfer}} = \frac{W_L}{2W_T} \quad \text{nepers/m} \qquad (5b\text{-}128)$$

$$W_L = \frac{R_s}{2}\oint_L [\|H_t\|^2 + \|H_z\|^2]\,dl \qquad (5b\text{-}129)$$

$$W_T = \frac{1}{2}\int_A \|E_t\|\|H_t\|\,dS \qquad (5b\text{-}130)$$

TM modes
$$\begin{cases} W_L \approx \frac{R_s}{8\pi^2\mu^2 f_c^2}\left(\frac{f}{f_c}\right)^2 \oint_L \left(\frac{\partial E_z}{\partial n}\right)^2 dl & (5b\text{-}131) \\[2ex] W_T \approx \frac{1}{2(\mu/\epsilon)^{\frac{1}{2}}}\left[1-\left(\frac{f_c}{f}\right)^2\right]^{\frac{1}{2}}\left(\frac{f}{f_c}\right)^2 \int_A E_z^2\,dS & (5b\text{-}132) \end{cases}$$

TE modes
$$\begin{cases} W_L \approx \frac{R_s}{2}\oint_L \left[H_z^2 + \left(\frac{f}{f_c}\right)^2\frac{1-(f_c/f)^2}{4\pi^2 f_c^2\mu\epsilon}\left(\frac{\partial H_z}{\partial l}\right)^2\right] dl & (5b\text{-}133) \\[2ex] W_T \approx \left(\frac{\mu}{\epsilon}\right)^{\frac{1}{2}}\left(\frac{f}{f_c}\right)^2\left[1-\left(\frac{f_c}{f}\right)^2\right]^{\frac{1}{2}}\int_A H_z^2\,dS & (5b\text{-}134) \end{cases}$$

where $|H_t|^2$ is the square of the total transverse magnetic field, $\partial/\partial n$ is the normal derivative at the boundary conducting wall, $\partial H_z/\partial l$ is the derivative of H_z tangent to curve L along the cross-sectional bounding wall, A is the cross-sectional area of the guide, $R_s = (\pi f \mu/\sigma_c)^{\frac{1}{2}}$ is the surface resistance, and σ_c is the conductivity of the boundary conductor.

ATTENUATION DUE TO IMPERFECT DIELECTRIC

$$\alpha_d \approx \frac{(\mu/\epsilon_r)^{\frac{1}{2}}\sigma_d}{2[1 - (f_c/f)^2]^{\frac{1}{2}}} \qquad \frac{\sigma_d}{\omega\epsilon_r} \ll 1 \qquad (5b\text{-}135)$$

where $\epsilon = \epsilon_r(1 - \sigma_d/i\omega\epsilon_r)$, σ_d is the conductivity of the dielectric in the guide, and ϵ_r is the real part of the dielectric constant ϵ. α_d is in nepers/meter.

The above approximate expressions for the attenuation constant are not valid for frequencies very close to the cutoff frequencies and for very high frequencies. It has also been assumed that the field configurations are not affected by the presence of small wall and dielectric losses.

Field components and propagation parameters for waves guided in rectangular and circular tubes are summarized in Table 5b-1. Table 5b-2 provides the field configurations for several lower-order modes in rectangular and circular waveguides. In a bounded waveguide, an arbitrary field $\mathbf{E}$ or $\mathbf{H}$ within the waveguide may be expanded in terms of the mode functions as follows:

$$\mathbf{E} = \sum_p A_p\mathbf{E}^{TM} + B_p\mathbf{E}^{TE}$$

$$\mathbf{H} = \sum_p B_p\mathbf{H}^{TE} + A_p\mathbf{H}^{TM}$$

i.e., the mode functions for TE and TM waves are a complete set.

For details concerning bounded waveguides of other simple shapes (such as elliptical, parabolic, triangular, etc.), the reader is referred to the literature.[1] For waveguides of arbitrary cross-sectional shape for which solutions in terms of known and tabulated eigenfunctions are not available, one must resort to numerical means[2] or to approximations based on variational techniques.[3] Numerically speaking, the problem reduces to finding the eigenvalue Γ which satisfies the Helmholz equation $(\nabla^2 + \Gamma^2)F = 0$ and the boundary condition $F = 0$ on C for TM waves and $\partial F/\partial n = 0$ on C for TE waves by the use of a computer. The well-known difference method has been used successfully for this type of problem.[3] The variational method offers a way to obtain a rather accurate value for the eigenvalue Γ which is related to the propagation constant γ by the relation $\gamma = \sqrt{\omega^2\mu\epsilon - \Gamma^2}$, from the knowledge of an approximate field configuration (i.e., a trial function). Specifically, for a TM modes, if a trial function $u(x,y)$ vanishes on the boundary and satisfies the conditions

$$\int_A uE_z^{(0)}\,dx\,dy = 0, \int_A uE_z^{(1)}\,dx\,dy = 0, \ldots, \int_A uE_z^{(n-1)}\,dx\,dy = 0 \quad (5b\text{-}136)$$

where $E_z^{(0)}, E_z^{(1)}, \ldots, E_z^{(n-1)}$ are the eigenfunctions for the equation

$$(\nabla^2 + \Gamma_n^2)E_z^{(n)} = 0$$

[1] F. E. Borgnis and C. H. Papas, Electromagnetic Waveguides and Resonators, "Handbuch der Physik," vol. 16, Springer-Verlag OHG, Berlin, 1958.
[2] R. F. Harrington, Field Computation by Moment Methods, the Macmillan Company, New York, 1968.
[3] F. E. Borgnis and C. H. Papas, "Randwertproblems der Mikrowellenphysik," Springer-Verlag OHG, Berlin, 1955.

TABLE 5b-1. FORMULAS FOR RECTANGULAR AND CIRCULAR WAVEGUIDES

Rectangular waveguides

TM waves

Field components — TM_{mn} modes

$$E_z = E_0 \sin\left(\frac{m\pi}{a}x\right)\sin\left(\frac{n\pi}{b}y\right)$$
$$H_x = \frac{-i\omega\epsilon}{4\pi^2 f_c^2}\frac{\partial E_z}{\partial y}$$
$$H_y = \frac{i\omega\epsilon}{4\pi^2 f_c^2}\frac{\partial E_z}{\partial x}$$
$$E_x = \frac{i\gamma}{4\pi^2 f_c^2}\frac{\partial E_z}{\partial x}$$
$$E_y = \frac{i\gamma}{4\pi^2 f_c^2}\frac{\partial E_z}{\partial y}$$
$$H_z = 0$$

Propagation parameters

$$(f_c)_{mn} = \frac{1}{2\sqrt{\mu\epsilon}}\left[\left(\frac{m}{a}\right)^2 + \left(\frac{n}{b}\right)^2\right]^{\frac{1}{2}}$$
$$(\lambda_w)_{mn} = \frac{1/(f\sqrt{\mu\epsilon})}{\{1 - [(f_c)_{mn}/f]^2\}^{\frac{1}{2}}}$$
$$\gamma = \frac{2\pi}{(\lambda_w)_{mn}}$$
$$(\alpha_w)_{mn} = \frac{2R_s}{b(\mu/\epsilon)^{\frac{1}{2}}[1-(f_c/f)^2]^{\frac{1}{2}}} \times \left[\frac{m^2(b/a)^3 + n^2}{m^2(b/a)^2 + n^2}\right]$$

TE waves

Field components — TE_{mn} modes

$$H_z = H_0 \cos\left(\frac{m\pi}{a}x\right)\cos\left(\frac{n\pi}{b}y\right)$$
$$E_x = \frac{i\omega\mu}{4\pi^2 f_c^2}\frac{\partial H_z}{\partial y}$$
$$E_y = \frac{-i\omega\mu}{4\pi^2 f_c^2}\frac{\partial H_z}{\partial x}$$
$$H_x = \frac{i\gamma}{4\pi^2 f_c^2}\frac{\partial H_z}{\partial x}$$
$$H_y = \frac{i\gamma}{4\pi^2 f_c^2}\frac{\partial H_z}{\partial y}$$
$$E_z = 0$$

Propagation parameters

$(f_c)_{mn}, (\lambda_w)_{mn}, \gamma$ for TE waves are the same as those for TM waves

$$(\alpha_w)_{mn} = \frac{2R_s}{b(\mu/\epsilon)^{\frac{1}{2}}[1-(f_c/f)^2]^{\frac{1}{2}}} \times \left\{\left(1+\frac{b}{a}\right)\left(\frac{f_c}{f}\right)^2 + \left[1-\left(\frac{f_c}{f}\right)^2\right]\frac{(b/a)[(b/a)m^2 + n^2]}{(b^2m^2/a^2) + n^2}\right\}$$

Notes: 1. a = length of the waveguide along x axis; b = length of the waveguide along y axis
2. m, n are integers; for TM waves $m \neq 0, n \neq 0$; For TE waves, either m or n may be zero, but not both.
3. Dominant mode is the TE_{10} wave having the lowest cutoff frequency.

Circular waveguides

TM waves

Field components — TM_{nl} modes

$$\left\{E_z = E_0 J_n(\Gamma_c\rho)\begin{Bmatrix}\cos \\ \sin\end{Bmatrix}n\phi\right\}$$
$$H_z = 0$$
$$H_\rho = \frac{-i\omega\epsilon}{4\pi^2 f_c^2\rho}\frac{\partial E_z}{\partial \phi}$$
$$H_\phi = \frac{i\omega\epsilon}{4\pi^2 f_c^2}\frac{\partial E_z}{\partial \rho}$$
$$E_\rho = \frac{i\gamma}{4\pi^2 f_c^2}\frac{\partial E_z}{\partial \rho}$$
$$E_\phi = \frac{i\gamma}{4\pi^2 f_c^2\rho}\frac{\partial E_z}{\partial \phi}$$

Propagation parameters

$$(\Gamma_c)_{nl} = \frac{p_{nl}}{a}; \quad p_{nl} = l\text{th root of } J_n(p_{nl}) = 0$$
$$(\lambda_c)_{nl} = \frac{2\pi a}{p_{nl}}$$
$$(f_c)_{nl} = \frac{p_{nl}}{2\pi a\sqrt{\mu\epsilon}}$$
$$(\lambda_w)_{nl} = \frac{1/(f\sqrt{\mu\epsilon})}{\{1 - [f_c/f]^2\}^{\frac{1}{2}}}$$
$$\gamma = \frac{2\pi}{(\lambda_w)_n}$$
$$(\alpha_w)_{nl} = \frac{R_s}{a(\mu/\epsilon)^{\frac{1}{2}}[1-(f_c/f)^2]^{\frac{1}{2}}}$$

TE waves

Field components — TE_{nl} modes

$$H_z = H_0 J_n(\Gamma_c\rho)\begin{Bmatrix}\cos \\ \sin\end{Bmatrix}n\phi$$
$$E_z = 0$$
$$E_\rho = \frac{i\omega\mu}{4\pi^2 f_c^2\rho}\frac{\partial H_z}{\partial \phi}$$
$$E_\phi = \frac{-i\omega\mu}{4\pi^2 f_c^2}\frac{\partial H_z}{\partial \rho}$$
$$H_\rho = \frac{i\gamma}{4\pi^2 f_c^2}\frac{\partial H_z}{\partial \rho}$$
$$H_\phi = \frac{i\gamma}{4\pi^2 f_c^2\rho}\frac{\partial H_z}{\partial \phi}$$

Propagation parameters

$$(\Gamma_c)_{nl} = \frac{p'_{nl}}{a}; \quad p'_{nl} = l\text{th roots of } J_n'(p'_{nl}) = 0$$
$$(\lambda_c)_{nl} = \frac{2\pi a}{p'_{nl}}$$
$$(f_c)_{nl} = \frac{p'_{nl}}{2\pi a\sqrt{\mu\epsilon}}$$
$$(\lambda_w)_{nl} = \frac{1/(f\sqrt{\mu\epsilon})}{\{1 - [f_c/f]^2\}^{\frac{1}{2}}}$$
$$\gamma = \frac{2\pi}{(\lambda_w)_{nl}}$$
$$(\alpha_w)_{nl} = \frac{R_s}{a(\mu/\epsilon)[1-(f_c/f)^2]^{\frac{1}{2}}} \times \left[\left(\frac{f_c}{f}\right)^2 + \frac{n^2}{(p'_{nl})^2 - n^2}\right]$$

Notes:
1. a = radius of the waveguide
2. n, l are integers, for both TE and TM Wave, n can be zero but not l.
3. Dominant mode is the TM_{11} wave having the lowest cutoff frequency; TE_{01} mode is the only mode whose α_w decreases monotonically as frequencies increase.

TABLE 5b-2. FIELD CONFIGURATIONS FOR SEVERAL LOWER-ORDER MODES
IN RECTANGULAR AND CIRCULAR GUIDES

TABLE 5b-2. FIELD CONFIGURATIONS FOR SEVERAL LOWER-ORDER MODES IN RECTANGULAR AND CIRCULAR GUIDES (*Continued*)

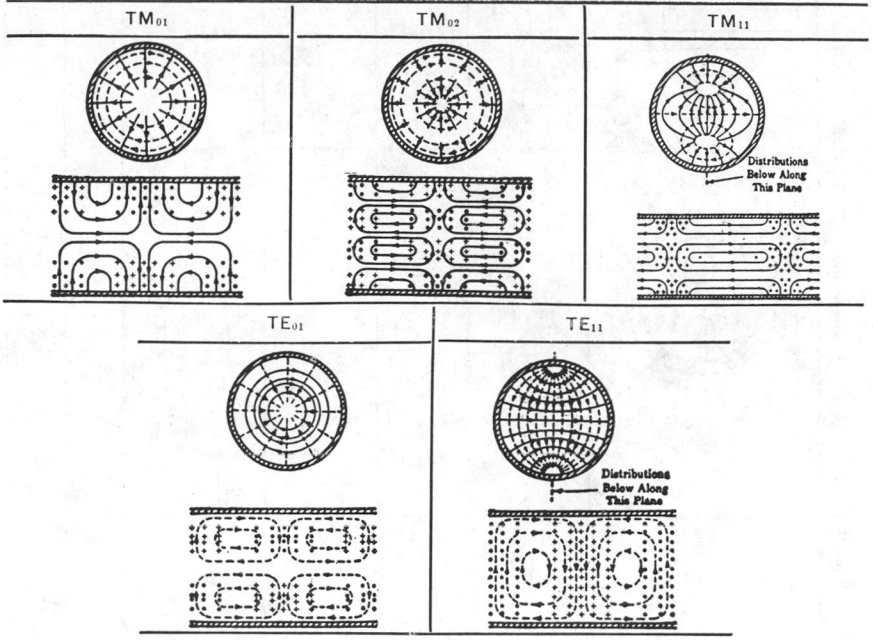

Note 1. The solid dots represent vectors coming out of the paper, and the crosses represent vectors going into the paper.

Note 2. The solid lines represent electric lines of force, and the dotted lines represent magnetic lines of force.

with $E_z^{(n)} = 0$ on the boundary, then, for $n > 0$,

$$\Gamma_n^2 \leq \frac{\int_A (\boldsymbol{\nabla} u)^2 \, dx \, dy}{\int_A u^2 \, dx \, dy} \qquad (5b\text{-}137)$$

for a *TE* mode, if a trial function $v(x,y)$ which are not subjected to any boundary conditions on the boundary satisfies the conditions

$$\int_A v H_z^{(0)} \, dx \, dy = 0, \int v H_z^{(1)} \, dx \, dy = 0, \ldots, \int v H_z^{(n-1)} \, dx \, dy = 0 \qquad (5b\text{-}138)$$

where $H_z^{(0)}, H_z^{(1)}, \ldots, H_z^{(n-1)}$ are the eigenfunctions for the equation

$$(\boldsymbol{\nabla}^2 + \Gamma_n^2) H_z^{(n)} = 0$$

with $\partial H_z^{(n)}/\partial n = 0$ on the boundary then, for $n > 0$,

$$\Gamma_n^2 \leq \frac{\int_A (\boldsymbol{\nabla} v)^2 \, dx \, dy}{\int_A v^2 \, dx \, dy} \qquad (5b\text{-}139)$$

Both $u(x,y)$ and $v(x,y)$ must be continuous within the bound with sectionally continuous derivatives.

TABLE 5b-3. SOME CONSTANTS OF COAXIAL, PARALLEL-WIRE, SHIELDED PAIRS AND PARALLEL-BAR TRANSMISSION LINES*

	Coaxial	Parallel-wire	Shielded pair ($p=\frac{s}{d}$, $q=\frac{s}{D}$)	Parallel-bar (Formulas for $a \ll b$)
Capacitance C, farads/m	$\dfrac{2\pi\epsilon}{\ln\left(\dfrac{r_0}{r_i}\right)}$	$\dfrac{\pi\epsilon}{\cosh^{-1}\left(\dfrac{s}{d}\right)}$	$\cdots\cdots\cdots$	$\dfrac{\epsilon b}{a}$
External inductance L, henrys/m	$\dfrac{\mu}{2\pi}\ln\left(\dfrac{r_0}{r_i}\right)$	$\dfrac{\mu}{\pi}\cosh^{-1}\left(\dfrac{s}{d}\right)$	$\cdots\cdots\cdots$	$\mu\dfrac{a}{b}$
Conductance G, mhos/m	$\dfrac{2\pi\sigma}{\ln\left(\dfrac{r_0}{r_i}\right)} = \dfrac{2\pi\omega\epsilon_v\epsilon_r''}{\ln\left(\dfrac{r_0}{r_i}\right)}$	$\dfrac{\pi\sigma}{\cosh^{-1}\left(\dfrac{s}{d}\right)} = \dfrac{\pi\omega\epsilon_v\epsilon_r''}{\cosh^{-1}\left(\dfrac{s}{d}\right)}$	$\cdots\cdots\cdots$	$\dfrac{\sigma b}{a} = \dfrac{\omega\epsilon_v\epsilon_r'b}{a}$
Resistance R, ohms/m	$\dfrac{R_s}{2\pi}\left(\dfrac{1}{r_0}+\dfrac{1}{r_i}\right)$	$\dfrac{2R_s}{\pi d}\left[\dfrac{s/d}{\sqrt{(s/d)^2-1}}\right]$	$\dfrac{2R_{e2}}{\pi d}\left[1+\dfrac{1+2p^2}{4p^4}\left(1-4q^2\right)\right] + \dfrac{8R_{e3}}{\pi D}q^2\left[1+q^2-\dfrac{1+4p^2}{8p^4}\right]$	$\dfrac{2R_s}{b}$
Internal inductance L_i, henrys/m (for high frequency)	$\longleftarrow$		$\dfrac{R}{\omega}$	$\longrightarrow$

Characteristic impedance at high frequency Z_0, ohms	$\dfrac{\eta}{2\pi}\ln\left(\dfrac{r_0}{r_i}\right)$	$\dfrac{\eta}{\pi}\cosh^{-1}\left(\dfrac{s}{d}\right)$	$\dfrac{\eta_1}{\pi}\left\{\ln\left[2p\left(\dfrac{1-q^2}{1+q^2}\right)\right] - \dfrac{1+4p^2}{16p^4}(1-4q^2)\right\}$	$\eta\dfrac{a}{b}$
Z_0 for air dielectric	$60\ln\left(\dfrac{r_0}{r_i}\right)$	$120\cosh^{-1}\left(\dfrac{s}{d}\right) \cong 120\ln\left(\dfrac{2s}{d}\right)$ if $s/d \gg 1$	$120\left\{\ln\left[2p\left(\dfrac{1-q^2}{1+q^2}\right)\right] - \dfrac{1+4p^2}{16p^4}(1-4q^2)\right\}$	$120\pi\dfrac{a}{b}$
Attenuation due to conductor α_c	$\dfrac{R}{2Z_0}$ $\longrightarrow$			
Attenuation due to dielectric α_d	$\dfrac{GZ_0}{2} = \dfrac{\sigma\eta}{2} = \pi\dfrac{\sqrt{\epsilon_r'\mu_r'}}{\lambda_0}\left(\dfrac{\epsilon_r''}{\epsilon_r'}\right)$ $\longrightarrow$			
Total attenuation, db/m	$8.686(\alpha_c + \alpha_d)$ $\longrightarrow$			
Phase constant for low-loss lines β	$\omega\sqrt{\mu\epsilon} = \dfrac{2\pi}{\lambda}$ $\longrightarrow$			

All units above are mks.
For the dielectric:

$\epsilon = \epsilon_r'\epsilon_0 =$ dielectric constant, farads/m
$\mu = \mu_r'\mu_0 =$ permeability, henrys/m
$\eta = \sqrt{\mu/\epsilon}$ ohms

$\epsilon_r'' =$ loss factor of dielectric $= \sigma/\omega\epsilon_0$
$R_s =$ skin-effect surface resistivity of conductor, ohms
$\lambda =$ wavelength in dielectric $= \lambda_0/\sqrt{\epsilon_r'\mu_r'}$

*Ramo and Whinnery, "Fields and Waves in Modern Radio," 2d ed., John Wiley & Sons, Inc., New York, 1953: Formulas for shielded pair obtained from Green, Leibe, and Curtis, *Bell System Tech. J.* **15**, 248–284 (April, 1936).

Conventional TEM Transmission Lines. For a two-conductor uniform line supporting the *TEM* waves, the differential equations for the voltage V and current I are

$$\frac{\partial V}{\partial z} = -L\frac{\partial I}{\partial t} - RI \qquad (5b\text{-}140)$$

$$\frac{\partial I}{\partial z} = -C\frac{\partial V}{\partial t} - GV \qquad (5b\text{-}141)$$

where L, C, R, and G are the inductance, capacitance, resistance, and conductance, respectively, all per unit length of the line.

If steady-state sinusoidal conditions of the form $e^{-i\omega t}$ are considered, then the equations become

$$\frac{\partial V}{\partial z} = -(R - i\omega L)I \qquad (5b\text{-}142)$$

$$\frac{\partial I}{\partial z} = -(G - i\omega C)V \qquad (5b\text{-}143)$$

Combining the above equations gives

$$\left(\frac{d^2}{dz^2} - \chi^2\right)\left\{\begin{matrix}I\\V\end{matrix}\right\} = 0 \qquad (5b\text{-}144)$$

where the propagation constant

$$\chi = \sqrt{(R - i\omega L)(G - i\omega C)} = \alpha + i\beta \qquad (5b\text{-}145)$$

The solution for Eq. (5b-144) is

$$V = Ae^{-\chi z} + Be^{\chi z} \qquad (5b\text{-}146)$$

$$I = \frac{1}{Z_0}(Ae^{-\chi z} - Be^{\chi z}) \qquad (5b\text{-}147)$$

where $Z_0 = \sqrt{(R - i\omega L)/(G - i\omega C)}$ and is called the characteristic impedance. A and B are constants to be determined according to the input and termination conditions. Tables 5b-3 and 5b-4 summarize constants for some common lines and some important formulas for transmission lines.

Another kind of quasi-*TEM* microwave transmission line is the strip line[1] which basically consists of two (or more) parallel metallic strips of generally different width separated by a dielectric medium. This structure cannot support a *TEM* wave although the dominant mode closely resembles the *TEM* wave of a simplified microstrip with dielectric material uniformly filling the entire region. Under this *TEM* wave approximation, the problem is essentially one of finding the electrostatic potential $\Phi(x,y)$ which satisfies the Laplace's equation $\nabla^2\Phi = 0$ and the boundary conditions $\Phi = \Phi_1$ on surface C_1 and $\Phi = \Phi_2$ on surface C_2. $\mathbf{E}_{\text{transverse}} = \nabla\Phi e^{\pm ikz}$ and $\mathbf{H}_{\text{transverse}} = \sqrt{\dfrac{\epsilon}{\mu}}(\mathbf{e}_z \times \nabla\Phi)e^{\pm ikz}$ with $k = \omega\sqrt{\mu\epsilon}$. The characteristic impedance of the line is

$$Z_c = \sqrt{\frac{\mu}{\epsilon}}\frac{(\Phi_1 - \Phi_2)^2}{\displaystyle\int_A \nabla\Phi \cdot \nabla\Phi \, dA} = \frac{\sqrt{\epsilon\mu}}{C}$$

where C is the capacitance of the structure per unit length.

Surface Waveguides. Another family of waveguides which is capable of guiding electromagnetic waves is the open-boundary structures. These structures consist of dielectric-coated planes and wires, corrugated planes and wires, interface between

---

[1] R. E. Collin, "Field Theory of Guided Waves," chap. 4, McGraw-Hill Book Company, New York, 1960.

TABLE 5b-4. SEVERAL IMPORTANT FORMULAS FOR SOME COMMON TRANSMISSION LINES*

Quantity	General line	Ideal line	Approximate results for low-loss lines
Propagation constant $\gamma = \alpha - i\beta$	$\sqrt{(R - i\omega L)(G - i\omega C)}$	$-i\omega\sqrt{LC}$	$\omega\sqrt{LC}\left(1 - \frac{RG}{4\omega^2 LC} + \frac{G^2}{8\omega^2 C^2} + \frac{R^2}{8\omega^2 L^2}\right)$ (See α and β below)
Phase constant β	$\mathrm{Im}\,(\gamma)$	$\omega\sqrt{LC} = \frac{\omega}{v} = \frac{2\pi}{\lambda}$	
Attenuation constant α	$\mathrm{Re}\,(\gamma)$	0	$\frac{R}{2Z_0} + \frac{GZ_0}{2}$
Characteristic impedance Z_0	$\sqrt{\dfrac{R - i\omega L}{G - i\omega C}}$	$\sqrt{\dfrac{L}{C}}$	$\sqrt{\dfrac{L}{C}}\left[1 - i\left(\frac{G}{2\omega C} - \frac{R}{2\omega L}\right)\right]$
Input impedance Z_i	$Z_0\left(\dfrac{Z_L \cosh \gamma l + Z_0 \sinh \gamma l}{Z_0 \cosh \gamma l + Z_L \sinh \gamma l}\right)$	$Z_0\left(\dfrac{Z_L \cos \beta l - iZ_0 \sin \beta l}{Z_0 \cos \beta l - iZ_L \sin \beta l}\right)$	
Impedance of shorted line	$Z_0 \tanh \gamma l$	$-iZ_0 \tan \beta l$	$Z_0\left(\dfrac{\alpha l \cos \beta l - i \sin \beta l}{\cos \beta l - i\alpha l \sin \beta l}\right)$
Impedance of open line	$Z_0 \coth \gamma l$	$+iZ_0 \cot \beta l$	$Z_0\left(\dfrac{\cos \beta l - i\alpha l \sin \beta l}{\alpha l \cos \beta l - i \sin \beta l}\right)$
Impedance of quarter-wave line	$Z_0\left(\dfrac{Z_L \sinh \alpha l + Z_0 \cosh \alpha l}{Z_0 \sinh \alpha l + Z_L \cosh \alpha l}\right)$	$\dfrac{Z_0^2}{Z_L}$	$Z_0\left(\dfrac{Z_0 + Z_L \alpha l}{Z_L + Z_0 \alpha l}\right)$
Impedance of half-wave line	$Z_0\left(\dfrac{Z_L \cosh \alpha l + Z_0 \sinh \alpha l}{Z_0 \cosh \alpha l + Z_L \sinh \alpha l}\right)$	Z_L	$Z_0\left(\dfrac{Z_L + Z_0 \alpha l}{Z_0 + Z_L \alpha l}\right)$
Voltage along line $V(z)$	$V_i \cosh \gamma z - I_i Z_0 \sinh \gamma z$	$V_i \cos \beta z + iI_i Z_0 \sin \beta z$	
Current along line $I(z)$	$I_i \cosh \gamma z - \dfrac{V_i}{Z_0} \sinh \gamma z$	$I_i \cos \beta z + i\dfrac{V_i}{Z_0} \sin \beta z$	
Reflection coefficient K_R	$\dfrac{Z_L - Z_0}{Z_L + Z_0}$	$\dfrac{Z_L - Z_0}{Z_L + Z_0}$	
Standing-wave ratio	$\dfrac{1 + |K_R|}{1 - |K_R|}$	$\dfrac{1 + |K_R|}{1 - |K_R|}$	

R, L, G, C = distributed resistance, inductance, conductance, capacitance per unit length

l = length of line

Subscript i denotes input end quantities.

Subscript L denotes load end quantities.

* Ramo and Whinnery, "Fields and Waves in Modern Radio," 2d ed., John Wiley & Sons, Inc., New York, 1953.

z = distance along line from input end

λ = wavelength measured along line

v = phase velocity of line equals velocity of light in dielectric of line for an ideal line

5-60 ELECTRICITY AND MAGNETISM

two different media. Special features of surface-wave modes having the usual propagation constant $e^{i\gamma z}$ along the axis and the structure are given in the following:

1. The field is characterized by an exponential decay away from the surface of the structure.

2. In most cases in which ϵ/ϵ_0, $\mu/\mu_0 > 1$, the phase velocities of the propagating surface-wave modes are less than the velocity of light in vacuum.

3. Below the cutoff frequency, a mode simply does not exist. In other words unlike the bounded waveguide case no evanescent mode exists.

4. The finite number of discrete surface-wave modes does not represent a complete set of solutions. In addition to the eigenfunction solutions there exists solutions with a continuous spectrum. (This property is in direct contrast to the mode property in bounded waveguides.)

5. Only TE, TM, or HE modes may exist on a surface-wave structure.

Detailed formulas are given for the circular dielectric waveguide as a representative surface-wave structure. It is understood that all fields vary as $e^{i\gamma z - i\omega t}$. The dielectric rod of radius a, having ϵ_1 and μ_0 as its permittivity and permeability, is assumed to be embedded in another dielectric medium with $\epsilon = \epsilon_0$ and $\mu = \mu_0$. Furthermore $\epsilon_1 > \epsilon_0$.

FIELD COMPONENTS

1. HE_{nm} modes with $n \neq 0$:

$$\begin{aligned}
E_z &= A_n J_n(s_1 r) \cos n\phi & r &\leq a & \text{(5b-148)}\\
&= B_n K_n(s_0 r) \cos n\phi & r &\geq a & \text{(5b-149)}\\
H_z &= C_n J_n(s_1 r) \sin n\phi & r &\leq a & \text{(5b-150)}\\
&= D_n K_n(s_0 r) \sin n\phi & r &\geq a & \text{(5b-151)}
\end{aligned}$$

2. TM_{om} modes:

$$\begin{aligned}
E_z &= A_o J_0(s_1 r) & r &\leq a\\
&= B_0 K_0(s_0 r) & r &\geq a\\
H_z &= 0 & &\text{for all } r
\end{aligned}$$

3. TE_{om} modes:

$$\begin{aligned}
E_z &= 0 & &\text{for all } r\\
E_z &= C_0 J_0(s_1 r) & r &\leq a\\
&= D_0 K_0(s_0 r) & r &\geq a
\end{aligned}$$

All other transverse field components may be found from Eqs. (5b-120) to (5b-123) with $\epsilon = \epsilon_1$, $\mu = \mu_0$ for $r \leq a$ and $\epsilon = \epsilon_0$, $\mu = \mu_0$ for $r \geq a$. A_n, B_n, C_n, D_n are amplitude coefficients. J_n and K_n are respectively the Bessel and modified Bessel functions.

PROPAGATION CONSTANT. The propagation constant γ is obtained by solving the following equations:

1. HE_{nm} modes ($n \neq 0$):

$$\left[\frac{(\epsilon_1/\epsilon_0) J_n'(s_1 a)}{s_1 a J_n(s_1 a)} + \frac{K_n'(s_0 a)}{s_0 a K_n(s_0 a)} \right] \left[\frac{J_n'(s_1 a)}{s_1 a J_n(s_1 a)} + \frac{K_n'(s_0 a)}{s_0 a K_n(s_0 a)} \right]$$

$$= n^2 \left(\frac{1}{s_0^2 a^2} + \frac{\epsilon_1/\epsilon_0}{s_1^2 a^2} \right) \left(\frac{1}{s_0^2 a^2} + \frac{1}{s_1^2 a^2} \right) \quad \text{(5b-152)}$$

$$s_0^2 a^2 + s_1^2 a^2 = \omega^2 \mu_0 \epsilon_0 a^2 \left(\frac{\epsilon_1}{\epsilon_0} - 1 \right) \quad \text{(5b-153)}$$

$$\gamma^2 = \omega^2 \mu_0 \epsilon_0 + s_0^2 = \omega^2 \mu_0 \epsilon_1 - s_1^2 \quad \text{(5b-154)}$$

2. TM_{om} modes:

$$\frac{(\epsilon_1/\epsilon_0) J_0'(s_1 a)}{s_1 a J_0(s_1 a)} + \frac{K_0'(s_0 a)}{s_0 a K_0(s_0 a)} = 0 \quad \text{(5b-155)}$$

with Eqs. (5b-153) and (5b-154)

3. TE_{om} modes:

$$\frac{J_0'(s_1 a)}{s_1 a J_0(s_1 a)} + \frac{K_0'(s_0 a)}{s_0 a K_0(s_0 a)} = 0 \tag{5b-156}$$

with Eqs. (5b-153) and (5b-154).

Numerical solutions of the above equations show that only the HE_{11} mode processes no cutoff frequency. The propagation constants as a function of frequency for the three lowest-order modes are given by Fig. 5b-1.

ATTENUATION. If the dielectric ϵ_1 for the cylindrical rod is imperfect and has a small conductivity σ_d, then there exists an attenuation constant α_d for all field components (i.e., all field components vary as $e^{-\alpha_d z}e^{i\gamma_z}e^{-i\omega t}$).

$$\alpha_d = 4.343\sigma_d \sqrt{\frac{\mu_0}{\epsilon_0}} \left| \frac{\int_{A_i} (\mathbf{E} \cdot \mathbf{E})\, dA}{\frac{\mu_0}{\epsilon_0} \int_{A_i + A_0} (\mathbf{E}_t \times \mathbf{H}_t^*) \cdot \mathbf{e}_z\, dA} \right| \tag{5b-157a}$$

where α_d is in db/meter, $\mathbf{E}_t$ and $\mathbf{H}_t$ are the transverse fields, A_i is the cross-sectional area of the dielectric rod, and $A_i + A_0$ is the total cross-sectional area. Figures 5b-1 and 5b-2 show respectively the propagation constant γ and the attenuation α_d as a function of frequency for the three lowest-order modes.

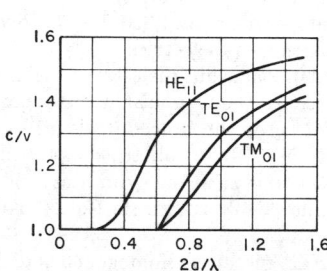

FIG. 5b-1. Velocity ratio c/v for polystyrene rod ($\epsilon_1 = 2.56$) embedded in free space. v is the phase velocity of the surface wave; c is the velocity of light in free space. a is the radius of the rod. λ is the free-space wavelength. [*From data obtained from W. Elsasser, J. Appl. Phys.* **20**, 1193 (1949).]

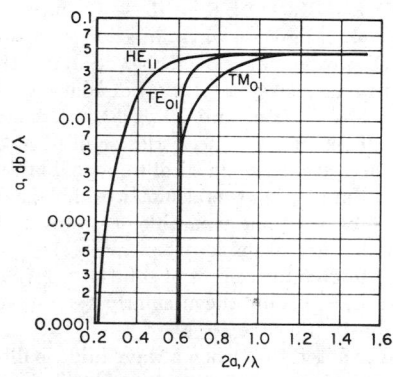

FIG. 5b-2. Attenuation for some surface wave modes along polystyrene rod of radius a ($\epsilon_1 = 2.56$ and $\tan \delta = 0.001$). Note that the attenuation α is in db/λ which is equal to 8.686 nepers/λ, where λ is the free-space wavelength. [*From data obtained from W. Elsasser, J. Appl. Phys.* **20**, 1193 (1949).]

If conductors are included as part of the surface-wave structures (such as the dielectric-coated wires), then in addition to the dielectric loss there is an attenuation constant α_c associated with the loss due to the finite conductivity of the conductors; i.e.,

$$\alpha_c = 4.343 \sqrt{\frac{\mu_0}{\epsilon_0}} \left| \frac{R_s/2 \oint_L |H_t|^2\, dl}{\frac{\mu_0}{\epsilon_0} \int_{A_i + A_0} (\mathbf{E}_t \times \mathbf{H}_t^*) \cdot \mathbf{e}_z\, dA} \right| \text{db/m} \tag{5b-157b}$$

where L is the cross-sectional curve around the conductor. In the case of the dielectric-coated wire, there are two dominate modes that have zero cutoff frequencies: the HE_{11} mode and the TM_{01} mode.

These formulas for the dielectric rod case are particularly useful in the study of fiber optics. Much more involved formulas for other types of surface-wave guides are also available; but for these the reader is referred to the literature.[1]

Inhomogeneously Filled Waveguides. Waveguides filled nonuniformly with homogeneous dielectrics or filled with inhomogeneous dielectrics offer many practical applications such as phase changers, matching transformers, etc. Previous formulas for homogeneously filled waveguides are not applicable for the present situation. Because of the complexity of the problem only several special cases have been treated. A procedure for deriving the electric and magnetic field components is given in the following for an inhomogeneously filled rectangular waveguide case[2]:

It is assumed that the nonuniformity is only in one of the three coordinate directions, say e_ξ where ξ may be x, y, or z. Then derive the electric and magnetic field components from the scalar potentials Ψ, Φ as follows:

$$E_1 = \nabla \times (e_\xi \Psi) \tag{5b-158}$$

$$H_1 = \frac{1}{i\omega\mu} \nabla \times \nabla \times (e_\xi \Psi) \tag{5b-159}$$

and

$$H_2 = \nabla \times (e_\xi \Phi) \tag{5b-160}$$

$$E_2 = \frac{i}{\omega\epsilon} \nabla \times \nabla \times (e_\xi \Phi) \tag{5b-161}$$

with $E = E_1 + E_2$, $H = H_1 + H_2$ in the general case. Substituting Eq. (5b-158) into Eq. (5b-58) gives a wave equation for Ψ and substituting Eq. (56-160) into Eq. (5b-59) gives a wave equation for Φ. Solving these differential equations for Φ and Ψ and satisfying the appropriate boundary conditions gives the solution of the problem. The above procedure is workable for rectangular waveguides filled nonuniformly with homogeneous dielectrics or filled with inhomogeneous dielectrics.

For a circular cylindrical waveguide filled with inhomogeneous dielectrics, the above procedure is, in general, not workable. This is because the resultant wave equation for Φ or Ψ is not separable. However, for special cases the procedure is still very useful. For example, when $\epsilon = \epsilon(z)$, $\mu = \mu_0$, Eqs. (5b-158) to (5b-161) may still be used to give the full set of solutions. When $\epsilon = \epsilon(r)$, $\mu = \mu_0$ in the cylindrical coordinates r, ϕ, z; only the circularly symmetric modes may be derived from Eqs. (5b-158) to (5b-161) with $e_\xi = e_r$.

It is noted that when a waveguide is filled by certain piecewise-homogeneous dielectrics (such as a circular waveguide filled with a concentric dielectric of different radius), the field components may still be derived for the region in which the dielectric is homogeneous and the complete solutions are obtained by matching the boundary conditions at the discontinuity.

Anisotropic Wave Propagation. A typical material having anisotropic electromagnetic property is the ferrite. In rectangular coordinates, the ferrite medium is characterized by the following relations:

$$B = \mu \cdot H \tag{5b-162}$$

$$\mu = \begin{bmatrix} \mu_1 & -i\mu_2 & 0 \\ i\mu_2 & \mu_1 & 0 \\ 0 & 0 & \mu_3 \end{bmatrix} \tag{5b-163}$$

$$\mu_1 = \mu_0 \left(1 + \frac{\omega_c \omega_M}{\omega_c^2 - \omega^2}\right) \qquad \omega_c = \left|\frac{e}{m}\right| B_0$$

$$\mu_2 = \mu_0 \frac{\omega \omega_M}{\omega_c^2 - \omega^2} \qquad \omega_M = \left|\frac{e}{m}\right| \mu_0 M_0 \tag{5b-164}$$

$$\mu_3 = \mu_0$$

[1] G. Goubau, *J. Appl. Phys.* **21**, 119 (1950); C. Yeh, *J. Appl. Phys.* **33**, 3235 (1962).
[2] C. Yeh and K. F. Casey, *IEEE Trans. Microwave Theory and Tech.* **MTT-13**, 297 (1965).

with the applied d-c magnetic field B_0 in the z direction. M_0 is the internal magnetization.

FARADAY ROTATIONS. Solutions for plane-wave propagating in the ferrite medium in a direction parallel to the applied static magnetic field $B_0 e_z$ are

$$\mathbf{E}^{\pm} = (\mathbf{e}_x \pm i\mathbf{e}_y)A^{\pm}e^{i\gamma^{\pm}z}e^{-i\omega t} \qquad (5b\text{-}165)$$

with
$$\gamma^{\pm} = \omega \sqrt{\epsilon(\mu_1 \mp \mu_2)} \qquad (5b\text{-}166)$$

Equations (5b-165) and (5b-166) are derived directly from Maxwell's equations and Eqs. (5b-162) and (5b-163). It is recognized that $\mathbf{E}^{+}$ is the right-handed circularly polarized plane wave of amplitude A^{+} and propagation constant γ^{+}, and $\mathbf{E}^{-}$ is the left-handed circularly polarized plane wave of amplitude A^{-} and propagation constant γ^{-}. A linearly polarized plane wave may be resolved into two counterrotating equal-amplitude circularly polarized waves, i.e., a linearly polarized wave $\mathbf{E} = \mathbf{E}^{+} + \mathbf{E}^{-}$ with $A^{+} = A^{-}$. Since the two circularly polarized component waves propagate at different velocities, the linearly polarized wave in the ferrite is rotated. When the wave is propagating in the $+z$ direction, the angle of rotation θ is given by

$$\theta = \tan^{-1}\frac{E_y}{E_x} = \tan^{-1} i\frac{e^{i\gamma^{-}z} - e^{i\gamma^{+}z}}{e^{i\gamma^{-}z} + e^{i\gamma^{+}z}}$$

$$= \frac{1}{2}(\gamma^{+} - \gamma^{-})z \qquad (5b\text{-}167)$$

$$= \frac{1}{2}\omega\sqrt{\mu_0\epsilon}\,z\left(\sqrt{1 + \frac{\omega_M}{\omega_c + \omega}} - \sqrt{1 + \frac{\omega_M}{\omega_c - \omega}}\right)$$

Reversing the direction of propagation (i.e., replacing i by $-i$) does not change the sense of θ. Hence, regardless of whether the waves are traveling in the direction of the static magnetic field $(+z)$ or in the opposite direction of the static magnetic field, their axes of polarization are rotated in the same sense with respect to the biasing magnetostatic field. This phenomenon is called the *Faraday rotation*. Reciprocity requires that the rotations be equal and opposite; thus the ferrite medium is non-reciprocal. Making use of the nonreciprocal nature of waves in ferrite medium, a number of very useful practical devices using ferrites have been invented: The ferrite gyrator, producing π phase shift in the $+z$ direction and zero phase shift in the $-z$ direction, the ferrite isolator, circulator, switch, etc.

FERRITE-LOADED WAVEGUIDES. Assuming that all field components in a waveguide containing anisotropic ferrites vary as $e^{i\gamma z - i\omega t}$, then the transverse components $\mathbf{E}_t$ and $\mathbf{H}_t$ can be derived from the longitudinal components E_z and H_z according to the following relations:

$$\mathbf{E}_t = \frac{1}{(\beta_1{}^2 - \gamma^2)^2 - \beta_2{}^4}\{\boldsymbol{\nabla}_t[(\beta_1{}^2 - \gamma^2)i\gamma E_z + \omega\gamma^2\mu_2 H_z]$$

$$- i\mathbf{e}_z \times \boldsymbol{\nabla}_t[\omega(\beta_1{}^2\mu_1 - \gamma^2\mu_1 + \beta_2{}^2\mu_2)H_z - i\gamma\beta_2{}^2 E_z]\} \qquad (5b\text{-}168)$$

$$\mathbf{H}_t = \frac{1}{(\beta_1{}^2 - \gamma^2) - \beta_2{}^4}\{\boldsymbol{\nabla}_t[\beta_2{}^2\omega\epsilon E_z + (\beta_1{}^2 - \gamma^2)i\gamma H_z]$$

$$+ i\mathbf{e}_z \times \boldsymbol{\nabla}_t[(\beta_1{}^2 - \gamma^2)\omega\epsilon E_z + \beta_2{}^2 i\gamma H_z]\} \qquad (5b\text{-}169)$$

with $\beta_1{}^2 = \omega^2\mu_1\epsilon$ and $\beta_2{}^2 = -\omega^2\mu_2\epsilon$. The longitudinal fields (E_z, H_z) satisfy

$$\boldsymbol{\nabla}_t{}^2 E_z + a_1 E_z + a_2 H_z = 0 \qquad (5b\text{-}170)$$

$$\boldsymbol{\nabla}_t{}^2 H_z + a_3 H_z + a_4 E_z = 0 \qquad (5b\text{-}171)$$

with

$$a_1 = (\beta_1{}^2 - \gamma^2) + \beta_2{}^2 \frac{\mu_2}{\mu_1} \qquad (5b\text{-}172)$$

$$a_2 = \omega\mu_3 i\gamma \frac{\mu_2}{\mu_1} \qquad (5b\text{-}173)$$

$$a_3 = (\beta_1{}^2 - \gamma^2) \frac{\mu_3}{\mu_1} \qquad (5b\text{-}174)$$

$$a_4 = -i\omega\gamma\epsilon \frac{\mu_2}{\mu_1} \qquad (5b\text{-}175)$$

According to Eqs. (5b-170) and (5b-171), a pure TE, TM, or TEM modes cannot exist in a waveguide filled with a ferrite, since if either of the longitudinal field E_z or H_z is zero, the entire longitudinal field vanishes and all transverse fields also vanish according to Eqs. (5b-168) and (5b-169).

Equations (5b-170) and (5b-171) may be decoupled by the introduction of functions Φ_1 and Φ_2 defined by

$$E_z = \Phi_1 + \Phi_2 \qquad (5b\text{-}176)$$

$$H_z = g_1\Phi_1 + g_2\Phi_2 \qquad (5b\text{-}177)$$

where

$$g_1 = \frac{p_1{}^2 - a_1}{a_2} = \frac{a_4}{p_1{}^2 - a_3} \qquad (5b\text{-}178)$$

$$g_2 = \frac{p_2{}^2 - a_1}{a_2} = \frac{a_4}{p_2{}^2 - a_3} \qquad (5b\text{-}179)$$

and $p_1{}^2$ and $p_2{}^2$ are the roots of the equation

$$p^4 - (a_1 + a_3)p^2 + a_1a_3 - a_2a_4 = 0 \qquad (5b\text{-}180)$$

Φ_1 and Φ_2 satisfy the equations

$$(\nabla_t{}^2 + p_1{}^2)\Phi_1 = 0 \qquad (\nabla_t{}^2 + p_2{}^2) = 0 \qquad (5b\text{-}181)$$

Rigorous theories of wave propagation in endless waveguides completely filled with a magnetized ferrite medium have been worked out for circular and rectangular waveguides. However, the solutions are too complicated to be included here.[1]

Periodic Structures. Any propagating mode in an empty perfectly conducting cylindrical tube has a phase velocity which is greater than the speed of light. From the need of electronic devices for the generation and amplification of microwaves and for the acceleration of charged elementary particles originated the demand for cylindrical waveguide structures in which the modes having a longitudinal component of the electric vector move with a phase velocity less than the velocity of light so as to enhance the energy exchange between the beam of particles and the wave fields. A practical slow-wave guide is the periodic structure.

Because of the periodicity of the structure, spatial harmonics of the modes must be taken into account. This is accomplished by the use of Floquet's theorem which states that for a given mode of propagation and at a given frequency, the wave functions at two points along the periodic guiding structure, separated by one spatial period of the structure, differ only by a complex constant. In other words, the general expression for the wave function, say E_z, should be of the form

$$E_z = \sum_{m=-\infty}^{\infty} A_m f_m(x,y) e^{i(\gamma_0 + 2m\pi/L)z} \qquad (5b\text{-}182)$$

[1] H. Suhl and R. C. Walker, *Bell System Tech. J.* **33**, 579 (1954); P. S. Epstein, *Revs. Modern Phys.* **28**, 3 (1956).

where γ_0 is the propagation constant of the fundamental wave $m = 0$, and L is the period of the guide.

As a representative example, a periodic disk-load circular waveguide will be considered. The radii of the tube and the circular apertures are b and a, respectively. The spatial period of the structure is L and the thickness of each disk is w. Only the circularly symmetrical TM waves will be considered. In accordance with Floquet's theorem, appropriate representation for $E_z{}^I$ in region I ($r \leq a$) is

$$E_z{}^I = \sum_{m = -\infty}^{\infty} A_m I_0(\chi_m r)e^{i\gamma_m z} \tag{5b-183}$$

where

$$\gamma_m{}^2 = k^2 + \chi_m{}^2 \qquad \gamma_m = \gamma_0 + \frac{2\pi m}{L} \tag{5b-184}$$

with $m = 0, \pm 1, \ldots$, $k = \omega \sqrt{\mu \epsilon}$. I_0 is the modified Bessel function. Appropriate series representation for $E_z{}^{II}$ in region II ($a \leq r \leq b$) may also be assumed. However, upon matching the tangential E_z and H_ϕ across the surface $r = a$, an infinite determinant for the propagation constant results. For the case of closely spaced disks ($\gamma_0 w \ll 1$), the infinite determinant reduces to

$$\frac{1}{\chi_0 a} \frac{I_1(x_0 a)}{I_0(s_0 a)} = \frac{1}{ka} \frac{J_1(ka)N_0(kb) - N_1(ka)J_0(kb)}{J_0(ka)N_0(kb) - N_0(ka)J_0(kb)} \tag{5b-185}$$

A typical ω vs. γ_0 diagram is shown in Fig. 5b-3.

Formulas for propagating modes along helix (tape helix, sheath helix, or wire helix) are also available in the literature.[1] Problems concerning wave propagation in a waveguide filled longitudinally with a sinusoidally varying dielectric material have also been solved.[2]

Waveguide Junctions. Strictly speaking, the problem of waveguide junctions should be analyzed as a boundary-value problem. However, extreme difficulties are encountered owing to the complex geometry of the junction region.[3] The problem is therefore reformulated in terms of lumped-circuit representations for the junctions and transmission-line representations for the waveguides. For an arbitrary junction of m waveguide arms, assuming that only one mode may propagate in each waveguide arm, the power flowing into the nth arm is

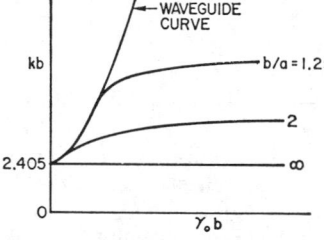

FIG. 5b-3. The ω-γ_0 diagram for the periodic disk-loaded circular waveguide with infinitesimal disk thickness. γ_0 is the propagation constant, b is the waveguide radius and a is the disk inner radius. [*From E. L. Chu and W. W. Hansen, J. Appl. Phys.* **18**, 996 (1947).]

$$\bar{P}_n = a_n a_n^* - b_n b_n^* \tag{5b-186}$$

where a_n is the incoming mode amplitude, and b_n is the reflected mode amplitude. The * represents the complex conjugate of the function. The voltage and current at the nth terminal are

$$v_n = \sum_m z_{nm} i_m \qquad i_n = \sum_m y_{nm} v_m \tag{5b-187}$$

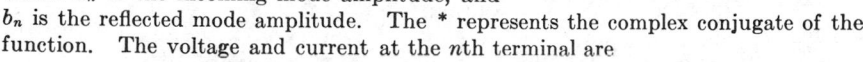

[1] D. A. Watkins, "Topics in Electromagnetic Theory," John Wiley & Sons, Inc., New York, 1958; R. M. Bevensee, "Electromagnetic Slow Wave Systems," John Wiley & Sons, Inc., New York, 1964.

[2] C. Yeh and K. F. Casey, *IEEE Trans. Microwave Theory and Tech.* **MTT-13**, 297 (1965).

[3] G. L. Matthaei, L. Young, and E. M. T. Jones, "Microwave Filters, Impedance-matching Networks and Coupling Structures," McGraw-Hill Book Company, New York, 1964.

where z_{nm} and y_{nm} are respectively the impedance and admittance matrix. The mode amplitudes are related to i_n and z_{nm} by the relations

$$a_n = \sqrt{\frac{z_n}{8}} \sum_{\lfloor m} \left(\frac{z_{nm}}{z_n} + \delta_{nm} \right) i_m \qquad \text{(5b-188)}$$

$$b_n = \sqrt{\frac{z_n}{8}} \sum_{m} \left(\frac{z_{nm}}{z_n} - \delta_{nm} \right) i_m \qquad \text{(5b-189)}$$

with δ_{nm} being the Kronecker delta. In matrix notation

$$\mathbf{A} = (\mathbf{Z}' + \mathbf{U})\mathbf{I}' \qquad \text{(5b-190)}$$
$$\mathbf{B} = (\mathbf{Z}' - \mathbf{U})\mathbf{I}' \qquad \text{(5b-191)}$$

where the elements of the matrices $\mathbf{Z}'$ and $\mathbf{I}'$ are respectively z_{nm}/z_n and $i_m\sqrt{z_n/8}$. $\mathbf{U}$ is the unit matrix. Combining Eqs. (5b-190) and (5b-191) gives

$$\mathbf{B} = \mathbf{SA} \qquad \text{(5b-192)}$$

with $\mathbf{S} = (\mathbf{Z}' - \mathbf{U})(\mathbf{Z}' + \mathbf{U})^{-1}$. $\mathbf{S}$ is called the *scattering matrix*. So the junction problem is reduced to an evaluation of the scattering matrix. For a reciprocal lossless junction, the scattering matrix possesses the following properties:

$$\begin{aligned} \tilde{\mathbf{S}} &= \mathbf{S} \qquad \text{(symmetry property of } \mathbf{S}) \\ \mathbf{S}^{-1} &= \tilde{\mathbf{S}}^* \qquad \text{(} \mathbf{S} \text{ is a unitary matrix)} \end{aligned} \qquad \text{(5b-193)}$$

where $\tilde{\mathbf{S}}$ is the transpose of $\mathbf{S}$ and $*$ indicates the complex conjugate.

Waveguide Discontinuities. A discontinuity in the structure of a waveguide results in the distortion of the nominal field distributions. It is assumed that only the dominant mode may propagate in this structure. Therefore, as far as the far-zone (away from the disturbance) dominant propagating mode is concerned, an equivalent circuit description of the disturbance will be adequate. Considering the discontinuity due to a post, an aperture or an abrupt change in the cross-sectional area of a waveguide, the disturbance may be represented by an admittance Y shunted across a uniform transmission line at the discontinuity. The shunt admittance is related to the reflection coefficient R of a dominant mode by the relation $R = Y/(2 - Y)$. For a time dependence of $e^{-i\omega t}$, and $Y = iB$, the admittance is inductive if B is positive and capacitive if B is negative.

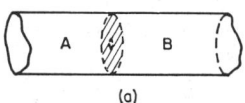

(a)

Fig. 5b-4. A sketch of an iris in a waveguide.

The variational expression for the shunt admittance of an iris in a waveguide is (see Fig. 5b-4)

$$Y_d = \frac{\int_{\text{Ap}} \int_{\text{Ap}} \mathbf{E}_{nA}(x',y') \cdot \mathbf{G}(x',y',x,y) \cdot \mathbf{E}_{nA}(x,y) \, dx' \, dy' \, dx \, dy}{\left(\iint_{\text{Ap}} \Phi_0 \cdot \mathbf{E}_{nA} \, dS \right)^2} \qquad \text{(5b-194)}$$

where

$$\mathbf{G} \; (Green's \; function) = \sum_{m,n} \sum_{p=A,B} Y_{mnp}\Phi_{mn}(x',y')\Phi_{mn}(x,y)$$

Y_{mnA}, Y_{mnB} = characteristic admittances for (m,n) modes in regions A and B
Φ_{mn} = transverse field vectors in the undisturbed guide
$\mathbf{E}_{nA}$ = normalized transverse aperture fields in region A
Ap = aperture area
Φ_0 = dominant mode transverse field vectors

The above expression for Y_d is variational in the sense that the variation of Y_d due to a small change of $\mathbf{E}_{nA}$ from its true value is vanishingly small.

For a rectangular waveguide of cross-sectional dimension $a \times b$, we have

$$\Phi_{mn} = A_{mn} \left[-\mathbf{e}_x \left(\frac{n}{b} \right) \cos \frac{m\pi x}{a} \sin \frac{n\pi y}{b} \right.$$
$$\left. + \mathbf{e}_y \left(\frac{m}{a} \right) \sin \frac{m\pi x}{a} \cos \frac{n\pi y}{b} \right] \quad \text{for } TE \text{ modes}$$
$$= A_{mn} \left[\mathbf{e}_x \left(\frac{m}{b} \right) \cos \frac{m\pi x}{a} \sin \frac{n\pi y}{b} \right.$$
$$\left. + \mathbf{e}_y \left(\frac{n}{b} \right) \sin \frac{m\pi x}{a} \cos \frac{n\pi y}{b} \right] \quad \text{for } TM \text{ modes}$$

(5b-195)

$$Y_{mn} = \sqrt{\frac{\epsilon}{\mu}} \left[1 - \left(\frac{k_{mn}}{k} \right)^2 \right]^{\frac{1}{2}} \quad \text{for } TE \text{ modes}$$
$$= \sqrt{\frac{\epsilon}{\mu}} \left[1 - \left(\frac{k_{mn}}{k} \right)^2 \right]^{-\frac{1}{2}} \quad \text{for } TM \text{ modes}$$

(5b-196)

$$\gamma^2 = k^2 - k_{mn}^2 \qquad k = \omega \sqrt{\mu\epsilon}$$

(5b-197)

$$k_{mn}^2 = \left(\frac{m\pi}{a} \right)^2 + \left(\frac{n\pi}{b} \right)^2$$

(5b-198)

$$A_{mn}^2 = 2 \left[\frac{n^2}{\epsilon_m} \left(\frac{a}{b} \right) + \frac{m^2}{\epsilon_n} \left(\frac{a}{b} \right) \right]^{-1}$$

(5b-199)

$$\epsilon_m = \begin{cases} 1 & \text{for } m = 0 \\ 2 & \text{for } m = 1, 2, \ldots \end{cases}$$

(5b-200)

For a circular waveguide of radius a, we have

$$\Phi_{mn} = A_{mn} \left[\mathbf{e}_r \frac{m J_m(k_{mn}r)}{k_{mn}r} \sin m\phi \right.$$
$$\left. + \mathbf{e}_\phi J'_m(k_{mn}r) \cos m\phi \right] \quad \text{for } TE \text{ modes} \quad \text{(5b-201)}$$
$$= B_{mn} \left[\mathbf{e}_r J'_m(k_{mn}r) \sin m\phi - \mathbf{e}_\phi \frac{m J_m(k_{mn}r)}{k_{mn}r} \cos m\phi \right] \quad \text{for } TM \text{ modes}$$

Y_{mn} is given by Eq. (5b-196), and γ is given by Eq. (5b-197).

$$J'_m(k_{mn}a) = 0 \qquad \text{for } TE \text{ modes} \quad \text{(5b-202)}$$
$$J_m(k_{mn}a) = 0 \qquad \text{for } TM \text{ modes} \quad \text{(5b-203)}$$

$$A_{mn} = \left[\frac{\pi}{\epsilon_m} \left(a^2 - \frac{m^2}{k_{mn}^2} \right) J_m^2(k_{mn}a) \right]^{-\frac{1}{2}}$$

(5b-204)

$$B_{mn} = \left[\frac{\pi}{\epsilon_m} (a J'_m(k_{mn}a))^2 \right]^{-\frac{1}{2}}$$

(5b-205)

ϵ_m is given by Eq. (5b-200). Some representative examples are given in Fig. 5b-5.

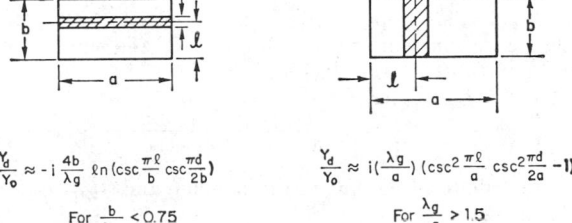

$$\frac{Y_d}{Y_0} \approx -i \frac{4b}{\lambda_g} \ell n \left(\csc \frac{\pi \ell}{b} \csc \frac{\pi d}{2b} \right)$$

$$\text{For } \frac{b}{\lambda_g} < 0.75$$

$$\frac{Y_d}{Y_0} \approx i \left(\frac{\lambda_g}{a} \right) \left(\csc^2 \frac{\pi \ell}{a} \csc^2 \frac{\pi d}{2a} - 1 \right)$$

$$\text{For } \frac{\lambda_g}{a} > 1.5$$

FIG. 5b-5. *Left:* capacitive iris   *Right:* inductive iris.

References

1. Marcuvitz, N., (ed.): "Waveguide Handbook," vol. 10 of MIT Rad. Lab. Ser., McGraw-Hill Book Company, New York, 1951.
2. Collin, R. E.: "Foundations for Microwave Engineering," McGraw-Hill Book Company, New York, 1966.
3. Ghose, R. N.: "Microwave Circuit Theory and Analysis," McGraw-Hill Book Company, New York, 1963.
4. Ramo, S., J. R. Whinnery, and T. Van Duzer: "Fields and Waves in Communication Electronics," John Wiley & Sons, Inc., New York, 1965.

5b-10. Cavity Resonators. Resonant cavities are used at high frequencies in place of lumped-circuit elements, primarily because they eliminate radiation and in general possess very low losses. Only eigenvalue solutions exist in a lossless cavity resonator completely enclosed by perfectly conducting walls. For a cavity filled with a homogeneous, isotropic dielectric, the pth eigenvector $\mathbf{E}_p$ satisfies

$$\begin{aligned}(\nabla^2 + k_p{}^2)\mathbf{E}_p &= 0 \quad \text{(everywhere within the cavity)}\\ \mathbf{n} \times \mathbf{E}_p &= 0 \quad \text{(on the enclosing wall)}\end{aligned} \qquad (5b\text{-}206)$$

where $k_p = \omega_p \sqrt{\mu\epsilon}$ ($p = 1, 2, 3, \ldots$) are the eigenvalues. ω_p is the resonant frequency for the pth mode.

The Q_p of a resonator for the pth mode is defined as follows:

$$Q_p = \omega_p \frac{\text{total time-average energy stored}}{\text{time-average power dissipated}} \qquad (5b\text{-}207)$$

$$= \frac{\Delta\omega}{\omega_p} \qquad (5b\text{-}208)$$

where $\Delta\omega$ is the bandwidth of the resonance curve. Hence Q_p is a measure of the amount of power dissipated for the pth mode. For an enclosed cavity with slightly lossy walls,

$$Q_p = \frac{2}{\delta_s} \frac{\int_V \mathbf{H}_p \cdot \mathbf{H}_p^* \, dV}{\oint_A \mathbf{H}_p \cdot \mathbf{H}_p^* \, dA} \qquad (5b\text{-}209)$$

where $\mathbf{H}_p$ is the magnetic field of the pth mode of the cavity without losses, and δ_s is the skin depth of the walls. A is the total surface enclosing the cavity region.

For a cavity composed of a uniform transmission line (which may support the TE, TM, TEM, or HE mode) with short-circuiting perfectly conducting ends, the Q_p of this cavity is related to the attenuation constant α_p of the transmission line by the relation[1]

$$Q_p = \frac{v_{\text{phase}}^p}{v_{\text{group}}^p} \frac{\gamma_p}{2\alpha_p} \qquad (5b\text{-}210)$$

where v_{phase}^p, v_{group}^p and γ_p are respectively the phase velocity ω_p/γ_p, the group velocity $\partial\omega_p/\partial\gamma_p$, and the phase constant of the pth mode. If the end plates possess a very small loss, then the total Q_T of this cavity is

$$\frac{1}{Q_T} = \frac{1}{Q_{\text{end plates}}} + \frac{1}{Q_{\text{trans. line}}}$$

where $Q_{\text{end plates}}$ is calculated according to Eq. (5b-209) and $Q_{\text{trans. line}}$ can be calculated according to Eq. (5b-210).

[1] C. Yeh, *Proc. IRE* **50**, 2145 (1962).

Simple Resonators. The mode functions for a cylindrical waveguide of simple cross section closed at both ends by short-circuiting plates are[1] (with d = length of the cavity):

For TM_{mnl} modes

$$E_{zmnl} = A_{mn}\Phi_{mn} \cos \frac{l\pi z}{d} \qquad (5\text{b-}211)$$

$$H_{zmnl} = 0 \qquad (5\text{b-}212)$$

$$\mathbf{E}_{tmnl} = -\frac{l\pi}{d} \frac{A_{mn}}{\Gamma_{mn}{}^{(TM)}} \nabla_t\Phi_{mn} \sin \frac{l\pi z}{d} \qquad (5\text{b-}213)$$

$$\mathbf{H}_{tmnl} = i\omega\epsilon \frac{A_{mn}}{\Gamma_{mn}{}^{(TM)}} (\mathbf{e}_z \times \nabla_t\Phi_{mn}) \cos \frac{l\pi z}{d} \qquad (5\text{b-}214)$$

with $(\nabla_t{}^2 + \Gamma_{mn}{}^{(TM)^2})\Phi_{mn} = 0$ and $\Phi_{mn} = 0$ on the cylindrical wall. The resonant frequency

$$\omega_{mnl}^{TM} = \frac{1}{\sqrt{\mu\epsilon}} \left[\Gamma_{mn}{}^{(TM)^2} + \left(\frac{l\pi}{d}\right)^2 \right]^{\frac{1}{2}}$$

For TE_{mnl} modes

$$E_{zmnl} = 0 \qquad (5\text{b-}215)$$

$$H_{zmnl} = B_{mn}\Psi_{mn} \sin \frac{l\pi z}{d} \qquad (5\text{b-}216)$$

$$\mathbf{E}_{tmnl} = \frac{-i\omega\mu}{\Gamma_{mn}{}^{(TE)}} B_{mn}(\mathbf{e}_z \times \nabla_t\Psi_{mn}) \sin \frac{l\pi z}{d} \qquad (5\text{b-}217)$$

$$\mathbf{H}_{tmnl} = \frac{l\pi}{d} \frac{1}{\Gamma_{mn}{}^{(TE)}} B_{mn}(\nabla_t\Psi_{mn}) \cos \frac{l\pi z}{d} \qquad (5\text{b-}218)$$

with $(\nabla_t{}^2 + \Gamma_{mn}{}^{(TE)^2})\Psi_{mn} = 0$ and $\partial\Psi_{mn}/\partial n = 0$ along the cylindrical wall. The resonant frequency

$$\omega_{mnl}^{TE} = \frac{1}{\sqrt{\mu\epsilon}} \left[\Gamma_{mn}{}^{(TE)^2} + \left(\frac{l\pi}{d}\right)^2 \right]^{\frac{1}{2}}$$

For a rectangular resonator with cross sections $a \times b$ we have

$$\Phi_{mn} = \sin \frac{m\pi x}{a} \cos \frac{n\pi y}{b} \qquad (5\text{b-}219)$$

$$\Psi_{mn} = \cos \frac{m\pi x}{a} \sin \frac{n\pi y}{b} \qquad (5\text{b-}220)$$

with $\qquad \Gamma_{mn}{}^{(TM)} = \Gamma_{mn}{}^{(TE)} = \left[\left(\frac{m\pi}{a}\right)^2 + \left(\frac{n\pi}{b}\right)^2 \right]^{\frac{1}{2}} \qquad (5\text{b-}221)$

For a circular cylindrical resonator of radius a, we have

$$\Phi_{mn} = J_m(\Gamma_{mn}{}^{(TM)}r) \begin{array}{c} \cos \\ \sin \end{array} m\phi \qquad (5\text{b-}222)$$

$$\Psi_{mn} = J_m(\Gamma_{mn}{}^{(TE)}r) \begin{array}{c} \cos \\ \sin \end{array} m\phi \qquad (5\text{b-}223)$$

where $\Gamma_{mn}{}^{(TM)}$ and $\Gamma_{mn}{}^{(TE)}$ satisfy the following equations:

$$J_m(\Gamma_{mn}{}^{(TM)}a) = 0 \qquad (5\text{b-}224)$$
$$J'_m(\Gamma_{mn}{}^{(TE)}a) = 0 \qquad (5\text{b-}225)$$

---

[1] Solutions are also available for resonators of more complex shapes, such as the ellipsoid-hyperbolid resonators [W. W. Hansen and R. D. Richtmeyer, *J. Appl. Phys.* **10**, 189 (1930)] and the reentrant cavities [D. C. Stinson, *Trans. IRE MTT-3*, 18 (1955)].

Solutions are also available for spherical cavity of radius a:

$$\mathbf{E}_{mnl}^{TE} = \nabla \times (\Phi_{mnl} r \mathbf{e}_r) \tag{5b-226}$$

$$\mathbf{H}_{mnl}^{TE} = -\frac{i}{\omega\mu} \nabla \times \nabla \times [\Phi_{mnl} r \mathbf{e}_r] \tag{5b-227}$$

$$\mathbf{H}_{mnl}^{TM} = \nabla \times [\Psi_{mnl} r \mathbf{e}_r] \tag{5b-228}$$

$$\mathbf{E}_{mnl}^{TM} = \frac{i}{\omega\epsilon} \nabla \times \nabla \times [\Psi_{mnl} r \mathbf{e}_r] \tag{5b-229}$$

$$\Phi_{mnl} = j_m(k_{mnl}^{(TE)} r) P_m{}^l (\cos \theta)_{\sin}^{\cos} l\phi \tag{5b-230}$$

$$\Psi_{mnl} = j_m(k_{mnl}^{(TM)} r) P_m{}^l (\cos \theta)_{\sin}^{\cos} l\phi \tag{5b-231}$$

where $j_m(x)$ is the spherical Bessel function. $k_{mnl}^{(TE)}$ and $k_{mnl}^{(TM)}$ satisfy

$$j_m(k_{mnl}^{(TE)} a) = 0, \quad j_m'(k_{mnl}^{(TM)} a) = 0 \tag{5b-232}$$

with

$$\omega_{mnl}^{(TE),(TM)} = \frac{k_{mnl}^{(TE),(TM)}}{\sqrt{\mu\epsilon}}$$

Field configurations for a few lower-order modes are given in Fig. 5b-6.

Small Perturbation Formula. The resonant frequency shift of a cavity due to the presence of a small foreign body having a dielectric constant ϵ_1 and a permeability μ_1 is

$$-\frac{\delta\omega}{\omega_p} = \frac{\int_{V_1} (\epsilon_1 - \epsilon_0)\mathbf{E}_1 \cdot \mathbf{E}_0^* \, dV + \int_{V_1} (\mu_1 - \mu_0)\mathbf{H}_1 \cdot \mathbf{H}_0^* \, dV}{\epsilon_0 \int_V \mathbf{E}_0 \cdot \mathbf{E}_0^* \, dV + \mu_0 \int_V \mathbf{H}_0 \cdot \mathbf{H}_0^* \, dV} \tag{5b-233}$$

where $\mathbf{E}_1$, $\mathbf{H}_1$ denote the resulting field vectors within the volume V_1 of the foreign body, and $\mathbf{E}_0$, $\mathbf{H}_0$ denote the undisturbed field vectors. V is the volume of the cavity. ω_p is the resonant frequency of the unperturbed cavity.

The resonant frequency shift of a cavity due to a small wall deformation is

$$-\frac{\delta\omega}{\omega_p} = \frac{\int_{\Delta V} (\mu_0 \mathbf{H}_0 \cdot \mathbf{H}_0^* - \epsilon_0 \mathbf{E}_0 \cdot \mathbf{E}_0^*) \, dV}{\mu_0 \int_V (\mathbf{H}_0 \cdot \mathbf{H}_0^*) \, dV + \epsilon_0 \int_V (\mathbf{E}_0 \cdot \mathbf{E}_0^*) \, dV} \tag{5b-234}$$

where ΔV is the small change in cavity volume.

Open Resonators. For very high frequency waves (such as light waves) any enclosed metallic cavity of reasonable dimensions for machining would have to operate on a very high order mode. The resonances of the mode would be so closely grouped that the natural bandwidths of the oscillating modes could not be separated, and the use as a resonant system would be impractical. By removing the sides from a closed cavity, a large number of modes can be eliminated owing to energy loss by radiation from the open sides; only the low-loss modes which are essentially *TEM* modes will remain. Assuming that z is the axis of the open resonator, and x, y are the transverse directions, one may obtain, from Maxwell's equations, the simple beam solutions which are characterized by a direction of propagation (the z axis) and by a unique plane phase front perpendicular to this axis:[1]

$$E_{mn}(z) = E_0 \frac{w_0}{w} H_m \left(\sqrt{2}\, \frac{x}{w} \right) H_n \left(\sqrt{2}\, \frac{y}{w} \right) \exp \left(-\frac{x^2 + y^2}{w^2} \right) \tag{5b-235}$$

---

[1] G. D. Boyd and J. P. Gordon, *Bell System Tech. J.* **40**, 489 (1961); A. G. Fox and T. Li, *ibid.* 453.

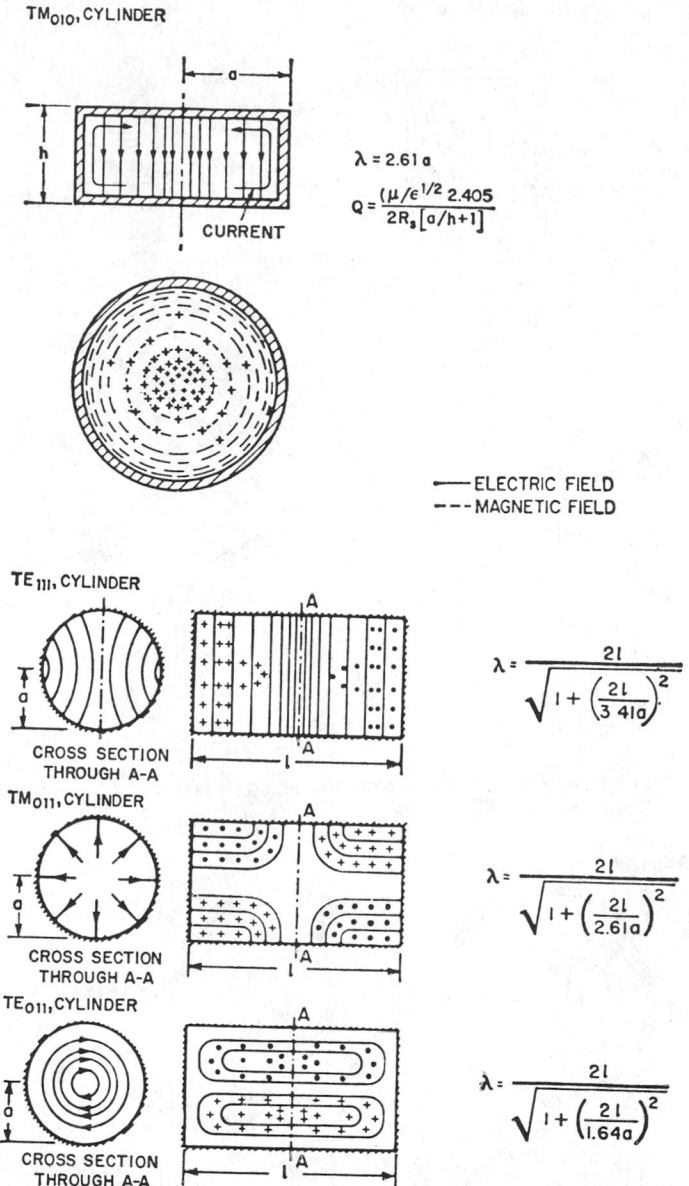

FIG. 5b-6. Field configurations for several selected modes. λ is the resonant wavelength.

where E_{mn} is a field component parallel to a wavefront for the (m,n) mode; H_m and H_n are Hermite polynomials of order m and n, respectively; $w_0 > \lambda$ is an arbitrary parameter with dimensions of length (w_0 may also be defined as the minimum spot size of the beam), $w(z) = w_0[1 + (z/z_0)^2]^{\frac{1}{2}}$, $z_0 = \pi w_0^2/\lambda$, and λ is the wavelength of a plane wave in the resonator medium. Possible positions of reflectors having radii of

TE$_{101}$, RECTANGULAR RESONATOR

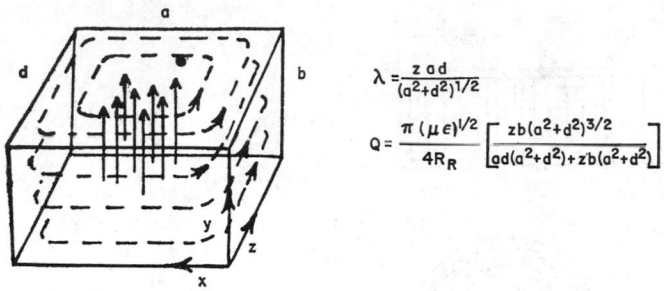

$$\lambda = \frac{z\,a\,d}{(a^2+d^2)^{1/2}}$$

$$Q = \frac{\pi\,(\mu\epsilon)^{1/2}}{4R_R}\left[\frac{zb(a^2+d^2)^{3/2}}{ad(a^2+d^2)+zb(a^2+d^2)}\right]$$

TM$_{101}$, SPHERE

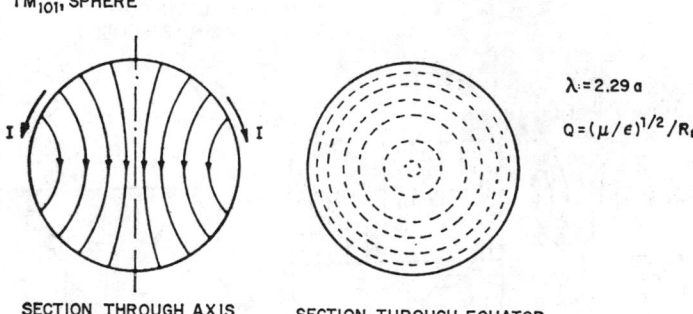

$\lambda = 2.29\,a$

$Q = (\mu/\epsilon)^{1/2}/R_s$

SECTION THROUGH AXIS   SECTION THROUGH EQUATOR

TE$_{101}$, SPHERE

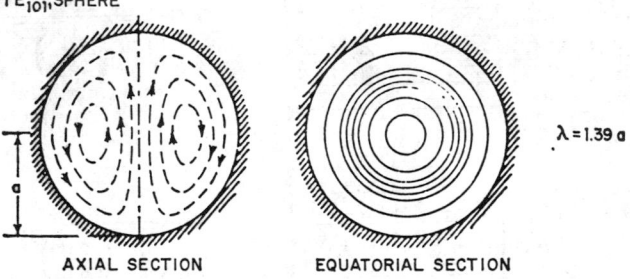

$\lambda = 1.39\,a$

AXIAL SECTION   EQUATORIAL SECTION

Fig. 5b-6 (*Continued*)

curvature R are given by the relation

$$R(z) = -w\,\frac{dz}{dw} = -\frac{1}{z}\,(z^2 + z_0^2) \qquad (5b\text{-}236)$$

The size of the reflector must be large enough to intercept substantially all the field for the mode of interest (say, $m = 0$, $n = 0$), so that energy loss due to diffraction may be acceptable. The modes with large m and n have fields extending farther out from the axis and so will suffer larger diffraction losses. In this way one can discriminate between the transverse modes and ensure that only a few will have low loss. As an example, let us design an optical resonator by using two reflectors having radii of curvature R_1 and R_2, and a mirror separation d. From Eq. (5b-236), we have

$$R_1 = -z_1 - \frac{z_0^2}{z_1} \qquad R_2 = -z_2 - \frac{z_0^2}{z_2}$$

with
$$z_2 - z_1 = d$$

Solving the above equations for z_0 gives

$$z_0 = \left[\frac{d(R_1 - d)(-R_2 - d)(R_1 - R_2 - d)}{(R_1 - R_2 - 2d)^2} \right]^{\frac{1}{2}}$$

which is the location of the minimum spot size $w_0 = (\lambda z_0/\pi)^{\frac{1}{2}}$.

The phase variation along the z axis for the (m,n) mode is

$$\beta z = -kz - (m + n + 1) \tan^{-1} \frac{z}{z_0} \qquad (5b\text{-}237)$$

where $k = 2\pi/\lambda$, and β is the propagation constant. The resonant condition requires

$$\beta d = q\pi \approx kd \qquad q = 1, 2, \ldots \qquad (5b\text{-}238)$$

where d is the minor separation. The frequency separation between longitudinal modes is $\Delta f = c/2d$; c = velocity of light in the resonator medium. Selected modal patterns are given in Fig. 5b-7.

The Q of an optical resonator is given by

$$Q = \frac{2\pi d}{\alpha \lambda} \qquad (5b\text{-}239)$$

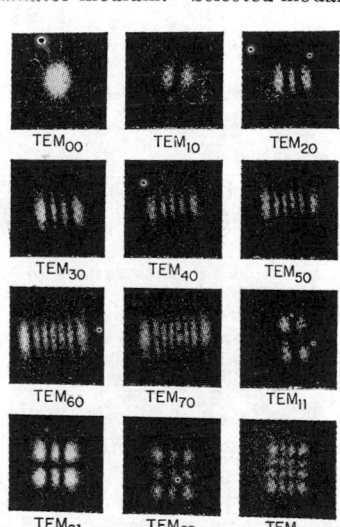

where α is the fractional power loss per bounce from a reflector and is the sum of diffraction and reflection losses. The diffraction loss is small only if the Fresnel number $N = a_1 a_2/\lambda d$, where a_1 and a_2 are radii of the mirrors, is much larger than unity.

General Considerations. DEGENERATE MODES. Modes with different field distributions but with the same resonant frequency.

EXCITATION OF CAVITY FIELDS. Excitation of cavity fields may be accomplished by the introduction of a conducting probe or antenna in the direction of the electric field lines, or by the introduction of a conducting loop with plane normal to the magnetic field lines, or by the introduction of a hole or iris between the cavity waveguide.[1] It is important to note that when the walls of the cavity have one or more apertures, the orthonormal sets $\mathbf{H}_p$ and $\mathbf{E}_p$, derived from the consideration of a completely enclosed cavity, are no longer adequate for an expansion of the cavity fields.[2] The electric vector $\mathbf{E}$ and the magnetic vector $\mathbf{H}$ of an electromagnetic field within a

FIG. 5b-7. Modal patterns in optical resonators. [*From H. Kogelnik and W. W. Rigrod, Proc. IRE* **50**, 220 (1962).]

[1] Smythe, W. R., "Static and Dynamic Electricity," 3d ed., McGraw-Hill Book Company, New York, 1968.

[2] K. Kurokawa, *IRE Trans.* **MTT-6**, 178 (1958).

cavity coupled to an outside source by means of a waveguide must be derived according to the relations

$$\mathbf{E} = \epsilon \sum_{p=1}^{\infty} \mathbf{E}_p \int_V \mathbf{E} \cdot \mathbf{E}_p \, dV \tag{5b-240}$$

$$\mathbf{H} = \sum_{p=1}^{\infty} \mathbf{H}_p \left[\frac{-i\omega\epsilon\mu}{k_p{}^2 - k^2} \int_A (\mathbf{n} \times \mathbf{E}) \cdot \mathbf{H}_p \, dA \right]$$

$$- \sum_{p=1}^{\infty} \mathbf{G}_p \left[\frac{i\omega\epsilon}{k^2} \int_A (\mathbf{n} \times \mathbf{E}) \cdot \mathbf{G}_p \, dA \right] \tag{5b-241}$$

where A consists of the perfectly conducting surface and the aperature surface, V is the volume of the cavity, $k^2 = \omega^2\mu\epsilon$, and

$$\left. \begin{array}{c} \nabla^2 \mathbf{E}_p + k_p{}^2 \mathbf{E}_p = 0 \\ \nabla \cdot \mathbf{E}_p = 0 \end{array} \right\} \quad \text{in } V \tag{5b-242}$$

$$\mathbf{n} \times \mathbf{E}_p = 0 \qquad \text{on } A$$

$$\left. \begin{array}{c} \nabla^2 \mathbf{H}_p + k_p{}^2 \mathbf{H}_p = 0 \\ \nabla \cdot \mathbf{H}_p = 0 \end{array} \right\} \quad \text{in } V \tag{5b-243}$$

$$\mathbf{n} \times (\nabla \times \mathbf{H}_p) = 0 \qquad \text{on } A$$

$$\left. \begin{array}{c} \nabla^2 \mathbf{G}_p + g_p{}^2 \mathbf{G}_p = 0 \\ \nabla \cdot \mathbf{G}_p = 0 \end{array} \right\} \quad \text{in } V \tag{5b-244}$$

$$\mathbf{n} \times (\nabla \times \mathbf{G}_p) = 0 \qquad \mathbf{n} \cdot \mathbf{G}_p = 0 \qquad \text{on } A$$

Hence $\mathbf{G}_p$ is derivable from scalar potential as follows:

$$\mathbf{G} = \nabla v_p \tag{5b-245}$$

References

1. Borgnis, F. E., and C. H. Papas: Electromagnetic Waveguides and Resonators, "Handbuch der Physik," vol. 16, Springer-Verlag OHG, Berlin, 1958.
2. Goubau, G.: "Electromagnetic Waveguides and Cavities," Pergamon Press, New York, 1961.
3. Collin, R. E.: "Foundations for Microwave Engineering," McGraw-Hill Book Company, New York, 1966.
4. Ramo, S., J. R. Whinnery, and T. Van Duzer: "Fields and Waves in Communication Electronics," John Wiley & Sons, Inc., New York, 1965.
5. Slater, J. C.: "Microwave Electronics," D. Van Nostrand Company, Inc., Princeton, N.J., 1950.

5b-11. Radiation. Solutions of radiation problems must satisfy not only Maxwell's equations and the appropriate boundary conditions but also Sommerfeld's radiation condition.

Radiation Field from Known Current Distributions. Given a distribution of electric and magnetic currents, specified by the density functions $\mathbf{J}(\mathbf{r})$ and $\mathbf{J}_m(\mathbf{r})$ occupying a finite region of space. Formal expressions for the electric vector $\mathbf{E}$ and the magnetic vector $\mathbf{H}$ in an unbounded space are given earlier by Eqs. (5b-81) through (5b-84). Consider a reference frame with its origin in the vicinity of the sources; let $\mathbf{r}$ be the coordinates of the observation point, and $\mathbf{r}'$ be the coordinates of the source point. In the far-zone region (i.e., $r \gg r'$ and $kr \gg 1$), the radiated fields which are purely transverse to the direction of propagation are

$$E_\theta = i\omega\mu \frac{e^{ikr}}{4\pi r} F_\theta(\theta,\phi) = \left(\frac{\mu}{\epsilon}\right)^{\frac{1}{2}} H_\phi \tag{5b-246}$$

$$E_\phi = i\omega\mu \frac{e^{ikr}}{4\pi r} F_\phi(\theta,\phi) = -\left(\frac{\mu}{\epsilon}\right)^{\frac{1}{2}} H_\theta \tag{5b-247}$$

with

$$F_\theta(\theta,\phi) = \int_{V_0} \left[\mathbf{J}(\mathbf{r}') \cdot \mathbf{e}_\theta + \left(\frac{\epsilon}{\mu}\right)^{\frac{1}{2}} \mathbf{J}_m(\mathbf{r}') \cdot \mathbf{e}_\phi \right] e^{-ik\mathbf{e}_r \cdot \mathbf{r}'} \, dV' \tag{5b-248}$$

$$F_\phi(\theta,\phi) = \int_{V_0} \left[\mathbf{J}(\mathbf{r}') \cdot \mathbf{e}_\phi - \left(\frac{\epsilon}{\mu}\right)^{\frac{1}{2}} \mathbf{J}_m(\mathbf{r}') \cdot \mathbf{e}_\theta \right] e^{-ik\mathbf{e}_r \cdot \mathbf{r}'} \, dV' \tag{5b-249}$$

where r, θ, ϕ are the spherical coordinates of the observation point and $\mathbf{e}_r$, $\mathbf{e}_\theta$, $\mathbf{e}_\phi$ are the corresponding unit vectors. The Poynting's vector (the time-average intensity of power flow) is

$$\mathbf{S} = \tfrac{1}{2}\mathrm{Re} \; (\mathbf{E} \times \mathbf{H}^*) = \frac{1}{8\lambda^2 r^2} \left(\frac{\mu}{\epsilon}\right)^{\frac{1}{2}} \{|F_\theta|^2 + |F_\phi|^2\} \mathbf{e}_r \tag{5b-250}$$

and the time-average power per unit solid angle is

$$p(\theta,\phi) = r^2 |\mathbf{S}| = \frac{1}{8\lambda^2} \left(\frac{\mu}{\epsilon}\right)^{\frac{1}{2}} \{|F_\theta|^2 + |F_\phi|^2\} \tag{5b-251}$$

The total time-average radiated power is

$$P = \int p(\theta,\phi) \, d\Omega = \int_0^{2\pi} \int_0^\pi p(\theta,\phi) \sin\theta \, d\theta \, d\phi \tag{5b-252}$$

The directivity characteristics of the radiating system are expressed by the gain function which is the ratio of the power radiated per unit solid angle in a direction (θ,ϕ) to average power radiated per unit solid angle. It is also referred to as the gain function with respect to an isotropic radiator radiating the same total power. Thus,

$$G(\theta,\phi) = \frac{p(\theta,\phi)}{\dfrac{1}{4\pi} \displaystyle\int p(\theta,\phi) \, d\Omega} = \frac{4\pi(|F_\theta|^2 + |F_\phi|^2)}{\displaystyle\int_0^{2\pi} \int_0^\pi (|F_\theta|^2 + |F_\phi|^2) \sin\theta \, d\theta \, d\phi} \tag{5b-253}$$

The absolute gain is the maximum value of the gain function, and directivity $=$ $10 \log_{10}[G(\theta,\phi)]_{\max}$ db.

THE ELECTRIC DIPOLE. The fields of an oscillating electric dipole of moment $p = p_0 \mathbf{e}_z = -(Id/i\omega)\mathbf{e}_z$, where $\mathbf{e}_z$ is a unit vector in the z direction, I is the uniform current, and d is the length of the structure, are

$$E_r = \frac{1}{2\pi\epsilon} \left(\frac{1}{r^3} - \frac{ik}{r^2}\right) \cos\theta \; p_0 e^{ikr} e^{-i\omega t} \tag{5b-254}$$

$$E_\theta = \frac{1}{4\pi\epsilon} \left(\frac{1}{r^3} - \frac{ik}{r^2} - \frac{k^2}{r}\right) \sin\theta \; p_0 e^{ikr} e^{-i\omega t} \tag{5b-255}$$

$$H_\phi = -\frac{i\omega}{4\pi} \left(\frac{1}{r^2} - \frac{ik}{r}\right) \sin\theta \; p_0 e^{ikr} e^{-i\omega t} \tag{5b-256}$$

The dipole is located at the origin of the spherical coordinates. The far-zone Poynting vector is

$$\mathbf{S} = \frac{\omega k^3}{32\pi^2 \epsilon} |p_0|^2 \frac{\sin^2\theta}{r^2} \mathbf{e}_r \tag{5b-257}$$

and the gain function is

$$G(\theta,\phi) = \tfrac{3}{2} \sin^2\theta \tag{5b-258}$$

THE MAGNETIC DIPOLE. The fields of an oscillating magnetic dipole of moment $\mathbf{m} = m_0 \mathbf{e}_z = IA\mathbf{e}_z$, where $\mathbf{e}_z$ is a unit vector in the z direction, I is the uniform current, and A is the area of the small current loops, are

$$E_\phi = \frac{k^2}{4\pi} \left(\frac{\mu}{\epsilon}\right)^{\frac{1}{2}} \left(\frac{1}{r} + \frac{i}{kr^2}\right) \sin \theta \, m_0 e^{ikr} e^{-i\omega t} \tag{5b-259}$$

$$H_r = \frac{1}{2\pi} \left(\frac{1}{r^3} - \frac{ik}{r^2}\right) \cos \theta \, m_0 e^{ikr} e^{-i\omega t} \tag{5b-260}$$

$$H_\theta = \frac{1}{4\pi} \left(\frac{1}{r^3} - \frac{ik}{r^2} - \frac{k^2}{r}\right) \sin \theta \, m_0 e^{ikr} e^{-i\omega t} \tag{5b-261}$$

$$A \ll \lambda^2$$

The dipole is located at the origin of the coordinates, and $\mathbf{e}_z$ is normal to the area A. The gain function is the same as that of an electric dipole.

THE LINEAR THIN-WIRE ANTENNA. The far-zone fields of a thin-wire center-driven antenna with the current distribution

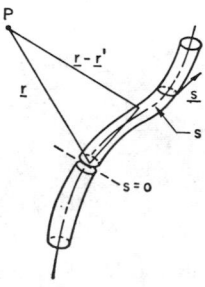

$$\mathbf{J}(\mathbf{r}) = \mathbf{e}_z I_0 \, \delta(x) \, \delta(y) \sin k(l - |z|) \tag{5b-262}$$

where I_0 is the current amplitude, $2l$ is the length of the wire, and the wire is centered at the origin, are

$$E_\theta = -i \sqrt{\frac{\mu}{\epsilon}} I_0 \frac{\cos (kl \cos \theta) - \cos kl}{\sin \theta} \cdot \frac{e^{ikr} e^{-i\omega t}}{2\pi r} \tag{5b-263}$$

$$H_\phi = \sqrt{\frac{\epsilon}{\mu}} E_\theta \tag{5b-264}$$

The far-zone Poynting vector is

FIG. 5b-8. Coordinates of a curved wire.

$$\mathbf{S} = \mathbf{e}_r \sqrt{\frac{\mu}{\epsilon}} \frac{I_0^2}{8\pi^2 r^2} \left[\frac{\cos (kl \cos \theta) - \cos kl}{\sin \theta}\right]^2 \tag{5b-265}$$

Here, the length l does not have to be much less than the wavelength λ.

Integral Equation of Thin-wire Antennas. Equations (5b-81) through (5b-84) show that when the current distribution is known, the radiation fields may be found by integration. Strictly speaking, the current distribution must be found by solving the boundary-value problem which is usually quite difficult except for some special cases.[1] For a thin curved wire antenna excited by a generator which produces an electric field $E_s^i(s)$ across a gap centered at $s = 0$, the current along the wire satisfies the following integral equation.[2] (Symmetric wire antenna)

$$\int_{L_s} J(s')\pi(s,s') \, ds' = c' \cos ks + \frac{i}{\sqrt{\mu/\epsilon}} \int_0^s E_\xi^i(\xi) \sin k(s - \xi) \, d\xi \tag{5b-266}$$

$$\pi(s,s') = G(s,s')\mathbf{s} \cdot \mathbf{s}' - \int_0^s \left[\frac{\partial G(\xi,s')}{\partial \xi} \boldsymbol{\xi} \cdot \mathbf{s}' + \frac{\partial G(\xi,s')}{\partial s'}\right.$$
$$\left. + G(\xi,s') \frac{\partial(\boldsymbol{\xi} \cdot \mathbf{s}')}{\partial \xi}\right] \cos k(s - \xi) \, d\xi \tag{5b-267}$$

$$G(s,s') = \frac{e^{ik|\mathbf{r}-\mathbf{r}'|}}{4\pi|\mathbf{r} - \mathbf{r}'|} \tag{5b-268}$$

where $\int_{L_s} ds'$ will represent the surface integral over the wire, s is the arc length measured from the center of the gap, and $\mathbf{s}$, $\mathbf{s}'$, $\boldsymbol{\xi}'$ are all unit vectors along the wire (see Fig. 5b-8). c' is determined by the condition that the current vanishes at both ends of the antenna. Equation (5b-266) may be solved numerically by reducing the integral equation to a finite set of algebraic equations.[2]

[1] S. A. Schelkunoff, "Advanced Antenna Theory," John Wiley & Sons, Inc., New York, 1952.
[2] K. K. Mei, *IEEE Trans.* **AP-13,** 374 (1965).

Knowing the exact current distribution, one may also find the input impedance at the gap. For example, for a delta gap source at $s = 0$—i.e., the last integral in Eq. (5b-266) becomes $[iV_0/(2\sqrt{\mu/\epsilon})]$ sin ks, where V_0 is the voltage across the gap— the input impedance is $Z_{in} = V_0/I(0)$, where $I(0)$ is the current at $s = 0$.

Radiation Field from Apertures. Given a surface S enclosing the sources and the values of **E** and **H** over the entire surface S. The field at a point P outside the region of the surface is given by

$$\mathbf{E}_p(\mathbf{r}) = \int_S \{i\omega\mu[\mathbf{n}' \times \mathbf{H}(\mathbf{r}')]G(\mathbf{r},\mathbf{r}') + [\mathbf{n}' \times \mathbf{E}(\mathbf{r}')] \times \nabla'G(\mathbf{r},\mathbf{r}')$$
$$+ [\mathbf{n}' \cdot \mathbf{E}(\mathbf{r}')]\nabla'G(\mathbf{r},\mathbf{r}')\} \, dS' \quad \text{(5b-269)}$$

$$\mathbf{H}_p(\mathbf{r}) = \int_S \{-i\omega\epsilon[\mathbf{n}' \times \mathbf{E}(\mathbf{r}')]G(\mathbf{r},\mathbf{r}') + [\mathbf{n}' \times \mathbf{H}(\mathbf{r}')] \times \nabla'G(\mathbf{r},\mathbf{r}')$$
$$+ [\mathbf{n}' \cdot \mathbf{H}(\mathbf{r}')]\nabla'G(\mathbf{r},\mathbf{r}')\} \, dS' \quad \text{(5b-270)}$$

$$G(\mathbf{r},\mathbf{r}') = \frac{e^{ik|\mathbf{r}-\mathbf{r}'|}}{4\pi|\mathbf{r}-\mathbf{r}'|} \quad \text{(5b-271)}$$

where $\mathbf{n}'$ is the unit vector normal to S directed outward from the region of the sources, $|\mathbf{r} - \mathbf{r}'|$ is the distance from dS' to P; the gradient operator ∇' is with respect to the primed coordinates on S. Another form of Eqs. (5b-269) and (5b-270) is

$$\mathbf{E}_p(\mathbf{r}) = -\int_S \left[G(\mathbf{r},\mathbf{r}') \frac{\partial \mathbf{E}(\mathbf{r}')}{\partial n'} - \mathbf{E}(\mathbf{r}') \frac{\partial G(\mathbf{r},\mathbf{r}')}{\partial n'} \right] dS' \quad \text{(5b-272)}$$

$$\mathbf{H}_p(\mathbf{r}) = -\int_S \left[G(\mathbf{r},\mathbf{r}') \frac{\partial \mathbf{H}(\mathbf{r}')}{\partial n'} - \mathbf{H}(\mathbf{r}') \frac{\partial G(\mathbf{r},\mathbf{r}')}{\partial n'} \right] dS' \quad \text{(5b-273)}$$

where $G(\mathbf{r},\mathbf{r}')$ is given by Eq. (5b-271). Equations (5b-269) and (5b-270) or Eqs. (5b-272) and (5b-273) may be regarded as an analytical formulation of the Huygens-Fresnel principle which states that each point on a given wavefront can be regarded as a secondary source which gives rise to a spherical wavelet; the wave at a field point is to be obtained by superposition of these elementary wavelets, with due regard to their phase differences when they reach the point in question.

Since the values of **E** and **H** over the entire surface S are not known for most antenna problems, it is therefore desirable to provide modified expressions for $\mathbf{E}_p(\mathbf{r})$ and $\mathbf{H}_p(\mathbf{r})$. For very high frequencies, we have

$$\mathbf{E}_p(\mathbf{r}) \approx \frac{i}{\omega\epsilon} \int_A \{k^2[\mathbf{n}' \times \mathbf{H}_a(\mathbf{r}')]G(\mathbf{r},\mathbf{r}') + [\mathbf{n}' \times \mathbf{H}_a(\mathbf{r}')] \cdot \nabla'[\nabla'G(\mathbf{r},\mathbf{r}')]$$
$$- i\omega\epsilon[\mathbf{n}' \times \mathbf{E}_a(\mathbf{r}')] \times \nabla'G(\mathbf{r},\mathbf{r}')\} \, dS' \quad \text{(5b-274)}$$

$$\mathbf{H}_p(\mathbf{r}) \approx \frac{-i}{\omega\mu} \int_A \{k^2[\mathbf{n}' \times \mathbf{E}_a(\mathbf{r}')]G(\mathbf{r},\mathbf{r}') + [\mathbf{n}' \times \mathbf{E}_a(\mathbf{r}')] \cdot \nabla'[\nabla'G(\mathbf{r},\mathbf{r}')]$$
$$+ i\omega\mu[\mathbf{n}' \times \mathbf{H}_a(\mathbf{r}')] \times \nabla'G(\mathbf{r},\mathbf{r}')\} \, dS' \quad A \gg \lambda^2 \quad \text{(5b-275)}$$

where $\mathbf{E}_a(\mathbf{r}')$ and $\mathbf{H}_a(\mathbf{r}')$ are the fields over the aperture. It is noted that in the present case the integration is carried out over an open surface A in contrast with that of the previous equations, i.e., Eqs. (5b-269) to (5b-273). Equations (5b-274) to (5b-275) may be used for the computation of fields for reflectors, lenses, and horns under the approximations that the field over the aperture is related in the most simple way possible to the primary sources—in the cases of lenses and reflectors, by the use of geometrical optics; in the case of horns, by considering the field distribution which would exist over the aperture plane of the horn extended to infinity. In the far zone ($r \gg r'$, $kr \gg 1$),

$$\mathbf{E}_p(\mathbf{r}) \approx \frac{ik}{4\pi r} e^{ikr} \mathbf{e}_r \times \int_A \left\{ \mathbf{n}' \times \mathbf{E}_a(\mathbf{r}') - \left(\frac{\mu}{\epsilon}\right)^{\frac{1}{2}} \mathbf{e}_r \times [\mathbf{n}' \times \mathbf{H}_a(\mathbf{r}')] \right\} e^{ik\mathbf{e}_r \cdot \mathbf{r}'} \, dS'$$
$$\text{(5b-276a)}$$

$$\mathbf{H}_p(\mathbf{r}) \approx \sqrt{\frac{\epsilon}{\mu}} [\mathbf{e}_r \times \mathbf{E}_p(\mathbf{r})] \quad A \gg \lambda^2 \quad \text{(5b-276b)}$$

When the aperture field is obtained by the simple considerations stated above, there is an elementary relation between the tangential components of the electric and magnetic vectors over the aperture:

$$H_a = \eta(\mathbf{s}' \times \mathbf{E}_a) \tag{5b-277}$$

and

$$\mathbf{E}_p^{(\text{far zone})}(\mathbf{r}) \simeq \frac{ike^{ikr}}{4\pi r} \, \mathbf{e}_r \times \int_A \left(\mathbf{n}' \times \mathbf{E}_a(\mathbf{r}') - \eta \left(\frac{\mu}{\epsilon}\right)^{\frac{1}{2}} \{\mathbf{e}_r \cdot [\mathbf{s}' \times \mathbf{E}_a(\mathbf{r}')]\mathbf{n}' \right.$$
$$\left. - [\mathbf{s}' \times \mathbf{E}_a(\mathbf{r}')](\mathbf{n}' \cdot \mathbf{e}_r)\} \right) e^{ik\mathbf{e}_r \cdot \mathbf{r}'} \, dS' \tag{5b-278a}$$

$$\mathbf{H}_p^{(\text{far zone})}(\mathbf{r}) \simeq \sqrt{\frac{\epsilon}{\mu}} \, [\mathbf{e}_r \times \mathbf{E}_p^{(\text{far zone})}(\mathbf{r})] \tag{5b-278b}$$

where $\mathbf{s}'$ is a unit vector along a ray through the aperture; $\eta = (\epsilon/\mu)^{\frac{1}{2}}$ for lenses and reflectors in free space; and for a horn, $\eta = (\Gamma_{mn}/\omega\mu)(1 - R)/(1 + R)$ for TE modes and $[\omega\epsilon/\Gamma_{mn}](1 - R)/(1 + R)$ for TM modes; with R being the reflection coefficient of the mode in the horn, and Γ_{mn} the eigenvalues of modes in an infinite horn.

Another approximate formula for the computation of radiated fields from a perfectly conducting reflector at very high frequencies is also available. The formula is based on the knowledge of the induced current distribution over the reflector; the induced current is obtained on the basis of geometrical optics. Let $(\mathbf{E}_i,\mathbf{H}_i)$ be the incident field, $\mathbf{s}_0$ a unit vector in the direction of the incident ray, and $\mathbf{n}$ a unit vector normal to the surface S at the point of incidence. The induced surface current density is

$$\mathbf{J} = 2(\mathbf{n} \times \mathbf{H}_i) = 2\sqrt{\frac{\epsilon}{\mu}} \, [\mathbf{n} \times (\mathbf{s}_0 \times \mathbf{E}_i)] \tag{5b-279a}$$

The induced surface charge density is

$$\rho = 2\epsilon(\mathbf{n} \cdot \mathbf{E}_i) \tag{5b-279b}$$

The radiated fields are

$$\mathbf{E}(\mathbf{r}) = \frac{2i}{\omega\epsilon} \int_S \{[\mathbf{n}' \times \mathbf{H}_i(\mathbf{r}')] \cdot \nabla'[\nabla'G(\mathbf{r},\mathbf{r}')] + k^2[\mathbf{n}' \times \mathbf{H}_i(\mathbf{r}')]G(\mathbf{r},\mathbf{r}')\} \, dS' \tag{5b-280a}$$

$$\mathbf{H}(\mathbf{r}) = 2\int_S \{[\mathbf{n}' \times \mathbf{H}_i(\mathbf{r}')] \times \nabla'G(\mathbf{r},\mathbf{r}')\} \, dS' \tag{5b-280b}$$

where S is the illuminated surface of the reflector and

$$G(\mathbf{r},\mathbf{r}') = \frac{e^{ik|\mathbf{r}-\mathbf{r}'|}}{4\pi|\mathbf{r}-\mathbf{r}'|}$$

The far-zone radiated fields are

$$\mathbf{E}(\mathbf{r}) \simeq \frac{i\omega\mu}{2\pi r} e^{ikr} \int_S (\mathbf{n}' \times \mathbf{H}_i(\mathbf{r}') - \{[\mathbf{n}' \times \mathbf{H}_i(\mathbf{r}')] \cdot \mathbf{e}_r\}\mathbf{e}_r)e^{-ikr'} \cdot \mathbf{e}_r \, dS \tag{5b-281a}$$

$$\mathbf{H}(\mathbf{r}) \simeq \left(\frac{\epsilon}{\mu}\right)^{\frac{1}{2}} [\mathbf{e}_r \times \mathbf{E}(\mathbf{r})] \tag{5b-281b}$$

Linear Arrays of Antennas. A great variety of radiation patterns can be realized by arranging in space a set of antennas operating at the same frequency. The linear array has been used quite successfully to synthesize certain desired radiation patterns.[1] A linear array is assumed to consist of n antennas with centers at the points x_p ($p = 0$, $1, \ldots, n - 1$) on the x axis. Each antenna is independently fed. Under the

---

[1] See the article by R. S. Elliott in "Microwave Scanning Antennas," edited by R. C. Hansen, Academic Press, Inc., New York, 1966.

approximation that the antennas do not interact with one another, the Poynting vector for the radiated field is

$$ \mathbf{S} = \mathbf{e}_r \sqrt{\frac{\mu}{\epsilon}} \frac{1}{8\pi^2 r^2} |G(\theta,\phi)A(\theta,\phi)|^2 \tag{5b-282} $$

where $G(\theta,\phi)$ is the radiation pattern of each individual antenna, and $A(\theta,\phi)$ is called the array factor. The radiation pattern of the entire array is

$$ u(\theta,\phi) = |G(\theta,\phi)A(\theta,\phi)| = G(\theta,\phi)|A(\theta,\phi)| \tag{5b-283} $$

For an array made up of center-fed half-wave dipoles ($kl = \pi/2$, $2l = $ length of the dipole) oriented parallel to the z axis,

$$ G(\theta,\phi) = \frac{\cos\,[(\pi/2)\cos\theta]}{\sin\theta} \tag{5b-284a} $$

$$ A(\theta,\phi) = \sum_{p=0}^{n-1} A_p e^{-ikx_p \sin\theta \cos\phi} \tag{5b-284b} $$

where A_p denotes the complex magnitude of the current. By the appropriate choice of A_p and x_p, many desired radiation patterns may be synthesized.

EQUALLY SPACED LINEAR ARRAY ($x_p = pd$; d is the uniform spacing). The array factor for an equally spaced linear array is

$$ A(\theta,\phi) = \sum_{p=0}^{n-1} a_p \xi^p \tag{5b-285} $$

where $\xi = e^{i\alpha}$, $\alpha = -kd\sin\theta\cos\phi - \gamma$, and $A_p = a_p e^{-ip\gamma}$. $e^{-ip\gamma}$ is the progressive phasing γ of the array currents, and a_p is the magnitude of the currents.

1. Uniform Array ($a_p = $ constant). The array factor of the uniform array is

$$ A(\theta,\phi) = \sum_{p=0}^{n-1} \xi^p = \frac{\xi^n - 1}{\xi - 1} \tag{5b-286} $$

Broadside Array ($\gamma = 0$, $kd < 2\pi$). Radiation is cast principally in the broadside direction; $\psi = \pi/2$. (ψ is the angle between the x axis and the line of observation, and $\cos\psi = \sin\theta\cos\phi$.)

End-fire Array ($kd = -\gamma$ or $+\gamma$). Radiation is cast principally in the direction of the line of sources.

Hansen-Woodyard Unilateral End-fire Array [$\gamma = -(kd + \pi/n)$ or $\gamma = +(kd + \pi/2)$]. Radiation is cast principally in the direction $\psi = \pi$ when $\gamma = +(kd + \pi/n)$, and in the direction $\psi = 0$ when $\gamma = -(kd + \pi/n)$.

Phase Array [n and kd ($<\pi$) are fixed]. By varying γ from 0 to kd, the major lobe rotates from the broadside direction to the end-fire direction.

2. Nonuniform Array ($a_p \neq $ constant). The array factor of a nonuniform array is given by Eq. (5b-285).

Binomial Array. a_p are chosen as the binomial coefficient: $a_p = (n-1)!/[(n-1-p)!p!]$. When $\gamma = 0$, $kd = \pi$, the binomial array yields a broadside pattern without side lobes.

Dolph-Chebyshev Array (n even, $d \geq \lambda/2$). By matching the polynomial $A(\theta,\phi)$ in Eq. (5b-285) to a Chebyshev polynomial, one may obtain an array of a given number of elements which gives the lowest side lobes for a prescribed antenna gain, or highest gain for a prescribed side-lobe level.

UNEQUALLY SPACED LINEAR ARRAYS. Although Eq. (5b-284) for unequally spaced array is considerably more difficult to handle than Eq. (5b-285) for equally spaced array, with the use of a computer numerical results can be obtained in a straightforward manner. An unequally spaced array is generally more "broadband" than an equally spaced array.

Radio-astronomical Antennas. Consideration must be given to the case where the incident wave from cosmic sources is partially polarized and polychromatic.[1] Assuming that the radio-astronomical antenna is conjugate-matched to the load, the power absorbed by the load is

$$P_{abs} = \frac{1}{2} \sqrt{\frac{\epsilon}{\mu}} A(\theta,\phi) Tr(p^{rad}p^{rad*}) \cdot (\widetilde{<E^{inc}E^{inc*}>}) \qquad (5b\text{-}287)$$

where $A(\theta,\phi)$ is the effective area of the receiving antenna, i.e., $A(\theta,\phi) = (\lambda^2/4\pi)G(\theta,\phi)$, $G(\theta,\phi)$ is the gain function of antenna in transmission. p^{rad} is the field polarization vector, .

$$p^{rad} = \frac{E^{rad}}{\sqrt{E^{rad} \cdot E^{rad*}}} \qquad (5b\text{-}288)$$

where E^{rad} is the electric vector of the far-zone field radiated by the antenna in transmission. $\widetilde{\langle E^{inc}E^{inc*}\rangle}$ is the transpose of $\langle E^{inc}E^{inc}\rangle$ and, $\langle W\rangle$ means

$$\langle W \rangle = \lim_{T \to \infty} \frac{1}{2T} \int_{-T}^{T} W \, dt \qquad (5b\text{-}289)$$

For example, if the incident polychromatic wave is narrow-band and has the form

$$E^{inc}(\mathbf{r},t) = [\mathbf{e}_\theta E_\theta(t) + \mathbf{e}_\phi E_\phi(t)]e^{-ikr}e^{-i\omega t} \qquad (5b\text{-}290)$$

where $E_\theta(t)$ and $E_\phi(t)$ are slowly varying functions of time, and ω is a mean frequency, then the time-average power absorbed by the conjugate-matched load is

$$P_{abs} = \frac{1}{2}(1 - m)A(\theta,\phi)\langle S^{inc}(\theta,\phi)\rangle + mA(\theta,\phi)\langle S^{inc}(\theta,\phi)\rangle \cos^2\frac{\gamma}{2} \qquad (5b\text{-}291)$$

with

$$\cos\gamma = \cos 2\chi' \cos 2\chi \cos (2\psi' - 2\psi) \sin 2\chi' \sin 2\chi \qquad (5b\text{-}292)$$

$$\langle S^{inc}(\theta,\phi)\rangle = \frac{1}{2}\sqrt{\frac{\epsilon}{\mu}} (\langle E_\theta E_\theta^*\rangle + \langle E_\phi E_\phi^*\rangle) \qquad (5b\text{-}293)$$

γ is the angle between the point $(2\psi, -2\chi)$ describing the polarization ellipse of the incident wave and the point $(2\psi', 2\chi')$ describing the polarization ellipse of the radiated wave, and m is the degree of polarization which is the ratio of the power density of the polarized part to the total power density.[2]

A way to measure the degree of coherence $|\gamma|$ of an incoming polychromatic signal by the use of a correlation interferometer which requires no phase-preserving link has been suggested by Brown and Twiss.[3] The correlation interferometer (which consists of two identical antennas) measures the correlation coefficient $|\gamma|$ which is defined as

$$|\gamma| = \left\{ \frac{\langle [M_1^2 - \langle M_1^2\rangle][M_2^2 - \langle M_2^2\rangle]\rangle}{\sigma(M_1^2)\sigma(M_2^2)} \right\}^{\frac{1}{2}} \qquad (5b\text{-}294)$$

where

$$\sigma^2(M_1^2) = \langle (M_1^2 - \langle M_1^2\rangle)^2\rangle \qquad (5b\text{-}295)$$

$$\sigma^2(M_2^2) = \langle (M_2^2 - \langle M_2^2\rangle)^2\rangle \qquad (5b\text{-}296)$$

$$M_1^2 = V_1 V_1^* \qquad M_2^2 = V_2 V_2^* \qquad (5b\text{-}297)$$

[1] H. C. Ko, *Proc. IRE* **49**, 1446 (1961).
[2] M. Born and E. Wolf, "Principles of Optics," 2d ed., Pergamon Press, New York, 1964.
[3] R. H. Brown and R. Q. Twiss, *Phil. Mag.* **45**, 663 (1954).

$V_1(t)V_1^*(t)$ is the power output of one antenna operating singly and $V_2(t)V_2^*(t)$ is the power output of the other antenna operating singly. The correlation interferometer of Brown and Twiss is an interferometer that measures $|\gamma(M_1{}^2,M_2{}^2)|$, while the conventional interferometer measures $\gamma(V_1,V_2)$. Hence, no phase-preserving link is necessary in the measurement of $|\gamma(M_1{}^2,M_2{}^2)|$, the antennas can be separated greatly, and thus high resolving powers can be realized. If the source of the polychromatic signal is a rectangular distribution of width $2w$, t' e correlation coefficient $|\gamma|$ is related to the width by the equation

$$|\gamma| = \left| \frac{\sin klw}{klw} \right| \qquad (5b\text{-}298)$$

where l is the separation of the interferometer, $k = \omega/c$, and ω is a mean frequency.

Lorentz Reciprocity Theorem. Let $(\mathbf{E}_a,\mathbf{H}_a)$ be the fields generated by sources $(\mathbf{J}_a,\mathbf{J}_{ma})$, and $(\mathbf{E}_b,\mathbf{H}_b)$ be the fields generated by sources $(\mathbf{J}_b,\mathbf{J}_{mb})$, operating at the same frequency. Then, Lorentz reciprocity theorem states that

$$\int_{\text{all space}} (\mathbf{E}_a \cdot \mathbf{J}_b - \mathbf{H}_a \cdot \mathbf{J}_{mb})\, dV = \int_{\text{all space}} (\mathbf{E}_b \cdot \mathbf{J}_a - \mathbf{H}_b \cdot \mathbf{J}_{ma})\, dV \qquad (5b\text{-}299)$$

With regard to antennas, the above theorem means that the receiving pattern of any antenna constructed of linear isotropic matter is identical to its transmitting pattern.

In general, reciprocity does not hold for an anisotropic medium. However, for the special case of an anisotropic plasma or ferrite, the concept of reciprocity can be generalized. This is based on the fact that the dielectric tensor of a magnetically biased plasma or the permeability tensor of a ferrite is symmetrical under a reversal of the biasing magnetostatic field: i.e., $\varepsilon(\mathbf{B}_0) = \tilde{\varepsilon}(-\mathbf{B}_0)$ or $\mu(\mathbf{B}_0) = \tilde{\mu}(-\mathbf{B}_0)$ where the tilde indicates the transpose dyadic. The reciprocity theorem then becomes

$$\int_{\text{all space}} (\mathbf{E}_b(-B_0) \cdot \mathbf{J}_a - \mathbf{H}_b(-B_0) \cdot \mathbf{J}_{ma})\, dV$$

$$= \int_{\text{all space}} (\mathbf{E}_a(B_0) \cdot \mathbf{J}_b - \mathbf{H}_a(B_0) \cdot \mathbf{J}_{mb})\, dV \qquad (5b\text{-}300)$$

Elementary Relations Concerning Antennas. Consider a transmitting antenna and a receiving antenna separated by a large distance r. The power absorbed by the receiving antenna is

$$P_r = P \frac{G_t G_r \lambda^2}{16\pi^2 r^2} \qquad (5b\text{-}301)$$

where P is the total power transmitted by the transmitting antenna, G_t and G_r are the respective gain functions of the two antennas for the direction of transmission, and λ is the wavelength of the radiated wave.

Now if an antenna is used for transmission as well as reception, such as for radar application, the power absorbed by the receiver from the scattered wave is

$$P_r = P \frac{\sigma\lambda^2 G_t{}^2}{(4\pi)^3 r^4} \qquad (5b\text{-}302)$$

where r is the distance from the antenna to the scatterer, and σ is the scattering cross section of the scatterer. The scattering cross section is defined as the actual cross section of a sphere that in the same position as the scatterer would scatter back to the receiver the same amount of energy as is returned by the scatterer.

Radiation from Charged Particles. Radiation results when a charged particle accelerates or decelerates (Bremsstrahlung), when a charged particle moves along a curved path at a constant velocity (cyclotron radiation), when a charged particle moves at a constant velocity which is faster than the phase velocity of light in the medium (Čerenkov radiation), when a charged particle moves at a uniform velocity along an uneven surface (Smith-Purcell radiation), or when a charged particle moves through two media with different electrical properties (transition radiation).[1]

POINT CHARGE IN ARBITRARY MOTION IN FREE SPACE.   The fields are:

$$E = \frac{q}{4\pi\epsilon_0 s^3}\left\{r_u\left(1 - \frac{u^2}{c^2}\right) + \frac{1}{c^2}[r \times (r_u \times \dot{u})]\right\} \qquad (5b\text{-}303)$$

$$B = \frac{1}{rc} r \times E \qquad (5b\text{-}304)$$

with

$$s = r - \frac{r \cdot u}{c} \qquad (5b\text{-}305)$$

$$r_u = r - \frac{ru}{c} \qquad (5b\text{-}306)$$

$$u = -\frac{dr}{dt'} \qquad \dot{u} = \frac{du}{dt'} \qquad (5b\text{-}307)$$

where r is the retarded radius vector which is the radius vector from the retarded position of the particle to the field point, u and $\dot{u}$ are respectively the velocity vector and the acceleration vector of the particle at the retarded position, t' is the time of emission, and c is the velocity of light in vacuum. q is the charge of the particle.   The second term in Eq. (5b-303) represents the radiated field. ϵ_0 is the free-space permittivity.   The directional rate of radiation is

$$-\frac{dU}{dt'}\, d\Omega = \frac{q^2 r}{16\pi^2\epsilon_0 s^5 c^3}\, |r \times (r_u \times \dot{u})|^2\, d\Omega \qquad (5b\text{-}308)$$

and the total rate of radiation is

$$-\frac{dU}{dt'} = \frac{q^2}{6\pi\epsilon_0 c^3}\, \frac{|\dot{u}|^2 - |u \times \dot{u}|^2/c^2}{(1 - u^2/c^2)^3} \qquad (5b\text{-}309)$$

$-dU/dt'$ is also the rate of energy loss by the particle.   Two useful special cases are listed in the following:

$u \| \dot{u}$ (Linear Motion)

$$-\frac{dU}{dt'}\, d\Omega = \frac{|\dot{u}|^2}{c^3}\left(\frac{q^2}{16\pi^2\epsilon_0}\right)\frac{\sin^2\theta}{(1 - (u/c)\cos\theta)^5}\, d\Omega \qquad (5b\text{-}310)$$

$$-\frac{dU}{dt'} = \frac{q^2|\dot{u}|^2}{6\pi\epsilon_0 c^3(1 - u^2/c^2)^3} \qquad (5b\text{-}311)$$

where θ is the angle between u and r.

$\dot{u} \perp u$ (Circular Motion)

$$-\frac{dU}{dt'}\, d\Omega = \frac{q^2|\dot{u}|^2}{16\pi^2\epsilon_0 c^3}\frac{(1 - u^2/c^2)\cos^2\alpha + (u/c - \sin\alpha\cos\phi)^2}{[1 - (u/c)\sin\alpha\cos\phi]^5}\, d\Omega \qquad (5b\text{-}312)$$

$$-\frac{dU}{dt'} = \frac{q^2|\dot{u}|^2}{6\pi\epsilon_0 c^3}\frac{1}{(1 - u^2/c^2)^2} \qquad (5b\text{-}313)$$

where $\sin\alpha = \cos\theta/\cos\phi$, θ is the angle between u and r, $\phi = \omega_0 t'$, $|u| = a\omega_0$, and $|\dot{u}| = a\omega_0^2$.   The charge is assumed to be moving in a circle of radius a with a constant angular velocity ω_0.

ČERENKOV RADIATION.   Čerenkov radiation occurs when a charged particle is moving in a material medium at a uniform speed u which is faster than the phase velocity of light in the medium.   If n is the index of refraction of the medium,

[1] J. V. Jelley, "Čerenkov Radiation," Pergamon Press, New York, 1958.

Čerenkov radiation occurs (when $nu > c$, $n > 1$) at a cone angle of $\theta = \cos^{-1}(c/nu)$ with respect to the direction of motion. The field components are singular in that direction. c is the speed of light in vacuum. Energy radiated per unit length of path per frequency interval $(dU/dl)\,d\omega$ is

$$\frac{dU}{dl}\,d\omega = \frac{q^2}{4\pi\epsilon_0 c^2}\left(1 - \frac{c^2}{n^2 u^2}\right)\omega\,d\omega \tag{5b-314}$$

and the total radiation rate is

$$\frac{dU}{dt'} = \frac{q^2 u}{4\pi\epsilon_0 c^2}\int\left(1 - \frac{c^2}{n^2 u^2}\right)\omega\,d\omega \tag{5b-315}$$

where the integration is carried out over ranges of ω where $c^2/n^2 u^2 < 1$.

TRANSITION RADIATION. A burst of radiation occurs when a charged particle, moving at constant speed u, passes through the boundary between two media having different optical properties. Unlike Čerenkov radiation, this transition radiation will occur at any velocity of the particle, though its intensity increases with the energy. Assuming that a charged particle enters normally into a half space of refractive index n from vacuum, the energy radiated per frequency interval per unit solid angle u for $u \ll c$ is

$$U\,d\Omega\,d\omega = \frac{q^2 u^2}{4\pi^3 c^3 \epsilon_0}\sin^2\theta\cos^2\theta\left(\frac{n^2 - 1}{n^2\cos\theta + \sqrt{n^2 - \sin^2\theta}}\right)^2 d\Omega\,d\omega \tag{5b-316}$$

where θ is the angle between the outward unit normal from the dielectric half space and the line connecting the observation point with the point that the charge particle enters into the dielectric half space. If the half space is a perfect conductor, we have

$$U\,d\Omega\,d\omega = \frac{q^2 u^2}{4\pi^3 c^3 \epsilon_0}\sin^2\theta\,d\Omega\,d\omega \qquad u \ll c \tag{5b-317}$$

The total energy spectral density per unit frequency interval for the perfectly conducting half-space case is

$$U\,d\omega = \frac{q^2 u^2}{3\pi^2 c^3 \epsilon_0}\,d\omega \qquad u \ll c \tag{5b-318}$$

SMITH-PURCELL RADIATION. Radiation occurs when a charged particle moves at a uniform velocity u along an uneven surface. Assuming that the uneven surface is a sinusoidal diffraction grating of period d and amplitude a, and the medium above the grating is vacuum, the power radiated per unit solid angle (for $u \ll c$) is

$$P\,d\Omega = \frac{2q^2 a^2\pi^2 u^4}{\epsilon_0 c^3 d^4}\frac{\{[1 - (u/c)\cos\theta]^2 - [1 - (u/c)^2]\sin^2\theta\cos^2\phi\}}{[1 - (u/c)\cos\theta]^5}\,d\Omega$$

and the total power radiated is $16a^2 q^2\pi^3 u^4/3d^4\epsilon_0 c^3$. θ is the angle between the axis of the grating and the line connecting the point of observation with the retarted position of the charged particle, and ϕ is the azimuthal angle.

References

1. Silver, S.: "Microwave Antenna Theory and Design," Boston Technical Lithographers, Boston, 1963.
2. Van Bladel, J.: "Electromagnetic Fields," McGraw-Hill Book Company, New York, 1964.
3. Harrington, R. F.: "Time-harmonic Electromagnetic Fields," McGraw-Hill Book Company, New York, 1961.
4. Panofsky, W. K. H. and M. Phillips: "Classical Electricity and Magnetism," 2d ed., Addison-Wesley Publishing Company, Reading, Mass., 1962,

5b-12. Scattering and Diffraction. Scattering occurs when a propagating wave is interrupted by an obstacle. In an unbounded region, the scattered fields must satisfy the appropriate boundary conditions as well as the radiation condition.

Scattering Cross Sections. The fundamental problem in diffraction is the determination of the total field in amplitude, phase, and polarization. However, in many cases it is not required to know the total field in complete detail at all points; it is often sufficient to know such quantities as the total scattered power, the total power absorbed by the obstacle, or the amplitude of the electric field in a specified direction and at a great distance from the obstacle.

The fields of a plane wave propagating in a direction **u** are given by

$$\mathbf{E}_i = A e^{i\mathbf{k}\cdot\mathbf{r}} \tag{5b-319}$$

$$\mathbf{H}_i = \left(\frac{\epsilon}{\mu}\right)^{\frac{1}{2}} (\mathbf{u}_3 \times \mathbf{A}) e^{i\mathbf{k}\cdot\mathbf{r}} \tag{5b-320}$$

where $\mathbf{k} = k\mathbf{u}_3 = \omega \sqrt{\mu\epsilon}\, \mathbf{u}_3$, and $\mathbf{A}$ may be $(A_1\mathbf{u}_1 + A_2\mathbf{u}_2)$. $\mathbf{u}_1$, $\mathbf{u}_2$, and $\mathbf{u}_3$ form an orthogonal set of unit vectors. At large distances, the scattered fields $(\mathbf{E}_{sc}, \mathbf{H}_{sc})$ resulting from this incident plane wave are

$$\mathbf{E}_{sc} = \frac{e^{ikr}}{r} \mathbf{F} \tag{5b-321}$$

$$\mathbf{H}_{sc} = \frac{e^{ikr}}{r} \left(\frac{\epsilon}{\mu}\right)^{\frac{1}{2}} (\mathbf{e}_r \times \mathbf{F}) \tag{5b-322}$$

where $\mathbf{F}$ is a complex vector which is transverse to the unit vector $\mathbf{e}_r$ in the radial direction. For example, in the spherical coordinates r, θ, ϕ,

$$\mathbf{F} = F_\theta(\theta,\phi)\mathbf{e}_\theta + F_\phi(\theta,\phi)\mathbf{e}_\phi \tag{5b-323}$$

The time-averaged power scattered by the obstacle is

$$P_{sc} = \frac{1}{2} \operatorname{Re}\left[\int_S (\mathbf{E}_{sc} \times \mathbf{H}_{sc}^*) \cdot d\mathbf{S} \right] = \frac{1}{2}\left(\frac{\epsilon}{\mu}\right)^{\frac{1}{2}} \int_{\substack{\text{unit} \\ \text{sphere}}} |\mathbf{F}|^2\, d\Omega \tag{5b-324}$$

where S is the surface of the scatterer, and Ω is the solid angle. If P_{abs} is the time averaged power dissipated in the scatterer, the following relationship can be shown

$$P_{abs} + P_{sc} = -\frac{2\pi}{\omega\mu} \operatorname{Im}\left[\mathbf{A}^* \cdot \mathbf{F}(\mathbf{u}_3)\right] \tag{5b-325}$$

where $\mathbf{F}(\mathbf{u}_3)$ is the radiation vector of the scattered wave in the direction of incidence (i.e., in the forward direction). The time-averaged incident power per unit area is

$$P_i = \frac{1}{2}\left(\frac{\epsilon}{\mu}\right)^{\frac{1}{2}} |A|^2 \tag{5b-326}$$

The following quantities are defined to characterize the reradiating, absorbing or transmitting properties of a three-dimensional obstacle in an incident plane-wave field:

TOTAL SCATTERING CROSS SECTION

$$\sigma_{sc} = \frac{P_{sc}}{P_i} = \frac{\int_\Omega |\mathbf{F}|^2\, d\Omega}{|A|^2} \tag{5b-327}$$

ABSORPTION CROSS SECTION

$$\sigma_a = \frac{P_{abs}}{P_i} \tag{5b-328}$$

EXTINCTION CROSS SECTION

$$\sigma_{ext} = \frac{P_{sc} + P_{abs}}{P_i} = \sigma_a + \sigma_{sc}$$

$$= -\frac{4\pi}{k} \operatorname{Im} \left[\frac{\mathbf{A}^* \cdot \mathbf{F}(\mathbf{u}_3)}{|\mathbf{A}|^2} \right] \tag{5b-329}$$

BISTATIC CROSS SECTION

$$\sigma_{bistatic} = \frac{4\pi |\mathbf{F}(\mathbf{u}')|^2}{|\mathbf{A}|^2} \tag{5b-330}$$

$\mathbf{u}'$ is the observation direction. The power scattered per unit solid angle in the $\mathbf{u}'$ direction is

$$\frac{dP_{sc}}{d\Omega} = \sigma_{bistatic} \frac{P_i}{4\pi} \tag{5b-331}$$

MONOSTATIC (RADAR) CROSS SECTION OR BACKSCATTERING CROSS SECTION

$$\sigma_{mono} = \frac{4\pi |\mathbf{F}(-\mathbf{u}_3)|^2}{|\mathbf{A}|^2} \tag{5b-332}$$

TRANSMISSION CROSS SECTION. A plane wave polarized in the $\mathbf{u}_p$ direction and propagating in the $\mathbf{u}$ direction is incident on a metallic screen provided with an aperture S. $\mathbf{F}(\mathbf{u})$ is the radiation vector of the transmitted wave with respect to the forward direction $\mathbf{u}$. The transmission cross section is

$$\sigma_t = \frac{2\pi}{k} \operatorname{Im} [\mathbf{u}_p \cdot \mathbf{F}(\mathbf{u})] \tag{5b-333}$$

Similar expressions to those given above are also available for two-dimensional scatterers. The far-zone scattered field due to an incident $\mathbf{E}$ wave whose electric vector is polarized in the z direction, which is parallel to the axis of the two-dimensional scatterer, is

$$E_z^{sc} \simeq \frac{e^{ikr}}{r^{\frac{1}{2}}} F(\phi)$$

where the cylindrical coordinates r, ϕ, z have been used. The scattering cross section, the extinction cross section, and the bistatic cross section are respectively

$$\sigma_{sc} = \frac{1}{|A|^2} \int_0^{2\pi} |F(\phi)|^2 \, d\phi$$

$$\sigma_{ext} = -\frac{4}{k} \operatorname{Re} \left[\left(\frac{\pi k}{-2i} \right)^{\frac{1}{2}} \frac{F(\mathbf{u})}{|A|} \right]$$

$$\sigma_{bistatic} = \frac{2\pi |F(\mathbf{u}')|^2}{|A|^2}$$

where $\mathbf{u}$ is the direction of incidence (i.e., the forward direction), $\mathbf{u}'$ is the observation direction, and A is the amplitude of the incident E wave. The formulas derived for an E wave can also be applied to an H wave provided that we replace E_z^{sc} by H_z^{sc}, $F(\phi)$ by $F(\phi)/(\mu/\epsilon)^{\frac{1}{2}}$, and A by $A/(\mu/\epsilon)^{\frac{1}{2}}$ in the above equations.

Integral Formulations. Scattering from objects of arbitrary shapes may be formulated in terms of integral equations. This formulation eliminates the necessity of separating the vector wave equations. Although in general the resultant integral equation is difficult to solve formally, with the help of high-speed computors numerical solutions may be readily obtained.[1]

[1] R. F. Harrington, "Field Computation by Moment Methods," The Macmillan Company, New York, 1968.

The surface current density on a perfectly conducting three-dimensional scatterer satisfies the following integral equation:[1]

$$J_s(\mathbf{r}_0) - 2\int_S \mathbf{n}(\mathbf{r}_0) \times [J_s(\mathbf{r}') \times \nabla'G(\mathbf{r}_0,\mathbf{r}')] \, dS' = 2\mathbf{n}(\mathbf{r}_0) \times H_i(\mathbf{r}_0) \quad (5\text{b-}334)$$

or

$$\mathbf{n}(\mathbf{r}_0) \times E_i(\mathbf{r}_0) = -i\omega\mu \int_S \mathbf{n}(\mathbf{r}_0) \times J_s(\mathbf{r}')G(\mathbf{r}',\mathbf{r}_0) \, dS'$$

$$- \frac{1}{i\omega\epsilon}\int_S [\nabla' \cdot J_s(\mathbf{r}')]\mathbf{n}(\mathbf{r}_0) \times \nabla'G(\mathbf{r}',\mathbf{r}_0) \, dS' \quad (5\text{b-}335)$$

where $E_i(\mathbf{r}_0)$ and $H_i(\mathbf{r}_0)$ are the known incident electric and magnetic fields at $\mathbf{r}_0$ on the scatterer, $\mathbf{n}(\mathbf{r}_0)$ is the unit outward normal on the scatterer at $\mathbf{r}_0$, $G(\mathbf{r}_0,\mathbf{r}') = e^{ik|\mathbf{r}_0-\mathbf{r}'|}/4\pi|\mathbf{r}_0 - \mathbf{r}'|$, and S is the surface of the scatterer. The scattered field may then be found from the relation

$$E^{sc}(\mathbf{r}) = i\omega\mu \int_S \left(u + \frac{1}{k^2}\nabla'\nabla'\right) G(\mathbf{r},\mathbf{r}') \cdot J_s(\mathbf{r}') \, dS' \quad (5\text{b-}336)$$

or

$$H^{sc}(\mathbf{r}) = \int_S J_s(\mathbf{r}') \times \nabla'G(\mathbf{r},\mathbf{r}') \, dS' \quad (5\text{b-}337)$$

in which $G(\mathbf{r},\mathbf{r}') = e^{ik|\mathbf{r}-\mathbf{r}'|}/4\pi|\mathbf{r} - \mathbf{r}'|$. J_s is obtained from Eq. (5b-334), and u is the unit dyadic.

The surface current density on a perfectly conducting two-dimensional cylindrical scatterer satisfies the following integral equation

$$E_z{}^i(\mathbf{r}_0) - i\omega\mu \int_l J_z(\mathbf{r}')G_c(\mathbf{r}_0,\mathbf{r}') \, dl' = 0 \quad (5\text{b-}338)$$

for an incident E wave, or

$$\frac{1}{2} J_t(\mathbf{r}_0) + \int_l J_t(\mathbf{r}') \frac{\partial}{\partial n(\mathbf{r}')} G_c(\mathbf{r}_0,\mathbf{r}') \, dl' = -H_z{}^i(\mathbf{r}_0) \quad (5\text{b-}339)$$

for an incident H wave. J_z is the current density along the axis of the cylindrical scatterer; J_t is the current density tangent to the boundary of the cylindrical scatterer but normal to the z axis. $\mathbf{n}$ is still the unit outward normal on the scatterer. $G_c(\mathbf{r},\mathbf{r}')$ is the two-dimensional Green's function:

$$G_c(\mathbf{r},\mathbf{r}') = -\frac{i}{4} H_0{}^{(1)}(k|\mathbf{r} - \mathbf{r}'|) \quad (5\text{b-}340)$$

where $H_0{}^{(1)}(p)$ is the Hankel function of the first kind of order zero and argument p. l is the cross-sectional bounding curve of the scatterer. The scattered field may then be found from the relation

$$E_z{}^{sc}(\mathbf{r}) = -i\omega\mu \int_l J_z(\mathbf{r}')G_c(\mathbf{r},\mathbf{r}') \, dl' \quad (5\text{b-}341)$$

or

$$H_z{}^{sc}(\mathbf{r}) = \int_l J_t(\mathbf{r}') \frac{\partial}{\partial n(\mathbf{r}')} G_c(\mathbf{r},\mathbf{r}') \, dl' \quad (5\text{b-}342)$$

Formulation of the problem of the scattering by dielectric obstacles in terms of integral equations is also possible. However, the results are too involved to be included here. The reader is referred to the literature.[2]

[1] A. W. Maue, *Z. Physik.* **126**, 601 (1949).
[2] P. C. Waterman, Scattering by Dielectric Obstacles, *Mitre Corp. Rept.* MTP-84, July, 1968.

Rayleigh Scattering (Low-frequency Scattering). Rather simple formulas for the scattered fields in the far zone are available when the wavelength of the incident wave is much greater than the largest linear dimension of the scatterer.[1]

DIELECTRIC SCATTERER. For three-dimensional dielectric scatterers,

$$\mathbf{E}^{sc} \simeq - \frac{k^2}{4\pi\epsilon} \, \mathbf{e}_r \times (\mathbf{e}_r \times \mathbf{p}^e) \, \frac{e^{ikr}}{r} \tag{5b-343}$$

$$\mathbf{H}^{sc} \simeq \frac{\omega k}{4\pi} \, \mathbf{e}_r \times \mathbf{p}^e \, \frac{e^{ikr}}{r} \tag{5b-344}$$

where $\mathbf{p}^e$ is the induced electric dipole moment which is orientated in the same direction as the electric vector of the incident field. $k = \omega \sqrt{\mu\epsilon}$ is the free-space wave number and $\mathbf{e}_r$ is a unit vector in the r direction. The induced electric dipole moment is the same as that for the static value for the dielectric sphere immersed in a static electric field orientated in the same direction as the incident electric vector. For a perfect dielectric sphere of radius a and a dielectric constant ϵ_1

$$\mathbf{p}^e = 4\pi\epsilon a^3 \, \frac{\epsilon_1/\epsilon - 1}{\epsilon_1/\epsilon + 2} \, \mathbf{e} \tag{5b-345}$$

$\mathbf{e}$ is a unit vector in the same direction as the electric vector of the incident field. The total scattering cross section is

$$\sigma_{sc} \simeq \frac{8}{3} \left(\frac{\epsilon_1/\epsilon - 1}{\epsilon_1/\epsilon + 2} \right)^2 (ka)^4 \pi a^2 \tag{5b-346}$$

The magnitude of the scattered field is

$$|F| = \left| \frac{\epsilon_1/\epsilon - 1}{\epsilon_1/\epsilon + 2} \, k^2 a^3 \sin \theta_s \right| \tag{5b-347}$$

where θ_s is the angle between the axis of the induced dipole and the point of observation.

For two-dimensional dielectric scatterers:

$$\left.
\begin{aligned}
E_z{}^{sc} &\simeq \frac{ik^2}{4} \left(\frac{-2i}{\pi k} \right)^{\frac{1}{2}} \frac{e^{ikr}}{r^{\frac{1}{2}}} E_z{}^i \left(\frac{\epsilon_1}{\epsilon} - 1 \right) S \\
\sigma_{sc}{}^E &= \left(\frac{\epsilon_1}{\epsilon} - 1 \right)^2 \frac{k^3 S^2}{4}
\end{aligned}
\right\} \text{ incident } E \text{ waves}$$

$$\sigma_{sc}{}^H = \frac{k^3}{8\epsilon\mu} \frac{|p_e|^2}{|H_z{}^i|^2} \text{ (incident } H \text{ wave)}$$

where S is the cross-section area of the cylinder, ϵ_1 is the dielectric constant of the cylinder, and p_e is the induced electric dipole moment. $E_z{}^i$ and $H_z{}^i$ are the magnitudes of the incident waves.

PERFECTLY CONDUCTING SCATTERER. The scattered wave for a small perfectly conducting obstacle is due not only to an induced electric dipole of moment $\mathbf{p}^e$ but also to an induced magnetic dipole of moment $\mathbf{p}^m$. For the case of a perfectly conducting sphere of radius a, the induced electric dipole moment is $4\pi\epsilon a^3 \mathbf{e}_x$, and the induced magnetic dipole moment is $(-2\pi\omega\mu\epsilon a^3/k)\mathbf{e}_y$. The incident electric vector is polarized in the $\mathbf{e}_x$ direction. The far-zone scattered electric fields are

$$E_\theta{}^{sc} \simeq \frac{e^{ikr}}{r} k^2 a^3 \cos \phi (\cos \theta - \tfrac{1}{2}) \tag{5b-348}$$

$$E_\phi{}^{sc} \simeq \frac{e^{ikr}}{r} k^2 a^3 \sin \phi (\tfrac{1}{2} \cos \theta - 1) \tag{5b-349}$$

[1] A. F. Stevenson, *J. Appl. Phys.* **24**, 1134, 1143 (1953).

The backscattering cross section is

$$\sigma_{\text{mono}} = 9\pi k^4 a^6 \tag{5b-350}$$

For two-dimensional perfectly conducting obstacles,

$$\sigma_{sc}{}^E = \frac{\pi^2}{k(\log kL)^2} \text{ (incident } E \text{ wave)}$$

$$\sigma_{sc}{}^H = \frac{1}{8} k^3 \left(2S^2 + \frac{|\mathbf{p}_e|^2}{\mu\epsilon|H^i|^2} \right) \text{ (incident } H \text{ wave)}$$

where L is the length of the contour, S is the cross-sectional area of the cylinder, and $\mathbf{p}_e$ is the induced electric dipole moment. For example, for a circular cylinder of radius a, $L = a$, $S = \pi a^2$, and $|\mathbf{p}_e| = 2\pi a^2 \sqrt{\mu\epsilon} |H^i|$.

Rayleigh-Gans Scattering or Born Approximation. Under the assumption that $|\epsilon_1/\epsilon - 1| \ll 1$, the scattered field by such a dielectric scatterer may be approximated by the following formula:

$$\mathbf{E}^{sc}(\mathbf{r}) = (\nabla^2 + k^2) \int_V \left(\frac{\epsilon_1}{\epsilon} - 1 \right) \mathbf{E}_0 G(\mathbf{r},\mathbf{r}') \, dV' \tag{5b-351}$$

where $G(\mathbf{r},\mathbf{r}') = e^{ik|\mathbf{r}-\mathbf{r}'|}/4\pi|\mathbf{r} - \mathbf{r}'|$, V is the volume of the scatterer, and $\mathbf{E}_0$ is the incident field. In the far zone of the scatterer and in the direction of the unit vector $\mathbf{e}$,

$$\mathbf{E}^{sc}(\mathbf{r}) \sim \frac{k^2 e^{ikr}}{4\pi r} \int_V [\mathbf{E}_0 - (\mathbf{E}_0 \cdot \mathbf{e})\mathbf{e}] \left(\frac{\epsilon_1}{\epsilon} - 1 \right) e^{-ik\mathbf{e}\cdot\mathbf{r}'} \, dV' \tag{5b-352}$$

For a dielectric sphere of radius a and $k^2 a^3 |\epsilon_1/\epsilon - 1| \ll 1$,

$$\mathbf{E}^{sc} \sim \frac{ka^2 e^{ikr}}{r} [\mathbf{E}_0 - (\mathbf{E}_0 \cdot \mathbf{e})\mathbf{e}] \frac{\epsilon_1/\epsilon - 1}{2 \sin \frac{1}{2}\theta} j_1(2ka \sin \tfrac{1}{2}\theta) \tag{5b-353}$$

where j_1 is the spherical Bessel's function of order 1. The total scattering cross section is

$$\sigma_{sc} = \frac{\pi a^2}{4} \left(\frac{\epsilon_1}{\epsilon} - 1 \right)^2 \left\{ \frac{5}{2} - \frac{\sin 4ka}{ka} + \frac{7}{16k^2 a^2} (\cos 4ka - 1) \right.$$
$$\left. + 2k^2 a^2 + \left(\frac{1}{2k^2 a^2} - 2 \right) [\gamma + \ln (4ka) - \text{Ci}(4ka)] \right\} \tag{5b-354}$$

where $\gamma = 0.5772 \ldots$ is the Euler's constant, and Ci is the cosine integral.

High-frequency Scattering. If the wavelength of an incident wave is much smaller than the smallest dimension of the scatterer, several approximation techniques for finding the scattered fields are available.

GEOMETRIC OPTICS APPROACH. Assume that a linearly polarized incident electric field in the direction $\mathbf{e}_p$ which is given by

$$\mathbf{E}^i = \mathbf{E}_0(\mathbf{e}_j) \frac{e^{ikR}}{R} \tag{5b-355}$$

impinges upon a perfectly conducting body (see Fig. 5b-9). The far-zone reflected electric field at the observation point is

$$\mathbf{E}^{rc} = D^{\frac{1}{2}} \{\mathbf{n}[\mathbf{n} \cdot \mathbf{E}^i] + \mathbf{n} \times [\mathbf{n} \times \mathbf{E}^i]\} \frac{R}{r} e^{ikr} \tag{5b-356}$$

where

$$D = \frac{R_1 R_2 \cos \theta}{(4R^2 + R_1 R_2) \cos \theta + 2R(R_1 \sin^2 \theta_1 + R_2 \sin^2 \theta_2)} \tag{5b-357}$$

and θ_1 and θ_2 are the angles between the incident ray and the directions of the principal radii of curvature R_1 and R_2. θ is the angle between the incident ray and **n**. **n** is the outward unit normal on the surface of the scatterer at the point where the incident ray intersects the scatterer. Equation (5b-356) is valid only if $r \gg R_1$ or R_2 and if diffraction effects are not important.

A slightly better approximation for the scattered field can be obtained by assuming that the scattered fields are due to the induced current density in the illuminated region and the induced line distribution of charge along the bounding curve between the illuminated and the shadow regions. The induced current density is found according to the geometric optics method. The scattered fields can then be obtained according to Eqs. (5b-280a) and (5b-280b).

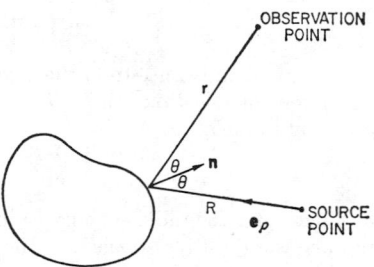

A useful expression for the high-frequency radar cross section of perfectly conducting convex scatterer is also available:[1]

$$\sigma_{\text{rad}} = \sigma_{\text{mono}} = \pi R_1 R_2 \qquad (5b\text{-}358)$$

FIG. 5b-9. Reflection from a conducting obstacle.

where R_1 and R_2 are the principal radii of curvature at the point at which the incident ray is perpendicular to the surface.

GEOMETRICAL THEORY OF DIFFRACTION. An extension of geometrical optics to account for diffraction phenomena has been proposed by Keller.[2] The main feature of the theory is the introduction of diffracted rays in addition to the usual rays of geometrical optics. These diffracted rays are produced by incident rays which hit edges, corners, or vertices of boundary surfaces, or which graze such surfaces. Some of these rays penetrate into the shadow regions and account for the existence of fields there. The initial value of the field on a diffracted ray is obtained by multiplying the field on the incident ray by a diffraction coefficient which takes different values for edge diffraction, vertex diffraction, etc. The value of the field along the diffracted ray is then obtained from its value at the diffraction point by the ordinary laws of geometrical optics. Several specific examples are given in the following.

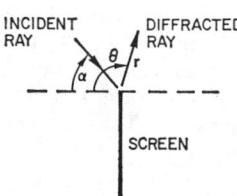

FIG. 5b-10. Diffracted rays from a straight conducting edge.

Fields Diffracted by Straight Edges. Let u_e be the field on a ray diffracted from an edge which is a straight line and the incident rays all lie in planes normal to the edge (see Fig. 5b-10). The diffracted field is

$$u_e = D u_i r^{-\frac{1}{2}} e^{ikr} \qquad (5b\text{-}359)$$

where D is the diffraction coefficient:

$$D = -\frac{e^{i\pi/4}}{2(2\pi k)^{\frac{1}{2}} \sin \beta} [\sec \tfrac{1}{2}(\theta - \alpha) \pm \csc \tfrac{1}{2}(\theta + \alpha)] \qquad (5b\text{-}360)$$

β is the angle between the incident ray and the edge, which is $\pi/2$ in the present normal incidence case. r is the distance from the edge. The angles between the incident

[1] R. G. Kouyoumjian, *Proc. IEEE* **53**, 864 (1965). High-frequency scattering by conducting ellipsoid has been treated by J. E. Burke and V. Twersky, *J. Acoust. Soc. Am.* **38**, 589 (1965).

[2] J. B. Keller, *J. Opt. Soc. Am.* **52**, 116 (1962).

and diffracted rays and the normal to the screen are θ and α, respectively. The upper sign applies when the boundary condition on the half-plane is $u = 0$ (i.e., $u_e = E_d$ where the incident E field is parallel to the edge), while the lower sign applies if it is $\partial u/\partial n = 0$.[1] (i.e., $u_e = H_d$ where the incident H field is parallel to the edge.) Equation (5b-359) is still valid for obliquely incident waves provided that θ and α are defined as above after the rays are first projected into the plane normal to the edge.

Fields Diffracted by Curved Edges. The diffracted field for a curved edge is

$$u_e = Du_i \left[r \left(1 + \frac{r}{\rho_1} \right) \right]^{-\frac{1}{2}} e^{ikr} \tag{5b-361}$$

where ρ_1 is the distance from the edge to the caustic of the diffracted rays, measured negatively in the direction of propagation. When the edge is a plane curve, ρ_1 is given by the relation

$$\frac{1}{\rho_1} = -\frac{\dot\beta}{\sin \beta} - \frac{\cos \delta}{\rho \sin^2 \beta} \tag{5b-362}$$

$\rho \geq 0$ denotes the radius of curvature of the edge, β is the angle between the incident ray and the (positive) tangent to the edge, $\dot\beta$ is the derivative of β with respect to arc length s along the edge, and δ is the angle between the diffracted ray and the normal to the edge.

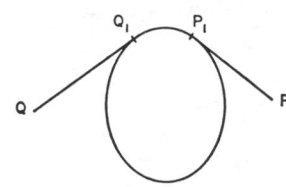

Fields of Vertex-diffracted Ray. The diffracted field from a vertex is

$$u = Cu_i \frac{e^{ikr}}{r} \tag{5b-363}$$

where C is the vertex diffraction coefficient which has been evaluated only for a circular cone.

Fig. 5b-11. Surface diffracted rays.

Fields of Surface-diffracted Rays. The diffracted rays are produced by incident rays which are tangent to the surface of the impenetrable body (see Fig. 5b-11). Each tangent ray splits at the point of tangency. One part continues along the path of the incident ray; another part travels along the surface of the body as a surface ray. This surface ray is a geodesic or shortest path on the body surface. Thus a single grazing incident ray gives rise to infinitely many diffracted rays.

The diffracted field is

$$u_d(P) = A_i(Q_1) \exp \{ik[\phi_i(Q_1) + t + s]\} \left[\frac{d\sigma(Q_1)}{d\sigma(P_1)} \right]^{\frac{1}{2}} \left[\frac{\rho_1}{s(\rho_1 + s)} \right]^{\frac{1}{2}} \cdot \sum_m D_m(P_1)D_m(Q_1) \exp \left[-\int_0^t \alpha_m(\tau)\, d\tau \right] \tag{5b-364}$$

where $A_i(Q_1)$ and $\phi_i(Q_1)$ are the amplitude and phase of the incident field at Q_1, t is the distance along the diffracted ray from Q_1 to P_1, s is the distance from P_1 to P, ρ_1 is the principal radius of curvature of the diffracted wavefront on the body, and $d\sigma(Q_1)/d\sigma(P_1)$ is the ratio of the width of a narrow strip of diffracted rays at Q_1 to that at P_1 on the surface of the body. The diffraction coefficients $D_m(P_1)$ and $D_m(Q_1)$ and the decay exponents α_m are obtained from a canonical problem with the appropriate boundary conditions. u_d corresponds to E_d with $u = 0$ on the cylindrical body when the incident E field is parallel to the axis of the cylinder while u_d corresponds to H_d with $\partial u/\partial n = 0$ on the cylindrical body when the incident H field is parallel to the

[1] For the special case of $\alpha = \pi/2$ with the boundary condition $\partial u/\partial n = 0$ on the screen, Eqs. (5b-359) and (5b-360) are not applicable. The revised form is

$$u_e = D' \frac{\partial u_i}{\partial n} r^{-\frac{1}{2}} e^{ikr} \qquad D' = -\frac{1}{ik} \left[\frac{\partial}{\partial \alpha} D(\theta,\alpha) \right] \Big|_{\alpha = \pi/2}$$

axis of the cylinder. Equation (5b-364) is not applicable without modification in the determination of the fields near the diffracting surface or near the shadow boundary. Application of Eq. (5b-364) to the problems of diffraction of waves by circular cylinders, spheres, parabolic cylinders, elliptic cylinders, etc., has been carried out successfully by Keller and his coworkers.[1]

Babinet's Principle. Consider three cases of a given source (1) radiating in free space, (2) radiating in the presence of an electrically conducting screen, and (3) radiating in the presence of a magnetically conducting screen. The electric and magnetic screens are said to be complementary if the two screens superimposed cover the entire $y = 0$ plane with no overlapping. Let the fields $y > 0$ be designated $(\mathbf{E}^i, \mathbf{H}^i)$, $(\mathbf{E}^e, \mathbf{H}^e)$, and $(\mathbf{E}^m, \mathbf{H}^m)$ for the cases 1, 2, and 3, respectively. Then Babinet's principle for complementary screens states that

$$\mathbf{E}^e + \mathbf{E}^m = \mathbf{E}^i \qquad \mathbf{H}^e + \mathbf{H}^m = \mathbf{H}^i \qquad (5\text{b-}365)$$

The above Babinet's principle allows replacement of the aperture problem with an equivalent "disk" problem. Consider a plane metallic obstacle (disk) at $y = 0$ immersed in an incident wave $(\mathbf{E}^i = \mathbf{E}_0, \mathbf{H}^i = \mathbf{H}_0)$. The scattered fields are $(\mathbf{E}^{sc}, \mathbf{H}^{sc})$. If one assumes that a wave $[\mathbf{E}^i = -\sqrt{(\mu/\epsilon)}\,\mathbf{E}_0, \mathbf{H}^i = \sqrt{(\epsilon/\mu)}\,\mathbf{H}_0]$ impinges on a metallic screen at $y = 0$ with an aperture of the same shape as the disk, the scattered fields on the shadow side of the aperture is $\mathbf{E}^{sc}_{\text{screen}} = \sqrt{(\mu/\epsilon)}\,\mathbf{H}^{sc}$, $\mathbf{H}^{sc}_{\text{screen}} = -\sqrt{(\epsilon/\mu)}\,\mathbf{E}^{sc}$ where $(\mathbf{E}^{sc}, \mathbf{H}^{sc})$ are the scattered fields on the $y > 0$ side of the disk.

Diffraction by Simple Objects. DIFFRACTION BY SPHERE. A plane wave in an infinite, homogeneous medium (ϵ, μ), whose electric vector is linearly polarized in the x direction, is incident upon a sphere of radius a and constitutive parameters ϵ_1, μ_1 from the negative z axis. The incident wave $(\mathbf{E}_i, \mathbf{H}_i)$, the penetrated wave $(\mathbf{E}_p, \mathbf{H}_p)$, and the scattered wave $(\mathbf{E}_{sc}, \mathbf{H}_{sc})$ are respectively

$$\mathbf{E}_i = \mathbf{e}_x E_0 e^{ikz} = E_0[\nabla \times \nabla \times (v_i \mathbf{r}\mathbf{e}_r) + i\omega\mu \nabla \times (w_i \mathbf{r}\mathbf{e}_r)] \qquad (5\text{b-}366a)$$

$$\mathbf{H}_i = \mathbf{e}_y \frac{k}{\mu\omega} E_0 e^{ikz} = E_0[-i\omega\epsilon \nabla \times (v_i \mathbf{r}\mathbf{e}_r) + \nabla \times \nabla \times (w_i \mathbf{r}\mathbf{e}_r)] \qquad (5\text{b-}366b)$$

$$\mathbf{E}_p = E_0[\nabla \times \nabla \times (v_p \mathbf{r}\mathbf{e}_r) + i\omega\mu_1 \nabla \times (w_p \mathbf{r}\mathbf{e}_r)] \qquad (5\text{b-}367a)$$

$$\mathbf{H}_p = E_0[-i\omega\epsilon_1 \nabla \times (v_p \mathbf{r}\mathbf{e}_r) + \nabla \times \nabla \times (w_p \mathbf{r}\mathbf{e}_r)] \qquad (5\text{b-}367b)$$

$$\mathbf{E}_{sc} = E_0[\nabla \times \nabla \times (v_s \mathbf{r}\mathbf{e}_r) + i\omega\mu \nabla \times (w_s \mathbf{r}\mathbf{e}_r)] \qquad (5\text{b-}368a)$$

$$\mathbf{H}_{sc} = E_0[-i\omega\epsilon \nabla \times (v_s \mathbf{r}\mathbf{e}_r) + \nabla \times \nabla \times (w_s \mathbf{r}\mathbf{e}_r)] \qquad (5\text{b-}369b)$$

with

$$v_i = \frac{-i}{k}\cos\phi \sum_{n=1}^{\infty} \frac{(i)^n 2n + 1}{n(n+1)} j_n(kr) P_n^1(\cos\theta) \qquad (5\text{b-}370a)$$

$$w_i = \frac{-i}{\omega\mu}\sin\phi \sum_{n=1}^{\infty} \frac{(i)^n 2n + 1}{n(n+1)} j_n(kr) P_n^1(\cos\theta) \qquad (5\text{b-}370b)$$

$$v_p = \frac{-i}{k_1}\cos\phi \sum_{n=1}^{\infty} c_n \frac{(i)^n 2n + 1}{n(n+1)} j_n(k_1 r) P_n^1(\cos\theta) \qquad (5\text{b-}371)$$

$$w_p = \frac{-i}{\omega\mu_1}\sin\phi \sum_{n=1}^{\infty} d_n \frac{(i)^n 2n + 1}{n(n+1)} j_n(k_1 r) P_n^1(\cos\theta) \qquad (5\text{b-}372)$$

$$v_s = \frac{-i}{k}\cos\phi \sum_{n=1}^{\infty} a_n \frac{(i)^n 2n + 1}{n(n+1)} h_n^{(1)}(kr) P_n^1(\cos\theta) \qquad (5\text{b-}373)$$

---

[1] B. R. Levy and J. B. Keller, *Communs. Pure Appl. Math.* **12**, 159 (1959).

$$w_s = \frac{-i}{\omega\mu} \sin\phi \sum_{n=1}^{\infty} b_n \frac{(i)^n 2n+1}{n(n+1)} h_n^{(1)}(kr) P_n^1(\cos\theta) \tag{5b-374}$$

$$a_n = \frac{(\epsilon_1/\epsilon) j_n(k_1a)[kaj_n(ka)]' - j_n(ka)[k_1aj_n(k_1a)]'}{(\epsilon_1/\epsilon) j_n(k_1a)[kah_n^{(1)}(ka)]' - h_n^{(1)}(ka)[k_1aj_n(k_1a)]'} \tag{5b-375a}$$

$$b_n = \frac{(\mu_1/\mu) j_n(k_1a)[kaj_n(ka)]' - j_n(ka)[k_1aj_n(k_1a)]'}{(\mu_1/\mu) j_n(k_1a)[kah_n^{(1)}(ka)]' - h_n^{(1)}(ka)[k_1aj_n(k_1a)]'} \tag{5b-375b}$$

$$c_n = \frac{(\mu_1\epsilon_1/\mu\epsilon)^{\frac{1}{2}}}{[k_1aj_n(k_1a)]'} \{[kaj_n(ka)]' - a_n[kah_n^{(1)}(ka)]'\} \tag{5b-375c}$$

$$d_n = \frac{1}{j_n(k_1a)} \{j_n(ka) - b_n h_n^{(1)}(ka)\} \tag{5b-375d}$$

The prime indicates the derivative of the function with respect to its argument, $k_1 = \omega\sqrt{\mu_1\epsilon_1}$, and $k = \omega\sqrt{\mu\epsilon}$. j_n and $h_n^{(1)}$ are respectively the Bessel and Hankel functions. P_n^1 is the associated Legendre function. For a perfectly conducting sphere, $a_n = [kaj_n(ka)]'/[kah_n^{(1)}(ka)]'$, $b_n = j_n(ka)/h_n^{(1)}(ka)$, $c_n = 0$, $d_n = 0$.

Far-zone Scattered Electric Field

$$\mathbf{E}_{sc}^{\text{far zone}} = \frac{e^{ikr}}{r}(F_\theta \mathbf{e}_\theta + F_\varphi \mathbf{e}_\phi) \tag{5b-376}$$

with

$$F_\theta = \frac{i}{k}\cos\phi \sum_{n=1}^{\infty} \frac{2n+1}{n(n+1)}\left[a_n \frac{d}{d\theta}P_n^1(\cos\theta) + b_n \frac{P_n^1(\cos\theta)}{\sin\theta}\right] \tag{5b-377}$$

$$F_\phi = -\frac{i}{k}\sin\phi \sum_{n=1}^{\infty} \frac{2n+1}{n(n+1)}\left[a_n \frac{P_n^1(\cos\theta)}{\sin\theta} + b_n \frac{d}{d\theta}P_n^1(\cos\theta)\right] \tag{5b-378}$$

Total Scattering Cross Section

$$\sigma_{sc} = \frac{2\pi}{k^2}\sum_{n=1}^{\infty}(2n+1)(|a_n|^2 + |b_n|^2) \tag{5b-379}$$

Extinction Cross Section

$$\sigma_{ext} = \frac{2\pi}{k^2}\text{Re}\left[\sum_{n=1}^{\infty}(2n+1)(a_n+b_n)\right] \tag{5b-380}$$

Radar Cross Section

$$\sigma_{rad} = \frac{\pi}{k^2}\left|\sum_{n=1}^{\infty}(2n+1)(-1)^n(a_n-b_n)\right|^2 \tag{5b-381}$$

Several typical curves for σ_{sc}, σ_{ext}, and σ_{rad} are given in Figs. 5b-12 and 5b-13.

High- and Low-frequency Limits

$$\sigma_{sc}^{(\text{conducting sphere})} \underset{ka\to\infty}{\simeq} 2\pi a^2[1 + 0.06595661(ka)^{-\frac{2}{3}} + 0.7797489(ka)^{-\frac{4}{3}}$$
$$- 2.8713350(ka)^{-2} - 0.3385447(ka)^{-\frac{8}{3}} + 0.058460(ka)^{-\frac{10}{3}} + \cdots] \tag{5b-382}$$

$$\sigma_{rad}^{(\text{conducting sphere})} \underset{ka\to\infty}{\simeq} \pi a^2 \tag{5b-383}$$

$$\sigma_{sc}^{(\text{conducting sphere})} \underset{ka\to 0}{\simeq} \frac{10\pi}{3}k^4a^6\left[1 + \frac{6}{25}(ka)^2\right] \tag{5b-384}$$

$$\sigma_{rad}^{(\text{conducting sphere})} \underset{ka\to 0}{\simeq} 9\pi k^4 a^6 \tag{5b-385}$$

$$\sigma_{sc}^{(\text{dielectric sphere})} \underset{ka\to 0}{\simeq} \frac{8\pi k^4 a^6}{3}\left[\left(\frac{\epsilon_1/\epsilon - 1}{2 + \epsilon_1/\epsilon}\right)^2 + \left(\frac{\mu_1/\mu - 1}{2 + \mu_1/\mu}\right)^2\right]$$

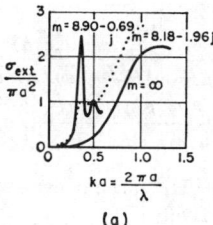 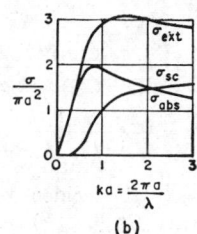

(a)   (b)

FIG. 5b-12. Typical cross sections for a sphere: (a) Typical extinction cross sections for various values of the index of refraction m [$= (\epsilon_1/\epsilon)^{\frac{1}{2}}$, where ϵ_1 is the complex dielectric constant of the sphere and ϵ is the free-space permittivity]. (b) Cross sections for an iron sphere. (At $\lambda = 0.42 \times 10^{-6}$ meter, the index of refraction for the iron sphere is $1.27 - j1.37$). a is the radius of the sphere, and λ is the free-space wavelength. (*From H. C. van der Hulst, "Light Scattering by Small Particles," John Wiley & Sons, Inc., New York, 1957.*)

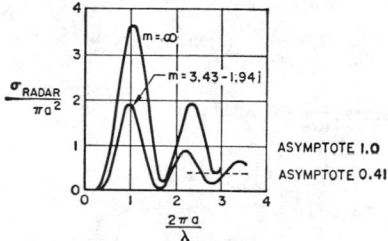

FIG. 5b-13. Typical radar cross section of a sphere with complex index of refraction, m. a is the radius of the sphere, and λ is the free-space wavelength. (*From H. C. van der Hulst, "Light Scattering by Small Particles," John Wiley & Sons, Inc., New York, 1957.*)

DIFFRACTION BY CIRCULAR CYLINDERS. A plane wave in an infinite, homogeneous medium (ϵ,μ) is incident upon a circular cylinder of radius a and constitutive parameters ϵ_1, μ_1 from the negative x axis. The axis of the cylinder is parallel to the z axis.[1] For an incident E wave, the incident wave $(\mathbf{E}_i{}^E, \mathbf{H}_i{}^E)$, the penetrated wave $(\mathbf{E}_p{}^E, \mathbf{H}_p{}^E)$ and the scattered wave $(\mathbf{E}_{sc}{}^E, \mathbf{H}_{sc}{}^E)$ are respectively

$$\mathbf{E}_i{}^E = E_0 e^{ikx}\mathbf{e}_z = E_0 \sum_{n=-\infty}^{\infty} (i)^n J_n(kr) e^{in\phi}\mathbf{e}_z \qquad (5b\text{-}386a)$$

$$\mathbf{H}_i{}^E = \frac{1}{i\omega\mu}(\boldsymbol{\nabla} \times \mathbf{E}_i{}^E) \qquad (5b\text{-}386b)$$

$$\mathbf{E}_p{}^E = E_0 \sum_{n=-\infty}^{\infty} (i)^n b_n{}^E J_n(k_1 r) e^{in\phi}\mathbf{e}_z \qquad (5b\text{-}387a)$$

$$\mathbf{H}_p{}^E = \frac{1}{i\omega\mu_1}(\boldsymbol{\nabla} \times \mathbf{E}_p{}^E) \qquad (5b\text{-}387b)$$

$$\mathbf{E}_{sc}{}^E = E_0 \sum_{n=-\infty}^{\infty} (i)^n a_n{}^E H_n{}^{(1)}(kr) e^{in\phi}\mathbf{e}_z \qquad (5b\text{-}388a)$$

$$\mathbf{H}_{sc}{}^E = \frac{1}{i\omega\mu}(\boldsymbol{\nabla} \times \mathbf{E}_{sc}{}^E) \qquad (5b\text{-}388b)$$

[1] The problem of the scattering by an elliptical dielectric cylinder has been treated [C. Yeh, *J. Math. Phys.* **4**, 65 (1963)]. Solution for the scattering by parabolic dielectric cylinder has also been obtained [C. Yeh, *J. Opt. Soc. Am.* **57**, 195 (1967)].

with

$$a_n{}^E = \frac{(\epsilon_1\mu/\mu_1\epsilon)^{\frac{1}{2}}J_n'(k_1a)J_n(ka) - J_n(k_1a)J_n'(ka)}{H_n{}^{(1)\prime}(ka)J_n(k_1a) - (\epsilon_1\mu/\mu_1\epsilon)^{\frac{1}{2}}H_n{}^{(1)}(ka)J_n'(k_1a)} \tag{5b-389a}$$

$$b_n{}^E = \frac{H_n{}^{(1)\prime}(ka)J_n(ka) - J_n'(ka)H_n{}^{(1)}(ka)}{H_n{}^{(1)\prime}(ka)J_n(k_1a) - (\epsilon_1\mu/\mu_1\epsilon)^{\frac{1}{2}}H_n{}^{(1)}(ka)J_n'(k_1a)} \tag{5b-389b}$$

For an incident H wave, the incident wave $(\mathbf{E}_i{}^H,\mathbf{H}_i{}^H)$, the penetrated wave $(\mathbf{E}_p{}^H,\mathbf{H}_p{}^H)$ and the scattered wave $(\mathbf{E}_{sc}{}^H,\mathbf{H}_{sc}{}^H)$ are respectively

$$\mathbf{E}_i{}^H = \frac{-1}{i\omega\epsilon}\,(\boldsymbol{\nabla}\times\mathbf{H}_i{}^H) \tag{5b-390a}$$

$$\mathbf{H}_i{}^H = H_0 e^{ikx}\mathbf{e}_z = H_0 \sum_{n=-\infty}^{\infty} (i)^n J_n(kr)e^{in\phi}\mathbf{e}_z \tag{5b-390b}$$

$$\mathbf{E}_p{}^H = \frac{-1}{i\omega\epsilon_1}\,(\boldsymbol{\nabla}\times\mathbf{H}_p{}^H) \tag{5b-391a}$$

$$\mathbf{H}_p{}^H = H_0 \sum_{n=-\infty}^{\infty} (i)^n b_n{}^H J_n(k_1r)e^{in\phi}\mathbf{e}_z \tag{5b-391b}$$

$$\mathbf{E}_{sc}{}^H = \frac{-1}{i\omega\epsilon}\,(\boldsymbol{\nabla}\times\mathbf{H}_{sc}{}^H) \tag{5b-392a}$$

$$\mathbf{H}_{sc}{}^H = H_0 \sum_{n=-\infty}^{\infty} (i)^n a_n{}^H H_n{}^{(1)}(kr)e^{in\phi}\mathbf{e}_z \tag{5b-392b}$$

with

$$a_n{}^H = \frac{(\epsilon\mu_1/\mu\epsilon_1)^{\frac{1}{2}}J_n'(k_1a)J_n(ka) - J_n(k_1a)J_n'(ka)}{H_n{}^{(1)\prime}(ka)J_n(k_1a) - (\epsilon\mu_1/\mu\epsilon_1)^{\frac{1}{2}}H_n{}^{(1)}(ka)J_n'(k_1a)} \tag{5b-393a}$$

$$b_n{}^H = \frac{H_n{}^{(1)\prime}(ka)J_n(ka) - J_n'(ka)H_n{}^{(1)}(ka)}{H_n{}^{(1)\prime}(ka)J_n(k_1a) - (\epsilon\mu_1/\mu\epsilon_1)^{\frac{1}{2}}H_n{}^{(1)}(ka)J_n'(k_1a)} \tag{5b-393b}$$

where the prime signifies the derivative of the function with respect to its argument, $k_1 = \omega\sqrt{\mu_1\epsilon_1}$ and $k = \omega\sqrt{\mu\epsilon}$. J_n and $H_n{}^{(1)}$ are respectively the Bessel and Hankel functions. For a perfectly conducting circular cylinder $a_n{}^E = -J_n(ka)/H_n{}^{(1)}(ka)$, $b_n{}^E = 0$ and $a_n{}^H = -J_n'(ka)/H_n{}^{(1)\prime}(ka)$, $b_n{}^H = 0$.

Far-zone Scattered Field

$$\mathbf{E}_{sc}{}^E \text{ (far zone)} \simeq \frac{e^{ikr}}{\sqrt{r}}\,\mathbf{e}_z F_z{}^E \tag{5b-394}$$

$$\mathbf{H}_{sc}{}^H \text{ (far zone)} \simeq \frac{e^{ikr}}{\sqrt{r}}\,\mathbf{e}_z F_z{}^H \tag{5b-395}$$

with

$$F_z{}^{E,H} = \left(\frac{-2i}{\pi k}\right)^{\frac{1}{2}} \sum_{n=-\infty}^{\infty} a_n{}^{E,H}e^{in\phi} \tag{5b-396}$$

Total Scattering Cross Section

$$\sigma_{sc}{}^{E,H} = -\frac{4}{k}\,\mathrm{Re} \sum_{n=-\infty}^{\infty} a_n{}^{E,H} \tag{5b-397}$$

Radar Cross Section

$$\sigma_{rad}^{E,H} = \frac{4}{k}\left| \sum_{n=0}^{\infty} d_n(-1)^n a_n{}^{E,H} \right|^2 \tag{5b-398}$$

with $d_0 = 1$, and $d_n = 2$ for $n \neq 0$.

High- and Low-frequency Limits

$$\sigma_{sc}{}^{E} \text{ (conducting cylinder)} \underset{ka \to \infty}{\simeq} 4a[1 + 0.49807659(ka)^{-\frac{2}{3}}$$

$$-0.01117656(ka)^{-\frac{4}{3}} + \cdots] \quad (5\text{b-}399)$$

$$\sigma_{sc}{}^{H} \text{ (conducting cylinder)} \underset{ka \to \infty}{\simeq} 4a[1 - 0.43211998(ka)^{-\frac{2}{3}}$$

$$-0.21371236(ka)^{-\frac{4}{3}} + \cdots] \quad (5\text{b-}400)$$

$$\sigma_{sc}{}^{E} \text{ (conducting cylinder)} \underset{ka \to 0}{\simeq} \frac{\pi^2}{k \ (\log ka)^2} \quad (5\text{b-}401)$$

$$\sigma_{sc}{}^{H} \text{ (conducting cylinder)} \underset{ka \to 0}{\simeq} \frac{3\pi^2}{4} a(ka)^3 \quad (5\text{b-}402)$$

A typical scattering cross section of a circular cylinder is given in Fig. 5b-14.

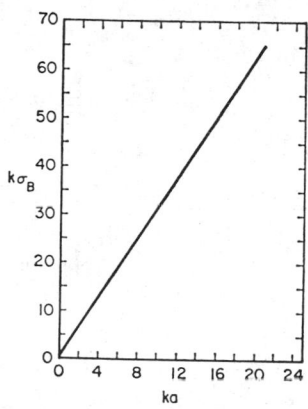

FIG. 5b-14. Backscattering cross section σ_B for a conducting cylinder of radius a. $k = \omega \sqrt{\mu\epsilon}$.

FIG. 5b-15. Diffraction by a semi-infinite conducting half plane.

DIFFRACTION BY A PERFECTLY CONDUCTING HALF PLANE. A perfectly conducting thin plane is located at the plane $x = 0$ and extends from $y = 0$ to $y = +\infty$ (see Fig. 5b-15). The solution for an incident plane wave

$$\left.\begin{matrix} E_z{}^i \\ H_z{}^i \end{matrix}\right\} = \begin{Bmatrix} E_0 \\ H_0 \end{Bmatrix} e^{-ikr \cos (\phi - \phi_0)} \quad (5\text{b-}403)$$

is

$$\left.\begin{matrix} E_z{}^i + E_z{}^{sc} \\ H_z{}^i + H_z{}^{sc} \end{matrix}\right\} = \begin{Bmatrix} E_0 \\ H_0 \end{Bmatrix} e^{-ikr \cos (\phi - \phi_0)} \frac{1 - i}{2} \int_{-\infty}^{a} e^{i\pi\tau^2/2} \, d\tau$$

$$\mp e^{-ikr \cos (\phi - \phi_0)} \frac{1 - i}{2} \int_{-\infty}^{b} e^{i\pi\tau^2/2} \, d\tau \quad (5\text{b-}404)$$

with

$$a = 2 \left(\frac{kr}{\pi}\right)^{\frac{1}{2}} \cos \left(\frac{\phi - \phi_0}{2}\right)$$

$$b = 2 \left(\frac{kr}{\pi}\right)^{\frac{1}{2}} \cos \left(\frac{\phi + \phi_0}{2}\right)$$

Numerical results may be obtained with the help of the tabulated values for the Fresnel integral which is $F(w) = \int_0^w e^{i(\pi/2)\tau^2} \, d\tau$.

DIFFRACTION BY AN APERTURE IN AN INFINITE CONDUCTING SCREEN. The total scattering cross section σ_{sc} of a strip is related to the transmission coefficient t of a slit by the relation

$$t^{H,E} = \frac{\sigma_{sc}{}^{E,H}}{2A} \tag{5b-405}$$

where A is the cross-sectional area of the aperture.[1]

Transmission through a Slit of Width δ

$$t^H = \frac{\sigma_{sc}{}^E}{2\delta} \underset{k\delta \to 0}{\sim} \frac{\pi^2/2k\delta}{[\gamma + \log(k\delta/8)]^2 + \pi^2/4}\left[1 + \frac{(k\delta)^2}{16} + \cdots\right] \tag{5b-405a}$$

$$t^E = \frac{\sigma_{sc}{}^H}{2\delta} \underset{k\delta \to 0}{\sim} \frac{\pi^2(k\delta)^3}{256}\left\{1 + \frac{5(k\delta)^2}{64}\left[1 - \frac{8}{5}\left(\gamma + \log\frac{k\delta}{8}\right)\right]\right\} \tag{5b-406}$$

$$t^H = \frac{\sigma_{sc}{}^E}{2\delta} \underset{k\delta \to \infty}{\sim} \left\{1 - \frac{\sin(k\delta - \pi/4)}{(2\pi)^{\frac{1}{2}}(k\delta)^{\frac{3}{2}}} + \frac{27\sin(k\delta + \pi/4)}{8(2\pi)^{\frac{1}{2}}(k\delta)^{\frac{7}{2}}}\right.$$
$$\left. + \frac{\sin 2(k\delta - \pi/4)}{8\pi(k\delta)^4} + \cdots\right\} \tag{5b-407}$$

$$t^E = \frac{\sigma_{sc}{}^H}{2\delta} \underset{k\delta \to \infty}{\sim} \left\{1 - 2\left(\frac{2}{\pi}\right)^{\frac{1}{2}}\frac{\cos(k\delta - \pi/4)}{(k\delta)^{\frac{3}{2}}} + \frac{2}{\pi}\frac{\cos 2k\delta}{(k\delta)^2}\right.$$
$$- \frac{1}{\pi}\left(\frac{2}{\pi}\right)^{\frac{1}{2}}\frac{\cos(3k\delta + \pi/4) - (7\pi/4)\cos(k\delta + \pi/4)}{(k\delta)^{\frac{5}{2}}}$$
$$\left. - \frac{1}{\pi^2}\frac{\sin 4k\delta - (5\pi/2)\sin 2k\delta}{(k\delta)^3} + \cdots\right\} \tag{5b-408}$$

with $\gamma = 0.5772$.

Transmission through a Circular Aperture of Radius a

$$t = \frac{\sigma_{sc}}{2\pi a^2} \underset{ka \to 0}{=} \frac{64}{27\pi^2}(ka)^4[1 + \tfrac{2}{2}\tfrac{2}{5}(ka)^2 + 0.3979(ka)^4 + \cdots] \tag{5b-409}$$

$$t = \frac{\sigma_{sc}}{2\pi a^2} \underset{ka \to \infty}{=} 1 - \frac{1}{\sqrt{\pi}}\frac{1}{(ka)^{\frac{3}{2}}}\sin\left(2ka - \frac{\pi}{4}\right)$$
$$+ \frac{1}{(ka)^2}\left[\frac{3}{4} + \frac{1}{2\pi}\sin 2\left(2ka - \frac{\pi}{4}\right)\right] - \cdots \tag{5b-410}$$

A typical transmission coefficient of a circular aperture is given in Fig. 5b-16.

Holography. Holography may be described as a method for recording and reconstructing the amplitude and phase information of a propagating field in a given plane.[2] Strictly speaking, rigorous electromagnetic theory of diffraction and polarization is required for an exact treatment of optical holography. Since the electromagnetic field under consideration is almost completely linearly polarized (i.e., only a small fraction of the energy is in the cross-polarization component of the field) and the wavelength of the field is much smaller than the smallest characteristic length of the scattering objects, a scalar physical optics description of the field is therefore adequate.

THE RECORDING PROCESS. The magnitude and the phase of a scattered wavefront can be recorded photographically by superposing a coherent reference wave on the field striking the photographic plate. One of the techniques for carrying out this

[1] H. Levine and J. Schwinger, *Communs. Pure Appl. Math.* **3**, 355 (1950).

[2] D. Gabor, *Proc. Roy. Soc. (London)*, ser. A, **197**, 454 (1949); ser. B, **64**, 449 (1951); E. N. Leith and J. Upatnieks, *J. Opt. Soc. Am.* **52**, 1123 (1962); G. W. Stroke, "An Introduction to Coherent Optics and Holography," Academic Press, Inc., New York, 1966.

superposition is illustrated in Fig. 5b-17 wherein a plane wave illuminates a region containing the scattering object and a triangular prism. The scattering object diffracts the incident radiation to generate a field with magnitude $A(x)$ and phase $\phi(x)$ at the recording photographic plate, while the prism turns the incident plane wave through a small angle θ to give a field with a uniform magnitude A_0 and a linear phase variation αx where $\alpha = 2\pi \sin \theta/\lambda \underset{\theta \text{ small}}{\simeq} 2\pi\theta/\lambda$ with λ = wavelength. The total field at the recording plate is

$$u_{\text{total}} = A_0 e^{-i\alpha x} + A(x)e^{i\phi(x)} \qquad (5b\text{-}411)$$

and the intensity to which the emulsion is sensitive is

$$I(x) = |u_{\text{total}}|^2 = A_0{}^2 + A^2(x) + 2A_0A(x)$$
$$\cos [\alpha x + \phi(x)] \qquad (5b\text{-}412)$$

Note that the intensity recorded by the photographic plate contains information concerning not only $A(x)$, the amplitude of the scattered wave, but also $\phi(x)$, the phase of the scattered wave.

THE RECONSTRUCTION PROCESS. Let us first consider the transmission characteristics of the recording photographic plate. The transmittance $T(x)$ of the resultant photographic plate, provided that the linear range of the Hurter-Driffield curve is used, is

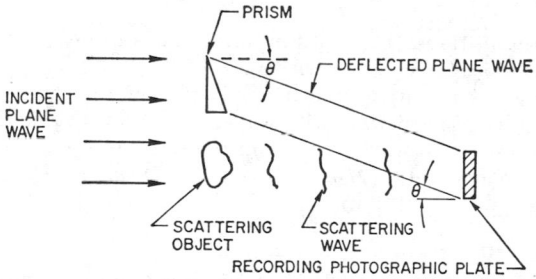

FIG. 5b-16. Transmission coefficient of a circular aperture of radius a. [*From C. Huang, R. D. Kodis, and H. Levine, J. Appl. Phys.* **26**, 151 (1955).]

$$T(x) \sim [I(x)]^{-\gamma/2} = \{A_0{}^2 + A^2(x) + 2A_0A(x) \cos [\alpha x + \phi(x)]\}^{-\gamma/2}$$
$$\sim 2A_0{}^2 - \gamma A^2(x) - \gamma A_0 A(x)e^{i\phi(x)+i\alpha x} - \gamma A_0A(x)e^{-i\phi(x)-i\alpha x} \qquad (5b\text{-}413)$$

where γ is the slope of the Hurter-Driffield curve. It has been assumed that the intensity of the reference wave is much greater than that of the radiation scattered

FIG. 5b-17. Schematic arrangement to illustrate recording of hologram.

by the object, so that the approximation made in dropping the higher-orders terms of the binomial expansion is justified. Note that neither the sign nor the exact magnitude of γ is of any consequence in the recording process; i.e., making a contact print of the photograph (hologram), which is equivalent to changing the sign of γ, serves only to shift the phase of the nonconstant portion of the transmittance an inconsequential 180°, whereas changing slightly the magnitude of γ serves only to enhance or to suppress the magnitude of this same portion of the transmittance.

To reconstruct the original wavefront it is only necessary to illuminate the hologram with a plane incident wave, as shown in Fig. 5b-18. As the plane wave passes through the photographic plate, it is multiplied by the transmittance $T(x)$, thereby producing four distinct components of radiation corresponding to four terms of Eq. (5b-413). The first term, being a constant, attenuates the parallel beam uniformly, but otherwise does not alter it. The second term also attenuates the beam, but not uniformly, so that the plane wave suffers some diffraction as it passes through the hologram. Recall that a common triangular prism shifts the phase of an incident ray by an amount proportional to its thickness at the point of incidence, a positive phase shift deflecting the ray upward and a negative one deflecting it downward. In the case of the third term in Eq. (5b-413), it represents an upward deflected beam multiplied by the scattered wave $A(x)e^{i\phi(x)}$; hence it is a reconstruction of the scattered wavefront. The fourth term represents a downward beam multiplied by the complex conjugate of the scattered wave. Hence, a copy of the scattered wavefront is con-

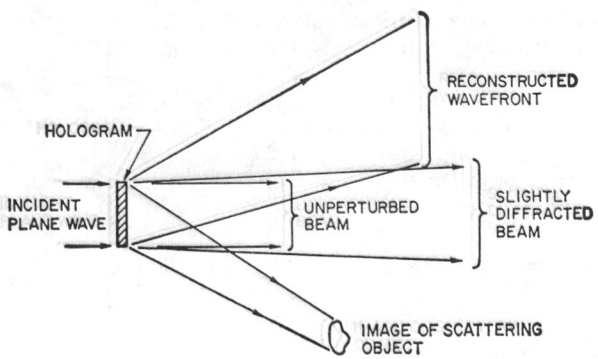

Fig. 5b-18. The reconstruction process—image formation from a hologram for the case of plane-wave illumination.

structed except that it travels backward in time. Consequently, a three-dimensional image of the scattering object is constructed.

Magnification. Magnification or demagnification of the image may be accomplished if one uses an incident wave with wavelength λ for making the hologram and uses an incident wave with wavelength λ' in the reconstruction of the image. The formula for linear magnification M is

$$M = \frac{\lambda'}{\lambda}\frac{q'}{q} \tag{5b-414}$$

where q is the distance of the original object from the hologram, and q' is the distance of the hologram from the final image plane.

Resolution. The ultimate resolution of the conventional Fresnel-transform projection wavefront-reconstruction technique described above is approximately one-half that of the recording media. However, higher resolutions may be obtained by the use of Fourier-transform holography.

References
1. Van Bladel, J.: "Electromagnetic Fields," McGraw-Hill Book Company, New York, 1964.
2. Born, M., and E. Wolf: "Principles of Optics," 2d ed., Pergamon Press, New York, 1964.
3. Jones, D. S.: "The Theory of Electromagnetism," Pergamon Press, New York, 1964.

4. King, R. W. P., and T. T. Wu: "The Scattering and Diffraction of Waves," Harvard University Press, Cambridge, Mass., 1959.
5. Hönl, H., A. W. Maue, and K. Westpfahl: Theory of Diffraction, "Handbuch der Physik," vol. 25, Springer-Verlag OHG, Berlin, 1961.
6. Some recent references on scattering and diffraction are given in the August, 1965, issue of *Proc. IEEE* on Radar Reflectivity.

5b-13. Waves in Plasma. Three basic features characterize plasmas and distinguish them from ordinary solids, liquids, or gases. The first feature is that at least some or all of the particles in a plasma are charged although the plasma as a whole is electrically neutral. The second feature is that Debye shielding effect must be present in plasmas. The third feature is that the product $\omega\tau$ must be large in order that plasma effects may be important. (ω = frequency of the wave in plasma, τ = the average time an electron travels between collisions with neutral molecules, or lattice ions, or impurities, etc.)

Basic Equations. The basic equations governing the waves in plasmas are the Boltzmann equation and Maxwell's equations:

$$\frac{\partial f_\alpha}{\partial t} + \mathbf{v} \cdot \frac{\partial f_\alpha}{\partial \mathbf{r}} + \mathbf{a} \cdot \frac{\partial f_\alpha}{\partial \mathbf{v}} = \left(\frac{\partial f_\alpha}{\partial t}\right)_c \tag{5b-415}$$

$$\nabla \times \mathbf{E} = -\frac{\partial \mathbf{B}}{\partial t} \qquad \mathbf{D} = \epsilon_0 \mathbf{E}$$

$$\nabla \times \mathbf{H} = \mathbf{J} + \frac{\partial \mathbf{D}}{\partial t} \qquad \mathbf{B} = \mu_0 \mathbf{H} \tag{5b-416}$$

$$\nabla \cdot \mathbf{D} = \rho_c$$

$$\nabla \cdot \mathbf{B} = 0$$

$$\rho_c = \sum_\alpha q_\alpha \iiint f_\alpha \, dv_x \, dv_y \, dv_z$$

$$\mathbf{J} = \sum_\alpha q_\alpha \iiint \mathbf{v} f_\alpha \, dv_x \, dv_y \, dv_z \tag{5b-417}$$

where $f_\alpha(\mathbf{x}, \mathbf{v}, t)$ is the distribution function for particles of type α, and $\mathbf{a}$ is the acceleration due to external forces, which for an electromagnetic field would be the Lorentz accleration $\mathbf{a} = (q_\alpha/m_\alpha)(\mathbf{E} + \mathbf{v} \times \mathbf{B})$. $(\partial f_\alpha/\partial t)_c$ is the time rate of change due to collisions. $\mathbf{E}, \mathbf{H}, \mathbf{B}, \mathbf{D}$ are the electromagnetic field vectors. ρ_c and $\mathbf{J}$ are respectively the charged density and the vector current density. q_α and m_α are respectively the charge and mass for particles of type α. $\mathbf{v}$ and $\mathbf{r}$ are the velocity and position vectors.

When collisions are neglected, we may set $(\partial f_\alpha/\partial t)_c = 0$ in Eq. (5b-415). This equation is called the collisionless Boltzmann equation or the Boltzmann-Vlasov equation.

HYDRODYNAMIC-CONTINUUM MODEL. Taking the appropriate moments of Eq. (5b-415) and making the assumption that (1) the mass density ρ_α for each species is unchanged, (2) the Lorentz force per unit mass for each species is $\langle \mathbf{a} \rangle_\alpha = (q_\alpha/m_\alpha)$ $(\mathbf{E} + \mathbf{u}_\alpha \times \mathbf{B})$, (3) viscous effects are negligible, i.e., the pressure is a scalar quantity, and (4) the flow-velocity difference among the various gas species is small and each gas has a maxwellian velocity distribution, one obtains the following equations for the hydrodynamic-continuum model:

$$\frac{\partial \rho_\alpha}{\partial t} + \nabla \cdot \rho_\alpha \mathbf{u}_\alpha = 0 \qquad \text{(mass conservation)} \tag{5b-418}$$

$$\frac{\partial \mathbf{u}_\alpha}{\partial t} + \mathbf{u}_\alpha \cdot \nabla \mathbf{u}_\alpha = \frac{q_\alpha}{m_\alpha}(\mathbf{E} + \mathbf{u}_\alpha \times \mathbf{B}) - \frac{\nabla p_\alpha}{\rho_\alpha} - \sum_\beta \nu_\alpha(\mathbf{u}_{\alpha\beta} - \mathbf{u}_\beta)$$

$$\text{(momentum conservation)} \tag{5b-419}$$

$$\nabla p_\alpha = U_\alpha^2 \nabla \rho_\alpha \qquad \text{(energy conservation)} \tag{5b-420}$$

with $\rho_\alpha = m_\alpha n_\alpha$, $\rho_c = \sum_\alpha q_\alpha n_\alpha$, and $\mathbf{J} = \sum_\alpha q_\alpha n_\alpha \mathbf{u}_\alpha$. The subscript α refers to particles of type α. ρ_α, m_α, n_α, ρ_c, $\mathbf{J}$, $\mathbf{u}_\alpha$, and p_α are respectively the mass density, mass, number density, charge density, current density, average velocity vector, and scalar pressure. U_α is the adiabatic or the isothermal sound speed, depending on the problem at hand. $\nu_{\alpha\beta}$ is the collision frequency for momentum transfer for particles of type α with those of type β.

Equations (5b-418) to (5b-420), together with Maxwell's equations (5b-416) provide a complete set of equations for the hydrodynamic model.

LINEARIZED MAGNETOHYDRODYNAMIC (MHD) MODEL. A set of linearized mhd equations may be obtained if we replace the above set of individual-species equations (5b-418) to (5b-420) by a set of equations for the gas as a whole:

$$\frac{\partial \rho}{\partial t} + \rho_0 \nabla \cdot \mathbf{u} = 0 \tag{5b-421}$$

$$\rho_0 \frac{\partial \mathbf{u}}{\partial t} = \mathbf{J} \times \mathbf{B}_0 - \nabla p \tag{5b-422}$$

$$\nabla p = U_s{}^2 \nabla \rho \tag{5b-423}$$
$$\mathbf{J} = \sigma_0 (\mathbf{E} + \mathbf{u} \times \mathbf{B}_0) \tag{5b-424}$$

where $\mathbf{B}_0$ is the applied magnetostatic field, σ_0 is the conductivity of the gas, U_s is the isothermal or adiabatic sound speed for the gas, ρ_0 is the equilibrium mass density of the gas. ρ, $\mathbf{u}$, p, $\mathbf{E}$, and $\mathbf{B}$ are all infinitesimal disturbances. A simplified Ohm's law [Eq. (5b-424) has been assumed. Equations (5b-421) to (5b-424), together with Maxwell's equations (5b-416)—with the assumption that the displacement vector term $\partial \mathbf{D}/\partial t$ is negligible—provide a complete set of equations for the linearized mhd model.

MAGNETOIONIC MODEL (COLD PLASMA MODEL). If we further assumed that the thermovelocity of electrons or ions is zero, (i.e., the term $\nabla p_\alpha / \rho_\alpha$ in Eq. (5b-419) is zero, and the inertial term $\mathbf{u}_\alpha \cdot \nabla \mathbf{u}_\alpha$ is omitted, then

$$\frac{\partial \mathbf{u}_\alpha}{\partial t} = \frac{q_\alpha}{m_\alpha} (\mathbf{E} + \mathbf{u}_\alpha \times \mathbf{B}) - \mathbf{u}_\alpha \sum_\beta \nu_{\alpha\beta} \tag{5b-425}$$

$$\mathbf{J} = \sum_\alpha q_\alpha n_\alpha \mathbf{u}_\alpha \tag{5b-426}$$

Equations (5b-425) and (5b-426), together with Maxwell's equations (5b-416), provide a complete set of equations for the cold plasma model.

Waves in Cold Plasmas. The linearized equations[1] (with harmonic time dependence $e^{-i\omega t}$) for waves in cold (electron)[2] plasmas are

$$\nabla \times \mathbf{B} = -i\omega \mu_0 \boldsymbol{\varepsilon} \cdot \mathbf{E} \tag{5b-427}$$
$$\nabla \times \mathbf{E} = i\omega \mathbf{B} \tag{5b-428}$$
$$\boldsymbol{\varepsilon} = \begin{bmatrix} \epsilon_{xx} & -i\epsilon_{xy} & 0 \\ i\epsilon_{xy} & \epsilon_{yy} & 0 \\ 0 & 0 & \epsilon_{zz} \end{bmatrix} \tag{5b-429}$$

[1] The linearization procedures are justified if the phase velocities of the waves under consideration are much greater than the average electron velocity.

[2] In an electron plasma, only the motion of electrons is important. The ions and the neutrons are assumed to be stationary. For very low frequency waves the motion of ions may be important. In that case the components of the dielectric tensor must be modified. See E. Astrom, *Arkiv Fysik.* **2**, 443 (1950).

$$\epsilon_{xx} = \epsilon_0 \left\{ 1 - \frac{\omega_p^2(\omega + i\nu)}{\omega[(\omega + i\nu)^2 - \omega_c^2]} \right\}$$

$$\epsilon_{xy} = \epsilon_0 \left[\frac{\omega_p^2 \omega_c}{\omega(\omega + i\nu + \omega_c)(\omega + i\nu - \omega_c)} \right]$$

$$\epsilon_{zz} = \epsilon_0 \left[1 - \frac{\omega_p^2}{\omega(\omega + i\nu)} \right]$$

with $\omega_p = (n_e e^2/m_e \epsilon_0)^{\frac{1}{2}}$ and $\omega_c = -(e/m_e)B_0$. ω_p, ω_c, and ν are respectively the plasma frequency for electrons, and the gyro frequency and collision frequency of electrons with all other heavy particles. $\mathbf{B}_0 = B_0 \mathbf{e}_z$ is the applied static magnetic field. $\mathbf{E}$ and $\mathbf{B}$ are the complex amplitudes of the electromagnetic fields.

For a plane wave propagating in the $\mathbf{n}$ direction, the electric vector has the form

$$\mathbf{E} = \mathbf{E}_0 e^{i\mathbf{k}\cdot\mathbf{r}} \qquad (5b\text{-}430)$$

where $\mathbf{E}_0$ is a constant vector, $\mathbf{r}$ is the position vector, $\mathbf{k} = \mathbf{n}\omega/v_{\mathrm{ph}}$ is the vector wave number, and v_{ph} is the phase velocity of the wave. The dispersion relation for the phase velocity, called the Appleton-Hartree equation, is obtained by substituting Eq. (5b-430) into Eqs. (5b-427) and (5b-428):

$$\Phi = U - \frac{Y^2 \sin^2 \gamma}{2(U - X)}$$
$$\pm \left[\frac{Y^4 \sin^4 \gamma}{4(U - X)^2} + Y^2 \cos^2 \gamma \right]^{\frac{1}{2}} \quad (5b\text{-}431)$$

$$k^2 c^2 = \omega^2 - \frac{\omega_p^2}{\Phi} \qquad (5b\text{-}432)$$

$$U = 1 + i\frac{\nu}{\omega} \qquad X = \frac{\omega_p^2}{\omega^2} \qquad Y^2 = \frac{\omega_c^2}{\omega^2}$$

$$c = \frac{1}{\sqrt{\mu_0 \epsilon_0}}$$

where γ is the angle between the direction of propagation and the direction of the static magnetic field $\mathbf{e}_z \cdot \mathbf{n}$ is in the yz plane. A sketch of the phase velocity vs. frequency for waves traveling in an arbitrary direction relative to $\mathbf{B}_0$ is given in Fig. 5b-19.

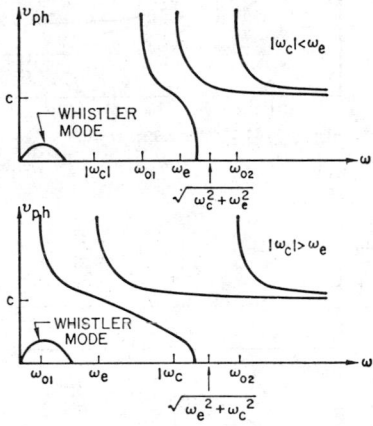

Fig. 5b-19. Phase velocity vs. frequency for waves traveling in an arbitrary direction relative to $\mathbf{B}_0$, the applied magnetic field, in an electron plasma. The above results are obtained according to the cold plasma model. $|\omega_c| = eB_0/m_e$,

$$\omega_e = (e^2 n_0/m_e \epsilon_0)^{\frac{1}{2}}$$

$\omega_{01} = [-|\omega_c| + (\omega_c^2 + 4\omega_e^2)^{\frac{1}{2}}]/2$, $\omega_{02} = \omega_{01} + |\omega_c|$, $n_0 =$ number density of electrons, $m_e =$ mass of electrons, and $c =$ velocity of light in vacuum.

A great deal of work on wave propagation in plasma filled guide[1] and on the scattering of waves by a plasma column[2] has also been carried out.

Alfvén Wave. Alfvén wave exists in a plasma at very low frequencies when the plasma can be adequately represented by the linearized mhd model. Assuming that ρ, $\mathbf{u}$, p, $\mathbf{E}$, and $\mathbf{B}$ in Eqs. (5b-416) and (5b-421) to (5b-424) are all proportional to exp $i(kx - \omega t)$ $\sigma_0 = \infty$; and the applied static magnetic field $\mathbf{B}_0$ lies in the xy plane and makes an angle γ with the positive x axis; one may obtain the following set of equations:

$$u_x(\omega^2 - k^2 V_a^2 \sin^2 \gamma - k^2 U^2) + u_y k^2 V_a^2 \sin \gamma \cos \gamma = 0 \qquad (5b\text{-}433)$$
$$u_x k^2 V_a^2 \sin \gamma \cos \gamma + u_y(\omega^2 - k^2 V_a^2 \cos^2 \gamma) = 0 \qquad (5b\text{-}434)$$
$$u_z(\omega^2 - k^2 V_a^2 \cos^2 \gamma) = 0 \qquad (5b\text{-}435)$$

[1] See, for example, A. W. Trivelpiece and R. W. Gould, *J. Appl. Phys.* **30**, 1784 (1959)
[2] See, for example, C. Yeh and W. V. T. Rusch, *J. Appl. Phys.* **36**, 2302 (1965).

where $V_a = B_0/(\mu_0\rho_0)^{\frac{1}{2}}$ is called the Alfvén velocity. According to Eq. (5b-435), we see that a wave linearly polarized in the z direction (the direction perpendicular to both $\mathbf{k}$ and $\mathbf{B}_0$) can exist if

$$v_{\text{ph}} = \frac{\omega}{k} = V_a \cos\gamma \tag{5b-436}$$

This wave is called the pure Alfvén wave. Solving of Eqs. (5b-433) and (4b-434) gives the phase velocity of mhd waves containing components u_x and u_y:

$$v_{\text{ph}} = \frac{\omega}{k} = \frac{1}{\sqrt{2}}\{(V_a{}^2 + U^2) \pm [(V_a{}^2 + U^2) - 4V_a{}^2 U^2 \cos^2\gamma]^{\frac{1}{2}}\}^{\frac{1}{2}} \tag{5b-437}$$

The plus and minus signs refer to fast and slow mhd waves. The above results are applicable only if $\omega \ll \omega_i$, $\omega \ll \omega_{ci}$ and $V_a \ll c$, where ω_i is the ion plasma frequency, ω_{ci} is the ion cyclotron frequency, and c is the velocity of light in vacuum. Hence, the dispersion characteristics of high-frequency waves must be found from the full set of equations for the hydrodynamic-continuum model.[1] A sketch of phase velocity vs. frequency for waves in a fully ionized plasma is given in Fig. 5b-20.

Longitudinal Electron Landau Waves. Let us now consider the problem of the propagation of small-amplitude longitudinal waves in an electron plasma with no uniform applied static magnetic field by the use of the Boltzmann-Vlasov equation. Assuming that

$$f = f_0(\mathbf{v}) + f_1(\mathbf{v})e^{ikz-i\omega t} \qquad |f_1| \ll f_0 \tag{5b-438}$$

$$\mathbf{E} = Ee_z e^{ikz-i\omega t} \tag{5b-439}$$

where f_0 is the equilibrium distribution function for electrons, and substituting Eqs. (5b-438) and (5b-439) into Eqs. (5b-415) and (5b-416), one has

$$\left[\frac{\omega_p{}^2}{n_0 k^2}\int\frac{(\partial f_0/\partial v_x)\,d^3v}{v_x - \omega/k} - 1\right]E = \mathbf{0} \tag{5b-440}$$

where n_0 is the equilibrium electron density.

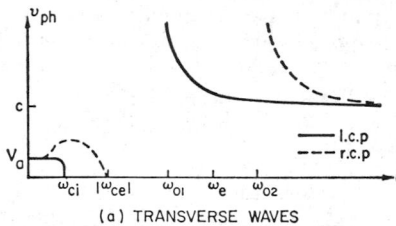

(a) TRANSVERSE WAVES

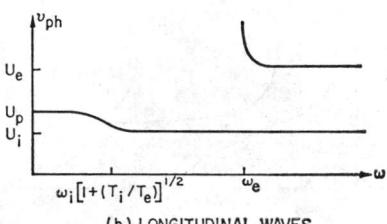

(b) LONGITUDINAL WAVES

FIG. 5b-20. Phase velocity vs. frequency for waves in a fully ionized plasma according to the hydrodynamic-continuum model. Waves are assumed to be propagating in the direction of $\mathbf{B}_0$, the applied magnetic field. $|\omega_{ce}| = eB_0/m_e$,

$$\omega_{ci} = eB_0/m_i, \quad \omega_e = (e^2n_0/m_e\epsilon_0)^{\frac{1}{2}}$$
$$\omega_i = (m_e/m_i)^{\frac{1}{2}}\omega_e$$

and $U_p = \gamma K(T_e + T_i)/m_i$ (the plasma sound speed). V_a is the Alfvén velocity. $U_e = \gamma KT_e/m_e$, $U_i = \gamma KT_i/m_i$, γ is the ratio of specific heats at constant pressure and constant volume, and K is the Boltzmann's constant. T_e, T_i, m_e, m_i are respectively the electron temperature, the ion temperature, the electron mass, and the ion mass. The above curves are valid only if $T_e \gg T_i$ and the phase velocity of the wave is not close to the thermovelocity of ions or electrons.

Setting the quantity in the square brackets to zero gives the dispersion equation for the longitudinal electron waves. The solution of this dispersion equation has been obtained for the case when f_0 is the maxwellian velocity distribution for a stationary plasma; i.e.,

$$f_0 = n_0 e^{-v^2/a^2}/\pi^{\frac{3}{2}}a^3$$

[1] The description of the propagation characteristics of waves according to the hydrodynamic-continuum model is not valid when the phase velocity of a particular mode of interest is close to the thermovelocity of ions or electrons. In that case, Boltzmann's equations must be used. See B. D. Fried and R. W. Gould, *Phys. Fluids* **4**, 139 (1961).

with $a^2 = 2KT/m_e$, K is the Boltzmann's constant, and T is the temperature:

$$k^2 = -k_D{}^2 \left(1 - 2C \int_0^C e^{z^2-C^2}\, dz + i\pi^{\frac12}Ce^{-C^2}\right) \tag{5b-441}$$

where $C = \omega/ka$, and $k_D{}^2 = 2\omega_p{}^2/a^2$ is the Debye wave number. The integral in the above equation is called the dispersion function and has been tabulated.[1] The last term, which is imaginary, is known as the Landau damping term. When $\omega/k \to \infty$, Eq. (5b-441) may be written as

$$\omega^2 = \omega_p{}^2 + \frac{3KTk^2}{m_e} + \cdots - \frac{2i\pi^{\frac12}\omega_p{}^5}{k^3a^3} e^{-\omega_p{}^2/k^2a^2} \tag{5b-442}$$

Hence the longitudinal waves will decay in a *collisionless* electron plasma. The Landau damping characteristics are also present for transverse waves.[2]

Motion of a Charged Particle in Electromagnetic Fields. The motion of a charged particle in electromagnetic fields is governed by the following equation:

$$m \frac{d\mathbf{v}}{dt} = q(\mathbf{E} + \mathbf{v} \times \mathbf{B}) \tag{5b-443}$$

where m, q, $\mathbf{v}$, $\mathbf{E}$, and $\mathbf{B}$ are respectively the mass of the particle, the charge, the velocity, the applied electric field, and the applied magnetic field. Some important behaviors of a charged particle in such an applied field are listed below:

IN CONSTANT AND UNIFORM $\mathbf{E}$ AND $\mathbf{B}$ FIELDS

1. The particle rotates about the $\mathbf{B}$ direction at a gyrofrequency $(\omega_c) = |qB/m|$ and with a radius $|V_0/\omega_c|$, where V_0 is the initial velocity of the particle in a plane normal to the $\mathbf{B}$ direction.

2. The particle possesses a drift velocity, $\mathbf{v}_D = \mathbf{E} \times \mathbf{B}/B^2 + m(\mathbf{g} \times \mathbf{B})/qB^2$, where $\mathbf{g}$ is the uniform gravitational field.

3. There is a constant acceleration in the $\mathbf{B}$ direction unless q and $\mathbf{E}$ are perpendicular to $\mathbf{B}$. (In the last case the particle drifts in the $\mathbf{B}$ direction with its initial velocity.)

IN A NONUNIFORM $\mathbf{B}$ FIELD. The particle possesses a drift velocity,

$$\mathbf{v}_D = \frac{\nabla_\perp B}{\omega_c B}\left(\frac12 V_\perp{}^2 + V_\parallel{}^2\right) \mathbf{e}_D$$

where $\nabla_\perp B$ is the gradient of the scalar B in the plane perpendicular to $\mathbf{B}$, $V_\perp$ and $V_\parallel$ are respectively the initial velocities perpendicular and parallel to the magnetic field $\mathbf{B}$, and $\mathbf{e}_D$ is a unit vector in the direction $\mathbf{B} \times \nabla B$.

ADIABATIC INVARIANCE OF $\boldsymbol{\mu}$. When the applied magnetic field changes slowly with space or time,

$$\frac{d\boldsymbol{\mu}}{dt} = 0$$

where $\boldsymbol{\mu}$ is the magnetic moment for the changed particle and $\boldsymbol{\mu} = -w_\perp \mathbf{B}/B^2$ with $w_\perp = \frac12 m V_\perp{}^2$, which is the kinetic energy of the motion perpendicular to $\mathbf{B}$.

ENERGY CONSERVATION IN A STATIONARY FIELD

$$\frac{d}{dt}(\tfrac12 mv^2 + q\Phi) = 0$$

[1] B. D. Fried and S. D. Conte, "The Plasma Dispersion Function," Academic Press, Inc., New York, 1961.

[2] See the treatment by Bernstein and Harris on waves in a hot plasma with an applied static magnetic field. [I. B. Bernstein, *Phys. Rev.* **109**, 10 (1958); E. G. Harris, *J. Nuclear Energy, Pt. C.* **2**, 138 (1961).] Also, for the treatment of waves in hot plasma-filled waveguides, see H. H. Kuehl, G. E. Stewart, and C. Yeh, *Phys. Fluids* **8**, 723 (1965).

where Φ is the potential energy per unit charge. The above equation indicates that the sum of kinetic and potential energies stays constant in a stationary field with $\mathbf{E} = -\nabla\Phi$.

References

1. Stix, T. H.: "The Theory of Plasma Waves," McGraw-Hill Book Company, New York, 1962.
2. Allis, W. P., S. J. Buchsbaum, and A. Bers: "Waves in Anisotropic Plasmas," The MIT Press, Cambridge, Mass., 1963.
3. Spitzer, L., Jr.: "Physics of Fully Ionized Gases," 2d ed., Interscience Publishers, a division of John Wiley & Sons, Inc., New York, 1962.
4. Heald, M. A., and C. B. Wharton: "Plasma Diagnostics with Microwaves," John Wiley & Sons, Inc., New York, 1964.
5. Huddlestone, R. H., and S. L. Leonard: "Plasma Diagnostics," Academic Press, Inc., New York, 1965.
6. Tanenbaum, B. S.: "Plasma Physics," McGraw-Hill Book Company, New York, 1967.

5b-14. Skin Effect. At high frequencies currents in a conductor tend to concentrate on the surface and decay approximately exponentially into the conductor. The concentration increases as frequency, conductivity, or permeability increases. The result is an increased resistance and decreased internal inductance at frequencies for which the effect is significant.

The basic equations governing the skin-effect phenomena are the Maxwell's equations applied to good conductors. A good conductor is defined by the following characteristics: the free-charge term is zero, i.e., $\rho = 0$; conduction current is given by Ohm's law, $\mathbf{J} = \sigma\mathbf{E}$, where σ is the conductivity; displacement current is negligible in comparison with conduction current, $\omega\epsilon \ll \sigma$. Under this assumption, Maxwell's equations are:

$$\nabla \times \mathbf{E} = i\omega\mu\mathbf{H} \qquad \nabla \cdot \mathbf{D} = 0$$
$$\nabla \times \mathbf{H} = \sigma\mathbf{E} \qquad \nabla \cdot \mathbf{B} = 0 \tag{5b-444}$$

with $\mathbf{B} = \mu\mathbf{H}$, $\mathbf{D} = \epsilon\mathbf{E}$, and $\mathbf{J} = \sigma\mathbf{E}$. A time dependence of $e^{-i\omega t}$ has been assumed for all field components and suppressed. Combining these equations and assuming that ϵ, μ, σ are independent of the position vector (i.e., a homogeneous medium), one has

$$\nabla^2\mathbf{P} - \tau^2\mathbf{P} = 0 \tag{5b-445}$$

where $\mathbf{P}$ may be $\mathbf{E}$, or $\mathbf{H}$, or $\mathbf{J}$; and $\tau^2 = -i\omega\mu\sigma = -2i/\delta^2$. $\delta = (2/\omega\mu\sigma)^{\frac{1}{2}}$ is called the skin depth; it is a measure of the decaying characteristics of fields within a conductor. The surface resistivity R_s is defined as $R_s = 1/\sigma\delta = (\omega\mu/2\sigma)^{\frac{1}{2}}$. Data for δ and R_s as functions of frequency are given for several common materials in Table 5b-5. The boundary conditions at the surface between a good dielectric and a good conductor are $\mathbf{n} \cdot \mathbf{J} = 0$ and $\mathbf{J} = \sigma\mathbf{E}_0$, where $\mathbf{n}$ is normal to the surface, and $\mathbf{E}_0$ is the applied field at the surface. The boundary conditions at the surface between two good conductors are the continuity of tangential electric and magnetic fields.

The internal impedance Z_i of a good conductor is defined as the ratio of the electric field at the surface to total current. The time-averaged power dissipated as Joules heat within the volume V is $\frac{1}{2}\int\sigma|E|^2\,dv$.

Formulas for Several Simple Conductors. PLANE SEMI-INFINITE CONDUCTOR. The plane conductor extends from $x = 0$ to $x = \infty$, and E_0 is an applied field in the z direction at $x = 0$.

$$J_z = \sigma E_0 e^{-x/\delta} e^{ix/\delta} \tag{5b-446}$$
$$Z_i = R_i - i\omega L_i = (1 - i)R_s \tag{5b-447}$$

TABLE 5b-5. SKIN-EFFECT QUANTITIES FOR CONDUCTORS

Metal	Resistivity* (ohm-m)10^8	Relative* permeability at 0.002 weber/m^2	$\delta\sqrt{\nu}$ δ = depth of penetration, m, ν = frequency, Hz	$10^7 R_s/\sqrt{\nu}$ R_s = surface resistivity, ohms/m^2
Aluminum	2.828	1	0.085	3.33
Brass (65.8 Cu, 34.2 Zn)	6.29†	1	0.126	4.99
Brass (90.9 Cu, 9.1 Zn)	3.65†	1	0.096	3.79
Graphite	1,000	1	1.592	62.81
Chromium	2.6†	1	0.081	3.21
Copper	1.724	1	0.066	2.61
Gold	2.22†	1	0.075	2.96
Lead	22	1	0.236	9.32
Magnesium	4.6	1	0.108	4.26
Mercury	95.8†	1	0.493	19.43
Nickel	7.8	100	0.014	55.71
Phosphor bronze	7.75†	1	0.140	5.54
Platinum	9.83†	1	0.158	6.22
Silver	1.629	1	0.064	2.55
Tin	11.5	1	0.171	6.73
Tungsten	5.51	1	0.118	4.67
Zinc	5.38†	1	0.117	4.60
Magnetic iron	10	200	0.011	90.9
Permalloy (78.5 Ni, 21.5 Fe)	16	8,000	0.0022	727
Supermalloy (5 Mo, 79 Ni, 16 Fe)	60	10^5	0.0012	4,880
Mumetal (75 Ni, 2 Cr, 5 Cu, 18 Fe)	62	20,000	0.0029	2,140

* Values from Pender and McIlwain, "Electrical Engineers' Handbook," 4th ed., John Wiley & Sons, Inc., New York, 1950.
† Values at 0°C; others at 20°C.

SOLID ROUND WIRE. For a solid round conductor of radius a with applied axial electric field E_0 at the surface, we have

$$J_z = \sigma E_0 \frac{J_0(i^{\frac12}r/\delta)}{J_0(i^{\frac12}a/\delta)} \tag{5b-448}$$

$$Z_i = R_i - i\omega L_i = \frac{-R_s}{\sqrt{2}a\pi}\frac{J_0(i^{\frac12}a/\delta)}{J_0'(i^{\frac12}a/\delta)} \tag{5b-449}$$

where $J_0(i^{\frac12}a/\delta)$ is a Bessel function of order zero with complex argument.[1] For $a/\delta \ll 1$,

$$Z_i \simeq \frac{1}{\pi a^2\sigma}\left[1 + \frac{1}{48}\left(\frac{a}{\delta}\right)^4\right] - i\frac{\omega\mu}{8\pi} \tag{5b-450}$$

for $a/\delta \gg 1$,

$$Z_i \simeq \frac{(1-i)R_s}{2\pi a} \tag{5b-451}$$

[1] S. Ramo, J. R. Whinnery, and T. Van Duzer, "Fields and Waves in Communication Electronics," chap. 5, John Wiley & Sons, Inc., New York, 1965: S. J. Haefner, *Proc. IRE* **25**, 434 (1937); H. A. Wheeler, *ibid.* **43**, 805 (1955).

Formulas are also available for tabular conductors and rectangular conductors as well as coated conductors.[1] An example of skin depth and high-frequency resistance of copper is given in Fig. 5b-21.

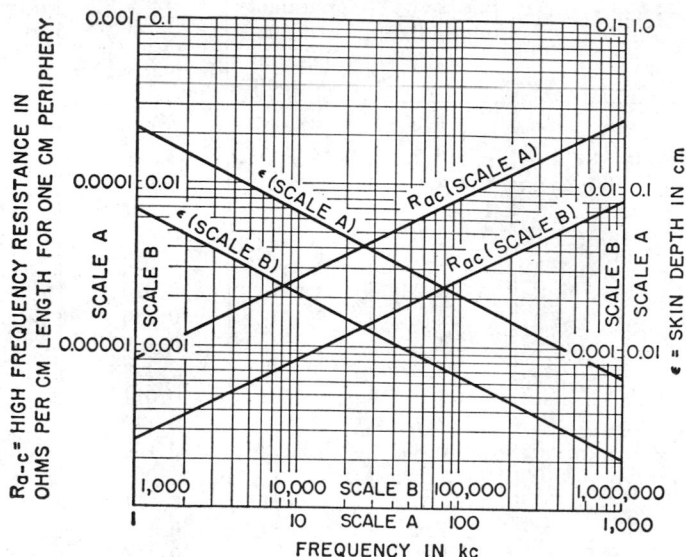

FIG. 5b-21. Skin depth and high-frequency resistance of copper. (*From F. E. Terman, "Radio Engineers' Handbook," p. 35, McGraw-Hill Book Company, New York, 1943.*)

Transient Penetration in the Plane Conductor. If a constant magnetic field H_0 is suddenly applied at time $t = 0$ to the surface of a semi-infinite plane conductor, field at depth x, time $t > 0$ is

$$H(x,t) = H_0 \left[1 - \text{erf} \left(\frac{x}{2} \sqrt{\frac{\mu\sigma}{t}} \right) \right] \tag{5b-452}$$

If the applied field increases linearly with time, $H(0,t) = Ct$ for $t > 0$:

$$H(x,t) = Ct \left\{ \left(1 + \frac{\mu\sigma x^2}{2t} \right) \left[1 - \text{erf} \left(\frac{x}{2} \sqrt{\frac{\mu\sigma}{t}} \right) \right] \right.$$
$$\left. - x \sqrt{\frac{\mu\sigma}{\pi t}} \exp \left(\frac{-\mu\sigma x^2}{4t} \right) \right\} \tag{5b-453}$$

Anomalous Skin Effect. At sufficiently low temperatures and high frequencies, the mean free path of the electrons in a good conductor becomes greater than the classically predicted skin depth, and the classical skin-effect equations break down. Thus, the radio-frequency skin conductivity is practically independent of bulk conductivity (measured at direct current) when the mean free path of the electrons is sufficiently long. Data for Na, Cu, Ag, Au, Pt, W, Al, Pb, and Sn have been given by Pippard, Chambers, and Dingle.[2]

[1] S. Ramo, J. R. Whinnery, and T. Van Duzer, "Fields and Waves in Communication Electronics," chap. 5, John Wiley & Sons, Inc., New York, 1965: S. J. Haefner, *Proc. IRE* **25**, 434 (1937); H. A. Wheeler, *ibid.* **43**, 805 (1955).
[2] R. B. Dingle, *Physica* **19**, 348 (1953); R. G. Chambers, *Nature* **165**, 239 (1950); A. B. Pippard, *Proc. Roy. Soc. (London),* ser. A, **191**, 385 (1947).

5c. Electrical Standards

F. K. HARRIS

The National Bureau of Standards

5c-1. Fundamental Considerations. The standards in terms of which electrical quantities are evaluated are derived from absolute measurements which serve to establish the magnitudes of the electrical units in terms of the base mechanical units. The relations between the fundamental mechanical units and the electrical units derived from them are required to satisfy two conditions: (1) the electrical watt should equal the mechanical watt; and (2) in a rationalized system the unit of resistance must be such as to make the wave impedance of free space numerically equal to $\mu_v c$, where μ_v is the conventionally assigned value of the permeability of free space, and c is the velocity of the electromagnetic wave.[1] The first condition fixes the *product* of the volt and the ampere (the watt), while the second fixes their *quotient* (the ohm).

Two types of *absolute* measurement have been used in assigning numerical values to the basic electrical standards in terms of mechanical units of length and time. In one measurement the ohm is evaluated in terms of the mechanical units of length and time; in the second the ampere is measured in units of length, mass, and time.

Most of the *absolute-ohm* determinations in the past have involved an inductor (either self or mutual) constructed in such a way that its inductance can be computed from its measured dimensions, together with the conventionally assigned permeability of the space around it. This inductor is then supplied with a periodically varying current, and its reactance at the known frequency is, in effect, compared with the resistance of a standard resistor.[2] Now, there is no reason why absolute-ohm determinations should not be made in terms of a computable capacitance and a frequency. Indeed, such a determination offers decided advantages in that the electric field of a capacitor can be confined by shields and the capacitance value made completely independent of neighboring objects, whereas the magnetic field of a computable inductor cannot be so limited as to be completely free from proximity influences. However, only since the recent discovery of a new theorem in electrostatics[3] has it been possible to design and construct capacitors whose value can be computed with sufficient accuracy from simple dimensional measurements to make attractive an absolute-ohm determination in terms of capacitance. In such a determination, the permittivity of the dielectric medium in the capacitor must be used in computing its value, and this is obtained through permeability—an assigned unit in the absolute mksa system— and an experimental value for the speed of light. Thus, the essential limitation in such an absolute-ohm determination is the uncertainty in the value used for the speed of light in vacuum. Two ohm determinations in terms of capacitance have been reported.[4]

---

[1] F. B. Silsbee, *Instruments* **26**, 1522 (1953).
[2] Thomas et al., *J. Research NBS* **43**, 291 (1949); Rayner, *Metrologia* **3**, 12 (1967).
[3] The Thompson-Lampard theorem is discussed in Sec. 5c-5.
[4] Cutkosky, *J. Research NBS* **65A**, 147 (1961); Thompson, *Metrologia* **4**, 1 (1968).

In an *absolute-ampere* experiment, a pair of coils is so arranged that the force or torque between them when they carry a current can be measured accurately. The arrangement is called a *current balance*. The current, thus measured in absolute amperes, is passed through a resistor whose value is known in absolute ohms. The resulting voltage drop is opposed to the electromotive force of a standard cell, and its emf is determined in *absolute* volts.[1]

Values having been assigned to physical standards of resistance and voltage on the basis of absolute measurements, the values of the other electrical units can be derived from them, using appropriate relationships. Thus the *ohm* and *volt* become the base units of electrical measurement, and their physical embodiments in resistance coils and standard cells become the fundamental electrical standards.

5c-2. History of Electrical Standards. The British Association Ohm (1864), resulting from the work of a committee under the leadership of Maxwell, represented the first concerted attempt by a responsible organization to realize an electrical standard based on absolute measurements correlating a mechanical and electrical system of units. At that time the Daniell cell was commonly used as the standard of emf. Later the Clark cell (1872) and its modification by Lord Rayleigh (1884) were used. Still more recently (by international agreement in 1908) the cadmium cell invented by Weston (1891) has entirely replaced the Clark cell and is in use today as the standard of emf.

Although the assignment of values to electrical standards on the basis of an absolute system of units has been generally recognized as desirable since the initial proposal of the British Association, the difficulties encountered in absolute measurements led to rather large uncertainties in the values of the standards. This resulted in the adoption (1894) of an auxiliary set known as the "international" units, which were a "reasonable approximation" of the absolute units and which could, it was hoped, be experimentally reproduced with sufficient accuracy for measurement purposes. These units were defined by the resistance of a uniform column of mercury of specified length and mass and by the current required for the deposition of silver at a specified rate from a silver nitrate solution. The units defined in terms of the "mercury" ohm and the "silver" ampere could be established easily within a few hundredths of a percent, but presently there was need for greater accuracy in measurements. Fortunately the techniques needed in absolute measurement also improved, and it became possible to establish values of the electrical units within about 10 ppm by absolute methods.

Accordingly, on January 1, 1948, the *international* units based on the "mercury" ohm and "silver" ampere were abandoned, and *absolute* units were adopted by international agreement. These absolute units are defined identically with the meter-kilogram-second-ampere (mksa) units of the Systéme International (SI) formalized by international agreement in 1960.

In the 1948 reassignment of the electrical units, the International Committee rounded to 10 ppm the factors needed to convert from "mean international" units to "absolute" units. These mean international units were based on intercomparisons at the International Bureau (BIPM) of *ohm* and *volt* standards supplied by several countries. The absolute ohm and ampere determinations performed at various national laboratories were reduced to this common basis in establishing conversion factors.[2] Table 5c-1 may be used to compute the value in "1948 absolute" units of a quantity that is known in the "international" units previously used in the United States. The table takes account of the difference between the *mean* international

[1] Driscoll, *J. Research NBS* **60**, 287 (1958); Driscoll and Cutkosky, *ibid.* 297; Vigoureux, *Metrologia* **1**, 3 (1965).
[2] Curtis, *J. Research NBS* **33**, 235 (1944).

ohm and volt established by the intercomparison of national units at BIPM, and the international ohm and volt maintained at NBS and used in the United States. Corresponding tables based on the "international" units maintained by other countries would be slightly different because each country has maintained its own standards independently, and small differences developed over the years between the units of one country and those of another.

In October, 1968, the International Committee reviewed the results of recent absolute-ohm and ampere determinations with the following results. The unit of resistance maintained by BIPM appeared to be within 0.2 ppm of its intended value in absolute ohms, and no change was made in the assignment of the standards with which this unit is maintained. Also, on the basis of the 1967 international comparison of standards, BIPM reported that $\Omega_{NBS} = \Omega_{BIPM} - 0.19 \ \mu\Omega$. Thus, it appeared that the unit of resistance maintained by NBS differs from its intended value by less than 0.4 ppm; and therefore the values of the standards with which the NBS ohm is maintained were not reassigned. On the other hand, from recent ampere determinations it was concluded that the unit of voltage maintained by BIPM was too large by 11 ppm[1]; and

TABLE 5c-1. UNITED STATES VALUES (1948)

1 international ohm	= 1.000495 absolute ohms
1 international volt	= 1.000330 absolute volts
1 international ampere	= 0.999835 absolute ampere
1 international coulomb	= 0.999835 absolute coulomb
1 international henry	= 1.000495 absolute henrys
1 international farad	= 0.999505 absolute farad
1 international watt	= 1.000165 absolute watts
1 international joule	= 1.000165 absolute joules

its assigned value was decreased by this amount, effective January 1, 1969. BIPM reported, as a result of the 1967 intercomparisons that $V_{NBS} = V_{BIPM} - 2.58 \ \mu V$. Thus the NBS unit of voltage was brought to equivalence with that maintained by BIPM (on January 1, 1969) by increasing the numerical assignment of emf of the standard cells of the National reference group that maintains the NBS volt, by 8.4 ppm.

5c-3. Maintenance of the Electrical Units. The National Bureau of Standards is assigned the duty of establishing and maintaining the electrical units defined by an Act of Congress,[2] and used by science and technology in the United States. It also measures and reports values of electrical standards for other laboratories[3] in terms of the NBS units—the legal units of the United States.

The NBS *ohm* and *volt* are maintained by groups of wire-wound resistors and of standard cells whose group averages are assumed to remain constant with time. These groups constitute the *primary* electrical standards of the United States, and, in effect, all values of the various electrical quantities are derived from them. The constancy of these primary standards is checked at intervals by appropriate absolute-ohm and ampere determinations; and the constancy of their ratio is monitored by annual determinations of proton precession frequency in a magnetic field that can be related directly to the NBS ampere. The ratio of the NBS volt to ohm, thus determined, has not varied by as much as 1 ppm in the eight years (1968) during which this

---

[1] It should not be supposed that this entire amount represents a drift since 1948 in the emf of the standard cells with which the unit had been maintained; a considerable part of the discrepancy may be attributable to incorrect assignment in 1948, on the basis of ohm and ampere determinations then available, together with the Committee's decision to round the assignment to a part in 10^5.

[2] Public Law 617—81st Congress.

[3] This service is voluntary on the part of the organization requesting the test, as NBS has no police powers. Also, with some exceptions, a fee is charged covering the cost of the test.

test has been available. Monitoring the stability of the electrical units by relating them to an invariant atomic constant (proton gyromagnetic ratio) marks a significant step toward better maintenance.

Values of the NBS volt and ohm are also compared with the corresponding units of other countries every three years by intercomparison of standards at BIPM. Values of the NBS standards have been reassigned on occasion[1] in the conformity with international action, to bring the NBS units into closer agreement with their intended absolute values.

5c-4. Standards of Resistance. The primary standard of resistance in the United States is a group of ten 1-Ω resistors of special construction. The present group comprising the primary standard are of the Thomas[2] type, made in 1933. They were wound of No. 12 Awg manganin wire, vacuum-annealed at 500°C, and sealed in air in double-walled containers. The individual members of the group are intercompared annually. The maximum net change in any member of the group, with respect to the group average, has been a little over 2 ppm and the average about 1 ppm over the last 30 years (to 1968). They can be intercompared or can be compared with other similar standards to about 1 part in 10^7. By suitable comparisons and by series and parallel combinations, ratios can be established to a few parts in 10^7, and the range of resistance can be extended from the primary standards to higher and lower values. Secondary standards can be assigned values stepwise from the primary group to a maximum of 10^8 Ω and to a minimum of 10^{-5} Ω or less. Manganin (a Cu-Ni-Mn alloy) is generally used in resistance standards as sheet material for resistors of low value (below 1 Ω) and wire for resistors of higher value. Evanohm (a Ni-Cr-Al-Cu alloy) is coming into use for stable wire-wound resistors, particularly in the higher resistance range. In this application, its higher resistivity is of advantage. Both alloys have low-temperature coefficients of resistance (a few ppm/°C at most) when appropriately heat-treated, and both have low thermoelectric power against copper (2 to 3 μV/°C).

The stability of a resistance standard depends on its construction, the extent to which initial strains have been removed by annealing, freedom from strain in use, and protection of the resistance element against air and moisture. The construction of the Thomas-type standards in the primary group probably represents the best approach yet made to the ideal: complete elimination of initial strain by a high-temperature anneal in an inert atmosphere, practically strain-free mounting in use, a reasonably large ratio of volume to surface area, and protection by sealing from atmospheric effects. The stability of these standards is better by a factor of 10 or even 100$\times$ than that of the usual resistance standard. No general statement is possible concerning the stability of standards, except that those of higher value are usually less stable and that hermetic sealing provides better protection from atmospheric effects than other forms of enclosure.

For resistors having values of 10^9 Ω or more, wire-wound construction is not practical, and a film of resistive material on an insulating base is regularly used. Such resistors are inherently less stable than wire-wound units, and in many instances their values depend on impressed voltage, humidity, and other factors. Thus, accurately known stable standards are not available in the very high resistance region. The usual techniques of measurement can be applied, but with increasing difficulty, only up to about 10^{12} Ω, and higher values cannot be determined stepwise from the primary standard. Stability of selected hermetically sealed resistors up to 10^{14} Ω has been studied over a period of years by a rate-of-charge method.[3] Under specified measurement conditions, the best of this group of resistors appear to be drifting at rates that range from 0.5 percent per year for 10^{14}-Ω units to 0.1 percent per year for 10^9-Ω units.

[1] See Sec. 5c-2.

[2] Thomas, *J. Research NBS* **5**, 295 (1930); **36**, 107 (1946).

[3] Scott, *J. Research NBS* **50**, 1947 (1953).

5c-5. Standards of Electromotive Force. The *primary* standard consists of a group of 40 saturated cadmium (Weston) cells[1] which are maintained at a temperature of 28°C, held constant within 0.01°C. Of these cells, 33 are of the acid type, sulfuric acid being present in the electrolyte at a concentration of 0.03 to 0.05 N. The remainder of the group are neutral in the sense that no acid has been added. The presence of acid prevents hydrolysis of the mercurous sulfate in the cell and decreases the solvent action of the electrolyte on the glass container. Thus it contributes to the constancy of emf of the cell. However, the emf of an acid cell is lower than that of a neutral cell by an amount proportional to the concentration of the acid (30 μV for 0.05 N acid).[2] Of the cells which make up the primary standard, 7 have been in the group since 1906. Of the remainder, 7 made in 1932 and 26 made in 1949 were added in 1955. New cells are made periodically, employing carefully purified materials, and are used to supplement the primary group.[3] The cells of the primary standard are intercompared periodically, and the group average is used as the standard of emf. An international comparison made in 1967 indicated that, at that time, the United States standard differed by 2.6 ppm from the standard maintained by the International Bureau. The report stated that $V_{USA} = V_{BIPM} - 2.58$ μV. On the basis of recent absolute measurements, it was decided that V_{BIPM} was in error, and that the emfs assigned to the BIPM reference bank of standard cells should be increased by 11 ppm; simultaneously the emfs assigned to the NBS reference bank were increased by 8.4 ppm to bring them into agreement with the international standard. These changes became effective January 1, 1969.

Secondary standards of emf may be cadmium cells of either the saturated or unsaturated type. Most modern cells, both saturated and unsaturated, are the "acid" type, containing sulfuric acid at a normality of about 0.05 N. Saturated cells maintain a more nearly constant emf over long periods of time than unsaturated cells, and are being used to an increasing extent as reference standards in many laboratories. The temperature coefficient of emf of a standard cell is the difference between rather large positive and negative coefficients of the positive and negative cell limbs, respectively. For a saturated cell this difference amounts to about -50 μV/°C at 28°C while the coefficients of the separate limbs are more than 300 μV/°C each. Thus it is not only essential that the cell temperature be held within close limits (<0.02°C for assignment to 1 μV), but in addition temperature differences between the cell limbs must be held within much closer limits. The coefficients of the limbs of an unsaturated cell match much more closely (usually within 5 μV), but the requirement of small temperature difference between the cell limbs is just as strict as for a saturated cell. Temperature control for saturated cells can be maintained by a regulated oil bath or air bath. If oil is used, it should be clear, of medium viscosity, acid-free, and without appreciable vapor pressure. Air baths are used in a number of laboratories to maintain standard cells at a constant temperature. In one such construction,[4] the cells are within a thick-walled aluminum box which is enclosed by and thermally insulated from an outer aluminum box. The latter is thermally insulated to protect it against ambient temperature changes and is maintained at a constant temperature (within a few hundredths of a degree) somewhat above ambient (usually at a temperature between 28 and 37°C). Temperature fluctuations within the inner compartment are attenuated, and the cells are in a nearly constant temperature environment with very low gradients.

[1] Hamer, *NBS Monograph* 84 (1965).
[2] The initial small decrease of emf usually observed in a neutral cell during the first few months after it is made is believed to result from the formation of acid in the electrolyte.
[3] These supplementary groups are kept under the same conditions as the primary group and are regularly compared with them. Thus, if a cell in the primary group should fail, another cell with a known history of constancy could be used to replace it.
[4] Mueller and Stimson, *J. Research NBS* **13**, 699 (1933).

Recently, saturated cells have been developed in which the solid materials are mechanically held in place. These cells can be shipped with little danger of injury, whereas the older type of saturated cell must be hand-carried in transport. When these "transportable" cells are mounted in an air bath, it is advantageous to arrange the temperature control circuitry so that it can be operated alternatively from power lines or from a 12-V battery. Thus the bath temperature can be held constant during transport as well as in the laboratory; and the cells need not be subjected to temperature shock from which they would recovery slowly.

The international formula (adopted in 1908) relating the emf of a saturated cadmium cell to its temperature is

$$E_t = E_{20} - 40 \times 10^{-6}(t - 20) - 0.95 \times 10^{-6}(t - 20)^2 + 0.01 \times 10^{-6}(t - 20)^3$$

where E_t is the emf at temperature t, and E_{20} is the emf at 20°C. This formula is stated to apply to either neutral or acid cells and to hold within about 1 μV for temperatures between 0 and 40°C. However, it was developed for cells in which a 12 percent amalgam was used. As a 10 percent amalgam is now used almost universally, the formula of Vigoureux and Watts[1] is to be preferred. This may be stated, for cells with 10 percent amalgam, as

$$E_t = E_{20} - 39.39 \times 10^{-6}(t - 20) - 0.903 \times 10^{-6}(t - 20)^2 \\ + 0.0066 \times 10^{-6}(t - 20)^3 - 0.00015 \times 10^{-6}(t - 20)^4$$

These formulas should be used with caution since, in manufacture, the concentration of the amalgam may vary slightly from its intended value. Thus for correction to better than 1 μV, the correction formula should not be used over an extended temperature range; i.e., where the highest accuracy is required, the cell emf should be assigned in terms of a reliable reference standard at a temperature within, say 1°C of the use temperature. Since cells are frequently maintained at 28°C, it is convenient to express the Vigoureux and Watts formula in terms of that temperature:

$$E_t = E_{28} - 52.90 \times 10^{-6}(t - 28) - 0.803 \times 10^{-6}(t - 28)^2 \\ + 0.0018 \times 10^{-6}(t - 28)^3 - 0.00015 \times 10^{-6}(t - 28)^4$$

Unsaturated cells (becoming saturated at 4°C) are used extensively as working standards of emf. Their temperature coefficient of emf is less by a factor of 10 (usually $< -5\mu$V/°C at room temperature) than that of a saturated cell; and they can be shipped by express or parcel post, since the electrode material is held in place by porous plugs. The emf of an unsaturated cell decreases slowly with time; for cells of recent manufacture this decrease usually amounts to 20 to 30 ppm/year. Because of this change of emf with time, it is advisable to check them periodically against a stable standard, and to discard them when their emf has dropped below 1.0183 V.

Certain precautions should be observed in using standard cells.

1. They should be protected from large or sudden temperature changes, which are accompanied by a large temporary change in emf; recovery time is slow and depends on the magnitude and duration of the temperature shock. In the case of a 10°C shock it has been observed that recovery within 1 ppm may require more than two months.

2. Cells should not be exposed to nearby sources of heat that may produce temperature inequality between the limbs; this could produce a large change in emf, since the large positive and negative temperature coefficients of the individual limbs tend to cancel each other only if their temperatures are equal.

3. Saturated cells must not be exposed to temperatures above 43°C or below -8°C; a metastable change in the cadmium sulfate crystals takes place at 43.4°C, and at -8°C the two-phase liquid-solid amalgam becomes a one-phase solid amalgam.

---

[1] Vigoureux and Watts, *Proc. Phys. Soc. (London)* **45**, 172 (1933).

4. Cells should not be exposed to strong light, as the mercurous sulfate is photosensitive; some cells have a band of black paint on the positive limb to protect the mercurous sulfate.

5. The internal resistance of an unsaturated cell is about 500 Ω in the high-resistance type and 100 Ω in the low-resistance type; the latter should be used with a deflection potentiometer. The resistance of a saturated cell may be as much as 750 Ω. Loss of sensitivity in potentiometer measurements, which is traced to the standard-cell circuit, may indicate the presence of a gas bubble in the negative limb of the cell. If the bubble cannot be removed by gentle tapping, the cell should be discarded. A convenient way of measuring cell resistance without injury to it is a potentiometer determination of the difference in its terminal voltage with and without a 1-MΩ resistor across its terminals; the voltage difference in microvolts is equal to the cell resistance in ohms. The resistor should be connected across the cell only long enough to make the measurement.

6. Current drawn from a cell which is used as an emf standard should be kept small and drawn for only a few seconds at most. Current should never exceed 100 μA. A cell which has been short-circuited may be assumed to be permanently damaged and should be discarded. Circuit arrangements which permit an alternating current through the cell should be avoided, as this results in a temporary change in emf. Also to be avoided are laboratory conditions that could produce moisture condensation on the cell walls, with resultant leakage current supplied by the cell.

7. Cells which have been exposed to temperature shock or roughly handled during shipment can suffer a substantial change in emf, with a recovery period that may be of several weeks' duration. Evidence is accumulating that, for transportable-type cells which have been held at constant temperature during shipment, the recovery period is quite short, and the cells may recover their normal emf within 1 to 2 days.

Zener diode networks are coming into use as reference voltage standards for potentiometer applications.[1] As of 1969 zener packages were commercially available which are adequate for 0.01 percent use, and which are stable within 0.001 percent over short periods.

5c-6. Capacitance Standards. Capacitors whose values can be computed accurately from measured dimensions are necessarily small—at most 100 pF (picofarads)[2] or so—and are limited to three-terminal types in which air, or another gas, or vacuum is the dielectric. A three-terminal design is required both to define precisely the geometry of the active electrode system and to permit the solid insulation that supports the electrode arrangement to be so located that it does not influence either the computed

[1] Eicke, *Trans. ISA* **3**, 93 (1964).

[2] A list of prefixes, to represent powers of 10, sponsored by the International Union of Physics and approved by the International Electrotechnical Commission (05-35-080 in *IEC Publ.* 80), has gained wide acceptance both abroad and in the United States. These prefixes and the corresponding powers of 10, adopted by the National Bureau of Standards, are given in the table below.

Prefix	Value	Prefix	Value
Tera	10^{12}	Deci	10^{-1}
Giga	10^{9}	Centi	10^{-2}
Mega	10^{6}	Milli	10^{-3}
Kilo	10^{3}	Micro	10^{-6}
Hecto	10^{2}	Nano	10^{-9}
Deka	10^{1}	Pico	10^{-12}

capacitance or the quadrature relation between current and voltage. Until recently such capacitors were usually built as a parallel-plate guard-ring arrangement, or as a system of coaxial guarded cylinders; and the accuracy of the assigned value was at best around 0.01 percent. A modification of the parallel-plate guard-ring design has been used to construct computable "guard-well" capacitors in the range below 1 pF (down to 0.001 pF). In this construction the working electrode is recessed behind the plane of the guard ring to take advantage of the concentration of field at an exposed sharp edge. The capacitance is a function of depth of the recess,[1] and can be as small as desired, while linear dimensions remain large enough for precise construction and measurement.

The situation as regards computable capacitors has changed completely since 1956, thanks to a new theorem in electrostatics.[2] The Thompson-Lampard theorem may be generalized as follows: "If four infinite cylindrical conductors of arbitrary sections are assembled with their generators parallel, to form a completely enclosed hollow cylinder in such a way that the internal cross-capacitances per unit length are equal, then in vacuum these cross-capacitances are ln $2/4\pi^2$ statfarads per cm." For small inequalities of cross-capacitances, it can be shown that the departure of their mean from the theoretical value is 0.087 $(\Delta C/C)^2$, where ΔC is their difference and C their mean. Thus, if the inequality is no more than 0.1 percent, the error in the computed value is less than 1 in 10^7. A practical realization of such a capacitor consists of four equal closely spaced round cylindrical rods with their axes parallel and at the corners of a square. Arranged as a three-terminal capacitor, the internal cross-capacitance amounts to 1 pF for approximately 50 cm length of the assembly and, if end effects are eliminated,[3] this capacitance can be computed as accurately as the length can be measured. In an early arrangement, made of an assembly of cylindrical gage rods, the uncertainty was stated as 2 ppm. In a later version of the Thompson-Lampard cross-capacitor, in which the active length is the distance between screen electrodes inserted in the central opening from both ends of the assembly,[4] the effective length is determined in a Fabry-Perot interferometer formed by optical flats on the opposed ends of the screen electrodes. Accuracy of length measurement and of capacitance assignment is an order of magnitude better than for the earlier gage-rod assembly. Bridges with transformer ratio arms[5] and high-resolution detectors[6] are now available with which capacitors can be intercompared with a precision approaching a part in 10^8 at the 1-pF level or can be stepped up by successive factors of 10 from this level without significant loss of precision. Thus the assignment of values to three-terminal capacitors in the range 1 to 10^6 pF is limited only by the stability of the unit. In addition to time dependence of stability, many capacitors have a value which is voltage-dependent. This voltage effect can vary within wide limits, depending on the construction of the unit. In some solid-dielectric capacitors, a change of several hundredths of a percent in value between low and rated voltage is observed; in others it can be substantially less than 1 ppm. In low-value air-dielectric capacitors whose dimensions are not altered by the electrostatic forces resulting from the applied voltage, the change in capacitance from low to rated voltage may be as small as a few parts in[7] 10^9. The best fixed-value solid-dielectric capacitors commercially available in 1969 showed drift rates that range upward from a few

[1] Snow, *NBS J. Research* **42**, 287 (1949).

[2] Lampard, *Proc. IEE* **104c**, 271 (1957); first announcement, *Nature*, **177**, 888 (1956).

[3] McGregor et al., *Trans. IRE* **7-I**, 253 (1958).

[4] Clothier, *Metrologia* **1**, 36 *LM17*, 232 (1965).

[5] Thompson, *Trans. IRE* **7-I**, 253 (1958); Cutkosky and Shields, *Trans. IRE* **9-I**, 243 (1960).

[6] Cutkosky, *IEEE Trans. IM*, 232 (1968).

[7] Shields, *J. Research NBS* **69C**, 265 (1965); Kusters and Petersens, *IEEE Trans.* **82,** (*Commun. & Electronics*) (1963).

ppm per year. A group of 10-pF reference-standard capacitors constructed at the National Bureau of Standards[1] have substantially less drift; over a three-year period their drift has been less than 1 in 10^7 per year. The voltage dependence of these capacitors is less than 1 in 10^8 up to 200 V.

Air capacitors (in both two- and three-terminal form up to 10^3 pF) are available as secondary standards. Adjustable standards in this range have a group of movable parallel plates that rotate with respect to a group of fixed interleaved parallel plates; the accuracy of adjustable air capacitors depends on the closeness with which the relative angular position of the fixed- and movable-plate systems can be set and reproduced, and on the quality of the bearings on which the electrode system rotates.

The phase-defect angles of air capacitors are always small, depending largely on the extent to which solid dielectric is present in the working field and to a much lesser extent on the presence of surface films[2] on the electrodes.

Solid-dielectric capacitors in which thin mica sheets are interleaved with metal foil are used as working standards up to 1 μF. The assembly is impregnated with wax to eliminate voids and air pockets, and is compressed through massive end plates to squeeze out excess wax. The quality and constancy of such a standard depend critically on the construction, being a function of the assembly pressure as well as the quality of the mica. In an alternative construction, mica sheets are coated with a thin layer of silver and stacked with interleaved metal foil. The assembly is not wax-impregnated but is sealed in a dry atmosphere. A general characteristic of mica capacitors is their increase in capacitance with increasing voltage. In the power and a-f range, this increase may be less than 10 ppm between low and rated voltage for the best wax-impregnated mica capacitors and has been observed to be in excess of 400 ppm in some silvered-mica capacitors. As the effect varies markedly from one unit to another, individual capacitors should be checked before they are used as working standards. Absorption and losses are always present in mica capacitors. The phase-defect angle of the best mica capacitors may amount to 1 to 2 minutes throughout the a-f range, and their capacitance value may be expected to remain constant within 0.01 and 0.02 percent over a period of many years. While mica is used in solid-dielectric capacitors of the best grade, polystyrene is used extensively in secondary standards. The phase-defect angle of such capacitors may be no larger than of mica capacitors of comparable size, but their temperature coefficient of capacitance is more than $4\times$ as great.

5c-7. Inductance Standards. Self- and mutual inductors, whose values can be computed from measured dimensions, have been built at the National Bureau of Standards and at other national laboratories for use in absolute-ohm measurements. Computable self-inductors are single-layer solenoids wound on marble, porcelain, low-expansion glass, or fused silica forms. In some instances an accurate screw thread has been cut into the cylindrical form to control the spacing of the winding.[3] Computable mutual inductors have been built following a design of Campbell[4] or Wenner's modification of it.[5] In each of these designs the primary consists of single-layer helical windings on a marble or porcelain cylinder, the sections being spaced in such a way that a relatively large annular space is available around the central portion of the cylinder, within which the field is very small. The multilayer secondary winding is located in this space, and since the field is small, the exact location of the secondary becomes relatively less critical. Such mutual inductors can be computed as accurately as can the self-inductance standards. However, both types of inductor, apart from being

[1] Cutkosky and Lee, *J. Research NBS* **69C**, 173 (1965).
[2] Koops, *Philips Tech. Rev.* **5**, 300 (1940).
[3] Curtis, Moon, and Sparks, *J. Research NBS* **21**, 371 (1938).
[4] Campbell, *Proc. Roy Soc. (London)*, ser. A, **79**, 428 (1907).
[5] Thomas, et al., *J. Research NBS* **43**, 325 (1949).

very difficult and expensive to build and compute, have relatively low time constants and are not generally useful for work outside the special field (absolute measurements) for which they are designed.

Self-inductance standards for laboratory work are usually multilayer coils of such shape that their inductance is maximum for a given size and length of wire.[1] Accurate computation of value from their measured dimensions is not possible, and their values are usually established from electrical measurements in terms of other inductors or a combination of resistance and capacitance. Laboratory mutual inductors also are usually designed to achieve a maximum time constant.

Higher inductance in a given volume or with a given amount of copper can be obtained if the winding is on a core of high-permeability material. Special ferromagnetic alloys are used for this purpose in sheet or strip form, or as a bonded granular or powder material. The gain in time constant is achieved at the expense of some nonlinearity in the inductor, since the core permeability is a function of current in the winding. Also, increased losses are to be expected from eddy currents and from hysteresis in the iron. By proper construction and the use of suitable core materials, these defects can be kept small, so that "iron-cored" inductance standards of moderate accuracy and stability are practicable.

Inductors wound as multilayer cylindrical coils of rectangular cross section, to achieve maximum time constant, set up an external field and, conversely are subject to pickup from stray fields in which they are placed. These effects are considerably reduced by dividing the coil into two equal sections wound in opposite directions so that the emfs induced in them by a changing external field tend to cancel. Such an arrangement is called *astatic*. A much greater degree of astaticism is attained when the coil is toroidal, with the winding uniformly distributed around the torus.

Adjustable standards of self- and mutual inductance are of two general kinds: the cross-coil type, in which the plane of a movable coil is turned to make various angles with the plane of the fixed coil; and the parallel-coil type, in which the plane of the movable coil is always parallel to the fixed coil. A familiar example of the cross-coil type is the Ayrton and Perry inductometer, in which the fixed and movable coils are zones of concentric spheres, with the movable coil pivoted on the common polar axis. Such an arrangement has serious faults. If the coils are nearly equal in radius to maximize their coupling, the rate of change of mutual inductance with angle is far from uniform; and the arrangement is not astatic. Probably the best example of the parallel-coil type is the Brooks inductometer with three pairs of link-shaped coils, designed to provide a uniform scale over most of its range. The coil dimensions are such that, at the maximum reading, the conditions for maximum time constant are approximately met. Also the system is arranged to be nearly astatic. The rotor, holding the movable pair of coils, turns on a shaft between pairs of fixed parallel coils. The coils are all connected in series when the instrument is to be used as a self-inductor; for use as a mutual inductor the circuits of the fixed and movable coils are separated.

All inductors are frequency-dependent as a result of distributed capacitance, eddy currents, and imperfect insulation between turns and layers of the winding. The effect of distributed capacitance is to increase both the effective resistance and inductance above their low-frequency values. At frequencies well below resonance the following formulas hold approximately:

$$R_{\text{eff}} = R_0(1 + 2\omega^2 L_0 C) \quad \text{and} \quad L_{\text{eff}} = L_0(1 + \omega^2 L_0 C)$$

where R_0 and L_0 are the values at zero frequency, and C is the equivalent capacitance considered to be connected across the terminals of the inductor. The effect of eddy

[1] Brooks, *J. Research NBS* **7**, 293 (1931).

currents is to increase the effective resistance and to decrease the effective inductance in accordance with the following formulas:

$$R_{\text{eff}} = R_0 + \frac{M^2\rho\omega^2}{\rho^2 + l^2\omega^2} \quad \text{and} \quad L_{\text{eff}} = L_0 - \frac{M^2l\omega^2}{\rho^2 + l^2\omega^2}$$

where ρ and l are, respectively, the equivalent resistance and self-inductance of the eddy-current circuit, and M is its coupling with the inductor. The effect of imperfect insulation (equivalent to a shunt resistance across the terminals of the inductor) is to decrease the effective inductance. However, it may increase or decrease the effective resistance,[1] depending on conditions. If the leakage resistance ρ is very high compared with the coil resistance, the following formulas hold:

$$L_{\text{eff}} = L_0 \left(1 - \frac{\omega^2L_0^2}{\rho^2}\right) \quad \text{and} \quad R_{\text{eff}} = R_0 \left(1 + \frac{\omega^2L_0^2}{R_0\rho}\right)$$

It must be borne in mind that ρ is the a-c resistance of the insulation and hence may be itself a function of frequency.

5c-8. Frequency Standards. The reciprocal relationship between frequency and time interval makes it desirable to examine first the definition of the *second*. Previous to 1956 the second was defined in terms of the earth's rotation—the second was 1/86,400 part of a mean solar day. As frequency standards improved, this definition became unacceptable because of small variations in the earth's rotational motion. Because the earth's orbital frequency around the sun is less subject to perturbations than its rotational frequency, the second was redefined in 1956 by international agreement, as the fraction 1/31,566,925.9747 of the tropical year 1900, beginning at 12 hr ephemeral time. Obviously, the precise determination of a frequency in terms of such a "second" requires a lengthy and involved set of observations, and, as atomic frequency standards improved greatly during the decade that followed, it became possible to measure frequencies in terms of these standards quickly and far more precisely. Thus, by international agreement in 1967 the *second* was redefined as the duration of 9,192,631,770 periods of the radiation corresponding to the transition between the two hyperfine levels of the ground state of the cesium 133 atom.

The National Bureau of Standards maintains standard frequency services from four transmitting stations: WWVH in Hawaii; WWV, WWVB, and WWVL in Colorado.[2] WWVH broadcasts on frequencies of 2.5, 5, 10, and 15 MHz; WWV on 2.5, 5, 10, 15, 20 and 25 MHz; WWVB on 60 kHz; and WWVL on 20 kHz. Frequencies transmitted by WWV (derived from cesium-controlled oscillators) are held stable to better than 2 parts in 10^{11} at all times, with deviations much less than 1 part in 10^{11} from day to day; frequencies at WWVH are derived from quartz oscillators, and frequency adjustments do not exceed 5 parts in 10^{10}. Both these stations broadcast standard audio frequencies of 440 and 600 Hz on each carrier frequency. WWVB and WWVL frequencies (cesium-controlled) are normally stable to better than 2 parts in 10^{11}, with day-to-day variations less than 1 part in 10^{11}. WWVB frequency has no offset; the others are all intentionally offset (-300 parts in 10^{10} in 1968) from standard frequency to reduce the departure of their broadcast time signals from UT2 astronomical time. Changes in the propagation medium (causing Doppler effect, diurnal shift, etc.) result in fluctuations of the order of a part in 10^7 in the HF carrier frequencies as received; the effects of the propagating medium on the received frequencies are much less at LF and VLF, and the full transmitted accuracy may be obtained with appropriate receiving techniques.

---

[1] Campbell and Childs "Measurement of Inductance, Capacitance, and Frequency" p. 191, D. Van Nostrand Company, Inc., Princeton, N.J., 1953.

[2] *NBS Spec. Publ.* 236.

Quartz crystals are used in vast numbers to control the frequencies of oscillators through much of the spectrum, in both measurement and communication applications. Their constancy depends on the closeness with which their temperature and pressure are controlled.

Tuning forks may be used as laboratory standard at power and audio frequencies. A precision fork, operating at constant temperature, may have a frequency that is stable to 10 ppm and, when corrected for barometric pressure, to 1 ppm. A battery-driven fork without temperature control may have a temperature coefficient less than -0.015 percent/°C and a voltage coefficient less than 0.01 percent/V. It should provide a frequency known to better than 0.1 percent under any specified laboratory condition.

The frequency of 60-Hz power in most localities affords a convenient reference point. However, even where power is supplied from a network that includes generating stations over an area of many hundreds of square miles, the frequency is not continuously held at precisely 60 Hz. It may depart by as much as 0.1 or 0.2 Hz, occasionally even more. Also, the frequency can be corrected only very slowly because of the large inertia of the system, perhaps as much as half an hour being required. The average frequency will be very close to 60 Hz over an extended time period, and synchronous clocks will usually keep time within a few seconds. However, a commercial power source cannot be employed reliably as a frequency standard to much better than 1 percent.

5c-9. Deflecting Instruments. Instruments used for the measurement of current, voltage, or power are made in a number of accuracy classes. The best grades, called "laboratory standards," may be in the $\frac{1}{10}$ (or $\frac{1}{20}$) percent class, meaning that, over the useful part of the scale, no marked point is in error by more than $\frac{1}{10}$ (or $\frac{1}{20}$) percent of the full-scale value. These are large instruments which must be carefully leveled to ensure good performance. Smaller portable instruments are made in 0.2, 0.5, and $\frac{3}{4}$ percent accuracy classes. The class of the instrument is generally stated in the maker's catalog. Switchboard instruments are generally in the 1 percent class, and panel instruments in the 1, 2, or even 5 percent class. Direct-current ammeters and voltmeters are almost universally permanent-magnet moving-coil instruments, whereas the construction of a-c instruments depends on the intended application. Moving-iron or electrodynamic instruments are used at power frequencies and, if suitably compensated, in the lower a-f range. Thermocouple ammeters are useful from low frequencies up to many megahertz, whereas thermocouple voltmeters are generally applicable only at power and audio frequencies unless they have special multipliers designed for high-frequency operation. Electrostatic voltmeters have no frequency limitation other than that imposed by low impedance at very high frequencies, and many electronic voltmeters are designed to operate from power frequencies up to many megahertz without serious error. Depending on operating principle and construction, deflecting instruments are subject to errors of various types: temperature, magnetic field, frequency, waveform, spring hysteresis, use in other than the intended position, and others.[1]

References

Campbell and Childs: "Measurement of Inductance, Capacitance, and Frequency," D. Van Nostrand Company, Inc., Princeton, N.J., 1935.

Curtis: "Electrical Measurements," McGraw-Hill Book Company, New York, 1937.

Drysdale and Jolley: "Electrical Measuring Instruments," 2d ed., John Wiley & Sons, Inc., New York, 1952.

[1] Standard C-39 of the USA Standards Institute contains performance specifications for deflecting instruments of various types and accuracy classes. Sections of C-39 also deal with electronic and digital instruments. A text on electrical instruments or measurements should be consulted for complete details on instrument performance.

Golding: "Electrical Measurements," Pitman Publishing Corporation, New York, 1946.
Harris: "Electrical Measurements," John Wiley & Sons, Inc., New York, 1952.
Historical Reports of the Committee on Electrical Standards Appointed by the British Association for the Advancement of Science, Cambridge University Press, London, 1913.
Keinath: "Die Technik elektrischer Messgerate," R. Oldenbourg Verlag, Munich, 1928.
NBS Handbook 77, vol. 1, Electricity and Electronics.
Procès Verbaux, Comité International des Poids et Mesures.
Public Law 619, 81st Congress, Ch. 486, 2d. Sess., Title 15, USCA221.
Vinal: "Primary Batteries," John Wiley & Sons, Inc., New York, 1950.

5d. Properties of Dielectrics

G. L. LINK AND D. B. HERRMANN

Bell Telephone Laboratories, Inc.

Dielectrics store electric energy and the intrinsic property of the material which measures this ability is the *dielectric constant* or *permittivity* ϵ'. (The term *constant* refers to the ϵ' independence of field strength; as shown below, ϵ' usually does depend on temperature, frequency, and other parameters.)   Part of the electric energy put into a dielectric is not recoverable from storage; i.e., it is "dissipated."   The intrinsic property which measures dissipation is the *loss factor* (index), ϵ''.   Other convenient parameters expressing this information include $\tan \delta = \epsilon''/\epsilon' = 1/Q$.   Typical values of $\tan \delta$ are of order 10^{-2}, with low-loss materials being of order 10^{-4} and high loss of order one or higher.

Since ϵ' measures motions at all higher frequencies, and ϵ'' measures motion at a particular frequency, it is not surprising to find a relation between ϵ' and the integral of ϵ'' over frequency.   This, the Kramers-Krönig relation, is

$$\epsilon'(f) - \epsilon_\infty = \frac{2}{\pi} \int_0^\infty \epsilon''(F) \frac{F \, dF}{F^2 - f^2}$$

$$\epsilon''(f) = -\frac{2}{\pi} \int_0^\infty (\epsilon'(F) - \epsilon_\infty) \frac{f \, dF}{F^2 - f^2}$$

$$\epsilon_\infty = \epsilon' \text{ at high frequency}$$

For the special case of $f = 0$ (i.e., very low frequency) this reduces to

$$\epsilon_0 - \epsilon_\infty = \frac{2}{\pi} \int_{-\infty}^\infty \epsilon'' d \ln F$$

The optical ϵ is determined by electronic polarizability and is the square of the (complex) refractive index.   Since this polarizability is really a measure of the size of the electron cloud around an atom, and since this size is about constant in most covalent compounds, it is possible to estimate ϵ_∞ for such compounds simply by adding individual contributions (provided we know the density and chemical structure of the compound), as illustrated in Table 5d-1.   The dependence of ϵ on density, ρ, is given by the Clausius-Mossotti (Lorentz-Lorenz) equation which is especially useful in estimating $\epsilon(t)$ when $\rho(t)$ is known (where $t°C$ is the temperature):

$$\frac{\epsilon - 1}{\epsilon + 2} = \frac{4\pi}{3} \frac{\rho}{M} \alpha$$

where ρ is density, M is molecular weight, and α is (electronic) polarizability. [The general validity of the $\epsilon - 1$ proportional to ρ relation is illustrated by comparing gas and liquids where typical ρ's differ by a factor of 10^3 as do the $(\epsilon - 1)$'s.] In practice, this relation holds to about 1 percent accuracy. More accurate relations require a third parameter; several are described by Brown in his lucid synopsis of dielectric theory.[1]

The electronic effects just described occur at such high (optical) frequencies because of the small mass of the electrons. These resonant dispersions are inertial, depending only on the electron's ability to rapidly respond to the imposed electric field variations. Any molecule with an asymmetric charge distribution constitutes a permanent dipole which also can respond to an alternating electric field, but because of the mass of the atoms the natural (inertial) period of motion will be much longer and hence dispersions occur at lower frequencies.

In the condensed phase, molecular motion is typically determined by viscous (not inertial) forces, and the corresponding relaxation (not resonant) dispersions occur at microwave, radio, and even lower frequencies. These dispersion processes result from the (slow) motion of the molecule being coupled to the alternating electric field by an attached permanent dipole. (Relaxation processes are characterized by ϵ' which monotonically decreases with increasing frequency throughout the entire dispersion range, unlike resonant dispersions.)

In addition to molecular motion in the condensed phase, it is also possible to have ionic motion. The d-c conduction caused by ions will cause ϵ'' to vary inversely with the frequency while ϵ' remains constant. If the ions are constrained, as frequently happens in inhomogeneous materials, the resulting induced "dipole" gives rise to a (usually low frequency) dispersion with appropriate changes in both ϵ' and ϵ''. (See Figs. 5d-2 and 3.)

The temperature dependence of ϵ' can often be estimated a priori. In the gas phase $\epsilon - 1$ is frequently proportional to the gas density and for nonpolar molecules is independent of frequency. Discernible frequency dispersions occur in polar molecules, e.g., water, in the microwave region.

In the condensed phase $\epsilon - 1$ is again proportional to density in nonpolar substances so that $d\epsilon/dt$ is small and negative except in the vicinity of phase transitions. For polar molecules, $d\epsilon/dt$ is usually negative, although it can be large and positive in that temperature range on the cool side of a dispersion. (In this case the $e^{-H/kT}$ dependence of dipolar motion dominates the smaller temperature dependence of the density, the large values of $d\epsilon/dt$ resulting from the high dipole concentration and consequent large values of $\epsilon_s - \epsilon_\infty$.) Phase transitions in polar molecules can be accompanied by dramatic drops in ϵ' as the dipole motion is frozen out. (These drops tend to be sharper when the molecule is more symmetrical, as illustrated by the nearly spherical camphor and its derivatives as contrasted to asymmetric derivatives of the planar benzene molecule, as illustrated in Figure 5d-1.)

Very high molecular weight materials (polymers) often form supercooled liquids rather than freeze into crystals, and consequently do not have discontinuities in $\epsilon(t)$. In those polymers that are partially crystalline, it is usually possible to discern two $\epsilon''(t)$ or $\epsilon''(f)$ loss peaks, one corresponding to each phase in which the relative ampli-

[1] W. F. Brown, Jr., "Handbuch der Physik," vol. 17, Springer-Verlag OHG, Berlin, 1956 (in English).

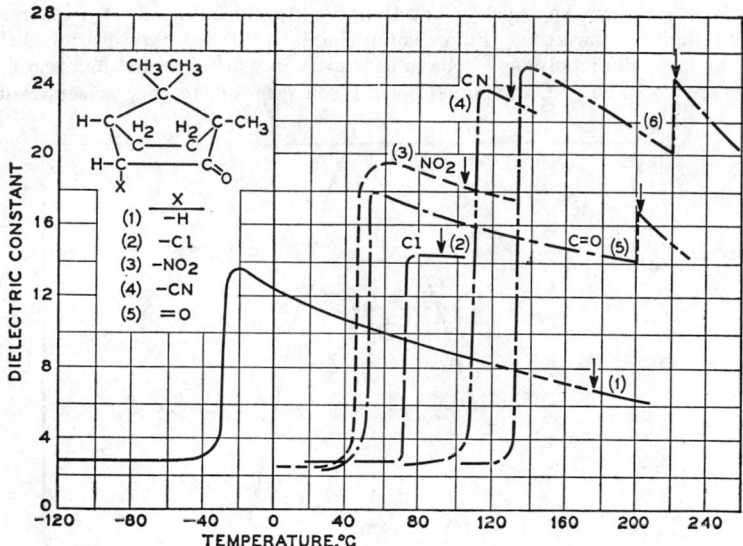

FIG. 5d-1a. Dielectric constant of camphor (1), chlorocamphor (2), nitrocamphor (3), cyanocamphor (4), camphor quinone (5), and camphoric anhydride (6). Heavy arrow indicates the melting point; values are independent of frequency below 100 kHz. [(*From Morgan and Lowry, J. Phys. Chem.* **34**, 2385 (1930).]

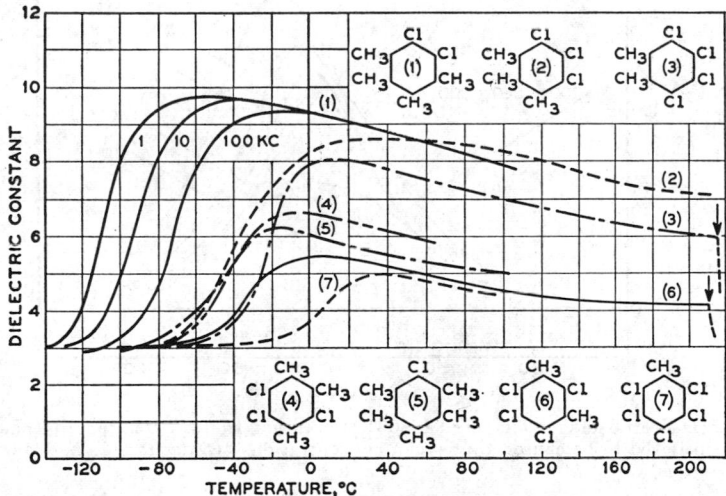

FIG. 5d-1b. Dielectric constant of polar hexasubstituted chloromethylbenzenes at 100 kHz: (1) dichlorophrenitene, (2) trichlorohemimellitene, (3) tetrachloro-*o*-xylene, (4) trichloro-pseudocumene, (5) pentamethylchlorobenzene, (6) tetrachloro-*m*-xylene, (7) pentachloro-toluene. [*From A. H. White and S. O. Morgan: J. Am. Chem. Soc.* **57**, 2078 (1935).]

tudes of ϵ'' depend on the degree of crystallinity (e.g., in Table 5d-7A, r_1 decreases as the crystallinity is lowered). Below some characteristic temperature for each substance, noncrystalline polymers behave as a glass in which all but the easiest dipole motions are frozen out. Commercial plastics sometimes have "plasticizers" added to

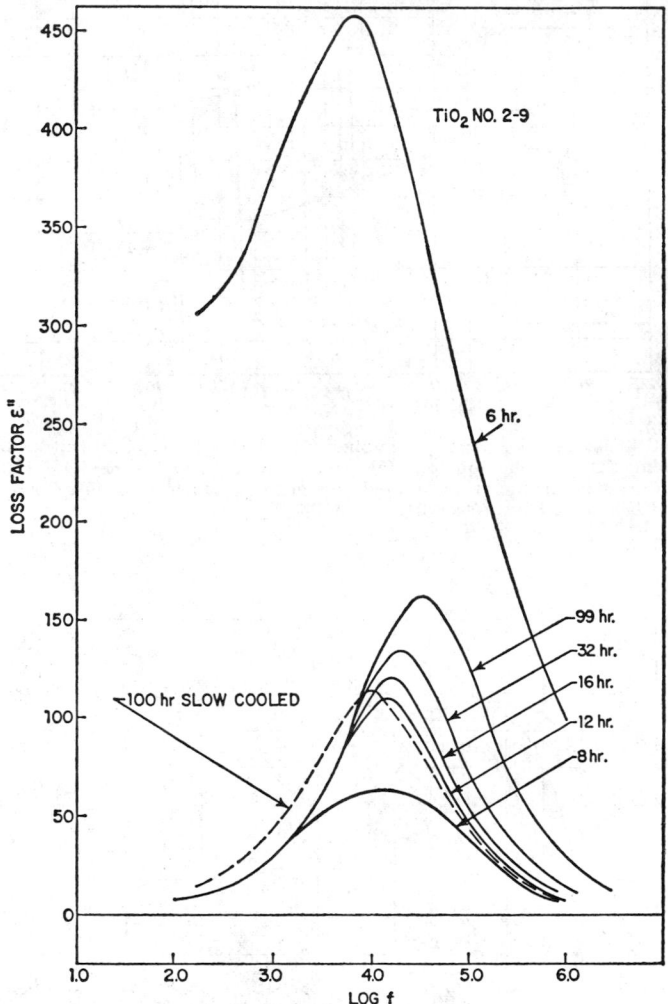

FIG. 5d-2. Dielectric loss for TiO_2 fired at 800°C for various times. Different loss characteristics are believed to be caused by oxygen concentration. (*Data from L. Egerton, private communication.*)

them to keep them less brittle to lower temperatures. Such additions usually raise the frequency for the peak $\epsilon''(f)$ by an amount dependent on the plasticizer concentration. Plastics also have other additives to stabilize them against chemical degradation and in the case of low-loss polymers these additives raise the value of ϵ'', giving considerable (greater than two times) variation in ϵ'' between different manufacturers of the "same" material. (Other differences result from different topological structures and impurities.)   Polar polymers have varying affinities for water, causing some to have dielectric properties which depend on the humidity of the environment.   Polymers that are chemically joined by a (loose) structure of three-dimensional bonds are not able to collapse to the glassy state and so remain "rubbery" even at low temperatures, with corresponding dielectric properties.

The last class of synthetic materials we consider are non-single-crystal metallic oxides.   These inorganic materials are characterized by high ϵ' (because of the high

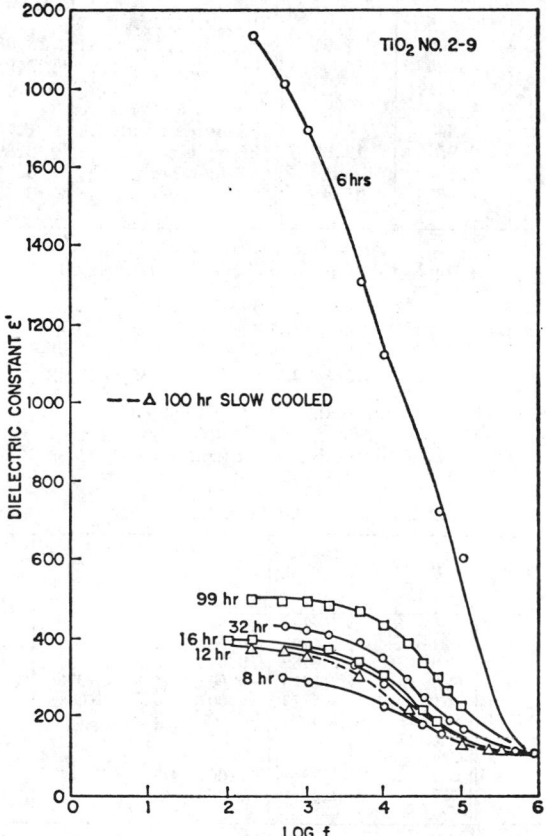

FIG. 5d-3. Dielectric constant corresponding to Fig. 5d-2.

polarizability of oxygen and the high density) and are stable at higher temperatures than organic materials.   One class of materials is known as (inorganic) "glasses"; a second class consists of small (about micrometer) crystallites joined together with binders or by sintering and is known as "ceramics."   Both are usually considered isotropic but under some conditions of formation are definitely anisotropic.   Both also are subject to dielectric dispersions caused by nonstoichiometric compositions (e.g., see Figs. 5d-2 and 5d-3), and at high temperatures the d-c conductivity may become important.

With this introduction it should be clear that the following listings are not intended to be exhaustive; rather, they are intended to be illustrative of the different types of

TABLE 5d-1. MOLAR REFRACTION—ESTIMATION OF OPTICAL FREQUENCY ϵ'

$$R = \frac{M}{\rho} \cdot \frac{\epsilon_\infty - 1}{\epsilon_\infty + 2} = \text{molar refraction}$$

$$\epsilon_\infty = \frac{1 + 2\rho R/M}{1 - \rho R/M} \quad \text{where } R = \Sigma r \text{ and } M = \Sigma m$$

Atom (structure)	r	m	Atom (structure)	r	m
C................	2.418	12.01	Br................	8.865	79.91
H................	1.100	1.01	I................	13.900	126.90
O (alcohol)........	1.525	16.00	S................	7.9	32.06
(carbonyl).......	2.211		N................	2.5–4.4	14.01
(ether)..........	1.643		Structure effects:		
(ester-OR).......	1.64		Double bond....	1.733	
F (one/carbon)	0.95	19.00	Triple bond.....	2.398	
($>$one/carbon) ..	1.1		3-member ring...	0.71	
Cl................	5.967	35.45	4-member ring...	0.48	

Illustration

Heptanol, $\rho(20°C) = 0.824$, $CH_3(CH_2)_5CH_2OH$

7 C	$7(2.418) = 16.926$	$7(12.01) =$	94.07
16 H	$16(1.100) = 17.600$	$16(1.01) =$	16.16
1 O$_{alc}$.	$1(1.525) = \underline{\;1.525\;}$	$1(16.00) =$	$\underline{16.00}$
	$R = 36.051$	$M =$	116.23

$$\rho R/M = 0.2556 \quad \epsilon_\infty = 2.03$$

Cf. $n^2 = 2.03 \quad \epsilon_\infty = 2.35$ (microwave)

n = refractive index (sodium D)

TABLE 5d-2. REFERENCE FLUIDS

Gases [1]	$(\epsilon - 1) \times 10^6$ (20°C, 1 atm)*	Nonpolar liquids	$\epsilon'(23°C)$	$a \equiv -d\epsilon'/dt$	Ref.	Polar liquids [3]	$\epsilon'(20°C)$	$\alpha \equiv -d \log_{10} \epsilon'/dt$
H$_2$........	253.8 ± 0.3	n-Hexane	1.8829	-0.0015	2	ϕCl	5.708	0.0013$_3$
He........	65.0 ± 0.4	Cyclohexane	2.0182	-0.0016	3	EtCl$_2$ª	10.65	0.0024$_0$
O$_2$........	494.7 ± 0.2	CCl$_4$	2.2315	-0.0020	4	MeOH^b	33.62	0.0026$_0$
N$_2$........	548.0 ± 0.5	ϕH†	2.2772	-0.0020	5	ϕNO$_2$	35.74	0.0022$_5$
Ar........	517.2 ± 0.4	Silicone	2.3000	-0.0028	5			
Air (dry, CO$_2$-free)	536.4 ± 0.3	(1 cs)‡ Cryogenic:						
		H$_2$§	1.228	-0.0034	3			
		O$_2$¶	1.507	-0.0024	3			

* $\epsilon(P,t)$: $\dfrac{(\epsilon - 1)(t,P)}{(\epsilon - 1)(20°C, 1 \text{ atm})} = \dfrac{(P)}{1 + 0.003411(t - 20)}$

P = atm, t = °C

0.02 % error for 0 to 30°C, and P = 1 to 0.1 atmo.

† ϕ represents a benzene ring minus one hydrogen. ϕH is benzene.

‡ Centistoke.

§ $\epsilon'(20.4$ K$)$

¶ $\epsilon'(80.0$ K$)$

ª Et = C$_2$H$_5$

^{b} Me = CH$_3$

References for Table 5d-2

1. *NBS Circ.* 537, 1953.
2. Mopsik, F. I.: *J. Research NBS* **71A**, 287 (1967).
3. *NBS Circ.* 514, 1951.
4. Hartmann, H., A. Neumann, and G. Rinck: *Z. phys. Chem.* **44**, 204 (1965).
5. Unpublished results of Bell Telephone Laboratories.

TABLE 5d-3. INORGANIC COMPOUNDS (SMALL MOLECULES)

Type/Name	ϵ'	t, °C	a (or α)	Range	Type/Name	ϵ'	t, °C	a (or α)	Range
Elements					*Hydrides*				
A	1.53_8	-191	0.0034	$-191, -184$	NH₃	25.	-77.7		
H₂	1.22_8	20.4°K	0.0034	14, 21K		22.4	-33.4		
D₂	1.277	20°K	0.004	18.8, 21.2K		18.9	5		
He	1.055_9	2.06°K				17.8	15		
	1.055_9	2.30				16.9	25		
	1.005_3	2.63				16.3	35		
	1.053_9	3.09			N₂H₄	52.9	20	$0.0021(\alpha)$	0, 25
	1.051_8	3.58			AsH₃	2.50	-100	0.0043	$-116, -72$
	1.048	4.19			SbH₃	$2.9_3{}^{B}$	-80		
O₂	1.507	-193	0.0024	$-218, -183$		$2.5_8{}^{B}$	-50		
N₂	1.454	-203	0.0029	$-210, -195$	PH₃	$2.5_5{}^{B}$	-60		
F₂	1.54	-202	0.0019	$-216, -190$		$2.7_1{}^{B}$	-25		
Cl₂	2.10_1	-50	0.0031	65, -33	*Halides*				
	1.9_1	14	0.0032	$-22, 14$	HF	$17._5$	-73		
	1.7_3	77				$13._4$	-42		
	1.5_4	142				$11._1$	-27		
Br₂	3.09	20	0.007	0, 50		8.4	0		
I₂	$11._1$	118			HCl	12.	-113		
	$11._7$	140				6.35	-15	$0.00288(\alpha)$	$-85, -15$
	$13._0$	168				4.6	28		
S	3.52	118			HBr	7.00	-85	$0.0026(\alpha)$	$-85, -70$
	3.48	231				3.8^{B}	$+25$		
Se	5.40	250	0.0025	237, 301	HI	3.39	-50	0.008	$-51, -37$
P	4.10	34				2.9^{B}	22		
	4.06	46			SbCl₃	$33.^{B}$	75.		
	3.86	85			SbCl₅	3.22	20	0.0046	2, 47
Oxides					SbBr₃	$20._9{}^{B}$	100		
H₂O	78.54	25	*	0, 100	SbI₃	$13._9{}^{B}$	175		
	34.59	200		100, 370	AsCl₃	12.6^{A}	20		
D₂O	78.25	25		0.4, 98	AsBr₃	9.0^{A}	35		
H₂O₂	84.2	0		$-30, 20$	GeCl₄	2.43_0	25	0.0024_0	0, 55
SO₂	17.6	-20	$0.00287(\alpha)$	$-65, -15$	SiCl₄	2.4_0	16		
	15.0_8	0			SrCl₄	2.87	20	0.0030	$-30, 20$
	$14._1$	20	0.077	14, 140	TiCl₄	2.80	20	0.0020	$-20, 20$
	2.1_0	154			PbCl₄	2.78	20		
SO₃	3.11	18			*Silicones*				
CO₂	1.60	20 (50 atm)			$(CH_3)_3Si[OSi(CH_3)_2]_nCH_3$				
N₂O	1.97	-90			$n = 1$	2.17	20		
	1.61	0	0.006	$-6, 14$	2	2.30	20		
N₂O₄	2.5_6	15			3	2.39	20		
Sulfides					4	2.46	20		
H₂S	9.26	-85.5			5	2.50	20		
	9.05	-78.5			$[OSi(CH_3)_2]_n$ (cyclo)				
CS₂	2.63	20	0.0013	0, 40	$n = 4$	2.39	20		
					5	2.50	20		
					6	2.59	20		
					7	2.68	20		
					8	2.74	20		

$$* \ \epsilon = 78.54[1 - 4.579 \times 10^{-3}(t - 25) + 1.19 \times 10^{-5}(t - 25)^2 - 2.8 \times 10^{-8}(t - 25)^3], \ \pm 0.03\%$$

materials and the different behavior in each type. In the first part we review some small-molecular-weight compounds which are well characterized when pure. In the second part we treat high-molecular-weight materials that are subject to some variation in dielectric properties caused by different conditions of formation (e.g., density variations) and different types and quantities of additives.

Data for Tables 5d-1-4 are taken largely from previous compilations;

"International Critical Tables," Washburn, ed., 1933.
Landolt-Börnstein Tabellen, vol. II, part 6, 6th ed., 1959.
Maryott and Smith: *NBS Circ.* 514, 1951.
Maryott and Buckley: *NBS Circ.* 537, 1953.
Buckley and Maryott: *NBS Circ.* 589, 1958.
Lange: "Handbook of Chemistry," 10th ed., 1967.

TABLE 5d-4. ORGANIC COMPOUNDS (SMALL MOLECULES)

Type/Name	Formula	t, °C	ϵ'	n_0^2	$10^2 a$	Range	Melting point	Boiling point
Hydrocarbons								
Straight chain	H(CH$_2$)$_n$H							
Methane	($n = 1$)	−173	1.70	....	0.2	−181, −159	−184	−161.5
Propane	($n = 3$)	0	1.61	....	0.2	−90, 15	−189.9	−42.17
Pentane	($n = 5$)	20	1.844	1.841(16°)	0.16	−50, 30	−131.5	36.2
Hexane	($n = 6$)	20	1.890	1.89	0.15	−10, 50	−94.3	69.0
Heptane	($n = 7$)	20	1.924	1.923(23°)	0.14	−50, 50	−90.5	98.5
Octane	($n = 8$)	20	1.948	1.95	0.13	−50, 50	−56.5	125.8
Nonane	($n = 9$)	20	1.972	1.97	0.13$_5$	−10, 90	−53.7	150.7
Decane	($n = 10$)	20	1.991	1.99	0.13	10, 110	−31	174
Undecane	($n = 11$)	20	2.005	1.99	0.12$_5$	10, 130	−26.5	195.8
Dodecane	($n = 12$)	20	2.014	2.01	0.12	10, 150	−12	214.5
Docosane	($n = 22$)	50	2.00	....			44.4	317.4
Octacosane	($n = 28$)	52.2	2.19					
Cyclo-	(CH$_2$)$_n$							
Pentane	($n = 5$)	20	1.965	1.97			−93.3	49.5
Hexane	($n = 6$)	20	2.023	2.042(15°)	0.16	10, 60	6.5	81.4
Aromatic								
Benzene	ϕH	20	2.284	2.25	0.20	10, 60	5.5	80.1
Toluene	ϕCH$_3$	25	2.379	2.25	0.24$_3$	0, 90	−95	110.8
Styrene	ϕCH$_2$CH$_3$	25	2.43	2.40				146
Diphenyl	$\phi\cdot\phi$	75	2.53	2.53(77°)	0.18	75, 155	70	254
		85	2.54	2.51(99°)				
Naphthalene		25	2.85				80.2	217.9

ϕ represents a benzene ring minus one hydrogen.

TABLE 5d-4. ORGANIC COMPOUNDS (SMALL MOLECULES) (*Continued*)

Type/Name	Formula	t, °C	ϵ_s	ϵ_∞	f_c, Hz	n_0^2	$10^2\alpha$	Range	Melting point	Boiling point
Alcohols										
Methanol	CH_3OH	0	37.98	6.1	1.87×10^9					
(wood alcohol)		20	33.64	5.7	3.00×10^9	1.77	0.264	5, 55	−97.8	64.6
		40	29.73	5.2	4.6×10^8					
Ethanol	$CH_3 \cdot CH_2OH$	0	28.39	4.45	6.17×10^8					
(grain alcohol)		20	25.07	4.2_6	1.11×10^9	1.85	0.270	−5, 70	−117.3	78.5
		40	22.14	4.18	1.8×10^9					
Glycol	$CH_2OH \cdot CH_2OH$	20	38.7	2.6_5	1.5×10^9	2.05	0.224	20, 100	−17.4	197.2
(ethylene glycol)		40	34.9	3.4_5	3.0×10^9					
1-Propanol	$CH_3 \cdot CH_2 \cdot CH_2OH$	0	25.0	6.0	1.56×10^8					
		20	20.8	2.65	$3.7_5 \times 10^8$		0.293	20, 90	−127	97.2
2-Propanol	$CH_3 \cdot CHOH \cdot CH_3$	0	24.4	6.7	1.3×10^8					
(isopropyl alcohol)		20	19.0	3.2	$5.4_5 \times 10^8$	1.90	0.310	20, 70	−89	82.3
1,2-Propanediol	$CH_3 \cdot CHOH \cdot CH_2OH$	20	32.0			2.08	0.27	20,		189
1,3-Propanediol	$CH_2OH \cdot CH_2 \cdot CH_2OH$	20	35.0			2.18	0.23	20,		214(*d*)
Glycerol	$CH_2OH \cdot CHOH \cdot CH_2OH$	25	42.5				0.208	0, 100		240
1-Butanol	$CH_3(CH_2)_2 \cdot CH_2OH$	25	17.1	2.95	$3.3_3 \times 10^8$	1.95	0.300	−40, 20	−89.2	117.7
2-Butanol	$CH_3CH_2CHOH \cdot CH_3$	20	15.8	3.5	3.16×10^8	1.95			−89	99.5
1-Pentanol	$CH_3(CH_2)_3 \cdot CH_2OH$	20	15.3	3.8	$2.1_4 \times 10^8$	1.99	0.23	15, 35	−78.5	138
1-Hexanol	$CH_3(CH_2)_n CH_2OH$	2	12.9	3.3	$1.5_2 \times 10^9$	2.00	0.35	15, 35	−51.6	157.2
	($n = 4$)	120	3.2	2.34	7.5×10^9					
1-Heptanol		2	11.7	3.10	$1.1_0 \times 10^8$	2.03			−34.6	176
	($n = 5$)	120	3.10	2.35	7.0×10^9					
1-Octanol		2	10.35	3.05	9.1×10^7	2.03	0.410	20, 60	−16.3	195
	($n = 6$)	120	3.05	2.35	$5.7_5 \times 10^9$					
1-Nonanol		2	9.05	3.05	8.0×10^7	2.05			−5	213
	($n = 7$)	120	3.05		3.5×10^9					
1-Decanol		2	7.75	3.10	9.6×10^7	2.07			−6	231
	($n = 8$)	120	3.10	2.40	3.0×10^9					

(*d*) = decomposes

TABLE 5d-4. ORGANIC COMPOUNDS (SMALL MOLECULES) (Continued)

Type/Name	Formula	t, °C	ϵ'	$10^2\alpha$ (or a)	Range	Melting point	Boiling point
Acids	R·COOH, $-\overset{\text{O}}{\underset{\parallel}{\text{C}}}-\text{O}-\text{H}$						
Formic	H·COOH	16	58.$_5$	...	...	8.4	100.7
Acetic	CH₃·COOH	20	6.15	...	...	16.6	118.1
Anhydride	CH₃CO·O·COCH₃	19	20.$_7$	...	...	−73.1	140.0
Propionic	CH₃·CH₂·COOH	10	3.30	...	...	−22	141.1
Butyric	CH₃(CH₂)₂·COOH	20	2.97	−0.23(a)	10, 70	−7.9	163.5
Isobutyric	CH₃·CHCOOH·CH₃	10	2.71	...	...	−47.0	154.4
Succinic	HOOC·(CH₂)₂·COOH	25	2.40	...	...	185	235(d)
Benzoic	φ·COOH	...	...	...	...	122	249
Esters	R·COOR', $-\overset{\text{O}}{\underset{\parallel}{\text{C}}}-\text{O}-$						
Methyl formate	H·COOCH₃	20	8.5	5(a)	0, 20	−99.0	31.8
Ethyl formate	H·COOCH₂CH₃	25	7.1$_6$	...	...	−79.4	54.2
Propyl formate	H·COO(CH₂)₂CH₃	19	7.7$_2$	2.2(a)	25, 40	−92.9	80.9
Methyl acetate	CH₃COOCH₃	25	6.02	1.5(a)	25	−98.7	57.8
Ethyl acetate	CH₃COOCH₂CH₃	25	6.02			−83.6	77.2
Aldehydes	R·CHO, $-\overset{\text{O}}{\underset{\parallel}{\text{C}}}-\text{H}$						
Formaldehyde	H·CHO	...	...	...	...	−92	−21
Acetaldehyde	CH₃·CHO	21	21.1	...	...	−123.5	21
Benzaldehyde	φ·CHO	20	17.$_8$	...	...	−26	179.5
Ketones	R·CO·R, $-\overset{\text{O}}{\underset{\parallel}{\text{C}}}-$						
Acetone	CH₃·CO·CH₃	25	20.7$_0$	0.205	−60, 40	−95	56.5
Acetophenone	φ—CO—CH₃	25	17.39	4(a)	25	19.7	202.3
Benzophenone	φ—CO—φ	50	11.4	...	...	48.5	306

Substance	Formula	t, °C	Dielectric constant			M.p., °C	B.p., °C
Quinone	CH=CH / C=O ⋯ C=O / CH=CH	23	2.66			115.7	
Ethers	R—O—R′						
Methyl ether	CH₃·O·CH₃	25	5.02	2.38(a)	25, 100	−138.5	−23.6
Ethyl ether	CH₃CH₂·O·CH₂CH₃	20	4.335	2.0(a)	20	−116.3	34.6
Propyl ether	CH₃CH₂CH₂·O·CH₂CH₂CH₃	26	3.3_9			−95.2	142
Vinyl ether	CH₂=CH·O·CH=CH₂	20	3.94				39
Phenyl ether	φ—O—φ	30	3.65	0.7(a)	30, 50	28	259
Nitrogen Derivatives							
Ethylamine	CH₃CH₂NH₂	10	6.94		−20, 10	−80.6	16.6
Aniline	φ·NH₂	20	7.07			−6.2	184.4
Pyridine	HC—CH / N ⋯ CH / CH=CH	24	12.3				
Ethane Derivatives							
Chloride	CH₃CH₂·Cl	170	6.29	0.196	−30, 30	−138.7	12.2
Bromide	CH₃CH₂·Br	20	9.39	0.150		−119	38.0
Iodide	CH₃CH₂·I	20	7.82		−20, 70	−109	72.2
Hydroxide (ethanol)	CH₃CH₂·OH	See Alcohols					
(acid) (acetic acid)	CH₃·COOH	See Acids					
Amine	CH₃·CH₂·NH₂	10	6.94	11.4(a)	−20, 10	−80.6	16.6
Nitrite (nitroethane)	CH₃CH₂·NO₂	30	28.0_6	9	30, 35		17
Nitrate	CH₃CH₂·NO₃	20	19.4		0, 50	−102	88.7
Zinc	CH₃CH₂·Zn·CH₂CH₃	20	2.5_5				
Thiol (ethanethiol)		15	6.9_1			−121	34.7
Thiocyanate	CH₃CH₂·CNS	21	$29._3$				
Benzene Derivatives							
Fluorobenzene	φF	25	5.42	0.130		−41.9	84.8
Chlorobenzene	φCl	20	5.71		0, 80	−55	132
Bromobenzene	φBr	25	5.40	0.115	70	−30.6	155

R represents an aliphatic hydrocarbon minus one hydrogen.
φ represents a benzene ring minus one hydrogen.
(d) = decomposes

TABLE 5d–4. ORGANIC COMPOUNDS (SMALL MOLECULES) (*Continued*)

Type/Name	Formula	t, °C	ϵ'	$10^2\alpha$ (or a)	Range	Melting point	Boiling point
Iodobenzene	ϕI	20	4.62	.....	.....	−31.4	188.6
Phenol	ϕOH	60	9.78	0.32	40, 70	+41	182
Benzoic acid	$\phi COOH$	See *Acids*					
Aniline	ϕNH_2	20	6.89	0.148	0, 50	−6.2	184.4
Nitrobenzene	ϕNO_2	25	34.82	0.225	10, 80	5.7	210.9
Toluene	ϕCH_3	25	2.379	0.243(a)	0, 90	−95	110.8
Structural Variations							
Orthodichlorobenzene	.....	25	9.93	0.194	0, 50	−17.5	180
Metadichlorobenzene	.....	25	5.04	0.120	0, 50	−24.8	172
Paradichlorobenzene	.....	50	2.41	0.18	50, 80	+53	173.4
1-Octanol	$CH_2OH \cdot (CH_2)_6 \cdot CH_3$	20	10.3_4	0.410	20, 60	−16.3	195
2-Octanol	$CH_3 \cdot CHOH \cdot (CH_2)_5 \cdot CH_3$	20	8.20	.....	.....	−38.6	179
3-Octanol	$CH_3CH_2 \cdot CHOH \cdot (CH_2)_4 \cdot CH_3$	20	7.03				
4-Octanol	$CH_3(CH_2)_2 \cdot CHOH \cdot (CH_2)_2 \cdot CH_3$	20	5.12				
d-Pinene	$C_{10}H_{16}$ (camphorlike structure)	25	2.64	.....	.....	−55	161
l-Pinene		20	2.76				
cis-1,2-Dichloroethylene	$\begin{array}{cc} Cl & Cl \\ | & | \\ CH{=}CH \end{array}$	25	9.20	.....	.....	−80.5	60.1
trans-1,2-Dichloroethylene	$\begin{array}{cc} Cl & \\ | & \\ CH{=}CH \\ & | \\ & Cl \end{array}$	25	2.14	.....	.....	−50	48.4

ϕ represents a benzene ring minus one hydrogen.

TABLE 5d-5. LOW-LOSS POLYMERS

Name	ϵ'	ρ, g/cc	$d\epsilon/dt$	t, °C	10^2	10^3	10^4	10^5	10^6	10^7	10^8	10^9	10^{10}	References
Polyethylene	.....	.....	.....	-80	30	40	60	80	:	:	:	:	:	1
				-20	50	50	50	50	55	:	:	:	:	1
	2.2860	0.9205	-0.0012	23	25	30	45	55	80	:	:	:	:	2
	2.355	0.966												2
Polypropylene	.....	.....	.....	-110	:	55	:	:	:	:	:	:	:	3
				-20	10	10	10	:	:	120	:	:	:	
				23	50	45	40	:	:	:	:	:	:	
Polyisobutylene	2.25	.....	.....	25	100	125	155	140	100	70	60	:	470	4
Polystyrene	2.23	.....	.....	25	400	100	100	<200	100	100	300	330	430	4, 5
Polystyrene-divinyl-benzene	2.56	.....	.....	25	<50	<50	<50	50	70	<200	<100	460	470	4
Polytetrafluoroethylene	2.55			25	210	110	100	110	130	200	380	240		
$d\epsilon/d\rho = 0.61$	2.0120	2.1000	-0.0004_4	23	<20	<20	20	30	60	130	240	240	140	2

References for Table 5d-5

1. Barrie, Buckingham, and Reddish: *Proc. IEE* **113**, 1849 (1966).
2. Unpublished results, Bell Telephone Laboratories.
3. Buckingham and Reddish: *Proc. IEE* **114**, 1810 (1967).
4. Von Hippel: "Dielectric Materials and Applications," John Wiley & Sons, Inc., New York, 1954.
5. Amrhein: *Kolloid-Z. u. Z. Polymere* **218-217**, 38 (1967).

TABLE 5d-6. POLAR POLYMERS

Name	t,°C	ε' 10^2	10^3	10^4	10^5	10^6	10^7	10^8	10^9	10^{10}	$10^4\varepsilon''$ 10^2	10^3	10^4	10^5	10^6	10^7	10^8	10^9	10^{10}	References
Polychlorotrifluoroethylene:																				
80% crystalline	+23	2.71_4	2.64_6	2.55_5	2.49_8	2.46_4	2.44_0	—	2.38_8	2.36_0		54.9	47.7	29.2	22.9	19.0	—	15.5	18.0	1
12% crystalline	22.7	2.59_5	2.49_2	2.39_8	2.34_2						62.2	69.9	52.6	31.0						
Polyvinyl chloride	20	3.1_8	3.1_0	3.0_2	2.9_6	2.8_8	2.8_7	2.8_5			41	56	68	62	46	33	23	16		2,3
	47	3.6_0	3.5_2	3.4_1	3.2_8	3.1_4	3.0_2	2.9_2	2.84		36	65	82	86	72	49	32	22		
	76	3.9_2	3.8_3	3.6_8	3.3	3.0	2.8_7	2.8	2.81		71	84	117	132	105	78	53	49	75	
	96	6.6_0	5.3_0	4.4_0	3.7	3.3	2.8	2.7	2.6		990	740	530	360	240	140	86	47		
Polyvinylidene chloride	110	9.9	8.6	6.8	5.6						1,020	1,140	1,210	1,060						2
	23	4.8_8	4.6_6	4.1_7	3.6_0	3.1_8	2.9_7	2.8_2	2.71	2.7_0	220	293	368	304	181	92	51	19	14	
	84	5.1_3	4.9_4	4.8_5	4.7_1	4.4_0	3.7_5	3.2	2.7_6	2.7_0	400	104	63	154	343	487	288	67		
Polyvinylidene fluoride	23	12.8	12.2	11.6	10.5	8.9	6.8	4.7	2.8_4		461	415	615	1,089	1,897	1,900	1,160	17		4,5
Polymethyl methacrylate	27	3.4_0	3.1_2	2.9_5	2.8_4	2.7_6	2.7_1	2.6_6	2.6_0	2.59	260	145	88	57	39	27	16	15		2
Polycarbonate	80	4.3_6	3.8_0	3.3_4	3.0_0	2.8_0	2.7_0	2.6_6	2.6_0	2.5_6	300	340	267	156	90	57	20			4,6
Polyphenylene oxide	23	—	2.9_2	2.9_1	2.8_0	2.8_0	2.7_7			2.5_6	—	1.5	4.5	15	30	35	27	20	15	4
	23		2.59		2.5_9							0.9			1.8					
Polysulfone	25	3.13	3.13		3.10				2.95			3.4	7.5							7
Polyamide	25	3.6_0	3.5_0	3.3_5	3.2_4	3.1_4	3.0_5	3.0	2.8_4	2.7_3	56	65	70	71	68	62	60			2,8
Epoxy	84	$13._5$	$11._2$	9.0	6.3	4.4	3.7	3.4	2.9_4	2.9_9	3,170	1,570	1,420	1,280	756	426	228	105	75	2
	25	3.9_6	3.9_0	3.8_8	3.6_7	3.5_4	3.4_2	3.2_4	3.01		27	44	79	95	96	91	98	88		
Rubbers:																				
Hevea (vulcanized)	27	2.9_4	2.9_4	2.9_3	2.8_8	2.7_4	2.5_2	2.4_2	2.3_6		14	7	18	63	122	103	43	11	14.4	2
Gutta-percha	25	2.6_1	2.6_0	2.5_8	2.5_5	2.5_3	2.5_0	2.4_7	2.4_0	2.3_6	1.3	1.04	2.3_2	5.3_5	10.6	20.0	29.6	27.0	11.9	
Neoprene	24	6.7_0	6.6_0	6.5_4	6.4_7	6.2_6	5.54	4.5	4.0	4.0	107	73	75	97	238	660	405	104	135	
Silicone	25	3.3_6	3.3_0	3.2_6	3.2_3	3.2_0	3.1_0	3.1_0	3.0_4	3.1_1	21	22	21	19	12	9	9	100	54	

1. Scott, A. H., D. J. Scheiber, A. J. Curtis, J. I. Lauritzen, Jr., and J. D. Hoffman: J. Research NBS, (66A(4), 269 (1962).
2. von Hippel, A.: "Dielectric Materials and Applications," John Wiley & Sons, Inc., New York, 1954.
3. Amrhein, E.: Kolloid-Z. u. Z. Polymere 216-217, 38 (1967).
4. Unpublished data of Bell Telephone Laboratories.
5. Peterlin, A., and J. H. Elwell: J. Materials Sci. 2, 1 (1967).
6. Matsuoka, S., and Y. Ishida: J. Polymer Sci. C4, 247 (1966).
7. Helbling, J. T.: 14th Annual Symposium on Wire and Cable, December, 1965.
8. McCall, D. W., and W. E. Anderson: J. Chem. Phys. 32(1), 237 (1960).

Additional topics covered by these references are: ferroelectrics, liquid crystals, solutions, and pressure dependence of ϵ. More recent data on pressure dependence can be found in:

Skinner, Cussler, and Fuoss: *J. Phys. Chem.* **72**, 1057 (1968).
Mopsik: *J. Research NBS* **71A**, 287 (1967).
Hartman, Neumann, and Rinck: *Z. Phys. Chem.* **44**, 204 (1965).

Data on dielectric breakdown are omitted. Scatter in reported values is about 10 percent in gases and several-fold in the condensed phases. Some trends are well established, however: Gases which can capture electrons and which have vibrational modes available to store and dissipate energy (e.g., SF_6) have higher breakdown strengths than air (by two or three times), and gases with no vibrational modes (e.g., He) have lower strengths than air. Condensed phases can have *intrinsic* breakdown strengths around 10^7 volts/cm, but practical values are 10 to 100 times lower than this. Discernible trends in breakdown field strength are: D-c values are 10 to 100 times higher than MHz values; values decrease with increasing thickness; and some polymers show a drop at elevated temperature.

TABLE 5d-7. DETAILED DESCRIPTION OF A PARTICULAR POLYMER

A. Loss Index "Map" for Polychlorotrifluoroethylene
Smoothed $10^4\epsilon''$, 80% Crystalline*

f, Hz	\(t\),°C										
	−50	−25	0	23	50	75	100	125	150	175	200
10^{-1}	360	187	100	53	26	61	58	58	80	102	532
2×10^{-1}	382	221	113	60	26	56	58	57	78	66	330
5×10^{-1}	389	279	139	75	30	50	58	55	78	42	175
10^0	378	326	160	88	34	45	58	54	80	35	121
2×10^0	355	370	189	105	38	40	59	53	83	30	79
5×10^0	312	421	239	133	49	35	60	53	85	29	38
10^1	279	449	298	162	59	33	61	52	85	33	23
2×10^1	243	457	364	197	72	35	63	54	83	42	15
5×10^1	198	426	445	257	100	44	65	58	79	58	11
10^2	167	382	488	315	130	54	66	61	76	70	11
2×10^2	145	335	511	387	175	67	68	65	74	83	13
5×10^2	126	269	487	490	267	95	70	71	73	99	20
10^3	113	224	444	549	348	131	76	77	72	111	28
2×10^3	104	192	391	572	445	188	90	82	74	117†	40
5×10^3	95	165	320	536	581	294	117	92	82	119	58
10^4	91	151	275	477	640	413	166	104	92	116	73
2×10^4	87	139	238	413	650	544	244	127	107	113	89
5×10^4	82	126	200	335	595	683	404	187	131	123	111
10^5	78	120	183	292	535	745	561	277	175	142	127
2×10^5	74	115	170	263	479	754	744	432	243	171	146
5×10^5	70	112	165	248	442	716	803	552	314	214	158
10^6	64	104	155	229	363	586	830	826	608	421	246
3.2×10^6	557	96	149	208	306	485	737	720	821	793	518
10^7	51	92	144	190	275	425	637	873	1,018	1,090	1,013
2.92×10^7	......	......	......	155	......	......	403	......	852		

B. Permittivity "Map" for Polchlorotrifluoroethylene‡
Smoothed ϵ', 80% Crystalline

Freq (Hz)											
10^{-1}	2.597	2.684	2.740	2.772	2.798	2.832	2.862	2.887	2.926	2.958	2.955
2×10^{-1}	2.582	2.674	2.735	2.770	2.797	2.831	2.860	2.886	2.923	2.956	2.946
5×10^{-1}	2.559	2.661	2.727	2.765	2.796	2.827	2.858	2.883	2.918	2.955	2.942
10^{0}	2.540	2.647	2.721	2.762	2.794	2.824	2.856	2.881	2.915	2.955	2.941
2×10^{0}	2.522	2.631	2.713	2.758	2.793	2.821	2.854	2.879	2.913	2.954	2.940
5×10^{0}	2.500	2.605	2.702	2.751	2.791	2.818	2.849	2.877	2.910	2.953	2.940
10^{1}	2.486	2.686	2.690	2.746	2.788	2.816	2.846	2.874	2.908	2.952	2.939
2×10^{1}	2.474	2.565	2.675	2.737	2.785	2.815	2.842	2.871	2.905	2.950	2.939
5×10^{1}	2.461	2.538	2.651	2.725	2.781	2.812	2.838	2.868	2.900	2.948	2.939
10^{2}	2.454	2.520	2.629	2.714	2.777	2.810	2.835	2.863	2.896	2.945	2.938
2×10^{2}	2.447	2.503	2.604	2.700	2.772	2.808	2.832	2.861	2.892	2.942	2.938
5×10^{2}	2.438	2.485	2.571	2.672	2.760	2.804	2.828	2.858	2.888	2.937	2.937
10^{3}	2.433	2.474	2.548	2.646	2.748	2.800	2.825	2.854	2.885	2.932	2.937
2×10^{3}	2.428	2.464	2.529	2.618	2.733	2.796	2.822	2.852	2.882	2.927	2.935
5×10^{3}	2.423	2.455	2.507	2.578	2.700	2.784	2.818	2.846	2.878	2.919	2.933
10^{4}	2.418	2.448	2.495	2.555	2.672	2.771	2.812	2.842	2.875	2.913	2.930
2×10^{4}	2.415	2.441	2.485	2.534	2.641	2.751	2.807	2.839	2.871	2.908	2.927
5×10^{4}	2.410	2.434	2.472	2.510	2.596	2.706	2.792	2.832	2.865	2.902	2.920
10^{5}	2.407	2.428	2.464	2.498	2.571	2.671	2.771	2.825	2.860	2.898	2.916
2×10^{5}	2.404	2.424	2.456	2.488	2.552	2.641	2.744	2.814	2.854	2.892	2.913
5×10^{5}	2.402	2.420	2.451	2.483	2.541	2.623	2.725	2.808	2.848	2.887	2.909
10^{6}	2.397	2.412	2.441	2.468	2.516	2.581	2.665	2.757	2.823	2.867	2.900
3.2×10^{6}	2.393	2.405	2.431	2.454	2.493	2.547	2.602	2.688	2.768	2.825	2.882
10^{7}	2.390	2.397	2.421	2.440	2.475	2.515	2.552	2.610	2.700	2.762	2.849
2.92×10^{7}	......	......	......	2.402	......	......	2.470	......	2.560	......	......

* Note ridges (r) and valleys (v) in ϵ'':
r_1 ($-50°$, 5×10^{-1} Hz; $23°$, 2×10^{3} Hz; $150°$, 10^{7} Hz),
r_2 ($75°$, 10^{-1} Hz; $100°$, 2×10^{1} Hz), r_3 ($150°$
  2×10^{-1} Hz; $175°$, 5×10^{3} Hz),
v_1 ($50°$, 10^{-1} Hz; $75°$, 2×10^{2} Hz),
v_2 ($100°$, 10^{-1} Hz; $125°$, 2×10^{1} Hz),
v_3 ($175°$, 2 Hz; $200°$, 10^{2} Hz)
† Adjusted value.
‡ Scott, Scheiber, Curtis, Lauritzen, and Hoffman, J. Research NBS **66A** 269 (1962). There is a typographical error in the data for specimen 0.80 at 1.75°C and 2×10^{3} Hz. The listed value of 137×10^{-4} for $f = 2.10^{3}$Hz, $t = 175°$C should be 117×10^{-4}.

TABLE 5d-8. CERAMICS AND GLASSES

Name and Ref. No.	t, °C	ε′ (f, Hz) 10²	10³	10⁴	10⁵	10⁶	10⁷	10⁸	10⁹	10¹⁰	10⁴ tan δ (f, Hz) 10²	10³	10⁴	10⁵	10⁶	10⁷	10⁸	10⁹	10¹⁰
Glass (Corning no.)																			
Soda lime																			
(0080)—high loss [1]	23	8.30	7.70	7.35	7.08	6.90	6.82	6.75	—	6.71	780	400	220	140	100	85	90	—	170
(7570)—high ε′	25	14.58	14.56	14.54	14.53	14.52	14.50	14.42	—	14.2	11.5	13.5	15.9	16.5	19.0	23.5	33	—	98
Soda, Pb, borosilicate																			
(7720) Pyrex*	24	4.74	4.70	4.67	4.64	4.62	4.61	—	—	4.59	78	42	29	22	20	23	—	43	85
Soda, borosilicate																			
(7740) Pyrex*	25	4.80	4.73	4.70	4.60	4.55	4.52	4.52	—	4.52	128	86	65	54	49	45	45	—	85
96% silica																			
(7900) Vycor [1]	20	3.85						3.85	—	3.82	6	6	6	6	6	6	—	10	9.4
	100	3.85						3.85	—	3.82		37	17	12	10	8.5	—	—	13
Fused silica																			
915c [1]	25	3.78							—	3.78	6.6	2.6	1.1	0.4	0.1	0.1	0.3	—	1.7
(7940) NBS [2]	RT		3.830	3.830	3.824	3.824	3.824		—	(3.83)		—	—	0.4	0.4	0.3	—	—	1.2
(7940) NRC [2]	RT		3.823	3.824	3.821	3.826	(3.824)		—	(3.83)		2	0.6	0.2	0.1	0.1	—	—	4
(7940) NPL [2]	RT		3.835	3.836	3.835	3.83	3.83		—	(3.833)		1.8	1.5	0.4	0.5	0.2	1	—	0.8₄
(7940) NBS [2]	RT		6.233	6.309	6.306	6.279	6.272		—	6.20		—	—	11.2	12.7	14.0	—	—	77
Ceramic																			
Steatite																			
AlSiMag A-196* [1]	25	5.90	5.88	5.84	5.80	5.70	5.65	5.60	—	5.24	30	59	79.5	55	30.5	19	16	—	26
	81	5.90	5.88	5.84	5.80	5.70	5.65	5.60	—	—	58	40	46.5	70.5	66	40.5	24	—	38
Forsterite																			
AlSiMag 243 [1]	85	6.37	6.37	6.37	6.36	6.32	6.28				21	13.7	8.0	<9	3.7	3.5			
Titania ceramic																			
NPOT 96 [1]	25	29.5						29.5		28.9	12	4.9	3.3	2.5	1.6	1.7	2		20
N 750 T 96	25	83.4								83.4	5.7	4.5	3.5	2.5	2.2	2.3	4.6		14.6
N 1400 T 110	25	131	130.8	130.7	130.5	130.2	130.2	130.0			6.7	5.5	3.3	1.4	3.0	5.5	7.0		
High alumina																			
85% [1]	25	8.22	8.18	8.17	8.17	8.16	8.16	8.16		8.08	20	13.4	11.4	10.5	9.0	7.5	9.0		27
96% [1]	25	8.83	8.83	8.82	8.80	8.80	8.80	8.80		8.79	14	5.7	4.8	3.8	3.3	3.2	3.0	9	14
99.5% [2]	23		9.43	(9.41)	9.43	9.43	9.43	9.43	9.55	9.41			7	4	1.6	2.0	0.4		1
99.5% [3]	23		9.55 (ρ = 3.83 g/cc)																
BeO (ρ = 2.88 g/cc) [3]	23			6.60					6.6				<1						
TiO₂—see Figs. 5d-2 and 3																			

Notes: (ρ = 3.83 g/cm); $d\varepsilon'/d\rho = 3.82\ \mathrm{cm^3/g}$; $d\varepsilon'/d\rho = 3.54\ \mathrm{cm^3/g}$

* American Lava Corporation.

References for Table 5d-8

1. Von Hippel, A. R.: "Dielectric Materials and Applications," John Wiley & Sons, Inc., New York, 1954.
2. Bussey, H. E., J. E. Gray, E. C. Bamberger, E. Rushton, G. Russell, B. W. Petley, and D. Morris: *IEEE Trans. Instr. Meas* IM13, 305 (1964).
3. Unpublished data of Bell Telephone Laboratories.

Name	$\log_{10} \rho(t,°C)$, $\Omega \cdot$ cm					
	25	100	250	350	500	900
Soda lime 0080.....................	12.4	....	6.4	5.1		
Soda, Pb, borosilicate 7720 Pyrex.....	16	....	8.8	7.2		
Soda, borosilicate 7740 Pyrex........	15	....	8.1	6.6		
96% silica 7900 Vycor..............	17	....	9.7	8.1		
Steatite AlSiMag A-196.............	>14	13	...	...	7.6	5.5
Forsterite AlSiMag 243.............	>14	13.7	...	...	10.1	6.5
High alumina 85%..................	>14					

TABLE 5d-9. NATURAL PRODUCTS

Type/Name	t, °C	f, Hz	ϵ'	$10^4 \tan \delta$	References
Waxes					
Bayberry...........	24	10^3	3.22–3.27	50–58	1
Beeswax, crude.....	24	10^3	2.87–2.88	28–30	1
yellow....	23	10^3	2.66	14	2
white.....	23	10^3	2.63	118	2
		10^6	2.43	84	2
		3×10^9	2.35	50	2
Spermaceti........	24	10^3	3.60–3.75	30–32	1
Microcrystalline....	25	10^3	2.5	4	4
Candelilla..........	24	10^3	2.38–2.49	46	1
Carnauba..........	24	10^3	2.66–2.83	34	1
Ozokerite.........	24	10^3	2.37–2.43	76–88	1
Resins and Pitches					
Canada balsam.....	22	2×10^3	3.22	130	4
Rosin.............	24	10^3	2.69–2.76	16–40	1
Manila copal......	24	10^3	3.05–3.09	44–87	1
Asphalt...........	25	10^3	2.6	1.200	4
Amber............	25	10^3	2.7	18	2
Shellac............	24	10^3	3.5–3.7	65–80	1, 2
Oils					
Castor	23	3×10^9	2.68	870	3
Tung.............	23	2×10^3	3.32	10	4
Miscellaneous					
Basalt............	RT	Low	12.		8
Granite...........	RT	Low	7.–9.		8
Sandstone.........	RT	Low	9.–11.		8
Diamond..........	26	1.6×10^6	5.7		5
Marble...........	RT	90–650	8.3	30–500	8
Plywood..........	23	2.4×10^{10}	1.7	200–700	7
Slate.............	RT	950	6.0–7.5	860	8
Mica (muscovite)...	25	10^6	6.6	2	6

References to Table 5d-9
1. Lee, J. A., and H. H. Lowry: *Ind. Eng. Chem.* **19**, 302–306 (1927).
2. Von Hippel, A. R.: "Dielectric Materials and Applications," John Wiley & Sons, Inc., New York, 1954.
3. Massachusetts Institute of Technology.
4. Bell Telephone Laboratories.
5. Rao, B., and N. Rao: *Nature* **161**, 729 (May 8, 1948).
6. Coutlee, K. G.: *ASTM Proc.* **46** (1946).
7. Surber, W. H., Jr., and G. E. Grouch, Jr.: *J. Appl. Phys.* **19**, 1130 (1948).
8. "International Critical Tables," vol. 6, McGraw-Hill Book Company, New York, 1927.

5e. Electrical Conductions in Gases

E. C. BEATY AND K. B. PERSSON

National Bureau of Standards
Boulder, Colorado

Under normal circumstances (thermal equilibrium at temperatures less than 500°C) gases are very good insulators. If free charges are introduced into the gas, electrical conduction can take place. For weak constant fields the charges have superimposed on their thermal motions a drift in the direction of the field. The drift velocity **v** is proportional to the electric field, the proportionality constant being called the mobility μ. The current density **J** is related to the number density n of particle with charge q, by $\mathbf{J} = qn\mathbf{v}$; and the conductivity is given by $\sigma = qn\mu$. Charges of different sign move in opposite directions, and so the currents contributed are all in the same direction. The total conductivity is just the scalar sum $\sigma = \Sigma q_i n_i \mu_i$.

Unfortunately these simple expressions do not have wide applicability. At field strengths commonly of interest, the drift velocity is not proportional to the electric field because of the effect of the field on the distribution of velocities of the free charges. A more consequential breakdown of the simple description occurs in the very common circumstance where it is the conduction current which is responsible (by direct or indirect means) for the production of the free charges. These processes frequently involve interactions with solid surfaces at the container walls or the electrodes. A description of the conduction processes must be concerned not only with the rate of transport of the charges, i.e., **v**, but also with the density of charges and the processes which produce and remove the charges.

For a-c fields an inductive effect is important, and the complex conductivity at radian frequency ω is approximately

$$\sigma = \frac{q^2 n}{m(j\omega + \nu_m)}$$

where ν_m, the momentum transfer collision frequency, is related to the low-frequency mobility by $\mu = q/m\nu_m$. This formula is more generally useful at high than at low frequencies.

Many of the data needed for a description of gaseous conductors are available in the literature; however, much of this information is not, and in many instances the accuracy of the data is questionable. The volume of such data is much too large to reproduce here; however, several collections are available.

References

1. Allen, C. W.: "Astrophysical Quantities," 2d ed., University of London, The Athlone Press, 1955.
2. Bekefi, G.: "Radiation Processes in Plasmas," John Wiley & Sons, Inc., New York, 1966.
3. Brown, Sanborn C.: "Basic Data of Plasma Physics," John Wiley & Sons, Inc., New York, 1959.
4. "Encyclopedia of Physics: Gas Discharges, I," vol. XXI, Springer-Verlag New York Inc., New York, 1968.

5. "Encyclopedia of Physics: Gas Discharges, II," vol. XXII, Springer-Verlag New York Inc., New York, 1968.
6. Griem, Hans R.: "Plasma Spectroscopy," McGraw-Hill Book Company, New York, 1964.
7. Holt, E. H., and R. E. Haskell: "Foundations of Plasma Dynamics," The Macmillan Company, New York, 1965.
8. Kieffer, L. J., and G. H. Dunn: *Revs. Modern Phys.* **38**, 1 (1966).
9. Kieffer, L. J.: *Atomic Data* **1**, 19 (1969).
10. Kieffer, L. J.: *Atomic Data* **1**, 120 (1969).
11. Loeb, Leonard B.: "Basic Processes of Gaseous Electronics," University of California Press, Berkeley, 1955.
12. Massey, H. S. W., and E. H. S. Burhop: "Electronic and Ionic Impact Phenomena," Oxford University Press, London, 1969. H. S. W. Massey and E. H. S. Burhop, "Electron Collisions with Atoms," vol. I; H. S. W. Massey, "Electron Collisions with Molecules—Photo-Ionization," vol. II.
13. McDaniel. Earl W.: "Collision Phenomena in Ionized Gases," John Wiley & Sons, Inc., New York, 1964.
14. Moiseiwitsch, B., and S. J. Smith: *Revs. Modern Phys.* **40**, 238 (1968).

5f. Magnetic Properties of Materials

R. M. BOZORTH

U.S. Naval Ordnance Laboratory

T. R. McGUIRE

IBM Thomas J. Watson Research Center

R. P. HUDSON[1]

National Bureau of Standards

5f-1. Types of Magnetism and Some Formulas.[2] *Diamagnetism.* Substances whose magnetic susceptibility

$$\chi = \frac{M}{H}$$

is negative are called diamagnetic. The Langevin-Pauli formula for the diamagnetic susceptibility of an atom is [1]

$$\chi = -\frac{Ne^2}{6mc^2} \sum \bar{r}^2$$

where $\bar{r}^2$ is the mean-square distance of the electron from the nucleus, and the summation is over all the electrons in the atom.

[1] Section on properties of paramagnetic salts.
[2] Contributed by D. F. Bleil, U.S. Naval Ordnance Laboratory.

Paramagnetism. Substances whose magnetic susceptibility is positive are called paramagnetic. Langevin made a classical statistical analysis of an ensemble of dipole moments in thermal equilibrium in a magnetic field. The magnetization is given by

$$M = N\mu L\left(\frac{\mu H}{kT}\right)$$

where N is the number of atoms per unit volume and μ is their dipole moment. The Langevin function for $\chi = \mu H/kT$ is

$$L(x) = \text{ctnh } x - \frac{1}{x}$$

If $\mu H \ll kT$, the Langevin formula reduces to the Curie law

$$\chi = \frac{N\mu^2}{3kT} = \frac{C}{T}$$

Introduction of the quantum theory into the statistics for atoms with total angular momentum quantum number J gives

$$M = NgJ\beta B_J\left(\frac{gJ\beta H}{kT}\right)$$

where g is the Landé factor, β (also μ_B) is the Bohr magneton, $eh/4\pi mc = 0.927 \times 10^{-20}$ erg/oersted, and the Brillouin function is

$$B_J(x) = \frac{2J+1}{2J}\text{ctnh }\frac{2J+1}{2J}x - \frac{1}{2J}\text{ctnh }\frac{x}{2J}$$

When the argument of the ctnh is much less than one, the susceptibility becomes

$$\chi = NJ(J+1)\frac{g^2\beta^2}{3kT}$$

Note. The above equations were derived on the assumption that the atoms are free and therefore they apply, in general, only to solids which are magnetically "dilute." For details, see Van Vleck [1].

Ferromagnetism. Ferromagnetic substances are characterized by the onset of a spontaneous magnetization (in a zero applied field) at temperatures for which $T < T_c$ where T_C is called the Curie temperature.

MOLECULAR FIELD (Modified Weiss). Consider the magnetic field applied to the dipoles in the Brillouin function to consist of the applied field H_a plus an internal field which is proportional to the magnetization. The effective field is

$$H_e = H_a + \gamma M$$

where γ is the molecular field coefficient. The magnetization is $M = NgJ\beta B_J(x) = M_0 B_J(x)$ where M_0 is the moment at 0 K, and when

$$x = \frac{gJ\beta H_e}{kT} \ll 1$$

$$M = \frac{Ng^2J(J+1)\beta^2H_e}{3kT}$$

A nonvanishing solution for M exists for $H_a = 0$ when $T \leq T_C$, where

$$T_C = \frac{Ng^2\beta^2\gamma J(J+1)}{3k}$$

Combining the above equations to get a temperature function for the argument of $B_J(x)$ for $H_a = 0$, we have

$$x = \frac{3J}{J+1}\frac{M/M_0}{T/T_C}$$

For $T > T_C$ the susceptibility is

$$\chi = \frac{M}{H_a} = \frac{g\beta M_0(J + 1)}{3k(T - T_C)} = \frac{C}{T - T_C}$$

This equation is called the *Curie-Weiss law*. It is usually written

$$\chi = \frac{C}{T - \theta}$$

where θ, called the *paramagnetic intercept* (Curie point), is found by experiment to be slightly larger than T_C when $T \gg T_C$ (see Sec. 5f-14). Many solids obey a Curie-Weiss law.

HEISENBERG EXCHANGE COUPLING. Heisenberg replaces[1] the molecular field assumption with the idea that the interaction between a pair of atoms i and j has the form

$$V_{ij} = -2\mathcal{J}\mathbf{S}_i \cdot \mathbf{S}_j$$

where $\mathbf{S}_i$ and $\mathbf{S}_j$ are quantum-mechanical spin operators, and $\mathcal{J}$ is the exchange energy. This problem has not been solved exactly; the most usual approximations are to consider interactions only between nearest neighbors and to assume that all states of the crystal with the same total spin have the same energy. For these approximations, the Heisenberg results can be taken over directly into the molecular field form from the preceding paragraph, with the following substitutions:

$$\mu \to g\beta S \qquad \mu^2 \to g^2\beta^2 S(S + 1)$$
$$\gamma \to \frac{2z|\mathcal{J}|}{Ng^2\beta^2}$$

where z is the number of nearest neighbors of a given atom. For those atoms for which $L \neq 0$, $S(S + 1)$ is replaced by $J(J + 1)$. These procedures and results usually go by the name of the "first Heisenberg approximation." The literature (see Van Vleck [1] and Smart [3]) should be consulted for information about other approximate solutions of the spin-operator problem.

Antiferromagnetism (Molecular Field). Antiferromagnetic substances are those in which the magnetic ions can be divided into equivalent sublattices which become spontaneously magnetized in an antiparallel arrangement below some temperature T_N. The antiparallel alignment occurs because of a large negative exchange integral. Van Vleck[2] considered two simple interpenetrating cubic lattices and nearest-neighbor interactions. Call one sublattice A and the other B. The effective field on an ion of lattice A is due to the ions of B; thus

$$H_{eA} = H_a - 2\gamma M_B$$
$$H_{eB} = H_a - 2\gamma M_A$$

where γ is the same as in the ferromagnetic case except that each sublattice has $N/2$ (see Smart [3]) and $\mathcal{J}$ is now negative. The susceptibility for $T > T_N$ is

$$\chi = \frac{Ng^2\beta^2 J(J + 1)}{3k(T + \theta)} = \frac{C}{T + \theta}$$

where $\theta = cT_N$, and $c = 1$ for the simple model.[3] The susceptibility below the Néel temperature for this simple model consists of two parts, the susceptibility parallel $(\chi_\parallel)$ and perpendiculare $(\chi_\perp)$ to the antiferromagnetic axis. $\chi_\parallel$ decreases and becomes

---

[1] W. Heisenberg, *Z. Physik* **49**, 619 (1928).

[2] J. H. Van Vleck, *J. Chem. Phys.* **9**, 85 (1941).

[3] For other models see J. Samuel Smart, *Phys. Rev.* **86**, 968 (1952); see also ref. 3.

zero as $T \to 0$; thus the susceptibility at absolute zero for a polycrystalline solid is

$$\chi_{T=0} = \tfrac{2}{3}\chi_{T=T_N}$$

Ferrimagnetism (Molecular Field).   Ferrimagnetic substances are those in which the magnetic ions can be divided into nonequivalent sublattices which become spontaneously magnetized in an antiparallel arrangement below some temperature T_C. A ferrite, i.e., $NiFe_2O_4$, is used as an example.   It is a spinel structure having a close-packed cubic oxygen lattice in which there are 8 tetrahedral and 16 octahedral sites occupied by magnetic ions.   The sites are labeled A and B, respectively.   Néel,[1] using the molecular field theory, gave the effective fields at the A and B sites as

$$H_A = H_a + \gamma_{AA}M_A - \gamma_{AB}M_B$$
$$H_B = H_a - \gamma_{AB}M_A + \gamma_{BB}M_B$$

where
$$\gamma_{ij} = \frac{2z_{ij}\mathcal{J}_{ij}}{N_j g^2 \beta^2}$$

z_{ij} is the number of nearest neighbors on the j sublattice to an atom on the i sublattice, $\mathcal{J}_{ij}$ is the exchange coupling between the electrons of those atoms, and N_j is the total number of magnetic ions on the j sublattice.
For $T > T_C$,
$$\chi = \frac{C}{T - T_C}\frac{T - \theta'}{T - T_{C'}}$$

where
$$C = \frac{Ng^2\beta^2 J(J+1)}{3k} \qquad \lambda = \frac{N_A}{N} \qquad \mu = \frac{N_B}{N}$$
$$T_C = \tfrac{1}{2}C[\lambda\gamma_{AA} + \mu\gamma_{BB} + \sqrt{(\lambda\gamma_{AA} - \mu\gamma_{BB})^2 + 4\lambda\mu\gamma_{AB}{}^2}]$$
$$T'_C = \tfrac{1}{2}C[\lambda\gamma_{AA} + \mu\gamma_{BB} - \sqrt{(\lambda\gamma_{AA} - \mu\gamma_{BB})^2 + 4\lambda\mu\gamma_{AB}{}^2}]$$
$$\theta' = \lambda\mu C(\gamma_{AA} + \gamma_{BB} + 2\gamma_{AB})$$

For $T < T_C$,
$$M_A = N_A g\beta J y_A \qquad M_B = N_B g\beta J y_B$$

where
$$y_A = B_J\left[\frac{Ng^2\beta^2 J^2}{3kT}(\lambda\gamma_{AA}y_A - \mu\gamma_{AB}y_B)\right]$$
$$y_B = B_J\left[\frac{Ng^2\beta^2 J^2}{3kT}(-\lambda\gamma_{AB}y_A + \mu\gamma_{BB}y_B)\right]$$

where $B_J(x)$ is the Brillouin function.
Gyromagnetic Ratio.   The magnetic moment of an amperian current loop is proportional to its angular momentum,

$$\mathbf{\mu} = \frac{g'e}{2mc}\mathbf{j} = \gamma'\mathbf{j}$$

or summed over an entire body,

$$\mathbf{M} = \gamma'\mathbf{J}$$

where $\mathbf{J}$ is the total angular momentum corresponding to the magnetic moment $\mathbf{M}$.
Both γ' and $g' = \dfrac{2mc}{e}\gamma'$ are called the "gyromagnetic ratio."   They are more properly called the "magnetomechanical ratio."   A change in either $\mathbf{J}$ or $\mathbf{M}$ produces a corresponding change in the other.

BARNETT[2] EFFECT.   Change of magnetization by rotation.
EINSTEIN–DE HAAS[3] EFFECT.   Change of rotation by magnetization.

[1] L. Néel, *Ann. Phys.* **3**, 137 (1948).
[2] S. J. Barnett, *Revs. Modern Phys.* **7**, 129 (1935).
[3] A. Einstein and W. J. de Haas, *Verhandl. deut. physik. Ges.* **17**, 152 (1915).

Measurements of many ferromagnetic materials by these methods yield values of $g' \le 2$, indicating that for them the electron spin is the predominant source of magnetism. For a free ion $g' = g$ (spectroscopic splitting factor), but in a crystalline field both g' and g may depart considerably from 2. When the orbital admixtures are not necessarily small, the relation[1]

$$g' = \frac{g}{g - \rho}$$

departs from the Kittel-Van Vleck relation for which $\rho = 1$. For substances where $\rho \ne 1$ see Smart [3] and Smit [5].

Spin Resonance. A substance with a magnetic moment in a static magnetic field H will absorb energy from an oscillating magnetic field of small intensity at right angles to the static field. The peak of the absorption curve occurs at the angular frequency

$$\omega = \frac{2\pi g\mu H}{h} = \gamma_r H$$

where

$$\gamma_r = \frac{ge}{2mc}$$

where μ is the appropriate unit for the magnetic moment, and g is the spectroscopic splitting factor.

PROTONS. μ is the nuclear magneton $\mu_P = eh/4\pi M_P c$, and $g = 5.58$.

$$\frac{\omega}{2\pi} = \nu(\text{kHz}) = 4.26H \text{ (oersteds)}$$

FREE ELECTRONS

$$\mu = \beta \quad \text{and} \quad g = 2$$
$$\nu \text{ (MHz)} = 2.80H \text{ (oersteds)}$$

PARAMAGNETIC SALTS.[2] The equation of motion, treating the body as a whole, may be obtained[3] by the use of $\mathbf{M} = \gamma_r \mathbf{J}$, and the torque $d\mathbf{J}/dt = \mathbf{M} \times \mathbf{H}$,

$$\frac{d\mathbf{M}}{dt} = \gamma_r(\mathbf{M} \times \mathbf{H})$$

where the components of $\mathbf{H}$ are

$$H_x = 2H_1 \cos \omega t \qquad H_y = 0 \qquad H_z = \text{static field}$$

The amplitude of the oscillatory field is small compared with that of the static field, and the resonance frequency is

$$\omega_0 = \gamma_r H_z$$

FERROMAGNETIC RESONANCE. Kittel[4] has shown that the above equations hold for ferromagnetic resonance if all demagnetizing effects are included. For example, the resonance frequency becomes

$$\omega = \gamma(BH)^{\frac{1}{2}}$$

for a specimen in the form of a thin disk with the static field parallel to the disk.

ANTIFERROMAGNETIC RESONANCE. Above the Curie temperature, paramagnetic resonance is found. Below the Curie temperature, the effective field[5] becomes

$$H_{\text{eff}} = [H_A(2H_E + H_A)]^{\frac{1}{2}}$$

where H_A is the effective anistropy field of one sublattice, and H_E is the exchange field.

[1] M. Blume, S. Geschwind, and Y. Yafet, Generalized Kittel–Van Vleck Relation between g and g'; Validity for Negative g-Factors, *Phys. Rev.* **181**, 478 (1969).
[2] For metals, see F. J. Dyson, *Phys. Rev.* **98**, 349 (1955).
[3] F. Bloch, *Phys. Rev.* **70**, 460 (1946).
[4] C. Kittel, *Phys. Rev.* **71**, 270 (1947); **73**, 155 (1948).
[5] C. Kittel, *Phys. Rev.* **82**, 565 (1951).

FERRIMAGNETIC RESONANCE. The individual sublattices must be considered in the resonance equation. An effective splitting factor[1] for the combined sublattices is given by

$$g_{eff} \frac{e}{2mc} = \frac{|\mathbf{M}|}{|\mathbf{S}|} = \frac{|\Sigma \mathbf{M} i|}{|\Sigma (\mathbf{M}_i/\gamma_i)|}$$

where $\mathbf{M}_i$ is the magnetization of the individual sublattice, and $\gamma_i = g_i(e/2mc)$ describes its magnetomechanical ratio.

References

1. Van Vleck, J. H.: "The Theory of Electric and Magnetic Susceptibilities," Oxford University Press, New York, 1932.
2. Kittel, C.: "Introduction to Solid State Physics," 3d ed., John Wiley & Sons, Inc., New York, 1967.
3. Smart, J. S.: "Effective Field Theories of Magnetism," W. B. Saunders Company, Philadelphia, 1966.
4. Bozorth, Richard M.: "Ferromagnetism," D. Van Nostrand Company, Inc., Princeton, N.J., 1951.
5. Smit, J., and H. P. J. Wijn: "Ferrites," John Wiley & Sons, Inc., New York, 1959.

5f-2. Magnetic Properties of Elements

TABLE 5f-1. SATURATION MAGNETIZATION AND CURIE POINTS
OF FERROMAGNETIC ELEMENTS*

Element	$\sigma_s(20°C)$	$M_s(20°C)$	$\sigma_0(0\ K)$	n_B	$T_{C'}\ K$	$\theta,\ K$	$T_N,\ K$	μ_{eff}	Ref.
Fe......	218.0	1,714	221.7	2.216	1043	1100	...	3.20	1,3
Co......	161.8	1,422	162.5	1.72	1404	1415	...	3.15	2
Ni......	54.39	484	58.57	0.616	631	650	...	1.61	3
Gd......	....	†	250	7	293	302	...	‡	
Tb.....	....	.....	330	9	222	238	229		
Dy.....	....	.....	350	10	85	159	179		
Ho.....	....	.....	345	10	20	87	131		
Er......	....	.....	300	9	20	40	84		
Tm.....	......	.....	230	7	25	....	56		
Cr......−	......	.....	....	.....	.....	.....	475		
Mn.....	......	.....	....	.....	.....	.....	100		

* σ_s and σ_0 = saturation moments per gram
   M_s = saturation moment per cm³
   n_B = number of Bohr magnetons per atom
T_C and θ = ferromagnetic and paramagnetic Curie points
   T_N = Néel temperature
   μ_{eff} = effective Bohr magneton number in the paramagnetic state
† Values of M_0 (M_s at 0 K) are 2,000 (Gd) to 3,000 (Ho) for the ferromagnetic rare earths, zero at 20°C; n_B is nearly the theoretical value of gJ (Table 5f-3) with an uncertain additional value of a few tenths of a unit.
‡ Values of μ_{eff} of the trivalent rare earths are nearly the theoretical ones given in Table 5f-3, except for Sm and Eu [4] and Yb.

References for Table 5f-1

1. Vogt, E.: "Landolt-Bornstein Tabellen," vol. II, part 9, p. 16, Springer-Verlag OHG, Berlin, 1962.
2. Myers, H. P., and W. Sucksmith: *Proc. Roy. Soc. (London)*, ser. A, **207**, 427 (1951).
3. Danan, H., A. Herr, and A. J. P. Meyer: *J. Appl. Phys.* **39**, 669 (1968); Crangle, J., and G. M. Goodman, *Bull. Am. Phys. Soc.* II, **15**, 269 (1970).
4. Van Vleck, J. H.: "Theory of Electric and Magnetic Susceptibilities," Clarendon Press, Oxford, 1932.

5f-3. Properties of Ferromagnetic Compounds. Tables 5f-4 and 5f-5 show respectively properties of binary compounds of iron group elements and of rare earth elements; Tables 5f-6 to 5f-8 list properties of pure spinel ferrites, of spinel ferrites containing $ZnFe_2O_4$, and of other ferrites; Table 5f-9 applies to garnet ferrites and Table 5f-10 to known weak ferromagnets of various compositions and structures.

[1] R. K. Wangsness, *Phys. Rev.* **93**, 68 (1954).

TABLE 5f-2. RELATIVE SATURATION MAGNETIZATION σ_s/σ_0,
AS DEPENDENT ON TEMPERATURE RELATIVE TO THE
CURIE POINT T/T_C

$\dfrac{T}{T_c}$	σ_s/σ_0 observed		Molecular field theory												
	Fe	Co, Ni	$J=\frac{1}{2}$	1	$\frac{3}{2}$	2	$\frac{5}{2}$	3	$\frac{7}{2}$	4	$\frac{9}{2}$	6	$\frac{15}{2}$	8	∞
0	1	1	1	1	1	1	1	1	1	1	1	1	1	1	1
0.1	0.996	0.996	1.000	1.000	1.000	1.000	1.000	1.000	1.000	0.999	0.999	0.998	0.996	0.995	0.965
0.2	0.99	0.99	1.000	0.999	0.998	0.997	0.994	0.992	0.989	0.986	0.984	0.977	0.971	0.969	0.928
0.3	0.975	0.98	0.997	0.993	0.987	0.980	0.974	0.967	0.962	0.957	0.952	0.941	0.933	0.931	0.887
0.4	0.95	0.96	0.986	0.973	0.960	0.949	0.938	0.929	0.922	0.915	0.910	0.897	0.888	0.885	0.841
0.45			0.974	0.957	0.941	0.927	0.915	0.905	0.897	0.890	0.884	0.871	0.862	0.860	0.816
0.5	0.93	0.94	0.958	0.937	0.918	0.901	0.889	0.878	0.870	0.862	0.856	0.843	0.834	0.831	0.789
0.55			0.936	0.911	0.889	0.872	0.858	0.848	0.839	0.831	0.825	0.812	0.803	0.800	0.759
0.6	0.90	0.90	0.907	0.879	0.856	0.838	0.824	0.813	0.804	0.796	0.790	0.777	0.768	0.766	0.726
0.65			0.872	0.841	0.817	0.798	0.784	0.773	0.764	0.757	0.751	0.738	0.729	0.727	0.689
0.7	0.85	0.83	0.829	0.796	0.771	0.753	0.739	0.728	0.719	0.712	0.706	0.694	0.686	0.684	0.647
0.75			0.776	0.742	0.717	0.699	0.686	0.675	0.667	0.660	0.655	0.643	0.635	0.633	0.600
0.8	0.77	0.73	0.710	0.678	0.654	0.636	0.624	0.614	0.606	0.600	0.595	0.584	0.577	0.575	0.545
0.85	0.70	0.66	0.630	0.599	0.576	0.561	0.549	0.540	0.533	0.528	0.523	0.514	0.507	0.506	0.479
0.9	0.61	0.56	0.525	0.498	0.479	0.465	0.454	0.448	0.442	0.438	0.434	0.426	0.420	0.419	0.397
0.95	0.40	0.40	0.379	0.359	0.344	0.334	0.327	0.322	0.317	0.314	0.311	0.305	0.302	0.301	0.285
1	0	0	0	0	0	0	0	0	0	0	0	0	0	0	0

Theoretical values as calculated by S. Smart, "Effective Field Theories of Magnetism," pp. 139–154, W. B. Saunders Company, Philadelphia, 1966; M. I. Darby, *Brit. J. Appl. Phys.* **18**, 1415 (1967); and private communication for $J > \frac{7}{2}$.

TABLE 5f-3. SOME ATOMIC CONSTANTS, AND PROPERTIES,
OF RARE-EARTH ELEMENTS*

Elements	S	L	J	g	gJ	μ_{eff}	G	C_m	d, g/cm³	m.p., °C	b.p., °C
(Y)	0	0	0	....	0	0	0	0	4.48	1509	
La	0	0	0	....	0	0	0	0	6.19	920	4200
Ce	0.5	3	2.5	6/7	15/7	2.535	5/28	0.804	6.77	795	2900
Pr	1	5	4	4/5	16/5	3.578	4/5	1.600	6.78	935	3020
Nd	1.5	6	4.5	8/11	36/11	3.618	81/44	1.636	7.00	1024	3180
Pm	2	6	4	3/5	12/5	2.683	16/5	0.900†	....	..	2700
Sm	2.5	5	2.5	2/7	5/7	0.845	125/28	0.089†	7.54	1072	1600
Eu	3	3	0	....	0	0	0	0†	5.26	826	1430
Gd	3.5	0	3.5	2	7	7.937	63/4	7.879	7.89	1312	2700
Tb	3	3	6	3/2	9	9.721	21/2	11.818	8.27	1356	2500
Dy	2.5	5	7.5	4/3	10	10.646	85/12	14.171	8.54	1407	2300
Ho	2	6	8	5/4	10	10.607	9/2	14.069	8.80	1461	2300
Er	1.5	6	7.5	6/5	9	9.581	51/20	11.481	9.05	1497	2600
Tm	1	5	6	7/6	7	7.561	7/6	9.149	9.33	1545	2100
Yb	0.5	3	3.5	8/7	4	4.536	9/28	2.573	6.98	824	1500
Lu	0	0	0	0	....	0	0	0	9.84	1652	1900

* S, L, and J = quantum numbers of trivalent rare-earth ions and usually apply to the elements
g = Landé factor
gJ = theoretical saturation in Bohr magnetons per atom
μ_{eff} = effective paramagnetic moment per atom
$G = (g-1)^2 J(J+1)$, DeGennes factor
C_m = Curie constant per mole (see Sec. 5f-1)
d = density

† C_m, the theoretical Curie constant for trivalent atoms, is usually observed in the metals and compounds except for Sm and Eu and Yb.

TABLE 5f-4. MAGNETIC MOMENT AND CURIE TEMPERATURE
OF SOME BINARY COMPOUNDS

Compound	Structure (type)	T_C, K	n_B per magnetic atom	Refs.
Au$_4$Mn.........	bc tetr. (Ni$_4$Mo)	363	4.15	1
Au$_4$V..........	bc tetr. (Ni$_4$Mo)	55	0.92	2, 3
CoB...........	orthorhombic (FeB)	477	0.28	4
Co$_2$B..........	tetragonal (CuAl$_2$)	429	0.76	5
Co$_3$B..........	orthorhombic (Fe$_3$C)	747	1.11	5
CoPt..........	tetragonal (AuCu)	813	0.17	6
CoS$_2$.	fcc pyrite (FeS$_2$)	122	0.84	7–9
		130	0.96	
CrBe$_{12}$.........	tetragonal (MoBe$_{12}$)	50	~0.2	10
CrBr$_3$..........	hexagonal (BiI$_3$)	37	3.0	11, 12
CrGe$_2$..........	....	98	~0.1	13, 14
CrI$_3$..........	hexagonal (BiI$_3$)	68	3.1	15
CrO$_2$..........	tetragonal (TiO$_2$)	378	2.07	16
		386		
Cr$_{1.2}$Pt$_{2.8}$.......	fcc (Cu$_3$Au)	>77	Cr = 2.56	17
			Pt = −0.47	
CrS$_{1.19}$.........	hexagonal (NiAs)	T_N = 160	0.11	18
		T_C = 305		
CrTe..........	hexagonal (NiAs)	239–334	2.45	19–21
Cr$_3$Te$_4$..........	monoclinic	T_N = 80	2.3	22
		T_C = 329		
FeAl..........	cubic (CsCl)	623	~1.0	23
Fe$_3$Al..........	bcc (CsCl superlattice)	773	FeI = 1.46	23, 24
			FeII = 2.14	
FeB...........	orthorhombic (FeB)	598	1.12	25
Fe$_2$B..........	tetragonal CuAl$_2$	1043	1.91	26
FeBe$_5$..........	fcc (MgCu$_2$)	75	~0.1	27
Fe$_3$C..........	orthorhombic (Fe$_3$C)	483	2.01	26, 28
Fe$_3$Cr..........	cubic (Cu$_3$Au)	993	~1.3	29
Fe$_3$Ge..........	hexagonal (Ni$_3$Sn)	365	1.90	48
FeP...........	orthorhombic (MnP)	215	0.36	30
Fe$_2$P..........	hexagonal (Fe$_2$P)	266	0.77	30, 31
		278	1.32	
Fe$_3$P..........	tetragonal (Ni$_3$P)	716	1.84	30, 32
FePd$_3$..........	fcc (Cu$_3$Au)	540	Fe = 2.7	17, 33
			Pd = 0.5	
FePt..........	tetragonal (AuCu)	743	~0.2	34
FeRh..........	cubic (CsCl)	T_N = 330	Fe = 3.0	35, 36
		T_C = 675	Rh = 0.9	
Fe$_3$Si..........	cubic (Cu$_2$MnAl)	808	FeI = 1.15	37, 38
			FeII = 2.15	
Fe$_3$Sn..........	hexagonal (Ni$_3$Sn)	743	1.9	39
MnAs..........	hexagonal (NiAs)	up 318	3.4	40, 41
		down 306		
MnB..........	orthorhombic (FeB)	578	1.92	42
MnB$_2$..........	hexagonal (AlB$_2$)	143	0.19	43, 44
		157	0.25	
MnBi..........	hexagonal (NiAs)	633	3.52	45, 46
Mn$_3$Ga.........	hexagonal	470	~0.02	47
Mn$_3$Ge.........	hexagonal (Ni$_3$Sn)	28	0.38	48
Mn$_5$Ge$_3$........	hexagonal (Mn$_5$Si$_3$)	320	2.5	49
Mn$_3$In..........	cubic (Cu$_5$Zn$_8$)γ brass	583	~0.1	50
MnPt$_3$..........	fcc (Cu$_3$Au)	<300	Mn = 3.60	51
			Pt = 0.17	
MnSb..........	hexagonal (NiAs)	583	3.53	52, 53
MnSi..........	cubic (FeSi)............	34	0.4	54
Mn$_5$Sn$_3$........	hexagonal (NiIn)	263	1.23	55
Mn$_5$Y..........	orthorhombic (GdMn$_5$)	490	2.2	56, 57
MnZn$_3$..........	hexagonal (Ni$_3$Sn)	>400	~1.0	58, 59
Ni$_3$Al..........	orthorhombic	75	~0.1	60
NiPt..........	tetragonal (AuCu)	136	0.06	61
Ni$_3$Y..........	rhombic (CeNi$_3$)	33	0.16	62
Sc$_3$In..........	hexagonal	7.5	0.06/Sc	63, 64
ZrZn$_2$..........	cubic (Cu$_2$Mg)	18	~0.2	65–68

References for Table 5f-4.

1. Meyer, A. J. P.: *Compt. rend.* **242**, 2315 (1965); **244**, 2028 (1957); *J. phys. radium* **20**, 430 (1959).
2. Creveling, L., H. L. Luo, and G. S. Knapp: *Phys. Rev. Letters* **18**, 851 (1967).
3. Cohen, R. L., R. C. Sherwood, and J. H. Wernick: *Phys. Letters* **26A**, 462 (1968).
4. Lundquist, N., H. P. Myers, and R. Westin: *Phil. Mag.* **7**, 1197 (1962).
5. Fruchart, R.: *Compt. rend.* **256**, 3304 (1963).
6. Velge, W. A., and K. J. DeVos: *Z. angew. Phys.* **21**, 115 (1966).
7. Morris, B., V. Johnson, and A. Wold: *J. Phys. Chem. Solids* **28**, 1565 (1967).
8. Miyakara, S., and T. Teranishi: *J. Appl. Phys.* **39**, 896 (1968).
9. Adachi, K., K. Sato, and M. Takeda: *J. Appl. Phys.* **39**, 900 (1968).
10. Wolcott, N. M., and R. L. Falge: *Bull. Am. Phys. Soc.* **13**, 572 (1968).
11. Tsubokawa, I.: *J. Phys. Soc. Japan* **15**, 1664 (1960).
12. Dillon, J. F.: *J. Phys. Soc. Japan* **19**, 1662 (1964).
13. Margolin, S. D., and I. G. Fakidov: *Phys. Metals Metallog.* **9**(6), 22 (1960).
14. Davidenko, N. I., and I. G. Fakidov: *Phys. Metals Metallog.* **24**(1), 194 (1967).
15. Dillon, J. F., and C. E. Olsen: *J. Appl. Phys.* **36**, 1259 (1965).
16. Swoboda, T. J., A. P. Cox, J. N. Ingraham, A. L. Oppegard, and M. S. Sadler: *J. Appl. Phys.* **32**, 3745 (1961).
17. Pickart, S. J., and R. Nathans: *J. Appl. Phys.* **33**, 1336 (1962).
18. Dwight, K., N. Menyuk, D. B. Rogers, and A. Wold: *J. Appl. Phys.* **33**, 1341 (1962).
19. Lotgering, F. K., and E. W. Gorter: *J. Phys. Chem. Solids* **3**, 238 (1957).
20. Aduchi, K.: *J. Phys. Soc. Japan* **16**, 2187 (1961).
21. Bertaut, E. F., G. Roult, R. Aleonard, R. Pauthenet, M. Chevreton, and R. Jansen: *Journal de Physique* **25**, 582 (1964).
22. Chevreton, M., and E. F. Bertaut: *Compt. rend.* **255**, 1275 (1962).
23. Dekhtyar, *Phys. Metals Metallog.* **23**(1), 36 (1967).
24. Nathans, R., M. T. Pigott, and C. G. Shull: *J. Phys. Chem. Solids* **6**, 38 (1958).
25. Lundquist, N., H. P. Myers, and R. Westin: *Phil. Mag.* **7**, 1187 (1962).
26. Bozorth, R. M.: "Ferromagnetism" D. Van Nostrand Company, Inc., Princeton, N.J., 1951.
27. Herr, A., and A. J. P. Meyer: *Compt. rend.* **265**, 1165 (1967).
28. Jannin, C., P. Lecocq, and A. Michel: *Compt. rend.* **257**, 1906 (1963).
29. Dekhtjar, M. V.: *Soviet Phys.–Solid State* **5**, 2297 (1963).
30. Meyer, A. J. P., and M. C. Cadeville: *J. Phys. Soc. Japan* **17B**, 223 (1962).
31. DeVos, K. J., W. A. Velge, M. G. Van der Steeg, and H. Zijlstra: *J. Appl. Phys.* **33**, 1320 (1962).
32. Gambino, R. J., T. R. McGuire, and Y. Nakamura: *J. Appl. Phys.* **38**, 1253 (1967).
33. Cable, J. W., E. O. Wollan, and W. C. Koehler: *Phys. Rev.* **138**, A755 (1965).
34. Velge, W. A., and K. J. DeVos: *Z. angew. Phys.* **21**, 115 (1966).
35. Kouvel, J. S., and C. C. Hartelius: *J. Appl. Phys.* **33**, 1343 (1962).
36. Shirane, G., C. W. Chen, P. A. Flinn, and R. Nathans: *J. Appl. Phys.* **34**, 1044 (1963).
37. Nakamura, Y.: *J. Phys. Soc. Japan* **18**, 797 (1963). (Ref. to A. Paoletti.)
38. Lecocq, P., and A. Michel: *Compt. rend.* **258**, 1817, (1964).
39. Janniri, C., P. Lecocq, and A. Michel: *Compt. rend.* **257**, 1906 (1963).
40. Guillaud, C.: *J. phys. radium* **12**, 223 (1951).
41. Goodenough, J. B. and J. A. Kafalas: *Phys. Rev.* **157**, 389 (1967).
42. Lundquist, N., H. P. Myers, and R. Westin: *Phil. Mag.* **7**, 1197 (1962).
43. Cadeville, M. C.: *J. Phys. Chem. Solids* **27**, 667 (1966).
44. Anderson, L., B. Bellby, and H. P. Myers: *Solid State Commun.* **4**, 77 (1966).
45. Guillaud, C.: *J. phys. radium* **12**, 223 (1951).
46. Adachi, K.: *J. Phys. Soc. Japan* **16**, 2187 (1961).
47. Tsuboya, I., and M. Sugihara: *J. Phys. Soc. Japan* **18**, 143 (1963).
48. Lecocq, Y., P. Lecocq, and A. Michel: *Compt. rend.* **256**, 4913 (1963).
49. Castelliz, L.: *Z. Metallk.* **46**, 198 (1955).
50. Aoyagi, K., and M. Sugihara: *J. Phys. Soc. Japan* **17**, 1072 (1962).
51. Pickart, S. J., and R. Nathans: *J. Appl. Phys.* **33**, 1336 (1962).
52. Guillaud, C.: *J. phys. radium* **12**, 223 (1951); 489 (1951).
53. Ido, H., T. Kameko, and K. Kamigaki: *J. Phys. Soc. Japan* **22**, 1418 (1967).
54. Williams, H. J., J. H. Wernick, R. C. Sherwood, and G. K. Wertheim: *J. Appl. Phys.* **37**, 1256 (1966).
55. Yasukochi, K., K. Kanematsu, and T. Ohoyama: *J. Phys. Soc. Japan* **16**, 429 (1961); 1123 (1961).
56. Cherry, L. V., and W. E. Wallace: *J. Appl. Phys.* **32**, 340 (1961).
57. Nassau, K., L. V. Cherry, and W. E. Wallace: *J. Phys. Chem. Solids* **16**, 123 (1960).

58. Tezuka, S., S. Sakai, and Y. Nakagawa: *J. Phys. Soc. Japan* **15**, 931 (1960).
59. Nakagawa, Y., S. Sakai, and T. Hori: *J. Phys. Soc. Japan* **17**, Suppl. B.1, 168 (1962).
60. deBoer, F. R., J. Biesterbos, and C. J. Schinkel: *Phys. Letters* **24A**, 355 (1967).
61. Watanabe, M., and S. Miyahara: *J. Phys. Soc. Japan* **23**, 451 (1967).
62. Paccard, D., and R. Pauthenet: *Compt. rend.* **264B**, 1056 (1967).
63. Matthias, B. T., A. M. Clogston, H. J. Williams, E. Corenzwit, and R. C. Sherwood: *Phys. Rev. Letters* **7**, 7 (1961).
64. Gardner, W. E., T. F. Smith, B. W. Howlett, C. W. Chu, and A. Sweedler: *Phys. Rev.* **166**, 577 (1968).
65. Matthias, B. T., and R. M. Bozorth: *Phys. Rev.* **169**, 604 (1958).
66. Pickart, S. J., H. A. Alperin, G. Shirane, and R. Nathans: *Phys. Rev. Letters* **12**, 444 (1964).
67. Ogawa, S., and N. Sakamoto: *J. Phys. Soc. Japan* **22**, 1214 (1967).
68. Foner, S., E. J. McNiff, and V. Sadagopan: *Phys. Rev. Letters* **19**, 1233 (1967); E. P. Wohlfarth: *Phys. Letters* **20**, 253 (1966).

TABLE 5f-5. T_C, CURIE POINTS OF FERROMAGNETIC BINARY COMPOUNDS OF RARE-EARTH ELEMENTS R IN KELVIN*

Part 1

R =	Ce	Pr	Nd	Sm	Gd	Tb	Dy	Ho	Er	Tm	Lu	Y	Refs.
R$_6$Mn$_{23}$	...	...	...	439	469	...	443	434	415	...	...	486	1, 2
RFe$_2$	221	...	...	675	793	695	640	603	587	613	610	...	3
RFe$_3$	...	...	...	651	728	648	600	567	550	539	529	...	3
R$_6$Fe$_{23}$	...	429?	...	...	659	574	524	651	491	475	485	471	3
R$_2$Fe$_{17}$	91	287	327	395	460	409	362	319	293	248	235	245	3
RCo$_2$	...	48	106	209	413	238	154	90	38	33	...	...	4
RCo$_3$	78	349	395	...	612	506	450	418	401	370	...	301	5
R$_2$Co$_7$	151	574	609	713	762	693	647	644	644	...	...	459	3, 23
RCo$_5$	737	912	910	1020	1008	980	966	1000	986	1020	...	977	6, 23
R$_2$Co$_{17}$	1083	1171	1150	1190	1209	1180	1152	1173	1186	1182	...	1167	7, 23
RNi	...	20	35	45	73	50	48	31	10	4	...	...	8
RNi$_2$	...	8	20	22	77	46	32	23	14	14	...	...	8
RNi$_3$	...	20	27	85	116	98	69	66	62	43	...	...	9
R$_2$Ni$_7$	48	85	87	...	118	101	81	70	67	...	...	58	10
RNi$_5$	...	...	9	25	36	27	15	10	13	7	...	...	11, 8
R$_2$Ni$_{17}$	...	...	...	641	623	615	604	611	602	603	...	621	24
RAl$_2$	8	33	63	122	176	119	51	27	20	5	...	...	12
R$_5$Si$_4$	...	...	...	336	225	140	76	25	...	...	...	...	13
RRu$_2$	...	39	29	...	85	...	...	...	8	...	...	...	14
RRh$_2$	...	8	7	2	73	39	28	17	7	...	...	...	15
ROs$_2$	...	28	22	36	67	34	15	9	3	...	...	...	14
RIr$_2$	...	15	12	37	89	44	23	12	4	1	...	...	14
RPt$_2$	...	6	4	6	37	16	14	9	3	...	...	...	15

Part 2

Compound	T_C	Ref.	Compound	T_C	Ref.	Compound	T_C	Ref.
RMn$_2$	...	11, 16	TbGa	155	20	DyN	17	29, 31
PrFe$_7$	283	17	Gd$_5$Pd$_2$	335	22	HoN	13	29
NdFe$_7$	327	17	Tb$_5$Pd$_2$	30	22	ErN	16	29
Tm$_3$Ni	12	17a	Dy$_5$Pd$_2$	25	22	DyP	5	32
EuB	8	18	Ho$_5$Pd$_2$	10	22	HoP	5	31, 32
PrSi$_2$	11	19	Gd$_2$AgIn	122	21	DyAs	2	33
CeGe$_2$	5	19	NdH$_2$	10	27	EuO	69	34
PrGe$_2$	19	19	EuH$_2$	25	25	EuS	17	34
NdGe$_2$	4	19	NdN	35	28, 29	EuSe	7	34, 35, 36
TbZn	160	20	GdN	69	30	EuI$_2$	5	26
GdCd	262	21	TbN	42	29, 31	Dy$_3$Al$_2$	76	37
TbHg	80	20						

* Data for compounds with nonmetallic elements compiled by F. Holtzberg and S. Methfessel, IBM Watson Research Center.

References for Table 5f-5 (R and metallic elements)

1. DeSavage, B. F., R. M. Bozorth, F. E. Wang, and E. R. Callen: *J. Appl. Phys.* **36**, 992 (1965).
2. Kirchmayer, H. R.: *IEEE Trans.* **MAG-2**, 493 (1966).
3. Salmans, L. R., K. Strnat, and G. I. Hoffer: Technical report, Air Force Materials Lab., Wright-Patterson Air Force Base, Ohio 45433 (1968).
4. Farrell, J., W. E. Wallace: *Inorg. Chem.* **5**, 105 (1966).
5. Lemaire, R., R. Pauthenet, J. Schweizer, and I. S. Silvera: *J. Phys. Chem. Solids* **28**, 2471 (1967).
6. Lemaire, R.: *Cobalt* **32**, 132 (1966).
7. Lemaire, R.: *Cobalt* **33**, 201 (1966).
8. Abrahams, S. C., R. C. Bernstein, J. H. Sherwood, J. H. Wernick, and H. J. Williams: *J. Phys. Chem. Solids* **25**, 1069 (1964).
9. Laforest, J., R. Lemaire, D. Paccard, and R. Pauthenet: *Compt. rend.* (*B*)**264**, 676 (1967).
10. Lemaire, R., D. Paccard and R. Pauthenet: *Compt. rend.* (B)**265**, 1280 (1967).
11. Nesbitt, E. A., H. J. Williams, J. H. Wernick, and R. C. Sherwood: *J. Appl. Phys.* **33**, 1674 (1962).
12. Williams, H. J., J. H. Wernick, E. A. Nesbitt, and R. C. Sherwood: *J. Phys. Soc. Japan* **17**(I), 91 (1962).
13. Holtzberg, F., R. J. Gambino, and T. R. McGuire: *J. Phys. Chem. Solids* **28**, 2283 (1967).
14. Bozorth, R. M., B. T. Matthias, H. Suhl, E. Corenzwit, and D. D. Davis: *Phys. Rev.* **115**, 1595 (1959).
15. Crangle, J., and J. W. Ross: *Proc. Intern. Conf. Magnetism, Nottingham*, p. 240, The Institute of Physics and the Physical Society, London, 1964.
16. Felcher, G. P., L. M. Corliss, and J. M. Hastings: *J. Appl. Phys.* **36**, 1001 (1965).
17. Strnat, K., G. Hoffer, and A. E. Ray: *IEEE Trans.* **MAG-2**, 489 (1966).
17a. Féron, J.-L., R. Lemaire, D. Paccard, and R. Pauthenet: *Compt. rend.* (B) **267**, 371 (1968).
18. Matthias, B. T., T. H. Geballe, K. Andres, E. Corenzwit, G. W. Hull, and J. P. Maita: *Science* **159**, 530 (1968).
19. Matthias, B. T., E. Corenzwit, and W. H. Zachariasen, *Phys. Rev.* **112**, 89 (1958).
20. Cable, J. W., W. C. Koehler, and E. O. Wollan: *Phys. Rev.* **136**, 240 (1964).
21. Sekizawa, K., and K. Yasukochi: *J. Phys. Soc. Japan* **21**, 684 (1966).
22. Berkowitz, A. E., F. Holtzberg, and S. Methfessel: *J. Appl. Phys.* **35**, 1030 (1964).
23. Buschow, K. H. J., J. F. Fast, and A. S. VanderGoot: *Phys. Status Solid* **29**, 719 (1968).
24. Carfagna, P. D., and W. E. Wallace: *J. Appl. Phys.* **39**, 5259 (1968).
25. Zanowick, R. L., and W. E. Wallace: *Phys. Rev.* **126**, 537 (1962).
26. McGuire, T. R., and M. W. Shafer: *J. Appl. Phys.* **35**, 984 (1964).
27. Henry, W. E.: *Phys. Rev.* **98**, 226 (1955).
28. Schumacher, D. P., and W. E. Wallace: *Inorg. Chem.* **5**, 1563 (1966).
29. Busch, G., P. Junod, F. Levy, A. Menth, and O. Vogt: *Phys. Letters* **14**, 264 (1965).
30. Schumacher, D. P., and W. E. Wallace: *J. Appl. Phys.* **36**, 984 (1965).
31. Child, H. R., M. H. Wilkinson, J. W. Cable, W. C. Koehler, and E. O. Wollan: *Phys. Rev.* **131**, 922 (1963).
32. Busch, G., P. Schwob, O. Vogt, and F. Hulliger: *Phys. Letters* **11**, 100 (1964).
33. Busch, G., O. Vogt, and F. Hulliger: *Phys. Letters* **15**, 301 (1965).
34. McGuire, T. R., and M. W. Shafer: *J. Appl. Phys.* **35**, 984 (1964).
35. McGuire, T. R., F. Holtzberg, and R. Joenk: *J. Phys. Chem. Solids* **29**, 410, (1967).
36. Busch, G., P. Junod, R. G. Morris, and J. Muheim: *Helv. Phys. Acta* **37**, 637 (1964).
37. Barbara, B., C. Bècle, J.-L. Féron, R. Lemaire, D. Paccard, and R. Pauthenet: *Compt. rend.* (B) **267**, 244 (1968).

Table 5f-6. Saturation Magnetization and Curie Points of Some Simple Ferrite Spinels[a]

Ferrite	X-ray density[b]	$4\pi M_s$ at room temperature	t_C, °C
$MnFe_2O_4$	5.00	4,900[c]	295–330[c,d]
Fe_3O_4	5.24	6,000[b]	585[b]
$CoFe_2O_4$	5.29	5,300[b]	520[b]
$NiFe_2O_4$	5.38	3,230[c]	580–600[c,d]
$CuFe_2O_4$	5.35	1,700[b,e]	455[b]
$MgFe_2O_4$	4.55	1,450[c,e]	320, 440[c,d,e]
$CdFe_2O_4$	....	0	
$ZnFe_2O_4$[f]	5.33	0	60
$Li_{0.5}Fe_{2.5}O_4$	4.75	3,240–3,900[c,d]	590–680[c,d]

[a] Prepared by F. G. Brockman, Philips Laboratories, Briarcliff Manor, N.Y.
[b] J. Smit and H. P. J. Wijn, "Ferrites," John Wiley & Sons, Inc., New York, 1959.
[c] Wilhelm H. von Aulock, ed., "Handbook of Microwave Ferrite Materials," Academic Press, Inc., New York, 1965.
[d] Range of values indicates extremes of reported values from various workers.
[e] Depends on heat treatment.
[f] $ZnFe_2O_4$ magnetic when quenched, otherwise nonmagnetic; t_C for rapid quench.

Table 5f-7. Bohr Magneton Numbers of Some Ferrite Spinels and of Corresponding Solid Solutions with $ZnFe_2O_4$[a]

Mol % $ZnFe_2O_4$	0	20	40	50	70
$MnFe_2O_4$[b]	4.5	5.6	6.7	7.0	6.3
$FeFe_2O_4$[c]	4.2	5.2	5.7	5.8	5.4
$CoFe_2O_4$[b]	3.7	5.0	6.1	6.3	5.2
$NiFe_2O_4$[b]	2.4	3.8	5.1	5.3	5.1
$MgFe_2O_4$[b]	1.8[d]	3.3[d]	4.2[d]	4.4[d]	4.2[d]
$(Li_{0.5}Fe_{0.5})Fe_2O_4$[c]	2.6	2.8	4.4	4.0	1.8
$CuFe_2O_4$[c]	1.3[d]	...	...	4.7[d]	

[a] Prepared by F. G. Brockman, Philips Laboratories, Briarcliff Manor, N.Y. Some values obtained by interpolation of data in references.
[b] C. Guillaud et al., from summary of E. W. Gorter.
[c] E. W. Gorter, *Philips Research Repts.* **9**, 295, 321, 403 (1954).
[d] Depends on heat treatment.

TABLE 5f-8. CURIE POINTS AND BOHR MAGNETON NUMBERS OF
OTHER COMPOUNDS WITH THE SPINEL STRUCTURE*

Composition	T_C, K	n_B per molecule	Ref.	Composition	T_C, K	n_B per molecule	Ref.
$CoCr_2O_4$...........	98	0.18	1	$LiCo_{0.5}Mn_{1.5}O_4$.....	50	0.33†	14
$CuCr_2O_4$...........	133	0.72	1	$LiMg_{0.5}Mn_{1.5}O_4$.....	38	2.97†	14
$FeCr_2O_4$...........	88	0.84	2	$LiNi_{0.5}Mn_{1.5}O_4$.....	130	3.28	14
$MnCr_2O_4$.........	43	1.20	3	$Mg_{0.5}Mn_{2.5}O_4$......	20	0.71	15
$MnFe_{0.5}Cr_{1.5}O_4$.....	224	0.77	4	$NiMn_2O_4$..........	113-160	1.75	16
$NiCr_2O_4$...........	78	0.33	1	$LiZn_{0.5}Mn_{1.5}O_4$.....	22	4.24	14
CoV_2O_4...........	145	1.20	3	$ZnNiMnO_4$.........	90	0.87	14
Co_2VO_4...........	160	1.33	5	$Zn_{0.5}Mn_{2.5}O_4$......	20	0.61	15
FeV_2O_4...........	109	1.06	5	Mn_2SnO_4..........	58	0.35	17
Fe_2VO_4...........	440	0.72	5	$CdCr_2S_4$..........	86	5.15	18
MnV_2O_4...........	56	2.05	6	$CoCr_2S_4$...........	238	2.55	2
$NiFeAlO_4$.........	444	0.57	7	$CuCr_2S_4$..........	420	4.58	19
$NiFeGaO_4$........	444	2.8-3.0	7	$CuCr_2S_3Cl$........	218	5.14	20
$NiFeInO_4$.........	313	2.5	8	$FeCr_2S_4$...........	193	1.5	2
$NiFeVO_4$.........	610	0.70	9	$HgCr_2S_4$..........	36	5.35	18
$Li_{0.5}Fe_{0.5}Rh_2O_4$.....	130	0.1	9	$MnCr_2S_4$..........	66	2.0	21
$LiFe_2O_3F$.........	903	2.1	10	$CdCr_2Se_4$.........	129.5	5.62	22
Fe_2TiO_4.........	142	0.36	11	$CuCr_2Se_4$.........	460	4.94	19
$MnCo_2O_4$.........	203	0.04	2	$CuCr_2Se_3Br$.......	274	2.74	23
Mn_3O_4...........	43	1.85	12	$HgCr_2Se_4$.........	106	5.64	24
$Co_{1.8}Mn_{1.2}O_4$......	191	1.1	13	$CuCr_2Te_4$.........	365	4.93	19
$CuCrMnO_4$.......	45	1.47†	14	$CuCr_2Te_3I$........	294	4.10	23
$Cu_{1.5}Mn_{1.5}O_4$......	80	3.35	14				
$CuNi_{0.5}Mn_{1.5}O_4$.....	150	3.15	14				
$CuRhMnO_4$........	35	2.35	14				

* Compiled by M. W. Shafer, IBM Research Center, Yorktown Heights, N.Y.
† Not completely saturated at 30 kOe.

References for Table 5f-8.

1. McGuire, T. R.: *Phys. Rev.* **86**, 599 (1952); *Brussels Intern. Conf. Solid State Phys.*, vol. 3, p. 50, 1958.
2. Lotgering, F. K.: Thesis, Utrecht, 1956.
3. Menyuk, N., et al.: *J. Appl. Phys.* **33**, (1962).
4. Gorter, E. W.: *Philips Research Repts.*, **9**, 295, 403 (1954).
5. Rodgers, D. B., et al.: *J. Phys. Chem. Solids* **24**, 347 (1963).
6. Villers, G., et al.: *Compt. rend.* **260**, 3017 (1965).
7. Maxwell, L. R., and S. J. Pickart: *Phys. Rev.* **92**, 1120 (1953).
8. Maxwell, L. R., and S. J. Pickart: *Phys. Rev.* **96**, 1501 (1954).
9. Blasse, G.: *Philips Research Repts.*, suppl. 3, 1964.
10. Okazaki, C., et al.: *J. Phys. Soc. Japan* **21**, 199 (1966).
11. Ishikawa, Y.: *Phys. Letters* **24A**, 725 (1967).
12. Dwight, K., and N. Menyuk: *Phys. Rev.* **119**, 1470 (1960).
13. Wickham, D. G., and W. J. Croft: *J. Phys. Chem. Solids* **7**, 351 (1958).
14. Blasse, G.: *J. Phys. Chem. Solids* **27**, 383 (1966).
15. Jacobs, I. S.: *J. Phys. Chem. Solids* **11**, 1 (1959).
16. Villers, G., and R. Buhl: *Compt. rend.* **26**, 3406 (1965).
17. Gilleo, M. A., and D. W. Mitchell: *J. Appl. Phys.* **305**, 20 (1959).
18. Baltzer, P. K., et al.: *Phys. Rev.* **151**, 367 (1966).
19. Lotgering, F. K.: *Proc. Intern. Conf. Magnetism, Nottingham*, p. 533, The Institute of Physics and the Physical Society, London, 1964.
20. Sleight, A. W., and H. S. Jarrett: *J. Phys. Chem. Solids* **29**, 868 (1968).
21. Menyuk, N., et al.: *J. Appl. Phys.* **36**, 1088 (1965).
22. Baltzer, P. K., et al.: *Phys. Rev.* **151**, 367 (1966).
23. Robbins, M., et al.: *J. Appl. Phys.* **39**, 662 (1968).
24. Baltzer, P. K., et al.: *Phys. Rev. Letters* **15**, 493 (1965).

TABLE 5f-9. SPONTANEOUS MAGNETIZATION AND COMPENSATION
POINTS OF SOME FERRITE GARNETS OF COMPOSITION $R_2Fe_5O_{12}$

R	n_B per $R_3Fe_5O_{12}$ (at 0 K)	Curie temperature,[†] K	Compensation temperature,[‡] K
Y.............	5.00	560	
Sm............	5.43	578	
Eu............	2.78	566	
Gd............	16.0	564	286
Tb............	18.2	568	246
Dy............	16.9	563	226
Ho............	15.2	567	137
Er............	10.2	556	83
Tm...........	1.2	549	$\begin{cases} 4 < T < 20\dagger \\ \text{None}\ddagger \end{cases}$
Yb............	0§	548	$\begin{cases} \text{None}\dagger\ddagger \\ 7.6\P \end{cases}$
Lu............	5.07	549	

* Compiled by B. A. Calhoun, IBM Research Center, Yorktown Heights, N.Y.
† R. Pauthenet, *Ann. phys.* [13] **3**, 424 (1958).
‡ S. Geller, J. P. Remeika, R. C. Sherwood, H. J. Williams, and G. P. Spinoza, *Phys. Rev.* **137A**, 1034 (1965).
§ Spontaneous moment exists at higher temperatures.
¶ J. W. Henderson and R. L. White, *Phys. Rev.* **123**, 1627 (1961).

WEAK FERROMAGNETISM[1] (Table 5f-10). Under certain magnetocrystallographic symmetry conditions, the magnetic sublattice vectors of an antiferromagnet can depart from strict collinearity and lower, rather than raise, the value of the thermodynamic energy or potential. The noncollinearity is induced by intrinsic anisotropic forces, and the canting of the sublattice moments causes a small spontaneous magnetic moment to exist in a nominally antiferromagnetic material. The presence of this weak ferromagnetism is characterized by an energy expression, which is antisymmetric with respect to an interchange of the sublattice moments, of the form

$$l_i m_j \pm l_j m_i \qquad i \neq j = x, y, z$$

where

$$l \equiv m_1 - m_2, \qquad m \equiv m_1 + m_2$$

and m_1, m_2 are the antiferromagnetic sublattice vectors. (Note that l and m are perpendicular because $|m_1| = |m_2|$.) The minus sign in the energy equation corresponds to a two-ion exchange energy (type E in Table 5f-10) first recognized by Dzialoshinski [21][2] and later explained by Moriya [22] in the case of $Fe_2O_3(\alpha)$; the plus sign in the energy equation characterizes a single-ion anisotropy energy (type A) initially described by Dzialoshinski [13] and by Moriya [14] as the source of canting in NiF_2.

5f-4. Saturation and Curie Points of Magnetic Alloys. These and some related properties of a number of alloy systems are presented in the form of curves (Figs. 5f-1 to 5f-10), and in Tables 5f-11 and 5f-12.

[1] Prepared by R. J. Joenk, IBM Corporation, Armonk, N.Y.
[2] The references in this paragraph are those for Table 5f-10. See also E. A. Turov, "Physical Properties of Magnetically Ordered Crystals," chaps. 5–8, Academic Press, Inc., New York, 1965; and T. Moriya, "Weak Ferromagnetism," in "Magnetism," G. T. Rado and H. Suhl, eds., Academic Press, Inc., New York, 1963, vol. 1, p. 85.

TABLE 5f-10. MAGNETIC PROPERTIES OF NOMINALLY
ANTIFERROMAGNETIC MATERIALS WITH A WEAK FERROMAGNETIC
MOMENT DUE TO SMALL-ANGLE CANTING

Compound	Non-magnetic space group*	Type	T_N, K	$\hat{\imath}$	$\hat{m}$	$\lvert m \rvert$, μ_B per atom	References
$BiCrO_3$	....	....	123	....	....	0.017	1
CuF_2	$C_{2h}{}^5$†	E	69	Near c	b	$\approx 2 \times 10^{-4}$	2
$Ni(IO_3)_2 \cdot 2HO(\beta)$	$D_{2h}{}^{15}$	....	3	....	....	≈ 0.1	3, 4
$NaMnF_3$	$D_{2h}{}^{16}$‡	E	60	[100]	[001]	....	5, 6
$NaNiF_3$	$D_{2h}{}^{16}$	E	156	[100]	[001]	0.058	7–9
$KMnF_3$	$D_{2h}{}^{16}$	A	88§	....	....	0.0034	10
$RFeO_3$¶	$D_{2h}{}^{16}$	E	620–750	[100]	[001]	0.05	11, 12
NiF_2	$D_{4h}{}^{14}$	A	73	⟨100⟩	⟨010⟩	0.029	13–15
PdF_2	$D_{4h}{}^{14}$	A	217	⟨100⟩	⟨010⟩	0.0031	16
CrF_3	$D_{3d}{}^6$**	E	80	(111)	(111)	0.056	17–19
FeF_3	$D_{3d}{}^6$	E	≈ 365	(111)	(111)	....	17, 20
$Fe_2O_3(\alpha)$	$D_{3d}{}^6$	E	960††	(100) near (111)	[100]	0.006	21–24
$MnCO_3$	$D_{3d}{}^6$	E	32	[010]	[100]	0.034	21, 25, 26
$CoCO_3$	$D_{3d}{}^6$	E	18	$\theta = 46°$	....	0.26	21, 25, 27, 28
$NiCO_3$	$D_{3d}{}^6$	E	25	$\theta = 63°$	....	0.33	29, 30
UO_2‡‡	$O_h{}^5$	....	31	{001}	....	0.011	31, 32

* The space group is usually identified in the paramagnetic temperature range of the crystal; in the ordered state the symmetry is generally lower, but often indistinguishably so for most purposes.

† The crystallographic axes are labeled such that $a > b > c$; b is the symmetry axis.

‡ The crystallographic axes are labeled such that $a < b < c$.

§ Weak ferromagnetism is observed below 81.5 K.

¶ Here $R = $ Y, La, and the rare earths. The data refer to the ordering of the Fe sublattices; the moment is temperature dependent because of spin reorientation and rare-earth ordering at various lower temperatures.

** The z axis is the threefold symmetry axis, and x is a twofold axis; θ is the polar angle.

†† There is a transition to an uncanted state at 260 K.

‡‡ More than two sublattices are probably required for a descriptive model.

References for Table 5f-10.

1. Sugawara, F., S. Iida, Y. Syono, and S. Akimoto: *J. Phys. Soc. Japan* **25**, 1553 (1968).
2. Joenk, R. J., and R. M. Bozorth: *J. Appl. Phys.* **36**, 1167 (1965).
3. Burgiel, J. C., V. Jaccarino, and A. L. Schalow: *Phys. Rev.* **122**, 429 (1961).
4. Meijer, H. C., and J. van den Handel: *Physica* **30**, 1633 (1964).
5. Shane, J. R., D. H. Lyons, and M. Kestigan: *J. Appl. Phys.* **38**, 1280 (1967).
6. Pickart, S. J., H. A. Alperin, and R. Nathans: *J. phys. radium* **20**, 565 (1964).
7. Yudin, V. M., and A. B. Sherman: *Phys. Status Solidi* **20**, 759 (1967).
8. Gurevich, A. G., E. I. Golovenchits, and V. A. Sanina: *J. Appl. Phys.* **39**, 1023 (1968).
9. Ogawa, S.: *J. Phys. Soc. Japan* **15**, 2361 (1960).
10. Heeger, A. J., O. Beckman, and A. M. Portis: *Phys. Rev.* **123**, 1652 (1961).
11. White, R. L.: *J. Appl. Phys.* **40**, 1061 (1969); this is a review paper and contains an extensive list of references; for $CeFeO_3$ see M. Robbins, G. K. Wertheim, A. Menth, and R. C. Sherwood: *J. Phys. Chem. Solids* **30**, 1823 (1969).
12. Turov, E. A., and V. E. Naish: *Phys. Metals Metallog.* **9**(1), 7 (1960); V. E. Naish and E. A. Turov: *ibid*, **11**(2), 1 and **11**(3), 1 (1961).
13. Dzialoshinski, I. E.: *Soviet Phys.–JETP* **6**, 1120 (1958).
14. Moriya, T.: *Phys. Rev.* **117**, 635 (1960).
15. Joenk, R. J., and R. M. Bozorth: *Proc. Intern. Conf. Magnetism, Nottingham*, p. 493, The Institute of Physics and the Physical Society, London 1964, p. 493.
16. Rao, R. P., R. C. Sherwood, and N. Bartlett: *J. Chem. Phys.* **40**, 3728 (1968).
17. Wollan, E. O., H. R. Child, W. C. Koehler, and M. K. Wilkinson: *Phys. Rev.* **112**, 1132 (1958).
18. Bozorth, R. M., and V. Kramer: *J. phys. radium* **20**, 393 (1959).
19. Hansen, W. N., and M. Griffel: *J. Chem. Phys.* **30**, 913 (1959).
20. Shane, J. R., and M. Kestigan: *J. Appl. Phys.* **39**, 1027 (1968).
21. Dzialoshinski, I. E.: *Soviet Phys.–JETP* **5**, 1259 (1957); *J. Phys. Chem. Solids* **4**, 241 (1958).
22. Moriya, T.: *Phys. Rev. Letters* **4**, 228 (1960); *Phys. Rev.* **120**, 91 (1960).
23. Tasaki, A., and S. Iida: *J. Phys. Soc. Japan* **18**, 1148 (1963).
24. Flanders, P. J., and W. J. Schuele: *Phil. Mag.* **9**, 485 (1964).
25. Borovik-Romanov, A. S., and M. P. Orlova: *Soviet Phys.–JETP* **4**, 531 (1957).
26. Borovik-Romanov, A. S.: *Soviet Phys.–JETP* **9**, 539 (1959).
27. Borovik-Romanov, A. S., and V. I. Ozhogin: *Soviet Phys.–JETP* **12**, 18 (1961).
28. Kaczer, J.: *Soviet Phys.–JETP* **16**, 1443 (1963).
29. Bizette, H., and B. Tsai: *Compt. rend.* **241**, 546 (1955).
30. Alikhanov, R. A.: *J. Phys. Soc. Japan* **17**, suppl. BIII, 58 (1962).
31. Hambourger, P. D., and J. A. Marcus: *Phys. Rev.* **158**, 438 (1967).
32. Cracknell, A. P.: *Phys. Letters* **27A**, 426 (1968).

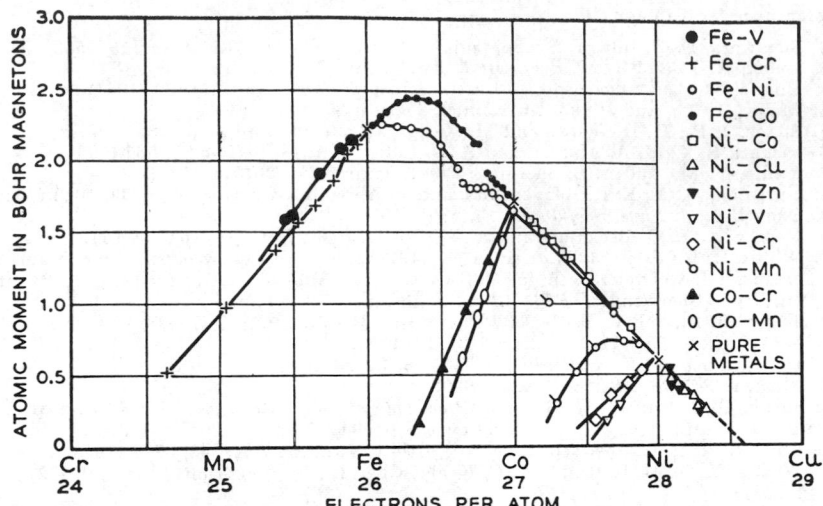

Fig. 5f-1. Saturation magnetization of intra-iron-group alloys as dependent on electron concentration. Data by Peschard (1925), Weiss, Forrer and Birch (1929), Forrer (1930), Sadron (1932), Fallot (1936, 1938), Farcas (1937), Marian (1937), and Guillaud (1944). [*R. M. Bozorth, Phys. Rev.*, **79**, 887 (1950).]

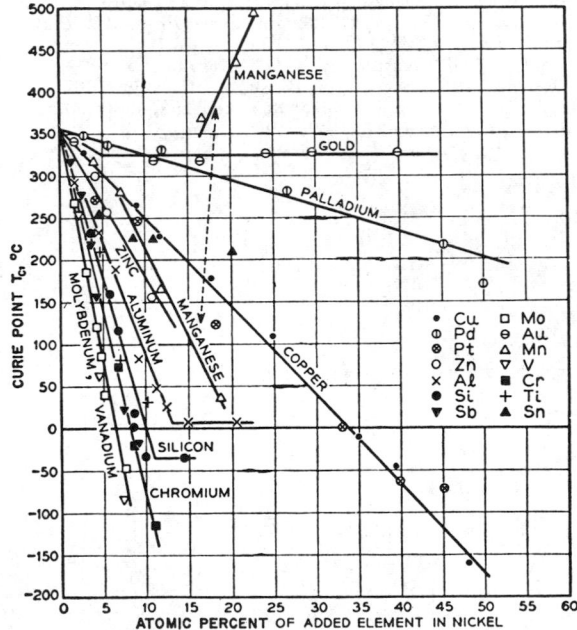

Fig. 5f-2. Change of Curie point with the composition of nickel alloys (atomic percent). Data by V. Marian, *Ann. physique* [11]7, 459 (1937). (*Bozorth, "Ferromagnetism," D. Van Nostrand Company, Inc., Princeton, N.J., p.* 721, 1951.)

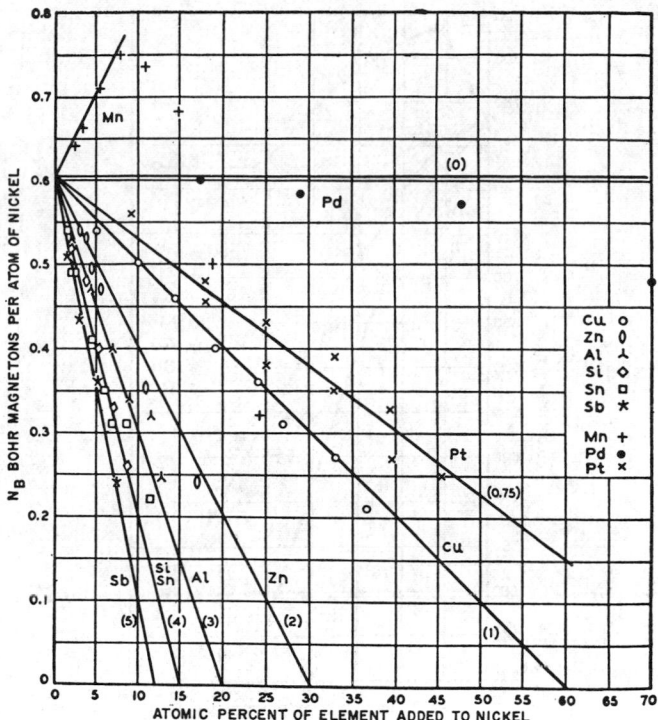

FIG. 5f-3. The saturation magnetization of nickel as affected by the addition of other elements having 1, 2, 3, . . . , electrons in the outermost shell. Data by Sadron, *Ann. physique* [10]**17**, 371 (1932). (*Bozworth*, *"Ferromagnetism,"* D. *Van Nostrand Company, Inc., Princeton, N.J., p.* 440, 1951.)

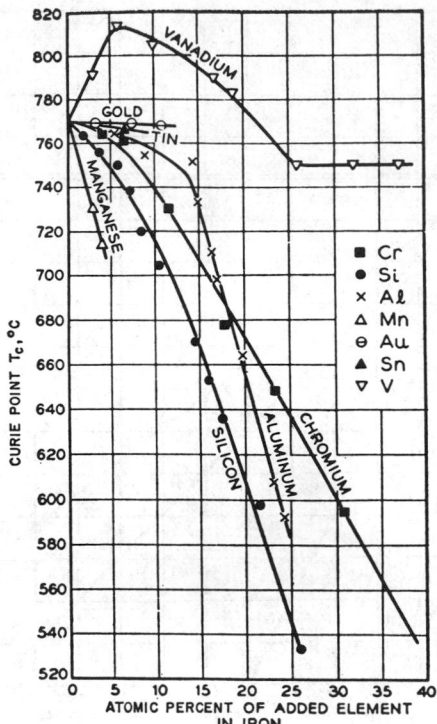

FIG. 5f-4. Change of Curie point of iron alloys with composition (atomic percent). Data by M. Fallot, *Ann. physique* [11]**6**, 305 (1936). (*Bozworth*, "*Ferromagnetism*," *D. Van Nostrand Company, Inc., Princeton, N.J.*, p. 722, 1951.)

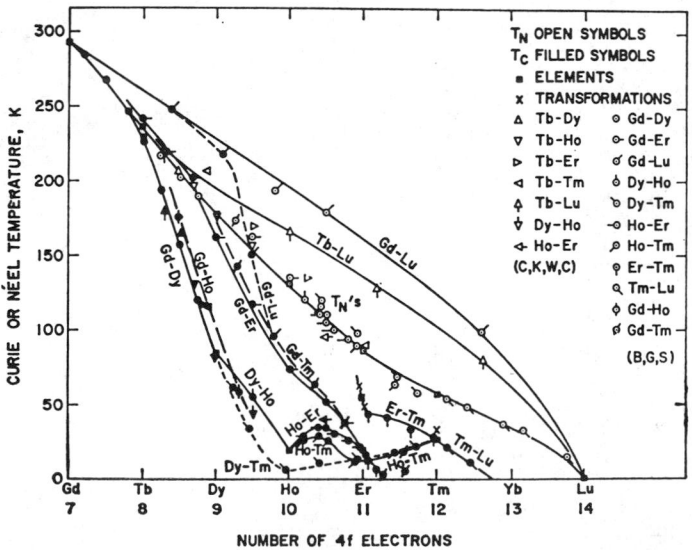

FIG. 5f-5. T_C, T_N, and T_z (change in magnetic structure) for heavy rare-earth alloys with one another. [1,2,3]

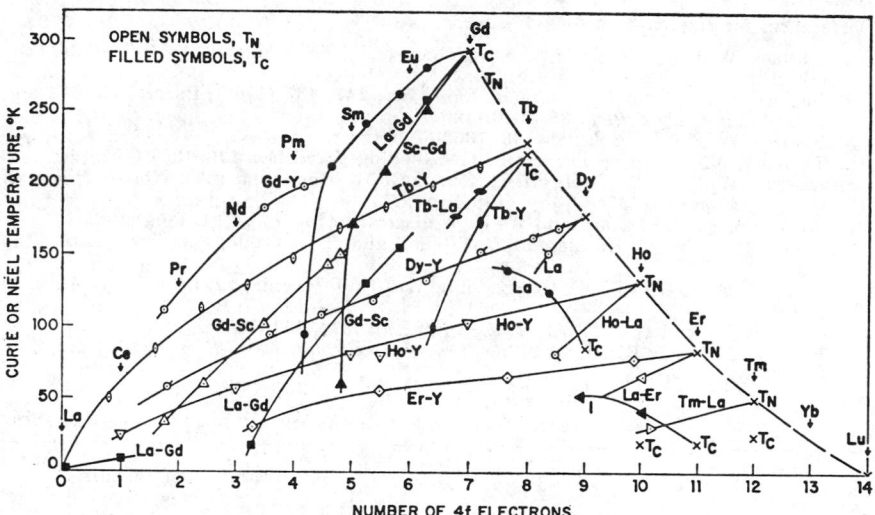

Fig. 5f-6. Dependence of Néel and Curie points of rare-earth metals on additions of non-magnetic La, Y, and Sc. Note that La often stabilizes the ferromagnetic phase, Y and Sc the antiferromagnetic phase. Data for La alloys [4,5]; for Y alloys [1,4,6,7]; and for Sc alloys [8,9].

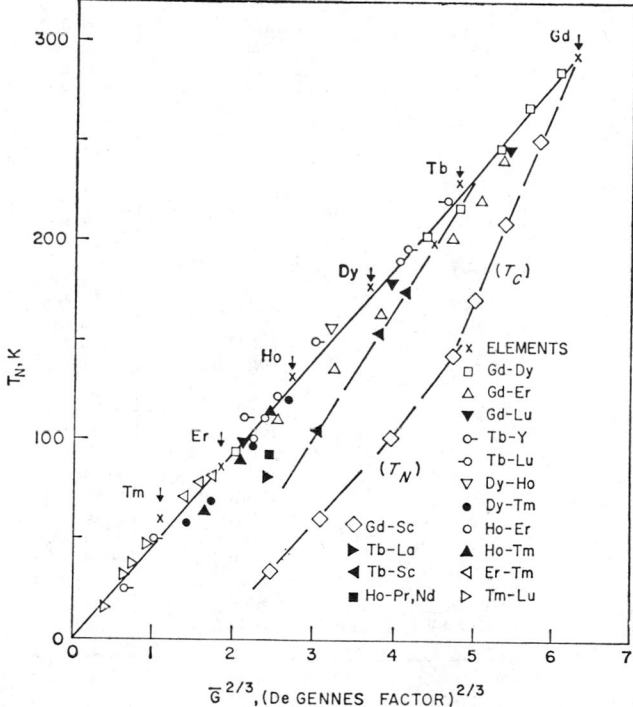

Fig. 5f-7. Néel points for rare-earth elements and various binary alloys, plotted against two-thirds power of average de Gennes factor, $G = (g - 1)^2 J(J + 1)$. Data as follows: Gd-Dy [4,5]; Gd-Er [1,5]; Gd-Lu [5]; Gd-Sc [6]; Tb-Sc [6,9]; Tb-Y, Lu [7]; Tb-La [8]; Ho-Pr, Nd [8]; and R-R [10]. [(de Gennes, P.: Compt. rendu, Sci. (Paris) **247**, 1836 (1968)).]

References for Figs. 5f-5, 5f-6, 5f-7.

1. Koehler, W. C., H. R. Child, E. O. Wollan, and J. W. Cable: *J. Appl. Phys.* **34**, 1335 (1963).
2. Bozorth, R. M., and R. Gambino: *Phys. Rev.* **147**, 487 (1966); Bozorth, R. M., and J. C. Suits: *J. Appl. Phys.* **35**, 1039 (1964).
3. Bozorth, R. M.: *J. Appl. Phys.* **38**, 1366 (1967).
4. Thoburn, W. C., S. Legvold, and F. H. Spedding: *Phys. Rev.* **110**, 1298 (1958).
5. Koehler, W. C., J. W. Cable, H. R. Child, R. M. Moon, and E. O. Wollan: *Proc. Int. Conf. on Magnetism, Nottingham* (1964), p. 271.
6. Weinstein, S., R. S. Craig, and W. E. Wallace: *J. Appl. Phys.* **34**, 1354 (1963).
7. Child, H. R., W. C. Koehler, E. O. Wollan, and J. W. Cable: *Phys. Rev.* **138A**, 1655 (1965).
8. Nigh, H. E., S. Legvold, F. H. Spedding, and B. J. Beaudry: *J. Chem. Phys.* **41**, 3799 (1964).
9. Child, H. R., and W. C. Koehler: *J. Appl. Phys.* **37**, 1353 (1966).
10. Bozorth, R. M., and R. J. Gambino, *Proc. Intern. Conf. Magnetism, Nottingham, England* (1964) p. 263.

TABLE 5f-11. PROPERTIES OF HIGH D-C PERMEABILITY

Material	Composition, % by wt. (remainder Fe)	Heat treatment,† °C	Permeability, gauss oersted	
			Initial μ_0	Maximum, μ_{max}
Iron (commercial)	0.2 impurity	950	200	5,000
Iron (purified)........	0.05 impurity	1480, 880	10,000	>200,000
0.5 Si-Fe............	0.5 Si	850	280	3,000
1.75 Si-Fe..........	1.75 Si	850	280	5,000
3.25 Si-Fe..........	3.25 Si	850	290	8,000
Oriented Si-Fe (cube on edge)......	3.25 Si	850, 1200	5,200	55,000
Oriented Si-Fe (cube on face)......	3.25 Si	850, 1200	......	116,000
6.5 Si-Fe............	6.5 Si	1000, 800(M)	1,390	67,000
Sendust (cast)........	9.6 Si, 5.4 Al	1000, 600(Q)	25,000	115,000
Sendust (sintered powder compact)...	9.6 Si, 5.4 Al	1275, 600(Q)	38,000	103,000
12-Alfenol...........	12 Al	1085	4,500	35,000
16-Alfenol...........	16 Al	900, 600(Q)	3,500	95,000
45 Permalloy........	45 Ni	1200	3,500	50,000
Monimax...........	3 Mo, 47 Ni	1125	3,000	60,000
48-50 Ni-Fe‡........	48-50 Ni	1200	5,000	70,000
50-50 Ni-Fe§ (oriented).........	50 Ni	1050	2,000	100,000
53 Ni-Fe............	53 Ni	1200, 450(M)	38,000	320,000
65 Ni-Fe............	60-65 Ni	1220, 675(M)	30,000	1,000,000
78 Permalloy........	78 Ni	1050, 600(Q)	15,000	100,000
4-79 Mo-Permalloy....	4 Mo, 79 Ni	1100(C)	25,000	200,000
Mumetal (U.K.)......	5 Cu, 4 Mo, 77 Ni	1100(C)	50,000	175,000
Mumetal (U.S.).......	5 Cu, 1.5 Cr, 77 Ni	1100	25,000	150,000
Supermalloy..........	5 Mo, 79 Ni	1300(C)	100,000	1,000,000
Hiperco-35..........	0.5 Cr, 35 Co	925	650	10,000
V-Permendur........	2 V, 49 Co	850	800	6,000
Supermendur........	2 V, 49 Co (purified)	850(M)	1,000	92,500

5f-5. Properties of Some Materials for Permanent Magnets. In the use of materials for permanent magnets, important quantities are the coercive force H_c, the residual induction B_r, and the energy product BH. The latter is the product of B and $-H$ for points on the demagnetization curve, the portion of the hysteresis loop that lies in the second quadrant. The maximum energy product $(BH)_m$ is the largest value of BH for points on the demagnetization curve, and this is the best single criterion for a material for use in permanent magnets. The point (B_d, H_d) corresponding to $(BH)_m$ is the desirable point[1] for operation (see Fig. 5f-11).

Demagnetization curves for several important materials are given in Fig. 5f-12, and constants for the commonly used materials in Table 5f-13.

[1] D. Hadfield, ed., "Permanent Magnets and Magnetism," John Wiley & Sons, Inc., New York, 1962.

MATERIALS IN SHEET FORM*

Coercive force H_c, oersteds	Saturation induction $(B_s = 4\pi M_s)$, gauss	Saturation hysteresis, W_h ergs/cm³	Curie temperature, t_C, °C	Electrical resistivity ρ, microhm-cm	Density d, g/cm³
1.00	21,580	2,500	770	10.7	7.87
0.05	21,580	300	770	9.7	7.87
0.90	21,400	2,300	765	18	7.83
0.80	20,900	2,100	750	34	7.75
0.70	20,200	1,600	740	49	7.67
0.08	20,200	700	740	48	7.67
0.07	20,200	. . . .	740	48	7.67
0.20	18,100	. . . .	690	80	7.45
0.035	10,000	100	480	81	6.96
0.030	~9,500	. . . .	480	106	6.78
0.080	14,800	270	650	100	6.74
0.030	8,100	62	450	153	6.53
0.07	16,000	1,200	440	50	8.17
0.06	14,500	800	400	65	8.27
0.05	15,800	300	480	45	8.21
0.10	15,800	390	500	45	8.22
0.025	15,600	. . . .	520	32	8.25
0.010	14,000	. . . .	610	20	8.40
0.05	10,800	200	580	16	8.60
0.015	8,700	200	460	55	8.76
0.014	7,340	. . . .	355	57	8.60
0.015	7,500	200	400	56	8.58
0.004	7,900	20	400	60	8.78
0.63	24,200	3,320	970	20	8.05
2.0	24,000	6,000	980	40	8.20
0.20	24,000	1,500	980	40	8.20

* Compiled by H. H. Helms, Jr., U.S. Naval Ordnance Laboratory.
† Dry hydrogen atmosphere:
   M Cooled in magnetic field.
   Q Quenched from indicated temperature.
   C Controlled cooling rate.
‡ Hipernik, AL 4750, high-permeability 49, Armco 48, etc.
§ Orthonol, Deltamax, Orthonik, Hipernik 5, Hy-Ra 49, Permenorm 5000 Z, etc.

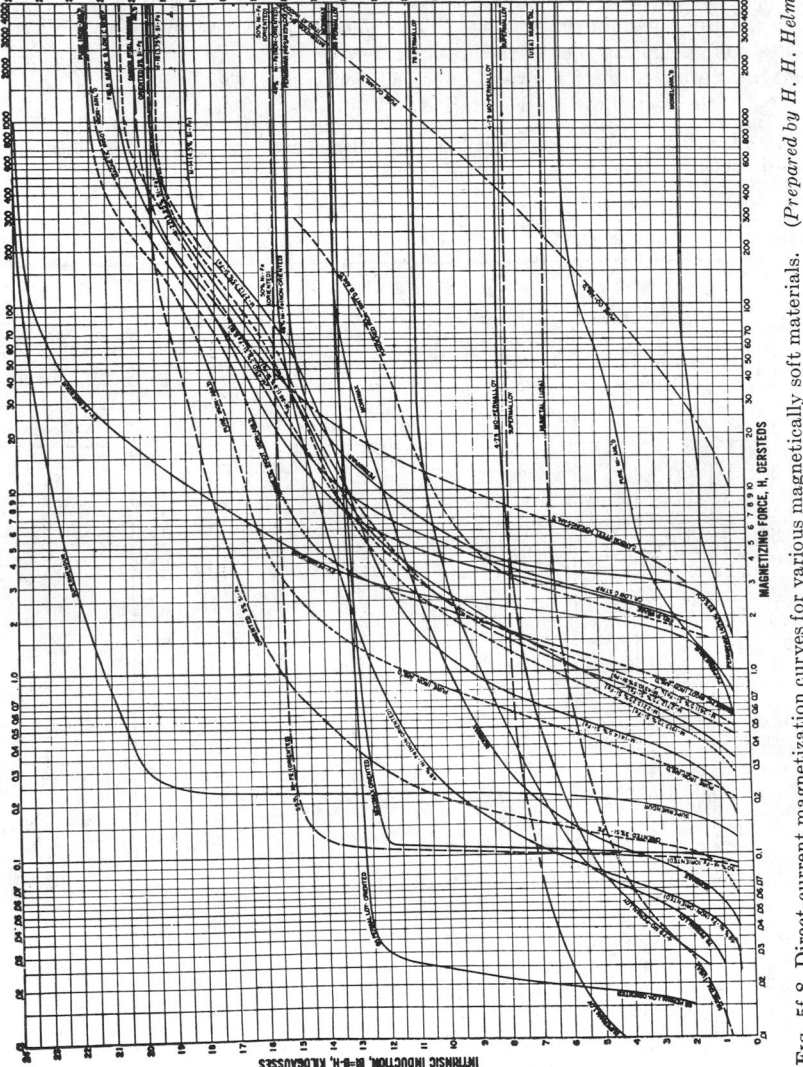

Fig. 5f-8. Direct-current magnetization curves for various magnetically soft materials. (*Prepared by H. H. Helms, U.S. Naval Ordnance Laboratory.*)

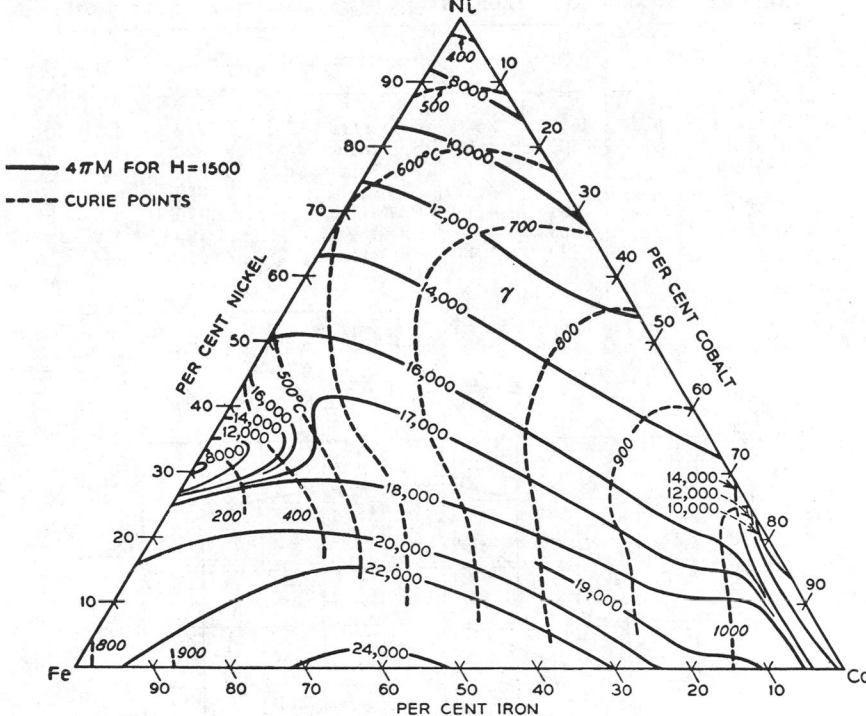

FIG. 5f-9. Approximate saturation ($4\pi M$ for $H = 1,500$) and Curie points of Fe-Co-Ni alloys. Temperature in °C. [*T. Kase, Sci. Repts. Tôhoku Imp. Univ.* **16**, 491 (1927).]

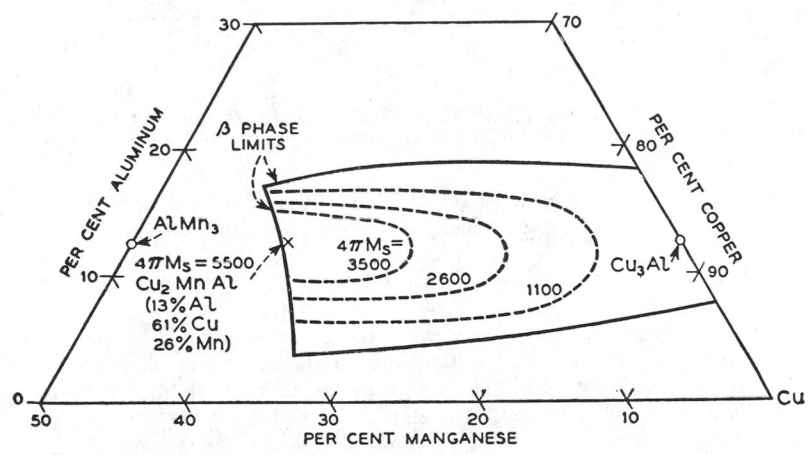

FIG. 5f-10. Saturation induction of Heusler Mn-Cu-Al alloys. Amounts in weight percent. [*Data by O. Heusler, Ann. Physik* [5]**19**, 155 (1934).]

TABLE 5f-12. MAGNETIC PROPERTIES OF HEUSLER-TYPE COMPOUNDS*

Compound	T_C	n_B	Compound	T_C	n_B
Cu₂MnAl.........	600	3.6	Ni₂MnGa.........	379	4.2
Cu₂MnIn.........	520	4.0	Ni₂MnIn.........	323	4.4
Cu₂MnSn.........	530	4.1	Ni₂MnSn.........	344	4.1
Co₂MnSi.........	985	5.1	Ni₂MnSb.........	360	3.3
Co₂MnGa.........	694	4.1	Pd₂MnGe.........	170	3.2
Co₂MnGe.........	905	5.1	Pd₂MnSn.........	189	4.2
Co₂MnSn.........	829	5.1	Pd₂MnSb.........	247	4.4
			Au₂MnAl.........	258	3.1

* As summarized by R. S. Tebble and D. J. Craik, "Magnetic Materials," Wiley Interscience, London 1969, p 152.   Data by Endo, Ohoyama, Kimura, Oxley, Tebble, Williams, and Webster.

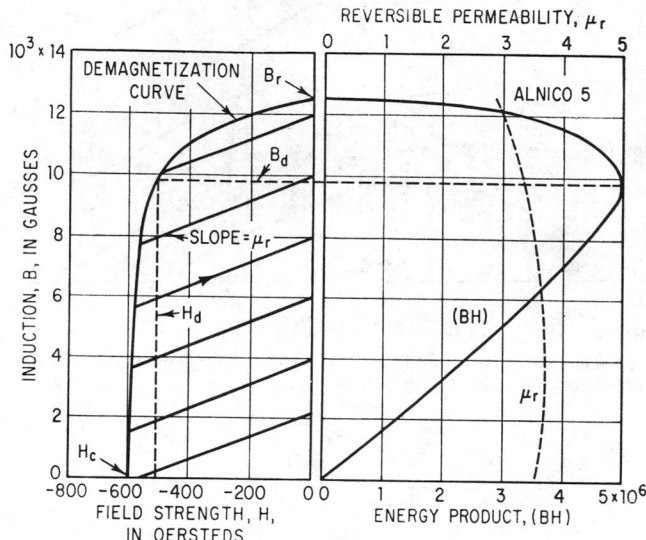

FIG. 5f-11. Demagnetization curve of Alnico 5, showing B_r, H_c, and optimum operating point (B_d, H_d).   Also energy-product curve and reversible permeability μ_r as function of B.

5f-6. Losses.   Losses in magnetic materials in alternating fields at low inductions (<100 gauss, approximately) are usually described by the following equation:[1]

$$\frac{R}{\mu L \nu} = aB + c + e\nu$$

R is in ohms (series) and L in henrys, as measured on an a-c bridge, μ the permeability, ν the frequency of alternating current in cps, B the maximum induction in gauss during the cycle, and a, c, and e the constants given in Table 5f-14.   The constant a is generally ascribed to hysteresis, c to lag, and e to eddy currents.

The loss angle δ is related to these constants and Q as follows:

$$\tan \delta = \frac{1}{Q} = \frac{R}{\omega L} = \frac{R}{2\pi \nu L}$$

C. D. Owens, *Proc. IRE* **41**. 359 (1953).

TABLE 5f-13. CONSTANTS OF PERMANENT-MAGNET MATERIALS*

Name	Composition, % by weight, remainder iron	H_c, coercive force, oersteds	B_r, residual flux density, gauss	Optimum point H_d	Optimum point B_d	(B_dH_d)max, energy product, 10^6 G-Oe	μ_r, reversible permeability at H_d, B_d	Density, g/cm³	Mechanical properties	Preparation
1% carbon steel	1C, 0.5 Mn	51	9,000	34	5,900	0.20	.....	7.8	Hard, strong	Hot-roll, machine, punch, heat-treat.
5% tungsten steel	5W, 0.7 C	70	10,560	47	7,000	0.33	30	8.1	Hard, strong	Hot-roll, machine, punch, heat-treat.
3½% chromium steel	3.5 Cr, 1 C, 0.5 Mn	66	9,500	45	6,500	0.29	35	7.8	Hard, strong	Hot-roll, machine, punch, heat-treat.
36% cobalt steel	36 Co, 3.75 W, 5.75 Cr, 0.8 C	240	9,750	147	6,300	0.93	12	8.2	Hard, strong	Cast, hot-roll, heat-treat.
Remalloy 2	12 Co, 20 Mo	340	8,550	220	5,400	1.2	.....	8.4	Hard, malleable	Hot-roll, machine, punch.
Vicalloy 2	52 Co, 13 V	415	9,000	325	7,000	2.3	...	8.2	Ductile	Cold-roll, draw.
Cunico 1	50 Cu, 21 Ni, 29 Co	680	3,400	400	2,000	0.8	3.2	8.3	Ductile	Cold-roll, machine.
Cunife 1	60 Cu, 20 Ni	500	5,400	320	4,000	1.3	1.7	8.6	Ductile	Draw, cold-roll, machine, punch.
Alnico 2	10 Al, 17 Ni, 12.5 Co, 6 Cu	550	7,250	350	4,600	1.60	6.4	7.1	Hard, brittle	Cast, ground.
Alnico 5	8 Al, 14 Ni, 24 Co, 3 Cu	620	12,500	525	10,200	5.30	4.3	7.3	Hard, brittle	Cast, ground.
Alnico 5-DG	8 Al, 14 Ni, 24 Co, 3 Cu	650	12,900	580	10,500	6.10	4.0	7.3	Hard, brittle	Chill-cast.
Alnico 5-7 (complete orientation)	8 Al, 14 Ni, 24 Co, 3 Cu	730	13,200	640	11,500	7.40	3.0	7.3	Hard, brittle	Chill-cast into hot mold, ground.
Alnico 6	8 Al, 16 Ni, 24 Co, 3 Cu, 2 Ti	750	10,500	525	7,100	3.70	4.8	7.4	Hard, brittle	Cast, ground.
Alnico 8	7 Al, 15 Ni, 35 Co, 4 Cu, 5 Ti	1,600	8,300	950	5,060	5.0	2.6	7.3	Hard, brittle	Cast, ground.
Alnico 9	7 Al, 15 Ni, 35 Co, 4 Cu, 5 Ti	1,450	10,500	1,100	7,700	8.5	1.3	7.3	Hard, brittle	Press in field at 280°C.
Bismanol	79.2 Bi, 20.8 Mn	3,650	4,800	2,000	2,640	5.3	1.1	8.1	Hard, brittle	Cold-work, draw, machine.
Platinum cobalt	76.7 Pt, 23.3 Co	4,300	6,450	2,800	3,400	9.5	1.2	15.7	Ductile	Press in field.
ESD Fe-Co (Lodex)	64 Fe-36 Co in Pb-Sb matrix	940	7,300	650	5,400	3.4	2.6	8.6	Malleable	Press in field.
Silmanal	86.75 Ag, 8.8 Mn, 4.45 Al	6,300 (H_{ci})	590	284	292	0.083	1.1	9.0	Ductile	Cold-roll, draw.
Barium ferrite	BaO·6 Fe₂O₃	1,850	2,200	900	1,100	1.0	1.15	4.7	Hard, brittle	Press, sinter.
Oriented	BaO·6 Fe₂O₃	2,200	4,000	1,750	2,000	3.5	1.05	5.0	Hard, brittle	Press in field.
Rubber bonded	BaO·6 Fe₂O₃	1,480	2,200	1,000	1,100	1.1	1.1	3.7	Flexible	Rolled sheet.
Samarium-cobalt†	SmCo₅	8,400	8,700	.....	.....	18.5	.....	95% theoretical	Hard, brittle	Press in field, uniaxial deformation.
Cast rare earth alloy‡	Co₀.₈Fe₀.₄Cu₁.₃₅Sm	4,000	6,400	.....	.....	8.8	.....	.....	Hard, brittle	Cast, age.

* Revised by H. H. Helms, Jr. and E. Adams, U.S. Naval Ordnance Laboratory.
† K. H. J. Buschow, W. Luiten, P. A. Naastepad, and F. F. Westendorp, *Philips Tech. Rev.* **29** (1968).
‡ E. A. Nesbitt, *J. Appl. Phys.* **40**, 1259 (1969).

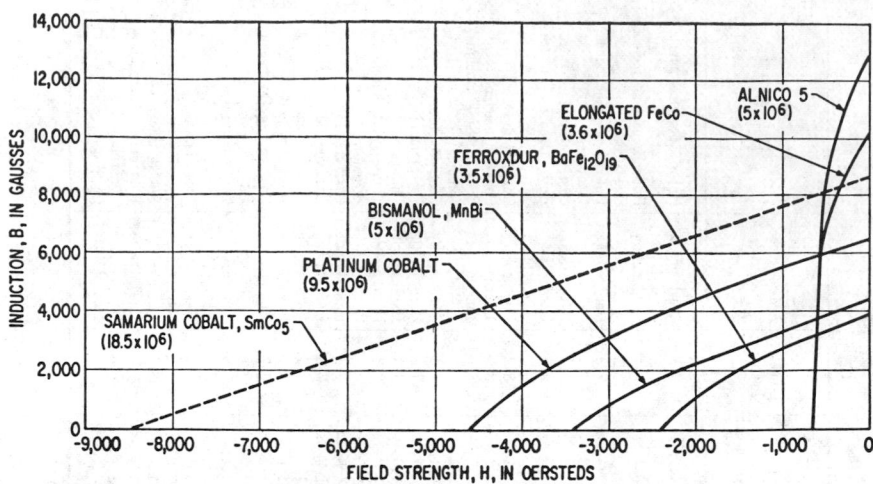

Fig. 5f-12. Demagnetization curves and maximum energy products, $(BH)_m$, of several types of permanent-magnet materials. (*Prepared by H. H. Helms and E. Adams, U.S. Naval Ordnance Laboratory.*)

This is valid only at low frequencies, when eddy-current shielding is negligible. Hysteresis losses at low inductions are described by the Raleigh relation

$$W_h = \frac{4\pi}{3} \frac{d\mu}{dH} H^3$$

per cycle, H being the maximum field strength during the cycle and $d\mu/dH$ the slope of the μ vs. H curve (near μ_0).

At high inductions, e.g., $B = 100$ to saturation, the relation often used to calculate hysteresis loss per cycle at maximum induction B is

$$W_h = \eta B^{1.6}$$

η being an empirical constant varying from 1 to 10^6.

TABLE 5f-14. MATERIAL CONSTANTS FOR LOSSES AT LOW INDUCTIONS
(*a* is hysteresis constant, *c* the "lag" constant, and *e* the eddy-current constant)

Material	Size	μ_0	$a \times 10^6$	$c \times 10^6$	$e \times 10^9$
Carbonyl iron........	5 μ	13	5	60	1
Mo Permalloy........	0.001-in. sheet	13,000	2	0.	10
Mo Permalloy........	120 mesh	125	1.6	30	19
Mo Permalloy........	400 mesh	14	11	140	7
Mn Zn ferrite........	....	1,500	1.6	4.8*	0.3
Ni Zn ferrite........	....	200	7	....	0.2

* $\nu < 1$ Mc/sec, higher values at higher frequencies.

5f-7. Antiferromagnetic Materials Studied by Neutron Diffraction.[1] *Introduction.* Since Table 5g-22 in the second edition of the Handbook was compiled, any magnetic structural distinction between ferro- (and ferri-) magnetic and antiferromagnetic materials has become increasingly arbitrary in view of the existence of many complex noncollinear or modulated structures with ferromagnetic components, which in some

[1] Compiled by D. E. Cox, Brookhaven National Laboratory, Upton, N.Y.

cases transform at some intermediate temperature to yet another structure. In general, such materials have been included in the table, and the only ones which have been systematically excluded are collinear ferromagnetic and ferrimagnetic materials. Even a few of the latter have been listed, however, if it is felt that the structural features are closely related to a basic antiferromagnetic arrangement.

Although the table was initially compiled in considerable detail with a format very similar to that of its predecessor, space limitations competing with an almost tenfold increase in the literature have necessitated the present highly abbreviated and concise form (Table 5f-15). A more detailed compilation is available on request from the author (Brookhaven National Laboratory Report No. 13822). For similar reasons, it has also not been possible to provide any structural details of a number of very interesting complex arrangements, in either the table or the accompanying figures. This is also true for solid solutions, where it is clearly impracticable to attempt to list all the relevant data. Reference to the original article is strongly urged in these cases.

Format and Abbreviations for Table 5f-15. COLUMN 1: MATERIAL. Materials have as far as possible been listed within structurally similar groups with the magnetic atoms in alphabetical order. There are, however, a number of departures from the latter scheme, for example, where compounds contain more than one such atom.

COLUMN 2: CRYSTAL CLASS AND NÉEL TEMPERATURE. The crystal classes have been abbreviated as follows: **C**(cubic), **T**(tetragonal), **H**(hexagonal), **R**(rhombohedral), **O**(orthorhombic), and **M**(monoclinic). The magnetic structures of all rhombohedral systems have been described in terms of the hexagonal unit cell. The crystal class is usually that cited in the neutron diffraction determination above the initial ordering temperature. The actual structure is sometimes known to be distorted from that assumed, and where there is a distortion associated with the (or one of the) magnetic transition(s), this has been denoted by *. A distortion apparently unconnected with any magnetic transition has been denoted by **. The Néel temperatures where listed are those cited in the neutron diffraction references, and are not necessarily determined by diffraction techniques. A second (or third) figure in parentheses indicates the temperature of a second (or third) transition, and † implies that the temperature in question corresponds to a Curie point (i.e., the appearance of a spontaneous moment).

A typical entry might be C: 64, which means a cubic lattice with a Néel point of 64 K.

COLUMN 3: MAGNETIC STRUCTURE. Most of the abbreviations used here can be found by reference to the figures and captions at the end. The description f. (or a.f.) sheets implies a structure with ferromagnetic (or antiferromagnetic) sheets which are coupled antiparallel to adjacent sheets. The symbol †† denotes that the structure described occurs over part of the composition range in solid solutions, and the symbol # (also used in column 4) denotes the presence of two magnetic phases. The use of braces indicates that the magnetic structure in question involves components from more than one type of mode and is therefore noncollinear. Changes in magnetic structure or additional magnetic ordering are entered opposite the appropriate transition temperature listed in parentheses in column 2.

It is to be noted that the magnetic unit cell is in many cases some multiple of the chemical cell; in order to save space this is not explicitly stated in the table but is very often obvious by reference to the appropriate figure.

COLUMN 4: MOMENT AND DIRECTION. A typical entry in this column lists first the magnetic moment in boldface type, followed by its direction. For example, the entry **1.7**: $\perp$[100]; 36°, [010] means a moment of $1.7\mu_B$ directed perpendicular to [100] and 36° from [010]. Where there are multiple entries of this sort, each moment and direction is that appropriate to the entry listed in column 3 on the same horizontal line. The moments which have been tabulated are for the most part those determined at the lowest temperature studied, which is 4.2 K in the majority of cases.

TABLE 5f-15. ANTIFERROMAGNETIC MATERIALS STUDIED BY NEUTRON DIFFRACTION

Material	Crystal class and Néel temperature, K	Magnetic* structure	Moment (in μ_B) and direction	References
NaCl and related structures, (see also Fig. 5f-13)				
CoO........	C*: 291	f2 (Fig. 5f-13)	**3.8**: 11.5°, [001]	290, 343, 363
		f2 or complex	**3.5**: 27.4°, [001]	247, 248
CrN........	C*: 273	f4A	**2.4**: [1̄10]	125
FeO........	C*: 198	f2	**3.3**: [111]	343, 345, 363
LiFeO₂.....	C	f2	**2.5-4.5**: ⊥[111]	138
MnO........	C*: 120	f2	**5**: ⊥[111]	343, 362, 363
(Mn,Co) O	C	f2		42
α-MnS......	C	f2	**5**: ⊥[111]	123, 363
Mn₀.₃₃Cr₀.₆₇S	C	f1	**4.1**: ⊥[001]	96
(Mn,Cr) S	C: 240			96
MnSe.......	C	f2	⊥[111]	363
Mn₀.₈Li₀.₁Se	C*: 71	f3	**4.6**: 45°, [001]	318
(Mn,Li) Se	C		⊥[001]	318
NiO........	C*: 530	f2	**1.8**: ⊥[111]	15, 16, 344, 346, 363
CeAs.......	C: 8	f1	**0.7**: [001]	341
CeSb.......	C: 16	complex		341
ErP........	C: 3.1	f2	**5.7**: ⊥[111]	112
ErSb.......	C: 3.7	f2	**7.0**: ⊥[111]	112
EuSe.......	C: 5.8	{f2 / sinusoidal}	⊥[111]? / ⊥[111]	323
EuTe.......	C: 7.8	f2	⊥[111]	410
GdBi.......	C	f2		266
GdS........	C	f2		266
GdSb.......	C	f2		266
GdSe.......	C	f2		266
HoP........	C: 5.5†	complex	**8.8**: ⊥[001]	112
HoSb.......	C: 9	f2	**9.3**: [100]	112
TbAs.......	C: 12	f2	**7.7**: [111]	112
TbP........	C: 9	f2	**6.2**: [111]	112
TbSb.......	C: 14	f2	**8.2**: [111]	112

Compound	Temp.	Structure	Moment: orientation	Refs.
UAs	C: 123, 128	f1	1.9, 2.1: [001]	260, 388, 412
	(>4.2)	f1A	2.2: [001]	
UN	C: 53	f1	0.8: [001]	148
UP	C: 125	f1	1.7, 1.9: [001]	149, 367
UP$_{0.95}$S$_{0.05}$	C: 122	f1	1.7: [001]	255
	(27)	f1 and f1A#	1.9: [001]	
U(P,S)	C	......	......	246
USb	C: 243	f1	>2.6, 2.2	260, 312

Perovskite and related structures (see also Fig. 5f-14)

Compound	Temp.	Structure	Moment: orientation	Refs.
DyAlO$_3$	O: 3.48	{ G / A }	4.7: [100] / 7.2: [010]	81, 195
TbAlO$_3$	O: 4.0	{ G / A }	6.9: [100] / 4.6: [010]	82, 267
Ba$_2$CoWO$_6$	C: 17	f2 (Fig. 5f-13b)	2.0: 23°, [111]	143
KCoF$_3$	C: 135	G	4.4	353
TbCoO$_3$	O: 3.31	{ A (Tb) / G (Tb) }	6.4: [100] / 4.6: [010]	267
DyCrO$_3$	O: 146 (2.16)	G(Cr) / complex(Dy)	2.8: [001] / 9.6: ⊥[001]	70, 77
ErCrO$_3$	O: 133 (16.8)	G(Cr) / C(Er)	2.9: varies / 5.2: [001]	70, 74
HoCrO$_3$	O: 140 (~12)	G(Cr) / { F / C } (Ho)	2.9: varies / 3.4: [100], 7.0: [010]	62, 70
KCrF$_3$	T	A	4.3: ⊥[001]	353
LaCrO$_3$	O: 282	G	2.5, 2.8: ⊥[001]	70, 216, 296
LuCrO$_3$	O: 112	G	2.5: ⊥[010]; 63°, [100]	70
NdCrO$_3$	O: 224 (~10)	G(Cr) / C(Nd)	2.6: varies / 1.3: [001]	70, 74
PbCrO$_3$	C: 240	G	1.9	349
PrCrO$_3$	O: 239 (>4.2)	G(Cr) / F(Pr)	2.5: varies / 0.5: [001]	70
TbCrO$_3$	O: 158 (4) (3.05)	G(Cr) / { F / C } (Tb) / complex(Tb)	2.9: [001] / 8.5: ⊥[001] / 2.6: varies	70, 76, 267
TmCrO$_4$	O: 124 (>4)	G(Cr) / F(Tm)	0.8: [001]	70

* The use of f1 and f2 are explained in Fig. 5f-13; G, A, F, and C in Fig. 5f-14.

TABLE 5-15. ANTIFERROMAGNETIC MATERIALS STUDIED BY NEUTRON DIFFRACTION (Continued)

Material	Crystal class and Néel temperature, K	Magnetic* structure	Moment (in μ_B) and direction	References
YCrO3	O: 141	G	3.0: [100]	70
YbCrO3	O: 118	G(Cr)	2.8: ⊥[010]; 68°, [100]	70
KCuF3	T	A	05: ⊥[001]	198
LaErO3	O: 2.4	{A / G / C}	1.1: [100] / 4.3: [010] / 4.5: [001]	283
BiFeO3	C**: 673	G	4.0	211, 324
ErFeO3	O: 620 (4.3)	G(Fe) / C(Er)	4.6: varies / 5.8: [001]	218
HoFeO3	O: 700 (6.5)	G(Fe) / {F / C}(Ho)	4.6: varies / 3.4: [100] / 6.7: [010]	218
KFeF3	C: 115	G	4.4	353
LaFeO3	O: 750	G	**4.4, 4.6**	216, 296
NdFeO3	O: 760	G(Fe)	4.6	218
Pb2FeNbO6	C	G	0.8	155
Pb3Fe2WO9	C: 363	G	3.3	324
RbFeF3	C: 105	G	4.3	396
Sr2FeMoO6	T: 450†	G	....	293
TbFeO3	O: 681	G(Fe)	4.8: varies	75, 267
	(8.4)	{F / C}(Tb)	2.4: [100] / 1.9: [010]	
	(3.1)	{A / G}(Tb)	6.6: [100] / 5.6: [010]	
TmFeO3	O	G(Fe)	varies	257
CaMnO3	C: 110	G	2.6	414
Ca0.85Bi0.15MnO3	O: 110	{C / G}		389
KMnF3	C**: 88-95	G	5.1	353
LaMnO3	O: 100	A	3.9: ⊥[001]	216, 414
(La,Ca)MnO3		C and F††		414
La(Mn,Cr)O3		....		57

Compound		Structure	Moment / direction	Ref
NH₄MnF₃	C: 84	G		321
NaMnF₃	C**: 60	G		321
NdMnO₃	O: 85	A(Mn)	1.7: ⊥[100]; 36°, [010]	340
PrMnO₃	O: 91	A(Mn)	1.8: [010]	340
RbMnF₃	C: 82	G		321
Mn₃GaN	C: 298	triangular		79
Ba₂NiWO₆	C	f2 (Fig. 5f-13b)	2.3: ⊥[111]	143
KNiF₃	C: 275	G	1.9: ⊥[111]	353
NaNiF₃	C: 149	G	2.2	159
EuTiO₃	O: 5.3	G	[100]	265
TbVO₃	O	C(V) / {C, F(Tb)}	1.3: [100]; 5.9: [100]; 4.8: [010]	267, 272
Spinel and related structures				
CoAl₂O₄	C: 4	(Fig. 5f-15a)?	3.6	347
MnAl₂O₄	C: 6.4	(Fig. 5f-15a)	3.3	347
Co₃O₄	C: 40	(Fig. 5f-15a)	3.4: ~60°, [111]	348
GeCo₂O₄	C: 20	complex / complex	3.2: [413] and [143]	62, 333, 270
CoCr₂O₄	C: 97† (31)	complex / ferrimagnetic spiral	2.5(Co), 2.7(Cr): ⊥[001] / ⊥[001]	334
CuCr₂O₄	T: 135†	complex (Fig. 5f-15b)		294, 338
FeCr₂O₄	C**: 80, 84† (35,42)	ferrimagnetic spiral	2.7: ⊥⟨100⟩ / ⊥[001]	37, 358
HgCr₂S₄	C: 60	spiral		190
MgCr₂O₄	C*: 15	complex		336
MnCr₂O₄	C: 43† (18)	ferrimagnetic spiral	4.3(Mn), 2.0(Cr₁), 3.1(Cr₁₁) / 3.1(Mn), 3.0(Cr): ⊥[001]	127, 157, 187
NiCr₂O₄	T: 65	complex	⊥[001]	335
ZnCr₂O₄	C: 15	unsolved	⊥[001]	339
ZnCr₂Se₄	C: 20	unsolved		184
GeFe₂O₄	C: 10	spiral		330, 331
Mn(Fe,Cr)₂O₄	C: 9	complex		332
ZnFe₂O₄	C: 33	complex		264
Mn₃a₂O₄		complex (Fig. 5f-15a)	3.6: [111]	183
CoMn₂O₄	T: 95-105† (70)	ferrimagnetic (Fig. 5f-15b)	3.2-3.4(tetr.): [110]; 2.9-2.1(oct.): ±58-39°, [110]	85, 88
(various degrees of inversion)				

TABLE 5f-15. ANTIFERROMAGNETIC MATERIALS STUDIED BY NEUTRON DIFFRACTION (*Continued*)

Material	Crystal class and Néel temperature, K	Magnetic* structure	Moment (in μ_B) and direction	References
$CrMn_2O_4$..........	T: 65† (>4.2)	ferrimagnetic (Fig. 5f-15b)	4.1(tetr.): [110]; 1.7(oct.): ±23°, [110]	89
$FeMn_2O_4$..........	T: 393† (55)	ferrimagnetic; complex	4.3(tetr.), 3.1(oct.)	90
$NiMn_2O_4$.......... (various degrees of inversion)	C: 116–164† (70)	ferrimagnetic	3.9(tetr.), 1.3(oct.)	86, 87
$GeNi_2O_4$..........	C: 15	complex	2.2: ⊥[111]	67
$\alpha\text{-}Mn_2TiO_4$.....	T: 62	complex	4.9(Mn_I), 4.8 (Mn_{II} and Mn_{III}): ⊥[010]	80
CoV_2O_4..........	C	complex?		156
MgV_2O_4..........	C: 45	complex	1.1: [001]	327
MnV_2O_4..........	T: 56† (52)	ferrimagnetic (Fig. 5f-15a)	4.4–4.5(Mn): [001]; 1.2(V): ±ca.50°, [001]	156, 326, 329, 337
NiAs and related structures				
CrAs........	O(Pnma): 300	spiral	....	400
Cr(Mn,As)....	O		....	400
Cr_2FeSe_4......	M	(Fig. 5f-16b)	2.6(Cr), 3.4(Fe): 55°, [10$\bar{1}$]	111
Cr_2NiS_4......	M	(Fig. 5f-16b)	2.0(Cr), 1.3(Ni): 45°, [10$\bar{1}$]	24
Cr_2TiS_4......	M	(Fig. 5f-16b)	....	253
CrS........	M: 460	complex	3.4	380
Cr_2S_3........	H: 125	spiral	2.1: ⊥[001]	250
Cr_3S_3........	R: 122	complex	2.4: 61°, [001]	78
Cr_3S_4........	M: 280	complex (Fig. 5f-16b)	2.3: 29°, [10$\bar{1}$]	25, 65
Cr_5S_6........	H: 303† (168)	ferrimagnetic	2.6–3.0: ⊥[001]	249
CrSb........	H: 705, 723	spiral (Fig. 5f-16a)	2.7, 2.8: [001]	376, 384
Cr(Sb,Te).....	H	canted††		139, 385
(Cr,Mn)Sb.....	H	canted††		139, 315, 384
CrSe........	H: 300	triangular (Fig. 5f-16b)	2.9: mainly ⊥[001] (10$\bar{1}$); 30°, [010]	126
Cr_3Se_4.......	M: 80			65

Substance	Structure / temp	Magnetic structure	Moment : direction	References
CrTe	H: 330†	ferromagnetic	....	385
	(150)	{ferromagnetic / a.f. (101) sheets}		
Cr_3Te_4	M: 329†	ferromagnetic	~1	65
	(80)	{ferromagnetic (Fig. 5f-16b)}		
FeS	H*: 600	(Fig. 5f-16a)	4: varies	18, 22
$Fe_{1-x}S$	M	(Fig. 5f-16a)	....	22, 365, 377, 378, 380
Fe_3Se_4	M: 483, 460†	(Fig. 5f-16a)	2.2(Fe_I), **1.4**(Fe_{II}): [110]	23
Fe_7Se_8	O(Pnma): 291†	(Fig. 5f-16a)	3.6(Fe_I), **4.5**(Fe_{II}): varies	19, 208
MnP	(50)	ferromagnetic / spiral	**1.3**, **1.6**: ⊥[001]	163,166
MnTe	H: 320, 323	(Fig. 5f-16a)	4.6: ⊥[001]	154, 244, 370
NiS	H: 263	(Fig. 5f-16a)	1.7: [001]	379, 380
Rutile and related structures				
CoF_2	T: 50	(Fig. 5f-17a)	3.0: [001]	161
$CrCl_2$	O: 20	complex	~3: along long Cr–Cl	98
CrF_2	M: 53	(Fig. 5f-17a)	~3: along long Cr–F	98
Cr_2TeO_6	T: 105	G (Fig. 5f-17b)	2.5: ⊥[001]	242, 280
Cr_2WO_6	T: 69	A (Fig. 5f-17b)	2.1: ⊥[001]	242, 278, 280
FeF_2	T: 90	(Fig. 5f-17a)	4.6: [001	161
FeOF	T: 315	(Fig. 5f-17a)	4.8: [001	109
Fe_2TeO_6	T: 219	G (Fig. 5f-17b)	**4.2**, **4.7**: [001]	242, 279
$Mn_{0.5}Cr_{0.5}O_2$	T: 390†	{ferromagnetic (Fig. 5f-17a)}	0.5: ⊥[001] / 0.4: ⊥[001]	395
MnF_2	T: 75	spiral	5.0: [001]	161, 295
MnO_2	T: 84	(Fig. 5f-17a)	5.0: 10°, [001]; ⊥[001	160, 422
NiF_2	T: 83	(Fig. 5f-17a)	⊥[001]	161
VF_2	T: 7	spiral	⊥[001]	8
V_2WO_6	T	A (Fig. 5f-17b)	⊥[001]	256
				242
Olivine and related structures (see also Fig. 5f-18)				
$CaCoSiO_4$	O(Pbnm): 16	complex	3.3: [001]	305
Co_2SiO_4	O(Pbnm): 49	C(Co_I) / C(Co_{II})	3.3: [001]	307
$LiCoPO_4$	O(Pbnm): 23	A	[001]	351
Cr_2BeO_4	O(Pbnm): 28	spiral	**1.6**(Cr_I), **2.8**(Cr_{II}): ⊥[010]	146

TABLE 5f-15. ANTIFERROMAGNETIC MATERIALS STUDIED BY NEUTRON DIFFRACTION (*Continued*)

Material	Crystal class and Néel temperature, K	Magnetic* structure	Moment (in μ_B) and direction	References
Fe₂SiO₄	O(Pbnm): 65	C(Fe_I)	[001]	350
		C(Fe_II)	[001]	
	(23)	{C	[001]	
		G(Fe_I)	[100]	
		A	[010]	
	65	C	3.2: [001]	141
		{A(Fe_I)	2.1: [100]	
		G}	1.1: [010]	
		C(Fe_II)	3.9: [001]	
LiFePO₄	O(Pbnm): 50	A	3.8: [001]	352
CaMnSiO₄	O(Pbnm): 9	G	[100]	108
LiMnPO₄	O(Pbnm): 35	A	5.2: [010]	303
Mn₂SiO₄	O(Pbnm): 50	C(Mn_I)	[010]	350
		C(Mn_II)	[010]	
	(13)	{C(Mn_I)	[010]	141
		G}	[100]	
		{C(Mn_II)	3.6: [010]	
		G}	3.6: [001]	
		C(Mn_II)	4.7: [010]	
LiNiPO₄	O(Pbnm): 23	A	[100]	351
Ni₂SiO₄	O(Pbnm): 34	complex	⊥[001]	302
PbFCl and related structures				
UAs₂	T: 283	(Fig. 5f-19b)	1.6: [001]	309
UBi₂	T: 183	(Fig. 5f-19a)	1.6: [001]	259
UOS	T: 55	(Fig. 5f-19c)	1.9: [001]	50
UOSe	T: 90	(Fig. 5f-19c)	2.2: [001]	287
UOTe	T: 160	(Fig. 5f-19a)	2.7: [001]	286
UP₂	T: 203	(Fig. 5f-19b)	1.0: [001]	387
USb₂	T: 206	(Fig. 5f-19b)	0.9: [001]	259
Corundum and related structures				
CoTiO₃	R: 38	(Fig. 5f-20c)	⊥[001]	300

Nb₂Co₄O₉	H: 30	f. [001] chains	~3: [001]	58
Cr₂O₃	R: 318	(Fig. 5f-20a)	2.8: [001]	91, 131
α-Fe₂O₃	R: 948	(Fig. 5f-20b)	~5: varies	122, 135, 147, 284, 297, 320, 342, 363
α-(Fe,Cr)₂O₃	R	spiral††		137
α-(Fe,Rh)₂O₃				235
α-(Fe,V)₂O₃				235
FeTiO₃	R: 68	(Fig. 5f-20c)	~4.0: [001]	355
α-Fe₂O₃–FeTiO₃	R	ferrimagnetic††		357
MnTiO₃	R: 41	a.f. (001) sheets	4.6: [001]	354
Nb₂Mn₄O₉	H: 125	f. [001] chains	~5: [001]	58
NiTiO₃	R	(Fig. 5f-20c)	2.3: ⊥[001]	354
CrVO₄ type structures (see also Fig. 5f-21)				
α-CoSO₄	O: 15.5	{C, G	2.9: [010], 1.4: [001]	144, 168
β-CrPO₄	O: 22	spiral	2.5: ⊥[001]	145
CrVO₄	O: 50	A	2.1: 27° [100]; 64°, [010]; 81°, [001]	168
TeSO₄	O: 21	C	4.1: [010]	168
MnSO₄		spiral	4.8	411
NiSO₄	O: 37	C	2.1: [010]	168
NiSeO₄	O	A	⊥[001]	171
CuSO₄ type structures (see also Fig. 5f-21)				
β-CoSO₄	O: 12	{A, G, C	2.3, 2.7: [100], 1.5, 1.9: [010], 1.7, 1.9: [001]	61, 93
CoSeO₄	O	{A, G, C		171
CuSO₄	O: 35	A	0.8: [100]	14, 271
MnSeO₄	O: 20	A	5.0: [100]	171
Calcite type structures				
CoCO₃	R: 20.4	(Fig. 5f-22)	46°, [001]	9
FeCO₃	R: 20	(Fig. 5f-22)	5.0: [001]	7, 317
MnCO₃	R: 32	(Fig. 5f-22)	⊥[001], [001]	7, 95, 316
NiCO₃	R: 25	(Fig. 5f-22)	63°, [001]	10

Table 5f-15. Antiferromagnetic Materials Studied by Neutron Diffraction (Continued)

Material	Crystal class and Néel temperature, K	Magnetic* structure	Moment (in μ_B) and direction	References
Garnet type structures				
$Dy_3Al_5O_{12}$	C: 2.49, 2.54	a.f. [100] chains	**9.0, 9.5**: $\langle 100 \rangle$	188, 195
$Er_3Ga_5O_{12}$	C: 0.79	a.f. [100] chains	5.9: $\langle 100 \rangle$	180
$Ho_3Al_5O_{12}$	C: 0.95	a.f. [100] chains	5.8: $\langle 100 \rangle$	181
$Nd_3Ga_5O_{12}$	C: 0.52	f. [100] chains	3.6: $\langle 100 \rangle$	179
$Tb_3Al_5O_{12}$	C: 1.35	a.f. [100] chains	5.7: $\langle 100 \rangle$	181
$YMnO_3$ type structures (see also Fig. 5f-23)				
$ErMnO_3$	H: 79	triangular	3.5(Mn): $\perp[001]$; 70° $\langle 100 \rangle$	221
$HoMnO_3$	H: 76	triangular	3.5(Mn): $\langle 100 \rangle$	221
$LuMnO_3$	H: 91	triangular	3.7(Mn): $\perp[001]$; 55°, $\langle 100 \rangle$	221
$ScMnO_3$	H: ~120	triangular	~4.0: $\perp[001]$; ~24°, $\langle 100 \rangle$	221
$TmMnO_3$	H: ~86	triangular	~3.8(Mn): $\perp[001]$; ~45°, $\langle 100 \rangle$	221
$YMnO_3$	H: 80	triangular	3.5: $\langle 100 \rangle$	62, 64, 69, 221
VF_3 and related structures				
CoF_3	R: 460	(Fig. 5f-22)	4.4: [001]	416
CrF_3	R: 80	(Fig. 5f-22)	3.0: $\perp[001]$	416
FeF_3	R: 394	(Fig. 5f-22)	~5.0: $\perp[001]$	416
MnF_3	M: 43	f. (101) sheets	4.0: (101)	416
MoF_3	R: 185	(Fig. 5f-22)	~3: $\perp[001]$	407
Miscellaneous anhydrous halides				
$CoBr_2$	H: 19	(Fig. 5f-24a)	2.8: $\perp[001]$	405
$CoCl_2$	R: 25	(Fig. 5f-24b)	3.0: $\perp[001]$	405
$CoCs_3Cl_5$	T: 0.52	f. (101) sheets	3.2: [001]	182
$CrCl_3$	H: 17	(Fig. 5f-24b, with vacancies)	2.7: $\perp[001]$	99
$FeBr_2$	H: 11	(Fig. 5f-24a)	4.4: [001]	405
$FeCl_2$	R: 24	(Fig. 5f-24b)	4.5: [001]	192, 405
$FeCl_3$	R: 15	spiral	4.3: $\perp[140]$	102
K_2IrCl_6	C: 3.05	f3A (Fig. 5f-13c)	[001]	197, 275
Cs_2MnCl_4	T	(Fig. 5f-24c)	[001]	262

Compound	Temp.	Structure	Direction	Refs.
CsMnF$_3$	H: 64	f. (001) sheets	⊥[001]	321
K$_2$MnF$_4$	T	(Fig. 5f-24c)	4.5: [001] [010]	263
MnBr$_2$	H: 2.16	complex		217, 415
MnI$_2$	H: 3.4	spiral	4.6: ⊥[307] [001]	101
K$_2$NiF$_4$	T: 97	(Fig. 5f-24c)		83, 325, 328
RbNiF$_3$	H	complex		322
K$_2$ReBr$_6$	C**: 15.3	f1 (Fig. 5f-13a)	⊥[001]?	275
K$_2$ReCl$_6$	C**: 11.9, 12.4	f1 (Fig. 5f-13a)	2.6, 2.7: ⊥[001]	275, 374
α-RuCl$_3$	H: ~30		~0.2	164
Miscellaneous oxides				
Co$_2$B$_2$O$_6$	O: 30	complex		301, 304
CoUO$_4$	O: 12	f. (111) sheets	4.2(Co): ⊥[010]	59
β-CaCr$_2$O$_4$	O	complex	2.4: [010]	133
CrUO$_4$	O	A(Cr) (Fig. 5f-25) A(U)	0.3: [010]	38
CuO	M: 230	f (10Ī) sheets		92
Er$_2$O$_3$	C: 3.36	complex	6.1(Er$_I$): ⟨111⟩ 5.4(Er$_{II}$): ⟨100⟩	71, 282
Ba$_{0.4}$Sr$_{1.6}$Zn$_2$Fe$_{12}$O$_{22}$	H: 400† (380)	ferrimagnetic	⊥[001]	314, 371, 372
BaCoFe$_{17}$O$_{27}$	H	spiral		419
BaSc$_{1.8}$Fe$_{10.2}$O$_{19}$	H	spiral		6
CaFe$_2$O$_4$	O: 200, 285 (120–170,140)	a.f. chains#	4.1–4.4: [001]	13, 72, 133, 399
Ca(Fe,Cr)$_2$O$_4$	O	complex	4.5, 4.9: [001]	133
Ca$_2$Fe$_2$O$_5$	O: 720, 730	complex	3.5: ⊥[001]	131, 170, 383
				175
FeSb$_2$O$_4$	T	A C (Fig. 5f-21) G		
FeTi$_2$O$_5$	O	a.f. [100] chains	4.4: ⊥[010]; 29°, [100]	273
FeWO$_4$	M: 66	f. (100) sheets	4.5: [001]	390
LiFeO$_2$	T: ~315	complex	4.2: [001]	138
β-NaFeO$_2$	O: 723	f. (101) sheets	[010]	63
FeUO$_4$	O: 55	C(Fe) C F (Fe) (Fig. 5f-25)	2: [100] 4: [001]	40, 41
	(42†)	C(U)	0.4: [100]	
Gd$_2$O$_3$	C: ~1.6	unsolved		115
BiMn$_2$O$_5$	O: 52	complex	⊥[001]	73

TABLE 5f-15. ANTIFERROMAGNETIC MATERIALS STUDIED BY NEUTRON DIFFRACTION (Continued)

Material	Crystal class and Néel temperature, K	Magnetic* structure	Moment (in μ_B) and direction	References
$CaMn_2O_4$	O: 225	complex	3.6: [100]	12
MnB_2O_6	O	spiral?	....	304
$\alpha\text{-}Mn_2O_3$	C: 90	unsolved	....	110
$MnUO_4$	O: ($\sim$50)12	f. (100) sheets	4.9: [010] [10$\bar{1}$]	39
$MnWO_4$	M: 16	complex	....	151
$Na_3B_6O_6$	O: 49	complex	....	304
CaV_2O_4	O	complex	1.1: [010]	189
UO_2	C: 30.8, 30.6	f1 (Fig. 5f-13a)	1.8: $\perp$[001]	169, 191, 413
Yb_2O_3	C: 2.25	complex	1.1(Yb_I): $\langle 111 \rangle$; 1.1(Yb_{II}): $\langle 111 \rangle$; 1.9(Yb_{II}): $\langle 110 \rangle$	282
Miscellaneous chalcogenides				
$AgCrSe_2$	R: 50	(Fig. 5f-24b)	2.7: $\perp$[111]	252
$NaCrSe_2$	R: 40	(Fig. 5f-24b)	2.3: $\perp$[111]	252
$CuFeS_2$	T	f. (001) sheets	3.9: [001]	153
$FeNb_3S_6$	H	complex	3.7: [001]	251
FeS_2	C	f2 (Fig. 5f-13b)	5: $\perp$[001]	84
$\beta\text{-}MnS$	C	f3A	5: $\perp$[011]	123
$\beta\text{-}MnS$	H	complex	5: [001]	123
MnS_2	C	f3A (Fig. 5f-13c)	5: [001]	185
$MnSe_2$	C	complex	5: [001]	130, 185
$MnTe_2$	C	f1 (Fig. 5f-13a)	5: $\perp$[001]	185
Miscellaneous hydrates				
$CoCl_2\cdot2D_2O$	M: 17.5	f. [001] chains	2.8: [010]	140, 142
$CoCl_2\cdot6H_2O$	M: 2.25	a.f. (001) planes	[001]	213
$CuCl_2\cdot2D_2O$	O: 4.3	{A (Fig. 5f-21) ; G (Fig. 5f-21)}	$\sim$1: [100] ; $\sim$0.1: [001]	360, 373, 392
$CuF_2\cdot2H_2O$	M: 10.9	(Fig. 5f-17a)	0.8: $\sim$[001]	1
$LiCuCl_3\cdot2H_2O$	M: 4.4	a.f. [100] chains	$\sim$1: $\perp$[010]; 49°, [001]; 158°, [100]	3
$Fe_3(PO_4)_2\cdot4H_2O$	M: 15†	complex	4.6: $\sim\perp$[100]; 10°, (101)	4
$FeSO_4\cdot H_2O$	M	f. (100) sheets	$\perp$[010]	261
$NiCl_2\cdot6H_2O$	M	f. (101) sheets	10°, [100]	214

CaC₂ and related structures

Compound	T / C	Structure	Moment : direction	Refs.
AlCr₂	T: 598	f. (001) sheets in sequence (+ − − +)	0.9 : 65°, [001]	29
MnAu₂	T: 363	spiral	3.0 : ⊥[001]	193, 194
CeC₂	T: 33	(Fig. 5f-17a)	1.7 : [001]	30
DyC₂	T: 59	sinusoidal	11.8 : [001]	33
HoC₂	T: 26	spiral	6.9 : ⊥[100]	30
NdC₂	T: 29	(Fig. 5f-17a)	3.0 : [001]	30
PrC₂	T: 15	(Fig. 5f-17a)	1.1 : [001]	30
TbAu₂	T: 55 (42.5)	sinusoidal	[001]	31
TbAg₂	T: 35	f. (100) sheets	9.0 : [001]	32
TbC₂	T: 66 (40)	f. (100) sheets / spiral	9.0 : [001] / 5.1 : ⊥[100]	30, 32
α-KO₂	T**: 7.1	f. (001) sheets	~1(O₂⁻); ⊥[001]	375

CsCl and related structures (see also Fig. 5f-14)

Compound	T / C	Structure	Moment : direction	Refs.
DyAg	C: 51	C	9.8 : [001]	26
FeRh	C: 678† (338)	F / G	3.3	359
Fe-Rh	C*: 515 (403)	A / A	⊥[001] / 4.1 : [100] and [001]‖	60, 233, 285, 359 / 44, 46
AuMn				
Au-Mn	C*: 147† (65)	F		46
Au₂MnAl (ordered)		spiral	4.4 : ⊥[001]	49
Au₂(Mn,Al)₂				49
MnHg	C*: 460	G	3.7, 3.9 : varies	291, 310, 311
β-MnZn	C	{F / G	1.7 / 2.9	196, 192
Pd₂MnAl (disordered)	C: 240	G	4.4	402
Pd₂MnIn (disordered)	C	G	4.3	401
Pd₂MnIn (ordered)	C: 142	f2 (Fig. 5f-13b)	4.3 : ⊥[111]	401
TbAg	C: 100	C	~9.0 : [001]	104
Tb(Ag,In)	C		⊥[001]	107
Tb(Ag,Pd)	C			107
TbCu	C: 115	C	~8.9 : [001]	104

TABLE 5f-15. ANTIFERROMAGNETIC MATERIALS STUDIED BY NEUTRON DIFFRACTION (Continued)

Material	Crystal class and Néel temperature, K	Magnetic* structure	Moment (in μ_B) and direction	References
CuAu-I and related structures				
CrPt	C	(Fig. 5f-26)	**2.2**: ⊥[001]	319
MnNi	T: 1073	(Fig. 5f-26)	**3.8, 4.0**: ⊥[001]	205, 313
Mn-Ni	T	....	....	313, 368, 369
MnPd	T: 813	(Fig. 5f-26)	**4.4**: ⊥[001]	212, 238, 313
Mn_2Pd_3	T: 643, 653	(Fig. 5f-26)	**4.1, 4.3**: ⊥(001)	176, 215, 239
$Mn_{11}Pd_{21}$	T	(Fig. 5f-26)	**4.0**: ⊥[001]	212
Mn-Pd	T	....	....	212, 313
MnPt	T: 973, 970	(Fig. 5f-26)	**4.3**: varies	20, 313
Mn-Pt	T	....	....	20, 21, 313
Cu₃Au and related structures				
Pt_3Fe	C: 170	C (Fig. 5f-14)	**3.4**: [001]	47, 209, 236
Pt-Fe		....	⊥[001]	47
$Pt_{3-x}(Fe,Mn)_{1+x}$		....	....	48
$(Pt,Pd)_3Fe$	C: 475	complex	[001]	210, 232
Mn_3Pt	(365)	triangular	**3.0**: ⟨112⟩	241, 416
	523	....	....	366
	(388)			
Mn-Pt	C	....	....	241
$Mn_3(Pt,Rh)$	C	....	....	241
$(Mn,Fe)_3Pt$	C	....	....	240, 241
Mn_3Rh	C: 855	triangular	**3.5**: ⟨112⟩	229, 241
Pd_3Mn	T: 170	complex	**4.0**(Mn), **0.2**(Pd): ⊥[001]	103
α-Zn₃Mn	C*: 150	A (Fig. 2)	**2.5**: ⊥[001]	292
Cu₂Sb and related structures				
CrAs	T: 393	complex	**1.1**(Cr$_I$), **1.2**(Cr$_{II}$): ⊥[001]	397
Fe_2As	T: 353	(Fig. 5f-27c)	**1.0**(Fe$_I$), **1.5**(Fe$_{II}$): ⊥[001]	207
FeMnAs	T	(Fig. 5f-27a)	**0.2**(Fe), **3.6**(Mn): ⊥[001]	421
Mn_2As	T: 573	(Fig. 5f-27a)	**3.7**(Mn$_I$), **3.5**(Mn$_{II}$): ⊥[001]	34
Mn_2Sb	T: 550†	(Fig. 5f-27b)	**2.1**(Mn$_I$), **3.9**(Mn$_{II}$): varies	17, 404
$Mn_2Sb_{0.7}As_{0.3}$	T (308–388)	(Fig. 5f-27a)	**2.3**(Mn$_I$), **2.8**(Mn$_{II}$): ⊥[001]	34

Material		Structure	Moment : Direction	Ref.
Mn₁.₉Cr₀.₁Sb	T	(Fig. 5f-27b)	[001]	120
Mn₁.₉₇Cr₀.₀₃Sb	(135) (115)	(Fig. 5f-27a) (Fig. 5f-27b)	1.4(MnI), 2.8(MnII):[001]; varies	35
Mn₁.₉Cr₀.₁Sb₀.₉₅In₀.₀₅	T	complex (Fig. 5f-27a) (Fig. 5f-27b)	⊥[001]; 1.8(MnI), 3.7(MnII): ⊥[001]	119
(Mn,Cr)₂Sb	T	(Fig. 5f-27a)	[001]; ⊥[001]	35
Rare-earth metals and alloys (see also Fig. 5f-28)				
Ce	H: 12.5	complex	0.6:[001]	408
Dy	H: 179 (90†)	spiral	9.5⊥[001]	409
Er	H: 80 (52) (20†)	ferromagnetic; sinusoidal	7.6:[001]; 9.0	100, 106
Eu	C: 91	spiral	5.9: ⊥[100]	298
Ho	H: 130 (~40) (~20†)	spiral; distorted spiral; distorted spiral	⊥[001]; ⊥[001]; 9.5	222, 225, 393
Nd	H: 19 (7.5)	sinusoidal (Nd₁); sinusoidal (Nd₂)	2.3:[10Ī0]; 1.8:30°,[10Ī0]	281
Pr	H: 25	sinusoidal	⊥[001]	105
Tb	H: 226, 229 (216,221†)	spiral; ferromagnetic	[001]; 7.0:[001]	152, 220, 393
Tm	H: 56 (40)	sinusoidal; antiphase	7: ⊥[001]	219
ErAl	O: 10	complex	8.4: ⊥[001]	55
HoAl	O: 26†	complex	8.8: ⊥[001]	54
TbAl	O: 72	complex	5.0–8.0(Tb), 1.1–2.5(Mn): ⊥[001]	56
TbMn₂	C: 40	spiral		129
TbNi₂	C: 46†	{F G}	7.2(Tb):[111]?; ⊥[111]?	162
Dy-X (X = Er,Y)				113
Er-X (X = Ho,Sc,Th,Y)				113, 116, 117, 224, 274, 361
Gd-X (X = Sc,Y)		spiral††		118

TABLE 5f-15. ANTIFERROMAGNETIC MATERIALS STUDIED BY NEUTRON DIFFRACTION (Continued)

Material	Crystal class and Néel temperature, K	Magnetic* structure	Moment (in μ_B) and direction	References
Ho-X .... (X = Sc,Tb,Th,Y)	....	....	....	113, 116, 117, 224, 258
Nd-Th	....	....	....	116
Pr-Th	....	....	....	116
Tb-X .... (X = La,Lu,Sc,Th,Y)	....	....	....	113, 114, 116, 117, 220, 226
Tm-Y	....	....	....	113
Transition metals and miscellaneous alloys and intermetallic compounds				
Cr	C: 311.5 (115)	sinusoidal; sinusoidal	$\perp\langle 100\rangle$; **0.6**: $\langle 100\rangle$	27, 45, 53, 94, 97, 124, 172, 174, 186, 276, 356, 364, 394, 403, 406
$Cr_{1-x}Co_x$ ($x \leq 0.053$)	C	(Fig. 5f-17a)††	....	158
$Cr_{1-x}Fe_x$ ($x \leq 0.047$)	C	(Fig. 5f-17a)††	....	28, 199
$Cr_{1-x}Mn_x$ ($x \leq 0.48$)	C	(Fig. 5f-17a)††	....	52, 173, 177, 178, 204, 243, 277
$Cr_{0.995-x}Mn_xV_{0.005}$ ($x \leq 0.025$)	C	(Fig. 5f-17a)††	....	228
$Cr_{1-x}Ni_x$ ($x \leq 0.01$)	C	....	....	158
$Cr_{1-x}Re_x$ ($x \leq 0.008$)	C	(Fig. 5f-17a)††	....	277
$Cr_{1-x}V_x$ ($x \leq 0.02$)	C	....	....	178, 227, 277
$Cr_{1-x}X$ (X = V,Mn,Nb,Mo,Tc,Ru,Rh, Ta,W,Re)	C	(Fig. 5f-17a) in a number of cases for $x >$ ca. 0.01	....	223
$CrAu_4$	T: ~400	....	....	386
γ-Cu-Mn_{1-x} ($0.05 \leq 0.31$)	C*	f1 (Fig. 5f-13a)	[001]	43, 269

Compound	Structure	Spin structure	Moment / direction	Refs
γ-Fe	C: 8	f1 (Fig. 5f-13a)	**0.7**: 19°, [001]	2
FeGe	H: 410	(Fig. 5f-29)	**1.7**: [001]	5, 398
Fe₁.₇₇Ge	H: 500†	complex	**1.3**(Fe$_I$), **1.1**(Fe$_{II}$): ⊥[001]	206
FeGe₂	T: 270, 315	$\{$ C / G (Fig. 5f-21) ; C (Fig. 5f-21)	⊥[001]; ⊥[001]; ⊥[001]	68, 234, 289
(Fe,Mn)₅Ge₃	H	C (Fig. 5f-21)		36, 381
Fe₃Mn	C: 425	pyramidal	**1.7**: ⟨111⟩	230
Fe₀.₆₉Mn₀.₃₁	C: 435	pyramidal	**1.5**: ⟨111⟩	391
γ-Fe-Mn	C	....	....	200, 201
(Fe,Ni)₃Mn	C	....	....	230
FeSn	H: 373	(Fig. 5f-29)	**1.5**: ⊥[001]	418
FeSn₂	T: 384	C (Fig. 5f-21)	**1.6**	202, 306
α-Mn	C: 95	complex	**0.1–1.8**	203, 245, 308, 364
Mn₃Ge	H: 360	triangular	**2**: varies [001]	229
Mn₃(Pt,Rh) (disordered)	C*	f1 (Fig. 5f-13a)		237
Mn₅Si₃	H: 68	sinusoidal	**0.4**(Mn$_I$), **1.2**(Mn$_{II}$): ⊥[001]	254
Mn₃Sn	H: 420 ~(270)	triangular	**2.5**: ⊥arbitrary [hkO]	229
(Mn,Fe)₃Sn	H	spiral	**2.3**: [110]	231
MnSn₂	T: ~325 (~74)	f. (110) sheetsᵃ ; f. (110) sheets in sequence + + – –	**2.3**: [110]	128, 134

Other compounds

Compound	Structure	Spin structure	Moment / direction	Refs
α-FeOOH	O(Pbnm): 362, 403	f. (001) sheets	**5**: [001]	167, 382
HoD₂	C: 8	unsolved		136
Mn₂N	O: 301	complex	**1.6**: [001]; 15°. [100]	268
MnOOH	O: ~40	spiral		150
Mn₂P	H: 103	sinusoidal	~**0**(Mn$_I$), **0.8**(Mn$_{II}$): ⊥⟨100⟩	420
α-O₂	M(C2/m): 24	f. (100) sheets ; f. (100) sheets	⊥⟨001⟩ [010]	51, 121 ; 11
TbD₂	C: 40	sinusoidal	**7.9**: [001]	136

ᵃ There is also a small sinusoidal component in the region 74 < T < 90 K.

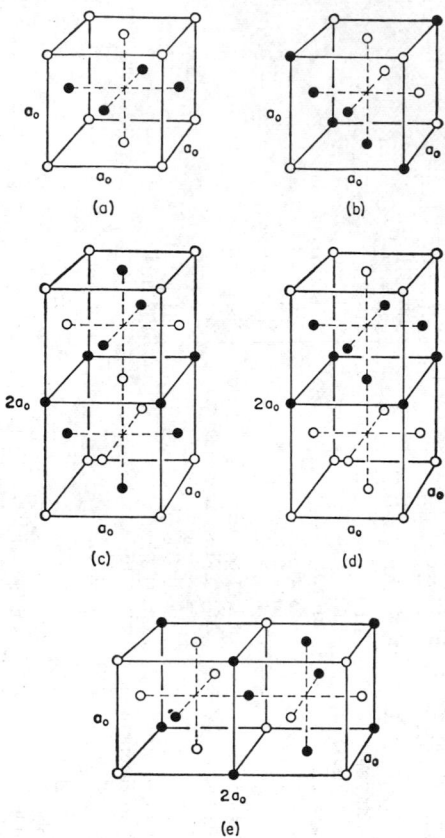

FIG. 5f-13. Ordering in f.c.c. structures: (a) type 1(f1), (b) type 2(f2), (c) type 3A (f3A), (d) type 1A(f1A), (e) type 4(f4).

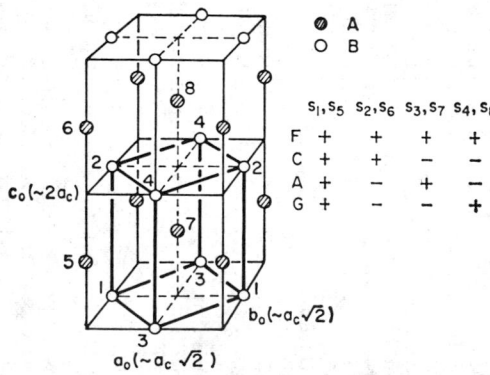

FIG. 5f-14. Ordering in orthorhombic perovskite (ABO_3) type structures. The ideal simple cubic cell (a_c) is shown in heavy outline.

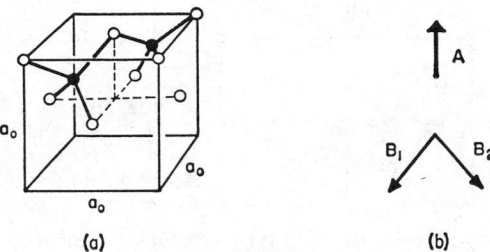

(a)   (b)

FIG. 5f-15. Ordering in spinel-type structures: (a) tetrahedral (A) nearest neighbors anti-parallel, (b) schematic Yefet-Kittel canting of octahedral (B) moments.

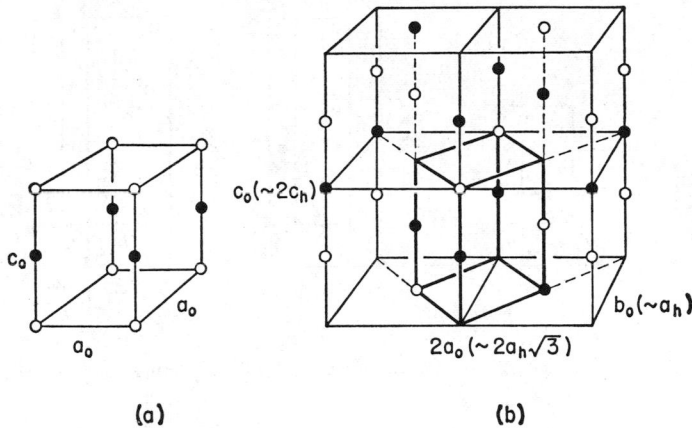

(a)   (b)

FIG. 5f-16. (a) CrSb and (b) Cr_3S_4 type magnetic structures.   The ideal hexagonal cell (a_h, c_h) is shown in heavy outline in the latter case.

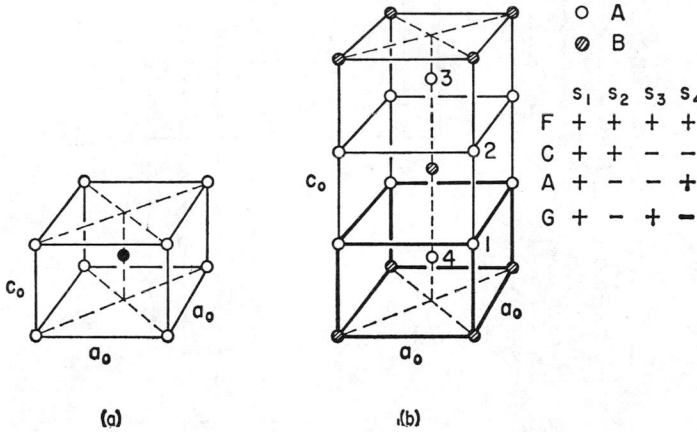

(a)   (b)

FIG. 5f-17. (a) b.c. tetragonal antiferromagnetic MnF_2 type structure; (b) ordering in trirutile (A_2BO_6) type structures.   B is a nonmagnetic ion.   The rutile cell is shown in heavy outline.

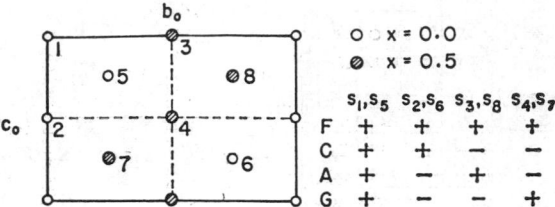

FIG. 5f-18. Ordering in olivine (Mg_2SiO_4) type structures projected on (100)-$Pbnm$ orientation.   Atoms 5 to 8 have been placed in idealized positions.

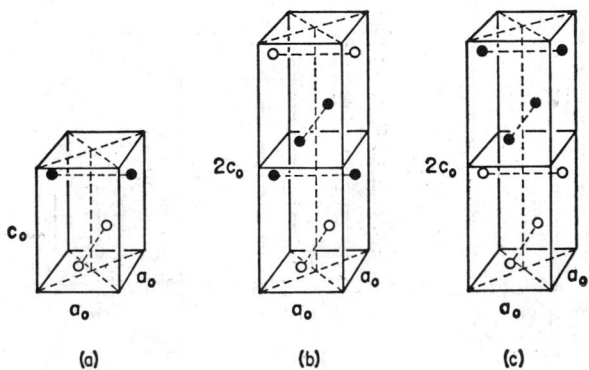

FIG. 5f-19. (a) UBi_2, (b) UAs_2, and (c) UOS type magnetic structures.

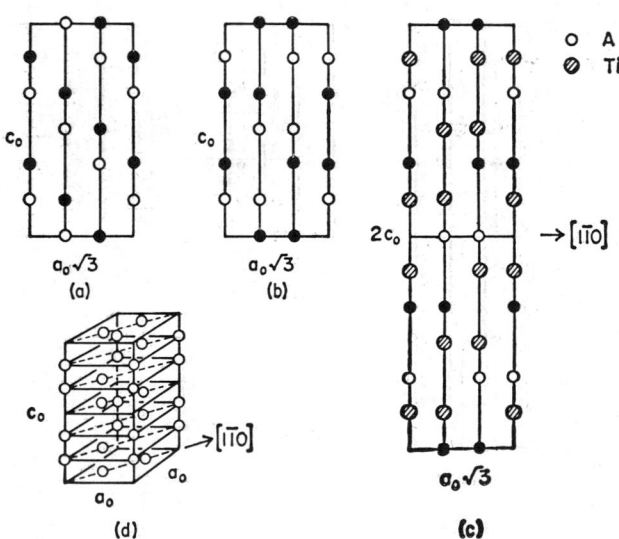

FIG. 5f-20. (a) Cr_2O_3, (b) Fe_2O_3, and (c) $FeTiO_3$ type magnetic structures projected on (110).   (d) cation positions in corundum and related structures.

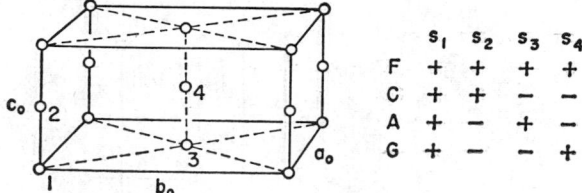

FIG. 5f-21. Ordering in $CrVO_4$ and $CuSO_4$ type structures.

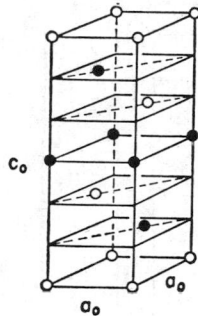

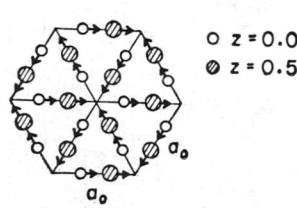

FIG. 5f-22. $MnCO_3$ (and CoF_3) type magnetic structure.

FIG. 5f-23. $YMnO_3$ type magnetic structure projected on (001).

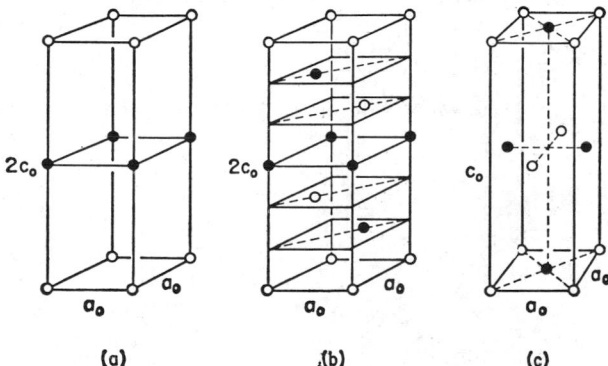

(a) (b) (c)

FIG. 5f-24. (a) $CoBr_2$, (b) $CoCl_2$, and (c) K_2NiF_4 type magnetic structures.

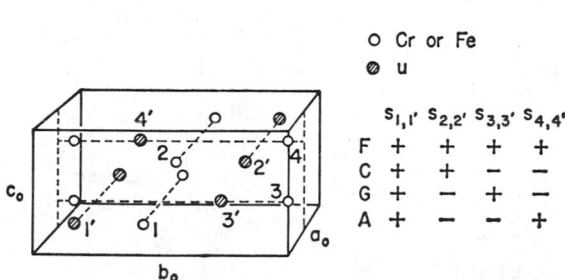

FIG. 5f-25. Ordering in $CrUO_4$ and $FeUO_4$.

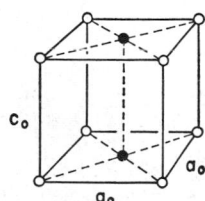

FIG. 5f-26. Ordering in CuAu-I type structures.

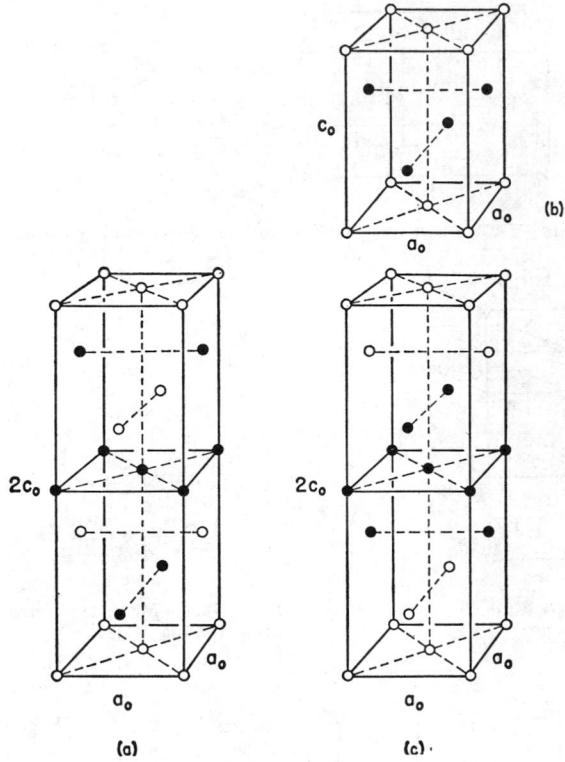

Fig. 5f-27. (a) Mn_2As, (b) Mn_2Sb, and (c) Fe_2As type magnetic structures.

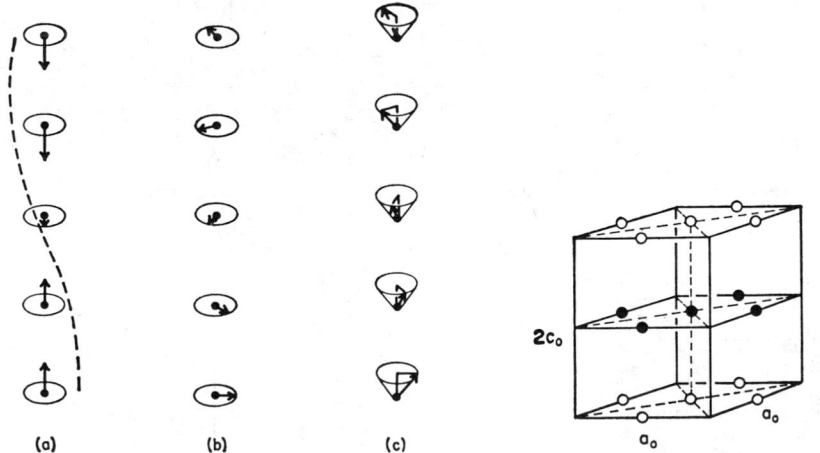

Fig. 5f-28. (a) Longitudinal sinusoidal, (b) screw spiral, and (c) ferromagnetic screw-cone spiral magnetic structures.

Fig. 5f-29 FeSn type magnetic structure.

References for Table 5f-15

1. Abrahams: *J. Chem. Phys.* **36**, 56 (1962).
2. Abrahams, Guttman, and Kasper: *Phys. Rev.* **127**, 2052 (1962).
3. Abrahams and Williams: *J. Chem. Phys.* **39**, 2923 (1963).
4. Abrahams: *J. Chem. Phys.* **44**, 2230 (1966).
5. Adelson and Austin: *J. Phys. Chem. Solids* **26**, 1795 (1965).
6. Aleshko-Ozhevskii, Sizov, Cheparin, and Yamzin: *Soviet Phys.–JETP Letters* **7**, 158 (1968).
7. Alikhanov: *Soviet Phys.–JETP* **9**, 1204 (1959).
8. Alikhanov: *Soviet Phys.–JETP* **10**, 814 (1960).
9. Alikhanov: *Soviet Phys.–JETP* **12**, 1029 (1961).
10. Alikhanov: *J. Phys. Soc. Japan* **17**, suppl. BIII, 58 (1962).
11. Alikhanov: *Soviet Phys.–JETP Letters* **5**, 349 (1967).
12. Allain and Boucher: *J. phys. radium* **26**, 789 (1965).
13. Allain, Boucher, Imbert, and Perrin: *Compt. rend.* **B263**, 9 (1966).
14. Almodovar, Frazer, Hurst, Cox, and Brown: *Phys. Rev.* **A138**, 153 (1965).
15. Alperin: *J. Appl. Phys.* **31**, 354S (1960).
16. Alperin: *J. Phys. Soc. Japan* **17**, suppl. BIII, 12 (1962).
17. Alperin, Brown, and Nathans: *J. Appl. Phys.* **34**, 1201 (1963).
18. Andresen: *Acta. Chem. Scand.* **14**, 919 (1960).
19. Andresen and Leciejewicz: *J. phys. radium* **25**, 574 (1964).
20. Andresen, Kjekshus, Mollerud, and Pearson: *Phil. Mag.* **11**, 1245 (1965).
21. Andresen, Kjekshus, Mollerud, and Pearson: *Acta. Chem. Scand.* **20**, 2529 (1966).
22. Andresen and Torbo: *Acta. Chem. Scand.* **21**, 2841 (1967).
23. Andresen: *Acta. Chem. Scand.* **22**, 827 (1968).
24. Andron and Bertaut: *J. phys. radium* **27**, 619 (1966).
25. Andron and Bertaut: *J. phys. radium* **27**, 626 (1966).
26. Arnold, Nereson, and Olsen: *J. Chem. Phys.* **46**, 4041 (1967).
27. Arrott, Werner, and Kendrick: *Phys. Rev. Letters* **14**, 1022 (1965).
28. Arrott, Werner, and Kendrick: *Phys. Rev.* **153**, 624 (1967).
29. Atoji: *J. Chem. Phys.* **43**, 222 (1965).
30. Atoji: *J. Chem. Phys.* **46**, 1891 (1967).
31. Atoji: *J. Chem. Phys.* **48**, 560 (1968).
32. Atoji: *J. Chem. Phys.* **48**, 3380 (1968).
33. Atoji: *J. Chem. Phys.* **48**, 3384 (1968).
34. Austin, Adelson, and Cloud: *J. Appl. Phys.* **33**, 1356 (1962).
35. Austin, Adelson, and Cloud: *Phys. Rev.* **131**, 1511 (1963).
36. Austin: *J. Appl. Phys.*, **40**, 1381 (1969).
37. Bacchella and Pinot: *J. phys. radium* **25**, 35 (1964).
38. Bacmann, Bertaut, and Bassi: *Bull. soc. franç. minéral. et crist.* **88**, 214 (1965).
39. Bacmann and Bertaut: *J. phys. radium* **27**, 726 (1966).
40. M. Bacmann, Bertaut, and Blaise: *Compt. rend.* **B266**, 45 (1968).
41. M. Bacmann, Chevalier, Bertaut, Roult, and Belakhovsky: *Compt. rend.* **267B**, 518 (1968).
42. Bacon, Street, and Tredgold: *Proc. Roy. Soc. (London)*, Ser. A, **217**, 252 (1953).
43. Bacon, Dunmur, Smith, and Street: *Proc. Roy. Soc. (London)*, Ser. A, **241**, 223 (1957).
44. Bacon and Street: *Proc. Phys. Soc. (London)* **72**, 470 (1958).
45. Bacon: *Acta. Cryst.* **14**, 823 (1961).
46. Bacon: *Proc. Phys. Soc. (London)* **79**, 938 (1962).
47. Bacon and Crangle: *Proc. Roy. Soc. (London)*, ser. A, **272**, 387 (1963).
48. Bacon and Mason: *Proc. Phys. Soc. (London)* **88**, 929 (1966).
49. Bacon and Mason: *Proc. Phys. Soc. (London)* **92**, 713 (1967).
50. Ballestracci, Bertaut, and Pauthenet: *J. Phys. Chem. Solids* **24**, 487 (1963).
51. Barrett, Meyer, and Wasserman: *J. Chem. Phys.* **47**, 592 (1967).
52. Bastow: *Proc. Phys. Soc. (London)* **88**, 935 (1966).
53. Bastow and Street: *Phys. Rev.* **141**, 510 (1966).
54. Becle, Lemaire, and Pauthenet: *Compt. rend.* **B266**, 994 (1968).
55. Becle and Lemaire: *Phys. Letters* **27A**, 541 (1968).
56. Becle, Lemaire, and Parthe: *Solid State Commun.* **6**, 115 (1968).
57. Bents: *Phys. Rev.* **106**, 225 (1957).
58. Bertaut, Corliss, Forratt, Aleonard, and Pauthenet: *J. Phys. Chem. Solids* **21**, 234 (1961).
59. Bertaut, Delapalme, Forratt, and Pauthenet: *J. phys. radium* **23**, 477 (1962).
60. Bertaut, de Bergevin, and Roult: *Compt. rend.* **256**, 1688 (1963).
61. Bertaut, Coing-Boyat, and Delapalme: *Phys. Letters* **3**, 178 (1963).

62. Bertaut: *Proc. Intern. Conf. Magnetism, Nottingham*, p. 275, The Institute of Physics and the Physical Society, London, 1964.
63. Bertaut, Delapalme, and Bassi: *J. phys. radium* **25**, 545 (1964).
64. Bertaut, Mercier, and Pauthenet: *J. phys. radium* **25**, 550 (1964).
65. Bertaut: *J. phys. radium.* **25**, 582 (1964).
66. Bertaut: *J. Appl. Phys.* **35**, 952 (1964).
67. Bertaut, Van Qui, Pauthenet, and Murasik: *J. phys. radium* **25**, 516 (1964).
68. Bertaut and Chenavas: *Solid State Commun.* **3**, 117 (1965).
69. Bertaut, Pauthenet, and Mercier: *Phys. Letters* **18**, 13 (1965).
70. Bertaut et al.: *J. Appl. Phys.* **37**, 1038 (1966).
71. Bertaut and Chevalier: *Compt. rend.* **262B**, 1707 (1966).
72. Bertaut, Chappert, Apostolov, and Semenov: *Bull. soc. franç. minéral. et crist.* **89**, 206 (1966).
73. Bertaut, Buisson, Quezel-Ambrunaz, and Quezel: *Solid State Commun.* **5**, 25 (1967).
74. Bertaut and Mareschal: *Solid State Commun.* **5**, 93 (1967).
75. Bertaut, Chappert, Mareschal, Rebouillat, and Sivardiere: *Solid State Commun.* **5**, 293 (1967).
76. Bertaut, Mareschal, and DeVries: *J. Phys. Chem. Solids* **28**, 2143 (1967).
77. Bertaut and Mareschal: *J. phys. radium* **29**, 67 (1968).
78. Bertaut, Cohen, Lambert-Andron, and Mollard: *J. phys. radium* **29**, 813 (1968).
79. Bertaut and Fruchart: *Solid State Commun.* **6**, 251 (1968).
80. Bertaut and Vincent: *Solid State Commun.* **6**, 269 (1968).
81. Bidaux and Meriel: *J. phys. radium* **29**, 220 (1968).
82. Bielen, Mareschal, and Sivardiere: *Z. angew. Phys.* **23**, 243 (1967).
83. Birgeneau, Guggenheim, and Shirane: *Phys. Rev. Letters*, **22**, 720 (1969).
84. Blinowski: *Nukleonika* **5**, 414 (1960).
85. Boucher and Oles: *J. phys. radium* **27**, 51 (1966).
86. Boucher, Buhl, and Perrin: *Compt. rend.* **B263**, 344 (1966).
87. Boucher, Buhl, and Perrin: *J. Appl. Phys.* **38**, 1109 (1967).
88. Boucher, Buhl, and Perrin: *J. Appl. Phys.* **39**, 632 (1968).
89. Boucher, Buhl, and Perrin: International Conference on Magnetic Oxides, Bucharest, 1963.
90. Boucher, Buhl, and Perrin: *J. Appl. Phys.*, **40**, 1126 (1969).
91. Brockhouse: *J. Chem. Phys.* **21**, 961 (1953).
92. Brockhouse: *Phys. Rev.* **94**, 781 (1954).
93. Brown and Frazer: *Phys. Rev.* **129**, 1145 (1963).
94. Brown, Wilkinson, Forsyth, and Nathans: *Proc. Phys. Soc. (London)* **85**, 1185 (1965).
95. Brown and Forsyth: *Proc. Phys. Soc. (London)* **92**, 125 (1967).
96. Burlet and Bertaut: *Compt. rend.* **264**, 323 (1967).
97. Bykov, Golovkin, Ageev, Levdik, and Vinogradov: *Soviet Phys.-Doklady* **4**, 1070 (1960).
98. Cable, Wilkinson, and Wollan: *Phys. Rev.* **118**, 950 (1960).
99. Cable, Wilkinson, and Wollan: *J. Phys. Chem. Solids* **19**, 29 (1961).
100. Cable, Wollan, Koehler, and Wilkinson: *J. Appl. Phys.* **32**, 49S (1961).
101. Cable, Wilkinson, Wollan, and Koehler: *Phys. Rev.* **125**, 1860 (1962).
102. Cable, Wilkinson, Wollan, and Koehler: *Phys. Rev.* **127**, 714 (1962).
103. Cable, Wollan, Koehler, and Child: *Phys. Rev.* **128**, 2118 (1962).
104. Cable, Koehler, and Wollan: *Phys. Rev.* **A136**, 240 (1964).
105. Cable, Moon, Koehler, and Wollan: *Phys. Rev. Letters* **12**, 553 (1964).
106. Cable, Wollan, and Koehler: *Phys. Rev.* **A140**, 1896 (1965).
107. Cable, Koehler, and Child: *J. Appl. Phys.* **36**, 1096 (1965).
108. Caron, Santoro, and Newnham: *J. Phys. Chem. Solids* **26**, 927 (1965).
109. Chappert and Portier: *Solid State Commun.* **4**, 395 (1966).
110. Chevalier, Roult, and Bertaut: *Solid State Commun.* **5**, 7 (1967).
111. Chevreton and Andron: *Compt. rend.* **264**, 316 (1967).
112. Child, Wilkinson, Cable, Koehler, and Wollan: *Phys. Rev.* **131**, 922 (1963).
113. Child, Koehler, Wollan, and Cable: *Phys. Rev.* **A138**, 1655 (1965).
114. Child and Koehler: *J. Appl. Phys.* **37**, 1353 (1966).
115. Child, Moon, Raubenheimer, and Koehler: *J. Appl. Phys.* **38**, 1381 (1967).
116. Child, Koehler, and Millhouse: *J. Appl. Phys.* **39**, 1329 (1968).
117. Child and Koehler: *Phys. Rev.* **174**, 562 (1968).
118. Child and Cable: *J. Appl. Phys.*, **40**, 1003 (1969).
119. Cloud, Jarrett, Austin, and Adelson: *Phys. Rev.* **120**, 1969 (1960).
120. Cloud, Bither, and Swoboda: *J. Appl. Phys.* **32**, 55S (1961).
121. Collins: *Proc. Phys. Soc. (London)* **89**, 415 (1966).
122. Corliss, Hastings, and Goldman: *Phys. Rev.* **93**, 893 (1954).

123. Corliss, Elliott, and Hastings: *Phys. Rev.* **104**, 924 (1956).
124. Corliss, Hastings, and Weiss: *Phys. Rev. Letters* **3**, 211 (1959).
125. Corliss, Elliott, and Hastings: *Phys. Rev.* **117**, 929 (1960).
126. Corliss, Elliott, Hastings, and Sass: *Phys. Rev.* **122**, 1402 (1961).
127. Corliss and Hastings: *J. Appl. Phys.* **33**, 1138 (1962).
128. Corliss and Hastings: *J. Appl. Phys.* **34**, 1192 (1963).
129. Corliss and Hastings: *J. Appl. Phys.* **35**, 1051 (1964).
130. Corliss and Hastings: *J. phys. radium* **25**, 557 (1964).
131. Corliss, Hastings, Nathans, and Shirane: *J. Appl. Phys.* **36**, 1099 (1965).
132. Corliss, Hastings, Kunnmann, and Banks: *I.U. Cr. 7th Intern. Congr. and Symp.*, Moscow, 1966.
133. Corliss, Hastings, and Kunnmann: *Phys. Rev.* **160**, 408 (1967).
134. Corliss and Hastings: *J. Appl. Phys.* **39**, 461 (1968).
135. Cox, Takei, Miller, and Shirane: *J. Phys. Chem. Solids* **23**, 863 (1962).
136. Cox, Shirane, Takei, and Wallace: *J. Appl. Phys.* **34**, 1352 (1963).
137. Cox, Takei, and Shirane: *J. Phys. Chem. Solids* **24**, 405 (1963).
138. Cox, Shirane, Flinn, Ruby, and Takei: *Phys. Rev.* **132**, 1547 (1963).
139. Cox, Shirane, and Takei: *Proc. Intern. Conf. Magnetism, Nottingham*, p. 291, The Institute of Physics and the Physical Society, London, 1964.
140. Cox, Frazer, and Shirane: *Phys. Letters* **17**, 103 (1965).
141. Cox, Frazer, Almodovar, and Kay: ACA Meeting, Gatlinburg, 1965.
142. Cox, Shirane, Frazer, and Narath: *J. Appl. Phys.* **37**, 1126 (1966).
143. Cox, Shirane, and Frazer: *J. Appl. Phys.* **38**, 1459 (1967).
144. Cox, Menzinger, and Frazer: ACA Meeting, Atlanta, 1967.
145. Cox and Williams: International Conference on Magnetic Oxides, Bucharest, 1968.
146. Cox, Frazer, Newnham, and Santoro: *J. Appl. Phys.*, **40**, 1124 (1969).
147. Curry, Johnston, and Besser: *Phil. Mag.* **12**, 221 (1965).
148. Curry: *Proc. Phys. Soc. (London)* **86**, 1193 (1965).
149. Curry: *Proc. Phys. Soc. (London)* **89**, 427 (1966).
150. Dachs: *J. Phys.* **25**, 563 (1964).
151. Dachs, Stoll, and Weitzel: *Z. Krist.* **125**, 120 (1967).
152. Dietrich and Als-Nielsen: *Phys. Rev.* **162**, 315 (1967).
153. Donnay, Corliss, Donnay, Elliott, and Hastings: *Phys. Rev.* **112**, 1917 (1958).
154. Doroshenko, Klyushin, Loshmanov, and Goman'kov: *Phys. Metals Metallog.* **12**(6), 119 (1961).
155. Drabkin, Mal'tsev, and Plakhtii: *Soviet Phys.–Solid State* **7**, 997 (1965).
156. Dwight, Menyuk, Rogers, and Wold: *Proc. Intern. Conf. Magnetism, Nottingham*, p. 538, The Institute of Physics and the Physical Society, London, 1964.
157. Dwight, Menyuk, Feinleib, and Wold: *J. Appl. Phys.* **37**, 962 (1966).
158. Endoh, Ishikawa, and Ohno: *J. Phys. Soc. Japan* **24**, 263 (1968).
159. Epstein, Makovsky, Melamud, and Shaked: *Phys. Rev.* **174**, 560 (1968).
160. Erickson: *Phys. Rev.* **85**, 745 (1952).
161. Erickson: *Phys. Rev.* **90**, 779 (1953).
162. Felcher, Corliss, and Hastings: *J. Appl. Phys.* **36**, 1001 (1965).
163. Felcher: *J. Appl. Phys.* **37**, 1056 (1966).
164. Fletcher, Gardner, Hooper, Hyde, Moore, and Woodhead: *Nature* **199**, 1089 (1963).
165. Forsyth, Johnson, and Brown: *Phil. Mag.* **10**, 713 (1964).
166. Forsyth, Pickart, and Brown: *Proc. Phys. Soc. (London)* **88**, 333 (1966).
167. Forsyth, Hedley, and Johnson: *J. Phys.* (C)**1**, 179 (1968).
168. Frazer and Brown: *Phys. Rev.* **125**, 1283 (1962).
169. Frazer, Shirane, Cox, and Olsen: *Phys. Rev.* **A140**, 1448 (1965).
170. Friedman, Shaked, and Shtrikman: *Phys. Letters* **25A**, 9 (1967).
171. Fuess and Will: *J. Appl. Phys.* **39**, 628 (1968).
172. Golovkin, Bykov, and Ledvik: *Soviet Phys.–JETP* **22**, 754 (1966).
173. Goman'kov and Loshmanov: *Soviet Phys.–Cryst.* **6**, 628 (1961).
174. Goman'kov, Litvin, Loshmanov, and Lyashchenko: *Soviet Phys.–Cryst.* **7**, 639 (1963).
175. Gonzalo, Cox, and Shirane: *Phys. Rev.* **147**, 415 (1966).
176. Gonzalo and Kay: *J. Phys. Soc. Japan* **21**, 1626 (1966).
177. Hamaguchi and Kunitomi: *J. Phys. Soc. Japan* **19**, 1849 (1964).
178. Hamaguchi, Wollan, and Koehler: *Phys. Rev.* **A138**, 737 (1965).
179. Hammann: *Phys. Letters* **26A**, 263 (1968).
180. Hammann: *J. phys. radium* **29**, 495 (1968).
181. Hammann: Private communication.
182. Hammann: *Physica*, **43**, 277 (1969).
183. Hastings and Corliss: *Phys. Rev.* **102**, 1460 (1956).
184. Hastings, Corliss, and Goldman: Pittsburgh Diffraction Conference, 1954.

185. Hastings, Elliott, and Corliss: *Phys. Rev.* **115**, 13 (1959).
186. Hastings: *Bull. Am. Phys. Soc.* **5**, 455 (1960).
187. Hastings and Corliss: *Phys. Rev.* **126**, 556 (1962).
188. Hastings, Corliss, and Windsor: *Phys. Rev.* **A138**, 176 (1965).
189. Hastings, Corliss, Kunnmann, and LaPlaca: *J. Phys. Chem. Solids* **28**, 1089 (1967).
190. Hastings and Corliss: *J. Phys. Chem. Solids* **29**, 9 (1968).
191. Henshaw and Brockhouse: *Bull. Am. Phys. Soc.* **2**, 9 (1957).
192. Herpin and Meriel: *Compt. rend.* **245**, 650 (1957).
193. Herpin, Meriel, and Villain: *Compt. rend.* **249**, 1334 (1959).
194. Herpin and Meriel: *J. phys. radium* **22**, 337 (1961).
195. Herpin and Meriel: *Compt. rend.* **259**, 2416 (1964).
196. Hori, Nakagawa, and Ishikawa: *J. Phys. Soc. Japan* **21**, 2080 (1966).
197. Hutchings and Windsor: *Proc. Phys. Soc. (London)* **91**, 928 (1967).
198. Hutchings, Samuelsen, Shirane, and Hirakawa: *Phys. Rev.*, **188**, 919 (1969).
199. Ishikawa, Hoshino, and Endoh: *J. Phys. Soc. Japan* **22**, 1221 (1967).
200. Ishikawa and Endoh: *J. Phys. Soc. Japan* **23**, 205 (1967).
201. Ishikawa and Endoh: *J. Appl. Phys.* **39**, 1318 (1968).
202. Iyengar, Dasannacharya, Vijayaraghavan, and Roy: *J. Phys. Soc. Japan* **17**, suppl. BIII, 41 (1962).
203. Kasper and Roberts: *Phys. Rev.* **101**, 537 (1956).
204. Kasper and Waterstrat: *Phys. Rev.* **109**, 1551 (1958).
205. Kasper and Kouvel: *J. Phys. Chem. Solids* **11**, 231 (1959).
206. Katsuraki: *J. Phys. Soc. Japan* **19**, 863 (1964).
207. Katsuraki and Achiwa: *J. Phys. Soc. Japan* **21**, 2238 (1966).
208. Kawaminami and Okazaki: *J. Phys. Soc. Japan* **22**, 924 (1967).
209. Kelarev, Klyushin, and Lyaschenko: *Phys. Metals Metallog.* **17**(5), 136 (1964).
210. Kelarev, Sidorov, Klyushin, and Abdulov: *Phys. Status Solidi* **24**, 385 (1967).
211. Kiselev, Ozerov, and Zhdanov: *Soviet Phys.–Doklady* **7**, 742 (1963).
212. Kjekshus, Mollerud, Andresen, and Pearson: *Phil. Mag.* **16**, 1063 (1967).
213. Kleinberg: *Bull. Am. Phys. Soc.* **11**, 759 (1966).
214. Kleinberg: *J. Appl. Phys.* **38**, 1453 (1967).
215. Klyushin and Ciszewski: *Phys. Metals Metallog.* **16**(5), 145 (1963).
216. Koehler and Wollan: *J. Phys. Chem. Solids* **2**, 100 (1957).
217. Koehler, Wilkinson, Cable, and Wollan: *J. phys. radium* **20**, 180 (1959).
218. Koehler, Wollan, and Wilkinson: *Phys. Rev.* **118**, 58 (1960).
219. Koehler, Cable, Wollan, and Wilkinson: *Phys. Rev.* **126**, 1672 (1962).
220. Koehler, Child, Wollan, and Cable: *J. Appl. Phys.* **34**, 1335 (1963).
221. Koehler, Yakel, Wollan, and Cable: *Phys. Letters* **9**, 93 (1964).
222. Koehler, Cable, and Wilkinson: *Phys. Rev.* **151**, 414 (1966).
223. Koehler, Moon, Trego, and Mackintosh: *Phys. Rev.* **151**, 405 (1966).
224. Koehler, Child, Cable, and Moon: *J. Appl. Phys.* **38**, 1384 (1967).
225. Koehler, Cable, Child, Wilkinson, and Wollan: *Phys. Rev.* **158**, 450 (1967).
226. Koehler, Child, Wollan, and Cable: *J. Appl. Phys.* **39**, 1331 (1968).
227. Komura and Kunitomi: *J. Phys. Soc. Japan* **20**, 103 (1965).
228. Komura, Hamaguchi, and Kunitomi: *Phys. Letters* **24A**, 299 (1967).
229. Kouvel and Kasper: *Proc. Intern. Conf. Magnetism, Nottingham*, p. 169, The Institute of Physics and the Physical Society, London, 1964.
230. Kouvel and Kasper: *J. Phys. Chem. Solids* **24**, 529 (1963).
231. Kouvel: *J. Appl. Phys.* **36**, 980 (1965).
232. Kouvel and Forsyth: *J. Appl. Phys.*, **40**, 1359 (1969).
233. Kren, Pal, and Szabo: *Phys. Letters* **9**, 297 (1964).
234. Kren and Szabo: *Phys. Letters* **11**, 215 (1964).
235. Kren, Szabo, and Konczos: *Phys. Letters* **19**, 103 (1965).
236. Kren, Szabo, and Tarnoczi: *Solid State Commun.* **4**, 31 (1966).
237. Kren: *Phys. Letters* **21**, 383 (1966).
238. Kren and Solyom: *Phys. Letters* **22**, 273 (1966).
239. Kren, Kadar, and Tarnoczi: *Phys. Letters* **25A**, 56 (1967).
240. Kren, Kadar, and Szabo: *Phys. Letters* **26A**, 556 (1968).
241. Kren, Kadar, Pal, Solyom, Szabo, and Tarnoczi: *Phys. Rev.* **171**, 574 (1968).
242. Kunnmann, LaPlaca, Corliss, Hastings, and Banks: *J. Phys. Chem. Solids* **29**, 1359 (1968).
243. Kunitomi, Hamaguchi, Sakamoto, Doi, and Komura: *J. phys. radium* **25**, 462 (1964).
244. Kunitomi, Hamaguchi, and Anzai: *J. phys. radium* **25**, 568 (1964).
245. Kunitomi, Yamada, Nakai, and Fujii: *J. Appl. Phys.*, **40**, 1265 (1969).
246. Kuznietz, Lander, and Baskin: *J. Appl. Phys.*, **40**, 1130 (1969).
247. van Laar: *Phys. Rev.* **A138**, 584 (1965).

248. van Laar: *Phys. Rev.* **141**, 538 (1966).
249. van Laar: *Phys. Rev.* **156**, 654 (1967).
250. van Laar: *Phys. Letters* **25A**, 27 (1967).
251. van Laar and Ijdo: Private communication.
252. van Laar and Bongers: Private communication.
253. Lambert-Andron, Berodias, and Chevreton: *Bull. soc. franç minéral. et crist.* **91**, 88 (1968).
254. Lander, Brown, Forsyth: *Proc. Phys. Soc. (London)* **91**, 332 (1967).
255. Lander, Kuznietz, and Baskin: *Solid State Commun.*, **6**, 877 (1968).
256. Lau, Stout, Koehler, and Child: *J. Appl. Phys.* **40**, 1136 (1969).
257. Leake, Shirane, and Remeika: *Solid State Commun.* **6**, 15 (1968).
258. Lebech: *Solid State Commun.* **6**, 761 (1968).
259. Leciejewicz, Troc, and Murasik: *Phys. Status Solidi* **22**, 517 (1967).
260. Leciejewicz, Murasik, and Troc: *Phys. Status Solidi* **30**, 157 (1968).
261. LeFur and Coing-Boyat: *I. U. Cr. 7th Intern. Congr. and Symp.*, Moscow, 1966.
262. Legrand and Verschuren: *J. phys. radium* **25**, 578 (1964).
263. Loopstra, van Laar, and Breed: *Phys. Letters* **26A**, 526 (1968).
264. McGuire and Pickart: *J. Phys. Chem. Solids* **24**, 1531 (1963).
265. McGuire, Shafer, Joenk, Alperin, and Pickart: *J. Appl. Phys.* **37**, 981 (1966).
266. McGuire, Gambino, Pickart, and Alperin: *J. Appl. Phys.* **40**, 1009 (1969).
267. Mareschal, Sivardiere, DeVries, and Bertaut: *J. Appl. Phys.* **39**, 1364 (1968).
268. Mekata, Haruna, and Takaki: *J. Phys. Soc. Japan* **25**, 234 (1968).
269. Meneghetti and Sidhu: *Phys. Rev.* **105**, 130 (1957).
270. Menyuk, Dwight, and Wold: *J. Phys. radium* **25**, 528 (1964).
271. Menzinger, Umebayashi, Frazer, and Cox: *Bull. Am. Phys. Soc.* **11**, 759 (1966).
272. Meriel: Private communication quoted by J. Mareschal et al. in ref. 267.
273. Miksic, Miller, and Cox: International Conference on Magnetic Oxides, Bucharest, 1968.
274. Millhouse, Koehler, and Child: *J. Appl. Phys.* **40**. 1006 (1969).
275. Minkiewicz, Shirane, Frazer, Wheeler, and Dorain: *J. Phys. Chem. Solids* **29**, 881 (1968).
276. Møller, Blinowski Mackintosh, and Brun: *Solid State Commun.* **2**, 109 (1964).
277. Møller, Trego, and Mackintosh: *Solid State Commun.* **3**, 137 (1965).
278. Montmory, Bertaut, and Mollard: *Solid State Commun.* **4**, 249 (1966).
279. Montmory, Belakhovsky, Chevalier, and Newnham: *Solid State Commun.* **6**, 317 (1968).
280. Montmory and Newnham: *Solid State Commun.* **6**, 323 (1968).
281. Moon, Cable, and Koehler: *J. Appl. Phys.* **35**, 1041 (1964).
282. Moon, Child, Koehler, and Raubenheimer: *J. Appl. Phys.* **38**, 1383 (1967).
283. Moreau, Mareschal, and Bertaut: *Solid State Commun.* **6**, 751 (1968).
284. Morrish, Johnston, and Curry: *Phys. Letters* **7**, 177 (1963).
285. Muldawer and de Bergevin: *J. Chem. Phys.* **35**, 1904 (1961).
286. Murasik and Niemiec: *Bull. acad. polon. sci. ser. sci. chim.* **13**, 291 (1965).
287. Murasik, Suski, Troc, and Leciejewicz: *Phys. Status Solidi* **30**, 61 (1968).
288. Murthy, Begum, Srinivasan, and Murthy: *Phys. Letters* **15**, 225 (1965).
289. Murthy, Begum, Somanathan, and Murthy: *Solid State Commun.* **3**, 113 (1965).
290. Nagamiya, Saito, Shimomura, and Uchida: *J. Phys. Soc. Japan* **20**, 1285 (1965).
291. Nakagawa, Watanabe, and Hori: *J. Phys. Soc. Japan* **19**, 2078 (1964).
292. Nakagawa and Hori: *J. Phys. Soc. Japan* **19**, 2082 (1964).
293. Nakayama, Nakagawa, and Nomura: *J. Phys. Soc. Japan* **24**, 219 (1968).
294. Nathans, Pickart, and Miller: *Bull. Am. Phys. Soc.* **6**, 54 (1961).
295. Nathans, Alperin, Pickart, and Brown: *J. Appl. Phys.* **34**, 1182 (1963).
296. Nathans, Will, and Cox: *Proc. Intern. Conf. Magnetism, Nottingham*, p. 237, The Institute of Physics and the Physical Society, London, 1964.
297. Nathans, Pickart, Alperin, and Brown: *Phys. Rev.* **A136**, 1641 (1964).
298. Nereson, Olsen, and Arnold: *Phys. Rev.* **A135**, 176 (1964).
299. Nereson: *J. Appl. Phys.* **37**, 4575 (1966).
300. Newnham, Fang, and Santoro: *Acta Cryst.* **17**, 240 (1964).
301. Newnham, Redman, and Santoro: *Z. Krist.* **121**, 6 (1965).
302. Newnham, Santoro, Fang, and Nomura: *Acta Cryst.* **19**, 147 (1965).
303. Newnham, Santoro, and Redman: *J. Phys. Chem. Solids* **26**, 445 (1965).
304. Newnham, Santoro, Seal, and Stallings: *Phys. Status Solidi* **16**, K17 (1966).
305. Newnham, Caron, and Santoro: *J. Am. Ceram. Soc.* **49**, 284 (1966).
306. Nicholson and Friedman: *Bull. Am. Phys. Soc.* **8**, 43 (1963).
307. Nomura, Santoro, Fang, and Newnham: *J. Phys. Chem. Solids* **25**, 901 (1964).
308. Oberteuffer, Marcus, Schwartz, and Felcher: *Phys. Letters* **28A**, 267 (1968).

309. Oles: *J. phys. radium* **26**, 561 (1965).
310. Oles: *Phys. Status Solidi* **8**, K167 (1965).
311. Oles: *Phys. Status Solidi* **14**, K39 (1966).
312. Olsen and Koehler: *J. Appl. Phys.* **40**, 1135 (1969).
313. Pal, Kren, Kadar, Szabo, and Tarnoczi: *J. Appl. Phys.* **39**, 538 (1968).
314. Perekalina, Sizov, Sizov, Yamzin, and Voskanyan: *Soviet Phys.–JETP* **25**, 266 (1967).
315. Pickart and Nathans: *J. Appl. Phys.* **30**, 280S (1959).
316. Pickart: *Bull. Am. Phys. Soc.* **5**, 59 (1960).
317. Pickart: *Bull. Am. Phys. Soc.* **5**, 357 (1960).
318. Pickart, Nathans, and Shirane: *Phys. Rev.* **121**, 707 (1961).
319. Pickart and Nathans: *J. Appl. Phys.* **34**,'1203 (1963).
320. Pickart, Nathans, and Alperin: *J. Phys.* **25**, 542 (1964).
321. Pickart, Alperin, and Nathans: *J. Phys.* **25**, 565 (1964).
322. Pickart and Alperin: *J. Appl. Phys.* **39**, 1332 (1968).
323. Pickart and Alperin: *J. Phys. Chem. Solids* **29**, 414 (1968).
324. Plakhtii, Mal'tsev, and Kaminker: *Bull. Acad. Sci. U.S.S.R.* **28**, 350 (1964).
325. Plumier and Legrand: *J. phys. radium* **23**, 474 (1962).
326. Plumier: *Compt. rend.* **255**, 2244 (1962).
327. Plumier and Tardieu: *Compt. rend.* **257**, 3858 (1963).
328. Plumier: *J. phys. radium* **24**, 741 (1963).
329. Plumier: *Proc. Intern. Conf. Magnetism, Nottingham*, p. 295, The Institute of Physics and the Physical Society, London, 1964.
330. Plumier: *Compt. rend.* **260**, 3348 (1965).
331. Plumier: *J. phys. radium* **27**, 213 (1966).
332. Plumier: *Compt. rend.* **B263**, 173 (1966).
333. Plumier: *Compt. rend.* **B264**, 278 (1967).
334. Plumier: *Compt. rend.* **B265**, 672 (1967).
335. Plumier: *Compt. rend.* **B265**, 726 (1967).
336. Plumier: *Compt. rend.* **B267**, 98 (1968).
337. Plumier: International Conference on Magnetic Oxides, Bucharest, 1968.
338. Prince: *Acta Cryst.* **10**, 554 (1957).
339. Prince: *J. Appl. Phys.* **32**, 68S (1961).
340. Quezel-Ambrunaz: *Bull. soc. franç. minéral. et crist.* **91**, 339 (1968).
341. Rainford, Turberfield, Busch, and Vogt: *J. Phys.* (C)**1**, 679 (1968).
342. Riste and Wanic: *J. Phys. Chem. Solids* **17**, 318 (1960).
343. Roth: *Phys. Rev.* **110**, 1333 (1958).
344. Roth: *Phys. Rev.* **111**, 772 (1958).
345. Roth: *Acta Cryst.* **13**, 140 (1960).
346. Roth and Slack: *J. Appl. Phys.* **31**, 352S (1960).
347. Roth: *J. phys. radium* **25**, 507 (1964).
348. Roth: *J. Phys. Chem. Solids* **25**, 1 (1964).
349. Roth and DeVries: *J. Appl. Phys.* **38**, 951 (1967).
350. Santoro, Newnham, and Nomura: *J. Phys. Chem. Solids* **27**, 655 (1966).
351. Santoro, Segal, and Newnham: *J. Phys. Chem. Solids* **27**, 1192 (1966).
352. Santoro and Newnham: *Acta Cryst.* **22**, 344 (1967).
353. Scatturin, Corliss, Elliott, and Hastings: *Acta Cryst.* **14**, 19 (1961).
354. Shirane, Pickart, and Ishikawa: *J. Phys. Soc. Japan* **14**, 1352 (1959).
355. Shirane, Pickart, Nathans, and Ishikawa: *J. Phys. Chem. Solids* **10**, 35 (1959).
356. Shirane and Takei: *J. Phys. Soc. Japan* **17**, suppl. BIII, 35 (1962).
357. Shirane, Cox, Takei, and Ruby: *J. Phys. Soc. Japan* **17**, 1598 (1962).
358. Shirane, Cox, and Pickart: *J. Appl. Phys.* **35**, 954 (1964).
359. Shirane, Nathans, and Chen: *Phys. Rev.* **A134**, 1547 (1964).
360. Shirane, Frazer, and Friedberg: *Phys. Letters* **17**, 95 (1965).
361. Shirane and Pickart: *J. Appl. Phys.* **37**, 1032 (1966).
362. Shull and Smart: *Phys. Rev.* **76**, 1256 (1949).
363. Shull, Strauser, and Wollan: *Phys. Rev.* **83**, 333 (1951).
364. Shull: *Revs. Modern Phys.* **25**, 100 (1953).
365. Sidhu, Heaton, and Mueller: *J. Appl. Phys.* **30**, 1323 (1959).
366. Sidhu, Anderson, and Zauberis: *Bull. Am. Phys. Soc.* **10**, 352 (1965).
367. Sidhu, Vogelsang, and Anderson: *J. Phys. Chem. Solids* **27**, 1197 (1966).
368. Sidorov and Doroshenko: *Phys. Metals Metallog.* **20**(6), 48 (1965).
369. Sidorov and Doroshenko: *Phys. Status Solidi* **16**, 737 (1966).
370. Sirota and Makovetskii: *Soviet Phys.-Doklady* **11**, 888 (1967).
371. Sizov, Sizov, and Yamzin: *Soviet Phys.–JETP Letters* **6**, 176 (1967).
372. Sizov, Sizov, and Yamzin: *Soviet Phys.–JETP* **26**, 736 (1968).
373. Skalyo, Cox, and Frazer: *Bull. Am. Phys. Soc.* **13**, 461 (1968).

374. Smith and Bacon: *J. Appl. Phys.* **37**, 979 (1966).
375. Smith, Niklow, Raubenheimer, and Wilkinson: *J. Appl. Phys.* **37**, 1047 (1966).
376. Snow: *Phys. Rev.* **85**, 365 (1952).
377. Sparks, Mead, Kirschbaum, and Marshall: *J. Appl. Phys.* **31**, 356S (1960).
378. Sparks, Mead, and Komoto: *J. Phys. Soc. Japan* **17**, suppl. BI, 249 (1962).
379. Sparks and Komoto: *J. Appl. Phys.* **34**, 1191 (1963).
380. Sparks and Komoto: *J. phys. radium* **25**, 567 (1964).
381. Suzuoka, Adelson, and Austin: *Acta Crysta.* **A24**, 513 (1968).
382. Szytula, Burewicz, Dimitrijevic, Krasnicki, Rzany, Todorovic, Wanic, and Wolski: *Phys. Status Solidi* **26**, 429 (1968).
383. Takeda, Yamaguchi, Tomiyoshi, Fukase, Sugimoto, and Watanabe: *J. Phys. Soc. Japan* **24**, 446 (1968).
384. Takei, Cox, and Shirane: *Phys. Rev.* **129**, 2008 (1963).
385. Takei, Cox, and Shirane: *J. Appl. Phys.* **37**, 973 (1966).
386. Toth, Arrott, Shinozaki, Werner, and Sato: *J. Appl. Phys.* **40**, 133 (1969).
387. Troc, Leciejewicz, and Ciszewski: *Phys. Status Solidi* **15**, 515 (1966).
388. Troc, Murasik, Zygmunt, and Leciejewicz: *Phys. Status Solidi* **23**, K123 (1967).
389. Turkevich and Plakhtii: *Soviet Phys.–Solid State* **10**, 754 (1968).
390. Ulki: *Z. Krist.* **124**, 192 (1967).
391. Umebayashi and Ishikawa: *J. Phys. Soc. Japan* **21**, 1281 (1966).
392. Umebayashi, Shirane, Frazer, and Cox: *Phys. Rev.* **167**, 519 (1968).
393. Umebayashi, Shirane, Frazer, and Daniels: *Phys. Rev.* **165**, 688 (1968).
394. Umebayashi, Shirane, Frazer, and Daniels: *J. Phys. Soc. Japan* **24**, 368 (1968).
395. Villers, Gibart, Druilhe, and Burlet: *J. Appl. Phys.* **39**, 590 (1968).
396. Wang, Cox, and Kestigian: *Bull. Am. Phys. Soc.* **13**, 468 (1968).
397. Watanabe, Nakagawa, and Sato: *J. Phys. Soc. Japan* **20**, 2244 (1965).
398. Watanabe and Kunitomi: *J. Phys. Soc. Japan* **21**, 1932 (1966).
399. Watanabe, Yamaguchi, Ohashi, Sugimoto, T. Okada, and Fukase: *J. Phys. Soc. Japan* **22**, 939 (1967).
400. Watanabe, Kazama, Yamaguchi, and Ohashi: *J. Appl. Phys.* **40**, 1128 (1969).
401. Webster and Tebble: *Phil. Mag.* **16**, 347 (1967).
402. Webster and Tebble: *J. Appl. Phys.* **39**, 471 (1968).
403. Werner and Arrott: *J. Appl. Phys.* **40**, 1447 (1969).
404. Wilkinson, Gingrich, and Shull: *J. Phys. Chem. Solids* **2**, 289 (1957).
405. Wilkinson, Cable, Wollan, and Koehler: *Phys. Rev.* **113**, 497 (1959).
406. Wilkinson, Wollan, and Koehler: *Bull. Am. Phys. Soc.* **5**, 456 (1960).
407. Wilkinson, Wollan, Child, and Cable: *Phys. Rev.* **121**, 74 (1961).
408. Wilkinson, Child, McHargue, Koehler, and Wollan: *Phys. Rev.* **122**, 1409 (1961).
409. Wilkinson, Koehler, Wollan, and Cable: *J. Appl. Phys.* **32**, 48S (1961).
410. Will, Pickart, Alperin, and Nathans: *J. Phys. Chem. Solids* **24**, 1679 (1963).
411. Will, Frazer, Shirane, Cox, and Brown: *Phys. Rev.* **A140**, 2139 (1965).
412. Williams, Heaton, and Campos: *J. Phys. Chem. Solids* **29**, 1702 (1968).
413. Willis and Taylor: *Phys. Letters* **17**, 188 (1965).
414. Wollan and Koehler: *Phys. Rev.* **100**, 545 (1955).
415. Wollan, Koehler, and Wilkinson: *Phys. Rev.* **110**, 638 (1958).
416. Wollan, Child, Koehler, and Wilkinson: *Phys. Rev.* **112**, 1132 (1958).
417. Wuttig: *Z. angew. Math. u. Phys.* **16**, 535 (1965).
418. Yamaguchi and Watanabe: *J. Phys. Soc. Japan* **22**, 1210 (1967).
419. Yamazin, Sizov, I. S. Zheludev, T. M. Perekalina, and Zalesskii: *Soviet Phys.–JETP* **23**, 395 (1966).
420. Yessik: *Phil. Mag.* **17**, 623 (1968).
421. Yoshii and Katsuraki: *J. Phys. Soc. Japan* **22**, 674 (1967).
422. Yoshimori: *J. Phys. Soc. Japan* **12**, 807 (1959).

5f-8. Gyromagnetic Ratios and Spectroscopic Splitting Factors. The magneto-mechanical ratio g' is defined by the relation

$$g' = \frac{M}{J}\frac{2mc}{e}$$

where m/e is the mass-to-charge ratio of the electron and c is the velocity of light. M/J is the ratio of the magnetic moment to the angular momentum of the electrons which contribute to the spontaneous magnetization as measured in an Einstein-de Haas or a Barnett-effect experiment (Sec. 5f-1).

The spectroscopic splitting factor g for ferromagnetic materials is defined as:

$$g = \frac{h\nu}{\mu_B H}$$

where ν is the Larmor precession frequency of the moment associated with a sample of the material in a field H as measured in a ferromagnetic-resonance experiment, h is Planck's constant, and μ_B is the Bohr magneton. See Table 5f-16.

The Kittel-Van Vleck relation between g and g' is usually written

$$1/g + 1/g' = 1$$

which reduces to $g = g' = 2$ when g and g' are equal. When g and g' differ considerably it has been proposed[1] that the relation is

$$g' = g/(g - \rho)$$

where ρ may be calculated from atomic structure when this is sufficiently well known. In some cases g' may then be negative.

TABLE 5f-16. VALUES OF g AND g'

Material	g'(obs) [1]	g(calc)	g(obs)†	Material [3]	g'(obs)	g(calc)
Fe..............	1.919	2.09	2.09 [2]	Nd₂O₃.........	0.77	3.3
Co.............	1.850	2.18	2.18 [2]	Gd₂O₃.........	2.12	1.9
Ni.............	1.837	2.19	2.19 [2]	Dy₂O₃.........	1.29	4.5
FeNi‡.........	1.908	2.10	2.10 [2]	Eu₂O₃.........	>4.5	<1.3
CoNi‡.........	1.846	2.18	2.17 [2]	CrCl₃.........	1.95	2.1
Cu₂MnAl.......	1.993	2.01	2.01 [4]	MnCO₃........	2.00	2.0
MnSb..........	1.978	2.02		MnSO₄........	2.29	1.8
NiFe₂O₄........	1.849	2.18	2.20 [5]	FeSO₄........	1.89	2.1
Fe-Ni-Mo§.....	1.905	2.10	2.10 [6]	CoCl₂.........	1.45	3.2
FeS₁.₂.........	1.9	2.1		CoSO₄........	1.57	2.8

Probable errors in g' are usually 0.002 to 0.004 in the metals, and several percent in the nonmetals.
† No values of g(obs) available for Nd₂O₃ to CoSO₄.
‡ Other compositions of alloys are given by Scott [1] and Asch [2].
§ Supermalloy: 17 Fe, 78 Ni, 5 Mo (wt. %).
In solid solutions of composition $A_x B_{1-x}$, g' is observed[1] to follow the relation

$$\frac{x + k(1 - x)}{x/g_1' + k(1 - x)/g_2'}$$

k being an empirical constant.

References for Table 5f-16

1. Scott, G. G.: *Revs. Modern Phys.* **34**, 102 (1962); *J. Phys. Soc. Japan* **17**, suppl. B1, 372 (1962).
2. Ash, G.: Thesis, Strasbourg, 1960; A. J. P. Meyer and G. Asch: *J. Appl. Phys.* **32**, 330S (1961); and articles by these authors in *Compt. rend.*, 1954–1959.
3. Sucksmith, W.: *Proc. Roy. Soc. (London)*, ser. A, **128**, 276 (1930); **133**, 179 (1931); **135**, 276 (1932); and *Helv. Phys. Acta* **8**, 205 (1933). Ni and Cu-Ni above T_C are included.
4. Yager, W. A., and F. R. Merritt: *Phys. Rev.* **75**, 318 (1949).
5. Yager, W. A., J. K. Galt, and F. R. Merritt: *Phys. Rev.* **99**, 1203 (1955).
9. Young, J. A., and E. A. Uehling: *Phys. Rev.* **94**, 544 (1954).
7. Blume, M., G. Geschwind, and Y. Yafet: *Phys. Rev.* **181**, 478 (1969).

5f-9. Change of Curie Point and Néel Point with Pressure. A hydrostatic pressure changes the Curie Point or Néel Point of magnetic materials, usually from about +5 K to −5 K per kilobar (987 atmospheres). See Table 5f-17.

[1] See Ref. 7 for Table 5f-16.

TABLE 5f-17. EFFECT OF PRESSURE ON CURIE AND NÉEL TEMPERATURES*[a,b]

Substance	dT/dP, K/kbar	Ref.	Substance	dT/dP, K/kbar	Ref.
Au_2Mn	AF 0.47	1	$Ni_{0.5}Zn_{0.5}Fe_2O_4$	FM 0.99	40
Au_4Mn	AF 2.7	2-4	$Ni_{0.8}Zn_{0.2}Fe_2O_4$	FM 0.73	40
	0.9 to 2.7^c		$NiFe_2O_4$	FM 1.16	40
$Ba_{0.1}Sr_{0.9}RuO_3$	FM -0.56	5	$Fe_{0.7}Mn_{0.3}$	AF -2.5	45
$CaCrO_3$	FM -0.23	6	$Fe_{0.70}Ni_{0.30}$	FM -5.8	9, 46
$CaMnO_3$	FM 0.41	5	$Fe_{0.64}Ni_{0.36}$	FM -4^c	9, 39
$Ca_{0.25}La_{0.75}MnO_3$	FM 1.07	7			47, 47a
$Ca_{0.85}Sr_{0.15}MnO_3$	FM 0.44	7a	$Fe_{0.32}Ni_{0.68}$	FM -0.1	9
$Ca_{0.75}Sr_{0.25}MnO_3$	FM 0.38	7a	$Fe_{0.47}Rh_{0.53}$	FM 5.1	1
$Ca_{0.5}Sr_{0.5}MnO_3$	FM 0.55	7a	$Fe_{0.96}Si_{0.04}$	FM -0.1	9
$CdCr_2S_4$	FM -0.58	8	$Fe_{0.90}Si_{0.10}$	FM 0.2	9
$CdCr_2Se_4$	FM -0.84	8, 5	Gd	FM -1.4	1, 9, 23
Co	FM 0.0	9, 10			27, 30
$CoCO_3$	FM 0.1	11			48
CoF_2	AF 0.0	12		1.2 to 1.6^c	49, 50
CoO	AF 0.60	13, 13a,	GdN	FM 0.08	51
		13b	$HgCr_2S_4$	FM 0.14	8
CoS_2	FM -0.64	5	$HgCr_2Se_4$	FM -0.95	8
Cr	AF -5.1^d	14, 15	Ho	AF -0.4	1, 27
	-5 to -6^c	16, 17			32
Cr_2O_3	AF -1.6	18		-0.3 to -0.5^c	52
CrN	AF 0.85	18a	$La_{0.75}Sr_{0.25}MnO_3$	FM 0.6	9
$CrS_{1.17}$	FM -2.6	19	$\gamma Mn_{0.95}Cu_{0.05}$	AF -3.5	52a
CrTe	FM -5.6	20, 21	MnAs	FM -12^e	53, 54
Cr_3Te_4	FM -7.35	21a			55
$CsNiF_3$	FM 0.53	22		AF 2.2	56
$CuCr_2S_4$	FM -1.39	5	MnBi	FM -0.7	55
$CuCr_2Se_4$	FM -0.44	5	$MnCO_3$	FM 0.4	11
Dy	AF -0.4	1, 23	$Mn_{1.2}In_{0.8}Cu_2$	AF 1.5^f	2
		24, 25	MnF_2	AF 0.3	57, 12
	-0.3 to -0.6^c	26, 27		0.3 to 0.7^c	
		28, 29	Mn_5Ge_3	FM 0.42	57a
Dy	FM -1.25	30	MnO	AF 0.3	58
$Dy_{0.58}Gd_{0.42}$	AF -0.85	31	MnP	FM -1.4	59, 60
	FM -1.01		MnS	AF 1.20	60a
Er	AF -0.25	24, 32	MnSb	FM -3	55, 20
	FM -0.8		MnTe	AF 2	61, 62
$ErCo_3$	FM -4.17	32a	$MnTe_2$	AF 0.2	63
Eu	AF 0.45	33, 34,	Ni	FM 0.32	1, 9, 64
		34a	Alumel-94 Ni	FM 0.04	9
	0 to 0.9^c		Monel	FM 0.07	9
EuO	FM 0.4	34, 35	$Ni_{0.80}Cu_{0.20}$	FM 0.14	1
		36	$Ni_{0.70}Cu_{0.30}$	FM 0.05	1
EuS	FM 0.28	37	NiS	AF -6.0	65
EuSe	AF 0.0	38	$Ni_{0.93}Si_{0.07}$	FM 0.16	1
EuTe	AF 0.0	38	$Ni_{0.91}Si_{0.09}$	FM 0.05	1
Fe	FM 0.0	9, 39	$Pd_{0.95}Co_{0.05}$	FM 0.1	65a
FeO	AF 0.65	13a	$Pd_{0.997}Fe_{0.003}$	FM -0.004	66
Garnets			$Pd_{0.97}Fe_{0.03}$	FM -0.1	66
$Gd_3Fe_5O_{12}$	FM 1.28	13	$RbFeCl_3$	FM 2.13	66a
$Gd_3AlFe_4O_{12}$	FM 0.58	40	$RbNiF_3$	FM 0.6	67
$Tb_3Fe_5O_{12}$	FM 1.23	13	Sc_3In	FM 0.2	68
$Dy_3Fe_5O_{12}$	FM 1.15	13	$SrRuO_3$	FM -0.63	5
$Er_3Fe_5O_{12}$	FM 1.22	13	Tb	AF -0.8	1, 23
$Yb_3Fe_5O_{12}$	FM 1.08	13		-0.76 to -1.1^c	26, 27
$Y_3Fe_5O_{12}$	FM 1.25	13		FM -1.2	52, 69
$Y_{1.5}Gd_{1.5}Fe_{4.5}$-				-1.06 to -1.24^c	70
$Al_{0.5}O_{12}$	FM 1.00	40	$Tb_{0.7}Y_{0.3}$	AF -0.5	27
Spinels			$Tb_{0.3}Y_{0.7}$	AF -0.3	27
Fe_3O_4	FM ~ 0	41	V_2O_3	AF -3.78	71
$Mn_{0.5}Zn_{0.5}Fe_2O_4$	FM 0.9	9, 42	$YCrO_3$	AF 0.38	5
$Ni_{0.3}Zn_{0.7}Fe_2O_4$	FM 0.83	40, 43	$ZrZn_2$	FM -1.2	72
		44			

* Compiled by D. B. McWhan, Bell Telephone Laboratories.
[a] AF—antiferromagnetic. FM—ferro or ferrimagnetic.
[b] 1 kbar $\approx$ 987 atm $\approx$ 1,020 kg/cm².
[c] Range of published values.
[d] $T_N(V) = T_N(V_0)\exp(-26.5\,\Delta V/V_0)$; see McWhan and Rice [14].
[e] Low-pressure phase FM; high-pressure phase AF.
[f] Heusler alloys: for series of alloys see Austin and Mishra [30].

References for Table 5f-17

1. Bloch, D.: *Ann. Phys.* [14] **1**, 93 (1966).
2. Hirone, T., T. Kaneko, and K. Kondo: *J. Phys. Soc. Japan* **18**, 65 (1963).
3. Matsumoto, M., T. Kaneko, and K. Kamigaki: *J. Phys. Soc. Japan* **24**, 953 (1968).
4. Tsuboi, T., T. Nakajima, and H. Takaki: *J. Phys. Soc. Japan* **19**, 768 (1964).
5. Menyuk, N., J. A. Kafalas, K. Dwight, and J. B. Goodenough: *J. Appl. Phys.* **40**, 1324 (1969).
6. Goodenough, J. B., J. M. Longo, and J. A. Kafalas: *Materials Research Bull.* **3**, 471 (1968).
7. Kafalas, J. A.: Private communication.
7a. Kafalas, J. A., N. Menyuk, K. Owight, and J. M. Longo: *J. Appl. Phys.* (1971) (to be published).
8. Srivastava, V. C.: *J. Appl. Phys.* **40**, 1017 (1969).
9. Patrick, L.: *Phys. Rev.* **93**, 384 (1954).
10. Leger, J. M., C. Susse, and B. Vodar: *Solid State Commun.* **5**, 755 (1967).
11. Astrov, D. N., G. A. Kytin, and M. P. Orlova: *Soviet Phys.- Solid State* **4**, 777 (1962).
12. Astrov, D. N., S. J. Novikova, and M. P. Orlova: *Soviet Phys.–JETP* **10**, 851 (1960).
13. Bloch, D., F. Chaissé, and R. Pauthenet: *J. Appl. Phys.* **37**, 1401 (1966).
13a. Okamoto, T., H. Fujii, Y. Hidaka, and E. Tatsumoto: *J. Phys. Soc. Japan* **23**, 1174 (1967).
13b. Holzappel, W. B., and H. G. Drickamer: *Phys. Rev.* **184**, 323 (1969).
14. McWhan, D. B., and T. M. Rice: *Phys. Rev. Letters* **19**, 846 (1967).
15. Mitsui, T., and C. T. Tomizuka: *Phys. Rev.* **137A**, 564 (1965).
16. Voronov, F. F.: *Soviet Phys.–JETP* **20**, 1342 (1965).
17. Litvin, D. F., and E. G. Ponyatovskii: *Soviet Phys.–Doklady* **9**, 388 (1964).
18. Worlton, T. G., R. M. Brugger, and R. B. Bennion: *J. Phys. Chem. Solids* **29**, 435 (1968).
18a. Bloch, D., P. Mallaro, and J. Voiron: *Compt. rend.* **269**, 553 (1969).
19. Kamigaichi, T., T. Okamoto, N. Iwata, and E. Tatsumoto: *J. Phys. Soc. Japan* **21**, 2730 (1966).
20. Ido, H., T. Kaneko, and K. Kamigaki: *J. Phys. Soc. Japan* **22**, 1418 (1967).
21. Grazhdankina, N. P., L. G. Gaidukov, K. P. Rodionov, M. I. Oleinik, and V. A. Schchipanov: *Soviet Phys.–JETP* **13**, 297 (1961).
21a. Bloch, D., and A. S. Pavlovic: "Advances in High Pressure Research," vol. III, Academic Press, Inc., New York, 1969.
22. Longo, J. M., and J. A. Kafalas: *J. Appl. Phys.* **40**, 160 (1969).
23. Bartholin, H., and D. Bloch: *J. Phys. Chem. Solids* **29**, 1063 (1968).
24. Milton, J. E., and T. A. Scott: *Phys. Rev.* **160**, 387 (1967).
25. Okamoto, T., N. Iwata, S. Ishida, and E. Tatsumato: *J. Phys. Soc. Japan* **21**, 2727 (1966).
26. Robinson, L. B., S. I. Tan, and K. F. Sterrett: *Phys. Rev.* **141**, 548 (1966).
27. McWhan, D. B., and A. L. Stevens: *Phys. Rev.* **154**, 438 (1967); **139**, A682 (1965).
28. Souers, P. C., and G. Jura: *Science* **145**, 575 (1964).
29. Landry, P., and R. Stevenson: *Can. J. Phys.* **41**, 1273 (1963).
30. Austin, I. G., and D. K. Mishra: *Phil. Mag.* **15**, 529 (1967).
31. Milstein, F., and L. B. Robinson: *Phys. Rev.* **159**, 466 (1967).
32. Okamoto, T., H. Fujii, Y. Hidaka, and E. Tatsumoto: *J. Phys. Soc. Japan* **24**, 951 (1968).
32a. Bloch, D., and F. Chaissé: *Compt. rend.* **268**, 660 (1969).
33. Grazhdankina, N. P.: *Soviet Phys.–JETP* **25**, 258 (1967).
34. McWhan, D. B., P. C. Souers, and G. Jura: *Phys. Rev.* **143**, 385 (1966).
34a. Menyuk, N., K. Owight, and J. A. Kafalas: *J. Appl. Phys.* (1971) (to be published).
35. Stevenson, R., and M. C. Robinson: *Can. J. Phys.* **43**, 1744 (1965).
36. Sokolova, G. K., K. M. Demchuk, K. P. Rodionov, and A. A. Samokhvalov: *Soviet Phys.–JETP* **22**, 317 (1966).
37. Schwob, P., and O. Vogt: *Phys. Letters* **24A**, 242 (1967).
38. Schwob, P., and O. Vogt: *J. Appl. Phys.* **40**, 1328 (1969).
39. Leger, J. M., C. Susse, and B. Vodar: *Solid State Commun.* **4**, 503 (1966).
40. Foiles, C. L., and C. T. Tomizuka: *J. Appl. Phys.* **36**, 3839 (1965).
41. Samara, G. A., and A. A. Giardini: *Phys. Rev.* **186**, 577 (1969).
42. Endo, S., S. Kume, M. Koizumi, C. Okazaki, and E. Hirota: *Phys. Letters* **18**, 232 (1965).
43. Werner, K.: *Ann. Physik* **7**, 403 (1959).
44. Adams, C. Q., and C. M. Davis, Jr.: *J. Appl. Phys.* **29**, 372 (1958).
45. Fujimora, H.: *J. Phys. Soc. Japan* **21**, 1860 (1966).
46. Graham, R. A., D. H. Anderson, and J. R. Holland: *J. Appl. Phys.* **38**, 223 (1967).

47. Livshits, L. D., and Yu. S. Genshaft: *Soviet Phys.–JETP* **19**, 560 (1964).
47a. Wayne, R. C., and L. C. Bartel: *Phys. Letters* **28A**, 96 (1968).
48. Robinson, L. B., F. Milstein, and A. Jayaraman: *Phys. Rev.* **134**, A187 (1964).
49. Livshitz, L. D., and Yu. S. Genshaft: *Soviet Phys.–JETP* **21**, 701 (1965).
50. Iwata, N., T. Okamoto, and E. Tatsumoto: *J. Phys. Soc. Japan* **24**, 948 (1968).
51. McWhan, D. B.: *J. Chem. Phys.* **44**, 3528 (1966).
52. Umebayashi, H., G. Shirane, B. C. Frazer, and W. B. Daniels: *Phys. Rev.* **165**, 688 (1968).
52a. Swaoka, A., T. Soma, S. Saito, and Y. Gndo: (to be published).
53. Menyuk, N., J. A. Kafalas, K. Dwight, and J. B. Goodenough: *Phys. Rev.* **177**, 942 (1969).
54. Grazhdankina, N. P., and Yu. S. Bersenyev: *Soviet Phys.–JETP* **24**, 707 (1967).
55. Samara, G. A., and A. A. Giardini: *Bull. Am. Phys. Soc.* **9**, 635 (1964) and "Physics of Solids at High Pressures," p. 308, C. T. Tomizuka and R. M. Emrick, eds., Academic Press, Inc., New York, 1965.
56. Bean, C. P., and D. S. Rodbell: *Phys. Rev.* **126**, 104 (1962).
57. Benedek, G. B., and T. Kushida: *Phys. Rev.* **118**, 46 (1960).
57a. Bloch, D., and R. Pauthenet: *Compt. rend.* **254**, 1222 (1962).
58. Bartholin, H., D. Bloch, and R. Georges: *Compt. rend.* **264**, 360 (1967).
59. Kawai, N., A. Sawaoka, and G. Kaji: *J. Phys. Soc. Japan* **23**, 896 (1967).
60. Kamigaichi, T., T. Okamoto, N. Iwata, and E. Tatsumoto: *J. Phys. Soc. Japan* **24**, 649 (1968).
60a. Georges, R.: *Compt. rend.* **268**, 16 (1969).
61. Grazhdankina, N. P.: *Soviet Phys.–JETP* **6**, 1178 (1958).
62. Ozawa, K., S. Anzai, and Y. Hamaguchi: *Phys. Letters* **20**, 132 (1966).
63. Sawaoka, A., S. Miyahara, and S. Minomura: *J. Phys. Soc. Japan* **21**, 1017 (1966).
64. Leger, J. M., C. Susse, R. Epain, and B. Vodar: *Solid State Commun.* **4**, 197 (1966).
65. Anzai, S., and K. Ozawa: *J. Phys. Soc. Japan* **24**, 271 (1968).
65a. Holzappel, W. B., J. A. Cohen, and H. G. Drickamer: *Phys. Rev.* **187**, 667 (1969).
66. Fawcett, E., D. B. Mcwhan, R. C. Sherwood, and M. P. Sarachik: *Solid State Commun.* **6**, 509 (1968).
66a. Longo, J. M., J. A. Kafalas, N. Menyuk, and K. Owight: *J. Appl. Phys.* (1971) (to be published).
67. Kafalas, J. A., and J. M. Longo: *Materials Research Bull.* **3**, 501 (1968).
68. Gardner, W. E., T. F. Smith, B. W. Howlett, C. W. Chu, and A. Sweedler: *Phys. Rev.* **166**, 577 (1968).
69. Wazzan, A. R., R. S. Vitt, and L. B. Robinson: *Phys. Rev.* **159**, 400 (1967).
70. Tatsumoto, E., H. Fujiwara, H. Fujii, N. Iwata, and T. Okamoto: *J. Appl. Phys.* **39**, 894 (1968).
71. Feinleib, J., and W. Paul: *Phys. Rev.* **155**, 841 (1967).
72. Wayne, R. C., and L. R. Edwards: *Phys. Rev.* **188**, 1042 (1969).

5f-10. Magnetic Anisotropy.[1] The magnetocrystalline anisotropy energy density E_K can be expanded in powers of the magnetization (**M**) components. Since **M** changes sign under time reversal, this expansion must contain only even[2] functions of the components of **M**. In addition, the number of terms in the expansion may be reduced by the requirement that E_K be invariant under the crystallographic symmetry operations. An equivalent phenomenological description of E_K can be obtained by expanding it in powers of the direction cosines of **M** with respect to the principal crystallographic axes.[3] Thus for cubic crystals, the anisotropy energy per unit volume is

$$E_K = K_1(\alpha_1{}^2\alpha_2{}^2 + \alpha_2{}^2\alpha_3{}^2 + \alpha_1{}^2\alpha_3{}^2) + K_2\alpha_1{}^2\alpha_2{}^2\alpha_3{}^2 \\ + K_3(\alpha_1{}^4\alpha_2{}^4 + \alpha_2{}^4\alpha_3{}^4 + \alpha_1{}^4\alpha_3{}^4) + \cdots$$

where the α's are the direction cosines of **M** with respect to the cubic axes. Usually $|K_2|$ is considerably smaller than $|K_1|$.

In hexagonal crystals,

$$E_K = K_1 \sin^2\theta + K_2 \sin^4\theta + K_3 \sin^6\theta + K_4 \sin^6\theta \cos 6\psi + \cdots$$

where θ is the polar angle of M with respect to the crystallographic c axis, and ψ is the azimuthal angle of M with respect to the a axis.

For tetragonal crystals,

$$E_K = K_1 \sin^2\theta + K_2 \sin^4\theta + K_3 \sin^4\theta \sin^2\psi \cos^2\psi + \cdots$$

where θ and ψ are defined as in the hexagonal case.

To obtain the angular dependence of the anisotropy energy for each of the 32 crystal classes the functions of the direction cosines of M tabulated by Döring[4] must be corrected[5] in order to satisfy the requirement of time-reversal invariance. This means omitting all terms which are of odd degree in the direction cosines.

Induced Anisotropy. Induced anisotropy is a nonintrinsic anisotropy which is produced by some external treatment, for example, by cold working or by annealing in a magnetic field or in the presence of an applied mechanical stress. The axes of such anisotropies do not, in general, correspond to principal crystallographic axes but are related to the directions of cold working, applied magnetic field, or applied mechanical stress. The induced anisotropy energy resulting from magnetic anneal is of the form[6]

$$E_{Ki} = -F(\alpha_1{}^2\alpha_1'^2 + \alpha_2{}^2\alpha_2'^2 + \alpha_3{}^2\alpha_3'^2) \\ -G(\alpha_1\alpha_2\alpha_1'\alpha_2' + \alpha_2\alpha_3\alpha_2'\alpha_3' + \alpha_1\alpha_3\alpha_1'\alpha_3') \cdots$$

where α_1, α_2, α_3 are the direction cosines of **M** with respect to the principal crystallographic axes and α_1', α_2', α_3' are the direction cosines of **M** with respect to the crystallographic axes at the time of the anneal.

[1] Compiled by V. J. Folen, Naval Research Laboratory.
[2] Except in cases where unidirectional anisotropy exists, as, for example, in Co-CoO interface materials. See W. H. Meiklejohn and C. P. Bean, *Phys. Rev.* **102**, 1413 (1956).
[3] N. S. Akulov, *Z. Physik* **57**, 249 (1929); **69**, 78 (1931).
[4] W. Döring, *Ann. Physik* **1**, 102 (1958).
[5] W. Döring and G. Simon, *Ann. Physik* **8**, 144 (1961).
[6] J. C. Slonczewski, "Magnetism," G. T Rado and H. Suhl, eds., Academic Press, Inc., New York, vol 1, 1963, p. 205.

TABLE 5f-18. MAGNETOCRYSTALLINE ANISOTROPY CONSTANTS OF SOME
CUBIC METALS [1]
(in ergs/cm³)

Material*	$10^{-4}K_1$	$10^{-4}K_2$	Material*	$10^{-4}K_1$	$10^{-4}K_2$
Fe [2] (Fig. 5f-30).	45	20	13.5% Si-Fe.....	6 [11]	
Fe [3]..........	48	0 ± 5	7% Al-Fe†.......	45 [12]	
Fe [4]..........	45	−3.5	16% Al-Fe†......	28 [12]	
77 K..........	52 [3]	∼9.5 [5]	27% Al-Fe†		
Ni (Fig. 5f-31)...	−5 [4, 6]	−2 [6]	Disordered....	−1 [12]	
77 K..........	−58 [6]	−21 [4,6]	Ordered.......	−12 [12]	
Fe-Ni (Fig.			13% Cu-Ni......	−2.3 [13]	
5f-32)........			24% Cu-Ni......	−0.5 [13]	
30% Fe-Co......	10.2 [7]	16.0 [7]	1.47% Cr-Ni†....	−1.1 [14]	
50% Fe-Co......	−7 [7]	−39 [7]	77 K..........	−27 [14]	
70% Fe-Co.....	−43 [7]	+5 [7]	4.08% Cr-Ni†....	0 [14]	
65% Co-Ni......	−26 [8]	15 [8]	77 K..........	−6.5 [14]	
20% Co-Ni.....	−0.4 [8]	0.8 [8]	1.28% V-Ni†....	−2.4 [14]	
25% Fe, 25% Co,			77 K..........	−36 [14]	
50% Ni.......	0.35 [9]	1.6 [9]	3.93% V-Ni†....	−0.28 [14]	
50% Fe, 10% Co,			77 K..........	−0.13 [14]	
40% Ni.......	6 [9]	16 [9]	2% Mo, 19% Fe,		
10% Fe, 40% Co,			79% Ni		
50% Ni.......	−7 [9]	−0.4 [9]	Quenched......	−0.14 [16]	
75% Ni-Mn.....	3.3 [15]	−7 [15]	Annealed......	−0.33 [16]	
77 K..........	5 [15]		6% Mo, 15% Fe,		
3.1% Si-Fe.....	36.5 [10]		79% Ni		
5.1% Si-Fe.....	28.5 [10]	−5 ± 5 [10]	Quenched......	0.53 [16]	
7.5% Si-Fe.....	17 [11]		Annealed......	0.77 [16]	

* All percentages are in weight percent except those indicated by †.

References for Table 5f-18

1. Unless specified, values are for room temperature. See also R. M. Bozorth, "Ferromagnetism," D. Van Nostrand Company, Inc., Princeton, N.J., 1951. Additional data are given in the original reports.
2. Bozorth, R. M.: *J. Appl. Phys.* **8**, 575 (1937). Constants determined from measurements by K. Honda, H. Masumoto and S. Kaya: *Sci. Repts. Tôhoku Univ.* **17**, 111 (1928).
3. Graham, C. D., Jr.: *J. Appl. Phys.* **30**, 317 (1959).
4. Hofmann, U.: *Z. angew. Phys.* **22**, 106 (1967).
5. Sato, H., and B. S. Chandrasekhar: *J. Phys. Chem. Solids* **1**, 228 (1957).
6. Aubert, G.: *J. Appl. Phys.* **39**, 504 (1968).
7. McKeehan, L. W.: *Phys. Rev.* **51**, 136 (1937) based on data in J. W. Shih: *Phys. Rev.* **46**, 139 (1934).
8. Shih, J. W.: *Phys. Rev.* **50**, 376 (1936).
9. McKeehan, L. W.: *Phys. Rev.*, **51**, 136 (1937).
10. Graham, C. D., Jr.: *J. Appl. Phys.* **30**, 317 S (1959).
11. Gengnagel, H., and H. Wagner: *Z. angew. Phys.* **13**, 174 (1961).
12. Hall, R. C.: *J. Appl. Phys.* **30**, 816 (1959).
13. Williams, H. J., and R. M. Bozorth: *Phys. Rev.* **55**, 673 (1939).
14. Wakiyama, T., and S. Chikazumi: *J. Phys. Soc. Japan* **15**, 1975 (1960).
15. Blanchard, A., and V. Tutovan: *Compt. rend.* **261**, 2852 (1965).
16. Aoyagi, K.: *J. Appl. Phys. (Japan)* **4**, 551 (1965).

TABLE 5f-19. MAGNETOCRYSTALLINE ANISOTROPY CONSTANTS
OF SOME CUBIC OXIDES AND OTHER CUBIC MATERIALS [1]
(in ergs/cm^3)

Material	$10^{-4}K_1$	$10^{-4}K_2$
$Y_3Fe_5O_{12}$(YIG) [2*]	-0.4	
4.2 K	-2.45	
$Y_3Ga_{1.47}Fe_{3.53}O_{12}$ [3*]	0	
0 K	-1.2	
$Y_3In_{0.20}Fe_{4.80}O_{12}$ [4]	-0.4	
4.2 K	-2.4	
$Y_3Sc_{0.87}Fe_{4.13}O_{12}$ [4]	~ -0.1	
4.2 K	-1.0	
$Y_{2.62}La_{0.38}Fe_5O_{12}$ [5*]	$K_1/M = -38 Oe$	
88 K	$K_1/M = -108 Oe$	
$Gd_3Fe_5O_{12}$ [2*,6]	0	
4.2 K	-22	
$Er_3Fe_5O_{12}$ [6]	-0.6	
100 K	-2.2	
$Yb_3Fe_5O_{12}$ [7]	-0.8	
1.5 K	-900	900
$Tb_3Fe_5O_{12}$ [6]	-1	
80 K	-76	-760
$Ho_3Fe_5O_{12}$ [6]	~ 0	
80 K	-80	-27
$Dy_3Fe_5O_{12}$ [6]	~ 0	
80 K	-97	21.4
$Sm_3Fe_5O_{12}$ [6]	-2	
80 K	-120	100
$Eu_3Fe_5O_{12}$ [8*]	-0.4	0
170 K	-2	1
$Tm_3Fe_5O_{12}$ [9*]	-1.1	~ 0
77 K	-21	10
$MnFe_2O_4$ [10]	-4	
92 K [11]	-19.1	-2.15
$Mn_{0.75}Fe_{2.25}O_4$ [12]	0.9	
90 K	0	-9
$Mn_{0.50}Fe_{2.50}O_4$ [12]	-1.6	
90 K	-10.7	
$Mn_{0.53}Mg_{0.75}Fe_{1.71}O_{3.96}$		
88 K [11]	-23.7	-1.6
$MnTi_{0.15}Fe_{1.85}O_4$		
4.2 K [13]	101	-230
$Mn_{0.78}Ni_{0.24}Fe_{1.98}O_4$ [12]	-4.1	
90 K	-13.5	-10
$Mn_{0.57}Ni_{0.20}Fe_{2.23}O_4$ [12]	-2.8	
90 K	-9.5	-8
$Mn_{0.53}Mg_{0.75}Co_{0.023}Fe_{1.71}O_{3.98}$ [14]	-2.8	
88 K	18.9	-147
$Mn_{0.53}Mg_{0.75}Co_{0.076}Fe_{1.71}O_{4.04}$ [14]	-0.7	
88 K	94	-423
$Mn_{0.99}Co_{0.009}Fe_{1.98}O_4$ 82 K [15]	19.8	-104
$Mn_{1.03}Co_{0.019}Fe_{1.94}O_4$ [16]	-22.2	
160 K	91.8	
$NiFe_2O_4$ [17*]	-6.9	0.2
4.2 K	-8.9	-0.06
$Ni_{0.76}Fe_{2.16}O_4$ [18]	-3.9	
77 K	-4.2	
$Ni_{0.20}Fe_{2.80}O_4$ [12]	-6.7	
90 K	12.3	-9
$Ni_{0.58}Mn_{0.17}Fe_{2.25}O_4$ [12]	-4.7	
90 K	-6.6	-9

TABLE 5f-19. MAGNETOCRYSTALLINE ANISOTROPY CONSTANTS
OF SOME CUBIC OXIDES AND OTHER CUBIC MATERIALS [1] (Continued)

Material	$10^{-4}K_1$	$10^{-4}K_2$
$Ni_{0.3}Zn_{0.45}Fe_{2.25}O_4$ [19]............	−1.7	
88 K.......................	−16	
$Ni_{0.7}Co_{0.004}Fe_{2.2}O_4$ [20]...........	−1.0	
77 K.......................	−19.6	
Fe_3O_4 [21]......................	−13.0	−4.4
150 K.......................	−4.0	−4.2
$Ga_{0.44}Fe_{0.93}{}^{2+}Fe_{1.61}{}^{3+}O_4$ [22]........	−80.9	
200 K.......................	−122	
$Ga_{0.76}Fe_{0.94}{}^{2+}Fe_{1.28}{}^{3+}O_4$ [22]........	−36.9	
200 K.......................	−75.7	
$Al_{0.21}Fe_{0.99}{}^{2+}Fe_{1.81}{}^{3+}O_4$ [23]........	−12	
120 K.......................	−10	
$Fe_{2.9}Ti_{0.10}O_4$ [21]................	−25.0	4.8
80 K.......................	2.3	−4.0
$Fe_{2.69}Ti_{0.31}O_4$ [21]................	−18.1	
80 K.......................	−1.0	
$Fe_{2.99}Co_{0.01}O_4$ [24]...............	0	
130 K.......................	69.2	
$CoFe_2O_4$ [25].....................	260	
$Co_{0.8}Fe_{2.2}O_4$ [20].................	290	
77 K.......................	440	
$Co_{1.1}Fe_{1.9}O_4$ [20].................	180	
$Co_{0.7}Zn_{0.3}Fe_{2.0}O_4$ [26]............	23.6	
$Co_{0.3}Zn_{0.2}Fe_{2.2}O_4$ [20]	150	
$Li_{0.55}Fe_{2.45}O_4$		
Ordered......................	−8.3 [27]	−0.2 [28*]
77 K.......................	−12.7 [27]	−9.7 [28*]
Disordered...................	−9.0 [27]	0.06 [28*]
77 K.......................	−16.2 [27]	−3.2 [28*]
$Mg_{0.94}Fe_{0.06}{}^{2+}Fe_2{}^{3+}O_4$		
Low M_s [29]...................	−4.6	
77 K.......................	−12.8	
High M_s.....................	−4.2	
77 K.......................	−15.3	
$Mg_{0.75}Fe_{0.25}{}^{2+}Fe_2{}^{3+}O_4$ [30]........	−4.8	
88 K.......................	−12.2	
$Mg_{0.50}Fe_{0.50}{}^{2+}Fe_2{}^{3+}O_4$ [30]........	−5.8	
88 K.......................	−6.0	
$Mg_{0.8}Fe_{2.2}O_4$ [11], 88 K..........	−10.9	−1.5
$CuFe_2O_4$ [31]....................	−0.04	
77 K.......................	−2.06	
EuO [32] 1.5 K................	$K_1/M = 190$ Oe*	
$CdCr_2Se_4$ [33], 4.4 K.............	0.22–1.8*	
$CdCr_2S$ [33], 4.4 K.............	0.38–4.4*	

* Values obtained from microwave measurements.

References for Table 5f-19

1. Unless specified, values are for room temperature. Additional data are given in the original reports.
2. Rodrigue, G. P., H. Meyer, and R. V. Jones: *J. Appl. Phys.* **31**, 376S (1960).
3. Lüthi, B., and T. Henningsen: *Proc. Intern. Conf. Magnetism, Nottingham,* The Institute of Physics and the Physical Society, London, 1964.
4. Pearson, R. F., and A. D. Annis: *J. Appl. Phys.* **39**, 1338 (1968).
5. Makram, H., and R. Krishnan: *J. phys. radium* **25**, 343 (1964).
6. Pearson, R. F.: *J. Appl. Phys.* **33**, 1236 (1962).
7. Pearson, R. F.: *Proc. Phys. Soc.* **86**, 1055 (1965).
8. Miyadai, T.: *J. Phys. Soc. Japan* **15**, 2205 (1960).
9. Miyadai, T.: *J. Phys. Soc. Japan* **17**, 1899 (1962).
10. Penoyer, R. F., and M. W. Shafer: *J. Appl. Phys.* **30**, 315S (1959).
11. Perthel, R., and G. Elbinger: *Z. angew. Phys.* **19**, 344 (1965).
12. Miyata, N.: *J. Phys. Soc. Japan* **16**, 1291 (1961).
13. Smit, J., F. K. Lotgering, and R. P. van Stapele: *J. Phys. Soc. Japan,* **17**, 268 (1962).
14. Elbinger, G.: *Phys. Status Solidi* **9**, 843 (1965).
15. Pearson, R. F., and R. W. Teale: *Proc. Phys. Soc. (London)* **75**, 314 (1960).
16. Pearson, R. F.: *Proc. Phys. Soc. (London)* **74**, 505 (1959).
17. Smith, A. B., and R. V. Jones: *J. Appl. Phys.* **37**, 1001 (1966).
18. Bozorth, R. M., B. B. Cetlin, J. K. Galt, F. R. Yager, and W. A. Merritt: *Phys. Rev.* **99**, 1898 (1955).
19. Ohta, K.: *Bull. Kobayashi Inst. Phys. Research* **10**, 149 (1960).
20. Bozorth, R. M., E. F. Tilden, and A. J. Williams: *Phys. Rev.* **99**, 1788 (1955).
21. Syono, Y.: *Japan, J. Geophys.* **4**, 71 (1965).
22. Pearson, R. F.: *J. Appl. Phys.* **31**, 160S (1960).
23. Pearson, R. F.: *J. phys. radium* **20**, 409 (1959).
24. Bickford, L. R., Jr., J. M. Brownlow, and R. F. Penoyer: *Proc. IEE* **104B**, 238 (1956).
25. Perthel, R., G. Elbinger, and W. Keilig: *Phys. Status Solidi* **17**, 151 (1966).
26. Okamura, T., and Y. Kojima: *Phys. Rev.* **85**, 690 (1952).
27. Folen, V. J.: *J. Appl. Phys.* **31**, 166S (1960).
28. Schnitzler, A. D., V. J. Folen, and G. T. Rado: *J. Appl. Phys.* **31**, 348S (1960).
29. Folen, V. J., and G. T. Rado: *J. Appl. Phys.* **29**, 438 (1958).
30. Elbinger, G.: *Naturwiss.* **48**, 498 (1961).
31. Okamura, T., and Y. Kojima: *Phys. Rev.* **86**, 1040 (1952).
32. Dillon, J. F., Jr., and C. E. Olsen: *Phys. Rev.* **135**, A434 (1964).
33. Berger, S. B., and H. L. Pinch: *J. Appl. Phys.* **38**, 949 (1967).

TABLE 5f-20. MAGNETOCRYSTALLINE ANISOTROPY CONSTANTS
OF SOME NONCUBIC MATERIALS*†
(in ergs/cm^3)

Material	$10^{-6}K_1$	$10^{-6}K_2$
Co [1] (Fig. 5f-33)...........	5.3	1.0
97 K.....................	7.9	1.0
Gd [2]		
10 K.....................	-0.85	2.5
Tb [3]		
205 K....................	$K_1 + \frac{8}{7}K_2 = -260$	
11 K.....................	$K_1 + \frac{8}{7}K_2 = -830$	
Dy [3]		
152 K....................	$K_1 + \frac{8}{7}K_2 = -260$	
22 K.....................	$K_1 + \frac{8}{7}K_2 = -750$	
1.8 wt. % Tb-Gd [4]		
190 K....................	-2.9	0.6
4.2 K....................	-13.1	6.6
1.3 wt. % Dy-Gd [4]		
180 K....................	-1.9	0.3
4.2 K....................	-8.8	4.2
MnBi [5]....................	$K_1 + K_2 = 12$	
88 K.....................	$K_1 + K_2 = 0$	
Mn$_2$Sb [6].................	0.30	0.13
200 K....................	-0.50	0.03
Mn$_{1.95}$Cr$_{0.05}$Sb [6]............	0.76	-0.12
200 K....................	0	
Mn$_5$Ge$_3$ [7].................	0.3	
77 K.....................	4.2	
Fe$_5$Ge$_3$ [8].................	-5	
77 K.....................	-8.5	
CrTe [9]		
283 K....................	-1.5	
0 K.....................	-5.5	
CrO$_2$ [10‡].................	$K_1/M = 450$ Oe	$K_2/M = 90$ Oe
4.2 K....................	$K_1/M = 690$ Oe	$K_2/M = 60$ Oe
Ga$_{0.89}$Fe$_{1.11}$O$_3$ [11], 4.2 K......	(a-c plane) 3.22	
	(b-c plane) 5.65	
CrBr$_3$ [12‡], 1.5 K...........	0.94	
Fe$_7$Se$_8$ [13]		
280 K....................	2.5	
4.2 K....................	35	
BaFe$_{12}$O$_{19}$..................	3.3 [14]	0 [15]
77 K.....................	4.3 [14]	0 [15]
BaCo$_{1.5}$Ti$_{1.5}$Fe$_9$O$_9$ [16]		
90 K.....................	-1.05	
BaNi^{2+}Ti^{4+}Fe$_{10}$O$_{19}$ [17]........	0.7	
BaZn^{2+}Ti^{4+}Fe$_{10}$O$_{19}$ [17]........	1.4	
PbFe$_{12}$O$_{19}$ [18]..............	2.2	0.030
2.3 K....................	2.8	0.062
Ba$_2$Co$_2$Fe$_{12}$O$_{22}$ [17]............	$K_1 + 2K_2 = -2.6$	
BaFe$_{18}$O$_{27}$ [19].............	3.41	0
77 K.....................	4.31	0
BaCo$_{0.5}$Fe$_{1.5}{}^{2+}$Fe$_{16}{}^{3+}$O$_{27}$ [19]....	2.38	0.11
77 K.....................	-0.04	2.14
BaCo$_{1.5}$Fe$_{0.5}{}^{2+}$Fe$_{16}{}^{3+}$O$_{27}$ [19]....	-0.66	0.63
77 K.....................	-4.98	3.61
BaCo$_{1.85}$Fe$_{16.2}$O$_{27}$ [20]........	$K_1 + \frac{8}{7}K_2 = -4.0$	
77 K.....................	$K_1 + \frac{8}{7}K_2 = -7.3$	
BaCoZnFe$_{16}$O$_{27}$ [16]		
77 K.....................	-3.4	
Ba$_4$Co$_2$Fe$_{24}$O$_{41}$ [14]............	$K_1 + 2K_2 = -1.7$	
Ba$_3$CoZnFe$_{24}$O$_{41}$ [14].........	$K_1 + 2K_2 = -0.8$	
77 K [16].................	$K_1 + 2K_2 = -0.8$	

* Unless otherwise specified, values are for room temperature.
† All materials are hexagonal except Ga$_{0.89}$Fe$_{1.11}$O$_3$ (orthorhombic), and Mn$_2$Sb, Mn$_{1.95}$Cr$_{0.05}$Sb, CrO$_2$ (tetragonal).
‡ Obtained from microwave measurements.

References for Table 5f-20

1. Sucksmith, W. and J. E. Thompson: *Proc. Roy. Soc. (London)*, ser A, **225**, 362 (1954).
2. Graham, C. D., Jr.: *J. Phys. Soc. Japan* **17**, 1310 (1962).
3. Rhyne, J. J., and A. E. Clark: *J. Appl. Phys.* **38**, 1379 (1967).
4. Tajima, K., and S. Chicazumi: *J. Phys. Soc. Japan* **23**, 1175 (1967).
5. Guillaud, C.: Thesis, Strasbourg, 1943.
6. Jarrett, H. S., P. E. Bierstedt, F. J. Darnell, and M. Sparks: *J. Appl. Phys.* **32**, 57S (1961).
7. Tawara, Y., and K. Sato: *J. Phys. Soc. Japan* **18**, 773 (1963).
8. Tawara, Y.: *J. Phys. Soc. Japan* **19**, 776 (1964).
9. Hirone, T., and S. Chiba: *J. Phys. Soc. Japan* **15**, 1991 (1960).
10. Rodbell, D. S.: *J. Phys. Soc. Japan* **21**, 1224 (1966).
11. Schelleng, J. H., and G. T. Rado: *Phys. Rev.* **179**, 541 (1969).
12. Dillon, J. F., Jr.: *J. Appl. Phys.* **33**, 1191 (1962).
13. Kamimura, T., K. Kamigaki, T. Hirone, and K. Sato: *J. Phys. Soc. Japan* **22**, 1235 (1967).
14. Casimir, H. B. G., J. Smit, U. Enz, J. F. Fast, H. P. J. Wijn, E. W. Gorter, A. J. W. Duyvesteyn, J. D. Fast, and J. J. deJong: *J. phys. radium* **20**, 360 (1959).
15. Private communication from E. W. Gorter.
16. Lotgering, F. K., U. Enz, and J. Smit: *Philips Research Repts.* **16**, 441 (1961).
17. Smit, J., and H. P. J. Wijn: "Ferrites," John Wiley & Sons, Inc., New York, 1959.
18. Pauthenet, R., and G. Rimet: *Compt. rend.* **249**, 1875 (1959).
19. Perekaline, T. M., and A. V. Zalesskii: *Soviet Phys.–JETP* **19**, 1337 (1954).
20. Bickford, L. R., Jr.: *J. Phys. Soc. Japan* **17**, suppl. BI, 272 (1962).

5f-11. Magnetostriction.[1] *Single Crystals.* When a cubic crystal is magnetized to saturation in a direction defined by the direction cosines α_1, α_2, α_3, the fractional change in length measured in the direction β_1, β_2, β_3 is given to the first approximation by

$$\left(\frac{\Delta l}{l}\right)_s = \lambda_s = \tfrac{3}{2}\lambda_{100}(\alpha_1{}^2\beta_1{}^2 + \alpha_2{}^2\beta_2{}^2 + \alpha_3{}^2\beta_3{}^2 - \tfrac{1}{3})$$
$$+ 3\lambda_{111}(\alpha_1\alpha_2\beta_1\beta_2 + \alpha_2\alpha_3\beta_2\beta_3 + \alpha_3\alpha_1\beta_3\beta_1)$$

[1] Prepared by A. E. Clark, U.S. Naval Ordnance Laboratory.

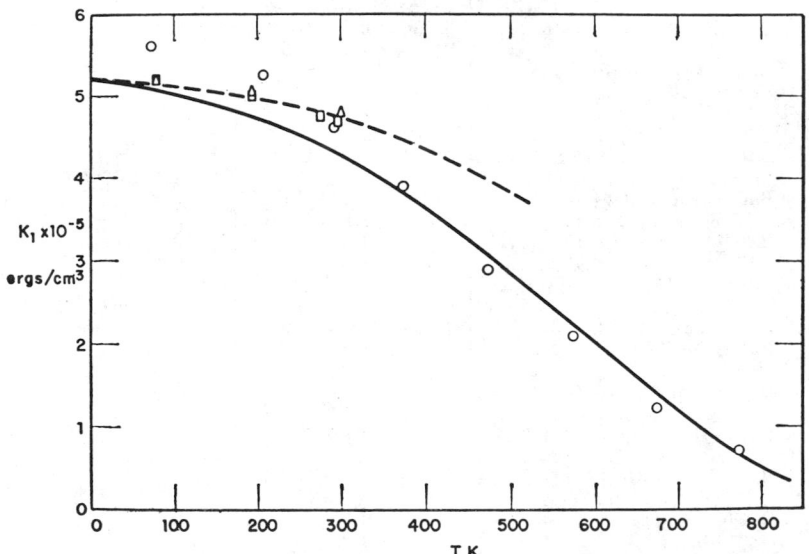

FIG. 5f-30. Measured values of K_1 for iron plotted against temperature. △—measurements of Graham. ☐—measurements of Bozorth as given by Graham. ○—older values of Bozorth as calculated from magnetization curves of Honda, Masumoto, and Kaya: --- $5.2 \times 10^5 \ (\sigma/\sigma_0)^5$. — $5.2 \times 10^5 \ (\sigma/\sigma_0)^{10}$. σ is the magnetization per unit mass. [*From W. J. Carr, Jr., J. Appl. Phys.* **31**, 69 (1960).]

provided that in the initial condition from which λ_s is measured, the domains are distributed equally among the easy directions of magnetization (six [100] directions in Fe, eight [111] directions in Ni). In any case, this equation gives the correct change in λ_s as the magnetization direction is varied. A five-constant expression which includes magnetization-direction cosines to the fourth power is sometimes used [1]. To lowest order there is no volume change associated with domain rotation. The constants λ_{100} and λ_{111} for some cubic materials are given in Tables 5f-21 and 5f-22 and in Fig. 5f-34.

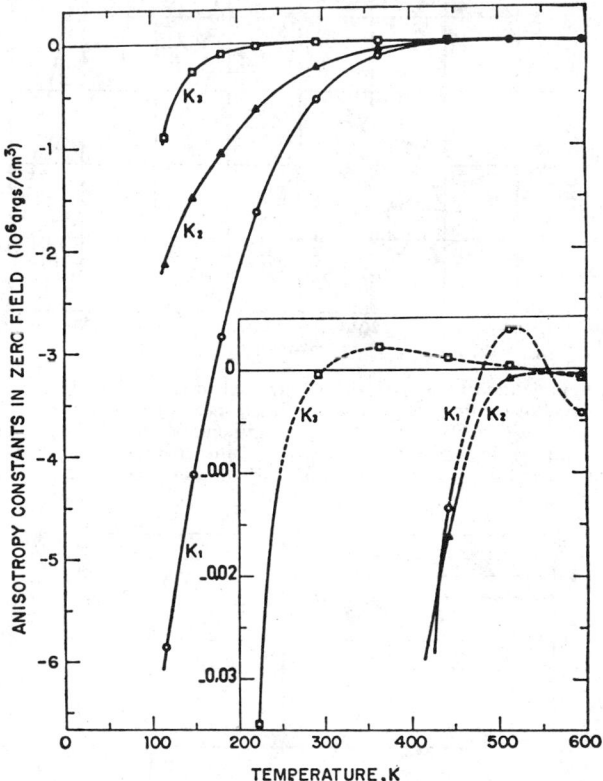

FIG. 5f-31. Temperature dependence of the anisotropy constants of nickel. [*From G. Aubert, J. Appl. Phys.* **39**, 504 (1968).]

The saturation magnetostriction of hexagonal crystals is described to lowest order by the following four-constant expression:

$$\lambda_s = \lambda_1{}^\alpha(\beta_1{}^2 + \beta_2{}^2)(\alpha_3{}^2 - \tfrac{1}{3}) + \lambda_2{}^\alpha\beta_3{}^2(\alpha_3{}^2 - \tfrac{1}{3})$$
$$+ \lambda^\gamma[\tfrac{1}{2}(\beta_1{}^2 - \beta_2{}^2)(\alpha_1{}^2 - \alpha_2{}^2) + 2\beta_1\beta_2\alpha_1\alpha_2]$$
$$+ 2\lambda^\epsilon(\beta_1\alpha_1 + \beta_2\alpha_2)\beta_3\alpha_3$$

In this equation λ^γ and λ^ϵ represent distortions which lower the symmetry of the crystal (as do λ_{100} and λ_{111} in cubic crystals), while $\lambda_1{}^\alpha$ and $\lambda_2{}^\alpha$ represent linear changes respectively along a direction in the basal plane and along the c axis [2]. The direction cosines are referred to rectangular axes so chosen that the 3 axis lies along the hexagonal c axis. The 1 and 2 axes are any two perpendicular directions in the basal plane

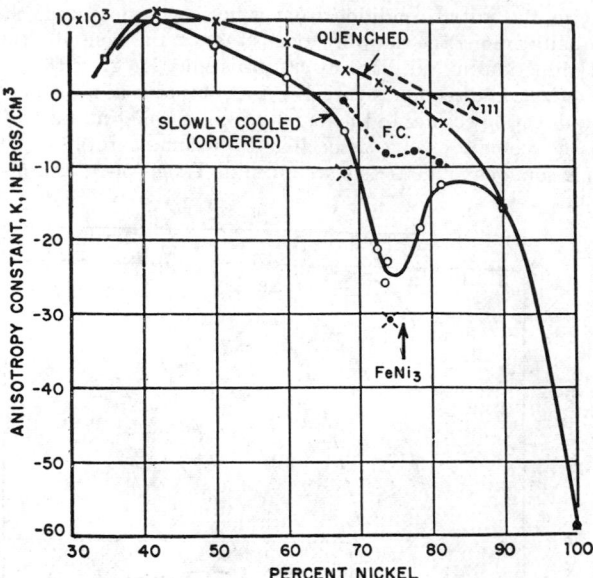

Fig. 5f-32. Magnetic anisotropy constants of quenched and of slowly cooled Fe-Ni alloys. Approximate rates of cooling, 10^5 and 2.5°C/hr, respectively, from 600 to 300°C.   Broken line F.C. shows values for 55°C/hr.   Line λ_{111} shows composition at which magnetostriction in [111] direction goes through zero.   Single low points at 68 and 74 percent nickel are for cooling rate of about 1.5°C/hr.   [*From R. M. Bozorth and J. G. Walker, Phys. Rev.* **89**, 624 (1953).]

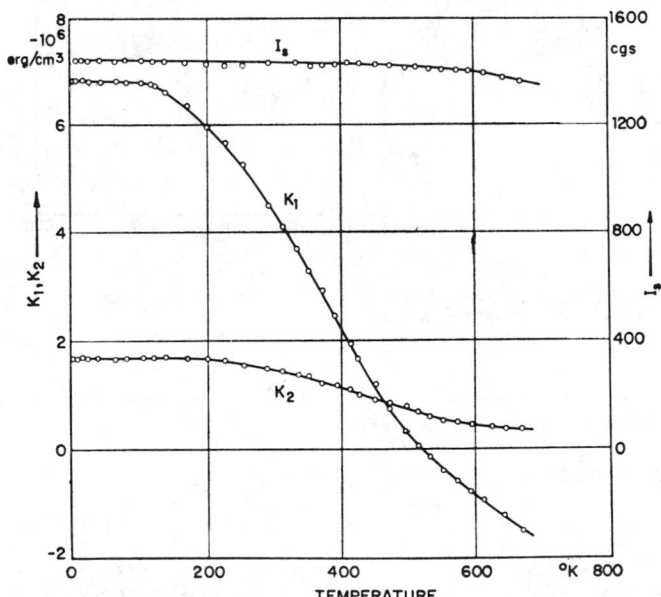

Fig. 5f-33. Temperature dependence of the spontaneous magnetization M_s and the anisotropy constants K_1 and K_2 for hexagonal cobalt. [*From R. Pauthenet, Y. Barnier, and G. Rimet, J. Phys. Soc. Japan* **17**, *suppl. BI*, 309 (1962).]

The magnetostriction constants of some hexagonal crystals are given in Table 5f-23. The volume change associated with magnetization rotation from the basal plane to the hexagonal axis is $2\lambda_1{}^\alpha + \lambda_2{}^\alpha$. In cobalt [3], the volume change is 26×10^{-6}.

Many coefficients are required to describe the magnetostriction of crystals of lower symmetry [4].

Polycrystalline Materials. If a polycrystal has randomly oriented crystallites and equally distributed domains, its fractional change in length upon application of a saturating magnetic field is given by

$$\lambda = \tfrac{3}{2}\,\bar{\lambda}_s(\cos^2\theta - \tfrac{1}{3})$$

θ is the angle between the direction of the magnetization and the direction in which the change is measured. In a cubic material the saturation magnetostriction $\bar{\lambda}_s$ can be calculated from

$$\bar{\lambda}_s = (\tfrac{2}{5} - K)\lambda_{100} + (\tfrac{3}{5} + K)\lambda_{111}$$

Here $K = \tfrac{1}{8}\log\,[2C_{44}/(C_{11} - C_{12})]$ where C_{11}, C_{12}, and C_{44} are the elastic moduli [5]. For an elastically isotropic material, $K = 0$.

Values of $\bar{\lambda}_s$ at room temperature for some alloys and compounds are given in Table 5f-24. (The specimens do not necessarily have randomly oriented crystallites.) The fractional change in length parallel to an applied magnetic field of a few hard magnetic materials is illustrated in Fig. 5f-35. The approach to saturation of the longitudinal magnetostriction of iron, cobalt, and nickel are shown in Fig. 5f-36.

Forced Magnetostriction. Above technical saturation, where domain effects are no longer important, the dimensions of a magnetic body continue to change with increasing magnetic field. In many materials, the distortion is essentially an isotropic volume change and is frequently referred to as the high-field "volume" magnetostriction ω. Unlike the shape magnetostriction which arises primarily from the strain dependence of the anisotropy energy, the isotropic volume magnetostriction results from the dependence of the exchange on strain. It is maximum near the Curie temperature and is linear with field at temperatures appreciably below. Room-temperature values of $\delta\omega/\delta H$ for some materials are listed in Table 5f-25. Although the anisotropic part of the forced magnetostriction is small in cubic crystals, it may be large in single crystals of lower symmetry. In Gd, for example, at 300 K the forced magnetostriction parallel to the hexagonal c axis is $+280 \times 10^{-10}/\text{Oe}$ whereas it is $-13 \times 10^{-10}/\text{Oe}$ for the magnetostriction parallel to the a axis [6].

In ferrites the forced magnetostriction changes sign at the magnetization compensation temperatures. Such reversals have been observed in the ferrimagnetic rare-earth iron garnets [7].

References

1. Becker, R., and W. Döring: "Ferromagnetismus," Springer-Verlag OHG, Berlin, 1939. See also: R. M. Bozorth, "Ferromagnetism," D. Van Nostrand Company, Inc., Princeton, N.J., 1951.
2. The arrangement of terms in our expression differ slightly from that of W. P. Mason: *Phys. Rev.* **96**, 302 (1954). In terms of Mason's coefficients: $\lambda_1{}^\alpha = -\tfrac{1}{2}(\lambda_A + \lambda_B)$, $\lambda_2{}^\alpha = -\lambda_C$, $\lambda^\gamma = \lambda_A - \lambda_B$, $\lambda^\epsilon = 2\lambda_D - \tfrac{1}{2}(\lambda_A + \lambda_C)$.
3. Bozorth, R. M.: *Phys. Rev.* **96**, 311 (1954).
4. Döring, W., and G. Simon: *Ann. Physik* **5**, 373 (1960). See also: E. R. Callen and H. B. Callen: *Phys. Rev.* **139**, A455 (1965); and W. J. Carr, Secondary Effects in Ferromagnetism, "Handbook der Physik," vol. 18/2, Springer-Verlag OHG, Berlin, 1966.
5. Callen, H. B., and N. Goldberg: *J. Appl. Phys.* **36**, 976 (1965).
6. Bozorth, R. M., and T. Wakiyama: *J. Phys. Soc. Japan* **17**, 1669 (1962).
7. Clark, A. E., B. F. DeSavage, N. Tsuya, and S. Kawakami: *J. Appl. Phys.* **37**, 1324 (1966).

TABLE 5f-21. MAGNETOSTRICTION CONSTANTS OF SOME CUBIC METALS

Metal	$\lambda_{100} \times 10^6$			$\lambda_{111} \times 10^6$			Reference
	4 K	77 K	~293 K	4 K	77 K	~293 K	
Fe..	....	....	20	....	....	−21	1
	23	23	24	−30	−30	−22	2
Ni	....	....	−51	....	....	−23	3
	....	−57	−54	....	−28	−23	4
Fe + 3.9 wt. % Si	16	16	25	−2	−2	−3	2
Fe + 5.6 wt. % Si	−8	−7	7	2	2	2	2
Fe + 4.6 wt. % Al	32	34	40	−20	−19	−14	2
Fe + 8.6 wt. % Al	....	85	92	....	−8	−6	2
Fe + 5.32 wt. % Ge	....	22	26	....	−20	−14	2
Fe + 6.07 wt. % V	....	....	28	....	....	−12	5
Fe + 14.4 wt. % V	....	....	43	....	....	−10	5
Fe + 5.72 wt. % Mo	....	....	33	....	....	−10	5
Fe + 7.24 wt. % Mo	....	....	39	....	....	−8	5
Fe + 14.7 wt. % Cr	....	....	51	....	....	−6	5
Fe + 19.9 wt. % Cr	....	....	52	....	....	−3	5
Fe + 1.39 wt. % Ti	....	....	18	....	....	−16	5
Fe + 2.09 wt. % Ti	....	....	15	....	....	−13	5
Fe + 2.47 wt. % Sn	....	....	13	....	....	−15	5
Fe + 3.76 wt. % Sn	....	....	12	....	....	−14	5

References for Table 5f-21

1. Carr, W. J., Jr., and R. Smoluchowski: *Phys. Rev.* **83,** 1236 (1951).
2. Gersdorf, R., J. H. M. Stoelinga, and G. W. Rathenau: *J. Phys. Soc. Japan* **17,** suppl. B-I, 342 (1962).
3. Bozorth, R. M., and R. W. Hamming: *Phys. Rev.* **89,** 865 (1953).
4. Corner, W. D., and F. Hutchinson: *Proc. Phys. Soc. (London)* **72,** 1049 (1958).
5. Hall, R. C.: *J. Appl. Phys.* **31,** 1037 (1960).

TABLE 5f-22. ROOM-TEMPERATURE MAGNETOSTRICTION
OF SOME MAGNETIC COMPOUNDS

Compound	$\lambda_{100} \times 10^6$	$\lambda_{111} \times 10^6$	Ref.	Compound	$\lambda_{100} \times 10^6$	$\lambda_{111} \times 10^6$	Ref.
Fe_3O_4	−20	80	1	$Er_3Fe_5O_{12}$	1	−5	5, 6
$Co_{0.8}Fe_{2.2}O_4$	−590	120	2	$Eu_3Fe_5O_{12}$	21	2	6
$NiFe_2O_4$	−46	−22	3	$Gd_3Fe_5O_{12}$	0	−3	6
$Ni_{0.8}Fe_{2.2}O_4$	−36	−4	2	$Tb_3Fe_5O_{12}$	−3	12	6
$MnFe_2O_4$	−35	−1	2	$Tm_3Fe_5O_{12}$	1	−5	6
$Y_3Fe_5O_{12}$	−1	−3	4, 6	$Yb_3Fe_5O_{12}$	1	−5	6
$Dy_3Fe_5O_{12}$	−14	−8	5, 6	CoO (10 kOe,			
$Ho_3Fe_5O_{12}$	−6	−4	5, 6	77 K)	9	..	7

References for Table 5f-22

1. Bickford, L. R., J. Pappis, and J. L. Stull: *Phys. Rev.* **99,** 1210 (1955).
2. Bozorth, R. M., E. F. Tilden, and A. J. Williams: *Phys. Rev.* **99,** 1788 (1955).
3. Smith, A. B., and R. V. Jones: *J. Appl. Phys.* **37,** 1001 (1966).
4. Clark, A. E., B. F. DeSavage, W. Coleman, E. R. Callen, and H. B. Callen: *J. Appl. Phys.* **34,** 1296 (1963).
5. Clark, A. E., B. F. DeSavage, N. Tsuya, and S. Kawakami: *J. Appl. Phys.* **37,** 1324 (1966).
6. Iida, S.: *J. Phys. Soc. Japan* **22,** 1201 (1967).
7. Nakamichi, J., and M. Yamamoto: *J. Phys. Soc. Japan,* **17,** suppl. B-I, 214 (1962).

TABLE 5f-23. MAGNETOSTRICTION OF SOME HEXAGONAL CRYSTALS

Crystal	T, K	$\lambda^\gamma \times 10^6$	$\lambda^\epsilon \times 10^6$	$\lambda_1{}^\alpha \times 10^6$	$\lambda_2{}^\alpha \times 10^6$	Ref.
Co............	RT	50	−233	70	−110	1
MnBi.........	RT	−45*	......	37*	−50*	2
Gd...........	0	105	34	143	−105	3,4
Gd...........	200	14	4	42	−110	3,4
Tb...........	0	8,500	15,000†	−2,600†	9,000†	5,6
Tb...........	150	4,700*	....	....	....	5
Dy...........	0	9,000	5,500†	....	....	7,8
Dy...........	200	24*	10*	....	....	7,8
Ho...........	0	2,300	....	....	....	9
Ho...........	150	3*	....	....	....	9
Er...........	0	−5,100†	....	....	....	10
Er...........	150	−1*	....	....	....	10

* H = 10 kOe   † Extrapolated from paramagnetic region.

References for Table 5f-23

1. Bozorth, R. M.: *Phys. Rev.* **96**, 311 (1954).
2. Williams, H. J., R. C. Sherwood, and D. L. Boothby: *J. Appl. Phys.* **28**, 445 (1957).
3. Bozorth, R. M.: *J. Phys. Soc. Japan* **17**, 1669 (1962).
4. Alstad, J., and S. Legvold: *J. Appl. Phys.* **35**, 1752 (1964).
5. Rhyne, J. J., and S. Legvold: *Phys. Rev.* **138**, A507 (1965).
6. DeSavage, B. F., and A. E. Clark: Fifth Rare Earth Research Conference, Ames, Iowa, 1965.
7. Clark, A. E., B. F. DeSavage, and R. M. Bozorth: *Phys. Rev.* **138**, A216 (1965).
8. Rhyne, J. J.: Ph.D. Thesis, Iowa State University, 1965.
9. Rhyne, J. J., S. Legvold, and E. T. Rodine: *Phys. Rev.* **154**, 266 (1967).
10. Rhyne, J. J., and S. Legvold: *Phys. Rev.* **140**, A2143 (1965).

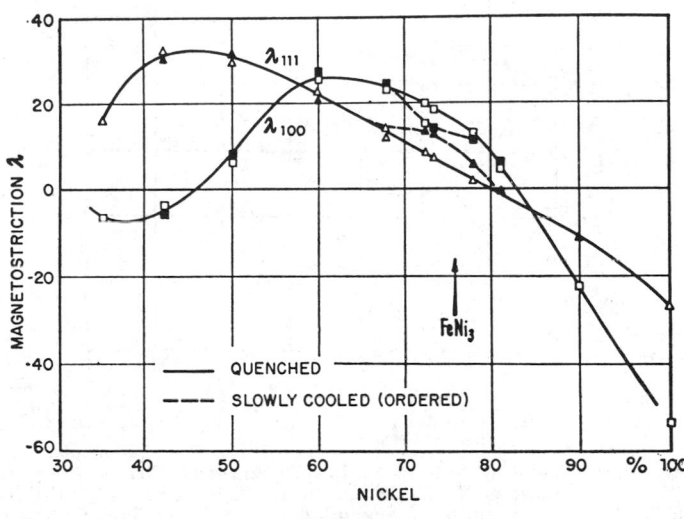

FIG. 5f-34a. See legend on page **5-212**.

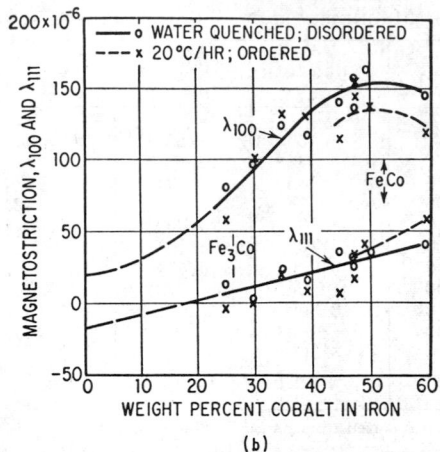

(b)

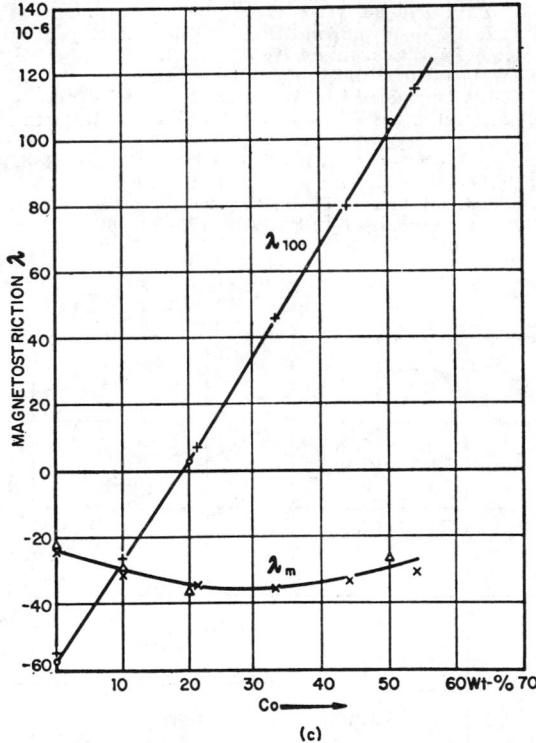

(c)

Fig. 5f-34. Saturation magnetostriction constants of iron-nickel, iron-cobalt, and nickel-cobalt alloys. (a) Iron-nickel, from R. M. Bozorth and J. G. Walker: *Phys. Rev.* **89,** 624 (1953). (b) Iron-cobalt from R. C. Hall: *J. Appl. Phys. Suppl.* **31,** 157S (1960). (c) Nickel-cobalt; +,× taken from M. Yamamoto and T. Nakamichi: *J. Phys. Soc. (Japan)* **13,** 228 (1958); ○, △ from R. C. Hall: *J. Appl. Phys.* **30,** 816 (1959). (*Figure taken from* "*Secondary Effects in Ferromagnetism,*" W. J. Carr. "*Handbuch der Physik,*" Vol. XVIII, 12 *Springer-Verlag, OHG, Berlin,* 1966.

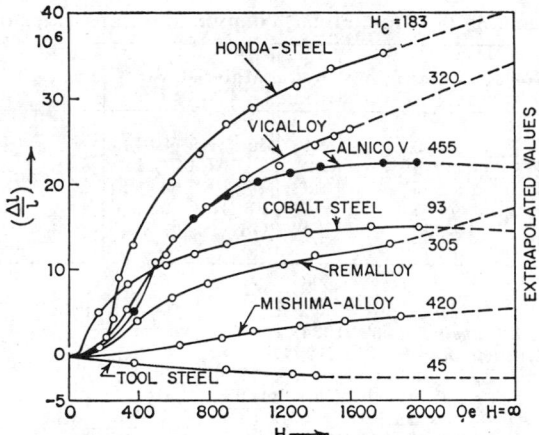

Fig. 5f-35. The longitudinal change in length of a few hard magnetic materials as a function of magnetic field. H_c is the coercive force. [From E. A. Nesbitt, J. Appl. Phys. **21**, 879 (1950).]

TABLE 5f-24. SATURATION MAGNETOSTRICTION OF SOME POLYCRYSTALLINE MATERIALS

Material, wt. %	$\lambda_s \times 10^6$	Ref.	Material, wt. %	$\lambda_s \times 10^6$	Ref.
Fe.....................	−9	1	20 % Co, 80 % Ni........	−21	2, 3
Co.....................	−62*	2	Gd.....................	−7‡	6
Ni.....................	−33	3	Fe_3O_4.................	40	7
80 % Fe, 20 % Co........	32*	4	$MnFe_2O_4$...............	−5	7
70 % Fe, 30 % Co........	35*	4	$CoFe_2O_4$...............	−110	7
40 % Fe, 60 % Co........	68*	4	$MgFe_2O_4$...............	−6	7
30 % Fe, 70 % Co........	57*	4	$Li_{0.5}Fe_{2.5}O_4$.........	−8	7
20 % Fe, 80 % Co........	32*	4	$NiFe_2O_4$...............	−26	7
80 % Fe, 20 % Ni........	30†	5	$CuFe_2O_4$...............	−10	8
70 % Fe, 30 % Ni........	10†	5	$Y_3Fe_5O_{12}$..............	−2	9
40 % Fe, 60 % Ni........	25†	5	$Gd_3Fe_5O_{12}$............	∼−1	10
20 % Fe, 80 % Ni........	2†	5	$Tb_3Fe_5O_{12}$............	+240§	11
10 % Fe, 90 % Ni........	−15†	5	$Dy_3Fe_5O_{12}$............	−100§	11
60 % Co, 40 % Ni........	6*, 31	2, 3	$Er_3Fe_5O_{12}$............	−2§	11
50 % Co, 50 % Ni........	∼0*, 19	2, 3			

* $H = 900$ Oe   † $H = 1,050$ Oe   ‡ $T = 200$ K   § $T = 100$ K, H = 5 kOe

References for Table 5f-24

1. Weil, L., and K. Reichel: J. phys. radium **15**, 72S (1954).
2. Yamamoto, M., and R. Miyasawa: Sci. Repts. Tôhoku Univ. **A5**, 22 (1953).
3. Went, J. J.: Physica **17**, 98 (1951).
4. Williams, S. R.: Rev. Sci. Instruments **3**, 675 (1932).
5. Masiyama, Y.: Sci. Repts. Tôhoku Univ. **20**, 574 (1931).
6. Corner, W. D., and F. Hutchinson: Proc. Phys. Soc. (London) **75**, 781 (1960).
7. See J. Smit and H. P. J. Wijn: "Ferrites," p. 169, John Wiley & Sons, Inc. New York, 1959.
8. Weisz, R. S: Phys. Rev. **96**, 800 (1954).
9. Clark, A. E., B. F. DeSavage, W. Coleman, E. R. Callen, and H. B. Callen: J. Appl. Phys. **34**, 1296 (1963).
10. Belov, K. P., and A. V. Pedko: J. Appl. Phys. **31**, suppl., 55S (1960).
11. Belov, K. P., and V. I. Sokolov: Soviet Physics–JETP **48**, 979 (1965).

TABLE 5f-25. HIGH FIELD VOLUME MAGNETOSTRICTION

Material, wt. %	$\delta\omega \times 10^{10}/\delta H$, Oe^{-1}	Ref.	Material, wt. %	$\delta\omega \times 10^{10}/\delta H$, Oe^{-1}	Ref.
Fe..............	6	1	40 % Ni, 60 % Fe	∼60	4
Co..............	∼6	2	3 % Al, 97 % Fe.	8	5
Ni..............	−0.6	3	15 % Al, 85 % Fe	14	5
20 % Ni, 80 % Fe	∼15	4	Fe$_3$O$_4$..........	−0.7	6
30 % Ni, 70 % Fe	∼290	4	Gd............	150	7

References for Table 5f-25

1. Kornetzki, M.: *Z. Physik* **87**, 560 (1934).
2. Bozorth, R. M.: *Phys. Rev.* **96**, 311 (1954).
3. Azumi, K., and J. E. Goldman: *Phys. Rev.* **93**, 630 (1954).
4. Masiyama, Y.: *Sci. Repts. Tôhoku Univ.* **20**, 574 (1931).
5. Hall, R. C: *J. Appl. Phys.* **28**, 707 (1957).
6. Kornetzki, M.: *Z. Physik* **97**, 662 (1935).
7. Coleman, W. E., and A. S. Pavlovic: *Phys. Rev.* **A135**, 426 (1964).

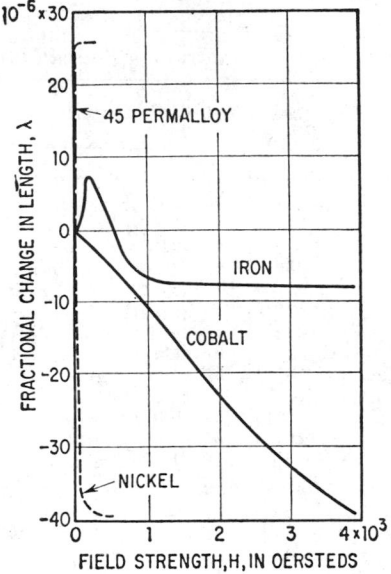

FIG. 5f-36. Longitudinal magnetostriction of iron, cobalt, nickel, and 45-permalloy as a function of field. (*Figure taken from R. M. Bozorth, "Ferromagnetism," D. Van Nostrand Company, Inc., Princeton, N.J.,* 1951.)

5f-12. Hall Constants of Ferromagnetic Elements and Alloys.[1] In ferromagnetic materials the Hall effect (see Sec 5a for definition) has its origin in two independent, experimentally distinguishable mechanisms. One is the result of the usual Lorentz force and is proportional to B, the magnetic induction in the material, while the second arises from spin-orbit coupling and is proportional to M, the macroscopic magnetiza-

---

[1] Compiled by A. C. Ehrlich and A. I. Schindler, U.S. Naval Research Laboratory.

tion. A conventional description of the Hall effect is given in terms of the Hall resistivity ρ_H by the following expression

$$\rho_H = R_0 B + R_s M$$

where ρ_H is the Hall electric field per unit current density, R_0 is the ordinary Hall coefficient, and R_s (sometimes designated by R_1) is the spontaneous (or extraordinary) Hall coefficient. Tables 5f-26 through 5f-30 list values of R_0 and R_s in cubic meters per coulomb (1 m^3/coul = 10^{-2} ohm-cm/Oe) at various temperatures. By convention, a negative (positive) sign for R_0 indicates electron (hole) type conduction.

TABLE 5f-26. HALL CONSTANTS OF POLYCRYSTALLINE ELEMENTS
(1 m^3/coul = 10^{-2} ohm-cm/Oe)

Element	T, K	$R_0 \times 10^{11}$, m^3/coul	$R_s \times 10^{11}$, m^3/coul	Ref.*	Element	T, K	$R_0 \times 10^{11}$, m^3/coul	$R_s \times 10^{11}$, m^3/coul	Ref.*
Fe.......	301	2.28	66.8	1†	Ni	296	−5.67	−49.2	3
	231	2.67	26.6			273	−5.55	−38.9	
	169	2.15	3.77			194	−5.00	−12.6	
	112	1.58	1.02			117	−4.10	−2.82	
	77	0.56	−0.08			96	−3.82	−1.96	
Co.......	293	−12.4	8.4	2‡		77	−3.53	−1.70	
	273	−13.6	7.2			4.15	−7.00	−0.6	4§
	193	−11.8	−0.9						
	77	−13.2	−4.2						

* After Table 5f-30.
† For higher-temperature data see Volkenshtein et al. [23].
‡ Contains additional data concerning dependence of R_0 and R_s on metallurgical treatment.
§ These data were obtained on a polycrystal in the "high field limit" ($\omega_c \tau \gg 1$).

In Tables 5f-26 through 5f-30 certain classes of data have been expressly omitted. These are (1) low-temperature (i.e., where the residual resistivity dominates the thermal resistivity) measurements on "pure" metals in the "low field limit," i.e., when $\omega_c \tau < 1$, ω_c being the cyclotron frequency, and τ the relaxation time; and (2) single-crystal measurements made primarily as studies of Fermi surface topology. Limited single-crystal studies have been made on both the $3d$ and the rare-earth $4f$ ferromagnetic metals. Both Hall coefficients for Fe are isotropic and are similar to polycrystalline data above 70 K. Some single-crystal rare-earth data and references to single-crystal work in other materials are given at the end of this compilation.

TABLE 5f-27. HALL CONSTANTS OF BINARY ALLOYS WITH NICKEL

Alloy, atomic %	Ref.*	T, K	$R_0 \times 10^{11}$, m³/coul	$R_s \times 10^{11}$, m³/coul	Alloy, atomic %	Ref.*	T, K	$R_0 \times 10^{11}$, m³/coul	$R_s \times 10^{11}$, m³/coul
Fe 0.07...	3†‡	292	−5.74	−50.8	Cu 1.30...	3†‡	293	−6.43	−75.0
		77	−3.66	−2.05			194	−6.14	−29.8
		4.15	−3.07	−1.72			77	−6.68	−9.0
Fe 1.08...	3†‡	293	−6.83	−69.8			4.15	−5.92	−19.3
		77	−3.85	−5.60	Cu 10.....	8†	293	−10.8	−225
		4.15	−0.83	−6.80			77	−14.4	−83.5
Fe 5.20...	3†‡	292	−10.3	−78.9	Cu 20.....	8†	301	−14.5	−512
		77	−7.70	−9.04			77	−19.8	−194
		4.15	−6.49	−10.3	Cu 30.....	8§†	293	−13	−783
Fe 10.5...	5	290	−17	−20			77	−19.5	−465
		77	−4.5	−12			14	−21.2	−414
Fe 16.....	5	290	−17	10.4	Cu 50.....	8§†	77	−14.8	152
		77	−22	6.6			20	−16.6	46.4
		20	−23.5	4.6	Pd 2......	9‡††	303	−6.50	−73.3
Fe 20.9...	6†	293	−20.6	34.0			77	−5.01	−8.40
Fe 25.3...	6†	293	−21.1	67.5			4	−3.80	−11.07
Fe 33.7...	6†	293	−21.3	63.5	Pd 5......	9‡††	306	−6.68	−104
Fe 80.....	1‡	313	−15.2	423			77	−6.53	−13.0
		77	−18.8	164			4	−5.83	−13.0
Fe 85.....	1‡	309	−10.9	454	Pd 10....	9†	301	−7.37	−132
		77	−11.0	195			77	−7.12	−23.6
Fe 90.....	1‡	303	−2.59	395			4	−6.81	−19.8
		77	2.55	159	Pd 30.....	9‡††	301	−9.69	−349
Fe 95.....	1‡	299	−0.90	279			77	−10.0	−93.8
		77	5.72	84.3			4	−10.4	−75.7
Fe 99.6...	1‡	303	3.15	85.0	Pd 60....	9‡††	300	−11.9	−891
		77	0.24	2.49			77	−12.8	−300
Co 0.51...	3†‡	293	−5.86	−58.4			4	−13.1	−242
		77	−3.17	−3.89	Al 8......	5	293	−10	−850
		4.15	−0.75	−1.50			77	−18	−410
Co 10.....	5	290	−22.5	−73.8			20	−18	−360
		77	−11	0.3	Si 3.......	5	293	−11	−390
		20	−12	−4.8			77	−10	−190
Co 11.....	7‡††	298	−11.3	−85.0			20	−7	−180
Co 20.....	5	290	−19	−6.5	Sn 3......	5	293	−20	−340
		77	−20	−1.5			77	−9	−180
		20	−21	−2.7			20	−8.5	−180
Co 30.....	5	295	−13	19	V 7........	10	293	−48	−2,200
		77	−28	5.4			77	−19.5	−2,550
		20	−29	−3.2			20	−16	−2,370
Co 53.....	7‡††	294	−19.6	52.3	Mo 3.....	10	293	−15	−1,070
Co 85.....	7‡††	294	−16.4	30.6			77	−10.5	−824
Co 98.0...	2†‡	293	−13.3	9.6			20	−11.5	−824
		77	−15.5	−6.1	W 1.6.....	10	293	−11	−395
		4.15	−15.9	−8.1			77	−10.5	−205
Co 99.77..	2†‡	293	−13.7	9.2			20	−7	−203
		193	−12.9	−0.3	Mn 25....	11‖¶‡	302	−12.6	298
		77	−14.0	−4.4			20	−24.0	112
		4.15	−15.0	−7.6	Mn 25....	11††#	231	−13	1,100
Cu 0.06...	3†‡	293	−5.70	−50.35			20	−17	110
		77	−3.52	−2.68					
		4.15	−3.40	−1.40					

* After Table 5f-30.
† This reference contains data for alloys with compositions intermediate to those given here.
‡ This reference contains data for this alloy for temperature(s) additional to those given here.
§ Values for R_0 are obtained from the slope $\partial \rho_H / \partial B$ at high fields. In these alloys $R_s(\partial M / \partial B)$ is great enough to cause errors in R_0 as large as 15 % since values of $\partial M / \partial B$ are not well known.
¶ Ordered.
# Disordered.
†† See also Volkenshtein et al. [25].

TABLE 5f-29. HALL CONSTANTS OF BINARY ALLOYS WITH IRON

Alloy, atomic %	Ref.*	T, K	$R_0 \times 10^{11}$, m³/coul	$R_s \times 10^{11}$, m³/coul
Co 0.5	15†	297	-2.66	146
		77	-3.43	27.3
Co 15	15†	299	-12.30	251
		77	-12.91	136
Co 35	15†	300	-21.03	33.9
		77	-24.89	16.1
Co 60	15†	300	-8.87	10.4
		77	-8.29	4.91
Co 75	15†	300	-9.94	15.2
		77	-9.69	4.92
Co 85	15†	298	-15.90	40.2
		77	-17.59	6.54
Cr 2.3	16†	315	6.1	220
		77	22	18
		4.2	18	79
Cr 5.1	16†	305	7.8	370
		77	24	42
		4.2	22	120
Cr 12.7	16†	308	6.7	640
		77	26	82
		4.2	41	79
Cr 25.1	16†	307	6.1	730
		77	15	98
		4.2	25	51
Al 25	17‡	295	0.0	4,700
Si 1.30	18§†	300	—	496
		77	—	235
Si 5.09	18§†	300	—	1,910
		77	—	1,600

* After Table 5f-30.
† This reference contains data for this alloy at temperature(s) additional to those given here and at intermediate compositions.
‡ Ordered.
§ See also Okamoto et al. [24].

TABLE 5f-28. HALL CONSTANTS OF TERNARY AND QUATERNARY ALLOYS WITH NICKEL

Alloy, atomic %	Ref.*	T, K	$R_0 \times 10^{11}$, m³/coul	$R_s \times 10^{11}$, m³/coul
Cu 2, Fe 1	12	300	-7.53	-120
		77	-8.53	-16.7
		20	-9.78	-11.9
Cu 4.5, Fe 2.5	12	300	-10.5	-150
		77	-14.0	-38.9
		20	-15.3	-27.8
Cu 10.7, Fe 20.2	13†	300	-21.1	68.1
		169	-24.0	32.2
		77	-28.2	20.2
		20	-30.5	16.6
Cu 14.1, Fe 16.5	13†	305	-20.1	78.7
		77	-26.4	29.3
		20	-28.2	25.6
Cu 19.8, Fe 10.0	13†	305	-18.8	64.0
		77	-26.1	28.9
		20	-28.2	25.9
Cu 25.6, Fe 3.3	13†	305	-15.1	-287
		77	-22.5	-163
		20	-24.7	-144
Fe 15, Mo 5, Mn 0.5	14†	300	-20	105
Fe 16, Cu 5, Cr 2	14†	300	-20	289
Mn4, Si 1	14†	300	—	337

* After Table 5f-30.
† This reference contains data for this alloy at temperature(s) additional to those given here and at intermediate compositions.

TABLE 5f-30. HALL CONSTANTS OF SINGLE CRYSTALS

Element	Ref.	T, K	B in basal plane		B along c axis*	
			$R_0 \times 10^{11}$, $\mathrm{m^3/coul}$	$R_s \times 10^{11}$, $\mathrm{m^3/coul}$	$R_0 \times 10^{11}$, $\mathrm{m^3/coul}$	$R_s \times 10^{11}$, $\mathrm{m^3/coul}$
Gd.........	19†‡	240	52.3	−872	−27.2	−4,080
		200	8.7	−367	−9.4	−3,200
		150	−30.1	−161	8.5	−1,900
		100	−40.0	−167	25.0	−890
		50	−34.7	−29	38.9	−170
Dy.........	20	148	—	117		
		119	—	210		
		78	—	130		
		39	—	22		
Tb.........	21	162	29.8	163		
		119	−3.5	200		
		79	−11.7	82.2		
		40	−15.8	5.7		
Fe..........	22	4 to 300	−10 to 1	−1 to 50		
Co..........	2, 23	4 to 300	10	−10 to 10		
Ni..........	23	4 to 300	5	50		

* Applies only to rare-earth elements.
† See also Volkenshtein [26].
‡ This reference contains data for this alloy for temperature(s) additional to those given here.

References for Tables 5f-26 through 5f-30

1. Soffer, S.: Thesis, Carnegie Institute of Technology, 1964; see also S. Soffer, J. A. Dreesen, and E. M. Pugh: *Phys. Rev.* **140**, A 668 (1965).
2. Dubois, J.: Thesis, Institute de Physique Expérimentale de l'Université de Lausanne; see also J. Dubois and D. Rivier: To be published.
3. Huguenin, R.: Thesis, Institute de Physique Expérimentale de l'Université de Lausanne, 1964; see also R. Huguenin, and D. Rivier: *Helv. Phys. Acta* **38**, 900 (1965).
4. Ehrlich, A. C., and D. Rivier: *J. Phys. Chem. Solids* **29**, 1293 (1968).
5. Smit, J., and J. Volger: *Phys. Rev.* **92**, 1576 (1953).
6. Jellinghaus, W., and M. P. de Andrés: *Ann. Physik* **5**, 187 (1960).
7. Foner, S., and E. M. Pugh: *Phys. Rev.* **91**, 20 (1953).
8. Cohen, P.: Thesis, Carnegie Institute of Technology, 1955; A. I. Schindler: Thesis, Carnegie Institute of Technology, 1950; see also A. I. Schindler and E. M. Pugh: *Phys. Rev.* **89**, 295 (1953); and E. M. Pugh: *Phys. Rev.* **97**, 647 (1955).
9. Dreesen, J. A., and E. M. Pugh: *Phys. Rev.* **120**, 1218 (1960).
10. Smit, J., and J. Volger: Private communication.
11. Dreesen, J. A.: *Phys. Rev.* **125**, 1215 (1962).
12. Sanford, E. R., A. C. Ehrlich, and E. M. Pugh: *Phys. Rev.* **123**, 1947 (1961).
13. Ehrlich, A. C., J. A. Dreesen, and E. M. Pugh: *Phys. Rev.* **133**, A 407 (1964).
14. Lavine, J. M.: *Phys. Rev.* **123**, 1273 (1961).
15. Beitel, F. P., and E. M. Pugh: *Phys. Rev.* **112**, 1516 (1958).
16. Carter, G. C., and E. M. Pugh: *Phys. Rev.* **152**, 498 (1966).
17. Volkenshtein, N. V., and G. V. Fedorov: *Soviet Phys.–JETP* **11**, 48 (1960).
18. Kooi, C.: *Phys. Rev.* **95**, 843 (1954).
19. Lee, R. S.: Private communication; see also R. S. Lee and S. Legvold: *Phys. Rev.* **162**, 431 (1967).
20. Rhyne, J. J.: *Phys. Rev.* **172**, 523 (1968).
21. Rhyne, J. J.: *J. Appl. Phys.* **40**, 1001 (1969).
22. Dheer, P. N.: *Phys. Rev.* **156**, 637 (1967).
23. Volkenshtein, N. V., G. V. Fedorov, and V. P. Shivokovskii: *Phys. Metals Metallog.* **11**, 151 (1961).
24. Okamoto, T., H. Tange, A. Nishomuva, and E. Tatsumoto: *J. Phys. Soc. Japan* **17**, 717 (1962).
25. Volkenshtein, N. V., and G. V. Fedorov: *Phys. Metals Metallog.* **9**, 21 (1960).
26. Volkenshtein, N. V., I. K. Grigovova, and G. V. Fedorova: *Soviet Physics–JETP* **23**, 1003 (1966).

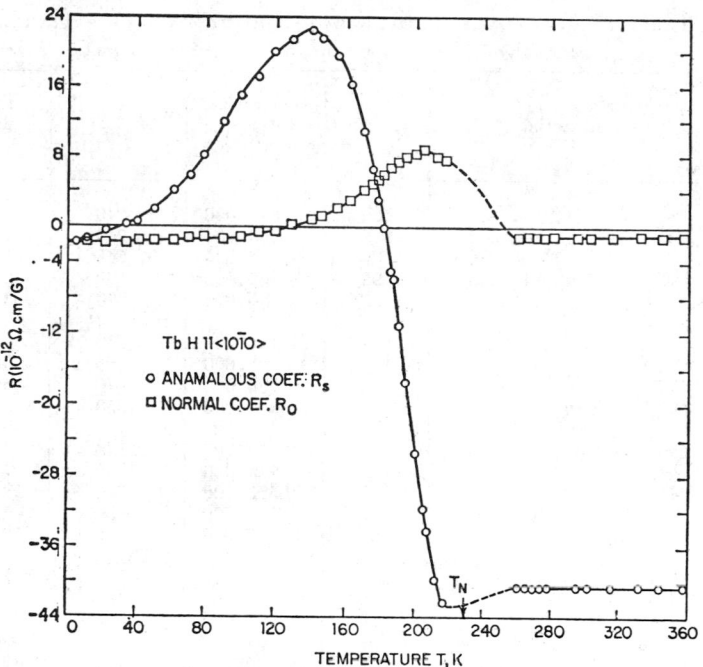

Fig. 5f-37. Temperature dependence of the anomalous and normal Hall coefficients for a single crystal of the heavy rare-earth element terbium. The sign change in R_s and the unique maximum occurring near $0.6T_N$ were found also in Dy but were not observed in the S-state ion Gd or in the iron-group elements. The different temperature dependence in Gd is attributed to the absence of orbital angular momentum of the 4f ion. The constant value of R_s in the paramagnetic region is a consequence of the dominance of spin-disorder scattering. The broad rise in R_0 below T_N, which remains after correction for the high field susceptibility, indicates a field dependence of the scattering. This effect is characteristic of Gd also. [*After J. J. Rhyne, J. Appl. Phys.* **40**, 1001 (1969)—*recent data included by the editors; see also Phys. Rev.* **172**, 523 (1968), *and R. S. Lee and S. Legvold; Phys. Rev.* **162**, 431 (1967).]

5f-13. Faraday Effect. *Magneto-optical Rotation.*[1] Linearly polarized light incident upon a magnetic material in which the magnetization is parallel to the light path emerges as elliptically polarized light. The major axis of the emergent elliptical light is rotated by the magnetization through an angle θ relative to the vibration direction of the incident light. θ is the Faraday rotation and is proportional to the magnetization **M** of the material and also to the path length L in the material,

$$\theta = KL\mathbf{M}$$

where the constant of proportionality K is known as *Kundt's constant* with the units deg/gauss-cm. The sign of the rotation is positive if the major axis of the ellipse is rotated in the same direction as the current flow in a solenoid used to create the magnetization **M**. Table 5f-31A gives values of saturation rotation

$$\frac{\theta_s}{L} = K\mathbf{M}_s$$

(**M**$_s$ is the saturation magnetization) for some ferromagnetic or ferrimagnetic materials below their Curie temperature.

[1] Prepared by James C. Suits, IBM Research Center, San Jose, Calif.

TABLE 5f-31. MAGNETO-OPTICAL ROTATION OF VARIOUS MATERIALS

A. Faraday Rotation: Ferromagnetic and Ferrimagnetic

Material	T, K	Saturation rotation, deg/cm	Wavelength, nm	Ref.
Fe.....................	RT	347,000	546	1
Co.....................	RT	363,000	546	1
Ni.....................	RT	98,000	546	1
Gd....................	93	$-325,000$	589	2
EuO...................	5	85,000	800	3
EuSe..................	4.2	140,000	750	4
$CrCl_3$.................	1.5	3,000	385	5
$CrBr_3$.................	1.5	500,000	470	5
CrI_3...................	1.5	150,000	950	5
$CdCr_2Se_4$...............	82	$-9,200$	1170	6
CrO_2...................	RT	135,000	1000	7
MnBi..................	RT	570,000	750	8
$RbNi_{0.75}Co_{0.25}F_3$.........	77	500	550	9
$RbFeF_3$...............	82	3,300	300	10
FeRh..................	348	90,500	700	7
$MgFe_2O_4$...............	RT	-700	1100	11
$Li_{0.5}Fe_{2.5}O_4$.............	RT	-970	1100	11
$NiFe_2O_4$...............	RT	27,000	330	12
$CoFe_2O_4$...............	RT	44,000	360	12
$YFeO_3$.................	RT	8000	600	13
$BaFe_{12}O_{19}$..............	RT	160	6000	11
$Ba_2Zn_2Fe_{12}O_{22}$..........	RT	80	8000	11
$Y_3Fe_5O_{12}$...............	RT	4,000	530	14
$Gd_3Fe_5O_{12}$..............	RT	95,000	330	15
$Ho_3Fe_5O_{12}$..............	RT	60	1100	16
$Er_3Fe_5O_{12}$...............	RT	65	3500	17
$Eu_3Fe_5O_{12}$..............	RT	-760	3100	18
$FeBO_3$.................	RT	4,800	480	19

B. Faraday Rotation: Paramagnetic and Diamagnetic

Material	Room-temperature Verdet constant V, min/Oe-cm	Wavelength, nm	Ref.
EuO....................	-10.0	1200	20
EuF_2...................	-6.6	435	20
TbAlG.................	-2.256	405	21
Eu glass...............	-2.55	450	22
Tb-Pr borate glass.......	-0.940	405	23
Corning 8363 (lead) glass..	0.10	600	24
Schott SFS-6 glass........	0.490	366	25
AO soda-lime glass.......	0.074	334	25
Quartz.................	0.01664	546	26
NaCl..................	0.0410	546	27
CaF_2..................	0.00883	589	28

For paramagnetic or diamagnetic materials or for ferromagnetic materials above their Curie temperature, the Faraday rotation is still proportional to **M** but is usually described in terms of the applied field **H**,

$$\theta = VL\mathbf{H},$$

where the constant of proportionality V is known as the *Verdet constant* with the units min/Oe-cm. Table 5f-31B gives values of Verdet constant at room temperature for a few representative materials.

References for Tables 5f-31A and 5f-31B

1. Breuer, W., and J. Jaumann: *Z. Physik* **173**, 117 (1963).
2. Lambeck, M., L. Michel, and M. Waldschmidt: *Z. angew. Phys.*, **15**, 369 (1963).
3. Ahn, K. Y., and J. C. Suits: *IEEE Trans.* **MAG-3**, 453 (1967); Suits, J. C.: *Proc. International Conf. on Ferrites*, Kyoto, Japan (1970).
4. Suits, J. C., B. E. Argyle, and M. J. Freiser: *J. Appl. Phys.* **37**, 1391 (1966).
5. Dillon, J. F., Jr., H. Kamimura, and J. P. Remeika: *J. Phys. Chem. Solids*, **27**, 1531 (1966).
6. Bongers, P. F., and G. Zanmarchi: *Solid State Commun.* **6**, 291 (1968).
7. Stoffel, A. M.: *J. Appl. Phys.* **40**, 1238 (1969).
8. Chen, D., J. F. Ready, and E. Bernal: *J. Appl. Phys.* **39**, 3916 (1968).
9. Suits, J. C., T. R. McGuire, and M. W. Shafer: *Appl. Phys. Letters*, **12**, 406 (1968).
10. Chen, F. S., H. J. Guggenheim, H. J. Levinstein, and S. Singh: *Phys. Rev. Letters* **19**, 948 (1967).
11. Zanmarchi, G., and P. F. Bongers: *J. Appl. Phys.* **40**, 1230 (1969).
12. Coren, R. L., and M. H. Francombe: *Journal de Physique* **25**, 233 (1964).
13. Tabor, W. J., A. W. Anderson, and L. G. Van Vitert: *J. Appl. Phys.* **41**, 3018 (1970).
14. Dillon, J. F., Jr.: *J. phys. radium* **20**, 374 (1959).
15. MacDonald, R. E., O. Voegeli, and C. D. Mee: *J. Appl. Phys.* **38**, 4101 (1967).
16. Krinchik, G. S., and M. V. Chetkin: *Soviet Phys.–JETP*, **13**, 509 (1961).
17. Krinchik, G. S., and M. V. Chetkin: *Soviet Phys.–JETP*, **14**, 485 (1962).
18. Krinchik, G. S., and G. K. Tyutneva: *Soviet Phys.–JETP*, **19**, 292 (1964).
19. Kurtzig, A. J., R. Wolfe, R. C. LeCraw, and J. W. Nielsen: *Appl. Phys. Letters* **14**, 350 (1969).
20. Suits, J. C.: Unpublished data.
21. Rubinstein, C. B., L. G. Van Uitert, and W. H. Grodkiewicz: *J. Appl. Phys.* **35**, 3069 (1964).
22. Shafer, M. W., and J. C. Suits: *J. Am. Ceram. Soc.* **49**, 261 (1966).
23. Rubinstein, C. B., S. B. Berger, L. G. Van Uitert, and W. A. Bonner: *J. Appl. Phys.* **35**, 2338 (1964).
24. Borelli, N. F.: *J. Chem. Phys.* **41**, 3289 (1964).
25. Robinson, C. C.: *Appl. Optics*, **3**, 1163 (1964).
26. Ramaseshan, S.: *Proc. Indian Acad. Sci.* **24**, 426 (1946).
27. Ramaseshan, S.: *Proc. Indian Acad. Sci.* **28**, 360 (1948).
28. Ramaseshan, S.: *Proc. Indian Acad. Sci.* **24**, 104 (1946).

Faraday Rotation at Microwave Frequencies.[1] The Faraday effect which occurs at microwave frequencies is described by the relation

$$\theta = \frac{\omega}{2c} \sqrt{\epsilon} \left(\sqrt{\mu + \kappa} - \sqrt{\mu - \kappa} \right) L$$

where θ = rotation, rad
  ω = angular frequency, rad/sec
  c = velocity of light
  L = path length, cm
  ϵ = dielectric constant

and μ and κ are components of a permeability tensor which describes the behavior of materials under the combined influence of a static and an orthogonal r-f magnetic field. When $\omega \gg 4\pi M\gamma$ and $\omega \gg \gamma H$, the tensor components are given approximately by

$$\mu \approx 1 \qquad \kappa \approx \frac{4\pi M\gamma}{\omega}$$

[1] Prepared by C. L. Hogan and H. Solt, Jr., Fairchild Camera & Instrument Corp.

TABLE 5f-32. FARADAY ROTATION IN FERRITE MATERIALS

A. Completely Filled Waveguide

Applied H, oersteds	Rotation, deg/cm			
	$Mn_{0.5}Zn_{0.5}Fe_2O_4$* ($4\pi M_o = 1,500$, $\lambda = 3.33$ cm)	$MgFe_2O_4$† ($4\pi M_s = 900$, $\lambda = 3.2$ cm)	$MgAl_{0.4}Fe_{1.6}O_4$† ($4\pi M_s = 540$, $\lambda = 3.2$ cm)	$MgAl_{0.8}Fe_{1.2}O_4$† ($4\pi M_s = 54$, $\lambda = 3.2$ cm)
0........	0	0	0	0
100......	....	....	....	1.1
200......	....	3	3	
400......	....	....	6	
500......	35	....	7.4	
600......	....	9		
1,000....	80	14.3	....	1.1
1,400....	....	14.3	7.4	
1,500....	120			
2,000....	123			
2,500....	123			

B. Waveguide Containing Slender Cylinders at Saturation

Composition	Frequency, GHz	$4\pi M_s$, gauss	Rotation, deg/cm	Loss, db/cm	Fig. of merit, deg/db
$Ni_{0.6}Zn_{0.4}Mn_{0.2}Fe_{1.8}O_4$‡......	4.0§	3,840	17.5	0.9	19.5
$Mg_{1.5}Mn_{0.2}Fe_{1.5}O_4$‡..........	4.0§	1,800	13.3	0.6	21.7
$Mg_{1.0}Mn_{0.1}Al_{0.2}Fe_{1.9}O_4$‡......	4.0§	1,600	10.5	0.026	410
$Ni_{0.4}Zn_{0.6}Mn_{0.02}Fe_{1.9}O_4$‡......	11.2¶	3,850	9.4	0.013	730
$Ni_{0.7}Zn_{0.2}Mn_{0.1}Fe_{1.5}O_4$‡......	11.2¶	2,800	5.6	......	2150
$Mg_{0.1}Mn_{0.02}Al_{0.2}Fe_{1.7}O_4$‡......	11.2¶	1,600	3.77	0.01+	370
Ferroxcube 4A**............	24.0††	3,360	13.8		
Ferroxcube 4B**............	24.0††	4,400	28.0		
Ferroxcube 4C**............	24.0††	4,365	20.0		
Ferroxcube 4D**............	24.0††	3,470	9.8		
Ferroxcube 4E**............	24.0††	2,315	5.8		

* C. L. Hogan, *Bell System Tech. J.* **31**, 1–30 (1952).
† Roberts, F. F.: *J. phys. radium* **12**, 305 (1951).
‡ Private communication from J. P. Schafer, Bell Telephone Laboratories.
¶ 1.35-cm-diameter rods supported in polystyrene in 5-cm-diameter waveguide.
§ 0.355-cm-diameter rods supported in polyfoam in 1.9-cm-diameter waveguide.
** A. A. T. M. van Trier, Thesis, Delft, 1953.
†† 1.0-mm-diameter rods.

where $\gamma = ge/2$ MHz $\approx 1.76 \times 10^7$ rad/sec-Oe, and **M** = intensity of magnetization of medium in cgs units. The rotation is then independent of frequency and field and is[1]

$$\theta = \frac{\sqrt{\epsilon}}{2c} 4\pi M\gamma$$

Table 5f-32 shows the Faraday rotation observed in a completely filled waveguide and in waveguides containing slender cylinders of ferrite along the waveguide axis.

[1] For further information, see C. L. Hogan, *Bell System Tech. J.* **31**, 1–30 (1952).

Measurements of completely filled waveguides are reliable only when the materials attenuate the wave appreciably because of the effects of internal reflections arising from the abrupt discontinuities at the ferrite-air interfaces. The data on the completely filled waveguide show the dependence of rotation upon magnetization as evidenced by the fact that the rotation approaches a limit as the applied field saturates the sample.

The data on the slender samples give the rotation at a field just sufficient to saturate the sample. The losses observed under these conditions are also shown along with the figure of merit given by the rotation in degrees per decibel of loss.

The dependence of Faraday rotation on magnetizing field is given[1] in Fig. 5f-38 for a slender sample.

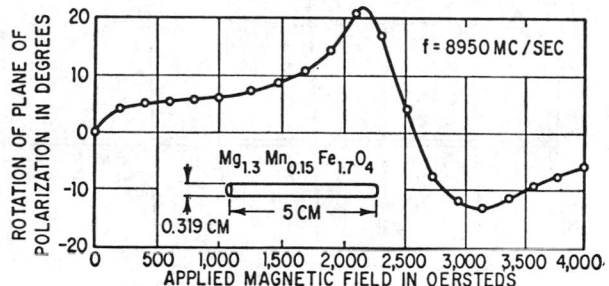

Fig. 5f-38. Faraday rotation in Mg-Mn ferrite as a function of the magnetic field strength Wavelength 3 cm; path length 5 cm.

Table 5f-33 giving data on semiconductors is included here because the phenomenon involved is closely related to the Faraday rotation in ferrites and can be described by an equation similar to that above when the tensor permeability is replaced by a tensor dielectric constant.

Additional data on solids, liquids, and gases will be found in Chap. 6.

5f-14. Susceptibility.[2] The atomic susceptibilities of the elements at room temperature are shown in Fig. 5f-39. Data are given in Table 5f-34 for materials which follow a Curie-Weiss law (see Sec. 5f-1) over a substantial temperature range.

The effective Bohr magnetons number per formula unit is given by the relation

$$\mu_{\text{eff}} = \left(3k\chi_{\text{mole}}\frac{T - \theta}{N\beta^2}\right)^{\frac{1}{2}} = 2.83 \sqrt{C}$$

where β is the Bohr magneton, and C and θ are constants of the Curie-Weiss law.

The chemical formulas as written are the simplest which include whole numbers only. In many cases, however, in order to make μ_{eff} per formula unit correspond to the magnetic moment of an actual paramagnetic ion, the magnetic data and calculations refer to the formula multiplied by $\frac{1}{2}$ or $\frac{1}{3}$ or $\frac{1}{4}$. When this is so, the multiplying factor is indicated immediately after the formula thus: $Dy_2O_2 \times \frac{1}{2}$. If μ_{eff} does correspond to the moment of a single dipole, then according to molecular field theory it is related to g and J by

$$\mu_{\text{eff}} = g \sqrt{J(J + 1)}$$

[1] Unpublished data by C. L. Hogan.
[2] Compiled by E. E. Anderson and A. Stelmach, Clarkson College of Technology.

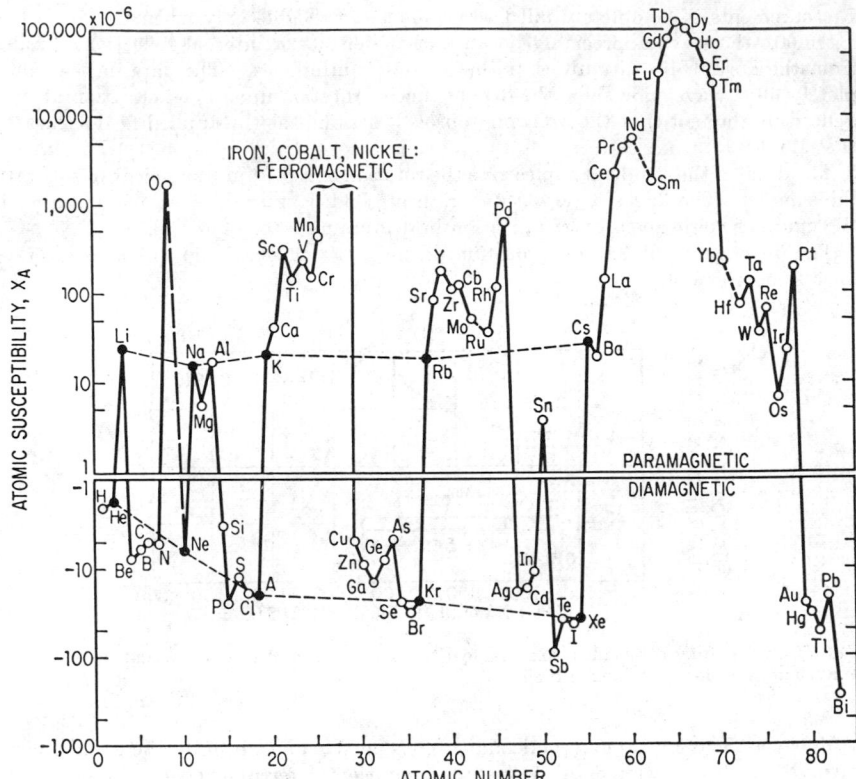

Fig. 5f-39. Atomic susceptibility of the elements at room temperature.   Gd to Tm are ferromagnetic at low temperatures.

All data in Table 5f-34 for which references are not given are taken from G. Foëx.[1] Paramagnetic properties of certain ferrites above their Curie temperatures have been discussed by Néel,[2] who gives references to experimental work in this field.   Also, further references to other materials which do not obey the Curie-Weiss law are to be found in Selwood[3] and in Staude.[4]

Electron-spin and orbital values for the rare-earth ions are given in Table 5f-3.

5f-15. Very Low Temperature Data. Properties of Paramagnetic Salts.   The properties of a paramagnetic ion embedded in a crystal lattice are determined by a number of factors, viz., the level structure of the free ion, the strength and symmetry of the crystalline electric field, and the dipole-dipole and/or exchange interactions with neighboring ions.   In general, ions with an odd number of unpaired ("magnetic") electrons will be found to have a number of doublets lying within ~ 1 cm^{-1} of the ground level and all other levels at much higher energies.   These latter are depopu-

[1] G. Foëx, "Constantes Sélectionnées: Diamagnétisme et Paramagnétisme," Tables de constantes et données numériques, Masson et Cie, Paris, 1957.

[2] L. Néel, *Ann. phys.* **3**(12), 137 (1948).

[3] P. W. Selwood, "Magnetochemistry," Interscience Publishers, Inc., New York, 1956.

[4] H. Staude, "Physikalisch-Chemisches Taschenbuch," vol. 2, p. 1624, Akademische Verlagsgesellschaft m.b.H., Leipzig, 1949.

TABLE 5f-33. FARADAY ROTATION IN SEMICONDUCTORS
A. Silicon:* Room Temperature, H in [100] Direction

Resistivity, ohm-cm	V, min/(Oe)(cm)			
	n-type, 9.6 GHz	n-type, 35 GHz	p-type, 9.6 GHz	p-type, 35 GHz
1.........	-0.25	-1.1	0.12	0.3
2.9.......	-0.35	-0.8	0.12	0.2
5.........	-0.25	-0.4	0.09	0.15
10.........	-0.15	-0.2	0.04	
22.........	....	-0.15	....	0.04
40.........	....	....	....	0.025

B. Germanium:† Room Temperature, H in [100] Direction, 24.9 GHz and Resistivity ρ, 3.8 ohm-cm

H, oersteds	Rotation, deg/cm
3,200.......	70
6,400.......	130
8,300.......	160
11,800.......	210

* Furdyna, J. K. and S. Broersma: *Phys. Rev.* **120**, 1995 (1960).
† Bouwknegt, A. and J. Volger: *Physica* **30**, 113 (1964).

lated at low temperatures, and one finds that Curie's law is followed in the liquid helium region unless the interactions are strong enough to modify this to the Curie-Weiss relation $\chi = C(T - \theta')^{-1}$. (Ions with an even number of unpaired electrons will usually have a singlet ground state and be nonmagnetic at low temperatures.)

Paramagnetic ions fall into separate groups within the Periodic Table, e.g., the $3d$, $4d$, and $5d$ transition series, the $4f$ (rare-earth) and $5f$ (transuranic) groups, wherein the general properties are determined by the relative strength of the electrostatic and spin-orbit interactions. Thus, in the case of the rare-earth ions the spin-orbit coupling $\zeta \mathbf{L} \cdot \mathbf{S}$ is dominant, $\mathbf{L}$ and $\mathbf{S}$ are not uncoupled to a first approximation, and J remains a good quantum number, giving rise to $(2J + 1)$-fold degenerate multiplets. The crystal field splits these into doublets characterized (roughly) by $\pm J_z$. The latter may then be split by an external magnetic field. If the electrostatic energy is large, as in the case of the iron-group $(3d)$ ions, for example, $\mathbf{L}$ and $\mathbf{S}$ precess separately about the electric field axis and the levels are characterized by quantum numbers L_z and S_z. The splittings are generally so large that only the lowest orbital level is populated, and the remaining $(2S + 1)$-fold degeneracy may be partially lifted by the combined action of the electric field and the spin-orbit interaction, with splittings ~ 1 cm^{-1}. The effect of crystalline fields of different symmetries on the iron-group ions is depicted in Fig. 5f-40.

A fictitious spin S is introduced to correspond to the multiplicity $(= 2S + 1)$ of the lowest lying group of levels, and the "spin Hamiltonian" takes the form

$$\mathcal{H} = D[S_z{}^2 - \tfrac{1}{3}S(S + 1)] + \mu_B[g_\parallel \mathfrak{B}_z S_z + g_\perp(\mathfrak{B}_x S_x + \mathfrak{B}_y S_y)]$$
$$+ A I_z S_z + B(I_x S_x + I_y S_y) \quad (5f\text{-}1)$$

TABLE 5f-34. MOLECULAR SUSCEPTIBILITIES, CURIE CONSTANTS, AND EFFECTIVE BOHR MAGNETON NUMBERS OF SOME PARAMAGNETIC MATERIALS*

(θ is constant in Curie-Weiss law)

Substance	$\chi_{mole} \times 10^6$ (20°C) (cgs units)	Range of validity of Curie-Weiss law, K	C	θ, K	μ_{eff}
$B_2O_3 \cdot Fe_2O_3 \cdot 2MgO(\times\frac{1}{2})$ [1]	....	....	2.7	−600	4.66
$B_2O_3 \cdot Fe_2O_3 \cdot 4CuO(\times\frac{1}{2})$ [1]	....	....	2.6	−635	4.58
$B_2O_3 \cdot Fe_2O_3 \cdot 4CoO(\times\frac{1}{2})$ [1]	....	....	5.0	−445	6.3
$B_2O_3 \cdot Fe_2O_3 \cdot 4NiO(\times\frac{1}{2})$ [1]	....	....	4.2	−832	5.8
$CeCl_3$	2,520	>80	0.787	−23	2.51
CeF_3	2,240	>80	0.794	−62	2.52
$Ce(NO_3)_3 \cdot 5H_2O$	2,335	290 ↔ 480	0.717	−17	2.39
$CoBr_2$ [24]	11,640	70 ↔ 300	3.43	−20	5.24
$Co(CN)_2$	~3,870	....	1.21	−9	3.11
$CoCl_2$ [23]	....	90–500	3.19	−28	5.16
	13,060	<400	3.56	18.5	5.33
$CoCr_2S_4(\times\frac{1}{2})$ [2]	4,700	>500	2.08	−410	4.08
CoF_2 [22]	8,660	50 ↔ 300	2.90	−50	5.15
$CoGeO_4(\times\frac{1}{2})$ [35]	13,100	>300	2.75	90	4.70
CoI_2 [24]	10,860	50–300	3.18	0	5.18
$Co(NO_3)_2 \cdot 6H_2O$ [3]	9,050	>8	2.58	8	4.58
CoO [4], [25]	5,235	>300	3.23	330	5.25
CoS_2 [26]	3,520	>155	0.49	161	1.85
$CoSO_4$ [27]	10,200	95–300	3.36	−47	5.65
$CoSO_4 \cdot 7H_2O$	9,780	77–350	2.94	−14	5.03
$CoTiO_3$ [5]	12,400	....	3.72	−9.2	5.64
$CrBr_3$ [6]	7,700	>100	1.94	51	3.94
$CrCl_2$ [28]	7,330	>225	2.97	−116	4.88
$CrCl_3$ [6]	6,860	210–690	1.82	24	3.82
CrF_3 [6]	4,450	>65	1.90	−124	3.90
CrI_3 [6]	6,700	>100	....	70	4.03
CrS [2]	1,610	>560	2.6	≈ −800	4.6
$Cr_2(SO_4)_3(\times\frac{1}{2})$ [3]	10,500	>11	2.94	11	4.89
$Cr_2(SO_4)_3 \cdot 18H_2O(\times\frac{1}{2})$ [3]	10,600	>19	2.88	19	4.85
$Cr(NO_3)_3 \cdot 9H_2O$ [3]	5,320	>20	1.41	20	3.43
$CrK(SO_4)_2 \cdot 12H_2O$	6,320	....	1.84	0	3.84
$CuCl_2$ [34]	1,340	155–670	0.457	−78	1.97
CuS [7]	....	<100	0.04	−25	0.58
$CuSO_4$ [34]	1,340	90 ↔ 500	0.50	−67	2.10
$CuSO_4 \cdot 5H_2O$	1,570	....	0.46	−0.7	1.92
$DyNi_2$ [36]	....	....	13.5	23	10.4
$Dy_2O_3(\times\frac{1}{2})$	43,200	....	13.6	−24	10.5
DyP [8]	....	....	13.0	9	10.22
$Dy_2(SO_4)_3 \cdot 8H_2O(\times\frac{1}{2})$	≈45,000	....	13.74	−5	10.5
$ErNi_2$ [36]	....	....	11.0	11	9.37
$Er_2O_3(\times\frac{1}{2})$	38,600	....	11.6	−8	9.65
$Er_2(SO_4)_3(\times\frac{1}{2})$	28,700	....	8.24	−2	8.12
$Er_2(SO_4)_3 \cdot 8H_2O(\times\frac{1}{2})$	36,500	....	11.18	−6	9.46
$EuCl_2$	26,600	....	7.80	−1	7.90
$Eu_2O_3(\times\frac{1}{2})$	5,550	....	3.26	−294	5.11
EuS	23,800	....	6.81	6	7.38
$EuSO_4$	25,800	....	7.64	−4	7.81
$FeCl_2$ [29]	13,200	90 ↔ 300	3.60	48	5.37
$FeCl_2 \cdot 4H_2O$	12,060	....	3.37	12	5.18
$FeCl_3$	13,900	....	3.93	10	5.6
$FeCr_2S_4(\times\frac{1}{2})$ [2]	4,100	>300	2.22	−240	4.22
FeF_2	9,460	....	3.88	−117	5.57
$Fe_2GeO_4(\times\frac{1}{2})$ [35]	11,500	>100	3.63	−15	5.40
$FeSO_4$ [30]	12,400	40 ↔ 300	3.60	−30.5	5.20
$FeSO_4 \cdot 7H_2O$ [31]	11,930	80 ↔ 300	3.52	−1.0	5.22
$Fe(NH_4)_2(SO_4)_2 \cdot 6H_2O$	13,100	....	3.78	2	5.49
$Fe_2(SO_4)_3(\times\frac{1}{2})$	12,100	....	4.3	−61	5.9
$Fe(NH_4)_2(SO_4)_2 \cdot 12H_2O$	14,900	>20	4.2	0	5.8
$Fe_4[Fe(CN)_6]_3 \cdot 14.5H_2O(\times\frac{1}{4})$ [9]	14,000	>77	3.92	14.7	5.6
$FeTiO_3$ [5], [32]	14,560	80 ↔ 290	3.37	23	5.46
$GdAs$ [10]	....	100 ↔ 500	7.8	....	≈7.9
GdB [10]	....	100 ↔ 500	7.8	....	≈7.9
$GdCl_3$	24,700	....	7.51	−11	7.75
$GdLa$ [11]	....	>20.4	30.0	−3 ± 2	....
$GdNi_2$ [36]	....	....	7.62	78	7.82
$Gd_2O_3(\times\frac{1}{2})$	24,500	....	7.61	−18	7.80
GdP [8]	28,700	100–500	7.45	39	7.72
GdS [10]	....	100 ↔ 500	7.8	....	≈7.9
$Gd_3S_4(\times\frac{1}{3})$ [12]	23,400	>77	7.02	0	≈7.50 ↔ 7.55
$GdSb$ [10]	....	100 ↔ 500	7.8	....	≈7.9
$GdSe$ [10]	....	100 ↔ 500	7.8	....	≈7.9
$Gd_3Se_4(\times\frac{1}{3})$ [12]	27,500	>77	8.24	0	≈8.14

TABLE 5f-34. MOLECULAR SUSCEPTIBILITIES, CURIE CONSTANTS, AND EFFECTIVE BOHR MAGNETON NUMBERS OF SOME PARAMAGNETIC MATERIALS* (*Continued*)

Substance	$\chi_{mole} \times 10^6$ (20°C) (cgs units)	Range of validity of Curie-Weiss law, K	C	θ, K	μ_{eff}
$Gd_2(SO_4)_3(\times\frac{1}{3})$	26,600	....	7.81	-0.4	7.90
$Gd_2(SO_4)_3 \cdot 8H_2O(\times\frac{1}{2})$	27,500	....	8.11	-2	8.06
GdTe [10]	....	100 ↔ 500	7.8	....	7.9
$HCrO_2$ [13]	3,300	....	1.934	-279.6	3.92
$HoNi_2(\times\frac{1}{3})$ [36]	....	....	13.7	12	10.5
$Ho_2O_3(\times\frac{1}{2})$	44,800	....	13.7	-14	10.5
$Ho_2(SO_4)_3(\times\frac{1}{3})$	45,900	....	13.8	-8	10.5
$Ho_2(SO_4)_3 \cdot 8H_2O(\times\frac{1}{2})$	44,300	....	13.6	-7	10.43
$KFe[Fe(CN)_6] \cdot 1.9H_2O$ [9]	....	>77	4.05	22	5.7
K_2MnO_4	1,270	....	0.383	-7	1.75
Li_2NiF_4 [12]	....	77 ↔ 473	....	....	3.18
$MgCr_2O_4(\times\frac{1}{2})$ [35]	2,800	>100	1.86	-350	3.84
$MgV_2O_4(\times\frac{1}{2})$ [35]	1,400	>50	1.47	-750	3.43
$MnBr_2$	14,000	70 ↔ 180	4.26	-2	5.84
$MnCO_3$	≈11,500	....	3.93	-40	5.61
$MnCl_2$ [23]	14,500	2 ↔ 800	4.17	3	5.78
$MnCo_2O_4(\times\frac{1}{3})$ [2]	2,820	>300	1.91	-380	3.91
$MnCr_2S_4(\times\frac{1}{3})$ [2]	8,200	>300	2.54	-10	4.51
MnF_2 [33]	10,730	76 ↔ 300	≈4.10	82	5.98
MnF_3 [14]	....	....	3.01	8	4.91
MnI_2	14,800	35 ↔ 200	4.21	-4	5.80
MnO	5,040	>120	4.90	-680	6.26
$Mn_2O_3(\times\frac{1}{2})$	7,080	....	3.40	-188	5.21
MnO_2	≈2,300	....	1.80	-480	3.78
$Mn(OH)_2$	≈13,700	200 ↔ 300	4.60	20	5.5
$Mn_2P_2O_7(\times\frac{1}{2})$	14,400	195 ↔ 770	4.58	-23	6.05
MnRh [15]	≈790 (4.2°K)	170 ↔ 700	....	≈ -260°K	
$MnSO_4$	13,960	77 ↔ 660	4.34	-22	5.88
$MnTiO_3$ [5]	....	....	4.36	-219	
$NdCl_3$ [16]	....	290 ↔ 570	1.861	-57.4	3.87
NdF_3	5,020	>155	1.76	-56	3.75
$NdNi_2(\times\frac{1}{3})$ [36]	....	....	1.75	10	3.74
$Nd_2O_3(\times\frac{1}{2})$	4,700	....	1.53	32	3.50
$Nd_2(SO_4)_3(\times\frac{1}{3})$	5,070	....	1.70	-42	3.69
$Nd_2(SO_4)_3 \cdot 8H_2O(\times\frac{1}{2})$	5,390	....	1.82	-44	3.81
$NiCl_2$	6,250	>540 / <510	1.50 / 1.37	28 / 71	3.47 / 3.32
NiF_2	3,450	100 ↔ 300	1.34	-97	3.27
$Ni_2GeO_4(\times\frac{1}{2})$ [35]	4,370	>100	1.31	0	3.24
$Ni(NO_3)_2 \cdot 6H_2O$ [3]	3,700	>21	1.01	21	2.86
$NiTiO_3$ [5]	....	....	1.24	-11	3.15
$PrCl_3$ [16]	....	285 ↔ 700	1.69	-29.4	3.69
$PrNi_2(\times\frac{1}{3})$ [36]	....	....	1.60	4	3.57
$Pr_2O_3(\times\frac{1}{2})$	4,450	....	1.62	-71	3.60
$Pr_2(SO_4)_3(\times\frac{1}{3})$	≈4,900	65 ↔ 370	1.64	-44	3.62
$TbNi_2(\times\frac{1}{3})$ [36]	....	....	12	35	9.82
TbP [8]	....	....	10.8	5	9.28
$Tb_2(SO_4)_3 \cdot 8H_2O(\times\frac{1}{2})$	37,500	....	11.86	-16	9.74
$TlMnF_3$ [17]	....	>150	4.87	148	6.25
TmAs [18]	....	>≈20	....	....	≈7.6
$TmNi_2(\times\frac{1}{3})$ [36]	....	....	6.64	0	7.28
$Tm_2O_3(\times\frac{1}{2})$ [19]	....	80 ↔ 980	7.2	....	7.56
TmP [18]	....	>≈20	7.2	....	≈7.0
TmSb [18]	....	>≈20	7.2	....	≈7.6
$Tm_2(SO_4)_3(\times\frac{1}{3})$	20,800	....	6.33	-11.7	7.11
UCl_4 [20]	....	....	1.35	-62	3.29
UBr_4 [20]	....	....	1.21	-35	3.12
UBr_3 [20]	....	....	1.35	25	3.29
UCl_3 [20]	....	....	1.15	-29	3.03
UF_4 [20]	....	....	1.36	-147	3.30
UI_3 [20]	....	....	1.36	5	3.31
KUF_5 [21]	....	....	1.30	-122	3.30
K_2UF_6 [21]	....	>198	1.47	-108	3.45
$CaUF_6$ [21]	....	....	1.31	-101	3.25
Na_3UF_7 [21]	....	....	1.45	-290	3.40
UO_2	2,240	....	1.06	-185	2.92
$U_3O_8(\times\frac{1}{3})$	525	....	0.24	-170	1.39
$U(SO_4)_2$	3,060	....	1.32	-140	3.25
$Yb_2O_3(\times\frac{1}{2})$	6,700	....	2.43	-68	4.40
$Yb_2(SO_4)_3 \cdot 8H_2O(\times\frac{1}{2})$	≈8,600	....	2.92	-42	4.83
$ZnCo_2O_4(\times\frac{1}{3})$ [2]	625	>100	0.21	-20	1.3
$ZnCr_2S_4(\times\frac{1}{3})$ [2]	5,750	>100	1.67	+10	3.66

* Compiled by E. E. Anderson and A. Stelmach, Clarkson College of Technology.

References for Table 5f-34

1. Benoit, R.: *Compt. rend.* **231**, 1216 (1950).
2. Lotgering, F. K.: *Philips Research Repts.* **11**, 190, 337 (1956).
3. Johnson, A. F., and H. Grayson-Smith: *Can. J. Research* **28A**, 229 (1950).
4. Elliott, N.: *J. Chem. Phys.* **22**, 1924 (1954).
5. Stickler, J. J., S. Kern, A. Wold, and G. S. Heller: *Phys. Rev.* **164**, 765 (1967).
6. Hansen, W. N., and M. Griffel: *J. Chem. Phys.* **30**, 913 (1959).
7. Munson, R. A., W. DeSorbo, and J. S. Kouvel: *J. Chem. Phys.* **47**, 1769 (1967).
8. Yaguchi, K.: *J. Phys. Soc. Japan* **21**, 1226 (1966).
9. Davidson, D., and L. A. Welo: *J. Phys. Chem.* **32**, 1191 (1928).
10. Iandelli, A.: *R. C. Accad. Naz. Lincei (Italy)* **30**, 201 (1961).
11. Thoburn, W. C., S. Legvold, and F. H. Spedding: *Phys. Rev.* **110**, 1298 (1968).
12. Yaguchi, K.: *J. Phys. Soc. Japan* **22**, 673 (1967).
13. Meisenheimer, R. G., and J. D. Swalen: *Phys. Rev.* **123**, 831 (1961).
14. Klemm, W., and E. Krose: *Z. anorg. Chem.* **253**, 226 (1947).
15. Kouvel, J. S., C. C. Hartelius, and L. M. Osika, *J. Appl. Phys.* **34**, 1095 (1963).
16. Sanchez, A. E.: *Rev. acad. cienc. exact., fis. y nat. Madrid* **34**, 202 (1940).
17. Kizhaev, S. A., A. G. Tutov, and V. A. Bokov: *Fiz. Tverd. Tela* **7**, 2868 (1965).
18. Busch, G., A. Menth, O. Vogt, and F. Hulliger: *Phys. Letters* **19**, 622 (1966).
19. Perakis, N., and F. Kern: *Phys. Kondens. Materie* **4**, 247 (1965).
20. Dawson, J. K.: *J. Chem. Soc.* **1951**, 429.
21. Elliott, N.: *Phys. Rev.* **76**, 431 (1949).
22. Schilt, A. A.: *J. Am. Chem. Soc.* **85**, 904 (1963).
23. Watanabe, T.: *J. Phys. Soc. Japan* **16**, 1131 (1961).
24. Bizette, H., C. Terrier and B. Tsai: *J. Phys. Radium* **20**, 421 (1959).
25. Singer, J. R.: *Phys. Rev.* **104**, 929 (1956).
26. Benoit, R.: *J. Chim. Phys.* **52**, 119 (1955).
27. Boravik-Romanov, A. S., V. R. Karasik, and N. M. Kreines: *Zh. Eksp. i Teor. Fiz.* **31**, 18 (1956).
28. Cable, J. W., M. K. Wilkinson, and E. O. Wollan: *Phys. Rev.* **118**, 950 (1960).
29. Wilkinson, M. K., J. W. Cable, E. O. Wollan, and W. C. Koehler: *Phys. Rev.* **113**, 497 (1959).
30. Frazer, B. C. and P. J. Brown: *Phys. Rev.* **125**, 1283 (1962).
31. Guha, B. C.: *Proc. Roy. Soc. (London)* **A206**, 353 (1951).
32. Ishakawa, Y., and S. Akimoto: *J. Phys. Soc. Japan* **13**, 1298 (1958).
33. Trapp, C., and J. W. Stout: *Phys. Rev. Letters* **10**, 157 (1963).
34. Escoffier, P., and J. Gauthier: *Compt. rend.* **252**, 271 (1961).
35. Blasse, G., and J. F. Fast: *Philips Res. Repts.* **18**, 393 (1963).
36. Farrell, J., and W. E. Wallace: *Inorg. Chem.* **5**, 105 (1966).

for axial symmetry. Here, D, A, and B are constants and I is the nuclear spin. D is determined by the crystalline electric field, and A and B by the hyperfine coupling. $g_{\parallel}$ and $g_{\perp}$ are the spectroscopic splitting factors for the z direction (parallel to the crystal-field symmetry axis) and in the xy plane, respectively. Terms representing the nuclear electric quadrupole interaction (for $I > \frac{1}{2}$) and the direct coupling of the nuclear magnetic moment with the external field have been omitted from Eq. (5f-1).

The parameters in the spin Hamiltonian are determined by electron paramagnetic resonance (epr) spectroscopy, and the correctness of the assumed crystal field symmetry can be checked by studying the angular dependence of the resonance pattern. Frequently the line width due to magnetic dipole interaction is comparable with the fine structure and hyperfine structure (hfs) separations. Then the established practice is to dilute the subject salt with an isomorphous diamagnetic salt. In most cases the electric field acting on the ion remains unaltered, but there are instances of drastic modifications occurring.

If all the ions in the crystal have the same symmetry axis, the susceptibility will be given by the formulas [1].

$$\chi_{\parallel} = Ng_{\parallel}^2\mu_B^2 \frac{S(S+1)}{3kT}\left[1 - \frac{D(2S-1)(2S+3)}{15kT}\right]$$

$$\chi_{\perp} = Ng_{\perp}^2\mu_B^2 \frac{S(S+1)}{3kT}\left[1 + \frac{D(2S-1)(2S+3)}{30kT}\right]$$

$$(5f\text{-}2)$$

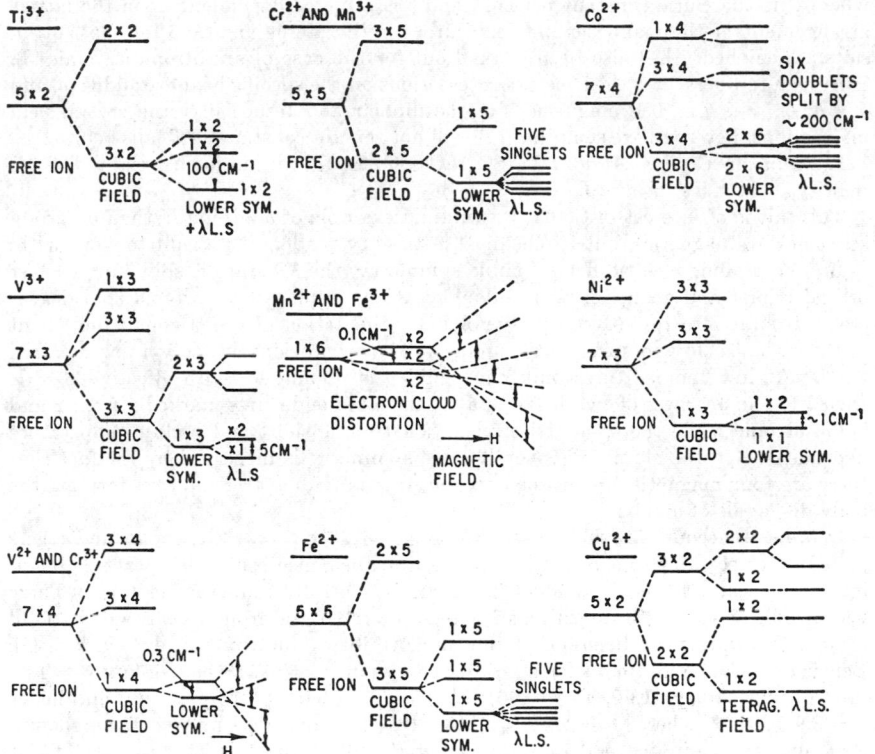

FIG. 5f-40. Iron transition group. Splitting of levels. (Not to scale.)

Here, N is Avogadro's number, k is the Boltzmann constant, and μ_B is the Bohr magneton value. D/k is measured in kelvins, and $N\mu_B{}^2/3k$ is very closely equal to $\frac{1}{8}$ for 1 g ion (0.12506). If there are groups of ions with different symmetry axes, these expressions can be used to calculate the susceptibility by averaging over the groups, and for the particular case of a powder the D term vanishes.

The Stark splitting introduces a contribution to the specific heat which, for $D \ll kT$, is given by

$$\frac{c_e T^2}{R} = b_e = \frac{D^2}{45} S(S+1)(2S-1)(2S+3) \tag{5f-3a}$$

and the hfs contribution is

$$\frac{c_n T^2}{R} = b_n = \frac{1}{9}(A^2 + 2B^2)S(S+1)I(I+1) \tag{5f-3b}$$

where A, B, and D are in kelvins, and R is the molar gas constant. If these parameters are given in cm^{-1}, the above expressions for b are to be multiplied by $(hc/k)^2 = 1.4388^2 = 2.070$.

Magnetic dipole interaction also gives rise to a term in the specific heat varying as T^{-2} in high-temperature approximation. For isotropic ions [2] this is

$$\frac{c_i T^2}{R} = b_i = fC^2$$

where C is the Curie constant per cm^3, and f is a factor dependent upon the lattice arrangement of the magnetic ions, e.g., 21.6 for the alums and 26.5 for the Tutton salts. The theory has also been worked out for the case of anisotropic crystals [3].

At high temperatures the effects of interactions on the specific heat are additive, and $b = b_e + b_n + b_i$. The magnitudes of the dipolar and (especially) the crystal field interactions, however, are such that cT^2 will not remain constant as T falls below 1 K.

1. *Titanium Cesium Alum.* $Ti_2(SO_4)_3 \cdot Cs_2SO_4 \cdot 24H_2O$; gram-ionic weight, 589.14; density, 2.019; Ti^{3+}; $3d^1$; $^2D_{\frac{3}{2}}$.

This salt is of interest in that it provides an example of a $3d^1$ ion in the iron-group series of paramagnetics, but its chemical instability renders it difficult to work with.

In a crystalline electric field of cubic symmetry, the ^2D state is split into a lower orbital triplet and an upper orbital doublet. The triplet, with twofold spin degeneracy, is split into three Kramers doublets by the action of the trigonal component of the crystal field and spin-orbit coupling. The separations are only a few hundred cm^{-1}. At low temperatures only one doublet is populated, with effective $S = \frac{1}{2}$. Owing to the nearness of two higher states, the g value departs markedly from 2 and is anisotropic while the spin-lattice relaxation time is short and highly temperature dependent. In the alums, the overall cubic symmetry is preserved by the fact that there are four magnetic ions in the unit cell, the individual trigonal axes forming the body diagonals of a cube.

In order to account for their observations $g_\| = 1.25$, $g_\perp = 1.14$, Bleaney and others [4] found it necessary to invoke the hypothesis of the d electron taking part in π bonding to the water molecules and to introduce an "orbital reduction factor." There will then be an excited state within a few tens of cm^{-1} of the ground state, which would explain the anomalous behavior of the susceptibility found by van den Handel [5]. Benzie and Cooke [6] found for a powder specimen: $C = 0.118$ emu/g-ion; $g = 1.12$ (and estimated $g_\| = 1.40$, $g_\perp = 0.96$); $bR/C = 2.7 \times 10^4$ (± 10 percent); and hence $b = 3.9 \times 10^{-5}$ where $b = b_e + b_n + b_i$. By dilution experiments they showed that nuclear hfs in the odd isotopes contributes 0.4×10^{-5} to b. Magnetic dipole interaction accounts for 0.3×10^{-5}, and the balance of 3.2×10^{-5} must be due to exchange. The spin-lattice relaxation time varied as $T^{-7.5}$ between 0.9 and 1.3 K, being 3.3×10^{-4} sec at 1 K. The excited doublets were estimated to lie at 174 and 364 cm^{-1}, respectively.

Adiabatic demagnetization experiments [7] gave $(S/R)_{T_c} = 0.22$, for $\mathcal{B}/T = 0.26$ T/K; here S is the entropy, ignoring the nuclear contribution. The odd isotopes Ti^{47} ($I = \frac{5}{2}$, 7.75 percent abundant) and Ti^{49} ($I = \frac{7}{2}$, 5.51 percent) will contribute $0.253R$ to the entropy.

2. *Chromic Potassium Alum.* $Cr_2(SO_4)_3 \cdot K_2SO_4 \cdot 24H_2O$; gram-ionic weight, 499.42; density, 1.83; Cr^{3+}; $3d^3$; $^4F_{\frac{3}{2}}$.

A cubic field leaves an orbital singlet lowest, some 10^4 cm^{-1} below the first triplet. This spin quadruplet splits into two Kramers doublets in any field of lower symmetry. $g = 1.98$. Paramagnetic resonance experiments by Bleaney [8] indicate two distinct values of the splitting δ below 160 K, a feature which is not understood. The spectrum intensities indicated roughly equal fractions for the two sets of ions, and Bleaney found it necessary to postulate the existence of a third set with a very small splitting to account for the observed value of b. Vilches and Wheatley [9] made calorimetric measurements down to 0.017 K—using a thermometer of CMN (cerous magnesium nitrate, q.v.)—and found that the specific heat down to 0.07 K could be accurately accounted for using the Bleaney assignments but somewhat different percentages: $\delta_1 = 0.27$ cm^{-1}, 28 percent; $\delta_2 = 0.15$ cm^{-1}, 28 percent; $\delta_3 = 0.035$ cm^{-1}, 44 percent.

The single odd isotope of chromium, Cr^{53} ($I = \frac{3}{2}$), has an abundance of 9.54 percent and a small nuclear moment, -0.47 nuclear magneton. Bleaney and Bowers [10] achieved resolution of the hfs in paramagnetic resonance studies of diluted (with aluminum) chromic potassium selenate alum enriched in Cr^{53}. Their value for A, 18.5×10^{-4} cm^{-1}, indicates that the associated entropy, $0.0954 \times R \ln 4$ or $0.132R$, will noticeably affect the properties of the chrome alums in the region 0.002 to 0.2 K, and this probably accounts for the fact that the dipolar specific-heat anomaly in this salt is found not in the region of $S/R \sim \ln 2$, but well below.

Adiabatic demagnetization experiments lead to an rms value of δ, and the results of various investigators [11–14] can be summarized by $\delta/k = 0.258 \pm 0.008$ K; hence $b_e = 0.017$. With $b_i = 0.001$, one obtains $b = 0.018$. From direct measurements Kapadnis [15] finds, for the range 1 to 4 K, $b = 0.0162$.

Daniels and Kurti [12] measured the course of entropy with absolute temperature down to 0.007 K; $T_c = 0.011_5$ K at $S/R = 0.40$ (neglecting hfs—see above); $\mathcal{B}/T = 1.6$ T/K. Further measurements by Beun et al. [14] are in quite good agreement with these results; hence it is probable that the unusually low temperatures estimated by deKlerk et al. [16] were in error. As Durieux et al. showed [17] by precise measurements in the liquid-helium region, this salt actually obeys the Curie-Weiss law, with $\theta' = -0.037$ K. They also show that a more accurate formula which takes account of the crystal field (and is valid down to ~ 0.3 K) is $\chi = C(T + 0.00304T^{-1} - \theta)^{-1}$, and here $\theta = -0.033$ K. For still lower temperatures a graphical function is given for calculating the diminution of θ (i.e., of the exchange interaction) with decreasing T.

3. *Chromic Methylammonium Alum.* $Cr_2(SO_4)_3 \cdot (CH_3NH_3)_2SO_4 \cdot 24H_2O$; gram-ionic weight, 492.39; density, 1.645; Cr^{3+}; $3d^3$; $^4F_{\frac{3}{2}}$.

The discussion is the same as for the potassium salt except that here there is a unique value for the splitting [8]. Adiabatic demagnetization experiments give $\delta/k = 0.270 \pm 0.005$ K [18-21] corresponding to $b_e = 0.0182$ K^2 which is significantly greater than the 0.245 K from the epr results. $b_i = 8.6 \times 10^{-4}$ K^2. Baker made a detailed reinvestigation [22] and showed that there exists a small rhombic distortion of the crystal field. This adds a term $+E(S_x^2 - S_y^2)$ to the Hamiltonian of Eq. (5f-1) and brings the derived value of δ/k [$\delta = 2(D^2 + 3E^2)^{\frac{1}{2}}$] up to 0.255 K, with $g = 1.976$. $D = 0.087$, $E = 0.009$ cm^{-1}. Thus there is still a discrepancy of some 5 percent in δ which remains unexplained.

Despite the greater atomic volume of this salt (299 cm^3) compared with that of the potassium alum (272 cm^3), its Curie temperature is 50 percent higher, viz., 0.016 K [21]. $(S/R)_{T_c} = 0.53$ for $\mathcal{B}/T \approx 1.34T$/K.

Durieux et al. [17] investigated departures from Curie's law for this salt as well as for the potassium alum (q.v.). They obtained $\theta' = -0.013$ and, in the formula $\chi = C[T + 0.00279T^{-1} - \theta]^{-1}$, $\theta = -0.010$ K.

4. *Vanadous Ammonium Sulfate.* $VSO_4 \cdot (NH_4)_2SO_4 \cdot 6H_2O$; gram-ionic weight, 387.26; density, ; V^{2+}; $3d^3$; $^4F_{\frac{3}{2}}$. (Blank space indicates density is not yet known.)

The discussion is the same as for Cr^{3+}. The V^{2+} ion oxidizes easily but is somewhat more stable in a zinc-diluted crystal such as is required for narrow epr lines. Preparation [23,24] is conducted in a chilled reducing solution and inert atmosphere. The spin Hamiltonian parameters are [25]: $g_{\parallel} = 1.951$ and probably isotropic; $D = 0.158$; $E = 0.04_9$, A $(= B) = 0.0088$ cm^{-1}. Vanadium is 99.8 percent V^{51}, with $I = \frac{7}{2}$. For $D \gg E$ and A it is shown [26] that $b = D^2 + 3E^2 + 105A^2/16$ (D, E, and A in kelvins) $= 0.056$.

5. *Ferric Ammonium Alum.* $Fe_2(SO_4)_3 \cdot (NH_4)_2SO_4 \cdot 24H_2O$; gram-ionic weight, 482.21; density, 1.71; Fe^{3+}; $3d^5$; $^6S_{\frac{5}{2}}$.

The free ion being in an S state, interaction with the crystalline electric field is small. The sixfold degenerate level is split by a cubic field into a doublet and a quadruplet. A distortion of trigonal symmetry will result in three doublets. Specific-heat data derived from paramagnetic relaxation measurements [27] suggested three equally spaced doublets, of overall splitting 0.23 K. This picture was substantiated by Meijer's analysis [28] of resonance measurements on aluminum-diluted iron alum [29]. Recent low-temperature specific-heat data [9] similarly analyzed [30] strongly suggest, however, that the trigonal component is negligibly small!

Elucidation of the properties by paramagnetic resonance studies has proved to be difficult. In the concentrated salt, magnetic interactions cause a large line width, comparable to the fine structure. Even upon dilution with the corresponding aluminum alum the resolution remains rather poor, probably owing to inhomogeneity as the splitting varies with degree of dilution.

Vilches and Wheatley [9] measured the specific heat down to 0.02 K and observed a pronounced λ-type anomaly at 0.0260 K. Cooke, Meyer, and Wolf [31] compared T^* with T as indicated by CMN, using a novel compound specimen technique, over the range 0.05 to 1 K. As noted above [9], these results when combined with the $S - T^*$ data of Kurti, Squire, and Simon reported in [32] agreed with the $S - T$ derivation of Vilches and Wheatley where overlap exists. $T_c = 0.026$ K at $S/R = 0.69$. Steenland et al. [33] found $(S/R)_{T_c} = 0.65$ at $\mathcal{B}/T = 1.15$ $T/$K.

There are also fairly wide variations in the values reported for b, viz., 0.0123 [34]. 0.0128 [11], 0.0135 [15,35], and 0.0143 [27]. The spread may be due to imperfect crystals in view of the sensitivity of the crystal field splitting to dilution. In the dilution range 3:1 to 6:1 (Al:Fe), b actually doubles; at 20:1 it is back to that of the pure ferric salt [35,36].

6. *Ferric Methylammonium Alum.* $Fe_2(SO_4)_3 \cdot (CH_3NH_3)_2SO_4 \cdot 24H_2O$; gram-ionic weight, 496.24; density,1.659; Fe^{3+}; $3d^5$; $^6S_{\frac{5}{2}}$.

The epr spectrum was found by Bleaney and Trenam [37] to be very complex, largely owing to the excitation of "forbidden" transitions as the measuring frequency was comparable in (energy) magnitude with the zero-field splittings. Their general Hamiltonian for the alums contains a cubic term and trigonal terms of second and fourth degree:

$$\mathcal{H} = g\mu_B\mathcal{B} \cdot \mathbf{S}$$
$$+ \frac{a}{6}\left[S_\xi^4 + S_\eta^4 + S_\zeta^4 - \frac{S}{5}(S+1)(3S^2+3S-1)\right] + D\left[S_z^2 - \frac{S}{3}(S+1)\right]$$
$$+ \frac{F}{180}[35S_z^4 - 30S(S+1)S_z^2 + 25S_z^2 - 6S(S+1) + 3S^2(S+1)^2]$$

Here, ξ, η, ζ form a set of mutually perpendicular axes, with respect to which the z axis is the (111) direction. The specific heat can be computed from

$$b = \frac{56}{9}D^2 + 2a^2 - \frac{14}{27}F(2a - F)$$

The term in F is always very small and detectable only in high-resolution circumstances. Zero-field splittings were found to be 0.393 cm^{-1} and 0.740 cm^{-1}. These are equal to $2D + \frac{5}{3}a$ and $4D - \frac{4}{3}a$, if one assumes that $a \ll D$. Then $D = -0.188$ cm^{-1} and $a = -0.010$ cm^{-1}. Cooke et al. [38] measured the specific heat between 0.17 and 20 K and showed that the $\pm\frac{1}{2}$ level lies lowest. $b = 0.455$. Their data were confirmed and extended to about 0.1 K by Croft and Exell [39]. This substance should be very useful as an inexpensive substitute for gadolinium sulfate (q.v.) as a refrigerant in the range 0.1 to 1 K.

7. *Manganous Ammonium Sulfate.* $MnSO_4 \cdot (NH_4)_2SO_4 \cdot 6H_2O$; gram-ionic weight, 391.24; density, 1.83; Mn^{2+}; $3d^5$; $^6S_{\frac{5}{2}}$.

The free ion being in an S state, the interaction with the crystalline electric field is small in this Tutton salt. Epr experiments on a zinc-diluted salt [40] showed that $g = 2.000$ and is isotropic to about 1 part in 10^4. The hfs is surprisingly large and has received an explanation from Abragam and Pryce [41] who postulate a distortion of the nominally spherical electron-cloud distribution by the electric field. D ($\sim$0.028 cm^{-1}) and E ($\sim$0.005 cm^{-1}) are sensitive to temperature and dilution, but A (= 0.009 cm^{-1}) is not. Mn^{55}, with $I = \frac{5}{2}$, is 100 percent abundant. The z axes of the two ions in the unit cell are inclined to each other at an angle 2α, where $\alpha = 32$ deg. Neglecting the effects of exchange, the susceptibility can be calculated from $\chi = \chi_{\parallel} \cos^2 \beta + \chi_{\perp} \sin^2 \beta$ and Eq. (5f-2) with $\beta = 32$ deg, 90 deg, and 58 deg for the K_1, K_2, and K_3 axes, respectively.

The low-temperature specific heat has been measured as a function of dilution [35] in order to separate the various contributions to $b = 0.032K^2$. Dipole-dipole interaction accounts for 0.0106, Stark splitting plus hfs for 0.016, and an appreciable exchange interaction must be present to account for the balance. Other values reported for b range from 0.0306 [31] to 0.034 [42].

A number of adiabatic demagnetization studies have been made, [31,32,34,43,45]. The entropy vs. temperature measurements of Miedema et al. [45] below the Curie point ($T_c = 0.14$ K at $S/R = 1.25$) can be added quite satisfactorily to those of Cooke and Hull [32] made largely above the Curie point ($T_c = 0.14$ K, $S/R = 1.27$). In this work, it was found that the sign of D is negative, contradicting the conclusions drawn in ref. 40. θ' should then range from $+0.035$ K (K_1) to -0.03 K (K_2) (and zero for a powder), modified by any exchange contribution. The results of Durieux et al. [17], however, from measurements in four (unspecified) directions range from $+0.079$ to -0.032 K. Recent specific heat measurements [9] between 0.1 and 1 K yield an entropy-temperature curve qualitatively in agreement with those mentioned above but displaced toward higher temperatures. The λ-type anomaly was located at 0.173 K and investigated in detail; $S_c/R = 1.27$.

8. *Cobaltous Ammonium Sulfate*. $CoSO_4 \cdot (NH_4)_2SO_4 \cdot 6H_2O$; gram-ionic weight, 395.25; density, 1.902; Co^{2+}; $3d^7$; $^4F_{\frac{9}{2}}$.

This is a Tutton salt. In a cubic field the sevenfold degenerate orbital level is split into two triplets and a singlet, with one triplet lying lowest. The latter, with fourfold spin degeneracy, is then split into Kramers doublets by the combined effect of the spin-orbit coupling and fields of lower symmetry. The separations are not very large (200 to 2,000 cm^{-1}); hence there is a large temperature-independent term in the susceptibility, the spin-lattice relaxation time is strongly temperature dependent ($\propto T^{-7}$) in the liquid-helium region, and paramagnetic resonance can be observed only below 20 K. Then only the lowest doublet is occupied, effective $S = \frac{1}{2}$, g is quite anisotropic (with accurately tetragonal symmetry; see Bleaney et al. [46]). $g_{\parallel} = 6.45$; $g_{\perp} = 3.05$ [47]. The z axes of the two ions in the unit cell are inclined to each other at an angle 2α, where $\alpha = 34$ deg. The susceptibility can be calculated from $\chi_1 = \chi_{\parallel}$ $\cos^2 \alpha + \chi_{\perp} \sin^2 \alpha$, for the K_1 axis. Co^{59}, $I = \frac{7}{2}$, is 100 percent abundant, and the nuclear contribution to the specific heat, b_n, is given [26] by $\frac{21}{16}(A_x{}^2 + A_y{}^2 + A_z{}^2)$, with the A's in kelvins. From $A_z = 0.0245$, $A_x = A_y = 0.002$ cm^{-1}, one finds that $b_n = 1.66 \times 10^{-3}$ K^2, which is to be compared with the experimental values [35,48,49] of 1.6 to 1.8×10^{-3}. The total specific heat $b \approx 4.2 \times 10^{-3}$, and b_i (calculated) $= 2.1 \times 10^{-3}$.

Adiabatic demagnetization experiments have been carried out by Garrett [50] and Miedema et al. [45]; $T_c = 0.084$ K; $S_c/R = 0.43$.

9. *Cobaltous Fluosilicate*. $CoSiF_6 \cdot 6H_2O$; gram-ionic weight, 309.10; density, 2.113; Co^{2+}; $3d^7$; $^4F_{\frac{9}{2}}$.

The discussion is the same as for cobaltous ammonium sulfate, except that there is only one ion in the unit cell. The "parallel" and "perpendicular" axes of the magnetic complex coincide with those of the crystal and with the susceptibility axes. $g_{\parallel} = 5.82$; $g_{\perp} = 3.44$; $A = 0.0184$, $B = 0.0047$ cm^{-1} [47]. Benzie, Cooke, and Whitley [35] find that b is large (18.5×10^{-3} K^2) and almost entirely due to exchange interaction. Isotropic exchange would give rise to a Curie-Weiss constant of 0.23 K, but none was detected. Van den Broek, van der Marel, and Gorter [49], however, obtained $bR/c_{\parallel}$ $= 0.45 \times 10^6$ and $bR/c_{\perp} = 1.74 \times 10^6$, values which would require the principal g-values to be 6.04 and 3.07 in order to yield the above mentioned figure for b.

10. *Nickel Fluosilicate.* NiSiF$_6 \cdot$6H$_2$O; gram-ionic weight, 308.85; density, 2.134; Ni^{2+}; $3d^8$; $^3F_4$.

A cubic field leaves an orbital singlet lowest, and this spin triplet is split by a trigonal or tetragonal field into a doublet and a singlet and by a rhombic field into three singlets. Penrose and Stevens [51] found that $g = 2.29$ and is isotropic; the doublet lies lowest (D negative) and D is constant at 0.12 cm^{-1} below 20 K but increases to 0.17 cm^{-1} at 90 K and 0 .32 cm^{-1} at 195 K. Nickel in zinc fluosilicate behaves similarly [52] except that at all temperatures from 20 to 300 K the splitting is about 20 percent greater in the dilute (1:4 and 1:16) salts.

$b = 0.013$ K^2 [35]; the contribution from the Stark splitting ($= 2D^2/9$) is 6.0×10^{-3} taking $D = 0.173$ K and from magnetic dipole interaction is 2×10^{-3}. For a 1:16 zinc-diluted specimen, $b = 0.012$ K^2. Investigations were carried out down to 0.02 K with a 15:85 Ni-Zn specimen by Hill, Meyer, and Milner [53]. They found that this salt obeyed the Curie law in the region of 1 K, and since the susceptibility is given by

$$\chi = \frac{C}{T} \left(1 - \frac{D}{3kT} + \frac{\theta}{T} \right)$$

then the Weiss constant $\theta \approx D/3k$, or about -0.07 K, taking $D = -0.14$ cm^{-1}. Thus the exchange is antiferromagnetic, in contrast to the experience of Benzie and Cooke [27] with the concentrated salt, where $\theta' = +0.1$ K, i.e., $\theta = +0.04$ K, using $D = -0.12$ cm^{-1}. From epr linewidth studies below 1 K, Svare and Seidel [54] showed that the exchange with the six nearest neighbor ions is ferromagnetic and isotropic, with the exchange constant $= -3.9 \times 10^{-18}$ erg. These authors give $g = 2.24$ and $D = -0.113$ cm^{-1}.

Extensive magnetothermodynamic investigations have been carried out by Giauque et al. [55] and used to derive absolute temperatures in the range 0.05 to 0.3 K.

11. *Cupric Potassium Sulfate.* CuSO$_4 \cdot$K$_2$SO$_4 \cdot$6H$_2$O; gram-ionic weight, 441.97; density, 2.22; Cu^{2+}; $3d^9$; $^2D_{\frac{3}{2}}$.

This is a Tutton salt. The orbital levels are split by a cubic field into an upper triplet and a lower doublet. In a tetragonal or rhombic field, the doublet is further split into two singlets separated by about 10^4 cm^{-1}. These are twofold degenerate in spin, and hence at low temperatures effectively $S = \frac{1}{2}$.

The symmetry is rhombic [46]: For a zinc-diluted specimen g_z, g_x, $g_y = 2.42$, 2.16, and 2.04, respectively; A_z, A_x, $A_y = -99$, <17, and $+61 \times 10^{-4}$ cm^{-1}, respectively (these are the mean values for the two isotopes Cu^{63} and Cu^{65}, weighted in proportion to their natural abundances, viz., 69.1 and 30.9 percent, respectively). The z axes of the two ions in the unit cell are inclined to each other at an angle 2α, where $\alpha = 43$ K. The theoretical value of $b_n = \frac{5}{16}(A_x{}^2 + A_y{}^2 + A_z{}^2) = 0.97 \times 10^{-4}$ K^2. Experimentally, $b = 5.7 \times 10^{-4}$, $b_n = 1.0 \times 10^{-4}$ K^2. The interaction contribution is thus 4.7×10^{-4}, of which dipole-dipole coupling accounts for 1.3×10^{-4} [35]. The nuclear hfs entropy $= R \ln 4$ or $1.386R$. Bleaney, Penrose, and Plumpton [56] observed that the epr lines of the undiluted salt showed "exchange narrowing." Their spectrum parameters, obtained at 90 K, differed from those cited above for the diluted salt in that $g_z = 2.36$, $g_x = 2.14$, and $\alpha = 42$ deg.

The magnetic cooling field is clouded by early entropy calculations being made using incorrect g values and by the questionable practice—for anisotropic substances —of working with powder specimens. $T_c \approx 0.05$ K [57].

12. *Cupric Sulfate.* $CuSO_4 \cdot 5H_2O$; gram-ionic weight, 249.69; density, 2.279; Cu^{2+}; $3d^9$; $^2D_{\frac{5}{2}}$.

The discussion is the same as for cupric potassium sulfate, except that the exchange energy between neighboring magnetic complexes is rather large (~ 0.15 cm^{-1}) and of the order of magnitude of a microwave quantum; hence the details of an epr spectrum will depend upon the microwave frequency used.

There are two ions in the unit cell. $g_{\parallel} = 2.47$ and $g_{\perp} = 2.06$ [58]. The angle between the tetragonal axes is approximately 90 deg, and, as a result, χ has an axis of symmetry normal to the plane of these two axes. The susceptibility along this symmetry axis χ_a is equal to $\chi_{\perp}$, while in any direction in the plane of the tetragonal axes the suscepti-bility χ_e has the value $\frac{1}{2}(\chi_{\parallel} + \chi_{\perp})$ [59]. The exchange interaction gives rise to a Weiss constant of -0.6 K [60]. The Curie constants per mole are $C_a = 0.407$ and $C_e = 0.480$. Benzie and Cooke [60] combined their results with a reinterpretation of the measure-ments of Krishnan and Mookherji [59] to show that the susceptibility contains a tem-perature-independent term, with anisotropy ($\chi_e - \chi_a$) of magnitude 65×10^{-6} per mole. This is in excellent agreement with the theoretical predictions of Polder [61]; viz.,

$$\chi_a = \frac{0.40}{T} + 20 \times 10^{-6} \qquad \chi_e = \frac{0.48}{T} + 78 \times 10^{-6}$$

The specific heat shows a rounded maximum at 1.4 K and a λ-type anomaly at 0.034 K [62]. Studies by Geballe and Giauque [63] led those authors to suggest that the two inequivalent Cu^{++} ions in the unit cell give rise to two systems of ions with different environments, and hence differing low-temperature behavior. This picture was supported by specific heat and susceptibility determinations by Miedema et al. [64] who found that below 1 K the Curie constant fell to one-half its high-temperature value and θ' changed to about $+0.02$ K. More detailed informa-tion on the interactions were furnished in proton magnetic resonance studies by Wittekoek, Poulis, and Miedema [65]. Extensive magnetothermodynamic measure-ments by Giauque et al. [62] enabled them to derive absolute temperatures in the range 0.02 to 0.4 K.

13. *Cerous Magnesium Nitrate.* $2Ce(NO_3)_3 \cdot 3Mg(NO_3)_2 \cdot 24H_2O$; gram-ionic weight, 764.85; density, 2.10; Ce^{3+}; $4f^1$; $^2F_{\frac{5}{2}}$.

The $J = \frac{7}{2}$ state lies about 2,500 cm^{-1} above the ground state. Under the com-bined action of the spin-orbit coupling and the electrostatic field (of C_{3v} symmetry) the $J = \frac{5}{2}$ multiplet is split into three doublets. For simplicity these may be referred to as the $J_z = \pm\frac{1}{2}$, $\pm\frac{3}{2}$, and $\pm\frac{5}{2}$ levels, although actually the $|\pm\frac{1}{2}\rangle$ and $|\pm\frac{5}{2}\rangle$ states are mixed and there are, furthermore, small admixtures from the $J = \frac{7}{2}$ manifold [66]. The salt is very anisotropic with $g_{\parallel} \leq 0.026$ [67] and $g_{\perp} = 1.840$ [68]. There is one magnetic ion in the unit cell.

The $J_z = \pm\frac{3}{2}$ level lies at 25.2 cm^{-1} [69] and causes the spin-lattice relaxation time to vary exponentially with T in the region of 2 K [70]. There is a relatively large tem-perature-independent term in the susceptibility (0.002 cgs/mole) [71]. Measurements of χ between 1.5 and 36 K [72] indicate that the third doublet lies at about 100 K.

The low-temperature specific heat is extremely small and is entirely accounted for by magnetic dipole interaction [73] plus the (in this instance not entirely negligible) lattice contribution; the Debye θ is ~ 60 K [74]; $b = 6 \times 10^{-6}$ K^2. The salt obeys Curie's law down to below 0.01 K [73], and hence is much used in magnetic cooling applications. Its extreme anisotropy and Curie law behavior have been made use of by Cooke, Meyer, and Wolf [31] to determine the χ-T relation for several isotropic substances (by a compound-specimen technique) down to 0.05 K. Absolute tem-perature determinations down to the region of 0.001 K by different methods [75,76] are in reasonably close agreement. Marked deviations from Curie's law set in below 6 mK, despite the fact that θ' is only -0.27 mK [77]. Wheatley [78] has obtained evidence that agglomerations of powdered CMN obey Curie's law down to as low as 2 mK.

14. *Cerous Ethyl Sulfate.* $Ce(C_2H_5SO_4)_3 \cdot 9H_2O$; gram-ionic weight, 677.66; density, 1.839; Ce^{3+}; $4f^1$; $^2F_{\frac{5}{2}}$.

The discussion is the same as for cerous magnesium nitrate, except that the crystal field has C_{3h} symmetry [79] and there are two ions in the unit cell although with a unique magnetic axis. The $J_z = +\frac{5}{2}$ level lies lowest, and the $J_z = \pm\frac{1}{2}$ level is at 6.6 K higher. In a crystal diluted 200:1 with lanthanum ethyl sulfate this order of levels was found to be inverted [80]. The third doublet ($J_z = \pm\frac{3}{2}$) is at some 200 cm^{-1} above the ground state. (J_z is not truly a good quantum number; compare cerous magnesium nitrate.) For the $\pm\frac{5}{2}$ level, $g_\parallel = 3.80$, $g_\perp = 0.22$; for the $\pm\frac{1}{2}$ level, $g_\parallel = 1.0$, $g_\perp = 2.2$ [81].

Between 0.3 and 1 K, the specific heat is given by [81]

$$\frac{cT^2}{R} = b_i + \delta^2 . \exp\left(-\frac{\delta}{T}\right)\left[1 + \exp\left(-\frac{\delta}{T}\right)\right]^{-2}$$

with $\delta = 6.6$ K. The magnitude of b_i, 11.2×10^{-4}, is six times that to be expected from magnetic dipole interaction [82] and is thought to arise mainly from electric quadrupole-quadrupole interaction between the cerium ions [83,84]. Supporting evidence was obtained from epr measurements below 1 K by Baker [85]. There is a temperature-independent term in $\chi_\perp$ of magnitude 4.7×10^{-4} emu/mole [80]. In the liquid-helium region the susceptibility along the crystal axis is given by

$$\chi_\parallel = \frac{C}{T}(1 - A)$$

where
$$C = \frac{N\mu_B{}^2}{4k} g_\parallel \left(\tfrac{5}{2}\right)^2 \qquad A = \frac{g_\parallel(\frac{5}{2})^2 - g_\parallel(\frac{1}{2})^2}{g_\parallel(\frac{5}{2})^2[1 + \exp(\delta/T)]}$$

Adiabatic demagnetization studies have been carried out down to 0.02 K by Johnson and Meyer [84]; $T_c = 0.05_3$ K at $S/R = 0.48$. Meyer and Smith [86] have measured the specific heat between 0.6 and 20 K. They find that $\delta = 6.7$ K and also observe an excess specific heat in the region of the maximum of the Schottky anomaly. This has been explained by Becker and Clover [87] in terms of an anomalous phonon spectrum arising from the spin-phonon interaction, taking account of the substantial broadening of the lowest energy levels.

15. *Neodymium Magnesium Nitrate.* $2Nd(NO_3)_3 \cdot 3Mg(NO_3)_2 \cdot 24H_2O$; gram-ionic weight, 768.99; density, 　; Nd^{3+}; $4f^3$; $^4I_{\frac{9}{2}}$.

The $^4I_{\frac{9}{2}}$ ground state is split by a crystalline electric field of C_{3v} symmetry [66] into five Kramers doublets. At liquid-helium temperatures only the lowest of these is populated. This is a mixture of states with $J_z = \pm\frac{7}{2}$, $\pm\frac{1}{2}$, $\pm\frac{5}{2}$. $g_\parallel = 0.45$; $g_\perp = 2.72$ [87]. The low-temperature thermal and magnetic behavior is modified by nuclear hyperfine coupling. Nd^{143} (12.20 percent) and Nd^{145} (8.30 percent) both have $I = \frac{7}{2}$, resulting in a contribution to the entropy of $0.205R \ln 8$ or $0.426R$. The hfs parameters in the spin Hamiltonian are: $B_{143} = 0.0312$, $B_{145} = 0.0194$, $A_{143} \sim 0.005$, $A_{145} \sim 0.003$ cm^{-1}.

Adiabatic demagnetization experiments down to 0.06 K have been made by Cooke, Meyer, and Wolf [31,34]. Deviations from Curie's law begin at 0.1 K, owing to the influence of the hfs. $b = 8.74 \times 10^{-4}$ K^2. Calculated values are: $b_i = 3.0 \times 10^{-5}$, $b_{143} = 6.53 \times 10^{-4}$, $b_{145} = 1.70 \times 10^{-4}$ K^2. Spin-lattice relaxation in the liquid-helium region has been investigated by Hudson and Mangum [88] and down to 0.3 K, in lanthanum-diluted crystals, by Jeffries et al. [89]. The Zeeman effect in the optical spectrum has been studied in detail by Dieke and Heroux [90]. Their values for $g_\parallel$ and $g_\perp$ for the ground state (0.420 and 2.629, respectively) are in reasonably good agreement with the epr data. They obtain 3.38 and zero for the g values of the $J_z = \pm \frac{3}{2}$ level, which is found to lie at 33.13 cm^{-1}.

16. *Neodymium Ethylsulfate.* $Nd(C_2H_5SO_4)_3 \cdot 9H_2O$; gram-ionic weight, 681.80; density, 1.872; Nd^{3+}; $4f^3$; $^4I_{\frac{9}{2}}$.

The ^4I ground state lies about 1,800 cm^{-1} below the next higher level. In a crystal field of C_{3h} symmetry it is split into five doublets and only the lowest is effectively occupied at 20 K and below. This is a mixture of the basic J_z states $|\pm \frac{7}{2}\rangle$ and $|\pm \frac{5}{2}\rangle$ [79c]. The first excited doublet, $|+\frac{1}{2}\rangle$, lies at 150 cm^{-1} (see below). There is one ion in the unit cell.

Detailed paramagnetic resonance investigations have been made by Bleaney, Scovil, and Trenam [91], who find discrepancies between the strong field and weak field results. The Hamiltonian of Eq. (5f-1) cannot be fitted to the latter. A possible explanation is that given by Stevens to account for the similar but much larger effects in gadolinium ethyl sulfate; viz., here one is not dealing with a static crystalline electric field, but the equilibrium positions of the surrounding dipoles may depend upon the state of the magnetic ion. $g_\parallel = 3.535$; $g_\perp = 2.072$. The hfs parameters in the spin Hamiltonian are: $A_{143} = 0.0380$, $A_{145} = 0.0236$, $B_{143} = 0.0199$, $B_{145} = 0.0124$ cm^{-1}. (See also Erickson [92]).

Optical transitions and the Zeeman effect have been studied by Dieke and Heroux [90], who obtain $g_\parallel = 3.50$ and $g_\perp = 2.06$ for the lowest doublet. The next doublet lies at 149.64 cm^{-1}.

Low-temperature studies, between 0.015 and 2 K, were made by Meyer [93] ($b = 1.13 \times 10^{-3}$ K^2), and in the liquid-helium region by Roberts, Sartain, and Borie [94] ($b = 1.09 \times 10^{-3}$). The theoretical value for b_e is 0.176×10^{-3} K^2, and from the epr data one calculates $b_{143} = 6.073 \times 10^{-3}$, $b_{145} = 2.348 \times 10^{-3}$ K^2; these together lead to a theoretical value of 1.11×10^{-3} K^2 for b. $T_c \approx 0.016$ K at $S/R \sim 0.45$. The theory of dipole-dipole interaction in anisotropic crystals and of the added effect on χ of the hfs developed by Daniels [82] has been applied to this salt by Meyer. The effect is much larger for the "parallel" direction ($2\frac{1}{2}$ percent correction at 0.5 K) than for the "perpendicular" direction (2 percent correction at 0.2 K). Blok, Shirley, and Stone [95] have determined the susceptibility temperature scale down to 0.01 K, using both gamma-ray heating and nuclear orientation methods. Their values of T agree with those of Meyer [93] down to 0.025 K. The specific heat between 1.5 and 20 K has been measured by Meyer and Smith [86]. The lattice contribution c_l showed the same features here as in several rare-earth ethyl sulfates; viz., c_l/T^3 fell rapidly between 1.5 and 3 K, then rose again, and passed through a broad maximum at 8 to 9 K, falling slowly with further increase of temperature.

17. *Gadolinium Sulfate.* $Gd_2(SO_4)_3 \cdot 8H_2O$; gram-ionic weight, 373.42; density, 3.010; Gd^{3+}; $4f^7$; ^8S.

The free ion having an ^8S ground state, the effect of the crystalline electric field is small and the splittings produced are only of the order of 1 cm^{-1}. Hebb and Purcell [96] showed that the eightfold degenerate ground level is split by a cubic field into two doublets and a quadruplet, and specific-heat data have always been interpreted on this basis. Epr studies at room temperature by Bogle and Heine [97] on a 200:1 samarium-diluted crystal showed, however, that there are four doublets lying at 0, 0.20, 0.48, and 0.82 cm^{-1}, respectively. There are two magnetically inequivalent ions in the unit cell, the crystal b axis being the twofold axis. $g = 1.99$ and is isotropic.

The spin Hamiltonian can be written most conveniently

$$\mathcal{H} = g\mu_B \mathcal{B} \cdot S + B_2^0 P_2^0 + B_2^2 P_2^2 + B_4^0 P_4^0$$

TABLE 5f-35. PROPERTIES OF PARAMAGNETIC SALTS

Paramagnetic salt	Gram-ionic weight M, g	Room-temp. density ρ, g/cm³	g values	Curie const. C, emu/g-ion	D, cm⁻¹	Spec. heat const b, K²
1. Titanium cesium alum $TiCs(SO_4)_2 \cdot 12H_2O$	589	2.019	$g_\| = 1.25$ $g_\perp = 1.14$	0.130[a]	....	3.9×10^{-5}
2. Chromic potassium alum $CrK(SO_4)_2 \cdot 12H_2O$	499	1.83	1.98	1.84	0.135 0.075[b]	1.8×10^{-2}
3. Chromic methylammonium alum $Cr(CH_3NH_3)(SO_4)_2 \cdot 12H_2O$	492	1.645	1.976	1.83	0.087	1.9×10^{-2c}
4. Vanadous ammonium sulfate $V(NH_4)_2(SO_4)_2 \cdot 6H_2O$	387	......	1.95	1.78	0.15_5[d]	5.6×10^{-2e}
5. Ferric ammonium alum $Fe(NH_4)(SO_4)_2 \cdot 12H_2O$	482	1.71	2.003	4.39	0.016[f]	$\sim 1.3 \times 10^{-2}$
6. Ferric methylammonium alum $Fe(CH_3NH_3)(SO_4)_2 \cdot 12H_2O$	496	1.659	2.003	4.39	0.188[g]	0.45
7. Manganous ammonium sulfate $Mn(NH_4)_2(SO_4)_2 \cdot 6H_2O$	391	1.83	2.00	4.38	0.027_7[h]	3.2×10^{-2}
8. Cobaltous ammonium sulfate $Co(NH_4)_2(SO_4)_2 \cdot 6H_2O$	395	1.902	$g_\| = 6.45$ $g_\perp = 3.05$	2.95[j]	........[i]	4.2×10^{-3}
9. Cobaltous fluosilicate $CoSiF_6 \cdot 6H_2O$	309	2.113	$g_\| = 5.82$ $g_\perp = 3.44$	$C_\| = 3.18$ $C_\perp = 1.11$	........[k]	(18.5×10^{-3})[l]
10. Nickel fluosilicate $NiSiF_6 \cdot 6H_2O$	309	2.134	2.29	1.31[m]	-0.12[n]	1.3×10^{-2}
11. Cupric potassium sulfate $CuK_2(SO_4)_2 \cdot 6H_2O$	442	2.22	$g_z = 2.14$ $g_y = 2.04$[o] $g_z = 2.36$	0.445[p]	....	5.7×10^{-4}
12. Cupric sulfate $CuSO_4 \cdot 5H_2O$	250	2.281	....	$C_\| = 0.480$ $C_\perp = 0.407$[q]		

13. Cerous magnesium nitrate $Ce_2Mg_3(NO_3)_{12} \cdot 24H_2O$	765	2.10	$g_\parallel \leq 0.026$, $g_\perp = 1.840$	$C_\perp = 0.317$	. . . .	6×10^{-6}
14. Cerous ethylsulfate $Ce(C_2H_5SO_4)_3 \cdot 9H_2O$	678	1.839	$g_\parallel = 3.80$, $g_\perp = 0.22^r$	$C_\parallel = 1.35$, $C_\perp = 0.0045$	. . . .	1.12×10^{-3}
15. Neodymium magnesium nitrate $Nd_2Mg_3(NO_3)_{12} \cdot 24H_2O$	769	2	$g_\parallel = 0.45$, $g_\perp = 2.72$	$C_\perp = 0.019$, $C_\perp = 0.694$	. . . .	8.74×10^{-4_s}
16. Neodymium ethyl sulfate $Nd(C_2H_5SO_4)_3 \cdot 9H_2O$	682	1.872	$g_\parallel = 3.535$, $g_\perp = 2.072$	$C_\parallel = 1.170$, $C_\perp = 0.402$	. . . .	11.1×10^{-4_t}
17. Gadolinium sulfate $Gd_2(SO_4)_3 \cdot 8H_2O$	373	3.010	1.99	7.80	0.0633^u	0.32 or 0.37^v

ᵃ $\tfrac{1}{3}\chi_\parallel + \tfrac{1}{3}\chi_\perp$.
ᵇ Two types of ion below 160 K.
ᶜ 1.7×10^{-2} using epr data.
ᵈ $E = 0.04_9$ cm⁻¹.
ᵉ δ_e/R, Stark contribution (calculated from epr data).
ᶠ $a = -0.0128$ cm⁻¹; splitting sensitive to temperature and dilution.
ᵍ $a = 0.010$ cm⁻¹; splitting little affected by dilution.
ʰ $E = 0.005$, $a = 0.0008$, $A = 0.0093$ cm⁻¹; D and E sensitive to temperature and dilution.
ⁱ Along K_1 axis.
ʲ $A = 0.0245$, $B = 0.002$ cm⁻¹. Tetragonal axes 68 deg apart.
ᵏ $A = 0.0184$, $B = 0.0047$ cm⁻¹.
ˡ Conflicting experimental data.
ᵐ Weiss $\theta \sim -0.07$ K.
ⁿ D varies with temperature and dilution.
ᵒ Rhombic symmetry; z axes 84 deg apart.
ᵖ For powder. Weiss $\theta \sim 0.035$ K.
q Tetragonal axes ~90 deg apart. Weiss $\theta \sim -0.6$ K.
ʳ Second doublet at 4.6 cm⁻¹ with $g_\parallel = 1.0$, $g_\perp = 2.2$; order of levels inverted in diluted salt.
ˢ Mostly due to hfs.
ᵗ Mostly due to hfs.
ᵘ $E = 0.013$, $F = 0.004$ cm⁻¹. Spectrum observed at room temperature.
ᵛ All experiments give one or the other of these two figures.

where the $P_n{}^m$ are operators [26,79c] which have the same transformation properties as the corresponding spherical harmonics $Y_n{}^m$, and $B_n{}^m$ are coefficients determined by fitting to the observed spectrum. The findings are: $b_2{}^0 = 3B_2{}^0 = 0.0633$, $b_2{}^2 = 3B_2{}^2 = 0.038$, $b_4{}^0 = 60B_4{}^0 = -0.0013$ cm^{-1}. The Stark specific-heat term b_e is calculated to be 0.195 K^2 from the expression $21(b_2{}^0)^2 + 7(b_2{}^2)^2 + 77(b_4{}^0)^2$, in kelvins, while $b_i = 0.10$ K^2; adding, one finds that $b = 0.30$ K^2. The last figure is notably smaller than that which is obtained from relaxation experiments [98] between 77 and 290 K. These give $bR/C = 3.9 \times 10^6$ Oe^2, and taking $C/R = 9.38 \times 10^{-8}$ K^{2}Oe^{-2}, $b = 0.37$ K^2. There are discrepancies among different measured values at low temperatures, viz., 0.32 from relaxation experiments [99–101], 0.37 from adiabatic demagnetization [102] and calorimetry [103]. The T^{-2} dependence of the specific heat breaks down below 3 K, owing to the large value of D ($= b_2{}^0$), and relaxation-type determinations involve rather large corrections for saturation effects owing to the large Curie constant.

Giauque and MacDougall [104] used this substance for their pioneering magnetic cooling experiments in 1933. Because of its large specific heat it is most useful as a cooling substance in the 0.1 to 1 K range (compare ferric methylammonium alum). Van Dijk [105] has measured the departures from the Curie Law down to 0.22 K.

References

1. Bleaney, B.: *Phys. Rev.* **78**, 214 (1950).
2. Van Vleck, J. H.: *J. Chem. Phys.* **5**, 320 (1937).
3. Daniels, J. M.: *Proc. Phys. Soc. (London)*, ser. A, **66**, 673 (1953).
4. Bleaney, B., G. S. Bogle, A. H. Cooke, H. J. Duffus, N. C. M. O'Brien, and K. W. H. Stevens: *Proc. Phys. Soc. (London)*, ser. A, **68**, 57 (1955).
5. van den Handel, J.: Thesis, Leiden University, 1940.
6. Benzie, R. J., and A. H. Cooke: *Proc. Roy. Soc. (London)*, ser A, **209**, 269 (1951).
7. Kurti, N., P. Laine, and F. Simon: *Compt. rend.* **204**, 675 (1937).
8. Bleaney, B.: *Proc. Roy. Soc. (London)*, ser. A, **204**, 203 (1950).
9. Vilches, O. E., and J. C. Wheatley: *Phys. Rev.* **148**, 509 (1966).
10. Bleaney, B., and K. D. Bowers: *Proc. Phys. Soc. (London)*, ser. A, **64**, 1135 (1951).
11. de Klerk, D.: Thesis, Leiden University, 1948.
12. Daniels, J. M., and N. Kurti: *Proc. Roy. Soc. (London)*, ser. A, **221**, 243 (1954).
13. Ambler, E., and R. P. Hudson: *Phys. Rev.* **95**, 1143 (1954).
14. Beun, J. A., A. R. Miedema, and M. J. Steenland: *Physica* **23**, 1 (1957).
15. Kapadnis, D. G.: *Physica* **22**, 159 (1956).
16. de Klerk, D., M. J. Steenland, and C. J. Gorter: *Physica* **15**, 649 (1949).
17. Durieux, M., H. van Dijk, H. ter Harmsel, and C. van Rijn: "Temperature: Its Measurement and Control in Science and Industry," vol. 3, part 1, p. 313, Reinhold Publishing Corporation, New York, 1962.
18. de Klerk, D., and R. P. Hudson: *Phys. Rev.* **91**, 278 (1953).
19. Gardner, W. E., and N. Kurti: *Proc. Roy. Soc. (London)*, ser. A, **223**, 542 (1954).
20. Hudson, R. P., and C. K. McLane: *Phys. Rev.* **95**, 932 (1954).
21. Ambler, E., and R. P. Hudson: *J. Chem. Phys.* **27**, 378 (1957).
22. Baker, J. M.: *Proc. Phys. Soc. (London)*, ser. B, **69**, 633 (1956).
23. Hutchison, C. A., and L. S. Singer: *Phys. Rev.* **89**, 256 (1953).
24. Kikuchi, C., H. M. Sirvetz, and V. W. Cohen: *Phys. Rev.* **92**, 109 (1953).
25. Bleaney, B., D. J. E. Ingram, and H. E. D. Scovil: *Proc. Phys. Soc. (London)*, ser. A, **64**, 601 (1951).
26. Bowers, K. D., and J. Owen: *Repts. Progr. Phys.* **18**, 304 (1955).
27. Benzie, R. J., and A. H. Cooke: *Proc. Phys. Soc. (London)*, ser. A, **63**, 213 (1950).
28. Meijer, P. H. E.: *Physica* **17**, 899 (1951).
29. Ubbink, J., J. A. Poulis, and C. J. Gorter: *Physica* **17**, 213 (1951).
30. Kimura, I., and N. Uryu: *J. Phys. Soc. Japan* **23**, 1204 (1967).
31. Cooke, A. H., H. Meyer, and W. P. Wolf: *Proc. Roy. Soc. (London)*, ser. A, **233**, 536 (1956).
32. Cooke, A. H.: *Proc. Phys. Soc. (London)*, ser. A, **62**, 269 (1949).
33. Steenland, M. J., D. de Klerk, M. L. Potters, and C. J. Gorter: *Physica* **17**, 149 (1951).
34. Cooke, A. H., H. Meyer, and W. P. Wolf: *Proc. Roy. Soc. (London)*, ser. A, **237**, 395 (1956).

35. Benzie, R. J., A. H. Cooke, and S. Whitley: *Proc. Roy. Soc. (London)*, ser. A, **232**, 277 (1955).
36. van der Marel, L. G., J. van den Broek, and C. J. Gorter: *Physica* **23**, 361 (1957).
37. Bleaney, B., and R. S. Trenam: *Proc. Roy. Soc. (London)*, ser. A, **223**, 1 (1954).
38. Cooke, A. H., H. Meyer, and W. P. Wolf: *Proc. Roy. Soc. (London)*, ser. A, **237**, 404 (1956).
39. Croft, A. J., and R. H. B. Exell: *Proc. Roy. Soc. (London)*, ser. A, **262**, 110 (1961).
40. Bleaney, B., and D. J. E. Ingram: *Proc. Roy. Soc. (London)*, ser. A, **205**, 336 (1951).
41. Abragam, A., and M. H. L. Pryce: *Proc. Roy. Soc. (London)*, ser. A, **206**, 173 (1951).
42. Bijl, D.: *Physica* **16**, 269 (1950).
43. Steenland, M. J., L. C. van der Marel, D. de Klerk, and C. J. Gorter: *Physica* **15**, 906 (1949).
44. Dabbs, J. W. T., and L. D. Roberts: *Phys. Rev.* **95**, 970 (1954).
45. Miedema, A. R., J. van den Broek, H. Postma, and W. J. Huiskamp: *Physica* **25**, 1177 (1959).
46. Bleaney, B., K. D. Bowers, and D. J. E. Ingram: *Proc. Roy. Soc. (London)*, ser. A, **228**, 147 (1955).
47. Bleaney, B., and D. J. E. Ingram: *Proc. Roy. Soc. (London)*, ser. A, **208**, 143 (1951).
48. Malaker, S. F.: *Phys. Rev.* **84**, 133 (1951).
49. van den Broek, J., L. C. van der Marel, and C. J. Gorter: *Physica* **25**, 371 (1959).
50. Garrett, C. G. B.: *Proc. Roy. Soc. (London)*, ser. A, **206**, 242 (1951).
51. Penrose, R. P., and K. W. H. Stevens: *Proc. Phys. Soc. (London)*, ser. A, **63**, 29 (1950).
52. Griffiths, J. H. E., and J. Owen: *Proc. Roy. Soc. (London)*, ser. A, **213**, 459 (1952).
53. Hill, J. S., H. Meyer, and J. H. Milner: *Cryogenics* **2**, 170 (1962).
54. Svare, I., and G. Seidel: *Phys. Rev.* **134**, A172 (1964).
55. Giauque, W. F., E. W. Hornung, G. E. Brodale, and R. A. Fisher: *J. Chem. Phys.* **46**, 1804 (1967); **47**, 2685 (1967).
56. Bleaney, B., R. P. Penrose, and B. I. Plumpton: *Proc. Roy. Soc. (London)*, ser. A, **198**, 406 (1949).
57. Steenland, M. J., D. de Klerk, J. A. Beun, and C. J. Gorter: *Physica* **17**, 161 (1951).
58. Bagguley, D. M. S., and J. H. E. Griffiths: *Proc. Roy. Soc. (London)*, ser. A, **201**, 366 (1950).
59. Krishnan, K. S., and A. Mookherji: *Phys. Rev.* **54**, 533, 841 (1938).
60. Benzie, R. J., and A. H. Cooke: *Proc. Phys. Soc. (London)*, ser. A, **64**, 124 (1951).
61. Polder, D.: *Physica* **9**, 709 (1942).
62. Giauque, W. F., R. A. Fisher, E. W. Hornung, and G. E. Brodale: *J. Chem. Phys.* **48**, 3728, 3906 (1968).
63. Geballe, T. H., and W. F. Giauque: *J. Am. Chem. Soc.* **74**, 3513 (1952).
64. Miedema, A. R., H. van Kempen, T. Haseda, and W. J. Huiskamp: *Physica* **28**, 119 (1962).
65. Wittekoek, S., N. J. Poulis, and A. R. Miedema: *Physica* **30**, 1051 (1964).
66. Judd, B. R.: *Proc. Roy. Soc. (London)*, ser. A, **232**, 458 (1955).
67. Estle, T. L., H. R. Hart, Jr., and J. C. Wheatley: *Phys. Rev.* **112**, 1576 (1958).
68. Williamson, S. J., H. C. Praddaude, R. F. O'Brien, and S. Foner, *Phys. Rev.* **181**, 642 (1969).
69. Thornley, J. H. M.: *Phys. Rev.* **132**, 1492 (1963).
70. Finn, C. B. P., R. Orbach, and W. P. Wolf: *Proc. Phys. Soc. (London)*, ser. A. **77**, 261 (1961).
71. Hudson, R. P., and W. R. Hosler: *Phys. Rev.* **122**, 1417 (1961).
72. Leask, M. J. M., R. Orbach, M. J. D. Powell, and W. P. Wolf: *Proc. Roy. Soc. (London)*, ser. A, **272**, 371 (1963).
73. Daniels, J. M., and F. N. H. Robinson: *Phil. Mag.* **44**(7), 630 (1953).
74. Bailey, C. A.: *Phil. Mag.* **4**, 833 (1959); *Proc. Phys. Soc. (London)*, ser. A, **83**, 369 (1964).
75. Frankel, R. B., D. A. Shirley, and N. J. Stone: *Phys. Rev.* **140**, A1020 (1965); **143**, 334 (1966).
76. Hudson, R. P., and R. S. Kaeser: *Physics* **3**, 95 (1967).
77. Peverley, J. R., and P. H. E. Meijer: *Phys. Status Solidi* **23**, 353 (1967).
78. Wheatley, J. C.: *Ann. Acad. Sci. Fennicae*, ser. A, **VI-210**, 15 (1966).
79. Elliott, R. J., and K. W. H. Stevens: *Proc. Roy. Soc. (London)*, ser. A, (*a*) **215**, 437 (1952); (*b*) **218**, 553 (1953); (*c*) **219**, 387 (1953).
80. Bogle, G. S., A. H. Cooke, and S. Whitley: *Proc. Phys. Soc. (London)*, ser. A, **64**, 931 (1951).
81. Cooke, A. H., S. Whitley, and W. P. Wolf: *Proc. Phys. Soc. (London)*, ser. B, **68**, 415 (1955).

82. Daniels, J. M.: *Proc. Phys. Soc. (London)*, ser. A, **66**, 673 (1953).
83. Finkelstein, R., and A. Mencher: *J. Chem. Phys.* **21**, 472 (1953).
84. Johnson, C. E., and H. Meyer: *Proc. Roy. Soc. (London)*, ser. A, **253**, 199 (1959).
85. Baker, J. M.: *Phys. Rev.* **136**, A1633 (1964).
86. Meyer, H., and P. L. Smith: *Phys. Chem. Solids* **9**, 285 (1959).
87. Cooke, A. H., and H. J. Duffus: *Proc. Roy. Soc. (London)*, ser. A, **229**, 407 (1955).
88. Hudson, R. P., and B. W. Mangum: "Magnetic and Electric Resonance and Relaxation," p. 135, North-Holland Publishing Company, Amsterdam, 1963.
89. Scott, P. L., and C. D. Jeffries: *Phys. Rev.* **127**, 32 (1962); R. H. Ruby, H. Benoit, and C. D. Jeffries: *ibid.* 51.
90. Dieke, G. H., and L. Heroux: *Phys. Rev.* **103**, 1227 (1956).
91. Bleaney, B., H. E. D. Scovil, and R. S. Trenam: *Proc. Roy. Soc. (London)*, ser. A, **223**, 15 (1954).
92. Erickson, L. E.: *Phys. Rev.* **143**, 295 (1966).
93. Meyer, H.: *Phil. Mag.* **2**(8), 521 (1957).
94. Roberts, L. D., C. C. Sartain, and B. Borie: *Revs. Modern Phys.* **25**, 170 (1953).
95. Blok, J., D. A. Shirley, and N. J. Stone: *Phys. Rev.* **143**, 78 (1966).
96. Hebb, M. H., and E. M. Purcell: *J. Chem. Phys.* **5**, 338 (1937).
97. Bogle, G. S., and V. Heine: *Proc. Phys. Soc. (London)*, ser. A, **67**, 734 (1954).
98. Broer, L. J. F., and C. J. Gorter: *Physica* **10**, 621 (1943).
99. Bijl, D.: *Physica* **16**, 269 (1950).
100. Benzie, R. J., and A. H. Cooke: *Proc. Phys. Soc. (London)*, ser. A, **63**, 201, 213 (1950).
101. deVries, A. J., D. A. Curtis, J. W. M. Livius, A. J. van Duyneveldt, and C. J. Gorter: *Physica* **36**, 91 (1967).
102. Giauque, W. F., and D. P. MacDougall: *J. Am. Chem. Soc.* **57**, 1175 (1935).
103. van Dijk, H., and W. V. Auer: *Physica* **9**, 785 (1942).
104. Giauque, W. F., and D. P. MacDougall: *Phys. Rev.* **43**, 768 (1933).
105. van Dijk, H.: *Physica* **9**, 720 (1942).

5f-16. Susceptibility in High Magnetic Fields.[1] The magnetic behavior of materials in high magnetic fields ($H \gtrsim 25$ kOe) can yield useful information whether the material is paramagnetic (P), ferromagnetic (F), ferrimagnetic (F$_i$), antiferromagnetic (A), weakly ferromagnetic (WF), i.e., canted antiferromagnetism, or some combination of these.   In Table 5f-36, an abbreviated notation is used to give a capsule view of the type of behavior encountered.

For usual paramagnetics, a comparison with a Brillouin curve reflecting the energy-level structure and the paramagnetic saturation moment (PS) are of interest.   Special cases occur with ions whose ground state is nonmagnetic and whose magnetization is induced rather than aligned.   Closely related behavior also occurs as a residual high-field susceptibility (χ_{HF}) of the Van Vleck type (χ_{VV}) after saturation by alignment. Other phenomena embraced by residual χ_{HF} are electron band susceptibility (band), magnetic sublattice rotation or a less well organized aligning process against antiferromagnetic interactions, magnetization rotation in monocrystals against large anisotropies (K), approach to ferro- or ferrimagnetic saturation (FS) in polycrystalline samples, and puzzling jumps in χ vs. H above the antiferromagnetic Néel point ($d\chi/dH$; $T > T_N$).

A large number of substances undergo first-order transitions as a function of temperature at $H = 0$ between two different ordered magnetic states below the paramagnetic region, e.g., A–F, A–F$_i$, A–WF, etc.   A related case is the first-order change from paramagnetic (P) to an ordered state.   In all these instances a high field can induce a transition from the less magnetic state to the more magnetic one, as manifested by the externally measurable total magnetic moment.   The critical fields for these transitions are distinctly temperature-dependent; i.e., $H_c(T - T_t)$, where T_t is the transition temperature in vanishingly small field.   These transitions lend themselves to thermodynamic analysis using the Clausius-Clapeyron equation.   One should also note the spiral or helical antiferromagnetic or ferromagnetic-like states (A$_h$,F$_h$) and their field-induced variants such as fan (Fan) or other intermediate states

[1] Prepared by I. S. Jacobs, General Electric Research and Development Laboratories.

TABLE 5f-36. MAGNETIZATION BEHAVIOR IN HIGH FIELDS

Substance	Behavior	H_c, kOe, or χ_v, emu/cc	T, K	Ref.
CeBi	$A \rightarrow F_i \rightarrow P$	11, 43	1.3	1
CeSb	$A \rightarrow F_i^I \rightarrow F_i^{II} \rightarrow P$	7, 22, 38 $\quad H \parallel [100]$	1.5	2
$CoBr_2 \cdot 2H_2O$	$A \rightarrow F_i \rightarrow P$	13.7, 29.8 $H \parallel b$	4.2	3
$CoCl_2$	$SF \rightarrow P, d\chi/dH$	34	4.2	4
$CoCl_2 \cdot 2H_2O$	$A \rightarrow F_i \rightarrow P$	31.6, 46.0 $H \parallel b$	4.2	3, 5
CoF_2	$A \rightarrow WF$	130 $\qquad H \perp c$	4.2	6, 7
$\beta\text{-}Co(OH)_2$	$SF \rightarrow P, d\chi/dH$	35 $\qquad H \perp c$	4.2	8
$\beta\text{-}CoSO_4$	$A \rightarrow WF$	12.5 $\qquad H \parallel c$	4.2	9
Co_5Y	K	$K = 5.7 \times 10^7$ erg/cc	300	10
$CoUO_4$	$A \rightarrow P$	55	4.2	11
Cr	$d\chi/dH$	$\Delta\chi/\chi < 0, \approx 300$ kOe	295	12
Cr_2BeO_4	$A_h \rightarrow$ Fan (?)	30, $H \parallel b$; 48, $H \parallel c$	4.2	13
Cr_2CuO_4	χ_{HF}	1.3×10^{-4}	4.2	14
Cr_2FeO_4	χ_{HF}	3.9×10^{-4}	4.2	14
$CrK(SO_4)_2 \cdot 12H_2O$	PS	$3.0\mu_B/Cr^{3+}$	1.3	15
Cr_2MnO_4	χ_{HF}	3.6×10^{-4}	4.2	14
$CrNaS_2$	$A \rightarrow SF \rightarrow P$	20, 138 $\quad H \perp c$	~4	16, 17
Cr_2O_3	$A \rightarrow SF$	59 $\qquad H \parallel c$	4.2	18
Cr_2ZnSe_4	$A_h \rightarrow P$	64	4.2	19
$Cu_{0.8}Cd_{0.2}Fe_2O_4$	χ_{HF}	$7 \times 10^{-4}, > 110$ kOe	300	20
$CuCl_2 \cdot 2H_2O$	$A \rightarrow SF \rightarrow P, d\chi/dH$	7, 150	1	21
$Cu(NO_3)_2 \cdot 2.5H_2O$	$A \rightarrow P$	35 $\qquad H \parallel b$	1.2	22
Dy	K	$K = 4.9 \times 10^8$ erg/cc	4.2	23
Dy_3Al_2	$A(?) \rightarrow F, H_c(T - T_t)$	21, $T_t = 20°$	4.2	23a
DyAu	$A \rightarrow P$	22	4.2	24
$DyCu_2$	$A_h \rightarrow P$	≈ 20	4.2	25
$DyEr_3$	$A_h \rightarrow X \rightarrow F$	22, 45 $\quad H \parallel a$	4.2	26
DySb	$A \rightarrow F_i \rightarrow P$	22, 40 $\quad H \parallel [100]$	1.5	27
Er	$F_A \rightarrow X \rightarrow F$	20, 140 $\quad H \perp c$	4.2	26
ErSb	$A \rightarrow P, d\chi/dH$	25	1.5	27
EuTe	$SF \rightarrow P, d\chi/dH$	75	2.1	4, 28
Fe	χ_{HF}, Band	$3.2 \times 10^{-5}, 4.3 \times 10^{-5}$	4.2	29–31
$Fe_{0.83}Al_{0.17}$	χ_{HF}, Band	3.0×10^{-5}	4.2	32
$FeBr_2$	$A \rightarrow P; \chi_{VV}$	31.5 $\qquad H \parallel c$	2	33
$FeCO_3$	$A \rightarrow P$	≈ 150 $\qquad H \parallel c$	4.2	34, 6
$FeCl_2$	$A_{II,\perp} \rightarrow P, \chi_{HF}$	10.6, 100	4.2	35, 36
$FeCl_2 \cdot 2H_2O$	$A \rightarrow F_i \rightarrow P$	39, 46 $\quad H \parallel a$	4.0	37
FeGe	$A \rightarrow SF$	67	4.2	38
$FeK_3(CN)_6$	PS	$1.0\mu_B/Fe^{3+}$	1.3	39
$FeNH_4(SO_4) \cdot 12H_2O$	PS	$5.0\mu_B/Fe^{3+}$	1.3	15, 39
$Fe_{0.5}Ni_{0.5}Br_2$	$A \rightarrow X \rightarrow P$	35, 60 $\quad H \perp c$	4.2	40
FeO	$d\chi/dH, T \gtrsim T_N$	$H > 90, 150$	150–400	41
$\alpha\text{-}Fe_2O_3$	$A \parallel \rightarrow SF$	68 $\qquad H \parallel c$	77	42–44
	$A \perp \rightarrow WF$	130, 162 $\quad H \perp c$	120, 77	45, 46
FeRh	$A \rightarrow F, H_c(T - T_t)$	270, 230; $T_t = 330°$	77	47, 48
Fe_7S_8	K	$K \approx 10^7$ erg/cc	1.2–300	49
$GdAlO_3$	$A \rightarrow SF \rightarrow P, d\chi/dH$	11, 34 $\quad H \parallel b$	1	50, 51
GdAs	$SF \rightarrow P$	165	1.6	52
$GdCu_2$	$A_h \rightarrow X \rightarrow P$	50, 100	4.2	53, 25
$Gd_3Fe_5O_{12}$	χ_{HF}	$1 \times 10^{-3}, H > 70$ kOe	300	54
GdP	$SF \rightarrow P$	90	1.6	28
$Gd_2(SO_4)_3 \cdot 8H_2O$	PS	$7.0\mu_B/Gd^{3+}$	1.3	15
Ho	$A \rightarrow F$ (?)	106 $\qquad H \parallel c$	40	55, 23
HoAl	FS	$7.1 \pm 0.2\mu_B/Ho$	4.2	56
HoSb	$A \rightarrow F_i \rightarrow P$	17, 23 $\quad H \parallel [100]$	1.6	57
MnAs	$P \rightarrow F, H_c(T,p)$	$29(p = 0), 110(p = 1$ kb$)$	307, 329	58
$MnAu_2$	$A \rightarrow SF$	≈ 47	4.2	59, 60
$MnCO_3$	$d\chi/dH, T > T_N$	$\Delta\chi/\chi \approx 0.14, 150$ kOe	300	61
$MnCl_2$	$A \rightarrow SF \rightarrow P, d\chi/dH$	9, 32	1.3	62
$MnCl_2 \cdot 4H_2O$	$A \rightarrow SF \rightarrow P; PS$	7.5, 20.6 $\quad H \parallel c'$	0.26; 1.3	63, 64
$Mn_{1.9}Cr_{0.1}Sb$	$A \rightarrow F_i, H_c(T - T_t)$	$\approx 100, T_t = 305°$	265	65
MnF_2	$A \rightarrow SF$	93 $\qquad H \parallel c$	4.2	66, 67
$Mn_{1.31}Fe_{0.69}As$	$A \rightarrow F, H_c(T - T_t)$	64, $T_t = 283°$	301	68
Mn_3GaC	$A \rightarrow F, H_c(T - T_t)$	150, $T_t = 150°$	100	69
Mn_3Ge_2	$A \rightarrow WF, H_c(T - T_t)$	190, $T_t = 160°$	77	70, 65

TABLE 5f-36. MAGNETIZATION BEHAVIOR IN HIGH FIELDS (*Continued*)

Substance	Behavior	H_c, kOe, or χ_v, emu/cc	T, K	Ref.
MnK_2F_4	A → SF	55 $H \parallel c$	4.2	71
$Mn_{1-y}LaO_3$	χ_{HF}	8×10^{-4}, $0 \leq y \lesssim 0.05$	77	72
$MnLiPO_4$	A → SF	40 $H \parallel a$	4.2	13
MnO	χ_{HF}	$d\chi/dH > 0$, <50 kOe	4.2	73
MnO_2	$d\chi/dH$, $T \gtrless T_N$	$\Delta\chi/\chi \approx 0.6$, ≈ 250 kOe	300	61
		$\Delta\chi/\chi \approx 0$, <300 kOe	84, 4	74
Mn_3O_4	χ_{HF}	3.0×10^{-4}	4.2	75, 76
$MnRb_2F_4$	A → SF	51 $H \parallel c$	4.2	71
$MnSO_4$	A_h → SF → P, $d\chi/dH$	40, 250	1.6	77
	$d\chi/dH$, $T > T_N$	$\Delta\chi/\chi \approx 0.19$, 200 kOe	300	61
$MnSO_4 \cdot H_2O$	A → SF → P, $d\chi/dH$	25, 320	1.6	77
	$d\chi/dH$, $T > T_N$	$\Delta\chi/\chi \approx 0.12$, ≈ 250 kOe	300	61
$MnSn_2$	A → WF, $H_c(T - T_t)$	120, $T_t = 73°$	4.2	78
Ni	χ_{HF}, Band	3×10^{-5}, 1.7×10^{-5}	4.2	29–31
Ni_3Al	χ_{HF}, Band	9.8×10^{-5}, 200 kOe	4.2	79
Ni_3Ga	χ_{HF}, Band	17×10^{-5}, 200 kOe	4.2	79
$Ni(OH)_2$	A → P	50 $H \parallel c$	4.2	80
Pd	χ_{HF}, Band	8×10^{-5}	4.2, 0.1	81, 82
$R_3Ga_5O_{12}$	PS	....	2.6	11
$[R = Gd,Er,Yb]Sc_2In$	χ_{HF}, Band	6.5×10^{-5}, 40 kOe	1.2	83
Tb	K	$K = 4.5 \times 10^8$ erg/cc	4.2	23
Tb_3Al_2	$F_A(?)$ → F, $H_c(T - T_t)$	30, $T_t = 10°$	4	23a
TbAs	A → P	≈ 30	1.6	52
$TbCu_2$	A_h → P	22	4.2	25
Tm	F_i → F	15, 28	4.2	84, 85
TmSb	Induced K_{HF}	$M[100] < [110] < [111]$	1.6	86
$Yb_3Fe_5O_{12}$	χ_{HF}	1.4×10^{-3}	4.2	87
$ZrZn_2$	χ_{HF}, Band	7.3×10^{-5}	4.2	88

References for Table 5f-36

1. Tsuchida, T., and Y. Nakamura: *J. Phys. Soc. Japan* **22**, 942 (1967).
2. Busch, G., and O. Vogt: *Phys. Letters* **25A**, 449 (1967).
3. Narath, A.: *J. Phys. Soc. Japan* **19**, 2244 (1964).
4. Jacobs, I. S., and S. D. Silverstein: *Phys. Rev. Letters* **13**, 272 (1964).
5. Kobayashi, H., and T. Haseda: *J. Phys. Soc. Japan* **19**, 765 (1964).
6. Ozhogin, V. I.: *Soviet Phys.–JETP* **18**, 1156 (1964).
7. Ozhogin, V. I.: *J. Appl. Phys.* **39**, 1029 (1968).
8. Takada, T., Y. Bando, M. Kiyama, H. Miyamoto, and T. Sato: *J. Phys. Soc. Japan* **21**, 2726 (1966).
9. Kreines, N. M.: *Soviet Phys.–JETP* **13**, 534 (1961).
10. Hoffer, G., and K. Strnat: *IEEE Trans.* **MAG-2**, 487 (1966).
11. Guillot, M., and R. Pauthenet: *J. Appl. Phys.* **36**, 1003 (1965).
12. Stevenson, R.: *Can. J. Phys.* **44**, 283 (1966).
13. Ranicar, J. H., and P. R. Elliston: *Phys. Letters* **25A**, 720 (1967).
14. Jacobs, I. S.: *J. Phys. Chem. Solids* **15**, 54 (1960).
15. Henry, W. E.: *Phys. Rev.* **88**, 559 (1952).
16. Blazey, K. W., and H. Rohrer: *Helv. Phys. Acta* **41**, 391 (1968).
17. Bongers, P. F., C. F. Van Bruggen, J. Koopstra, W. Omloo, G. A. Wiegers, and F. Jellinek: *J. Phys. Chem. Solids* **29**, 977 (1968).
18. Foner, S., and S. L. Hou: *J. Appl. Phys.* **33**, 1289 (1962).
19. Plumier, R.: *J. phys. radium* **27**, 213 (1966).
20. Rode, V. E., A. V. Vedyaev, and B. N. Krainov: *Soviet Phys.–Solid State* **5**, 1277 (1963).
21. van der Sluijs, J. C. A., B. A. Zweers, and D. de Klerk: *Phys. Letters* **24A**, 637 (1967).
22. Myers, B. E., L. Berger, and S. A. Friedberg: *J. Appl. Phys.* **40**, 1149 (1969).
23. Rhyne, J. J., S. Foner, E. J. McNiff, Jr., and R. Doclo: *J. Appl. Phys.* **39**, 892 (1968).
23a. Barbara, B., C. Bècle, J. L. Féron, R. Lemaire, and R. Pauthenet: *Compt. rend.* **267**, B244 (1968).
24. Kaneko, T., *J. Phys. Soc. Japan* **25**, 905 (1968).
25. Sherwood, R. C., H. J. Williams, and J. H. Wernick: *J. Appl. Phys.* **35**, 1049 (1964).
26. Bozorth, R. M., R. J. Gambino, and A. E. Clark: *J. Appl. Phys.* **39**, 883 (1968).
27. Busch, G., and O. Vogt: *J. Appl. Phys.* **39**, 1334 (1968).

28. Busch, G., P. Junod, P. Schwob, O. Vogt, and F. Hulliger: *Phys. Letters* **9**, 7 (1964).
29. Freeman, A. J., N. A. Blum, S. Foner, R. B. Frankel, and E. J. McNiff, Jr.: *J. Appl. Phys.* **37**, 1338 (1966).
30. Stoelinga, J. H. M., and R. Gersdorf: *Phys. Letters* **19**, 640 (1966).
31. Foner, S., A. J. Freeman, N. A. Blum, R. B. Frankel, E. J. McNiff, Jr., and H. C. Praddaude: *Phys. Rev.* **181**, 863 (1969).
32. Wakiyama, T., and J. P. Rebouillat: Physical Society of Japan, April 1968 Meeting.
33. Jacobs, I. S., and P. E. Lawrence: *J. Appl. Phys.* **35**, 996 (1964).
34. Jacobs, I. S.: *J. Appl. Phys.* **34**, 1106 (1963); *USAF Tech. Document. Rept. No.* ML-TDR-64-58; and unpublished work.
35. Jacobs, I. S., and P. E. Lawrence: *Phys. Rev.* **164**, 866 (1967); and earlier refs. cited therein.
36. Carrara, P., J. de Gunzbourg, and Y. Allain: *J. Appl. Phys.* **40**, 1035 (1969).
37. Narath, A.: *Phys. Rev.* **139**, A1221 (1965).
38. Beckman, O., H. Schwartz, and K. Å. Blom: *Bull. Am. Phys. Soc.* **13**, 461 (1968).
39. Henry, W. E.: *Phys. Rev.* **106**, 465 (1957).
40. Hirone, T., T. Kamigaki, and H. Yoshida: Physical Society of Japan, April 1968 Meeting.
41. Zavadskii, E. A., I. G. Fakidov, and N. Ya. Samarin: *Soviet Phys.–JETP* **20**, 558 (1965).
42. Foner, S.: *Proc. Intern. Conf. Magnetism, Nottingham*, p. 438, The Institute of Physics and the Physical Society, London, 1964.
43. Besser, P. J., and A. H. Morrish: *Phys. Letters* **13**, 389 (1964).
44. Kaneko, T., and S. Abe: *J. Phys. Soc. Japan* **20**, 2001 (1965).
45. Voskanyan, R. A., R. Z. Levitin, and V. A. Shchurov: *Soviet Phys.–JETP* **26**, 302 (1968).
46. Beyerlein, R. A., and I. S. Jacobs: *Bull. Am. Phys. Soc.* **14**, 349 (1969).
47. Zavadskii, E. A., and I. G. Fakidov: *Soviet Phys.–Solid State* **9**, 103 (1967); H_c linear with $|T - T_c|$.
48. McKinnon, J. B., D. A. Melville, and E. W. Lee: IPPS Solid State Physics Conference, Manchester, January, 1968; H_c parabolic with $|T - T_c|$.
49. Adachi, K., and K. Sato: *J. Appl. Phys.* **39**, 1343 (1968); earlier refs. cited therein.
50. Cashion, J. D., A. H. Cooke, J. F. B. Hawkes, M. J. M. Leask, T. L. Thorp, and M. R. Wells: *J. Appl. Phys.* **39**, 1360 (1968).
51. Blazey, K. W., and H. Rohrer: *Phys. Rev.* **173**, 574 (1968).
52. Busch, G., O. Vogt, and F. Hulliger: *Phys. Letters* **15**, 301 (1965).
53. Jacobs, I. S., and J. S. Kouvel: Unpublished.
54. Rode, V. E., and A. V. Vedyaev: *Soviet Phys.–JETP* **18**, 286 (1964).
55. Flippen, R. B.: *J. Appl. Phys.* **35**, 1047 (1964).
56. Bècle, C., R. Lemaire, and R. Pauthenet: *Compt. rend.* **266**, B994 (1968).
57. Busch, G., P. Schwob, and O. Vogt: *Phys. Letters* **23**, 636 (1966).
58. DeBlois, R. W., and D. S. Rodbell: *Phys. Rev.* **130**, 1347 (1963).
59. Jacobs, I. S., J. S. Kouvel, and P. E. Lawrence: *J. Phys. Soc. Japan* **17**, suppl. BI, 157 (1962).
60. Sato, K., T. Hirone, H. Watanabe, S. Maeda, and K. Adachi: *J. Phys. Soc. Japan* **17**, suppl. BI, 160 (1962).
61. Stevenson, R.: *Can. J. Phys.* **40**, 1385 (1962).
62. Giauque, W. F., G. E. Brodale, R. A. Fisher, and E. W. Hornung: *J. Chem. Phys.* **42**, 1 (1965); W. F. Giauque, R. A. Fisher, E. W. Hornung, R. A. Butera, and G. E. Brodale: *ibid.*, 9.
63. Rives, J. E.: *Phys. Rev.* **162**, 491 (1967).
64. Henry, W. E.: *Phys. Rev.* **94**, 1146 (1954).
65. Flippen, R. B., and F. J. Darnell: *J. Appl. Phys.* **34**, 1094 (1963).
66. Jacobs, I. S.: *J. Appl. Phys.* **32**, 61S (1961).
67. deGunzbourg, J., and J. P. Krebs: *J. phys. radium* **29**, 42 (1968).
68. Flippen, R. B.: *Phys. Rev. Letters* **21**, 1079 (1968).
69. Bouchaud, J. P., R. Fruchart, R. Pauthenet, M. Guillot, H. Bartholin, and F. Chaissé: *J. Appl. Phys.* **37**, 971 (1966).
70. Zavadskii, E. A., and I. G. Fakidov: *Soviet Phys.–JETP* **24**, 887 (1967).
71. Breed, D. J.: *Physica* **37**, 35 (1967).
72. Jacobs, I. S., and W. L. Roth: *Bull. Am. Phys. Soc.* **8**, 213 (1963); *USAF ASD Tech. Rept.* 61-630, February, 1962.
73. Bloch, D., J. L. Féron, R. Georges, and I. S. Jacobs: *J. Appl. Phys.* **38**, 1474 (1967).
74. van der Sluijs, J. C. A.: Thesis, Leiden University, 1967.
75. Jacobs, I. S.: *J. Phys. Chem. Solids* **11**, 1 (1959).
76. Moruzzi, V. L.: *J. Appl. Phys.* **32**, 59S (1961).

77. Allain, Y., J. P. Krebs, and J. de Gunzbourg: *J. Appl. Phys.* **39**, 1124 (1968).
78. Kouvel, J. S., and I. S. Jacobs: *J. Appl. Phys.* **39**, 467 (1968).
79. de Boer, F. R., C. J. Schinkel, J. Biesterbos, and S. Proost: *J. Appl. Phys.* **40**, 1049 (1969).
80. Takada, T., Y. Bando, M. Kiyama, H. Miyamoto, and T. Sato: *J. Phys. Soc. Japan* **21**, 2745 (1966).
81. Foner, S., and E. J. McNiff, Jr.: *Phys. Rev. Letters* **19**, 1438 (1967).
82. Chouteau, G., R. Fourneaux, K. Gobrecht, and R. Tournier: *Phys. Rev. Letters* **20**, 193 (1968).
83. Gardner, W. E., T. F. Smith, B. W. Howlett, C. W. Chu, and A. Sweedler: *Phys. Rev.* **166**, 577 (1968).
84. Foner, S., M. Schieber, and E. J. McNiff, Jr.: *Phys. Rev. Letters*, **25**, 321 (1967).
85. Legvold, S., and D. B. Richards: *Bull. Am. Phys. Soc.* **13**, 440 (1968).
86. Vogt, O., and B. R. Cooper: *J. Appl. Phys.* **39**, 1202 (1968).
87. Clark, A. E., and E. Callen: *J. Appl. Phys.* **39**, 5972 (1968).
88. Foner, S., E. J. McNiff, Jr., and V. Sadagopan: *Phys. Rev. Letters* **19**, 1233 (1967).

(X). The notation (X) is also invoked when the nature of a state is not known, but its field region is well described.

Lastly, there is the particularly rich area for high-field studies in mapping out the magnetic phase diagrams of antiferromagnetics. As the field along the moment axis increases, there is often a transition from the antiferromagnetic state (A) to the transverse spin-flopped state (SF). This critical field is a measure of the anisotropy. For a further increase of the field, this state gives way to the nearly saturated P state. Sometimes this SF → P behavior is nonlinear, i.e., $(d\chi/dH)_{H<H_c} = f(H)$, arising from one or more higher-order interactions. In other cases, either in the presence of strong anisotropy or nearer to the Néel point, the transition A → P is found, but with differing steepness during the transition, e.g., nearly first order, or distinctly second or higher order, respectively.

Workers in this rapidly expanded area are cautioned that pulsed field measurements tend to be isentropic rather than isothermal, which condition can seriously modify the results obtained.

5f-17. Demagnetizing and Form Factors.[1] The magnetic field strength **H** in a magnetized body is given by

$$\mathbf{H} = \mathbf{H}_a + \mathbf{H}_D$$

where $\mathbf{H}_a$ is the applied field due to outside sources and $\mathbf{H}_D$, called the demagnetizing field, is due to the effective magnetic poles in the body itself, which occur at points where the intensity of magnetization **M** has nonvanishing divergence.

In a uniformly magnetized ellipsoid the effective poles are all on the surface, and they contribute a uniform demagnetizing field, with components along the principal axes, $i = 1, 2,$ and 3, given by

$$(H_D)_i = -N_i M_i$$

where N_i is a constant, known as the demagnetizing factor for magnetization along ith principal axis. For any ellipsoid

$$N_1 + N_2 + N_3 = 4\pi$$

In terms of a set of axes not along the principal axes of the ellipsoid, the demagnetizing field is described by

$$(H_D)_k = -\sum_j N_{kj} M_j$$

[1] Prepared by W. J. Carr, Jr., Westinghouse Research and Development Center.

where the tensor components N_{kj} are easily calculated from the three demagnetizing factors.

In the general case of a nonellipsoidal body, or for nonuniform magnetization, the demagnetizing field varies from point to point, and the above description does not apply; however, many shapes can be approximated by some particular ellipsoid, and in this way if the magnetization is nearly uniform, approximate demagnetizing factors can be defined. For example, a rod is approximated by an elongated prolate ellipsoid of revolution, and a disk by a flattened oblate ellipsoid.

For ellipsoids of revolution, i.e., spheroids, in which the ratio of the long to the short axis is m, the demagnetizing factor N for magnetization along the long axis is as follows:

Prolate Spheroid

$$N = \frac{4\pi}{m^2 - 1} \left[\frac{m}{\sqrt{m^2 - 1}} \log_e (m + \sqrt{m^2 - 1}) - 1 \right]$$

Oblate Spheroid

$$N = \frac{2\pi}{m^2 - 1} \left(\frac{m^2}{\sqrt{m^2 - 1}} \arcsin \frac{\sqrt{m^2 - 1}}{m} - 1 \right)$$

Since for the prolate spheroid the two short axes are equal, and for the oblate case the two long axes are equal, then for either case the demagnetizing factor along the short axis is easily obtained by using the above formulas and the expression for the sum. Formulas and tables for the general ellipsoid have been given by Osborn[1] and Stoner.[2]

For rods magnetized in a uniform field in their long direction, the effective demagnetizing factor at the middle of the rod depends to some extent upon permeability, since permeability affects the distribution of magnetic flux along the rod. Demagnetizing factors for rods have been given by Bozorth and Chapin[3] (see Table 5f-37).

TABLE 5f-37. DEMAGNETIZING FACTORS $N/4\pi$ FOR RODS* AND ELLIPSOIDS
MAGNETIZED PARALLEL TO THE LONG AXIS

Dimensional ratio (length/diameter)	Rod	Prolate ellipsoid	Oblate ellipsoid
0	1.0	1.0	1.0
1	0.27	0.3333	0.3333
2	0.14	0.1735	0.2364
5	0.040	0.0558	0.1248
10	0.0172	0.0203	0.0696
20	0.00617	0.00675	0.0369
50	0.00129	0.00144	0.01472
100	0.00036	0.000430	0.00776
200	0.000090	0.000125	0.00390
500	0.000014	0.0000236	0.001567
1000	0.0000036	0.0000066	0.000784
2000	0.0000009	0.0000019	0.000392

* Values for the rods were obtained empirically for high-permeability materials.

Demagnetizing fields also play a role in magnetostriction, where the strain in a magnetized body is to some extent dependent upon its shape, the effect being known as the magnetostriction form effect. As calculated by Becker,[4] in an elastically isotropic

[1] J. A. Osborn, *Phys. Rev.* **67**, 351 (1945).
[2] E. C. Stoner, *Phil. Mag.* **36** [7], 803 (1945).
[3] R. M. Bozorth and D. M. Chapin, *J. Appl. Phys.* **13**, 320 (1942).
[4] R. Becker, *Z. Physik* **87**, 547 (1934).

prolate spheriod magnetized parallel to the long axis, the observed longitudinal magnetostriction is greater than that of an infinite rod by the form effect strain,

$$\Delta\lambda = \frac{1}{2} M^2 N \left(\frac{1}{3k} + \frac{a}{2G}\right)$$

where, again, M is the intensity of magnetization, N the demagnetizing factor in the long direction, k the elastic compression modulus, G the shear modulus, and

$$a = -\frac{1}{N}\frac{\partial N}{\partial e_{11}}$$

e_{11} being the strain in the longitudinal direction. Values for a are given in Table 5f-38. $\Delta\lambda$ is usually important only for large values of $M^2 N$.

TABLE 5f-38. MAGNETOSTRICTION FORM FACTORS
(See equation for $\Delta\lambda$)

m = length/diameter	N	a
1	4.19	0.80
2	2.18	1.07
3	1.37	1.23
4	0.95	1.31
5	0.70	1.38
10	0.255	1.53
15	0.135	1.60
20	0.085	1.63
30	0.043	1.68

The volume form effect strain in the above case is

$$\Delta\omega = \frac{M^2 N}{2k}$$

and the transverse strain along the two equivalent short axes is simply $(\Delta\omega - \Delta\lambda)/2$. Other cases have been considered by Gersdorf[1] and Carr.[2]

Brown[3] has shown that actually the strains calculated by Becker are only average strains, and in reality the form effect is not a uniform strain but changes from point to point within the ellipsoid. As calculated by Gersdorf,[4] the form-effect strain at the center of a sphere is about twice as great as the average value.

In the case of an elastically anisotropic single crystal, similar but somewhat more complicated expressions apply for the form effect. Of particular interest is the form effect in a disklike oblate single-crystal spheroid magnetized in the equatorial plane. In general, the crystal axes and the principal axes of the ellipsoid will not coincide. For a cubic crystal the average form-effect components to the strain tensor are,[5] for $i = 1, 2, 3$,

$$(e_{ii})_{fe} = \text{const} + (h_1)_{fe}\,\alpha_i{}^2$$
$$(e_{ij})_{fe} = \text{const} + (h_2)_{fe}\,\alpha_i\alpha_j$$

[1] R. Gersdorf, *Physica* **26**, 553 (1960).
[2] W. J. Carr, Jr., "Handbuch der Physik," vol. 18/2, S. Flügge, ed., Springer-Verlag OHG, Berlin, 1966.
[3] W. F. Brown, Jr., *Revs. Modern Phys.* **25**, 131 (1953).
[4] R. Gersdorf, *Physica* **26**, 553 (1960).
[5] W. J. Carr, Jr.: "Handbuch der Physik," vol. 18/2, S. Flügge, ed., Springer-Verlag OHG, Berlin, 1966.

where the components are referred to the cubic crystal axes, and α_i is the direction cosine of magnetization with respect to the ith axis,

$$(h_1)_{fe} = -\frac{F}{3}\frac{M^2}{c_{11} - c_{12}}$$

$$(h_2)_{fe} = -\frac{F}{6}\frac{M^2}{c_{44}}$$

with c_{11}, c_{12}, and c_{44} the elastic constants; and F defined by

$$F = \frac{3\pi}{2\epsilon^4}\left[(1 - \epsilon^2)(3 + 2\epsilon^2) - \frac{3}{\epsilon}(1 - \epsilon^2)^{\frac{1}{2}} \arcsin \epsilon \right]$$

ϵ being the eccentricity, defined in terms of the principal axes a, b, c of the ellipsoid by

$$a = b = \frac{c}{(1 - \epsilon^2)^{\frac{1}{2}}}$$

The constants in the equations for the form-effect strains depend upon the orientation of the ellipsoid with respect to crystal axes, but are independent of the direction of magnetization, which takes any direction within the plane of the disk.

5g. Electrochemical Information

GORDON ATKINSON

University of Maryland

Editor's note.   The symbols used in this section are essentially those recommended by the International Union of Pure and Applied Chemistry (cf. Christiansen, 1960).

Conductance data and transference numbers are taken from Harned and Owen (1958), Kortüm, and Robinson and Stokes (1959). Additional data can be found in these books, in Conway (1952) and in Parsons (1955).

Much information concerning a wide variety of electrochemical phenomena including conductance can be found in the general tables: "International Critical Tables," 1926–1933; "International Annual Tables," 1950——; and Landolt-Börnstein 1950——.

Additional information may be derived from tables of polarographic data compiled by Kolthoff and Lingane (1952) and by von Stackelberg (1950).

Standard electromotive forces of half cells were taken from Latimer (1952).   Many additional data are available in his tables.   Note especially his table on page 345 for alkaline solutions.   Other values of E^o may be calculated from the free-energy data of Rossini et al. (1952).

Activity coefficients were selected from extensive tables in Harned and Owen (1958), Kortüm and Bockris (1951), and Robinson and Stokes (1959).   Additional data can be found in these sources and in Robinson and Stokes (1949) and in Conway (1952).

Dissociation constants are from Harned and Owen (1958) and Hood, Redlich, and Reilly (1954).   Constants for many other equilibria can be found in Harned and Owen

(1958), Redlich (1946), Bjerrum et al. (1957), Scudder (1914), and can be derived from thermodynamic data of Rossini et al. (1952) and of Latimer (1952).

The molal heat content (enthalpy) data were taken from Harned and Owen (1958).

Standard entropies of ions were taken from Latimer (1952) and from Powell and Latimer (1951). Additional values can be found in those sources and in Robinson and Stokes (1959) and in Kortüm and Bockris (1951).

Electrochemical data of many other kinds have been tabulated by Robinson and Stokes (1959), Harned and Owen (1958), and Kortüm and Bockris (1951). Information especially useful for the electrometric determination of pH has been assembled by Bates (1954). Polarographic data have been collected by Kolthoff and Lingane (1952) and by von Stackelberg (1950).

Further data on the rates of fast equilibria can be found in Caldin (1963) and Stuehr and Yeager (1965).

Notes on Abbreviations, Symbols, and Terminology Used in Table 5g-3 and in the Discussion Which Follows. The letters (g), (l), (s), and (aq) denote *gas, liquid, solid*, and *aqueous solution*, respectively. These symbols are often omitted for substances which are in their most familiar states.

Pt. Many authors writing symbols for electrodes include the symbol "Pt" whenever no solid conducting element appears elsewhere in the formulation of the electrode. Its purpose is to remind the reader that some connection (not necessarily platinum) to the external portion of the circuit must be provided. The symbol is not essential and has been omitted in Table 5g-3.

Cathode and *Anode.* The words *cathode* and *anode* are not essential for a discussion of electrochemical cells. They are not used in the explanation which follows. Because some writers use the words frequently, their meanings must be understood. At the cathode reduction occurs; at the anode oxidation occurs. In the external portion of the circuit electrons flow from anode to cathode, whereas the "positive current" is said to flow in the external conductor from cathode to anode. Within the cell the "positive current" flows from anode to cathode, thus completing the circuit. The current within the cell consists of both positive ions moving from anode to cathode and negative ions moving from cathode to anode. Note that in an electrochemical cell operating spontaneously the anode is the negative pole and the cathode is the positive pole. For a somewhat more detailed discussion of the words, see Daniels and Alberty (1955).

E denotes the electromotive force (emf) of a cell or half cell.

E^o denotes the standard emf defined below.

ΔG denotes the increase in Gibbs free energy for the reaction specified.

ΔG^o denotes the standard increase in free energy. It is related to E^o by an equation similar to Eq. (5g-1).

TABLE 5g-1. EQUIVALENT CONDUCTANCES AND CATION TRANSFERENCE
NUMBERS OF ELECTROLYTES IN AQUEOUS SOLUTIONS AT 25°C
(Λ in cm^2 ohm^{-1} equivalent^{-1}; N in equivalent liter^{-1})

	N	0	0.001	0.01	0.02	0.05	0.1
HCl	Λ	426.16	421.36	412.00	407.24	399.09	391.32
	t_+	0.8209	......	0.8251	0.8266	0.8292	0.8314
LiCl	Λ	115.03	112.40	107.32	104.65	100.11	95.86
	t_+	0.3364	......	0.3289	0.3261	0.3211	0.3168
NaCl	Λ	126.45	123.74	118.51	115.76	111.06	106.74
	t_+	0.3963	......	0.3918	0.3902	0.3876	0.3854
KCl	Λ	149.86	146.95	141.27	138.34	133.37	128.96
	t_+	0.4906	......	0.4902	0.4901	0.4899	0.4898
NH$_4$Cl	Λ	149.7	......	141.28	138.33	133.29	128.75
	t_+	0.4909	......	0.4907	0.4906	0.4905	0.4907
KBr	Λ	151.9	......	143.43	140.48	135.68	131.39
	t_+	0.4849	......	0.4833	0.4832	0.4831	0.4833
NaI	Λ	126.94	124.25	119.24	116.70	112.79	108.78
KI	Λ	150.38	......	142.18	139.45	134.97	131.11
	t_+	0.4892	......	0.4884	0.4883	0.4882	0.4883
KNO$_3$	Λ	144.96	141.84	132.82	132.41	126.31	120.40
	t_+	0.5072	......	0.5084	0.5087	0.5093	0.5103
KHCO$_3$	Λ	118.00	115.34	110.08	107.22		
NaO$_2$C$_2$H$_3$	Λ	91.0	88.5	83.76	81.24	76.92	72.80
	t_+	0.5507	......	0.5537	0.5550	0.5573	0.5594
NaO$_2$C(CH$_3$)$_2$CH$_3$	Λ	82.70	80.31	75.76	73.39	69.32	65.27
NaOH	Λ	247.8	244.7	238.0			
AgNO$_3$	Λ	133.36	130.51	124.76	121.41	115.24	109.14
	t_+	0.4643	......	0.4648	0.4652	0.4664	0.4682
$\frac{1}{2}$MgCl$_2$	Λ	129.40	124.11	114.55	110.04	103.08	97.10
$\frac{1}{2}$CaCl$_2$	Λ	135.84	130.36	120.36	115.65	108.47	102.46
	t_+	0.4380	......	0.4264	0.4220	0.4140	0.4060
$\frac{1}{2}$SrCl$_2$	Λ	135.80	130.33	120.29	115.54	108.25	102.19
$\frac{1}{2}$BaCl$_2$	Λ	139.98	134.34	123.94	119.09	111.48	105.19
$\frac{1}{2}$Na$_2$SO$_4$	Λ	129.9	124.15	112.44	106.78	97.75	89.98
	t_+	0.386	......	0.3848	0.3836	0.3829	0.3828
$\frac{1}{2}$CuSO$_4$	Λ	133.6	115.26	83.12	72.20	59.05	50.58
$\frac{1}{2}$ZnSO$_4$	Λ	132.8	115.53	84.91	74.24	61.20	52.64
$\frac{1}{3}$LaCl$_3$	Λ	145.8	137.0	121.8	115.3	106.2	99.1
	t_+	0.477	......	0.4625	0.4576	0.4482	0.4375
$\frac{1}{3}$K$_3$Fe(CN)$_6$	Λ	174.5	163.1				
$\frac{1}{4}$K$_4$Fe(CN)$_6$	Λ	184.5	167.24	134.83	122.82	107.70	97.87

TABLE 5g-2. LIMITING EQUIVALENT CONDUCTANCES OF IONS IN WATER IN INFINITELY DILUTE SOLUTION
(cm^2 ohm^{-1} $equivalent^{-1}$)

Ion	$t °C$	Λ_0	Ion	$t °C$	Λ_0
H^+..................	15	300.6	OH^-...............	25	197.6
	25	349.8	Cl^-.................	15	61.42
	35	397.0		25	76.34
Li^+..................	25	38.69		35	92.21
Na^+.................	15	39.75	Br^-................	15	63.3
	25	50.11		25	78.3
	35	61.53		35	94.2
K^+..................	15	59.66	I^-..................	25	76.8
	25	73.50	NO_3^-..............	25	71.4
	35	88.21	ClO_4^-.............	25	68.0
NH_4^+..............	25	73.4	HCO_3^-.............	25	44.5
Ag^+................	25	61.92	$CH_3CO_2^-$..........	25	40.9
Tl^+.................	25	74.7	$ClCH_2CO_2^-$........	25	39.8
$\frac{1}{2} Mg^{++}$.............	25	53.06	$CH_3CH_2CO_2^-$.......	25	35.8
$\frac{1}{2} Ca^{++}$.............	25	59.50	$CH_3(CH_2)_2CO_2^-$....	25	32.6
$\frac{1}{2} Sr^{++}$.............	25	59.46	$C_6H_5CO_2^-$.........	25	32.3
$\frac{1}{2} Ba^{++}$.............	25	63.64	$HC_2O_4^-$...........	25	40.2
$\frac{1}{2} Cu^{++}$.............	25	54	$\frac{1}{2} C_2O_4^-$............	25	74.2
$\frac{1}{2} Zn^{++}$.............	25	53	$\frac{1}{2} SO_4^-$.............	25	80
$\frac{1}{3} La^{3+}$.............	25	69.5	$\frac{1}{3} Fe(CN)_6^{3-}$........	25	101
$\frac{1}{3} Co(NH_3)_6^{3+}$.......	25	102	$\frac{1}{4} Fe(CN)_6^{4-}$........	25	111

N denotes the number of Faradays (F) of electricity. N may have any positive value. For simplicity it is arbitrarily chosen as unity for all of Table 5g-3 and for each example of its use.

Significance of Table 5g-3 and Conventions. When current passes through a reversible electrolytic cell, oxidation occurs at one electrode and reduction at the other. When the direction of the current is reversed, the chemical reaction is reversed and oxidation and reduction exchange places. While no current is passing through the cell, a reversible emf can be measured with a potentiometer. Electromotive forces of cells are important thermodynamic data since

$$\Delta G = -NFE \qquad (5g\text{-}1)$$

It is conventional to associate ΔG with the reaction which occurs when N Faradays, i.e., *ca.* N 96,500 coulombs, of positive electricity is passed through the cell from left to right. It is conventional to write E as positive if this current flows spontaneously from left to right through the cell, i.e., if electrons are caused by the cell reaction to move in the external part of the circuit from left to right. According to this convention E of the cell is positive if the right-hand electrode is positive with respect to the left-hand electrode. If the cell is rewritten in the reverse order, the algebraic sign of its emf is changed. [The negative sign in Eq. (5g-1) is a consequence of these two conventions.] Examples:

$$\text{H}_2,\ \text{HCl(aq)},\ \text{Cl}_2 \qquad E^o = 1.3595 \text{ volts at } 25°\text{C} \qquad (5g\text{-}2)$$
$$\text{Cl}_2,\ \text{HCl(aq)},\ \text{H}_2 \qquad E^o = -1.3595 \text{ volts at } 25°\text{C} \qquad (5g\text{-}3)$$

In these equations the symbol ^{o} (read "standard") indicates that all the cell reactants and products are in their standard states, i.e., each is at unit activity. Actually there are no criteria for the decision that the activity of any single ion (a_+ of H^+ or a_- of Cl^-, in this example) is unity. The emf of the cell is completely determined, however, by a product of ion activities, in this example by

$$a_+ a_- = a_2 \qquad (5g\text{-}4)$$

The activity a_2 of the solute, e.g., HCl, can be measured and is known for many electrolytes as functions of their concentration.

The emf of a cell may be regarded as the net result of two opposing half-cell reactions, one at each electrode. Each of these two half reactions may be thought of as having a tendency to liberate electrons, or each may be considered to possess a tendency to consume electrons. The half reaction having the greater tendency to acquire electrons forces the other half reaction to surrender them, or according to the alternative point of view, the half reaction having the greater tendency to liberate electrons forces the other to accept them. These two points of view are designated below as plan A and plan B, respectively. Either plan is quite correct and general. Example: Consider the cell of Eq. (5g-2), H_2, HCl(aq), Cl_2. At the left-hand electrode the half reaction, for $N = 1$, may be considered to be either (a) or (b); thus

$$\begin{array}{cc} \text{Plan } A & \text{Plan } B \\ \tfrac{1}{2}\text{H}_2 \to \text{H}^+ + \Theta \quad (a) & \Theta + \text{H}^+ \to \tfrac{1}{2}\text{H}_2 \quad (b) \end{array} \qquad (5g\text{-}5a, 5b)$$

The opposing half-cell reaction (at the other electrode) is written

$$\text{Cl}^- \to \tfrac{1}{2}\text{Cl}_2 + \Theta \quad (c) \qquad \Theta + \tfrac{1}{2}\text{Cl}_2 \to \text{Cl}^- \quad (d) \qquad (5g\text{-}5c, 5d)$$

Since E^o of cell (2) is positive it is obvious that half reaction (c) has less tendency to proceed than half reaction (a), and that (d) has more tendency to proceed than (b). The difference in each case is 1.3595 volts.

TABLE 5g-3. STANDARD ELECTROMOTIVE FORCES OF HALF CELLS IN WATER AT 25°C

(E^o in absolute volts relative to the standard hydrogen electrode)

Acid solution

Half-cell reaction	Electrode	E^o electrode potential
$\ominus + Li^+ \rightarrow Li$	Li^+, Li	-3.045
$\ominus + K^+ \rightarrow K$	K^+, K	-2.925
$\ominus + Rb^+ \rightarrow Rb$	Rb^+, Rb	-2.925
$\ominus + \frac{1}{2}Ba^{++} \rightarrow \frac{1}{2}Ba$	Ba^{++}, Ba	-2.90
$\ominus + \frac{1}{2}Sr^{++} \rightarrow \frac{1}{2}Sr$	Sr^{++}, Sr	-2.89
$\ominus + \frac{1}{2}Ca^{++} \rightarrow \frac{1}{2}Ca$	Ca^{++}, Ca	-2.87
$\ominus + Na^+ \rightarrow Na$	Na^+, Na	-2.714
$\ominus + \frac{1}{2}Mg^{++} \rightarrow \frac{1}{2}Mg$	Mg^{++}, Mg	-2.37
$\ominus + \frac{1}{3}Am^{3+} \rightarrow \frac{1}{3}Am$	Am^{3+}, Am	-2.32
$\ominus + \frac{1}{3}Pu^{3+} \rightarrow \frac{1}{3}Pu$	Pu^{3+}, Pu	-2.07
$\ominus + \frac{1}{4}Th^{4+} \rightarrow \frac{1}{4}Th$	Th^{4+}, Th	-1.90
$\ominus + \frac{1}{3}Np^{3+} \rightarrow \frac{1}{3}Np$	Np^{3+}, Np	-1.86
$\ominus + \frac{1}{2}Be^{++} \rightarrow \frac{1}{2}Be$	Be^{++}, Be	-1.85
$\ominus + \frac{1}{3}U^{3+} \rightarrow \frac{1}{3}U$	U^{3+}, U	-1.80
$\ominus + \frac{1}{3}Al^{3+} \rightarrow \frac{1}{3}Al$	Al^{3+}, Al	-1.66
$\ominus + \frac{1}{2}Ti^{++} \rightarrow \frac{1}{2}Ti$	Ti^{++}, Ti	-1.63
$\ominus + \frac{1}{2}Zn^{++} \rightarrow \frac{1}{2}Zn$	Zn^{++}, Zn	-0.763
$\ominus + U^{4+} \rightarrow U^{3+}$	U^{4+}, U^{3+}	-0.61
$\ominus + \frac{1}{2}Fe^{++} \rightarrow \frac{1}{2}Fe$	Fe^{++}, Fe	-0.440
$\ominus + Eu^{3+} \rightarrow Eu^{++}$	Eu^{3+}, Eu^{++}	-0.43
$\ominus + \frac{1}{2}Cd^{++} \rightarrow \frac{1}{2}Cd$	Cd^{++}, Cd	-0.403
$\ominus + Tl^+ \rightarrow Tl$	Tl^+, Tl	-0.3363
$\ominus + \frac{1}{2}Ni^{++} \rightarrow \frac{1}{2}Ni$	Ni^{++}, Ni	-0.250
$\ominus + AgI \rightarrow Ag + I^-$	I^-, AgI, Ag	-0.151

Acid solution

Half-cell reaction	Electrode	E^o electrode potential
$\ominus + \frac{1}{2}Cl_2(g) \rightarrow Cl^-$	$Cl^-, Cl_2(g)$	1.3595
$\ominus + \frac{1}{3}Au^{3+} \rightarrow \frac{1}{3}Au$	Au^{3+}, Au	1.50
$\ominus + \frac{8}{5}H^+ + \frac{1}{5}MnO_4^- \rightarrow \frac{4}{5}H_2O(l) + \frac{1}{5}Mn^{++}$	H^+, MnO_4^-, Mn^{++}	1.51
$\ominus + Bk^{4+} \rightarrow Bk^{3+}$	Bk^{4+}, Bk^{3+}	1.6
$\ominus + Ce^{4+} \rightarrow Ce^{3+}$	Ce^{4+}, Ce^{3+}	1.61
$\ominus + AmO_2^{++} \rightarrow AmO_2^+$	AmO_2^{++}, AmO_2^+	1.64
$\ominus + Au^+ \rightarrow Au$	Au^+, Au	1.68
$\ominus + Am^{4+} \rightarrow Am^{3+}$	Am^{4+}, Am^{3+}	2.18
$\ominus + H^+ + \frac{1}{2}F_2(g) \rightarrow HF(aq)$	$H^+, HF(aq), F_2(g)$	3.06

Basic solution

Half-cell reaction	Electrode	E^o electrode potential
$\ominus + \frac{1}{2}Ca(OH)_2 \rightarrow \frac{1}{2}Ca + OH^-$	$Ca(OH)_2, Ca$	-3.03
$\ominus + \frac{1}{2}Sr(OH)_2 \cdot 8H_2O \rightarrow \frac{1}{2}Sr + HO^- + 4H_2O$	$Sr(OH)_2 \cdot 8H_2O, Sr$	-2.99
$\ominus + \frac{1}{2}Ba(OH)_2 \cdot 8H_2O \rightarrow \frac{1}{2}Ba + OH^- + 4H_2O$	$Ba(OH)_2 \cdot 8H_2O, Ba$	-2.97
$\ominus + H_2O \rightarrow \frac{1}{2}H_2(g) + OH^-$	$H_2O, H_2(g)$	-2.93
$\ominus + \frac{1}{3}La(OH)_3 \rightarrow \frac{1}{3}La + OH^-$	$La(OH)_3, La$	-2.90
$\ominus + \frac{1}{2}Mg(OH)_2 \rightarrow \frac{1}{2}Mg + OH^-$	$Mg(OH)_2, Mg$	-2.69
$\ominus + \frac{1}{2}Be_2O_3^- + \frac{3}{2}H_2O \rightarrow \frac{1}{2}Be + \frac{3}{2}OH^-$	$Be_2O_3^-, Be$	-2.62
$\ominus + \frac{1}{4}Th(OH)_4 \rightarrow \frac{1}{4}Th + OH^-$	$Th(OH)_4, Th$	-2.48
$\ominus + \frac{1}{3}H_2AlO_3 + \frac{1}{3}H_2O \rightarrow \frac{1}{3}Al + \frac{4}{3}OH^-$	$H_2AlO_3^-, Al$	-2.35
$\ominus + \frac{1}{4}SiO_3^- + \frac{3}{4}H_2O \rightarrow \frac{1}{4}Si + \frac{1}{2}OH^-$	SiO_3^{--}, Si	-1.70
$\ominus + \frac{1}{2}Mn(OH)_2 \rightarrow \frac{1}{2}Mn + OH^-$	$Mn(OH)_2, Mn$	-1.55
$\ominus + \frac{1}{3}Cr(OH)_3 \rightarrow \frac{1}{3}Cr + OH^-$	$Cr(OH)_3, Cr$	-1.3

E	Couple	Electrode reaction
-1.245	$Zn(OH)_2, Zn$	$\Theta + \frac{1}{2}Zn(OH)_2 \rightarrow \frac{1}{2}Zn + OH^-$
-1.216	ZnO_2^{--}, Zn	$\Theta + \frac{1}{2}ZnO_2^{--} + H_2O \rightarrow \frac{1}{2}Zn + 2OH^-$
-1.14	Te, Te^{--}	$\Theta + \frac{1}{2}Te \rightarrow \frac{1}{2}Te^{--}$
-1.12	$SO_3^{--}, S_2O_4^{--}$	$\Theta + SO_3^{--} + H_2O \rightarrow \frac{1}{2}S_2O_4^{--} + 2OH^-$
-1.0	$In(OH)_3, In$	$\Theta + \frac{1}{3}In(OH)_3 \rightarrow \frac{1}{3}In + OH^-$
-0.93	SO_4^{--}, SO_3^{--}	$\Theta + \frac{1}{2}SO_4^{--} + \frac{1}{2}H_2O \rightarrow \frac{1}{2}SO_3^{--} + OH^-$
-0.92	Se, Se^{--}	$\Theta + \frac{1}{2}Se \rightarrow \frac{1}{2}Se^{--}$
-0.89	$P, PH_3(g)$	$\Theta + \frac{1}{3}P + H_2O \rightarrow \frac{1}{3}PH_3(g) + OH^-$
-0.877	$Fe(OH)_2, Fe$	$\Theta + \frac{1}{2}Fe(OH)_2 \rightarrow \frac{1}{2}Fe + OH^-$
-0.828	$H_2O, H_2(g)$	$\Theta + H_2O \rightarrow \frac{1}{2}H_2(g) + OH^-$
-0.809	$Cd(OH)_2, Cd$	$\Theta + \frac{1}{2}Cd(OH)_2 \rightarrow \frac{1}{2}Cd + OH^-$
-0.73	$Co(OH)_2, Co$	$\Theta + \frac{1}{2}Co(OH)_2 \rightarrow \frac{1}{2}Co + OH^-$
-0.72	$Ni(OH)_2, Ni$	$\Theta + \frac{1}{2}Ni(OH)_2 \rightarrow \frac{1}{2}Ni + OH^-$
-0.58	$SO_3^{--}, S_2O_3^{--}$	$\Theta + \frac{1}{2}SO_3^{--} + \frac{3}{4}H_2O \rightarrow \frac{1}{4}S_2O_3^{--} + \frac{3}{2}OH^-$
-0.56	$Fe(OH)_3, Fe(OH)_2$	$\Theta + Fe(OH)_3 \rightarrow Fe(OH)_2 + OH^-$
-0.56	$O_2(g), O_2^-$	$\Theta + O_2(g) \rightarrow O_2^-$
-0.48	S, S^{--}	$\Theta + \frac{1}{2}S \rightarrow \frac{1}{2}S^{--}$
-0.44	Bi_2O_3, Bi	$\Theta + \frac{1}{6}Bi_2O_3 + \frac{1}{2}H_2O \rightarrow \frac{1}{3}Bi + OH^-$
-0.358	Cu_2O, Cu	$\Theta + \frac{1}{2}Cu_2O + \frac{1}{2}H_2O \rightarrow Cu + OH^-$
-0.345	$TlOH, Tl$	$\Theta + TlOH \rightarrow Tl + OH^-$
-0.24	HO_2^-, OH	$\Theta + HO_2^- + H_2O \rightarrow OH + 2OH^-$
-0.13	$CrO_4^{--}, Cr(OH)_3$	$\Theta + \frac{1}{3}CrO_4^{--} + \frac{4}{3}H_2O \rightarrow \frac{1}{3}Cr(OH)_3 + \frac{5}{3}OH^-$
-0.080	$Cu(OH)_2, Cu_2O$	$\Theta + Cu(OH)_2 \rightarrow \frac{1}{2}Cu_2O + OH^-$
-0.076	$O_2(g), HO_2^-$	$\Theta + \frac{1}{2}O_2(g) + \frac{1}{2}H_2O \rightarrow \frac{1}{2}HO_2^- + \frac{1}{2}OH^-$
-0.05	$Tl(OH)_3, TlOH$	$\Theta + \frac{1}{2}Tl(OH)_3 \rightarrow \frac{1}{2}Tl(OH) + OH^-$
-0.05	$MnO_2, Mn(OH)_2$	$\Theta + \frac{1}{2}MnO_2 + H_2O \rightarrow \frac{1}{2}Mn(OH)_2 + OH^-$
0.01	NO_3^-, NO_2^-	$\Theta + \frac{1}{2}NO_3^- + \frac{1}{2}H_2O \rightarrow \frac{1}{2}NO_2^- + OH^-$
~ 0.136	Sn^{++}, Sn	$\Theta + \frac{1}{2}Sn^{++} \rightarrow \frac{1}{2}Sn$
∓ 0.126	Pb^{++}, Pb	$\Theta + \frac{1}{2}Pb^{++} \rightarrow \frac{1}{2}Pb$
± 0.0000	$H^+, H_2(g)$	$\Theta + H^+ \rightarrow \frac{1}{2}H_2(g)$
0.05	UO_2^{++}, UO_2^+	$\Theta + UO_2^{++} \rightarrow UO_2^+$
0.095	$Br^-, AgBr, Ag$	$\Theta + AgBr \rightarrow Ag + Br^-$
0.147	Np^{4+}, Np^{3+}	$\Theta + Np^{4+} \rightarrow Np^{3+}$
0.15	Sn^{4+}, Sn^{++}	$\Theta + \frac{1}{2}Sn^{4+} \rightarrow \frac{1}{2}Sn^{++}$
0.153	Cu^{++}, Cu^+	$\Theta + Cu^{++} \rightarrow Cu^+$
0.2223	$Cl^-, AgCl, Ag$	$\Theta + AgCl \rightarrow Ag + Cl^-$
0.337	Cu^{++}, Cu	$\Theta + \frac{1}{2}Cu^{++} \rightarrow \frac{1}{2}Cu$
0.52	$H^+, C_2H_4(g), C_2H_6(g)$	$\Theta + H^+ + \frac{1}{2}C_2H_4(g) \rightarrow \frac{1}{2}C_2H_6(g)$
0.521	Cu^+, Cu	$\Theta + Cu^+ \rightarrow Cu$
0.5355	I^-, I_2	$\Theta + \frac{1}{2}I_2 \rightarrow I^-$
0.536	I^-, I_3^-	$\Theta + \frac{1}{3}I_3^- \rightarrow I^-$
0.62	H^+, UO_2^{++}, U^{4+}	$\Theta + 2H^+ + \frac{1}{2}UO_2^{++} \rightarrow \frac{1}{2}U^{4+} + H_2O(l)$
0.75	H^+, NpO_2^+, Np^{4+}	$\Theta + 4H^+ + NpO_2^+ \rightarrow Np^{4+} + 2H_2O(l)$
0.771	Fe^{3+}, Fe^{++}	$\Theta + Fe^{3+} \rightarrow Fe^{++}$
0.789	$Hg_2^{++}, Hg(l)$	$\Theta + \frac{1}{2}Hg_2^{++} \rightarrow Hg(l)$
0.7991	Ag^+, Ag	$\Theta + Ag^+ \rightarrow Ag$
0.920	Hg^{++}, Hg_2^{++}	$\Theta + Hg^{++} \rightarrow \frac{1}{2}Hg_2^{++}$
0.93	PuO_2^{++}, PuO_2^+	$\Theta + PuO_2^{++} \rightarrow PuO_2^+$
0.97	Pu^{4+}, Pu^{3+}	$\Theta + Pu^{4+} \rightarrow Pu^{3+}$
0.987	Pd^{++}, Pd	$\Theta + \frac{1}{2}Pd^{++} \rightarrow \frac{1}{2}Pd$
1.0652	$Br^-, Br_2(l)$	$\Theta + \frac{1}{2}Br_2(l) \rightarrow Br^-$
1.15	NpO_2^{++}, NpO_2^+	$\Theta + NpO_2^{++} + NpO_2^+$
1.15	H^+, PuO_2^+, Pu^{4+}	$\Theta + 4H^+ + PuO_2^+ \rightarrow 2H_2O(l) + Pu^{4+}$
1.229	$H^+, O_2(g)$	$\Theta + H^+ + \frac{1}{4}O_2(g) \rightarrow \frac{1}{2}H_2O(l)$
1.33	$H^+, Cr_2O_7^{--}, Cr^{3+}$	$\Theta + \frac{7}{6}H^+ + \frac{1}{6}Cr_2O_7^{--} \rightarrow \frac{7}{6}H_2O(l) + \frac{1}{3}Cr^{3+}$

TABLE 5g-3. STANDARD ELECTROMOTIVE FORCES OF HALF CELLS IN WATER AT 25°C (*Continued*)

(E^o in absolute volts relative to the standard hydrogen electrode)

Half-cell reaction	Electrode	E^o electrode potential	Half-cell reaction	Electrode	E^o electrode potential
Basic solution			Basic solution		
$\ominus + \frac{1}{2}SeO_4^{--} + \frac{1}{2}H_2O \rightarrow \frac{1}{2}SeO_3^{--} + OH^-$	SeO_4^{--}, SeO_3^{--}	0.05	$\ominus + \frac{1}{4}O_2(g) + \frac{1}{2}H_2O \rightarrow OH^-$	$O_2(g)$, OH^-	0.401
$\ominus + \frac{1}{2}HgO(r) + \frac{1}{2}H_2O \rightarrow \frac{1}{2}Hg + OH^-$	$HgO(r)$, Hg	0.098	$\ominus + \frac{1}{2}NiO_2 + H_2O \rightarrow \frac{1}{2}Ni(OH)_2 + OH^-$	NiO_2, $Ni(OH)_2$	0.49
$\ominus + \frac{2}{3}Mn(OH)_3 \rightarrow Mn(OH)_2 + OH^-$	$Mn(OH)_3$, $Mn(OH)_2$	0.1	$\ominus + AgO + \frac{1}{2}H_2O \rightarrow \frac{1}{2}Ag_2O + OH^-$	AgO, Ag_2O	0.57
			$\ominus + \frac{1}{2}MnO_4^{--} + H_2O \rightarrow \frac{1}{2}MnO_2 + 2OH^-$	MnO_4^{--}, MnO_2	0.60
$\ominus + Co(OH)_3 \rightarrow Co(OH)_2 + OH^-$	$Co(OH)_3$, $Co(OH)_2$	0.17	$\ominus + \frac{1}{2}ClO_2 + \frac{1}{2}H_2O \rightarrow \frac{1}{2}ClO^- + OH^-$	ClO_2^-, ClO^-	0.66
$\ominus + \frac{1}{2}PbO_2 + \frac{1}{2}H_2O \rightarrow \frac{1}{2}PbO(r) + OH^-$	PbO_2, $PbO(r)$	0.248	$\ominus + \frac{1}{2}HO_2^- + \frac{1}{2}H_2O \rightarrow \frac{3}{2}OH^-$	HO_2^-, OH^-	0.88
$\ominus + \frac{1}{2}ClO_3^- + \frac{1}{2}H_2O \rightarrow \frac{1}{2}ClO_2^- + OH^-$	ClO_3^-, ClO_2^-	0.33	$\ominus + \frac{1}{2}ClO^- + \frac{1}{2}H_2O \rightarrow \frac{1}{2}Cl^- + OH^-$	ClO^-, Cl^-	0.89
$\ominus + \frac{1}{2}Ag_2O + \frac{1}{2}H_2O \rightarrow Ag + OH^-$	Ag_2O, Ag	0.344	$\ominus + \frac{1}{2}O_3(g) + \frac{1}{2}H_2O \rightarrow \frac{1}{2}O_2(g) + OH^-$	$O_3(g)$, $O_2(g)$	1.24

TABLE 5g-4. THEORETICAL VOLTAGES FOR BATTERY REACTIONS AT 25° IN H_2O

Reaction	ΔG^o, kcal	E^o, volts
$MnO_2(s) + Zn(s) + 4H^+ \rightarrow Mn^{++} + Zn^{++} + 2H_2O$	-92.054	1.996
$HgO(s) + Zn(s) \rightarrow ZnO(s) + Hg(l)$	-62.100	1.346
$Pb(s) + PbO_2(s) + 2H_2SO_4 \rightarrow 2PbSO_4(s) + 2H_2O$	-94.204	2.042
$Pb(s) + PbO_2(s) + 4H^+ \rightarrow 2Pb^{++} + 2H_2O$	-73.084	1.584
$Mg(s) + 2AgCl(s) \rightarrow Mg^{++} + 2Cl^-$	-119.246	2.585
$Pb(s) + Ag_2O(s) \rightarrow PbO(s) + 2Ag(s)$	-42.230	0.916
$2NiOOH(s) + Cd(s) + 2H_2O \rightarrow 2Ni(OH)_2(s) + Cd(OH)_2(s)$	-60.056	1.302
$H_2(g) + O_2(g) + OH^- \rightarrow HO_2^- + H_2O$	-35.193	0.763
$Zn(s) + Cl_2(g) \rightarrow Zn^{++} + 2Cl^-$	-97.884	2.122
$2AgO(s) + Zn(s) \rightarrow ZnO(s) + Ag_2O(s)$	-85.500	1.854
$Zn(s) + 2AgCl(s) \rightarrow Zn^{++} + 2Cl^- + 2Ag(s)$	-45.396	0.984
$2MnO_2(s) + Zn(s) + 3H_2O + 2OH^- \rightarrow ZnO(s) + 2Mn(OH)_4^-$	-52.531	1.139

Similarly the cell

$$Tl,\ TlCl(aq),\ Cl_2(g) \qquad E^o = 1.6958\ volts \qquad (5g\text{-}6)$$

involves two opposing half reactions which are

	Plan A	Plan B	
	$Tl \rightarrow Tl^+ + \ominus$ (e)	$\ominus + Tl^+ \rightarrow Tl$ (f)	$(5g\text{-}6e,\ 6f)$
and	$Cl^- \rightarrow \frac{1}{2}Cl_2 + \ominus$ (c)	$\ominus + \frac{1}{2}Cl_2 \rightarrow Cl^-$ (d)	$(5g\text{-}5c,\ 5d)$

Since E^o of the cell is 1.6958 volts, the tendency of (e) is 1.6958 *greater* than that of (c) and the tendency of (f) is 1.6958 volts *less* than that of (d). To simplify the tabulation of relative half-cell emfs it has long been the custom to compare all reactions with (a) in plan A or (b) in plan B. In the same sense that the altitude of sea level is arbitrarily set equal to zero, the half-cell emfs of (a) and (b) are called zero and the emfs of all other half cells are listed relatively to (a) or to (b) depending upon the "plan" used by an author. Since the tendency of (e) is 1.6958 volts greater than that of (c), which, in turn, is 1.3595 volts less than that of (a), the appropriate entries for the table are, respectively,

Plan A	Plan B
$Tl \rightarrow Tl^+ + \ominus$ $\quad E^o = 0.3363$ volt	$\ominus + Tl^+ \rightarrow Tl$ $\quad E^o = -0.3363$ volt $\qquad (5g\text{-}6e,\ 6f)$

It should be clearly understood that all the standard emfs of Table 5g-3 are equilibrium values and are valid strictly only when no current is passing or when the current passing is so small that resulting changes in the cell are negligible. The reversal of such a current would not affect the magnitude and, of course, could not alter the algebraic sign of the emf of a cell or half cell. The choice of plan A or plan B is an arbitrary one and has nothing to do with the direction in which current is actually passed through a given cell.

Only plan B emfs are listed here. Most electrochemists have preferred this approach, and it has been overwhelmingly recommended by the international commissions concerned with such matters. It has also been adopted in many of the most recent physical chemistry texts. In the past American physical chemists have preferred plan A and it was promoted by the extensive treatise of Professor W. B. Latimer. To use his tables one must note that Latimer's standard half-cell emfs are the negatives of the respective "electrode potentials" and that his equations must be written

TABLE 5g-5. SELECTED MEAN-IONIC-ACTIVITY COEFFICIENTS $\gamma_\pm$ OF ELECTROLYTES IN AQUEOUS SOLUTIONS AT 25°C

(m in mole kg^{-1})

m	$HClO_4$	HNO_3	$LiCl$	$NaCl$	$NaClO_3$	$NaClO_4$	$NaBrO_3$	$NaNO_3$	KCl	KNO_3	$RbCl$	$CsCl$	$AgNO_3$	$TlClO_4$
0.1	0.803	0.791	0.790	0.778	0.772	0.775	0.758	0.762	0.770	0.739	0.764	0.756	0.734	0.730
0.2	0.778	0.754	0.757	0.735	0.720	0.729	0.696	0.703	0.718	0.663	0.709	0.694	0.657	0.652
0.5	0.769	0.720	0.739	0.681	0.645	0.668	0.605	0.617	0.649	0.545	0.634	0.606	0.536	0.527
1.0	0.823	0.724	0.774	0.657	0.589	0.629	0.528	0.548	0.604	0.443	0.583	0.544	0.429	
2.0	1.055	0.793	0.921	0.668	0.538	0.609	0.450	0.478	0.573	0.333	0.546	0.495	0.316	
4.0	2.08	.....	1.510	0.783	.....	0.626	.....	0.408	0.577	.....	0.538	0.473	0.210	

m	$MgCl_2$	$Mg(ClO_4)_2$	$CaCl_2$	$CaBr_2$	CaI_2	$SrCl_2$	BaI_2	$Sr(NO_3)_2$	$ZnCl_2$	$Zn(ClO_4)_2$	$UO_2(ClO_4)_2$	H_2SO_4	Na_2SO_4	Cs_2SO_4
0.1	0.529	0.590	0.518	0.532	0.560	0.511	0.542	0.478	0.515	0.581	0.626	0.2655	0.445	0.456
0.2	0.489	0.578	0.472	0.492	0.531	0.462	0.509	0.410	0.462	0.564	0.634	0.2090	0.365	0.382
0.5	0.481	0.647	0.448	0.491	0.561	0.430	0.523	0.329	0.394	0.629	0.790	0.1557	0.268	0.291
1.0	0.570	0.946	0.500	0.597	0.741	0.461	0.649	0.275	0.339	0.929	1.390	0.1316	0.204	0.235
2.0	1.053	2.65	0.792	1.121	1.640	0.670	1.221	0.232	0.289	2.74	5.91	0.1276	.....	.....
4.0	5.54	34.1	2.934	6.28	.....	1.977	.....	.....	0.307	38.8	160.2	0.1700	.....	.....

m	$CuSO_4$	$MgSO_4$	$ZnSO_4$	$CdSO_4$	UO_2SO_4	$AlCl_3$	$LaCl_3$	$EuCl_3$	$K_3Fe(CN)_6$	$K_4Fe(CN)_6$	$Al_2(SO_4)_3$	$Th(NO_3)_4$
0.1	(0.150)	(0.150)	(0.150)	(0.150)	(0.150)	0.337	0.314	0.318	0.268	0.139	0.0350	0.279
0.2	0.104	0.108	0.104	0.102	0.102	0.305	0.274	0.282	0.212	0.100	0.0225	0.225
0.5	0.062	0.068	0.063	0.061	0.0611	0.331	0.266	0.276	0.155	0.062	0.0143	0.189
1.0	0.043	0.049	0.043	0.041	0.0439	0.539	0.342	0.371	0.128	.....	0.0175	0.207
2.0	.....	0.042	0.035	0.032	0.0367	.....	0.825	0.995	.....	.....	.....	0.326
4.0	.....	.....	.....	.....	0.0433	.....	.....	.....	.....	.....	.....	0.647

TABLE 5g-6. MEAN-ACTIVITY COEFFICIENTS $\gamma_\pm$ OF HCl IN AQUEOUS SOLUTION
(m in mole kg^{-1})

m	0°	10°	20°	25°	40°	50°	60°
0.0001	0.9890	0.9890	0.9892	0.9891	0.9885	0.9879	0.9879
0.0002	0.9848	0.9846	0.9844	0.9842	0.9833	0.9831	0.9831
0.0005	0.9756	0.9756	0.9759	0.9752	0.9741	0.9738	0.9734
0.001	0.9668	0.9666	0.9661	0.9656	0.9643	0.9639	0.9632
0.002	0.9541	0.9544	0.9527	0.9521	0.9505	0.9500	0.9491
0.005	0.9303	0.9300	0.9294	0.9285	0.9265	0.9250	0.9235
0.01	0.9065	0.9055	0.9052	0.9048	0.9016	0.9000	0.8987
0.02	0.8774	0.8773	0.8768	0.8755	0.8715	0.8690	0.8666
0.05	0.8346	0.8338	0.8317	0.8304	0.8246	0.8211	0.8168
0.1	0.8027	0.8016	0.7985	0.7964	0.7891	0.7850	0.7813
0.2	0.7756	0.7740	0.7694	0.7667	0.7569	0.7508	0.7437
0.5	0.7761	0.7694	0.7616	0.7571	0.7432	0.7344	0.7237
1.0	0.8419	0.8295	0.8162	0.8090	0.7865	0.7697	0.7541
2.0	1.078	1.053	1.024	1.009	0.9602	0.9327	0.9072
4.0	2.006	1.911	1.812	1.762			

in the reverse direction to fit plan B. It should also be noted that his equations are written for integral values of N but not necessarily for $N = 1$.

The Use of Table 5g-3. To calculate E^o of any cell, e.g.,

$$\text{Tl, TlCl(aq), AgCl(s), Ag} \tag{5g-7}$$

write the half-cell reaction and E^o for the *right*-hand electrode

$$\theta + \text{AgCl} \to \text{Ag} + \text{Cl}^- \qquad E^o = 0.2223 \text{ volt}$$

Subtract both the half-cell reaction of E^o of the *left*-hand electrode

$$-[\theta + \text{Tl}^+ \to \text{Tl}] \qquad -E^o = -(0.3363 \text{ volt})$$

to give

$$\text{Tl} + \text{AgCl} \to \text{Ag} + \text{TlCl(aq)} \qquad E^o = 0.5568 \text{ volt} \tag{5g-8}$$

or the completely equivalent form

$$\text{Tl} + \text{AgCl} \to \text{Ag} + \text{Tl}^+ + \text{Cl}^- \qquad E^o = 0.5568 \text{ volt}$$

Since E^o is positive, ΔG^o is negative for the reaction. Equation (5g-8) is therefore the equation for the reaction actually taking place in the cell when all activities are unity. If the cell had been written Ag, AgCl, TlCl(aq), Tl the indicated reaction would have been

$$\text{Ag} + \text{TlCl(aq)} \to \text{Tl} + \text{AgCl} \qquad E^o = -0.5586 \text{ volt} \tag{5g-9}$$

The conclusions about the actual reaction and the absolute values of E^o and ΔG^o are unchanged.

TABLE 5g-7. DISSOCIATION CONSTANTS OF WATER AND OF ELECTROLYTES IN AQUEOUS SOLUTIONS
(Constants* are on the molality scale. *Italics* indicate maximum values)

Material	°C	0°	5°	10°	15°	20°	25°	30°	35°	40°	45°	50°
Water	$K \times 10^{14}$	0.1139	0.1846	0.2920	0.4505	0.6809	1.008	1.469	2.089	2.919	4.018	5.474
Formic acid	$K_A \times 10^4$	1.638	1.691	1.728	1.749	1.765	*1.772*	1.768	1.747	1.716	1.685	1.650
Acetic acid	$K_A \times 10^5$	1.657	1.700	1.729	1.745	1.753	*1.754*	1.750	1.728	1.703	1.670	1.633
Propionic acid	$K_A \times 10^5$	1.274	1.305	1.326	1.336	*1.338*	1.336	1.326	1.310	1.280	1.257	1.229
n-Butyric acid	$K_A \times 10^5$	1.563	1.574	*1.576*	1.569	1.542	1.515	1.484	1.439	1.395	1.347	1.302
Chloroacetic acid	$K_A \times 10^3$	1.528	......	1.488	......	......	1.379	......	......	1.230	......	......
Lactic acid	$K_A \times 10^4$	1.287	......	......	......	......	*1.374*	......	......	......	......	1.270
Glycolic acid	$K_A \times 10^4$	1.334	......	......	......	......	*1.475*	......	......	......	......	1.415
Sulfuric acid	$K_{2A} \times 10^2$	......	1.80	......	1.36	......	1.01	......	0.75	......	0.56	......
Carbonic acid	$K_{1A} \times 10^7$	2.64	3.04	3.44	3.81	4.16	4.45	4.71	4.90	5.04	5.13	5.19
	$K_{2A} \times 10^{11}$	2.36	2.77	3.24	3.71	4.20	4.69	5.13	5.62	6.03	6.38	6.73
Phosphoric acid	$K_{1A} \times 10^3$	8.97	......	......	......	......	7.52	......	......	......	......	5.50
	$K_{2A} \times 10^8$	4.85	5.24	5.57	5.89	6.12	6.34	6.46	6.53	6.58	*6.59*	6.55
Nitric acid	K_A	......	......	......	......	......	21	......	......	......	......	......
Glycine	$K_A \times 10^3$	......	......	3.94	......	4.31	4.47	4.59	......	4.81	......	......
	$K_B \times 10^5$	......	......	4.68	5.12	5.57	6.04	6.52	6.98	7.43	7.87	......
Alanine	$K_A \times 10^3$	......	......	......	......	4.47	4.57	4.66	4.71	4.74	4.76	......
	$K_B \times 10^6$	......	......	......	......	6.90	7.47	8.08	8.61	9.10	9.60	......

* Letter subscripts on K indicate dissociation as acid or base, respectively; number subscripts indicate first or second dissociation.

TABLE 5g-8. RATE CONSTANTS FOR PROTON TRANSFER REACTIONS IN H_2O AT $25°$

Reaction: $A + B \underset{k_b}{\overset{k_f}{\rightleftharpoons}} C + D$	k_f, M^{-1} sec^{-1}	k_b, sec^{-1}
$H^+ - OH^- \rightleftharpoons H_2O$	1.4×10^{11}	2.5×10^{-5}
$D^+ + OD^- \rightleftharpoons D_2O$	8.4×10^{10}	2.5×10^{-6}
$H^+ + SO_4^{--} \rightleftharpoons HSO_4^-$	1×10^{11}	1×10^9
$H^+ + F^- \rightleftharpoons HF$	1.0×10^{11}	7×10^7
$H^+ + HS^- \rightleftharpoons H_2S$	7.5×10^{10}	4.3×10^3
$H^+ + HCO_3^- \rightleftharpoons H_2CO_3$	4.7×10^{10}	8×10^6
$H^+ + HCOO^- \rightleftharpoons HCOOH$	5×10^{10}	8.6×10^6
$H^+ + CH_3COO^- \rightleftharpoons CH_3COOH$	4.5×10^{10}	7.8×10^5
$OH^- + NH_4^+ \rightleftharpoons NH_3 \cdot H_2O$	3.4×10^{10}	6×10^5
$OH^- + C_6H_5OH \rightleftharpoons H_2O + C_6H_5O^-$	1.4×10^{10}	1.3×10^6
$OH^- + HCN \rightleftharpoons H_2O + CN^-$	3.7×10^9	5.2×10^4
$OH^- + HPO_4^{--} \rightleftharpoons H_2O + PO_4^{3-}$	2×10^9	2×10^7

TABLE 5g-9. RELATIVE APPARENT MOLAL HEAT CONTENT φL AND PARTIAL
MOLAL HEAT CONTENT $\bar{L}_2$ OF SOLUTES IN DILUTE
AQUEOUS SOLUTIONS AT 25°C
(cal mole^{-1})

m		0.0001	0.0004	0.0016	0.0064	0.0100	0.0400	0.0900
NaCl	φL	4.5	8.5	17.0	33	40	67	83
	$\bar{L}_2$	6.5	12.5	24.0	46	57	92	104
NaIO$_3$	φL	4.0	7.5	14.0	21	21	0	
	$\bar{L}_2$	5.8	11.0	19.8	24	20	-41	
KCl	φL	4.5	8.5	16.0	31	38	65	77
	$\bar{L}_2$	6.5	12.5	24.0	46	55	82	91
KClO$_4$	φL	4.3	8.0	13.0	16	14	-28	
	$\bar{L}_2$	6.2	11.3	16.6	13	4	-86	
Li$_2$SO$_4$	φL	24	47	91	177	218	377	488
	$\bar{L}_2$	35	69	135	260	317	508	620
Cs$_2$SO$_4$	φL	20	39	71	121	139	161	137
	$\bar{L}_2$	29	57	102	161	176	152	87
SrCl$_2$	φL	23	46	86	161	195	332	420
	$\bar{L}_2$	34	66	125	232	277	443	528
SrBr$_2$	φL	23	44	82	152	182	293	366
	$\bar{L}_2$	33	64	119	216	254	383	452
Ba(NO$_3$)$_2$	φL	19	36	59	72	66	-46	-223
	$\bar{L}_2$	27	51	75	68	37	-195	-528

TABLE 5g-10. STANDARD ENTROPIES OF MONATOMIC IONS
IN AQUEOUS SOLUTIONS AT 25°C
(Referred* to $H_2 \rightarrow 2H^+ + 2\theta$; $\Delta S^o = 0$; cal mole^{-1} deg^{-1})

Ion	$\bar{S}^o$	Ion	$\bar{S}^o$	Ion	$\bar{S}^o$
Cs^+	31.8	Ca^{++}	−13.2	Cr^{3+}	−73.5
Tl^+	30.4	Cd^{++}	−14.6	Al^{3+}	−74.9
Rb^+	29.7	Mn^{++}	−20	Ga^{3+}	−83
K^+	24.5	Cu^{++}	−23.6	U^{4+}	−78
Ag^+	17.67	Zn^{++}	−25.45	Pu^{4+}	−87
Na^+	14.4	Fe^{++}	−27.1	I^-	26.14
Li^+	3.4	Mg^{++}	−28.2	Br^-	19.25
Pb^{++}	5.1	U^{3+}	−36		
Ba^{++}	3.0	Pu^{3+}	−39	Cl^-	13.17
Hg^{++}	−5.4	Gd^{3+}	−43	F^-	−2.3
Sn^{++}	−5.9	In^{3+}	−62		
Sr^{++}	−9.4	Fe^{3+}	−70.1	S^-	−6.4

* This is not equivalent to the setting of S^o of H^+ equal to zero; cf. Klotz (1950).

TABLE 5g-11. STANDARD ENTROPIES OF POLYATOMIC IONS
IN AQUEOUS SOLUTIONS AT 25°C
(Referred* to $H_2 \rightarrow 2H^+ + 2\theta$; $\Delta S^o = 0$; cal mole^{-1} deg^{-1})

Ion	$\bar{S}^o$	Ion	$\bar{S}^o$	Ion	$\bar{S}^o$
OH^-	−2.5	HSO_4^-	30.3	PO_4^{3-}	−52
ClO^-	10.0			AsO_4^{3-}	−34.6
HCO_2^-	21.9	$H_2AsO_4^-$	28	HF_2^-	0.5
ClO_2^-	24.1	$H_2PO_4^-$	21.3	BF_4^-	40
NO_2^-	29.9	$HN_2O_4^-$	34	SiF_6^-	−12
NO_3^-	35.0	BeO_2^-	−27	$CuCl_2^-$	49.2
ClO_3^-	39.0	CO_3^-	−12.7	$AuCl_4^-$	61
BrO_3^-	38.5	SO_3^-	−7	$PdCl_4^-$	36
IO_3^-	28.0	SO_4^-	4.1	$PtCl_4^-$	42
ClO_4^-	43.2	SeO_4^-	5.7	$PtCl_6^-$	52.6
MnO_4^-	45.4	$N_2O_2^-$	6.6	I_3^-	41.5
HCO_3^-	22.7	$C_2O_4^-$	10.6	$Ag(CN)_2^-$	49
HSO_3^-	26	$Cr_2O_7^-$	51.1	$Ni(CN)_4^-$	33
SH^-	14.9	HPO_4^-	−8.6	$FeCl^{++}$	−22
		$HAsO_4^-$	0.9		

* This is not equivalent to the setting of S^o of H^+ equal to zero; cf. Klotz (1950).

Effect of Concentration. For the case where the reactants are not at unit activity, the Nernst equation is used:

$$E = E^o - \frac{RT}{NF} \log \pi a_i^{\nu_i} \tag{5g-10}$$

For the example above ($N = 1$, 25°C) this reduces to

$$E = 0.5586 - 0.059 \log \frac{a_{Ag}a_{TlCl}}{a_{Tl}a_{AgCl}} \tag{5g-11}$$

The activities of pure solids and liquids are conventionally taken as unity so that

$$a_{Ag} = a_{AgCl} = a_{Tl} = 1$$

and $\qquad\qquad E = 0.5586 - 0.05915 \log a_{TlCl}$

In many cases the activities of the dissolved species (e.g., TlCl above) are approximated by their molar concentrations.

References

1. Bates, R. G.: "Electrometric pH Determinations," John Wiley & Sons, Inc., New York, 1954.
2. Bjerrum, J., G. Schwarzenbach, and L. G. Sillen: "Stability Constants of Metal-Ion Complexes with Solubility Products of Inorganic Substances," 2d ed., Chemical Society, London, 1964.
3. Caldin, E. F.: "Fast Reactions in Solution," Blackwell Scientific Publications, Ltd., Oxford, 1964.
4. Christiansen, J. A.: *J. Am. Chem. Soc.* **82**, 5517 (1960).
5. Conway, B. E.: "Electrochemical Data," Elsevier Publishing Company, Amsterdam, 1952.
6. Daniels, F., and R. A. Alberty: "Physical Chemistry," 2d ed., John Wiley & Sons, Inc., New York, 1961.
7. Harned, H. S.: in Electrochemical Constants, *Natl. Bur. Standards (U.S.) Circ.* 524, 1953.
8. Harned, H. S., and B. B. Owen: "The Physical Chemistry of Electrolytic Solutions," 3d ed., Reinhold Publishing Corporation, New York, 1958.
9. Hood, G. C., O. Redlich, and C. A. Reilly: *J. Chem. Phys.* **22**, 2067 (1954).
10. "International Critical Tables of the Numerical Data of Physics, Chemistry and Technology," McGraw-Hill Book Company, New York, 1926–1933.
11. Klotz, I. M.: "Chemical Thermodynamics," revised ed., W. A. Benjamin, Inc., New York, 1964.
12. Kolthoff, I. M., and J. J. Lingane: "Polarography," 2d ed., Interscience Publishers, Inc., New York, 1952.
13. Kortüm, G.: "Treatise on Electrochemistry," Elsevier Publishing Company, Amsterdam, 1965.
14. Kortüm, G., and J. O'M. Bockris: "Textbook of Electrochemistry," vol. II, Elsevier Publishing Company, Amsterdam, 1951.
15. Landolt-Börnstein: "Physikalische-Chemische Tabellen," 5th ed., Springer-Verlag OHG, Berlin, 1927–1935; 6th ed., 1950–?.
16. Latimer, W. M.: "The Oxidation States of the Elements and Their Potentials in Aqueous Solution," 2d ed., Prentice-Hall, Inc., Englewood Cliffs, N.J., 1952.
17. Parsons, R.: "Handbook of Electrochemical Constants," Butterworth Scientific Publishers, London, 1955.
18. Powell, R. F., and W. M. Latimer: *J. Chem. Phys.*, **19**, 1139 (1951).
19. Redlich, O.: *Chem. Revs.*, **39**, 333 (1946).
20. Robinson, R. A., and R. H. Stokes: *Trans. Faraday Soc.*, **45**, 612 (1949).
21. Robinson, R. A., and R. H. Stokes: "Electrolyte Solutions," 2d ed., Butterworth Scientific Publications, London, 1959.
22. Rossini, F. D., D. D. Wagman, W. H. Evans, S. Levine, and I. Jaffe: Selected Values of Chemical Thermodynamic Properties, *Natl. Bur. Standards (U.S.) Circ.* 500, 1952.
23. Scudder, H.: "The Electrical Conductivity and Ionization Constants of Organic Compounds," D. Van Nostrand Company, Inc., Princeton, N.J., 1914.
24. Stuehr, J., and E. Yeager: Chap. 6 in "Physical Acoustics," vol. IIA, Academic Press, Inc., New York, 1965.
25. van Stackelberg, M.: "Polarographische Arbeitsmethoden," Walter De Gruyter & Co., Berlin, 1950.

5h. Electric and Magnetic Fields in the Earth's Environment

M. SUGIURA AND J. P. HEPPNER

NASA-Goddard Space Flight Center

EARTH'S MAIN MAGNETIC FIELD

5h-1. Sources of the Magnetic Field at the Earth's Surface. The largest contribution to the magnetic field observed on the surface of the earth comes from its main field. The *main field* is considered to be produced in the earth's fluid core where the magnetic field and the convective motion of thermal origin are coupled to constitute a self-maintaining dynamo. The main field (Secs. 5h-4 and 5h-5) at the earth's surface is distorted to varying degrees by the earth's crustal anomalies (Sec. 5h-7) and by the magnetic fields from sources external to the solid earth. The latter include those from ionospheric currents (Secs. 5h-8 to 5h-10), effects of plasmas in the magnetosphere (Secs. 5h-11 and 5h-21), and the distortion of the magnetospheric magnetic field by the solar wind (Secs. 5h-12 and 5h-20).

5h-2. Units Conventionally Used in Geomagnetism [1]. Except in dealing with magnetic material, the magnetic permeability is unity in cgs units, and thus the distinction between magnetic induction **B** and magnetic field intensity **H** can be dispensed with without confusion. Hence, although the cgs unit for **H** is *oersted*, the unit for **B**, namely *gauss* (Γ), is conventionally used for both **B** and **H**. For weak magnetic fields and for describing time variations of the field, the unit *gamma* (γ) is used, where $1\ \gamma = 10^{-5}\ \Gamma$. Declination and inclination are expressed in angular measure, normally in degrees and minutes.

5h-3. Component Nomenclature [1,2]. The magnetic field vector is described by three orthogonal components, by the magnitude and two angles specifying the direction, or by a mixture of these. The standard nomenclature in geomagnetism is as follows:

$X, Y,$ and Z three components measured positively northward, eastward, and vertically downward, respectively

B (or F)   the magnitude of the field vector, (also called total force or intensity) $F = (X^2 + Y^2 + Z^2)^{\frac{1}{2}}$

H the magnitude of the horizontal intensity, $H = (X^2 + Y^2)^{\frac{1}{2}}$

D declination: the angular deviation of the horizontal projection of the field vector from geographic north, taken positive when eastward, $D = \tan^{-1}(Y/X)$

I inclination, or dip angle: the inclination of the field vector from the horizontal plane, taken positive when downward, $I = \tan^{-1}(Z/H)$

5h-4. Dipole Description of the Main Field [1–3]. The main field is described, to a first approximation, by the field of a magnetic dipole placed at the earth's center (i.e., a *centered dipole*) with the following equations and parameters [4]:

$$H = H_0 \left(\frac{a}{r}\right)^3 \sin \Theta$$

$$Z = 2H_0 \left(\frac{a}{r}\right)^3 \cos \Theta$$

$$M = H_0 a^3$$

where $M = 8.0052 \times 10^{25}$ gauss cm³, the dipole moment

$H_0 = 0.30953$ gauss, the field intensity at the magnetic equator $(I = 0°,$
$\Theta = 90°)$ for $r = a$

$a = 6371.2$ km, the reference radius for a spherical earth

$r =$ radial distance, km, from the earth's center

$\Theta =$ geomagnetic colatitude, from the geomagnetic coordinate system, below

Geomagnetic coordinates (Θ and Λ, respectively, for geomagnetic colatitude and east longitude) are given in terms of geographic coordinates (θ and λ, respectively, for geographic colatitude and east longitude) by

$$\cos \Theta = \cos \theta \cos \theta_0 + \sin \theta \sin \theta_0 \cos (\lambda - \lambda_0)$$
$$\sin \Lambda = \sin \theta \sin (\lambda - \lambda_0)/\sin \Theta$$

where [4] $\theta_0 = 11.44°$, the geographic colatitude of the north *geomagnetic pole* (also called the north *dipole pole*);

$\lambda_0 = 290.24°$, the geographic east longitude of the north geomagnetic pole

The errors in describing the magnetic field at the earth's surface by means of the above centered dipole are as great as 10 percent in some regions. For a somewhat more accurate dipole description some use has been made of an *eccentric dipole* [5] representation in which a dipole is located at a position displaced from the earth's center. This is equivalent to adding the potential terms for $g_2{}^1$, $h_2{}^1$, $g_2{}^2$, and $h_2{}^2$ from the spherical harmonic description (Sec. 5h-5) to the potential term $H_0 \cos \theta$ for the centered dipole. As indicated by the $(a/r)^{n+1}$ dependence of the potential in the spherical harmonic description (Sec. 5h-5), the errors in describing the main field by means of a dipole decrease rapidly with increasing distance r from the earth.

The study of phenomena related to the interaction between the geomagnetic field and the solar wind (Secs. 5h-10 to 5h-12), including many of the indirect effects of this interaction, is often facilitated by use of the parameter geomagnetic local time in place of either local mean or standard time. This is a consequence of the magnetic field being symmetrical, to a first approximation, about the magnetic axis rather than the geographic axis. *Geomagnetic Local Time* [1,2] is defined analogous to the conventional local time system by replacing the geographic axis with the geomagnetic axis and defining *geomagnetic noon* for a location P as being the time when the subsolar point lies on the geomagnetic meridian of P. Geomagnetic Local Time for any location P is thus given in angular measure by $180° + \Lambda_p - \Lambda_S$, where Λ_p and Λ_S are the geomagnetic longitudes of the location P and the subsolar point, respectively.

An additional coordinate parameter which has been effective in organizing the data on geomagnetic phenomena related to particle motions in space is the *invariant latitude*, $\Psi - \cos^{-1} 1/L^{\frac{1}{2}}$. Here, L is the shell parameter defined in Sec. 5h-22 that gives the distance in the geomagnetic equatorial plane (in units of earth radii R_e) to a field line having its intercept at the earth's surface at the invariant latitude Ψ. Although L, and thus Ψ, is usually computed from a spherical harmonic field description (Sec. 5h-5), it is noted here because invariant latitude is frequently used in place of the dipole geomagnetic latitude.

5h-5. Spherical Harmonic Description of the Main Field [1,2,5]. The main field can be described, outside its source region, by a scalar magnetic potential V satisfying Laplace's equation $\nabla^2 V = 0$. $\mathbf{B} = -\nabla V$ then defines the vector magnetic field. Expressed in a spherical harmonic series,

$$V = a \sum_{n=1}^{\infty} \left(\frac{a}{r}\right)^{n+1} \sum_{m=0}^{n} P_n{}^m (\cos \theta) (g_n{}^m \cos m\lambda + h_n{}^m \sin m\lambda)$$

where θ = geographic colatitude (see Sec. 5h-4 for other coordinate symbols).

$P_n{}^m$ (cos θ) = associated Legendre function of degree n and order m. The normalization introduced by Schmidt is most frequently used.

$g_n{}^m$, $h_n{}^m$ = constants called the *Gauss coefficients*.

The vector components of the magnetic field (Sec. 5h-3) are given by

$$X = \frac{1}{r}\frac{\partial V}{\partial \theta}$$

$$Y = -\frac{1}{r \sin \theta}\frac{\partial V}{\partial \lambda}$$

$$Z = \frac{\partial V}{\partial r}$$

X and Z above are applicable to a spherical earth; for accurate evaluations of X and Z on the surface of the actual earth, its oblateness should be taken into account.

TABLE 5h-1. GAUSS COEFFICIENTS FOR THE INTERNATIONAL GEOMAGNETIC REFERENCE FIELD [4] FOR EPOCH 1965.0*

(Units: gammas)

n	m	$g_n{}^m$	$h_n{}^m$	n	m	$g_n{}^m$	$h_n{}^m$
1	0	−30,339	0	6	2	4	106
1	1	−2,123	5,758	6	3	−229	68
2	0	−1,654	0	6	4	3	−32
2	1	2,994	−2,006	6	5	−4	−10
2	2	1,567	130	6	6	−112	−13
3	0	1,297	0	7	0	71	0
3	1	−2,036	−403	7	1	−54	−57
3	2	1,289	242	7	2	0	−27
3	3	843	−176	7	3	12	−8
4	0	958	0	7	4	−25	9
4	1	805	149	7	5	−9	23
4	2	492	−280	7	6	13	−19
4	3	−392	8	7	7	−2	−17
4	4	256	−265	8	0	10	0
5	0	−223	0	8	1	9	3
5	1	357	16	8	2	−3	−13
5	2	246	125	8	3	−12	5
5	3	−26	−123	8	4	−4	−17
5	4	−161	−107	8	5	7	4
5	5	−51	77	8	6	−5	22
6	0	47	0	8	7	12	−3
6	1	60	−14	8	8	6	−16

* Courtesy of J. C. Cain and S. J. Cain.

Table 5h-1 gives the Gauss coefficients for the International Geomagnetic Reference Field [4] for epoch 1965.0. Table 5h-2 gives a second set of coefficients determined directly from measurements by the *OGO*-2 and *OGO*-4 satellites [6] taken between October, 1965, and December, 1967, but referenced through use of secular change terms (Sec. 5h-6) to the same date, $t = 0$ at time 1965.0. Comparison of different sets of coefficients gives an indication of the accuracy of spherical harmonic descriptions of the main field. The relatively uniform latitude and longitude distribution of data from the polar orbiting *OGO* satellites selected for magnetically quiet periods ($K_p = 0$, Sec. 5h-13) provides a data set suitable for examining differ-

TABLE 5h-2. GAUSS COEFFICIENTS BASED ON THE *OGO*-2 AND 4 SATELLITE
DATA [6] AND REFERENCED TO EPOCH 1965.0*
(Units: gammas)

n	m	$g_n{}^m$	$h_n{}^m$	n	m	$g_n{}^m$	$h_n{}^m$
1	0	−30,338	0	8	3	−11	5
1	1	−2,114	5,768	8	4	−2	−18
2	0	−1,661	0	8	5	5	6
2	1	2,994	−2,013	8	6	−6	20
2	2	1,597	103	8	7	12	−3
3	0	1,298	0	8	8	10	−22
3	1	−2,041	−403	9	0	10	0
3	2	1,295	237	9	1	8	−22
3	3	856	−169	9	2	2	15
4	0	956	0	9	3	−12	4
4	1	807	153	9	4	14	−2
4	2	480	−269	9	5	0	−3
4	3	−387	16	9	6	0	10
4	4	260	−274	9	7	2	13
5	0	−221	0	9	8	4	1
5	1	360	17	9	9	0	4
5	2	250	128	10	0	−3	0
5	3	−33	−127	10	1	−2	2
5	4	−152	−100	10	2	2	1
5	5	−61	99	10	3	−5	2
6	0	46	0	10	4	−2	6
6	1	60	−11	10	5	7	−4
6	2	8	104	10	6	6	1
6	3	−229	70	10	7	0	−2
6	4	−3	−31	10	8	0	4
6	5	4	−12	10	9	2	3
6	6	−100	0	10	10	0	−10
7	0	71	0	11	0	2	0
7	1	−53	−62	11	1	−1	1
7	2	3	−27	11	2	−2	3
7	3	14	−6	11	3	4	−1
7	4	−31	7	11	4	−1	−3
7	5	−3	19	11	5	1	0
7	6	12	−24	11	6	−1	−1
7	7	3	−26	11	7	1	−2
8	0	10	0	11	8	2	−1
8	1	5	9	11	9	−1	−4
8	2	−4	−13	11	10	4	−1
				11	11	3	3

* Courtesy of J. C. Cain and S. J. Cain.

ences between calculated and measured values—particularly because the data, at altitudes >400 km, is not significantly affected by the crustal anomalies described in Sec. 5h-7. Using the *OGO* set of coefficients, the root-mean-square values of the residuals in scalar B for the difference (measured minus computed) for *OGO*-2 and *OGO*-4 data are 7 and 9 γ, respectively. The corresponding rms residuals, using the coefficients adopted for the International Geophysical Reference Field, are 39 and 57 γ, respectively.

Figure 5h-1 illustrates the form of the main magnetic field at the earth's surface in terms of the total scalar intensity B. The lack of dipole symmetry and the possible ambiguity of using the term magnetic pole without further definition are evident features. The preferred terminology is to use the term *dip pole* when referring to the location where $I = 0°$ and the term *geomagnetic pole*, or *dipole pole* (Sec. 5h-4), when referring to the surface intercepts of the axis of the centered dipole determined from the coefficients $g_1{}^0$, $g_1{}^1$ and $h_1{}^1$. It is further evident that pole positions do not correspond to locations of maximum field intensity.

5h-6. Secular Variations of the Main Field [1,2,7–9]. The main magnetic field as observed at the earth's surface varies noticeably on a time scale of several or more years. These changes are called the *secular variations*. The secular variations can be represented by maps with contours of equal rate of change for various quantities.

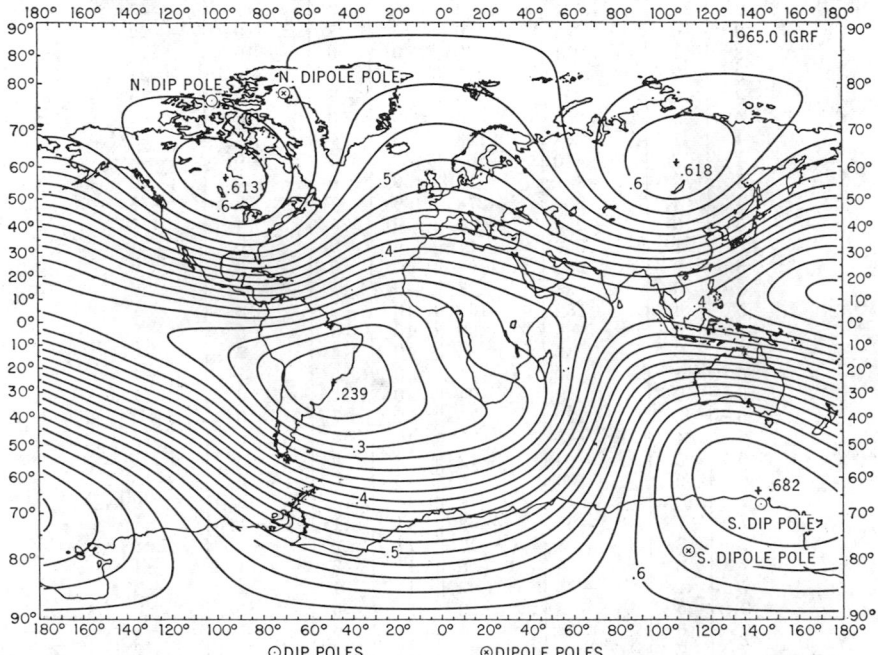

FIG. 5h-1. The earth's main magnetic field (Sec. 5h-5): the distribution of the magnitude B on the earth's surface as calculated from the 1965.0 International Geomagnetic Reference Field (Table 5h-1); B measured in gauss. (*Courtesy of S. J. Cain.*)

Such contours for the total force B are referred to as *isopors;* an example of an isopor chart for the epoch of 1965 is shown in Fig. 5h-2.

The secular variation is also expressed by regarding the Gauss coefficients (Sec. 5h-5) as being time-dependent [10]. The coefficients $g_n{}^m$ and $h_n{}^m$ are then replaced, respectively, by $g_n + \dot{g}_n{}^m t$ and $h_n{}^m + \dot{h}_n{}^m t$, where t is time measured in years from a specific epoch date. The coefficients $\dot{g}_n{}^m$ and $\dot{h}_n{}^m$ are given in Table 5h-3 for epoch 1965.0 for the International Geomagnetic Reference Field.

The nondipole part of the main field as expressed by the spherical harmonics for different epochs appears to indicate the presence of two types of gross features: (1) regions over which the pattern of deviations from a dipole field has remained relatively constant through the history of accurate measurements, and (2) regions over which the pattern of deviations appears to be drifting westward at rates of 0.2 to 0.3 deg/year. The analyses also suggest a poleward shifting of the dipole field as indicated by a change of 24 γ/year in the term $dg_2{}^0/dt$ (Table 5h-3) of the harmonic analysis which is the largest change observed in any single coefficient. There are also indications that the dipole moment (Sec. 5h-4) has decreased at a relatively uniform rate from about 8.55×10^{25} gauss cm^3 in 1835 [5] to 8.005×10^{25} in 1965.

5h-7. Crustal Anomalies [8,11,12]. The main magnetic field is locally distorted at the earth's surface by differences in the degree of magnetization of various rock formations. This magnetization is primarily dependent on the magnetite, Fe_3O_4, content and exists to the depth at which the Curie temperature (575°C for magnetite) is reached. Over many regions of the earth this depth is estimated to be about 20 km,

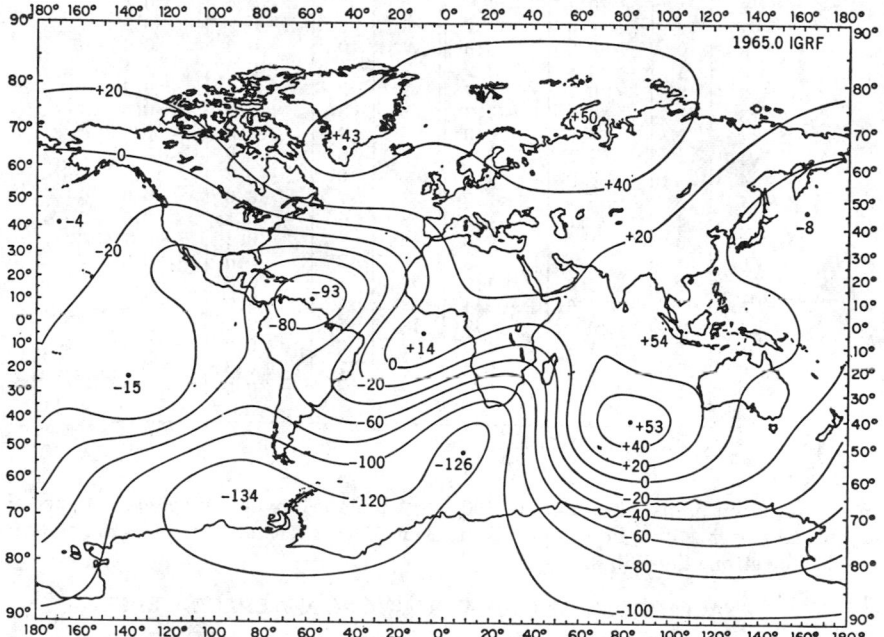

Fig. 5h-2. The secular variations (Sec. 5h-6) in the magnitude B of the main magnetic field: calculated from the secular variation terms in the 1965.0 International Geomagnetic Reference Field (Table 5h-3); units are gammas. (*Courtesy of S. J. Cain.*)

but variations in estimates range from 10 to 30 km. The magnetization is of two types: induced (proportional to the rock susceptibility and the main field B), and permanent (the remnant magnetization from a geologically earlier state of magnetization). Extensive mapping of magnetic anomalies has been carried out over many regions of the earth's surface to aid in inferring the subsurface geology. The distribution, intensity, and number of anomalies as a function of their dimensions are highly variable for different regions. However, considering dimensions >1 km, statistics indicate: (1) that the number of anomalies sharply decreases between anomaly dimensions of 25 and 50 km, (2) that between anomaly dimensions of 50 and 300 km the number of anomalies gradually decreases with increasing dimensions, and (3) that the comparatively small number of anomalies that have a·maximum dimension >300 km usually appear in a lineation pattern that marks a major geologic boundary resulting from large-scale tectonic activity. On a similar scale (resolution about 1 km), anomaly intensities are statistically indicated by noting that contour

TABLE 5h-3. SECULAR VARIATION TERMS FOR EPOCH 1965.0 IN THE SPHERICAL
HARMONIC EXPRESSION: BASED ON THE INTERNATIONAL GEOMAGNETIC
REFERENCE FIELD [4]*
(Units: gammas/year)

n	m	$\dot{g}_n{}^m$	$\dot{h}_n{}^m$	n	m	$\dot{g}_n{}^m$	$\dot{h}_n{}^m$
1	0	15.3	0.0	6	2	1.1	−0.4
1	1	8.7	−2.3	6	3	1.9	2.0
2	0	−24.4	0.0	6	4	−0.4	−1.1
2	1	0.3	−11.8	6	5	−0.4	0.1
2	2	−1.6	−16.7	6	6	−0.2	0.9
3	0	0.2	0.0	7	0	−0.5	0.0
3	1	−10.8	4.2	7	1	−0.3	−1.1
3	2	0.7	0.7	7	2	−0.7	0.3
3	3	−3.8	−7.7	7	3	−0.5	0.4
4	0	−0.7	0.0	7	4	0.3	0.2
4	1	0.2	−0.1	7	5	−0.0	0.4
4	2	−3.0	1.6	7	6	−0.2	0.2
4	3	−0.1	2.9	7	7	−0.6	0.3
4	4	−2.1	−4.2	8	0	0.1	0.0
5	0	1.9	0.0	8	1	0.4	0.1
5	1	1.1	2.3	8	2	0.6	−0.2
5	2	2.9	1.7	8	3	0.0	−0.3
5	3	0.6	−2.4	8	4	−0.0	−0.2
5	4	0.0	0.8	8	5	−0.1	−0.3
5	5	1.3	−0.3	8	6	0.3	−0.4
6	0	−0.1	0.0	8	7	−0.3	−0.3
6	1	−0.3	−0.9	8	8	−0.5	−0.3

* Courtesy of J. C. Cain and S. J. Cain.

maps are often prepared in units of 100 γ. Anomalies involving changes of several
thousand gammas over horizontal distances of several tens of kilometers are considered to be strong anomalies.

SURFACE GEOMAGNETIC VARIATIONS OF SPACE ORIGIN

5h-8. Solar Daily Variation on Quiet Days [1–3]. The regular magnetic variation
observed on an average magnetically quiet day is called the *Sq variation*. The source
of *Sq* is a system of electric currents in the *E* region of the ionosphere (Sec. 5h-15).
Systematic winds produced mainly by the solar heating of the atmosphere generate
the current through a dynamo action which is effective between roughly 95 and 130
km where the electric conductivity is high. The relevant conductivity elements are
given in Sec. 5h-18.

Examples of magnetograms on a typical quiet day are shown in Fig. 5h-3 for observatories at low and middle latitudes and near the dip equator. An illustrative picture
of the *Sq* current system is shown in Fig. 5h-4. Within about 200 km of the dip
equator, $I = 0°$ (Sec. 5h-3), the *Sq* range in *H* is very large, as indicated by the
Huancayo, Peru, magnetogram in Fig. 5h-3. The concentrated ionospheric current
along the dip equator on the day side of the earth which produces this large change in
H is referred to as the *equatorial electrojet*. It results from an enhanced effective conductivity (Cowling conductivity, Sec. 5h-18) in the east-west direction in a narrow
belt at the dip equator. Typical ranges in the magnitude of *Sq* at the earth's surface
are roughly 50 to 250 γ at the dip equator and 30 to 60 γ at magnetic latitudes 10 to
60°. Both the form and magnitude of *Sq* can vary considerably from one day to the
next and also with season and solar activity.

5h-9. Lunar Geomagnetic Variation [1,2].   The tidal oscillation of the atmosphere due to the gravitational force of the moon generates an electric current in the ionosphere which gives rise to the *lunar geomagnetic variation,* conventionally denoted by *L*.   Amplitudes of *L* are about one-tenth of those of *Sq*, and are not generally recognizable in the magnetograms by the eye.   An exception to this is *L* near the dip equator, where its amplitude is enhanced over the corresponding amplitudes at higher latitudes by a rate greater than that for the similar enhancement in *Sq* (Sec. 5h-8).   The current system for *L* varies greatly with lunar phase and season.   (*Note.* The use of *L* for lunar variations is not related to the *L* shell parameter described in Sec. 5h-22.)

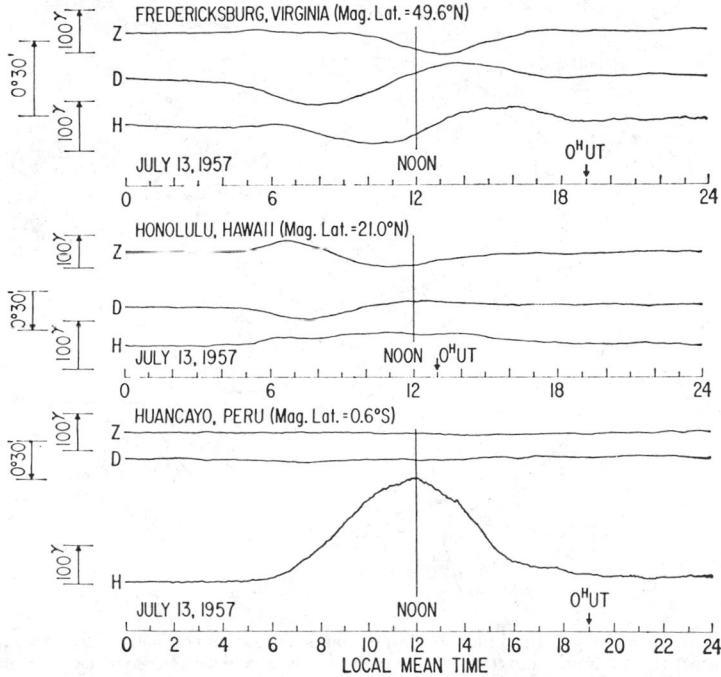

Fig. 5h-3. Typical magnetograms on a quiet day, showing the solar daily variation *Sq* (Sec. 5h-8). *H* is the horizontal component, *D* the declination, and *Z* the vertical component. The *H* trace for Huancayo shows the effect of the equatorial electrojet.

5h-10. High-latitude Magnetic Disturbances [2,3,13,14].   At high latitudes the magnetic field is disturbed more frequently and more severely than at middle and low latitudes.   Disturbances are statistically greatest along oval-shaped belts encircling the dipole poles in each hemisphere roughly at Θ (or $90° - \Psi$) = 15 to 20° near noon, 17 to 23° near the twilight meridians, and 19 to 26° near midnight [15–17]. (*Note.* Θ is used in this section as geomagnetic colatitude for each hemisphere.) As the ionospheric electric currents causing the magnetic disturbances in these belts are associated with the occurrence of aurora, they are usually referred to as the *auroral belts* or *auroral ovals* [18].   Within these belts maximum disturbance intensities are most commonly encountered at Θ = 20 to 25° in the nighttime geomagnetic local time range (Sec. 5h-4) 20^h to 04^h.   The frequency of occurrence of aurora is also a maximum in these nighttime strips, and this statistic has led to the designation of the strips centered near $\Theta = 23°$ as the *auroral zones*. The colatitudes Θ indicated above decrease by several degrees during times of weak activity and increase during times of intense activity and magnetic storms. During intense storms values of Θ as great as 35° are not uncommon in the midnight sector.

Figure 5h-5 is an idealized representation of the distribution of ionospheric electric currents for a disturbance of moderate intensity (*Note.* Because current may also flow along magnetic field lines to magnetospheric regions, currents are not drawn as being continuous within the ionosphere.) Several major features to note are: (1) The most intense current flows westward in the auroral belt toward the 22^h to 23^h region from the morning sector. This concentration of current is often called the

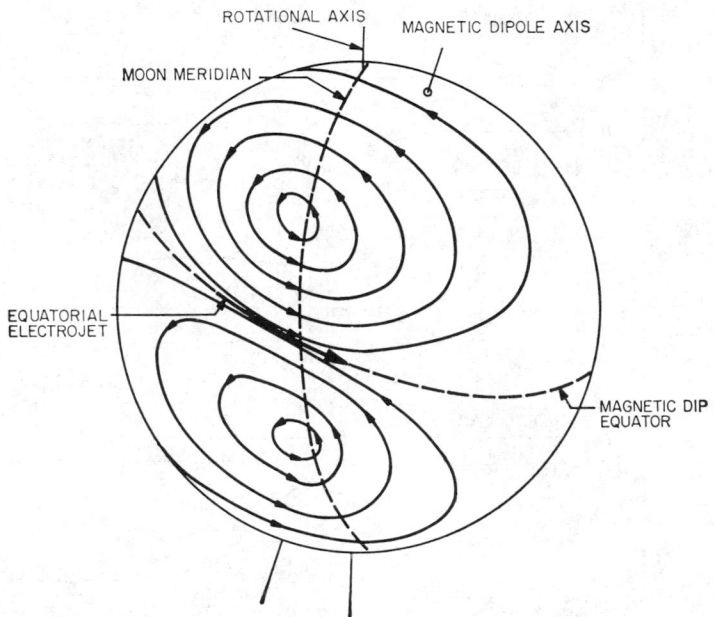

FIG. 5h-4. Illustration of the ionospheric current system for the solar daily variation Sq (Sec. 5h-8); the stronger current vortex represents the average condition in northern summer.

auroral electrojet. (2) A weaker auroral electrojet current flows eastward toward the 22^h to 23^h region from the evening sector. (3) A comparatively uniform current flows across the *polar cap* (the region poleward from minimum Θ for the auroral belt) from the late evening sector to the late morning sector. (4) Much weaker currents spread outward to the middle-latitude regions. Although shown only for one hemisphere, the same distribution occurs simultaneously in the other hemisphere, indicating that the magnetic field lines connecting the two auroral belts are also lines of electrical equipotentials in the electric field system (Sec. 5h-28) driving the electrojet currents. Sounding rocket measurements have shown that the auroral electrojets are primarily Hall currents (Sec. 5h-18) driven by horizontal electric fields (Sec. 5h-28) in the region of the ionosphere where energetic electrons precipitate from the magnetosphere and increase the ionization and hence the conductivity.

Enhancements of these disturbances typically last one to several hours and are often repeated at intervals of several hours. Disturbances of this type are sometimes referred to as *magnetic bays* (a term originating from the resemblance of the H trace on magnetograms during such events to the coastline of a bay on a map). Directly beneath the auroral electrojets, variations are largest in the H component and usually lie within the range 100 to 1,000 γ, but may on occasions reach values greater than 2,000 γ. Large bay disturbances often occur during magnetic storms (Sec. 5h-11). The magnetic activity indices K_p and AE (Sec. 5h-13) are essentially measures of the severity of magnetic disturbances in the auroral belt.

5h-11. Magnetic Storms [1,2,19]. Worldwide magnetic disturbances lasting one to several days are called *magnetic storms*. Magnetic storms often, but not always, begin with a sudden, worldwide increase in the magnetic field. This sudden field increase, called a *sudden commencement* and denoted by *SC*, is due to a compression of

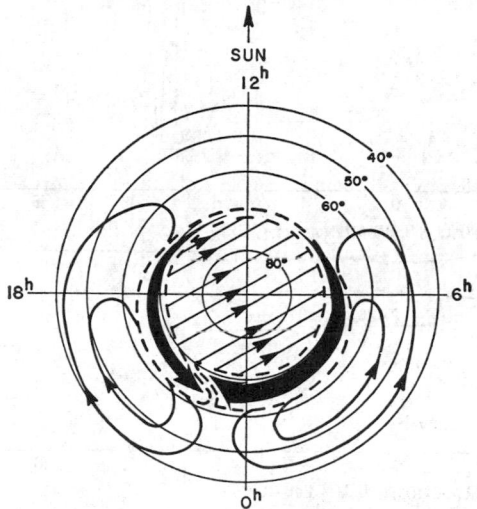

FIG. 5h-5. Illustration of ionospheric currents for a high-latitude disturbance. Concentric circles represent geomagnetic latitude $(90° - \Theta)$ circles.

the magnetosphere by a sudden increase in the pressure on the magnetosphere boundary exerted by the solar wind; magnetic variations of this type are discussed in Sec. 5h-12. Examples of magnetograms taken during a magnetic storm are shown in Fig. 5h-6.

Typically, one-half to a few hours after an *SC* the magnetic field begins to decrease all over the globe, and after reaching a minimum in several hours recovers slowly toward the prestorm level. The field decrease, often called a *Dst* decrease (Sec. 5h-13) is due to a creation of a plasma belt at a distance of 2.5 to 6 earth radii [20]. The plasma belt exerts magnetic stress such that the magnetic field in the magnetosphere is inflated, resulting in a field decrease inside the plasma belt and an increase outside it [21]. In the developing phase of a magnetic storm the plasma belt grows most rapidly in the late afternoon sector of the magnetosphere, and consequently the field decrease observed on the earth is also larger in this sector than in other sectors. The magnetic stress of the plasma belt mainly comes from its protons in the energy range 10 to 100 kev.

During magnetic storms high-latitude disturbances of the type discussed in Sec. 5h-10 occur in greater intensity than under normal conditions (e.g., see College, Alaska, in Fig. 5h-6). Field variations produced by an auroral electrojet may reach 3,000

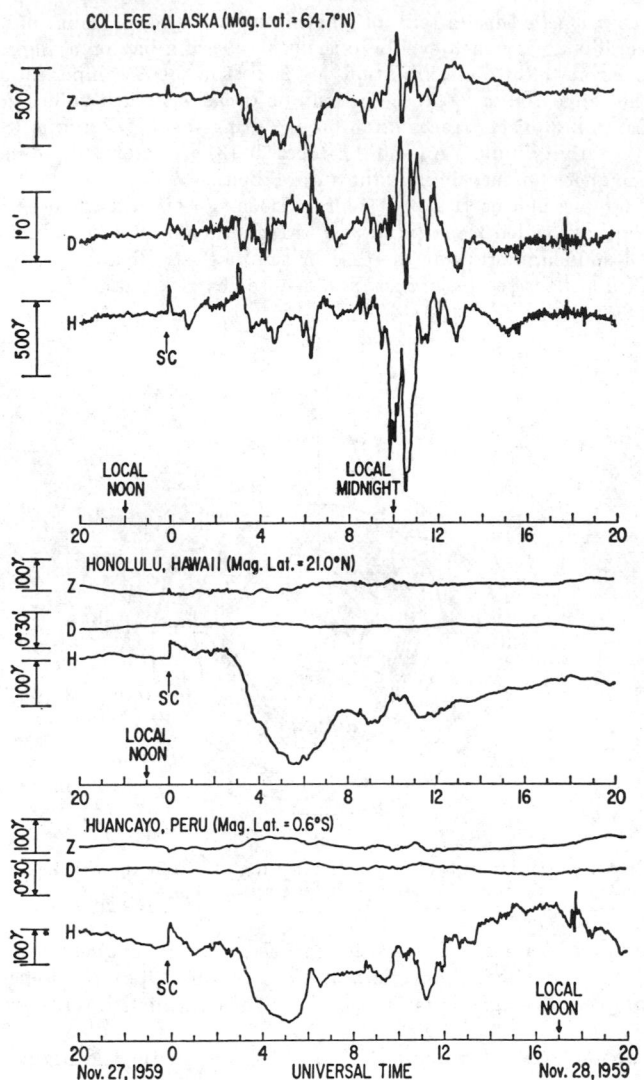

COLLEGE, ALASKA (Mag. Lat. = 64.7°N)

HONOLULU, HAWAII (Mag. Lat. = 21.0°N)

HUANCAYO, PERU (Mag. Lat. = 0.6°S)

UNIVERSAL TIME

Nov. 27, 1959   Nov. 28, 1959

FIG. 5h-6. Magnetograms taken during a magnetic storm (Sec. 5h-11): examples from high-latitude (top) and low-latitude (middle and bottom) stations.

γ during an exceptionally large storm. In such a storm the Dst decrease may exceed 400 γ.

The frequency of occurrence and the average intensity of magnetic storms are statistically correlated with solar activity. However, individual magnetic storms can by no means be traced to active regions on the sun, and conversely solar flares do not always produce magnetic storms on the earth. During the International Geophysical Year, July, 1957, to December, 1958, representing a period of maximum solar activity, magnetic storms with Dst decreases exceeding 40 γ occurred at a rate of 55 storms per year, and the average Dst decrease was 110 γ, with a maximum of

434 γ observed on September 13, 1957, which was still the largest (as of November 1, 1968) since 1957. In 1964 a year of solar activity minimum, there were only 15 magnetic storms (as defined above), and the average Dst decrease of these storms was 61 γ, with a maximum of 109 γ [22,23].

5h-12. Magnetic Variations Caused by Compressions and Expansions of the Magnetosphere. Compression or expansion of the magnetosphere responding to an increase or decrease of the solar wind pressure on its boundary surface produces a class of magnetic variations of considerable interest, in spite of their generally small magnitudes. Sudden commencements (SC) of magnetic storms (Sec. 5h-11) and *sudden impulses* belong to this class. A sudden impulse, often denoted by SI, is a sudden increase or decrease in the magnetic field observed simultaneously over the world [2].

The theoretical magnetic field change ΔB at the ground is related to changes in the proton density n and the velocity v of the solar wind approximately by the formula [24]

$$\Delta B = 0.03\Delta(n^{\frac{1}{2}}v)$$

where ΔB is in γ, n in protons/cm^3, and v in km/sec. For sudden commencements ΔB ranges from several to 100 γ, whereas ΔB for sudden impulses is normally less than 20 γ. The effect of an abrupt compression of the magnetosphere boundary is transmitted inward hydromagnetically [25]. Presence of continuous time variations in Dst (Sec. 5h-13) with appreciable amplitudes suggests that occurrences of more gradual, as compared with sudden, compression or expansion of the magnetosphere are not infrequent, but their significance has not as yet been explored.

5h-13. Magnetic Indices [1,26,27]. *Planetary 3-hourly Index Kp.* The Kp index expresses the intensity of geomagnetic activity mainly at high latitudes (Sec. 5h-10) for each 3-hr interval of the Greenwich day, in a scale of thirds in the order:

0o 0+ 1− 1o 1+ 2− 2o 2+ · · · 8− 8o 8+ 9− 9o

which may be condensed to a scale of integers from 0 to 9. Kp is based on magnetic records from 12 selected observatories lying between 47.7 and 62.5° dipole latitude with the average of 56°. The Kp index is published regularly by the International Association of Geomagnetism and Aeronomy in the No. 12 series of its Bulletin. Indices ap, Ap, Ci, Cp, etc., are also found in the same publication.

Three-hourly Equivalent Planetary Amplitudes, ap. The conversion from Kp to ap is made according to the following table:

$Kp =$	0o	0+	1−	1o	1+	2−	2o	2+	3−	3o	3+	4−	4o	4+
$ap =$	0	2	3	4	5	6	7	9	12	15	18	22	27	32
$Kp =$	5−	5o	5+	6−	6o	6+	7−	7o	7+	8−	8o	8+	9−	9o
$ap =$	39	48	56	67	80	94	111	132	154	179	207	236	300	400

At a standard station the average range of the most disturbed of the three magnetic components is 2 ap (γ); for instance, if $Kp = 6+$, the range is 188 γ. The scale for ap is linear, while that for Kp is quasi-logarithmic.

Daily Equivalent Planetary Amplitude, Ap. The average of the eight values of ap for each day is the daily equivalent planetary amplitude Ap. Hence Ap is also expressed in units of 2 γ for a standard station.

Daily International Character Figure Ci, and Daily Planetary Character Figure Cp. The daily international character figure Ci is the average of the daily character figure C for all collaborating observatories; Ci varies from 0.0 to 2.0. Ci is available for every day since 1884.

A more recently introduced substitute index Cp is derived from the daily sum of ap according to the following scheme.

ap sum up to:

22	34	44	55	66	78	90	104	120	139	164	190	228
$Cp = 0.0$	0.1	0.2	0.3	0.4	0.5	0.6	0.7	0.8	0.9	1.0	1.1	1.2

ap sum up to:

273	320	379	453	561	729	1119	1399	1699	1999	2399	3199	3200
$Cp = 1.3$	1.4	1.5	1.6	1.7	1.8	1.9	2.0	2.1	2.2	2.3	2.4	2.5

Though values of Ci and Cp are found to be nearly the same, Cp is more reliable and should be used in preference to Ci.

Auroral Electrojet Index AE [28]. The AE index is designed to be a measure of global auroral electrojet activity with higher time resolution than Kp. The index is normally given at 2.5-min intervals, but hourly averages are also used. AE represents the instantaneous range of disturbance of the horizontal component H from a set of observatories that are relatively uniformly distributed in longitude between magnetic colatitudes θ of 30 and 19°. $AE = \Delta H$ (maximum) $+ |\Delta H$ (minimum)$|$, where ΔH (maximum) is taken from the observatory showing the maximum positive deviation of H, and $|\Delta H$ (minimum)$|$ is taken from the observatory showing the largest negative value in H.

Dst Index [22]. The component of disturbance magnetic field that is axially symmetric with respect to the geomagnetic dipole axis is called Dst. The Dst index, computed at hourly or 2.5-min intervals, is useful for studies of magnetic storm phenomena (Sec. 5h-11). $Dst = 1/n \ (\Delta H_1 + \Delta H_2 + \cdots + \Delta H_n)$ represents the average deviation of the horizontal component from quiet-day values for a set of n observatories that are relatively uniformly distributed in longitude and located at low ($<35°$) magnetic latitudes.

5h-14. Induced Currents in the Earth [1,29–31]. Since the solid earth is a good conductor, currents are induced within the earth by any time-varying magnetic field changes imposed on it; and these induced currents in turn modify the field changes observed on and above the earth. For magnetic fields applied from sources external to the earth, the longer the periods of the changes, the greater the depth of the penetration of the fields. This property is used to estimate the conductivity distribution within the earth; a method frequently used is to assume a model with a set of parameters which are to be determined so as to fit the observed variations on the earth. In the case of slow variations such as Sq, approximately two-thirds of the variation observed on the earth is directly from the currents in the ionosphere, and one-third from the currents induced in the earth. A simple conductivity model often used assumes a uniform, conducting sphere surrounded by a less conducting shell. Using Sq and disturbance fields of various periods, the conductivity in this sphere in such a model has been estimated to be of the order of 10^{-1} mho meter^{-1}; values of this magnitude may be considered as giving conductivities for the region roughly from 500 to 1,000 km depth. Estimates of conductivity in the lower mantle are based on studies of secular variations (Sec. 5h-6) or on direct calculations of conduction processes in rocks and minerals; though the reliability of these estimates is uncertain, values of the order of 10^2 mho meter^{-1} have been given. In the earth's molten core (below about 2,900 km depth) where the dynamo mechanism is thought to be operative maintaining the earth's field, conductivity estimates of the order of 10^5 to 10^6 mho meter^{-1} have been suggested. Upper mantle conductivity seems to have regional anomalies which can make various types of transient geomagnetic variations appear quite different at stations close together. Near the surface of the earth the crustal conductivity can vary from one region to another by a large factor, since the conductivity depends on the type of rocks, mineral contents, temperature pressure, and water content. The most significant variables are the water content and its salinity. Such factors as regional variations in the thickness of the crust and the dependence of the conductivity on the frequency of incident electromagnetic waves further make a general statement on the crustal conductivity virtually impracticable. Effects of the electromagnetic induction in the oceans are not negligible in certain types of geomagnetic variations.

IONOSPHERE

5h-15. General [2,32–34]. The *ionosphere* extends from about 50 km above the ground to altitudes at which hydrogen ions become the main ion constituent. However, the definitions of the upper and lower boundaries of the ionosphere and its divisions into several regions, C, D, E, and F, are made for conventional, and not necessarily physically compelling, reasons.

The C region, extending to about 70 km height, is produced mainly by cosmic rays, and its peak electron density near 65 km height is of the order of 10^2 electrons cm^{-3}. Above the C region the ionization is due to the solar radiation with wavelengths from 1215.7 Å (for Lyman α of neutral hydrogen) to those of X rays. Altitudes from approximately 70 to 90 km are referred to as the D region, where the main ion constituents are NO^+ with O_2^+ and N_2^+ as secondary members. Hydrogen Lyman α is the main ionizing agency for NO^+. The peak electron density of the D region is of the order of 10^3 electrons cm^{-3} at 75 to 80 km height. Negative ions, particularly O_2^-, NO_2^-, and NO_3^-, are abundant in the nighttime D region. Their concentration decreases greatly in daytime as a consequence of photodetachment of the electrons in sunlight. The E region extends from about 90 to 140 km height. The ions there are mainly NO^+ and O_2^+, with N_2^+ ions as a secondary constituent. The steep increase in the electron density in the lower part of the E region is due to the ionization by X rays. Hydrogen Lyman β at 1025.6 Å and CIII at 976 Å are also important ionizing agencies. The maximum electron density in the E region is of the order of 10^5 electrons/cm^3 (see Sec. 5h-16 for electron density profiles). The F region, beginning from the top of the E region near 140 km height, has its peak electron density of the order of 10^6 electrons cm^{-3} between approximately 200 and 400 km height. The region above this F peak is often referred to as the *topside ionosphere*. The main ion constituent in the F region is O^+, mainly produced by EUV radiation in the wavelength range 300 to 800 Å. Although there is no agreed upper boundary for the F region, this may be taken to be the altitude beyond which hydrogen ions become the predominant ion constituent. This altitude depends upon the temperature, but is typically near 1,000 km; the region above this height is often called the *protonosphere*. In the transition from the region where O^+ ions are predominant to the region where H^+ ions take over, He^+ ions may become the predominant constituent.

The F region is sometimes divided into F_1 and F_2 layers because of the apparent ledge in the electron density profile, e.g., as seen in Fig. 5h-7. However, it is questionable if this separation is physically definable.

5h-16. Electron Density Profiles. The electron density profile in the ionosphere is variable to a great extent with latitude, local time, season, and solar activity. Several examples of these variations are shown in Fig. 5h-7. Variations in the ionospheric electron density involve considerable complexity; for instance, its diurnal variation is generally not symmetric with respect to local noon. The height distribution of the electron density in the F region is determined by a combination of photochemical equilibrium and diffusive equilibrium. The global distribution of the F-region electron

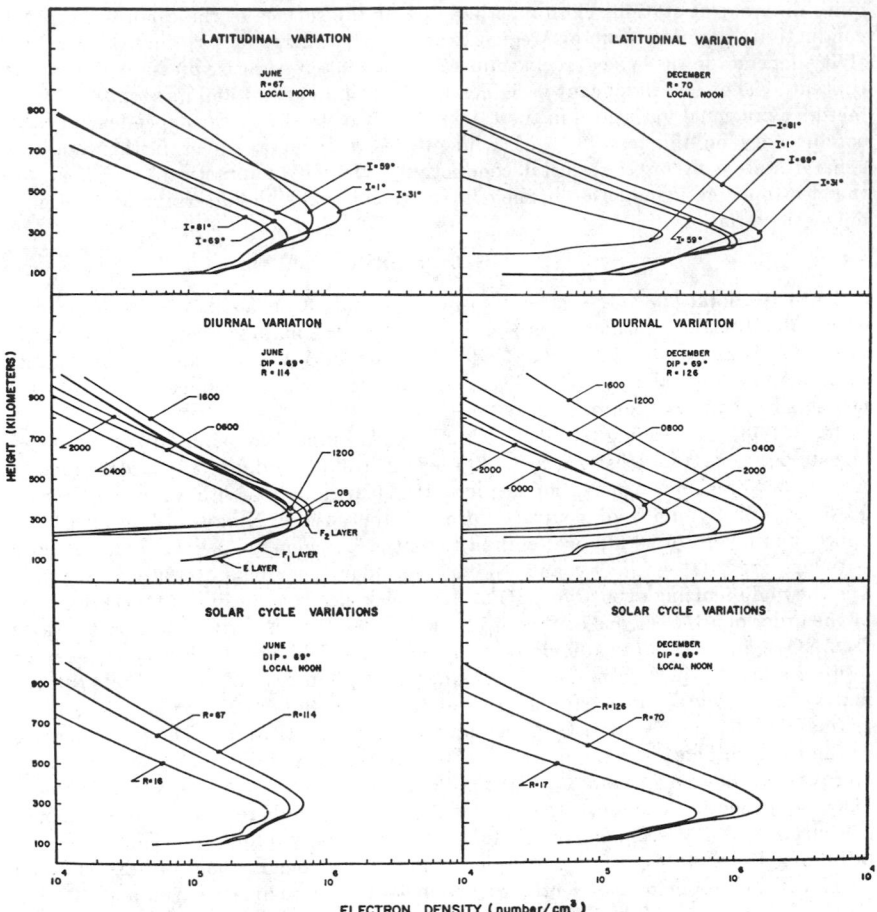

FIG. 5h-7. Monthly median electron density profiles (Sec. 5h-16): latitudinal, diurnal, and solar cycle variations. Below the peak of the F region the profiles were derived from ionograms, and above this altitude they were extrapolated exponentially. The stations used for the magnetic dip angles of 1, 31, 59, 69, and 81° are Huancayo, Bogota, Grand Bahama, Wallops Island, and Godhaven, respectively. (*Courtesy of T. E. VanZandt and A. R. Laird.*)

density is controlled to an appreciable degree by the geomagnetic field through its influence on the mass transport of ionization. An example of this nature is the "equatorial anomaly" exhibiting a minimum electron density over the magnetic equator and two maxima about 20° off, and on both sides of, the equator. Other outstanding anomalous characteristics of the F-region electron density include the high electron densities at middle latitudes near December in both hemispheres, i.e., the "December anomaly," and the larger electron densities in local winter than in summer, i.e., the "winter anomaly."

5h-17. Ionospheric Irregularities [35–37]. A variety of irregularities in the electron density distribution have been recognized as resulting from dynamical behavior involving ionospheric wind shears, particle precipitation from the magnetosphere, electrostatic wave instabilities, ionization by meteoritic impact, etc. Several of the most common types are noted below.

Thin horizontal layers of abnormally high electron density and sharp ledges in the electron density altitude profile are frequently found within the regular E region. These layers are called *sporadic E* [38,39] and denoted by E_S. The layers are often only several hundred meters to a few kilometers in thickness at temperate latitudes where they are thought to be produced primarily by wind shears. In auroral regions sporadic E occurs in many forms involving patches and magnetic field aligned irregularities of highly variable thickness. The combined effect of nonuniform particle precipitation from the magnetosphere and drift-wave instabilities in the ionosphere is largely responsible for the complexity of the auroral ionization which is called *auroral sporadic E*. In association with visible aurora the electron density is found to be roughly proportional to the square root of the optical brightness of the most prominent visible auroral emissions. Over the magnetic dip equator, embedded in the equatorial electrojet, sporadic-E ionization occurs in the form of magnetic field aligned irregularities of small dimensions (<1 meter to hundreds of meters) and small density contrast (approximately 1 percent). These *equatorial irregularities*, which are effective as low-frequency radar reflectors, are believed to be the result of a plasma instability which is closely related to the intensity of the equatorial electrojet current (Sec. 5h-8).

In, and above, the F region, *magnetic field aligned irregularities* are particularly prevalent over the polar caps and auroral belts. Although broken up into filamentary columns with cross-section dimensions of roughly 0.1 to 2 km, the individual filaments are probably >100 km in length. Density contrasts between adjacent filaments can involve changes by factors of 2, but contrasts of the order of 1 to 10 percent are much more common. When observed from the ground using radio sounding techniques, these irregularities give the normal F region the appearance of being spread upward. The term *spread F* has been commonly used to describe this appearance of the F region. Spread F is also observed frequently over the magnetic dip equator but appears to be more confined to F-region altitudes.

5h-18. Electric Conductivity in the Ionosphere [2,40,41]. Because of the low collision frequencies and of the presence of the strong magnetic field, the conductivity in the ionosphere above 70 km height is highly anisotropic for low-frequency electromagnetic waves and for d-c electric fields. Writing the current density **j** as

$$\mathbf{j} = \sigma_0\,\mathbf{E}_\| + \sigma_1\,\mathbf{E}_\perp + \sigma_2\,(\hat{\mathbf{B}} \times \mathbf{E})$$

where **B** = magnetic induction
    E = electric field intensity
    $\hat{\mathbf{B}} = \mathbf{B}/|\mathbf{B}|$
    $\mathbf{E}_\| = (\mathbf{E}\cdot\hat{\mathbf{B}})\hat{\mathbf{B}}$ = electric field intensity parallel to **B**
    $\mathbf{E}_\perp = \mathbf{E} - \mathbf{E}_\|$ = electric field intensity perpendicular to **B**

The conductivity elements σ_0, σ_1, and σ_2 are called the *direct* (or *longitudinal*), *Pedersen*, and *Hall conductivities*, respectively. For electromagnetic waves of frequency ω, conductivities σ_0, σ_1, and σ_2 are respectively given, in mks units, by

$$\sigma_0 = ne^2\left[\frac{1}{m_e(\nu_e - i\omega)} + \frac{1}{m_i(\nu_i - i\omega)}\right]$$

$$\sigma_1 = ne^2\left\{\frac{\nu_e - i\omega}{m_e[(\nu_e - i\omega)^2 + \omega_e{}^2]} + \frac{\nu_i - i\omega}{m_i[(\nu_i - i\omega)^2 + \omega_i{}^2]}\right\}$$

$$\sigma_2 = ne^2\left\{\frac{\omega_e}{m_e[(\nu_e - i\omega)^2 + \omega_e{}^2]} - \frac{\omega_i}{m_i[(\nu_i - i\omega)^2 + \omega_i{}^2]}\right\}$$

where n = electron density in electron-ion pairs/m³
    e = electronic charge in coulombs
    $m_{e,i}$ = electron, or ion, mass in kilograms
    $\omega_{e,i} = eB/m_{e,i}$ = cyclotron (or gyro) frequency in radians/sec

The d-c conductivity is obtained by setting $\omega = 0$. In the case when the electric field is perpendicular to the magnetic field, the energy loss per unit volume due to joule heating can be expressed as $\mathbf{j} \cdot \mathbf{j}/\sigma_3$ ($= \mathbf{j} \cdot \mathbf{E}$), where

$$\sigma_3 = \sigma_1 + \frac{\sigma_2{}^2}{\sigma_1}$$

The effective conductivity σ_3 is called the *Cowling conductivity*. If the electric field is parallel to the magnetic field, the energy loss is $\mathbf{j} \cdot \mathbf{j}/\sigma_0$ as in the absence of the magnetic field. The Cowling conductivity plays an important role in the dynamo

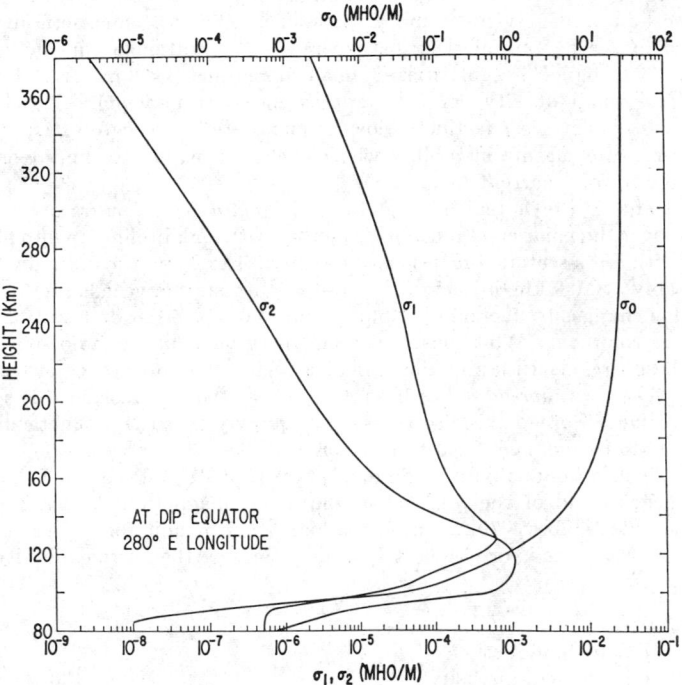

FIG. 5h-8. Typical conductivity profile (Sec. 5h-18); the upper scale for σ_0, the direct conductivity, and the lower scale for σ_1 and σ_2, the Pedersen and Hall conductivities, respectively.

region of the ionosphere. For instance, over the dip equator, where the magnetic field is horizontal, the effective conductivity for a horizontal electric field directed perpendicular to the magnetic field becomes σ_3 when the vertical Hall current is inhibited by polarization of the medium. The value of σ_3 is larger over the dip equator, and this is the reason for the existence of the equatorial electrojet mentioned in Sec. 5h-8.

The conductivities are latitude-dependent through the magnetic field, and also vary with the electron density distribution and hence with local time, latitude, season, and solar activity. The following rules generally hold: (1) The direct conductivity σ_0 is much larger than σ_1 and σ_2 throughout the ionosphere above about 70 km height; at this height σ_0 and σ_1 are approximately equal, indicating that the conductivity is nearly isotropic. (2) In the 90- to 130-km region, where the Sq (and L) current flows (Sec. 5h-8), σ_2 is larger than σ_1. (3) Above this region σ_1 becomes greater than σ_2. Figure 5h-8 shows a typical conductivity profile for noon over the dip equator.

5h-19. Refractive Index in the Ionosphere [42]. For a plane electromagnetic wave propagating in a homogeneous magneto-ionic medium permeated by a uniform magnetic field, the complex refractive index n in the Appleton-Hartree approximation is given (in the standard notations and mks units) by

$$n^2 = 1 - \frac{X}{1 - iZ - \frac{1}{2}Y_T{}^2/(1 - X - iZ) \pm [\frac{1}{4}Y_T{}^4/(1 - X - iZ)^2 + Y_L{}^2]^{\frac{1}{2}}}$$

where $\omega_0{}^{\cdot} = Ne^2/\epsilon_0 m$

$\omega_H = \mu_0 H_0 |e|/m$

$\omega_L = (\mu_0 H_0 e/m) \cos\theta$

$\omega_T = (\mu_0 H_0 e/m) \sin\theta$

$X = \omega_0{}^2/\omega^2$

$Y_L = \omega_L/\omega$

$Y_T = \omega_T/\omega$

$Z = \nu/\omega$

ω = angular frequency of the wave

ϵ_0 = electric permittivity of free space

μ_0 = magnetic permittivity of free space

e = electron charge, numerically negative

m = electron mass

H_0 = magnitude of the ambient magnetic field $\mathbf{H_0}$

N = number density of electrons

θ = angle between $\mathbf{H_0}$ and the direction of propagation of the plane wave

ν = frequency of collision of electrons with heavy particles

The upper and lower signs in the above equation correspond to the *ordinary* and the *extraordinary* wave, respectively. For ionospheric propagation there are two useful approximations: the quasi-longitudinal (QL) and the quasi-transverse (QT) approximations.

For QL:

$$n^2 \doteq 1 - \frac{X}{1 - iZ \pm |Y_L|}$$

where the upper and lower signs are for the ordinary and the extraordinary wave, respectively.

For QT:

$$n^2 \doteq 1 - \frac{X}{1 - iZ + (1 - X - iZ)\cot^2\theta}$$

for the ordinary wave, and

$$n^2 \doteq 1 - \frac{X}{1 - iZ - Y_T{}^2/(1 - X - iZ)}$$

for the extraordinary wave.

These approximations are valid if the following conditions are satisfied:

$$\frac{Y_T{}^4}{4Y_L{}^2} \ll |(1 - X - iZ)^2| \qquad \text{for } QL$$

$$\frac{Y_T{}^4}{4Y_L{}^2} \gg |(1 - X - iZ)^2| \qquad \text{for } QT$$

The existence of the two modes, the ordinary and the extraordinary, is due to the presence of the magnetic field. In the absence of the latter the refractive index n would be given, ignoring the collision effects and the ion motions, by

$$n^2 = 1 - \frac{Ne^2}{\epsilon_0 m\omega^2}$$

A plane wave of angular frequency ω propagating vertically upward in a medium with electron density N varying with height will be reflected back at the height at which N reaches a value that makes n vanish; this value of N is given by

$$N = 1.24 \times 10^{-8} f^2$$

in terms of the wave frequency f ($= \omega/2\pi$) in hertz. Given N, the angular frequency ω_0 ($= \sqrt{Ne^2/\epsilon_0 m}$) is called the *electron plasma frequency*. If there is a peak in the electron density, say N_m, the plasma frequency is also a maximum, say ω_m, at the same height. Then radio waves with angular frequency ω less than ω_m will all be reflected from varying heights, and those with ω greater than ω_m will penetrate through the medium. The frequency f_m ($= \omega_m/2\pi$) is called the *critical* or *penetration* frequency of the medium. The electron collisions with heavy particles result in an absorption of the wave energy and a modification of the refractive index. In the ionosphere the latter modification is generally slight, and reflection still occurs near the level where the plasma frequency is equal to the wave frequency. The presence of the geomagnetic field introduces another complication, namely, that an incident wave is split into two waves, ordinary and extraordinary, with different polarizations, which propagate in general independently. The penetration frequencies for these two waves are different. For the ordinary wave this frequency is, under normal conditions, the maximum plasma frequency mentioned above.

MAGNETOSPHERE

5h-20. The Magnetic Field Configuration [2,43,44]. The earth's magnetic field is confined to a bounded region of space by the solar wind plasma (Sec. 5h-21); this region is called the *magnetosphere*, and its outer boundary, the *magnetopause*. Its inner boundary is not well-defined, but may be regarded as the altitude above which the convection of the plasma and the motion of the energetic charged particles are predominantly controlled by the geomagnetic field. If this altitude is taken to be that at which the cyclotron frequency of the main constituent ions becomes comparable with their collision frequency, this is about 150 km height or near the base of the ionospheric F region.

The magnetopause along the sun-earth line is at about $10R_e$ ($1R_e$ = earth's radius = 6371.2 km, Fig. 5h-9), and flares out toward its flanks to about $14R_e$ in the meridian plane through 6 and 18 hr local time [45,46]. Beyond about $20R_e$ behind the earth the magnetopause is nearly a cylindrical surface of radius 15 to $20R_e$, enclosing the *magnetosphere* tail [46]. The surface current on the magnetopause (which is a consequence of the solar wind particles being reflected away from the magnetopause) increases the magnetic field inside it; the effect of the current is such that the magnetic field intensity just inside the magnetopause is made approximately twice as large as it would be if there were no confinement of the field. Roughly speaking, the surface current flows from the morning to the evening side on the sunlit part of the magnetopause and closes its circuit by flowing around two foci, one in each hemisphere, so as to form two vortices around them. The magnetic field immediately inside the magnetopause is everywhere parallel to the magnetopause except directly behind the foci, where the magnetic field just inside is normal to the magnetopause. These two singular points are called the null points. The current on the nearly cylindrical part of the magnetopause enclosing the magnetosphere tail flows from the evening toward the morning side both on the northern and southern halves; how the current closes is not as yet understood, though there have been models in which the current returns to the evening side along the equatorial plane where the magnetic field is weak. The general picture depicted above is necessarily highly idealized and serves merely as a guide toward a better understanding of the real magnetopause, which must have a much more complex structure.

The magnetic field configuration in the magnetosphere is schematically shown in Fig. 5h-10. Let the polar angle of the points at which the two magnetic field lines through the null points intersect the earth's surface be denoted by Θ_1, assuming, for simplicity, a symmetry with respect to the dipole equator. Then, Θ_1 is near 15° (i.e., near, but not necessarily always identified with, the poleward boundary of the auroral electrojet belt near noon, as shown in Fig. 5h-5) [47]. In the (geomagnetic) noon meridian half plane containing the null points, the field lines intersecting the

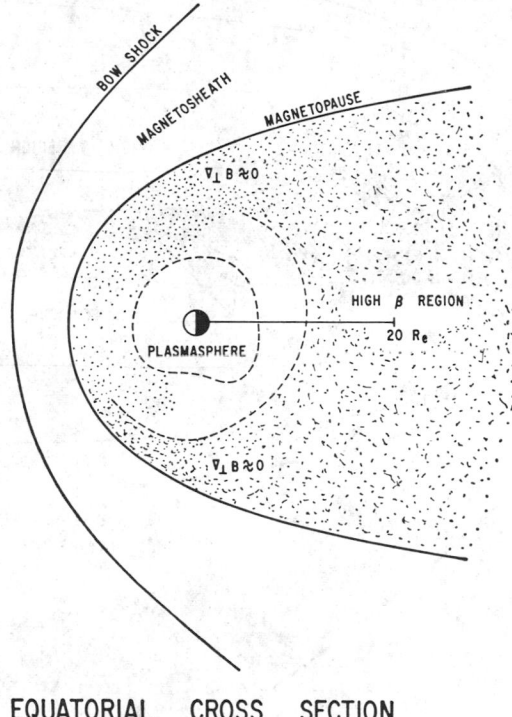

EQUATORIAL CROSS SECTION

Fig. 5h-9. Illustrations of the magnetosphere (Secs. 5h-20 and 5h-21): equatorial cross section; dots represent presence of plasmas; the sun is to the left.

earth's surface at polar angles greater than Θ_1 are dipolar in their gross character, though they are deformed by the compression from the front. The field lines intersecting the earth's surface at polar angles less than Θ_1 are drawn back over the pole toward the tail. In the midnight meridian half plane all field lines extend toward the tail, but there is a somewhat analogous polar angle, say Θ_2, where the field lines leaving the earth at angles greater than Θ_2 maintain some characteristics of a dipolar field at large distances. This angle, Θ_2, is frequently near 19° and corresponds roughly to the poleward boundary of the auroral electrojet belt near midnight, as shown in Fig. 5h-5. The magnetic lines leaving the earth at polar angles less than Θ_2 extend to great distances in the geomagnetic tail (Fig. 5h-10) where they become nearly parallel to the direction of the solar wind flow outside the magnetosphere. The magnetospheric tail

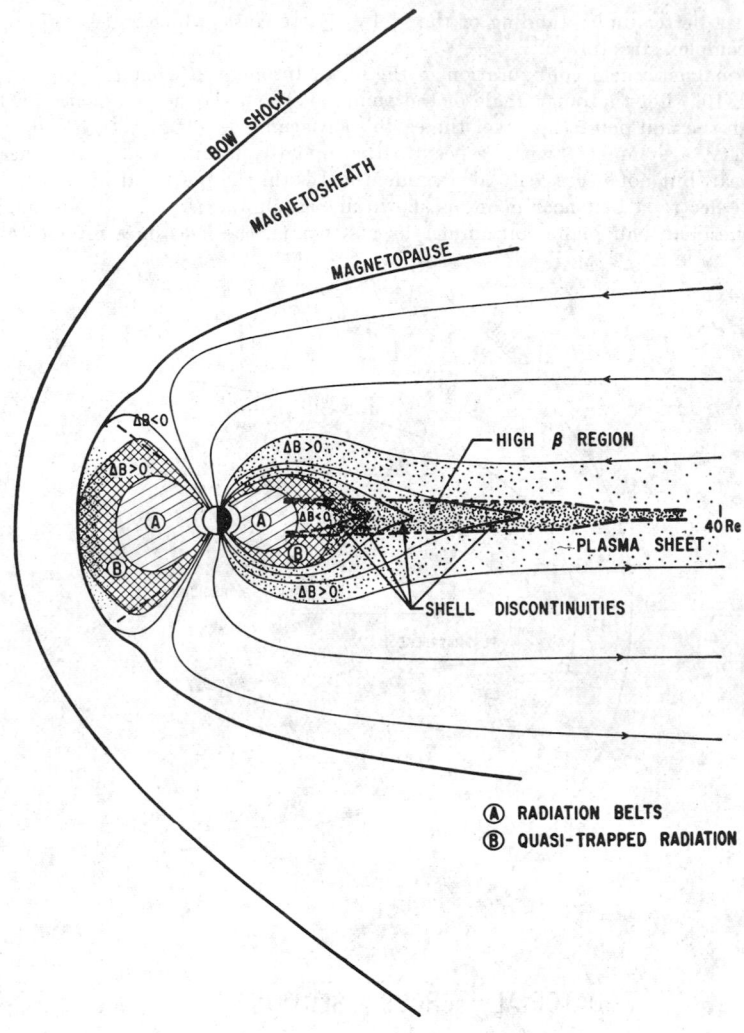

NOON-MIDNIGHT CROSS SECTION

F$_{IG}$. 5h-10. Illustrations of the magnetosphere (Secs. 5h-20 and 5h-21): noon-midnight meridianal cross section; dots represent presence of plasmas; the sun is to the left, and the tail extends to the right.

extends to at least $80R_e$ [48] and may still exist at $1,000R_e$ in a less well-defined form [49,50]. In the midnight meridian a third polar angle, Θ_3, can be designated corresponding to the lowest latitude of the east-west auroral electrojet, shown schematically in Fig. 5h-5. The radiation belts (region A of Fig. 5h-10) lie on magnetic shells emanating from the earth at $\Theta > \Theta_3$. The magnetic field in the radiation-belt region is represented fairly accurately by the spherical harmonic description given in Sec. 5h-5. A transitory region (region B of Fig. 5h-10) where trapping of energetic (e.g., >40 kev) particles is highly variable usually bounds the outermost magnetic shell of the radiation belts.   Within the shells of this transitory region the spherical harmonic and dipole descriptions of the field become less accurate, and large deviations commonly occur.   It has not been clearly established whether nightside field lines in region B (Fig. 5h-10) emanate from the earth at $\Theta > \Theta_3$ or $\Theta < \Theta_3$; Θ_3 field lines could lie within region B or at the boundary between regions A and B.   The uncertainty is greatly influenced by the fact that Θ_3 (as noted in Sec. 5h-10) varies greatly with the level of activity.   Close to the nightside equatorial plane beyond region B (Fig. 5h-10), in what can be called the *near-tail* region, the magnetic field maintains a shell-like structure, but is highly deformed by internal plasma pressures that inflate and stretch the field shells toward the antisolar direction.   These shells emanate from the earth at $\Theta_2 \leq \Theta < \Theta_3$.

For accurate mathematical description of the field in the outer magnetosphere expressions for the distortions created by the solar wind compression and the stresses exerted by plasmas within the magnetosphere and magnetospheric tail need to be added to the spherical harmonic description (Sec. 5h-5).   However, as of 1970, expressions which adequately represent these effects analytically throughout the magnetosphere had not been derived.   The time variability is one of the principal problems that is illustrated by considering some of the effects observed during storms, below, but present to some degree at all times: (1) a high solar wind pressure compresses the magnetosphere inward (Sec. 5h-12) such that Θ_1 is increased and more magnetic flux is pushed back into the tail; (2) an enhanced plasma belt is created deep in the magnetosphere and exerts stresses that inflate the magnetosphere (Sec. 5h-11); (3) complex changes in the plasma behavior in the near-tail region locally disturb the field [51]; and (4) at the times of magnetic bays (Sec. 5h-10), or substorms, the plasma stress is partially released and the near-tail field suddenly relaxes, or collapses, toward a more dipolar configuration [45].

5h-21. Charged Particle Content.   *Plasmas* [20,52,53].   The presence of plasmas in the magnetosphere and their influences on the magnetic field have been mentioned in Sec. 5h-20.   High fluxes of plasma are observed in the *plasma sheet* that lies approximately on the equatorial plane separating the earthward magnetic field in the northern half of the tail from the oppositely directed field in the southern half (Fig. 5h-10).   The thickness of the sheet is 4 to $6R_e$ at the distance from the earth of about $17R_e$.   The electrons in the plasma sheet have a broad, quasi-thermal energy spectrum peaked anywhere between a few hundred and a few thousand electron volts with a non-Maxwellian high-energy tail.   In the midnight sector the plasma sheet reaches, under quiet conditions, distances of about $10R_e$ from the earth and comes closer when disturbed.   The sheet extends to the flanks and toward the front side, enveloping the magnetosphere, as shown in Figs. 5h-9 and 10.   The inner boundary of the plasma sheet is well-defined on the evening to the afternoon side, but appears to be more diffuse on the morning side; however, detailed plasma behaviors on the dayside of the magnetosphere are as yet not known.   On the nightside the plasma and the magnetic field show shell-like structures often with distinct discontinuities between neighboring shells.   Whenever the ratio β of the plasma kinetic energy density $(\frac{1}{2}nmv^2)$ to the magnetic field energy density $(B^2/8\pi)$ exceeds unity, the magnetic field is disturbed by the diamagnetic effect of the plasma, and the dipolar characteristics of the field are appreciably modified or completely lost.   An example of the latter is seen in the outer skirts of the magnetosphere near the dawn and dusk meridians and near the geomagnetic equator where the magnetic field gradient is almost zero from about $10R_e$ to the magnetopause.

From ground-based observations of whistlers (Sec. 5h-33) and direct satellite measurements, it has been found that sudden decreases in electron densities occur near the magnetic equatorial plane at distances from the earth's center that vary with local time and disturbance from 3.5 to $7R_e$.   The region of higher electron densities, about

10^2 cm^{-3} is called the *plasmasphere*, as shown in Fig. 5h-9. The outer boundary of the plasmasphere is referred to as the *plasmapause*, and its position moves toward the earth during periods of high magnetic activity [54]. The plasmas inside the plasmasphere are of much lower energy than those in the plasma sheet, and unlike the latter, the plasma within the plasmapause is not known to play any major role in causing magnetic disturbances except that their presence makes the conductivity along the magnetic field lines very large.

Energetic Particles [19,55,56]. The magnetosphere is populated with charged particles of a wide range of energies. These particles are distinguished from such transients as galactic and solar cosmic rays by their being "trapped" in the geomagnetic field for varying lengths of time. Hence their motion in the geomagnetic field is of fundamental importance; basic properties of their motion are discussed in Sec. 5h-22. Energies of the plasmas discussed above partially overlap with the low-energy part of particles described here, and the division of the charged-particle content of the magnetosphere into plasmas and energetic particles becomes arbitrary in some cases. Grouping of the particle populations is not definitively settled and is subject to future revisions. However, roughly speaking, the energetic particles in the magnetosphere can be divided into two groups: *trapped* and *quasi-trapped* particles. The definition of "being trapped" is by no means unambiguous; if a particle drifts around the earth repeatedly, it is considered to be trapped; see Sec. 5h-22 for particle drifts.

The region of trapped particles is from $L \sim 1.2$ to 6; see Sec. 5h-22 for the definition of L. This population includes the so-called inner and outer radiation zones or belts that were discovered by the early probes. The inner zone contains protons of energies, $E > 30$ Mev, with fluxes of about 10^4 cm^{-2} sec^{-1}, and peak intensities near $L \sim 1.5$; these particles are relatively stable with a time constant of the order of 1 year. The outer zone with electrons of $E > 1.5$ Mev is at $L \sim 3$ to 4, and its flux is highly variable, particularly during magnetic storms. The time constant for these electrons is roughly hours to weeks. These early observations of penetrating radiation with heavily shielded Geiger tubes led to the concept of the inner- and outer-zone structure with a "slot" near $L \sim 2$ where the count rate was a minimum [57]. However, later observations of particles in wider energy ranges have revealed that the structure described above was, to some extent, due to instrumental factors, and that different group of particles have grossly different spatial distributions and time variations, some with double peaks and a minimum near $L \sim 2$, and others without such a structure. Nevertheless the terms "inner and outer zones" are used to refer to the regions approximately $L < 2$ and $L > 2$, respectively, without implying that the two regions are separated by a sharply defined boundary. Protons in the energy range 0.1 Mev $< E <$ 5 Mev occupy the outer zone from $L \sim 2$ to 6 and are found to be relatively stable with small fluctuations on time scales of days; this population of protons has high fluxes with a peak of the order of 10^8 cm^{-2} sec^{-1} near $L \sim 3.5$ and carries most of the energy content of the trapped particles. Low-energy protons of E from a few kev to about 50 kev occupy a broad region from 3 to $10R_e$, and their fluxes vary greatly during magnetic storms. These low-energy protons are mainly responsible for the inflation of the magnetosphere and the *Dst* decrease on the earth's surface during magnetic storms, described in Sec. 5h-11. The contribution to the storm effects from electrons in the energy range from a few hundred ev to 50 kev is likely to be appreciable, but probably only about one-fourth of that from the protons [20]. Electrons with higher energies (e.g., $E \gtrsim 300$ kev) occupy the outer zone, and their fluxes suddenly increase at the time of magnetic storms and decay to an equilibrium level with a time constant of days to weeks [58,59]. The sources of the trapped particles are not as yet well understood. The radioactive decay of albedo neutrons produced by nuclear collisions of galactic (and solar) cosmic-ray protons with oxygen and nitrogen nuclei in the atmosphere has been proposed to be the main source for the high-energy protons in the inner zone.   For the outer-zone protons, several processes involving an inward diffusion and acceleration have been proposed; but there are no definitive proofs as to how efficient these mechanisms are.   One suggestion is that the diffusion is caused by violation of the third adiabatic invariant (Sec. 5h-22), due to such magnetic perturbations as sudden impulses (Sec. 5h-12), and that conservation of the first and second invariants (Sec. 5h-22) leads to energization of the particles as they diffuse inward. A similar process resulting from fluctuations in the electric field of the magnetospheric convection system (Sec. 5h-23) has also been considered.

The series of high-altitude nuclear explosions of 1958 (i.e., the Argus experiment) was designed to test the possibility of trapping energetic particles by the geomagnetic field [60].   Electrons mainly from the β decay of the fission fragments were injected into a thin shell near $L \sim 2$, and were found to be stable in position during their lifetime of a few weeks.   The Starfish explosion of July 9, 1962 [61] created a more intense and extensive artificial radiation belt; initially, a maximum flux ($\sim 10^9$ electrons cm^{-2} sec^{-1}) was near $L \sim 1.3$ with large fluxes extending to beyond $L \sim 4$.   For $L < 1.5$ the electron decay was slow, with lifetimes of the order of years, and was in agreement with the theoretical expectation for decay from Coulomb scattering of the electrons in the atmosphere.   Beyond $L \sim 1.7$ the decay was considerably faster, with lifetimes of months to a week; the rapid decay was thought to be due to resonant interaction of electrons with electromagnetic waves such as whistlers (Sec. 5h-33) or due to processes related to magnetic disturbances.   Interactions between particles and electromagnetic waves of various frequencies from VLF to ELF or ULF (Secs. 5h-29 to 5h-33) constitute an important subject concerning the particle behaviors in the magnetosphere [62], but direct observational evidence for these interactions is in most cases still lacking.

The quasi-trapped particles occupy regions between the trapping region and the magnetopause on the front to the flanks of the magnetosphere, and the near-earth tail region on the nightside.   These particles mainly consist of low-energy protons and electrons with $E \lesssim 50$ kev.   Their fluxes are highly variable with geomagnetic activity; this population and its extension into the near-tail plasma sheet is probably the reservoir for the particles precipitating into the atmosphere during high-latitude disturbances.   A major supply of particles for the trapping region may also come from the quasi-trapping region.   Occasional high fluxes of particles have been observed in different regions; for instance, sudden flux increases of electrons with $E \gtrsim 40$ kev and with omnidirectional fluxes up to about 10^7 electrons cm^{-2} sec^{-1} observed in the tail [63] (often referred to as electron "islands") and "spikes" of directional electron fluxes up to 10^9 electrons cm^{-2} sec^{-1} sterad^{-1} encountered at low altitudes ($\sim 1,000$ km) [64] are examples of intensified particle activity in the quasi-trapping region.

5h-22. Energetic Particle Motion [2,56,65,66].   There are three fundamental characteristics in the motion of a charged particle in a dipole-type magnetic field such as the earth's field: (1) gyration about a line of force, (2) oscillation between mirror points along lines of force, and (3) longitudinal drift around the center of the dipole. Corresponding to these three periodicities there are three *adiabatic invariants* that are conserved.

For the gyrating motion the *first*, or *magnetic moment*, *invariant* μ is given by

$$\mu = \frac{P_\perp{}^2}{2m_0 B}$$

where $P_\perp$ = the component of the (relativistic) momentum perpendicular to the magnetic field vector **B**

m_0 = the particle rest mass

$B = |\mathbf{B}|$

The motion of a charged particle can be investigated with high accuracies by an approximation in which the center of the gyrating motion, referred to as the *guiding center*, is followed. For the oscillatory motion of the guiding center between mirror points, the *second*, or *longitudinal, invariant* is conserved. This invariant, J, is defined by

$$J = \oint P_{\parallel} \, ds$$

where $P_{\parallel}$ is the guiding-center momentum parallel to the line of force, and the integral is taken over a complete oscillation. For J to be conserved, the drift velocity perpendicular to the lines of force along which the guiding center oscillates must be small compared with its velocity along the line of force.

The guiding center drifts from a line of force to an adjacent line of force such that J is constant, and thus the lines of force, along which the guiding center moves, form a surface on which J is constant. If this surface is closed, namely, if the guiding-center returns to a line of force which it previously traversed, there is a third periodicity. For this precessional motion of the guiding center, the *third*, or *flux, invariant* is conserved. The invariant is the magnetic flux Φ, enclosed by the closed surface just described. Obviously, if the magnetic field is static, this invariant is trivial, but it remains constant even when the field varies in time, provided that the time of precession is small compared with the time scale for the field variations. Thus the third invariant Φ is the first to be violated when the time scale of field fluctuations becomes short; the next is the second invariant J; and the first invariant μ is violated only by very rapid field variations with periods less than the particle's gyroperiod.

From the constancy of μ, the relation between the magnetic field strength B and the *pitch angle* α (= the angle between the velocity vector and the line of force about which the particle gyrates) is

$$\frac{\sin^2 \alpha}{B} = \text{constant}$$

In particular, if $B = B_m$ at the mirror point (where $\alpha = \pi/2$), the pitch angle at the equator α_0 is given by

$$\sin \alpha_0 = \left(\frac{B_0}{B_m}\right)^{\frac{1}{2}}$$

where B_0 is B at the equator.

If μ is constant, and if there is no electric field,

$$J = 2P \int_A^{A'} \left(1 - \frac{B}{B_m}\right)^{\frac{1}{2}} ds$$

where P is the linear momentum, and the integral is taken over a half oscillation from one mirror point A to its conjugate point A'. The quantity I defined by

$$I = \frac{J}{2P}$$

is a function of the mirror point A and the magnetic field configuration, and is independent of the particle properties. I can be considered as a function of position only, if the mirror point A is taken to be coincident with the point for which I is being evaluated. For a dipole field of moment M the quantity $L_d{}^3 B/M$ can be expressed as a function, say F, with an argument $I^3 B/M$,

$$\frac{L_d{}^3 B}{M} = F\left(\frac{I^3 B}{M}\right)$$

where L_d is the distance from the dipole to the equatorial crossing of the line of force that passes the point at which I is being evaluated. For the earth's magnetic field the *magnetic shell parameter* [67], L, is defined by the formula

$$\frac{L^3B}{M} = F\left(\frac{I^3B}{M}\right)$$

where the functional form of F is taken to be identical with the corresponding function for the dipole field, but B and I are now computed using the spherical harmonic expression as given in Sec. 5h-5; M is the earth's magnetic dipole moment. L is usually measured in units of earth radii. Data for the trapped particles are normally organized in the (B,L) coordinate system. Because of the distortions of the geomagnetic field by the solar wind (Sec. 5h-20) and the charged particles inside the magnetosphere (Sec. 5h-21), the B, L coordinates as computed from the spherical harmonic expression for the surface field are usually accurate only to $L \sim 5$. Beyond this distance the B, L coordinates computed in this way become gradually meaningless with increasing distance. At present there is no adequate analytical expression for the distorted magnetosphere field.

So far the motion of a charged particle has been considered in the presence of a magnetic field $\mathbf{B}$. In the presence of an electric field $\mathbf{E}$ in addition to $\mathbf{B}$, the particle drifts in a direction transverse to $\mathbf{B}$ with a velocity $\mathbf{v}_t$ given in a guiding-center approximation by

$$\mathbf{v}_t = \mathbf{E} \times \frac{\mathbf{B}}{B^2} + \epsilon_\perp \mathbf{B} \times \frac{\nabla B}{qB^3} + 2\epsilon_\parallel \mathbf{B} \times \frac{\mathbf{R}}{qB^2R^2}$$

where $\epsilon_\perp$ and $\epsilon_\parallel$ are kinetic energy perpendicular and parallel to $\mathbf{B}$, respectively; q is the charge; $\mathbf{R}$ is the vector radius of curvature of the field line about which the particle gyrates: $R = |\mathbf{R}|$, and in the absence of currents, $1/R = \nabla_\perp B/B$. The three terms are often referred to as the $\mathbf{E} \times \mathbf{B}$ drift, the gradient drift, and the curvature drift, respectively. The longitudinal drift mentioned earlier corresponds to the combined effects of the gradient and curvature drifts in the geomagnetic field.

5h-23. Plasma Convection [2,68]. If a plasma in a magnetic field is taken to be a perfect conductor, Ohm's law is expressed by

$$\mathbf{E} + \mathbf{v} \times \mathbf{B} = 0$$

where $\mathbf{E}$ and $\mathbf{B}$ are the electric and magnetic field intensities, respectively, and $\mathbf{v}$ is the bulk velocity of the plasma. Under these conditions the magnetic field is referred to as a *frozen-in* field [66,69,70], and magnetic field lines can be identified by the plasma. $\mathbf{E} + \mathbf{v} \times \mathbf{B} = 0$ is also referred to as the *hydromagnetic approximation*, and as an approximation it has been useful in explaining the generation of electric fields in the magnetosphere and the consequences of these electric fields in causing electric currents in the ionosphere and electric field drifts (Sec. 5h-22) (proportional to $\mathbf{E} \times \mathbf{B}/B^2$) of energetic particles (i.e., particles of higher energy in populations of lower-energy density than the plasma particles moving in bulk with velocity $\mathbf{v}$). In the frozen-in state, motions of plasma and magnetic field lines leave the magnetic field configuration unchanged and are said to interchange the lines of force. This has led to the concept that there may be a continuous convection of plasma in the magnetosphere. The geomagnetic field lines are also tightly "frozen" into the earth because of its high conductivity; however, the presence of a nonconducting neutral atmosphere between the earth and the ionosphere decouples the two frozen-in regions and makes magnetospheric *interchange motions* [71] possible. If the magnetic field configuration is unchanged, $\partial\mathbf{B}/\partial t = 0$, and consequently curl $\mathbf{E} = $ curl $(-\mathbf{v} \times \mathbf{B}) = 0$.

This implies that magnetic lines of force and the streamlines of the flow are equipotentials of the electric field. Also, because magnetospheric magnetic fields are not uniform, interchange motions are accompanied by changes in plasma density and by adiabatic heating and cooling of the plasma during compressions and expansions.

The gross pattern of convection in the magnetosphere and magnetospheric tail has been largely inferred from the distribution of currents associated with the auroral electrojets (Sec. 5h-10), using two assumptions: (1) that the Hall conductivity σ_2 is much greater than the Pedersen conductivity σ_1 in the lower ionosphere (Sec. 5h-18), and (2) that the magnetic field lines in and above the lower ionosphere are lines of infinite conductivity and thus equipotential lines [47,72-74]. The electric field $E = -v \times B$ perpendicular to the electric current in the direction $-v$ (i.e., opposite v as a consequence of the electron motion in the region 100 to 130 km where the ion's transverse motion is inhibited by collisions) then maps directly into the magnetosphere from hemisphere to hemisphere. In the magnetospheric equatorial plane this produces a flow pattern in which plasma adjacent to the magnetopause on the nightside and near the twilight meridians flows, in general, away from the sun, and plasma in the central regions of the near-earth tail flows toward the sun. This picture is frequently interpreted as indicating that viscous interaction between the solar wind and magnetosphere pushes the outermost magnetospheric field and plasma roughly parallel to the direction of the solar wind flow. The flow from the tail toward the earth and sun is then regarded as the return flow. There are substantial reasons for believing that some such convection pattern exists associated with the auroral electrojects. In detail, however, the hydromagnetic approximation cannot strictly hold in that the ionospheric Pedersen conductivity is not negligible and tends to short-circuit the convective electric field. Deviations from the hydromagnetic approximation involving electrical loading in the ionosphere may, in fact, be responsible for many of the dynamical changes (e.g., the collapse of the near-tail magnetic field with the onset of a magnetic bay (Sec. 5h-20) can be viewed as being caused by an ionospheric release of the plasma pressure in the distant magnetosphere).

Other effects, in addition to Pedersen current short circuiting, can cause a breakdown of the hydromagnetic approximation. Lack of sufficient plasma density to justify the infinite conductivity assumption is one possibility that could be important along the polar cap field lines extending deep into the magnetospheric tail. The hydromagnetic approximation also loses validity in regions where the electron and ion pressures are large; in these cases the expression becomes $E + v \times B = (ne)^{-1} \operatorname{grad} p$, where n, e, and p are the number density, charge, and pressure. Thus, in association with field shell discontinuities (Sec. 5h-20) in the near-tail and other large irregularities, significant deviations from the approximation are expected. Many outstanding problems in magnetospheric dynamics are closely related to determining the degree of applicability of the hydromagnetic approximation. Inasmuch as the problems relate to having information on electric fields in space, additional discussion is given in Sec. 5h-28.

INTERPLANETARY MEDIUM

5h-24. The Solar Winds [43,44,75,76]. One of the basic dynamical properties of the solar corona is its continuous expansion outward. The resulting plasma flow is referred to as the *solar wind*. The solar wind velocity varies with activity on the sun, and ranges from 300 to about 850 km sec^{-1}. The azimuthal direction of the average flow of the ions is 2 or 3 deg east of the sun after aberration due to the earth's orbital motion is corrected for, but the deviations in the arrival angle from the above average can be as large as 15 deg. Deviations of the flow direction from the solar ecliptic plane by as much as 5 deg have also been observed. Statistically the daily mean solar wind velocity v is related to the daily sum of three-hourly geomagnetic activity index K_p (Sec. 5h-13) by a formula [77]

$$v \text{ km sec}^{-1} = 8.44 \Sigma Kp + 330$$

However, on a finer time scale geomagnetic activity is not related to the solar wind velocity in any simple manner. The solar wind gas is mainly composed of electrons and protons. It also contains alpha particles in considerably variable quantity, from less than 1 to nearly 20 percent, with an average of 3 to 6 percent. A typical ion density in the solar wind is 1 to 10 protons cm^{-3}, but low values of less than 0.1 proton cm^{-3} and a maximum exceeding 80 protons cm^{-3} have been observed. A transverse "temperature" (for the direction normal to the bulk velocity) has been determined from the angular distribution of the ion velocities. The proton temperature so defined varies from 10^4 to 10^6 K. The temperature determined for the direction parallel to the bulk velocity appears to be greater than the transverse temperature by a factor typically about 5 [78]. Data on the electron temperature in the solar wind are scarce, but a characteristic temperature of 8×10^5 K has been reported for low-energy electrons in the energy range between 33 and 1,000 ev; however, the electron temperature is likely to change appreciably with solar activity, as does the bulk velocity of the ions [79].

5h-25. Interplanetary Magnetic Field [43,44,75,80]. Solar magnetic fields that extend beyond the solar corona are swept outward into interplanetary space by the solar wind (Sec. 5h-24). Thus, interplanetary space is permeated by magnetic fields of solar origin to distances that are thought to be of the order 100 AU (1 AU = mean sun-earth distance, $1.496 \cdot 10^{13}$ cm). The intensity of the interplanetary magnetic field near the earth's orbit is usually within the range 2 to 10 γ, with an average near 5 γ [81], but values as high as 25 γ are occasionally encountered. Although highly variable in direction, the statistical average direction in the ecliptic plane is closely approximated, considering only the rotation of the sun and the solar wind velocity. Thus, taking α as the angle between a radial line from the sun and the magnetic field in the ecliptic plane: $\alpha = \tan^{-1}(\omega r/v) = \tan^{-1}(428/v)$ at 1 AU where r = distance from the sun, ω = angular velocity of the sun's rotation, and v = solar wind velocity in km sec^{-1}.

In the ecliptic plane the magnetic field pattern sometimes indicates a sector structure with the field direction alternately pointing toward or away from the sun (on an average at angle α) in neighboring sectors. This pattern then corotates with the sun at its rotational period of approximately 27 days [81]. Although the magnetic field is clearly dominated by the solar wind, in that its energy density is only of the order of 1 percent of the kinetic energy density of the solar wind, it significantly modulates the flux of solar cosmic rays whose energy density is considerably less than that of the field. The flux of solar cosmic rays is thus anisotropic and roughly follows the filamentary magnetic flux tubes that constitute the prominent fine structure of the field [82]. In addition to this fine structure, hydromagnetic discontinuities are caused by the transient and nonuniform emission of solar wind plasma from the sun. These discontinuities develop into shock waves or form stationary contact surfaces, in the frame of reference of the plasma [83]. Effects of interplanetary discontinuities on the magnetosphere are discussed in Sec. 5h-26.

5h-26. Solar Wind–Magnetosphere Interaction [2,44]. The Alfvén speed v_A (Sec. 5h-29) in the solar wind is typically 60 km sec^{-1} (with density $n = 3$ protons cm^{-3}, and $B = 5 \gamma$); and hence the solar wind is hydromagnetically supersonic with a representative Alfvén Mach number, $M_A (= v/v_A)$, of 7. Because of the variability in n and B, M_A may be as low as 1.5 or may exceed 10. A standing bow shock is created on the upstream side of the magnetosphere boundary (Sec. 5h-20), analogous to that of an object placed in a supersonic flow of fluid. Whereas in the latter case sound waves communicate the presence of the object upstream to divert the flow around it, magnetoacoustic waves, which have a group velocity of the order of the Alfvén speed, convey the message upstream in the solar wind to the shock. Since the mean free path in the interplanetary medium is of the order of 1 AU, the magnetosphere bow shock is a collision-free shock.

In the subsolar region the flow is subsonic behind the shock with ion velocities considerably randomized. As the gas flows toward the flanks of the magnetosphere, the flow becomes more ordered, and beyond a sonic line the velocity is again supersonic. The flow pattern appears to be in agreement with theoretical results obtained using continuum models for the solar wind.

According to gasdynamical calculations, the distance D from the earth's center to the magnetosphere boundary along the sun-earth line is approximately given by [84]

$$D = aH_0^{\frac{1}{3}}(2\pi K\rho v^2)^{-\frac{1}{6}}$$

where a = earth's radius = 6.3712 × 10^8 cm
     H_0 = dipole field intensity on earth's surface at the equator = 0.30953 gauss
     ρ = solar wind density
     v = solar wind velocity
     K = an adjustable constant which depends on the ratio of the specific heats, $\gamma(= C_p/C_v)$

For large Mach numbers K approaches 0.844 for $\gamma = 2$ and 0.881 for $\gamma = \frac{5}{3}$. For a Newtonian gas model K is 2 if the particle reflection at the boundary is assumed to be specular or "elastic," or unity if this is assumed to be "inelastic." For large Mach numbers the standoff distance Δ at the nose of the magnetosphere can be expressed approximately by [84]

$$\frac{\Delta}{D} = \frac{1.1[(\gamma - 1)M_A{}^2 + 2]}{(\gamma + 1)M_A{}^2}$$

These theoretical results roughly agree with observations [45]. Representative observational values of D and Δ are 10 and 3 earth radii, respectively. As a result of the solar wind–magnetosphere interaction, the earth's magnetic field is drawn out to large distances forming a magnetosphere tail (Sec. 5h-20).

ELECTRIC FIELDS

5h-27. Atmospheric Electricity [85,86]. Electrical phenomena between the earth's surface and an altitude of 30 km (roughly the peak altitude for airplane and balloon observations) are usually referred to as being atmospheric. At altitudes >80 km, measurements of "electric fields in space" (Sec. 5h-28) have only recently been initiated. Between 30 and 80 km, measurements are essentially nonexistent, and theory is at most only an extrapolation from the lower "atmospheric" regions.

Within the atmospheric region the ionizing agents are primarily cosmic rays and radiation from radioactive material in the earth and dust (both natural and from bomb debris). The vertical atmospheric electric field under clear weather conditions decreases rapidly with altitude from roughly 100 to 200 volts meter^{-1} at the surface to 10 to 30 volts meter^{-1} at an altitude of 6 km. Above roughly 10 km the potential gradient is often too weak to measure reliably, but appears typically to decrease to about 5 and 1 volts meter^{-1}, at 10 and 20 km, respectively. The conductivity (which depends primarily on ion mobilities) increases over these altitudes and ranges roughly from 1 to 10 to 30 × 10^{-14} ohm^{-1} meter^{-1} at 0, 6, and 10 km, respectively. Field intensity and conductivity values, such as those above, are subject to many variations, even under clear weather conditions. Both seasonal and diurnal variations are generally recognized. Secular changes are also reported, which appear to be related to artificial factors such as changes in air pollution and periods of bomb testing. The fair weather electric field is negative upward such that positive ions move toward the earth. This fair weather electric current is, in general, considered to be a return current in that the principal generators for air-earth currents lie in regions where the atmosphere is disturbed by thunderstorms and precipitation. Models of the electrical structure of thunderstorms have been constructed for particular cases, but there is little agreement on the generality of such models when lightning and precipitation are included. Electric fields of several thousand volts per meter are commonly associated with disturbed conditions. The total potential drop associated with lightning reaches values of several million volts.

5h-28. Electric Fields in Space [14,19]. Direct space measurements of the electric field **E** have only recently become feasible by measuring differences in floating potential between identical probes, symmetrically oriented relative to the spacecraft motion and the sun, and separated by long base lines for each axis of measurement [87]. The measurements have been successfully conducted in the ionosphere between 80 and 300 km at middle latitudes and in the auroral belt. Measurements have also been attempted in traverses through the magnetosphere and into the interplanetary medium, but these, to date, have involved errors, due to insufficient base line relative to plasma sheath phenomena, that prohibit accurate d-c values. Limits on the electric field and valid a-c electric field measurements (Secs. 5h-32 and 5h-33) have, however, come from these early efforts. Electric fields have also been indirectly measured by observing the motion of artificially created barium ion clouds above the region (approximately below 180 km) where collisions affect the cloud motion. The motion **v** in the known magnetic field **B** gives the electric field **E** from $\mathbf{E} + \mathbf{v} \times \mathbf{B} = 0$ (Sec. 5h-23) [88].

The principal source of electric fields in the outer magnetosphere and high-latitude ionosphere appears to be the convection of plasma (Sec. 5h-23) driven indirectly by interaction between the solar wind and the earth's magnetic field. Both direct probe measurements [89] of **E** and motion measurements of **v** for artificial Ba^+ clouds [90] have demonstrated that auroral electrojet currents are mainly Hall currents (in the direction $-\mathbf{v}$ as shown by the simultaneous magnetic disturbance field of the current) in that the quantities **v** and **E** have both been found to be at right angles to **B** in a number of cases, including examples of both eastward and westward electrojet flow (Fig. 5h-5). These measurements essentially show that the hydromagnetic approximation (Sec. 5h-23) $\mathbf{E} + \mathbf{v} \times \mathbf{B} = 0$ gives a reasonable picture of the convection dynamics in the auroral belt (Sec. 5h-10) region. The lack of a **v** component along **B** and one example [89] of a constant value for **E** between roughly 90 and 130 km when a probe trajectory fell along a magnetic field line during that section of a rocket flight provide additional support for treating the magnetic field lines as equipotentials. However, while verifying the proper vector relationships for $\mathbf{E} + \mathbf{v} \times \mathbf{B} = 0$ the measurements have shown that the electric field is highly variable in magnitude and can drop to extremely low values in narrow shells where the ionospheric conductivity is exceptionally high. These narrow shells contain discrete auroral forms. Over most of the belt within which aurora is occurring **E** is found to be > 0.01 volt meter^{-1} with values between 0.02 and 0.05 volt meter^{-1} being fairly common [89,90]. Values as high as 0.13 volt meter^{-1} have been observed, and as the total time of sampling has been small, it is highly probable that greater intensities are not uncommon. In contrast, directly on shells of discrete auroral forms **E** has been observed to drop below 0.005 volt meter^{-1}, which suggests that the **E** field is partially short-circuited in local strips of exceptionally high ionospheric conductivity. Observations that visual auroral structures show identical details when observed simultaneously at conjugate points in the northern and southern auroral belts [91] and that the magnetic disturbances are correlated in detail [92] illustrate that the potential distribution maps from hemisphere to hemisphere along magnetic field lines. Corresponding field intensities in the equatorial plane of the magnetosphere, taking into account the magnetic field geometry and its distortion in the near-tail region, are roughly 20 to 100 times less than in the ionosphere along the same magnetic shells.

Over the polar cap (Sec. 5h-10) there have not been definitive electric field measurements. If the polar-cap ionospheric currents are Hall currents, it can be estimated that the electric field has properties similar to those observed in the auroral belt. It is also not known whether or not magnetic field lines emanating from the polar cap can be treated as electrical equipotential lines in that plasma densities at large distances along these shells outside the equatorial plasma sheet in the tail have been below measurement thresholds.

In middle- and low-latitude regions E and v measurements in and above the ionosphere have given values in the range 10^{-4} to 2×10^{-3} volt meter^{-1}, which in most cases is close to the magnitude of possible errors in measurement [88–90]. Measurements have not been made during periods of large disturbances, or storms, when one might expect the high-latitude convection system to penetrate more deeply into the magnetosphere. E magnitudes of the order 10^{-3} volt meter^{-1} can be expected from the Sq (Sec. 5h-8) dynamo (i.e., winds) in the ionospheric E region [93].

In addition to the convection process, a variety of mechanisms have been proposed for generating electrostatic fields relevant to specific problems in the ionosphere and magnetosphere [74,94,95]. Other than those cases where the indirect evidence is convincing (e.g., the polarization field required for the equatorial electrojet, Sec. 5h-8), their existence and/or importance is unproved. Electrostatic electric fields also play an essential role in a number of theoretical treatments of the magnetopause boundary and the bow shock (Sec. 5h-26). In the case of the bow shock, fields as strong as 5 volts meter^{-1} have been postulated as existing in a thin region at the shock front [96]. However, the limited measurements available indicate that the fields are more of the order $-v \times B$, where v is the plasma bulk velocity as in the hydromagnetic approximation (Sec. 5h-23) [89]. Similarly, in the interplanetary medium, E, as indicated in the spacecraft frame of reference, is compatible with $-v \times B$ for the solar wind velocity v (*Note.* This field does not exist in the frame of reference of the solar wind which is moving with velocity v.) For typical ranges of solar wind–interplanetary magnetic field parameters, the range of E (in spacecraft coordinates) is roughly 10^{-3} to 10^{-2} volt meter^{-1} [89].

Relative to accelerating charged particles that enter the earth's (magnetospheric) frame of reference from inertial space, an additional electric field exists which is caused by charge separation induced by the rotation of the ionosphere with the earth [72,73]. The potential of this *corotational* field in the ionosphere is $V = 90$ $\sin^2 \Theta$ in kilovolts, where Θ is the colatitude. As the corotational field is zero in the earth's reference frame, it does not create currents.

WAVE PHENOMENA

5h-29. Magnetohydrodynamic Waves [66,69,97]. *Alfvén Waves.* In a perfectly conducting fluid permeated by a uniform magnetic field B_0, Alfvén waves propagate in the direction of B_0 with the Alfvén velocity V_A given by

$$V_A = \frac{B_0}{\sqrt{4\pi\rho}}$$

where $B_0 = |B_0|$ (in emu, i.e., in gauss) and ρ = density (in cgs units, i.e., g cm^{-3}). The magnetic perturbation b and the fluid velocity v are both transverse to B_0; and $b^2/8\pi = \frac{1}{2}\rho v^2$; i.e., an equipartition of energy holds.

Magnetoacoustic Waves. In a perfectly conducting, compressible fluid permeated by a uniform magnetic field $\mathbf{B}_0$, magnetohydrodynamic waves propagate with a phase velocity V that satisfies

$$V^4 + (V_A^2 + C_0^2)V^2 + V^2 C_0^2 \cos^2 \theta = 0$$

where V_A = Alfvén velocity ($= B_0/\sqrt{4\pi\rho}$)

C_0 = sound velocity ($= \sqrt{\gamma p/\rho}$; γ = ratio of specific heats, p = pressure, and ρ = density)

θ = angle between $\mathbf{B}_0$ and direction of propagation

When $\theta = 0$, the roots of the above equation are $\pm C_0$ and $\pm V_A$, corresponding to a pure acoustic wave and a pure Alfvén wave propagation along $\mathbf{B}_0$. When $\theta = \pi/2$, there is only one mode propagating with phase velocity, $V = \pm(C_0^2 + V_A^2)^{\frac{1}{2}}$; in this wave the fluid velocity $\mathbf{v}$ is perpendicular to $\mathbf{B}_0$, and the magnetic perturbation $\mathbf{b}$ is parallel to $\mathbf{B}_0$. The latter velocity is called the *magnetoacoustic* velocity. Between $\theta = 0$ and $\pi/2$ the two modes are coupled.

5h-30. Plasma Waves [98,99]. *General.* For a two-component (electron-ion) cold plasma in a uniform magnetic field B_0 the dispersion equation is given by

$$An^4 - Bn^2 + C = 0$$

where n = refractive index ($= |\mathbf{k}|c/\omega$; $\mathbf{k}$ = angular wave number, c = velocity of light)

$A = S \sin^2 \theta + P \cos^2 \theta$

$B = RL \sin^2 \theta + PS (1 + \cos^2 \theta)$

$C = PRL$

$S = \frac{1}{2}(R + L)$

$D = \frac{1}{2}(R - L)$

$R = 1 - \alpha\omega^2/[(\omega + \Omega_i)(\omega - \Omega_e)]$

$L = 1 - \alpha\omega^2/[(\omega - \Omega_i)(\omega + \Omega_e)]$

$P = 1 - \alpha$

$\alpha = (\Pi_e^2 + \Pi_i^2)/\omega^2$

$\Pi_e^2 = 4\pi n_e e^2/m_e$

$\Pi_i^2 = 4\pi n_i Z^2 e^2/m_i$

$\Omega_e = e B_0/m_e c$ = electron cyclotron frequency

$\Omega_i = Z e B_0/m_i c$ = ion cyclotron frequency

ω = angular frequency of the wave

$B_0 = |\mathbf{B}_0|$

e(or Ze) = magnitude of electron (or ion) charge in esu units

$m_{e,i}$ = electron or ion mass in grams

$n_{e,i}$ = electron or ion number density, cm^{-3}

θ = angle between the propagation direction (i.e., that of $\mathbf{k}$) and $\mathbf{B}_0$

The dispersion equation is quadratic in n^2, and hence there are in general two modes. In particular, for $\theta = 0$: $n^2 = R$ (with right-handed circular polarization), and $n^2 = L$ (with left-handed circular polarization); and for $\theta = \pi/2$: $n^2 = RL/S$ (the extraordinary mode), and $n^2 = P$ (the ordinary mode). Polarization, left- or right-handed, is defined, for positive ω, with respect to the direction of the ambient magnetic field $\mathbf{B}_0$. *Resonances* occur when $n^2 = \pm \infty$, and *cutoffs*, when $n^2 = 0$. Resonances: (1) $\theta = 0$: the *electron cyclotron resonance* at $\omega = \Omega_e$ ($R \rightarrow \pm \infty$); and the *ion cyclotron resonance* at $\omega = \Omega_i$ ($L \rightarrow \pm \infty$). (2) $\theta = \pi/2$: the *lower hybrid resonance* at ω_{LH}, where ω_{LH} approximately satisfies

$$\omega_{LH}^{-2} = (\Omega_i^2 + \Pi_i^2)^{-1} + (\Omega_i \Omega_e)^{-1}$$

and the *upper hybrid resonance* at ω_{UH}, where ω_{UH} approximately satisfies

$$\omega_{UH}{}^2 = \Omega_e{}^2 + \Pi_e{}^2$$

Numerous modes of plasma waves have been investigated extensively, but only those that are frequently encountered in the geophysical environment are mentioned below. All these modes are defined in a homogeneous plasma. The plasma in the magnetosphere cannot necessarily be regarded as being homogeneous, and drift waves may play an important role in inducing instabilities, but these and other instabilities are not discussed here since they are not as yet adequately understood under geophysical conditions.

Magnetohydrodynamic Waves $(\omega \ll \Omega_i)$

$$n^2 = 1 + \gamma \qquad \text{for the fast mode (compressional wave)} \qquad (5h\text{-}1)$$
$$n^2 \cos^2 \theta = 1 + \gamma \qquad \text{for the slow mode (Alfvén wave)} \qquad (5h\text{-}2)$$

where
$$\gamma = 4\pi(n_i m_i + n_e m_e)\,\frac{c^2}{B_0{}^2}$$

$$\approx \frac{4\pi\rho c^2}{B_0{}^2} \qquad \rho = \text{plasma density}$$

the characteristic velocity $c/\sqrt{1+\gamma}$ is called the Alfvén velocity. When $\gamma \gg 1$, this reduces to $B_0/\sqrt{4\pi\rho}$, which is V_A given in Sec. 5h-29. The slow mode disappears at $\omega = \Omega_i$. The fast mode exists at frequencies above Ω_i and continues on to the whistler mode. The phase velocity of the fast mode is isotropic.

Ion Cyclotron Waves $(\omega \lesssim \Omega_i)$. The slow wave in Eq. (5h-2) has a characteristic dispersion relation just below the ion cyclotron frequency. In the neighborhood of $\omega = \Omega_i$, n^2 for the two modes are approximately

$$n^2 \approx \frac{\gamma}{1 + \cos^2 \theta} \qquad \text{for the fast mode} \qquad (5h\text{-}3)$$

$$n^2 \cos^2 \theta \approx \gamma(1 + \cos^2 \theta)\,\frac{\Omega_i{}^2}{\Omega_i{}^2 - \omega^2} \qquad \text{for the slow mode} \qquad (5h\text{-}4)$$

The ion cyclotron resonance appears in the slow mode; for this mode the dispersion relation can be rewritten as

$$\omega^2 \approx \Omega_i{}^2 \left(1 + \frac{\Pi_i{}^2}{k_\parallel{}^2 c^2} + \frac{\Pi_i{}^2}{k_\parallel{}^2 c^2 + k_\perp c^2}\right)^{-1}$$

where $k_\parallel$ and $k_\perp$ denote the components of **k** parallel and perpendicular to **B**$_0$, respectively. Waves in this mode are called ion cyclotron waves.

The "Whistler" Mode $(\Omega_i < \omega < \Omega_e)$. The branch that is an extension of the fast mode in Eq. (5h-1) is called the *whistler mode*. The term "whistler mode" originates from the circumstance that whistlers (Sec. 5h-33) propagate in this mode; however, the use of this term is not limited to the propagation of natural whistlers. The refractive index for this mode is approximately given by

$$n^2 = 1 - \frac{\alpha\omega}{\omega - \Omega_e \cos \theta}$$

provided that

$$\Omega_e \sin^4 \theta \ll 4\omega^2(1 - \alpha)^2 \cos^2 \theta \qquad \text{(quasi-longitudinal propagation)}$$

and that

$$\Omega_e{}^2 \sin^2 \theta \ll |2\omega^2(1 - \alpha)|$$

where $\alpha = \Pi_e{}^2/\omega^2$ in the present approximation. For $\theta = 0$ the wave has a right-handed, circular polarization, and exhibits a resonance at $\omega = \Omega_e$, i.e., the electron cyclotron frequency; the wave becomes evanescent above Ω_e.

Ion Acoustic and Electrostatic Ion Cyclotron Waves.   In a plasma with finite electron and ion temperatures T_e and T_i, respectively, *ion acoustic waves* propagate in the direction parallel to $\mathbf{B}_0$ with a dispersion relation

$$\frac{1}{\omega^2} = \frac{1}{k^2}\frac{m_i}{Z\kappa T_e} + \frac{1}{\Pi_i{}^2}$$

if β_e ($\equiv 8\pi n_e\kappa T_e/B_0{}^2$) is small and if $T_i \ll T_e$, where κ is the Boltzmann constant. If T_i is comparable to T_e, the ion thermal velocity becomes comparable to the wave phase velocity, and the wave will be strongly Landau-damped.

For $\theta \neq 0$, and for frequencies above Ω_i but close to it, the *electrostatic ion cyclotron wave* can propagate, under certain conditions, with a dispersion relation

$$\omega^2 \approx \Omega_i{}^2 + k_\perp{}^2\frac{Z\kappa T_e}{m_i}$$

where $k_\perp$ is the component of $\mathbf{k}$ perpendicular to $\mathbf{B}_0$.

Electrostatic Plasma Waves.   In the absence of a magnetic field, a plasma resonates electrostatically with the frequency Π_e ($\approx \sqrt{4\pi n_e e^2/m_e}$, ignoring the ion motion). This frequency is called the *Langmuir* or *plasma frequency*. The reflection of radio waves from the ionosphere is due to this resonance (Sec. 5h-19).

5h-31. Geomagnetic Pulsations [100–102].   Rapid geomagnetic fluctuations with periods approximately from 0.2 sec to 10 min (or roughly 5 to 0.001 Hz in frequency) are generally referred to as *pulsations* or *micropulsations*. Fluctuations (or signals) in the frequency range 3,000 to 3 Hz are often grouped under ELF (extra low frequency) waves and those with frequencies <3 Hz under ULF (ultrahigh frequency); in this scheme of nomenclature pulsations may be regarded as being ULF phenomena. However, various names and corresponding abbreviations have been used for different groups of pulsations; these include: "transient" (*pt*), "continuous" (*pc*), "giant" (*pg*) pulsations and "pearls" (*pp*). The International Association of Geomagnetism and Aeronomy recommended in 1963 adoption of a scheme in which pulsations are divided into two groups: regular (and continuous) and irregular pulsations, denoted by *pc* and *pi*, respectively. These groups are divided further into several subgroups according to the following scheme:

Type	Range of period, sec
Regular (continuous) pulsations	
pc 1	0.2– 5
pc 2	5 – 10
pc 3	10 – 45
pc 4	45 –150
pc 5	150 –600
Irregular pulsations	
pi 1	1 – 40
pi 2	40 –150

However, the above classification is solely based on periods and should be considered as a tentative scheme, which is likely to be modified as the sources of different types of pulsations become identified. The morphological characteristics of pulsations are extremely complex, and only a few types of pulsations are described below.

Quasi-sinusoidal continuous oscillations with frequencies about 0.1 to 5 Hz and with "beating" characteristics have been interpreted as hydromagnetic emissions from the magnetosphere. Because of the resemblance of the appearance of their amplitude variations to pearls on a necklace they are often called "pearls." In a frequency-time display (i.e., sonagram) these pulsations often consist of discrete repetitive emissions of rising frequency, but those of irregular structure or of other types are also observed. In the case of repetitive emissions, wave packets are observed alternatingly at magnetically conjugate points (i.e., at the northern and southern feet of a line of magnetic force), indicating that the hydromagnetic waves propagate along the lines of force back and forth between two hemispheres. These wave packets are likely to be generated by cyclotron instabilities involving protons in the magnetosphere.

There has been much discussion of the possibility of eigenoscillations of the magnetosphere, but such oscillations have not as yet been positively identified. Because of the inhomogeneities of the plasma density and of the axial asymmetry of the magnetic field, the oscillation of the magnetosphere is likely to be extremely complex. Nevertheless some types of pulsations appear to be caused by regional oscillations of the magnetosphere. In particular, large-amplitude damped oscillations observed in the auroral zones (Sec. 5h-10) appear to be due to oscillations of the magnetic field lines embedded in the auroral zones; small-amplitude (<1 γ) regular, continuous oscillations belonging to pc 2, 3 may be due to oscillations of a large volume of the magnetosphere predominantly on the dayside. Transverse, regular oscillations with amplitudes of 1 to 10 γ and with periods from 1 to several minutes have been observed by a satellite at a distance of about 7 earth radii from the earth, mostly on geomagnetically quiet days and in the dayside magnetosphere.

Another group of magnetic oscillations observed in the magnetosphere has periods from a fraction of one second to several seconds and amplitudes less than 1 γ; these are probably the same oscillations as those observed at high latitudes and grouped under hydromagnetic emissions or pc 1. Many other groups of magnetic oscillations will probably be found in the magnetosphere, but satellite observations of pulsations have just begun, and classifications and studies of generation mechanisms are still left for the future.

At high latitudes irregular pulsations (pi 1 and 2) with wide ranges of periods and amplitudes are observed in association with magnetic and auroral disturbances; some of these pulsations are closely related to particle precipitations. Hence these pulsations can be regarded as part of the disturbance phenomena discussed in Secs. 5h-10 and 5h-11.

5h-32. ELF (Extra Low Frequency) Waves [100,101]. The main sources for natural electromagnetic signals or emissions observed on the earth in the ELF range, 3 to 300 Hz, are *"sferics"* (or *atmospherics*), *earth-ionosphere cavity resonances*, and *ELF emissions*.

Sferics [85]. Emissions from lightning, when VLF signals are filtered, have slow tails which contain frequencies below a few hundred hertz. These signals propagate in a "cavity" between the earth and the ionosphere, as if in a waveguide, with low attenuation and characteristic dispersion.

Cavity Resonances [103]. Wavelengths of *ELF* signals can be of the order of several thousand kilometers, and such long-wavelength signals can cause resonances in the earth-ionosphere cavity; these resonances are sometimes called the *Schumann resonances*. Typical resonance frequencies for the first five modes are near 8, 14, 20, 27, and 32 Hz. These resonance frequencies vary with conditions in the lower ionosphere (the lower D and C regions), and exhibit diurnal variations normally of the order of a fraction of one hertz. Harmonics of the resonance as high as the 11th ($\sim$70 Hz) have been observed. Appreciable shifts in the resonance frequencies are seen during severe ionospheric disturbances and at the times of high-altitude nuclear explosions.

ELF Emissions [104,105]. Electromagnetic emissions of natural origin, normally classed under VLF phenomena (Sec. 5h-33), often extend into the ELF range; thus the division between VLF and ELF is somewhat artificial. Particularly in cases where the propagation is in the whistler mode, the designation ELF is often only a convenient way of indicating the low-frequency portion of the VLF phenomena. In the magnetosphere the most prominent emission in the total ELF-VLF range, in the sense of integrated occurrence and intensity in space and time, falls in the range 300 to 3,000 Hz and has been called *ELF hiss*. Satellite observations indicate that this signal is almost continuously present at altitudes >300 km in the late morning hours along L shells (Sec. 5h-22), corresponding to invariant latitudes of 55 to 65° [106,89]. The percentage of time of occurrence as a function of signal intensity decreases markedly for E between 60 and 180 μv meter^{-1} (rms) at an altitude of 700 km [89]. Similarly, for samples distributed between 240 and 2,700 km altitude, occurrences decrease markedly for B between 2 and 6 milligammas (rms) [106]. Although occurrence is most frequent near 60° invariant latitude in the late morning, the total region of frequent occurrence (e.g., >10 percent of the time) extends throughout the dayside hours 6h to 18h local time and invariant latitudes 50 to 70°. Although the ELF hiss signal is relatively steady in the sense that rapid changes in intensity are absent, it is frequently accompanied by a second signal, called *ELF chorus*. The chorus signals consist of a long series of wave packets, each having a duration of the order of one second, and the characteristic that the frequency rises with time within each packet. The time-space distribution of ELF hiss and chorus observed from satellites, and the fact that their occurrence at the earth's surface is less common and more erratic, suggest that these signals are repeatedly reflected from hemisphere to hemisphere from ionospheric levels. There is evidence that this reflection occurs roughly at the altitude where the signal frequency equals the proton gyrofrequency (but is presumably affected by the presence of heavier ions), and that the effective gyrofrequency also acts as a low-frequency cutoff [107]. Some signal apparently reaches the earth's surface through mode-coupling mechanisms.

ELF signals of a more transient nature than the ELF hiss, noted above, are encountered in the auroral-belt and polar-cap regions. These are frequently associated with irregularities (Sec. 5h-17) in electron density and electric fields when observed by satellites [89]. Although most ELF emissions propagating in the whistler mode are believed to be generated in the magnetosphere, it has been suggested that a strong signal near 700 Hz in the auroral zone might be caused by proton cyclotron radiation in the ionosphere [108].

Part of the energy of the ELF (and VLF) emission from a lightning impulse propagates upward into the ionosphere and sometimes triggers a *proton whistler* [109]. In a frequency-time display, such as in a sonogram, a proton whistler has a dispersion characteristic of slowly rising frequency that asymtotically approaches the proton cyclotron frequency at the point of observation by a satellite. The frequency at which this proton whistler originates in the frequency-time display is an extension of the trace that corresponds to the "electron whistler," to be discussed in Sec. 5h-33. Proton whistlers are thought to be ion cyclotron waves (Sec. 5h-30).

5h-33. Whistlers and VLF Emissions [100,104]. *Whistlers* are electromagnetic signals in audio frequencies originating from lightning strokes. They are called whistlers because of their whistling sound when converted into audio signals. Whistlers typically have a descending tone from above 10 to 1 kHz; however, the upper limit can be as high as 30 kHz or even higher, and the lowest may extend to the ELF or even ULF range. The duration of a whistler is about one second, but some whistlers last only for a fraction of a second and others for two or three seconds. Whistlers propagate in the whistler mode (Sec. 5h-30), which is roughly a guided mode along the magnetic field lines. Only a slight electron density gradient is required to make tubes of magnetic force act like ducts for whistler propagation. Ducted whistlers often propagate back and forth between the two hemispheres repeatedly. The group velocity, with which wave energy propagates, has a maximum at a frequency of say, f_1, and decreases for frequencies above and below f_1. Therefore, in a frequency-time display (e.g., in a sonogram) a signal trace for a whistler that traveled a long distance shows the earliest arrival at f_1 and a gradual delay in arrival time as f departs from f_1 above and below. A whistler exhibiting such a dispersion characteristic is called a *nose whistler,* and the frequency f_1 of the minimum delay, the *nose frequency.* In a homogeneous medium f_1 is $\frac{1}{4}f_H$, where f_H is the electron cyclotron frequency ($= \Omega_e/2\pi$). The group delay time, say t, for a whistler that has traversed a one-hop path from one hemisphere to the other can be approximately expressed by

$$t = Df^{-\frac{1}{2}}$$

for f well below the nose frequency; the constant D, called the *dispersion constant* or simply the *dispersion,* is given approximately by

$$D = (2c)^{-1} \int_{\text{path}} (f_p/f_H^{\frac{1}{2}}) \, ds \qquad \text{sec}^{\frac{1}{2}}$$

where c is the velocity of light, and f_p the electron plasma frequency ($= \Pi_e/2\pi$). Apart from a constant, the integrand reduces to $(n_e/B)^{\frac{1}{2}}$ where n_e is the electron density, and B the magnetic field intensity. Thus various models for the electron density in the magnetosphere can be tested by comparing the observed D with the calculated values. Studies of electron densities in the magnetosphere by means of whistlers have shown that there is a "knee" in the electron density profile at several earth radii and that the electron density drops substantially beyond this distance (Sec. 5h-21). Whistlers have been detected by satellites at various altitudes in the magnetosphere, and their behaviors are now being investigated in detail.

In addition to whistlers, there are other types of emissions in the VLF range; these are called *VLF emissions.* Several types of these emissions are observed in close association with whistlers, suggesting that they are triggered by the latter. VLF emissions may last steadily for minutes, or even hours, or may occur in bursts; converted to sound waves, they may produce a hissing sound or show a musical tone. (A division of hiss into ELF and VLF groups is entirely artificial.) VLF emissions are most frequently observed at middle and high latitudes, and indicate similarity in occurrence and form at magnetically conjugate areas in the northern and southern hemispheres. Such mechanisms as electron cyclotron radiation and Čerenkov radiation have been suggested for the origin of VLF emissions. The possibility of these VLF waves having significant interaction with energetic particles in the radiation belt has been extensively investigated [56,62].

References

1. Chapman, S., and J. Bartels: "Geomagnetism," Clarendon Press, Oxford, 1940, 1951, 1962 (corrected).
2. Matsushita, S., and W. H. Campbell, eds.: "Physics of Geomagnetic Phenomena," Academic Press, Inc., New York, 1967.
3. Vestine, E. H., L. LaPorte, I. Lange, and W. E. Scott: The Geomagnetic Field, Its Description and Analysis, *Carnegie Inst. Wash. Publ.* 580, 1947.
4. Adopted at the International Association of Geomagnetism and Aeronomy Symposium on Description of the Earth's Magnetic Field, Washington, D.C., Oct. 22–25, 1968.
5. Vestine, E. H.: Chap. II-2, p. 181, in ref. 2.
6. Cain, J. C.: Personal communication December, 1968.
7. Vestine, E. H., L. Laporte, C. Cooper, I. Lange, W. C. Hendrix: Description of the Earth's Main Magnetic Field and Its Secular Change, *Carnegie Inst. Wash. Publ.* 578, 1947.
8. Symposium on Magnetism of the Earth's Interior: *J. Geomag. Geoel.,* **17** (3–4) (1965).
9. Cain, J. C., and S. J. Hendricks: The Geomagnetic Secular Variation 1900–1965, *NASA Tech. Note* TN-D-4527, 1968.

10. Cain, J. C., S. J. Hendricks, R. A. Langel, and W. V. Hudson: *J. Geomag. Geoel.* **19**, 335 (1967).
11. Beloussov, V. V., P. J. Hart, B. C. Heezen, H. Kuno, V. A. Magnitsky, T. Nagata, A. R. Ritsema, and G. P. Woollard, eds.: The Earth's Crust and Upper Mantle, *Am. Geophys. Union Geophys. Monograph* 13, Washington, D.C., 1969; in particular, chap. 5.
12. Serson, P. H., and W. L. W. Hannaford: *J. Geophys. Research* **62**, 1 (1957).
13. McCormac, B. M., ed.: Aurora and Airglow, *Proc. NATO Study Inst.*, 1966, Reinhold Book Corporation, New York, 1967.
14. McCormac, B. M., ed.: Aurora and Airglow, *Proc. NATO Study Inst.*, 1968, Reinhold Book Corp., New York, 1969.
15. Harang, L.: *Terrest. Magnetism and Atmospheric Elec.*, **51**, 353 (1946).
16. Fukushima, N.: *J. Fac. Sci. Univ., Tokyo*, **8**, 293 (1953).
17. Heppner, J. P.: Ref. 13, p. 75.
18. Akasofu, S.-I.: *Space Sci. Rev.* **4**, 498 (1965).
19. McCormac, B. M., ed.: Earth's Particles and Fields, *Proc. NATA Advanced Study Inst.*, 1967, Reinhold Book Corporation, New York, 1968.
20. Frank, L. A., Ref. 19, p. 67.
21. Hoffman, R. A., and L. J. Cahill, Jr.: *J. Geophys. Research* **73**, 6711 (1968).
22. Sugiura, M.: *Ann. IGY* **35**, 9, Pergamon Press, New York, 1964.
23. Sugiura, M., and S. J. Hendricks: *NASA Tech. Note*, NASA TN D-5748, 1970.
24. Mead, G. D.: *J. Geophys. Research* **69**, 1181 (1964).
25. Sugiura, M.: *J. Geophys. Research* **70**, 4151 (1965).
26. Bartels, J.: *Ann. IGY* **4**, 227, Pergamon Press, New York, 1957.
27. Lincoln, J. V.: Chap. I-3, p. 67, in ref. 2.
28. Davis, T. N., and M. Sugiura: *J. Geophys. Research* **71**, 785 (1966).
29. Rikitake, T.: "Electromagnetism and the Earth's Interior," Elsevier Publishing Company, Amsterdam, 1966.
30. Price, A. T.: Chap. II-3, p. 235, in ref. 2.
31. Madden, T. R., and C. M. Swift, Jr.: In ref. 11.
32. Ratcliffe, J. A., ed.: "Physics of the Upper Atmosphere," Academic Press, Inc., New York, 1960.
33. Rishbeth, H.: *Rev. Geophys.* **6**, 33 (1968).
34. Donahue, T. M.: *Science*, **159**, 489 (1968).
35. Cohen, R.: Chap. III-4, p. 561, in ref. 2.
36. Herman, J. R.: *Rev. Geophys.* **4**, 255 (1966).
37. Symposium on Upper Atmospheric Winds, Waves, and Ionospheric Drifts, IAGA Assembly, 1967; *J. Atmospheric and Terrest. Phys.* **30**(5), (1968).
38. Smith, E. K., and S. Matsushita, eds.: "Ionospheric Sporadic E," Pergamon Press, Oxford, 1962.
39. Smith, E. K., Jr.: Chap. III-5, p. 615, in ref. 2.
40. Baker, W. G., and D. F. Martyn: *Phil. Trans. Roy. Soc. London*, Ser. A, **246**, 281 (1953).
41. Chapman, S.: *Nuovo Cimento* **4** (suppl.), 1385 (1956).
42. Ratcliffe, J. A.: "The Magneto-ionic Theory and its Applications to the Ionosphere," Cambridge University Press, London, 1959.
43. Hess, W. N., and G. D. Mead, eds.: "Introduction to Space Science," 2d ed., Gordon and Breach, Science Publishers, Inc., New York, 1968.
44. King, J. W., and W. S. Newman, eds.: "Solar-Terrestrial Physics," Academic Press, Inc., New York, 1967.
45. Heppner, J. P., M. Sugiura, T. L. Skillman, B. G. Ledley, and M. Campbell: *J. Geophys. Research* **72**, 5417 (1967).
46. Ness, N. F.: Ref. 44, p. 57.
47. Heppner, J. P.: In ref. 14.
48. Ness, N. F., K. W. Behannon, S. C. Cantarano, and C. S. Scearce: *J. Geophys. Research* **72**, 927 (1967).
49. Ness, N. F., C. S. Scearce, and S. C. Cantarano: *J. Geophys. Research* **72**, 3769 (1967).
50. Wolfe, J. H., R. W. Silva, D. D. McKibbin, and R. H. Mason: *J. Geophys. Research* **72**, 4577 (1967).
51. Sugiura, M., T. L. Skillman, B. G. Ledley, and J. P. Heppner: Presented at International Symposium on the Physics of the Magnetosphere, Washington, D.C., September, 1968. Sugiura, M.: In "The World Magnetic Survey 1957–1969," IAGA Bulletin No. 28.
52. Bame, S. J.: Ref. 19, p. 373.
53. Vasyliunas, V. M.: *J. Geophys. Research* **73**, 2839 (1968).
54. Carpenter. D. L.: *J. Geophys. Research* **71**, 693 (1966).

55. McCormac, B. M., ed.: Radiation Trapped in the Earth's Magnetic Field, *Proc. Advanced Study Inst.*, Bergen, 1965, D. Reidel Publishing Co., Dordrecht, Holland, 1966.
56. Hess, W. N.: "The Radiation Belt and Magnetosphere," Blaisdell Publishing Company, a division of Ginn and Company, Waltham, Mass., 1968.
57. Van Allen, J. A., and L. A. Frank: *Nature* **183**, 430 (1959).
58. Brown, W. L., L. J. Cahill, L. R. Davis, C. E. McIlwain, and C. S. Roberts: *J. Geophys. Research* **73**, 153 (1968).
59. Williams, D. J., J. F. Arens, and L. J. Lanzerotti: *J. Geophys. Research* **73**, 5673 (1968).
60. Symposium on Scientific Effects of Artificially Introduced Radiations at High Altitudes, *J. Geophys. Research* **64**(8), 865 (1959).
61. Collected Papers on the Artificial Radiation Belt from the July 9, 1962, Nuclear Detonation, W. N. Hess, ed., *J. Geophys Research* **68**(3), 605 (1963).
62. Kennel, C. F., and H. E. Petschek: *J. Geophys. Research* **71**, 1 (1966).
63. Anderson, K. A.: *J. Geophys. Research* **70**, 4741 (1965); P. Serlemitsos: *ibid.* **71**, 61 (1966); A. Konradi: *ibid.* 2317; E. W. Hones, Jr., S. Singer, and C. S. R. Rao: *ibid.* **73**, 7339 (1968).
64. McDiarmid, I. B., and J. R. Burrows: *J. Geophys. Research* **70**, 3031 (1965).
65. Northrop, T. G.: "The Adiabatic Motion of Charged Particles," Interscience Publishers, a division of John Wiley & Sons, Inc., New York, 1963.
66. Alfvén, H., and C.-G. Fälthammar: "Cosmical Electrodynamics," Clarendon Press, Oxford, 1963.
67. McIlwain, C. E.: Ref. 55, p. 45.
68. Hines, C. O.: *Space Sci. Rev.* **3**, 342 (1964).
69. Cowling, T. G.: "Magnetohydrodynamics," Interscience Publishers, Inc., New York, 1957.
70. Dungey, J. W.: "Cosmic Electrodynamics," Cambridge University Press, London, 1958.
71. Gold, T.: *J. Geophys. Research* **64**, 1219 (1959).
72. Axford, W. I., and C. O. Hines: *Can. J. Phys.* **39**, 1433 (1961).
73. Taylor, H. E., and E. H. Hones: *J. Geophys. Research* **70**, 3605 (1965).
74. Obayshi, T., and A. Nishida: *Space Sci. Rev.* **8**, 3 (1968).
75. Parker, E. N.: "Interplanetary Dynamical Processes," Interscience Publishers, a division of John Wiley & Sons, Inc., New York, 1963.
76. Axford, W. I.: *Space Sci. Rev.*, **8**, 331 (1968).
77. Snyder, C. W., M. Neugebauer, and U. R. Rao: *J. Geophys. Research* **68**, 6361 (1963).
78. Hundhausen, A. J., J. R. Asbridge. S. J. Bame, H. E. Gilbert, and I. B. Strong: *J. Geophys. Research* **72**, 87 (1967).
79. Frank, L. A.: (abstract) *Trans. Am. Geophys. Union* **49**, 262 (1968).
80. Wilcox, J. M.: *Space Sci. Rev.* **8**, 258 (1968).
81. Ness, N. F., C. S. Scearce, J. B. Seek, and J. M. Wilcox: "Space Research," vol. VI, p. 581, R. L. Smith-Rose, ed., Spartan Books, Washington, D.C., 1966.
82. McCracken, K. G., and N. F. Ness: *J. Geophys. Research* **71**, 3315 (1966).
83. Colburn, D. S., and C. P. Sonett: *Space Sci. Rev.* **5**, 439 (1966).
84. Spreiter, J. R., A. L. Summers, and A. Y. Alksne: *Planet. Space Sci.* **14**, 223 (1966).
85. Coroniti, S. C., ed.: Problems of Atmospheric and Space Electricity, *Proc. 3d Intern. Conf. Atmospheric and Space Elec.*, 1963, Elsevier Publishing Company, Amsterdam, 1965.
86. Smith, L. G., ed.: Recent Advances in Atmospheric Electricity, *Proc. 2d Conf. Atmospheric Elec.*, Pergamon Press, New York, 1958.
87. Fahleson, U.: *Space Sci. Rev.* **7**, 238 (1967).
88. Haerendel, G., R. Lüst, and E. Rieger: *Planet. Space Sci.* **15**, 1 (1967).
89. Aggson, T. L., J. P. Heppner, N. C. Maynard, and D. S. Evans: Personal Communications; presentations at International Symposium on the Physics of the Magnetosphere, September, 1968.
90. Wescott, E. M., J. Stolarik, and J. P. Heppner: *Trans. Am. Geophys. Union* **49**, 155 (1968).
91. Davis. T. N.: In ref. 14.
92. Wescott, E. M., and K. B. Mather: *J. Geophys. Research* **70**, 29 (1965).
93. Maeda, H.: *J. Geomag. Geoelec.* **7**, 121 (1955).
94. Kern, J. W.: In ref. 2.
95. Chamberlain, J. W.: "Physics of the Aurora and Airglow," Academic Press, Inc., New York, 1961.
96. Tidman, D. A.: *J. Geophys. Research* **72**, 1799 (1967).
97. Ferraro, V. C. A., and C. Plumpton: "An Introduction to Magneto-fluid Mechanics," Oxford University Press, London, 1961.

98. Stix, T. H.: "The Theory of Plasma Waves," McGraw-Hill Book Company, New York, 1962.
99. Akhiezer, A. I., I. A. Akhiezer, R. V. Polovin, A. G. Sitenko, and K. N. Stepanov: "Collective Oscillations in a Plasma," tr. H. S. H. Massey, tr. ed. R. J. Tayler, The MIT Press, Cambridge, Mass., 1967.
100. Bleil, D. F., ed.: "Natural Electromagnetic Phenomena below 30 kc/s," Plenum Press, Plenum Publishing Corporation, New York, 1964.
101. Campbell, W. H.: Ref. 2, p. 822.
102. Troitskaya, V. A.: Ref. 44, p. 213.
103. Madden, T., and W. Thompson: *Rev. Geophys.* **3**, 211 (1965).
104. Helliwell, R. A.: "Whistlers and Related Ionospheric Phenomena," Stanford University Press, Stanford, Calif., 1965.
105. Gurnett, D. A.: Ref. 19, p. 349.
106. Taylor, W. W. L., and D. A. Gurnett: *J. Geophys. Research* **73**, 5615 (1968).
107. Gurnett, D. A., and T. B. Burns: *Univ. Iowa Preprint* 68-28, Department of Physics and Astronomy, 1968.
108. Egeland, A., G. Gustafsson, S. Olsen, J. Aarons, and W. Barron: *J. Geophys. Research* **70**, 1079 (1965).
109. Gurnett, D. A., S. D. Shawhan, N. M. Brice, and R. L. Smith: *J. Geophys. Research* **70**, 1665 (1965).

5i. Lunar, Planetary, Solar, Stellar, and Galactic Magnetic Fields

M. SUGIURA,[1] J. P. HEPPNER,[1] AND E. BOLDT[2]

NASA—Goddard Space Flight Center

H. W. BABCOCK[3] AND ROBERT HOWARD[4]

Hale Observatories
Carnegie Institution of Washington
California Institute of Technology

LUNAR AND PLANETARY MAGNETIC FIELDS

5i-1. Moon.[5,6] According to the measurements made aboard the satellite *Explorer* 35, there appeared to be no magnetic field attributable to the moon at the distance of 800 km from the lunar surface. On the basis of the *Explorer* 35 observations the magnetic moment of the moon, even if the moon is magnetized, must be less than 4×10^{20} cgs units, which is less than 10^{-5} times the earth's magnetic moment. The conductivity of the moon seems to be sufficiently low to allow the interplanetary

[1] Lunar and planetary fields.
[2] Galactic fields.
[3] Stellar fields.
[4] Solar fields.
[5] N. F. Ness, K. W. Behannon, C. S. Scearce, and S. C. Cantarano, *J. Geophys. Research* **72**, 5769 (1967).
[6] C. P. Sonett, D. S. Colburn, and R. G. Currie, *J. Geophys. Research* **72**, 5503 (1967).

magnetic field to be convected through it without noticeable change; the upper limit to the effective average conductivity has been estimated to be 10^{-5} mho meter^{-1}.

5i-2. Venus.[1] *Mariner V* detected a bow shock around Venus; the bow shock appeared to be similar to, but much smaller in dimension than, that of the earth (Sec. 5h-20). The creation of the bow shock has been attributed to the presence of a dense ionosphere which prevents rapid penetration of the solar wind magnetic field and plasma into the atmosphere. The standoff distance of the bow shock at the time of the *Mariner V* traversal appeared to be about 4,000 km (or about 0.7 Venus radii) from the surface of the plane. No planetary magnetic field was detected at this distance. The upper limit to the magnetic dipole moment of Venus was estimated to be about 10^{-3} times that of the earth. The observation that trapped charged particles (electrons with $E_e > 45$ kev and protons with $E_p > 320$ kev) were absent in the vicinity of Venus is in agreement with the above estimate.

SOLAR FIELD

5i-3. General Magnetic Field of the Sun. Magnetic fields on the solar surface are measured by means of the Zeeman effect in solar spectrum lines. Since 1952 measurements of magnetic fields outside sunspots have been made with the solar magnetograph.[2] Tables 5i-1 and 5i-2 summarize data on magnetic fields in polar regions.

TABLE 5i-1. THE POLAR MAGNETIC FIELDS OF THE SUN: 1912–1954

Investigator	Field intensity at North Pole^a	Years of measurement	Remarks
Hale, Langer^b.........	-4 gauss	1912–1932	Reanalysis in 1935 of early data
Nicholson, Ellerman,	$+3 \pm 1.7$	1933–1934	$\pm 45°$ solar latitude
and Hickox^c	-2.0 ± 2.8	1948–1949	Visual
von Klüber^d..........	<1–2	1949–1950	$\pm 45°$, photographic
Thiessen^e.............	$+1.5 \pm 3.5$	1947–1948	$\pm 45°$, photoelectric
	$+1.5 \pm 0.75$	1949	
	$+2.4 \pm 0.5$	1951	
Kiepenheuer^f..........	<1	1951	Full disk, photoelectric
H. D. and H. W. Babcock^g..........	$+2$–4	1952–1954	$\pm 70°$, full disk, photoelectric

^{a} Polarity definition: magnetic vector toward observer is $+$.
^{b} G. E. Hale, *Nature* **136**, 703 (1935).
^{c} *Ann. Rept. Mt. Wilson Observ., C.I.W. Yearbook*, p. 12, 1934; p. 138, 1949.
^{d} H. von Klüber, *Monthly Notices Roy. Astron. Soc.* **111**, 2 (1951); **114**, 242 (1954).
^{e} G. Thiessen, *Z. Astrophys.* **26**, 16(1949); **30**, 185(1952); *Nature* **169**, 147(1952); *Ann. astrophys.* **9**, 101 (1946).
^{f} K. O. Kiepenheuer, *Astrophys. J.* **117**, 447 (1953).
^{g} H. W. Babcock, *Astrophys. J.* **118**, 387 (1953); **119**, 687 (1954); H. W. Babcock and H. D. Babcock, *Publ. Astron. Soc. Pacific* **64**, 282 (1952); *Astrophys. J.* **121**, 349 (1955).

Although field strengths of the polar fields are measured to be small, this is strictly an effect of integration over a relatively large area. At least a large fraction of the polar magnetic fields is confined to small regions where the magnetic field is some tens of gauss in strength.[3] The polar fields are somewhat variable,[4] and may be thought of as the result of the poleward drift of following portions of old low-latitude active

[1] C. W. Snyder et al., Collection of *Mariner V* Papers, *Science* **158**, 1665 (1967).
[2] H. W. Babcock, *Astrophys. J.* **118**, 387 (1953).
[3] R. Howard, "Stellar and Solar Magnetic Fields," R. Lüst, ed., p. 129, North-Holland Publishing Co., Amsterdam, 1965.
[4] H. W. and H. D. Babcock, *Astrophys. J.* **121**, 349 (1955).

regions, rather than as a relatively permanent dipole field.[1] During the maximum of solar cycle 19 (1957), the polar fields of the sun reversed polarity.[2]

5i-4. Sunspot Fields. Sunspots[3] vary greatly in both size and magnetic field strength, although area and field show a positive correlation. Sunspot areas vary from as small as one to as large as 5,500 millionths of a solar hemisphere. Sunspot magnetic fields vary from about 100 to nearly 4,000 gauss. Spots generally consist of an inner dark region called the *umbra*, and a surrounding annular region, brighter than the umbra, called the *penumbra*. The magnetic fields are measured from the Zeeman effect in absorption lines in sunspot spectra. Within a sunspot the magnetic field strength varies with distance r from the center of the spot approximately as Broxon's[4] formula: $H = H_m(1 - r^2/b^2)$, where H_m is the maximum field strength, and b is the

TABLE 5i-2. THREE-MONTH AVERAGES OF NORTH AND SOUTH POLAR MAGNETIC FIELDS OF THE SUN (GAUSS)*

Year	Jan.–Mar.		Apr.–June		July–Sept.		Oct.–Dec.	
	N	S	N	S	N	S	N	S
1956	+2	−2	+1.5	−1	+1	−1	+1	−1.5
1957	−0.5	−1	0	0	0	+0.5	+1	+2
1958	0	+1	+2	+2	+1	+1	−2	+1.5
1959	−2	+2	.....	.....	−1	+0.5	−1	+1
1960	−1	+0.5	−1	+0.5	−1.5	+1	−1	+1
1961	−1	+1	−1	0	−0.5	0	−0.5	0
1962	−0.5	0	−1	−0.5	−1	+0.5	−0.5	+0.5
1963	.....	.....	−0.5	+0.5	−1	+0.5	−0.5	0
1964	−0.5	+1	−0.5	+0.5	−0.5	+0.5	−0.5	+0.5
1965	+0.5	+1	−0.5	+0.5	−1.5	+0.5	−1	+0.5
1966	−1	+0.5	0	+0.5	−0.5	0	−0.5	−0.5
1967	0	+0.5	−0.5	0	−0.5	+0.5	+1	+0.5
1968	+1	+1.5	−0.5	+0.5	0	0	0	0

* Taken from : H. D. Babcock, *Astrophys. J.* **130,** 364 (1959); R. Howard, "Stellar and Solar Magnetic Fields," p. 129, R. Lüst, ed., North-Holland Publishing Company, Amsterdam, 1965; D. W. Beggs and H. von Klüker, *Monthly Notices Roy. Astron. Soc.* **127,** 134 (1964); R. Howard, *Bull. Am. Astron. Soc.,* **1,** 280 (1969); and unpublished data.

outer radius of the penumbra. Sunspots generally occur in groups consisting of an eastward and westward spot or group of spots. These are called the preceding or following spots according to the direction of solar rotation. In most cases the magnetic polarities of the preceding sunspots are all the same and opposite to that of the following spots. In any cycle the preceding spots in the north hemisphere of the sun are nearly all of the same magnetic polarity, and this is opposite to the polarity of the preceding spots in the southern hemisphere.[5] During the next 11-year activity cycle all spot polarities are reversed. Thus the solar activity cycle is actually a 22-cycle. Solar magnetic field data covering a period of many years have been published in the

[1] H. W. Babcock, *Astrophys. J.* **133,** 572 (1961); V. Bumba and R. Howard, *ibid.* **141,** 1502 (1965).

[2] Harold D. Babcock, *Astrophys. J.* **130,** 364 (1959).

[3] R. J. Bray and R. E. Loughhead, "Sunspots," John Wiley & Sons, Inc., New York, 1964.

[4] J. W. Broxon, *Phys. Rev.* **62,** 508 (1942).

[5] G. E. Hale and S. B. Nicholson, *Astrophys. J.* **62,** 270 (1925).

TABLE 5i-3. MAGNETIC STAR DATA (AS OF 1958)

No.	Star or HD	R.A.*	Dec.*	m_v	Sp.	$w\dagger$	No. of obs.‡	H_e extremes§	Per.‖	Remarks#
1	2453	0h25m50s	+32°09'	6.7	A2p	0.14	6/15	− 425 − 710	....	Like HD 188041
2	4174	0 41 53	+40 24	7.5	M2e	....	11/41	−1200 +1100	....	[Ne III], [O III], H
3	8441	1 21 23	+42 53	6.6	A2p	0.08	13/31	− 750 + 400	2.96	Sr, Gd; sp. binary
4	9996	1 35 30	+45 09	6.3	A0	0.1:	2/6	− 990 + 135	....	Sp. binary
5	43 Cas	1 38 36	+67 47	5.5	A0p	0.7:	1/2	−1200	....	Si, Sr
6	10783	1 43 04	+8 18	6.6	A2p	0.3:	18/53	−1200 +2200	4.16	Si, Sr, Cr, (Eu, Gd)
7	11187	1 48 10	+54 40	7.1	A4p	0.27	7/10	70 +1250	....	Si, λ4201
8	HR 710	2 23 37	−15 34	5.8	A0p	0.15	39/50	−1080 − 320	....	Sr, Eu, Cr; sp. binary
9	21 Per	2 54 15	+31 44	5.2	A0p	0.6:	13/44	−1270 +1350	1.73	Si, Mn, Sr, Eu, λ4200; variable profiles
10	19445	3 05 28	+26 09	8.0	F6	....	1/3		....	High-velocity subdwarf
11	20210	3 12 53	+34 30	6.4	A7	0.4	1/4	260 + 415	....	Ba; sp. binary
12	9 Tau	3 34 01	+23 03	6.7	A2p	0.15	1/7	+ 140	....	Si
13	HR 1105	3 37 48	+63 03	5.3	S	....	2/9	+ 450	....	Heavy elements
14	25354	3 59 52	+37 55	7.9	A0p	0.2v	4/8	− 380x 0:	....	Eu, Cr, Mn, λ4201; var. profiles
15	41 Tau	4 03 32	+27 28	5.3	A0p	0.43	3/8	− 530 + 700	....	Sr, Si, λ4201; pec. profiles; sp. binary
16	68 Tau	4 22 36	+17 49	4.2	A2V	0.2	1/4	− 400	....	Metallic-line star
17	30466	4 46 06	+29 29	7.2	A0p	0.5v	2/8	? +2320	....	Si, λ4201; var. profiles
18	32633	5 02 51	+33 51	6.9	B9p	0.4:	24/24	−3960x +2220	6.43	Si, Cr; rapid reversal
19	16 Ori	5 06 34	+9 46	5.4	F2	0.3	2/4	− 420 + 375	....	Metallic-line star
20	μ Lep	5 10 41	−16 16	3.3	B9p	0.3	5/28	− 170 + 325	....	Si, Mn, Y
21	WY Gem	6 08 54	+23 14	7.4	M3p	....	1/8	+ 540	....	[Fe II]
22	42616	6 10 10	+41 43	6.9	A2p	0.4:	4/12	− 840	....	Sr; K-profile pec.; varies
23	45677	6 25 59	−13 01	7.5	B2e	0.3:	1/7	−1600 0:	....	H, Fe II, [Fe II], [S II] in emission; K pec.; varies
24	49976	6 48 18	− 7 59	6.2	A0p	0.4v	1/14	− 810 +2120	....	Sr, Ca; profiles diverse, vary irregularly
25	50169	6 49 25	− 1 35	8.9	A4p	0.12	6/11	+ 670	....	Sr, (Eu); resembles HD 188041
26	R Gem	7 04 21	+22 47	6+	Se	....	2/2	+ 370 + 400	....	
27	56495	7 14 33	− 7 26	7.5	A3p	0.4v	2/7	? + 570	....	Sr, Cr, Mg I
28	53 Cam	7 57 27	+60 28	6.0	A2p	1/4–1v	20/20	−5120x +3700	....	Ti, Sr, Cr, Eu, Mg II
29	15 Cnc	8 10 03	+29 48	5.6	A0p	0.7:	0/9	0: str	....	Si; profile of K varies
30	71866	8 27 52	+40 24	6.7	A0p	0.26	61/65	−1700x +2000	....	Eu, Gd, Sr
31	3 Hya	8 33 02	− 7 48	5.6	A2p	0.31	16/22	− 480 + 740	....	Sr; velocity varies; sp. binary?
32	49 Cnc	8 42 02	+10 16	5.6	A0p	0.26	9/26	− 200 +1450	....	Si, Sr, (Eu?)

No.	Name	α	δ	m	Sp					P	Remarks
33	ν Cnc	8 59 49	+24 39	5.4	B9p	0.5:	2/2	+ 105	+ 470	.....	Si, Sr, Cr; pec. profiles
34	κ Cnc	9 05 02	+10 52	5.1	B8p	0.13	8/17	− 640	+ 460	.....	Mn, Si; sp. binary (6ᵈ4)
35	30 UMa	10 20 33	+65 49	4.9	A0p	0.08	2/3	− 290	+ 340	.....	Si, Sr, (Mn); sp. binary (11ᵈ6)
36	45 Leo	10 25 01	+10 01	5.9	A2p	0.2v	5/14	−1000	+ 400	.....	Many profiles peculiar and variable; ☓
37	98088	11 14 26	− 6 52	6.0	A0p	0.4:	12/15	−1150	+ 800	5.905	Sr, Ba, Ti; no. ☓; sp. binary
38	17 Com A	12 26 25	+26 11	5.4	A3	0.4:	9/21	.....	+ 360	.....	Sr, Cr, (Eu); profiles-vary
39	17 Com B	12 26 25	+26 11	6.7	A4p	.....	0/2	.....	+mod.	.....	Metallic-line star
40	110066	12 36 14	+36 14	6.3	B8p	0.1:	5/7	− 55	+ 300	.....	Sr, Cr; λ 4210 wide
41	l Cen	12 37 10	−39 43	4.8	.....	0.2:	1/1	.....	+ 580	.....	Mn, Si
42	ν Vir	12 39 07	− 1 10	2.9	F0V	.....	1/3	− 390	.....	.....	Standard F0
43	111133	12 44 30	+ 6 13	6.4	A4p	0.14	1/2	− 990	.....	.....	Sr, Cr
44	α² CVn	12 53 42	+38 35	2.9	A0p	0.3v	28/96	−1400☓	+1600	5.469	Eu, Cr, Sr; profiles vary
45	115708	13 16 11	+26 38	8.3	A2p	0.2:	1/3	.....	+ 740	.....	Sr, Eu
46	78 Vir	13 31 35	+ 3 55	4.9	A2p	0.2	50/76	−1680☓	− 140	3.77	Sr, Cr, Eu
47	BD1913	13 53 50	+45 59	9.7	Ap	0.2:	/1	.....	+ 500:	.....	BD+46°
48	125248	14 15 52	−18 29	5.7	A0p	0.2v	33/40	−1900☓	+2100	9.29	Eu, Cr; long-period sp. binary
49	126515	14 23 23	+ 1 13	7.0	A2p	0.1+	1/4	.....	+1310	.....	Cr, Si, Sr; (☓); pec. profiles
50	π Boo A	14 38 22	+16 37	4.9	B8p	0.4	2/12	+ 75	+ 190	.....	Si, Mn, Y, Sc
51	μ Lib A	14 46 34	−13 56	5.4	A0p	0.3:	7/13	−1300	− 200	.....	Sr, Cr
52	133029	14 58 56	+47 28	6.2	A0p	0.4	50/74	+1150	+3270	.....	Si, Cr, λ 4201
53	134793	15 09 05	+ 8 43	8.2	A3p	0.3 + v	4/11	− 530	+ 450	.....	Eu, Sr, Cr; widths vary
54	135297	15 11 48	+ 0 33	8.0	A0p	0.3:	1/3	−1110	.....	.....	Sr, Cr
55	β CrB	15 25 46	+29 17	3.7	F0p	0.15	61/89	− 960	+1020	18.50	Sr, Eu: sp. binary
56	33 Lib	15 26 45	−17 16	7.2	F0p	0.15	1/2	.....	+1120	.....	Sr, Eu
57	ι CrB	15 59 26	+29 59	4.6	A0p	0.07	6/10	− 340	+ 75	.....	Mn, Si, Sr, Zr, Y
58	ω Oph	16 29 10	−21 21	5.3	A7p	0.6	0/19	.....	.....	.....	Sr, Cr; pec. profiles
59	45 Her	16 45 19	+ 5 20	4.9	A0p	0.4 + v	0/6	+ 840	+1430	.....	Eu, Sr, Si; profiles vary
60	52 Her	16 47 46	+46 04	6.9	A4p	0.4	2/19	− 500	.....	.....	Sr, (Eu)
61	153286	16 54 41	+47 26	6.2	F	.....	2/3	−1200☓	+1440	6.01	Sr; metallic-line star
62	165474	16 59 16	+15 01	7.4	A7p	0.4:	32/38	− 740	+ 900	.....	(Sr, Cr, Mn); profiles vary
63	171586	18 03 25	+12 00	6.7	A2p	0.15	1/3	− 540	.....	.....	Eu, Sr
64	173650	18 33 08	+ 4 54	6.4	A0p	0.8:	1/4	.....	+ 700	.....	Sr, Cr
65	.....	18 43 28	+21 55	.....	A4p	0.2+	20/43	− 540	.....	10.1	Sr, Eu, Si, Mn, Cr, Gd, λ 4201; pec. variable profiles
66	10 Aql	18 56 29	+13 50	5.9	A4p	0.1:	5/10	− 315	+ 440	.....	Sr, Eu, Mn
67	21 Aql	19 11 11	+ 2 12	5.2	B8	0.4	4/6	− 590	+ 173	.....	Si
68	RR Lyr	19 23 52	+42 41	7.8	F	.....	18/47	−1580	+1170	.....	
69	51 Sgr	19 33 00	−24 50	5.7	F	.....	1/3	− 230	.....	.....	Sr, Eu. Metallic-line star; sp. binary
70	184905	19 33 09	+43 50	6.6	A0p	1:v	0/26	.....	.....	.....	Si, Eu, Ca; profiles vary

TABLE 5i-3. MAGNETIC STAR DATA (AS OF 1958) (Continued)

No.	Star or HD	R.A.*	Dec.*	m_v	Sp.	w†	No. of obs.‡	H_e extremes§		Per.‖	Remarks#
71	187474	19h 48m 27s	−40°01′	5.4	A0p	0.1+	5/5	−1870	+1700	2350	Eu, Si, Ti, Fe, (Mn, Al)
72	188041	19 50 42	− 3 15	5.6	A5p	0.11	75/84	− 230	+1470	226	Gd, Eu, Sr; secular changes; variable amplitude
73	190073	20 00 31	+ 5 36	7.9	Aep	0.2+	1/12	+ 120		·····	Ca
74	191742	20 08 04	+42 24	7.8	A7p	0.12	2/5	− 510	− 175	·····	Sr, (Si, Eu)
75	192678	20 12 18	+53 30	7.1	A4p	0.2:	0/1		+2000:	·····	Cr
76	192913	20 14 23	+27 37	6.7	A0p	0.2:	4/10	− 670	+ 380	·····	Si, λ 4201
77	73 Dra	20 32 11	+74 47	5.2	A2p	0.13	9/14	− 700	+ 200:	·····	Ti, Eu, Sr; sp. variations periodic?
78	ν Equ	21 07 55	+ 9 56	4.8	A7p	0.09	21/31	+ 180	+ 880	·····	Eu, Mg, Sr, (Si)
79	θ¹ Mic	21 17 34	−41 01	4.9	A2p	0.6−	1/3	− 650		·····	Eu, Sr, Cr; diverse profiles
80	AG Peg	21 48 37	+12 23	7.6	B+M	·····	14/30	−1000	+ 500	·····	Sp. binary
81	VV Cep	21 55 14	+63 23	5–6	M+B	0.8:	5/17	− 360	+ 850	·····	Si, λ 4201
82	215038	22 38 18	+75 24	8.0	A0p	0.15	0/2	−3000:		·····	Sr, Cr, Eu, (Si)
83	216533	22 50 36	+58 33	4.9	A2p	0.8v	5/6	− 650	0	·····	Sr, Ca, Eu, Cr
84	κ Psc	23 24 22	+ 5 58	4.5	A2p	0.3:	0/17		+ 660	·····	Mn, Si; Y has neg. polarity
85	β Scl	23 30 18	−38 06	4.8	B9p	0.4:	1/3			·····	Sr, Cr; pec. profiles
86	ι Phe	23 32 23	−42 54	5.3	A2p	0.8:	0/12			·····	Sr, Ca, Eu, Si, λ 4201; pec.
87	108 Aqr	23 48 46	−19 11	6.2	A0p	0.8:	2/22		+2300	·····	Eu, Si, Sr, λ 4201
88	224801	23 58 10	+44 58	6.1	A0p	½–1	0/4			·····	Eu, Mg, Sr, Cr
89	4778	0 47 30	+44 44		A0p						

* Position for 1950.
† Index of line width, w.
‡ Number of plates measured/number of plates taken.

§ H_e = effective field intensity in gauss; crossover effect indicated by x.
‖ Period in days, or irregular.
Elements showing abnormal line intensity, italicized if variable.

ESSA Research Laboratory monthly series, *IER-FB Solar Geophysical Data* (Superintendent of Documents, Government Printing Office, Washington, D.C.).

STELLAR MAGNETIC FIELDS

5i-5. Spectral Observations. Many stars have strong magnetic fields that can be detected and measured by means of the Zeeman effect. This method requires that the spectrum lines be relatively sharp, i.e., not much broadened by stellar rotation, and that the magnetic field be largely coherent as to polarity over the visible hemisphere of the star. The presence of numerous lines of the metals and of the rare-earth elements, showing predictable Zeeman splitting and polarization in a magnetic field, facilitates measurement. Instrumentation includes a rather large telescope for light-gathering power, a differential optical analyzer for polarization, and spectrographic equipment of high dispersion and high resolution. Most of the results to date have been obtained with the 100-, 120-, and 200-in. telescopes and coudé spectrographs of the Mount Wilson, Lick, and Palomar Observatories, respectively. Results have been limited to stars brighter than 8.5 magnitude (photographic). Brighter stars can be observed at higher dispersion (4.5 Å/mm) and with better precision.

Except in a very few instances (e.g., HD215441), the components of Zeeman patterns are not individually resolved, but the use of a differential analyzer for right-hand and left-hand circular polarization permits measurement of the displacement of the centroid of the blended Zeeman pattern when the two modes of polarization are compared. Results are expressed in terms of the effective field H_e. This is the uniform longitudinal magnetic field in gauss that would produce the measured displacement. It has been shown that a uniformly magnetized spherical star, with limb-darkening, viewed pole-on, would have a field strength at the pole equal to 3.3 H_e. By convention, the polarity is taken to be positive when the field vector points toward the observer.

Stars showing strong magnetic fields are mostly of spectral type late B, A, and early F.[1,2] The most outstanding are the stars previously classified as the peculiar stars and spectrum variables of type A, practically all of which show fields in the range of several hundred to a few thousand gauss. All stellar fields adequately tested are found to be variable; many of the variations are periodic. Among the spectrum variables, the magnetic variations, roughly sinusoidal, are synchronous with periodic variations in the intensity of lines of various groups of elements such as the rare earths, chromium, and strontium. These variations are generally attributed to axial rotation of a star carrying an asymmetric distribution of magnetic areas.

The periods of variation are characteristically a few days, but range up to 226 days for HD188041 and 2,350 days for HD187474. Preston[3] has tabulated the periodic magnetic variables as identified in 1967. Of these, 15 show reversals of magnetic polarity; only 3—HD188041, 78 Virginis, and HD215441—show always the same polarity.

The strongest magnetic field yet measured in nature is that of the AOp star HD215441; for this the field at maximum has been measured at 35,700 gauss.

Table 5i-3 summarizes data for 89 magnetic stars as of 1958,[1] except that recently determined periods have been added for several stars from the work of Preston, Renson, Steinitz, and Wehlau.

Table 5i-4 provides data for 38 additional magnetic stars discovered between 1958 and 1966.

Much of the observational and interpretive work on the subject is reviewed by various authors in the Proceedings of the American Astronomical Society–National

---

[1] H. W. Babcock, *Astrophys. J.* **128**, 228 (1958).
[2] H. W. Babcock, *Astrophys. J. Supp.* **3** (30), (1958).
[3] G. W. Preston, *Astrophys. J.* **150**, 547 (1967).

TABLE 5i-4. MAGNETIC STAR DATA (FOR STARS DISCOVERED 1958–1966)

Star or HD	R.A.*	Dec.*	m_v	Sp	w	No. of obs.	H_e extremes	
2837	0ʰ29ᵐ59ˢ	+43°29'	9.1	A0		1/1		+700 ± 127
5797	0 58 6	+60 14	8.8	A0p		3/3	0 ± 148	+1420 ± 120
9393	1 30 53	+43 41	8.5	A0p		4/4	−1960 ± 272	+2790 ± 170
12288	2 0 14	+69 23	8.0	A0p		4/5	−1345 ± 95	−195 ± 109
16778	2 40 51	+59 40	7.7	B9p(?)		3/6	+21 ± 153	+1620 ± 141
17775	2 50 48	+61 43	8.8	A0p		1/1		+1290 ± 111
18078	2 53 34	+56 1	8.0	A2p	2	3/3	+700 ± 90	+1075 ± 115
24712	3 53 23	− 12 13	5.9	A5, F0		3/4	+575 ± 60	+1000 ± 125
50729	6 52 19	− 4 51	9.1	A5p		1/1	−540 ± 88	
51106	6 53 52	− 1 30	7.7	A3p		1/3		+890 ± 190
E Pup / 55719	7 10 56	−40 26	5.4	A2		1/1		+1215 ± 150
59435	7 27 42	− 9 10	7.9	A5p		2/2	−430 ± 88	+848 ± 103
89069	10 17 42	+78 59	8.1	(A0p) A3p		6/6	−440 ± 114	+445 ± 112
94660	10 53 12	−42 2	6.3	A0	≈0.1	7/12	−1960 ± 87	−1020 ± 108
115606	13 16 4	+13 13	8.3	A2	≈0.3	2/4	−810 ± 139	−60 ± 143
133652	15 4 7	−30 46	6.0	A0p	≈0.5	1/4	−2080 ± 320	
141988	15 47 53	+62 28	8.3	A2p		4/5	−810 ± 122	+1235 ± 129
143939	16 2 3	−39 20	7.0	B9p	≈0.5	2/3	+690 ± 236	+730 ± 260
162950	17 50 57	+27 12	7.8	A3		1/1	−565 ± 87	
170973	18 30 7	+ 3 38	6.3	A0p		8/8	−1140 ± 71	+755 ± 52
171782	18 34 31	+ 5 15	7.9	A0p		11/16	−1380 ± 130	+1190 ± 181
177984	19 4 45	+ 7 37	9.1	A2p		1/1	−785 ± 110	
179259	19 8 36	+44 30	8.9	A5p		2/3	−540 ± 77	+40 ± 118
183806	19 30 27	−45 18	5.9	A0p	≈0.5	1/3	−720 ± 271	
186343	19 41 17	+22 12	8.2	A2p		1/1	−430 ± 60	
190145	19 58 48	+67 22	7.4	A2p		1/2	−580 ± 77	
190068	20 0 52	+15 15	8.0	A0p	0.4	4/4	+990 ± 192	+1780 ± 183
189932	20 1 3	−33 54	6.9	F0p		1/4		+525 ± 86
355163	20 10 44	+13 52	8.7	A0p		1/1		+790 ± 228
192687	20 13 47	+13 43	8.6	A2	≈0.3	1/2		+1120 ± 264
+29°4202	20 49 3	+29 39	8.8	A0p		4/4	−1520 ± 90	+1500 ± 134
200311	20 59 47	+43 54	7.9	(A0p) B9p		7/13	−1900 ± 159	+760 ± 139
201174	21 4 56	+45 6	8.5	A0p		27/32	−1825 ± 143	+1765 ± 177
204411	21 25 26	+48 40	5.3	(A3p) F0?		5/6	−515 ± 41	+665 ± 70
212385	22 22 16	−39 20	6.9	A2p		1/2	−1260 ± 319	
215441	22 42 42	+55 22	8.6	A0p		28/37	+4100 ± 370	+35,700
220147	23 19 3	+62 11	7.6	B9p		4/5	−835 ± 151	+735 ± 138
221568	23 30 55	+57 41	8.0	A0p		6/8	−225 ± 172	+470 ± 69

* Position for 1960.

Aeronautics and Space Administration Symposium held at Greenbelt, Maryland, in 1965.[1] The book is replete with references.

GALACTIC MAGNETIC FIELD

5i-6. Summary. Some of the gross features of the galactic magnetic field have been inferred from information related to cosmic rays (cf. Ginzburg and Syrovatskii, 1964). A comparison of the observed cosmic-ray electron spectrum with the non-thermal radio spectrum arising from galactic synchrotron radiation indicates (Okuda and Tanaka, 1968) that the magnetic field is 10 to 20 microgauss near the galactic center, 5 to 10 microgauss near the solar system, and $\gtrsim 2.5$ microgauss for the halo. Dynamical considerations (Parker, 1968) of the cosmic-ray pressure, due mainly to energetic protons, suggest that the average field of the disk is about 5 microgauss.

[1] "The Magnetic and Related Stars," Robert C. Cameron, ed., Mono Book Corporation, Baltimore, 1967.

Polarization measurements (cf. van de Hulst, 1967) of galactic nonthermal radio emission indicate that the coherence scale of the magnetic field of the disk is about 10^2 light years. The Faraday rotation measure for the polarization of distant discrete radio sources varies quite smoothly with galactic coordinates (Morris and Berge, 1964; Gardner and Davies, 1966) and corresponds to a field whose lines of force run parallel to the galactic plane in the direction $l^{II} \approx 70°$ for $b^{II} > 0$, while below the plane ($b^{II} < 0$) the direction of the field is opposite. These directions are in general agreement with the studies of the polarization of starlight by magnetically aligned interstellar grains (Smith, 1956; Behr 1959) and with the direction of the local Orion spiral arm (Sharpless, 1965). A search (Verschuur, 1968) for the Zeeman splitting of the 21-cm-absorption line by the atomic hydrogen of this local arm yields a limit to this HI-associated field as 0.6 ± 0.9 microgauss. A relatively strong magnetic field of 20 microgauss in the Perseus spiral arm, in the direction of Cassiopeia A, was clearly detected by the Zeeman effect in the course of the same observations. This measurement of a strong HI-associated magnetic field suggests that the search for detectable Zeeman effects in other absorption or emission spectra throughout the galactic disk should yield much new information.

References

Behr, A.: *Nachr. Akad. Wiss. Göttingen Math.-physik. Kl.* IIa 185 (1959).

Gardner, F. F., and R. D. Davies: *Australian J. Phys.* **19**, 129, 441 (1966).

Ginzburg, V. L., and S. I. Syrovatskii: "Origin of Cosmic Rays," Pergamon Press, New York, 1964.

Morris, D., and G. L. Berge: *Astrophys. J.* **139**, 1388 (1964).

Okuda, H., and Y. Tanaka: *Can. J. Phys.* **46**, S642 (1968).

Parker, E. N.: "Stars and Stellar Systems," vol. 7., "Nebulae and Interstellar Matter," B. Middlehurst and L. Aller, eds., University of Chicago Press, Chicago, 1968.

Sharpless, S.: "Stars and Stellar Systems," vol. 5, "Galactic Structure," A. Blaauw and M. Schmidt, eds., University of Chicago Press, Chicago, 1965.

Smith, E. van P.: *Astrophys. J.* **124**, 43 (1956).

van de Hulst, H. C.: "Annual Review of Astronomy and Astrophysics," vol. 5, L. Goldberg, ed., Annual Reviews, Inc., Palo Alto, 1967.

Verschuur, G. L.: *Phys. Rev. Letters* **21**, 775 (1968).

Section 6

OPTICS

BRUCE H. BILLINGS, Editor

Joint Commission on Rural Reconstruction, Taipei, Taiwan.

CONTENTS

6a. Fundamental Definitions, Standards, and Photometric Units

6a-1. Fundamental Definitions

Absorptance. The ratio of the radiant flux lost by absorption to the incident radiant flux. If I_0 represents the incident flux, I_r the reflected flux, I_t the transmitted flux, the absorptance is given by the expression

$$\frac{I_0 - (I_t + I_r)}{I_0}$$

Absorption, Bouger's Law. If I_0 is the incident flux, I the flux passing through a thickness x of a material whose *absorption coefficient* is α,

$$I = I_0 e^{-\alpha x}$$

where it is implied that I and I_0 are measured within the material.

The *extinction coefficient* k is given by the relation $k = \alpha\lambda/4\pi$, where λ is the wavelength *in vacuo* and α is the absorption coefficient. The mass absorption is given by k/d, where d is the density. The transmittance is given by I/I_0.

Absorption Spectrum. The spectrum obtained by the examination of light from a source that gives a continuous spectrum, after this light has passed through an absorbing medium. The absorption spectrum will be marked by dark lines or bands; in the case of gases these will be the reverse of many of the features of the emission spectrum. When the absorbing medium is in the solid or liquid state, the spectrum of the transmitted light shows broad, dark regions which are not resolvable into lines and usually have no sharp or distinct edges.

Achromatic. A term applied to a lens, signifying that its focal length is the same for two quite different wavelengths.

Angular Aperture. The arc sine of the ratio of radius to focal length of a lens.

Apochromat. A term applied to a photographic or microscopic objective indicating that its focal length is the same for three quite different wavelengths.

Astigmatism. An error characteristic of the formation of images oblique to the axis of axially symmetric optical systems. When astigmatism is present, the sharpest image of a radial line will be formed at a distance from the lens different from the sharpest image of a tangential line.

Balmer Series of Spectral Lines. The wavelengths of a series of lines in the spectrum of hydrogen given in nanometers by the equation

$$\lambda = 364.6 \frac{N^2}{N^2 - 4}$$

where N is an integer having values greater than 2.

Beer's Law (1852). If two solutions of the same salt are made in the same solvent, one of which is, say, twice the concentration of the other, the absorptance of a given thickness of the first solution should be equal to that of twice the thickness of the second.

Blackbody. An almost completely enclosed cavity in a material of constant temperature. A small hole in the cavity completely absorbs all wavelengths of incident radiant energy.

Brewster's Law (1811). The tangent of the polarizing angle for a nonabsorbing substance is equal to the index of refraction. The polarizing angle is that angle of incidence for which the completely polarized reflected ray is at right angles to the refracted ray. If n is the index of refraction, and θ the polarizing angle, $n = \tan\theta$.

Candela. Symbol cd. International unit of luminous intensity. It is $\frac{1}{60}$ of the intensity of one square centimeter of a blackbody at the temperature of solidification of platinum (2045 K).

Chemiluminescence. Emission of light during a chemical reaction.

Christiansen Effect. When a clean, finely powdered, homogeneous substance such as glass or quartz is immersed in a liquid of the same index of refraction, nearly complete transparency can be obtained, but only for substantially monochromatic light. If white light is incident, the transmitted color corresponds to the particular wavelength band for which the two substances, solid and liquid, have very nearly equal indices of refraction. Because of differences of dispersion, the indices of refraction will sufficiently match for only a narrow band of the spectrum.

Chromatic Aberration. Because of the differences in the indices of refraction for different wavelengths, light of various wavelengths from the same source cannot be sharply focused at the same distance by any lens. The differences of focus are called chromatic aberration.

Coma. An aberration characteristic of the formation of images oblique to the axis of axially symmetric optical systems. The image of a point consists of a family of eccentric circles all tangent to two intersecting, nearly straight lines in the focal plane.

Conjugate Foci. Rays close to the axis, divergent from a point source on the axis of an axially symmetric optical system, converge on another point on the axis. The point of convergence and the position of the source are interchangeable and are called conjugate foci.

Diffraction. Deviation of light from the paths and foci prescribed by rectilinear propagation (geometrical optics) consequent on the wave nature of light. Thus, even with a very small, distant source, some light, in the form of bright and dark bands, is found within a geometrical shadow because of the diffraction of the light at the edge of the object forming the shadow.

Diffraction Grating. An array of fine, parallel, equally spaced reflecting or transmitting lines which mutually enhance the effects of diffraction at the edges of each so as to concentrate the diffracted light very close to a few directions characteristic of the spacing of the lines and the wavelength of the diffracted light. If i is the angle of incidence, d the angle of diffraction, s the center-to-center distance between successive rulings, n the order of the spectrum, the wavelength is

$$\lambda = \frac{s}{n}(\sin i + \sin d)$$

Dispersion. In any spectrum-forming device, the difference of position along the spectrum per unit of wavelength difference, e.g., 1 millimeter per nanometer.

Dispersion Equations. It is convenient and sometimes necessary to obtain equations relating refractive index to wavelength so that one can interpolate or perhaps even extrapolate with considerable accuracy and also obtain the most accurate values for $dn/d\lambda$. The equation due to Hartmann is $n = n_0 + (C/\lambda - \lambda_0)$. That due to Cauchy is $n = A + (B/\lambda^2) + (C/\lambda^4)$, and a more complicated one derived by

Sellmeier is $n^2 = 1 + (A_0\lambda^2/\lambda^2 - \lambda_0^2)$. An extension of the Sellmeier equation that is useful for covering more than one absorption region is $n^2 = 1 + \displaystyle\sum_{i=0}^{m} \frac{A_i\lambda^2}{\lambda^2 - \lambda_i^2}$. Finally, the Helmholtz expression, which includes an additional term $B_i/(\lambda^2 - \lambda_i^2)$ is useful within absorption regions as well. Usually, some of the terms of the summation are replaced by a constant. In practice, one of the above expressions is often used, and then a more accurate fit is found by an appropriate curve-fitting technique such as the method of least squares. A formula developed by Herzberger, which in some respects resembles Helmholtz's, is employed in Sec. 6d-2 to generalize the data in a condensed glass table.

Dispersive Power. If n_1 and n_2 are the indices of refraction of a substance for wavelengths λ_1 and λ_2, and n is the mean index, or that for sodium light, the dispersive power of that substance for the specified wavelengths is

$$\omega = \frac{n_2 - n_1}{n - 1}$$

This is also called *Mean Dispersion*. See also *Reciprocal Dispersion*.

Doppler Effect (Light). Change of wavelength of the light observed which arises from any change of relative velocity of the observer with respect to the source of light.

Emissive Power. See *Radiation Formula, Planck's*.

Emissivity. Ratio of flux radiated by a hot substance to the flux radiated by a blackbody at the same temperature. Emissivity is usually a function of wavelength.

Extinction Coefficient. See *Absorption*.

Faraday Effect. Rotation of the plane of polarization produced when linearly polarized light is passed through certain substances in a magnetic field, the light traveling in a direction parallel to the magnetic field. For a given substance, the rotation is proportional to the thickness traversed by the light and to the magnetic field strength.

Fermat's Principle of Least Time. The path followed by a ray between two points is that along which light can be propagated in less time than for any neighboring path.

Fraunhofer's Lines. When sunlight is examined through a spectroscope, an enormous number of dark lines parallel to the length of the slit are seen against a bright continuous spectrum. The dark lines are Fraunhofer's lines. They are caused by resonance absorption by all the elements in the layers of vapors by which the sun is surrounded. The continuous spectrum from which those resonant frequencies are absorbed is produced by the extremely hot, highly compressed substances in the body of the sun proper. Many of the reversed or dark lines have the same wavelengths as bright lines found in the emission spectrum of the absorbing elements.

Huygens' Theory of Light. Light is a continuous, cyclical disturbance propagated through space with constant velocity, frequency, and wavelength in any homogeneous transparent substance. Every point (subjected to that disturbance) acts as the center of a new disturbance having the same frequency, velocity, and wavelength radiating from it equally in all directions. The secondary disturbances from the neighboring points which were simultaneously disturbed by the initial wave coalesce to produce a net effect only along a surface which is the envelope of all the simultaneous neighboring secondary disturbances. This surface forms a new wavefront, which is further propagated in the same manner.

Illuminance. Luminous flux incident per unit area. Metric units are the lux, one lumen per square meter, and the phot, one lumen per square centimeter. In the United States the lumen per square foot is commonly used. Unit illuminance is produced by a unit source at unit distance; hence the older names meter-candle for the metric unit the lux and the foot-candle, which is the same as the lumen per square foot.

Index of Refraction. For any substance this is the ratio of the velocity of propagation of waves of light in a vacuum to its velocity of propagation in the substance. It is also the ratio of the sine of the angle of incidence to the sine of the angle of refraction. The index of refraction for any substance varies with the wavelength of the refracted light.

Irradiance. Radiant power incident per unit area of a surface. The preferred symbol for this quantity is E; it is expressed in watts/m^2.

Kirchhoff's Laws of Radiation. For each wavelength and temperature the emittance of any substance is equal to its absorptance.

Lambert's Law of Absorption. Each layer of equal thickness of any homogeneous substance absorbs an equal fraction of the light which is incident upon that layer.

Lambert's Law of Illumination. The illuminance of a surface on which light falls normally from a source small compared with its distance is inversely proportional to the square of the distance of the surface from the source. If the normal to the surface is inclined at an angle with the direction of the rays, the illuminance is proportional to the cosine of that angle.

Lens Combination. If f_1 and f_2 are the focal lengths of two thin lenses separated by a distance d, the focal length of the system is

$$F = \frac{f_1 f_2}{f_1 + f_2 - d}$$

Lens Formulas. The focal length F and distances p and q of pairs of conjugate foci (positive if convex) for a single thin lens which has index of refraction n and whose surfaces have radii of curvature r_1 and r_2 are connected by

$$\frac{1}{F} = \frac{1}{p} + \frac{1}{q} = (n - 1) \left(\frac{1}{r_1} + \frac{1}{r_2} \right)$$

If that lens has thickness t,

$$F = \frac{n r_1 r_2}{(n - 1)[n(r_1 + r_2) - t(n - 1)]}$$

Lumen. See *Luminous Flux.*

Luminance. The luminous flux per unit solid angle emitted per unit area as projected on a plane normal to the line of sight. The unit of luminance is that of a perfectly diffusing surface giving out one lumen per square centimeter and is called the *lambert.* The millilambert (0.001 lambert) is a more convenient unit. The candela per square centimeter is the luminance of a surface which has, in the direction considered, a luminous intensity of one candela per square centimeter of projected area.

Luminosity. Ratio of the luminous flux in lumens to the total radiant flux in watts.

Luminosity Maximum. 673 lumens per watt for 555 nm.

Luminous Flux. The total amount of light emitted by a source per unit time is called the luminous flux from the the source. The unit of luminous flux, the *lumen* (symbol lm), is the flux emitted in a unit solid angle by a point source that has one candela luminous intensity. A one-candela point source that radiates uniformly in all directions thus emits 4π lumens into all space.

Luminous Intensity. Luminous flux emitted per unit solid angle. The unit of luminous intensity is the candela. The symbol for luminous intensity is I.

The mean horizontal intensity is the average intensity measured in a horizontal plane passing through the source. The mean spherical intensity is the average intensity measured in all directions; it is equal to the total luminous flux in lumens divided by 4π.

Magnifying Power. In an optical instrument this is the ratio of the visual angle subtended by the image of the object seen through the instrument to the visual angle subtended by the object when observed by the unaided eye. In the case of the microscope or simple magnifier the object when viewed by the unaided eye is supposed to be at a distance of 25 cm.

Minimum Deviation. The deviation, or change of direction, of light passing through a prism is a minimum when the angle of incidence is equal to the angle of emergence. If D is the angle of minimum deviation and A the angle of the prism, the index of refraction of the prism for the wavelength used is

$$n = \frac{\sin \tfrac{1}{2}(A + D)}{\sin \tfrac{1}{2}A}$$

Molecular Refraction. The molecular refraction of a substance can be computed by the following relation:

$$N = \frac{M(n^2 - 1)}{d(n^2 + 2)}$$

where N is the molecular refraction for a specified wavelength and temperature, M the molecular weight, d the density, and n the refractive index for the specified conditions.

Nodal Points. Two points on the axis of a lens such that a ray entering the lens in the direction of one leaves as if from the other and parallel to the original direction.

Optical Density. The common logarithm of the reciprocal of transmittance

$$D = \log \frac{1}{t}$$

Polarized Light. Light which exhibits different properties in different directions at right angles to the line of propagation is said to be polarized. Specific rotation is the power of materials to rotate the plane of polarization. It is stated in terms of the rotation in degrees per decimeter per unit density or concentration.

Principal Focus. For a lens or spherical mirror, this is the point of convergence of light coming from a source at an infinite distance.

Radiance. The radiant power (flux) emitted in a specified direction, per unit projected area of surface, per unit solid angle. The preferred symbol for this quantity is L; it is expressed in watts per steradian per square meter.

Radiant Energy. When a substance is excited—e.g., because it has a temperature above 0 K—it radiates energy, called radiant energy. This may be the amount of energy emitted during the entire radiating lifetime of the body, it may be the amount of energy for a given time period, or it may be the amount in a given volume of space. The preferred symbol for this quantity is Q; it is expressed in joules.

Radiant Density. The radiant energy per unit volume of space is sometimes a useful quantity; it is called radiant density. The preferred symbol for this quantity is w, and it is expressed in joules/m^3.

Radiant Exitance. Radiant power emitted into a full sphere (4π steradians) by a unit area of source. The preferred symbol for this quantity is M; it is expressed in watts/m^2.

Radiant Flux. The rate at which energy is radiated is called radiant power or flux. Radiant energy is the time integral of radiant flux. The preferred symbol for this quantity is Φ; it is expressed in watts = joules/sec.

Radiant Intensity. Radiant flux per unit solid angle, expressed in watts per steradian. The preferred symbol for radiant intensity is I.

Radiation Formula, Planck's. The spectral exitance of a blackbody at wavelength λ and in a spectral range $\Delta\lambda$ can be written

$$M_\lambda = \frac{c_1 \lambda^{-5} \Delta\lambda}{e^{c_2/\lambda T} - 1}$$

where M_λ is watts/m²; c_1 and c_2 are constants with numerical values 3.7415×10^{-16} watt $\cdot$ m² and 0.014388 m $\cdot$ K, respectively; λ is the wavelength in meters; and T is Kelvin temperature.

Radius of Curvature from Spherometer Readings. If l is the mean length of the sides of the nearly equilateral triangle formed by the points of the three legs, and d is the normal distance from the mid-point of the triangle to the spherical surface on which the points rest, then the radius of curvature of the surface is

$$R = \frac{l^2}{6d} + \frac{d}{2}$$

Reciprocal Dispersion. $\nu = (n_D - 1)/(n_F - n_C)$, where n_C, n_D, and n_F are indices of refraction for the Fraunhofer lines C, D, F. The index n_d for the Fraunhofer d lines is sometimes used instead of n_D. The ν value is sometimes called the *Abbe Number.*

Reflectance. The ratio of the reflected flux to the flux incident on a surface is called the reflectance; it may refer to diffuse or to specular reflection. In general, it varies with the angle of incidence and with the wavelength of the light. Symbol ρ.

Reflection of Light at a Smooth Boundary between an Absorbing Medium and a Transparent Medium. At normal incidence, if n_0 is the index of the transparent medium, and n_1 and k_1 are the index and extinction coefficients of the absorbing medium, the reflectance is

$$\rho = \frac{(n_0 - n_1)^2 + k_1^2}{(n_0 + n_1)^2 + k_1^2}$$

Reflection of Light by a Smooth Surface between Two Transparent Media (Fresnel's Formulas). If i is the angle of incidence, r the angle of refraction, n_1 the index of refraction of the medium from which the light is incident, n_2 the index of refraction of the other, then for unpolarized incident light the reflectance is

$$R = \frac{1}{2}\left[\frac{\sin^2(i-r)}{\sin^2(i+r)} + \frac{\tan^2(i-r)}{\tan^2(i+r)}\right] \qquad \text{where } \frac{\sin r}{\sin i} = \frac{n_1}{n_2}$$

If $i = 0$ (normal incidence),

$$R = \left(\frac{n_2 - n_1}{n_2 + n_1}\right)^2$$

Refraction at a Spherical Surface. If u is the distance of a point object from a spherical surface separating two media, v is the distance of the point image or the intersection of a nearby refracted ray with the line defined by the object and the center of curvature, n_1 and n_2 are the indices of refraction of the first and second media, and r is the radius of curvature of the separating surface, then

$$\frac{n_2}{v} + \frac{n_1}{u} = \frac{n_2 - n_1}{r}$$

Refractivity. This is given by $n - 1$, where n is the index of refraction; the *specific refractivity* is given by $(n - 1)/d$, where d is the density; *molecular refractivity* is the product of the specific refractivity and the molecular weight.

Relationships between Radiometric Units. In a nonabsorbing medium, for the geometrical arrangement in which the source and receiver areas are both perpendicular to the line joining their centers, the radiation quantities above are related as follows:

$$\Phi = \frac{\partial Q}{\partial t}, \quad M = \frac{\Phi}{A_s}, \quad L = \frac{M}{\omega_r}, \quad E = L\omega_s, \quad I = Er^2, \quad \Phi = I\omega_r$$

where A_s is source area, ω_s is the solid angle subtended by the source area at the receiver, ω_r is the solid angle subtended by the receiver area at the source, and r is the distance between the centers of the source and receiver areas. The corresponding relations connect the corresponding photometric quantities, which are distinguished by the root "lumi-" in place of "radi-." The same symbols are preferred for the photometric quantities as for the corresponding radiometric quantities. When symbols are needed for both photometric and radiometric quantities in the same context, the symbols for photometric quantities should be followed by the subscript v. The symbols of radiometric quantities should, in such cases, be followed by the subscript e.

Resolving Power. For a telescope or microscope this is the minimum separation of two objects for which they appear distinct and separate when viewed through the instrument. Resolving power is often specified by the reciprocal, e.g., lines per millimeter.

Rotatory Power, Molecular or Atomic. This is the product of the specific rotatory power by the molecular or atomic weight. Magnetic rotatory power is given by

$$\frac{\theta}{t} G \cos \alpha$$

where G is the magnetic field strength, t is the thickness traversed, θ is the rotation of the plane of polarization by the Faraday effect, and α is the angle between the field and the direction of the light.

Snell's Law of Refraction. If i is the angle of incidence, r the angle of refraction, v the velocity of light waves in the first medium, and v' the velocity in the second medium, the relative index of refraction n is

$$n = \frac{\sin i}{\sin r} = \frac{v}{v'}$$

Specific Rotation. If there is n g of active substance in v cm³ of solution, and the light passes through 1 cm, r being the observed rotation of the plane of polarization in degrees, the specific rotation (for 1 cm) is

$$[\alpha] = \frac{rv}{n1}$$

Spectral Irradiance. Irradiance per unit wavelength interval. The preferred symbol for this quantity is E_λ; it is measured in units of watts per square meter per micrometer.

Spectral Radiance. Radiance per unit wavelength interval. The preferred symbol for this quantity is L_λ; it is measured in units of watts per steradian per square meter per micrometer.

Spectral Series. Spectral lines or groups of lines which occur in an orderly sequence in the spectrum of an element.

Spherical Aberration. When large surfaces of spherical mirrors or lenses are used, the light divergent from a point source is not focused exactly at a point. The phenomenon is known as spherical aberration. For oblique pencils it produces coma.

Spherical Mirrors. If R is the radius of curvature, F is the principal focus, and f_1 and f_2 are any two conjugate focal distances,

$$\frac{1}{f_1} + \frac{1}{f_2} = \frac{1}{F} = \frac{2}{R}$$

If the transverse dimensions of the object and the image are O and I, respectively, and u and v their distances from the mirror,

$$\frac{O}{I} = \frac{u}{v}$$

Total Reflection. When light passes from a denser medium to one in which the velocity is greater, refraction ceases and total reflection begins, at a certain critical angle of incidence such that

$$\sin \theta = \frac{1}{n}$$

where n is the index of the denser medium relative to that of the less dense.

Transmissivity. The internal transmittance for unit thickness of a nondiffusing substance.

Transmittance. If Φ_0 and Φ are the incident and transmitted luminous flux, respectively, the transmittance is given by Φ/Φ_0.

Transmittance, External. The external transmittance is the ratio of the flux that is transmitted through a sample to that which is incident on it. This is the quantity that is usually measured. The greater the losses by reflections at the surfaces, the smaller is the external transmittance; the greater the absorption, the smaller is the external transmittance.

Transmittance, Internal. The ratio of the flux incident internally on the second internal surface of a sample to that leaving the first surface is the internal transmittance. This is not a measurable quantity but is obtained from measurements of external transmittance corrected for reflection losses. Internal transmittance is related to sample thickness and absorption coefficient by Bouguer's law:

$$\text{Internal transmittance} = \exp{(-\alpha x)}$$

Transmittance, Luminous. External transmittance, when flux is measured in photometric units (lumens).

Transmittance, Radiant. External transmittance, when flux is measured in radiometric (powers) units.

Transmittancy. The transmittancy is the ratio of the transmittance of a solution to that of a solvent.

Wein's Displacement Law. When the temperature of a radiating blackbody increases, the wavelength corresponding to maximum radiance decreases in such a way that the product of the absolute temperature and wavelength is constant.

$$\lambda_{max}T = 0.0028978 \text{ m·K}$$

Zeeman Effect. The splitting of a spectrum line into several symmetrically disposed polarized components, which occurs when the source of light is placed in a strong magnetic field. The directions of polarization and the appearance of the effect depend on the direction from which the source is viewed relative to the lines of force.

6a-2. Fundamental Standards

Candela. The international standard unit of luminous intensity. It is $\frac{1}{60}$ of the intensity of one square centimeter of a blackbody radiator at the temperature of solidification of platinum (2045 K).

Primary Standard of Wavelength. The krypton[86] line whose vacuum wavelength is $6.057802106 \times 10^{-7}$ m. This is the unperturbed $2P_{10}$-$5d_5$ transition of [86]Kr. The actual definition was given by defining the standard meter as 1,650,763.73 vacuum wavelengths of the krypton line.

Velocity of Light. An acceptable present value is $2.9979250 \pm 10 \times 10^8$ meters per second (in SI units). This figure is taken from a paper by Taylor, Parker, and Langenberg, *Rev. Mod. Phys.*, **41**, 375 (1969). It should be considered an interim value pending completion of the work of the Task Group on Fundamental Constants, of the Committee on Data for Science and Technology, International Council of Scientific Unions. It is expected that the Task Group's recommended figure will be available in 1973.

6a-3. Photometric Quantities, Units, and Standards

Apostilb. Unit of luminance. $1/\pi$ cd/m². Symbol, asb.

Blondel. Alternate name for apostilb.

Candela. Unit of luminous intensity. It is $\frac{1}{60}$ of the intensity of one square centimeter of a blackbody radiator at the temperature of solidification of platinum (2045 K). Symbol, cd.

Efficiency of a Source of Light. The efficiency of a source is the ratio of the total luminous flux to the total power consumed. It is expressed in lumens per watt.

Foot-Lambert. Unit of luminance equal to $1/\pi$ candela per square foot. Symbol, fL.

Lambert. Unit of luminance equal to $1/\pi$ candela per square centimeter. Symbol, L. $1/\pi$ sb.

Least Mechanical Equivalent of Light. One lumen at the wavelength of maximum luminosity (555 nm) equals 0.00147 watt; 1 watt at the same wavelength equals 680 lumens.

Lumen. The lumen is the unit of luminous flux. Symbol, lm. It is equal to the flux through a unit solid angle (steradian) from a one-candela point source or to the flux on a unit surface all points of which are at unit distance from a one-candela uniform point source. Luminous power of 1 talbot per second.

Lux. Unit of illuminance. 1 lm/m². Symbol, lx.

Nit. Unit of luminance. 1 cd/m². Symbol, nt.

Phot. Unit of illuminance. 1 lm/cm². Symbol, ph.

Relative Luminosity. The relative luminosity for a particular wavelength is the ratio of the luminosity for that wavelength to the maximum luminosity. Values of the relative luminosity are given in Sec. 6j, Colorimetry.

Spherical Candlepower. The spherical candlepower of a lamp is the average intensity (candela) of the lamp in all directions. It is equal to the total luminous flux from the lamp, in lumens, divided by 4π.

Stilb. Unit of luminance. 1 cd/cm². Symbol, sb.

Talbot. Unit of luminous energy, the product of lumens times seconds, 1 lm·sec.

Troland. Unit of retinal illuminance. Illuminance produced on the retina of the human eye when a surface having luminance of 1 cd/m² is viewed through a pupil whose area is 1 mm². 0.4 times the illuminance produced on the retina when a surface having 1 millilambert luminance is viewed through a pupil having 1 millimeter diameter. Symbol, td.

6a-4. Photometric Equivalents

Candela per Square Centimeter (Luminance). 1 stilb, 10,000 nit, π apostilbs, 3.1416 lamberts, 3,141.6 millilamberts.

Candela per Square Inch (Luminance). 0.48695 lambert; 486.95 millilamberts.

Foot-Candle (Illuminance). 1 lumen incident per square foot, 1.0764 milliphots, 10.764 lumens per square meter, 10.764 lux.

Foot-Lambert (Luminance). 1.0764 millilambert.

Lambert (Luminance). 0.3183 candela per square centimeter; 2.054 candela per square inch. One lumen is emitted per square centimeter of a perfectly diffusing surface having a luminance of 1 lambert.

Lumen (Luminous Flux). Emitted by 0.07958 candela spherical intensity. A source of one candela spherical intensity emits $4\pi = 12.566$ lumens.

Lumen per Square Centimeter per Steradian (Luminance). 3.1416 lamberts.

Lumen per Square Foot (Illuminance). One foot-candle, 10.764 lumens per square meter; 10.764 lux.

Lumen per Square Foot per Steradian (Luminance). 3.3816 millilamberts.

Lumen per Square Meter (Illuminance). 1×10^{-4} phot, 0.092902 foot-candle or lumen per square foot; 1.0 lux.

Lux. 1×10^{-4} phot, 0.1 milliphot, 0.092902 foot-candle.

Meter-Candle (Illuminance). 1 lux, or 0.0929 lumen emitted per square foot (perfect diffusion).

Millilambert (Luminance). 0.929 foot-lambert.

Milliphot (Illuminance). 0.001 phot, 0.929 foot-candle; 10 lux.

Phot. 1,000 milliphots, 1.000×10^4 lm/m², 10^{-4} lx; 929 candles.

Stilb. 10,000 nit, 3.1416 lamberts, 31,416 apostilbs.

6b. Refractive Index of Special Crystals and Certain Glasses

STANLEY S. BALLARD

JAMES STEVE BROWDER

JOHN F. EBERSOLE

University of Florida

Refractive indices for the following materials are given in this section:

Ammonium dihydrogen phosphate (ADP) and Potassium dihydrogen phosphate (KDP)
Barium fluoride
Barium titanate
Cadmium fluoride
Cadmium iodide
Cadmium selenide
Cadmium sulfide
Calcite
Calcium fluoride
Cesium bromide
Cesium iodide
Crystal quartz
Cuprous chloride
Diamond
Fused silica
Germanium
Irtrans 1 to 6
Lanthanum fluoride
Lead bromide
Lead chloride
Lead fluoride
Lead selenide
Lead sulfide
Lead telluride
Lithium fluoride
Magnesium fluoride
Magnesium oxide
Muscovite mica
Potassium bromide
Potassium chloride
Potassium iodide
Rubidium bromide
Rubidium chloride

Rubidium iodide
Ruby
Sapphire
Selenium
Silicon
Silver chloride
Sodium chloride
Sodium fluoride
Sodium nitrate
Spinel
Strontium titanate
T-12
Tellurium
Thallium bromide
Thallium chloride
Thallium bromide-chloride (KRS-6)
Thallium bromide-iodide (KRS-5)
Titanium dioxide
Zinc sulfide
Group III–Group V compounds:
    Gallium antimonide
    Gallium arsenide
    Gallium phosphide
    Indium antimonide
    Indium arsenide
    Indium phosphide
Nonoxide chalcogenic glasses:
    Arsenic-modified selenium glass
    Arsenic triselenide glass
    Arsenic trisulfide glass
    A telluride glass
    Texas Instruments Glass No. 1173
Special glasses:
    Cer-Vit
    Corning Vycor

Refractive index, or index of refraction, can be defined in a number of ways. The complex refractive index, often written as $n + ik$, is defined and described in Sec. 6g. The real part of the refractive index of a substance can be defined as the ratio of the velocity of light in vacuo to the phase velocity of the light in the substance. Usually this quantity is called the absolute refractive index. Often the relative index—the ratio of the absolute refractive index of one substance to the absolute refractive index of another—is the more useful quantity. The refractive index relative to air is expecially useful, since most optical systems have air for both the initial and the final

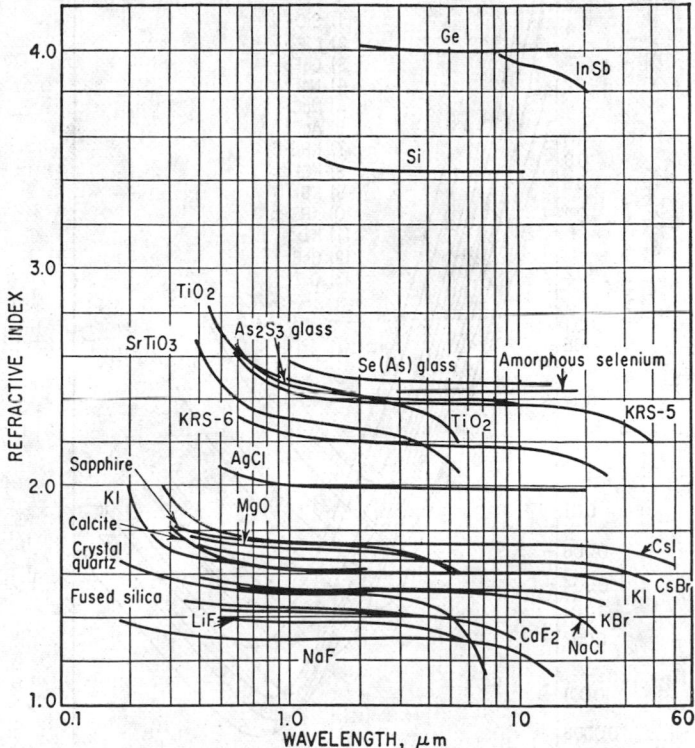

FIG. 6b-1. Refractive index vs. wavelength for several optical materials.

media and since the refractive index of air does not vary appreciably from that of a vacuum (see Table 6e-5). Unless stated otherwise, the refractive-index data in this section are values relative to air for the experimental conditions described. The absolute refractive index can be obtained from these data by multiplying by the appropriate values of the absolute refractive index of air.

The index of refraction varies throughout the entire electromagnetic spectrum in a manner described by theory. Between two absorption bands, the region of primary interest for most materials, the index decreases with increasing wavelength. Near the first absorption band the decrease is rapid; then the decrease becomes quite slow until an inflection point is reached; the decrease of index with wavelength then becomes more rapid again as the second absorption band is approached. This behavior can be seen in Fig. 6b-1. The figure also suggests the large range of values of refractive index covered by optical materials. Some of the more recently available semiconductors, silicon, germanium, and tellurium, have very large refractive indices—about 3.4, 4, and 6, respectively. In Fig. 6b-2 the slopes of many of the curves shown in Fig. 6b-1 are plotted, following the usual practice of plotting $-dn/d\lambda$ rather than $dn/d\lambda$. Regions of smallest dispersion ($d^2n/d\lambda^2 = 0$) and values of $|dn/d\lambda|$ in various spectral regions can be obtained directly from this figure.

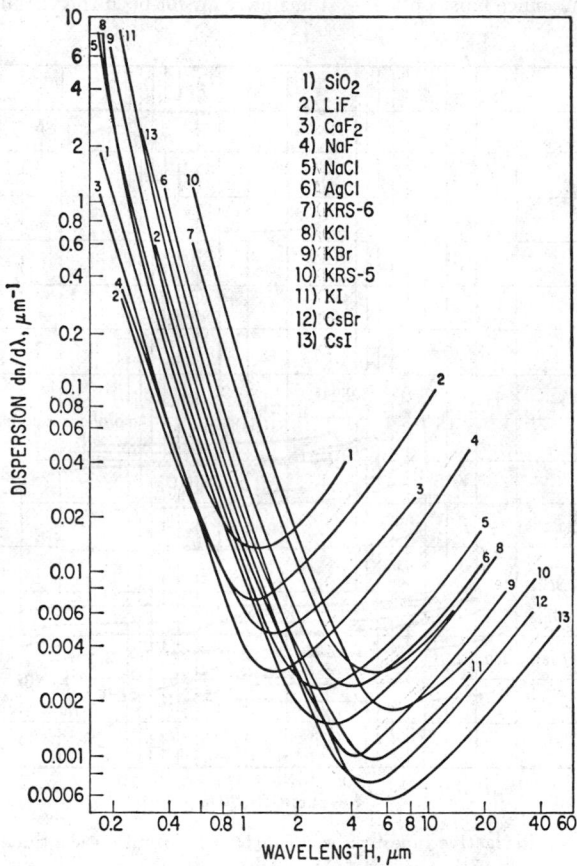

FIG. 6b-2. Dispersion vs. wavelength for several optical materials. [*From A. Smakula, Opt. Acta* **9**, 205 (1962).]

The method usually employed for the precise measurement of refractive index is that of minimum deviation. A prism is made from the material; it is mounted on a prism table, and the prism angle is measured; then the angles of minimum deviation are determined for a series of spectral lines. For some measurements, a monochromator is used to obtain light of the desired wavelength, but where possible, spectral lines provide radiation of a more accurately known wavelength. Thus, precise measurements are usually reported as values of refractive index for a number of irregularly spaced wavelengths of light. However, intermediate values may often be desired by the optical designer. Thus, interpolation techniques, some based on physical principles and others which are strictly mathematical or graphical, have been developed. The equations so derived, relating refractive index to wavelength, are called dispersion formulas. The formulas reported in the literature are included here. They not only provide the interpolation mechanism described above, but they can be used to evaluate the experimental data. The formulas also contribute to a better understanding of the properties of the material they describe.

For some applications the change of refractive index with temperature (dn/dT), the so-called temperature coefficient, is an important consideration. The relative temperature coefficient, $(1/n)(dn/dT)$, is also used. The available data for these properties of each material are included. A complete description includes telling how the temperature coefficient changes for each wavelength, as illustrated for several materials in Fig. 6b-3, and also how it varies with temperature. The variations of the temperature coefficient with both wavelength and temperature are presented when available, but it is usually sufficient to indicate a single value for a given wavelength or temperature region. The second-order effects d^2n/dT^2 and

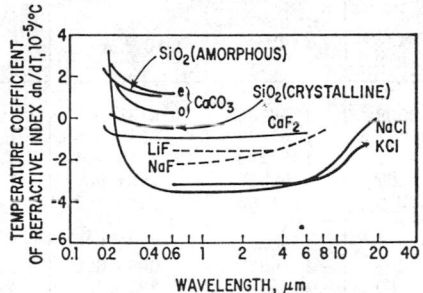

FIG. 6b-3. The temperature coefficient of refractive index of several materials. [*Adapted from A. Smakula, Opt. Acta* **9**, 205 (1962).]

$d^2n/d\lambda\,dT$ have not been reported. They will occasionally be found in the references for some of the older, more throughly measured materials, such as rock salt.

If at all possible, the reference describing the original work has been cited; for some materials additional references to compendia and collections of data are given.

A number of conventions have been assumed for this section. Unless stated otherwise, wavelengths—even those in dispersion formulas—are given in micrometers (μm), and temperatures T are in kelvins (K), employing the new Système International set of units. Many sets of data were recorded using degrees Celsius (°C), and no attempt has been made to convert them to kelvins. The index data listed in the tables are generally the measured rather than computed values. Where the data are computed, the table is immediately followed by a dispersion formula. *Note:* For some materials where the data were measured, a dispersion equation is given as a supplement, but these instances are clearly indicated. If the temperature at which measurements were made is not reported, it can be assumed to be room temperature, or about 20°C. The refractivity $(n - 1)$ of several materials is reported rather than the refractive index n. For anisotropic media, both the refractive index for the ordinary ray (n_o) and the refractive index for the extraordinary ray (n_e) are usually given. Finally, a table entitled Miscellaneous Refractive-index Data has been included; it covers those materials for which only a very few data are available.

Comments on any important physical or chemical characteristics of the materials are presented in Sec. 6c.

The individual materials are treated in the same order as listed at the beginning of this section, with the exception of those included in Table 6b-64, Miscellaneous Refractive-index Data.

(The literature search extended back to January, 1959. It was restricted primarily to "optical" journals, i.e., *Journal of the Optical Society of America*, *Applied Optics*, *Optics & Spectroscopy*, *Optica Acta*, and *Infrared Physics*. It is realized that optical data on semiconductor materials are to be found also through the literature of solid-state physics.)

Ammonium Dihydrogen Phosphate (ADP) and Potassium Dihydrogen Phosphate (KDP)

TABLE 6b-1. REFRACTIVE INDICES OF ADP AND KDP IN AIR AT 24.8°C

ADP			KDP		
λ, 10^{-4} μm	n_o	n_e	λ, 10^{-4} μm	n_o	n_e
2138.560	1.62598	1.56738	2138.560	1.60177	1.54615
2288.018	1.60785	1.55138	2288.018	1.58546	
2536.519	1.58688	1.53289	2446.905	1.57228	
2967.278	1.56462	1.51339	2464.068	1.57105	
3021.499	1.56270	1.51163	2536.519	1.56631	1.51586
3125.663	1.55917	1.50853	2800.869	1.55263	1.50416
3131.545	1.55897	1.50832	2980.628	1.54618	1.49824
3341.478	1.55300	1.50313	3021.499	1.54433	1.49708
3650.146	1.54615	1.49720	3035.781	....	1.49667
3654.833	1.54608	1.49712	3125.663	1.54117	1.49434
3662.878	1.54592	1.49698	3131.545	1.54098	1.49419
3906.410	1.54174	....	3341.478	....	1.48954
4046.561	1.53969	1.49159	3650.146	1.52932	1.48432
4077.811	1.53925	1.49123	3654.833	1.52923	1.48423
4358.350	1.53578	1.48831	3662.878	1.52909	1.48409
4916.036	....	1.48390	3906.410	....	1.48089
5460.740	1.52662	1.48079	4046.561	1.52341	1.47927
5769.590	1.52478	1.47939	4077.811	1.52301	1.47898
5790.654	1.52466	1.47930	4358.350	1.51990	1.47640
6328.160	1.52195	1.47727	4916.036	....	1.47254
10 139.75	1.50835	1.46895	5460.740	1.51152	1.46982
11 287.04	1.50446	1.46704	5769.580	1.50987	
11 522.76	1.50364	1.46666	5790.654	1.50977	1.46856
			6328.160	1.50737	1.46685
			10 139.75	1.49535	1.46041
			11 287.04	1.49205	1.45917
			11 522.76	1.49135	1.45893
			13 570.70	1.48455	
			15 231.00	....	1.45521
			15 295.25	....	1.45512

The accuracy of the data is believed to be ±0.00003 or better.   The index values for ADP are adapted from F. Zernike, Jr., *J. Opt. Soc. Am.* **54**, 1215 (1964), and **55**, 210E (1965).   The indexes for KDP are from **54**, 1215 (1964).   Zernike also reports several absolute index values.   Index values of ADP and KDP for 0.4860, 0.5890, and 0.6560 μm are given in the "International Critical Tables," vol. VII, pp. 19, 27, McGraw-Hill Book Company, New York, 1930.   In addition, Zernike lists many computed index values in air, using the following dispersion equation:

$$ n^2 = \frac{A + B\nu^2}{1 - \nu^2/C} + \frac{D}{E - \nu^2} $$

where $\nu = 1/\lambda$ in cm^{-1}.

TABLE 6b-2. CONSTANTS OF THE DISPERSION EQUATION FOR ADP

	Extraordinary ray	Ordinary ray
A	2.163510	2.302842
B	9.616676 × 10^{-11}	1.1125165 × 10^{-10}
C	7.698751 × 10^9	7.5450861 × 10^9
D	1.479974 × 10^6	3.775616 × 10^6
E	2.500000 × 10^5	2.500000 × 10^5

From F. Zernike, Jr.: *J. Opt. Soc. Am.* **55**, 210E (1965).

TABLE 6b-3. CONSTANTS OF THE DISPERSION EQUATION FOR KDP

	Extraordinary ray	Ordinary ray
A	2.132668	2.259276
B	8.637494 × 10^{-11}	1.008956 × 10^{-10}
C	8.142631 × 10^9	7.726408 × 10^9
D	8.069981 × 10^5	3.251305 × 10^6
E	2.500000 × 10^5	2.500000 × 10^5

From F. Zernike, Jr., *J. Opt. Soc. Am.* **54**, 1215 (1964).

Temperature dependences of n_o and n_e for ADP and KDP are given by R. A. Phillips: *J. Opt. Soc. Am.* **56**, 629 (1966); M. Yamazaki and T. Ogawa, *ibid.*, 1407; and by V. N. Vishnevskii and I. V. Stefanski: *Opt. Spectr. U.S.S.R.* **20**, 195 (1966).

Barium Fluoride

TABLE 6b-4. REFRACTIVE INDEX OF BARIUM FLUORIDE AT 25°C

λ, μm	n	λ, μm	n	λ, μm	n
0.2652	1.51217	0.589262	1.47443	2.5766	1.46262
0.28035	1.50668	0.643847	1.47302	2.6738	1.46234
0.28936	1.50390	0.656279	1.47274	3.2434	1.46018
0.296728	1.50186	0.706519	1.47177	3.422	1.45940
0.30215	1.50044	0.85211	1.46984	5.138	1.45012
0.3130	1.49782	0.89435	1.46942	5.3034	1.44904
0.32546	1.49521	1.01398	1.46847	5.343	1.44878
0.334148	1.49363	1.12866	1.46779	5.549	1.44732
0.340365	1.49257	1.36728	1.46673	6.238	1.44216
0.34662	1.49158	1.52952	1.46613	6.6331	1.43899
0.361051	1.48939	1.681	1.46561	6.8559	1.43694
0.366328	1.48869	1.7012	1.46554	7.0442	1.43529
0.404656	1.48438	1.97009	1.46472	7.268	1.43314
0.435835	1.48173	2.1526	1.46410	9.724	1.40514
0.486133	1.47855	2.32542	1.46356	10.346	1.39636
0.546074	1.47586				

Adapted from I. H. Malitson: *J. Opt. Soc. Am.* **54**, 628 (1964).

Malitson also reports several computed index values, using the following dispersion equation:

$$n^2 - 1 = \frac{0.643356\lambda^2}{\lambda^2 - (0.057789)^2} + \frac{0.506762\lambda^2}{\lambda^2 - (0.10968)^2} + \frac{3.8261\lambda^2}{\lambda^2 - (46.3864)^2}$$

TABLE 6b-5. TEMPERATURE COEFFICIENT OF REFRACTIVE INDEX

λ, μm	Refractive index			dn/dT, (10⁻⁶/°C)
	15°C	35°C	55°C	
0.4046563	1.484054	1.483753	1.483452	−15.05
0.4358342	1.481416	1.481116	1.480816	−15.00
0.4861327	1.478234	1.477930	1.477628	−15.15
0.5460740	1.475559	1.475255	1.474951	−15.20
0.589262*	1.474124	1.473820	1.473515	−15.22
0.6562793	1.472439	1.472135	1.471830	−15.23
0.6678149	1.472196	1.471892	1.471586	−15.25
0.7065188	1.471474	1.471167	1.470863	−15.28
0.767858*	1.470538	1.470230	1.469920	−15.45

* Intensity weighed mean of doublet.
Adapted from I. H. Malitson, *loc. cit.*

The low-temperature dependence of refractive index is given by T. W. Houston, L. F. Johnson, P. Kisliuk, and D. J. Walsh: *J. Opt. Soc. Am.* **53**, 1286 (1963).

Cadmium Sulfide

TABLE 6b-6. REFRACTIVE INDEX OF CADMIUM SULFIDE

λ, μm	n_e	n_o
0.5120	2.751	
0.5130	2.743	
0.5140	2.737	
0.5150	2.726	2.743
0.5160	2.720	2.735
0.5170	2.714	2.727
0.5180	2.706	2.718
0.5190	2.702	2.709
0.5200	2.698	2.702
0.5210	2.694	2.700
0.5220	2.689	2.694
0.5230	2.685	2.687
0.5240	2.680	2.681
0.5250	2.675	2.674
0.5275	2.665	2.661
0.5300	2.654	2.649
0.5325	2.644	2.638
0.5350	2.637	2.628
0.5375	2.628	2.617
0.5400	2.622	2.609
0.5425	2.612	2.602
0.5450	2.606	2.594
0.5475	2.600	2.587
0.5500	2.593	2.580
0.5750	2.545	2.528
0.6000	2.511	2.493
0.6250	2.484	2.467
0.6500	2.463	2.446
0.6750	2.446	2.427
0.7000	2.432	2.414
0.7500	2.409	2.390
0.8000	2.392	2.374
0.8500	2.378	2.361
0.9000	2.368	2.350
0.9500	2.359	2.341
1.0000	2.352	2.334
1.0500	2.346	2.328
1.1000	2.340	2.324
1.1500	2.336	2.320
1.2000	2.332	2.316
1.2500	2.329	2.312
1.3000	2.326	2.309
1.3500	2.323	2.306
1.4000	2.321	2.304

From T. M. Bieniewski and S. J. Czyzak: *J. Opt. Soc. Am.* **53**, 496 (1963).

For computational purposes, dispersion equations are given by S. J. Czyzak, W. M. Baker, R. C. Crane, and J. B. Howe: *J. Opt. Soc. Am.* **47**, 240–243 (1957):

Ordinary ray:

$$n_o{}^2 = 5.235 + \frac{1.819 \times 10^7}{\lambda^2 - 1.651 \times 10^7}$$

Extraordinary ray:

$$n_e{}^2 = 5.239 + \frac{2.076 \times 10^7}{\lambda^2 - 1.651 \times 10^7}$$

Calcite

TABLE 6b-7. REFRACTIVE INDEX OF CALCITE

λ, μm	n_o	n_e	λ, μm	n_o	n_e	λ, μm	n_o	n_e
0.198	....	1.57796	0.410	1.68014	1.49640	1.229	1.63926	1.47870
0.200	1.90284	1.57649	0.434	1.67552	1.49430	1.273	1.63849	
0.204	1.88242	1.57081	0.441	1.67423	1.49373	1.307	1.63789	1.47831
0.208	1.86733	1.56640	0.508	1.66527	1.48956	1.320	1.63767	
0.211	1.85692	1.56327	0.533	1.66277	1.48841	1.369	1.63681	
0.214	1.84558	1.55976	0.560	1.66046	1.48736	1.396	1.63637	1.47789
0.219	1.83075	1.55496	0.589	1.65835	1.48640	1.422	1.63590	
0.226	1.81309	1.54921	0.643	1.65504	1.48490	1.479	1.63490	
0.231	1.80233	1.54541	0.656	1.65437	1.48459	1.497	1.63457	1.47744
0.242	1.78111	1.53782	0.670	1.65367	1.48426	1.541	1.63381	
0.257	1.76038	1.53005	0.706	1.65207	1.48353	1.609	1.63261	
0.263	1.75343	1.52736	0.768	1.64974	1.48259	1.615	....	1.47695
0.267	1.74864	1.52547	0.795	1.64886	1.48216	1.682	1.63127	
0.274	1.74139	1.52261	0.801	1.64869	1.48216	1.749	....	1.47638
0.291	1.72774	1.51705	0.833	1.64772	1.48176	1.761	1.62974	
0.303	1.71959	1.51365	0.867	1.64676	1.48137	1.849	1.62800	
0.312	1.71425	1.51140	0.905	1.64578	1.48098	1.909	....	1.47573
0.330	1.70515	1.50746	0.946	1.64480	1.48060	1.946	1.62602	
0.340	1.70078	1.50562	0.991	1.64380	1.48022	2.053	1.62372	
0.346	1.69833	1.50450	1.042	1.64276	1.47985	2.100	....	1.47492
0.361	1.69316	1.50224	1.097	1.64167	1.47948	2.172	1.62099	
0.394	1.68374	1.49810	1.159	1.64051	1.47910	3.324	....	1.47392

From F. F. Martens, *Ann. Physik* **6**, 603 (1901), for 0.198 to 0.768 μm; W. Gifford, *Proc. Roy. Soc.* (*London*) **70**, 329 (1902), for 0.226, 0.303, 0.330, 0.361, 0.706, and 0.795 μm; A. Carvallo, *Compt. Rend.* **126**, 950 (1898), and *J. Phys. Radium*, ser. 3, **9**, 465 (1900), for 0.346, and 0.801 to 2.172 μm; A. Smakula, *Office Tech. Serv.* (*OTS*) *Rept.* 111053, 1952, for 3.324 μm.

The data fit together to within a few parts in the fifth decimal. Data for the ordinary ray at about 0.21 μm appear a little out of line.

TABLE 6b-8. TEMPERATURE COEFFICIENTS OF REFRACTIVE INDEX

λ, μm	dn_o/dT $(10^{-5}/°C)$	dn_e/dT $(10^{-5}/°C)$
0.211	2.150	
0.231	1.397	2.198
0.298	0.604	1.641
0.361	0.360	1.449
0.441	0.325	1.318
0.467	0.319	
0.480	0.305	1.287
0.508	0.287	1.234
0.589	0.240	1.213
0.643	0.208	1.185

From F. J. Micheli, *Ann. Physik* **4**, 7, 772 (1902).

Calcium Fluoride

TABLE 6b-9. REFRACTIVE INDEX OF CALCIUM FLUORIDE AT 24°C,
AND TEMPERATURE COEFFICIENT OF INDEX FOR MEAN
TEMPERATURE OF 19°C

λ, μm	Computed index	Measured difference		dn/dT $(10^{-6}/°C)$
		Synthetic	Natural	
0.228803	1.47635	−2	+1	−6.2
0.24827	1.46793	+3	+5	−7.0
0.2537	1.46602	+9	+12	−7.5
0.26520	1.46233	−1	0	−8.1
0.28035	1.45828	−1	+1	−8.4
0.296728	1.45467	−2	0	−8.8
0.334148	1.44852	−1	+2	−9.2
0.34662	1.44694	−3	0	−9.4
0.365015	1.44490	−4	0	−9.6
0.4046563	1.44151	−3	+1	−9.8
0.4358342	1.43949	−3	0	−10.0
0.4861327	1.43703	−4	0	−10.2
0.546074	1.43494	−3	+1	−10.4
0.589262	1.43381	−2	+2	−10.4
0.643847	1.43268	−2	+3	−10.4
0.6562793	1.43246	−2	+1	−10.4
0.6678149	1.43226	−1	+1	−10.5
0.7065188	1.43167	−2	+2	−10.5
0.767858	1.43088	−2	+2	−10.6
0.85212	1.43002	−1	+4	−10.6
0.8944	1.42966	0	+1	−10.6
1.01398	1.42879	−2	0	−10.5
1.3622	1.42691	+1	+8	−10.0
1.39506	1.42675	+1	+6	−9.9
1.52952	1.42612	+4	+4	−9.6
1.7012	1.42531	+2	+4	−9.4
1.81307	1.42478	0	+9	−9.1
1.97009	1.42401	+3	+3	−8.9
2.1526	1.42306	−1	+1	−8.7
2.32542	1.42212	+3	+4	−8.5
2.4374	1.42147	0	+2	−8.5
3.3026	1.41561	0	+3	−8.2
3.422	1.41467	+2	+2	−8.1
3.5070	1.41398	−1	+2	−8.0
3.7067	1.41229	+2	+2	−7.8
4.258	1.40713	+4	+4	−7.5
5.01882	1.39873	+1	+5	−7.3
5.3034	1.39520	+3	+3	−7.2
6.0140	1.38539	+5	+5	−7.0
6.238	1.38200	−6	0	−7.0
6.63306	1.37565	0	+1	−6.9
6.8559	1.37186	−8	+2	−6.7
7.268	1.36443	+2	+7	−6.5
7.4644	1.36070	+5	+6	−6.4
8.662	1.33500	−4	+3	−6.0
9.724	1.30756	+1	+5	−5.8

The third and fourth columns give the difference between the computed and measured values of refractive index for synthetic calcium fluoride and natural calcium fluoride (fluorite).

Dispersion equation:

$$n^2 - 1 = \sum_j \frac{A_j \lambda^2}{\lambda^2 - \lambda_j^2}$$

TABLE 6b-10. CONSTANTS OF THE DISPERSION EQUATION AT 24°C

$\lambda_1 = 0.050263605$	$\lambda_1{}^2 = 0.002526430$	$A_1 = 0.5675888$
$\lambda_2 = 0.1003909$	$\lambda_2{}^2 = 0.01007833$	$A_2 = 0.4710914$
$\lambda_3 = 34.649040$	$\lambda_3{}^2 = 1200.5560$	$A_3 = 3.8484723$

Adapted from I. H. Malitson, *Appl. Opt.* **2**, 1103 (1963).

Cesium Bromide

TABLE 6b-11. REFRACTIVITY OF CESIUM BROMIDE AT 27°C

λ. μm	0.0	0.1	0.2	0.3	0.4	0.5	0.6	0.7	0.8	0.9
0	......	......	......	......	73,519	70,896	69,583	68,825	68,345	68,022
1	67,793	67,624	67,496	67,397	67,318	67,254	67,201	67,157	67,120	67,088
2	67,061	67,036	67,015	66,996	66,979	66,963	66,948	66,935	66,923	66,911
3	66,901	66,890	66,881	66,871	66,862	66,853	66,845	66,837	66,829	66,821
4	66,813	66,805	66,798	66,790	66,782	66,775	66,767	66,760	66,752	66,745
5	66,737	66,730	66,722	66,715	66,707	66,699	66,691	66,683	66,675	66,667
6	66,659	66,651	66,643	66,634	66,626	66,617	66,609	66,600	66,591	66,582
7	66,573	66,564	66,555	66,545	66,536	66,526	66,517	66,507	66,497	66,487
8	66,477	66,467	66,457	66,446	66,436	66,425	66,414	66,403	66,392	66,381
9	66,370	66,359	66,347	66,335	66,324	66,312	66,300	66,288	66,276	66,263
10	66,251	66,238	66,226	66,213	66,200	66,187	66,174	66,160	66,147	66,134
11	66,120	66,106	66,092	66,078	66,064	66,050	66,035	66,021	66,006	65,991
12	65,976	65,961	65,946	65,931	65,915	65,900	65,884	65,868	65,852	65,836
13	65,820	65,804	65,787	65,770	65,754	65,737	65,720	65,703	65,685	65,668
14	65,651	65,633	65,615	65,597	65,579	65,561	65,543	65,524	65,505	65,487
15	65,468	65,449	65,430	65,411	65,391	65,372	65,352	65,332	65,312	65,292
16	65,272	65,251	65,231	65,210	65,190	65,169	65,148	65,126	65,105	65,084
17	65,062	65,040	65,018	64,996	64,974	64,952	64,929	64,907	64,884	64,861
18	64,838	64,815	64,792	64,768	64,745	64,721	64,697	64,673	64,649	64,625
19	64,600	64,576	64,551	64,526	64,501	64,476	64,450	64,425	64,399	64,374
20	64,348	64,322	64,295	64,269	64,243	64,216	64,189	64,162	64,135	64,108
21	64,080	64,053	64,025	63,997	63,969	63,941	63,913	63,884	63,856	63,827
22	63,798	63,769	63,739	63,710	63,681	63,651	63,621	63,591	63,561	63,530
23	63,500	63,469	63,438	63,407	63,376	63,345	63,313	63,282	63,250	63,218
24	63,186	63,154	63,121	63,089	63,056	63,023	62,990	62,957	62,923	62,890
25	62,856	62,822	62,788	62,754	62,719	62,685	62,650	62,615	62,580	62,545
26	62,509	62,474	62,438	62,402	62,366	62,330	62,293	62,256	62,220	62,183
27	62,146	62,108	62,071	62,033	61,995	61,957	61,919	61,881	61,842	61,803
28	61,764	61,725	61,686	61,646	61,607	61,567	61,527	61,487	61,446	61,406
29	61,365	61,324	61,283	61,242	61,200	61,158	61,116	61,074	61,032	60,990
30	60,947	60,904	60,861	60,818	60,775	60,731	60,687	60,643	60,599	60,555
31	60,510	60,465	60,420	60,375	60,330	60,284	60,238	60,192	60,146	60,100
32	60,053	60,007	59,960	59,912	59,865	59,817	59,770	59,722	59,673	59,625
33	59,576	59,527	59,478	59,429	59,380	59,330	59,280	59,230	59,179	59,129
34	59,078	59,027	58,976	58,924	58,873	58,821	58,769	58,717	58,664	58,611
35	58,558	58,505	58,452	58,398	58,344	58,290	58,236	58,181	58,126	58,071
36	58,016	57,960	57,905	57,849	57,792	57,736	57,679	57,622	57,565	57,508
37	57,450	57,392	57,334	57,276	57,217	57,158	57,099	57,040	56,980	56,920
38	56,860	56,800	56,739	56,678	56,617	56,556	56,494	56,432	56,370	56,308
39	56,245	56,182	56,119							

From W. S. Rodney and R. J. Spindler, *J. Res. NBS* **51**, 123–126 (1953).

Dispersion equation:

$$n^2 = 5.640752 - 0.000003338\lambda^2 + \frac{0.0018612}{\lambda^2} + \frac{41,110.49}{\lambda^2 - 14,390.4} + \frac{0.0290764}{\lambda^2 - 0.024964}$$

The average temperature coefficient of refractive index for two samples of different origin is given by Rodney and Spindler as 7.9×10^{-5} per °C.

TABLE 6b-12. REFRACTIVITY OF CESIUM IODIDE AT 24°C

Cesium Iodide

λ, μm	n	λ, μm	n	λ, μm	n	λ, μm	n	λ, μm	n	λ, μm	n	λ, μm	n	λ, μm	n	λ, μm	n	λ, μm	n
0.280	103 939	0.840	76 352	1.80	74 702	10.8	73 852	16.4	73 267	22.0	72 435	27.6	71 334	33.2	69 941	38.8	68 227	44.4	66 151
0.300	97 872	0.860	76 252	1.84	74 683	11.0	73 835	16.6	73 242	22.2	72 400	27.8	71 289	33.4	69 886	39.0	68 159	44.6	66 070
0.320	93 700	0.880	76 159	1.88	74 664	11.2	73 818	16.8	73 216	22.4	72 365	28.0	71 244	33.6	69 830	39.2	68 091	44.8	65 988
0.340	90 649	0.900	76 074	1.92	74 647	11.4	73 800	17.0	73 190	22.6	72 330	28.2	71 199	33.8	69 774	39.4	68 023	45.0	65 905
0.360	88 324	0.920	75 993	1.96	74 631	11.6	73 783	17.2	73 164	22.8	72 294	28.4	71 153	34.0	69 717	39.6	67 954	45.2	65 822
0.380	86 497	0.940	75 918	2.00	74 616	11.8	73 765	17.4	73 137	23.0	72 258	28.6	71 107	34.2	69 660	39.8	67 884	45.4	65 739
0.400	85 027	0.960	75 848	2.20	74 551	12.0	73 746	17.6	73 111	23.2	72 222	28.8	71 061	34.4	69 602	40.0	67 814	45.6	65 655
0.420	83 823	0.980	75 782	2.40	74 500	12.2	73 728	17.8	73 083	23.4	72 185	29.0	71 014	34.6	69 544	40.2	67 744	45.8	65 570
0.440	82 820	1.00	75 721	2.60	74 460	12.4	73 709	18.0	73 056	23.6	72 148	29.2	70 967	34.8	69 486	40.4	67 673	46.0	65 485
0.460	81 975	1.04	75 608	2.80	74 427	12.6	73 690	18.2	73 028	23.8	72 111	29.4	70 919	35.0	69 427	40.6	67 601	46.2	65 399
0.480	81 255	1.08	75 508	3.00	74 400	12.8	73 670	18.4	72 999	24.0	72 073	29.6	70 871	35.2	69 368	40.8	67 530	46.4	65 313
0.500	80 635	1.12	75 419	3.50	74 346	13.0	73 650	18.6	72 971	24.2	72 035	29.8	70 823	35.4	69 308	41.0	67 457	46.6	65 226
0.520	80 097	1.16	75 339	4.00	74 305	13.2	73 630	18.8	72 942	24.4	71 997	30.0	70 774	35.6	69 248	41.2	67 384	46.8	65 138
0.540	79 626	1.20	75 268	4.50	74 270	13.4	73 610	19.0	72 913	24.6	71 958	30.2	70 725	35.8	69 188	41.4	67 311	47.0	65 051
0.560	79 213	1.24	75 203	5.00	74 239	13.6	73 589	19.2	72 883	24.8	71 919	30.4	70 676	36.0	69 127	41.6	67 237	47.2	64 962
0.580	78 846	1.28	75 144	5.50	74 210	13.8	73 568	19.4	72 853	25.0	71 880	30.6	70 626	36.2	69 065	41.8	67 163	47.4	64 873
0.600	78 520	1.32	75 091	6.00	74 181	14.0	73 547	19.6	72 823	25.2	71 840	30.8	70 576	36.4	69 004	42.0	67 088	47.6	64 783
0.620	78 229	1.36	75 042	6.50	74 152	14.2	73 525	19.8	72 793	25.4	71 800	31.0	70 525	36.6	68 941	42.2	67 013	47.8	64 693
0.640	77 967	1.40	74 997	7.00	74 122	14.4	73 504	20.0	72 762	25.6	71 759	31.2	70 474	36.8	68 879	42.4	66 937	48.0	64 602
0.660	77 731	1.44	74 956	7.50	74 091	14.6	73 481	20.2	72 731	25.8	71 718	31.4	70 422	37.0	68 815	42.6	66 861	48.2	64 511
0.680	77 517	1.48	74 919	8.00	74 059	14.8	73 459	20.4	72 699	26.0	71 677	31.6	70 371	37.2	68 752	42.8	66 784	48.4	64 419
0.700	77 323	1.52	74 884	8.50	74 026	15.0	73 436	20.6	72 667	26.2	71 635	31.8	70 318	37.4	68 688	43.0	66 707	48.6	64 326
0.720	77 146	1.56	74 852	9.00	73 991	15.2	73 413	20.8	72 635	26.4	71 593	32.0	70 266	37.6	68 623	43.2	66 629	48.8	64 233
0.740	76 985	1.60	74 822	9.50	73 954	15.4	73 389	21.0	72 602	26.6	71 551	32.2	70 213	37.8	68 558	43.4	66 551	49.0	64 139
0.760	76 836	1.64	74 795	10.0	73 916	15.6	73 366	21.2	72 570	26.8	71 508	32.4	70 159	38.0	68 493	43.6	66 472	49.2	64 045
0.780	76 700	1.68	74 769	10.2	73 901	15.8	73 342	21.4	72 536	27.0	71 465	32.6	70 105	38.2	68 427	43.8	66 392	49.4	63 950
0.800	76 575	1.72	74 745	10.4	73 885	16.0	73 317	21.6	72 503	27.2	71 422	32.8	70 051	38.4	68 361	44.0	66 312	49.6	63 855
0.820	76 459	1.76	74 723	10.6	73 868	16.2	73 292	21.8	72 469	27.4	71 378	33.0	69 996	38.6	68 294	44.2	66 232	49.8	63 759
																		50.0	63 662

OPTICS

Dispersion equation:

$$n^2 - 1 = \sum_i \frac{K_i \lambda^2}{\lambda^2 - \lambda_i^2}$$

TABLE 6b-13. CONSTANTS OF THE DISPERSION EQUATION

i	λ_i^2	K_i
1	0.00052701	0.34617251
2	0.02149156	1.0080886
3	0.032761	0.28551800
4	0.044944	0.39743178
5	25,921.0	3.3605359

From W. S. Rodney, *J. Opt. Soc. Am.* **45**, 987 (1955).

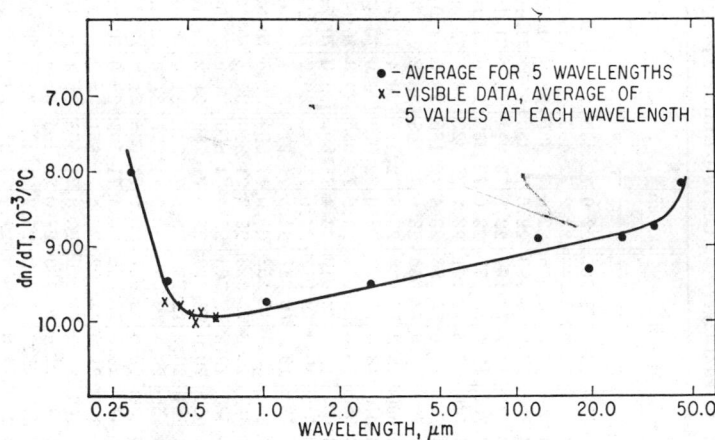

FIG. 6b-4. The temperature coefficient of refractive index of cesium iodide. [*From W. S. Rodney, J. Opt. Soc. Am.* **45**, 987 (1955).]

The data for the region 0.185 to 0.76 μm were taken at 23°C by F. F. Martens, *Ann. Physik* **6**, 603 (1901). Similar values are given by J. W. Gifford, *Proc. Phys. Soc. (London)* **70**, 329 (1902), and by H. Trommsdorff, *Z. Physik* **2**, 576 (1901). R. B. Sosman, "The Properties of Silica," Chemical Catalog Company, Inc., New York, 1927, gives a collation of the above data. The data for the wavelength region from 0.8325 to 2.30 μm (at 20°C) are taken from A. Carvallo, *Compt. Rend.* **126**, 728 (1898). For the region from 2.60 to 7.0 μm (at 18°C), the data are taken from H. Rubens, *Wied. Ann.* **54**, 488 (1895). The values fit together well (to a few parts in the fifth decimal place). They are, however, questionable at the extreme ends of the range (0.185, 5.00, 6.45, and 7.0 μm).

Crystal Quartz

TABLE 6b-14. REFRACTIVE INDEX OF CRYSTAL QUARTZ

λ, μm	n_o	n_e	λ, μm	n_o	n_e
0.185	1.65751	1.68988	1.5414	1.52781	1.53630
0.198	1.65087	1.66394	1.6815	1.52583	1.53422
0.231	1.61395	1.62555	1.7614	1.52468	1.53301
0.340	1.56747	1.57737	1.9457	1.52184	1.53004
0.394	1.55846	1.56805	2.0531	1.52005	1.52823
0.434	1.55396	1.56339	2.30	1.51561	
0.508	1.54822	1.55746	2.60	1.50986	
0.5893	1.54424	1.55335	3.00	1.49953	
0.768	1.53903	1.54794	3.50	1.48451	
0.8325	1.53773	1.54661	4.00	1.46617	
0.9914	1.53514	1.54392	4.20	1.4569	
1.1592	1.53283	1.54152	5.00	1.417	
1.3070	1.53090	1.53951	6.45	1.274	
1.3958	1.52977	1.53832	7.0	1.167	
1.4792	1.52865	1.53716			

TABLE 6b-15. TEMPERATURE COEFFICIENTS OF REFRACTIVE INDEX

λ, μm	dn_o/dT $(10^{-5}/°C)$	dn_e/dT $(10^{-5}/°C)$	λ, μm	dn_o/dT $(10^{-5}/°C)$	dn_e/dT $(10^{-5}/°C)$
0.202	+0.321	+0.267	0.298	−0.311	−0.415
0.206	0.253	0.198	0.313	−0.348	−0.450
0.210	0.193	0.143	0.325	−0.352	−0.469
0.214	0.124	0.083	0.340	−0.393	−0.501
0.219	0.074	0.027	0.361	−0.418	−0.521
0.224	0.017	−0.048	0.441	−0.475	−0.593
0.226	−0.008	−0.075	0.467	−0.485	−0.601
0.228	−0.027	−0.093	0.480	−0.499	−0.610
0.231	−0.052	−0.112	0.508	−0.514	−0.616
0.257	−0.186	−0.265	0.589	−0.539	−0.642
0.274	−0.235	−0.323	0.643	−0.549	−0.653
0.288	−0.279	−0.385			

From F. J. Micheli, *Ann. Physik* **4,** 7 (1902).

Diamond

TABLE 6b-16. REFRACTIVE INDEX OF DIAMOND

λ, μm	n	λ, μm	n
0.480	2.4368	0.589	2.4175
0.486	2.4354	0.644	2.4114
0.546	2.4235	0.656	2.4104

From von S. Rösch, *Opt. Acta* **12,** 253 (1965).

TABLE 6b-17. REFRACTIVE INDEX AT 20°C FOR THREE SPECIMENS OF FUSED SILICA

λ, µm	Computed index	Measured difference			
		C-D-G.E.*	Corning	Dynasil	General Electric
0.213856	1.534307	−27	−29	−42	−31
0.214438	1.533722	− 2	−11	−21	−22
0.226747	1.522750	+70	+71	+68	+73
0.230209	1.520081	−21	−28	−31	−23
0.237833	1.514729	+ 1	+13	+23	+19
0.239938	1.513367	+ 3	+ 6	+ 2	+ 9
0.248272	1.508398	+ 2	+ 6	− 1	+ 7
0.265204	1.500029	−29	−32	−25	−13
0.269885	1.498047	+ 3	+ 7	− 4	+11
0.275278	1.495913	− 3	+ 2	+ 8	+12
0.280347	1.494039	+ 1	− 4	− 9	−11
0.289360	1.490990	+20	+18	+22	+20
0.296728	1.488734	−14	− 7	−12	− 4
0.302150	1.487194	− 4	− 9	− 2	+ 4
0.330259	1.480539	− 9	+ 1	+10	+ 3
0.334148	1.479763	− 3	− 8	− 1	+ 9
0.340365	1.478584	+ 6	+ 9	+ 2	− 8
0.346620	1.477468	+ 2	−17	−12	−14
0.361051	1.475129	+ 1	+ 3	− 9	− 8
0.365015	1.474539	−19	−11	−15	−21
0.404656	1.469618	+ 2	+ 1	− 1	+ 2
0.435835	1.466693	− 3	+ 5	+ 1	+ 3
0.467816	1.464292	+ 8	+ 5	+ 3	+ 6
0.486133	1.463126	+ 4	+ 6	+ 5	+ 7
0.508582	1.461863	+ 7	+ 4	+ 1	+ 5
0.546074	1.460078	+ 2	+ 4	+ 1	− 5
0.576959	1.458846	+ 4	+ 5	+ 3	+ 4
0.579065	1.458769	+ 1	+ 6	+ 6	+ 6
0.587561	1.458464	+ 6	+ 3	− 2	+ 1
0.589262	1.458404	− 4	+ 6	+ 3	+ 7
0.643847	1.456704	+ 6	+ 9	+ 4	+ 7
0.656272	1.456367	+ 3	+ 7	+ 5	+ 7
0.667815	1.456067	+ 3	+ 8	+ 6	+ 3
0.706519	1.455145	+ 5	+10	+12	+ 7
0.852111	1.452465	+ 5	+ 8	+ 3	+ 5
0.894350	1.451835	+ 5	+11	+ 5	+10
1.01398	1.450242	+ 8	+ 6	+ 3	+ 6
1.08297	1.449405	− 5	+ 8	+ 1	+ 9
1.12866	1.448869	+ 1	+ 7	+ 8	+ 9
1.3622	1.446212	−12	− 6	−14	−12
1.39506	1.445836	+ 4	− 1	+ 4	− 3
1.4695	1.444975	− 5	+ 3	+ 9	+10
1.52952	1.444268	+ 2	+ 8	+ 6	0
1.6606	1.442670	−20	−14	−19	−11
1.681	1.442414	+ 6	− 2	−10	+ 8
1.6932	1.442260	0	+ 7	− 6	+ 1
1.70913	1.442057	+ 3	0	+ 3	− 1
1.81307	1.440699	+21	− 7	− 7	+ 6
1.97009	1.438519	+ 1	+ 6	+12	+12
2.0581	1.437224	− 4	− 3	− 9	−11
2.1526	1.435769	−29	−22	−25	−24
2.32542	1.432928	−18	−10	− 3	− 6
2.4374	1.430954	−24	−23	−21	−14
3.2439	1.413118	+32	+21	+29	+25
3.2668	1.412505	+25	+20	+30	+25
3.3026	1.411535	+25	+32	+30	+28
3.422	1.408180	+20	+40	+42	+37
3.5070	1.405676	−16	−26	−20	−10
3.5564	1.404174	−24	−27	−29	−18
3.7067	1.399389	−19	−22	−14	− 9
Average of absolute values of measured differences..........................		10.5	11.9	12.2	11.7

* Difference for arithmetical mean table of values compiled from experimental data of Corning (C), Dynasil (D), and General Electric (G.E.).

Adapted from I. H. Malitson, *J. Opt. Soc. Am.* **55**, 1205 (1965).

Dispersion equation:

$$n^2 - 1 = \frac{0.6961663\lambda^2}{\lambda^2 - (0.0684043)^2} + \frac{0.4079426\lambda^2}{\lambda^2 - (0.1162414)^2} + \frac{0.8974794\lambda^2}{\lambda^2 - (9.896161)^2}$$

The companies that submitted material for interspecimen comparison are Corning Glass Works, Dynasil Corporation of America, and the General Electric Company. They identify their brands as Corning code 7940 fused silica, Dynasil high-purity synthetic fused silica, and General Electric type 151. Each company submitted material from four different production runs. All specimens are considered to be of comparable optical quality, produced according to the highest standards of purity and uniformity.

TABLE 6b-18. TEMPERATURE COEFFICIENT OF REFRACTIVE INDEX

λ, μm	n, 26°C	n, 471°C	dn/dT (10^{-6}/°C)	n, 828°C	dn/dT (10^{-6}/°C)
0.23021	1.52034	1.52908	+19.6	1.53584	+19.3
0.23783	1.51496	1.52332	+18.8	1.52985	+18.6
0.2407	1.51361	1.52201	+18.9	1.52832	+18.3
0.2465	1.50970	1.51774	+18.1	1.52391	+17.7
0.24827	1.50865	1.51665	+18.0	1.52289	+17.8
0.26520	1.50023	1.50763	+16.6	1.51351	+16.5
0.27528	1.49615	1.50327	+16.0	1.50899	+16.0
0.28035	1.49425	1.50143	+16.2	1.50691	+15.8
0.28936	1.49121	1.49818	+15.7	1.50358	+15.4
0.29673	1.48892	1.49584	+15.6	1.50112	+15.2
0.30215	1.48738	1.49407	+15.1	1.49942	+15.0
0.3130	1.48462	1.49126	+14.9	1.49641	+14.7
0.33415	1.48000	1.48633	+14.2	1.49135	+14.1
0.36502	1.47469	1.48089	+14.0	1.48563	+13.6
0.40466	1.46978	1.47575	+13.4	1.48033	+13.2
0.43584	1.46685	1.47248	+12.7	1.47716	+12.9
0.54607	1.46028	1.46575	+12.3	1.47004	+12.2
0.5780	1.45899	1.46429	+11.9	1.46870	+12.1
1.01398	1.45039	1.45562	+11.8	1.45960	+11.5
1.12866	1.44903	1.45426	+11.8	1.45820	+11.4
1.254*	1.44772	1.45283	+11.5	1.45700	+11.6
1.36728	1.44635	1.45140	+11.4	1.45549	+11.4
1.470*	1.44524	1.45031	+11.4	1.45440	+11.4
1.52952	1.44444	1.44961	+11.6	1.45352	+11.3
1.660*	1.44307	1.44799	+11.1	1.45174	+10.8
1.701	1.44230	1.44733	+11.3	1.45140	+11.3
1.981*	1.43863	1.44361	+11.2	1.44734	+10.9
2.262*	1.43430	1.43933	+11.3	1.44306	+10.9
2.553*	1.42949	1.43450	+11.3	1.43854	+11.3
3.00*	1.41995	1.42495	+11.2	1.42877	+11.0
3.245*	1.41353	1.41893	+12.2	1.42243	+11.1
3.37*	1.40990	1.41501	+11.5	1.41915	+11.5

* Wavelength determination by narrow-bandwidth interference filters.
From J. H. Wray and J. T. Neu, *J. Opt. Soc. Am.* **59**, 774 (1969).

These values of dn/dT are for Corning No. 7940, ultraviolet grade.

Germanium

TABLE 6b-19. REFRACTIVE INDEX OF GERMANIUM AT 27°C

λ, μm	Single-crystal n	Polycrystal n
2.0581	4.1016	4.1018
2.1526	4.0919	4.0919
2.3126	4.0786	4.0785
2.4374	4.0708	4.0709
2.577	4.0609	4.0608
2.7144	4.0552	4.0554
2.998	4.0452	4.0452
3.3033	4.0369	4.0372
3.4188	4.0334	4.0339
4.258	4.0216	4.0217
4.866	4.0170	4.0167
6.238	4.0094	4.0095
8.66	4.0043	4.0043
9.72	4.0034	4.0033
11.04	4.0026	4.0025
12.20	4.0023	4.0020
13.02	4.0021	4.0018

Adapted from C. D. Salzberg and J. J. Villa, *J. Opt. Soc. Am.* **48**, 579 (1958).

Resistivity is about 50 ohm-cm.

TABLE 6b-20. TEMPERATURE COEFFICIENT OF REFRACTIVE INDEX AT 24.5°C
AND RELATIVE TEMPERATURE COEFFICIENT

λ, μm	dn/dT $(10^{-4}/°C)$	$(1/n)(dn/dT)$ $(10^{-4}/°C)$
1.934	5.919	1.432
2.174	5.285	1.287
2.246	5.251	1.280
2.401	5.037	1.231

From D. H. Rank, H. E. Bennett, and D. C. Cronemeyer, *J. Opt. Soc. Am.* **44**, 13 (1954).

For computational purposes, a dispersion equation for the wavelength region 0.5 to 6.0 μm is given by M. Herzberger and C. D. Salzberg, *J. Opt. Soc. Am.* **52**, 420 (1962):

$$n = A + BL + CL^2 + D\lambda^2 + E\lambda^4$$

where $L = \dfrac{1}{\lambda^2 - 0.028}$

$A = 3.99931$
$B = 0.391707$
$C = +0.163492$
$D = -0.0000060$
$E = +0.000000053$

Irtrans 1 to 6

TABLE 6b-21. REFRACTIVE INDEX OF IRTRANS 1 TO 6

λ, μm	Irtran 1	Irtran 2	Irtran 3	Irtran 4	Irtran 5	Irtran 6
1.0000	1.3778	2.2907	1.4289	2.485	1.7227	2.838
1.2500	1.3763	2.2777	1.4275	2.466	1.7188	2.773
1.5000	1.3749	2.2706	1.4263	2.456	1.7156	2.742
1.7500	1.3735	2.2662	1.4251	2.450	1.7123	2.725
2.0000	1.3720	2.2631	1.4239	2.447	1.7089	2.714
2.2500	1.3702	2.2608	1.4226	2.444	1.7052	2.707
2.5000	1.3683	2.2589	1.4211	2.442	1.7012	2.702
2.7500	1.3663	2.2573	1.4196	2.441	1.6968	2.698
3.0000	1.3640	2.2558	1.4179	2.440	1.6920	2.695
3.2500	1.3614	2.2544	1.4161	2.438	1.6868	2.693
3.5000	1.3587	2.2531	1.4141	2.437	1.6811	2.691
3.7500	1.3558	2.2518	1.4120	2.436	1.6750	2.689
4.0000	1.3526	2.2504	1.4097	2.435	1.6684	2.688
4.2500	1.3492	2.2491	1.4072	2.434	1.6612	2.687
4.5000	1.3455	2.2477	1.4047	2.433	1.6536	2.686
4.7500	1.3416	2.2462	1.4019	2.433	1.6455	2.685
5.0000	1.3374	2.2447	1.3990	2.432	1.6368	2.684
5.2500	1.3329	2.2432	1.3959	2.431	1.6275	2.683
5.5000	1.3282	2.2416	1.3926	2.430	1.6177	2.683
5.7500	1.3232	2.2399	1.3892	2.429	1.6072	2.682
6.0000	1.3179	2.2381	1.3856	2.428	1.5962	2.681
6.2500	1.3122	2.2363	1.3818	2.426	1.5845	2.681
6.5000	1.3063	2.2344	1.3778	2.425	1.5721	2.680
6.7500	1.3000	2.2324	1.3737	2.424	1.5590	2.680

λ, μm	Irtran 1	Irtran 2	Irtran 3	Irtran 4	Irtran 5	Irtran 6
7.0000	1.2934	2.2304	1.3693	2.423	1.5452	2.679
7.2500	1.2865	2.2282	1.3648	2.422	1.5307	2.678
7.5000	1.2792	2.2260	1.3600	2.421	1.5154	2.678
7.7500	1.2715	2.2237	1.3550	2.419	1.4993	2.677
8.0000	1.2634	2.2213	1.3498	2.418	1.4824	2.677
8.2500	1.2549	2.2188	1.3445	2.417	1.4646	2.676
8.5000	1.2460	2.2162	1.3388	2.416	1.4460	2.675
8.7500	1.2367	2.2135	1.3330	2.415	1.4265	2.675
9.0000	1.2269	2.2107	1.3269	2.413	1.4060	2.674
9.2500	......	2.2078	1.3206	2.411	......	2.674
9.5000	......	2.2048	1.3141	2.410	......	2.673
9.7500	......	2.2018	1.3073	2.409	......	2.672
10.0000	......	2.1986	1.3002	2.407	......	2.672
11.0000	......	2.1846	1.2694	2.401	......	2.669
12.0000	......	2.1688	......	2.394	......	2.666
13.0000	......	2.1508	......	2.386	......	2.663
14.0000	......	......	......	2.378	......	2.660
15.0000	......	......	......	2.370	......	2.657
16.0000	......	......	......	2.361	......	2.655
17.0000	......	......	......	2.352	......	......
18.0000	......	......	......	2.343	......	......
19.0000	......	......	......	2.333	......	......
20.0000	......	......	......	2.323	......	......

From *Kodak Pamphlet U-71*, 1968.

Index of refraction values were experimentally determined at selected wavelengths between 1 and 10 μm. Coefficients of an interpolation formula were established and reduced by least-square methods, and the values computed. All values beyond 10 μm are extrapolated.

Other references give a dispersion equation:

$$n = A + BL + CL^2 + D\lambda^2 + E\lambda^4$$

$$L = \frac{1}{\lambda^2 - 0.028}$$

TABLE 6b-22. CONSTANTS FOR THE DISPERSION EQUATION

Material	Wavelength range, μm		A	$B \times 10^2$	$C \times 10^4$	$D \times 10^5$	$E \times 10^8$
	From	To					
Irtran 1......	1.0	6.7	1.37770	0.1348	$+2.16$	-150.41	-441
Irtran 2......	1.0	13.5	2.25698	3.2586	$+6.79$	-52.72	-60.4
Irtran 4......	8.0	14.0	2.43508	5.156757	$+24.90192$	-27.24521	-9.85413
Irtran 6......	8.0	14.0	2.68238	11.80290	$+327.680$	-12.0298	$+21.773$

Constants for Irtrans 1 and 2 are adapted from M. Herzberger and C. D. Salzberg, *J. Opt. Soc. Am.* **52**, 420 (1962). Constants for Irtrans 4 and 6 are from A. I. Funai, *Lockheed Missiles & Space Co. Rept.* LMSC/6-78-68-34, pp. 7–8, 1968.

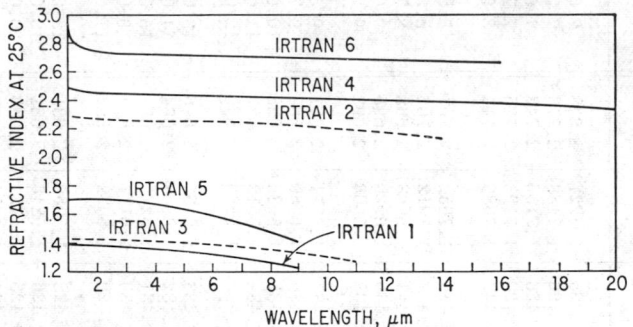

FIG. 6b-5. Refractive index vs. wavelength for Irtrans 1 to 6. (*From Kodak Pamphlet U-71*, 1968.)

The temperature coefficient of refractive index dn/dT of Irtran 4, from 198 to 295 K, has essentially a constant value of 4.8×10^{-5} per K from 3 to 13 μm [adapted from A. R. Hilton and C. E. Jones, *Appl. Opt.* **6**, 1513 (1967)].

REFRACTIVE INDEX OF CRYSTALS AND GLASSES

6-33

Lanthanum Fluoride

TABLE 6b-23. REFRACTIVE INDEX OF LANTHANUM FLUORIDE

λ, μm	n_e (observed)	n_e (computed)	n_o (observed)	n_o (computed)
0.25365	1.64866	1.64866	1.65587	1.65587
0.31315	1.61803	1.61694		1.63639
0.36633	1.61803	1.61694		1.62520
0.40465	1.61184	1.61216	1.61797	1.61733
0.43583	1.60950	1.60916	1.61664	1.61546
0.54607	1.60223	1.60223	1.60597	1.60597

Adapted from M. P. Wirick, *Appl. Opt.* **5**, 1966 (1966).

Dispersion equations:

$$n_e = 1.58330 + \frac{77.850}{\lambda - 1346.5} \qquad n_o = 1.57376 + \frac{153.137}{\lambda - 686.2}$$

A mean value between n_e and n_o of about 1.58 between 0.8 and 2.0 μm is reported by J. B. Mooney, *Infrared Phys.* **6**, 153 (1966).

Lead Fluoride

TABLE 6b-24. REFRACTIVE INDEX OF LEAD FLUORIDE

λ, μm	n	λ, μm	n
0.3088	1.915	0.5711	1.771
0.3188	1.887	0.6259	1.764
0.3338	1.882	0.6930	1.758
0.3462	1.854	0.7775	1.753
0.3645	1.849	0.8861	1.748
0.3810	1.826	1.031	1.744
0.4045	1.824	1.237	1.743
0.4266	1.804	1.545	1.742
0.4565	1.801	12	1.62
0.4876	1.787	15	1.52
0.5277	1.785	18	1.40

The index values from 0.3088 to 1.545 μm are for a 0.8869-μm film of lead fluoride on fused quartz, and were adapted from J. M. Bennett, E. J. Ashley, and H. E. Bennett, *Appl. Opt.* **4**, 961 (1965). The last three values are adapted from B. Welber, *Appl. Opt.* **6**, 925 (1967).

Lithium Fluoride

TABLE 6b-25. REFRACTIVE INDEX OF LITHIUM FLUORIDE

λ, μm	n	λ, μm	n	λ, μm	n
0.1935	1.4450	0.366	1.40121	4.50	1.33875
0.1990	1.4413	0.391	1.39937	5.00	1.32661
0.2026	1.4390	0.4861	1.39480	5.50	1.31287
0.2063	1.4367	0.50	1.39430	6.00	1.29745
0.2100	1.4346	0.80	1.38896	6.91	1.260
0.2144	1.4319	1.00	1.38711	7.53	1.239
0.2194	1.4300	1.50	1.38320	8.05	1.215
0.2265	1.4268	2.00	1.37875	8.60	1.190
0.231	1.4244	2.50	1.37327	9.18	1.155
0.254	1.41792	3.00	1.36660	9.79	1.109
0.280	1.41188	3.50	1.35868		
0.302	1.40818	4.00	1.34942		

Data at a temperature of 20°C for wavelengths 0.193 to 0.231 μm are taken from Z. Gyulai, *Z. Physik* **46**, 84 (1927); at 20°C for 0.254 to 0.486 μm are taken from H. Harting, *Sitzber. Deut. Akad. Wiss. Berlin* **IV**, 1–25 (1948); at 23.6°C for 0.50 to 6.0 μm from L. W. Tilton and E. K. Plyler, *J. Res. NBS* **47**, 25 (1951); at 18°C for 6.91 to 9.79 μm from H. W. Hohls, *Ann. Physik* **29**, 433 (1937). The data for the four spectral regions reported here fit together to within a few parts in the fourth decimal place.

For computational purposes, a dispersion equation for the wavelength range 0.5 to 6.0 μm is given by M. Herzberger and C. D. Salzberg, *J. Opt. Soc. Am.* **52**, 420 (1962):

$$n = A + BL + CL^2 + D\lambda^2 + E\lambda^4$$

where $L = \dfrac{1}{\lambda^2 - 0.028}$

$A = 1.38761$
$B = 0.001796$
$C = -0.000041$
$D = -0.0023045$
$E = -0.00000557$

Magnesium Fluoride

TABLE 6b-26. REFRACTIVE INDEX OF MAGNESIUM FLUORIDE AT 21°C

λ $(10^{-4}\,\mu m)$	n_o	n_e	λ $(10^{-4}\,\mu m)$	n_o	n_e
1,780 ± 2	1.43975	1.45365	5,015.68	1.37972	1.39163
1,849.68	1.43424	1.44797	5,085.82	1.37953	1.39142
2,536.5	1.40208	1.41483	5,460.74	1.37859	1.39043
2,893.59	1.39485	1.4073	5,875.62	1.37774	1.38954
4,046.56	1.38359	1.39566	5,893.7	1.37770	1.38950
4,340.465	1.38215	1.39415	6,234.37	1.37713	1.38889
4,358.35	1.38207	1.39407	6,438.47	1.37681	1.38858
4,471.48	1.38160	1.39357	6,562.79	1.37662	1.38838
4,678.16	1.38082	1.39275	6,678.15	1.37647	1.38822
4,799.92	1.38039	1.39231	6,907.16	1.37618	1.38790
4,921.93	1.37001	1.39192	7,065.25	1.37599	1.38771

The index values from 0.1780 to 0.289359 μm are from D. L. Steinmetz, W. G. Phillips, M. Wirick, and F. F. Forbes, *Appl. Opt.* **6**, 1001 (1967). The values from 0.404656 to 0.706525 are those of A. Duncanson and R. W. H. Stevenson, *Proc. Phys. Soc. (London)* **72**, 1001 (1958). The birefringence of magnesium fluoride in the vacuum ultraviolet is discussed by V. Chandrasekharan and H. Damany, *Appl. Opt.* **7**, 939 (1968), and **8**, 671 (1969), and many values are listed.

For computational purposes Duncanson and Stevenson also give two dispersion equations:

$$n_o = 1.36957 + \frac{35.821}{\lambda - 1492.5}$$

$$n_e = 1.38100 + \frac{37.415}{\lambda - 1494.7}$$

TABLE 6b-27. TEMPERATURE COEFFICIENT OF REFRACTIVE INDEX

λ, μm	dn_o/dT $(10^{-5}/°C)$	dn_e/dT $(10^{-5}/°C)$
0.4047	+0.23	+0.17
0.7065	+0.19	+0.10

From A. Duncanson and R. W. H. Stevenson, *loc. cit.*

Magnesium Oxide

TABLE 6b-28. REFRACTIVE INDEX OF MAGNESIUM OXIDE AT 23.3°C

λ, μm	n	λ, μm	n
0.36117	1.77318	1.97009	1.70885
0.365015	1.77186	2.24929	1.70470
1.01398	1.72259	2.32542	1.70350
1.12866	1.72059	3.3033	1.68526
1.36728	1.71715	3.5078	1.68055
1.52952	1.71496	4.258	1.66039
1.6932	1.71281	5.138	1.63138
1.7092	1.71258	5.35	1.62404
1.81307	1.71108		

Dispersion equation:

$$n^2 = 2.956362 - 0.01062387\lambda^2 - 0.0000204968\lambda^4 - \frac{0.02195770}{\lambda^2 - 0.01428322}$$

From R. E. Stephens and I. H. Malitson, *J. Res. NBS* **49**, 249 (1952). The earlier measurements of J. Strong and R. T. Brice, *J. Opt. Soc. Am.* **25**, 207 (1935), are discussed therein. A systematic difference in values of n appears to exist, the data of Strong and Brice being about 37×10^{-5} higher.

TABLE 6b-29. TEMPERATURE COEFFICIENT OF REFRACTIVE INDEX

λ, μm	dn/dT $(10^{-6}/°C)$				
	20°C	25°C	30°C	35°C	40°C
7.679	13.6	13.7	13.8	13.9	14.0
7.065	14.1	14.2	14.3	14.4	14.5
6.678	14.4	14.5	14.6	14.7	14.8
6.563	14.5	14.6	14.7	14.8	14.9
5.893	15.3	15.4	15.5	15.6	15.7
5.461	15.9	16.0	16.1	16.2	16.3
4.861	16.9	17.0	17.1	17.2	17.3
4.358	18.0	18.1	18.2	18.3	18.4
4.047	18.9	19.0	19.1	19.2	19.3

From R. E. Stephens and I. H. Malitson, *loc. cit.*

Muscovite Mica

TABLE 6b-30. REFRACTIVE INDICES OF MUSCOVITE MICA

λ_F, μm	λ_S, μm	n_F	n_S	λ_F, μm	λ_S, μm	n_F	n_S
	$t = 5.24$ μm thick				$t = 20.82$ μm thick		
0.6665	0.6675	1.590	1.592	0.6960	0.6985	1.598	1.594
0.6188	0.6210	1.594	1.600	0.6316	0.6336	1.593	1.598
0.5573	0.5590	1.595	1.600	0.5539	0.5555	1.596	1.601
0.5082	0.5090	1.600	1.603	0.4935	0.4950	1.600	1.605
0.4538	0.4555	1.602	1.608	0.4600	0.4615	1.062	1.607
0.4320	0.4330	1.608	1.611	0.4310	0.4326	1.604	1.610
	$t = 13.91$ μm thick				$t = 48.68$ μm thick		
0.6910	0.6930	1.590	1.594	0.6914	0.6935	1.591	1.596
0.5995	0.6010	1.595	1.599	0.6110	0.6125	1.594	1.598
0.5293	0.5308	1.598	1.603	0.5470	0.5487	1.596	1.601
0.4740	0.4754	1.602	1.606	0.4958	0.4971	1.599	1.603
0.4300	0.4308	1.607	1.611	0.4667	0.4680	1.601	1.606
				0.4408	0.4425	1.603	1.609

Adapted from M. A. Jeppeson and A. M. Taylor, *J. Opt. Soc. Am.* **56**, 451 (1966).

The values for the fast and slow rays are for four different thicknesses t.

Potassium Bromide

TABLE 6b-31. REFRACTIVE INDEX OF POTASSIUM BROMIDE AT 22°C

λ, μm	n	λ, μm	n	λ, μm	n	λ, μm	n
0.404656	1.589752	1.01398	1.54408	6.238	1.53288	17.40	1.50390
0.435835	1.581479	1.12866	1.54258	6.692	1.53225	18.16	1.50076
0.486133	1.571791	1.36728	1.54061	8.662	1.52903	19.01	1.49703
0.508582	1.568475	1.7012	1.53901	9.724	1.52695	19.91	1.49288
0.546074	1.563928	2.44	1.53733	11.035	1.52404	21.18	1.48655
0.587562	1.559965	2.73	1.53693	11.862	1.52200	21.83	1.48311
0.643847	1.555858	3.419	1.53612	14.29	1.51505	23.86	1.47140
0.706520	1.552447	4.258	1.53523	14.98	1.51280	25.14	1.46324

From R. E. Stephens, E. K. Plyler, W. S. Rodney, and R. J. Spindler, *J. Opt. Soc. Am.* **43**, 111–112 (1953).

Dispersion equation:

$$n^2 = 2.361323 - 0.000311497\lambda^2 - 0.000000058613\lambda^4 + \frac{0.007676}{\lambda^2} + \frac{0.0156569}{\lambda^2 - 0.0324}$$

The average value of the temperature coefficient of refractive index is given as 4.0×10^{-5} per °C.

Potassium Chloride

TABLE 6b-32. REFRACTIVE INDEX OF POTASSIUM CHLORIDE

λ, μm	n	λ, μm	n	λ, μm	n
0.185409	1.82710	0.410185	1.50907	5.3039	1.470013
0.186220	1.81853	0.434066	1.50503	5.8932	1.468804
0.197760	1.73120	0.441587	1.50390	8.2505	1.462726
0.198990	1.72438	0.467832	1.50044	8.8398	1.460858
0.200090	1.71870	0.486149	1.49841	10.0184	1.45672
0.204470	1.69817	0.508606	1.49620	11.786	1.44919
0.208216	1.68308	0.53383	1.49410	12.965	1.44346
0.211078	1.67281	0.54610	1.49319	14.144	1.43722
0.21445	1.66188	0.56070	1.49218	15.912	1.42617
0.21946	1.64745	0.58931	1.49044	17.680	1.41403
0.22400	1.63612	0.58932	1.490443	18.2	1.409
0.23129	1.62043	0.62784	1.48847	18.8	1.401
0.242810	1.60047	0.64388	1.48777	19.7	1.398
0.250833	1.58979	0.656304	1.48727	20.4	1.389
0.257317	1.58125	0.67082	1.48669	21.1	1.379
0.263200	1.57483	0.76824	1.48377	22.2	1.374
0.267610	1.57044	0.78576	1.483282	23.1	1.363
0.274871	1.56386	0.88398	1.481422	24.1	1.352
0.281640	1.55836	0.98220	1.480084	24.9	1.336
0.291368	1.55140	1.1786	1.478311	25.7	1.317
0.308227	1.54136	1.7680	1.475890	26.7	1.300
0.312280	1.53926	2.3573	1.474751	27.2	1.275
0.340358	1.52726	2.9466	1.473834	28.2	1.254
0.358702	1.52115	3.5359	1.473049	28.8	1.226
0.394415	1.51219	4.7146	1.471122		

Dispersion equations (for the ultraviolet and visible, respectively):

$$n^2 = a^2 + \frac{M_1}{\lambda^2 - \lambda_1{}^2} + \frac{M_2}{\lambda^2 - \lambda_2{}^2} - k\lambda^2 - h\lambda^4$$

$$n^2 = b^2 + \frac{M_1}{\lambda^2 - \lambda_1{}^2} + \frac{M_2}{\lambda^2 - \lambda_2{}^2} - \frac{M_3}{\lambda_3{}^2 - \lambda^2}$$

$$a^2 = 2.174967 \qquad k = 0.000513495$$
$$M_1 = 0.008344206 \qquad h = 0.06167587$$
$$\lambda_1{}^2 = 0.0119082 \qquad b^2 = 3.866619$$
$$M_2 = 0.00698382 \qquad M_3 = 5,569.715$$
$$\lambda_2{}^2 = 0.0255550 \qquad \lambda_3{}^2 = 3,292.47$$

Refractive-index data for the wavelength ranges indicated are from the following sources: (1) 0.185409 to 0.76824 μm at 18°C, F. F. Martens, *Ann. Physik* **6**, 619 (1901); (2) 18.2 to 28.8 μm, H. W. Hohls, *Ann. Physik* **29**, 433 (1937); (3) 0.58932 to 17.680 μm at 15°C, F. Paschen, *Ann. Physik* **26**, 120 (1908). Note that the data of Paschen and of Martens overlap in a small region, and both sets are presented. There is less spread between Hohls's data and Paschen's data than there is among Paschen's data in the region where they join. The fit in this region is good (~0.0005). Paschen also presents two dispersion curves which fit the data of Martens to about five parts in the fifth decimal.

Temperature coefficient of refractive index:

$$n = 1.490443 - (T - 15)0.000034$$

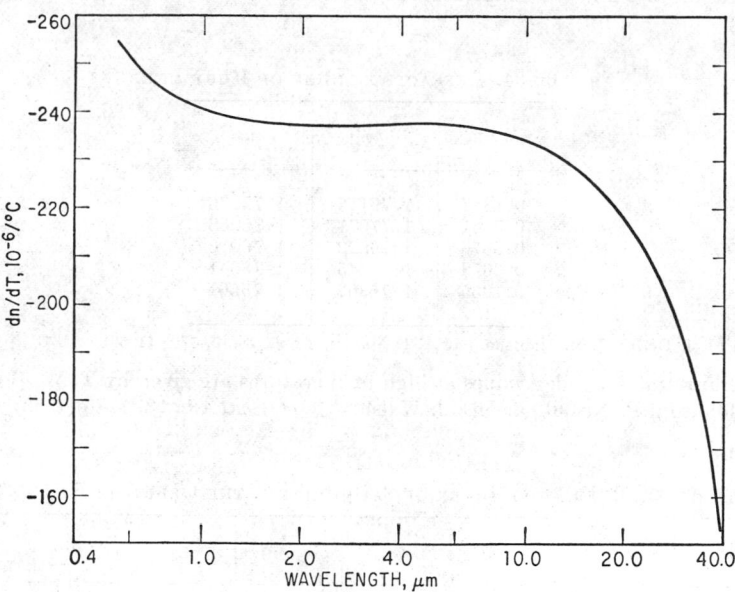

FIG. 6b-6. Average temperature coefficient of refractive index dn/dT of potassium chloride near room temperature.   [*From F. Paschen, Ann. Physik* **26**, 120 (1908).]

Potassium Iodide

TABLE 6b-33. REFRACTIVE INDEX OF POTASSIUM IODIDE

λ, μm	n	λ, μm	n	λ, μm	n
0.248	2.0548	0.656	1.65809	10.02	1.6201
0.254	2.0105	0.707	1.6537	11.79	1.6172
0.265	1.9424	0.728	1.6520	12.97	1.6150
0.270	1.9221	0.768	1.6494	14.14	1.6127
0.280	1.8837	0.811	1.6471	15.91	1.6085
0.289	1.85746	0.842	1.6456	18.10	1.6030
0.297	1.83967	0.912	1.6427	19	1.5997
0.302	1.82769	1.014	1.6396	20	1.5964
0.313	1.80707	1.083	1.6381	21	1.5930
0.334	1.77664	1.18	1.6366	22	1.5895
0.366	1.74416	1.77	1.6313	23	1.5858
0.391	1.72671	2.36	1.6295	24	1.5819
0.405	1.71843	3.54	1.6275	25	1.5775
0.436	1.70350	4.13	1.6268	26	1.5729
0.486	1.68664	5.89	1.6252	27	1.5681
0.546	1.67310	7.66	1.6235	28	1.5629
0.588	1.66654	8.84	1.6218	29	1.5571
0.589	1.66643				

The values for the wavelengths 0.248 through 1.083 μm are from H. Harting, *Sitzber. Deut. Akad. Wiss. Berlin* **IV**, 1 (1948); for 1.18 through 29 μm the values are from K. Korth, *Z. Physik* **84**, 677 (1933). The temperature coefficient of refractive index for 0.546 μm in the temperature region 38 to 90°C is -5.0×10^{-5} per °C.

Ruby

TABLE 6b-34. REFRACTIVE INDEX OF RUBY AT 22°C

λ, μm	n_0	n_e
0.4358	1.78115	1.77276
0.5461	1.77071	1.76258
0.5876	1.76822	1.76010
0.6678	1.76445	1.75641
0.7065	1.76302	1.75501

From M. J. Dodge, I. H. Malitson, and A. I. Mahan, *Appl. Opt.* **8**, 1703 (1969).

Several refractive-index values at high temperatures are given by T. W. Houston, L. F. Johnson, P. Kisliuk, and D. J. Walsh, *J. Opt. Soc. Am.* **53**, 1286 (1963).

Sapphire

TABLE 6b-35. REFRACTIVE INDEX OF SAPPHIRE FOR THE ORDINARY RAY AT 24°C

λ, μm	n_0	λ, μm	n_0	λ, μm	n_0	λ, μm	n_0
0.26520	1.83360	0.435834	1.78120	1.52952	1.74660	3.3026	1.70231
0.28035	1.82427	0.546071	1.77078	1.6932	1.74368	3.3303	1.70140
0.28936	1.81949	0.576960	1.76884	1.70913	1.74340	3.422	1.69818
0.29673	1.81595	0.579066	1.76871	1.81307	1.74144	3.5070	1.69504
0.30215	1.81351	0.64385	1.76547	1.9701	1.73833	3.7067	1.68746
0.3130	1.80906	0.706519	1.76303	2.1526	1.73444	4.2553	1.66371
0.33415	1.80184	0.85212	1.75885	2.24929	1.73231	4.954	1.62665
0.34662	1.79815	0.89440	1.75796	2.32542	1.73057	5.1456	1.61514
0.361051	1.79450	1.01398	1.75547	2.4374	1.72783	5.349	1.60202
0.365015	1.79358	1.12866	1.75339	3.2439	1.70437	5.419	1.59735
0.39064	1.78826	1.36728	1.74936	3.2668	1.70356	5.577	1.58638
0.404656	1.78582	1.39506	1.74888				

Adapted from I. H. Malitson, *J. Opt. Soc. Am.* **52**, 1377 (1962).

The refractive index n_e for the extraordinary ray at 1.014 μm was also measured by Malitson and determined to be 1.74794. The birefringence of sapphire in the vacuum ultraviolet is discussed by V. Chandrasekharan and H. Damany, *Appl. Opt.* **7**, 939 (1968), and **8**, 671 (1969); and many values are listed.

Temperature coefficients of index dn/dT were determined from differences between indices measured at 19°C and those at 24°C. The results indicate that the coefficient is positive and decreases from about 20×10^{-6} per °C at the short wavelengths to about 10×10^{-6} per °C near 4μm. An average value of 13×10^{-6} per °C for the visible region was determined from additional measurements made at 17, 24, and 31°C. (From I. H. Malitson, *loc. cit.*)

For computational purposes, Malitson gives a dispersion equation for the wavelength region 0.270 to 5.60 μm:

$$n^2 - 1 = \frac{A_1\lambda^2}{\lambda^2 - \lambda_1^2} + \frac{A_2\lambda^2}{\lambda^2 - \lambda_2^2} + \frac{A_3\lambda^2}{\lambda^2 - \lambda_3^2}$$

TABLE 6b-36. CONSTANTS OF THE DISPERSION EQUATION AT 24°C

$\lambda_1 = 0.06144821$	$\lambda_1{}^2 = 0.00377588$	$A_1 = 1.023798$
$\lambda_2 = 0.1106997$	$\lambda_2{}^2 = 0.0122544$	$A_2 = 1.058264$
$\lambda_3 = 17.92656$	$\lambda_3{}^2 = 321.3616$	$A_3 = 5.280792$

Selenium

TABLE 6b-37. REFRACTIVE INDEX OF SELENIUM AT 23°C (± 2°C)

λ, μm	n_0	n_e
1.06	2.790 ± 0.008	3.608 ± 0.008
1.15	2.737 ± 0.008	3.573 ± 0.008
3.39	2.65 ± 0.01	3.46 ± 0.01
10.6	2.64 ± 0.01	3.41 ± 0.01

These values are for single-crystal selenium [from L. Gampel and F. M. Johnson, *J. Opt. Soc. Am.* **59**, 72 (1969)]. For the region 9 to 23 μm, the index values are 2.78 ± 0.02 for the ordinary and 3.58 ± 0.02 for the extraordinary ray with no appreciable variation [from R. S. Caldwell and H. Y. Fan, *Phys. Rev.* **114**, 664 (1959)]. For amorphous selenium in the region 2.5 to 15 μm, index values of 2.46 to 2.38 are referenced by Caldwell and Fan.

Silicon

TABLE 6b-38. REFRACTIVE INDEX OF SILICON AT 26°C

λ, μm	n	λ, μm	n	λ, μm	n	λ, μm	n
1.3570	3.4975	2.1526	3.4476	4.00	3.4255	7.50	3.4186
1.3673	3.4962	2.3254	3.4430	4.258	3.4242	8.00	3.4184
1.3951	3.4929	2.4373	3.4408	4.50	3.4236	8.50	3.4182
1.5295	3.4795	2.7144	3.4358	5.00	3.4223	10.00	3.4179
1.6606	3.4696	3.00	3.4320	5.50	3.4213	10.50	3.4178
1.7092	3.4664	3.3033	3.4297	6.00	3.4202	11.04	3.4176
1.8131	3.4608	3.4188	3.4286	6.50	3.4195		
1.9701	3.4537	3.50	3.4284	7.00	3.4189		

From C. D. Salzberg and J. J. Villa, *J. Opt. Soc. Am.* **47**, 244 (1957).

The purity of the silicon sample is not specified. These data are about five parts in the third decimal place lower than those reported by H. B. Briggs, *Phys. Rev.* **77**, 287 (1950). The refractive index of adequately pure (30 ohm-cm) cast polycrystal silicon should have refractive-index values very near those of single crystals.

The relative temperature coefficient of refractive index is $(1/n) \, dn/dT = (3.9 \pm 0.4) \times 10^{-5}$ per °C in a temperature range from 77 to 400 K [from M. Cardona, W. Paul, and H. Brooks, *J. Phys. Chem. Solids* **8**, 204 (1959)].

For computational purposes, a dispersion equation for the wavelength region 1.3 to 11.0 μm is given by M. Herzberger and C. D. Salzberg, *J. Opt. Soc. Am.* **52**, 420 (1962):

$$n = 3.41696 + 0.138497L + 0.013924L^2 - 0.0000209\lambda^2 + 0.000000148\lambda^4$$

where $L = 1/(\lambda^2 - 0.028)$.

Silver Chloride

TABLE 6b-39. REFRACTIVE INDEX OF SILVER CHLORIDE AT 23.9°C

λ, μm	n	λ, μm	n	λ, μm	n	λ, μm	n
0.5	2.09648	2.3	2.00465	4.5	1.99866	13.5	1.96133
0.6	2.06385	2.4	2.00424	5.0	1.99745	14.0	1.95807
0.7	2.04590	2.5	2.00386	5.5	1.99618	14.5	1.95467
0.8	2.03485	2.6	2.00351	6.0	1.99483	15.0	1.95113
0.9	2.02752	2.7	2.00318	6.5	1.99339	15.5	1.94743
1.0	2.02239	2.8	2.00287	7.0	1.99185	16.0	1.94358
1.1	2.01865	2.9	2.00258	7.5	1.99021	16.5	1.93958
1.2	2.01582	3.0	2.00230	8.0	1.99847	17.0	1.93542
1.3	2.01363	3.1	2.00203	8.5	1.98661	17.5	1.93109
1.4	2.01189	3.2	2.00177	9.0	1.98464	18.0	1.92660
1.5	2.01047	3.3	2.00151	9.5	1.98255	18.5	1.92194
1.6	2.00931	3.4	2.00126	10.0	1.98034	19.0	1.91710
1.7	2.00833	3.5	2.00102	10.5	1.97801	19.5	1.91208
1.8	2.00750	3.6	2.00078	11.0	1.97556	20.0	1.90688
1.9	2.00678	3.7	2.00054	11.5	1.97297	20.5	1.90149
2.0	2.00615	3.8	2.00030	12.0	1.97026		
2.1	2.00559	3.9	2.00007	12.5	1.96742		
2.2	2.00510	4.0	1.99983	13.0	1.96444		

From L. W. Tilton, E. K. Plyler, and R. E. Stephens, *J. Opt. Soc. Am.* **40**, 540 (1950).

Dispersion equation:

$$n^2 = 4.00804 - 0.00085111\lambda^2 - 0.00000019762\lambda^4 + \frac{0.079086}{\lambda^2 - 0.04584}$$

The temperature coefficient of refractive index is given as approximately 6.1×10^{-5} per °C at 0.61 μm.

Sodium Chloride

TABLE 6b-40. REFRACTIVE INDEX OF SODIUM CHLORIDE

λ, μm	n	λ, μm	n	λ, μm	n	λ, μm	n
0.19	1.85343	1.1786	1.53031	4.0	1.52190	12.50	1.47568
0.20	1.79073	1.2016	1.53014	4.1230	1.52156	12.9650	1.47160
0.22	1.71591	1.2604	1.52971	4.7120	1.51979	13.0	1.47141
0.24	1.67197	1.3126	1.52937	5.0	1.51899	14.0	1.46189
0.26	1.64294	1.4	1.52888	5.0092	1.51883	14.1436	1.46044
0.28	1.62239	1.4874	1.52845	5.3009	1.51790	14.7330	1.45427
0.30	1.60714	1.5552	1.52815	5.8932	1.51593	15.0	1.45145
0.35	1.58232	1.6	1.52798	6.0	1.51548	15.3223	1.44743
0.40	1.56769	1.6368	1.52781	6.4825	1.51347	15.9116	1.44090
0.50	1.55175	1.6848	1.52764	6.80	1.51200	16.0	1.44001
0.589	1.54427	1.7670	1.52736	7.0	1.51136	17.0	1.42753
0.6400	1.54141	1.8	1.52728	7.0718	1.51093	17.93	1.4149
0.6874	1.53930	2.0	1.52670	7.22	1.51020	18.0	1.41393
0.70	1.53881	2.0736	1.52649	7.59	1.50850	19.0	1.39914
0.7604	1.53682	2.1824	1.52621	7.6611	1.50822	20.0	1.38307
0.7858	1.53607	2.2464	1.52606	7.9558	1.50665	20.57	1.3735
0.80	1.53575	2.3	1.52594	8.0	1.50655	21.0	1.36563
0.8835	1.53395	2.3560	1.52579	8.04	1.5064	21.3	1.352
0.90	1.53366	2.6	1.52525	8.8398	1.50192	22.3	1.3403
0.9033	1.53361	2.6505	1.52512	9.0	1.50105	22.8	1.318
0.9724	1.53253	2.9466	1.52466	9.00	1.50100	23.6	1.299
1.0	1.53216	3.0	1.52434	9.50	1.49980	24.2	1.278
1.0084	1.53206	3.2736	1.52371	10.0	1.49482	25.0	1.254
1.0540	1.53153	3.5	1.52317	10.0184	1.49462	25.8	1.229
1.0810	1.53123	3.5359	1.52312	11.0	1.48783	26.6	1.203
1.1058	1.53098	3.6288	1.52286	11.7864	1.48171	27.3	1.175
1.1420	1.53063	3.8192	1.52238	12.0	1.48004		

Refractive-index data for rock salt are from the following sources for the indicated wavelengths: the data for wavelengths given with two-figure accuracy (0.19, 0.50, . . .) or three-figure accuracy (10.0, 11.0, . . .) are reported for 20°C by F. Kohlrausch, "Praktische Physik," vol. II, p. 528, B. G. Teubner, Leipzig, 1943; the data reported to four figures (1.299) in index at 18°C (even though they are three figures in wavelength) are from H. W. Hohls, *Ann. Physik* **29**, 433 (1937); other data at 20°C are from W. W. Coblentz, *J. Opt. Soc. Am.* **4**, 443 (1914). Still more data have been published by Langley, Martens, Paschen, Rubens, Trowbridge, Nichols, and others, but all have apparently measured natural crystals of undetermined purity; the data all agree to the fifth decimal place.

TABLE 6b-41. TEMPERATURE COEFFICIENT OF REFRACTIVE INDEX AT ABOUT 60°C

λ, μm	dn/dT $(10^{-5}/°C)$	λ, μm	dn/dT $(10^{-5}/°C)$
0.202	3.134	0.589	−3.622
0.206	2.229	0.643	−3.636
0.210	1.570	0.656	−3.652
0.214	0.861	1.1	−3.642
0.219	0.235	1.6	−3.557
0.224	−0.187	2.7	−3.427
0.226	−0.382	3.96	−3.286
0.229	−0.598	4.96	−3.172
0.231	−0.757	6.4	−3.149
0.257	−1.979	8.85	−2.405
0.274	−2.396	10.02	−2.2
0.288	−2.602	11.79	−1.6
0.298	−2.727	12.97	−1.4
0.313	−2.862	14.14	−1.2
0.325	−2.987	14.73	−1.0
0.340	−3.068	15.32	−0.8
0.361	−3.194	15.91	−0.7
0.441	−3.425	17.93	−0.5
0.467	−3.454	20.57	0
0.480	−3.468	22.3	0
0.508	−3.517		

From F. J. Micheli, *Ann. Physik* **4,** 7 (1902) for the region 0.202 through 0.643 μm; from E. Liebreich, *Verhandl. Deut. Physik. Ges.* **13,** 709 (1911) for the region 0.656 through 15.91 μm; and from H. Rubens and E. F. Nichols, *Weid. Ann.* **60,** 454 (1897) for the region 17.93 to 22.3 μm.

Sodium Fluoride

TABLE 6b-42. REFRACTIVE INDEX OF SODIUM FLUORIDE

λ, μm	n	λ, μm	n	λ, μm	n	λ, μm	n
0.186	1.3930	0.486	1.32818	3.9	1.309	9.4	1.251
0.193	1.3854	0.546	1.32640	4.1	1.308	9.8	1.241
0.199	1.3805	0.588	1.32552	4.5	1.305	10.3	1.233
0.203	1.3772	0.589	1.32549	4.7	1.303	10.8	1.222
0.206	1.3745	0.656	1.32436	4.9	1.302	11.3	1.209
0.210	1.3718	0.707	1.32372	5.1	1.301	11.7	1.193
0.214	1.3691	0.720	1.32349	5.3	1.299	12.5	1.180
0.219	1.3665	0.768	1.32307	5.5	1.297	13.2	1.163
0.227	1.3630	0.811	1.32272	5.7	1.295	13.8	1.142
0.231	1.3606	0.842	1.32247	5.9	1.294	14.3	1.118
0.237	1.3586	0.912	1.32198	6.1	1.292	15.1	1.093
0.240	1.35793	1.014	1.32150	6.3	1.290	15.9	1.065
0.248	1.35500	1.083	1.32125	6.5	1.288	16.7	1.034
0.254	1.35325	1.27	1.320	6.7	1.286	17.3	1.000
0.265	1.34999	1.48	1.319	6.9	1.284	18.1	0.963
0.270	1.34881	1.67	1.318	7.1	1.281	18.6	0.924
0.280	1.34645	1.83	1.318	7.3	1.279	19.3	0.881
0.289	1.34462	2.0	1.317	7.5	1.277	19.7	0.838
0.297	1.34328	2.2	1.317	7.7	1.274	20.0	0.82
0.302	1.34232	2.4	1.316	7.9	1.272	20.5	0.75
0.313	1.34062	2.6	1.315	8.1	1.269	21.0	0.70
0.334	1.33795	2.8	1.314	8.3	1.266	21.5	0.65
0.366	1.33482	3.1	1.313	8.5	1.263	22.0	0.55
0.391	1.33290	3.3	1.312	8.7	1.261	22.5	0.45
0.405	1.33194	3.5	1.311	8.9	1.258	23.0	0.33
0.436	1.33025	3.7	1.309	9.1	1.252	23.5	0.25
						24.0	0.24

From Alexander Smakula, *U.S. Dept. Comm. Office Tech. Serv. Doc.* 111,052, pp. 88–89, October, 1952, who references these values.

The data fit together well (within a few parts in the fifth decimal place).

TABLE 6b-43. TEMPERATURE COEFFICIENT OF REFRACTIVE INDEX

λ, μm	dn/dT (10^{-5} per °C)	T, °C
0.546	−1.6	18–80
3.5	−1.6	18–80
8.5	−0.7	18–80

From H. W. Hohls, *Ann. Physik* **29**, 433 (1937).

Sodium Nitrate

TABLE 6b-44. REFRACTIVE INDEX OF SODIUM NITRATE

λ, μm	n_o	n_e	λ, μm	n_o	n_e
0.434	1.6126	1.340	0.578	1.5860	1.336
0.436	1.6121	1.340	0.589	1.5840	1.336
0.486	1.5998	1.338	0.656	1.5791	1.334
0.501	1.5968	1.337	0.668	1.5783	1.334
0.546	1.5899	1.336			

From *International Critical Tables*, vol. VII, p. 26, McGraw-Hill Book Company, New York, 1930.

Spinel

TABLE 6b-45. REFRACTIVE INDEX OF SPINEL

λ, μm	n
0.4861	1.736
0.5893	1.727
0.6563	1.724

From Linde Air Products Company Technical Data Sheets.

Strontium Titanate

TABLE 6b-46. REFRACTIVE INDEX OF STRONTIUM TITANATE

λ, μm	n	λ, μm	n
0.404657	2.6481	1.52952	2.2848
0.435834	2.5680	1.7012	2.2783
0.486132	2.4897	1.81307	2.2744
0.546074	2.4346	1.871	2.2710
0.576960	2.4149	1.918	2.2704
0.579066	2.4137	2.1526	2.2624
0.587562	2.4090	2.3126	2.2564
0.589262	2.4081	2.4374	2.2525
0.643847	2.3837	2.5628	2.2466
0.656279	2.3790	2.6707	2.2438
0.667815	2.3750	2.7248	2.2404
0.706519	2.3634	3.2434	2.2211
0.767858	2.3488	3.3026	2.2181
0.85212	2.3337	3.4226	2.2124
0.89440	2.3276	3.5070	2.2088
1.01398	2.3147	3.5564	2.2063
1.12866	2.3055	3.7067	2.1990
1.3622	2.2921	4.2553	2.1680
1.39506	2.2906	5.138	2.1119
1.517	2.2859	5.3034	2.1004

From I. H. Malitson, National Bureau of Standards, private communication, 1960.

Tellurium

TABLE 6b-47. REFRACTIVE INDEX OF TELLURIUM

λ, μm	n_o	n_e	λ, μm	n_o	n_e
4.0	4.929	6.372	9.0	4.802	6.253
5.0	4.864	6.316	10.0	4.796	6.246
6.0	4.838	6.286	12.0	4.789	6.237
7.0	4.821	6.270	14.0	4.785	6.230
8.0	4.809	6.257			

The data are for single-crystal tellurium [from R. S. Caldwell and H. Y. Fan, *Phys. Rev.* **114**, 664 (1959)]. The data of P. A. Hartig and J. J. Loferski, *J. Opt. Soc. Am.* **44**, 17 (1954), may be compared: The latter are probably in error, owing to an averaging effect of the two indices. (Hartig's data are lower for the high index and higher for the low index.) The 8-μm value reported here (the datum of Caldwell) is probably about 0.002 too low.

Thallium Bromide

TABLE 6b-48. REFRACTIVE INDEX OF THALLIUM BROMIDE

λ, μm	n	λ, μm	n
0.438	2.652	0.750	2.350
0.546	2.452	9.98	2.338
0.578	2.424	13.95	2.321
0.589	2.418	19.76	2.321
0.650	2.384	24.39	2.321

The indices in the visible region are from Tom F. W. Barth, *Am. Mineralogist* **14**, 358 (1929). The indices in the infrared region were measured at 45°C by D. E. McCarthy, *Appl. Opt.* **4**, 878 (1965).

Thallium Chloride

TABLE 6b-49. REFRACTIVE INDEX OF THALLIUM CHLORIDE

λ, μm	n	λ, μm	n
0.436	2.400	0.750	2.198
0.546	2.270	10.0	2.193
0.578	2.253	12.47	2.191
0.589	2.247	18.35	2.182
0.650	2.223		

The indices in the visible region are from Tom F. W. Barth, *Am. Mineralogist* **14**, 358 (1929). The indices in the infrared region were measured at 45°C by D. E. McCarthy, *Appl. Opt.* **4**, 878 (1965).

Thallium Bromide-Chloride (KRS-6)

TABLE 6b-50. REFRACTIVE INDEX OF THALLIUM BROMIDE-CHLORIDE

λ, μm	n	λ, μm	n	λ, μm	n
Na-D	2.3367	2.0	2.2059	11.0	2.1723
0.6	2.3294	2.2	2.2039	12.0	2.1674
0.7	2.2982	2.4	2.2024	13.0	2.1620
0.8	2.2660	2.6	2.2011	14.0	2.1563
0.9	2.2510	2.8	2.2001	15.0	2.1504
1.0	2.2404	3.0	2.1990	16.0	2.1442
1.1	2.2321	3.5	2.1972	17.0	2.1377
1.2	2.2255	4.0	2.1956	18.0	2.1309
1.3	2.2212	4.5	2.1942	19.0	2.1236
1.4	2.2176	5.0	2.1928	20.0	2.1154
1.5	2.2148	6.0	2.1900	21.0	2.1067
1.6	2.2124	7.0	2.1870	22.0	2.0976
1.7	2.2103	8.0	2.1839	23.0	2.0869
1.8	2.2086	9.0	2.1805	24.0	2.0752
1.9	2.2071	10.0	2.1767		

From G. Hettner and G. Leisegang, *Optik* **3**, 305 (1948).

The composition of the KRS-6 used was 44 percent thallium bromide and 56 percent thallium chloride.

Thallium Bromide-Iodide (KRS-5)

TABLE 6b-51. REFRACTIVE INDEX OF THALLIUM BROMIDE-IODIDE AT 25°C

λ, µm	n	λ, µm	n	λ, µm	n	λ, µm	n	λ, µm	n	λ, µm	n	λ, µm	n	λ, µm	n
0.540	2.68059	1.46	2.40938	5.80	2.37832	15.0	2.35812	22.1	2.33161	26.7	2.30844	31.3	2.28011	35.9	2.24609
0.560	2.64959	1.48	2.40854	6.00	2.37797	15.2	2.35751	22.2	2.33116	26.8	2.30789	31.4	2.27943	36.0	2.24528
0.580	2.62390	1.50	2.40774	6.20	2.37763	15.4	2.35690	22.3	2.33070	26.9	2.30732	31.5	2.27875	36.1	2.24447
0.600	2.60221	1.52	2.40697	6.40	2.37729	15.6	2.35629	22.4	2.33025	27.0	2.30676	31.6	2.27807	36.2	2.24366
0.620	2.58261	1.54	2.40623	6.60	2.37695	15.8	2.35566	22.5	2.32979	27.1	2.30619	31.7	2.27738	36.3	2.24284
0.640	2.56748	1.56	2.40552	6.80	2.37661	16.0	2.35502	22.6	2.32933	27.2	2.30562	31.8	2.27669	36.4	2.24202
0.660	2.55337	1.58	2.40484	7.00	2.37627	16.2	2.35438	22.7	2.32887	27.3	2.30505	31.9	2.27600	36.5	2.24120
0.680	2.54092	1.60	2.40419	7.20	2.37592	16.4	2.35373	22.8	2.32840	27.4	2.30448	32.0	2.27531	36.6	2.24038
0.700	2.52986	1.62	2.40355	7.40	2.37558	16.6	2.35307	22.9	2.32793	27.5	2.30390	32.1	2.27461	36.7	2.23955
0.720	2.51998	1.64	2.40295	7.60	2.37523	16.8	2.35240	23.0	2.32746	27.6	2.30332	32.2	2.27391	36.8	2.23872
0.740	2.51110	1.66	2.40236	7.80	2.37488	17.0	2.35173	23.1	2.32699	27.7	2.30274	32.3	2.27321	36.9	2.23788
0.760	2.50309	1.68	2.40180	8.00	2.37452	17.2	2.35104	23.2	2.32652	27.8	2.30216	32.4	2.27251	37.0	2.23705
0.780	2.49583	1.70	2.40125	8.20	2.37416	17.4	2.35035	23.3	1.32604	27.9	2.30157	32.5	2.27180	37.1	2.23621
0.800	2.48922	1.72	2.40073	8.40	2.37380	17.6	2.34965	23.4	2.32556	28.0	2.30098	32.6	2.27109	37.2	2.23536
0.820	2.48318	1.74	2.40022	8.60	2.37343	17.8	2.34894	23.5	2.32508	28.1	2.30039	32.7	2.27038	37.3	2.23452
0.840	2.47766	1.76	2.39974	8.80	2.37305	18.0	2.34822	23.6	2.32460	28.2	2.29979	32.8	2.26966	37.4	2.23367
0.860	2.47258	1.78	2.39926	9.00	2.37267	18.2	2.34750	23.7	2.32411	28.3	2.29920	32.9	2.26895	37.5	2.23281
0.880	2.46790	1.80	2.39881	9.20	2.37229	18.4	2.34676	23.8	2.32362	28.4	2.29860	33.0	2.26823	37.6	2.23196
0.900	2.46358	1.82	2.39837	9.40	2.37190	18.6	2.34602	23.9	2.32313	28.5	2.29800	33.1	2.26750	37.7	2.23110
0.920	2.45958	1.84	2.39794	9.60	2.37150	18.8	2.34527	24.0	2.32264	28.6	2.29739	33.2	2.26678	37.8	2.23024
0.940	2.45587	1.86	2.39753	9.80	2.37110	19.0	2.34451	24.1	2.32215	28.7	2.29679	33.3	2.26605	37.9	2.22937
0.960	2.45242	1.88	2.39713	10.0	2.37069	19.2	2.34374	24.2	2.32165	28.8	2.29618	33.4	2.26532	38.0	2.22850
0.980	2.44920	1.90	2.39674	10.2	2.37027	19.4	2.34296	24.3	2.32115	28.9	2.29556	33.5	2.26458	38.1	2.22763
1.00	2.44620	1.92	2.39637	10.4	2.36985	19.6	2.34217	24.4	2.32065	29.0	2.29495	33.6	2.26384	38.2	2.22676
1.02	2.44339	1.94	2.39600	10.6	2.36942	19.8	2.34138	24.5	2.32014	29.1	2.29433	33.7	2.26310	38.3	2.22588
1.04	2.44076	1.96	2.39565	10.8	2.36898	20.0	2.34058	24.6	2.31964	29.2	2.29371	33.8	2.26236	38.4	2.22500
1.06	2.43830	1.98	2.39531	11.0	2.36854	20.1	2.34017	24.7	2.31913	29.3	2.29309	33.9	2.26161	38.5	2.22412
1.08	2.43598	2.00	2.39498	11.2	2.36809	20.2	2.33976	24.8	2.31861	29.4	2.29247	34.0	2.26087	38.6	2.22323
1.10	2.43380	2.20	2.39214	11.4	2.36763	20.3	2.33935	24.9	2.31810	29.5	2.29184	34.1	2.26011	38.7	2.22234
1.12	2.43175	2.40	2.38997	11.6	2.36717	20.4	2.33894	25.0	2.31758	29.6	2.29121	34.2	2.25936	38.8	2.22145
1.14	2.42981	2.60	2.38826	11.8	2.36669	20.5	2.33853	25.1	2.31707	29.7	2.29058	34.3	2.25860	38.9	2.22055
1.16	2.42798	2.80	2.38688	12.0	2.36622	20.6	2.33811	25.2	2.31655	29.8	2.28994	34.4	2.25784	39.0	2.21965
1.18	2.42625	3.00	2.38574	12.2	2.36573	20.7	2.33770	25.3	2.31602	29.9	2.28931	34.5	2.25708	39.1	2.21875
1.20	2.42462	3.20	2.38478	12.4	2.36523	20.8	2.33727	25.4	2.31550	30.0	2.28867	34.6	2.25631	39.2	2.21784
1.22	2.42307	3.40	2.38396	12.6	2.36473	20.9	2.33685	25.5	2.31497	30.1	2.28802	34.7	2.25554	39.3	2.21693
1.24	2.42159	3.60	2.38325	12.8	2.36422	21.0	2.33643	25.6	2.31444	30.2	2.28738	34.8	2.25477	39.4	2.21602
1.26	2.42020	3.80	2.38261	13.0	2.36371	21.1	2.33600	25.7	2.31391	30.3	2.28673	34.9	2.25400	39.5	2.21510
1.28	2.41887	4.00	2.38204	13.2	2.36318	21.2	2.33557	25.8	2.31337	30.4	2.28608	35.0	2.25322	39.6	2.21418
1.30	2.41760	4.20	2.38153	13.4	2.36265	21.3	2.33514	25.9	2.31283	30.5	2.28543	35.1	2.25244	39.7	2.21326
1.32	2.41640	4.40	2.38105	13.6	2.36211	21.4	2.33471	26.0	2.31229	30.6	2.28477	35.2	2.25166	39.8	2.21233
1.34	2.41525	4.60	2.38061	13.8	2.36157	21.5	2.33427	26.1	2.31175	30.7	2.28411	35.3	2.25087	39.9	2.21140
1.36	2.41416	4.80	2.38019	14.0	2.36101	21.6	2.33383	26.2	2.31121	30.8	2.28345	35.4	2.25008	40.0	2.21047
1.38	2.41312	5.00	2.37979	14.2	2.36045	21.7	2.33339	26.3	2.31066	30.9	2.28279	35.5	2.24929		
1.40	2.41212	5.20	2.37940	14.4	2.35988	21.8	2.33295	26.4	2.31011	31.0	2.28212	35.6	2.24849		
1.42	2.41117	5.40	2.37903	14.6	2.35930	21.9	2.33251	26.5	2.30956	31.1	2.28145	35.7	2.24769		
1.44	2.41025	5.60	2.37867	14.8	2.35871	22.0	2.33206	26.6	2.30900	31.2	2.28078	35.8	2.24689		

Dispersion equation:

$$n^2 - 1 = \sum_i \frac{K_i \lambda^2}{\lambda^2 - \lambda_i^2}$$

TABLE 6b-52. CONSTANTS OF THE DISPERSION EQUATION AT 25°C

i	λ_i^2	K_i
1	0.0225	1.8293958
2	0.0625	1.6675593
3	0.1225	1.1210424
4	0.2025	0.04513366
5	27,089.737	12.380234

From W. S. Rodney and I. H. Malitson, *J. Opt. Soc. Am.* **46**, 956 (1956).

Data are also given in the reference for temperatures of 19 and 31°C. A comparison of these data, taken with the 45.7 to 54.3 mole percent mixed crystal, is made with the older data taken with 42 to 58 crystal material by L. W. Tilton, E. K. Plyler, and R. E. Stephens, *J. Res. NBS* **43**, 81 (1949). The 45.7 to 54.3 composition, which has the lowest freezing temperature of the binary system, should give the best optical homogeneity.

TABLE 6b-53. TEMPERATURE COEFFICIENT OF REFRACTIVE INDEX

λ, μm	dn/dT $(10^{-6}/\mathrm{K})$	λ, μm	dn/dT $(10^{-6}/\mathrm{K})$
0.577	-254	14	-228
1.1	-240	16	-225
2	-238	20	-217
4	-237	25	-207
6	-237	30	-195
8	-236	35	-175
10	-235	40	-152
12	-232		

From A. I. Funai, *Lockheed Missiles & Space Co. Rept.* LMSC/6-78-68-34, p. 46, (1968) who references Rodney and Malitson.

The composition of KRS-5 in Table 6b-52 was 45.7 percent thallium bromide and 54.3 percent thallium iodide.

Titanium Dioxide

TABLE 6b-54. REFRACTIVE INDEX OF RUTILE

λ, μm	n_o	n_e	λ, μm	n_o
0.4358	2.853	3.216	2.0000	2.399
0.4916	2.725	3.051	2.5000	2.387
0.4960	2.718	3.042	3.0000	2.380
0.5461	2.652	2.958	3.5000	2.367
0.5770	2.623	2.921	4.0000	2.350
0.5791	2.621	2.919	4.5000	2.322
0.6907	2.555	2.836	5.0000	2.290
0.7082	2.548	2.826	5.5000	2.200
1.0140	2.484	2.747		
1.5296	2.454	2.710		

Dispersion equation—ordinary ray:

$$n_o^2 = 5.913 + \frac{2.441 \times 10^7}{\lambda^2 - 0.803 \times 10^7}$$

Dispersion equation—extraordinary ray:

$$n_e^2 = 7.197 + \frac{3.322 \times 10^7}{\lambda^2 - 0.843 \times 10^7}$$

For wavelengths between 0.4358 and 1.5296 μm, the calculated refractive-index data from J. R. DeVore, *J. Opt. Soc. Am.* **41**, 418 (1951), are given. The dispersion equations are also due to DeVore; note that wavelength must be specified in angstroms in these equations. The other data are the observed values of W. F. Parsons, private communication. It should be noted that the data of Parsons and of DeVore do not fit together smoothly, probably indicating that there are differences from sample to sample. Data for dn/dT are listed by Alexander Smakula, *U.S. Dept. Comm. Office Tech. Serv. Doc.* 111,052, October, 1952; the original data of Z. Schroeder, *Z. Krist.* **67**, 509 (1928), are quoted herein.

The refractive index of the anastase form of titanium dioxide is discussed by T. N. Krylova and G. O. Bagdyk'yants, *Opt. Spectr. U.S.S.R.* **9**, 339 (1960), and several other references are listed.

Zinc Sulfide

TABLE 6b-55. REFRACTIVE INDICES OF HEXAGONAL ZINC SULFIDE

λ, μm	n_e	n_o	λ, μm	n_e	n_o
0.3600	2.709	2.705	0.5000	2.425	2.421
0.3750	2.640	2.637	0.5250	2.407	2.402
0.4000	2.564	2.560	0.5500	2.392	2.386
0.4100	2.544	2.539	0.5750	2.378	2.375
0.4200	2.525	2.522	0.6000	2.368	2.363
0.4250	2.514	2.511	0.6250	2.358	2.354
0.4300	2.505	2.502	0.6500	2.350	2.346
0.4400	2.488	2.486	0.6750	2.343	2.339
0.4500	2.477	2.473	0.7000	2.337	2.332
0.4600	2.463	2.459	0.8000	2.328	2.324
0.4700	2.453	2.448	0.9000	2.315	2.310
0.4750	2.449	2.445	1.0000	2.303	2.301
0.4800	2.443	2.438	1.2000	2.294	2.290
0.4900	2.433	2.428	1.4000	2.288	2.285

From T. M. Bieniewski and S. J. Czyzak, *J. Opt. Soc. Am.* **53**, 496 (1963).

TABLE 6b-56. REFRACTIVE INDEX OF CUBIC ZINC SULFIDE

λ, μm	n	λ, μm	n
0.4400	2.488	0.6000	2.359
0.4600	2.458	0.6500	2.346
0.4800	2.435	0.7000	2.334
0.5000	2.414	0.9000	2.306
0.5250	2.395	1.0500	2.293
0.5500	2.384	1.2000	2.282
0.5750	2.375	1.4000	2.280

These values are for synthetic cubic zinc sulfide [from S. J. Czyzak, W. M. Baker, R. C. Crane, and J. B. Howe, *J. Opt. Soc. Am.* **47**, 240 (1957)]. Refractive-index values in the wavelength region 0.365 to 1.53 μm for natural cubic zinc sulfide are give by J. R. DeVore, *J. Opt. Soc. Am.* **41**, 416 (1951).

For computational purposes, Czyzak, Baker, Crane, and Howe and give a dispersion equation:

$$n^2 = 5.131 + \frac{1.275 \times 10^7}{\lambda^2 - 0.732 \times 10^7}$$

The hexagonal form of zinc sulfide is called wurtzite, and the cubic form is called sphalerite. Natural-crystal zinc sulfide occurs in the cubic form only and is called sphalerite or zincblende. Note that the birefringence of hexagonal zinc sulfide is very small but fairly constant. Cubic zinc sulfide evidences electro-optic properties.

Group III—Group V Compounds

Gallium Antimonide

TABLE 6b-57. REFRACTIVE INDEX OF GALLIUM ANTIMONIDE

λ, μm	n_D	n_R
1.8	3.820	3.61
1.9	3.802	3.59
2.0	3.789	3.57
2.1	3.780	3.55
2.2	3.764	3.54
2.3	3.758	3.53
2.4	3.755	3.52
2.5	3.749	3.49

From D. F. Edwards and G. S. Haynes, *J. Opt. Soc. Am.* **59**, 414 (1959).

These values are for p-type single-crystal gallium antimonide which has a purity corresponding to 7.5×10^{16} carriers/cm^3 measured at room temperature. The refractive index n_D was calculated from the relation for the angle of minimum deviation, with an estimated error of 0.3 percent. The refractive index n_R was deduced from reflectivity measurements, with an error of less than 2 percent. The discrepancy between the values of n_D and n_R is not explained.

Gallium Arsenide

TABLE 6b-58. REFRACTIVE INDEX OF GALLIUM ARSENIDE

λ, μm	n	λ, μm	n
0.78 $\pm$ 0.01	3.34 $\pm$ 0.04	14.5 $\pm$ 0.05	2.82 $\pm$ 0.04
8.0 $\pm$ 0.05	3.34 $\pm$ 0.04	15.0 $\pm$ 0.05	2.73 $\pm$ 0.04
10.0 $\pm$ 0.05	3.135 $\pm$ 0.04	17.0 $\pm$ 0.05	2.59 $\pm$ 0.04
11.0 $\pm$ 0.05	3.045 $\pm$ 0.04	19.0 $\pm$ 0.05	2.41 $\pm$ 0.04
13.0 $\pm$ 0.05	2.97 $\pm$ 0.04	21.9 $\pm$ 0.1	2.12 $\pm$ 0.04
13.7 $\pm$ 0.05	2.895 $\pm$ 0.04		

The experimental data seem to be somewhat more scattered than the reported experimental errors indicate. The data are from L. C. Barcus, *Phys. Rev.* **111,** 167 (1958), and L. C. Barcus, Lowell Institute of Technology, private communication. The index is approximately 3.34 from about 2 to 7 μm.

Indium Antimonide

TABLE 6b-59. REFRACTIVE INDEX OF INDIUM ANTIMONIDE

λ, μm	n	λ, μm	n
7.87	4.00	15.13	3.88
8.00	3.99	15.79	3.87
9.01	3.96	16.96	3.86
10.06	3.95	17.85	3.85
11.01	3.93	18.85	3.84
12.06	3.92	19.98	3.82
12.98	3.91	21.15	3.81
13.90	3.90	22.20	3.80

From R. G. Breckenridge, *Phys. Rev.* **96,** 571 (1954).

These values are for a sample of indium antimonide which has a purity corresponding to 2.0×10^{16} carriers/cm³; measured at room temperature. These data are in agreement with values reported by T. S. Moss, *Proc. Phys. Soc. (London),* ser. B, **70,** 776 (1954). The temperature dependence of index of refraction for three different temperatures is given by R. F. Potter, *Appl. Opt.* **5,** 35 (1966).

Nonoxide Chalcogenic Glasses

Arsenic-modified Selenium Glass

TABLE 6b-60. REFRACTIVE INDEX OF ARSENIC-MODIFIED SELEN IUM GLASS AT 27°C

λ, μm	n, prism A	n, prism B	λ, μm	n, prism A	n, prism B
1.0140	2.5774	2.5783	7.00	2.4778	2.4787
1.1286	2.5554	2.5565	7.50	......	2.4784
1.3622	2.5285	2.5294	8.10	2.4772	2.4778
1.5295	2.5173	2.5183	8.50	......	2.4775
1.7012	2.5089	2.5100	9.10	2.4765	2.4771
2.1526	2.4950	2.4973	9.50	......	2.4768
3.00	2.4861	2.4882	10.00	2.4756	2.4767
3.4188	2.4841	2.4858	10.50	......	2.4759
4.00	2.4825	2.4835	11.00	2.4752	2.4758
4.50	......	2.4822	11.50	......	2.4753
5.00	2.4803	2.4811	12.00	......	2.4749
5.50	......	2.4804	13.00	......	2.4760[*sic*]
6.00	2.4789	2.4798	13.50	......	2.4748
6.50	......	2.4792	14.00	......	2.4743

From C. D. Salzberg and J. J. Villa, *J. Opt. Soc. Am.* **47,** 244 (1957).

Arsenic Trisulfide Glass

TABLE 6b-61. REFRACTIVE INDEX OF ARSENIC TRISULFIDE GLASS AT 25°C

λ, μm	n	λ, μm	n	λ, μm	n
0.560	2.68689	1.800	2.43009	7.000	2.39899
0.580	2.65934	2.000	2.42615	7.200	2.39806
0.600	2.63646	2.200	2.42318	7.400	2.39709
0.620	2.61708	2.400	2.42086	7.600	2.39610
0.640	2.60043	2.600	2.41898	7.800	2.39508
0.660	2.58594	2.800	2.41742	8.000	2.39403
0.680	2.57323	3.000	2.41608	8.200	2.39294
0.700	2.56198	3.200	2.41491	8.400	2.39183
0.720	2.55195	3.400	2.41386	8.600	2.39068
0.740	2.54297	3.600	2.41290	8.800	2.38949
0.760	2.53488	3.800	2.41200	9.000	2.38827
0.780	2.52756	4.000	2.41116	9.200	2.38700
0.800	2.52090	4.200	2.41035	9.400	2.38570
0.820	2.51483	4.400	2.40956	9.600	2.38436
0.840	2.50928	4.600	2.40878	9.800	2.38298
0.860	2.50418	4.800	2.40802	10.000	2.38155
0.880	2.49949	5.000	2.40725	10.200	2.38007
0.900	2.49515	5.200	2.40649	10.400	2.37855
0.920	2.49114	5.400	2.40571	10.600	2.37698
0.940	2.48742	5.600	2.40493	10.800	2.37536
0.960	2.48396	5.800	2.40414	11.000	2.37369
0.980	2.48074	6.000	2.40333	11.200	2.37196
1.000	2.47773	6.200	2.40250	11.400	2.37018
1.200	2.45612	6.400	2.40166	11.600	2.36833
1.400	2.44357	6.600	2.40079	11.800	2.36643
1.600	2.43556	6.800	2.39991	12.000	2.36446

Dispersion equation:
$$n^2 - 1 = \sum_{i=1}^{i=5} \frac{K_i \lambda^2}{\lambda^2 - \lambda_i^2}$$

TABLE 6b-62. CONSTANTS FOR THE DISPERSION EQUATION FOR 25°C

i	λ_i^2	K_i
1	0.0225	1.8983678
2	0.0625	1.9222979
3	0.1225	0.8765134
4	0.2025	0.1188704
5	750	0.9569903

From I. H. Malitson, W. S. Rodney, and T. A. King, *J. Opt. Soc. Am.* **48**, 633 (1958).

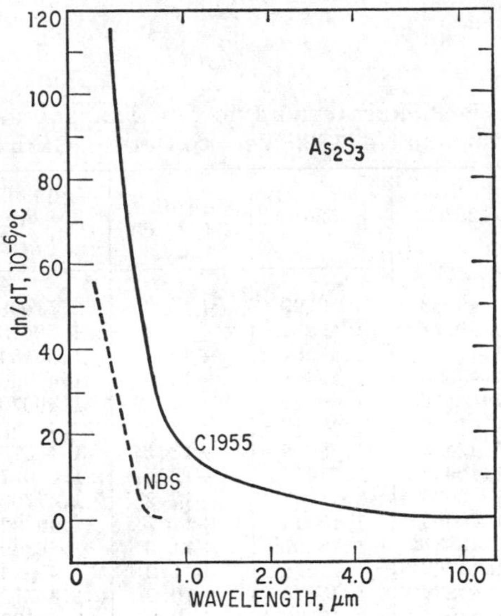

FIG. 6b-7. The temperature coefficient of refractive index dn/dT for two types of arsenic trisulfide glass. [*From I. H. Malitson, W. S. Rodney, and T. A. King, J. Opt. Soc. Am.* **48,** 633 (1958).]

It should be noted that arsenic and sulfur form an entire glass system, with varying properties. Therefore, one should expect different values of transmission, refractive index, and so forth, for samples from different batches.

Special Glasses

Corning Vycor

TABLE 6b-63. REFRACTIVE INDEX VS. TEMPERATURE FOR
CORNING NO. 7913 (VYCOR), OPTICAL GRADE

λ, μm	n, 28°C	n, 526°C	dn/dT (10⁻⁶/°C)	n, 826°C	dn/dT (10⁻⁶/°C)
0.26520	1.49988	1.50799	$+16.3$	1.51438	$+18.2$
0.28936	1.49074	1.49831	$+15.2$	1.50418	$+16.8$
0.29673	1.48851	1.49587	$+14.8$	1.50164	$+16.5$
0.30215	1.48694	1.49423	$+14.6$	1.49990	$+16.2$
0.3130	1.48416	1.49121	$+14.2$	1.49679	$+15.8$
0.33415	1.47949	1.48622	$+13.5$	1.49158	$+15.2$
0.36502	1.47415	1.48065	$+13.1$	1.48570	$+14.5$
0.40466	1.46925	1.47547	$+12.5$	1.48027	$+13.8$
0.43584	1.46628	1.47234	$+12.2$	1.47708	$+13.5$
0.54607	1.45960	1.46544	$+11.7$	1.46992	$+12.9$
0.5780	1.45831	1.46407	$+11.6$	1.46849	$+12.8$
1.01398	1.44968	1.45526	$+11.2$	1.45924	$+12.0$
1.12866	1.44831	1.45373	$+10.9$	1.45779	$+11.9$
1.254*	1.44677	1.45222	$+10.9$	1.45627	$+11.9$
1.36728	1.44554	1.45095	$+10.9$	1.45504	$+11.9$
1.470*	1.44422	1.44965	$+10.9$	1.45370	$+11.9$
1.52952	1.44356	1.44896	$+10.8$	1.45306	$+11.9$
1.660*	1.44206	1.44750	$+11.0$	1.45157	$+11.9$
1.701	1.44137	1.44677	$+10.8$	1.45088	$+11.9$
1.981*	1.43750	1.44291	$+10.9$	1.44702	$+11.9$
2.262*	1.43298	1.43839	$+10.9$	1.44258	$+12.0$
2.553*	1.42825	1.43373	$+11.0$	1.43824	$+12.5$

* Wavelength determination by narrow-bandwidth interference filters.
From J. H. Wray and J. T. Neu, *J. Opt. Soc. Am.* **59**, 774 (1969).

TABLE 6b-64. MISCELLANEOUS REFRACTIVE-INDEX DATA

Material	Wavelength Region, μm	Refractive index	Reference
Barium titanate....................	Visible and infrared	2.40	
Cadmium fluoride....................	2	1.63	1
	5.5	1.53	1
	11	1.45	1
Cadmium iodide......................	0.6	2.7	2
Cadmium selenide...................	1–8	2.45	3
	>8	2.42	3
Cuprous chloride....................	0.4–20.5	1.93	4
Lead bromide........................	"White light"	2.53	5
Lead chloride.......................	"Yellow light"	2.2	6
Lead selenide.......................	1–3.5	3.5–4.6	7
	5	4.6	8
Lead sulfide........................	3.0	4.10 ± 0.06	7
	6.0	4.19 ± 0.06	9
	$dn/dT = 6 \times 10^{-4}$ per °C in the temperature range 20–300°C		9
Lead telluride......................	1–3.5	4.1–5.3	7
	3.9–20	5.10	10
Rubidium bromide....................	1–8	1.53	11
Rubidium chloride...................	1–8	1.48	11
Rubidium iodide.....................	1–8	1.62	11
T-12................................	"Near infrared"	1.41	12
Gallium phosphide...................	<1	3.5	13
	1–8	3.2–2.9	13
	>8	2.8	13
Indium arsenide....................	4–15	3–3.5	14
Indium phosphide....................	2–15	3–3.5	14
Arsenic triselenide glass.............	4	2.796	15
	2–16	2.812–2.768	15
Telluride glass (Ge-As-Te)...........	5	3.5	16
Texas Instruments Glass No. 1173.....	3.3	2.63	16
	5	2.62	17
	$dn/dT = 79 \times 10^{-6}$ per °C in the region 3–13 μm		17
Cer-Vit material C-101..............	0.589 at 25°C	1.540	18

References for Table 6b-64
1. Adapted from B. Welber: *Appl. Opt.* **6**, 925 (1967).
2. Adapted from P. O. Nilsson: *Appl. Opt.* **7**, 435 (1968).
3. Vitrikhovsky, N. I., L. F. Gudymenko, A. F. Maznichenko, V. N. Malinko, E. V. Pidlinsu, and S. F. Terekhova: *Ukr. Fiz. Zh.* **12**,796 (1967).
4. Heilmeier, G. H.: *Appl. Opt.* **3**, 1281 (1964).
5. Moss, T. S., and A. G. Peacock: *Infrared Phys.* **1**, 104 (1961).
6. "Handbook of Chemistry and Physics," Chemical Rubber Publishing Co., Cleveland, Ohio, 1960.
7. Avery, D. G.: *Proc. Phys. Soc.* (*London*), ser. B, **66**, 134 (1953).
8. Smakula, A.: *Opt. Acta* **9**, 205 (1962).
9. Avery, D. G.: *Proc. Phys. Soc.* (*London*), ser. B, **67**, 2 (1954).
10. Smakula, A., J. Kalnajs, and M. J. Redman: *Appl. Opt.* **3**, 323 (1964).
11. Mott, N. F., and R. W. Gurney: "Electronic Processes in Ionic Crystals," p. 12, Oxford University Press, New York, 1950. (From the values given for the dielectric constants.)
12. Ballard, S. S.: *Japan. J. Appl. Phys.* **4**, suppl. 1, 23 (1965).
13. Willardson, R. K., and A. C. Beer: "Semiconductors and Semimetals," vol. 3, Academic Press, Inc., New York, 1967.
14. Oswald, F., and R. Schade: *Z. Naturforsch.* **9a**, 611 (1954).
15. Savage, J. A., and S. Nielson: *Infrared Phys.* **5**, 195 (1965).
16. Ballard, S. S., and J. S. Browder: *Appl. Opt.* **5**, 1873 (1966).
17. Hilton, A. R., and C. E. Jones: *Appl. Opt.* **6**, 1513 (1967).
18. Monnier, R. C.: *Appl. Opt.* **6**, 1437 (1967).

6c. Transmission and Absorption of Special Crystals and Certain Glasses

STANLEY S. BALLARD

JAMES STEVE BROWDER

JOHN F. EBERSOLE

University of Florida

The transmittances of the following materials are discussed in this section:

Ammonium dihydrogen phosphate
(ADP) and Potassium dihydrogen
phosphate (KDP)
Barium fluoride
Barium titanate
Cadmium selenide
Cadmium sulfide
Calcite
Calcium fluoride
Cesium bromide
Cesium iodide
Crystal quartz
Cuprous chloride
Diamond
Fused silica
Germanium
Irtrans 1 to 6
Lanthanum fluoride
Lead bromide
Lead chloride
Lead fluoride
Lead selenide
Lead sulfide
Lead telluride
Lithium fluoride
Magnesium fluoride
Magnesium oxide
Mica
Potassium bromide
Potassium chloride
Potassium iodide
Rubidium bromide
Rubidium chloride
Rubidium iodide

Ruby
Sapphire
Selenium
Silicon
Silver chloride
Sodium chloride
Sodium fluoride
Spinel
Strontium titanate
Sulfur
T-12
Tellurium
Thallium bromide
Thallium chloride
Thallium bromide-chloride (KRS-6)
Thallium bromide-iodide (KRS-5)
Titanium dioxide
Group III–Group V compounds:
    Gallium antimonide
    Gallium arsenide
    Gallium phosphide
    Indium antimonide
    Indium arsenide
    Indium phosphide
Nonoxide chalcogenic glasses:
    Arsenic-modified selenium glass
    Arsenic triselenide glass
    Arsenic trisulfide glass
    A telluride glass
    Texas Instruments Glass No. 1173
Special glasses:
    Cer-Vit
    Corning glasses

The materials listed above can be used in the infrared, visible, and/or ultraviolet regions of the spectrum for prisms, lenses, windows, and other components of optical systems. "Standard" glasses, of which there are many kinds, are not discussed here, but certain unusual glasses are included.

When transmission data are given, external or internal transmittance is specified. The external transmittance is the fraction of the incident intensity that is transmitted; it is determined by both the absorbing and the reflecting properties of the material. Internal transmittance is descriptive of the result of absorption processes only (if scattering can be neglected).

Transmittance data, usually in the form of curves, over the ultraviolet, visible, and/or infrared regions of the spectrum are included when available for each material. The transmission curves are often given with just the long- and short-wavelength extremes shown. Unless the accompanying text indicates otherwise, interpolation between the short- and long-wavelength curves can be made by a straight line with reasonable accuracy.

Since the transmission of a material depends upon its temperature, instances in which the temperature dependence for a material is known to be appreciable are pointed out.

Notes are included to give information of possible interest and practical value on most of the materials.

Unless stated otherwise, wavelengths are given in micrometers (μm), and temperatures are in kelvins (K), employing the new Système International units. Many sets of data were recorded using degrees Celsius (°C), and no attempt has been made to convert these to kelvins.

An overall concept of the spectral regions of transparency of the materials is given in Fig. 6c-1. This figure indicates the wavelength range over which a sample 2 mm thick has an external transmittance of 10 percent or more.

When light is incident on a sample, part of it is reflected, part is absorbed, and part is transmitted through the sample. The absorption of a material is expressed by the Lambert-Beer-Bouguer law, which can be written as

$$I = I_1 e^{-\alpha x}$$

where I is the intensity of light at a distance x in the material, I_1 is the intensity just inside the front surface, and α is the absorption coefficient.

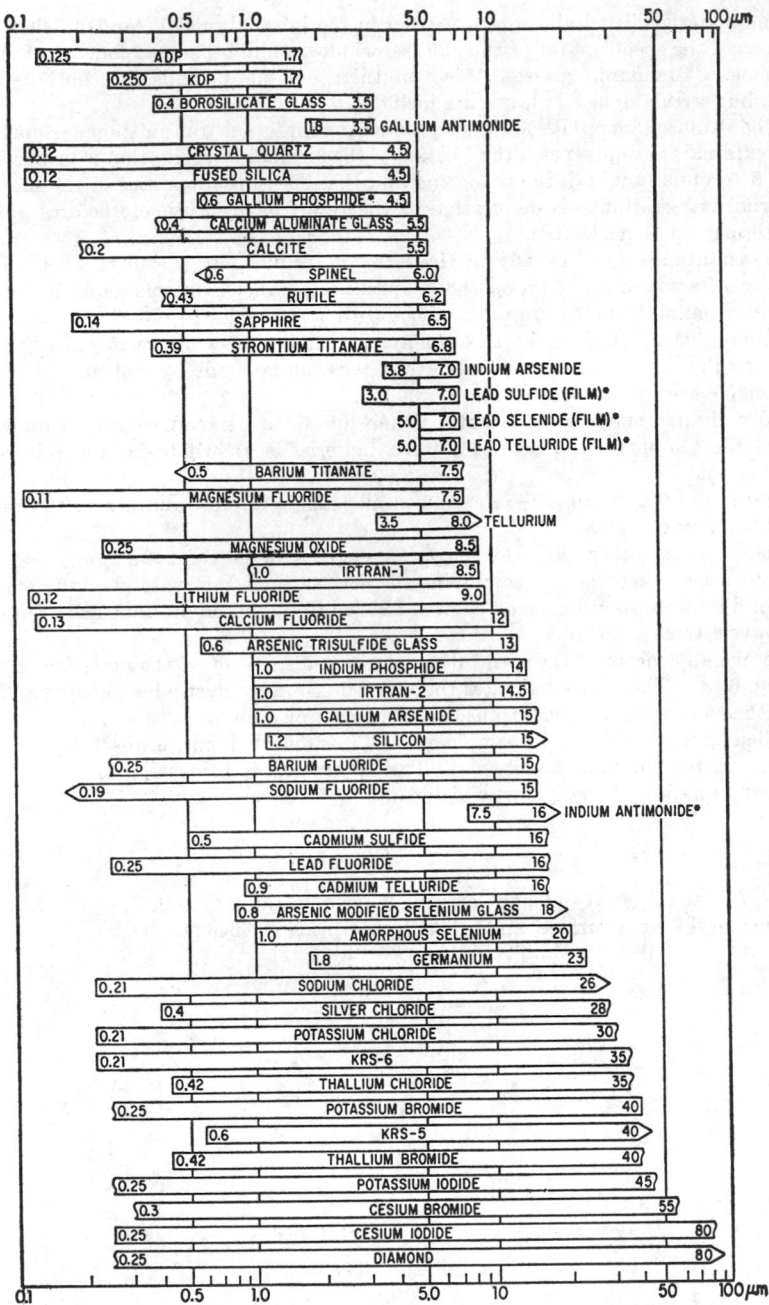

Fig. 6c-1. Transmission regions.   The limiting wavelengths, for both long and short cutoff, have been chosen as those wavelengths at which a sample 2 mm thick has 10 percent transmittance.   Materials marked with an asterisk (*) have a maximum external transmittance less than 10 percent.

For comparison purposes the short- and long-wavelength absorption edges of several materials are included in Figs. 6c-2 to 6c-4.

(The literature search extended back to January, 1959.   It was restricted primarily to "optical" journals, i.e., *Journal of the Optical Society of America, Applied Optics, Optics & Spectroscopy, Optica Acta, and Infrared Physics.*   It is realized that optical data on semiconductor materials are to be found also through the literature of solid-state physics.)

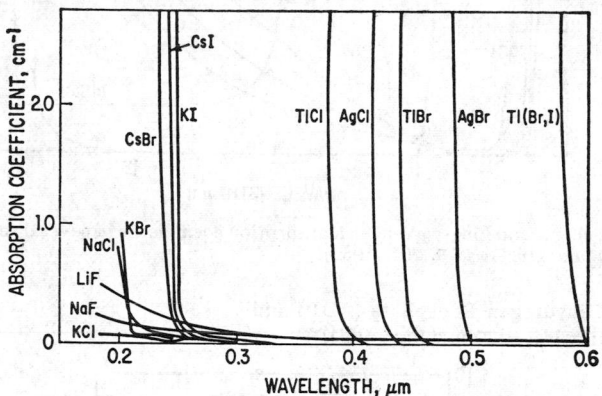

FIG. 6c-2. The short-wavelength absorption edges of several alkali, silver, and thallium halides.   [*From A. Smakula, Opt. Acta* **9**, 205 (1962).]

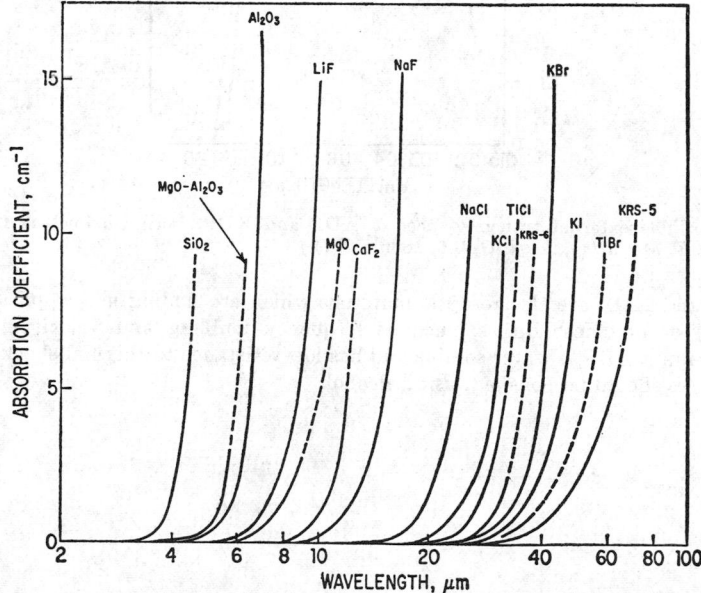

FIG. 6c-3. The long-wavelength absorption edges of several ionic crystals.   [*From A. Smakula, Opt. Acta* **9**, 205 (1962).]

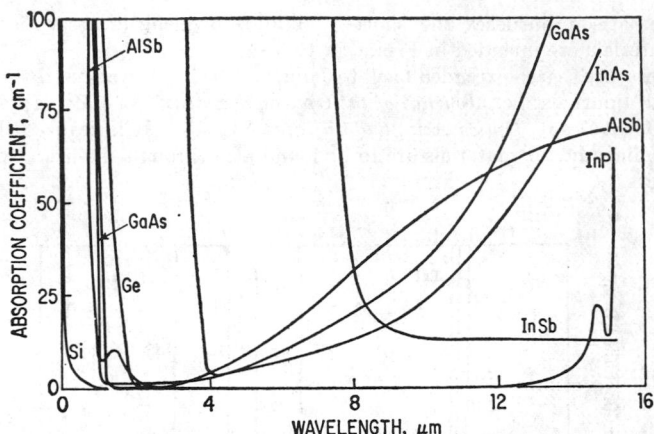

Fig. 6c-4. The short- and long-wavelength absorption coefficients of several semiconductors. [*From A. Smakula, Opt. Acta* **9,** 205 (1962).]

Ammonium Dihydrogen Phosphate (ADP) and
Potassium Dihydrogen Phosphate (KDP)

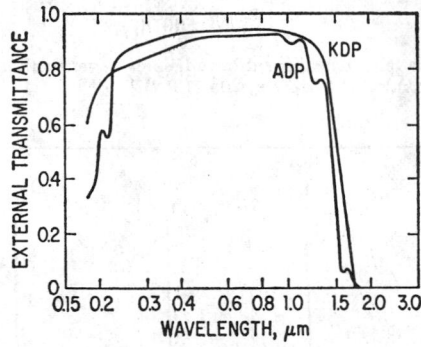

Fig. 6c-5. The external transmittances of ADP and KDP, both thicknesses 1.65 mm. [*From D. E. McCarthy, Appl. Opt.* **6,** 1896 (1967).]

ADP and KDP are electro-optic materials which are finding new applications in the field of nonlinear optics, such as frequency-doubling and velocity-matching experiments. ADP is water-soluble and has low resistance to thermal shock. Optical surfaces should be polished, using alcohol.

Barium Fluoride

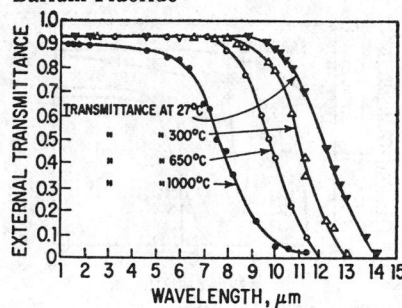

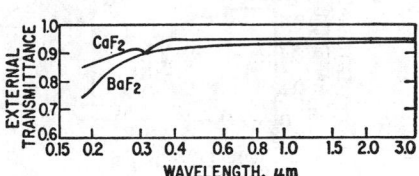

FIG. 6c-6. The external transmittance of barium fluoride at 27, 300, and 650°C, thickness 8.0 mm; and at 1000°C, thickness 7.6 mm. [*Adapted from U. P. Oppenheim and A. Goldman, J. Opt. Soc. Am.* **54**, 127 (1964).]

FIG. 6c-7. The short-wavelength external transmittance of barium fluoride, thickness 12.0 mm; and calcium fluoride, thickness 5.0 mm. [*Adapted from D. E. McCarthy, Appl. Opt.* **6**, 1896 (1967).]

Barium fluoride can be obtained in cylinders of diameters up to 6 in. The material can be cut with a Norton diamond wheel at about 4,000 rpm, but very, very light pressure should be applied in order to prevent cleavage. It has a high melting point (1280°C). It is much less hard than magnesium oxide, and because of its brittleness, it is less suitable in applications in which it is subjected to mechanical stress. Above room temperature, the ultraviolet absorption edge shifts from approximately 0.135 μm to longer wavelengths.

Barium Titanate

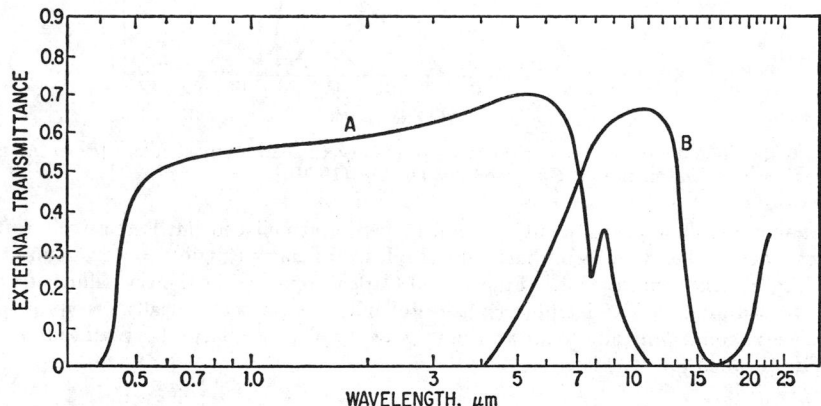

FIG. 6c-8. The external transmittance of barium titanate: (*A*) single crystal, thickness 0.25 mm; (*B*) powder, thickness about 10 μm. [*From A. F. Iatsenko, Soviet Phys.—Tech. Phys.* **2**, 2257 (1957).]

Barium titanate is well known for its electrical properties. Because of its transmission properties, it has applications in immersion lenses for infrared detectors.

Cadmium Selenide

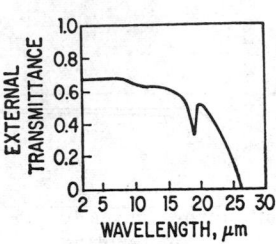

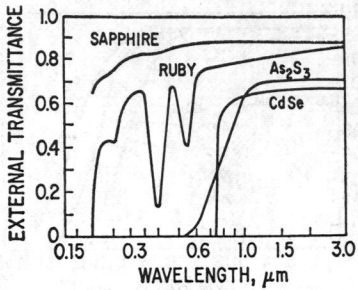

FIG. 6c-9. The infrared external transmittance of cadmium selenide, thickness 1.67 mm. [*Adapted from D. E. McCarthy, Appl. Opt.* **4**, 317 (1965).]

FIG. 6c-10. The short-wavelength external transmittances of cadmium selenide, thickness 1.67 mm; ruby, thickness 6.10 mm; sapphire, thickness 3.0 mm; and arsenic trisulfide glass, thickness 5.0 mm. [*From D. E. McCarthy, Appl. Opt.* **6**, 1896 (1967).]

The surface of cadmium selenide takes a mirror-like polish but scratches easily.

Cadmium Sulfide

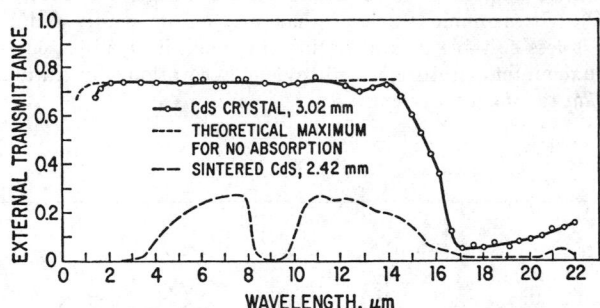

FIG. 6c-11. The external transmittance of single-crystal cadmium sulfide. [*From A. B. Francis and A. L. Carlson, J. Opt. Soc. Am.* **50**, 118 (1960).]

Cadmium sulfide is easily cut, ground, lapped, and polished but is relatively soft. It can be produced in 1-in.-diameter samples. It has negligible water solubility but can be dissolved in acids. Francis and Carlson report that the crystalline structure of cadmium sulfide is cubic (sphalerite) if it has been chemically precipitated, and is hexagonal (wurtzite) if grown from the vapor phase. Natural-crystal cadmium sulfide is called greenockite.

Calcite

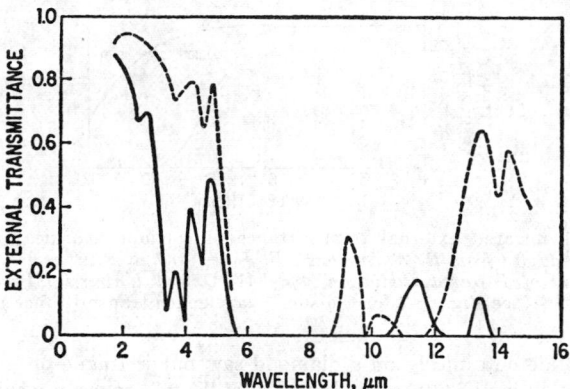

FIG. 6c-12. The infrared external transmittance of calcite for the ordinary ray (solid curve) and for the extraordinary ray (dashed curve); thickness 1 mm. [*From R. E. Nysander, Phys. Rev.* **28**, 291 (1909).]

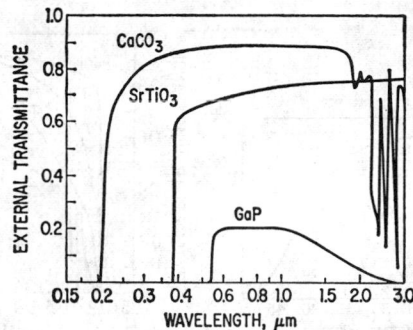

FIG. 6c-13. The short-wavelength external transmittance of calcite, thickness 33.3 mm; strontium titanate, thickness 1.0 mm; and gallium phosphide, thickness 1.0 mm. [*From D. E. McCarthy, Appl. Opt.* **6**, 1896 (1967).]

Calcite is important historically. Its birefringent properties are well known, and although they are important academically and in polarizing prisms, they are disadvantageous in many instrument applications. Calcite, which is a form of calcium carbonate, is known to exist in many varieties, such as Iceland spar, oriental alabaster, and onyx.

Calcium Fluoride

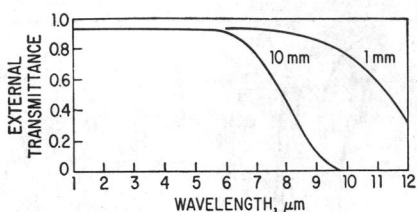

Fig. 6c-14. The infrared external transmittance of calcium fluoride for two different thicknesses. (*Adapted from R. A. Smith, F. E. Jones, and R. P. Chasmar, "The Detection and Measurement of Infrared Radiation," p. 341, Oxford University Press, London and New York, 1957.*)   (See Fig. 6c-7 for the short-wavelength transmittance.)

Calcium fluoride cuts nicely on a diamond saw but is fragile on a diamond fine-grinding wheel.   It is difficult to grind; very light "cuts" are recommended.   Crystals of 9-in. diameter can be produced.   It is practically insoluble in water but is soluble in ammonium solutions.

Cesium Bromide

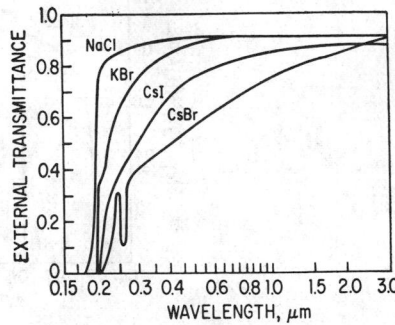

Fig. 6c-15. The infrared external transmittance of cesium bromide, thickness 10 mm. [*Adapted from D. E. McCarthy, Appl. Opt.* **2**, 591 (1963).]

Fig. 6c-16. The short-wavelength external transmittances of cesium bromide and cesium iodide, both thicknesses 10.0 mm; and potassium bromide and sodium chloride, both thicknesses 5.0 mm. [*From D. E. McCarthy, Appl. Opt.* **6**, 1896 (1967).]

Crystalline cesium bromide is available in 7.5-in.-diameter ingots.   Like other alkali halide materials, it is hygroscopic and must be used in a dry atmosphere.   It is also soluble in alcohol and in many acids.

Cesium Iodide

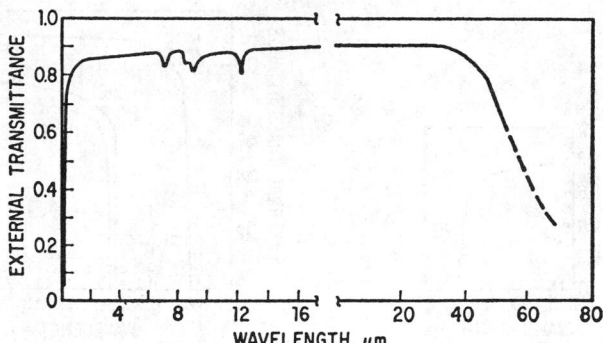

FIG. 6c-17. The external transmittance of cesium iodide; short-wavelength portion, thickness 3 mm. [*From E. K. Plyler and F. R. Phelps, J. Opt. Soc. Am.* **42,** 432 (1952)]; and long-wavelength portion, thickness 5 mm. [*From E. K. Plyler and N. Acquista, J. Opt. Soc. Am.* **43,** 978 (1953), and **48,** 668 (1958).] (See Fig. 6c-16 for the short-wavelength transmittance.)

Crystalline cesium iodide is available in ingots of diameter up to 5.5 in. It is hygroscopic. It is mechanically stable and evidences negligible flow or change of shape with time [D. E. McCarthy, *Appl. Opt.* **2,** 591 (1963)].

Crystal Quartz

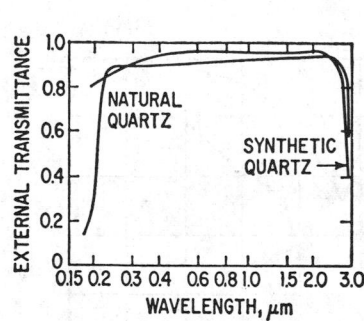

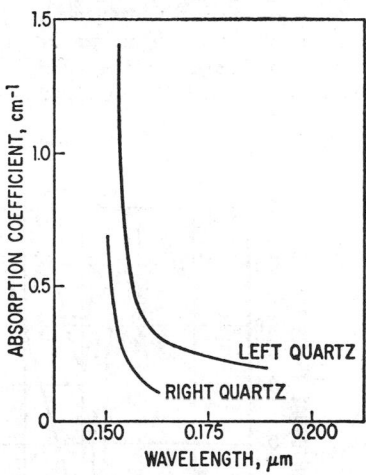

FIG. 6c-18. The external transmittances of synthetic crystal quartz, thickness 6.35 mm, and natural crystal quartz, thickness 1.0 mm. [*From D. E. McCarthy, Appl. Opt.* **6,** 1896 (1967).]

FIG. 6c-19. The ultraviolet absorption edges of natural crystal quartz. Right-handed quartz seems to be more transparent below 0.2 μm than left-handed. This puzzling effect, however, might be accidental. [*From A. Smakula: Optica Acta* **9,** 205 (1962).]

Crystal quartz is of little interest in modern infrared techniques. Historically, the material was of interest until synthetic fused silica was available. Quartz crystals crack easily when heated, in contrast to fused silica. See Table 6c-1, Fused Silica.

Cuprous Chloride

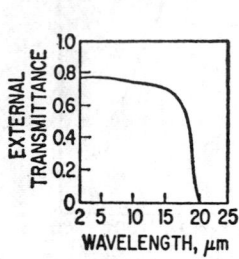

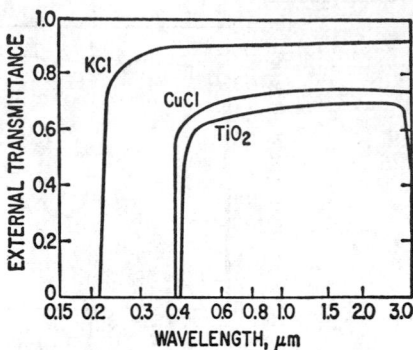

Fig. 6c-20. The infrared external transmittance of cuprous chloride, thickness 9.1 mm. [Adapted from D. E. McCarthy, Appl. Opt. 4, 316 (1965).]

Fig. 6c-21. The short-wavelength external transmittance of cuprous chloride, thickness 9.1 mm; potassium chloride, thickness 10.0 mm; and rutile, thickness 5.0 mm. [From D. E. McCarthy, Appl. Opt. 6, 1896 (1967).]

Cuprous chloride is hygroscopic and is not a common material. However, cubic-structure crystals with a volume of a few cubic centimeters have been grown at the RCA Semiconductor and Materials Division. Cuprous chloride is transparent from 0.4 to 20 μm. The dihydrogen phosphates, by contrast, are transparent only in the range 0.2 to 1.5 μm. Thus, unlike cuprous chloride, they cannot be used to electro-optically modulate far-infrared radiation. [From F. Sterzer, D. Blattner, and S. Miniter, J. Opt. Soc. Am. 54, 62 (1964).]

Diamond

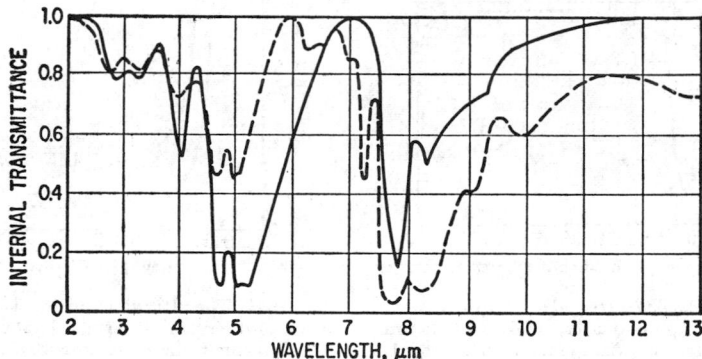

Fig. 6c-22. The infrared internal transmittances of two types of diamond.

The ultraviolet absorption edge of diamond varies from 0.23 to 0.30 μm, depending on the type. This subject has been studied by W. G. Simeral in a dissertation written at the University of Michigan at Ann Arbor (1953), and by R. L. Hauser in a thesis written at Ohio State University at Columbus (1952), from which the data are taken.

Fused Silica

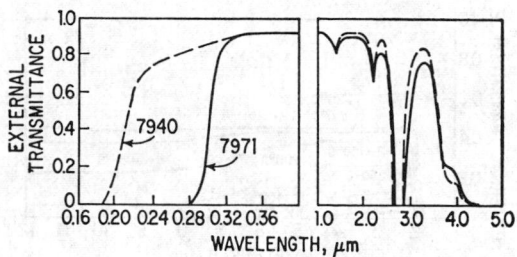

FIG. 6c-23. The external transmittances of fused silica Corning No. 7940 and U. L. E. modified fused silica Corning No. 7971, both thicknesses 10 mm. [*From C. L. Rathmann, G. H. Mann, and M. E. Nordberg, Appl. Opt.* **7**, 819 (1968).]

Fused silica cuts and grinds very well and is otherwise suitable for many applications. Blanks up to 156 in. (396 cm) in diameter have been successfully formed. Fused silica has zero cold-water solubility but can be dissolved in hydrofluoric acid.

TABLE 6c-1. TERMS USED TO DESCRIBE FIVE TYPES OF
SILICON DIOXIDE MATERIALS

Natural quartz	*Transparent vitreous silica*	*Translucent fused silica*
Quartz	Quartz	Translucent fused quartz
Crystalline quartz	Quartz glass	(Quarzgut, in German)
Quartz crystal	(Quarzglas, in German)	Translucent fused silica
Natural quartz	Fused quartz	Fused silica
Rock crystal	Vitreous quartz	Translucent vitreous silica
	Fused quartz glass	Vitreous silica
	Optical quartz glass	
	Optical fused quartz	
Cultured quartz	Clear fused quartz	
	Transparent fused quartz	*High-silica glass*
Cultured quartz	Fused silica	
Synthetic quartz	Synthetic fused silica	Silica glass
Synthetic quartz crystal	Transparent fused silica	Vitreous silica
	Clear fused silica	High-silica glass
	Silica glass	
	Transparent vitreous silica	
	Clear vitreous silica	
	Vitreous silica	

From J. S. Laufer, *J. Opt. Soc. Am.* **55**, 458 (1965).

Laufer also gives a lucid discussion concerning properties of these materials, their structures, and how they are formed.

There is a large variety of high-homogeneity fused silica now available from such companies as Corning Glass Works, Dynasil Corporation of America, and General Electric Company. Amersil, Inc., has a special line of fused silica products under trade names such as Suprasil, Ultrasil, Infrasil, Homosil, and Optosil. Transmission properties of several of these are shown in Fig. 6c-83 in the special glasses section.

Germanium

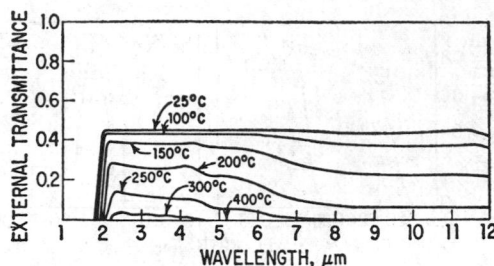

FIG. 6c-24. The external transmittance of single-crystal, p-type, 30-ohm-cm-resistivity germanium, thickness 2.80 mm. [*From D. T. Gillespie, A. L. Olsen, and L. W. Nichols, Appl. Opt.* **4**, 1488 (1965).]

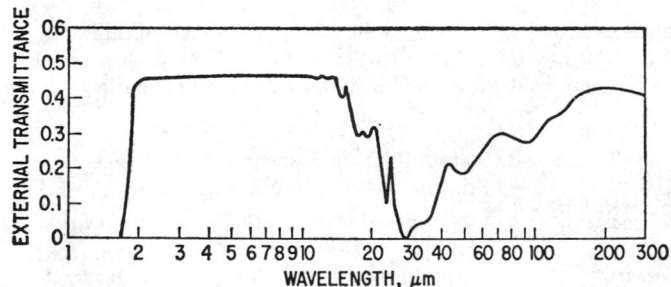

FIG. 6c-25. The external transmittance of polycrystalline, n-type germanium, thickness 2 mm. [*From Exotic Materials, Inc., Infrared Phys.* **5**, ii (1965).]

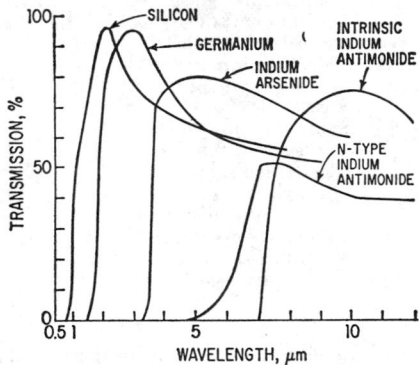

FIG. 6c-26. External transmission curves of materials antireflected with zinc sulfide, showing some characteristics of semiconductor filters. [*From J. S. Seeley and S. D. Smith, Appl. Opt.* **5**, 81 (1966).] (Germanium can also be antireflection-coated with silicon monoxide.)

Germanium can be used as an optical material both as a single crystal and in polycrystalline form. It is hard and brittle at room temperature and tends to fracture during fabrication. Its chemical inertness makes it useful for optical applications, although its electrical properties are affected by moisture.

Optical-quality polycrystalline germanium is available in n-type only and can be produced in difficult shapes, including domes. 24-in.-diameter polycrystalline pieces are available.

Germanium has zero cold-water solubility but can be dissolved in aqua regia and hot sulfuric acid. Germanium becomes opaque at high temperatures (Fig. 6c-24).

A summary of the short-wavelength absorption in germanium is given by T. S. Moss, "Optical Properties of Semi-conductors," pp. 133–151, Academic Press, Inc., New York, 1959.

Irtrans 1 to 6

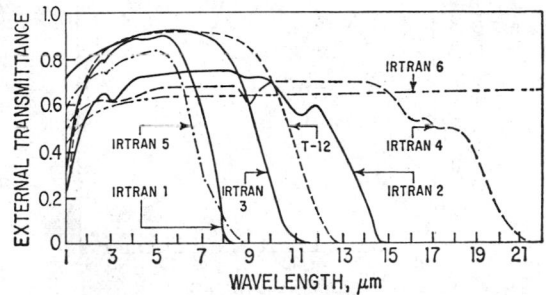

FIG. 6c-27. The external transmittances of Irtran 1 and single-crystal magnesium fluoride, both thicknesses 0.110 in. [*From A. L. Olsen and W. R. McBride, J. Opt. Soc. Am.* **53**, 1003 (1963).]

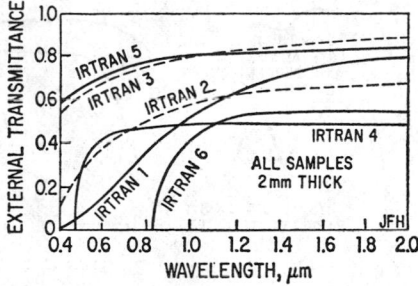

FIG. 6c-28. The infrared external transmittances of Irtrans 1 to 5 and of T-12, thicknesses 6.2 mm. [*Adapted from S. S. Ballard, Japan. J. Appl. Phys.* **4**, suppl. 1, 23 (1965).] Also Irtran 6, thickness 2 mm. (*Adapted from Kodak Pamphlet* U-71, 1968.) (Irtran 6 transmits to beyond 30 μm.)

FIG. 6c-29. The short-wavelength external transmittances of Irtrans 1 to 6. (*From Kodak Pamphlet* U-71, 1968.)

The Kodak Irtrans (*Infrared-trans*mitting) are hot-pressed, compacted materials. The chemical compositions of Irtrans 1, 2, 3, 4, 5, and 6 are respectively polycrystalline magnesium fluoride, zinc sulfide, calcium fluoride, zinc selenide, magnesium oxide, and cadmium telluride. The pure powder is heated and compacted in a high-pressure apparatus to produce a blank of roughly the desired size and shape. The blanks are subsequently ground and polished to give windows, meniscus lenses, or hemispherical domes.

The physical and chemical properties of the Irtrans are, in general, favorable for the user. Because they are polycrystalline, there cannot be cleavage along any crystal planes; they do not exhibit cold flow. Their thermal shock resistance is high. Their melting points are all well above 1000 K, although their maximum useful temperatures are somewhat lower. Irtrans 1, 2, 4, 5, and 6 are insoluble in water; and Irtran 3 is practically insoluble. Irtrans 4 and 6 are slightly soluble in acids. All the Irtrans are available as very thin windows. Disks of Irtrans 3 through 6 can be made 6 in. in diameter, while samples of Irtrans 1 and 2 can be produced as large as 8 in.

Lanthanum Fluoride

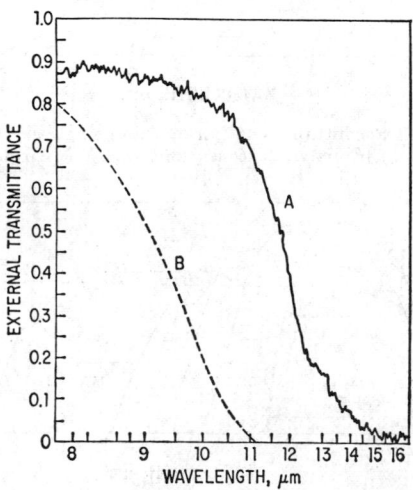

Fig. 6c-30. The infrared external transmittance of lanthanum fluoride, thicknesses 0.43 mm (curve *A*) and 11.7 mm (curve *B*). [*Adapted from J. B. Mooney, Infrared Phys.* **6**, 153 (1966).]

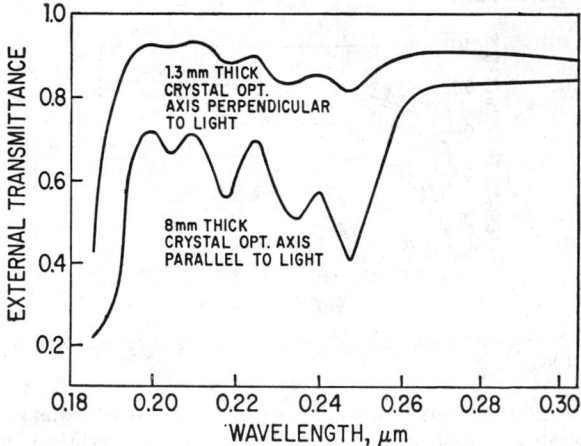

Fig. 6c-31. The ultraviolet external transmittance of lanthanum fluoride for two thicknesses. [*From M. P. Wirick, Appl. Opt.* **5**, 1966 (1966).]

In additional to its broad transmission range (0.13 to 13 μm for 0.5-mm samples), lanthanum fluoride exhibits a high degree of thermal and chemical stability. In a 100°C oven for 100 hr, it is insoluble in water and sodium hydroxide, and has negligible (only a fraction of a percent) solubility in sulfuric, hydrochloric, and nitric acids.

Lead Bromide

Lead Chloride

Lead Fluoride

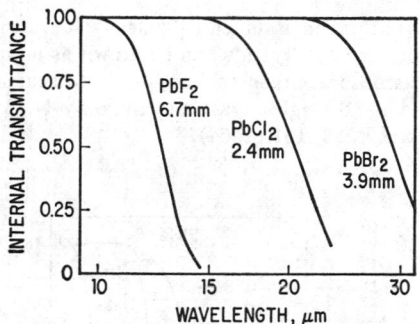

FIG. 6c-32. The internal transmittances of three lead halides, thicknesses shown. [*From T. S. Moss and A. G. Peacock, Infrared Phys.* **1**, 104 (1961).]   The crystals were grown from the melt in a Stockbarger-type vacuum furnace.

Lead bromide and lead chloride are 40 times less soluble in water than sodium chloride or potassium bromide, and lead fluoride is 500 times less soluble.

Lead Selenide

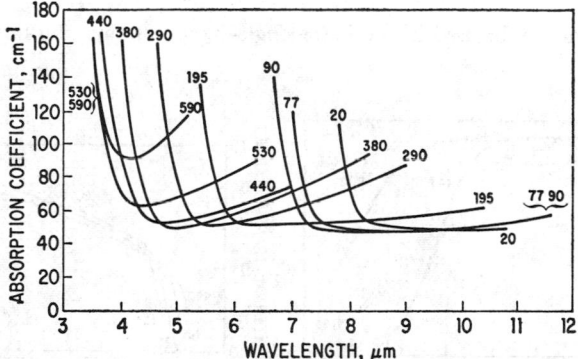

FIG. 6c-33. The absorption coefficient of lead selenide for several temperatures (in kelvins), thickness 0.68 mm, purity unknown. [*Adapted from A. F. Gibson, Proc. Phys. Soc.* **65B**, 378 (1952).]

Lead selenide can be used in single-crystal or thin-film form and can be grown by the Bridgman-Stockbarger technique in the absence of oxygen.

Lead Sulfide

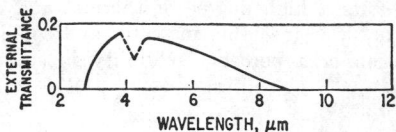

FIG. 6c-34. The external transmittance of natural crystalline lead sulfide, thickness unknown; no detectable impurity by X-ray analysis. (*From S. S. Ballard, K. A. McCarthy, and W. L. Wolfe, Optical Materials for Infrared Instrumentation, Univ. Michigan Rept. 2389-11S, 101, 1959.*)

Lead sulfide occurs as natural crystals (called galena); it is also available as synthetic crystals, and in thin-film form. Crystals can be grown as large as a few centimeters in size by the Bridgman-Stockbarger technique. A discussion of the differences between lead sulfide films and single crystals is given by N. A. Vlasenko and M. P. Lisitsa, *Opt. Spectr. U.S.S.R.* **16**, 161 (1964).

Lead Telluride

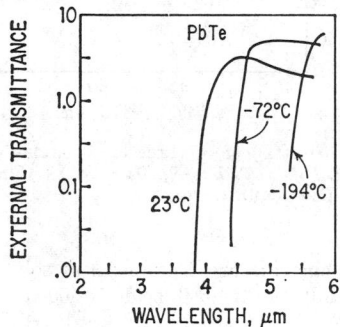

FIG. 6c-35. The external transmittance of lead telluride for three temperatures, thickness 0.11 mm. [*From M. A. Clark and R. J. Cashman, Phys. Rev.* **85**, 1043 (1952).]

Lead telluride can be used in synthetic-single-crystal or thin-film form.

Lithium Fluoride

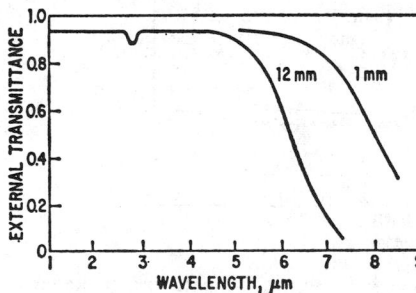 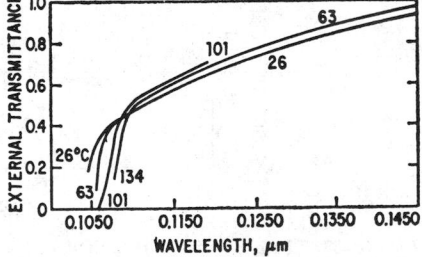

FIG. 6c-36. The infrared external transmittance of lithium fluoride for two thicknesses. (*Adapted from R. A. Smith, F. E. Jones, and R. P. Chasmar, "The Detection and Measurement of Infrared Radiation," Oxford University Press, London and New York, 1957.*)

FIG. 6c-37. The short-wavelength external transmittance of lithium fluoride for four temperatures, thickness 1.55 mm. [*From A. H. Laufer, J. A. Pirog, and J. R. McNesby, J. Opt. Soc. Am.* **55**, 64 (1965).]

When lithium fluoride is grown in a vacuum, the absorption at $2.8\mu m$ attributed to the H-F band disappears.   Cylindrical castings of diameter 6 in. are available.   It is only slightly soluble in water but can be dissolved in acids.   Lithium fluoride and arsenic trisulfide glass should combine to give a satisfactory achromatic lens for the 2- to 5-μm region.

Magnesium Fluoride

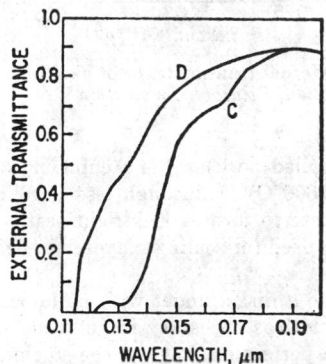

FIG. 6c-38. The ultraviolet external transmittance of single-crystal magnesium fluoride, thicknesses 14.82 mm (curve C) and 1.40 (curve D).   [*Adapted from D. L. Steinmetz, W. G. Phillips, M. Wirick, and F. F. Forbes, Appl. Opt.* **6**, 1001 (1967).]   (See Fig. 6c-27 for the infrared transmittance.)

Although good cleavages have occasionally been obtained, conchoidal fracture is more common.   Ground surfaces have been polished on pitch laps with rouge, but beeswax with putty powder appears to give a higher polish more rapidly.

Magnesium Oxide

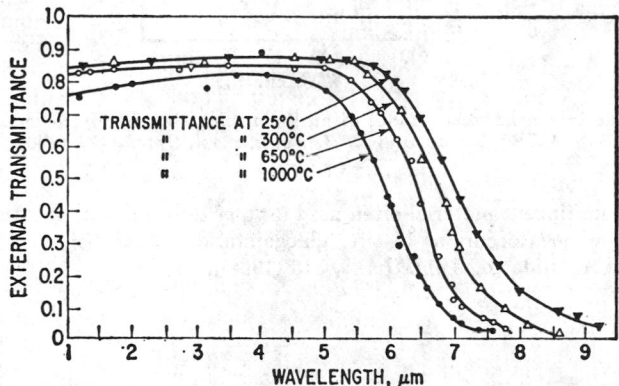

FIG. 6c-39. The infrared external transmittance of magnesium oxide at several temperatures, thickness 5.5 mm.   [*Adapted from U. P. Oppenheim and A. Goldman, J. Opt. Soc. Am.* **54**, 127 (1964).]

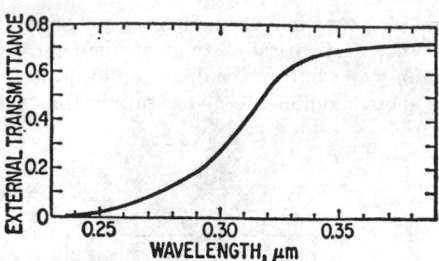

Fig. 6c-40. The ultraviolet external transmittance of magnesium oxide, thickness 0.49 mm.
[*From measurements by A. Sutton, Polaroid Corporation.*]

Magnesium oxide, also called periclase, is a cubic crystal of fairly high hardness and high melting point (2800°C). Although less hard than sapphire, it may be pressed against metal gaskets to form a leak-tight seal. This method, though not always successful, has been used for sealing magnesium oxide windows to absorption cells at high temperatures.

The crystal can be cut on a disk grinder with no lubricant. Hard work with an aluminum oxide finishing cloth is necessary to get a smooth finish. It can also be used without polishing if a perfect cleavage of the single crystal has been obtained. Some specimens show a little O-H absorption, probably due to water. The polished surfaces of optical components of magnesium oxide can be protected from attack by atmospheric moisture with evaporated coatings of silicon monoxide. Magnesium oxide has a slippage plane that may affect the mechanical strength of certain optical components.

Mica

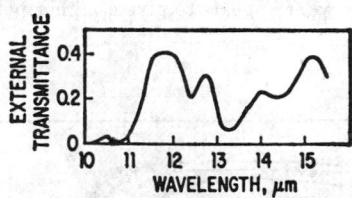

Fig. 6c-41. The external transmittance of an 8-μm-thick cleaved film of muscovite mica.
[*From S. Ruthberg, M. W. Barnes, and R. H. Noyce, Appl. Opt.* **2**, 177 (1963).]

Mica is a birefringent material often used for quarter-wave plates. In addition to muscovite, mica also occurs as biotite, phlogopite, and fluorphogopite [from J. M. Serratosa and A. Hildalgo, *Appl. Opt.* **3**, 315 (1964)].

Potassium Bromide

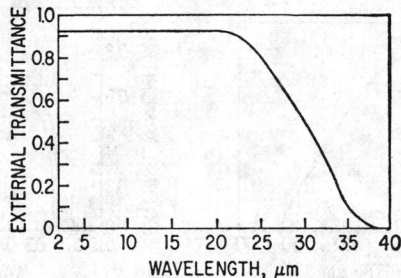

FIG. 6c-42. The infrared external transmittance of potassium bromide, thickness 5 mm. [*Adapted from D. E. McCarthy, Appl. Opt.* **2**, 591 (1963).]   (See Fig. 6c-16 for the short-wavelength transmittance.)

Potassium bromide is grown in the same manner as sodium chloride and is hygroscopic.   It is also soluble in alcohol and glycerin.   It is available in 12-in.-diameter ingots.   Very pure samples have been obtained, but they cleave.

Potassium Chloride

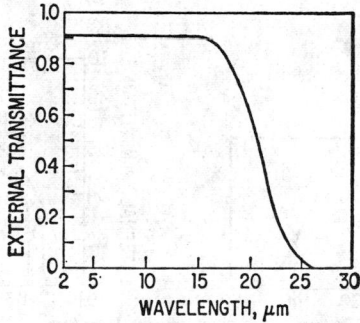

FIG. 6c-43. The infrared external transmittance of potassium chloride, thickness 10 mm. [*Adapted from D. E. McCarthy, Appl. Opt.* **4**, 317 (1965).]   (See Fig. 6c-21 for the short-wavelength transmittance.)

Potassium chloride is grown in the same way as sodium chloride, but sometimes multiple crystals instead of single-crystal ingots result; therefore, the large-size prisms are somewhat rare and more expensive.   Crystals of 12-in. diameter are available.   Its water solubility is only about half that of potassium bromide; it is soluble in alkalies, ether, and glycerin.

Potassium Iodide

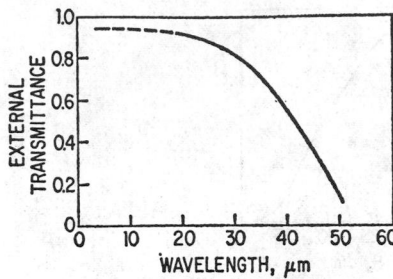

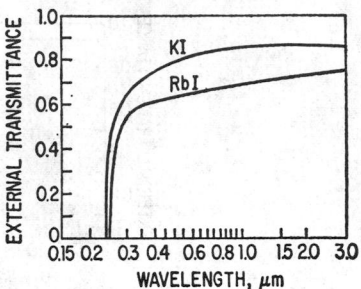

Fig. 6c-44. The infrared external transmittance of potassium iodide, thickness 0.83 mm. [*From J. Strong, Phys. Rev.* **38**, 1818 (1931).]

Fig. 6c-45. The short-wavelength transmittances of potassium iodide, thickness 4.3 mm; and rubidium iodide, thickness 1.36 mm. [*Adapted from D. E. McCarthy, Appl. Opt.* **7**, 1243 (1968).]

Potassium iodide is valuable as a prism material, but it is too hygroscopic (being about twice as soluble in water as potassium bromide) and too soft for field use. It is also soluble in alcohol and in ammonia. Ingots of 7.5-in. diameter are available.

Rubidium Bromide

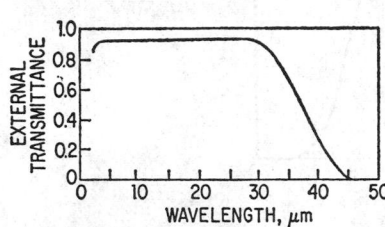

Fig. 6c-46. The infrared external transmittance of rubidium bromide, thickness 5.3 mm. [*Adapted from D. E. McCarthy, Appl. Opt.* **7**, 1997 (1968).]

Fig. 6c-47. The short-wavelength external transmittances of rubidium bromide, thickness 5.31 mm; rubidium chloride, thickness 2.1 mm; and gallium arsenide, thickness 0.25 mm. [*From D. E. McCarthy, Appl. Opt.* **7**, 1243 (1968).]

Rubidium bromide, chloride, and iodide are hygroscopic; and care must be used in handling them to preserve the surface polish. The gradual decrease in transmittance at shorter wavelengths is due to surface scattering that is caused by the roughness of the surface and not by absorption or scattering within the material itself [D. E. McCarthy, *Appl. Opt.* **7**, 1243 (1968)].

Rubidium Chloride

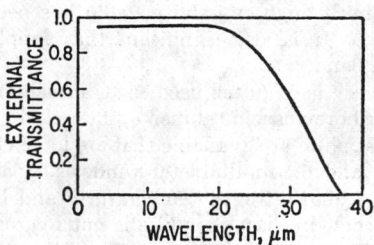

FIG. 6c-48. The infrared external transmittance of rubidium chloride, thickness 3.0 mm. [*Adapted from D. E. McCarthy, Appl. Opt.* **7**, 1997 (1968).]   (See Fig. 6c-47 for the short-wavelength transmittance and the discussion following it.)

Rubidium Iodide

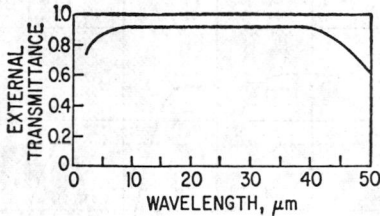

FIG. 6c-49. The infrared external transmittance of rubidium iodide, thickness 3.91 mm. [*Adapted from D. E. McCarthy, Appl. Opt.* **7**, 1997 (1968).]   (See Fig. 6c-45 for the short-wavelength transmittance and the discussion following Fig. 6c-47.)

Ruby

See Fig. 6c-10 for the short-wavelength transmittance.   Ruby is essentially sapphire (aluminum oxide) with a 0.05 percent by weight chromium impurity.   The absorption of ruby at room temperature is discussed by D. C. Cronemeyer, *J. Opt. Soc. Am.* **56**, 1703 (1966), and at high temperatures by S. V. Grum-Grzhimailo and G. V. Klimusheva, *Opt. Spectr. U.S.S.R.* **8**, 179 (1960).   Ruby is mechanically rugged and is not hygroscopic.   A 6-mm sample transmits out to 5 μm.

Sapphire

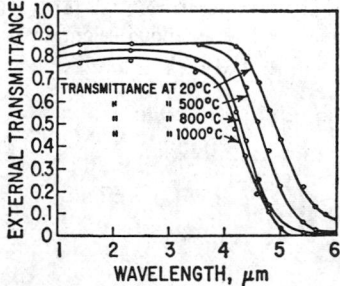

FIG. 6c-50. The infrared external transmittance of sapphire for several temperatures, thickness 8 mm.   [*Adapted from U. P. Oppenheim and U. Even, J. Opt. Soc. Am.* **52**, 1078 (1962).] (See Fig. 6c-10 for the short-wavelength transmittance.)

A transmission loss of not more than 3 percent is encountered when the material is heated to 440°C. A fairly high emissivity (0.05) has been reported and may be significant in reradiation. It is also significant that sapphire does not show the 2.8-μm water absorption band.

Since this material is very hard (often used as an abrasive), it must be ground and polished with diamond or boron carbide abrasive; the techniques are therefore difficult and costly. Blanks of sapphire are available that are large enough for 3-in.-diameter domes and hemispheres and 5.5-in.-diameter windows. Sapphire has a very high thermal conductivity at liquid-nitrogen temperatures and below, and so it can be used as a substrate for cooled cells. Sapphire is not hygroscopic, but it is slightly soluble in acids and alkalies.

Selenium

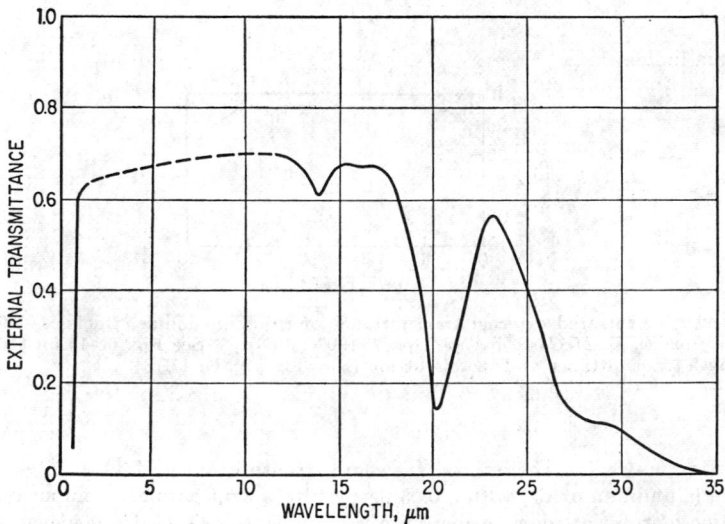

Fig. 6c-51. The external transmittance of amorphous selenium, thickness 169 mm. [*From R. S. Caldwell, Special Report on Contract DA* 36-039-SC-71131, *Purdue University, January,* 1958.]

Selenium can exist in various forms. The trigonal (or crystalline) form is most stable. The amorphous (or vitreous) form is fairly stable below 50°C but converts to the trigonal form at higher temperatures. A fairly thorough discussion of the optical properties of trigonal and amorphous selenium is given by R. S. Caldwell and H. Y. Fan, *Phys. Rev.* **114,** 664 (1959).

Silicon

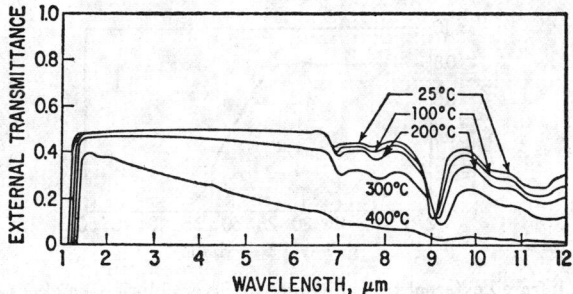

FIG. 6c-52. The external transmittance of single-crystal, n-type, 5-ohm-cm-resistivity silicon, thickness 2.80 mm. [*From D. T. Gillespie, A. L. Olsen, and L. W. Nichols, Appl. Opt.* **4**, 1488 (1965).]

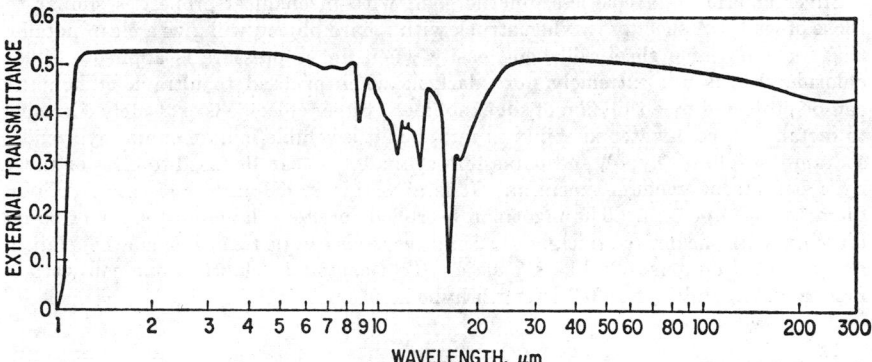

FIG. 6c-53. The external transmittance of polycrystalline n- or p-type silicon, thickness 2 mm. [*From Exotic Materials, Inc., Infrared Phys.* **5**, ii (1965).]

The physical and chemical properties of silicon are very similar to those of germanium. Optical-grade silicon (impurity content less than $10^8/cm^3$) has high resistance to thermal and mechanical shocks. One sample which was heated to 400°C and quenched in ice water remained unaffected. It can be used as an irdome if aerodynamic heating does not take the skin temperature above 300°C. Optical-quality polycrystalline silicon is available in n and p types and can be produced in 16-in.-diameter sizes.

Silicon has zero cold-water solubility but can be dissolved in a mixture of hydrofluoric acid and nitric acid. Silicon can be ground on Blanchards; diamond curve generators can be used; and normal pitches and polishing compounds are acceptable, although operations take 1.5 times as long as those with quartz. Silicon monoxide can be used as a low-reflectance coating at relatively short wavelengths. Silicon dioxide usually overcoats, and the combination is stable. At longer wavelengths zinc sulfide can be used as a coating (see Fig. 6c-26). Both should be vacuum-evaporated.

An absorption band usually occurs at 9 μm, but this is due to a Si-O stretching (from the coating or "poisoning") superimposed on a weak lattice vibration. A summary of the short-wavelength absorption in silicon is given by T. S. Moss, "Optical Properties of Semi-conductors," pp. 116–126, Academic Press, Inc., New York, 1959.

Silver Chloride

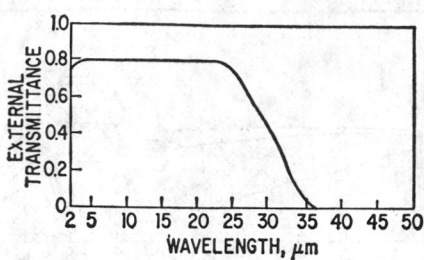

FIG. 6c-54. The infrared external transmittance of silver chloride, thickness 0.5 mm. (Silver chloride has a sharp short-wavelength absorption edge at about 0.42 μm.) [*Adapted from D. E. McCarthy, Appl. Opt.* **2**, 591 (1963).]

Silver chloride is a colorless, ductile solid with mechanical properties similar to those of lead. A sheet of it, when struck with a hard object, will give a clear metallic ring, but the same sheet will bend easily when finger pressure is applied. Silver chloride that is not extremely pure darkens when exposed to ultraviolet light; it can be protected by a thin film of silver sulfide. Silver chloride is extremely corrosive to metal. Its cold-water solubility is zero, but it is soluble in ammonium hydroxide, sodium thiosulfate (hypo), and potassium cyanide. It can be fused to glass or silver by a permanent vacuum-type seal. It is available in cylindrical pieces of 3.75-in. diameter by 5 or 6 in. The ingot can be rolled to give a large-area sheet of 20-in. diameter with the desired thickness. Single crystals can be turned on a lathe, planed, and operated on generally like a plastic. To clean silver chloride, one can wash it first in water, and then in 0.2 strength hypo solution.

Sodium Chloride

FIG. 6c-55. The external transmittance of sodium chloride for two thicknesses. (*Adapted from R. A. Smith, F. E. Jones, and R. P. Chasmar,* "The Detection and Measurement of Infrared Radiation," *Oxford University Press, London and New York,* 1957.) (See Fig. 6c-16 for the short-wavelength transmittance.)

Sodium chloride, the natural form of which is ordinary rock salt, polishes easily, and, although hygroscopic, can be protected by evaporated coatings and plastics; selenium films have been used successfully. It is soluble in glycerin. Synthetic single-crystal sodium chloride is available in 12-in.-diameter sizes.

Sodium Fluoride

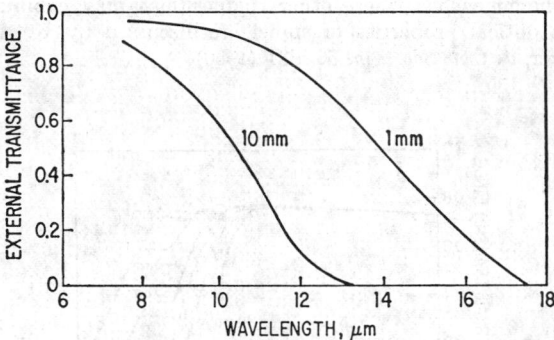

FIG. 6c-56. The infrared external transmittance of sodium fluoride for two thicknesses. (*Data of S. S. Ballard, private communication.*)

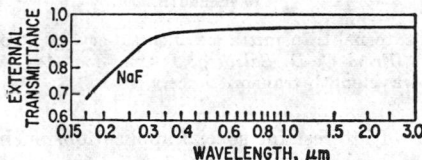

FIG. 6c-57. The short-wavelength external transmittance of sodium fluoride, thickness 2.16 mm. [*Adapted from D. E. McCarthy, Appl. Opt.* **6**, 1896 (1967).]

Sodium fluoride transmits in about the same region as calcium fluoride and lithium fluoride, but it is less satisfactory mechanically.   It has some significance, however, in cases where a remarkably low refractive index is needed.   It can also be easily evaporated as a thin film and can be used for reflection-reducing coatings.

Spinel

FIG. 6c-58. The external transmittance of spinel, thickness 5.4 mm. [*From Linde Air Products Co., Technical Data Sheets.*]

Spinel can be grown by the flame fusion (Verneuil) process.   It is somewhat softer than sapphire and is thus more easily worked for optical purposes.   For applications where optical isotropy is desired, the advantages of cubic spinel over trigonal sapphire are obvious.   The drawback of spinel is that it does not transmit as far into either the infrared or the ultraviolet.   The absorption band at 2.8 μm may be due to entrapped water.   The mixture ratio 1:3.5, magnesium oxide to aluminum oxide, is not unique; other ratios which exist seem to give essentially the same physical properties.　The 1:3.5 mixture has been an easy one from which to grow gem stones and is usually homogeneous where other compositions may exhibit concentration gradients.　The optical properties of spinel are discussed by K. A. Wickersheim and R. A. Lefever, *J. Opt. Soc. Am.* **50**, 831 (1960).

Strontium Titanate

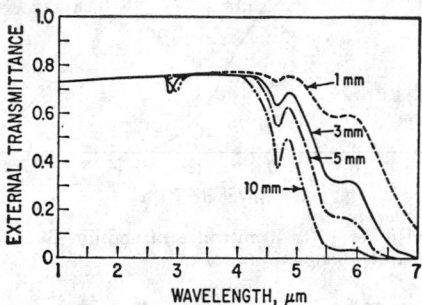

FIG. 6c-59. The infrared external transmittance of single-crystal strontium titanate at 26°C for several thicknesses.　[*From C. D. Salzberg, J. Opt. Soc. Am.* **51**, 1149 (1961).]　(See Fig. 6c-13 for the short-wavelength transmittance.)

Strontium titanate is of interest for special applications such as immersion lenses. The single crystals can be ground with 220 carborundum on a lead lap, then by finer compounds like Linde "A" on a 50-50 tin-lead lap.　Salzberg reports that strontium titanate does not suffer any appreciable loss of transmittance when cooled below 187°C.

Sulfur

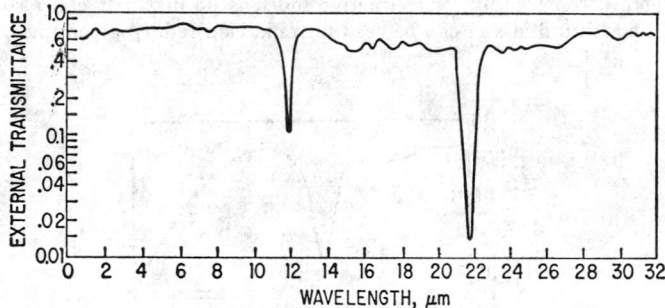

FIG. 6c-60. The external transmittance of rhombic sulfur, thickness 0.4 mm.　[*From C. MacNeill, J. Opt. Soc. Am.* **53**, 398 (1963).]

Unlike sodium chloride and potassium bromide, rhombic sulfur is not hygroscopic. It is easy to prepare in virtually any rectangular size needed, and is inexpensive.　It can be useful for windows in infrared equipment and in surveillance systems.

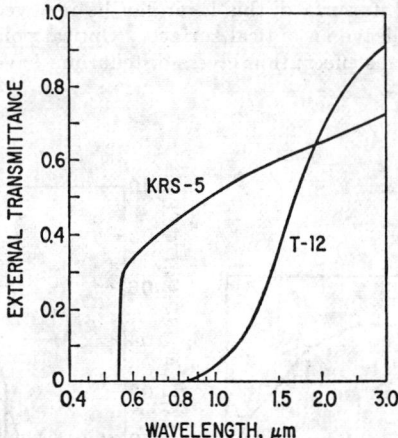

FIG. 6c-61. The short-wavelength external transmittances of T-12, thickness 4.0 mm; and KRS-5, thickness 2.0 mm. [*From D. E. McCarthy, Appl. Opt.* **6**, 1896 (1967).]   (See Fig. 6c-28 for the infrared transmittance of T-12.)

T-12 is a development of the Harshaw Chemical Company of Cleveland, Ohio. It is an optically·integral two-phase polycrystalline body consisting of barium fluoride and calcium fluoride in nearly equal molar proportions.   Translucent, almost marble-like in appearance, it has high resistance to thermal shock.   The maximum size of cylindrical blanks that have been produced to date is about 7.5-in. diameter and several inches thick; larger sizes are feasible.   Its useful temperature limit in air is around 600°C.

Tellurium

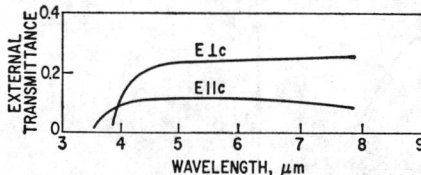

FIG. 6c-62. The external transmittance of tellurium for two polarizations, thickness 0.85 mm.   [*From J. J. Loferski, Phys. Rev.* **93**, 707 (1954).]   R. S. Caldwell and H. V. Fan, *Phys. Rev.* **114**, 664 (1954), report a strong absorption band at 11 μm for the E||c radiation only.

FIG. 6c-63. The external transmittance of polycrystalline tellurium, thickness 3.5 mm. The cut-on wavelength (1% transmittance point) for this sample was 3.85 μm. [*Adapted from D. E. McCarthy, Appl. Opt.* **7**, 1997 (1968).]

Tellurium is an interesting anisotropic crystal.   Loferski gives more data on the polarization effects, and Caldwell and Fan have also given extensive data.   Optical activity has been observed by K. C. Nomura, *Phys. Rev. Letters* **5**, 500 (1960).   Single crystals of tellurium are hard to grow; a polycrystalline ingot (which is brittle and difficult to polish) usually forms from the melt.   2- by 2- by 10-mm crystals can be grown rather well from the vapor phase.   A hexagonal crystal symmetry results and cleavage occurs parallel to the c axis.   It is difficult to cleave or cut in other directions because the crystal will fracture. When tellurium is ground, a conducting layer tends to form on its surface; this layer can be removed by chemical etching, although this does not leave an optical surface. Optical polishing can probably be accomplished without this effect; thin layers of tellurium have been made that have a good optical finish.

Thallium Bromide

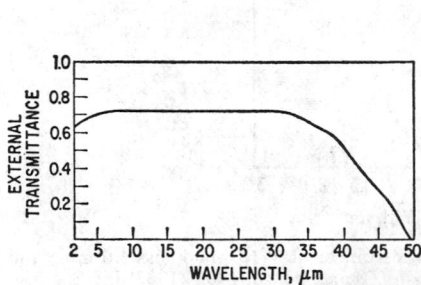

FIG. 6c-64. The infrared external transmittance of thallium bromide, thickness 1.65 mm. [*Adapted from D. E. McCarthy, Appl. Opt.* **4**, 317 (1965).]

FIG. 6c-65. The short-wavelength external transmittances of thallium bromide, thallium bromide-chloride (KRS-6), and thallium chloride, all thicknesses 1.65 mm. [*From D. E. McCarthy, Appl. Opt.* **6**, 1896 (1967).]

Thallium bromide can be ground a very small amount at a time without cracking or chipping. It bends like lead and is only slightly soluble in water.

Thallium Chloride

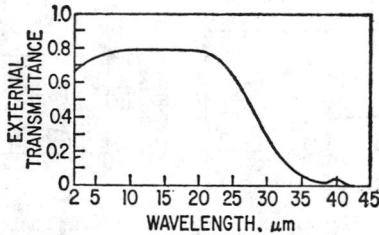

FIG. 6c-66. The infrared external transmittance of thallium chloride, thickness 1.65 mm. [*Adapted from D. E. McCarthy, Appl. Opt.* **4**, 317 (1965).] (See Fig. 6c-65 for the short-wavelength transmittance.)

Thallium chloride can be cut easily on a diamond saw (melted beeswax is used as a "lubricant" while it is being sawed). It is difficult to grind to dimension since the material flows to the edge and fills the grinding wheel. Thin strips bend like lead. It is soluble in water.

Thallium Bromide-Chloride (KRS-6)

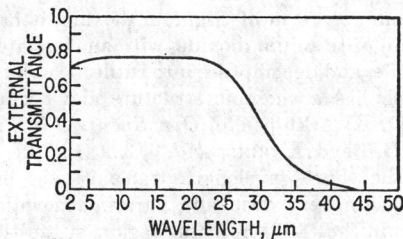

FIG. 6c-67. The infrared external transmittance of thallium bromide-chloride (KRS-6), thickness 1.65 mm. [*Adapted from D. E. McCarthy, Appl. Opt.* **4**, 317 (1965).] (See Fig. 6c-65 for the short-wavelength transmittance.)

KRS-6 has not been given very much attention because of the greater usefulness of KRS-5, which covers about the same transmission range. KRS-6 is almost unavailable.

Thallium Bromide-Iodide (KRS-5)

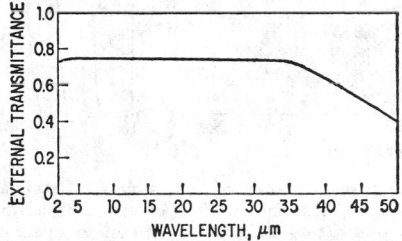

FIG. 6c-68. The infrared external transmittance of thallium bromide-iodide (KRS-5), thickness 2 mm. [*Adapted from D. E. McCarthy, Appl. Opt.* **2**, 591 (1963).] (See Fig. 6c-61 for the short-wavelength transmittance.)

KRS-5 has a waxy quality somewhat similar to that of silver chloride. It has often shown polarization properties due to strain birefringence. The modern compositions, however, show little, if any, of this effect. The thallium salts are toxic, and so KRS-5 should be handled with care. It has a serious tendency to cold-flow and change its shape with time; this is a result of plastic memory, or the gradual relief of strains. The proper annealing technique is not yet well understood, and even that would not completely eliminate the problem. It can be cut with a diamond saw very slowly. Then aloxite and crocus polishing cloths can be used. The refractive index of KRS-5 is not always uniform and sometimes appears as a gradient. If the composition corresponding to the minimum melting point is chosen, the change in refractive index should be less than 10^{-5}. Because of its higher reflection losses and shorter wavelength range, it is not as good a prism material as cesium iodide. KRS-5 is only slightly soluble in water but can be dissolved in alcohol, nitric acid, and aqua regia. It can be produced in ingots of diameter up to 5 in.

Titanium Dioxide

Rutile is the most common form of titanium dioxide; it has tetragonal structure. Anastase is another form of titanium dioxide, with an elongated tetragonal structure. It is less stable than rutile, and decomposes into rutile at high temperatures. Finally, there is brookite, which has a rhombic structure and occurs only rarely. [From Ya. S. Bobovich and D. K. Arkhipenko, *Opt. Spectr. U.S.S.R.* **17**, 407 (1964); and T. N. Krylova and G. O. Bagdyk'yants, *ibid.* **9**, 339 (1960).]

See Fig. 6c-21 for the short-wavelength transmittance of rutile. It transmits out to 5 μm. Rutile has a very deep and narrow absorption band at 3277 cm⁻¹ (about 3 μm) which lies in the O-H stretching region, suggesting the presence of O-H groups in the crystals. [From B. H. Soffer, *J. Chem. Phys.* **35**, 940 (1961).]

Group III—Group V Compounds

Gallium Antimonide

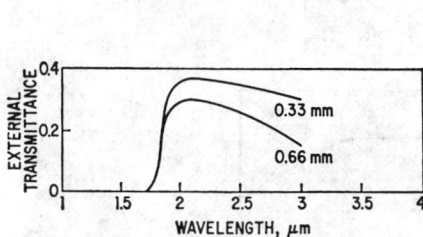

Fig. 6c-69. The external transmittance of gallium antimonide for two thicknesses. (*From D. F. Edwards, University of Michigan, Willow Run Laboratories, private communication, 1959.*)

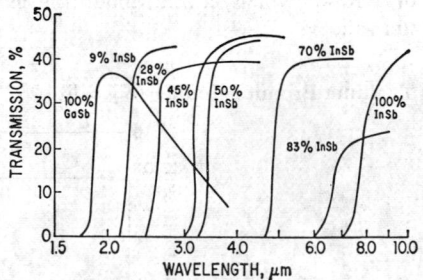

Fig. 6c-70. External transmission curves for various compositions of a gallium-indium-antimonide system. [*From J. S. Wrobel and H. Levinstein, Infrared Phys.* **7**, 201 (1967).]

Gallium Arsenide

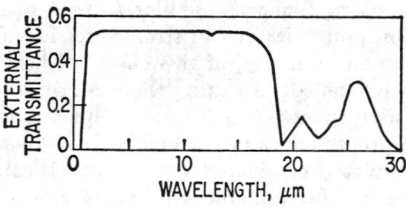

Fig. 6c-71. The external transmittance of gallium arsenide, thickness 0.5 mm. [*Adapted from D. E. McCarthy, Appl. Opt.* **7**, 1997 (1968).] (See Fig. 6c-47 for the short-wavelength transmittance.)

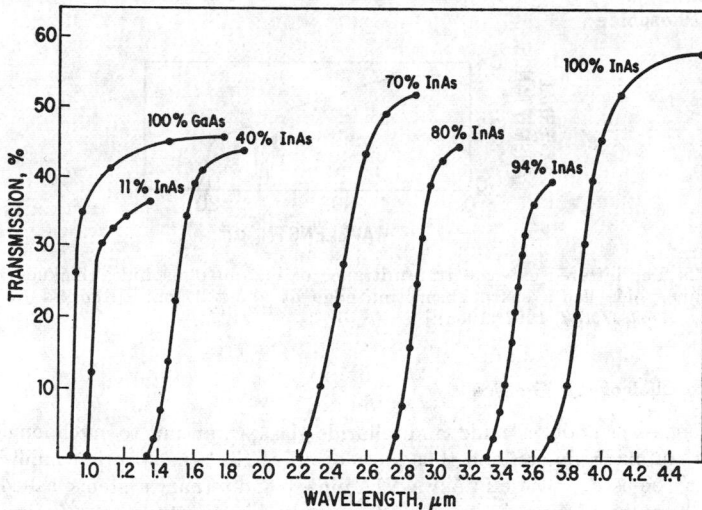

FIG. 6c-72. External transmission curves for various compositions of a gallium-indium-arsenide system. [*From J. S. Wrobel and H. Levinstein, Infrared Phys.* **7**, 201 (1967).]

Gallium Phosphide

See Fig. 6c-13 for the transmittance of gallium phosphide.

Indium Antimonide

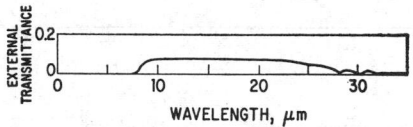

FIG. 6c-73. The external transmittance of indium antimonide, thickness 1.0 mm. [*Adapted from D. E. McCarthy, Appl. Opt.* **7**, 1997 (1968).]   (See Fig. 6c-70.)

Indium antimonide is soft and brittle.   To cut it, one may use a diamond wheel with care.   To polish, lap in a manner very similar to that used for silicon and germanium.   Samples of indium antimonide are available in sizes up to 0.125 by 0.375 in. with thicknesses from 0.03 to 0.01 in.   Samples as large as 1.25 by 0.875 in. have been grown.   Indium antimonide has a large Faraday effect.   It can be used for filters (particularly in atmospheric heat measurements); it "cuts on" at about 7 μm; it can be antireflection-coated with lead chloride or zinc sulfide (see Fig. 6c-26).

Indium Arsenide

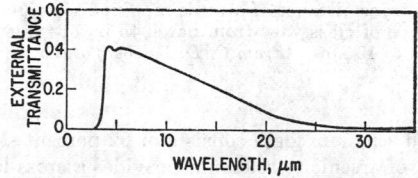

FIG. 6c-74. The external transmittance of indium arsenide, thickness 0.17 mm. [*Adapted from D. E. McCarthy, Appl. Opt.* **7**, 1997 (1968).]   (See Fig. 6c-26 and 6c-72.)

Indium Phosphide

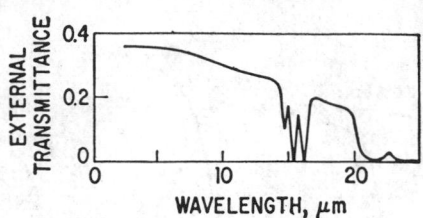

Fig. 6c-75. The infrared external transmittance of indium phosphide, thickness 1.0 mm. Indium phosphide has a sharp absorption edge at about 1 μm. [*Adapted from D. E. McCarthy, Appl. Opt.* **7**, 1997 (1968).]

Nonoxide Chalcogenic Glasses

The nonoxide sulfide, selenide, and telluride glasses transmit to much longer wavelengths than oxide glasses, and thus most or all of the atmospheric window at 8 to 14 μm may be used. These IVA-VA-VIA binary and ternary systems, called chalcogenides, have been extensively studied, notably by Savage et al., *Infrared Phys.* **5**, 195 (1965); Hilton et al., *Appl. Opt.* **5**, 1877 (1966); and Worrall, *Infrared Phys.* **8**, 49 (1968). The glasses are soft, weak, very poorly durable, and severely limited in maximum-use temperature compared to the oxides. The sulfide glasses have limited transparency in the visible, and the selenides and tellurides are essentially opaque.

Arsenic-modified Selenium Glass

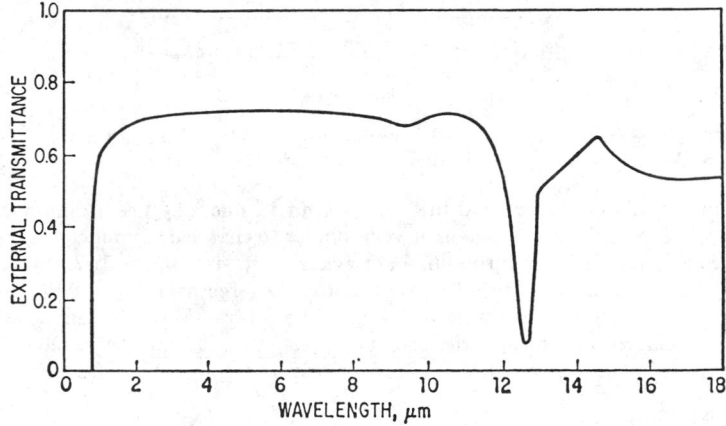

Fig. 6c-76. The external transmittance of arsenic-modified selenium glass, thickness 2 mm. The properties of this type of glass vary from batch to batch; this curve may be regarded as typical. It transmits to 19 μm. [*From C. D. Salzberg and J. J. Villa, J. Opt. Soc. Am.* **47**, 244 (1957).]

This arsenic-modified selenium glass consists of 92 percent selenium and 8 percent arsenic. The addition of arsenic to selenium provides a cross-linked structure which causes the material to have a higher softening point and reduces the tendency of the material to crystallize. At one time the material was formed from 80 percent selenium and 20 percent arsenic triselenide.   This is equivalent to the 92-8 percent selenium-arsenic mixture and explains why the material is often called 80-20 arsenic-selenium glass.

For a glass, this material shows unusual deformation properties; it tends to soften and flow because of the relative motion of the chains of selenium atoms.   An unloaded glass sample is free from viscous flow up to temperatures of 70°C.   Above 70°C, the glass flows quite easily.   Thus, particular care must be taken in machining.   This glass shows a strong absorption at 12.7 μm.   It makes good optical contact with lead sulfide, although the thermal conductivities are different, and they will separate unless "potted" plastically.   An 8-in. 210-deg hyperhemispherical dome $\frac{1}{4}$ in. thick has been made.

Arsenic Triselenide Glass

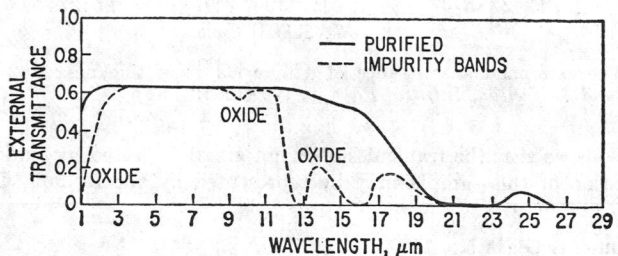

FIG. 6c-77. The external transmittance of arsenic triselenide, thickness 1.75 mm, showing the effect of small amounts (estimated 0.05 atomic percent) of oxide impurities on transmission.   [*Adapted from J. A. Savage and S. Nielson, Infrared Phys.* **5**, 195 (1965).]

Arsenic triselenide has a melting temperature of about 400°C.

Arsenic Trisulfide Glass

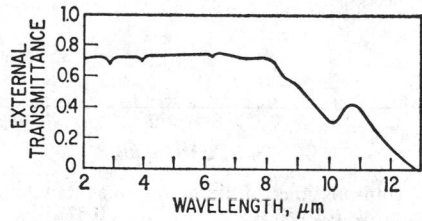

FIG. 6c-78. The infrared external transmittance of arsenic trisulfide, thickness 6.4 mm. [*From J. A. Savage and S. Nielson, Infrared Phys.* **5**, 195 (1965).]   (See Fig. 6c-10 for the short-wavelength transmittance.)

The properties of arsenic trisulfide glass vary with different batches.   This difference is probably caused by various quantities of arsenic di- and penta-sulfide as well as free sulfur.   Although it cracks rather easily under thermal strain, a sample window of arsenic trisulfide was subjected to a 500°C air blast on one side for more than 30 sec and did not crack.   A 4.5-in. dome has traveled successfully at Mach 2.8 and has withstood 160g without damage.   Arsenic trisulfide is quite soft and brittle (although experiments are in progress to harden it somewhat) and can be pressed, sawed, ground, and polished rather easily.   The coefficient of expansion of arsenic trisulfide is similar to that of aluminum, so it can be used in aluminum mounts quite satisfactorily. Silicon monoxide is often used as a coating for reducing reflection losses. Arsenic trisulfide has zero cold-water solubility but can be dissolved in alkalies. Samples of 20-in. diameter are available.

A Telluride Glass

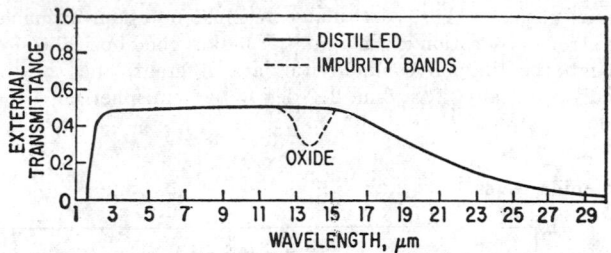

Fig. 6c-79. The external transmittance of a telluride glass, thickness 1.62 mm. [*From J. A. Savage and S. Nielson, Infrared Phys.* **5**, 195 (1965).]

This curve shows that the transmission is not greatly affected by oxide impurities. The composition of this sample may be represented by the formula $Ge_{10}As_{50}Te_{40}$.

Texas Instruments Glass No. 1173

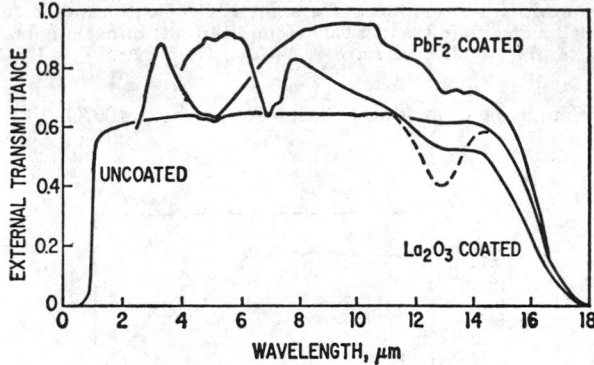

Fig. 6c-80. The external transmittance of Texas Instruments Glass No. 1173, thickness 3 mm. [*From Texas Instruments technical data literature* (1967).]

The composition of this glass can be represented by the formula $Ge_{28}Sb_{12}Se_{60}$. It is a relatively new material produced by Texas Instruments, Inc., Dallas, Texas. Samples of 12-in. diameter are available. The softening point is 370°C; it is insoluble in water but can be dissolved in alkalies.

Special Glasses

Cer-Vit

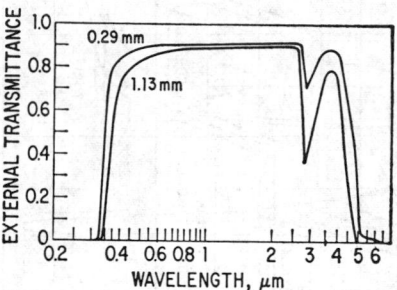

FIG. 6c-81. The external transmittance of Cer-Vit material type C-101, for two thicknesses. [*From R. M. Fuller, D. G. Rathburn, and Robert J. Bell, Appl. Opt.* **7**, 1243 (1968).]

Cer-Vit is the trade name of a material made by Owens-Illinois, Inc., Toledo, Ohio, which belongs to the family known as "glass ceramics." Their principal uniqueness is their tailorability to yield specific properties and combinations of properties not previously available for specific end uses. In the process of manufacturing Cer-Vit material products, a special glass is first melted, and then formed into monolithic structures of the desired shape. Finally, a controlled heat treatment converts the glass to a nonporous, polycrystalline material exhibiting a high degree of isotropy in all its properties.

The main advantage of Cer-Vit material is that it has an essentially zero thermal-expansion coefficient in a broad temperature range. Thus its principal application is for use as mirror blanks.

Corning Glasses

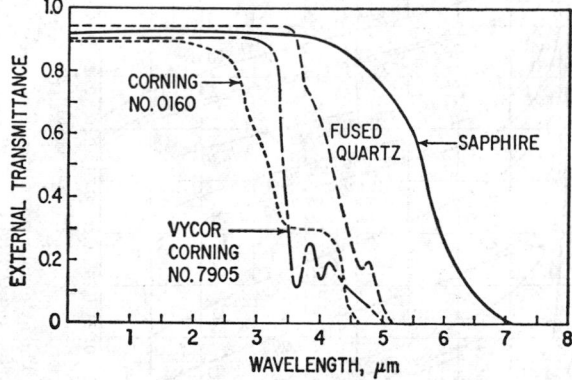

FIG. 6c-82. The external transmittances of Corning No. 0160 and No. 7905 (Vycor) glasses, as compared with common samples of fused quartz and sapphire, all thicknesses 2 mm. (*From Kodak Pamphlet* U-73, 1968.)

These two Corning glasses have been used extensively as substitutes for fused silica in appropriate applications.

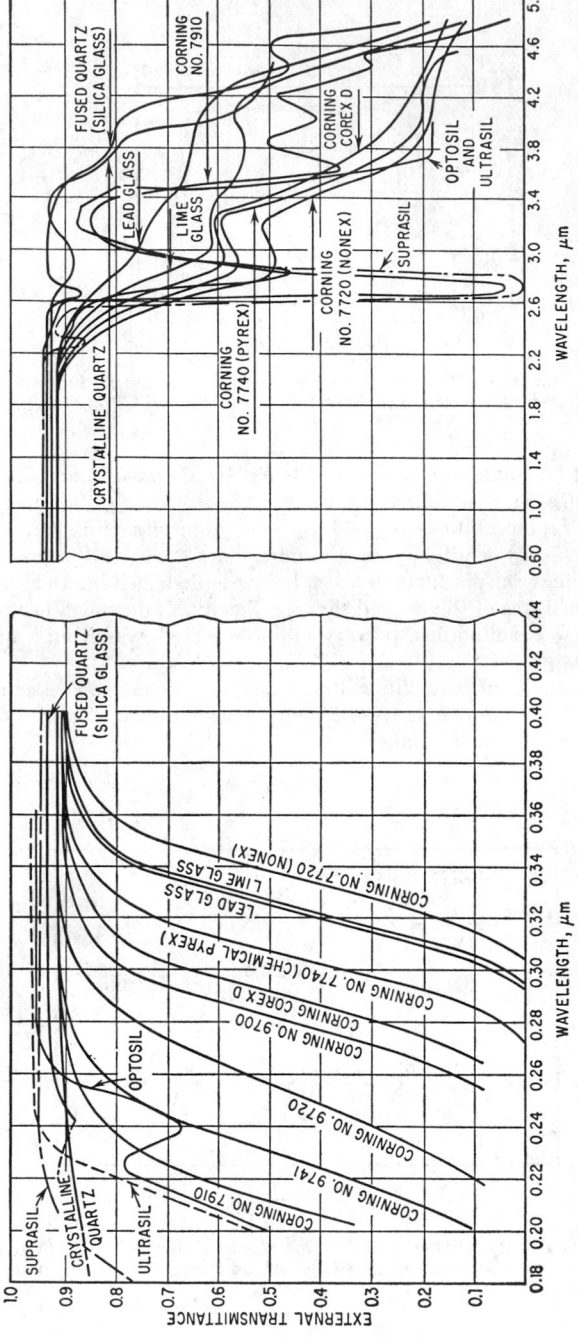

FIG. 6c–83. The external transmittance of several samples of Corning and Amersil glasses. *(From Fused Quartz Catalog, Q-6, General Electric Company, 1957, and Optical Fused Quartz, a catalog of the Amersil Quartz Division, Engelhard Industries, Inc.)*

6d. Geometrical Optics and the Index of Refraction of Various Optical Glasses

WILLEM BROUWER
Diffraction Limited, Inc.

NANCY R. McCLURE
Eastman Kodak Co.

6d-1. Geometrical Optics. Geometrical optics ignores the wave nature of light. The concept of a light ray is introduced in order to make the optical calculations easier. If a light ray traverses the boundary between two homogeneous media, the deviation from the normal is determined by *Snell's Law*, which states: The refracted ray, the normal to the surface at the point of incidence, and the incoming ray are in one plane; they obey the relationship

$$n \sin \phi = n' \sin \phi' \tag{6d-1}$$

in which n is the refractive index of the first medium, n' the refractive index of the second medium, ϕ and ϕ' the angles between the normal and the ray before and after refraction. This is the fundamental relationship in geometrical optical calculations.

Since spherical surfaces are most common in optics, only these will be described. An optical system always possesses an axis of symmetry, so that all centers of curvature lie on this axis.

To investigate the behavior of a light ray traversing the boundary between two optical media with refractive indices n and n', we select a coordinate system with the Z axis along the axis of symmetry. The positive direction is the direction in which the light travels. The origin is selected in such a way that the XY plane goes through the point where the ray intersects the surface. The ray can then be defined by its direction cosines L, M and its intersection point x, y. We can derive the following relation, with the help of Snell's Law, written in matrix form:

and

$$\begin{bmatrix} n'L' \\ x' \end{bmatrix} = \begin{bmatrix} 1 & -(n'\cos\phi' - n\cos\phi)/r \\ 0 & 1 \end{bmatrix} \begin{bmatrix} nL \\ x \end{bmatrix}$$

$$\begin{bmatrix} n'M' \\ y' \end{bmatrix} = \begin{bmatrix} 1 & -(n'\cos\phi' - n\cos\phi)/r \\ 0 & 1 \end{bmatrix} \begin{bmatrix} nM \\ y \end{bmatrix} \tag{6d-2}$$

in which r is the radius of curvature of the refracting surface (r is positive if the direction from the vertex to the center of curvature measured along the z axis is the direction in which the light travels) and in which primes refer to conditions after refraction.

If it is desired to follow the light ray through more refracting surfaces, we translate the coordinate system to the incidence point on the next surface without rotation. The relation governing this translation is given by

and

$$\begin{bmatrix} n'L'' \\ x'' \end{bmatrix} = \begin{bmatrix} 1 & 0 \\ d'/n' & 1 \end{bmatrix} \begin{bmatrix} n'L' \\ x' \end{bmatrix}$$

$$\begin{bmatrix} n'M'' \\ y'' \end{bmatrix} = \begin{bmatrix} 1 & 0 \\ d'/n' & 1 \end{bmatrix} \begin{bmatrix} n'M' \\ y' \end{bmatrix} \tag{6d-3}$$

in which d' is the distance between the two incidence points measured along the ray.

By multiplying the proper matrices for a complete system, we can find the relationship between the incoming ray and the outgoing ray. This relationship can always be written in the form

$$\begin{bmatrix} n'L' \\ x' \end{bmatrix} = \begin{bmatrix} B & -A \\ -D & C \end{bmatrix} \begin{bmatrix} nL \\ x \end{bmatrix} \quad \text{and} \quad \begin{bmatrix} n'M' \\ y' \end{bmatrix} = \begin{bmatrix} B & -A \\ -D & C \end{bmatrix} \begin{bmatrix} nM \\ y \end{bmatrix} \quad (6d\text{-}4)$$

A is called the *power of the system*. The value of all matrices, calculated as determinants, is $+1$. This gives us a good check on the final calculations in the form of

$$BC - AD = +1 \tag{6d-5}$$

In general, the values of A, B, C, and D are different for different rays.

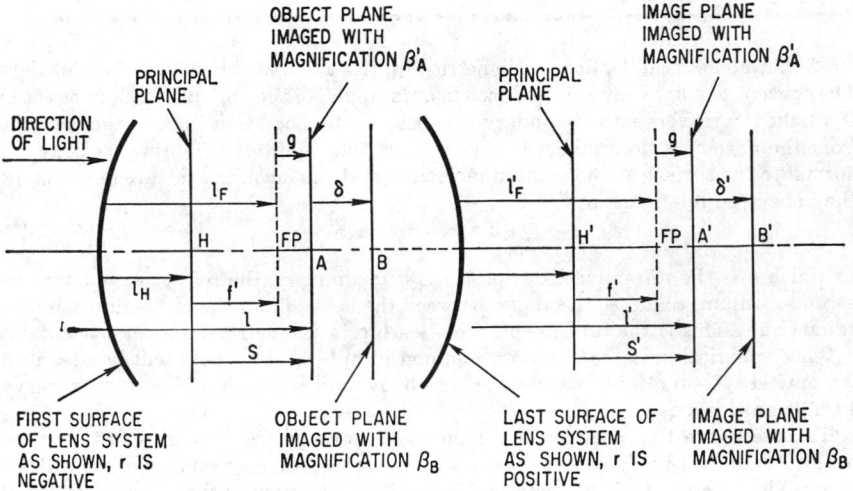

ALL DISTANCES SHOWN IN THIS FIGURE ARE POSITIVE. FOR THIS REASON ALL PLANES SHOWN IN THE OBJECT SPACE ARE VIRTUAL

FIG. 6d-1. Significant planes in a lens system.

For image forming, as opposed to analyzing, optical systems require that all rays coming from a certain object point should go through its corresponding image point. To investigate this we select an object point at a distance l, measured along the ray, from the point of incidence with the first surface of the system and an image point on the same ray at a distance l' from the point where this ray leaves the last surface of the system. This is shown in Fig. 6d-1. With the help of matrices of the form (6d-3) we find for the transformation matrix

$$\begin{bmatrix} B + (l/n)A & -A \\ (l'/n')(l/n)A + (l'/n')B - (l/n)C - D & C - (l'/n')A \end{bmatrix} \tag{6d-6}$$

An examination of the matrix elements shows that, in general, it will be impossible to find an image point which satisfies these conditions. If, however, we define the image point of the ray as the point where this ray intersects, in the image space, the plane through the axis of the system and the object point, we shall have the condition

$$\frac{x'}{x} = \frac{y'}{y} = \beta' \tag{6d-7}$$

(β' is called the linear magnification), and we get the following relations:

$$\begin{bmatrix} n'L' \\ x' \end{bmatrix} = \begin{bmatrix} 1/\beta' & -A \\ 0 & \beta' \end{bmatrix} \begin{bmatrix} nL \\ x \end{bmatrix} \quad \text{and} \quad \begin{bmatrix} n'M' \\ y' \end{bmatrix} = \begin{bmatrix} 1/\beta' & -A \\ 0 & \beta' \end{bmatrix} \begin{bmatrix} nM \\ y \end{bmatrix} \quad \text{(6d-8)}$$

and

$$\beta' = C - \frac{l'}{n'} A = \frac{1}{B + (l/n)A} \quad \text{(6d-9)}$$

$$\frac{l'}{n'} \frac{l}{n} A + \frac{l'}{n'} B - \left(\frac{l}{n}\right) C - D = 0 \quad \text{(6d-10)}$$

The task of the designer is to find a system for which Eqs. (6d-8) hold true for every object point lying in a given object plane. This, however, is impossible, and in practice a good lens will be a solution for which the deviations are within the tolerances allowed by the purpose for which the lens will be used. These deviations are called *aberrations*.

In order to get an insight into the properties of an optical system, the *paraxial laws* are often used. To arrive at these laws we develop all the quantities involved in power series and use only the first term of every series. The quantities L and L' can now be replaced by α and α', being the angles between the axis and the ray. The origin of the coordinate system can now be chosen at the intersection point of the surface and the axis. The distances can all be measured along the axis instead of along the ray. Equations (6d-2) and (6d-3) now become

$$\begin{bmatrix} n'\alpha' \\ x' \end{bmatrix} = \begin{bmatrix} 1 & -(n'-n)/r \\ 0 & 1 \end{bmatrix} \begin{bmatrix} n\alpha \\ x \end{bmatrix} \quad \text{(6d-11)}$$

and

$$\begin{bmatrix} n'\alpha'' \\ x'' \end{bmatrix} = \begin{bmatrix} 1 & 0 \\ t'/n' & 1 \end{bmatrix} \begin{bmatrix} n'\alpha' \\ x' \end{bmatrix} \quad \text{(6d-12)}$$

Similar to Eqs. (6d-8), (6d-9), and (6d-10) we find for the paraxial relationship of object and image points that

$$\begin{bmatrix} n'\alpha' \\ x' \end{bmatrix} = \begin{bmatrix} 1/\beta'_p & -A_p \\ 0 & \beta'_p \end{bmatrix} \begin{bmatrix} n\alpha \\ x \end{bmatrix} \quad \text{(6d-13)}$$

$$\beta'_p = C_p - \frac{l'}{n'} A_p = \frac{1}{B_p + (l/n)A_p} \quad \text{(6d-14)}$$

$$\frac{l'}{n'} \frac{l}{n} A_p + \frac{l'}{n'} B_p - \frac{l}{n} C_p - D_p = 0 \quad \text{(6d-15)}$$

in which the subscript p denotes that we deal with paraxial quantities.

It is customary, however, to measure the distances l and l' not from the outer surfaces of the lens but from its *principal points*. The principal points are defined as the object and image points on the axis with a magnification of $+1$. These points follow from Eq. (6d-14):

$$\frac{l_H}{n} = \frac{1 - B_p}{A_p} \quad \text{and} \quad \frac{l'_H}{n'} = \frac{C_p - 1}{A_p} \quad \text{(6d-16)}$$

If we measure the object distance s and image distance s' from the principal points, it is easily shown that the following relationship holds:

$$\frac{n'}{s'} = \frac{n}{s} + A_p \quad \text{(6d-17)}$$

If we move the object point to infinity, we call its image the *focal point*. For the distance between this focal point and its corresponding principal point, which is called the *focal length f'*, we find, with the help of Eq. (6d-17), that

$$s'_F = f' = \frac{n'}{A_p} \tag{6d-18}$$

By making $s' = \infty$, we find that

$$s_F = f = \frac{-n}{A_p} \tag{6d-19}$$

Table 6d-1 gives a selection of useful paraxial formulas. From this table we see that all the image distances can be calculated if we know the four quantities A_p,

TABLE 6d-1. PARAXIAL FORMULAS

$$\frac{l_H}{n} = \frac{1 - B_p}{A_p} \qquad \frac{l'_H}{n'} = \frac{C_p - 1}{A_p}$$

$$\frac{l_F}{n} = -\frac{B_p}{A_p} \qquad \frac{l'_F}{n'} = \frac{C_p}{A_p}$$

$$\frac{f}{n} = -\frac{1}{A_p} \qquad \frac{f'}{n'} = \frac{1}{A_p}$$

$$\frac{n'}{s'} = \frac{n}{s} + A_p$$

$$\beta' = \frac{ns'}{n's} = 1 - A_p \frac{s'}{n'} = \frac{1}{1 + A_p(s/n)} = -A_p \frac{g'}{n'} = \frac{1}{A_p(g/n)}$$

$$\beta'^2 = -\frac{g'n}{gn'} = \frac{gg'}{nn'} = -\frac{1}{A_p{}^2}$$

$$\frac{\delta}{n} = -\frac{\beta'_B - \beta'_A}{\beta'_A \beta'_B A_p} \qquad \frac{\delta'}{n'} = \frac{\beta'_B - \beta'_A}{A_p}$$

$$\frac{\delta'}{n'} = \beta'_A \beta'_B \frac{\delta}{n}$$

B_p, C_p, and D_p of our lens system. These are readily found with the help of Eqs. (6d-11) and (6d-12). For example, we find for a lens in air with radii r_1 and r_2, refractive index n, and thickness t, the following:

$$\begin{bmatrix} B_p & -A_p \\ -D_p & C_p \end{bmatrix} = \begin{bmatrix} 1 & -(1-n)/r_2 \\ 0 & 1 \end{bmatrix} \begin{bmatrix} 1 & 0 \\ t/n & 1 \end{bmatrix} \begin{bmatrix} 1 & -(n-1)/r_1 \\ 0 & 1 \end{bmatrix} =$$

$$\begin{bmatrix} 1 - (t/n)(1-n)/r_2 & -\{(n-1)/r_1 + (1-n)/r_2 - (t/n)[(n-1)(1-n)/r_1 r_2]\} \\ t/n & 1 - (t/n)(n-1)/r_1 \end{bmatrix} \tag{6d-20}$$

If the thickness of a single lens is small compared with its focal length, the thickness is often assumed to be zero in order to simplify the calculation. Such a lens is referred to as *thin*. For a thin lens we find from Eq. (6d-20) that

$$A_p = (n-1)\left(\frac{1}{r_1} - \frac{1}{r_2}\right) \tag{6d-21}$$

$$B_p = C_p = 1 \qquad D_p = 0$$

The principal points of such a thin lens are now located in the lens, according to Eqs. (6d-16).

All these formulas can be used for *mirrors* if the following sign rules are adopted: For the first mirror, n' is to be taken negative; i.e., for a mirror in air $n' = -1$. All refractive indices after the first reflection but before the second reflection are negative. After the second reflection the refractive indices are again positive until the third reflection, where they become negative, and so on. All distances remain positive when they are in the direction the light was traveling before it entered the system.

The deviations of actual optical systems from ideal optical systems are called *aberrations*. There are two kinds of aberrations:

1. Deviations of the rays from the paths given by the ideal laws as given by the paraxial equations, called *monochromatic aberrations*.

2. Changes in image position and size due to changes of refractive index with wavelength, called *chromatic aberrations*. These already exist in the paraxial approximation.

To find the aberrations for a given system we have to follow several rays step by step through the system. This is called *ray tracing*.

To get an insight into aberrations, the next term in the power series which gave us the paraxial law is often used. The deviations of the rays found in this way are called *third-order aberrations*.

In an aberration-free system β' and A, in Eqs. (6d-8), are constants. For every ray in such a system

$$n'L' = \left(\frac{1}{\beta'}\right) nL - Ax \qquad (6\text{d-}22)$$

For a point on the axis of the system we can replace L and L' by $\sin U$ and $\sin U'$, where U and U' represent the angle between the optical axis and the ray before and after passing through the system. Now Eq. (6d-22) becomes

$$n' \sin U' = \frac{n \sin U}{\beta'} \qquad (6\text{d-}23)$$

This is called *Abbe's sine condition*. Sometimes β' is replaced by x'/x, where x is a small distance perpendicular to the axis in the object plane and imaged as x' perpendicular to the axis in the image plane. The sine condition now reads

$$n'x' \sin U' = nx \sin U \qquad (6\text{d-}24)$$

For a system in air without chromatic aberrations it is seen from Eq. (6d-13) that β' and A_p should be constants for light of different wavelengths, while for a thin lens this reduces to $A_p = \text{const}$. In general, however, using Eqs. (6d-21) we find after differentiation with respect to n that

$$\Delta A_p = \Delta \frac{1}{f} = \frac{\Delta n}{n-1} \frac{1}{f} = \frac{\Delta n}{n-1} A_p \qquad (6\text{d-}25)$$

For calculations of chromatic aberrations the ν-value is often used. The ν-value is defined as

$$\nu = \frac{n_d - 1}{n_F - n_C} \qquad (6\text{d-}26)$$

in which n_d is the refractive index of the glass at the d-line in a helium spectrum ($\lambda = 0.58756$ μm); n_F is the refractive index of the glass at the F-line in a mercury spectrum ($\lambda = 0.48613$ μm); n_C is the refractive index of the glass at the C-line in a mercury spectrum ($\lambda = 0.65628$ μm). Using the ν-value we find for Eq. (6d-26) that

$$\frac{\Delta A_p}{A_p} = -\frac{\Delta f'}{f'} = \frac{1}{\nu} \tag{6d-27}$$

The *illuminance of image* can be calculated for an optical system. Let dS be an element of area of the object. The flux dF into the element of solid angle $d\omega$, defined by a double cone of aperture α to $\alpha + d\alpha$, is

$$dF = B\, dS \cos \alpha\, d\omega$$

in which B is the *luminance* of the object. Now, from the way $d\omega$ is defined, it follows that

$$d\omega = 2\pi \sin \alpha\, d\alpha$$

so $\qquad\qquad dF = 2\pi B\, dS\, d\alpha \cos \alpha \sin \alpha$

If the cone admitted by the optical system has an angle U, we can calculate the total light flux going into the system. We find that

$$F_U = \int_0^U 2\pi B\, dS\, d\alpha \cos \alpha \sin \alpha = \pi B\, dS \sin^2 U \tag{6d-28}$$

A similar relation holds for an area dS' in the image space, where the total flux is given by

$$F_{U'} = \pi B'\, dS' \sin^2 U' \tag{6d-29}$$

In the case where dS' is the image of dS and the cone given by U' is the same cone as given by U after passing through the optical system with a transmittance τ, $F_{U'} = \tau F_U$ and the relation between B and B' becomes

$$B'\, dS' \sin^2 U' = \tau B\, dS \sin^2 U$$

If the optical system is aberration-free, we can use the sine condition and we find that

$$B' = \tau \left(\frac{n'}{n}\right)^2 B \tag{6d-30}$$

In applications where the total energy in the image is important, we can use these results. In other applications, however, the total energy is not so important as the *light flux density* or *illuminance E*. Here we find on the image side that

$$E' = \frac{F'}{dS'} = \pi B' \sin^2 U'$$

If the system is again aberration-free, we can use the other form of the sine condition and Eq. (6d-30) to arrive at

$$E' = \tau \pi B \frac{\sin^2 U}{G^2} \tag{6d-31}$$

It is also important to find the maximum value of $\sin U$ which is passed by an optical system. For a single thin lens, this is determined by the diameter of the lens.

In a multiple system the situation is more complicated. In order to find what cone is passed, for instance, by the nth surface, we calculate where a system consisting of the first $(n - 1)$ surfaces images the nth surface in the object space and with what magnification. If this surface has a radius ρ_n, then it will be imaged with a magnification $G_n\rho_n$. Now every ray going through a point in this image in the object space will go through the nth surface if this ray is passed by all the other surfaces. So if we image all surfaces and diaphragms in this way in the object space, we can determine which rays are passed by the whole system. The smallest of these images as seen from the axial point of the object is called the *entrance pupil*. The diameter of the entrance pupil determines U for the whole system. The image of the entrance pupil in the image space of the whole system is called the *exit pupil*. A ray through the center of the entrance pupil is called a *chief ray* or *principal ray*. It follows from the definition of an image that a chief ray after passing through the system goes through the center of the exit pupil.

A special case arises when the object is at infinity. The entrance pupil is determined in the same way. In this case we can write $\sin U = \rho/p$, in which ρ is the radius of the entrance pupil and p the object distance. For G we write p'/p and arrive at the illuminance in this case by

$$E' = \tau\pi B \left(\frac{\rho/p}{p'/p}\right)^2 = \tau\pi B \left(\frac{\rho}{p'}\right)^2 = \tau\pi B \left(\frac{\rho}{f'}\right)^2$$

If we now introduce the *f-number* N, defined by $N = f/2\rho$, we find that

$$E' = \frac{1}{4}\tau\pi B \left(\frac{1}{N}\right)^2 \tag{6d-32}$$

The *field of view* is another important property of an optical system. There are two things which can limit the field of view. Aberrations have in general a tendency to increase if the bundles travel more obliquely through the system. If the aberrations become so large that the image is no longer useful, this is a limiting factor. On the other hand, the bundles may become so oblique that they are no longer passed by the system. In order to find this limit, we consider a cone of rays all going through the center of the entrance pupil. This cone will be limited to an angle V by one of the other images in the object space of the surfaces and diaphragms. We call V the *field angle* and the limiting boundary the *field stop*. The chief rays with an angle V will come from certain points of the object and so define the size of the object which can be imaged usefully by the optical system. In general, the cones from the edge of the field will be smaller than the one in the center of the field. This effect is called *vignetting*.

6d-2. Index of Refraction of Various Glasses. Table 6d-2 includes the index of refraction for the wavelengths for the d, A', C, F, and h lines for each glass. For Schott glasses, the values of reciprocal dispersion ν are based on n_d; for B & L and EK glasses, they are based on n_D. To interpolate the index for any other wavelength, obtain the universal functions[1] a_i for the desired wavelength from Table 6d-3 and substitute in

$$n(\lambda) = a_1 n_{A'} + a_2 n_C + a_3 n_F + a_4 n_h \tag{6d-33}$$

The matrix C_{ij} given in Table 6d-4 can be used to compute the universal functions for other wavelengths in the range 0.4047 to 0.7682 μm.

$$a_j(\lambda) = C_{1j} + C_{2j}\lambda^2 + C_{3j}L + C_{4j}L^2 \tag{6d-34}$$

where λ is the wavelength in micrometers, and $L = 1/(\lambda^2 - 0.028)$.

[1] M. Herzberger, *Opt. Acta* **6**, 197 (1959).

TABLE 6d-2. A REPRESENTATIVE SELECTION OF AVAILABLE OPTICAL GLASSES

Manufacturer*	Type	ν	n_d, 0.5876 μm	$n_{A'}$, 0.7682 μm	n_C, 0.6563 μm	n_F, 0.4861 μm	n_h, 0.4047 μm
Schott	FKS01	81.61	1.48523	1.48135	1.48342	1.48936	1.49520
Schott	FK05	70.34	1.48749	1.48282	1.48534	1.49227	1.49893
Schott	PKS01	69.69	1.52054	1.51556	1.51824	1.52571	1.53295
Schott	FK6	67.28	1.44628	1.44188	1.44424	1.45088	1.45737
Schott	PSKS01	67.25	1.55753	1.55205	1.55498	1.56327	1.57137
B & L	BSC	67.0	1.49808	1.49316	1.49577	1.50320	1.51048
Schott	PSKS6	65.41	1.60310	1.59704	1.60028	1.60950	1.61857
B & L	BSC	63.5	1.51107	1.50578	1.50860	1.51665	1.52454
Schott	PSK1	62.88	1.54771	1.54198	1.54505	1.55376	1.56230
Schott	K11	61.59	1.50013	1.49479	1.49765	1.50577	1.51379
B & L	DBC	61.2	1.58811	1.58184	1.58513	1.59474	1.60424
B & L	C	60.5	1.51258	1.50708	1.50999	1.51846	1.52685
B & L	DBC	60.3	1.62011	1.61342	1.61696	1.62724	1.63748
Schott	BaK2	59.66	1.53996	1.53407	1.53720	1.54625	1.55528
B & L	C	58.6	1.52307	1.51729	1.52036	1.52929	1.53819
B & L	LBC	57.4	1.57497	1.56619	1.56956	1.57953	1.58951
B & L	DBC	57.2	1.61109	1.60423	1.60785	1.61853	1.62923
B & L	EDBC	57.2	1.65709	1.64972	1.65362	1.66510	1.67649
Schott	K10	56.46	1.50137	1.49560	1.49867	1.50755	1.51647
EK	110	56.15	......	1.68877	1.69313	1.70554	1.71786
B & L	DBC	55.5	1.63810	1.63074	1.63461	1.64611	1.65772
B & L	LaC	54.8	1.69111	1.68305	1.68730	1.69910	1.71248
Schott	LaK18	54.67	1.72875	1.72004	1.72469	1.73802	1.75126
B & L	CF	54.6	1.52568	1.51954	1.52277	1.53239	1.54215
B & L	EDBC	53.9	1.61711	1.60980	1.61368	1.62512	1.63670
Schott	BaLF5	53.61	1.54739	1.54086	1.54432	1.55453	1.56494
B & L	LBF	53.4	1.58809	1.58110	1.58479	1.59580	1.60697
Schott	LaK19	53.24	1.75496	1.74574	1.75065	1.76483	1.77902
Schott	KF6	52.16	1.51742	1.51105	1.51443	1.52435	1.53446
EK	210	51.18	......	1.72483	1.72978	1.74417	1.75877
B & L	CF	51.0	1.52408	1.51759	1.52100	1.53127	1.54185
B & L	LBF	51.0	1.56210	1.55518	1.55879	1.56982	1.58115
B & L	EDBC	50.9	1.65714	1.64894	1.65323	1.66614	1.67927
Schott	LaK17	50.48	1.78847	1.77841	1.78375	1.79937	1.81514
Schott	SSK7	50.36	1.61847	1.61069	1.61479	1.62707	1.63972
Schott	BaF5	49.25	1.60729	1.59949	1.60359	1.61592	1.62871
B & L	LaF	48.0	1.70012	1.69098	1.69576	1.71033	1.72536
B & L	LaF	47.5	1.72013	1.71063	1.71561	1.73077	1.74645
Schott	LaF21	47.37	1.78831	1.77767	1.78330	1.79994	1.81694
B & L	ELF	47.3	1.54110	1.53386	1.53768	1.54912	1.56102
B & L	BF	47.2	1.67008	1.66123	1.66585	1.68004	1.69472
Schott	BaF8	47.04	1.62374	1.61541	1.61980	1.63306	1.64692
Schott	LaSF1	46.76	1.80279	1.79184	1.79763	1.81480	1.83239
EK	310	46.42	......	1.73491	1.74033	1.75638	1.77301
B & L	BF	46.0	1.58391	1.57598	1.58013	1.59282	1.60615
B & L	ELF	45.5	1.55860	1.55086	1.55495	1.56722	1.58010
EK	430	44.69	......	1.76582	1.77164	1.78902	1.80706
B & L	BF	43.6	1.60542	1.59682	1.60130	1.61518	1.62987
B & L	LF	42.5	1.57262	1.56425	1.56861	1.58208	1.59637
Schott	LFS1	42.19	1.54765	1.53959	1.54382	1.55680	1.57091
B & L	LaF	42.0	1.72016	1.70957	1.71508	1.73220	1.74965
Schott	BaSF6	41.88	1.66755	1.65766	1.66284	1.67878	1.69579
EK	450	41.80	......	1.79180	1.79814	1.81738	1.83767
Schott	LaSF5	41.00	1.88069	1.86722	1.87430	1.89578	1.91825
Schott	BaSFS6	40.99	1.70181	1.69108	1.69672	1.71384	1.73200
Schott	BaSF10	39.31	1.65016	1.63999	1.64529	1.66183	1.67968
B & L	DF	38.0	1.60514	1.59538	1.60045	1.61638	1.63358
Schott	BaSFS2	37.99	1.72340	1.71160	1.71779	1.73683	1.75729
Schott	LaF13	37.83	1.77551	1.76283	1.76946	1.78996	1.81201
Schott	FS1	37.06	1.58407	1.57440	1.57945	1.59521	1.61274
B & L	DBF	36.6	1.65715	1.64616	1.65189	1.66984	1.68943
B & L	DF	36.2	1.62114	1.61066	1.61610	1.63325	1.65197
Schott	BaSF14	34.95	1.69968	1.68749	1.69384	1.71388	1.73599
B & L	EDF	33.8	1.64916	1.63754	1.64355	1.66275	1.68397
B & L	EDF	32.2	1.67269	1.66012	1.66663	1.68751	1.71084
Schott	BaSFS7	32.15	1.73627	1.72243	1.72961	1.75251	1.77817
Schott	SFS08	31.30	1.61339	1.60174	1.60777	1.62737	1.65071
B & L	LaF	31.0	1.86725	1.85036	1.85910	1.88706	1.91874
B & L	EDF	29.3	1.72022	1.70555	1.71309	1.73766	1.76542
Schott	LaF9	28.39	1.79504	1.77827	1.78695	1.81495	1.84675
B & L	EDF	27.8	1.75084	1.73473	1.74303	1.77005	1.80089
Schott	SFS3	26.10	1.78470	1.76688	1.77607	1.80613	1.84085
Schott	SFS09	23.83	1.84666	1.82578	1.83651	1.87204	1.91363
Schott	SFS06	20.36	1.95250	1.92545	1.93928	1.98606	2.04280

* Schott   Jenaer Glaswerk Schott & Geb.
  B & L   Bausch & Lomb.
  EK   Eastman Kodak.

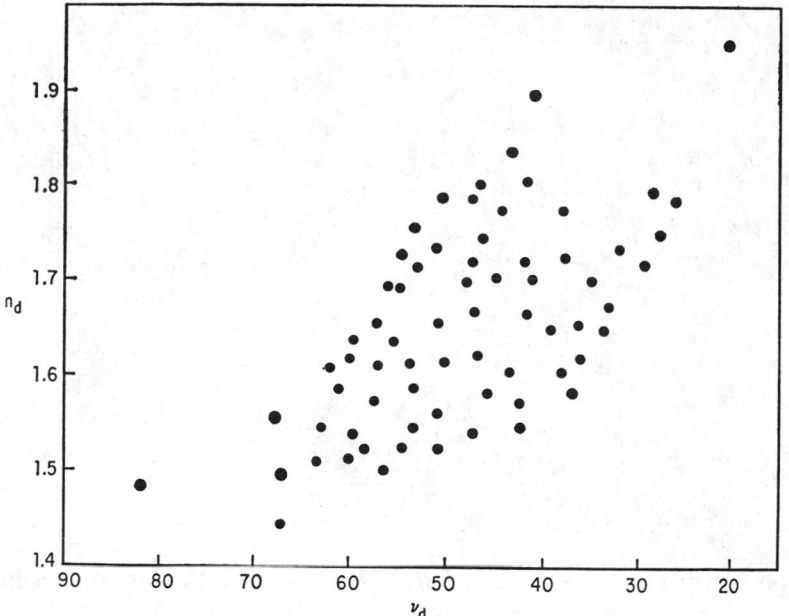

Fig. 6d-2. Graphical representation of n_d vs. v_d of the glasses shown in Table 6d-2.

TABLE 6d-3. UNIVERSAL FUNCTIONS FOR USE IN EQ. (6d-33)

Line	λ, μm	a_1	a_2	a_3	a_4
A'	0.76820	+1.000000	0.000000	0.000000	0.000000
C	0.65630	0.000000	1.000000	0.000000	0.000000
D	0.58930	−0.219082	0.951088	0.317290	−0.049296
d	0.58760	−0.220644	0.943152	0.328101	−0.050609
e	0.54610	−0.198539	0.652493	0.619332	−0.073287
F	0.48610	0.000000	0.000000	1.000000	0.000000
g	0.43580	0.146293	−0.362709	0.835623	0.380793
G'	0.43410	0.145947	−0.360970	0.810864	0.404160
h	0.40470	0.000000	0.000000	0.000000	1.000000

TABLE 6d-4. MATRIX OF COEFFICIENTS FOR DETERMINING UNIVERSAL FUNCTIONS

j	C_{1j}	C_{2j}	C_{3j}	C_{4j}
1	−10.181350	18.313606	−9.815670	2.683413
2	13.140533	−19.603381	8.696877	−2.234029
3	2.192521	−4.346029	3.108244	−0.954736
4	−0.149673	0.311663	−0.267335	0.105345

6e. Index of Refraction for Visible Light of Various Solids, Gases, and Liquids

TABLE 6e-1. INDEX OF REFRACTION OF SOME LIQUIDS RELATIVE TO AIR*

Substance	Density	Temp., °C	Indices of refraction				
			0397 μm H	0.434 μm G'	0.486 μm F	0.589 μm D	0.656 μm C
Acetaldehyde, CH₃CHO.........	0.780	20	......	1.3394	1.3359	1.3316	1.3298
Acetone, CH₃COCH₃............	0.791	20	......	1.3678	1.3639	1.3593	1.3573
Aniline, C₆H₅·NH₂..............	1.022	20	......	1.6204	1.6041	1.5863	1.5793
Alcohol, methyl, CH₃·OH........	0.794	20	1.3399	1.3362	1.3331	1.3290	1.3277
Alcohol, ethyl, C₂H₅·OH.........	0.808	0	......	1.3773	1.3739	1.3695	1.3677
Alcohol, ethyl..................	0.800	20	......	1.3700	1.3666	1.3618	1.3605
Alcohol, ethyl, dn/dt...........	.....	20	......	−0.0004	−0.0004	−0.0004	−0.0004
Alcohol, n-propyl, C₃H₇·OH.....	0.804	20	......	1.3938	1.3901	1.3854	1.3834
Benzene, C₆H₆.................	0.880	20	......	1.5236	1.5132	1.5012	1.4965
Benzene, C₆H₆ dn/dt...........	.....	20	......	−0.0007	−0.0006	−0.0006	−0.0006
Bromnaphthalene, C₁₀H₇Br......	1.487	20	1.7289	1.7041	1.6819	1.6582	1.6495
Carbon disulfide, CS₂..........	1.293	0	1.7175	1.6920	1.6688	1.6433	1.6336
Carbon disulfide...............	1.263	20	1.6994	1.6748	1.6523	1.6276	1.6182
Carbon tetrachloride, CCl₄.......	1.591	20	......	1.4729	1.4676	1.4607	1.4579
Chinolin, C₉H₇N...............	1.090	20	......	1.6679	1.6470	1.6245	1.6161
Chloral, CCl₃·CHO.............	1.512	20	......	1.4679	1.4624	1.4557	1.4530
Chloroform, CHCl₃.............	1.489	20	1.463	1.458	1.4530	1.4467	1.4443
Decane, C₁₀H₂₂.................	0.728	14.9	......	1.4200	1.4160	1.4108	1.4088
Ether, ethyl, C₂H₅·O·C₂H₅.......	0.715	20	......	1.3607	1.3576	1.3538	1.3515
Ether, ethyl, dn/dt.............	.....	20	......	−0.0006	−0.0006	−0.0006	−0.0006
Ethyl nitrate, C₂H₅·O·NO₂.......	1.109	20	......	1.395	1.392	1.3853	1.3830
Formic acid, H·CO₂H...........	1.219	20	......	1.3804	1.3764	1.3714	1.3693
Glycerine, C₃H₈O₃.............	1.260	20	......	1.4828	1.4784	1.4730	1.4706
Hexane, CH₃(CH₂)₄CH₃.........	0.660	20	......	1.3836	1.3799	1.3754	1.3734
Hexylene, CH₃(CH₂)₃CH·CH₂....	0.679	23.3	......	1.4059	1.4007	1.3945	1.3920
Methylene iodide, CH₂I₂........	3.318	20	1.8027	....	1.7692	1.7417	1.7320
Methylene iodide dn/dt.........	.....	20	......	....	−0.0007	−0.0007	−0.0006
Naphthalene, C₁₀H₈............	0.962	98.4	......	....	1.6031	1.5823	1.5746
Nicotine, C₁₀H₁₄N₂.............	1.012	22.4	......	1.5439	....	1.5239	1.5198
Octane, CH₃(CH₂)₆CH₃.........	0.707	15.1	......	1.4097	1.4046	1.4007	1.3987
Oil:							
Almond.....................	0.92	0	......	....	1.4847	1.4782	1.4755
Anise seed..................	0.99	15.1	1.6084	....	1.5743	1.5572	1.5508
Anise......................	0.99	21.4	......	....	1.5647	1.5475	1.5410
Bitter almond...............	1.06	20	......	1.5775	1.5623	....	1.5391
Cassia.....................	.....	10	1.7039	....	1.6389	1.6104	1.6007
Cassia.....................	.....	22.5	1.6985	....	1.6314	1.6026	1.5930
Cinnamon...................	1.05	23.5	......	....	1.6508	1.6188	1.6077
Olive......................	0.92	0	......	....	1.4825	1.4763	1.4738
Rock.......................	.....	0	......	....	1.4644	1.4573	1.4545
Turpentine..................	0.87	10.6	1.4939	....	1.4817	1.4744	1.4715
Turpentine..................	0.87	20.7	1.4913	....	1.4793	1.4721	1.4692
Pentane, CH₃(CH₂)₃CH₃........	0.625	15.7	......	1.3645	1.3610	1.3581	1.3570
Phenol, C₆H₅OH...............	1.060	40.6	......	1.5684	1.5558	1.5425	1.5369
Phenol........................	1.021	82.7	......	......	1.5356	....	1.5174
Styrene, C₆H₅CH·CH₂..........	0.910	16.6	......	1.5816	1.5659	1.5485	1.5419
Thymol, C₁₀H₁₄O..............	0.982	....	......	....	1.5386	....	1.5228
Toluene, CH₃·C₆H₅.............	0.86	20	......	1.5170	1.5070	1.4955	1.4911
Water, H₂O....................	.....	20	1.3435	1.3404	1.3372	1.3330	1.3312
Water.........................	.....	0	1.3444	1.3413	1.3380	1.3338	1.3319
Water.........................	.....	40	1.3411	1.3380	1.3349	1.3307	1.3290
Water.........................	.....	80	1.3332	1.3302	1.3270	1.3230	1.3313

* "Smithsonian Physical Tables," 1954, Table 551.

TABLE 6e-2. INDEX OF REFRACTION FOR SOLUTIONS OF SALTS AND ACIDS RELATIVE TO AIR*

Substance	Density	Temp., °C	Indices of refraction for spectrum lines				
			C	D	F	$H\gamma$	H

Solutions in Water

Substance	Density	Temp., °C	C	D	F	$H\gamma$	H
Ammonium chloride	1.067	27.05	1.37703	1.37936	1.38473	....	1.39336
Ammonium chloride	1.025	29.75	1.34850	1.35050	1.35515	....	1.36243
Calcium chloride	1.398	25.65	1.44000	1.44279	1.44938	....	1.46001
Calcium chloride	1.215	22.9	1.39411	1.39652	1.40206	....	1.41078
Calcium chloride	1.143	25.8	1.37152	1.37369	1.37876	....	1.38666
Hydrochloric acid	1.166	20.75	1.40817	1.41109	1.41774	....	1.42816
Nitric acid	1.359	18.75	1.39893	1.40181	1.40857	....	1.41961
Potash (caustic)	1.416	11.0	1.40052	1.40281	1.40808	....	1.41637
Potassium chloride	Normal solution		1.34087	1.34278	1.34719	1.35049	
Potassium chloride	Double normal		1.34982	1.35179	1.35645	1.35994	
Potassium chloride	Triple normal		1.35831	1.36029	1.36512	1.36890	
Soda (caustic)	1.376	21.6	1.41071	1.41334	1.41936	....	1.42872
Sodium chloride	1.189	18.07	1.37562	1.37789	1.38322	1.38746	
Sodium chloride	1.109	18.07	1.35751	1.35959	1.36442	1.36823	
Sodium chloride	1.035	18.07	1.34000	1.34191	1.34628	1.34969	
Sodium nitrate	1.358	22.8	1.38283	1.38535	1.39134	....	1.40121
Sulfuric acid	1.811	18.3	1.43444	1.43669	1.44168	....	1.44883
Sulfuric acid	1.632	18.3	1.42227	1.42466	1.42967	....	1.43694
Sulfuric acid	1.221	18.3	1.36793	1.37009	1.37468	....	1.38158
Sulfuric acid	1.028	18.3	1.33663	1.33862	1.34285	....	1.34938
Zinc chloride	1.359	26.6	1.39977	1.40222	1.40797	....	1.41738
Zinc chloride	1.209	26.4	1.37292	1.37515	1.38026	....	1.38845

Solutions in Ethyl Alcohol

Substance	Density	Temp., °C	C	D	F	$H\gamma$	H
Ethyl alcohol	.....	25.5	1.35971	1.35971	1.36395	....	1.37094
Ethyl alcohol	.....	27.6	1.35372	1.35556	1.35986	....	1.36662
Fuchsin (nearly saturated)	.....	16.0	1.3918	1.398	1.361		1.3759
Cyanin (saturated)	.....	16.0	1.3831	....	1.3705	....	1.3821

Note: Cyanin in chloroform also acts anomalously; for example, Sieben gives for a 4.5 % solution $\mu_A = 1.4593$, $\mu_B = 1.4695$, μ_F (green) = 1.4514, μ_G (blue) = 1.4554. For a 9.9 % solution he gives $\mu_A = 1.4902$, μ_F (green) = 1.4497, μ_G (blue) = 1.4597.

Solutions of Potassium Permanganate in Water

Wave-length, μm	Spectrum line	Index for 1 % sol	Index for 2 % sol	Index for 3 % sol	Index for 4 % sol	Wave-length, μm	Spectrum line	Index for 1 % sol	Index for 2 % sol	Index for 3 % sol	Index for 4 % sol
0.687	B	1.3328	1.3342	......	1.3382	0.516	...	1.3368	1.3385		
0.656	C	1 3335	1.3348	1.3365	1.3391	0.500	...	1.3374	1.3383	1.3386	1.3404
0.617	...	1.3343	1.3365	1.3381	1.3410	0.486	F	1.3377	......	......	1.3408
0.594	...	1.3354	1.3373	1.3393	1.3426	0.480	...	1.3381	1.3395	1.3398	1.3413
0.589	D	1.3353	1.3372	......	1.3426	0.464	...	1.3397	1.3402	1.3414	1.3423
0.568	...	1.3362	1.3387	1.3412	1.3445	0.447	...	1.3407	1.3421	1.3426	1.3439
0.553	...	1.3366	1.3395	1.3417	1.3438	0.434	...	1.3417	......		1.3452
0.527	E	1.3363				0.423	...	1.3431	1.3442	1.3457	1.3468
0.522	...	1.3362	1.3377	1.3388							

* "Smithsonian Physical Tables," 1954, Table 552.

Table 6e-3. Liquids Used for Determining Refractive Index
by Transmission Method*

Liquid	N_D, 24°C
Trimethylene chloride	1.446
Cineole	1.456
Hexahydrophenol	1.466
Decahydronaphthalene	1.477
Isoamylphthalate	1.486
Tetrachloroethane	1.492
Pentachloroethane	1.501
Trimethylene bromide	1.513
Chlorobenzene	1.523
Ethylene bromide + chlorobenzene	1.533
o-Nitrotoluene	1.544
Xylidine	1.557
o-Toluidine	1.570
Aniline	1.584
Bromoform	1.595
Iodobenzene + bromobenzene	1.603
Iodobenzene + bromobenzene	1.613
Quinoline	1.622
α-Chloronaphthalene	1.633
α-Bromonaphthalene + α-chloronaphthalene	1.640–1.650
α-Bromonaphthalene + α-iodonaphthalene	1.660–1.690
Methylene iodide + iodobenzene	1.700–1.730
Methylene iodide	1.738
Methylene iodide saturated with sulfur	1.78
Yellow phosphorus, sulfur, and methylene iodide† (8:1:1 by weight)	2.06

* "Handbook of Chemistry and Physics," 36th ed., p. 2669, Chemical Rubber Publishing Company, 1954–1955.

† Can be diluted with methylene iodide to cover range 1.74–2.06. For precautions in use, cf. West, *Am. Mineralogist* **21**, 245–249 (1936).

TABLE 6e-4. INDEX OF REFRACTION OF PLASTICS*
Optical Plastics

Name of monomer	Refractive index (N^{20})	Reciprocal dispersive power	Name of monomer	Refractive index (N^{20})	Reciprocal dispersive power
p-Methoxy styrene	1.5967	28	Allyl methacrylate	1.5196	49.0
β-Amino-ethyl methacrylate	1.537	52.5	Benzhydryl methacrylate	1.5933	31.0
Methyl α-bromoacrylate	1.5672	46.5	Benzyl methacrylate	1.5680	36.5
Vinyl benzoate	1.5775	30.7	n-Butyl methacrylate	1.483	49
Phenyl vinyl ketone	1.586	26.0	Tert-butyl methacrylate	1.4638	51
Vinyl carbazole	1.683	18.8	o-Chlorobenzhydryl methacrylate	1.6040	30
Lead methacrylate	1.645	28	α-(o-Chlorophenyl)-ethyl methacrylate	1.5624	37.5
2-Chlorocyclohexyl methacrylate	1.5179	56	Cyclohexyl-cyclohexyl methacrylate	1.5250	53
1-Phenyl-cyclohexyl methacrylate	1.5645	40	Cyclohexyl methacrylate	1.5064	56.9
Triethoxy-silicol methacrylate	1.436	53	p-Cyclohexyl-phenyl methacrylate	1.5575	39.0
p-Bromophenyl methacrylate	1.5964	33	α-β-Diphenyl-ethyl methacrylate	1.5816	30.5
2-3 Dibromopropyl methacrylate	1.5739	44	Menthyl methacrylate	1.4890	54.5
Diethyl-amino-ethyl methacrylate	1.5174	54	Ethylene dimethacrylate	1.5063	53.4
1-Methyl-cyclohexyl methacrylate	1.5111	54	Hexamethylene glycol dimethacrylate	1.5066	56
n-Hexyl methacrylate	1.4813	57	Methacrylic anhydride	1.5228	48.5
2-6-Dichlorostyrene	1.6248	31.3	Methyl methacrylate	1.4913	57.8
β-Bromo-ethyl methacrylate	1.5426	40	m-Nitro-benzyl methacrylate	1.5845	27.4
μ-Polychloroprene	1.5540	36	2-Nitro-2-methyl-propyl methacrylate	1.4868	48
Methyl α-chloracrylate	1.5172	57	α-Phenyl-allyl methacrylate	1.5573	34.8
β-Naphthyl methacrylate	1.6298	24	α-Phenyl-n-amyl methacrylate	1.5396	40
Vinyl phenyl sulfide	1.6568	27.5	α-Phenyl-ethyl methacrylate	1.5487	37.5
Methacryl methyl salicylate	1.5707	34	β-Phenyl-ethyl methacrylate	1.5592	36.5
Methyl isopropenyl ketone	1.5200	54.5	Tetrahydrofurfuryl methacrylate	1.5096	54
Ethylene glycol mono-methacrylate	1.5119	56	Vinyl methacrylate	1.5129	46
N-Benzyl methacrylamide	1.5965	34.5	Styrene	1.5907	30.8
β-Phenyl-sulfone ethyl methacrylate	1.5682	39	Vinyl formate	1.4757	55
N-Methyl methacrylamide	1.5398	47.5	Phenyl cellosolve methacrylate	1.5624	36.2
N-Allyl methacrylamide	1.5476	47	p-Methoxy-benzyl methacrylate	1.552	32.5
Methacryl-phenyl salicylate	1.6006	36	Ethylene chlorohydrin methacrylate	1.517	54
N-β-Methoxyethyl methacrylamide	1.5246	53	o-Chlorostyrene	1.6098	31
N-β-Phenylethyl methacrylamide	1.5857	37	Pentachlorophenyl methacrylate	1.608	22.5
Cyclohexyl α-ethoxyacrylate	1.4969	58	Phenyl methacrylate	1.5706	35.0
1-3-Dichloropropyl-2-methacrylate	1.5270	56	Vinyl naphthalene	1.6818	20.9
2-Methyl-cyclohexyl methacrylate	1.5028	53	Vinyl thiophene	1.6376	29
3-Methyl-cyclohexyl methacrylate	1.4947	55	Eugenol methacrylate	1.5714	33
3-3-5-Trimethyl-cyclohexyl methacrylate	1.485	54	m-Cresyl methacrylate	1.5683	36.8
N-Vinyl phthalimide	1.6200	24.1	o-Methyl-p-methoxy styrene	1.5868	30.3
Fluorenyl methacrylate	1.6319	23.1	o-Methoxy styrene	1.5932	29.7
α-Naphthyl-carbinyl methacrylate	1.63	25	o-Methyl styrene	1.5874	32
p-p'-Xylylenyl dimethacrylate	1.5559	37	Ethyl sulfide dimethacrylate	1.547	44
Cyclohexanediol-1-4 dimethacrylate	1.5067	54.3	Allyl cinnamate	1.57	30
Ethylidene dimethacrylate	1.4831	52.9	Diacetin methacrylate	1.4855	50
p-Divinyl benzene	1.6150	28.1	Ethylene glycol benzoate methacrylate	1.555	36.8
Decamethylene glycol dimethacrylate	1.4990	56.3	Ethyl glycolate methacrylate	1.4903	55
Vinyl cyclohexene dioxide	1.5303	56.4	p-Isopropyl styrene	1.554	35
Methyl α-methylene butyrolactone	1.5118	53.9	Bornyl methacrylate	1.5059	54.6
α-Methylene butyrolactone	1.5412	56.4	Triethyl carbinyl methacrylate	1.4889	57
4-Dioxolylmethyl methacrylate	1.5084	59.7	Butyl mercaptyl methacrylate	1.5390	41.8
Methylene-α-valerolactone	1.5431	47.8	o-Chlorobenzyl methacrylate	1.5823	37
o-Methoxy-phenyl methacrylate	1.5705	33.4	α-Methallyl methacrylate	1.4917	49
Isopropyl methacrylate	1.4728	57.9	β-Methallyl methacrylate	1.5110	47
Trifluoroisopropyl methacrylate	1.4177	65.3	α-Naphthyl methacrylate	1.6411	20.5
β-Ethoxy-ethyl methacrylate	1.4833	32.0	Ethyl acrylate	1.4685	58
Name of polymer			Cinnamyl methacrylate	1.5951	26.5
Condensation resin from di- (p-aminocyclohexyl) methane and sebacic acid	1.5199	52.0	Methyl acrylate	1.4793	59
Columbia resin 39	1.5001	58.8	Terpineyl methacrylate	1.514	50
			Furfuryl methacrylate	1.5381	39.2

* H. C. Raine, Plastic Glasses, *Proc. London Conf. Opt. Instruments* **1950**, 243.

TABLE 6e-4. INDEX OF REFRACTION OF PLASTICS* (*Continued*)

Polystyrene

Spectral line	Wavelength, Å	Refractive index at		
		15°C	35°C	55°C
A	7,679	1.581^2	1.578^5	1.575^8
C	6,563	1.587^0	1.584^3	1.581^6
D$_1$	5,896	1.592^3	1.589^7	1.586^9
F	4,861	1.606^2	1.603^4	1.600^6
g	4,358	1.617^6	1.614^8	1.612^0

Polycyclohexyl Methacrylate

Spectral line	Wavelength, Å	Refractive index at		
		15°C	35°C	55°C
A	7,679	1.501^6	1.499^2	1.496^4
C	6,563	1.504^4	1.502^1	1.499^2
D$_1$	5,896	1.507^1	1.504^6	1.501^8
F	4,861	1.513^4	1.501^0	1.508^1
g	4,358	1.518^4	1.516^0	1.513^1

Polymethyl Methacrylate

Spectral line	Wavelength, Å	Refractive index at 20°C
C	6,563	1.489^0
D	5,896	1.491^3
e	5,461	1.493^2
F	4,861	1.497^5
g	4,358	1.501^9

TABLE 6e-5. INDEX OF REFRACTION OF GASES AND VAPORS*

Wave-length, μm	$(n-1)10^3$†				Wave-length, μm	$(n-1)10^3$‡			
	Air	O_2	N_2	H_2		Air	O_2	CO_2	H_2
0.4861	0.2951	0.2734	0.3012	0.1406	0.4360	0.2971	0.2743	0.4563	0.1418
0.5461	0.2936	0.2717	0.2998	0.1397	0.5461	0.2937	0.2704	0.4506	0.1397
0.5790	0.2930	0.2710	.....	0.1393	0.6709	0.2918	0.2683	0.4471	0.1385
0.6563	0.2919	0.2698	0.2982	0.1387	6.709	0.3881	0.2643	0.4804	0.1361
					8.678	0.2888	0.2650	0.4579	0.1361

* The values are for 0°C and 760 mm Hg.
† Cuthbertson, 1910.
‡ Koch, 1909.

Substance	Kind of light	Indices of refraction	Substance	Kind of light	Indices of refraction
Acetone.............	D	1.001079–1.001100	Hydrogen..........	White	1.000138–1.000143
Ammonia...........	White	1.000381–1.000385	Hydrogen..........	D	1.000132
Ammonia...........	D	1.000373–1.000379	Hydrogen sulfide....	D	1.000644
Argon..............	D	1.000281		D	1.000623
Benzene............	D	1.001700–1.001823	Methane...........	White	1.000443
Bromine...........	D	1.001132	Methane...........	D	1.000444
Carbon dioxide......	White	1.000449–1.000450	Methyl alcohol......	D	1.000549–1.000623
Carbon dioxide......	D	1.000448–1.000454	Methyl ether........	D	1.000891
Carbon disulfide.....	White	1.001500	Nitric oxide........	White	1.000303
	D	1.001478–1.001485	Nitric oxide........	D	1.000297
Carbon monoxide....	White	1.000340	Nitrogen...........	White	1.000295–1.000300
	White	1.000335	Nitrogen...........	D	1.000271–1.000298
Chlorine...........	White	1.000772	Nitrous oxide.......	White	1.000503–1.000507
Chlorine...........	D	1.000773	Nitrous oxide.......	D	1.000516
Chloroform.........	D	1.001436–1.001464	Oxygen.............	White	1.000272–1.000280
Cyanogen..........	White	1.000834	Oxygen.............	D	1.000271–1.000272
Cyanogen..........	D	1.000784–1.000825	Pentane............	D	1.001711
Ethyl alcohol........	D	1.000871–1.000885	Sulfur dioxide.......	White	1.000665
Ethyl ether.........	D	1.001521–1.001544	Sulfur dioxide.......	D	1.000686
Helium.............	D	1.000036	Water..............	White	1.000261
Hydrochloric acid....	White	1.000449	Water..............	D	1.000249–1.000259
	D	1.000447			

* "Smithsonian Physical Tables," 1954, Table 554. A formula was given by Biot and Arago expressing the dependence of the index of refraction of a gas on pressure and temperature. More recent experiments confirm their conclusions. The formula is $n_t - 1 = \dfrac{n_0 - 1}{1 + at}\dfrac{p}{760}$, where n_t is the index of refraction for temperature t, n_0 for temperature zero, a the coefficient of expansion of the gas with temperature, and p the pressure of the gas in millimeters of mercury.

AIR

D. H. Rank, in an article in "Advances in Spectroscopy," vol. I, indicates that the Edlen formula for the dispersion of air agrees with experimental data from the ultraviolet (2,000 Å) to the near infrared (2.06 μm).

The equations are for air free of carbon dioxide at standard conditions and for standard air (dry, 0.03 percent carbon dioxide), respectively:

$$(n - 1)10^8 = 6,431.8 + \frac{2,949,330}{146 - \sigma^2} + \frac{25,536}{41 - \sigma^2}$$

$$(n - 1)10^8 = 6,432.8 + \frac{2,949,810}{146 - \sigma^2} + \frac{25,540}{41 - \sigma^2}$$

σ is the wave number in μm^{-1}.

Furthermore, Rank reports the equation of Barrell and Sears for the change of index with temperature and pressure:

$$n_{Tp} - 1 = (n_{15,760} - 1) \frac{p(1 + \beta_{Tp})(1 + 15\alpha)}{760(1 + 760\beta_{15})(1 + \alpha T)}$$

where T = temperature
p = pressure
α = 0.00366
β_T = $(1.049 - 0.015^7 T)10^{-6}$
β_{15} = $0.813^5 \times 10^{-6}$

6f. Optical Characteristics of Various Uniaxial and Biaxial Crystals

TABLE 6f-1. INDEX OF REFRACTION OF SELECTED UNIAXIAL MINERALS*

Mineral	Formula	Index of refraction	
		Ordinary ray	Extraordinary ray

Uniaxial Positive Minerals

Mineral	Formula	Ordinary ray	Extraordinary ray
Ice	H_2O	1.309	1.313
Sellaite	MgF_2	1.378	1.390
Chrysocolla	$CuO \cdot SiO_2 \cdot 2H_2O$	$1.460 \pm$	$1.570 \pm$
Laubanite	$2CaO \cdot Al_2O_3 \cdot 5SiO_2 \cdot 6H_2O$	1.475	1.486
Chabazite	$(Ca, Na_2)O \cdot Al_2O_3 \cdot 4SiO_2 \cdot 6H_2O$	$1.480 \pm$	$1.482 \pm$
Douglasite	$2KCl \cdot FeCl_2 \cdot 2H_2O$	1.488	1.500
Hydronephelite	$2Na_2O \cdot 3Al_2O_3 \cdot 6SiO_2 \cdot 7H_2O$	1.490	1.502
Apophyllite	$K_2O \cdot 8CaO \cdot 16SiO_2 \cdot 16H_2O$	$1.535 \pm$	$1.537 \pm$
Quartz	SiO_2	1.544	1.553
Coquimbite	$Fe_2O_3 \cdot 3SO_3 \cdot 9H_2O$	1.550	1.556
Brucite	$MgO \cdot H_2O$	1.559	1.580
Alunite	$K_2O \cdot 3Al_2O_3 \cdot 4SO_3 \cdot 6H_2O$	1.572	1.592
Penninite	$5(Mg, Fe)O \cdot Al_2O_3 \cdot 3SiO_2 \cdot 4H_2O$	1.576	1.579
Cacoxenite	$2Fe_2O_3 \cdot P_2O_5 \cdot 12H_2O$	1.582	1.645
Eudialite	$6Na_2O \cdot 6(Ca, Fe)O \cdot 20(Si, Zr)O_2 \cdot NaCl$	1.606	1.611
Dioptase	$CuO \cdot SiO_2 \cdot H_2O$	1.654	1.707
Phenacite	$2BeO \cdot SiO_2$	1.654	1.670
Parisite	$2CeOF \cdot CaO \cdot 3CO_2$	$1.676 \pm$	1.757
Willemite	$2ZnO \cdot SiO_2$	1.691	1.719
Vesuvianite	$2(Ca, Mn, Fe)O \cdot (Al, Fe)$ $(OH, F)O \cdot 2SiO_2$	$1.716 \pm$	1.721
Xenotime	$Y_2O_3 \cdot P_2O_5$	1.721	1.816
Connellite	$20CuO \cdot SO_3 \cdot 2CuCl_2 \cdot 20H_2O$	1.724	1.746
Benitoite	$BaO \cdot TiO_2 \cdot 3SiO_2$	1.757	1.804
Ganomalite	$6PbO \cdot 4(Ca, Mn)O \cdot 6SiO_2 \cdot H_2O$	1.910	1.945
Scheelite	$CaO \cdot WO_3$	1.918	1.934
Zircon	$ZrO_2 \cdot SiO_2$	$1.923 \pm$	$1.968 \pm$
Powellite	$CaO \cdot MoO_3$	1.974	1.978
Calomel	$HgCl$	1.973	2.650
Cassiterite	SnO_2	1.997	2.093
Zincite	ZnO	2.013	2.029
Phosgenite	$PbO \cdot PbCl_2 \cdot CO_2$	2.114	2.140
Penfieldite	$PbO \cdot PbCl_2$	2.130	2.210
Iodyrite	AgI	2.210	2.220
Tapiolite	$FeO \cdot (Ta, Nb)_2O_5$	2.270	2.420 (Li line)
Wurtzite	ZnS	2.356	2.378
Derbylite	$6FeO \cdot Sb_2O_3 \cdot 5TiO_2$	2.450	2.510 (Li line)
Greenockite	CdS	2.506	2.529
Rutile	TiO_2	2.616	2.903
Moissanite	CSi	2.654	2.697
Cinnabar	HgS	2.854	3.201

TABLE 6f-1. INDEX OF REFRACTION OF SELECTED UNIAXIAL
MINERALS* (*Continued*)

Mineral	Formula	Index of refraction	
		Ordinary ray	Extraordinary ray
Uniaxial Negative Minerals			
Chiolite...........	$2NaF \cdot AlF_3$	1.349	1.342
Hanksite..........	$11Na_2O \cdot 9SO_3 \cdot 2CO_2 \cdot KCl$	1.481	1.461
Thaumasite.......	$3CaO \cdot CO_2 \cdot SiO_2 \cdot SO_3 \cdot 15H_2O$	1.507	1.468
Hydrotalcite......	$6MgO \cdot Al_2O_3 \cdot CO_2 \cdot 15H_2O$	1.512	1.498
Cancrinite.......	$4Na_2O \cdot CaO \cdot 4Al_2O_3 \cdot 2CO_2 \cdot 9SiO_2 \cdot 3H_2O$	1.524	1.496
Milarite..........	$K_2O \cdot 4CaO \cdot 2Al_2O_3 \cdot 24SiO_2 \cdot H_2O$	1.532	1.529
Kaliophilite.......	$K_2O \cdot Al_2O_3 \cdot 2SiO_2$	1.537	1.533
Mellite...........	$Al_2O_3 \cdot C_{12}O_9 \cdot 18H_2O$	1.539	1.511
Marialite........	"Ma" $= 3Na_2O \cdot 3Al_2O_3 \cdot 18SiO_2 \cdot 2NaCl$	1.539	1.537
Nephelite........	$Na_2O \cdot Al_2O_3 \cdot 2SiO_2$	1.542	1.538
Wernerite........	$Me_1Ma_1 \pm$	1.578	1.551
Beryl............	$3BeO \cdot Al_2O_3 \cdot 6SiO_2$	1.581 ±	1.575 ±
Torbernite.......	$CuO \cdot 2UO_3 \cdot P_2O_5 \cdot 8H_2O$	1.592	1.582
Meionite..........	"Me" $= 4CaO \cdot 3Al_2O_3 \cdot 6SiO_2$	1.597	1.560
Melilite..........	Contains Na_2O, CaO, Al_2O_3, SiO_2, etc.	1.634	1.629
Apatite..........	$9CaO \cdot 3P_2O_5 \cdot Ca(F, Cl)_2$	1.634	1.631
Calcite..........	$CaO \cdot CO_2$	1.658	1.486
Gehlenite........	$2CaO \cdot Al_2O_3 \cdot SiO_2$	1.669	1.658
Tourmaline.......	Contains Na_2O, FeO, Al_2O_3, B_2O_3, SiO_2, etc.	1.669 ±	1.638 ±
Dolomite.........	$CaO \cdot MgO \cdot 2CO_2$	1.681	1.500
Magnesite........	$MgO \cdot CO_2$	1.700	1.509
Pyrochroite.......	$MnO \cdot H_2O$	1.723	1.681
Corundum........	Al_2O_3	1.768	1.760
Smithsonite.......	$ZnO \cdot CO_2$	1.818	1.618
Rhodochrosite.....	$MnO \cdot CO_2$	1.818	1.595
Jarosite..........	$K_2O \cdot 3Fe_2O_3 \cdot 4SO_3 \cdot 6H_2O$	1.820	1.715
Siderite..........	$FeO \cdot CO_2$	1.875	1.635
Pyromorphite.....	$9PbO \cdot 3P_2O_5 \cdot PbCl_2$	2.050	2.042
Barysilite........	$3PbO \cdot 2SiO_2$	2.070	2.050
Mimetite.........	$9PbO \cdot 3As_2O_5 \cdot PbCl_2$	2.135	2.118
Matlockite........	$PbO \cdot PbCl_2$	2.150	2.040
Stolzite..........	$PbO \cdot WO_3$	2.269	2.182
Geikielite........	$(Mg, Fe)O \cdot TiO_2$	2.310	1.950
Vanadinite........	$9PbO \cdot 3V_2O_5 \cdot PbCl_2$	2.354	2.299
Wulfenite........	$PbO \cdot MoO_3$	2.402	2.304 (Li line)
Octahedrite......	TiO_2	2.554	2.493
Massicotite.......	PbO	2.665	2.535 (Li line)
Proustite.........	$3Ag_2S \cdot As_2S_3$	2.979	2.711 (Li line)
Pryargyrite.......	$3Ag_2S \cdot Sb_2S_3$	3.084	2.881 (Li line)
Hematite.........	Fe_2O_3	3.220	2.940 (Li line)

* "Smithsonian Physical Tables," 1954, Table 546.   Selected by Edgar T. Wherry from a private compilation of Esper S. Larsen, of the U.S. Geological Survey.

TABLE 6f-2. INDEX OF REFRACTION OF SELECTED BIAXIAL MINERALS*

Mineral	Formula	Index of refraction		
		n_α	n_β	n_γ
Biaxial Positive Minerals				
Stercorite........	$NaO \cdot (NH_4)_2O \cdot P_2O_5 \cdot 9H_2O$	1.439	1.441	1.469
Aluminite.......	$Al_2O_3 \cdot SO_3 \cdot 9H_2O$	1.459	1.464	1.470
Tridymite.......	SiO_2	1.469	1.470	1.473
Thenardite......	$Na_2O \cdot SO_3$	1.464	1.474	1.485
Carnallite......	$KCl \cdot MgCl_2 \cdot 6H_2O$	1.466	1.475	1.494
Alunogen........	$Al_2O_3 \cdot 3SO_3 \cdot 16H_2O$	1.474	1.476	1.483
Melanterite.....	$FeO \cdot SO_3 \cdot 7H_2O$	1.471	1.478	1.486
Natrolite........	$Na_2O \cdot Al_2O_3 \cdot 3SiO_2 \cdot 2H_2O$	1.480	1.482	1.493
Arcanite.........	$K_2O \cdot SO_3$	1.494	1.495	1.497
Struvite.........	$(NH_4)_2O \cdot 2MgO \cdot P_2O_5 \cdot 12H_2O$	1.495	1.496	1.500
Heulandite......	$CaO \cdot Al_2O_3 \cdot 6SiO_2 \cdot 3H_2O$	1.498	1.499	1.505
Thomsonite.....	$(Na_2, Ca)O \cdot Al_2O_3 \cdot 2SiO_2 \cdot 3H_2O$	1.497	1.503	1.525
Harmotome.....	$(K_2, Ba)O \cdot Al_2O_3 \cdot 5SiO_2 \cdot 5H_2O$	1.503	1.505	1.508
Petalite.........	$Li_2O \cdot Al_2O_3 \cdot 8SiO_2$	1.504	1.510	1.516
Monetite........	$2CaO \cdot P_2O_5 \cdot H_2O$	1.515	1.518	1.525
Newberyite.....	$2MgO \cdot P_2O_5 \cdot 7H_2O$	1.514	1.519	1.533
Gypsum.........	$CaO \cdot SO_3 \cdot 2H_2O$	1.520	1.523	1.530
Mascagnite.....	$(NH_4)_2O \cdot SO_3$	1.521	1.523	1.533
Albite..........	"Ab" $= Na_2O \cdot Al_2O_3 \cdot 6SiO_2$	1.525	1.529	1.536
Hydromagnesite..	$4MgO \cdot 3CO_2 \cdot 4H_2O$	1.527	1.530	1.540
Wavellite.......	$3Al_2O_3 \cdot 2P_2O_5 \cdot 12(H_2O, 2HF)$	1.525	1.534	1.552
Kieserite........	$MgO \cdot SO_3 \cdot H_2O$	1.523	1.535	1.586
Copiapite........	$2Fe_2O_3 \cdot 5SO_3 \cdot 18H_2O$	1.530	1.550	1.592
Whewellite......	$CaO \cdot C_2O_3 \cdot H_2O$	1.491	1.555	1.650
Variscite........	$Al_2O_3 \cdot P_2O_5 \cdot 4H_2O$	1.551	1.558	1.582
Labradorite......	Ab_2An_3	1.559	1.563	1.568
Gibbsite.........	$Al_2O_3 \cdot 3H_2O$	1.566	1.566	1.587
Wagnerite.......	$3MgO \cdot P_2O_5 \cdot MgF_2$	1.569	1.570	1.582
Anhydrite.......	$CaO \cdot SO_3$	1.571	1.576	1.614
Colemanite......	$2CaO \cdot 3B_2O_3 \cdot 5H_2O$	1.586	1.592	1.614
Fremontite......	$Na_2O \cdot Al_2O_3 \cdot P_2O_5 \cdot (H_2O, 2HF)$	1.594	1.603	1.615
Vivianite.......	$3FeO \cdot P_2O_5 \cdot 8H_2O$	1.579	1.603	1.633
Pectolite........	$Na_2O \cdot 4CaO \cdot 6SiO_2 \cdot H_2O$	1.595	1.604	1.633
Calamine........	$2ZnO \cdot SiO_2 \cdot H_2O$	1.614	1.617	1.636
Chondrodite.....	$4MgO \cdot SiO_2 \cdot Mg(F, OH)_2$	1.604	1.617	1.636
Turquoise.......	$CuO \cdot 3Al_2O_3 \cdot 2P_2O_5 \cdot 9H_2O$	1.610	1.620	1.650
Topaz...........	$2AlOF \cdot SiO_2$	1.619	1.620	1.627
Celestite........	$SrO \cdot SO_3$	1.622	1.624	1.631
Prehnite.........	$2CaO \cdot Al_2O_3 \cdot 3SiO_2 \cdot H_2O$	1.616	1.626	1.649
Barite..........	$BaO \cdot SO_3$	1.636	1.637	1.648
Anthophyllite....	$MgO \cdot SiO_2$	1.633	1.642	1.657
Sillimanite.......	$Al_2O_3 \cdot SiO_2$	1.638	1.642	1.653
Forsterite.......	$2MgO \cdot SiO_2$	1.635	1.651	1.669

TABLE 6f-2. INDEX OF REFRACTION OF SELECTED BIAXIAL
MINERALS* (Continued)

Mineral	Formula	Index of refraction		
		n_α	n_β	n_γ

Biaxial Positive Minerals

Mineral	Formula	n_α	n_β	n_γ
Enstatite	$MgO \cdot SiO_2$	1.650	1.653	1.658
Euclase	$2BeO \cdot Al_2O_3 \cdot 2SiO_2 \cdot H_2O$	1.653	1.656	1.673
Triplite	$3MnO \cdot P_2O_5 \cdot MnF_2$	1.650	1.660	1.672
Spodumene	$Li_2O \cdot Al_2O_3 \cdot 4SiO_2$	1.660	1.666	1.676
Diopside	$CaO \cdot MgO \cdot 2SiO_2$	1.664	1.671	1.694
Olivine	$2(Mg, Fe)O \cdot SiO_2$	1.662	1.680	1.699
Triphylite	$Li_2O \cdot 2(Fe, Mn)O \cdot P_2O_5$	1.688	1.688	1.692
Zoisite	$4CaO \cdot 3Al_2O_3 \cdot 6SiO_2 \cdot H_2O$	1.700	1.702	1.706
Strengite	$Fe_2O_3 \cdot P_2O_5 \cdot 4H_2O$	1.708	1.708	1.745
Diaspore	$Al_2O_3 \cdot H_2O$	1.702	1.722	1.750
Staurolite	$2FeO \cdot 5Al_2O_3 \cdot 4SiO_2 \cdot H_2O$	1.736	1.741	1.746
Chrysoberyl	$BeO \cdot Al_2O_3$	1.747	1.748	1.757
Azurite	$3CuO \cdot 2CO_2 \cdot H_2O$	1.730	1.758	1.838
Scorodite	$Fe_2O_3 \cdot As_2O_5 \cdot 4H_2O$	1.765	1.774	1.797
Olivenite	$4CuO \cdot As_2O_5 \cdot H_2O$	1.772	1.810	1.863
Anglesite	$PbO \cdot SO_3$	1.877	1.882	1.894
Titanite	$CaO \cdot TiO_2 \cdot SiO_2$	1.900	1.907	2.034
Claudetite	As_2O_3	1.871	1.920	2.010
Sulfur	S	1.950	2.043	2.240
Cotunnite	$PbCl_2$	2.200	2.217	2.260
Huebnerite	$MnO \cdot WO_3$	2.170	2.220	2.320
Manganite	$Mn_2O_3 \cdot H_2O$	2.240	2.240	2.530 (Li)
Raspite	$PbO \cdot WO_3$	2.270	2.270	2.300
Mendipite	$2PbO \cdot PbCl_2$	2.240	2.270	2.310
Tantalite	$(Fe, Mn)O \cdot Ta_2O_5$	2.260	2.320	2.430 (Li)
Wolframite	$(Fe, Mn)O \cdot WO_3$	2.310	2.360	2.460 (Li)
Crocoite	$PbO \cdot CrO_3$	2.310	2.370	2.660 (Li)
Pseudobrookite	$2Fe_2O_3 \cdot 3TiO_2$	2.380	2.390	2.420 (Li)
Stibiotantalite	$Sb_2O_3 \cdot Ta_2O_5$	2.374	2.404	2.457
Montroydite	HgO	2.370	2.500	2.650 (Li)
Brookite	TiO_2	2.583	2.586	2.741
Massicot	PbO	2.510	2.610	2.710

Biaxial Negative Minerals

Mineral	Formula	n_α	n_β	n_γ
Mirabilite	$Na_2O \cdot SO_3 \cdot 10H_2O$	1.394	1.396	1.398
Thomsenolite	$NaF \cdot CaF_2 \cdot AlF_3 \cdot H_2O$	1.407	1.414	1.415
Natron	$Na_2O \cdot CO_2 \cdot 10H_2O$	1.405	1.425	1.440
Kalinite	$K_2O \cdot Al_2O_3 \cdot 4SO_3 \cdot 24H_2O$	1.430	1.452	1.458
Epsomite	$MgO \cdot SO_3 \cdot 7H_2O$	1.433	1.455	1.461
Sassolite	$B_2O_3 \cdot H_2O$	1.340	1.456	1.459
Borax	$Na_2O \cdot 2B_2O_3 \cdot 10H_2O$	1.447	1.470	1.472

Table 6f-2. Index of Refraction of Selected Biaxial Minerals* (Continued)

Mineral	Formula	Index of refraction		
		n_α	n_β	n_γ
Biaxial Negative Minerals				
Goslarite	$ZnO \cdot SO_3 \cdot 7H_2O$	1.457	1.480	1.484
Pickeringite	$MgO \cdot Al_2O_3 \cdot 4SO_3 \cdot 22H_2O$	1.476	1.480	1.483
Bloedite	$Na_2O \cdot MgO \cdot 2SO_3 \cdot 4H_2O$	1.483	1.487	1.486
Trona	$3Na_2O \cdot 4CO_2 \cdot 5H_2O$	1.410	1.492	1.542
Thermonatrite	$Na_2O \cdot CO_2 \cdot H_2O$	1.420	1.495	1.518
Stilbite	$(Ca, Na_2)O \cdot Al_2O_3 \cdot 6SiO_2 \cdot 5H_2O$	1.494	1.498	1.500
Niter	$K_2O \cdot N_2O_5$	1.334	1.505	1.506
Kainite	$MgO \cdot SO_3 \cdot KCl \cdot 3H_2O$	1.494	1.505	1.516
Gaylussite	$Na_2O \cdot CaO \cdot 2CO_2 \cdot 5H_2O$	1.444	1.516	1.523
Scolecite	$CaO \cdot Al_2O_3 \cdot 3SiO_2 \cdot 3H_2O$	1.512	1.519	1.519
Laumontite	$CaO \cdot Al_2O_3 \cdot 4SiO_2 \cdot H_2O$	1.513	1.524	1.525
Orthoclase	$K_2O \cdot Al_2O_3 \cdot 6SiO_2$	1.518	1.524	1.526
Microcline	Same as preceding	1.522	1.526	1.530
Anorthoclase	$(Na, K)_2O \cdot Al_2O_3 \cdot 6SiO_2$	1.523	1.529	1.531
Glauberite	$Na_2O \cdot CaO \cdot 2SO_3$	1.515	1.532	1.536
Cordierite	$4(Mg, Fe)O \cdot 4Al_2O_3 \cdot 10SiO_2 \cdot H_2O$	1.534	1.538	1.540
Chalcanthite	$CuO \cdot SO_3 \cdot 5H_2O$	1.516	1.539	1.546
Oligoclase	Ab_4An	1.539	1.543	1.547
Beryllonite	$Na_2O \cdot 2BeO \cdot P_2O_5$	1.552	1.558	1.561
Kaolinite	$Al_2O_3 \cdot 2SiO_2 \cdot 2H_2O$	1.561	1.563	1.565
Biotite	$K_2O \cdot 4(Mg, Fe)O \cdot 2Al_2O_3 \cdot 6SiO_2 \cdot H_2O$	1.541	1.574	1.574
Autunite	$CaO \cdot 2UO_3 \cdot P_2O_5 \cdot 8H_2O$	1.553	1.575	1.577
Anorthite	"An" = $CaO \cdot Al_2O_3 \cdot 2SiO_2$	1.576	1.584	1.588
Lanthanite	$La_2O_3 \cdot 3CO_2 \cdot 9H_2O$	1.520	1.587	1.613
Pyrophyllite	$Al_2O_3 \cdot 4SiO_2 \cdot H_2O$	1.552	1.588	1.600
Talc	$3MgO \cdot 4SiO_2 \cdot H_2O$	1.539	1.589	1.589
Hopeite	$3ZnO \cdot P_2O_5 \cdot 4H_2O$	1.572	1.590	1.590
Muscovite	$K_2O \cdot Al_2O_3 \cdot 6SiO_2 \cdot 2H_2O$	1.561	1.590	1.594
Amblygonite	$Al_2O_3 \cdot P_2O_5 \cdot 2LiF$	1.579	1.593	1.597
Lepidolite	$Al_2O_3 \cdot 3SiO_2 \cdot 2(K, Li)F$	1.560	1.598	1.605
Phlogopite	$K_2O \cdot 6MgO \cdot Al_2O_3 \cdot 6SiO_2 \cdot 2H_2O$	1.562	1.606	1.606
Tremolite	$CaO \cdot 3MgO \cdot 4SiO_2$	1.600	1.616	1.627
Actinolite	$CaO \cdot 3(Mg, Fe)O \cdot 4SiO_2$	1.614	1.630	1.641
Wollastonite	$CaO \cdot SiO_2$	1.620	1.632	1.634
Lazulite	$(Fe, Mg)O \cdot Al_2O_3 \cdot P_2O_5 \cdot H_2O$	1.612	1.634	1.643
Danburite	$CaO \cdot B_2O_3 \cdot 2SiO_2$	1.632	1.634	1.636
Glaucophane	$Na_2O \cdot 2FeO \cdot Al_2O_3 \cdot 6SiO_2$	1.621	1.638	1.638
Andalusite	$Al_2O_3 \cdot SiO_2$	1.632	1.638	1.643
Hornblende	Contains Na_2O, MgO, FeO, SiO_2, etc.	1.634	1.647	1.652
Datolite	$2CaO \cdot 2SiO_2 \cdot B_2O_3 \cdot H_2O$	1.625	1.653	1.669
Erythrite	$3CoO \cdot As_2O_5 \cdot 8H_2O$	1.626	1.661	1.699
Monticellite	$CaO \cdot MgO \cdot SiO_2$	1.651	1.662	1.668

TABLE 6f-2. INDEX OF REFRACTION OF SELECTED BIAXIAL
MINERALS* (Continued)

Mineral	Formula	Index of refraction		
		n_α	n_β	n_γ
Biaxial Negative Minerals				
Strontianite.....	$SrO \cdot CO_2$	1.520	1.667	1.667
Witherite........	$BaO \cdot CO_2$	1.529	1.676	1.677
Aragonite.......	$CaO \cdot CO_2$	1.531	1.682	1.686
Axinite..........	$6(Ca, Mn)O \cdot 2Al_2O_3 \cdot B_2O_3 \cdot 8SiO_2 \cdot H_2O$	1.678	1.685	1.688
Dumortierite....	$8Al_2O_3 \cdot B_2O_3 \cdot 6SiO_2 \cdot H_2O$	1.678	1.686	1.689
Cyanite.........	$Al_2O_3 \cdot SiO_2$	1.712	1.720	1.728
Epidote.........	$4CaO \cdot 3(Al, Fe)_2O_3 \cdot 6SiO_2 \cdot H_2O$	1.729	1.763	1.780
Atacamite.......	$3CuO \cdot CuCl_2 \cdot 3H_2O$	1.831	1.861	1.880
Fayalite.........	$2FeO \cdot SiO_2$	1.824	1.864	1.874
Caledonite.......	$2(Pb, Cu)O \cdot SO_3 \cdot H_2O$	1.818	1.866	1.909
Malachite.......	$2CuO \cdot CO_2 \cdot H_2O$	1.655	1.875	1.909
Lanarkite........	$2PbO \cdot SO_3$	1.930	1.990	2.020
Leadhillite......	$4PbO \cdot SO_3 \cdot 2CO_2 \cdot H_2O$	1.870	2.000	2.010
Cerusite........	$PbO \cdot CO_2$	1.804	2.076	2.078
Laurionite......	$PbCl_2 \cdot PbO \cdot H_2O$	2.077	2.116	2.158
Matlockite......	$PbO \cdot PbCl_2$	2.040	2.150	2.150
Baddeleyite.....	ZrO_2	2.130	2.190	2.200
Lepidocrocite....	$Fe_2O_3 \cdot H_2O$	1.930	2.210	2.510
Limonite........	$2Fe_2O_3 \cdot 3H_2O$ in part	2.170	2.290	2.310
Goethite........	$Fe_2O_3 \cdot H_2O$	2.210	2.350	2.350 (Li)
Valentinite......	Sb_2O_3	2.180	2.350	2.350
Turgite.........	$2Fe_2O_3 \cdot H_2O$ in part	2.450	2.550	2.550 (Li)
Realgar.........	AsS	2.460	2.590	2.610 (Li)
Terlinguaite.....	Hg_2OCl	2.350	2.640	2.660 (Li)
Hutchinsonite...	$(Tl, Ag)_2S \cdot PbS \cdot 2As_2S_3$	3.078	3.176	3.188
Stibnite........	Sb_2S_3	3.194	4.303	4.460

* "Smithsonian Physical Tables," 1954, Table 548. The values are arranged in the order of increasing β index of refraction and are for the sodium D line except where noted. Selected by Edgar T. Wherry from private compilation of Esper S. Larsen, of the U.S. Geological Survey.

6g. Optical Properties of Metals

GEORG HASS

U.S. Army Electronics Command, Night Vision Laboratory

LAWRENCE HADLEY

Colorado State University

The optical properties of metals are usually characterized by two parameters, the index of refraction n and the extinction coefficient k. The refractive index is defined as the ratio of the phase velocity of light in vacuum to the phase velocity of light in the material. The extinction coefficient is related to the exponential decay of the wave as it passes through the medium. Both of these parameters are contained in the equation for the propagation of a wave in an absorbing medium:

$$E = E_0 e^{-2\pi kx/\lambda_0} e^{-2\pi i(nx/\lambda_0 - \nu t)}$$

where E_0 is the amplitude of the wave measured at the point $x = 0$ in the medium, E is the instantaneous value of the electric vector measured at a distance x from the first point and at some time t, ν is the frequency of the source, and λ_0 is the wavelength in vacuum. The two parameters n and k (called the optical "constants," even though they vary strongly with frequency) can be combined to give a complex index of refraction $N = n - ik$. It should be noted that in much of the older literature the complex index of refraction is written as $N = n(1 - i\kappa)$. Consequently, the k used here will equal the $n\kappa$ which is found tabulated in many places elsewhere. This κ is called the absorption index.

In much of the current literature the real and imaginary parts of the complex dielectric constant are given instead of the index of refraction and the extinction coefficient. They are related through the following equations

$$\epsilon = \epsilon_1 - i\epsilon_2 = N^2 = n^2 - k^2 - 2ink$$

A second point on which some confusion has arisen in the literature is that of the absorption coefficient α, which appears in the familiar equation $I = I_0 e^{-\alpha x}$. The absorption coefficient α is related to the extinction coefficient by $\alpha = 4\pi k/\lambda_0$. The use of the above absorption equation implies, however, that the intensities I and I_0 are to be measured *within* the absorbing medium and that the total thickness of the medium is sufficiently great that there are no interference effects arising from multiple reflection.

When light is reflected from a metal surface, it experiences a phase shift which is a function of the angle of incidence and the state of polarization of the incident light. If r_p and r_s represent respectively the amplitude ratios of the reflected electric vector to the incident electric vector for light polarized parallel and perpendicular to the plane of incidence, then

$$\frac{r_p}{r_s} = \frac{|r_p| e^{i\beta_p}}{|r_s| e^{i\beta_s}} = e^{i\Delta} \tan \psi$$

6-118

It may be shown that the phase angle Δ and the azimuth angle ψ are related to the refractive index and the extinction coefficient for a particular angle of incidence ϕ_1 by the following equations

$$\frac{r_{1p}}{r_{1s}} = \frac{|r_{1p}|e^{i\beta_p}}{|r_{1s}|e^{i\beta_s}} = e^{i\Delta} \tan \psi$$

$$n^2 - k^2 = n_1{}^2 \sin^2 \phi_1 \left\{ 1 + \frac{\tan^2 \phi_1(\cos^2 2\psi - \sin^2 2\psi \sin^2 \Delta)}{(1 + \sin 2\psi \cos \Delta)^2} \right\}$$

$$nk = \frac{n_1{}^2 \sin^2 \phi_1 \tan^2 \phi_1 \sin 2\psi \cos 2\psi \sin \Delta}{(1 + \sin 2\psi \cos \Delta)^2}$$

where n_1 is the refractive index of the incident medium. Since these angles are relatively easily measured quantities, these equations form the basis of several of the methods used to determine the optical constants of metals. Reference 1 also lists a number of other methods for these determinations.

Since reflection methods are used in determining the constants, they are strongly dependent on the characteristics of the metallic surface. These characteristics vary considerably with the chemical and mechanical treatment. Accordingly, there has always been a certain degree of controversy on the subject of the optical constants of metals. Since the oldest measurements were made, there has been considerable development in the preparation of metallic surfaces by evaporation in a vacuum. The properties of such surfaces are frequently quite different from those of surfaces of bulk metals prepared by polishing. By no means all the metallic constants have been determined on such freshly prepared surfaces.

It is also well known that the presence of an extremely thin surface film on a metal will significantly alter the values of the phase and azimuth angles, making ellipsometric measurements subject to some difficulties. The appropriate corrections to be made in the presence of such surface films are given in ref. 2.

The relationships existing among n and k and the reflectance, transmittance, and phase shift are given here for several cases of interest. Since the properties of an absorbing dielectric material can also be expressed by a complex index $N = n - ik$, the following equations have general application.

CASE I.   Reflection at the boundary between two homogeneous, isotropic massive media, the one a dielectric of refractive index n_0, which is assumed to be the medium of incidence, and the other an opaque absorbing medium whose complex refractive index will be denoted by $N_1 = n_1 - ik_1$:

$$n_0 \quad | \quad N_1$$

Incident
light
$\longrightarrow$

a. *Intensity reflectance* (normal incidence):

$$R = \frac{(n_0 - n_1)^2 + k_1{}^2}{(n_0 + n_1)^2 + k_1{}^2} \tag{6g-1}$$

b. *Phase change on reflection* (normal incidence):

$$\rho = \tan^{-1} \frac{2n_0 k_1}{n_0{}^2 - n_1{}^2 - k_1{}^2} \tag{6g-2}$$

CASE II. Reflection, transmission, and absorption of light by a thin absorbing film N_1 of true thickness h_1 surrounded by homogeneous, isotropic, massive media, the incident medium being a dielectric of refractive index n_0 and the emergent medium being an absorbing medium whose complex refractive index is given by $N_2 = n_2 - ik_2$:

$$n_0 \quad \bigg| \quad N_1 \quad \bigg| \quad N_2$$

Incident
light
$\longrightarrow$ $\leftarrow h_1 \rightarrow$

a. *Intensity reflectance* (normal incidence):

$$R = \frac{a_1 e^{\sigma} + a_2 e^{-\sigma} + a_3 \cos \nu + a_4 \sin \nu}{b_1 e^{\sigma} + b_2 e^{-\sigma} + b_3 \cos \nu + b_4 \sin \nu} \tag{6g-3}$$

where:

$$a_1 = [(n_0 - n_1)^2 + k_1^2][(n_1 + n_2)^2 + (k_1 + k_2)^2]$$
$$a_2 = [(n_0 + n_1)^2 + k_1^2][(n_1 - n_2)^2 + (k_1 - k_2)^2]$$
$$a_3 = 2\{[n_0^2 - (n_1^2 + k_1^2)][(n_1^2 + k_1^2) - (n_2^2 + k_2^2)] + 4n_0 k_1(n_1 k_2 - n_2 k_1)\}$$
$$a_4 = 4\{[n_0^2 - (n_1^2 + k_1^2)](n_1 k_2 - n_2 k_1) - n_0 k_1[(n_1^2 + k_1^2) - (n_2^2 + k_2^2)]\}$$
$$\sigma = \frac{4\pi k_1 h_1}{\lambda_0} \qquad \lambda_0 = \text{vacuum wavelength}$$
$$\nu = \frac{4\pi n_1 h_1}{\lambda_0}$$
$$b_1 = [(n_0 + n_1)^2 + k_1^2][(n_1 + n_2)^2 + (k_1 + k_2)^2]$$
$$b_2 = [(n_0 - n_1)^2 + k_1^2][(n_1 - n_2)^2 + (k_1 - k_2)^2]$$
$$b_3 = 2\{[n_0^2 - (n_1^2 + k_1^2)][(n_1^2 + k_1^2) - (n_2^2 + k_2^2)] - 4n_0 k_1(n_1 k_2 - n_2 k_1)\}$$
$$b_4 = 4\{[n_0^2 - (n_1^2 + k_1^2)](n_1 k_2 - n_2 k_1) + n_0 k_1[(n_1^2 + k_1^2) - (n_2^2 + k_2^2)]\}$$

b. *Phase change on reflection* (normal incidence):

$$\rho = \tan^{-1} \frac{c_1 e^{\sigma} + c_2 e^{-\sigma} + c_3 \cos \nu + c_4 \sin \nu}{d_1 e^{\sigma} + d_2 e^{-\sigma} + d_3 \cos \nu + d_4 \sin \nu} \tag{6g-4}$$

where

$$c_1 = 2n_0 k_1[(n_1 + n_2)^2 + (k_1 + k_2)^2]$$
$$c_2 = -2n_0 k_1[(n_1 - n_2)^2 + (k_1 - k_2)^2]$$
$$c_3 = 8n_0 n_1[n_1 k_2 - n_2 k_1]$$
$$c_4 = -4n_0 n_1[(n_1^2 + k_1^2) - (n_2^2 + k_2^2)]$$
$$d_1 = [n_0^2 - (n_1^2 + k_1^2)][(n_1 + n_2)^2 + (k_1 + k_2)^2]$$
$$d_2 = [n_0^2 - (n_1^2 + k_1^2)][(n_1 - n_2)^2 + (k_1 - k_2)^2]$$
$$d_3 = 2[n_0^2 + (n_1^2 + k_1^2)][(n_1^2 + k_1^2) - (n_2^2 + k_2^2)]$$
$$d_4 = 4[n_0^2 + (n_1^2 + k_1^2)](n_1 k_2 - n_2 k_1)$$

The symbols σ and ν have the same definitions as in Eq. (6g-3).

c. *Intensity transmittance* (normal incidence): This denotes the percentage of incident intensity which is transmitted into the final medium.

$$T = \frac{16 n_0 n_2 (n_1^2 + k_1^2)}{b_1 e^{\sigma} + b_2 e^{-\sigma} + b_3 \cos \nu + b_4 \sin \nu} \tag{6g-5}$$

where b_1, b_2, b_3, b_4, σ, and ν are defined as in Eq. (6g-3). Alternatively, one can write

$$T = (1 - R)\Psi \tag{6g-6}$$

where
$$\Psi = \frac{4n_2(n_1{}^2 + k_1{}^2)}{g_1 e^{\sigma} + g_2 e^{-\sigma} + g_3 \cos \nu + g_4 \sin \nu}$$

$$g_1 = \frac{b_1 - a_1}{4n_0} = n_1[(n_1 + n_2)^2 + (k_1 + k_2)^2]$$

$$g_2 = \frac{b_2 - a_2}{4n_0} = -n_1[(n_1 - n_2)^2 + (k_1 - k_2)^2]$$

$$g_3 = \frac{b_3 - a_3}{4n_0} = -4k_1(n_1 k_2 - n_2 k_1)$$

$$g_4 = \frac{b_4 - a_4}{4n_0} = 2k_1[(n_1{}^2 + k_1{}^2) - (n_2{}^2 + k_2{}^2)]$$

d. Intensity absorptance (normal incidence):

$$A = (1 - R - T) = (1 - R)(1 - \Psi) \tag{6g-7}$$

This is the percentage of incident intensity which is absorbed by the film. In the simpler case where the emergent medium is a nonabsorbing material of refractive index n_2, the formulas for R, T, and Ψ can be obtained from Eqs. (6g-3) and (6g-6) by setting $k_2 = 0$.

CASE III.   The effect of a nonabsorbing surface film of refractive index n_1 and thickness h_1 on the reflectance of an opaque metal of complex index $N_2 = n_2 - ik_2$, where n_0 is the index (real) of the incident medium:

$$n_0 \quad | \quad n_1 \quad | \quad N_2$$

Incident light $\longrightarrow$ $\leftarrow h_1 \rightarrow$

a. Intensity reflectance (normal incidence):

$$R = \frac{r_1{}^2 + r_2{}^2 - 2r_1 r_2 \cos (\nu - \delta)}{1 + r_1{}^2 r_2{}^2 - 2r_1 r_2 \cos (\nu - \delta)} \tag{6g-8}$$

where

$$r_1 = \frac{n_1 - n_0}{n_1 + n_0} \qquad r_2 = \left[\frac{(n_2 - n_1)^2 + k_2{}^2}{(n_2 + n_1)^2 + k_2{}^2}\right]^{\frac{1}{2}}$$

$$\nu = \frac{4\pi n_1 h_1}{\lambda_0} \qquad \delta = \tan^{-1}\left(\frac{2n_1 k_2}{n_1{}^2 - n_2{}^2 - k_2{}^2}\right)$$

δ is the absolute phase change at the dielectric-metal boundary.

A minimum value of reflectance occurs when $\nu - \delta = 2m\pi$, where m is an integer:

$$R_{\min} = \frac{(r_1 - r_2)^2}{(1 - r_1 r_2)^2} \tag{6g-9}$$

A maximum value of reflectance occurs when $\nu - \delta = (2m - 1)\pi$:

$$R_{\max} = \frac{(r_1 + r_2)^2}{(1 + r_1 r_2)^2} \tag{6g-10}$$

CASE IV.   The reflectance of a metal can be increased by the addition of a pair of dielectric layers to its surface (see ref. 3):

By using pairs of dielectric films with alternately low and high indices of refraction, mirror protection and reflectance enhancement over a rather broad spectral region can be achieved. To obtain the maximum reflectance increase with a low-index–high-index film pair, the metal surface must first be coated with low-index material until its reflectance decreases to a minimum at the wavelength at which highest reflectance is desired. Then the high-index material must be applied until the reflectance reaches a maximum. For further reflectance increase, more film pairs must be added in the same sequence. Under these conditions, the low-index film adjacent to the metal is effectively $\lambda/4$, and all other films are truly $\lambda/4$ thick. The optical thickness of the effectively $\lambda/4$-thick film on the metal surface can be determined from the following equation:

$$n_1 t_1 = \frac{\lambda}{4}\frac{\delta}{180°} \tag{6g-11}$$

where δ is the absolute phase change at the dielectric-metal boundary, as given by

$$\tan \delta = \frac{2n_1 k}{n_1{}^2 - n^2 - k^2} \tag{6g-12}$$

where n_1 is the refractive index of the low-index dielectric film, and n and k are the constants for the mirror material.

For normal incidence the maximum reflectance of a metal with optical constants n and k when coated with low-index n_L–high-index n_H film pairs is given by the following expression:

$$R = \left|\frac{1 - Y^{2x}Z}{1 + Y^{2x}Z}\right|^2 \tag{6g-13}$$

where

$$Y = \frac{n_H}{n_L} \qquad Z = n_L\left|\frac{1 + r_3}{1 - r_3}\right|$$

x is the number of film pairs, and

$$r_3 = \left(\frac{(n_L - n)^2 + k^2}{(n_L + n)^2 + k^2}\right)^{\frac{1}{2}} \tag{6g-14}$$

For opaque coatings, the reflectances R_s and R_p and their dependence on angle of incidence i are given below. As before, the subscripts s and p refer to light polarized perpendicular and parallel to the plane of incidence. Here, the incident medium has refractive incidence of unity, and n and k are the values for the coating material. Normal incidence:

$$R_s = R_p = \frac{(n - 1)^2 + k^2}{(n + 1)^2 + k^2} \tag{6g-15}$$

Reflectances as a function of angle of incidence i:

$$R_s = \frac{a^2 + b^2 - 2a\cos i + \cos^2 i}{a^2 + b^2 + 2a\cos i + \cos^2 i} \tag{6g-16}$$

$$R_p = R_s \cdot \frac{a^2 + b^2 - 2a\sin i \tan i + \sin^2 i \tan^2 i}{a^2 + b^2 + 2a\sin i \tan i + \sin^2 i \tan^2 i} \tag{6g-17}$$

where
$$2a^2 = [(n^2 - k^2 - \sin^2 i)^2 + 4n^2k^2]^{\frac{1}{2}} + (n^2 - k^2 - \sin^2 i)$$
$$2b^2 = [(n^2 - k^2 - \sin^2 i)^2 + 4n^2k^2]^{\frac{1}{2}} - (n^2 - k^2 - \sin^2 i)$$

For unpolarized light with equal amplitudes of perpendicular and parallel components, the reflectance is

$$R = \tfrac{1}{2}(R_s + R_p) \tag{6g-18}$$

A great deal of work remains to be done in this area. The following tables and graph include both old and new data on the optical constants and reflectance of metals as a function of wavelength. In recent years many of these values have been extended into the vacuum ultraviolet region, and in some cases, further into the infrared region.

Many of the values of refractive index and extinction coefficient given in these tables have been calculated from graphical values of the real and imaginary parts of the dielectric coefficient. In order to facilitate these calculations a computer program was written. Because of this it was not possible to maintain a uniform standard of usable significant digits. Where the computer was used, the data were computed to four digits beyond the decimal place. If some question exists as to the reliability of a particular datum, the original reference should be consulted.

References for Sec. 6g

1. Heavens, O. S.: "Physics of Thin Films," G. Hass and R. Thun, eds., vol. 2, pp. 193–238, Academic Press, Inc., New York, 1964.
2. Burge, D. K., and H. E. Bennett: *J. Opt. Soc. Am.* **54**, 1428 (1964).
3. Hass, G., and A. P. Bradford: *J. Opt. Soc. Am.* **44**, 810 (1954).

OPTICS

Table 6g-1. Optical Constants of Metals

Metal	Wave-length, μm	Index of refraction n	Extinction coefficient k	Reflec-tance calculated	Ref.
Aluminum, evaporated..........	0.010	0.99	0.0041	0.00003	1
	0.011	....	0.0051		
	0.012	0.99	0.0068		
	0.013	0.99	0.0079		
	0.014	....	0.0087		
	0.015	0.98	0.0076		
	0.016	....	0.0084		
	0.017	0.99	0.0038		
	0.018	0.98	0.0043		
	0.019	....	0.0044		
	0.020	0.97	0.0048		
	0.021	....	0.0048		
	0.022	....	0.0052		
	0.023	....	0.0060		
	0.024	0.96	0.0064		
	0.025	....	0.0067		
	0.026	....	0.0074		
	0.027	....	0.0079		
	0.028	....	0.0084		
	0.029	....	0.0088		
	0.030	0.93	0.0096		
	0.0344	0.96	0.0095	....	2
	0.0376	0.943	0.0110		
	0.0413	0.912	0.0125		
	0.0443	0.880	0.0141		
	0.0477	0.838	0.0159		
	0.0516	0.785	0.0182		
	0.0563	0.718	0.0213		
	0.0620	0.635	0.0267		
	0.0652	0.580	0.0307		
	0.0689	0.520	0.0355		
	0.0729	0.445	0.0424		
	0.0775	0.345	0.0632		
	0.0826	0.225	0.22		
	0.0920	0.104	0.39		
	0.1032	0.033	0.58		
	0.0584	0.71	0.018	....	3
	0.735	0.455	0.043		
	0.120	0.057	1.15	0.9019	4
	0.140	0.065	1.43	0.9122	
	0.160	0.080	1.73	0.9231	
	0.180	0.095	1.97	0.9290	
	0.200	0.110	2.20	0.9275	
	0.220	0.130	2.40	0.9261	
	0.240	0.160	2.53	0.9174	
	0.2200	0.14	2.35	0.918	5
	0.240	0.16	2.60	0.921	
	0.260	0.19	2.85	0.920	
	0.280	0.22	3.13	0.922	
	0.300	0.25	3.33	0.921	
	0.320	0.28	3.56	0.922	
	0.340	0.31	3.80	0.923	
	0.360	0.34	4.01	0.924	
	0.380	0.37	4.25	0.926	
	0.400	0.40	4.45	0.926	
	0.436	0.47	4.84	0.927	

TABLE 6g-1. OPTICAL CONSTANTS OF METALS (*Continued*)

Metal	Wave-length, μm	Index of refraction n	Extinction coefficient k	Reflectance calculated	Ref.
Aluminum evaporated (*Cont.*)....	0.450	0.51	5.00	0.925	
	0.492	0.64	5.50	0.922	
	0.546	0.82	5.99	0.916	
	0.578	0.93	6.33	0.915	
	0.650	1.30	7.11	0.907	
	0.700	1.55	7.00	0.888	6, 7
	0.750	1.80	7.12	0.877	
	0.800	1.99	7.05	0.864	
	0.850	2.08	7.15	0.863	
	0.900	1.96	7.70	0.885	
	0.950	1.75	8.50	0.912	
	2.0	2.3	16.5	0.968	8
	4.0	6.1	30.4	0.975	
	6.0	10.8	42.6	0.978	
	8.0	17.9	55.3	0.979	
	10.0	26.0	67.3	0.980	
	12.0	33.1	78.0	0.982	
Antimony, evaporated..........	0.0310	1.0291	0.2429	0.0143	9
	0.0400	1.0055	0.3332	0.0269	
	0.0500	0.8194	0.1464	0.0162	
	0.0620	0.6976	0.1290	0.0373	
	0.0830	0.4989	0.4109	0.1739	
	0.1000	0.6013	0.8316	0.2612	
	0.1140	0.6167	1.0297	0.3286	
	0.1240	0.5141	1.0697	0.4016	
	0.1650	0.6694	1.3594	0.4223	
	0.1950	0.6246	1.2728	0.4134	
	0.2180	0.6805	1.4328	0.4419	
	0.2420	0.6000	1.2083	0.4030	
	0.2820	0.4602	1.5754	0.6010	
Antimony, single crystal, optic axis	0.4000	1.1297	2.8768	0.6473	10
	0.4500	1.4972	2.9054	0.5920	
	0.5000	1.6031	3.4308	0.6542	
	0.5500	2.0620	3.8797	0.6623	
	0.6000	2.6008	4.1910	0.6592	
	0.6500	2.7296	4.2131	0.6551	
	0.7000	2.7698	3.9714	0.6305	
	0.8000	2.9409	4.0803	0.6344	
	0.8500	3.0463	4.1689	0.6390	
	0.9000	3.2938	4.3415	0.6466	
	0.9500	3.5726	4.2266	0.6314	
	1.0000	3.8139	4.1165	0.6197	
	1.1000	3.9268	4.1255	0.6196	
	1.2000	4.0621	4.0989	0.6170	
	1.3000	4.4216	4.1049	0.6175	
	1.4000	4.6672	3.9853	0.6111	
	1.5000	4.8446	3.9838	0.6127	
	1.6000	4.9338	3.9928	0.6142	
	1.8000	5.1815	4.0432	0.6201	
	2.0000	5.4148	4.1737	0.6302	
	2.2000	5.8958	4.3082	0.6433	
	2.5000	6.3294	4.2263	0.6463	
	3.0000	7.0784	4.3936	0.6652	
	3.5000	7.8361	4.2432	0.6738	
	4.0000	7.9222	3.9194	0.6663	
	4.5000	7.3412	3.0649	0.6281	

TABLE 6g-1. OPTICAL CONSTANTS OF METALS (*Continued*)

Metal	Wave-length, μm	Index of refraction n	Extinction coefficient k	Reflec-tance calculated	Ref.
Antimony, single crystal, optic axis (*Cont.*)	5.0000	6.8470	2.0447	0.5835	
	5.5000	6.5086	4.0368	0.6028	
	6.0000	6.2512	3.3433	0.6078	
	6.5000	6.0593	3.5235	0.6106	
	7.0000	5.4416	3.9511	0.6188	
	7.5000	5.3852	4.3360	0.6384	
	8.0000	5.0797	5.0003	0.6721	
	8.5000	5.6760	5.7849	0.7152	
	9.0000	5.2135	6.5407	0.7438	
	9.5000	5.0092	7.1967	0.7721	
	10.0000	5.0755	7.4673	0.7809	
Antimony, basal plane............	0.4000	0.9898	2.5258	0.6171	
	0.4500	1.1146	2.8710	0.6493	
	0.5000	1.3788	3.0824	0.6362	
	0.5500	1.6696	3.5338	0.6595	
	0.6000	1.9499	3.7950	0.6624	
	0.6500	2.0179	3.7910	0.6562	
	0.7000	2.1127	3.8812	0.6586	
	0.8000	2.5606	4.5888	0.6964	
	0.8500	2.7946	4.7233	0.6955	
	0.9000	2.8059	5.0964	0.7226	
	0.9500	3.1582	5.1453	0.7113	
	1.0000	3.3243	5.1138	0.7035	
	1.1000	3.5400	5.3800	0.7140	
	1.2000	3.8542	5.5005	0.7135	
	1.3000	4.3282	5.5797	0.7091	
	1.4000	4.5459	5.6094	0.7078	
	1.5000	4.6786	5.7175	0.7118	
	1.6000	5.1888	5.6854	0.7061	
	1.8000	5.6637	5.8972	0.7139	
	2.0000	6.0180	6.1981	0.7254	
	2.2000	6.5269	6.5728	0.7385	
	2.5000	7.3291	6.7539	0.7451	
	3.0000	8.8769	6.9281	0.7560	
	3.5000	10.2871	6.3186	0.7541	
	4.0000	11.1722	5.2362	0.7455	
	4.5000	11.2230	4.5888	0.7366	
	5.0000	11.0380	4.0542	0.7264	
	5.5000	10.5985	3.9911	0.7182	
	6.0000	10.4824	3.6442	0.7111	
	6.5000	10.2814	3.2875	0.7022	
	7.0000	10.2090	3.4675	0.7034	
	7.5000	9.8931	3.7248	0.7014	
	8.0000	9.8153	3.4843	0.6959	
	8.5000	9.4542	3.8448	0.6952	
	9.0000	9.0686	3.7602	0.6860	
	9.5000	8.7140	3.8903	0.6817	
	10.0000	8.1891	3.9321	0.6721	
Antimony, evaporated..........	1.0	3.4	4.1	0.624	11
	1.5	4.5	4.4	0.637	
	2.0	5.4	4.6	0.652	
	2.5	6.2	4.8	0.669	
	3.0	6.8	4.9	0.679	
	4.0	7.8	5.0	0.694	
	5.0	7.6	4.5	0.677	
	6.0	7.3	4.0	0.656	

TABLE 6g-1. OPTICAL CONSTANTS OF METALS (*Continued*)

Metal	Wave-length, μm	Index of refraction n	Extinction coefficient k	Reflec-tance calculated	Ref.
Antimony, evaporated (*Cont.*)....	7.0	7.0	4.0	0.650	
	8.0	6.6	4.5	0.662	
	9.0	6.2	4.9	0.673	
	10.0	5.7	5.6	0.701	
	11.0	5.1	6.3	0.735	
	12.0	4.3	7.0	0.777	
Antimony, bulk................	1.0000	2.8000	4.5000	0.6771	12
	2.0000	3.4000	4.4000	0.6488	
	3.0000	4.0000	4.3000	0.6321	
	4.0000	4.4000	4.1000	0.6171	
	5.0000	4.8000	4.0000	0.6132	
	6.0000	5.0000	4.0000	0.6154	
	7.0000	5.1000	3.9000	0.6108	
	8.0000	4.9000	3.9000	0.6082	
	9.0000	4.4000	3.8000	0.5963	
	10.0000	2.0000	3.9000	0.6696	
	11.0000	2.0000	5.0000	0.7647	
	12.0000	6.0000	9.0000	0.8154	
Barium, evaporated............	0.1440	0.7400	0.1100	0.0262	13
	0.1550	0.6300	0.2000	0.0656	
	0.1650	0.5700	0.2900	0.1055	
	0.1770	0.5300	0.4300	0.1607	
	0.1900	0.5300	0.5300	0.1914	
	0.2070	0.5400	0.6300	0.2198	
	0.2250	0.5900	0.7300	0.2290	
	0.2480	0.6100	0.8700	0.2714	
	0.2750	0.6900	0.9300	0.2583	
	0.3100	0.7700	1.0900	0.2872	
Barium, evaporated............	0.4046	0.82	1.07	0.264	14
	0.4358	0.78	1.10	0.287	
	0.4916	0.86	1.26	0.318	
	0.5461	0.89	1.51	0.392	
	0.5780	0.88	1.52	0.398	
	0.3130	0.76	7.84	0.288	15
	0.3650	0.72	7.10	0.347	
	0.4047	0.69	7.12	0.425	
	0.4358	0.72	2.12	0.454	
	0.4916	0.81	2.19	0.471	
	0.5461	0.90	2.19	0.520	
	0.5780	0.90	2.32	0.548	
Beryllium, evaporated..........	0.4046	2.48	2.20	0.415	14
	0.4358	2.56	2.23	0.420	
	0.4916	2.64	2.25	0.423	
	0.5461	2.66	2.36	0.439	
	0.5780	2.64	2.27	0.426	
	0.8	2.7	2.8	0.498	16
	0.9	2.7	2.9	0.511	
	1.2	2.4	3.5	0.597	
	1.5	2.4	4.7	0.715	
	2.0	2.5	5.8	0.782	
	2.5	2.75	7.25	0.835	
	3	3.45	9.0	0.863	
	3.5	3.9	10.8	0.889	
	4	5.0	12.2	0.892	
	5	6.85	14.3	0.897	
	6	8.0	15.4	0.899	

TABLE 6g-1. OPTICAL CONSTANTS OF METALS (*Continued*)

Metal	Wavelength, μm	Index of refraction n	Extinction coefficient k	Reflectance calculated	Ref.
Beryllium, evaporated (*Cont.*)....	7	9.1	17.1	0.908	
	8	11.0	19.4	0.915	
	9	11.4	18.6	0.909	
	10	11.9	20.0	0.916	
	11	11.9	21.1	0.922	
Bismuth, cleaved, single crystal...	0.350	0.82	2.57	0.669	17
	0.370	0.87	2.78	0.690	
	0.390	0.93	3.00	0.708	
	0.410	0.99	3.17	0.717	
	0.430	1.09	3.01	0.675	
	0.440	1.17	3.30	0.700	
	0.450	1.28	3.38	0.692	
	0.460	1.25	3.41	0.700	
	0.470	1.1	2.87	0.652	
	0.490	1.11	2.94	0.661	
	0.510	1.18	2.93	0.646	
	0.530	1.19	3.03	0.659	
	0.550	1.24	3.17	0.671	
	0.570	1.28	3.27	0.678	
	0.589	1.35	3.36	0.679	
	0.610	1.37	3.52	0.696	
	0.630	1.42	3.60	0.698	
	0.650	1.46	3.71	0.705	
	0.670	1.52	3.65	0.691	
Bismuth, single crystal, optic axis.	0.4000	1.4477	2.2794	0.4824	10
	0.4500	1.6082	2.4872	0.5048	
	0.5000	1.7412	2.6705	0.5245	
	0.5500	2.1015	2.8313	0.5234	
	0.6000	2.0230	3.0647	0.5633	
	0.6500	2.0542	3.3346	0.5982	
	0.7000	2.2682	3.9679	0.6567	
	0.7500	2.6717	4.4540	0.6793	
	0.8000	2.8324	4.5191	0.6773	
	0.8500	3.2570	4.3598	0.6491	
	0.9000	3.4014	4.4687	0.6542	
	0.9500	3.7072	4.3294	0.6374	
	1.0000	4.0801	4.3872	0.6378	
	1.1000	4.1849	4.5402	0.6476	
	1.2000	4.6152	4.6368	0.6519	
	1.3000	5.0398	4.5934	0.6499	
	1.4000	5.4377	4.5791	0.6515	
	1.5000	5.8069	4.6066	0.6562	
	1.6000	6.1789	4.4021	0.6515	
	1.8000	6.6576	4.1381	0.6485	
	2.0000	7.0147	3.8348	0.6446	
	2.2000	8.0032	3.3987	0.6543	
	2.5000	7.6241	2.8331	0.6299	
	3.0000	7.6813	2.1220	0.6153	
	3.5000	7.5552	1.4427	0.5985	
	4.0000	7.3769	1.0574	0.5861	
	4.5000	7.3832	1.5847	0.5943	
	5.0000	7.4842	2.1245	0.6086	
	5.5000	7.6034	1.8741	0.6077	
	6.0000	7.5045	1.7656	0.6021	
	6.5000	7.6564	2.5338	0.6235	
	7.0000	7.5650	2.2868	0.6150	

TABLE 6g-1. OPTICAL CONSTANTS OF METALS (*Continued*)

Metal	Wave-length, μm	Index of refraction n	Extinction coefficient k	Reflec-tance calculated	Ref.
Bismuth, single crystal optic axis. (*Cont.*)	7.5000	7.4913	2.1492	0.6094	
	8.0000	7.5900	2.3254	0.6166	
	8.5000	7.6510	2.4572	0.6216	
	9.0000	7.7011	2.1231	0.6160	
	9.5000	7.6220	1.8696	0.6083	
	10.0000	7.4470	2.0141	0.6050	
Bismuth, basal plane............	0.4000	1.2015	2.5384	0.5743	
	0.4500	1.4371	2.7139	0.5679	
	0.5000	1.5195	2.8300	0.5766	
	0.5500	1.7009	3.1453	0.6042	
	0.6000	1.7896	3.4645	0.6382	
	0.6500	1.7655	3.8232	0.6828	
	0.7000	2.1909	4.3589	0.6997	
	0.7500	2.1788	4.7273	0.7314	
	0.8000	2.6054	4.0088	0.7264	
	0.8500	3.0036	5.2270	0.7229	
	0.9000	3.2281	5.3592	0.7229	
	0.9500	3.7049	5.4522	0.7143	
	1.0000	3.9147	5.2750	0.6987	
	1.1000	4.2638	5.6991	0.7166	
	1.2000	4.8698	5.7805	0.7130	
	1.3000	5.3944	5.8394	0.7123	
	1.4000	5.9624	5.9624	0.7162	
	1.5000	6.4591	5.8754	0.7134	
	1.6000	7.0329	5.7586	0.7120	
	1.8000	7.8028	5.2994	0.7044	
	2.0000	8.2569	4.8142	0.6966	
	2.2000	8.5265	4.5388	0.6937	
	2.5000	8.7956	3.7633	0.6805	
	3.0000	8.7609	3.0419	0.6647	
	3.5000	8.7156	2.8742	0.6604	
	4.0000	8.5067	2.0161	0.6397	
	4.5000	8.8043	2.3909	0.6542	
	5.0000	8.7850	2.1855	0.6504	
	5.5000	8.7449	2.0183	0.6468	
	6.0000	8.9090	2.6210	0.6608	
	6.5000	8.7959	2.2511	0.6517	
	7.0000	8.8615	1.3880	0.6426	
	7.5000	8.8662	1.0997	0.6401	
	8.0000	8.8560	1.9309	0.6488	
	8.5000	8.8718	2.0514	0.6509	
	9.0000	8.6748	1.8848	0.6428	
	9.5000	8.6008	2.2965	0.6470	
	10.0000	8.6907	2.3934	0.6511	
Bismuth, polycrystalline bulk....	1.0000	1.700	3.3000	0.6260	12
	2.0000	2.0000	3.2000	0.5842	
	3.0000	2.1000	3.3000	0.5902	
	4.0000	2.2000	3.4000	0.5963	
	5.0000	2.3000	3.4000	0.5902	
	6.0000	2.4000	3.5000	0.5968	
	7.0000	2.4000	3.6000	0.6085	
	8.0000	2.5000	3.7000	0.6145	
	9.0000	2.5000	3.9000	0.6358	
	10.0000	2.6000	4.0000	0.6409	
	11.0000	2.9000	4.1000	0.6377	
	12.0000	3.8000	4.3000	0.6340	
	13.0000	5.6000	4.4000	0.6440	

TABLE 6g-1. OPTICAL CONSTANTS OF METALS (*Continued*)

Metal	Wave-length, μm	Index of refraction n	Extinction coefficient k	Reflec-tance calculated	Ref.
Bismuth, bulk.................	1.00	4.5	5.0	0.674	18
	1.15	5.0	4.9	0.667	
	1.41	5.6	4.6	0.654	
	1.88	6.6	3.7	0.631	
	2.76	7.9	2.7	0.635	
	3.55	8.6	1.2	0.632	
	5.01	8.2	1.5	0.623	
	5.68	7.6	2.2	0.614	
Cadmium, single crystal, optic axis	0.2400	0.6084	1.7917	0.5802	19
	0.2600	0.3856	1.4521	0.6171	
	0.2800	0.3679	1.4406	0.6271	
	0.3000	0.3598	1.5425	0.6596	
	0.3200	0.3990	1.6791	0.6658	
	0.3400	0.4393	1.7757	0.6637	
	0.3600	0.5085	1.9075	0.6561	
	0.3800	0.6132	2.0140	0.6316	
	0.4000	0.7135	2.1585	0.6242	
	0.4200	0.7846	2.2750	0.6246	
	0.4400	0.9026	2.3653	0.6082	
	0.4600	0.9969	2.5127	0.6129	
	0.4800	1.1035	2.6959	0.6225	
	0.5000	1.1887	2.9023	0.6401	
	0.5200	1.2001	3.0332	0.6581	
	0.5400	1.3023	3.1521	0.6581	
	0.5600	1.4213	3.4088	0.6748	
	0.5800	1.5736	3.7049	0.6907	
	0.6000	1.9777	4.0350	0.6854	
	0.6100	2.0728	4.0910	0.6833	
	0.6500	2.3704	4.6743	0.7145	
	0.7000	1.9733	4.9663	0.7644	
	0.7500	2.0713	4.9245	0.7540	
	0.8000	2.3097	4.8492	0.7320	
	0.8500	2.2939	4.7605	0.7262	
	0.9000	1.9543	4.4440	0.7255	
	0.9500	2.0371	4.1725	0.6941	
	1.0000	2.5603	4.5503	0.6932	
	1.0500	3.0715	4.8202	0.6914	
	1.1000	3.2299	4.5466	0.6650	
	1.1500	3.2022	4.2736	0.6434	
	1.2000	3.4700	3.8731	0.6032	
	1.2500	2.7996	3.9291	0.6252	
	1.3000	2.8810	4.2190	0.6493	
	1.5000	2.1558	4.0009	0.6679	
	1.6000	1.8758	5.2885	0.7930	
	1.7000	1.5254	6.1852	0.8633	
	1.8000	1.7290	6.1424	0.8469	
	2.0000	1.7071	6.7952	0.8724	
Cadmium, basal plane..........	0.2400	0.5397	1.6677	0.5810	
	0.2600	0.2966	1.5678	0.7134	
	0.2800	0.2992	1.5873	0.7155	
	0.3000	0.2810	1.6550	0.7434	
	0.3200	0.3059	1.7982	0.7523	
	0.3400	0.4123	1.6492	0.6502	
	0.3600	0.4206	2.0801	0.7348	
	0.3800	0.4885	2.2313	0.7284	
	0.4000	0.5323	2.3671	0.7322	

TABLE 6g-1. OPTICAL CONSTANTS OF METALS (*Continued*)

Metal	Wave-length, μm	Index of refraction n	Extinction coefficient k	Reflec-tance calculated	Ref.
Cadmium, basal plane (*Cont.*)....	0.4200	0.6001	2.5080	0.7288	
	0.4400	0.6647	2.6929	0.7347	
	0.4600	0.7664	2.9099	0.7354	
	0.4800	0.8792	3.1564	0.7394	
	0.5000	0.9053	3.3689	0.7583	
	0.5200	1.0188	3.6122	0.7620	
	0.5400	1.1062	3.8825	0.7732	
	0.5600	1.3257	4.1337	0.7643	
	0.5800	1.6193	4.5080	0.7616	
	0.6000	1.9007	4.8982	0.7654	
	0.6100	1.9076	5.1768	0.7836	
	0.6500	2.1409	5.3669	0.7785	
	0.7000	2.2464	5.0414	0.7501	
	0.7500	2.2127	4.9735	0.7475	
	0.8000	2.3403	4.9545	0.7378	
	0.8500	2.5598	4.7405	0.7087	
	0.9000	2.7066	4.8400	0.7087	
	0.9500	2.7465	4.9044	0.7116	
	1.0000	2.4292	4.4767	0.6944	
	1.0500	3.2890	4.7537	0.6791	
	1.1000	4.2983	4.0725	0.6150	
	1.1500	4.4419	4.4610	0.6412	
	1.2000	4.3324	4.1085	0.6176	
	1.2500	4.2414	3.9610	0.6069	
	1.3000	3.7442	4.1411	0.6223	
	1.5000	2.0119	4.5330	0.7283	
	1.6000	1.8016	5.1677	0.7915	
	1.7000	1.5862	5.3487	0.8208	
	1.8000	1.6502	5.8175	0.8385	
	2.0000	1.5222	6.6975	0.8811	
Cadmium, evaporated..........	0.0500	0.9800	0.2000	0.0102	20
	0.0600	0.9900	0.2200	0.0121	
	0.0700	1.0000	0.2500	0.0154	
	0.0800	1.0000	0.2800	0.0192	
	0.0900	1.0000	0.3000	0.0220	
	0.0950	1.0000	0.3100	0.0235	
	0.1000	1.0100	0.2900	0.0204	
	0.1050	1.0200	0.2600	0.0164	
	0.1100	1.0300	0.2200	0.0118	
	0.1150	1.0400	0.1900	0.0090	
	0.1200	1.0100	0.1300	0.0042	
	0.1220	0.9800	0.1200	0.0038	
	0.1240	0.9300	0.1700	0.0090	
	0.1260	0.8900	0.1600	0.0105	
	0.1280	0.8500	0.1900	0.0169	
	0.1300	0.8100	0.2900	0.0358	
	0.1320	0.7800	0.3600	0.0540	
	0.1340	0.7400	0.4300	0.0786	
	0.1360	0.7100	0.4800	0.0997	
	0.1380	0.6700	0.5100	0.1210	
	0.1400	0.6400	0.5300	0.1382	
	0.1500	0.5100	0.6200	0.2344	
	0.1600	0.4400	0.7200	0.3210	
	0.1700	0.4200	0.8100	0.3714	
	0.1800	0.4100	0.9100	0.4177	
	0.1900	0.4000	1.0000	0.4595	

TABLE 6g-1. OPTICAL CONSTANTS OF METALS (*Continued*)

Metal	Wave-length, μm	Index of refraction n	Extinction coefficient k	Reflectance calculated	Ref.
Cadmium, evaporated (*Cont.*)....	0.2000	0.4000	1.1000	0.4953	
	0.2100	0.4000	1.1900	0.5261	
	0.2200	0.4000	1.2500	0.5458	
	0.2300	0.4000	1.3200	0.5678	
	0.2400	0.4000	1.4200	0.5976	
	0.2500	0.4000	1.5400	0.6306	
	0.2600	0.4100	1.6900	0.6615	
	0.2700	0.4100	1.8000	0.6863	
	0.2800	0.4100	2.0000	0.7261	
Cadmium, bulk.................	0.589	1.13	5.01	0.847	21
Calcium, evaporated...........	0.4046	0.34	1.56	0.678	24
	0.4358	0.29	1.64	0.734	
	0.4916	0.29	1.92	0.783	
	0.5461	0.27	2.16	0.828	
	0.5780	0.29	2.31	0.834	
Cerium, evaporated............	0.4358	1.41	1.97	0.418	14
	0.5461	1.74	2.39	0.474	
	0.5780	1.91	2.58	0.495	
Cesium, evaporated............	0.2536	0.916	0.143	0.007	22
	0.3126	0.827	0.174	0.018	
	0.3650	0.671	0.233	0.057	
	0.4047	0.540	0.320	0.127	
	0.4358	0.425	0.438	0.235	
	0.5461	0.278	0.950	0.561	
	0.5780	0.264	1.123	0.631	
Chromium, evaporated.........	0.133	0.83	0.35	0.044	23
	0.145	0.84	0.50	0.076	
	0.156	0.90	0.54	0.077	
	0.169	1.22	0.75	0.111	
	0.178	1.68	0.92	0.163	
	0.193	2.23	1.17	0.244	
	0.205	2.46	1.37	0.289	
Chromium, bulk................	0.579	2.97	4.85	0.698	24
Cobalt, evaporated.............	0.1130	1.0748	0.8094	0.1332	25
	0.1240	1.0592	0.8497	0.1462	
	0.1380	1.0512	0.9513	0.1775	
	0.1550	1.0252	1.0729	0.2193	
	0.1770	1.0169	1.2783	0.2866	
	0.2070	1.1542	1.5595	0.3472	
	0.2480	1.2683	1.7346	0.3778	
	0.3100	1.3477	2.0776	0.4515	
	0.3870	1.5693	2.6763	0.5439	
	0.6200	2.1726	4.0274	0.6694	
	1.2400	3.8513	6.2316	0.7530	
	2.4800	4.8379	9.5082	0.8445	
	0.4400	1.9000	3.1800	0.5897	26
	0.5400	2.5000	3.7600	0.6210	
	0.6600	3.0000	4.1200	0.6361	
	0.8100	3.5400	4.5900	0.6603	
	1.0300	3.8500	5.2700	0.6998	
	1.3100	4.0000	5.7200	0.7156	
	1.6700	4.6100	5.8600	0.7198	
	2.1600	5.0000	6.0600	0.7250	
Cobalt, polycrystalline..........	2.5000	5.1000	7.8000	0.7919	27
	3.0000	4.8800	8.4600	0.8161	
	4.0000	4.7000	11.0000	0.8775	

TABLE 6g-1. OPTICAL CONSTANTS OF METALS (*Continued*)

Metal	Wave-length, μm	Index of refraction n	Extinction coefficient k	Reflec-tance calculated	Ref.
Cobalt, polycrystalline (*Cont.*)....	4.5000	4.7800	12.6000	0.9005	
	5.0000	4.7000	14.7000	0.9244	
	5.5000	4.7600	15.2000	0.9356	
	6.0000	5.0000	17.5000	0.9416	
	6.5000	5.2000	19.3000	0.9494	
	7.0000	5.4000	20.9000	0.9548	
	8.0000	5.8000	24.0000	0.9627	
	9.0000	6.56	27.200	0.968	
	10.0000	7.1000	29.5000	0.9697	
	11.000	8.1000	32.6000	0.9717	
	12.0000	9.0000	34.7000	0.9724	
	14.0000	10.2000	38.0000	0.9740	
	15.0000	11.2000	40.5000	0.9750	
	17.0000	13.5000	45.0000	0.9758	
	19.0000	14.9000	49.0000	0.9775	
	20.0000	15.2000	51.7000	0.9793	
Copper, bulk.................	0.3650	1.0719	2.0710	0.5004	28
	0.4050	1.0769	2.2890	0.5491	
	0.4360	1.0707	2.4610	0.5860	
	0.5000	1.0308	2.7843	0.6528	
	0.5500	0.7911	2.7177	0.7013	
	0.5780	0.3250	2.8923	0.8716	
	0.6000	0.1491	3.2867	0.9508	
	0.6500	0.1074	3.9104	0.9740	
	0.7500	0.1034	4.8847	0.9835	
	1.0000	0.1471	6.9334	0.9881	
Copper, single crystal..........	0.4400	1.1070	2.5565	0.5965	29
	0.4600	1.0942	2.6320	0.6131	
	0.4800	1.0618	2.7124	0.6341	
	0.5000	1.0836	2.7684	0.6390	
	0.5200	1.0438	2.7784	0.6490	
	0.5400	0.9324	2.7348	0.6674	
	0.5600	0.6470	2.7200	0.7440	
	0.5800	0.2805	2.9764	0.8931	
	0.6000	0.1360	3.3464	0.9565	
	0.6200	0.1040	3.6525	0.9714	
	0.6400	0.0972	3.9114	0.9765	
	0.6600	0.0897	4.0692	0.9798	
Copper, evaporated............	0.450	0.87	2.20	0.583	6, 7
	0.500	0.88	2.42	0.625	
	0.550	0.756	2.462	0.669	31
	0.600	0.186	2.980	0.928	
	0.650	0.142	3.570	0.960	
	0.700	0.150	4.049	0.966	
	0.750	0.157	4.463	0.970	
	0.800	0.170	4.840	0.973	
	0.850	0.182	5.222	0.975	
	0.900	0.190	5.569	0.977	
	0.950	0.197	5.900	0.978	
	1.000	0.197	6.272	0.981	
	5.0	2.92	27.45	0.985	32
	1.35	0.45	7.81	0.971	33
	1.69	0.58	9.96	0.977	
	2.28	0.82	13.0	0.981	
	3.00	1.22	17.1	0.984	
	3.4	1.53	20.3	0.985	

TABLE 6g-1. OPTICAL CONSTANTS OF METALS (*Continued*)

Metal	Wave-length, μm	Index of refraction n	Extinction coefficient k	Reflec-tance calculated	Ref.
Copper, evaporated (*Cont.*).......	3.97	1.94	23.1	0.986	
	4.87	2.86	28.9	0.987	
	5.8	3.71	34.6	0.988	
	7.0	5.25	40.7	0.988	
	7.3	5.79	43.2	0.988	
	8.35	7.28	49.2	0.988	
	9.6	9.76	57.2	0.988	
	10.25	11.0	60.6	0.988	
	10.8	12.6	64.3	0.988	
	12.25	15.5	71.9	0.989	
	0.1025	1.05	0.70	0.098	84
	0.1113	0.95	0.73	0.115	
	0.1215	0.95	0.78	0.137	
	0.1306	0.96	0.83	0.148	
	0.1392	1.00	0.91	0.165	
	0.1500	1.02	1.02	0.192	
	0.1603	0.98	1.04	0.219	
	0.1700	0.94	1.12	0.254	
	0.1800	0.90	1.21	0.296	
	0.1900	0.88	1.36	0.335	
	0.2000	0.94	1.51	0.378	
Gallium, liquid................	0.40	0.59	4.50	0.896	34
	0.50	0.89	5.60	0.898	
	0.60	1.25	6.60	0.897	
	0.70	1.65	7.60	0.898	
	0.80	2.09	8.50	0.898	
	0.87	2.40	9.20	0.900	
Gallium, thin solid film..........	0.4200	0.9555	1.7897	0.4561	35
	0.4400	1.0775	1.9675	0.4736	
	0.4600	1.1045	2.0688	0.4927	
	0.4800	1.1020	2.1643	0.5158	
	0.5000	0.9737	2.2645	0.5684	
	0.5200	0.8608	2.3582	0.6184	
	0.5400	1.0281	2.4366	0.5908	
	0.5600	1.3351	2.5578	0.5548	
	0.5800	1.5585	2.6436	0.5394	
	0.6000	1.6796	2.6346	0.5242	
	0.6200	1.7059	2.4678	0.4912	
	0.6400	1.5447	2.0943	0.4311	
Germanium, evaporated........	0.0490	0.8100	0.0300	0.0113	36
	0.0550	0.7600	0.0400	0.0191	
	0.0580	0.7200	0.500	0.0273	
	0.0610	0.6800	0.0700	0.0380	
	0.0670	0.5700	0.1200	0.0804	
	0.0690	0.5300	0.1600	0.1042	
	0.0720	0.4800	0.1800	0.1362	
	0.0740	0.4500	0.2600	0.1705	
	0.0760	0.4000	0.3600	0.2343	
	0.0800	0.3800	0.4200	0.2695	
	0.0840	0.3400	0.4800	0.3287	
	0.0870	0.3200	0.5900	0.3877	
	0.0920	0.3100	0.6500	0.4202	
	0.1050	0.3500	0.9100	0.4718	
	0.1220	0.4200	1.0400	0.4577	
	0.1610	0.5300	1.4600	0.5260	

TABLE 6g-1. OPTICAL CONSTANTS OF METALS (*Continued*)

Metal	Wave-length, μm	Index of refraction n	Extinction coefficient k	Reflec-tance calculated	Ref.
Germanium, single crystal.......	0.0540	0.8400	0.0900	0.0099	37
	0.0560	0.8300	0.0900	0.0110	
	0.0590	0.8100	0.0900	0.0135	
	0.0620	0.7800	0.1100	0.0190	
	0.0650	0.7300	0.1300	0.0298	
	0.0690	0.6700	0.2000	0.0526	
	0.0730	0.6300	0.3500	0.0933	
	0.0770	0.6200	0.4400	0.1199	
	0.0830	0.6300	0.6300	0.1748	
	0.0880	0.6800	0.7600	0.1964	
	0.0950	0.7300	0.8800	0.2249	
	0.1030	0.7900	0.9600	0.2341	
	0.1133	0.8300	1.1200	0.2789	
	0.1240	0.8800	1.2900	0.3229	
Germanium, bulk..............	0.365	4.2	2.6	0.503	39
	0.405	4.25	2.2	0.475	
	0.430	4.1	2.2	0.468	
	0.465	4.15	2.27	0.476	
	0.49	4.5	2.3	0.494	
	0.52	4.8	2.25	0.504	
	0.545	5.15	2.15	0.515	
	0.58	5.5	1.8	0.416	
	0.60	5.7	1.25	0.509	
	0.63	5.45	0.85	0.485	
	0.655	5.3	0.70	0.472	
	0.68	5.0	0.55	0.449	
	0.250	1.7	1.35	0.245	40
	0.310	2.47	1.58	0.320	
	0.370	2.63	1.35	0.299	
	0.400	2.71	1.20	0.287	
	0.430	3.32	1.99	0.413	
	0.490	4.19	2.57	0.501	
	0.540	4.28	2.40	0.492	
	0.620	4.66	1.65	0.465	
	0.700	4.63	0.95	0.432	
	0.900	4.33	0.47	0.396	
	1.100	4.17	0.43	0.379	
	1.300	4.12	0.36	0.375	
	0.124	0.94	0.87	0.168	41
	0.138	0.92	1.10	0.219	
	0.155	0.94	1.37	0.219	
	0.177	1.00	1.74	0.287	
	0.190	1.07	2.00	0.321	
	0.207	1.27	2.38	0.354	
	0.255	1.54	2.47	0.417	
	0.247	1.63	2.88	0.419	
	0.258	1.74	3.15	0.448	
	0.269	2.06	3.61	0.501	
	0.281	3.18	4.26	0.526	
	0.295	3.94	3.45	0.508	
	0.310	3.94	2.88	0.544	
	0.354	4.00	2.42	0.598	
	0.413	4.15	1.88	0.647	
	0.442	4.12	1.87	0.513	
	0.477	4.27	2.00	0.484	
	0.516	4.71	2.00	0.466	

TABLE 6g-1. OPTICAL CONSTANTS OF METALS (*Continued*)

Metal	Wave-length, μm	Index of refraction n	Extinction coefficient k	Reflec-tance calculated	Ref.
Germanium, bulk (*Cont.*)........	0.562	5.15	1.83	0.446	
	0.619	5.30	0.90	0.474	
	0.826	4.64	0.20	0.477	
	1.24	4.27	0.05	0.422	
	2.48	4.08	0.00	0.388	
Germanium, evaporated........	1.33	......	0.131	......	42
	1.43	......	0.085		
	1.54	4.50	0.061	0.405	
	1.67	4.45	0.040	0.401	
	1.82	4.40	0.036	0.396	
	2.00	4.35	0.03	0.392	
	2.22	4.29	0.02	0.387	
	2.50	4.26	0.02	0.384	
	0.4	2.3	2.8	0.509	43
	0.5	3.4	2.25	0.443	
	0.6	4.5	1.7	0.457	
	0.7	5.15	1.3	0.479	
	0.8	5.27	0.9	0.475	
	0.9	5.2	0.6	0.464	
	1.0	5.1	0.45	0.455	
	2.0	4.6			
	3.0	4.4			
	4.0	4.35			
	5.0	4.3			
	6.0	4.3			
	7.0	4.3			
	8.0	4.3			
	9.0	4.3			
	10.0	4.3			
Gold, evaporated..............	0.025	0.890	0.386	0.0433	38
	0.026	0.900	0.390	0.0431	
	0.027	0.906	0.392	0.0429	
	0.028	0.910	0.396	0.0433	
	0.029	0.910	0.400	0.0441	
	0.030	0.906	0.407	0.0459	
	0.031	0.900	0.416	0.0484	
	0.032	0.893	0.426	0.0512	
	0.033	0.882	0.440	0.0556	
	0.034	0.867	0.453	0.0604	
	0.035	0.855	0.470	0.0661	
	0.036	0.849	0.490	0.0719	
	0.037	0.846	0.512	0.0779	
	0.038	0.850	0.535	0.0832	
	0.039	0.865	0.555	0.0862	
	0.040	0.894	0.570	0.0859	
	0.041	0.925	0.572	0.0825	
	0.042	0.940	0.570	0.0803	
	0.043	0.942	0.562	0.0781	
	0.044	0.935	0.550	0.0758	
	0.045	0.910	0.542	0.0766	
	0.046	0.870	0.540	0.0814	
	0.047	0.855	0.548	0.0859	
	0.048	0.846	0.565	0.0920	
	0.049	0.846	0.600	0.1018	
	0.050	0.850	0.645	0.1142	
	0.051	0.860	0.695	0.1275	

TABLE 6g-1. OPTICAL CONSTANTS OF METALS (*Continued*)

Metal	Wave-length, μm	Index of refraction n	Extinction coefficient k	Reflec-tance calculated	Ref.
Gold, evaporated (*Cont.*)	0.052	0.872	0.740	0.1392	
	0.053	0.890	0.795	0.1532	
	0.054	0.915	0.825	0.1582	
	0.055	0.950	0.840	0.1571	
	0.056	0.985	0.848	0.1544	
	0.057	1.022	0.850	0.1503	
	0.058	1.055	0.842	0.1444	
	0.059	1.085	0.830	0.1382	
	0.060	1.113	0.813	0.1314	
	0.061	1.134	0.795	0.1253	
	0.062	1.146	0.770	0.1182	
	0.063	1.153	0.750	0.1127	
	0.064	1.157	0.730	0.1075	
	0.065	1.155	0.710	0.1026	
	0.066	1.140	0.700	0.1005	
	0.067	1.125	0.694	0.0995	
	0.068	1.107	0.687	0.0984	
	0.069	1.088	0.680	0.0975	
	0.070	1.075	0.678	0.0976	
	0.071	1.060	0.680	0.0990	
	0.072	1.050	0.685	0.1010	
	0.073	1.042	0.690	0.1029	
	0.074	1.038	0.697	0.1050	
	0.075	1.033	0.704	0.1073	
	0.076	1.030	0.713	0.1100	
	0.077	1.029	0.720	0.1120	
	0.078	1.028	0.730	0.1149	
	0.079	1.028	0.739	0.1174	
	0.080	1.029	0.745	0.1190	
	0.081	1.030	0.752	0.1200	
	0.082	1.033	0.759	0.1226	
	0.083	1.037	0.765	0.1239	
	0.084	1.041	0.770	0.1249	
	0.085	1.048	0.775	0.1257	
	0.086	1.053	0.780	0.1267	
	0.087	1.061	0.784	0.1272	
	0.088	1.070	0.789	0.1279	
	0.089	1.080	0.793	0.1282	
	0.090	1.090	0.798	0.1289	
	0.091	1.100	0.801	0.1290	
	0.092	1.110	0.806	0.1297	
	0.093	1.121	0.809	0.1298	
	0.094	1.133	0.812	0.1300	
	0.095	1.146	0.815	0.1301	
	0.096	1.159	0.819	0.1305	
	0.097	1.170	0.823	0.1311	
	0.098	1.180	0.826	0.1315	
	0.099	1.190	0.831	0.1324	
	0.100	1.200	0.836	0.1334	
	0.105	1.215	0.862	0.1397	
	0.110	1.218	0.896	0.1486	
	0.115	1.232	0.930	0.1571	
	0.120	1.258	0.963	0.1649	
	0.125	1.282	0.992	0.1718	
	0.130	1.307	1.020	0.1783	
	0.135	1.330	1.048	0.1849	

TABLE 6g-1. OPTICAL CONSTANTS OF METALS (*Continued*)

Metal	Wave-length, μm	Index of refraction n	Extinction coefficient k	Reflectance calculated	Ref.
Gold, evaporated (*Cont.*)	0.140	1.357	1.070	0.1899	
	0.145	1.386	1.089	0.1941	
	0.150	1.419	1.102	0.1967	
	0.155	1.450	1.108	0.1978	
	0.160	1.483	1.106	0.1971	
	0.165	1.512	1.093	0.1941	
	0.170	1.519	1.070	0.1888	
	0.175	1.500	1.070	0.1886	
	0.180	1.470	1.085	0.1921	
	0.185	1.442	1.107	0.1976	
	0.190	1.427	1.135	0.2049	
	0.195	1.424	1.170	0.2138	
	0.200	1.427	1.215	0.2251	
	0.450	1.40	1.88	0.397	6, 7
	0.500	0.84	1.84	0.504	
	0.550	0.331	2.324	0.815	31
	0.600	0.200	2.897	0.919	
	0.650	0.142	3.374	0.955	
	0.700	0.131	3.842	0.967	
	0.750	0.140	4.266	0.971	
	0.800	0.149	4.654	0.974	
	0.850	0.157	4.993	0.976	
	0.900	0.166	5.335	0.978	
	0.950	0.174	5.691	0.979	
	1.000	0.179	6.044	0.981	
Gold, crystalline	0.4400	1.5778	1.9077	0.3863	29
	0.4600	1.4843	1.8257	0.3754	
	0.4800	1.2543	1.7301	0.3787	
	0.5000	0.8031	1.8180	0.5100	
	0.5200	0.5264	2.1277	0.6929	
	0.5400	0.3772	2.4520	0.8092	
	0.5600	0.3054	2.7501	0.8682	
	0.5800	0.2524	3.0106	0.9050	
	0.6000	0.2113	3.2411	0.9294	
	0.6200	0.1906	3.4621	0.9431	
	0.6400	0.1667	3.6902	0.9555	
	1.0000	0.2200	6.7100	0.9811	44
	1.5000	0.3600	10.4000	0.9869	
	2.0000	0.5500	13.9000	0.9888	
	2.5000	0.8200	17.3000	0.9892	
	3.0000	1.1700	21.0000	0.9895	
	4.0000	2.0400	27.9000	0.9896	
	5.0000	3.2700	35.2000	0.9896	
	6.0000	4.7000	35.2000	0.9896	
	8.0000	7.8200	54.60000	0.9898	
	10.0000	11.5000	67.5000	0.9902	
	12.0000	15.4000	80.5000	0.9909	
	1.0000	0.3100	5.5800	0.9623	45
	2.0000	0.5400	11.2000	0.9831	
	3.0000	0.9300	16.7000	0.9868	
	4.0000	1.4900	22.2000	0.9881	
	5.0000	2.1900	27.7000	0.9887	
	6.0000	3.0100	33.0000	0.9891	
	7.0000	3.9700	38.3000	0.9894	
	8.0000	5.0500	43.5000	0.9895	
	9.0000	6.2100	48.6000	0.9897	

TABLE 6g-1. OPTICAL CONSTANTS OF METALS (*Continued*)

Metal	Wave-length, μm	Index of refraction n	Extinction coefficient k	Reflectance calculated	Ref.
Gold, crystalline (*Cont.*)	10.0000	7.4100	53.4000	0.9899	
	11.0000	8.7100	58.2000	0.9900	
	2.5000	0.6900	14.4000	0.9869	87
	3.0000	1.2500	17.6000	0.9841	
	4.0000	1.8800	23.2000	0.9862	
	4.5000	2.2800	25.3000	0.9860	
	5.0000	2.7100	28.5000	0.9869	
	6.0000	4.7100	34.5000	0.9846	
	6.5000	5.4800	37.4000	0.9848	
	7.0000	6.6200	39.4000	0.9836	
	8.0000	7.9000	44.0000	0.9843	
	8.5000	9.7200	45.6000	0.9823	
	9.0000	10.0000	47.9000	0.9834	
	9.5000	10.9000	50.6000	0.9839	
Indium, evaporated	0.4200	0.6505	1.8448	0.5753	35
	0.4400	0.8128	1.8085	0.5041	
	0.4600	0.8676	1.8902	0.5085	
	0.4800	0.8103	1.9252	0.5359	
	0.5000	1.0190	2.0805	0.5150	
	0.5200	1.0536	2.2068	0.5362	
	0.5400	1.0778	2.3242	0.5564	
	0.5600	1.1743	2.4185	0.5559	
	0.5800	1.2039	2.4919	0.5648	
	0.6000	1.2915	2.5900	0.5680	
	0.6200	1.4285	2.7406	0.5738	
	0.6400	1.4502	2.8307	0.5861	
	0.7100	1.3800	6.2400	0.8762	46
	1.0500	1.8300	7.9400	0.8970	
	1.5600	2.3100	11.3000	0.9334	
	2.2000	3.5300	15.8000	0.9477	
	2.6800	4.4300	18.2000	0.9509	
	3.1400	5.5000	21.2000	0.9553	
	4.0000	7.6000	26.1000	0.9597	
	5.9500	13.4000	35.6000	0.9637	
	8.0000	19.2000	42.2000	0.9649	
	10.0000	23.8000	51.7000	0.9710	
Iridium, evaporated	0.0500	0.65	0.88	0.255	47
	0.0550	0.76	0.99	0.255	
	0.0600	0.88	1.08	0.251	
	0.0650	1.02	1.08	0.221	
	0.0700	1.13	0.97	0.175	
	0.0750	1.15	0.90	0.153	
	0.0800	1.14	0.90	0.154	
	0.0850	1.10	0.93	0.166	
	0.0900	1.09	0.98	0.182	
	0.0950	1.11	1.06	0.204	
	0.1000	1.14	1.13	0.220	
	0.1100	1.27	1.23	0.238	
	0.1200	1.36	1.21	0.227	
	0.1300	1.38	1.16	0.213	
	0.1400	1.35	1.14	0.205	
	0.1500	1.28	1.18	0.222	
	0.1600	1.17	1.29	0.275	
	0.17	1.07	1.45	0.340	
	0.18	1.01	1.64	0.400	
	0.19	0.95	1.81	0.460	

TABLE 6g-1. OPTICAL CONSTANTS OF METALS (*Continued*)

Metal	Wave-length, μm	Index of refraction n	Extinction coefficient k	Reflec-tance calculated	Ref.
Iridium, evaporated (*Cont.*)......	0.20	0.89	1.93	0.510	
	0.21	0.82	1.98	0.555	
	0.22	0.74	2.01	0.585	
Iron, single crystal.............	0.3670	1.9500	3.5300	0.6314	50
	0.3730	2.0200	3.5700	0.6305	
	0.3800	2.1100	3.5600	0.6223	
	0.3930	2.2000	3.6800	0.6300	
	0.4060	2.3200	3.7100	0.6256	
	0.4190	2.4600	3.7600	0.6231	
	0.4320	2.5800	3.8000	0.6214	
	0.4450	2.6900	3.8500	0.6216	
	0.4580	2.8300	3.8400	0.6152	
	0.4710	2.9200	3.8800	0.6161	
	0.4840	3.0400	3.8600	0.6105	
	0.4970	3.1200	3.8700	0.6094	
	0.5100	3.1900	3.8600	0.6068	
	0.5230	3.2500	3.8500	0.6047	
	0.5460	3.3500	3.8400	0.6020	
	0.5490	3.3600	3.8400	0.6018	
	0.5620	3.4200	3.8500	0.6018	
	0.5750	3.4400	3.8700	0.6033	
	0.5880	3.4600	3.8800	0.6040	
	0.6010	3.4900	3.8900	0.6044	
	0.6140	3.5000	3.8800	0.6034	
	0.6270	3.5300	3.9300	0.6074	
	0.6400	3.4900	3.9700	0.6114	
	0.6530	3.5600	4.0400	0.6163	
	0.6660	3.5700	4.0200	0.6145	
	0.6790	3.5800	4.1000	0.6210	
	0.6920	3.5800	4.1700	0.6267	
Iron, evaporated...............	0.4400	2.9400	3.3400	0.5592	26
	0.5400	3.1100	3.6200	0.5853	
	0.6600	3.3100	3.7500	0.5943	
	0.8100	3.6900	3.9400	0.6066	
	1.0300	3.8100	4.4400	0.6443	
	1.3100	4.1200	5.3100	0.6971	
	1.6700	4.0600	5.9400	0.7333	
	2.1600	3.8100	6.3800	0.7613	
Iron, bulk...................	0.4800	2.9057	3.8201	0.6106	51
	0.5000	3.0222	3.8449	0.6096	
	0.5200	3.0931	3.8246	0.6057	
	0.5400	3.2151	3.8485	0.6062	
	0.5600	3.2972	3.3563	0.6044	
	0.5800	3.3629	3.8509	0.6028	
	0.6000	3.3975	3.8410	0.6014	
	0.6200	3.4396	3.8537	0.6019	
	0.6400	3.4558	3.8500	0.6014	
	0.6600	3.4801	3.8563	0.6016	
	0.589	2.36	3.20	0.561	21
Lanthanum, evaporated.........	0.4046	1.34	2.33	0.508	14
	0.4358	1.35	2.49	0.539	
	0.5461	1.79	3.43	0.634	
	0.5780	1.74	3.47	0.644	
Lead, bulk....................	0.589	2.01	3.48	0.620	21
Lead, evaporated..............	0.7000	1.6800	3.6700	0.6746	52
	0.8000	1.5100	4.2400	0.7512	

TABLE 6g-1. OPTICAL CONSTANTS OF METALS (*Continued*)

Metal	Wave-length, μm	Index of refraction n	Extinction coefficient k	Reflectance calculated	Ref.
Lead, evaporated (*Cont.*).........	0.9000	1.4400	4.8500	0.8046	
	1.0000	1.4100	5.4000	0.8387	
	1.1000	1.4200	5.9700	0.8631	
	1.2000	1.4600	6.3500	0.8741	
	1.3000	1.5100	7.1200	0.8940	
	1.4000	1.5900	7.6700	0.9030	
	1.5000	1.6700	8.2400	0.9110	
	1.7000	1.9000	9.3700	0.9210	
	2.0000	2.2800	11.1000	0.9319	
	2.5000	3.2000	13.7000	0.9377	
	3.0000	4.2700	16.4000	0.9424	
	3.5000	5.3900	18.6000	0.9443	
	4.0000	6.5800	20.8000	0.9463	
	5.0000	9.0400	24.8000	0.9495	
	6.0000	11.7000	28.1000	0.9508	
	7.0000	14.1000	30.9000	0.9523	
	8.0000	16.4000	33.6000	0.9542	
	9.0000	18.7000	35.8000	0.9552	
	10.0000	21.0000	37.4000	0.9554	
	11.0000	24.6000	40.5000	0.9571	
	12.0000	24.6000	40.5000	0.9571	
Magnesium, evaporated........	0.1200	0.2500	0.4000	0.4194	53
	0.1400	0.1500	0.9500	0.7303	
	0.1600	0.2000	1.2000	0.7222	
	0.1800	0.2500	1.3000	0.6925	
	0.2000	0.2000	1.4000	0.7647	
	0.2200	0.1500	1.5000	0.8321	
	0.2400	0.1000	1.6000	0.8939	
	0.4046	0.52	2.05	0.681	14
	0.4358	0.52	2.65	0.777	
	0.4916	0.53	2.92	0.805	
	0.5461	0.57	3.47	0.843	
	0.5780	0.48	3.71	0.880	
Magnesium, bulk..............	0.589	0.37	4.42	0.931	21
Manganese, evaporated.........	0.4358	2.08	2.62	0.491	14
	0.5461	2.46	3.07	0.540	
	0.5780	2.59	3.04	0.532	
Manganese, bulk..............	0.4600	1.97	3.43	0.617	54
	0.5000	1.92	3.42	0.620	
	0.5408	2.10	3.53	0.619	
	0.5890	2.26	3.71	0.629	
	0.6410	2.61	3.97	0.637	
	0.6800	2.85	4.05	0.635	
Mercury, liquid...............	0.40	0.73	3.01	0.758	34
	0.50	1.04	3.70	0.767	
	0.60	1.39	4.32	0.772	
	0.70	1.76	4.83	0.773	
	0.80	2.14	5.33	0.776	
	0.87	2.40	5.63	0.778	
	0.3022	0.55	2.25	0.705	55
	0.3130	0.44	2.53	0.792	
	0.3650	0.64	2.97	0.778	
	0.4047	0.79	3.40	0.786	
	0.4358	0.88	3.47	0.774	
Molybdenum, bulk.............	0.0550	0.4900	0.6300	0.2511	56
	0.0580	0.5200	0.9600	0.3564	

TABLE 6g-1. OPTICAL CONSTANTS OF METALS (*Continued*)

Metal	Wave-length, μm	Index of refraction n	Extinction coefficient k	Reflectance calculated	Ref.
Molybdenum, bulk (*Cont.*).......	0.0730	1.1500	1.3500	0.2863	
	0.0740	1.0300	1.2200	0.2655	
	0.0830	1.1700	1.2000	0.2389	
	0.0870	1.1700	1.2000	0.2389	
	0.0880	1.3200	1.3300	0.2617	
	0.0890	1.3300	1.3500	0.2663	
	0.0910	1.3800	1.3800	0.2707	
	0.0970	1.5500	1.3100	0.2456	
	0.1030	1.6700	1.2000	0.2204	
	0.1070	1.6400	1.0800	0.1937	
	0.1100	1.3300	1.0500	0.1855	
	0.1200	0.8300	1.4200	0.3812	
	0.1210	0.8100	1.4200	0.3878	
	0.1260	1.0000	1.9000	0.4744	
	0.1290	0.9200	1.9200	0.5009	
	0.1340	0.9500	2.0200	0.5179	
	0.1400	1.0300	2.4200	0.5871	
	0.1450	0.9700	2.4500	0.6074	
	0.1600	1.1000	2.7700	0.6358	
	0.1750	1.6700	3.0200	0.5889	
	0.2480	1.2300	2.7300	0.6040	
	0.2540	1.3000	2.7300	0.5919	
	0.2650	1.4000	2.9300	0.6096	
	0.2800	1.6300	3.0000	0.5904	
	0.3120	2.0200	3.0100	0.5556	
	0.3660	2.4300	2.9700	0.5278	
	0.4050	2.5000	3.0000	0.5294	
	0.4360	2.4800	3.0100	0.5314	
	0.4920	2.7500	3.4500	0.5764	
	0.5460	3.0000	3.4200	0.5667	
	0.5780	3.1800	3.4100	0.5629	
	0.4720	2.8600	3.0000	0.5213	48
	0.5010	3.1700	3.0000	0.5195	
	0.5610	3.4300	3.0000	0.5207	
	0.6220	3.5600	3.0100	0.5230	
	0.8000	3.6900	3.0200	0.5257	
	1.0000	3.8300	3.5500	0.5736	
	1.2000	4.0000	4.0400	0.6128	
	1.4000	4.3100	4.5200	0.6455	
	1.6000	4.6000	5.0000	0.6735	
	1.8000	4.8300	5.4900	0.6987	
	2.0000	5.0700	5.8600	0.7151	
	0.436	2.95	3.283	0.553	57
	0.546	3.59	3.403	0.560	
	0.578	3.65	3.274	0.549	
Neodymium, evaporated.........	0.3950	0.89	1.20	0.290	58
	0.4060	0.84	1.22	0.311	
	0.4170	0.79	1.28	0.347	
	0.4280	0.67	1.36	0.422	
	0.4420	0.56	1.42	0.497	
	0.4550	0.46	1.47	0.571	
	0.4690	0.40	1.46	0.609	
	0.4840	0.37	1.42	0.620	
	0.5000	0.38	1.41	0.610	
	0.5170	0.43	1.44	0.582	
	0.5360	0.43	1.44	0.582	

TABLE 6g-1. OPTICAL CONSTANTS OF METALS (*Continued*)

Metal	Wave-length, μm	Index of refraction n	Extinction coefficient k	Reflectance calculated	Ref.
Neodymium, evaporated (*Cont.*).	0.5560	0.42	1.42	0.583	
	0.5770	0.40	1.38	0.586	
	0.6000	0.37	1.37	0.606	
	0.6250	0.35	1.36	0.619	
	0.6520	0.34	1.35	0.624	
	0.6820	0.32	1.32	0.633	
	0.7140	0.28	1.30	0.664	
	0.7500	0.27	1.30	0.673	
	0.7900	0.28	1.32	0.669	
	0.8340	0.28	1.42	0.694	
	0.8830	0.30	1.50	0.695	
Nickel, evaporated............	0.4400	1.5600	2.6800	0.5457	26
	0.5400	1.8500	3.2700	0.6067	
	0.6600	2.0600	3.8900	0.6636	
	0.8100	2.3700	4.2100	0.6740	
	1.0300	2.8700	4.8700	0.7033	
	1.3100	3.3600	5.6500	0.7361	
	1.6700	3.6200	6.1600	0.7558	
	2.1600	4.2500	6.2500	0.7448	
Nickel, bulk..................	0.4800	1.7763	3.2765	0.6147	51
	0.5000	1.8282	3.3886	0.6246	
	0.5200	1.8796	3.5061	0.6348	
	0.5400	1.9245	3.6268	0.6454	
	0.5600	1.9670	3.7469	0.6556	
	0.5800	1.9830	3.9208	0.6732	
	0.6000	2.0663	3.9950	0.6741	
	0.6200	2.1278	4.0887	0.6788	
	0.6400	2.1890	4.2074	0.6858	
	0.6600	2.2498	4.3338	0.6933	
	0.420	1.41	2.53	0.538	59
	0.589	1.79	3.33	0.621	21
	0.750	2.19	4.36	0.700	60
	1.000	2.63	5.26	0.742	
	2.25	3.95	9.20	0.855	
Nickel, evaporated.............	2.0	3.74	8.80	0.850	32
	1.12	2.63	4.28	0.666	33
	1.58	2.89	5.08	0.718	
	2.18	3.18	6.13	0.769	
	2.72	3.44	7.15	0.806	
	3.4	3.72	8.49	0.842	
	4.4	4.35	10.59	0.876	
	5.4	4.92	12.4	0.896	
	6.75	5.86	15.2	0.916	
	8.7	7.31	19.2	0.933	
	10.5	8.86	22.5	0.941	
	12.5	10.2	26.2	0.950	
Niobium, bulk.................	0.4720	2.2600	2.2600	0.4255	48
	0.5010	2.3900	2.2800	0.4272	
	0.5610	2.5700	2.3400	0.4358	
	0.6220	2.5200	2.5200	0.4621	
	0.8000	2.2300	3.0400	0.5466	
	1.0000	2.0300	4.0000	0.6775	
	1.2000	2.0000	5.0400	0.7675	
	1.4000	2.1800	6.1800	0.8195	
	1.6000	2.4200	7.1300	0.8452	
	1.8000	2.7800	8.3600	0.8679	
	2.0000	3.1300	9.1300	0.8753	

TABLE 6g-1. OPTICAL CONSTANTS OF METALS (*Continued*)

Metal	Wave-length, μm	Index of refraction n	Extinction coefficient k	Reflec-tance calculated	Ref.
Niobium, crystalline, bulk......	1.0000	1.5200	4.2900	0.7544	61
	1.5000	1.5600	7.1200	0.8910	
	2.0000	1.8500	8.7000	0.9117	
	2.5000	2.5000	11.0000	0.9250	
	3.0000	2.9000	12.2000	0.9293	
	3.5000	3.8000	15.6000	0.9429	
	4.0000	4.2000	17.4000	0.9491	
	4.5000	4.7000	19.0000	0.9522	
	5.0000	5.5000	21.6000	0.9568	
	5.5000	6.4000	23.0000	0.9561	
	6.0000	7.2000	25.0000	0.9584	
	6.5000	7.8000	25.9000	0.9583	
	7.0000	9.0000	27.7000	0.9585	
	7.5000	9.9000	29.4000	0.9597	
	8.0000	10.6000	31.0000	0.9613	
	8.5000	11.4000	33.2000	0.9637	
	9.0000	12.5000	34.9000	0.9643	
	9.5000	14.0000	36.4000	0.9639	
	10.0000	15.6000	38.7000	0.9648	
	12.0000	19.1000	42.0000	0.9648	
	14.0000	24.8000	45.4000	0.9636	
	15.0000	26.1000	48.8000	0.9665	
Niobium, bulk.................	0.579	1.80	2.11	0.414	24
Palladium, evaporated..........	0.3021	1.5	2.0	0.415	62
	0.3404	1.5	2.1	0.437	
	0.4358	1.8	2.4	0.471	
	0.5085	1.9	2.7	0.516	
	0.5461	2.3	2.7	0.494	
Platinum, evaporated..........	0.0580	0.9700	1.0300	0.2149	63
	0.0730	1.0800	0.7900	0.1274	
	0.1220	1.2800	1.1600	0.2176	
Platinum, bulk.................	0.589	2.06	4.26	0.701	21
Platinum, electrolytic..........	0.257	1.17	1.93	0.445	64
	0.441	1.94	3.16	0.584	
	0.589	2.63	3.54	0.591	
	0.668	2.91	3.66	0.594	
Platinum, sputtered............	1.00	3.42	6.3	0.770	49
	1.97	5.92	9.8	0.830	
	3.29	7.50	12.2	0.860	
	4.65	10.9	15.5	0.890	
Plutonium, bulk...............	0.5461	2.0700	3.6100	0.6313	65
Potassium, evaporated..........	0.1270	0.9600	0	0.0004	66
	0.1390	0.9700	0	0.0002	
	0.1420	0.9800	0	0.0001	
	0.1480	0.9800	0	0.0001	
	0.1560	0.9600	0	0.0004	
	0.1590	0.9600	0	0.0004	
	0.1720	0.9300	0	0.0013	
	0.1780	0.9200	0	0.0017	
	0.1830	0.9100	0	0.0022	
	0.1880	0.8900	0	0.0034	
	0.1980	0.8800	0	0.0041	
	0.2080	0.8400	0	0.0076	
	0.2180	0.8200	0	0.0098	
	0.2280	0.7900	0	0.0138	
	0.2520	0.7200	0	0.0265	

TABLE 6g-1. OPTICAL CONSTANTS OF METALS (*Continued*)

Metal	Wave-length, μm	Index of refraction n	Extinction coefficient k	Reflectance calculated	Ref.
Potassium, evaporated (*Cont.*)....	0.3120	0.3900	0	0.1926	
	0.2536	0.744	0.049	0.022	67
	0.3126	0.410	0.080	0.178	
	0.3650	0.150	0.443	0.605	
	0.4047	0.105	0.710	0.757	
	0.4358	0.121	0.978	0.781	
	0.5461	0.091	1.42	0.886	
	0.5780	0.094	1.57	0.897	
	0.3126	0.51	0.07	0.107	68
	0.3341	0.30	0.21	0.308	
	0.3650	0.21	0.42	0.488	
	0.4047	0.12	0.56	0.694	
	0.4358	0.08	0.68	0.804	
	0.4916	0.07	1.22	0.894	
	0.5461	0.05	1.41	0.935	
	0.5780	0.05	1.60	0.938	
Potassium, bulk behind glass.....	0.472	0.070	1.00	0.869	69
	0.589	0.068	1.50	0.920	
	0.605	0.066	1.77	0.938	
	0.546	0.06	1.29	0.914	70
Potassium, bulk behind KBr.....	2.5000	0.3500	7.5500	0.9762	71
	3.2000	0.6500	9.6500	0.9729	
	4.5000	1.3800	14.3000	0.9737	
	5.5000	1.6600	17.1000	0.9778	
	7.2000	2.8900	21.6000	0.9768	
	10.1500	4.77	28.2000	0.9770	
Rhenium, bulk................	0.436	2.62	2.97	0.510	72
	0.589	3.18	3.55	0.576	
	0.4720	3.0000	3.0200	0.5223	48
	0.5010	3.3000	3.0300	0.5230	
	0.5610	3.4600	3.0800	0.5289	
	0.6220	3.4100	3.1300	0.5336	
	0.8000	3.3800	4.1700	0.6303	
	1.0000	3.3300	5.3700	0.7201	
	1.2000	3.5600	6.6000	0.7787	
	1.4000	4.1700	7.7900	0.8092	
	1.6000	4.7800	8.7300	0.8256	
	1.8000	5.6300	9.5200	0.8327	
	2.0000	6.0000	10.0200	0.8394	
Rhodium, bulk................	0.4000	0.8400	3.9100	0.8201	73
	0.5000	1.0900	4.1700	0.7996	
	0.6000	1.4300	4.6200	0.7901	
	0.7000	1.6800	5.6700	0.8291	
	0.8000	2.0300	6.3600	0.8364	
	0.9000	2.2700	6.5000	0.8285	
	1.0000	2.3300	6.8000	0.8374	
	1.5000	3.1000	8.5200	0.8613	
	2.0000	3.2800	9.8700	0.8866	
	2.5000	4.4500	12.2000	0.9003	
	3.0000	5.0800	14.3000	0.9158	
	3.5000	5.6500	16.9900	0.9321	
	4.0000	5.9900	17.4600	0.9323	
	4.5000	6.6900	19.0900	0.9368	
	5.0000	6.7900	19.8800	0.9404	
	5.5000	7.6300	23.7100	0.9521	
	6.0000	8.0000	24.5400	0.9532	

TABLE 6g-1. OPTICAL CONSTANTS OF METALS (*Continued*)

Metal	Wave-length, μm	Index of refraction n	Extinction coefficient k	Reflec-tance calculated	Ref.
Rhodium, bulk (*Cont.*)...........	6.5000	8.9100	25.5000	0.9541	
	7.0000	8.9100	28.9100	0.9618	
	7.5000	9.5900	31.4200	0.9651	
	8.0000	9.1200	33.3100	0.9699	
	8.5000	10.6900	35.6200	0.9696	
	9.0000	10.9700	37.5400	0.9717	
	9.5000	11.3200	38.8900	0.9728	
	10.0000	12.0700	41.7600	0.9748	
	10.5000	12.8800	43.9500	0.9757	
	11.0000	13.8300	56.2100	0.9765	
	0.579	1.54	5.67	0.782	24
Rhodium, evaporated...........	0.546	1.62	4.63	0.771	74
Rubidium, vacuum deposited.....	0.2536	1.031	0.056	0.001	22
	0.3022	0.833	0.071	0.010	
	0.3126	0.814	0.078	0.012	
	0.3341	0.745	0.090	0.024	
	0.3650	0.496	0.135	0.121	
	0.4047	0.275	0.373	0.377	
	0.4358	0.181	0.636	0.598	
	0.5461	0.157	1.05	0.742	
	0.5780	0.164	1.19	0.763	
Selenium, single crystal, ⊥ to *C* axis	0.2810	1.9000	1.7263	0.3328	75
	0.2940	2.1800	1.8715	0.3595	
	0.3100	2.4965	1.8145	0.3565	
	0.3260	2.7947	1.6764	0.3504	
	0.3440	3.0381	1.4318	0.3380	
	0.3640	3.1990	1.2222	0.3309	
	0.3860	3.3009	0.9467	0.3192	
	0.4100	3.2483	0.7219	0.3003	
	0.4400	3.1415	0.5825	0.2816	
	0.4750	3.0063	0.5455	0.2644	
	0.5150	2.9317	0.6123	0.2594	
	0.5600	3.0460	0.7337	0.2794	
	0.6100	3.2439	0.7229	0.2999	
	0.6200	3.3708	0.6497	0.3095	
	0.6500	3.4611	0.3149	0.3078	
	0.6900	3.3106	0.1420	0.2881	
Selenium, ∥ to *C* axis...........	0.3440	3.3908	3.0509	0.5255	
	0.3640	3.6899	2.7304	0.4988	
	0.3860	3.8083	2.7282	0.5016	
	0.3960	4.2412	2.4875	0.4960	
	0.4100	4.4447	2.3207	0.4925	
	0.4400	4.5964	1.7340	0.4644	
	0.4750	4.3729	1.3572	0.4304	
	0.5150	4.2640	1.2090	0.4153	
	0.5600	4.4567	1.1926	0.4286	
	0.6200	4.7644	0.7210	0.4353	
	0.6350	4.7660	0.5245	0.4313	
	0.6900	4.4723	0.0347	0.4026	
Selenium, evaporated amorphous..	0.2400	1.881	1.131	0.215	76
	0.2600	2.069	1.257	0.248	
	0.2800	2.280	1.285	0.265	
	0.3000	2.453	1.240	0.271	
	0.3200	2.570	1.157	0.270	
	0.3400	2.661	1.060	0.267	
	0.3625	2.734	0.965	0.265	

TABLE 6g-1. OPTICAL CONSTANTS OF METALS (*Continued*)

Metal	Wave-length, μm	Index of refraction n	Extinction coefficient k	Reflectance calculated	Ref.
Selenium, evaporated amorphous (*Cont.*)	0.3875	2.792	0.877	0.263	
	0.4000	2.820	0.838	0.262	
	0.4250	2.871	0.756	0.262	
	0.4500	2.917	0.679	0.262	
	0.4750	2.963	0.600	0.262	
	0.5000	3.003	0.515	0.263	
	0.5250	3.041	0.410	0.263	
	0.5500	3.051	0.282	0.260	
	0.5750	3.005	0.147	0.252	
	0.6000	2.922	0.061	0.240	
	0.6297	2.810	0.000		
	0.6766	2.710	0.000		
	0.7429	2.633	0.000		
	0.8349	2.580	0.000		
	0.9643	2.539	0.000		
	1.1470	2.494	0.000		
	1.4350	2.464	0.000		
	1.6490	2.454	0.000		
	1.9410	2.445	0.000		
	2.3630	2.435	0.00		
Silicon, single crystal............	0.0650	0.5000	0.1300	0.1177	77
	0.0690	0.4500	0.1700	0.1555	
	0.0730	0.3700	0.3600	0.2624	
	0.0770	0.3700	0.3700	0.2651	
	0.0820	0.4100	0.5400	0.2806	
	0.0880	0.4300	0.5900	0.2812	
	0.0950	0.4100	0.6600	0.3233	
	0.1030	0.3600	0.7700	0.4104	
	0.1130	0.4600	1.0400	0.4274	
	0.1240	0.4800	1.1800	0.4641	
	0.1240	0.5200	1.2600	0.4664	78
	0.1300	0.5700	1.3300	0.4615	
	0.1380	0.5800	1.5000	0.5112	
	0.1460	0.6500	1.6400	0.5196	
	0.1550	0.6700	1.7700	0.5474	
	0.1650	0.6800	1.9500	0.5894	
	0.1770	0.7500	2.2500	0.6308	
	0.1990	0.8000	2.5300	0.6681	
	0.2060	1.1400	2.8300	0.6378	
	0.2140	1.2700	3.0000	0.6411	
	0.2210	1.5000	3.1700	0.6319	
	9.2290	1.6600	3.1500	0.6094	
	0.2380	1.7500	3.2500	0.6138	
	0.2480	1.7000	3.3800	0.6366	
	0.2580	1.6700	3.6700	0.6757	
	0.2690	2.0900	4.3800	0.7098	
	0.2810	3.3300	5.1300	0.7044	
	0.2940	4.8300	3.9500	0.6104	
	0.3100	4.9000	3.5500	0.5866	
	0.3260	5.1000	3.0000	0.5585	
	0.3440	5.2000	3.0900	0.5666	
	0.3640	7.0000	2.1600	0.5922	
	0.3860	6.0000	0.4200	0.5120	
	0.4100	5.1100	0.1700	0.4529	
	0.4400	4.6700	0.1300	0.4193	
	0.4750	4.3300	0.1100	0.3906	

TABLE 6g-1. OPTICAL CONSTANTS OF METALS (*Continued*)

Metal	Wave-length, μm	Index of refraction n	Extinction coefficient k	Reflec-tance calculated	Ref.
Silicon, single crystal (*Cont.*).....	0.5150	4.1600	0.1000	0.3753	
	0.5600	4.0000	0.1000	0.3603	
	0.6200	3.9200	0.0500	0.3523	
Silicon, evaporated.............	0.5000	4.3000	0.7400	0.3994	79
	0.5500	4.4000	0.6300	0.4045	
	0.6000	4.3500	0.5900	0.3994	
	0.6500	4.2300	0.5700	0.3887	
	0.7000	4.1900	0.4000	0.3815	
	0.7500	4.1700	0.3700	0.3791	
	0.8000	4.0600	0.2100	0.3668	
Silicon, bulk (low purity)........	0.589	4.18	0.376	0.380	60
	1.25	3.67	0.294	0.330	
	2.25	3.53	0.282	0.315	
Silver, bulk...................	0.1030	1.6500	0.4100	0.0821	80
	0.1100	1.5900	0.3200	0.0661	
	0.1240	1.4000	0.2700	0.0399	
	0.1300	1.3000	0.2900	0.0324	
	0.1340	1.2500	0.3200	0.0319	
	0.1370	1.2300	0.3700	0.0371	
	0.1430	1.1500	0.4200	0.0414	
	0.1550	1.0700	0.5800	0.0739	
	0.1620	1.0600	0.6900	0.1016	
	0.1710	1.0700	0.7800	0.1253	
	0.1790	1.0900	0.8700	0.1493	
	0.1960	1.1500	1.0100	0.1848	
	0.2160	1.2300	1.1000	0.2043	
	0.2360	1.2800	1.1800	0.2232	
	0.2620	1.3900	1.2800	0.2436	
	0.2810	1.5800	1.2400	0.2287	
	0.2920	1.7100	1.0800	0.1963	
	0.3010	1.8100	0.8500	0.1600	
	0.3070	1.7400	0.5400	0.1076	
	0.3140	1.3500	0.2300	0.0315	
	0.3180	1.1500	0.1900	0.0216	
	0.3200	1.0700	0.1800	0.0086	
	0.3210	0.9800	0.1600	0.0066	
	0.3230	0.8900	0.1600	0.0105	
	0.3260	0.6800	0.1700	0.0460	
	0.3290	0.4900	0.2100	0.1344	
	0.3320	0.2300	0.4000	0.4501	
	0.3350	0.1700	0.5900	0.6040	
	0.3440	0.1400	0.9400	0.7435	
	0.3620	0.1000	1.3400	0.8669	
	0.3760	0.0900	1.5700	0.9015	
Silver, evaporated..............	0.1025	1.19	0.57	0.078	84
	0.1113	1.09	0.56	0.067	
	0.1215	1.10	0.57	0.063	
	0.1306	1.14	0.57	0.063	
	0.1392	1.04	0.54	0.074	
	0.1500	0.96	0.66	0.100	
	0.1603	0.94	0.86	0.149	
	0.1700	0.95	0.91	0.195	
	0.1800	0.99	1.07	0.226	
	0.1900	1.02	1.11	0.250	
	0.2000	1.13	1.23	0.263	
	0.2200	1.3200	1.2900	0.2507	81

TABLE 6g-1. OPTICAL CONSTANTS OF METALS (*Continued*)

Metal	Wave-length, μm	Index of refraction n	Extinction coefficient k	Reflectance calculated	Ref.
Silver, evaporated (*Cont.*)........	0.2300	1.3800	1.3100	0.2521	
	0.2400	1.3700	1.3300	0.2580	
	0.2500	1.3900	1.3400	0.2594	
	0.2550	1.3900	1.3400	0.2594	
	0.2600	1.4500	1.3500	0.2588	
	0.2650	1.4700	1.3400	0.2554	
	0.2700	1.5100	1.3300	0.2515	
	0.2750	1.5100	1.3100	0.2465	
	0.2800	1.5700	1.2700	0.2358	
	0.2850	1.6100	1.2400	0.2287	
	0.2900	1.6000	1.1700	0.2127	
	0.2950	1.6400	1.0800	0.1937	
	0.3000	1.6700	0.9600	0.1702	
	0.3020	1.6500	0.8200	0.1423	
	0.3040	1.6400	0.7500	0.1291	
	0.3060	1.6200	0.6800	0.1156	
	0.3080	1.5800	0.6100	0.1008	
	0.3100	1.5400	0.5400	0.0865	
	0.3120	1.4700	0.4800	0.0713	
	0.3140	1.4000	0.4300	0.0580	
	0.3160	1.3000	0.3800	0.0431	
	0.3180	1.1900	0.3400	0.0309	
	0.3200	1.0700	0.3200	0.0244	
	0.3220	0.9200	0.3000	0.0255	
	0.3240	0.7900	0.3000	0.0407	
	0.3260	0.6400	0.3500	0.0896	
	0.3280	0.4800	0.4400	0.1946	
	0.3300	0.3000	0.5500	0.3977	
	0.3320	0.2300	0.6800	0.5342	
	0.3340	0.2000	0.7900	0.6124	
	0.3360	0.1900	0.9200	0.6641	
	0.3380	0.1800	1.0500	0.7114	
	0.3400	0.1600	1.1400	0.7581	
	0.3450	0.1400	1.2700	0.8077	
	0.3500	0.1200	1.3500	0.8440	
	0.3550	0.1000	1.4200	0.8760	
	0.3600	0.0900	1.5200	0.8971	
	0.3650	0.0700	1.6000	0.9244	
	0.3700	0.0600	1.7000	0.9402	
Silver, bulk...................	0.226	1.41	1.11	0.199	82
	0.293	1.57	0.97	0.168	
	0.316	1.13	0.43	0.043	
	0.332	0.41	0.65	0.320	
	0.395	0.16	1.91	0.872	
	0.500	0.17	2.94	0.932	
	0.589	0.18	3.64	0.951	
Silver, evaporated..............	0.3021	1.2	0.8	0.124	83
	0.3261	0.5	0.5	0.200	
	0.3404	0.22	1.0	0.646	
	0.40	0.075	1.93	0.939	6, 7
	0.45	0.055	2.42	0.968	
	0.50	0.050	2.87	0.979	
	0.55	0.055	3.32	0.982	
	0.60	0.060	3.75	0.984	
	0.65	0.070	4.20	0.985	
	0.70	0.075	4.62	0.987	

TABLE 6g-1. OPTICAL CONSTANTS OF METALS (*Continued*)

Metal	Wave-length, μm	Index of refraction n	Extinction coefficient k	Reflectance calculated	Ref.
Silver, evaporated (*Cont.*)........	0.75	0.080	5.05	0.988	
	0.80	0.090	5.45	0.988	
	0.85	0.100	5.85	0.989	
	0.90	0.105	6.22	0.989	
	0.95	0.110	6.56	0.989	
Silver, crystalline..............	0.4400	0.0462	2.5985	0.9765	29
	0.4600	0.0410	2.8039	0.9817	
	0.4800	0.0415	3.0136	0.9837	
	0.5000	0.0468	3.2019	0.9835	
	0.5200	0.0427	3.3988	0.9865	
	0.5400	0.0448	3.5696	0.9870	
	0.5600	0.0453	3.7499	0.9880	
	0.5800	0.0496	3.9335	0.9880	
	0.6000	0.0489	4.0881	0.9890	
	0.6200	0.0552	4.2595	0.9885	
	0.6400	0.0542	4.4320	0.9896	
Silver, chemically deposited......	0.750	0.17	5.16	0.976	60
	1.00	0.24	6.96	0.981	
	1.50	0.45	10.7	0.985	
	2.25	0.77	15.4	0.987	
	2.89	1.39	19.0	0.985	49
	4.37	4.34	32.6	0.985	
Silver, evaporated..............	1.2500	0.3700	7.700	0.9785	30
	1.5000	0.4500	9.000	0.9783	
	2.0000	0.6500	12.2000	0.9828	
	3.000	1.3000	18.2000	0.9845	
	4.000	2.3000	24.3000	0.9847	
	5.000	3.5000	30.4000	0.9852	
	6.000	5.0000	36.0000	0.9850	
	7.000	6.9000	41.0000	0.9842	
	8.000	8.9000	46.000	0.9839	
	9.0000	11.0000	50.0000	0.9834	
	10.0000	13.0000	54.0000	0.9830	
	1.000	0.2500	6.8100	0.9791	45
	2.000	0.6800	13.6000	0.9855	
	3.0000	1.3800	20.3000	0.9868	
	4.0000	2.3400	26.9000	0.9873	
	5.0000	3.5200	33.2000	0.9875	
	6.0000	4.8700	39.4000	0.9877	
	7.0000	6.3100	45.300	0.9880	
	8.0000	7.8600	50.9000	0.9882	
	9.0000	9.3600	56.0000	0.9885	
	10.0000	10.8000	60.7000	0.9887	
	11.0000	12.0000	64.8000	0.9890	
	12.0000	12.8000	67.8000	0.9893	
	1.0	0.129	6.83	0.989	31
	2.0	0.48	14.4	0.991	8
	4.0	1.89	28.7	0.991	
	6.0	4.15	42.6	0.991	
	8.0	7.14	56.1	0.991	
	10.00	10.69	69.0	0.991	
	12.00	14.50	81.4	0.992	
Sodium, vacuum deposited.......	0.2536	0.026	0.621	0.928	7
	0.2652	0.028	0.735	0.930	
	0.3126	0.040	1.02	0.925	
	0.3650	0.042	1.44	0.947	

TABLE 6g-1. OPTICAL CONSTANTS OF METALS (*Continued*)

Metal	Wave-length, μm	Index of refraction n	Extinction coefficient k	Reflectance calculated	Ref.
Sodium, vacuum deposited (*Cont.*)	0.4047	0.048	1.56	0.946	
	0.4358	0.048	1.80	0.956	
	0.5461	0.029	2.32	0.982	
	0.5780	0.027	2.59	0.986	
Sodium, bulk behind NaCl.......	2.2500	0.4100	11.6000	0.9880	71
	2.5000	0.5000	12.5000	0.9874	
	3.0000	0.6100	14.7000	0.9888	
	3.5000	0.8100	17.2000	0.9892	
	4.0000	1.0200	19.4000	0.9893	
	4.7500	1.7100	22.3000	0.9864	
	7.2000	3.6600	32.9000	0.9867	
	10.1500	6.4300	44.5000	0.9874	
Strontium, evaporated...........	0.4046	0.55	1.28	0.456	14
	0.4358	0.57	1.50	0.516	
	0.4916	0.58	1.61	0.544	
	0.5461	0.63	1.99	0.619	
	0.5780	0.61	2.13	0.658	
Tantalum, bulk................	0.4720	2.5200	2.9600	0.5234	48
	0.5010	2.5500	2.9600	0.5226	
	0.5610	2.4700	2.8100	0.5044	
	0.6220	2.1300	2.8900	0.5306	
	0.8000	1.3400	3.6400	0.7138	
	1.0000	1.3000	4.4200	0.7905	
	1.2000	1.4100	5.1000	0.8227	
	1.4000	1.6900	5.7800	0.8337	
	1.6000	2.0000	6.3900	0.8395	
	1.8000	2.5000	6.9400	0.8345	
	2.0000	2.9800	7.5400	0.8360	
Tin (gray), single crystal........	1.0000	4.7000	1.6000	0.4636	85
	3.0000	4.6000	1.3000	0.4433	
	4.0000	4.6000	1.1000	0.4351	
	5.0000	4.5000	0.9000	0.4205	
	6.0000	4.4000	1.0000	0.4164	
	7.0000	4.4000	0.9000	0.4127	
	8.0000	4.3000	0.8000	0.4013	
	9.0000	4.2000	0.8000	0.3931	
	9.5000	4.2000	0.8000	0.3931	
	10.0000	4.1000	0.8000	0.3846	
	10.5000	4.0000	0.9000	0.3801	
	11.0000	4.0000	1.0000	0.3846	
	11.500	3.900	1.000	0.3762	
	12.000	3.800	1.000	0.3677	
	12.500	3.7000	0.900	0.3537	
	13.000	3.6000	0.900	0.3446	
	13.500	3.6000	0.900	0.3446	
	14.000	3.6000	0.800	0.3394	
	14.500	3.5000	0.900	0.3352	
	15.000	3.3000	0.800	0.3100	
	15.500	3.2000	0.800	0.2998	
	16.000	3.2000	0.800	0.2998	
	16.500	3.2000	0.900	0.3062	
	17.000	3.000	0.900	0.2861	
	17.500	2.900	1.000	0.2844	
	18.000	2.800	1.100	0.2843	
	18.500	2.700	1.3000	0.2978	
	19.000	2.500	1.500	0.3103	

TABLE 6g-1. OPTICAL CONSTANTS OF METALS (*Continued*)

Metal	Wave-length, μm	Index of refraction n	Extinction coefficient k	Reflec-tance calculated	Ref.
Tin (gray), single crystal (*Cont.*)..	19.500	2.400	1.6000	0.3201	
	20.000	2.100	2.0000	0.3828	
Titanium, evaporated..........	0.436	2.04	2.85	0.530	86
	0.546	2.53	3.33	0.570	
	0.578	2.64	3.42	0.577	
	0.650	3.03	3.65	0.590	
Titanium, bulk...............	2.000	4.3800	4.8400	0.6655	87
	2.500	4.5700	5.3900	0.6957	
	3.000	4.5700	5.8300	0.7188	
	3.500	4.5600	6.5800	0.7542	
	4.000	4.6600	7.2700	0.7804	
	4.500	4.6600	8.0700	0.8082	
	5.000	4.8700	9.1800	0.8359	
	5.500	5.0700	10.3000	0.8581	
	6.000	5.3800	11.3000	0.8722	
	6.500	5.6300	12.2000	0.8832	
	7.000	5.9900	13.2000	0.8926	
	7.500	6.3100	13.9000	0.8977	
	8.000	6.5600	14.8000	0.9050	
	8.500	6.9600	16.1000	0.9137	
	9.000	7.5600	16.6000	0.9133	
	9.5000	8.5600	17.1000	0.9108	
	10.0000	9.0100	17.8000	0.9136	
Thallium, evaporated..........	0.0800	1.3500	0.1000	0.0239	20
	0.0900	1.1200	0.1900	0.0111	
	0.0950	1.1200	0.2400	0.0158	
	0.1000	1.2000	0.2800	0.0241	
	0.1020	1.1300	0.1900	0.0116	
	0.1040	0.9400	0.1500	0.0069	
	0.1060	0.8600	0.1900	0.0159	
	0.1080	0.8100	0.2500	0.0295	
	0.1100	0.7700	0.3300	0.0499	
	0.1150	0.6700	0.4600	0.1068	
	0.1200	0.5900	0.5500	0.1663	
	0.1300	0.500	0.7700	0.2965	
	0.1400	0.4800	0.9600	0.3830	
	0.1500	0.5400	1.1200	0.4043	
	0.1600	0.6100	1.2800	0.4232	
	0.1700	0.6700	1.3900	0.4323	
	0.1800	0.7300	1.4900	0.4399	
	0.1900	0.7800	1.5900	0.4523	
	0.2000	0.8400	1.7000	0.4646	
	0.2100	0.8800	1.8000	0.4804	
	0.2200	0.9100	1.9000	0.4985	
	0.2300	0.9700	2.0000	0.5077	
	0.2400	1.0300	2.1100	0.5194	
	0.2500	1.1100	2.2800	0.5399	
	0.2600	1.1900	2.4000	0.5491	
	0.2700	1.2600	2.5400	0.5640	
	0.2800	1.3500	2.7000	0.5785	
	0.4200	0.8099	2.0312	0.5623	35
	0.4400	0.9407	2.1154	0.5434	
	0.4600	1.0414	2.1942	0.5362	
	0.4800	1.1057	2.2655	0.5377	
	0.5000	1.1336	2.3377	0.5473	
	0.5200	1.1588	2.3860	0.5523	

TABLE 6g-1. OPTICAL CONSTANTS OF METALS (*Continued*)

Metal	Wavelength, μm	Index of refraction n	Extinction coefficient k	Reflectance calculated	Ref.
Thallium, evaporated (*Cont.*).....	0.5400	1.1540	2.4437	0.5650	
	0.5600	1.1622	2.5081	0.5761	
	0.5800	1.1135	2.5729	0.5983	
	0.6000	1.0681	2.6497	0.6218	
	0.6200	1.0405	2.7005	0.6367	
	0.6400	0.9717	2.8043	0.6693	
Tungsten, bulk................	0.4720	2.9900	2.2600	0.4312	48
	0.5010	3.0400	2.3400	0.4421	
	0.5610	3.2800	2.5200	0.4682	
	0.6220	3.4100	2.6300	0.4826	
	0.8000	3.5400	2.7600	0.4984	
	1.0000	3.0400	3.5200	0.5765	
	1.2000	3.0400	4.2800	0.6490	
	1.4000	2.9400	4.5700	0.6770	
	1.6000	2.4700	5.1300	0.7424	
	1.8000	2.1300	6.4900	0.8359	
	2.0000	2.0000	7.0200	0.8627	
	0.579	2.76	2.71	0.486	24
	0.589	3.46	3.25	0.545	88
Vanadium, bulk................	2.000	2.0800	6.4300	0.8363	87
	2.500	2.1800	7.3700	0.8647	
	3.000	2.4400	8.8100	0.8909	
	3.500	2.9500	10.6000	0.9078	
	4.000	3.2400	11.5000	0.9137	
	4.500	3.7500	12.8000	0.9195	
	5.000	4.1500	14.3000	0.9281	
	5.500	4.7200	14.6000	0.9316	
	6.000	5.2000	16.8000	0.9351	
	7.000	6.1800	18.5000	0.9372	
	7.500	6.7400	19.8000	0.9403	
	8.000	7.0000	20.2000	0.9407	
	9.000	8.1000	22.8000	0.9462	
Zinc.........................	0.2573	0.554	0.612	0.206	64
	0.2749	0.456	1.167	0.476	
	0.2981	0.469	1.598	0.602	
	0.3255	0.599	2.229	0.682	
	0.3611	0.720	2.610	0.705	
	0.3982	0.846	2.917	0.716	
	0.4413	0.934	3.178	0.730	
	0.4678	1.049	3.485	0.743	
	0.508	1.406	4.101	0.751	
	0.5893	1.932	4.661	0.745	
	0.668	2.618	4.083	0.731	
Zinc, single crystal, optic axis....	0.2650	0.2354	1.6357	0.7759	19
	0.3050	0.2510	1.8528	0.7991	
	0.3450	0.2737	2.1737	0.8275	
	0.3850	0.3069	2.5088	0.8466	
	0.4250	0.3589	2.8140	0.8530	
	0.4650	0.4430	3.1379	0.8515	
	0.5050	0.6395	3.4013	0.8206	
	0.5450	0.7737	3.9129	0.8250	
	0.5850	1.0017	3.8683	0.7888	
	0.5920	1.2525	3.9961	0.7619	
	0.6000	1.4856	4.0555	0.7374	
	0.6250	1.8562	3.9706	0.6896	
	0.6400	3.0132	3.9974	0.6243	

TABLE 6g-1. OPTICAL CONSTANTS OF METALS (*Continued*)

Metal	Wave-length, μm	Index of refraction n	Extinction coefficient k	Reflectance calculated	Ref.
Zinc, single crystal, optic axis (*Cont.*)	0.6800	3.4234	4.3232	0.6421	
	0.7200	3.5908	4.4614	0.6495	
	0.7500	3.7577	4.6239	0.6585	
	0.8000	3.8086	4.6212	0.6575	
	0.8500	3.2523	4.2447	0.6396	
	0.9000	2.9459	3.5761	0.5845	
	0.9500	3.2039	3.0042	0.5200	
	1.0000	2.8821	3.4766	0.5755	
	1.0500	1.9808	4.2004	0.7013	
	1.1000	1.7768	4.5307	0.7483	
	1.1500	1.5853	4.9013	0.7935	
	1.2000	1.5407	5.3192	0.8227	
	1.2500	1.5762	5.8843	0.8472	
	1.3000	1.4824	6.2296	0.8681	
	1.4000	1.5571	6.7753	0.8812	
	1.5000	1.7921	6.9973	0.8737	
	1.6500	1.9241	7.5619	0.8829	
Zinc, basal plane...............	0.2650	0.2806	1.7997	0.7699	
	0.3050	0.3013	2.0077	0.7894	
	0.3450	0.3147	2.3041	0.8212	
	0.3850	0.3911	2.7463	0.8250	
	0.4250	0.4774	3.0476	0.8335	
	0.4650	0.5470	3.4277	0.8453	
	0.5050	0.7568	3.7627	0.8264	
	0.5450	0.9725	4.2879	0.8254	
	0.5850	1.3329	4.4751	0.7907	
	0.5920	1.7048	4.7923	0.7748	
	0.6000	2.0802	4.7231	0.7383	
	0.6250	3.2515	4.2980	0.6441	
	0.6400	3.4512	4.1942	0.6309	
	0.6800	3.7549	4.3042	0.6417	
	0.7200	3.9369	4.6356	0.6566	
	0.7500	4.0269	4.8027	0.6668	
	0.8000	4.1241	4.7768	0.6638	
	0.8500	3.5064	4.1994	0.6303	
	0.9000	3.1807	3.4709	0.5691	
	0.9500	3.3991	2.7684	0.4967	
	1.0000	2.8717	3.2873	0.5547	
	1.0500	1.9701	4.0176	0.6843	
	1.1000	1.6897	4.4062	0.7464	
	1.1500	1.3095	4.9025	0.8216	
	1.2000	1.2889	5.4001	0.8501	
	1.2500	1.3835	5.8910	0.8630	
	1.3000	1.3165	6.2212	0.8805	
	1.4000	1.3628	6.6886	0.8917	
	1.5000	1.4744	6.9688	0.8922	
	1.6500	1.4469	7.4158	0.9051	
Zirconium, polycrystalline........	2.5000	3.8000	6.0500	0.7451	61
	3.0000	3.9500	6.4600	0.7615	
	3.5000	3.4500	7.5500	0.8203	
	4.0000	3.5700	8.7100	0.8524	
	4.5000	3.7500	9.8000	0.8735	
	5.0000	3.9900	11.5000	0.8984	
	5.5000	4.3500	12.8000	0.9096	
	6.0000	4.5200	14.0000	0.9202	
	6.5000	5.0000	15.3000	0.9260	

OPTICAL PROPERTIES OF METALS **6-155**

TABLE 6g-1. OPTICAL CONSTANTS OF METALS (*Continued*)

Metal	Wave-length, μm	Index of refraction n	Extinction coefficient k	Reflec-tance calculated	Ref.
Zirconium, polycrystalline (*Cont.*).	7.0000	5.5000	16.6000	0.9308	
	8.0000	6.4000	18.3000	0.9343	
	9.0000	7.3000	21.0000	0.9427	
	10.0000	8.2000	23.0000	0.9465	
	11.0000	9.0500	25.0000	0.9501	
	12.0000	10.0000	26.40000	0.9511	
	15.0000	12.4000	32.50000	0.9599	
	16.0000	12.6000	34.60000	0.9635	
	17.0000	13.3000	36.60000	0.9655	

References for Table 6g-1

1. Philipp, H. R., and H. Ehrenreich: *J. App. Phys.* **35**, 1416 (1964).
2. Ditchburn, R. W., and G. H. C. Freeman: *Proc. Roy. Soc. (London)*, ser. A, **294**, 20 (1966).
3. Madden, R. R., L. R. Canfield, and G. Hass: *J. Opt. Soc. Am.* **53**, 620 (1963).
4. Daudae, A., M. Priol, and S. Robin: *Compt. Rend.* **263**, 1178 (1966).
5. Hass, G., and J. E. Waylonis: *J. Opt. Soc. Am.* **50**, 1133 (TB15)(1960).
6. Schulz, L. G., and F. R. Tangherlini: *J. Opt. Soc. Am.* **44**, 362 (1954).
7. Schulz, L. G.: *J. Opt. Soc. Am.* **44**, 357 (1954).
8. Beattie, J. R.: *Physica* **23**, 898 (1957).
9. Lemonnier, J., Y. LeCalvez, G. Stepahan, and S. Robin: *Compt. Rend.* **264**, 1599 (1967).
10. Lenham, A. P., D. M. Treherne, and P. J. Metcalfe: *J. Opt. Soc. Am.* **55**, 1072 (1965).
11. Shkliarevskii, Avdeenko, and Padalka: *Opt. Specktroskopiya* **6**, 528 (1959).
12. Potapov, E. V.: *Soviet Phys.—JETP* **20**, 307 (1965).
13. Fisher, E. I., I. Fugita, and G. L. Weissler: *J. Opt. Soc. Am.* **56**, 1560 (1966).
14. O'Bryan, H. M.: *J. Opt. Soc. Am.* **26**, 122 (1936).
15. Maurer, R. J.: *Phys. Rev.* **57**, 653 (1940).
16. Shkliarevskii, I. N., and R. G. Yarovaya: *Opt. Spectr. U.S.S.R.* **11**, 355 (1961).
17. Dix, F. E., and L. H. Rowse: *J. Opt. Soc. Am.* **14**, 306 (1927).
18. Hodgson, J. N.: *Proc. Phys. Soc. (London)*, ser B, **67**, 269 (1954).
19. Lenham, A. P., and D. M. Treherne: *Proc. Phys. Soc.* **83**, 1059 (1964).
20. Jelinek, T. M., R. N. Hamm, E. T. Arakawa, and R. H. Huebner: *J. Opt. Soc. Am.* **36**, 185 (1966).
21. Drude, P.: *Ann. Physik* **39**, 481 (1890).
22. Ives, H. E., and N. B. Briggs: *J. Opt. Soc. Am.* **27**, 395 (1937).
23. Robin, S.: *Compt. Rend.* **236**, 674 (1953).
24. Wartenberg, H. V.: *Verhandl. Deut. Physik. Deut. Ges.* **12**, 105 (1910).
25. Yu, A. Y. C., T. M. Donovan, and W. E. Spicer: *Phys. Rev.* **167**, 670 (1968).
26. Clemens, K. H., and J. Jaumann:*Z. Physik* **173**, 135 (1963).
27. Kirillova, M. M., and R. A. Charikov: *Opt. Spectr. U.S.S.R.* **17**, 134 (1965).
28. Roberts, S.: *Phys. Rev.* **118**, 1509 (1960).
29. Otter, M.: *Z. Physik* **161**, 163 (1961).
30. Dold, B., and R. Mecke: *Optik* **22**, 435 (1965).
31. Weiss, K.: *Z. Naturforsch.* **3a**, 143 (1948).
32. Beattie, J. R., and G. K. T. Conn: *Phil. Mag.* **46**, 989 (1955).
33. Shkliarevskii, I. N., and V. G. Padalka: *Opt. Specktroskopiya* **6**, 78 (1959).
34. Schulz, L. G.: *J. Opt. Soc. Am.* **47**, 64 (1957).
35. Bor, J., and C. Bartholomew: *Proc. Phys. Soc. (London)* **90**, 1153 (1966).
36. Marton, L., and J. Toots: *Phys. Rev.* **160**, 602 (1967).
37. Sasaki, T.: *J. Phys. Soc. Japan* **18**, 701 (1963).
38. Canfield, L. R., G. Hass, and W. R. Hunter: *J. Phys.* **25**, 124 (1964).
39. Archer, R. J.: *Phys. Rev.* **110**, 354 (1958).
40. Avery, D. G., and P. L. Clegg: *Proc. Phys. Soc. (London)*, ser. B, **66**, 512 (1953).
41. Philipp, H. R., and E. A. Taft: *Phys. Rev.* **113**, 1002 (1959).
42. Huldt, L., and T. Staflin: *Opt. Acta* **6**, 27 (1959).

43. Brattain, W. H., and H. B. Briggs: *Phys. Rev.* **75**, 1705 (1949).
44. Motulevich, G. P., and A. A. Shubin: *Soviet Phys.—JETP* **20**, 560 (1965).
45. Padalka, V. G., and I. N. Shklyarevskii: *Opt. Spectr. U.S.S.R.* **11**, 285 (1961).
46. Motulevich, G. P., and A. A. Shubin: *Soviet Phys.—JETP* **17**, 33 (1963).
47. Hass, G., G. F. Jacobus, and W. R. Hunter: *J. Opt. Soc. Am.* **57**, 758 (1967).
48. Barnes, B. T.: *J. Opt. Soc. Am.* **56**, 1546 (1966).
49. Forsterling, K., and V. Freedericksz: *Ann. Physik* **40**, 201 (1913).
50. Yolken, H. T., and J. Kruger: *J. Opt. Soc. Am.* **55**, 842 (1965).
51. Menzel, E., and J. Gebhart: *Z. Physik* **168**, 392 (1962).
52. Golovashkin, A. I.: *Soviet Phys. JETP* **21**, 548 (1965).
53. Priol, M., A. Daudé, and S. Robin: *Compt. Rend.* **264b** 935 (1967).
54. Nathanson, J. B.: *J. Opt. Soc. Am.* **20**, 594 (1930).
55. O'Brien, Brian: *Phys. Rev.* **27**, 93 (1926).
56. Juenker, D. W., L. J. LeBlanc, and C. R. Martin: *J. Opt. Soc. Am.* **58**, 164 (1968).
57. Summers, R. D.: *J. Opt. Soc. Am.* **24**, 262 (1934).
58. Kern, E.: *Z. Physik* **148**, 38 (1957).
59. Tool, A. Q.: *Phys. Rev.* **31**, 1 (1910).
60. Ingersoll, L. R.: *Astrophys. J.* **32**, 282 (1910).
61. Kirillova, M. M., and B. A. Charikov: *Phys. Metals Metallog.* (GB) **16**, 41 (1964).
62. Malé, D., and J. Trompette: *J. Phys. Radium* **18**, 128 (1957).
63. Jacobus, G. F., R. P. Madden, and L. R. Canfield: *J. Opt. Soc. Am.* **53**, 1084 (1963).
64. Meier, W.: *Ann. Physik* **31**, 1017 (1910).
65. Larson, D. T., and D. L. Cash: *J. Nucl. Mater.* **24**, 2232 (1967).
66. Sutherland, J. C., and L. T. Arakawa: *J. Opt. Soc. Am.* **58**, 1080 (1968).
67. Ives, H. E., and H. B. Briggs: *J. Opt. Soc. Am.* **26**, 238 (1936); **27**, 182 (1937).
68. Bolle, H. J.: *Z. Physik* **143**, 538 (1956).
69. Duncan, R. W.: *Phys. Rev.* **1**, 306 (1913).
70. Morgan, R.: *Phys. Rev.* **20**, 203 (1922).
71. Althoff, R., and J. H. Hertz: *Infrared Phys.* **7**, 11 (1967).
72. Lange, H.: *Z. Physik* **94**, 650 (1935).
73. Bolotin, G. A., and T. P. Chukina: *Opt. Spectr. U.S.S.R.* **23**, 333 (1967).
74. Hass, Schroeder, and Turner: *J. Opt. Soc. Am.* **46**, 31 (1956).
75. Tutihasi, S., and I. Chen: *Phys. Rev.* **158**, 623 (1967).
76. Koehler, Odencrantz, and White: *J. Opt. Soc. Am.* **49**, 109 (1959).
77. Sasaki, T., and K. Ishiguro: *Phys. Rev.* **127**, 1091 (1962).
78. Philipp, H. R., and E. A. Taft: *Phys. Rev.* **120**, 37 (1960).
79. Pulker, H., and E. Ritter: *Optik* **21**, 21 (1964).
80. Taft, E. A., and H. R. Phillip: *Phys. Rev.* **121**, 1100 (1961).
81. Huebner, R. H., E. T. Arakawa, R. A. MacRae, and R. N. Hamm: *J. Opt. Soc. Am.* **54**, 1434 (1964).
82. Minor, R. S.: *Ann. Physik* **10**, 581 (1903).
83. Philip, R., and J. Trompette: *Compt. Rend.* **241**, 627 (1955).
84. Canfield, L. R., and G. Hass: *J. Opt. Soc. Am.* **55**, 61 (1965).
85. Lindquist, R. E., and A. A. Ewald: *Phys. Rev.* **135**, A191 (1964).
86. Hass, G., and A. P. Bradford: *J. Opt. Soc. Am.* **47**, 125 (1957).
87. Bolotin, G. A., et al.: *Fiz. Metal. Metalloved.* **13**, 823 (1962).
88. Littleton, J. T.: *Phys. Rev.* **35**, 306 (1912).

TABLE 6g-2. PERCENT NORMAL-INCIDENCE REFLECTANCE OF FRESHLY
EVAPORATED MIRROR COATINGS OF ALUMINUM, SILVER, GOLD,
COPPER, RHODIUM, AND PLATINUM, FROM THE
ULTRAVIOLET TO THE INFRARED*†

λ, μm	Al	Ag	Au	Cu	Rh	Pt
0.220	91.5	28.0	27.5	40.4	57.8	40.5
0.240	91.9	29.5	31.6	39.0	63.2	46.9
0.260	92.2	29.2	35.6	35.5	67.7	51.5
0.280	92.3	25.2	37.8	33.0	70.7	54.9
0.300	92.3	17.6	37.7	33.6	73.4	57.6
0.315	92.4	5.5	37.3	35.5	75.0	59.4
0.320	92.4	8.9	37.1	36.3	75.5	60.0
0.340	92.5	72.9	36.1	38.5	76.9	62.0
0.360	92.5	88.2	36.3	41.5	78.0	63.4
0.380	92.5	92.8	37.8	44.5	78.1	64.9
0.400	92.4	95.6	38.7	47.5	77.4	66.3
0.450	92.2	97.1	38.7	55.2	76.0	69.1
0.500	91.8	97.9	47.7	60.0	76.6	71.4
0.550	91.5	98.3	81.7	66.9	78.2	73.4
0.600	91.1	98.6	91.9	93.3	79.7	75.2
0.650	90.5	98.8	95.5	96.6	81.1	76.4
0.700	89.7	98.9	97.0	97.5	82.0	77.2
0.750	88.6	99.1	97.4	07.0	82.6	77.9
0.800	86.7	99.2	98.0	98.1	83.1	78.5
0.850	86.7	99.2	98.2	98.3	83.4	79.5
0.900	89.1	99.3	98.4	98.4	83.6	80.5
0.950	92.4	99.3	98.5	98.4	83.9	80.6
1.0	94.0	99.4	98.6	98.5	84.2	80.7
1.5	97.4	99.4	99.0	98.5	87.7	81.8
2.0	97.8	99.4	99.1	98.6	91.4	81.8
3.0	98.0	99.4	99.3	98.6	95.0	90.6
4.0	98.2	99.4	99.4	98.7	95.8	93.7
5.0	98.4	99.5	99.4	98.7	96.4	94.9
6.0	98.5	99.5	99.4	98.7	96.8	95.6
7.0	98.6	99.5	99.4	98.7	97.0	95.9
8.0	98.7	99.5	99.4	98.8	97.2	96.0
9.0	98.7	99.5	99.4	98.8	97.4	96.1
10.0	98.7	99.5	99.4	98.9	97.6	96.2
15.0	98.9	99.6	99.4	99.0	98.1	96.5
20.0	99.0	99.6	99.4			
30.0	99.2	99.6	99.4			

* The reflectance of a good evaporated mirror coating is always higher than that of a polished or electroplated surface of the same material.
† G. Hass, in R. Kingslake, ed., "Applied Optics and Optical Engineering," vol. III, pp. 309–330, Academic Press, Inc., New York, 1965.

TABLE 6g-3. PERCENT REFLECTANCE OF VARIOUS POLISHED METALS
AT CLOSE TO NORMAL INCIDENCE

Wavelength λ, μm	Au (1)	Be (2)	Cu (3)	Mo (8)	Ni	Pd (7)	Rh (8)	Ag	Ta (8)
0.25	....	56	25.9	....	47.5 (2)	....	....	25 (6)	
0.30	....	50	25.3	....	41.5 (2)	....	....	13 (6)	
0.35	....	....	27.5	....	45.0 (2)	....	....	68 (6)	
0.40	36.0	48	30.0	44.0	53.3 (2)	....	....	87.5 (6)	
0.50	41.5	46	43.7	45.5	59.7 (1)	....	76	95.2 (6)	38.0
0.60	87.0	....	71.8	47.6	64.5 (1)	....	....	......	45.0
0.70	93.0	....	83.1	49.8	67.6 (1)	....	79	96.1 (3)	56.0
0.80	....	50	88.6	52.3	......	....	81	96.2 (3)	64.5
1.0	....	54.5	90.1	58.2	74.1 (4)	74.8	84	96.4 (3)	78.5
2.0	....	....	95.5	81.6	84.4 (4)	....	91	97.3 (3)	90.5
4.0	....	....	97.3	90.5	......	88.1	92.5	97.7 (3)	93.0
6.0	....	....	98.0	93.0	......	....	93.5	98.0 (3)	93.2
8.0	....	....	98.3	93.7	96.0 (5)	94.7	94	98.7 (3)	93.8
10.0	....	....	98.4	94.5	......	96.5	95	98.9 (3)	94.5
12.0	....	....	98.4	95.2	......	96.5	....	98.9 (3)	95.0

Wavelength λ, μm	Al (7)	Sb (8)	Cd (7)	Cr (8)	Fe (8)	Ir (7)	Co (7)	Mg (8)	W (9)
0.6	....	53	....	55.5	57.5	....	....	....	53.1
1.0	73.3	55	71.0	57.0	65.0	79.4	67.5	74.0	57.6
2.0	82.0	60	....	63.0	78.0	....	....	77.0	90.0
3.0	88.3	65	93	70.0	84.5	91.4	76.7	80.5	94.3
4.0	91.4	68	....	76.0	89.5	93.3	80.7	83.5	94.8
5.0	93.7	...	95.9	81.0	91.5	94.0	86.0	86.0	95.3
6.0	....	70	....	85.0	93.0	94.5	....	88.0	95.8
7.0	95.0	...	....	....	94.0	94.7	93.0	91.0	
8.0	96.9	...	97.2	89.0	94.0	94.8	95.8	93.0	
9.0	....	72	98.0	92.0	94.0	95.5	96.4	93.0	
10.0	97.0	...	98.0	93.0	....	95.8	96.8		
12.0	97.3	...	98.2	....	....	96.1	96.6		

Numbers in parentheses refer to the references which follow.

References for Table 6g-3
1. Tool, A. Q.: *Phys. Rev.* **31**, 1 (1910).
2. Coblentz, W. W., and R. Stair: *Natl. Bur. Standards J. Research* **2**, 343 (1929).
3. Hagen, E., and H. Rubens: *Ann. Physik* **8**, 16 (1902), and **11**, 873 (1903).
4. Ingersoll, L. R.: *Astrophys. J.* **32**, 282 (1910).
5. Hagen, E., and H. Rubens: *Berliner Sitzber.* 491 (1909).
6. Minor, R. S.: *Ann. Physik* **10**, 581 (1903).
7. Coblentz, W. W.: *Natl. Bur. Standards Bull.* **2**, 457 (1906).
8. Coblentz, W. W.: *Natl. Bur. Standards Bull.* **7**, 198 (1911).
9. Coblentz, W. W., and W. B. Emerson: *Natl. Bur. Standards Bull.* **14**, 306 (1918–1919).

TABLE 6g-4. CALCULATED REFLECTANCE AND TRANSMITTANCE OF AL FILMS ON TRANSPARENT SUBSTRATES OF $n = 1.5$ FOR VARIOUS WAVELENGTHS AS A FUNCTION OF FILM THICKNESS
(Calculated values agree with directly measured ones for film thicknesses $t > 80$ Å; back surface antireflected)*

Film thickness, angstroms	Wavelength, nm									
	220		300		400		546		650	
	$R\%$	$T\%$	$R\%$	$T\%$	$R\%$	$T\%$	$R\%$	$T\%$	$R\%$	$T\%$
40	14	82	19	74	25	65	33	51	38	42
80	33	60	43	47	52	36	60	24	63	18
120	52	40	62	27	70	19	74	12	75	9
160	67	25	74	16	79	11	81	7	82	5
200	76.3	15.2	81.5	9.1	84.9	5.9	85.6	3.5	85.4	2.6
240	82.4	9.1	86.0	5.1	88.1	3.3	88.1	2.0	87.5	1.4
280	86.2	5.4	88.4	3.1	90.0	1.9	89.5	1.1	88.8	0.8
320	88.5	3.2	90.0	1.8	91.1	1.1	90.4	0.5	89.6	0.4
360	89.8	1.9	90.9	1.0	91.7	0.6	90.9	0.4	90.0	0.3
400	90.6	1.1	91.4	0.5	92.1	0.4	91.2	0.2	90.3	0.2
500	91.5	0.3	92.0	0.1	92.5	<0.1	91.5	<0.1	90.6	<0.1

* From measurements by G. Hass and J. E. Waylonis; for similar tables of other metals see H. Mayer "Physik Duenner Schichten," vol. I, Wissenschaftliche Verlagsgesellschaft, Stuttgart, 1950.

TABLE 6g-5. THICKNESS OF AL FILMS FOR WHICH THE TRANSMITTANCE IS 0.5 PERCENT AT VARIOUS WAVELENGTHS

Wavelength, angstroms	Film thickness, angstroms
5,460	320
3,000	390
2,200	450
1,216	700
735	~7,000
585	~12,000

TABLE 6g-6. THE EXTREME ULTRAVIOLET REFLECTANCE OF EVAPORATED PLATINUM AND RHODIUM IN THE WAVELENGTH REGION FROM 585 TO 2,000 Å*
(Both film materials show very little aging during exposure to air)

λ, angstroms		584	735	900	1,000	1,105	1,216	1,360	1,486	1,606	2,000
$R\%$	Pt	20.9	12.9	16.1	17.5	20.6	23.0	24.3	25.7	26.0	30.0
	Rh	20.9	13.5	12.5	11.0	9.5	9.2	12.5	24.8	35.0	49.0

* G. Hass and R. Tousey, *J. Opt. Soc. Am.* **49**, 593 (1959).

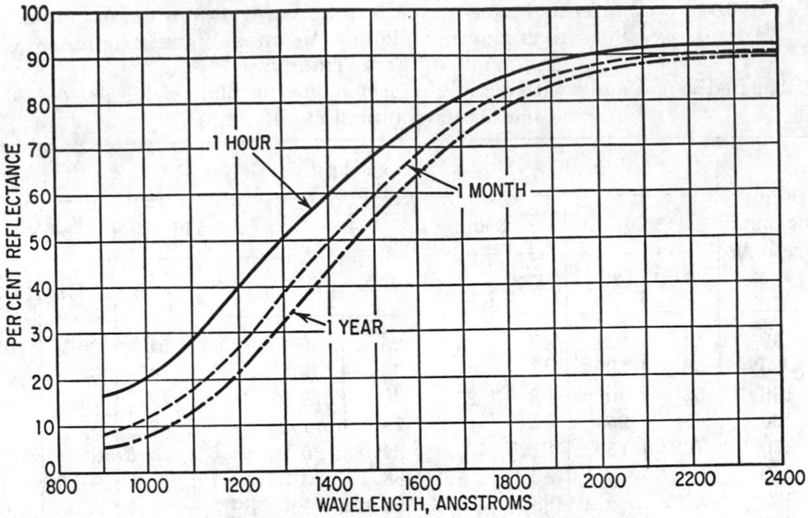

FIG. 6g-1. Extreme ultraviolet reflectance of best-quality aluminum films after 1 hour, 1 month, and 1 year exposure to air; wavelength region 900 to 2,400 Å.   [*From G. Hass and R. Tousey, J. Opt. Soc. Am.* **49**, 593 (1959).]

6h. Reflection[1]

TABLE 6h-1. INFRARED DIFFUSE PERCENTAGE REFLECTING FACTORS OF DRY PIGMENTS*

Wavelength, μm	Co_2O_3	CuO	Cr_2O_3	PbO	Fe_2O_3	Y_2O_3	$PbCrO_4$	Al_2O_3	ThO_2	ZnO	MgO	CaO	ZrO_2	$PbCO_3$	$MgCO_3$	White lead paint	Zn oxide paint
0.60†	3	...	27	52	26	74	70	84	86	82	86	85	86	88	85	76	68
0.95†	4	24	45	...	41	...	...	88	...	86	...	...	84	93	89	79	72
4.4	14	15	33	51	30	34	41	21	47	8	16	22	23	29	11		
8.8	13	...	5	26	4	11	5	20	7	3	2	4	5	10	4		
24.0	6	4	8	10	9	10	7	6	10	5	9	6	5	7	9		

A surface of plate glass, ground uniformly with the finest emery and then silvered, used at an angle of 75 deg, reflected 90 percent at ¼ μm, approached 100 for longer waves, only 10 at 1 μm, less than 5 in the visible red and approached 0 for shorter waves. Similar results were obtained with a plate of rock salt for transmitted energy when roughened merely by breathing on it. In both cases the finer the surface, the more suddenly it cuts off the short waves.

* "Smithsonian Physical Tables," 1954, Table 581.

† Nonmonochromatic means from Coblentz.

[1] Metallic reflections are discussed in Sec. 6g.

TABLE 6h-2. REFLECTION COEFFICIENTS FOR VISIBLE
MONOCHROMATIC RADIATION*

Material	Wavelengths, μm			
	0.400	0.500	0.600	0.700
Carbon black in oil....................	0.003	0.003	0.003	0.003
Clay:				
Kaolin (treated).....................	0.82	0.81	0.82	0.82
Kaolin (untreated).................	0.75	0.79	0.85	0.86
White Georgia......................	0.94	0.92	0.93	0.94
$MgCO_3$.............................	....	0.98	0.99	
Magnesium oxide.....................	0.97	0.98	0.99	0.98
Paint:				
Lithopone..........................	0.95	0.98	0.98	0.98
$MgCO_3$-vinyl acetate lacquer........	0.90	0.88	0.88	0.88
ZnO-milk..........................	0.74	0.84	0.85	0.86
Paper:				
Blotting...........................	0.64	0.72	0.79	0.79
Calendered........................	0.64	0.69	0.73	0.76
Crepe, green.......................	0.23	0.49	0.19	0.48
Crepe, red.........................	0.03	0.02	0.21	0.69
Crepe, yellow......................	0.17	0.44	0.75	0.79
Newsprint stock....................	0.38	0.61	0.63	0.78
Peach:				
Green.............................	0.18	0.17	0.62	0.63
Ripe...............................	0.10	0.10	0.41	0.42
Pear:				
Green.............................	0.04	0.12	0.29	0.41
Ripe..............................	0.08	0.19	0.46	0.53
Pigment:				
Chrome yellow.....................	0.05	0.13	0.70	0.77
French ochre......................	0.06	0.14	0.50	0.56
Porcelain enamel:				
Blue..............................	0.44	0.10	0.05	0.23
Orange............................	0.09	0.09	0.59	0.69
Red...............................	0.05	0.03	0.08	0.62
White.............................	0.77	0.73	0.72	0.70
Yellow............................	0.11	0.46	0.62	0.62
Talcum, Italian.....................	0.94	0.89	0.88	0.88
Wheat flour........................	0.75	0.87	0.94	0.97

* J. L. Michaelson, in "Handbook of Chemistry and Physics," 36th ed., p. 2689, Chemical Rubber
Publishing Company, 1954–1955.

TABLE 6h-3. REFLECTION COEFFICIENTS FOR INCANDESCENT LIGHT*

Material	Nature of surface	Coefficient	Authority
Aluminum, "Alzak"	Diffusing	0.77–0.81	3
"Alzak"	Specular	0.79–0.83	3
On glass	First surface	0.82–0.86	4
Polished	Specular	0.69	3
Black paper	Diffusing	0.05–0.06	4
Chromium	Specular	0.62	4
Copper	Specular	0.63	4
Gold	Specular	0.75	1
Magnesium oxide	Diffusing	0.98	5
Nickel	Specular	0.62–0.64	1, 3
Platinum	Specular	0.62	1
Porcelain enamel	Glossy	0.76–0.79	3
Porcelain enamel	Ground	0.81	3
Porcelain enamel	Mat	0.72–0.76	3
Silver	Polished	0.93	1
Silvered glass	Second surface	0.88–0.93	3
Snow	Diffusing	0.93	2
Steel	Specular	0.55	1
Stellite	Specular	0.58–0.65	4

1. Hagen and Rubens. 2. Nutting, Jones, and Elliot. 3. J. E. Bock. 4. Frank Benford. 5. J. L. Michaelson.

* "Handbook of Chemistry and Physics," 36th ed., p. 2689, Chemical Rubber Publishing Company. 1954–1955.

6h-1. Residual Rays and Crystal Reflectivity. Over certain narrow wavelength regions in the infrared various crystals have reflectivities over 60 percent. Since their reflectivities are much lower in other wavelength regions, multiple reflections of a broad band of wavelengths will result in isolation of the narrow region. For wavelengths over 50 μm this technique is still much used as a simple method of producing a narrow band of wavelengths from a blackbody source such as a globar or Nernst filament.

The high reflectivity which is sometimes referred to as "metallic" is associated with a particularly strong absorption band. The formula for reflectivity of an absorbing medium at normal incidence can be written

$$R = \frac{(n_0 - n_1)^2 + k_1^2}{(n_0 - n_1)^2 + k_1^2}$$

where k_1 is the extinction coefficient and n_1 is the index of refraction. For very large k the reflectivity approaches 100 percent. However, in such regions of high absorption the refractive index is usually changing rapidly and the region of highest reflectivity is thus slightly displaced from the wavelength of peak absorption. The residual-ray wavelength value is useful in crystal physics. It represents the wavelength of strong coupling between acoustic vibrations in the crystal and an electromagnetic disturbance. At this wavelength a photon is converted directly into a phonon.

Table 6h-4 lists the peak wavelengths which can be isolated by multiple reflection from a particular crystal. Table 6h-5 shows the reflectivity of a few specific crystals as a function of wavelength.

TABLE 6h-4. WAVELENGTHS OF RESIDUAL RAYS
[After many reflections from a given material the reflected radiation contains only
a few ("residual") wavelengths in the range 18 to 150 μm. Unit of
$\lambda = 10^4 \overset{\circ}{A} = 10^{-4}$ cm]

Substance	λ			λ_{mean}	Authority
NH₄Cl...............	46.3	.....	54.0	51.5	Nichols and Day, 1908
					Rubens, 1914
NH₄Br..............	55.3	.....	62.3	59.3	Rubens, 1914
PbCl₂..............	74	92	114	91	Rubens, 1913
TlCl................	....	91.6	.....	....	Rubens, 1914
TlBr...............	....	117	.....	....	Rubens, 1914
TlI.................	....	151.8	.....	....	Rubens, 1914
Hg₂Cl₂.............	91.6	.....	117.8	98.8	Rubens, 1913
AgCl...............	74	.....	90	81.5	Rubens, 1913
AgBr...............	....	112.7	.....	....	Rubens, 1913
AgCN..............	....	93	.....	....	Rubens, 1914
MgCO₃.............	....	30.2	.....	....	Coblentz, 1908
CaF₂...............	22.0	33.0	.....	....	Czerny, 1923
CaCO₃..............	....	29.4	.....	....	Aschkinass, 1900
CaCO₃..............	93	.....	116	98.7	Rubens, 1913
CaCO₃..............	....	39	.....	....	Liebisch and Rubens, 1919
SrCO₃..............	43.2	.....	.....	....	Nichols and Day, 1908
BaCO₃.............	....	46	.....	....	Nichols and Day, 1908
NaCl...............	47	.....	54	52	Nichols and Day, 1908
					Rubens, 1913
					Rubens and Aschkinass, 1898
					Rubens and Hollnagel, 1923
NaCl...............	....	52	.....	....	Czerny, 1923
NaBr...............	50–55		.....	....	Aschkinass, 1900
KCl................	62.3	.....	70.6	63.4	Rubens, 1913
					Rubens and Aschkinass, 1898
					Rubens and Hollnagel, 1910
KCl................	....	63	.....	....	Czerny, 1923
KBr...............	74.0	.....	86.0	83.3	Aschkinass, 1900
					Rubens, 1913
					Rubens and Hollnagel, 1910
KI.................	....	94	.....	....	Czerny, 1923
H₂KAl₂(SiO₄)₃........	18.4	.....	21.5	....	Coblentz, 1908
					Rubens and Nichols, 1897

Quartz residual rays (ordinary).........	8.4	9.0	12.5	21	26	....	Liebisch and Rubens, 1919
Quartz residual rays (extraordinary).....	8.4	9.0	12.9	19.7	27	....	Liebisch and Rubens, 1919

TABLE 6h-5. REFLECTIVITY AS A FUNCTION OF WAVELENGTH FROM
MULTIPLE PLATES OF CsBr*

Wavelength, μm	1 reflection from CsBr	2 reflections from CsBr	3 reflections from CsBr	4 reflections from CsBr
80	0.8			
82	1.5			
84	2.7	0.1		
86	7.2	0.5		
88	16.8	2.8	0.5	0.1
90	28.1	7.9	2.2	0.6
92	36.7	13.5	4.9	1.8
94	42.0	17.6	7.4	3.1
96	45.5	20.7	9.4	4.3
98	46.2	21.3	9.9	4.6
100	47.9	22.9	11.0	5.3
102	53.0	28.1	14.9	7.9
104	58.6	34.3	20.1	11.8
106	64.1	41.1	26.3	16.9
108	71.6	51.3	36.7	26.3
110	77.2	59.6	46.0	35.5
112	81.5	66.4	54.2	44.1
114	84.3	71.1	59.9	50.5
116	85.4	72.9	62.3	53.2
118	85.6	73.3	62.7	53.7
120	86.2	74.3	64.0	55.2
122	86.2	74.3	64.0	55.2
124	85.6	73.3	62.7	53.7
126	85.6	73.3	62.7	53.7
128	85.1	72.4	61.6	52.4
130	83.8	70.2	58.8	49.3
132	81.2	66.0	53.5	43.5
134	77.7	60.4	46.9	36.4
136	71.9	51.7	37.2	26.7
138	65.8	43.3	28.5	18.7
140	59.6	35.5	21.2	12.6
145	47.4	22.5	10.7	5.1
150	42.1	17.7	7.5	3.1
155	37.6	14.1	5.3	2.0
160	35.0	12.3	4.3	1.5
165	32.4	10.5	3.4	1.1
170	30.3	9.2	2.8	0.8
175	28.9	8.4	2.4	0.7
180	28.1	7.9	2.2	0.6
185	26.9	7.2	2.0	0.5
190	26.1	6.8	1.8	0.5
195	25.7	6.6	1.7	0.4
200	24.8	6.2	1.6	0.4
210	24.4	6.0	1.5	0.4
220	23.9	5.7	1.4	0.3
230	23.3	5.4	1.3	0.3
240	23.0	5.2	1.2	0.3
250	22.4	5.0	1.1	0.3
260	22.1	4.9	1.1	0.2
270	21.7	4.7	1.0	0.2

* Personal communication from M. Czerny.

TABLE 6h-6. REFLECTIVITY OF LiF*

λ, μm	R, %	λ, μm	R, %
64.5	24.0	22.4	84.6
57.8	24.0	21.8	82.3
50.0	24.0	21.1	77.8
42.9	25.2	20.8	75.6
39.7	29.8	20.5	73.7
38.5	31.6	20.2	72.2
37.5	35.1	20.0	71.7
37.2	35.9	19.7	71.4
36.2	45.1	19.4	72.2
35.7	50.0	19.2	72.9
34.6	55.6	18.9	73.7
33.8	61.5	18.5	74.6
33.3	64.4	18.0	76.5
32.8	69.6	17.7	77.0
32.2	73.9	17.4	77.3
31.1	79.1	17.1	77.1
29.9	83.2	16.9	76.6
29.4	84.2	16.6	75.8
29.1	85.5	16.3	73.1
28.5	86.7	16.1	70.1
27.9	87.8	15.9	65.8
27.0	88.6	15.7	60.1
26.2	89.6	15.5	48.8
25.6	90.0	15.2	28.1
25.0	89.2	14.2	12.4
24.6	88.5	13.9	9.5
24.1	88.2	13.5	5.9
23.1	86.7	12.6	1.6

* Personal communication from M. Czerny.

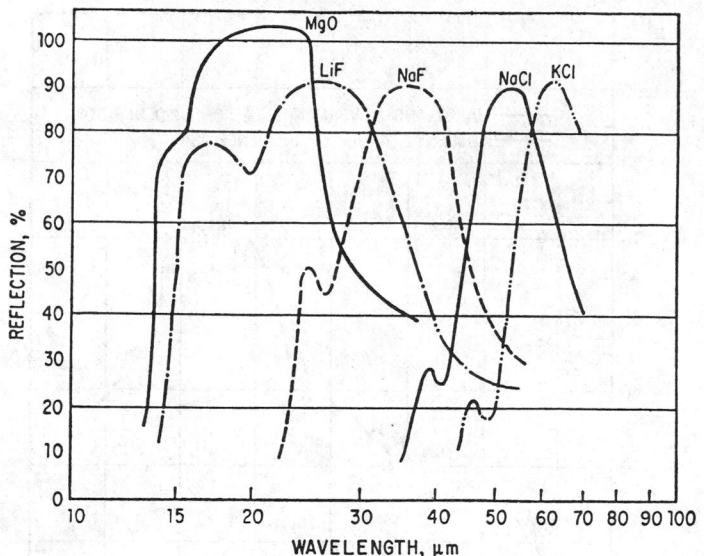

FIG. 6h-1. Reflection of MgO and of some alkali halide crystals relative to an aluminum mirror.   [*MgO data from Burstein, Oberly, and Plyler, Proc. Indian Acad. Sci.* **28,** 388 (1948). *Other data from Hohls,* 1937.]

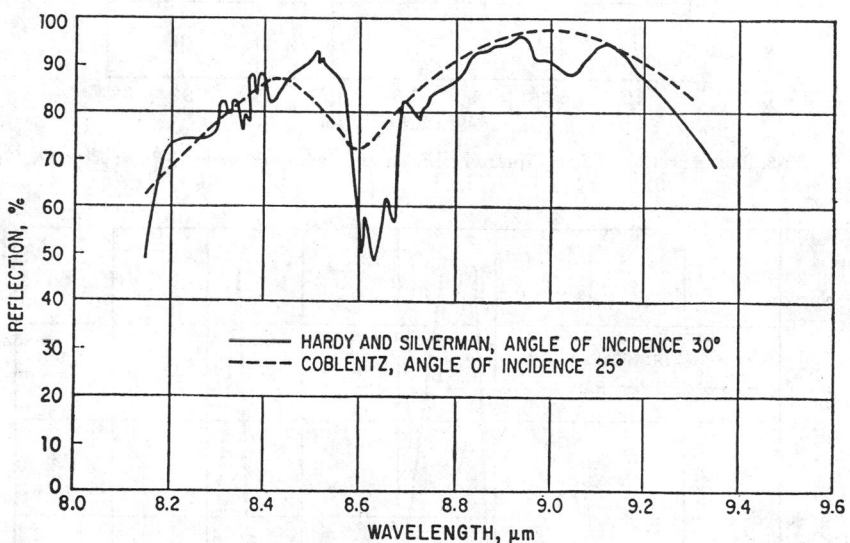

FIG. 6h-2. The reflection of crystalline quartz.   [*From Hardy and Silverman, Phys. Rev.* **37,** 176 (1931).]

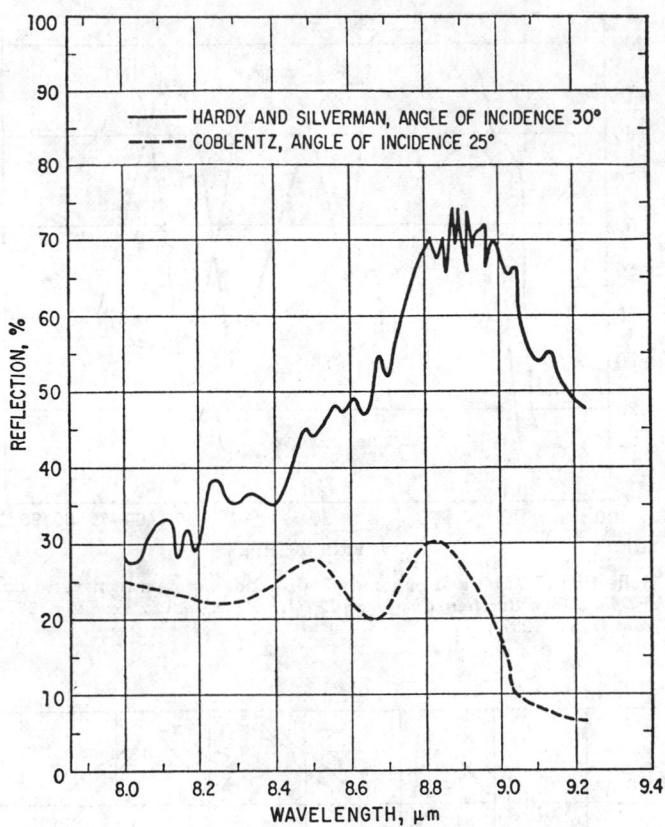

FIG. 6h-3. The reflection of fused quartz.   [*From Hardy and Silverman, Phys. Rev.* **37**, 176 (1931).]

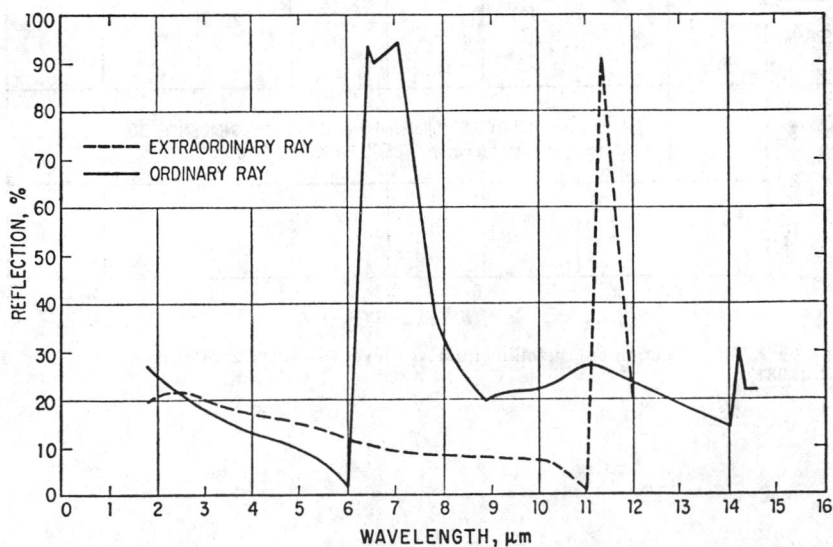

FIG. 6h-4. The reflection of calcite (CaCO₃).   [*From Nyswander, Phys. Rev.* **28**, 291 (1909).]

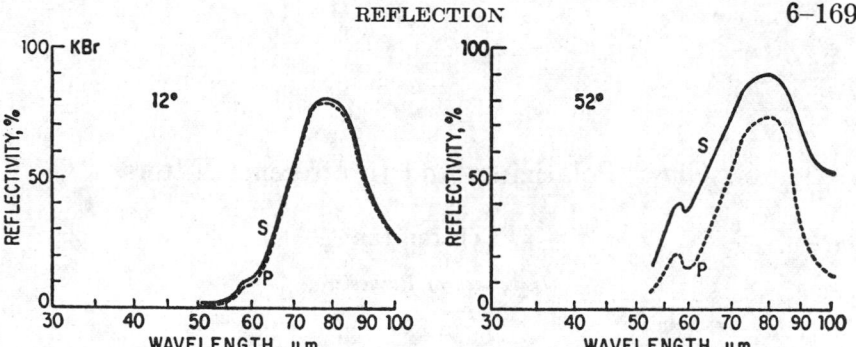

REFLECTIVITY OF PLANE POLARIZED LIGHT BY KBr CRYSTAL AT INCIDENT ANGLES OF 12° AND 52°

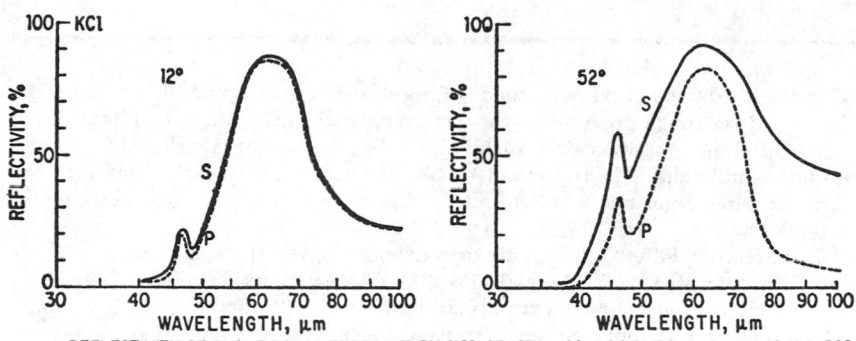

REFLECTIVITY OF PLANE POLARIZED LIGHT BY KCl CRYSTAL AT INCIDENT ANGLES OF 12° AND 52°

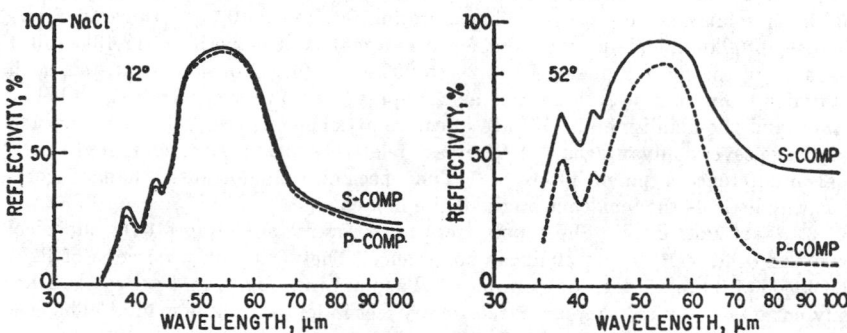

REFLECTIVITY OF PLANE POLARIZED LIGHT BY NaCl CRYSTAL AT INCIDENT ANGLES OF 12° AND 52°

Fig. 6h-5. The reflectivity of various crystals for different states of polarization. [*From A. Mitsuishi, J. Opt. Soc. Am. 50, 433 (1960).*]

6i. Glass, Polarizing, and Interference Filters

P. BAUMEISTER

University of Rochester

J. EVANS

Air Force Cambridge Research Laboratories
Sacramento Peak Observatory

This chapter briefly surveys methods of spectral filtering, by which we mean the technique of isolating a portion of the electromagnetic spectrum with filters which function in either reflection or transmission. Several survey articles [1,2] discuss these filters in detail. The important classes of filters which are discussed here are absorption filters, polarizers, mesh filters, interference devices, and polarization interference filters.

6i-1. Absorption Filters. There are many types of these filters, such as: (1) Glass-doped with impurities such as metal ions [3]. The commonly available filters [4,5] are useful in the spectral region from 0.25 to 2.5 μm. (2) Crystals, such as alkali halides or semiconductors. Spectral transmittance data are tabulated by several authors [6,7] and also by manufactures [8]. (3) Gelatin sheets impregnated with organic dyes [9] are inexpensive filters for the region 0.3 to 1.5 μm. (4) Gas cells and liquid solutions are often excellent filtering devices [1,10,11]. Infrared filters consisting of alkali halide powders dispersed in a matrix of polyethene [12,13] exhibit passbands in the spectral region from 20 to 200 μm. (6) Thin films of metals such as aluminum and indium [14] are used as bandpass filters in the spectral region below 0.1 μm, and the alkali metals [15] are effective at longer wavelengths. Absorption filters have several advantages: (1) They are relatively inexpensive, compared to the usual interference type of filter. (2) The spectral transmittance changes comparatively little as the incidence angle of the flux changes.

6i-2. Sheet Polarizers. Sheet polarizers have several advantages over the nicol prism and other early types of linear polarizers. They accept a wide cone of light (half angle of 30 to 45 deg, for example). They are thin, light, and rugged and are easily cut to any desired shape. Pieces many feet in length can be made. The cost is almost negligible compared with that of a nicol prism.

If a sheet polarizer is mounted perpendicular to a beam of 100 percent linearly polarized radiation, and if the polarizer is slowly turned in its own plane, the transmittance k varies between a maximum value k_1 and a minimum value k_2 according to the following law:

$$k = (k_1 - k_2)(\cos^2 \theta) + k_2 \qquad (6i\text{-}1)$$

When such a polarizer is placed in a beam of unpolarized radiation, the transmittance is $\frac{1}{2}(k_1 + k_2)$. When two identical polarizers are mounted in the bean with their axes crossed, the transmittance is $k_1 k_2$.

The principal transmittance values k_1 and k_2 vary with wavelength, the variation being different for different types of polarizers. Table 6i-1 presents data for several

TABLE 6i-1. SPECTRAL PRINCIPAL TRANSMITTANCE OF SHEET POLARIZERS*

Wave-length, μm	HN-22 sheet		HN-32 sheet		HN-38 sheet		KN-36 sheet		HR sheet	
	k_1	k_2	k_1	k_2	k_1	k_2	k_1	k_2	k_1	k_2
0.375	.11	.000,005	.33	.001	.54	.02	.42	.002	.00	.00
0.40	.21	.000,01	.47	.003	.67	.04	.51	.001	.00	.00
0.45	.45	.000,003	.68	.000,5	.81	.02	.65	.000,3	.00	.00
0.50	.55	.000,002	.75	.000,05	.86	.005	.71	.000,05	.00	.00
0.55	.48	.000,002	.70	.000,02	.82	.000,7	.74	.000,04	.00	.00
0.60	.43	.000,002	.67	.000,02	.79	.000,3	.79	.000,03	.01	.00
0.65	.47	.000,002	.70	.000,02	.82	.000,3	.83	.000,08	.05	.00
0.7	.59	.000,003	.77	.000,03	.86	.000,7	.88	.02	.10	.00
1.0									.55	.05
1.5									.65	.00
2.0									.70	.00
2.5									.10	.02

* Data supplied by R. C. Jones, Polaroid Corporation, Cambridge, Mass.   For each type of polarizer, the transmittance values near the ends of the useful range depend on the type of supporting sheet or lamination used.   Also some variation from lot to lot must be expected.

well-known types, produced by Polaroid Corporation, Cambridge, Mass.   H sheet, perhaps the most widely used sheet polarizer, is effective throughout the visual range; it is produced in three modifications having total luminous transmittance (for C.I.E. Illuminant C light) of 22 percent (Type HN-22), 32 percent (Type HN-32), and 38 percent (HN-38).   Type HN-22 provides the best extinction, Type HN-38 provides the highest transmittance, and Type HN-32 represents a compromise that is preferred in many applications.   K sheet, also useful throughout the visual range, is particularly intended for applications involving very high temperature.   Its transmittance is 35 to 40 percent.   HR sheet is effective in the infrared range from 0.7 to 2.2 μm.

Table 6i-2 presents data for a Polaroid Corporation ultraviolet light-polarizing filter HNP'B.   The characteristics for wavelengths longer than 0.400 μm are the same as for HN-32 in Table 6i-1.   In Fig. 6i-1 are curves showing a range of k values which can be achieved with this class of ultraviolet polarizer.   Absorbing polarizers are also made by Polacoat, Inc., Blue Ash, Ohio.

TABLE 6i-2. SPECTRAL PRINCIPAL TRANSMITTANCE OF
ULTRAVIOLET SHEET POLARIZER HNP'B (3.5)

λ	k_1	k_2	λ	k_1	k_2
(275)*	(0.250)	(0.0126)	340	0.602	0.0002
280	0.328	0.0110	350	0.568	0.0001
290	0.340	0.0040	360	0.550	0.0003
300	0.372	0.0017	370	0.568	0.0007
310	0.448	0.0009	380	0.604	0.0009
320	0.546	0.0006	390	0.644	0.0008
330	0.611	0.0003	400	0.688	0.0005

* This is the effective cutoff wavelength of the supporting plastic layer.   The HNP'B foil itself transmits to about 250 nm.   In this region the foil has much lower dichroism.

6i-3. Mesh Filters and Interference Devices. *Metal mesh filters* consist of an array of thin metal strips, rectangles, disks, apertures, etc., which are either unsupported or deposited on a thin plastic sheet. They have an effect similar to an iris in a microwave guide, with the exception that the mesh array functions in free space. They have been used principally in the spectral region from 100 μm to one millimeter

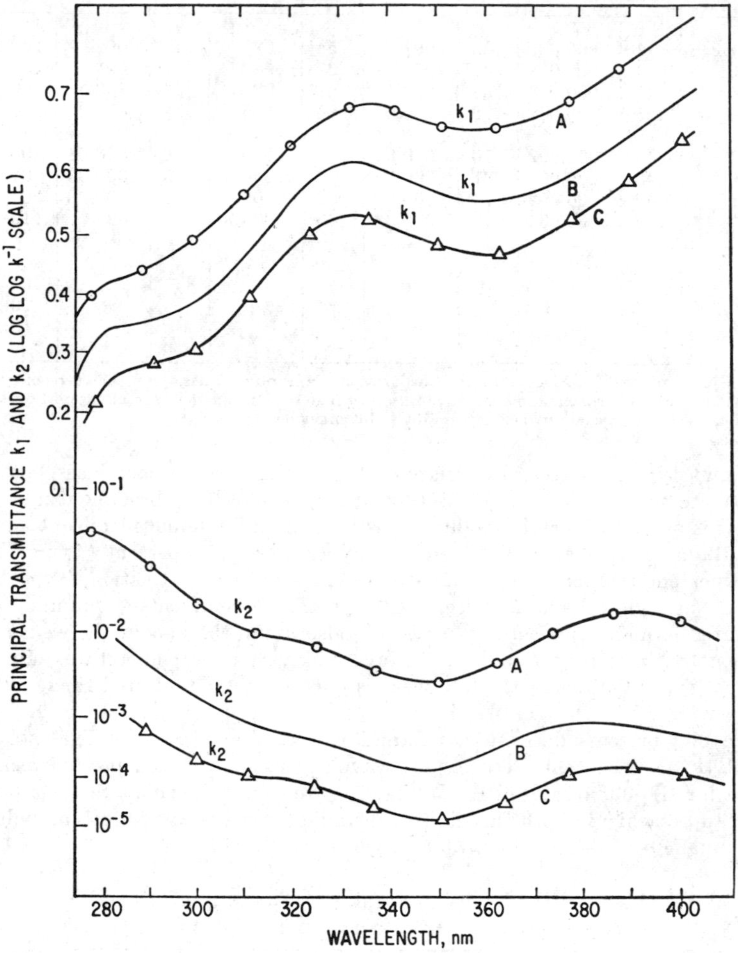

FIG. 6i-1. Polarized spectral transmittances of Polariod type HNP′B film. (Sample *B* represents standard concentration.)

as high-frequency pass filters [16], low-frequency pass filters [17], and bandpass filters [18].

Optical interference devices utilize the optical interference between reflecting surfaces and usually consist of a stack of thin films deposited on a suitable substrate. The term *multilayer* is often used generically to describe such coatings. They are used as mirrors and bandpass filters in the spectral region from the far infrared to 0.12 μm, which is the transmission limit of the lithium fluoride substrate. The wavelength at which they function is determined by the thickness of layers.

At the risk of oversimplification, such filters can be classed in two broad categories: (1) long-wave pass or short-wave pass filters, which consist of a periodic structure; (2) bandpass filters, which consist of a single optical cavity or several optical cavities. The terms "optical cavity" and "periodic structure" are explained below.

The term *periodic structure* refers to a group of two or more layers which is repeated many times. For example, the simplest type is a quarter-wave stack [19], which consists of a stack of layers of equal optical thickness and alternating between a low-index **L** and a high-index **H**. The stack is then of the design:

$$air\ \textbf{H L H L}\cdots\textbf{H L H}\ substrate$$

where *air* is the incident medium. If such a stack were fabricated to reflect in the visible portion of the spectrum, then the **H** layer would typically consist of titanium dioxide (index 2.35) and **L** is silicon dioxide (index 1.55). Such multilayers are used

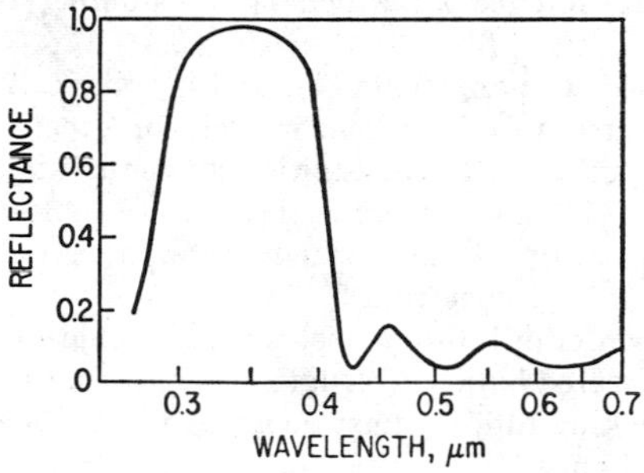

FIG. 6i-2. The measured spectral reflectance R of a multilayer mirror. Its transmittance is approximately $l - R$. (*Courtesy of Schott Glaswork, Mainz.*)

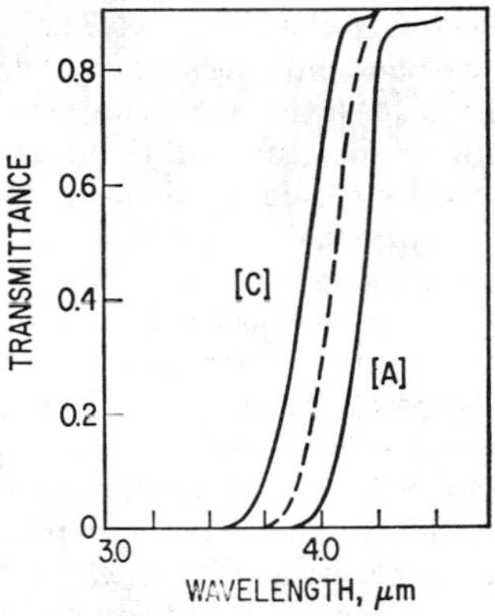

FIG. 6i-3. Measured spectral transmittance of a multilayer type of long-wave pass filter at normal incidence (curve *a*), at $\phi = 47$ deg incidence (dashed curve), and cooled to liquid nitrogen temperature at $\phi = 47$ deg (curve *c*). (*Courtesy of Optical Coating Laboratory, Inc.*)

because they have one or more useful *stop bands*—a region of the spectrum in which they are strongly reflecting and hence have a low transmittance. They have a substantial transmittance in the spectral region outside their stop bands. For example, Fig. 6i-2 depicts the spectral reflectance of a multilayer whose stop band extends from approximately 0.3 to 0.4 μm and which has a passband in the visible portion of the spectrum. Similar types of stacks with periodic structures are used as long-wave pass and short-wave pass filters, particularly in the infrared portion of the spectrum. As is true of all optical-interference devices, their transmittance changes with the angle of incidence, as shown in Fig. 6i-3. Since these periodic stacks are strongly reflecting in their stop bands, they are used as laser mirrors, reflection filters, and also as components of the optical-cavity type filters described in the next paragraph.

Filters with a relatively narrow spectral passband width are constructed by separating two reflectors with a "spacer" to form an *optical cavity*. More complex filters are produced by combining several optical cavities. There are several ways in which such optical-cavity filters can be fabricated. The reflectors can be composed of (1) semireflecting metal layers, such as silver or aluminum, or (2) a multilayer mirror, such as the quarter-wave stack described previously.

The optical cavity can be formed in several ways: (1) The reflectors are deposited on ultraflat fused-quartz plates and are held extremely parallel by a mechanical structure. Thus the cavity contains either a vacuum or some gas. This instrument is called the "Fabry-Perôt interferometer." (2) The cavity can contain a solid material, as when the reflectors are deposited on a solid slab of fused quartz or a sheet of mica. (3) The spacer is a thin solid film, which is deposited by evaporation in a vacuum when the filter is fabricated.

Before surveying the bandpass filters which can be fabricated with such optical cavities, we must define the filter's important attributes, which are: (1) The maximum transmittance T_{max} in its passband. (2) The wavelength λ_0 at which T_{max} occurs. (3) The spectral width of the passband, which is expressed in terms of its half width $\Delta\lambda_{\frac{1}{2}}$ at $0.5T_{max}$ or its deciwidth $\Delta\lambda_{\frac{1}{10}}$ at $0.1T_{max}$. (4) The resolution $Q = \lambda_0/\Delta\lambda_{\frac{1}{2}}$. (5) The off-band rejection, which is nebulously defined as the attenuation outside the passband. (6) The angle shift, which is the change of λ_0 at the angle of the incidence flux changes. (7) The angular field, which is related to the angle shift. The passband broadens when the filter is illuminated with convergent flux. The angular field is the maximum permissible cone angle of this flux. (8) The shift of λ_0 with temperature. (9) The order of interference m, which is the number of half wavelengths in the resonant cavity.

The Fabry-Perôt interferometer is described in many texts [20] and is principally used to measure with high resolution the spectral profile of lines in emission spectra. It can also be used as a narrow-band filter, especially if a regulator is used to maintain the parallelism of the reflectors [21]. It has the following properties:

1. It has a high resolution. A Q of 10^5 is easily attained, which corresponds to a bandwidth of 0.005 nm in the visible portion of the spectrum.

2. Since this Q is attained by using a high order of interference, there are undesired passbands adjacent to the main passband separated by a wavelength of approximately λ_0/m. For most applications, auxiliary "blocking filters" must be ganged in tandem with the FP to suppress these undesired passbands.

3. Its off-band rejection is rather poor, compared to that of a multicavity filter or a polarization-interference filter. The spectral shape of the passband is given to a good approximation by $T(\lambda) = T_{max}[1 + (\lambda - \lambda_0)^2 a]^{-1}$ where the constant a is related to $\Delta\lambda_{\frac{1}{2}}$. This line shape is called "Lorentzian" and has the property that the off-band transmittance decreases quite slowly. The ratio $\Delta\lambda_{\frac{1}{2}}/\Delta\lambda_{\frac{1}{10}}$ is independent of the Q.

4. As the angle of incidence ϕ increases, the angle θ of the flux in the cavity also increases, and the wavelength λ of the passband shifts to shorter wavelengths, as given by

$$2nd \cos \theta = \lambda m \qquad (6i\text{-}2)$$

where n and d are respectively the refractive index and physical thickness of the spacer. Although this equation neglects any effects of the phase shift upon reflection from the mirrors, it is a good approximation for $m > 10$. For small angles Eq. (6i-2) is combined with Snell's law, and the fractional change in wavelength is expressed as

$$\frac{\lambda_0 - \lambda}{\lambda_0} \cong \frac{n_0^2\phi^2}{2n^2} \qquad (6i\text{-}3)$$

where n_0 is the index of the incident medium in which ϕ is measured. As the fractional bandwidth of the filter is decreased, the flux must be more nearly perfectly collimated in order to maintain the fractional "angle shift" of the passband comparable to the Q^{-1} of the filter.

5. Although it is theoretically possible to attain a high resolution by using cavity reflectors with a reflectance very close to 100 percent, this is usually undesirable for

several reasons: First, the effect of any absorption in cavity spacer or in the reflectors is greatly enhanced, and this drastically reduces T_{max}. Second, the passband width does not decrease below a certain limit owing to a lack of planeness in the surface of the reflectors [22].

Single-cavity bandpass filters have also been constructed using a solid material for the cavity, as, for example, a slab of fused quartz [23] or a sheet of mica [24]. Compared with the conventional Fabry-Perôt, these have the advantages of greater mechanical and thermal stability. They also have a smaller angle shift, as can be seen from the effect of n^2 in the denominator of Eq. (6i-3). However, the T_{max} is smaller, because of the absorption of the spacer. A typical mica filter [24] has a T_{max} of 60 percent, $\Delta\lambda_{\frac{1}{2}}$ of 0.15 nm, and an order number of 150 at λ_0 of 600 nm. Multilayer mirrors are usually used for the reflectors in both the "air-spaced" and the "solid spacer" types of Fabry-Perôt filters.

The bandpass filters which are fabricated entirely of thin films have several distinct features: **(1)** Multiple-cavity filters [25] can be constructed, which have the advantage of superior off-band rejection. **(2)** Metal films can be used to advantage in the

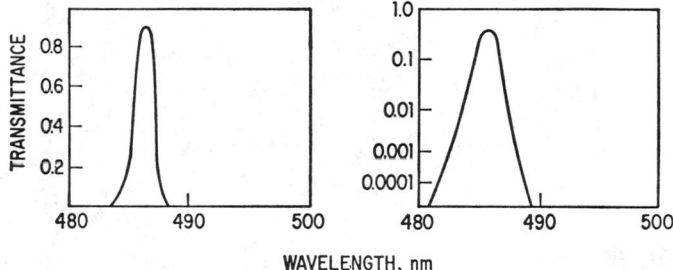

WAVELENGTH, nm

Fig. 6i-4. The measured spectral transmittance of a multiple cavity bandpass filter composed of thin films. (*Courtesy of Spectrolab, Inc.*)

Fig. 6i-5. The measured spectral transmittance (on linear and logarithmic scales) of a multilayer bandpass filter of the multiple-cavity type. (*Courtesy of Spectrolab, Inc.*)

filters. This provides another means of achieving a large off-band rejection. **(3)** The passbands can be located at wavelengths from the vacuum ultraviolet [26] to the infrared [2].

In the selection of such filters, the problem of blocking unwanted transmission bands should be considered. For example, Fig. 6i-4 shows the spectral transmittance of a single-cavity filter composed of dielectric layers. In addition to the passband in the green portion of the spectrum, there are broad regions of transmittance in both the blue and red. In most applications, auxiliary filters must be added in tandem with the multilayer filter to suppress these unwanted bands. Such a filter is said to be "completely blocked." Of course, the addition of these auxiliary filters reduces the T_{max}. Thus it is important to note whether a manufacturer is supplying the transmission curve of a "blocked" or an "unblocked" filter.

Compared with the single-cavity filter, the multiple-cavity filter has a passband shape which is more nearly rectangular, and it also has a superior off-band rejection. Figure 6i-5 depicts the spectral transmittance of a multicavity filter. Another method of obtaining an improved off-band rejection is to use filters which contain metal films [27]. The transmittance of such a filter is shown in Fig. 6i-6, and although its peak transmittance is rather small, its off-band transmittance is more than 40 db below T_{max}. Filters with a bandwidth $\Delta\lambda_{\frac{1}{2}}$ of a few angstroms have been fabricated [28] for use in the spectral region from 0.5 to 0.9 μm.

Reflection Filters. In certain regions of the spectrum, and especially in the ultraviolet, the reflection filter offers the simplest method of achieving a substantial T_{max} and a large off-band rejection. Figure 6i-7 depicts the spectral transmittance of an ultraviolet filter in which the incident flux reflects from four mirrors. The residual-ray type of reflection filter [2] has been used for many years to isolate portions of the infrared spectrum.

Choice of Filter. There are so many varied applications of bandpass filters that no single criterion can be used to judge their relative merit. Consider, for example, the selection of a filter in terms of its passband width. The half width is not a useful criterion of its performance if substantial off-band rejection is required. For many years solar physicists have relied upon the Lyot-Öhman filter described in Sec. 6i-4 to photograph absorption lines on the sun, because of its superior off-band rejection.

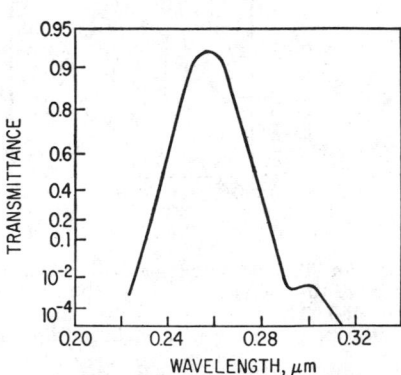

Fig. 6i-6. The measured spectral transmittance of an ultraviolet bandpass filter which contains metal films. (*Courtesy of Infrared Industries, Waltham, Mass.*)

Fig. 6i-7. The measured spectral transmittance of a reflection filter. (*Courtesy of Schott Glaswork, Mainz.*)

The angular field is an attribute which must be related to the filter's physical size and to the optical system in which it is used. Suppose that a filter is combined with an objective lens which collects radiant flux from a distant source and focuses it upon a detector, as shown in Fig. 6i-8. Although a filter of small physical dimensions is required if it is placed at its focus, the highly convergent flux broadens its passband. Clearly the flux is least convergent at A in front of the objective lens, but this requires a filter of large diameter D_1. A smaller filter could be placed at B, although the angular subtense of the flux is magnified by the ratio D_1/D_2.

It is often necessary to compute the flux transmitted by the filter placed at the focus. If a uniform, collimated irradiance H_0 impinges upon the objective lens, then a flux $dF = H_0 2\pi y \, dy$ is contained in an annulus of width dy at a radius y from the center. Since $y = f \sin \phi$ for an aplanatic lens [29] of focal length f, the total flux is

$$2\pi H_0 f^2 \int_0^{\frac{1}{2}\phi_m} T(\phi) \cos \phi \sin \phi \, d\phi \qquad (6i\text{-}4)$$

where ϕ_m is the total cone angle.   Several authors [30,31] have obtained approximate solutions to this equation, for specific types of filters.

Combinations of Filters.   Often two or more filters are ganged in a tandem array to sharpen the passband or to increase the off-band rejection.   Each filter has a reflectance R_i and transmittance T_i, and the tandem array has reflectance R and transmittance T.   The largest attenuation is attained when the filters are either: (1) nonreflecting, $R_i = 0$, or (2) arranged so that the flux reflected at each surface leaves the optical system and is not collected by the detector.   Only in this case is it true that T is the product: $T = T_1 T_2 \cdots T_i$.

The poorest attenuation is obtained when the multiply reflected beams are collected by the detector and all the filters are nonabsorbing, $R_i + T_i = 1.0$.   In this case it can be shown [32] that $R/T = R_1/T_1 + R_2/T_2 + \cdots + R_i/T_i$.

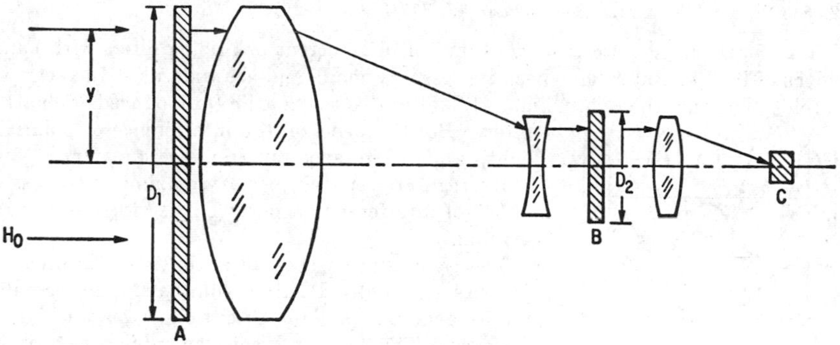

FIG. 6i-8. Showing the various positions at which a filter may be placed in a flux-collecting optical system.

If a single nonabsorbing filter transmits 1 percent, a tandem array of two identical filters transmits 0.503 percent.   In the more general case of tandem array of semireflecting, absorbing filters, we must measure the reflectance from both sides of each filter and use more complex equations [32] to compute the transmittance of the array.

References for Sec. 6i-3

1. Geffcken, W.: "Landolt-Börnstein Zahlenwerte und Functionen, Technik," Teil 3, pp. 925–962, Springer-Verlag OHG, Berlin, 1957.
2. Greenler, R.: "Concepts of Classical Optics," John Strong, ed., pp. 580–596, W. H. Freeman and Company, San Francisco, 1958.
3. Weyl, W. A.: "Coloured Glasses," Society of Glass Technology, Sheffield, England, 1951.
4. Catalogue of the Jenaer Glaswerk, Schott und Gen., Mainz, Germany.
5. Catalogue of the Corning Glass Company, Corning, N.Y.
6. Hellwege, K., and A. Hellwege, eds.: "Landolt-Börnstein Zahlenwerte und Functionen, Eigenschaften der Materie in ihren Aggregatzuständen," Teil 8, Springer-Verlag OHG, Berlin, 1962.
7. Ballard, S.: Section 6c of this Handbook.
8. Catalogue of the Harshaw Chemical Company, Cleveland, Ohio, 1967.
9. Catalogue of Wratten Filters, Eastman Kodak Company, Rochester, N.Y.
10. Pellicori, S., C. Johnson, and F. King: *Appl. Opt.* **5**, 1916 (1966).
11. Pellicore, S.: *Appl. Opt.* **3**, 361 (1964).
12. Yamada, Y., A. Mitsuishi, and H. Yoshinaga, *J. Opt. Soc. Am.* **52**, 17 (1962).
13. Mitsuishi, A., Y. Otsuka, S. Fujita, and H. Yoshinaga: *Japan. J. Appl. Phys.* **2**, 574 (1963).
14. Hunter, W., and R. Tousey: *J. Phys. Radium* **25**, 148 (1964).
15. Movikov, V., and G. Vasni: *Soviet J. Opt. Tech.* **34**, 639 (1967).
16. Ulrich, R.: *Infrared Phys.* **7**, 37 (1967).
17. Ulrich, R.: *Infrared Phys.* **7**, 65 (1967).

18. Rawcliffe, R., and C. Randall: *Appl. Opt.* **6**, 1353 (1967).
19. Heavens, O. S.: "Optical Properties of Thin Solid Films," p. 217, Dover Publications, Inc., New York, 1963.
20. Strong, J.: Ref. 2, chaps. 11 and 12.
21. Ramsey, J.: *Appl. Opt.* **5**, 1297 (1966).
22. Davis, S.: *Appl. Opt.* **2**, 727 (1963).
23. Herriot, D.: Bell Telephone Laboratories, Murray Hill, N.J.
24. Dobrowolski, J.: *J. Opt. Soc. Am.* **49**, 794 (1959).
25. Thelen, A.: *J. Opt. Soc. Am.* **56**, 1533 (1966).
26. Harrison, D.: *Appl. Opt.* **7**, 210 (1968).
27. Schroeder, D.: *J. Opt. Soc. Am.* **52**, 1380 (1962).
28. Thin Film Products Company—Now a division of Infrared Industries, Waltham, Mass.
29. Kingslake, R.: "Applied Optics and Optical Engineering," vol. 2, p. 202, Academic Press, Inc., New York, 1965.
30. Linder, S.: *Appl. Opt.* **6**, 1201 (1967).
31. Pidgeon, C., and S. Smith: *J. Opt. Soc. Am.* **54**, 1459 (1964).
32. Smith, T.: *Trans. Opt. Soc. London* **27**, 317 (1925–1926).

6i-4. Birefringent Filters. Two types of birefringent polarizing filters with bandwidths of 0.12 to 100 Å have been successfully made and are widely used in solar research. They are the Lyot-Öhman [1,2] filter and the Solc [3] (pronounced "Scholtz") filter. Both depend on the interference of polarized light, and have spectral transmission curves with sharp regularly spaced primary maxima at successive orders of interference, separated by series of secondary maxima.

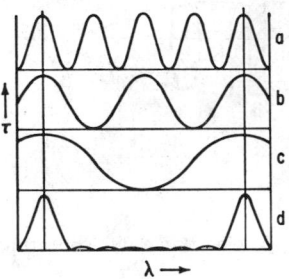

FIG. 6i-9. Spectral transmission of the single *b* elements of a three-element Lyot-Ohman filter (*a, b, c*) and the assembled filter *d*.

The Lyot-Öhman (LO) filter consists of alternating polarizers (usually Polaroid film) and plane-parallel birefringent plates cut with optic axes parallel to the surfaces. The first and last elements are polarizers. In the simplest configuration, the polarizers are parallel, and the birefringent elements are set with optic axes either alternating between plus and minus 45 deg to the plane of polarization or all parallel to either angle. The thicknesses of the birefringent elements (*b* elements) increase in powers of 2, but may be arranged in any order.

The light entering a given *b* element is divided into ordinary and extraordinary beams which traverse the element at different velocities and are combined by the following polarizer to interfere. The transmission of a single *b* element between a pair of polarizers is

$$\tau = \cos^2 \pi\gamma \tag{6i-5}$$

where $\gamma = (d/\lambda)\,(\epsilon - \omega)$, ϵ and ω being the extraordinary and ordinary indices of refraction, and d the thickness of the *b* element. Since the thickness of the ith element is $d_i = 2^{i-1}d_1$, the transmission of the whole filter is

$$\tau = \cos^2 \pi\gamma_1 \cos^2 \pi\gamma_2 \cdots \cos^2 \pi\gamma_n = \cos^2 \pi\gamma_1 \cos^2 2\lambda\gamma_1 \cdots \cos^2 2^{n-1}\pi\gamma_1 \tag{6i-6}$$

where n is the total number of *b* elements. Figure 6i-9 shows the spectral transmission curves for the individual *b* elements of a three-element filter and the curve of their product, i.e., the transmission curve for the assembled filter. All but two of the transmission peaks of the thickest element in Fig. 6i-9a coincide with transmission minima of one of the thinner elements and are thereby suppressed. Lyot [1] showed that Eq. (6i-6) is identical with the expression for the spectral transmission of a diffraction grating of 2^n rulings:

$$\tau = \frac{\sin 2^n \gamma_1}{2^n \sin \gamma_1} \tag{6i-7}$$

The bandwidth at half intensity, $\delta\lambda$, is very nearly the half width between the transmission peak at $\gamma_1 =$ integer, and the first zero at $\gamma_n =$ integer $+ \frac{1}{2}$, given approximately by

$$\delta\lambda = \frac{\sigma\lambda}{2\gamma_n} = \frac{\sigma\lambda}{2^n\gamma_1} \tag{6i-8}$$

The band separation, $\Delta\lambda$, is the interval between successive transmission peaks for the thinnest b element, approximately:

$$\Delta\lambda = \frac{\sigma\lambda}{\gamma_1} \tag{6i-9}$$

The mysterious factor σ allows for the effect of dispersion in $\epsilon - \omega$. It is about 0.9, and varies by a few percent for different crystals and wavelength regions. The finesse is obviously

$$F = \frac{\Delta\lambda}{\delta\lambda} = 2^n \tag{6i-10}$$

A most important property of any narrow-band filter is its suppression ratio, $S = T_m/T_p$, where T_m is the integrated transmission over the whole principal band between the adjacent zeros (of width $2\delta\lambda$), and T_p is the integrated parasitic transmission between successive principal bands. For the LO filter, $S = 0.11$.

If $\epsilon - \omega$ is known in the neighborhood of a desired passband, Eqs. (6i-2), (6i-4), and (6i-5) are sufficient to calculate the γ_1 and n required for a filter of any desired bandwidth and band separation. Normally one provides for $\Delta\lambda$ sufficiently large to allow unwanted passbands to be eliminated with glass filters or an interference filter. In making the calculation, one must allow for the dependence of γ on temperature. The net effect is a shift in the wavelength of the passband by $\Delta\lambda_T = k\,\Delta T$. For quartz and calcite, $k = -0.66$ and -0.42 Å per °C, respectively. It is necessary, therefore, to control the temperature of the filter to keep $\Delta\lambda_T$ within an acceptable tolerance (usually about 0.2 bandwidth). However, the temperature dependence provides a convenient fine tuning of sufficient range to correct for small deviations from an assumed $\epsilon - \omega$.

This is the basic Lyot-Öhman filter. It has limitations. The wavelength of the passband for light traversing it at an angle to the instrumental axis is shifted. Let ϕ be the angle in air between the light ray and the axis, and θ the azimuth of the incident plane measured from the crystal optic axis of the first b element. Then

$$\Delta\lambda(\phi,\theta) = \pm\lambda\sigma\frac{\phi^2}{2\omega^2}(\cos^2\theta - \sin^2\theta) \tag{6i-11}$$

The plus and minus signs apply, respectively, for positive and negative crystals. Let $\Delta\lambda(\phi)$ be the maximum acceptable shift. The corresponding ϕ at $\theta = 0$ or $\pi/2$ is

$$\phi = \left[2\Delta\lambda(\phi)\frac{\omega^2}{\sigma\lambda}\right]^{\frac{1}{2}} \approx 2.2\left[\frac{\Delta\lambda(\phi)}{\lambda}\right]^{\frac{1}{2}} \tag{6i-12}$$

This is a fairly stringent restriction. For example, in a filter for $\lambda = 6,563$ ($H\alpha$) with $\delta\lambda = 0.5$ Å, $\Delta\lambda(\phi) = 0.1$ Å is a reasonable tolerance. Then $\phi = 9 \times 10^{-3}$ radian.

Lyot invented a wide field elaboration of the simple filter in which each b element is made of two equal pieces of half the calculated thickness. The two are rotated 90 deg with respect to each other, and a 90-deg polarization rotator (usually a $\lambda/2$ plate) is mounted between them. Then for a given light ray, θ in the first half element becomes $\theta + \pi/2$ in the second half element, and by Eq. (6i-11), the $\Delta\lambda(\phi,\theta)$'s compensate. Since Eq. (6i-11) neglects higher-order terms, the compensation is not perfect, but Lyot's device does increase the radius of the useful field by factors of 26 and 6 for b elements of quartz and calcite.

Another problem is the loss of light by absorption in the polarizers of filters with large finesse. A Polaroid film usually transmits about 80 percent of the desired light, which means an optical density of $0.091n + 0.39$ for the filter as a whole. In some uses the loss of 95 percent of the light is serious. To alleviate the pain, Evans [2] devised a "split element" filter. Here half the b elements are cut into two equal halves and crossed as in Lyot's wide-field filter. The remaining elements are inserted between these equal halves with axes at 45 deg, and a unit of two birefringent elements can then be placed between successive polarizers. This reduces the optical density to $0.091n/2 + 0.39$, a very considerable improvement when n approaches 8 or 10. A better solution is to use more transparent polarizers like Rochon prisms, which may be no more expensive than the split-element construction. The prism polarizers are the only presently practical approach at wavelengths less than about 4,200 Å, where the absorption of Polaroid film begins to become excessive.

Beyond the limited slow tuning by temperature variation, each transmission band of a birefringent filter has a fixed central wavelength. It is feasible, however, to tune to any desired wavelength by adding phase shifters to the b elements. The condition for a transmission band at a given wavelength, λ_1, is that γ be integral at λ_1 in all b elements. If each b element has a phase shifter which adds $\Delta\gamma$, adjustable from 0 to 1, this condition can be satisfied.

The elegant approach is a filter with b elements of adjustable thickness. Each element is a pair of wedges, one of which slides with respect to the other to vary the total thickness in the optical path. The $\Delta\gamma$ is then a function of the wedge position and wavelength, but at every wavelength the required $\Delta\gamma$ can be achieved. Hence the variable thickness tuning works at all wavelengths for which the polarizers are effective. So far, the mechanical problems of control in filters of 8 or 10 elements have prevented use of this method. However, one of the modern small control computers could deal with these problems quite easily.

The second approach to filter tuning is relatively simple mechanically, but is effective only over a limited spectral range. Before entering the following polarizer, the light emerging from a b element is elliptically polarized. As λ varies and γ goes through a range of 1, the elliptical figure goes through the cycle from vertical linear to right circular, horizontal linear, left circular, and back to vertical linear. If now we add a $\lambda/4$ plate with its axis at 45 deg to the axis of the b element, elliptical polarization is converted to plane polarization rotated at an angle $\Psi = \frac{1}{2}\gamma$ to the axis of the $\lambda/4$ plate. If now we rotate the following polarizer, we can adjust it to transmit any $\Delta\gamma$, regardless of the wavelength. In a simple filter the Ψ's are simply proportional to the thicknesses of the preceding b elements, which progress in powers of two. The wavelength limitation is imposed by the fact that a simple $\lambda/4$ plate is $\lambda/4$ at only one wavelength. Light leaks develop in the wavelength intervals between principal bands as we depart from that wavelength. However, the leaks are tolerable for most purposes over a range of several hundred angstroms.

The tuning range of the rotating polarizer filter can be greatly extended by the use of achromatic $\lambda/4$ plates, which can now be constructed by known principles [4]. One such filter with a 0.25 Å passband at $\lambda 6,563$, tunable from 4,200 to 7,000 Å, is presently under construction.

All commercially available LO filters have b elements of quartz with $\epsilon - \omega \sim 0.009$, or calcite with $\epsilon - \omega \sim 0.18$, or both. The use of other materials has been confined to a few experiments. A typical example has an aperture of 30 mm, a 0.5 Å passband at 6,563, with high-order calcite elements of wide field construction. The two thickest elements are tunable over a range of ± 2 Å by rotating polarizers. The total length is about 16 cm.

The Solc filter [3] consists of a pile of N plane-parallel birefringent plates placed between two polarizers.   The plates are identical in thickness, and are cut with the optic axes parallel to the surfaces.   Two basic arrangements give identical filtering characteristics.   The folded filter has crossed polarizers, and the optic axes of the plates are set successively at angles of plus or minus $\pi/4N$ to the electric vector of light from the first polarizer.   The fan filter has parallel polarizers, and the orientations of successive b elements increase monotonically, $\pi/4N$, $3\pi/4N$, $5\pi/4N$, etc.   The action of the pile on polarized light is not readily apparent, but can be understood qualitatively if one thinks of the pile as a device for producing $N + 1$ different pathlengths, among which the light is distributed somewhat unevenly.   The resulting transmission curve resembles that of a grating of $N + 1$ rulings, but has some significant differences.   The transmission is

$$\tau = \frac{\sin N\chi}{\sin \chi} \cos \chi \tan \frac{\pi}{2N} \tag{6i-13}$$

where χ is defined by

$$\cos \chi = \cos \pi\gamma \cos \frac{\pi}{2N}$$

and γ is the retardation of a single plate.

The separation of successive transmission bands is approximately

$$\Delta\lambda = o\,\frac{\lambda}{\gamma} \tag{6i-14}$$

The bandwidth approaches

$$\delta\lambda = \frac{\sqrt{3}}{2}\,\frac{o\lambda}{N\gamma} \tag{6i-15}$$

as N increases.   Equation (6i-15) is accurate within one percent when $N \geq 5$.   The Solc and LO filters have the same band separation and bandwidth if $\gamma = \gamma_1$ and $N = \sqrt{3} \times 2^{n-1}$.   In its basic form, however, the Solc filter suppresses parasitic light much less effectively than the Lyot filter. S (Solc) increases with N.   It is 0.22 for $N = 16$ and approaches an upper limit of 0.27.   However, Solc [5] showed that S can be reduced to <0.05 by simply altering the orientations of the plates slightly at the cost of some increase in $\delta\lambda$ (which can be compensated by increasing N about 20 percent).   This is the only practical elaboration of the Solc filter so far proposed.

The effect of inclination of the light rays to the instrumental axis is about the same as that for the simple LO filter.   No one has devised an analytic expression for $\Delta\lambda(\phi,\theta)$, but Beckers and Dunn [6] calculated it numerically.   The field could be enlarged by using plates of Lyot's compound form, but in a filter of large finesse this would be very expensive.

Similarly, wavelength tuning is possible only by varying the thickness of all N individual plates equally.

The great virtue of the Solc filter is its transparency.   It has only two polarizers.   Against this advantage we must consider a very expensive form of construction.   There are more optical elements to be worked than for an LO filter, and the tolerances in thickness are very much smaller.   The thickness tolerances for all b elements in an LO filter are about 3 μm for quartz and 0.15 μm for calcite.   Beckers and Dunn found that the tolerances in a Solc filter are smaller by a factor of the order of $1/N$.   They note, however, that Solc has made a successful filter with $N = 80$, which speaks well for the skill of his opticians.

On the whole, it is probably far less expensive to build an LO filter with prism polarizers than a Solc filter of equivalent performance.   If a wide angular field is required, or wavelength tuning, the Solc filter is not a realistic competitor.

References for Sec. 6i-4

1. Lyot: *Ann. Astrophys.* **761**, 2 (1944).
2. Evans: *J. Opt. Soc. Am.* **39**, 229 (1949).
3. Evans: *J. Opt. Soc. Am.* **48**, 142 (1958).
4. Pancharatnan: *Proc. Indian Acad. Sci.*, sec. B, **41**, 137 (1955).
5. Solc: *Cesk. Casopis Fys.* **10**(1), 16 (1960).
6. Beckers and Dunn: *Air Force Rept.* AFCRL-65-605, Instrumentation Paper 75. (Obtainable from Clearing House for Federal Scientific and Technical Information, 5285 Port Royal Road, Springfield, Va. 22151.)

6j. Colorimetry

D. L. MACADAM

Eastman Kodak Company

6j-1. Luminosity. *Photopic Luminosity.* RELATIVE PHOTOPIC LUMINOSITY ($\bar{y}$): Adopted in 1931 by International Commission on Illumination (C.I.E.)(intended to represent normal eyes, for fields subtending about 2 deg, having about 1 millilambert luminance).

ABSOLUTE PHOTOPIC LUMINOSITY (K_λ lumens per watt): 680 times photopic luminosities given in Table 6j-1.

LUMINOUS FLUX (lumens):

$$F = \sum_{\lambda=380}^{770} P_\lambda K_\lambda$$

for spectral distribution of radiant energy, P_λ (watts per 5-nm-wavelength band).

LUMINOUS TRANSMITTANCE:

$$t = \frac{\displaystyle\sum_{\lambda=380}^{\lambda=770} \tau_\lambda P_\lambda K_\lambda}{\displaystyle\sum_{\lambda=380}^{\lambda=770} P_\lambda K_\lambda}$$

or

$$t = \frac{\displaystyle\sum_{\lambda=380}^{\lambda=770} \tau_\lambda P_\lambda \bar{y}}{\displaystyle\sum_{\lambda=380}^{\lambda=770} P_\lambda \bar{y}}$$

for material with spectral transmittance τ_λ irradiated with spectral distribution P_λ.

LUMINOUS REFLECTANCE r: Substitute spectral reflectance ρ_λ for τ_λ in either of above. Revisions of photopic relative luminosity data, recommended in 1951 by the United States Technical Committee on Colorimetry of C.I.E.:

370 nm 0.0001	380 0.0004	390 0.0015	400 0.0045	410 0.0093	420 0.0175	430 0.0273	440 0.0379	450 0.0468

These revisions have not been adopted by C.I.E.

Scotopic Luminosity. RELATIVE VALUES V' (Table 6j-1): Adopted in 1951 by C.I.E. (intended to represent normal eyes of young subjects, age ≤ 30, when observing at angles of not less than 5 deg from foveal center, under conditions of complete dark adaptation).

INTERNATIONAL PHOTOMETRIC STANDARD: Blackbody at temperature (2045 K) of solidification of platinum has luminance of 60 candela/cm² for both scotopic and photopic conditions.

ABSOLUTE SCOTOPIC LUMINOSITY K_λ': 1,725 times scotopic luminosities given in Table 6j-1.

SCOTOPIC LUMENS, SCOTOPIC, LUMINOUS TRANSMITTANCE, AND SCOTOPIC LUMINOUS REFLECTANCE: Substitute K_λ', V', for scotopic luminosity in formulas for corresponding photopic quantities.

6j-2. Standard Illuminants for Colorimetry. The C.I.E. makes a distinction between *illuminant* and *source*. *Source* refers to a physical emitter of light, such as a lamp or the sun and sky. *Illuminant* refers to a specific spectral power distribution, not necessarily provided directly by a source. The definitions of the standard sources are considered secondary, as it is conceivable that new developments in lamps and filters will bring about improved standard sources that represent the standard illuminants more accurately. STANDARD ILLUMINANT A is the spectral power distribution from a full (planckian, blackbody) radiator at 2858.7 K (IPTS, 1968, $c_2 = 0.014388$ m · K). STANDARD SOURCE A is a gas-filled, coiled-coil tungsten-filament lamp operating at a correlated color temperature of 2855.5 K. A lamp with a fused-quartz envelope or window is recommended if the spectral power distribution of illuminant A must be closely approximated in the ultraviolet. No recommendation has been made for a standard source representing standard illuminant D_{6500}.

STANDARD SOURCES B AND C and their spectral power distributions are omitted here because the sources are obsolete and their distributions are obsolescent. These sources, distributions, and colorimetric computation data based on them were published in the first and second editions of this Handbook (Tables 6j-4, 6j-5, and 6j-6). For general use in colorimetry, illuminants A and D_{6500} given in the new table 6j-2 should suffice.

STANDARD ILLUMINANT D_{6500} represents a phase of daylight with a correlated color temperature of approximately 6504 K. (The odd 4 K resulted from the change of the International Practical Temperature Scale, made by the Comité International des Poids et Mesures (CIPM) in 1968, after all details of illuminant D_{6500} had been adopted by the C.I.E., which had used the value of $c_2 = 0.014380$ m · K promulgated by the CIPM in 1960.)

OTHER ILLUMINANTS D: Whenever a phase of daylight other than D_{6500} is desired, the following rules may be used to define it:

1. *Chromaticity.* The C.I.E. 1931 x, y chromaticity coordinates of the daylight to be defined must satisfy $y_D = -3x_D^2 + 2.87x_D - 0.275$ and $0.25 \leq x_D \leq 0.38$.

TABLE 6j-1. PHOTOPIC AND SCOTOPIC LUMINOSITY DATA*

Wavelength, nm	Photopic $\bar{y}$	Scotopic V'	Wavelength, nm	Photopic $\bar{y}$	Scotopic V'
380	0.0000	0.00059	580	0.8700	0.1212
385	0.0001	0.00111	585	0.8163	0.0899
390	0.0001	0.00221	590	0.7570	0.0655
395	0.0002	0.00453	595	0.6949	0.0469
400	0.0004	0.00929	600	0.6310	0.03325
405	0.0006	0.01850	605	0.5668	0.02312
410	0.0012	0.03484	610	0.5030	0.01593
415	0.0022	0.0604	615	0.4412	0.01088
420	0.0040	0.0966	620	0.3810	0.00737
425	0.0073	0.1436	625	0.3210	0.00497
430	0.0116	0.1998	630	0.2650	0.003335
435	0.0168	0.2625	635	0.2170	0.002235
440	0.0230	0.3281	640	0.1750	0.001497
445	0.0298	0.3931	645	0.1382	0.001005
450	0.0380	0.4550	650	0.1070	0.000677
455	0.0480	0.5129	655	0.0816	0.000459
460	0.0600	0.5672	660	0.0610	0.0003129
465	0.0739	0.6205	665	0.0446	0.0002146
470	0.0910	0.6756	670	0.0320	0.0001480
475	0.1126	0.7337	675	0.0232	0.0001026
480	0.1390	0.7930	680	0.0170	0.0000716
485	0.1693	0.8509	685	0.0119	0.0000502
490	0.2080	0.9043	690	0.0082	0.00003533
495	0.2586	0.9491	695	0.0057	0.00002502
500	0.3230	0.9817	700	0.0041	0.00001780
505	0.4073	0.9984	705	0.0029	0.00001273
510	0.5030	0.9966	710	0.0021	0.00000914
515	0.6082	0.9750	715	0.0015	0.00000660
520	0.7100	0.9352	720	0.0010	0.00000478
525	0.7932	0.8796	725	0.0007	0.000003482
530	0.8620	0.8110	730	0.0005	0.000002546
535	0.9149	0.7332	735	0.0004	0.000001870
540	0.9540	0.6497	740	0.0003	0.000001379
545	0.9803	0.5644	745	0.0002	0.000001022
550	0.9950	0.4808	750	0.0001	0.000000760
555	1.0002	0.4015	755	0.0001	0.000000567
560	0.9950	0.3288	760	0.0001	0.000000425
565	0.9786	0.2639	765	0.0000	0.000000320
570	0.9520	0.2076	770	0.0000	0.000000241
575	0.9154	0.1602	775	0.0000	0.000000183
			780	......	0.000000139

* "The Science of Color." Available only from Optical Society of America, 2100 Pennsylvania Avenue NW, Washington, D.C. 20037.

TABLE 6j-2a. SPECTRAL POWER DISTRIBUTIONS OF
C.I.E. ILLUMINANTS A AND D_{6500}

λ, nm	A, $S(\lambda)$	D_{6500}, $S(\lambda)$	λ, nm	A, $S(\lambda)$	D_{6500}, $S(\lambda)$
300	0.93	0.03	575	110.80	96.1
05	1.13	1.7	80	114.44	95.8
10	1.36	3.3	85	118.08	92.2
15	1.62	11.8	90	121.73	88.7
20	1.93	20.2	95	125.39	89.3
325	2.27	28.6	600	129.04	90.0
30	2.66	37.1	05	132.70	89.8
35	3.10	38.5	10	136.34	89.6
40	3.59	39.9	15	139.99	88.6
45	4.14	42.4	20	143.62	87.7
350	4.74	44.9	625	147.23	85.5
55	5.41	45.8	30	150.83	83.3
60	6.15	46.6	35	154.42	83.5
65	6.95	49.4	40	157.98	83.7
70	7.82	52.1	45	161.51	81.9
375	8.77	51.0	650	165.02	80.0
80	9.80	50.0	55	168.51	80.1
85	10.90	52.3	60	171.96	80.2
90	12.09	54.6	65	175.38	81.2
95	13.36	68.7	70	178.76	82.3
400	14.71	82.8	675	182.11	80.3
05	16.15	87.1	80	185.42	78.3
10	17.68	91.5	85	188.70	74.0
15	19.29	92.5	90	191.93	69.7
20	21.00	93.4	95	195.11	70.7
425	22.79	90.1	700	198.26	71.6
30	24.67	86.7	05	201.35	73.0
35	26.64	95.8	10	204.40	74.3
40	28.70	104.9	15	207.40	68.0
45	30.85	110.9	20	210.36	61.6
450	33.09	117.0	725	213.26	65.7
55	35.41	117.4	30	216.11	69.9
60	37.81	117.8	35	218.91	72.5
65	40.30	116.3	40	221.66	75.1
70	42.87	114.9	45	224.35	69.3
475	45.52	115.4	750	226.99	63.6
80	48.24	115.9	55	229.58	55.0
85	51.04	112.4	60	232.11	46.4
90	53.91	108.8	65	234.58	56.6
95	56.85	109.1	70	237.00	66.8
500	59.86	109.4	775	239.36	65.1
05	62.93	108.6	80	241.67	63.4
10	66.06	107.8	85	243.91	63.8
15	69.25	106.3	90	246.11	64.3
20	72.50	104.8	95	248.24	61.9
525	75.79	106.2	800	250.32	59.5
30	79.13	107.7	05	252.34	55.7
35	82.52	106.0	10	254.30	52.0
40	85.95	104.4	15	256.21	54.7
45	89.41	104.2	20	258.06	57.4
550	92.91	104.0	825	259.85	58.9
55	96.44	102.0	30	261.59	60.3
60	100.00	100.0			
65	103.58	98.2			
70	107.18	96.3			

TABLE 6j-2b. SPECTRAL POWER DISTRIBUTIONS OF ILLUMINANTS D_{5500} AND D_{7500}

λ nm	$D_{5500},$ $S(\lambda)$	$D_{7500},$ $S(\lambda)$	λ nm	$D_{5500},$ $S(\lambda)$	$D_{7500},$ $S(\lambda)$
300	0.02	0.04	600	94.4	87.2
310	2.1	5.1	610	95.1	86.1
320	11.2	29.8	620	94.2	83.6
330	20.6	54.9	630	90.4	78.7
340	23.9	57.3	640	92.3	78.4
350	27.8	62.7	650	88.9	74.8
360	30.6	63.0	660	90.3	74.3
370	34.3	70.3	670	93.9	75.4
380	32.6	66.7	680	90.0	71.6
390	38.1	70.0	690	79.7	63.9
400	61.0	101.9	700	82.8	65.1
410	68.6	111.9	710	84.8	68.1
420	71.6	112.8	720	70.2	56.4
430	67.9	103.1	730	79.3	64.2
440	85.6	121.2	740	85.0	69.2
450	98.0	133.0	750	71.9	58.6
460	100.5	132.4	760	52.8	42.6
470	99.9	127.3	770	75.9	61.4
480	102.7	126.8	780	71.8	58.3
490	98.1	117.8	790	72.9	59.1
500	100.7	116.6	800	67.3	54.7
510	100.7	113.7	810	58.7	47.9
520	100.0	108.7	820	65.0	52.9
530	104.2	110.4	830	68.3	55.5
540	102.1	106.3			
550	103.0	104.9			
560	100.0	100.0			
570	97.2	95.6			
580	97.7	94.2			
590	91.4	87.0			

TABLE 6j-3. SPECTRAL POWER FUNCTIONS FOR COMPOSING ILLUMINANTS D

λ, nm	$P_0(\lambda)$	$P_1(\lambda)$	$P_2(\lambda)$	λ, nm	$P_0(\lambda)$	$P_1(\lambda)$	$P_2(\lambda)$
300	0.04	0.02	0.0	600	90.5	− 5.8	3.2
310	6.0	4.5	2.0	610	90.3	− 7.2	4.1
320	29.6	22.4	4.0	620	88.4	− 8.6	4.7
330	55.3	42.0	8.5	630	84.0	− 9.5	5.1
340	57.3	40.6	7.8	640	85.1	−10.9	6.7
350	61.8	41.6	6.7	650	81.9	−10.7	7.3
360	61.5	38.0	5.3	660	82.6	−12.0	8.6
370	68.8	42.4	6.1	670	84.9	−14.0	9.8
380	63.4	38.5	3.0	680	81.3	−13.6	10.2
390	65.8	35.0	1.2	690	71.9	−12.0	8.3
400	94.8	43.4	−1.1	700	74.3	−13.3	9.6
410	104.8	46.3	−0.5	710	76.4	−12.9	8.5
420	105.9	43.9	−0.7	720	63.3	−10.6	7.0
430	96.8	37.1	−1.2	730	71.7	−11.6	7.6
440	113.9	36.7	−2.6	740	77.0	−12.2	8.0
450	125.6	35.9	−2.9	750	65.2	−10.2	6.7
460	125.5	32.6	−2.8	760	47.7	− 7.8	5.2
470	121.3	27.9	−2.6	770	68.6	−11.2	7.4
480	121.3	24.3	−2.6	780	65.0	−10.4	6.8
490	113.5	20.1	−1.8	790	66.0	−10.6	7.0
500	113.1	16.2	−1.5	800	61.0	− 9.7	6.4
510	110.8	13.2	−1.3	810	53.3	− 8.3	5.5
520	106.5	8.6	−1.2	820	58.9	− 9.3	6.1
530	108.8	6.1	−1.0	830	61.9	− 9.8	6.5
540	105.3	4.2	−0.5				
550	104.4	1.9	−0.3				
560	100.0	0.0	0.0				
570	96.0	− 1.6	0.2				
580	95.1	− 3.5	0.5				
590	89.1	− 3.5	2.1				

TABLE 6j-4. Chromaticities and Multipliers for Composing Illuminants D

T_c*	x_D	y_D	M_1	M_2
4000	0.3823	0.3838	−1.505	2.827
4100	0.3779	0.3812	−1.464	2.460
4200	0.3737	0.3786	−1.422	2.127
4300	0.3697	0.3760	−1.378	1.825
4400	0.3658	0.3734	−1.333	1.550
4500	0.3621	0.3709	−1.286	1.302
4600	0.3585	0.3684	−1.238	1.076
4700	0.3551	0.3659	−1.190	0.871
4800	0.3519	0.3634	−1.140	0.686
4900	0.3487	0.3610	−1.090	0.518
5000	0.3457	0.3587	−1.040	0.367
5100	0.3429	0.3564	−0.989	0.230
5200	0.3401	0.3541	−0.939	0.106
5300	0.3375	0.3519	−0.888	−0.005
5400	0.3349	0.3497	−0.837	−0.105
5500	0.3325	0.3476	−0.786	−0.195
5600	0.3302	0.3455	−0.736	−0.276
5700	0.3279	0.3435	−0.685	−0.348
5800	0.3258	0.3416	−0.635	−0.412
5900	0.3237	0.3397	−0.586	−0.469
6000	0.3217	0.3378	−0.536	−0.519
6100	0.3198	0.3360	−0.487	−0.563
6200	0.3179	0.3342	−0.439	−0.602
6300	0.3161	0.3325	−0.391	−0.635
6400	0.3144	0.3308	−0.343	−0.664
6500	0.3128	0.3292	−0.296	−0.688
6600	0.3112	0.3276	−0.250	−0.709
6700	0.3097	0.3260	−0.204	−0.726
6800	0.3082	0.3245	−0.159	−0.739
6900	0.3067	0.3231	−0.114	−0.749
7000	0.3054	0.3216	−0.070	−0.757
7100	0.3040	0.3202	−0.026	−0.762
7200	0.3027	0.3189	0.017	−0.765
7300	0.3015	0.3176	0.060	−0.765
7400	0.3003	0.3163	0.102	−0.763
7500	0.2991	0.3150	0.144	−0.760
7600	0.2980	0.3138	0.184	−0.755
7700	0.2969	0.3126	0.225	−0.748
7800	0.2958	0.3115	0.264	−0.740
7900	0.2948	0.3103	0.303	−0.730
8000	0.2938	0.3092	0.342	−0.720
8100	0.2928	0.3081	0.380	−0.708
8200	0.2919	0.3071	0.417	−0.695
8300	0.2910	0.3061	0.454	−0.682
8400	0.2901	0.3051	0.490	−0.667
8500	0.2892	0.3041	0.526	−0.652
9000	0.2853	0.2996	0.697	−0.566
9500	0.2818	0.2956	0.856	−0.471
10000	0.2788	0.2920	1.003	−0.369
10500	0.2761	0.2887	1.139	−0.265
11000	0.2737	0.2858	1.266	−0.160
12000	0.2697	0.2808	1.495	0.045
13000	0.2664	0.2767	1.693	0.239
14000	0.2637	0.2732	1.868	0.419
15000	0.2614	0.2702	2.021	0.586
17000	0.2578	0.2655	2.278	0.878
20000	0.2539	0.2603	2.571	1.231
25000	0.2499	0.2548	2.907	1.655
5503†	0.3324	0.3475	−0.785	−0.198
6504‡	0.3127	0.3291	−0.295	−0.689
7504§	0.2990	0.3150	0.145	−0.760

* All correlated color temperatures T_c are based on $c_2 = 0.014388$ m·K.
† Standard illuminant D_{5500}; $T_c = 5500 \times 1.4388/1.4380 = 5503$ K (approximately).
‡ Standard illuminant D_{6500}; $T_c = 6500 \times 1.4388/1.4380 = 6504$ K (approximately).
§ Standard illuminant D_{7500}; $T_c = 7500 \times 1.4388/1.4380 = 7504$ K (approximately).

TABLE 6j-5. COLOR-MIXTURE FUNCTIONS FOR SMALL AND LARGE FIELDS

λ, nm	$\bar{x}(\lambda)$	$\bar{y}(\lambda)$	$\bar{z}(\lambda)$	$\bar{x}_{10}(\lambda)$	$\bar{y}_{10}(\lambda)$	$\bar{z}_{10}(\lambda)$
380	0.0014	0.0000	0.0065	0.0002	0.0000	0.0007
385	0.0022	0.0001	0.0105	0.0007	0.0001	0.0029
390	0.0042	0.0001	0.0201	0.0024	0.0003	0.0105
395	0.0076	0.0002	0.0362	0.0072	0.0008	0.0323
400	0.0143	0.0004	0.0679	0.0191	0.0020	0.0860
405	0.0232	0.0006	0.1102	0.0434	0.0045	0.1971
410	0.0435	0.0012	0.2074	0.0847	0.0088	0.3894
415	0.0776	0.0022	0.3713	0.1406	0.0145	0.6568
420	0.1344	0.0040	0.6456	0.2045	0.0214	0.9725
425	0.2148	0.0073	1.0391	0.2647	0.0295	1.2825
430	0.2839	0.0116	1.3856	0.3147	0.0387	1.5535
435	0.3285	0.0168	1.6230	0.3577	0.0496	1.7985
440	0.3483	0.0230	1.7471	0.3837	0.0621	1.9673
445	0.3481	0.0298	1.7826	0.3867	0.0747	2.0273
450	0.3362	0.0380	1.7721	0.3707	0.0895	1.9948
455	0.3187	0.0480	1.7441	0.3430	0.1063	1.9007
460	0.2908	0.0600	1.6692	0.3023	0.1282	1.7454
465	0.2511	0.0739	1.5281	0.2541	0.1528	1.5549
470	0.1954	0.0910	1.2876	0.1956	0.1852	1.3176
475	0.1421	0.1126	1.0419	0.1323	0.2199	1.0302
480	0.0956	0.1390	0.8130	0.0805	0.2536	0.7721
485	0.0580	0.1693	0.6162	0.0411	0.2977	0.5701
490	0.0320	0.2080	0.4652	0.0162	0.3391	0.4153
495	0.0147	0.2586	0.3533	0.0051	0.3954	0.3024
500	0.0049	0.3230	0.2720	0.0038	0.4608	0.2185
505	0.0024	0.4073	0.2123	0.0154	0.5314	0.1592
510	0.0093	0.5030	0.1582	0.0375	0.6067	0.1120
515	0.0291	0.6082	0.1117	0.0714	0.6857	0.0822
520	0.0633	0.7100	0.0782	0.1177	0.7618	0.0607
525	0.1096	0.7932	0.0573	0.1730	0.8233	0.0431
530	0.1655	0.8620	0.0422	0.2365	0.8752	0.0305
535	0.2257	0.9149	0.0298	0.3042	0.9238	0.0206
540	0.2904	0.9540	0.0203	0.3768	0.9620	0.0137
545	0.3597	0.9803	0.0134	0.4516	0.9822	0.0079
550	0.4334	0.9950	0.0087	0.5298	0.9918	0.0040
555	0.5121	1.0000	0.0057	0.6161	0.9991	0.0011
560	0.5945	0.9950	0.0039	0.7052	0.9973	0.0000
565	0.6784	0.9786	0.0027	0.7938	0.9824	0.0000
570	0.7621	0.9520	0.0021	0.8787	0.9556	0.0000
575	0.8425	0.9154	0.0018	0.9512	0.9152	0.0000
580	0.9163	0.8700	0.0017	1.0142	0.8689	0.0000
585	0.9786	0.8163	0.0014	1.0743	0.8256	0.0000
590	1.0263	0.7570	0.0011	1.1185	0.7774	0.0000
595	1.0567	0.6949	0.0010	1.1343	0.7204	0.0000
600	1.0622	0.6310	0.0008	1.1240	0.6583	0.0000
605	1.0456	0.5668	0.0006	1.0891	0.5939	0.0000
610	1.0026	0.5030	0.0003	1.0305	0.5280	0.0000
615	0.9384	0.4412	0.0002	0.9507	0.4618	0.0000
620	0.8544	0.3810	0.0002	0.8563	0.3981	0.0000
625	0.7514	0.3210	0.0001	0.7549	0.3396	0.0000

TABLE 6j-5. COLOR-MIXTURE FUNCTIONS FOR
SMALL AND LARGE FIELDS (*Continued*)

λ, nm	$\bar{x}(\lambda)$	$\bar{y}(\lambda)$	$\bar{z}(\lambda)$	$\bar{x}_{10}(\lambda)$	$\bar{y}_{10}(\lambda)$	$\bar{z}_{10}(\lambda)$
630	0.6424	0.2650	0.0000	0.6475	0.2835	0.0000
635	0.5419	0.2170	0.0000	0.5351	0.2283	0.0000
640	0.4479	0.1750	0.0000	0.4316	0.1798	0.0000
645	0.3608	0.1382	0.0000	0.3437	0.1402	0.0000
650	0.2835	0.1070	0.0000	0.2683	0.1076	0.0000
655	0.2187	0.0816	0.0000	0.2043	0.0812	0.0000
660	0.1649	0.0610	0.0000	0.1526	0.0603	0.0000
665	0.1212	0.0446	0.0000	0.1122	0.0441	0.0000
670	0.0874	0.0320	0.0000	0.0813	0.0318	0.0000
675	0.0636	0.0232	0.0000	0.0579	0.0226	0.0000
680	0.0468	0.0170	0.0000	0.0409	0.0159	0.0000
685	0.0329	0.0119	0.0000	0.0286	0.0111	0.0000
690	0.0227	0.0082	0.0000	0.0199	0.0077	0.0000
695	0.0158	0.0057	0.0000	0.0138	0.0054	0.0000
700	0.0114	0.0041	0.0000	0.0096	0.0037	0.0000
705	0.0081	0.0029	0.0000	0.0066	0.0026	0.0000
710	0.0058	0.0021	0.0000	0.0046	0.0018	0.0000
715	0.0041	0.0015	0.0000	0.0031	0.0012	0.0000
720	0.0029	0.0010	0.0000	0.0022	0.0008	0.0000
725	0.0020	0.0007	0.0000	0.0015	0.0006	0.0000
730	0.0014	0.0005	0.0000	0.0010	0.0004	0.0000
735	0.0010	0.0004	0.0000	0.0007	0.0003	0.0000
740	0.0007	0.0002	0.0000	0.0005	0.0002	0.0000
745	0.0005	0.0002	0.0000	0.0004	0.0001	0.0000
750	0.0003	0.0001	0.0000	0.0003	0.0001	0.0000
755	0.0002	0.0001	0.0000	0.0002	0.0001	0.0000
760	0.0002	0.0001	0.0000	0.0001	0.0000	0.0000
765	0.0001	0.0000	0.0000	0.0001	0.0000	0.0000
770	0.0001	0.0000	0.0000	0.0001	0.0000	0.0000
775	0.0001	0.0000	0.0000	0.0000	0.0000	0.0000
780	0.0000	0.0000	0.0000	0.0000	0.0000	0.0000

TABLE 6j-6. MAXIMUM POSSIBLE LUMINOUS EFFICIENCY K_m
(In lumens per watt of sources having indicated chromaticities)

y	$x = 0.1$	0.2	0.3	0.4	0.5	0.6	0.7
0.7	475	590	677*	\(680 at $x = 0.337$, $y = 0.659$\)			
0.6	425	548	620	670*			
0.5	375	500	553	590	610*		
0.4	310	430	480	505	500	480*	
0.3	245	350	380	385	370	320	226*
0.2	155	250	270	255	185		
0.1	80*	138	130				

* Entries marked with asterisks are for chromaticities beyond the domain of real colors. They are useful for interpolations only. Approximate values for intermediate chromaticities may be determined by tabular interpolation. A contour diagram permitting more accurate interpolation than Table 6j-8 was published in "The Science of Color," p. 308 (citation at bottom of Table 6j-1).

TABLE 6j-7. MAXIMUM POSSIBLE LUMINOUS REFLECTANCE
[For samples having indicated chromaticities when illuminated by standard
illuminant D_{6500} (top) and A (bottom)]

y	$x = 0.1$	0.2	0.3	0.4	0.5	0.6	0.7
			r_D (or t_D), %				
0.7	32	53					
0.6	34	62	77	0*			
0.5	33	66	85	97	76*		
0.4	31	69	93	85	59	43*	
0.3	26	65	81	64	39	23	12*
0.2	14	36	43	35	14		
0.1	...	15	13				
			r_A (or t_A), %				
0.7	21	38					
0.6	22	42	61	0*			
0.5	20	45	67	83	77*		
0.4	18	36	48	71	86	64*	
0.3	14	22	27	36	52	38	21*
0.2	7	12	15	17	19		
0.1	0*	5	4				

* See footnote following Table 6j-6. Contour diagrams permitting more accurate interpolation for illuminants A and C than Table 6j-7 were published in "The Science of Color," pp. 310, 311 (citation at bottom of Table 6j-1). Similar diagrams for illuminants A, B, C, D_{6500}; blackbodies at 1600, 1800, 2000, 2200, 2360, 3000, 3200 K; "equal energy," two different neon lamps, two different mercury lamps, and a Lucalox 400-watt alkali-metal-vapor lamp are available in *ELECLAB Tech. Note 200/68*, which may be obtained from the Naval Ship Research and Development Center, Annapolis Division, Annapolis, Md. 21402.

The correlated color temperature T_c ($c_2 = 0.014388$ m · K) of daylight D_{T_c} determines

$$x_D = -\frac{4.607 \times 10^9}{T_c^3} + \frac{2.9678 \times 10^6}{T_c^2} + \frac{99.11}{T_c} + 0.244063$$

when $4000 \text{ K} \le T_c \le 7000 \text{ K}$, or

$$x_D = -\frac{2.00064 \times 10^9}{T_c^3} + \frac{1.9018 \times 10^6}{T_c^2} + \frac{247.48}{T_c} + 0.23704$$

when $7000 \text{ K} \le T_c \le 25,000 \text{ K}$.

2. *Relative Spectral Power Distribution*

$$P(\lambda) = P_0(\lambda) + M_1 P_1(\lambda) + M_2 P_2(\lambda)$$

where $P_0(\lambda)$, $P_1(\lambda)$, and $P_2(\lambda)$ are functions of wavelength λ, given in Table 6j-3. And

$$M_1 = \frac{-1.3515 - 1.7703 x_D + 5.9114 y_D}{0.0241 + 0.2562 x_D - 0.7341 y_D}$$

$$M_2 = \frac{0.03 - 31.4424 x_D + 30.0717 y_D}{0.0241 + 0.2562 x_D - 0.7341 y_D}$$

To facilitate the use of this recommendation, values of x_D, y_D, M_1, and M_2 for correlated color temperatures in the range from 4000 to 25,000 K are given in Table 6j-4. Although these formulas may be used to calculate the relative spectral power distributions for daylight of any desired correlated color temperature, the use of D_{6500} is recommended whenever that is possible. When D_{6500} cannot be used, one of the two other distributions D_{5500} or D_{7500} given in Table 6j-2 should be used, if possible.

TABLE 6j-8. CHROMATICITY COORDINATES OF BLACKBODIES

$$(c_2 = 0.014388 \text{ m} \cdot \text{K})$$

T, K	x	y	u	v
1000	0.6525	0.3447	0.4476	0.3547
1500	0.5855	0.3932	0.3577	0.3603
1600	0.5730	0.3993	0.3449	0.3605
1700	0.5608	0.4043	0.3333	0.3604
1800	0.5490	0.4083	0.3229	0.3602
1900	0.5376	0.4112	0.3135	0.3597
2000	0.5265	0.4133	0.3049	0.3590
2100	0.5157	0.4146	0.2971	0.3582
2200	0.5054	0.4152	0.2900	0.3573
2300	0.4955	0.4152	0.2835	0.3563
2400	0.4860	0.4147	0.2775	0.3552
2500	0.4768	0.4137	0.2720	0.3540
2600	0.4680	0.4123	0.2670	0.3528
2700	0.4597	0.4106	0.2624	0.3516
2800	0.4517	0.4086	0.2581	0.3502
2900	0.4441	0.4064	0.2542	0.3489
3000	0.4368	0.4041	0.2505	0.3476
3100	0.4298	0.4015	0.2471	0.3462
3200	0.4232	0.3989	0.2439	0.3448
3300	0.4170	0.3962	0.2410	0.3435
3400	0.4109	0.3935	0.2382	0.3422
3500	0.4053	0.3906	0.2358	0.3408
3600	0.3997	0.3879	0.2332	0.3395
3700	0.3945	0.3849	0.2310	0.3381
3800	0.3896	0.3823	0.2289	0.3369
3900	0.3847	0.3794	0.2268	0.3356
4000	0.3804	0.3767	0.2251	0.3344
4500	0.3607	0.3635	0.2173	0.3284
5000	0.3450	0.3516	0.2114	0.3231
5500	0.3324	0.3410	0.2069	0.3183
6000	0.3220	0.3317	0.2033	0.3141
6500	0.3135	0.3236	0.2004	0.3103
7000	0.3063	0.3165	0.1981	0.3070
7500	0.3003	0.3103	0.1962	0.3041
8000	0.2952	0.3048	0.1946	0.3014
8500	0.2907	0.2999	0.1932	0.2990
9000	0.2869	0.2956	0.1921	0.2969
9500	0.2836	0.2918	0.1912	0.2950
10000	0.2806	0.2883	0.1903	0.2933
20000	0.2565	0.2577	0.1839	0.2771
∞	0.2399	0.2342	0.1800	0.2636

The spectral power distributions of daylight vary seasonally, particularly in the ultraviolet (λ <400 nm), but these formulas and tabulated data should be used pending recommendations of the C.I.E. concerning such variations. The formulas and data are believed to be sufficiently accurate for colorimetric uses, but they should not be used for other purposes if high accuracy is needed. Direct spectroradiometric measurements should be made in such cases.

6j-3. Colorimetry *Standard Color-mixture Data.* C.I.E. standard observer for color measurement is determined by the specifications for the equal-energy spectrum, as given in Table 6j-5. The *tristimulus values* are the amounts of three colors necessary to match equal energies of the indicated wavelengths. The value of $\bar{y}$ given in Table 6j-5 is the standard luminosity function or relative luminosity.

TRISTIMULUS VALUES:

$$X = 680 \sum_{\lambda=380}^{\lambda=770} P_\lambda \bar{x}$$

$$Y = 680 \sum_{\lambda=380}^{\lambda=770} P_\lambda \bar{y} = F \text{ (lumens)}$$

$$Z = 680 \sum_{\lambda=380}^{\lambda=770} P_\lambda \bar{z}$$

for spectral distribution of radiant energy P_λ (watts per 5-nm-wavelength band). For material with spectral transmittance τ_λ:

$$X = \frac{\displaystyle\sum_{\lambda=380}^{\lambda=770} \tau_\lambda P_\lambda \bar{x}}{\displaystyle\sum_{\lambda=380}^{\lambda=770} P_\lambda \bar{y}}$$

$$Y = \frac{\displaystyle\sum_{\lambda=380}^{\lambda=770} \tau_\lambda P_\lambda \bar{x}}{\displaystyle\sum_{\lambda=380}^{\lambda=770} P_\lambda \bar{y}}$$

$$Z = \frac{\displaystyle\sum_{\lambda=380}^{\lambda=770} \tau_\lambda P_\lambda \bar{z}}{\displaystyle\sum_{\lambda=380}^{\lambda=770} P_\lambda \bar{y}}$$

Relative values of P_λ are sufficient for determining tristimulus values X, Y, Z of material. For reflecting materials, substitute ρ_λ for τ_λ in above formulas.

For samples subtending more than 5-deg visual angle, the values of $\bar{x}_{10}$, $\bar{y}_{10}$, $\bar{z}_{10}$ in Table 6j-5 are probably more appropriate than $\bar{x}$, $\bar{y}$, $\bar{z}$, for colorimetry, but Y based on $\bar{y}_{10}$ has no photometric significance.

Data designed to facilitate manual computation of tristimulus values based on $\bar{x}$, $\bar{y}$, $\bar{z}$ for illuminants A, B, and C; blackbody sources of 1000, 1500, 1900, 2360, 3000, 3500, 4800, 6000, 6500, 7000, 8000, 10,000, 24,000 K; and infinite temperature; for five phases of natural daylight and for three commercial sources of artificial daylight, are tabulated in "The Science of Color."[1]

[1] Citation at bottom of Table 6j-1 (p. 6–184).

Chromaticity Coordinates. Horizontal coordinate $x = X/(X + Y + Z)$. Vertical coordinate $y = Y/(X + Y + Z)$.

Illuminant	x	y
C.I.E. standard A..............	0.4476	0.4074
C.I.E. standard B..............	0.3484	0.3516
C.I.E. standard C..............	0.3101	0.3162
C.I.E. standard D$_{6500}$..........	0.3127	0.3290
D$_{5500}$.........................	0.3324	0.3475
D$_{7500}$.........................	0.2990	0.3150

c.i.e. 1931 (x,y) diagram: Produced by plotting the chromaticity coordinates, x horizontally, y vertically, to equal scales.

c.i.e. 1960 (u,v) diagram: Provisionally recommended for use whenever a projective transformation of the (x,y) diagram yielding more nearly uniform chromaticity spacing is desired; it is formed by plotting u horizontally and v vertically, to equal scales, where

$$u = \frac{4x}{3 - 2x + 12y}, \qquad v = \frac{6y}{3 - 2x + 12y}$$

Alternatively,

$$u = \frac{4X}{X + 15Y + 3Z}, \qquad v = \frac{6Y}{X + 15Y + 3Z}$$

c.i.e. 1964 U^*, V^*, W^*, coordinate system: Provisionally recommended for use whenever a three-dimensional color-coordinate system perceptually more nearly uniformly spaced than the (X,Y,Z) system is desired. It is formed by plotting U^*, V^*, and W^* to equal scales along mutually orthogonal axes, where, with $1 \leq Y \leq 100$ and u_0, v_0 representing light that appears achromatic under the conditions prevailing in the application of interest (usually that is the illuminant),

$$W^* = 25Y^{\frac{1}{3}} - 17 \qquad U^* = 13W^*(u - u_0) \qquad V^* = 13W^*(v - v_0)$$

spectrum locus: Curve obtained by plotting chromaticity coordinates x, y or u, v for all wavelengths listed in Table 6j-5.

dominant wavelength: Wavelength corresponding to the intersection of the spectrum locus with the straight line drawn from the point representing the light source or illuminant, through the point representing the light reflected from (or transmitted by) the sample.

complementary wavelength: Wavelength corresponding to the intersection of the spectrum locus with the straight line drawn from the point representing the light from the sample through the point representing the light source or illuminant (used when dominant wavelength is not determinate).

purity: Ratio of distance from source point to sample point, compared with distance from source point to point on the spectrum locus representing the dominant wavelength (or, in case in which dominant wavelength is not determinate, ratio of distance from source point to sample point compared with distance from source point to collinear point on line joining extremities of the spectrum locus).

planckian locus: Curve produced when x, y in Table 6j-8 or the corresponding values of u, v, are plotted.

Correlated color temperature of an illuminant is the temperature corresponding to the point on the planckian locus which is at the foot of the perpendicular to that locus, from the point representing the illuminant in the C.I.E. 1960 (u,v) diagram.

c.i.e. 1964 color-difference formula: For evaluating difference between two closely similar colors specified by U_1^*, V_1^*, W_1^*, and U_2^*, V_2^*, W_2^*,

$$\Delta E = [(U_1^* - U_2^*)^2 + (V_1^* - V_2^*)^2 + (W_1^* - W_2^*)^2]^{\frac{1}{2}}$$

This and three other formulas proposed for test for the same purpose were published in *J. Opt. Soc. Am.* **58**, 291 (1968), which should be consulted for details.

The following formulas, to the end of the section on colorimetry, are not recommended by the C.I.E. or by any other organization. They are presented for trial and use by anyone who finds them to be applicable to his problems.

GEODESIC TRANSFORMATION OF (x,y) CHROMATICITY DIAGRAM: This nonlinear transformation of the (x,y) diagram provides the most nearly uniform plane representation of small-color-difference data for 14 observers.[1]

$$\xi = 3{,}751a^2 - 10a^4 - 520b^2 + 13{,}295b^3 + 32{,}327ab - 25{,}491a^2b - 41{,}672ab^2 + 10a^3b - 5{,}227\sqrt{a} + 2{,}952\sqrt[4]{a}$$

in which $a = 10x/(2.4x + 34y + 1)$ and $b = 10y/(2.4x + 34y + 1)$.

$$\eta = 404b - 185b^2 + 52b^3 + 69a(1 - b^2) - 3a^2b + 30ab^3$$

in which $a = 10x/(4.2y - x + 1)$ and $b = 10y/(4.2y - x + 1)$. ξ and η are given in units of root-mean-square errors of color matching by the 14 normal observers. All straight lines in the (ξ,η) diagram represent paths (geodesics) of minimum accumulated color differences, as evaluated according to the observer data.

According to the Schrödinger hypothesis,[2] such geodesics [straight lines in the (ξ,η) diagram] drawn outward from the achromatic point should represent series of colors of constant hue. The point representing C.I.E. standard illuminant D_{65} is at $\xi = 861.2$, $\eta = 395.7$.

CHROMATICITY DIFFERENCE between any two colors specified by (ξ_1,η_1) and (ξ_2,η_2) is

$$\Delta c = [(\xi_1 - \xi_2)^2 + (\eta_1 - \eta_2)^2]^{\frac{1}{2}}$$

SATURATION of color specified by ξ, η, Y: $s = w_1 [(\xi - \xi_a)^2 + (\eta - \eta_a)^2]^{\frac{1}{2}}$, where $w_1 = 0.054 + 0.46Y^{\frac{1}{3}}$ ($1 < Y < 80$), and ξ_a and η_a are the specifications of the achromatic color, usually the illuminant.

HUE (h) expressed as an angle clockwise from the vertical drawn downward from the point representing the achromatic color,

$$h = \tan^{-1}\frac{\xi_a - \xi}{\eta_a - \eta}$$

where $0 < h < 90°$ when $\xi_a - \xi$ and $\eta_a - \eta$ are both positive
           $h = 90°$ when $\xi_a - \xi$ is positive and $\eta = \eta_a$
   $90° < h < 180°$ when $\xi_a - \xi$ is positive and $\eta_a - \eta$ is negative
$180° < h < 270°$ when $\xi_a - \xi$ and $\eta_a - \eta$ are both negative
         $h = 270°$ when $\xi_a - \xi$ is negative and $\eta = \eta_a$
$270° < h < 360°$ when $\xi_a - \xi$ is negative and $\eta_a - \eta$ is positive.

COLOR DIFFERENCE between any two colors specified by (ξ_1,η_1), (ξ_2,η_2) and luminous reflectances Y_1 and Y_2 (in percent):

$$\Delta c = w_1 \left[(\xi_1 - \xi_2)^2 + (\eta_1 - \eta_2)^2 + \frac{w_2^2(Y_1 - Y_2)^2}{Y^2} \right]^{\frac{1}{2}}$$

where $w_2 = 1$ for sharply juxtaposed samples subtending about 2 deg, and Y is the average of Y_1 and Y_2. For a less well-defined dividing line, w_2 may be considerably less, e.g., for 5-deg separation between large samples $w_2 \cong 0.07$. For extremely small samples, to be distinguished from a uniform-background color, $w_2 \cong 7$, and w_1 should be multiplied by about 0.1.

LIGHTNESS:[3] $L = 2.468Y^{\frac{1}{3}} - 1.636$, where $1.2 < Y < 79$.

[1] D. L. MacAdam, *J. Opt. Soc. Am.* **32**, 247 (1942); L. Silberstein and D. L. MacAdam, *ibid.* **35**, 32 (1945); W. R. J. Brown and D. L. MacAdam, *ibid.* **39**, 808 (1949); W. R. J. Brown, *ibid.* **47**, 137 (1957); D. L. MacAdam, Appl. Opt. **11** (January, 1971).
[2] E. Schrödinger, *Ann. Physik* **63**, 515 (1920).
[3] J. H. Ladd and J. E. Pinney, *Proc. Inst. Radio Engrs.* **43**, 1137 (1955).

CHROMATIC ADAPTATION: Colors that are viewed with eyes adapted to a chromaticity S but appear the same as other colors viewed with eyes adapted to some other chromaticity T may be calculated in terms of tristimulus values, R, G, B based on the set of primaries

$$X_R = 0.747 \qquad Y_R = 0.253$$
$$X_G = 1.75 \qquad Y_G = -0.75$$
$$X_B = 0.1785 \qquad Y_B = 0.0$$

by multiplication of constant *adaptation factors*, K_ρ, K_γ, K_β times

and
$$\rho = a_1 + b_1 R^{p(R,T)}$$
$$\gamma = a_2 + b_2 G^{p(G,T)}$$
$$\beta = a_3 + b_3 B^{p(B,T)}$$

and solution of similar equations for R', G', B' for the corresponding colors

$$R' = \frac{(K_\rho \rho - a_4)^{1/p(R,S)}}{b_4}$$

$$G' = \frac{(K_\gamma \gamma - a_5)^{1/p(G,S)}}{b_5}$$

$$B' = \frac{(K_\beta \beta - a_6)^{1/p(B,S)}}{b_6}$$

For complete adaptation, such that S and T look alike, the adaptation factors are $K_\rho = \sigma_T/\sigma_S$, $K_\gamma = \gamma_T/\gamma_S$, $K_\beta = \beta_T/\beta_S$. Incomplete adaptation can be provided for by using, instead, the ratios determined from any observed pair of corresponding colors, preferably not very different from S or T.

The formulas for R, G, B should be normalized so that the maximum values used in these formulas do not exceed 80. For values of X, Y, Z not exceeding 80, i.e., for ordinary colored materials in daylight (in illuminants C and D), whose spectral reflectance p_λ (and Y) are expressed in percent, the formulas for R, G, B based on the primaries specified above are

$$R = 0.32X + 0.74Y - 0.069Z$$
$$G = -0.46X + 1.36Y + 0.10Z$$
$$B = Z$$

For other illuminants, only the normalizations need be changed. Values of R, G, B, R', G', B' less than 1.0 should be avoided, so far as possible, by adjusting the normalizations. Any that are nevertheless less than 1 should immediately be set arbitrarily at 1. For data based on the normalization given above, the tristimulus values X', Y', Z' of the corresponding colors for eyes adapted to T may be computed from the values of R', G', B' by use of

$$X' = 1.75R' - 0.15G' + 0.216B'$$
$$Y' = 0.59R' + 0.41G'$$
$$Z' = B'$$

For other normalizations, the coefficients of R', G', B' in these formulas should be divided by the ratios of the corresponding adjusted formulas to the formulas given above for R, G, B.

Exponents: The exponents $p(R,S)$ and $p(R,T)$ of R and R' may be computed from the formula

$$p(R,i) = 12.82x^2 + 0.53y^2 - 7.26x + 3.75y - 11.05xy + 1.21$$

where x and y are the chromaticity coordinates of the color, S or T, to which the eyes are adapted in each instance. Whenever a value of p greater than 1.0 results from the use of this or the two following equations, the value 1 should be used instead. The exponents $p(G,S)$ and $p(G,T)$ of G and G' are computed from the formula $p(G,i)$ $= 1 - 23(x - y)^2 + 2.3(x - y)$. The exponents $p(B,S)$ and $p(B,T)$ of B and B' are computed from the formula $p(B,i) = 1.8 - x - 2y$.

Constants: For each exponent p, the corresponding coefficients a and b, for use in the formulas for ρ, γ, β, R', G', B' are

$$b = \frac{6.1}{55.63^p - 2.422^p}$$
$$a = 7.8 - b(55.63)^p$$

e.g., for $p = 1$, $a = 1.42$, $b = 0.115$; for $p = \frac{1}{2}$, $a = 0.065$, $b = 1.046$.

TRANSFORMATIONS OF COLOR-MIXTURE DATA. In general, the tristimulus values R, G, B are the amounts (in a mixture matching a color specified by X, Y, Z) of a set of primaries located at (x_r,y_r), (x_g,y_g), and (x_b,y_b) in the (x,y) chromaticity diagram,

$$R = a_{11}X + a_{12}Y + a_{13}Z$$
$$G = a_{21}X + a_{22}Y + a_{23}Z$$
$$B = a_{31}X + a_{32}Y + a_{33}Z$$

Note that the color-mixture functions for the spectrum $\bar{x}$, $\bar{y}$, $\bar{z}$, $\bar{r}$, $\bar{g}$, $\bar{b}$ are simply special cases; their symbols may be substituted for X, Y, Z, R, G, B in these formulas. Let m_1 represent the slope, and let b_1 represent the y axis intercept of the line $y = m_1 x + b_1$ drawn between the points (x_g,y_g), (x_b,y_b) representing the G and B primaries. Then

$$a_{12} = a_{11}\frac{b_1 - 1}{b_1 + m_1} \quad \text{and} \quad a_{13} = \frac{a_{11}b_1}{b_1 + m_1}$$

where a_{11} is arbitrary and may be used to normalize the R function. If, as is often the case, the line joining the G and B primaries is more than 45 deg from horizontal, accuracy is best preserved if it is specified in terms of its reciprocal slope M_1 and x-axis intercept B_1, $x = M_1 y + B_1$. Then

$$a_{12} = a_{11}\frac{B_1 + M_1}{B_1 - 1} \quad \text{and} \quad a_{13} = \frac{a_{11}B_1}{B_1 - 1}$$

Similarly, in terms of the slope m_2 and y-axis intercept b_2 of the line through the R and B primaries, the coefficients in the formula for G are

$$a_{21} = a_{22}\frac{b_2 + m_2}{b_2 - 1} \quad a_{23} = \frac{a_{22}b_2}{b_2 - 1}$$

where a_{22} is the normalization constant. Finally, in terms of the slope m_3 and y-axis intercept b_3 of the line through the R and G primaries the coefficients in the formula for B are

$$a_{31} = a_{33}\frac{b_3 + m_3}{b_3} \quad a_{32} = a_{33}\frac{b_3 - 1}{b_3}$$

where a_{33} is the normalization constant. In terms of the two axis intercepts,

$$a_{31} = a_{33}(1 - B_3^{-1}) \quad a_{32} = a_{33}(1 - b_3^{-1})$$

The reverse transformations are

$$X = A_1[(b_2 - b_3)R + (b_3 - b_1)G + (b_1 - b_2)B]$$
$$Y = A_2[(m_3b_2 - m_2b_3)R + (m_1b_3 - m_3b_1)G + (m_1b_2 - m_2b_1)B]$$
$$Z = A_3\left[(m_3 - m_2)R + (m_1 - m_3)G + (m_2 - m_1)W - \frac{X}{A_1} - \frac{Y}{A_2}\right]$$

where A_1, A_2, A_3 are constants that depend on the normalizations of R, G, B. They can be determined by calculating R, G, B and the corresponding values of the expressions in the brackets for some one color, e.g., the illuminant, and dividing those results into the original values of X, Y, Z. Note that the last two terms in the brackets in Z are simply the quantities that appear in brackets in the formulas for X and Y.

6k. Radiometry

J. KASPAR

The Aerospace Corporation

6k-1. Blackbody Radiation. These tables contain various radiation functions derived from the Planck function

$$W(\lambda, T) = \frac{c_1}{\lambda^5(e^{c_2/\lambda T} - 1)}$$

where $W(\lambda,T)$ is defined as the power radiated per unit wavelength interval at wavelength λ by unit area of a blackbody at temperature T K. c_2 was taken to be 1.438 cm K. The constant c_1 does not enter into the functions here tabulated.

The maximum value of $W(\lambda,T)$ is given by

$$W_{max}(T) = 1.290 \times 10^{-15} T^5 \qquad \text{W/(cm}^2 \cdot \mu\text{m)}$$

while the Stefan-Boltzmann function is given by

$$\int_0^\infty W \, d\lambda = 5.679 \times 10^{-12} T^4 \qquad \text{W/cm}^2$$

6k-2. Optical Pyrometry (Narrow-band Radiation). When an optical pyrometer which has been calibrated to read the true temperature of a blackbody source is sighted on a nonblack source, it reads values of "brightness temperature" $T_{br}(\lambda,T)$ lower than the true temperature T K. Brightness temperature is related to true temperature through the following formula, which is derived from Planck's formula:

$$\ln \epsilon(\lambda, T) = \frac{c_2}{\lambda} \left(\frac{1}{T} - \frac{1}{T_{br}} \right)$$

where $c_2 = 1.4350$ cm $\cdot$ K (international temperature scale of 1948)

$\epsilon(\lambda,T)$ = emittance of the source at wavelength λ and temperature T

Commercial radiation pyrometers, although broad-band, do not utilize the complete spectrum of radiant energy. Hence there is no simple formula for precise calculation of the effect on temperature readings of varying emittance of the source. Table 6k-10 was calculated using the relation

$$T(\text{K}) = \frac{T_{\text{apparent}}(\text{K})}{\epsilon_t^{\frac{1}{4}}}$$

where ϵ_t is the total emittance. It may be used to estimate approximate corrections in radiation pyrometry.

TABLE 6k-1. BLACKBODY RADIATION FUNCTIONS*

λT, cm·deg	$\dfrac{W(\lambda, T)}{W_{max}(T)}$	$\dfrac{\int_0^\lambda W\,d\lambda}{\int_0^\infty W\,d\lambda}$	λT, cm·deg	$\dfrac{W(\lambda, T)}{W_{max}(T)}$	$\dfrac{\int_0^\lambda W\,d\lambda}{\int_0^\infty W\,d\lambda}$
0.050	2.999×10^{-7}	1.316×10^{-9}	0.155	3.032×10^{-1}	1.610×10^{-2}
0.051	4.775×10^{-7}	2.184×10^{-9}	0.160	3.457×10^{-1}	1.979×10^{-2}
0.052	7.452×10^{-7}	3.552×10^{-9}	0.165	3.892×10^{-1}	2.396×10^{-2}
0.053	1.142×10^{-6}	5.665×10^{-9}	0.170	4.332×10^{-1}	2.862×10^{-2}
0.054	1.718×10^{-6}	8.871×10^{-9}	0.175	4.772×10^{-1}	3.379×10^{-2}
0.055	2.545×10^{-6}	1.366×10^{-8}	0.180	5.208×10^{-1}	3.946×10^{-2}
0.056	3.709×10^{-6}	2.068×10^{-8}	0.185	5.636×10^{-1}	4.561×10^{-2}
0.057	5.326×10^{-6}	3.084×10^{-8}	0.190	6.053×10^{-1}	5.225×10^{-2}
0.058	7.544×10^{-6}	4.532×10^{-8}	0.195	6.455×10^{-1}	5.935×10^{-2}
0.059	1.054×10^{-5}	6.568×10^{-8}	0.200	6.840×10^{-1}	6.690×10^{-2}
0.060	1.455×10^{-5}	9.395×10^{-8}	0.22	8.169×10^{-1}	1.011×10^{-1}
0.061	1.985×10^{-5}	1.327×10^{-7}	0.24	9.126×10^{-1}	1.405×10^{-1}
0.062	2.676×10^{-5}	1.853×10^{-7}	0.26	9.712×10^{-1}	1.834×10^{-1}
0.063	3.570×10^{-5}	2.558×10^{-7}	0.28	9.972×10^{-1}	2.282×10^{-1}
0.064	4.713×10^{-5}	3.493×10^{-7}	0.30	9.971×10^{-1}	2.736×10^{-1}
0.065	6.613×10^{-5}	4.721×10^{-7}	0.32	9.771×10^{-1}	3.185×10^{-1}
0.066	7.984×10^{-5}	6.319×10^{-7}	0.34	9.432×10^{-1}	3.621×10^{-1}
0.067	1.025×10^{-4}	8.380×10^{-7}	0.36	8.999×10^{-1}	4.040×10^{-1}
0.068	1.305×10^{-4}	1.101×10^{-6}	0.38	8.512×10^{-1}	4.438×10^{-1}
0.069	1.649×10^{-4}	1.435×10^{-6}	0.40	7.997×10^{-1}	4.813×10^{-1}
0.070	2.066×10^{-4}	1.856×10^{-6}	0.42	7.475×10^{-1}	5.164×10^{-1}
0.071	2.571×10^{-4}	2.380×10^{-6}	0.44	6.961×10^{-1}	5.492×10^{-1}
0.072	3.176×10^{-4}	3.030×10^{-6}	0.46	6.464×10^{-1}	5.796×10^{-1}
0.073	3.897×10^{-4}	3.831×10^{-6}	0.48	5.990×10^{-1}	6.079×10^{-1}
0.074	4.751×10^{-4}	4.810×10^{-6}	0.50	5.543×10^{-1}	6.341×10^{-1}
0.075	5.757×10^{-4}	5.999×10^{-6}	0.52	5.125×10^{-1}	6.583×10^{-1}
0.076	6.934×10^{-4}	7.436×10^{-6}	0.54	4.735×10^{-1}	6.807×10^{-1}
0.077	8.304×10^{-4}	9.162×10^{-6}	0.56	4.375×10^{-1}	7.013×10^{-1}
0.078	9.891×10^{-4}	1.122×10^{-5}	0.58	4.042×10^{-1}	7.204×10^{-1}
0.079	1.172×10^{-3}	1.367×10^{-5}	0.60	3.735×10^{-1}	7.381×10^{-1}
0.080	1.382×10^{-3}	1.657×10^{-5}	0.62	3.453×10^{-1}	7.544×10^{-1}
0.081	1.621×10^{-3}	1.997×10^{-5}	0.64	3.193×10^{-1}	7.694×10^{-1}
0.082	1.893×10^{-3}	2.395×10^{-5}	0.66	2.956×10^{-1}	7.834×10^{-1}
0.083	2.201×10^{-3}	2.859×10^{-5}	0.68	2.737×10^{-1}	7.963×10^{-1}
0.084	2.548×10^{-3}	3.398×10^{-5}	0.70	2.537×10^{-1}	8.083×10^{-1}
0.085	2.938×10^{-3}	4.020×10^{-5}	0.72	2.354×10^{-1}	8.194×10^{-1}
0.086	3.373×10^{-3}	4.735×10^{-5}	0.74	2.185×10^{-1}	8.297×10^{-1}
0.087	3.859×10^{-3}	5.555×10^{-5}	0.76	2.030×10^{-1}	8.392×10^{-1}
0.088	4.397×10^{-3}	6.491×10^{-5}	0.78	1.888×10^{-1}	8.481×10^{-1}
0.089	4.993×10^{-3}	7.556×10^{-5}	0.80	1.758×10^{-1}	8.564×10^{-1}
0.090	5.651×10^{-3}	8.763×10^{-5}	0.82	1.638×10^{-1}	8.641×10^{-1}
0.091	6.373×10^{-3}	1.013×10^{-4}	0.84	1.528×10^{-1}	8.713×10^{-1}
0.092	7.165×10^{-3}	1.166×10^{-4}	0.86	1.426×10^{-1}	8.780×10^{-1}
0.093	8.030×10^{-3}	1.339×10^{-4}	0.88	1.332×10^{-1}	8.843×10^{-1}
0.094	8.973×10^{-3}	1.532×10^{-4}	0.90	1.246×10^{-1}	8.901×10^{-1}
0.095	9.998×10^{-3}	1.747×10^{-4}	0.92	1.166×10^{-1}	8.956×10^{-1}
0.096	1.111×10^{-2}	1.986×10^{-4}	0.94	1.093×10^{-1}	9.007×10^{-1}
0.097	1.231×10^{-2}	2.252×10^{-4}	0.96	1.024×10^{-1}	9.055×10^{-1}
0.098	1.360×10^{-2}	2.546×10^{-4}	0.98	9.613×10^{-2}	9.100×10^{-1}
0.099	1.500×10^{-2}	2.870×10^{-4}	1.0	9.029×10^{-2}	9.143×10^{-1}
0.100	1.649×10^{-2}	3.228×10^{-4}	1.1	6.679×10^{-2}	9.319×10^{-1}
0.105	2.563×10^{-2}	5.591×10^{-4}	1.2	5.035×10^{-2}	9.451×10^{-1}
0.110	3.785×10^{-2}	9.162×10^{-4}	1.3	3.862×10^{-2}	9.551×10^{-1}
0.115	5.350×10^{-2}	1.431×10^{-3}	1.4	3.007×10^{-2}	9.629×10^{-1}
0.120	7.281×10^{-2}	2.145×10^{-3}	1.5	2.375×19^{-2}	9.690×10^{-1}
0.125	9.588×10^{-2}	3.099×10^{-3}	1.6	1.899×10^{-2}	9.738×10^{-1}
0.130	1.227×10^{-1}	4.336×10^{-3}	1.7	1.536×10^{-2}	9.777×10^{-1}
0.135	1.530×10^{-1}	5.897×10^{-3}	1.8	1.255×10^{-2}	9.808×10^{-1}
0.140	1.866×10^{-1}	7.822×10^{-3}	1.9	1.035×10^{-2}	9.834×10^{-1}
0.145	2.232×10^{-1}	1.015×10^{-2}	2.0	8.612×10^{-3}	9.856×10^{-1}
0.150	2.622×10^{-1}	1.290×10^{-2}			

* Table by Reynolds et al., ref. 4.

Table 6k-2. Total Blackbody Radiation[*]

T, K	$\int_0^\infty W\,d\lambda$, W/cm²	$W_{max}(T)$, W/(cm²·μm)	T, K	$\int_0^\infty W\,d\lambda$, W/cm²	$W_{max}(T)$, W/(cm²·μm)
1	5.679×10^{-12}	1.290×10^{-15}	1060	7.170	1.726
5	3.549×10^{-9}	4.030×10^{-12}	1080	7.726	1.895
10	5.679×10^{-8}	1.290×10^{-10}	1100	8.315	2.077
15	2.875×10^{-7}	9.794×10^{-10}	1120	8.937	2.273
20	9.086×10^{-7}	4.127×10^{-9}	1140	9.591	2.483
30	4.600×10^{-6}	3.134×10^{-8}	1160	10.29	2.709
40	1.454×10^{-5}	1.321×10^{-7}	1180	11.01	2.951
50	3.549×10^{-5}	4.030×10^{-7}	1200	11.78	3.209
60	7.360×10^{-5}	1.003×10^{-6}	1220	12.58	3.486
70	1.364×10^{-4}	2.168×10^{-6}	1240	13.43	3.781
80	2.326×10^{-4}	4.226×10^{-6}	1260	14.32	4.096
90	3.726×10^{-4}	7.616×10^{-6}	1280	15.25	4.431
100	5.679×10^{-4}	1.290×10^{-5}	1300	16.22	4.789
110	8.315×10^{-4}	2.077×10^{-5}	1320	17.25	5.169
120	1.178×10^{-3}	3.209×10^{-5}	1340	18.32	5.572
130	1.622×10^{-3}	4.789×10^{-5}	1360	19.43	6.001
140	2.181×10^{-3}	6.936×10^{-5}	1380	20.59	6.455
150	2.875×10^{-3}	9.794×10^{-5}	1400	21.81	6.936
160	3.722×10^{-3}	1.352×10^{-4}	1420	23.09	7.446
170	4.743×10^{-3}	1.831×10^{-4}	1440	24.42	7.986
180	5.961×10^{-3}	2.437×10^{-4}	1460	25.80	8.556
190	7.401×10^{-3}	3.194×10^{-4}	1480	27.24	9.158
200	9.086×10^{-3}	4.127×10^{-4}	1500	28.75	9.794
210	1.105×10^{-2}	5.267×10^{-4}	1520	30.31	10.46
220	1.331×10^{-2}	6.647×10^{-4}	1540	31.94	11.17
230	1.590×10^{-2}	8.301×10^{-4}	1560	33.63	11.92
240	1.885×10^{-2}	1.027×10^{-3}	1580	35.39	12.70
250	2.218×10^{-2}	1.260×10^{-3}	1600	37.22	13.52
260	2.595×10^{-2}	1.532×10^{-3}	1620	39.12	14.39
270	3.018×10^{-2}	1.851×10^{-3}	1640	41.08	15.30
280	3.491×10^{-2}	2.220×10^{-3}	1660	43.12	16.26
290	4.017×10^{-2}	2.645×10^{-3}	1680	45.24	17.26
300	4.600×10^{-2}	3.134×10^{-3}	1700	47.43	18.31
310	5.245×10^{-2}	3.692×10^{-3}	1720	49.71	19.42
320	5.955×10^{-2}	4.328×10^{-3}	1740	52.06	20.57
330	6.735×10^{-2}	5.047×10^{-3}	1760	54.50	21.78
340	7.589×10^{-2}	5.860×10^{-3}	1780	57.01	23.05
350	8.522×10^{-2}	6.774×10^{-3}	1800	59.61	24.37
360	9.538×10^{-2}	7.799×10^{-3}	1820	62.31	25.75
370	1.065×10^{-1}	8.944×10^{-3}	1840	65.09	27.20
380	1.184×10^{-1}	1.022×10^{-2}	1860	67.97	28.71
390	1.314×10^{-1}	1.164×10^{-2}	1880	70.94	30.29
400	1.454×10^{-1}	1.321×10^{-2}	1900	74.01	31.94
410	1.605×10^{-1}	1.494×10^{-2}	1920	77.18	33.65
420	1.768×10^{-1}	1.686×10^{-2}	1940	80.44	35.44
430	1.942×10^{-1}	1.896×10^{-2}	1960	83.81	37.31
440	2.128×10^{-1}	2.127×10^{-2}	1980	87.29	39.25
450	2.328×10^{-1}	2.380×10^{-2}	2000	90.86	41.27
460	2.542×10^{-1}	2.656×10^{-2}	2100	110.5	52.67
470	2.771×10^{-1}	2.958×10^{-2}	2200	133.1	66.47
480	3.015×10^{-1}	3.286×10^{-2}	2300	159.0	83.01
490	3.274×10^{-1}	3.643×10^{-2}	2400	188.5	102.7
500	3.549×10^{-1}	4.030×10^{-2}	2500	221.8	126.0
520	4.152×10^{-1}	4.904×10^{-2}	2600	259.5	153.2
540	4.829×10^{-1}	5.922×10^{-2}	2700	301.8	185.1
560	5.585×10^{-1}	7.103×10^{-2}	2800	349.1	222.0
580	6.426×10^{-1}	8.465×10^{-2}	2900	401.7	264.5
600	7.360×10^{-1}	1.003×10^{-1}	3000	460.0	313.4
620	8.392×10^{-1}	1.182×10^{-1}	3100	524.5	369.2
640	9.527×10^{-1}	1.385×10^{-1}	3200	595.5	432.8
660	1.087	1.615×10^{-1}	3300	673.5	504.7
680	1.215	1.875×10^{-1}	3400	758.9	586.0
700	1.364	2.168×10^{-1}	3500	852.2	677.4
720	1.527	2.496×10^{-1}	3600	953.8	779.9
740	1.703	2.862×10^{-1}	3700	1065	894.4
760	1.895	3.270×10^{-1}	3800	1184	1022
780	2.102	3.724×10^{-1}	3900	1314	1164
800	2.326	4.226×10^{-1}	4000	1454	1321
820	2.567	4.782×10^{-1}	4500	2328	2380
840	2.827	5.394×10^{-1}	5000	3549	4030
860	3.106	0.667×10^{-1}	5500	5197	6491
880	3.406	6.806×10^{-1}	6000	7360	10030
900	3.726	7.616×10^{-1}	6500	10140	14960
920	4.069	8.500×10^{-1}	7000	13640	21680
940	4.434	9.465×10^{-1}	7500	17970	30610
960	4.824	1.052	8000	23260	42260
980	5.239	1.166	8500	29640	57230
1000	5.679	1.290	9000	37260	76160
1020	6.147	1.424	9500	46260	99800
1040	6.644	1.569	10000	56790	12900

[*] Table by Reynolds et al., ref. 4.

TABLE 6k-3. TOTAL EMITTANCE

Material	Temperature, K	Type*	Total† emittance	References‡
Aluminum:				
Polished.................	370–630	h	0.04–0.06	1
Heavily oxidized.........	360–800	h	0.2–0.33	1
Electrolytically oxid., 4–10 μm thick..............	310	h	0.72–0.83	2
Aluminum oxide............	80–500	n	0.76–	1
	1200–1750		0.45–0.41	
Aluminum oxide layer:				
0.25 μm thick.............	311	n	0.06	4
0.50 μm thick...........	311	n	0.11	
1.0 μm thick.............	311	n	0.30	
2.0 μm thick.............	311	n	0.65	
3.0 μm thick.............	311	n	0.70	
4.0 μm thick.............	311	n	0.70	
7.0 μm thick.............	311	n	0.75	
Antimony:				
Polished.................	300–350	n	0.28–0.31	3
Beryllium.................	1100–1300–1480	n	0.41–0.57–0.87	1
Bismuth...................	350		0.34	3, 2
Brass:				
Highly polished...........	500–610	h	0.02	2
Polished................	373	n	0.06	1
Oxidized................	450–590	n	0.56–0.64	1
Bronze, 4–7 aluminum:				
Polished................	450–1270	n	0.03–0.06	1
Oxidized................	450–1270	n	0.08–0.16	1
Cadmium..................	80–300	h	0.03	1
Carbon:				
Rough...................	1200–2000	n	0.84–0.81	1
Polished................	1200–2000	n	0.82–0.79	1
Ceria (cerium dioxide):				
Powder coating...........	670–1070–1350–2250	n	0.53–0.30–0.90–0.93	1
Heat treated.............	1300–1700–1900	h	0.65–0.40–0.50	1
Chromium.................	370–600–750–1220	n	0.06–0.10–0.42	1
	80	h	0.07	1
Cobalt...................	350–600–1030	n	0.20–0.28–0.74	1
Copper:				
Polished.................	80–800	h	0.02–0.03	1
Polished pure.............	300–700–970–1410	h	0.04–0.07–0.19–0.15	2
Oxidized.................	300–600–800–1100	h	0.38–0.47–0.59–0.87	1
Polished.................	80–380–1160	n	0.02–0.01–0.02	1
Oxidized.................	80–540–700–1078	n	0.66–0.78–0.90–0.93	1
Gold:				
Polished.................	80–1100	h	0.01–0.07	1
Graphite:				
ATJ.....................	700–1400–2700	n	0.81–0.74–0.90	1
Pyrolytic, basal **plane**......	1600–1800–2700	n	0.67–0.49–0.35	1
Pyrolytic, c plane..........	1570–1900–1900–2500	n	1.0–1.0–0.82	1
Pyrolytic film on ATJ.....	2000–2600	h	0.65–0.73	1
Iron and steel:				
Armco and pure, polished...	160–1100	h	0.05–0.25	1
	600–1100	n	0.2–0.56	
Cast, polished.............	300–915–1355	h	0.21–0.21–0.28	1
Cast, oxidized.............	360–800–1350	h	0.62–0.73–0.73	2
Wrought, smooth..........	300–1800	h	0.27	2
Smooth sheet, rolled.......	800–1350	h	0.48–0.60	2
Electrolytic, oxidized......	310–700	h	0.78–	2
	1025–1800		0.89–0.94	

TABLE 6k-3. TOTAL EMITTANCE (*Continued*)

Material	Temperature, K	Type*	Total† emittance	References‡
Stainless steel 310.........	800–1400	h	0.25	1
Stainless steel 310 (grid blasted, oxidized).......	400–1050	h	0.74–0.84	1
Stainless steel 347 (stably (oxidized).............	600–1400	h	0.86–0.91	1
Stainless steel 303 (stably oxidized)..............	600–1400	h	0.75–0.87	1
Stainless steel 18-8, polished oxidized..............	350–650	n	0.15–0.20	1
	350–650		0.84	
Stainless steel, AISI 316, polished	80–900 1100–1300–1420	n	0.19–0.35–0.62–0.32	1
Iron oxide, Fe_2O_3...........	310–1350	h	0.82–0.89	
Lead:				
Polished.................	310–530	h	0.04–0.08	3
Gray, oxidized...........	270–470	h	0.28	2
Oxidized at 473 K........	473	h	0.63	4
Magnesium, polished........	410–490	h	0.12	1
Manganin, bright, rolled.....	391	n	0.048	4
Mercury, pure, clean........	273–373		0.09–0.12	2
Molybdenum:				
Vapor blasted.............	400–1800	h	0.07–0.21	1
Oxidized.................	600–810	h	0.81	1
Polished.................	1600–2300–2900	n	0.13–0.28–0.28	1
Oxidized.................	600–800	n	0.83	1
Nickel:				
Polished.................	80–1100	h	0.02–0.17	1
Polished.................	80–550–1450	n	0.07–0.04–0.19	1
Oxidized.................	420–700– 980	n	0.07–0.39– 0.47	1
Nickel-chromium Alloys:				
Inconel X, stably oxidized..	580–1370	h	0.89–0.93	1
Inconel X, polished........	80–800–1200	h	0.06–0.12–0.23	1
René 41, oxidized	550–1350	n	0.78–0.87	1
Inconel-NBSA-418 coating.	750–1250	n	0.68–0.64	1
Nichrome 80-20, oxidized...	480–900–1200	h	0.62–0.67–0.78	1
Nickel-copper Alloys:				
K Monel 5700............	80–900–1300	n	0.14–0.20–0.30	1
Monel 400...............	300	h	0.12	
Nickel-molybdenum alloys:				
Haynes Alloy C, oxidized...	580–1370	h	0.90–0.96	1
Inor-8, polished...........	340–1240	n	0.15–0.25	1
Niobium-tungsten alloys:				
90-10, polished............	1970–2580	h	0.27	1
85-15, polished............	1970–2580	h	0.26	1
Palladium.................	400–1520	n	0.02–0.17	1
Platinum, cold rolled........	420–1500	n	0.02–0.16	
	690–1700	h	0.09–0.17	1
Rhodium:				
As received..............	490–1520	n	0.02–0.09	1
Polished.................	420–1080–1250	n	0.012–0.068–0.034	2
Plated on stainless steel....	80	h	0.075	1
Silver:				
Polished.................	650–1100	h	0.03–0.04	1
Thermally etched........	460–1100	h	0.04–0.08	1
Tantalum:				
Polished, gas adsorbed.....	480–1480	h	0.1–0.25	1
Polished.................	1600–2920	h	0.17–0.30	1

TABLE 6k-3. TOTAL EMITTANCE (*Continued*)

Material	Temperature, K	Type*	Total† emittance	References‡
Tantalum carbide, polished...	1600–3000	n	0.2–0.33	1
Tantalum nitride, polished....	800–1500	n	0.74–0.80–0.60	1
Tellurium.................	295	n	0.22	4
Tin, polished..............	310–360	h	0.05	2
Titanium:				
Polished.................	900	h	0.24	1
Electropolished...........	250–370	h	0.1–0.13	1
Titanium:				
Oxidized.................	640–950–1100	h	0.54–0.60	1
Oxidized, 306 hr at 813 K..	360–700	n	0.35–0.48	1
Oxidized, 306 hr at 580 K..	360–700	n	0.11–0.20	1
Titanium-aluminum alloy:				
A-110-AT, polished........	80–600–1100	n	0.12–0.20–0.52	1
Titanium-manganese Alloys:				
C-110M, AMS 4908,				
polished...............	300–900	n	0.05–0.17	1
oxidized...............	300–900	n	0.50–0.61	1
RS-120, oxidized..........	600–1100	h	0.67–0.72	1
Tungsten:				
Polished.................	400–2000–3400	h	0.04–0.24–0.34	1
Polished.................	1400–3000	n	0.15–0.32	1
Filament.................	300–2300	n	0.032–0.28	4
Uranium...................	1200	h	0.35	1
Zinc:				
Pure, polished............	300–530	n	0.02–0.06	2
Oxidized.................	300–470–800	n	0.28–0.14–0.11	2
Zirconium, as received......	1150–1250–1450–1860	n	0.33–0.39–0.26–0.32	1
Water.....................	273–373	n	0.92–0.96	4
Ice:				
Smooth, H₂O.............	273	n	0.96	4
Rough crystals...........	273	n	0.985	4
Glass.....................	293	n	0.94	4
Lacquer:				
White....................	373	n	0.925	4
Black matte..............	373	n	0.97	4
Oil paints, all colors........	273–373	n	0.92–0.96	4
Enamel...................	295	n	0.90–0.95	4
Candle soot...............	273–373	n	0.952	4
Plaster...................	273–373	n	0.91	4
Paper.....................	373	n	0.92	4
Rubber, hard, glossy plate....	297	n	0.945	4
Quartz (fused).............	295	n	0.932	4

* n = normal (emittance), h = hemispherical (emittance).
† The emittances correspond to the given temperatures. Linear interpolation between points fairly accurate.
‡ References are on p. 6-204 following Table 6k-5.

TABLE 6k-4. LOW-TEMPERATURE TOTAL HEMISPHERICAL EMITTANCES*

Material	80 K	150 K	300 K
Gold, polished.........................	0.018	0.020	0.023
Platinum black on gold†.................	0.95	1.00	1.0
Parson's optical black, heavy coat........	0.68	0.88	1.0
3M velvet, 9564 black, heavy coat........	0.80	0.80	1.0
Gold black on gold, heavy coat...........	0.75	0.90	0.95
Fuller black 3811......................	0.60	0.70	0.82
Midland sicon black 7 × 942.............	0.70	0.70	0.70
Anodized aluminum, 28 μm thick........	0.60	0.82	0.86

* Taken from graphs in ref. 7.
† Special preparation, see ref. 7.

6-204 OPTICS

TABLE 6k-5. NORMAL SPECTRAL EMITTANCE OF METALS AND ALLOYS AT λ = 0.665 μm

Material	Temperature, K	Normal spectral emittance*	Ref.
Cobalt	1180–1530	0.39–0.37	1
Copper:			
Polished	1080	0.15	2
Oxidized	1080	0.15	2
As received	1080	0.25	2
Germanium	1000–2000	0.50–0.53†	1
Gold	1220–1330	0.50–0.53‡	1
Graphite:			
GBE	1080–1905	0.80–0.71	2
GBH	1080–1905	0.86–0.77	2
7087	1080–1800	0.89–0.79	2
Iron, Armco, 2 μm rough	1130–1430	0.38–0.35	1
Molybdenum	1000–1800–2900	0.4–0.32–0.31†	1
Nickel:			
Polished	1080–1500	0.36–0.32	1
Oxidized	1100–1500	0.86–0.82	1
Osmium	1200–1800–2500	0.55–0.38–0.39	1
Palladium	1100–1550	0.40–0.33	1
Platinum:			
Polished	1100–1900	0.28–0.29†	1
Cold rolled	1100–1500	0.32–0.42†	1
Pyroceram 9608	1135–1465	0.48	5
Rhenium	1800–3000	0.41†	1
Rhodium	1120–1820	0.24–0.17	1
Silicon	1000–1700	0.64–0.46†	1
Tantalum:			
Polished	2300–3300	0.36†	1
Aged	1100–1600–2800	0.49–0.44–0.41†	1
Thorium, heat treated	1300–1650	0.38	1
Titanium, polished	1250–1650	0.48–0.47†	1
Zirconium	1000–2000	0.43–0.41†	1
Zirconium oxide	1155–1800	0.4–0.55	5
Stainless steels:			
321, bright	1080–1465	0.38–0.30	2
321, dull oxidized	1080–1465	0.67–0.52	2
AM 350, bright	1080–1465	0.38–0.33	2
AM 350, oxidized	1135–1465	0.75–0.58	2
PH 15-7 Mo, 5–40 μm rough	1080–1465	0.40–0.36	2
Cobalt alloy N-155:			
Polished	1080–1465	0.36–0.28	2
Oxidized	1080–1465	0.72–0.70	2
Inconel X:			
Polished	1080–1465	0.44–0.39	2
Oxidized	1080–1465	0.89–0.66	2
Bronze:			
4–7 Al	1080–1245	0.65	2
6–8 Al	1080–1245	0.72–0.70	2

* The emittances correspond to the given temperatures. Linear interpolation between points is fairly accurate.
† At λ = 0.65 μm.
‡ At λ = 0.64 μm.

References for Tables 6k-3 to 6k-5

1. Touloukian, Y. S., ed.: "Thermophysical Properties of High Temperature Solid Materials," The Macmillan Company, New York, N.Y., 1967.
2. Gubareff, G. G., J. E. Janson, and R. H. Tosberg: "Thermal Radiation Properties Survey," Honeywell Research Center, Minneapolis, Minn., 1960.

3. Sparrow, E. M., and R. D. Cess: "Radiation Heat Transfer," Brooks/Cole Publishing Company, Belmont, Calif., 1966.
4. "American Institute of Physics Handbook," 2d ed. McGraw-Hill Book Company, New York, 1963.
5. Rolling, R. E., and A. F. Funai: Investigation of the Effect of Surface Condition on the Radiant Properties of Metals, *Air Force Rept.* AFML-TR-64-363; part 1, 1964; part 2, 1967. (Directional polarized spectral and total emittances.)
6. Schatz, E. A.: Emittance and Reflectance of Intermetallic Compounds, *Am. Machine and Foundry Co. Progr. Rept.* 2, Contract AF 33 (657)-8877, 1962.
7. Jenkins, R. T., C. P. Butler, and W. J. Parker: Total Hemispherical Emittance Measurements over the Temperature Range 77°K to 300°K, *U.S. Navy Rept.* USNRDL-TR-663, SSD-TDR-62-189, 1963.
8. Wood, W. D., H. W. Deem, and C. F. Lucks: Thermal Radiative Properties of Selected Materials, *DMIC Rept.* 177, 1962.
9. Weber, D.: Bibliography of Emissivity of Solids and Liquids, *Hughes Aircraft Co. Res. Labs. Doc.* 3W15-34, 1957.
10. Hottel, H. C.: Normal Total Emissivity of Various Surfaces, table A-23 in "Heat Transmission," W. H. McAdams, ed., McGraw-Hill Book Company, New York, 1954.
11. Singham, J. R.: "Tables of Emissivities of Surfaces," *Intern. J. Heat Mass Transfer,* **5**, 67–76 (1962).
12. "Temperature: Its Measurement and Control in Science and Industry," Reinhold Publishing Corporation, New York, 1941.
13. Landolt-Börnstein, "Zahlenwerte und Functionen," 6th ed., vol. IV, p. 4, Springer-Verlag, New York, 1967.

6k-3. Emittance of Solids

ϵ_n = total normal emittance (emission of radiant energy of all wavelengths normal to the specified surface divided by the corresponding emission from a blackbody)

ϵ_h = total hemispherical emittance (emittance for radiation into hemisphere above emitting surface)

For metals: $\dfrac{\epsilon_h}{\epsilon_n}$ = (1.05–1.33)   (most metals ~ 1.2)

For nonmetals: $\dfrac{\epsilon_h}{\epsilon_n}$ = (0.95–1.05)   (most nonmetals ~ 0.98)

These relations are strictly valid only for specularly reflecting surfaces.

Emittances depend on such factors as surface roughness, work hardening, impurity content, and surface contamination. Tabulated values, while critically selected, can therefore only serve as a guide.

ϵ_λ = spectral emittance (emission of radiant energy within a small wavelength increment at wavelength λ, divided by the corresponding emission from a blackbody). The quantity depends on the angle of emission. For pyrometric temperature determination, the normal spectral emittance at about 0.665 μm is of importance.

r_λ = spectral reflectance (fraction of incident unpolarized radiation of wavelength λ reflected into the hemisphere above the reflecting surface). The quantity depends on the angle of incidence.

In consequence of Kirchhoff's law and the Helmholtz reciprocity relations:

$$\epsilon_\lambda = 1 - r_\lambda$$

if the angles of emission and incidence are the same. The relation is valid for specular and diffuse reflection and may be used to determine emittances from reflectance measurements. A corresponding relation does not in general exist for total emittances and reflectances. Off-normal emitted radiation is in general partially polarized.

TABLE 6k-6. SPECTRAL EMITTANCE OF OXIDES FOR $\lambda = 0.65$ μm*

Material	Range of observed values	Probable value for the oxide formed on smooth metal
Aluminum oxide	0.22–0.40	0.30
Beryllium oxide	0.07–0.37	0.35
Cerium oxide	0.58–0.80	
Chromium oxide	0.60–0.80	0.70
Cobalt oxide	....	0.75
Columbium oxide	0.55–0.71	0.70
Copper oxide	0.60–0.80	0.70
Iron oxide	0.63–0.98	0.70
Magnesium oxide	0.10–0.43	0.20
Nickel oxide	0.85–0.96	0.90
Thorium oxide	0.20–0.57	0.50
Tin oxide	0.32–0.60	
Titanium oxide	....	0.50
Uranium oxide	....	0.30
Vanadium oxide	....	0.70
Yttrium oxide	....	0.60
Zirconium oxide	0.18–0.43	0.40
Alumel (oxidized)	....	0.87
Cast iron (oxidized)	....	0.70
Chromel P (90 Ni, 10 Cr) (oxidized)	....	0.87
80 Ni, 20 Cr (oxidized)	....	0.90
60 Ni, 24 Fe, 16 Cr (oxidized)	....	0.83
55 Fe, 37.5 Cr, 7.5 Al (oxidized)	....	0.78
70 Fe, 23 Cr, 5 Al, 2 Co (oxidized)	....	0.75
Constantan (55 Cu, 45 Ni) (oxidized)	....	0.84
Carbon steel (oxidized)	....	0.80
Stainless steel (18-8) (oxidized)	....	0.85
Porcelain	0.25–0.50	

The emittance of oxides and oxidized metals depends to a large extent upon the roughness of the surface. In general, higher values of emissivity are obtained on the rougher surfaces.

* American Institute of Physics, "Temperature, Its Measurement and Control in Science and Industry," p. 1313, Reinhold Publishing Corporation, New York, 1941.

TABLE 6k-7. SPECTRAL EMITTANCE OF OXIDES FOR $\lambda = 0.665\ \mu m$*

Material	Temperature, °C	Emittance	Year
Aluminum oxide, 99.5%.....................	1000–1600	0.175	1952
Beryllium oxide, white (hot pressed)...........	927–1063	0.21	1948
Magnesium oxide...........................	1000–1470	0.18	1957
Nickel oxide, NiO...........................	816–1204	0.87–0.82	1957
Silicon dioxide.............................	1000–1600	0.18	1952
Tantalum oxide.............................	816–1204	0.78	1957
Thorium oxide.............................	1268–1800	0.40	1952

The emittance of white oxides depends strongly on purity. Only low values are shown.
* See ref. 13, p. 6-205.

TABLE 6k-8. THERMAL-CONTROL MATERIALS*

Material	α_s/ϵ	Absorptance† α_s at 70°F	Emittance ϵ at 70°F
Paints:			
White silicate on Al......................	0.15	0.13	0.85
White epoxy...........................	0.25	0.22	0.89
Al-silicone.............................	0.92	0.22	0.24
Al-acrylic..............................	0.85	0.41	0.48
Black acrylic...........................	1.06	0.93	0.88
Black silicone..........................	1.15	0.89	0.77
Optical solar reflector (second surface mirror, Ag)......................................	0.0625	0.05	0.80
Stainless steel, sandblasted (AI SI 410).......	0.88	0.75	0.85
6061 Al, rolled, chemically cleaned..........	2.7	0.16	0.06
Al foil, annealed..........................	2.4	0.12	0.04
Al, sandblasted...........................	2.0	0.42	0.21
Al, Reynolds wrap foil:			
Dull side...............................	5.0	0.2	0.04
Shiny side..............................	6.3	0.18	0.03
Inconel quilted...........................	3.2	0.38	0.12
Inconel, X-foil...........................	6.6	0.66	0.10
Hanovia gold on René 41..................	6.0	0.53	0.09
Gold, high purity on Al...................	9.0	0.27	0.03

* Space Materials Handbook, Air Force Rept. AFML-TR 60-40, suppl. 2, 1966. See reference for additional materials, details on composition, and stability.
† α_s = absorptance for solar radiation.

TABLE 6k-9. RELATION BETWEEN BRIGHTNESS TEMPERATURE AND TRUE
TEMPERATURE FOR VARIOUS VALUES OF SPECTRAL
EMISSIVITY AT $\lambda = 0.65$ μm*

Brightness temp., °C....	800	1000	1200	1400	1600	1800	2000
Emissivity ϵ (0.65 μm)	True temp., °C						
0.05	982	1265	1567	1846	2236	2609	3011
0.10	934	1194	1467	1752	2054	2370	2704
0.15	909	1156	1413	1681	1958	2248	2549
0.20	891	1130	1377	1632	1895	2168	2451
0.30	867	1095	1329	1567	1813	2064	2320
0.40	850	1071	1296	1525	1757	1995	2236
0.50	837	1053	1272	1493	1717	1944	2174
0.60	827	1039	1252	1467	1685	1905	2125
0.70	819	1027	1236	1447	1659	1872	2087
0.80	812	1017	1222	1429	1636	1844	2054
0.90	805	1008	1210	1413	1617	1821	2025

* American Institute of Physics, "Temperature, Its Measurement and Control in Science and Industry," Reinhold Publishing Corporation, New York, 1941.

TABLE 6k-10. RELATION BETWEEN APPARENT AND TRUE TEMPERATURE
FOR VARIOUS VALUES OF THE TOTAL EMISSIVITY*

Apparent temp., °C	100	200	400	600	800	1000	1200	1400	1600	1800
Total emissivity ϵ_t	True temp., °C									
0.05	422	686	1137	1567	1993	2317	2841	3264	3687	4110
0.10	316	536	913	1275	1632	1989	2345	2701	3057	3413
0.15	264	460	799	1126	1449	1771	2093	2415	2736	3058
0.20	231	410	725	1029	1330	1629	1929	2228	2527	2827
0.30	189	347	630	904	1175	1446	1717	1987	2258	2528
0.40	164	307	568	823	1075	1327	1579	1830	2082	2333
0.50	146	278	523	763	1002	1240	1478	1716	1954	2192
0.60	132	255	489	718	945	1173	1400	1628	1855	2082
0.70	121	238	461	680	900	1119	1337	1556	1775	1993
0.80	113	223	437	649	861	1073	1284	1496	1707	1919
0.90	106	211	417	623	828	1034	1239	1445	1650	1855

* American Institute of Physics, "Temperature, Its Measurement and Control in Science and Industry," Reinhold Publishing Corporation, New York, 1941.

TABLE 6k-11. EFFICIENCIES OF ILLUMINANTS*

Lamp	Rating, or specification	Eff.	Ab. eff.
Acetylene...............	1.0 liters/hr	0.67	0.0010
Arc, electric:			
Carbon, enclosed d-c....	6.6 amp opal globe and reflector	5.9	0.0087
Carbon, open d-c.......	9.6 amp clear globe	11.8	0.0173
High intensity..........	150 amp bare arc	18.5	0.0272
Magnetite d-c..........	6.6 amp	21.6	0.0318
Gas burner, open flame....	Bray high pressure	0.22	0.00032
Gas mantle, incandescent:			
High pressure..........	0.578 lumen/(Btu. hr)	2.0	0.0030
Low pressure...........	0.350 lumen/(Btu. hr)	1.2	0.0018
Incandescent electric carbon filament:			
First commercial........	....	1.6	0.0023
Squirted cellulose.......	....	3.3	0.0048
Metalized..............	....	4.0	0.0059
Tungsten filaments:			
Vacuum...............	25 watt 120 volt (1,000 hr life)	10.6	0.0156
Gas-filled..............	40 watt 120 volt (1,000 hr life)	11.6	0.0171
Gas-filled..............	60 watt 120 volt (1,000 hr life)	13.9	0.0204
Gas-filled..............	100 watt 120 volt (750 hr life)	16.3	0.0239
Gas-filled..............	1,000 watt 120 volt (1,000 hr life)	21.6	0.0318
Gas-filled..............	5,000 watt 120 volt (75 hr life)	32.8	0.0482
Fluorescent lamps:			
General line...........	20 watt standard warm white (T12)	50.0	0.0735
General line...........	40 watt standard warm white (T12)	64.0	0.0940
General line...........	90 watt standard warm white (T17)	58.0	0.0850
Slimline...............	96T8 (120 ma) standard warm white	76.0	0.1115
Slimline...............	96T12 (425 ma) standard warm white	69.0	0.1015
General line...........	40 W daylight (T12)	54.0	0.0795
General line...........	40 W green (T12)	84.0	0.1235
General line...........	40 W blue (T12)	33.0	0.0485
General line...........	40 W red (T12)	3.6	0.0053
Mercury lamps...........	400 W (E1)	50.0	0.0735
	1,000 W (A6)	65.0	0.0955
Sodium.................	10,000 lumen	55.0	0.0808

The rating listed is the commercial rating of the lamp. The absolute efficiency is the equivalent power in light flux (0.556 μm) per watt input. Efficiency is given in lumens per watt input.

* "Handbook of Chemistry and Physics," 49th ed., p. E-196, Chemical Rubber Publishing Company, 1968–1969. Compiled by J. M. Smith and C. E. Weitz.

TABLE 6k-12. APPROXIMATE BRIGHTNESS OF VARIOUS LIGHT SOURCES*

Source		Lamberts†
Natural sources:		
Clear sky	Average brightness	2.5
Sun (as observed from earth's surface)	At meridian	519,000
Sun (as observed from earth's surface)	Near horizon	1,885
Moon (as observed from earth's surface)	Bright spot	0.8
Combustion sources:		
Candle flame (sperm)	Bright spot	3.1
Kerosene flame (flat wick)	Bright spot	3.8
Illuminating-gas flame	Fishtail burner	1.3
Welsbach mantle	Bright spot	20.0
Acetylene flame	Mees burner	34.0
Incandescent electric lamps:		
Carbon filament		165
Metalized carbon filament (Gem)		300
Tungsten filament	Vacuum lamp, 10 lumens per watt	650
Tungsten filament	Gas-filled lamp, 20 lumens per watt	3,800
Tungsten filament	750-watt projector lamp, 26 lumens per watt	7,500
Fluorescent lamps:		
20 watt T12 standard warm white		1.67
40 watt T12 standard warm white		2.10
96T12 standard warm white		2.052
Electric-arc lamps:		
Plain carbon arc	Positive crater 7 mm non-rotating	55,000
High-intensity carbon arc	Positive crater 8 mm non-rotating	125,000
High-intensity carbon arc	Positive crater 13.6 mm non-rotating	220,000
High-intensity carbon arc	Positive crater	314,000
Mercury lamps:		
Low-pressure mercury arc	50-in. a-c rectified tube	6.6
400 W (H1)		440
1,000 W (A6)	Water-cooled	94,000
Sodium lamps	10,000 lumens	18

* "Handbook of Chemistry and Physics," 49th ed., pp. E-196, 197, Chemical Rubber Publishing Company, 1968–1969. Compiled by J. M. Smith and C. E. Weitz.
† To convert lamberts to foot-lamberts multiply by 929. To convert lamberts to candelas/cm² divide by π.

TABLE 6k-13. PROPERTIES OF TUNGSTEN*

Temp., K	Normal brightness new candela/cm²	Spectral emissivity		Color emissivity	Total emissivity	Brightness temp. 0.65 μm	Color temp.
		0.65 μm	0.467 μm				
300	....	0.472	0.505	....	0.032		
400	....	.....	.....	....	0.042		
500	....	.....	.....	....	0.053		
600	....	.....	.....	....	0.064		
700	....	.....	.....	....	0.076		
800	....	.....	.....	....	0.088		
900	....	.....	.....	....	0.101		
1000	0.0001	0.458	0.486	0.395	0.114	966	1007
1100	0.001	0.456	0.484	0.392	0.128	1059	1108
1200	0.006	0.454	0.482	0.390	0.143	1151	1210
1300	0.029	0.452	0.480	0.387	0.158	1242	1312
1400	0.11	0.450	0.478	0.385	0.175	1332	1414
1500	0.33	0.448	0.476	0.382	0.192	1422	1516
1600	0.92	0.446	0.475	0.380	0.207	1511	1619
1700	2.3	0.444	0.473	0.377	0.222	1599	1722
1800	5.1	0.442	0.472	0.374	0.236	1687	1825
1900	10.4	0.440	0.470	0.371	0.249	1774	1928
2000	20.0	0.438	0.469	0.368	0.260	1861	2032
2100	36	0.436	0.467	0.365	0.270	1946	2136
2200	61	0.434	0.466	0.362	0.279	2031	2241
2300	101	0.432	0.464	0.359	0.288	2115	2345
2400	157	0.430	0.463	0.356	0.296	2198	2451
2500	240	0.428	0.462	0.353	0.303	2280	2556
2600	350	0.426	0.460	0.349	0.311	2362	2662
2700	500	0.424	0.459	0.346	0.318	2443	2769
2800	690	0.422	0.458	0.343	0.323	2523	2876
2900	950	0.420	0.456	0.340	0.329	2602	2984
3000	1260	0.418	0.455	0.336	0.334	2681	3092
3100	1650	0.416	0.454	0.333	0.337	2759	3200
3200	2100	0.414	0.452	0.330	0.341	2837	3310
3300	2700	0.412	0.451	0.326	0.344	2913	3420
3400	3400	0.410	0.450	0.323	0.348	2989	3530
3500	4200	0.408	0.449	0.320	0.351	3063	3642
3600	5200	0.406	0.447	0.317	0.354	3137	3754

* "Handbook of Chemistry and Physics," 49th ed., p. E-228, Chemical Rubber Publishing Company, 1968-1969. Roeser and Wensel, National Bureau of Standards.

TABLE 6k-14. THE EMITTANCE OF WELL-DEFINED TUNGSTEN RIBBON AS A
FUNCTION OF WAVELENGTH AT TEMPERATURES BETWEEN 1600
AND 2800 K*

Wavelength, μm	Emissivity						
	1600 K	1800 K	2000 K	2200 K	2400 K	2600 K	2800 K
0.25	0.447	0.442	0.437	0.430	0.424	0.416	0.410
0.30	0.482	0.478	0.474	0.470	0.465	0.461	0.456
0.35	0.479	0.476	0.473	0.470	0.467	0.464	0.461
0.40	0.481	0.477	0.475	0.471	0.468	0.464	0.461
0.50	0.469	0.465	0.462	0.458	0.455	0.451	0.448
0.60	0.455	0.451	0.448	0.444	0.441	0.438	0.434
0.70	0.444	0.440	0.436	0.432	0.428	0.423	0.419
0.80	0.431	0.426	0.420	0.414	0.409	0.404	0.400
0.90	0.413	0.407	0.401	0.396	0.390	0.386	0.383
1.0	0.390	0.386	0.382	0.376	0.373	0.371	0.368
1.1	0.367	0.364	0.361	0.358	0.355	0.353	0.352
1.2	0.344	0.343	0.342	0.341	0.340	0.339	0.338
1.3	0.322	0.322	0.323	0.323	0.324	0.324	0.325
1.4	0.300	0.302	0.306	0.308	0.310	0.311	0.313
1.5	0.281	0.284	0.288	0.292	0.296	0.299	0.302
1.6	0.264	0.268	0.273	0.278	0.283	0.288	0.292
1.8	0.234	0.241	0.247	0.255	0.262	0.268	0.275
2.0	0.210	0.219	0.227	0.235	0.243	0.251	0.259
2.2	0.190	0.201	0.210	0.218	0.228	0.236	0.245
2.4	0.176	0.187	0.196	0.206	0.215	0.224	0.233
2.6	0.164	0.175	0.185	0.194	0.205	0.214	0.224

* J. C. De Vos, *Physica* **20**, 690 (1954).

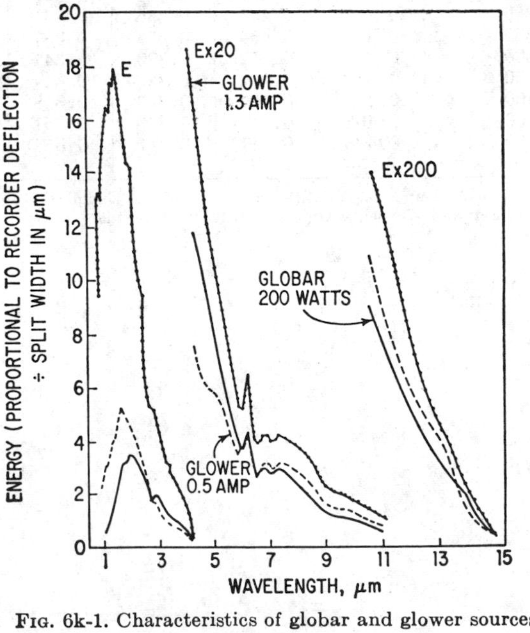

FIG. 6k-1. Characteristics of globar and glower sources.

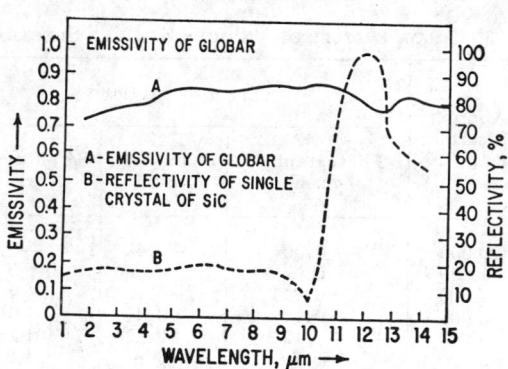

F<small>IG</small>. 6k-2. Emissivity of globar.

6k-4. Stellar Radiation. Brightness of stars as seen by any photoreceiver may be expressed as a stellar magnitude, related to the effective irradiance I in watts/cm² received from the star:

$$\text{Stellar magnitude } m = -2.5 \log_{10} \frac{I}{I_0}$$

The effective irradiance I from the star as seen by a photoreceiver is

$$I = \int_0^\infty J(\lambda)\sigma(\lambda)\, d\lambda$$

where $J\lambda$ = spectral distribution of radiation received from the star, in watts/cm² per wavelength increment $d\lambda$. $J(\lambda)$ for stars approximates blackbody distribution for the assumed surface temperatures.

    $\sigma(\lambda)$ = photoreceiver's spectral-response function normalized at the response peak.

For visual magnitude

$$I_0 = \tfrac{1}{685} \times 10^{-(24.18/2.5)} = 3.1 \times 10^{-13}\ \text{W/cm}^2$$

(Cf. definition of lumen, page 6–5; definition of stellar magnitude, "Smithsonian Tables," 8th ed., Table 798.)

Star brightness as seen by photoreceivers other than the eye is also expressed as a stellar magnitude (e.g., bolometric magnitude, photographic magnitude). The magnitude scales are generally adjusted by setting I_0 so that a class A0 star (surface temperature 11,000 K) appears of the same magnitude to each photoreceiver. For stars at other temperatures the effective-irradiance integral can be evaluated to obtain an index, which when added to visual magnitude gives the star's magnitude as seen by other receivers. Early stellar photometry used the non-color-sensitized (blue-sensitive) photographic plate; the difference between photographic and visual magnitude was called color index. Difference between bolometric and visual magnitude was called heat index. Indices for the principal spectral classes of stars and for several photoreceivers are given in Table 6k-15.

TABLE 6k-15. COLOR INDICES OF VARIOUS STELLAR SPECTRAL CLASSES

Spectral class	Approx eff. surface temp., K	Index			
		Photographic, visual	Bolometric, visual	S4 photosurface, visual	PbS, visual
B0..........	20,000	−0.30	−1.4	−0.15	+0.2
A0..........	11,000	0	0	0	0
F0..........	7,500	+0.33	+0.6	+0.30	−0.4
gG0.........	5,000	+0.70	+0.4	+0.7	−1.0
gK0.........	4,200	+1.12	+0.1	+1.0	−1.5
gM0.........	3,400	+1.70	−0.8	+1.1	−2.6

Effective temperature: Kuiper, *Astrophys. J.* **88**, 464 (1938).
S4 index: computed from manufacturers' data on 1P21 photomultiplier.
Bolometric index: Kuiper, *Astrophys. J.* **88**, 452 (1938).
Photographic index: "Smithsonian Tables," 8th ed.
PbS index: computed from manufacturers' data.

6k-5. Luminance of a Blackbody and Tungsten.[1] The luminance of a blackbody and of tungsten ribbon can be represented as a function of temperature by the following formulas:

$$\log L = 7.2010 - \frac{1.1376 \times 10^4}{T} + \frac{0.00613 \times 10^8}{T^2} \qquad \text{for a blackbody}$$

$$\log L = 6.8045 - \frac{1.1236 \times 10^4}{T} + \frac{0.00538 \times 10^8}{T^2} \qquad \text{for a tungsten ribbon}$$

where L is the luminance and T is the temperature.

[1] J. C. De Vos, *Physica* **20**, 715 (1954).

TABLE 6k-16. BRIGHTNESS OF STARS AS SEEN BY VARIOUS PHOTORECEIVERS

Star	Spectral type	Visual magnitude	S4 photosurface magnitude	Lead sulfide magnitude
Sirius	A0	−1.58	−1.6	−1.6
Canopus	F0	−0.86	−0.6	−1.3
α Centauri	G0	0.06	0.8	−0.9
Vega	A0	0.14	0.1	0.1
Capella	G0	0.21	0.9	−0.8
Arcturus	K0	0.24	1.3	−1.3
Rigel	B8p	0.34	0.3	0.3
Procyon	F5	0.48	1.0	−0.2
Achernar	B5	0.60	0.5	0.7
Betelgeuse (var.)	M0	0.7 ± 0.5	1.8 ± 0.5	−1.9 ± 0.5
β Centauri	B1	0.86	0.7	1.1
Altair	A5	0.89	1.0	0.7
α Crucis	B1	1.05	0.9	1.3
Aldebaran	K5	1.06	2.1	−0.8
Pollux	K0	1.21	2.2	−0.3
Spica	B2	1.21	1.1	1.4
Antares	M0	1.22	2.3	−1.4
Fomalhaut	A3	1.29	1.4	1.2
Deneb	A2p	1.33	1.4	1.2
Regulus	B8	1.34	1.3	1.4
β Crucis	B1	1.50	1.4	1.7
Castor	A0	1.58	1.6	1.6
γ Crucis	M3	1.61	2.7	−1.4
ε Canis Majoris	B1	1.63	1.5	1.8
ε Ursa Majoris	A0p	1.68	1.7	1.7
γ Orionis	B2	1.70	1.6	1.9
λ Scorpii	B2	1.71	1.6	1.9
ε Carniae	K0	1.74	2.7	0.2
ε Orionis	B0	1.75	1.6	2.0
β Tauri	B8	1.78	1.7	1.8
β Carniae	A0	1.80	1.8	1.8
α Triang. Aust.	K2	1.88	2.9	0.2
α Persei	F5	1.90	2.4	1.2
η Ursa Majoris	B3	1.91	1.8	2.1
γ Geminorum	A0	1.93	1.9	1.9
α Ursa Majoris	K0	1.95	3.0	0.5
ε Sagitarii	A0	1.95	2.0	2.0
δ Canis Majoris	F8p	1.98	2.6	1.1
β Canis Majoris	B1	1.99	1.9	2.2

TABLE 6k-17. SOLAR SPECTRAL IRRADIANCE*

λ	P_λ	D_λ	λ	P_λ	D_λ
0.140	0.0000048	0.00050	0.420	0.1758	11.19
0.150	0.0000176	0.00059	0.425	0.1705	11.83
0.160	0.000059	0.00087	0.430	0.1651	12.45
0.170	0.00015	0.00164	0.435	0.1675	13.06
0.180	0.00035	0.00349	0.440	0.1823	13.71
0.190	0.00076	0.00760	0.445	0.1936	14.41
0.200	0.00130	0.0152	0.450	0.2020	15.14
0.205	0.00167	0.0207	0.455	0.2070	15.90
0.210	0.00269	0.0288	0.460	0.2080	16.66
0.215	0.00445	0.0420	0.465	0.2060	17.43
0.220	0.00575	0.0609	0.470	0.2045	18.19
0.225	0.00649	0.0835	0.475	0.2055	18.95
0.230	0.00667	0.1079	0.480	0.2085	19.72
0.235	0.00593	0.1312	0.485	0.1986	20.47
0.240	0.00630	0.1534	0.490	0.1959	21.20
0.245	0.00723	0.1788	0.495	0.1966	21.92
0.250	0.00704	0.2053	0.500	0.1946	22.65
0.255	0.0104	0.2375	0.505	0.1922	23.36
0.260	0.0130	0.2808	0.510	0.1882	24.07
0.265	0.0185	0.3391	0.515	0.1833	24.76
0.270	0.0232	0.4163	0.520	0.1833	25.43
0.275	0.0204	0.4960	0.525	0.1852	26.12
0.280	0.0222	0.5758	0.530	0.1842	26.80
0.285	0.0315	0.6752	0.535	0.1818	27.48
0.290	0.0482	0.8225	0.540	0.1783	28.14
0.295	0.0584	1.020	0.545	0.1754	28.80
0.300	0.0514	1.223	0.550	0.1725	29.44
0.305	0.0602	1.430	0.555	0.1720	30.08
0.310	0.0686	1.668	0.560	0.1695	30.71
0.315	0.0757	1.935	0.565	0.1700	31.34
0.320	0.0819	2.227	0.570	0.1705	31.97
0.325	0.0958	2.555	0.575	0.1710	32.60
0.330	0.1037	2.925	0.580	0.1705	33.23
0.325	0.1057	3.312	0.585	0.1700	33.86
0.340	0.1050	3.702	0.590	0.1685	34.49
0.345	0.1047	4.090	0.595	0.1665	35.11
0.350	0.1074	4.483	0.600	0.1646	35.72
0.355	0.1067	4.879	0.605	0.1626	36.33
0.360	0.1055	5.271	0.610	0.1611	36.93
0.365	0.1122	5.674	0.620	0.1576	38.11
0.370	0.1173	6.099	0.630	0.1542	39.26
0.375	0.1152	6.529	0.640	0.1517	40.39
0.380	0.1117	6.949	0.650	0.1487	41.50
0.385	0.1097	7.359	0.660	0.1468	42.00
0.390	0.1099	7.765	0.670	0.1443	43.67
0.395	0.1191	8.189	0.680	0.1418	44.73
0.400	0.1433	8.675	0.690	0.1398	45.78
0.405	0.1651	9.245	0.700	0.1369	46.80
0.410	0.1759	9.876	0.710	0.1344	47.80
0.415	0.1783	10.53	0.720	0.1314	48.79

TABLE 6k-17. SOLAR SPECTRAL IRRADIANCE* (Continued)

λ	P_λ	D_λ	λ	P_λ	D_λ
0.730	0.1290	49.75	3.6	0.00135	98.720
0.740	0.1260	50.69	3.7	0.00123	98.816
0.750	0.1235	51.62			
0.800	0.1107	55.95	3.8	0.00111	98.902
0.850	0.0988	59.83	3.9	0.00103	98.982
			4.0	0.00095	99.055
0.900	0.0889	63.30	4.1	0.00087	99.122
0.950	0.0835	66.49	4.2	0.00078	99.182
1.000	0.0746	69.42			
1.1	0.0592	74.37	4.3	0.00071	99.238
1.2	0.0484	78.35	4.4	0.00065	99.289
			4.5	0.00059	99.335
1.3	0.0396	81.61	4.6	0.00053	99.376
1.4	0.0336	84.32	4.7	0.00048	99.414
1.5	0.0287	86.62			
1.6	0.0244	88.59	4.8	0.00045	99.448
1.7	0.0202	90.24	4.9	0.00041	99.480
			5.0	0.000383	99.509
1.8	0.0159	91.58	6.0	0.000175	99.716
1.9	0.0120	92.63	7.0	0.000099	99.817
2.0	0.0103	93.48			
2.1	0.0090	94.19	8.0	0.000060	99.876
2.2	0.0079	94.82	9.0	0.000038	99.912
			10.0	0.000025	99.935
2.3	0.0068	95.36	11.0	0.0000170	99.951
2.4	0.0064	95.85	12.0	0.0000120	99.962
2.5	0.0054	96.287			
2.6	0.0048	96.664	13.0	0.0000087	99.969
2.7	0.0043	97.001	14.0	0.0000055	99.975
			15.0	0.0000049	99.9785
2.8	0.0039	97.305	16.0	0.0000038	99.9817
2.9	0.0035	97.579	17.0	0.0000031	99.9843
3.0	0.0031	97.823			
3.1	0.0026	98.034	18.0	0.0000024	99.9863
3.2	0.00226	98.214	19.0	0.0000020	99.9879
			20.0	0.0000016	99.9893
3.3	0.00192	98.368	λ_∞		100.0
3.4	0.00166	98.501			
3.5	0.00146	98.616			

* *NASA Rept.* X-322-68-304, August, 1968. Based on measurements on board NASA-711 *Galileo* at 38,000 ft.

λ   Wavelength, μm
P_λ Solar spectral irradiance averaged over small bandwidth centered at λ, W/(cm².μm).
D_λ Percentage of the solar constant associated with wavelengths shorter than λ
Solar constant   0.013510 W/cm².

TABLE 6k-18. ENERGY DISTRIBUTION IN THE SPECTRA OF THE SELECTED STARS IN CGS UNITS*

No.	λ†	$E(\lambda)$, erg/(cm²·sec) per unit $\Delta\lambda$							
		β Ari	ζ Per	β Ori	ϵ Ori	β Tau	ϵ Ori	ζ Ori	α Leo
1	2	3	4	5	6	7	8	9	10
1	3,300	0.024_5	0.060_2	0.71_4	0.31_1	0.15_6	0.31_3	0.31_4	0.18_0
2	3,400	0.0244	0.0577	0.695	0.284	0.154	0.301	0.288	0.172
3	3,500	0.0243	0.0552	0.670	0.263	0.148	0.281	0.278	0.164
4	3,600	0.0244	0.0528	0.648	0.246	0.141	0.261	0.259	0.157
5	3,700	0.0251	0.0502	0.671	0.226	0.131	0.242	0.238	0.148
6	3,800	0.035_5	0.051_0	0.74_4	0.23_6	0.17_9	0.22_1	0.21_7	0.21_9
7	3,929	0.0539	0.0487	0.710	0.233	0.199	0.199	0.195	0.259
8	3,970	0.0586	0.0475	0.696	0.230	0.198	0.197	0.192	0.258
9	4,036	0.0600	0.0461	0.673	0.220	0.195	0.190	0.184	0.251
10	4,102	0.0581	0.0442	0.649	0.208	0.186	0.179	0.174	0.238
11	4,221	0.0550	0.0410	0.603	0.189	0.170	0.163	0.158	0.219
12	4,340	0.0527	0.0388	0.559	0.172	0.156	0.148	0.143	0.199
13	4,500	0.0495	0.0364	0.510	0.153	0.141	0.133	0.129	0.179
14	4,600	0.0475	0.0350	0.478	0.143	0.132	0.124	0.121	0.168
15	4,700	0.0455	0.0335	0.448	0.134	0.123	0.117	0.113	0.158
16	4,861	0.0418	0.0315	0.413	0.120	0.111	0.106	0.103	0.142
17	5,000	0.0384	0.0300	0.386	0.110	0.102	0.096_4	0.095_0	0.130
18	5,150	0.0355	0.0287	0.356	0.099_0	0.093_7	0.087_3	0.086_5	0.118
19	5,300	0.0333	0.0274	0.327	0.089_3	0.085_7	0.079_1	0.079_5	0.107
20	5,500	0.0307	0.0255	0.290	0.079_5	0.075_4	0.069_5	0.070_0	0.094_1
21	5,700	0.0283	0.0233	0.261	0.070_5	0.066_6	0.062_4	0.060_2	0.083_6
22	5,850	0.0263	0.0219	0.243	0.064_3	0.060_8	0.058_0	0.055_0	0.077_6
23	6,000	0.0246	0.0208	0.230	0.058_7	0.056_4	0.054_2	0.050_0	0.073_0
24	6,200	0.0225	0.0195	0.218	0.053_0	0.051_6	0.049_5	0.046_0	0.067_9
25	6,400	0.0209	0.0186	0.206	0.051_0	0.048_7	0.047_8	0.042_5	0.063_9
26	6,500	0.0202	0.0181	0.198	0.048_5	0.047_0	0.047_0	0.040_5	0.061_4
27	6,563	0.0198	0.0176	0.192	0.046_0	0.045_6	0.045_3	0.039_3	0.059_5
28	6,600	0.0195	0.0173	0.188	0.045_0	0.044_8	0.044_9	0.038_0	0.058_8
29	6,700	0.0189	0.0165	0.176	0.042_1	0.042_0	0.041_2	0.035_5	0.056_0
30	6,800	0.0180	0.0156	0.165	0.039_0	0.039_2	0.038_8	0.033_2	0.052_2
31	7,000	0.0160	0.0139	0.144	0.034	0.033_9	0.033_7	0.028_7	0.044_8
32	7,100	......	......	0.134	0.032	0.031_1	0.032_1	0.026_6	0.041_5
33	7,200	......	......	0.125	0.030	0.028_5	0.030_0	0.024_3	0.036_5

TABLE 6k-18. ENERGY DISTRIBUTION IN THE SPECTRA OF THE
SELECTED STARS IN CGS UNITS* (*Continued*)

No.	λ†	$E(\lambda)$, erg/(cm²·sec) per unit $\Delta\lambda$							
		γ UMa	η UMa	α Oph	α Lyr	δ Cyg	α Aql	α Cyg	α Peg
1	2	11	12	13	14	15	16	17	18
1	3,300	0.029_4	0.16_5	0.033	0.33_9	0.024_9	0.12_3	0.10_3	0.033_4
2	3,400	0.0296	0.156	0.0344	0.320	0.0249	0.124	0.106	0.0336
3	3,500	0.0298	0.147	0.0349	0.314	0.0248	0.126	0.109	0.0340
4	3,600	0.0300	0.139	0.0354	0.308	0.0247	0.128	0.112	0.0339
5	3,700	0.0302	0.129	0.0383	0.306	0.0246	0.135	0.158	0.0342
6	3,800	0.051_8	0.14_3	0.059_0	0.50_0	0.045_2	0.19_7	0.21_7	0.055_2
7	3,929	0.0784	0.161	0.0750	0.778	0.0620	0.232	0.214	0.0874
8	3,970	0.0804	0.163	......	0.798	0.0612	......	0.212	0.0906
9	4,036	0.0806	0.157	0.0896	0.795	0.0596	0.294	0.208	0.0873
10	4,102	0.0770	0.150	0.0866	0.765	0.0571	0.288	0.201	0.0831
11	4,221	0.0710	0.137	0.0830	0.709	0.0525	0.276	0.187	0.0770
12	4,340	0.0667	0.127	0.0796	0.655	0.0484	0.268	0.176	0.0707
13	4,500	0.0615	0.114	0.0758	0.598	0.0438	0.258	0.165	0.0642
14	4,600	0.0584	0.107	0.0739	0.564	0.0413	0.250	0.159	0.0607
15	4,700	0.0552	0.099_0	0.0712	0.531	0.0389	0.243	0.153	0.0570
16	4,861	0.0504	0.089_0	0.0650	0.484	0.0356	0.226	0.143	0.0520
17	5,000	0.0471	0.082_0	0.0609	0.449	0.0332	0.212	0.135	0.0482
18	5,150	0.0438	0.074_5	0.0571	0.413	0.0306	0.198	0.127	0.0443
19	5,300	0.0406	0.068_0	0.0535	0.382	0.0285	0.186	0.119	0.0411
20	5,500	0.0368	0.060_0	0.0491	0.345	0.0256	0.174	0.110	0.0371
21	5,700	0.0329	0.052_6	0.0453	0.313	0.0230	0.162	0.102	0.0333
22	5,850	0.0304	0.048_0	0.0425	0.290	0.0212	0.154	0.095_2	0.0388
23	6,000	0.0286	0.044_3	0.0401	0.272	0.0196	0.146	0.090_0	0.0286
24	6,200	0.0266	0.040_1	0.0379	0.248	0.0180	0.137	0.083_3	0.0261
25	6,400	0.0246	0.037_7	0.0354	0.230	0.0170	0.128	0.077_9	0.0245
26	6,500	0.0234	0.036_2	0.0336	0.220	0.0166	0.122	0.074_5	0.0234
27	6,563	0.0225	0.035_0	0.0324	0.209	0.0161	0.119	0.072_5	0.0227
28	6,600	0.0221	0.034_0	0.0318	0.204	0.0156	0.117	0.071_0	0.0221
29	6,700	0.0208	0.032_0	0.0300	0.190	0.0144	0.112	0.067_5	0.0205
30	6,800	0.0199	0.029_5	0.0282	0.178	0.0133	0.107	0.064_5	0.0190
31	7,000	......	0.026	0.0245	0.154	0.0115	0.095	0.056_6	0.0162
32	7,100	......	......	0.023	0.145	0.0105	0.090	0.052_5	0.0146
33	7,200	......	......	0.021	0.136	......	0.086	0.047_5	0.0131

* Kharitonov, A. V., *Soviet Astron.—AJ* **7**, 258 (1963).
† Wavelength in angstroms.

TABLE 6k-19. SOLAR ULTRAVIOLET FLUX INCIDENT ON EARTH'S ATMOSPHERE*

λ, Å	$\log f$,[†] W/(cm^2·Å)	$\log (f/E)$,[‡] photons/ (cm^2·sec·Å)	λ, Å	$\log f$,[†] W/(cm^2·Å)	$\log (f/E)$,[‡] photons/ (cm^2·sec·Å)
10	−11	4.7	900	−9.4	8.2
20	−10.2	5.8	1,000	−9.3	8.4
50	−9.6	6.8	1,100	−9.8	7.9
100	−9.5	7.2	1,200	−9.7	8.1
200	−9.4	7.6	1,400	−9.3	8.5
400	−9.8	7.5	1,600	−8.3	9.6
600	−9.8	7.7	1,800	−7.6	10.4
800	−9.8	7.8	2,000	−7.1	10.9

* Compiled by G. R. Cook, The Aerospace Corp.
[†] C. W. Allen, "Astrophysical Quantities," 2d ed., p. 173, Athlone Press, University of London, London, 1963.
Mean solar intensity with spectrum lines smoothed less the dominant resonance lines:

$$\text{HI } 1216 \text{ Å} \ldots \ldots 6 \times 10^{-7} \text{ W/cm}^2$$
$$\text{HeI } 584 \text{ Å} \ldots \ldots 0.1 \times 10^{-7} \text{ W/cm}^2$$
$$\text{HeII } 303 \text{ Å} \ldots \ldots 0.3 \times 10^{-7} \text{ W/cm}^2$$

[‡] Photon energy $E = hc/\lambda$.

TABLE 6k-20. LABORATORY VACUUM ULTRAVIOLET SOURCES*[a]

Name	Gas	Pressure, torrs	Wave-length, Å	Excitation method	Flux, photons/ (cm^2·sec·Å)
Continua					
Hopfield................	He	50–200	600–1,000	Condensed spark	10^{10}–10^{11}[b]
Argon..................	Ar	50–200	1,060–1,500	Condensed spark	10^{10}–10^{11}[b]
Krypton................	Kr	50–200	1,250–1,800	Condensed spark	10^9–10^{10}[b]
Xenon..................	Xe	50–200	1,500–1,800	Condensed spark	10^9–10^{10}[b]
Hydrogen...............	H$_2$	1–2	1,600–5,000	A-c or d-c glow	10^7–10^{8}[c]
Lyman/90% He + 10% air	...	0.02–0.05	300 ~ 5,000	Condensed spark	[d]
Synchrotron............	...	....	100–5,000	180 MeV	10^8–10^{9}[e]
X-ray fluorescence.......	...	....	10–100	Soft X-ray tube	[f]
Line Emission					
Hydrogen...............	H$_2$	1–2	850–1,600	A-c or d-c glow	~10^{11}[g]
Resonance line/He + 10%	Ar	~1	1,165, 1,236 1,470, 1,295	Microwave	10^{14}[h]
Resonance line/Ar + 10%.	H$_2$	~1	1,216	Microwave	10^{13}[h]
Resonance line/Ar + 10%.	O$_2$	~1	1,302–1,306	Microwave	10^{12}[h]
Resonance line/Ar + 10%.	N$_2$	~1	1,743–1,745	Microwave	10^{12}[h]
Spark spectra He + 10%..	Air	0.05	200–1,500	Condensed a-c	[i]
Hollow cathode..........	He	0.1	231–1,640	D-c glow	10^6–10^{7}[i]

* Compiled by G. R. Cook, The Aerospace Corp.

Notes for Table 6k-20

^a An account of this subject may be found in J. A. R. Samson, "Vacuum Ultraviolet Spectroscopy," chap. 5, John Wiley & Sons, Inc., New York, 1967.

^b Fluxes are approximate, and represent values that one may expect to obtain at the maximum of the continuum with a 1- or 2-m normal-incidence monochromator with a 600- or 1,200-line/mm grating. Absolute flux measurements have been reported by Metzger and Cook, *J. Opt. Soc. Am.* **55**, 516 (1965), and by R. E. Huffman, J. C. Larabee, and Y. Tanaka, *Appl. Opt.* **4**, 1581 (1965). The Ar, Kr, and Xe continua may also be excited with less intensity by microwaves. See P. G. Wilkenson and E. T. Byran, *Appl. Opt.* **4**, 581 (1965). Greater intensity may be obtained in high-energy single-flash technique. See J. A. Golden and A. L. Myerson, *J. Opt. Soc. Am.* **48**, 548 (1958).

^c At about 1,850 Å. See D. M. Packer and C. Lock, *J. Opt. Soc. Am.* **41**, 699 (1951).

^d This source requires current densities of 30,000 Å/cm² or more in the light-source capillary tubes. Flash tubes have been designed which produce a well-developed photographic spectrum after two or three flashes. See W. R. S. Garton, *J. Sci. Instr.* **36**, 11 (1959), and M. Nakamura, *Sci. Light (Tokyo)* **16**, 179 (1967). For wavelengths shorter than about 1,000 Å the continuum contains numerous emission lines.

^e These values are for the NBS 180-MeV, $R = 83$ cm, electron synchrotron at a distance of about 2 m along the tangent to the orbit before entering the spectrograph with $\lambda = 304$ Å. See K. Codling and P. Madden, *J. Appl. Phys.* **36**, 380 (1956). For 6-GeV electrons in a 31.7-m orbit see R. Haensel and C. Kunz, *Z. Angew. Phys.* **23**, 276 (1967). The wavelength of the maximum of the continuum decreases according to $\lambda = 2.35R/E^3$, where λ is in Å, R is in meters, and E in GeV. For 1 GeV and $R = 31.7$ m, the maximum of the continuum is at about 75 Å.

^f Fluorescence in the 10- to 100-Å region is detected with proportional counters containing P-10 or methane gas. For analysis of the light elements Mg to Be typical counting rates vary from 30 to 7,200 per sec, with peak to background ratios between 4 and 55. See B. L. Henke in "Advances in X-ray Analysis," vol. 8, p. 269, Plenum Press, Plenum Publishing Corporation, New York, 1965.

^g This is the flux observed at $\lambda = 1215.6$ with a 1-m monochromator with the light source operated 400 mA. See D. M. Packer and C. Lock, *J. Opt. Soc. Am.* **41**, 699 (1951). A wavelength table of the H_2 and many line spectra with relative intensities has been prepared by K. E. Schubert and R. D. Hudson, ATN-64 (9233)-2, October, 1963, The Aerospace Corp., P. O. Box 95085, Los Angeles, Calif. 90045.

^h About 50-W microwave power at 2450 MH coupled to the gas in a 13-mm OD capillary. See H. Okabe, *J. Opt. Soc. Am.* **54**, 478 (1964). A table of wavelengths of emission lines from neutral and ionized atoms in the 6 to 2,000 Å range has been prepared by R. L. Kelly, UCRL 5612, University of California, Lawrence Radiation Laboratory, Livermore, Calif. For each line there are one or more references to the original literature.

ⁱ Current densities less than for the Lyman discharge allow pulse rates in the 50 to 400 per sec region. These rates are convenient for photoelectric detection. Details of this source have been published by P. Lee and G. E. Weissler, *J. Opt. Soc. Am.* **42**, 80 (1952).

^j These are photon fluxes at the entrance slit of a 1-m grazing incident monochromator necessary to produce an output current of 10^{-9} amp from a Bendix magnetic-type multiplier. See E. Hinnov and F. Hofmann, *J. Opt. Soc. Am.* **53**, 1259 (1963).

61. Wavelengths for Spectrographic Calibration[1]

TABLE 6l-1. WAVELENGTH STANDARDS FOR THE VACUUM ULTRAVIOLET*

Wavelength, Å	Intensity	Spectrum	Estimated relative error ($\pm$mÅ)	Wavelength, Å	Intensity	Spectrum	Estimated relative error ($\pm$mÅ)
1,942.273	20	Hg II	2	1,774.941[a]	20	Si I	4
1,930.902	10	C I	2	1,769.658[a]	1	Si I	4
1,900.284	5	Hg II	2	1,753.113[a]	2	Si I	3
1,880.969	5	Si I	2	1,749.771[a]	1	Si I	5
1,870.547	20	Hg II	4	1,745.246	30	N I	3
1,869.548	8	Hg II	2	1,743.322	10	N II	4
1,867.590	1	N II	3	1,742.724	60	N I	3
1,864.742	5	N II	2	1,740.327	15	N II	3
1,862.806	2	N II	5	1,736.582	8	Si I	4
1,861.750[a]	1	Si I	2	1,732.142	15	Hg II	4
1,859.406	3	Ni I	2	1,730.874	2	N I	3
1,857.956	8	Ni I	4	1,727.332[a]	4	Si I	3
1,853.260	3	Si I	4	1,721.081	20	C II	3
1,850.665	5	Si I	5	1,720.158	18	C II	4
1,849.497	50R[b]	Hg I	4	1,707.397	4	Hg II	4
1,849.380	5	Ni I	4	1,704.558[a]	4	Si I	4
1,848.237	5	Si I	4	1,702.805[a]	8	Si I	4
1,846.014	8	N II	4	1,702.733	8	Hg II	4
1,844.304	10	N II	4	1,700.522	3	Si I	4
1,842.066	1	N II	5	1,693.756	15	Si I	4
1,839.995	4	Si I	4	1,676.913	5	Si I	4
1,833.264	1	C	5	1,672.405	2	Hg II	3
1,831.973	5	N II	4	1,658.117[c]	20	C I	1
1,830.458	4	N II	4	1,657.899[c]	15	C I	4
1,820.336	20	Hg II	4	1,657.541	1	C I	5
1,816.921	8	Si II	2	1,657.374[c]	10	C I	1
1,808.003	5	Si II	4	1,657.243	1	C I	5
1,807.303	30	N II	5	1,657.001[c]	30	C I	1
1,803.888	2	Hg II	2	1,656.923[c]	15	C I	1
1,796.897	15	Hg II	4	1,656.454	4	C I	4
1,787.805[a]	10	Si I	2	1,656.259	15	C I	1
1,782.817	15	Na III	4	1,654.055	5	C I	3
1,775.677	1	Hg I	4	1,653.644	2	Hg II	3

[1] This section presents calibration standards in the ultraviolet and infrared wavelength regions. For corresponding data on visible wavelengths, see Sec. 7.

TABLE 6l-1. WAVELENGTH STANDARDS FOR THE VACUUM ULTRAVIOLET* (Continued)

Wavelength, Å	Intensity	Spectrum	Estimated relative error (±mÅ)	Wavelength, Å	Intensity	Spectrum	Estimated relative error (±mÅ
1,649.932	10	Hg II	4	1,329.590	40	C I	1
1,640.474	80^d	He II	4	1,329.108	40	C I	2
1,640.342	100^d	He II	2	1,328.836^d	15	C I	10
1,630.180	2	Si I	3	1,327.927	10	N I	2
1,629.931	4	Si I	4	1,326.572	15	N I	4
1,629.830	4	N II	4	1,321.712	20	Hg II	3
1,629.366	4	Si I	4	1,319.684	30	N I	4
1,613.251	4	He II	4	1,319.003	20	N I	2
1,605.321	1	He II	3	1,316.287	1	N I	1
1,602.598	15	C I	3	1,311.365	20	C I	3
1,592.245	4	Si I	3	1,310.952	25	N I	1
1,589.607	2	Si I	3	1,310.548	25	N I	3
1,574.035	1	N II	3	1,309.278	3	Si II	5
1,561.433	20	C I	2	1,307.928	10	Hg II	3
1,561.339	5	C I	4	1,306.036	25	O I	3
1,560.687^d	15	C I	12	1,304.872	30	O I	5
1,560.301	2	C I	5	1,302.173	30	O I	1
1,504.474	5	Hg III	4	1,288.430	5	C I	3
1,494.673	60	N I	4	1,280.852^e	10	C I	1
1,492.824	30	N I	4	1,280.604^e	8	C I	3
1,492.624	80	N I	5	1,280.403^e	5	C I	4
1,485.600	8	Si II	2	1,280.340^e	15	C I	1
1,481.760	30	C I	3	1,280.140^e	8	C I	1
1,470.082	5	C I	3	1,279.897^e	10	C I	1
1,469.844	15	C I	4	1,279.230	8	C I	3
1,467.405	20	C I	3	1,277.727	20	C I	1
1,466.723	5	N I	4	1,277.551	50	C I	4
1,463.838	40	C	3	1,277.282	40	C I	1
1,463.346	40	C I	2	1,276.754	3	N II	1
1,459.034	20	C I	4	1,265.001	1	Si II	1
1,439.094	10	Si II	2	1,261.559^f	15	C I	1
1,411.948	30	N I	3	1,261.430^f	8	C I	4
1,393.322	1	Hg III	2	1,261.128^f	8	C I	1
1,364.165	8	C I	4	1,261.000^f	8	C I	1
1,361.267	8	Hg II	4	1,260.930^f	8	C I	2
1,357.140	5	C I	2	1,260.738^f	8	C I	1
1,355.598	2	O I	3	1,259.523	10	C I	3
1,354.292	8	C I	3	1,253.816	5	C I	1
1,350.074	4	Hg II	2	1,251.164	8	Si II	4
1,335.692	80	C II	5	1,250.586	4	Hg I	4
1,335.184	8	Hg	3	1,248.426	5	Si II	4
1,334.520	60	C II	5	1,246.738	1	Si II	3
1,331.737	20	Hg II	4	1,243.309	15	N I	4

TABLE 6l-1. WAVELENGTH STANDARDS FOR THE VACUUM ULTRAVIOLET* (Continued)

Wavelength, Å	Intensity	Spectrum	Estimated relative error (±mÅ)	Wavelength, Å	Intensity	Spectrum	Estimated relative error (±mÅ)
1,243.179	20	N I	1	1,098.103	40	N I	5
1,229.172	1	N I	1	1,097.990	25	N I	4
1,228.790	10	N I	4	1,097.245	50	N I	4
1,228.410	5	N I	4	1,096.749	35	N I	4
1,225.372	10	N I	1	1,096.322	35	N I	2
1,225.028	15	N I	4	1,095.940	35	N I	3
1,215.662	100R^b	H	5	1,085.707	50	N II	3
1,215.167	5	He II	5	1,085.546	3	N II	5
1,215.086	5	He II	4	1,085.442	3	N II	3
1,200.708^g	30	N I	2	1,084.970	2	He II	4
1,200.226^g	40	N I	1	1,084.910	2	He II	5
1,199.718^g	2	N I	4	1,084.579	30	N II	3
1,199.551^g	50	N I, C I	5	1,083.990	20	N II	4
1,194.496	5	Si I	1	1,070.821	0	N I	5
1,194.060	3	C I	3	1,069.984	30	N I	1
1,193.674	3	C I	3	1,068.476	35	N I	4
1,193.388^d	3	C I	8	1,067.607	35	N I	4
1,193.243	15	C I	2	1,041.688	1	O I	4
1,193.013	15	C I	4	1,040.941	15	O I	4
1,189.628	5	N I	4	1,039.233	20	O I	4
1,189.244	3	N I	3	1,037.627	0	O	3
1,188.972	5	N I	1	1,037.020	0	C II	1
1,177.694	15	N I	3	1,028.162	8	O I	3
1,176.626	3	N I	5	1,027.433	20	O I	3
1,176.508	15	N I	1	1,025.728	60	H	3
1,170.276	1	N I	3	1,025.298	2^i	He II	5
1,169.692	1	N I	1	990.805^h	2	O I	4
1,168.537	20	N I	4	990.210^h	8	O I	4
1,168.334	8	N I	4	990.132^h	1	O I	4
1,167.450	25	N I	4	988.776^h	15	O I	4
1,164.322	8	N I	3	988.661^{h,d}	2	O I	4
1,163.884	12	N I	4	977.967	1	O I	4
1,158.138	1	C I	5	964.626	1	N I	4
1,158.030	8	C I	4	963.991	5	N I	4
1,152.149	2	O I	5	953.658	15	N I	4
1,134.988	25	N I	4	953.415	15	N I	3
1,134.426	25	N I	4	952.522	4	N I	4
1,134.176	20	N I	4	952.414	8	O I	4
1,101.293	40	N I	5	952.304	8	N I	4
1,100.362	30	N I	4	950.114	0	O I	4
1,099.259	40	Hg II	3	949.742	25	H	4
1,099.153	25	N I	5	910.279	0	N I	5
1,098.264	40	N I	5	909.692	0	N I	5

TABLE 6l-1. WAVELENGTH STANDARDS FOR THE VACUUM ULTRAVIOLET* (*Continued*)

Wavelength, Å	Intensity	Spectrum	Estimated relative error (±mÅ)	Wavelength, Å	Intensity	Spectrum	Estimated relative error (±mÅ)
906.722	1	N I	2	893.079	0	Hg II	2
906.426	15	N I	4	888.363	0	N I	2
906.202	10	N I	3	888.019	0	N I	4
905.829	5	N I	4	875.092	5	N I	5

* *J. Opt. Soc. Am.* **45**, 10 (1955).
 [a] Identification: A. Fowler, *Proc. Roy. Soc.* (*London*), ser. A, **123**, 422 (1929); J. C. Boyce and H. A. Robinson, *J. Opt. Soc. Am.* **26**, 133 (1936).
 [b] Self-reversed resonance line.
 [c] Resolved $2p^2\,^3P - 3s\,^3P^0$ multiplet.
 [d] Blended line.
 [e] Completely resolved $2p^2\,^3P - 4s\,^3P^0$ multiplet.
 [f] Completely resolved $2p^2\,^3P - 3d\,^3P^0$ multiplet.
 [g] Resolved $2p^3\,^4S^0 - 3s\,^4P$ multiplet.
 [h] $2p^4\,^3P - 3s'\,^3D^0$ multiplet.
 [i] Diffuse line.

TABLE 6l-2. PROPOSED INTERNATIONAL WAVELENGTH STANDARDS
IN THE VACUUM ULTRAVIOLET

Wave-length, Å, this research	Spectrum	Wave-length, Å, More and Rieke[a]	Wave-length, Å, Boyce and Rieke[b]	Wave-length, Å, Weber and Watson[c]	Wave-length, Å, other observers	Wave-length, Å, mean value
1,930.902	C I	.....	0.900	0.889	......	1,930.897
1,745.246	N I	.....	0.246	0.255	......	1,745.249
1,742.724	N I	.....	0.734	0.733	......	1,742.730
1,740.327	N II	.....	.....	0.320	0.315[d]	1,740.321
1,658.117	C I	.....	0.126	0.127	......	1,658.123
1,657.899	C I	0.909	.....	.....	0.891[e]	1,657.900
1,657.374	C I	.....	0.380	0.381	......	1,657.378
1,657.001	C I	.....	0.005	.....	6.998[e]	1,657.001
1,656.259	C I	0.266	.....	.....	0.255[e]	1,656.260
1,560.301	C I	0.308	0.316	.....	......	1,560.308
1,494.673	N I	0.672	0.669	0.668	......	1,494.670
1,492.624	N I	.....	0.630	0.634	......	1,492.630
1,481.760	C I	0.771	.....	.....	0.750[f]	1,481.760
1,335.692	C II	0.700	.....	.....	0.684[g]	1,335.692
1,329.590	C I	0.587	.....	.....	0.583[h]	1,329.587
1,329.108	C I	0.102	0.101	.....	......	1,329.104
1,277.282	C I	.....	0.274	.....	0.280[h]	1,277.279
1,261.559	C I	.....	0.560	.....	0.565[h]	1,261.561
1,200.708	N I	0.719	0.706	0.693	......	1,200.708
1,200.226	N I	0.217	0.220	0.215	......	1,200.219
1,199.551	N I	0.552	0.547	0.557	......	1,199.552
1,177.694	N I	0.701	.....	0.677	......	1,177.691
1,176.508	N I	0.506	.....	0.498	......	1,176.504
1,167.450	N I	0.442	.....	0.454	......	1,167.449
1,134.988	N I	0.977	0.980	0.980	......	1,134.981
1,134.426	N I	.....	0.419	0.416	......	1,134.420
1,134.176	N I	.....	0.171	0.169	......	1,134.172
1,085.546	N II	.....	0.546	0.546	......	1,085.546
1,084.579	N II	0.584	0.579	0.582	......	1,084.580
1,083.990	N II	.....	0.991	0.990	......	1,083.990
990.805	C I	0.790	0.797	.....	......	990.797
990.210	C I	0.198	0.213	.....	......	990.207

[a] K. R. More and C. A. Rieke, *Phys. Rev.* **50,** 1054 (1936).
[b] J. C. Boyce and C. A. Rieke, *Phys. Rev.* **47,** 653 (1935).
[c] R. L. Weber and W. W. Watson, *J. Opt. Soc. Am.* **26,** 307 (1936).
[d] A. Fowler, *Proc. Roy. Soc. (London),* ser. A, **123,** 422 (1929).
[e] A. G. Shenstone, *Phys. Rev.* **72,** 411 (1947).
[f] E. Ekefors, *Z. Physik* **63,** 437 (1930).
[g] B. Edlén, *Z. Physik* **98,** 561 (1936); *Nature* **159,** 129 (1947).
[h] F. Paschen and G. Kruger, *Ann. Phys.* **7,** 1 (1930).

TABLE 6l-3. INFRARED STANDARD WAVELENGTHS

Wave-length, μm	State	Description	Substance	Ref.
0.54607	Emission	AH-4 lamp	Mercury	9
0.57696	Emission	AH-4 lamp	Mercury	9
0.57907	Emission	AH-4 lamp	Mercury	9
1.01398	Emission	AH-4 lamp	Mercury	9
1.12866	Emission	AH-4 lamp	Mercury	9
1.140	Liquid	....	Benzene	6
1.35703	Emission	AH-4 lamp	Mercury	9
1.36728	Emission	AH-4 lamp	Mercury	9
1.39506	Emission	AH-4 lamp	Mercury	9
1.52452	Emission	AH-4 lamp	Mercury	9
1.6606	Liquid	0.5-mm cell	1,2,4-Trichlorobenzene	9
1.671	Liquid	....	Benzene	6
1.69202	Emission	AH-4 lamp	Mercury	9
1.69419	Emission	AH-4 lamp	Mercury	9
1.70727	Emission	AH-4 lamp	Mercury	9
1.71090	Emission	AH-4 lamp	Mercury	9
1.81307	Emission	AH-4 lamp	Mercury	9
1.97009	Emission	AH-4 lamp	Mercury	9
2.008	Gas	....	Carbon dioxide	
2.150	Liquid	....	Benzene	
2.1526	Liquid	0.5-mm cell	1,2,4-Trichlorobenzene	9
2.22	Liquid	....	Carbon disulfide	9
2.24929	Emission	AH-4 lamp	Mercury	9
2.3126	Liquid	0.5-mm cell	1,2,4-Trichlorobenzene	9
2.32542	Emission	AH-4 lamp	Mercury	9
2.37	Solid	25-μm film	Polystyrene	Wright
2.4030	Liquid	0.5-mm cell	1,2,4-Trichlorobenzene	9
2.4374	Liquid	0.5-mm cell	1,2,4-Trichlorobenzene	9
2.439	Gas	....	Carbon oxysulfide central min	8
2.464	Liquid	....	Benzene	5
2.4944	Liquid	0.5-mm cell	1,2,4-Trichlorobenzene	9
2.5434	Liquid	0.5-mm cell	1,2,4-Trichlorobenzene	9
2.688	Gas	....	Carbon dioxide	Barker and Wu
2.7144	Vapor	5.0-cm cell	Methanol	9
2.765	Gas	....	Carbon dioxide	Barker and Wu
2.79	Solid		Lithium fluoride	9
2.996	Gas	200-mm 5.0-cm cell	Ammonia-zero branch	2
3.2204	Solid	25-μm film	Polystyrene	9
3.230	Gas	....	Carbon oxysulfide central min	8
3.2432	Solid	25-μm film	Polystyrene	9
3.2666	Solid	25-μm film	Polystyrene	9
3.3033	Solid	25-μm film	Polystyrene	9
3.3101	Solid	25-μm film	Polystyrene	9

TABLE 6l-3. INFRARED STANDARD WAVELENGTHS (*Continued*)

Wavelength, μ	State	Description	Substance	Ref.
3.320	Gas	....	Methane-zero branch	7
3.3293	Gas	5.0-cm cell	Methane	9
3.4188	Solid	25-μm film	Polystyrene	9
3.426	Gas	....	Carbon oxysulfide central min	8
3.465	Gas	....	Hydrogen chloride central min	
3.5078	Solid	25-μm film	Polystyrene	9
4.258	Gas	Atmospheric	Carbon dioxide	9
4.613	Vapor	....	Carbon disulfide central min	5
4.866	Vapor	5.0-cm cell	Methanol	9
4.875	Gas	....	Carbon oxysulfide central min	8
5.138	Solid	50-μm film	Polystyrene	9
5.284	Gas	....	Carbon oxysulfide central min	8
5.292	Gas	....	Ethylene central min	5
5.549	Solid	50-μm film	Polystyrene	9
5.847	Gas	....	Carbon oxysulfide central min	8
6.154	Gas	200 mm 5.0-cm cell	Ammonia-zero branch	2
6.238	Solid	50-μm film	Polystyrene	9
6.692	Solid	50-μm film	Polystyrene	9
6.753	Liquid	....	Benzene	S. Silverman
6.925	Gas	....	Ethylene-zero branch	5
7.268	Liquid	0.05-mm cell	Methylcyclohexane	9
7.681	Gas	....	Methane-zero branch	3
8.241	Gas	200-mm 5.0-cm cell	Ammonia	2
8.362	Gas	200-mm 5.0-cm cell	Ammonia	2
8.490	Gas	200-mm 5.0-cm cell	Ammonia	2
8.623	Gas	200-mm 5.0-cm cell	Ammonia	2
8.762	Gas	200-mm 5.0-cm cell	Ammonia	2
9.057	Gas	200-mm 5.0-cm cell	Ammonia	2
9.216	Gas	200-mm 5.0-cm cell	Ammonia	2
9.295	Gas	200-mm 5.0-cm cell	Ammonia	2
9.378	Gas	200-mm 5.0-cm cell	Ammonia	2
9.548	Gas	....	Carbon oxysulfide central min	8
9.608	Vapor	....	Methyl chloride	4
9.672	Vapor	5-cm cell	Methanol	9
9.673	Gas	....	Ammonia	Wright
9.724	Solid	50-μm film	Polystyrene	9
9.807	Vapor	....	Methyl chloride	4
9.85	Gas	....	Ammonia	Wright

TABLE 6l-3. INFRARED STANDARD WAVELENGTHS (*Continued*)

Wave-length, μm	State	Description	Substance	Ref.
10.073	Gas	200-mm 5.0-cm cell	Ammonia	2
10.53	Gas	....	Ethylene-zero branch	5
11.008	Gas	200-mm 5.0-cm cell	Ammonia	2
11.035	Solid	50-μm film	Polystyrene	9
11.26	Gas	200-mm 5.0-cm cell	Ammonia	J. Opt. Soc. Am.
11.475	Liquid	0.05-mm cell	Methylcyclohexane	9
11.793	Gas	200-mm 5.0-cm cell	Ammonia	2
11.862	Liquid	0.05-mm cell	Methylcyclohexane	9
12.075	Gas	200-mm 5.0-cm cell	Ammonia	2
12.381	Gas	200-mm 5.0-cm cell	Ammonia	2
12.732	Gas	....	Acetylene	1
12.809	Gas	....	Acetylene	1
12.885	Gas	....	Acetylene	1
12.961	Gas	....	Acetylene	1
12.99	Gas	....	Ammonia	Wright
13.69	Gas	....	Acetylene	1
13.883	Gas	Atmospheric	Carbon dioxide	9
14.29*	Solid	50-μm film	Polystyrene	9
14.42	Liquid	....	Toluene 1% in carbon disulfide	9
14.98	Gas	Atmospheric	Carbon dioxide	9
15.48	Liquid	0.05 mm (1:4 CS_2)	Unknown in technical grade of 1,2,4-trichlorobenzene	9
17.40*	Liquid	0.025-mm cell	1,2,4-Trichlorobenzene	9
18.16	Liquid	0.025-mm cell	1,2,4-Trichlorobenzene	9
20.56	Liquid	0.05-mm cell	1,2,4-Trichlorobenzene (sat. sol. in CS_2)	9
21.52	Liquid	0.05-mm cell	Toluene	9
21.80	Liquid	0.025-mm cell	1,2,4-Trichlorobenzene	9
22.76*	Liquid	0.025-mm cell	1,2,4-Trichlorobenzene	9
23.85	Vapor	Atmospheric	Water	9

* Broad bands.

References

1. Levin and Meyer: *J. Opt. Soc. Am.* **16**, 137 (1928); Meyer and Levin: *Phys. Rev.* **29**(2), 293 (1927).
2. Oetjen, Kao, and Randall: *Rev. Sci. Instr.* **13**, 515 (1942).
3. Cooley: *Astrophys. J.* **62**, 73 (1925).
4. Bennett and Meyer: *Phys. Rev.* **32**, 888 (1927).
5. McKinney, Leberknight, and Warner: *J. Am. Chem. Soc.* **59**, 481 (1937).
6. Liddel and Kaspar: *J. Research Natl. Bur. Standards* **11**, 599 (1933).
7. Nielsen and Nielsen: *Phys. Rev.* **48**, 864 (1935).
8. Bartunek and Baker: *Phys. Rev.* **48**, 516 (1935).
9. Plyler: *J. Research Natl. Bur. Standards* **45**, 463.

6m. Magneto-, Electro-, and Elasto-optic Constants

WILLIAM R. COOK, JR. AND HANS JAFFE

Gould, Inc.

6m-1. Magnetic Rotation (Faraday Effect). The most important interaction between magnetic field and light wave propagation is a rotation of the plane of polarization of a light wave traveling parallel to a magnetic field component

$$\alpha = VHl \tag{6m-1}$$

where H is magnetic field strength, and l the path length. This is the Faraday effect. The coefficient V is known as the Verdet constant. The Faraday effect results from a difference in propagation velocity for left and right circular polarized light. For a constant value of this difference, the Verdet constant is inversely proportional to wavelength. The tables give V in angular minutes/oersted·cm. Positive sign indicates rotation of the polarization plane in the same sense as a positive current in a coil producing the field.

TABLE 6m-1a. VERDET CONSTANTS OF GASES AND LIQUIDS[a]

Gas[b]	$(n_D{}^0 - 1) \times 10^{3}{}_{c}$	$10^6 V_0$	Liquid	λ, μm	t, °C	$n_D{}^{20\ c,d}$	$10^2 V$
He	0.036	+ 0.40	P	0.589	33		+13.3
Ar	2.81	+ 9.36	S	0.589	114	$1.929^{110°}$	+ 8.1
H$_2$		+ 6.2$_9$	H$_2$O[e]	0.546	25		+ 1.547
N$_2$	0.297	+ 6.4$_6$	H$_2$O	0.589	20	1.3330	+ 1.309
O$_2$	0.272	+ 5.69	D$_2$O	0.589	19.7	$1.3384^{20°}$	+ 1.257
Air	0.293	+ 6.27	H$_3$PO$_4$	0.578	97.4		+ 1.35
Cl$_2$	0.773 ·	+31.9	CS$_2$	0.589	20	1.6255	+ 4.255
HCl	0.447	+21.5	CCl$_4$	0.578–0.589	25.1	$1.463^{15°}$	+ 1.60
H$_2$S	0.63	+41.5	SbCl$_5$	0.578	18	$1.601^{14°}$	+ 7.45
NH$_3$	0.37$_6$	+19.0	TiCl$_4$	0.578	17	1.61	− 1.65
CO	0.34	+11.0	TiBr$_4$[f]	0.578	46		− 5.3
CO$_2$	0.45	+ 9.39	Methanol	0.589	18.7	1.3289	+ 0.958
NO	0.297	−58	Acetone	0.578–0.589	20.0	1.3585	+ 1.116
CH$_4$	0.444	+17.4	Toluene	0.578–0.589	15.0	1.4950	+ 2.71
n-C$_4$H$_{10}$		+44.0	Benzene	0.578–0.589	15.0	1.5005	+ 3.00
			Chlorobenzene	0.589	15	1.5246	+ 2.92
			Nitrobenzene	0.589	15	1.5523	+ 2.17
			Bromoform	0.589	17.9	1.5960	+ 3.13

[a] Selected except as noted from R. de Malleman, "Tables des constantes selectionées, pouvoir rotatoire magnétique (effet Faraday)," Hermann & Cie, Paris, 1951.

[b] V_0 for $\lambda = 0.578$ μm as reduced to 0°C and 760 mm Hg.

[c] "Handbook of Chemistry and Physics," Chemical Rubber Publishing Co., Cleveland, Ohio.

[d] Indices of refraction for organic chemicals from Eastman Kodak Co. Organic Chemicals List No. 39, 1954.

[e] V. Sivaramakrishnan, *Proc. Indian Acad. Sci.* **39**, 31 (1954); *J. Indian Inst. Sci.* **36**, 193 (1954).

[f] P. Fritsch, *Compt. Rend.* **217**, 447 (1943).

TABLE 6m-1b. VERDET CONSTANTS OF SOLIDS
(At room temperature except as noted)

Solid	V 0.633 μm	V 0.700 μm	Ref.
Oxide glasses			
$39Tl_2O\cdot61SiO_2$(moles)	0.12	0.10	8
$20TeO_2\cdot80PbO$	0.14	0.127	8
$24Pr_2O_3\cdot76B_2O_3$	−0.26	−0.22	8
$24Nd_2O_3\cdot76B_2O_3$	−0.14	−0.105	8
$85Bi_2O_3\cdot15B_2O_3$	0.10	0.085	8
$85PbO\cdot15B_2O_3$	0.115	0.093	8
$85Tl_2O\cdot15B_2O_3$	0.122	0.092	8
$2.67Ce_2O_3\cdot P_2O_5$	−0.174	−0.132	10
As_2S_3	0.26	0.194	9

Solid	n 0.5461 μm	V 0.5461 μm	n 0.5893 μm	V 0.5893 μm	Ref.
Oxide glasses					
SiO_2	1.4601	0.01664	1.4585	0.01421	3b
Dense flint 18	1.8999	0.1180	1.8900	0.0969	3b
Lead glass (Corning 8363)	.....	0.133	.....	0.107	9
Oxide crystals					
$NH_4Al(SO_4)_2\cdot12H_2O$	.....	0.0151	1.4594	0.0128	3c
$KAl(SO_4)_2\cdot12H_2O$	.....	0.0144	1.4564	0.0124	3c
$NH_4Fe(SO_4)_2\cdot12H_2O$ at 26°C	.....	−0.00145	1.4848	−0.00058	1
Same at −111°C	.....	−0.0145	.....	−0.0111	1
$NiSO_4\cdot6H_2O$ at 24°C	.....	0.0256	$\omega = 1.5109$	0.0221	4
Same at 1.36°K	.....	0.419	.....	.....	2
$MgAl_2O_4$ (spinel)	.....	....	1.7181	0.021	6
$CaCO_3$ (calcite)	.....	....	$\omega = 1.6585$	0.019	7
$NaClO_3$	.....	0.0105	1.5151	0.0081	3c
SiO_2 (quartz)	$\omega = 1.5462$	0.01952	1.5443	0.01664	3b
Al_2O_3 (corundum)	$\omega = 1.7712$	0.0240	1.7685	0.0210	3d
Cubic halide crystals					
$NaCl$	.....	0.0410	1.5443	0.0345	3c
$NaBr$	.....	0.0621	1.6412	....	3c
KCl	.....	0.0328	1.4904	0.0275	3c
KBr	1.5641	0.0500	1.5600	0.0425	3c
KI	1.6731	0.083	1.6664	0.070	3c
NH_4Cl	.....	0.0430	1.6426	0.0362	3c
NH_4Br	.....	0.0601	1.7108	0.0504	3c
CaF_2	.....	....	1.4338	0.00883	3a
Tetrahedral cubic crystals					
C, diamond	.....	0.0278	2.4172	0.0233	3a
$CuCl$	.....	0.20 ± 0.03	1.793	....	5
ZnS	.....	0.287	2.3683	0.226	3a

References to Table 6m-1b

1. Kaufmann, H.: *Ann. Physik* **18**, 251 (1933).   (Paramagnetic rotation.)
2. Levy and van den Handel: *Physica* **15**, 717 (1951).   (Paramagnetic rotation.)
3. Ramaseshan, S.: *Proc. Indian Acad. Sci.*: (a) **24**, 104 (1946); (b) **24**, 426 (1946); (c) **28**, 360 (1948); (d) **34**, 97 (1951); (e) *Current Sci.* (*India*) **20**, 150 (1951).
4. O'Connor, Beck, and Underwood: *Phys. Rev.* **60**, 443 (1941).
5. Gassmann, G.: *Ann. Physik* **35**, 638 (1939).   A volume of 23.9 cm³/mole is assumed to derive this value of V.
6. DuBois: *Ann. Physik* **51**, 537 (1894).

7. Chauvin: *J. Phys.* **9**, 5 (1890).
8. Borrelli, N. F.: Personal communication; also, *J. Chem. Phys.* **41**, 3289 (1964).
9. Robinson, C. C.: *Appl. Opt.* **3**, 1163 (1964).
10. Berger, S. B., C. B. Rubinstein, C. R. Kurkjian, and A. W. Treptow: *Phys. Rev.* **133A**, 723 (1964).

6m-2. The Kerr Effect. The lowest-order effect of an electric field on the refractive index of an isotropic material permitted by symmetry is quadratic in the electric field. The observed effect is an induced birefringence, the Kerr effect. It has substantial magnitude in polar liquids. (See also Sec. 6m-6 for ferroelectric crystals.) The Kerr constant K is defined by the relation

$$\Gamma = \frac{(n_p - n_s)l}{\lambda} = lKE^2 \qquad (6m-2)$$

where Γ is the retardation (path difference in fractions of the wavelength λ), n_p and n_s are the refractive indices parallel and normal to the applied field E, and l is the path length. As customary, Table 6m-2 gives K in electrostatic units.

TABLE 6m-2. TABULATED CHARACTERISTICS OF LIQUIDS WITH KNOWN LARGE KERR CONSTANTS*

Liquid	Symbol†	Kerr constant K, 10^{-7} esu ($\lambda = 0.589$ μm)	Static dielectric constant‡ ϵ'	Melting point, °C / Boiling point, °C	$n_D{}^{20}/H\beta - H\alpha$	Short-wavelength cutoff λ_{c0}, nm§
Carbon disulfide	CS₂	+3.23	2.6	−108.6/+46.3	1.6295/0.0343	
Acetone	C₃H₆O	+16.3	21.9	−94.3/+56.1	1.3591/0.0068	
Methyl ethyl ketone	C₄H₈O	+13.6	18.5	−86.4/+79.6	1.3791/0.0071	
Pyridine	C₅H₅N	+20.4	12.5	−42/+115.3	1.509/0.0163	3,000
Ethyl cyanoacetate	C₅H₇NO₂	+38.8	27.7	−22.5/+206	1.4179/0.0044	3,100
o-Dichlorobenzene	C₆H₄Cl₂	+42.6	7.5	−17.6/+179	1.549/0.0176	3,000
Benzenesulfonyl chloride	C₆H₅ClO₂S	+89.9	....	+14.5/+247	....	3,000
Nitrobenzene	C₆H₅NO₂	+326	36.1	+5.7/210.9	1.5529/0.0252	4,600
Ethyl B-aminocrotonate	C₆H₁₁NO₂	+31.0	....	+33.9/210	....	
Paraldehyde	C₆H₁₂O₃	−23.0	14.5 12.0	+10.5/+124	1.4198/0.0081	3,200
Benzaldehyde	C₇H₆O	+80.8	18.0 14.1	−56.0/+179.5	1.5464/0.0232	4,000
p-Chlorotoluene	C₇H₇Cl	+23.0	6.4	+7.8/+162.5	1.521/0.0164	3,200
o-Nitrotoluene	C₇H₇NO₂	+174	27.4	−4.1/+222.3	1.5462/	
m-Nitrotoluene	C₇H₇NO₂	+177	23.8	+15.5/+231	1.5475/	4,600
p-Nitrotoluene	C₇H₇NO₂	+222	18.7	+51.3/+238	1.5346/	
Benzyl alcohol	C₇H₈O	−15.4	13.0 10.8	−15.3/+205.8	1.5399/0.0173	3,200
m-Cresol	C₇H₈O	+21.2	13.0 5.0	+10/+202.8	1.540/0.0181	3,400
m-Chloroacetophenone	C₈H₇ClO	+69.1				
Acetophenone	C₈H₈O	+66.6	18.3 15.8	+19.7/+202.3	1.5339/0.0217	3,800
Quinoline	C₉H₇N	+15.0	9.0	−19.5/+239.7	1.6283/0.0312	3,600
Ethyl salicylate	C₉H₁₀O₃	+19.6	8.6	+1.3/+231.5	1.5226/0.0206	3,400
Carvone	C₁₀H₁₄O	+23.6	11.2	<0/+230	....	4,000
Ethyl benzoylacetate	C₁₁H₁₂O₃	+16.0	12.8	<0/+270	1.5338/0.0202	
Water	H₂O	+4.0	81	0/100	1.3330/	

* The data reported here are from ICT except where noted.
† The chemical symbol as shown here is used in the ICT for locating or reference purposes only.
‡ Several values for each liquid are listed in the ICT for several temperatures and frequencies. A single value in this table means that the approximate value is valid at both audio and radio frequencies (10^8 to 10^9) and at temperatures near 20°C. If two values are quoted, the first refers to audio, the second to radio frequencies.
§ Low-resolution measurements.

6m-3. Elasto-optic Effects in Isotropic Bodies. Elastic stresses generally cause changes of the refractive index proportional to the stress. This index change is different (and usually of opposite sign) for the electric vector of the light wave parallel and normal to the direction of an applied uniaxial stress. Isotropic bodies therefore develop induced uniaxial birefringence, with a retardation

$$\Gamma = \frac{(n_p - n_s)l}{\lambda} = \frac{CTl}{\lambda} \tag{6m-3}$$

where l is the light-path length in the body, λ the wavelength, and T the applied tensile stress. Positive value of C means that the refractive index for the electric vector parallel to an applied tension is the higher one. The customary unit for the stress-optical coefficient C is 1 Brewster $= 10^{-13}$ cm^2/dyne $= 10^{-12}$ m^2/N. See Sec. 6m-4 for absolute values of refractive-index change in some isotropic materials.

TABLE 6m-3. ENGINEERING STRESS-OPTICAL COEFFICIENT C FOR
VARIOUS ISOTROPIC MATERIALS

Material	C,. Brewsters $(10^{-12}$ m^2/N)	t, °C	λ, μm	Ref.
Polystyrene (glassy)	+8 to +10	27	0.546	1
Polymethyl methacrylate:				
Plexiglas II	−3.8	27	0.546	1
Lucite 130	−2.7	27	0.546	1
Polyphenyl methacrylate	+39.8	27	0.546	1
Polycyclohexyl methacrylate	+5.9	27	0.546	1
p-Cl Polystyrene	+23	27	0.546	1
Benzyl methacrylate	+45	20–25	.....	1
Natural rubber	2,000	20	.....	2
Gutta-percha	3,080	85	.....	2
Polyethylene	~2,000	130	.....	2
Polymethylene	~1,700	180	.....	2
Celluloid	11			
Bakelite	53			
Gelatin	1,700–14,000			
Fused silica (code 7940)*	−3.36	.....	0.633	3
Borosilicate (code 7070)*	−4.6	.....	0.633	3
Lead silicate (code 8363)*	+1.1	.....	0.633	3

* Codes refer to Corning designation.

References for Table 6m-3

1. Rudd, J. F., and R. D. Andrews: *J. Appl. Phys.* **31**, 818 (1960).
2. Saunders, D. W.: *Trans. Faraday Soc.* **52**, 1414, 1425 (1956).
3. Borelli, N. F., and R. A. Miller: *Appl. Opt.* **7**, 743 (1966).

6m-4. Elasto-optic Effects in Crystals. The optic effects of stress in crystals are in general dependent on the orientation of the stress tensor with respect to the crystal axes. The changes in refraction caused by stress are superimposed on permanent birefringence in all crystal systems except the cubic. The refractive properties of the crystal may be represented by the index ellipsoid. The changes of refraction due to applied stress are then expressed in additive terms in the coefficients of the index ellipsoid equation. Related to orthogonal crystal axes, the ellipsoid equation is

$$a_1x^2 + a_2y^2 + a_3z^2 + 2a_4yz + 2a_5zx + 2a_6xy = 1 \tag{6m-4}$$

For all systems except the monoclinic and triclinic $a_4 = a_5 = a_6 = 0$, and a_1, a_2, a_3 are reciprocal to the square of the refractive indices for vibration in the respective axial directions. The additive terms in the index ellipsoid coefficients are related to the stress components by

$$\Delta a_i = q_{ij}T_j \qquad (i, j \text{ from 1 to 6, summation over repeated index}) \qquad (6\text{m-}5)$$

$\Delta a_{1...3}$ determine changes of magnitude of the refractive indices, while $\Delta a_{4...6}$ cause either rotation of the index ellipsoid or a reduction of its symmetry from cubic or uniaxial to biaxial. The q_{ij} are the stress-optic coefficients; there are 36 independent q_{ij} for an asymmetric crystal. Crystal symmetries introduce equations between some coefficients and make others zero. The matrices of coefficients for all crystal classes have been derived.[1,2] Earlier derivations [quoted for instance by W. P. Mason, *Bell System Tech. J.* **29**, 161–188 (1950)] show additional equalities which are erroneous.

For orthorhombic crystals the matrix is:

$$
\begin{matrix}
q_{11} & q_{12} & q_{13} & 0 & 0 & 0 \\
q_{21} & q_{22} & q_{23} & 0 & 0 & 0 \\
q_{31} & q_{32} & q_{33} & 0 & 0 & 0 \\
0 & 0 & 0 & q_{44} & 0 & 0 \\
0 & 0 & 0 & 0 & q_{55} & 0 \\
0 & 0 & 0 & 0 & 0 & q_{66}
\end{matrix}
$$

For crystals of the tetragonal classes $\bar{4}2\,m$ and $4/mmm$ the same matrix holds with the equalities

$$q_{11} = q_{22},\ q_{12} = q_{21},\ q_{13} = q_{23},\ q_{31} = q_{32},\ q_{44} = q_{55}$$

For the cubic crystal classes $\bar{4}3m$ and $m3m$ one finds the additional equalities $q_{33} = q_{11}$, $q_{13} = q_{31} = q_{12}$, $q_{66} = q_{44}$, so that only three independent coefficients remain: q_{11}, q_{12}, q_{44}. In the less symmetric cubic classes $m3$ and 23, however, there are four independent coefficients, with $q_{12} = q_{23} = q_{31}$ and $q_{13} = q_{21} = q_{32}$, but $q_{12} \neq q_{21}$. The matrix for isotropic materials differs from that for the most symmetric cubic class by the additional relation $q_{44} = q_{11} - q_{12}$.

The matrix for the trigonal classes 32 and $3m$ is:

$$
\begin{matrix}
q_{11} & q_{12} & q_{13} & q_{14} & 0 & 0 \\
q_{12} & q_{11} & q_{13} & -q_{14} & 0 & 0 \\
q_{31} & q_{31} & q_{33} & 0 & 0 & 0 \\
q_{41} & -q_{41} & 0 & q_{44} & 0 & 0 \\
0 & 0 & 0 & 0 & q_{44} & 2q_{41} \\
0 & 0 & 0 & 0 & q_{14} & q_{11} - q_{12}
\end{matrix}
$$

The matrix for the hexagonal classes $6mm$ and $6/mm$ follows from the preceding by setting $q_{41} = q_{14} = 0$.

The changes of refraction caused by elastic deformation may also be expressed as a function of the strain tensor. The resulting strain-optic coefficients p_{ij} are related to the stress-optic coefficients by

$$p_{ij} = q_{ij}c_{kj} \qquad (6\text{m-}6)$$

[1] J. F. Nye, "Physical Properties of Crystals," 2d ed., Clarendon Press, Oxford, 1960.
[2] S. Bhagavantam: "Crystal Symmetry and Physical Properties," Academic Press, Inc., New York and London, 1966.

The p_{ij} are dimensionless coefficients generally of magnitude in the 10^{-2} to 1 range. The relations introduced by symmetry are the same as for the stress-optic coefficients except that for the trigonal crystals $p_{56} = p_{41}$; for both trigonal and hexagonal crystals $p_{66} = (p_{11} - p_{12})/2$; also for isotropic bodies $p_{44} = (p_{11} - p_{12})/2$.

In piezoelectric crystals the values of some q_{ik} and p_{ik} depend on the electric boundary conditions. This effect is more pronounced for the q_{ik} than p_{ik}. The listed values may be assumed to relate to the short-circuit (constant E) condition.

The optic path difference is found by solving the equation of the modified index ellipsoid for the desired wave propagation direction. The solutions are of the form $\Gamma = n^3(\Delta a_i - \Delta a_k)\gamma l$, where n is the appropriate refractive index, and γ a numerical constant in the order of unity.

TABLE 6m-4a. STRESS-OPTIC CONSTANTS OF CRYSTALS* (10^{-12} m^2/N)

Crystal	Symmetry	q_{11}	q_{12}	q_{44}	q_{13}	q_{31}	q_{33}	q_{66}	q_{14}	q_{41}
KCl	$m3m$	4.75	2.87	−4.32	q_{12}	q_{12}	q_{11}	q_{44}	0	0
NaCl	$m3m$	1.27	2.58	−0.84						
LiF	$m3m$	−0.40	1.12	−0.83						
CaF$_2$	$m3m$	−0.29	1.16	0.698						
Diamond	$m3m$	−0.43	0.37	−0.27						
ZnS	$\bar{4}3m$	......	......	−3.16	q_{12}	q_{12}	q_{11}	q_{44}	0	0
Ammonium alum	$m3$	5.5	11.6	−1.15	10.9	q_{13}	q_{11}	q_{44}	0	0
Potassium alum.	$m3$	3.7	9.1	−0.63	8.5					
Ba(NO$_3$)$_2$	$m3$	$q_{11}-q_{12} = $ −23.84		−1.69	$q_{11}-q_{13} = $ −17.13					
Pb(NO$_3$)$_2$	$m3$	$q_{11}-q_{12} - $ −19.13		−1.39	$q_{11}-q_{13} = $ −11.84					
NH$_4$H$_2$PO$_4$	$\bar{4}2m$	8.6	7.9	−5.8	−37.3	12.3	−35.7	−12.2	0	0
Calcite	$32/m$	−0.61	0.92	0.35	2.53	1.58	−0.45	$(q_{11}-q_{12})$	−1.11	0.67
α-Quartz	32	1.11	2.50	−1.010	1.97	2.77	0.183	$(q_{11}-q_{12})$	−0.097	−0.320
Beryl	$6/mmm$	−0.356	0.592	−2.32	0.739	0.931	−0.426	$(q_{11}-q_{12})$	0	0

Crystal	Symmetry	q_{11}	q_{22}	q_{33}	q_{44}	q_{55}	q_{66}	q_{12}	q_{21}	q_{13}	q_{31}	q_{23}	q_{32}
Barite	mmm	−0.079	1.77	2.56	0.20	0.435	1.39	3.81	1.49	1.64	0.44	1.23	1.23
Topaz	mmm	−0.514	−0.594	−0.456	−0.852	−0.233	−0.750	0.326	0.521	0.226	0.376	0.251	0.222
Rochelle salt	222	......	......	......	−0.9	1.7	−1.7						

* R. Bechmann, "Landolt-Börnstein," New Series, Group III, vol. 1, K.-H. Hellwege, ed., 1966; vol. 2, 1969. Most of data determined at 0.589 μm.

TABLE 6m-4b. STRAIN-OPTIC CONSTANTS OF CUBIC CRYSTALS AND SOME GLASSES

Material	Symmetry	p_{11}	p_{12}	p_{44}	p_{13}	Wavelength, μm	Refs.
Fused silica	Glass	+0.121	+0.270	$\frac{1}{2}(p_{11}-p_{12})$	p_{12}	0.633	1
Code 7070*	Glass	+0.113	+0.23	....	....	0.633	2
Code 8363*	Glass	+0.196	+0.185	....	....	0.633	2
As$_2$S$_3$	Glass	+0.277	+0.272	....	....	0.633	1
As$_2$S$_3$	Glass	+0.308	+0.299	....	....	1.15	1
KCl	$m3m$	0.215	0.159	−0.024	p_{12}	0.589	3, 4
NaCl	$m3m$	0.137	0.178	−0.011	....	.....	3, 4
LiF	$m3m$	0.02	0.130	−0.045	....	0.589	3, 4
Tl(Br,I) = KRS-5	$m3m$	$p_{11}-p_{12}=0.08$		0.157	....	0.61	
MgO	$m3m$	−0.32	−0.08	−0.096	....	0.560	4
CaF$_2$	$m3m$	0.056	0.228	0.024	....	0.589	3, 4
Y$_3$Al$_5$O$_{12}$	$m3m$	−0.029	+0.009	−0.062	....	0.633	1
Y$_3$Fe$_5$O$_{12}$	$m3m$	0.025	0.073	0.041	....	0.633	1
Y$_3$Ga$_5$O$_{12}$	$m3m$	0.091	0.019	0.079	....	0.633	5
SrTiO$_3$	$m3m$	0.15	0.095	0.072	....	0.633	5
Diamond	$m3m$	−0.31	+0.09	−0.12	....	0.589	3, 4
GaAs	$\bar{4}3m$	−0.165	−0.140	−0.072	p_{12}	1.15	1
GaP	$\bar{4}3m$	−0.151	−0.082	−0.074	....	0.633	1
Ammonium alum	$m3$	0.378	0.465	−0.009	0.454	0.589	3, 4
Potassium alum	$m3$	0.275	0.354	−0.005	0.345	0.589	3
Barium nitrate	$m3$	$p_{11}-p_{12}=0.992$		−0.0205	$p_{11}-p_{13}=0.713$	0.589	3
Lead nitrate	$m3$	$p_{11}-p_{12}=0.281$		−0.039	$p_{11}-p_{13}=0.174$	0.589	3, 4

Crystal	Symmetry	p_{11}	p_{12}	p_{44}	p_{13}	p_{31}	p_{33}	p_{66}	p_{14}	p_{41}	Wavelength, μm	Ref.
NH$_4$H$_2$PO$_4$	$\bar{4}2m$	−0.11	−0.15	−0.056	−0.93	0.20	−0.71	−0.077	0	0	.....	3
		0.302	0.246	....	0.236	0.195	0.263	0.075			0.633	1
KH$_2$PO$_4$	$\bar{4}2m$	.....	.....	....	0.246	0.225	0.221	−0.0685	....	....	0.560	3
		0.251	0.249	....	0.246	0.225	0.221	0.058	....	....	0.633	1
TiO$_2$	$4/mmm$	0.011	0.172	....	0.168	0.0965	0.058	....	0	0	0.633	1
Calcite	$\bar{3}2/m$	0.095	0.189	−0.090	0.215	0.309	0.178	$\frac{1}{2}(p_{11}-p_{12})$	−0.006	0.010	0.589	3, 4
αAl$_2$O$_3$	$\bar{3}2/m$	0.25	0.038	0.10	≤0.005	0.032	0.23	....	<0.02	≤0.01	0.633	5
α-Quartz	32	0.138	0.250	−0.0685	0.259	0.258	0.098	$\frac{1}{2}(p_{11}-p_{12})$	−0.029	−0.042	0.589	3, 4
LiNbO$_3$	$3m$	0.036	0.072	....	0.092	0.178	0.088	$\frac{1}{2}(p_{11}-p_{12})$	0.070	0.155	0.633	1, 5
LiTaO$_3$	$3m$	0.0804	0.0804	0.022	0.094	0.086	0.150	....	0.031	0.024	0.633	1
CdS	$6mm$	0.142	0.066	∼0.054	....	0.041	....	$\frac{1}{2}(p_{11}-p_{12})$	0	0	0.633	1
Beryl	$6/mmm$	0.010	0.175	−0.152	0.191	0.313	0.023	$\frac{1}{2}(p_{11}-p_{12})$	0	0	0.589	3

Measured at $\lambda = 0.589\ \mu m$

Crystal	Symmetry	p_{11}	p_{22}	p_{33}	p_{44}	p_{55}	p_{66}	p_{12}	p_{21}	p_{13}	p_{31}	p_{23}	p_{32}	Ref.
Barite	mmm	0.21	0.24	0.31	0.002	−0.012	0.037	0.25	0.34	0.16	0.275	0.19	0.22	3, 4
Topaz	mmm	−0.085	−0.120	−0.083	−0.095	−0.031	−0.098	−0.069	0.093	0.052	0.095	0.065	0.085	3
Rochelle salt	222	.....	.....	.....	−0.006	−0.005₄	−0.016	.....	.....	.....	.....	.....	.....	3

* Corning Glass Works.
The full matrix is given for the first example of each symmetry. Those labeled 0 are identically zero by symmetry; ... indicates no value available.

References for Table 6m-4b

1. Dixon, R. W.: *J. Appl. Phys.* **38**, 5149 (1967).
2. Borelli, N. F., and R. A. Miller: *Appl. Opt.* **7**, 745 (1968).
3. Bechmann, R.: "Landolt-Börnstein," New Series, Group III, vol. 1, K.-H. Hellwege, ed., 1966; vol. 2, 1969.
4. Krishnan, R. S.: *Progr. Cryst. Phys.* **1** (1958).
5. Reintjes, J., and M. B. Schulz: *J. Appl. Phys.* **39**, 5254 (1968).

6m-5. Linear Electro-optic Effect (Pockels Effect). A linear interaction between the vectorial quantities, electric field or displacement, and the tensor quantities, stress or strain, is permitted by symmetry in the 20 piezoelectric crystal classes. The symmetry relations are exactly the same as for the piezoelectric effects, specifically the converse piezoelectric effect giving strain as function of applied field. The linear electro-optic effect has in recent years been termed Pockels effect in recognition of F. Pockels who made the first systematic studies in the 1890s. The defining set of equations is

$$\Delta a_i = r_{ij} E_j \tag{6m-7}$$

where the Δa_i are increments of the coefficient of the index ellipsoid as in Sec. 6m-4, E_j are the components of the electric field vector, and r_{ij} the electro-optic coefficients. There are 18 in the absence of any symmetry. The values of all electro-optic coefficients depend on the elastic boundary conditions. If the superscripts T and S denote respectively the conditions of zero stress (free crystal) and zero strain (clamped crystal), one finds

$$r_{ij}^T = r_{ij}^S + q_{ik}^E e_{jk} = r_{ij}^S + p_{ik}^E d_{jk} \tag{6m-8}$$

Here $e_{jk} = (\partial T_k / \partial E_j)_S$ and $d_{jk} = (\partial S_k / \partial E_j)_T$ are the customary piezoelectric coefficients (see Sec. 9f). Note that the order of subscripts in the electro-optic tensor is the reverse of the customary order in piezoelectric coefficients, where i indicates the electric and j the elastic variable. The r_{ij}^S are sometimes referred to as the direct electro-optic effect. It is generally observed at frequencies above the principal elastic resonances of the crystal which are typically in the 100-kHz range. The additional term in Eq. (6m-8) describes the "indirect" electro-optic effect which may have the same order of magnitude as the direct effect. The r_{ij}^T as well as the r_{ij}^S may be regarded as special values of a second-order dielectric constant relating dielectric displacement to two electric field components at different frequencies (see Sec. 6n, Nonlinear Optic Effects). One may also express the Pockels effect with dielectric displacement or polarization as independent variable. The coefficients so defined show less temperature dependence than the r_{ij}, especially in ferroelectric crystals.

The effect of symmetry on the electro-optic matrix for the crystal classes represented in Table 6m-5 is as follows:

Orthorhombic						Tetragonal						Trigonal						Cubic		
222			mm^2			$4mm$			$\bar{4}2m$			32			$3m$			$\bar{4}3m$		
0	0	0	0	0	r_{13}	0	0	r_{13}	0	0	0	r_{11}	0	0	0	$-r_{22}$	r_{13}	0	0	0
0	0	0	0	0	r_{23}	0	0	r_{13}	0	0	0	$-r_{11}$	0	0	0	r_{22}	r_{13}	0	0	0
0	0	0	0	0	r_{33}	0	0	r_{33}	0	0	0	0	0	0	0	0	r_{33}	0	0	0
r_{41}	0	0	0	r_{42}	0	0	r_{42}	0	r_{41}	0	0	r_{41}	0	0	0	r_{42}	0	r_{41}	0	0
0	r_{52}	0	r_{51}	0	0	r_{42}	0	0	0	r_{41}	0	0	$-r_{41}$	0	r_{42}	0	0	0	r_{41}	0
0	0	r_{63}	0	0	0	0	0	0	0	0	r_{63}	0	$-2r_{11}$	0	$-2r_{22}$	0	0	0	0	r_{41}

The hexagonal class $6mm$ has the same matrix as $4mm$.

As in the elasto-optic case, optic path differences result from the Δa_i of Eq. (6m-7) after multiplication with n^3. The utility of crystals with about equal Pockels coefficients is therefore strongly dependent on the value of the refractive index. The specific equations for preferred orientation in all crystal classes have been tabulated.[1]

[1] O. G. Vlokh and I. S. Zheludev, *Soviet Phys.—Cryst.* **5**, 368–380 (1960).

TABLE 6m-5. LINEAR ELECTRO-OPTIC (POCKELS) EFFECT

Crystal	Symmetry	Wavelength, μm	Electro-optic coefficients r_{ij}, 10^{-12} m/V at constant stress unless noted				Indices of refraction n_o, n_e	Relative dielectric constants ϵ_1, ϵ_3	Refs.
			r_{11}	r_{41}	r_{52}	r_{63}			
CuCl	$\bar{4}3m$	0.548	0	6.5	r_{41}	r_{41}	1.991	7.7*	1–3
ZnS	$\bar{4}3m$	0.546	0	2.0	r_{41}	r_{41}	2.384	8.37	4, 5
ZnTe	$\bar{4}3m$	0.569	0	4.50	r_{41}	r_{41}	3.111	10.1	6, 5
ZnTe	$\bar{4}3m$	10.6	0	1.4	r_{41}	r_{41}	2.7	.....	40
CdTe	$\bar{4}3m$	1.0	0	2.24	r_{41}	r_{41}	2.845	9.65	7, 8, 5
CdTe	$\bar{4}3m$	27.95	0	5.0	r_{41}	r_{41}	2.53	.....	36
GaAs	$\bar{4}3m$	3.39, 10.6*	0	1.6	r_{41}	r_{41}	3.34	11.5	9, 40
Hexamethylenetetramine	$\bar{4}3m$	0.546	0	0.8 ± 0.1	r_{41}	r_{41}	1.591	3.2	11–13a
GaP*	$\bar{4}3m$	0.683	0	−0.97	r_{41}	r_{41}	3.310	8.5	10
KH₂PO₄	$\bar{4}2m$	0.546	0	$+8.7_7$	r_{41}	−10.3†	1.512, 1.470	42, 21	14, 15
KD₂PO₄	$\bar{4}2m$	0.546	0	+8.8	r_{41}	−26.4†	1.508, 1.468	58*, 50	16, 14, 41, 3
KH₂AsO₄	$\bar{4}2m$	0.546	0	+12.5	r_{41}	−10.9	1.571, 1.521	54, 21	14, 17
NH₄H₂PO₄	$\bar{4}2m$	0.546	0	+24.5	r_{41}	−8.5†	1.527, 1.481	56, 15.4	14, 17
HgS	32	0.633	3.1	1.5	$-r_{41}$	0	2.885, 3.232	.....	39
HgS	32	3.39	4.2	2.7	$-r_{41}$	0	2.637, 2.900	.....	39
Quartz	32	0.589	-0.4_7	0.19_5	$-r_{41}$	0	1.544, 1.553	4.6	18, 2, 19
Rochelle salt	222	0.589	0	−2.0	−1.7	+0.32	1.495–1.490	.....	18, 13a

TABLE 6m-5. LINEAR ELECTRO-OPTIC (POCKELS) EFFECT (Continued)

Crystal	Symmetry	Wavelength, μm	r_{22}	r_{33}	r_{13}	r_{23}	r_{12}	$r_{33} - (n_o/n_e)^3 r_{13} = r_2$	Indices of refraction n_o, n_e	Relative dielectric constants ε_1, ε_3	Refs.
Tourmaline	3m	0.589	0.3	......	....	r_{13}	......	......	1.70–1.63	8.2, 7.5	18
LiNbO3	3m	3.39	3.1	......	....	r_{13}	......	16	2.147, 2.081	84, 30	20–22
LiNbO3	3m	0.633	6.7	31.0	9.7	r_{13}	32.6	20.1	2.287, 2.200	84, 30	21–23
LiNbO3*	3m	0.633	3.4	30.8	8.6	r_{13}	28	21.1	2.287, 2.200	44, 29	24, 21, 22
LiTaO3	3m	0.633	~0.1	30	7.0	r_{13}		23.5	2.175, 2.180	51, 45	25, 22
CdS	6mm	0.633	0	2.4*	1.1*	r_{13}	3.7‡	5.0	2.501, 2.519‡	9.35, 10.3	26, 40, 5
Ba0.8Na0.4Nb2O6*	mm2	0.633	0	29.5	7.2	8.2	95	(r_{42} = 79)		......	37
Ba0.8Na0.4Nb2O6	mm2	0.633	0	62	18	13.5	42	(r_{42} N.A.)	2.326–2.221	245, 50	27
Sr0.75Ba0.25Nb2O6	4mm	0.633	0	1,340	67	r_{13}			2.312, 2.299	~3400	25
K1.2Li0.8Nb2O6	4mm	0.633	0	78	9	r_{13}		130	2.277, 2.163	309, 100	28
K0.4Sr0.8Nb2O6	4mm?	0.633	0	....	....	r_{13}		108	~2.25	~800	29
BaTiO3	4mm	0.546	0	....	....	r_{13}	1,640		2.436, 2.365	2920, 168	30–32
BaTiO3	4mm	0.633	0	28*	8*	r_{13}	840*§		2.386, ~2.325	1970, 110*	31–34
K(Ta,Nb)O3[KTN]¶	4mm	0.633	0	....	....	r_{13}	14,000	1,120	2.318, 2.275	......	35

* Values at constant strain.
† r_{63} at constant strain is (from top to bottom) 8.8, 21, and 5.5×10^{-12} m/V.
‡ Measured at 0.589 μm.
§ Measured at 0.546 μm.
¶ Curie point = 60°C. Constants vary sharply with composition.

References for Table 6m-5

1. Walsh, T. E.: Personal communication, September, 1968.
2. Winchell, A. N.: "Microscopic Characters of Artificial Minerals," John Wiley & Sons, New York, 1931.
3. Kaminow, I. P., and E. H. Turner: *Proc. IEEE* **54**, 1374 (1966).
4. Namba, S.: *J. Opt. Soc. Am.* **51**, 76 (1961).
5. Berlincourt, D., H. Jaffe, and L. R. Shiozawa: *Phys. Rev.* **129**, 1009 (1963).
6. Sliker, T. R., and J. M. Jost: *J. Opt. Soc. Am.* **56**, 130 (1966).
7. Stafsudd, O. M., F. A. Haak, and K. Radisavljevic: *Appl. Opt.* **6**, 1276 (1967).
8. Marple, D. T. F.: *J. Appl. Phys.* **35**, 539 (1964).
9. Yariv, A., C. A. Mead, and J. V. Parker: *IEEE J. Quantum Electron.* **QE-2**, 243 (1966).
10. Nelson, D. F., and E. H. Turner: *J. Appl. Phys.* **39**, 3337 (1968).
11. Sliker, T. R.: *J. Opt. Soc. Am.* **54**, 1348 (1964).
12. Buhrer, C. F., and L. Ho: *Appl. Opt.* **3**, 1500 (1964).
13. Heilmeier, G. H.: *Appl. Opt.* **3**, 1281 (1964). (Reports higher r_{41} than 11 and 12.)
13a. Winchell, A. N.: "The Optic Properties of Organic Compounds," Academic Press, Inc., New York, 1954.
14. Ott, J. H., and T. R. Sliker: *J. Opt. Soc. Am.* **54**, 1442 (1964).
15. Carpenter, R. O'B.: *J. Acoust. Soc. Am.* **26**, 1145 (1953).
16. Sliker, T. R., and S. R. Burlage: *J. Appl. Phys.* **34**, 1837 (1963).
17. Berlincourt, D. A., D. R. Curran, and H. Jaffe: *Phys. Acoust.* **1**, 181 (1964).
18. Pockels, F.: *Abhandl. Ges. Wiss. Göttingen Math.-Physik. Kl.* **39**, 1 (1894).
19. Mason, W. P.: "Piezoelectric Crystals and Their Applications to Ultrasonics," D. Van Nostrand Company, Inc., Princeton, N.J., 1950.
20. Smakula, P. H., and P. C. Claspy: *Trans. Met. Soc. AIME* **239**, 421 (1967).
21. Boyd, G. D., R. C. Miller, K. Nassau, W. L. Bond, and A. Savage: *Appl. Phys. Letters* **5**, 234 (1964).
22. Warner, A. W., M. Onoe, and G. A. Coquin: *J. Acoust. Soc. Am.* **42**, 1223 (1967).
23. Zook, J. D., D. Chen, and G. N. Otto: *Appl. Phys. Letters* **11**, 159 (1967).
24. Turner, E. H.: *Appl. Phys. Letters* **8**, 303 (1966).
25. Spencer, E. G., P. V. Lenzo, and A. A. Ballman: *Proc. IEEE* **55**, 2074 (1967).
26. Gainon, D. A.: *J. Opt. Soc. Am.* **54**, 270 (1964).
27. Geusic, J. E., H. J. Levinstein, J. J. Rubin, S. Singh, and L. G. Van Uitert: *Appl. Phys. Letters* **11**, 269 (1967).
28. Van Uitert, L G., S. Singh, H. J. Levinstein, J. E. Geusic, and W. A. Bonner: *Appl. Phys. Letters* **11**, 161 (1967).
29. Giess, E. A., G. Burns, D. F. O'Kane, and A. W. Smith: *Appl. Phys. Letters* **11**, 233 (1967).
30. Johnson, A. R., and J. M. Weingart: *J. Opt. Soc. Am.* **55**, 828 (1965).
31. Shumate, M. S.: *Appl. Phys. Letters* **5**, 178 (1964).
32. Berlincourt, D. A., and H. Jaffe: *Phys. Rev.* **111**, 143 (1958).
33. Kaminow, I. P.: *Appl. Phys. Letters* **8**, 54, 305 (1966).
34. Johnston, A. R.: *Appl. Phys. Letters* **7**, 195 (1965).
35. van Raalte, J. A.: *J. Opt. Soc. Am.* **57**, 671 (1967).
36. Johnson, C. J.: *Proc. IEEE* **56**, 1719 (1968).
37. Turner, E. H.: Personal communication, November, 1968.
38. Turner, E. H., I. P. Kaminow, and E. D. Kolb: *IEEE J. Quantum Electron.* **QE-4**, 234 (1968).
39. Turner, E. H.: *IEEE J. Quantum Electron.* **QE-3**, 695 (1967).
40. Kaminow, I. P.: *IEEE J. Quantum Electron.* **QE-4**, 23 (1968).
41. Phillips, R. A.: *J. Opt. Soc. Am.* **56**, 629 (1966).

6m-6. Electro-optic Effects in Ferroelectric Crystals. Ferroelectric crystals with high dielectric constants show higher-order electro-optic effects of significant magnitude. In oxide ferroelectrics, it has been found that these effects can be adequately described as quadratic (Kerr effect) if electric polarization P, instead of field E, is taken as the independent variable. Since the products of two electric polarization components form a symmetric tensor, the quadratic electro-optic coefficients g_{ik} have the same symmetry relation as the elasto-optic coefficients, and we have

$$\Delta a_i = g_{ij}(P^2)_j \qquad (6m\text{-}9)$$

where the index $j - 1, \ldots, 6$ gives the usual contracted tensor components.   For the cubic class $m3m$, there are three independent coefficients g_{11}, g_{12}, g_{44}.   Above the Curie point (excluding its immediate vicinity), P is a linear function of E.   In rationalized units, for cubic crystals

$$P = (\epsilon - 1)\epsilon_0 E \approx \epsilon\epsilon_0 E \qquad \text{(6m-10)}$$

the dielectric constant ϵ follows the law

$$\epsilon = \frac{C}{T - T_c} \qquad \text{(6m-11)}$$

where C is the Curie constant, and T_c the Curie temperature.   Equations (6m-9) to (6m-11) combine to

$$\Delta a_i = \frac{g_{ij} C^2 \epsilon_0^2 E^2}{(T - T_c)^2} \qquad \text{(6m-12)}$$

The conventional Kerr constants are obtained by

$$K_{[100]} = \frac{2n^3(g_{11} - g_{12})\epsilon^2}{\lambda} \qquad \text{(6m-13)}$$

or

$$K_{[110]} = \frac{2n^3 g_{44} \epsilon^2}{\lambda} \qquad \text{(6m-14)}$$

for electric field in the crystallographic [100] and [110] directions, respectively.   The values for crystals which are not cubic above the Curie point refer to quasi-cubic axes [4].   The g_{ik} show little temperature dependence, and apply also below the Curie point if the resultant of spontaneous polarization P_s and polarization induced by an applied field is inserted into Eq. (6m-9).   The relation to the linear Pockels coefficients for an applied field small compared to the spontaneous polarization is of the form $r_{33} = 2\epsilon_3\epsilon_0 g_{11} P_s$.

TABLE 6m-6.   KERR CONSTANTS OF FERROELECTRIC CRYSTALS

Crystal	Symmetry	Wave-length, μm	Kerr constants, m^4/C^2				n	Curie temperature, K	Curie constant, K	Refs.
			$g_{11}-g_{12}$	g_{11}	g_{12}	g_{44}				
BaTiO$_3$	$m3m$, $4mm$	0 633	+0.13	......	......	......	2.29	370	170,000	1, 2
SrTiO$_3$	$m3m$	0.633	+0.14	......	......	......	2.38	....	......	1
KTa$_{0.65}$Nb$_{0.35}$O$_3$	$m3m$	0.623	+0.174	+0.136	−0.038	+0.147	2.29	271	145,000	1, 3
KTaO$_3$	$m3m$	0.623	+0.16	......	......	+0.12	2.24	~4	......	1, 3
LiNbO$_3$	$3m$	0.633?	0.07	0.00$_4$	0.02$_5$	0.06$_4$	2.291-2.201	1468	......	4, 6
LiTaO$_3$	$3m$	0.633?	0.08	0.10	0.01$_7$	0.07	2.175,2.180	883	......	4, 5
Ba$_{0.8}$Na$_{0.4}$Nb$_2$O$_6$	$mm2$	......	0.110	0.155	0.044	......	2.326-2.221	833	......	4

References for Table 6m-6

1. Geusic, J. E., S. K. Kurtz, L. G. Van Uitert, and S. H. Wemple: *Appl. Phys. Letters* **4**, 141 (1964).
2. Nakamura, E., T. Mitsui, and J. Furuichi, *J. Phys. Soc. Japan* **18**, 1477 (1963).
3. Chen, F. S., J. E. Geusic, S. K. Kurtz, J. G. Skinner, and S. H. Wemple: *J. Appl. Phys.* **37**, 388 (1966).
4. Wemple, S. H., M. Di Domenico Jr., and I. Camlibel: *Appl. Phys. Letters* **12**, 209 (1968).
5. Spencer, E. G., P. V. Lenzo, and A. A. Ballman, *Proc. IEEE* **55**, 2074 (1967).
6. Boyd, G. D., W. L. Bond, and H. L. Carter: *J. Appl. Phys.* **38**, 1941 (1967).

6n. Nonlinear Optical Coefficients

F. ZERNIKE

The Perkin-Elmer Corporation

For frequency mixing in the optical region one normally uses the fact that the polarization of a material is a nonlinear function of the electric field:

$$P = xE(1 + a_1E + a_2E^2 + \cdots)$$

In the limit of small electric fields this reduces to $P = xE$ the term which is responsible for the linear refractive index.

Although second-harmonic generation and other mixing experiments were first done using lasers, it is not necessary that the light be coherent. Indeed, other effects such as the Kerr, Pockels, and Raman effects, are manifestations of the same nonlinearity and have been well known for some time.

The effects treated here are those due to the second-order nonlinear polarization $P^{NL} = xa_1E^2$. They occur in acentric materials only. Taking account of the fact that nonlinear polarizability is a tensor, the nonlinear polarization is written as[1] $P_i{}^{NL} = d_{ijk}E_jE_k$.

Assuming that the two interacting fields are sinusoidal traveling waves with frequencies ω_1 and ω_2 and wavevectors k_1 and k_2, application of basic trigonometry shows that the nonlinear polarization in general has five components: one d-c component and four components with frequencies and corresponding wave-vectors:

$$\omega_1 + \omega_2, \mathbf{k}_1 + \mathbf{k}_2 \qquad \omega_1 - \omega_2, \mathbf{k}_1 - \mathbf{k}_2$$
$$2\omega_1, 2\mathbf{k}_1 \qquad 2\omega_2, 2\mathbf{k}_2 \qquad \text{respectively}$$

By considering the nonlinear polarization as a perturbation to the linear source term in Maxwell's equation, it can be shown that each frequency component of the nonlinear polarization generates an electromagnetic wave with the same frequency but 90 deg out of phase [1]. Thus, if $\omega_1 = \omega_2$, the nonlinearity will generate a d-c electric field (optical rectification) and a wave at twice the frequency of the input (usually called the second harmonic). Similarly if either ω_1 or ω_2 is zero, the effect is the linear electro-optic effect (see Sec. 6m-5). The nonlinear polarizability in this case is related to the normally used electro-optical coefficient as $d_{ijk}(\omega,0,\omega) = (n^4/4\pi)r_{ijk}$, where n is the refractive index at frequency ω.

In a matter analogous to the one in which Fresnel's equations are derived in the linear optics case, it can be shown that the nonlinear source term also generates a reflected component at the mixed frequency [2].

6n-1. Phase Matching. The interaction is said to be phase-matched if the wave vectors of the polarization wave and the accompanying electromagnetic wave are equal. In this case both energy and momentum are conserved: $\omega_3 = \omega_1 + \omega_2$,

---

[1] Here the dummy suffix notation is adopted. Summation is implied whenever suffices are repeated on one side of an equation.

$k_3 = k_1 + k_2$, and efficient energy transfer occurs. In general, because of the dispersion in the mixing crystal, one had

$$k_3 = {}_1k + k_2 + \Delta k \tag{6n-1}$$

In a small-signal approximation, i.e., no significant depletion of the waves at ω_1 and ω_2, it can be shown that the signal at ω_3 depends on Δk and on l, the length of the crystal, as

$$S(\omega_3) \sim \left[\frac{l \sin (\Delta kl/2)}{\Delta kl/2} \right]^2$$

This function has a maximum for $\Delta kl = \pi$. The crystal length for which this maximum occurs is called the coherence length:

$$l_{\text{coh}} = \frac{\pi}{\Delta k}$$

An often-employed method of phase matching [3] utilizes the fact that in a uniaxial birefringent crystal the index of refraction for an extraordinary ray, n^{ext}, can be made to vary between the extraordinary index n_e and the ordinary index n_0, by varying the angle θ between the wave normal and the optical axis:

$$n^{\text{ext}} = \frac{n_o n_e}{(n_e{}^2 \sin^2 \theta + n_o{}^2 \cos^2 \theta)^{\frac{1}{2}}}$$

In this method all three waves are propagated with parallel-wave normals. This reduces Eq. (6n-1) to the form

$$\omega_3 n_3 = \omega_1 n_1 + \omega_2 n_2 + \omega_3 \Delta n$$

Now one or two of the waves are polarized in a plane parallel to the optical axis (extraordinary polarization), and the remaining one(s) are polarized as ordinary rays. By choosing the correct value of θ the refractive indices of the extraordinary rays are adjusted to give $\Delta n = 0$. In this arrangement the direction of the extraordinary ray is not parallel to its wave normal, unless $\theta = 90$ deg, thus causing the length in which all three waves overlap to be finite. For a more detailed treatment, including the effects of crystal symmetry, see Midwinter and Warner [4].

In a modification of this method all three waves are propagated in a direction perpendicular to the optical axis, and the temperature of the crystal is varied to adjust the refractive indices to the values necessary to give $\Delta n = 0$. This method is often referred to as temperature tuning. It has the advantage that the ray directions of all the waves remain parallel [5]. Also, the variation of index with angle of propagation is smallest when $\theta = 90$ deg, allowing for sharper focusing of the beams. A d-c electric field can be applied to "fine tune" the phase-matching condition, using the electro-optic effect [6].

Phase matching in biaxial crystal has been treated by Hobden [7].

6n-2. Symmetry and Contraction of d_{ijk}. Armstrong [8] et al. have shown that $d_{ijk}(\omega_1, \omega_2, \omega_3) = d_{kji} (\omega_3, \omega_2, \omega_1) = d_{jik}(\omega_2, \omega_1, \omega_3)$. This reduces the number of independent coefficients from 81 to 27. It also shows that the last two indices are interchangeable: $d_{ijk} = d_{ikj}$. It is therefore possible to write the tensor in a contracted form: d_{il}, with l running from 1 to 6. Now $d_{il} = d_{ijk}$ for $k = j$ and $d_{il} = \frac{1}{2}(d_{ijk} + d_{ikj})$ for $k \neq j$. Equation (6n-2) shows the normally used, contracted matrix with the column matrix on which it operates.

$$\begin{pmatrix} P_x \\ P_y \\ P_z \end{pmatrix} = \begin{pmatrix} d_{11} & d_{12} & d_{13} & d_{14} & d_{15} & d_{16} \\ d_{21} & d_{22} & d_{23} & d_{24} & d_{25} & d_{26} \\ d_{31} & d_{32} & d_{33} & d_{34} & d_{35} & d_{36} \end{pmatrix} \begin{Bmatrix} E_x{}^2 \\ E_y{}^2 \\ E_z{}^2 \\ 2E_y E_z \\ 2E_x E_z \\ 2E_x E_y \end{Bmatrix} \tag{6n-2}$$

Note that the form of the matrix is very much the same as the form of the piezo-electric matrix, except that here the 4, 5, 6 columns operate on $2E_yE_z$, etc. In other words, the usual factor of 2 is included in the field upon which the matrix operates, instead of in the matrix itself. This definition is not uniformly accepted in the literature, and caution should be exercised.

6n-3. Symmetry of d_{ik}, Kleinman's Conjecture. The nonlinearity of the polarization must be invariant to those symmetry operations which transform the crystal into itself. For a specific crystal class the matrix of the second-order susceptibility is homologous with the piezoelectric matrix, except for the factor 2 mentioned above. The values of the matrix elements are of course not related to those of the piezoelectric matrix.

For second-harmonic generation Kleinman has suggested that in nondispersive nonabsorbing media a second symmetry condition obtains [9]. This condition allows all three indices i, j, and k in the coefficient d_{ijk} to be freely interchanged. In the absence of any other symmetry conditions the number of independent coefficients is then reduced to ten:

$$
\begin{array}{cccccc}
d_{11} & d_{12} & d_{13} & d_{14} & d_{15} & d_{16} \\
d_{16} & d_{22} & d_{23} & d_{24} & d_{14} & d_{12} \\
d_{15} & d_{24} & d_{33} & d_{23} & d_{13} & d_{14}
\end{array}
$$

Combined with the symmetry conditions which govern the piezoelectric tensor, this condition reduces the number of independent coefficients even further than in the piezoelectric matrix. For example, in quartz the normal symmetry conditions give two independent coefficients, $d_{11} = -d_{12} = -d_{26}$, and $d_{14} = -d_{25}$, with all the other coefficients $\equiv 0$. Kleinman's condition requires $d_{14} = d_{25}$, and so $d_{14} \equiv 0$.

The allowed values of the nonlinear optical coefficients for the crystal classes listed in Table 6n-1 are:

Crystal Class	Coefficients
4mm, 6mm	$d_{31} = d_{32}$; d_{33}; $d_{24} = d_{15}$
3m	$d_{21} = -d_{22} = d_{16}$; $d_{24} = d_{15}$; $d_{31} = d_{32}$; d_{36}
mm2	d_{31}; d_{32}; d_{33}; d_{24}; d_{15}
42m	$d_{14} = d_{25}$; d_{36}
43m	$d_{14} = d_{25} = d_{36}$
222	d_{14}; d_{25}; d_{36}
32	$d_{11} = -d_{12} = -d_{26}$; $d_{14} = -d_{25}$

6n-4. Output Power. In a small-signal approximation the output power is given by [10]

$$
S(\omega_3) = S(\omega_1)S(\omega_2) \frac{32\pi^3\omega_3{}^2 l^2}{c^3 n_1 n_2 n_3} d_{\text{eff}}{}^2 \frac{\sin \Delta kl/2^2}{\Delta kl/2} \tag{6n-3}
$$

Here $S(\omega_n)$ is the power at the frequency ω_n in ergs cm^{-2}, d_{eff} is the nonlinear coefficient multiplied by the terms introduced by the matrix because of the direction of propagation in the medium, l is the length of the crystal, and n_i is the index at the frequency ω_i.

Since d_{eff} is most often given in esu, but intensities are usually expressed in watts cm^{-2}, a more convenient equation is [11]

$$
P(\omega_3) = \frac{13.04 \, P(\omega_1)P(\omega_2)d_{\text{eff}}{}^2 l^2}{n_1 n_2 n_3 \lambda_3{}^2} \left(\frac{\sin \Delta kl/2}{\Delta kl/2} \right)^2 \tag{6n-4}
$$

where l is the crystal length, and λ the free-space wavelength, both in cm. $P(\omega)$ is in watts cm^{-2}, and d_{eff} is in esu. Equations (6n-3) and (6n-4) are for single-mode inputs.

In esu the dimension of the nonlinear coefficient is cm/stat. volt. In mks units this becomes meters/volt. The conversion from esu to mks units is given by $d(\text{mks}) =$

TABLE 6n-1. NONLINEAR OPTICAL COEFFICIENTS

Material	Class	λ	Coefficients in 10^{-9} esu			Ref.	Index data
			d_{15}	d_{31}	d_{33}		
BaTiO₃.........	$4mm$	1.06	57 ± 19	60 ± 20	22 ± 7	13	
K₆LiNbO₃......	$4mm$	1.06	....	15 ± 2	20 ± 2	15	15
ZnO...........	$6mm$	1.06	7.6 ± 2.5	7 ± 2	16 ± 7	13	
CdS...........	$6mm$	1.06	57 ± 19	53 ± 9	102 ± 32	13	
		10.6	69 ± 17	63 ± 15	105 ± 30	16	
ZnS...........	$6mm$	1.06	....	......	25 ± 8	17	
		10.6	51 ± 20	45 ± 15	89 ± 30	16	
CdSe..........	$6mm$	1.06	....	......	147 ± 49	17	
		10.6	74 ± 18	68 ± 15	130 ± 30	16	
			d_{31}	d_{22}	d_{33}		
LiNbO₃*........	$3m$	1.06	22 ± 2	9 ± 5	145 ± 27	18, 19, 36	20, 18
Ag₃AsS₃ (proustite)..........	$3m$	1.15	48 ± 12	80 ± 20	....	21	21
			d_{15}	d_{24}	d_{33}		
Ba₂NaNb₅O₁₅...	$mm2$	1.06	25 ± 8	22 ± 2	31 ± 2	22	22

Material	Class	λ	Coefficients in 10^{-9} esu		Ref.	Index data
			d_{14}	d_{36}		
KH₂PO₄(KDP).......	$\bar{4}2m$	1.06	1.6 ± 0.5	....	13	23
		1.15	....	1.6 ± 0.4	35	
KD₂PO₄..............	$\bar{4}2m$	1.06	1.5 ± 0.4	1.5 ± 0.4	13	
NH₄H₂PO₄...........	$\bar{4}2m$	1.06	1.36 ± 0.16	....	13	23
		0.6328	....	1.36 ± 0.16	24	
KH₂AsO₄.............	$\bar{4}2m$	1.06	1.8 ± 0.5	1.7 ± 0.5	13	
InAs................	$\bar{4}3m$	10.6	1000 ± 300	....	16	
CdTe...............	$\bar{4}3m$	10.6	400 ± 150	....	16	31
ZnS................	$\bar{4}3m$	1.06	....	80 ± 24	17	
		10.6	73 ± 20	....	16	
ZnSe...............	$\bar{4}3m$	1.06	....	105 ± 32	17	31
		10.6	187 ± 70	....	16	
ZnTe...............	$\bar{4}3m$	1.06	....	353 ± 111	17	31
		10.6	220 ± 80	....	16	
GaP................	$\bar{4}3m$	1.06	238 ± 40	136 ± 40	13, 17	
GaAs...............	$\bar{4}3m$	1.06	760 ± 190	827 ± 240	13, 17	32
		10.6	880 ± 300			
N₄(CH₂)₆............	$\bar{4}3m$	1.06	16 ± 5	....	25	
			d_{14}	d_{36}		
Ammonium oxalate....	222	0.6943	1.25	....	26	
C₆H₅COHN—CH₂CO₂H......... (hippuric acid)	222	0.6943	6.8	....	27	
HIO₃ (α-iodic acid)....	222	1.15	15 ± 6		35	28
			d_{11}			
SiO₂ (quartz).........	32	1.06	1.33 ± 0.4	....	13	33
Se..................	32	10.6	190 ± 100	....	16	
Te..................	32	10.6	12,700	....	29	29
AlPO₄...............	32	1.06	1.37 ± 0.42		13	
HgS (cinnabar)......	32	10.6	150 ± 50	....	30	30
K₂S₂O₆..............	32	0.6943	0.50	....	34	34

*The values of d_{31} and d_{22} in LiNbO₃ have opposite signs.

$d(\text{esu})/3 \times 10^4$. It is sometimes given as $d(\text{mks}) = 4\pi d(\text{esu})/3 \times 10^4$. In the latter case the $d(\text{mks})$ is the nonlinear susceptibility, and the nonlinear polarization is given by

$$P = \epsilon_0 dEE$$

where ϵ_0 is the permittivity of free space.

Some authors include ϵ_0 in the coefficient. Then the conversion from esu to mks units becomes

$$d(\text{mks}) = 3.68 \times 10^{-15} d(\text{esu})$$

6n-5. Coefficients for Second-harmonic Generation. A number of nonlinear coefficients for second-harmonic generation are listed in Table 6n-1. In all cases these have been measured in experiments generating the second harmonic of the wavelength listed. Most of the reported measurements were made relative to a "known" crystal. The values given in Table 6n-1 are absolute values. They were all calculated from these relative measurements, using the listed coefficient d_{36} for ADP. The reference for each crystal is given in the first reference column. The second reference column gives available index-of-refraction data.

In selecting a crystal for a particular application, it should be borne in mind that a large nonlinear coefficient is not the only requirement for efficient generation. It should also be possible to grow crystals of optical quality to the required size, and the material should be transparent at all frequencies involved. Another danger is that the crystal may suffer optical damage from the large incident intensities. This type of damage was first observed in $LiNbO_3$ [12].

6n-6. Miller's Rule. Miller has found empirically that if $d_{ijk}^{\omega_1 \omega_2 \omega_3}$ is written as

$$d_{ijk}^{\omega_1 \omega_2 \omega_3} = x_{ii}^{\omega_1} x_{jj}^{\omega_2} x_{kk}^{\omega_3} \Delta_{ijk}$$

where $x_{ii}^{\omega_1}$ is the ii component of the linear optical susceptibility at frequency ω_1, then the allowed components of Δ_{ijk} for all materials have the same magnitude [13,14]. This provides a helpful pointer to good new materials.

6n-7. Material Evaluation. A useful technique for evaluating materials in powder form has been developed by Kurtz and Perry [37].

References

1. Minck, R. W., R. W. Terhune, and C. C. Wang: *Proc. IEEE* **54**, 1357 (1966). Rather than list the multitude of papers on this effect we list this one review paper. References to other papers will be found in it.
2. Bloembergen, N., and P. S. Pershan: *Phys. Rev.* **128**, 606 (1962).
3. Giordmaine, J. A.: *Phys. Rev. Letters* **8**, 19 (1962); P. D. Maker, R. W. Terhune, M. Nisenoff, and C. M. Savage: *ibid.*, 21.
4. Midwinter, J. E., and J. Warner: *Brit. J. Appl. Phys.* **16**, 1135 (1965).
5. Miller, R. C., G. D. Boyd, and A. Savage: *Appl. Phys. Letters* **6**, 77 (1965).
6. Adams, N. I., III, and J. J. Barrett: *IEEE J. Quantum Electron.* **QE-2**, 21 (1966).
7. Hobden, M. V.: *J. Appl. Phys.*: **38**, 4365 (1967).
8. Armstrong, J. A., N. Bloembergen, J. Ducuing, and P. S. Pershan: *Phys. Rev.* **127**, 1918 (1962).
9. Kleinman, D. A.: *Phys. Rev.* **126**, 1977 (1962).
10. Kleinman, D. A.: *Phys. Rev.* **128**, 1761 (1962).
11. Midwinter, J. E., and Frits Zernike: *IEEE J. Quantum Electron.* **QE-5**, (February, 1969).
12. Ashkin, A., G. D. Boyd, J. M. Dziedzic, R. G. Smith, A. A. Ballman, J. J. Levinstein and K. Nassau: *Appl. Phys. Letters* **9**, 72 (1966).
13. Miller, R. C.: *Appl. Phys. Letters* **5**, 17 (1964).
14. Robinson, F. N. H.: *Bell System Tech. J.* **46**, 913, 1967.
15. van Uitert, L. G., S. Singh, M. J. Levinstein, J. E. Geusic and W. A. Bonner: *Appl. Phys. Letters* **11**, 161, 1967.
16. Patel, C. K. N.: *Phys. Rev. Letters* **16**, 613 (1966).
17. Soreff, R. A., and H. W. Moos, *J. Appl. Phys.* **35**, 2152, 1964.

18. Boyd, G. D., Robert C. Miller, K. Nassau, W. L. Bond, and A. Savage: *Appl. Phys. Letters* **5**, 234 (1964).
19. Bjorkholm, J. E.: *Appl. Phys. Letters* **13**, 36 (1968).
20. Fay, Homer, W. J. Alford, and H. M. Dess: *Appl. Phys. Letters* **12**, 89 (1968); G. D. Boyd, W. L. Bond, and H. L. Carter: *J. Appl. Phys.* **38**, 1941, (1967); M. V. Hobden and J. Warner: *Phys. Letters* **22**, 243 (1966).
21. Hulme, K. F., O. Jones, P. H. Davies, and M. V. Hobden: *Appl. Phys. Letters* **10**, 133 (1967).
22. Geusic, J. E., H. J. Levinstein, S. Singh, R. G. Smith, and L. G. van Uitert: *Appl. Phys. Letters* **12**, 306 (1968).
23. Zernike, F., Jr.: *J. Opt. Soc. Am.* **54**, 1215, 1964.
24. Francois, G. E.: *Phys. Rev.* **143**, 597 (1966).
25. Heilmeier, G. H., N. Ockman, R. Braunstein, and D. A. Kramer: *Appl. Phys. Letters* **5**, 229 (1964).
26. Izrailenko, A. N., R. Yu Orlov, and V. A. Kopsik: *Kristallografiya* **13**, 171 (1968); translation, *Soviet Phys.—Cryst.* **13**, 136 (1968).
27. Orlov, R. Yu: *Kristallografiya* **11**, 410 (1966); translation, *Soviet Phys.—Cryst.* **11**, 410 (1966).
28. Kurtz, S. K., T. T. Perry, and J. G. Bergman, Jr.: *Appl. Phys. Letters* **12**, 186 (1968).
29. Patel, C. K. N.: *Phys. Rev. Letters* **15**, 1027 (1965).
30. Jerphagnon, J., E. Batifol, G. Tsoucaris, and M. Sourbe: *Compt. Rend.* **265B**, 495 (1967).
31. Marple, D. T. F.: *J. Appl. Phys.* **35**, 539 (1964).
32. Marple, D. T. F.: *J. Appl. Phys.* **35**, 1241 (1964).
33. Radakrishnan, T.: *Proc. Indian Acad. Sci*, sec. A, **25**, 260, 1947.
34. Hobden, M. V., D. S. Robertson, P. H. Davies, K. F. Hulme, J. Warner, and J. Midwinter: *Phys. Letters* **22**, 65 (1966).
35. Bjorkholm, J. E.: *IEEE J. Quantum Electron.* **QE-4**, 970 (1968).
36. Byer, R. L., and S. E. Harris: *Phys. Rev.* **168**, 1065 (1968).
37. Kurtz, S. K.: *IEEE J. Quantum Electron.* **QE-4**, 578 (1968).

60. Specific Rotation

Table 60-1. SPECIFIC ROTATION*
Solids

Substance	Wave-length, μm	Rota-tion, deg/min	Substance	Wave-length, μm	Rota-tion, deg/min
Cinnabar (HgS)	D	+32.5	Quartz	0.3726	+58.894
Lead hyposulfate	D	5.5		0.3609	63.628
Potassium hyposulfate	D	8.4		0.3582	64.459
Quartz	3.676	0.34		0.3466	69.454
	1.342	3.89		0.3441	70.587
	0.7604	12.668		0.3402	72.448
	0.7184	14.304		0.3360	74.571
	0.6867	15.746		0.3286	78.579
	0.6562	17.318		0.3247	80.459
	0.5895932	21.7010		0.3180	84.972
	0.5895	21.684		0.2747	121.052
	0.5892617	21.729		0.2571	143.266
	0.5889965	21.7492		0.2313	190.426
	0.5889	21.727		0.2265	201.824
	0.5460741	25.538		0.2194	220.731
	0.5269	27.543		0.21740	229.96
	0.4861	32.773		0.2143	235.972
	0.4307	42.604		0.1750	453.5
	0.4101	47.481		0.1525	776.0
	0.3968	51.193	Sodium bromate	D	2.8
	0.3933	52.155	Sodium chlorate	D	3.13
	0.3820	55.625			

Specific rotation or rotatory power is given in degrees per decimeter for liquids and solutions and in degrees per millimeter for solids; + signifies right-handed rotation, − left. Specific rotation varies with the wavelength of light used, with temperature and, in the case of solutions, with the concentration. When sodium light is used, indicated by D in the wavelength column, a value of $\lambda = 0.5893$ may be assumed.

Optical rotatory power for a large number of organic compounds will be found in the "International Critical Tables," vol. VII; for sugars, vol. II.

*Most of the data taken from "Handbook of Chemistry and Physics," 36th ed., pp. 2752, 2753, 2754, Chemical Rubber Publishing Company, 1954–1955.

TABLE 6o-1. SPECIFIC ROTATION (*Continued*)
Liquid

Liquid	Temp., °C	Wave-length, μm	Specific rotation, deg/dm
Amyl alcohol........................	......	D	$-$ 5.7
Camphor............................	204	D	$+$ 70.33
Cedar oil...........................	15	D	$-$ 30 to -40
Citron oil..........................	15	D	$+$ 62
Ethyl malate $(C_2H_5)_2C_4H_4O_5$..........	11	D	$-$ 10.3 to -12.4
Menthol............................	35.2	D	-49.7
Nicotine $C_{10}H_{14}N_2$.................	10–30	D	-162
	20	0.6563	-126
	20	0.5351	-207.5
	20	0.4861	-253.5
Turpentine $C_{10}H_6$..................	20	D	$-$ 37
	20	0.6563	$-$ 29.5
	20	0.5351	$-$ 45
	20	0.4861	$-$ 54.5

Specific rotation or rotatory power is given in degrees per decimeter for liquids and solutions and in degrees per millimeter for solids; + signifies right-handed rotation, − left. Specific rotation varies with the wavelength of light used, with temperature and, in the case of solutions, with the concentration. When sodium light is used, indicated by D in the wavelength column, a value of λ = 0.5893 may be assumed.

Optical rotatory power for a large number of organic compounds will be found in the "International Critical Tables," vol. VII; for sugars, vol. II.

* Most of the data taken from "Handbook of Chemistry and Physics," 36th ed., pp. 2752, 2753, 2754, Chemical Rubber Publishing Company, 1954–1955.

TABLE 6o-1. SPECIFIC ROTATION* (Continued)
Solutions†

Substance	Solvent	Temp., °C	Wave-length, μm	Specific rotation, deg/dm	Correction for concentration or temp.
Albumen.............	Water	...	D	− 25 to −38	
Arabinose...........	Water	20	D	− 105.0	
Camphor............	Alcohol	20	D	+ 54.4 − 0.135d for	
					d = 45–91
	Benzene	20	D	+ 56 − 0.166d for	
					d = 47–90
	Ether	...	D	+ 57	
Dextrose d-glucose	Water	20	D	+ 52.5 + 0.025d for	
$C_6H_{12}O_6$					d = 1–18
			0.5461	+ 62.03 + 0.04257c for	
					c = 6–32
Galactose............	Water	...	D	+ 83.9 + 0.078d −	
				0.21t for d = 4–36 and	
					t = 10–30°C
l-Glucose (β).........	Water	20	D	− 51.4	
Invert sugar $C_6H_{12}O_6$...	Water	20	D	− 19.7 − 0.036c for	
					c = 9–35
				$\alpha_t = \alpha_{20} + 0.304(t - 20)$	
				+ 0.00165	
				$(t - 20)^2$ for t = 3–30°C	
		25	0.5461	− 21.5	
Lactose.............	Water	20	D	+ 52.4 + 0.072	
				(20° − t) for c = 5	
			0.5461	+ 61.9 + 0.085	
				(20° − t) for c = 5	
Levulose fruit sugar....	Water	25	D	− 88.5 − 0.145d for	
				d = 2.6–18.6	
		25	0.5461	− 105.30	
Maltose.............	Water	20	D	+ 138.48 − 0.01837d for	
				d = 5–35	
		25	0.5461	+ 153.75	
Mannose............	Water	20	D	+ 14.1 c = 10.2	
Nicotine............	Water	20	D	− 77 for d = 1–16	
	Benzene	20	D	− 164 for d = 8–100	
Potassium tartrate.....	Water	20	D	+ 27.14 + 0.0992c −	
				0.00094c² for c = 8–50	
Quinine sulfate........	Water	17	D	− 214	
Santonin.............	Alcohol	20	D	− 161.0 c = 1.78	
		20	D	+ 693 c = 4.05	
	Chloroform	20	D	− 202.7 + 0.309d for	
				d = 75–96.5	
	Alcohol	20	0.6867	+ 442 c = 4.05	
			0.5269	+ 991 c = 4.05	
			0.4861	+1,323 c = 4.05	

† Corrections for values of the specific rotation for concentration are given in the last column. c indicates concentration in grams per 100 ml of solution; d indicates the concentration in grams per 100 g of solution.

TABLE 6o-1. SPECIFIC ROTATION (*Continued*)
Solutions†

Substance	Solvent	Temp., °C	Wave-length, μm	Specific rotation, deg/dm	Correction for concentration or temp.
Sodium potassium tar-trate (rochelle salt)	Water	20	D	+ 29.75 − 0.0078c	
Sucrose (cane sugar) $C_{12}H_{22}O_{11}$	Water	20	D	+ 66.412 + 0.01267d − 0.000376d² for d = 0–50 $\alpha_t = \alpha_{20}[1 − 0.00037 (t − 20)]$ for t = 14–30°C	

Solutions†

Sucrose dissolved in water, 20°C

μm	Spec. rot.	μm	Spec. rot.	μm	Spec. rot.
670.8 (Li)	+50.51	510.6 (Cu)	+90.46	435.3 (Fe)	+128.5
643.8 (Cd)	55.04	508.6 (Cd)	91.16	433.7 (Fe)	129.8
636.2 (Zn)	56.51	481.1 (Zn)	103.07	431.5 (Fe)	130.7
589.3 (Na)	66.45	480.0 (Cd)	103.62	428.2 (Fe)	133.6
578.2 (Cu)	69.10	472.2 (Zn)	107.38	427.2 (Fe)	134.2
578.0 (Hg)	69.22	468.0 (Zn)	109.49	426.1 (Fe)	134.9
570.0 (Cu)	71.24	467.8 (Cd)	109.69	419.1 (Fe)	140.0
546.1 (Hg)	78.16	438.4 (Fe)	126.5	414.4 (Fe)	144.2
521.8 (Cu)	86.21	437.6 (Fe)	127.2	388.9 (Fe)	166.7
515.3 (Cu)	88.68	435.8 (Hg)	128.49	383.3 (Fe)	171.8
				382.6 (Fe)	173.1

Solutions†

Substance	Solvent	°C	μm	Spec. rot.	Correction
Tartaric acid (ord.).....	Water	20	D	+15.06 − 0.131c	
		20	0.6563	7.75	
		20	D	8.86	for d = 41
		20	0.5351	9.65	
		20	0.4861	9.37	
Turpentine............	Alcohol	20	D	−37 − 0.00482d − 0.00013d² for d = 0–90	
	Benzene	20	D	−37 − 0.0265d for d = 0–91	
Xylose..............	Water	20	D	+19.13	d = 2.7

† Corrections for values of the specific rotation for concentration are given in the last column. c indicates concentration in grams per 100 ml of solution; d indicates the concentration in grams per 100 g of solution.

6p. Radiation Detection

R. M. BURLEY

Baird-Atomic, Inc.

6p-1. General. Radiation detectors can be classed as either thermal detectors or quantum detectors. In the former the radiation is absorbed and transformed into heat in the detector, producing a temperature rise in the device. Some characteristic of the detector changes as a function of temperature, and this characteristic can be measured to determine the quantity of radiation striking the detector. In this type of receiver, then, the quantity actually measured is the temperature change. In the quantum detector, on the other hand, the incident photons change the detector characteristic directly.

There can be as many thermal detectors as there are material characteristics which change with temperature. Table 6p-1 lists some of the fundamental processes that are used in thermal detectors.

TABLE 6p-1. THERMAL DETECTORS

Device	*Measured characteristic*
Bolometer...................	Change of electrical resistance with temperature
Thermocouple...............	Peltier effect or change of contact potential at a junction as a function of temperature
Pneumatic detector...........	Change of gas pressure in an enclosed chamber as a function of temperature

Various kinds of quantum detectors are mentioned and described briefly in Table 6p-2.

TABLE 6p-2. QUANTUM DETECTORS

Device	*Measured characteristic*
Photoelectric cell.............	The emission of an electron from a surface when struck by sufficiently energetic photons
Photoconductor cell..........	The resistance of the cell changes directly as a result of photon absorption
Photovoltaic cell.............	A voltage is generated directly as a result of the absorption of a photon
Photographic plate...........	A silver halide is reduced to silver by photon absorption

Both types—thermal and quantum detectors—are manufactured in single elements, multiple elements, and image detectors. The important characteristics of radiation detectors are:

Spectral Response. The reciprocal of the amount of monochromatic incident radiant power that it takes to produce a given detector output response (e.g., electrical signal or photographic density), plotted against radiation wavelength. For thermal detectors, the response is generally independent of wavelength over a range from the ultraviolet to wavelengths which approach the dimensions of the detector. Quantum detectors show a long-wavelength cutoff, related to the energy gap of the

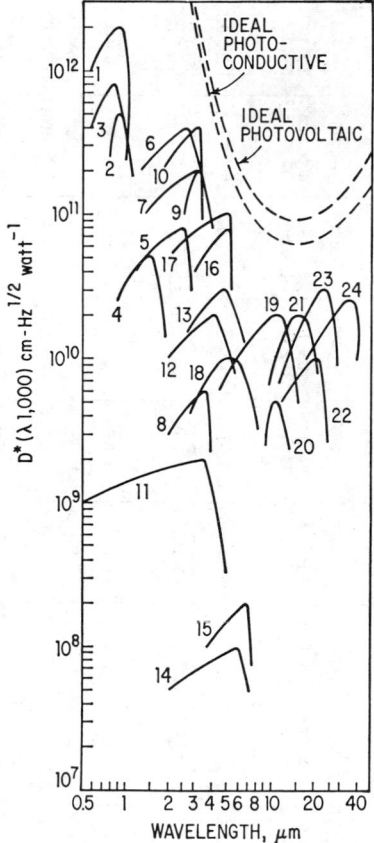

FIG. 6p-1. Characteristics of commercially available semiconductor detectors.

detector material (Fig. 6p-1). Table 6p-3 shows the characteristics of detectors in Fig. 6p-1. The spectral defectivities in Fig. 6p-1 were measured under a 60-deg field of view and with a 295 K background temperature. The theoretical values of peak D^* lie on the dashed curves.

Noise. The fluctuation in the output of the detector when the incident radiation is steady.

JOHNSON NOISE, due to thermal agitation, is present in any electrical circuit, as follows:

$$\text{rms noise voltage} = \sqrt{4kTR(f_2 - f_1)}$$

(6p-1)

where k = Boltzmann's constant (1.374 $\times 10^{-23}$ joule/K)

T = Kelvin temperature

R = resistive component of circuit element

f_2, f_1 = bandwidth limits

SHOT NOISE, arising from statistical fluctuation in electron tube current, is given by

$$\bar{\imath} = \sqrt{2eI(f_2 - f_1)}$$

where $\bar{\imath}$ = rms noise current

e = electronic charge (1.59 $\times 10^{-19}$ coulomb)

I = direct electron current, amp

Key to Detector Manufacturers in Table 6p-3

a.	Aerojet-General Corp.	Azusa, California
b.	Avco Corporation, Electronics Div.	Cincinnati, Ohio
c.	Barnes Engineering Co.	Stamford, Conn.
d.	Block Engineering, Inc.	Cambridge, Mass.
e.	Catron Electronic Corp.	Geneva, Ill.
f.	E. G. and G.	Boston, Mass.
g.	Electronic Corp. of America	Cambridge, Mass.
h.	Electro-Nuclear Laboratories, Inc.	Menlo Park, Calif.
i.	Honeywell Radiation Center	Boston, Mass.
j.	Infrared Industries, Inc.	Waltham, Mass.
k.	Mithras, Inc.	Cambridge, Mass.
l.	Networks Electronic Corp.	Chatsworth, Calif.
m.	Philco-Ford Corp.	Spring City, Penn.
n.	Raytheon Co.	Waltham, Mass.
o.	Santa Barbara Research Center	Goleta, Calif.
p.	SAT—Paris, France; (U.S. Representative: Elteck Corp.)	Larchmont, N.Y.
q.	Sensor Precision Ind.	Medfield, Mass.
r.	Texas Instruments Inc.	Dallas, Texas
s.	United Detector Technology	Santa Monica, Calif.

TABLE 6p-3. CHARACTERISTICS OF AVAILABLE SEMICONDUCTOR DETECTORS

Detector material:	Si Silicon		GaAs Gallium arsenide	Ge Germanium		PbS Lead sulfide		InAs Indium arsenide			PbSe Lead selenide	
	(1)	(2)	(3)	(4)	(5)	(6)	(7)	(8)	(9)	(10)	(11)	(12)
Operating mode	pv	pc	pv	pv	pc	pc	pc	pv	pv	pv	pc	pc
Typical peak D^* at 1,000-Hz modulation frequency, cm Hz$^{\frac{1}{2}}$ watt^{-1}	2×10^{12}	5×10^{11}	8×10^{11}	5×10^{10}	8×10^{10}	4×10^{11}	3×10^{11}	6×10^{9}	2×10^{11}	4×10^{11}	2×10^{9}	2×10^{10}
Wavelength, μm	0.9	0.9	0.85	1.5	2.5	2.7	3.1	3.5	3.2	3.1	3.4	4.1
Field of view, deg	....	....	....	....	....	....	60	....	60	60	....	60
Background temperature, K	....	....	....	....	295	295	295	295	295	295	295	295
Best measured peak D^*, cm Hz$^{\frac{1}{2}}$ watt^{-1} (conditions as above)	1×10^{13}	1×10^{12}	....	....	1.5×10^{11}	7×10^{11}	4×10^{11}	1×10^{10}	3.5×10^{11}	7×10^{11}	2×10^{10}	5×10^{10}
Spectral range exhibiting greater than 50% relative response, μm	0.6–1.0	0.8–1.06	0.6–0.95	0.9–1.7	1.2–2.8	1.3–3.2	1.4–3.8	2.0–3.8	2.5–3.4	1.8–3.3	0.5–4.2	2.0–5.3
Normal operating temperature, K	295	295	295	295	295	195	77	295	195	77	295	195
Operating temperature limits, K (50% peak D^* degradation points)	—, 320	—, 350	....	....	—, 310	130, 250	—, 160	—, 320	—, 210	—, 180	—, 310	—, 230
Typical time constant, seconds	5×10^{-7}	5×10^{-6}	1×10^{-6}	1×10^{-7}	3×10^{-4}	5×10^{-3}	3×10^{-3}	$<1 \times 10^{-6}$	$<1 \times 10^{-6}$	5×10^{-7}	2×10^{-6}	3×10^{-5}
Nominal resistance, ohms	1×10^{6}	1×10^{6}	1×10^{6}	2×10^{5}	1×10	1×10^{6}	2×10^{6}	3×10^{4}	5×10^{4}	5×10^{5}	2×10^{6}	5×10^{6}
Area Configuration												
Single detectors:												
Size range, min. to max., inches	0.004–0.5	0.004–0.7	0.004–0.060	0.004–0.5	0.001–1.0	0.001–1.0	0.001–1.0	0.004–0.1	0.004–0.1	0.004–0.1	0.003–0.5	0.003–0.5
Shape (round, square, or rectangular)	□	□	□	—	□	□	□	○	○	○	□	□
Typical package	Any TO-5/18	Flat mount	TO 18	Any TO 5/18, BNC	Flat mount	Glass Dewar	Glass Dewar	TO 5/18	Glass Dewar	Glass Dewar	Flat mount	Glass Dewar
Detector arrays:												
Minimum size per detector, inches	0.004	0.004	....	....	0.001	0.001	0.001	0.003	0.003	0.003	0.003	0.003
Minimum size per space, inches	0.002	0.001	....	....	0.001	0.001	0.001	0.002	0.002	0.002	0.002	0.002
Dimensions—see code	0.002"	0.020"			0.001"	0.001"	0.001"	0.002"	0.002"	0.002"	0.001"	0.001"
Manufacturer(s)—see key	f,h,m,r,s	k	m	h,m,r	e,g,j,o,q	g,j,o,q	g,j,o,q	c,h,m,r	c,m,r	c,o,r	j,o	j,o

TABLE 6p-3. CHARACTERISTICS OF AVAILABLE SEMICONDUCTOR DETECTORS (*Continued*)

	(13) PbSe Lead selenide	(14)	(15) InSb Indium antimonide	(16) InSb Indium antimonide	(17) InSb Indium antimonide	(18) Ge:Au Gold-doped germanium	(19) Ge:Hg Mercury-doped germanium	(20) (Hg-Cd)Te Mercury-cadmium telluride	(21) Ge:Cd Cadmium-doped germanium	(22) Si:Sb Antimony-doped silicon	(23) Copper-doped germanium	(24) Zinc-doped germanium
Detector material / Operating mode	pc	pem	pc	pc	pv	pc	pc	pv	pc	pc	pc	pc
Typical peak D^* at 1,000-Hz modulation frequency, cm Hz$^{\frac{1}{2}}$ watt^{-1}	3×10^{10}	1×10^{8}	2×10^{8}	8×10^{10}	1×10^{11}	1×10^{10}	2×10^{10}	5×10^{9}	2×10^{10}	1×10^{10}	3×10^{10}	2.5×10^{10}
Wavelength, μm	4.8	6.0	6.8	5.3	5.1	5.0	10.5	10.6	16	20	23	36
Field of view, deg	60			60	60	60	60	60	60	60	60	60
Background temperature, K	295			295	295	295	295	295	295	295	295	295
Best measured peak D^*, cm Hz$^{\frac{1}{2}}$ watt^{-1} (conditions as above)	5×10^{10}	3×10^{8}	......	1×10^{11}	2×10^{11}	2×10^{10}	5×10^{10}	2×10^{10}	4×10^{10}	2×10^{10}	5×10^{10}	5×10^{10}
Spectral range exhibiting greater than 50% relative response, μm	2.7-6.3	2.0-7.0	3.6-7.3	3.0-5.4	2.0-5.4	3.0-7.5	6-14	9-13	11-20	12-23	15-27	20-40
Normal operating temperature, K	77	295	295	77	77	60	27	77	4.2	4.2	4.2	4.2
Operating temperature limits K (50% peak D^* degradation points)	-, 160			-, 95	-, 105	-, 80	-, 40	-, 100	-, 26	-, 10	-, 20	-, 6
Typical time constant, seconds	4×10^{-5}	2×10^{-7}	1×10^{-6}	6×10^{-6}	$<1 \times 10^{-6}$	1×10^{-7}	2×10^{-7}	$<1 \times 10^{-8}$	1×10^{-7}	7×10^{-7}	5×10^{-7}	2×10^{-8}
Nominal resistance, ohms	5×10^{6}	1×10^{1}	2×10^{1}	1×10^{4}	1×10^{5}	1×10^{5}	1×10^{5}	2.5×10^{1}	1×10^{5}	7×10^{6}	1×10^{5}	2.5×10^{5}
Area Configuration Single detectors: Size range, min. to max., inches	0.003-0.5	0.015-0.040	0.040-0.1	0.003-0.1	0.003-0.1	0.003-0.1	0.003-0.1	0.020-0.080	0.003-0.1	0.004-0.1	0.003-0 1	0.003-0.1
Shape (round, square, or rectangular)	□	□	□	□ □	□ ○	□ □	□		□ □	□ □	□ □	□ □
Typical package	Glass Dewar	Metal container	Flat mount	Glass Dewar	Glass Dewar	Glass Dewar	Metal Dewar	Glass or metal Dewar	Metal Dewar	Metal Dewar	Metal Dewar	Metal Dewar
Detector arrays: Minimum size per space, inches	0.003			0.003	0.003	0.003	0.003	Developmental	0.003	0.004	0.003	0.003
Minimum size per space, inches	0.002			0.002	0.002	0.002	0.002	....	0.002	0.004	0.002	0.002
Dimensions—see code	0.001"			0.001"	0.002"	0.002"	0.002"		0.002"	0.004"	0.002"	0.002"
Manufacturer(s)—see key	j,o	h,i,j	d,j	h,i,l,m,n,o	b,c,l,m,o	c,m,n,o	a,m,n,o,r	p	n,o,r	a	a,m,n,o,r	o

pc Photoconductive mode   pv Photovoltaic mode   pem Photoelectromagnetic mode   Array dimension code: linear ⟶; 2-dimensional ⟂; linear staggered - -; space between rows ⟂.

Other sources of noise are current noise, photon noise, and flicker noise. Since noise limits the detection of weak signals when unlimited amplification is available, sensitivity of a detector is often expressed as noise equivalent power.

Noise Equivalent Power (NEP). The incident radiation that it takes to produce a detector output signal equal to detector noise in a specified bandwidth (generally 1 Hz). Generally the incident radiation is chopped and expressed in watts as the rms value of the fundamental component of the chopped radiation.

Specific detectivity D^* is the reciprocal of the noise equivalent power of the detector referred to unit area and 1-Hz bandwidth.

$$D^* = \frac{\sqrt{A \, \Delta f}}{\text{NEP}} \qquad \text{watt}^{-1} \text{ cm sec}^{-\frac{1}{2}} \tag{6p-3}$$

where A = detector area, cm²

Δf = effective noise bandwidth, Hz

Time Constant. Dynamic response of many detectors to a step function can be approximated by a single exponential of the form $(1 - \epsilon^{-t/\tau})$, in which t is time and τ is the time constant. The frequency response is then approximately

$$\frac{1}{\sqrt{1 + (2\pi f \tau)^2}} \tag{6p-4}$$

where f is the electrical frequency of the signal.

The signal and noise as a function of frequency for a typical lead sulfide detector and a typical lead selenide detector are shown in Figs. 6p-2 and 6p-3.

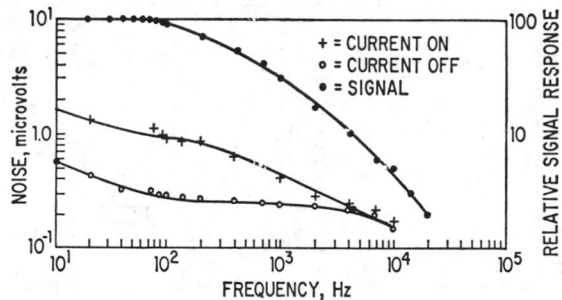

Fig. 6p-2. PbS detector at 25°C. Signal vs. chopping frequency, and noise vs. frequency.

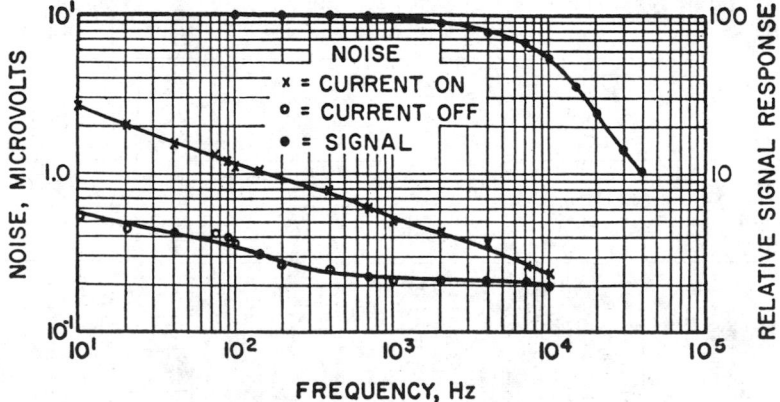

Fig. 6p-3. PbSe detector at −195°C. Signal vs. chopping frequency, and noise vs. frequency.

TABLE 6p-4. PHOTOTUBE TYPES, PHOTOSENSITIVE-DEVICE CLASSIFICATION CHART
Phototubes

Response	Single-unit		Twin-unit		Multiplier
	Vacuum	Gas	Vacuum	Gas	
S-1	917 919 922 925 6570	1P40 1P41 868 918 921 923 927 928 930 6405/1640 6953	. . . .	920	7120
S-3	926	1P29			
S-4	1P39 929 934 5653 7043	1P37 5581 5582 5583	5652	5584	1P21 931-A 6328 6472 7117
S-5	935	. . . .	. . . .	. . . .	1P28
S-8	. . . .	. . . .	. . . .	. . . .	1P22
S-9	1P42				
S-10	. . . .	. . . .	. . . .	. . . .	6217
S-11	. . . .	. . . .	. . . .	. . . .	2020 5819 6199 6342-A 6655-A 6810-A 7264
Extended S-11	. . . .	. . . .	. . . .	. . . .	7046
S-13	. . . .	. . . .	. . . .	. . . .	6903
S-17	. . . .	. . . .	. . . .	. . . .	7029
S-19	. . . .	. . . .	. . . .	. . . .	7200
S-20	. . . .	. . . .	. . . .	. . . .	7265 7326

TABLE 6p-4. PHOTOTUBE TYPES, PHOTOSENSITIVE-DEVICE CLASSIFICATION
CHART (*Continued*)

Camera Tubes

Image orthicons	Vidicons	Iconoscopes
5820	6198	1850-A
6474	6326	
6849	7038	
7198	7262	
7513	7263	

Image-converter Tubes

Response	Infrared-sensitive types	Ultraviolet-sensitive types
S-1	6032 6032-A 6914 6914-A 6929	
S-21	. . . . . .	7404

6p-2. Photoemissive Detectors. A tabulation of photoemissive detector tubes of single, twin, and electron multiplier types is shown in Table 6p-4. The multiplier types have sensitive surfaces ranging in diameter from about 1 to 11 cm with 6, 10, or 14 stages. Table 6p-5 shows dark noise equivalent power data of some of the more sensitive types (ref. 4) with various spectral characteristics. The spectral characteristic curves of various photo surfaces are shown in Fig. 6p-4.

TABLE 6p-5. THE DARK NOISE EQUIVALENT POWER
OF VARIOUS PHOTOEMISSIVE DETECTOR TUBES

Type No.	Spectral response	Dark NEP at response peak		
		At +25°C, watts	At temp. shown, watts	at °C
1P21	S-4	5×10^{-16}	5×10^{-17}	-55
7102†	S-1	1.7×10^{-12}	5×10^{-14}	-60
7200	S-19	5×10^{-16}	3×10^{-17}	-78
7264	S-11	4×10^{-15}	1.2×10^{-15}	$-?0$
7265	S-20	1.6×10^{-15}	2.3×10^{-16}	-80

† NEP figures are for long-wavelength response peak at 0.8 μm.

The dark NEP data apply only if there is no substantial amount of unchopped effective radiation reaching the photocathode. Otherwise the phototube becomes dominated by shot noise, and the NEP is then given by

$$\text{NEP} = \sqrt{\frac{2eW_f}{C(f_2 - f_1)}}$$

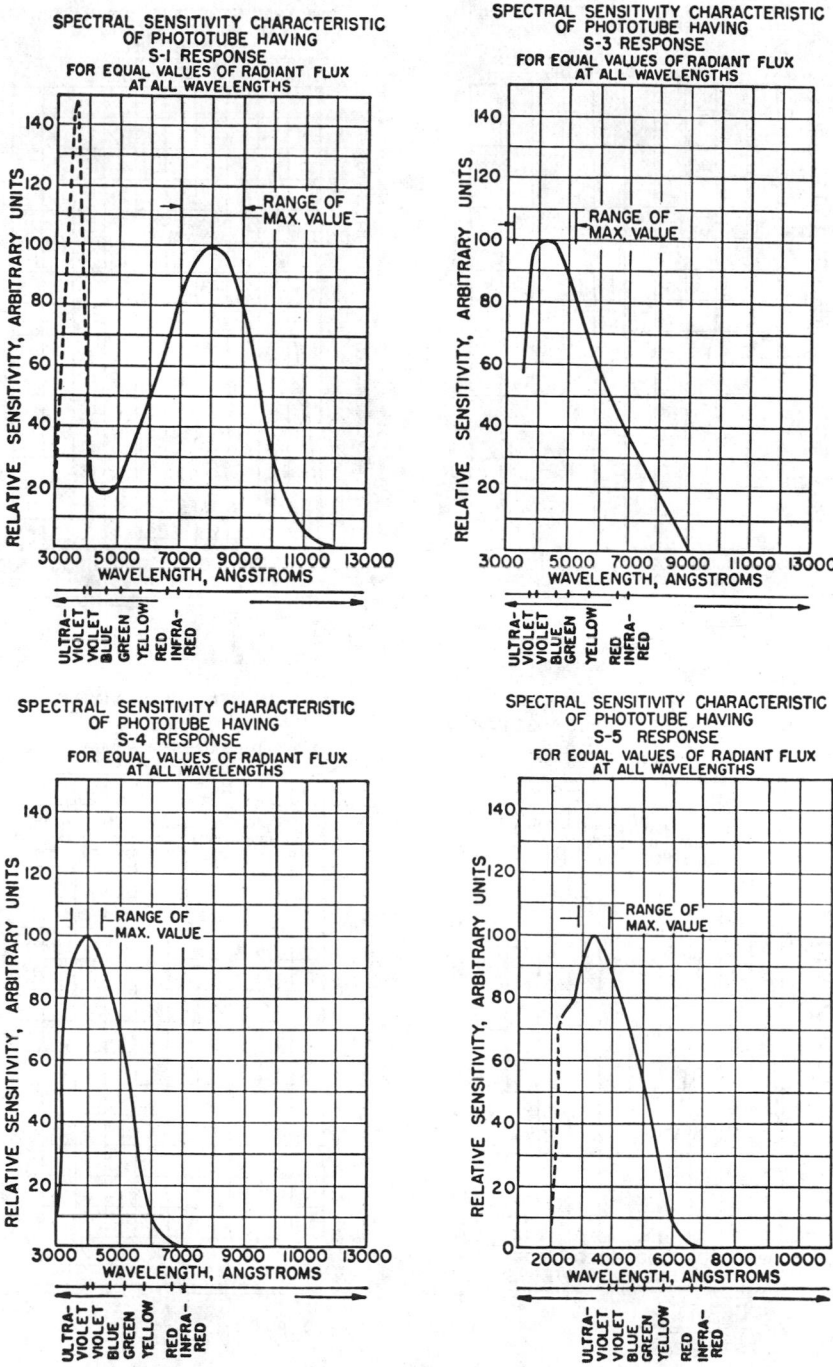

Fig. 6p-4. Spectral sensitivities of commercially available phototubes.

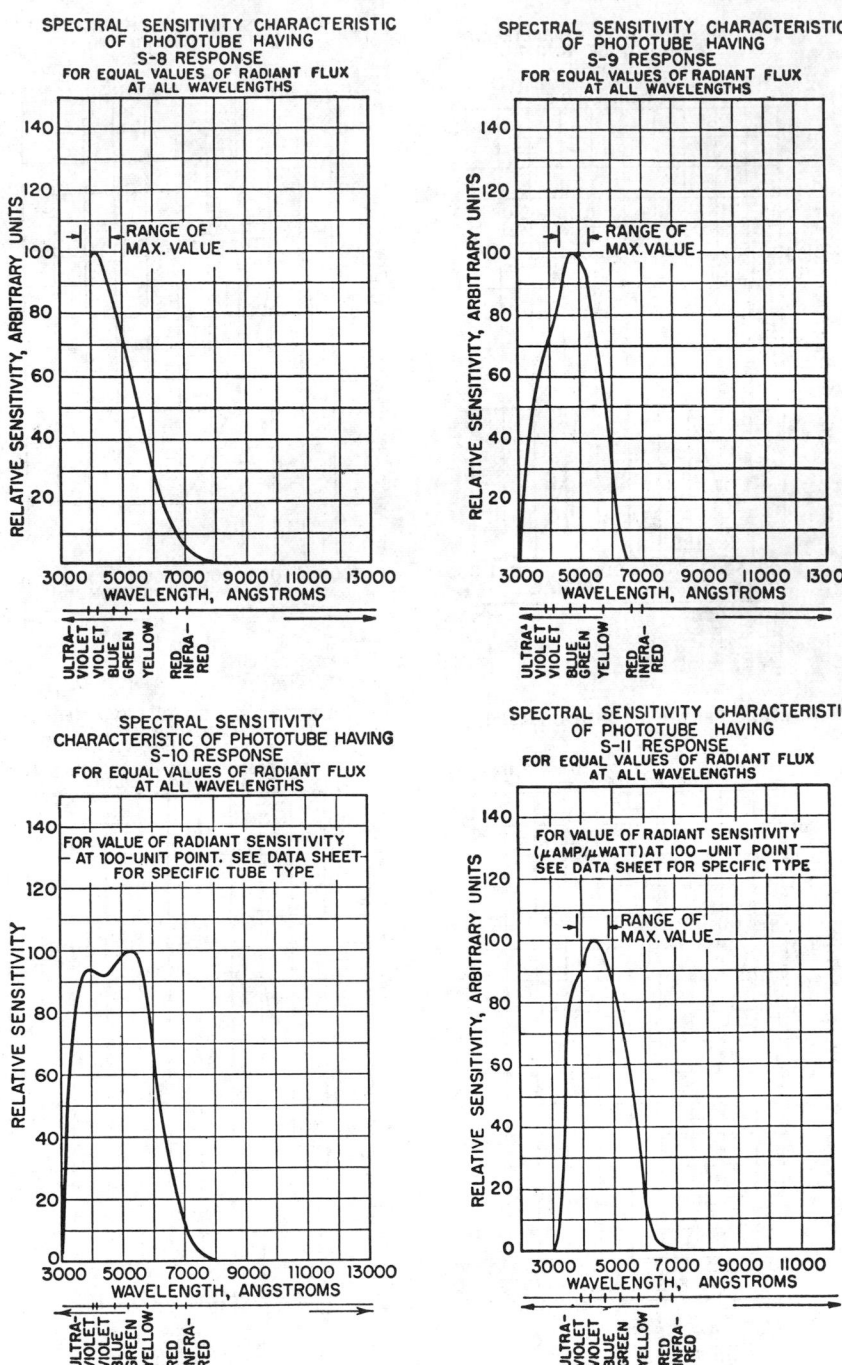

FIG. 6p-4 (*Continued*)

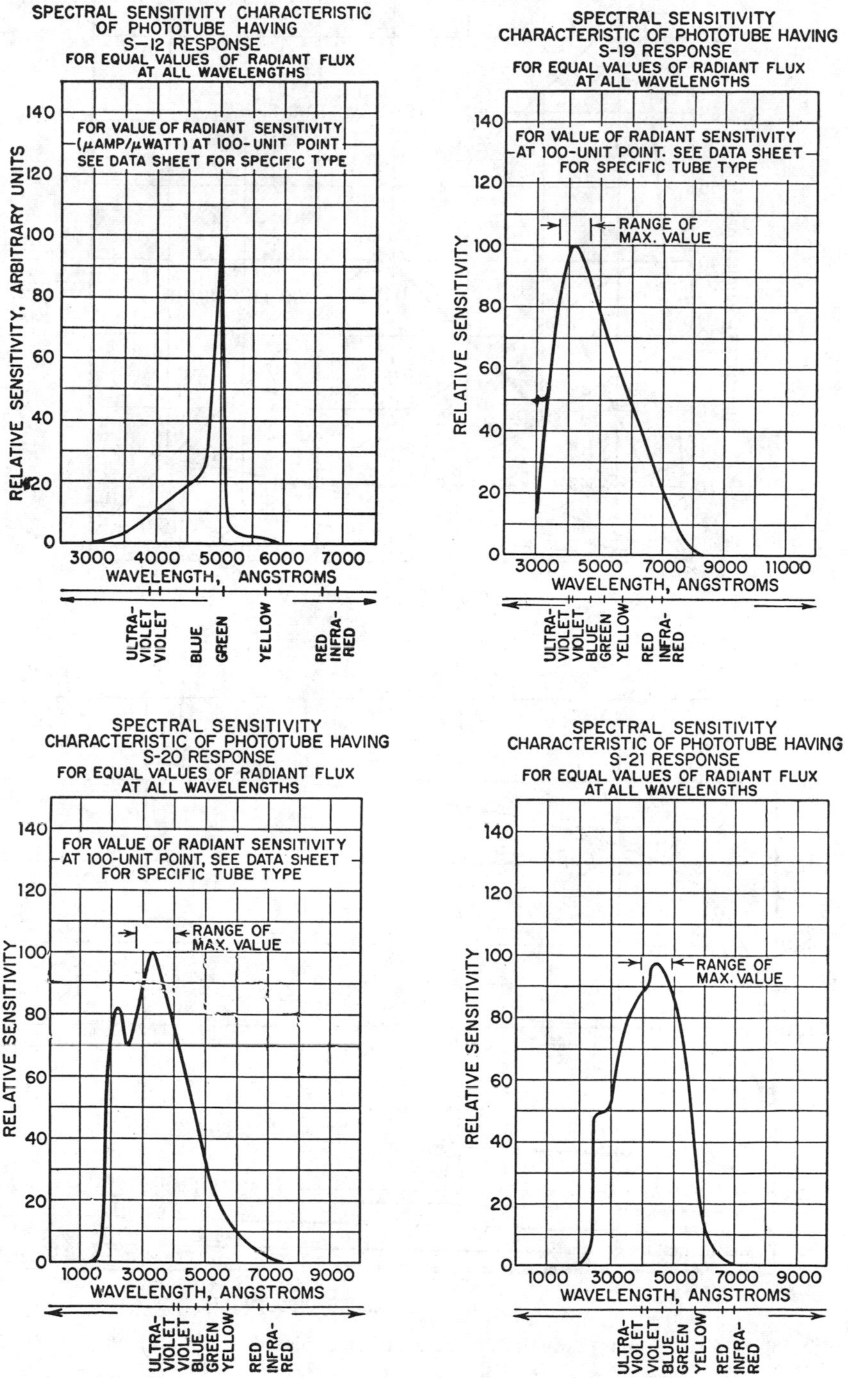
SPECTRAL SENSITIVITY CHARACTERISTIC OF PHOTOTUBE HAVING S—12 RESPONSE
FOR EQUAL VALUES OF RADIANT FLUX AT ALL WAVELENGTHS
FOR VALUE OF RADIANT SENSITIVITY (μAMP/μWATT) AT 100-UNIT POINT SEE DATA SHEET FOR SPECIFIC TYPE
RELATIVE SENSITIVITY, ARBITRARY UNITS
140
120
100
80
60
40
20
0
3000
4000
5000
6000
7000
WAVELENGTH, ANGSTROMS
ULTRA-VIOLET
VIOLET
BLUE
GREEN
YELLOW
RED
INFRA-RED
SPECTRAL SENSITIVITY CHARACTERISTIC OF PHOTOTUBE HAVING S-19 RESPONSE
FOR EQUAL VALUES OF RADIANT FLUX AT ALL WAVELENGTHS
FOR VALUE OF RADIANT SENSITIVITY AT 100-UNIT POINT. SEE DATA SHEET FOR SPECIFIC TUBE TYPE
RANGE OF MAX. VALUE
RELATIVE SENSITIVITY
140
120
100
80
60
40
20
0
3000
5000
7000
9000
11000
WAVELENGTH, ANGSTROMS
ULTRA-VIOLET
VIOLET
BLUE
GREEN
YELLOW
RED
INFRA-RED
SPECTRAL SENSITIVITY CHARACTERISTIC OF PHOTOTUBE HAVING S-20 RESPONSE
FOR EQUAL VALUES OF RADIANT FLUX AT ALL WAVELENGTHS
FOR VALUE OF RADIANT SENSITIVITY AT 100-UNIT POINT, SEE DATA SHEET FOR SPECIFIC TUBE TYPE
RANGE OF MAX. VALUE
RELATIVE SENSITIVITY
140
120
100
80
60
40
20
0
1000
3000
5000
7000
9000
WAVELENGTH, ANGSTROMS
ULTRA-VIOLET
VIOLET
BLUE
GREEN
YELLOW
RED
INFRA-RED
SPECTRAL SENSITIVITY CHARACTERISTIC OF PHOTOTUBE HAVING S-21 RESPONSE
FOR EQUAL VALUES OF RADIANT FLUX AT ALL WAVELENGTHS
RANGE OF MAX. VALUE
RELATIVE SENSITIVITY
140
120
100
80
60
40
20
0
1000
3000
5000
7000
9000
WAVELENGTH, ANGSTROMS
ULTRA-VIOLET
VIOLET
BLUE
GREEN
YELLOW
RED
INFRA-RED

FIG. 6p-4 (*Continued*)

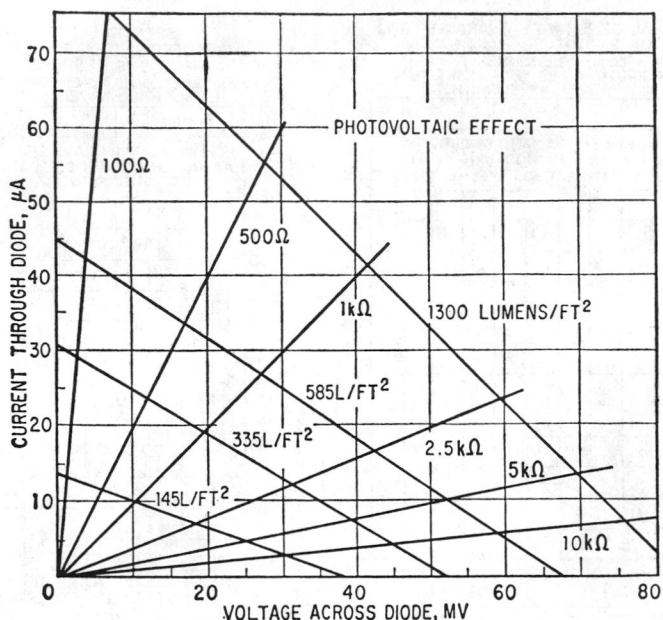

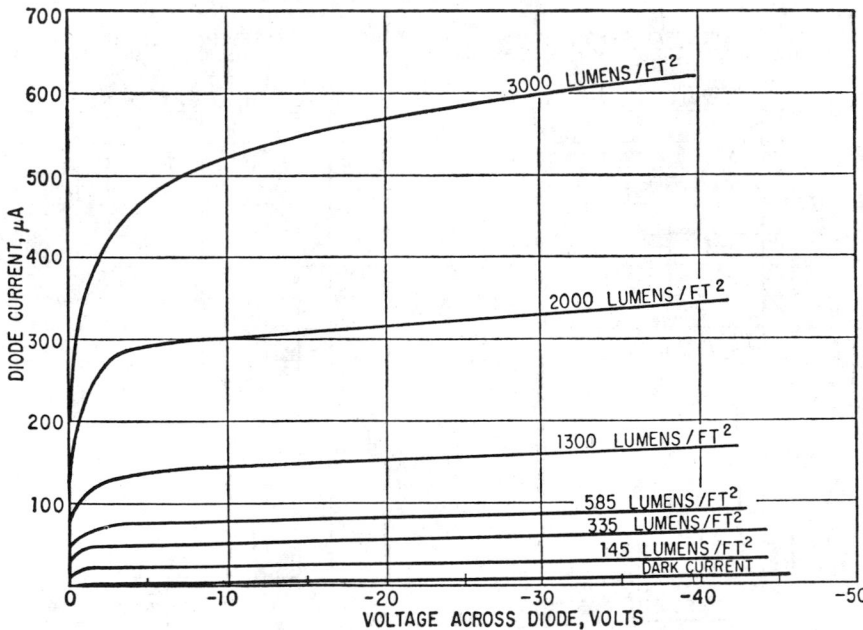

FIG. 6p-5. Germanium photodiode curves showing biased and unbiased photovoltaic characteristics.

where e, f_2, f_1 = quantities defined under shot noise

W_f = unchopped background radiation reaching the photocathode, effective watts

C = cathode sensitivity, amp per effective watt

6p-3. Germanium Photodiodes and Silicon Cells. The characteristics of germanium photodiodes are shown in Fig. 6p-5. Silicon photovoltaic cells (ref. 2) are used largely for the conversion of solar energy into electrical energy. Typical data are shown in Fig. 6p-6.

6p-4. Cadmium Sulfide, Cadium Selenide, and Selenium Detectors. CdS and CdSe cells (ref. 3), listed in Table 6p-6, are available in photoconductive surfaces, areas 1 to 100 mm², potted in transparent resin or sealed in glass envelope. Some CdSe cells have very low dark conductance: $<10^{-4}$ μmho per square surface. Low dark resistance equivalent illumination $<10^{-5}$ ft-c makes CdS and CdSe suitable for detecting low light levels without light chopping. Both have long time constants at low light levels. Comb-type electrodes (series 500 and 500L) provide greater conductance values. Series 400, 400L, and 600L have a rectangular sensitive surface of approximately $\frac{1}{16}$ by $\frac{3}{16}$ in. Spectral performance curves are given in Fig. 6p-7. Conductances at various illumination levels are given in Fig. 6p-8. The spectral response of a typical selenium cell is shown in Fig. 6p-9.

6p-5. Image Converters or Image Intensifiers. Available types (ref. 4), shown in Table 6p-7, have a semitransparent photoemissive cathode at the input end of an evacuated glass envelope (ref. 5). Electrons emitted in a pattern corresponding to the image falling on the surface are accelerated and focused onto an output surface of electroluminescent phosphor at high potential to produce a bright picture. Types 6914 and 6929, having a silver-oxygen-cesium photocathode, can be used to convert an image formed in near-infrared radiation to a visible image. If a fast optical system is used to image a visible scene on the photosurface, the output phosphor image can be made brighter than the scene. For 2870 K tungsten illumination of noncolor selective objects, the screen brightness/scene brightness ratio is given by

$$\frac{\text{Conversion index}}{4f^2M^2}$$

where f = f/ratio of optical system

M = magnification (output/input image)

For all three types listed: Magnification (output/input image size) = 0.8; screen phosphor = P20.

6p-6. Vidicons. Available types (ref. 6) have a photoconducting layer on the inside of a window at the input end of an evacuated glass envelope (ref. 7). The charge pattern developed when an image falls on this surface is scanned by an electron beam to produce a video signal. Vidicons are used in compact television pickup equipment.

Other types 6326, 7038, 7263, 7290, 7325 have similar resolution, sensitive area, relative spectral response, and generally similar characteristics. Vidicons have also been made with infrared sensitive PbS photocathodes (ref. 13). Figures 6p-10 and 6p-11 and Table 6p-8 show characteristics of Vidicon Type 7262.

6p-7. Image Orthicon. The image orthicon (ref. 8) is a television pickup tube that has a semitransparent photocathode, a mesh screen, and a thin dielectric target of high resistivity. Electrons emitted by the cathode are focused on the target, causing secondary emission which leaves the rear face of the target with a positive charge pattern. This is scanned by an electron beam whose signal currents are amplified by electron multiplier stages to provide a high-level output current.

TABLE 6p-6. CHARACTERISTICS OF CdS AND CdSe CELLS

Refer to Fig. 6p-8

Material†	2 (CdS)					3/3A (CdSe)					4 (CdSe)				5 (CdS)					7 (CdS)				
Light level, ft-c	0.01	0.1	1	10	100	0.01	0.1	1	10	100	0.1	1	10	100	0.01	0.1	1	10	100	0.01	0.1	1	10	100
Time constant (to $1 - 1/e$ of final reading):																								
Rise, sec.	2.5	0.59	0.14	0.037	0.008	0.29	0.069	0.017	0.004	0.001	0.25	0.047	0.010	0.002	2.8	0.30	0.074	0.021	0.002	1.1	0.25	0.047	0.010	0.002
Decay, sec.	0.57	0.17	0.053	0.016	0.005	0.030	0.014	0.007	0.003	0.002	0.053	0.023	0.010	0.005	1.3	0.22	0.058	0.021	0.005	0.12	0.053	0.023	0.010	0.005
Temp. characteristic $100 \times \frac{G_t}{G_{25^\circ C}}$:																								
−25°C	46	110	116	110	92	600	340	200	130	120	130	99	99	98	77	80	95	95	98	110	94	88	86	86
0°C	52	100	110	104	95	250	240	160	120	110	110	100	100	101	78	82	96	99	101	110	98	94	93	93
25°C	100	100	100	100	100	100	100	100	100	100	100	100	100	100	100	100	100	100	100	100	100	100	100	100
50°C	250	103	97	98	99	78	28	51	79	73	84	89	95	96	160	120	110	110	96	89	94	99	102	107
75°C	370	120	93	94	94	95	48	31	44	48	60	65	80	88	230	120	120	110	88	50	69	87	97	120

† Last digit of CL cell number.

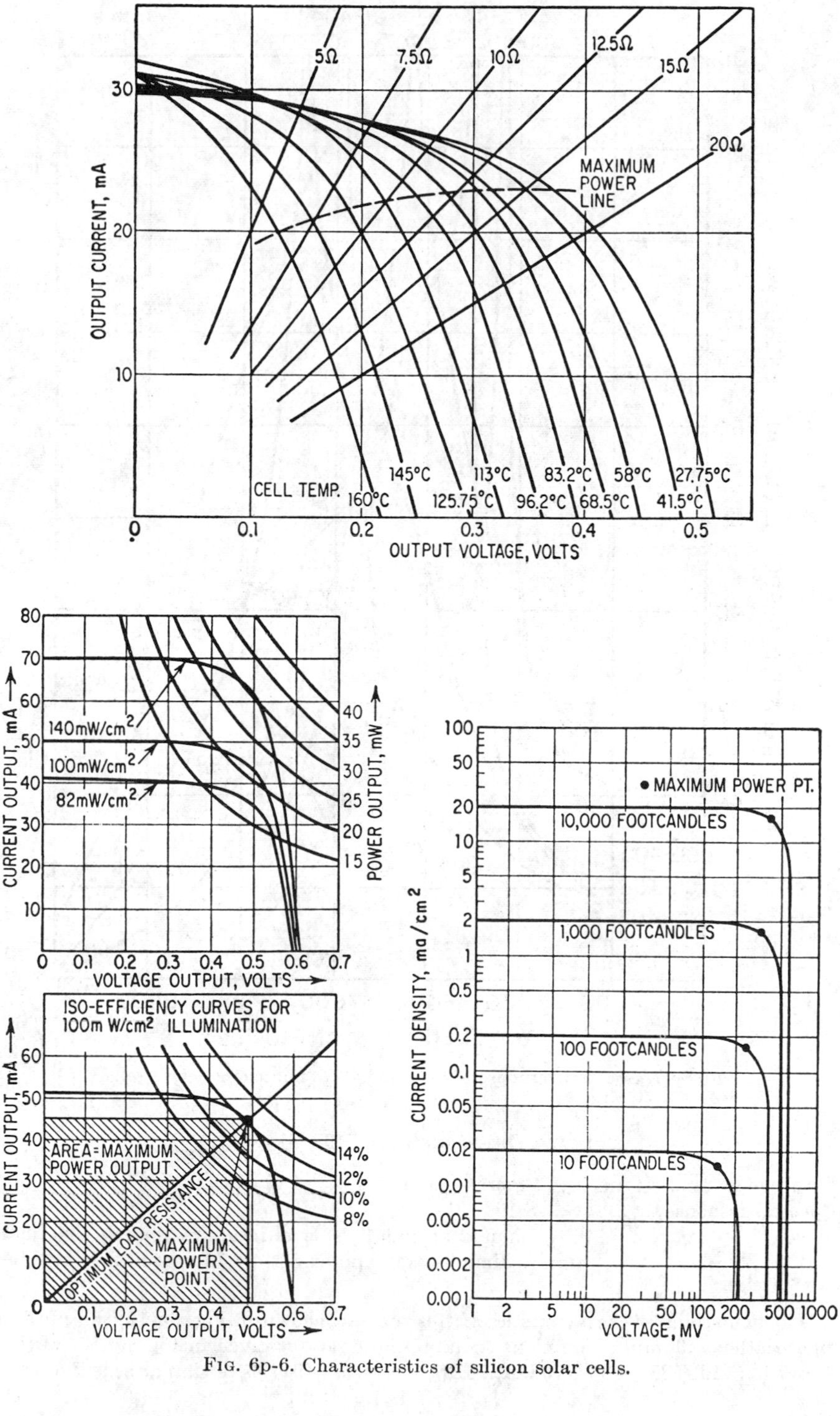

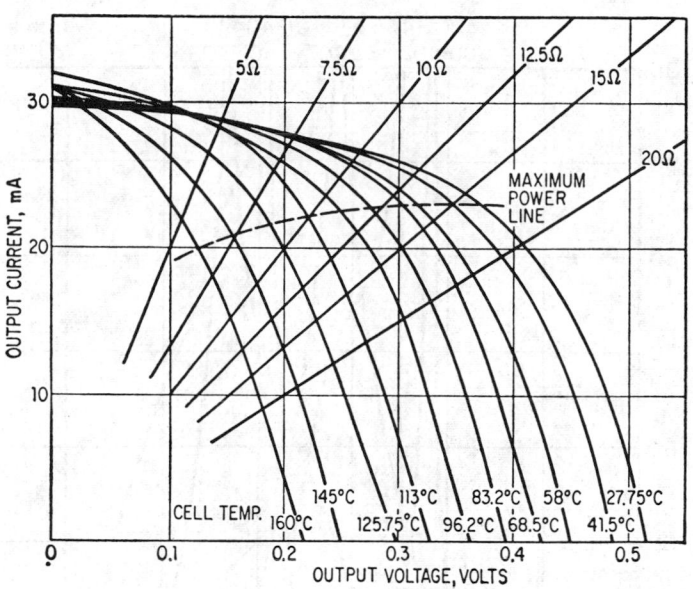

FIG. 6p-6. Characteristics of silicon solar cells.

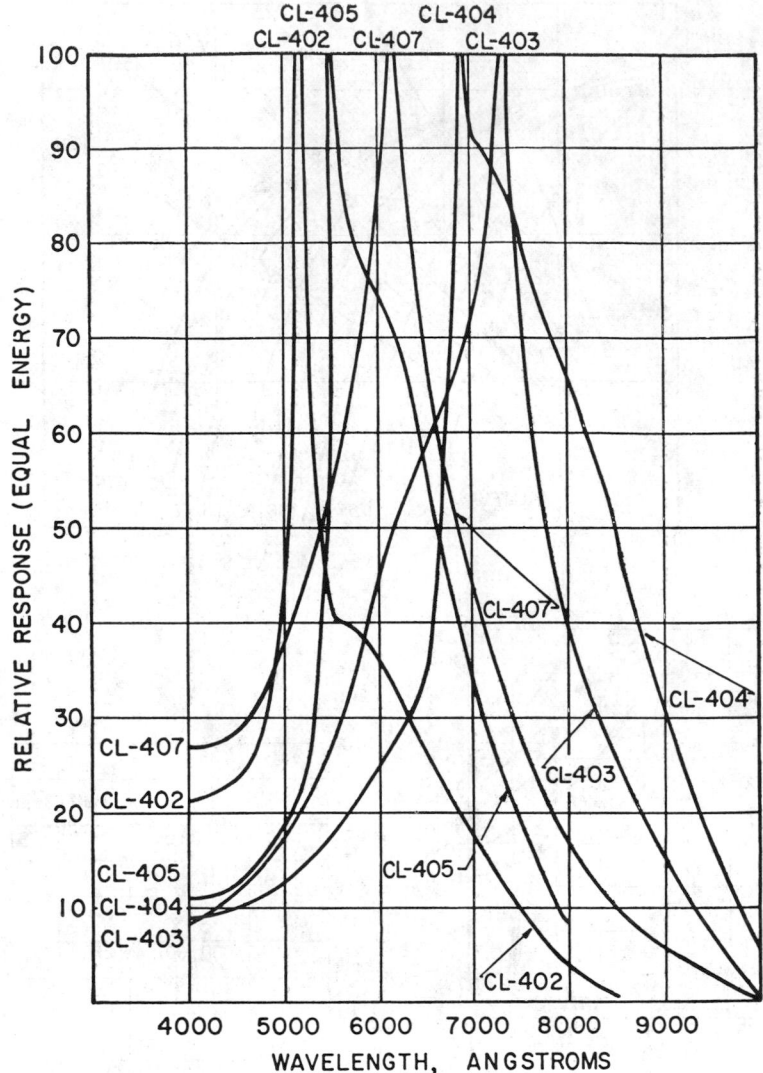

FIG. 6p-7. Spectral response of CdS and CdSe photoconductive cells.

IMAGE ORTHICON TYPE 7198 DATA†

Sensitive area.............. Approximately 1.1 × 1.4 in.
Spectral response.......... S-10
Resolution................ Limited at high light level to approximately 600 lines;
                            diminishes to approximately 75 lines at 2×10^{-5} ft-c

† See ref. 6,

The light-transfer characteristics of this tube are shown in Fig. 6p-12; the effect of photocathode illumination on the signal-to-noise ratio is given in Fig. 6p-13. Other types (ref. 10)(5820, 6849, 7389A, 7513, 7611, and 4401) have similar sensitive area,

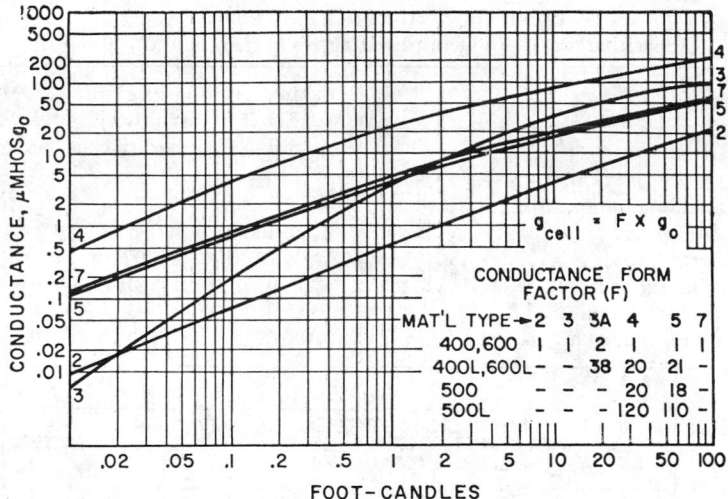

FIG. 6p-8. Characteristics of CdS and CdSe cells.

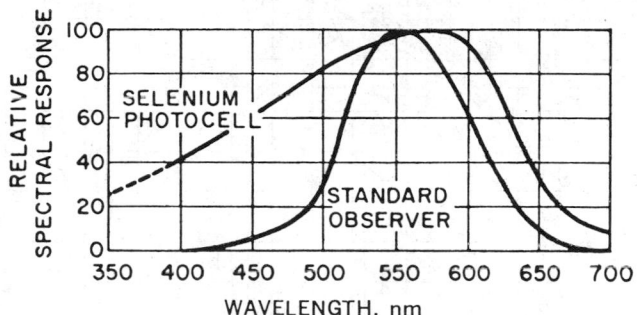

FIG. 6p-9. Spectral response of selenium photocell.

TABLE 6p-7. IMAGE CONVERTERS

Type no.	Spectral response	Screen volts	Cathode diameter, in.	Conversion index	Resolution (Note 3)		Screen background
					At center	At 0.3 in. from center	
6914	S-1	16,000	1.00	15 (Note 1)	28	13	0.16 (Note 4)
6929	S-1	12,000	0.75	10 (Note 1)	33	9	0.21 (Note 4)
7404	S-21	12,000	0.75	6,000 (Note 2)	33	9	10^{-10} (Note 5)

Notes:
1. Ratio of output lumens to incident lumens at photocathode (2870 K).
2. Number of output lumens produced by one incident watt at 2,537 Å.
3. Resolution in line pairs per millimeter at photocathode.
4. Equivalent screen background input in incident microlumens/cm^2.
5. Equivalent screen background input in incident watts/cm^2.

TABLE 6p-8. VIDICON TYPE 7262 DATA

(Sensitive area, $\frac{3}{8}$ by $\frac{1}{2}$ in.; limiting resolution, 600 to 750 lines)

	Max sensitivity operation	Avg sensitivity operation	Min lag operation
Highlight illumination, ft-c	2	15	100
Target voltage	60–100	30–50	15–25
Dark current, μa	0.2	0.02	0.004
Highlight target current, μa	0.4–0.5	0.3–0.4	0.3–0.4
Signal current, peak μa	0.2–0.3	0.3–0.4	0.3–0.4
Signal current, avg μa	0.08–0.1	0.1–0.2	0.1–0.2

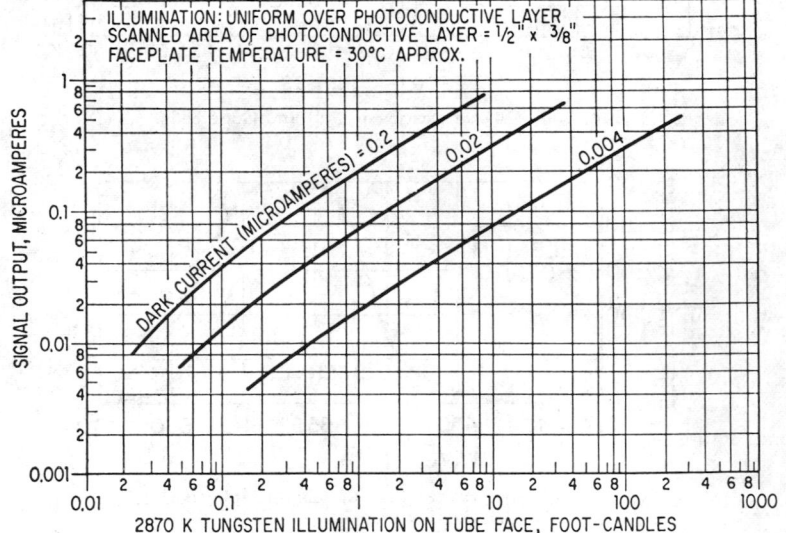

FIG. 6p-10. Typical light-transfer characteristics of Vidicon Type 7262.

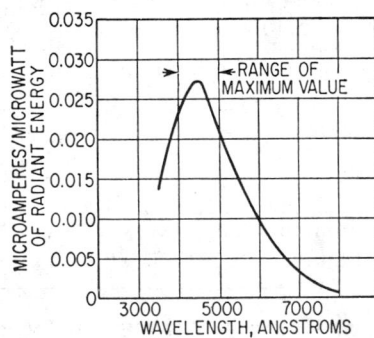

FIG. 6p-11. For Vidicon Type 7262, curve for equal values of signal-output current at all wavelengths. Signal-output microamperes from scanned area of $\frac{1}{2}$ by $\frac{3}{8}$ in. = 0.02, dark current (μa) = 0.02.

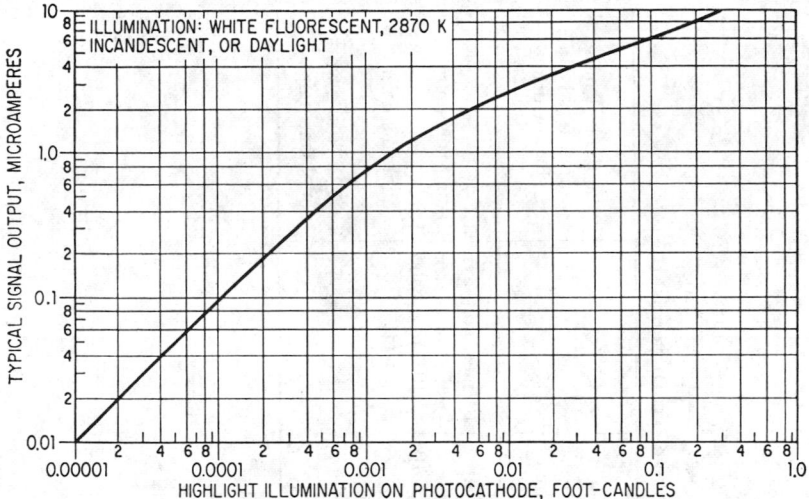

FIG. 6p-12. Basic light-transfer characteristics of Type 7198.

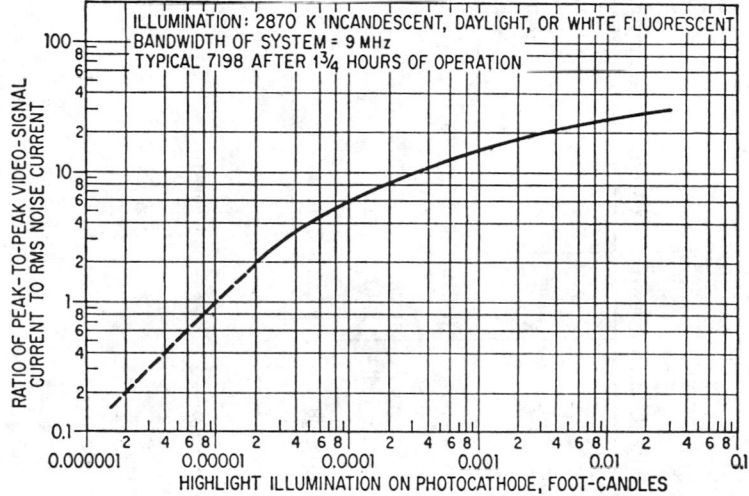

FIG. 6p-13. Effect of photocathode illumination on signal-to-noise ratio of typical 7198.

resolution, and relative spectral response. Some later types (Z5294 and C74037) have a thin film target, giving better than 300-line resolution at 10^{-5} ft-c. Type Z5395 has an S-1 photocathode. The developmental Type C73477 has an image-intensifier stage.

6p-8. Photographic Emulsions. Characteristics of various photographic materials manufactured by Eastman Kodak Company are shown in ref. 11.

6p-9. Commercially Available Thermal Detectors. Characteristics of some available thermal detectors are shown in Table 6p-9. Another thermal detector is the evaporograph (ref. 12). The sensitive element here is a transparent membrane

Table 6p-9. Characteristics of Commercially Available Thermal Detectors†

Detectors	Material	Time const, sec	Area	Frequency of measurement, Hz	Resistance, ohms	V/W	Equivalent noise input for 1-Hz bandwidth
Bolometer‡	Platinum	0.016	6.5 × 0.25 mm	10	40	10 rms volt/avg watt	1.7×10^{-10} avg watt equal rms noise
Bolometer§	Mixture manganese, nickel, and cobalt oxide	0.20 0.40	2.5 × 0.2 mm	15	3×10^6	1,210 rms volt/avg watt	1.8×10^{-10} avg watt equal rms noise
Golay pneumatic cell¶	Gas-filled cavity	0.015	3-mm circle	10	....	....	6×10^{-11} avg watt equal rms noise

† See ref. 13.
‡ Made by Baird-Atomic, Inc.
§ Made by Servo Corporation of America and Barnes Engineering Company.
¶ Made by Eppley Laboratory, Inc.

.about 2 cm in diameter on which is formed an oil film. When infrared radiation falls on this film, the oil evaporates. The difference in thickness of the film can then be seen with white-light interference colors. 'The room-temperature device can detect temperature differences of the order of 1°C.

References

1. "RCA Tube Handbook," HB-3, General Electric Data Sheet.
2. Hoffman Electronic Corp., Data Sheets T1B 32-58, HSD5-1-60; International Recti-fier Corp., "Engineering Handbook."
3. Clairex Corp. Data Sheet 400, 500, 600; "RCA Tube Handbook," HB-3; NAVORD Report 4649, U.S. Naval Ordnance Laboratory, Corona, Calif.
4. "RCA Tube Handbook," HB-3.
5. *Proc. IRE* **47**, 1467, 1605, 1618 (1959).
6. "RCA Tube Handbook," HB-3.
7. *Proc. IRE* **47**, 1607 (1959).
8. The Image Orthicon, *Proc. IRE*, July, 1946.
9. "Smithsonian Physical Tables," 9th ed.
10. General Electric, Westinghouse, and RCA Data Sheets.
11. "Kodak Photographic Plates for Scientific and Technical Use," 7th ed., Eastman Kodak Company, Rochester, N.Y.
12. Baird-Atomic, Inc., Data Sheet.
13. *Proc. IRE* **47**, 1503 (1959).

6q. Radio Astronomy

EUGENE E. EPSTEIN

The Aerospace Corporation

Radio astronomy involves the study of that emission from astronomical objects that falls in the radio portion of the electromagnetic spectrum [arbitrarily defined here as the range from 1 mm (300 GHz) to 1 km (300 kHz)]. Because of absorption by the earth's ionosphere, observations at $\lesssim 10$ MHz are usually made by space-borne equipment. At $\gtrsim 20$ GHz, absorption by molecular constituents in the earth's atmosphere (principally water and oxygen) becomes increasingly severe with fre-quency; however, valuable observations at 220 GHz can still be made at sea level [1]. Large radio telescopes achieve resolutions of the order of 2 arc minutes; however, interferometers and arrays can achieve much finer resolution. Intercontinental interferometers have recently achieved the equivalent of 0.001 arc second resolution [2].

The basic quantity measured by radio astronomers is flux S defined as

$$S = \iint B(\theta,\phi) \, d\Omega$$

where B is the brightness as a function of position on the sky, in $W/(m^2 \cdot Hz \cdot sr)$. For convenience, radio astronomers often equate B with the Planck blackbody radiation function; hence, signal strengths are frequently expressed in terms of the brightness temperature of an equivalent blackbody. Since $h\nu \ll kT$ for most, but not all, radio astronomical cases of interest, the Rayleigh-Jeans approximation to the Planck function is often used;[1] it is,

$$B = \frac{2h\nu^3}{c^2}\frac{kT}{h\nu} \approx \frac{2kT}{\lambda^2}$$

Typical fluxes of astronomical objects range from 10^4 to 10^{-3} flux units [1 flux unit = 10^{-26} W/(m^2 · Hz)]. Flux densities at 400 MHz and spectra of a number of representative radio sources are given in Table 6q-1 and Fig. 6q-1, respectively.

TABLE 6q-1. FLUX DENSITIES OF SOME NONTHERMAL RADIO SOURCES*

Source	Flux density at 400 MHz†
Cassiopeia A	6,100
Cygnus A	4,500
Hydra A	133
Taurus A	1,230
Virgo A	580
3C 28	66
3C 48	36
3C 98	25
3C 147	52
3C 273	59
3C 286	23
3C 295	52
3C 298	24
3C 310	25
3C 452	29
CTA 21	9
CTA 102	6

* Adapted from J. D. Kraus, "Radio Astronomy," McGraw-Hill Book Company, New York, 1966.
† In flux units [10^{-26} W/(m^2 · Hz)].

Sources of radio emission can be considered in three broad categories: solar system objects, galactic objects, and extragalactic objects. A brief overview follows; most of the references cited here are review articles.

6q-1. Solar System Objects. Solar radio emission can be described as originating from both a "quiet Sun" and an "active Sun" [3]. The active Sun emission is time-varying on a scale ranging from fractions of a second at wavelengths in the decameter range to minutes, hours, or days at shorter wavelengths. This short-term activity frequently originates in the solar corona and is often closely associated with optically observed sunspot and flare activity. The quiet Sun emission, as the appellation implies, is not generally associated with variable phenomena of the Sun, except the 11-year solar cycle.

Radio emission from the Moon [4,6], Mercury [5,6], and Mars [6] (objects with little or no atmosphere) arises from their surface layers and reveals information about the thermal and electrical properties of these layers. Emission from Venus [6,7], Jupiter [6], Saturn [6], Uranus [6], and Neptune [6] provides information on the thermal properties of their atmospheres and important constraints on models of their atmospheres. Observations of Jupiter [8] have also revealed the presence of an extensive magnetosphere filled with charged particles in a fashion similar to the Earth's Van

---

[1] For a more complete discussion of radiation measurements, see J. D. Kraus, "Radio Astronomy," McGraw-Hill Book Company, New York, 1966. This is an excellent general treatise on the subject and includes a historical introduction and thorough discussions of antennas and receivers. Another important reference is A. D. Kuz'min and A. E. Salomonovich, "Radioastronomical Methods of Antenna Measurements," translated by E. Jacobs, Academic Press, Inc., New York, 1966. Also, for a general handbook on astronomy, see C. W. Allen, "Astrophysical Quantities," The Athlone Press, University of London, 1963.

Allen belts.   Radar studies [6,7,9] of the Moon and planets have yielded information on surface roughness, dielectric constants, topographic features, orientations of rotation axes, and rotation rates (revealing, for example, that Venus is in retrograde rotation with a period of 243 days).

6q-2. Galactic Objects.   The broad patterns of the spiral structure of our galaxy have been traced out by radio spectroscopic studies of the 21-cm (1,420-MHz) emission line of neutral hydrogen (HI) gas [10].   The Doppler shifts and angular distributions of the line emission give information on galactic dynamics.   The line intensities give information on the densities of the HI gas concentrations.   Studies

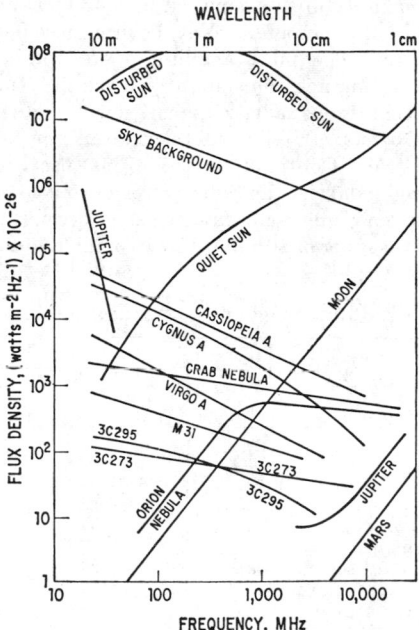

FIG. 6q-1. Spectra of typical radio sources. [*From J. D. Kraus, "Radio Astronomy," McGraw-Hill Book Company, New York*, 1966.]

of continuum [11] radiation reveal the concentration of emission to the galactic plane and to the galactic center (just as the 21-cm line studies do).   The continuum radiation is of two kinds: thermal emission (also known as free-free emission and as coulomb bremsstrahlung) from regions of ionized hydrogen (HII) gas and synchrotron radiation (also known as magnetobremsstrahlung) from electrons with relativistic velocities moving in weak magnetic fields [12].

Thermal emission data have led to the identification of many dense HII regions. These regions also emit recombination lines [13] ($n \rightarrow n-1$ transitions, where $n \sim 50$ to ~ 150).   Also found originating near these regions are 18-cm lines from OH radicals [14] that are often highly polarized and variable in intensity.   The emission regions are usually quite small (a fraction of an arc second), and the emission intensity occasionally corresponds to brightness temperatures as high as 10^9 K! Such high brightness temperatures are clearly not thermal in origin, and "masering" action has been proposed as the explanation, with pumping resulting from intense UV or IR radiation coming from stars embedded in the same cloud.

Emission lines from NH_3 (1.25 cm, 24.0 GHz), H_2O (1.35 cm, 22.2 GHz), and CH_2O (6.21 cm, 4.83 GHz) have recently been detected[1] from these regions of dense gas; in some cases, the H_2O may be "masering." This kind of observation provides data on the structure and dynamics of these regions and, with varieties of other data, has pointed to the suggestion that star and planetary formation is occurring in these regions. With C, H, N, and O, all the ingredients for life are present. It has been further suggested that perhaps life originated not in the primordial oceans on the surfaces of planets, but rather in the primordial clouds out of which the stars and planetary systems condensed.

A recently discovered category of galactic objects of great interest is pulsars.[2] The radio emission from these objects is pulsed in character, with interpulse periods ranging from 30 msec to 4 sec. In a few cases, the interpulse periods have been found to be slowly lengthening. The pulse amplitudes are not constant. Pulsars are thought to be rapidly rotating neutron stars, which are the remnants of supernovae. However, the exact mechanism of radiation is unknown.

6q-3. Extragalactic Objects. HI line emission at 21 cm has been detected from many nearby galaxies [15]. Continued emission has also been detected at many wavelengths from normal galaxies; the radio power output is of the order of 10^{30} to 10^{32} W. There are also many *radio* galaxies, so called because their radio power output is much higher—of the order of 10^{35} to 10^{38} W (see Table 6q-2.) Other strong emit-

TABLE 6q-2. POWER OUTPUT OF ASTRONOMICAL OBJECTS

Object	Optical power, watts	Radio power, watts
White dwarf star	10^{23}	?
Sun	4×10^{26}	10^{12}
Supergiant star	10^{31}	?
Flare star	10^{25}	10^{16}
Supernova remnant	10^{29}	10^{28}
Normal galaxy (10^{11} solar masses)	10^{37}	10^{30}–10^{32}
Radio galaxy	10^{37}	10^{35}–10^{38}
Quasi-stellar radio source	10^{39}	10^{37}–10^{38}

From J. D. Kraus, "Radio Astronomy," McGraw-Hill Book Company, New York, 1966.

ters of radio (and optical) energy are the quasi-stellar radio sources [16] (abbreviated as either quasar, QSO, or QSS)—10^{37} to 10^{38} W. QSS are the most rapidly moving and most distant objects of which we are aware—velocities up to ≈ 0.8 c, and distances up to a few billion light years. One school of thought holds them to be galaxies in very early stages of formation when nonequilibrium conditions obtain. Their angular sizes seem to be a function of radio wavelength, being as small as $\sim 10^{-3}$ arc seconds for some objects [2] at $\lambda \approx 10$ cm. Many radio sources are variable in flux [17] (see Fig. 6q-2), and in some sources repeated outbursts have been measured [18]. The sequence of events depicted in Fig. 6q-2 is thought to be the result of an adiabatically expanding cloud of relativistic electrons in a weak magnetic field (10^{-3} to 10^{-5} gauss); however, the origin of the relativistic electrons is unknown. Seyfert galaxies [19] constitute a class of objects thought to be an intermediate evolutionary stage between QSS and normal galaxies because some of their radio characteristics (spectra and time

---

[1] See the current literature for information on this new and rapidly growing area of radio astronomy.

[2] The current literature contains information on this exciting new field.

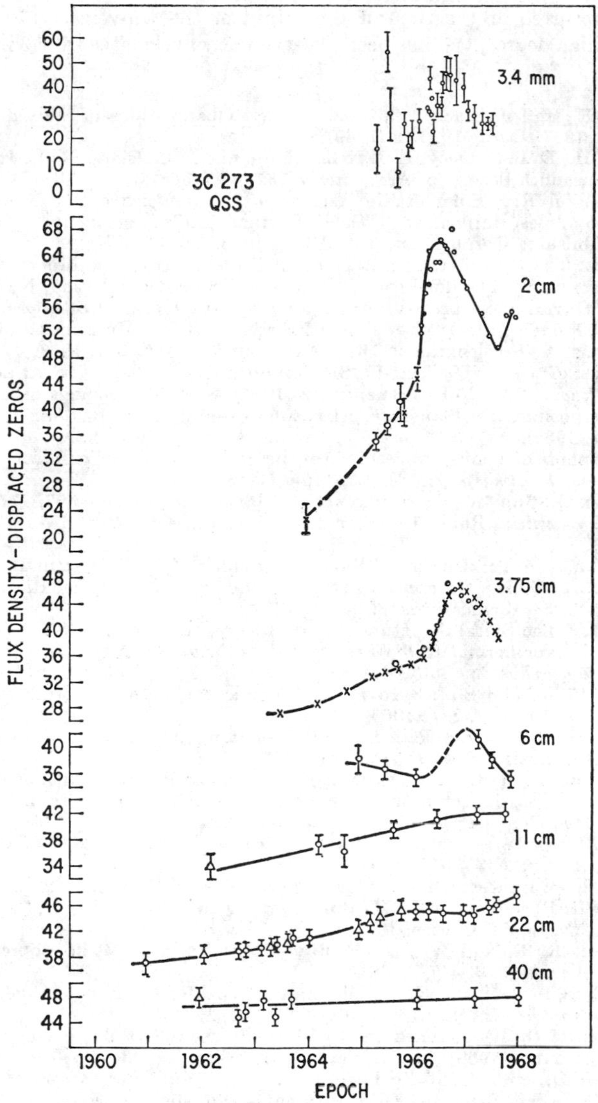

FIG. 6q-2. Variations with time of the flux at several wavelengths from the quasi-stellar radio source 3C273. [*From K. I. Kellermann and I. I. K. Pauliny-Toth, Ann, Rev. Astron. & Astrophys.* **6,** 417 (1968).]

variability) and optical characteristics (excited, turbulent nuclei) resemble those of QSS; yet, in general appearance, they resemble more-or-less normal galaxies. Extensive statistical analyses, important in cosmological studies [20], have been made of the numbers of radio sources as a function of flux [21].

Another feature of the radio sky that has contributed to cosmological investigations is the microwave background [22]. This radiation, corresponding to that from a 2.7 K blackbody, is postulated to be the much diluted radiation left over from the fireball that occurred at the time of the origin of the Universe. It is apparently isotropic to a high degree, and has been observed at wavelengths from 3 mm to 20 cm.

References

1. Reber, E. E., and J. Stacey: 3.4- and 1.4-mm Observations of the Lunar Eclipse on 18 October 1967, *Icarus* **10**, 171 (1969).
2. Cohen, M. H., D. L. Jauncey, K. I. Kellermann, and B. G. Clark: Radio Interferometry at One-thousandth Second of Arc, *Science* **162**, 88 (1968).
3. Kundu, Mukul R.: "Solar Radio Astronomy," Interscience Publishers, Inc., New York, 1965; Jules Aarons, ed.: "Solar System Radio Astronomy," Plenum Press, Plenum Publishing Corporation, New York, 1965.
4. Weaver, H.: The Interpretation of Thermal Emission from the Moon, "Solar System Radio Astronomy," Plenum Press, Plenum Publishing Corporation, New York, 1965.
5. Morrison, David, and Carl Sargan: The Microwave Phase Effect of Mercury, *Astrophys. J.* **150**, 1105 (1967); also, David Morrison: On the Interpretation of Mercury Observations at Wavelengths of 3.4 and 19 mm, *ibid.* **152**, 661, 1968.
6. Proceedings of the 1969 IAU-URSI Symposium on Planetary Atmospheres and Surfaces, *Radio Sci.*, Vol. 5 (February, 1970); also, Proceedings of the 1965 IAU-URSI Symposium on Planetary Atmospheres and Surfaces, *Radio Science* **69D**, (December, 1965).
7. The Atmosphere of Venus, Proceedings of the Second Arizona Conference on Planetary Atmospheres, *J. Atmospheric Sci.* **25**, (July, 1968).
8. Smith, Alex G.: Jupiter, The Radio-active Planet, *Am. Scientist* **57**, 177 (1969); also, James W. Warwick: Radio Emission from Jupiter, *Ann. Rev. Astron. & Astrophys.* **2**, 1 (1964).
9. Evans, John, and Tor Hagfors: "Radar Astronomy," McGraw-Hill Book Company, New York, 1968; also, Gordon H. Pettengill and Irwin I. Shapiro: Radar Astronomy, *Ann. Rev. Astron. & Astrophys.* **3**, 377 (1965).
10. Bok, Bart J.: The Spiral Structure of Our Galaxy, *Am. Scientist* **55**, 375 (1967).
11. Mills, B. Y.: Northermal Radio Frequency Radiation from the Galaxy, *Ann. Rev. of Astron. & Astrophys.* **2**, 185 (1964).
12. Ginzburg, V. L., and S. I. Syrovatskii: Cosmic Magnetobremsstrahlung, *Ann. Rev. Astron. & Astrophys.* **3**, 297 (1965).
13. Mezger, P. G., and Patrick Palmer: Radio Recombination Lines: A New Observational Tool in Astrophysics, *Science*, **160**, 29 (1968).
14. Barrett, Alan H.: Radio Observations of Interstellar Hydroxyl Radicals, *Science* **157**, 881, 1967; also, B. J. Robinson and R. X. McGee: OH Molecules in the Interstellar Medium, *Ann. Rev. Astron. Astrophys.* **5**, 183 (1967).
15. Roberts, Morton S.: Neutral Atomic Hydrogen in 32 Galaxies of Small Angular Diameter, *Astron. J.* **73**, 945 (1968).
16. Burbidge, E. Margaret: Quasi-stellar Objects, *Ann. Rev. Astron. & Astrophys.* **5**, 399 (1967); G. R. Burbidge and E. M. Burbidge: "Quasi-stellar Objects," W. H. Freeman and Company, San Francisco, 1967.
17. Kellermann, K. I., and I. I. K. Pauliny-Toth: Variable Radio Sources, *Ann. Rev. Astron. & Astrophys.* **6**, 417 (1968).
18. Pauliny-Toth, I. I. K., and K. I. Kellermann: Repeated Outbursts in the Radio Galaxy 3C 120, *Astrophys. J.* **152**, L169 (1968).
19. Proceedings of the Conference on Seyfert Galaxies and Related Objects, *Astron. J.* **73**, 836 (November, 1968); Ray J. Weymann: Seyfert Galaxies, *Sci. Am.* **220**, 28 (1968).
20. Novikov, I. D., and Ya. B. Zeldovic: Cosmology, *Ann. Rev. Astron. & Astrophys.* **5**, 627 (1967); Allan Sandage: Observational Cosmology, *Observatory* **88**, 91 (1968).
21. Ryle, M.: The Counts of Radio Sources, *Ann. Rev. Astron. & Astrophys.* **6**, 249 (1968).
22. Partridge, R. B.: The Primeval Fireball Today, *Am. Scientist* **57**, 37 (1969).

6r. Far Infrared

ERNEST V. LOEWENSTEIN

DONALD R. SMITH

Air Force Cambridge Research Laboratories

6r-1. Sources. Only two broad-band sources are bright enough for use in the far-infrared region. The first is a silicon carbide source of the type used in commercial infrared spectrometers. This source has a temperature of about 1200 K, but transmission measurements indicate that its emissivity decreases at low wave numbers. Therefore, the apparent temperature of such a source decreases with decreasing wave number, and partly for this reason globar sources are rarely if ever used at wave numbers lower than 100 cm^{-1}. The mercury-discharge lamp is the source almost universally used for the far-infrared region. The radiant energy at wave numbers greater than about 140 cm^{-1} comes almost exclusively from the hot envelope. The apparent temperature of this fused-quartz envelope is about 900 to 1200 K [1,2]. Below 140 cm^{-1}, as the envelope becomes progressively more transparent, the radiation from the mercury plasma becomes more important. This radiation follows the same ν^2 wave-number dependence as the envelope emission, but with an effective temperature of the order of 5000 to 7000 K in the central part of the tube [3-6]. Therefore, the apparent temperature of the mercury-lamp source increases rapidly with decreasing wave number below 140 cm^{-1}. The emission of the lamp in this region closely approximates a ν wave-number dependence as observed by McCubbin [7]. Mercury lamps need not be water- or air-cooled, but should be operated with the outer envelope removed since it is highly absorbing in the far infrared, even if it is made of fused quartz. There are many different types of high-pressure mercury-discharge lamps commercially available which are satisfactory as far-infrared sources; the 100-W and 85-W G.E. lamps are the most widely used. Lamps of higher wattage are not, in general, better sources unless a large source area is required.

References for Sec. 6r-1

1. Plyler, E. K., D. J. Yates, and H. A. Gebbie: *J. Opt. Soc. Am.* **52,** 859 (1962).
2. Louden, W. C., and K. Schmidt: *Illum. Engr.* **60,** 696 (1965).
3. Cano, R., and M. Mattioli: *Infrared Phys.* **7,** 25 (1967).
4. Smith, D. R., R. L. Morgan, and E. V. Loewenstein: *J. Opt. Soc. Am.* **58,** 433 (1968).
5. Elenbaas, W.: "The High Pressure Mercury Vapour Discharge," Interscience Publishers, Inc., New York, 1951.
6. Filippov, O. K., and V. M. Pivovarov: *Opt. Spectr.* **16,** 282 (1964).
7. McCubbin, T. King, Jr.: Doctoral Dissertation, The Johns Hopkins University, (1951).

6r-2. Detectors. A large variety of detectors is available for the far infrared, and their characteristics are summarized in Table 6r-1. The room-temperature detectors are slower and less sensitive than the cooled ones, all of which must be cooled to 4.2 K or lower.

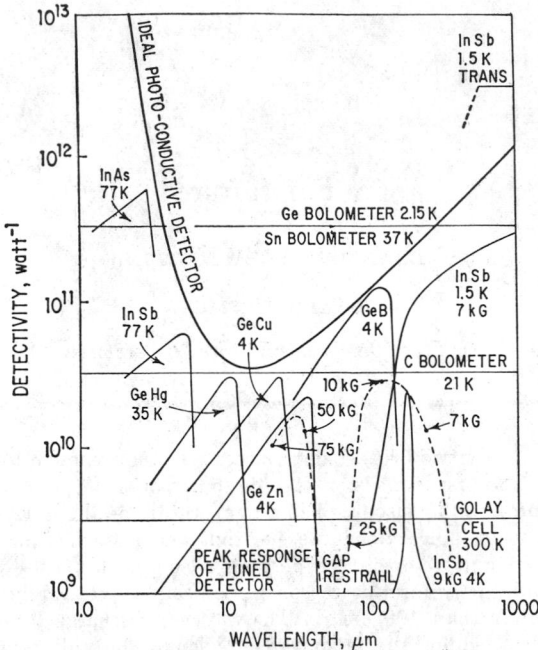

FIG. 6r-1. Performance of far infrared detectors. [Ref. E. Putley, in "Spectroscopic Techniques for Far Infrared and Submillimetre Waves," D. H. Martin, ed., John Wiley & Sons, New York, 1967.]

TABLE 6r-1. CHARACTERISTICS OF FAR-INFRARED DETECTORS

Detector	NEP (for 1 Hz bandwidth), watt	τ sec	Wavelength range	Refs.	Remarks
Golay cell.	10^{-10}	10^{-2}	Visible to several mm	1	Room-temperature pneumatic detector
Thermopile.	0.2–1×10^{-10}	0.150 in vacuum 0.02 in xenon	Visible to several mm	2	Room temperature
Modified thermistor bolometer. . . .	7×10^{-10}	6×10^{-3}	5 μm to 4 mm	3	Room temperature
Carbon bolometer.	3×10^{-11}	10^{-3}	Entire infrared	4	<2 K
Germanium bolometer.	3×10^{-12}	10^{-3}	1.2 μm to several mm	5, 6	<2 K
InSb (low field).	5×10^{-12}	2×10^{-7}	200 μm to several mm	7, 8	~1.5 K; weak magnetic field
InSb (no field).	3×10^{-13}	10^{-7}	300 μm to several mm	14	~1.8 K; no field, cooled transformer
InSb (high field).	5×10^{-11}	$<10^{-6}$	Tuned, narrow band	7, 8	~4 K; wavelength of peak response, fn. of field
Germanium photoconductor.	~10^{-12}	10^{-7}	30 to 135 μm	9	~4 K; Ga doped, compensated
Germanium bolometer.	3×10^{-14}	10^{-2}	Not given, probably $\lambda\lambda > 100$ μm	15	0.4 K, liquid ^{3}He bath, closed-cycle refrigeration
Silicon bolometer.	~10^{-13}	~10^{-3}	Near IR to several mm	10	1 K; silicon doped with phosphorus and boron
Josephson detector.	~5×10^{-13}	$<10^{-9}$	See remarks	11	4 K; wavelength range narrow dependent on material used
Superconductor Sn bolometer. . . .	10^{-2}	10^{-2}	10 μm to several mm	12	3.7 K
GaAs photoconductor.	10^{-12}	$<10^{-6}$	<195 μm to >1 mm	13	4 K

References for Table 6r-1

1. Golay, M. J. E.: *Rev. Sci. Instr.* **20**, 816 (1949).
2. Stafsudd, O., and N. Stevens: *Appl. Opt.* **7**, 2320 (1968).
3. Allen, C., F. Arams, M. Wang, and C. C. Bradley: *Appl. Opt* **8**, 813 (1969).
4. Boyle, W. S., and K. F. Rodgers, Jr.: *J. Opt. Soc. Am.* **49**, 66 (1959).
5. Low, Frank J.: *J. Opt. Soc. Am.* **51**, 1300 (1961).
6. Zwerdling, S., R. A. Smith, and J. P. Theriault: *Infrared Phys.* **8**, 271 (1968).
7. Putley, E.: *J. Sci. Instr.* **43**, 857 (1966).
8. Martin, D. H., ed.: "Spectroscopic Techniques for Far Infrared and Submillimetre Waves," chap. 4. North-Holland Publishing Company, Amsterdam, 1967.
9. Moore, W. J., and H. Shenker: *Infrared Phys.* **5**, 99 (1965).
10. Silvera, I.: Private communication.
11. Grimes, C. C., P. L. Richards, and S. Shapiro: *Phys. Rev. Letters,* **17**, 431 (1966).
12. Martin, D. H., and D. Bloor, *Cryogenics* **1**, 159 (1961).
13. Stillman, G. E., C. M. Wolfe, I. Melngailis, C. D. Parker, P. E. Tannenwald, and J. O. Dimmock: *Appl. Phys. Letters* **13**, 83 (1968).
14. Kinch, M. A., and B. V. Rollin: *Brit. J. Appl. Phys.* **14**, 672 (1963).
15. Drew, H. D., and A. J. Sievers: *App. Opt.* **8**, 2067 (1969).

Most of the detectors are broad-band devices, but some, such as the InSb detector in a magnetic field, are narrow band. The Josephson detector is also a narrow-band detector, and is the fastest and most sensitive yet reported. Difficulties of manufacture are formidable, however, and no general use has been made of this detector.

The material for the germanium and silicon bolometers must be compensated, and no accurate work has been published on the required donor and acceptor concentrations. Ordinarily an ingot of the material is grown, and a search is made for a volume which produces good detectors. Figure 6r-1 gives a graphical comparison of some of the far-infrared detectors.

6r-3. Filters. *Transmission Filters.* The major advantage of transmission filters is that they may be placed anywhere in the optical beam of the instrument. The most widely used filters in the far infrared for the elimination of high-wave-number radiation are absorptive-type transmission filters. These include crystal quartz and black polyethylene, whose transmittance is shown in Figs. 6r-2 and 6r-3. Other materials such as sapphire, fused quartz, and mica can also be used.

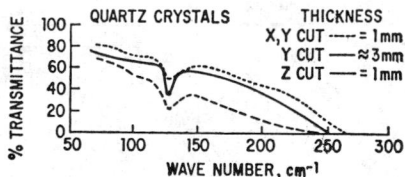

Fig. 6r-2. The transmittance of natural Brazilian crystal quartz. [*From R. V. McKnight and K. D. Möller, J. Opt. Soc. Am.* **54**, 132 (1964).]

Transmission filters with steeper cut-on slopes have been described by Yamada, et al. [1]. They consist of alkali halide crystals suspended in polyethylene and have since come into wide use as low-pass filters for the 400 to 50 cm⁻¹ region. By varying the combination of crystal powders used, the cut-on point can be shifted over a wide range of wave numbers. Several of these filters are shown in Fig. 6r-4. If carbon black is mixed with the crystal powders in these filters, the need for a separate black polyethylene filter is eliminated.

Polyethylene filter gratings [2,3] are useful below 50 cm⁻¹ where few absorption filters exist. These are constructed by pressing a sheet of polyethylene on a heated metal reflection grating. The position of the transmission minimum is a function of groove shape and spacing; for a symmetric groove of 90-deg apex angle it occurs at $\lambda/d \approx 0.3$ To give a satisfactory band stop two such filters should be used. T in Fig. 6r-5 represents the transmission for a typical filter.   Since these filters work by diffraction, their performance depends upon their location and orientation in the optical beam [3].   Making filter gratings out of black polyethylene eliminates the need for a separate black polyethylene filter.

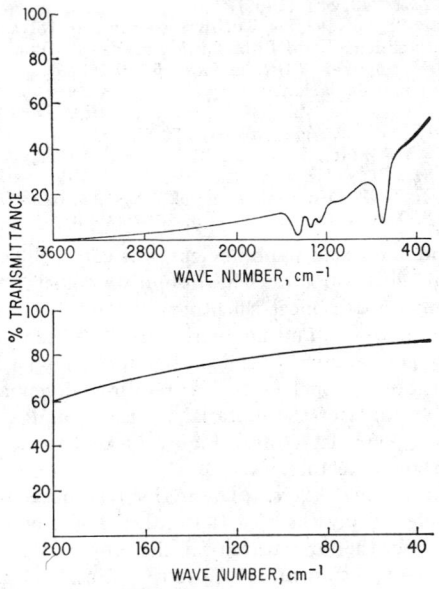

Fig. 6r-3. The transmittance of black polyethylene; thickness, 0.1 mm.   [*From K. D. Möller et al., Appl. Opt.* **5**, 403 (1966).]

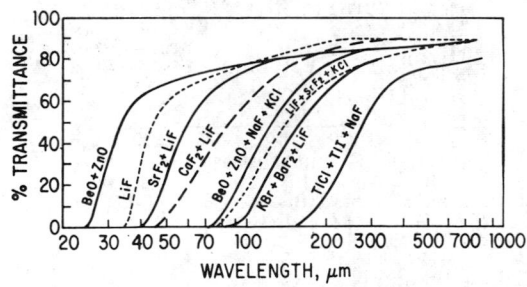

Fig. 6r-4. The transmittance of several typical crystal powder filters.   [*From Y. Yamada et al., J. Opt. Soc. Am.* **52**, 17 (1962).]

Thick plates (4.0 to 10.0 mm) of the alkali halide crystals used in the filters described by Yamada et al. are called reststrahlen plates.   These are usually single crystals, but plates made of pressed or bonded powders can also be used.   In such thicknesses these crystals have extremely strong absorption in the far infrared, but are quite transparent in the near infrared where they are used for prisms and windows.   This high-pass characteristic makes these plates useful as chopper blades in far-infrared spectrometers [4].   The chopper used in this manner becomes a low-pass filter since

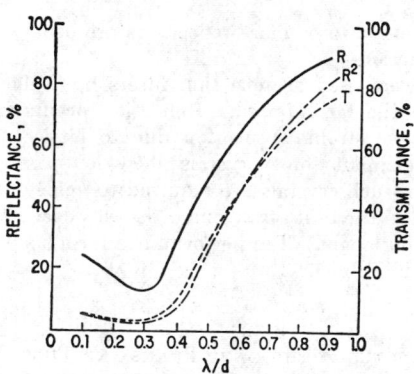

Fɪɢ. 6r-5. Comparison of the efficiency of reflection and transmission gratings. R and R^2 represent the characteristics of the reflection from one and two gratings, respectively, and T is the transmittance of a single-transmission filter grating. [*From K. D. Möller and R. V. McKnight, J. Opt. Soc. Am.* **55**, 1075 (1965).]

most of the near-infrared radiation passes unchopped. These plates are also useful for checking the spectral purity of a grating spectrometer. By choosing the proper material and thickness, the first-order radiation of a grating can be eliminated while the overlapping higher orders at shorter wavelengths are allowed to pass. Since only unwanted radiation is recorded, a good estimate of the spectral purity is obtained. The cutoff wave numbers for some typical reststrahlen plates are summarized in Table 6r-2.

TABLE 6r-2. WAVE NUMBERS CORRESPONDING TO A TRANSMITTANCE
OF 10% FOR RESTSTRAHLEN CRYSTAL PLATES*

Material	Thickness, mm	Cutoff wave number (10 % T)
NaF	10.0	830
BaF₂	9.1	770
NaCl	6.0	490
KCl	5.3	380
KRS-6	6.0	340
KBr	4.0	260
KRS-5	5.1	210
CsBr	5.1	200
CsI	5.0	130

* Data from E. K. Plyler and L. R. Blaine, *J. Res. NBS* **64**, 55 (1960), and S. S. Ballard et al., State-of-the-Art Report: Optical Materials for Infrared Instrumentation, University of Michigan, 1959.

Reflection Filters. Reststrahlen crystals, scatter plates, gratings in the zero order and wire meshes have been employed as reflection filters. They are a rigid part of the optical system, and a filter change must be performed carefully so that the optical alignment is not disturbed. To allow for quick interchanges without the need for realignment, filter-wheel assemblies with three or four positions are generally used, but at a cost of a certain amount of instrument space. Further complications in the optical path result because two or more reflections are usually necessary to achieve a satisfactory filtering action.

Reststrahlen plates were used as reflection filters by Rubens and others in the earliest experiments in the far infrared. Reflection occurs in the spectral range where these crystals have strong absorption due to lattice vibration. These are basically bandpass filters, but some crystals show low-pass characteristics. Two or more reflections from such crystals are required to achieve adequate attenuation in the band-stop region. The best arrangement is two reflection plates set in crossed positions at the polarizing angle. The performance for a single reflection from such crystals is summarized in Table 6r-3.

TABLE 6r-3. WAVELENGTHS OF THE BAND PEAKS AND THOSE CORRESPONDING TO
THE REFLECTIVITY OF 50% FOR VARIOUS RESTSTRAHLEN CRYSTALS*

Reststrahlen crystal	Peak wavelength, μm	Wavelength at 50% level, μm
NaF	34	27 – 42.5
CaF$_2$	34	21.3– 41.2
BaF$_2$	45	31 – 60
NaCl	53	44.3– 65
KCl	63	54 – 72
KBr	79	68.5– 88
KI	92	82 –101
CsBr	122	95 –143
CsI	145	124 –170
TlCl	130	63
TlBr	170	95
KSR-6	155	75
KRS-5	170	112

* From A. Mitsuishi et al., *J. Opt. Soc. Am.* **52**, 14 (1962).

Scatter plates are metal mirrors with an abraded surface which scatter wavelengths shorter than the dimensions of the abrasions in all directions. Such plates ground with carborundum of grades 120, 220, and 320 give cut-on points at about 70, 125, and 180 cm^{-1}, respectively [5,6].

In the case of reflection grating, the wavelengths of interest are reflected in zero order while shorter wavelengths are diffracted out of the beam. Their performance depends on the groove shape and spacing and the angle of incidence of the radiation. Two reflections from such gratings are required to duplicate the performance of a single transmission filter grating and four such reflections to achieve an adequate band-stop attenuation [7,8]. R^2 in Fig. 6r-5 shows the performance of a set of two such filters, while that of a single filter is given by R.

Wire-cloth meshes [9] are the most efficient filters presently in use in the far infrared. These scatter or transmit the unwanted short-wavelength radiation while reflecting the longer wavelengths specularly. Their filter characteristics depend on the wire diameter and spacing as well as on the angle of incidence. The reflectance for five different meshes appears in Fig. 6r-6. In all cases the ratio of the wire diameter to wire spacing was between 0.35 and 0.46 with an angle of incidence of 15 deg. As with other reflection filters, two or more reflections are generally used. The characteristics of the reflectivity of a single mesh and of a set of two and three meshes are

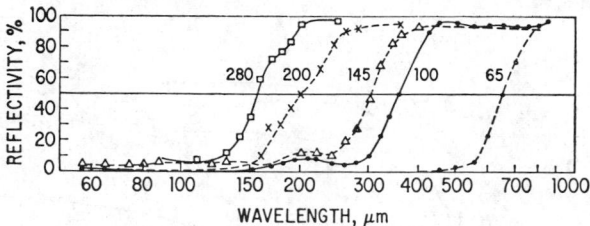

FIG. 6r-6. Reflectance of wire cloth meshes of various mesh number at 15 deg angle of incidence. [*From A. Mitsiushi et al., Japan J. Appl. Phys.* **2**, 574 (1963).]

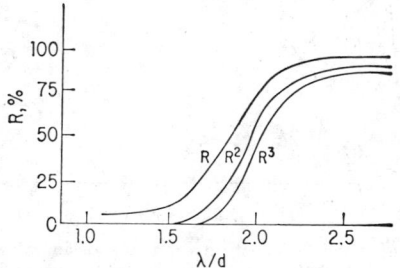

FIG. 6r-7. Reflectance of a typical wire-cloth mesh as a function of λ/d. R, R^2, R^3 represent the effect of one, two, and three reflections, respectively. [*From K. D. Möller et al., J. Opt. Soc. Am.* **55**, 1233 (1965).]

shown in Fig. 6r-7 for a typical mesh. Self-supporting electroformed metallic meshes may also be used as reflection filters, but are not as effective as wire-cloth meshes [10].

Interference Filters. Interference filters for the far infrared making use of the interference between two electroformed metallic meshes have been described by several workers [11–13]. These two-grid filters operating in a high order of interference produce a series of extremely narrow transmission peaks. The complementary structure of metallic meshes for the far infrared was investigated by Ulrich [14]. This complementary structure is made by depositing copper in the form of squares on a Mylar substrate and is referred to as a "capacitive" grid, in contrast to the usual metallic mesh or "inductive" grid. Interference filters consisting of two capacitive

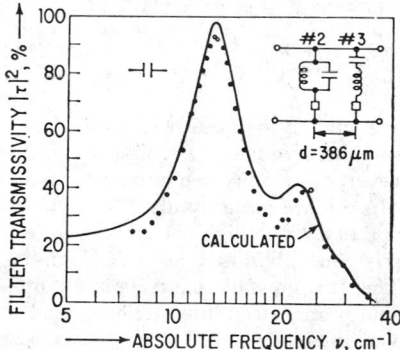

FIG. 6r-8. The transmittance of an interference filter consisting of one inductive and one capacitive grid of grid constant 368 and 250 μm, respectively. The separation of the grids is 386 μm. [*From R. Ulrich, Infrared Phys.* **7**, 37 (1967).]

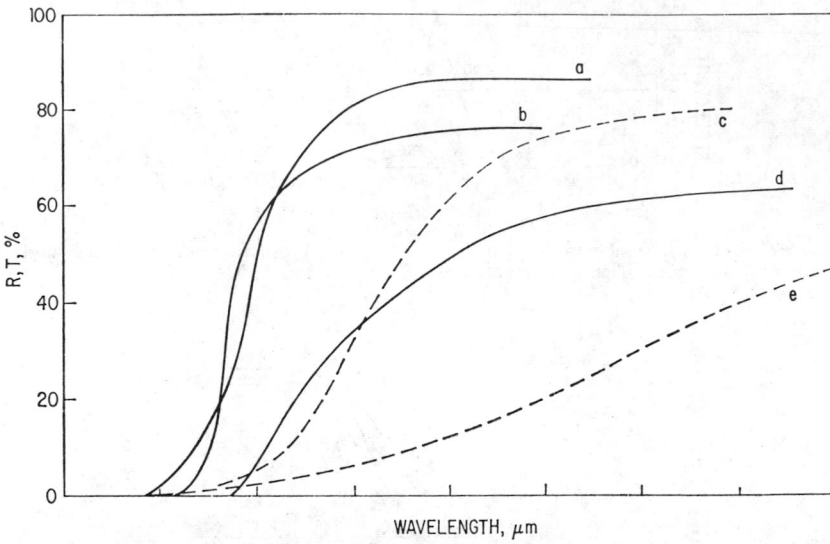

FIG. 6r-9. Characteristics of various reflection and transmission filters. (a) Three reflections from a typical wire-cloth mesh. (b) A crystal powder filter at 4.2 K containing NaCl, KCl, KBr, KI, CsBr, and CsI. (c) A four-grid interference filter of nonidentical grids. (d) The same filter as curve b at room temperature. (e) A set-of-two transmission filter gratings.

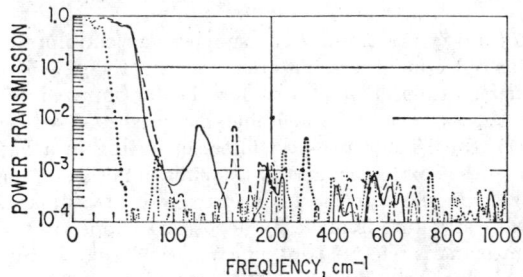

FIG. 6r-10. The transmittance of four-grid filters consisting of capacitive grids of different grid constants. (Grid constants $g_1 = g_2 = g_3 = 102$ μm, $g_4 = 51$ μm, and the spacers $s_1 = s_2 = 50$ μm, $s_3 = 40$ μm. $--- g_1 = g_4 = 25$ μm, $g_2 = g_3 = 51$ μm, and $s_1 = s_3 = 28$ μm, $s_2 = 20$ μm. $--- g_1 = g_4 = 25$ μm, $g_2 = g_3 = 51$ μm, and $s_1 = s_2 = s_3 = 20$ μm.) [From R. Ulrich, Appl. Opt. **7**, 1987 (1968).]

grids have characteristics similar to those of the two inductive grid filters except that the finesse increases with frequency. A filter consisting of one inductive and one capacitive grid, however, shows only one interference maximum. One such filter is shown in Fig. 6r-8. In a more recent study Ulrich [15] emphasized the need for using more than two grids to achieve high-performance filters in the far infrared and described multigrid interference filters with low-pass, high-pass, bandpass, and band-stop characteristics. The transmission characteristics of several low-pass filters with extremely wide band stops, steep cut-on slopes, and good attenuation in the band-stop region are given in Fig. 6r-10. These filters consisted of four nonidentical capacitive grids with varied spacers. Although in principle these filters may be scaled to any desired frequency range, in practice they are presently limited to frequencies below 100 cm^{-1}.

Low-temperature Filters. Cooled filters are necessary to minimize heating of a cooled sample under investigation and of the detector itself by radiation from room-temperature surroundings. Cooling increases the cut-on slope and transmittance of many filters. In addition, some materials which are opaque in the far infrared at room temperature become transparent at liquid helium temperature with good transmission characteristics. Many alkali halide crystals [16] and heavily doped silicon [17] are such examples. These filters are usually placed somewhere in the detector dewar for operation at low temperatures. The crystal powder filters described by Yamada et al. show improved performance at low temperatures [18], and crystal quartz, black polyethylene, and sapphire have also been found to be effective filters at these temperatures [19], with improved performance relative to room-temperature operation.

References for Sec. 6r-3

1. Yamada, Y., A. Mitsuishi, and H. Yoshinaga: *J. Opt. Soc. Am.* **52**, 17 (1962).
2. Möller, K. D., and R. V. McKnight: *J. Opt. Soc. Am.* **53**, 760 (1963).
3. Möller, K. D., and R. V. McKnight, *J. Opt. Soc. Am.* **55**, 1075 (1965).
4. Plyler, E. K., and L. R. Blaine: *J. Res. NBS* **64**, 55 (1960).
5. Bloor, D., T. J. Dean, G. O. Jones, D. H. Martin, P. A. Mawer, and C. H. Perry: *Proc. Roy. Soc. (London)*, ser. A, **260**, 510 (1961).
6. Robinson, D. W.: *J. Opt. Soc. Am.* **49**, 966 (1959).
7. White, J. U.: *J. Opt. Soc. Am.* **37**, 713 (1947).
8. Möller, K. D., V. P. Tomaselli, L. R. Skube and B. K. McKenna: *J. Opt. Soc. Am.* **55**, 1233 (1965).
9. Mitsuishi, A., Y. Otsuka, S. Fujita, and H. Yoshinaga: *Japan. J. Appl. Phys.* **2**, 574 (1963).
10. Ressler, G. M., and K. D. Möller: *Appl. Opt.* **6**, 893 (1967).
11. Rawcliffe, R. D., and C. M. Randall: *Appl. Opt.* **6**, 1353 (1967).
12. Ulrich, R., K. F. Renk, and L. Genzel: *IEEE Trans. on MTT* **11**, 363 (1963).
13. Renk, K. F., and L. Genzel: *Appl. Opt.* **1**, 643 (1962).
14. Ulrich, R.: *Infrared Phys.* **7**, 37 (1937).
15. Ulrich, R.: *Appl. Opt.* **7**, 1987 (1968).
16. Hadni, A., J. Claudel, X. Gerbaux, G. Morlot, and J. M. Munier: *Appl. Opt.* **4**, 487 (1965).
17. Neuringer, L. J., and R. C. Milward: *Appl. Opt.* **6**, 978 (1967).
18. Zwerdling, S., and J. P. Theriault: *Appl. Opt.* **7**, 209 (1968).
19. Wheeler, R. G., and J. C. Hill: *J. Opt. Soc. Am.* **56**, 657 (1966).

6r-4. Calibration. The pure rotation spectra of simple gases can be used to calibrate the far-infrared region. A number of such gases and the regions in which they are useful for calibration are given in Table 6r-4. The wave numbers of the rotational lines for each gas appear in Tables 6r-5 to 6r-8. Calibration can also be accomplished by using the higher orders of diffraction of a strong visible or near-infrared line, such as the mercury green line or a laser line.

TABLE 6r-4. CALIBRATION GASES FOR THE FAR-INFRARED REGION

Calibration gas	Useful calibration region, cm⁻¹	Number of lines	Dipole moment,* debyes	Average spacing, cm⁻¹	J_{max}
H_2O	18–400	~250	1.85	Irregular	
CO	4–120	30	0.112	3.8	7
HCN	3–120	40	2.98	2.9	8
N_2O	1–40	60	0.167	0.8	15
CH_3Cl	1–40	45	1.87	0.9	14
HCl	20–300	15	1.08	20.9	3

* From R. D. Nelson, Jr. et al., *Selected Values of Electric Dipole Moments for Molecules in the Gas Phase*, National Bureau of Standards Report *NSRDS-NBS* 10 (September 1967).

Table 6r-5. Calculated Wave Numbers of the Pure Rotational Spectrum of CO, HCN, N$_2$O Molecules*

J	C^{12}O^{16}	N$_2$^{14}O^{16}	HC^{12}N^{14}	J	C^{12}O^{16}	N$_2$^{14}O^{16}	HC^{12}N^{14}
0	3.84$_5$	0.83$_8$	2.95$_6$	25	99.54$_1$	21.77$_6$	76.66$_3$
1	7.69$_0$	1.67$_6$	5.91$_3$	26	103.33$_4$	22.61$_3$	79.59$_5$
2	11.53$_4$	2.51$_4$	8.86$_9$	27	107.12$_4$	23.44$_9$	82.52$_4$
3	15.37$_9$	3.35$_2$	11.82$_5$	28	110.90$_9$	24.28$_5$	85.45$_3$
4	19.22$_2$	4.19$_0$	14.78$_1$	29	114.69$_0$	25.12$_2$	88.37$_9$
5	23.05$_5$	5.02$_8$	17.73$_6$	30	118.46$_7$	25.95$_8$	91.30
6	26.90$_7$	5.86$_6$	20.69$_1$	31		26.79$_4$	94.22
7	30.74$_8$	6.70$_4$	23.64$_6$	32		27.62$_9$	97.14
8	34.58$_8$	7.54$_2$	26.59$_9$	33		28.46$_5$	100.06
9	38.42$_6$	8.38$_0$	29.55$_3$	34		29.30$_1$	102.98
10	42.26$_3$	9.21$_7$	32.50$_5$	35		30.13$_6$	105.89
11	46.09$_8$	10.05$_5$	35.45$_7$	36		30.97$_1$	108.80
12	49.93$_2$	10.89$_3$	38.40$_8$	37		31.80$_6$	111.71
13	53.76$_3$	11.73$_0$	41.35$_8$	38		32.64$_1$	114.61
14	57.59$_3$	12.56$_8$	44.30$_7$	39		33.47$_6$	117.51
15	61.42$_0$	13.40$_5$	47.25$_5$	40		34.31$_0$	120.41
16	65.24$_5$	14.24$_3$	50.20$_2$	41		35.14$_5$	
17	69.06$_8$	15.08$_0$	53.14$_8$	42		35.97$_9$	
18	72.88$_8$	15.91$_8$	56.09$_2$	43		36.81$_3$	
19	76.70$_5$	16.75$_5$	59.03$_6$	44		37.64$_7$	
20	80.51$_9$	17.59$_2$	61.97$_7$	45		38.48$_1$	
21	84.33$_0$	18.42$_9$	64.91$_8$	46		39.31$_4$	
22	88.13$_8$	19.26$_6$	67.85$_6$	47		40.14$_7$	
23	91.94$_3$	20.10$_3$	70.79$_3$	48		40.98$_0$	
24	95.74$_4$	20.94$_0$	73.72$_9$	49		41.81$_3$	
				50		42.64$_6$	

* From K. N. Rao et al., *J. Res. NBS* **67A**, 351 (1963).
Path length 40 cm at pressures 2 to 3 cm of Hg for HCN, and 40 to 60 cm of Hg for CO and N$_2$O.

TABLE 6r-6. CALCULATED WAVE NUMBERS OF THE PURE
ROTATIONAL SPECTRUM OF METHYL CHLORIDE*

$$\nu = 2B(J + 1) - 4D(J + 1)^3$$

J	CH_3Cl^{35} $2B = 0.8868$ cm^{-1} $4D = 2.4\cdot10^{-6}$ cm^{-1} ν, cm^{-1}	CH_3Cl^{37} $2B = 0.8731$ cm^{-1} $4D = 3.6\cdot10^{-6}$ cm^{-1} ν, cm^{-1}	J	CH_3Cl^{35} $2B = 0.8868$ cm^{-1} $4D = 2.4\cdot10^{-6}$ cm^{-1} ν, cm^{-1}	CH_3Cl^{37} $2B = 0.8731$ cm^{-1} $4D = 3.6\cdot10^{-6}$ cm^{-1} ν, cm^{-1}
0	0.886_8	0.8731	11	10.6_4	10.47
1	1.77_4	1.746	12	11.52	11.34
2	2.660	2.619	13	12.4_1	12.21
3	3.547	3.492	14	13.29	13.0_9
4	4.43_4	4.365	15	14.1_8	13.9_6
5	5.320	5.238	16	15.06	14.8_3
6	6.207	6.11_1	17	15.9_5	15.7_0
7	7.093	6.983	18	16.83	16.5_7
8	7.979	7.85_6	19	17.7_2	17.43
9	8.86_6	8.72_8	20	18.60	18.30
10	9.75_2	9.60_0			

* From K. D. Möller et al., *J. Opt. Soc. Am.* **55**, 1233 (1965).

TABLE 6r-7. CALCULATED WAVE NUMBERS OF THE PURE
ROTATIONAL SPECTRUM OF HYDROGEN CHLORIDE*

$J'' \to J'$	HCl^{35}	HCl^{37}	$J'' \to J'$	HCl^{35}	HCl^{37}
$0 \to 1$	20.878_2	20.846_8	$8 \to 9$	186.38_7	186.10_8
$1 \to 2$	41.743_7	41.681_0	$9 \to 10$	206.69_8	206.38_8
$2 \to 3$	65.583_9	62.489_9	$10 \to 11$	226.88_3	226.54_3
$3 \to 4$	83.386_1	83.260_8	$11 \to 12$	246.93_0	246.56_0
$4 \to 5$	104.137_7	103.981_3	$12 \to 13$	266.82_8	266.42_9
$5 \to 6$	124.82_6	124.63_9	$13 \to 14$	286.56_4	286.13_6
$6 \to 7$	145.43_9	145.22_1	$14 \to 15$	306.12_7	305.66_9
$7 \to 8$	165.96_3	165.71_4			

* D. H. Martin, "Spectroscopic Techniques," North-Holland Publishing Company, Amsterdam, 1967.

TABLE 6r-8. PURE ROTATIONAL WATER-VAPOR ABSORPTION LINES*

Wave number, cm⁻¹	Intensity,† grams/cm²	Assignment						Wave number, cm⁻¹	Intensity,† grams/cm²	Assignment					
0.742	0.01	6	1	6	5	2	3	73.262	7,170	3	3	0	3	2	1
6.115	2.63	3	1	3	2	2	0	74.109	7,350	5	1	4	5	0	5
10.846	3.06	5	1	5	4	2	2	74.881	113	8	3	6	7	4	3
12.683	28.0	4	1	4	3	2	1	75.523	10,100	4	2	3	4	1	4
14.645	2.37	6	4	3	5	5	0	77.322	280	9	4	5	9	3	6
14.944	29.2	4	2	3	3	3	0	78.200	2,690	7	2	5	7	1	6
15.834	3.67	5	3	3	4	4	0	78.918	2,670	3	3	1	3	2	2
18.577	1,790	1	1	0	1	0	1	79.774	10,100	4	0	4	3	1	3
20.705	19.1	5	3	2	4	4	1	80.999	404	9	3	6	9	2	7
25.085	1,180	2	1	1	2	0	2	81.622	254	8	4	4	8	3	5
30.561	48.4	4	2	2	3	3	1	82.155	9,700	4	3	2	4	2	3
32.367	54.3	5	2	4	4	3	1	85.636	351	7	3	4	6	4	3
32.953	858	2	0	2	1	1	1	87.760	2,900	5	3	3	5	2	4
36.604	5,590	3	1	2	3	0	3	88.076	39,600	4	1	4	3	0	3
37.137	1,710	1	1	1	0	0	0	88.882	1,840	7	4	3	7	3	4
38.245	3.65	7	2	5	8	1	8	89.583	3,060	5	2	4	5	1	5
38.464	862	3	1	2	2	2	1	92.528	34,200	2	2	1	1	1	0
38.640	82.0	6	3	4	5	4	1	96.070	6,000	6	3	4	6	2	5
38.790	6,090	3	2	1	3	1	2	96.208	2,110	6	1	5	6	0	6
38.965	3.66	8		4	7	6	1	96.231	1,230	6	4	2	6	3	3
39.113	6.66	7	4	4	6	5	1	98.808	793	6	2	4	5	3	3
39.715	1.27	8	5	3	7	6	2	99.025	10,900	2	2	0	1	1	1
40.283	1,900	4	2	2	4	1	3	99.095	12,100	5	1	4	4	2	3
40.988	1,640	2	2	0	2	1	1	100.026	555	8	2	6	8	1	7
42.640	23.9	7	4	3	6	5	2	100.509	42,600	5	0	5	4	1	4
43.240	23.1	8	2	7	7	3	4	101.529	5,780	5	4	1	5	3	2
43.639	1.65	8	4	5	9	1	8	104.293	1,960	4	4	0	4	3	1
44.099	192	6	2	5	5	3	2	104.573	15,700	5	1	5	4	0	4
44.859	1.22	7	4	4	8	1	7	105.592	6,020	4	4	1	4	3	2
47.055	4,860	5	2	3	5	1	4	105.659	7,150	6	2	5	6	1	6
48.058	31	7	2	6	6	3	3	106.147	2,090	5	4	2	5	3	3
53.444	2,360	4	1	3	4	0	4	107.091	1,250	7	3	5	7	2	6
55.405	6,190	2	2	1	2	1	2	107.747	4,500	6	4	3	6	3	4
55.701	14,700	2	1	2	1	0	1	111.051	880	7	4	4	7	3	5
57.265	13,800	3	0	3	2	1	2	111.124	13,500	3	2	2	2	1	1
58.777	1,040	6	3	3	6	2	4	116.596	1,340	8	4	5	8	3	6
59.871	1,270	6	2	4	6	1	5	117.066	288	9	5	4	9	4	5
59.950	1,670	7	3	4	7	2	5	117.969	4,560	7	1	6	7	0	7
62.301	5,330	5	3	2	5	2	3	120.072	15,000	6	0	6	5	1	5
63.996	989	5	2	3	4	3	2	120.523	2,070	8	3	6	8	2	7
64.022	3,090	3	2	2	3	1	3	121.905	46,700	6	1	6	5	0	5
67.249	281	8	3	5	8	2	6	122.415	155	8	3	5		4	4
68.062	2,500	4	3	1	4	2	2	122.847	895	9	2	7	9	1	8
69.196	1,700	4	1	3	3	2	2	123.128	1,610	7	2	6	7	1	7
72.187	9,130	3	1	3	2	0	2	124.137	256	8	5	3	8	4	4

TABLE 6r-8. PURE ROTATIONAL WATER-VAPOR ABSORPTION LINES* (Continued)

Wave number, cm⁻¹	Intensity,† grams/cm	Assignment						Wave number, cm⁻¹	Intensity,† grams/cm²	Assignment					
124.659	200	9	4	6	9	3	7	173.282	4,300	8	1	7	7	2	6
126.697	5,740	6	1	5	5	2	4	173.500	8,370	4	2	2	3	1	3
126.995	41,500	4	2	3	3	1	2	176.010	18,200	9	0	9	8	1	8
128.599	1,650	7	5	2	7	4	3	176.151	6,080	9	1	9	8	0	8
130.856	909	6	5	1	6	4	2	177.540	22,600	4	3	1	3	2	2
131.742	2,990	5	5	0	5	4	1	178.474	235	0	1	9	0	0	0
131.877	2,770	6	5	2	6	4	3	178.663	271	7	7	0	7	6	1
131.904	578	7	5	3	7	4	4	179.073	709	0	2	9	0	1	0
131.966	1,000	5	5	1	5	4	2	179.106	200	8	7	2	8	6	3
132.459	871	8	5	4	8	4	5	181.389	14,500	8	2	7	7	1	6
132.658	33,000	3	2	1	2	1	2	183.465	1,120	5	3	2	5	0	5
133.433	3,450	7	2	5	6	3	4	188.189	15,700	5	3	3	4	2	2
134.097	124	9	5	5	9	4	6	193.480	7,990	9	1	8		2	7
135.213	241	0	4	7	0	3	8	194.328	3,410	0	0	0	9	1	9
135.855	338	9	3	7	9	2	8	194.387	10,200	0	1	0	9	0	9
137.385	139	0	5	6	0	4	7	195.804	2,650	9	2	7	8	3	6
138.823	935	8	1	7	8	0	8	197.256	108	0	3	7	9	4	6
138.993	38,800	7	0	7	6	1	6	197.498	2,810	9	2	8	8	1	7
139.785	13,100	7	1	7	6	0	6	197.719	297	1	1	0	1	0	1
140.711	12,500	5	2	4	4	1	3	202.470	29,700	6	3	4	5	2	3
141.435	2,880	8	2	7	8	1	8	202.690	89,800	4	4	1	3	3	0
144.958	139	0	2	8	0	1	9	202.915	30,000	4	4	0	3	3	1
148.655	535	3	3	0	3	0	3	208.451	47,400	5	3	2	4	2	3
149.054	26,400	3	3	1	2	2	0	210.884	476	5	4	1	5	1	4
150.515	80,000	3	3	0	2	2	1	212.566	5,100	1	0	1	0	1	0
151.303	17,000	7	1	2	6	2	5	212.591	1,700	1	1	1	0	0	0
152.507	444	0	3	8	0	2	9	212.633	1,410	0	1	9	9	2	8
153.455	30,300	6	2	5	5	1	4	213.924	5,690	7	3	5	6	2	4
154.088	208	9	6	3	9	5	4	214.556	4,330	0	2	9	9	1	8
155.736	159	8	6	2	8	5	3	214.855	218	6	3	3	6	0	6
156.372	867	7	6	1	7	5	2	214.878	217	6	4	2	6	1	5
156.447	350	6	6	0	6	5	1	215.126	362	3	3	1	2	0	2
156.451	483	8	6	3	8	5	4	216.876	112	2	2	1	2	1	2
156.480	1,050	6	6	1	6	5	2	221.673	15,800	5	2	3	4	1	4
156.556	290	7	6	2	7	5	3	221.735	518	0	2	8	9	3	7
157.588	9,510	8	0	8	7	1	7	223.712	9,060	8	3	6	7	2	5
157.923	28,700	8	1	8	7	0	7	226.273	21,200	5	4	2	4	3	1
158.904	1,500	9	1	8	9	0	9	227.030	523	7	4	3	7	1	6
160.169	505	9	2	8	9	1	9	227.825	64,300	5	4	1	4	3	2
160.207	443	9	3	6	8	4	5	230.732	756	2	0	2	1	1	1
161.789	385	4	3	1	4	0	4	230.743	2,270	2	1	2	1	0	1
165.829	169	1	2	9	1	1	0	231.213	1,960	1	1	0	0	2	9
166.217	1,170	8	2	6	7	3	5	231.874	188	1	3	8	0	4	7
166.704	7,410	7	2	6	6	1	5	232.118	659	1	2	0	0	1	9
170.359	66,200	4	3	2	3	2	1	233.327	1,490	9	3	7	8	2	6

TABLE 6r-8. PURE ROTATIONAL WATER-VAPOR ABSORPTION LINES* (*Continued*)

Wave number, cm^{-1}	Intensity,† grams/cm^2	Assignment						Wave number, cm^{-1}	Intensity,† grams/cm^2	Assignment					
244.216	2,040	0	3	8	9	2	7	311.744	146	2	4	9	1	3	8
244.535	737	1	2	9	0	3	8	314.741	382	4	4	1	3	1	2
245.344	8,920	6	3	3	5	2	4	315.088	3,060	8	4	4	7	3	5
245.753	3,690	4	3	2	3	0	3	323.633	5,330	6	3	4	5	0	5
247.915	38,500	6	4	3	5	3	2	323.935	9,280	8	5	4	7	4	3
248.826	904	3	0	3	2	1	2	327.571	5,060	7	6	2	6	5	1
248.831	301	3	1	3	2	0	2	327.610	15,200	7	6	1	6	5	2
249.477	268	2	1	1	1	2	0	328.173	3,160	8	5	3	7	4	4
249.900	808	2	2	1	1	1	0	334.617	435	5	4	2	4	1	3
253.814	13,200	6	4	2	5	3	3	335.160	4,700	7	2	5	6	1	6
253.946	20,500	5	5	1	4	4	0	340.556	1,710	8	3	5	7	2	6
253.975	61,600	5	5	0	4	4	1	343.212	1,220	9	5	5	8	4	4
256.117	272	7	3	4	7	0	7	349.792	3,160	7	7	1	6	6	0
257.109	281	1	3	9	0	2	8	349.792	9,500	7	7	0	6	6	1
266.199	6,600	7	4	4	6	3	3	351.786	7,120	8	6	3	7	5	2
266.843	108	4	0	4	3	1	3	352.006	2,380	8	6	2	7	5	3
266.845	325	4	1	4	3	0	3	354.125	3,390	9	4	5	8	3	6
267.552	295	3	1	2	2	2	1	354.595	3,850	9	5	4	8	4	5
271.851	316	2	3	0	1	2	9	357.270	2,370	6	4	3	5	1	4
276.150	2,960	6	2	4		1	5	358.492	1,250	0	5	6	9	4	5
277.430	123	7	5	2	7	2	5	369.343	124	1	5	7	0	4	6
278.263	37,700	6	5	2	5	4	1	370.002	1,250	7	3	5	6	0	6
278.523	12,600	6	5	1	5	4	2	374.521	4,420	8	7	2	7	6	1
280.358	8,840	8	4	5	7	3	4	374.527	1,470	8	7	1	7	6	2
281.168	101	9	4	5	9	1	8	375.342	976	9	6	4	8	5	3
281.915	1,870	5	3	3	4	0	4	376.224	2,940	9	6	3	8	5	4
282.263	20,800	7	4	3		3	4	376.377	100	2	5	8	1	4	7
284.381	103	3	2	1	2	3	0	383.826	447	0	5	5	9	4	6
284.778	105	5	0	5	4	1	4	384.845	904	7	4	4	6	1	5
289.451	12,500	7	3	4	6	2	5	385.502	400	5	4	1	4	1	4
290.737	1,180	9	4	6	8	3	5	394.272	2,360	8	8	1	7	7	0
298.430	1,310	0	4	7	9	3	6	394.272	786	8	8	0	7	7	1
301.871	6,710	7	5	3	6	4	2	396.435	791	8	2	6	7	1	7
303.001	28,500	6	6	1	5	5	0	397.325	1,950	9	3	6	8	2	7
303.005	9,510	6	6	0	5	5	1	397.681	1,050	0	6	5	9	5	4
303.116	20,300	7	5	2	6	4	3	398.959	606	9	7	3	8	6	2
304.895	151	1	4	8	0	3	7	398.994	1,820	9	7	2	8	6	3
309.474	116	6	4	3	6	1	6								

* Only the following lines are included in the table:

$$50 < \nu < 400 \text{ cm}^{-1} \quad \text{and} \quad I > 100$$
$$1 < \nu < 50 \text{ cm}^{-1} \quad \text{and} \quad I > 1$$
$$\nu < 1 \text{ cm}^{-1} \quad \text{all}$$

Private Communication from Clough, S. A. and W. S. Benedict.
† Intensity values are good at best to three significant figures.

6r-5. Far-infrared Polarizers. Polarizers for the far infrared have been made of stacks of dielectric plates at the Brewster angle, wire grids, and pyrolitic graphite. In addition, a Michelson interferometer acts as a polarizer when the radiant flux is incident on the beam splitter at the Brewster angle (cf. section on beam splitters).

Pile-of-plates polarizers have been discussed by Bird and Schurcliff [1], and a polarizer using polyethylene sheets has been reported by Mitsuishi et al [2]. The light is incident on the plates at the Brewster angle, and the polarizance of the device is

$$P = \frac{1 - \left(\dfrac{2n}{n^2 + 1}\right)^{4m}}{1 + \left(\dfrac{2n}{n^2 + 1}\right)^{4m}} \tag{6r-1}$$

where n is the refractive index and m the number of plates. (Polarizance is defined as the percent polarization of the output beam when the input is completely unpolarized.) For far-infrared polarizers two different plate thicknesses must be used to avoid interference effects which seriously reduce the polarizance at certain wavelengths. This polarizer is more easily built in the laboratory than the grating polarizer, but occupies more instrument space. Figure 6r-11 illustrates the polarizance of pile-of-plates polarizers using various combinations of polyethylene ($n = 1.5$) sheets. Equation (6r-1) gives $P = 88$ percent with 10 sheets and 97.5 percent with 15 sheets.

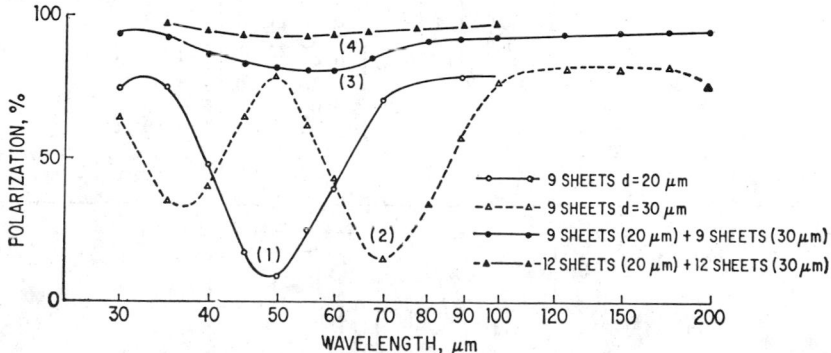

FIG. 6r-11. Degree of polarization with different numbers and thicknesses of polyethylene sheets.

The wire grid operates on the principle (discovered by Hertz) that radiation polarized parallel to the grids is reflected, while that polarized perpendicular is transmitted, for wavelengths larger than the grid constant. These polarizers have been made by evaporating metal at a large angle of incidence onto transmission gratings of the appropriate spacing so that one side of each groove is coated while the other remains transparent. The results obtained by Hass and O'Hara [3] are summarized in Tables 6r-9 and 6r-10, and the transmittance of their polarizers is shown in Fig. 6r-12.

TABLE 6r-9. DESCRIPTIONS OF POLARIZERS

Designation	Source of grating	Substrate and thickness	Conductor	Periodicity
DP1	Diffraction products	Polymethyl methacrylate, 0.051 mm	Aluminum (lightly coated)	2,160 grooves/mm* = 0.463 μm/groove
DP2	Diffraction products	Polymethyl methacrylate, 0.051 mm	Aluminum (heavily coated)	2,160 grooves/mm* = 0.463 μm/groove
NRL	Naval Research Lab.	Polyethylene, 0.152 mm	Aluminum (medium coat)	600 grooves/mm* = 1.69 μm/groove
BM	Buckbee Mears	Mylar sheet, 0.038 mm	Gold strips 0.01 mm wide	39.3 lines/mm = 25.4 μm/line

* The blaze angle is about 20°.

TABLE 6r-10. TRANSMITTANCE AND DEGREE OF POLARIZATION

Wave number, cm^{-1}	Degree of polarization P, %				Transmittance T_1			
	DP1	DP2	NRL	BM	DP1	DP2	NRL	BM
2.5	99.0	....	....	>99.5	0.985	....	....	>0.995
49.5	97.8	....	96.4	98.4	0.86	....	0.87	0.81
83	98.8	....	97.9	98.4	0.86	....	0.88	0.86
160	98.9	99.0	98.0	89.0	0.86	0.80	0.83	0.67
300	98.1	....	96.6	......	0.65	....	0.84	
600	98.2	....	96.0	......	0.94	....	0.81	
1,025	96.3	99.4	89.0	......	0.86	0.53	0.57	
2,000	88.0	99.5	63.0	......	0.90	0.65	0.43	
3,500	71.0	98.4	33.0	......	0.90	0.54	0.35	
5,710	....	95.0	....	......	....	0.39		
10,000	....	84.1	....	......	.....	0.27		

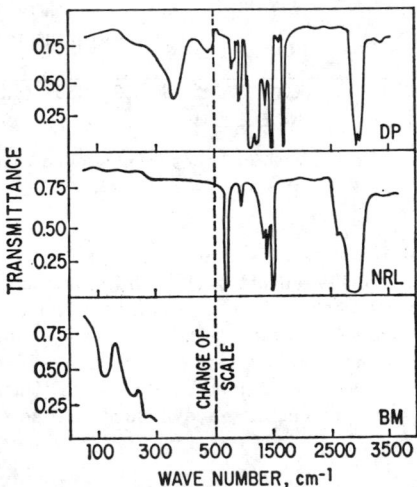

FIG. 6r-12. Transmittance of gratings. The DP polymethyl methacrylate grating and the NRL polyethylene grating were unaluminized and measured in unpolarized radiation. The BM metal-strip grating was measured in the high-transmission direction in polarized radiation.

TABLE 6r-11. TRANSMITTANCE OF PYROGRAPHITE POLARIZER PGP1 FOR
RADIATION WITH ELECTRIC FIELD IN THE c DIRECTION

Wave number, cm^{-1}	$T_1 \pm 2\%$
17.1	0.519
22.7	0.504
33.3	0.512
42.0	0.487
51.0	0.495
58.8	0.519
66.2	0.520
71.0	0.505
77.0	0.507
81.5	0.494

TABLE 6r-12. PERCENTAGE POLARIZATION OF PYROGRAPHITE POLARIZER PGP1

Wave number, cm^{-1}	$T_2 \times 10^{3}$*	Percentage polarization
16.7	1.7	99.65 ± 0.35
21.7	4.5	99.11 ± 0.21
28.6	2.4	99.53 ± 0.06
500	7.5	98.03
666.7	10.7	97.45
1,000	11.4	96.20
2,000	7.7	93.58

* T_2 is the transmittance for the unwanted direction of polarization.

A thin foil of pyrolitic graphite, which has a layered crystal structure, acts as a polarizer [4] in both the far and the near infrared. The transmittance for the desired polarization is rather low (about 50 percent), but the polarizance is above 99 percent. The results obtained by Rupprecht et al. [4] are summarized in Tables 6r-11 and 6r-12.

References for Sec. 6r-5

1. Bird, G. R., and W. A. Schurcl ff: *J. Opt. Soc. Am.* **49**, 235 (1959).
2. Mitsuishi, A., Y. Yamada, S. Fujita, and H. Yoshinaga: *J. Opt. Soc. Am.* **50**, 433 (1960).
3. Hass, M., and M. O'Hara: *Appl. Opt.* **4**, 1027 (1965).
4. Rupprecht, G., D. M. Ginsberg, and J. D. Leslie: *J. Opt. Soc. Am.* **52**, 665 (1962).

6r-6. Optical Constants of Far-infrared Materials. Precise values of refractive index and reasonably good values of absorption coefficient have been determined for far-infrared materials by two techniques. Both are basically interferometric: one is the use of a Michelson Fourier spectrometer with the sample in one arm [1,2], referred to as an "asymmetric Michelson"; the other is the analysis of the channel-spectrum fringes (fringes of equal chromatic order) resulting from interference between the multiple beams produced by internal reflections in a plane-parallel sample of material [3]. In the asymmetric Michelson method, the sample is placed in one arm, and an interferogram is taken; the amplitude of the resulting spectrum gives the absorption coefficient while the phase gives the refractive index. The analysis of the channel spectra is based on the fact that the fringe position depends on the index only, whereas the amplitude depends on both index and absorption coefficient. The channel-spectrum fringes are revealed by spectra, which may be taken with either a conventional or a Fourier spectrometer.

In spite of the fact that the absorption coefficient can in theory be derived by the above methods, in most of the data given below it is derived from analysis of a low-resolution transmission spectrum, using the refractive index found in the inter-ferometric method. This is so because discrepancies between absorption coefficients calculated from the asymmetric Michelson or channel spectrum and those calculated from the transmission measurements are always resolved in favor of the latter.

The tables and graphs below list the optical constants for the following materials:

Crystal quartz	Mylar (polyethylene terephthalate)
Sapphire	Irtran VI (hot-pressed CdTe)
Germanium	Teflon (polytetrafluoroethylene)
Silicon	CdTe (crystalline)
Fused quartz	GaAs (crystalline)

The quantites given are index and absorption coefficient α which is related to k, the imaginary part of the complex refractive index, by

$$\alpha = 4\pi k\sigma$$

where $\sigma =$ wave number of the radiation. Units of σ and α are cm^{-1} in all cases. Except where noted, measurements were made at room temperature.

The optical constants of Mylar are labeled with subscripts 1 and 2. *If* Mylar is uniaxial, 1 denotes the ordinary optical constants, 2 the extraordinary. Mylar is probably biaxial, but it is difficult to determine this for the far infrared. The samples were aligned by using a polarizing microscope.

TABLE 6r-13. OPTICAL CONSTANTS OF CRYSTAL QUARTZ FROM 20 TO 200 CM⁻¹*

Wave number, σ, cm⁻¹	Refractive indices†			Absorption coefficients,‡ cm⁻¹	
	n_0	n_e	$n_e - n_0$	α_0	α_e
20.2	2.1073	2.1541	0.0468		
25.2	2.1076	2.1561	0.0485		
30.2	2.1076	2.1560	0.0484	0.10	0.10
35.3	2.1083	2.1564	0.0481		
40.3	2.1093	2.1573	0.0480		
45.4	2.1105	2.1580	0.0475	0.15	0.12
50.4	2.1114	2.1590	0.0476		
55.4	2.1124	2.1602	0.0478		
60.5	2.1134	2.1615	0.0481	0.32	0.21
65.5	2.1147	2.1629	0.0482		
70.6	2.1159	2.1644	0.0485		
75.6	2.1175	2.1662	0.0487	0.47	0.37
80.6	2.1190	2.1679	0.0489		
85.7	2.1209	2.1699	0.0490		
90.7	2.1228	2.1718	0.0490	0.61	0.56
95.8	2.1248	2.1739	0.0491		
100.8	2.1269	2.1762	0.0493		
105.8	2.1291	2.1787	0.0496	0.90	0.83
110.9	2.1316	2.1815	0.0499		
115.9	2.1343	2.1842	0.0499		
120.9	2.1376	2.1872	0.0496	1.2	1.1
122.0	2.1383	(2.1877)§	0.0494	1.3	
123.0	2.1393	(2.1882)	0.0489	1.3	
124.0	2.1400	(2.1888)	0.0488	1.9	
125.0	2.1413	(2.1895)	0.0482	2.5	
126.0	2.1421	2.1902	0.0481	4.3	
127.0	2.1426	(2.1909)	0.0483	6.0	
128.0	2.1419	(2.1916)	0.0497	8.5	
129.0	2.1408	(2.1923)	0.0515	8.0	
130.0	2.1403	(2.1930)	0.0527	7.1	
131.0	2.1406	2.1937	0.0531	5.3	
132.0	2.1413	(2.1944)	0.0531	4.7	
133.0	2.1419	(2.1950)	0.0531	3.8	
134.0	2.1428	(2.1957)	0.0529	3.6	
135.0	2.1434	(2.1964)	0.0530	3.1	
136.1	2.1441	2.1971	0.0530	2.8	1.3
141.1	2.1478	2.2009	0.0531		
146.1	2.1515	2.2049	0.0534		
151.2	2.1553	2.2089	0.0536	2.8	1.9
156.2	2.1592	2.2131	0.0539		
161.3	2.1635	2.2177	0.0542		
166.3	2.1678	2.2222	0.0544	3.3	2.4
171.3	2.1725	2.2273	0.0548		
176.4	2.1773	2.2325	0.0552		
181.4	2.1826	2.2381	0.0555	4.1	3.2
186.5	2.1882	2.2440	0.0558		
191.5	2.1941	2.2502	0.0561		
196.5	2.2005				
201.6	2.2072				

* E. E. Russell and E. E. Bell, *J. Opt. Soc. Am.* **57**, 341 (1967).

† The total, estimated probable error in the measured values of the refractive indices is ±0.001 except at wave numbers less than 25 cm⁻¹ and greater than 175 cm⁻¹, where the error can be somewhat greater.

‡ The estimated probable error in the measured absorption coefficients is approximately ±100 % for $\alpha < 0.2$; ±50 % for $0.2 < \alpha < 0.4$; ±20 % for $0.4 < \alpha < 0.8$; and ±10 % for $\alpha > 0.8$.

§ The bracketed values of the extraordinary-ray refractive indices were interpolated from the neighboring values.

TABLE 6r-14. OPTICAL CONSTANTS OF SAPPHIRE FROM 20 TO 175 CM^{-1}*

Wave number, σ, cm^{-1}	Refractive indices†			Absorption coefficients,‡ cm^{-1}	
	n_o	n_e	$n_e - n_o$	α_o	α_e
20.2	3.0688	3.4111	0.3423		
25.2	3.0698	3.4129	0.3436		
30.2	3.0704	3.4134	0.3430	0.4	0.5
35.3	3.0720	3.4163	0.3443		
40.3	3.0740	3.4187	0.3447		
45.4	3.0752	3.4232	0.3480	1.7	2.2
50.4	3.0770	3.4260	0.3490		
55.4	3.0795	3.4294	0.3499		
60.5	3.0822	3.4334	0.3512	3.6	4.0
65.5	3.0843	3.4391	0.3548		
70.6	3.0870	3.4444	0.3574		
75.6	3.0906	3.4510	0.3604	4.9	7.6
80.6	3.0941	3.4569	0.3628		
85.7	3.0982	3.4625	0.3643		
90.7	3.1019	3.4689	0.3670	7.2	12.7
95.8	3.1060	3.4766	0.3706		
100.8	3.1103	3.4836	0.3733		
105.8	3.1147	3.4908	0.3761	9.9	17.8
110.9	3.1198	3.4993	0.3795		
115.9	3.1249	3.5081	0.3832		
120.9	3.1304	3.5185	0.3881	12.9	24.0
126.0	3.1357	3.5279	0.3922		
131.0	3.1422	3.5375	0.3953		
136.1	3.1485	3.5508	0.4023	15.7	29.6
141.1	3.1549	3.5612	0.4063		
146.1	3.1623	3.5746	0.4123		
151.2	3.1696	3.5856	0.4160	19.7	35.9
156.2	3.1765	3.6042	0.4277		
161.3	3.1854				
166.3	3.1921			26.2	
171.3	3.2018				
176.4	3.2113				

* E. E. Russell and E. E. Bell, *J. Opt. Soc. Am.* **57,** 543 (1967).

† The total estimated probable error of the measured values of the refractive indices is ±0.002 except at wave numbers less than 25 cm^{-1} and greater than 150 cm^{-1} where the error may be somewhat greater.

‡ The estimated probable error of the measured absorption coefficients is approximately 50 % for α < 1.0 cm^{-1}; ±20 % for 1.0 cm^{-1} < α < 20 cm^{-1}; ±30 % for α > 20 cm^{-1}.

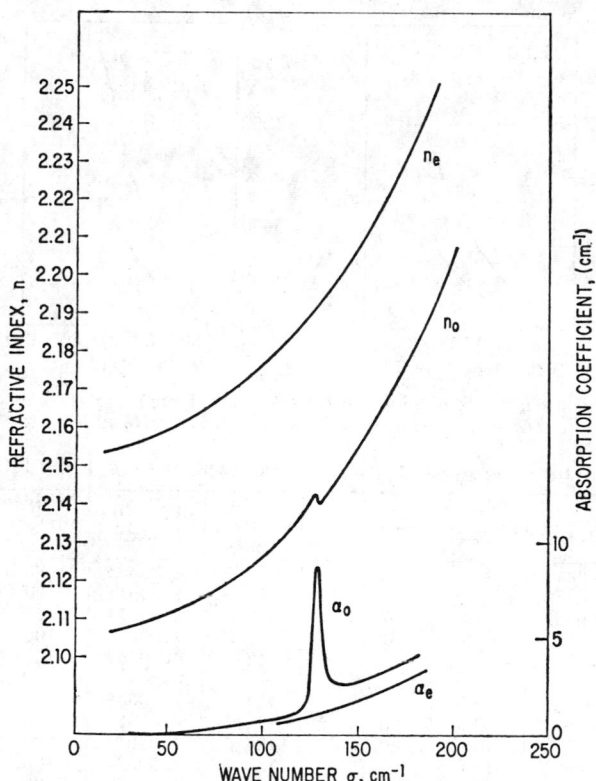

FIG. 6r-13a. Optical constants of crystal quartz from 20 to 200 cm⁻¹. [*E. E. Russell and E. E. Bell, J. Opt. Soc. Am.* **57**, 341 (1967).]

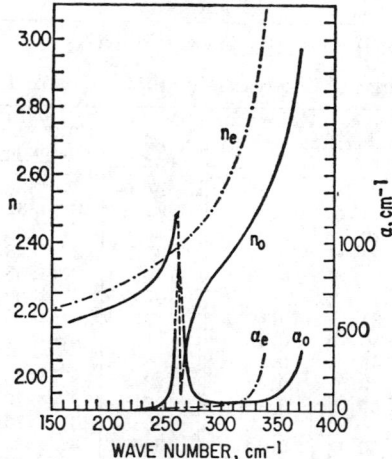

FIG. 6r-13b. Optical constants of crystal quartz from 150 to 370 cm⁻¹. [*E. E. Russell and E. E. Bell, J. Opt. Soc. Am.* **57**, 341 (1967).]

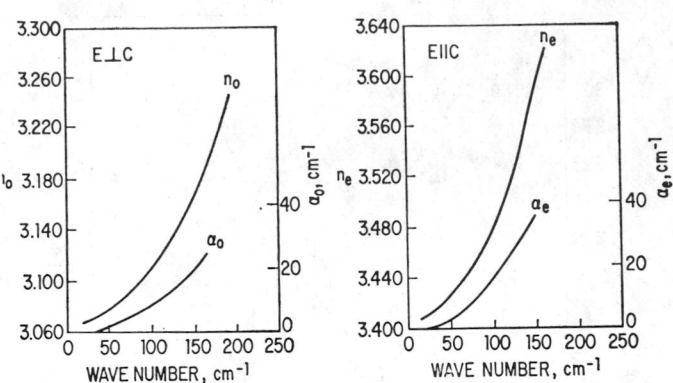

Fig. 6r-14. Optical constants of sapphire from 20 to 175 cm⁻¹: (a) ordinary ray and (b) extraordinary ray. [E. E. Russell and E. E. Bell, J. Opt. Soc. Am. **57**, 543 (1967).]

TABLE 6r-15. OPTICAL CONSTANTS OF GERMANIUM FROM 10 TO 140 cm⁻¹*

Wave number, cm⁻¹	Refractive index n	Wave number, cm⁻¹	Absorption coefficient, cm⁻¹	Wave number, cm⁻¹	Refractive index n	Wave number, cm⁻¹	Absorption coefficient, cm⁻¹
17.5	4.0041	10	1.05	92.5	4.00606	85	1.76
22.5	4.00452	15	1.02	97.5	4.00611	90	2.04
27.5	4.00480	20	1.14	102.5	4.00571	95	2.43
32.5	4.00509	25	1.08	107.5	4.00572	100	3.08
37.5	4.00535	30	1.05	112.5	4.00553	105	3.03
42.5	4.00540	35	0.99	117.5	4.00538	110	2.81
47.5	4.00567	40	0.84	122.5	4.00519	115	2.70
52.5	4.00570	45	0.73	127.5	4.0061	120	2.63
57.5	4.00590	50	0.73	132.5	4.0056	125	2.45
62.5	4.00610	55	0.81	137.5	4.0058	130	2.0
67.5	4.00627	60	0.87	142.5	4.0066	135	1.9
72.5	4.00631	65	1.00			140	1.9
77.5	4.00616	70	1.22				
82.5	4.00608	75	1.64				
87.5	4.00619	80	1.59				

* C. M. Randall and R. D. Rawcliffe, Appl. Opt. **6**, 1889 (1967).

TABLE 6r-16. OPTICAL CONSTANTS OF SILICON FROM 10 TO 140 cm⁻¹*

Wave number, cm⁻¹	Refractive index n	Wave number, cm⁻¹	Absorption coefficient, cm⁻¹	Wave number, cm⁻¹	Refractive index n	Wave number, cm⁻¹	Absorption coefficient, cm⁻¹
22.5	3.4172	10	0.54	87.5	3.41828	75	0.32
27.5	3.4167	15	0.52	92.5	3.41832	80	0.33
32.5	3.41753	20	0.54				
37.5	3.41756	25	0.56	97.5	3.41848	85	0.28
42.5	3.41779	30	0.58	102.5	3.41858	90	0.34
				107.5	3.41860	95	0.43
47.5	3.41790	35	0.60	112.5	3.41866	100	0.44
52.5	3.41791	40	0.48	117.5	3.41852	105	0.44
57.5	3.41796	45	0.43				
62.5	3.41807	50	0.44	122.5	3.41860	110	0.54
67.5	3.41818	55	0.46	127.5	3.41873		
				132.5	3.41873		
72.5	3.41824	60	0.34	137.5	3.4184		
77.5	3.41825	65	0.32	142.5	3.4188		
82.5	3.41824	70	0.33				

* C. M. Randall and R. D. Rawcliffe, Appl. Opt. **6**, 1889 (1967).

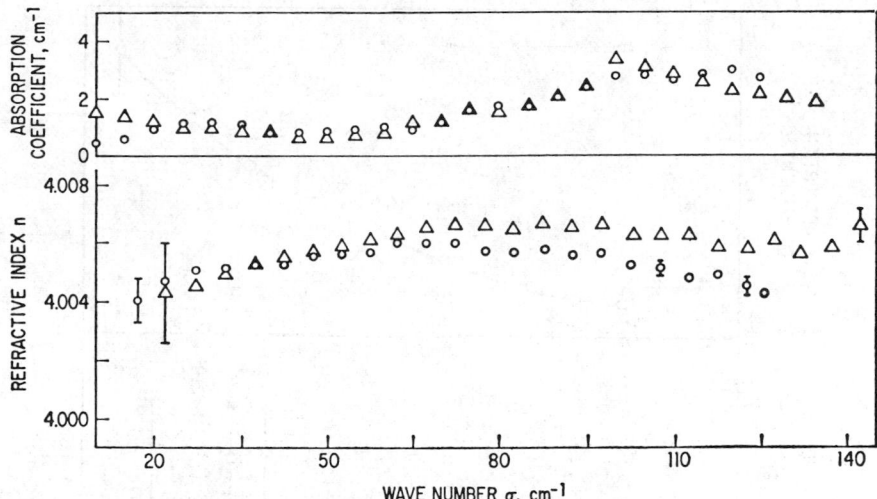

Fig. 6r-15. Optical constants of germanium from 10 to 140 cm⁻¹. Circles: 6-mm sample, triangles: 2-mm sample. [*C. M. Randall and R. D. Rawcliffe, Appl. Opt.* **7,** 213 (1968).]

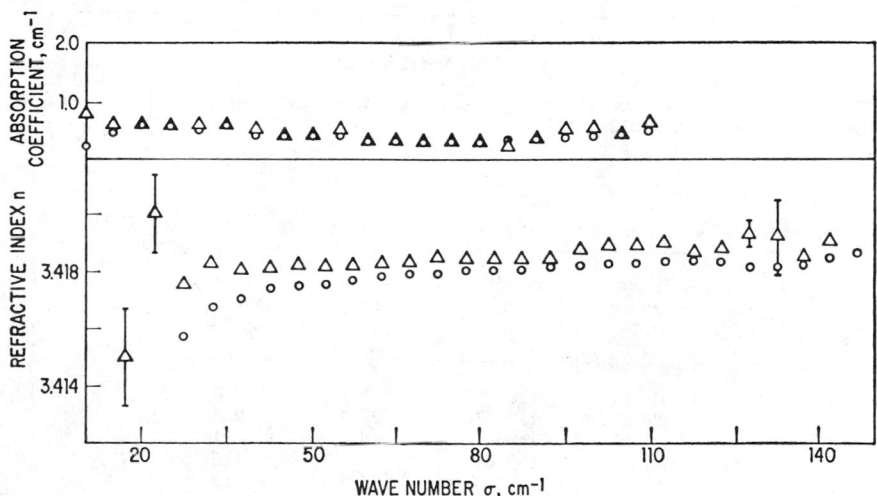

Fig. 6r-16. Optical constants of silicon from 10 to 140 cm⁻¹. Circles: 6-mm sample, triangles: 2-mm sample. [*C. M. Randall and R. D. Rawcliffe, Appl. Opt.* **7,** 213 (1968).]

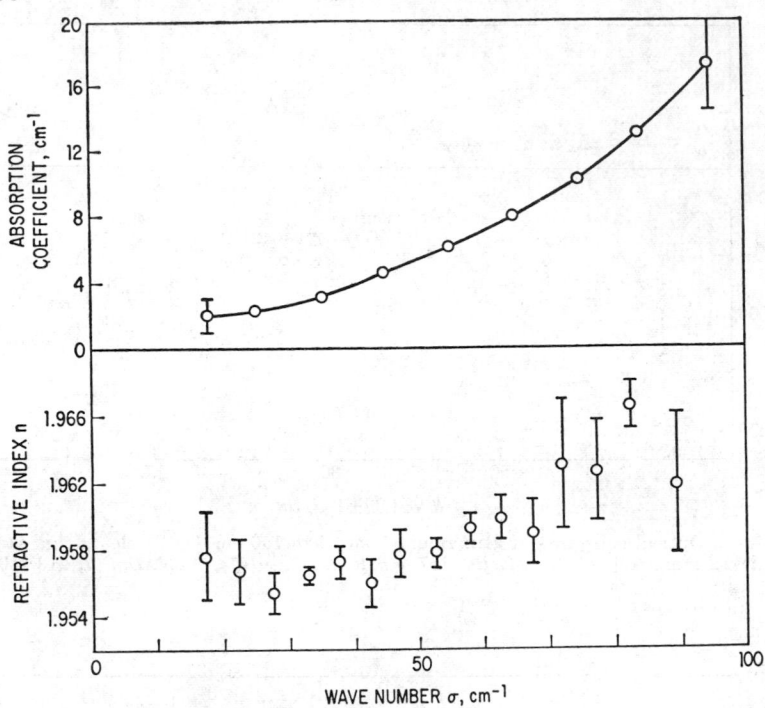

FIG. 6r-17. Optical constants of fused quartz from 15 to 95 cm⁻¹.   [*C. M. Randall and R. D. Rawcliffe, Appl. Opt.* **7,** 213 (1968).]

TABLE 6r-17. OPTICAL CONSTANTS OF FUSED QUARTZ FROM 15 TO 95 CM⁻¹*

Wave number, cm⁻¹	Refractive index n	Wave number, cm⁻¹	Absorption coefficient, cm⁻¹
17.32	1.9576	15	2.0
22.11	1.9567	25	2.3
27.48	1.9554	35	3.1
32.86	1.9565	45	4.6
37.63	1.9573	55	6.2
42.43	1.9560	65	8.0
46.72	1.9576	75	10.2
52.54	1.9576	85	13.0
57.89	1.9592	95	17.2
62.65	1.9598		
67.49	1.9589		
72.08	1.9630		
77.42	1.9626		
82.66	1.9665		
89.32	1.9618		

C. M. Randall and R. D. Rawcliffe, *Appl. Opt.* **6,** 1889 (1967).

TABLE 6r-18. OPTICAL CONSTANTS OF MYLAR (POLYETHYLENE TEREPHTHALATE) FROM 40 TO 350 cm^{-1}*

σ	n_1	σ	n_2	σ	n_1	σ	n_2
55.29	1.7171	54.16	1.7525	209.84	1.6907	204.98	1.7305
61.17	1.7153	59.74	1.7561	215.57	1.6922	210.52	1.7324
67.06	1.7137	65.31	1.7593	221.37	1.6930	216.05	1.7343
72.94	1.7127	70.89	1.7618	227.34	1.6925	221.60	1.7360
78.84	1.7111	76.53	1.7624	233.61	1.6898	227.21	1.7371
84.85	1.7078	82.31	1.7602	239.98	1.6867	232.91	1.7375
91.01	1.7020	88.27	1.7546	246.07	1.6855	238.76	1.7368
97.25	1.6956	94.33	1.7478	251.92	1.6860	244.61	1.7361
103.33	1.6925	100.24	1.7443	257.69	1.6871	250.37	1.7361
109.23	1.6926	105.97	1.7444	263.45	1.6881	256.00	1.7369
115.01	1.6945	111.57	1.7464	269.26	1.6888	261.54	1.7383
120.76	1.6965	117.10	1.7492	275.18	1.6888	267.04	1.7399
126.59	1.6973	122.61	1.7521	281.24	1.6880	272.55	1.7414
132.63	1.6954	128.15	1.7543	287.25	1.6874	278.14	1.7423
139.05	1.6891	133.87	1.7540	293.03	1.6882	283.85	1.7425
145.45	1.6834	141.88	1.7255	298.58	1.6903	289.62	1.7423
151.49	1.6824	148.52	1.7158	303.95	1.6933	295.31	1.7425
157.31	1.6837	154.18	1.7176	309.20	1.6968	300.83	1.7437
163.04	1.6857	159.68	1.7209	314.34	1.7008	306.26	1.7455
168.77	1.6877	165.20	1.7239	319.42	1.7050	311.64	1.7474
174.58	1.6887	170.73	1.7265	324.48	1.7093	317.01	1.7493
180.50	1.6887	176.36	1.7281	329.52	1.7134	322.40	1.7510
186.51	1.6879	182.08	1.7287	334.57	1.7174	327.82	1.7525
192.49	1.6874	187.87	1.7286	339.62	1.7212	333.43	1.7530
198.36	1.6878	193.66	1.7285	344.63	1.7252	339.20	1.7526
204.13	1.6891	199.37	1.7291				

σ	α_1	α_2	σ	α_1	α_2
50.0	12.9	12.7	175.0	55.2	61.1
55.0	21.0	17.7	180.0	60.2	64.8
60.0	27.7	20.9	185.0	62.8	67.7
65.0	31.9	25.0	190.0	63.6	68.8
70.0	35.7	26.9	195.0	63.2	68.2
75.0	40.2	32.9	200.0	62.4	66.4
80.0	45.4	39.9	205.0	62.7	64.7
85.0	51.2	46.0	210.0	64.6	64.1
90.0	55.1	50.0	215.0	68.9	65.6
95.0	55.2	49.9	220.0	75.3	68.3
100.0	52.5	45.4	225.0	83.0	72.3
105.0	48.7	42.4	230.0	90.5	77.4
110.0	45.8	41.6	235.0	95.8	82.0
115.0	45.0	43.4	240.0	95.2	86.7
120.0	46.8	47.7	245.0	92.0	88.5
122.5	48.6	50.8	250.0	91.3	87.2
125.0	51.4	55.4	255.0	92.3	88.4
127.5	53.1	63.6	260.0	95.4	90.4
130.0	57.5	68.2	265.0	100.2	93.7
132.0	61.3	76.5	270.0	106.4	98.9
134.0	61.9	80.7	275.0	114.3	104.6
136.0	64.7	88.4	280.0	120.2	110.5
138.0	66.4	109.2	285.0	122.0	115.8
140.0	66.8	123.0	290.0	121.7	118.1
142.0	65.0	117.8	295.0	121.1	114.4
144.0	61.8	104.5	300.0	121.7	108.8
146.0	62.3	86.8	305.0	123.8	111.0
148.0	58.3	79.7	310.0	126.7	111.8
150.0	54.8	74.7	315.0	130.0	114.1
152.5	53.9	67.0	320.0	133.0	118.8
155.0	51.0	63.3	325.0	135.4	124.0
157.5	49.5	60.0	330.0	137.7	132.4
160.0	49.0	59.4	335.0	140.5	140.7
165.0	49.3	57.1	340.0	144.3	151.6
170.0	51.7	57.8			

* Unpublished work of the subsection authors.

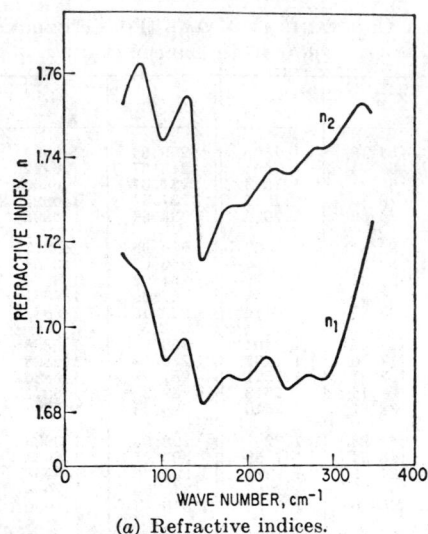

(a) Refractive indices.

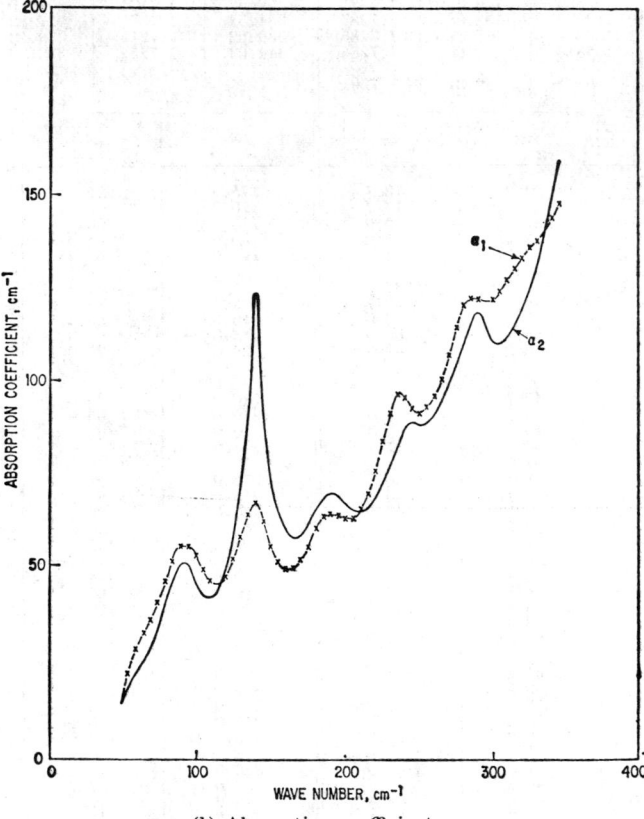

(b) Absorption coefficients.

FIG. 6r-18. Optical constants of Mylar from 40 to 350 cm⁻¹. [*E. V. Loewenstein and D. R. Smith, Appl. Opt.* **10,** 577 (1971).]

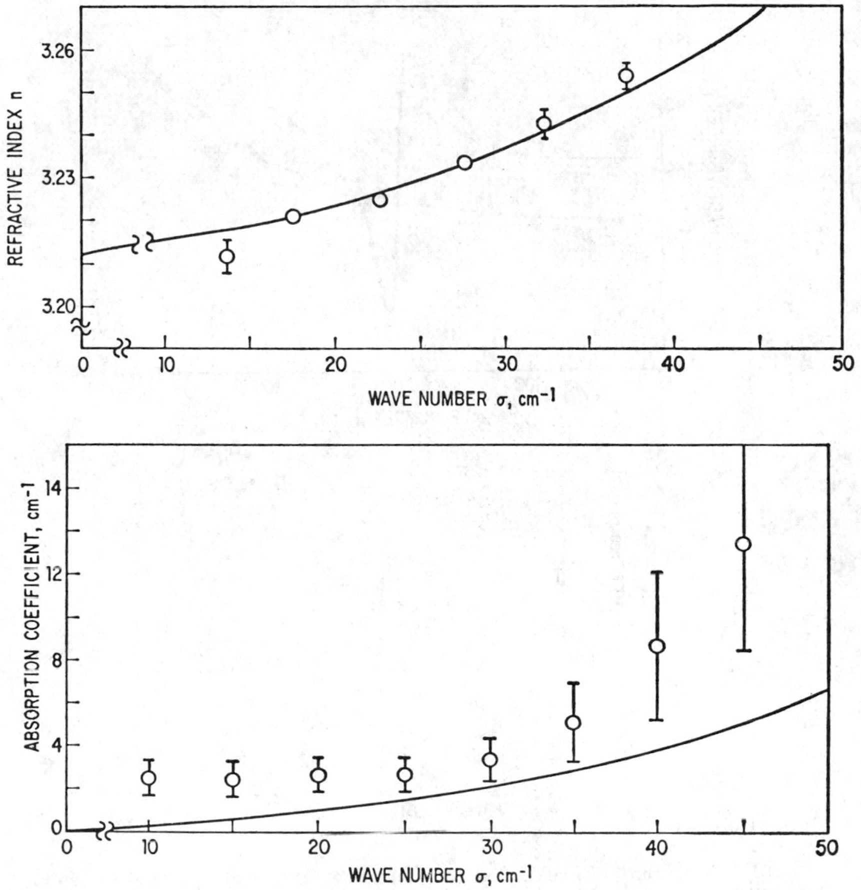

FIG. 6r-19. Optical constants of Irtran VI (hot pressed CdTe) from 10 to 45 cm⁻¹. The solid line on the absorption coefficient graph was calculated using a Lorentz oscillator model based on the index measurements. [*C. M. Randall and R. D. Rawcliffe, Appl. Opt.* **7,** 213 (1968).]

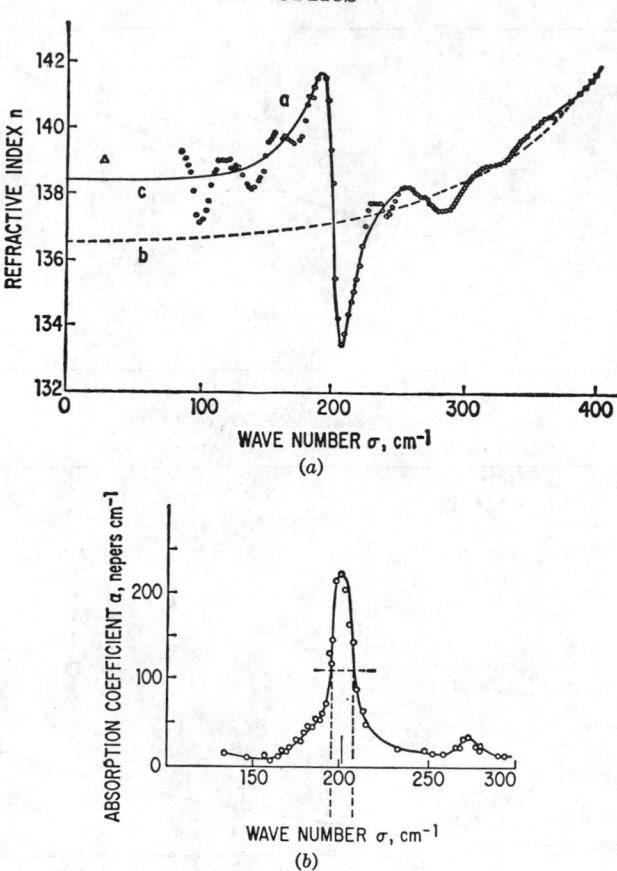

FIG. 6r-20. Optical constants of Teflon (polytetrafluoroethylene) from 100 to 350 cm⁻¹.
(a) Refractive index—circles are the experimental points including channel-spectrum
fringes. Solid line is drawn through the average value of the experimental points. Dashed
line represents the calculated contribution to the index from absorption at higher wave
numbers. (b) Absorption coefficient. [*J. E. Chamberlain and H. A. Gebbie, Appl. Opt.*
5, 393 (1966).]

TABLE 6r-19. OPTICAL CONSTANTS OF IRTRAN VI (HOT-PRESSED CdTe)
FROM 10 TO 45 CM⁻¹*

Wave number, cm⁻¹	Refractive index n	Wave number, cm⁻¹	Absorption coefficient, cm⁻¹
13.73	3.212	10	2.5
17.68	3.221	15	2.4
22.63	3.225	20	2.7
27.53	3.233	25	2.7
32.40	3.243	30	3.4
37.23	3.254	35	5.1
		40	8.7
		45	13.4

* C. M. Randall and R. D. Rawcliffe, *Appl. Opt.* **7,** 213 (1968).

TABLE 6r-20. OPTICAL CONSTANTS OF CdTe FROM 33 TO 833 CM^{-1}, AT 300 K AND FROM 33 TO 400 CM^{-1} AT 8 K*

Temperature, 300 K			Temperature, 8 K		
Wave number, cm^{-1}	Index n	Absorption coefficient	Wave number, cm^{-1}	Index n	Absorption coefficient
833	2.57	0.3	400	2.52	0.5
769	2.57	0.3	370	2.51	0.6
667	2.56	0.3	345	2.49	0.9
500	2.54	0.4	323	2.47	1.8
454	2.53	0.4	313	2.46	6.4
417	2.52	0.5	300	2.45	17.0
385	2.51	0.7	286	2.43	5.2
357	2.49	0.9	278	2.42	3.6
333	2.47	2.5	270	2.40	5.7
323	2.46	4.6	263	2.39	8.0
313	2.45	10.0	256	2.37	8.5
286	2.42	Very large	244	2.33	4.1
250	2.34	Very large	233	2.29	3.4
222	2.22	Very large	222	2.24	3.0
200	2.03	Very large	213	2.17	5.3
100	3.49	Very large	200	2.04	17.0
90	3.37	Very large	192	1.92	Very large
77	3.25	Very large	132	4.43	Very large
67	3.19	Very large	128	4.15	12.0
59	3.16	Very large	125	3.96	7.0
53	3.14	14.0	119	3.71	3.4
50	3.13	10.0	113	3.56	2.4
45	3.12	6.6	108	3.44	2.3
40	3.11	4.2	100	3.33	2.8
37	3.10	3.5	89	3.23	3.7
36	3.10	3.3	83	3.18	3.2
33	3.09	3.0	80	3.16	2.9
			77	3.15	2.9
			74	3.13	2.8
			71	3.12	2.0
			67	3.10	0.7
			63	3.09	0.3
			59	3.08	0.1
			56	3.07	<0.1
			53	3.06	<0.1
			50	3.05	<0.1
			40	3.03	<0.1
			33	3.02	<0.1

* C. J. Johnson, G. H. Sherman, and R. Weil, *Appl. Opt.* 8, 1667 (1969).

TABLE 6r-21. OPTICAL CONSTANTS OF GaAs FROM 33 TO 833 cm^{-1} AT 300 K AND FROM 33 TO 400 cm^{-1} AT 8 K

Temperature, 300 K						Temperature, 8 K		
Wave number, cm^{-1}	Index n	Absorption coefficient	Wave number, cm^{-1}	Index n	Absorption coefficient	Wave number, cm^{-1}	Index n	Absorption coefficient
833	3.27	0.1	333	2.71	Very large	426	3.12	1.4
813	3.27	0.2	238	4.44	Very large	400	3.07	1.9
769	3.26	0.3	233	4.30	Very large	370	2.98	4.0
714	3.25	0.2	222	4.11	Very large	357	2.92	8.2
667	3.25	0.2	213	4.00	Very large	345	2.83	21.0
625	3.24	0.5	204	3.92	Very large	333	2.72	Very large
588	3.23	0.7	196	3.87	Very large	263	5.86	Very large
556	3.21	4.3	172	3.76	17.0	256	5.01	18.0
526	3.20	21.0	167	3.74	14.0	250	4.62	16.0
500	3.18	Very large	143	3.68	6.7	222	3.99	4.5
476	3.16	Very large	125	3.65	4.9	200	3.82	2.9
454	3.14	Very large	111	3.63	4.1	167	3.69	1.4
435	3.12	15.0	100	3.62	3.6	143	3.64	0.5
426	3.10	12.0	83	3.61	2.8	125	3.61	0.2
413	3.08	13.0	74	3.60	2.4	111	3.60	0.1
394	3.04	9.0	67	3.60	1.8	100	3.59	<0.1
385	3.01	13.0	59	3.59	1.2	67	3.57	<0.1
370	2.96	21.0	50	3.59	0.5	33	3.55	<0.1
357	2.90	Very large	40	3.58	0.1			
345	2.82	Very large	33	3.58	<0.1			

References for Sec. 6r-6

1. Bell, E. E.: *Infrared Phys.* **6,** 57 (1966).
2. Chamberlain, J. E., J. E. Gibbs, and H. A. Gebbie: *Nature* **198,** 874 (1963).
3. Randall, C. M., and R. D. Rawcliffe: *Appl. Opt.* **6,** 1889 (1967).
4. Russell, E. E., and E. E. Bell: *J. Opt. Soc. Am.* **57,** 341 (1967).
5. Russell, E. E., and E. E. Bell: *J. Opt. Soc. Am.* **57,** 543, (1967).
6. Randall, C. M., and R. D. Rawcliffe: *Appl. Opt.* **7,** 213 (1968).
7. Unpublished work of the subsection authors.
8. Chamberlain, J. E., and H. A. Gebbie: *Appl. Opt.* **5,** 393 (1966).
9. Johnson, C. J., G. H. Sherman, and R. Weil: *Appl. Opt.* **8,** 1667 (1969).

6r-7. Beam Splitters for the Far Infrared. Michelson interferometers for the far infrared may employ as beam splitters either a metal mesh or an uncoated Mylar (polyethylene terephthalate) film. For maximum efficiency, any beam splitter should have $R \approx T \approx \frac{1}{2}$, where R and T are the intensity reflection and transmission coefficients.

For the dielectric film beam splitter the efficiency may be calculated directly in terms of the optical constants, since R and T are given by the Fresnel coefficients. The efficiency is [1]

$$\mathcal{E} = \frac{8RT^2(1 - \cos(4\pi n\sigma d \cos\theta')}{[1 + R^2 - 2R\cos(4\pi n\sigma d\cos\theta')]^2}$$

where σ = wave number of the radiation

n = refractive index (at the wave number σ)

d = beam splitter thickness

θ' = angle beteween a ray and the surface normal *inside* the beam splitter film

Since R and T depend upon polarization, the beam splitter will affect the state of polarization of the emergent radiation. When the radiation is incident upon the

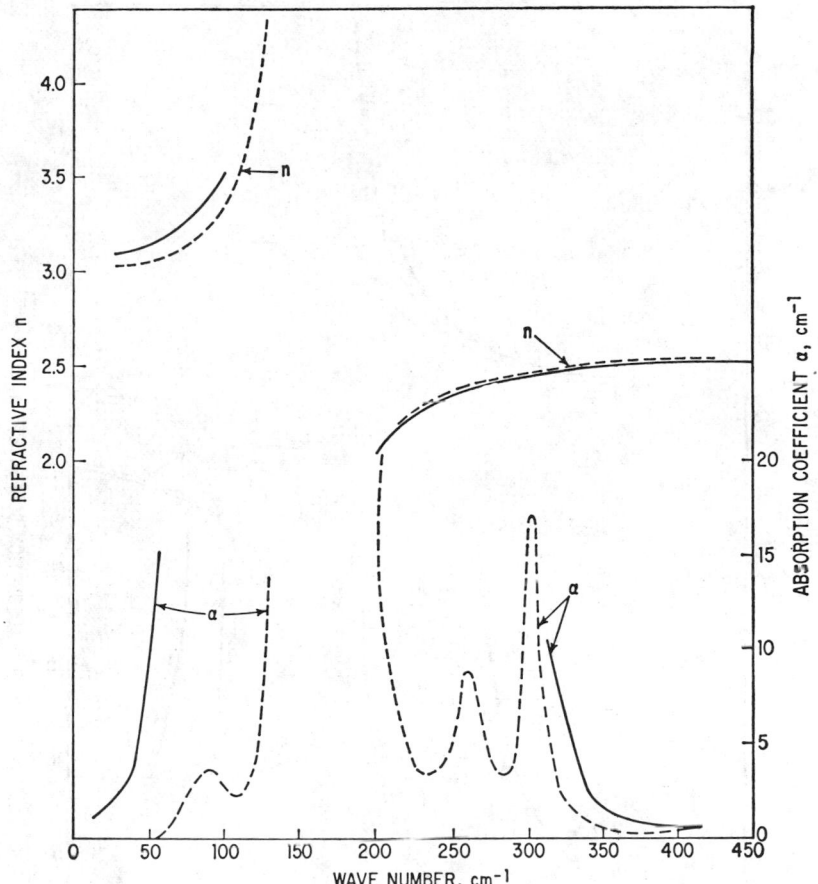

Fig. 6r-21. Optical constants of CdTe from 33 to 450 cm⁻¹. Solid line, 300 K; dashed line, 8 K. [*C. J. Johnson, G. H. Sherman, and R. Weil, Appl. Opt.* **8**, 1667 (1969).]

beam splitter at the Brewster angle (60 deg for Mylar) instead of the usual 45 deg, the interferometer produces 100 percent polarized radiation. The efficiency of a Mylar beam splitter 6 μm thick at 45 and 60 deg of angle incidence is illustrated in Fig. 6r-23.

The optical characteristics of a metal mesh depend upon λ/d, the ratio of wavelength to spacing. For $\lambda/d \gg 1$ the mesh acts as a mirror, whereas $\lambda/d \ll 1$ gives some transmission. The details of the reflection and transmission depend in a complicated way upon the details of the mesh geometry [2,3]. In the vicinity of $\lambda/d = 2$, however, the mesh acts as an acceptable beam splitter [4]. A variety of meshes are compared with a 12-μm Mylar beam splitter in Fig. 6r-24.

References for Sec. 6r-7

1. Loewenstein, E. V., and A. Engelsrath: *J. Phys. Suppl.* 3–4, **28**, 153 (1967).
2. Wessel, F.: *Hochfrequenztechnik* **54**, 62 (1939).
3. Ulrich, R.: *Infrared Phys.* **7**, 37 (1967).
4. Russell, E. E., and E. E. Bell: *Infrared Phys.* **6**, 75 (1966).

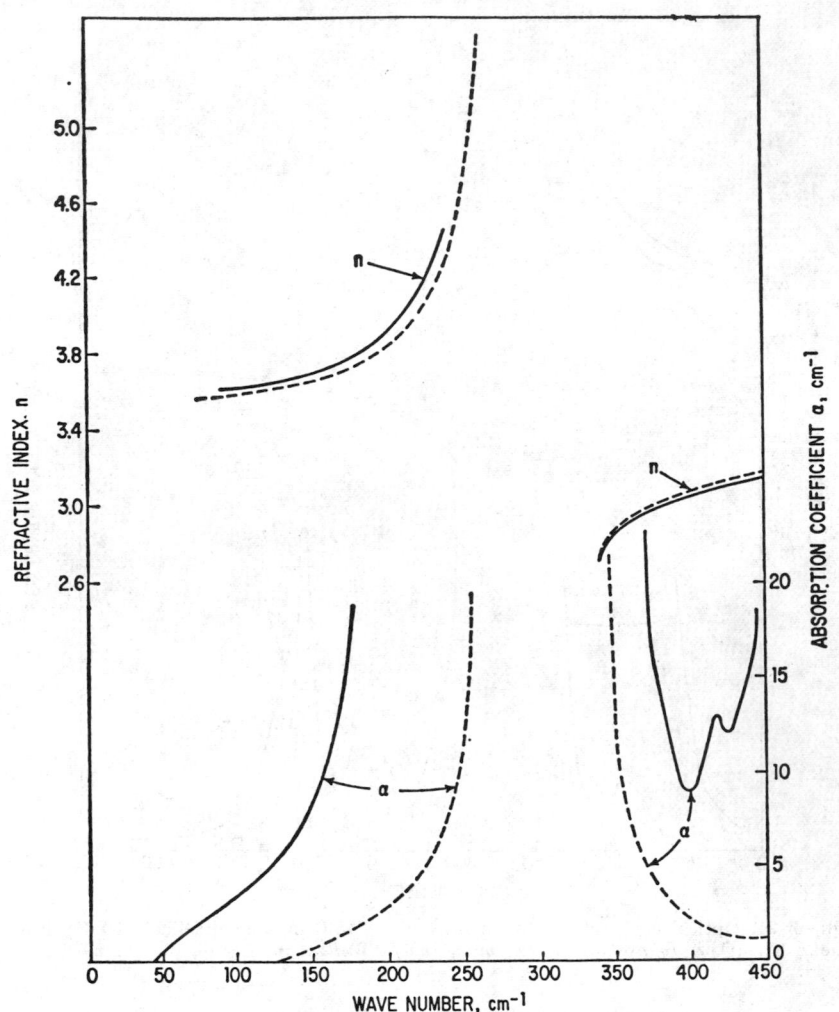

FIG. 6r-22. Optical constants of GaAs from 33 to 450 cm^{-1}. Solid line, 300 K; dashed line, 8 K. [*C. J. Johnson, G. H. Sherman, and R. Weil, Appl. Opt.* **8,** 1667 (1969).]

6r-8. Far-infrared Lasers. Listed in the following tables are the laser lines of wavelength greater than 50 μm reported through September, 1968. The power levels given are reported by most authors to be uncertain by a factor of 3. Even then, greater variations occur from one experimenter to another because the power depends on such factors as excitation conditions, gas pressure, and cleanliness of the discharge tube. The 337-μm line of HCN is the best example of the discrepancies to be found. In the pulsed mode it is reported in ref. 1 to produce 0.6-W peak, whereas ref. 19 reports 10-W peak power. In the continuous mode ref. 13 reports 0.1 W, whereas ref. 20 gives 0.6 W. All power levels listed in the tables are peak power in the pulsed mode except where specifically labeled otherwise.

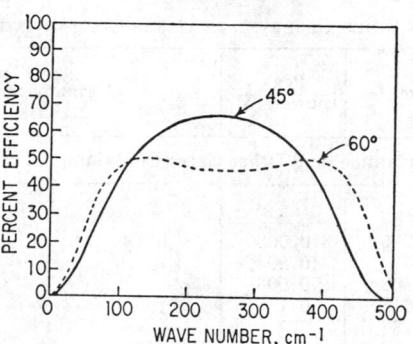

FIG. 6r-23. Calculated efficiency for Mylar beam splitters used at 45- and 60-deg angles of incidence. [*E. V. Loewenstein and A. Engelsrath, J. Phys. Suppl. 3–4,* **28,**153 (1967).]

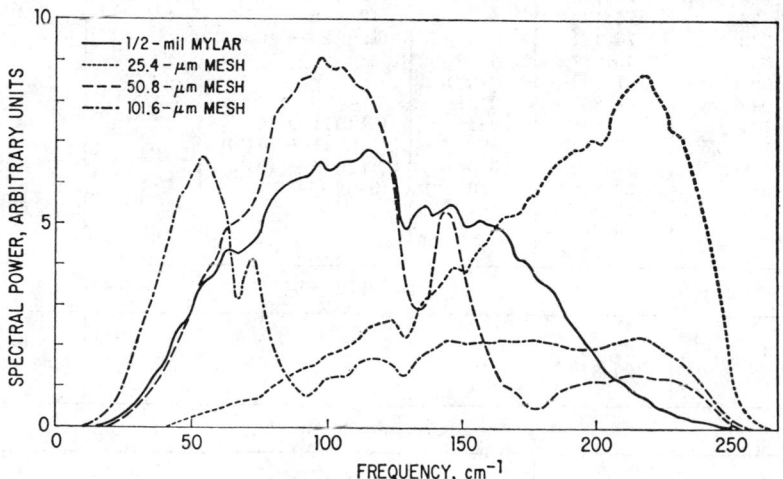

FIG. 6r-24. Background spectra obtained with various beam splitters under otherwise identical experimental conditions. [*E. E. Russell and E. E. Bell, Infrared Phys.* **6,** 75, (1966).]

Except in Table 6r-26 the wavelengths are the measured quantities, and the wave numbers given are simply $\sigma = 10^4/\lambda$.

In Table 6r-26 the frequencies were measured by comparison with a klystron, and the wavelengths were calculated using $C = 2.997925 \times 10^8$ m/sec.

TABLE 6r-22. LASER LINES OBSERVED IN GASES CONTAINING N, C, AND H OR D

λ, μm	σ, cm^{-1}	Peak power, W	Assignment	Refs.
		Dimethylamine, and Other Gases Containing C, H, and N		
71.899	139.084	0.3		1
73.101	136.796	0.008		
76.093	131.418	0.005		
77.001	129.868	0.003		
81.554	122.618	0.1		
96.401	103.733	0.2		
98.693	101.325	0.8		
101.257	98.759	0.2		
112.066	89.233	0.2		
116.132	86.109	0.5		
126.164	79.262	3	$(12^{2d}0)27 \rightarrow (05^{1d}0)26$	14
128.629(CW)	77.743	9	$(12^{2d}0)26 \rightarrow (05^{1d}0)25$ } HCN	
130.838	76.430	4	$(12^{0}0)26 \rightarrow (05^{1c}0)25$	
134.932	74.111	0.8	$(12^{0}0)26 \rightarrow (05^{1c}0)24$	
201.059	49.737	0.05		
211.001(CW)	47.393	0.2		
222.949	44.853	0.08		
309.731	32.2861	0.4	$(11'0)11 \rightarrow 10$	2
310.908(CW)	32.1639	1	$(11'0)11 \rightarrow (04^{0}0)$ } HCN	
336.579(CW)	29.7107	0.6	$(11'0)10 \rightarrow (04^{0}0)9$	
372.547	26.8422	0.6	$(04^{0}0)9 \rightarrow 8$	
537.7	18.60			11
538.2	18.58			
		CH_3CN		
334.4	29.90			18
334.8	29.87			
		CH_4 and $^{15}NH_3$		
110.240	90.711			1
113.311	88.253			
138.768	72.063			
165.150	60.551			
		$CD_4 + ND_3$		
181.789	55.009		$(22^{0}0)23 \rightarrow (22^{0}0)22$	1, 8, and 14
189.948(CW)	52.646	2×10^{-4}	$(22^{0}0)22 \rightarrow (09^{1c}0)21$ } DCN	
194.706(CW)	51.359	$3 \times 10^{-}{}_{4}$	$(22^{0}0)21 \rightarrow (09^{1c}0)20$	
204.387	48.927		$(09^{1c}0)20 \rightarrow (09^{1c}0)19$	

TABLE 6r-23. LASER LINES OBSERVED IN H_2O AND D_2O

λ, μm	σ, cm^{-1}	Peak power, W	Assignment	Refs.
		H_2O		
53.906	185.51	0.0008		3
55.077(CW)	181.56	0.06	$(020)5_{50} \rightarrow (020)5_{41}$	4, 5, 6
57.660	173.43	0.02	$(100)9_{19} \rightarrow (020)8_{44}$	
67.177	148.86	0.01	$\left\{\begin{array}{l}(100)6_{25} \rightarrow (020)5_{50} \\ (020)4_{41} \rightarrow (020)4_{32}\end{array}\right\}$	
73.402	136.24	0.002		
78.455(CW)	127.46	0.007	$(100)8_{08} \rightarrow (020)8_{35}$	
79.106(CW)	126.41	0.006	$(020)8_{44} \rightarrow (020)8_{35}$	
89.775	111.39	0.006		
115.32(CW)	86.64	0.0007	$(020)8_{35} \rightarrow (020)8_{26}$	
118.65(CW)	84.28	0.001	$(001)6_{42} \rightarrow (020)6_{61}$	
120.08	83.28	......	$(001)6_{42} \rightarrow (001)6_{33}$	
220.23(CW)	45.407	......	$(100)5_{23} \rightarrow (020)5_{50}$	8
		D_2O		
56.845	175.92			3
71.965	138.96			
72.429	138.07			
72.747	137.46			
73.337	136.36			
74.545	134.15			
76.305	131.05			
84.111	118.89			
84.291(CW)	118.64			7
107.71(CW)	92.84			

TABLE 6r-24. LASER LINES OBSERVED IN NEON

λ, μm	σ, cm^{-1}	Continuous power, W	Assignment	Ref.
50.70(CW)	197.2		$7p[3/2]_2 - 6d[3/2]_2^0$	9
52.39(CW)	190.9		$7p'[1/2]_1 - 6d'[3/2]_2^0$	
55.68(CW)	179.6		$7p[3/2]_1 - 6d[7/2]_4^0$	
72.15(CW)	138.6		$8p'[1/2]_0 - 7d'[3/2]_1^0$	
86.9(CW)	115.1		$8p'[3/2]_2 - 7d'[5/2]_2^0$	
88.46(CW)	113.0	$> 10^{-9}$	$8p[3/2]_1 - 7d[5/2]_2^0$	
89.93(CW)	111.20		$8p[5/2]_3 - 7d[7/2]_3^0$	
93.02(CW)	107.50			
106.02(CW)	94.322	$\sim 10^{-9}$	$10p[1/2]_0 - 9d[3/2]_1^0$	
124.4(CW)	80.39	$\sim 10^{-9}$	$9p[3/2]_1 - 8d[5/2]_2^0$	
			$9p[3/2]_2 - 8d[5/2]_3^0$	
126.1(CW)	79.30	$\sim 10^{-9}$		
132.8(CW)	75.30	$\sim 10^{-9}$		

TABLE 6r-25. LASER LINES OBSERVED IN MISCELLANEOUS GASES

Gas	λ, μm	σ, cm^{-1}	Peak power, W	Assignment	Ref.
ICN	773.5	12.928			12
He	95.788	109.94	3	$3p^1P_1{}^0 - 3d^1D_2$	10
CH$_3$CN and (CH$_3$)$_2$SO$_4$	119.0	84.0			18

TABLE 6r-26. LASER LINES WHOSE FREQUENCIES HAVE BEEN DETERMINED
BY DIRECT COMPARISON WITH A KLYSTRON
(The wavelengths are calculated using $C = 2.997925 \times 10^8$ m/sec.)

Gas	Frequency, GHz	λ, μm	σ, cm^{-1}	Ref.
DCN	1,466.787	204.3872	48.92674	15
	1,539.257	194.7644	51.34409	
	1,539.745	194.7027	51.36035	
	1,577.789	190.0080	52.62937	
	1,578.279	189.9490	52.64571	
D$_2$O	1,578.279	189.9490	52.64571	16
C$_2$N$_2$	1,539.756	194.7013	51.36072	16
HCN	964.3123	310.8874	32.16599	17
	890.7595	336.5583	29.71253	

References for Sec. 6r-8

1. Mathias, L. E. S., A. Crocker, and M. S. Wills: *IEEE J. Quantum Electron.* **QE-4**, 205 (1968).
2. Lide, D. R., Jr., and A. G. Maki: *Appl. Phys. Letters* **11**, 62 (1967).
3. Mathias, L. E. S., and A. Crocker: *Phys. Letters* **13**, 35 (1964).
4. Hartman, B., and B. Kleman: *Appl. Phys. Letters* **12**, 168 (1968).
5. Benedict, W. S.: *Appl. Phys. Letters* **12**, 170 (1968).
6. Pollack, M. A., and W. J. Tomlinson: *Appl. Phys. Letters* **12**, 173 (1968).
7. Müller, W. M. and G. T. Flesher: *Appl. Phys. Letters* **8**, 217 (1966).
8. Müller, W. M., and G. T. Flesher: *Appl. Phys. Letters* **10**, 93 (1967).
9. Patel, C. K. N., W. L. Faust, R. A. McFarlane, and C. G. B. Garrett: *Proc. IEEE* **52**, 713 (1964).
10. Mathias, L. E. S., A. Crocker, and M. S. Wills: *IEEE, J. Quantum Electron.* **QE-3**, 170 (1967).
11. Steffen, H., J. Steffen, J. F. Moser, and F. K. Kneubühl: *Phys. Letters* **20**, 20 (1966).
12. Steffen, H., J. Steffen, J. F. Moser and F. K. Kneubühl: *Phys. Letters* **21**, 425 (1966).
13. Gebbie, H. A., N. W. B. Stone, W. Slough, J. E. Chamberlain, and W. A. Sheraton: *Nature* **211**, 62 (1966).
14. Maki, Arthur G.: *Appl. Phys. Letters* **12**, 122 (1968).
15. Hocker, L. O., and A. Javan: *Appl. Phys. Letters* **12**, 124 (1968).
16. Hocker, L. O., D. Ramachandra Rao, and A. Javan: *Phys. Letters* **24A**, 690 (1967).
17. Hocker, L. O., A. Javan, D. Ramachandra Rao, L. Frenkel, and T. Sullivan: *Appl. Phys. Letters* **10**, 147 (1967).
18. Prettl, W., and L. Genzel: *Phys. Letters* **23**, 443 (1966).
19. Gebbie, H. A., N. W. B. Stone, and F. D. Findlay, *Nature* **202**, 685 (1964).
20. Kotthaus, J. P.: *Appl. Opt.* **7**, 2422 (1968).

6s. Optical Masers

ROBERT J. COLLINS

University of Minnesota

6s-1. Introduction. In the short time since the first explicit proposals [1] that stimulated emission be used as an amplifying mechanism, devices employing this principle have become common in the microwave and optical regions of the spectrum. Less than three years passed between the proposal and the observation by Zweiger and Townes [2] of gain ammonia gas at 23.879 kmc. After these initial experiments, it was clear that stimulated emission could be used to build either amplifiers or oscillators. The original work led to the construction of an amplifier using ammonia gas, in which the inverted system was prepared by the electromagnetic separation of the excited ammonia molecules. The device was called a *maser*, which is an acronym for Microwave Amplification by Stimulated Emission of Radiation. In 1960, when Maiman [3] first reported stimulated emission in the optical region of the spectrum, an additional acronym came into use—"laser" for Light Amplification by Stimulated Emission of Radiation. The extension from the microwave region to the optical portion of the spectrum of the use of stimulated emission as an amplifying mechanism followed an explicit proposal to use a 3-level energy system for a maser. In this suggestion, pumping or inversion was to be accomplished by an external energy source and stimulated emission was to occur between two of the three levels. This Bloembergen proposal was first successfully carried out by Scovil, Feher, and Seidel [5].

In the construction of oscillators the active material must be contained in a cavity with means to control the mode of oscillation. At frequencies $<10^{11}$ Hz, cavities with all dimensions comparable to wavelengths can be built, making mode selection straightforward. This approach is not convenient at optical frequencies. The solution to the problem of control by using a multimode cavity in the form of an interferometer and the natural line width of the transition was suggested by Schawlow and Townes [6]. This suggestion is used in operating lasers. As a consequence of the form of the cavity used, the output of a laser is a beam with a well-defined phase front. The angular divergence of the beam should be diffraction-limited by the cavity aperture. Although clearly predicted, the output radiation of an optical maser in the form of a well-collimated beam of radiation is one of its most striking properties. The isolation of single oscillating modes from lasers is now common practice. Accompanying the appearance of the directional beam, the spectral line width decreases. Within the narrow beam width of the output, the high radiation level occurs. Using Nd^{3+} doped glasses, lasers have been constructed with peak power outputs of $\sim 10^{12}$ watts at 1.06 μm for $\sim 3 \times 10^{-12}$ sec [7].

In addition to the requirement of mode selection [30], oscillations can be sustained only if the rate of supplying energy to the mode through emission exceeds the rate of loss of energy from the cavity. This statement can be simply expressed in terms of the absorption coefficient. From thermodynamic considerations of an atom containing only two energy levels, each of statistical weight unity, the probability that a light wave incident on an atom in the ground state will be absorbed is equal to the probability that the light wave will stimulate the emission of radiation from an atom in the upper state. It should be noted that the radiation produced by the stimulated emission will have the same phase and direction as the original radiation. Light passing through a crystal l cm long, containing N_0 atoms/cm³ all in the ground state, is attenuated by an amount $e^{-\sigma N_0 l}$, where σ is the cross section for absorption (or stimulated emission) at the wavelength of the radiation. On the other hand, if all the atoms were in the upper state, the light would be amplified by $e^{+\sigma N_0 l}$. To sustain oscillations in a cavity the gain must exceed the loss, and thus, with end reflectivities R_1 and R_2,

$$e^{2\sigma \, \Delta n l} \geq \frac{1}{R_1 R_2} \qquad (6s\text{-}1)$$

where $\sigma \, \Delta n l$ is the gain/transversal, and Δn is the excess population in the upper level over the lower level. Through the relation between absorption coefficients and line widths, (Δn), the excess population needed to satisfy the above condition, for a gaussian line shape, is

$$\Delta n \geq \frac{4\pi^2}{\sqrt{\pi \ln 2}} \frac{\tau \, \Delta \nu}{\lambda_1{}^2} \ln \frac{1}{\sqrt{R_1 R_2}} \qquad (6s\text{-}2)$$

where λ = wavelength in the medium
    τ = radiative lifetime in the upper level
    $\Delta \nu$ = full width of the half maximum of the transition
A consideration of Eq. (6s-2) and its implications will indicate some of the properties of a material suitable for a laser. Initially, of course, energy levels connected by a radiative transition must be found in which an inversion of population can be created.

The red fluorescence (6,924 Å) of the Cr^{3+} ion of an Al_2O_3 lattice was the first used to satisfy the conditions of small $\Delta \nu$, high quantum efficiency, reasonable τ, and convenient absorption bands for pumping. Figure 6s-1 gives the energy-level diagram of Cr^{3+} in pink ruby, showing the green and violet absorption bands and the two red fluorescent transitions $R_1 \{E \rightarrow 4A_2\}$ and $R_2 \{2A \rightarrow 4A\}$. One serious drawback of ruby as a laser material is evident in Fig. 6s-1. The R_1 emission line originates in the transition from an excited state to the ground state. Hence, before any optical gain by stimulated emission can occur, the ground state must be depopulated. To depopulate the ground state in ruby, the pumping radiation must have ~10^3 watts/cm² in the green absorption band.

Optical pumping [3,8] of ruby was first accomplished using millisecond-long pulses from Xe flash lamps. In later work, Hg-Xe vapor lamps have been used to produce CW operation.

The first experimental [9] use of stimulated emission in a gas mixture to observe oscillation was in the He-Ne mixture. Excitation of the neon was achieved through collisions of the second kind between neon and excited atoms of helium.

The long chain of events leading to the eventual development of lasers began with the explicit introduction by Einstein of stimulated emission. This was followed by the theoretical work of Tolman [10], experimental observations by Ladenburg [11], and recognition by Fabrikant [12], that gain was possible. The immediate activity, however, began in 1951 with several groups actively engaged in attempts to produce gain through stimulated emission.

At present, approximately 1,000 different laser wavelengths have been identified. The range of wavelengths extends from 2,358 Å in Ne IV to 372.80 μm in diethylamine. The distribution of the 350 discrete wavelengths from 2,600 Å to 1 μm is such that almost every increment of 100 Å has several laser lines. The coverage is not as dense from 1 to 10 μm; and from 10 to 370 μm only 70 lines in H_2O, D_2O, NH_2, Ne, CN, BrCN, and diethylamine have been observed.

The output of a laser generally consists of more than one axial mode with frequencies separated by $c/2l$. The simultaneous existence of more than one frequency in the output affects the properties of the beam. When a single-frequency axial and transition are emitted from a well-controlled cavity, the amplitude and frequency are stable [13] to one part in 10^9. Fluctuations or noise occur largely as phase noise. With more than one frequency the output shows random fluctuations due to phase modulation between the separate frequencies. Through various techniques [14] of mode locking it is possible to produce pulses separated by $\sim 10^{-9}$ and 10^{-12} sec long. In Table 6s-1 is a list of the wavelengths at which laser oscillation has been reported, together with material in which it was observed. For the purpose of this handbook it is not reasonable to discuss each separately. Only the commonly used lasers are treated below.

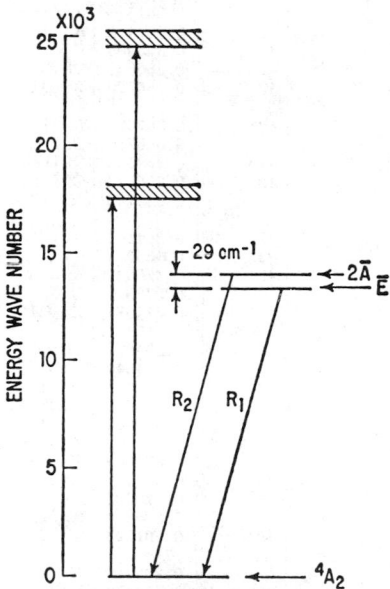

FIG. 6s-1. Energy level Cr^{3+} in Al_2O_3 showing absorption bands and the R_1R_2 emission lines.

In its simplest form a laser consists of a gain medium, a feedback mechanism, a source of input energy, a method of coupling between the input energy and the actual gain medium, and a method of extracting power. The forms of input energy are:

1. Optical energy from gas discharge or incandescent sources (pump lamps). The first pump lamps used were the Xe flash lamps. Later lamps included mercury vapor, tungsten, ribbons, the sun, other lasers, light-emitting diodes, and shock waves.

2. Electric discharges in the gaseous material itself.

3. Thermal excitation of one species followed by excitation of the actual gain material.

4. Direct electrical injection of carriers in semiconductor junctions.

5. Electron-beam excitation of solids. In these cases the incident electron-beam energy varies from $\sim 50,000 \rightarrow 300,000$ electron volts.

6. Chemical dissociation into excited states.

7. Exothermic chemical reaction producing molecules in excited states.

8. Thermodynamic processes—e.g., rapid expansion of gases.

OPTICS

TABLE 6s-1. OBSERVED LASER OSCILLATIONS*

Wave-length, μm	Material	Reference	Wave-length, μm	Material	Reference
0.23580	Ne IV	42	0.338554	O IV	42
0.247350	Ne	42	0.339286	Ne II	42
0.247718	Xe	42	0.339287	Cl III	43
0.262490	Ar IV	42	0.339340	Ne II	42
0.263270	Cl	43	0.339345	Cl III	43
0.264941	Kr	42	0.345423	Xe III	42
0.266450	Kr II	42	0.347870	N IV	42
0.267798	Ne III	42	0.348296	Xe	42
0.267868	Ne III	42	0.348302	N IV	42
0.269182	Xe	42	0.350742	Kr III	42
0.274151	Kr	42	0.351113	Ar III	42
0.275391	Ar III	42	0.351415	Ar III	42
0.275959	F III	43	0.353003	Cl III	43
0.27775	Ne III	42	0.356069	Cl III	43
0.282608	F IV	43	0.356420	Kr III	42
0.28688	Ne	42	0.35769	Ar II	42
0.288424	Ar III	42	0.363786	Cl III	43
0.291292	Ar IV	42	0.364546	Cl III	43
0.292624	Ar IV	42	0.366920	Cl III	43
0.29837	Xe III	42	0.363786	Ar III	42
0.298386	O III	42	0.364546	Xe	42
0.300264	Ar	42	0.366920	Xe	42
0.3024	Ar III	42	0.37052	Ar	42
0.304715	O III	42	0.372046	Cl III	43
0.304974	Kr	42	0.374573	Xe III	42
0.30548	Ar III	42	0.374878	Cl III	43
0.306346	O IV	42	0.374949	OH	42
0.307978	Xe	42	0.375468	O III	42
0.312156	F III	43	0.3757	ZnO	50
0.312443	Kr III	42	0.375986	O III	42
0.3125	Gd³⁺(glass)	42	0.378099	Xe III	42
0.217418	F III	43	0.379528	Ar III	42
0.319143	Cl III	43	0.380327	Xe	42
0.320274	F II	43	0.385826	Ar III	42
0.323943	Kr III	43	0.397293	Xe	42
0.324694	Xe III	43	0.399499	N II	43
0.325	ZnS	49	0.402478	F II	43
0.330592	Xe	42	0.406048	Xe III	42
0.331984	Ne II	42	0.406736	Kr III	42
0.332379	Ne II	42	0.409729	N III	42
0.332437	Ne	42	0.410326	N III	42
0.33275	Ne II	42	0.413128	Kr III	42
0.332902	Ne II	42	0.41325	Cl II	43
0.333082	Xe	42	0.414660	Ar III	42
0.333107	Ne III	42	0.415445	Kr III	42
0.33614	Ar III	42	0.417181	Kr III	42
0.334478	Ar III	42	0.418292	Ar	42
0.334550	Ne II	42	0.421405	Xe III	42
0.334776	P IV	43	0.422225	P III	43
0.335004	Xe	42	0.422651	Kr III	42
0.335852	Ar III	42	0.424026	Xe III	42
0.336732	N III	42	0.427460	Xe III	42
0.33750	Kr III	42	0.428592	Xe III	42
0.337833	Ne II	42	0.430577	Xe III	42
0.338134	O IV	42	0.4318	Kr II	42

* The references for this table have been selected, not for historical purposes, but on the grounds of availability in modest libraries and to lead the user into the appropriate literature.

TABLE 6s-1. OBSERVED LASER OSCILLATIONS (*Continued*)

Wave-length, μm	Material	Refer-ence	Wave-length, μm	Material	Refer-ence
0.434738	O II	42	0.491	CdS	27
0.435126	O KK	42	0.491766	Cl II	43
0.437073	Ar II	42	0.492404	Zn II	59
0.4387	Kr II	42	0.495410	Xe III	42
0.441493	O II	42	0.496500	Xe II	42
0.44156	Cd	59	0.496509	Ar II	42
0.441697	O II	42	0.4986	I II	42
0.443422	Xe III	42	0.499255	Ar	42
0.444328	Kr III	42	0.500772	Xe	42
0.4482	Ar II	42	0.50164	Kr II	42
0.451045	N III	42	0.501717	Ar II	42
0.451441	N III	42	0.5022	Kr II	42
0.454504	Ar II	42	0.504489	Xe II	42
0.455259	F	43	0.505463	Br II	64
0.456784	F	43	0.507830	Cl II	43
0.457720	Kr II	42	0.5097	Se II	65
0.457936	Ar II	42	0.51031	Cl II	44
0.4583	Kr II	42	0.5126	Kr II	42
0.460302	Xe II	42	0.513175	Ge II	59
0.460552	O	42	0.51418	Ar II	42
0.460957	Ar II	42	0.514533	Ar II	42
0.46152	Kr II	42	0.515904	Xe	42
0.461917	Kr II	42	0.517865	Ge II	59
0.463031	N II	42	0.518238	Br II	64
0.463392	Kr II	42	0.5185	Br II	65
0.464740	C III	47	0.520832	Kr II	42
0.464908	O II	42	0.5216	I II	42
0.465011	C III	47	0.521790	Cl II	43
0.465016	Kr II	42	0.522130	Cl II	43
0.465795	Ar II	42	0.5228	Se II	65
0.467373	Xe III	42	0.523826	Br II	64
0.468045	Kr II	42	0.523889	Xe III	42
0.46805	In II	59	0.526017	Xe II	42
0.468357	Xe III	42	0.52615	Xe II	42
0.4695	Kr II	42	0.5287	Ar II	42
0.47231	Xe III	42	0.530868	Kr II	42
0.472689	Ar II	42	0.5314	Xe II	42
0.47404	Cl II	43	0.53207	S II	59
0.474266	Br II	64	0.533203	Br II	64
0.474892	Xe	48	0.5334	Br II	65
0.476244	Kr II	42	0.533749	Cd II	60
0.476488	Ar II	42	0.53410	Mn I	61
0.476571	Kr II	42	0.53457	S II	59
0.476874	Cl II	42	0.535289	Xe	42
0.478134	Cl II	42	0.53721	Pb II	59
0.4797	Hg III	42	0.537804	Cd II	60
0.482518	Kr II	42	0.539215	Cl II	43
0.484666	Kr II	42	0.539459	Xe	42
0.4862	Xe II	42	0.54009	Xe III	55
0.486948	Xe III	42	0.5407	I II	42
0.487986	Ar II	42	0.5419	I	42
0.4887	Xe II	42	0.541916	Xe II	42
0.488906	Ar II	42	0.54204	Mn I	61
0.489688	Cl II	43	0.54287	S II	59
0.490473	Cl II	43	0.54328	S II	59

TABLE 6s-1. OBSERVED LASER OSCILLATIONS (*Continued*)

Wave-length, μm	Material	Reference	Wave-length, μm	Material	Reference
0.54538	S II	59	0.6130	Eu^{3+}(chelate)	69
0.54703	Mn I	61	0.6150	Hg II	42
0.54736	S II	59	0.616574	P II	43
0.5498	As II	65	0.616880	Kr II	48
0.55022	Ar III	42	0.6171	As II	65
0.55164	Mn I	61	0.627090	Xe II	42
0.552439	Xe III	52	0.6293	He-Ne	95
0.55375	Mn I	61	0.6328	He-Ne	95
0.5559	As II	65	0.634724	F	43
0.55906	CO	62	0.637148	F	43
0.559237	O III	42	0.6401	He-Ne	95
0.55934	CO	62	0.65530	Sn II	59
0.55960	CO	62	0.64710	Kr II	42, 48
0.55983	CO	62	0.652865	Xe II	52
0.56004	CO	62	0.65700	Kr II	42, 48
0.56025	CO	62	0.6585	I II	42
0.56038	CO	62	0.65955	CO	62
0.5625	I II	42	0.65995	CO	62
0.56400	S II	59	0.66031	CO	62
0.56470	S II	59	0.66064	CO	62
0.5652	As II	65	0.66091	CO	62
0.5659	Xe II	42	0.66115	CO	62
0.566662	N II	42	0.66135	CO	62
0.567603	N II	42	0.667193	P	43
0.5678	Hg II	42	0.672138	O II	42
0.5678	I II	42	0.676457	Kr II	42, 48
0.567953	N II	42	0.68400	Sn II	59
0.568192	Kr II	42	0.685	Cd Se	27
0.5727	Xe II	42	0.687096	Kr II	42, 48
0.5751	Xe II	42	0.690	CdSSe	27
0.5753	Kr II	42	0.6904	I II	42
0.5760	I II	42	0.6943	$Cr^{3+}(Al_2O_3)$	3, 8, 72
0.57987	Sn II	59	0.6969	$Sm^{2+}(SrF_2)$	74
0.5939	He-Ne	95	0.7010	$Cr^{3+}(Al_2O_3)$	71
0.594	GaSe	27	0.7032	I II	42
0.595573	Xe	42	0.7040	$Cr^{3+}(Al_2O_3)$	71
0.597112	Xe II	42	0.7083	$Sm^{2+}(CaF_2)$	74
0.5985	$Pr^{3+}(LaF_3)$	91	0.7346	Hg II	42
0.602427	P II	43	0.74783	Zn II	59
0.603419	P II	43	0.75875	Zn II	59
0.604312	P II	59	0.76118	Zn II	59
0.6046	He-Ne	95	0.775786	Zn II	59
0.60629	CO	62	0.7828	Xe II	42
0.60657	CO	62	0.784563	P II	59
0.60682	CO	62	0.786	CdTe	27
0.60705	CO	62	0.7989	Xe II	42
0.60725	CO	62	0.79930	Kr II	42
0.60742	CO	62	0.8250	I II	42
0.608804	P II	43	0.8330	Xe II	42
0.6094	Xe II	42	0.834961	H_2	75
0.609474	Cl II	43	0.84	Ga AsP	27
0.61028	Zn II	59	0.8408	Xe	42
0.6113	$Eu^{3+}(Y_2O_3)$	68	0.8443	Xe II	42
0.6118	He-Ne	95	0.8446	O I	42
0.6127	I II	42	0.8446	Ne-O_2	41

TABLE 6s-1. OBSERVED LASER OSCILLATIONS (*Continued*)

Wavelength, μm	Material	Reference	Wavelength, μm	Material	Reference
0.84463	Br I	41	1.0461	N_2	66
0.84464	Br I	41	1.0468	$Pr^{3+}(CaWO_4)$	78
0.84467	Br I	41	1.0472	N_2	66
0.84468	Br I	41	1.0480	N_2	66
0.845	Ga As	27, 70	1.0491	N_2	66
0.8547	Hg II	42	1.0495	N_2	66
0.8569	Xe	42	1.0505	N_2	66
0.8582	Xe II	42	1.057	Nd^{3+}(borate glass)	26
0.8589	Kr II	42	1.0574	$Nd^{3+}(SrWO_4)$	73
0.9628	Hg II	42	1.0576	$Nd^{3+}(CaWO_4)$	73
0.8677	Hg	42	1.0576	$Nd^{3+}(SrMoO_4)$	73
0.8684	N_2	66	1.0586	Hg II	42
0.8691	N_2	66	1.0586	$Nd^{3+}(PbMoO_4)$	73
0.8698	N_2	66	1.059	$Nd^{3+}(SrMoO_4)$	73
0.8704	N_2	66	1.06	$Nd^{3+}(CaWO_4)$	73
0.87099	N_2	66	1.06	$Nd^{3+}(SeOCl_2)$	77
0.8714	Xe II	42	1.06	Nd^{3+}(silicate glass)	26
0.8780	Ar II	42	1.06	$Nd^{3+}(YAG)$	76
0.8800	I II	42	1.060	$Nd^{3+}(BaF_2)$	73
0.8844	N_2	66	1.0607	$Nd^{3+}(SrWO_4)$	73
0.8847	N_2	66	1.061	$Nd^{3+}(CaMoO_4)$	73
0.8852	N_2	66	1.061	Nd^{3+}(barium crown glass)	26
0.8856	N_2	66	1.0611	$Nd^{3+}(SrMoO_4)$	73
0.8862	N_2	66	1.0618	S I	41
0.8871	N_2	66	1.0621	He-Ne	95
0.887625	H_2	75	1.0623	$Nd^{3+}(Gd_3Ga_5O_{12})$	76
0.8879	N_2	66	1.0627	$Nd^{3+}(SrMoO_4)$	73
0.8886	N_2	66	1.0631	$Nd^{3+}(LaF_3)$	73
0.8892	N_2	66	1.0633	$Nd^{3+}(CaWO_4)$	73
0.8898	N_2	66	1.0633	$Nd^{3+}(Y_3Ga_5O_{12})$	76
0.889884	H_2	75	1.0634	Xe II	
0.8909	N_2	66	1.0638	$Nd^{3+}(CeF_3)$	94
0.9063	Xe II	42	1.0640	$Nd^{3+}(SrMoO_4)$	73
0.907	InP	27	1.0641	$Nd^{3+}(CaWO_4)$	73
0.9145	$Nd^{3+}(CaWO_4)$	73	1.0650	$Nd^{3+}(CaWO_4)$	73
0.9180	Nd^{3+}(glass)	26	1.0652	$Nd^{3+}(SrMoO_4)$	73
0.9265	Xe II	42	1.066	$Nd^{3+}(CaWO_4)$	73
0.9287	Xe II	42	1.067	$Nd^{3+}(CaMoO_4)$	73
0.9396	Hg II	42	1.0689	C I	41
0.9451	Cl I	92	1.073	$Nd^{3+}p(Y_2O_3)$	78
0.9697	Xe II	42	1.0741	$Nd^{3+}(Gd_2O_3)$	78
0.98	I	45	1.078	$Nd^{3+}(Y_2O_3)$	78
1.01	I	45	1.0789	$Nd^{3+}(Gd_2O_3)$	78
1.015	Yb^{3+}(silicate glass)	26	1.0798	He-Ne	95
1.0295	Ne I	87	1.0844	He-Ne	95
1.0296	$Yb^{3+}(YAG)$	88	1.0923	Ar II	56
1.03	I	45	1.0935	Ar	56
1.037	$Nd^{3+}(SrF_2)$	73	1.0950	Xe	42
1.0399	$Nd^{3+}(LAF_3)$	78	1.1181	Hg II	42
1.04	I	45	1.1143	He-Ne	95
1.04	$Pr^{3+}(Ca(NbO_3)_2)$	93	1.116	$Tm^{2+}(CaF_2)$	73
1.0437	$Nd^{3+}(SrF_2)$	73	1.116214	H_2	73
1.0449	N_2	66	1.1177	He-Ne	95
1.0455	S I	41	1.1180	He-Ne	95
1.0457	$Nd^{3+}(CaF_2)$	41	1.1181	Hg II	42

TABLE 6s-1. Observed Laser Oscillations (Continued)

Wavelength, μm	Material	Reference	Wavelength, μm	Material	Reference
1.1222	H₂	75	1.5550	Hg II	42
1.1390	He-Ne	95	1.60	InPAs	27
1.1409	He-Ne	95	1.61	Er³⁺(Ca(NbO₃)₂)	78
1.1523	He-Ne	95	1.612	Er³⁺(CaWO₄)	78
1.1525	He-Ne	95	1.618	Ar I	53
1.1601	He-Ne	95	1.622	Ni²⁺	90
1.1614	He-Ne	95	1.6452	Er³⁺(YAG)	73
1.1617	He-Ne	95	1.6602	Er³⁺(YAG)	73
1.1767	He-Ne	95	1.690	Kr I	53
1.1770	He-Ne	95	1.6918	Hg I	18
1.1788	Ne I	87	1.6921	Hg I	18
1.1790	Ne I	87	1.694	Ar I	53
1.1985	He-Ne	95	1.694	Kr I	53
1.1988	He-Ne	95	1.6942	Hg I	18
1.2066	He-Ne	95	1.7073	Hg I	18
1.2069	He-Ne	95	1.7110	Hg I	18
1.2303	N₂	66	1.7162	Ne I	87
1.2312	N₂	66	1.750	Co²⁺	78
1.2319	N₂	66	1.77	InGaAs	27
1.2334	N₂	66	1.784	Kr	53
1.2347	N₂	66	1.793	Ar	53
1.2459	Ne I	87	1.803	Co²⁺	78
1.2545	Hg	42	1.8130	Hg I	53
1.2689	He-Ne	95	1.819	Kr	53
1.2887	He-Ne	95	1.8210	Ne I	87
1.28991	Mn I	61	1.8281	He-Ne	95
1.2912	He-Ne	95	1.8287	He-Ne	95
1.2981	Hg	42	1.8309	He-Ne	95
1.30578	H₂	75	1.8408	He-Ne	95
1.31623	H₂	75	1.8596	He-Ne	95
1.32932	Mn I	61	1.8602	He-Ne	95
1.33174	Mn I	61	1.8751	H I	75
1.3372	Nd³⁺(CaWO₄)	78	1.880	Tm³⁺(YAG)	88
1.345	Nd³⁺(CaWO₄)	78	1.884	Tm³⁺(YAG)	88
1.3472	Ar I	56	1.91	Tm³⁺(Ca(NbO₃)₂)	78
1.3585	N I	41	1.911	Tm³⁺(CaWO₄)	73
1.36246	Mn I	61	1.916	Tm³⁺(CaWO₄)	73
1.3655	Hg I	42	1.921	Kr I	53
1.37	Nd³⁺(glass)	26	1.934	Tm³⁺(Er₂O₃)	78
1.3859	Cl I	43	1.9574	Ne I	87
1.38641	Mn I	61	1.9577	Ne I	87
1.387	Nd³⁺(CaWO₄)	78	1.972	Tm³⁺(SrF₂)	73
1.3891	Cl I	43	1.9755	Cl I	86
1.39956	Mn I	61	1.992	Co²⁺	78
1.401	Nd³⁺(borate glass)	26	2.014	Tm³⁺(YAG)	88
1.4539	C I	41	2.0199	Cl I	86
1.4542	II	46	2.0267	Xe I	53
1.4544	N I	41	2.0350	Ne I	87
1.48	Er³⁺(YAl₅O₁₂)	78	2.0354	Ne I	87
1.5	Er³⁺(YAl₅O₁₂)	78	2.046	Ho³⁺(CaWO₄)	73
1.51	GaSb	27	2.047	Ho³⁺(Ca(NbO₃)₂)	78
1.5231	He-Ne	95	2.053	Co²⁺	78
1.5235	He-Ne	95	2.059	Ho³⁺(CaWO₄)	73
1.5295	Hg I	57	2.0616	Ar I	53
1.5426	Er³⁺(glass, Er³⁺ − Yb³⁺)	89	2.0650	C I	82

TABLE 6s-1. OBSERVED LASER OSCILLATIONS (*Continued*)

Wave-length, μm	Material	Reference	Wave-length, μm	Material	Reference
2.0650	C I	82	2.967	Ne I	53
2.07	InGaAs	40	2.9788	Ar I	53
2.092	Ho^{3+}(CaF$_2$)	73	2.9813	Ne I	53
2.092	Ho^{3+}(YAG)	78	2.9845	Kr I	53
2.098	Ho^{3+}(YAG)	78	2.9878	Kr I	53
2.10119	Ne I	53	3.0268	Ne I	53
2.1041	He-Ne	95	3.0278	Ne I	53
2.116	Kr I	53	3.0453	Ar I	53
2.123	Ho^{3+}(YAG)	78	3.0536	Kr	53
2.1339	Ar I	53	3.0672	Kr I	53
2.1708	He-Ne	95	3.0996	Ar I	53
2.189	Kr I	53	3.1	In As	27
2.2045	Ar I	53	3.0078	Xe I	53
2.2	U^{3+}	73	3.1333	Ar I	53
2.2845	Br I	85	3.1346	Ar I	53
2.3139	Ar I	53	3.1515	Kr I	53
2.3200	Xe I	53	3.20	Cs	54
2.3260	Ne I	53	3.236	I I	84
2.3511	Br I	85	3.24804	N$_2$	17
2.36	Dy^{2+}(CaF$_2$)	78	3.2739	Xe I	53
2.3951	Ne I	53	3.29522	N$_2$	67
2.396	He-Ne	95	3.30756	N$_2$	67
2.3973	Ar I	53	3.31240	N$_2$	67
2.4466	Cl I	86	3.31616	N$_2$	67
2.5	U^{3+}	78	3.3179		53
2.5014	Ar I	53	3.3182	Ne I	53
2.523	Kr I	53	3.31892	N$_2$	67
2.5400	Ne I	53	3.3342	Ne I	53
2.5421	Ar I	53	3.3362	Ne I	53
2.5494	Ar I	41	3.3409	Kr I	53
2.5512	Ar I	41	3.3419	Kr I	53
2.5524	Ne I	53	3.3676	Xe I	53
2.5364	Ar I	41	3.3813	Ne I	53
2.5668	Ar I	41	3.3849	Ne I	53
2.5987	I I	84	3.39	He-Ne	95
2.6267	Kr I	53	3.3912	Ne I	53
2.6276	Xe I	53	3.3922	Ne I	53
2.6288	Kr I	53	3.4060	C I	82
2.6518	Xe I	53	3.431	I I	84
2.660	Xe I	53	3.4345	Xe I	53
2.6843	Ar I	53	3.440	Xe I	53
2.7356	Ar I	53	3.4481	Ne I	53
2.7571	I I	84	3.4487		53
2.7581	Ne I	53	3.45212	N$_2$	67
2.784	Ne I	53	3.45852	N$_2$	67
2.8202	Ar I	53	3.45377	N$_2$	67
2.8238	Ar I	53	3.4680	Kr I	53
2.8375	Br I	85	3.46804	N$_2$	67
2.8618	Kr I	53	3.47127	N$_2$	67
2.864	Ne I	53	3.4883	Kr I	53
2.8663	Ar I	53	3.4895	Kr I	53
2.8783	Ar I	53	3.5080	Xe I	53
2.8843	Ar I	53	3.5160	C I	82
2.9272	Ar I	53	3.5845	Ne I	53
2.9496	Ne I	53	3.6219	Xe I	53

TABLE 6s-1. OBSERVED LASER OSCILLATIONS (*Continued*)

Wave-length, μm	Material	Refer-ence	Wave-length, μm	Material	Refer-ence
3.6518	Xe I	53	5.27380	CO	63
3.6798	Xe I	53	5.28465	CO	63
3.6859	Xe I	53	5.29570	CO	63
3.700	HC I	83	5.3000	Kr I	53
3.734	HC I	83	5.3919	Kr I	53
3.7746	Ne I	53	5.30695	CO	63
3.808	HCl	83	5.31820	CO	63
3.810	HCl	83	5.32415	CO	63
3.840	HCl	83	5.33490	CO	63
3.869	Xe I	53	5.34590	CO	63
3.894	Xe I	53	5.35695	CO	63
3.93	Hg I	58	5.36820	CO	63
3.9817	Ne I	53, 41	5.37950	CO	63
3.9966	Xe I	53	5.54041		53
4.14	HgCdTe	27	5.4048	Ne I	53
4.160	Xe I	53	5.40801	CO	63
4.28	PbS	80	5.41961	CO	63
4.3748	Kr I	53	5.42682	CO	63
4.602	Xe I	53	5.42991	CO	63
4.7330	Ar I	53	5.44247	CO	63
4.8773	Kr I	53	5.4680	Ar I	53
4.8832	Kr I	53	5.4770	Ar I	53
4.9160	Ar I	53	5.5700	Kr I	53
4.9213	Ar I	53	5.5754	Xe I	53
5.03755	CO	63	5.5863	Kr I	53
5.04750	CO	63	5.5970	C I	82
5.05755	CO	63	5.6306	Kr I	53
5.06773	CO	63	5.662	Ne I	53
5.07807	CO	63	5.8477	Ar I	53
5.08845	CO	63	5.88	Hg I	58
5.09905	CO	63	5.8666	Ar I	53
5.10410	CO	63	5.9550	NO	81
5.10985	CO	63	5.9639	NO	8
5.11418	CO	63	5.9750	NO	81
5.1216	Ar I	53	6.0008	NO	81
5.1218	Ar I	53	6.0266	NO	81
5.12445	CO	63	6.0327	NO	81
5.13485	CO	63	6.0386	NO	81
5.14530	CO	63	6.0403	NO	81
5.15595	CO	63	6.050	Ar I	53
5.16666	CO	63	6.0629	NO	81
5.17220	CO	63	6.0885	NO	81
5.17765	CO	63	6.0934	NO	81
5.18250	CO	63	6.1417	NO	81
5.18865	CO	63	6.1545	NO	81
5.19290	CO	63	6.1972	NO	81
5.20	InSb	27	6.2752	NO	81
5.20345	CO	63	6.49	Hg I	58
5.21410	CO	63	6.565	PbTe	80
5.22498	CO	63	6.9429	Ar I	41
5.23600	CO	63	6.9448	Ar I	41
5.24195	CO	63	7.0581	Kr I	41
5.24710	CO	63	7.18	Cs	54
5.25250	CO	63	7.2150	Ar I	53
5.26310	CO	63	7.314	Xe I	51

TABLE 6s-1. OBSERVED LASER OSCILLATIONS (*Continued*)

Wave-length, μm	Material	Reference	Wave-length, μm	Material	Reference
7.33	Ne I	41	10.8416	N₂O	33
7.4221	Ne I	41	10.8523	N₂O	33
7.4237	Ne I	41	10.8628	N₂O	33
7.4799	Ne I	41	10.8736	N₂O	33
7.4994	Ne I	41	10.8843	N₂O	33
7.6163	Ne I	41	10.8951	N₂O	33
7.6461	Ne I	41	10.9061	N₂O	33
7.6510	Ne I	41	10.9170	N₂O	33
7.6925	Ne I	41	10.9280	N₂O	33
7.7015	Ne I	41	10.9390	N₂O	33
7.7407	Ne I	41	10.9510	N₂O	33
7.7655	Ne I	41	10.9612	N₂O	33
7.78	Ne I	41	10.9724	N₂O	33
7.8063	Ar I	41	10.96	Ne	33, 53
7.8368	Ne I	41	10.981	Ne I	33
8.0088	Ne I	41	10.9838	N₂O	33
8.0621	Ne I	41	10.9951	N₂O	33
8.18384	N₂	67	11.0067	N₂O	33
8.337		53	11.0184	N₂O	33
8.21106	N₂	67	11.0296	N₂O	33
8.57	PbSe	27	11.0416	N₂O	33
8.83370	Ne I	41	11.29	Xe I	53
8.8413	Ne I	41	11.4823	N₂-Cs₂	36
9.004	Xe I	51	11.4893	N₂-Cs₂	36
9.0890	Ne I	41	11.5031	N₂-Cs₂	36
9.5691	CO₂	32	11.5099	N₂-Cs₂	36
9.5862	CO₂	32	11.5166	N₂-Cs₂	36
9.6063	CO₂	32	11.5237	N₂-Cs₂	36
9.6211	CO₂	32	11.5307	N₂-Cs₂	36
9.6391	CO₂	32	11.5376	N₂-Cs₂	36
9.657	CO₂	32	11.5446	N₂-Cs₂	36
9.6762	CO₂	32	11.5962	N₂-Cs₂	36
9.702	Xe I	51	11.861	Ne I	41
10.063	Ne I	41	11.902	Ne I	41
10.5135	CO₂	32	12.141	Ar I	41
10.5326	CO₂	32	12.266	Xe I	41
10.5518	CO₂	32	12.835	Ne I	41
10.5713	CO₂	32	12.972	Xe I	41
10.5912	CO₂	32	13.739	Ne I	41
10.6118	CO₂	32	13.759	Ne I	41
10.6324	CO₂	32	14.93	Ne	53
10.6534	CO₂	32	15.037	Ar I	41
10.6748	CO₂	32	16.56	PbSnTe	27
10.6965	CO₂	32	16.638	Ne I	41
10.7194	CO₂	32	16.668	Ne I	41
10.7415	CO₂	32	16.893	Ne I	41
10.7698	N₂O	33	16.931	H₂O	37
10.7748	CO₂	32	16.947	Ne I	41
10.7799	N₂O	33	17.158	Ne I	41
10.7880	CO₂	33	17.189	Ne I	41
10.7901	N₂O	33	17.802	Ne I	41
10.8005	N₂O	33	17.837		53
10.8107	N₂O	33	17.841	Ne I	41
10.8212	N₂O	33	17.888	Ne I	41
10.8312	N₂O	33	18.396	Ne I	41

TABLE 6s-1. OBSERVED LASER OSCILLATIONS (*Continued*)

Wave-length, μm	Material	Refer-ence	Wave-length, μm	Material	Refer-ence
18.506	Xe I	53	55.077	H_2O	37
19.472		53	55.68	Ne I	41
20.480	Ne I	41	56.845	D_2O	35
21.471	NH_4	38	57.355	Ne I	41
21.752	Ne I	41	57.660	H_2O	37
22.542	NH_4	38	67.177	H_2O	37
22.563	NH_4	38	68.329	Ne I	41
22.675	NH_4	38	71.965	D_2O	35
22.836	Ne I	41	72.15	Ne I	41
23.365	H_2O	37	72.429	D_2O	35
24.918	NH_4	M38	72.747	D_2O	35
25.423	Ne I	41	73.337	D_2O	35
26.282	NH_4	38	73.402	H_2O	37
26.666	H_2O	37	74.545	D_2O	35
26.944	Ar I	53	76.305	D_2O	35
27.974	H_2O	37	78.455	H_2O	37
28.053	Ne I	41	79.106	H_2O	37
28.054	H_2O	37	84.111	D_2O	35
28.273	H_2O	37	84.291	D_2O	35
28.356	H_2O	37	84.047	Ne I	41
31.55		53	86.9	Ne I	41
31.928	Ne I	41	88.46	Ne I	41
31.951	NH_4	38	89.775	H_2O	37
32.02		53	89.93	Ne I	41
32.52		53	93.02	Ne I	41
32.929	H_2O	37	106.02	Ne I	41
33.033	H_2O	37	107.71	D_2O	36, 35
33.896	D_2O	35	115.42	H_2O	37
34.679	Ne I	41	118.65	H_2O	37
35.000	H_2O	37	120.08	H_2O	37
35.090	D_2O	35	124.4	Ne I	41
35.602	Ne I	41	126.1	Ne I	41
35.841	H_2O	37	126.24	Diethylamine	40
36.319	D_2O	35	128.75	Diethylamine	40
36.524	D_2O	35	130.95	Diethylamine	40
36.619	H_2O	37	132.8	Ne I	41
37.231	Ne I	41	135.03	Diethylamine	40
37.791	D_2O	35	171.6	D_2O	36
37.859	H_2O	37	181.90	Deuterium and BrCN	40
38.094	H_2O	37	190.08	Deuterium and BrCN	40
39.698	H_2O	37	194.83	Deuterium and BrCN	40
40.629	H_2O	37	201.19	Diethylamine	40
40.994	D_2O	35	204.53	Deuterium and BrCN	40
41.741	Ne I	41	211.14	Diethylamine	40
45.523	H_2O	37	223.25	Diethylamine	40
47.251	H_2O	37	309.94	Diethylamine	40
47.469	H_2O	37	311.08	Diethylamine	40
47.693	H_2O	37	311.3	CN	39
48.677	H_2O	37	336.7	CN	39
50.70	Ne I	41	336.83	Diethylamine	40
52.39	Ne I	41	372.30	Diethylamine	40
53.486	Ne I	41			
53.906	H_2O	37			
54.019	Ne I	41			
54.117	Ne I	41			

Although the feedback mechanism is reflection, the metallic mirrors used in the first lasers have been replaced by multilayer dielectrics with the reflectivity maximum matched to the wavelength of interest. Another device commonly used for reflectors is the multiple reflections from an etalon of transparent plates.

6s-2. Noble Gas Lasers. One important class of lasers consists of the electrically excited noble gas lasers, and of these the most common is the He-Ne. In the He-Ne

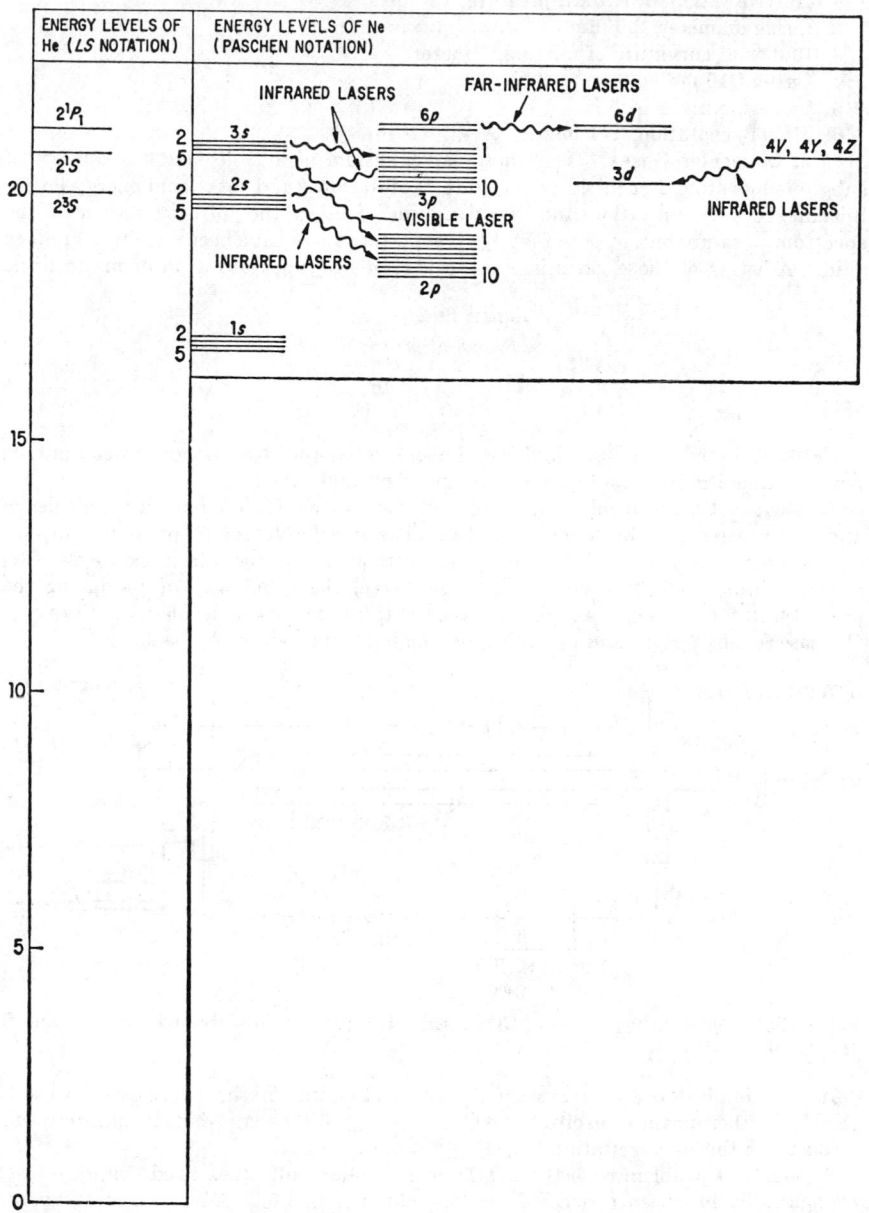

FIG. 6s-2. The energy levels to the He-Ne system.

laser excitation of the Ne occurs through collisions with excited He atoms; the appropriate energy levels are given in Fig. 6s-2. Specifically identified are only three of the more than 100 laser lines originating from the neon atom covering the range 0.5939 to 120 μm. The details of construction of a laser operating at these wavelengths involve pressures, discharge currents, tube diameter, and choice of cavity optics; however, typical parameters for 6,328 Å or 1.15 μm are:

1. He³:Ne ratio, 10:1; total pressure, 2.5 torrs
2. Inside diameter, 2.0 mm
3. Radius of curvature of mirrors, 1 meter
4. Current 10 ma voltage, ~3 kv
5. Output, 5 mw
6. Mirror separation, ~1 meter

6s-3. Molecular Gases. The energy-level system of a molecular gas consists of subgroupings around each electronic state of vibrational and rotational energy levels. Spacing between vibration and rotation levels falls in the infrared region of the spectrum. At present, approximately 15 molecular gases have been shown to produce gain. A listing of those gases is given below, identifying the region of major lines.

TABLE 6s-2

Gas	Region of principal lines, μm
CO	5
CO₂	10
H₂O	10–100

Until 1969, the CO_2 laser [32] has demonstrated the highest continuous output power. Reports of output powers in excess of 1,000 watts are common. A very convenient method of reaching high output power with a reasonable overall efficiency and size has recently been reported [19]. The specific details of pressure, current, etc., are dependent upon the dimensions. Although it is possible to excite the CO_2 directly through electron impact in a discharge, the usual way of producing the excited CO_2 for lasers is through a collision with excited N_2 molecules. A "typical" CO_2 laser configuration and operating parameters are given in Fig. 6s-3.

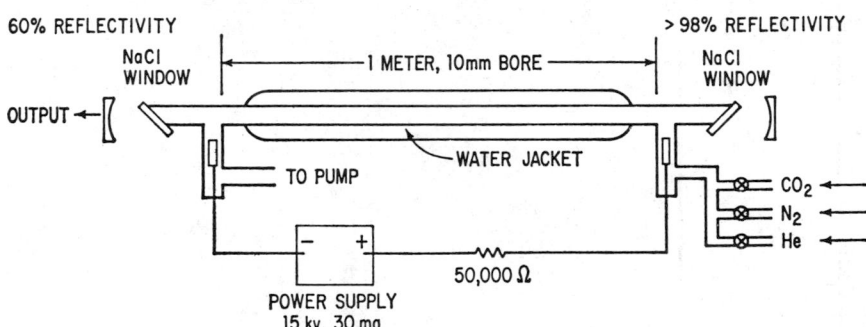

FIG. 6s-3. Schematic diagram of a CO_2 laser. The pressure maintained by the pump is approximately 10 torrs.

In the simple low-power system described electrical discharge is used to excite the N_2. Other means of exciting the CO_2 have been discussed. The most interesting seems to be thermal excitation [20,21] in flowing mixtures.

A persistent problem with the CO_2 laser in a "sealed-off" (i.e., fixed volume of gas) configuration has been a decay in output power with life. Advances in technique and understanding of the failure mechanism should alleviate this problem shortly.

For CO_2 lasers the appropriate energy levels are identified in Fig. 6s-4. Excitation of the CO_2 molecule from the ground state to the "001" level results from a collision with an excited N_2 molecule. This is followed by laser decay from the "001" level to the "100" level. Before recycling can occur, the CO_2 must decay back to the ground state.

6s-4. Ion Lasers. In an intense discharge it is possible to produce inverted populations among levels of the ions, and thus observe laser action. Most of the noble gases, the halogens, Group II and IV elements make up the two dozen ions that have been reported. Probably the most important is the argon or argon-krypton mixtures. The principal reason has been its convenient spectral region and relatively high output power. Continuous powers up to $\sim$10 watts in mixtures of Ar and Kr have been

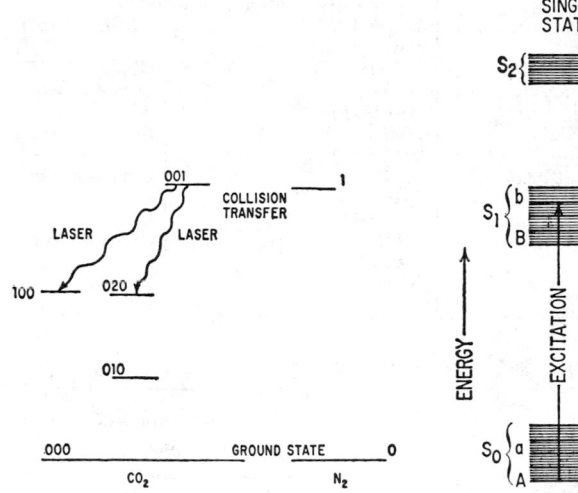

FIG. 6s-4. Appropriate energy levels for the $CO_2:N_2:He$ laser system.

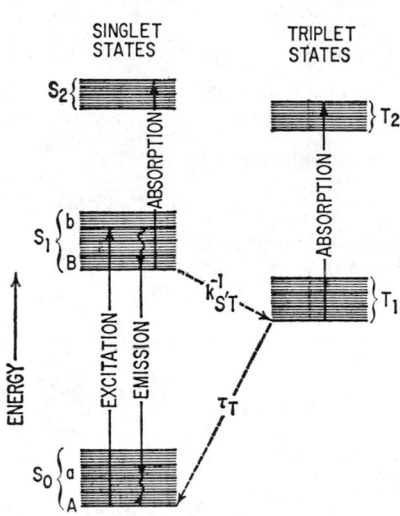

FIG. 6s-5. Energy levels for organic dye lasers.

reported. In this combination outputs of $\sim\frac{1}{4}$ watt are available at a series of lines from 4,800 to 6,471 Å.

6s-5. Organic Dye Lasers. The unique property of the organic dye lasers [25] that makes them of interest is their ability to be tuned [22] in frequency over portions of the visible spectrum. The materials used are given in Table 6s-3, along with the central wavelengths of laser emission. By using various frequency-selective devices within the cavity (e.g., etalon, gratings, or prisms) oscillation can be tuned over $\sim$500 Å. These organic materials, in general, have two sets of electronic states, the singlet and triplet states, with the lower level a singlet (see Fig. 6s-5). The optical properties are determined by these levels. In transition between vibrational sublevels, the single states have been observed to "lase." A "quenching action" is built in by the absorption of the triplet levels. Pumping has been accomplished with either other lasers or specially designed short flash lamps.

6s-6. Glass Lasers. The glass laser [26,29,30] in particular the Nd^{3+} doped glasses, is of particular interest because as a class they produce the highest peak powers and the shortest pulses. (See Table 6s-4).

In a glass rod one meter by 38 mm containing 3 percent by weight of Nd^{3+} doping an output of 5,000 joules in 3 ms has been reported [26]. Operation in a cw mode

TABLE 6s-3. STRUCTURE, WAVELENGTH, AND SOLVENTS FOR LASER DYES

Dye	Solvent	Wavelength, nm
Acridine Red	E ↑ OH	Red 600–630
Pyronin B	MeOH H_2O	Yellow
Rhodamine 6G	E ↑ OH MeOH H_2O DMSO Polymethyl methacrylate	Yellow 570–610
Rhodamine B	E ↑ OH MeOH Polymethyl methacrylate	Red 605–635
Na-fluorescein	E ↑ OH H_2O	Green 530–560
2,7-Dichlorofluorescein	E ↑ OH	Green 530–560
7-Hydroxycoumarin	H_2O (pH ~ 9)	Blue 450–470
4-Methylumbelliferone	H_2O (pH ~ 9)	Blue 450–470
Esculin	H_2O (pH ~ 9)	Blue 450–470
7-Diethylamino-4-methylcoumarin	E ↑ OH	Blue
Acetamidopyrene trisulfonate	MeOH H_2O	Green-yellow
Pyrylium salt	MeOH	Green

TABLE 6s-4. LASER IONS IN GLASS

Ion	Host	Wavelength, μm	Inversion for 1% gain/cm, cm^3
Nd^{3+}	K-Ba-Si La-Ba-Th-B Na-Ca-Si	1.06 1.37 0.92	0.7×10^{18} 3.5×10^{18}
Nd^{3+}	YAG	1.065	1.1×10^{16}
Yb^{3+}	Li-Mg-Al-Si K-Ba-Si	1.015 1.06	2.8×10^{18} 11.0×10^{18}
Ho^{3+}	Li-Mg-Al-Si	2.1	
Er^{3+}	Yb-Na-K-Ba-Si Li-Mg-Al-Si Yb-Al-Zn-P_2O_5 Yb-fluorophosphate	1.543 1.55 1.536 1.54	1.8×10^{18} 9×10^{17}
Tm^{3+}	Li-Mg-Al-Si Yb-Li-Mg-Al-Si	1.85 2.015	

or at high pulse repution rate is limited by the low thermal conductivity of glass. However, when the laser material is segmented as discs or slabs some relief from this problem is possible. One configuration using slabs has operated at 100 pulses/sec.

The wide inhomogeneous fluorescent line width (180 Å) of Nd in glass makes it possible through a combination of Q-switching and mode locking to generate very short pulses. To date the shortest pulse reported is 2.5×10^{-13} sec. In many applications not only is the pulse width important but the brightness watts/steradian $\cdot$ cm^2 and the energy available. A recent report [7] is 51 joules in 3×10^{-12} sec or 17×10^{12} watts.

In designing to produce maximum brightness an essential feature is control of the cavity modes. The control in the form of a pinhole at the focus of a lens has been

TABLE 6s-5. SEMICONDUCTOR LASERS

Material	Photon energy, eV	Method of excitation
ZnS	3.82	
ZnO	3.30	Electron beam
CdS	2.50	
		Optical
GaSe	2.09	
CdS_xSe_{1-x}	1.80–2.50	Electron beam
CdSe	1.82	
CdTe	1.58	
$Ga(As_xP_{1-x})$	1.41–1.95	p-n junction
GaAs	1.47	
		Electron beam
		Optical
		Avalanche
InP	1.37	
$In_xGa_{1-x}AS$	1.5	p-n junction
GaSb	0.82	
		Electron beam
InP_xAs_{1-x}	1.40	p-n junction
InAs	0.40	
		Electron beam
		Optical
InSb	0.23	p-n junction
		Electron beam
		Optica
Te	0.34	Electron beam
PbS	0.29	p-n junction
		Electron beam
PbTe	0.19	p-n junction
		Electron beam
		Optical
PbSe	0.145	p-n junction
		Electron beam
$Hg_xCd_{1-x}Te$	0.30–0.33	Optical
$Pb_xSn_{1-x}Te$	0.075–0.19	

used in a Pockels cell Q-switched oscillator-amplifier chain to generate 2×10^{17} w/cm^2. The overall efficiency of pulsed glass lasers is generally given as the ratio of the incremental pump energy in excess of threshold to the output power with glass lasers operating in the long pulse mode (pulse times $\sim 10^{-3}$ sec) efficiencies as high as 8 percent have been reported.

6s-7. Crystalline Solids. The first laser $Cr^{3+}:Al_2O_3$ possessed all the general properties of this large class of lasers. They are characterized by impurities in low concentrations ($\sim 10^{-3} \to 10^{-2}$) and fluorescent emissions between $0.6 \to 2$ μm. Excitation of the fluorescent levels is through absorption bands lying at higher photon energies. At this time the two most commonly encountered are the $Cr^{3+}:Al_2O_3$ and the $Nd^{3+}:YAG$.

6s-8. Semiconductor Lasers. The first observations [28] of stimulated emission were obtained from forward-biased GaAs diodes. After this initial work other schemes of pumping were employed (e.g., optical pumping, electron beam, and avalanche breakdown). The range of materials is given in Table 6s-5, along with the approximate wavelengths of emission and the pumping systems. Several reviews have been written [27].

6s-9. Chemical Lasers. Various proposals have been to utilize the energy of exothermic chemical reactions to produce inverted populations of molecules. Examples of such reactions are the combination of H and F to form HF. Papers have been written reporting observations of laser action near 2.8 μm in HF [96] and in the transfer of energy obtained from the exothermic reaction of H and F forming HF to the CO_2 molecule [97].

References

1. Although the concepts of stimulated emission follow from even the simplest treatments of radiation interacting with an energy-level system and are treated in many tests, the first explicit proposal to use stimulated emission in an actual device that the author is aware of followed the demonstration of negative temperatures [E. M. Purcell and R. V. Pound, *Phys. Rev.* **81**, 279 (1951)].
   Weber, J.: *IRE Trans. Electron Devices* **3**, 1 (1953).
   Gordon, Zeiger, and Townes: *Phys. Rev.* **95**, 282 (1954).
   Basov, N. G., and A. M. Prokhorov: *J. Exptl. Theoret. Phys. U.S.S.R.* **27**, 431 (1954).
   Chang, W. S. C.: "Principles of Quantum Electronics," Addison-Wesley Publishing Company, Inc., Reading, Mass., (1969).
   In addition to the published papers, oral unpublished comments at conferences were made by Weber (June, 1952), Basov and Prokhorov (May, 1952), Nethercot and Townes (May, 1951), applicable to the microwave frequency region. In the optical region of the spectrum the first observation of stimulated emission in solids was by T. H. Maimen, *Nature* **18**, 492 (1960).
2. Gordon, J. P., H. S. Zweiger, and C. H. Townes: *Phys. Rev.* **99**, 1264 (1955).
3. Maimen, T.: *Nature* **187**, 493 (1960).
4. Bloembergen, N. B.: *Phys. Rev.* **104**, 324 (1956).
5. Scovil, Feher, and Seider: *Phys. Rev.* **105**, 762 (1957).
6. Schawlow, A. L., and C. H. Townes: *Phys. Rev.* **112**, 1940 (1958).
7. Gobeli, G.: *Electron. News* **14**, 72 (Mar. 17, 1969: N. Basov et al.: *IEEE J. Quantum Electron.* **QE-4**, 604 (October, 1968).
8. Collins, Nelson, Schawlow, Bond, Garrett, and Kaiser: *Phys. Rev. Letters* **5**, 303 (1960).
9. Javan, Bennett, and Herriott: *Phys. Rev. Letters* **6**, (1961).
10. Tolman, R. C.: *Phys. Rev.* **23**, 693 (1924).
11. Ladenburg, R.: *Physik* **48**, 15 (1949).
12. Fabrikant, V. A.: *Tr. Vses Elektrotekhn. Inst.* **41**, 254 (1940); see especially p. 273.
13. Tomlinson, W. J., and R. L. Fork: *Appl. Opt.* **8**, 121–129 (January, 1969).
14. For a review see A. J. DeMaria, *Proc. IEEE* **57**, 2–25 (January, 1969).
15. Patel, C. K. N., and R. J. Kerl: *Appl. Phys. Letters* **5**, 81 (1964).
16. Patel, C. K. N.: *Appl. Phys. Letters* **7**, 246 (1965).
17. Patel, C. K. N.: *Appl. Phys. Letters* **6**, 12 (1965); W. R. Bennett, Jr.: *Appl. Opt. Suppl.* **2**, 3 (1965).
18. Bennett, W. R.: *Appl. Opt. Suppl.* **2**, 3 (1965); A. Crocker, H. A. Gebbie, M. F. Kimmitt, and L. E. S. Mathias: *Nature* **201**, 250 (1964); W. W. Muller and G. T. Flesher: *Appl. Phys. Letters* **8**, 217 (1966).
19. Tiffany, W. B., R. Targ, and J. D. Foster: *Appl. Phys. Letters* **3** (1969).
20. Basov, N. G., A. N. Oraevskie, and V. A. Scheglov: *Soviet Phys.—Tech. Phys.* **12**, 243 (1969).
21. Fern, M. E., J. Vergeyen, and B. E. Cherrington: *Appl. Phys. Letters* **14**, 337 (1969).
22. Soffer and McFarland: *Appl. Phys. Letters* **10**, 266 (1967).
23. Sorokin, P., and J. Lankard: *IBM J. Res. & Develop.* **10**, 162 (1969); F. Schaefer, W. Schmidt, and J. W. Volze: *Appl. Phys. Letters* **9**, 306 (1966); M. Spaeth and Bartfield: *ibid*, 179.
24. Bridges, W. B., and A. N. Chester: *J. Quantum Electron.* **QE-1**, 66 (1965).
25. Snaveley, B. B.: *Proc. IEEE* **57**, 1374–1390 (1969).
26. Young, G.: *Proc. IEEE* **57**, 1274 (1969).
27. Nathan, M.: *Proc. IEEE* **54**, 1276–1290 (October, 1966).

28. Hall, R. N., G. E. Fenner, J. D. Kingsley, T. J. Soltys, and R. O. Carlson: *Phys. Rev. Letters* **9**, 366–378 (November, 1962); M. I. Nathan, W. P. Dumke, G. Burns, F. H. Dill, Jr., and G. J. Lasher: *Appl. Phys. Letters* **1**, 62–64 (November, 1962); T. M. Quist, R. H. Rediker, R. J. Keyes, W. E. Krag, B. Lax, A. L. McWhorter, and H. J. Zeiger: *ibid.*, 91–92 (December, 1962).
29. Snitzer, E.: *Phys. Rev. Letters* **7**, 444–446 (December, 1961).
30. Snitzer, E.: *Proc. IEEE* **54**(10), 1249–1261 (October, 1966).
31. Boyd, G. D., and J. P. Gordon: *Bell System Tech. J.* **40**, 496 (1961).
32. Patel, C. K. N.: *Phys. Rev.* **136**, A1187 (1964).
33. Patel, C. K. N.: *Appl. Phys. Letters* **6**, 12 (1965).
34. Patel, C. K. N. *Appl. Phys. Letters* **7**, 273 (1965).
35. Mathias, L. E. S., and A. Crocker: *Phys. Letters* **13**, 35 (1964).
36. Muller, W. W., and G. T. Flesher: *Appl. Phys. Letters* **8**, 217 (1966).
37. Crocker, A., H. A. Gebbie, M. F. Kimmitt, and L. E. S. Mathias: *Nature* **201**, 250 (1964).
38. Mathias, L. E. S., A. Crocker, and M. S. Wills: *Phys. Letters* **14**, 33 (1965).
39. Muller, W. W., and G. T. Flesher: *Appl. Phys. Letters* **8**, 217 (1966).
40. Mathias, L. E. S., A. Crocker, and M. S. Wills: *Elec. Letters* **1**, 45 (1965).
41. Flesher, G. T., and W. M. Muller: *Proc. IEEE* **54**, 543 (1966); R. A. McFarland, W. L. Faust, C. K. N. Patel, and C. G. B. Garrett: *Quantum Electron.* **III**, 561 (1964); see also ref. 53 for the above.
42. Bridges, W. B. *Proc. IEEE* **52**, 843 (1964); W. B. Bridges and A. N. Chester: *IEEE J. Quant. Electron.* QE-1, 66 (1965); *Appl. Opt.* **4**, 573 (1965).
43. Cheo, P. K., and H. G. Cooper: *Appl. Phys. Letters* **7**, 202 (1965)
44. Zarowin, C. B.: *Appl. Phys. Letters* **9**, 241 (1966).
45. Javan, A.: *Proc. IEEE* **52**, 1350 (1964).
46. Jarrett, S. M., J. Nunez, and G. Gould: *Appl. Phys. Letters* **8**, 1950 (1966).
47. McFarland, R. A.: *Appl. Opt.* **3**, 1196 (1964).
48. Bridges, W. B., and A. S. Halstead: *IEEE J. Quantum Electron.* QE-2, 84 (1966).
49. Hurwitz, C. E.: *Appl. Phys. Letters* **9**, 116 (1966).
50. Nicoll, F. H.: *Appl. Phys. Letters* **9**, 13 (1966).
51. Faust, W. L., R. A. McFarland, C. K. N. Patel, and C. G. B. Garrett: *Appl. Phys. Letters* **1**, 85 (1962).
52. Bridges, W. B., and A. S. Halstead: *IEEE J. Quantum. Electron.* QE-2, 84 (1966).
53. Faust, W. L., R. A. McFarland, C. K. N. Patel, and C. G. B. Garrett: *Phys. Rev.* **133**, A1476 (March, 1963).
54. Rabinowitz, P., S. Jacobs, and G. Gould: *Appl. Opt.* **1**, 513 (1962).
55. Neusel, R. H.: *IEEE J. Quantum Electron.* QE-2, 70 (1966).
56. Horrigan, F. A., S. H. Koozekanani, and R. A. Paananen: *Appl. Phys. Letters* **6**, 41 (1965).
57. Paananen, R. A., C. L. Tang, F. A. Horrigan, and H. Statz: *J. Appl. Phys.* **34**, 3148 (1963).
58. Bockasten, K., M. Gararaglia, B. A. Lengyel, and T. Lundholm: *J. Opt. Soc. Am.* **55**, 1051 (1965).
59. Silfvast, W. T., G. R. Fowles, and B. D. Hopkins: *Appl. Phys. Letters* **8**, 318 (1966); G. R. Fowles, W. T. Silfvast, and R. C. Jensen: *IEEE J. Quantum Electron.* QE-1, 183 (1965).
60. Fowles, G. R., and W. T. Silfvast: *IEEE J. Quantum Electron.* QE-1, 131 (1965).
61. Piltch, M. W. T. Walter, N. Solimene, G. Gould, and W. R. Bennett, Jr.: *Appl Phys. Letters* **7**, 309 (1965); M. Piltch, W. T. Walter, N. Solimene, and C. Gould: *ibid.* **9**, 253 (1966).
62. Mathias, L. E. S., and J. T. Parker: *Phys. Letters* **7**, 194 (1963).
63. Patel, C. K. N., and R. J. Kerl: *Appl. Phys. Letters* **5**, 81 (1964).
64. Keeffe, William M., and Walter J. Graham: *Appl. Phys. Letters* **7**, 263 (1965).
65. Bell, W. E., and A. L. Bloom: *IEEE J. Quant. Electron.* QE-1, 400 (1965).
66. Mathias, L. E. S., and J. T. Parker: *Appl. Phys. Letters* 3, 16 (1963).
67. McFarland, R. A.: *IEEE J. Quantum Electron.* QE-2, 229 (1966).
68. Chang, N. C.: *J. Appl. Phys.* **34**, 3500 (1963).
69. Wolff, Hoskins: *Appl. Phys. Letters* **4**, 113 (1964); E. Schimitschek: *ibid.* **3**, 117 (1963).
70. Hall, R. N., et al.: *Phys. Rev. Letters* **9**, 366 (1962); M. Nathan et al.: *Appl. Phys. Letters* **1**, 62 (1962); T. M. Quist et al.: *ibid.*, 91.
71. Schawlow, A. L., and G. Devlin: *Phys. Rev. Letters* **6**, 95 (1961).
72. Abella, I. D., and H. J. Cummins: *J. Appl. Phys.* **32**, 1177 (1961).
73. Johnson, L.: *J. Appl. Phys.* **34**, 897 (1963).
74. Kaiser, W., et al.: *Phys. Rev.* **123**, 766 (1961); **126**, 2079 (1962).
75. Bockasten, K.: *J. Opt. Soc. Am.* **56**, 1260 (1966).

76. Geusic, J., et al.: *Appl. Phys. Letters* **4**, 182 (1964).
77. Heller, A.: *Appl. Phys. Letters* **9**, 108 (1966).
78. Kiss, Z. J., and R. J. Pressley: *Proc. IEEE* **54**, 1236–1247 (1966).   (This is a general review article of crystalline solid lasers.)
79. Harris, S. E.: *Proc. IEEE* **57**, 2096 (1969).
80. Hurwitz, C., A. R. Calawa, and R. H. Rediker: *IEEE J. Quant. Electron.* (correspondence) **QE-1**, 102–104 (May, 1965): J. F. Butler and A. R. Calawa: *J. Electrochem. Soc.* **54**, 1056–1057 (October, 1965).
81. Pollack, M. A.: *Appl. Phys. Letters* **9**, 94 (1966).
82. McFarland, R. A.: *Appl. Phys. Letters* **5**, 91 (1964).
83. Kasper, J. V. V., and G. C. Pimentel: *Phys. Rev. Letters* **14**, 352 (1965).
84. Rigden, J. D., and A. D. White: *Nature* **198**, 774 (1963).
85. Jarrett, S. M., J. Nunez, and G. Gould: *Appl. Phys. Letters* **8**, 240 (1966); **7**, 294 (1965); **8**, 150 (1966).
86. Paananen, R. A., C. L. Tang, and F. A. Horrigan: *Appl. Phys. Letters* **3**, 154 (1963).
87. Zitter, R. N.: *J. Appl. Phys.* **35**, 3070 (1964).
88. Johnson, L. F., J. E. Geusic, and L. G. Vanuitert: *Appl. Phys. Letters* **7**, 127 (1965).
89. Snitzer, E., and R. Woodcock, *Appl. Phys. Letters* **6**, 45 (1965).   For a review see Elias Snitzer, *Proc. IEEE* **54**, 1249 (1966).
90. Johnson, L. F., R. E. Dietz, and H. J. Guggenheim: *Phys. Rev. Letters* **11**, 318 (1963).
91. Soloman, R., and L. Mueller: *Appl. Phys. Letters* **3**, 135 (1963).
92. Bockasten, K.: *Appl. Phys. Letters* **4**, 118 (1964).
93. Ballman, A. A., S. P. S. Porto, and A. Yariv: *J. Appl. Phys.* **34**, 3155 (1963).
94. O'Connor, J. R., and W. A. Hargreaves, *Appl. Phys. Letters* **4**, 208 (1964).
95. The large number of papers concerning He-Ne lasers makes it difficult to reference only one.   For details see refs. 9, 41, 53, and for reviews: A. L. Bloom, *Proc. IEEE* **54**, 1263 (October, 1966); K. Tomiyasu, *IEE J. Quantum Electron.* **QE-1**, 199–219 (August, 1965); Laser Bibliography III, *ibid.* **QE-2**, 124–151 (June, 1966).
96. Spencer, D. F., T. A. Jacobs, H. Mirels, and R. Gross, *Int. J. Chem. Kinetics*, **1**, (Sept., 1969).
97. Cook, T., T. Falk, and R. Stephens: *Appl. Phys. Letters* **15**, 318 (1969).

Section 7

ATOMIC AND MOLECULAR PHYSICS[1]

H. M. CROSSWHITE, Editor

The Johns Hopkins University

CONTENTS

[1] Sections 7a through 7d were originally prepared by the previous editor, the late G. H. Dieke. Where no contributor is specifically mentioned, the material was compiled by the section editor.

7a. The Periodic System

TABLE 7a-1. ALPHABETICAL LIST OF THE ELEMENTS

In later tables the elements are arranged according to increasing order number Z. This table gives in alphabetical order the names of the elements in English, French, and German, together with the chemical symbol, year of discovery, and order number of each. (A dash means that the name of the element in French or German is the same as in English.)

English	Name in French	Name in German	Year of discovery	Symbol	Z
Actinium...............	—	—	1899	Ac	89
Alabamine*.............	....	....	....	(Ab)	(85)
Alumin(i)um...........	Aluminium	Aluminium	1827	Al	13
Americium.............	Américium	—	1945	Am	95
Antimony..............	Antimoine	Antimon	Old	Sb	51
Argentum*.............	....	....	....	Ag	47
Argon..................	—	—	1894	A	18
Arsenic................	—	Arsen	Old	As	33
Astatine...............	—	—	1940	At	85
Barium................	—	Baryum	1808	Ba	56
Berkelium.............	—	—	1950	Bk	97
Beryllium.............	Béryllium	—	1798	Be	4
Bismuth...............	—	Wismut	1753	Bi	83
Boron.................	Bore	Bor	1808	B	5
Bromine...............	Brome	Brom	1826	Br	35
Cadmium..............	—	—	1817	Cd	48
Calcium...............	—	—	1808	Ca	20
Californium...........	—	—	1950	Cf	98
Carbon................	Carbone	Kohlenstoff	Old	C	6
Cassiopeium*..........	....	....	....	Lu	71
Celtium*..............	....	....	....	(Ct)	(72)
Cerium................	Cérium	Cer	1803	Ce	58
Cesium................	Césium	Caesium	1860	Cs	55
Chlorine...............	Chlore	Chlor	1774	Cl	17
Chromium.............	Chrome	Chrom	1797	Cr	24
Cobalt................	—	—	1735	Co	27
Columbium*...........	....	....	....	(Cb)	41
Copper................	Cuivre	Kupfer	Old	Cu	29
Curium................	—	—	1944	Cm	96
Deuterium.............	—	—	1930	D	1
Dysprosium...........	—	—	1886	Dy	66
Einsteinium...........	—	—	1955	Es	99
Emanation*...........	....	....	....	Rn	86
Erbium................	—	—	1843	Er	68

Table 7a-1. Alphabetical List of the Elements (*Continued*)

English	Name in French	Name in German	Year of discovery	Symbol	Z
Europium...............	—	—	1896	Eu	63
Fermium................	—	—	1955	Fm	100
Ferrum*.................	....	....	....	Fe	26
Fluorine................	Fluor	Fluor	1771	F	9
Francium...............	—	—	1939	Fr	87
Gadolinium.............	—	—	1880	Gd	64
Gallium................	—	—	1875	Ga	31
Germanium.............	—	—	1886	Ge	32
Gold...................	Or	—	Old	Au	79
Hafnium................	—	—	1923	Hf	72
Helium.................	Hélium	—	1895	He	2
Holmium................	—	—	1879	Ho	67
Hydrogen...............	Hydrogéne	Wasserstoff	1766	H	1
Illinium*...............	....	....	....	(Il)	(61)
Indium.................	—	—	1863	In	49
Iodine..................	Iode	Jod	1811	I	53
Iridium.................	—	—	1803	Ir	77
Iron....................	Fer	Eisen	Old	Fe	26
Kalium*................	....	....	....	K	19
Krypton................	—	—	1898	Kr	36
Lanthanum.............	Lanthane	Lanthan	1839	La	57
Lawrencium............	—	—	1961	Lw	103
Lead...................	Plomb	Blei	Old	Pb	82
Lithium................	—	—	1817	Li	3
Lutetium...............	Lutétium	—	1907	Lu	71
Magnesium.............	Magnésium	—	1755	Mg	12
Manganese.............	Manganèse	Mangan	1774	Mn	25
Masurium*.............	....	....	....	(Ma)	(43)
Mendelevium...........	—	—	1955	Md	101
Mercury...............	Mercure	Quecksilber	Old	Hg	80
Molybdenum...........	Molybdène	Molybdän	1778	Mo	42
Natrium*...............	....	....	....	Na	11
Nebulium*..............					
Neodymium.............	Néodyme	Neodym	1885	Nd	60
Neon...................	Néon	—	1898	Ne	10
Neptunium.............	—	—	1940	Np	93
Nickel..................	—	—	1751	Ni	28
Niobium................	—	—	1801	Nb	41
Niton..................	....	....	....	Rn	86
Nitrogen...............	Nitrogène	Stickstoff	1772	N	7
Nobelium...............	—	—	1958	No	102
Osmium................	—	—	1803	Os	76
Oxygen................	Oxygène	Sauerstoff	1774	O	8
Palladium..............	—	—	1803	Pd	46
Phosphorus.............	Phosphore	Phosphor	1669	P	15
Platinum...............	Platine	Platin	1735	Pt	78
Plumbum*..............	....	....	....	Pb	82
Plutonium..............	—	—	1940	Pu	94
Polonium...............	—	—	1898	Po	84
Potassium..............	—	Kalium	1807	K	19
Praseodymium..........	Praséodyme	Praseodym	1879	Pr	59
Prometheum............	Prométheum	—	1947	Pm	61
Protactinium...........	—	—	1917	Pa	91
Radium................	—	—	1898	Ra	88
Radon.................	—	—	1900	Rn	86
Rhenium...............	—	—	1925	Re	75
Rhodium...............	....	....	1803	Rh	45

TABLE 7a-1. ALPHABETICAL LIST OF THE ELEMENTS (*Continued*)

English	Name in French	Name in German	Year of discovery	Symbol	Z
Rubidium	—	—	1861	Rb	37
Ruthenium	Ruthenium	—	1844	Ru	44
Samarium	—	—	1879	Sm	62
Scandium	—	—	1879	Sc	21
Selenium	Sélénium	Selen	1817	Se	34
Silicon	Silicium	Silicium	1823	Si	14
Silver	Argent	Silber	Old	Ag	47
Sodium	—	Natrium	1807	Na	11
Stannum*	....	....	....	Sn	50
Stibium*	....	....	....	Sb	51
Strontium	—	—	1790	Sr	38
Sulfur	Soufre	Schwefel	Old	S	16
Tantalum	Tantale	Tantal	1802	Ta	73
Technetium	—	—	1937	Tc	43
Tellurium	Tellure	Tellur	1782	Te	52
Terbium	—	—	1843	Tb	65
Thallium	—	—	1861	Tl	81
Thorium	—	—	1828	Th	90
Thulium	—	—	1879	Tm	69
Tin	Etain	Zinn	Old	Sn	50
Titanium	Titane	Titan	1791	Ti	22
Tritium	—		....	T	1
Tungsten	Tungstène	Wolfram	1781	W	74
Uranium	—	Uran	1789	U	92
Vanadium	—	—	1830	V	23
Virginium*	....	....	....	(Vi)	(87)
Wolfram*	....	....		W	74
Xenon	Xénon	—	1898	Xe	54
Ytterbium	—	—	1878	Yb	70
Yttrium	—	—	1794	Y	39
Zinc	—	Zink	1746	Zn	30
Zirconium	—	Zircon	1789	Zr	40

* Alternate or obsolete names. An order number between parentheses means that the discovery of the element was an error and another element has taken its place. Element symbols between parentheses have been given up.

TABLE 7a-2. PERIODIC SYSTEM OF THE ELEMENTS

1 H $\underline{2}$																	2 He $\underline{1}$, 3
3 Li $\underline{2}$	4 Be $\underline{1}$, 3											5 B $\underline{2}$	6 C 1, $\underline{3}$	7 N 2, $\underline{4}$	8 O 1, $\underline{3}$, 5	9 F $\underline{2}$, 4	10 Ne $\underline{1}$, 3
11 Na $\underline{2}$	12 Mg $\underline{1}$, 3											13 Al $\underline{2}$	14 Si 1, $\underline{3}$	15 P 2, $\underline{4}$	16 S $\underline{3}$, 5	17 Cl $\underline{2}$, 4	18 A $\underline{1}$, 3
19 K $\underline{2}$	20 Ca $\underline{1}$, 3	21 Sc $\underline{2}$, 4	22 Ti 1, $\underline{3}$, 5	23 V 2, 4, $\underline{6}$	24 Cr 1, 3, 5, $\underline{7}$	25 Mn 2, 4, $\underline{6}$, 8	26 Fe 1, 3, $\underline{5}$, 7	27 Co 2, $\underline{4}$, 6	28 Ni 1, $\underline{3}$, 5	29 Cu $\underline{2}$, 4	30 Zn $\underline{1}$, 3	31 Ga $\underline{2}$, 4	32 Ge 1, $\underline{3}$	33 As 2, $\underline{4}$	34 Se $\underline{3}$, 5	35 Br $\underline{2}$, 4	36 Kr $\underline{1}$, 3
37 Rb $\underline{2}$	38 Sr $\underline{1}$, 3	39 Y $\underline{2}$, 4	40 Zr 1, $\underline{3}$, 5	41 Nb 2, 4, $\underline{6}$	42 Mo 1, 3, 5, $\underline{7}$	43 Tc 4, $\underline{6}$, 8	44 Ru 3, $\underline{5}$, 7	45 Rh 2, $\underline{4}$	46 Pd $\underline{1}$, 3, 5	47 Ag $\underline{2}$, 4	48 Cd $\underline{1}$, 3	49 In $\underline{2}$, 4	50 Sn 1, $\underline{3}$	51 Sb 2, $\underline{4}$	52 Te 1, $\underline{3}$, 5	53 I $\underline{2}$, 4	54 Xe $\underline{1}$, 3
55 Cs $\underline{2}$	56 Ba $\underline{1}$, 3	57 La $\underline{2}$, 4	* 72 Hf 1, $\underline{3}$, 5	73 Ta $\underline{4}$, 6	74 W $\underline{5}$, 7	75 Re 4, $\underline{6}$, 8	76 Os 3, $\underline{5}$, 7	77 Ir $\underline{4}$, 6	78 Pt 1, $\underline{3}$, 5	79 Au $\underline{2}$	80 Hg $\underline{1}$, 3	81 Tl $\underline{2}$	82 Pb 1, $\underline{3}$	83 Bi 2, $\underline{4}$	84 Po 3	85 At	86 Rn $\underline{1}$, 3
87 Fr $\underline{2}$	88 Ra $\underline{1}$, 3	89 Ac $\underline{2}$	†														

*Lanthanides (rare earths)	58 Ce $\underline{1}$, 3, 5	59 Pr $\underline{4}$	60 Nd $\underline{5}$	61 Pm $\underline{6}$	62 Sm $\underline{7}$, 9	63 Eu 6, $\underline{8}$, 10	64 Gd 7, $\underline{9}$, 11	65 Tb 6, $\underline{8}$	66 Dy $\underline{5}$	67 Ho $\underline{4}$	68 Er 1, $\underline{3}$, 5	69 Tm $\underline{2}$	70 Yb $\underline{1}$, 3	71 Lu $\underline{2}$, 4
†Actinides	90 Th $\underline{3}$, 5	91 Pa $\underline{4}$	92 U $\underline{5}$, 7	93 Np $\underline{6}$	94 Pu $\underline{7}$	95 Am $\underline{8}$	96 Cm $\underline{9}$	97 Bk	98 Cf	99 Es	100 Fm	101 Md	102 No	103 Lw

The numbers under the elements indicate the observed multiplicity in the first spectrum. The value for the ground state is underlined.

TABLE 7a-3. PROPERTIES OF ELEMENTS*

Z	Symbol	Element	Atomic wt.†	Valency	Atomic diam in Å	Mass No. and (abundance)
(1)	(2)	(3)	(4)	(5)	(6)	(7)
1	H	Hydrogen	1.00797	1	3.0	1(99.985), 2(0.0146)
2	He	Helium	4.0026	0	.....	4(100), 3(1.3 × 10⁻⁴)
3	Li	Lithium	6.939	1	3.13	7(92.48), 6(7.52)
4	Be	Beryllium	9.0122	2	2.25	9(100)
5	B	Boron	10.811	3	.....	11(81.17), 10(18.83)
6	C	Carbon	12.01115	±4, 2	1.54	12(98.9), 13(1.1)
7	N	Nitrogen	14.0067	−3, 5, 2	1.06	14(99.635), 15(0.365)
8	O	Oxygen	15.9994	−2	.....	16(99.76), 18(0.204), 17(0.039)
9	F	Fluorine	18.9984	−1	1.36	10(100)
10	Ne	Neon	20.183	0	3.20	20(90.92), 22(8.82), 21(0.257)
11	Na	Sodium	22.9898	1	3.83	23(100)
12	Mg	Magnesium	24.312	2	3.20	24(78.60), 26(11.29), 25(10.11)
13	Al	Aluminum	26.9815	3	2.82	27(100)
14	Si	Silicon	28.086	4	2.34	28(92.28), 29(4.67), 30(3.05)
15	P	Phosphorus	30.9738	5, ±3	2.16	31(100)
16	S	Sulfur	32.064	6, 4, −2	2.12	32(95.018), 34(4.215), 33(0.74), 36(0.016)
17	Cl	Chlorine	35.453	±1, 7, 5	1.94	35(75.4), 37(24.6)
18	A	Argon	39.948	0	3.82	40(99.60), 36(0.337), 38(0.060)
19	K	Potassium	39.102	1	4.76	39(93.1), 41(6.9), 40‡(0.012)
20	Ca	Calcium	40.08	2	3.93	40(96.96), 44(2.06), 42(0.64), 48(0.19), 43(0.15), 46(0.0033)
21	Sc	Scandium	44.956	3	3.20	45(100)
22	Ti	Titanium	47.90	4, 3	2.93	48(73.45), 46(7.95), 47(7.75), 49(5.51), 50(5.34)
23	V	Vanadium	50.942	5, 4, 2	2.71	51(99.76), 50(0.24)
24	Cr	Chromium	51.996	6, 3, 2	2.57	52(83.76), 53(9.55), 50(4.31), 54(2.38)
25	Mn	Manganese	54.9380	7, 4, 2, 6, 3	2.5	55(100)
26	Fe	Iron	55.847	3, 2	2.52	56(91.64), 54(5.81), 57(2.21), 58(0.34)
27	Co	Cobalt	58.9332	3, 2	2.50	59(100)
28	Ni	Nickel	58.71	2, 3	2.49	58(67.76), 60(26,16), 62(3.66), 61(1.25), 64(1.16)
29	Cu	Copper	63.54	2, 1	2.551	63(69.09), 65(30.91)
30	Zn	Zinc	65.37	2	2.748	64(48.89), 66(27.81), 68(18.61), 67(4.07), 70(0.620)
31	Ga	Gallium	69.72	3	2.7	69(60.2), 71(39.8)
32	Ge	Germanium	72.59	4	2.788	74(36.74), 72(27.37), 70(20.55), 76(7.67), 73(7.61)
33	As	Arsenic	74.9216	5, ±3	2.50	75(100)
34	Se	Selenium	78.96	6, 4, −2	2.32	80(49.82), 78(23.52), 82(9.19), 76(9.02), 77(7.58), 74(0.87)
35	Br	Bromine	79.909	±1, 5	2.26	79(50.5), 81(49.5)
36	Kr	Krypton	83.80	0	4.0	84(56.90), 86(17.37), 82(11.55), 83(11.56), 80(2.27), 78(0.354)
37	Rb	Rubidium	85.47	1	5.02	85(72.15), 87(27.85)
38	Sr	Strontium	87.62	2	4.29	88(82.56), 86(9.86), 87(7.02), 84(0.56)
39	Y	Yttrium	88.905	3	3.62	89(100)
40	Zr	Zirconium	91.22	4	3.19	90(51.46), 94(17.40), 92(17.11), 91(11.23), 96(2.80)
41	Nb	Niobium	92.906	5, 3	2.94	93(100)
42	Mo	Molybdenum	95.94	6, 3, 5	2.80	98(23.75), 96(16.5), 92(15.86), 95(15.7), 100(9.62), 97(9.45), 94(9.12)
43	Tc	Technetium	(99)	7		
44	Ru	Ruthenium	101.07	3, 4, 6, 8	2.67	102(31.34), 104(18.27), 101(16.98), 99(12.81), 100(12.70), 96(5.7), 98(2.22)
45	Rh	Rhodium	102.905	3, 4	2.7	103(100)
46	Pd	Palladium	106.4	2, 4	2.745	106(27.2), 109(26.8), 105(22.6), 110(13.5), 104(9.3), 102(0.8)
47	Ag	Silver	107.870	1	2.883	107(51.35), 109(48.65)

TABLE 7a-3. PROPERTIES OF ELEMENTS* (Continued)

Z	Symbol	Element	Atomic wt.†	Valency	Atomic diam in Å	Mass No. and (abundance)
(1)	(2)	(3)	(4)	(5)	(6)	(7)
48	Cd	Cadmium	112.40	2	3.042	114(28.86), 112(24.07), 111(12.75), 110(12.39), 113(12.26), 116(7.58), 106(1.215), 108(0.875)
49	In	Indium	114.82	3	3.14	115(95.77), 113(4.23)
50	Sn	Tin	118.69	4, 2	3.164	120(33.03), 118(23.98), 116(14.07), 119(8.62), 117(7.54), 124(6.11), 122(4.78), 112(0.90), 114(0.61), 115(0.35)
51	Sb	Antimony	121.75	3, 5	3.228	121(57.25), 123(42.75)
52	Te	Tellurium	127.60	4, 6, −2	2.9	130(34.46), 128(31.72), 126(18.72), 125(7.01), 124(4.63), 122(2.49), 123(0.89), 120(0.091)
53	I	Iodine	126.9044	−1, 5, 7	2.7	127(100)
54	Xe	Xenon	131.30	0	4.4	132(26.96), 129(26.44), 131(21.17), 134(10.44), 136(8.95), 130(4.07), 128(1.90), 124(0.094), 126(0.088)
55	Cs	Cesium	132.905	1	5.40	133(100)
56	Ba	Barium	137.34	2	4.48	138(71.66), 137(11.32), 136(7.81), 135(6.59), 134(2.42), 130(0.101), 132(0.097)
57	La	Lanthanum	138.91	3	3.741	139(99.91) 138(0.089)
58	Ce	Cerium	140.12	3, 4	3.64	140(88.48), 142(11.07), 138(0.250), 136(0.193)
59	Pr	Praesodymium	140.907	3	3.65	141(100)
60	Nd	Neodymium	144.24	3	3.63	142(27.13), 144(23.87), 146(17.18), 143(12.20), 145(8.30), 148(5.72), 150(5.60)
61	Pm	Promethium	(145)	3		
62	Sm	Samarium	150.35	3	.....	152(26.63), 154(22.53), 147(15.07), 149(13.84), 148(11.27), 150(7.47), 144(3.16)
63	Eu	Europium	151.96	3, 2	4.08	153(52.23), 151(47.77)
64	Gd	Gadolinium	157.25	3	3.59	158(24.78), 160(21.79), 156(20.59), 157(15.71), 155(14.78), 154(2.15), 152(0.20)
65	Tb	Terbium	158.924	3	3.54	159(100)
66	Dy	Dysprosium	162.50	3	3.54	164(28.18), 162(25.53), 163(24.97), 161(18.88), 160(2.294), 158(0.0902), 156(0.0524)
67	Ho	Holmium	164.930	3	3.52	165(100)
68	Er	Erbium	167.26	3	3.50	166(33.41), 168(27.07), 167(22.94), 170(14.88), 164(1.56), 162(0.1)
69	Tm	Thulium	168.934	3	3.48	169(100)
70	Yb	Ytterbium	173.04	3, 2	3.87	174(31.84), 172(21.82), 173(16.13), 171(14.26), 176(12.73), 170(3.03), 168(0.14)
71	Lu	Lutetium	174.97	3	3.47	175(97.5), 176(2.5)
72	Hf	Hafnium	178.49	4	3.17	180(35.11), 178(27.10), 177(18.47), 179(13.85), 176(5.30), 174(0.18)
73	Ta	Tantalum	180.948	5	2.94	181(100)
74	W	Wolfram	183.85	6	2.82	184(30.68), 186(29.17), 182(25.77), 183(14.24), 180(0.122)
75	Re	Rhenium	186.2	7, 4, −1	2.75	187(62.93), 185(37.07)
76	Os	Osmium	190.2	4, 6, 8	2.70	192(41.0), 190(26.4), 189(16.1), 188(13.3), 187(1.64), 186(1.59), 184(0.018)
77	Ir	Iridium	192.2	3, 4, 6	2.709	193(61.5), 191(38.5)
78	Pt	Platinum	195.09	4, 2	2.769	195(33.7), 194(32.8), 196(25.4), 198(7.23), 192(0.78), 190(0.012)
79	Au	Gold	196.967	3, 1	2.878	197(100)

TABLE 7a-3. PROPERTIES OF ELEMENTS* (Continued)

Z	Symbol	Element	Atomic wt.†	Valency	Atomic diam in Å	Mass No. and (abundance)
(1)	(2)	(3)	(4)	(5)	(6)	(7)
80	Hg	Mercury	200.59	2, 1	3.10	202(29.80), 200(23.13), 199(16.84), 201(13.2), 198(10.02), 204(6.85), 196(0.15)
81	Tl	Thallium	204.37	1, 3	3.42	205(70.5), 203(29.5)
82	Pb	Lead	207.19	2, 4	3.49	208(52.3), 206(23.6), 207(22.6), 204(1.5), 202(<0.0004)
83	Bi	Bismuth	208.980	3, 5	3.64	209(100)
84	Po	Polonium	(210)	2, 4	.....	210‡
85	At	Astatine	(210)	....	.....	206‡, 215‡
86	Rn	Radon	(222)	0	.....	222‡
87	Fr	Francium	(223)	1	.....	223‡
88	Ra	Radium	(226.05)	2	.....	226‡, 228‡, 224‡, 223‡
89	Ac	Actinium	(227)	3	.....	227‡, 228‡
90	Th	Thorium	232.038	4	3.6	232‡(100)
91	Pa	Protactinium	(231)	5	.....	231‡
92	U	Uranium	238.03	6, 5, 4, 3	3.0	238‡(99.28), 235‡(0.715), 234‡(0.0058)
93	Np	Neptunium	(237)	6, 5, 4, 3	.....	237‡, 239‡
94	Pu	Plutonium	(242)	6, 5, 4, 3	.....	238‡, 239‡, 242‡
95	Am	Americium	(243)	3	.....	243‡
96	Cm	Curium	(247)	3	.....	247‡
97	Bk	Berkelium	(249)	4, 3	.....	249‡
98	Cf	Californium	(251)	3	.....	251‡
99	Es	Einsteinium	....	....	.....	254‡
100	Fm	Fermium	....	....	.....	255‡
101	Md	Mendelevium	....	....	.....	256‡
102	No	Nobelium	....	....	.....	255‡
103	Lw	Lawrencium	....	....	.....	257‡

* Much of the material in this table was taken from Henry D. Hubbard and William F. Meggers, "Key to Periodic Chart of the Atoms," 1950. Courtesy of W. M. Welch Manufacturing Company, Chicago.

† Official 1961 values based on carbon-12; courtesy of the International Union of Pure and Applied Chemistry and Butterworth Publications. The atomic weight of some elements varies because of natural variations in the isotope composition. The observed ranges are B, ±0.003; C, ±0.00005; H, ±0.00001; O, ±0.0001; Si, ±0.001; S, ±0.003. In order to convert the atomic weights given in the table for those based on oxygen-16, multiply by 1.0003203.

‡ Radioactive element; mass number of the most abundant or most stable isotope.

7b. The Electronic Structure of Atoms

Explanation of Table 7b-1. COLUMN (3): Electronic structure of the ground state. Rare-gas shells and similar closed shells are indicated by appropriate symbols and only the electrons outside them given explicitly. All structures are based on spectroscopic evidence except in a few cases (e.g., Fr, At) where there is no reasonable doubt about predictions.

The electron printed in boldface when removed produces the ground state of the ion. Where the other electrons are rearranged in the ion this is indicated in a footnote.

COLUMN (4): Ground state of atom.

COLUMN (5): First ionization potential of atom (in electron volts).[1]

COLUMN (6): Ground state of ion. For electron configuration of ion, see column (3).

COLUMN (7): Second ionization potential (ionization potential of singly ionized atom) in electron volts.

COLUMN (8): Resonance potentials (see below).

COLUMN (9): Resonance lines (see below).

RESONANCE POTENTIALS AND RESONANCE LINES: The resonance potential is the energy (in electron volts) required to raise an atom from the ground state to the lowest excited state. The resonance line is the spectrum line absorbed or emitted in this or the reverse transition. There is a clear and unambiguous situation with regard to resonance lines and potentials for atoms with simple structure such as the alkalies. For more complicated atoms the matter needs further amplification.

A line is not considered a resonance line if the excited state has the same parity as the ground state and thus the transition is forbidden as a dipole transition. If the line is allowed as a dipole transition but very weak, i.e., if it violates an approximate dipole-selection rule (usually the spin-selection rule $\Delta S = 0$), it is called subresonance line r. The resonance line R proper is the first line allowed by all the selection rules. Both R and r then are given in such cases. For the heavy elements r may be very strong.

The resonance potentials are in general those corresponding to the lines, with one exception. If there is a lower state than that of the first resonance line for which transitions to the ground state are forbidden by the J-selection rule (but allowed by the parity rule) this state is metastable. It may, however, often be excited by direct electron collisions, and the excitation potential for this state is given as first resonance potential followed by a letter m. There is no observed resonance line corresponding to this transition. An asterisk on the second resonance potential indicates that the corresponding line is that also marked with an asterisk.

A C preceding column (8) means that there are states of the same parity as the ground state between it and the first resonance state. These often belong to the electron configuration of the ground state. A C is *not* listed if these states are merely additional levels of the ground-state multiplet.

[1] For conversion from wave numbers into electron volts or vice versa, see Table 7a-2.

TABLE 7b-1. ELECTRONIC STRUCTURE OF ATOMS[a]

Z	El.	Ground state	Ground state	IP	Ion ground state	IP	Resonance potentials	Resonance lines	
(1)	(2)	(3)	(4)	(5)	(6)	(7)	(8)	(9)	
1	H	$1s$	^2S	13.599	.....	.....	10.15	1,215.67(^2P)	
2	He	$1s^2$	^1S	24.588	^2S	54.418	20.96m 21.13	591.43($^3P_1$)	584.35(^1P)
3	Li	[He]$2s$	^2S	5.392	^1S	75.641	1.84	6,707.85($^2P_{1/2}$)	
4	Be	$-2s^2$	^1S	9.323	^2S	18.211	2.71 5.25	4,548.3($^3P_1$)	2,348.61($^1P_1$)
5	B	$-2s^22p$	$^2P_{1/2}$	8.298	^1S	25.156	3.57 4.94	3,470.6($^4P_{1/2}$)	2,497.72(^2S)
6	C	$-2s^22p^2$	$^3P_0$	11.260	$^2P_{1/2}$	24.383	C, 4.16 7.46	2,967.22(^5S)	1,656.998(^3P)
7	N	$-2s^22p^3$	^4S	14.53	$^3P_0$	29.602	C, 10.28	1,200.71(^4P)	
8	O	$-2s^22p^4$	$^3P_2$	13.618	^4S	35.118	C, 9.11 9.48	1,355.60(^5S)	1,302.17(^3S)
9	F	$-2s^22p^5$	$^2P_{3/2}$	17.423	$^3P_2$	34.98	12.69 12.98	976.50(^4P)	954.82(^2P)
10	Ne	$-2s^22p^6$	^1S	21.565	$^2P_{3/2}$	40.964	16.62m 16.84	743.71($^3P_1$)	735.89($^1P_1$)
11	Na	[Ne]$3s$	^2S	5.139	^1S	47.29	2.10	5,889.95($^2P_{1/2}$)	
12	Mg	$-3s^2$	^1S	7.646	^2S	15.035	2.71m 4.33	4,571.10($^3P_1$)	2,852.12($^1P_1$)
13	Al	$-3s^23p$	$^2P_{1/2}$	5.986	^1S	18.828	3.13	3,961.52(^2S)	
14	Si	$-3s^23p^2$	$^3P_0$	8.152	$^2P_{1/2}$	16.346	C, 4.93		2,516.11($^3P_1$)
15	P	$-3s^23p^3$	^4S	10.487	$^3P_0$	19.72	C, 6.93	1,787.65($^4P_{1/2}$)	1,774.94($^4P_{3/2}$)
16	S	$-3s^23p^4$	$^3P_2$	10.360	^4S	23.4	C, 6.50 6.83	1,900.27(^5S)	1,807.31(^3S)
17	Cl	$-3s^23p^5$	$^2P_{3/2}$	12.967	$^3P_2$	23.80	8.88 9.16	1,389.78($^4P_{3/2}$)	1,347.32($^2P_{3/2}$)
18	A	$-3s^23p^6$	^1S	15.760	$^2P_{3/2}$	27.62	11.55m 11.83	1,066.66($^3P_1$)	1,049.22($^1P_1$)
19	K	[A]$4s$	^2S	4.341	^1S	31.81	1.61	7,664.91($^2P_{1/2}$)	
20	Ca	$-4s^2$	^1S	6.113	^2S	11.872	1.88 2.92	6,572.78($^3P_1$)	4,226.73($^1P_1$)
21	Sc	$-3d4s^2$	$^2D_{3/2}$	6.54	$^3D_1$	12.80	1.94m 1.98	6,378.82($^4F_{3/2}$)	6,305.67($^2D_{3/2}$)
22	Ti	$-3d^24s^2$	$^3F_2$	6.82	$^2F_{3/2}$	13.57	C, 1.96 2.39	6,296.65($^5G_2$)	5,173.74($^3D_1$)
23	V	$-3d^44s^2$	$^4F_{3/2}$	6.74[b]	$^5D_0$	14.65	C, 2.23 2.54	5,537.72($^6G_{1/2}$)	4,851.48($^4D_{1/2}$)
24	Cr	$-3d^54s$	^7S	6.765	^6S	16.49	C, 2.90	4,289.72($^7P_2$)	
25	Mn	$-3d^54s^2$	^6S	7.435	^7S	15.640	C, 2.27 3.06	5,432.55($^8P_{3/2}$)	4,034.49($^6P_{3/2}$)
26	Fe	$-3d^64s^2$	$^5D_4$	7.87	$^6D_{9/2}$	16.18	C, 2.39 3.20	5,166.29($^7D_5$)	3,859.91($^5D_4$)
27	Co	$-3d^74s^2$	$^4F_{9/2}$	7.864[c]	$^3F_4$	17.05	C, 2.91 3.50	4,233.99($^6F_{11/2}$)	3,526.85($^4F_{9/2}$)
28	Ni	$-3d^84s^2$	$^3F_4$	7.635[d]	$^2D_{5/2}$	18.15	C, 3.18 3.64	3,884.58($^5D_4$)	3,670.43($^3P_2$)
29	Cu	$-3d^{10}4s$	^2S	7.726	^1S	20.292	C, 3.79	3,273.96($^2P_{1/2}$)	
30	Zn	$-3d^{10}4s^2$	^1S	9.394	^2S	17.964	4.01m 5.77	3,075.90($^3P_1$)	2,138.56($^1P_1$)
31	Ga	$-3d^{10}4s^24p$	$^2P_{1/2}$	5.999	^1S	20.51	3.06	4,032.98(^2S)	
32	Ge	$-3d^{10}4s^24p^2$	$^3P_0$	7.900	$^2P_{1/2}$	15.935	C, 4.64m	2,651.58($^3P_1$)	
33	As	$-3d^{10}4s^24p^3$	^4S	9.81	$^3P_0$	18.63	C, 6.26	1,972.62($^4P_{1/2}$)	
34	Se	$-3d^{10}4s^24p^4$	$^3P_2$	9.75	^4S	21.5	C, 5.95 6.30	2,074.79(^5S)	1,960.90($^3S_4$)
35	Br	$-3d^{10}4s^24p^5$	$^2P_{3/2}$	11.814	$^3P_2$	21.6	7.83 8.29	1,576.5($^4P_{3/2}$)	1,488.6($^2P_{3/2}$)
36	Kr	$-3d^{10}4s^24p^6$	^1S	14.000	$^2P_{3/2}$	24.56	9.91m 9.99*	1,235.82*($^3P_1$)	1,164.86(^1P)
37	Rb	[Kr]$5s$	^2S	4.177	^1S	27.5	1.56	7,947.64($^2P_{1/2}$)	
38	Sr	$-5s^2$	^1S	5.696	^2S	11.030	1.78m 2.68*	6,892.58($^3P_1$)	4,607.33*(^1P)
39	Y	$-4d5s^2$	$^2D_{3/2}$	6.379	^1S	12.236	1.31	9,494.81($^2P_{1/2}$)	
40	Zr	$-4d^25s^2$	$^3F_2$	6.837	$^4F_{3/2}$	13.13	C, 1.83 2.71	6,762.38($^5G_2$)	4,575.52($^3G_3$)
41	Nb	$-4d^45s$	$^6D_{1/2}$	6.883	$^5D_0$	14.32	C, 2.07m 2.97*	5,320.21($^6F_{1/2}$)	4,168.12*($^6F_{1/2}$)
42	Mo	$-4d^55s$	^7S	7.10	^6S	16.15	C, 3.18	3,902.96($^7P_{3/2}$)	
43	Tc	$-4d^55s^2$	^6S	7.28	^7S	15.26	2.09 2.88	5,924.57($^8P_{3/2}$)	4,297.06($^6P_{3/2}$)
44	Ru	$-4d^75s$	$^5F_5$	7.366	$^4F_{9/2}$	16.76	C, 3.13 3.26*	3,964.90($^7D_5$)	3,799.35*($^5D_4$)
45	Rh	$-4d^85s$	$^4F_{9/2}$	7.464	$^3F_4$	18.07	C, 3.36	3,692.36($^4D_{3/2}$)	
46	Pd	$-4d^{10}$	^1S	8.33	$^2D_{5/2}$	19.42	C, 4.22m 5.01*	2,763.09($^3P_1$)	2,447.91*($^1P_1$)
47	Ag	$-4d^{10}5s$	^2S	7.576	^1S	21.48	3.66	3,382.89($^2P_{1/2}$)	
48	Cd	$-4d^{10}5s^2$	^1S	8.994	^2S	16.908	3.73m 5.29	3,261.04($^3P_1$)	2,288.02($^1P_1$)
49	In	$-4d^{10}5s^25p$	$^2P_{1/2}$	5.786	^1S	18.833	3.02	4,101.76(^2S)	
50	Sn	$-4d^{10}5s^25p^2$	$^3P_0$	7.344	$^2P_{1/2}$	14.632	C, 4.29m 4.33*	2,863.32*($^3P_1$)	
51	Sb	$-4d^{10}5s^25p^3$	^4S	8.642	$^3P_0$	16.5	C, 5.36	2,311.47($^4P_{1/2}$)	
52	Te	$-4d^{10}5s^25p^4$	$^3P_2$	9.01	^4S	18.6	C, 5.49 5.78	2,259.02($^5S_2$)	2,142.75($^3S_1$)
53	I	$-4d^{10}5s^25p^5$	$^2P_{3/2}$	10.451	$^3P_2$	19.135	6.77 7.66	2,062.1($^4P_{3/2}$)	1,617.7($^2P_{3/2}$)
54	Xe	$-4d^{10}5s^25p^6$	^1S	12.130	$^2P_{3/2}$	21.21	8.31m 8.44*	1,469.62*($^3P_1$)	1,295.56($^1P_1$)
55	Cs	[Xe]$6s$	^2S	3.894	^1S	25.1	1.38	8,943.46($^2P_{1/2}$)	8,521.10($^2P_{3/2}$)

TABLE 7b-1. ELECTRONIC STRUCTURE OF ATOMS (*Continued*)

Z	El.	Ground state	Ground state	IP	Ion ground state	IP	Resonance potentials	Resonance lines	
(1)	(2)	(3)	(4)	(5)	(6)	(7)	(8)	(9)	
56	Ba	$-6s^2$	^1S	5.212	^2S	10.004	$C,\ 1.52m \quad 2.24^*$	$7,911.36(^1P_1)$	$5,535.53^*(^1P)$
57	La	$-5d6s^2$	$^2D_{\frac{3}{2}}$	5.61e	$^3F_2$	11.06	$C,\ 1.64 \quad 1.84^*$	$7,539.24(^4F_{\frac{3}{2}})$	$6,753.05(^2D_{\frac{3}{2}})$
58	Ce	$-4f5d6s^2$	$^1G_4$	5.65f	$^4H_{\frac{7}{2}}$	10.85			
59	Pr	$-4f^36s^2$	$^4I_{\frac{9}{2}}$	5.42	$^5I_4$	10.55			
60	Nd	$-4f^46s^2$	$^5I_4$	5.49	$^6I_{\frac{7}{2}}$	10.73			
61	Pm	$-4f^56s^2$	$^6H_{\frac{5}{2}}$	5.55	$^7H_2$	10.90			
62	Sm	$-4f^66s^2$	$^7F_0$	5.63	$^8F_{\frac{1}{2}}$	11.07	$C,\ 1.71m \quad 1.74^*$	$7,141.13^*(^9F_1)$	$6,725.88(^9G_1)$
63	Eu	$-4f^76s^2$	^8S	5.68	^9S	11.25	$C,\ 1.74 \quad 2.66^*$	$7,106.48(^{10}P_{\frac{7}{2}})$	$4,661.88^*(^8P_{\frac{7}{2}})$
64	Gd	$-4f^75d6s^2$	$^9D_2$	6.16	$^{10}D_{\frac{5}{2}}$	12.1	....		4,225.85
65	Tb	$-4f^96s^2$	$^6H_{\frac{15}{2}}$	5.98					
66	Dy	$-4f^{10}6s^2$	I_8	5.93	.....	11.67			
67	Ho	$-4f^{11}6s^2$	$^4I_{\frac{15}{2}}$	6.02	.....	11.80			
68	Er	$-4f^{12}6s^2$	$^3H_6$	6.10	$^4H_{\frac{13}{2}}$	11.93			
69	Tm	$-4f^{13}6s^2$	$^2F_{\frac{7}{2}}$	6.18	$^3F_4$	12.05	....		5,675.83
70	Yb	$-4f^{14}6s^2$	^1S	6.25	^2S	12.17	....		3,987.99
71	Lu	$-4f^{14}5d6s^2$	$^2D_{\frac{3}{2}}$	6.15	^1S	13.9			
72	Hf	$-4f^{14}5d^26s^2$	$^3F_2$	7.0	$^2D_{\frac{3}{2}}$	14.9	$C,\ 2.19$		
73	Ta	$-4f^{14}5d^36s^2$	$^4F_{\frac{3}{2}}$	7.88	$^5F_1$	16.2	$C,\ 2.90$	4,280.47	
74	W	$-4f^{14}5d^46s^2$	$^5D_0$	7.98	$^6D_{\frac{1}{2}}$	17.7	$C,\ 2.40m$	$4,982.16(^7F_1)$	
75	Re	$-4f^{14}5d^56s^2$	^6S	7.87	^7S	16.6	$C,\ 2.35 \quad 3.58$	$5,275.53(^8P_{\frac{5}{2}})$	$4,008.75(^6P_{\frac{7}{2}})$
76	Os	$-4f^{14}5d^66s^2$	$^5D_4$	8.7	$^6D_{\frac{9}{2}}$	17	$C,\ 2.80$	$4,420.67(^7D_4)$	3,464.72
77	Ir	$-4f^{14}5d^76s^2$	$^4F_{\frac{9}{2}}$	9.2	.....	.....	$C,\ 3.26$	$3,800.12(^4D_{\frac{9}{2}})$	
78	Pt	$-4f^{14}5d^96s$	$^3D_3$	9.0	$^2D_{\frac{5}{2}}$	18.56	$C,\ 3.74 \quad 4.04$	$3,315.05(^3D_4)$	$3,064.71(^3P_2)$
79	Au	$[g]6s$	^2S	9.22	^1S	20.5	$C,\ 4.63$	$2,675.95(^2P_{\frac{1}{2}})$	$2,427.95(^2P_{\frac{3}{2}})$
80	Hg	$-6s^2$	^1S	10.437	^2S	18.757	$4.67m \quad 6.70$	$2,536.52(^3P_1)$	$1,849.57(^1P_1)$
81	Tl	$-6s^26p$	$^2P_{\frac{1}{2}}$	6.108	^1S	20.42	3.29	$3,775.72(^2S)$	
82	Pb	$-6s^26p^2$	$^3P_0$	7.415	$^2P_{\frac{1}{2}}$	15.032	$C,\ 4.33m \quad 4.37$	$2,833.07^*(^3P_1)$	
83	Bi	$-6s^26p^3$	^4S	7.287	$^3P_0$	16.68	$C,\ 4.04$	$3,067.72(^4P_{\frac{1}{2}})$	
84	Po	$-6s^26p^4$	$^3P_2$	8.43	.....	.....	....		2,449.99
85	At	$-6s^26p^5$	$^2P_{\frac{3}{2}}^*$						
86	Rn	$-6s^26p^6$	^1S	10.745	$^2P_{\frac{3}{2}}^*$	.....	$6.77m \quad 6.94^*$	$1,786.07^*(^3P_1)$	$1,451.56(^1P_1)$
87	Fr	$[Rn]7s$	$^2S^*$	......	$^1S^*$				
88	Ra	$-7s^2$	^1S	5.277	^2S	10.14	$1.62m \quad 2.57^*$	7,141.21	$4,825.91^*$
89	Ac	$-6d7s^2$	$^2D_{\frac{3}{2}}$	6.9	^1S	12.1			
90	Th	$-6d^27s^2$	$^3F_2$	......	$^4F_{\frac{3}{2}}$	11.5			
91	Pa	$-5f^26d7s^2$	$^4K_{\frac{11}{2}}$						
92	U	$-5f^36d7s^2$	$^5L_6$	6.08	$^4I_{\frac{9}{2}}$			5,915,40	
93	Np	$-5f^46d7s^2$	$^6L_{\frac{11}{2}}$	5.8					
94	Pu	$-5f^67s^2$	$^7F_0$	5.8	$^8F_{\frac{1}{2}}$				
95	Am	$-5f^77s^2$	^8S	6.05	^9S				
96	Cm	$-5f^76d7s^2$	$^9D_2$						

^{a} Data taken from current literature. Use has been made of Moore, "Atomic Energy Levels," vols. I–III, and "Smithsonian Physical Tables," 9th ed.

^{b} Normal state of ion $-3d^4$.

^{c} Normal state of ion $-3d^8$.

^{d} Normal state of ion $-3d^9$.

^{e} Normal state of ion $-5d^2$.

^{f} Normal state of ion $-4f5d^2$.

^{g} Structure of closed shells $[Xe]4f^{14}5d^{10}$.

7c. Energy-level Diagrams of Atoms

A number of energy-level diagrams are represented in Figs. 7c-1 through 7c-15. An attempt has been made to select typical cases which show characteristic features derived from optical spectra. The following comments may be helpful.

In almost all cases the energy levels have been arranged according to the Russell-Saunders scheme, also called L, S coupling. This means that each level is characterized by the total orbital angular momentum L, the resultant spin S or rather the multiplicity $2S + 1$ [1 for singlets ($S = 0$), 2 for doublets ($S = \frac{1}{2}$), 3 for triplets ($S = 1$), etc.], and its total angular momentum quantum number J. The scale of the figures usually does not permit showing the individual multiplet components. However, the total width is indicated unless it is no greater than the thickness of the line.

Even levels are shown by entire lines or blocks, odd levels by broken ones. When an entire column has the same parity as in the simple spectra, the odd parity is indicated by the term symbol at the bottom of the column in the usual way, e.g., $^3F^0$.

The horizontal line across the whole width of the diagram is at the first ionization potential. This is indicated by the term symbol for the ground state of the ion. In some cases higher ionization potentials are also indicated.

The electron configuration is given by symbols explained with each individual diagram.

Transitions which correspond to spectrum lines are left out in order to avoid confusion except for the important lowest transitions which often give rise to the strongest lines. The resonance line R is the lowest transition to the ground state allowed by the selection rules of L, S coupling, which are change of parity, no change in multiplicity ($\Delta S = 0$), and $\Delta J = 0 \pm 1$. The subresonance line r is a line from a lower level than that responsible for the resonance line; it obeys the same selection rules except $\Delta S = 0$. It is usually very weak for the lighter atoms but may be quite strong for the heavier elements (e.g., 2,537 of Hg).

The spectra represented in the figures are given in Table 7c-1.

TABLE 7c-1. SPECTRA REPRESENTED BY FIGS. 7c-1 THROUGH 7c-15

Z	Element	Figure
2	He I	7c-1
6	C I	7c-2
7	N I	7c-3
8	O I	7c-4
11	Na I	7c-5
13	Al I	7c-6
17	Cl I	7c-7
18	A I	7c-8
20	Ca I	7c-9
25	Mn I	7c-10
26	Fe II	7c-11
	Fe I	7c-12
29	Cu I	7c-13
80	Hg I	7c-14
57–70	Ce IV-Yb IV	7c-15

Further diagrams of simple spectra are found in Grotrian.[1]

[1] W. Grotrian, "Graphische Darstellung der Spektren von Atomen und Ionen mit ein, zwei, und drei Valenzelektronen," vol. 1, 245 pp.; vol. 2, 268 pp.; Springer-Verlag OHG, Berlin, 1928.

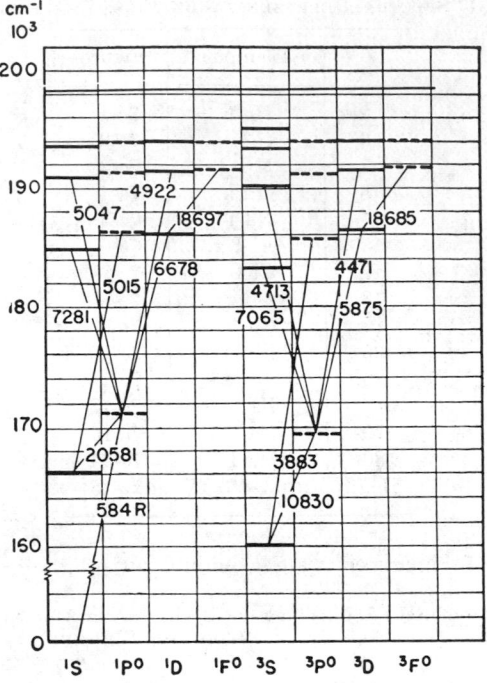

FIG. 7c-1. Energy-level diagram of He I—simplest atom with two valence electrons.   The wavelengths of the principal lines are indicated.

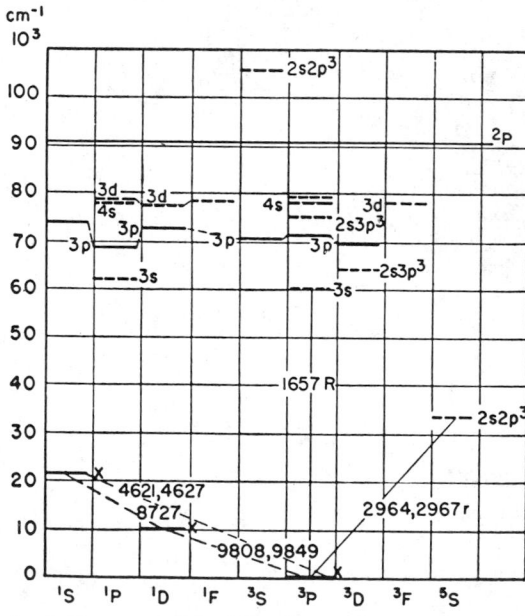

FIG. 7c-2. Energy-level diagram of C I—four valence electrons, lowest state $2s^2 2p^2$.   Most excited states are $2s^2 2p \cdot nx$.   The orbit nx of the last electron only is indicated in the figure except where one of the $2s$ electrons is excited, as, for instance, $2s2p^3$.   The important forbidden lines are indicated.

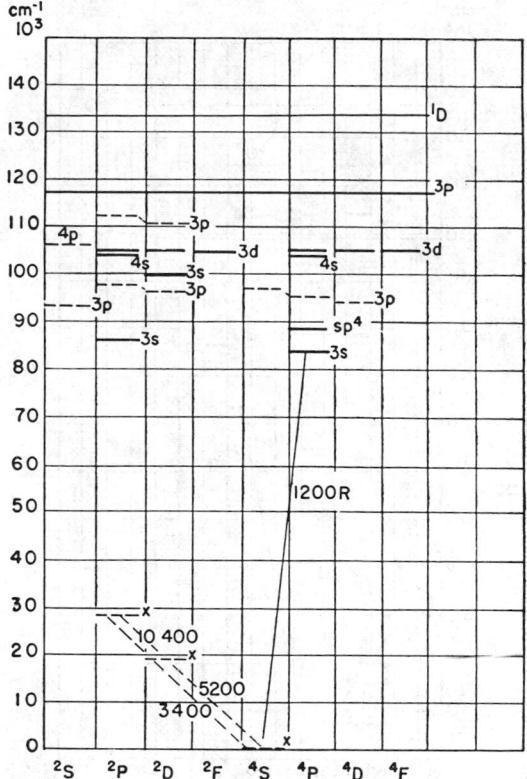

FIG. 7c-3. Energy-level diagram of N I—five valence electrons, normal state $2s^2 2p^3$. Excited states are $2s^2 2p^2 \cdot nx$, nx being indicated in the figure. When the $2s$ electron is excited, the full configuration is given, e.g., $2s 2p^4$. The important forbidden lines are indicated.

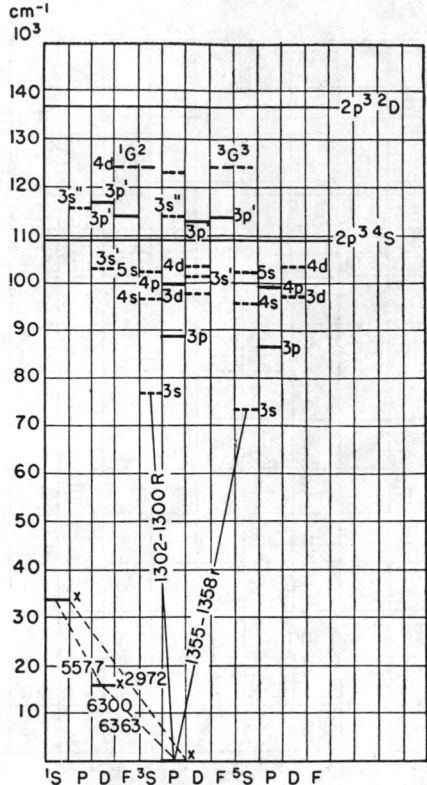

FIG. 7c-4. Energy-level diagram of O I—six valence electrons. Normal state is $2s^2 2p^4$; excited states are $2s^2 2p^3 \cdot nx$, nx being indicated in the figure. The important forbidden lines are indicated.

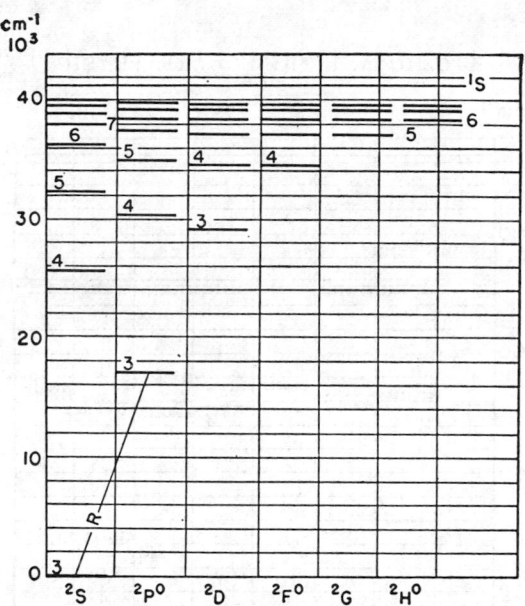

FIG. 7c-5. Energy-level diagram of Na I. Simple diagram typical for elements with one valence electron. The other alkalies have essentially the same scheme.

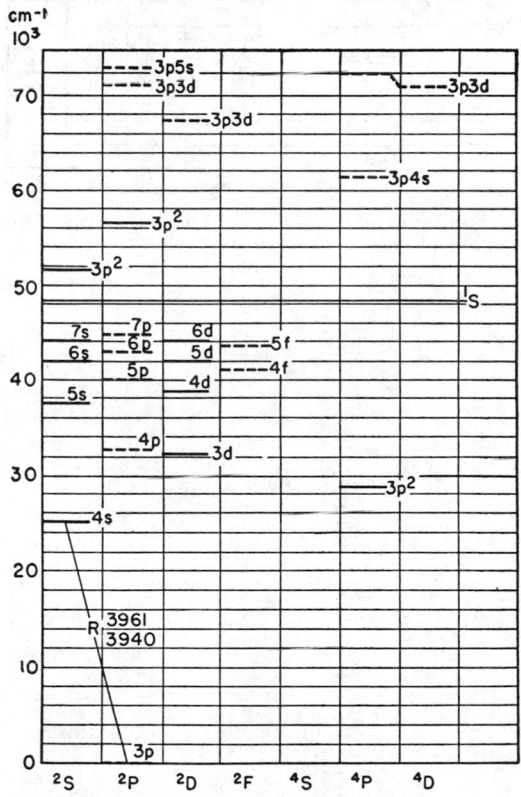

FIG. 7c-6. Energy-level diagram of Al I—three valence electrons. Normal state is $3s^23p$, excited states $3s^2 \cdot nx$ (nx is indicated) or $3s3pnx'$ (nx' is indicated with primed letters). The primed levels converge to a higher ionization limit.

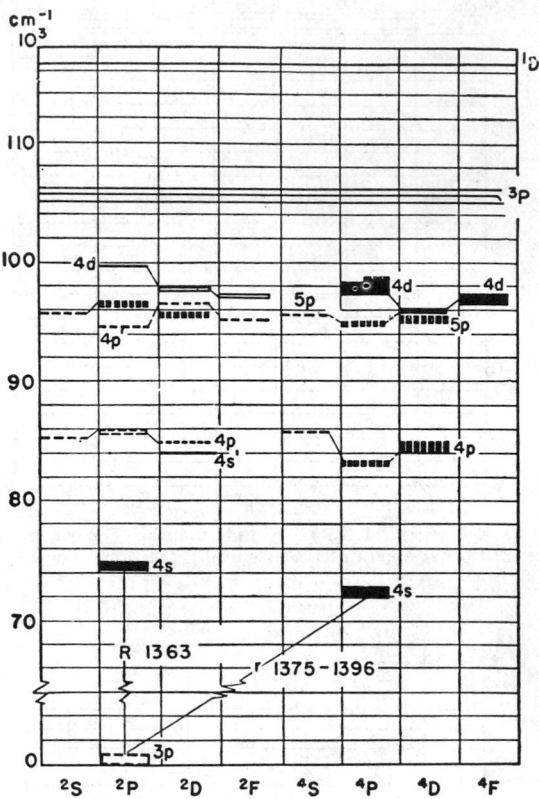

FIG. 7c-7. Energy-level diagram of Cl I—seven valence electrons.   Ground state is $3s^2 3p^5$, excited states $3s^2 3p^4 nx$, nx being indicated in the figure.

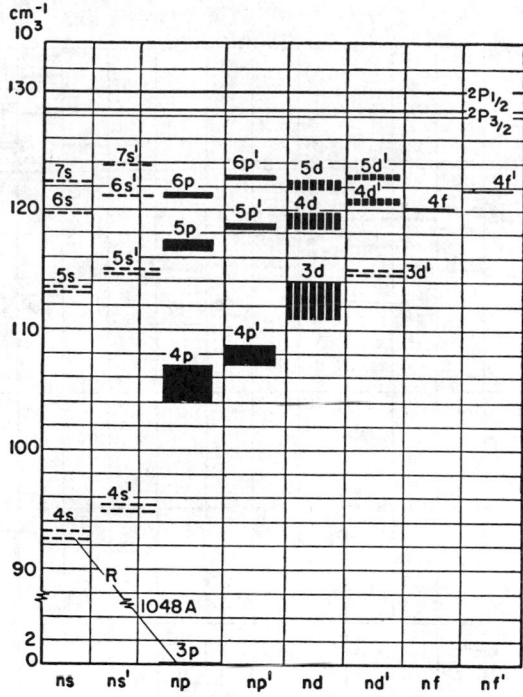

Fig. 7c-8. Energy-level diagram of A I—typical for the rare gases except helium. Levels $3s^2 3p^5 \cdot nx$ with nx indicated. L,S coupling is not appropriate here and therefore symbols like ^3P, etc., have no meaning. The primed levels converge to the higher ionization potential. See also Table 7e-3.

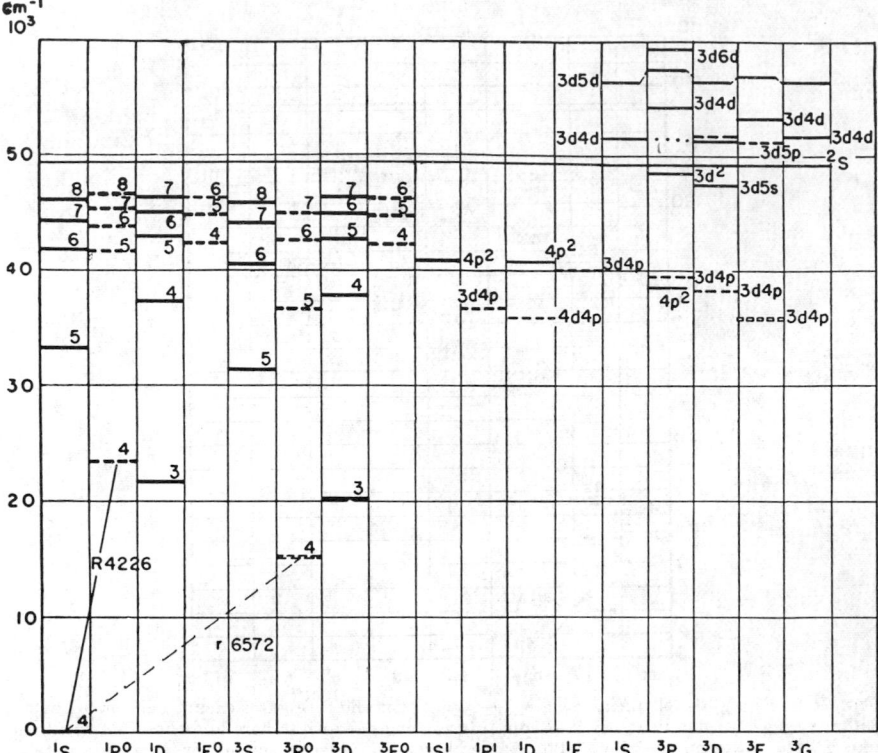

FIG. 7c-9. Energy-level diagram of Ca I. Characteristic for the elements in the second column of the periodic system. Ground state $4s^2$ and regular excited states $4s\,nx$ are indicated only by the value of n in the appropriate column. Levels with both electrons excited are given at the right.

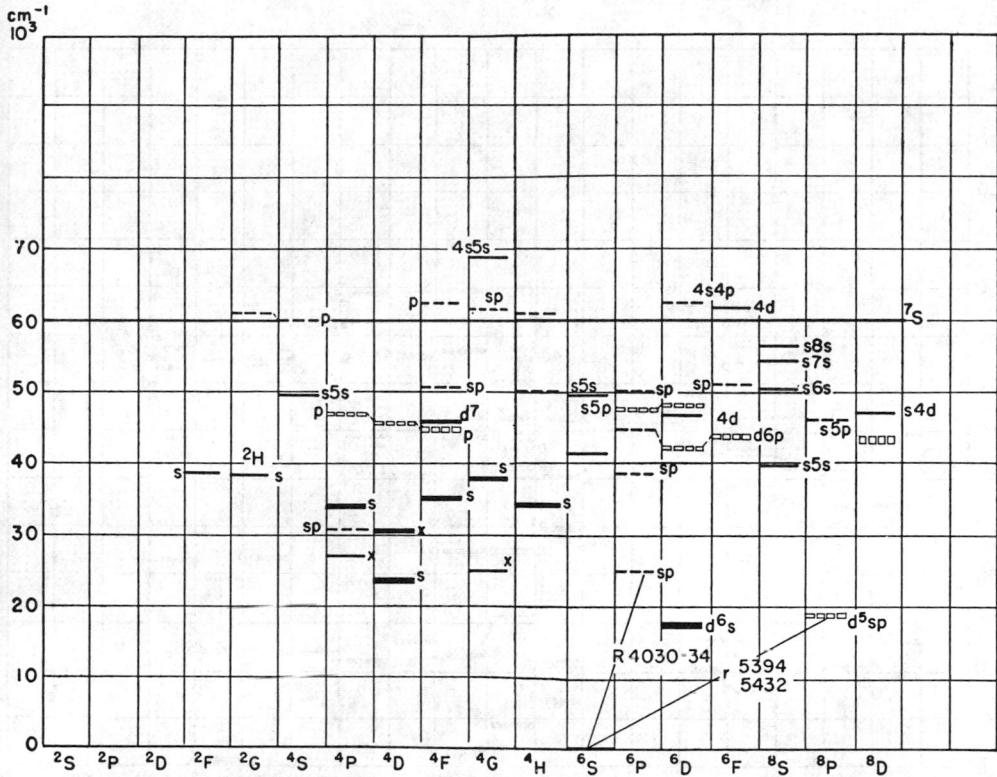

FIG. 7c-10. Energy-level diagram of Mn I. A typical element of the transition group. Seven valence electrons. Ground state $3d^5 4s^2$. This produces 16 multiplet levels of which only four (^6S, ^4P, ^4D, ^4G) are known. They are marked by an x. The other low states are $3d^6 4s$ (s), $3d^6 4p$ (p), $3d^5 4s 4p$ (sp), $3d^5 4p^2$ (p^2), $3d^7$ (d^7). The symbol between parentheses indicates how the level is marked in the figure. If higher than $3d$, $4s$, $4p$, electrons are involved, the value of n is marked, e.g., $3d^5 4s 5p$ ($s5p$) or $3d^5 4s 4d$ ($s4d$). In general, the number of $3d$ electrons is left out in the figure (except for $3d^7$). Compare Mn I with Fe II, which has the same number of electrons (Fig. 7c-11).

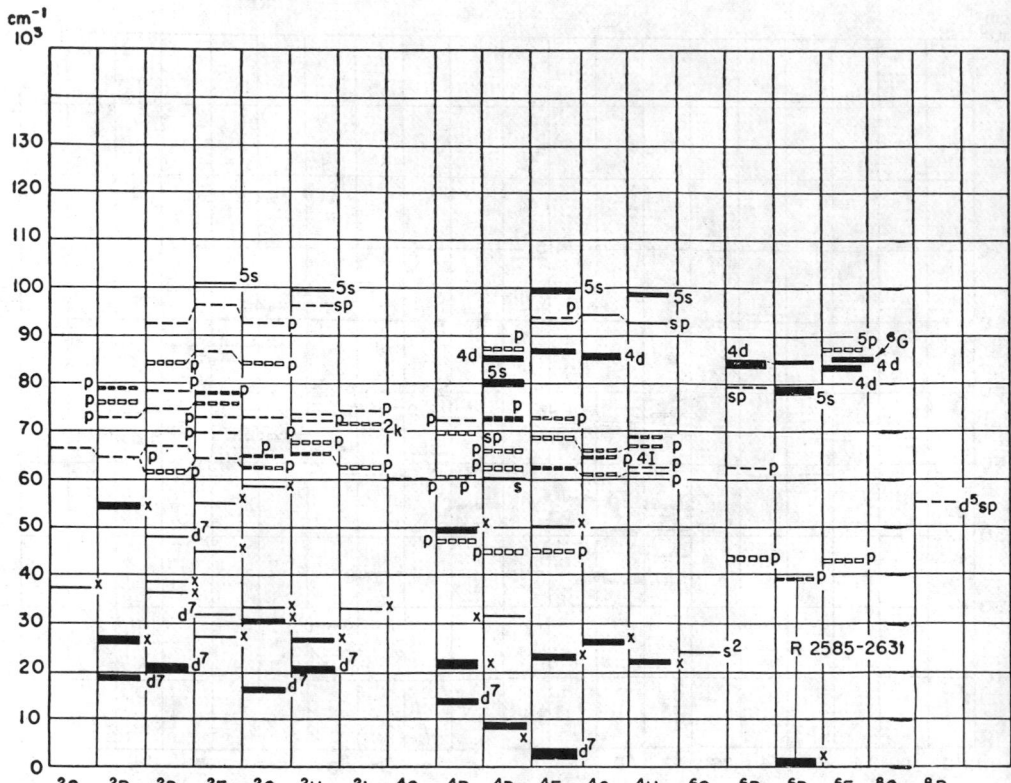

FIG. 7c-11. Energy-level diagram of Fe II. Fe II has the same number of electrons as Mn I and therefore the same type of levels. The relative position of the levels is, however, greatly changed by the increase in the nuclear charge. In general, there is a tendency for levels containing $3d$ electrons to be lower than those with $4s$ or $4p$ electrons. The ground state is $3d^64s$. There are 24 multiplet levels of this configuration, of which 23 are known (marked with x). The excited levels are marked as for Mn I.

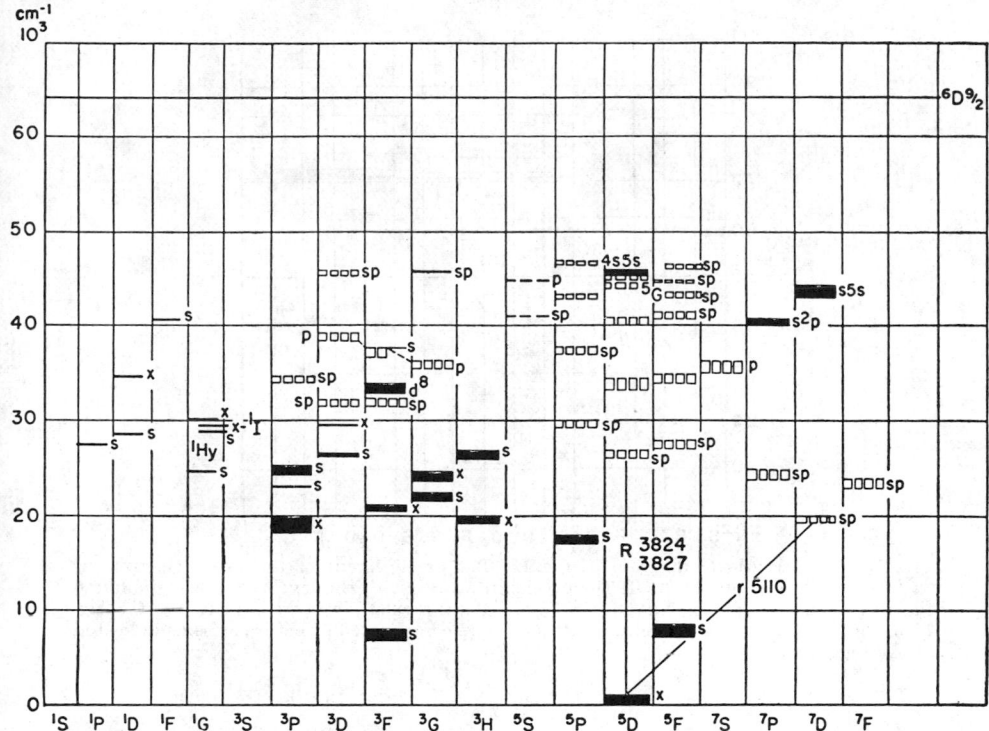

FIG. 7c-12. Energy-level diagram of Fe I.   The spectrum of Fe I is one of the best studied and is of particular importance because of the use of iron lines for wavelength standards and other applications (see Table 7e-6).   Eight valence electrons, ground-state configuration $3d^6 4s^2$, which gives 16 multiplet levels, of which 9 are known (marked x in the figure). Other configurations leading to low-lying levels are $3d^7 4s$ (s), $3d^6 4s 4p$ (sp), $3d^8$ (d^8), $3d^7 4p$ (p), $3d^5 4s^2 4p$ ($s^2 p$).   If n values higher than for $3d$, $4s$, $4p$ are involved, they are indicated as, e.g., $3d^6 4s 5s$ ($s5s$).

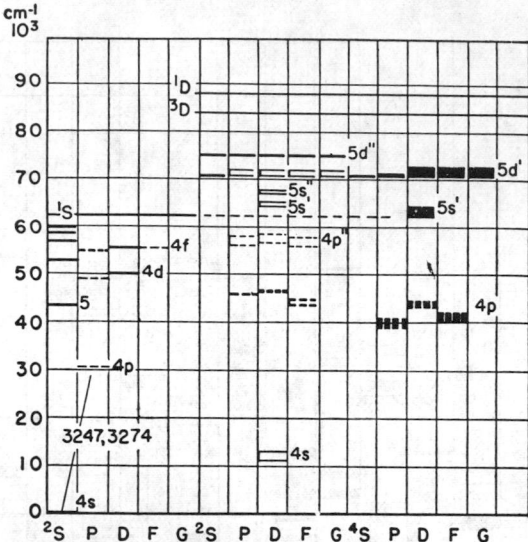

FIG. 7c-13. Energy-level diagram of Cu I. The arrangement of the outer electrons is $3d^{10}4s\,{}^2S$ in the ground state. If the $4s$ electron is excited, the levels are very similar to those of an alkali as shown, e.g., in Fig. 7c-5. These regular levels are indicated at the left. If one of the $3d$ electrons is excited, levels of more complicated structure arise as indicated at the right.

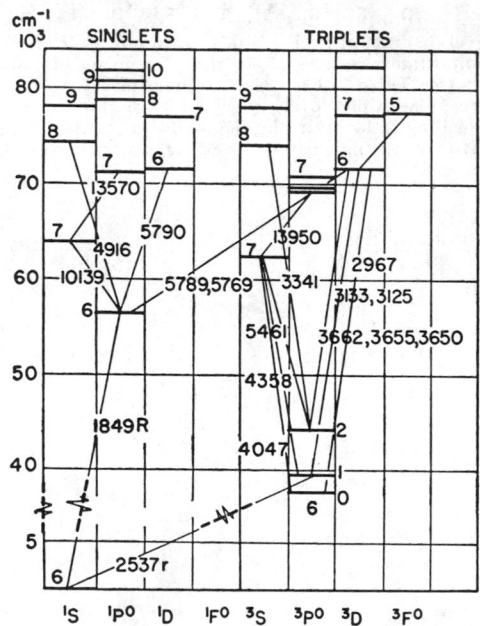

FIG. 7c-14. Energy-level diagram of Hg I. This is the diagram of a typical two-electron spectrum with singlets and triplets. Because of the wide use of the mercury spectrum in many applications, the wavelengths of many transitions are indicated. Single triplet transitions are relatively strong. See also Table 7e-7 and Figs. 7e-5 and 7e-6.

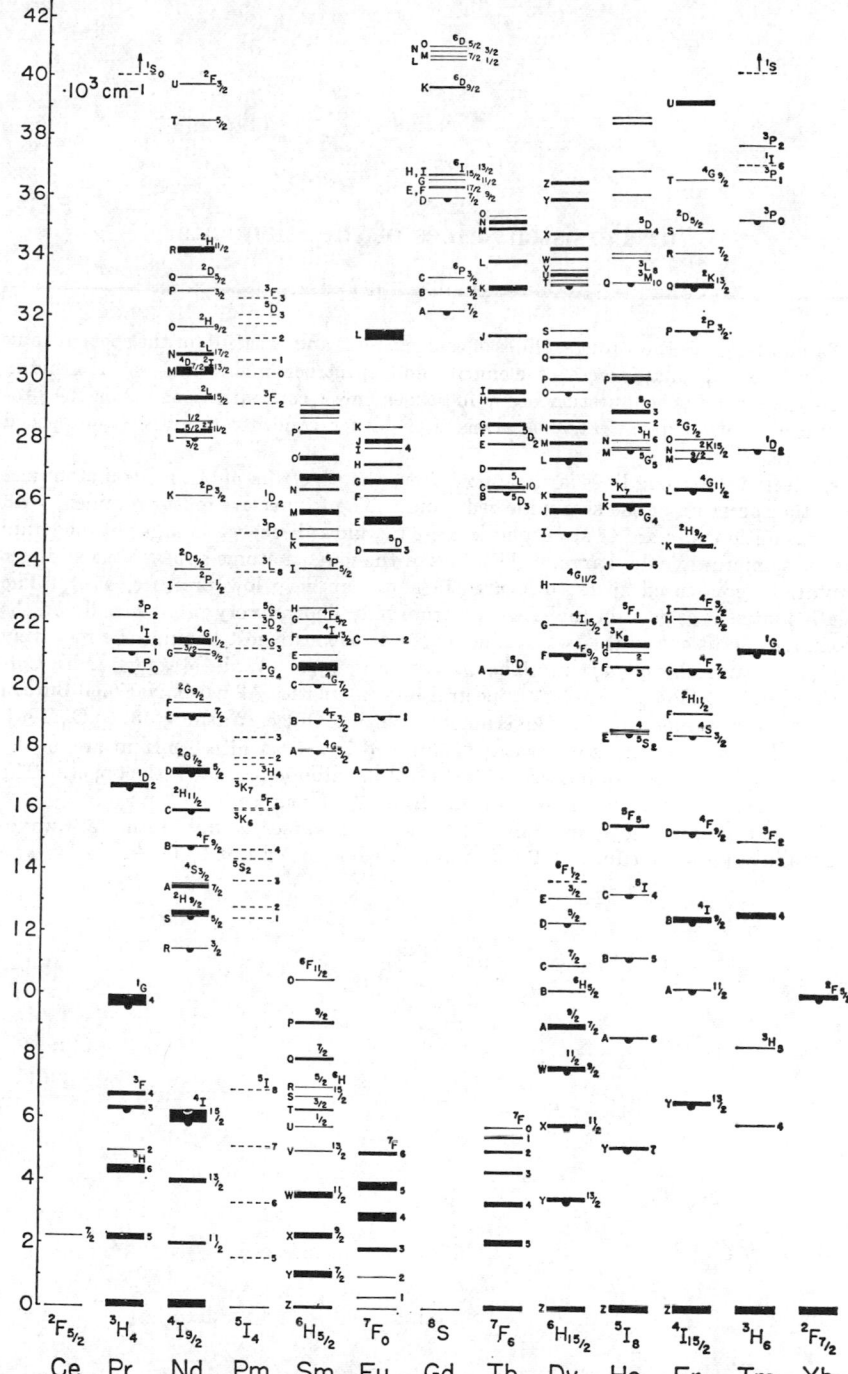

Fig. 7c-15. Lowest energy levels of the three-valent ions as determined from the crystal absorption and fluorescence spectra. (In most cases the data are for anhydrous chloride.) The thickness of the levels indicates the amount of splitting of the free ion level in the crystal field. The ground state is indicated on the bottom; the excited states are designated by empirical letters A, B, C, etc., or by the equivalent L,S coupling symbols like $^5D_2$. The J values are given at the sides. The data for Pm are tentative. The scheme is complete to about 25,000 cm.$^{-1}$ A semicircle under a level shows that it is the upper state of fluorescence transitions. (*Source of data, The Johns Hopkins University.*)

7d. Persistent Lines of the Elements

Table 7d-1 gives the strongest lines of each element and is useful for the spectroscopic identification of small traces of elements and spectrochemical analysis in general, when the elements in question occur in rather small concentrations. For the procedure of routine quantitative analysis with larger concentrations, see the special literature.

A selection of strong lines is given both from the spectrum of the neutral atom and from the spectrum of the singly ionized atom. The former are most prominent with mild excitation (d-c arc at atmospheric pressure, glow discharge in a gas at moderate pressure, microwave discharge). The lines of the ionized atoms appear with stronger excitation (condensed spark, discharge in a gas at very low pressure, etc.). The relative intensities even in the same spectrum may depend very pronouncedly on the discharge conditions so that what is indicated as the strongest line may be relatively weak at a particular condition. The data are taken from W. F. Meggers, C. H. Corliss, and B. F. Scribner, "Tables of Spectral-line Intensities," 2 parts, National Bureau of Standards Monograph 32, Government Printing Office, Washington, D.C., 1961. These tables list the relative intensities, obtained in a 10-Å direct-current arc, of the lines of 70 elements mixed in a concentration of 0.1 atomic percent with copper. The lines of gaseous and unstable elements are from older sources.

In general, wavelengths in Table 7d-1 and other tables of this section are wavelengths in standard air for $\lambda > 2,000$ Å and in vacuum for $\lambda < 2,000$ Å.

TABLE 7d-1. PERSISTENT LINES OF THE ELEMENTS

Z	Symbol	Neutral atoms				Singly ionized			
		Strongest line	Other strong lines			Strongest line	Other strong lines		
1	H	1,215.66	6,562.85	4,861.33					
2	He	584.33	5,875.62	3,888.65	....	303.78			
3	Li	6,707.85	6,103.64	....	....	199.26			
4	Be	2,348.61	2,650.47	3,321.34	....	3,130.42	3,131.07		
5	B	2,497.73	2,496.78	....	....	1,362.46	3,451.41		
6	C	1,657.01	2,478.57	....	....	1,335.71	4,267.27	2,836.71	
7	N	1,134.98	4,109.98	4,099.94	....	1,085.74	5,679.56	5,666.64	
8	O	1,302.19	7,771.93	7,774.14	7,775.43	834.47			
9	F	954.80	6,856.02	6,902.46	....	606.81			
10	Ne	735.89	5,852.49	6,402.25	5,400.56	460.73			
11	Na	5,889.95	5,895.92	8,194.81	3,302.32	372.04			
12	Mg	2,852.13	3,838.26	5,183.62	3,832.31	2,795.53	2,802.70		
13	Al	3,961.53	3,092.78	3,944.03	3,082.16	1,670.81	2,669.17	2,816.18	
14	Si	2,516.11	2,881.60	2,524.11	2,528.51	1,817.0			
15	P	1,774.94	2,535.65	2,553.28	....	1,542.32			
16	S	1,807.31	9,212.91	9,228.11	4,694.13	1,259.53			
17	Cl	1,347.2	....	....	....	1,071.05	4,794.54	4,810.06	4,819.46
18	A	1,048.22	8,115.31	7,067.22	6,965.43	919.78			
19	K	7,604.91	7,698.98	4,044.14	4,047.20	600.77			
20	Ca	4,226.73	4,454.78	6,162.17	4,434.96	3,933.67	3,968.47	3,179.33	8,542.09
21	Sc	3,911.81	3,907.49	4,020.40	5,081.56	3,613.84	3,630.74	4,246.83	3,572.53
22	Ti	3,998.64	3,653.50	3,642.68	4,981.73	3,349.41	3,234.52	3,372.80	3,383.76
23	V	4,379.24	3,183.98	4,111.78	4,384.72	3,093.11	3,102.30	3,110.71	2,908.82
24	Cr	3,578.69	3,593.49	4,254.35	3,605.33	2,835.63	2,677.16	3,843.25	2,849.84
25	Mn	4,030.76	4,033.07	2,794.82	4,034.49	2,576.10	2,593.73	2,605.69	2,949.20
26	Fe	3,734.87	3,581.20	3,719.94	4,045.82	2,599.40	2,611.87	2,598.37	2,404.88
27	Co	3,453.50	3,405.12	3,502.28	3,569.38	2,388.92	2,528.62		
28	Ni	3,414.76	3,524.34	3,515.05	3,619.39	2,394.52	2,216.47		
29	Cu	3,247.54	3,273.96	5,218.20	5,105.54	2,135.98	2,700.96	2,192.26	
30	Zn	2,138.56	3,345.02	4,810.53	4,722.16	2,061.91	2,025.51		
31	Ga	4,172.06	4,032.98	2,943.64	2,874.24	1,414.44			
32	Ge	2,651.18	2,709.63	3,039.06	2,754.59	1,649.26			
33	As	1,890.43	2,780.22	2,860.44	2,349.84	1,266.36			
34	Se	1,960.91	2,039.85	2,062.79	8,918.80	1,192.29			
35	Br	1,488.4	....	....	....	1,015.42	4,704.86	4,785.50	4,816.71
36	Kr	1,235.82	5,870.92	5,570.29	....	917.43			
37	Rb	7,800.23	7,947.60	4,201.85	4,215.56	741.4			
38	Sr	4,607.33	6,408.47	4,962.26	5,480.84	4,077.71	4,215.52	3,464.46	
39	Y	4,102.38	4,077.38	3,620.94	4,643.70	3,710.30	3,600.73	3,774.33	4,374.94
40	Zr	3,601.19	3,519.60	3,835.96	4,687.80	3,391.98	3,438 23	3,496.21	3,572.47
41	Nb	4,058.94	4,079.73	4,100.92	3,580 27	3,094.18	3,130.79	2,927.81	2,950.88
42	Mo	3,798.25	3,864.11	3,132.59	3,902.96	2,775.40	2,816.15	2,848.23	2,871.51
43	Tc	3,636.10	4,297.06	4,262.26	....	2,543.24	2,610.00	3,237.02(?)‡	
44	Ru	3,728.13	3,498.94	3,726.93	4,080.60	2,678.76	2,402.72	2,456.57	
45	Rh	3,692.36	3,528.02	3,434.89	3,657.99	2,520.53	2,490.77	2,715.31	
46	Pd	3,404.58	3,609.55	3,634.70	3,421.24	2,488.92			
47	Ag	3,280.68	3,382.89	5,209.07	5,465.49	2,413.18	2,437.79	2,246.41	
48	Cd	2,288.02	3,610.51	5,085.82	3,466.20	2,265.02	2,144.38		
49	In	4,511.32	4,101.77	3,256.09	3,039.36	1,586.4			
50	Sn	2,839.99	2,863.33	3,034.12	2,706.41	2,152.22			
51	Sb	2,598.05	2,528.52	2,877.92	3,232.52	1,606.98			
52	Te	2,385.76	2,383.25	2,142.75	....	1,161.52			
53	I	1,830.4	....	....	....	1,233.97	2,062.38	5,464.61	
54	Xe	1,469.62	4,671.23	4,624.28	....	1,100.42			

TABLE 7d-1. PERSISTENT LINES OF THE ELEMENTS (Continued)

Z	Symbol	Neutral atoms — Strongest line	Neutral atoms — Other strong lines			Singly ionized — Strongest line	Singly ionized — Other strong lines		
55	Cs	8,521.10	8,943.50	4,555.36	6,723.28	926.75			
56	Ba	5,535.55	6,110.78	6,498.76	7,059.94	4,554.03	4,934.09	6,141.72	6,496.90
57	La	6,249.93	5,177.31	5,234.27	5,501.34	3,949.10	4,086.72	3,794.78	4,333.74
58	Ce	5,699.23	5,159.69	5,161.48	5,245.92	4,186.60	3,952.54	3,801.53	3,999.24
59	Pr	4,951.36	4,939.74	5,045.53	4,695.77	4,179.42	4,222.98	4,225.33	3,908.43
60	Nd	4,924.53	4,883.81	4,634.24	5,620.54	4,303.58	4,061.09	3,863.36	4,012.25
61	Pm	....	....	....	....	3,892.16	3,910.26	3,998.96†	
62	Sm	4,296.74	5,071.20	5,175.42	4,336.14	3,568.27	3,592.60	3,885.29	4,424.34
63	Eu	4,205.05	3,819.67	3,930.48	3,907.10	4,594.03	4,627.22	4,661.88	3,212.81
64	Gd	4,225.85	3,783.05	4,078.70	4,053.64	3,768.39	3,422.47	3,646.19	3,340.47
65	Tb	4,326.47	4,318.85	3,765.14	4,338.45	3,509.17	3,702.85	3,568.51	3,324.40
66	Dy	4,211.72	4,045.99	4,186.78	4,194.85	3,531.70	3,968.42	3,645.41	3,944.70
67	Ho	3,796.75	3,810.73	4,103.84	4,053.93	3,456.00	3,891.02	3,398.98	3,484.84
68	Er	4,007.97	3,862.82	4,151.10	3,892.69	3,906.34	3,372.76	3,692.64	3,499.11
69	Tm	4,094.19	4,105.84	3,717.92	4,187.62	3,462.20	3,848.02	3,131.26	3,425.08
70	Yb	3,987.98	3,464.36	5,556.48	7,699.49	3,694.19	3,289.37	2,891.38	2,970.56
71	Lu	3,281.74	3,359.56	3,312.11	3,376.50	2,615.42	2,911.39	3,077.60	3,507.39
72	Hf	2,866.37	3,072.88	2,916.48	2,940.77	3,399.80	3,561.66	2,820.22	3,505.23
73	Ta	2,653.27	2,714.67	2,647.47	2,656.61	3,012.54	2,685.17	2,400.63	2,635.58
74	W	4,008.75	4,074.36	4,294.61	2,724.35	2,555.09	2,571.44	2,658.04	2,764.27
75	Re	3,460.46	3,464.73	3,424.62	2,999.60	3,580.15	2,461.84	2,608.50	2,733.04
76	Os	2,909.06	3,058.66	3,301.56	4,260.85	....	2,538.00	2,486.24	
77	Ir	3,220.78	2,543.97	3,133.32	3,800.12	....	3,731.36	2,242.68	
78	Pt	3,064.71	2,659.45	2,702.40	2,733.96	1,777.09	2,488.74		
79	Au	2,675.95	2,427.95	3,122.78	2,748.26	1,740.47	2,802.19		
80	Hg	1,849.68	2,536.52	4,358.35	5,460.74	1,649.96			
81	Tl	3,519.24	5,350.46	3,775.72	3,429.43	1,908.64			
82	Pb	4,057.83	3,683.48	2,801.99	2,833.06	1,726.75	2,203.51	5,608.8	
83	Bi	3,067.72	3,897.98	2,938.30	2,989.03	1,902.41			
84	Po	2,449.99							
85	At								
86	Rn	1,786.07	7,450.00	7,055.42					
87	Fr								
88	Ra	4,825.91	....	....	....	3,814.42	4,682.28	3,649.55	4,340.64
89	Ac	....	....	....	....	4,168.40	4,088.40	3,863.12	
90	Th	3,719.44	3,803.07	3,304.24	3,967.39	4,019.13	2,837.30	3,469.92	3,392.03
91	Pa	....	2,743	....	....	....	2,743.9	3,054.6	3,957.8¶
92	U	3,812.00	3,854.88	3,871.04	3,566.60	3,859.58	3,854.66	3,670.07	3,890.36
93	Np	....	....	....	....	....	2,956.6	3,829.2	4,290.9¶
94	Pu	....	....	....	....	....	2,835.5	3,907.1	3,989.7¶
95	Am	....	....	....	....	....	2,832.3	3,926.2	4,188.2¶
96	Cm								
97	Bk								
98	Cf								
99	Es								
100	Fm								
102	No								
103	Lw								

† Scribner, Bozman, Meggers, J. Research Natl. Bur. Standards 46, 85 (1951) (Pm).
‡ Scribner, Bozman, Meggers, J. Research Natl. Bur. Standards 45, 476 (1950).
¶ Fred, Tomkins, J. Opt. Soc. Am. 39, 357 (1949).

7e. Important Atomic Spectra

H. M. CROSSWHITE AND G. H. DIEKE

The Johns Hopkins University

7e-1. General. The tables and figures of this section furnish data on spectra which are often used for reference. These are chiefly the spectra of the rare gases which can easily be obtained with simple discharge tubes (a neon advertising sign, for instance, is a good source for the neon spectrum); the iron spectrum which is the best source of standard lines for a spectrograph of moderate to high dispersion; and the mercury spectrum which, like that of helium, is particularly useful for spectrographs of low dispersion.

Data on other spectra of varying degrees of accuracy and completeness can be found in the MIT tables;[1] Kayser, "Handbuch der Spectroscopie," vols. 5–8; Paschen und Götze (1922); Fowler (1922); C. E. Moore, "Multiplet Tables" (1945); and Brode, "Chemical Spectroscopy" (1943).

An atlas of spectra is Gatterer and Junkes (1937 and 1945). For the solar spectrum, Minnaert, Mulders, and Houtgast (1940) is recommended.

The various tables of spectra and figures presented in this section are as follows:

Spectrum	Table	Figure
Helium.................	7e-1	
Neon Ne I.............	7e-2	7e-1
Argon A I..............	7e-3	7e-2
Krypton Kr I..........	7e-4	7e-3
Xenon Xe I............	7e-5	7e-4
Iron Fe I...............	7e-6	7e-5
Mercury Hg I..........	7e-7	7e-6, 7

The wavelengths and intensities are listed as completely as space permits. Special attention has been paid to lines which can be used as standards for wavelength measurements of high accuracy.

The figures, which are direct photoelectric traces obtained at The Johns Hopkins University, will help to orient the reader in the particular spectra. The traces were made with a logarithmic amplifier and calibrated to compensate for variations in sensitivity of spectrograph and measuring devices. Furthermore, the intensity scale is the same for all spectra so that the values indicate relative brightnesses of the light sources. Intensities as read from the charts, however, are not meant for high accuracy.

In a number of spectra numerical intensity values are given on a logarithmic scale. Also the conditions under which the spectra were produced are shown in each case.

[1] See the references on p. 7-96a.

Without the knowledge of such conditions intensity tables have little meaning because the intensities vary greatly with the discharge conditions.

In both figures and tables (except for helium) the intensities are standardized to give the energy flux from 100 cm² of the light source per unit solid angle in ergs per second.

In Figs. 7e-1 through 7e-5, only whole numbers are given in the wavelength designations. Values accurate to several decimal places appear for many of these lines in Tables 7e-2 through 7e-7.

7e-2. Standard Wavelengths. Since October, 1960, the international standard of length is officially defined in terms of the orange line of the krypton isotope with mass 86. The anstrom unit (Å) is exactly 10^{-10} meter. The meter is defined as exactly 1,650,763.73 wavelengths *in vacuo* of the Kr^{86} line, which has

$$\lambda_{vac} = 6,057.80211 \text{ Å}$$
$$\lambda_{air} = 6,056.12525 \text{ Å}$$

This line has the indicated wavelength when the atoms are free from interactions. If a lamp meets the following specifications, the wavelength is within 10^{-4} Å of the nominal value.

1. Purity of Kr^{86} not less than 99 percent.
2. Temperature of the coldest point of the lamp not higher than 63 K (triple point of nitrogen). The Kr pressure is then about 0.03 mm of Hg.
3. The current density must not exceed 4 ma/mm².
4. For a hot-cathode d-c lamp the anodes should be toward the observer.

Wavelengths of Kr^{84}, which is the predominant constituent of natural krypton, are approximately 0.001 Å larger in the visible than the Kr^{86} wavelengths.

For accurate spectroscopic wavelength measurements wavelength standards should be used as follows: (1) For interferometric measurements of the highest accuracy, the primary standard. (2) For other interferometric measurements and grating measurements of exceptional accuracy, the primary standard and secondary standards of Kr^{86} or natural Kr*, Ne*, A*, Hg^{198}*, Fe* (in a low-pressure source), and Th determined to four decimals. The values for the elements marked by an asterisk will be found in Tables 7e-2 to 7e-7 of this section. (3) For other grating measurements, in general those listed under (2) and many other lines produced by stable low-pressure light sources and measured reliability to at least three decimals.

Note. Using lines of one order of the grating as standards for different overlapping orders may or may not lead to errors, depending on the properties of the particular grating.

Helium I. The He I spectrum (Table 7e-1) consists of singlets and triplets. The latter appear as double lines except under the most favorable conditions. This is because the 2^3P_2 and 2^3P_1 levels almost coincide, whereas the 2^3P_0 level is about 1 cm^{-1} removed. The wavelengths are taken from the literature [see especially W. C. Martin, *J. Research NBS* **64**, 19 (1960)]. The intensities I_1 and I_2 are quantitative measurements at the following conditions: I_1, discharge with external electrodes, frequency 15 MHz, pressure 7.5 mm; I_2, same, pressure 0.25 mm; I_0, estimates from the literature.

TABLE 7e-1. THE SPECTRUM OF HELIUM I AND II

λ	Classification				He II	I_0	I_1	I_2
	Singlets		Triplets					
243.027	...	...	...	...	$4 \to 1$			
256.317	...	...	...	...	$3 \to 1$			
303.781	...	...	...	...	$2 \to 1$			
522.2128	1S	4P						
537.0296	1S	3P						
584.331	1S	2P						
591.4117	1S	...	...	2p				
1,084.975	...	...	...	...	$5 \to 2$			
1,215.171	...	...	...	...	$4 \to 2$			
1,640.474	...	...	...	...	$3 \to 2$			
2,696.119	...	...	2s	9p	....	1		
2,723.191	...	...	2s	8p	....	1		
2,763.804	...	...	2s	7p	....	2		
2,829.076	...	...	2s	6p	....	4		
2,945.106	...	...	2s	5p	....	6		
3,187.745	...	...	2s	4p	....	8		
3,203.14	...	...	...	...	$5 \to 3$			
3,354.550	2S	7P	...	...	....	2		
3,447.586	2S	6P	...	...	....	2		
3,587.270	...	...	2p	9d	....	2		
3,587.405	...	...	2p	9d	....	1		
3,590.314	...	...	2p	9s	....	1		
3,599.448	...	...	2p	9s	....	1		
3,613.643	2S	5P	...	...	....	3	19	260
3,634.232	...	...	2p	8d	....	2		
3,634.369	...	...	2p	8d	....	1		
3,651.990	...	...	2p	8s	....	1		
3,652.130	...	...	2p	8s	....	1		
3,705.005	...	...	2p	7d	....	3	28	260
3,705.148	...	...	2p	7d	....	1		
3,732.865	...	...	2p	7s	....	1		
3,733.010	...	...	2p	7s	....	1		
3,819.6072	...	...	2p	6d	....	4	84	680
3,819.758	...	...	2p	6d	....	1		
3,867.475	...	...	2p	6s	....	2	23	160
3,867.630	...	...	2p	6s	....	1		
3,888.648	...	...	2s	3p	....	10	10,000	10,000
3,964.7289	2S	4P	...	...	....	4	140	2,100
4,009.268	2P	7D	...	...	....	1	5	89
4,023.973	2P	7S	...	...	....	1		
4,026.1912	...	...	2p	5d	....	5	370	1,450
4,026.359	...	...	2p	5d	....	1		
4,120.812	...	...	2p	5s	....	3	90	480
4,120.993	...	...	2p	5s	....	1		

TABLE 7e-1. THE SPECTRUM OF HELIUM I AND II (Continued)

λ	Classification				He II	I_0	I_1	I_2
	Singlets		Triplets					
4,143.761	2P	6D	...	...	....	2	19	210
4,168.967	2P	6S	...	...	....	1	3	36
4,387.9294	2P	5D	...	...	....	3	83	590
4,437.551	2P	5S	...	...	....	1	17	290
4,471.479	...	...	2p	4d	....	6	2,300	2,220
4,471.682	...	...	2p	4d	....	1		
4,685.75	...	...	...	...	$4 \to 3$			
4,713.1455	...	...	2p	4s	....	3	350	370
4,713.376	...	...	2p	4s	....	1		
4,921.9310	2P	4D	...	...	....	4	57	1,800
5,015.6799	2S	3P	...	...	....	6	710	3,106
5,047.738	2P	4S	...	...	....	2	120	860
5,411.551	...	...	...	...	$7 \to 4$			
5,875.621	...	...	2p	3d	....	10	18,200	7,100
5,875.966	...	...	2p	3d	....	1		
6,559.71	...	...	...	...	$6 \to 4$			
6,678.151	2P	3D	...	...	....	6	2,400	1,850
7,065.190	...	...	2p	3s	....	5	7,100	1,450
7,065.707	...	...	2p	3s	....	1		
7,281.349	2P	3S	...	...	....	3*	1,450	
10,123.77	...	...	...	...	$5 \to 4$			
10,829.088	...	...	2s	$2p_0$	....	500	105,000	6,950
10,830.248	...	...	2s	$2p_1$	....	1,500		
10,830.337	...	...	2s	$2p_2$	....	2,500		
12,784.79†	...	...	3d	5f	....	10†		
12,790.27	3D	5F	...	...	....	1·		
17,003.11	...	...	3p	4d	....	20		
18,685.12	...	...	3d	4f	....	70		
18,697.00	3D	4F	...	...	....	10		
20,580.9	2S	2P	...	...	....	5,000		

* Change in the I_0 scale. From here on National Bureau of Standards values.
† Wavelengths and intensities from here on from Humphreys and Kostkowski, *J. Research Natl. Bur. Standards* **49**, 73 (1952).

The classification is indicated by capital letters for singlets, lower-case letters for triplets. A few of the He II lines are also listed. They have elaborate fine structures.

Neon I. The neon spectrum is moderately rich in lines and may serve, like the other rare-gas spectra, as an easily obtained comparison spectrum. Any neon-sign manufacturer can produce a satisfactory tube. The wavelengths of the strong lines have been measured with great accuracy and have been adopted as international secondary standards,[1] often replacing the primary standard for interferometric measurements.

Table 7e-2 lists the principal neon lines. The wavelengths are interferometric wavelengths when followed by a capital letter.

B, Burns, Adams, Longwell, *J. Opt. Soc. Am.* **40**, 339 (1950)

H, Humphreys, *J. Research Natl. Bur. Standards* **20**, 17(1938)

[1] *Trans. Intern. Astron. Union* **5**, 86 (1935); **9**, 204 (1957); **10**, 229 (1958).

TABLE 7e-2. THE SPECTRUM OF NEON I

Wavelength	Classification				I_0	$\log I_1$	$\log I_2$	$\log I_3$
	System.		Paschen					
2,647.42	$3s_{12}$	$8p_1$	$1s_5$	$7p_{6,7}$	8			
2,675.24	$3s_{11}$	$7p'_{12}$	$1s_4$	$6p_4$	8			
2,675.64	$3s_{11}$	$7p'_{11}$	$1s_4$	$6p_5$	8			
2,872.663	$3s'_{00}$	$6p'_{00}$	$1s_2$	$5p_1$	5	.....	.....	2.73
2,913.168	$3s_{12}$	$5p'_{01}$	$1s_5$	$4p_2$	8	.....	.....	3.16
2,932.721	$3s'_{01}$	$6p_{00}$	$1s_2$	$5p_3$	7	.....	.....	3.30
2,947.297	$3s_{11}$	$5p_{12}$	$1s_4$	$4p_4$	8	.....	.....	3.2?
2,974.714	$3s_{12}$	$5p_{12}$	$1s_5$	$4p_6$	9	.....	.....	3.6?
2,980.642	$3s'_{00}$	$5p'_{01}$	$1s_2$	$4p_2$	5.5	.....	.....	2.7
2,980.922	$3s'_{00}$	$5p'_{11}$	$1s_2$	$4p_5$	6	.....	.....	2.80
2,982.663	$3s_{12}$	$5p_{23}$	$1s_5$	$4p_9$	9	.....	.....	3.52
2,992.420	$3s_{11}$	$5p_{00}$	$1s_4$	$4p_3$	8}	.....	.....	3.32
2,992.438	$3s_{12}$	$5p_{01}$	$1s_5$	$4p_{10}$	8}			
3,012.129	$3s_{11}$	$5p_{12}$	$1s_4$	$4p_6$	6	.....	.....	2.93
3,012.955	$3s_{11}$	$5p_{11}$	$1s_4$	$4p_7$	6	.....	.....	2.98
3,017.348	$3s_{11}$	$5p_{22}$	$1s_4$	$4p_8$	6	.....	.....	3.12
3,057.388	$3s'_{01}$	$5p'_{00}$	$1s_2$	$4p_1$	9	.....	.....	2.7
3,076.971	$3s'_{01}$	$5p'_{12}$	$1s_2$	$4p_4$	8	.....	.....	2.80
3,126.1986 B	$3s'_{01}$	$5p_{00}$	$1s_2$	$4p_3$	8	.....	.....	3.61
3,148.6107 B	$3s'_{01}$	$5p_{11}$	$1s_2$	$4p_7$	7	.....	.....	2.44
3,153.4107 B	$3s'_{01}$	$5p_{22}$	$1s_2$	$4p_8$	6	.....	.....	2.4?
3,167.5762 B	$3s'_{01}$	$5p_{01}$	$1s_2$	$4p_{10}$	6	.....	.....	2.21
3,369.8076 B	$3s_{12}$	$4p'_{12}$	$1s_5$	$3p_4$	10	.....	.....	3.90
3,369.9069 B	$3s_{12}$	$4p'_{01}$	$1s_5$	$3p_2$	15	.....	.....	4.36
3,375.6489 B	$3s_{12}$	$4p'_{11}$	$1s_5$	$3p_5$	6	.....	.....	2.98
3,417.9031 B	$3s_{11}$	$4p'_{12}$	$1s_4$	$3p_4$	10	.....	.....	4.62
3,418.0066 H	$3s_{11}$	$4p'_{01}$	$1s_4$	$3p_2$	6	.....	.....	4.14
3,423.9120 B	$3s_{11}$	$4p'_{11}$	$1s_4$	$3p_5$	6	.....	.....	3.57
3,447.7022 B	$3s_{12}$	$4p_{12}$	$1s_5$	$3p_6$	8	.....	.....	4.91
3,450.7641 B	$3s_{12}$	$4p_{11}$	$1s_5$	$3p_7$	6	.....	.....	4.18
3,454.1942 B	$3s_{11}$	$4p_{00}$	$1s_4$	$3p_3$	7	.....	.....	4.72
3,460.5235 B	$3s'_{00}$	$4p'_{01}$	$1s_2$	$3p_2$	7	.....	.....	4.37
3,464.3385 B	$3s_{12}$	$4p_{22}$	$1s_5$	$3p_8$	7	.....	.....	4.27
3,466.5781 B	$3s'_{00}$	$4p'_{11}$	$1s_2$	$3p_5$	8	.....	.....	4.64
3,472.5706 B	$3s_{12}$	$4p_{23}$	$1s_5$	$3p_9$	10	.....	.....	4.90
3,498.0632 B	$3s_{11}$	$4p_{12}$	$1s_4$	$3p_6$	7	.....	.....	4.45
3,501.2154 B	$3s_{11}$	$4p_{11}$	$1s_4$	$3p_7$	8	.....	.....	4.53
3,510.7207 B	$3s_{12}$	$4p_{01}$	$1s_5$	$3p_{10}$	6	.....	.....	3.85
3,515.1900 B	$3s_{11}$	$4p_{22}$	$1s_4$	$3p_8$	8	.....	.....	4.55
3,520.4714 B	$3s'_{01}$	$4p'_{00}$	$1s_2$	$3p_1$	20	.....	.....	5.32

TABLE 7e-2. THE SPECTRUM OF NEON I (*Continued*)

Wavelength	System.		Paschen		I_0	$\log I_1$	$\log I_2$	$\log I_3$
3,562.9551 *B*	$3s_{11}$	$4p_{01}$	$1s_4$	$3p_{10}$	3			
3,593.5263 *B*	$3s'_{01}$	$4p'_{12}$	$1s_2$	$3p_4$	10	.....	.....	4.70
3,593.639 *B*	$3s'_{01}$	$4p'_{01}$	$1s_2$	$3p_2$	9	.....	.....	4.50
3,600.1694 *B*	$3s'_{01}$	$4p'_{11}$	$1s_2$	$3p_5$	7	.....	.....	4.17
3,609.1787 *B*	$3s'_{00}$	$4p_{01}$	$1s_2$	$3p_{10}$	6	.....	.....	3.26
3,633.6643 *B*	$3s'_{01}$	$4p_{00}$	$1s_2$	$3p_3$	7	.....	.....	4.28
3,682.2421 *B*	$3s'_{01}$	$4p_{12}$	$1s_2$	$3p_6$	7	.....	.....	4.21
3,685.7351 *B*	$3s'_{01}$	$4p_{11}$	$1s_2$	$3p_7$	7	.....	.....	4.08
3,701.2247 *B*	$3s'_{01}$	$4p_{22}$	$1s_2$	$3p_8$	7	.....	.....	4.06
3,754.2148 *B*	$3s'_{01}$	$4p_{01}$	$1s_2$	$3p_{10}$	6	.....	.....	3.42
4,270.2674 *B*	$3p_{01}$	$7d_{00}$	$2p_{10}$	$7d_6$	4	2.460		
4,275.5598 *B*	$3p_{01}$	$6d'_{22}$	$2p_{10}$	$6s_1''''$	5	2.70	2.61	
4,306.2625 *B*	$3p_{01}$	$8s_{12}$	$2p_{10}$	$6s_5$	5			
4,334.1267 *B*	$3p_{01}$	$7s'_{01}$	$2p_{10}$	$5s_2$	5			
4,363.524 *M*	$3p_{23}$	$9d_{34}$	$2p_9$	$9d'_4$	5			
4,381.220 *M*	$3p_{23}$	$10s_{12}$	$2p_9$	$8s_5$	3			
4,395.556 *M*	$3p_{22}$	$9d_{33}$	$2p_8$	$9d_4$	4			
4,422.5205 *B*	$3p_{01}$	$6d_{12}$	$2p_{10}$	$6d_3$	8	2.97	2.90	
4,424.8096 *B*	$3p_{01}$	$6d_{01}$	$2p_{10}$	$6d_5$	8	2.89	2.81	
4,425.400 *M*	$3p_{01}$	$6d_{00}$	$2p_{10}$	$6d_6$	7			
4,433.7239 *B*	$3p_{23}$	$8d_{34}$	$2p_9$	$8d'_4$	5	2.34	2.19	
4,460.175 *M*	$3p_{23}$	$9s_{12}$	$2p_9$	$7s_5$	6			
4,466.8120 *B*	$3p_{22}$	$8d_{33}$	$2p_8$	$8d_4$	5	2.02	1.81	
4,475.656 *M*	$3p_{11}$	$7d'_{12}$	$2p_7$	$7s_1''$	6			
4,483.199 *B*	$3p_{01}$	$7s_{11}$	$2p_{10}$	$5s_4$	7	2.098		
4.488.0926 *B*	$3p_{01}$	$7s_{12}$	$2p_{10}$	$5s_5$	8	2.811	2.673	
4,500.182 *M*	$3p'_{11}$	$8d'_{12}$	$2p_5$	$8s_1''$	4			
4,517.736 *M*	$3p'_{12}$	$8d'_{23}$	$2p_4$	$8s_1'''$	6			
4,525.764 *M*	$3p_{11}$	$8d_{22}$	$2p_7$	$8d_1''$	5			
4,536.312	$3p_{01}$	$5d'_{11}$	$2p_{10}$	$5s_1'$	7	2.694	2.699	
4,537.7545 *B*	$3p_{01}$	$5d'_{22}$	$2p_{10}$	$5s_1''''$	10	3.3	3.4	
4,538.2927 *B*	$3p_{23}$	$7d_{23}$	$2p_9$	$7d'_1$	8			
4,540.3801 *B*	$3p_{23}$	$7d_{34}$	$2p_9$	$7d'_4$	10	2.964	2.854	
4,552.598 *M*	$3p_{11}$	$9s_{11}$	$2p_7$	$7s_4$	3			
4,565.888 *M*	$3p_{12}$	$8d_{23}$	$2p_6$	$8d'_1$	4.5			
4,575.0620 *B*	$3p_{22}$	$7d_{33}$	$2p_8$	$7d_4$	8	2.714	2.569	
4,582.035 *M*	$3p_{22}$	$6d'_{23}$	$2p_8$	$6s_1'''$	7	2.4	2.3	
4,582.4521 *B*	$3p_{23}$	$8s_{12}$	$2p_9$	$6s_5$	7	2.4	2.3	
4,609.910 *M*	$3p'_{11}$	$7d'_{12}$	$2p_5$	$7s_1''$	7	2.19		
4,614.391 *M*	$3p_{22}$	$8s_{11}$	$2p_8$	$6s_4$	6	2.204		

TABLE 7e-2. THE SPECTRUM OF NEON I (*Continued*)

Wavelength	System.		Paschen		I_0	$\log I_1$	$\log I_2$	$\log I_3$
4,617.837 M	$3p_{22}$	$8s_{12}$	$2p_8$	$6s_5$	5			
4,628.3113 B	$3p'_{12}$	$7d'_{23}$	$2p_4$	$7s'''_1$	7	2.49	2.39	
4,636.125 M	$3p_{11}$	$7d_{22}$	$2p_7$	$7d''_1$	5	2.0		
4,636.630	$3p_{11}$	$7d_{11}$	$2p_7$	$7d_2$	5	2.0		
4,645.4180 B	$3p_{11}$	$6d'_{12}$	$2p_7$	$6s''_1$	8	2.672	2.607	
4,649.904 M	$3p_{22}$	$7s'_{01}$	$2p_8$	$5s_2$	5			
4,656.3936 B	$3p_{01}$	$6s'_{01}$	$2p_{10}$	$4s_2$	8	2.916	2.828	2.799
4,661.1054 B	$3p_{01}$	$6s'_{00}$	$2p_{10}$	$4s_3$	7	2.634	2.559	
4,670.884 M	$3p'_{12}$	$8s'_{01}$	$2p_4$	$6s_2$	5			
4,678.218 M	$3p_{12}$	$7d_{23}$	$2p_6$	$7d'_1$	8	2.4	2.3	
4,679.135 M	$3p_{12}$	$7d_{12}$	$2p_6$	$7d_3$	7	2.2	2.1	
4,687.6724 B	$3p_{12}$	$6d'_{23}$	$2p_6$	$6s'''_1$	6	2.410	2.340	
4,702.526	$3p_{01}$	$5d_{11}$	$2p_{10}$	$5d_2$	7	2.472	2.427	
4,704.3949 B	$3p_{01}$	$5d_{12}$	$2p_{10}$	$5d_3$	15	3.701	3.729	3.437
4,708.8619 B	$3p_{01}$	$5d_{01}$	$2p_{10}$	$5d_5$	12	3.688	3.693	3.459
4,710.0669 B	$3p_{01}$	$5d_{00}$	$2p_{10}$	$5d_6$	10	3.33	3.33	3.34
4,712.0661 B	$3p_{23}$	$6d_{23}$	$2p_9$	$6d'_1$	10	2.96	2.90	2.55
4,715.3466 B	$3p_{23}$	$6d_{34}$	$2p_9$	$6d_4$	15	3.57	3.50	3.17
4,725.145 M	$3p_{12}$	$8s_{12}$	$2p_6$	$6s_5$	5			
4,749.5754 B	$3p_{22}$	$6d_{22}$	$2p_8$	$6d'_1$	8	2.78	2.68	
4,752.7320 B	$3p_{22}$	$6d_{33}$	$2p_8$	$6d_4$	10	3.329	3.243	2.974
4,788.9270 B	$3p_{23}$	$7s_{12}$	$2p_9$	$5s_5$	12	3.16	3.05	
4,790.217 B	$3p'_{11}$	$6d_{22}$	$2p_5$	$6s''_1$	10	2.84	2.77	
4,800.100 B	$3p_{12}$	$7d_{23}$	$2p_4$	$7d'_1$	5			
4,810.0640 B	$3p'_{12}$	$6d'_{23}$	$2p_4$	$6s'''_1$	7	3.07	3.01	2.70
4,817.6386 B	$3p_{11}$	$6d_{22}$	$2p_7$	$6d''_1$	8	2.861	2.775	2.597
4,818.748	$3p_{11}$	$6d_{11}$	$2p_7$	$6d_2$	7	2.599	2.499	2.335
4,821.9236 B	$3p_{22}$	$7s_{11}$	$2p_8$	$5s_4$	8	2.864	2.646	2.693
4,823.174	$3p_{00}$	$6d'_{11}$	$2p_3$	$6s'_1$	6	2.3	2.2	
4,827.3444 B	$3p_{01}$	$6s_{11}$	$2p_{10}$	$4s_4$	10	2.9	2.8	
4,827.587 B	$3p_{22}$	$7s_{12}$	$2p_8$	$5s_5$	8			
4,837.3139 B	$3p_{01}$	$6s_{12}$	$2p_{10}$	$4s_5$	9	3.442	3.402	3.177
4,852.6571 B	$3p'_{01}$	$6d'_{22}$	$2p_2$	$6s''''_1$	6	2.731	2.632	
4,863.0800 B	$3p_{12}$	$6d_{23}$	$2p_6$	$6d'_1$	6	3.131	3.064	
4,865.5009 B	$3p_{12}$	$6d_{12}$	$2p_6$	$6d_3$	6			
4,866.477 B	$3p_{12}$	$6d_{33}$	$2p_6$	$6d_4$	5 5	2.61	2.53	
4,867.010	$3p'_{11}$	$7s'_{00}$	$2p_5$	$5s_3$	5	2.4	2.3	
4,884.9170 B	$3p'_{12}$	$7s'_{01}$	$2p_4$	$5s_2$	10	3.2	3.2	3.0
4,892.1007 B	$3p_{11}$	$7s_{11}$	$2p_7$	$5s_4$	9	2.58	2.38	
4,928.241 B	$3p'_{01}$	$7s'_{01}$	$2p_2$	$5s_2$	5			

TABLE 7e-2. THE SPECTRUM OF NEON I (*Continued*)

Wavelength	Classification System.		Paschen		I_0	$\log I_1$	$\log I_2$	$\log I_3$
4,939.0457 B	$3p_{12}$	$7s_{12}$	$2p_6$	$5s_4$	6	2.626	2.462	
4,944.9899 B	$3p_{12}$	$7s_{12}$	$2p_6$	$5s_5$	6	2.641	2.517	
4,957.0335 B	$3p_{11}$	$5d'_{12}$	$2p_7$	$5s''_1$	10	3.3	3.4	
4,957.123 B	$3p_{11}$	$5d'_{22}$	$2p_7$	$5s''''_1$	7			
4,973.538	$3p'_{11}$	$6d_{22}$	$2p_5$	$6d''_1$	6	2.496	2.406	2.89
4,994.913 B	$3p'_{12}$	$6d_{23}$	$2p_4$	$6d'_1$	7ur	2.451	2.365	
5,005.1587 B	$3p_{12}$	$5d'_{23}$	$2p_6$	$5s'''_1$	10	3.10	3.13	3.58
5,011.003 M	$3p_{00}$	$6d_{11}$	$2p_3$	$6d_2$	4	2.279	2.208	
5,022.864 B	$3p_{22}$	$6s'_{01}$	$2p_8$	$4s_2$	4	2.592	2.506	
5,031.3504 B	$3p_{23}$	$5d_{23}$	$2p_9$	$5d'_1$	9	3.634	3.665	3.374
5,035.989	$3p_{23}$	$5d_{12}$	$2p_9$	$5d_3$	5	2.818	2.823	
5,037.7512 B	$3p_{23}$	$5d_{34}$	$2p_9$	$5d'_4$	10	4.27	4.29	4.01
5,074.2007 B	$3p_{22}$	$5d_{22}$	$2p_8$	$5d'_1$	5	3.53	3.54	3.27
5,080.3852 B	$3p_{22}$	$5d_{33}$	$2p_8$	$5d_4$	8	4.038	4.061	3.803
5,104.7011 B	$3p_{11}$	$6s'_{00}$	$2p_7$	$4s_3$	5	2.798	2.745	
5,113.6724 B	$3p_{01}$	$4d'_{11}$	$2p_{10}$	$4s'_1$	7	3.475	3.654	3.326
5,116.5032 B	$3p_{01}$	$4d'_{12}$	$2p_{10}$	$4s''_1$	8	4.11	4.36	3.92
5,122.2565 B	$3p'_{11}$	$5d'_{12}$	$2p_5$	$5s''_1$	8	3.6	3.6	
5,144.9384 B	$3p'_{12}$	$5d'_{23}$	$2p_4$	$5s'''_1$	10	3.9	4.0	
5,150.077	$3p_{12}$	$6s'_{01}$	$2p_6$	$4s_2$	5	2.9	2.9	
5,151.9610 B	$3p_{11}$	$5d_{22}$	$2p_7$	$5d''_1$	7	3.595	3.597	3.352
5,154.4271 B	$3p_{11}$	$5d_{11}$	$2p_7$	$5d_2$	6	3.292	3.286	
5,156.6672 B	$3p_{11}$	$5d_{12}$	$2p_7$	$5d_3$	6	2.5	2.5	
5,158.9018 B	$3p_{00}$	$5d'_{11}$	$2p_3$	$5s'_1$	6	3.087	3.094	
5,188.6122 B	$3p_{23}$	$6s_{12}$	$2p_9$	$4s_5$	8	3.813	3.898	3.519
5,191.3223 B	$3p'_{01}$	$5d'_{11}$	$2p_2$	$5s'_1$	5			
5,193.1302 B	$3p'_{01}$	$5d'_{12}$	$2p_2$	$5s''_1$	8⎫	3.6	3.6	
5,193.2227 B	$3p'_{01}$	$5d'_{22}$	$2p_2$	$5s''''_1$	8⎭			
5,203.8962 B	$3p_{12}$	$5d_{23}$	$2p_6$	$5d'_1$	8	3.837	3.884	3.515
5,208.8648 B	$3p_{12}$	$5d_{12}$	$2p_6$	$5d_3$	7	3.584	3.585	
5,210.5672 B	$3p_{12}$	$5d_{33}$	$2p_6$	$5d_4$	6	.....	.....	2.860
5,214.3389 B	$3p_{12}$	$5d_{01}$	$2p_6$	$5d_5$	5	2.777	2.745	
5,222.3517 B	$3p_{22}$	$6s_{11}$	$2p_8$	$4s_4$	6	3.549	3.431	3.592
5,234.0271 B	$3p_{22}$	$6s_{12}$	$2p_8$	$4s_5$	6	3.161	3.125	
5,274.0393 B	$3p'_{11}$	$6s'_{01}$	$2p_5$	$4s_2$	5.5	2.767	2.649	
5,280.0853 B	$3p'_{11}$	$6s'_{00}$	$2p_5$	$4s_3$	6	2.962	2.899	2.660
5,298.1891 B	$3p'_{12}$	$6s'_{01}$	$2p_4$	$4s_2$	8	3.492	3.396	3.300
5,304.7580 B	$3p_{11}$	$6s_{11}$	$2p_7$	$4s_4$	7	3.255	3.154	3.088
5,326.3968 B	$3p_{01}$	$4d_{11}$	$2p_{10}$	$4d_2$	7	3.388	3.540	
5,330.7775 B	$3p_{01}$	$4d_{12}$	$2p_{10}$	$4d_3$	12	4.547	4.771	4.360

TABLE 7e-2. THE SPECTRUM OF NEON I (*Continued*)

Wavelength	Classification System.	Classification Paschen	I_0	$\log I_1$	$\log I_2$	$\log I_3$
5,341.0938 B	$3p_{01}$ $4d_{01}$	$2p_{10}$ $4d_5$	20	4.537	4.732	
5,343.2834 B	$3p_{01}$ $4d_{00}$	$2p_{10}$ $4d_6$	12	4.3	4.5	3.936
5,349.2038 B	$3p'_{01}$ $6s'_{01}$	$2p_2$ $4s_2$	8	3.072	3.004	2.810
5,360.0121 B	$3p_{12}$ $6s_{11}$	$2p_6$ $4s_4$	8	3.392	3.297	3.129
5,372.3110 B	$3p_{12}$ $6s_{12}$	$2p_6$ $4s_5$	7	3.318	3.282	2.196
5,374.9774 B	$3p_{00}$ $5d_{11}$	$2p_3$ $5d_2$	6	3.002	2.984	
5,383.2503 B	$3p_{00}$ $5d_{01}$	$2p_3$ $5d_5$	4	2.487	2.525	
5,400.5616 B	$3s_{11}$ $3p'_{00}$	$1s_4$ $2p_1$	50	4.735	5.079	4.832
5,412.6490 B	$3p'_{01}$ $5d_{12}$	$2p_2$ $5d_3$	9	2.948	3.015	
5,418.5584 B	$3p'_{01}$ $5d_{01}$	$2p_2$ $5d_5$	8	2.88	2.85	
5,433.6513 B	$3p_{01}$ $5s'_{01}$	$2p_{10}$ $3s_2$	9	3.349	3.377	3.223
5,448.5091 B	$3p_{01}$ $5s'_{00}$	$2p_{10}$ $3s_3$	8	3.077	3.169	
5,494.4158 B	$3p'_{11}$ $6s_{11}$	$2p_5$ $4s_4$	6	2.843	2.745	
5,533.6788 B	$3p'_{12}$ $6s_{12}$	$2p_4$ $4s_5$	7	2.738	2.720	
5,538.6510 B	$3p_{00}$ $6s_{11}$	$2p_3$ $4s_4$	6	2.625	2.532	
5,562.7662 B	$3p_{22}$ $4d'_{23}$	$2p_8$ $4s_1'''$	10	3.9	4.1	3.7
5,652.5664 B	$3p_{11}$ $4d'_{11}$	$2p_7$ $4s_1'$	7	3.400	3.562	3.240
5,656.6588 B	$3p_{11}$ $4d'_{22}$	$2p_7$ $4s_1''''$	10	4.20	4.40	3.96
5,662.5489 B	$3p_{01}$ $5s_{11}$	$2p_{10}$ $3s_4$	7	3.438	3.665	
5,689.8163 B	$3p_{01}$ $5s_{12}$	$2p_{10}$ $3s_5$	8	4.179	4.305	3.949
5,719.2248 B	$3p_{12}$ $4d'_{23}$	$2p_6$ $4s_1'''$	10	3.9	4.1	3.7
5,748.2985 B	$3p_{23}$ $4d_{23}$	$2p_9$ $4d'_1$	10	4.4	4.6	4.1
5,760.5885 B	$3p_{23}$ $4d_{12}$	$2p_9$ $4d_3$	7	3.603	3.800	
5,764.4188 B	$3p_{23}$ $4d_{24}$	$2p_9$ $4d'_4$	15	5.080	5.312	4.868
5,804.4496 B	$3p_{22}$ $4d_{22}$	$2p_8$ $4d_1''$	10	4.374	4.585	4.121
5,811.4066 B	$3p_{22}$ $4d_{11}$	$2p_8$ $4d_2$	8	3.53	3.69	
5,820.1558 B	$3p_{22}$ $4d_{33}$	$2p_8$ $4d_4$	10	4.870	5.080	4.638
5,852.4878 S	$3s'_{01}$ $3p'_{00}$	$1s_2$ $2p_1$	50	5.904	6.268	6.442
5,868.4183 B	$3p'_{11}$ $4d'_{11}$	$2p_5$ $4s_1$	7	3.659	4.341	
5,872.8275 B	$3p'_{11}$ $4d'_{22}$	$2p_5$ $4s_1''''$	10	4.47	4.74	4.27
5,881.8950 S	$3s_{12}$ $3p'_{01}$	$1s_5$ $2p_2$	20	5.235	6.300	5.974
5,902.4623 B	$3p'_{12}$ $4d'_{23}$	$2p_4$ $4s_1'''$	6	4.82	5.05	4.626
5,902.7835 B	$3p'_{12}$ $4d'_{22}$	$2p_4$ $4s_1''''$	1.5			
5,906.4294 B	$3p_{11}$ $4d_{22}$	$2p_7$ $4d_1''$	6	4.448	4.671	4.185
5,913.6327 B	$3p_{11}$ $4d_{11}$	$2p_7$ $4d_2$	9	4.133	4.303	3.927
5,918.9068 B	$3p_{00}$ $4d'_{11}$	$2p_3$ $4s_1'$	9	4.09	4.28	3.860
5,944.8342 S	$3s_{12}$ $3p'_{12}$	$1s_5$ $2p_4$	10	5.365	6.380	6.104
5,961.6228 B	$3p'_{01}$ $4d'_{11}$	$2p_2$ $4s_1'$	7	3.903	4.198	3.717
5,965.4710 B	$3p'_{01}$ $4d'_{12}$	$2p_2$ $4s_1''$	10	4.54	4.75	4.25
5,974.6273 B	$3p_{12}$ $4d_{23}$	$2p_6$ $4d'_1$	10	4.7	5.6	

TABLE 7e-2. THE SPECTRUM OF NEON I (*Continued*)

Wavelength	Classification System.		Paschen		I_0	$\log I_1$	$\log I_2$	$\log I_3$
5,975.5340 S	$3s_{12}$	$3p'_{11}$	$1s_5$	$2p_5$	*12	5.14	6.05	
5,987.9074 B	$3p_{12}$	$4d_{12}$	$2p_6$	$4d_3$	8	4.373	4.601	4.058
5,991.6532 B	$3p_{12}$	$4d_{23}$	$2p_6$	$4d_4$	7	4.049	4.237	3.729
6,000.9275 B	$3p_{12}$	$4d_{01}$	$2p_6$	$4d_5$	6	3.725	3.925	
6,029.9971 S	$3s_{11}$	$3p'_{01}$	$1s_4$	$2p_2$	10	5.200	6.266	5.748
6,046.1348 B	$3p_{11}$	$5s'_{01}$	$2p_7$	$3s_2$	4	3.249	3.961	
6,064.5359 B	$3p_{11}$	$5s'_{00}$	$2p_7$	$3s_3$	4	3.613	3.995	
6,074.3377 S	$3s_{11}$	$3p_{00}$	$1s_4$	$2p_3$	10	5.411	6.490	6.093
6,096.1630 S	$3s_{11}$	$3p_{12}$	$1s_4$	$2p_4$	8	5.428	6.550	6.161
6,128.4498 B	$3s_{11}$	$3p'_{11}$	$1s_4$	$2p_5$	6	4.908	5.580	5.024
6,143.0623 S	$3s_{12}$	$3p_{12}$	$1s_5$	$2p_6$	10	5.48	6.63	6.198
6,163.5939 S	$3s'_{00}$	$3p'_{01}$	$1s_3$	$2p_2$	12	5.231	6.488	6.010
6,174.8829 B	$3p_{12}$	$4d_{23}$	$2p_4$	$4d_1$	5	3.9	4.3	
6,182.1460 B	$3p_{23}$	$5s_{12}$	$2p_9$	$3s_5$	7	3.610	4.737	4.334
6,189.0649 B	$3p_{12}$	$4d_{12}$	$2p_4$	$4d_3$	5	3.544	3.846	
6,193.0663 B	$3p_{12}$	$4d_{23}$	$2p_4$	$4d_4$	4	.....	3.498	
6,205.7775 B	$3p_{00}$	$4d_{11}$	$2p_3$	$4d_2$	6	3.785	4.043	
6,213.8758 B	$3p_{22}$	$5s_{11}$	$2p_8$	$3s_4$	7	4.376	4.473	
6,217.2813 S	$3s_{12}$	$3p_{11}$	$1s_5$	$2p_7$	15	5.359	6.436	5.962
6,246.7294 B	$3p_{22}$	$5s_{12}$	$2p_8$	$3s_5$	6	3.929	4.129	
6,266.4950 S	$3s'_{00}$	$3p'_{11}$	$1s_3$	$2p_5$	15	5.336	6.606	6.156
6,293.7447 B	$3p'_{00}$	$5s'_{01}$	$2p_5$	$3s_2$	6	3.683	3.900	
6,304.7892 S	$3s_{11}$	$3p_{12}$	$1s_4$	$2p_6$	6	5.422	6.391	6.009
6,313.6921 B	$3p'_{00}$	$5s'_{00}$	$2p_5$	$3s_3$	7	3.899	4.151	
6,328.1646 B	$3p'_{12}$	$5s'_{01}$	$2p_4$	$3s_2$	8	4.424	4.546	
6,334.4279 S	$3s_{12}$	$3p_{22}$	$1s_5$	$2p_8$	10	5.567	6.679	6.281
6,351.8618 B	$3p_{00}$	$5s'_{01}$	$2p_3$	$3s_2$	6			
6,382.9914 S	$3s_{11}$	$3p_{11}$	$1s_4$	$2p_7$	12	5.503	6.684	6.221
6,402.2460 B	$3s_{12}$	$3p_{23}$	$1s_5$	$2p_9$	20	5.93	6.83	6.389
6,421.7108 B	$3p'_{01}$	$5s_{12}$	$2p_2$	$3s_5$	6	3.701	3.893	
6,444.7118 B	$3p_{12}$	$5s_{12}$	$2p_6$	$3s_5$	7	4.094	4.191	3.823
6,506.5279 S	$3s_{11}$	$3p_{22}$	$1s_4$	$2p_8$	15	5.635	6.709	6.287
6,532.8824 S	$3s'_{00}$	$3p_{11}$	$1s_3$	$2p_7$	6	5.381	6.531	6.094
6,598.9529 S	$3s'_{01}$	$3p'_{01}$	$1s_2$	$2p_2$	15	5.736	6.691	6.213
6,652.0925 B	$3s'_{01}$	$3p_{00}$	$1s_2$	$2p_3$	7	4.279	4.681	4.203
6,666.8967 B	$3p_{00}$	$5s_{11}$	$2p_3$	$3s_4$	6			
6,678.2764 S	$3s'_{01}$	$3p'_{12}$	$1s_2$	$2p_4$	9	5.840	6.806	6.393
6,717.0428 S	$3s'_{01}$	$3p'_{11}$	$1s_2$	$2p_5$	2	5.765	6.712	6.286
6,929.4672 B	$3s'_{01}$	$3p_{12}$	$1s_2$	$2p_6$	10	5.965	6.783	6.421
7,024.0500 B	$3s'_{01}$	$3p_{11}$	$1s_2$	$2p_7$	9	5.436	6.068	5.568

TABLE 7e-2. THE SPECTRUM OF NEON I (*Continued*)

Wavelength	Classification System.		Paschen		I_0	$\log I_1$	$\log I_2$	$\log I_3$
7,032.4127 S	$3s_{12}$	$3p_{01}$	$1s_5$	$2p_{10}$	10	5.732	6.917	6.362
7,051.2923 S	$3p_{01}$	$3d'_{11}$	$2p_{10}$	$3s'_1$	5	4.286	.....	4.281
7,059.1074 S	$3p_{01}$	$3d'_{12}$	$2p_{10}$	$3s''_i$	7.5	4.868	5.534	4.904
7,173.9380 B	$3s'_{01}$	$3p_{22}$	$1s_2$	$2p_8$	10	5.793	6.411	6.022
7,245.1665 B	$3s_{11}$	$3p_{01}$	$1s_4$	$2p_{10}$	10	5.751	6.756	6.289
7,438.8981 B	$3s'_{00}$	$3p_{01}$	$1s_3$	$2p_{10}$	8	5.510	6.424	
7,472.4386 S	$3p_{01}$	$3d_{11}$	$2p_{10}$	$3d_2$	4	4.432	5.021	4.441
7,488.8712 S	$3p_{01}$	$3d_{12}$	$2p_{10}$	$3d_3$	9	5.398	6.052	5.424
7,535.7741 S	$3p_{01}$	$3d_{01}$	$2p_{10}$	$3d_5$	8	5.352	5.978	5.387
7,544.0443 S	$3p_{01}$	$3d_{00}$	$2p_{10}$	$3d_6$	6	4.962	5.667	4.956
7,724.6281 B	$3p'_{00}$	$5s_{11}$	$2p_1$	$3s_4$	10			
7,839.0546 S	$3p_{23}$	$3d'_{23}$	$2p_9$	$3s'''_1$	30	3.303	3.939	3.19
7,927.1177 S	$3p_{22}$	$3d'_{11}$	$2p_8$	$3s'_1$	40	.....	.....	3.48
7,936.9961 S	$3p_{22}$	$3d'_{12}$	$2p_8$	$3s''_1$	70	3.487	4.043	4.040
7,943.1814 S	$3p_{22}$	$3d'_{23}$	$2p_8$	$3s'''_1$	200	4.718	5.412	4.725
8,082.4576 B	$3s'_{01}$	$3p_{01}$	$1s_2$	$2p_{10}$	200	4.676	5.203	4.629
8,118.5492 S	$3p_{11}$	$3d'_{11}$	$2p_7$	$3s'_1$	100	4.452	5.030	4.419
8,128.9108 S	$3p_{11}$	$3d'_{12}$	$2p_7$	$3s'_1$	60	3.916	4.633	3.85
8,136.4057 S	$3p_{11}$	$3d'_{22}$	$2p_7$	$3s''''_1$	300	5.047	5.718	5.029
8,248.6824 S	$3p_{12}$	$3d'_{11}$	$2p_6$	$3s'_1$	30	3.467	4.038	3.34
8,259.3790 S	$3p_{12}$	$3d'_{12}$	$2p_6$	$3s'_1$	150	4.327	.....	4.280
8,266.0772 S	$3p_{12}$	$3d'_{23}$	$2p_6$	$3s'''_1$	250	.....	5.387	4.691
8,267.1166 S	$3p_{12}$	$3d'_{22}$	$2p_6$	$3s''''_1$	80			
8,300.3263 S	$3p_{23}$	$3d_{23}$	$2p_9$	$3d'_1$	600	5.31	5.97	5.316
8,365.7486 S	$3p_{23}$	$3d_{12}$	$2p_9$	$3d_3$	150	4.439	.....	4.415
8,377.6065 S	$3p_{23}$	$3d_{34}$	$2p_9$	$3d'_4$	800	.....	.....	5.957
8,417.1591 S	$3p_{22}$	$3d_{23}$	$2p_8$	$3d'_1$	100	4.2	4.9	
8,418.4274 S	$3p_{22}$	$3d_{22}$	$2p_8$	$3d'_1$	400	5.15	5.87	5.244
8,463.3575 S	$3p_{22}$	$3d_{11}$	$2p_8$	$3d_2$	150	4.433	5.039	4.452
8,484.4435 S	$3p_{22}$	$3d_{12}$	$2p_8$	$3d_3$	80	3.930	4.678	3.90
8,495.3598 S	$3p_{22}$	$3d_{33}$	$2p_8$	$3d_4$	500	5.703	6.324	5.764
8,544.6959 S	$3p_{22}$	$3d_{01}$	$2p_8$	$3d_5$	60	4.014	4.752	3.98
8,571.3524 S	$3p'_{11}$	$3d'_{11}$	$2p_5$	$3s'_1$	100	4.332	5.012	4.330
8,591.2587 S	$3p'_{11}$	$3d'_{22}$	$2p_5$	$3s''''_1$	400	5.436	6.057	5.450
8,634.6470 S	$3p_{11}$	$3d_{22}$	$2p_7$	$3d'_1$	600	5.3	6.0	5.386
8,647.0411 S	$3p_{12}$	$3d'_{12}$	$2p_4$	$3s''_1$	300	4.709	5.235	
8,654.3831 S	$3p'_{12}$	$3d'_{23}$	$2p_4$	$3s'''_1$	1,500	5.56	6.26	5.747
8,655.5224 S	$3p'_{12}$	$3d'_{22}$	$2p_4$	$3s''''_1$	400			
8,679.4925 S	$3p_{00}$	$3d'_{11}$	$2p_3$	$3s'_1$	500⎫	5.2	5.8	5.016
8,681.9211 S	$3p_{11}$	$3d_{11}$	$2p_7$	$3d_2$	500⎭			5.075

ATOMIC AND MOLECULAR PHYSICS

TABLE 7e-2. THE SPECTRUM OF NEON I (*Continued*)

Wavelength	Classification				I_0	$\log I_1$	$\log I_2$	$\log I_3$
	System.		Paschen					
8,704.1116 S	$3p_{11}$	$3d_{12}$	$2p_7$	$3d_3$	200	4.243	4.992	4.201
8,771.6563 S	$3p'_{01}$	$3d'_{11}$	$2p_2$	$3s'_1$	400	4.845	5.467	4.888
8,780.6210 S	$3p_{12}$	$3d_{23}$	$2p_6$	$3d'_1$	1,200	.....	.....	5.642
8,783.7533 S	$3p'_{01}$	$3d'_{23}$	$2p_2$	$3s'''_1$	1,000	.....	.....	5.488
8,830.9072 S	$3p_{12}$	$3d_{11}$	$2p_6$	$3d_2$	50	3.606	4.258	3.61
8,853.8669 S	$3p_{12}$	$3d_{12}$	$2p_6$	$3d_3$	700	5.233	5.805	5.246
8,865.3060 S	$3p_{12}$	$3d_{33}$	$2p_6$	$3d_4$	100⎫	5.0	5.6	5.0
8,865.7552 S	$3p_{01}$	$4s'_{01}$	$2p_{10}$	$2s_2$	500⎭			
8,919.5007 S	$3p_{12}$	$3d_{01}$	$2p_6$	$3d_5$	300	4.623	5.290	4.624
8,988.57	$3p_{01}$	$4s'_{00}$	$2p_{10}$	$2s_3$	200	4.310	4.712	4.12
9,148.672 S	$3p'_{11}$	$3d_{22}$	$2p_5$	$3d''_1$	600	4.809	5.501	4.808
9,201.759 S	$3p'_{11}$	$3d_{11}$	$2p_5$	$3d_2$	600	4.786	5.381	4.826
9,220.058 S	$3p'_{12}$	$3d_{23}$	$2p_4$	$3d'_1$	400	4.54	5.23	4.624
9,221.580 S	$3p'_{12}$	$3d_{22}$	$2p_4$	$3d''_1$	200	4.0	4.7	
9,226.690 S	$3p'_{11}$	$3d_{12}$	$2p_5$	$3d_3$	200	4.040	4.785	4.01
9,275.520 S	$3p'_{12}$	$3d_{11}$	$2p_4$	$3d_2$	100	.....	4.466	3.83
9,300.853 S	$3p'_{12}$	$3d_{12}$	$2p_4$	$3d_3$	600	4.650	5.261	4.639
9,310.584 S	$3p'_{11}$	$3d_{00}$	$2p_5$	$3d_6$	150	4.213	4.966	3.60
9,313.973 S	$3p'_{12}$	$3d_{33}$	$2p_4$	$3d_4$	300	4.224	4.947	4.23
9,326.507 S	$3p_{00}$	$3d_{11}$	$2p_3$	$3d_2$	600	4.682	5.285	4.710
9,373.308 S	$3p'_{12}$	$3d_{01}$	$2p_4$	$3d_5$	200	4.008	4.712	3.96
9,425.379 S	$3p_{00}$	$3d_{01}$	$2p_3$	$3d_5$	500	4.472	5.225	4.47
9,459.210 S	$3p'_{01}$	$3d_{12}$	$2p_2$	$3d_3$	300	4.211	4.969	4.15
9,486.68	$3p_{01}$	$4s_{11}$	$2p_{10}$	$2s_4$	500	4.793	5.280	4.76
9,534.163 S	$3p'_{01}$	$3d_{01}$	$2p_2$	$3d_5$	500	4.555	5.319	4.567
9,547.405 S	$3p'_{01}$	$3d_{00}$	$2p_2$	$3d_6$	300	4.241	4.986	4.15
9,665.424 S	$3p_{01}$	$4s_{12}$	$2p_{10}$	$2s_5$	1,000	5.207	5.552	5.155

TABLE 7e-2. THE SPECTRUM OF NEON I (*Continued*)

Wavelength	Classification				I_0
	System.		Paschen		
10,295.417	$3p_{22}$	$4s'_{01}$	$2p_8$	$2s_2$	80
562.408	$3p'_{00}$	$3d'_{11}$	$2p_1$	$3s'_1$	200
620.664	$3p_{11}$	$4s_{01}$	$2p'_7$	$2s_2$	40
798.07	$3p_{11}$	$4s_{00}$	$2p_7$	$2s_3$	150
844.477	$3p_{12}$	$4s_{01}$	$2p_6$	$2s_2$	200
11,143.02	$3p_{22}$	$4s_{11}$	$2p_8$	$2s_4$	300
177.533	$3p_{22}$	$4s_{12}$	$2p_9$	$2s_5$	300
390.439	$3p_{22}$	$4s_{12}$	$2p_8$	$2s_5$	110
409.134	$3p'_{11}$	$4s_{01}$	$2p_5$	$2s_2$	100
522.745	$3p'_{12}$	$4s'_{01}$	$2p_4$	$2s_2$	150
11,525.02	$3p_{11}$	$4s_{11}$	$2p_7$	$2s_4$	90
536.345	$3p'_{00}$	$3d_{11}$	$2p$	$3d_2$	50
601.536	$3p_{00}$	$4s'_{01}$	$2p_3$	$2s_2$	25
614.11	$3p'_{11}$	$4s_{00}$	$2p_5$	$2s_3$	80
688.002	$3p'_{00}$	$3d_{01}$	$2p_1$	$3d_5$	10
11,766.792	$3p'_{01}$	$4s'_{01}$	$2p_2$	$2s_2$	60
789.05	$3p_{12}$	$4s_{11}$	$2p_6$	$2s_4$	50
789.895	$3p_{11}$	$4s_{12}$	$2p_7$	$2s_5$	10
984.94	$3p'_{01}$	$4s_{00}$	$2p_2$	$2s_3$	10
12,066.340	$3p_{12}$	$4s_{12}$	$2p_6$	$2s_5$	15
12,459.39	$3p'_{11}$	$4s_{11}$	$2p_5$	$2s_4$	2
595.01	$3p'_{12}$	$4s_{11}$	$2p_4$	$2s_4$	
689.21	$3p_{00}$	$4s_{11}$	$2p_3$	$2s_4$	1
769.532	$3p'_{11}$	$4s_{12}$	$2p_5$	$2s_5$	
887.16	$3p'_{01}$	$4s_{11}$	$2p_2$	$2s_4$	
12,912.021	$3p'_{12}$	$4s_{12}$	$2p_4$	$2s_5$	
13,219.248	$3p'_{01}$	$4s_{12}$	$2p_2$	$2s_5$	
15,230.713	$3p'_{00}$	$4s'_{01}$	$2p_1$	$2s_2$	
17,161.94	$3p'_{00}$	$4s_{11}$	$2p_1$	$2s_4$	

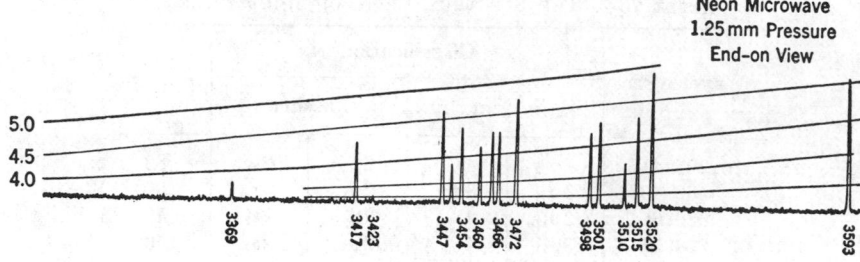

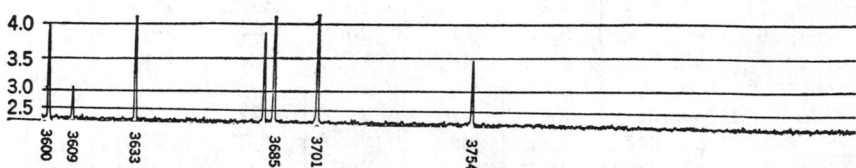

FIG. 7e-1. Photoelectric traces of the neon spectrum, microwave discharge at 1.25 mm. Wavelength range is 3,000 to 10,000 Å.

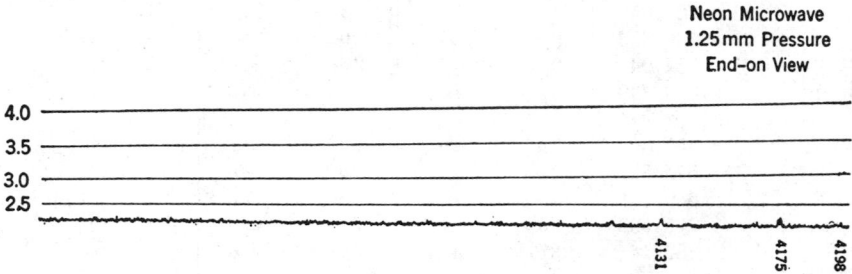

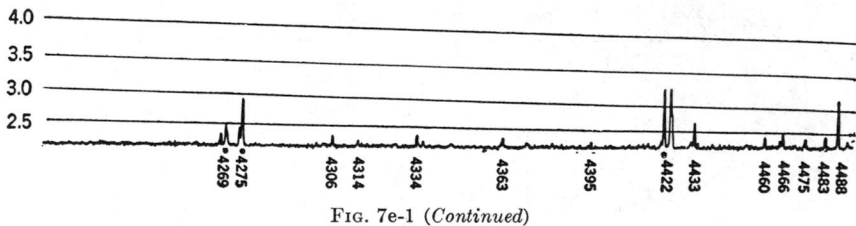

FIG. 7e-1 (Continued)

Neon Microwave
1.25 mm Pressure
End-on View

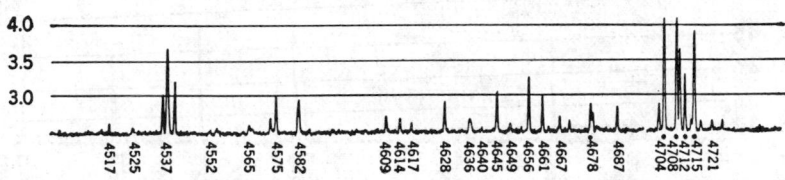

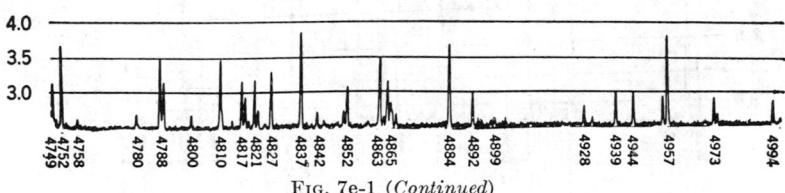

FIG. 7e-1 (Continued)

Neon Microwave
1.25 mm Pressure
End-on View

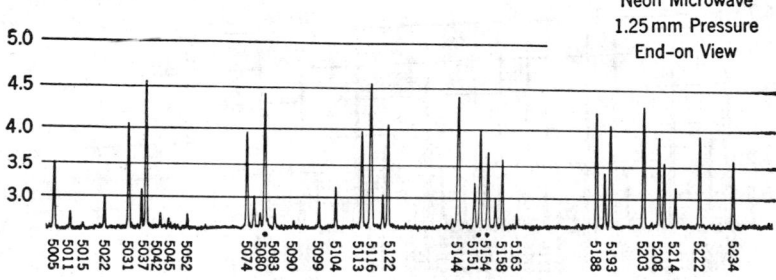

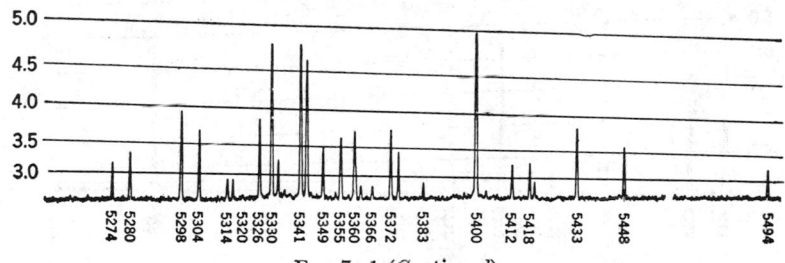

FIG. 7e-1 (Continued)

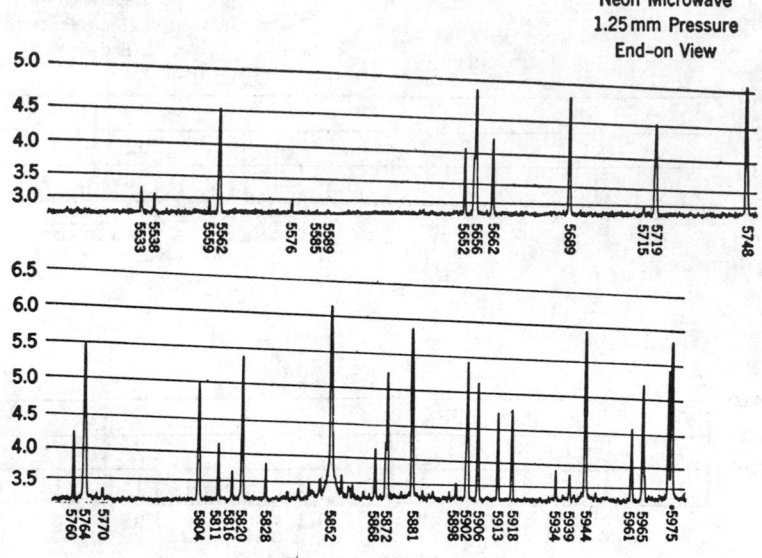

FIG. 7e-1 (*Continued*)

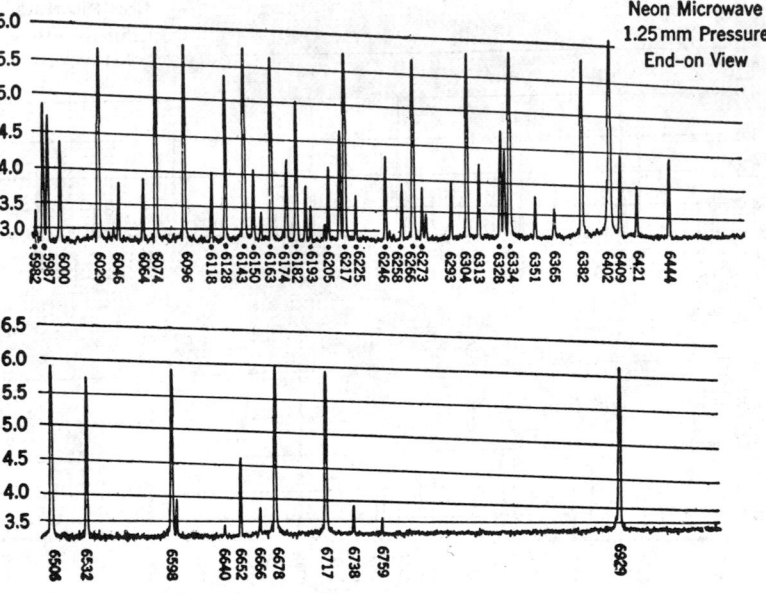

FIG. 7e-1 (*Continued*)

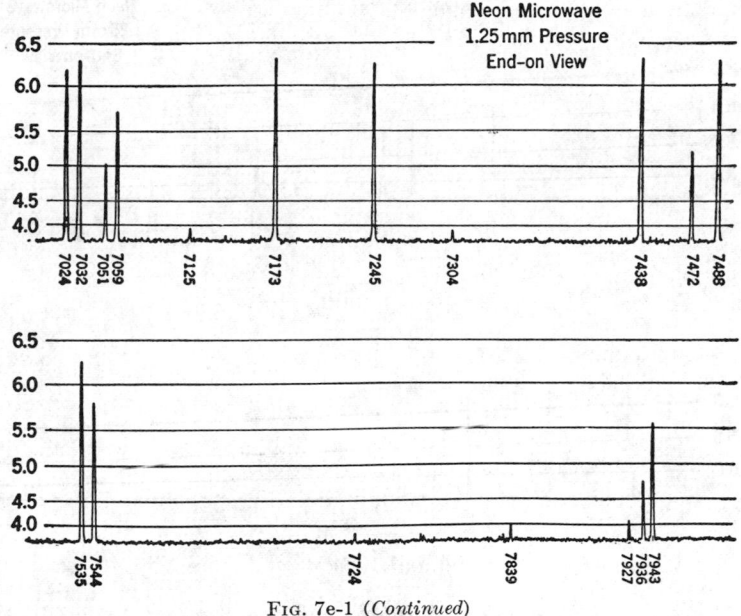

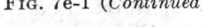

FIG. 7e-1 (*Continued*)

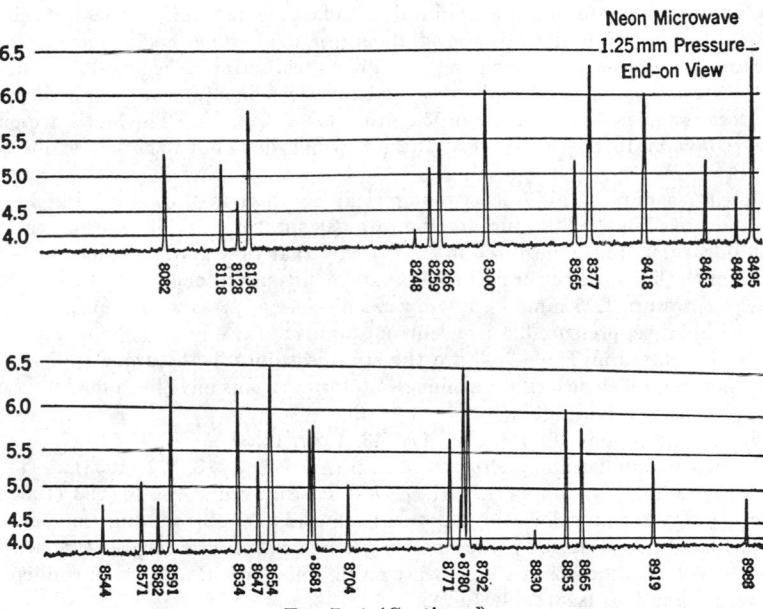

FIG. 7e-1 (*Continued*)

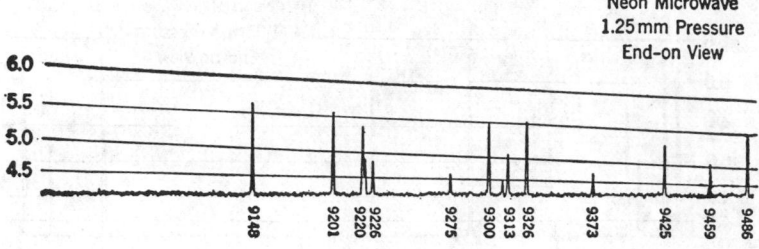

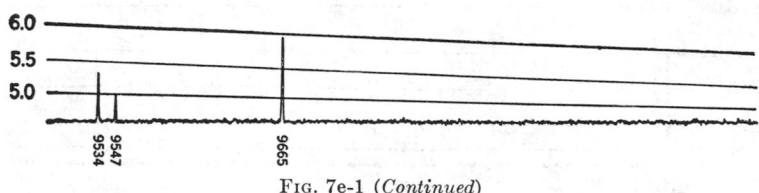

Fig. 7e-1 (*Continued*)

M, Meggers and Humphreys, *J. Research Natl. Bur. Standards* **13,** 293 (1934)
S, International secondary standard[1]

The *classification* is expressed in two notations:

Systematic (Modified Racah). Orbital angular momentum of the last electron (valence electron) is specified by the symbols *s, p, d,* etc. (*not* the angular momentum of the *configuration* as in *L, S* coupling). The first subscript is the angular momentum *K* of the atom exclusive of the spin of the valence electron minus $\frac{1}{2}$. The second index is the total angular momentum *J* of the atom ($J = K \pm \frac{1}{2}$). The levels are primed if they converge to the $^2P_\frac{1}{2}$ level of the ion which lies above the lowest ionization limit $^2P_\frac{3}{2}$.

Paschen Notation. This is a semiempirical notation first used by Paschen and extensively used in the literature for the rare-gas spectra. It is now obsolete.

The *intensities* are standardized in such a way that they give the energy flux from 100 cm² of the light source per unit solid angle in ergs per second. I_1, glow discharge, 60 cycles, pressure 1.25 mm; I_2, microwave discharge; pressure 10 mm; I_3, hollow-cathode discharge, pressure 3.5 mm, current 90 ma.

Argon I. Listed in Table 7e-3 are the strongest lines in the argon spectrum and some others for which accurate wavelength determinations have been made. Letters indicate origin of wavelengths:

B, Burns and Adams, *J. Opt. Soc. Am.* **43,** 1020 (1953)
L, Littlefield and Turnbull, *Proc. Roy. Soc. (London)* **A218,** 577 (1953)
M, Meggers and Humphreys, *J. Research Natl. Bur. Standards* **13,** 293 (1934)

There are systematic deviations between the wavelengths of different observers, and care should be exercised if the lines are to be used as wavelength standards.

COLUMNS 2 TO 5: Classification, systematic (modified Racah) and conventional Paschen designations (see Table 7e-2).

COLUMNS 6 AND 7: Intensities (logarithmic scale): I_1, intensity in 60-cycle a-c glow discharge; current 60 ma, argon pressure 3 mm; I_2, hollow-cathode discharge with iron electrodes, current 150 ma, argon pressure 1 mm.

[1] *Trans. Intern. Astron. Union* **5,** 86 (1935).

TABLE 7e-3. THE SPECTRUM OF ARGON I

λ	Classification				Intensities	
	System.		Paschen		$\log I_1$	$\log I_2$
3,319.3446 B	$4s_{12}$	$7p_{12}$	$1s_5$	$5p_6$		
3,373.4823 B	$4s_{11}$	$7p_{00}$	$1s_4$	$5p_5$		
3,554.3048 L	$4s_{12}$	$6p_{12}$	$1s_5$	$4p_6$		
3,567.6550 L	$4s_{12}$	$6p_{23}$	$1s_5$	$4p_9$		
3,572.2960 B	$4s'_{01}$	$7p_{00}$	$1s_2$	$5p_5$		
3,606.5207 L	$4s_{11}$	$6p_{00}$	$1s_4$	$4p_5$		
3,649.8310 L	$4s'_{01}$	$6p'_{00}$	$1s_2$	$4p_1$		
3,834.6775 L	$4s'_{01}$	$6p_{00}$	$1s_2$	$4p_5$	2.18	
3,894.6609 L	$4s'_{01}$	$6p_{01}$	$1s_2$	$4p_{10}$	1.75	
3,947.5046 L	$4s_{12}$	$5p'_{12}$	$1s_5$	$3p_2$	1.54	
3,948.9785 L	$4s_{12}$	$5p'_{01}$	$1s_5$	$3p_2$	3.09	2.65
4,044.4176 L	$4s_{11}$	$5p'_{12}$	$1s_4$	$3p_2$	3.16	
4,045.9645 L	$4s_{11}$	$5p'_{01}$	$1s_4$	$3p_2$	2.17	
4,054.5259 L	$4s_{11}$	$5p'_{11}$	$1s_4$	$3p_4$	1.92	
4,158.5906 L	$4s_{12}$	$5p_{12}$	$1s_5$	$3p_6$	3.80	3.56
4,164.1794 L	$4s_{12}$	$5p_{11}$	$1s_5$	$3p_7$	3.03	2.62
4,181.8833 L	$4s'_{00}$	$5p'_{01}$	$1s_3$	$3p_2$	3.13	2.56
4,190.7126 L	$4s_{12}$	$5p_{22}$	$1s_5$	$3p_8$	....	3.11
4,191.0292 L	$4s'_{00}$	$5p'_{11}$	$1s_3$	$3p_4$		
4,198.3174 L	$4s_{11}$	$5p_{00}$	$1s_4$	$3p_5$	3.53	
4,200.6745 L	$4s_{12}$	$5p_{23}$	$1s_5$	$3p_9$	3.83	
4,251.1848 L	$4s_{12}$	$5p_{01}$	$1s_5$	$3p_{10}$	2.73	
4,259.3615 L	$4s'_{01}$	$5p'_{00}$	$1s_2$	$3p_1$	3.40	
4,266.2865 L	$4s_{11}$	$5p_{12}$	$1s_4$	$3p_6$	3.29	3.11
4,272.1688 L	$4s_{11}$	$5p_{11}$	$1s_4$	$3p_7$	3.54	
4,300.1005 L	$4s_{11}$	$5p_{00}$	$1s_4$	$3p_5$	3.40	
4,333.5611 L	$4s'_{01}$	$5p'_{12}$	$1s_2$	$3p_2$	3.32	3.00
4,335.3374 L	$4s'_{01}$	$5p'_{01}$	$1s_2$	$3p_2$	2.95	2.52
4,345.1679 L	$4s'_{01}$	$5p'_{11}$	$1s_2$	$3p_4$	2.91	2.59
4,363.7944 L	$4s_{11}$	$5p_{01}$	$1s_4$	$3p_{10}$	1.89	2.30
4,510.7332 L	$4s'_{01}$	$5p_{00}$	$1s_2$	$3p_5$	3.13	2.92
4,522.3231 L	$4s'_{00}$	$5p_{01}$	$1s_3$	$3p_{10}$	2.62	2.19
4,596.0963 L	$4s'_{01}$	$5p_{11}$	$1s_2$	$3p_7$	2.65	2.20
4,628.4406 L	$4s'_{01}$	$5p_{22}$	$1s_2$	$3p_8$	2.42	
4,702.3160 L	$4s'_{01}$	$5p_{01}$	$1s_2$	$3p_{10}$	2.74	2.27
4,768.6750 B	$4p_{01}$	$6d'_{12}$	$2p_{10}$	$6s''_1$	1.63	
4,876.2610 L	$4p_{01}$	$7d_{12}$	$2p_{10}$	$7d_3$	1.80	
4,887.9478 B	$4p_{01}$	$7d_{01}$	$2p_{10}$	$7d_5$	1.77	
5,060.0793 B	$4p_{23}$	$8d_{34}$	$2p_9$	$8d_4$	1.65	
5,151.3943 B	$4p_{01}$	$6d_{00}$	$2p_{10}$	$6d_6$	2.00	

TABLE 7e-3. THE SPECTRUM OF ARGON I (*Continued*)

λ	Classification				Intensities	
	System.		Paschen		$\log I_1$	$\log I_2$
5,162.2847 L	$4p_{01}$	$6d_{01}$	$2p_{10}$	$6d_5$	2.47	
5,187.7467 L	$4p_{01}$	$5d'_{12}$	$2p_{10}$	$5s''_1$	2.53	2.01
5,221.2690 L	$4p_{23}$	$7d_{34}$	$2p_9$	$7d'_4$	2.17	
5,252.7857 L	$4p_{22}$	$7d_{33}$	$2p_8$	$7d_4$	1.85	
5,373.4951 B	$4p_{11}$	$7d_{22}$	$2p_7$	$7d''_1$	1.45	
5,410.4750 B	$4p_{12}$	$7d_{23}$	$2p_6$	$7d'_1$	2.49	
5,421.3492 L	$4p_{23}$	$8s_{12}$	$2p_9$	$5s_5$	2.00	
5,439.9903 B	$4p_{01}$	$7s_{11}$	$2p_{10}$	$4s_4$	1.67	
5,451.6506 L	$4p_{01}$	$7s_{12}$	$2p_{10}$	$4s_5$	2.42	2.00
5,457.4158 B	$4p_{22}$	$8s_{11}$	$2p_8$	$5s_4$	1.09	
5,467.1626 B	$4p_{22}$	$8s_{12}$	$2p_8$	$5s_5$	1.28	
5,473.455 B	$4p_{22}$	$7s'_{01}$	$2p_8$	$4s_2$	1.45	
5,495.8728 L	$4p_{23}$	$6d_{34}$	$2p_9$	$6d'_4$	2.72	2.39
5,506.1105 L	$4p_{22}$	$6d_{33}$	$2p_8$	$6d_4$	2.00	1.98
5,524.9576 L	$4p_{23}$	$5d'_{23}$	$2p_9$	$5s'''_1$	1.70	1.43
5,558.7015 L	$4p_{01}$	$5d_{12}$	$2p_{10}$	$5d_3$	2.84	2.48
5,572.5406 L	$4p_{22}$	$5d'_{23}$	$2p_8$	$5s'''_1$	2.35	2.09
5,588.7213 B	$4p_{22}$	$5d'_{22}$	$2p_8$	$5s''''_1$	1.55	
5,597.4783 B	$4p'_{12}$	$6d'_{23}$	$2p_3$	$6s'''_1$	1.58	
5,606.7328 L	$4p_{01}$	$5d_{01}$	$2p_{10}$	$5d_5$	2.84	2.56
5,650.7042 L	$4p_{01}$	$5d_{00}$	$2p_{10}$	$5d_6$	2.54	2.21
5,659.1278 B	$4p_{12}$	$8s_{12}$	$2p_6$	$5s_5$	1.61	
5,681.8976 L	$4p_{12}$	$6d_{23}$	$2p_6$	$6d'_1$	1.78	1.43
5,739.5191 L	$4p_{11}$	$5d'_{22}$	$2p_7$	$5s'''_1$	2.25	1.93
5,772.1143 L	$4p_{12}$	$5d'_{23}$	$2p_6$	$5s'''_1$	1.83	1.71
5,802.0802 L	$4p_{12}$	$6d_{01}$	$2p_6$	$6d_5$	1.69	
5,834.2640 L	$4p_{12}$	$5d'_{12}$	$2p_6$	$5s''_1$	2.01	1.75
5,860.3098 L	$4p_{01}$	$6s'_{01}$	$2p_{10}$	$3s_2$	2.19	2.05
5,882.6245 L	$4p_{01}$	$6s'_{00}$	$2p_{10}$	$3s_3$	2.41	1.98
5,888.5830 L	$4p_{23}$	$7s_{12}$	$2p_9$	$4s_5$	2.78	2.34
5,912.0848 L	$4p_{01}$	$4d'_{11}$	$2p_{10}$	$4s'_1$	2.82	2.62
5,928.8119 L	$4p_{22}$	$7s_{11}$	$2p_8$	$4s_4$	2.43	2.17
5,942.6676 L	$4p_{22}$	$7s_{12}$	$2p_8$	$4s_5$	1.96	1.84
5,987.3027 B	$4p_{23}$	$5d_{33}$	$2p_9$	$5d_4$	2.10	1.75
5,999.0004 B	$4p_{22}$	$5d_{22}$	$2p_8$	$5d''_1$	1.90	
6,005.7246 B	$4p'_{12}$	$8s_{11}$	$2p_3$	$5s_4$	1.33	
6,013.6790 B	$4p_{23}$	$5d_{12}$	$2p_9$	$5d_3$	1.75	
6,025.1515 B	$4p'_{12}$	$7s'_{01}$	$2p_3$	$4s_2$	1.97	
6,032.1273 L	$4p_{23}$	$5d_{34}$	$2p_9$	$5d'_4$	3.33	2.91
6,043.2232 L	$4p_{22}$	$5d_{33}$	$2p_8$	$5d_4$	2.88	2.46

TABLE 7e-3. THE SPECTRUM OF ARGON I (*Continued*)

λ	Classification				Intensities	
	System.		Paschen		$\log I_1$	$\log I_2$
6,052.7230 L	$4p_{01}$	$4d'_{22}$	$2p_{10}$	$4s_1''''$	2.28	1.84
6,059.3723 L	$4p_{01}$	$4d'_{12}$	$2p_{10}$	$4s_1''$	2.59	2.25
6,098.8046 B	$4p_{11}$	$7s_{11}$	$2p_7$	$4s_4$	2.10	2.05
6,105.6346 L	$4p'_{11}$	$5d'_{22}$	$2p_4$	$5s_1''''$	2.28	2.81
6,145.4406 L	$4p'_{12}$	$4d'_{23}$	$2p_3$	$5s_1'''$	2.25	1.93
6,155.2393 B	$4p_{12}$	$7s_{11}$	$2p_6$	$4s_4$	1.93	
	$4p'_{11}$	$5d'_{12}$	$2p_4$	$5s_1''$		
6,170.1734 L	$4p_{12}$	$7s_{12}$	$2p_6$	$4s_5$	2.25	
6,173.0949 L	$4p_{11}$	$5d_{22}$	$2p_7$	$5d_1''$	2.30	2.71
6,212.5015 L	$4p_{12}$	$5d_{23}$	$2p_6$	$5d_1'$	2.26	1.97
6,215.9423 B	$4p'_{12}$	$5d'_{12}$	$2p_3$	$5s_1''$	2.01	
6,296.8739 L	$4p'_{01}$	$5d'_{12}$	$2p_2$	$5s_1''$	2.18	
6,307.6561 L	$4p_{12}$	$5d_{12}$	$2p_6$	$5d_3$	2.36	2.09
6,364.8940 L	$4p_{11}$	$5d_{00}$	$2p_7$	$5d_6$	1.75	
6,369.5756 L	$4p_{12}$	$5d_{01}$	$2p_6$	$5d_5$	2.05	
6,384.7160 L	$4p_{01}$	$6s_{11}$	$2p_{10}$	$3s_4$	2.60	2.34
6,416.3064 L	$4p_{01}$	$6s_{12}$	$2p_{10}$	$3s_5$	3.36	2.87
6,431.5553 L	$4p_{22}$	$6s'_{01}$	$2p_8$	$3s_2$	1.60	
6,466.5498 L	$4p_{00}$	$5d_{11}$	$2p_5$	$5d_2$	1.64	
6,538.1118 L	$4p_{23}$	$4d'_{23}$	$2p_9$	$4s_1'''$	2.18	
6,604.8542 B	$4p_{22}$	$4d'_{23}$	$2p_8$	$4s_1'''$	2.43	
6,660.6784 B	$4p_{11}$	$6s'_{01}$	$2p_7$	$3s_3$	2 12	
6,664.0533 B	$4p_{22}$	$4d'_{12}$	$2p_8$	$4s_1''''$	2.16	
6,677.2812 B	$4s_{11}$	$4p'_{00}$	$1s_4$	$2p_1$	3.40	3.01
6,698.8752 B	$4p_{12}$	$6s'_{01}$	$2p_6$	$3s_2$	1.97	
6,719.2193 B	$4p_{00}$	$5d_{01}$	$2p_5$	$5d_5$	1.92	
6,752.8347 B	$4p_{01}$	$4d_{12}$	$2p_{10}$	$4d_3$	3.60	3.26
6,766.6134 B	$4p_{12}$	$4d'_{11}$	$2p_6$	$4s_1'$	2.27	
6,827.2529 B	$4p'_{12}$	$5d_{01}$	$2p_3$	$5d_5$	1.89	
6,871.2898 B	$4p_{01}$	$4d_{01}$	$2p_{10}$	$4d_5$	3.53	3.26
6,888.1704 B	$4p_{11}$	$4d'_{12}$	$2p_7$	$4s_1''$	2.45	
6,937.6658 B	$4p_{01}$	$4d_{00}$	$2p_{10}$	$4d_6$	3.15	2.86
6,965.4304 B	$4s_{12}$	$4p'_{01}$	$1s_5$	$2p_2$	5.06	4.75
7,030.2519 B	$4p_{23}$	$6s_{12}$	$2p_9$	$3s_5$	3.57	3.19
7,067.2175 B	$4s_{12}$	$4p'_{12}$	$1s_5$	$2p_3$	5.01	4.75
7,107.4777 B	$4p_{22}$	$6s_{12}$	$2p_8$	$3s_5$	2.79	
7,125.825 B	$4p'_{11}$	$6s'_{01}$	$2p_4$	$3s_2$	2.47	
7,147.0408 B	$4s_{12}$	$4p'_{11}$	$1s_5$	$2p_4$	4.42	3.83
7,206.9812 B	$4p'_{12}$	$6s'_{01}$	$2p_3$	$3s_2$	2.93	
7,272.9349 B	$4s_{11}$	$4p'_{01}$	$1s_4$	$2p_2$	4.71	4.23
7,311.724 B	$4p_{11}$	$6s_{11}$	$2p_7$	$3s_4$	2.89	

TABLE 7e-3. THE SPECTRUM OF ARGON I (*Continued*)

λ	Classification				Intensities	
	System.		Paschen		$\log I_1$	$\log I_2$
7,353.316	$4p_{22}$	$4d_{33}$	$2p_8$	$4d_4$	3.32	
7,372.1189 B	$4p_{23}$	$4d_{34}$	$2p_9$	$4d_4'$	3.76	3.44
7,383.9796 B	$4s_{11}$	$4p_{12}'$	$1s_4$	$2p_3$	5.02	5.03
7,412.334 B	$4p_{11}'$	$4d_{22}'$	$2p_4$	$4s_1''''$	2.55	
7,425.290 B	$4p_{12}'$	$4d_{23}'$	$2p_3$	$4s_1'''$	2.48	
7,471.1676 B	$4s_{11}$	$4p_{11}'$	$1s_4$	$2p_4$	2.86	
7,503.8685 B	$4s_{01}'$	$4p_{00}'$	$1s_2$	$2p_1$	5.35	5.28
7,514.6514 B	$4s_{11}$	$4p_{00}$	$1s_4$	$2p_5$	5.22	5.07
7,635.1056 B	$4s_{12}$	$4p_{12}$	$1s_5$	$2p_6$	5.53	5.36
7,723.7599 B	$4s_{12}$	$4p_{11}$	$1s_5$	$2p_7$	5.44	5.19
7,891.0777 B	$4p_{12}$	$4d_{12}$	$2p_6$	$4d_3$	3.60	
7,948.1755 B	$4s_{00}'$	$4p_{11}'$	$1s_3$	$2p_4$	5.13	5.13
8,006.1566 B	$4s_{11}$	$4p_{12}$	$1s_4$	$2p_6$	5.23	5.06
8,014.7853 B	$4s_{12}$	$4p_{22}$	$1s_5$	$2p_8$	5.30	5.29
8,103.6920 B	$4s_{11}$	$4p_{11}$	$1s_4$	$2p_7$	5.31	5.30
8,115.3108 B	$4s_{12}$	$4p_{23}$	$1s_5$	$2p_9$	5.58	5.59
8,264.5221 B	$4s_{01}'$	$4p_{01}'$	$1s_2$	$2p_2$	5.28	5.07
8,408.2094 B	$4s_{01}'$	$4p_{12}'$	$1s_2$	$2p_3$	5.36	5.35
8,424.6473 B	$4s_{11}$	$4p_{22}$	$1s_4$	$2p_8$	5.35	5.48
8,521.4428 B	$4s_{01}'$	$4p_{11}'$	$1s_2$	$2p_4$	5.18	5.09
8,605.7790 B	$4p_{12}'$	$4d_{12}$	$2p_3$	$4d_3$		
8,620.4602 B	$4p_{00}$	$4d_{01}$	$2p_5$	$4d_5$		
8,667.9438 B	$4s_{00}'$	$4p_{11}$	$1s_3$	$2p_7$	4.52	4.64
8,761.6907 B	$4p_{01}'$	$4d_{12}$	$2p_2$	$4d_3$		
8,799.082 B	$4p_{12}'$	$4d_{01}$	$2p_3$	$4d_5$		
9,122.9660 B	$4s_{12}$	$4p_{01}$	$1s_5$	$2p_{10}$	....	5.58
9,194.636 B	$4p_{01}$	$5s_{00}'$	$2p_{10}$	$2s_2$		
9,224.4955 B	$4s_{01}'$	$4p_{12}$	$1s_2$	$2p_6$	....	5.19
9,354.218 M	$4s_{01}'$	$4p_{11}$	$1s_2$	$2p_7$	....	4.18
9,657.7841 M	$4s_{11}$	$4p_{01}$	$1s_4$	$2p_{10}$	....	5.36
9,784.5010 M	$4s_{01}'$	$4p_{22}$	$1s_2$	$2p_8$	....	4.72
10,470.051 M	$4s_{00}'$	$4p_{01}$	$1s_3$	$2p_{10}$		

TABLE 7e-3. THE SPECTRUM OF ARGON I (*Continued*)
Vacuum Argon Wavelengths in the Near Infrared*

λ	Classification				I_0
	System.		Paschen		
10,676.489 H	$4p_{01}$	$5s_{12}$	$2p_{10}$	$2s_5$	500
684.698 H	$4p_{23}$	$3d'_{23}$	$2p_9$	$3s'''_1$	200
11,081.901 H	$4p_{22}$	$3d'_{22}$	$2p_8$	$3s''''_1$	200
671.903 H	$4p_{12}$	$3d'_{12}$	$2p_6$	$3s''_1$	100
12,115.639 H	$4p_{23}$	$3d_{23}$	$2p_9$	$3d'_1$	300
12,346.770 H	$4p_{22}$	$3d_{23}$	$2p_8$	$3d'_1$	150
406.2184 R	$4p_{11}$	$3d_{11}$	$2p_7$	$3d_2$	400
442.724 H	$4p_{01}$	$3d_{12}$	$2p_{10}$	$3d_3$	500
459.523 H	$4p_{22}$	$5s_{11}$	$2p_8$	$2s_4$	400
491.0793 R	$4p_{23}$	$5s_{12}$	$2p_5$	$2s_5$	700
12,705.755 H	$4p'_{01}$	$3d'_{11}$	$2p_2$	$3s'_1$	150
806.2474 R	$4p_{22}$	$3d_{22}$	$2p_8$	$3d''_1$	300
960.2029 R	$4p_{01}$	$3d_{01}$	$2p_{10}$	$3d_5$	250
13,011.8209 R	$4p'_{12}$	$5s'_{01}$	$2p_3$	$2s_2$	200
217.606 H	$4p_{01}$	$3d_{00}$	$2p_{10}$	$3d_6$	150
13,231.727 H	$4p_{23}$	$3d_{33}$	$2p_9$	$3d_4$	200
276.2656 R	$4p'_{12}$	$3d'_{23}$	$2p_3$	$3s'''_1$	750
316.8552 R	$4p'_{11}$	$3d'_{22}$	$2p_4$	$3s''''_1$	600
370.7679 R	$4p_{12}$	$3d_{23}$	$2p_6$	$3d'_1$	800
507.8818 R	$4p_{22}$	$3d_{33}$	$2p_8$	$3d_4$	850
13,603.051 H	$4p'_{12}$	$3d'_{22}$	$2p_3$	$3s''''_1$	55
626.3909 R	$4p_{11}$	$3d_{22}$	$2p_7$	$3d''_1$	500
682.2918 R	$4p'_{01}$	$3d'_{12}$	$2p_2$	$3s''_1$	300
722.3286 R	$4p_{23}$	$3d_{34}$	$2p_9$	$3d'_4$	1000
16,945.2129 R	$4p_{12}$	$3d_{12}$	$2p_6$	$3d_3$	100

* From Report of Commission 14 of the International Union, December, 1960. H measured by Humphreys and Paul, *J. phys.* **19,** 424 (1958); R measured by Littlefield and Rowley in the above-mentioned report.

Argon Microwave
6.5 mm Pressure
End-on View

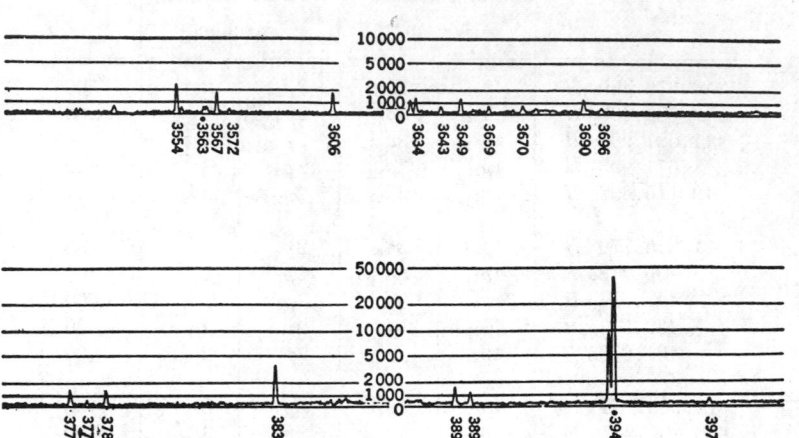

FIG. 7e-2. Photoelectric traces of the argon spectrum, microwave discharge at 6.5 mm pressure.   Wavelength range is 3,500 to 10,000 Å.

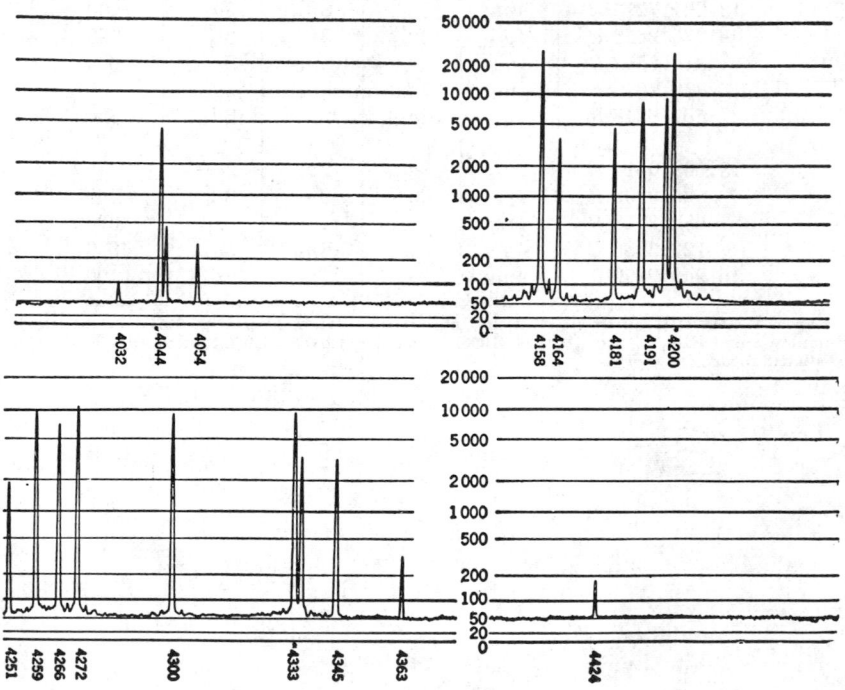

FIG. 7e-2 (*Continued*)

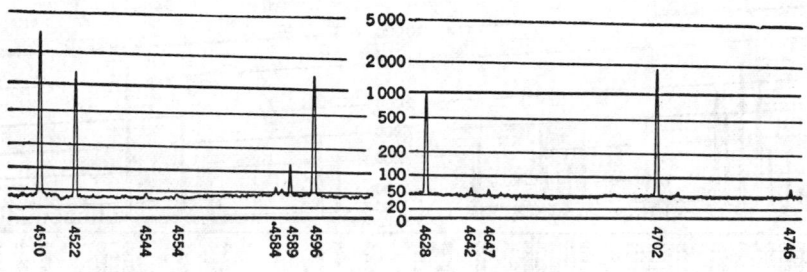

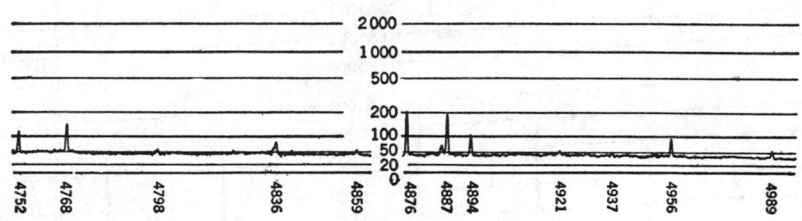

FIG. 7e-2 (*Continued*)

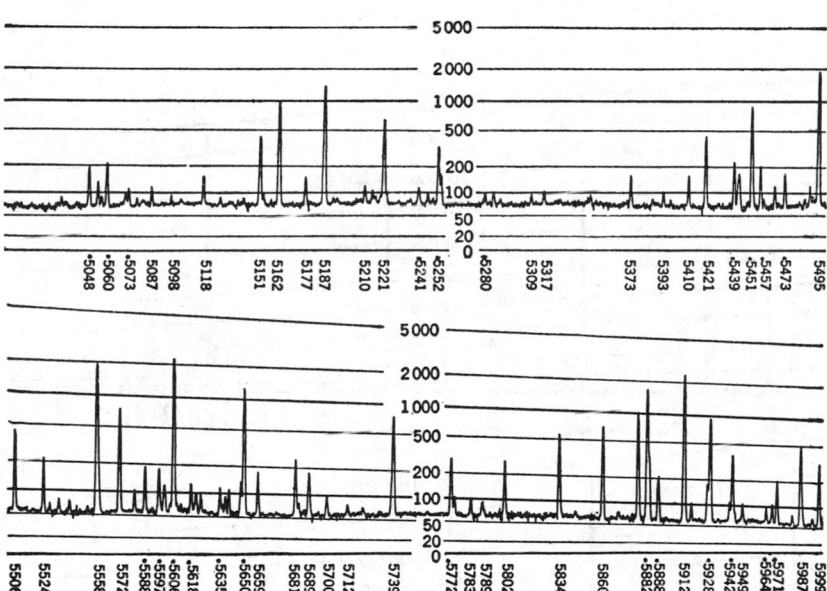

FIG. 7e-2 (*Continued*)

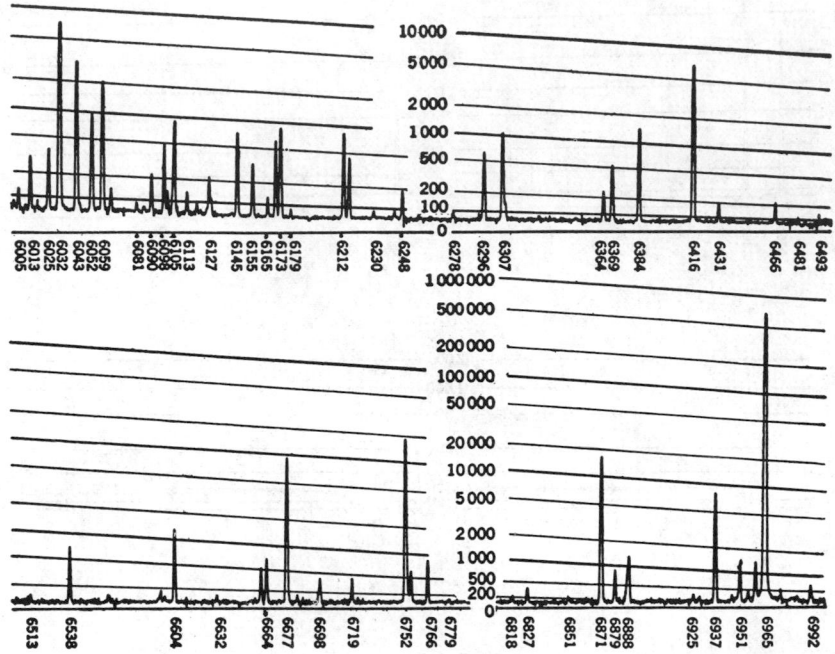

FIG. 7e-2 (*Continued*)

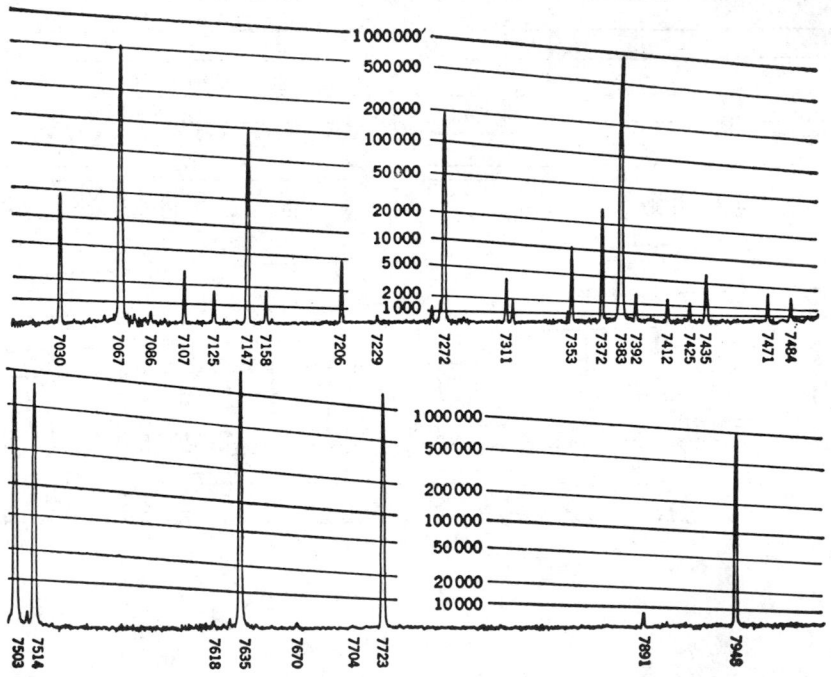

FIG. 7e-2 (*Continued*)

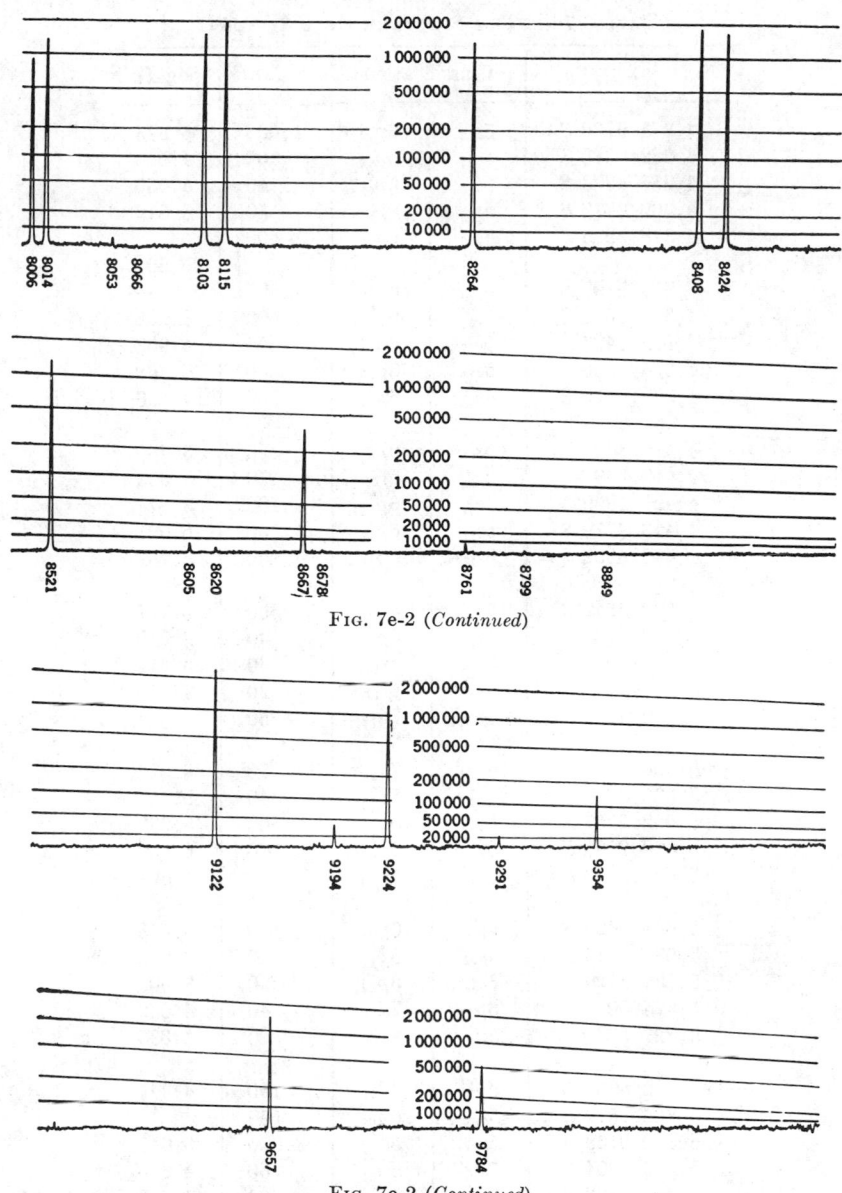

FIG. 7e-2 (*Continued*)

FIG. 7e-2 (*Continued*)

Krypton I. Notation and arrangement of Table 7e-4 are similar to those of the Tables 7e-2 and 7e-3.

Wavelengths

All values given to 8 significant figures are interferometric values.

    S, International secondary standard[1]

    L, Littlefield, *Proc. Roy. Soc.* (*London*), ser. A, **187**, 220 (1946)

[1] *Trans. Intern. Astron. Union* **5**, 87 (1935).

TABLE 7e-4. THE SPECTRUM OF KRYPTON I

λ	Classification		I_0	$\log I_1$
4,273.9700 S	$5s_{12}$	$6p_{12}$	1,000	5.573
4,282.9683 S	$5s_{12}$	$6p_{11}$	100	4.540
4,286.4873 S	$5s'_{00}$	$6p'_{01}$	40	4.039
4,300.4877 S	$5s'_{00}$	$6p'_{11}$	50	3.812
4,318.5525 S	$5s_{12}$	$6p_{22}$	400⎫	
				5.66
4,319.5797 S	$5s_{12}$	$6p_{23}$	1,000⎭	
4,351.3607 S	$5s'_{01}$	$6p'_{00}$	100	3.938
4,362.6423 S	$5s_{12}$	$6p_{01}$	500	4.958
4,376.1220 S	$5s_{11}$	$6p_{00}$	800	5.208
4,399.9670 S	$5s'_{01}$	$6p'_{12}$	200	4.430
4,410.369	$5s'_{01}$	$6p'_{01}$	50	3.440
4,418.769	$5s'_{01}$	$5f_{22}$	50	3.391
4,425.1909	$5s'_{01}$	$6p'_{11}$	100	3.874
4,453.9179 S	$5s_{11}$	$6p_{12}$	600	5.027
4,463.6902 S	$5s_{11}$	$6p_{11}$	800	5.252
4,502.3547 S	$5s_{11}$	$6p_{22}$	600	5.117
4,550.298	$5s_{11}$	$6p_{01}$	40	3.210
4,812.607	$5s'_{00}$	$4f_{11}$	40	3.611
4,969.08	$5s'_{01}$	$4f_{12}$	20	3.560
5,490.94	$5p_{01}$	$7d_{12}$	50	3.903
5,500.71	$5p_{01}$	$7d_{01}$	50	3.924
5,520.52	$5p_{23}$	$8d_{34}$	40	3.757
5,562.2257 S	$5s_{12}$	$5p'_{12}$	500	5.338
5,570.2895 S	$5s_{12}$	$5p'_{01}$	2,000	5.937
5,580.3890 L	$5s'_{01}$	$6p_{00}$	80	4.399
5,649.5629 S	$5s'_{00}$	$6p_{01}$	100	4.518
5,672.4514 L	$5s_{12}$	$5p'_{11}$	50	3.993
5,707.5128 L	$5s'_{01}$	$6p_{12}$	40	3.800
5,824.50	$5p_{22}$	$7d_{33}$	40	4.032
5,827.07	$5p_{01}$	$8s_{12}$	20	3.833
5,832.8600 L	$5p_{23}$	$7d_{34}$	100	4.345
5,866.7514 L	$5s'_{01}$	$6p_{01}$	50	
5,870.9158 S	$5s_{11}$	$5p'_{12}$	3,000	6.040
5,879.9004 L	$5s_{11}$	$5p'_{01}$	50	4.696
5,993.8503 S	$5s_{11}$	$5p'_{11}$	60	4.618
6,012.1570 L	$5p_{01}$	$6d_{12}$	50	4.550
	$5p_{12}$	$9s_{12}$		
6,035.82	$5p_{11}$	$7d_{22}$	15	3.707
6,056.1274 L *	$5p_{01}$	$6d_{01}$	60	4.617
6,075.24	$5p_{12}$	$7d_{23}$	20	3.780
6,082.8630 L	$5p_{01}$	$6d_{00}$	40	4.292

* The vacuum wavelength 6,057.80211 Å of this line of Kr^{86} defines the international standard of length.

TABLE 7e-4. THE SPECTRUM OF KRYPTON I (*Continued*)

λ	Classification		I_0	log I_1
6,151.38	$5p_{12}$	$7d_{12}$	20	3.798
6,222.71	$5p_{22}$	$8s_{11}$	20	3.865
6,236.3520 L	$5p_{23}$	$8s_{12}$	30	4.140
6,346.66	$5p_{23}$	$6d_{23}$	20	3.795
6,373.58	$5p_{22}$	$6d_{22}$	30	4.027
6,421.0285 L	$5p_{22}$	$6d_{33}$	100	4.900
6,456.2910 L	$5p_{23}$	$6d_{34}$	200	5.103
6,576.42	$5p_{12}$	$8s_{12}$	20	3.799
6,652.24	$5p_{11}$	$6d_{22}$	40	4.351
6,699.23	$5p_{12}$	$6d_{23}$	60	4.474
6,740.10	$5p_{11}$	$6d_{12}$	20	3.75
6,813.10	$5p_{12}$	$6d_{12}$	50	4.466
6,846.40	$5p_{01}$	$7s_{11}$	20	3.83
6,869.63	$5p_{12}$	$6d_{01}$	20	4.025
6,904.68	$5p_{01}$	$7s_{12}$	100	5.029
7,224.109	$5p_{01}$	$5d_{12}$	100	5.090
7,287.262	$5p_{01}$	$6s'_{01}$	80	4.966
7,425.54	$5p_{22}$	$7s_{11}$	60	4.707
7,486.850	$5p_{01}$	$6s'_{00}$ ⎫	100	5.119
	$5p_{23}$	$7s_{12}$ ⎭		
7,493.58	$5p_{22}$	$5d_{11}$	20⎫	
				4.692
7,494.15	$5p_{22}$	$7s_{12}$	30⎭	
7,587.4135	$5s_{11}$	$5p_{00}$	1,000	6.357
7,601.5465	$5s_{12}$	$5p_{12}$	2,000	6.908
7,685.2472	$5s'_{01}$	$5p'_{00}$	1,000	6.369
7,694.5401	$5s_{12}$	$5p_{11}$	1,200	6.507
7,741.39	$5p_{23}$	$5d_{23}$	40	4.340
7,746.831	$5p_{01}$	$5d_{00}$	150	5.317
7,776.28	$5p_{22}$	$5d_{22}$	40	4.509
7,806.52	$5p_{11}$	$7s_{11}$	50	4.536
7,854.823	$5s'_{00}$	$5p'_{01}$	800	6.448
7,863.91	$5p_{23}$	$5d_{12}$	20	4.250
7,881.76	$5p_{11}$	$5d_{11}$	30	4.318
7,904.62	$5p_{12}$	$7s_{11}$	30	4.17
7,913.443	$5p_{01}$	$5d_{01}$	200	5.536
7,920.47	$5p_{23}$	$5d_{33}$	40	4.38
7,928.602	$5p_{22}$	$5d_{33}$	180	5.458
7,946.99	$5p_{22}$	$6s'_{01}$	20	4.05
7,982.42	$5p_{12}$	$7s_{12}$	100	4.826
8,059.5053	$5s'_{00}$	$5p'_{11}$	1,500	6.422
8,104.3660	$5s_{12}$	$5p_{22}$	4,000	6.813

ATOMIC AND MOLECULAR PHYSICS

TABLE 7e-4. THE SPECTRUM OF KRYPTON I (*Continued*)

λ	Classification		I_0	$\log I_1$
8,112.9023	$5s_{12}$	$5p_{23}$	6,000	6.994
8,190.0570	$5s_{11}$	$5p_{12}$	3,000	6.682
8,218.40	$4d_{12}$	$6f_{22}$	80	3.99
8,263.2412	$5s'_{01}$	$5p'_{12}$	3,000	6.764
8,272.36	$5p_{12}$	$5d_{23}$	100	5.171
8,281.05	$5s'_{01}$	$5p'_{01}$	1,500	6.450
8,298.1091	$5s_{11}$	$5p_{11}$	5,000	6.857
8,412.45	$5p_{10}$	$5d_{12}$	100	4.746
8,498.21	$5p_{10}$	$6s'_{01}$	30	4.16
8,508.8736	$5s'_{01}$	$5p'_{11}$	3,000	6.537
8,537.93	$4d_{00}$	$5f_{11}$	40	4.17
8,560.89	$5p_{00}$	$7s_{11}$	50	4.22
8,569.02	$4d_{00}$	$6p'_{11}$	20	3.85
8,605.85	$4d_{33}$	$6f_{45}$	40	4.16
8,697.50	$5p_{22}$	$5d_{01}$	40	4.341
8,755.20	$4d_{01}$	$5f_{22}$	30	4.13
8,764.09	$5p_{23}$	$4d'_{23}$	150	5.149
8,776.7498	$5s_{11}$	$5p_{22}$	6,000	6.941
8,805.78	$4d_{01}$	$6p'_{11}$	20	3.78
8,928.6934	$5s_{12}$	$5p_{01}$	2,000	6.893
8,967.53	$5p_{23}$	$4d'_{22}$	10	3.95
8,977.99	$5p_{22}$	$4d'_{22}$	50	4.925
8,999.19	$5p_{11}$	$5d_{00}$	30	4.528
9,094.33	$4d_{22}$	$6f_3$	4h	3.94
9,111.69	$5p_{23}$	$4d'_{12}$	20	4.27
9,122.49	$5p_{22}$	$4d'_{12}$	20	4.32
9,243.54	....	....	30	4.783
9,270.96	$4d_{12}$	$5f_{12}$	10	4.38
9,326.03	$4d_{34}$	$5f_3$	10	4.17
9,352.23	$4d_{34}$	$5f_4$	100	5.122
9,362.03	$5p_{12}$	$5d_{01}$	100	5.181
9,450.88	$5p_{12}$	$4d'_{23}$	20	4.44
9,540.89	$5p_{11}$	$4d'_{22}$	30	4.72
9,687.83	$5p_{12}$	$4d'_{22}$	10	4.06
9,704.22	$5p_{11}$	$4d'_{12}$	50	5.00
9,714.85	$4d_{33}$	$5f_3$	15	2.26
9,743.11	$4d_{33}$	$5f_{44}$	50	4.990
9,751.74	$5s_{11}$	$5p_{01}$	2,000	6.545
9,856.24	$5p_{12}$	$4d'_{12}$	500	5.677
11,819.43	$5p_{01}$	$6s_{12}$	2,000	

TABLE 7e-4. THE SPECTRUM OF KRYPTON I (*Continued*)

λ	Classification		I_0	log I_1
12,204.39	$4d_{34}$	$4f_{45}$	700	
12,879.00	$4d_{33}$	$4f_{44}$	500	
13,177.38	$5p_{22}$	$6s_{11}$	850	
13,622.28	$5p_{22}$	$4d_{11}$	800	
13,634.22	$5p_{23}$	$6s_{12}$	1,700	
14,426.93	$5p_{11}$	$6s_{11}$	1,100	
14,734.46	$5p_{23}$	$4d_{23}$	900	
15,239.85	$5p_{22}$	$4d_{22}$	900	
15,335.29	$5p_{01}$	$4d_{12}$	850	
16,784.65	$5p_{12}$	$4d_{23}$	950	
16,890.40	$5p_{22}$	$4d_{33}$	1,000	
16,896.58	$5p_{01}$	$4d_{01}$	700	
16,935.71	$5p_{11}$	$4d_{22}$	800	
18,167.12	$5p_{23}$	$4d_{34}$	1,500	

Wavelengths not followed by a capital letter and all I_0 values are taken from the three following sources:

4,273 to 7,601 A: Meggers, deBruin, and Humphreys, *J. Research Natl. Bur. Standards* **7**, 643 (1931)

7,685 to 9,856 A: Meggers and Humphreys, *J. Research Natl. Bur. Standards* **10**, 443 (1933)

11,819 to 18,167 A: Humphreys and Kostkowski, *J. Research Natl. Bur. Standards* **49**, 73 (1952).

I_1, intensity in a microwave discharge at 1.6 mm pressure. This is approximately the vapor pressure of krypton at the temperature of liquid nitrogen (77 K). Immersing a discharge tube with krypton at a room-temperature pressure of more than 7 mm in liquid nitrogen will keep the pressure very steady at about 1.6 mm and therefore will produce very constant intensities.

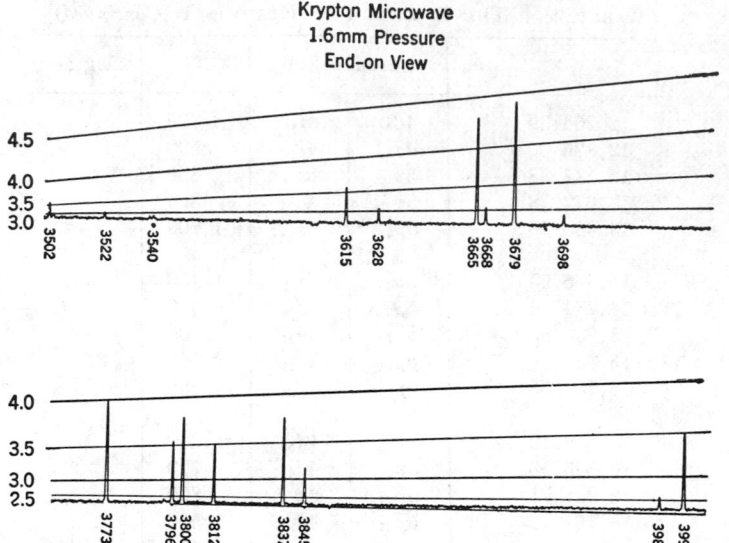

Fig. 7e-3. Photoelectric traces of the krypton spectrum, microwave discharge at 1.6 mm pressure.   Wavelength range is 3,500 to 10,000 Å.

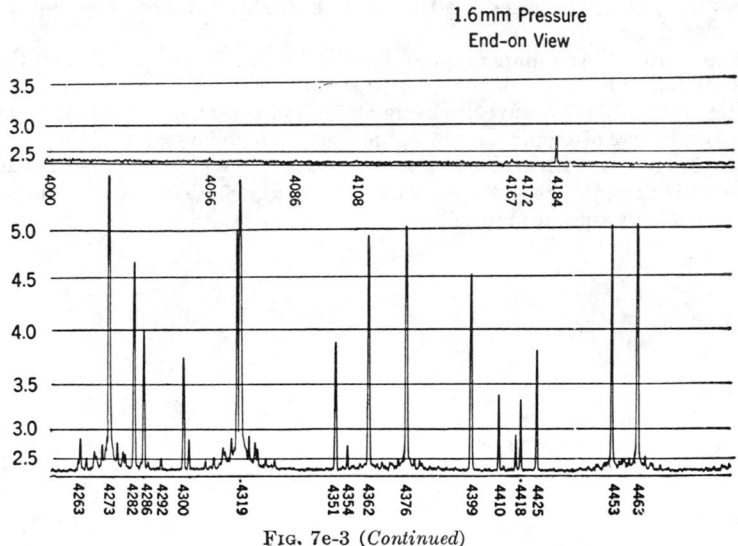

Fig. 7e-3 (*Continued*)

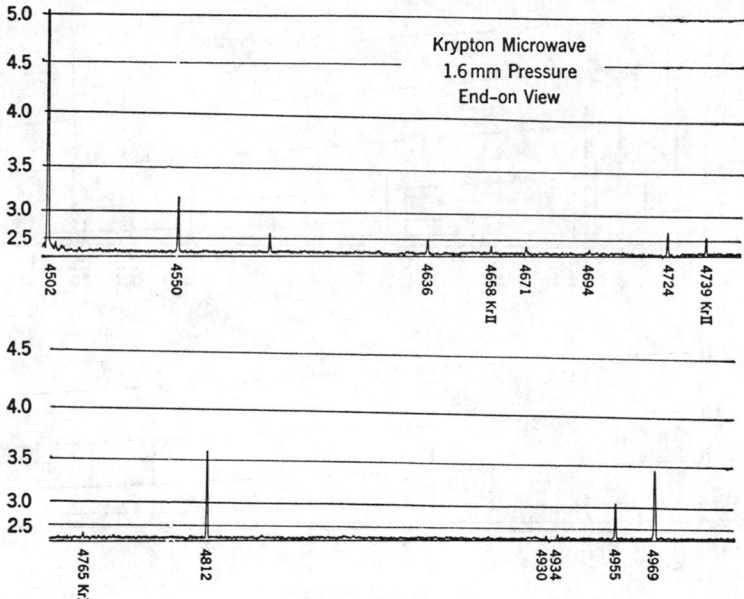

Fig. 7e-3 (*Continued*)

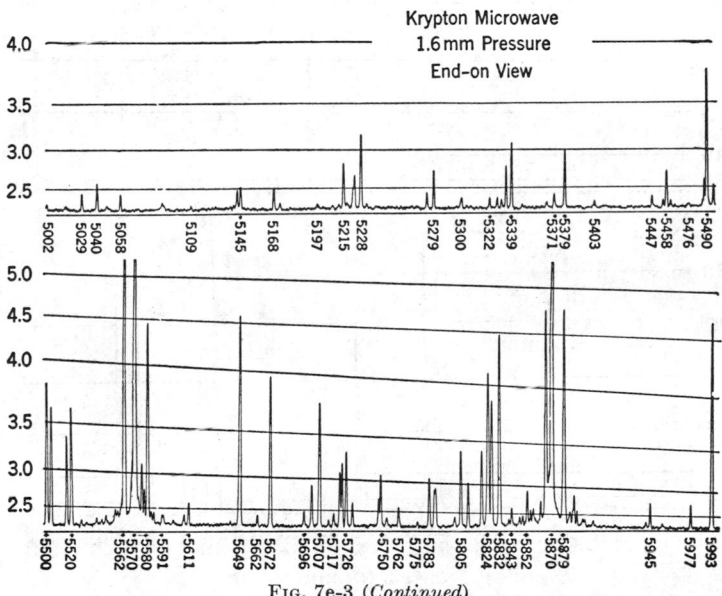

Fig. 7e-3 (*Continued*)

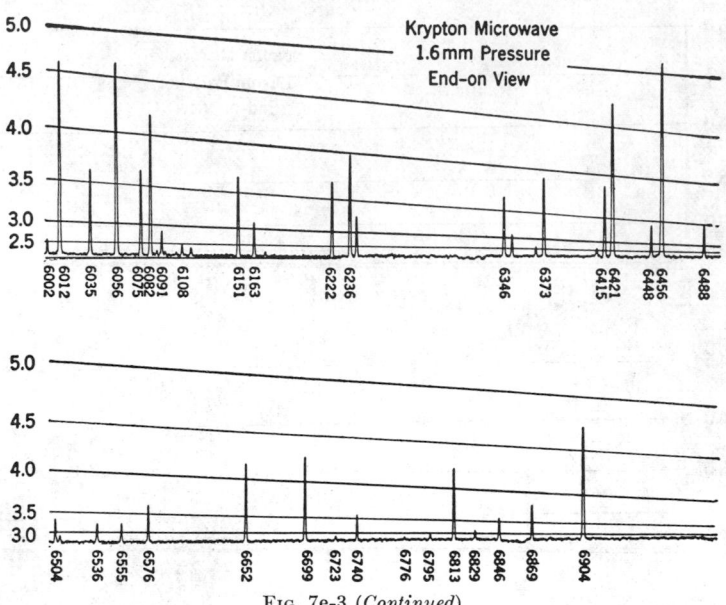

FIG. 7e-3 (*Continued*)

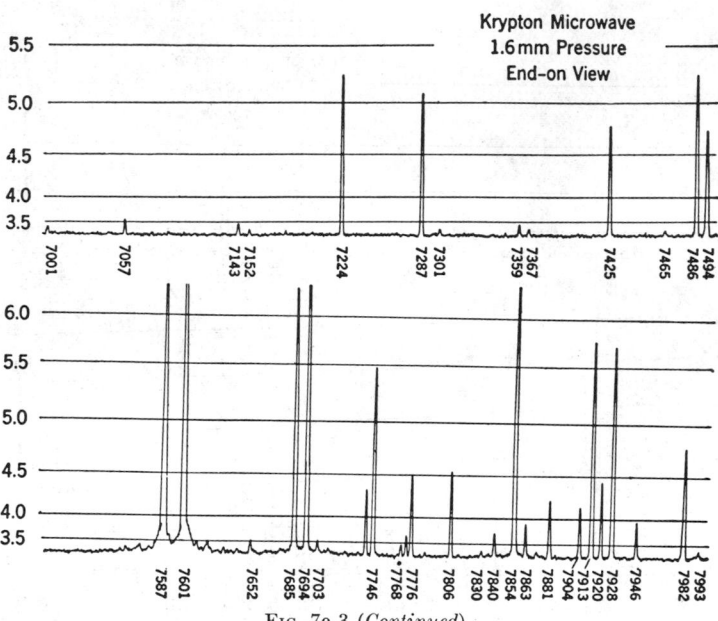

FIG. 7e-3 (*Continued*)

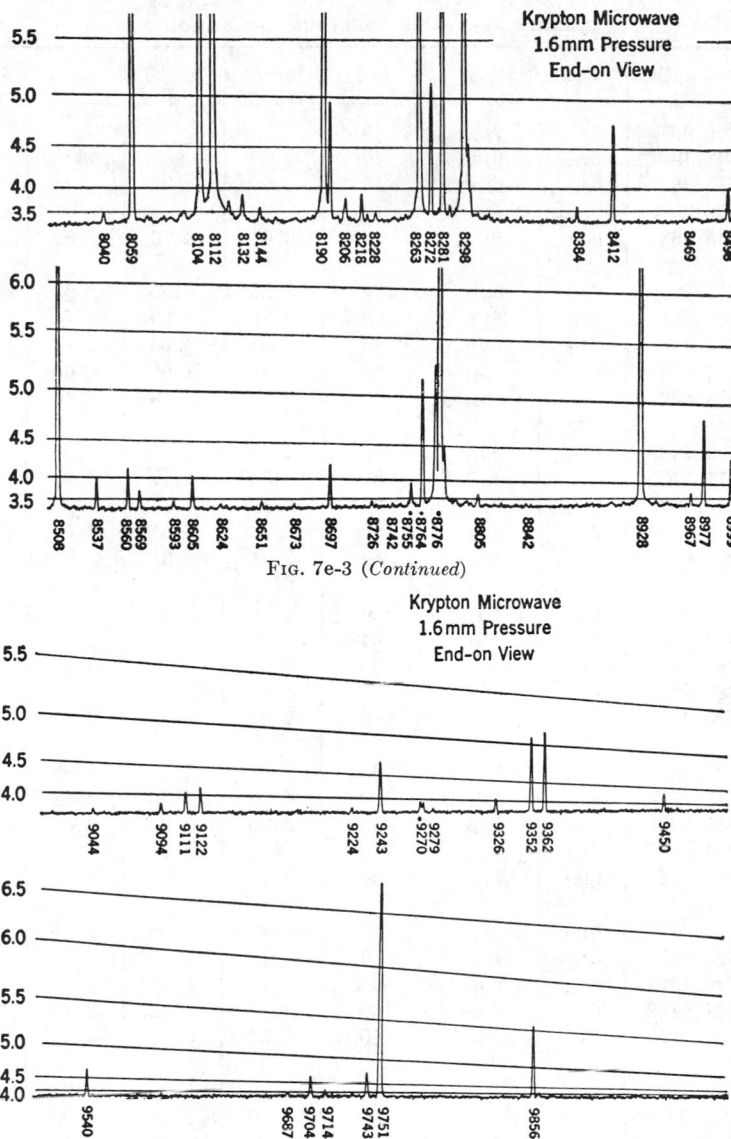

FIG. 7e-3 (Continued)

FIG. 7e-3 (Continued)

Xenon I. Wavelengths in Table 7e-5 are from Humphreys and Meggers[1] and Humphreys and Kostkowski[2] (above 11,000 Å). Notation is the same as for Ne I and A I.

Intensities are as follows: I_0, conventional estimates quoted from the literature; I_1, microwave discharge, pressure of 0.002 mm; I_2, same, $p = 0.07$ mm; I_3, same, $p = 16$ mm; I_4, d-c glow discharge, $p = 4.1$ mm.[3]

For significance of the intensity scale, see Table 7e-2.

[1] C. V. Humphreys and W. F. Meggers, *J. Research Natl. Bur. Standards* **10**, 139 (1933).
[2] Humphreys and Kostkowski, *J. Research Natl. Bur. Standards* **49**, 73 (1952).
[3] The I_1 to I_4 intensities were measured by M. Thekaekara, S.J.

TABLE 7e-5. THE SPECTRUM OF XENON I

Wavelength	Classification		I_0	$\log I_1$	$\log I_2$	$\log I_3$	$\log I_4$
3,685.90	$6s_{12}$	$9p_{12}$	40				
3,693.49	$6s_{12}$	$9p_{23}$	40				
3,745.38	$6s_{11}$	$6f_{12}$	10				
3,796.30	$6s_{12}$	$5f_{23}$	40				
3,948.163	$6s_{11}$	$5f_{12}$	60	3.06	3.70	2.89	2.32
3,950.925	$6s_{12}$	$8p_{12}$	120	3.86	4.55	3.62	3.21
3,967.541	$6s_{12}$	$8p_{23}$	200	3.94	4.66	3.74	3.34
3,974.417	$6s_{12}$	$8p_{22}$	40	3.02	3.70	2.71	2.34
3,985.202	$6s_{12}$	$8p_{01}$	30	2.91	3.60	2.65	2.26
4,078.8207	$6s_{11}$	$8p_{00}$	100	4.06	4.32	3.40	2.76
4,109.7093	$6s_{11}$	$8p_{12}$	60	3.33	4.00	3.05	2.66
4,116.1151	$6s_{11}$	$8p_{11}$	80	3.56	4.17	3.23	2.71
4,135.1337	$6s_{11}$	$8p_{22}$	20	2.66	3.31		
4,193.5296	$6s_{12}$	$4f_{23}$	150	3.62	4.51	3.54	3.25
4,203.6945	$6s_{12}$	$4f_{12}$	50	2.91	4.01		
4,205.404	$6s_{12}$	$4f_{11}$	10	....	3.02		
4,372.287	$6s_{11}$	$4f_{22}$	20				
4,383.9092	$6s_{11}$	$4f_{12}$	100	3.08	4.13	3.12	2.83
4,385.7693	$6s_{11}$	$4f_{11}$	70	....	2.80	2.82	2.55
4,500.9772	$6s_{12}$	$6p'_{01}$	500	4.06	5.13	4.23	2.98
4,524.6805	$6s_{12}$	$6p'_{12}$	400	3.97	4.85	3.96	3.64
4,582.7474	$6s_{11}$	$6p'_{00}$	300	4.16	4.66	3.68	3.42
4,611.8896	$6s_{12}$	$7p_{11}$	100	2.86	3.86	2.84	2.61
4,624.2757	$6s_{12}$	$7p_{12}$	1,000	4.76	5.61	4.72	4.44
4,671.226	$6s_{12}$	$7p_{23}$	2,000	4.98	5.81	4.99	4.70
4,690.9711	$6s_{12}$	$6p'_{11}$	100	3.29	4.46	3.43	3.25
4,697.020	$6s_{12}$	$7p_{22}$	300	4.21	5.17	4.13	3.92
4,734.1524	$6s_{11}$	$6p'_{12}$	600	4.25	5.27	4.39	4.10
4,792.6192	$6s_{12}$	$7p_{01}$	150	3.48	4.32	3.29	3.12
4,807.019	$6s_{11}$	$7p_{00}$	500	4.52	5.31	4.35	4.12
4,829.709	$6s_{11}$	$7p_{11}$	400	4.27	5.19	4.21	3.97
4,843.294	$6s_{11}$	$7p_{12}$	300	4.50	5.06	4.07	3.84
4,916.508	$6s_{11}$	$6p'_{11}$	500	4.04	5.15	4.16	3.95
4,923.1522	$6s_{11}$	$7p_{22}$	500	4.30	5.22	4.21	3.99
5,028.2796	$6s_{11}$	$7p_{01}$	200	3.54	4.52	3.42	3.25
5,162.711	$6s'_{00}$	$7f_{11}$	10	2.86	3.30	3.12	
5,362.244	$6p_{01}$	$10d_{01}$	15 }	2.97	3.24	3.20	
5,364.626	$6p_{01}$	$10d_{12}$	30 }				
5,392.795	$6s'_{00}$	$6f_{11}$	100 }	3.31	3.86	3.35	2.46
5,394.738	$6p_{01}$	$7s'_{01}$	20 }				

TABLE 7e-5. THE SPECTRUM OF XENON I (*Continued*)

Wavelength	Classification		I_0	log I_1	log I_2	log I_3	log I_4
5,439.923	$6s'_{01}$	$7f_{12}$	30	3.65	3.49	3.21	2.03
5,460.037	$6p_{01}$	$11s_{12}$	15	3.23	3.12	2.81	
5,488.555	$6p_{22}$	$11d_{33}$	20h	2.85	3.22	3.56	
5,552.385	$6p_{01}$	$9d_{12}$	80	3.32	3.78	3.48	2.42
5,566.615	$6p_{01}$	$9d_{01}$	100	3.41	3.86	3.52	2.10
5,581.784	$6p_{01}$	$9d_{00}$	50	3.13	3.53	3.52	
5,618.878	$6p_{22}$	$10d_{33}$	80	3.21	3.60	3.61	2.21
5,688.373	$6s'_{01}$	$6f_{22}$	40	2.97	3.41	2.84	
5,695.750	$6s'_{01}$	$6f_{12}$	100⎫				
5,696.479	$6s'_{01}$	$6f_{11}$	80⎭	3.61	4.06	3.50	2.62
5,715.716	$6p_{01}$	$10s_{12}$	70⎫				
5,716.252	$6p_{23}$	$10d_{34}$	80⎭	3.56	4.00	3.83	2.57
5,807.311	$6p_{22}$	$9d_{23}$	15	2.39	2.93	2.67	
5,814.505	$6p_{22}$	$9d_{22}$	60	3.16	3.58	3.31	2.16
5,823.890	$6s'_{00}$	$5f_{11}$	300	3.96	4.65	4.08	3.23
5,824.800	$6p_{22}$	$9d_{33}$	150				
5,856.509	$6p_{01}$	$8d_{22}$	15	2.61	3.21	2.81	
5,875.018	$6p_{01}$	$8d_{12}$	100	4.03	5.41	3.77	2.98
5,894.988	$6p_{01}$	$8d_{01}$	100	3.92	4.44	3.85	3.02
5,904.462	$6p_{23}$	$9d_{23}$	20	3.15	3.42	3.16	
5,922.550	$6p_{23}$	$9d_{33}$	20	3.02	3.52	3.23	
5,931.241	$6p_{01}$	$8d_{00}$	80⎫				
5,934.172	$6p_{23}$	$9d_{34}$	100⎭	3.83	4.32	4.05	2.95
5,974.152	$6p_{12}$	$10d_{23}$	40	3.50	3.42	3.57	
5,989.18	$6p_{12}$	$10d_{12}$	20	2.90	3.19	3.17	
5,998.115	$6p_{22}$	$10s_{11}$	30	3.17	3.51	3.12	
6,007.909	$6p_{22}$	$10s_{12}$	15	2.87	3.20	2.79	
6,111.759	$6p_{11}$	$9d_{22}$	30⎫				
6,111.951	$6p_{23}$	$10s_{12}$	40⎭	3.63	....	3.72	2.56
6,152.069	$6p_{22}$	$8d_{23}$	20	....	3.46		
6,163.660	$6p_{22}$	$8d_{22}$	90⎫				
6,163.935	$6s'_{01}$	$5f_{22}$	80⎭	3.95	....	3.85	3.07
6,178.302	$6s'_{01}$	$5f_{12}$	150⎫				
6,179.665	$6s'_{01}$	$5f_{11}$	120⎭	3.99	....	3.95	3.28
6,182.420	$6p_{22}$	$8d_{33}$	300	4.19	....	4.19	3.42
6,189.10	$6p_{01}$	$9s_{11}$	20	2.89	3.43	3.16	
6,198.260	$6p_{01}$	$9s_{12}$	100⎫				
6,200.890	$6p_{12}$	$9d_{23}$	60⎭	3.72	3.64	3.72	3.01
6,206.297	$6p_{22}$	$8d_{01}$	20	3.18	....	3.27	
6,224.169	$6p_{12}$	$9d_{12}$	40	....	3.67	3.39	

TABLE 7e-5. THE SPECTRUM OF XENON I (*Continued*)

Wavelength	Classification		I_0	$\log I_1$	$\log I_2$	$\log I_3$	$\log I_4$
6,261.212	$6p_{23}$	$8d_{23}$	50	3.39	4.03	3.45	
6,265.301	$6s'_{00}$	$8p_{01}$	40	3.18	3.87	2.96	
6,286.011	$5d_{34}$	$8f_{45}$	100	3.34	3.82	3.84	
6,292.649	$6p_{23}$	$8d_{33}$	50	3.43	4.06	3.47	
6,318.062	$6p_{23}$	$8d_{34}$	500	4.34	4.93	4.42	3.66
6,430.155	$6p_{12}$	$10s_{12}$	20	....	3.44		
6,469.705	$6p_{01}$	$7d_{12}$	300	4.15	4.92	4.05	3.56
6,472.841	$6p_{01}$	$7d_{11}$	150	3.92	4.57	3.70	3.20
6,487.765	$6p_{01}$	$7d_{22}$	120	3.90	4.59	3.72	3.22
6,497.43	$5d_{34}$	$7f_{33}$	30hl				
6,498.718	$6p_{11}$	$8d_{22}$	100	3.90	4.44	3.89	3.09
6,504.18	$6s'_{01}$	$8p_{00}$	200h	3.82	4.37	4.16	3.05
6,521.508	$6p_{11}$	$8d_{12}$	40	3.30	3.88	3.25	
6,533.159	$6p_{22}$	$9s_{11}$	100		4.32	3.56	
6,543.360	$6p_{22}$	$9s_{12}$	40	3.78	3.95		
6,554.196	$5d_{12}$	$7f_{23}$	50hl	3.54	4.02	3.78	
6,595.561	$6p_{12}$	$8d_{23}$	100	4.08	4.61	4.05	3.20
6,632.464	$6p_{12}$	$8d_{12}$	50	3.76	4.32	3.73	
6,666.965	$6p_{23}$	$9s_{12}$	60⎫	4.26	5.03	4.19	3.69
6,668.920	$6p_{01}$	$7d_{00}$	150⎭				
6,678.972	$6s'_{01}$	$8p_{01}$	25	3.49	4.12		
6,681.036	$5d_{00}$	$6f_{11}$	20				
6,728.008	$6p_{01}$	$7d_{01}$	200	4.48	5.22	4.34	3.85
6,777.57	$5d_{01}$	$6f_{12}$	50⎫	3.86	4.32	3.85	2.96
6,778.60	$5d_{01}$	$6f_{11}$	40⎭				
6,827.315	$6s'_{00}$	$4f_{11}$	200	3.91	4.12	4.27	3.83
6,846.613	$6p_{22}$	$7d_{12}$	60	3.95	4.72	4.03	3.45
6,866.838	$6p_{22}$	$7d_{22}$	50	3.87	4.56		
6,872.107	$5d_{34}$	$6f_{45}$	100	4.19	4.84	4.52	3.58
6,882.155	$6p_{22}$	$7d_{33}$	300	4.77	5.41	4.68	4.14
6,925.53	$5d_{12}$	$6f_{23}$	100	3.97	4.51	3.88	3.25
6,976.182	$6p_{23}$	$7d_{23}$	100	4.07	4.93	3.99	3.52
7,119.598	$6p_{23}$	$7d_{34}$	500	4.91	5.62	4.92	4.43
7,257.94	$5d_{33}$	$6f_{44}$	60	4.07	4.73	4.07	3.39
7,262.54	$6p_{11}$	$7d_{12}$	20	4.02	4.70	3.83	3.26
7,266.49	$6p_{11}$	$7d_{11}$	25	....	4.60		
7,283.961	$6s'_{01}$	$4f_{22}$	40⎫	4.61	5.33	4.50	4.00
7,285.301	$6p_{11}$	$7d_{22}$	60⎭				
7,316.272	$6s'_{01}$	$4f_{12}$	70	4.09	5.07	4.35	3.83
7,321.452	$6s'_{01}$	$4f_{11}$	80	....	5.00		

TABLE 7e-5. THE SPECTRUM OF XENON I (*Continued*)

Wavelength	Classification		I_0	$\log I_1$	$\log I_2$	$\log I_3$	$\log I_4$
7,336.480	$6p_{22}$	$5d'_{23}$	50	4.57	5.02	3.97	3.56
7,355.58	$5d_{00}$	$5f_{11}$	40	3.80	5.63	3.79	3.26
7,386.002	$6p_{01}$	$8s_{12}$	100	4.26	5.16	4.27	3.85
7,393.793	$6p_{12}$	$7d_{23}$	150	4.49	5.30	4.46	3.96
7,400.41	$6p_{12}$	$7d_{12}$	30	4.05	4.80	3.89	3.46
7,451.00	$5d_{01}$	$5f_{22}$	25	3.69	4.46	....	3.05
7,472.01	$5d_{01}$	$5f_{12}$	40	4.37	4.94	4.19	3.65
7,474.01	$5d_{01}$	$5f_{11}$	25				
7,492.23	$6p_{23}$	$5d'_{23}$	20	4.18	4.65	3.64	3.27
7,559.79	$5d_{34}$	$5f_{33}$	40	3.76	4.72	3.88	3.35
7,584.680	$5d_{34}$	$5f_{45}$	200	4.59	5.42	4.86	4.28
7,642.025	$6s'_{00}$	$6p'_{01}$	500⎱	4.98	5.92	5.36	4.88
7,643.91	$5d_{12}$	$5f_{33}$	100⎰				
7,664.56	$5d_{12}$	$5f_{12}$	30	4.26	4.83	4.00	3.47
7,740.31	$6p_{12}$	$7d_{01}$	40	3.87	4.59	3.67	3.17
7,783.66	$5d_{22}$	$6f_{33}$	50	3.90	4.55	3.84	3.17
7,802.651	$6p_{22}$	$8s_{11}$	100	4.31	5.19	4.33	3.89
7,881.320	$6p_{22}$	$8s_{12}$	100		4.73		3.45
7,887.395	$6s'_{01}$	$6p'_{00}$	300	5.20	5.66	4.90	4.45
7,937.41	$6p_{00}$	$7d_{11}$	40	3.75	4.42	3.50	3.05
7,967.341	$6s'_{00}$	$7p_{11}$	500	4.82	5.45	4.97	4.53
8,029.67	$5d_{33}$	$5f_{33}$	100	3.95	4.85	3.79	3.46
8,057.258	$5d_{33}$	$5f_{44}$	200	4.55	5.33	4.67	4.10
8,061.340	$6p_{23}$	$8s_{12}$	150	4.53	5.38	4.55	4.12
8,101.98	$5d_{23}$	$6f_{33}$	100	3.92	4.71	3.93	3.25
8,171.02	$5d_{01}$	$8p_{22}$	100	4.52	4.97	4.01	3.55
8,206.341	$6s'_{00}$	$6p'_{11}$	700	4.85	6.01	5.20	4.85
8,231.6348	$6s_{12}$	$6p_{12}$	10,000	5.66	7.16	6.87	6.37
8,266.519	$6s'_{01}$	$6p'_{01}$	500	4.75	5.93	5.20	4.72
8,280.1163	$6s_{11}$	$6p_{00}$	7,000	5.99	6.73	6.71	6.21
8,346.823	$6s'_{01}$	$6p'_{12}$	2,000	5.50	6.36	5.82	5.29
8,409.190	$6s_{12}$	$6p_{11}$	2,000	4.96	6.60	6.01	5.63
8,522.55	$6s'_{00}$	$7p_{00}$	30	3.69	4.72	3.69	3.28
8,530.10	$6p_{12}$	$8s_{11}$	30	3.79	4.74	3.83	3.39
8,576.01	$6s'_{01}$	$7p_{00}$	200	4.38	5.26	4.42	3.98
8,624.24	$6p_{12}$	$8s_{12}$	80	4.07	5.00	4.10	3.65
8,648.54	$6s'_{01}$	$7p_{11}$	250	4.65	5.56	4.77	4.32
8,692.20	$6s'_{01}$	$7p_{12}$	100⎱	4.47	5.13	4.31	3.87
8,696.86	$5d_{22}$	$5f_{33}$	200⎰		5.19	4.46	3.86
8,709.64	$5d_{22}$	$5f_{22}$	40	3.93	....	3.84	

TABLE 7e-5. THE SPECTRUM OF XENON I (*Continued*)

Wavelength	Classification		I_0	$\log I_1$	$\log I_2$	$\log I_3$	$\log I_4$
8,739.39	$6p_{01}$	$6d_{12}$	300	4.99	6.03	5.22	4.80
8,758.20	$6p_{22}$	$6d_{23}$	100	4.13	5.35	4.8	4.01
8,819.412	$6s_{12}$	$6p_{23}$	5,000	5.75	....	7.02	6.51
8,862.32	$6p_{01}$	$6d_{01}$	300	5.10	6.17	5.44	4.99
8,908.73	$6p_{01}$	$6d_{00}$	200	4.76	5.94	5.12	4.71
8,930.83	$6s'_{01}$	$6p'_{11}$	200	4.93	6.02	5.25	4.74
8,952.254	$6s_{11}$	$6p_{12}$	1,000	5.92	6.76	6.72	6.23
8,981.05	$6p_{23}$	$6d_{23}$	100	4.34	5.61	4.61	4.23
8,987.57	$6p_{22}$	$6d_{22}$	200	4.73	5.82	5.00	4.55
9,025.98	$6p_{11}$	$6d_{11}$	30	4.58	5.25	4.38	3.87
9,032.18	$5d_{00}$	$4f_{11}$	50	4.49	5.36	4.69	4.14
9,045.446	$6s_{12}$	$6p_{22}$	400	5.60	6.00	5.73	5.28
9,096.13	$5d_{23}$	$5f_{33}$	50	4.39	5.32	4.53	3.98
9,152.12	$5d_{01}$	$4f_{22}$	20	4.16	5.30		
9,162.654	$6s_{11}$	$6p_{11}$	500	5.97	6.93	6.94	6.39
9,167.52	$6p_{22}$	$6d_{33}$	100	....	6.22		
9,203.20	$5d_{01}$	$4f_{12}$	30	4.60	5.67	4.88	4.36
9,211.38	$5d_{01}$	$4f_{11}$	25	4.21	5.40	4.73	4.06
9,301.95	$5d_{34}$	$4f_{33}$	30	....	5.46	4.73	4.20
9,306.64	$6s'_{01}$	$7p_{01}$	40	4.74	5.59	4.75	4.33
9,374.76	$5d_{34}$	$4f_{45}$	100	4.86	5.66	5.61	5.08
9,412.01	$6p_{23}$	$6d_{33}$	60	4.66	5.10	5.05	4.56
9,445.34	$5d_{12}$	$4f_{23}$	80	4.81	5.86	5.31	4.77
9,497.07	$5d_{12}$	$4f_{12}$	40	4.40	5.50	4.71	4.19
9,513.379	$6p_{23}$	$6d_{34}$	200	5.48	6.30	5.91	5.41
9,585.14	$6p_{22}$	$6d_{01}$	20	3.95	....	4.27	3.77
9,685.32	$6p_{12}$	$6d_{23}$	150	5.04	6.04	5.40	4.88
9,700.99	$6p_{23}$	$6d_{12}$	20	4.14	6.00	4.31	3.82
9,718.16	$6p_{11}$	$6d_{22}$	100	5.04	6.95	5.31	4.80
9,799.699	$6s_{12}$	$6p_{01}$	2,000	5.79	6.78	7.00	6.49
9,923.192	$6s_{11}$	$6p_{22}$	3,000	6.19	....	7.03	6.51
10,023.72	$5d_{12}$	$4f_{33}$	50	4.49	....	4.85	4.39
10,107.34	$5d_{12}$	$4f_{44}$	80				
10,838.34	$6s_{11}$	$6p_{01}$	1,000				
11,742.26	$5d_{23}$	$4f_{34}$	90				
12,623.32	$6p_{01}$	$7s_{12}$	300				
13,656.48	$6p_{22}$	$7s_{11}$	150				
14,142.09	$6p_{22}$	$7s_{12}$	80				
14,732.38	$6p_{23}$	$7s_{12}$	200				
15,418.01	$6p_{11}$	$7s_{11}$	110				

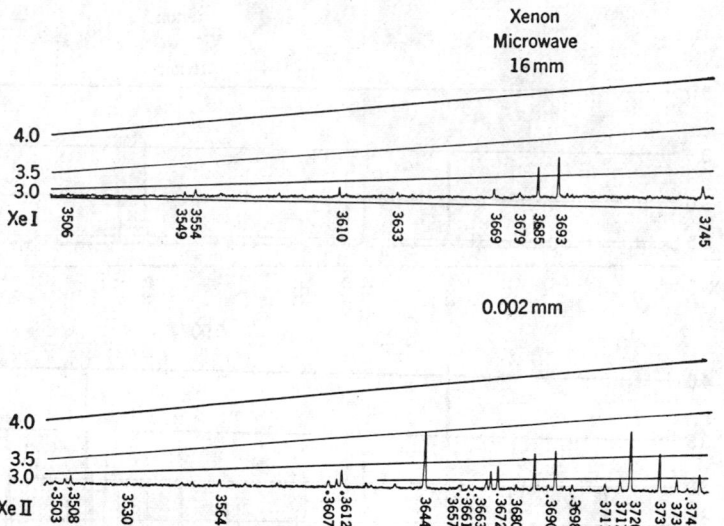

Fig. 7e-4. Photoelectric traces of the xenon spectrum, microwave discharges at 16 mm (upper traces) and 0.002 mm (lower traces). Wavelength range is 3,500 to 10,000 Å. The 16-mm trace shows the Xe I spectrum with the lines broadened. The strongest lines in the 0.002-mm trace are those for Xe II.

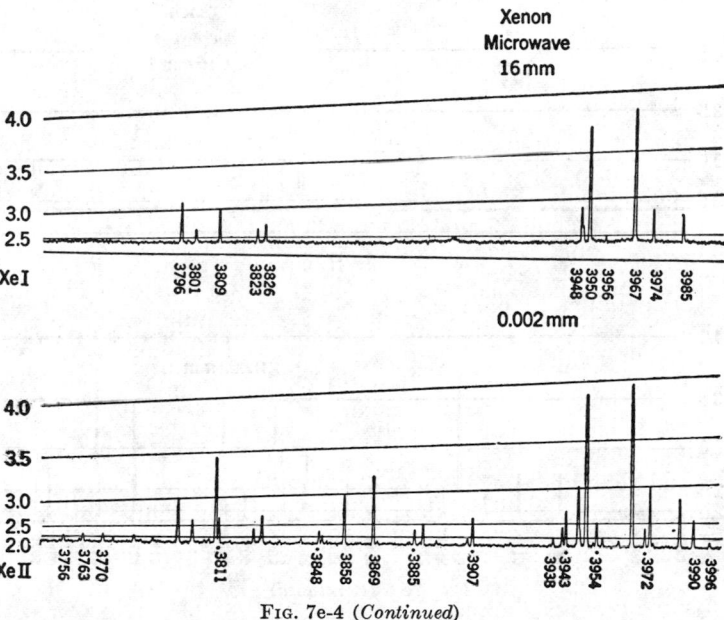

Fig. 7e-4 (Continued)

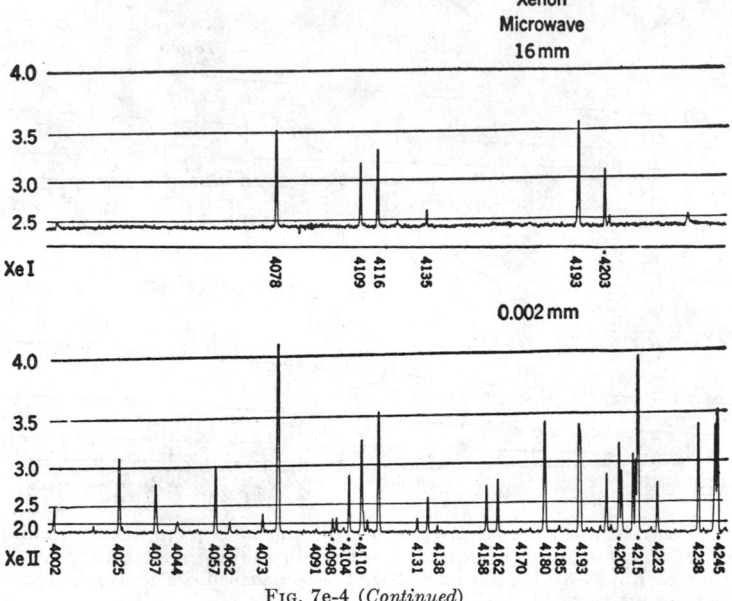

FIG. 7e-4 (*Continued*)

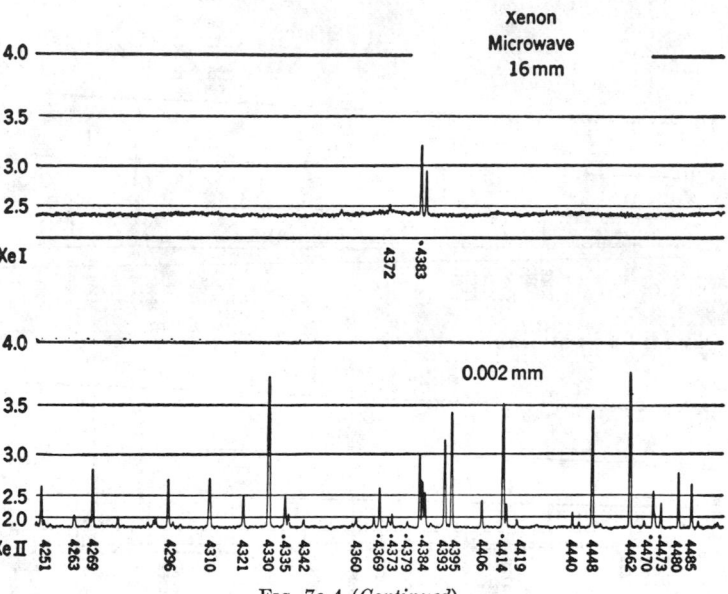

FIG. 7e-4 (*Continued*)

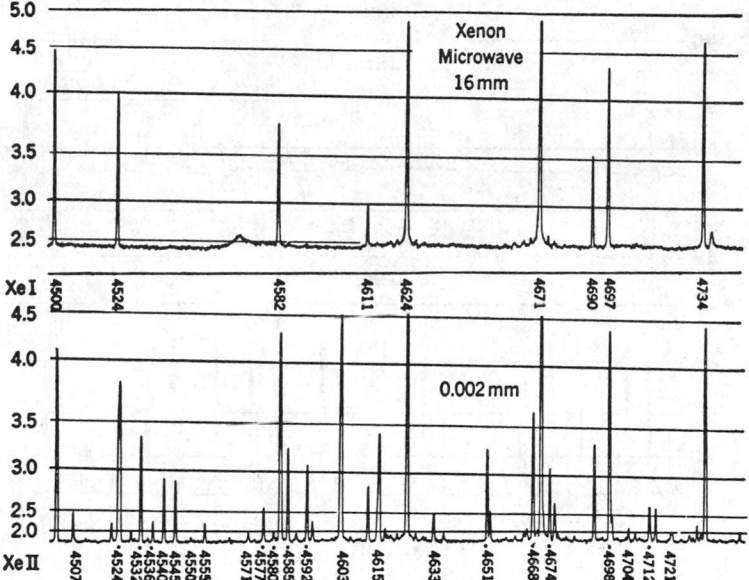

FIG. 7e-4 (*Continued*)

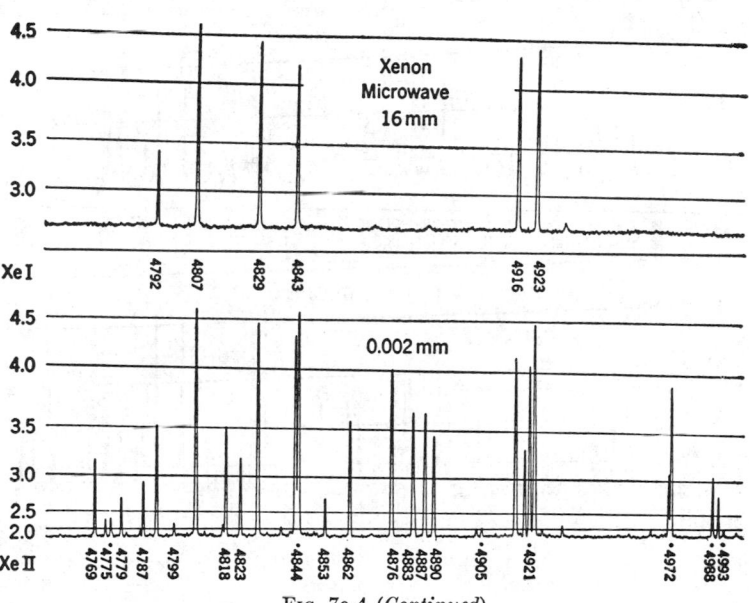

FIG. 7e-4 (*Continued*)

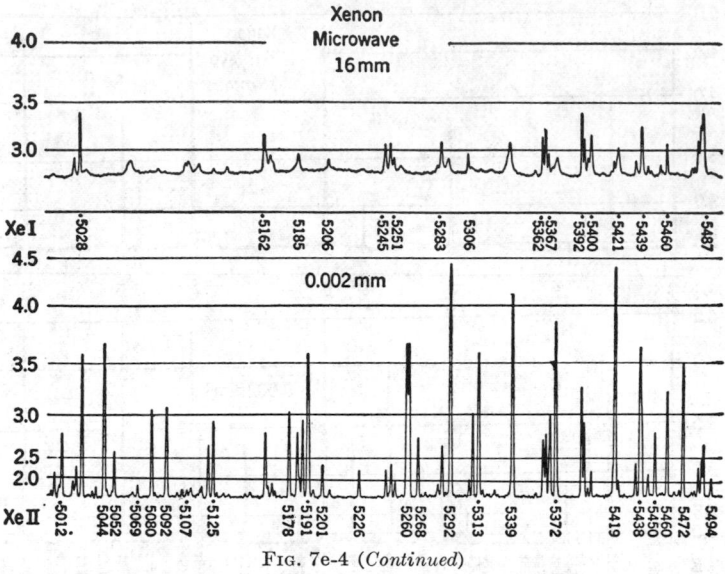

FIG. 7e-4 (Continued)

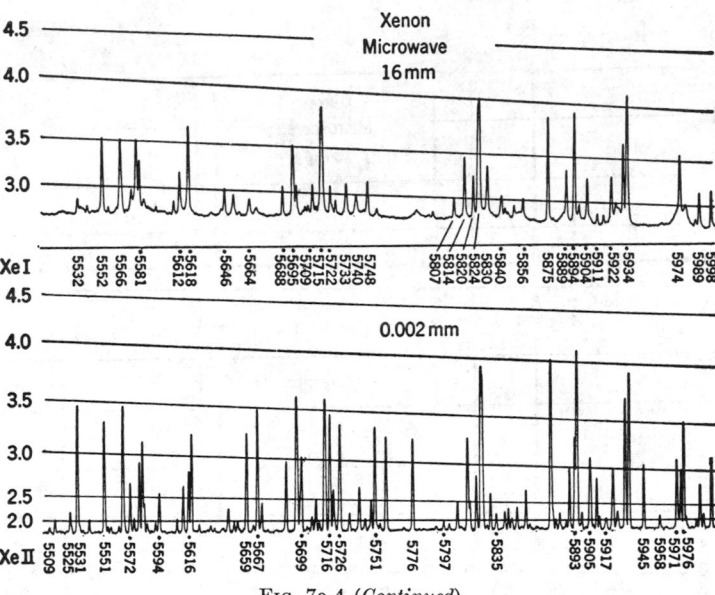

FIG. 7e-4 (Continued)

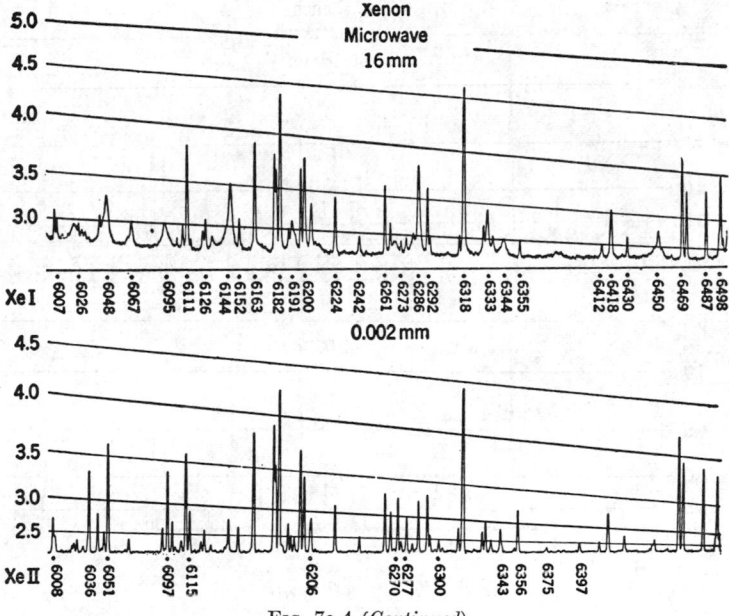

FIG. 7e-4 (Continued)

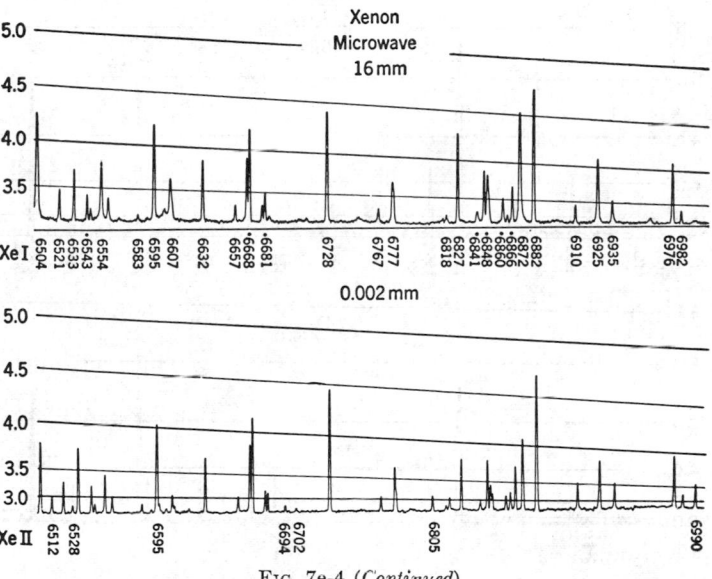

FIG. 7e-4 (Continued)

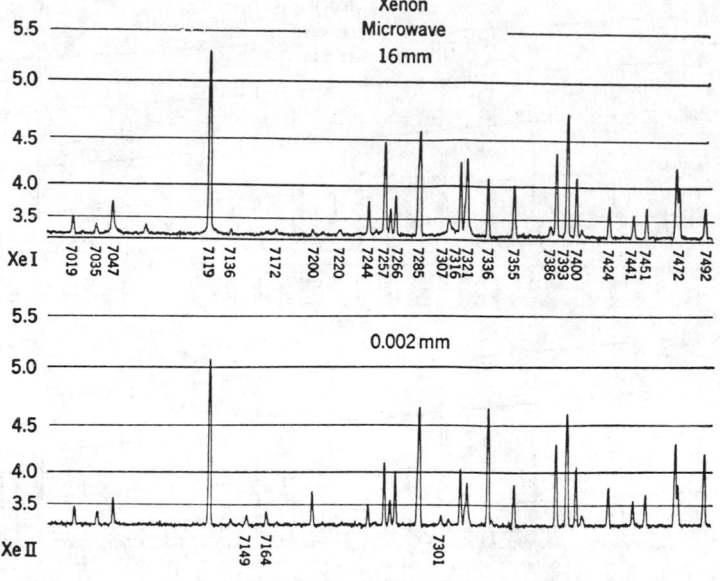

Fig. 7e-4 (Continued)

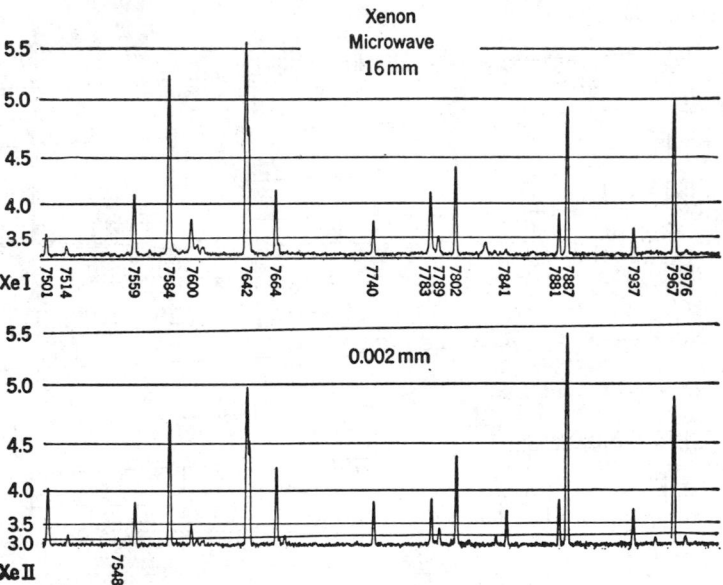

Fig. 7e-4 (Continued)

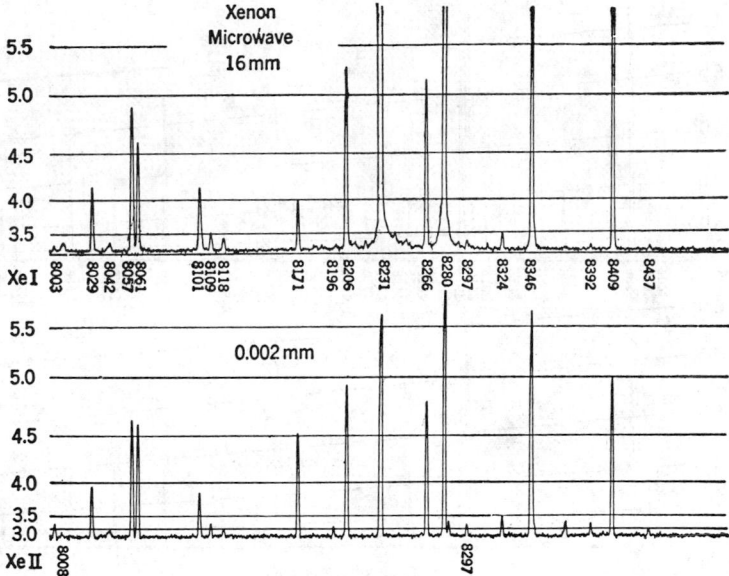

FIG. 7e-4 (*Continued*)

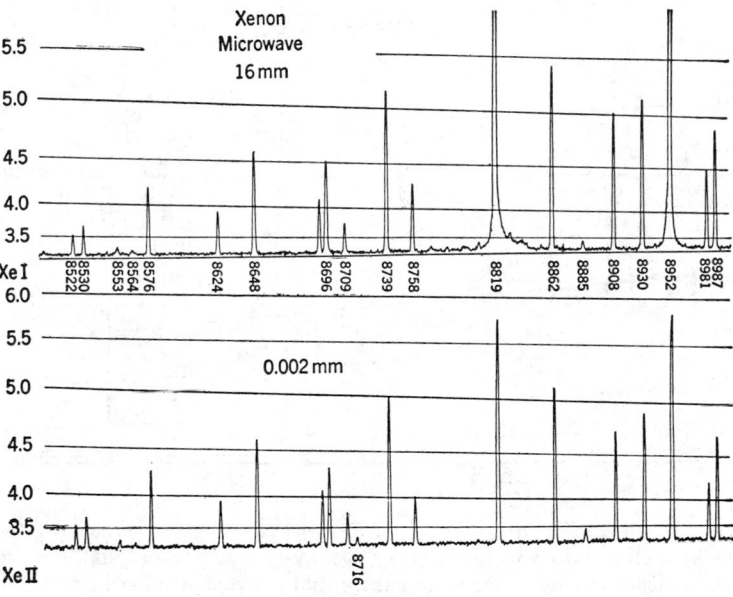

FIG. 7e-4 (*Continued*)

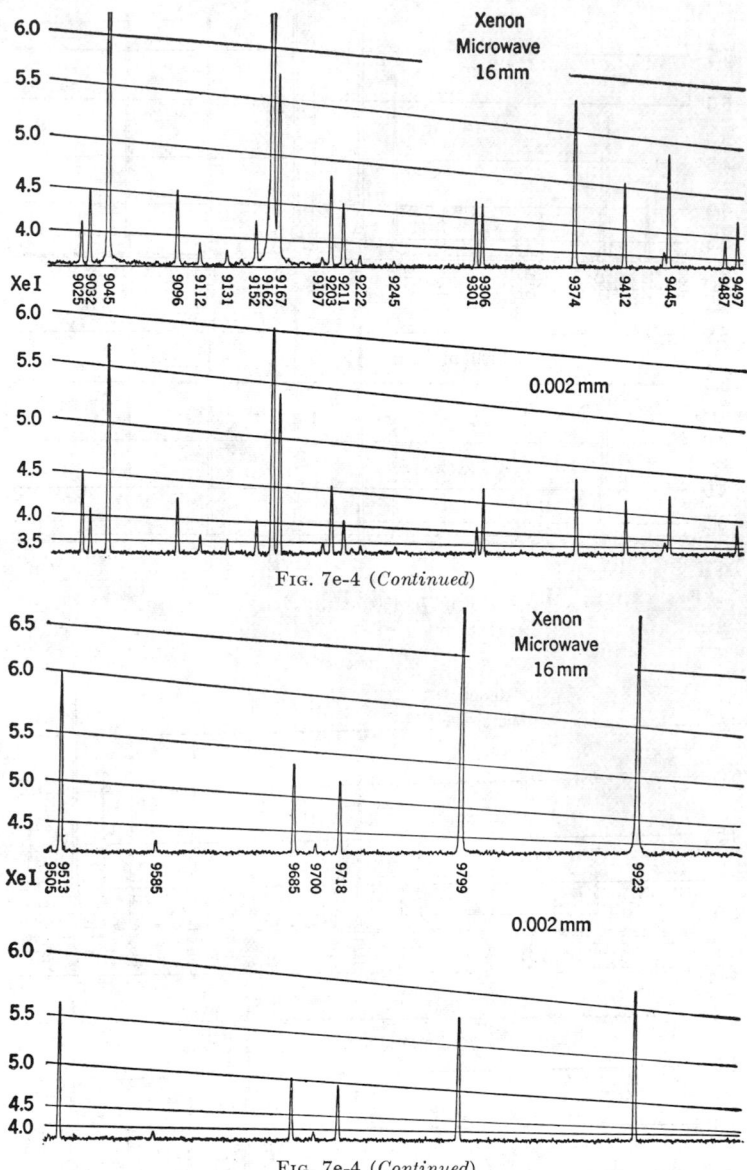

FIG. 7e-4 (*Continued*)

FIG. 7e-4 (*Continued*)

Iron I. The lines of the iron spectrum are extensively used as wavelength standards and may be used equally well as intensity standards. The traditional iron arc in air no longer satisfies the demands on accuracy and convenience because the lines are relatively broad, the wavelengths are not constant, and the arc cannot be made to burn steadily. A hollow-cathode discharge[1] with iron electrodes and neon at about 3 mm pressure is much superior. Microwave discharges[2] with volatile iron salts in a rare gas also give very sharp lines but are less suitable for providing intensity standards.

[1] Crosswhite, Dieke, and Legagneur, *J. Opt. Soc. Am.* **45**, 270 (1955).
[2] W. F. Meggers and F. O. Westfall, *J. Research Natl. Bur. Standards* **44**, 447 (1950)

Accurate wavelength measurements by several independent investigators are available for many iron lines. From measurements on the atmospheric arc Edlen has determined many well-defined energy levels and from these has computed a set of self-consistent wavelengths, which in 1955 were adopted as international secondary standards.[1] Many other lines are unsuitable as standards because their wavelengths are not constant. These ordinarily come from high-lying excited states.

No such difficulties are encountered with the hollow-cathode or other low-pressure discharges. At the present time, however, not enough measurements have been made to qualify any line as an international standard and there are some contradictory results. The values given in column λ_2 may be considered accurate in general to better than 0.001. As there is a systematic shift between the wavelengths of the low-pressure discharges and those of the arc in air, the arc wavelengths should not be used for the low-pressure discharge and vice versa. More and improved wavelengths for column λ_2 may be expected in the near future.

Explanation of Table 7e-6. COLUMN λ_1. The wavelengths of iron arc in air given to four decimal places are international standards.[1] The rest are taken from the compilation of Russell and Moore.[2]

COLUMN λ_2. Wavelengths of the hollow-cathode discharge: The four-decimal figures without letter designation are derived from the international standards of column λ_1 by applying observed pressure-shift corrections to the levels involved. The letters signify: L, Blackie and Littlefield,[3] measured with the reflecting echelon; H, J, at The Johns Hopkins University with the Fabry-Perot interferometer;[4,5] N, Stanley and Meggers;[6] W, Williams and Middleton[7] with the vacuum echelon. Values quoted to three decimal places are grating measurements made by interpolation between the above standards.[5]

CLASSIFICATION. Standard L, S coupling notation is used. E' is the energy of the upper state above the ground state in wave numbers. For more accurate values see Moore.[8]

INTENSITY COLUMNS. log I_2, quantitative intensities of a standard hollow-cathode discharge in neon at 3.5 mm pressure, current 90 ma.[5] Values with three decimals are photoelectric measurements; those with two decimals, photographic measurement with photoelectric calibration. Sensitivity calibration above 3,150 Å and standard tungsten ribbon-filament lamp calibrated by the National Bureau of Standards; between 2,700 and 3,150 Å, indirect calibration through self-absorption behavior; below 2,700, extrapolated. The scale in the log I_2 column is the same as for neon and argon (see Tables 7e-2 and 7e-3).

log I_3, iron arc in air, current 1 A, photographic measurements on arbitrary scale. Sensitivity correction as for I_2. r, self-reversal between 10 and 30 percent; R, same, larger than 30 percent.

log I_4, iron arc in air, current 2.2 A photoelectric measurement; otherwise same as I_3.

log νA_ν, absolute line emissive power in units of microergs per second per excited atom. Derived from measurements of Crosswhite,[5] King,[9] King and King,[10] and Carter.[11]

[1] *Trans. Intern. Astron. Union* **9**, 216 (1957).
[2] Russell and Moore, *Trans. Am. Phil. Soc.* **34**, 113 (1944).
[3] J. Blackie and T. A. Littlefield, *Proc. Roy. Soc. (London)*, ser. A, **234**, 398 (1956).
[4] (H) R. W. Stanley and G. H. Dieke, *J. Opt. Soc. Am.* **45**, 280 (1955).
[5] (J) H. M. Crosswhite, *Johns Hopkins Spectroscopic Rept.* 13, 1958.
[6] R. W. Stanley and W. F. Meggers, *Natl. Bur. Standards J. Research* **58**, 41 (1957).
[7] W. E. Williams and A. Middleton, *Proc. Roy. Soc. (London)*, ser. A, **172**, 159 (1939).
[8] C. E. Moore, Atomic Energy Levels, *Natl. Bur. Standards Circ.* 467, vol. 2, 1952.
[9] R. B. King, *Astrophys. J.* **95**, 78 (1942).
[10] R. B. King and A. S. King, *Astrophys. J.* **87**, 24 (1938).
[11] W. W. Carter, *Phys. Rev.* **76**, 962 (1949).

TABLE 7e-6. THE SPECTRUM OF IRON I

λ_1	λ_2	Classification		E'	$\log I_2$	$\log I_3$	$\log I_4$	$\log \nu A_\nu$
2,440.106	.109	a^2H_4	t^3H_4	60,758	4.2			
2,442.567	.568	a^3H_5	t^3H_5	60,549	4.0			
2,443.8728	.8718	a^5F_5	x^3G_5	47,835	4.36			
2,457.5980	.5975 L	a^5F_5	v^5F_5	47,606	5.09			
2,462.6483	.6474	a^5D_4	x^5F_4	40,594	5.08			
2,465.1500	.1490	a^5F_4	v^5F_4	47,930	4.85			
2,468.8803	.8793	a^5F_5	w^5G_5	47,420	4.86			
2,472.343	.345	a^5F_4	x_2G_4	47,812	4.87			
		a^5F_5	w^5G_6	47,363				
2,472.8962	.8953	a^5D_3	x^5F_3	40,842	5.13			
2,473.156	.155	a^5D_4	y^7P_4	40,422	5.00			
2,474.8151	.8141	a^5F_3	v^5F_2	48,123	4.69			
2,479.7774	.7765	a^5D_2	x^5F_2	41,018	5.07			
2,483.2718	.2709	a^5D_4	x^5F_5	40,257	5.75			
2,483.531	.533	a^5F_2	v^5F_2	48,239	4.54			
2,484.186	.185	a^5D_1	x^5F_1	41,131	4.97			
2,486.372	.367	a^5D_4	y^7P_3	40,207	4.90			
2,488.1437	.1428	a^5D_3	x^5F_4	40,594	5.59			
2,489.751	.750	a^5D_0	x^5F_1	41,131	4.98			
2,490.6454	.6445	a^5D_2	x^5F_3	40,842	5.45			
2,491.1562	.1553	a^5D_1	x^5F_2	41,018	5.20			
2,496.5343	.5332	a^5F_4	w^5G^5	47,420	4.78			
2,501.1332	.1326 L	a^5D_4	x^5D_3	39,970	5.03			
2,507.899	.902	a^5F_3	w^5G_4	47,590	4.73			
2,510.8362	.8353	a^5D_3	x^5D_2	40,231	5.04			
2,512.361	.367	a^5D_3	y^7P_2	40,207	4.63			
2,517.658	.661	a^5F_2	w^5G_3	47,693	4.58			
2,518.1029	.1020	a^5D_2	x^5D_1	40,405	4.92			
2,522.8505	.8496	a^5D_4	x^5D_4	39,626	5.54			
2,524.2939	.2928	a^5D_1	x^5D_0	40,491	4.65			
2,527.4358	.4346	a^5D_3	x^5D_3	39,970	5.30			
2,529.1361	.1351	a^5D_2	x^5D_2	40,231	4.86			
2,535.6086	.6076	a^5D_0	x^5D_1	40,405	4.60			
2,540.9734	.9719 L	a^5D_1	x^5D_2	40,231	4.85			
2,542.101	.100	b^3F_2	r^3G_3	60,365	4.46			
2,543.920	.921	b^3F_3	r^3G_4	60,172	4.40			
2,545.9795	.9789 L	a^5D_2	x^5D_3	39,970	4.92			
2,549.6142	.6140 L	a^5D_3	x^5D_4	39,626	4.87			
2,576.6916	.6907 L	a^5F_5	x^5G_5	45,726	4.50			
2,584.5370	.5364 L	a^5F_5	x^5G_6	45,608	5.17			
2,599.565	. . . . . .	a^5F_4	x^5G_4	45,833	4.50			
2,605.6578	.6566	a^5F_5	y^3G_5	45,295	3.86			
2,606.8280	.8270 L	a^5F_4	x^5G_5	45,726	4.56			
2,618.0191	.0179	a^5F_3	x^5G_3	45,914	4.3			
2,623.532	.533	a^5F_3	x^5G_4	45,833	4.65			
2,635.8100	.8096 L	a^5F_2	x^5G_3	45,914	4.48			
2,643.9992	.9980	a^5F_1	x^5G_2	45,965	4.32			
2,666.8133	.8122	a^5F_5	v^5D_4	44,415	4.45			
2,679.0626	.0622 L	a^5F_5	w^5F_5	44,244	4.79			
2,689.2130	.2131 L	a^5F_4	v^5D_3	44,551	4.63			
2,699.1075	.1063	a^5F_4	v^5D_4	44,415	4.20			

TABLE 7e-6. THE SPECTRUM OF IRON I (*Continued*)

λ_1	λ_2	Classification		E'	$\log I_2$	$\log I_3$	$\log I_4$	$\log \nu A_\nu$
2,706.5829	.5829 L	a^5F_3	v^5D_2	44,664	4.59	.....	.....	4.59
2,711.6560	.6555 L	a^5F_4	w^5F_5	44,244	4.29			
2,719.027	.020	a^5D_4	y^5P_3	36,767	5.44			
		(b^3F_3)	(t^3F_3)	(57,641)				
2,720.9035	.9024	a^5D_3	y^5P_2	37,158	5.08			
2,723.5786	.5776 L	a^5D_2	y^5P_1	37,410	4.61			
2,733.5816	.5810 L	a^5F_5	w^5D_4	43,500	4.96			
2,735.4762	.4750	a^5F_4	w^5D_3	43,923	4.71	4.70R	.....	3.56
2,737.3108	.3099 L	a^5D_1	y^5P_1	37,410	4.74	4.70R	.....	2.88
2,742.2554	.2542	a^5F_3	w^5D_2	44,184	4.4	4.50R	.....	3.28
2,742.4064	.4060 L	a^5D_2	y^5P_2	37,158	5.02	4.64R	.....	2.97
2,744.0691	.0680	a^5D_0	y^5P_1	37,410	4.33	4.66R	.....	2.65
2,750.1415	.1404	a^5D_3	y^5P_3	36,767	5.02	4.66R	.....	3.04
2,756.3295	.3284	a^5D_1	y^5P_2	37,158	4.36	.....	.....	2.3
2,761.7810	.7798	a^5F_2	w^5D_2	44,184	4.14	4.32R	.....	3.03
2,762.0275	.0263	a^5F_3	w^5D_3	43,923	4.09	4.40R	.....	3.06
2,767.5232	.5220	a^5F_4	w^5D_4	43,500	4.39	4.44R	.....	3.00
2,772.0748	.0736	a^5F_5	z^5H_5	42,992	4.47	4.64R	.....	2.38
2,778.2214	.2205 L	a^5F_5	y^5G_5	42,912	4.70	4.49R	.....	2.98
2,788.106	.104	a^5F_5	y^5G_6	42,784	5.60	4.65R	.....	3.89
2,797.7765	.7752	a^5F_4	z^5H_4	43,109	4.24	4.16R	.....	2.63
2,804.5212	.5207 L	a^5F_4	y^5G_4	43,023	4.65	4.48R	.....	2.99
2,806.9852	.9845 L	a^5F_4	z^5H_5	42,992	5.02	4.56R	.....	3.11
2,813.2877	.2867 L	a^5F_4	y^5G_5	42,912	5.37	4.02R	.....	3.53
2,823.2767	.2763 L	a^5F_3	y^5G_3	43,138	4.56	4.50R	.....	3.10
2,825.5569	.5559 L	a^5F_3	z^5H_4	43,109	4.81	4.52R	.....	3.17
2,828.8094	.8081	a^5F_2	z^5H_3	43,326	4.16	3.69	.....	2.15
2,832.4364	.4357 L	a^5F_3	y^5G_4	43,023	4.90	4.65R	.....	3.35
2,838.1205	.1187	a^5F_2	y^5G_2	43,210	4.36	4.33r	.....	2.85
2,843.6314	.6311	a^5F_4	x^5P_3	42,860	4.27	4.3R	.....	2.74
2,843.9775	.9762	a^5F_2	y^5G_3	43,138	4.96	4.61R	.....	3.36
2,851.7979	.7973 L	a^5F_1	y^5G_2	43,210	4.80	4.60R	.....	3.25
2,863.8644	.8634	a^5D_2	z^5G_3	35,612	3.23	4.10r	.....	1.48
2,869.3083	.3073	a^5D_3	z^5G_4	35,259	3.76	4.41R	.....	1.78
2,874.1733	.1723	a^5D_4	z^5G_5	34,782	3.92	4.48R	.....	1.84
2,912.1589	.1575	a^5D_4	y^5F_3	34,329	4.08	4.55R	.....	2.111
2,929.0085	.0073	a^5D_3	y^5F_2	34,547	4.20	4.50R	.....	2.215
2,936.9049	.9034	a^5D_4	y^5F_4	34,040	5.02	4.55R	.....	2.89
2,941.3438	.3423	a^5D_2	y^5F_1	34,692	3.82	4.55R	.....	2.009
2,947.8773	.8758	a^5D_3	y^5F_3	34,329	4.95	4.54R	.....	2.89
2,953.9411	.9400 N	a^5D_2	y^5F_2	34,547	4.75	4.52R	.....	2.766
2,957.3660	.3646 N	a^5D_1	y^5F_1	34,692	4.45	4.55R	.....	2.519
2,965.2561	.2545 N	a^5D_0	y^5F_1	34,692	4.01	4.54R	.....	2.27
		(a^3G_5)	(v^3H_5)	(55,430)				
2,966.8997	.8982	a^5D_4	y^5F_5	33,695	5.48	4.47R	.....	3.269
2,970.106	.110	a^5D_2	z^3P_1	34,363	4.80	4.52R	.....	1.40
		a^5D_1	y^5F_2	34,547				
2,973.1336	.1324	a^5D_2	y^5F_3	34,329	5.13	4.5R	....	3.067
2,973.2368	.2356	a^5D_3	y^5F_4	34,040	4.7	4.5R	.....	2.713
2,981.4459	.4450 N	a^5D_3	z^3P_2	33,947	4.83	4.52R	.....	2.41
2,983.5714	.5699 N	a^5D_4	y^5D_3	33,507	5.11	4.55R	.....	3.040
2,986.4569	.4558	a^5D_1	z^3P_1	34,362	3.26	3.57	.....	0.59
2,987.2923	.2904 N	a^5F_4	x^5F_3	40,842	3.52	4.44r	.....	2.60

ATOMIC AND MOLECULAR PHYSICS

TABLE 7e-6. THE SPECTRUM OF IRON I (*Continued*)

λ_1	λ_2	Classification		E'	$\log I_2$	$\log I_3$	$\log I_4$	$\log \nu A_\nu$
2,994.4281	.4274 N	a^5D_3	y^5D_2	33,082	5.1	4.56R	.....	3.065
2,994.5033	.5022	a^5D_0	z^3P_1	34,363	4.4			
2,999.5125	.5118 N	a^5F_5	x^5F_5	40,527	4.73	4.63R	.....	3.29
3,000.4527	.4513	a^3F_4	y^3G_5	45,295	4.11	3.98	.....	2.86
3,000.9489	.9481 N	a^5D_2	y^5D_1	34,017	4.94	4.53R	.....	3.068
3,003.0323	.0304 N	a^5F_3	x^5F_2	41,018	3.31	4.39r	.....	2.66
3,007.2832	.2824 N	a^5D_2	z^3P_2	33,947	4.36	4.5R	.....	2.01
3,008.1399	.1390 N	a^5D_1	y^5D_0	34,122	4.62	4.49R	.....	2.827
3,009.5707	.5693 N	a^5F_4	x^5F_4	40,594	4.27	4.61R	.....	3.09
3,016.186	.185	a^5F_2	x^5F_1	41,131	3.19	4.20r	.....	2.48
3,017.6288	.6271 N	a^5D_1	y^5D_1	34,017	3.80	4.51R	.....	2.119
3,018.9848	.9827 N	a^5F_3	x^5F_3	40,842	3.72	4.48R	.....	2.94
3,020.4918	.4909 N	a^5D_2	y^5D_2	33,802	4.7	4.5R	.....	2.796
3,020.6405	.6392	a^5D_4	y^5D_4	33,096	5.64	4.4R	.....	3.473
3,021.0743	.0727 N	a^5D_3	y^5D_3	33,507	5.24	4.46R	.....	3.241
3,024.0337	.0328 N	a^5D_1	z^3P_2	33,947	4.59	4.54R	.....	2.27
3,025.638	.636	a^3H_6	w^3H_6	52,431	4.17	4.12r	.....	4.15
3,025.8442	.8423 N	a^5D_0	y^5D_1	34,017	4.69	4.54R	.....	2.780
3,026.4637	.4612 N	a^5F_2	x^5F_2	41,018	3.48	4.43R	.....	2.76
3,030.1494	.1477	a^3H_5	w^3H_5	52,613	3.92	4.04	.....	4.00
3,031.213	.213	a^3H_4	w^3H_4	52,769	3.81	3.96	.....	3.93
3,031.638	.636	a^5F_1	x^5F_1	41,131	3.36	4.39R	.....	2.71
3,037.3901	.3885 N	a^5D_1	y^5D_2	33,802	4.89	4.56R	.....	2.986
3,040.4281	.4272 N	a^5F_4	x^5F_5	40,257	3.74	4.34r	.....	2.57
3,041.6386	.6370	a^3F_3	y^3G_4	45,428	3.9	3.8		
3,041.7401	.7381 N	a^5F_3	x^5F_4	40,594	3.9	4.3r	.....	2.6
3,042.0215	.0198	a^5F_1	x^5F_2	41,018	3.12	4.16r	.....	2.45
3,042.6667	.6643 N	a^5F_2	x^5F_3	40,842	3.39	4.36r	.....	2.63
3,047.6060	.6039 N	a^5D_2	y^5D_3	33,507	5.02	4.56R	.....	3.044
3,057.4471	.4457 N	a^5F_5	x^5D_4	39,626	4.82	4.7R	.....	3.4
3,059.0871	.0859 N	a^5D_3	y^5D_4	33,096	5.06	4.5R	.....	2.901
3,067.2457	.2437 N	a^5F_4	x^5D_3	39,970	4.66	4.7R	.....	3.2
3,075.7214	.7193 N	a^5F_3	x^5D_2	40,231	4.09	4.7R	.....	3.2
3,083.7430	.7409 N	a^5F_2	x^5D_1	40,405	3.78	4.6R	.....	3.0
3,091.5786	.5768	a^5F_1	x^5D_0	40,491	3.41	4.5r	.....	2.69
3,099.8968	.8950	a^5F_1	x^5D_1	40,405	4.1	4.5R	.....	2.8
3,099.9695	.9678 N	a^5F_4	x^5D_4	39,626	4.1	4.6R	.....	2.9
3,100.3054	3032 N	a^5F_2	x^5D_2	40,231	3.80	4.5R	.....	2.9
3,100.6667	.6649 N	a^5F_3	x^5D_3	39,970	4.01	4.6R	.....	3.0
3,116.6337	.6319	a^5F_1	x^5D_2	40,231	2.86	3.99	.....	2.22
3,125.653	.651	a^5F_2	x^5D_3	39,970	3.49	4.18	.....	2.30
3,134.1115	.1099 N	a^5F_3	x^5D_4	39,626	3.05	3.97	.....	2.04
3,142.453	.454	z^7D_3	e^7S_3	51,570	2.84	3.25	.....	2.98
3,142.8908	.8890	a^3P_2	w^3P_2	50,187	2.90	3.23	.....	2.86
3,143.2434	.2426	a^5D_4	z^3F_3	31,805	2.88	3.05	.....	-0.09
3,143.990	.989	z^5D_4	i^5D_4	57,698	3.14	3.19	.....	3.68
3,151.353	.352	a^3G_4	y^1H_5	53,722	3.39	3.45	3.880	3.42
3,153.200	.200	z^7D_3	f^5F_4	51,462	3.03	3.44	3.76	3.14
3,157.040	.037	z^7D_4	e^7G_5	51,229	3.24	3.66	3.974	3.70
3,157.88	.886	z^7D_2	e^7S_3	51,570	3.06	3.48	3.830	3.24

TABLE 7e-6. THE SPECTRUM OF IRON I (Continued)

λ_1	λ_2	Classification		E'	$\log I_2$	$\log I_3$	$\log I_4$	$\log \nu A_\nu$
3,160.658	.659	z^7D_4	e^7F_4	51,192	3.33	3.73	4.06	3.44
3,161.949	.947	z^7D_5	e^7G_6	50,968	3.08	3.56	3.766	3.16
3,165.860	.859	z^7D_3	e^7G_4	51,335	2.87	3.24	3.543	2.96
3,166.435	.436	b^3F_4	t^3D_3	52,213	3.08	3.44	3.772	3.35
3,175.447	.445	z^7D_5	e^7F_5	51,192	3.44	3.77	4.072	3.42
3,178.015	.014	z^7D_5	f^7D_4	50,808	3.29	3.65	3.948	
3,180.223	.223	z^7D_3	e^7F_4	51,192	3.81	4.07	4.37r	3.87
3,180.7562	.7556	a^5D_2	z^3F_2	32,134	3.79	3.73	3.81R	0.68
3,182.9798	.9781	a^5P_2	v^3D_3	49,135	3.07	3.30	3.521	2.63
3,184.8955	.8941	a^5D_3	z^3F_3	31,805	4.29	4.15R	3.97R	1.1
3,188.567	.570	z^7D_5	e^5G_5	50,704	3.06	3.31	3.81	2.95
3,188.819	.820	z^7D_1	e^5G_2	51,370	3.48	3.58	3.95	3.34
3,191.6599	.6594	a^5D_4	z^3D_3	31,323	4.42	4.21R	4.00R	1.13
3,192.799	.799	z^7D_1	e^7F_2	51,331	3.57	3.86	4.06	3.66
		(b^3G_4)	(v^3H_5)	(55,430)				
3,193.228	.2245	a^5D_4	z^3F_4	31,307	4.86	4.44R	4.66R	1.4
3,196.930	.926	z^7D_4	e^7F_5	50,833	4.41	4.4r	4.67r	4.0
3,196.977	.9868	a^5D_3	z^3D_2	31,686	.....	4.1r	.....	0.49
3,199.530	.500	z^7D_4	f^7D_4	50,808	4.03	4.08	3.53	
		(a^5D_1)	(z^3F_2)	32,134				
3,200.475	.470	z^7D_2	e^7F_3	51,149	3.97	4.09	4.407	
		z^7D_2	e^5S_2	51,149				
3,200.7854	.7849	a^5D_2	z^3D_1	31,937	3.10	3.22	.....	0.17
3,205.400	.3959 N	z^7D_1	e^7F_1	51,208	3.68	4.00	4.308	3.77
3,209.297	.295	z^5F_2	g^5G_3	58,710	3.76	3.48	3.887	
		z^7F_6	g^7D_5	53,801				
3,210.230	.228	z^7D_4	e^5G_5	50,704	3.56	3.64	4.05	3.24
3,210.830	.829	z^7D_2	f^7D_1	51,048	3.65	3.89	4.25	3.57
3,211.487	.486	z^7D_1	e^5S_2	51,149	3.0	3.34	.....	3.02
3,211.683	.678	z^5F_5	g^5G_5	58,002	3.81	3.56	4.11	4.12
3,211.989	.988	z^7D_5	e^7P_4	50,475	3.2	4.13	4.65r	3.68
3,214.044	.017	z^5F_4	g^5G_5	58,271	4.38	4.07	4.78r	
		z^7D_3	f^7D_3	50,862				
		(z^7D_3)	(e^7P_2)	(50,861)				
3,214.3964	.3950	a^5D_2	z^3F_3	31,805	4.39	4.07	.....	0.95
3,215.940	.938	z^7D_2	f^7D_2	50,999	3.84	4.00	4.346	3.72
3,217.380	.377	z^7D_5	f^5D_4	50,423	3.75	3.87	4.162	3.46
3,219.581	.582	z^7D_3	f^7D_4	50,808	3.93	4.12	4.41	3.81
3,219.806	.804	z^7D_4	e^7P_3	50,611	3.87	3.95	4.33	
		(a^5D_1)	(z^3D_1)	(31,937)				
3,222.069	.066	z^7D_5	f^7D_5	50,378	4.53	4.52r	4.79R	
3,225.789	.785	z^7D_5	e^7F_6	50,342	4.76	4.65R	4.89R	4.3
3,227.798	.795	z^7D_4	f^5D_3	50,534	4.04	4.11	4.48	3.78
3,229.1221	.1216	a^5D_0	z^3D_1	31,937	3.70	3.62	.....	0.54
3,230.210	.207	z^7D_2	e^7P_2	50,861	3.32	3.54	.....	3.16
3,230.963	.963	z^7D_3	f^5D_2	50,699	3.66	3.83	4.156	3.50
3,233.053	.051	b^3H_6	x^3I_7	57,028	3.87	3.52	4.060	4.28
3,233.967	.968	z^7D_4	e^7P_4	50,475	3.76	3.82	4.149	3.44
3,234.6133	.6133	a^5D_3	z^3D_3	31,323	4.09	3.91	3.75	0.75
3,236.2231	.2219 N	a^5D_3	z^3F_4	31,307	4.55	4.18	3.98r	1.03
3,239.436	.432	z^7D_4	f^5D_4	50,423	4.12	4.10	4.427	
3,244.190	.187	z^7D_4	f^7D_5	50,378	4.22	4.07	4.368	3.77

TABLE 7e-6. THE SPECTRUM OF IRON I (*Continued*)

λ_1	λ_2	Classification		E'	log I_2	log I_3	log I_4	log νA_2
3,246.0054	.0049	a^5D_1	z^3D_2	31,686	4.02	.....	3.72	0.48
3,246.962	.964	a^5P_2	x^3P_1	48,516	2.92	3.47	.....	3.80
3,248.206	.204	z^7D_3	f^5D_3	50,534	3.1	3.75	3.964	3.45
3,254.3628	.3608	b^3H_5	x^3I_6	57,070	3.80	3.56	3.99	4.23
3,257.5940	.5935 N	a^5P_3	v^5F_2	48,239	3.11	3.53	3.71	2.92
3,265.0473	.0468	a^5D_2	z^3D_3	31,323	4.03	3.83	3.69	0.69
3,265.6182	.6164	a^5P_3	v^5P_2	48,163	3.78	4.03	4.293	3.36
3,271.0014	.9996	a^5P_2	v^5P_1	48,290	3.74	4.05	4.279	3.46
3,280.2613	.2593	b^3H_4	x^3I_5	57,104	3.78	3.46	3.89	4.16
3,284.5888	.5870	a^5P_2	v^5P_2	48,163	2.81	3.40	3.552	2.81
3,286.7541	.7508 W	a^5P_3	v^5P_3	47,967	4.42	4.38	4.62r	3.76
3,292.022	.020	a^3D_3	u^3F_4	56,593	3.70	3.49	3.861	4.13
3,292.5910	.5892	a^5P_1	v^5P_1	48,290	3.41	3.76	4.008	3.19
3,298.1331	.1313	a^5P_1	v^5F_2	48,239	3.27	3.51	3.687	2.93
3,305.9719	.9700	a^5P_2	v^5P_3	47,967	4.09	4.25	4.44	3.65
3,306.356	.352	a^5P_1	v^5P_2	48,163	4.20	4.29	4.48	3.72
3,314.7420	.7399	a^3D_2	u^3F_3	56,783	3.67	3.41	3.80	4.03
3,323.7375	.7355	b^3P_2	v^3P_2	52,916	.....	3.41	3.723	3.71
3,328.8667	.8646	b^3H_5	u^3H_5	56,383	3.50	.....	3.666	4.11
3,337.6664	.6644	a^3G_5	u^3G_4	51,668	3.25	.....	3.433	3.25
3,340.5666	.5647	a^3P_2	x^3P_2	48,305	3.16	.....	3.395	2.79
3,341.906	.906	a^3G_5	6_5	51,630	3.22			
3,342.2163	.2144	a^3P_2	v^5P_1	48,290	2.86	.....	3.53	
3,342.298	.292	b^3P_1	8_1	52,858	3.31			
3,347.9271	.9252	a^3P_2	v^5F_2	48,239	3.06	.....	3.331	2.71
3,355.2287	.2266	b^3H_4	u^3H_4	56,423	3.43	.....	3.615	4.08
3,369.549	.549	a^3G_4	u^3G_4	51,668	3.82	3.74	3.964	
3,370.7852	.7832	a^3G_5	u^3G_5	51,374	4.07	3.99	4.196	4.00
3,378.676	.676	a^3G_5	v^3F_4	51,305	3.59	3.41	3.70	3.30
3,379.0206	.0187	a^5P_3	w^3D_2	47,136	3.38	3.48	3.74	2.76
3,380.1117	.1097	a^3G_3	u^3G_3	51,826	3.71	3.63	3.86	3.61
3,383.9808	.9789	a^5P_3	x^3F_3	47,093	3.82	3.81	3.99	3.08
3,389.748	.741	a^5P_1	1_2	47,420	3.05	.....	3.832	3.12
3,392.3058	.3038	a^5P_2	x^3F_2	47,197	3.72	3.72	.....	3.02
3,392.6540	.6520	a^5P_3	w^3D_3	47,107	4.14	4.20	4.32	3.49
3,394.5854	.5834	a^5P_2	u^5D_1	47,177	3.41	3.54	3.683	2.94
3,396.9774	.9757	a^5F_3	y^5P_2	37,158	2.74	3.62	3.47	1.46
3,399.3356	.3337	a^5P_2	w^3D_2	47,136	4.13	4.22	4.301	3.55
3,401.5200	.5180	a^5F_4	y^5P_3	42,967	3.15	3.92	3.79r	2.61
3,402.256	.255	b^3H_6	v^3H_6	55,490	3.67	3.51	3.770	4.15
3,404.3557	.3537	a^5P_2	x^3F_3	47,093	4.06	4.11	4.270	3.37
3,406.8021	.8001	a^5P_1	w^3D_1	47,272	3.75	3.86	3.95	3.16
3,407.4611	.4573 W	a^5P_3	x^3F_4	46,889	4.63	4.68	4.67r	3.94
3,413.1339	.1295 W	a^5P_2	w^3D_3	47,017	4.44	4.39	4.42r	3.67
3,417.8428	.8408	a^5P_1	u^5D_1	47,177	4.62?	4.19	4.241	3.51
3,418.507	.507	a^5P_1	u^5D_0	47,172	3.88	4.09	4.173	3.44
3,422.6583	.6563	a^5P_1	w^3D_2	47,136	3.84	3.96	4.12	3.25
3,424.2861	.2841	a^5P_3	u^5D_3	46,745	4.04	4.17	4.228	3.43
3,426.383	.381	a^5P_3	y^3P_2	46,727	3.59	3.94	4.14	
3,426.637	.630	a^5P_2	y^3P_1	46,902	3.73	3.85	.....	3.23

TABLE 7e-6. THE SPECTRUM OF IRON I (*Continued*)

λ_1	λ_2	Classification		E'	$\log I_2$	$\log I_3$	$\log I_4$	$\log \nu A_\nu$
3,427.1213	.1193	a^5P_3	u^5D_4	46,721	4.57	4.63	4.61r	2.96
3,428.1948	.1928	a^5P_2	u^5D_2	46,889	3.98	4.06	4.127	3.36
3,440.6069	.6053	a^5D_4	z^5P_3	29,056	5.76	5.46R	4.6R	2.30
3,440.9899	.9888 N	a^5D_3	z^5P_2	29,469	5.39	5.22R	4.5R	1.91
3,443.8775	.8761 N	a^5D_2	z^5P_1	29,733	5.02	4.89r	4.32R	1.728
3,445.1508	.1493	a^5P_2	u^5D_3	46,745	4.28	4.32	4.34r	3.55
3,450.3304	.3284	a^5P_1	y^3P_1	46,902	3.75	3.93	3.922	3.18
3,451.9166	.9146	a^5P_1	u^5D_2	46,889	3.76	3.93	4.13	3.39
3,452.2760	.2746	a^5F_3	y^3F_4	36,686	3.69	4.14	4.13	2.04
3,465.8621	.8602 N	a^5D_1	z^5P_1	29,733	5.13	5.02r	4.36R	1.898
3,475.4511	.4497 N	a^5D_2	z^5P_2	29,469	5.32	5.13R	4.48R	2.031
3,476.7036	.7020 N	a^5D_0	z^5P_1	29,733	4.80	4.74r	4.32R	1.578
3,490.5749	.5740 N	a^5D_3	z^5P_3	29,056	5.38	5.06R	4.43R	1.971
3,497.110	....	a^5P_3	w^5P_3	46,137	3.58	3.99	4.152	3.30
3,497.8420	.8407 N	a^5D_1	z^5P_2	29,469	4.82	4.62r	4.30R	1.537
3,513.8196	.8177 N	a^5F_5	z^3G_5	35,379	4.55	4.48	4.48R	2.16
3,521.2630	.2610 N	a^5F_4	z^3G_4	35,768	4.45	4.52	4.51	2.30
3,526.0415	.0397 N	a^5D_2	z^5P_3	29,056	4.65	} 4.6	} 4.7R	0.83
3,526.1676	.1657	a^5F_3	z^3G_3	36,079	4.15			2.00
3,533.201	.196	z^7F_1	e^7G_2	51,540	3.96	3.98	4.20	4.07
3,536.556	.554	z^7F_2	e^7G_3	51,461	4.15	4.15	4.425	4.29
3,541.083	.083	z^7F_4	e^7G_5	51,229	4.34	4.29	4.56r	4.30
3,542.076	.076	z^7F_3	e^7G_4	51,335	4.29	4.24	4.52r	4.26
3,554.1196	.1181	a^5F_3	z^5G_2	35,856	3.16	3.85	4.04	1.85
3,554.922	.9245 N	z^7F_6	e^7G_6	50,968	4.29	4.53	4.79r	4.50
3,556.877	.877	z^7F_4	f^5F_5	51,103	4.22	4.10	4.326	4.14
3,558.5170	.5149	a^5F_2	z^3G_1	36,079	4.54	4.73	4.59R	2.55
3,565.3807	.3789 N	a^5F_3	z^3G_4	35,768	4.98	5.22	4.80R	2.99
3,570.0996	.0963 H	a^5F_4	z^3G_5	35,379	5.13	5.51R }	} 5.11R	3.14
3,570.243	.247	z^7F_6	e^7G_7	50,652	4.91	.....		
3,571.995	.995	z^7F_5	e^7F_5	50,833	3.94	3.87	4.124	3.89
3,573.896	.886	b^3H_4	t^3G_3	54,600	3.81	3.79	4.00	4.28
3,581.195	.1925 H	a^5F_5	z^5G_6	34,844	5.56	5.73R	4.98R	3.7
3,582.201	.201	b^3H_6	12_5	54,014	4.05	.....	4.01	4.21
3,584.6627	.6605	a^3G_5	y^3H_5	49,604	4.14	4.09	4.32	3.94
3,585.3206	.3191	a^5F_3	z^5G_3	35,612	4.60	4.72	5.02	2.79
3,585.7068	.7053	a^5F_4	z^5G_4	35,257	4.35	4.47	4.74	2.47
3,586.114	.112	b^3H_6	l^3G_5	53,983	4.40	4.02	4.23	
3,586.9861	.9836 N	a^5F_2	z^5G_2	35,856	4.60	4.71	4.64R	2.46
3,589.1063	.1048	a^5F_5	z^5G_5	34,782	3.66	4.11	4.34	2.00
3,594.632	.631	z^7F_4	f^7D_4	50,808	3.91	3.91	4.068	3.84
3,603.2068	.2046	a^3G_5	v^3G_5	49,461	4.117	4.08	4.274	3.88
3,605.450	.454	a^3G_4	y^3H_4	49,727	4.386	4.22	4.56r	
		(z^7F_6)	(f^7D_5)	50,378				}
3,606.6821	.6799	a^3G_5	y^3H_6	49,434	4.38	4.52	4.65r	
3,608.8609	.8591 H	a^5F_1	z^5G_2	35,856	5.239	5.27r	4.78R	3.02
3,610.159	.149	z^7F_6	e^7F_6	50,342	4.353	4.26	4.53r	
3,617.788	.784	c^3P_2	u^3D_3	51,969	4.137	4.01	4.26	4.19
3,618.7694	.7675 H	a^5F_2	z^5G_3	35,612	5.364	5.35r	4.83R	3.18
3,621.4640	.4618	a^3G_4	y^3H_5	49,604	4.33	4.30	4.48r	
3,622.001	.004	a^3G_3	v^3G_2	49,851	4.16	4.11	4.36	4.02

TABLE 7e-6. THE SPECTRUM OF IRON I (*Continued*)

λ_1	λ_2	Classification		E'	$\log I_2$	$\log I_3$	$\log I_4$	$\log \nu A_\nu$
3,623.1878	.1856	a^3H_6	z^3H_6	46,982	4.141	3.88	4.013	3.28
3,631.4646	.4630 H	a^5F_3	z^5G_4	35,257	5.441	5.38	4.85R	3.01
3,634.326	.325	z^7P_4	e^5G_3	51,219	3.53	.....	3.83	3.66
3,638.2998	.2976	a^3G_3	y^3H_4	49,727	3.96	3.95	4.15	3.78
3,640.3918	.3896	a^3G_4	v^3G_5	49,461	4.253	4.21	4.390	4.00
3,645.822	.818	c^3P_0	u^3D_1	52,512	3.56	.....	3.83	3.83
3,647.8439	.8422	a^5F_4	z^5G_5	34,782	5.411	5.30r	4.80R	2.91
3,649.3045	.3025	a^5D_4	z^5F_3	27,395	3.58	4.00		
3,649.5090	.5067	a^3G_5	w^3F_4	49,109	4.397	4.23	3.44	3.99
3,650.031	.026	z^7P_3	e^7S_3	51,570	3.54	4.05	} 3.0	
3,650.2811	.2788	a^3H_5	z^3H_5	47,008	4.141	3.86	}	
3,651.4699	.4676	a^3G_3	v^3G_4	49,628	4.361	4.21	3.471	4.09
3,659.5188	.5165	a^3H_4	z^3H_4	47,107	3.899	3.78	3.843	3.12
3,669.5229	.5206	a^3G_4	w^3F_3	49,243	4.101	3.95	4.19	3.76
3,676.3135	.3112	b^3F_4	x^3G_5	47,835	3.934	3.72	3.844	3.22
3,677.6309	.6286	a^3G_3	w^3F_2	49,433	4.15	4.16	4.38	3.97
3,679.9152	.9129 H	a^5D_4	z^5F_4	27,167	5.071	4.88r	4.36R	1.449
3,682.226	....	a^1D_2	w^1D_2	55,754	4.175	3.97	4.260	4.69
3,683.0562	.0541 N	a^5D_3	z^5F_2	27,560	3.945	3.89	4.10R	0.496
3,684.1102	.1079	a^3G_4	v^3D_3	49,135	4.156	4.04	4.210	3.76
3,685.998	.995	z^7P_4	e^7F_5	50,833	4.01	4.00	4.22	3.99
3,687.4589	.4560 H	a^5F_5	y^5F_4	34,040	4.663	5.11r	4.63R	2.378
3,689.457	.457	z^7P_4	f^7D_4	50,808	3.876	3.97	4.196	
		b^3P_1	w^3P_1	50,043				
3,694.005	.005	z^7P_2	e^7S_3	51,570	4.11	4.16	4.333	4.21
3,695.054	.050	b^3F_3	v^5F_4	47,930	4.014	3.80	3.998	
3,697.426	.424	z^7P_3	e^5G_3	51,219	3.485	3.73	3.837	3.67
3,701.086	.085	z^7P_3	e^7F_4	51,192	4.10	4.11	4.330	4.16
3,703.556	.546	a^3G_3	w^3F_3	49,243	3.47	.....	3.93	
3,704.4635	.4612	a^3G_5	y^1G_4	48,703	3.971	4.12	4.00	3.50
3,705.5674	.5658 H	a^5D_3	z^5F_3	27,395	5.249	4.04r	5.45R	1.698
3,707.048	.041	z^7P_3	e^7F_3	51,149	3.79	3.83	4.040	3.85
3,707.8231	.8214	a^5D_2	z^5F_1	27,666	4.17 }	4.56	4.65R	
3,707.9216	.9200	a^5P_3	y^5S_2	44,512	4.42 }			
3,709.2484	.2458 N	a^5F_4	y^5F_3	34,329	4.758	5.00r	4.66R	2.540
3,716.442	.439	z^7P_4	e^7P_3	50,611	3.877	3.87	4.083	
3,719.9367	.9345 H	a^5D_4	z^5F_5	26,875	5.954	5.73R	4.76R	2.541
3,722.5642	.5629 H	a^5D_2	z^5F_2	27,560	5.10	5.06r	4.45R	1.747
3,724.3796	.3774	a^3P_2	x^3D_3	45,221	4.04	3.99	4.162	3.20
3,727.6211	.6187 N	a^5F_3	y^5F_2	34,547	4.69	4.97r	4.68R	2.543
3,730.3884	.3859	a^1G_4	u^3G_5	51,826	3.64	.....	3.804	3.70
3,732.399	.396	a^5P_2	y^5S_2	44,512	4.29	4.22	4.43r	
3,733.3191	.3168 H	a^5D_1	z^5F_1	27,666	5.00	4.96r	4.46R	1.624
3,734.8659	.8643 N	a^5F_5	y^5F_5	33,695	5.57	5.76R	5.03R	3.475
3,737.1333	.1317 H	a^5D_3	z^5F_4	27,167	5.89	5.57R	4.79R	2.408
3,738.3078	.3053	b^3H_5	z^1I_6	53,094	4.31	3.86	4.19	4.27
3,743.3640	.3614 N	a^5F_2	y^5F_1	34,692	4.53	4.77	4.75R	2.392
3,745.5623	.5602	a^5D_2	z^5F_3	27,395	5.66	5.38R	4.7R	2.25
3,745.9013	.8988 N	a^5D_0	z^5F_1	27,666	5.09	4.96r	.....	1.71
3,748.2639	.2618 N	a^5D_1	z^5F_2	27,560	5.41	5.19R	4.61R	1.990
3,749.4875	.4852 N	a^5F_4	y^5F_4	34,040	5.43	5.57R	4.98R	3.310

TABLE 7e-6. THE SPECTRUM OF IRON I (*Continued*)

λ_1	λ_2	Classification		E'	$\log I_2$	$\log I_3$	$\log I_4$	$\log \nu A_\nu$
3,753.6134	.6111	a^5P_3	w^5D_2	44,184	3.75	3.92	4.134	3.04
3,758.2350	.2326 H	a^5F_3	y^5F_3	34,329	5.25	5.35r	4.90R	3.139
3,760.052	.0491 H	a^3H_6	z^3I_7	45,978	4.51	3.88	4.130	3.27
3,763.7910	.7887 H	a^5F_2	y^5F_2	34,547	5.01	5.17r	4.81R	2.926
3,765.5414	.5385 H	b^3H_6	y^3I_7	52,655	4.52	4.25	4.60r	
3,767.1939	.1912 H	a^5F_1	y^5F_1	34,692	4.75	5.03r	4.89R	2.785
3,785.950	.950	a^3H_5	z^3I_6	46,027	4.36	3.86	4.04	3.11
3,786.6781	.6764	a^5F_1	z^3P_0	34,556	3.91	3.86	3.93r	1.44
3,787.8825	.8800 H	a^5F_1	y^5F_2	34,547	4.450	4.76	4.63R	2.290
3,790.0943	.0923 H	a^5F_2	z^3P_1	34,363	4.345	4.22	4.32R	1.62
3,794.340	....	a^3H_4	z^3I_5	46,136	4.226	3.74	3.936	3.09
3,795.0045	.0017 N	a^5F_2	y^5F_2	34,329	4.580	4.89	4.69R	2.384
3,797.517	.514	b^3H_6	w^3H_6	52,431	4.091	4.01	4.344	4.32
3,798.5134	.5110 N	a^5F_4	y^5F_5	33,695	4.421	4.66	4.61R	2.028
3,799.5498	.5468 H	a^5F_3	y^5F_4	34,040	4.577	4.82	4.69R	2.306
3,805.3450	.3424 H	b^3H_4	y^3I_5	52,889	4.304	4.18	4.440	4.50
3,806.6992	.6966	b^3H_5	w^3H_5	52,613	3.945	3.98	4.24	
		(b^3F_3)	(w^3D_2)	47,136				
3,807.5392	.5369	a^5P_1	w^5D_2	44,184	3.59	3.90	4.076	2.99
3,812.9658	.9638 H	a^5F_3	z^3P_2	33,947	4.784	4.70	4.68R	2.16
3,814.5247	.5230	a^5F_1	z^3P_1	34,363	3.74	3.80	3.90R	1.35
3,815.8430	.8401 H	a^3F_4	y^3D_3	38,175	5.291	5.19r	4.98R	3.36
3,820.4274	.4251 H	a^5F_5	y^5D_4	33,096	5.444	5.36r	4.98R	3.233
3,821.1807	.1781	b^3H_5	y^3I_6	52,514	4.21	.....	4.48	4.48
3,824.4455	.4432 H	a^5D_4	z^5D_3	26,140	5.357	5.04r	4.65R	1.634
3,825.8834	.8808 H	a^5F_4	y^5D_3	33,507	5.240	5.42r	4.99R	3.094
3,827.8256	.8227 H	a^3F_3	y^3D_2	38,678	5.091	5.06r	4.96R	3.31
3,834.2244	.2219 H	a^5F_3	y^5D_2	33,802	4.973	5.11r	4.83R	2.846
3,839.2584	.2537	a^1G_4	x^1G_4	50,614	4.114	3.98	4.15	3.90
3,840.4397	.4376 N	a^5F_2	y^5D_1	34,017	4.697	5.02r	4.72R	2.609
3,841.0499	.0476 H	a^3F_2	y^3D_1	38,996	4.942	4.98	4.86R	3.19
3,843.2596	.2567 H	a^1G_4	z^1F_3	50,587	4.160			
3,846.8023	.8003 H	a^3D_3	t^3D_3	52,213	3.938	3.95	4.22	4.18
3,849.9694	.9591 H	a^5F_1	y^5D_0	34,122	4.326	4.80	4.65R	2.34
3,850.8193	.8175	a^5F_2	z^3P_2	33,947	4.083	4.25	4.34R	1.63
3,852.5752	.5728	a^5P_3	w^5D_4	43,500	3.381	3.78	3.938	2.76
3,856.3731	.3713 H	a^5D_3	z^5D_2	26,340	5.365	5.08r	4.25R	1.691
3,859.2143	.2110	a^3H_6	y^3G_5	45,295	4.17	.....	4.31	3.36
3,859.9132	.9121 H	a^5D_4	z^5D_4	25,900	5.978	5.52R	4.76R	2.244
3,865.5256	.5228 N	a^5F_1	y^5D_1	34,017	4.250	4.72	4.64R	2.25
3,867.2184	.2156 H	c^3P_2	w^3P_2	50,817	3.801	3.82	4.004	3.77
3,872.5032	.5007 H	a^5F_2	y^5D_2	33,802	4.366	4.77	4.63R	2.159
3,873.7624	.7607 H	a^3H_5	y^3G_4	45,428	4.158	3.91	4.11	3.17
3,878.0206	.0179 N	a^5F_3	y^5D_3	33,501	4.36	4.79	4.66R	2.184
3,878.5745	.5731 H	a^5D_2	z^5D_1	26,479	5.257	5.00r	4.68R	1.694
3,885.5121	.5098	a^3P_1	x^3D_2	45,282	3.57	.....	3.92	2.97
3,886.2839	.2820 H	a^5D_3	z^5D_3	26,140	5.619	5.11r	4.60R	1.865
3,887.0504	.0474 N	a^5F_4	y^5D_4	33,096	4.303	4.63	4.59R	2.075
3,888.5165	.5134 H	a^3F_2	y^3D_2	38,678	4.459	4.57	4.78R	2.70
3,893.3935	.3909	b^3G_5	v^3G_5	49,461	3.71	3.80	4.04	3.65
3,895.6579	.6562 H	a^5D_1	z^5D_0	26,550	4.907	4.81r	4.43R	1.266

TABLE 7e-6. THE SPECTRUM OF IRON I (*Continued*)

λ_1	λ_2	Classification		E'	$\log I_2$	$\log I_3$	$\log I_4$	$\log \nu A_\nu$
3,897.896	.8898 H	a^3G_5	w^5G_6	47,363	2.54 }	4.09	4.35	
3,898.0111	.0105 N	a^5F_1	y^5D_2	33,802	3.33			
3,899.7086	.7076 H	a^5D_2	z^5D_2	26,340	5.112	4.99r	4.43R	1.402
3,902.9484	.9452 H	a^3F_3	y^3D_3	38,175	4.624	4.72	4.78R	2.87
3,903.9011	.8984	b^3G_4	y^3H_4	49,727	3.20	3.60	3.794	3.43
3,906.4814	.4792 H	a^5D_1	z^5D_1	26,479	4.371	4.40	4.28R	0.816
3,907.464	....	a^3G_3	x^3G_3	47,834	3.02			
3,907.9371	.9345	a^3G_3	w^5G_2	47,831	3.51	3.55	3.669	3.05
3,916.733	.731	b^3H_6	6_5	51,630	3.732	3.56	3.79	3.67
3,917.1834	.1814	a^5F_2	y^5D_3	33,507	3.19	3.94	4.01	0.695
3,918.644	....	b^3G_3	v^3G_3	49,851	3.52	.....	3.87	3.53
3,920.2601	.2577 H	a^5D_0	z^5D_1	26,479	4.848	4.74r	4.34R	1.324
3,922.9134	.9113 H	a^5D_3	z^5D_4	25,900	5.084	4.91r	4.42R	1.300
3,925.946	....	b^3P_0	x^3P_1	48,516	3.40	3.63	3.81	3.29
3,927.9216	.9197 H	a^5D_1	z^5D_2	26,340	5.107	4.96	4.51R	1.391
3,930.2981	.2963 H	a^5D_2	z^5D_3	26,140	5.161	5.00r	4.49R	1.389
3,935.8143	.8123 H	b^3P_2	v^5F_2	48,239	3.41	3.62	3.764	
3,940.8797	.8777	a^5F_3	y^5D_4	33,096	2.91	3.66	3.66	0.500
3,942.4418	.4391	b^3P_1	x^3P_2	48,305	3.14	3.54	3.688	3.14
3,948.7778	.7750	b^3H_5	u^3G_4	51,668	3.773	3.72	3.955	3.85
3,949.9558	.9524 H	a^5P_3	x^5P_2	42,860	3.804	3.92	3.996	2.75
3,951.164	.1634 H	a^3D_1	y^1D_2	51,708	3.715	3.66	3.879	3.78
3,952.6045	.6013 H	a^3G_5	z^3H_5	47,008	3.680	3.59	3.802	3.07
3,956.6796	.6769 H	a^3G_5	z^3H_6	46,982	4.428	4.03	4.49	3.76
3,966.0645	.0620	a^3F_2	y^3D_2	38,175	3.41	3.85	4.04r	
3,966.630	.627	z^5D_4	f^5F_5	51,103	3.781	3.79	4.055	
3,967.4234	.4206	b^3H_4	u^3G_3	51,826	3.07	3.59	3.836	3.74
3,969.2595	.2567 H	a^3F_4	y^3F_3	37,163	4.796	4.81	4.85R	2.77
3,971.3250	.3223	a^3G_5	x^3F_4	46,889	3.54	3.64	3.865	3.13
3,977.7437	.7411 H	a^5P_2	x^5P_2	42,860	3.932	3.90	4.121	2.88
3,981.7743	.7710 H	a^3G_4	z^3H_4	47,107	3.593	3.55	3.686	2.95
3,983.9593	.9568 H	a^3G_4	x^3F_3	47,197	3.677	3.72	3.880	3.18
3,997.3952	.3921 H	a^3G_4	z^3H_5	47,008	4.300	4.10	4.290	3.56
3,998.0554	.0527	a^3G_5	u^5D_4	46,721	3.613	3.78	3.981	3.21
4,005.2440	.2415 H	a^3F_3	y^3F_2	37,521	4.591	4.64	4.76R	2.66
4,009.7154	.7128 H	a^5P_1	x^5P_2	42,860	3.772	3.78	3.994	2.73
4,014.534	.5308 H	a^1H_5	y^1H_5	53,722	3.934	3.69	3.962	4.13
4,021.8696	.8663 H	a^3G_3	z^3H_4	47,107	3.990	3.75	4.033	3.31
4,045.8147	.8139 H	a^3F_4	y^3F_4	36,686	5.565	5.39r	5.08R	3.34
4,062.4440	.4409 H	b^3P_1	y^3S_1	47,556	3.716	4.04	3.90	3.25
4,063.5963	.5942 H	a^3F_3	y^3F_3	37,163	5.247	5.20r	4.96R	3.19
4,066.979	.974	b^3P_2	1_2	47,420	3.686	3.49	3.66	2.97
4,067.2738	.2711	b^3F_4	x^3D_3	45,221	3.37			
4,067.984	.978	z^5D_4	e^7P_4	50,475	3.720	3.66	3.89	3.63
4,071.7399	.7371 H	a^3F_2	y^3F_2	37,521	5.114	4.99r	4.98R	3.14
4,076.636	.6294 H	z^5D_4	f^5D_4	50,423	3.641	3.63	3.940	3.66
4,100.738	.7374 H	a^5F_5	z^3F_4	31,307	3.627	3.38	3.279	0.48
4,107.4917	.4880 H	b^3P_2	u^5D_1	47,177	3.638	3.72	3.957	3.26
4,109.8053	.8016 H	b^3P_1	w^3D_1	47,272	3.544	3.56	3.784	3.09
4,114.4485	.4456	b^3P_2	w^3D_2	47,136	3.25	3.37	3.478	2.78

TABLE 7e-6. THE SPECTRUM OF IRON I (*Continued*)

λ_1	λ_2	Classification		E'	$\log I_2$	$\log I_3$	$\log I_4$	$\log \nu A_\nu$
4,118.5484	.5446 H	a^1H_5	z^1I_6	53,094	4.225	3.93	4.30	4.38
4,120.2087	.2061 H	b^3G_4	z^1H_5	48,383	3.302	3.30	3.393	2.85
4,121.8050	.8021	b^3P_2	x^3F_3	47,093	3.281	3.34	3.420	2.70
4,127.6113	.6083 H	b^3P_0	w^3D_1	47,272	3.581	3.55	3.81	3.12
4,132.0603	.0576 H	a^3F_2	y^3F_3	37,163	4.581	4.53	4.81R	2.58
4,132.9024	.8995	b^3P_1	w^3D_2	47,136	3.512	3.63	3.86	
		(a^3F_2)	(y^5P_2)	37,158				
4,134.6798	.6770 H	b^3P_2	w^3D_3	47,017	3.929	3.86	4.17	3.44
4,137.002	.9974 H	a^1P_1	y^1D_2	51,708	3.566	3.45	3.677	3.57
4,143.4174	.4145	a^1G_4	y^1G_4	48,703	4.298	3.97		
4,143.8703	.8680 H	a^3F_3	y^3F_4	36,686	4.862	4.70	4.86R	2.68
4,147.6719	.6687 H	a^3F_4	z^3G_3	36,079	3.399	3.65	3.81	1.64
4,149.372	.3658 H	z^5F_5	e^7G_6	50,968	3.15	3.31	3.446	3.25
4,152.1704	.1693 H	a^5F_3	z^3F_3	31,805	3.474	3.33	3.26	0.52
4,153.906	.901	z^5F_3	f^5F_4	51,462	3.616	3.74	3.909	3.78
4,154.5021	.4992	b^3P_2	y^3P_1	46,902	3.749	3.53	4.15	
4,156.8021	.7992	b^3P_2	u^5D_2	46,889	3.781	3.76	4.064	3.32
4,157.788	.781	z^5F_2	f^5F_3	51,604	3.448	3.57	3.726	3.61
4,170.9044	.9014	c^3P_2	x^3P_2	48,305	3.300	3.40	3.575	3.03
4,172.126	....	a^3D_3	w^3P_2	50,187	3.323	3.37	3.57	3.27
4,172.7454	.7445	a^5F_3	z^3D_2	31,686	3.678	3.45	3.519	0.77
4,174.9137	.9128	a^5F_4	z^3D_3	31,323	3.783	3.48	3.452	0.65
4,175.6386	.6356	b^3P_1	u^5D_2	46,889	3.705	3.74	4.004	3.26
4,176.571	.566	z^5F_4	f^5F_5	51,103	3.358	3.49	3.638	
		(z^5F_3)	(e^7F_2)	51,331				
4,177.5949	.5932 H	a^5F_4	z^3F_4	31,307	3.747	3.44	3.393	1.13
4,181.7571	.7542 H	b^3P_2	u^5D_3	46,745	4.125	4.11	4.427	3.66
4,184.8941	.8914 H	b^3P_2	y^3P_2	46,727	3.665	3.66	3.904	3.13
4,187.0436	.0371	z^7D_3	e^7D_2	43,634	4.110	4.12	4.48r	
4,187.8015	.7950	z^7D_4	e^7D_3	43,435	4.146	4.12	4.49r	
4,191.4358	.4297 H	z^7D_2	e^7D_1	43,764	3.923	4.04	4.336	3.19
4,195.337	....	z^5F_5	e^5G_5	50,704	3.551	3.63	3.80	3.56
4,196.218	.209	z^5F_3	e^5G_3	51,219	3.30	3.37	3.54	3.37
4,198.3098	.3036 H	z^7D_5	e^7D_4	43,163	4.161	4.11	4.46r	
4,199.0981	.0948 H	a^1G_4	z^7H_5	55,526	4.620	4.23	4.64r	
4,202.0320	.0282 H	a^3F_4	z^3G_4	35,768	4.540	4.66	4.81R	2.47
4,203.9867	.9878	b^3P_1	y^3P_2	46,727	3.619	3.60	3.852	3.08
4,206.6985	.6953 H	a^5D_3	z^7P_3	24,181	3.85?	3.35	3.30	
4,207.1208	.1208	b^3P_2	z^3S_1	46,601	3.03	3.26		
4,210.3497	.3431	z^7D_1	e^7D_1	43,764	3.87	3.86	4.124	
4,213.650	.647	b^3P_1	y^3P_0	46,673	3.30	3.33	3.425	2.98
4,216.1854	.1826 H	a^5D_4	z^7P_4	23,711	4.636	3.83	3.83r	−0.16
4,217.551	.545	z^5F_1	e^5G_2	51,370	3.180	3.51	3.698	3.55
4,219.3641	.3597 H	a^1H_5	y^1I_6	52,514	4.019	3.80	4.124	4.12
4,222.2181	.2128 H	z^7D_3	e^7D_3	43,435	3.717	3.86	4.097	2.90
4,224.176	.171	z^5F_4	e^7F_5	50,833	3.400	3.57	3.91	3.81
4,225.460	.454	z^5F_2	e^5G_3	51,219	3.347	3.55	3.756	3.59
4,227.434	.4257 H	z^5F_5	e^5G_6	50,523	4.268	4.15	4.520	3.86
4,231.525	....	a^3D_3	v^3G_3	49,851	2.84	.....	.....	3.55
4,232.732	.7261	a^5D_2	z^7P_2	24,507	3.02	.....	.....	−1.13
4,233.6089	.6019 H	z^7D_1	e^7D_2	43,634	4.021	4.06	4.42r	2.95
4,235.9433	.9361 H	z^7D_4	e^7D_4	43,163	4.432	4.27	4.67r	3.17

ATOMIC AND MOLECULAR PHYSICS

TABLE 7e-6. THE SPECTRUM OF IRON I (*Continued*)

λ_1	λ_2	Classification		E'	$\log I_2$	$\log I_3$	$\log I_4$	$\log \nu A_\nu$
4,238.816	.8087 H	z^5F_3	e^5G_4	50,980	3.661	3.81	3.982	3.78
4,239.847	....	a^3G_5	y^3G_5	45,295	2.67	....	....	−0.80
4,245.2594	.2564 H	b^3P_0	z^3S_1	46,661	3.191	3.43	3.570	2.80
4,247.432	.4246 H	z^5F_4	e^5G_5	50,704	3.749	3.75	4.008	3.77
4,248.2275	.2244	c^3P_1	x^3P_2	48,305	2.969	3.18	3.267	2.72
4,250.1248	.1181	z^7D_2	e^7D_3	43,435	4.278	4.22	4.59r	3.01
4,250.7896	.7867	a^3F_3	z^3G_3	36,079	4.508	4.59	4.76R	2.45
4,258.3174	.3150 H	a^5D_2	z^7P_3	24,181	3.573	.....	2.99	−0.84
4,260.4794	.4733 H	z^7D_5	e^7D_5	42,816	4.894	4.62	4.95r	3.41
4,267.830	.826	c^3P_0	x^3P_1	48,516	3.14	3.33	3.417	2.90
4,271.1589	.1521	z^7D_3	e^7D_4	43,163	4.40	4.25	4.67r	3.12
4,271.7634	.7601 H	a^3F_4	z^3G_5	35,379	5.088	4.96r	4.95R	2.88
4,282.4057	.4026 H	a^5P_3	z^5S_2	40,895	4.391	4.12	4.48r	2.85
4,285.4453	.4422	b^3H_6	y^3H_6	49,434	3.08	3.23	3.276	2.87
4,291.466	.4627 H	a^3F_3	z^5G_2	35,856	3.881	3.36	3.215	
		a^5D_3	z^7P_4	23,711	.....	.....	.....	−0.86
4,294.1271	.1240 H	a^3F_4	z^5G_4	35,257	4.148	4.35	4.65R	2.07
4,298.0403	.0371	a^1G_4	x^3G_5	47,835	3.313	3.23	3.255	2.64
4,299.2409	.2338 H	z^7D_4	e^7D_5	42,816	4.394	4.23	4.66r	2.82
		(b^3H_5)	(y^3H_5)	49,604				
4,305.4545	.4513	c^3P_2	y^3S_1	47,556	3.20	3.29	3.344	2.69
4,307.9048	.9014 H	a^3F_3	z^3G_4	35,768	5.129	4.91r	4.93R	3.01
4,309.3771	.3739	b^3G_5	z^3H_6	46,982	3.524	3.44	3.60	2.87
4,315.0872	.0837 H	a^5P_2	z^5S_2	40,895	4.212	4.03	4.31	2.78
4,325.7647	.7615 H	a^3F_2	z^3G_3	36,079	5.181	4.96r	4.95R	3.06
		(a^5D_4)	(z^7F_3)	23,111				
4,327.100	....	a^1D_2	y^1D_2	51,708	3.310	.....	3.38	3.15
4,337.0484	.0459 H	a^3F_3	z^5G_5	35,612	3.471	3.98	4.15r	1.59
4,347.239	....	a^5D_4	z^7F_4	22,997	2.53	.....	.....	−1.16
4,352.7371	.7337 H	a^5P_1	z^5S_2	40,895	3.9	3.82	3.998	2.47
4,367.5811	.5774 H	b^3G_4	z^3H_5	47,008	3.400	3.32		
4,369.7745	.7711 H	a^1G_4	z^1G_4	47,453	3.910	3.55	3.699	3.04
4,375.9318	.9290 H	a^5D_4	z^7F_5	22,846	4.945	4.04	4.11R	0.11
4,383.5473	.5449 H	a^3F_4	z^5G_5	34,782	5.472	4.99r	5.08R	3.23
4,388.412	.407	z^5P_3	e^5P_2	51,837	3.200	3.36	3.441	3.34
4,390.9542	.9509	b^3G_3	z^3H_4	47,107	3.217	3.20	3.110	2.39
4,404.7525	.7503 H	a^3F_3	z^5G_4	35,257	5.068	4.91	4.95R	2.93
4,408.4176	.4147	a^5P_2	x^5D_1	40,405	2.62	3.53	3.599	2.02
4,415.1250	.1222 H	a^3F_2	z^5G_3	35,612	4.528	4.71	4.81R	2.45
4,422.5703	.5675 P_1	b^3P_1	x^3D_1	45,552	3.483	3.53	3.669	2.76
4,427.3118	.3093 H	a^5D_3	z^7F_4	23,193	4.823	3.99	4.08R	0.09
4,430.6175	.6145	a^5P_1	x^5D_0	40,491	2.67	3.57	3.66	2.08
4,433.223	.220	z^5P_2	e^5P_1	52,020	3.12	3.25	3.328	3.26
4,442.3428	.3398	a^5P_2	x^5D_2	40,231	2.33	3.87	4.06	2.44
4,443.1963	.1929	b^3P_0	x^3D_1	45,552	3.509	3.57	3.72	2.81
4,445.48	.4699	a^5D_2	z^7F_2	23,193	2.39	.....	.....	1.43
4,447.7212	.7182	a^5P_1	x^5D_1	40,405	3.017	3.78	3.958	2.36
4,450.320	.316	c^3P_0	y^3S_1	47,556	3.39	.....	2.432	1.78
4,454.3835	.3803	b^3P_2	x^3D_2	45,282	3.364	3.41	3.484	2.53
4,459.1213	.1183	a^5P_3	x^5D_3	39,970	3.24	3.89	4.072	2.42
4,461.6544	.6523 H	a^5D_2	z^7F_3	23,111	4.576	3.94	3.88R	−0.10
4,466.5542	.5501 H	b^3P_2	x^3D_3	45,221	4.057	3.93	4.164	
		(a^5D_1)	(z^7F_0)	23,270	.....	.....	.....	−1.07
4,469.381	.3742 H	z^5P_2	e^5P_3	51,837	3.391	3.47	3.614	3.51

TABLE 7e-6. THE SPECTRUM OF IRON I (*Continued*)

λ_1	λ_2	Classification		E'	$\log I_2$	$\log I_3$	$\log I_4$	$\log \nu A_\nu$
4,476.0206	.0168 H	b^3P_1	x^3D_2	45,282	3.895	3.85	4.086	3.14
4,482.1720	.1684 H	a^5D_1	z^7F_2	23,193 ⎫				
4,482.2563	.2533	a^5P_1	x^5D_2	40,231 ⎭	4.4	3.9	4.0	−0.36
4,489.7416	.7391 H	a^5D_0	z^7F_1	23,245	3.741	3.41	.....	−0.77
4,494.5669	.5627 H	a^5P_2	x^5D_3	39,970	3.353	3.98	4.182	2.53
4,517.5289	.5254	c^3P_1	y^3P_1	46,902	2.40	.....	2.724	2.65
4,528.6175	.6132 H	a^5P_3	x^5D_4	39,626	3.747	4.17	4.46r	2.74
4,531.1520	.1485	a^3F_4	y^5F_4	34,040	3.050	.....	3.804	1.36
4,547.8505	.8468	a^1D_2	z^1F_3	50,587	3.409	.....	3.425	3.35
4,592.6547	.6511	a^3F_3	y^5F_3	34,329	2.77	.....	3.500	1.10
4,602.9446	.9388 J	a^3F_4	y^5F_5	33,695	3.123	.....	3.774	1.10
4,647.4370	.4327 J	b^3G_5	y^3G_5	45,295	3.532	.....	3.473	2.51
4,667.459	.4519 J	z^5P_3	e^7P_4	50,475	3.231	.....	3.455	3.20
4,678.852	.8440 J	z^5P_3	f^5D_4	50,423	3.254	.....	3.556	3.28
4,691.4144	.4101 J	b^3G_4	y^3G_4	45,428	3.345	.....	3.330	2.39
4,707.2807	.2727 J	z^5D_3	e^5F_4	47,378	3.342	.....	3.525	2.86
4,710.2864	.2825 J	b^3G_3	y^3G_3	45,563	3.26	.....	3.127	2.22
4,733.5955	.5926	a^3F_4	y^5D_4	33,096	2.42	.....	3.025	0.47
4,736.7807	.7717 J	z^5D_4	e^5F_5	47,006	3.517	.....	3.798	3.07
4,741.5321	.5205	b^3P_2	w^5D_3	43,923	2.44	.....	2.87	1.74
4,745.806 ⎫	.801	z^5P_2	f^5D_3	50,534	2.605	.....	2.86	
⎭		y^5D_4	f^5G_3	54,161				
4,772.817 ⎫	.815	c^3P_2	x^3D_2	45,282	2.602	.....	2.84	
⎭		a^3F_3	y^5D_3	33,507				
4,786.8106	.8069	c^3P_2	x^3D_3	45,221	2.888	.....	3.161	2.19
4,789.6537	.6499 J	a^1D_2	z^1D_2	49,477	3.415	.....	3.301	2.92
4,859.7480	.7399 J	z^7F_2	e^7D_1	43,764	3.654	.....	4.017	2.87
4,871.3244	.3177 J	z^7F_3	e^7D_2	43,634	4.096	.....	4.529	3.36
4,872.1444	.1369 J	z^7F_1	e^7D_1	43,764	3.790	.....	4.207	3.05
4,878.2182	.2094 J	z^7F_0	e^7D_1	43,764	3.527	.....	3.894	2.74
4,890.7616	.7541 J	z^7F_2	e^7D_2	43,634	4.049	.....	4.352	3.18
4,891.4989	.4915 J	z^7F_4	e^7D_3	43,435	4.404	.....	4.64r	3.44
4,903.3169	.3085 J	z^7F_1	e^7D_2	43,634	3.513	.....	3.852	2.68
4,919.0003	.9922 J	z^7F_3	e^7D_3	43,435	4.178	.....	4.410	3.21
4,920.5096	.5016 H	z^7F_5	e^7D_4	43,163	4.681	.....	4.80r	
4,924.7753	.7717	a^3P_2	y^3D_2	38,678	2.75	.....	3.030	1.28
4,938.8206	.8223 J	z^7F_2	e^7D_3	43,435	3.438	.....	3.74	2.54
4,939.6896	.6859	a^5F_5	z^5F_4	27,167	3.024	.....	3.350	0.00
4,957.3054	.2988 J	z^7F_4	e^7D_4	43,163	3.14			
4,957.6059	.5952 H	z^7F_6	e^7D_5	42,816	5.16	.....	5.0R	
4,966.0968	.0871 J	z^5F_5	e^5F_5	47,005	3.400	.....	3.614	2.88
4,982.507	.4977 J	y^5D_4	f^5P_3	53,161	3.430	.....	3.714	3.81
4,994.1323	.1284 J	a^5F_4	z^5F_3	27,395	3.191	.....	3.410	0.09
5,001.871	.8616 J	z^3F_4	e^3D_3	51,294	3.861	.....	3.895	
5,006.1254	.1172 J	z^7F_5	e^7D_5	42,816	4.051	.....	4.176	2.89
5,012.0712	.0672 J	a^5F_5	z^5F_5	26,875	3.791	.....	3.887	0.48
5,014.950	.9413 J	z^3F_3	e^3D_2	51,740	3.538	.....	3.682	3.57
5,041.7585	.7544 J	a_3F_4	z^3F_3	31,806	4.241	.....	3.748	1.01
5,049.8253	.8187 J	a^3P_2	y^3D_3	38,175	3.506	.....	3.979	2.10
5,051.6379	.6337 J	a^5F_4	z^5F_4	27,167	3.523	.....	3.690	0.34
5,079.2279	.2240	a^5P_2	z^5P_1	37,410	3.732	.....	3.557	1.56
5,083.3413	.3374	a^5F_3	z^5F_3	26,875	3.278	.....	3.492	0.10

TABLE 7e-6. THE SPECTRUM OF IRON I (*Continued*)

λ_1	λ_2	Classification		E'	log I_2	log I_3	log I_4	log νA_ν
5,110.4139	.4120 J	a^5D_4	z^7D_4	19,562	4.238	.....	3.613	−0.85
		(a^1H_5)	(z^1H_5)	48,383				
5,123.7231	.7192	a^5F_1	z^5F_1	27,666	3.323	.....	3.415	0.18
5,127.3624	.3585	a^5F_4	z^5F_5	26,875	3.002	.....	3.212	−0.14
5,133.692	.6885 J	y^5F_5	f^5G_6	53,169	3.577	.....	3.786	3.89
5,150.8425	.8385	a^5F_2	z^5F_3	27,395	2.506	.....	3.322	0.50
5,166.2841	.2814 J	a^5D_4	z^7D_5	19,351	3.901	.....	3.190	−1.50
5,167.4905	.4878 H	a^3F_4	z^3D_3	31,323	5.37	.....	4.71R	−1.67
5,168.9003	.8974 J	a^5D_2	z^7D_3	19,757	3.926	.....	3.48r	−1.03
5,171.5987	.5955 H	a^3F_4	z^3F_4	31,307	4.651	.....	4.23R	1.25
5,191.4615	.4544 J	z^7P_2	e^7D_1	43,764	3.701	.....	4.080	2.93
5,192.3509	.3437 J	z^7P_3	e^7D_2	43,435	3.914	.....	3.250	2.05
5,194.9441	.9410 J	a^3F_2	z^3F_3	31,805	4.275	.....	3.88r	0.96
5,198.7149	.7108	a^5P_1	y^5P_2	37,158	2.39	.....	3.32	1.30
5,202.3395	.3364 J	a^5P_3	y^5P_3	36,767	2.85	.....	3.725	1.77
5,204.5840	.5822 J	a^5D_2	z^7D_2	19,913	3.464	.....	2.86	−1.74
5,216.2770	.2738 J	a^3F_2	z^3F_2	32,134	4.171	.....	3.78	1.08
5,225.531	.5253 J	a^5D_1	z^7D_1	20,020	2.90	.....	.....	−1.78
5,227.1911	.1876 H	a^3F_3	z^3D_2	31,686	5.02	.....	4.93	
5,232.9474	.9400 H	z^7P_4	e^7D_5	42,816	4.436	.....	4.61r	2.95
5,235.392	.3858 J	b^3F_3	x^5D_3	39,970	2.73	.....	2.96	
		c^3F_4	u^3D_3	51,969				
5,236.204	....	c^3F_2	8_1	52,858	1.83			
5,242.4955	.4903 J	a^1I_6	z^1H_5	48,383	3.20	.....	3.326	3.73
5,247.065	.0494 J	a^5D_2	z^7D_3	19,757	2.89	.....	.....	−2.00
5,250.211	.216	a^5D_0	z^7D_1	20,020	2.44	.....	.....	−1.93
5,250.6490	.6449	a^5P_2	y^5P_3	36,767	2.78	.....	3.402	1.32
5,263.3134	.3038 J	z^5D_2	e^5D_2	45,334	3.195	.....	3.60	2.65
5,266.5626	.5548 J	z^7P_3	e^7D_4	43,163	4.033	.....	4.281	3.06
5,269.5402	.5363 J	a^5F_5	z^5D_4	25,900	5.058	.....	4.68r	1.45
5,270.3602	.3557 J	a^3F_2	z^3D_1	31,937	4.914	.....	.....	1.48
5,281.7970	.7896 J	z^7P_2	e^7D_3	43,435	3.477	.....	3.832	2.63
5,283.6283	.6203 J	z^5D_3	e^5D_3	45,061	3.811	.....	4.045	3.07
5,302.3073	.2991 H	z^5D_1	e^5D_2	45,335	3.423	.....	3.736	2.79
5,307.3633	.3604 H	a^3F_2	z^3F_3	31,805	3.337	.....	3.00	0.26
5,324.1864	.1784 H	z^5D_4	e^5D_4	44,677	4.182	.....	4.393	3.36
5,328.0418	.0386 J	a^5F_4	z^5D_3	26,140	4.867	.....	4.70R	1.29
5,328.5336	.5309 J	a^3F_3	z^3D_3	31,323	4.507	.....	4.20r	1.19
5,332.9020	.8987 H	a^3F_3	z^3F_4	31,307	3.951	.....	3.155	0.36
5,339.9371	.9286 H	z^5D_2	e^5D_3	45,061	3.874	.....	3.846	2.87
5,341.0255	.0236 H	a^3F_2	z^3D_2	31,686	4.65	.....	4.00r	1.11
5,364.874	.8717 J	z^5G_2	e^5H_3	54,491	3.384	.....	3.64	3.91
5,367.470	.4671 H	z^5G_3	e^5H_4	54,237	3.564	.....	3.79	4.02
5,369.965	.9621 H	z^5G_4	e^5H_5	53,874	3.725	.....	3.91	4.10
5,371.4926	.4892 H	a^5F_3	z^5D_2	26,340	4.622	.....	4.61R	1.10
		(z^3G_4)	(c^3G_3)	54,379				
5,383.374	.3689 H	z^5G_5	e^5H_6	53,353	3.844	.....	4.11	4.23
5,397.1311	.1272 H	a^5F_4	z^5D_4	25,900	4.459	.....	4.43R	0.81
5,404.144	.1185 J	z^3G_4	e^3H_5	54,267	3.819	.....	4.08	
5,405.7781	.7744 H	a^5F_2	z^5D_1	26,479	4.353	.....	4.49R	0.86
5,424.072	.0686 H	z^5G_6	e^5H_1	53,275	3.842	.....	4.08	4.19
5,429.6999	.6963 H	a^5F_3	z^5D_3	26,140	4.414	.....	4.48R	0.89
5,434.5268	.5237 H	a^5F_1	z^5D_0	26,550	4.048	.....	4.28R	0.72

TABLE 7e-6. THE SPECTRUM OF IRON I (*Continued*)

λ_1	λ_2	Classification		E'	$\log I_2$	$\log I_3$	$\log I_4$	$\log \nu A_\nu$
5,446.9197	.9168 H	a^5F_2	z^5D_2	26,340	4.337	.....	4.42R	0.82
5,455.6131	.6093 H	a^5F_1	z^5D_1	26,479	4.144	.....	4.42R	0.72
5,497.5196	.5159 H	a^5F_1	z^5D_2	26,340	3.374	.....	3.60	0.12
5,501.4686	.4633 H	a^5F_3	z^5D_4	25,900	3.299	.....	3.46	−0.06
5,506.7824	.7785 H	a^5F_2	z^5D_3	26,140	3.494	.....	3.68	0.17
5,569.6256	.6174 H	z^5F_2	e^5D_1	45,509	3.541	.....	3.807	2.89
5,572.8501	.8419 H	z^5F_3	e^5D_2	45,334	3.806	.....	4.06	3.11
5,586.7634	.7555 H	z^5F_4	e^5D_3	45,061	4.074	.....	4.43	3.26
5,615.6521	.6434 H	z^5F_5	e^5D_4	44,677	4.262	.....	4.375	3.35
5,624.5501	.5417 H	z^5F_2	e^5D_2	45,334	3.319	.....	3.574	2.62
5,658.8247	.8156 H	z^5F_3	e^5D_3	45,061	3.22	.....	3.597	2.62
5,662.525	.516	y^5F_5	g^5D_4	51,351	3.661	.....	3.241	3.09
7,187.341	....	y^5D_4	e^5F_5	47,006	3.53			
7,445.776	....	y^5F_3	e^5F_3	47,756	3.48			
7,495.088	....	y^5F_4	e^5F_4	47,378	3.53			
7,511.045	....	y^5F_5	e^5F_2	47,006	3.66			
7,586.044	....	z^5G_5 .	e^3F_4	47,961	3.39			
7,780.586	....	z^3G_3	e^3F_2	48,928	3.28			
7,937.166	....	z^5G_5	e^5F_4	47,378	4.040			
7,998.972	....	z^5G_4	e^5F_3	47,756	3.26			
8,046.073	....	z^5G_3	e^3F_2	48,532	3.36			
8,220.406	....	z^5G_6	e^5F_5	47,006	3.69			
8,248.151	....	z^5G_4	e^5F_4	47,378	3.34			
8,327.063	....	a^5P_2	z^5P_1	29,773	3.61			
8,331.941	....	z^3G_5	e^5F_4	47,378	3.11			
8,387.781	....	a^5P_3	z^5P_2	29,469	3.79			
8,661.908	....	a^5P_1	z^5P_2	29,469	3.75			
8,688.633	....	a^5P_3	z^5P_3	29,056	4.161			
8,824.227	....	a^5P_2	z^5P_3	29,056	3.76			

Mercury I.　This spectrum is very useful because of the ease with which it can be obtained.　Any low-pressure mercury tube gives sharp lines; for example, a commercial so-called bactericidal lamp is suitable.　High-pressure lamps give broader lines and very high pressure lamps (commercial type H6) a continuous spectrum.　The mercury spectrum is useful as a general reference spectrum.　Under high dispersion most lines show elaborate isotopic and hyperfine structure because there are six isotopes with considerable abundance: 196 (0.15 percent), 198 (10.12 percent), 199 (16.84 percent), 200 (23.13 percent), 201 (13.2 percent), 202 (29.80 percent), and 204 (6.85 percent).　The two odd ones have lines with hyperfine structure.　The structure of the lines is sometimes useful for obtaining the resolving power of spectrographs (for details of structure, see Schüler and Burns and Adams[1]).　An example is shown in Fig. 7e-5.

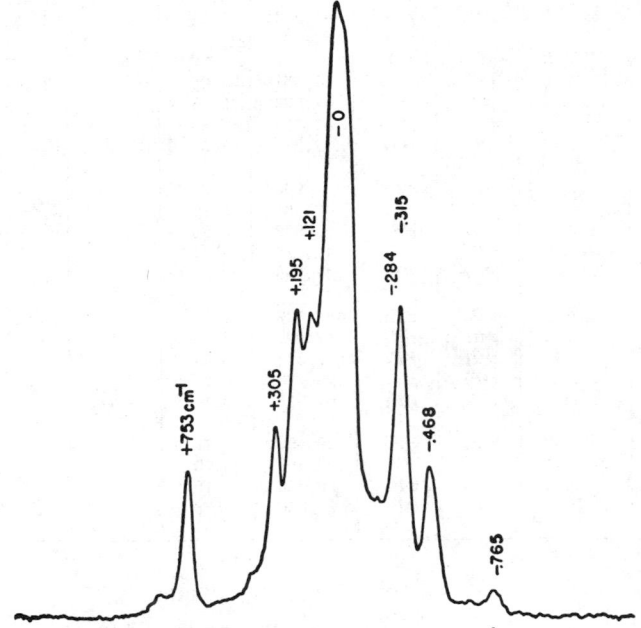

Fig. 7e-5. High-dispersion photoelectric trace of the 5461-Å line of ordinary mercury showing isotope and hyperfine structure.　Resolving power was 400,000.

Pure Hg[198] can be obtained by irradiation of gold with neutrons.　Lamps with this isotope are now commercially available and the spectrum shows very sharp single lines.　Meggers has proposed to adopt the wavelength of the green line (5,461) of Hg[198] as a primary standard of length.　International adoption of this proposal, however, awaits investigation of the variability of the wavelength with discharge conditions.　In the meantime most of the strong lines of Hg[198], particularly those marked S in Table 7e-7, may be used as standards for interferometric wavelength measurements.

Hg[202] is the most abundant isotope in natural mercury.　Tubes with nearly pure Hg[202] are also available and their wavelengths may also be used as standards.

Table 7e-7 gives the wavelengths of natural mercury, Hg[198] and Hg[202].　All values listed between 2,300 and 6,900 Å are recent interferometric wavelengths; those outside this interval are known with much less accuracy.

[1] Schüler and Keyston, *Z. Physik* **72**, 423 (1931); Schüler and Jones, *Z. Physik* **79**, 631 (1932); Burns and Adams, *J. Opt. Soc. Am.* **42**, 716 (1952).

TABLE 7e-7. THE SPECTRUM OF MERCURY I

Classification		λ (Hg nat.)		λ Hg[198]	λ Hg[202]	log I
6^1S	6^1P	1,402.72	O	. . . .	. . . . . .	(4)
6^1S	7^1P	1,849.52	O	. . . .	. . . . . .	(20)
6^1S	7^3P_2	2,296.97	O	. . . .	. . . . . .	
6^3P_0	10^3S	2,345.433	O	45.4400	45.4369	5.33
6^3P_0	8^3D_1	2,378.316	O	78.3246	78.3224	6.60
6^3P_1	10^3S	2,446.895		46.8998	46.8974	4.44
6^3P_0	9^3S	2,464.057		64.0636	64.0614	4.31
6^3P_1	8^3D_2	2,481.996		81.9993	81.9971	5.43
6^3P_1	8^3D_1	2,482.710		82.7131	82.7112	4.94
6^3P_1	8^1D_2	2,483.815		83.8215	83.8196	5.23
6^3P_0	7^3D_1	2,534.764		34.7691	34.7662	6.35
6^1S	6^3P_1	2,536.517		36.5063	36.5277	8.95
6^3P_1	9^1S	. . . .		63.8610	63.8584	
6^3P_1	9^3S	2,576.285		76.2904	76.2882	5.00
6^3P_1	7^3D_2	2,652.039		52.0425	52.0399	6.20
6^3P_1	7^3D_1	2,653.679		53.6827	53.6809	6.75
6^3P_1	7^1D	2,655.127		55.1305	55.1284	5.63
6^3P_2	9^3D_3	2,698.828		98.8314	98.8293	5.35
6^3P_0	8^3S	2,752.778		52.7828	52.7801	5.58
6^3P_2	10^3S	2,759.706		59.7103	59.7077	4.0
6^3P_2	8^3D_3	2,803.465		03.4706	03.4678	5.25
6^3P_2	8^3D_2	2,804.434		04.4378	04.4357	4.56
6^3P_2	8^3D_1	2,805.344		05.347	05.3474	3.49
6^3P_2	8^1D	2,806.759		06.765	06.7630	3.52
6^3P_1	8^1S	2,856.935		56.9389	56.9357	4.30
6^3P_1	8^3S	2,893.594		93.5982	93.5952	5.88
6^3P_2	9^3S	2,925.410		25.4135	25.4104	4.82
6^3P_0	6^3D_1	2,967.280		67.2832	67.2819	6.52
6^3P_0	6^1D	2,967.543				
6^3P_2	7^3D_3	3,021.498		21.4996	21.4973	6.09
6^3P_2	7^3D_2	3,023.475		23.4764	23.4739	5.45
6^3P_2	7^3D_1	3,025.606		25.6080	25.6056	4.43
6^3P_2	7^1D	3,027.487		27.4896	27.4874	4.76
6^3P_1	6^3D_2	3,125.6681		25.6698	25.6675	6.62
6^3P_1	6^3D_1	3,131.5485		31.5513	31.5480	6.48
6^3P_1	6^1D	3,131.8391		31.8423	31.8394	6.56
6^3P_2	8^3S	3,341.4766		41.4814	41.4766	5.85
6^3P_2	6^3D_3	3,650.1533		50.1564*	50.1532	6.94
6^3P_2	6^3D_2	3,654.8363		54.8392	54.8361	6.51
6^3P_2	6^3D_1	3,662.879		62.8826	62.8801	5.70

TABLE 7e-7. THE SPECTRUM OF MERCURY I (*Continued*)

Classification		λ (Hg nat.)	λ Hg^{198}	λ Hg^{202}	log I
6^3P_2	6^1D	3,663.2793	63.2808	63.2778	6.35
6^1P	9^1D	3,704.1655	04.1698	04.1712	3.94
6^1P	8^1D	3,906.371	06.3715	06.3715	4.56
6^3P_0	7^3S	4,046.5630	46.5712*	46.5619	7.09
6^3P_1	7^1S	4,077.8314	77.8379	77.8284	6.00
6^1P	9^1S	4,108.054	08.0574	08.0572	
6^1P	7^3D_2	4,339.2232	39.2244	39.2251	4.74
6^1P	7^1D	4,347.4945	47.4958	47.4967	5.17
6^3P_1	7^3S	4,358.3277	58.3375	58.3257	7.07
6^1P	8^1S	4,916.068	16.0681	16.0677	4.35
6^3P_2	7^3S	5,460.7348	60.7532 S	60.7355	6.76
6^1P_1	6^3D_2	5,769.5982	69.5984 S	69.6000	6.02
6^1P_1	6^3D_1	5,789.664	89.669	89.671	4.41
6^1P	6^1D_2	5,790.6630	90.6628 S	90.6648	5.97
7^3S	8^1P	....	6,072.7128	72.6260	
7^1S	9^1P	....	6,234.4020	34.3776	
7^1S	8^1P	....	6,716.4289	16.3253	
7^3S	8^3P_2	6,907.52 O	07.4612	07.4675	
7^3S	8^3P_1	7,082.01 O			
7^3S	8^3P_0	7,092.20 O			
6^1P	7^1S	10,139.75 O	....	....	6.20
7^3S	7^3P_2	11,287.04 O	....	....	5.98
7^1S	7^1P	13,570.70 O	....	....	5.36
7^3S	7^3P_1	13,673.09 O	....	....	5.53
7^3S	7^3P_0	13,950.75 O	....	....	5.26
		15,295.25 O	....	....	5.78
6^1D	5^1F	16,918.3 O			
7^3P_2	7^3D_3	16,920.97 O			
6^3D_1	5^4F_2	16,942.33 O	....	....	4.72
7^3P_2	7^1D	17,072.67 O	....	....	4.90
6^3D_2	5^3F_3	17,109.57 O	....	....	4.74
6^3D_3	5^3F_4	17,202.08 O			
7^3P_0	8^3S	22,499.29 O			
7^3P_1	8^3S	23,253.47 O	....	... ...	4.49
7^3P_2	8^3S	36,261 O			

Values obtained by Blank[1] for Hg^{198} are 3,650.1569, 4,046.5716, and 4,358.3376.

Intensities are rough photoelectric values obtained at The Johns Hopkins University with a low-pressure neon-mercury discharge. The scale is the same as for neon (Table 7e-2). Intensities may be considerably different for other discharge conditions.

Mercury Tube

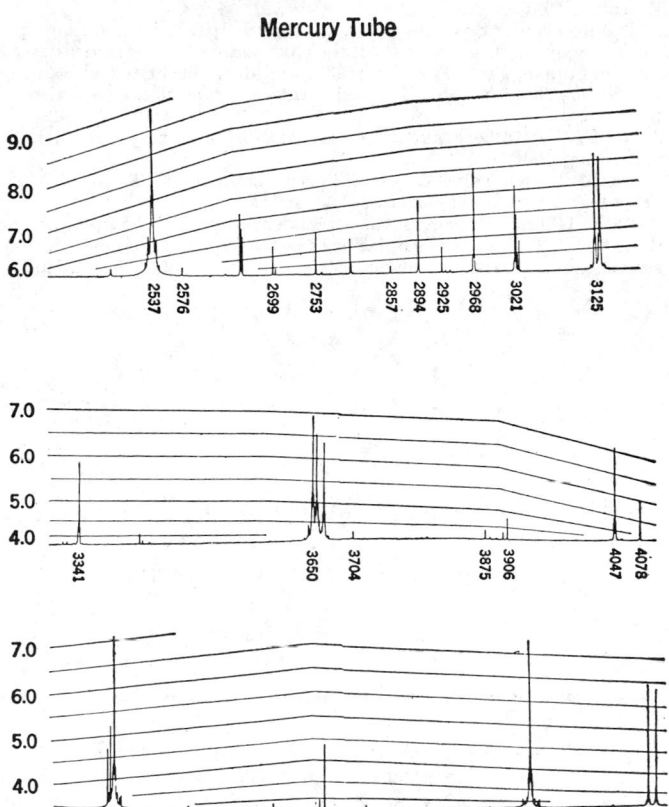

Fig. 7e-6. Photoelectric traces of the mercury spectrum, low-pressure mercury tube, 60 Hz discharge. Wavelength range 2400 to 5800 Å. In order to bring out the weaker lines, the sensitivity was increased so that the ghosts of the strong lines show.

Notes on Table 7e-7. All wavelengths are interferometric values by Burns,[2] except where otherwise noted.

Those marked *O* (natural mercury) are older values, sometimes of questionable accuracy. The values of Hg^{198} marked by * or *S* are averages, the latter proposed for international standards.

[1] Blank, *J. Opt. Soc. Am.* **40,** 345 (1950).
[2] Burns, Adams, and Longwell, *J. Opt. Soc. Am.* **40,** 339 (1950); Burns and Adams, *J. Opt. Soc. Am.* **42,** 56 (1952); **42,** 716 (1952).

References

Brode, W. R.: "Chemical Spectroscopy," 2d ed., 677 pp. John Wiley & Sons, Inc., New York; 1943.
Fowler, A.: "Report on Series in Line Spectra," Fleetway Press, London, 1922.
Gatterer, A., and J. Junkes: "Spektren der seltenen Erden," 2 vols., text and plates, Specolar Vaticana, 1945.
Kayser, H.: "Handbuch der Spektroscopie," 8 vols., S. Hirzel Verlag, Leipzig. Volumes 1–6 (1912) are now out of date but are still the chief source for the earlier developments in spectroscopy. Volumes 7 (1934) and 8 (1932) are more recent but also not quite up to modern standards. This handbook contains the most detailed compilation of spectroscopic data.
Minnaert, M., G. F. W. Mulders, and J. Houtgast: "Photometric Atlas of the Solar Spectrum," Amsterdam, 1940.
Moore, C. E.: A Multiplet Table of Astrophysical Interest: I, Table of Multiplets; II, Finding List, *Contrib. Princeton Univ. Obs.* 20, 1945.
Paschen, F., and R. Götze: "Seriengesetze der Liniensprektren," 154 pp., Springer-Verlag OHG, Berlin, 1922. This and the similar book by Fowler, though now largely out of date nevertheless contain material on the simpler spectra not found conveniently anywhere else.

7f. X-Ray Wavelengths and Atomic Energy Levels

J. A. BEARDEN AND J. S. THOMSEN

The Johns Hopkins University

Tables 7f-1 and 2 list the wavelengths of virtually all experimentally observed X-ray emission lines (excluding satellites) and absorption edges. These are taken from a review by J. A. Bearden [1] and are expressed in terms of the Å* unit, which is defined [2] in terms of the W $K\alpha_1$ line by setting $\lambda_{W\ K\alpha_1} = 0.209\ 0100$ Å*. This figure was chosen to make the Å* unit as close to unity as possible; the difference was then estimated to be zero with a probable error of 5 ppm (parts per million). However the conversion factor must remain an *experimentally determined quantity*.

A number of prominent X-ray reference lines (Cr $K\alpha_2$, Cu $K\alpha_1$, Mo $K\alpha_1$, and Ag $K\alpha_1$) were carefully remeasured in terms of the W $K\alpha_1$ standard [3]. An extensive survey of all experimental X-ray wavelength measurements was made, and the necessary corrections have been applied, as far as possible, to put each one on a basis consistent with the above set of reference values. When two or more values of comparable accuracy were available for the same wavelength, appropriate weighted averages were taken. The same procedure was followed with absorption edges. In all cases estimated probable errors were included.

A thorough recheck of the five reference lines is currently in progress. It now appears that at least one of the crystals used in the original work [3] contained significant imperfections. As a result of this redetermination some of the reference wavelength ratios may be shifted by as much as 10 ppm. Furthermore, a recent reevaluation of the atomic constants by Taylor et al. [4], which includes the highly precise a-c Josephson effect work, indicates that the conversion factor between Å* units and angstroms may differ from unity by about 20 ppm. As a result of these two developments most of the wavelengths listed below must be considered to have probable errors of no less than 10 ppm in terms of Å* units and perhaps somewhat larger errors in terms of angstroms.

Bearden and Burr [5] combined the emission line data of this table with the photoelectron measurements of Hagström, Nordling, and Siegbahn [6] to obtain a revised set of X-ray atomic energy levels. A separate least-squares adjustment was carried out for each element to obtain the values most consistent with all available data. The results are given below in Table 7f-3.

While the reviews [1,5] cited above outline the general principles of the wavelength and energy-level evaluations, full details are found only in the original separately published reports [7,8]. These must be consulted for references to all of the original papers used and for details of the procedures employed in the evaluations.

References

1. Bearden, J. A.: Rev. Mod. Phys. 39, 78 (1967).
2. Bearden, J. A.: Phys. Rev. 137, B455 (1965).
3. Bearden, J. A., A. Henins, J. G. Marzolf, W. C. Sauder, and J. S. Thomsen: Phys. Rev. 135, A899 (1964).
4. Taylor, B. N., W. H. Parker, and D. N. Langenberg: Rev. Mod. Phys. 41, 375 (1969).
5. Bearden, J. A., and A. F. Burr: Rev. Mod. Phys. 39, 125 (1967).
6. Hagström, S., C. Nordling, and K. Siegbahn: "Alpha-, Beta-, and Gamma-ray Spectroscopy," vol. 1, p. 845, K. Siegbahn, ed., North-Holland Publishing Company, Amsterdam, 1965.
7. Bearden, J. A., and collaborators: X-ray Wavelengths, AEC Rept. NYO-10586, 1964. Price $8.45. (Available from Clearinghouse for Federal Scientific and Technical Information, National Bureau of Standards, U.S. Department of Commerce, Springfield, Va. 22151.)
8. Bearden, J. A., and A. F. Burr: Atomic Energy Levels, AEC Rept. NYO-2543-1, 1965. Price $6.00. (Available from Clearinghouse for Federal Scientific and Technical Information, National Bureau of Standards, U.S. Department of Commerce, Springfield, Va. 22151.)

TABLE 7f-1. X-RAY WAVELENGTHS IN Å* UNITS AND IN keV

Designation	Wave-length, Å*	p.e.†	keV	Wave-length, Å*	p.e.†	keV
	3 Lithium			*4 Beryllium*		
$\alpha\ KL$	228	1	0.0543	114	1	0.1085
	5 Boron			*6 Carbon*		
$\alpha\ KL$	67.6	3	0.1833	44.7	3	0.277
	7 Nitrogen			*8 Oxygen*		
$\alpha\ KL$	31.6	4	0.3924	23.62	3	0.5249
	9 Fluorine			*10 Neon*		
$\alpha\ KL$	18.32	2	0.6768			
$\alpha_{1,2}\ KL_{II,III}$				14.610	3	0.8486
$\beta\ KM$				14.452	5	0.8579
	11 Sodium			*12 Magnesium*		
$\alpha_{1,2}\ KL_{II,III}$	11.9101	9	1.0410	9.8900	2	1.25360
$\beta\ KM$	11.575	2	1.0711	9.521	2	1.3022
$L_{II,III}M$	407.1	5	0.03045	251.5	5	0.0493
$L_I L_{II,III}$	376	1	0.0330	317	1	0.0392
	13 Aluminum			*14 Silicon*		
$\alpha_2\ KL_{II}$	8.34173	9	1.48627	7.12791	9	1.73938
$\alpha_1\ KL_{III}$	8.33934	9	1.48670	7.12542	9	1.73998
$\beta\ KM$	7.960	2	1.5574	6.753	1	1.8359
$L_{II,III}$	171.4	5	0.0724	135.5	4	0.0915
$L_I L_{II,III}$	290	1	0.0428			

Designation	Wave-length, Å*	p.e.†	keV	Wave-length, Å*	p.e.†	keV
	21 Scandium (Cont.)			*22 Titanium (Cont.)*		
$\beta_1\ L_{II}M_{IV}$	31.02	2	0.3996	27.05	2	0.4584
$l\ L_{III}M_I$	35.59	3	0.3483	31.36	2	0.3953
$\alpha_{1,2}\ L_{III}M_{IV,V}$	31.35	3	0.3954	27.42	2	0.4522
	23 Vanadium			*24 Chromium*		
$\alpha_2\ KL_{II}$	2.50738	2	4.94464	2.293606	3	5.40551
$\alpha_1\ KL_{III}$	2.50356	2	4.95220	2.28970	2	5.41472
$\beta_{1,3}\ KM_{II,III}$	2.28440	2	5.42729	2.08487	2	5.94671
$\beta_5\ KM_{IV,V}$	2.26951	6	5.4629	2.07087	6	5.9869
$\beta_{3,4}\ L_I M_{II,III}$	21.19‡	9	0.585	18.96	2	0.654
$\eta\ L_{II}M_I$	27.34	3	0.4535	24.30	3	0.5102
$\beta_1\ L_{II}M_{IV}$	23.88	4	0.5192	21.27	1	0.5828
$l\ L_{III}M_I$	27.77	1	0.4465	24.78	3	0.5003
$\alpha_{1,2}\ L_{III}M_{IV,V}$	24.25	3	0.5113	21.64	3	0.5728
$M_{II,III}M_{IV,V}$	337	9	0.037	309	9	0.040
	25 Manganese			*26 Iron*		
$\alpha_2\ KL_{II}$	2.10578	2	5.88765	1.939980	9	6.39084
$\alpha_1\ KL_{III}$	2.101820	9	5.89875	1.936042	9	6.40384
$\beta_{1,3}\ KM_{II,III}$	1.91021	2	6.49045	1.75661	2	7.05798
$\beta_5\ KM_{IV,V}$	1.8971	1	6.5352	1.7442	1	7.1081
$\beta_{3,4}\ L_I M_{II,III}$	17.19	2	0.721	15.65	2	0.792
$\eta\ L_{II}M_I$	21.85	2	0.5675	19.75	4	0.628
$\beta_1\ L_{II}M_{IV}$	19.11	2	0.6488	17.26	1	0.7185
$l\ L_{III}M_I$	22.29	1	0.5563	20.15	1	0.6152
$\alpha_{1,2}\ L_{III}M_{IV,V}$	19.45	6	0.6374	17.59	2	0.7050
$M_{II,III}M_{IV,V}$	273	6	0.045	243	5	0.051

Each element entry gives: wavelength λ (Å), probable error (p.e.), and energy (keV).

15 Phosphorus / 16 Sulfur

Designation	15 Phosphorus λ (Å)	p.e.	keV	16 Sulfur λ (Å)	p.e.	keV
$\alpha_2\ KL_{II}$	6.160‡	1	2.0127	5.37496	8	2.30664
$\alpha_1\ KL_{III}$	6.157‡	1	2.0137	5.37216	7	2.30784
$\beta\ KM$	5.796	2	2.1390	5.0316	2	2.4640
$\beta_2\ KM$				5.0233	3	2.4681
$L_{II,III}M$						
$l,\eta\ L_{II,III}M_I$	103.8	4	0.1194	83.4	3	0.1487

17 Chlorine / 18 Argon

Designation	17 Chlorine λ (Å)	p.e.	keV	18 Argon λ (Å)	p.e.	keV
$\alpha_2\ KL_{II}$	4.7307	1	2.62078	4.19474	5	2.95563
$\alpha_1\ KL_{III}$	4.7278	1	2.62239	4.19180	5	2.95770
$\beta_{1,3}\ KM_{II,III}$	4.4034	3	2.8156	3.8860	2	3.1905
$\eta\ L_{II}M_I$	67.33	9	0.1841	55.9‡	1	0.2217
$l\ L_{III}M_I$	67.90	9	0.1826	56.3‡	1	0.2201

19 Potassium / 20 Calcium

Designation	19 Potassium λ (Å)	p.e.	keV	20 Calcium λ (Å)	p.e.	keV
$\alpha_2\ KL_{II}$	3.7445	2	3.3111	3.36166	3	3.68809
$\alpha_1\ KL_{III}$	3.7414	2	3.3138	3.35539	3	3.69168
$\beta_{1,3}\ KM_{II,III}$	3.4539	2	3.5896	3.0897	2	4.0127
$\beta_5\ KM_{IV,V}$	3.4413	4	3.6027	3.0746	3	4.0325
$\eta\ L_{II}M_I$	47.24	2	0.2625	40.46	2	0.3064
$\beta_1\ L_{II}M_{IV}$				35.94	2	0.3449
$l\ L_{III}M_I$	47.74	1	0.25971	40.96	2	0.3027
$\alpha_{1,2}\ L_{III}M_{IV,V}$				36.33		0.3413
$M_{II,III}N_I$	692	9	0.0179	525	9	0.0236

21 Scandium / 22 Titanium

Designation	21 Scandium λ (Å)	p.e.	keV	22 Titanium λ (Å)	p.e.	keV
$\alpha_2\ KL_{II}$	3.0342	1	4.0861	2.75216	2	4.50486
$\alpha_1\ KL_{III}$	3.0309‡	1	4.0906	2.74851	2	4.51084
$\beta_{1,3}\ KM_{II,III}$	2.7796	2	4.4605	2.51391	2	4.93181
$\beta_5\ KM_{IV,V}$	2.7634	3	4.4865	2.4985	2	4.9623
$\gamma\ L_{II}M_I$	35.13	2	0.3529	30.89	3	0.4013

27 Cobalt / 28 Nickel

Designation	27 Cobalt λ (Å)	p.e.	keV	28 Nickel λ (Å)	p.e.	keV
$\alpha_2\ KL_{II}$	1.792850	9	6.91530	1.661747	8	7.46089
$\alpha_1\ KL_{III}$	1.788965	9	6.93032	1.657910	8	7.47815
$\beta_{1,3}\ KM_{II,III}$	1.62079	3	7.64943	1.500135	4	8.26466
$\beta_5\ KM_{IV,V}$	1.60891	3	7.7059	1.48862	1	8.3286
$\beta_{3,4}\ L_I M_{II,III}$	14.31	3	0.870	13.18	3	0.941
$\eta\ L_{II}M_I$	17.87	3	0.694	16.27	6	0.762
$\beta_1\ L_{II}M_{IV}$	15.666	8	0.7914	14.271	9	0.8688
$l\ L_{III}M_I$	18.292	6	0.6778	16.693	3	0.7427
$\alpha_{1,2}\ L_{III}M_{IV,V}$	15.972	6	0.7762	14.561	3	0.8515
$M_{II,III}M_{IV,V}$	214		0.058	190	2	0.0651

29 Copper / 30 Zinc

Designation	29 Copper λ (Å)	p.e.	keV	30 Zinc λ (Å)	p.e.	keV
$\alpha_2\ KL_{II}$	1.544390	2	8.02783	1.439000	8	8.61578
$\alpha_1\ KL_{III}$	1.540562	2	8.04778	1.435155	7	8.63886
$\beta_3\ KM_{II}$	1.3926	1	8.9029			
$\beta_{1,3}\ KM_{II,III}$	1.392218	9	8.90529	1.29525	2	9.5720
$\beta_2\ KN_{II,III}$	1.38109	3	8.9770	1.28372	2	9.6580
$\beta_5\ KM_{IV,V}$				1.2848	1	9.6501
$\beta_{3,4}\ L_I M_{II,III}$	12.122	8	1.0228	11.200	7	1.1070
$\eta\ L_{II}M_I$	14.90		0.832	13.68	2	0.906
$\beta_1\ L_{II}M_{IV}$	13.053	3	0.9498	11.983	2	1.0347
$l\ L_{III}M_I$	15.286	9	0.8111	14.02	2	0.884
$\alpha_{1,2}\ L_{III}M_{IV,V}$	13.336	3	0.9297	12.254	3	1.0117
$M_{II,III}M_{IV,V}$	173		0.072	157	3	0.079

31 Gallium / 32 Germanium

Designation	31 Gallium λ (Å)	p.e.	keV	32 Germanium λ (Å)	p.e.	keV
$\alpha_2\ KL_{II}$	1.34399	1	9.22482	1.258011	9	9.85532
$\alpha_1\ KL_{III}$	1.340083	9	9.25174	1.254054	9	9.88642
$\beta_3\ KM_{II}$	1.20835	5	10.2603	1.12936	2	10.9780
$\beta_1\ KM_{III}$	1.20789	2	10.2642	1.12894	2	10.9821
$\beta_2\ KN_{II,III}$	1.19600	2	10.3663	1.11686	2	11.1008
$\beta_5\ KM_{IV,V}$	1.1981	2	10.348	1.1195	1	11.0745
$\beta_{3,4}\ L_I M_{II,III}$	10.359‡	8	1.197	9.640		1.2861
$\gamma\ L_{II}M_I$				9.581		1.2941

† The probable error (p.e.) is the error in the last digit of wavelength.   Designation indicates both conventional Siegbahn notation (if applicable) and transition, e.g., $\beta_1\ L_{II}M_{IV}$ denotes a transition between L_{II} and M_{IV} levels, which is the $L\beta_1$ line in Siegbahn notation.

‡ This is an interpolated value. In some instances, no experimental values were available; in others, experimental measurements appeared clearly inconsistent with other data, as indicated by a Moseley diagram.

TABLE 7f-1. X-RAY WAVELENGTHS IN Å* UNITS AND IN keV (Continued)

Elements 31–36

Designation	Wave-length Å*	p.e.†	keV	Wave-length Å*	p.e.†	keV
	31 Gallium (Cont.)			*32 Germanium (Cont.)*		
$\eta\,L_{II}M_I$	12.597	2	0.9842	11.609	2	1.0680
$\beta_1\,L_{III}M_{IV}$	11.023	2	1.1248	10.175	1	1.2185
$l\,L_{III}M_I$	12.953	2	0.9572	11.965	4	1.0362
$\alpha_{1,2}\,L_{III}M_{IV,V}$	11.292	1	1.09792	10.4361	8	1.18800
	33 Arsenic			*34 Selenium*		
$\alpha_2\,KL_{II}$	1.17987	1	10.50799	1.10882	2	11.1814
$\alpha_1\,KL_{III}$	1.17588	1	10.54372	1.10477	2	11.2224
$\beta_3\,KM_{II}$	1.05783	5	11.7203	0.99268	5	12.4896
$\beta_1\,KM_{III}$	1.05730	3	11.7262	0.99218	3	12.4959
$\beta_2\,KN_{II,III}$	1.04500	1	11.8642	0.97992	5	12.6522
$\beta_5\,KM_{IV,V}$	1.0488	1	11.822	0.9843	1	12.595
$\beta_{3,4}\,L_IM_{II,III}$	8.929		1.3884	8.321†	9	1.490
$\eta\,L_{II}M_I$	10.734	1	1.1550	9.962	1	1.2446
$\beta_1\,L_{III}M_{IV}$	9.4141	8	1.3170	8.7358	5	1.41923
$l\,L_{III}M_I$	11.072	1	1.1198	10.294	1	1.2044
$\alpha_{1,2}\,L_{III}M_{IV,V}$	9.6709	8	1.2820	8.9900	5	1.37910
$M_V\text{-}N_{III}$				230	2	0.0538
	35 Bromine			*36 Krypton*		
$\alpha_2\,KL_{II}$	1.04382	2	11.8776	0.9841	1	12.598
$\alpha_1\,KL_{III}$	1.03974		11.9242	0.9801	1	12.649
$\beta_3\,KM_{II}$	0.93327	5	13.2845	0.8790	1	14.104
$\beta_1\,KM_{III}$	0.93279	2	13.2914	0.8785	1	14.112
$\beta_2\,KN_{II,III}$	0.92046	2	13.4695	0.8661	1	14.315
$\beta_5\,KM_{IV,V}$	0.9255	1	13.396	0.8708	2	14.238
$\beta_4\,KN_{IV,V}$				0.8653	2	14.328

Elements 37–40

Designation	Wave-length Å*	p.e.†	keV	Wave-length Å*	p.e.†	keV
	37 Rubidium (Cont.)			*38 Strontium (Cont.)*		
M_IM_{III}	144.4	3	0.0859	85.7	2	0.1447
$M_{II}M_{IV}$	91.5	2	0.1355			
$M_{II}N_I$	57.0	2	0.2174	51.3	1	0.2416
$M_{III}M_{IV,V}$	96.7	2	0.1282	91.4	2	0.1357
$M_{III}N_I$	59.5	2	0.2083	53.6	1	0.2313
$\zeta_2\,M_{IV}N_{II}$	127.8	2	0.0970			
$M_{IV}N_{III}$	126.8	2	0.0978			
$\zeta_1\,M_{IV}N_{II,III}$				108.0	2	0.1148
$\zeta_1\,M_VN_{III}$	128.7	2	0.0964	108.7	1	0.1140
	39 Yttrium			*40 Zirconium*		
$\alpha_2\,KL_{II}$	0.83305	1	14.8829	0.79015	1	15.6909
$\alpha_1\,KL_{III}$	0.82884	1	14.9584	0.78593	1	15.7751
$\beta_3\,KM_{II}$	0.74126	3	16.7258	0.70228	4	17.654
$\beta_1\,KM_{III}$	0.74072	2	16.7378	0.70173	3	17.6678
$\beta_2\,KN_{II,III}$	0.72864	4	17.0154	0.68993	4	17.970
$\beta_5\,KM_{IV,V}$	0.7345		16.879	0.6959		17.815
$\beta_4\,KN_{IV,V}$	0.72776	5	17.036	0.68901	5	17.994
$\beta_3\,L_IM_{III}$	6.0186	3	2.0600	5.6681	3	2.1873
$\beta_4\,L_IM_{II}$	5.9832	3	2.0722	5.6330	3	2.2010
$\gamma_{2,3}\,L_IN_{II,III}$	5.2830	3	2.3468	4.9536	3	2.5029
$\eta\,L_{II}M_I$	7.0406	3	1.76095	6.6069	3	1.87654
$\beta_1\,L_{II}M_{IV}$	6.2120	3	1.99584	5.8360	3	2.1244
$\gamma_5\,L_{II}N_I$	5.8754	3	2.1102	5.4977	3	2.2551
$\gamma_1\,L_{II}N_{IV}$				5.3843	3	2.3027
$l\,L_{III}M_I$	7.3563	3	1.68536	6.9185	3	1.79201
$\alpha_2\,L_{III}M_{IV}$	6.4558	3	1.92047	6.0778	3	2.0399
$\alpha_1\,L_{III}M_V$	6.4488	2	1.92256	6.0705	2	2.04236
$\beta_6\,L_{III}N_I$				5.7101	3	2.1712
$\beta_{2,15}$	6.0942	3	2.0344	5.5863	3	2.2194

Note: this is a rotated, high-density numeric reference table. Each element is given with wavelength λ (Å), relative intensity, and energy (keV).

37 Rubidium / 38 Strontium

Designation	Rb λ (Å)	Rb I	Rb E (keV)	Sr λ (Å)	Sr I	Sr E (keV)
$\alpha_2\, KL_{II}$	0.92969	1	13.3358	0.87943	1	14.0979
$\alpha_1\, KL_{III}$	0.925553	9	13.3953	0.87526	9	14.1650
$\beta_3\, KM_{II}$	0.82921	3	14.9517	0.78345	3	15.8249
$\beta_1\, KM_{III}$	0.82868	2	14.9613	0.78292	2	15.8357
$\beta_2\, KN_{II,III}$	0.81645	3	15.1854	0.77081	3	16.0846
$\beta_5\, KM_{IV,V}$	0.8219	1	15.085	0.7764	1	15.969
$\beta_4\, KN_{IV,V}$	0.8154	1	15.205	0.76989	5	16.104
$\beta_3\, L_I M_{III}$	6.8207	3	1.81771	6.4026	3	1.93643
$\gamma_{2,3}\, L_I N_{II,III}$	6.7876	3	1.82659	6.3672	3	1.94719
$\eta\, L_{II} M_I$	6.0458	3	2.0507	5.6445	3	2.1965
$\beta_1\, L_{II} M_{IV}$	8.0415	4	1.54177	7.5171	3	1.64933
$\gamma_5\, L_{II} N_I$	7.0759	3	1.75217	6.6239	3	1.87172
$l\, L_{III} M_I$	6.7553	3	1.83532	6.2961	3	1.96916
$\alpha_2\, L_{III} M_{IV}$	8.3636	4	1.48238	7.8362	3	1.55215
$\alpha_1\, L_{III} M_V$	7.3251	3	1.69256	6.8697	3	1.80474
$\beta_6\, L_{III} N_I$	7.3183	2	1.69413	6.8628	2	1.80656
$\beta_{2,15}\, L_{III} N_{IV,V}$	6.9842	3	1.77517	6.5191	3	1.90181

Rb / Sr soft L and M lines

Designation	Rb λ (Å)	Rb I	Rb E (keV)	Sr λ (Å)	Sr I	Sr E (keV)
$\beta_4\, L_I M_{II}$	7.767‡	9	1.596	7.304	5	1.697
$\beta_3\, L_I M_{III}$	9.255	1	1.3396	7.264	5	1.707
$\beta_{2,4}\, L_I M_{II,III}$	8.1251	5	1.52590	7.576‡	3	1.6366
$\eta\, L_{II} M_I$	9.585	1	1.2935	7.279	5	1.703
$\beta_1\, L_{III} M_{IV}$	8.3746	5	1.48043	7.817‡	3	1.5860
γ_5				7.510	4	1.6510
$\alpha_{1,2}\, L_{III} M_{IV,V}$				7.250	5	1.710
$L_{II} N_{III}$	184.6	3	0.0672			
$M_I M_{III}$	164.7	3	0.0753			
$M_{II} M_{III}$	109.4	3	0.1133			
$M_{II} M_{IV}$	76.9	2	0.1613			
$M_{II} N_I$	113.8	3	0.1089			
$M_{III} M_{IV,V}$	79.8	3	0.1554			
$\zeta_2\, M_{IV} N_{II}$	191.1	2	0.06488			
$M_{IV} N_{III}$	189.5	3	0.0654			
$\xi_1\, M_V N_{III}$	192.6	2	0.06437			

41 Niobium / 42 Molybdenum

Designation	Nb λ (Å)	Nb I	Nb E (keV)	Mo λ (Å)	Mo I	Mo E (keV)
$\alpha_2\, KL_{II}$	0.75044	1	16.5210	0.713590	6	17.3743
$\alpha_1\, KL_{III}$	0.74620	1	16.6151	0.709300	1	17.47934
$\beta_3\, KM_{II}$	0.66634	3	18.6063	0.632872	9	19.5903
$\beta_1\, KM_{III}$	0.66576	2	18.6225	0.632288	9	19.6083
$\beta_5^{\,I}\, KM_{IV}$				0.62708	5	19.771
$\beta_5^{\,I}\, KM_V$				0.62692	5	19.776
$\beta_2^{\,II}$	0.65416	4	18.953	0.62107	5	19.963
$\beta_2\, KN_{II,III}$	0.65318	5	18.981	0.62099	2	19.9652
$\beta_4\, KN_{IV,V}$				0.62001	9	19.996
$\beta_3\, L_I M_{III}$	5.3455	3	2.3194	5.0488	3	2.4557
$\gamma_{2,3}\, L_I N_{II,III}$	5.3102	3	2.3348	5.0133	3	2.4730
$\eta\, L_{II} M_I$	4.6542	2	2.6638	4.3800	2	2.8306
$\beta_1\, L_{II} M_{IV}$	6.2109	3	1.99620	5.8475	3	2.1202
$\gamma_5\, L_{II} N_I$	5.4923	3	2.2574	5.17708	8	2.39481
$\gamma_1\, L_{II} N_{IV}$	5.1517	3	2.4066	4.8369	2	2.5632
$l\, L_{III} M_I$	5.0361	3	2.4618	4.7258	2	2.6235
$\alpha_2\, L_{III} M_{IV}$	6.5176	3	1.90225	6.1508	3	2.01568
$\alpha_1\, L_{III} M_V$	5.7319	3	2.1630	5.41437	8	2.28985
$\beta_6\, L_{III} N_I$	5.7243	3	2.16589	5.40655	8	2.29316
$\beta_{2,15}\, L_{III} N_{IV,V}$	5.3613	3	2.3125	5.0488	5	2.4557
$\gamma\, L_{III}\,\ldots$	5.2379	3	2.3670	4.9232	2	2.5183
$M_{II} M_{IV}$	81.5	2	0.1522	76.7	2	0.1617
$M_{II} N_I$	46.48	9	0.267	80.9	3	0.1533
$M_{III} M_V$	48.5	2	0.256			
$M_{III} N_I$	86.5	2	0.1434	82.1	2	0.1511
$M_{III} M_{IV,V}$	93.4	2	0.1328	70.0	4	0.177
$M_{II} M_{IV}$	72.1	3	0.1718	68.9	3	0.1798
$M_{II} N_I$	38.4	2	0.323	35.3	3	0.351
$M_{III} M_V$	33.1	2	0.375	74.9	1	0.1656
$\gamma\, M_{III} N_{IV,V}$	78.4	2	0.1582	37.5	2	0.331
$M_{III} N_I$	40.7	2	0.305			
$M_{III} M_V$	34.9	2	0.356			
$\zeta\, M_{IV,V} N_{II,III}$	72.19	9	0.1717	64.38	7	0.1926
$M_{IV,V} O_{II,III}$	61.9	2	0.2002	54.8	2	0.2262

TABLE 7f-1. X-RAY WAVELENGTHS IN Å* UNITS AND IN keV (Continued)

Designation	43 Technetium / 45 Rhodium λ (Å*)	p.e.†	keV	44 Ruthenium / 46 Palladium λ (Å*)	p.e.†	keV
$\alpha_2\ KL_{II}$	0.67932‡	3	18.2508	0.647408	5	19.1504
$\alpha_1\ KL_{III}$	0.67502‡	3	18.3671	0.643083	4	19.2792
$\beta_1\ KM_{II}$	0.60188‡	4	20.599	0.573067	4	21.6346
$\beta_1\ KM_{III}$	0.60130‡	4	20.619	0.572482	4	21.6568
$\beta_2\ KN_{II,III}$	0.59024‡	5	21.005	0.56166	3	22.074
$\beta_5^{I}\ KM_{IV}$				0.5680	2	21.829
$\beta_5^{I}\ KM_{V}$				0.56785	9	21.834
β_4				0.56089	9	22.104
$\beta_4\ L_{II}M_{IV}$	4.8873‡	8	2.5368	4.5230	2	2.7411
$\beta_3\ L_{I}M_{III}$				4.4866	2	2.7634
$\gamma_{2,3}\ L_{I}N_{II,III}$				3.8977	2	3.1809
$\eta\ L_{II}M_{I}$				5.2050	2	2.38197
$\beta_1\ L_{II}M_{IV}$				4.62058	3	2.68323
$\gamma_5\ L_{II}N_{I}$				4.2873	2	2.8918
$\gamma_1\ L_{II}N_{IV}$				4.1822	2	2.9645
$l\ L_{III}M_{I}$				5.5035	3	2.2528
$\alpha_2\ L_{III}M_{IV}$				4.85381	7	2.55431
$\alpha_1\ L_{III}M_{V}$	5.1148‡	3	2.4240	4.84575	5	2.55855
$\beta_6\ L_{III}N_{I}$				4.4866	3	2.7634
$\beta_{2,15}\ L_{III}N_{IV,V}$				4.3718	2	2.8360
$M_{II}M_{IV}$				62.2	1	0.1992
$M_{II}N_{I}$				32.3	2	0.384
$M_{III}M_{V}$				25.50	9	0.486
$M_{III}N_{I}$				68.3	1	0.1814
$M_{III}N_{IV}$				26.9	1	0.462
$\gamma\ M_{III}N_{IV,V}$				52.34	7	0.2369
$M_{IV,V}O_{II,III}$				44.8	1	0.2768

45 Rhodium

Designation	λ (Å*)	p.e.†	keV
$\alpha_2\ KL_{II}$	0.617630	4	20.0737
$\alpha_1\ KL_{III}$	0.613279	4	20.2161

46 Palladium

Designation	λ (Å*)	p.e.†	keV
$\alpha_2\ KL_{II}$	0.589821	3	21.0201
$\alpha_1\ KL_{III}$	0.585448	3	21.1771

Designation	47 Silver (Cont.) / 49 Indium λ (Å*)	p.e.†	keV	48 Cadmium (Cont.) / 50 Tin λ (Å*)	p.e.†	keV
$\beta_4\ KN_{IV,V}$	0.48598	3	25.512	3.68203	9	3.36719
$\beta_4\ L_{I}M_{II}$	3.87023	5	3.20346	3.64495	9	3.40145
$\beta_3\ L_{I}M_{III}$	3.83313	9	3.23446	3.1377	2	3.9513
$\gamma_2\ L_{I}N_{II}$	3.31216	9	3.7432	4.19315	9	2.95675
$\gamma_3\ L_{I}N_{III}$	3.30635	9	3.7498	3.73823	4	3.31657
$\eta\ L_{II}M_{I}$	4.4183	3	2.8061	3.42551	9	3.61935
$\beta_1\ L_{II}M_{IV}$	3.93473	3	3.15094	3.33564	6	3.71686
$\gamma_5\ L_{II}N_{I}$	3.61638	9	3.42832	4.48014	9	2.76735
$\gamma_1\ L_{II}N_{IV}$	3.52260	4	3.51959	3.96496	6	3.12691
$l\ L_{III}M_{I}$	4.7076	2	2.6337	3.95635	4	3.13373
$\alpha_2\ L_{III}M_{IV}$	4.16294	5	2.97821	3.61467	9	3.42994
$\alpha_1\ L_{III}M_{V}$	4.15443	3	2.98431	3.51408	4	3.52812
$\beta_6\ L_{III}N_{I}$	3.70774	9	3.25603	3.4367	2	3.6075
$\beta_{2,15}\ L_{III}N_{IV,V}$	3.70335	9	3.34781	3.43015	9	3.61445
$\beta_{10}\ L_{I}M_{IV}$	3.61158	9	3.43287			
$\beta_9\ L_{I}M_{V}$	3.60497	9	3.43917			
$M_{I}N_{II,III}$	18.8	2	0.658	52.0	2	0.2384
$M_{II}M_{IV}$	54.0	1	0.2295	22.9	2	0.540
$M_{II}N_{I}$	20.66	7	0.600	19.40	7	0.639
$M_{III}M_{IV}$	60.5	1	0.2048	58.7	2	0.2111
$M_{III}M_{V}$	26.0	7	0.478	24.5	1	0.507
$\gamma\ M_{III}N_{IV,V}$	21.82	7	0.568	20.47	7	0.606
$\zeta\ M_{IV,V}N_{II,III}$	39.77	7	0.3117	30.4	1	0.408
$M_{V}N_{I}$	24.4	2	0.509	36.8	1	0.3371
$M_{IV,V}O_{II,III}$	33.5	3	0.370	30.8	1	0.403

49 Indium

Designation	λ (Å*)	p.e.†	keV
$\alpha_2\ KL_{II}$	0.516544	3	24.0020
$\alpha_1\ KL_{III}$	0.512113	3	24.2097

50 Tin

Designation	λ (Å*)	p.e.†	keV
$\alpha_2\ KL_{II}$	0.495053	3	25.0440
$\alpha_1\ KL_{III}$	0.490599	3	25.2713

Wavelengths are given in Å (ångström), energies in keV; the small integer is the relative intensity.

Transition	47 Silver λ(Å)	I	keV	48 Cadmium λ(Å)	I	keV	51 Antimony λ(Å)	I	keV	52 Tellurium λ(Å)	I	keV
$\beta_3\ KM_{II}$	0.546200	4	22.0089	0.521123	4	23.7911	0.455181	4	27.2377	0.435877	5	28.4440
$\beta_1\ KM_{III}$	0.545605	4	22.7236	0.520520	4	23.8187	0.454545	4	27.2759	0.435236	5	28.4860
$\beta_2^{I}\ KN_{II}$	0.53513	5	23.168				0.44500	1	27.8608	0.425915	8	29.1093
$\beta_2\ KN_{II,III}$	0.53503	2	23.1728	0.510228	4	24.2991	0.44374	3	27.940	0.42467	3	29.195
$\beta_5^{I}\ KM_{IV}$	0.54118	9	22.909	0.5093	2	24.346	0.45098	2	27.491	0.43184	3	28.710
$\beta_5^{I}\ KM_{V}$	0.54101	9	22.917	0.51670	9	23.995	0.45086	2	27.499	0.43175	3	28.716
$\beta_4\ KN_{IV,V}$												
$\beta_5\ KM_{IV,V}$	0.53431	9	23.217				0.44393	4	27.928	0.42495	9	29.175
$\beta_4\ L_1M_{III}$	4.2888	2	2.8908	4.0711	2	3.0454	3.50697	9	3.5353	3.34335	3	3.7083
$\beta_3\ L_1M_{II}$	4.2522	2	2.9157	4.0346	2	3.0730	3.46984	5	3.5731	3.30585	3	3.7500
$\gamma_{2,3}\ L_1N_{II,III}$	3.6855	2	3.3640	3.4892	2	3.5533	2.9800	5	4.1605	2.8327	2	4.3768
$\gamma_4\ L_1O_{II,III}$							2.9264	2	4.2367	2.7775	2	4.4638
$\eta\ L_{II}M_I$	4.9217	2	2.5191	4.6605	2	2.6603	3.98327	9	3.11254	3.78876	9	3.27234
$\beta_1\ L_{II}M_{IV}$	4.37414	4	2.83441	4.14622	5	2.99022	3.55531	3	3.48721	3.38487	3	3.66280
$\gamma_5\ L_{II}N_I$	4.0451	2	3.0650	3.8222	2	3.2437	3.24907	9	3.8159	3.08475	9	4.0192
$\gamma_1\ L_{II}N_{IV}$	3.9437	2	3.1438	3.7246	2	3.3287	3.16213	4	3.92081	3.00115	3	4.13112
$l\ L_{III}M_I$	5.2169	3	2.3765	4.9525	3	2.5034	4.26873	9	2.90440	4.07165	9	3.04499
$\alpha_2\ L_{III}M_{IV}$	4.60545	9	2.69205	4.37588	7	2.83329	3.78073	6	3.27929	3.60891	4	3.43542
$\alpha_1\ L_{III}M_V$	4.59743	9	2.69674	4.36767	5	2.83861	3.77192	5	3.28694	3.59994	3	3.44398
$\beta_6\ L_{III}N_I$	4.2417	2	2.9229	4.0162	2	3.0870	3.43606	9	3.60823	3.26901	3	3.7926
$\beta_{2,15}\ L_{III}N_{IV,V}$	4.1310	2	3.0013	3.90887	4	3.17179	3.33838	3	3.71381	3.17505	3	3.90486
$\beta_7\ L_{III}O_I$							3.324	4	3.730	3.1564	3	3.9279
$\beta_{10}\ L_{III}M_{IV}$				3.7988	2	3.2637	3.27404	9	3.7868	3.12170	9	3.9716
$\beta_9\ L_{III}M_V$				3.7920	2	3.2696	3.26763	9	3.7942	3.11513	9	3.9800
$M_I N_{II,III}$	59.3	1	0.2090	20.1	2	0.616				47.3	1	0.2621
$M_{II}M_{IV}$	28.1	2	0.442	56.5	1	0.2194				20.0	1	0.619
$M_{II}N_{IV}$				26.2	2	0.474				16.93	5	0.733
$M_{III}N_{IV}$	65.5	1	0.1892	22.1	1	0.560				54.2	1	0.2287
$M_{III}M_V$	29.8	1	0.417	62.9	1	0.1970				21.5	5	0.575
$M_{III}N_I$	25.01	9	0.496	27.9	1	0.445				17.94	1	0.691
$\gamma\ M_{III}N_{IV,V}$	47.67	9	0.2601	23.3‡	1	0.531				25.3	5	0.491
$M_{IV}O_{II,III}$	40.9	2	0.303	43.6	1	0.2844				31.24	9	0.397
$\zeta\ M_{IV,V}N_{II,III}$				37.4	2	0.332				25.7	1	0.483
$M_V O_{III}$												

Transition	47 Silver λ(Å)	I	keV	48 Cadmium λ(Å)	I	keV	51 Antimony λ(Å)	I	keV	52 Tellurium λ(Å)	I	keV
$\alpha_2\ KL_{II}$	0.563798	4	21.9903	0.539422	3	22.9903	0.474827	3	26.1108	0.455784	3	27.2017
$\alpha_1\ KL_{III}$	0.5594075	6	22.16292	0.535010	3	23.1736	0.470354	3	26.3591	0.451295	3	27.4723
$\beta_3\ KM_{II}$	0.497685	4	24.9115	0.475730	5	26.0612	0.417737	4	29.6792	0.400659	4	30.9443
$\beta_1\ KM_{III}$	0.497069	4	24.9424	0.475105	6	26.0955	0.417085	5	29.7256	0.399995	5	30.9957
$\beta_2\ KN_{II,III}$	0.487032	4	25.4564	0.465328	7	26.6438	0.407973	5	30.3895	0.391102	6	31.7004
$\beta_5\ KM_{IV,V}$	0.49306	2	25.145									

TABLE 7f-1. X-RAY WAVELENGTHS IN Å* UNITS AND IN keV (Continued)

Left panel

Designation	51 Antimony (Cont.) Wavelength, Å*	p.e.†	keV	52 Tellurium (Cont.) Wavelength, Å*	p.e.†	keV
$KO_{II,III}$	0.40666	1	30.4875	0.38974	1	31.8114
$\beta_5^{II} KM_{IV}$	0.41388	1	29.9560			
$\beta_5^{I} KM_V$	0.41378	1	29.9632			
$\beta_1 KN_{IV,V}$	0.40702	1	30.4604			
$\beta_4 L_I M_{II}$	3.19014	9	3.8864	3.04661	9	4.0695
$\beta_3 L_I M_{III}$	3.15258	9	3.9327	3.00893	9	4.1204
$\gamma_{2,3} L_I N_{II,III}$	2.6953	2	4.5999	2.5674	2	4.8290
$\gamma_4 L_I O_{II,III}$	2.6398	2	4.6967	2.5113	2	4.9369
$\eta L_{II} M_I$	3.60765	9	3.43661	3.43832	9	3.60586
$\beta_1 L_{II} M_{IV}$	3.22567	4	3.84357	3.07677	6	4.02958
$\gamma_5 L_{II} N_I$	2.93187	9	4.2287	2.79007	9	4.4437
$\gamma_1 L_{II} N_{IV}$	2.85159	3	4.34779	2.71241	6	4.5709
$l L_{III} M_I$	3.88826	9	3.18860	3.71696	9	3.33555
$\alpha_2 L_{III} M_{IV}$	3.44840	6	3.59532	3.29846	6	3.7588
$\alpha_1 L_{III} M_V$	3.43941	4	3.60472	3.28920	6	3.76933
$\beta_6 L_{III} N_I$	3.11513	9	3.9800	2.97088	9	4.1732
$\beta_{2,15} L_{III} N_{IV,V}$	3.02335	3	4.10078	2.88217	8	4.3017
$\beta_7 L_{III} O_I$	3.0052	3	4.1255	2.8634	3	4.3298
$\beta_{10} L_I M_{IV}$	2.97917	9	4.1616	2.84679	9	4.3551
$\beta_9 L_{III} M_V$	2.97261	9	4.1708	2.83897	9	4.3671
$M_{II} N_I$	45.2		0.2743			
$M_{II} N_{IV}$	18.8	1	0.658	17.6	1	0.703
$M_{III} M_V$	15.98	5	0.776			
$M_{III} N_I$	52.2	1	0.2375	50.3	1	0.2465
$\gamma M_{III} N_{IV,V}$	20.2	1	0.612	19.1	1	0.648
$M_{IV} O_{II,III}$	16.92	4	0.733	15.93	4	0.778
				21.34	5	0.581
$\zeta M_{IV,V} N_{II,III}$	28.88	8	0.429	26.72	9	0.464
$M_V O_{III}$				21.78	5	0.569

Right panel

Designation	55 Cesium (Cont.) Wavelength, Å*	p.e.†	keV	56 Barium (Cont.) Wavelength, Å*	p.e.†	keV
$\gamma_4 L_I O_{II,III}$	2.1741	2	5.7026	2.0756	3	5.9733
$\eta L_{II} M_I$	2.9932	2	4.1421	2.8627	3	4.3309
$\beta_1 L_{II} M_{IV}$	2.6837	2	4.6198	2.56821	5	4.82753
$\gamma_5 L_{II} N_I$	2.4174	2	5.1287	2.3085	3	5.3707
$\gamma_1 L_{II} N_{IV}$	2.3480	2	5.2804	2.2415	2	5.5311
$l L_{III} M_I$	3.2670	2	3.7950	3.1355	2	3.9541
$\alpha_2 L_{III} M_{IV}$	2.9020	2	4.2722	2.78553	5	4.45090
$\alpha_1 L_{III} M_V$	2.8924	2	4.2865	2.77595	5	4.46626
$\beta_6 L_{III} N_I$	2.5932	2	4.7811	2.4826	2	4.9939
$\beta_{2,15} L_{III} N_{IV,V}$	2.5118	2	4.9359	2.40435	6	5.1565
$\beta_7 L_{III} O_I$	2.4849	2	4.9893	2.3806	2	5.2079
$\beta_{10} L_{III} M_{IV}$	2.4920	2	4.9752	2.3869	2	5.1941
$\beta_9 L_{III} M_V$	2.4783	2	5.0026	2.3764	2	5.2171
$\gamma M_{III} N_{IV,V}$				12.75	3	0.973
$M_{IV} O_{II}$				15.91	5	0.779
$M_{IV} O_{III}$				15.72	9	0.789
				20.64	4	0.601
$M_V O_{III}$				16.20	5	0.765
$N_{IV} O_{II}$	188.6	1	0.06574	163.3	2	0.07590
$N_{IV} O_{III}$	183.8	1	0.06746	159.0	2	0.07796
$N_V O_{III}$	190.3	1	0.06515	164.6	2	0.07530

57 Lanthanum (in Cesium column) and **58 Cerium** (in Barium column)

Designation	57 Lanthanum Wavelength, Å*	p.e.†	keV	58 Cerium Wavelength, Å*	p.e.†	keV
$\alpha_2 KL_{II}$	0.375313	2	33.0341	0.361683	2	34.2789
$\alpha_1 KL_{III}$	0.370737	2	33.4418	0.357092	2	34.7197
$\beta_3 KM_{II}$	0.328686	4	37.7202	0.316520	4	39.1701
$\beta_1 KM_{III}$	0.327983	3	37.8010	0.315816	2	39.2573

	53 Iodine			54 Xenon		
$\alpha_2\ KL_{II}$	0.437829	7	28.3172	0.42087‡	2	29.458
$\alpha_1\ KL_{III}$	0.433318	5	28.6120	0.41634‡	2	29.779
$\beta_3\ KM_{II}$	0.384564	4	32.2394	0.36941‡	2	33.562
$\beta_1\ KM_{III}$	0.383905	4	32.2947	0.36872‡	2	33.624
$\beta_2\ KN_{II,III}$	0.37523‡	2	33.042	0.36026‡	3	34.415
$\beta_4\ L_IM_{II}$	2.91207	9	4.2575			
$\beta_3\ L_IM_{III}$	2.87429	9	4.3134			
$\gamma_{2,3}\ L_IN_{II,III}$	2.4475	2	5.0657			
$\gamma_4\ L_IO_{II,III}$	2.3913	2	5.1848			
$\eta\ L_{II}M_I$	3.27979	9	3.7801			
$\beta_1\ L_{II}M_{IV}$	2.93744	6	4.22072			
$\gamma_5\ L_{II}N_I$	2.65710	9	4.6660			
$\gamma_1\ L_{II}N_{IV}$	2.58244	8	4.8009			
$l\ L_{III}M_I$	3.55754	9	3.48502			
$\alpha_2\ L_{III}M_{IV}$	3.15791	6	3.92604			
$\alpha_1\ L_{III}M_V$	3.14860	6	3.93765	3.0166‡	2	4.1099
$\beta_6\ L_{III}N_I$	2.83672	9	4.3706			
$\beta_{2,15}\ L_{III}N_{IV,V}$	2.75053	8	4.5075			
$\beta_7\ L_{III}O_I$	2.7288	3	4.5435			
$\beta_{10}\ L_IM_{IV}$	2.72104	9	4.5564			
$\beta_9\ L_IM_V$	2.71352	9	4.5690			

	55 Cesium			56 Barium		
$\alpha_2\ KL_{II}$	0.404835	4	30.6251	0.389668	5	31.8171
$\alpha_1\ KL_{III}$	0.400290	4	30.9728	0.385111	4	32.1936
$\beta_3\ KM_{II}$	0.355050	4	34.9194	0.341507	4	36.3040
$\beta_1\ KM_{III}$	0.354364	7	34.9869	0.340811	3	36.3782
$\beta_2\ KN_{II,III}$	0.34611	2	35.822	0.33277	1	37.257
$KO_{II,III}$				0.33127	2	37.426
$\beta_5^{II}\ KM_{IV}$				0.33835	2	36.643
$\beta_5^{I}\ KM_V$				0.33814	2	36.666
$\beta_4\ KN_{IV,V}$				0.33229	2	37.311
$\beta_4\ L_IM_{II}$	2.6666	2	4.6494	2.5553	2	4.8519
$\beta_3\ L_IM_{III}$	2.6285	2	4.7167	2.5164	2	4.9269
$\gamma_2\ L_IN_{II}$	2.2371	2	5.5420	2.1387	2	5.7969
$\gamma_3\ L_IM_{III}$	2.2328	2	5.5527	2.1342	2	5.8092

	(continued)			(continued)		
$\beta_2\ KN_{II,III}$	0.320117	7	38.7299	0.30816	1	40.233
$KO_{II,III}$	0.31864	2	38.909	0.30668	2	40.427
$\beta_5^{II}\ KM_{IV}$	0.32563	2	38.074	0.31357	2	39.539
$\beta_5^{I}\ KM_V$	0.32546	2	38.094	0.31342	2	39.558
$\beta_4\ KN_{IV,V}$	0.31931	2	38.828	0.30737	2	40.337
$\beta_4\ L_IM_{II}$	2.4493	3	5.0620	2.3497	4	5.2765
$\beta_3\ L_IM_{III}$	2.4105	3	5.1434	2.3109	3	5.3651
$\gamma_2\ L_IN_{II}$	2.0460	4	6.060	1.9602	3	6.3250
$\gamma_3\ L_IN_{III}$	2.0410	4	6.074	1.9553	3	6.3409
$\gamma_4\ L_IO_{II,III}$	1.9830	4	6.252	1.8991	4	6.528
$\eta\ L_{II}M_I$	2.740	3	4.525	2.6203	4	4.7315
$\beta_1\ L_{II}M_{IV}$	2.45891	5	5.0421	2.3561	3	5.2622
$\gamma_5\ L_{II}N_I$	2.2056	4	5.621	2.1103	3	5.8751
$\gamma_1\ L_{II}N_{IV}$	2.1418	3	5.7885	2.0487	4	6.052
$\gamma_8\ L_{II}O_I$				2.0237	4	6.126
$l\ L_{III}M_I$	3.006	3	4.124	2.8917	4	4.2875
$\alpha_2\ L_{III}M_{IV}$	2.67533	5	4.63423	2.5706	3	4.8230
$\alpha_1\ L_{III}M_V$	2.66570	5	4.65097	2.5615	2	4.8402
$\beta_6\ L_{III}N_I$	2.3790	4	5.2114	2.2818	3	5.4334
$\beta_{2,15}\ L_{III}N_{IV,V}$	2.3030	3	5.3835	2.2087	2	5.6134
$\beta_7\ L_{III}O_I$	2.275	3	5.450	2.1701	2	5.7132
$\beta_{10}\ L_IM_{IV}$	2.290	3	5.415	2.1958	5	5.646
$\beta_9\ L_IM_V$	2.282	3	5.434	2.1885	3	5.6650
$\gamma\ M_{III}N_{IV,V}$	12.08	4	1.027	11.53	1	1.0749
$\beta\ M_{IV}N_{VI}$	14.51	5	0.854	13.75	4	0.902
$\zeta\ M_VN_{III}$	19.44	5	0.638	18.35	4	0.676
$\alpha\ M_VN_{VI,VII}$	14.88	5	0.833	14.04	2	0.883
$M_VO_{II,III}$				14.39	5	0.862
$N_{IV,V}O_{II,III}$	152.6	6	0.0812	144.4	6	0.0859

	59 Praseodymium			60 Neodymium		
$\alpha_2\ KL_{II}$	0.348749	2	35.5502	0.336472	2	36.8474
$\alpha_1\ KL_{III}$	0.344140	2	36.0263	0.331846	2	37.3610
$\beta_3\ KM_{II}$	0.304975	5	40.6529	0.294027	3	42.1665
$\beta_1\ KM_{III}$	0.304261	4	40.7482	0.293299	2	42.2713
$\beta_2\ KN_{II,III}$	0.29679	2	41.773	0.2861‡	1	43.33
$\beta_4\ L_IM_{II}$	2.2550	4	5.4981	2.1669	3	5.7216
$\beta_3\ L_IM_{III}$	2.2172	3	5.5918	2.1268	2	5.8294

TABLE 7f-1. X-RAY WAVELENGTHS IN Å* UNITS AND IN KEV (Continued)

Designation	59 Praseodymium (Cont.) λ, Å*	p.e.†	keV	60 Neodymium (Cont.) λ, Å*	p.e.†	keV	Designation	63 Europium (Cont.) λ, Å*	p.e.†	keV	64 Gadolinium (Cont.) λ, Å*	p.e.†	keV
$\gamma_2\,L_I N_{II}$	1.8791	4	6.598	1.8013	4	6.883	$\gamma_2\,L_I N_{II}$	1.5961	2	7.7677	1.5331	2	8.087
$\gamma_3\,L_I N_{III}$	1.8740	4	6.616	1.7964	4	6.902	$\gamma_3\,L_I N_{III}$	1.5903	2	7.7961	1.5297	2	8.105
$\gamma_4\,L_I O_{II,III}$	1.8193	4	6.815	1.7445	4	7.107	$\gamma_4\,L_I O_{II,III}$	1.5439	1	8.0304	1.4839	2	8.355
$\eta\,L_{II} M_I$	2.512	3	4.935	2.4094	4	5.1457	$\eta\,L_{II} M_I$	2.1315	2	5.8166	2.0494	1	6.0495
$\beta_1\,L_{II} M_{IV}$	2.2588	3	5.4889	2.1669	2	5.7216	$\beta_1\,L_{II} M_{IV}$	1.9203	2	6.4564	1.8468	2	6.7132
$\gamma_5\,L_{II} N_I$	2.0205	4	6.136	1.9355	4	6.406	$\gamma_5\,L_{II} N_I$	1.7085	2	7.2566	1.6412	2	7.5543
$\gamma_1\,L_{II} N_{IV}$	1.9611	3	6.3221	1.8779	2	6.6021	$\gamma_1\,L_{II} N_{IV}$	1.6574	2	7.4803	1.5924	2	7.7858
$\gamma_8\,L_{II} O_I$	1.9362	4	6.403	1.8552	5	6.683	$\gamma_8\,L_{II} O_I$	1.6346	2	7.5849	1.5707	2	7.894
$\gamma_6\,L_{II} O_{IV}$							$\gamma_6\,L_{II} O_{IV}$	1.6282	2	7.6147	1.5644	2	7.925
$l\,L_{III} M_I$	2.7841	4	4.4532	2.6760	4	4.6330	$l\,L_{III} M_I$	2.3948	2	5.1772	2.3122	2	5.3621
$\alpha_2\,L_{III} M_{IV}$	2.4729	4	5.0135	2.3807	3	5.2077	$\alpha_2\,L_{III} M_{IV}$	2.1315	2	5.8166	2.0578	2	6.0250
$\alpha_1\,L_{III} M_V$	2.4630	3	5.0337	2.3704	2	5.2304	$\alpha_1\,L_{III} M_V$	2.1209	2	5.8457	2.0468	2	6.0572
$\beta_6\,L_{III} N_I$	2.1906	2	5.660	2.1039	3	5.8930	$\beta_6\,L_{III} N_I$	1.8737	2	6.6170	1.8054	2	6.8671
$\beta_{2,15}\,L_{III} N_{IV,V}$	2.1194	4	5.850	2.0360	3	6.0894	$\beta_{2,15}\,L_{III} N_{IV,V}$	1.8118	2	6.8432	1.7455	2	7.1028
$\beta_7\,L_{III} O_I$	2.0919	4	5.927	2.0092	3	6.1708	$\beta_7\,L_{III} O_I$	1.7851	2	6.9453	1.7203	2	7.2071
$\beta_5\,L_{III} O_{IV,V}$							$\beta_5\,L_{III} O_{IV,V}$	1.7772	2	6.9763	1.7130	2	7.2374
$\beta_{10}\,L_I M_{IV}$	2.1071	4	5.884	2.0237	3	6.1265	$\beta_{10}\,L_I M_{IV}$	1.7993	3	6.890	1.7315	3	7.160
$\beta_9\,L_I M_V$	2.1004	4	5.903	2.0165	3	6.1484	$\beta_9\,L_I M_V$	1.7916	3	6.920	1.7240	3	7.192
$L_I O_{IV,V}$							$L_I O_{IV,V}$				1.4807	3	8.373
$\gamma\,M_{III} N_{IV,V}$	10.998	9	1.1273	10.505	9	1.180	$\gamma\,M_{III} N_{IV,V}$	9.211	9	1.346	8.844	9	1.402
$\beta\,M_{IV} N_{VI}$	13.06	2	0.950	12.44	2	0.997	$\beta\,M_{IV} N_{VI}$	10.750	7	1.1533	10.254	6	1.2091
$\zeta\,M_V N_{III}$	17.38	4	0.714	16.46	4	0.753	$\zeta\,M_V N_{III}$	14.22	2	0.872	13.57	2	0.914
$\alpha\,M_V N_{VI,VII}$	13.343	5	0.9292	12.68	2	0.978	$\alpha\,M_V N_{VI,VII}$	10.96	3	1.131	10.46	3	1.185
$N_{IV,V} O_{II,III}$	113	1	0.1095	107	7	0.116	$N_{IV,V} O_{II,III}$	112.0	6	0.1107			
	136.5	4	0.0908	128.9		0.0962							

Designation	61 Promethium λ, Å*	p.e.†	keV	62 Samarium λ, Å*	p.e.†	keV	Designation	65 Terbium λ, Å*	p.e.†	keV	66 Dysprosium λ, Å*	p.e.†	keV
$\alpha_2\,KL_{II}$	0.324803	4	38.1712	0.313698	2	39.5224	$\alpha_2\,KL_{II}$	0.283423	2	43.7441	0.274247	2	45.2078
$\alpha_1\,KL_{III}$	0.320160	4	38.7247	0.309040	2	40.1181	$\alpha_1\,KL_{III}$	0.278724	2	44.4816	0.269533	2	45.9984
$\beta_3\,KM_{II}$	0.28363‡	4	43.713	0.27376	2	45.289	$\beta\,KM_{II}$	0.24683	2	50.229	0.23862	2	51.957
$\beta_1\,KM_{III}$	0.28290‡	3	43.826	0.27301	2	45.413							
$\beta_2\,KN_{II,III}$	0.2759‡	1	44.94	0.2662	1	46.58							
$KO_{II,III}$				0.26491	3	46.801							

This dense rotated table lists X-ray line wavelengths (Å), relative intensities, and photon energies (keV) for the elements 63 Europium, 64 Gadolinium, 67 Holmium, and 68 Erbium. Each element is given as three sub-columns. Values are transcribed below to the best reading of this dense source; "—" marks an absent entry.

Upper block (L and M series, with leading K-region lines)

Line	Eu λ (Å)	Eu I	Eu keV	Gd λ (Å)	Gd I	Gd keV	Ho keV	Ho I	Ho λ (Å)	Er keV	Er I	Er λ (Å)
$\beta_1\,KM_{III}$	—	—	—	0.27111	3	45.731	50.382	2	0.24608	52.119	2	0.23788
$\beta_2\,KN_{II,III}$	—	—	—	—	—	—	51.72	2	0.2397‡	53.51	2	0.2317‡
$KO_{II,III}$	—	—	—	—	—	—	51.965	3	0.23358	53.774	3	0.23056
$\beta_5\,KM_{IV,V}$	—	—	—	—	—	—	—	—	—	52.494	7	0.23618
$\beta_4\,L_I M_{II}$	2.0797	4	5.961	2.00095	6	6.1963	6.9403	2	1.7864	7.2039	2	1.72103
$\beta_3\,L_I M_{III}$	2.0421	4	6.071	1.96241	3	6.3180	7.0959	2	1.7472	7.3702	7	1.6822
$\gamma_2\,L_I N_{II}$	—	—	—	1.66044	6	7.4668	8.398	2	1.4764	8.7140	7	1.42278
$\gamma_3\,L_I N_{III}$	—	—	—	1.65601	3	7.4867	8.423	2	1.4718	8.7532	7	1.41640
$\gamma_4\,L_I O_{II,III}$	—	—	—	1.60728	3	7.7137	8.685	2	1.4276	9.0195	7	1.37459
$\eta\,L_{II} M_I$	—	—	—	2.21824	3	5.5892	6.2839	2	1.9730	6.5342	7	1.89743
$\beta_1\,L_{II} M_{IV}$	1.9559	6	6.339	1.99806	3	6.2051	6.978	3	1.7768	7.2477	7	1.71062
$\gamma_5\,L_{II} N_I$	—	—	—	1.77934	3	6.9678	7.8535	2	1.5787	8.1661	7	1.51824
$\gamma_1\,L_{II} N_{IV}$	1.7989	9	6.892	1.72724	3	7.1780	8.102	2	1.5303	8.4188	7	1.47266
$\gamma_6\,L_{II} O_{IV}$	—	—	—	1.6966	9	7.3076	8.212	2	1.5097	8.5753	7	1.44579
$l\,L_{III} M_I$	—	—	—	2.4823	4	4.9945	5.5467	2	2.2352	5.7431	7	2.15877
$\alpha_2\,L_{III} M_{IV}$	2.2926	4	5.4078	2.21062	3	5.6084	6.2380	2	1.9875	6.4577	3	1.91991
$\alpha_1\,L_{III} M_V$	2.2822	3	5.4325	2.1998	2	5.6361	6.2728	2	1.9765	6.4952	3	1.90881
$\beta_6\,L_{III} N_I$	—	—	—	1.94643	3	6.3697	7.3667	2	1.6830	7.3705	3	1.68213
$\beta_{2,15}\,L_{III} N_{IV,V}$	—	—	—	1.88221	3	6.5870	7.4753	2	1.6585	7.6357	7	1.62369
$\beta_7\,L_{III} O_I$	—	—	—	1.85626	3	6.6791	7.5094	2	1.6510	7.7272	7	1.60447
$\beta_5\,L_{III} O_{IV,V}$	—	—	—	1.84700	9	6.7126	7.436	2	1.6673	7.8055	7	1.58837
$\beta_{10}\,L_{III} M_{IV}$	—	—	—	1.86990	3	6.6304	8.714	3	1.4228	7.7130	9	1.60743
$\beta_9\,L_{III} M_V$	—	—	—	1.86166	3	6.6597	8.246	3	1.5035	7.7501	9	1.59973
$\gamma\,M_{III} N_{IV,V}$	—	—	—	9.600	9	1.291	1.461	9	8.486	1.522	9	8.144
$\beta\,M_{IV} N_{VI}$	—	—	—	11.27	1	1.0998	1.2661	6	9.792	1.3250	6	9.357
$\zeta\,M_V N_{III}$	—	—	—	14.91	4	0.831	0.955	2	12.98	0.998	2	12.43
$\alpha\,M_V N_{VI,VII}$	—	—	—	11.47	3	1.081	1.240	2	10.00	1.293	2	9.59
$N_{IV,V} O_{VI,VII}$	—	—	—	98	4	0.126	0.144	1	86	0.149	1	83
$N_{IV,V} O_{II,III}$	—	—	—	117.4	4	0.1056	0.1213	4	102.2	0.128	8	97.2

| | 63 Europium | | 64 Gadolinium | | 67 Holmium | | 68 Erbium | |

Lower block (K series)

Line	Eu λ (Å)	Eu I	Eu keV	Gd λ (Å)	Gd I	Gd keV	Ho λ (Å)	Ho I	Ho keV	Er λ (Å)	Er I	Er keV
$\alpha_2\,KL_{II}$	0.303118	2	40.9019	0.293038	2	42.3089	0.265486	2	46.6997	0.257110	2	48.2211
$\alpha_1\,KL_{III}$	0.298446	2	41.5422	0.288353	2	42.9962	0.260756	2	47.5467	0.252365	2	49.1277
$\beta_3\,KM_{II}$	0.264332	5	46.9036	0.25534	2	48.555	0.23083	2	53.711	0.22341	2	55.494
$\beta_1\,KM_{III}$	0.263577	5	47.0379	0.25460	2	48.697	0.23012	2	53.877	0.22266	2	55.681
$\beta_2\,KN_{II,III}$	0.256923	8	48.256	0.24816	2	49.959						
$KO_{II,III}$	0.255645	7	48.497	0.24687	3	50.221						
$\beta_5\,KM_{IV,V}$	—	—	—	0.25275	3	49.052						
$\beta_4\,L_I M_{II}$	1.9255	2	6.4389	1.8540	2	6.6871						
$\beta_3\,L_I M_{III}$	1.8867	2	6.5713	1.8150	2	6.8311						

TABLE 7f-1. X-RAY WAVELENGTHS IN Å* UNITS AND IN keV (Continued)

Left group: *67 Holmium (Cont.)* and *68 Erbium (Cont.)*. Right group: *69 Thulium (Cont.)* → *71 Lutetium* and *70 Ytterbium (Cont.)* → *72 Hafnium*.

Designation	Wave-length, Å*	p.e.†	keV	Wave-length, Å*	p.e.†	keV	Designation	Wave-length, Å*	p.e.†	keV	Wave-length, Å*	p.e.†	keV
	67 Holmium (Cont.)			*68 Erbium (Cont.)*				*69 Thulium (Cont.) / 71 Lutetium*			*70 Ytterbium (Cont.) / 72 Hafnium*		
$\beta_2\,KN_{II,III}$	0.2241‡	2	55.32	0.2167‡	2	57.21	$\alpha\,M_V N_{VI,VII}$	8.48	1	1.462	8.149	5	1.5214
$KO_{II,III}$	0.22305	2	55.584	0.21581	3	57.450	$N_{IV} N_{VI}$				65.1	7	0.190
$\beta_5\,KM_{IV,V}$	0.22855	3	54.246	0.22124	3	56.040	$N_V N_{VI,VII}$				69.3	5	0.179
$\beta_4\,L_I M_{III}$	1.6595	2	7.4708	1.6007	1	7.7453	$\alpha_2\,KL_{II}$	0.234081	2	52.9650	0.227024	3	54.6114
$\beta_3\,L_I M_{II}$	1.6203	2	7.6519	1.5616	1	7.9392	$\alpha_1\,KL_{III}$	0.229298	2	54.0698	0.222227	3	55.7902
$\gamma_2\,L_I N_{II}$	1.3698	2	9.051	1.3210	2	9.385	$\beta_3\,KM_{II}$	0.20309‡	4	61.05	0.19686‡	4	62.98
$\gamma_3\,L_I N_{III}$	1.3643	2	9.087	1.3146	2	9.4309	$\beta_1\,KM_{III}$	0.20231‡	3	61.283	0.19607‡	3	63.234
$\gamma_4\,L_I O_{II,III}$	1.3225	2	9.374	1.2752	2	9.722	$\beta_2\,KN_{II,III}$	0.1969‡	2	62.97	0.1908‡	2	64.98
$\eta\,L_{II} M_I$	1.8264	2	6.7883	1.7566	1	7.0579	$KO_{II,III}$	0.19589	2	63.293			
$\beta_1\,L_{II} M_{IV}$	1.6475	2	7.5253	1.5873	1	7.8109	$\beta_5\,KM_{IV,V}$	0.20084	2	61.732			
$\gamma_5\,L_{II} N_I$	1.4618	2	8.481	1.4067	3	8.814	$\beta_4\,L_I M_{III}$	1.44056	5	8.6064	1.39220	5	8.9054
$\gamma_1\,L_{II} N_{IV}$	1.4174	2	8.747	1.3641	2	9.089	$\beta_3\,L_I M_{II}$	1.40140	5	8.8469	1.35300	5	9.1634
$\nu_8\,L_{II} O_I$	1.3983	2	8.867				$\gamma_2\,L_I N_{II}$	1.1853	2	10.460	1.14442	5	10.8335
$\gamma_6\,L_{II} O_{IV}$	1.3923	2	8.905	1.3397	3	9.255	$\gamma_3\,L_I N_{III}$				1.13841	5	10.8907
$l\,L_{III} M_I$	2.0860	2	5.9434	2.015	1	6.152	$\gamma_4\,L_I O_{II,III}$	1.17953	4	10.5110	1.10376	5	11.2326
$\alpha_2\,L_{III} M_{IV}$	1.8561	2	6.6795	1.7955	2	6.9050	$\eta\,L_{II} M_I$	1.1435	1	10.8425	1.10303	5	11.2401
$\alpha_1\,L_{III} M_V$	1.8450	2	6.7198	1.78425	9	6.9487	$\beta_1\,L_{II} M_{IV}$	1.5779	1	7.8575	1.52325	5	8.1393
$\beta_6\,L_{III} N_I$	1.6237	2	7.6359	1.5675	2	7.909	$\gamma_5\,L_{II} N_I$	1.42359	3	8.7090	1.37410	5	9.0227
$\beta_{2,15}\,L_{III} N_{IV,V}$	1.5671	2	7.911	1.51399	9	8.1890	$\gamma_1\,L_{II} N_{IV}$	1.2596	1	9.8428	1.21537	5	10.2011
$\beta_7\,L_{III} O_I$				1.4941	3	8.298	$\gamma_8\,L_{II} O_I$	1.22228	4	10.1434	1.17900	5	10.5158
$\beta_5\,L_{III} O_{IV,V}$	1.5378	2	8.062	1.4848	3	8.350	$\gamma_6\,L_{II} O_{IV}$	1.2047	1	10.2915	1.16138	5	10.6754
$\beta_{10}\,L_I M_{IV}$	1.5486	3	8.006	1.4941	3	8.298	$l\,L_{III} M_I$	1.1987	1	10.3431	1.15519	5	10.7325
$L_I O_{IV,V}$	1.3208	3	9.387				$\alpha_2\,L_{III} M_{IV}$	1.8360	1	6.7528	1.78145	5	6.9596
$\beta_9\,L_I M_V$				1.4855	5	8.346	$\alpha_1\,L_{III} M_V$	1.63029	5	7.6049	1.58046	5	7.8446
$M_{III} N_{IV}$	7.865	9	1.576	7.60	1	1.632	$\beta_6\,L_{III} N_I$	1.61951	3	7.6655	1.56958	5	7.8990
$\gamma\,M_{III} N_{IV,V}$				7.546	8	1.643	$\beta_{15}\,L_{III} N_{IV}$	1.4189	1	8.7376	1.37410	5	9.0227
$\gamma\,M_{III} N_V$				8.592	3	1.4430	$\beta_2\,L_{III} N_V$	1.3715	1	9.0395	1.32783	5	9.3371
$\beta\,M_{IV} N_{VI}$	8.965	4	1.3830	11.37	1	1.0901	$\beta_7\,L_{III} O_I$	1.37012	3	9.0489	1.32639	5	9.3473
$\zeta\,M_V N_{III}$	11.86	1	1.0450	8.82	1	1.406	$\beta_5\,L_{III} O_{IV,V}$	1.34949	5	9.1873	1.30564	5	9.4958
$\alpha\,M_V N_{VI,VII}$	9.20	2	1.348	72.7	9	0.171							
$N_{IV} N_{VI}$				76.3	7	0.163							
$N_V N_{VI,VII}$													

69 Thulium and 70 Ytterbium

Line	Tm λ (Å)	I	Tm E (keV)	Yb λ (Å)	I	Yb E (keV)
α₂ KL_II	0.249095	2	49.7726	0.241424	2	51.3540
α₁ KL_III	0.244338	2	50.7416	0.236655	2	52.3889
β₃ KM_II	0.21636	2	57.304	0.2096‡	1	59.14
β₁ KM_III	0.21556	2	57.517	0.20884	8	59.37
β₂ KN_{II,III}	0.2098‡	2	59.09	0.2033‡	2	60.98
KO_{II,III}	0.20591	2	59.346	0.20226	2	61.298
β₅ KM_{IV,V}	0.21404	2	57.923	0.20739	2	59.782
β₄ L_I M_II	1.5448	2	8.026	1.49138	3	8.3132
β₃ L_I M_III	1.5063	2	8.231	1.45233	5	8.5367
γ₂ L_I N_II	1.2742	2	9.730	1.22879	7	10.0897
γ₃ L_I N_III	1.2678	2	9.779	1.2232	5	10.1431
γ₄ L_I O_{II,III}	1.2294	2	10.084	1.1853	1	10.4603
η L_II M_I	1.6963	2	7.3088	1.63560	5	7.5802
β₁ L_II M_IV	1.5304	2	8.101	1.47565	5	8.4018
γ₅ L_II N_I	1.3558	2	9.144	1.3063	1	9.4910
γ₁ L_II N_IV	1.3153	2	9.426	1.26769	5	9.8701
γ₈ L_II O_I	1.2905	2	9.607	1.24923	5	9.9246
γ₆ L_II O_IV				1.24271	3	9.9766
l L_III M_I	1.9550	2	6.3419	1.89415	5	6.5455
α₂ L_III M_IV	1.7351	2	7.1331	1.68285	5	7.3673
α₁ L_III M_V	1.7268‡	2	7.1799	1.67189	4	7.4156
β₆ L_III N_I	1.5162	2	8.177	1.4661	1	8.4563
β₂,₁₅ L_III N_{IV,V}	1.4640	2	8.468	1.41550	5	8.7588
β₇ L_III O_I				1.3948	1	8.8889
β₅ L_III O_{IV,V}	1.4349	2	8.641	1.38696	7	8.9390
β₁₀ L_I M_IV	1.4410	3	8.604	1.3915	1	8.9100
β₉ L_I M_V	1.4336	3	8.648	1.3838	1	8.9597
L_I O_I				1.1886	1	10.4312
L_I O_{IV,V}	1.2263	3	10.110	1.1827	1	10.4833
L_II M_II				1.58844	9	7.8052
t L_II O_{II,III}				1.2453	1	9.9561
L_III O_{II,III}				1.83091	9	6.7715
M_III N_I				1.3898	1	8.9209
γ M_III N_V				8.470	9	1.464
β M_IV N_VI				7.024	8	1.765
ζ M_V N_III	8.249	7	1.503	7.909	2	1.5675
				10.48	1	1.183

73 Tantalum and 74 Tungsten (main lines)

Line	Ta λ (Å)	I	Ta E (keV)	W (col 1)	I	W (col 3)
β₅ L_III O_{IV,V}	1.34183	7	9.2397	9.5546	5	1.29761
L_I M_I	1.3430	2	9.232	8.6685	9	1.43025
β₁₀ L_I M_IV	1.3358	1	9.2816	9.5503	9	1.29819
β₉ L_I M_V	1.16227	9	10.6672	9.6090	9	1.29025
γ₁₁ L_I N_V	1.16107	9	10.6782	11.0451	9	1.12250
L_I O_I	1.53333	9	8.0858	11.0553	9	1.12146
L_II O_IV	1.2014	1	10.3198	11.2034	9	1.10664
β₁₇ L_II M_III	1.7760	1	6.9810	11.2622	9	1.10086
L_II N_V	1.34524	9	9.2163	8.3735	9	1.48064
υ L_III N_VI				8.6312	9	1.43643
t L_III O_{II,III}				10.5258	9	1.17788
s L_III M_III				10.7037	9	1.15830
L_III N_II				7.1954	9	1.72305
ξ L_III N_II				7.4532	9	1.66346
u L_III N_{VI,VII}				9.1239	9	1.35887
L_III O_{II,III}				9.1802	9	1.35053
M_III N_I				9.5249	9	1.30165
γ M_III N_V	6.768	6	1.832	1.572	9	7.887
ζ₂				1.895	4	6.544
β M_IV N_VI	7.601	2	1.6312	1.2800	7	9.686
ζ₁				1.6976	1	7.303
α M_V N_{VI,VII}	7.840	2	1.5813	1.2800	7	9.686
M_IV N_VI	63.0	5	0.197			
N_V N_{VI,VII}	65.7	2	0.1886	1.6446	1	7.539

73 Tantalum and 74 Tungsten (K lines)

Line	Ta λ (Å)	I	Ta E (keV)	W λ (Å)	I	W E (keV)
α₂ KL_II	0.220305	8	56.277	0.213828	2	57.9817
α₁ KL_III	0.215497	4	57.532	0.2090100	Std	59.31824
β₃ KM_II	0.190890	9	64.9488	0.185181	2	66.9514
β₁ KM_III	0.190089	4	65.223	0.184374	2	67.2443
β₂^{II} KN_III	0.185188	9	66.949	0.17960	1	69.031
β₂^{I} KN_III	0.185011	8	67.013	0.179421	7	69.101
KO_{II,III}	0.184031	7	67.370	0.178444	5	69.479
KL_I				0.21592	4	57.42
β₅^{II} KM_IV	0.188920	6	65.626	0.183264	5	67.652

TABLE 7f-1. X-RAY WAVELENGTHS IN Å* UNITS AND IN keV (Continued)

Designation	73 Tantalum (Cont.) Wave-length, Å*	p.e.†	keV	74 Tungsten (Cont.) Wave-length, Å*	p.e.†	keV	Designation	75 Rhenium Wave-length, Å*	p.e.†	keV	76 Osmium Wave-length, Å*	p.e.†	keV
$\beta_5^I\ KM_V$	0.188757	6	65.683	0.183092	7	67.715	$\alpha_2\ KL_{II}$	0.207611	1	59.7179	0.201639	2	61.4867
$\beta_4\ KN_{IV,V}$	0.18451	1	67.194	0.17892	2	69.294	$\alpha_1\ KL_{III}$	0.202781	2	61.1403	0.196794	2	63.0005
$\beta_4\ L_IM_{II}$	1.34581	3	9.2124	1.30162	5	9.5252	$\beta_3\ KM_{II}$	0.179697	3	68.994	0.174431	3	71.077
$\beta_3\ L_IM_{III}$	1.30678	3	9.4875	1.26269	5	9.8188	$\beta_1\ KM_{III}$	0.178880	3	69.310	0.173611	3	71.413
$\gamma_2\ L_IN_{II}$	1.1053	1	11.217	1.06806	3	11.6080	$\beta_2^{II}\ KN_{II}$	0.17425	1	71.151	0.16910	1	73.318
$\gamma_3\ L_IN_{III}$	1.09936	4	11.2776	1.06200	6	11.6743	$\beta_2^I\ KN_{III}$	0.174054	6	71.232	0.168906	6	73.402
$\gamma'_4\ L_IO_{II}$	1.06544	3	11.6366	1.02863	3	12.0530	$KO_{II,III}$	0.17308	1	71.633	0.16798	1	73.808
$\gamma_4\ L_IO_{III}$	1.06467	3	11.6451	1.02775	3	12.0634	$\beta_5^{II}\ KM_{IV}$	0.17783	1	69.719	0.17262	1	71.824
$\eta\ L_{II}M_I$	1.47106	5	8.4280	1.42110	3	8.7243	$\beta_5^I\ KM_V$	0.17766	1	69.786	0.17245	1	71.895
$\beta_1\ L_{III}M_{IV}$	1.32698	3	9.3431	1.281809	9	9.67235	$\beta_4\ KN_{IV,V}$	0.17362	2	71.410	0.16842	2	73.615
$\gamma_5\ L_{II}N_I$	1.1729	3	10.5702	1.13235	3	10.9490	$\beta_4\ L_IM_{II}$	1.25917	5	9.8463	1.21844	5	10.1754
$\nu_1\ L_{II}N_{IV}$	1.13794	3	10.8952	1.09855	3	11.2859	$\beta_3\ L_IM_{III}$	1.22031	5	10.1598	1.17955	7	10.5108
$\gamma_8\ L_{II}O_I$	1.1205	1	11.0646	1.08113	4	11.4677	$\gamma_2\ L_IN_{II}$	1.03233	5	12.0098	0.99805	5	12.4224
$\gamma_6\ L_{II}O_{IV}$	1.11388	3	11.1306	1.07448	5	11.5387	$\gamma_3\ L_IN_{III}$	1.02613	7	12.0824	0.99186	5	12.4998
$l\ L_{III}M_I$	1.72841	5	7.1731	1.6782	1	7.3878	$\gamma'_4\ L_IO_{II}$	0.99334	5	12.4813	0.96033	8	12.910
$\alpha_2\ L_{III}M_{IV}$	1.53293	2	8.0879	1.48743	2	8.3352	$\gamma_4\ L_IO_{III}$	0.99249	5	12.4920	0.95938	8	12.923
$\alpha_1\ L_{III}M_V$	1.52197	2	8.1461	1.47639	2	8.3976	$\eta\ L_{II}M_I$	1.37342	5	9.0272	1.32785	7	9.3370
$\beta_6\ L_{III}N_I$	1.33094	8	9.3153	1.29989	7	9.6117	$\beta_1\ L_{III}M_{IV}$	1.23858	2	10.0100	1.19727	7	10.3553
$\beta_{15}\ L_{III}N_{IV}$	1.28619	5	9.6394	1.24631	3	9.9478	$\gamma_5\ L_{II}N_I$	1.09388	5	11.3341	1.05693	5	11.7303
$\beta_2\ L_{III}N_V$	1.28454	5	9.6518	1.24460	3	9.9615	$\gamma_1\ L_{II}N_{IV}$	1.06099	5	11.6854	1.02503	5	12.0953
$\beta_7\ L_{III}O_I$	1.26385	5	9.8098	1.22400	4	10.1292	$\gamma_8\ L_{II}O_I$	1.04398	5	11.8758	1.00788	5	12.3012
$\beta_5\ L_{III}O_{IV,V}$	1.2555	1	9.8750	1.21545	3	10.2004	$\gamma_6\ L_{II}O_{IV}$	1.03699	9	11.956	1.00107	7	12.3848
$L_I\ M_I$				1.3365	3	9.277	$l\ L_{III}M_I$	1.63056	5	7.6036	1.58498	7	7.8222
$\beta_{10}\ L_IM_{IV}$	1.2537	2	9.889	1.21218	3	10.2279	$\alpha_2\ L_{III}M_{IV}$	1.44396	5	8.5862	1.40234	5	8.8410
$\beta_9\ L_IM_V$	1.2466	2	9.946	1.20479	7	10.2907	$\alpha_1\ L_{III}M_V$	1.43290	4	8.6525	1.39121	5	8.9117
$L_I\ N_I$	1.11521	9	11.1173				$\beta_6\ L_{III}N_I$	1.25100	5	9.9105	1.21349	5	10.2169
$L_I\ N_{IV}$	1.08377	7	11.4398	1.0468	2	11.844	$\beta_{15}\ L_{III}N_{IV}$	1.20819	5	10.2617	1.17167	5	10.5816
$\gamma_{11}\ L_IN_V$	1.08205	7	11.4580	1.0458	1	11.856	$\beta_2\ L_{III}N_V$	1.20660	4	10.2752	1.16979	8	10.5985
$L_IN_{VI,VII}$	1.06305	9	11.6570				$\beta_7\ L_{III}O_I$	1.18610	5	10.4529	1.14933	8	10.7872
L_IO_I	1.06771	9	11.6118	1.0317	3	12.017	$\beta_5\ L_{III}O_{IV,V}$	1.17721	5	10.5318	1.1405	1	10.8711

Left-hand element block (wavelength λ in Å, relative intensity, energy in keV):

Line	λ	Int.	E (keV)
$L_I O_{IV,V}$	1.06192	9	11.6752
$L_{II} M_{II}$	1.43048	1	8.6671
$\beta_{17} L_{II} M_{III}$	1.3864	1	8.9428
$L_{II} M_V$	1.31897	9	9.3998
$L_{II} N_{II}$	1.1600	2	10.688
$L_{II} N_{III}$	1.1553	1	10.7316
$L_{II} N_V$	1.13687	9	10.9055
$v L_{II} N_{VI}$	1.1158	1	11.1113
$L_{II} O_{III}$	1.11789	9	11.0907
$L_{II} O_{III}$	1.11693	9	11.1001
$t L_{III} M_{II}$	1.67265	9	7.4123
$s L_{III} M_{III}$	1.61264	1	7.6881
$L_{III} N_{II}$	1.3167	1	9.4158
$L_{III} N_{III}$	1.3086	4	9.4742
$u L_{III} N_{VI,VII}$	1.25778	1	9.8572
$L_{III} O_{II,III}$	1.2601	3	9.839
$M_I N_{II,III}$	5.40	2	2.295
$M_{II} N_{IV}$	5.570	4	2.226
$M_{III} N_I$	7.612	9	1.629
$M_{III} N_{IV}$	6.353	5	1.951
$\gamma M_{III} N_V$	6.312	4	1.964
$M_{III} O_{IV,V}$	5.83	2	2.126
$\zeta_2 M_{IV} N_{II}$	5.67	3	2.19
$M_{IV} N_{III}$	9.330	5	1.3288
$\beta M_{IV} N_{VI}$	8.90	2	1.393
$M_{IV} O_{II}$	7.023	1	1.7655
$\zeta_1 M_V N_{III}$	7.09	2	1.748
$\alpha M_V N_{VI,VII}$	9.316	4	1.3308
$\alpha_2 M_V N_{VI}$	7.252	1	1.7096
$M_V O_{III}$	7.30	2	1.700
$N_V N_{VI}$	58.2	1	0.2130
$N_V N_{VII}$	61.1	2	0.2028

Right-hand blocks — middle element (E in keV, Int., λ in Å), 77 Iridium (λ, Int., E), 78 Platinum (λ, Int., E):

Line	E (keV)	Int.	λ (Å)	77 Ir λ	Int.	77 Ir E	78 Pt λ	Int.	78 Pt E
$\beta_{10} L_{II} M_{IV}$	12.095	2	1.0250	1.17218	5	10.5770	1.13353	5	10.9376
$\beta_9 L_{II} M_V$	9.261	2	1.3387	1.16487	4	10.6433	1.12637	6	11.0071
$L_I N_I$	9.741	2	1.2728	1.0420	1	11.899	0.9772	3	12.687
$L_I N_{IV}$	11.052	3	1.1218	1.0119	1	12.252	0.9765	3	12.696
$\gamma_{11} L_I N_V$	11.120	2	1.1149	1.0108	1	12.266	0.96318	7	12.8721
$L_I O_I$				0.9965	1	12.442	0.95603	5	12.9683
$L_I O_{IV,V}$	11.510	1	1.0771	0.9900	1	12.524	1.2934	2	9.586
$L_{II} M_{II}$	11.488	2	1.0792	1.3366	1	9.2761	1.2480	2	9.934
$\beta_{17} L_{II} M_{III}$	7.632	3	1.6244	1.2927	1	9.5910	1.18977	7	10.4205
$L_{II} M_V$	7.926	3	1.5642	1.2305	1	10.0753	1.03973	5	11.9243
$L_{II} N_{II}$	9.712	2	1.2765	1.0839	1	11.438	1.0050	2	12.337
$L_{II} N_{III}$	9.784	2	1.2672	1.0767	1	11.515	1.0047	2	12.340
$v L_{II} N_{VI}$	10.1733	5	1.21868	1.0404	1	11.917	1.5347	2	8.079
$L_{II} O_{III}$	10.153	2	1.2211	1.0397	1	11.925	1.4735	2	8.414
$t L_{III} M_{II}$	2.397	9	5.172	1.5789	1	7.8525	1.20086	7	10.3244
$s L_{III} M_{III}$	2.79	2	4.44	1.5178	1	8.1682	1.14537	7	10.8245
$L_{III} N_{II}$	1.973	2	6.28	1.2283	1	10.0933	4.79	2	2.59
$L_{III} N_{III}$	2.314	4	5.357	1.1815	1	10.4931	5.81	2	2.133
$u L_{III} N_{VI,VII}$	1.684	8	7.360				4.955	4	2.502
$M_I N_{II}$	2.021	4	6.134				6.89	2	1.798
$M_I N_{III}$	2.035	3	6.092	5.931	5	2.090	5.724	5	2.166
$M_{II} N_I$	2.203	8	5.628	5.885	2	2.1067	5.682	4	2.182
$M_{III} N_I$	1.3787	5	8.993	8.664	5	1.4310	8.359	5	1.4831
$M_{III} N_{IV}$	1.446	8	8.573	8.239	8	1.505			
$M_{III} N_V$	1.8349	1	6.757	6.504	4	1.9061	6.267	1	1.9783
$\gamma M_{IV} N_V$	1.822	9	6.806				8.310	4	1.4919
$\zeta_2 M_{IV} N_{II}$	1.3835	4	8.962	8.629	4	1.4368	6.490	1	1.9102
$M_{IV} N_{III}$				6.729	1	1.8425			
$M_{IV} O_{III}$	1.7731	2	6.992						
$\alpha M_V N_{VI,VII}$	1.7754	1	6.983				51.9	1	0.2388
$N_{IV} N_{VI}$	1.770	9	7.005				54.7	2	0.2266
$N_V N_{VI,VII}$	0.2295	2	54.0						
	0.2221	1	55.8						
	0.208	3	59.5						
	0.2122	1	58.4						
$\alpha_2 KL_{II}$				0.195904	2	63.2867	0.190381	4	65.122
$\alpha_1 KL_{III}$				0.191047	2	64.8956	0.185511	4	66.832
$\beta_3 KM_{II}$				0.169367	2	73.2027	0.164501	3	75.368
$\beta_1 KM_{III}$				0.168542	2	73.5608	0.163675	3	75.748

Table 7f-1. X-ray Wavelengths in Å* Units and in keV (Continued)

77 Iridium (Cont.) and 78 Platinum (Cont.)

Designation	λ (Å*)	p.e.†	keV	λ (Å*)	p.e.†	keV
	77 Iridium (Cont.)			*78 Platinum (Cont.)*		
$\beta_2^{II}\ KN_{II}$	0.16415	1	75.529	0.15939	1	77.785
$\beta_2^{I}\ KN_{III}$	0.163956	7	75.619	0.15920	1	77.878
$KO_{II,III}$	0.163019	5	76.053	0.15826	1	78.341
$\beta_5^{II}\ KM_{IV}$	0.16759	2	73.980	0.16271	2	76.199
$\beta_5^{I}\ KM_V$	0.167373	2	74.075	0.16255	3	76.27
$\beta_4\ KN_{IV,V}$	0.16352	9	75.821	0.15881	2	78.069
$\beta_4\ L_I M_{II}$	1.17958	3	10.5106	1.14223	5	10.8543
$\beta_3\ L_I M_{III}$	1.14085	3	10.8674	1.10394	5	11.2308
$\gamma_2\ L_I N_{II}$	0.96545	3	12.8418	0.93427	5	13.2704
$\gamma_3\ L_I N_{III}$	0.95931	5	12.9240	0.92791	5	13.3613
$\gamma'_4\ L_I O_{III}$	0.92831	3	13.3555	0.89747	4	13.8145
$\gamma_4\ L_I O_{III}$	0.92744	3	13.3681	0.89659	4	13.8281
$\eta\ L_{II} M_I$	1.28448	3	9.6522	1.2429	2	9.975
$\gamma_5\ L_{II} M_{IV}$	1.15781	3	10.7083	1.11990	2	11.0707
$\gamma_5\ L_{II} N_I$	1.02175	5	12.1342	0.9877	2	12.552
$\gamma_1\ L_{II} N_{IV}$	0.99085	3	12.5126	0.95797	3	12.9420
$\gamma_8\ L_{II} O_I$	0.97409	3	12.7279	0.9411	1	13.173
$\gamma_6\ L_{II} O_{IV}$	0.96708	4	12.8201	0.9342	2	13.271
$l\ L_{III} M_I$	1.54094	3	8.0458	1.4995	2	8.268
$\alpha_2\ L_{III} M_{IV}$	1.36250	5	9.0995	1.32432	2	9.3618
$\alpha_1\ L_{III} M_V$	1.35128	3	9.1751	1.31304	3	9.4423
$\beta_6\ L_{III} N_I$	1.17796	3	10.5251	1.14355	5	10.8418
$\beta_{15}\ L_{III} N_{IV}$	1.13707	3	10.9036	1.10200	3	11.2505
$\beta_2\ L_{III} N_V$	1.13532	3	10.9203	1.08168	3	11.4619
$\beta_7\ L_{III} O_I$	1.11489	3	11.1205	1.0724	2	11.561
$\beta_5\ L_{III} O_{IV,V}$	1.10585	3	11.2114			
$L_I M_I$	1.2102	2	10.245	1.16962	9	10.6001
$\beta_{10}\ L_I M_{IV}$	1.09702	4	11.3016	1.06183	7	11.6762
$\beta_9\ L_I M_V$	1.08975	5	11.3770	1.05446	5	11.7577
$L_I N_I$	0.9766	2	12.695			
$L_I N_{IV}$	0.9459	2	13.108	0.9455	2	13.113

79 Gold (Cont.) and 80 Mercury (Cont.)

Designation	λ (Å*)	p.e.†	keV	λ (Å*)	p.e.†	keV
	79 Gold (Cont.)			*80 Mercury (Cont.)*		
$\beta_1\ KM_{III}$	0.158982	3	77.984	0.154487	3	80.253
$\beta_2^{II}\ KN_{III}$	0.15483	2	80.08	0.15040	2	82.43
$\beta_2^{I}\ KN_{III}$	0.154618	9	80.185	0.15020	2	82.54
$KO_{II,III}$	0.15694	4	80.667	0.14931	2	83.04
KL_I	0.18672	7	66.40			
$\beta_5^{II}\ KM_{IV}$	0.158062	5	78.438	0.15353	2	80.75
$\beta_5^{I}\ KM_V$	0.157880	5	78.529			
$\beta_5\ KM_{IV.v}$						
$\beta_4\ KN_{IV.v}$	0.154224	5	80.391	0.14978	2	82.78
$\beta_4\ L_I M_{II}$	1.10651	3	11.2047	1.07222	7	11.5630
$\beta_3\ L_I M_{III}$	1.06785	9	11.6103	1.03358	7	11.9953
$\gamma_2\ L_I N_{III}$	0.90434	3	13.7095	0.87544	7	14.162
$\gamma_3\ L_I N_{III}$	0.89783	5	13.8090	0.86915	7	14.265
$\gamma'_4\ L_I O_{II}$	0.86816	4	14.2809	0.84013	7	14.757
$\gamma_4\ L_I O_{III}$	0.86703	4	14.2996	0.83894	7	14.778
$\eta\ L_{II} M_I$	1.20273	3	10.3083	1.1640	1	10.6512
$\beta_1\ L_{II} M_{IV}$	1.08353	3	11.4423	1.04868	5	11.8226
$\gamma_5\ L_{II} N_I$	0.95559	3	12.9743	0.92453	7	13.410
$\gamma_1\ L_{II} N_{IV}$	0.92650	3	13.3817	0.89646	5	13.8301
$\gamma_8\ L_{II} O_I$	0.90989	5	13.6260	0.87995	7	14.090
$\gamma_6\ L_{II} O_{IV}$	0.90297	3	13.7304	0.87319	7	14.199
$l\ L_{III} M_I$	1.45964	9	8.4939	1.4216	7	8.7210
$\alpha_2\ L_{III} M_{IV}$	1.28772	3	9.6280	1.25264	5	9.8976
$\alpha_1\ L_{III} M_V$	1.27640	3	9.7133	1.24120	7	9.9888
$\beta_6\ L_{III} N_I$	1.11092	3	11.1602	1.07975	7	11.4824
$\beta_2\ L_{III} N_V$	1.07188	5	11.5667	1.04151	7	11.9040
$\beta_7\ L_{III} O_I$	1.07022	5	11.5847	1.03975	7	11.9241
$\beta_5\ L_{III} O_{IV.v}$	1.04974	8	11.8106	1.01937	7	12.1625
$\beta_{10}\ L_I M_{IV}$	1.04044	5	11.9163	1.00987	7	12.2769
$L_I M_I$	1.13525	7	10.9210	1.0999	2	11.272
$L_I N_{IV}$	1.02789	7	12.0617	0.9962	2	12.446

$\gamma_{11}\ L_IN_V$	0.9446	2	13.126	0.9143	2	13.560
$L_IO_{IV,V}$	0.9243	3	13.413			
L_IO_I				0.8995	2	13.784
L_IO_{IV}				0.8943	1	13.864
L_IO_V				0.8934	1	13.878
$L_{II}M_{II}$	1.2502	3	9.917	1.213	1	10.225
$\beta_{17}\ L_{II}M_{III}$	1.2069	2	10.273	1.1667	1	10.6265
$L_{II}M_V$	1.1489	2	10.791	1.1129	2	11.140
$L_{II}N_{II}$	1.0120	2	12.251	0.9792	2	12.661
$L_{II}N_{III}$	1.0054	3	12.332	0.97173	4	12.7588
$v\ L_{II}N_{VI}$	0.97161	6	12.7603	0.93931	5	13.1992
$L_{II}O_{III}$	0.96979	5	12.7843			
$t\ L_{III}M_{II}$	1.4930	3	8.304	1.4530	2	8.533
$s\ L_{III}M_{III}$	1.4318	2	8.659	1.3895	2	8.923
$L_{III}N_{II}$	1.16545	5	10.6380	1.1310	2	10.962
$L_{III}N_{III}$	1.1560	3	10.725	1.1226	2	11.044
$u\ L_{III}N_{VI,VII}$	1.11145	4	11.1549	1.07896	5	11.4908
$L_{III}O_{II,III}$	1.10923	6	11.1772	1.0761	3	11.521
M_IN_{III}	4.631‡	9	2.677	4.460	9	2.780
$M_{II}N_{IV}$	4.780	4	2.594	4.601	4	2.695
$M_{III}N_I$	6.669	9	1.859	6.455	9	1.921
$M_{III}N_{IV}$	5.540	5	2.238	5.357	5	2.314
$\gamma\ M_{III}N_V$	5.500	4	2.254	5.319	4	2.331
$M_{III}O_I$				4.876	9	2.543
$M_{III}O_{IV,V}$	4.869	9	2.546	4.694	8	2.641
$\zeta_2\ M_{IV}N_{II}$	8.065	5	1.5373	7.790	5	1.592
$M_{IV}N_{III}$	7.645	8	1.622	7.371	8	1.682
$\beta\ M_{IV}N_{VI}$	6.038	1	2.0535	5.828	1	2.1273
$\zeta_1\ M_VN_{III}$	8.021	4	1.5458	7.738	4	1.6022
$\alpha_2\ M_VN_{VI}$	6.275	3	1.9758	6.058	3	2.047
$\alpha_1\ M_VN_{VII}$	6.262	1	1.9799	6.047	1	2.0505
M_VO_{III}				5.987	9	2.071
$N_{IV}N_{VI}$	50.2	1	0.2470	48.1	2	0.258
$N_VN_{VI,VII}$	52.8	1	0.2348	50.9	1	0.2436
	79 Gold			*80 Mercury*		
$\alpha_2\ KL_{II}$	0.185075	2	66.9895	0.179958	3	68.895
$\alpha_1\ KL_{III}$	0.180195	2	68.8037	0.175068	3	70.819
$\beta_3\ KM_{II}$	0.159810	2	77.580	0.155321	3	79.822

$\beta_9\ L_IM_V$	1.02063	7	12.1474	0.9871	2	12.560
L_IN_I	0.9131	1	13.578	0.8827	2	14.045
L_IN_{IV}	0.88563	7	13.999			
$\gamma_{11}\ L_IN_V$	0.88433	7	14.020	0.85657	7	14.474
L_IO_I	0.87074	5	14.2385	0.8452	2	14.670
$L_IO_{IV,V}$	0.86400	5	14.3497	0.8350	2	14.847
$L_{II}M_{II}$	1.1708	1	10.5892	1.1387	5	10.888
$\beta_{17}\ L_{II}M_{III}$	1.12798	5	10.9915	1.0916	5	11.358
$L_{II}M_V$	1.0756	2	11.526			
$L_{II}N_{III}$	0.9402	2	13.186	0.90894	7	13.640
$v\ L_{II}N_{VI}$	0.90837	5	13.6487	0.87885	7	14.107
$L_{II}O_{II}$	0.90746	7	13.662	0.8784	1	14.114
$L_{II}O_{III}$	0.90638	7	13.679	0.8758	1	14.156
$t\ L_{III}M_{II}$	1.41366	7	8.7702	1.3746	2	9.019
$s\ L_{III}M_{III}$	1.35131	7	9.1749	1.3112	2	9.455
$L_{III}N_{II}$	1.09968	7	11.2743	1.0649	2	11.642
$L_{III}N_{III}$	1.09026	7	11.3717	1.0585	1	11.713
$u\ L_{III}N_{VI,VII}$	1.04752	5	11.8357			
$u'\ L_{III}N_{VI}$				1.01769	7	12.1826
$u\ L_{III}N_{VII}$				1.01674	7	12.1940
$L_{III}O_{II,III}$	1.0450	2	11.865			
$L_{III}O_{II}$				1.01558	7	12.2079
$L_{III}O_{III}$				1.01404	7	12.2264
$L_{III}P_{II,III}$	1.03876	7	11.9355			
M_IN_{III}	4.300	9	2.883			
$M_{II}N_{IV}$	4.432	4	2.797			
$M_{III}N_I$	6.259	9	1.981	6.09	2	2.036
$M_{III}N_{IV}$	5.186	5	2.391			
$\gamma\ M_{III}N_V$	5.145	4	2.410	4.984‡	2	2.4875
$M_{III}O_I$	4.703	9	2.636			
$M_{III}O_{IV,V}$	4.522	6	2.742			
$\zeta_2\ M_{IV}N_{II}$	7.523	5	1.648			
$M_{IV}N_{III}$	7.101	8	1.746	6.87	2	1.805
$\beta\ M_{IV}N_{VI}$	5.624	1	2.2046	5.4318‡	9	2.2825
$\zeta_1\ M_VN_{III}$	7.466	4	1.6605			
$\alpha_2\ M_VN_{VI}$	5.854	3	2.118			
$\alpha_1\ M_VN_{VII}$	5.840	1	2.1229	5.6476‡	9	2.1953
M_VO_{III}	5.767	9	2.150			
$N_{IV}N_{VI}$	46.8	2	0.265	45.2‡	3	0.274
$N_VN_{VI,VII}$	49.4	1	0.2510	47.9‡	3	0.259

TABLE 7f-1. X-RAY WAVELENGTHS IN Å* UNITS AND IN keV (Continued)

Left block: **81 Thallium** and **82 Lead**. Right block: **81 Thallium (Cont.)** / **83 Bismuth** and **82 Lead (Cont.)** / **84 Polonium**.

Designation	λ, Å* (81 Tl)	p.e.†	keV	λ, Å* (82 Pb)	p.e.†	keV	Designation	λ, Å* (81 Tl Cont. / 83 Bi)	p.e.†	keV	λ, Å* (82 Pb Cont. / 84 Po)	p.e.†	keV
$\alpha_2\ KL_{II}$	0.175036	2	70.8319	0.170294	2	72.8042	$\alpha_2\ M_V N_{VI}$	5.472	2	2.2656	5.299	2	2.3397
$\alpha_1\ KL_{III}$	0.170136	2	72.8715	0.165376	2	74.9694	$\alpha_1\ M_V N_{VII}$	5.460	1	2.2706	5.286	1	2.3455
$\beta_3\ KM_{II}$	0.150980	6	82.118	0.146810	4	84.450	$M_V O_{III}$				5.168	9	2.399
$\beta_1\ KM_{III}$	0.150142	5	82.576	0.145970	6	84.936	$N_{IV} N_{VI}$				42.3	2	0.293
$\beta_2^{II}\ KN_{II}$	0.14614	1	84.836	0.14212	2	87.23	$N_V N_{VI,VII}$	46.5	2	0.267	45.0	1	0.2756
$\beta_2^{I}\ KN_{III}$	0.14595	1	84.946	0.14191	1	87.364	$N_{VI} O_{IV}$	115.3	2	0.1075	102.4	1	0.1211
$KO_{II,III}$	0.14509	1	85.451	0.141012	8	87.922	$N_{VI} O_V$	113.0	1	0.10968	100.2	2	0.1237
KP	0.14917	1	83.114	0.1408	1	88.06	$N_{VII} O_V$	117.7	1	0.10530	104.3	1	0.1189
$\beta_5\ KM_{IV,V}$	0.14553	2	85.19	0.14512	2	85.43	$\alpha_2\ KL_{II}$	0.165717	2	74.8148	0.16130‡	1	76.862
$\beta_5^{II}\ KM_{IV}$				0.14495	3	85.53	$\alpha_1\ KL_{III}$	0.160789	7	77.1079	0.15636‡	1	79.290
$\beta_5^{I}\ KM_V$				0.14155	3	87.59	$\beta_3\ KM_{II}$	0.142779	7	86.834	0.13892‡	2	89.25
$\beta_4\ KN_{IV,V}$							$\beta_1\ KM_{III}$	0.141948	3	87.343	0.13807‡	2	89.80
$\beta_4\ L_I M_{II}$	1.03918	3	11.9306	1.0075	1	12.306	$\beta_2^{II}\ KN_{II}$	0.13818	1	89.733	0.13438‡	2	92.26
$\beta_3\ L_I M_{III}$	1.00062	3	12.3904	0.96911	7	12.7933	$\beta_2^{I}\ KN_{III}$	0.13797	1	89.864	0.13418‡	2	92.40
$\gamma_2\ L_I N_{II}$	0.84773	5	14.6251	0.8210	2	15.101	$KO_{II,III}$	0.13709	1	90.435			
$\gamma_3\ L_I N_{III}$	0.84130	4	14.7368	0.8147		15.218	$\beta_5\ KM_{IV,V}$	0.14111	1	87.860			
$\gamma'_4\ L_I O_{II}$	0.81308	5	15.2482	0.78706	7	15.752	$\beta_4\ KN_{IV,V}$	0.13759	2	90.11			
$\gamma_4\ L_I O_{III}$	0.81184	5	15.2716	0.7858	1	15.777	$\beta_4\ L_I M_{II}$	0.97690	4	12.6912	0.9475	3	13.086
$\eta\ L_{II} M_I$	1.12769	3	10.9943	1.09241	7	11.3493	$\beta_3\ L_I M_{III}$	0.93855	3	13.2098	0.9091	3	13.638
$\beta_1\ L_{II} M_{IV}$	1.01513	4	12.2133	0.98291	3	12.6137	$\gamma_2\ L_I N_{II}$	0.79565	3	15.5824			
$\gamma_5\ L_{II} N_I$	0.89500	4	13.8526	0.86655	5	14.3075	$\gamma_3\ L_I N_{III}$	0.78917	5	15.7102			
$\gamma_1\ L_{II} N_{IV}$	0.86752	3	14.2915	0.83973	3	14.7644	$\gamma'_4\ L_I O_{II}$	0.76198	3	16.2709	0.772	1	16.07
$\gamma_8\ L_{II} O_I$	0.8513	2	14.564	0.82365	5	15.0527	$\gamma_4\ L_I O_{III}$	0.76087	3	16.2947			
$\gamma_6\ L_{II} O_{IV}$	0.8442	2	14.685	0.81683	5	15.1783	$\gamma_{13}\ L_I P_{II,III}$	0.75690	3	16.3802			
$L_{II} P_I$				0.81583	5	15.1969	$\eta\ L_{II} M_I$	1.05856	3	11.7122			
$l\ L_{III} M_I$	1.38477	3	8.9532	1.34990	7	9.1845	$\beta_1\ L_{II} M_{IV}$	0.951978	9	13.0235	0.9220	2	13.447
$\alpha_2\ L_{III} M_{IV}$	1.21875	3	10.1728	1.18648	5	10.4495	$\gamma_5\ L_{II} N_I$	0.83923	5	14.7732			
$\alpha_1\ L_{III} M_V$	1.20739	4	10.2685	1.17501	5	10.5515	$\gamma_1\ L_{II} N_{IV}$	0.81311	2	15.2477			
$\beta_6\ L_{III} N_I$	1.04963	5	11.8118	1.0210	2	12.143	$\gamma_8\ L_{II} O_I$	0.7973	1	15.551	0.7848	9	15.744
$\beta_{15}\ L_{III} N_{IV}$	1.01201	3	12.2510	0.98389	1	12.6011							
$\beta_2\ L_{III} N_V$	1.01031	3	12.2715	0.98221	7	12.6226							
$\beta_7\ L_{III} O_I$	0.99017	5	12.5212	0.9620	7	12.888							

Designation	λ (Å)		keV	λ (Å)		keV	Designation	λ (Å)		keV	λ (Å)		keV
$\beta_8 L_{III}O_{IV,V}$	0.98058	3	12.6436	0.9526	1	13.015	$\gamma_6 L_{II}O_{IV}$	0.79043	3	15.6853	0.7645	2	16.218
$L_I M_I$	1.0644	2	11.648	1.0323	2	12.010	$l\ L_{III}M_I$	1.31610	7	9.4204	1.2829	5	9.664
$\beta_{10} L_I M_{IV}$	0.96389	7	12.8626	0.9339	2	13.275	$\alpha_2 L_{III}M_{IV}$	1.15536	1	10.73091	1.12548†	5	11.0158
$\beta_9 L_I M_V$	0.95675	7	12.9585	0.9268	1	13.377	$\alpha_1 L_{III}M_V$	1.14386	2	10.8388	1.11386	4	11.1308
$L_I N_I$	0.8549	1	14.503	0.82859	7	14.963	$\beta_6 L_{III}N_I$	0.99331	3	12.4816	0.9672	2	12.819
$L_I N_{IV}$	0.83001	7	14.937	0.80364	7	15.427	$\beta_{15} L_{III}N_{IV}$	0.95702	5	12.9549	0.9312	5	13.314
$\gamma_{11} L_I N_V$	0.82879	5	14.9593	0.80233	9	15.453	$\beta_2 L_{III}N_V$	0.95518	4	12.9799	0.92937		13.3404
$L_I N_{VI,VII}$	0.8158	1	15.198	0.7884	1	15.725	$\beta_7 L_{III}O_I$	0.93505	3	13.2593			
$L_I O_I$	0.80861	5	15.3327	0.7897	7	15.699	$\beta_5 L_{III}O_{IV,V}$	0.92556	9	13.3953			
$L_I O_{IV,V}$	1.0997	1	11.274	0.78257	7	15.843	$L_I M_I$	1.0005	4	12.39	0.8996	2	13.782
$L_{II}M_{II}$	1.05609	7	11.7397	1.0644	2	11.648	$\beta_{10} L_I M_{IV}$	0.90495	3	13.7002			
$\beta_{17} L_{II}M_{III}$	1.00722	5	12.3093	1.0223	1	12.127	$\beta_9 L_I M_V$	0.89791	1	13.8077			
$L_{II}M_V$	0.882	2	14.057	0.9747	1	12.720	$L_I N_I$	0.8022	5	15.456			
$L_{II}N_{II}$	0.87996	5	14.0893	0.8585	3	14.442	$L_I N_V$	0.7795	5	15.904			
$L_{II}N_{III}$	0.85048	5	14.5777	0.85192	7	14.553	$\gamma_{11} L_I N_V$	0.77728	5	15.951			
$L_{II}N_V$	0.8490	1	14.604	0.8382	2	14.791	$L_I N_{VI,VII}$	0.7641	5	16.23			
$v\ L_{II}N_{VI}$	1.34154	5	9.2417	0.82327	7	15.060	$L_I O_{IV,V}$	0.75791	9	16.358			
$L_{II}O_{II}$	1.27807	5	9.7007	0.8200	1	15.120	$L_{II}M_{II}$	1.0346	5	11.98			
$L_{II}O_{III}$	1.0286	1	12.053	1.30767	7	9.4811	$\beta_{17} L_{II}M_{III}$	0.98913	5	12.5344			
$t\ L_{III}M_{II}$	0.9888	1	12.538	1.24385	7	9.9675	$L_{II}M_V$	0.94419	9	13.1310			
$s\ L_{III}M_{III}$	0.98738	5	12.5566	1.01040	7	12.2705	$L_{II}N_I$	0.8344	1	14.86			
$L_{III}N_{II}$	0.98538	5	12.5820	0.0005	1	12.392	$L_{II}N_{III}$	0.8248	9	15.031			
$L_{III}N_{III}$	0.97926	5	12.6607	0.96133	7	12.8968	$v\ L_{II}N_{VI}$	0.79721	5	15.552			
$u\ L_{III}N_{VI,VII}$				0.9586	1	12.934	$L_{II}O_{III}$	0.79384	1	15.6178			
$L_{III}O_{II}$				0.9578	1	12.945	$t\ L_{III}M_{II}$	1.2748	5	9.7252			
$L_{III}O_{I,I}$				0.95118	7	13.0344	$s\ L_{III}M_{III}$	1.2105	5	10.2421			
$L_{III}P_{II,III}$				3.872	9	3.202	$L_{III}N_{II}$	0.98280	5	12.6151			
$M_I N_I$	4.013	9	3.089	4.655	8	2.664	$L_{III}N_{III}$	0.97321	2	12.7394			
$M_I N_{II}$	4.116	4	3.013	3.968	5	3.124	$u\ L_{III}N_{VI,VII}$	0.93505	2	13.2593			
$M_{II}N_{IV}$	5.834	8	2.107	5.704	8	2.174	$L_{III}O_{II}$	0.9323	4	13.298			
$M_{III}N_I$	4.865	5	2.548	4.715	3	2.630	$L_{III}O_{III}$	0.9302	9	13.328			
$M_{III}N_{IV}$	4.823	4	2.571	4.674	1	2.6527	$L_{III}P_{II,III}$	0.92418	9	13.4159			
$\gamma\ M_{III}N_V$	4.216	6	2.941	4.244	9	2.921	$M_I N_{II}$	3.892	4	3.185			
$M_{III}O_I$	7.032	5	1.763	4.069	6	3.047	$M_I N_{III}$	3.740	8	3.315			
$M_{III}O_{IV,V}$	5.249	1	2.3621	6.802	5	1.823	$M_{II}N_{IV}$	3.834	5	3.234			
$\xi_2 M_{IV}N_{II}$	5.196	9	2.386	6.384	7	1.942	$M_{III}N_{IV}$	5.537	2	2.239			
$M_{IV}N_{III}$	6.974	4	1.778	5.076	1	2.4427	$M_{III}N_V$	4.571	9	2.712			
$\beta\ M_{IV}N_{VI}$				5.004	9	2.477	$\gamma\ M_{III}N_V$	4.532	6	2.735			
$M_{IV}O_{II}$				6.740	3	1.8395	$M_{III}O_I$	4.105		3.021			
$\xi_1 M_V N_{III}$							$M_{III}O_{IV,V}$	3.932		3.153			

TABLE 7f-1. X-RAY WAVELENGTHS IN Å* UNITS AND IN keV (Continued)

83 Bismuth (Cont.)

Designation	Wave-length, Å*	p.e.†	keV
$\zeta_2\ M_{IV}N_{II}$	6.585	5	1.883
$M_{IV}N_{III}$	6.162	8	2.012
$\beta\ M_{IV}N_{VI}$	4.909	1	2.5255
$M_{IV}O_{II}$	4.823	3	2.571
$M_{IV}P_{II,III}$	4.59	2	2.70
$\zeta_1\ M_V N_{IV}$	6.521	4	1.901
$\alpha_2\ M_V N_{VI}$	5.130	2	2.4170
$\alpha_1\ M_V N_{VII}$	5.118	1	2.4226
$N_I P_{II,III}$	13.30	6	0.932
$N_{VI}O_{IV}$	91.6	1	0.1354
$N_{VII}O_V$	93.2	1	0.1330

85 Astatine

Designation	Wave-length, Å*	p.e.†	keV
$\alpha_2\ KL_{II}$	0.15705‡	2	78.95
$\alpha_1\ KL_{III}$	0.15210‡	2	81.52
$\beta_3\ KM_{II}$	0.13517‡	4	91.72
$\beta_1\ KM_{III}$	0.13432‡	4	92.30
$\beta_2^{II}\ KN_{II}$	0.13072‡	4	94.84
$\beta_2^{I}\ KN_{III}$	0.13052‡	4	94.99
$\beta_3\ L_I M_{III}$	0.88135‡	9	14.067
$\beta_1\ L_{II}M_{IV}$	0.89349‡	9	13.876
$\gamma_1\ L_{II}N_{IV}$	0.76289‡	9	16.251
$\alpha_2\ L_{III}M_{IV}$	1.09671‡	5	11.3048
$\alpha_1\ L_{III}M_V$	1.08500‡	5	11.4268

87 Francium

Designation	Wave-length, Å*	p.e.†	keV
$\alpha_2\ KL_{II}$	0.14896‡	3	83.23
$\alpha_1\ KL_{III}$	0.14399‡	3	86.10
$\beta_3\ KM_{II}$	0.12807‡	5	96.81
$\beta_1\ KM_{III}$	0.12719‡	5	97.47

84 Polonium (Cont.)

Designation	Wave-length, Å*	p.e.†	keV

86 Radon

Designation	Wave-length, Å*	p.e.†	keV
$\alpha_2\ KL_{II}$	0.15294‡	3	81.07
$\alpha_1\ KL_{III}$	0.14798‡	3	83.78
$\beta_3\ KM_{II}$	0.13155‡	5	94.24
$\beta_1\ KM_{III}$	0.13069‡	5	94.87
$\beta_2^{II}\ KN_{II}$	0.12719‡	5	97.47
$\beta_2^{I}\ KN_{III}$	0.12698‡	5	97.64
$\beta_3\ L_I M_{III}$	0.85436‡	9	14.512
$\beta_1\ L_{II}M_{IV}$	0.86605‡	9	14.316
$\gamma_1\ L_{II}N_{IV}$	0.73928‡	9	16.770
$\alpha_2\ L_{III}M_{IV}$	1.06899‡	5	11.5979
$\alpha_1\ L_{III}M_V$	1.05723‡	5	11.7270

88 Radium

Designation	Wave-length, Å*	p.e.†	keV
$\alpha_2\ KL_{II}$	0.14512‡	2	85.43
$\alpha_1\ KL_{III}$	0.14014‡	2	88.47
$\beta_3\ KM_{II}$	0.12469‡	3	99.43
$\beta_1\ KM_{III}$	0.12382‡	3	100.13

89 Actinium and 90 Thorium

Designation	89 Actinium Wave-length, Å*	p.e.†	keV	90 Thorium Wave-length, Å*	p.e.†	keV
$\alpha_2\ KL_{II}$	0.14141‡	2	87.67	0.137829	2	89.953
$\alpha_1\ KL_{III}$	0.136417‡	8	90.884	0.132813	2	93.350
$\beta_3\ KM_{II}$	0.12143‡	2	102.10	0.118268	3	104.831
$\beta_1\ KM_{III}$	0.12055‡	2	102.85	0.117396	9	105.609
$\beta_2^{II}\ KN_{II}$	0.11732‡	2	105.67	0.11426	1	108.511
$\beta_2^{I}\ KN_{III}$	0.11711‡	2	105.86	0.114040	9	108.717
$KO_{II,III}$				0.11322	1	109.500
$\beta_5\ KM_{IV,V}$				0.116667	9	106.269
$\beta_4\ KN_{IV,V}$				0.11366	2	109.08
$\beta_4\ L_I M_{II}$				0.79257	3	15.6429
$\beta_3\ L_I M_{III}$	0.77822	9	15.931	0.75479	3	16.4258
$\gamma_2\ L_I N_{II}$				0.64221	4	19.305
$\gamma_3\ L_I N_{III}$				0.63559	4	19.507
$\gamma_4\ L_I O_{II}$				0.61251	4	20.242
$\gamma_{13}\ L_I P_{II,III}$				0.61098	4	20.292
				0.60705	8	20.424
$\eta\ L_{II}M_I$				0.85446	4	14.5099
$\beta_1\ L_{II}M_{IV}$	0.78903	9	15.713	0.765210	4	16.2022
$\gamma_5\ L_{II}N_I$				0.67491	4	18.370
$\gamma_1\ L_{II}N_{IV}$	0.67351	9	18.408	0.65313	3	18.9825
$\gamma_8\ L_{II}O_I$				0.63898	5	19.403
$\gamma_6\ L_{II}O_{IV}$				0.63258	4	19.599
$L_{II}P_I$				0.6316	1	19.629
$L_{II}P_{IV}$				0.62991	9	19.682
$u\ L_{III}M_I$				1.11508	4	11.1186
$\alpha_2\ L_{III}M_{IV}$	0.99178	5	12.5008	0.96788	3	12.8096
$\alpha_1\ L_{III}M_V$	0.97993	5	12.6520	0.95600	3	12.9687
$\beta_6\ L_{III}N_I$				0.82790	8	14.975
$\beta_{15}\ L_{III}N_{IV,V}$				0.79539	5	15.5875
$\beta_2\ L_{III}N_V$				0.79354	3	15.6237
$\beta_7\ L_{III}O_I$				0.77437	4	16.0105
$\beta_5\ L_{III}O_{IV,V}$				0.76468	5	16.213

Table 1

Line	λ	I	E	I
β_2^{II} KN_{II}	0.12379‡	5	100.16	5
β_2^{I} KN_{III}	0.12355‡	5	100.33	5
β_4 $L_I M_{II}$	0.82789‡	9	14.976	9
β_3 $L_I M_{III}$				
γ_2 $L_I N_{II}$				
γ_3 $L_I N_{III}$				
γ'_4 $L_I O_{I,II}$				
γ_4 $L_I O_{II,III}$				
γ_{13} $L_I P_{II,III}$	0.83940‡	9	14.770	9
η $L_{II} M_I$	0.71652‡	9	17.303	9
β_1 $L_{II} M_{IV}$				
γ_5 $L_{II} N_I$				
γ_1 $L_{II} N_{IV}$				
γ_8				
γ_6 $L_{II} O_{IV}$				
$L_{II} P_I$				
l $L_{III} M_I$				
α_2 $L_{III} M_{IV}$	1.04230	5	11.8950	5
α_1 $L_{III} M_V$	1.03049	5	12.0313	5
β_6 $L_{III} N_I$				
β_{15} $L_{III} N_{IV}$				
β_2 $L_{III} N_V$	0.858	2	14.45	2
β_7 $L_{III} O_I$				
β_5 $L_{III} O_{IV,V}$				
$L_{III} P_I$				

Table 2

Line	E	I	λ	λ	I	E
β_2^{II} KN_{II}	102.89	3	0.12050‡	0.76338	5	16.241
β_2^{I} KN_{III}	103.07	3	0.12029‡	0.76087	9	16.295
β_4 $L_I M_{II}$	14.7472	5	0.84071	0.7301	1	16.981
β_3 $L_I M_{III}$	15.4449	5	0.80273	0.7234	1	17.139
γ_2 $L_I N_{II}$	18.179	5	0.68199	0.64755	5	19.146
γ_3 $L_I N_{III}$	18.357	5	0.67538	0.6276	5	19.755
γ'_4 $L_I O_{I,II}$	19.036	5	0.65131	0.62636	9	19.794
γ_4 $L_I O_{II,III}$	19.084	5	0.64965	0.6160	1	20.128
γ_{13} $L_I P_{II,III}$	19.218	5	0.64513	0.6146	1	20.174
η $L_{II} M_I$	13.6630	5	0.90742	0.6083	1	20.383
β_1 $L_{II} M_{IV}$	15.2358	5	0.81375	0.8338	1	14.869
γ_5 $L_{II} N_I$	17.274	5	0.71774	0.79257	4	15.6429
γ_1 $L_{II} N_{IV}$	17.849	5	0.69463	0.7579	1	16.359
γ_8	18.230	5	0.6801	0.6620	1	18.729
γ_6 $L_{II} O_{IV}$	18.414	5	0.67328	0.6521	1	19.014
$L_{II} P_I$	18.439	5	0.6724	0.64064	9	19.353
l $L_{III} M_I$	10.6222	5	1.16719	0.6369	1	19.466
α_2 $L_{III} M_{IV}$	12.1962	5	1.01656	0.6356	1	19.506
α_1 $L_{III} M_V$	12.3397	5	1.00473	0.6312	1	19.642
β_6 $L_{III} N_I$	14.2362	5	0.87088	1.08009	9	11.4788
β_{15} $L_{III} N_{IV}$	14.8086	5	0.83722	1.0112	1	12.261
β_2 $L_{III} N_V$	14.8414	5	0.83537	0.8190	2	15.138
β_7 $L_{III} O_I$	15.190	5	0.8162	0.8082	1	15.341
β_5 $L_{III} O_{IV,V}$	15.3771	5	0.80627	0.77661	5	15.964
$L_{III} P_I$	15.402	5	0.8050	0.7713	1	16.074
β_{10} $L_I M_{IV}$	15.988	5	0.77546	0.7690	1	16.123
β_9 $L_I M_V$	16.131	5	0.76857	0.7625	2	16.260
$L_I N_I$	18.036	1	0.6874	2.934	8	4.23
$L_I N_{IV}$	18.600	1	0.6666	2.442	9	5.08
γ_{11} $L_I N_V$	18.633	1	0.6654	3.537	9	3.505
$L_I O_{IV,V}$	19.167	1	0.6468	3.011	2	4.117
β_{17} $L_{II} M_{III}$	14.692	1	0.8438	2.618	5	4.735
$L_{II} N_{III}$	17.604	1	0.7043	4.568	5	2.714
$L_{II} N_V$	17.884	1	0.6932	3.718	3	3.335
$L_{II} O_{II}$	18.286	1	0.6780	3.679	3	3.370
L_2 $L_{II} O_{III}$	18.330	1	0.6764	3.283	2	3.78
$L_{III} P_{II,III}$	18.466	1	0.6714	3.131	9	3.959
$L_{III} N_{II}$	14.387	1	0.8618	5.340	3	2.322
$L_{III} N_{III}$	14.566	1	0.8512	4.911	5	2.524
u $L_{III} N_{VI,VII}$	15.146	1	0.8186	3.941	1	3.1458
$L_{III} P_{II,III}$	15.425	1	0.8038	3.808	4	3.256

TABLE 7f-1. X-RAY WAVELENGTHS IN Å* UNITS AND IN KEV (*Continued*)

Designation	Wave-length, Å*	p.e.†	keV	Wave-length, Å*	p.e.†	keV
	89 Actinium (Cont.)			*90 Thorium (Cont.)*		
$\zeta_1\ M_VN_{III}$				5.245	5	2.364
$\alpha_2\ M_VN_{VI}$				4.151	2	2.987
$\alpha_1\ M_VN_{VII}$				4.1381	9	2.9961
M_VP_{III}				3.760	9	3.298
N_IP_{II}				9.44	7	1.313
N_IP_{III}				9.40	7	1.1319
$N_{II}O_{IV}$				11.56	5	1.072
$N_{II}P_I$				11.07	7	1.120
$N_{III}O_V$				13.8	1	0.897
$N_{IV}N_{VI}$				33.57	9	0.3693
$N_VN_{VI,VII}$				36.32	9	0.3414
$N_{VI}O_{IV}$				49.5	1	0.2505
$N_{VI}O_V$				48.2	1	0.2572
$N_{VII}O_V$				50.0	1	0.2479
$O_{III}P_{IV,V}$				68.2	3	0.1817
$O_{IV,V}Q_{II,III}$				181	5	0.068
	91 Protactinium			*92 Uranium*		
$\alpha_2\ KL_{II}$	0.134343‡	9	92.287	0.130968	4	94.665
$\alpha_1\ KL_{III}$	0.129325‡	3	95.868	0.125947	3	98.439
$\beta_3\ KM_{II}$	0.11523‡	2	107.60	0.112296	4	110.406
$\beta_1\ KM_{III}$	0.114345‡	8	108.427	0.111394	5	111.300
$\beta_2^{II}\ KN_{II}$	0.11129‡	2	111.40	0.10837	1	114.40
$\beta_2^{I}\ KN_{III}$	0.11107‡	2	111.62	0.10818	1	114.60
$KO_{II,III}$				0.10744	1	115.39
$\beta_5\ KM_{IV,V}$				0.11069	1	112.01
$\beta_4\ KN_{IV,V}$				0.10780	2	115.01
$\beta_4\ L_IM_{II}$	0.7699	1	16.104	0.747985	9	16.5753
$\beta_3\ L_IM_{III}$	0.73230	5	16.930	0.71029	2	17.4550

Designation	Wave-length, Å*	p.e.†	keV	Wave-length, Å*	p.e.†	keV
	91 Protactinium (Cont.)			*92 Uranium (Cont.)*		
M_IN_{II}				2.92	2	4.25
M_IN_{III}				2.753	8	4.50
M_IO_{III}				2.304	7	5.38
M_IP_{III}				2.253	6	5.50
$M_{II}N_I$	3.441	5	3.603	3.329	4	3.724
$M_{II}N_{IV}$	2.910	2	4.260	2.817	2	4.401
$M_{II}O_{IV}$	2.527	4	4.906	2.443	4	5.075
$M_{III}N_I$	4.450	4	2.786	4.330	2	2.863
$M_{III}N_{IV}$	3.614	2	3.430	3.521	2	3.521
$\gamma\ M_{III}N_V$	3.577	1	3.4657	3.479	1	3.563
$M_{III}O_I$	3.245	9	3.82	3.115	7	3.980
$M_{III}O_{IV,V}$	3.038	2	4.081	2.948	2	4.205
$\zeta_2\ M_{IV}N_{II}$	5.193	2	2.3876	5.050	2	2.4548
$M_{IV}N_{III}$				4.625	5	2.681
$\beta\ M_{IV}N_{VI}$	3.827	1	3.2397	3.716	1	3.3367
$M_{IV}O_{II}$	3.691	2	3.359	3.576	1	3.4666
$\zeta_1\ M_VN_{III}$	5.092	2	2.4350	4.946	2	2.507
$\alpha_2\ M_VN_{VI}$	4.035	3	3.072	3.924	1	3.1595
$\alpha_1\ M_VN_{VII}$	4.022	1	3.0823	3.910	1	3.1708
N_IO_{III}				10.09	7	1.229
N_IP_{II}				8.81	7	1.41
N_IP_{III}				8.76	7	1.42
$N_{II}P_I$				10.40	7	1.192
$N_{III}O_V$				12.90	9	0.961
$N_{IV}N_{VI}$				31.8	1	0.390
$N_VN_{VI,VII}$				34.8	1	0.357
$N_{IV}O_{IV}$				43.3	2	0.286
$N_{VI}O_V$				42.1	2	0.295
$N_IP_{IV,V}$				8.60	7	1.44

$\gamma_2\ L_IN_{II}$	0.6239	1	19.872	0.605237	9	20.4847
$\gamma_3\ L_IN_{III}$	0.6169	1	20.098	0.598574	9	20.7127
$\gamma'_4\ L_IO_{II}$				0.576700	9	21.4984
$\gamma_4\ L_IO_{II,III}$	0.5937	1	20.882	0.57499	9	21.562
γ_{13}				0.5706	1	21.729
$\eta\ L_{II}M_I$	0.8295	1	14.946	0.80509	2	15.3997
$\beta_1\ L_{II}M_{IV}$	0.74232	5	16.702	0.719984	8	17.2200
$\gamma_5\ L_{II}N_I$	0.6550	1	18.930	0.63557	2	19.5072
$\gamma_1\ L_{II}N_{IV}$	0.63358	9	19.568	0.614770	9	20.1671
$\gamma_8\ L_{II}O_I$				0.60125	5	20.621
$\gamma_6\ L_{II}O_{IV}$	0.6133	1	20.216	0.594845	9	20.8426
$L_{II}P_{IV}$				0.59203	5	20.942
$l\ L_{III}M_I$	1.0908	1	11.366	1.06712	2	11.6183
$\alpha_2\ L_{III}M_{IV}$	0.94482‡	5	13.1222	0.922558	9	13.4388
$\alpha_1\ L_{III}M_V$	0.93284	5	13.2907	0.910639	9	13.6147
$\beta_6\ L_{III}N_I$	0.8079	1	15.347	0.78838	2	15.7260
$\beta_{15}\ L_{III}N_{IV}$				0.756642	9	16.3857
$\beta_2\ L_{III}N_V$	0.7737	1	16.024	0.754681	9	16.4283
$\beta_7\ L_{III}O_I$	0.7546	2	16.431	0.73602	6	16.845
$\beta_5\ L_{III}O_{IV,V}$	0.7452	2	16.636	0.726305	9	17.0701
$L_{III}P_I$				0.72521	5	17.096
$L_{III}P_{IV,V}$				0.72240	5	17.162
$\beta_{10}\ L_IM_{IV}$	0.7088	2	17.492	0.68760	5	18.031
$\beta_9\ L_IM_V$	0.7018	1	17.667	0.681014	8	18.2054
L_IN_{IV}				0.59096	5	20.979
$\gamma_{11}\ L_IN_V$				0.58986	5	21.019
$L_IO_{IV,V}$				0.5725	1	21.657
$\beta_{17}\ L_{II}M_{III}$				0.74503	5	16.641
$L_{II}N_{III}$				0.6228	1	19.907
$v\ L_{II}N_{VI}$				0.6031	1	20.556
$L_{III}O_{III}$				0.59728	5	20.758
$L_{II}P_{II,III}$				0.5930	2	20.906
$t\ L_{III}M_{II}$				1.0347	1	11.982
$s\ L_{III}M_{III}$				0.9636	1	12.866
$L_{III}N_{II}$				0.78017	9	15.892
$L_{III}N_{III}$				0.7691	1	16.120
$u\ L_{III}N_{VI,VII}$				0.738603	9	16.7859
$L_{III}O_{II}$				0.7333	1	16.907
$L_{III}O_{III}$				0.7309	1	16.962
$L_{III}P_{II,III}$				0.72426	5	17.118

	93 Neptunium			*94 Plutonium*		
$\beta_4\ L_IM_{II}$	0.72671	2	17.0607	0.70620	2	17.5560
$\beta_3\ L_IM_{III}$	0.68920‡	9	17.989	0.66871	2	18.5405
$\gamma_2\ L_IN_{II}$	0.5873	5	21.11	0.57068	2	21.7251
$\gamma_3\ L_IN_{III}$	0.5810	5	21.34	0.564001	9	21.9824
$\gamma'_4\ L_IO_{II}$				0.5432	1	22.823
$\gamma_4\ L_IO_{II,III}$	0.5585	5	22.20	0.5416	1	22.891
$\eta\ L_{II}M_I$	0.7809	2	15.876	0.7591	1	16.333
$\beta_1\ L_{II}M_{IV}$	0.698478	9	17.7502	0.67772	2	18.2937
$\gamma_5\ L_{II}N_I$	0.616	1	20.12	0.5988	1	20.704
$\gamma_1\ L_{II}N_{IV}$	0.596498	9	20.7848	0.578882	9	21.4173
γ_8				0.5658	1	21.914
$\gamma_6\ L_{II}O_{IV}$	0.57699	5	21.488	0.55973	2	22.1502
$l\ L_{III}M_I$	1.0428	6	11.890	1.0226	1	12.124
$\alpha_2\ L_{III}M_{IV}$	0.901045	9	13.7597	0.88028	2	14.0842
$\alpha_1\ L_{III}M_V$	0.889128	9	13.9441	0.86830	2	14.2786
$\beta_6\ L_{III}N_I$	0.769	1	16.13	0.75148	2	16.4983
$\beta_{15}\ L_{III}N_{IV}$				0.7205	1	17.208
$\beta_2\ L_{III}N_V$	0.736230	9	16.8400	0.71851	2	17.2553
$\beta_7\ L_{III}O_I$				0.7003	1	17.705
$\beta_5\ L_{III}O_{IV,V}$	0.70814	2	17.5081	0.69068	2	17.9506
$\beta_{10}\ L_IM_{IV}$				0.6482	1	19.126
$\beta_9\ L_IM_V$				0.6416	1	19.323
$u\ L_{III}N_{VI,VII}$				0.7031	1	17.635

	95 Americium		
$\beta_4\ L_IM_{II}$	0.68639	2	18.0627
$\beta_3\ L_IM_{III}$	0.64891	2	19.1059
$\gamma_2\ L_IN_{II}$	0.5544	2	22.361
$\beta_1\ L_{II}M_{IV}$	0.657655	9	18.8520
$\gamma_1\ L_{II}N_{IV}$	0.561886	9	22.0652
$\gamma_6\ L_{II}O_{IV}$	0.54311	2	22.8282
$l\ L_{III}M_I$	1.0012	6	12.384
$\alpha_2\ L_{III}M_{IV}$	0.860266	9	14.4119
$\alpha_1\ L_{III}M_V$	0.848187	9	14.6172
$\beta_6\ L_{III}N_I$	0.73418	2	16.8870
$\beta_{15}\ L_{III}N_{IV}$	0.70341	2	17.6258
$\beta_2\ L_{III}N_V$	0.701390	9	17.6765
$\beta_5\ L_{III}O_{IV,V}$	0.67383	2	18.3996

TABLE 7f-2. WAVELENGTHS OF X-RAY EMISSION LINES AND ABSORPTION EDGES: IN NUMERICAL ORDER

Wavelength, Å*	p.e.†	Element	Designation	keV
0.10723	1	92 U	K / Abs. edge	115.62
0.10744	1	92 U	$KO_{II,III}$	115.39
0.10780	2	92 U	$K\beta_4$ / $KN_{IV,V}$	115.01
0.10818	1	92 U	$K\beta_2^{I}$ / KN_{III}	114.60
0.10837	1	92 U	$K\beta_2^{II}$ / KN_{II}	114.40
0.11069	1	92 U	$K\beta_5$ / $KM_{IV,V}$	112.01
0.11107	2	91 Pa	$K\beta_2^{I}$ / KN_{III}	111.62
0.11129	2	91 Pa	$K\beta_2^{II}$ / KN_{II}	111.40
0.111394	5	92 U	$K\beta_1$ / KM_{III}	111.300
0.112296	4	92 U	$K\beta_3$ / KM_{II}	110.406
0.11307	1	90 Th	K / Abs. edge	109.646
0.11322	2	90 Th	$KO_{II,III}$	109.500
0.11366	2	90 Th	$K\beta_4$ / $KN_{IV,V}$	109.08
0.114040	9	90 Th	$K\beta_2^{I}$ / KN_{III}	108.717
0.11426	1	90 Th	$K\beta_2^{II}$ / KN_{II}	108.511
0.114345	8	91 Pa	$K\beta_1$ / KM_{III}	108.427
0.11523	2	91 Pa	$K\beta_3$ / KM_{II}	107.60
0.116667	9	90 Th	$K\beta_5$ / $KM_{IV,V}$	106.269
0.11711	2	89 Ac	$K\beta_2^{I}$ / KN_{III}	105.86
0.11732	2	89 Ac	$K\beta_2^{II}$ / KN_{II}	105.67
0.117396	9	90 Th	$K\beta_1$ / KM_{III}	105.609
0.118268	3	90 Th	$K\beta_3$ / KM_{II}	104.831
0.12029	3	88 Ra	$K\beta_2^{I}$ / KN_{III}	103.07
0.12050	3	88 Ra	$K\beta_2^{II}$ / KN_{II}	102.89
0.12055	2	89 Ac	$K\beta_1$ / KM_{III}	102.85
0.12143	2	89 Ac	$K\beta_3$ / KM_{II}	102.10
0.12358	5	87 Fr	$K\beta_2^{I}$ / KN_{III}	100.33
0.12379	5	87 Fr	$K\beta_2^{II}$ / KN_{II}	100.16
0.12382	3	88 Ra	$K\beta_1$ / KM_{III}	100.13
0.12469	3	88 Ra	$K\beta_3$ / KM_{II}	99.43
0.125947	3	92 U	$K\alpha_1$ / KL_{III}	98.439
0.12698	5	86 Rn	$K\beta_2^{I}$ / KN_{III}	97.64
0.12719	5	86 Rn	$K\beta_2^{II}$ / KN_{II}	97.47
0.12719	5	87 Fr	$K\beta_1$ / KM_{III}	97.47
0.14512	2	88 Ra	$K\alpha_2$ / KL_{II}	85.43
0.14553	2	81 Tl	$K\beta_4$ / $KN_{IV,V}$	85.19
0.14595	1	81 Tl	$K\beta_2^{I}$ / KN_{III}	84.946
0.145970	6	82 Pb	$K\beta_1$ / KM_{III}	84.936
0.14614	1	81 Tl	$K\beta_2^{II}$ / KN_{II}	84.836
0.146810	4	82 Pb	$K\beta_3$ / KM_{II}	84.450
0.14798	3	86 Rn	$K\alpha_1$ / KL_{III}	83.78
0.14896	3	87 Fr	$K\alpha_2$ / KL_{II}	83.23
0.14917	1	81 Tl	$K\beta_5$ / $KM_{IV,V}$	83.114
0.14918	1	80 Hg	K / Abs. edge	83.109
0.14931	2	80 Hg	$KO_{I,III}$	83.04
0.14978	2	80 Hg	$K\beta_4$ / $KN_{IV,V}$	82.78
0.150142	5	81 Tl	$K\beta_1$ / KM_{III}	82.576
0.15020	2	80 Hg	$K\beta_2^{I}$ / KN_{III}	82.54
0.15040	2	80 Hg	$K\beta_2^{II}$ / KN_{II}	82.43
0.150980	6	81 Tl	$K\beta_3$ / KM_{II}	82.118
0.15210	2	85 At	$K\alpha_1$ / KL_{III}	81.52
0.15294	3	86 Rn	$K\alpha_2$ / KL_{II}	81.07
0.15353	2	80 Hg	$K\beta_5$ / $KM_{IV,V}$	80.75
0.153593	5	79 Au	K / Abs. edge	80.720
0.153694	7	79 Au	$KO_{I,III}$	80.667
0.154224	5	80 Hg	$K\beta_4$ / $KN_{IV,V}$	80.391
0.154487	3	79 Au	$K\beta_1$ / KM_{III}	80.253
0.154618	9	79 Au	$K\beta_2^{I}$ / KN_{III}	80.185
0.15483	2	80 Hg	$K\beta_2^{II}$ / KN_{II}	80.08
0.155321	3	80 Hg	$K\beta_3$ / KM_{II}	79.822
0.15636	1	84 Po	$K\alpha_1$ / KL_{III}	79.290
0.15705	2	85 At	$K\alpha_2$ / KL_{II}	78.95
0.157880	5	79 Au	$K\beta_5^{II}$ / KM_{V}	78.529
0.158062	7	79 Au	$K\beta_5^{II}$ / KM_{IV}	78.438
0.15818	1	78 Pt	K / Abs. edge	78.381
0.15826	1	78 Pt	$KO_{I,III}$	78.341
0.15881	2	78 Pt	$K\beta_4$ / $KN_{IV,V}$	78.069
0.158982	3	79 Au	$K\beta_1$ / KM_{III}	77.984

Energy (keV)	Level	Line	Element	Intensity	Wavelength (Å)
77.878	KN_{III}	$K\beta_2^{I}$	78 Pt	1	0.15920
77.785	KN_{II}	$K\beta_2^{II}$	78 Pt	1	0.15939
77.580	KM_{III}	$K\beta_3$	79 Au	2	0.159810
77.1079	KL_{III}	$K\alpha_1$	83 Bi	2	0.160789
76.862	KL_{II}	$K\alpha_2$	84 Po	1	0.16130
76.27	KM_{V}	$K\beta_5^{I}$	78 Pt	3	0.16255
76.199	KM_{IV}	$K\beta_5^{II}$	78 Pt	2	0.16271
76.101	Abs. edge	K	78 Pt	1	0.16292
76.053	$KO_{II,III}$	$K\beta_4$	77 Ir	5	0.163019
75.821	$KN_{IV,V}$	$K\beta_1$	77 Ir	2	0.16352
75.748	KM_{III}	$K\beta_2^{I}$	78 Pt	3	0.163675
75.619	KN_{III}	$K\beta_2^{II}$	77 Ir	7	0.163956
75.529	KN_{II}	$K\beta_3$	77 Ir	3	0.16415
75.368	KM_{III}	$K\alpha_1$	78 Pt	2	0.164501
74.9694	KL_{II}	$K\alpha_2$	82 Pb	9	0.165376
74.8148	KL_{III}	$K\beta_5^{I}$	83 Bi	2	0.165717
74.075	KL_{II}	$K\beta_5^{II}$	77 Ir	1	0.167373
73.980	KM_{V}	$K\beta_3$	77 Ir	1	0.16759
73.856	KM_{IV}	$K\beta_2^{I}$	77 Ir	2	0.16787
73.808	Abs. edge	K	76 Os	6	0.16798
73.615	$KO_{II,III}$	$K\beta_4$	76 Os	1	0.16842
73.5608	$KN_{IV,V}$	$K\beta_1$	76 Os	1	0.168542
73.402	KM_{III}	$K\beta_2^{I}$	77 Ir	2	0.168906
73.318	KN_{III}	$K\beta_2^{II}$	76 Os	2	0.16910
73.2027	KN_{II}	$K\beta_3$	77 Ir	2	0.169367
72.8715	KM_{II}	$K\beta_4$	81 Tl	2	0.170136
72.8042	KL_{III}	$K\beta_2^{I}$	82 Pb	2	0.170294
71.895	KL_{II}	$K\beta_2^{II}$	76 Os	1	0.17245
71.824	KM_{V}	$K\beta_3$	76 OS	2	0.17262
71.658	KM_{IV}	$K\alpha_1$	76 OS	1	0.17302
71.633	Abs. edge	K	75 Re	1	0.17308
71.413	$KO_{II,III}$	$K\beta_1$	75 Re	3	0.173611
71.410	KM_{III}	$K\beta_4$	76 Os	3	0.17362
71.232	KN_{III}	$K\beta_2^{I}$	75 Re	8	0.174054
71.151	KN_{II}	$K\beta_2^{II}$	75 Re	1	0.17425
71.077	KM_{II}	$K\beta_3$	76 Os	3	0.174431
70.8319	KL_{II}	$K\alpha_2$	81 Tl	2	0.175036
70.819	KL_{III}	$K\alpha_1$	80 Hg	3	0.175068

Energy (keV)	Level	Line	Element	Intensity	Wavelength (Å)
96.81	KM_{II}	$K\beta_3$	87 Fr	5	0.12807
95.868	KL_{III}	$K\alpha_1$	91 Pa	3	0.129325
94.99	KN_{III}	$K\beta_2^{I}$	85 At	4	0.13052
94.87	KM_{III}	$K\beta_1$	86 Rn	5	0.13069
94.84	KN_{II}	$K\beta_2^{II}$	85 At	4	0.13072
94.665	KL_{II}	$K\alpha_2$	92 U	4	0.130968
94.24	KM_{II}	$K\beta_3$	86 Rn	5	0.13155
93.350	KL_{III}	$K\alpha_1$	90 Th	2	0.132813
92.40	KN_{III}	$K\beta_2^{I}$	84 Po	4	0.13418
92.30	KM_{III}	$K\beta_1$	85 At	9	0.13432
92.287	KL_{II}	$K\alpha_2$	91 Pa	2	0.134343
92.26	KM_{II}	$K\beta_3$	84 Po	4	0.13438
91.72	KL_{II}	$K\alpha_2$	85 At	8	0.13517
90.884	KL_{III}	$K\alpha_1$	89 Ac	1	0.136417
90.534	Abs. edge	K	83 Bi	1	0.13694
90.435	$KO_{II,III}$	$K\beta_4$	83 Bi	1	0.13709
90.11	KL_{II}	$K\alpha_2$	83 Bi	2	0.13759
89.953	KN_{III}	$K\beta_2^{I}$	90 Th	1	0.137829
89.864	KM_{II}	$K\beta_3$	83 Bi	2	0.13797
89.80	KN_{II}	$K\beta_2^{II}$	84 Po	2	0.13807
89.733	KM_{II}	$K\beta_3$	83 Bi	2	0.13817
89.25	KL_{III}	$K\alpha_1$	84 Po	2	0.13892
88.47	KL_{III}	$K\alpha_1$	88 Ra	1	0.14014
88.06	Abs. edge	K	82 Pb	5	0.1408
88.005	$KN_{IV,V}$	$K\beta_5$	82 Pb	8	0.140880
87.922	KL_{II}	$K\alpha_2$	82 Pb	1	0.141012
87.860	$KO_{II,III}$	$K\beta_4$	83 Bi	2	0.14111
87.67	KN_{III}	$K\beta_2^{I}$	89 Ac	3	0.14141
87.59	KM_{III}	$K\beta_1$	82 Pb	3	0.14155
87.364	KN_{II}	$K\beta_2^{II}$	82 Pb	3	0.14191
87.343	KM_{II}	$K\beta_3$	83 Bi	2	0.141948
87.23	KL_{III}	$K\alpha_1$	82 Pb	7	0.14212
86.834	KM_{II}	$K\beta_3$	83 Bi	3	0.142779
86.10	KL_{III}	$K\alpha_1$	87 Fr	1	0.14399
85.533	Abs. edge	K	81 Tl	3	0.14495
85.53	KM_{V}	$K\beta_5^{I}$	82 Pb	1	0.14495
85.451	KL_{II}	$K\alpha_2$	81 Tl	3	0.14509
85.43	KM_{IV}	$K\beta_5^{II}$	82 Pb	2	0.14512

TABLE 7f-2. WAVELENGTHS OF X-RAY EMISSION LINES AND ABSORPTION EDGES: IN NUMERICAL ORDER (Continued)

Wavelength, Å*	p.e.†	Element	Designation	keV
0.17766	1	75 Re	$K\beta_5^{I}$ KM_V	69.786
0.17783	1	75 Re	$K\beta_5^{II}$ KM_{IV}	69.719
0.17837	2	74 W	K Abs. edge	69.508
0.178444	5	74 W	$KO_{II,III}$	69.479
0.178880	3	75 Re	$KM_{II,III}$	69.310
0.17892	2	74 W	$K\beta_1$ $KN_{IV,V}$	69.294
0.179421	7	74 W	$K\beta_4$ KN_{III}	69.101
0.17960	1	74 W	$K\beta_2^{I}$ KN_{II}	69.031
0.179697	3	75 Re	$K\beta_3$ KM_{II}	68.994
0.179958	3	80 Hg	$K\alpha_2$ KL_{II}	68.895
0.180195	2	79 Au	$K\alpha_1$ KL_{III}	68.8037
0.183092	7	74 W	$K\beta_5^{I}$ KM_V	67.715
0.183264	5	74 W	$K\beta_5^{II}$ KM_{IV}	67.652
0.18394	1	73 Ta	K Abs. edge	67.403
0.184031	7	73 Ta	$KO_{II,III}$	67.370
0.184374	7	74 W	$K\beta_1$ KM_{III}	67.2443
0.18451	1	73 Ta	$K\beta_4$ $KN_{IV,V}$	67.194
0.185011	8	73 Ta	$K\beta_2^{I}$ KN_{III}	67.013
0.185075	2	79 Au	$K\alpha_2$ KL_{II}	66.9895
0.185181	9	74 W	$K\beta_3$ KM_{II}	66.9514
0.185188	4	73 Ta	$K\beta_2^{II}$ KN_{II}	66.949
0.185511	4	78 Pt	$K\alpha_1$ KL_{III}	66.832
0.18672	4	79 Au	KL_I	66.40
0.188757	6	73 Ta	$K\beta_5^{I}$ KM_V	65.683
0.188920	6	73 Ta	$K\beta_5^{II}$ KM_{IV}	65.626
0.18982	5	72 Hf	K Abs. edge	65.31
0.190089	4	73 Ta	$K\beta_1$ KM_{III}	65.223
0.190381	4	78 Pt	$K\alpha_2$ KL_{II}	65.122
0.1908	1	72 Hf	$K\beta_2$ $KN_{II,III}$	64.98
0.190890	2	73 Ta	$K\beta_3$ KM_{II}	64.9488
0.191047	2	77 Ir	$K\alpha_1$ KL_{III}	64.8956
0.19585	5	71 Lu	K Abs. edge	63.31
0.19589	2	71 Lu	$KO_{II,III}$	63.293
0.22855	3	67 Ho	$K\beta_5$ $KM_{IV,V}$	54.246
0.229298	2	71 Lu	$K\alpha_1$ KL_{III}	54.0698
0.23012	2	67 Ho	$K\beta_1$ KM_{III}	53.877
0.23048	1	66 Dy	K Abs. edge	53.793
0.23056	3	66 Dy	$KO_{II,III}$	53.774
0.23083	2	67 Ho	$K\beta_3$ KM_{II}	53.711
0.2317	2	66 Dy	$K\beta_2$ $KN_{II,III}$	53.47
0.234081	2	71 Lu	$K\alpha_2$ KL_{II}	52.9650
0.23618	3	66 Dy	$K\beta_5$ $KM_{IV,V}$	52.494
0.236655	2	70 Yb	$K\alpha_1$ KL_{III}	52.3889
0.23788	2	66 Dy	$K\beta_1$ KM_{III}	52.119
0.23841	1	65 Tb	K Abs. edge	52.002
0.23858	3	65 Tb	$KO_{II,III}$	51.965
0.23862	2	66 Dy	$K\beta_3$ KM_{II}	51.957
0.2397	2	65 Tb	$K\beta_2$ $KN_{II,III}$	51.68
0.241424	2	70 Yb	$K\alpha_2$ KL_{II}	51.3540
0.244338	2	69 Tm	$K\alpha_1$ KL_{III}	50.7416
0.24608	2	65 Tb	$K\beta_1$ KM_{III}	50.382
0.24681	1	64 Gd	K Abs. edge	50.233
0.24683	3	65 Tb	$K\beta_3$ KM_{II}	50.229
0.24687	3	64 Gd	$KO_{II,III}$	50.221
0.24816	3	64 Gd	$K\beta_2$ $KN_{II,III}$	49.959
0.249095	2	69 Tm	$K\alpha_2$ KL_{II}	49.7726
0.252365	2	68 Er	$K\alpha_1$ KL_{III}	49.1277
0.25275	3	64 Gd	$K\beta_5$ $KM_{IV,V}$	49.052
0.25460	2	64 Gd	$K\beta_1$ KM_{III}	48.697
0.25534	1	64 Gd	$K\beta_3$ KM_{II}	48.555
0.25553	1	63 Eu	K Abs. edge	48.519
0.255645	7	63 Eu	$KO_{II,III}$	48.497
0.256923	8	63 Eu	$K\beta_2^{I}$ $KN_{II,III}$	48.256
0.257110	2	68 Er	$K\alpha_2$ KL_{II}	48.2211
0.260756	2	67 Ho	$K\alpha_1$ KL_{III}	47.5467
0.263577	5	63 Eu	$K\beta_1$ KM_{III}	47.0379

Wavelength (nm)	Int.	Element	Line	Level	Energy (keV)
0.195904	2	77 Ir	$K\alpha_2$	KL_{II}	63.2867
0.19607	3	72 Hf	$K\beta_1$	KM_{III}	63.234
0.196794	2	76 Os	$K\alpha_1$	KL_{III}	63.0005
0.19686	4	72 Hf	$K\beta_3$	KM_{II}	62.98
0.1969	2	71 Lu	$K\beta_2$	$KN_{II,III}$	62.97
0.20084	2	71 Lu	$K\beta_5$	$KM_{IV,V}$	61.732
0.201639	5	76 Os	$K\alpha_2$	KL_{II}	61.4867
0.20224	2	70 Yb	K	Abs. edge	61.30
0.20226	3	70 Yb		$KO_{II,III}$	61.298
0.20231	3	71 Lu	$K\beta_1$	KM_{III}	61.283
0.202781	4	75 Re	$K\alpha_1$	KL_{III}	61.1403
0.20309	2	71 Lu	$K\beta_3$	KM_{II}	61.05
0.2033	2	70 Yb	$K\beta_2$	$KN_{II,III}$	60.89
0.20739	1	70 Yb	$K\beta_5$	$KM_{IV,V}$	59.782
0.207611	5	75 Re	$K\alpha_2$	KL_{II}	59.7179
0.20880	8	69 Tm	K	Abs. edge	59.38
0.20884	2	70 Yb	$K\beta_1$	KM_{III}	59.37
0.20891	Std.	69 Tm		$KO_{II,III}$	59.346
0.2090100	1	74 W	$K\alpha_1$	KL_{III}	59.31824
0.2096	2	70 Yb	$K\beta_3$	KM_{II}	59.14
0.2098	2	69 Tm	$K\beta_2$	$KN_{II,III}$	59.09
0.213828	4	74 W	$K\alpha_2$	KL_{II}	57.9817
0.21404	2	69 Tm	$K\beta_5$	$KM_{IV,V}$	57.923
0.215497	1	73 Ta	$K\alpha_1$	KL_{III}	57.532
0.21556	3	69 Tm	$K\beta_1$	KM_{III}	57.517
0.21567	4	68 Er	K	Abs. edge	57.487
0.21581	2	68 Er		$KO_{II,III}$	57.450
0.21592	2	74 W	$K\alpha_2$	KL_{II}	57.42
0.21636	8	69 Tm	$K\beta_3$	KM_{II}	57.304
0.2167	3	68 Er	$K\beta_2$	$KN_{II,III}$	57.21
0.220305	2	73 Ta	$K\alpha_2$	KL_{II}	56.277
0.22124	1	68 Er	$K\beta_5$	$KM_{IV,V}$	56.040
0.222227	2	72 Hf	$K\alpha_1$	KL_{III}	55.7902
0.22266	2	68 Er	$K\beta_1$	KM_{III}	55.681
0.22291	1	67 Ho	K	Abs. edge	55.619
0.22305	3	67 Ho		$KO_{II,III}$	55.584
0.22341	2	68 Er	$K\beta_3$	KM_{II}	55.494
0.2241	2	67 Ho	$K\beta_2$	$KN_{II,III}$	55.32
0.227024	3	72 Hf	$K\alpha_2$	KL_{II}	54.6114
0.264332	5	63 Eu	$K\beta_3$	KM_{II}	46.9036
0.26464	5	62 Sm	K	Abs. edge	46.849
0.26491	3	62 Sm		$KO_{II,III}$	46.801
0.265486	2	67 Ho	$K\alpha_2$	KL_{II}	46.6997
0.2662	1	62 Sm	$K\beta_2$	$KN_{II,III}$	46.57
0.269533	2	66 Dy	$K\alpha_1$	KL_{III}	45.9984
0.27111	3	62 Sm	$K\beta_5$	$KM_{IV,V}$	45.731
0.27301	2	62 Sm	$K\beta_1$	KM_{III}	45.413
0.27376	2	62 Sm	$K\beta_3$	KM_{II}	45.289
0.274247	5	66 Dy	$K\alpha_2$	KL_{II}	45.2078
0.27431	1	61 Pm	K	Abs. edge	45.198
0.2759	2	61 Pm	$K\beta_2$	$KN_{II,III}$	44.93
0.278724	3	65 Tb	$K\alpha_1$	KL_{III}	44.4816
0.28290	2	61 Pm	$K\beta_1$	KM_{III}	43.826
0.283423	4	65 Tb	$K\alpha_2$	KL_{II}	43.7441
0.28363	5	61 Pm	$K\beta_3$	KM_{II}	43.713
0.28453	1	60 Nd	K	Abs. edge	43.574
0.2861	2	60 Nd	$K\beta_2$	$KN_{II,III}$	43.32
0.288353	2	60 Nd	$K\alpha_1$	KL_{III}	42.9962
0.293038	3	64 Gd	$K\alpha_2$	KL_{II}	42.3089
0.293399	2	64 Gd	$K\beta_3$	KM_{II}	42.2713
0.294027	3	60 Nd	$K\beta_1$	KM_{III}	42.1665
0.29518	5	60 Nd	K	Abs. edge	42.002
0.29679	2	59 Pr	$K\beta_2$	$KN_{II,III}$	41.773
0.298446	2	59 Pr	$K\alpha_1$	KL_{III}	41.5422
0.303118	4	63 Eu	$K\alpha_2$	KL_{II}	40.9019
0.304261	5	63 Eu	$K\beta_1$	KM_{III}	40.7482
0.304975	2	59 Pr	$K\beta_3$	KM_{II}	40.6529
0.30648	2	59 Pr	K	Abs. edge	40.453
0.30668	5	58 Ce		$KO_{II,III}$	40.427
0.30737	2	58 Ce	$K\beta_4{}^{I}$	$KN_{IV,V}$	40.337
0.30816	1	58 Ce	$K\beta_2$	$KN_{II,III}$	40.233
0.309040	2	58 Ce	$K\alpha_1$	KL_{III}	40.1181
0.31342	2	62 Sm	$K\beta_5{}^{I}$	KM_{V}	39.558
0.31357	1	58 Ce	$K\beta_5{}^{II}$	KM_{IV}	39.539
0.313698	2	58 Ce	$K\alpha_2$	KL_{II}	39.5224
0.315816	2	62 Sm	$K\beta_1$	KM_{III}	39.2573
0.316520	2	58 Ce	$K\beta_3$	KM_{II}	39.1701
0.31844	5	57 La	K	Abs. edge	38.934

TABLE 7f-2. WAVELENGTHS OF X-RAY EMISSION LINES AND ABSORPTION EDGES: IN NUMERICAL ORDER (Continued)

Left column group:

Wavelength, Å*	p.e.†	Element	Designation		keV
0.31864	2	57 La	$K\beta_4^{\mathrm{I}}$	$KO_{\mathrm{II,III}}$	38.909
0.31931	2	57 La	$K\beta_2$	$KN_{\mathrm{IV,V}}$	38.828
0.320117	7	57 La	$K\beta_2$	$KN_{\mathrm{I,III}}$	38.7299
0.320160	4	61 Pm	$K\alpha_1$	KL_{III}	38.7247
0.324803	4	61 Pm	$K\alpha_2$	KL_{II}	38.1712
0.32546	2	57 La	$K\beta_5^{\mathrm{I}}$	KM_{IV}	38.094
0.32563	2	57 La	$K\beta_5^{\mathrm{II}}$	KM_{V}	38.074
0.327983	2	57 La	$K\beta_1$	KM_{III}	37.8010
0.328686	4	57 La	$K\beta_3$	KM_{II}	37.7202
0.33104	1	56 Ba	K	Abs. edge	37.452
0.33127	2	56 Ba	$K\alpha_1$	$KO_{\mathrm{II,III}}$	37.426
0.331846	2	60 Nd	$K\alpha_1$	KL_{III}	37.3610
0.33229	2	56 Ba	$K\beta_4^{\mathrm{II}}$	KN_{IV}	37.311
0.33277	1	56 Ba	$K\beta_2$	$KN_{\mathrm{I,III}}$	37.257
0.336472	2	60 Nd	$K\alpha_2$	KL_{II}	36.8474
0.33814	2	56 Ba	$K\beta_5^{\mathrm{I}}$	KM_{V}	36.666
0.33835	2	56 Ba	$K\beta_5^{\mathrm{II}}$	KM_{IV}	36.643
0.340811	3	56 Ba	$K\beta_1$	KM_{III}	36.3782
0.341507	4	56 Ba	$K\beta_3$	KM_{II}	36.3040
0.344140	1	59 Pr	$K\alpha_1$	KL_{III}	36.0263
0.34451	2	55 Cs	K	Abs. edge	35.987
0.34611	2	55 Cs	$K\beta_2$	$KN_{\mathrm{II,III}}$	35.822
0.348749	2	59 Pr	$K\alpha_2$	KL_{II}	35.5502
0.354364	7	55 Cs	$K\beta_1$	KM_{III}	34.9869
0.355050	2	55 Cs	$K\beta_3$	KM_{II}	34.9194
0.357092	2	58 Ce	$K\alpha_1$	KL_{III}	34.7197
0.3584	5	54 Xe	K	Abs. edge	34.59
0.36026	3	54 Xe	$K\beta_2$	$KN_{\mathrm{II,III}}$	34.415
0.361683	2	58 Ce	$K\alpha_2$	KL_{II}	34.2789
0.36872	2	54 Xe	$K\beta_1$	KM_{III}	33.624
0.36941	2	54 Xe	$K\beta_3$	KM_{II}	33.562
0.370737	2	57 La	$K\alpha_1$	KL_{III}	33.4418
0.37381	1	53 I	K	Abs. edge	33.1665

Right column group:

keV	Designation		Element	p.e.†	Wavelength, Å*
27.4723	$K\alpha_1$	KL_{III}	52 Te	3	0.451295
27.2759	$K\beta_1$	KM_{III}	49 In	4	0.454545
27.2377	$K\beta_3$	KM_{II}	49 In	4	0.455181
27.2017	$K\alpha_2$	KL_{II}	52 Te	3	0.455784
26.7159	K	Abs. edge	48 Cd	1	0.46407
26.6438	$K\beta_2$	$KN_{\mathrm{II,III}}$	48 Cd	7	0.465328
26.3591	$K\alpha_1$	KL_{III}	51 Sb	3	0.470354
26.1108	$K\alpha_2$	KL_{II}	51 Sb	3	0.474827
26.0955	$K\beta_1$	KM_{III}	48 Cd	6	0.475105
26.0612	$K\beta_3$	KM_{II}	48 Cd	5	0.475730
25.5165	K	Abs. edge	47 Ag	1	0.48589
25.512	$K\beta_4$	$KN_{\mathrm{IV,V}}$	47 Ag	9	0.4859
25.4564	$K\beta_2$	$KN_{\mathrm{II,III}}$	47 Ag	4	0.487032
25.2713	$K\alpha_1$	KL_{III}	50 Sn	3	0.490599
25.145	$K\beta_5$	$KM_{\mathrm{IV,V}}$	47 Ag	2	0.49306
25.0440	$K\alpha_2$	KL_{II}	50 Sn	3	0.495053
24.9424	$K\beta_1$	KM_{III}	47 Ag	4	0.497069
24.9115	$K\beta_3$	KM_{II}	47 Ag	4	0.497685
24.348	K	Abs. edge	46 Pd	1	0.5092
24.346	$K\beta_4$	$KN_{\mathrm{IV,V}}$	46 Pd	2	0.5093
24.2991	$K\beta_2$	$KN_{\mathrm{II,III}}$	46 Pd	4	0.510228
24.2097	$K\alpha_1$	KL_{III}	49 In	3	0.512113
24.0020	$K\alpha_2$	KL_{II}	49 In	3	0.516544
23.995	$K\beta_5$	$KM_{\mathrm{IV,V}}$	46 Pd	9	0.51670
23.8187	$K\beta_1$	KM_{III}	46 Pd	4	0.520520
23.7911	$K\beta_3$	KM_{II}	46 Pd	4	0.521123
23.2198	K	Abs. edge	45 Rh	1	0.53395
23.217	$K\beta_4^{\mathrm{I}}$	$KN_{\mathrm{IV,V}}$	45 Rh	9	0.53401
23.1736	$K\alpha_1$	KL_{III}	48 Cd	3	0.535010
23.1728	$K\beta_2^{\mathrm{I}}$	$KN_{\mathrm{II,III}}$	45 Rh	2	0.53503
23.168	$K\beta_2^{\mathrm{II}}$	KN_{II}	45 Rh	5	0.53513
23.109	L_{I}	Abs. edge	94 Pu	1	0.5365
22.9841	$K\alpha_2$	KL_{II}	48 Cd	3	0.539422

Energy (keV)	Level	Line	Element	Intensity	λ (Å)
22.917	KM_V	$K\beta_5^I$	45 Rh	9	0.54101
22.909	KM_{IV}	$K\beta_5^{II}$	45 Rh	9	0.541_8
22.891	L_IO_{III}	$L\gamma_4$	94 Pu	1	0.5416
22.8282	$L_{II}O_{IV}$	$L\gamma_4'$	95 Am	2	0.54311
22.823	L_IO_{IV}	$K\beta_1$	94 Pu	1	0.5432
22.7236	KM_{III}	$K\beta_3$	45 Rh	4	0.545605
22.6989	KM_{II}	$L\gamma_2$	45 Rh	4	0.546290
22.361	L_IN_{II}	L_{II}	95 Am	2	0.5544
22.253	Abs. edge	$L\gamma_4$	94 Pu	1	0.5572
22.20	$L_IO_{II,III}$	$L\gamma_4$	93 Np	5	0.5585
22.16292	KL_{III}	$K\alpha_1$	47 Ag	6	0.5594075
22.1502	$L_{II}O_{IV}$	$L\gamma_6$	94 Pu	2	0.55973
22.1193	Abs. edge	K	44 Ru	1	0.56051
22.104	$KN_{IV,V}$	$K\beta_4$	44 Ru	9	0.56089
22.074	$KN_{II,III}$	$K\beta_2$	44 Ru	3	0.56166
22.0652	$L_{II}N_{IV}$	$L\gamma_1$	95 Am	9	0.561883
21.9903	KL_{II}	$L\alpha_2$	47 Ag	4	0.563793
21.9824	L_IN_{III}	$L\gamma_3$	94 Pu	9	0.56400
21.914	$L_{II}O_I$	$L\gamma_8$	94 Pu	1	0.5658
21.834	KM_V	$K\beta_5^I$	44 Ru	9	0.56785
21.829	KM_{IV}	$K\beta_5^{II}$	44 Ru	9	0.5680
21.771	Abs. edge	L_I	92 U	1	0.5695
21.729	$L_IP_{II,III}$	$L\gamma_{13}$	92 U	1	0.5706
21.1251	L_IN_{II}	$L\gamma_2$	94 Pu	2	0.57068
21.6568	KM_{III}	$K\beta_1$	44 Ru	4	0.572482
21.657	$L_IO_{IV,V}$		92 U	1	0.5725
21.6346	KM_{II}	$K\beta_3$	44 Ru	4	0.573067
21.562	L_IO_{III}	$L\gamma_4$	92 U	9	0.57499
21.4984	L_IO_{II}	$L\gamma_4'$	92 U	9	0.576700
21.488	$L_{II}O_{IV}$	$L\gamma_6$	93 Np	5	0.57699
21.4173	$L_{II}N_{IV}$	$L\gamma_1$	94 Pu	9	0.578882
21.34	L_IN_{III}	$L\gamma_3$	93 Np	9	0.5810
21.1771	KL_{III}	$K\alpha_1$	46 Pd	5	0.585448
21.11	L_IN_{II}	$L\gamma_2$	93 Np	3	0.5873
21.0473	Abs. edge	K	43 Tc	5	0.58906
21.0201	KL_{II}	$K\alpha_2$	46 Pd	1	0.589821
21.019	L_IN_V	$L\gamma_{11}$	92 U	3	0.58986
21.005	$KN_{II,III}$	$K\beta_5^I$	43 Tc	5	0.59024
20.979	L_IN_{IV}	$K\beta_2$	92 U	5	0.59096

λ (Å)	Element	Intensity	Line	Level	Energy (keV)
0.37523	53 I	2	$K\beta_2$	$KN_{II,III}$	33.042
0.375313	57 La	2	$K\alpha_2$	KL_{II}	33.0341
0.383905	53 I	4	$K\beta_1$	KM_{III}	32.2947
0.384564	53 I	4	$K\beta_3$	KM_{II}	32.2394
0.385111	56 Ba	4	$K\alpha_1$	KL_{II}	32.1936
0.389668	56 Ba	5	$K\alpha_2$	KL_{II}	31.8171
0.38974	52 Te	1	K	$KO_{II,III}$	31.8114
0.38974	52 Te			Abs. edge	31.8114
0.391102	52 Te	6	$K\beta_2$	$KN_{II,III}$	31.7004
0.399995	52 Te	5	$K\beta_1$	KM_{III}	30.9957
0.400290	55 Cs	4	$K\alpha_1$	KL_{III}	30.9728
0.400659	52 Te	4	$K\beta_3$	KM_{II}	30.9443
0.404835	55 Cs	4	$K\alpha_2$	KL_{II}	30.6251
0.40666	51 Sb	1	K	$KO_{II,III}$	30.4875
0.40668	51 Sb			Abs. edge	30.4860
0.40702	51 Sb	1	$K\beta_4^I$	$KN_{IV,V}$	30.4604
0.407973	51 Sb	5	$K\beta_2$	$KN_{II,III}$	30.3895
0.41378	51 Sb	1	$K\beta_5^I$	KM_V	29.9632
0.41388	51 Sb	1	$K\beta_5^{II}$	KM_{IV}	29.9560
0.41634	54 Xe	2	$K\alpha_1$	KL_{III}	29.779
0.417085	51 Sb	3	$K\beta_1$	KM_{III}	29.7256
0.417737	51 Sb	4	$K\beta_3$	KM_{II}	29.6792
0.42087	54 Xe	2	$K\alpha_2$	KL_{II}	29.458
0.42467	50 Sn	3	K	$KO_{II,III}$	29.195
0.42467	50 Sn			Abs. edge	29.1947
0.42495	50 Sn	1	$K\beta_4^I$	$KN_{IV,V}$	29.175
0.425915	50 Sn	3	$K\beta_2$	$KN_{II,III}$	29.1093
0.43175	50 Sn	8	$K\beta_5^I$	KM_V	28.716
0.43184	50 Sn	3	$K\beta_5^{II}$	KM_{IV}	28.710
0.433318	53 I	3	$K\alpha_1$	KL_{III}	28.6120
0.435236	50 Sn	5	$K\beta_1$	KM_{III}	28.4860
0.435877	50 Sn	5	$K\beta_3$	KM_{II}	28.4440
0.437829	53 I	7	$K\alpha_2$	KL_{II}	28.3172
0.44371	49 In	1	K	Abs. edge	27.940
0.44374	49 In	3		$KO_{II,III}$	27.9420
0.44393	49 In	1	$K\beta_4^I$	$KN_{IV,V}$	27.928
0.44500	49 In	3	$K\beta_2$	$KN_{II,III}$	27.8608
0.45086	49 In	2	$K\beta_5^I$	KM_V	27.499
0.45098	49 In	2	$K\beta_5^{II}$	KM_{IV}	27.491

TABLE 7f-2. WAVELENGTHS OF X-RAY EMISSION LINES AND ABSORPTION EDGES: IN NUMERICAL ORDER (*Continued*)

Wavelength, Å*	p.e.†	Element	Designation		keV
0.5919	1	92 U	L_{II}	Abs. edge	20.945
0.59203	5	92 U	$L\gamma_4$	$L_{II}P_{IV}$	20.942
0.5930	2	92 U	$L\gamma_6$	$L_{II}P_{II,III}$	20.906
0.5937	1	91 Pa	$L\gamma_1$	$L_{II}O_{IV}$	20.882
0.594845	9	92 U	$L\gamma_6$	$L_{II}N_{IV}$	20.8426
0.596498	9	93 Np	$L\gamma_1$	$L_{II}O_{III}$	20.7848
0.59728	5	92 U	$L\gamma_3$	L_IN_{III}	20.758
0.598574	9	92 U	$L\gamma_5$	$L_{II}N_I$	20.7127
0.5988	1	94 Pu	$L\gamma_8$	$L_{II}O_I$	20.704
0.60125	5	92 U	$K\beta_1$	KM_{III}	20.621
0.60130	4	43 Tc	$K\beta_3$	KM_{III}	20.619
0.60188	4	43 Tc	$L\nu$	$L_{II}N_{VI}$	20.599
0.6031	1	92 U	$L\gamma_2$	L_IN_{II}	20.556
0.605237	9	92 U	L_I	Abs. edge	20.4847
0.6059	1	90 Th	$L\gamma_{13}$	$L_IP_{II,III}$	20.464
0.60705	8	90 Th	$L\gamma_4$	$L_IO_{IV,V}$	20.424
0.6083	1	90 Th	$L\gamma_4'$	L_IO_{III}	20.383
0.61098	4	90 Th	$L\gamma_6$	L_IO_{III}	20.292
0.61251	4	90 Th	$K\alpha_1$	$L_{II}O_{IV}$	20.242
0.6133	1	91 Pa	$L\gamma_1$	KL_{III}	20.216
0.613279	4	45 Rh	$L\gamma_5$	L_IO_I	20.2161
0.6146	9	90 Th	$L\gamma_3$	$L_{II}N_{IV}$	20.174
0.614770	1	92 U	$K\alpha_2$	$L_IN_{VI,VII}$	20.1671
0.6160	1	90 Th	K	$L_{II}N_I$	20.128
0.616	1	93 Np	$K\beta_4$	$L_{II}N_{III}$	20.12
0.6169	4	91 Pa	$K\beta_2$	KL_{II}	20.098
0.617630	1	45 Rh	$K\beta_2^{II}$	Abs. edge	20.0737
0.61978	1	42 Mo	$L\gamma_2$	$KN_{IV,V}$	20.0039
0.62001	9	42 Mo		$KN_{II,III}$	19.996
0.62099	2	42 Mo		KN_{II}	19.9652
0.62107	5	42 Mo		$L_{II}N_{III}$	19.963
0.6228	1	92 U		L_IN_{VII}	19.907
0.6239	1	91 Pa		L_IN_{II}	19.872
0.66576	2	41 Nb	$K\beta_1$	KM_{III}	18.6225
0.66634	3	41 Nb	$K\beta_3$	KM_{II}	18.6063
0.6666	1	88 Ra		L_IN_{VI}	18.600
0.66871	2	94 Pu	$L\beta_3$	L_IM_{III}	18.5405
0.6707	1	88 Ra	L_{II}	Abs. edge	18.486
0.6714	1	88 Ra	$L\beta_3$	$L_IP_{II,III}$	18.466
0.6724	1	88 Ra		$L_{II}P_I$	18.439
0.67328	5	88 Ra	$L\gamma_6$	$L_{II}O_{IV}$	18.414
0.67351	9	89 Ac	$L\gamma_1$	$L_{II}N_{IV}$	18.408
0.67383	2	95 Am	$L\beta_5$	$L_{III}O_{IV,V}$	18.3996
0.67491	4	90 Th	$L\gamma_5$	$L_{II}N_I$	18.370
0.67502	3	43 Tc	$K\alpha_1$	KL_{III}	18.3671
0.67538	5	88 Ra	$L\gamma_3$	L_IN_{III}	18.357
0.6764	1	88 Ra		$L_{II}O_{III}$	18.330
0.67772	2	94 Pu	$L\beta_1$	$L_{II}M_{IV}$	18.2937
0.6780	1	88 Ra		$L_{II}O_{II}$	18.286
0.67932	3	43 Tc	$K\alpha_2$	KL_{II}	18.2508
0.6801	1	88 Ra	$L\gamma_8$	$L_{II}O_{III}$	18.230
0.681014	8	92 U	$L\beta_9$	L_IM_V	18.2054
0.68199	5	88 Ra	$L\gamma_2$	L_IN_{II}	18.179
0.68639	2	95 Am	$L\beta_4$	L_IM_{II}	18.0627
0.6867	1	94 Pu	L_{III}	Abs. edge	18.054
0.6874	1	88 Ra	$L\gamma_1$	$L_{II}N_{IV}$	18.036
0.68760	5	92 U	$L\beta_{10}$	L_IM_{IV}	18.031
0.68883	1	40 Zr	K	Abs. edge	17.9989
0.68901	5	40 Zr	$K\beta_4$	$KN_{IV,V}$	17.994
0.68920	9	93 Np	$L\beta_3$	L_IM_{III}	17.989
0.68993	4	40 Zr	$K\beta_2$	$KN_{II,III}$	17.970
0.69068	2	94 Pu	$L\beta_2$	$L_{III}O_{IV,V}$	17.9506
0.6932	1	88 Ra	$L\beta_5$	$L_{III}O_{IV,V}$	17.884
0.69463	5	88 Ra	$L\gamma_1$	$L_{II}N_{IV}$	17.849
0.6959	1	40 Zr	$K\beta_5$	$KM_{IV,V}$	17.815
0.698478	9	93 Np	$L\beta_1$	$L_{II}M_{IV}$	17.7502

Wavelength (Å)	Intensity	Element	Line	Transition	Energy (keV)
0.62636	9	90 Th	$L\gamma_{11}$	$L_{I}N_{V}$	19.794
0.62692	5	42 Mo	$K\beta_5^{I}$	KM_{V}	19.776
0.62708	5	42 Mo	$K\beta_5^{II}$	KM_{IV}	19.771
0.6276	1	90 Th	L_{II}	$L_{I}N_{IV}$	19.755
0.6299	—	90 Th	Abs. edge	L_{II}	19.683
0.62991	9	90 Th	$K\beta_1$	$L_{II}P_{IV}$	19.682
0.6312	1	90 Th	$L\gamma_6$	$L_{II}P_{II,III}$	19.642
0.6316	1	90 Th	$K\beta_3$	$L_{II}P_{I}$	19.629
0.632288	9	42 Mo	$L\gamma_1$	KM_{III}	19.6083
0.63258	4	90 Th	$L\gamma_5$	$L_{II}O_{IV}$	19.599
0.632872	2	42 Mo	$L\gamma_3$	KM_{II}	19.5903
0.63358	9	91 Pa	$L\gamma_8$	$L_{II}N_{IV}$	19.568
0.63557	2	92 U	Lv	$L_{II}N_{I}$	19.5072
0.63559	4	90 Th	$L\beta_9$	$L_{I}N_{III}$	19.507
0.6356	1	90 Th	$L\gamma_2$	$L_{II}O_{III}$	19.506
0.6369	4	90 Th	$K\alpha_1$	$L_{II}O_{II}$	19.466
0.63898	1	90 Th		$L_{II}O_{I}$	19.403
0.64064	5	90 Th	L_{I}	$L_{II}N_{VI}$	19.353
0.6416	5	94 Pu	$L\gamma_{13}$	$L_{I}M_{V}$	19.323
0.64221	5	90 Th	$K\alpha_2$	$L_{II}N_{II}$	19.305
0.643083	2	44 Ru	KL_{III}	KL_{III}	19.2792
0.6445	—	88 Ra	Abs. edge		19.236
0.64513	5	88 Ra	$L\beta_{10}$	$L_{I}P_{II,III}$	19.218
0.6468	4	88 Ra	$L\beta_3$	$L_{I}O_{IV,V}$	19.167
0.647408	1	44 Ru	$K\alpha_2$	KL_{II}	19.1504
0.64755	1	90 Th	$L\gamma_4$	$L_{I}N_{I}$	19.146
0.6482	3	94 Pu	$L\gamma_4'$	$L_{I}M_{IV}$	19.126
0.64891	5	95 Am	$L\beta_{10}$	$L_{I}M_{III}$	19.1059
0.64965	4	88 Ra	$L\beta_3$	$L_{I}O_{III}$	19.084
0.65131	1	88 Ra	$L\gamma_4$	$L_{I}O_{II}$	19.036
0.6521	9	90 Th	$L\gamma_4'$	$L_{II}N_{V}$	19.014
0.65298	—	41 Nb	Abs. edge	K	18.9869
0.65313	1	90 Th	$K\gamma_1$	$KN_{IV,V}$	18.9825
0.65416	5	41 Nb	$K\beta_4$	$KN_{II,III}$	18.981
0.6550	4	41 Nb	$K\beta_2$	$L_{II}N_{I}$	18.953
0.657655	1	91 Pa	$L\gamma_5$	$L_{II}M_{IV}$	18.930
0.6620	9	95 Am	$L\beta_1$	$L_{II}M_{III}$	18.8520
0.66...	1	90 Th		$L_{I}N_{III}$	18.729
0.6654	1	88 Ra	$L\gamma_{11}$	$L_{I}N_{V}$	18.633

Wavelength (Å)	Intensity	Element	Line	Transition	Energy (keV)
0.7003	1	94 Pu	$L\beta_7$	$L_{III}O_{I}$	17.705
0.701390	9	95 Am	$L\beta_2$	$L_{III}N_{V}$	17.6765
0.70173	3	40 Zr	$L\beta_1$	KM_{III}	17.6678
0.7018	1	91 Pa	$L\beta_9$	$L_{I}M_{V}$	17.667
0.70228	4	40 Zr	$K\beta_3$	KM_{II}	17.654
0.7031	1	94 Pu	Lv	$L_{III}N_{VI,VII}$	17.635
0.70341	2	95 Am	$L\beta_{15}$	$L_{III}N_{IV}$	17.6258
0.7043	1	88 Ra	$L\beta_4$	$L_{II}N_{III}$	17.604
0.70620	2	94 Pu	$L\beta_5$	$L_{I}M_{II}$	17.5560
0.70814	2	93 Np	$L\beta_{10}$	$L_{III}O_{IV,V}$	17.5081
0.7088	2	91 Pa	$K\alpha_1$	$L_{I}M_{IV}$	17.492
0.709300	1	42 Mo	$L\beta_3$	KL_{III}	17.47934
0.71029	2	92 U	$K\alpha_2$	$L_{I}M_{III}$	17.4550
0.713590	6	42 Mo	$L\gamma_1$	KL_{II}	17.3743
0.71652	9	87 Fr	$L\gamma_5$	$L_{II}N_{VI,VII}$	17.303
0.71774	5	88 Ra	$L\gamma_1$	$L_{III}N_{IV}$	17.274
0.71851	2	94 Pu	$L\beta_2$	$L_{II}N_{I}$	17.2553
0.719984	8	92 U	$L\beta_1$	$L_{III}N_{V}$	17.2200
0.7205	1	94 Pu	$L\beta_{15}$	$L_{I}N_{I}$	17.208
0.7223	5	92 U	L_{III}	Abs. edge	17.165
0.72240	1	92 U	$L\beta_9$	$L_{III}O_{III}$	17.162
0.7234	5	90 Th	$L\beta_5$	$L_{II}O_{II}$	17.139
0.72426	5	92 U	$L\beta_4$	$L_{III}P_{II,III}$	17.118
0.72521	9	92 U	$L\beta_7$	$L_{III}P_{I}$	17.096
0.726305	2	92 U	$L\beta_{10}$	$L_{III}O_{IV,V}$	17.0701
0.72671	5	93 Np	$L\beta_4$	$L_{I}M_{II}$	17.0607
0.72766	5	39 Y	K	Abs. edge	17.038
0.72776	4	39 Y	$K\beta_4$	$KN_{IV,V}$	17.036
0.72864	1	90 Th	$K\beta_2$	$KN_{II,III}$	17.0154
0.7301	1	92 U	$L\beta_{10}$	$L_{I}M_{IV}$	16.981
0.7309	5	91 Pa		$L_{III}O_{III}$	16.962
0.73230	1	92 U	$L\beta_3$	$L_{I}M_{III}$	16.930
0.7333	2	95 Am		$L_{III}O_{II}$	16.907
0.73418	1	39 Y	$L\beta_6$	$L_{III}N_{I}$	16.8870
0.7345	6	92 U	$K\beta_5$	$KM_{IV,V}$	16.879
0.73602	9	93 Np	$L\beta_7$	$L_{III}O_{I}$	16.845
0.736230	9	92 U	$L\beta_2$	$L_{III}N_{VI,VII}$	16.8400
0.738603	1	92 U	Lv	$L_{III}N_{VI,VII}$	16.7859
0.73928	1	86 Rn	$L\gamma_1$	$L_{II}N_{IV}$	16.770

TABLE 7f-2. WAVELENGTHS OF X-RAY EMISSION LINES AND ABSORPTION EDGES: IN NUMERICAL ORDER (Continued)

Wavelength, Å*	p.e.†	Element		Designation	keV
0.74072	2	39 Y	$K\beta_1$	KM_{III}	16.7378
0.74126	3	39 Y	$K\beta_3$	KM_{II}	16.7258
0.74232	5	91 Pa	$L\beta_1$	$L_{II}M_{IV}$	16.702
0.74503	5	92 U	$L\beta_{17}$	$L_{II}M_{III}$	16.641
0.7452	2	91 Pa	$L\beta_5$	$L_{III}O_{IV,V}$	16.636
0.74620	1	41 Nb	$K\alpha_1$	KL_{III}	16.6151
0.747985	9	92 U	$L\beta_4$	$L_{II}M_{II}$	16.5753
0.75044	1	41 Nb	$K\alpha_2$	KL_{II}	16.5210
0.75148	2	94 Pu	$L\beta_6$	$L_{III}N_I$	16.4983
0.7546	2	91 Pa	$L\beta_7$	$L_{III}O_I$	16.431
0.754681	9	92 U	$L\beta_2$	$L_{III}N_V$	16.4283
0.75479	3	90 Th	$L\beta_3$	$L_I M_{III}$	16.4258
0.756642	9	92 U	$L\beta_{15}$	$L_{III}N_{IV}$	16.3857
0.75690	3	83 Bi	$L\gamma_{13}$	$L_I P_{II,III}$	16.3802
0.7571	1	83 Bi	L_I	Abs. edge	16.376
0.7579	1	90 Th		$L_{II}M_V$	16.359
0.75791	5	83 Bi		$L_I O_{IV,V}$	16.358
0.7591	1	94 Pu	$L\eta$	$L_{II}M_I$	16.333
0.7607	1	90 Th	L_{III}	Abs. edge	16.299
0.76087	9	90 Th		$L_{III}P_{IV,V}$	16.295
0.76087	3	83 Bi	$L\gamma_4$	$L_I O_{III}$	16.2947
0.76198	3	83 Bi	$L\gamma_4'$	$L_I O_{II}$	16.2709
0.7625	2	90 Th		$L_{III}P_{II,III}$	16.260
0.76289	9	85 At	$L\gamma_1$	$L_{III}N_{IV}$	16.251
0.76338	5	90 Th		$L_{III}P_I$	16.241
0.7641	5	83 Bi		$L_I N_{VI,VII}$	16.23
0.7645	2	84 Po	$L\gamma_6$	$L_{II}O_{IV}$	16.218
0.76468	5	90 Th	$L\beta_1$	$L_{II}M_{IV}$	16.213
0.765210	9	90 Th	$L\beta_9$	$L_I M_V$	16.2022
0.76857	5	88 Ra	$L\beta_6$	$L_{III}N_I$	16.131
0.769	1	93 Np		$L_{III}O_{III}$	16.13
0.7690	1	90 Th		$L_{III}N_{II}$	16.123
0.7691	1	92 U		$L_{III}N_{III}$	16.120
0.7973	1	83 Bi	$L\gamma_8$	$L_{II}O_I$	15.551
0.8022	1	83 Bi		$L_I N_I$	15.456
0.80233	9	82 Pb	$L\gamma_{11}$	$L_I N_V$	15.453
0.80273	5	88 Ra	$L\beta_3$	$L_I M_{III}$	15.4449
0.8028	1	88 Ra	L_{III}	Abs. edge	15.444
0.80364	7	82 Pb		$L_I N_{IV}$	15.427
0.8038	1	88 Ra		$L_{III}P_{II,III}$	15.425
0.8050	1	88 Ra		$L_{III}P_I$	15.402
0.80509	2	92 U	$L\eta$	$L_{II}M_I$	15.3997
0.80627	5	88 Ra	$L\beta_5$	$L_{III}O_{IV,V}$	15.347
0.8079	1	91 Pa	$L\beta_6$	$L_{III}O_I$	15.3371
0.8081	1	81 Tl	L_I	Abs. edge	15.343
0.8082	1	90 Th		$L_{III}N_{III}$	15.341
0.80861	5	81 Tl		$L_I O_{IV,V}$	15.3327
0.81163	9	90 Th		$L_I M_I$	15.276
0.81184	5	81 Tl	$L\gamma_4$	$L_I O_{III}$	15.2716
0.81308	5	81 Tl	$L\gamma_4'$	$L_I O_{II}$	15.2482
0.81311	2	83 Bi	$L\gamma_1$	$L_I N_{IV}$	15.2477
0.81375	5	88 Ra	$L\beta_1$	$L_{II}M_{IV}$	15.2358
0.8147	1	82 Pb	$L\gamma_3$	$L_{II}N_{III}$	15.218
0.81538	5	82 Pb	L_{II}	Abs. edge	15.2053
0.8154	2	37 Rb	$K\beta_4$	$K N_{IV,V}$	15.205
0.81554	5	37 Rb	K	Abs. edge	15.2023
0.8158	1	81 Tl		$L_I O_I$	15.198
0.81583	5	82 Pb		$L_{II}P_I$	15.1969
0.8162	1	88 Ra	$L\beta_7$	$L_{III}O_I$	15.190
0.81645	3	37 Rb	$K\beta_2$	$K N_{II,III}$	15.1854
0.81683	5	82 Pb	$L\gamma_6$	$L_{III}O_{IV}$	15.1783
0.8186	1	88 Ra	Lu	$L_{III}N_{VI,VII}$	15.146
0.8190	2	90 Th		$L_{III}N_{II}$	15.138
0.8200	1	82 Pb		$L_{II}O_{III}$	15.120
0.8210	2	82 Pb	$L\gamma_2$	$L_I N_{II}$	15.101
0.8219	1	37 Rb	$K\beta_5$	$K M_{IV,V}$	15.085

0.76973	5	38 Sr	K	Abs. edge	16.107	0.82327	7	82 Pb	Lv	$L_{II}N_{VI}$	15.060
0.7699	1	91 Pa	$L\beta_4$	L_IM_{II}	16.104	0.82365	5	82 Pb	$L\gamma_8$	$L_{II}O_I$	15.0527
0.76989	5	38 Sr	$K\beta_4$	$KN_{IV,V}$	16.104	0.8248	1	83 Bi		$L_{II}N_{III}$	15.031
0.77081	3	38 Sr	$K\beta_2$	$KN_{II,III}$	16.0846	0.82789	9	87 Fr	$L\beta_3$	L_IM_{III}	14.976
0.7713	1	90 Th		$L_{III}O_{II}$	16.074	0.82790	8	90 Th	$L\beta_6$	$L_{III}N_I$	14.975
0.772	1	84 Po	$L\gamma_2$	L_IN_{II}	16.07	0.82859	7	82 Pb		L_IN_I	14.963
0.7737	1	91 Pa	$L\beta_2$	$L_{III}N_V$	16.024	0.82868	2	37 Rb	$K\beta_1$	KM_{III}	14.9613
0.77437	4	90 Th	$L\beta_7$	$L_{III}O_I$	16.0105	0.82879	5	81 Tl	$L\gamma_{11}$	L_IN_V	14.9593
0.77546	5	88 Ra	$L\beta_{10}$	L_IM_{IV}	15.988	0.82884	1	39 Y	$K\alpha_1$	KL_{III}	14.9584
0.7764	1	38 Sr	$K\beta_5$	$KM_{IV,V}$	15.969	0.82921	3	37 Rb	$K\beta_3$	KM_{II}	14.9517
0.77661	5	90 Th	Lu	$L_{III}N_{VI,VII}$	15.964	0.8295	1	91 Pa	$L\eta$	$L_{II}M_I$	14.946
0.77728	5	83 Bi	$L\gamma_{11}$	L_IN_V	15.951	0.83001	7	81 Tl		L_IN_{IV}	14.937
0.77822	9	89 Ac	$L\beta_3$	L_IM_{III}	15.931	0.83305	1	39 Y	$K\alpha_2$	KL_{II}	14.8829
0.77954	5	83 Bi		L_IN_{IV}	15.904	0.8338	1	90 Th		$L_{II}M_{II}$	14.869
0.78017	9	92 U		$L_{III}N_{II}$	15.892	0.8344	9	83 Bi		$L_{II}N_{II}$	14.86
0.7809	2	93 Np	$L\eta$	$L_{II}M_I$	15.876	0.8350	2	80 Hg		$L_IO_{IV,V}$	14.847
0.78196	5	82 Pb	L_I	Abs. edge	15.855	0.8353	1	80 Hg	L_I	Abs. edge	14.842
0.78257	7	82 Pb		$L_IO_{IV,V}$	15.843	0.83537	5	88 Ra	$L\beta_2$	$L_{III}N_V$	14.8414
0.78292	2	38 Sr	$K\beta_1$	KM_{III}	15.8357	0.83722	5	88 Ra	$L\beta_{15}$	$L_{III}N_{IV}$	14.8086
0.78345	3	38 Sr	$K\beta_3$	KM_{II}	15.8249	0.8382	2	82 Pb		$L_{II}N_V$	14.791
0.7858	1	82 Pb	$L\gamma_4$	L_IO_{III}	15.777	0.83894	7	80 Hg	$L\gamma_4$	L_IO_{III}	14.778
0.78593	1	40 Zr	$K\alpha_1$	KL_{III}	15.7751	0.83923	5	83 Bi	$L\gamma_5$	$L_{II}N_I$	14.7732
0.78706	7	82 Pb	$L\gamma_4'$	L_IO_{II}	15.752	0.83940	9	87 Fr	$L\beta_1$	$L_{II}M_{IV}$	14.770
0.78748	9	84 Po	$L\gamma_1$	$L_{II}N_{IV}$	15.744	0.83973	3	82 Pb	$L\gamma_1$	$L_{II}N_{IV}$	14.7644
0.78838	2	92 U	$L\beta_5$	$L_{III}N_I$	15.7260	0.84013	7	80 Hg	$L\gamma_4'$	L_IO_{II}	14.757
0.7884	1	82 Pb		$L_IN_{VI,VII}$	15.725	0.84071	5	88 Ra	$L\beta_4$	L_IM_{II}	14.7472
0.7887	1	83 Bi	L_{II}	Abs. edge	15.719	0.84130	4	81 Tl	$L\gamma_3$	L_IN_{III}	14.7368
0.78903	9	89 Ac	$L\beta_1$	$L_{II}M_{IV}$	15.713	0.8434	1	81 Tl	L_{II}	Abs. edge	14.699
0.78917	5	83 Bi	$L\gamma_3$	L_IN_{III}	15.7102	0.8438	1	88 Ra	$L\beta_{17}$	$L_{II}M_{III}$	14.692
0.7897	1	82 Pb		L_IO_I	15.699	0.8442	2	81 Tl	$L\gamma_6$	$L_{II}O_{IV}$	14.685
0.79015	1	40 Zr	$K\alpha_2$	KL_{II}	15.6909	0.8452	2	80 Hg		L_IO_I	14.670
0.79043	3	83 Bi	$L\gamma_6$	$L_{II}O_{IV}$	15.6853	0.84773	5	81 Tl	$L\gamma_2$	L_IN_{II}	14.6251
0.79257	4	90 Th	$L\beta_4$	L_IM_{II}	15.6429	0.848187	9	95 Am	$L\alpha_1$	$L_{III}M_V$	14.6172
0.79257	4	90 Th	$L\beta_{17}$	$L_{II}M_{III}$	15.6429	0.8490	1	81 Tl		$L_{II}O_{II}$	14.604
0.79354	3	90 Th	$L\beta_2$	$L_{III}M_V$	15.6237	0.85048	5	81 Tl	Lv	$L_{II}N_{VI}$	14.5777
0.79384	5	83 Bi		$L_{II}O_{III}$	15.6178	0.8512	1	88 Ra		$L_{III}N_{III}$	14.566
0.79539	5	90 Th	$L\beta_{15}$	$L_{III}N_{IV}$	15.5875	0.8513	2	81 Tl	$L\gamma_8$	$L_{II}O_I$	14.564
0.79565	3	83 Bi	$L\gamma_2$	L_IN_{II}	15.5824	0.85192	7	82 Pb		$L_{II}N_{III}$	14.553
0.79721	9	83 Bi	Lv	$L_{II}N_{VI}$	15.552	0.85436	9	86 Rn	$L\beta_3$	L_IM_{III}	14.512

Table 7f-2. Wavelengths of X-ray Emission Lines and Absorption Edges: In Numerical Order (Continued)

Wavelength, Å*	p.e.†	Element	Designation	keV
0.85446	4	90 Th	$L\eta$ $L_{II}M_I$	14.5099
0.8549	1	81 Tl	L_IN_I	14.503
0.85657	7	80 Hg	$L\gamma_{11}$ L_IN_V	14.474
0.858	2	87 Fr	$L\beta_2$ $L_{II}N_V$	14.45
0.8585	3	82 Pb	L_IN_{II}	14.442
0.860266	9	95 Am	$L\alpha_2$ $L_{III}M_{IV}$	14.4119
0.8618	1	88 Ra	$L_{III}N_{II}$	14.387
0.86376	5	79 Au	L_I Abs. edge	14.3537
0.86400	5	79 Au	$L_IO_{IV,V}$	14.3497
0.8653	2	36 Kr	$K\beta_4$ $KN_{IV,V}$	14.328
0.86552	1	36 Kr	K Abs. edge	14.3244
0.86605	9	86 Rn	$L\beta_1$ $L_{II}M_{IV}$	14.316
0.8661	1	36 Kr	$K\beta_2$ $KN_{II,III}$	14.315
0.86655	5	82 Pb	$L\gamma_5$ $L_{II}N_I$	14.3075
0.86703	4	79 Au	$L\gamma_4$ L_IO_{III}	14.2996
0.86752	3	81 Tl	$L\gamma_1$ $L_{II}N_{IV}$	14.2915
0.86816	4	79 Au	$L\gamma_4'$ L_IO_{II}	14.2809
0.86830	2	94 Pu	$L\alpha_1$ $L_{III}M_V$	14.2786
0.86915	7	80 Hg	$L\gamma_3$ L_IN_{III}	14.265
0.87074	5	79 Au	L_IO_I	14.2385
0.8708	2	88 Ra	$K\beta_5$ $KM_{IV,V}$	14.238
0.87088	5	80 Hg	$L\beta_6$ $L_{III}N_I$	14.2362
0.8722	1	80 Hg	L_{II} Abs. edge	14.215
0.87319	7	80 Hg	$L\gamma_6$ $L_{II}O_{IV}$	14.199
0.87526	1	38 Sr	$K\alpha_1$ KL_{III}	14.1650
0.87544	7	80 Hg	$L\gamma_2$ L_IN_{II}	14.162
0.8758	1	80 Hg	$L_{II}O_I$	14.156
0.8784	1	36 Kr	$K\beta_1$ KM_{III}	14.114
0.8785	1	80 Hg	$L\nu$ $L_{II}N_{VI}$	14.112
0.87885	7	36 Kr	$K\beta_3$ KM_{II}	14.107
0.8790	1	38 Sr	$K\alpha_2$ KL_{II}	14.104
0.87943	1	80 Hg	$L\gamma_8$ $L_{II}O_I$	14.0979
0.87995	7	80 Hg	$L_{II}O_I$	14.090
0.9234	1	83 Bi	L_{III} Abs. edge	13.426
0.9236	1	77 Ir	L_I Abs. edge	13.423
0.92413	4	83 Bi	$L_{III}P_{II,III}$	13.4159
0.9243	3	87 Fr	$L_IO_{IV,V}$	13.413
0.92453	7	80 Hg	$L\gamma_5$ $L_{II}N_I$	13.410
0.9255	1	35 Br	$K\beta_5$ $KM_{IV,V}$	13.396
0.925553	9	37 Rb	$K\alpha_1$ KL_{III}	13.3953
0.92556	3	83 Bi	$L\beta_5$ $L_{III}O_{IV,V}$	13.3953
0.92650	3	79 Au	$L\gamma_1$ $L_{II}N_{IV}$	13.3817
0.9268	1	82 Pb	$L\beta_9$ L_IM_V	13.377
0.92744	3	77 Ir	$L\gamma_4$ L_IO_{III}	13.3681
0.92791	5	78 Pt	$L\gamma_3$ L_IN_{III}	13.3613
0.92831	3	77 Ir	$L\gamma_4'$ L_IO_{II}	13.3555
0.92937	5	84 Po	$L\beta_2$ $L_{III}N_V$	13.3404
0.92969	1	37 Rb	$K\alpha_2$ KL_{II}	13.3358
0.9302	2	83 Bi	$L\beta_{1,5}$ $L_{III}O_{III}$	13.328
0.9312	2	84 Po	$L_{III}N_{IV}$	13.314
0.9323	2	83 Bi	$L_{III}O_{II}$	13.298
0.93279	5	35 Br	$K\beta_1$ KM_{III}	13.2914
0.93284	5	91 Pa	$L_{III}N_{IV}$	13.2907
0.93327	2	35 Br	$K\beta_3$ KM_{II}	13.2845
0.9339	5	82 Pb	$L\beta_{10}$ L_IM_{IV}	13.275
0.93414	2	78 Pt	L_{II} Abs. edge	13.2723
0.9342	5	78 Pt	$L\gamma_6$ $L_{II}O_{IV}$	13.271
0.93427	5	78 Pt	$L\beta_6$ $L_{III}N_I$	13.2704
0.93505	5	83 Bi	$L\beta_7$ $L_{III}O_I$	13.2593
0.93505	5	83 Bi	Lu $L_{III}N_{VI,VII}$	13.2593
0.93855	3	83 Bi	$L\beta_3$ L_IM_{III}	13.2098
0.93931	3	83 Bi	$L\nu$ $L_{II}N_{VI}$	13.1992
0.9402	5	78 Pt	$L_{II}N_{III}$	13.186
0.9411	2	79 Au	$L\gamma_8$ $L_{II}O_I$	13.173
0.94419	1	78 Pt	$L\beta_9$ L_IM_V	13.1310
0.9446	2	77 Ir	$L\gamma_{11}$ L_IN_V	13.126

keV	Designation	Line	Element	Intensity	λ (Å)
14.0893	$L_{II}N_{III}$		81 Tl	5	0.87996
14.0842	$L_{III}M_{IV}$	$L\alpha_2$	94 Pu	2	0.88028
14.067	$L_I M_{III}$	$L\beta_3$	85 At	9	0.88135
14.045	$L_I N_I$		80 Hg	2	0.8827
14.020	$L_I N_V$	$L\gamma_{11}$	79 Au	7	0.88433
13.999	$L_{II}M_{II}$		81 Tl	2	0.88563
13.959	$L_{II}M_V$		93 Np	9	0.8882
13.9441	$L_{III}M_V$	$L\alpha_1$	78 Pt	1	0.889128
13.883	Abs. edge	L_{III}	78 Pt	1	0.8931
13.878	$L_{II}O_V$		85 At	9	0.8934
13.876	$L_{III}M_{IV}$	$L\alpha_2$	78 Pt	1	0.89349
13.864	$L_{II}O_{IV}$	$L\gamma_6$	81 Tl	4	0.8943
13.8526	$L_{II}N_I$	$L\gamma_5$	80 Hg	5	0.89500
13.8301	$L_{II}N_{IV}$	$L\gamma_1$	78 Pt	4	0.89646
13.8281	$L_{II}O_{III}$		78 Pt	4	0.89659
13.8145	$L_{II}O_{II}$		79 Au	5	0.89747
13.8090	$L_I N_{III}$	$L\gamma_3$	83 Bi	3	0.89783
13.8077	$L_{II}M_V$		78 Pt	2	0.89791
13.784	$L_{III}O_I$	$L\beta_7$	84 Po	2	0.8995
13.782	$L_{III}O_{IV,V}$	$L\beta_5$	93 Np	9	0.8996
13.7597	$L_{III}M_{IV}$	$L\alpha_2$	79 Au	5	0.901045
13.7361	Abs. edge	L_{III}	79 Au	3	0.90259
13.7304	$L_{II}O_{IV}$	$L\gamma_6$	79 Au	3	0.90297
13.7095	$L_I N_{III}$	$L\gamma_3$	83 Bi	4	0.90434
13.7002	$L_{II}M_{IV}$	$L\beta_1$	79 Au	5	0.90495
13.679	$L_{II}O_{III}$		79 Au	7	0.90638
13.6630	$L_{II}M_I$	$L\eta$	88 Ra	5	0.90742
13.662	$L_{II}O_{II}$		79 Au	7	0.90746
13.6487	$L_{II}N_{VI}$		79 Au	5	0.90837
13.640	$L_{II}N_{III}$		79 Au	7	0.90894
13.638	$L_{II}M_{III}$	$L\beta_3$	80 Hg	1	0.9091
13.6260	$L_{III}O_I$	$L\beta_7$	84 Po	2	0.90989
13.6147	$L_{III}M_V$	$L\alpha_1$	79 Au	1	0.910639
13.578	$L_I N_I$		92 U	7	0.9131
13.560	$L_I N_V$	$L\gamma_{11}$	79 Au	9	0.9143
13.470	Abs. edge	K	35 Br	1	0.9204
13.4695	$K N_{II,III}$	$K\beta_2$	35 Br	2	0.92046
13.447	$L_{II}M_{IV}$	$L\beta_1$	84 Po	2	0.9220
13.4388	$L_{III}M_{IV}$	$L\alpha_2$	92 U	9	0.922558

keV	Designation	Line	Element	Intensity	λ (Å)
13.1222	$L_{III}M_{IV}$	$L\alpha_2$	91 Pa	5	0.94482
13.113	$L_I N_I$		78 Pt	2	0.9455
13.108	$L_I N_{IV}$		77 Ir	2	0.9459
13.086	$L_I M_{II}$	$L\beta_4$	84 Po	3	0.9475
13.0406	Abs. edge	L_{III}	82 Pb	5	0.95073
13.0344	$L_{III}P_{II,III}$		82 Pb	7	0.95118
13.0235	$L_{II}M_{IV}$	$L\beta_1$	83 Bi	9	0.951978
13.015	$L_{III}O_{IV,V}$	$L\beta_5$	82 Pb	1	0.9526
12.9799	$L_{III}N_V$	$L\beta_2$	83 Bi	4	0.95518
12.9743	$L_{II}N_I$	$L\gamma_5$	79 Au	3	0.95559
12.972	Abs. edge	L_{II}	76 Os	1	0.9558
12.9687	$L_{III}M_V$	$L\alpha_1$	90 Th	3	0.95600
12.9683	$L_{II}O_{IV,V}$	$L\gamma_6$	76 Os	5	0.95603
12.9585	$L_I M_{IV}$	$L\beta_{10}$	81 Tl	7	0.95675
12.9549	$L_{II}N_{IV}$	$L\gamma_1$	83 Bi	5	0.95702
12.945	$L_{III}O_{III}$		82 Pb	1	0.9578
12.9420	$L_{III}O_{II}$		78 Pt	3	0.95797
12.934	$L_{II}N_{III}$		82 Pb	1	0.9586
12.9240	$L_I N_{III}$	$L\gamma_3$	77 Ir	5	0.95931
12.923	$L_I O_{III}$	$L\gamma_4$	76 Os	8	0.95938
12.910	$L_{II}O_{II}$		76 Os	8	0.96033
12.8968	$L_{III}N_{VI,VII}$	Lu	82 Pb	7	0.96133
12.888	$L_{III}O_I$	$L\beta_7$	82 Pb	1	0.9620
12.8721	$L_I O_I$		76 Os	7	0.96318
12.866	$L_{III}M_{III}$	Ls	92 U	1	0.9636
12.8626	$L_I M_{IV}$	$L\beta_{10}$	81 Tl	7	0.96389
12.8418	$L_I N_{III}$		77 Ir	3	0.96545
12.8201	$L_{II}O_{IV}$		77 Ir	4	0.96708
12.820	Abs. edge		77 Ir	1	0.9671
12.819	$L_{III}N_I$	$L\beta_6$	84 Po	2	0.9672
12.8096	$L_{III}M_{IV}$	$L\alpha_2$	90 Th	2	0.96788
12.7933	$L_{II}M_{III}$	$L\beta_3$	82 Pb	7	0.96911
12.7843	$L_{II}O_{III}$		77 Ir	5	0.96979
12.7603	$L_I N_{VI}$		77 Ir	6	0.97161
12.7588	$L_{II}N_{III}$		78 Pt	4	0.97173
12.7394	$L_{III}N_{VIII}$		83 Bi	5	0.97321
12.7279	$L_{II}O_I$	$L\gamma_8$	77 Ir	3	0.97409
12.720	$L_{II}M_V$		82 Pb	1	0.9747
12.696	$L_I N_V$	$L\gamma_{11}$	76 Os	3	0.9765

TABLE 7f-2. WAVELENGTHS OF X-RAY EMISSION LINES AND ABSORPTION EDGES: IN NUMERICAL ORDER (Continued)

Wavelength, Å*	p.e.†	Element	Designation	keV
0.9766	2	77 Ir	$L_I N_I$	12.695
0.97690	4	83 Bi	$L\beta_4$ $L_I M_{II}$	12.6912
0.9772	3	76 Os	$L_I N_{IV}$	12.687
0.9792	2	78 Pt	$L_{II} N_{II}$	12.661
0.97926	5	81 Tl	L_{III} $L_{III} P_{II,III}$	12.6607
0.9793	1	81 Tl	K Abs. edge	12.660
0.97974	1	34 Se	$K\beta_2$ Abs. edge	12.6545
0.97974	5	34 Se	$K N_{II,III}$	12.6522
0.97993	5	89 Ac	$L\alpha_1$ $L_{III} M_V$	12.6520
0.9801	1	36 Kr	$K\alpha_1$ $K L_{III}$	12.649
0.98058	3	81 Tl	$L\beta_5$ $L_{III} O_{IV,V}$	12.6436
0.98221	7	82 Pb	$L\beta_2$ $L_{III} N_V$	12.6226
0.98280	5	83 Bi	$L_{III} N_{II}$	12.6151
0.98291	3	82 Pb	$L\beta_1$ $L_{II} M_{IV}$	12.6137
0.98389	7	82 Pb	$L\beta_{15}$ $L_{III} N_{IV}$	12.6011
0.9841	1	36 Kr	$K\alpha_2$ $K L_{II}$	12.598
0.9843	1	34 Se	$K\beta_5$ $K M_{IV,V}$	12.595
0.98538	5	81 Tl	$L\beta_9$ $L_{III} O_{III}$	12.5820
0.9871	2	80 Hg	$L_{III} O_{II}$	12.561
0.98738	5	81 Tl	$L_{III} O_{II}$	12.5566
0.9877	2	78 Pt	$L\gamma_5$ $L_I N_I$	12.552
0.9888	5	81 Tl	Lu $L_{III} N_{VI,VII}$	12.538
0.98913	1	83 Bi	$L\beta_{17}$ $L_{II} M_{III}$	12.5344
0.9894	1	75 Re	L_I Abs. edge	12.530
0.9900	1	75 Re	$L_I O_{IV,V}$	12.524
0.99017	5	81 Tl	$L\beta_7$ $L_{III} O_I$	12.5212
0.99085	3	77 Ir	$L\gamma_1$ $L_{II} N_{IV}$	12.5126
0.99178	5	89 Ac	$L\alpha_2$ $L_{III} M_{IV}$	12.5008
0.99186	5	76 Os	$L\gamma_3$ $L_I N_{III}$	12.4998
0.99218	3	34 Se	$K\beta_1$ $K M_{III}$	12.4959
0.99249	5	75 Re	$L\gamma_4$ $L_{III} O_{III}$	12.4920
0.99268	5	34 Se	$K\beta_3$ $K M_{II}$	12.4896
0.99331	3	83 Bi	$L\beta_6$ $L_{III} N_I$	12.4816
1.0250	2	74 W	$L_I O_{IV,V}$	12.095
1.02503	5	76 Os	$L_{II} N_{IV}$	12.0953
1.02613	7	75 Re	$L_I N_{III}$	12.0824
1.02775	3	74 W	$L_I O_{III}$	12.0634
1.02789	7	79 Au	$L\beta_{10}$ $L_I M_{IV}$	12.0617
1.0286	1	81 Tl	$L_{III} N_{III}$	12.053
1.02863	3	74 W	$L\gamma_4'$ $L_I O_{II}$	12.0530
1.03049	5	87 Fr	$L\alpha_1$ $L_{III} M_V$	12.0313
1.0317	3	74 W	$L_I O_I$	12.017
1.03233	5	75 Re	$L_I N_{II}$	12.0098
1.0323	2	82 Pb	$L\gamma_2$ $L_I M_{II}$	12.010
1.03358	7	80 Hg	$L_I M_{III}$	11.9953
1.0346	9	83 Bi	$L\beta_3$ $L_I M_{II}$	11.98
1.0347	1	92 U	Lt $L_{III} M_{II}$	11.982
1.03699	9	75 Re	$L\gamma_6$ $L_{III} O_{IV}$	11.956
1.0371	1	75 Re	L_{II} L_{II}	11.954
1.03876	7	79 Au	Abs. edge	11.9355
1.03918	3	81 Tl	$L\beta_4$ $L_{III} P_{II,III}$	11.9306
1.0397	1	75 Re	$L_{II} M_{II}$	11.925
1.03973	5	76 Os	$L_{III} O_{III}$	11.925
1.03974	2	35 Br	$K\alpha_1$ $K L_{III}$	11.9243
1.03975	7	80 Hg	$L\beta_2$ $L_{III} N_V$	11.9242
1.03976	5	79 Au	$L_{III} N_V$	11.9241
1.04000	1	79 Au	Abs. edge	11.9212
1.0404	3	75 Re	$L\nu$ $L_{II} N_{VI}$	11.917
1.04044	7	79 Au	$L\beta_5$ $L_{III} O_{IV,V}$	11.9163
1.04151	1	80 Hg	$L\beta_{15}$ $L_{III} N_{IV}$	11.9040
1.0420	5	75 Re	$L_I N_I$	11.899
1.04230	6	87 Fr	$L\alpha_2$ $L_{III} M_{IV}$	11.8950
1.0428	3	93 Np	Ll $L_{III} M_I$	11.890
1.04382	2	35 Br	$K\alpha_2$ $K L_{II}$	11.8776
1.04398	5	75 Re	$L\gamma_8$ $L_I O_I$	11.8758
1.0450	2	79 Au	$L_{III} O_{II,III}$	11.865
1.0450	1	33 As	K Abs. edge	11.865

Block 1

λ (Å)	Designation	Siegbahn	Element	Int.	Energy (keV)
11.8642	$KN_{II,III}$	$K\beta_2$	33 As	3	1.04500
11.856	$L_I N_V$	$L\gamma_{11}$	74 W	1	1.0458
11.844	$L_I N_{IV}$		74 W	2	1.0468
11.8357	$L_{III} N_{VI,VII}$		79 Au	5	1.04752
11.8226	$L_{II} M_{IV,V}$		80 Hg	5	1.04868
11.822	$K M_{IV,V}$	Lu	33 As	1	1.0488
11.8118	$L_{III} N_I$	$L\beta_1$	81 Tl	5	1.04963
11.8106	$L_{III} O_I$	$K\beta_5$	79 Au	8	1.04974
11.7577	$L_I M_V$	$L\beta_6$	78 Pt	5	1.05446
11.7397	$L_{III} M_{III}$	$L\beta_7$	81 Tl	7	1.05609
11.7303	$L_I N_I$	$L\beta_9$	76 Os	5	1.05693
11.7270	$L_I M_{III}$	$L\beta_{17}$	86 Rn	5	1.05723
11.7262	$K M_{III}$	$L\gamma_5$	33 As	2	1.05730
11.7203	$K M_{III}$	$L\alpha_1$	33 As	5	1.05783
11.713	$L_{III} N_{III}$	$K\beta_1$	80 Hg	1	1.0585
11.7122	$L_{II} M_I$	$K\beta_3$	83 Bi	3	1.05856
11.6854	$L_{II} N_{IV}$	$L\eta$	75 Re	5	1.06099
11.682	Abs. edge	$L\gamma_1$	73 Ta	1	1.0613
11.6762	$L_{II} M_{IV}$	L_I	78 Pt	7	1.06183
11.6752	$L_I O_{IV,V}$	$L\beta_{10}$	73 Ta	9	1.06192
11.6743	$L_I N_{III}$		74 W	6	1.06200
11.6570	$L_{II} N_{VI,VII}$	$L\gamma_3$	73 Ta	9	1.06357
11.648	$L_{II} M_{III}$		82 Pb	2	1.0644
11.648	$L_I M_I$		81 Tl	2	1.0644
11.6451	$L_I O_{III}$		73 Ta	3	1.06467
11.642	$L_{III} N_{II}$	$L\gamma_4$	80 Hg	2	1.0649
11.6366	$L_I O_{II}$		73 Ta	3	1.06544
11.6183	$L_{II} M_I$	$L\gamma_4'$	92 U	2	1.06712
11.6118	$L_I O_I$	Ll	73 Ta	9	1.06771
11.6103	$L_I M_{III}$		79 Au	9	1.06785
11.6080	$L_I N_{II}$	$L\beta_3$	74 W	3	1.06806
11.5979	$L_{III} M_{IV}$	$L\gamma_2$	86 Rn	5	1.06899
11.5847	$L_{III} N_V$	$L\alpha_2$	79 Au	3	1.07022
11.5667	$L_{III} N_{IV}$	$L\beta_2$	79 Au	5	1.07188
11.5630	$L_I M_{III}$	$L\beta_{15}$	80 Hg	7	1.07222
11.562	Abs. edge	$L\beta_4$	78 Pt	1	1.0723
11.561	$L_{III} O_{IV,V}$	L_{III}	78 Pt	2	1.0724
11.5387	$L_{II} O_{IV}$	$L\beta_5$	74 W	5	1.07448
11.538	Abs. edge	L_{II}	74 W	1	1.0745

Block 2

Energy (keV)	Int.	Element	Siegbahn	Designation	λ (Å)
0.99334	5	75 Re	$L\gamma_4'$	$L_I O_{II}$	12.4813
0.9962	2	80 Hg	$L\beta_{10}$	$L_I M_{IV}$	12.446
0.9965	1	75 Re		$L_I O_I$	12.442
0.99805	5	76 Os	$L\gamma_2$	$L_I N_{II}$	12.4224
1.0005	1	82 Pb		$L_{III} N_{III}$	12.392
1.0005	9	83 Bi	$L\beta_3$	$L_I M_I$	12.39
1.00062	3	81 Tl	$L\gamma_3$	$L_I M_{III}$	12.3904
1.00107	5	76 Os	Ll	$L_I O_{IV}$	12.3848
1.0012	6	95 Am	L_{II}	$L_{III} M_I$	12.384
1.0014	1	76 Os		Abs. edge	12.381
1.0047	2	76 Os		$L_{II} O_{III}$	12.340
1.00473	5	88 Ra	$L\alpha_1$	$L_{III} M_V$	12.3397
1.0050	2	76 Os	Lv	$L_{II} N_{VII}$	12.337
1.0054	3	77 Ir		$L_{II} N_{VII}$	12.332
1.00722	5	81 Tl	$L\beta_4$	$L_I M_V$	12.3093
1.0075	1	82 Pb		$L_{II} M_I$	12.306
1.00788	5	76 Os	$L\gamma_8$	$L_I O_{III}$	12.3012
1.0091	1	80 Hg	L_{III}	Abs. edge	12.286
1.00987	7	80 Hg	$L\beta_5$	$L_{III} O_{IV,V}$	12.2769
1.01031	3	81 Tl	$L\beta_2$	$L_{III} N_V$	12.2715
1.01040	7	82 Pb		$L_{III} N_{II}$	12.2705
1.0108	1	75 Re	$L\gamma_{11}$	$L_I N_V$	12.266
1.0112	1	90 Th	Ls	$L_{III} M_{III}$	12.261
1.0119	1	75 Re		$L_I N_{IV}$	12.252
1.0120	2	77 Ir		$L_I N_{III}$	12.251
1.01201	3	81 Tl	$L\beta_{15}$	$L_{III} N_{IV}$	12.2510
1.01404	7	80 Hg		$L_{III} O_{III}$	12.2264
1.01513	4	81 Tl	$L\beta_1$	$L_{II} M_{IV}$	12.2133
1.01558	7	80 Hg		$L_{III} O_{II}$	12.2079
1.01556	5	88 Ra	$L\alpha_2$	$L_{III} M_{IV}$	12.1962
1.01674	7	80 Hg	Lu	$L_{III} N_{VII}$	12.1940
1.01769	7	80 Hg	Lu'	$L_{III} N_{VI}$	12.1826
1.01937	7	80 Hg	$L\beta_7$	$L_{III} N_{VI}$	12.1625
1.02063	7	79 Au	$L\beta_9$	$L_I O_I$	12.1474
1.0210	7	82 Pb	$L\beta_6$	$L_{III} N_I$	12.143
1.02175	1	77 Ir	$L\gamma_5$	$L_{II} N_I$	12.1342
1.0223	5	82 Pb	$L\beta_{17}$	$L_{II} M_{III}$	12.127
1.0226	1	94 Pu	Ll	$L_{III} M_I$	12.124
1.02467	5	74 W	L_I	Abs. edge	12.0996

Table 7f-2. Wavelengths of X-ray Emission Lines and Absorption Edges: in Numerical Order (*Continued*)

Wavelength, Å*	p.e.†	Element	Designation	Designation	keV
1.0756	2	79 Au		$L_{II}M_V$	11.526
1.0761	3	78 Pt		$L_{III}O_{II,III}$	11.521
1.0767	1	75 Re	$L\nu$	$L_{II}N_{III}$	11.515
1.0771	5	74 W	Lu	$L_{II}N_{VI}$	11.510
1.07896	2	78 Pt		$L_{III}N_{VI,VII}$	11.4908
1.0792	7	74 W	$L\beta_6$	$L_{I}O_{III}$	11.488
1.07975	9	80 Hg	Lt	$L_{III}N_I$	11.4824
1.08009	4	90 Th	$L\gamma_8$	$L_{III}M_{II}$	11.4788
1.08113	3	74 W	$L\beta_7$	$L_{I}O_I$	11.4677
1.08168	7	78 Pt	$L\gamma_{11}$	$L_{III}O_I$	11.4619
1.08205	3	73 Ta	$L\beta_1$	$L_{I}N_V$	11.4580
1.08353	7	79 Au		$L_{I}M_{IV}$	11.4423
1.08377	1	73 Ta		$L_{I}N_{IV}$	11.4398
1.0839	5	75 Re	$L\alpha_1$	$L_{II}N_{II}$	11.438
1.08500	5	85 At	$L\beta_9$	$L_{II}M_V$	11.4268
1.08975	7	77 Ir	Ll	$L_{I}M_V$	11.3770
1.09026	1	79 Au	$L\beta_{17}$	$L_{III}N_{III}$	11.3717
1.0908	5	91 Pa	$L\eta$	$L_{III}M_I$	11.366
1.0916	5	80 Hg	$L\gamma_5$	$L_{II}M_I$	11.358
1.09241	7	82 Pb	$L\alpha_2$	$L_{II}N_I$	11.3493
1.09388	1	75 Re	$L\beta_{10}$	$L_{I}N_I$	11.3341
1.09671	5	85 At	$L\gamma_1$	$L_{III}M_{IV}$	11.3048
1.09702	4	77 Ir	$L\gamma_3$	$L_{II}N_{IV}$	11.3016
1.09855	1	74 W		$L_{II}N_{III}$	11.2859
1.09936	7	73 Ta		$L_{II}N_{III}$	11.2776
1.0997	1	81 Tl		$L_{II}M_{II}$	11.274
1.0997	7	79 Au	L_I	Abs. edge	11.274
1.09968	2	80 Hg		$L_{III}N_{II}$	11.2743
1.0999	9	72 Hf		$L_{I}M_I$	11.272
1.10086	2	80 Hg	$L\beta_2$	$L_{I}O_{IV}$	11.2622
1.10200	9	78 Pt		$L_{III}N_V$	11.2505
1.10303	3	72 Hf	$L\gamma_4$	$L_{I}O_{III}$	11.2401
1.10376	5	72 Hf	$L\gamma_4'$	$L_{I}O_{II}$	11.2326
1.13353	5	76 Os	$L\beta_{10}$	$L_{I}M_{IV}$	10.9376
1.13525	5	79 Au		$L_{I}M_I$	10.9210
1.13532	3	77 Ir	$L\beta_2$	$L_{III}N_V$	10.9203
1.13687	9	73 Ta		$L_{II}N_V$	10.9055
1.13707	3	77 Ir	$L\beta_{15}$	$L_{III}N_{IV}$	10.9036
1.13794	3	73 Ta	$L\gamma_1$	$L_{II}N_{IV}$	10.8952
1.13841	5	72 Hf	$L\gamma_3$	$L_{I}N_{III}$	10.8907
1.1387	5	80 Hg		$L_{III}M_{II}$	10.888
1.1402	1	71 Lu	Lt	Abs. edge	10.8740
1.1405	1	76 Os	$L\beta_5$	$L_{III}O_{IV,V}$	10.8711
1.1408	1	76 Os		Abs. edge	10.8683
1.14085	3	77 Ir	$L\beta_3$	$L_{I}M_{III}$	10.8674
1.14223	5	78 Pt	$L\beta_4$	$L_{I}M_{II}$	10.8543
1.1435	1	71 Lu		$L_{I}O_{II,III}$	10.8425
1.14355	5	78 Pt	$L\gamma_4$	$L_{I}N_I$	10.8418
1.14386	2	83 Bi	$L\alpha_1$	$L_{III}M_V$	10.8388
1.14442	5	72 Hf	$L\gamma_2$	$L_{I}N_{II}$	10.8335
1.14537	7	76 Os	Lu	$L_{III}N_{VI,VII}$	10.8245
1.1489	2	77 Ir	$L\beta_7$	$L_{III}O_I$	10.791
1.14933	8	76 Os	$L\gamma_6$	Abs. edge	10.7872
1.1548	1	72 Hf	$L\beta_7$	$L_{II}O_{IV}$	10.7362
1.15519	5	72 Hf	$L\gamma_6$	$L_{III}M_{III}$	10.7325
1.1553	1	73 Ta		$L_{II}M_{IV}$	10.7316
1.15536	1	83 Bi	$L\alpha_2$	$L_{III}M_{IV}$	10.73091
1.1560	3	77 Ir		$L_{III}N_{III}$	10.725
1.15781	3	77 Ir	$L\beta_1$	$L_{III}N_{III}$	10.7083
1.15830	9	72 Hf	Lv	$L_{II}N_{VI}$	10.7037
1.1600	2	73 Ta		$L_{II}N_{II}$	10.688
1.16107	9	71 Lu	$L\gamma_{11}$	$L_{I}N_V$	10.6782
1.16138	5	72 Hf	$L\gamma_8$	$L_{I}O_I$	10.6754
1.16227	9	71 Lu		$L_{I}N_{IV}$	10.6672
1.1640	1	80 Hg	$L\eta$	$L_{II}M_I$	10.6512
1.16487	4	75 Re	$L\beta_9$	$L_{I}M_V$	10.6433

Upper table

Energy (keV)	Transition	Line	Element	Order	Wavelength
10.6380	$L_{II}N_{II}$		77 Ir	5	1.16545
10.6265	$L_{II}M_{III}$	$L\beta_{17}$	78 Pt	1	1.1667
10.6222	$L_{III}M_{I}$	Ll	88 Ra	5	1.16719
10.6001	$L_{I}M_{I}$		78 Pt	9	1.16962
10.5985	$L_{III}N_{V}$	$L\beta_{2}$	76 Os	8	1.16979
10.5892	$L_{I}M_{II}$		79 Au	1	1.1708
10.5816	$L_{III}N_{IV}$	$L\beta_{15}$	76 Os	5	1.17167
10.5770	$L_{II}N_{IV}$	$L\beta_{10}$	75 Re	5	1.17218
10.5702	$L_{II}N_{I}$	$L\gamma_{5}$	73 Ta	2	1.1729
10.5515	$L_{II}M_{V}$	$L\alpha_{1}$	82 Pb	1	1.17501
10.54372	KL_{III}	$K\alpha_{1}$	33 As	2	1.17588
10.5318	$L_{III}O_{IV,V}$	$L\beta_{5}$	75 Re	5	1.17721
10.5306	Abs. edge	L_{III}	75 Re	1	1.173
10.5258	$L_{II}N_{V}$	$L\beta_{6}$	72 Hf	9	1.17788
10.5251	$L_{III}N_{I}$	$L\gamma_{1}$	77 Ir	3	1.17796
10.5158	$L_{II}N_{IV}$	$L\gamma_{3}$	72 Hf	5	1.17900
10.5110	$L_{I}N_{III}$	$L\beta_{3}$	71 Lu	4	1.17953
10.5108	$L_{I}M_{III}$	$L\beta_{4}$	76 Os	7	1.17955
10.5106	$L_{I}M_{II}$	$K\alpha_{2}$	77 Ir	3	1.17958
10.50799	KL_{II}	Lu	33 As	1	1.17987
10.4931	$L_{III}N_{VI,VII}$	L_{I}	75 Re	1	1.1815
10.4904	Abs. edge		70 Yb	1	1.1827
10.4833	$L_{I}O_{IV,V}$	$L\gamma_{4}$	70 Yb	1	1.1853
10.4603	$L_{II}O_{II,III}$	$L\gamma_{2}$	70 Yb	1	1.1853
10.460	$L_{II}N_{I}$	$L\beta_{7}$	71 Lu	2	1.18610
10.4529	$L_{III}O_{I}$	$L\alpha_{2}$	75 Re	5	1.18648
10.4495	$L_{III}M_{IV}$		82 Pb	5	1.1886
10.4312	$L_{II}O_{I}$		70 Yb	1	1.18977
10.4205	$L_{II}M_{V}$	K	76 Os	7	1.1958
10.3682	Abs. edge	$K\beta_{2}$	31 Ga	1	1.19600
10.3663	$KN_{II,III}$	$L\beta_{1}$	31 Ga	2	1.19727
10.3553	$L_{II}M_{IV}$	$K\beta_{5}$	76 Os	7	1.1981
10.348	$KM_{IV,V}$		31 Ga	2	1.1985
10.3448	Abs. edge	L_{II}	71 Lu	1	1.1987
10.3431	$L_{III}O_{IV}$	$L\gamma_{6}$	71 Lu	1	1.20086
10.3244	$L_{III}N_{II}$		76 Os	7	1.2014
10.3198	$L_{III}O_{II,III}$		71 Lu	1	1.20273
10.3083	$L_{II}N_{II}$	$L\eta$	79 Au	2	1.2047
10.2915	$L_{II}N_{I}$	$L\gamma_{8}$	71 Lu	3	—

Lower table

Energy (keV)	Transition	Line	Element	Order	Wavelength
11.2308	$L_{I}M_{III}$	$L\beta_{3}$	78 Pt	5	1.10394
11.2224	KL_{III}	$K\alpha_{1}$	34 Se	2	1.10477
11.217	$L_{I}N_{I}$	$L\gamma_{2}$	73 Ta	1	1.1053
11.212	Abs. edge	L_{III}	77 Ir	1	1.1058
11.2114	$L_{III}O_{IV,V}$	$L\beta_{5}$	77 Ir	3	1.10585
11.2047	$L_{I}M_{II}$	$L\beta_{4}$	79 Au	3	1.10651
11.2034	KL_{II}	$K\alpha_{2}$	72 Hf	9	1.10664
11.1814	$L_{III}O_{II,III}$	$L\beta_{6}$	34 Se	2	1.10882
11.1772	$L_{III}N_{I}$	Lu	77 Ir	6	1.10923
11.1602	$L_{III}N_{VI,VII}$	L_{II}	79 Au	3	1.11092
11.1549	$L_{II}M_{V}$	$L\alpha_{1}$	77 Ir	4	1.11145
11.140	Abs. edge	$L\gamma_{6}$	78 Pt	2	1.1129
11.132	$L_{III}M_{V}$	$L\beta_{7}$	73 Ta	1	1.1137
11.1308	$L_{II}O_{IV}$	Ll	84 Po	4	1.11386
11.1306	$L_{III}O_{I}$	Lv	73 Ta	3	1.11388
11.1205	$L_{III}M_{I}$	K	77 Ir	3	1.11489
11.120	$L_{I}N_{III}$	$K\beta_{2}$	74 W	2	1.1149
11.1186	$L_{I}N_{I}$	$K\beta_{5}$	90 Th	4	1.11508
11.1173	$L_{I}N_{VI}$	$L\beta_{1}$	73 Ta	9	1.11521
11.1113	Abs. edge	$L\gamma_{8}$	73 Ta	1	1.1158
11.1036	$KN_{II,III}$	$L\gamma_{11}$	32 Ge	5	1.11658
11.1008	$L_{II}O_{III}$	$L\alpha_{2}$	32 Ge	2	1.11686
11.1001	$L_{II}O_{II}$	$L\beta_{9}$	73 Ta	9	1.11693
11.0907	$KM_{IV,V}$	$L\eta$	73 Ta	9	1.11789
11.0745	$L_{II}M_{IV}$	$L\beta_{17}$	32 Ge	1	1.1195
11.0707	$L_{I}O_{I}$	$K\beta_{1}$	78 Pt	2	1.11990
11.0646	$L_{I}N_{V}$	$K\beta_{3}$	73 Ta	1	1.1205
11.0553	$L_{I}N_{IV}$		72 Hf	9	1.12146
11.052	$L_{I}N_{IV}$		74 W	3	1.1218
11.0451	$L_{III}N_{II,III}$		72 Hf	5	1.12250
11.044	$L_{III}M_{IV}$		78 Pt	2	1.1226
11.0158	$L_{I}M_{V}$		84 Po	5	1.12548
11.0071	$L_{II}M_{I}$		76 Os	6	1.12637
10.9943	$L_{I}M_{III}$		81 Tl	3	1.12769
10.9915	KM_{III}		79 Au	5	1.12798
10.9821	KM_{II}		32 Ge	2	1.12894
10.9780	$L_{II}N_{II}$	$L\gamma_{5}$	32 Ge	9	1.12936
10.962	$L_{II}N_{I}$		78 Pt	2	1.1310
10.9490			74 W	3	1.13235

TABLE 7f-2. WAVELENGTHS OF X-RAY EMISSION LINES AND ABSORPTION EDGES: IN NUMERICAL ORDER (Continued)

Wavelength, Å*	p.e.†	Element	Designation	keV
1.20479	7	74 W	$L\beta_9$ L_IM_V	10.2907
1.20660	4	75 Re	$L\beta_2$ $L_{III}N_V$	10.2752
1.2069	2	77 Ir	$L\beta_{17}$ $L_{II}M_{III}$	10.273
1.20739	4	81 Tl	$L\alpha_1$ $L_{III}M_V$	10.2685
1.20789	2	31 Ga	$K\beta_1$ KM_{III}	10.2642
1.20819	5	75 Re	$L\beta_{15}$ $L_{III}N_{IV}$	10.2617
1.20835	5	31 Ga	$K\beta_3$ KM_{II}	10.2603
1.2102	2	77 Ir	Ls L_IM_I	10.245
1.2105	1	83 Bi	$L\beta_{10}$ $L_{III}M_{III}$	10.2421
1.21218	3	74 W	$L\beta_{10}$ L_IM_{IV}	10.2279
1.213	1	78 Pt	$L_{III}M_{II}$	10.225
1.21349	5	76 Os	$L\beta_6$ L_IN_I	10.2169
1.21537	5	72 Hf	$L\gamma_5$ $L_{III}O_{IV,V}$	10.2011
1.21545	3	74 W	$L\beta_5$ L_IN_I	10.2004
1.2155	1	74 W	L_{III} Abs. edge	10.1999
1.21844	5	76 Os	$L\beta_4$ L_IM_{II}	10.1754
1.21868	5	74 W	Lu $L_{III}N_{VI,VII}$	10.1733
1.21875	3	81 Tl	$L\alpha_2$ $L_{III}M_{IV}$	10.1728
1.22031	5	75 Re	$L\beta_3$ L_IM_{III}	10.1598
1.2211	2	74 W	$L_{III}O_{II,III}$	10.153
1.22228	4	71 Lu	$L\gamma_1$ $L_{II}N_{IV}$	10.1434
1.22232	5	70 Yb	$L\gamma_3$ L_IN_{III}	10.1431
1.22400	4	74 W	$L\beta_7$ $L_{III}O_I$	10.1292
1.2250	1	69 Tm	L_I Abs. edge	10.1206
1.2263	3	75 Re	$L_{III}N_{III}$	10.110
1.2283	1	69 Tm	$L_{II}N_{II}$	10.0933
1.22879	7	70 Yb	$L\gamma_2$ L_IN_{II}	10.0897
1.2294	2	69 Tm	$L\gamma_4$ $L_IO_{II,III}$	10.084
1.2305	1	75 Re	$L_{II}M_V$	10.0753
1.23858	2	75 Re	$L\beta_1$ $L_{II}M_{IV}$	10.0100
1.24120	5	80 Hg	$L\alpha_1$ $L_{III}M_V$	9.9888
1.24271	3	70 Yb	$L\gamma_6$ $L_{II}O_{IV}$	9.9766
1.2428	1	70 Yb	L_{II} Abs. edge	9.9761
1.2848	1	30 Zn	$K\beta_5$ $KM_{IV,V}$	9.6501
1.28619	5	73 Ta	$L\beta_{15}$ $L_{III}N_{IV}$	9.6394
1.28772	3	79 Au	$L\alpha_2$ $L_{III}M_{IV}$	9.6280
1.2892	1	69 Tm	Abs. edge	9.6171
1.28989	7	74 W	$L\beta_6$ $L_{III}N_I$	9.6117
1.29025	9	72 Hf	$L\beta_9$ L_IO_{IV}	9.6090
1.2905	2	69 Tm	$L\gamma_6$ $L_{II}O_{IV}$	9.607
1.2927	1	75 Re	$L\beta_{17}$ $L_{II}M_{III}$	9.5910
1.2934	2	76 Os	$K\beta_{1,3}$ KM_{II}	9.586
1.29525	2	30 Zn	L_{III} $KM_{II,III}$	9.5720
1.2972	1	72 Hf	Abs. edge	9.5577
1.29761	5	72 Hf	$L\beta_5$ $L_{III}O_{IV,V}$	9.5546
1.29819	9	72 Hf	$L\beta_{10}$ L_IM_{IV}	9.5503
1.30162	5	74 W	$L\beta_4$ L_IM_{II}	9.5252
1.30165	9	72 Hf	Lu $L_{III}N_{VI,VII}$	9.5249
1.30564	5	72 Hf	$L\beta_7$ $L_{III}O_I$	9.4958
1.3063	1	72 Hf	$L\gamma_5$ $L_{II}N_I$	9.4910
1.30678	3	70 Yb	$L\beta_3$ L_IM_{III}	9.4875
1.30767	7	73 Ta	Lt $L_{III}M_{II}$	9.4811
1.3086	1	82 Pb	$L_{III}M_{III}$	9.4742
1.3112	2	73 Ta	$L_{III}N_{III}$	9.455
1.31304	3	80 Hg	$L\alpha_1$ $L_{III}M_V$	9.4423
1.3146	1	78 Pt	$L\gamma_3$ L_IN_{III}	9.4309
1.3153	2	68 Er	$L\gamma_1$ $L_{II}N_{IV}$	9.426
1.31610	7	69 Tm	Ll $L_{III}M_I$	9.4204
1.3167	1	83 Bi	$L_{II}N_{II}$	9.4158
1.31897	9	73 Ta	L_IN_{II}	9.3998
1.3190	1	73 Ta	$L\alpha_1$ $L_{II}M_V$	9.3994
1.3208	3	67 Ho	L_I Abs. edge	9.387
1.3210	2	67 Ho	$L\gamma_2$ L_IN_{II}	9.385
1.3225	2	68 Er	$L\gamma_4$ $L_IO_{II,III}$	9.374
1.32432	2	67 Ho	$L\alpha_2$ $L_{III}M_{IV}$	9.3618
1.32639	5	78 Pt	$L\beta_2$ $L_{III}N_V$	9.3473

	Level	Line	Element		
9.3431	$L_{II}M_{IV}$	$L\beta_1$	73 Ta	3	1.32698
9.3371	$L_{III}N_{IV}$	$L\beta_{15}$	72 Hf	5	1.32783
9.3370	$L_{II}M_I$	$L\eta$	76 Os	7	1.32785
9.3153	$L_{III}M_I$	$L\beta_6$	73 Ta	8	1.33094
9.2816	$L_I M_V$	$L\beta_9$	71 Lu	1	1.3358
9.277	$L_I M_I$		74 W	3	1.3365
9.2761	$L_{II}M_{III}$		75 Re		1.3366
9.2622	Abs. edge		68 Er	1	1.3386
9.261	$L_{II}M_{III}$	L_{II}	74 W	2	1.3387
9.255	$L_{II}O_{IV}$	$L\beta_{17}$	68 Er	3	1.3397
9.25174	KL_{III}	$L\gamma_6$	31 Ga	9	1.340083
9.2490	L_{III}	$K\alpha_1$	71 Lu	1	1.3405
9.2417	$L_{III}M_{III}$	L_{III}	81 Tl	5	1.34154
9.2397	$L_{III}O_{IV,V}$	Lt	71 Lu	7	1.34183
9.232	$L_I M_{IV}$	$L\beta_5$	71 Lu	2	1.3430
9.22482	KL_{II}	$L\beta_{10}$	31 Ga	1	1.34399
9.2163	$L_{III}O_{II,III}$	$K\alpha_2$	71 Lu	9	1.34524
9.2124	$L_I M_{II}$	$L\beta_4$	73 Ta	3	1.34581
9.1873	$L_{II}O_I$	$L\beta_7$	71 Lu	5	1.34949
9.1845	$L_{III}M_{II}$	Ll	82 Pb	7	1.34990
9.1802	$L_{III}N_{III}$		72 Hf	9	1.35053
9.1751	$L_{II}M_V$	$L\alpha_1$	77 Ir	3	1.35128
9.1749	$L_{III}M_{III}$	Ls	79 Au	7	1.35131
9.1634	$L_{II}M_{IV}$	$L\beta_3$	72 Hf	5	1.35300
9.144	$L_I N_I$	$L\gamma_5$	69 Tm	2	1.3558
9.1239	$L_{III}N_{II}$		72 Hf	9	1.35887
9.0995	$L_{III}M_{IV}$	$L\alpha_2$	77 Ir	5	1.36250
9.089	$L_{II}N_{IV}$	$L\gamma_1$	68 Er	2	1.3641
9.087	$L_{IV}N_{IV}$	$L\gamma_3$	67 Ho	2	1.3643
9.0548	Abs. edge	Lt	66 Dy	1	1.3692
9.051	$L_I N_{II}$	$L\gamma_2$	67 Ho	2	1.3698
9.0489	$L_{III}N_V$	$L\beta_2$	71 Lu	3	1.37012
9.0395	$L_{III}N_{III}$	$L\delta_{15}$	71 Lu	1	1.3715
9.0272	$L_{II}M_I$	$L\eta$	75 Re	5	1.37342
9.0227	$L_{II}M_{IV}$	$L\beta_1$	72 Hf	5	1.37410
9.0227	$L_{III}M_I$	$L\beta_6$	72 Hf	5	1.37410
9.0195	$L_I O_{II,III}$	$L\gamma_4$	66 Dy	7	1.37459
9.019	$L_{III}M_{II}$	Lt	80 Hg	2	1.3746
8.9803	Abs. edge	K	29 Cu	5	1.38059

	Level	Line	Element		
9.975	$L_{II}M_I$	$L\eta$	78 Pt	2	1.2429
9.9675	$L_{III}M_{III}$	Ls	82 Pb	7	1.24385
9.9615	$L_{III}N_{IV}$	$L\beta_2$	74 W	3	1.24460
9.9561	$L_{II}O_{II,III}$		70 Yb	1	1.2453
9.9478	$L_{III}N_{IV}$	$L\beta_{15}$	74 W	3	1.24631
9.946	$L_I M_V$	$L\beta_9$	73 Ta	2	1.2466
9.934	$L_{II}M_{III}$	$L\beta_{17}$	76 Os	2	1.2480
9.9246	$L_{II}O_I$	$L\gamma_8$	70 Yb	5	1.24923
9.917	$L_{II}M_{II}$		77 Ir	3	1.2502
9.9105	$L_{III}N_I$	$L\beta_6$	75 Re	5	1.25100
9.8976	$L_{III}M_{IV}$	$L\alpha_2$	80 Hg	7	1.25264
9.889	$L_I M_{IV}$	$L\beta_{10}$	73 Ta	2	1.2537
9.88642	KL_{III}	$K\alpha_1$	32 Ge	9	1.254054
9.8766	Abs. edge	L_{III}	73 Ta	1	1.2553
9.8750	$L_{III}O_{IV,V}$	$L\beta_5$	73 Ta	3	1.2555
9.8572	$L_{III}N_{VI,VII}$	Lu	73 Ta	4	1.25778
9.85532	KL_{II}	$K\alpha_2$	32 Ge	9	1.258011
9.8463	$L_I M_{II}$	$L\beta_4$	75 Re	5	1.25917
9.8428	$L_{II}N_I$	$L\gamma_5$	71 Lu	1	1.2596
9.839	$L_{III}O_{II,III}$		73 Ta	3	1.2601
9.8188	$L_I M_{III}$	$L\beta_3$	74 W	5	1.26269
9.8098	$L_{III}O_I$	$L\beta_7$	73 Ta	5	1.26385
9.784	$L_{III}N_{III}$		74 W	2	1.2672
9.7801	$L_{II}N_{IV}$	$L\gamma_1$	70 Yb	5	1.26769
9.779	$L_{II}N_{II}$	$L\gamma_3$	69 Tm	2	1.2678
9.7574	Abs. edge	L_I	68 Er	1	1.2706
9.741	$L_{II}M_V$	$L\gamma_2$	74 W	2	1.2728
9.730	$L_{II}N_I$	Lt	69 Tm	2	1.2742
9.7252	$L_{III}M_{II}$	$L\gamma_4$	83 Bi	1	1.2748
9.722	$L_I O_{II,III}$	$L\alpha_1$	68 Er	2	1.2752
9.7133	$L_{III}M_V$		79 Au	3	1.27640
9.712	$L_{III}N_{II}$		74 W	2	1.2765
9.7007	$L_{III}M_{III}$	Ls	81 Tl	5	1.27807
9.67235	$L_{II}M_{IV}$	$L\beta_1$	74 W	9	1.281809
9.664	$L_{III}M_I$	Ll	84 Po	5	1.2829
9.6607	Abs. edge	K	30 Zn	1	1.2834
9.6580	$KN_{II,III}$	$K\beta_2$	30 Zn	2	1.28372
9.6522	$L_{II}M_I$	$L\eta$	77 Ir	3	1.28448
9.6518	$L_{III}N_V$	$L\beta_2$	73 Ta	2	1.28454

TABLE 7f-2. WAVELENGTHS OF X-RAY EMISSION LINES AND ABSORPTION EDGES: IN NUMERICAL ORDER *(Continued)*

Wavelength, Å*	p.e.†	Element		Designation	keV
1.38109	3	29 Cu	$K\beta_2$	$KM_{IV,V}$	8.9770
1.3838	1	70 Yb	$L\beta_9$	L_IM_V	8.9597
1.38477	3	81 Tl	Ll	$L_{III}M_I$	8.9532
1.3862	1	70 Yb	L_{III}	Abs. edge	8.9441
1.3864	1	73 Ta	$L\beta_{17}$	$L_{II}M_{III}$	8.9428
1.38696	7	70 Yb	$L\beta_5$	$L_{III}O_{IV,V}$	8.9390
1.3895	2	78 Pt	Ls	$L_{III}O_{II,III}$	8.923
1.3898	1	70 Yb	L_{II}	Abs. edge	8.9209
1.3905	1	67 Ho	$L\alpha_1$	$L_{III}M_V$	8.9164
1.39121	5	76 Os	$L\beta_{10}$	L_IM_V	8.9117
1.3915	1	70 Yb	$L\beta_4$	L_IM_{II}	8.9100
1.39220	5	72 Hf	$K\beta_{1,3}$	$KM_{II,III}$	8.9054
1.392218	9	29 Cu	$L\gamma_6$	$L_{II}O_{IV}$	8.90529
1.3923	2	67 Ho	$K\beta_3$	KM_{III}	8.905
1.3926	1	29 Cu	$K\beta_3$	KM_I	8.9029
1.3948	1	70 Yb	$L\gamma_{\varepsilon}$	$L_{III}O_I$	8.8889
1.3983	2	67 Ho	$L\beta_3$	$L_{III}O_I$	8.867
1.40140	5	71 Lu	$L\alpha_2$	$L_{III}M_{III}$	8.8469
1.40234	5	76 Os	$L\gamma_5$	$L_{III}M_{IV}$	8.8410
1.4067	3	68 Er	Lt	$L_{II}N_I$	8.814
1.41366	7	79 Au	$L\beta_{2,15}$	$L_{III}M_{II}$	8.7702
1.41550	5	70 Yb	$L\gamma_3$	$L_{III}N_{IV,V}$	8.7588
1.41640	7	66 Dy	$L\gamma_1$	$L_{II}N_{IV}$	8.7532
1.4174	2	67 Ho	$L\beta_6$	L_IN_{IV}	8.747
1.4189	2	71 Lu	$L\eta$	L_IN_V	8.7376
1.42110	3	74 W	Ll	$L_{II}M_I$	8.7243
1.4216	1	80 Hg	L_I	$L_{III}M_I$	8.7210
1.4223	1	65 Tb	Ll	Abs. edge	8.7167
1.42278	7	66 Dy	$L\gamma_2$	$L_IO_{IV,V}$	8.7140
1.4228	3	65 Tb	$L\beta_1$	L_IM_{IV}	8.714
1.42359	3	71 Lu	$L\beta_1$	L_IM_{III}	8.7090
1.4276	2	65 Tb	$L\gamma_4$	$L_IO_{II,III}$	8.685
1.43025	9	72 Hf	L_IM_I	L_IM_I	8.6685
1.4941	3	68 Er	$L\beta_7$	$L_{III}O_I$	8.298
1.4941	3	68 Er	$L\beta_{10}$	L_IM_{IV}	8.298
1.4995	2	78 Pt	Ll	$L_{III}M_I$	8.268
1.500135	8	28 Ni	$K\beta_{1,3}$	$KM_{II,III}$	8.26466
1.5023	1	65 Tb	L_{II}	Abs. edge	8.2527
1.5035	2	65 Tb	$L\gamma_6$	$L_{II}O_{IV}$	8.246
1.5063	2	69 Tm	$L\beta_3$	L_IM_{III}	8.231
1.5097	2	65 Tb	$L\gamma_8$	$L_{III}O_I$	8.212
1.51399	9	68 Er	$L\beta_{2,15}$	$L_{III}N_{IV,V}$	8.1890
1.5162	2	69 Tm	$L\beta_6$	$L_{III}N_I$	8.177
1.5178	1	75 Re	Ls	$L_{III}M_{III}$	8.1682
1.51824	7	66 Dy	$L\gamma_5$	$L_{III}N_I$	8.1661
1.52197	2	73 Ta	$L\alpha_1$	$L_{III}M_V$	8.1461
1.52325	5	72 Hf	$L\eta$	$L_{III}M_I$	8.1393
1.5297	2	64 Gd	$L\gamma_3$	L_IN_{III}	8.105
1.5303	2	65 Tb	$L\gamma_1$	$L_{II}N_{IV}$	8.102
1.5304	2	69 Tm	$L\beta_1$	$L_{II}M_{IV}$	8.101
1.53293	2	73 Ta	$L\alpha_2$	$L_{III}M_{IV}$	8.0879
1.5331	2	64 Gd	$L\gamma_2$	L_IN_{II}	8.087
1.53333	9	71 Lu	Lt	$L_{III}M_{II}$	8.0858
1.5347	2	76 Os	L_{III}	Abs. edge	8.079
1.5368	1	67 Ho	$L\beta_5$	$L_{III}O_{IV,V}$	8.0676
1.5378	2	67 Ho	L_I	Abs. edge	8.062
1.5381	1	63 Eu	Lt	$L_{III}M_{II}$	8.0607
1.540562	2	29 Cu	$K\alpha_1$	KL_{III}	8.04778
1.54094	3	77 Ir	Ll	$L_{III}M_I$	8.0458
1.5439	1	63 Eu	$L\gamma_4$	$L_IO_{II,III}$	8.0304
1.544390	9	29 Cu	$K\alpha_2$	KL_{II}	8.02783
1.5448	2	69 Tm	$L\beta_4$	L_IM_{II}	8.026
1.5486	3	67 Ho	$L\beta_{10}$	L_IM_{IV}	8.006
1.5616	1	68 Er	$L\beta_3$	L_IM_{III}	7.9392
1.5632	1	64 Gd	L_{II}	Abs. edge	7.9310
1.5642	3	74 W	Ls	$L_{III}M_{III}$	7.926

(This is a rotated, dense X-ray wavelength reference table. Columns, in image order, are: Energy (keV) · Level transition · Line (Siegbahn) · Element (Z, symbol) · Intensity · Wavelength λ (Å). The page holds two such tables.)

Table (upper block)

E (keV)	Level	Line	Element	Int.	λ (Å)
7.925	$L_{II}O_{IV}$	$L\gamma_6$	64 Gd	2	1.5644
7.911	$L_{III}N_{IV,V}$	$L\beta_{2,15}$	67 Ho	2	1.5671
7.909	$L_{III}N_I$	$L\beta_6$	68 Er	2	1.5675
7.8990	$L_{III}M_V$	$L\alpha_1$	72 Hf	5	1.56958
7.894	$L_{II}O_I$	$L\gamma_8$	64 Gd	2	1.5707
7.8575	$L_{II}M_I$	$L\eta$	71 Lu	1	1.5779
7.8535	$L_{II}N_I$	$L\gamma_5$	65 Tb	2	1.5787
7.8525	$L_{III}M_{III}$	Lt	75 Re	5	1.5789
7.8446	$L_{III}M_{IV}$	$L\alpha_2$	72 Hf	7	1.58046
7.8222	$L_{III}M_I$	Ll	76 Os	7	1.58498
7.8109	$L_{II}M_{IV}$	$L\beta_1$	68 Er	9	1.5873
7.8055	$L_{III}O_{IV,V}$	$L\beta_5$	66 Dy	2	1.58837
7.8052	$L_{II}M_{II}$	$L\gamma_3$	70 Yb	1	1.58844
7.7961	$L_I N_{III}$	L_{III}	63 Eu	2	1.5903
7.7897	Abs. edge	$L\gamma_1$	66 Dy		1.5916
7.7858	$L_{II}N_{IV}$	$L\gamma_2$	64 Gd	2	1.5924
7.7677	$L_{II}N_{II}$	$L\beta_9$	63 Eu	9	1.5961
7.7501	$L_I M_V$	L_l	66 Dy	1	1.59973
7.7478	Abs. edge	$L\beta_4$	62 Sm	7	1.6002
7.7453	$L_I M_{II}$	$L\beta_7$	68 Er	3	1.6007
7.7272	$L_{III}O_I$	$L\gamma_4$	66 Dy	9	1.60447
7.714	$L_I O_{II,III}$	$L\beta_{10}$	62 Sm	3	1.60728
7.7130	$L_I M_{IV}$	K	66 Dy	9	1.60743
7.70954	Abs. edge	$K\beta_5$	27 Co	3	1.60815
7.7059	$KM_{IV,V}$	L_s	27 Co	2	1.60891
7.6881	$L_{III}M_{III}$	$L\alpha_1$	73 Ta	2	1.61264
7.6555	$L_{II}M_V$	$L\beta_3$	71 Lu	7	1.61951
7.6519	$L_{III}M_{III}$	$K\beta_{1,3}$	67 Ho	2	1.6203
7.64943	$KM_{II,III}$	$L\beta_6$	27 Co	1	1.62079
7.6359	$L_{III}N_I$	$L\beta_{2,15}$	67 Ho	3	1.6237
7.6357	$L_{III}N_{IV,V}$	L_t	66 Dy	2	1.62369
7.6324	$L_{III}M_{II}$	L_{II}	74 W	5	1.6244
7.6199	Abs. edge	$L\gamma_6$	63 Eu		1.6271
7.6147	$L_{III}O_I$	$L\alpha_2$	63 Eu	5	1.6282
7.6049	$L_{III}M_{IV}$	Ll	71 Lu	5	1.63029
7.6036	$L_{III}M_I$	$L\gamma_8$	75 Re	5	1.63056
7.5849	$L_{II}O_I$	$L\eta$	63 Eu	5	1.6346
7.5802	$L_{II}M_I$	$L\gamma_5$	70 Yb	2	1.63560
7.5543	$L_{II}N_I$		64 Gd	2	1.6412

Table (lower block)

E (keV)	Level	Line	Element	Int.	λ (Å)
8.6671	$L_{II}M_{II}$		73 Ta	9	1.43048
8.659	$L_{III}M_{III}$	L_s	77 Ir	2	1.4318
8.6525	$L_{III}M_V$	$L\alpha$	75 Re	4	1.43290
8.6496	Abs. edge	L_{III}	69 Tm	1	1.4334
8.641	$L_I M_V$	$L\beta_9$	69 Tm	3	1.4336
8.63886	$L_{III}O_{IV,V}$	$L\beta_5$	69 Tm	2	1.4349
8.6312	KL_{III}	$K\alpha_1$	30 Zn	7	1.435155
8.61578	$L_{II}M_{III}$	$L\beta_{17}$	72 Hf	9	1.43643
8.6064	KL_{II}	$K\alpha_2$	30 Zn	8	1.439000
8.604	$L_I M_{II}$	$L\beta_4$	71 Lu	5	1.44056
8.5862	$L_I M_{IV}$	$L\beta_{10}$	69 Tm	3	1.4410
8.5830	$L_{III}M_{IV}$	$L\alpha_2$	75 Re	5	1.44396
8.5753	Abs. edge	L_{II}	66 Dy	1	1.4445
8.5367	$L_{III}O_{IV}$	$L\gamma_6$	66 Dy	7	1.44579
8.533	$L_I M_{III}$	$L\beta_3$	70 Yb	5	1.45233
8.4939	$L_{III}M_{II}$	Lt	78 Pt	2	1.4530
8.481	$L_{III}M_I$	Ll	79 Au	9	1.45964
8.468	$L_I N_I$	$L\gamma_5$	67 Ho	2	1.4618
8.4563	$L_{III}N_{IV,V}$	$L\beta_{2,15}$	69 Tm	2	1.4640
8.4280	$L_{III}N_I$	$L\beta_6$	70 Yb	1	1.4661
8.423	$L_{II}M_I$	$L\eta$	73 Ta	5	1.47106
8.4188	$L_I N_{III}$	$L\gamma_3$	65 Tb	2	1.47266
8.414	$L_I N_{IV}$	$L\gamma_1$	66 Dy	7	1.4735
8.4018	$L_{III}M_{III}$	L_s	76 Os	2	1.47565
8.398	$L_I M_V$	$L\beta_1$	70 Yb	5	1.4764
8.3976	$L_I N_{II}$	$L\gamma_2$	65 Tb	2	1.47639
8.3864	$L_{III}N_V$	$L\alpha_1$	74 W	1	1.4784
8.3735	$L_{II}O_I$	L_t	64 Gd	9	1.48064
8.373	Abs. edge	L_t	72 Hf	3	1.4807
8.3575	Abs. edge		64 Gd	1	1.4835
8.355	$L_I O_{II,III}$	L_{III}	68 Er	2	1.4839
8.350	$L_{II}O_{IV,V}$	$L\gamma_4$	64 Gd	3	1.4848
8.346	$L_I M_V$	$L\beta_5$	68 Er	5	1.4855
8.3352	$L_{III}M_{IV}$	$L\beta_9$	74 W	2	1.48743
8.33165	Abs. edge	$L\alpha_2$	28 Ni	1	1.48807
8.3286	$KM_{IV,V}$	K	28 Ni	4	1.48862
8.3132	$L_I M_{II}$	$K\beta_5$	70 Yb	3	1.49138
8.304	$L_{III}M_{II}$	$L\beta_4$	77 Ir	3	1.4930
		Lt			

TABLE 7f-2. WAVELENGTHS OF X-RAY EMISSION LINES AND ABSORPTION EDGES: IN NUMERICAL ORDER (Continued)

Wavelength, Å*	p.e.†	Element	Designation	Designation	keV
1.6475	2	67 Ho	$L\beta_1$	$L_{II}M_{IV}$	7.5253
1.6497	1	65 Tb	L_{III}	Abs. edge	7.5153
1.6510	2	65 Tb	$L\beta_5$	$L_{III}O_{IV,III}$	7.5094
1.65601	3	62 Sm	$L\gamma_3$	L_IN_{III}	7.487
1.6574	2	63 Eu	$L\gamma_1$	$L_{II}N_{IV}$	7.4803
1.657910	8	28 Ni	$K\alpha_1$	KL_{III}	7.47815
1.6585	2	65 Tb	$L\beta_7$	$L_{III}O_I$	7.4753
1.6595	2	67 Ho	$L\beta_4$	L_IM_{II}	7.4708
1.66044	6	62 Sm	$L\gamma_2$	L_IN_{II}	7.467
1.661747	8	28 Ni	$K\alpha_2$	KL_{II}	7.46089
1.66346	9	72 Hf	Ls	$L_{III}M_{III}$	7.4532
1.6673	3	65 Tb	$L\beta_{10}$	L_IM_{IV}	7.436
1.6674	5	61 Pm	Lt	$L_{III}M_{II}$	7.436
1.67189	4	70 Yb	$L\alpha_1$	$L_{III}M_V$	7.4156
1.67265	9	73 Ta	Lt	$L_{III}M_{II}$	7.4123
1.6782	1	74 W	Ll	$L_{III}M_I$	7.3878
1.68213	7	66 Dy	$L\beta_6$	$L_{III}N_I$	7.3705
1.6822	2	66 Dy	$L\beta_3$	L_IM_{III}	7.3702
1.68285	5	70 Yb	$L\alpha_2$	$L_{III}M_{IV}$	7.3673
1.6830	2	65 Tb	$L\beta_{2,15}$	$L_{III}N_{IV,V}$	7.3667
1.6953	1	62 Sm	L_{II}	Abs. edge	7.3132
1.6963	2	69 Tm	$L\eta$	$L_{II}M_I$	7.3088
1.6966	9	62 Sm	$L\gamma_6$	$L_{II}O_{IV}$	7.308
1.7085	7	63 Eu	$L\gamma_5$	L_IN_I	7.2566
1.71062	2	66 Dy	$L\beta_1$	$L_{II}M_{IV}$	7.2477
1.7117	1	64 Gd	L_{III}	Abs. edge	7.2430
1.7130	3	64 Gd	$L\beta_5$	$L_{III}O_{IV,V}$	7.2374
1.7203	2	64 Gd	$L\beta_7$	$L_{III}O_I$	7.2071
1.72103	2	64 Gd	$L\beta_4$	L_IM_{II}	7.2039
1.72305	7	66 Dy	Lt	$L_{III}M_{II}$	7.1954
1.7240	9	72 Hf	$L\beta_9$	L_IM_V	7.192
1.72724	3	64 Gd	$L\gamma_1$	$L_{II}N_{IV}$	7.178
1.7268	2	62 Sm	$L\alpha_1$	$L_{III}M_V$	7.1799

Wavelength, Å*	p.e.†	Element	Designation	Designation	keV
1.8450	2	67 Ho	$L\alpha_1$	$L_{III}M_V$	6.7198
1.8457	1	62 Sm	L_{III}	Abs. edge	6.7172
1.8468	2	64 Gd	$L\beta_1$	$L_{II}M_{IV}$	6.7132
1.84700	9	62 Sm	$L\beta_5$	$L_{III}O_{IV,V}$	6.7126
1.8540	2	64 Gd	$L\beta_4$	L_IM_{II}	6.6871
1.8552	5	60 Nd	$L\gamma_8$	$L_{II}O_I$	6.683
1.8561	2	67 Ho	L_{II}	Abs. edge	6.6795
1.85626	3	62 Sm	$L\alpha_2$	$L_{III}M_{IV}$	6.679
1.86166	3	62 Sm	$L\beta_7$	$L_{III}O_I$	6.660
1.86990	3	62 Sm	$L\beta_9$	L_IM_V	6.634
1.8737	2	63 Eu	$L\beta_6$	$L_{III}N_I$	6.6170
1.8740	4	59 Pr	$L\gamma_3$	L_IN_{III}	6.616
1.8779	4	60 Nd	$L\gamma_2$	L_IN_{II}	6.6021
1.8791	3	59 Pr	$L\gamma_1$	$L_{II}N_{IV}$	6.598
1.8821	2	62 Sm	$L\beta_{2,15}$	$L_{III}N_{IV,V}$	6.586
1.8867	5	63 Eu	$L\beta_3$	L_IM_{III}	6.5713
1.8934	5	58 Ce	Ll	$L_{III}M_I$	6.548
1.89415	5	70 Yb	Ll	Abs. edge	6.5455
1.89643	1	25 Mn	K	Abs. edge	6.5376
1.8971	7	25 Mn	$K\beta_5$	$KM_{IV,V}$	6.5352
1.89743	4	66 Dy	$L\eta$	$L_{II}M_I$	6.5342
1.8991	3	58 Ce	$L\gamma_4$	$L_IO_{II,III}$	6.528
1.90881	9	66 Dy	$L\alpha_1$	$L_{III}M_V$	6.4952
1.91021	2	25 Mn	$K\beta_{1,3}$	$KM_{II,III}$	6.49045
1.9191	1	61 Pm	L_{III}	Abs. edge	6.4605
1.91991	3	66 Dy	$L\alpha_2$	$L_{III}M_{IV}$	6.4577
1.9203	2	63 Eu	$L\beta_1$	$L_{II}M_{IV}$	6.4564
1.9255	2	63 Eu	$L\beta_4$	L_IM_{II}	6.4389
1.9255	5	59 Pr	L_I	Abs. edge	6.439
1.9355	4	60 Nd	$L\gamma_5$	L_IN_I	6.406
1.936042	9	26 Fe	$K\alpha_1$	KL_{III}	6.40384
1.9362	3	59 Pr	$L\gamma_8$	$L_{II}O_I$	6.403
1.939980	9	26 Fe	$K\alpha_2$	KL_{II}	6.39084

Wavelength (Å)	Element	Line	Level	Ref	Energy (keV)
1.72841	73 Ta	Ll	$L_{III}M_I$	5	7.1731
1.7315	64 Gd	$L\beta_{10}$	$L_I M_{IV}$	3	7.160
1.7381	69 Tm	$L\alpha_2$	$L_{III}M_{IV}$	2	7.1331
1.7390	60 Nd	L_I	Abs. edge	1	7.1294
1.7422	65 Tb	$L\beta_6$	$L_{III}N_I$	2	7.1163
1.74346	26 Fe	K	Abs. edge	1	7.11120
1.7442	26 Fe	$K\beta_5$	$K M_{IV,V}$	1	7.1081
1.7445	60 Nd	$L\beta_{2,15}$	$L_I O_{II,III}$	4	7.107
1.7455	64 Gd	$L\beta_3$	$L_{III}N_{IV,V}$	2	7.1028
1.7472	65 Tb	$K\beta_{1,3}$	$L_I M_{III}$	2	7.0959
1.75661	26 Fe	$L\eta$	$K M_{II,III}$	2	7.05798
1.7566	61 Pm	L_{II}	$L_I M_I$	1	7.0579
1.7676	71 Lu	L_{III}	Abs. edge	5	7.014
1.7760	63 Eu	$L\beta_1$	$L_{III}M_{II}$	1	6.9810
1.7761	65 Tb	$L\beta_5$	Abs. edge	1	6.9806
1.7768	63 Eu	$L\gamma_5$	$L_{II}M_{IV}$	3	6.978
1.7772	62 Sm	Ll	$L_{II}O_{IV,V}$	2	6.9763
1.77934	72 Hf	$L\alpha_1$	$L_{II}N_{IV}$	3	6.968
1.78145	68 Er	$L\beta_7$	$L_{III}M_I$	5	6.9596
1.78425	63 Eu	$L\beta_3$	$L_{III}M_V$	9	6.9487
1.7851	65 Tb	$K\alpha_1$	$L_{III}O_I$	2	6.9453
1.7864	27 Co	$L\beta_9$	$L_I M_{IV}$	2	6.9403
1.788965	63 Eu	$K\alpha_2$	$K L_{III}$	2	6.93032
1.7916	27 Co	$L\alpha_2$	$L_I M_V$	9	6.920
1.792850	68 Er	$L\gamma_3$	$K L_{II}$	3	6.91530
1.7955	60 Nd	$L\gamma_1$	$L_{III}M_{IV}$	9	6.9050
1.7964	61 Pm	$L\beta_{10}$	$L_{II}N_{III}$	3	6.902
1.7989	63 Eu	$L\gamma_2$	$L_{II}N_{IV}$	2	6.892
1.7993	60 Nd	$L\beta_6$	$L_{II}M_{IV}$	4	6.890
1.8013	64 Gd	$L\beta_{2,15}$	$L_I N_{II}$	9	6.883
1.8054	63 Eu	L_I	$L_{III}N_I$	3	6.8671
1.8118	59 Pr	$L\beta_3$	$L_{II}N_{IV,V}$	4	6.8432
1.8141	64 Gd	$L\gamma_4$	Abs. edge	2	6.834
1.8150	59 Pr	$L\eta$	$L_I M_{III}$	2	6.8311
1.8193	67 Ho	Lt	$L_I O_{II,III}$	5	6.815
1.8264	70 Yb	Ll	$L_{II}M_I$	2	6.7883
1.83091	71 Lu	L_{II}	$L_{III}M_{II}$	2	6.7715
1.8360	60 Nd	L_I	$L_{III}M_I$	4	6.7528
1.8440	60 Nd	L_{II}	Abs. edge	2	6.7234

Wavelength (Å)	Element	Line	Level	Ref	Energy (keV)
1.94643	62 Sm	$L\beta_6$	$L_{III}N_I$	3	6.3693
1.9550	69 Tm	Ll	$L_{II}M_I$	2	6.3419
1.9553	58 Ce	$L\gamma_3$	$L_I N_{III}$	3	6.3409
1.9559	61 Pm	$L\beta_{2,15}$	$L_{III}N_{IV,V}$	6	6.339
1.9602	58 Ce	$L\gamma_2$	$L_I N_I$	3	6.3250
1.9611	59 Pr	$L\gamma_1$	$L_{II}N_{IV}$	3	6.3221
1.96241	62 Sm	$L\beta_3$	$L_I M_{III}$	3	6.318
1.9730	65 Tb	$L\eta$	$L_{II}M_{III}$	2	6.2839
1.9765	65 Tb	$L\alpha_1$	$L_{III}M_V$	2	6.2728
1.9780	57 La	L_I	Abs. edge	5	6.268
1.9830	57 La	$L\gamma_4$	$L_I O_{II,III}$	4	6.252
1.9875	65 Tb	$L\alpha_2$	$L_{III}M_{IV}$	2	6.2380
1.9967	60 Nd	L_{III}	Abs. edge	1	6.2092
1.99806	62 Sm	$L\beta_1$	$L_{III}M_{IV}$	3	6.2051
2.00095	62 Sm	$L\beta_4$	$L_I M_{II}$	6	6.196
2.0092	60 Nd	$L\beta_7$	$L_{II}O_I$	3	6.1708
2.0124	58 Ce	L_{II}	Abs. edge	5	6.161
2.015	68 Er	Ll	$L_{III}M_I$	1	6.152
2.0165	60 Nd	$L\beta_9$	$L_I M_V$	3	6.1484
2.0205	59 Pr	$L\gamma_5$	$L_{II}N_I$	4	6.136
2.0237	58 Ce	$L\beta_{10}$	$L_{II}O_I$	4	6.126
2.0237	60 Nd	$L\beta_{2,15}$	$L_I M_{IV}$	3	6.1265
2.0360	60 Nd	$L\gamma_3$	$L_{III}N_{IV,V}$	3	6.0894
2.0410	57 La	$L\beta_3$	$L_I N_{III}$	4	6.074
2.0421	61 Pm	$L\gamma_2$	$L_I M_{III}$	4	6.071
2.0460	57 La	$L\alpha_1$	$L_{II}N_{II}$	2	6.060
2.0468	64 Gd	$L\gamma_1$	$L_{III}M_{IV}$	4	6.0572
2.0487	58 Ce	$L\eta$	$L_{II}N_{IV}$	2	6.052
2.0494	64 Gd	$L\alpha_2$	$L_{III}M_I$	4	6.0495
2.0578	64 Gd	L_I	$L_{II}N_I$	1	6.0250
2.0678	56 Ba	K	Abs. edge	2	5.996
2.07020	24 Cr	$K\beta_5$	Abs. edge	5	5.9888
2.07087	24 Cr	$L\gamma_4$	$K M_{IV,V}$	5	5.9869
2.0756	56 Ba	L_{III}	$L_I O_{II,III}$	6	5.9733
2.0791	59 Pr	$L\beta_1$	Abs. edge	6	5.963
2.0797	61 Pm	$K\beta_1$	$L_{III}M_{IV}$	3	5.961
2.08487	24 Cr	$K\beta_{1,3}$	$K M_{II,III}$	5	5.94671
2.0860	67 Ho	Ll	$L_{II}M_I$	4	5.9434
2.0919	59 Pr	$L\beta_7$	$L_{III}O_I$	2	5.927

TABLE 7f-2. WAVELENGTHS OF X-RAY EMISSION LINES AND ABSORPTION EDGES: IN NUMERICAL ORDER (Continued)

Wavelength, Å*	p.e.†	Element	Designation	keV
2.1004	4	59 Pr	Lβ9 — LIIMV	5.903
2.101820	9	25 Mn	Kα1 — KLIII	5.89875
2.1039	3	60 Nd	Lβ6 — LIIINI	5.8930
2.1053	5	57 La	LII — Abs. edge	5.889
2.10578	2	25 Mn	Kα2 — KLII	5.88765
2.1071	2	59 Pr	Lβ10 — LIMIV	5.884
2.1103	3	58 Ce	Lγ5 — LIINI	5.8751
2.1194	4	59 Pr	Lβ2,15 — LIIINIV,V	5.850
2.1209	2	63 Eu	Lα1 — LIIIMV	5.8457
2.1268	2	60 Nd	Lβ3 — LIMIII	5.8294
2.1315	2	63 Eu	Lη — LIIMI	5.8166
2.1315	2	63 Eu	Lα2 — LIIIMIV	5.8166
2.1342	2	56 Ba	Lγ3 — LINIII	5.8092
2.1387	2	56 Ba	Lγ2 — LINII	5.7969
2.1418	3	57 La	Lγ1 — LIINIV	5.7885
2.15877	7	66 Dy	Ll — LIIIMI	5.7431
2.166	1	58 Ce	LIII — Abs. edge	5.723
2.1669	3	60 Nd	Lβ4 — LIMII	5.7216
2.1669	2	60 Nd	Lβ1 — LIIMIV	5.7216
2.1673	5	55 Cs	LI — Abs. edge	5.721
2.1701	2	58 Ce	Lβ7 — LIIIOI	5.7132
2.1741	2	55 Cs	Lγ4 — LIOII,III	5.7026
2.1885	2	58 Ce	Lβ9 — LIMV	5.6650
2.1906	4	59 Pr	Lβ6 — LIIINI	5.660
2.1958	5	58 Ce	Lβ10 — LIMIV	5.646
2.1998	2	62 Sm	Lα1 — LIIIMV	5.6361
2.2048	1	56 Ba	LII — Abs. edge	5.6233
2.2056	4	57 La	Lγ5 — LIINI	5.621
2.2087	2	58 Ce	Lβ2,15 — LIIINIV,V	5.6134
2.21062	3	62 Sm	Lα2 — LIIIMIV	5.6090
2.2172	3	59 Pr	Lβ3 — LIMIII	5.5918
2.21824	3	62 Sm	Lη — LIIMI	5.589
2.2328	2	55 Cs	Lγ3 — LINIII	5.5527
2.4094	4	60 Nd	Lη — LIIMI	5.1457
2.4105	3	57 La	Lβ3 — LIMIII	5.1434
2.4174	2	55 Cs	Lγ5 — LIINI	5.1287
2.4292	1	54 Xe	LII — Abs. edge	5.1037
2.442	9	90 Th	MIOIII	5.08
2.443	4	92 U	MIIOIV	5.075
2.4475	2	53 I	Lγ2,3 — LINII,III	5.0657
2.4493	3	57 La	Lβ4 — LIMII	5.0620
2.45891	5	57 La	Lβ1 — LIIMIV	5.0421
2.4630	2	59 Pr	Lα1 — LIIIMV	5.0337
2.4729	3	59 Pr	Lα2 — LIIIMIV	5.0135
2.4740	1	55 Cs	LIII — Abs. edge	5.0113
2.4783	2	55 Cs	Lβ9 — LIMV	5.0026
2.4823	4	62 Sm	Ll — LIIIMI	4.9945
2.4826	2	56 Ba	Lβ6 — LIIINI	4.9939
2.4849	2	55 Cs	Lβ7 — LIIIOI	4.9893
2.4920	2	55 Cs	Lβ10 — LIMIV	4.9752
2.49734	5	22 Ti	K — Abs. edge	4.96452
2.4985	2	22 Ti	Kβ5 — KMIV,V	4.9623
2.50356	2	23 V	Kα1 — KLIII	4.95220
2.50738	2	23 V	Kα2 — KLII	4.94464
2.5099	1	52 Te	LI — Abs. edge	4.9397
2.5113	2	52 Te	Lγ4 — LIOII,III	4.9369
2.5118	2	55 Cs	Lβ2,15 — LIIINIV,V	4.9359
2.512	3	59 Pr	Lη — LIIMI	4.935
2.51391	2	22 Ti	Kβ1,3 — KMII,III	4.93181
2.5164	2	56 Ba	Lβ3 — LIMIII	4.9269
2.527	4	91 Pa	MIIOIV	4.906
2.5542	5	53 I	LII — Abs. edge	4.8540
2.5553	2	56 Ba	Lβ4 — LIMII	4.8519
2.5615	3	58 Ce	Lα1 — LIIIMV	4.8402
2.5674	2	52 Te	Lγ2,3 — LINII,III	4.8290
2.56821	5	56 Ba	Lβ1 — LIIMIV	4.82753

Upper block

λ (Å)	Level	Line	Element	n	λ (Å)
4.8230	$L_{III}M_{IV}$	$L\alpha_2$	58 Ce	3	2.5706
4.8009	$L_{II}N_{IV}$	$L\gamma_1$	53 I	8	2.58244
4.7822	Abs. edge	L_{III}	54 Xe	1	2.5926
4.7811	$L_{III}N_I$	$L\beta_6$	55 Cs	2	2.5932
4.735	$M_{II}O_{IV}$	$L\eta$	90 Th	5	2.618
4.7315	$L_{II}M_I$	$L\beta_3$	58 Ce	4	2.6203
4.7167	$L_I M_{III}$	L_I	55 Cs	2	2.6285
4.6984	Abs. edge	$L\gamma_4$	51 Sb	1	2.6388
4.6967	$L_I O_{II,III}$	$L\gamma_5$	51 Sb	2	2.6398
4.6660	$L_{II}N_I$	$L\alpha_1$	53 I	9	2.65710
4.65097	$L_{III}M_V$	$L\beta_4$	57 La	5	2.56570
4.6494	$L_I M_{II}$	$L\alpha_2$	55 Cs	2	2.6666
4.63423	$L_{III}M_{IV}$	Ll	57 La	5	2.67533
4.6330	$L_{II}M_I$	$L\beta_1$	60 Nd	4	2.6760
4.6198	Abs. edge	L_{II}	55 Cs	2	2.6837
4.6126	$L_I N_{II,III}$	$L\gamma_{2,3}$	52 Te	1	2.6879
4.5999	$L_{II}N_{IV}$	$L\gamma_1$	51 Sb	2	2.6953
4.5709	$L_I M_V$	$L\beta_9$	52 Te	6	2.71241
4.5690	Abs. edge	L_{III}	53 I	9	2.71352
4.5587	$L_I M_{IV}$	$L\beta_{10}$	53 I	5	2.7196
4.5564	$L_{III}O_I$	$L\beta_7$	53 I	9	2.72104
4.5435	$L_{II}M_I$	$K\alpha_1$	57 La	3	2.7288
4.525	KL_{III}	$L\beta_{2,15}$	22 Ti	3	2.740
4.51084	$L_{III}N_{IV,V}$	$K\alpha_2$	53 I	2	2.74851
4.5075	KL_{II}	K	22 Ti	8	2.75053
4.50486	$M_I N_{III}$	$K\beta_5$	92 U	2	2.75216
4.50	Abs. edge	$L\alpha_1$	21 Sc	8	2.753
4.489	$KM_{IV,V}$	L_I	21 Sc	1	2.762
4.4865	$L_{III}M_V$	$L\gamma_4$	56 Ba	3	2.7634
4.46626	Abs. edge	$K\beta_{1,3}$	50 Sn	5	2.77595
4.4648	$L_I O_{II,III}$	Ll	50 Sn	1	2.7769
4.4638	$KM_{II,III}$	$L\alpha_2$	21 Sc	2	2.7775
4.4605	$L_{III}M_I$	$L\gamma_5$	59 Pr	2	2.7796
4.4532	$L_{III}M_{IV}$	L_{II}	56 Ba	4	2.7841
4.45090	$L_{II}N_I$	$L\gamma_{2,3}$	52 Te	5	2.78553
4.4437	$L_{II}N_I$		92 U	9	2.79007
4.401	$M_{II}N_{IV}$		51 Sb	2	2.817
4.3819	Abs. edge		50 Sn	5	2.8294
4.3768	$L_I N_{II,III}$			2	2.8327

Lower block

λ (Å)	Level	Line	Element	n	λ (Å)
5.5467	$L_{III}M_I$	Ll	65 Tb	2	2.2352
5.5420	$L_I N_I$	$L\gamma_2$	55 Cs	2	2.2371
5.5311	$L_{II}N_{IV}$	$L\gamma_1$	56 Ba	2	2.2415
5.50	$M_I P_{III}$	$L\beta_4$	92 U	6	2.253
5.4981	$L_I M_{II}$	$L\beta_1$	59 Pr	4	2.2550
5.4889	$L_{II}M_{IV}$	L_{III}	59 Pr	3	2.2588
5.484	Abs. edge	K	57 La	1	2.261
5.4639	$KM_{IV,V}$	$K\beta_5$	23 V	1	2.2691
5.4629	Abs. edge	L_I	23 V	6	2.26951
5.4528	$L_{III}O_I$	$L\beta_9$	54 Xe	1	2.2737
5.450	$L_I M_V$	$L\beta_6$	57 La	3	2.275
5.434	$L_{III}N_I$	$L\alpha_1$	57 La	3	2.282
5.4334	$L_{III}M_V$	$K\beta_{1,3}$	58 Ce	3	2.2818
5.4325	$KM_{II,III}$	$K\alpha_1$	61 Pm	3	2.2822
5.42729	KL_{III}	$L\beta_{10}$	23 V	2	2.28440
5.41472	$L_I M_{IV}$	$L\alpha_2$	24 Cr	2	2.28970
5.415	$L_{III}M_{IV}$	$K\alpha_2$	57 La	3	2.290
5.4078	KL_{II}	$L\beta_{2,15}$	61 Pm	4	2.2926
5.405509	$L_{III}N_{IV,V}$	$L\gamma_5$	24 Cr	3	2.293606
5.3835	$M_I O_{III}$	$L\beta_3$	57 La	3	2.3030
5.38	$L_{II}N_I$	Ll	92 U	7	2.304
5.3707	$L_I M_{III}$	L_{II}	56 Ba	3	2.3085
5.3651	$L_{III}M_I$	$L\gamma_1$	58 Ce	3	2.3109
5.3621	Abs. edge	$L\beta_4$	64 Gd	2	2.3122
5.3581	$L_{II}N_{IV}$	$L\beta_1$	55 Cs	1	2.3139
5.2804	$L_I M_{II}$	L_{III}	55 Cs	2	2.3480
5.2765	$L_{III}M_{IV}$	$L\alpha_1$	58 Ce	4	2.3497
5.2622	Abs. edge	$L\beta_9$	58 Ce	3	2.3561
5.2470	$L_{III}M_V$	$L\beta_6$	56 Ba	1	2.3629
5.2304	$L_I M_V$	$L\beta_7$	60 Nd	2	2.3704
5.2171	$L_{III}N_I$	$L\alpha_2$	56 Ba	2	2.3764
5.2114	$L_{III}O_I$	$L\beta_{10}$	57 La	4	2.3790
5.2079	$L_{III}M_{IV}$	L_I	56 Ba	2	2.3806
5.2077	$L_I M_{IV}$	$L\gamma_4$	60 Nd	3	2.3807
5.1941	$L_{II}N_I$	Ll	56 Ba	2	2.3869
5.192	Abs. edge	$L\beta_{2,15}$	53 I	5	2.3880
5.1848	$L_I O_{II,III}$		53 I	2	2.3913
5.1772	$L_{III}M_I$		63 Eu	2	2.3948
5.1565	$L_{III}N_{IV,V}$		56 Ba	6	2.40435

TABLE 7f-2. WAVELENGTHS OF X-RAY EMISSION LINES AND ABSORPTION EDGES: IN NUMERICAL ORDER (Continued)

Wavelength, Å*	p.e.†	Element	Designation		keV
2.83672	9	53 I	$L\beta_6$	$L_{III}N_I$	4.3706
2.83897	9	52 Te	$L\beta_9$	$L_I M_V$	4.3671
2.84679	9	52 Te	$L\beta_{10}$	$L_I M_{IV}$	4.3551
2.85159	3	51 Sb	$L\gamma_1$	$L_{II}N_{IV}$	4.34779
2.8555	1	52 Te	L_{III}	Abs. edge	4.3418
2.8627	3	56 Ba	$L\eta$	$L_{II}M_I$	4.3309
2.8634	3	52 Te	$L\beta_7$	$L_{II}O_I$	4.3298
2.87429	9	53 I	$L\beta_3$	$L_I M_{III}$	4.3134
2.88217	8	52 Te	$L\beta_{2,15}$	$L_{III}N_{IV,V}$	4.3017
2.884	5	92 U	M_{III}	Abs. edge	4.299
2.8917	4	58 Ce	Ll	$L_{III}M_I$	4.2875
2.8924	2	55 Cs	$L\alpha_1$	$L_{III}M_V$	4.2865
2.9020	2	55 Cs	$L\alpha_2$	$L_{III}M_{IV}$	4.2722
2.910	2	91 Pa		$M_{II}N_{IV}$	4.260
2.91207	9	53 I	$L\beta_4$	$L_I M_{II}$	4.2575
2.92	2	92 U		$M_I N_{II}$	4.25
2.9260	1	49 In	L_I	Abs. edge	4.2373
2.9264	2	49 In	$L\gamma_4$	$L_I O_{II,III}$	4.2367
2.93187	2	51 Sb	$L\gamma_5$	$L_{II}N_I$	4.2287
2.934	8	90 Th		$M_I N_{III}$	4.23
2.93744	6	53 I	$L\beta_1$	$L_{II}M_{IV}$	4.22072
2.948	2	92 U		$M_{III}O_{IV,V}$	4.205
2.97088	9	52 Te	$L\beta_6$	$L_{III}N_I$	4.1732
2.97261	9	51 Sb	$L\beta_9$	$L_I M_V$	4.1708
2.97917	9	51 Sb	$L\beta_{10}$	$L_I M_{IV}$	4.1616
2.9800	2	49 In	$L\gamma_{2,3}$	$L_I N_{II,III}$	4.1605
2.9823	1	50 Sn	L_{II}	Abs. edge	4.1573
2.9932	2	55 Cs	$L\eta$	$L_{II}M_I$	4.1421
3.0003	1	51 Sb	L_{III}	Abs. edge	4.1323
3.00115	3	50 Sn	$L\gamma_1$	$L_{II}N_{IV}$	4.13112
3.0052	3	51 Sb	$L\beta_7$	$L_{III}O_I$	4.1255
3.006	3	57 La	Ll	$L_{III}M_I$	4.124
3.00893	9	52 Te	$L\beta_3$	$L_I M_{III}$	4.1204
3.27404	9	49 In	$L\beta_{10}$	$L_I M_{IV}$	3.7868
3.27979	9	53 I	$L\eta$	$L_{II}M_I$	3.7801
3.283	9	90 Th		$M_{III}O_I$	3.78
3.28920	6	52 Te	$L\alpha_1$	$L_{III}M_V$	3.76933
3.29846	9	52 Te	$L\alpha_2$	$L_{III}M_{IV}$	3.7588
3.30585	3	50 Sn	$L\beta_3$	$L_I M_{III}$	3.7500
3.30635	9	47 Ag	$L\gamma_3$	$L_I N_{III}$	3.7498
3.31216	9	47 Ag	$L\gamma_2$	$L_I N_{II}$	3.7432
3.3237	1	49 In	L_{III}	Abs. edge	3.7302
3.324	4	49 In	$L\beta_7$	$L_{III}O_I$	3.730
3.3257	1	48 Cd	L_{II}	Abs. edge	3.7280
3.329	4	92 U		$M_{II}N_I$	3.724
3.333	5	92 U	M_{IV}	Abs. edge	3.720
3.33564	6	48 Cd	$L\gamma_1$	$L_{II}N_{IV}$	3.71686
3.33838	3	49 In	$L\beta_{2,15}$	$L_{III}N_{IV,V}$	3.71381
3.34335	9	50 Sn	$L\beta_4$	$L_I M_{II}$	3.7083
3.346	5	81 Tl	M_I	Abs. edge	3.705
3.35839	3	20 Ca	$K\alpha_1$	$K L_{III}$	3.69168
3.359	5	83 Bi	M_{II}	Abs. edge	3.691
3.36166	3	20 Ca	$K\alpha_2$	$K L_{II}$	3.68809
3.38487	3	50 Sn	$L\beta_1$	$L_{II}M_{IV}$	3.66280
3.42551	9	48 Cd	$L\gamma_5$	$L_{II}N_I$	3.61935
3.43015	9	48 Cd	$L\beta_9$	$L_I M_V$	3.61445
3.43606	9	49 In	$L\beta_6$	$L_{III}N_I$	3.60823
3.4365	1	19 K	K	Abs. edge	3.6078
3.4367	2	48 Cd	$L\beta_{10}$	$L_I M_{IV}$	3.6075
3.437	1	46 Pd	L_I	Abs. edge	3.607
3.43832	9	52 Te	$L\eta$	$L_{II}M_I$	3.60586
3.43941	4	51 Sb	$L\alpha_1$	$L_{III}M_V$	3.60472
3.441	5	91 Pa		$M_{II}N_I$	3.603
3.4413	4	19 K	$K\beta_5$	$K M_{IV,V}$	3.6027
3.44840	6	51 Sb	$L\alpha_2$	$L_{III}M_{IV}$	3.59532
3.4539	2	19 K	$K\beta_{1,3}$	$K M_{II,III}$	3.5896

Upper section

λ (Å)	Level	Designation	Element	Intensity	λ (Å)
3.57311	$L_I M_{III}$	$L\beta_3$	49 In	9	3.46984
3.565	Abs. edge	M_I	80 Hg	5	3.478
3.563	$M_{III} N_V$	$M\gamma$	92 U	1	3.479
3.5533	$L_I N_{II,III}$	$L\gamma_{2,3}$	46 Pd	2	3.4892
3.550	Abs. edge	M_{II}	82 Pb	5	3.492
3.545	Abs. edge	M_V	92 U	5	3.497
3.5376	Abs. edge	L_{III}	48 Cd	1	3.5047
3.53528	$L_I M_I$	$L\beta_4$	49 In	9	3.50697
3.52812	$L_{III} N_{IV,V}$	$L\beta_{2,15}$	48 Cd	4	3.51408
3.5258	Abs. edge	L_{II}	47 Ag	1	3.5164
3.521	$M_{III} N_{IV}$	$M\gamma$	92 U	2	3.521
3.51959	$L_{II} N_{IV}$	$L\gamma_1$	47 Ag	4	3.52260
3.505	$M_{II} N_I$	M_{IV}	90 Th	9	3.537
3.48721	$L_{II} M_{IV}$	$L\beta_1$	49 In	4	3.55531
3.485	Abs. edge	M_{IV}	90 Th	5	3.557
3.48502	$L_{III} M_I$	Ll	53 I	9	3.55754
3.4666	$M_{IV} O_{II}$	$M\gamma$	92 U	1	3.576
3.4657	$M_{III} N_V$	$M\gamma$	91 Pa	1	3.577
3.44398	$L_{III} M_V$	$L\alpha_1$	50 Sn	3	3.59994
3.43917	$L_{III} M_V$	$L\beta_9$	47 Ag	9	3.60497
3.43661	$L_I M_V$	$L\eta$	51 Sb	9	3.60765
3.43542	$L_{III} M_{IV}$	$L\alpha_2$	50 Sn	4	3.60891
3.43287	$L_{II} M_V$	$L\beta_{10}$	47 Ag	9	3.61158
3.430	$M_{III} N_{IV}$	$M_{III} N_{IV}$	91 Pa	2	3.61467
3.42994	$L_{II} N_I$	$L\beta_6$	48 Cd	9	3.61638
3.42832	$L_{II} N_I$	$L\gamma_5$	47 Ag	9	3.616
3.428	Abs. edge	M_I	79 Au	5	3.629
3.417	Abs. edge	L_I	45 Rh	5	3.634
3.412	Abs. edge	M_{II}	81 Tl	5	3.64495
3.40145	$L_I M_{III}$	$L\beta_3$	48 Cd	9	3.679
3.370	$M_{III} N_{IV}$	$M\gamma$	90 Th	2	3.68203
3.36719	$L_I M_{II}$	$L\beta_4$	48 Cd	9	3.6855
3.3640	$L_I N_{II,III}$	$L\gamma_{2,3}$	45 Rh	2	3.691
3.359	$M_{IV} O_{II}$	$M\gamma$	91 Pa	2	3.6999
3.35096	Abs. edge	L_{III}	47 Ag	1	3.70335
3.34781	$L_{III} N_{IV,V}$	$L\beta_{2,15}$	47 Ag	3	3.716
3.3367	$M_{IV} N_{VI}$	$M\beta$	92 U	1	3.71696
3.33555	$L_{II} M_I$	Ll	52 Te	9	3.718
3.335	$M_{III} N_{IV}$		90 Th	3	

Lower section

λ (Å)	Level	Designation	Element	Intensity	λ (Å)
4.117	$M_{II} N_{IV}$	$L\alpha_1$	90 Th	2	3.011
4.1099	$L_{III} M_V$	$L\beta_{2,15}$	54 Xe	2	3.0166
4.10078	$L_{III} N_{IV,V}$	$K\alpha_1$	51 Sb	3	3.02235
4.0906	$K L_{II}$	$K\alpha_2$	21 Sc	1	3.0309
4.0861	$K L_{II}$		21 Sc	1	3.0342
4.081	$M_{III} O_{IV,V}$		91 Pa	2	3.038
4.0695	$L_I M_{II}$	$L\beta_4$	52 Te	9	3.04661
4.041	Abs. edge	M_{III}	90 Th	5	3.068
4.0381	Abs. edge	K	20 Ca	1	3.0703
4.0325	$K M_{IV,V}$	$K\beta_5$	20 Ca	3	3.0746
4.02958	$L_{II} M_{IV}$	$L\beta_1$	52 Te	6	3.07677
4.0192	$L_{II} M_{IV}$	$L\gamma_5$	50 Sn	9	3.08475
4.0190	$L_{II} N_I$	L_I	48 Cd	9	3.0849
4.0127	$K M_{II,III}$	$K\beta_{1,3}$	20 Ca	1	3.0897
4.007	Abs. edge	M_I	83 Bi	5	3.094
3.9800	$L_I N_I$	$L\beta_3$	50 Sn	9	3.11513
3.9800	$L_{III} N_I$	$L\beta_6$	51 Sb	9	3.11513
3.980	$M_{III} O_I$	Ll	92 U	7	3.115
3.9716	$L_{II} M_{IV}$	$L\gamma_2$	50 Sn	9	3.12170
3.959	$M_{III} O_{IV,V}$	L_{II}	90 Th	3	3.131
3.9541	$L_I M_I$	Ll	56 Ba	2	3.1355
3.9513	Abs. edge	L_{III}	48 Cd	2	3.1377
3.9393	$L_{III} O_I$	$L\gamma_2$	49 In	1	3.1473
3.93765	$L_{III} M_{IV}$	$L\alpha_2$	53 I	6	3.14860
3.9327	$L_{III} N_{IV,V}$	$L\beta_3$	51 Sb	9	3.15258
3.9288	Abs. edge	L_{III}	50 Sn	9	3.1557
3.9279	$L_{III} O_I$	$L\beta_7$	53 I	3	3.1564
3.92604	$L_{III} M_{IV}$	$L\alpha_2$	49 In	6	3.15791
3.92081	$L_{II} N_{IV}$	$L\gamma_1$	50 Sn	4	3.16213
3.90486	$L_{III} N_{IV,V}$	$L\beta_{2,15}$	51 Sb	3	3.17505
3.8364	$L_I M_{II}$	$L\beta_4$	82 Pb	9	3.19014
3.854	Abs. edge	M_I	51 Sb	5	3.217
3.84357	$L_{II} N_{IV}$	$L\beta_1$	91 Pa	4	3.22567
3.82	$M_{III} O_{II}$	$L\gamma_5$	49 In	9	3.245
3.8159	$L_{II} N_I$	Ll	47 Ag	1	3.24907
3.8072	Abs. edge	L_{III}	55 Cs	2	3.2564
3.7950	$L_{III} N_I$	$L\beta_9$	49 In	9	3.2670
3.7942	$L_I M_V$	$L\beta_6$	50 Sn	9	3.26763
3.7926	$L_{III} N_I$				3.26901

TABLE 7f-2. WAVELENGTHS OF X-RAY EMISSION LINES AND ABSORPTION EDGES: IN NUMERICAL ORDER (Continued)

Wavelength, Å*	p.e.†	Element	Designation		keV
3.7228	1	46 Pd	L_{II}	Abs. edge	3.33031
3.7246	2	46 Pd	$L\gamma_1$	$L_{II}N_{IV}$	3.3287
3.729	5	90 Th	M_V	Abs. edge	3.325
3.73823	4	48 Cd	$L\beta_1$	$L_{II}M_{IV}$	3.31657
3.740	9	83 Bi		$M_I N_{III}$	3.315
3.7414	2	19 K	$K\alpha_1$	KL_{III}	3.3138
3.7445	2	19 K	$K\alpha_2$	KL_{II}	3.3111
3.760	5	90 Th		$M_V P_{III}$	3.298
3.762	9	78 Pt	M_I	Abs. edge	3.296
3.77192	4	49 In	$L\alpha_1$	$L_{III}M_V$	3.28694
3.78073	6	49 In	$L\alpha_2$	$L_{III}M_{IV}$	3.27929
3.783	5	80 Hg	M_{II}	Abs. edge	3.277
3.78876	9	50 Sn	$L\eta$	$L_{II}M_I$	3.27234
3.7920	2	46 Pd	$L\beta_9$	$L_I M_V$	3.2696
3.7988	2	46 Pd	$L\beta_{10}$	$L_I M_{IV}$	3.2637
3.80774	9	47 Ag	$L\beta_6$	$L_{III}N_I$	3.25603
3.808	4	90 Th		$M_{IV}O_{II}$	3.256
3.8222	2	46 Pd	$L\gamma_5$	$L_{II}N_I$	3.2437
3.827	1	91 Pa	$M\beta$	$M_{IV}N_{VI}$	3.2397
3.83313	9	47 Ag	$L\beta_3$	$L_I M_{III}$	3.23446
3.834	4	83 Bi		$M_{II}N_{IV}$	3.234
3.835	5	44 Ru	L_I	Abs. edge	3.233
3.87023	5	47 Ag	$L\beta_4$	$L_I M_{II}$	3.20346
3.87090	5	18 A	K	Abs. edge	3.20290
3.872	9	82 Pb		$M_I N_{III}$	3.202
3.8860	2	18 A	$K\beta_{1,3}$	$KM_{II,III}$	3.1905
3.88826	9	51 Sb	Ll	$L_{III}M_I$	3.18860
3.892	9	83 Bi		$M_I N_{II}$	3.185
3.8977	9	44 Ru		$L_I N_{II,III}$	3.1809
3.904	2	83 Bi	M_{III}	Abs. edge	3.176
3.9074	5	46 Pd	L_{II}	Abs. edge	3.17298
3.90887	1	46 Pd	$L\beta_{2,15}$	$L_{III}N_{IV,V}$	3.17179
3.910	4	92 U	$M\alpha_1$	$M_V N_{VII}$	3.1708
4.198	1	81 Tl	M_{III}	Abs. edge	2.9535
4.216	6	81 Tl		$M_{III}O_{IV,V}$	2.941
4.236	5	75 Re	M_I	Abs. edge	2.927
4.2417	2	45 Rh	$L\beta_6$	$L_{II}N_I$	2.9229
4.244	9	82 Pb		$M_{III}O_I$	2.921
4.2522	2	45 Rh		$L_I M_{III}$	2.9157
4.260	5	77 Ir	$L\beta_3$	$L_I M_{III}$	2.910
4.26873	9	49 In	Ll	$L_{III}M_I$	2.90440
4.2873	2	44 Ru	Ll	$L_{III}M_I$	2.8918
4.2888	2	45 Rh	$L\gamma_5$	$L_{II}N_I$	2.8908
4.300	9	79 Au	$L\beta_4$	$M_I N_{III}$	2.883
4.304	5	42 Mo		$M_{III}N_{III}$	2.881-
4.330	2	92 U	L_I	$M_{III}N_I$	2.863
4.355	1	80 Hg	M_{III}	Abs. edge	2.8469
4.36767	5	46 Pd	$L\alpha_1$	$L_{III}M_V$	2.83861
4.369	1	44 Ru	L_{III}	Abs. edge	2.8377
4.3718	2	44 Ru	$L\beta_{2,15}$	$L_{III}N_{IV,V}$	2.8360
4.37414	4	45 Rh	$L\beta_1$	$L_{II}M_{IV}$	2.83441
4.37588	7	46 Pd	$L\alpha_2$	$L_{III}M_{IV}$	2.83329
4.3800	2	42 Mo	$L\gamma_{2,3}$	$L_I N_{II,III}$	2.8306
4.3971	1	17 Cl	K	Abs. edge	2.81960
4.4034	3	17 Cl	$K\beta$	KM	2.8156
4.407	5	74 W	M_I	Abs. edge	2.813
4.4183	2	47 Ag	$L\eta$	$L_{II}M_I$	2.8061
4.432	4	79 Au		$M_{II}N_{IV}$	2.797
4.433	5	76 Os	M_{II}	Abs. edge	2.797
4.436	1	43 Te	L_{II}	Abs. edge	2.7948
4.44	9	74 W		$M_I O_{II,III}$	2.79
4.450	4	91 Pa		$M_I N_{III}$	2.786
4.460	9	78 Pt		$M_I N_{II}$	2.780
4.48014	9	48 Cd	Ll	$L_{III}M_I$	2.76735
4.4866	9	44 Ru	$L\beta_3$	$L_I M_{III}$	2.7634
4.4866	3	44 Ru	$L\beta_6$	$L_{III}N_I$	2.7634

3.915	5	77 Ir	M_{I}	Abs. edge	3.167	4.518	1	79 Au	M_{III}	Abs. edge	2.7439
3.924	1	92 U	$M\alpha_{2}$	$M_{V}N_{VI}$	3.1595	4.522	6	79 Au		$M_{III}O_{IV,V}$	2.742
3.932	6	83 Bi		$M_{III}O_{IV,V}$	3.153	4.5230	2	44 Ru	$L\beta_{4}$	$L_{I}M_{II}$	2.7411
3.93473	3	47 Ag	$L\beta_{1}$	$L_{II}M_{IV}$	3.15094	4.532	2	83 Bi	$M\gamma$	$M_{III}N_{V}$	2.735
3.936	5	79 Au	M_{II}	Abs. edge	3.150	4.568	5	90 Th		$M_{III}N_{I}$	2.714
3.941	1	90 Th	$M\beta$	$M_{IV}N_{VI}$	3.1458	4.571	5	83 Bi		$M_{III}N_{IV}$	2.712
3.9425	5	45 Rh	L_{II}	Abs. edge	3.1448	4.572	5	83 Bi	M_{IV}	Abs. edge	2.711
3.9437	2	45 Rh	$L\gamma_{1}$	$L_{II}N_{IV}$	3.1438	4.575	5	41 Nb	L_{I}	Abs. edge	2.710
3.95635	4	48 Cd	$L\alpha_{1}$	$L_{III}M_{V}$	3.13373	4.585	5	73 Ta	M_{I}	Abs. edge	2.704
3.96496	6	48 Cd	$L\alpha_{2}$	$L_{III}M_{IV}$	3.12691	4.59	2	83 Bi		$M_{IV}P_{II,III}$	2.70
3.968	5	82 Pb		$M_{II}N_{IV}$	3.124	4.59743	9	45 Rh	$L\alpha_{1}$	$L_{III}M_{V}$	2.69674
3.98327	9	49 In	$L\eta$	$L_{II}M_{I}$	3.11254	4.601	4	78 Pt		$M_{II}N_{IV}$	2.695
4.013	9	81 Tl		$M_{I}N_{III}$	3.089	4.60545	9	45 Rh	$L\alpha_{2}$	$L_{III}M_{IV}$	2.69205
4.0162	2	46 Pd	$L\beta_{6}$	$L_{III}N_{I}$	3.0870	4.620	5	75 Re	M_{II}	Abs. edge	2.684
4.022	1	91 Pa	$M\alpha_{1}$	$M_{V}N_{VII}$	3.0823	4.62058	3	44 Ru	$L\beta_{1}$	$L_{II}M_{IV}$	2.68323
4.0346	2	46 Pd	$L\beta_{3}$	$L_{I}M_{III}$	3.0730	4.625	5	92 U		$M_{IV}N_{III}$	2.681
4.035	3	91 Pa	$M\alpha_{2}$	$M_{V}N_{VI}$	3.072	4.630	1	43 Tc	L_{III}	Abs. edge	2.6780
4.0451	2	45 Rh	$L\gamma_{5}$	$L_{II}N_{I}$	3.0650	4.631	9	77 Ir		$M_{I}N_{III}$	2.677
4.047	1	82 Pb	M_{III}	Abs. edge	3.0632	4.6542	2	41 Nb	$L\gamma_{2,3}$	$L_{I}N_{II,III}$	2.6638
4.058	5	43 Tc	L_{I}	Abs. edge	3.055	4.655	8	82 Pb		$M_{II}N_{I}$	2.664
4.069	6	82 Pb		$M_{III}O_{IV,V}$	3.047	4.6605	2	46 Pd	$L\eta$	$L_{II}M_{I}$	2.6603
4.0711	2	46 Pd	$L\beta_{4}$	$L_{I}M_{II}$	3.0454	4.674	1	82 Pb	$M\gamma$	$M_{III}N_{V}$	2.6527
4.071	5	76 Os	M_{I}	Abs. edge	3.045	4.686	1	78 Pt	M_{III}	Abs. edge	2.6459
4.07165	9	50 Sn	Ll	$L_{III}M_{I}$	3.04499	4.694	8	78 Pt		$M_{III}O_{IV,V}$	2.641
4.093	5	78 Pt	M_{II}	Abs. edge	3.029	4.703	9	79 Au		$M_{III}O_{I}$	2.636
4.105	9	83 Bi		$M_{III}O_{I}$	3.021	4.7076	2	47 Ag	Ll	$L_{III}M_{I}$	2.6337
4.116	4	81 Tl		$M_{II}N_{IV}$	3.013	4.715	3	82 Pb		$M_{III}N_{IV}$	2.630
4.1299	5	45 Rh	L_{III}	Abs. edge	3.0021	4.719	1	42 Mo	L_{II}	Abs. edge	2.6274
4.1310	2	45 Rh	$L\beta_{2,15}$	$L_{III}N_{IV,V}$	3.0013	4.7258	2	42 Mo	$L\gamma_{1}$	$L_{II}N_{IV}$	2.6235
4.1381	9	90 Th	$M\alpha_{1}$	$M_{V}N_{VII}$	2.9961	4.7278	1	17 Cl	$K\alpha_{1}$	KL_{III}	2.62239
4.14622	5	46 Pd	$L\beta_{1}$	$L_{II}M_{IV}$	2.99022	4.7307	1	17 Cl	$K\alpha_{2}$	KL_{II}	2.62078
4.151	2	90 Th	$M\alpha_{2}$	$M_{V}N_{VI}$	2.987	4.757	5	82 Pb	M_{IV}	Abs. edge	2.606
4.15443	3	47 Ag	$L\alpha_{1}$	$L_{III}M_{V}$	2.98431	4.764	5	83 Bi	M_{V}	Abs. edge	2.603
4.16294	5	47 Ag	$L\alpha_{2}$	$L_{III}M_{IV}$	2.97821	4.780	4	77 Ir		$M_{II}N_{IV}$	2.594
4.180	1	44 Ru	L_{II}	Abs. edge	2.9663	4.79	2	76 Os		$M_{I}N_{III}$	2.59
4.1822	2	44 Ru	$L\gamma_{1}$	$L_{II}N_{IV}$	2.9645	4.815	5	74 W	M_{II}	Abs. edge	2.575
4.19180	5	18 A	$K\alpha_{1}$	KL_{III}	2.95770	4.823	3	83 Bi		$M_{IV}O_{II}$	2.571
4.19315	9	48 Cd	$L\eta$	$L_{II}M_{I}$	2.95675	4.823	4	81 Tl	$M\gamma$	$M_{III}N_{V}$	2.571
4.19474	5	18 A	$K\alpha_{2}$	KL_{II}	2.95563	4.8369	2	42 Mo	$L\gamma_{5}$	$L_{II}N_{I}$	2.5632

TABLE 7f-2. WAVELENGTHS OF X-RAY EMISSION LINES AND ABSORPTION EDGES: IN NUMERICAL ORDER (Continued)

Wavelength, Å*	p.e.†	Element	Designation		keV
4.84575	5	44 Ru	$L\alpha_1$	$L_{III}M_V$	2.55855
4.85381	7	44 Ru	$L\alpha_2$	$L_{III}M_{IV}$	2.55431
4.861	1	77 Ir	M_{III}	Abs. edge	2.5505
4.865	5	81 Tl		$M_{III}N_{IV}$	2.548
4.869	9	77 Ir		$M_{III}O_{IV,V}$	2.546
4.876	9	78 Pt		$M_{III}O_I$	2.543
4.879	5	40 Zr	L_I	Abs. edge	2.541
4.8873	8	43 Tc	$L\beta_1$	$L_{II}M_{IV}$	2.5368
4.909	1	83 Bi	$M\beta$	$M_{IV}N_{VI}$	2.5255
4.911	5	90 Th		$M_{IV}N_{III}$	2.524
4.913	1	42 Mo	L_{III}	Abs. edge	2.5234
4.9217	2	45 Rh	$L\eta$	$L_{II}M_I$	2.5191
4.9232	2	42 Mo	$L\beta_{2,15}$	$L_{III}N_{IV,V}$	2.5183
4.946	2	92 U		$M_V N_{III}$	2.507
4.952	5	81 Tl	M_{IV}	Abs. edge	2.504
4.9525	3	46 Pd	Ll	$L_{III}M_I$	2.5034
4.9536	3	40 Zr	$L\gamma_{2,3}$	$L_I N_{II,III}$	2.5029
4.955	4	76 Os		$M_{II}N_{IV}$	2.502
4.955	5	82 Pb	M_V	Abs. edge	2.502
4.984	2	80 Hg	$M\gamma$	$M_{III}N_V$	2.4875
5.004	9	82 Pb		$M_{IV}O_{II}$	2.477
5.0133	3	42 Mo	$L\beta_3$	$L_I M_{III}$	2.4730
5.0185	1	16 S	K	Abs. edge	2.47048
5.020	3	73 Ta	M_{II}	Abs. edge	2.470
5.0233	3	16 S	$K\beta_z$	KM	2.4681
5.031	1	41 Nb	L_{II}	Abs. edge	2.4641
5.0316	2	16 S	$K\beta_1$	KM	2.46404
5.0361	3	41 Nb	$L\gamma_1$	$L_{II}N_{IV}$	2.4618
5.043	5	76 Os	M_{III}	Abs. edge	2.458
5.0488	3	42 Mo	$L\beta_4$	$L_I M_{II}$	2.4557
5.0488	5	42 Mo	$L\beta_6$	$L_{III}N_I$	2.4557
5.050	2	92 U	$M\zeta_2$	$M_{IV}N_{VI}$	2.4548
5.076	1	82 Pb	$M\beta$	$M_{IV}N_{VI}$	2.4427
5.40655	8	42 Mo	$L\alpha_1$	$L_{III}M_V$	2.29316
5.41437	8	42 Mo	$L\alpha_2$	$L_{III}M_{IV}$	2.28985
5.4318	9	80 Hg	$M\beta$	$M_{IV}N_{VI}$	2.2825
5.435	1	74 W	M_{III}	Abs. edge	2.2811
5.460	1	81 Tl	$M\alpha_1$	$M_V N_{VII}$	2.2706
5.472	2	81 Tl	$M\alpha_2$	$M_V N_{VI}$	2.2656
5.4923	3	41 Nb	$L\beta_1$	$L_{II}M_{IV}$	2.2574
5.4977	3	40 Zr	$L\gamma_5$	$L_{II}N_I$	2.2551
5.500	4	77 Ir	$M\gamma$	$M_{III}N_V$	2.254
5.5035	3	44 Ru	Ll	$L_{III}M_I$	2.2528
5.537	8	83 Bi		$M_{III}N_I$	2.239
5.540	5	77 Ir		$M_{III}N_{IV}$	2.238
5.570	4	73 Ta		$M_{III}N_{III}$	2.226
5.579	1	40 Zr	L_{III}	Abs. edge	2.2225
5.584	5	79 Au	M_V	Abs. edge	2.220
5.5863	3	40 Zr	$L\beta_{2,15}$	$L_{III}N_{IV,V}$	2.2194
5.59	1	78 Pt	M_{IV}	Abs. edge	2.217
5.592	5	38 Sr	L_I	Abs. edge	2.217
5.624	1	79 Au	$M\beta$	$M_{IV}N_{VI}$	2.2046
5.628	8	74 W		$M_{III}O_I$	2.203
5.6330	3	40 Zr	$L\beta_3$	$L_I M_{III}$	2.2010
5.6445	3	38 Sr	$L\gamma_{2,3}$	$L_I N_{II,III}$	2.1965
5.6476	9	80 Hg	$M\alpha_1$	$M_V N_{VII}$	2.1953
5.650	5	73 Ta	M_{III}	Abs. edge	2.194
5.6681	3	40 Zr	$L\beta_4$	$L_I M_{II}$	2.1873
5.67	3	73 Ta		$M_{III}O_{IV,V}$	2.19
5.682	4	76 Os	$M\gamma$	$M_{III}N_V$	2.182
5.704	8	82 Pb		$M_{III}N_I$	2.174
5.7101	3	40 Zr		$M_{III}N_I$	2.1712
5.724	5	76 Os		$M_{III}N_{IV}$	2.166
5.7243	2	41 Nb	$L\alpha_1$	$L_{III}M_V$	2.16589
5.7319	3	41 Nb	$L\alpha_2$	$L_{III}M_{IV}$	2.1630
5.756	1	39 Y	L_{II}	Abs. edge	2.1540

Upper table

Energy (keV)	Level	Line	Element	Int.	λ (Å)
2.150	$M_V\,O_{III}$		79 Au	9	5.767
2.1435	Abs. edge	K	15 P	1	5.784
2.1391	$K\,M$	$K\beta$	15 P	2	5.796
2.133	$M_{II}\,N_I$		76 Os	2	5.81
2.133	Abs. edge	M_V	78 Pt	1	5.81
2.1273	$M_{IV}\,N_{VI}$	$M\beta$	78 Pt	1	5.828
2.126	$M_{III}\,O_I$		73 Ta	2	5.83
2.126	Abs. edge	M_{IV}	77 Ir	1	5.83
2.1244	$L_{II}\,M_{IV}$	$L\beta_1$	40 Zr	3	5.8360
2.1229	$M_V\,N_{VII}$	$M\alpha_1$	79 Au	1	5.840
2.1202	$L_{II}\,M_I$	$L\eta$	42 Mo	3	5.8475
2.118	$M_V\,N_{VI}$	$M\alpha_2$	79 Au	3	5.854
2.1102	$L_{II}\,N_I$	$L\gamma_5$	39 Y	3	5.8754
2.107	$M_{III}\,N_I$		81 Tl	8	5.884
2.1067	$M_{III}\,N_V$	$M\gamma$	75 Re	2	5.885
2.090	$M_{III}\,N_{IV}$		75 Re	5	5.931
2.0794	Abs. edge	L_{III}	39 Y	1	5.962
2.0722	$L_I\,M_{III}$	$L\beta_3$	39 Y	3	5.9832
2.071	$M_V\,O_{III}$		78 Pt	9	5.987
2.063	Abs. edge	L_I	37 Rb	5	6.008
2.0600	$L_I\,M_{II}$	$L\beta_4$	39 Y	3	6.0186
2.0535	$M_{IV}\,N_{VI}$	$M\beta$	77 Ir	1	6.038
2.0507	$L_I\,N_{II,III}$	$L\gamma_{2,3}$	37 Rb	3	6.0458
2.0505	$M_V\,N_{VII}$	$M\alpha_1$	78 Pt	1	6.047
2.048	Abs. edge	M_V	77 Ir	1	6.05
2.047	$M_V\,N_{VI}$	$M\alpha_2$	78 Pt	3	6.058
2.04236	$L_{III}\,M_V$	$L\alpha_1$	40 Zr	2	6.0705
2.042	Abs. edge	M_{IV}	76 Os	5	6.073
2.0399	$L_{II}\,N_{IV}$	$L\gamma_1$	40 Zr	3	6.0778
2.036	$M_{III}\,N_I$		80 Hg	2	6.09
2.035	$M_{III}\,N_V$	$M\gamma$	74 W	3	6.092
2.0344	$L_{III}\,N_I$	$L\beta_6$	39 Y	3	6.0942
2.021	$M_{III}\,N_{IV}$		74 W	4	6.134
2.01568	$L_{III}\,M_I$	Ll	42 Mo	3	6.1508
2.0137	$K\,L_{III}$	$K\alpha_1$	15 P	1	6.157
2.0127	$K\,L_{II}$	$K\alpha_2$	15 P	1	6.160
2.012	$M_{IV}\,N_{III}$		83 Bi	8	6.162
2.0085	Abs. edge	L_{II}	38 Sr	1	6.173
1.99620	$L_{II}\,M_I$	$L\eta$	41 Nb	3	6.2109

Lower table

Energy (keV)	Level	Line	Element	Int.	λ (Å)
2.4350	$M_V\,N_{III}$	$M\zeta_1$	91 Pa	2	5.092
2.4240	$L_{III}\,M_V$	$L\alpha_1$	43 Tc	3	5.1148
2.4226	$M_V\,N_{VII}$	$M\alpha_1$	83 Bi	1	5.118
2.4170	$M_V\,N_{VI}$	$M\alpha_2$	83 Bi	2	5.130
2.410	$M_{III}\,N_V$	$M\gamma$	79 Au	4	5.145
2.4066	$L_{II}\,N_I$	$L\gamma_5$	41 Nb	3	5.1517
2.406	Abs. edge	M_V	81 Tl	5	5.153
2.404	Abs. edge	M_{IV}	80 Hg	5	5.157
2.399	$M_I\,N_{III}$		82 Pb	9	5.168
2.397	$M_I\,N_{III}$		74 W	9	5.172
2.39481	$L_{II}\,M_{IV}$	$L\beta_1$	42 Mo	8	5.17708
2.391	$M_{III}\,N_{IV}$		79 Au	5	5.186
2.3876	$M_{IV}\,N_{VII}$		91 Pa	2	5.193
2.386	$M_{IV}\,N_{VI}$	$M\beta$	81 Tl	9	5.196
2.38197	$M_{IV}\,O_{II}$	$M\zeta_2$	44 Ru	5	5.2050
2.377	$L_{II}\,M_I$	$L\eta$	39 Y	2	5.217
2.3765	Abs. edge	L_I	45 Rh	5	5.2169
2.3706	$L_{III}\,M_I$	Ll	41 Nb	3	5.230
2.369	Abs. edge	L_{III}	75 Re	1	5.234
2.3670	$L_{III}\,N_{IV,V}$	$L\beta_{2,15}$	41 Nb	5	5.2379
2.364	$M_V\,N_{VI}$	$M\zeta_2$	90 Th	3	5.245
2.3621	$M_{IV}\,N_{VI}$	$M\beta$	81 Tl	1	5.249
2.3468	$L_I\,N_{II,III}$	$L\gamma_{2,3}$	39 Y	3	5.2830
2.3455	$M_V\,N_{VII}$	$M\alpha_1$	82 Pb	1	5.286
2.3397	$M_V\,N_{VI}$	$M\alpha_2$	82 Pb	3	5.299
2.3348	$L_I\,M_{III}$	$L\beta_3$	41 Nb	1	5.3102
2.331	$M_{III}\,N_V$	$M\gamma$	78 Pt	2	5.319
2.322	$M_{III}\,N_{II}$	$M\zeta_2$	90 Th	5	5.340
2.3194	$L_I\,M_{II}$	$L\beta_4$	41 Nb	3	5.3455
2.314	$M_{II}\,N_{IV}$	M_V	74 W	4	5.357
2.314	$M_{III}\,N_{IV}$		78 Pt	5	5.357
2.313	Abs. edge	M_{IV}	80 Hg	1	5.36
2.3125	$L_{III}\,N_{III}$	$L\beta_6$	41 Nb	7	5.3613
2.30784	$K\,L_{III}$	$K\alpha_1$	16 S	8	5.37216
2.307	Abs. edge		79 Au	1	5.374
2.30664	$K\,L_{II}$	$K\alpha_2$	16 S	8	5.37496
2.3053	Abs. edge	L_{II}	83 Bi	1	5.378
2.3027	$L_{II}\,N_{IV}$	$L\gamma_1$	40 Zr	3	5.3843
2.295	$M_I\,N_{III}$		73 Ta	2	5.40

TABLE 7f-2. WAVELENGTHS OF X-RAY EMISSION LINES AND ABSORPTION EDGES: IN NUMERICAL ORDER (Continued)

Wavelength, Å*	p.e.†	Element	Designation		keV
6.2120	3	39 Y	$L\beta_1$	$L_{II}M_{IV}$	1.99584
6.259	9	79 Au		$M_{III}N_I$	1.981
6.262	1	77 Ir	$M\alpha_1$	M_VN_{VII}	1.9799
6.267	3	76 Os	$M\beta$	$M_{IV}N_{VI}$	1.9783
6.275	2	77 Ir	$M\alpha_2$	M_VN_{VI}	1.9758
6.28	3	74 W		$M_{II}N_I$	1.973
6.2961	3	38 Sr	$L\gamma_5$	$L_{II}N_I$	1.96916
6.30	1	76 Os	M_V	Abs. edge	1.967
6.312	4	73 Ta	$M\gamma$	$M_{III}N_{III}$	1.964
6.33	1	75 Re	M_{IV}	Abs. edge	1.958
6.353	5	73 Ta		$M_{III}N_{IV}$	1.951
6.3672	7	38 Sr	$L\beta_3$	L_IM_{III}	1.94719
6.384	1	82 Pb		$M_{IV}N_{III}$	1.942
6.387	3	38 Sr	L_{III}	Abs. edge	1.9411
6.4026	3	38 Sr	$L\beta_4$	L_IM_{II}	1.93643
6.4488	9	39 Y	$L\alpha_1$	$L_{III}M_V$	1.92256
6.455	3	78 Pt		$M_{III}N_I$	1.921
6.4558	3	39 Y	$L\alpha_2$	$L_{III}M_{IV}$	1.92047
6.47	1	36 Kr	L_I	Abs. edge	1.915
6.490	1	76 Os	$M\alpha$	$M_VN_{VI,VII}$	1.9102
6.504	1	75 Re	$M\beta$	$M_{IV}N_{VI}$	1.9061
6.5176	3	41 Nb	Ll	$L_{II}M_I$	1.90225
6.5191	3	38 Sr	L_{III}	$L_{III}N_I$	1.90181
6.521	4	83 Bi	$L\beta_6$	M_VN_{III}	1.901
6.544	4	72 Hf	$M\xi_1$	$M_{III}N_V$	1.895
6.560	5	75 Re	$M\gamma$	$M_{III}N_V$	1.890
6.585	5	83 Bi	$M\xi_2$	$M_{IV}N_{II}$	1.883
6.59	1	74 W	M_{IV}	Abs. edge	1.880
6.6069	3	40 Zr	$L\eta$	$L_{II}M_I$	1.87654
6.6239	3	38 Sr	$L\beta_1$	$L_{II}M_{IV}$	1.87172
6.644	1	37 Rb	L_{II}	Abs. edge	1.8661
6.669	9	77 Ir		$M_{III}N_I$	1.859
6.729	1	75 Re	$M\alpha$	$M_VN_{VI,VII}$	1.8425
7.30	2	73 Ta	$M\beta$	M_VO_{III}	1.700
7.303	1	72 Hf	$L\beta_4$	$M_{IV}N_{VI}$	1.6976
7.304	5	36 Kr	$L\alpha_1$	L_IM_{II}	1.697
7.3183	2	37 Rb	$L\alpha_2$	$L_{III}M_V$	1.69413
7.3251	3	37 Rb	Ll	$L_{III}M_{IV}$	1.69256
7.3563	3	39 Y		$L_{II}M_I$	1.68536
7.360	8	74 W		$M_{III}N_I$	1.684
7.371	8	78 Pt		$M_{IV}N_{III}$	1.682
7.392	1	36 Kr	L_{III}	Abs. edge	1.6772
7.466	4	79 Au	$M\xi_1$	M_VN_{III}	1.6605
7.503	1	34 Se	L_I	Abs. edge	1.6525
7.510	4	36 Kr	L_{II}	$L_{II}N_I$	1.6510
7.5171	3	38 Sr	$L\eta$	$L_{II}M_I$	1.64933
7.523	5	79 Au	$M\xi_2$	$M_{IV}N_{II}$	1.648
7.539	1	72 Hf	$M\alpha$	$M_VN_{VI,VII}$	1.6446
7.546	8	68 Er	$M\gamma$	$M_{III}N_V$	1.643
7.576	3	36 Kr	$L\beta_1$	$L_{II}M_{IV}$	1.6366
7.60	1	68 Er		$M_{III}N_{IV}$	1.632
7.601	2	71 Lu		$M_{IV}N_{VI}$	1.6312
7.612	9	73 Ta	$M\beta$	$M_{III}N_I$	1.629
7.645	8	77 Ir	$M\gamma$	$M_{IV}N_{III}$	1.622
7.738	4	78 Pt		M_VN_{III}	1.6022
7.753	5	35 Br	L_{II}	Abs. edge	1.599
7.767	9	35 Br	$L\beta_{3,4}$	$L_IM_{II,III}$	1.596
7.790	5	78 Pt	$M\xi_2$	$M_{IV}N_{II}$	1.592
7.817	3	36 Kr	$L\alpha_{1,2}$	$L_{III}M_{IV,V}$	1.5860
7.8362	3	38 Sr	Ll	$L_{III}M_I$	1.58215
7.840	2	71 Lu	$M\alpha$	$M_VN_{VI,VII}$	1.5813
7.865	9	67 Ho	$M\gamma$	$M_{III}N_{IV,V}$	1.576
7.887	9	72 Hf		$M_{III}N_I$	1.572
7.909	2	70 Yb	$M\alpha$	$M_{IV}N_{VI}$	1.5675
7.94813	5	13 Al	K	Abs. edge	1.55988
7.960	2	13 Al	$K\beta$	KM	1.55745

Upper panel

Wavelength	Level (transition)	Line	Element	Int.
1.5530	Abs. edge	L III	35 Br	5
1.5458	M V N III	M ζ1	77 Ir	4
1.54177	L II M I	L η	37 Rb	4
1.5373	M IV N II	M ζ2	77 Ir	5
1.5293	Abs. edge	L I	33 As	1
1.52590	L II N IV	L β1	35 Br	5
1.522	M III N IV,V	M γ	66 Dy	9
1.5214	M V N VI,VII	M α	70 Yb	5
1.505	M IV N III		75 Re	8
1.503	M IV N VI		69 Tm	7
1.4919	M V N VI	M β	76 Os	4
1.490	L I M II,III	M ζ1	34 Se	9
1.48670	K L II	L β3,4	13 Al	9
1.48627	K L III	K α2	13 Al	9
1.4831	M IV N II	K α2	76 Os	5
1.48238	L III M I	M ζ2	37 Rb	5
1.48043	L III M IV,V	L l	35 Br	1
1.4747	Abs. edge	L α1,2	34 Se	9
1.464	M III N I	L II	70 Yb	1
1.462	M V N VI,VII	M α	69 Tm	9
1.461	M III N IV,V	M γ	65 Tb	5
1.4609	Abs. edge	M v	69 Tm	8
1.446	M IV N III	M β	74 W	3
1.4430	M IV N VI		68 Er	7
1.44	N I P V,V		92 U	5
1.4415	Abs. edge	M IV	68 Er	4
1.4368	M V N III	M ζ1	75 Re	1
1.4340	Abs. edge	L III	34 Se	5
1.4310	M IV N II	M ζ2	75 Re	5
1.41923	L II M IV	L β1	34 Se	7
1.42	N I P III		92 U	1
1.4132	Abs. edge	L I	32 Ge	7
1.41	N I P II		92 U	1
1.406	M V N VI,VII	M α	68 Er	9
1.402	M III N IV,V	M γ	64 Gd	5
1.4013	Abs. edge	M v	68 Er	2
1.393	M IV N III		73 Ta	1
1.3884	L I M II,III	L β3,4	33 As	5
1.3835	M V N III	M ζ1	74 W	4

Lower panel

Wavelength	Int.	Element	Line	Level (transition)
6.738	1	14 Si	K	Abs. edge
6.740	3	82 Pb	M ζ1	M V N III
6.7530	1	14 Si	K β	K M
6.755	3	37 Rb	L γ5	L II N IV
6.757	1	74 W	M β	M IV N VI
6.768	6	71 Lu	M γ	M III N V
6.7876	3	37 Rb	L β3	L I M III
6.802	5	82 Pb	M ζ2	M IV N II
6.806	9	74 W	L β4	M IV O II
6.8207	9	74 W	M v	L I M II
6.83	3	37 Rb	L III	Abs. edge
6.862	1	38 Sr	L α1	L III M V
6.8628	2	38 Sr	L α2	L III M IV
6.8697	3	73 Ta	M IV	Abs. edge
6.87	1	80 Hg	δ	M IV N III
6.87	2	76 Os	L l	M III N I
6.89	2	40 Zr	L I	L III M I
6.9185	3	35 Br	M ζ1	Abs. edge
6.959	5	81 Tl	M α1	M V N III
6.974	4	74 W	L β6	M V N VII
6.983	1	37 Rb	M α2	L III N I
6.9842	3	74 W	M β	M IV N VI
6.992	2	74 W	M γ	M V O III
7.005	9	73 Ta	M ζ2	M IV N VI
7.023	1	70 Yb	L η	M III N V
7.024	8	81 Tl	L β1	M IV N II
7.032	5	39 Y		L II M I
7.0406	3	37 Rb	M v	L II M IV
7.0759	3	73 Ta	K α1	M IV O II,III
7.09	2	79 Au	K α2	M IV N III
7.101	8	73 Ta	L II	Abs. edge
7.11	1	14 Si	M α	K L III
7.12542	9	14 Si	K α1	K L II
7.12791	9	14 Si	K α2	Abs. edge
7.168	1	36 Kr	L II	L II N III
7.250	5	73 Ta	M α	M V N VI,VII
7.252	1	36 Kr	L β3	L I M III
7.264	5	73 Ta	L γ5	L II M III
7.279	5	36 Kr	L γ5	L II N I

TABLE 7f-2. WAVELENGTHS OF X-RAY EMISSION LINES AND ABSORPTION EDGES: IN NUMERICAL ORDER (*Continued*)

Wavelength, Å*	p.e.†	Element	Designation		keV
8.965	4	67 Ho	$M\beta$	$M_{IV}N_{VI}$	1.3830
8.9900	5	34 Se	$L\alpha_{1,2}$	$L_{III}M_{IV,V}$	1.37910
8.993	5	74 W	$M\zeta_2$	$M_{IV}N_{II}$	1.3787
9.125	1	33 As	L_{II}	Abs. edge	1.3587
9.20	2	67 Ho	$M\alpha$	$M_V N_{VI,VII}$	1.348
9.211	9	63 Eu	$M\gamma$	$M_{III}N_{IV,V}$	1.346
9.255	1	35 Br	$L\eta$	$L_{II}M_I$	1.3396
9.316	4	73 Ta	$M\zeta_1$	$M_V N_{III}$	1.3308
9.330	5	73 Ta	$M\zeta_2$	$M_{IV}N_{II}$	1.3288
9.357	6	66 Dy	$M\beta$	$M_{IV}N_{VI}$	1.3250
9.367	7	33 As	L_{III}	Abs. edge	1.3235
9.40	7	90 Th	$N_I P_{III}$	$N_I P_{III}$	1.319
9.4141	8	33 As	$L\beta_1$	$L_{II}M_{IV}$	1.3170
9.44	7	90 Th	$N_I P_{II}$	$N_I P_{II}$	1.313
9.5122	5	12 Mg	K	Abs. edge	1.30339
9.517	5	31 Ga	L_I	Abs. edge	1.3028
9.521	2	12 Mg	$K\beta$	KM	1.3022
9.581	2	32 Ge	$L\beta_3$	$L_I M_{III}$	1.2941
9.585	1	35 Br	Ll	$L_{III}M_I$	1.2935
9.59	2	66 Dy	$M\alpha$	$M_V N_{VI,VII}$	1.293
9.600	9	62 Sm	$M\gamma$	$M_{III}N_{IV,V}$	1.291
9.640	8	32 Ge	$L\beta_4$	$L_I M_{II}$	1.2861
9.6709	8	33 As	$L\alpha_{1,2}$	$L_{III}M_{IV,V}$	1.2820
9.686	7	72 Hf	$M\zeta_2$	$M_{IV}N_{II}$	1.2800
9.686	7	72 Hf	$M\zeta_1$	$M_V N_{VI}$	1.2800
9.792	6	65 Tb	$M\beta$	$M_{IV}N_{VI}$	1.2661
9.8900	2	12 Mg	$K\alpha_{1,2}$	$KL_{II,III}$	1.25360
9.924	1	32 Ge	L_{II}	Abs. edge	1.2494
9.962	1	34 Se	$L\eta$	$L_{II}M_I$	1.2446
10.00	2	65 Tb	$M\alpha$	$M_V N_{VI,VII}$	1.240
10.09	7	92 U	$N_I O_{III}$	$N_I O_{III}$	1.229
10.175	1	32 Ge	$L\beta_1$	$L_{II}M_{IV}$	1.2185
10.187	1	32 Ge	L_{III}	Abs. edge	1.2170

Wavelength, Å*	p.e.†	Element	Designation		keV
12.131	1	30 Zn	L_{III}	Abs. edge	1.02201
12.254	3	30 Zn	$L\alpha_{1,2}$	$L_{III}M_{IV,V}$	1.0117
12.43	2	66 Dy	$M\zeta$	$M_{IV}N_{III}$	0.998
12.44	2	60 Nd	$M\beta$	$M_{IV}N_{VI}$	0.997
12.459	5	60 Nd	M_{IV}	Abs. edge	0.9951
12.597	2	31 Ga	$L_{II}M_I$	$L_{II}M_I$	0.9842
12.68	2	60 Nd	$M\alpha$	$M_V N_{VI,VII}$	0.978
12.737	5	60 Nd	M_V	Abs. edge	0.9734
12.75	3	56 Ba	$M\gamma$	$M_{III}N_{IV,V}$	0.973
12.90	9	92 U	$N_{III}O_V$	$N_{III}O_V$	0.961
12.953	2	31 Ga	Ll	$L_{III}M_I$	0.9572
12.98	2	65 Tb	$M\zeta$	$M_V N_{III}$	0.955
13.014	1	29 Cu	L_{II}	Abs. edge	0.95268
13.053	2	29 Cu	$L\beta_1$	$L_{II}M_{IV}$	0.9498
13.06	2	59 Pr	$M\beta$	$M_{IV}N_{VI}$	0.950
13.122	5	59 Pr	M_{IV}	Abs. edge	0.9448
13.18	2	28 Ni	$L\beta_{3,4}$	$L_I M_{II,III}$	0.941
13.288	1	29 Cu	L_{III}	Abs. edge	0.93306
13.30	6	83 Bi	$N_I P_{II,III}$	$N_I P_{II,III}$	0.932
13.336	3	29 Cu	$L\alpha_{1,2}$	$L_{III}M_{IV,V}$	0.9297
13.343	5	59 Pr	$M\alpha$	$M_V N_{VI,VII}$	0.9292
13.394	5	59 Pr	M_V	Abs. edge	0.9257
13.57	2	64 Gd	$M\zeta$	$M_{IV}N_{III}$	0.914
13.68	2	30 Zn	$L\eta$	$L_{II}M_I$	0.906
13.75	4	58 Ce	$M\beta$	$M_{IV}N_{VI}$	0.902
13.8	1	90 Th	$N_{III}O_V$	$N_{III}O_V$	0.897
14.02	2	30 Zn	Ll	$L_{III}M_I$	0.884
14.04	2	58 Ce	$M\alpha$	$M_V N_{VI,VII}$	0.883
14.22	2	63 Eu	$M\zeta$	$M_V N_{III}$	0.872
14.242	5	28 Ni	L_{II}	Abs. edge	0.8706
14.271	6	28 Ni	$L\beta_1$	$L_{II}M_{IV}$	0.8688
14.3018	1	10 Ne	K	Abs. edge	0.866889
14.31	3	27 Co	$L\beta_{3,4}$	$L_I M_{II,III}$	0.870

Element	I	Line	Level designation	λ (Å)	E (keV)
58 Ce	5		$M_V O_{II,III}$	14.39	0.862
10 Ne	5	$K\beta$	KM	14.452	0.8579
57 La	5	$M\beta$	$M_{IV} N_{VI}$	14.51	0.854
28 Ni	5	L_{III}	Abs. edge	14.525	0.8536
28 Ni	3	$L\alpha_{1,2}$	$L_{II} M_{IV,V}$	14.561	0.8515
10 Ne	3	$K\alpha_{1,2}$	$KL_{II,III}$	14.610	0.8486
57 La	5	$M\alpha$	$M_V N_{VI,VII}$	14.88	0.833
29 Cu	2	$L\eta$	$L_{II} M_I$	14.90	0.832
62 Sm	4	$M\zeta$	$M_V N_{III}$	14.91	0.831
29 Cu	9	Ll	$L_{III} M_I$	15.286	0.8111
56 Ba	1	M_{IV}	Abs. edge	15.56	0.7967
27 Co	5	L_{II}	Abs. edge	15.618	0.7938
26 Fe	4	$L\beta_{3,4}$	$L_I M_{II,III}$	15.65	0.792
27 Co	8	$L\beta_1$	$L_{II} M_{IV}$	15.666	0.7914
56 Ba	9		$M_{IV} O_{III}$	15.72	0.789
56 Ba	1	M_V	Abs. edge	15.89	0.7801
56 Ba	5		$M_{IV} O_{II}$	15.91	0.779
27 Co	5	L_{III}	Abs. edge	15.915	0.7790
52 Te	4	$M\gamma$	$M_{III} N_{IV,V}$	15.93	0.778
27 Co	6	$L\alpha_{1,2}$	$L_{III} M_{IV,V}$	15.972	0.7762
51 Sb	5		$M_{II} N_{IV}$	15.98	0.776
56 Ba	5		$M_V O_{III}$	16.20	0.765
28 Ni	3	$L\eta$	$L_{II} M_I$	16.27	0.762
60 Nd	4	Ll	$L_{III} M_I$	16.46	0.753
28 Ni	9	L_I	Abs. edge	16.693	0.7427
24 Cr	1	$M\gamma$	$M_{III} N_{IV,III}$	16.7	0.741
51 Sb	4		$M_{II} N_V$	16.92	0.733
50 Sn	4	$L\beta_{3,4}$	$L_I M_{II,III}$	16.93	0.733
25 Mn	5	L_{II}	Abs. edge	17.19	0.721
26 Fe	4	$L\beta_1$	$L_{II} M_{IV}$	17.202	0.7208
26 Fe	5	$M\zeta$	$M_V N_{III}$	17.26	0.7185
59 Pr	1	L_{III}	Abs. edge	17.38	0.714
26 Fe	4	$L\alpha_{1,2}$	$L_{II} M_{IV,V}$	17.525	0.7074
52 Te	5	$L\eta$	$L_{II} M_I$	17.59	0.7050
27 Co	2	$M\gamma$	$M_{II} N_I$	17.6	0.703
50 Sn	1		$L_{II} M_I$	17.87	0.694
24 Cr	3		$M_{III} N_{IV,III}$	17.94	0.691
57 La	5		Abs. edge	17.9	0.691
29 Cu	8	Ll	$L_{III} M_I$	18.292	0.6778

Element	I	Line	Level designation	λ (Å)	E (keV)
64 Gd	6	$M\beta$	$M_{IV} N_{VI}$	10.254	1.2091
34 Se	1	Ll	$L_{III} M_I$	10.294	1.2044
30 Zn	1	L_I	Abs. edge	10.31	1.197
31 Ga	9	$L\beta_{3,4}$	$L_I M_{II,III}$	10.359	1.197
92 U	7	$N_{II} P_I$	$N_{II} P_I$	10.40	1.192
32 Ge	8	$L\alpha_{1,2}$	$L_{II} M_{IV,V}$	10.4361	1.18800
64 Gd	3	$M\alpha$	$M_V N_{VI,VII}$	10.46	1.185
70 Yb	1	$M\zeta$	$M_V N_{III}$	10.48	1.183
60 Nd	9	$M\gamma$	$M_{III} N_{IV,V}$	10.505	1.180
63 Eu	5	M_{IV}	Abs. edge	10.711	1.1575
33 As	1	$L\eta$	$L_{II} M_I$	10.734	1.1550
63 Eu	7	$M\beta$	$M_{IV} N_{VI}$	10.750	1.1533
31 Ga	5	L_{II}	Abs. edge	10.828	1.1450
63 Eu	3	$M\alpha$	$M_V N_{VI,VII}$	10.96	1.131
59 Pr	9	$M\gamma$	$M_{III} N_{IV,V}$	10.998	1.1273
63 Eu	5	M_V	Abs. edge	11.013	1.1258
31 Ga	2	$L\beta_1$	$L_{II} M_{IV}$	11.023	1.1248
33 As	1	Ll	$L_{III} M_I$	11.072	1.1198
90 Th	7	$N_{II} P_I$	$N_{II} P_I$	11.07	1.120
31 Ga	1	L_{III}	Abs. edge	11.100	1.1169
30 Zn	7	$L\beta_{3,4}$	$L_I M_{II,III}$	11.200	1.1070
62 Sm	5	$M\beta$	$M_{IV} N_{VI}$	11.27	1.0998
31 Ga	1	M_{IV}	Abs. edge	11.288	1.0983
68 Er	3	$L\alpha_{1,2}$	$L_{III} M_{IV,V}$	11.292	1.09792
62 Sm	1	$M\zeta$	$M_V N_{III}$	11.37	1.0901
58 Ce	5	$M\alpha$	$M_V N_{VI,VII}$	11.47	1.081
62 Sm	5	$M\gamma$	$M_{III} N_{IV,V}$	11.53	1.0749
90 Th	5	M_V	Abs. edge	11.552	1.0732
11 Na	1	$N_{II} O_{IV}$	$N_{II} O_{IV}$	11.56	1.072
11 Na	2	K	Abs. edge	11.569	1.07167
32 Ge	2	$K\beta$	KM	11.575	1.0711
30 Zn	1	$L\eta$	$L_{III} M_I$	11.609	1.0680
67 Ho	9	L_{II}	Abs. edge	11.862	1.04523
11 Na	2	$M\zeta$	$M_V N_{III}$	11.86	1.0450
32 Ge	3	$K\alpha_{1,2}$	$KL_{II,III}$	11.9101	1.04098
30 Zn	5	Ll	$L_{III} M_I$	11.965	1.0362
57 La	3	$L\beta_1$	$L_{II} M_{IV,V}$	11.983	1.0347
30 Zn	5	$M\gamma$	$M_{III} N_{IV,III}$	12.08	1.027
29 Cu	8	$L\beta_{3,4}$	$L_I M_{II,III}$	12.122	1.0228

TABLE 7f-2. WAVELENGTHS OF X-RAY EMISSION LINES AND ABSORPTION EDGES: IN NUMERICAL ORDER (Continued)

Wavelength, Å*	p.e.†	Element	Line	Designation	keV
18.32	2	9 F	$K\alpha$	KL	0.6768
18.35	4	58 Ce	$M\zeta$	M_V-N_{III}	0.676
18.8	1	51 Sb		$M_{II}N_I$	0.658
18.8	2	47 Ag		$M_I N_{II,III}$	0.658
18.96	4	24 Cr	$L\beta_{3,4}$	$L_I M_{III}$	0.654
19.11	2	25 Mn	$L\beta_1$	$L_{II}M_{IV}$	0.6488
19.1	1	52 Te		$M_{III}N_I$	0.648
19.40	7	48 Cd		$M_{II}N_{IV}$	0.639
19.44	5	57 La	$M\zeta$	M_V-N_{III}	0.638
19.45	1	25 Mn	$L\alpha_{1,2}$	$L_{III}M_{IV,V}$	0.6374
19.66	5	53 I	$M_{IV,V}$	Abs. edge	0.631
19.75	4	26 Fe	$L\eta$	$L_{II}N_I$	0.628
20.0	1	50 Sn		$M_I N_{II,III}$	0.619
20.1	2	46 Pd	Ll	$L_{III}M_I$	0.616
20.2	1	26 Fe		$M_{III}N_{IV,V}$	0.6152
20.47	7	48 Cd	$M\gamma$	$M_V N_{III}$	0.612
20.64	4	56 Ba	$M\zeta$	$M_{II}N_{IV}$	0.606
20.66	1	47 Ag	L_{III}	$M_{II}N_I$	0.601
20.7	5	24 Cr	$L\beta_{3,4}$	$L_I M_{II,III}$	0.600
21.19	1	23 Va	$L\beta_1$	$L_{II}M_{IV}$	0.598
21.27	5	24 Cr		$L_{II}M_{IV}$	0.5828
21.34	3	52 Te		$M_{IV}O_{II,III}$	0.581
21.5	5	50 Sn		$M_{III}N_I$	0.5728
21.64	7	24 Cr	$L\alpha_{1,2}$	$L_{III}M_{IV,V}$	0.569
21.78	2	52 Te	$M\gamma$	$M_V O_{III}$	0.568
21.82	1	47 Ag	$L\eta$	$M_{III}N_{IV,V}$	0.5675
21.85	1	25 Mn	Ll	$L_{II}N_I$	0.560
22.1	2	46 Pd		$M_{II}N_I$	0.5563
22.29	1	25 Mn		$L_{III}N_I$	0.540
22.9	1	48 Cd		$M_{II}N_I$	0.5317
23.32	2	8 O	K	Abs. edge	0.531
23.3	1	46 Pd	$M\gamma$	$M_{III}N_{IV,V}$	
33.1	2	41 Nb		$M_{II}N_{IV}$	0.375
33.5	3	47 Ag		$M_{IV,V}O_{II,III}$	0.370
33.57	9	90 Th		$N_{IV}N_{VI}$	0.3693
34.8	1	92 U		N_V-$N_{VI,VII}$	0.357
34.9	2	41 Nb		$M_{III}N_{IV,V}$	0.356
35.13	2	21 Sc	$M\gamma$	$L_{II}M_I$	0.3529
35.13	1	20 Ca	$L\eta$	L_{II}	0.3529
35.3	3	42 Mo		$M_{II}N_I$	0.351
35.49	1	20 Ca		Abs. edge	0.34931
35.59	3	21 Sc	L_{III}	$L_{III}M_I$	0.3483
35.63	1	20 Ca	Ll	$L_{II,III}$	0.34793
35.94	2	20 Ca	$L\beta_1$	$L_{II}M_{IV}$	0.3449
36.32	9	90 Th		N_V-$N_{VI,VII}$	0.3414
36.33	2	20 Ca	$L\alpha_{1,2}$	$L_{III}M_{IV,V}$	0.3413
36.8	1	48 Cd	$M\zeta$	$M_{IV,V}$-$N_{II,III}$	0.3371
37.4	2	46 Pd		$M_{IV,V}O_{II,III}$	0.332
37.5	2	42 Mo		$M_{III}N_I$	0.331
38.4	3	41 Nb		$M_{II}N_I$	0.323
39.77	7	47 Ag	$M\zeta$	$M_{IV,V}$-$N_{II,III}$	0.3117
40.46	2	20 Ca	$L\eta$	$L_{II}M_I$	0.3064
40.7	2	41 Nb		$M_{III}N_I$	0.305
40.9	2	45 Rh		$M_{IV,V}O_{II,III}$	0.303
40.96	2	20 Ca	Ll	$L_{II}M_I$	0.3027
42.1	2	92 U		$N_{VI}O_V$	0.295
42.1	1	19 K	$L_{II,III}$	Abs. edge	0.2946
42.3	2	82 Pb		$N_{IV}O_V$	0.293
42.3	2	92 U		$N_{VI}O_{IV}$	0.286
43.3	1	46 Pd	$M\zeta$	$M_{IV,V}$-$N_{II,III}$	0.2844
43.6	3	6 C	K	Abs. edge	0.28384
43.68	1	6 C	$K\alpha$	KL	0.277
44.7	1	44 Ru		$M_{IV,V}O_{II,III}$	0.2768
45.0	2	82 Pb		N_V-$N_{VI,VII}$	0.2756
45.2	3	80 Hg		$N_{IV}N_{VI}$	0.274

λ (Å)	Transition	Designation	Element	Int.	E
0.2743	$M_{II}M_{IV}$		51 Sb	1	45.2
0.267	$M_{II}N_{I}$		39 Y	9	46.48
0.267	$N_{V}N_{VI,VII}$		81 Tl	2	46.5
0.265	$N_{IV}N_{VI}$		79 Au	2	46.8
0.2625	$L_{II}M_{I}$	$L\eta$	19 K	2	47.24
0.2621	$M_{II}M_{IV}$		50 Sn	9	47.3
0.2601	$M_{IV,V}N_{II,III}$	$M\zeta$	45 Rh	1	47.67
0.25971	$L_{III}M_{I}$	Ll	19 K	3	47.74
0.259	$N_{V}N_{VI,VII}$		80 Hg	2	47.9
0.258	$N_{IV}N_{VI}$		78 Pt	1	48.1
0.2572	$N_{IV}O_{V}$		90 Th	2	48.2
0.256	$M_{III}N_{I}$		39 Y	1	48.5
0.2510	$N_{V}N_{VI,VII}$		79 Au	1	49.4
0.2505	$N_{VI}O_{IV}$		90 Th	1	49.5
0.2479	$N_{VI}O_{V}$		90 Th	1	50.0
0.2470	$N_{IV}N_{VI}$		77 Ir	1	50.2
0.2465	$M_{III}N_{V}$		52 Te	1	50.3
0.2436	$N_{V}N_{VI,VII}$		78 Pt	1	50.9
0.2416	$M_{II}N_{I}$		38 Sr	1	51.3
0.2388	$N_{IV}N_{VI}$		76 Os	2	51.9
0.2384	$M_{II}N_{VI}$		48 Cd	1	52.0
0.2375	$M_{III}M_{V}$		51 Sb	7	52.2
0.2369	$M_{IV,V}N_{II,III}$	$M\zeta$	44 Ru	1	52.34
0.2348	$N_{V}N_{VI,VII}$		77 Ir	1	52.8
0.2313	$M_{II}N_{I}$		38 Sr	2	53.6
0.2295	$N_{II}N_{IV}$		74 W	1	54.0
0.2295	$M_{II}M_{IV}$		47 Ag	1	54.0
0.2287	$M_{III}M_{V}$		50 Sn	2	54.2
0.2266	$N_{V}N_{VI,VII}$		76 Os	2	54.7
0.2262	$M_{IV,V}O_{II,III}$		42 Mo	1	54.8
0.2221	$N_{IV}N_{VI}$		74 W	1	55.8
0.2217	$L_{II}M_{I}$	$L\eta$	18 A	1	55.9
0.2201	$L_{III}M_{I}$	Ll	18 A	2	56.3
0.2194	$M_{II}M_{IV}$		46 Pd	1	56.5
0.2174	$M_{II}N_{I}$		37 Rb	1	57.0
0.2130	$N_{IV}N_{VI}$		73 Ta	1	58.2
0.2122	$N_{V}N_{VII}$		74 W	2	58.4
0.2111	$M_{III}M_{V}$		48 Cd	2	58.7
0.2090	$M_{II}M_{IV}$		45 Rh	1	59.3

λ (Å)	Transition	Designation	Element	Int.	E
0.5249	KL	$K\alpha$	8 O	3	23.62
0.5192	$L_{II}M_{IV}$	$L\beta_1$	23 Va	4	23.88
0.5113	$L_{III}M_{IV,V}$	$L\alpha_{1,2}$	23 Va	3	24.25
0.511	Abs. edge	$M_{IV,V}$	50 Sn	5	24.28
0.5102	$L_{II}M_{I}$	$L\eta$	24 Cr	3	24.30
0.509	$M_{V}N_{I}$		47 Ag	2	24.4
0.507	$M_{III}N_{I}$		48 Cd	1	24.5
0.5003	$L_{III}M_{I}$	Ll	24 Cr	1	24.78
0.496	$M_{III}N_{IV,V}$	$M\gamma$	45 Rh	9	25.01
0.491	$M_{IV}O_{II,III}$		50 Sn	3	25.3
0.486	$M_{II}N_{IV}$		44 Ru	9	25.50
0.483	$M_{V}O_{III}$		50 Sn	1	25.7
0.478	$M_{III}N_{I}$	$M\gamma$	47 Ag	1	26.0
0.474	$M_{II}N_{I}$		46 Pd	2	26.2
0.464	$M_{IV,V}N_{II,III}$	$M\zeta$	52 Te	9	26.72
0.462	$M_{III}N_{IV,V}$	$M\gamma$	44 Ru	1	26.9
0.4584	$L_{II}M_{IV}$	$L\beta_1$	22 Ti	2	27.05
0.4544	Abs. edge	$L_{II,III}$	22 Ti	3	27.29
0.4535	$L_{II}M_{I}$	$L\eta$	22 Ti	2	27.34
0.4522	$L_{II}M_{IV,V}$	$L\alpha_{1,2}$	23 Va	1	27.42
0.4465	$L_{III}M_{I}$	Ll	22 Ti	2	27.77
0.445	$M_{III}N_{I}$		23 Va	5	27.9
0.442	$M_{II}N_{I}$		46 Pd	8	28.1
0.4408	Abs. edge	$M_{IV,V}$	45 Rh	1	28.13
0.429	$M_{IV,V}N_{II,III}$	$M\zeta$	48 Cd	1	28.88
0.417	$M_{III}N_{I}$		51 Sb	1	29.8
0.408	$M_{V}O_{III}$	M_{IV}	45 Rh	5	30.4
0.403	$M_{V}O_{III}$		48 Cd	3	30.8
0.4022	Abs. edge	M_{IV}	48 Cd	1	30.82
0.4013	$L_{II}M_{I}$	$L\eta$	47 Ag	2	30.89
0.4000	Abs. edge	K	22 Ti	5	30.99
0.3996	$L_{II}M_{IV}$	$L\beta_1$	7 N	9	31.02
0.3981	Abs. edge	M_{V}	21 Sc	3	31.14
0.397	$M_{IV,V}N_{II,III}$	$M\zeta$	47 Ag	2	31.24
0.3954	$L_{III}M_{IV,V}$	$L\alpha_{1,2}$	50 Sn	4	31.35
0.3953	$L_{III}M_{I}$	Ll	21 Sc	1	31.36
0.3924	KL	$K\alpha$	22 Ti	2	31.60
0.390	$N_{IV}N_{VI}$		7 N	1	31.8
0.384	$M_{II}N_{I}$		92 U	2	32.3
			44 Ru	1	

TABLE 7f-2. WAVELENGTHS OF X-RAY EMISSION LINES AND ABSORPTION EDGES: IN NUMERICAL ORDER (Continued)

Wavelength, Å*	p.e.†	Element	Designation	keV
59.5	3	74 W	$N_V N_{VI}$	0.208
59.5	2	37 Rb	$M_{III}N_I$	0.2083
60.5	1	47 Ag	$M_{III}M_V$	0.2048
61.1	2	73 Ta	$N_V N_{VI,VII}$	0.2028
61.9	2	41 Nb	$M_{IV,V}O_{II,III}$	0.2002
62.2	1	44 Ru	$M_{II}M_{IV}$	0.1992
62.9	1	46 Pd	$M_{III}M_V$	0.1970
63.0	5	71 Lu	$N_{IV}N_{VI}$	0.197
64.38	7	42 Mo	$M_{IV,V}N_{II,III}$ $M\zeta$	0.1926
65.1	7	70 Yb	$N_{IV}N_{VI}$	0.190
65.5	1	45 Rh	$M_{III}M_V$	0.1892
65.7	2	71 Lu	$N_V N_{VI,VII}$	0.1886
67.33	9	17 Cl	$L_{II}M_I$ $L\eta$	0.1841
67.6	3	5 B	KL $K\alpha$	0.1833
67.90	9	17 Cl	$L_{III}M_I$ Ll	0.1826
68.2	3	90 Th	$O_{III}P_{IV,V}$	0.1817
68.3	1	44 Ru	$M_{III}M_V$	0.1814
68.9	2	42 Mo	$M_{II}M_{IV}$	0.1798
69.3	5	70 Yb	$N_V N_{VI,VII}$	0.179
70.0	4	40 Zr	$M_{IV,V}O_{II,III}$	0.177
72.1	3	41 Nb	$M_I M_{III}$	0.1718
72.19	9	41 Nb	$M_{IV,V}N_{II,III}$ $M\zeta$	0.1717
72.7	9	68 Er	$N_{IV}N_{VI}$	0.171
74.9	1	42 Mo	$M_{III}M_V$	0.1656
76.3	7	68 Er	$N_V N_{VI,VII}$	0.163
76.7	2	40 Zr	$M_{II}M_{IV}$	0.1617
76.9	2	35 Br	$M_{II}N_I$	0.1613
78.4	3	41 Nb	$M_{III}M_V$	0.1582
79.8	3	35 Br	$M_{III}N_I$	0.1554
80.9	2	40 Zr	$M_{III}M_V$	0.1533
81.5	2	39 Y	$M_{III}M_V$	0.1522
82.1	2	40 Zr	$M_{IV,V}N_{II,III}$ $M\zeta$	0.1511
83.	1	66 Dy	$N_{IV,V}N_{VI,VII}$	0.149
117.4	4	62 Sm	$N_{IV,V}O_{II,III}$	0.1056
117.7	1	81 Tl	$N_{VII}O_V$	0.10530
123.	1	14 Si	Abs. edge $L_{II,III}$	0.1006
126.8	2	37 Rb	$M_{IV}N_{III}$	0.0978
127.8	2	37 Rb	$M_V N_{II}$ $M\zeta_2$	0.0970
128.7	2	37 Rb	$M_V N_{II}$ $M\zeta_1$	0.0964
128.9	7	60 Nd	$N_{IV,V}O_{II,III}$	0.0962
135.5	4	14 Si	$L_{II,III}M$	0.0915
136.5	5	59 Pr	$N_{IV,V}O_{II,III}$	0.0908
137.0	5	30 Zn	Abs. edge M_{II}	0.0905
142.5	1	13 Al	Abs. edge L_I	0.08701
143.9	5	30 Zn	Abs. edge M_{III}	0.0862
144.4	6	58 Ce	$N_{IV,V}O_{II,III}$	0.0859
144.4	6	37 Rb	$M_I M_{III}$	0.0859
152.6	6	57 La	$N_{IV,V}O_{II,III}$	0.0812
157.	3	30 Zn	$M_{II,III}M_{IV,V}$	0.079
159.0	2	56 Ba	$N_{IV}O_{III}$	0.07796
159.5	5	29 Cu	Abs. edge M_{II}	0.0777
163.3	2	56 Ba	$N_{IV}O_{III}$	0.07590
164.6	2	56 Ba	$N_V O_{III}$	0.07530
164.7	5	35 Br	$M_I M_{III}$	0.0753
166.0	1	29 Cu	Abs. edge M_{III}	0.0747
170.4	5	13 Al	Abs. edge $L_{II,III}$	0.07278
171.4	5	13 Al	$L_{II,III}M$	0.0724
173.	3	29 Cu	$M_{II,III}M_{IV,V}$	0.072
181.	5	90 Th	$O_{IV,V}Q_{II,III}$	0.068
183.8	1	55 Cs	$N_{IV}O_{III}$	0.06746
184.6	3	35 Br	$M_I M_{II}$	0.0672
188.4	1	28 Ni	Abs. edge M_{III}	0.06581
188.6	3	55 Cs	$N_{IV}O_{II}$	0.06574
189.5	3	35 Br	$M_{IV}N_{III}$	0.0654
190.3	1	55 Cs	$N_V O_{III}$	0.06515
190.	2	28 Ni	$M_{II,III}M_{IV,V}$	0.0651

Int.	Element	Designation	Transition	λ (Å)	E (keV)
3	16 S	Ll, η	$L_{II,III}\,M_I$	83.4	0.1487
2	38 Sr		$M_{II}\,M_{IV}$	85.7	0.1447
1	65 Tb		$N_{IV,V}\,N_{VI,VII}$	86.	0.144
2	39 Y		$M_{III}\,M_{IV,V}$	86.5	0.1434
2	38 Sr		$M_{II}\,M_{IV}$	91.4	0.1357
2	37 Rb		$M_{II}\,M_{IV}$	91.5	0.1355
1	83 Bi		$N_{VII}\,O_{IV}$	91.6	0.1354
1	83 Bi		$N_{VI,V}\,N_{II,III}$	93.2	0.1330
2	39 Y	Mζ	Abs. edge	93.4	0.1328
1	15 P	$L_{II,III}$	$M_{III}\,M_{IV,V}$	94.	0.132
2	37 Rb		$N_{IV,V}\,O_{II,III}$	96.7	0.1282
8	66 Dy		$N_{VI,V}\,N_{VI,VII}$	97.2	0.128
1	62 Sm		$N_{VI}\,O_V$	98.	0.126
2	82 Pb		$N_{IV,V}\,O_{II,III}$	100.2	0.1237
4	65 Tb		$N_{VI}\,O_{IV}$	102.2	0.1213
1	82 Pb		$L_{II,III}\,M$	102.4	0.1211
4	15 P		$N_{VII}\,O_V$	103.8	0.1194
1	82 Pb		$N_{IV,V}\,N_{VI,VII}$	104.3	0.1189
1	60 Nd		$M_{IV}\,N_{II,III}$	107.	0.116
2	38 Sr	Mζ2	$M_V\,N_{III}$	108.0	0.1148
1	38 Sr	Mζ1	$M_{II}\,M_{IV}$	108.7	0.1140
3	35 Br	M_I	Abs. edge	109.4	0.1133
5	29 Cu	K	Abs. edge	110.6	0.1121
1	4 Be		$N_{IV,V}\,O_{II,III}$	111.	0.111
6	63 Eu		$N_{VI}\,O_V$	112.0	0.1107
1	81 Tl		$N_{IV,V}\,N_{VI,VII}$	113.0	0.10968
1	59 Pr		$M_{III}\,M_{IV,V}$	113.	0.1095
3	35 Br	Kα	KL	113.8	0.1089
1	4 Be		$M_{III}\,M_{IV,V}$	114.	0.1085
2	81 Tl		$N_{VI}\,O_{IV}$	115.3	0.1075

Int.	Element	Designation	Transition	λ (Å)	E (keV)
2	35 Br	Mζ2	$M_{IV}\,N_{II}$	191.1	0.06488
2	35 Br	Mζ1	$M_V\,N_{III}$	192.6	0.06437
1	12 Mg	L_I	Abs. edge	197.3	0.06284
5	27 Co	$M_{II,III}$	Abs. edge	202.	0.061
1	16 S		$L_I\,L_{II,III}$	203.	0.061
6	27 Co		$M_{II,III}\,M_{IV,V}$	214.	0.058
1	53 I	$N_{IV,V}$	Abs. edge	224.	0.0552
1	3 Li	K	Abs. edge	226.5	0.05475
1	34 Se	M_V	Abs. edge	227.8	0.05443
1	3 Li	Kα	KL	228.	0.0543
2	34 Se		$M_V\,N_{III}$	230.	0.0538
1	26 Fe	$M_{II,III}$	Abs. edge	230.	0.0538
5	26 Fe		$M_{II,III}\,M_{IV,V}$	243.	0.051
1	12 Mg	L_{II}	Abs. edge	249.3	0.04973
1	12 Mg	L_{III}	Abs. edge	250.7	0.04945
5	12 Mg		$L_{II,III}\,M$	251.5	0.04929
6	25 Mn		$M_{II,III}\,M_{IV,V}$	273.	0.045
1	13 Al		$M_{II,III}\,M_{IV,V}$	290.	0.0428
9	24 Cr		$L_I\,L_{II,III}$	309.	0.040
1	12 Mg		$M_{II,III}\,M_{IV,V}$	317.	0.0392
9	23 V		$L_I\,L_{II,III}$	337.	0.0368
1	11 Na	N_I	Abs. edge	376.	0.03299
5	35 Br	$L_{II,III}$	Abs. edge	399.	0.0311
5	11 Na		$L_{II,III}\,M$	405.	0.0306
5	11 Na	M_I	Abs. edge	407.1	0.03045
5	17 Cl	O_I	Abs. edge	417.	0.0297
5	53 I		$M_{II,III}\,N_I$	444.	0.0279
9	20 Ca		$M_{II,III}\,N_I$	525.	0.0236
9	19 K		$M_{II,III}\,N_I$	692.	0.0179

TABLE 7f-3. RECOMMENDED VALUES OF THE ATOMIC ENERGY LEVELS, AND PROBABLE ERRORS IN eV*

Level	1 H	2 He	3 Li	4 Be	5 B	6 C	7 N	8 O
K	13.59811[a]	24.58678[b]	54.75 ± 0.02 (54.75)	111.0 ± 1.0 (111.0)	188.0 ± 0.4 [188.0][c]	283.8 ± 0.4 [283.8][c] (283.8)	401.6 ± 0.4 [401.6][c]	532.0 ± 0.4 [532.0][c]
L_I	. . . .	. . . .	. . . .	. . . .	. . . .	. . . .	. . . .	23.7 ± 0.4 [23.7][d]
L_{II,III}	. . . .	. . . .	. . . .	. . . .	4.7 ± 0.9	6.4 ± 1.9	9.2 ± 0.6	7.1 ± 0.8

Level	9 F	10 Ne	11 Na	12 Mg	13 Al	14 Si	15 P	16 S
K	685.4 ± 0.4 [685.4][c]	866.9 ± 0.3 (866.9)	1072.1 ± 0.4 [1072.1][c] (1072.)	1305.0 ± 0.4 [1305.0][c] (1303.)	1559.6 ± 0.4 [1559.6][c] (1559.8)	1838.9 ± 0.4 [1838.9][c]	2145.5 ± 0.4 [2145.5][d]	2472.0 ± 0.4 [2472.0][c] (2470.)
L_I	⟨31.⟩	⟨45.⟩	63.3 ± 0.4 [63.3][d]	89.4 ± 0.4 [89.4][d] (63.)	117.7 ± 0.4 [117.7][d] (87.)	148.7 ± 0.4 [148.7][d]	189.3 ± 0.4 [189.3][d]	229.2 ± 0.4 [229.2][d]
L_{II,III}	8.6 ± 0.8	18.3 ± 0.4	31.1 ± 0.4 (31.)	51.4 ± 0.5 (50.)	73.1 ± 0.5 (72.8)	99.2 ± 0.5 (100.6)	132.2 ± 0.5 (132.)	164.8 ± 0.7

* Where applicable, photoelectron direct measurements are listed in square brackets [ ] immediately under the recommended values. The measured values of the X-ray absorption energies[g] are shown in parentheses ( ). Interpolated values are enclosed in angular brackets ⟨ ⟩.

TABLE 7f-3. RECOMMENDED VALUES OF THE ATOMIC ENERGY LEVELS, AND PROBABLE ERRORS IN eV* (Continued)

Level	17 Cl	18 Ar	19 K	20 Ca	21 Sc	22 Ti	23 V	24 Cr
K	2822.4 ± 0.3 [2822.4]c (2020.)	3202.9 ± 0.3 (3202.9)	3607.4 ± 0.4 [3607.4]c (3607.8)	4038.1 ± 0.4 [4038.1]c (4038.1)	4492.8 ± 0.4 [4492.8]c	4966.4 ± 0.4 [4966.4]d (4964.5)	5465.1 ± 0.3 [5465.1]c (5464.)	5989.2 ± 0.3 [5989.2]c (5989.)
L_I	270.2 ± 0.4 [270.2]d	320. (320.)d	377.1 ± 0.4 [377.1]d	437.8 ± 0.4 [437.8]d	500.4 ± 0.4 [500.4]d	563.7 ± 0.4 [563.7]d	628.2 ± 0.4 [628.2]d	694.6 ± 0.4 [694.6]d
L_{II}	201.6 ± 0.3	247.3 ± 0.3	296.3 ± 0.4	350.0 ± 0.4	406.7 ± 0.4	461.5 ± 0.4	520.5 ± 0.3	583.7 ± 0.3
L_{III}	200.0 ± 0.3	245.2 ± 0.3	293.6 ± 0.4	346.4 ± 0.4	402.2 ± 0.4	455.5 ± 0.4	512.9 ± 0.3	574.5 ± 0.3
M_I	17.5 ± 0.4	25.3 ± 0.4	33.9 ± 0.4	43.7 ± 0.4	53.8 ± 0.4	60.3 ± 0.4	66.5 ± 0.4	74.1 ± 0.4
$M_{II,III}$	6.8 ± 0.4	12.4 ± 0.3	17.8 ± 0.4	25.4 ± 0.4	32.3 ± 0.5	34.6 ± 0.4	37.8 ± 0.3	42.5 ± 0.3
$M_{IV,V}$	......	......	......	......	6.6 ± 0.5	3.7	2.2 ± 0.3	2.3 ± 0.4

Level	25 Mn	26 Fe	27 Co	28 Ni	29 Cu	30 Zn	31 Ga	32 Ge
K	6539.0 ± 0.4 [6539.0]c (6538.)	7112.0 ± 0.9 [7111.3]e,f (7111.2)	7708.9 ± 0.3 [7708.9]e (7709.5)	8332.8 ± 0.4 [8332.8]e (8331.6)	8978.9 ± 0.4 [8978.9]e,a (8980.3)	9658.6 ± 0.6 [9658.6]v (9660.7)	10367.1 ± 0.5 [10367.1]v (10368.2)	11103.1 ± 0.7 [11103.8]v (11103.6)
L_I	769.0 ± 0.4 [769.0]d	846.1 ± 0.4 [846.1]d	925.6 ± 0.4 [925.6]d	1008.1 ± 0.4 [1008.1]d	1096.1 ± 0.4 [1096.0]d	1193.6 ± 0.9	1297.7 ± 1.1	1414.3 ± 0.7 [1413.6]v
L_{II}	651.4 ± 0.4	721.1 ± 0.9 (720.8)	793.6 ± 0.3 (793.8)	871.9 ± 0.4 (870.6)	951.0 ± 0.4 [950.0]A (953.)	1042.8 ± 0.6 (1045.)	1142.3 ± 0.5	1247.8 ± 0.7 (1249.)
L_{III}	640.3 ± 0.4	708.1 ± 0.9 (707.4)	778.6 ± 0.3 (779.0)	854.7 ± 0.4 (853.6)	931.1 ± 0.4 [931.4]A (933.)	1019.7 ± 0.6 (1022.)	1115.4 ± 0.5 (1117.)	1216.7 ± 0.7 (1217.0)
M_I	83.9 ± 0.5	92.9 ± 0.9	100.7 ± 0.4	111.8 ± 0.6	119.8 ± 0.6	135.9 ± 1.1	158.1 ± 0.5	180.0 ± 0.8
M_{II}, M_{III}	48.6 ± 0.4	54.0 ± 0.9 (54.)	59.5 ± 0.3 (61.)	68.1 ± 0.4 (66.)	73.6 ± 0.4 (75.)	86.6 ± 0.6 (86.)	106.8 ± 0.7 / 102.9 ± 0.5	127.9 ± 0.9 / 120.8 ± 0.7
$M_{IV,V}$	3.3 ± 0.5	3.6 ± 0.9	2.9 ± 0.3	3.6 ± 0.4	1.6 ± 0.4	8.1 ± 0.6	17.4 ± 0.5	28.7 ± 0.7

TABLE 7f-3. RECOMMENDED VALUES OF THE ATOMIC ENERGY LEVELS, AND PROBABLE ERRORS IN eV* (Continued)

Level	33 As	34 Se	35 Br	36 Kr	37 Rb	38 Sr	39 Y	40 Zr
K	11866.7 ± 0.7 [11866.7]i (11865.)	12657.8 ± 0.7 [12657.8]p (12654.5)	13473.7 ± 0.4 (13470.)	14325.6 ± 0.8 (14324.4)	15199.7 ± 0.3 (15202.)	16104.6 ± 0.3 (16107.)	17038.4 ± 0.3 (17038.)	17997.6 ± 0.4 (17999.)
L_I	1526.5 ± 0.8	1653.9 ± 3.5	1782.0 ± 0.4 [1782.0]j	1921.0 ± 0.6 [1921.2]k	2065.1 ± 0.3 [2065.4]j	2216.3 ± 0.3 [2216.2]j	2372.5 ± 0.3 [2372.7]l	2531.6 ± 0.3 [2531.6]l
L_II	(1529.) 1358.6 ± 0.7	(1652.5) 1476.2 ± 0.7	1596.0 ± 0.4 [1596.2]j	1727.2 ± 0.5 [1727.2]k (1730.)	1863.9 ± 0.3 [1863.4]j	2006.8 ± 0.3 [2006.6]j (2008.5)	2155.5 ± 0.3 [2155.0]l (2154.0)	2306.7 ± 0.3 [2306.5]l (2305.3)
L_III	(1358.7) 1323.1 ± 0.7	(1474.7) 1435.8 ± 0.7	1549.9 ± 0.4 [1549.7]j	1674.9 ± 0.5 [1674.8]k (1677.)	1804.4 ± 0.3 [1804.6]j	1939.6 ± 0.3 [1939.9]j (1941.)	2080.0 ± 0.3 [2080.2]l (2079.4)	2222.3 ± 0.3 [2222.5]l (2222.5)
M_I	(1323.5) 203.5 ± 0.7	(1434.0) 231.5 ± 0.7	256.5 ± 0.4	....	322.1 ± 0.3	357.5 ± 0.3	393.6 ± 0.3	430.3 ± 0.3
M_II	146.4 ± 1.2	168.2 ± 1.3	189.3 ± 0.4	222.7 ± 1.1	247.4 ± 0.3	279.8 ± 0.3	312.4 ± 0.4	344.2 ± 0.4
M_III	140.5 ± 0.8	161.9 ± 1.0	181.5 ± 0.4	213.8 ± 1.1	238.5 ± 0.3	269.1 ± 0.3	300.3 ± 0.4	330.5 ± 0.4
M_IV	{ 41.2 ± 0.7	{ 56.7 ± 0.8	70.1 ± 0.4	{ 88.9 ± 0.8	111.8 ± 0.3	135.0 ± 0.3	159.6 ± 0.3	182.4 ± 0.3
M_V			69.0 ± 0.4		110.3 ± 0.3	133.1 ± 0.3	157.4 ± 0.3	180.0 ± 0.3
N_I			27.3 ± 0.5	24.0 ± 0.8	29.3 ± 0.3	37.7 ± 0.3	45.4 ± 0.3	51.3 ± 0.3
N_II	{ 2.5 ± 1.0	{ 5.6 ± 1.3	5.2 ± 0.4	{ 10.6 ± 1.9	{ 14.8 ± 0.4	19.9 ± 0.3	25.6 ± 0.4	28.7 ± 0.4
N_III			4.6 ± 0.4		14.0 ± 0.3			

Level	41 Nb	42 Mo	43 Tc	44 Ru	45 Rh	46 Pd	47 Ag	48 Cd
K	18985.6 ± 0.4 (18987.)	19999.5 ± 0.3 (20004.)	21044.0 ± 0.7	22117.2 ± 0.3 (22119.)	23219.9 ± 0.3 (23219.8)	24350.3 ± 0.3 (24348.)	25514.0 ± 0.3 (25516.)	26711.2 ± 0.3 (26716.)
L_I	2697.7 ± 0.3 [2697.7]l	2865.5 ± 0.3 [2866.0]l	3042.5 ± 0.4 [3042.5]l	3224.0 ± 0.3 [3224.3]l	3411.9 ± 0.3 [3412.0]l (3417.)	3604.3 ± 0.3 [3604.6]l (3607.)	3805.8 ± 0.3 [3806.2]m (3807.)	4018.0 ± 0.3 [4018.1]m (4019.)
L_II	2464.7 ± 0.3 [2464.7]l	2625.1 ± 0.3 [2624.5]l (2627.)	2793.2 ± 0.4 [2973.2]l	2966.9 ± 0.3 [2966.8]l (2966.3)	3146.1 ± 0.3 [3146.3]l (3145.)	3330.3 ± 0.3 [3330.3]l (3330.3)	3523.7 ± 0.3 [3523.6]v,m (3526.)	3727.0 ± 0.3 [3727.1]m (3728.)
L_III	2370.5 ± 0.3 [2370.6]l	2520.2 ± 0.3 [2520.2]l (2523.2)	2676.9 ± 0.4 [2676.9]l	2837.9 ± 0.3 [2837.7]l (2837.7)	3003.8 ± 0.3 [3003.5]q,i (3002.)	3173.3 ± 0.3 [3173.0]q,i (3173.0)	3351.1 ± 0.3 [3350.8]p (3351.0)	3537.5 ± 0.3 [3537.3]p (3537.6)

Level	49 In	50 Sn	51 Sb	52 Te	53 I	54 Xe	55 Cs	56 Ba
M_I	468.4 ± 0.3	504.6 ± 0.3	....	585.0 ± 0.3	627.1 ± 0.3	669.9 ± 0.3	717.5 ± 0.3	770.2 ± 0.3
M_{II}	378.4 ± 0.4	409.7 ± 0.4	444.9 ± 1.5	482.8 ± 0.3	521.0 ± 0.3	559.1 ± 0.3	602.4 ± 0.3	650.7 ± 0.3
M_{III}	363.0 ± 0.4	392.3 ± 0.3	425.0 ± 1.5	460.6 ± 0.3	496.2 ± 0.3	531.5 ± 0.3	571.4 ± 0.3	616.5 ± 0.3
M_{IV}	207.4 ± 0.3	230.3 ± 0.3	256.4 ± 0.5	283.6 ± 0.3	311.7 ± 0.3	340.0 ± 0.3	372.8 ± 0.3	410.5 ± 0.3
M_V	204.4 ± 0.3	227.0 ± 0.3	252.9 ± 0.4	279.4 ± 0.3	307.0 ± 0.3	334.7 ± 0.3	366.7 ± 0.3	403.7 ± 0.3
N_I	58.1 ± 0.3	61.8 ± 0.3	....	74.9 ± 0.3	81.0 ± 0.3	86.4 ± 0.3	95.2 ± 0.3	107.6 ± 0.3
$N_{II}\}$, $N_{III}\}$	33.9 ± 0.4	34.8 ± 0.4	38.9 ± 1.9	43.1 ± 0.4	47.9 ± 0.4	51.1 ± 0.4	{62.6 ± 0.3 / 55.9 ± 0.3}	66.9 ± 0.4
$N_{IV,V}$	3.2 ± 0.3	1.8 ± 0.3	....	2.0 ± 0.3	2.5 ± 0.4	1.5 ± 0.3	3.3 ± 0.3	9.3 ± 0.3

Level	49 In	50 Sn	51 Sb	52 Te	53 I	54 Xe	55 Cs	56 Ba
K	27939.9 ± 0.3	29200.1 ± 0.3 (29195.)	30491.2 ± 0.3 (30486.)	31813.8 ± 0.3 (31811.)	33169.4 ± 0.4 (33167.)	34561.4 ± 1.1 (34590.)	35984.6 ± 0.4 (35987.)	37440.6 ± 0.4 (37452.)
L_I	4237.5 ± 0.3 [4237.7]^m (4237.3)	4464.7 ± 0.3 [4464.5]^v (4464.8)	4698.3 ± 0.3 [4698.3]^m (4698.4)	4939.2 ± 0.3 [4939.3]^m (4939.7)	5188.1 ± 0.3 [5188.1]^j	5452.8 ± 0.4 (5452.8)	5714.3 ± 0.4 [5712.7]^j (5721.)	5988.8 ± 0.4 [5986.8] (5996.)
L_{II}	3938.0 ± 0.3 [3937.8]^m (3939.3)	4156.1 ± 0.3 [4156.2]^v (4157.)	4380.4 ± 0.3 [4380.6]^m (4382.)	4612.0 ± 0.3 [4612.0]^m (4612.6)	4852.1 ± 0.3 [4852.0]^j	5103.7 ± 0.4 (5103.7)	5359.4 ± 0.3 [5359.5]^j (5358.)	5623.6 ± 0.3 [5623.6] (5623.3)
L_{III}	3730.1 ± 0.3 [3730.1]^v (3730.2)	3928.8 ± 0.3 [3928.8] (3928.8)	4132.2 ± 0.3 [4132.2]^v (4132.3)	4341.4 ± 0.3 [4341.2]^v (4341.8)	4557.1 ± 0.3 [4557.1]^j	4782.2 ± 0.4 (4782.2)	5011.9 ± 0.3 [5012.0]^j (5011.3)	5247.0 ± 0.3 [5247.3] (5247.0)
M_I	825.6 ± 0.3	883.8 ± 0.3	943.7 ± 0.3	1006.0 ± 0.3	1072.1 ± 0.3	....	1217.1 ± 0.4	1292.8 ± 0.4
M_{II}	702.2 ± 0.3	756.4 ± 0.4	811.9 ± 0.3	869.7 ± 0.3	930.5 ± 0.3	999.0 ± 2.1	1065.0 ± 0.5	1136.7 ± 0.5
M_{III}	664.3 ± 0.3	714.4 ± 0.3	765.6 ± 0.3	818.7 ± 0.3	874.6 ± 0.3	937.0 ± 2.1	997.6 ± 0.5	1062.2 ± 0.5
M_{IV}	450.8 ± 0.3	493.3 ± 0.3	536.9 ± 0.3	582.5 ± 0.3	631.3 ± 0.3	....	739.5 ± 0.5	796.1 ± 0.3
M_V	443.1 ± 0.3	484.8 ± 0.3	527.5 ± 0.3	572.1 ± 0.3	619.4 ± 0.3	672.3 ± 0.5	725.5 ± 0.5	780.7 ± 0.5
N_I	121.9 ± 0.3	136.5 ± 0.4	152.0 ± 0.3	168.3 ± 0.3	186.4 ± 0.3	....	230.8 ± 0.4	253.0 ± 0.5
$N_{II}\}$, $N_{III}\}$	77.4 ± 0.4	88.6 ± 0.4	98.4 ± 0.5	110.2 ± 0.5	122.7 ± 0.5	146.7 ± 3.1	{172.3 ± 0.6 / 161.6 ± 0.6}	{191.8 ± 0.7 / 179.7 ± 0.6}
$N_{IV}\}$, $N_V\}$	16.2 ± 0.3	23.9 ± 0.3	31.4 ± 0.3	39.8 ± 0.3	49.6 ± 0.3	....	{78.8 ± 0.5 / 76.5 ± 0.5}	{92.5 ± 0.5 / 89.9 ± 0.5}
$O_I\}$, O_{II}	0.1 ± 4.5	0.9 ± 0.5	6.7 ± 0.5	11.6 ± 0.6	13.6 ± 0.6	....	{22.7 / 13.1 ± 0.5}	{39.1 ± 0.6 / 16.6 ± 0.5}
O_{III}	0.8 ± 0.4	1.1 ± 0.5	2.1 ± 0.4	2.3 ± 0.5	3.3 ± 0.5	....	11.4 ± 0.5	14.6 ± 0.5

TABLE 7f-3. RECOMMENDED VALUES OF THE ATOMIC ENERGY LEVELS, AND PROBABLE ERRORS IN eV* (Continued)

Level	57 La	58 Ce	59 Pr	60 Nd	61 Pm	62 Sm	63 Eu	64 Gd
K	38924.6 ± 0.4 (38934.)	40443.0 ± 0.4 (40453.)	41990.6 ± 0.5 (42002.)	43568.9 ± 0.4 (43574.)	45184.0 ± 0.7 (45198.)	46834.2 ± 0.5 (46849.)	48519.0 ± 0.4 (48519.)	50239.1 ± 0.5 (50233.)
L_I	6266.3 ± 0.5 [6266.3]^n	6548.8 ± 0.5 [6548.5]^n	6834.8 ± 0.5 [6834.9]^n	7126.0 ± 0.4 [7125.8]^n (7129.)	7427.9 ± 0.8 [7427.9]^o	7736.8 ± 0.5 [7736.2]^n (7748.)	8052.0 ± 0.4 [8051.7]^n (8061.)	8375.6 ± 0.5 [8375.4]^n (8386.)
L_{II}	5890.6 ± 0.4 [5890.7]^n	6164.2 ± 0.4 [6164.3]^n	6440.4 ± 0.5 [6440.2]^n	6721.5 ± 0.4 [6721.8]^o	7012.8 ± 0.6 [7012.8]^o	7311.8 ± 0.4 [7312.0]^n (7313.)	7617.1 ± 0.4 [7617.6]^n (7620.)	7930.3 ± 0.4 [7930.5]^n (7931.)
L_{III}	5482.7 ± 0.4 [5482.6]^n	5723.4 ± 0.4 [5723.6]^n	5964.3 ± 0.4 [5964.3]^n	6207.9 ± 0.4 [6208.0]^n (6209.)	6459.3 ± 0.6 [6459.4]^o	6716.2 ± 0.5 [6716.8]^n (6717.)	6976.9 ± 0.4 [6976.7]^n (6981.)	7242.8 ± 0.4 [7242.8]^n (7243.)
M_I	1361.3 ± 0.3	1434.6 ± 0.6	1511.0 ± 0.8	1575.3 ± 0.7	.....	1722.8 ± 0.8	1800.0 ± 0.5	1880.8 ± 0.5
M_{II}	1204.4 ± 0.6	1272.8 ± 0.6	1337.4 ± 0.7	1402.8 ± 0.6	1471.4 ± 6.2	1540.7 ± 1.2	1613.9 ± 0.7	1688.3 ± 0.7
M_{III}	1123.4 ± 0.5	1185.4 ± 0.5	1242.2 ± 0.6	1297.4 ± 0.5	1356.9 ± 1.4	1419.8 ± 1.1	1480.6 ± 0.6	1544.0 ± 0.8
M_{IV}	848.5 ± 0.4	901.3 ± 0.6	951.1 ± 0.6	999.9 ± 0.6	1051.5 ± 0.9	1106.0 ± 0.9	1160.6 ± 0.6	1217.2 ± 0.6
M_V	831.7 ± 0.4	883.3 ± 0.5	931.0 ± 0.6	977.7 ± 0.6	1026.9 ± 1.0	1080.2 ± 0.6	1130.9 ± 0.6	1185.2 ± 0.6
N_I	270.4 ± 0.8	289.6 ± 0.7	304.5 ± 0.9	315.2 ± 0.8	.....	345.7 ± 0.9	360.2 ± 0.7	375.8 ± 0.7
N_{II}	205.8 ± 1.2	223.3 ± 1.1	236.3 ± 1.5	243.3 ± 1.6	242. ± 16.	265.6 ± 1.9	283.9 ± 1.0	288.5 ± 1.2
N_{III}	191.4 ± 0.9	207.2 ± 0.9	217.6 ± 1.1	224.6 ± 1.3		247.4 ± 1.5	256.6 ± 0.8	270.9 ± 0.9
$N_{IV,V}$	98.9 ± 0.8	110.0 ± 0.6	113.2 ± 0.7	117.5 ± 0.7	120.4 ± 2.0	129.0 ± 1.2	133.2 ± 0.6	140.5 ± 0.8
$N_{VI,VII}$	.....	0.1 ± 1.2	2.0 ± 0.6	1.5 ± 0.9	.....	5.5 ± 1.1	0.0 ± 3.2	0.1 ± 3.5
O_I	32.3 ± 7.2	37.8 ± 1.3	37.4 ± 1.0	37.5 ± 0.9	.....	37.4 ± 1.5	31.8 ± 0.7	36.1 ± 0.8
$O_{II,III}$	14.4 ± 1.2	19.8 ± 1.2	22.3 ± 0.7	21.1 ± 0.8	.....	21.3 ± 1.5	22.0 ± 0.6	20.3 ± 1.2

Level	65 Tb	66 Dy	67 Ho	68 Er	69 Tm	70 Yb	71 Lu	72 Hf
K	51995.7 ± 0.5 (52002.)	53788.5 ± 0.5 (53793.)	55617.7 ± 0.5 (55619.)	57485.5 ± 0.5 (57487.)	59389.6 ± 0.5	61332.3 ± 0.5 (61300.)	63313.8 ± 0.5 (63310.)	65350.8 ± 0.6 (65310.)
L_I	8708.0 ± 0.5 [8707.6]^n (8717.)	9045.8 ± 0.5 [9046.5]^n	9394.2 ± 0.4 [9394.3]^n (9399.)	9751.3 ± 0.4 [9751.5]^n (9757.)	10115.7 ± 0.4 [10115.6]^n (10121.)	10486.4 ± 0.4 [10487.3]^n (10490.)	10870.4 ± 0.4 [10870.1]^n (10874.)	11270.7 ± 0.4 [11271.6]^o (11274.)
L_{II}	8251.6 ± 0.4 [8251.8]^n (8253.)	8580.6 ± 0.4 [8580.4]^n (8583.)	8917.8 ± 0.4 [8918.2]^n (8916.)	9264.3 ± 0.4 [9264.3]^n (9262.)	9616.9 ± 0.4 [9617.1]^n (9617.1)	9978.2 ± 0.4 [9977.9]^n (9976.)	10348.6 ± 0.4 [10349.0]^n (10345.)	10739.4 ± 0.4 [10738.9]^o (10736.)

Level	73 Ta	74 W	75 Re	76 Os	77 Ir	78 Pt	79 Au	80 Hg
L_{III}	7514.0 ± 0.4 [7514.2][n] (7515.)	7790.1 ± 0.4 [7789.6][n] (7789.7)	8071.1 ± 0.4 [8070.6][n] (8068.)	8357.9 ± 0.4 [8357.6][n] (8357.5)	8648.0 ± 0.4 [8647.8][n] (8649.6)	8943.6 ± 0.4 [8942.6][n] (8944.1)	9244.1 ± 0.4 [9243.8][n]	9560.7 ± 0.4 [9560.4][g] (9558.)
M_I	1967.5 ± 0.6	2046.8 ± 0.4	2128.3 ± 0.6	2206.5 ± 0.6	2306.8 ± 0.7	2398.1 ± 0.4	2491.2 ± 0.5	2600.9 ± 0.4
M_{II}	1767.7 ± 0.9	1841.8 ± 0.5	1922.8 ± 1.0	2005.8 ± 0.6	2089.8 ± 1.1	2173.0 ± 0.4	2263.5 ± 0.4	2365.4 ± 0.4
M_{III}	1611.3 ± 0.8	1675.6 ± 0.9	1741.2 ± 0.9	1811.8 ± 0.5	1884.5 ± 1.1	1949.8 ± 0.4	2023.6 ± 0.5	2107.6 ± 0.4
M_{IV}	1275.0 ± 0.6	1332.5 ± 0.4	1391.5 ± 0.7	1453.3 ± 0.5	1514.6 ± 0.7	1576.3 ± 0.4	1639.4 ± 0.4	1716.4 ± 0.4
M_V	1241.2 ± 0.7	1294.9 ± 0.4	1351.4 ± 0.8	1409.3 ± 0.6	1467.7 ± 0.9	1527.8 ± 0.4	1588.5 ± 0.4	1661.7 ± 0.4
N_I	397.9 ± 0.8	416.3 ± 0.5	435.7 ± 0.8	449.1 ± 1.0	471.7 ± 0.9	487.2 ± 0.6	506.2 ± 0.6	538.1 ± 0.4
N_{II}	310.2 ± 1.2	331.8 ± 0.6	343.5 ± 1.4	366.2 ± 1.5	385.9 ± 1.6	396.7 ± 0.7	410.1 ± 1.8	437.0 ± 0.5
N_{III}	385.0 ± 1.0	292.9 ± 0.6	306.6 ± 0.9	320.0 ± 0.7	336.6 ± 1.6	343.5 ± 0.5	359.3 ± 0.5	380.4 ± 0.5
N_{IV}}	147.0 ± 0.8	154.2 ± 0.5	161.0 ± 1.0	{176.7 ± 1.2	179.6 ± 1.2	{198.1	204.8 ± 0.5	223.7 ± 0.5
N_V}				167.6 ± 1.5		184.9 ± 1.3	195.0 ± 0.4	213.7 ± 0.5
$N_{VI,VII}$	2.6 ± 1.5	4.2 ± 1.6	3.7 ± 3.0	4.3 ± 1.4	5.3 ± 1.9	6.3 ± 1.0	6.9 ± 0.5	17.1 ± 0.5
O_I	39.0 ± 0.8	62.9 ± 0.5	51.2 ± 1.3	59.8 ± 1.7	53.2 ± 3.0	54.1 ± 0.5	56.8 ± 0.5	64.9 ± 0.6
O_{II}}	25.4 ± 0.8	26.3 ± 0.6	20.3 ± 1.5	29.4 ± 1.6	32.3 ± 1.6	23.4 ± 0.6	28.0 ± 0.6	{38.1 ± 0.6
O_{III}}								30.6 ± 0.6

Level	73 Ta	74 W	75 Re	76 Os	77 Ir	78 Pt	79 Au	80 Hg
K	67416.4 ± 0.6 (67403.)	69525.0 ± 0.3 (69508[e])	71676.4 ± 0.4 (71658.)	73870.8 ± 0.5	76111.0 ± 0.5	78394.8 ± 0.7 (78381.)	80724.9 ± 0.5 (80720.)	83102.3 ± 0.8
L_I	11681.5 ± 0.3 [11680.2][p] (11682.)	12099.8 ± 0.3 [12098.2][p] (12099.6)	12526.7 ± 0.4	12968.0 ± 0.4 (12972.)	13418.5 ± 0.3 (13423.)	13879.9 ± 0.4 (13883.)	14352.8 ± 0.4 (14353.7)	14839.3 ± 1.0 (14842.)
L_{II}	11136.1 ± 0.3 [11136.1][p] (11132.)	11544.0 ± 0.3 [11541.4][p] (11538.)	11958.7 ± 0.3 [11956.9][p] (11954.)	12385.0 ± 0.4 (12381.)	12824.1 ± 0.3 [12824.0][c,p] (12820.)	13272.6 ± 0.3 [13272.5][c,p] (13272.3)	13733.6 ± 0.3 [13733.5][c,p] (13736.)	14208.7 ± 0.7 (14215.)
L_{III}	9881.1 ± 0.3 [9880.3][p] (9877.7)	10206.8 ± 0.3 [10204.2][p] (10200.)	10535.3 ± 0.3 [10534.2][p] (10531.)	10870.9 ± 0.3 [10870.7][p] (10868.)	11215.2 ± 0.3 [11215.1][c,p] (11212.)	11563.7 ± 0.3 [11563.7][c,p] (11562.)	11918.7 ± 0.3 [11918.2][c,p] (11921.)	12283.9 ± 0.4 [12284.0][c,p] (12286.)
M_I	2708.0 ± 0.4	2819.6 ± 0.4	2931.7 ± 0.4	3048.5 ± 0.4	3173.7 ± 1.7	3296.0 ± 0.9	3424.9 ± 0.3 [3424.8][p]	3561.6 ± 1.1
M_{II}	2468.7 ± 0.3 [2468.6][p]	2574.9 ± 0.3 [2575.0][p]	2681.6 ± 0.4	2792.2 ± 0.3 [2791.9][p]	2908.7 ± 0.3 [2909.1][p]	3026.5 ± 0.4 [3026.5][p] (3029.)	3147.8 ± 0.4 [3149.5][p]	3278.5 ± 1.3
M_{III}	2194.0 ± 0.3 [2194.1][p]	2281.0 ± 0.3 [2281.0][p]	2367.3 ± 0.3 [2367.3][p]	2457.2 ± 0.4 [2457.4][p]	2550.7 ± 0.3 [2550.5][p] (2550.5)	2645.4 ± 0.4 [2645.5][p] (2645.9)	2743.0 ± 0.3 [2743.1][p] (2744.0)	2847.1 ± 0.4 [2847.1][p]

TABLE 7f-3. RECOMMENDED VALUES OF THE ATOMIC ENERGY LEVELS, AND PROBABLE ERRORS IN eV* (Continued)

Level	73 Ta	74 W	75 Re	76 Os	77 Ir	78 Pt	79 Au	80 Hg
M_{IV}	1793.2 ± 0.3 [1793.1]^p	1871.6 ± 0.3 [1871.4]^p	1948.9 ± 0.3 [1948.9]^p	2030.8 ± 0.3 [2031.0]^p	2116.1 ± 0.3 [2116.1]^p	2201.9 ± 0.3 [2201.9]^p	2291.1 ± 0.3 [2291.2]^p (2307.)	2384.9 ± 0.3 [2384.9]^p
M_V	1735.1 ± 0.3 [1735.2]^p	1809.2 ± 0.3 [1809.3]^p	1882.9 ± 0.3 [1882.9]^p	1960.1 ± 0.3 [1960.2]^p	2040.4 ± 0.3 [2040.5]^p	2121.6 ± 0.3 [2121.6]^p	2205.7 ± 0.3 [2206.1]^p (2220.)	2294.9 ± 0.3 [2294.9]^p
N_I	565.5 ± 0.5	595.0 ± 0.4	625.0 ± 0.4	654.3 ± 0.5	690.1 ± 0.4	722.0 ± 0.6	758.8 ± 0.4	800.3 ± 1.0
N_{II}	464.8 ± 0.5	491.6 ± 0.4	517.9 ± 0.5	546.5 ± 0.5	577.1 ± 0.4	609.2 ± 0.6	643.7 ± 0.5	676.9 ± 2.4
N_{III}	404.5 ± 0.4	425.3 ± 0.4	444.4 ± 0.5	468.2 ± 0.6	494.3 ± 0.4	519.0 ± 0.6	545.4 ± 0.5	571.0 ± 1.4
N_{IV}	241.3 ± 0.4	258.8 ± 0.4	273.7 ± 0.4	289.4 ± 0.5	311.4 ± 0.4	330.8 ± 0.5	352.0 ± 0.4	378.3 ± 1.0
N_V	229.3 ± 0.3	245.4 ± 0.4	260.2 ± 0.4	272.8 ± 0.6	294.9 ± 0.4	313.3 ± 0.4	333.9 ± 0.4	359.8 ± 1.2
N_{VI}	25.0 ± 0.4 {}	36.5 ± 0.4 {}	40.6 ± 0.4	46.3 ± 0.6	63.4 ± 0.4 {}	74.3 ± 0.4 {}	86.4 ± 0.4 {}	102.2 ± 0.5
N_{VII}		33.6 ± 0.4			60.5 ± 0.4	71.1 ± 0.5	82.8 ± 0.5	98.5 ± 0.5
O_I	71.1 ± 0.5	77.1 ± 0.4	82.8 ± 0.5	83.7 ± 0.6	95.2 ± 0.4	101.7 ± 0.4	107.8 ± 0.7	120.3 ± 1.3
O_{II}	44.9 ± 0.4	46.8 ± 0.5	45.6 ± 0.7	58.0 ± 1.1	63.0 ± 0.6	65.3 ± 0.7	71.7 ± 0.7	80.5 ± 1.3
O_{III}	36.4 ± 0.4	35.6 ± 0.5	34.6 ± 0.6	45.4 ± 1.0	50.5 ± 0.6	51.7 ± 0.7	53.7 ± 0.7	57.6 ± 1.3
$O_{IV,V}$	5.7 ± 0.4	6.1 ± 0.4	3.5 ± 0.5	....	3.8 ± 0.4	2.2 ± 1.3	2.5 ± 0.5	6.4 ± 1.4

Level	81 Tl	82 Pb	83 Bi	84 Po	85 At	86 Rn	87 Fr	88 Ra
K	85530.4 ± 0.6	88004.5 ± 0.7 (88005.)	90525.9 ± 0.7 (90534.)	93105.0 ± 3.8	95729.9 ± 7.7	98404. ± 12.	101137. ± 13.	103921.9 ± 7.2
L_I	15346.7 ± 0.4 (15343.)	15860.8 ± 0.5 (15855.)	16387.5 ± 0.4 (16376.)	16939.3 ± 9.8	17493. ± 29.	18049. ± 38.	18639. ± 40.	19236.7 ± 1.5 (19236.0)
L_{II}	14697.9 ± 0.3 [14697.3]^p (14699.)	15200.0 ± 0.4 (15205.)	15711.1 ± 0.3 [15708.4]^p (15719.)	16244.3 ± 2.4	16784.7 ± 2.5	17337.1 ± 3.4	17906.5 ± 3.5	18484.3 ± 1.5
L_{III}	12657.5 ± 0.3 [12656.3]^{c,p} (12660.)	13035.2 ± 0.3 [13034.9]^{c,p} (13041.)	13418.6 ± 0.3 [13418.3]^{c,p} (13426.)	13813.8 ± 1.0 ⟨13813.8⟩	14213.5 ± 2.0 ⟨14213.5⟩	14619.4 ± 3.0 ⟨14619.4⟩	15031.2 ± 3.0 ⟨15031.2⟩	15444.4 ± 1.5 (15444.0)
M_I	3704.1 ± 0.4	3850.7 ± 0.5	3999.1 ± 0.3 [3999.1]^p	4149.4 ± 3.9	⟨4317.⟩	⟨4482.⟩	⟨4652.⟩	4822.0 ± 1.5
M_{II}	3415.7 ± 0.3 [3415.7]^p	3554.2 ± 0.3 [3554.2]^p	3696.3 ± 0.3 [3696.4]^p	3854.1 ± 9.8	4008. ± 28.	4159. ± 38.	4327. ± 40.	4489.5 ± 1.8

Level	89 Ac	90 Th	91 Pa	92 U	93 Np	94 Pu	95 Am	96 Cm
M_{III}	2956.6 ± 0.3 [2956.5]p	3066.4 ± 0.4 [3066.5]p	3176.9 ± 0.3 [3176.8]p	3301.9 ± 9.9	3426. ± 29.	3538. ± 38.	3663. ± 40.	3791.8 ± 1.7
M_{IV}	2485.1 ± 0.3 [2485.2]p	2585.6 ± 0.3 [2585.5]p (2606.)	2687.6 ± 0.3 [2687.4]p	2798.0 ± 1.2	2908.7 ± 2.1	3021.5 ± 3.1	3136.2 ± 3.1	3248.4 ± 1.6
M_{V}	2389.3 ± 0.3 [2389.4]p	2484.0 ± 0.3 [2484.2]p (2502.)	2579.6 ± 0.3 [2579.5]p	2683.0 ± 1.1	2786.7 ± 2.1	2892.4 ± 3.1	2999.9 ± 3.1	3104.9 ± 1.6
N_{I}	845.5 ± 0.5	893.6 ± 0.7	938.2 ± 0.3 [938.7]p	995.3 ± 2.9	⟨1042.⟩	⟨1097.⟩	⟨1153.⟩	1208.4 ± 1.6
N_{II}	721.3 ± 0.8	763.9 ± 0.8	805.3 ± 0.3 [805.3]p	851. ± 12.	886. ± 30.	929. ± 40.	980. ± 42.	1057.6 ± 1.8
N_{III}	609.0 ± 0.5	644.5 ± 0.6	678.9 ± 0.3 [678.9]p	705. ± 14.	740. ± 30.	768. ± 40.	810. ± 43.	879.1 ± 1.8
N_{IV}	406.6 ± 0.4	435.2 ± 0.5	463.6 ± 0.3 [463.6]p	500.2 ± 2.4	533.2 ± 3.2	566.6 ± 4.0	603.3 ± 4.1	635.9 ± 1.6
N_{V}	386.2 ± 0.5	412.9 ± 0.6	440.0 ± 0.3 [440.1]p	473.4 ± 1.3	....	....	577. ± 34.	602.7 ± 1.7
N_{VI}	122.8 ± 0.4	142.9 ± 0.4	161.9 ± 0.5	....	....	....	....	298.9 ± 2.4
N_{VII}	118.5 ± 0.4	138.1 ± 0.4	157.4 ± 0.6	....	....	....	....	254.4 ± 2.1
O_{I}	136.3 ± 0.7	147.3 ± 0.8	159.3 ± 0.7	....	....	....	....	200.4 ± 2.0
O_{II}	99.6 ± 0.6	104.8 ± 1.0	116.8 ± 0.7	....	....	....	....	152.8 ± 2.0
O_{III}	75.4 ± 0.6	86.0 ± 0.6	92.8 ± 0.6	....	....	....	....	....
O_{IV}	15.3 ± 0.4	21.8 ± 0.4	26.5 ± 0.5	31.4 ± 3.2	....	....	....	67.2 ± 1.7
O_{V}	13.1 ± 0.4	19.2 ± 0.4	24.4 ± 0.6	....	....	....	....	43.5 ± 2.2
P_{I}	....	3.1 ± 1.0	....	....	....	....	....	18.8 ± 1.8
$P_{II,III}$	....	0.7	2.7 ± 0.7	....	....	....	....	....
Level	**89 Ac**	**90 Th**	**91 Pa**	**92 U**	**93 Np**	**94 Pu**	**95 Am**	**96 Cm**
K	106755.3 ± 5.3	109650.9 ± 0.9	112601.4 ± 2.4	115606.1 ± 1.6	118678. ± 33.	121818. ± 44.	125027. ± 55.	128220
L_{I}	19840. ± 18.	20472.1 ± 0.5 (20464.)	21104.6 ± 1.8 (21128.)	21757.4 ± 0.3 (21771.)	22426.8 ± 0.9	23097.2 ± 0.9 (23109.)	23772.9 ± 2.0 (23772.9)	24460
L_{II}	19083.2 ± 2.8	19693.2 ± 0.4 (19683.)	20313.7 ± 1.5 (20319.)	20947.6 ± 0.3 (20945.)	21600.5 ± 0.4	22266.2 ± 0.7 (22253.)	22944.0 ± 1.0	23779
L_{III}	15871.0 ± 2.0 ⟨15871.0⟩	16300.3 ± 0.4 [16299.6]q (16299.)	16733.1 ± 1.4 (16733.)	17166.3 ± 0.3 [17168.5]r (17165.)	17610.0 ± 0.4 (17606.2)	18056.8 ± 0.6 (18053.1)	18504.1 ± 0.9 (18504.1)	18930
M_{I}	⟨5002.⟩	5182.2 ± 0.3 [5182.3]q	5366.9 ± 1.6	5548.0 ± 0.4	5723.2 ± 3.6	5932.9 ± 1.4	6120.5 ± 7.5	6288
M_{II}	4656. ± 18.	4830.4 ± 0.4 [4830.6]q	5000.9 ± 2.3	5182.2 ± 0.4 [5180.9]r	5366.2 ± 0.7 [5366.4]s	5541.2 ± 1.7	5710.2 ± 2.1	5895

TABLE 7f-3. RECOMMENDED VALUES OF THE ATOMIC ENERGY LEVELS, AND PROBABLE ERRORS IN eV* (Continued)

Level	89 Ac	90 Th	91 Pa	92 U	93 Np	94 Pu	95 Am	96 Cm
M_{III}	3909. ± 18.	4046.1 ± 0.4 [4046.1]q (4041.)	4173.8 ± 1.8	4303.4 ± 0.3 [4303.6]r (4299.)	4434.7 ± 0.5 [4434.6]s	4556.6 ± 1.5	4667.0 ± 2.1	4797
M_{IV}	3370.2 ± 2.1	3490.8 ± 0.3 [3490.7]q (3485.)	3611.2 ± 1.4 (3608.)	3727.6 ± 0.3 [3728.1]r (3720.)	3850.3 ± 0.4 [3849.8]s	3972.6 ± 0.6 [3972.7]t	4092.1 ± 1.0	4227
M_V	3219.0 ± 2.1	3332.0 ± 0.3 [3332.1]q (3325.)	3441.8 ± 1.4 (3436.)	3551.7 ± 0.3 [3551.7]r (3545.)	3665.8 ± 0.4 [3664.2]s	3778.1 ± 0.6 [3778.0]t	3886.9 ± 1.0	3971
N_I	⟨1269.⟩	1329.5 ± 0.4 [1329.8]q	1387.1 ± 1.9	1440.8 ± 0.4 [1441.3]r	1500.7 ± 0.8 [1500.7]s	1558.6 ± 0.8	1617.1 ± 1.1	1643
N_{II}	1080. ± 19.	1168.2 ± 0.4 [1168.3]q	1224.3 ± 1.6	1272.6 ± 0.3 [1272.5]r	1327.7 ± 0.8 [1327.7]s	1372.1 ± 1.8	1411.8 ± 8.3	1440
N_{III}	890. ± 19.	967.3 ± 0.4 [967.6]q	1006.7 ± 1.7	1044.9 ± 0.3 [1044.9]r	1086.8 ± 0.7 [1086.8]s	1114.8 ± 1.6	⟨1135.7⟩	1154
N_{IV}	674.9 ± 3.7	714.1 ± 0.4 [714.4]q	743.4 ± 2.1	780.4 ± 0.3 [779.7]r	815.9 ± 0.5 [817.1]s	848.9 ± 0.6 [848.9]t	878.7 ± 1.0	
N_V		676.4 ± 0.4 [676.4]q	708.2 ± 1.8	737.7 ± 0.3 [737.6]r	770.3 ± 0.4 [773.2]s	801.4 ± 0.6 [801.4]t	827.6 ± 1.0	
N_{VI}		344.4 ± 0.3 [344.2]q	371.2 ± 1.6	391.3 ± 0.6	415.0 ± 0.8 [415.0]s	445.8 ± 1.7		
N_{VII}		335.2 ± 0.4 [335.0]q	359.5 ± 1.6	380.9 ± 0.9	404.4 ± 0.5 [404.4]s	432.4 ± 2.1		385
O_I		290.2 ± 0.8	309.6 ± 4.3	323.7 ± 1.1	....	351.9 ± 2.4	....	
O_{II}		229.4 ± 1.1		259.3 ± 0.5	283.4 ± 0.8 [283.4]s	274.1 ± 4.7	....	
O_{III}		181.8 ± 0.4 [181.8]q	222.9 ± 3.9	195.1 ± 1.3	206.1 ± 0.7 [206.1]s	206.5 ± 4.7		
O_{IV}		94.3 ± 0.4 [94.4]q		105.0 ± 0.5	109.3 ± 0.7 [108.8]s	116.0 ± 1.2	115.8 ± 1.3	
O_V		87.9 ± 0.3 [88.1]q	94.1 ± 2.8	96.3 ± 1.4	101.3 ± 0.5 [101.4]s	105.4 ± 1.0	103.3 ± 1.1	
P_I		59.5 ± 1.1	....	70.7 ± 1.2				
P_{II}		49.0 ± 2.5	....	42.3 ± 9.0				
P_{III}		43.0 ± 2.5	....	32.3 ± 9.0				

TABLE 7f-3. RECOMMENDED VALUES OF THE ATOMIC ENERGY LEVELS, AND PROBABLE ERRORS IN EV* *(Continued)*

Level	97 Bk	98 Cf	99 Es	100 Fm	101 Md	102 No	103 Lw
K	[131590 ± 40]u	135960	139490	143090	146780	150540	154380
LI	[25275 ± 17]u	26110	26900	27700	28530	29380	30240
LII	[24385 ± 17]u	25250	26020	26810	27610	28440	29280
LIII	[19452 ± 20]u	19930	20410	20900	21390	21880	22360
MI	[6556 ± 21]u	6754	6977	7205	7441	7675	7900
MII	[6147 ± 31]u	6359	6574	6793	7019	7245	7460
MIII	[4977 ± 31]u	5109	5252	5397	5546	5688	5710
MIV	4366	4497	4630	4766	4903	5037	5150
MV	4132	4253	4374	4498	4622	4741	4860
NI	[1755 ± 22]u	1799	1868	1937	2010	2078	2140
NII	1554	1616	1680	1747	1814	1876	1930
NIII	1235	1279	1321	1366	1410	1448	1480
OI	[398 ± 22]u	419	435	454	472	484	490

a J. E. Mack, 1949, as given in C. E. Moore, "Atomic Energy Levels" (U.S. National Bureau of Standards, Washington, D.C., 1949), vol. 1, p. 1.

b G. Herzberg, 1957, as given in C. E. Moore, "Atomic Energy Levels" (U.S. National Bureau of Standards, Washington, D.C., 1958), vol. 3, p. 238.

c S. Hagström and S. E. Karlsson, Arkiv Fysik 26, 451 (1964); and S. Hagström, Z. Physik 178, 82 (1964).

d A. Fahlman, D. Hamrin, R. Nordberg, C. Nordling, and K. Siegbahn, Phys. Rev. Letters 14, 127 (1965); R. Nordberg, K. Hamrin, A. Fahlman, C. Nordling, and K. Siegbahn, Z. Physik 192, 462 (1966).

e E. Sokolowski, Arkiv Fysik 15, 1 (1959).

f S. Hagstrom, C. Nordling, and K. Siegbahn, Alpha-, Beta-, and Gamma-Ray Spectroscopy, K. Siegbahn, Ed. (North-Holland Publ. Co. Amsterdam, 1965), Vol. 1, p. 845.

g C. Nordling, Arkiv Fysik 15, 397 (1959).

h E. Sokolowski, C. Nordling, and K. Siegbahn, Arkiv Fysik 12, 301 (1957).

i C. Nordling and S. Hagström, Arkiv Fysik 16, 515 (1960).

j I. Andersson and S. Hagström, Arkiv Fysik 27, 161 (1964).

k M. O. Krause, Phys. Rev. 140, A1845 (1965).

l A. Fahlman, O. Hörnfeldt, and C. Nordling, Arkiv Fysik 23, 75 (1962).

m P. Bergvall, O. Hörnfeldt, and C. Nordling, Arkiv Fysik 17, 113 (1960).

n P. Bergvall and S. Hagström, Arkiv Fysik 17, 61 (1960).

o S. Hagström, Z. Physik 178, 82 (1964).

p A. Fahlman and S. Hagström, Arkiv Fysik 27, 69 (1964).

q C. Nordling and S. Hagström, Z. Physik 178, 418 (1964).

r C. Nordling and S. Hagström, Arkiv Fysik 15, 431 (1959).

s S. Hagström, Bull. Am. Phys. Soc. 11, 389 (1966).

t A. Fahlman, K. Hamrin, R. Nordberg, C. Nordling, K. Siegbahn, and L. W. Holm, Phys. Letters 19, 643 (1966).

u J. M. Hollander, M. D. Holtz, T. Novakov, and R. L. Graham, Arkiv Fysik. 28, 375 (1965).

v J. A. Bearden, Rev. Mod. Phys. 39, 78 (1967). J. A. Bearden, X-Ray Wavelengths, NYO 10586 (National Technical Information Service, U.S. Dept. of Commerce, Springfield, Va. 22151).

7g. Constants of Diatomic Molecules

K. P. HUBER

National Research Council of Canada

Explanation of Columns in Table 7g-1

(1) Identification of molecule.

(2) Mass numbers of the constituent atoms to which the data refer. If, in the original paper, the mass numbers are not clearly specified, or, if the data refer to the normal isotopic mixture, the mass numbers for the most abundant isotope are given in parentheses.

(3) Reduced mass μ in unified atomic mass units ($^{12}C = 12.0000000$). Precise atomic masses were taken from the 1961 nuclidic mass table [L. A. König, J. H. E. Mattauch, and A. H. Wapstra, *Nucl. Phys.* **31**, 18 (1962)].

(4) Designation of the ground state of the molecule. For multiplet Π, Δ, $\ldots$ states the spin-orbit coupling constant A has been added.

(5) (6) (8) (9) Rotational constant B_e.
  Rotation-vibration interaction constant α_e
  (from $B_v = B_e - \alpha_e(v + \frac{1}{2}) + \cdots$).
  Vibrational frequency ω_e.
  Anharmonic constant $\omega_e x_e$ (from $G(v) =$
    $\omega_e(v + \frac{1}{2}) - \omega_e x_e(v + \frac{1}{2})^2 + \cdots$).

All constants in cm^{-1}. They are derived from the analyses of molecular spectra in the microwave, infrared, visible, and vacuum uv region. For $^1\Sigma$ states, the constants in these columns correspond to the coefficients Y_{01}, $-Y_{11}$, Y_{10}, and $-Y_{20}$, respectively, in the Dunham series expansion for the term values

$$T_{vJ} = \sum_{lm} Y_{lm}(v + \tfrac{1}{2})^l J^m (J + 1)^m$$

(7) Equilibrium internuclear distance r_e in Å, calculated without correction from

$$r_e = \left(\frac{h}{8\pi^2 c \mu B_e}\right)^{\frac{1}{2}}$$

(10) Dissociation energy $D_0{}^\circ$ in electron-volts (eV). Data obtained by a large variety of both spectroscopic and thermochemical methods have been included.

Uncertain quantities are enclosed in parentheses ( ). Quantities in square brackets [ ] in columns (5) and (8) refer to B_0 and $\Delta G(\frac{1}{2})$ respectively. * after ω_e and $\omega_e x_e$ indicates that these numbers are for the natural isotopic mixture rather than for the isotope specified in column (2).

The physical constants and conversion factors given in Appendix VII of the following book have been used throughout: G. Herzberg, "Electronic Spectra and Electronic

Structure of Polyatomic Molecules," D. Van Nostrand Company, Inc., Princeton, N.J., 1966.

The data included in the table are taken from a new compilation of vibrational and rotational constants for the electronic states of all known diatomic molecules. This compilation is presently being prepared by G. Herzberg and K. P. Huber and will provide further details and the literature references. A critical table of dissociation energies has recently been published by A. G. Gaydon in his book "Dissociation Energies and Spectra of Diatomic Molecules," 3d edition, Chapman & Hall, Ltd., London, 1968.

TABLE 7g-1. CONSTANTS OF DIATOMIC MOLECULES

	m_1	m_2	μ	Ground state	B_e	α_e	r_e	ω_e	$\omega_e x_e$	D_0^0
(1)	(2)	(2)	(3)	(4)	(5)	(6)	(7)	(8)	(9)	(10)
Ag$_2$	107	109	53.94779					192.4	0.643	1.63
AgAl	107	27	21.544070		0.12796	0.00076	2.4728	256.60	1.13	(1.7$_9$)
AgAu	(107)	197	(69.29476)					(200)		2.0$_6$
AgBr	107	79	45.40207$_2$	$^1\Sigma^+$	0.0648337$_8$	0.0002359$_7$	2.3931	250.49	0.6871	3.0
AgCl	107	35	26.349782	$^1\Sigma^+$	0.122983860	0.00059540$_5$	2.2808	343.49	1.17	3.2$_4$
AgCu	(107)	(63)	(39.61199$_8$)					229.5		1.7$_6$
AgF	107	19	16.131608	$^1\Sigma^+$	0.2657	0.0019	1.983	513.45	2.59	3.6$_3$
AgGa	(107)	(69)	(41.90678$_9$)					184.7	0.65	
AgH	107	1	0.9984128$_8$	$^1\Sigma^+$	6.449	0.201	1.618$_1$	1759.9	34.06	2.4$_1$
	107	2	1.9768579$_5$		3.2572	0.0722	1.6180	1250.7	17.17	
AgI	107	127	58.02466	$^1\Sigma^+$	0.044876	0.0001473	2.5444	206.52	0.445	2.4$_4$
AgIn	107	115	55.38014					155.8	0.42	2.6$_2$
AgO	107	16	13.913242$_5$	$^2\Pi(r)$ $A = (+135)$	0.3028	0.0025	2.000	490.4*	3.0*	(2.4)
AgSe	(107)	(80)	(45.73067$_4$)							
AgSn	(107)	(120)	(56.51557)					(233)		
AgTe	(107)	(130)	(58.64437)	$(^2\Pi)$				195.3	0.30	1.3$_7$
Al$_2$	27	27	13.490767$_5$	$^3\Sigma_g^-$	0.2054	0.0012	2.466$_5$	350.01	2.022	1.8$_6$
AlAu	27	197	23.730767$_1$	$^1\Sigma^+(0^+)$	0.12991$_3$	0.00066$_8$	2.3384	333.00	1.16$_3$	3.3$_4$
AlBr	27	79	20.107087$_0$	$^1\Sigma^+$	0.1591	0.000853	2.296	378.0	1.28	4.4$_3$
AlCl	27	35	15.230144$_3$	$^1\Sigma^+$	0.243926$_7$	0.0016002$_1$	2.1302	481.30	1.95	5.0$_3$
AlF	27	19	11.1484731	$^1\Sigma^+$	0.552468	0.004950	1.6544	801.9$_3$	4.7$_0$	6.8$_9$
AlH	27	1	0.97153601	$^1\Sigma^+$	6.3906$_6$	0.1858$_1$	1.6478	1682.56$_3$	29.09$_0$	2.91
	27	2	1.8741981$_2$		3.3186	0.0697	1.6463	1211.95	15.138	
AlH$^+$	27	1		$^2\Sigma^+$	6.763	0.398	1.601$_8$			2.94
AlI	27	127	22.250735$_7$	$^1\Sigma^+$				316.1	1.0	3.7$_7$
AlO	27	16	10.0419499	$^2\Sigma^+$	0.6413$_6$	0.0058$_6$	1.6178	979.23	6.97	4.9$_9$
AlP	27	31	14.4200738							2.2$_0$
AlS	27	32	14.632787$_4$	$^2\Sigma^+$	0.2799	0.0018	2.028$_8$	617.12	3.33	3.7$_0$
AlSe	27	(80)	(20.171277$_4$)	$(^2\Sigma)$				467.6	2.08	3.4$_5$
AlTe	27	(130)	(22.341268)	$(^2\Sigma)$						3.3$_7$
Ar$_2$	40	40	19.9811919	$(^1\Sigma_g^+)$				30.7	2.64	0.0096$_3$
Ar$_2^+$	40	40								$\geq$1.04$_6$
As$_2$	75	75	37.46079$_0$	$^1\Sigma_g^+$	0.10165	0.00034	2.104$_0$	429.4	1.12	3.94
As$_2^+$	75	75		$(^2\Sigma_u^+)$				314.8	1.25	
AsCl	75	(35)	(23.841218)	$(^3\Sigma^-)$				443	2	2.7
AsF	75	19	15.153352$_7$	$(^3\Sigma^-)$				($\geq$680)	(2.7)	

Molecule	z_1	z_2	μ	State	B_e	α_e	r_e	ω_e	$\omega_e x_e$	D_0^0
AsH	75	1	0.99444817	$^3\Sigma^-$	$[7.199_8]$		$[1.5344]$			<3.6
AsD	75	2	1.9613749_3	$^3\Sigma^-$	$[3.669_8]$		$[1.5304]$			2.7_6
AsN	75	14	11.797993_1	$^1\Sigma^+$	0.5457	0.0038	1.618_1	1068.0	5.36	(6.5)
AsO	75	16	13.180933_8	$^2\Pi_i$; $A=+1027$	0.48519	0.003320	1.6236	966.5_5	4.88	$\le 4.980_4$
AsO+	75	16		$^1\Sigma^+$	0.5199	0.0031	1.568_4	[1091.32]	(5)	(7.7)
Au2	197	197	98.483276	$^1\Sigma_g^+$	0.028013	0.0000723	2.4719	190.8	0.420	2.2
AuBa	197	(138)	(81.11371)	$(^2\Sigma^+)$				128.8	0.18	(2.8)
AuBe	197	9	8.6178757	$^2\Sigma^+$	0.46074	0.00400	2.0605	607.68	3.53	(3.2)
AuBi	197	209	101.39785_6	$(^2\Sigma^+)$				157.7	0.25	(2.6)
AuCa	197	40	33.2221413	$(^1\Sigma^+)$				220.0	0.62	(2.4)
AuCl	197	35	29.696607_4					382.8	1.30	(3.5)
AuCr	197	(52)	(41.101862_1)							2.1_0
AuCu	197	(63)	(47.692224)	(O^+)				250	0.7	2.3_6
AuGa	197	69	51.05848_2	$(^2\Pi_i)$; $A=+(1550)$				225.62	0.567	(2.0)
AuGe	197	(74)	(53.74919)					[251.7]		3.2_5
AuH	197	1	1.00269470	$^1\Sigma^+$	7.2401	0.2136	1.5238	2305.01	43.12	3.2_6
AuD	197	2	1.9937152_5	$^1\Sigma^+$	3.6415	0.07614	1.5238	1634.98	21.655	
AuMg	197	24	21.381386_7	$^2\Sigma^+$	0.13214	0.00073	2.4427	307.9	1.1	(2.6)
AuPb	197	(208)	(101.160960)	$(^2\Pi_i)$				158.6	0.6	(1.4)
AuPd	197	(106)	(58.87247)	$(^2\Pi)$						1.44
AuSe	197	(80)	(56.85028_0)	$(^2\Pi_i)$; $A=+(1070)$				270		3.2_6
AuSi	197	28	24.497348_8	$(^2\Pi_i)$; $A=+(2550)$				[391.2]	(1.35)	2.4_9
AuSn	197	(120)	(74.53153)	$(^2\Sigma^+)$				190.4	(1.26)	(2.6)
AuSr	197	(88)	(60.77977)	$(^2\Pi)$				153.33	0.19	(2.8)
AuTe	197	(130)	(78.27889)					212.5	0.50	3.2_5
AuU	197	238	107.78430_2							
B2	11	11	5.5046525_5	$^2\Sigma_g^-$	1.212	0.014	1.590	1051.3	9.35	2.9
BaBr	(138)	79	(50.19401_6)	$^2\Sigma^+$				193.8	0.42	(4.3)
BaCl	138	35	27.895369	$^2\Sigma^+$				279.3	0.89	4.5_6
BaF	(138)	19	(16.6990125)	$^2\Sigma^+$	$[0.2158_5]$		$[2.162_7]$	468.9	1.79	6.05
BaH	(138)	1	(1.00051336)	$^2\Sigma^+$	3.38285	0.06599	2.2317	1168.31	14.50	1.95
BaD	138	2	(1.9851096_7)		1.70712	0.02369	2.2303	829.89	7.35	1.97
BaO	(138)	16	14.332553_5	$^1\Sigma^+$	0.3126146	0.0013935	1.9397	669.8_1	2.05_4	5.74
BaS	138	(32)	(25.954702)	$^1\Sigma^+$	[0.10308]	0.0035	[2.5102]	377.1	1.22	4.3_7
BBr	11	79	9.6615017	$^1\Sigma^+$	0.490		1.88_7	684.31	3.52	4.49
BC	(11)	12	(5.7416623_6)	$^1\Sigma^+$						4.6_0
BCl	11	35	8.3731666	$^1\Sigma^+$	0.6838	0.00646	1.715_9	839.12	5.11	5.5_1
BeBr	9	(79)	(8.0885079)	$(^2\Sigma^+)$				713.8	3.5	(4.3)
BeCl	9	35	7.1634925	$(^2\Sigma^+)$				846.4	5.14	3.9_9

TABLE 7g-1. CONSTANTS OF DIATOMIC MOLECULES (Continued)

(1)	m_1 (2)	m_2 (2)	μ (3)	Ground state (4)	B_e (5)	α_e (6)	r_e (7)	ω_e (8)	$\omega_e x_e$ (9)	D_0^0 (10)
BeF	9	19	6.1125863	$^2\Sigma^+$	1.4877	0.01685	1.3615	1265.6	9.12	5.8_5
BeH	9	1	0.90645690	$^2\Sigma^+$	10.308	0.300	1.3432	2058.6	35.5	(2.1)
	9	2	1.6461989_0		5.6807	0.1218	1.3426			
BeH⁺	9	1		$^2\Sigma^+$	10.7996	0.2935	1.3123	2221.7	39.79	(3.2)
	9	2			5.9546	0.1233	1.3114	1647.64	21.85	
BeO	9	16	5.7643286	$^1\Sigma^+$	1.6510	0.0190	1.3309	1487.323	11.8297	4.6_4
BeS	9	32	7.0304618	$^1\Sigma^+$	0.79059	0.00664	1.7415	997.94	6.137	(4.1)
BF	11	19	6.9701832	$^1\Sigma^+$	1.516_5	0.019_5	1.2629	1402.1_3	11.8_4	7.8_1
BH	11	1	0.92330324	$^1\Sigma^+$	12.021	0.412	1.2324	2366.9_0	49.39_5	3.42
	11	2	1.70261630		6.54_2	0.171	1.230_2	1759	(28)	
BH⁺	11	1		$^2\Sigma^+$	$[12.374]$		$[1.2147]$			3.46
Bi₂	209	209	104.49020_9	$^1\Sigma_g^+$	0.022806	0.000050	2.6597	172.71	0.3227	1.95
BiBr	209	79	57.285377	(O^+)	(0.115)		(2.21)	209.50	0.466	2.04
BiCl	209	35	29.956251_8	O^+	0.231_5	0.0015	2.04_5	308.0	0.96	2.74
BiF	209	19	17.4151901	O^+				512.6	2.3_4	3.08
BiH	209	1	1.00298823	$^2\Pi$	5.137	0.148	1.808_8	1698.9	31.6	(3.4)
	209	2	1.9948760_6		2.592	0.054	1.805_6	1205.5	16.1	
BiI	209	127	78.95721_0					163.9	0.31	(2.7)
BiO	209	16	14.8577356	$^2\Pi$	0.3034	0.0022	1.934	692.4	4.34	(2.7)
BiPb	209	(208)	(104.23866_1)		$[0.112764]$		$[2.3219]$			3.4_7
BiS	209	32	27.729687_7					408.71	1.46	1.4_3
BiSe	209	80	57.80949_6					$[264.8]$	(0.4)	3.1_7
BiTe	209	(130)	(80.10914)					208.5	0.52	2.80
BN	11	14	6.1635126_8	$^3\Pi$	1.666	0.025	1.281	1514.6	12.3	(5.1)
BO	11	16	6.5093398_8	$^2\Sigma^+$	1.7820	0.0166	1.2044	1885.69	11.810	8.2_9
Br₂	79	81	39.95242_9	$^1\Sigma_g^+$	0.081101	0.000321	2.2809	323.33	1.081	1.9704_5
Br₂⁺	79	81		$^2\Pi_{g,3/2}$ $A = -3150$				(376)	(1.25)	3.25
BrCl	79	35	24.231733	$^1\Sigma^+$	0.152469_5	0.000775	2.1361	443.5^*	1.8^*	2.233
BrF	79	19	15.312218_4	$^1\Sigma^+$	0.355843	0.002612	1.7589	672.6	4.5	2.384
BrO	79	16	13.2994297	$^2\Pi_{3/2}$	$[0.427789]$		$[1.7213]$	$778._7^*$	6.8_2^*	2.39_7
BS	11	32	8.1893677	$^2\Sigma^+$	0.7948_9	0.0060_5	1.6092	1180.17	6.31	6.2_0
BSi	11	(28)	(7.9003922)							2.9_5
C₂	12	12	6.0000000	$^1\Sigma_g^+$	1.8198_4	0.0176_5	1.2425	1854.71	13.34_0	6.2_5
C₂(⁻)	12	12		$(^2\Sigma_g^+)$	1.7468_5	0.0167	1.2682	1781.04	11.58_5	(8.4)
CaBr	40	79	(26.528909)	$^2\Sigma^+$				285.3	0.86	(3.9)
CaCl	40	35	(18.649660_2)	$^2\Sigma^+$	0.15195	0.00078_3	2.4390	369.8	1.31	4.0_8
CaF	40	19	12.8767409	$^2\Sigma^+$	$[(0.3510)]$		$[(1.931)]$	587.1	2.74	5.4_8

Molecule	Ref	Ref	μ (amu)	State (A in cm⁻¹)	B_e	α_e	r_e (Å)	ω_e	$\omega_e x_e$	D_0° (eV)
CaH	(40)	1	(0.98303388)	$^2\Sigma^+$	4.2778	0.0963	2.0022	1299	19.5	$\le$1.70
	(40)	2	(1.9174626_6)	$^2\Sigma^+$	2.196	0.035	2.001	.....	.....	.....
CaI	(40)	127	(30.392038)	$(^2\Sigma^+)$	.....	.....	.....	242.0	0.64	(3.3)
CaO	(40)	16	(11.4229222)	$^1\Sigma^+$	0.4444	0.0034	1.822_3	733.4	5.28	4.0_3
CaS	40	32	17.761768_8	$^1\Sigma$	0.17667	0.00083_7	2.3178	462.23	1.78	3.4_5
CBr	12	(79)	(10.4161613)	$^2\Pi_r$, A = +466	[0.4872]	.....	$[1.822_6]$	.....	.....	$\le$4.11
CCl	12	35	8.9341385_5	$^2\Pi_r$, $A_0 = +135$	0.6970	0.0067	1.645_3	[866.1]	(6.2)	(3.8)
Cd₂	(112)	(114)	(56.44717)	$(^1\Sigma_g^+)$	.....	.....	.....	.....	.....	0.08
CdBr	(114)	(79)	(46.61856_7)	$^2\Sigma$	.....	.....	.....	230.0	0.50	(1.6)
CdCl	(114)	35	26.754971	$(^2\Sigma)$	.....	.....	.....	330.5	1.2	2.1
CdF	(114)	19	(16.282573_0)	$^2\Sigma^+$	.....	.....	.....	(535)	.....	0.678
CdH	(114)	1	(0.99989613)	$^2\Sigma^+$	5.437	0.218	1.761_7	1430.7	46.3	0.704
	(114)	2	(1.9791066_0)	$^2\Sigma^+$	2.788	0.168	1.748	.....	.....	.....
CdH⁺	(114)	1	.....	$^1\Sigma^+$	6.071	0.189	1.667_2	1775.4	37.3	(2.0)
	(114)	2	.....	$^1\Sigma^+$	3.075	0.0682	1.664	1262.5	19.01	.....
CdI	(114)	127	(60.02651)	.....	.....	.....	.....	178.5	0.625	(1.4)
CdO	(114)	16	(14.025397_7)	.....	.....	.....	.....	.....	.....	<3.8
CdS	(114)	(32)	(24.964643)	.....	.....	.....	.....	.....	.....	<2.0
CeO	(140)	16	(14.3538823)	(constants for Ω = 3 component)	[0.35687]	.....	[1.8141]	(830)	.....	8.1_8
CeS	(140)	(32)	(26.024731_2)	.....	.....	.....	.....	.....	.....	5.9_0
CF	12	19	7.3545996_3	$^2\Pi_r$, $A_0 = +77$	1.4172_0	0.0184_0	1.2717	1308.1	11.10	5.4_2
CH	12	1	0.92974056	$^2\Pi_r$, A = +28	14.457	0.534	1.1199	2859.1	63.3	3.47
	12	2	1.72463608	$^2\Pi_r$	7.808	0.212	.....	.....	.....	3.52
CH⁺	12	1	.....	$^1\Sigma^+$	14.177_6	0.4917	1.1309	[2739.70]	34.1	4.09
Cl₂	35	35	17.484427_3	$^1\Sigma_g^+$	0.24407	0.00153	1.9875	559.71	2.70	2.4795_0
Cl₂⁺	35	35	.....	$^2\Pi_{g,i}$	0.2696	0.00166	1.891	645.2	3.00_2	4.01
ClF	35	19	12.3102870	$^1\Sigma^+$	0.5164802	0.004358_5	1.6283	786.34	6.23	2.56
ClO	35	16	10.9749310	$^2\Pi_i$, A = −282	[0.620525]	.....	[1.5733]	868	7.5	2.746
CN	12	14	6.46219328	$^2\Sigma^+$	1.8992	0.01701	1.1720	2068.745	13.134	7.91
CN⁺	12	14	.....	$^1\Sigma$	1.8964	0.0188	1.1728	2033.05	16.14	4.9_5
CO	12	16	6.85520870	$^1\Sigma^+$	1.931271	0.017513	1.1283	2169.8233	13.2939	11.09_1
CO⁺	12	16	.....	$^2\Sigma^+$	1.97720	0.01896	1.1151	2214.24	15.164	8.34
Co₂	59	59	29.466594_6	.....	.....	.....	.....	(320)	.....	1.6_9
CoBr	59	(79)	(33.738542)	(constants for lowest observed $^1\Sigma$ state)	.....	.....	.....	.....	.....	.....
CoCl	59	(35)	(21.946552_3)	.....	.....	.....	.....	(420)	.....	.....
CoH	59	1	0.99088003	$(^3\Phi)_4$	[7.151]	.....	[1.5424]	(1890)	.....	.....
	59	2	1.9475429_1	$(^3\Phi)_4$	[3.722]	.....	$[1.525_0]$	.....	.....	.....

TABLE 7g-1. CONSTANTS OF DIATOMIC MOLECULES (*Continued*)

(1)	m_1 (2)	m_2 (2)	μ (3)	Ground state (4)	B_e (5)	α_e (6)	r_e (7)	ω_e (8)	$\omega_e x_e$ (9)	$D_0{}^\circ$ (10)
CoO	59	16	12.5804778	$^2\Sigma^+$				(850)	(6)	3.7_5
CP	12	31	8.6491182_4		0.7986	0.00597	1.5622	1239.67	6.86	(6.9)
Cr$_2$	(52)	(52)	(25.970256_0)							1.6
CrBr	(52)	(79)	(31.324279)							3.3_6
CrCl	(52)	(35)	(20.898785_6)	$(^6\Sigma^+)$				(292)		3.7_5
CrF	(52)	19	(13.9103741)	$(^6\Sigma^+)$				(536)		4.5_7
CrH	(52)	1	(0.98864215)	$^6\Sigma^+$	6.220	0.179	1.655_7	[1581]		
	(52)	2	(1.9389166_4)		[3.142]		$[1.663_5]$			
CrI	(52)	127	(36.855837)							2.9_4
CrO	(52)	16	(12.2290256)	$^5\Pi$ A = 110	0.5286	0.0050	1.614_9	898.8	6.5	4.3_8
CS	12	32	8.72519425	$^1\Sigma^+$	0.82005	0.00592	1.5349	1285.08	6.46	7.8_6
				(constants for lowest observed state)						
Cs$_2$	133	133	66.45255	$^1\Sigma_g^+$	(0.0127)	0.000026_1	(4.4_7)	42.0164	0.08102	0.45_0
CsBr	133	79	49.51600_1	$^1\Sigma^+$	0.03606939	0.00012410_3	3.0722	149.5	0.36	4.1_0
CsCl	133	35	27.684694	$^1\Sigma^+$	0.07209140	0.00033753	2.9063	209*	0.75	4.5_5
CSe	12	(80)	(10.4333609_1)	$^1\Sigma^+$	0.58_5	0.004	1.67	1035.9	4.8_8	(6.8)
CsF	133	19	16.622294_8	$^1\Sigma^+$	0.18436969	0.00117571	2.3453	353	1.7	5.1_6
CsH	133	1	1.00024035	$^1\Sigma^+$	2.708_5	0.057_0	2.494_5	891.2_9	12.79	(1.9)
CsHg	133	(202)	(80.15787)	$(^2\Sigma^+)$						$D_e = 0.048_2$
CsI	133	127	64.91771	$^1\Sigma^+$	0.02362732	0.00006825_9	3.3152	119.2	0.25	3.4_0
CsRb	133	(85)	(51.81051)	$^1\Sigma^+$				49.4		
Cu$_2$	63	63	31.464797	$^1\Sigma_g^+$	0.10874	0.000614	2.2197	266.1	1.025	1.97
CuBr	63	81	35.399211	$^1\Sigma^+$	0.1008	0.0004_7	2.17_4	313.4	1.05	3.4_3
CuCl	63	35	22.478148_0	$^1\Sigma^+$	0.17818_5	0.00101	2.0515	417.74	1.65	3.8_2
CuF	63	19	14.5928365	$^1\Sigma^+$	0.3794029	0.0032298	1.7449	622.6_5	3.95	4.4_2
CuH	63	1	0.99193919	$^1\Sigma^+$	7.938	0.249	1.4632	1940.4	37.0	2.73
	63	2	1.9516387_4		4.0375	0.09140	1.4627	1384.38	19.14	2.76
CuI	63	127	42.068553	$^1\Sigma^+$	0.0735	0.0003	2.33_5	264.50	0.60	<3.27
CuO	63	16	12.7533705	$^2\Pi_{(?)}$ 272 A = (+)272	[0.4429]		$[1.727_6]$	633*	$4._5$*	2.7_2
CuS	(63)	(32)	(21.200782_6)	$(^2\Pi)$ A = (433)				414.2	1.70	2.9_0
CuSe	(63)	(80)	(35.206516)	$(^2\Pi)$ A = (1590)				300.0	0.50	
CuSn	(63)	(120)	(41.26960_1)							1.8_6
CuTe	(63)	(130)	(42.39334_7)	$(^2\Pi)$				254.0	1.01	
DyO	(164)	16	(14.572994_2)							6.5_2
ErO	(166)	16	(14.586638_4)							6.5_8

EuO..........	(153)	16	(14.480329_9)	....	....	....	....	....	....	5.8_0
F_2..........	19	19	9.4992023	$^1\Sigma_g^+$	[0.8828]	....	[1.4178]	[891.8_5]	(15.6)	1.60
F_2^+..........	19	19	....	$^2\Pi_{g,i}$	[≥1.010]	(0.01)	[≤1.326]	≥1073	9	3.2_9
FeBr..........	(56)	(79)	(32.734038)	....	....	....	....	(300)		
FeCl..........	(56)	35	(21.517041_0)	($^6\Sigma$)	....	....	....	406.6	1.2	
				or ($^4\Sigma$)	....	....	....	397.0	1.6	
FeO..........	(56)	16	(12.4381535)	$^{5(7)}\Sigma_0$	0.51271	0.00376	1.625_9	880.53	4.63	3.9_2
FeS..........	(56)	(32)	(20.343723_0)	....	....	....	....	....	....	3.3_1
Ga_2..........	(69)	(71)	(34.95548_6)	($^3\Sigma$)	....	....	....	....	....	(1.2)
GaBr..........	69	81	37.22062_7	$^1\Sigma^+$	0.081839_3	0.0003207	2.3525	263.1	0.81	4.31
GaCl..........	69	35	23.199027	$^1\Sigma^+$	0.149895	0.000776	2.2018	365.3	1.2	4.92
GaF..........	69	19	14.893279_4	$^1\Sigma^+$	0.3590_5	0.0028_2	1.7755	622.2	3.2	5.9_8
GaH..........	(69)	1	(0.99330126)	$^1\Sigma^+$	6.137	0.181	1.662_9	1604.52	28.77	2.87
	(69)	2	(1.9569183_7)	$^1\Sigma^+$	[3.083]	....	[1.671_6]			
GaI..........	69	127	44.66612_6	$^1\Sigma^+$	0.056934_7	0.000189	2.5747	216.6	0.5	3.47
GaO..........	69	16	12.982250_0	$^2\Sigma$	[(0.4271)]	....	[(1.743_6)]	767.5	6.24	3.9_1
GaTe..........	(69)	(130)	(45.03244_3)	....	....	....	....	....	....	2.6
GdO..........	(158)	16	(14.523901_0)	....	....	....	....	841.0	3.70	7.5_0
								or 830.0	2.25	
Ge_2..........	(74)	(72)	(36.45387_0)	....	....	....	....	....	....	2.8_1
GeBr..........	74	79	38.16902_8	$^2\Pi_r$	....	....	....	$295._4$	0.7_2	≤3.5
				$A = +1150$						(2.6?)
GeCl..........	74	35	23.738983	$^2\Pi_r$	....	....	....	406.6	1.30	≤4.4
				$A = +975$						(3.5?)
GeCo..........	(74)	59	(32.79086_8)	....	....	....	....	....	....	2.4_3
GeCr..........	(74)	(52)	(30.50573_5)	....	....	....	....	....	....	1.7_2
GeCu..........	(74)	(63)	(33.99198_1)	....	....	....	....	....	....	2.0_7
GeF..........	(74)	19	(15.113976_0)	$^2\Pi_r$	....	....	....	665.2	2.80	5.0_0
				$A = +935$						
GeFe..........	(74)	(56)	(31.84120_8)	....	....	....	....	....	....	2.1_4
GeH..........	72	1	0.99389792	$^2\Pi_r$	6.727	0.193	1.587_9	1908	37	2.9_5
				$A = +891$						
	72	2	1.9592356_0	$^2\Pi_r$	3.413	0.070	1.587_8	1359	19	
GeI..........	(74)	127	(46.71177_5)	$^2\Pi_{\frac{1}{2}}$	....	....	....	246.1	0.5_0	
GeNi..........	(74)	(58)	(32.47960_7)	....	....	....	....	....	....	2.8_8
GeO..........	74	16	13.149624_7	$^1\Sigma^+$	0.4856961	0.0030756	1.6246	$985._5$*	4.29*	6.78
GeS..........	74	32	22.318826	$^1\Sigma^+$	0.1865653	0.0007486	2.0121	575.8	1.80	5.676
GeSe..........	(74)	(80)	(38.40100_3)	$^1\Sigma^+$	....	....	....	408.7	1.36	4.9_1
GeSi..........	(74)	(28)	(20.295639)	....	....	....	....	....	....	3.0_8
GeTe..........	74	130	47.11256	$^1\Sigma^+$	0.06533820	0.00017246	2.3402	323.9*	0.75*	4.2_2
H_2..........	1	1	0.50391261	$^1\Sigma_g^+$	60.864	3.07638	0.74138	4400.39	120.815	4.47800
	1	2	0.67171136	$^1\Sigma_g^+$	45.6378	1.95004	0.74155	3811.924	90.7113	4.51369
	2	2	1.00705110	$^1\Sigma_g^+$	30.442	1.0623	0.74154	3118.46	64.097	4.55618
H_2^+..........	1	1	....	$^2\Sigma_g^+$	29.8	1.4	1.06_0	2321.8	66.3	2.65056

TABLE 7g-1. CONSTANTS OF DIATOMIC MOLECULES (Continued)

(1)	m_1 (2)	m_2 (2)	μ (3)	Ground state (4)	B_e (5)	α_e (6)	r_e (7)	ω_e (8)	$\omega_e x_e$ (9)	D_0^0 (10)
HBr	1	81	0.99542703	$^1\Sigma^+$	8.46488	0.23328	1.4144	2648.975	45.217_5	3.759
	2	81	1.96518641	$^1\Sigma^+$	4.287_6	0.083_8	1.4144	1884.75	22.71_8	3.806
	3	81	2.90766698_6	$^1\Sigma^+$	2.8974	0.0459	1.4146	[1519.26]	.....	3.828
HBr+	1			$^2\Pi_i$ $A_0 = -2652$	8.072_0	0.236_3	1.4484	$[2348.2_3]$	(45.4)	3.93_5
HCl	1	35	0.97959272	$^1\Sigma^+$	10.59341_6	0.30718_1	1.2745	2990.946_3	52.8186	4.4333
	2	35	1.9044136_2	$^1\Sigma^+$	5.448794	0.113291_1	1.2746	2145.163	27.1825	4.4849
	3	35	2.7765712_6	$^1\Sigma^+$	3.7458	0.0611	1.2731	$[1735.5_1]$	.....	4.509_5
HCl+	2	35		$^2\Pi_i$ $A_0 = -643$	9.9463	0.3183	1.3154	2675.4	53.5	4.75_7
				$A_0 = -643$	5.1158	0.1170		[1863.96]	.....	
He2	4	4	2.00130180_5							$D_e = 0.00097$
He2+	4	4		$^2\Sigma_u^+$	7.211	0.224	1.080_8	$1698._5$	$35._3$	2.2_0
HF	1	19	0.95705545	$^1\Sigma^+$	20.9557	0.798_0	0.91680	4138.32	89.88	5.85_9
	2	19	1.82104538	$^1\Sigma^+$	11.000	0.2907	0.9174	2998.19_2	45.76_1	5.92_9
	3	19	2.6028411_5	$^1\Sigma^+$	7.692	0.1757	0.9176	2508.54	32.54	5.95_9
HF+	1	19		$^2\Pi_i$				3140	90	3.40 (H⁺ + F)
HfO	(180)	16	(14.689234_4)	$^1\Sigma^+$	[0.3860]		[1.724]	(968)		7.9
Hg2	(200)	(202)	(100.482250)	$^1\Sigma_g^+$			3.3	(36)		0.07_1
HgBr	(202)	79	(56.74551_1)	$(^2\Sigma^+)$			[2.23]	188.13	0.9665	0.71
HgCl	(202)	35	(29.807955_3)	$^2\Sigma^+$				292.61	1.6025	1.0_4
HgF	(202)	19	(17.3649659)	$(^2\Sigma^+)$				490	3.8	(1.9)
HgH	(202)	1	(1.00282118)	$^2\Sigma^+$	5.549_0	0.312	1.7405	1387.09	83.01	0.376
	(202)	2	(1.9942153_7)		2.7989	0.1133	1.7379	995.15	49.93	0.398
HgH+	(202)	1		$^1\Sigma^+$	6.613	0.206	1.594_4	2033.87	46.16	(2.3)
	(202)	2		$(^2\Sigma^+)$	3.328	0.0736	1.593_7	1442.15	23.24	(2.4)
HgI	(202)	127	(77.93524_1)					125.0	1.0	0.35
HgS	(202)	(32)	(27.602570_1)							$<2.1_7$
HgTl	(202)	(205)	(101.73073_0)					26.9	0.69	(0.031)
HI	1	127	0.99988452	$^1\Sigma^+$	6.5108_3	0.1686_4	1.6092	2308.09_1	38.981_0	3.056_8
	2	127	1.9826357_3	$^1\Sigma^+$	3.28414	0.06129		1639.939	20.087	3.097_6
	3	127	2.9460330_4	$^1\Sigma^+$	[2.19326]				.....	
HI+	1	127		$^2\Pi_{\frac{3}{2}}$ $A = -13470$			1.6690	(2170)	.....	3.13
HoO	165	16	14.580864_9				[1.6152]			6.6_2
I2	127	127	63.45217_6	$^1\Sigma_g^+$	0.037389	0.0001210	2.6656	214.51886	0.6738	1.5424_4

Molecule	A	B	μ	State	B_e	α_e	r_e	ω_e	$\omega_e x_e$	D_0°
I_2^+	127	127		$^2\Pi_{1/2}$ $A=-5080$						2.72
IBr	127	79	48.65878₂	$^1\Sigma^+$	0.056788	0.000199	2.4699	268.71	0.83	1.817
ICl	127	35	27.414665₆	$^1\Sigma^+$	0.1141556	0.0005357	2.3209	384.293	1.501	2.15306
IF	127	19	16.5245694	$^1\Sigma^+$	0.2799₅	0.00188₇	1.908₅	608.1₉	2.4₅	2.88
In₂	(115)	(115)	(57.45204)							0.8₅
InBr	115	81	47.48032₇	$^1\Sigma^+$	0.0548943	0.0001862	2.5432	221.0	0.65	3.9₉
InCl	115	35	26.809804	$^1\Sigma^+$	0.1090580₃	0.0005177₉	2.4012	317.4	1.01	4.44
InF	115	19	16.302865₃	$^1\Sigma^+$	0.2624₇	0.0018₆	1.9848	535.3₅	2.6₄	5.2₅
InH	115	1	0.99906243	$^1\Sigma^+$	4.994₅	0.142₈	1.8380	1476.0₄	25.61	2.49
(InD)	115	2	1.9794060₆		2.523	0.051	1.837	1048	12.4	
InI	115	127	60.30322	$^1\Sigma^+$	0.036585₇	0.0001040	2.7540	177.1	0.4	2.52
InO	115	16	14.040451₃	$(^2\Sigma)$				703.09	3.71	3.4₃
InS	115	(32)	(25.012376)							≤3.2₅
InSb	(115)	(121)	(58.91379)							2.9₄
InSe	(115)	(80)	(47.13430₁)							1.5₄
InTe	(115)	(130)	(60.97284)							2.5₀
IO	127	16	14.2045817	$^2\Pi_{3/2}$	0.34026	0.002696₆	1.8676	681.4₇	4.2₉	2.1₉
IrC	(193)	12	(11.2974351₃)							1.9
IrO	(193)	16	(14.7705681)							6.4
K₂	39	39	19.481857₀	$^1\Sigma_g^+$	0.056743	0.000165	3.9050	92.021	0.2829	0.514
KBr	39	79	26.084986	$^1\Sigma^+$	0.08122108	0.00040481	2.8208	213*	0.80	3.94
KCl	39	35	18.429177₇	$^1\Sigma^+$	0.1286347	0.0007899	2.6666	281*	1.3₀	4.35
KF	39	19	12.7712448	$^1\Sigma^+$	0.2793873₉	0.0023349₁	2.1714	426	2.4	5.07
KH	39	1	0.98241435	$^1\Sigma^+$	3.412₁	0.081₇	2.242₅	983.6	14.3₂	1.8₆
(KD)	39	2	1.9151069₈		1.753₈	0.0318	2.240₃	706.₆	7.7₅	1.8₉
KHg	(39)	(202)	(32.6625326)	$(^2\Sigma^+)$						0.046₆
KI	39	127	29.810831	$^1\Sigma^+$	0.06087472	0.00026776	3.0478	187	0.57	3.4₀
Kr₂	(84)	(84)	(41.955752₀)							(0.015)
Kr₂⁺	(84)	(84)								≥0.995
La₂	139	139	69.45303							2.5₀
LaF	139	19	16.712596₀	$^1\Sigma^+$	[0.2458]		[2.026]	(550)		(3.6₁)
LaO	139	16	14.343296₅	$^2\Sigma^+$	0.3526	0.0014	1.826	811.6	2.23	8.2₅
LaS	139	(32)	(25.989954)	$^2\Sigma^+$						5.9₆
Li₂	7	7	3.5080027	$^1\Sigma_g^+$	0.6727₂	0.00704	2.6727	351.43₅	2.592	1.117
LiBr	7	79	6.443192₁	$^1\Sigma^+$	0.555394₁	0.005640₂	2.1704	563.2*	3.5₁*	4.3₄
LiCl	7	35	5.8435748	$^1\Sigma^+$	0.706525	0.008012	2.0207	641*	4.2*	4.8₄
LiCs	7	133	6.664204₇	$^1\Sigma^+$				(167)		
LiF	7	19	5.1238105	$^1\Sigma^+$	1.345261	0.020300	1.5639	910.34	7.929	5.9₅
LiH	7	1	0.88123834	$^1\Sigma^+$	7.5131	0.2132	1.5957	1405.649	23.200	2.4289
(LiD)	7	2	1.5648708₁		4.233107	0.091550	1.5951	1055.12	13.228	2.451₀
LiI	7	127	6.648441₁	$^1\Sigma^+$	0.443181₉	0.004090	2.3919	498.1₆	3.3₉	3.5₇
LiK	7	(39)	(5.9454392)	$^1\Sigma^+$				(207)		1.8₉
LiO	(7)	16	4.8765327	$^2\Pi_i$			(1.62)	(745)		3.5₃

TABLE 7g-1. CONSTANTS OF DIATOMIC MOLECULES (Continued)

(1)	m_1 (2)	m_2 (2)	μ (3)	Ground state (4)	B_e (5)	α_e (6)	r_e (7)	ω_e (8)	$\omega_e x_e$ (9)	D_0° (10)
LiRb	(7)	(85)	(6.480537_5)	$^1\Sigma^+$				(185)	3.30	7.2_3
LuO	(175)	16	(14.655002_3)	$^1\Sigma^+$		0.00378	3.891	841.6	1.64_5	0.0495
Mg$_2$	24	24	11.9925223	$(^1\Sigma_g^+)$	0.0928_7			51.12	1.34	≤3.35
MgBr	24	79	18.3945354	$^2\Sigma^+$				373.8	(2.05)	3.2_6
MgCl	24	35	14.226871_3	$^2\Sigma^+$	0.24502	0.0015_8	2.1991	[462.12]	(4.94)	4.7_5
MgF	24	19	10.6012335	$^2\Sigma^+$	0.51922	0.00470	1.7500	[711.69]		≤2.49
MgH	(24)	1	(0.96718516)	$^2\Sigma^+$	5.818	0.1668	1.730_4	1497.0	32.4	(2.1)
	(24)	2	(1.8580737_0)		3.0307	0.0654	1.7302	1077.76	16.09	
MgH$^+$	(24)	1		$^1\Sigma^+$	6.411	0.206	1.648_8	1695.3	30.2	
	(24)	2		$^1\Sigma^+$	3.321	0.064	1.652_8	1226.6	16.30	
MgI	(24)	127	(20.172435_0)	$(^2\Sigma^+)$				[312]		(2.4)
MgO	24	16	9.5957763	$^1\Sigma^+$	0.5743	0.0050	1.749_0	785.0_6	5.1_8	3.5_3
MgS	(24)	(32)	(13.7042728)					525.2	2.93	<2.4
Mn.	55	55	27.469026_3	$(^9\Sigma_g^+)$	(constants for lowest observed state)					0.3_3
MnBr	55	(79)	(32.390087)	$(^7\Sigma^+)$				288	(0.9)	3.2_2
MnCl	55	(35)	(21.367888_7)	$(^7\Sigma^+)$				383	(0.7)	3.7_0
MnF	55	19	14.1166536	$(^7\Sigma^+)$				618.5	2.6	4.3_5
MnH	55	1	0.98966997	$^7\Sigma^+$	5.6841	0.1570	1.7311	1548.0	28.8	(2.5)
	55	2	1.9428738_5		2.8956	0.0514	1.7310	1102.5	13.9	
MnI	55	127	38.340221	$(^2\Sigma^+)$				(240)	(1.5)	2.8_9
MnO	55	16	12.3881675					839.5	4.7	3.7_0
MnS	55	(32)	(20.210343_2)							2.9_3
MoO	(98)	16	(13.748766)							5.0
N$_2$	14	14	7.00153719	$^1\Sigma_g^+$	1.9987	0.01781	1.0976	2358.07	14.188	9.760
N$_2^+$	14	14		$^2\Sigma_g^+$	1.932	0.020	1.116	2207.19	16.14	8.72
N$_2^{++}$	14	14	Data for lowest $^1\Sigma_g^+$		[1.8801]		[1.1316]	(1960)		
Na$_2$	23	23	11.4948863	$^1\Sigma_g^+$	0.15471	0.00079	3.0788	159.23	0.726	0.760
NaBr	23	79	17.803437_7	$^1\Sigma^+$	0.1512532_9	0.0009409_5	2.5020	302*	1.5_0*	3.80
NaCl	23	35	13.870687_6	$^1\Sigma^+$	0.2180630_6	0.0016248	2.3608	366*	2.0_5*	4.25
NaCs	23	133	19.599477	$^1\Sigma^+$			2.6	(98)		0.69
NaF	23	19	10.4021901	$^1\Sigma^+$	0.43690005_6	0.0045571	1.9259	536	3.4	4.9_5
NaH	23	1	0.96549966	$^1\Sigma^+$	4.9012	0.1353	1.8874	1172.2	19.72	2.0_5
	23	2	1.8518630_3		2.5575	0.0520	1.8866	$[826.1_0]$		
NaI	23	127	19.463752_8	$^1\Sigma^+$	0.1178055_9	0.0006477_7	2.7114	258	1.0_8	3.0_5
NaK	23	(39)	(14.458700_8)	$^1\Sigma^+$				123.29	0.400	0.62_1
NaRb	23	(85)	(18.091511)	$^1\Sigma^+$				106.64	0.455	0.57

Note: this page is a large table printed sideways (landscape). Column headings are not printed on this page; the symbols below are the standard constants to which each column corresponds. Values in ( ) or [ ] are as printed (estimated / uncertain). "·····" indicates no entry.

Molecule	A_1	A_2	μ	State	B_e	α_e	r_e (Å)	ω_e	$\omega_e x_e$	D_0^0 (eV)
NbO	93	16	13.645648_6	$(^2\Delta)$, $\|A\| \approx 15$	0.4321	0.0021	1.690_0	989.0_3	3.8_3	7.8_1
NBr	14	79	11.8928388	$^3\Sigma^-$	0.444	0.0040	1.78_7	691.75	4.720	2.9_0
NCl	14	35	9.9990236	$^3\Sigma^-$	$[0.6408_5]$	·····	$[1.6144]$	827.0	5.1	(4.1)
NdF	(142)	19	(16.755233_3)	·····	·····	·····	·····	·····	·····	5.8_7
NdO	(142)	16	(14.3749906)	·····	·····	·····	·····	·····	·····	7.4_8
NF	14	19	8.0613378_8	$^3\Sigma^-$	1.2056_8	0.01492	1.3170	1141.37	8.99	(4.4)
NH	14	1	0.94016028	$^3\Sigma^-$	16.667_6	0.645_7	1.0372	$[3125.5_3]$	$[79]$	3.5_9
ND	14	2	1.76083610	$^3\Sigma^-$	8.9074	0.2530	1.0367	$[2418]$	(45)	3.6_4
NH^+	14	1	·····	$^2\Pi_r$, $A = +78$	$[15.35]$	·····	$[1.081]$	$[2922]$	·····	4.1
ND^+	14	2	·····	·····	$[8.244]$	·····	$[1.0776]$	$[2143.04]$	·····	·····
Ni_2	(58)	(58)	(28.967671)	·····	·····	·····	·····	·····	·····	2.3_6
NiBr	(58)	(79)	(33.409121)	·····	·····	·····	·····	(315)	·····	3.6_9
NiCl	(58)	(35)	(21.806685_0)	·····	·····	·····	·····	(410)	·····	3.8_2
NiF	(58)	19	(14.3068434)	·····	·····	·····	·····	(740)	·····	·····
NiH	58	1	0.99059317	$^2\Delta$	7.815	0.231	1.476	2000	40	2.6
NiD	58	2	1.9464350_5	$^2\Delta_{5/2}$, $A_0 = -490$	4.037	0.090	1.465	1430	20	·····
NiI	(58)	127	(39.776343)	·····	·····	·····	·····	·····	·····	2.9_9
NiO	(58)	16	(12.5343926)	·····	·····	·····	·····	·····	·····	3.7_5
NO	14	16	7.4664332_0	$^2\Pi_r$, $A = +123$	1.7048_5	0.0176_5	1.1508	1904.12	14.088	6.507
NO^+	14	16	·····	$^1\Sigma^+$	2.002	0.0202	1.062_0	2377.1	16.35	10.858
NS	14	32	9.7380290_2	$^2\Pi_r$, $A = +223$	0.7754_5	0.0061	1.4941	1219.6	7.6	(6.0)
NS^+	14	32	·····	$^1\Sigma^+$	·····	·····	·····	≥ 944	·····	·····
NSe	14	(80)	(11.9159661)	·····	(1.09)	·····	(1.26)	·····	·····	(5.1)
O_2	16	16	7.99745747	$^3\Sigma_g^-$	1.44567	0.01579	1.2075	1580.361_3	12.0730	5.115_5
O_2^+	16	16	·····	$^2\Pi_{g,i}$, $A = +195$	1.6920	0.01984	1.1161	1903.85	16.18	6.674
O_2^-	16	16	·····	$^2\Pi_{g,i}$	·····	·····	·····	1108	9	4.1_0
OF	16	19	8.6838822_3	$^2\Pi_i$	·····	·····	(1.30)	$[1028.5]$ in Ar matrix	·····	·····
OH	16	1	0.94808710	$^2\Pi_i$, $A_0 = -140$	18.867	0.708	0.9707_8	3739.94	86.350	4.39_5
OD	16	2	1.78854794	$^2\Pi_i$	9.991	0.258	0.9712	2716.1	42.15	4.45_7
OH^+	16	1	·····	$^3\Sigma^-$	16.781	0.724	1.0294	$[2955]$	$[85]$	4.8_1
OD^+	16	2	·····	$^3\Sigma^-$	8.900	0.274_6	1.0290	$[2187]$	$[51]$	4.8_4
OH^-	16	1	·····	$(^1\Sigma^+)$	·····	·····	·····	·····	·····	4.76_7
P_2	31	31	15.4868817	$^1\Sigma_g^+$	0.30348	0.00143	1.8939	780.89	2.820	5.031
P_2^+	31	31	·····	$^2\Sigma_g^+$	≥ 0.3038	0.0021	$\leq 1.892_9$	$[\geq 733.7]$	·····	(3.7)
Pb_2	(208)	(208)	(103.988322)	·····	·····	·····	·····	256.5	2.96	1.0_0
PbBr	(208)	79	(57.209068_9)	$^2\Pi_{1/2}(\tfrac{1}{2})$	·····	·····	·····	207.5	0.50	2.5

(not certain that this is the ground state)

TABLE 7g-1. CONSTANTS OF DIATOMIC MOLECULES (*Continued*)

(1)	m_1 (2)	m_2 (2)	μ (3)	Ground state (4)	B_e (5)	α_e (6)	r_e (7)	ω_e (8)	$\omega_e x_e$ (9)	D_0^0 (10)
PbCl	(208)	35	(29.9355411$_3$)	$^2\Pi_{1/2}(\tfrac12)$				303.8	0.88	3.1
PbF	(208)	19	(17.4081885)	$^2\Pi_{1/2}$	0.2414	0.0018	2.00$_3$	507.2	2.30	3.6$_4$
PbH	(208)	1	(1.00296499)	$A = +8266$; $^2\Pi_{1/2}$	4.971	0.144	1.838$_5$	1564.1	29.75	≤1.59
PbI	(208)	127	(78.81349$_4$)	$A = +(8200)$; $^2\Pi_{1/2}(\tfrac12)$				160.5	0.25	(2.0)
PbO	208	16	14.8526391	$^1\Sigma^+(0^+)$	0.3073056	0.0019148	1.9218	721.8	3.70	3.83
PbS	208	32	27.711940$_6$	0^+	0.1163195	0.0004365	2.2868	429.40	1.30	3.4$_3$
PbSe	208	80	57.732418	0^+	0.05059952	0.00012993	2.4022	277.6*	0.51*	2.9$_5$
PbTe	(208)	(130)	(79.96121)	0^+				211.96	0.43	2.63
Pd$_2$	(106)	(108)	(53.44710)							1.4$_3$ (VI)
PdC	(106)	12	(10.778659$_1$)							4.7$_7$ (VI)
PdH	(106)	2	(1.9765122$_2$)	$^2\Sigma^+$	[3.613]		[1.536$_4$]		vibr. numbering unknown	2.9
PdO	(106)	16	(13.896135$_1$)							(4.9)
PF	31	19	11.7755965	$A = +324$; $^3\Sigma^-$	0.5665	0.00456	1.589$_7$	846.75	4.489	(6.7)
PF$^+$	31	19		$^3\Sigma^-$	0.6360	0.0048	1.500$_3$	1053.25	5.047	
PH	31	1	0.97606596	$^2\Pi_r$	[8.412], [4.363]	0.240$_1$	[1.4329], [1.4294]	[2299.60], [1666]		(3.0$_5$)
PH$^+$	31	2	1.8911294$_7$	$A = +296$; $^2\Pi_r$	8.505$_1$, [4.350$_3$]		1.4250, [1.4314]			3.0$_6$
PN	31	14	9.6433616$_5$	$^1\Sigma^+$	0.7862$_1$	0.00557	1.4911	1337.24	6.983	6.3$_6$
PO	31	16	10.5479381	$A = +224$; $^2\Pi_r$	0.7337	0.0056	1.4759	1233.3$_3$	6.57	≤6.15
PO$^+$	31	16		$^1\Sigma^+$				1405	(5)	7.8$_8$
PrO	141	16	14.3643633		(0.29)		(1.9$_2$)	(820)		(5.6)
PS	31	32	15.7325010	$A = +321$; $^2\Pi_r$				739.1	2.96	6.6
PS$^+$	31	32		$^1\Sigma^+$				844.6	3.3	4.9$_1$
PtB	(195)	(11)	(10.4208584)							
PtC	195	12	11.3042295$_7$	$^1\Sigma$	0.5303	0.0032	1.676$_9$	1051.18	4.87	6.30
PtH	(195)	1	(1.00264229)	$^2\Delta_{5/2}$	7.198	0.198	1.5283	[2293.50]		3.6$_6$
PtO	(195)	16	(14.7821844)							3.8$_2$
RaCl	(226)	(35)	(30.283614)	$^2\Sigma^+$				256.2	0.71	
Rb$_2$	(85)	(85)	(42.45585$_5$)	$^1\Sigma_g^+$				57.2$_8$	0.096	0.49$_0$
RbBr	85	79	40.90270$_1$	$^1\Sigma^+$	0.04752795	0.00018600	2.9447	169	0.46	4.0$_0$
RbCl	85	35	24.768529	$^1\Sigma^+$	0.0876403$_8$	0.0004536$_1$	2.7867	228*	0.92	4.4$_0$
RbF	85	19	15.524831$_5$	$^1\Sigma^+$	0.2106639$_5$	0.0015227$_6$	2.2703	376*	1.9	5.1$_0$

Molecule	Z	Mass	μ	State	B_e	α_e	r_e	ω_e	$\omega_e x_e$	D_0 (eV)
RbH	(85)	1	(0.99600356)	$^1\Sigma^+$	3.020	0.072	2.367	936.94	14.21	(1.9)
RbI	85	127	50.8727₅	$^1\Sigma^+$	0.03283293	0.0001094₆	3.1769	139	0.34	3.4₇
RhC	103	12	10.746788₆	$^2\Sigma$	0.6027	0.00396	1.613₃	1049.87	4.94	6.01
RhO	103	16	13.843208							4.3
RuC	(102)	12	(10.735774₄)							6.5₅
RuO	(102)	16	(13.824938)					880.8	13.1	5.3
S₂	32	32	15.9850369	$^3\Sigma_g^-$	0.2954₇	0.001570	1.8892	725.64₇	2.844	4.38
Sb₂	(121)	(123)	60.94787	$^1\Sigma_g^+$				269.85	0.59	3.06
SbBi	(121)	209	76.59208	$^1\Sigma^+$				220.0	0.50	(3.0)
SbBr	(121)	(79)	(47.75009₄)					(242.1)	(0.56)	(3.2)
SbCl	(121)	35	(27.123853)					(369.0)	(0.92)	(4.6)
SbF	(121)	19	16.418463					(614.2)	(2.77)	(4.2)
SbH	(121)	1	(0.99949368)	$^3\Sigma^-$	[5.87]		[1.69₃]			
SbH (²H)	(121)	2	(1.9810996₀)	$^3\Sigma^-$	[2.94]		[1.70.]			
SbN	(121)	14	(12.549581₆)	$^1\Sigma^+$				942.0	5.6	(4.8)
SbO	(121)	16	(14.126107₁)	$^2\Pi_i$ $A=+2272$				817	5.₀	(4.1)
SbSe	(121)	(80)	(48.11370₁)	$(^2\Pi_1)$				326.1	1.04	(3.1)
SbTe	(121)	(130)	(62.62182)	$(^2\Pi_{1/2})$				(284)	(0.2)	2.78
Sc₂	45	45	22.477959₅	$^1\Sigma^+$						1.6₅
ScF	45	19	13.35₆989	$^1\Sigma^+$	0.3950	0.00266	1.787₆	735.6	3.8	6.0₆
ScO	45	16	11.7974776	$^2\Sigma^+$	[0.51340]		[1.6683]	971.55	3.95	6.9₆
ScS	45	(32)	(18.684147₄)	$(^2\Sigma^+)$						4.9₂
Se₂	80	80	39.958256	$^3\Sigma_g^-$	0.08992	0.000288	2.1660	385.3028	0.96363	3.410
SeF	(80)	19	(15.349416₃)	$^2\Pi_i$ $A=-560$	[0.363]		[1.73₁]			
SeH	(80)	1	(0.99527385)	$^2\Pi_i$ $A_0=-1600$	[7.98]			(2400)		3.2
SeO	80	16	13.3274820	$^2\Sigma^-$	0.4704 (constants for F₂ levels)	0.0032₆	1.639₈	915.43	4.52	4.3₄
SeS	(80)	(32)	(22.836079₁)	$^2\Pi_i$ $A_0=-377$						(3.9₀)
SH	32	1	0.97702732	$^2\Pi_i$	[9.461₁]	(0.30₀)	[1.3504]	(2702)	(60)	3.53
SH (²H)	32	2	1.8947416₇	$^2\Pi_i$	[4.900₁]	(0.11₁)	[1.3474]	(1940)	(31)	
SH⁺	32	1		$^3\Sigma^-$	[9.134₀]		[1.3744]			3.57₅
Si₂	28	28	13.988463₆	$^2\Sigma_g^-$	0.2390	0.0013	2.245	510.98	2.02	3.1₀
SiBr	28	(79)	(20.654728)	$^2\Pi_r$ $A=+418$				425.4	1.5	(3.7)
SiBr⁺	28	(79)		$^1\Sigma^+$						(5.5)
SiC	28	12	(8.3979222₄)					535.8	1.6	4.6₄
SiCl	28	35	15.542282₁	$^2\Pi_r$ $A=+207$	0.25619	0.00163	2.0576	535.89	2.29	(4.5)
SiF	28	19	11.3148106	$^2\Pi_r$ $A=+162$	0.5813₈	0.00490	1.6008	857.2₀	4.74	5.5₇

TABLE 7g-1. CONSTANTS OF DIATOMIC MOLECULES (*Continued*)

(1)	m_1 (2)	m_2 (2)	μ (3)	Ground state (4)	B_e (5)	α_e (6)	r_e (7)	ω_e (8)	$\omega_e x_e$ (9)	D_0^0 (10)
SiH.......	(28)	1	(0.97278225)	$^2\Pi_r$ $A = +143$	7.4979	0.2149	1.5203	2045	36	3.06
SiI.......	(28)	2	(1.8788414$_8$)	$^2\Pi_r$	3.8849	0.0801	1.5197	1471	19	3.09
	28	127	22.9233244	$^2\Pi_{1/2}$ $A = +(757)$				360.5	1.0$_1$	
SiN.......	.28	14	9.3321338	$^2\Sigma^+$	0.7310	0.00567	1.5720	1151.68$_0$	6.560$_0$	(6.2)
SiO.......	28	16	10.1767074	$^1\Sigma^+$	0.7267514	0.005038$_5$	1.5097	1241.4$_4$	5.92	8.26
SiO+......			Spectrum previously attributed to SiO+ now known to be due to SiN.							
SiS.......	.28	.32	14.920888$_7$	$^1\Sigma^+$	0.3035290	0.0014736	1.9293	749.6$_9$	2.58	6.37
SiSe......	28	80	20.722468$_8$	$^1\Sigma^+$	0.1920116	0.0007767	2.0583	580.0*	1.78*	5.4$_5$
SiTe......	28	(130)	(23.019425)					481.2	1.30	4.60
SmO.......	(152)	16	(14.471297$_6$)							6.1$_6$
Sn$_2$......	(120)	(118)	(59.44677)							1.9$_9$
SnBr......	(120)	(79)	(47.59307$_6$)	$^2\Pi_r$ $A = +2467$				247.7	0.62	(3.0)
SnCl......	(120)	35	(27.073116)	$^2\Pi_r$ $A = +2361$	0.2733	0.0011	1.942	354.0	1.1	≤4.25
SnF.......	118	19	16.361890	$^2\Pi_r$ $A = +2317$	5.383	0.137	1.770	586.1*	2.76*	4.7$_9$
SnH.......	(120)	1	(0.99942466)	$^2\Pi_r$ $A = +2178$	2.7195	0.049	1.7690	1715	30	2.6$_0$
SnI.......	(120)	2	(1.9808284$_6$)	$^2\Pi$				1218	15	(2.2)
	(120)	127	(61.65195)					199.0	0.5$_5$	5.49
SnO.......	120	16	14.112333$_1$	$^1\Sigma^+$	0.3557190	0.002142$_9$	1.8325	822.1	3.73	4.7$_7$
SnS.......	120	32	25.241415	$^1\Sigma^+$	0.136861$_9$	0.000506$_3$	2.2090	487.26	1.358	4.09
SnSe......	120	80	47.95428$_5$	$^1\Sigma^+$	0.0649977$_6$	0.0001704$_8$	2.3256	331.2*	0.736*	3.65
SnTe......	(120)	(130)	(62.35204)	$^1\Sigma$				259.5	0.50	5.35$_5$
SO........	32	16	10.6613029$_9$	$^3\Sigma^-$	0.720817	0.005736	1.4811	1148.19	6.116	(3.9)
SrBr......	(88)	79	(41.58494$_7$)	$^2\Sigma^+$				216.5	0.51	4.2$_5$
SrCl......	(88)	35	(25.017065)	$^2\Sigma^+$	0.25045	0.00148	2.0757	302.3	0.95	5.5$_5$
SrF.......	(88)	19	(15.622111)	$^2\Sigma^+$	3.6751	0.0814	2.1456	500.1	2.21	≤1.68
SrH.......	(88)	1	(0.99640162)	$^2\Sigma^+$	1.8609	0.0292	2.1449	1206.2	17.0	≤1.70
SrI.......	(88)	2	(1.9689885$_3$)	$(^2\Sigma^+)$				173.9	0.42	(3.5)
	(88)	127	(51.93244)							
SrO.......	88	16	13.532585$_7$	$^1\Sigma^+$	0.33798	0.00219	1.9198	653.2*	3.92*	4.4$_2$
SrS.......	(88)	(32)	(23.444936)	$(^1\Sigma)$						3.4$_7$

SrSe	(88)	(80)	(41.86045_1)	$(^1\Sigma)$	....	....	....	....	....	2.8_9
TaO	181	16	14.6958719	$^2\Delta_{\frac{3}{2}}$	0.40288	0.00182	1.6874	1028.69	3.51	8.4
				$A = +1753$						
TbO	159	16	14.532317_7	....	....	....	....	....	....	7.5_2
Te_2	(130)	(128)	(64.44897)	$(O_g{}^+)$	....	....	[2.59]	251	0.55	2.65
TeH	(130)	1	(1.00006663)	$^2\Pi_i$	[5.56]	....	[1.741]			
				$A_0 = -2250$						
TeO	128	16	14.217027_7	O^+	0.3560	0.002375	1.825_0	797.69	4.00	3.9_0
TeS	(130)	(32)	(25.657389)	(O^+)	....	....	....	470.0	1.4	(3.4_7)
TeSe	(130)	(80)	(49.47827_4)	(O^+)	....	....	....	318.0	1.0	2.8_9
ThN	232	14	13.2061101_9	....	....	....	....	....	....	5.9
ThO	232	16	14.9634507	$(^1\Sigma^+)$	0.332644	0.001302	1.8403	895.77	2.39	8.5
					(constants for lowest $^1\Sigma$ state)					
ThP	232	31	27.326119_5	....	....	....	....	....	....	4.0
Ti_2	(48)	(48)	(23.973973_9)	....	....	....	....	....	....	1.3_0
TiC	(48)	12	(9.5979161_0)	....	....	....	....	....	....	≤5.5
TiCl	(48)	(35)	(20.221291_3)	$(^4\Sigma)$	....	....	....	379.7	3.41	
TiO	(48)	16	(11.9938851)	$^3\Delta_r$	0.5355	0.0030	1.6200	1008.26	4.13	6.9_5
				$A = +(50)$						
TiS	(48)	(32)	(19.181618_0)	....	....	....	....	....	....	4.7
Tl_2	(205)	(205)	(102.48723_1)	....	....	....	....	....	....	(0.6)
TlBr	205	81	58.01440_2	$^1\Sigma^+$	0.0423896	0.0001278	2.6182	192.10	0.39	3.42_1
TlCl	205	35	29.872564_2	$^1\Sigma^+$	0.09139701_0	0.00039793_2	2.4848	287.47	1.24	3.82_0
TlF	205	19	17.3868729	$^1\Sigma^+$	0.2231501_5	0.0015038_7	2.0844	477.3	2.3	4.57_5
TlH	(205)	1	(1.00289416)	$^1\Sigma^+$	4.806	0.154	1.870_2	1390.7	22.7	1.97_5
	(205)	2	(1.9945039_7)	$^1\Sigma^+$	2.419	0.057	1.869	987.7	12.04	2.00_0
TlI	205	127	78.37846_3	$^1\Sigma^+$	0.0271674_6	0.0000663_5	2.8137	(150)	....	2.77_0
TmO	169	16	14.611481_6	....	....	....	....	....	....	6.0
UN	238	14	13.2251212	....	....	....	....	....	....	5.4_6
UO	238	16	14.9878626	....	....	....	....	....	....	7.88
US	238	(32)	(28.186418_0)	....	....	....	....	....	....	5.8_0
VO	51	16	12.1729619	$^4\Sigma^-$	0.5480	0.0034	1.589_7	1011.56	4.97	6.4_0
WO	(184)	16	(14.7153823)	$(^3\Sigma^-)$	....	....	....	[(1055)]	....	6.7
Xe_2	(132)	(129)	(65.193613)	$^1\Sigma_g{}^+$	....	....	....	....	....	(0.019)
$Xe_2{}^+$	(132)	(129)	....		....	....	....	....	....	≥0.967
Y_2	89	89	44.452715	....	....	....	....	....	....	1.6_2
YbCl	(174)	(35)	(29.115457)	$(^2\Sigma)$	(constants for lowest observed state)			293.6	1.23	(2.1)
YbH	(174)	1	(1.00201939)	$^2\Sigma^+$	3.995	0.0986	2.052_1	1249.54	21.055	1.6_2
	(174)	2	(1.9910471_4)	$^2\Sigma^+$	2.0121	0.0352	2.0513	886.38	10.475	1.6_4
YbO	(174)	16	(14.647934_5)	....	....	....	....	....	....	≤(3.8)
YCl	89	35	25.097389	$^1\Sigma^+$	0.1162	0.0003	2.404	380.7	1.3	(3.4)
					(constants for lowest observed $^1\Sigma$ state)					

TABLE 7g-1. CONSTANTS OF DIATOMIC MOLECULES (*Continued*)

(1)	m_1 (2)	m_2 (2)	μ (3)	Ground state (4)	B_e (5)	α_e (6)	r_e (7)	ω_e (8)	$\omega_e x_e$ (9)	D_0^0 (10)
YF	89	19	15.653395	$^1\Sigma^+$	0.29041	0.00163	1.9257	636.3	2.50	6.2
YLa	89	139	54.20931							2.0_5
YO	89	16	13.556054_5	$^2\Sigma^+$	[0.3881]	(0.0016)	$[1.790_0]$	852.5	2.45	7.3_9
YS	89	(32)	(23.515467)	$(^2\Sigma^+)$						5.4_5
ZnCl	(64)	35	(22.604390_2)	$^2\Sigma$				390.5	1.55	2.1
ZnF	(64)	19	(14.6459382)	$^2\Sigma$				(630)	(3.5)	
ZnH	(64)	1	(0.99218372)	$^2\Sigma^+$	6.6794	0.2500	1.5949	1607.6	55.14	0.84_4
	(64)	2	(1.9525855_1)	$^2\Sigma^+$	[3.3497]		[1.6054]	(1150.0)	(28.02)	0.87_3
ZnH+	(64)	1		$^1\Sigma^+$	7.403	0.236	1.5149	1916	39	(2.5)
	(64)	2		$^1\Sigma^+$	3.767	0.107	1.513_9	1364.8	19.8	
ZnI	64	127	42.512907	$(^2\Sigma)$				223.4	0.75	1.4
ZnO	(64)	16	(12.7939101)							2.8
ZnS	(64)	(32)	(21.313048_7)							2.0_8
ZnSe	(64)	(80)	(35.517195)							1.3_7
ZrN	(90)	14	(12.115950_3)							5.8_1
ZrO	90	16	13.579058_9	$^1\Sigma^+$	0.4241	0.0023	1.711	978.0_7	5.04	7.8_4

7h. Constants of Polyatomic Molecules

G. HERZBERG

National Research Council of Canada

L. HERZBERG

Communications Research Center, Ottawa, Canada

7h-1. Introduction. The following tables present some of the more important data on simple polyatomic molecules derived from infrared, Raman, and microwave spectra. Tables 7h-1 through 7h-4 give the fundamental vibrational frequencies (in cm^{-1}) of most triatomic and four-atomic molecules for which these quantities are available and for a few important five- and six-atomic molecules. The point groups to which the molecules belong are indicated in the last column. The numbering of the vibrations is in accordance with the practice followed by many authors in recent years[1] and now established by international agreement.[2]

For most molecules listed the fundamentals are active in both the infrared and the Raman spectrum. However, for molecules of high symmetry, certain vibrations cannot occur in the Raman spectrum, others cannot occur in the infrared spectrum, and a few in neither one: for triatomic linear symmetric molecules ($D_{\infty h}$), ν_1 is Raman active and ν_2 and ν_3 infrared active; for four-atomic linear symmetric molecules ($D_{\infty h}$), ν_1, ν_2, and ν_4 are Raman active and ν_3 and ν_5 infrared active; for four-atomic planar molecules with a threefold axis (D_{3h}), ν_1 is Raman active, ν_2 infrared active, and ν_3 and ν_4 are both Raman and infrared active; for five-atomic tetrahedral molecules (T_d) all vibrations are Raman active but only ν_3 and ν_4 are infrared active; for linear symmetric six-atomic molecules, the vibrations ν_1, ν_2, ν_3, ν_6, ν_7 are Raman active and the remaining ones are infrared active; for six-atomic molecules with three mutually perpendicular planes of symmetry (V_h), the vibrations ν_7, ν_9, ν_{10}, ν_{11}, ν_{12} are infrared active and all others, except ν_4, are Raman active; for six-atomic molecules of C_{2h} symmetry, ν_1, ν_2, ν_3, ν_4, ν_5, and ν_8 are Raman active, and the others are infrared active.

Tables 7h-5 through 7h-15 give the rotational constants $A_{[0]}$, $B_{[0]}$, $C_{[0]}$ of selected triatomic, four-atomic, five-atomic, and six-atomic molecules. These rotational constants are, apart from the factor $h/8\pi^2 c$, the reciprocal moments of inertia, and therefore from them the geometrical parameters of the molecule can be determined if a sufficient number of isotopes have been investigated. The geometrical parameters thus obtained are also listed in Tables 7h-5 through 7h-15.

The constants $A_{[0]}$, $B_{[0]}$, $C_{[0]}$ refer to the lowest vibrational level which still includes the zero-point vibration. In the few cases in which these constants have been determined for the true equilibrium positions, the equilibrium contants A_e, B_e, C_e are also listed.

Microwave spectra give the constants in megahertz while infrared and Raman

---

[1] G. Herzberg, "Molecular Spectra and Molecular Structure. II. Infrared and Raman Spectra of Polyatomic Molecules," D. Van Nostrand Company, Inc., Princeton, N.J., 1945.
[2] R. S. Mulliken, *JCP* **23**, 1997 (1955).

spectra give them in cm^{-1}. Here all microwave values have been converted to cm^{-1} by dividing by $c = 2.997925 \times 10^{10}$ cm/sec.

In the alphabetical order used, D is counted as an H in order to have the deuterated molecules appear with the corresponding nondeuterated ones. Element symbols without mass numbers refer to the most abundant isotope.

Many of the data have been taken from the books by Herzberg[1]; by Gordy, Smith, and Trambarulo[2]; by Townes and Schawlow[3]; and the more recent compilations of Shimanouchi[4] and Starck[5]. In addition, some of the literature up to 1968 has been included. For detailed tables of microwave data reference should be made to the Microwave Spectral Tables prepared by Cord, Petersen, Lojko, and Haas.[6]

7h-2. Fundamental Vibrations

TABLE 7h-1. TRIATOMIC MOLECULES

Molecule	ν_1 cm^{-1}	ν_2 cm^{-1}	ν_3 cm^{-1}	Point group	Molecule	ν_1 cm^{-1}	ν_2 cm^{-1}	ν_3 cm^{-1}	Point group
BO₂.....	1070	464	1322	$D_{\infty h}$	HPO....	....	1187	985	C_s
BO₂⁻....	1070‡	610‡	1970‡	$D_{\infty h}$	H₂S.....	2614.6	1182.7	2627	C_{2v}
BrCN...	575	341.5	2198.3	$C_{\infty v}$	HDS....	....	1090	(2684)	C_s
C₃......	(1230)‡	63.1	2040	$D_{\infty h}$	D₂S.....	1896.4	855.5	1910	C_{2v}
CF₂.....	1102	667	(1222)	C_{2v}	H₂Se....	2344.5	1034.2	2357.8	C_{2v}
ClCN....	714	378.4	2215.6	$C_{\infty v}$	HDSe...	1691	912	2352	C_s
Cl₂O....	640	(300)	686	C_{2v}	D₂Se....	1686.7	741.4	1697.4	C_{2v}
ClO₂.....	945.5	447.4	1110.5	C_{2v}	HSiBr...	1547.s	771.4	408.0	C_s
CNC....	....	321	......	$D_{\infty h}$	HSiCl...	....	805.s	522.4	C_s
CO₂.....	$\begin{cases}1388.2\\1285.5\dagger\end{cases}$	667.4	2349.2	$D_{\infty h}$	ICN....	470*	321*	2158*	$C_{\infty v}$
					KrF₂....	449	232.6	588	$D_{\infty h}$
CO₂⁺....	1280	......	(1469)	$D_{\infty h}$	N₃⁻.....	1350*	630*	2080*	$D_{\infty h}$
CS₂......	658.0	396.7	1533	$D_{\infty h}$	NF₂.....	1074.3	(573)‡	(931)‡	C_{2v}
FCN....	(2294)	451.3	1076.5	$C_{\infty v}$	NH₂.....	....	1497.2	......	C_{2v}
F₂O.....	928	461	831	C_{2v}	N₂O.....	1284.9	588.8	2223.8	$C_{\infty v}$
HCF....	....	1403	......	C_s	N₂O⁺....	1736.6	461.2	1126.4	$C_{\infty v}$
HCN....	2096.9	712.0	3311.5	$C_{\infty v}$	NO₂.....	1318	749.8	1617.8	C_{2v}
DCN....	1925.3	569.0	2630.3	$C_{\infty v}$	NO₂⁻....	1345*	816*	1236*	C_{2v}
HCO....	(2700)	1820.2	1083.0	C_s	NO₂⁺....	1400*	538*	2358*	$(D_{\infty h})$
HCP....	3216.9	674.3	1278.2	$C_{\infty v}$	NOCl...	1799	596	332	C_s
HF₂⁻....	(595)‡	1240‡	1500‡	$C_{\infty v}$(?)	NOF....	1844.0	765.9	521	C_s
HgBr₂...	225	41	293	$D_{\infty h}$	O₃.....	1110	705	1042.2	C_{2v}
HgCl₂...	360	70	413	$D_{\infty h}$	OCN⁻...	2180*	870*	......	$C_{\infty v}$
HgI₂.....	156	33	(235)	$D_{\infty h}$	OCS.....	2062.2	520.4	859.0	$C_{\infty v}$
HNO....	3596	1562	1110	C_s	SCN⁻....	2066*	483*	750*	$C_{\infty v}$
H₂O.....	3656.7	1594.8	3755.8	C_{2v}	SeCN⁻...	2051.5*	575*	......	$C_{\infty v}$
HDO....	2726.7	1402.2	3707.5	C_s	SiCC....	1742	......	591	$C_{\infty v}$
D₂O....	2671.5	1178.3	2788.0	C_{2v}	SO₂.....	1151.4	517.7	1361.8	C_{2v}
HOCl....	3609.2	1242	739	C_s	UO₂²⁺...	860‡	(252)‡	930‡	C_{2v}
DOCl....	2666.0	911	739	C_s	XeF₂....	515	213.2	558	$D_{\infty h}$

( ) Values in parentheses are uncertain or have been obtained indirectly.

* Observed in liquid.

† Fermi resonance between ν_1 and $2\nu_2$.

‡ Observed in crystal or solid matrix.

[1] G. Herzberg, "Molecular Spectra and Molecular Structure," vol. II, "Infrared and Raman Spectra of Polyatomic Molecules," 1945, vol. III. "Electronic Spectra and Electronic Structure of Polyatomic Molecules," D. Van Nostrand Company, Inc., Princeton, N.J., 1945, 1966.

[2] W. Gordy, W. V. Smith, and R. F. Trambarulo, "Microwave Spectroscopy," John Wiley & Sons, Inc., New York, 1953.

[3] C. H. Townes and A. L. Schawlow, "Microwave Spectroscopy," McGraw-Hill Book Company, New York, 1955.

[4] T. Shimanouchi, Tables of Molecular Vibrational Frequencies, parts 1–3, *Natl. Standard Ref. Data Ser. NBS* 6, pp. 11, 17, 1967–1968.

[5] B. Starck, in Landolt-Börnstein New Series Group II, vol. 4, 1967.

[6] M. S. Cord, J. D. Petersen, M. S. Lojko, and R. H. Haas, Microwave Spectral Tables, *NBS Monograph* 70, vols. 3, 4, and 5, 1968.

TABLE 7h-2. FOUR-ATOMIC MOLECULES

Molecule	ν_1 cm^{-1}	ν_2 cm^{-1}	ν_3 cm^{-1}	ν_4 cm^{-1}	ν_5 cm^{-1}	ν_6 cm^{-1}	Point group
AsCl₃...........	410	193	370	159	......	......	C_{3v}
AsF₃...........	740.3	336.5	702.2	262.3	......	......	C_{3v}
AsH₃...........	2116.1	906	2123	1003	......	......	C_{3v}
AsD₃...........	1523.1	660.0	1529.3	714	......	......	C_{3v}
BBr₃...........	279	372	802	151	......	......	D_{3h}
BCl₃...........	471	460	956	243	......	......	D_{3h}
BI₃...........	189*	305*	692*	......	......	......	D_{3h}
BF₃...........	888	691.4	1453.7	480.4	......	......	D_{3h}
BiCl₃...........	288	130	242	96	......	......	C_{3v}
BrO₃⁻...........	803*	428*	828*	350*	......	......	C_{3v}
CH₃...........	......	611‡	......	......	......	......	D_{3h}
CD₃...........	......	463‡	......	......	......	......	D_{3h}
C₂H₂...........	3372.7	1973.7	3294.9	611.7	729.2	......	$D_{\infty h}$
C₂HD...........	3335.6	1853.8	2583.6	518.4	677.8	......	$C_{\infty v}$
C₂D₂...........	2703.8	1763.8	2439.2	510.7	536.4	......	$D_{\infty h}$
C₂I₂...........	2113	191	718	307	(115)	......	$D_{\infty h}$
C₂N₂...........	2329.9	854.2	2157.8	507.2	233.1	......	$D_{\infty h}$
Cl₂CO...........	1827	567	285	580	849	440	C_{2v}
Cl₂CS...........	1139	503	288	471	818	292	C_{2v}
ClF₃...........	752	527	326	703	434	364	C_s
ClO₃⁻...........	940*	617*	988*	479*	......	......	C_{3v}
CO₃⁻⁻...........	1063*	878*	1415*	680*	......	......	D_{3h}
F₂BO...........	1369	856.0	491.0	......	......	......	C_{2v}
FClCO...........	1868	1095	776	501	415	667	C_s
F₂CO...........	1942	965	584	774	1249	626	C_{2v}
HC₂Br........	3325	2085	618	618	295	......	$C_{\infty v}$
HC₂Cl........	3340	2110	756	604	326	......	$C_{\infty v}$
DC₂Cl........	2612	1980	742	472	(312)	......	$C_{\infty v}$
HC₂F........	3355	2255	1055	578	367	......	$C_{\infty v}$
HCO₂⁻........	2825*	1584*	1386*	1352*	773*	1069*	C_{2v}
H₂CO...........	2766.4	1746.1	1500.6	2843.4	1251.2	1167	C_{2v}
HDCO...........	2845	2120.5	1723.3	1399	1041	1074	C_s
D₂CO...........	2055.8	1700	1105.7	2159.7	990	938	C_{2v}
HFCO...........	2981.0	1836.9	1342.5	1064.8	662.5	(1175)	C_s
HN₃...........	3335.6	2139.8	1263.7	1150.5	534.2	607.0	C_s
HNCO...........	3531	2274	1527	777.1	659.8	577.5	C_s
HNCS...........	3537.9	1973	999	615	467	834	C_s
cis-HNO₂.......	3426	1639	(1370)	856	620	638	C_s
trans-HNO₂.....	3590	1698	1264	793	598	544	C_s
cis-DNO₂.......	2530	......	......	816	(591)	508	C_s
trans-DNO₂.....	2650	1690	1018	739	591	416	C_s
H₂O₂...........	3599	(1380)	880	309	3608	1266	C_2
D₂O₂...........	2510*	1009*	878*	229	2482*	1004*	C_2
H₂S₂...........	2513*	882*	510*	416	2577	886	C_2
IO₃⁻...........	779*	390‡	826*	330*	......	......	C_{3v}
NCl₃...........	535*	347*	637*	254*	......	......	C_{3v}
NF₃...........	1032	647	905	493	......	......	C_{3v}
trans-N₂F₂......	(1636)	(1010)	(592)	360	989	421	C_{2h}
NH₃...........	3336.7	950.4	3443.8	1626.8	......	......	C_{3v}
NH₂D...........	......	......	1592	884	......	......	C_s
NHD₂...........	......	2418	1234	813	2556	1464	C_s
ND₃...........	2420.4	747.5	2564.0	1191.2	......	......	C_{3v}
N₂H₂...........	......	1481*	......	3120.1	1286*	......	C_{2h}
NO₃⁻...........	1048*	824*	1357*	720*	......	......	D_{3h}
P₄.............	606	363	465	......	......	......	T_d
PBr₃...........	380	162	400	116	......	......	C_{3v}
PCl₃...........	507	260	493	189	......	......	C_{3v}

TABLE 7h-2. FOUR-ATOMIC MOLECULES (*Continued*)

Molecule	ν_1 cm^{-1}	ν_2 cm^{-1}	ν_3 cm^{-1}	ν_4 cm^{-1}	ν_5 cm^{-1}	ν_6 cm^{-1}	Point group
PF$_3$............	892	487	860	344	......	......	C_{3v}
PFBr$_2$..........	817	421	......	......	393	......	C_s
PFCl$_2$..........	827	524	......	......	496	......	C_s
PF$_2$Cl..........	860	527	......	......	833	......	C_s
PFClBr........	822	503	415	......	......	......	C_1
PH$_3$............	2322.9	992.0	2327.7	1118.3	......	......	C_{3v}
PH$_2$D..........	......	1700	1097	892	......	......	C_s
PHD$_2$..........	2320	......	906	......	......	980	C_s
PD$_3$............	1694	730	1700	806	......	......	C_{3v}
SbCl$_3$..........	360	165	320	134	......	......	C_{3v}
SbH$_3$..........	1890.9	781.5	1894.2	830.9	......	......	C_{3v}
SbD$_3$..........	1358.8	561.1	1362.0	592.5	......	......	C_{3v}
S$_2$Cl$_2$..........	448	438	206	102	538	242	(C_2)
SO$_3$............	1067	498	1391.2	531	......	......	D_{3h}
SOBr$_2$..........	1121	405	267	120	379	223	C_s
SOCl$_2$..........	1230	490	344	194	445	284	C_s
SOF$_2$..........	1333	808	530	(410)	748	390	C_s

( ) Values in parentheses are uncertain or have been obtained indirectly.
* Observed in liquid or solution.
‡ Observed in crystal or solid matrix.

TABLE 7h-3. SOME FIVE-ATOMIC MOLECULES

Molecule	ν_1 cm^{-1}	ν_2 cm^{-1}	ν_3 cm^{-1}	ν_4 cm^{-1}	ν_5 cm^{-1}	ν_6 cm^{-1}	Point group
CH$_4$	2916.7	1533.6	3018.9	1306.2	......	......	T_d
CD$_4$	2108.9	1091.9	2259.3	995.6	......	......	T_d
CH$_3$D	2973 2914 } †	2200.0	1300	3016.9	1471	1155	C_{3v}
CHD$_3$	2993	2142	1003	2263	1291	1036	C_{3v}
CF$_4$	908.5	435.0	1283.0	631.7	......	......	T_d
CCl$_4$	459.0	221	794.3 756 } †	310.0	......	......	T_d
CBr$_4$	267*	122*	671*	182*	......	......	T_d
CI$_4$	178‡	90‡	555‡	123‡	......	......	(T_d)
SiH$_4$	2187.0	974.6	2190.6	914.2	......	......	T_d
SiF$_4$	800	268	1031.8	389.4	......	......	T_d
SiCl$_4$	425	149	619.0	221.3	......	......	T_d
SiBr$_4$	249*	90*	487*	137*	......	......	T_d
SiI$_4$	168	63	405	94	......	......	T_d
GeH$_4$	2106	930.9	2113.6	819.3	......	......	T_d
GeF$_4$	(740)	(200)	800	260	......	......	T_d
GeCl$_4$	396*	134*	453*	172*	......	......	T_d
GeBr$_4$	235*	79*	327*	112*	......	......	T_d
GeI$_4$	159	60	264	80	......	......	T_d
SnH$_4$	......	758	1901.1	677	......	......	T_d
SnCl$_4$	366	104	403	134	......	......	T_d
SnBr$_4$	220	64	279	88	......	......	T_d
CH$_3$F	2964.5	1460.5	1048.6	3005.8	1466.5	1182.4	C_{3v}
CH$_3$Cl	2966.7	1354.9	732.1	3042.4	1452.1	1017.3	C_{3v}
CH$_3$Br	2972	1305.9	611.1	3056.6	1442.7	954.7	C_{3v}
CH$_3$I	2953.2	1250.8	533.2	3060.3	1437.4	882.4	C_{3v}
CHF$_3$	3034.5	1139.5	697.0	1377.5	1152	508	C_{3v}
CHCl$_3$	3032.0	671.1	364.8	1218	768	256	C_{3v}
CHBr$_3$	3042	541	222	1149	669	155	C_{3v}
CHI$_3$	(3040)*	385*	145*	1064*	581*	92*	C_{3v}
CF$_3$Cl	1104	782	475	1217	559	351	C_{3v}
CF$_3$Br	1083	762	350	1208	549	305	C_{3v}
CF$_3$I	1073	741	285	1185	540	267*	C_{3v}
SiH$_3$F	2206	990	872	2196	(956)	728.1	C_{3v}
SiH$_3$Cl	2201	949	551	2195	954.4	664.0	C_{3v}
SiH$_3$Br	2200	930	430	2196	950.4	632.6	C_{3v}
SiH$_3$I	2191.8	903	(355)	2205.6	941.0	592.4	C_{3v}
SiHCl$_3$	2274	497	250*	810	600	179*	C_{3v}
SiHBr$_3$	2232*	358*	169*	769*	473*	111*	C_{3v}
GeH$_3$F	2120.6	859.0	689.1	2131.8	874.2	642.5	C_{3v}
GeH$_3$Cl	2121	848	423	2129.1	874.7	603.9	C_{3v}
GeH$_3$Br	2116	833	305	2127.0	871.4	578.1	C_{3v}
GeH$_3$I	2112	812	248	2120.6	853.0	558.7	C_{3v}

( ) Values in parentheses are uncertain or have been obtained indirectly.
* Observed in liquid or solution.
† Fermi resonance.
‡ Observed in crystal or solid matrix.

TABLE 7h-4. SOME SIX-ATOMIC MOLECULES

Molecule	ν_1 cm^{-1}	ν_2 cm^{-1}	ν_3 cm^{-1}	ν_4 cm^{-1}	ν_5 cm^{-1}	ν_6 cm^{-1}	ν_7 cm^{-1}	ν_8 cm^{-1}	ν_9 cm^{-1}	ν_{10} cm^{-1}	ν_{11} cm^{-1}	ν_{12} cm^{-1}	Point group
HC≡C—C≡CH	(3293)*	2184	874	3329	2020	627	482	630	(220)	.....	.....	.....	$D_{\infty h}$
C$_2$H$_4$	3026.4	1622.6	1342.2	1023	3102.5	1222	949.3	943	3105.5	826.0	2988.7	1443.5	V_h
C$_2$D$_4$	2251*	1515*	981*	(726)	2305	(1009)	720.0	780	2345	(586)	2201.0	1077.9	V_h
C$_2$F$_4$	1872	778	394	(190)	1340	551	407	510	1337	218	1186	558	V_h
C$_2$Cl$_4$	1571*	447*	237*	(110)	1000*	347*	288*	512*	908*	176*	777*	310*	V_h
C$_2$Br$_4$	1546*	266*	144*	(66)	886*	211*	245‡	463*	766*	119*	635*	188*	V_h
C$_2$I$_4$	1448*	181*	106*	.....	780*	146*	225‡	(418)	638*	94‡	525*	129‡	V_h
H$_2$C:CF$_2$	3058.3	1728.5	1410	925.3	550	590	3099.8	1302	955	438	801	613	C_{2v}
cis-C$_2$H$_2$F$_2$	3135	1715	1266	1014	(255)	(866)	(482)	3135	1376	1127	768	756	C_{2v}
trans-C$_2$H$_2$F$_2$	.....	.....	.....	.....	.....	874	325	774	3115	1274	1159	(410)	C_{2h}
H$_2$C:CCl$_2$	3035*	1627	1400	603	299	686*	3130*	1095	800	372	875	460	C_{2v}
cis-C$_2$H$_2$Cl$_2$	3086	1591	1179*	711*	173*	876*	406*	3072	1303	857	571	697	C_{2v}
trans-C$_2$H$_2$Cl$_2$	3071*	1576*	1270*	844*	349*	898	(192)	758*	3090	1200	827	265*	C_{2h}
H$_2$C:CBr$_2$	3023*	1593*	1379*	467*	184*	668*	3108*	1065*	696*	322*	886*	405*	C_{2v}
cis-C$_2$H$_2$Br$_2$	3084	1584*	1150*	580*	109*	866	372*	3059*	1264	757	466	670	C_{2v}
trans-C$_2$H$_2$Br$_2$	3089*	1581*	1250*	745*	217*	899	.....	736*	3099	1163	688	(192)	C_{2h}
N$_2$O$_4$	1360	813	283	.....	1724*	500*	680	.....	1749	380	1265	752	V_h
N$_2$H$_4$	3325	3261*	1493	1098	873*	780	377	3350	3314	1628	1275	950	C_2
CH$_3$CN	2965.3	2267.3	1400.0	919.9	3009.0	1454.0	1041.0	361.0	.....	.....	.....	.....	C_{3v}
CH$_3$NC	2965.8	2166.0	1410.0	944.6	3014.3	1466.9	1129.3	263	.....	.....	.....	.....	C_{3v}
CH$_3$OH	3682	2977	2844	1477	1455	1340	1056	1034	(2977)	1477	1171*	270	C_s
CH$_3$SH	2946	2869	2607	1475	1335	1070	803	704	2999	1430	955	(600)	C_s

( ) Values in parentheses are uncertain or have been obtained indirectly.
* Observed in liquid only.
‡ Observed in crystal or solid matrix.

7h-3. Rotational Constants and Geometrical Parameters

Molecule	$B_{[0]}$, cm^{-1}	Point group	Geometrical parameters
Br^{79}C^{12}N^{14}	0.1374348		
Br^{79}C^{13}N^{14}	0.1358729		
Br^{79}C^{12}N^{15}	0.1315857	$C_{\infty v}$	$r_0(CBr) = 1.790$ Å
Br^{81}C^{12}N^{14}	0.1366539		$r_0(CN) = 1.159$ Å
Br^{81}C^{13}N^{14}	0.1350802		
Br^{81}C^{12}N^{15}	0.1308165		
Cl^{35}C^{12}N	0.199164$_8$		
Cl^{35}C^{13}N	0.198129$_4$		
Cl^{36}C^{12}N	0.19707	$C_{\infty v}$	$r_0(CCl) = 1.631$ Å
Cl^{37}C^{12}N	0.195043$_3$		$r_0(CN) = 1.159$ Å
Cl^{37}C^{13}N	0.193957$_6$		
C^{12}O$_2$	$\begin{cases} 0.3902_0 \\ B_e = 0.3916_3 \end{cases}$	$D_{\infty h}$	$r_0(CO) = 1.1621$ Å; $r_e(CO) = 1.1601$ Å
C^{13}O$_2$	0.39025	$D_{\infty h}$	$r_0(CO) = 1.1618$ Å
CO$_2{}^+$	0.3804	$D_{\infty h}$	$r_0(CO) = 1.177$ Å
CS$_2$	0.1092	$D_{\infty h}$	$r_0(CS) = 1.554$ Å
FC^{12}N^{14}	0.3520502		$r_0(CF) = 1.262$ Å
FC^{13}N^{14}	0.3518367	$C_{\infty v}$	$r_0(CN) = 1.159$ Å
FC^{12}N^{15}	0.3397823		
HC^{12}N	$\begin{cases} 1.47822 \\ B_c = 1.4849 \end{cases}$		
HC^{13}N	1.43999	$C_{\infty v}$	$r_0(CH) = 1.064$ Å; $r_e(CH) = 1.0657$ Å
DC^{12}N	$\begin{cases} 1.20775 \\ B_e = 1.2118 \end{cases}$		$r_0(CN) = 1.156$ Å; $r_e(CN) = 1.1530$ Å
DC^{13}N	1.18707		
HC^{12}P	0.6663292		
HC^{13}P	0.6384179	$C_{\infty v}$	$r_0(CH) = 1.0667$ Å
DC^{12}P	0.5665385		$r_0(CP) = 1.542$ Å
DC^{13}P	0.5479633		
I^{127}C^{12}N	0.1075931	$C_{\infty v}$	$r_0(CI) = 1.995$ Å
I^{127}C^{13}N	0.105974		$r_0(CN) = 1.159$ Å
N$_2{}^{14}$O	$\begin{cases} 0.4190113 \\ B_e = 0.42118_1 \end{cases}$		
N^{14}N^{15}O	0.4189825	$C_{\infty v}$	$r_0(NN) = 1.126$ Å; $r_e(NN) = 1.126$ Å
N^{15}N^{14}O	$\begin{cases} 0.404856_7 \\ B_e = 0.406935_5 \end{cases}$		$r_0(NO) = 1.191$ Å; $r_e(NO) = 1.186$ Å
N$_2{}^{15}$O	0.404859$_2$		
O^{16}C^{12}S^{32}	0.202857		
O^{16}C^{13}S^{32}	0.2022025		
O^{16}C^{12}S^{33}	0.2003016		
O^{16}C^{12}S^{34}	0.1978974		
O^{16}C^{12}S^{35}	0.19564		
O^{16}C^{12}S^{36}	0.193456	$C_{\infty v}$	$r_0(CO) = 1.1637$ Å
O^{16}C^{13}S^{34}	0.197194		$r_0(CS) = 1.5584$ Å
O^{16}C^{14}S^{32}	0.201581		
O^{17}C^{12}S^{32}	0.196258		
O^{18}C^{12}S^{32}	0.190292		
O^{18}C^{12}S^{34}	0.185458		
O^{18}C^{13}S^{32}	0.189829		
O^{16}C^{12}Se^{74}	0.1366207		
O^{16}C^{12}Se^{76}	0.1357085		
O^{16}C^{12}Se^{77}	0.1352681		
O^{16}C^{12}Se^{78}	0.1348404		
O^{16}C^{12}Se^{79}	0.1344213	$C_{\infty v}$	$r_0(CO) = 1.1588$ Å
O^{16}C^{12}Se^{80}	0.1340143		$r_0(CSe) = 1.7090$ Å
O^{16}C^{12}Se^{82}	0.1332276		
O^{16}C^{13}Se^{78}	0.1335960		
O^{16}C^{13}Se^{80}	0.1327598		
Te^{122}C^{12}S^{32}	0.05284063		
Te^{123}C^{12}S^{32}	0.05273401		
Te^{124}C^{12}S^{32}	0.05262940		
Te^{125}C^{12}S^{32}	0.05252608	$C_{\infty v}$	$r_0(TeC) = 1.904$ Å
Te^{126}C^{12}S^{32}	0.05242467		$r_0(CS) = 1.557$ Å
Te^{128}C^{12}S^{32}	0.05222620		
Te^{130}C^{12}S^{32}	0.05203367		

TABLE 7h-6. TRIATOMIC, ASYMMETRIC TOP MOLECULES

Molecule	$A_{[0]}$ (cm^{-1})	$B_{[0]}$ (cm^{-1})	$C_{[0]}$ (cm^{-1})	Point group	Geometrical parameters
H$_2$O......	27.8778	14.5092	9.2869	C_{2v}	r_0(OH) = 0.9568 Å; $\sphericalangle_0$(HOH) = 105.05°
HDO.....	23.3786	9.1020	6.4173	C_s	r_e(OH) = 0.9572 Å; $\sphericalangle_e$(HOH) = 104.52°
D$_2$O......	15.3846	7.2716	4.8458	C_{2v}	
H$_2$S^{32}.....	10.3599	9.0156	4.7315	C_{2v}	r_e(HS) = 1.335$_6$ Å; $\sphericalangle_e$HSH = 92.1°
HDS.....	9.683	4.843	3.140	C_s	
H$_2$Se^{80}...	8.1703	7.7272	3.9013	C_{2v}	r_e(HSe) = 1.460$_5$ Å; $\sphericalangle_e$HSeH = 90.9°
D$_2$Se^{80}....	4.1905$_5$	3.8662$_1$	1.9861$_4$		
H$_2$Te.....	6.248$_6$	6.097$_0$	3.036$_1$	C_{2v}	r_0(HTe) = 1.653 Å; $\sphericalangle$HTeH = 90.2$_5$°
BH$_2$......	41.64	7.24$_8$	6.00$_8$	C_{2v}	r_0(BH) = 1.18 Å; $\sphericalangle$HBH = 131°
NH$_2$......	23.72$_8$	12.94$_2$	8.16$_9$	C_{2v}	r_0(NH) = 1.024 Å; $\sphericalangle$HNH = 103.4°
HCO.....	22.36$_5$	1.494$_4$	1.400$_8$	C_s	$\{r_0$(CH) = 1.08 Å (assumed); $\sphericalangle$HCO = 119.5° r_0(CO) = 1.19$_8$ Å
DCO.....	13.64$_1$	1.281$_2$	1.171$_2$		
HNO.....	18.4792	1.4115	1.3071	C_s	$\{r_0$(NH) = 1.063 Å r_0(NO) = 1.212 Å; $\sphericalangle$HNO = 108.6°
DNO.....	10.5222	1.2920	1.1462		
HPO.....	8.855	0.7024	0.6488	C_s	
HCF.....	15.5$_5$	1.221	1.126	C_s	
HCCl....	15.75	0.6054	0.5882	C_s	
NO$_2$......	8.00251	0.433665	0.410493	C_{2v}	r_0(NO) = 1.193 Å; $\sphericalangle$ONO = 134.1°
CF$_2$......	2.94736	0.41719	0.36469	C_{2v}	r_0(CF) = 1.300 Å; $\sphericalangle$FCF = 104.94°
SiF$_2$......	1.02076	0.29433	0.22784	C_{2v}	r_0(SiF) = 1.591 Å; $\sphericalangle$FSiF = 101.0°
O$_3$.......	3.55345	0.445276	0.394749	C_{2v}	r_0(O'O) = 1.278 Å; $\sphericalangle$OO'O = 116.8°
SO$_2$......	2.02736	0.34417	0.293535	C_{2v}	r_0(SO) = 1.432 Å; $\sphericalangle$OSO = 119.5$_5$°
S$_2$O......	1.39811	0.16875$_3$	0.15034$_2$	C_s	r_0(SO) = 1.46$_5$ Å; r(SS) = 1.88$_4$; $\sphericalangle$SSO = 118.0°
NOF.....	3.175189	0.395077	0.350519	C_s	$\{r_0$(NO) = 1.13 Å; r_0(NF) = 1.52 Å; $\sphericalangle$ONF = 110°
NOCl^{35}...	2.8493	0.191383	0.179343	C_s	r_0(NCl) = 1.975 Å; r_0(NO) = 1.139 Å; $\sphericalangle$ClNO = 113.3°
NOCl^{37}...	2.8486	0.186825	0.175327		
NOBr^{79}...	2.7799	0.12499	0.11962	C_s	r_0(NBr) = 2.14 Å; r_0(NO) = 1.15 Å; $\sphericalangle$BrNO = 114°
NOBr^{81}...	2.7799	0.12417	0.11886		
NS^{32}F....	1.65841	0.290615	0.246607	C_s	r_0(SF) = 1.646 Å; r_0(SN) = 1.446 Å; $\sphericalangle$NSF = 116.9°
NS^{34}F....	1.61101	0.290245	0.245262		
Cl^{35}O$_2$....	1.73718	0.331971	0.277992	C_{2v}	r_0(ClO) = 1.473 Å; $\sphericalangle$OClO = 117.6°
F$_2$O......	1.960777	0.363466	0.305792	C_{2v}	r_0(OF) = 1.409 Å; $\sphericalangle$FOF = 103.3°

TABLE 7h-7. FOUR-ATOMIC LINEAR MOLECULES

Molecule	$B_{[0]}$ cm^{-1}	Point group	Geometrical parameters
C_2H_2	$\begin{cases} 1.1766 \\ B_e = 1.1817 \end{cases}$	$D_{\infty h}$	$\left.\begin{array}{l} r_0(CC) = 1.208 \text{ Å}; \ r_e(CC) = 1.204 \text{ Å} \\ r_0(CH) = 1.057 \text{ Å}; \ r_e(CH) = 1.059 \text{ Å} \end{array}\right.$
C_2HD	$\begin{cases} 0.9910 \\ B_e = 0.9948 \end{cases}$	$C_{\infty v}$	
C_2D_2	$\begin{cases} 0.84787 \\ B_e = 0.8507_5 \end{cases}$	$D_{\infty h}$	
$C_2N_2{}^{14}$	0.1571_2	$D_{\infty h}$	$r_0(C-C) = 1.389 \text{ Å}; \ r_0(C\equiv N) = 1.154 \text{ Å}$
$C_2N_2{}^{15}$	0.1477_4		(assumed)
$HC_2{}^{12}Cl^{35}$	0.189606		$\begin{cases} r_0(CH) = 1.052 \text{ Å} \\ r_0(CC) = 1.211 \text{ Å} \\ r_0(CCl) = 1.632 \text{ Å} \end{cases}$
$HC_2{}^{12}Cl^{37}$	0.185874	$C_{\infty v}$	
$DC_2{}^{12}Cl^{35}$	0.173020		
$DC_2{}^{12}Cl^{37}$	0.169592		
$FCCH$	0.323764		$\begin{cases} r_0(CH) = 1.053 \text{ Å} \\ r_0(CC) = 1.198 \text{ Å} \\ r_0(CF) = 1.279 \text{ Å} \end{cases}$
$FC^{13}CH$	0.323579		
$FCC^{13}H$	0.312681	$C_{\infty v}$	
$FCCD$	0.291403		
$FC^{13}CD$	0.291332		
$FCC^{13}D$	0.283071		

TABLE 7h-8. FOUR-ATOMIC SYMMETRIC TOP MOLECULES

Molecule	$A_{[0]}$ or $C_{[0]}$ cm^{-1}	$B_{[0]}$ cm^{-1}	Point group	Geometrical parameters
$AsCl_3{}^{35}$		0.071623	C_{3v}	$r_0(AsCl) = 2.161 \text{ Å}; \ \sphericalangle ClAsCl = 98.4°$
$AsCl_3{}^{37}$		0.068204		
AsF_3		0.1961013	C_{3v}	$r_0(AsF) = 1.712 \text{ Å}; \ \sphericalangle FAsF = 102°$ (assumed)
AsH_3		3.75154	C_{3v}	$r_0(AsH) = 1.517 \text{ Å}; \ \sphericalangle HAsH = 91.7°$
AsD_3		1.91723		
BF_3	0.17635	0.3527	D_{3h}	$r_0(BF) = 1.295 \text{ Å}$
$N^{14}F_3$		0.356261	C_{3v}	$r_0(NF) = 1.371 \text{ Å}; \ \sphericalangle FNF = 102.1_5°$
$N^{15}F_3$		0.354557		
NH_3	6.196	9.9443	C_{3v}	$r_0(NH) = 1.017 \text{ Å}; \ \sphericalangle HNH = 107.8°$
ND_3	3.117	5.1423		
$PCl_3{}^{35}$		0.087305	C_{3v}	$r_0(PCl) = 2.043 \text{ Å}; \ \sphericalangle ClPCl = 100.1°$
$PCl_3{}^{37}$		0.082974		
PF_3		0.260847	C_{3v}	$r_0(PF) = 1.535 \text{ Å}; \ \sphericalangle FPF = 100.1°$ (assumed)
PH_3		4.45236	C_{3v}	$r_0(PH) = 1.419 \text{ Å}; \ \sphericalangle HPH = 93.3°$
PD_3		2.31728		
$Sb^{121}Cl_3$		0.05850	C_{3v}	$r_0(SbCl) = 2.325 \text{ Å}; \ \sphericalangle ClSbCl = 99.5°$
$Sb^{123}Cl_3$		0.05840		
$Sb^{121}H_3$		2.93643		
$Sb^{123}H_3$		2.93588	C_{3v}	$r_0(SbH) = 1.704 \text{ Å}; \ \sphericalangle HSbH = 91.1°$
$Sb^{121}D_3$		1.49081		
$Sb^{123}D_3$		1.49027		

TABLE 7h-9. FOUR-ATOMIC ASYMMETRIC TOP MOLECULES

Molecule	$A_{[0]}$ cm^{-1}	$B_{[0]}$ cm^{-1}	$C_{[0]}$ cm^{-1}	Point group	Geometrical parameters
CCl$_2$^{36}O.....	0.264141	0.115913	0.0804639	C_{2v}	r_0(CO) = 1.166 Å; r_0(CCl) = 1.746 Å
CCl^{35}Cl^{37}O..	0.262440	0.112743	0.0787704	C_s	$\angle$ClCCl = 111.3°
C^{12}F$_2$O^{16}....	0.394054	0.392037	0.196166		r_0(CF) = 1.312 Å
C^{13}F$_2$O^{16}....	0.394095	0.391847	0.196129	C_{2v}	r_0(CO) = 1.174 Å
C^{13}F$_2$O^{18}....	0.394055	0.362869	0.188574		$\angle$FCF = 108.0°
CH$_2$O.......	9.4053	1.29536	1.13425	C_{2v}	r_0(CH) = 1.102 Å; r_0(CO) = 1.210 Å; $\angle$HCH = 121.1°
C^{12}HFO^{16}...	3.04056	0.39227	0.34680		r_0(CF) = 1.338 Å; r_0(CO) = 1.181 Å
C^{13}HFO^{16}...	2.95221	0.39211	0.34548	C_s	r_0(CH) = 1.095 Å;
C^{12}DFO^{16}...	2.17117	0.39233	0.33162		$\angle$FCO = 122.8°; $\angle$HCO = 127.3°
C^{12}HFO^{18}...	2.99439	0.37035	0.32901		
Cl^{35}F$_3$.......	0.458573	0.153830	0.115039	C_s	ClF$_2$F': $\begin{cases} r_0$(ClF) = 1.698 Å; $\angle$FClF = 175.0° \\ r_0$(ClF') = 1.598 Å; $\angle$F'ClF = 87.5° \end{cases}$
Cl^{37}F$_3$.......	0.455421	0.153836	0.114840		
HN$_3$^{14}.......	20.34	0.401416	0.392988		HN'N''N''': r_0(N'H) = 1.00 Å; $\angle$N'N''N''' = (180°) (assumed)
HN^{14}N$_2$^{14}...	(20.58)	0.389187	0.381192	C_s	r_0(N'N'') = 1.237 Å
HN$_2$^{14}N^{15}...	(20.58)	0.388327	0.380432		r_0(N''N''') = 1.133 Å; $\angle$HN'N'' = 114.1°
DN$_3$^{14}.......	11.47	0.378603	0.365769		
HNCO......	30.5$_9$	0.369289	0.363938	C_s	r_0(HN) = 0.987 Å; $\angle$HNC = 128.1°: r_0(NC) = 1.207 Å; r_0(CO) = 1.171 Å
DNCO......	17.3$_4$	0.344025	0.336221		$\angle$NCO = (180°) (assumed)
HNC^{12}S^{32}...	44.9$_0$	0.196250	0.194989		r_0(NH) = 0.989 Å; r_0(NC) = 1.216 Å;
HNC^{12}S^{34}...	....	0.19162$_6$	0.19042$_3$	C_s	r_0(CS) = 1.561 Å; $\angle$HNC = 135.0°
DNC^{12}S^{32}..	23.5$_8$	0.183477	0.181634		$\angle$NCS = 180° (assumed)
H$_2$O$_2$	10.068	0.8740	0.8384	C_2	r_0(OH) = 0.950 Å; r_0(OO) = 1.475 Å; $\angle$OOH = 94.8°; dihedral angle = 119.8°
cis N$_2$F$_2$.....	0.656682	0.265075	0.188510	C_{2v}	r_0(NF) = 1.384 Å; r_0(NN) = 1.214 Å
cis N^{14}N^{15}F$_2$.	0.643874	0.263556	0.186677		$\angle$FNN = 114.5°

TABLE 7h-10. FIVE-ATOMIC LINEAR MOLECULES

Molecule	$B_{[0]}$, cm^{-1}	Point group	Geometrical parameters
HC^{12}C^{12}C^{12}N^{14}.........	0.151740		
HC^{12}C^{12}C^{13}N^{14}.........	0.151112		
HC^{12}C^{13}C^{12}N^{14}.........	0.151099		
HC^{13}C^{12}C^{12}N^{14}.........	0.147050		
HC^{12}C^{12}C^{12}N^{15}.........	0.147332	$C_{\infty v}$	r_0(CH) = 1.057 Å; r_0(C$\equiv$C) = 1.203 Å; r_0(C—C) = 1.382 Å; r_0(CN) = 1.157 Å
DC^{12}C^{12}C^{12}N^{14}.........	0.140817		
DC^{12}C^{12}C^{13}N^{14}.........	0.140181		
DC^{12}C^{13}C^{12}N^{14}.........	0.140350		
DC^{13}C^{12}C^{12}N^{14}.........	0.137002		
DC^{12}C^{12}C^{12}N^{15}.........	0.136775		
C$_3$O$_2$.................	0.07321	$D_{\infty h}$	r_0(CO) = 1.160 Å (assumed); r_0(CC) = 1.28$_0$ Å

TABLE 7h-11. FIVE-ATOMIC SYMMETRIC AND SPHERICAL TOP MOLECULES

Molecule	$A_{[0]}$ or $C_{[0]}$, cm^{-1}	$B_{[0]}$, cm^{-1}	Point group	Geometrical parameters
CF$_3$Br^{79}......	....	0.069984	C_{3v}	r_0(CF) = 1.33 Å; $\angle$(FCF) = 108° (assumed)
CF$_3$Br^{81}......	....	0.069331		r_0(CBr) = 1.908 Å
CF$_3$Cl^{35}......	....	0.111262	C_{3v}	r_0(CF) = 1.328 Å; $\angle$(FCF) = 108° (assumed)
CF$_3$Cl^{37}......	....	0.108458		r_0(CCl) = 1.740 Å
CF$_3$I.........	0.1910	0.050809	C_{3v}	r_0(CF) = 1.33 Å (assumed); $\angle$FCF = 108° (assumed)
				r_0(CI) = 2.134 Å
CH$_4$........	....	5.2412	T_d	r_0(CH) = 1.0940 Å
CH$_3$D........	5.243	3.878	C_{3v}	r_e(CH) = 1.085 Å
CHD$_3$........	....	3.2795	C_{3v}	r_0(CD) = 1.091$_8$ Å
CD$_4$........	....	2.6329	T_d	
CHBr$_3$^{79}......	....	0.041616		
CHBr$_3$^{81}......	....	0.040605	C_{3v}	r_0(CH) = 1.06$_8$ Å; $\angle$(BrCBr) = 110°48′
CDBr$_3$^{79}......	....	0.041344		r_0(CBr) = 1.930 Å
CDBr$_3$^{81}......	....	0.040345		
C^{12}H$_3$Br^{79}....	}5.129{	0.319160		
C^{12}H$_3$Br^{81}....		0.317947	C_{3v}	r_0(CBr) = 1.939 Å; $\angle$(HCH) = 110°58′
C^{13}H$_3$Br^{79}....	....	0.304194		r_0(CH) = 1.096 Å
C^{13}H$_3$Br^{81}....	....	0.302971		
C^{12}D$_3$Br^{79}....	}2.591{	0.257332	C_{3v}	r_0(CBr) = 1.9391 Å; $\angle$(DCD) = 111°26′
C^{12}D$_3$Br^{81}....		0.256218		r_0(CD) = 1.104 Å
CHCl$_3$^{35}.....	....	0.110146		r_0(CH) = 1.073 Å; $\angle$(ClCCl) = 110°24′
CHCl$_3$^{37}......	....	0.104389	C_{3v}	r_0(CCl) = 1.767 Å
CDCl$_3$^{35}......	....	0.108414		
C^{12}H$_3$Cl^{35}....	(5.14)	0.443402		
C^{12}H$_3$Cl^{37}....	....	0.436574	C_{3v}	r_0(CCl) = 1.7810 Å; $\angle$(HCH) = 110°31′
C^{13}H$_3$Cl^{35}....	5.124	0.426835		r_0(CH) = 1.113 Å
C^{13}H$_3$Cl^{37}....	....	0.419957		
C^{12}D$_3$Cl^{35}....	....	0.361647	C_{3v}	r_0(CCl) = 1.7810 Å; $\angle$(DCD) = 110°43′
C^{12}D$_3$Cl^{37}....	....	0.355528		r_0(CD) = 1.104 Å
C^{12}H$_3$Cl^{36}....	....	0.439892	C_{3v}	
C^{12}HF$_3$......	....	0.345196		r_0(CH) = 1.098 Å; $\angle$(FCF) = 108°48′
C^{12}DF$_3$......	....	0.330940	C_{3v}	r_0(CF) = 1.332 Å
C^{13}HF$_3$......	....	0.347640		
C^{12}H$_3$F......	5.08$_1$	0.851794		r_0(CH) = 1.10$_9$ Å; $\angle$(HCN) = 110°0′
C^{13}H$_3$F......	....	0.829318	C_{3v}	r_0(CF) = 1.385 Å
C^{12}D$_3$F......	....	0.682132		
C^{12}H$_3$I......	5.134	0.250215	C_{3v}	r_0(CH) = 1.106 Å; $\angle$(HCH) = 111°10′
C^{13}H$_3$I......	....	0.237465		r_0(CI) = 2.1396 Å
C^{12}D$_3$I......	....	0.201482	C_{3v}	r_0(CI) = 2.1392 Å; r_0(CD) = 1.104; $\angle$(DCD) = 111°37′
Ge^{70}F$_3$^{19}Cl^{35}...	....	0.072334		
Ge^{70}F$_3$^{19}Cl^{37}...	....	0.070320		
Ge^{72}F$_3$^{19}Cl^{35}...	....	0.072301	C_{3v}	r_0(GeF) = 1.688 Å; $\angle$(FGeF) = 107°42′
Ge^{72}F$_3$^{19}Cl^{37}...	....	0.070283		r_0(GeCl) = 2.067 Å
Ge^{74}F$_3$^{19}Cl^{35}...	....	0.072270		
Ge^{74}F$_3$^{19}Cl^{37}...	....	0.070248		
GeH$_4$........	....	2.70$_3$	T_d	
GeD$_4$........	....	1.3512	T_d	r_0(GeH) = 1.524 Å
GeHD$_3$........	....	1.669	C_{3v}	
GeH$_3$D......	....	1.969	C_{3v}	
Ge^{70}H$_3$Br^{79}....	....	0.081342		
Ge^{70}H$_3$Br^{81}....	....	0.080395		
Ge^{72}H$_3$Br^{79}....	....	0.080269		
Ge^{72}H$_3$Br^{81}....	....	0.079322	C_{3v}	r_0(GeH) = 1.55 Å; $\angle$(HGeH) = 112°0′
Ge^{74}H$_3$Br^{79}....	....	0.079251		r_0(GeBr) = 2.297 Å
Ge^{74}H$_3$Br^{81}....	....	0.078303		
Ge^{76}H$_3$Br^{79}....	....	0.078282		
Ge^{76}H$_3$Br^{81}....	....	0.077332		
Ge^{70}HCl$_3$^{35}....	....	0.072475		
Ge^{72}HCl$_3$^{35}....	....	0.0723586		
Ge^{74}HCl$_3$^{35}....	....	0.0722445	C_{3v}	r_0(GeCl) = 2.1139 Å; $\angle$(ClGeCl) = 108°17′
Ge^{70}HCl$_3$^{37}....	....	0.0688389		r_0(GeH) = 1.55 Å
Ge^{72}HCl$_3$^{37}....	....	0.0687284		
Ge^{74}HCl$_3$^{37}....	....	0.0686207		

TABLE 7h-11. FIVE-ATOMIC SYMMETRIC AND SPHERICAL TOP MOLECULES
(Continued)

Molecule	$A_{[0]}$ or $C_{[0]}$, cm^{-1}	$B_{[0]}$, cm^{-1}	Point group	Geometrical parameters
Ge^{70}H$_3$Cl^{35}....	....	0.146825		
Ge^{74}H$_3$Cl^{35}....	2.603	0.144563	C_{3v}	r_0(GeH) = 1.52 Å; $\not\subset$(HGeH) = 111°4′
Ge^{74}H$_3$Cl^{37}....	....	0.139359		r_0(GeCl) = 2.147 Å
Ge^{76}H$_3$Cl^{37}....	....	0.13831		
Ge^{70}H$_3$F.....	....	0.33699	C_{3v}	
MnO$_3$F......	....	0.137732	C_{3v}	
POCl$_3$^{35}......	....	0.067220	C_{3v}	r_0(PCl) = 1.99 Å; $\not\subset$(ClPCl) = 103°36′
POCl$_3$^{37}......	....	0.064457		r_0(PO) = 1.45 Å
PO^{16}F$_3$........	....	0.153248	C_{3v}	r_0(PF) = 1.52 Å; $\not\subset$(FPF) = 102°30′
PO^{18}F$_3$........	....	0.146610		r_0(PO) = 1.45 Å
PS^{32}Cl$_3$^{35}......	....	0.046787		r_0(PCl) = 2.02 Å; $\not\subset$(ClPCl) = 100°30′
PS^{32}Cl$_3$^{37}......	....	0.045222	C_{3v}	r_0(PS) = 1.85 Å
PS^{34}Cl$_3$^{35}......	....	0.045702		
PS^{32}F$_3$........	....	0.088650		r_0(PF) = 1.53 Å; $\not\subset$(FPF) = 100°18′
PS^{33}F$_3$........	....	0.087218	C_{3v}	r_0(PS) = 1.87 Å
PS^{34}F$_3$........	....	0.086052		
Re^{185}O$_3$Cl^{35}....	....	0.069856		
Re^{185}O$_3$Cl^{37}....	....	0.067547	C_{3v}	r_0(ReO) = 1.761 Å; $\not\subset$(OReO) = 108°20′
Re^{187}O$_3$Cl^{35}....	....	0.069834		r_0(ReCl) = 2.230 Å
Re^{187}O$_3$Cl^{37}....	....	0.067525		
SiF$_3$Br^{79}......	....	0.051702	C_{3v}	r_0(SiF) = 1.560 Å; r_0(SiBr) = 2.153 Å
SiF$_3$Br^{81}......	....	0.051173		$\not\subset$(FSiF) = 108°30′ (assumed)
SiF$_3$Cl^{35}......	....	0.082650	C_{3v}	r_0(SiF) = 1.560 Å; r_0(SiCl) = 1.989 Å
SiF$_3$Cl^{37}......	....	0.080491		$\not\subset$(FSiF) = 108°30′ (assumed)
Si^{28}H$_4$........	....	2.864	T_d	r_0(SiH) = 1.480 Å
SiHD$_3$........	....	1.7755	C_{3v}	
Si^{28}H$_3$Br^{79}....	....	0.144159		
Si^{28}H$_3$Br^{81}....	....	0.143187		
Si^{29}H$_3$Br^{79}....	....	0.141196	C_{3v}	r_0(SiH) = 1.57 Å; $\not\subset$(HSiH) = 111°20′
Si^{29}H$_3$Br^{81}....	....	0.140220		r_0(SiBr) = 2.209 Å
Si^{30}H$_3$Br^{79}....	....	0.138409		
Si^{30}H$_3$Br^{81}....	....	0.137431		
SiHCl$_3$^{35}......	....	0.0824732	C_{3v}	r_0(SiH) = 1.47 Å; $\not\subset$(ClSiCl) = 109°22′
SiHCl$_3$^{37}......	....	0.0782564		r_0(SiCl) = 2.021 Å
Si^{28}H$_3$Cl^{35}....	....	0.22261		
Si^{30}H$_3$Cl^{35}....	....	0.21634		
Si^{28}H$_3$Cl^{37}....	....	0.21723		
Si^{28}D$_3$Cl^{35}....	....	0.19739	C_{3v}	r_0(SiCl) = 2.048 Å; $\not\subset$(HSiH) = 110°57′
Si^{29}D$_3$Cl^{35}....	....	0.19515		r_0(SiH) = 1.50 Å
Si^{30}D$_3$Cl^{35}....	....	0.19303		
Si^{28}D$_3$Cl^{37}....	....	0.19256		
Si^{28}HF$_3$......	....	0.240435		
Si^{29}HF$_3$......	....	0.240021	C_{3v}	r_0(SiF) = 1.565 Å; $\not\subset$(FSiF) = 108°17′
Si^{30}HF$_3$......	....	0.239622		r_0(SiH) = 1.455 Å (assumed)
Si^{28}H$_3$F......	....	0.477927		
Si^{29}H$_3$F......	....	0.473550		
Si^{30}H$_3$F......	....	0.469411	C_{3v}	r_0(SiH) = 1.503 Å; r_0(SiF) = 1.593 Å
Si^{28}D$_3$F......	....	0.408732		$\not\subset$(HSiH) = (111°) (assumed)
Si^{29}D$_3$F......	....	0.406120		
Si^{30}D$_3$F......	....	0.403678		
SiH$_3$I......	....	0.10726	C_{3v}	
SnH$_4$........	....	2.16$_9$	T_d	r_0(SnH) = 1.701 Å
HnSD$_3$.......	....	1.3573	C_{3v}	

TABLE 7h-12. FIVE-ATOMIC ASYMMETRIC TOP MOLECULES

Molecule	$A_{[0]}$, cm^{-1}	$B_{[0]}$, cm^{-1}	$C_{[0]}$, cm^{-1}	Point group	Geometrical parameters
CH₂Br₂	$\left[A - \dfrac{B+C}{2}\right] = 0.821$			C_{2v}	$r_0(\text{CBr}) = 1.907$ Å; $\sphericalangle(\text{HCH}) = 112°$ (elec. diffr.)
CH₂CO	9.37	0.343347	0.330758	C_{2v}	
CHDCO		0.321790	0.306032	C_s	$r_0(\text{CH}) = 1.075$ Å; $\sphericalangle(\text{HCH}) = 122.0°$
CD₂CO		0.304237	0.285286	C_{2v}	$r_0(\text{CO}) = 1.16$ Å (assumed); $r_0(\text{CC}) \doteq 1.31_4$ Å
CH₂Cl₂³⁵	1.06746	0.11076	0.10224	C_{2v}	
CH₂Cl³⁵Cl³⁷	1.063344	0.10779	0.099677	C_s	
CH₂Cl₂³⁷	1.0592	0.1048	0.09713	C_{2v}	$r_0(\text{CCl}) = 1.7724$ Å; $\sphericalangle(\text{ClCCl}) = 111°47'$
CHDCl₂³⁵	0.9072	0.1102	0.1010	C_s	$r_0(\text{CH}) = 1.068$ Å; $\sphericalangle(\text{HCH}) = 112°0'$
CHDCl³⁵Cl³⁷	0.90364	0.10732	0.09845	C_1	
CD₂Cl₂³⁵	0.78976	0.1095	0.09985	C_{2v}	
CD₂Cl³⁵Cl³⁷	0.78661	0.10666	0.09740	C_s	
CH₂ClBr	$\left[A - \dfrac{B+C}{2}\right] = 0.897_5$			C_s	$r_0(\text{CBr}) = 1.911$ Å; $r_0(\text{CCl}) = 1.766$ Å (assumed) $\sphericalangle(\text{HCH}) = 112°$ (elec. diffr.)
CH₂F₂	1.6391	0.3537	0.3085	C_{2v}	$r_0(\text{CH}) = 1.09$ Å; $\sphericalangle(\text{HCH}) = 112°$ $r_0(\text{CF}) = 1.36$ Å; $\sphericalangle(\text{FCF}) = 108°$
CH₂N₂	9.112	0.377109	0.361759	C_{2v}	$r_0(\text{NN}) = 1.12$ Å; $r_0(\text{CN}) = 1.32$ Å;
CD₂N₂		0.334984	0.311764		$r_0(\text{CH}) = 1.08$ Å; $\sphericalangle\text{HCH} = 127°$
HCO₂H	2.58548	0.402112	0.347447		$r_0(\text{CH}) = 1.097$ Å; $r_0(\text{CO}) = 1.202, 1.343$ Å
DCO₂H	1.9250	0.402138	0.332087	C_s	$r_0(\text{OH}) = 0.972$ Å; $\sphericalangle\text{OCO} = 124.9°$;
HCO₂D	2.2052	0.392356	0.332565		$\sphericalangle\text{HCO} = 124.1°$; $\sphericalangle\text{COH} = 106.3°$
HNO₃	0.434005	0.403610	0.208831	C_s	$r_0(\text{NO}) = 1.199, 1.211, 1.406$ Å; $r_0(\text{OH}) = 0.964$ Å;
DNO₃	0.432656	0.377355	0.201301		$\sphericalangle\text{ONO} = 130.3°, 113.8_5°$; $\sphericalangle\text{NOH} = 102.1_5°$
S³²O₂F₂	0.171261	0.169218	0.168685	C_{2v}	$r_0(\text{SO}) = 1.405$ Å; $\sphericalangle\text{OSO} = 124.0°$
S³⁴O₂F₂	0.171243	0.169117	0.168586		$r_0(\text{SF}) = 1.530$ Å; $\sphericalangle\text{FSF} = 96.1°$
Si²⁸H₂F₂	0.82359₃	0.26024	0.21272	C_{2v}	$r_0(\text{SiH}) \doteq 1.470$ Å; $r_0(\text{SiF}) = 1.577$ Å;
Si²⁸D₂F₂	0.62992₅	0.24842	0.20435		$\sphericalangle\text{FSiF} = 107.9°$

TABLE 7h-13. SIX-ATOMIC LINEAR MOLECULES

Molecule	$B_{[0]}$, cm^{-1}	Point group	Geometrical parameters
C_4H_2	0.14689	$\left.\begin{array}{c}\\\\\end{array}\right\} D_{\infty h}$	$\begin{cases} r_0(C—C) = 1.376 \text{ Å, assuming } r_0(C\equiv C) = 1.205 \text{ Å} \\ r_0(CH) = 1.046 \text{ Å} \end{cases}$
C_4D_2	0.12767		
$C_4H_2^+$	0.14013	$D_{\infty h}$	

TABLE 7h-14. SIX-ATOMIC SYMMETRIC TOP MOLECULES

Molecule	$B_{[0]}$, cm^{-1}	Point group	Geometrical parameters
$B^{10}H_3CO$	0.299544	$\left.\begin{array}{c}\\\\\\\end{array}\right\} C_{3v}$	$\begin{cases} r_0(BH) = 1.194 \text{ Å} \\ r_0(BC) = 1.540 \text{ Å}; \ \measuredangle(HBH) = 113°52' \\ r_0(CO) = 1.131 \text{ Å} \end{cases}$
$B^{11}H_3CO$	0.288773		
$B^{10}D_3CO$	0.251185		
$B^{11}D_3CO$	0.244721		
CF_3CN^{14}	0.0982523	$\left.\begin{array}{c}\\\\\end{array}\right\} C_{3v}$	$\begin{cases} r_0(CF) = 1.335 \text{ Å} \\ r_0(CN) = 1.158 \text{ Å (assumed)}; \\ \measuredangle(FCF) = 108° \text{ (assumed)} \\ r_0(CC) = 1.464 \text{ Å} \end{cases}$
CF_3CN^{15}	0.0952611		
$C^{12}H_3C^{12}N^{14}$	0.306842	C_{3v}	$\begin{cases} r_0(CH) = 1.092 \text{ Å} \\ r_0(CC) = 1.460 \text{ Å}; \ \measuredangle(HCH) = 109°8' \\ r_0(CN) = 1.158 \text{ Å} \end{cases}$
$C^{12}H_3C^{12}N^{15}$	0.297599		
$C^{12}H_3C^{13}N^{14}$	0.306686		
$C^{13}H_3C^{12}N^{14}$	0.297977		
$C^{12}D_3C^{12}N^{14}$	0.262119		
$C^{13}D_3C^{13}N^{14}$	0.261798		
$CH_3Hg^{198}Br^{81}$	0.03754	$\left.\begin{array}{c}\\\\\end{array}\right\} C_{3v}$	$\begin{cases} r_0(HgBr) = 2.406 \text{ Å} \\ r_0(CH) = 1.092 \text{ Å (assumed)}; \\ \measuredangle(HCH) = 109°7' \text{ (assumed)} \\ r_0(CHg) = 2.07 \text{ Å} \end{cases}$
$CH_3Hg^{202}Br^{79}$	0.03802		
$CH_3Hg^{202}Br^{81}$	0.03743		
$CH_3Hg^{198}Cl^{35}$	0.069296	$\left.\begin{array}{c}\\\\\\\\\\\\\\\end{array}\right\} C_{3v}$	$\begin{cases} r_0(CH) = 1.092 \text{ Å (assumed)}; \ \measuredangle(HCH) = 109°7' \\ r_0(CHg) = 2.059 \text{ Å} \\ r_0(HgCl) = 2.282 \text{ Å} \end{cases}$
$CH_3Hg^{198}Cl^{37}$	0.066918		
$CH_3Hg^{199}Cl^{35}$	0.069286		
$CH_3Hg^{199}Cl^{37}$	0.066906		
$CH_3Hg^{200}Cl^{35}$	0.069275		
$CH_3Hg^{200}Cl^{37}$	0.066895		
$CH_3Hg^{202}Cl^{35}$	0.069255		
$CH_3Hg^{202}Cl^{37}$	0.066872		
$CH_3Hg^{204}Cl^{35}$	0.069234		
$CH_3Hg^{204}Cl^{37}$	0.066849		
CH_3NC^{12}	0.335328	$\left.\begin{array}{c}\\\\\end{array}\right\} C_{3v}$	$\begin{cases} r_0(CH) = 1.094 \text{ Å} \\ r_0(C—N) = 1.427 \text{ Å}; \ \measuredangle(HCH) = 109°46' \\ r_0(N\equiv C) = 1.167 \text{ Å} \end{cases}$
CH_3NC^{13}	0.323420		
CD_3NC^{12}	0.286266		
CD_3NC^{13}	0.276150		
SiH_3CN	0.1658$_7$	$\left.\begin{array}{c}\\\\\end{array}\right\} C_{3v}$	$\begin{cases} r_0(SiH) = 1.47 \text{ Å (assumed)}; \ \measuredangle HSiC = 108° \\ \text{(assumed)}; \\ r_0(SiC) = 1.848, \ r_0(CN) = 1.156 \text{ Å} \end{cases}$
SiD_3CN	0.1512$_7$		

TABLE 7h-15. SIX-ATOMIC ASYMMETRIC TOP MOLECULES

Molecule	$A_{[0]}$ cm^{-1}	$B_{[0]}$ cm^{-1}	$C_{[0]}$ cm^{-1}	Point group	Geometrical parameters
C$_2$H$_4$.............	4.828	1.0012	0.8282	V_h	r_0(CH) = 1.086 Å; $\not\!\!\triangleleft$HCH = 117.6°
C$_2$D$_4$............	2.432	0.7369	0.5630		r_0(C=C) = 1.339 Å
CH$_2$CF$_2$.........	0.367003	0.347873	0.17830$_5$	C_{2v}	r_0(CH) = (1.080 Å); $\not\!\!\triangleleft$HCH = 121.8°
CD$_2$CF$_2$.........	0.35324	0.29998	0.16199		r_0(CF) = (1.325 Å); $\not\!\!\triangleleft$FCF = 108.9°
					r_0(CC) = (1.313 Å)
CH$_2$CFCl^{35}......	0.35630	0.17019	0.11503	C_s	
CH$_2$CFCl^{37}......	0.35629	0.16528	0.11276		
CH$_2$CHBr^{79}.....	....	0.13886	0.12885	C_s	
CH$_2$CHBr^{81}.....	....	0.13804	0.12816		
CH$_2$CHCl^{35}.....	....	0.201138	0.181635	C_s	r_0(CH) = 1.07$_9$ Å; r_0(CH') = 1.090 Å; r_0(CC) = 1.33$_2$ Å
CH$_2$CHCl^{37}.....	....	0.196922	0.178165		r_0(CCl) = 1.72$_6$ Å; $\not\!\!\triangleleft$CCH = 119.5°, 123.8°, 121.0°;
CH$_2$CHI........	....	0.10870	0.10229	C_s	$\not\!\!\triangleleft$CCCl = 122.3°
CH$_3$OH.........	....	0 82299	0.79263	C_s	r_0(C—O) = 1.425 Å; $\not\!\!\triangleleft$COH = 108.5°
CD$_3$OH........	....	0.66186	0.64272		r_0(CH) = 1.094 Å; $\not\!\!\triangleleft$HCH = 108.6°
CH$_3$OD........	....	0.78273	0 73309		r_0(OH) = 0.945 Å
					methyl tilt 3.3°
CH$_3$SH........	(5.68)	0.4305$_4$	0.4130$_9$	C_s	r_0(CH) = 1.104 Å; r_0(SH) = 1.329 Å
CH$_3$SD........	(3.04)	0.4222$_7$	0.3911$_9$		r_0(CS) = 1.818 Å; $\not\!\!\triangleleft$HCH = 110.3°
CD$_3$SH........	(4.03)	0.3516$_3$	0.3398$_9$		$\not\!\!\triangleleft$CSH = 100.3°
HC$_2$CHO.......	2.20912	0.160985	0.150091	C_s	r_0(C=C) = 1.215 Å; r_0(C≡C) = 1.209 Å
DC$_2$CHO.......	2.22715	0.148895	0.139359		r_0(C—C) = 1.445 Å; r_0(CH) = 1.106 Å, 1.055 Å
HC$_2$CDO.......	1.72668	0.159825	0.146060		$\not\!\!\triangleleft$CCO = 123.7°, $\not\!\!\triangleleft$CCC = 178.4°
DC$_2$CDO.......	1.70368	0.117739	0.135747		
N$_2$H$_4$...........	3.981$_4$	$B_0 + C_0 = 1.60633$		C_2	r_0(NH) = 1.02 Å (assumed); r_0(NN) = 1.45 Å (assumed)
					$\not\!\!\triangleleft$HNH = 106°; $\not\!\!\triangleleft$HNN = 112°
					Dihedral angle = 90.0°
NH$_2$CHO.......	2.42555	0.37939	0.32802	C_1	r_0(NH') = 1.014 Å; r_0(NH'') = 1.002 Å;
NH$_2$CDO.......	1.83288	0.37936	0.31421		r_0(CN) = 1.376 Å; r_0(CH) = 1.102 Å
ND$_2$CHO.......	1.99191	0.34002	0.29055		r_0(CO) = 1.193 Å; $\not\!\!\triangleleft$H'NH'' = 118.9°;
					$\not\!\!\triangleleft$H'NC = 120.6°
					$\not\!\!\triangleleft$H'NC = 117.1$_5$°; $\not\!\!\triangleleft$NCO = 123.8°;
					$\not\!\!\triangleleft$NCH = 113.2°
					$\not\!\!\triangleleft$OCH = 123.0°; $\not\!\!\triangleleft$H'NC − NCO = 7°;
					$\not\!\!\triangleleft$H''NC − NCH = 12°

7i. Atomic Transition Probabilities[1]

W. L. WIESE AND B. M. GLENNON

National Bureau of Standards

In the following tables, we present *selected* critically evaluated atomic transition probabilities for the 20 lightest elements. For this group of elements many data of moderate or sometimes even high accuracy are available from various experimental and theoretical sources. The material selected here is obtained principally from Hartree-Fock calculations (which partly include the effects of configuration interaction), from the Coulomb approximation, from the nuclear charge expansion method, and, experimentally, from emission measurements with stabilized arcs, from lifetime experiments (with delayed coincidence techniques as well as with the Hanle effect method) and from anomalous dispersion measurements.

1. *Guideposts for the Selection of Data.* The listed data are mostly the same as those chosen by us for two recent *comprehensive* critical data compilations [1,2] which are several times larger than the present table. For the inclusion of data into this much more compact table we have used the following guideposts: Only lines with uncertainties estimated not to exceed 50 percent are included; only the more prominent lines of a spectrum, that is, the lines of at least moderate strength, are listed (even if reliable data are known for weak lines); and normally only those lines are included which have been observed before, i.e., which are listed in multiplet or other spectral line tables [3–6]. However, we have not been too rigid about the last requirement, especially for spectra of higher stages of ionization. These spectra have recently come into prominence, but are as yet rather incompletely represented in present multiplet tables. For these spectra we have thus listed the most prominent lines—when good *f*-value data are available—even in cases when we had only calculated wavelengths at our disposal. (In order to indicate that the calculated wavelengths are normally much more uncertain than the measured ones, the former are given in square brackets.) We believe that with the greatly expanded scale of research in plasma physics and astrophysics it will be only a short time before many of these lines are observed and may be needed for diagnostic studies.[2]

As stated above, most of the data for this tabulation have been taken from two recent comprehensive compilations published in 1966 (H through Ne [1]) and in 1969 (Na through Ca [2]). But, in addition, we have also evaluated and included the most recent material through early 1970. Especially for the spectra of He, Li, Be, B, C, Ne, and Si we have found quite a bit of newer, more accurate data. In such cases we present the new data, list the individual references and indicate there which particular experimental or theoretical method the author has used.

2. *Definitions, Units, and Conversion Factors.* In the current literature several equivalent expressions for the atomic transition probability have found widespread

---

[1] Contribution of the National Bureau of Standards, not subject to copyright.
[2] We have usually not listed any data for stages of ionization beyond six. Some material for still higher stages of ionization is found in Wiese et al. [1,2].

acceptance. Not only the transition probability (per second) for spontaneous emission A_{ki} from upper atomic state k to lower state i, but also the (absorption) oscillator strength or f value and the line strength S are widely used. In addition, the $\log gf$ is often employed in the astrophysical literature (g is the statistical weight). For the present tables, where we have to restrict ourselves to one quantity to achieve a compact presentation, we have chosen to list A_{ki}. Quantum theory yields for it the expression

$$A_{ki} = \frac{64\pi^4 e^2 \nu_{ik}^3}{3hc^3} \frac{1}{g_k} |<i| \sum_p \mathbf{r}|k>|^2 \tag{7i-1}$$

where the summation in the squared matrix element is over the position vectors $\mathbf{r}$ of all p electrons of the atom, and ν_{ik} is the frequency.

TABLE 7i-1. NUMERICAL CONVERSION FACTORS FOR ALLOWED LINES
The transition probability is listed in units s^{-1}, and the f value is dimensionless. The wavelength λ must be used in angstroms, and g_i and g_k are the statistical weights of the lower and upper states, respectively. (Note that in the tables, with the exception of hydrogen, A_{ki} is given in units 10^8 s^{-1}).

	Transition probability	Oscillator strength	Line strength
Transition probability $A_{ki} =$	—	$\dfrac{6.670_2 \times 10^{15}}{\lambda^2} \dfrac{g_i}{g_k} f_{ik}$	$\dfrac{2.026_1 \times 10^{18}}{g_k \lambda^3} S$
Oscillator strength $f_{ik} =$	$1.4992 \times 10^{-16} \lambda^2 \dfrac{g_k}{g_i} A_{ki}$	—	$\dfrac{303.7_5}{g_i \lambda} S$
Line strength $S =$	$4.935_6 \times 10^{-19} g_k \lambda^3 A_{ki}$	$3.292_1 \times 10^{-3} g_i \lambda f_{ik}$	—

The f value and the line strength S are numerically related to A_{ki} by the formulas given in Table 7i-1 (see also [2]). The line strength is as usual given in atomic units, which are for allowed (or electric dipole) transitions

$$a_0^2 e^2 = 7.187_3 \times 10^{-59} \ m^2 C^2.$$

The statistical weights, which are listed for all presented lines, are related to the inner or total angular momentum quantum number J by $g = 2J + 1$.

Aside from the quantities listed in Table 7i-1, the transition probability for induced or stimulated emission B_{ki} and the transition probability for absorption B_{ik} may become important in special fields, for example, in laser research. These quantities are numerically related to the transition probability for spontaneous emission by

$$B_{ki} = 6.01\lambda^3 A_{ki} \tag{7i-2}$$
and
$$B_{ik} = 6.01\lambda^3 \frac{g_k}{g_i} A_{ki} \tag{7i-3}$$

where λ is the wavelength in angstroms.

Occasionally the emission oscillator strength f_{ki} has been employed.   This quantity is related to the normally used (absorption) oscillator strength by

$$f_{ki} = - \frac{g_i}{g_k} f_{ik} \qquad (7\text{i-}4)$$

3. *Discussion of Data Tables.*   In this compilation we list the transition probabilities of *individual* spectral lines, whenever the nearest known neighboring lines differ by at least a few parts in 10^4 in wavelength.[1]   We often present several lines of a multiplet, usually the stronger ones, but omit the weaker ones.   In the relatively few cases where the lines of a multiplet are all so closely grouped together that they are difficult or impossible to resolve, we list the *multiplet* value (as well as the *multiplet* statistical weights) instead of the individual line data.   These data are marked by a dagger.   If just a portion of the lines in a multiplet (or lines from different multiplets) differs in wavelength by less than one part in 10^4, we have omitted these lines, since they would overlap completely under most experimental conditions so that they might be mistaken for a single line.

For hydrogen, we list "average" transition probabilities A_{ki}^{*}, which are needed for most practical applications.   These (calculated) transition probabilities are exact values for the number of digits given.   For hydrogen, all states with the same principal quantum number are degenerate, so that only a single line having an "average" transition probability is observed for all possible combinations involving the principal quantum numbers i and k.   The only assumption entering into the application of average transition probabilities is that the atomic substates must be occupied according to their statistical weights [1,7], which is the case for any plasma which is not too dilute.

The spectra of hydrogen-like ions are not included in this tabulation, since their transition probabilities may be obtained simply by scaling the hydrogen values A_H according to

$$A_Z = Z^4 A_H \qquad (7\text{i-}5)$$

where Z is the nuclear charge.

For all other tabulated spectra we give accuracy estimates for the transition probabilities and present for purposes of identification all available multiplet numbers as given by Moore [3–5].   The evaluation of the accuracy of the presented material is the most crucial (and normally the most time-consuming) part of a critical data compilation.   We have therefore discussed our evaluation procedures extensively in the general introductions to our larger compilations [1,2], from which—as was mentioned before—we have extracted most of the data presented here.   Because of limitations of space we have to refer here to these discussions and may also state that we have used in this compilation exactly the same procedures for the evaluation of all newer material.

In addition to the allowed lines, we also list transition probabilities for some prominent forbidden lines because they are of interest in astrophysics and atmospheric physics.   We always present *total* transition probabilities, i.e., the sum of the magnetic dipole and electric quadrupole values for a given line (in [1,2], on the other hand, we have listed the separate values).

For a number of *magnetic dipole* lines the line strengths are essentially given by straight numbers.   In some of these cases, furthermore, the transition probabilities of the respective *electric quadrupole* lines at the same wavelength are smaller by several orders of magnitude for all the ions covered in this table.   The principal reason for this is that the wavelengths are relatively large (detailed estimates are given by Naqvi

---

[1] In cases where only moderate spectral resolution is achieved, the multiplet tables [3–6] or [1,2] should be checked for the existence of other nearby lines.

as well as Shortley, Aller, Baker, and Menzel [8]). The *total* transition probabilities in such instances, if the wavelength λ is known, may thus simply be obtained from

$$A = 2.697 \times 10^{13} g_k^{-1} \lambda^{-3} S_m \qquad (7\text{i-}6)$$

where λ is in angstroms, and the magnetic dipole line strength S_m is in atomic units. The S_m values for these lines are tabulated in Table 7i-2.

TABLE 7i-2. LINE STRENGTHS FOR SOME FORBIDDEN TRANSITIONS

Configuration	Line	S, atomic units
$nsnp*$	$^3P_0{}^\circ - {}^3P_1{}^\circ$	2.00
	$^3P_1{}^\circ - {}^3P_2{}^\circ$	2.50
np	$^2P_{\frac{1}{2}}{}^\circ - {}^2P_{\frac{3}{2}}{}^\circ$	1.33
np^2	$^3P_1 - {}^3P_2$	2.50
np^3	$^2D_{\frac{3}{2}}{}^\circ - {}^2D_{\frac{5}{2}}{}^\circ$	2.40
	$^2P_{\frac{1}{2}}{}^\circ - {}^2P_{\frac{3}{2}}{}^\circ$	1.33
np^4	$^3P_2 - {}^3P_1$	2.50
np^5	$^2P_{\frac{1}{2}}{}^\circ - {}^2P_{\frac{3}{2}}{}^\circ$	1.33

* Complete shells, like $1s^2 2s^2$, are omitted. The principal quantum number n has the values $n = 2,3$.

4. *Availability of Data for Heavier Elements.* For most other elements not included in this table, with the exception of the alkalies and some selected lines for elements of the iron group and the alkaline earths, the accuracy and reliability of atomic transition probabilities—if there are any available at all—are still rather poor and at the present time hardly worth a detailed critical compilation such as this. Thus, until more and especially more accurate material becomes available, we have to refer to the following sources: (1) Bibliography on Atomic Transition Probabilities, *NBS Special Publ.* 320, B. M. Miles and W. L. Wiese, 1970. This is an annotated bibliography which lists literature references ordered by elements and stages of ionization and indicates the various experimental or theoretical methods that have been employed. (2) Experimental Transition Probabilities, *NBS Monograph 53*, by C. Corliss and W. Bozman, 1962. This tabulation lists about 25,000 atomic oscillator strengths, mostly for heavier elements, obtained from arc intensity measurements which are generally of moderate or rather poor quality, as many comparisons with other data have shown. The data of Corliss and Bozman show many large discrepancies with other material, especially for the alkalies and alkaline earths and for lines from higher excited levels of the iron group elements. Thus great caution should be exercised when employing these data. (3) A special critical evaluation of transition probability data is available for the spectra of Ba I and II, *NBS Tech. Note* 474, by B. M. Miles and W. L. Wiese, 1968.

5. *Regularities and Systematic Trends.* Some remarks are in order about the recently detected regularities in atomic oscillator strengths, because these are of great value for evaluating the reliability of existing data as well as for determining additional numerical values by simple interpolation techniques. Three principal regularities have been detected (for detailed discussions see [9–11]), which may be briefly stated in the following way:

DEPENDENCE OF f VALUES ON NUCLEAR CHARGE Z. This dependence may be readily derived from conventional perturbation theory, with the result that f may be represented by a power series in Z^{-1}:

$$f = a_0 + a_1 Z^{-1} + a_2 Z^{-2} + \cdots$$

where the first term a_0 is a hydrogenic f value [9,10] which vanishes for all transitions which do not involve a change in the principal quantum number.   Three graphical examples exhibiting this systematic trend for different physical situations are given in Figs. 7i-1 to 7i-3, where the f value is plotted against $1/Z$.

SYSTEMATIC TRENDS OF f VALUES WITHIN SPECTRAL SERIES.   Within a spectral series, the dependence of f on the principal quantum number n (or the effective quantum number n^*) is found to be always a smooth one, in an analogous fashion as for hydrogen.   For lower values of n the f value is not always monotonically decreasing

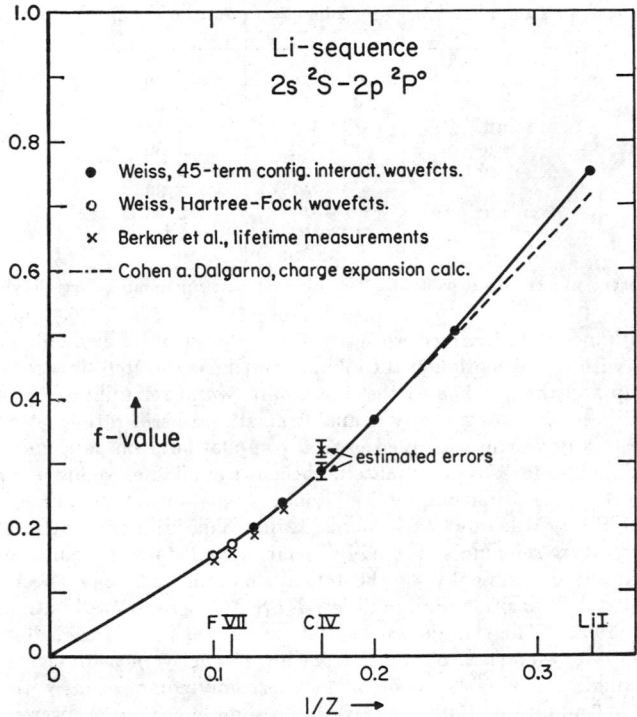

FIG. 7i-1. Oscillator strengths vs. $1/Z$ for the $2s$–$2p$ transition of the lithium isoelectronic sequence.   (*From Ref.* [10], *where the quoted authors and methods are discussed in detail.*)

(see Fig. 7i-4), but for higher n the f values gradually tend to obey the hydrogenic dependence $f \sim (n^*)^{-3}$.   Two examples for these trends are given in graphical form (Figs. 7i-4 and 5), where $n^{*3}f$ is plotted against n^*.

HOMOLOGOUS ATOMS.   The third principal regularity concerns homologous atoms, i.e., atoms with the same outer electron structure.   Here we have found that for certain analogous groups of spectral lines the f values remain approximately constant throughout a family of homologous atoms.   For example, the principal resonance lines of the alkalies, i.e., $2s$–$2p$ for Li, $3s$–$3p$ for Na, $4s$–$4p$ for K, etc., are all close to unity.   This behavior is readily understood on the basis of the Wigner-Kirkwood partial f-sum rule.   If it is assumed that most of the strength of a spectral series is concentrated in its leading transition (for example, $3s$–$3p$ has the dominant strength in a $3s$–np series), then it follows that for this dominant transition array the mean f value is approximately given by the value obtained from the partial f-sum rule.   Further-

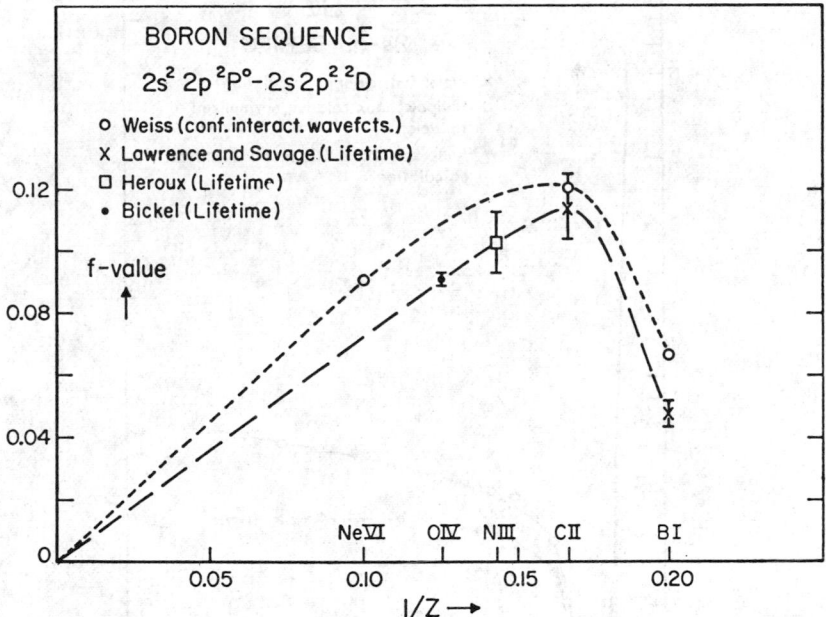

FIG. 7i-2. Oscillator strengths vs. $1/Z$ for the $2s^2\,2p\ ^2P^\circ\text{-}2s2p^2\ ^2D$ transition of the boron isoelectronic sequence. (*From Ref.* [9], *where the quoted authors and methods are discussed in detail.*)

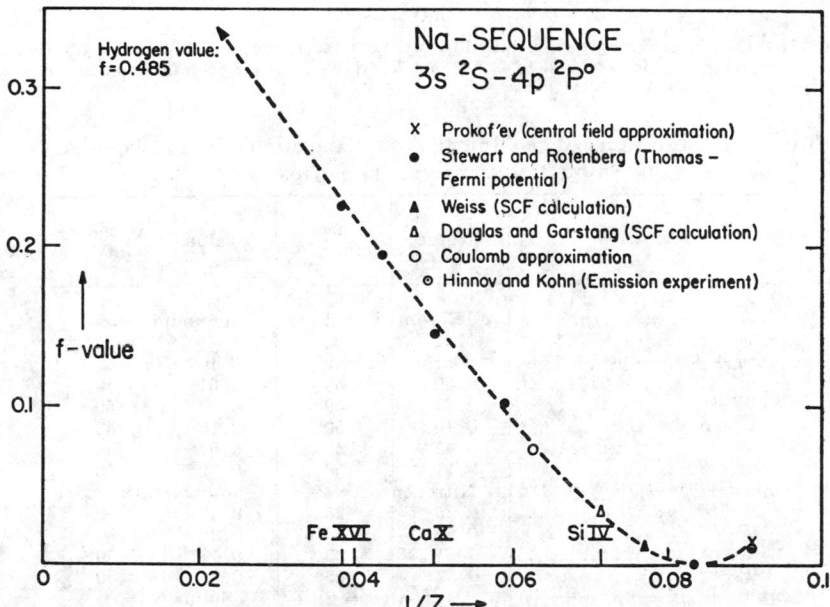

FIG. 7i-3. Oscillator strengths vs. $1/Z$ for the $3s\ ^2S\text{-}4p\ ^2p^\circ$ transition of the sodium isoelectronic sequence. (*From Ref.* [10], *where the quoted authors and methods are discussed in detail.*)

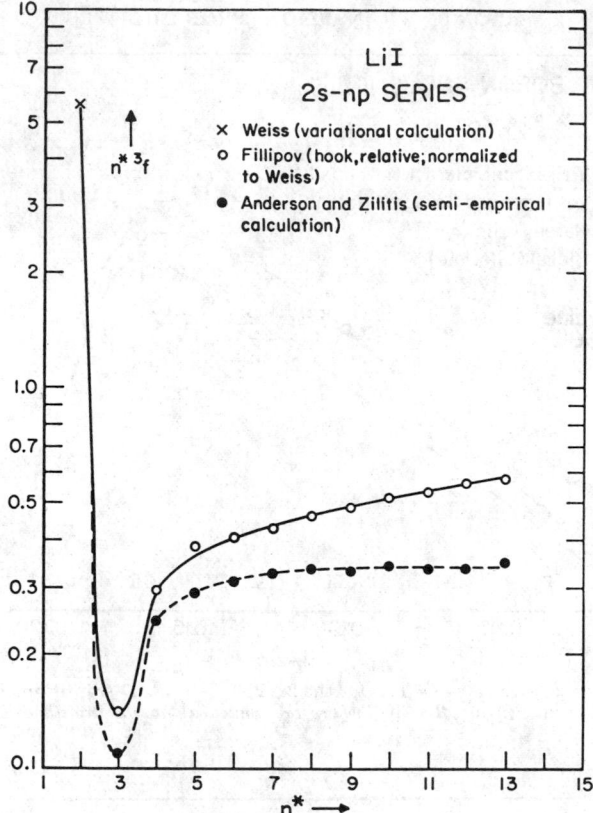

FIG. 7i-4. Oscillator strengths multiplied by n^{*3} vs. effective principal quantum number n^* for the resonance series $2s$-np of Li I. *(From Ref. [10], where the quoted authors and methods are discussed in detail.)*

TABLE 7i-3. COMPARISON OF MULTIPLET f VALUES FOR HOMOLOGOUS ATOMS IN SOME DOMINANT s-p TRANSITION ARRAYS*

Transition	f value	Uncertainty, %	f value	Uncertainty, %
$(n + 1)s - (n + 1)p$......	Boron ($n = 2$)		Aluminum ($n = 3$)	
$^2S-^2P^\circ$...............	1.21	25	1.41	25
$np(n + 1)s - np(n + 1)p$..	Carbon ($n = 2$)		Silicon ($n = 3$)	
$^3P^\circ-^3D$...............	0.50	50	0.61	50
$^3P^\circ-^3P$...............	0.31	50	0.39	50
$^3P^\circ-^3S$...............	0.10	50	0.13	50
$^1P^\circ-^1D$...............	0.42	50	0.67	50
$^1P^\circ-^1S$...............	0.11	50	0.12	50
$np^2(n + 1)s - np^2(n + 1)p$	Nitrogen ($n = 2$)		Phosphorus ($n = 3$)	
$^4P-^4D^\circ$...............	*0.36*	25	0.57	50
$^4P-^4P^\circ$...............	*0.23*	25	0.36	50
$^4P-^4S^\circ$...............	*0.088*	25	0.13	50
$^2P-^2P^\circ$...............	*0.318*	25	0.39	50
$np^3(n + 1)s - np^3(n + 1)p$	Oxygen ($n = 2$)		Sulfur ($n = 3$)	
$^5S^\circ-^5P$...............	*0.922*	10	*1.1*	50
$^3S^\circ-^3P$...............	*0.898*	10	*1.1*	50

* The data are the adopted "best" values. Data from experimental sources are in italics.

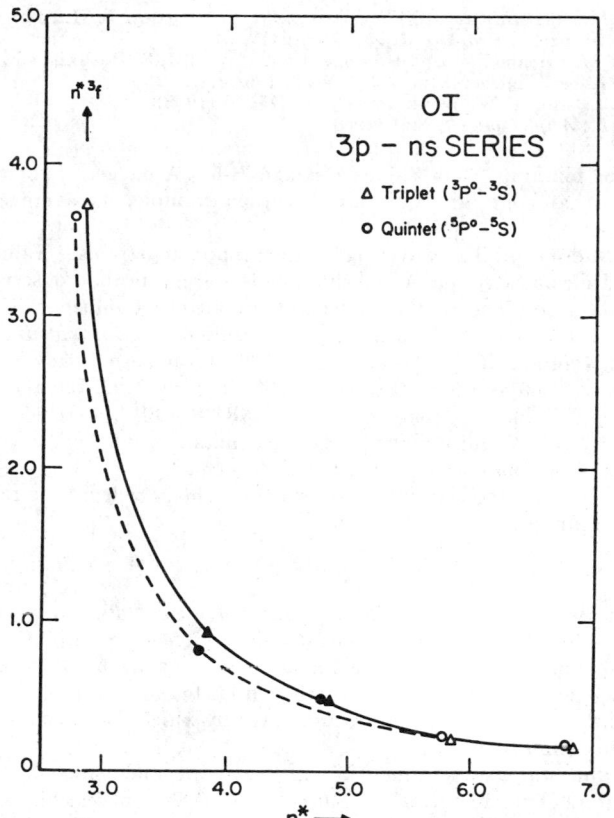

FIG. 7i-5. Oscillator strengths multiplied by n^{*3} vs. effective principal quantum number n^* for the $3p$-ns series of O I. The solid circles and triangles indicate that experimental values are involved in the data. *(From Ref. [10], where the quoted authors and methods are discussed in detail.)*

more, in all homologous atoms the breakdown of the total strength of a transition array into multiplets and individual lines remains the same as long as the coupling scheme remains constant. It follows therefore that for all lines of dominant transition arrays in homologous atoms the f values should stay approximately constant. An example is given in Table 7i-3. More extensive comparisons are found in [10].

References

1. Wiese, W. L., M. W. Smith, and B. M. Glennon: Atomic Transition Probabilities, vol. 1, Hydrogen through Neon, *Natl. Standard Ref. Data Ser. NBS* 4, 1966.
2. Wiese, W. L., M. W. Smith, and B. M. Miles: Atomic Transition Probabilities, vol. 2, Sodium through Calcium, *Natl. Standard Ref. Data Ser. NBS* 22, 1969.
3. Moore, C. E.: A Multiplet Table of Astrophysical Interest, rev. ed., *NBS Tech. Note* 36, 1959.
4. Moore, C. E.: An Ultraviolet Multiplet Table, *NBS Circ.* 488, sec. 1, 1950.
5. Moore, C. E.: Selected Tables of Atomic Spectra, secs. 1 and 2, Si I, II, III, IV, *Natl. Standard Ref. Data Ser. NBS* 3, 1965, 1967.
6. Kelly, R. L.: "Atomic Emission Lines Below 2000 Å," U.S. Government Printing Office, Washington, D.C., 1968.
7. Bethe, H. A., and E. E. Salpeter: "Quantum Mechanics of One- and Two-electron Atoms," Academic Press, Inc., New York, 1957.

8. Naqvi, A. M.: Thesis, Harvard University, 1951; G. Shortley, L. H. Aller, J. E. Baker, and D. H. Menzel: *Astrophys. J.* **93**, 178 (1941).
9. Wiese, W. L.: "Beam Foil Spectroscopy," vol. 2, p. 385, S. Bashkin, ed., Gordon and Breach, Science Publishers, Inc., New York, 1968.
10. Wiese, W. L., and A. W. Weiss: *Phys. Rev.* **175**, 50 (1968).
11. Wiese, W. L.: *Appl. Optics* **7**, 2361 (1968).

Explanations for Main Data Tables 7i-4 and 7i-5. A dagger (†) before a row of data indicates that *multiplet* values are given, for example, the averaged multiplet wavelength.

WAVELENGTH COLUMN: The wavelengths are given in angstroms. Values in square brackets [ ] are calculated and are likely to be less accurate than observed ones.

MULTIPLET COLUMN: The numbers refer to the multiplet numbers of C. E. Moore, "A Multiplet Table of Astrophysical Interest," revised edition, *Nat. Bur. Standards Tech. Note* 36, 1959; or, if "uv" is added, to C. E. Moore, An Ultraviolet Multiplet Table, *Natl. Bur. Standards Circ.* 488, sec. 1, 1950; or, for Si I, II, III, and IV, to C. E. Moore, "Selected Tables of Atomic Spectra," NSRDS-NBS 3, secs. 1 and 2. (Preceded by "UV," if in the ultraviolet.) All are available from the U.S. Government Printing Office, Washington, D.C. 20402.

STATISTICAL WEIGHTS COLUMN: The statistical weight g_k of level k is related to the inner quantum number J by

$$g_k = 2J_k + 1$$

The J's are listed in C. E. Moore, Atomic Energy Levels, *Natl. Bur. Standards Circ.* 467, vol. III, 1958, U.S. Government Printing Office, Washington, D.C. 20402.

TRANSITION PROBABILITY COLUMN: Normally, the A_{ki}'s are listed in units $10^8\ s^{-1}$. But for hydrogen and the forbidden lines, they are listed in units s^{-1} and the number given in parentheses ( ) indicates the power of ten by which the transition probability values have to be multiplied.

ACCURACY COLUMN: The accuracy ratings are to be understood in the sense of "estimated extent of possible errors." Since it is at present not feasible to give specific numerical error limits for each evaluated f value, the data are assigned to one of several levels of accuracy which differ by about factors of three. Further details are found in [1,2].

SOURCE COLUMN: The numbers refer to the references given below. n indicates normalization to an absolute scale different from the one in the listed reference.

References for Tables 7i-4 and 7i-5

1. Wiese, W. L., M. W. Smith, and B. M. Glennon: Atomic Transition Probabilities, vol. 1, Hydrogen through Neon, *Natl. Standard Ref. Data Ser. NBS* 4, 1966.
2. Wiese, W. L., M. W. Smith, and B. M. Miles: Atomic Transition Probabilities, vol. 2, Sodium through Calcium, *Natl. Standard Ref. Data Ser. NBS* 22, 1969.
3. Green, L. C., N. C. Johnson, and E. K. Kolchin: *Astrophys. J.* **144**, 369 (1966). (Central field approximation with exchange and configuration mixing.)
4. Cohen, M., and P. S. Kelly: *Can. J. Phys.* **45**, 1661 (1967). (Self-consistent field calculation.)
5. Cohen, M., and P. S. Kelly: *Can. J. Phys.* **45**, 2079 (1967). (Self-consistent field calculation.)
6. Weiss, A. W.: *Phys. Rev.* **188**, 119 (1969) and to be published. (Self-consistent field calculation with configuration mixing.)
7. Bergström, I., J. Bromander, R. Buchta, L. Lundin, and I. Martinson: *Physics Letters* **28A**, 721 (1969). (Lifetime measurement.)
8. Froese, C.: *J. Chem. Phys.* **47**, 4010 (1967). (Self-consistent field calculation.)
9. Pfennig, H., P. Steele, and E. Trefftz: *J. Quant. Spectr. & Radiative Transfer* **5**, 355 (1965). (Self-consistent field calculation.)
10. Lawrence, G. M., and B. D. Savage: *Phys. Rev.* **141**, 67 (1966). (Lifetime measurement.)

11. Hese, A., and H. P. Weise: *Z. Physik* **215**, 95 (1968). (Lifetime measurement.)
12. Warner, B.: *Monthly Notices Roy. Astron. Soc.* **139**, 1 (1968). (Scaled Thomas-Fermi approximation with limited configuration mixing.)
13. Weiss, A. W.: *Phys. Rev.* **162**, 71 (1967). (Self-consistent field calculation with configuration mixing.)
14. Roberts, J. R., and K. L. Eckerle: *Phys. Rev.* **153**, 87 (1967). (Relative emission measurement.)
15. Steele, R., and E. Trefftz: *J. Quant. Spectr. & Radiative Transfer* **6**, 833 (1966). (Self-consistent field calculation with configuration mixing.)
16. Curnette, B., W. S. Bickel, R. Girardeau, and S. Bashkin: *Phys. Letters* **27A**, 680 (1968). (Lifetime measurement.)
17. Warner, B.: *Monthly Notices Roy. Astron. Soc.* **141**, 273 (1968). (Scaled Thomas-Fermi approximation.)
18. Heroux, L.: *Phys. Rev.* **153**, 156 (1967). (Lifetime measurement.)
19. Bickel, W. S., R. Girardeau, and S. Bashkin: *Phys. Letters* **28A**, 154 (1968). (Lifetime measurement.)
20. Lewis, M. R., T. Marshall, E. H. Carnevale, F. S. Zimoch, and G. W. Wares: *Phys. Rev.* **164**, 94 (1967). (Lifetime measurement.)
21. Gaillard, M., and J. E. Hesser: *Astrophys. J.* **152**, 695 (1968). (Lifetime measurement.)
22. Lawrence, G. M.: *Bull. Am. Phys. Soc. II*, **13**, 424 (1968). (Lifetime measurement.)
23. Bickel, W. S.: *Phys. Rev.* **162**, 7 (1967). (Lifetime measurement.)
24. Bickel, W. S., and S. Bashkin: *Phys. Letters*, **20**, 488 (1966). (Lifetime measurement.)
25. Bridges, J. M., and W. L. Wiese: *Phys. Rev.* **A2**, 285 (1970). (Emission measurement.)
26. Lilly, R. A., and J. R. Holmes: *J. Opt. Soc. Am.* **58**, 1406 (1968). (Relative emission measurement.)
27. Hesser, J. E.: *Phys. Rev.* **174**, 68 (1968). (Lifetime measurement.)
28. Hofmann, W.: *Z. Naturforsch*, **24a**, 990 (1969). (Emission measurement.)

7-210   ATOMIC AND MOLECULAR PHYSICS

TABLE 7-4. TRANSITION PROBABILITIES FOR ALLOWED LINES

Hydrogen

Wavelength, Å	Transition	g_i	g_k	Average transition probability A_{ki}^*, s^{-1}	Source*
914.039	1–20	2	800	3.928(+3)	1
914.286	1–19	2	722	5.077(+3)	1
914.576	1–18	2	648	6.654(+3)	1
914.919	1–17	2	578	8.858(+3)	1
915.329	1–16	2	512	1.200(+4)	1
915.824	1–15	2	450	1.657(+4)	1
916.429	1–14	2	392	2.341(+4)	1
917.181	1–13	2	338	3.393(+4)	1
918.129	1–12	2	288	5.066(+4)	1
919.352	1–11	2	242	7.834(+4)	1
920.963	1–10	2	200	1.263(+5)	1
923.150	1– 9	2	162	2.143(+5)	1
926.226	1– 8	2	128	3.869(+5)	1
930.748	1– 7	2	98	7.568(+5)	
937.803	1– 6(L_e)	2	72	1.644(+6)	1
949.743	1– 5(L_δ)	2	50	4.125(+6)	1
972.537	1– 4(L_γ)	2	32	1.278(+7)	1
1025.72	1– 3(L_β)	2	18	5.575(+7)	1
1215.67	1– 2(L_α)	2	8	4.699(+8)	1
3682.81	2–20	8	800	2.172(+3)	1

Hydrogen (Continued)

Wavelength, Å	Transition	g_i	g_k	Average transition probability A_{ki}^*, s^{-1}	Source*
8467.26	3–17	18	578	3.444(+3)	1
8502.49	3–16	18	512	4.680(+3)	1
8545.39	3–15	18	450	6.490(+3)	1
8598.39	3–14	18	392	9.211(+3)	1
8665.02	3–13	18	338	1.343(+4)	1
8750.47	3–12	18	288	2.021(+4)	1
8862.79	3–11	18	242	3.156(+4)	1
9014.91	3–10	18	200	5.156(+4)	1
9229.02	3– 9	18	162	8.905(+4)	1
9545.98	3– 8(P_e)	18	128	1.651(+5)	1
10049.4	3– 7(P_δ)	18	98	3.358(+5)	1
10938.1	3– 6(P_γ)	18	72	7.783(+5)	1
12818.1	3– 5(P_β)	18	50	2.201(+6)	1
16407.2	4–12	32	288	1.620(+4)	1
16806.5	4–11	32	242	2.556(+4)	1
17362.1	4–10	32	200	4.235(+4)	1
18174.1	4– 9	32	162	7.459(+4)	1
18751.0	3– 4(P_α)	18	32	8.986(+6)	1
19445.6	4– 8	32	128	1.424(+5)	1
21655.0	4– 7	32	98	3.041(+5)	1

λ (Å)	Transition	g_i	g_k	A	Acc.
3686.83	2– 19	8	722	2.809(+3)	1
3691.55	2– 18	8	648	3.685(+3)	1
3697.15	2– 17	8	578	4.910(+3)	1
3703.85	2– 16	8	512	6.658(+3)	1
3711.97	2– 15	8	450	9.210(+3)	1
3721.94	2– 14	8	392	1.303(+4)	1
3734.37	2– 13	8	338	1.893(+4)	1
3750.15	2– 12	8	288	2.834(+4)	1
3770.63	2– 11	8	242	4.397(+4)	1
3797.90	2– 10	8	200	7.122(+4)	1
3835.38	2– 9	8	162	1.216(+5)	1
3889.05	2– 8	8	128	2.215(+5)	1
3970.07	2– 7 (Hε)	8	98	4.389(+5)	1
4101.73	2– 6 (Hδ)	8	72	9.732(+5)	1
4340.46	2– 5 (Hγ)	8	50	2.530(+6)	1
4861.32	2– 4 (Hβ)	8	32	8.419(+6)	1
6562.80	2– 3 (Hα)	8	18	4.410(+7)	1
8392.40	3–20	18	800	1.517(+3)	1
8413.32	3–19	18	722	1.964(+3)	1
8437.96	3–18	18	648	2.580(+3)	1

λ (Å)	Transition	g_i	g_k	A	Acc.
26252.0	4– 6	32	72	7.711(+5)	1
27575	5– 12	50	288	1.402(+4)	1
28722	5– 11	50	242	2.246(+4)	1
30384	5– 10	50	200	3.800(+4)	1
32961	5– 9	50	162	6.908(+4)	1
37395	5– 8	50	128	1.388(+5)	1
40512.0	4– 5	32	50	2.699(+6)	1
43753	6– 12	72	288	1.288(+4)	1
46525	5– 7	50	98	3.253(+5)	1
46712	6– 11	72	242	2.110(+4)	1
51273	6– 10	72	200	3.688(+4)	1
59066	6– 9	72	162	7.065(+4)	1
74578	5– 6	50	72	1.025(+6)	1
75005	6– 8	72	128	1.561(+5)	1
123680	6– 7	72	98	4.561(+5)	1

TABLE 7i-4. TRANSITION PROBABILITIES FOR ALLOWED LINES (*Continued*)

Helium

Wavelength, Å	Multiplet no.	g_i	g_k	Transition probability A_{ki}, 10^8 s^{-1}	Accuracy, %	Source*
508.643	8 uv	1	3	0.306	10	3
509.998	7 uv	1	3	0.454	10	3
512.098	6 uv	1	3	0.722	10	3
515.617	5 uv	1	3	1.25	10	3
522.213	4 uv	1	3	2.46	3	1
537.030	3 uv	1	3	5.66	1	1
584.334	2 uv	1	3	17.99	1	1
†2763.8	—	3	9	0.0132	10	3
†2829.07	12 uv	3	9	0.0204	10	3
†2945.10	11 uv	3	9	0.0339	10	3
†3187.74	3	3	9	0.0505	10	1
3296.77	9	1	9	0.0102	10	3
3354.55	8	1	3	0.0150	10	3
3447.59	7	1	3	0.0233	10	3
3613.64	6	1	3	0.0393	10	3
†3634.2	28	9	15	0.0273	10	3
†3705.0	25	9	15	0.0415	10	3
†3819.6	22	9	15	0.0671	10	3
†3888.65	2	9	9	0.09478	1	1
3926.53	58	3	5	0.0194	10	3
3964.73	5	1	3	0.0717	3	1
4009.27	55	3	5	0.0296	10	3
†4026.2	18	9	15	0.121	10	3
4120.8	16	9	3	0.0436	10	3
4143.76	53	3	5	0.0488	10	3

Helium (*Continued*)

Wavelength, Å	Multiplet no.	g_i	g_k	Transition probability A_{ki}, 10^8 s^{-1}	Accuracy, %	Source*
18555.6	—	5	3	0.00277	10	1
†18686	—	15	21	0.139	10	1
18696.9	—	5	7	0.138	10	1
19089.4	—	3	5	0.0713	10	3
†19543	—	15	9	0.00597	10	1
20581.3	—	1	3	0.01976	1	1
21120	—	9	3	0.0656	10	3
21132.0	—	3	1	0.0459	10	3
[33299]	—	1	3	0.00290	10	3
†[37026]	—	9	15	0.0127	10	3
†[40365]	—	15	21	0.0260	10	3
[40396]	—	5	7	0.0259	10	3
[41216]	—	3	5	0.0153	10	3
†[42947]	—	3	9	0.0108	3	1
[46053]	—	3	1	0.0150	10	3
†[46936]	—	9	3	0.0204	10	3
†[108800]	—	3	9	0.00231	10	3

Lithium

Wavelength, Å	Multiplet no.	g_i	g_k	Transition probability A_{ki}, 10^8 s^{-1}	Accuracy, %	Source*
Li I:						
†2394.36	5 uv	2	6	0.00355	10	1
†2425.41	4 uv	2	6	0.00484	10	1
†2475.06	3 uv	2	6	0.00697	10	1
†2562.31	2 uv	2	6	0.0107	10	1
†2741.19	1 uv	2	6	0.0142	10	1

λ (Å)	Ref.	g_i	g_k	A	Acc.	Ref.
4168.97	52	3	1	0.0181	10	3
4387.93	51	3	5	0.0899	10	3
4437.55	50	3	1	0.0322	10	3
†4471.5	14	3	15	0.257	10	3
4713.2	12	9	3	0.0934	10	3
4921.93	48	3	5	0.199	10	3
5015.68	4	1	3	0.1338	1	1
5047.74	47	3	1	0.0670	10	3
†5875.7	11	9	15	0.706	3	1
6678.15	46	3	5	0.638	3	1
†7065.3	10	9	3	0.278	3	1
7281.35	45	3	1	0.181	3	1
†9463.57	67	3	9	0.00561	10	3
9603.42	71	1	3	0.00586	10	3
†9702.66	75	9	3	0.00871	10	3
†10311	74	9	15	0.0201	10	3
10667.6	73	9	3	0.0145	10	3
10830.3	1	3	5	0.1022	1	1
†10912.9	79	15	21	0.0212	10	1
10917.0	84	5	7	0.0212	10	1
11013.1	70	1	3	0.00928	10	3
11045.0	88	3	5	0.0184	10	3
11225.9	87	3	1	0.0113	10	3
†11969.1	72	9	15	0.0349	10	3
†12528	—	3	9	0.00608	10	1
†12785	—	15	21	0.0462	10	1
12790.3	—	5	7	0.0461	10	1
†12846	—	9	3	0.0274	10	3
12968.4	—	3	5	0.0336	10	3
[13411.8]	—	3	1	0.0205	10	3
15083.7	—	1	3	0.0137	10	1
†17002	—	9	15	0.0664	10	3

λ (Å)	Ref.	g_i	g_k	A	Acc.	Ref.
†3232.63	2	2	6	0.0117	10	1
†3985.5	—	6	2	0.0250	10	1
†4132.6	—	6	10	0.106	10	1
†4273.1	—	6	2	0.0460	10	1
†4602.9	6	6	10	0.236	10	4
†4971.7	5	6	2	0.106	10	4
†6103.6	4	6	10	0.716	10	1
†6707.8	1	2	6	0.372	3	1
8126.4	3	6	2	0.349	10	1
†10510.6	—	6	10	0.0194	10	1
†11032.1	—	6	2	0.0144	10	1
†12237.7	—	6	10	0.0341	10	1
†12793.3	—	10	14	0.0463	10	1
†13557.8	—	6	2	0.0276	10	1
†17546.1	—	6	10	0.0719	10	4
†18703.1	—	10	14	0.138	10	1
†19274.8	—	10	6	0.00481	10	4
†24464.7	—	6	2	0.0774	10	4
†[25197]	—	6	10	0.00819	10	1
†26877.8	—	2	6	0.0377	10	1
†[28417]	—	6	2	0.00922	10	1
†[38081]	—	6	10	0.0136	10	1
†[41791]	—	10	6	0.00286	10	1
†[54633]	—	6	2	0.0225	10	1
†[68592]	—	2	6	0.00778	10	4
Li II:						
178.015	2 uv	1	3	77.9	3	1
199.282	1 uv	1	3	256	3	1
†[944.72]	—	3	9	1.39	10	5
[1093.2]	—	1	3	1.38	10	5
†1132.1	—	9	15	3.90	10	5

* For references see pp. 7-208 and 7-209.

TABLE 7i-4. TRANSITION PROBABILITIES FOR ALLOWED LINES (*Continued*)

Lithium (Continued)

Wavelength, Å	Multiplet no.	g_i	g_k	Transition probability A_{ki}, 10^8 s^{-1}	Accuracy, %	Source*
†1166.4	—	9	3	1.07	10	5
†1198.09	—	3	9	2.88	3	1
[1237.4]	—	3	5	3.16	10	5
1253.3	—	3	1	0.784	10	5
1420.89	—	1	3	2.82	3	1
†1493.0	—	9	15	11.2	3	1
1653.1	—	9	3	2.96	10	5
1681.66	—	3	5	10.1	3	1
1755.33	—	3	1	2.03	10	5
†2674.43	4 uv	3	9	0.192	10	1
[2952.5]	—	1	3	0.202	10	1
†[3029.1]	—	9	15	0.549	10	1
†[3155.4]	—	9	3	0.318	10	1
†[3195.8]	—	15	21	0.739	10	1
[3199.4]	—	5	7	0.736	10	1
[3250.1]	—	3	5	0.528	10	1
[3306.5]	—	3	1	0.252	10	1
†3684.1	2	3	9	0.295	10	5
4156.3	3	1	3	0.351	10	5
†4325.7	5	9	15	1.11	10	5
†[4671.8]	—	15	21	2.21	10	1
[4678.4]	—	5	7	2.21	10	1
[4787.5]	—	3	5	1.17	10	1
†[4840.8]	—	15	9	0.0895	10	5
†4881.3	4	9	3	0.738	10	5

Boron

Wavelength, Å	Multiplet no.	g_i	g_k	Transition probability A_{ki}, 10^8 s^{-1}	Accuracy, %	Source*
B I:						
†1826.2	3 uv	6	10	2.23	25	6, 10
†2089.3	2 uv	6	10	0.494	25	6, 10, 11
2496.77	1 uv	2	2	0.65	25	6, 7
2497.72	1 uv	4	2	1.30	25	6, 7
8667.2	—	2	2	0.0162	25	1
8668.6	—	4	2	0.0324	25	1
†11661	—	2	6	0.196	25	6
15625	—	2	2	0.051	25	1
15629	—	4	2	0.103	25	1
†16243	—	6	10	0.119	25	6
B II:						
1362.46	1 uv	1	3	11.1	25	10
1624.0	3 uv	9	9	10.0	25	10
[1842.8]	—	3	1	6.8	50	1
3451.41	1	3	5	0.64	25	7
†4121.95	2	15	21	2.11	25	7
B III:						
†518.25	—	2	6	12.5	10	1
†677.09	—	6	10	56.8	10	1
†758.60	—	2	2	16.3	10	1
†2066.3	—	2	6	1.91	3	1
†4243.60	1	6	10	1.11	10	1
†4487.46	2	10	14	2.10	10	1
†7838.5	—	2	6	0.222	10	1

λ(Å)	Mult.	g_i	g_k	A_{ki}	Acc.(%)	Ref.
[5038.7]	—	3	1	0.533	10	5
†5484.8	1	3	9	0.228	3	1
[9562.2]	—	1	3	0.0518	3	1
†[57324]	—	9	15	0.00110	3	1

Beryllium

λ(Å)	Mult.	g_i	g_k	A_{ki}	Acc.(%)	Ref.
Be I:						
2348.61	1 uv	1	3	5.3	25	6,7
†2494.6	3 uv	9	15	1.4	50	9
†2650.6	2 uv	9	9	4.29	25	6,7
†3321.2	1	9	3	1.6	50	7,8
[3455.2]	—	3	1	2.09	25	1,7
3515.54	7	3	5	0.13	50	1
3813.40	5	3	5	0.23	50	1
Be II:						
†1036.31	1 uv	2	6	1.66	10	1
†1512.4	4 uv	6	10	11.4	10	1
†1776.2	3 uv	6	2	4.22	10	1
†3130.6	1	2	6	1.15	3	1
†3247.7	5	6	2	0.410	10	1
†3274.64	2	2	6	0.133	10	4
†4360.9	4	6	10	1.12	10	4
†4673.46	6	10	14	2.21	10	1
†5270.7	3	6	2	1.00	10	4
†12094	—	2	6	0.128	10	1
Be III:						
88.314	—	1	3	362	3	1
100.254	—	1	3	1220	3	1
[398.19]	—	1	3	42.8	3	1
†[583.01]	—	3	9	16.5	3	1
†[3721.8]	—	3	9	0.342	3	1
[6141.2]	—	1	3	0.0877	3	1

λ(Å)	Mult.	g_i	g_k	A_{ki}	Acc.(%)	Ref.
B IV:						
52.682	—	1	3	1080	3	1
60.313	—	1	3	3720	3	1
†[344.19]	—	3	9	54.6	3	1
[381.13]	—	1	3	51.0	3	1
†2823.4	—	3	9	0.455	3	1
[4499.4]	—	1	3	0.125	3	1

Carbon

λ(Å)	Mult.	g_i	g_k	A_{ki}	Acc.(%)	Ref.
C I:						
†1261.3	9 uv	9	9	1.2	50	1
†1277.5	7 uv	9	15	1.6	50	1
1279.25	6 uv	5	7	0.11	50	1
†1280.4	5 uv	9	9	0.82	50	1
†1329.3	4 uv	9	9	1.4	50	1
1431.60	65 uv	5	7	1.5	50	1
1432.12	65 uv	5	5	1.4	50	1
1432.54	65 uv	5	3	1.3	50	10
1459.05	38 uv	5	3	0.37	50	10
1463.33	37 uv	5	7	2.1	50	1
1467.45	36 uv	5	3	0.46	50	10,13
1481.77	34 uv	5	5	0.33	50	10,13
†1561.0	3 uv	9	15	1.25	25	1
†1657.2	2 uv	9	9	3.20	25	1
1751.9	62 uv	1	3	0.87	50	1
1930.93	33 uv	5	3	3.1	50	1
2478.56	61 uv	1	3	0.33	50	1
4268.99	16	3	5	0.0032	50	1
4371.33	14	3	3	0.0097	50	1
4932.00	13	3	1	0.046	50	1
5052.12	12	3	5	0.017	50	1
5380.24	11	3	3	0.016	50	1

* For references see pp. 7-208 and 7-209.

TABLE 7i-4. TRANSITION PROBABILITIES FOR ALLOWED LINES (Continued)

Carbon (Continued)

Wavelength, Å	Multiplet no.	g_i	g_k	Transition probability A_{ki}, 10^8 s^{-1}	Accuracy, %	Source*
6587.75	22	3	3	0.024	50	1
8335.19	10	3	1	0.32	50	1
9061.48	3	3	5	0.065	50	1
9062.53	3	1	3	0.083	50	1
9078.32	3	3	3	0.062	50	1
9088.57	3	3	1	0.25	50	1
9094.89	3	5	5	0.19	50	1
9111.85	3	5	3	0.11	50	1
9603.09	2	1	3	0.024	50	1
9620.86	2	3	3	0.074	50	1
9658.49	2	5	3	0.12	50	1
10124	—	3	3	0.171	25	1
10548.0	20	3	3	0.010	50	1
10683.1	1	3	5	0.13	50	1
10685.3	1	1	3	0.10	50	1
10691.2	1	5	7	0.18	50	1
10707.3	1	3	3	0.072	50	1
10729.5	1	5	5	0.043	50	1
11602.9	25	3	5	0.0099	25	1
11609.9	25	3	3	0.0492	25	1
11619.0	25	5	7	0.0073	25	1
11631.6	25	5	5	0.0453	25	1
11638.6	25	5	3	0.0163	25	1

Carbon (Continued)

Wavelength, Å	Multiplet no.	g_i	g_k	Transition probability A_{ki}, 10^8 s^{-1}	Accuracy, %	Source*
4371.59	45	2	4	0.83	25	1
4372.49	45	4	6	1.40	25	1
4374.28	45	6	8	1.99	25	1
†4411.4	39	10	14	2.11	25	1
5143.49	16	4	2	0.72	25	1
5145.16	16	6	6	0.60	25	1
5151.08	16	6	4	0.385	25	1
5640.50	15	2	4	0.109	25	1
5648.08	15	4	4	0.217	25	1
5662.51	15	6	4	0.325	25	1
†5890.4	5	10	6	0.337	25	13
6578.03	2	2	4	0.361	25	13
6582.85	2	2	2	0.361	25	13
6783.75	14	6	8	0.370	25	1
6787.09	14	4	2	0.307	25	1
6791.30	14	6	4	0.195	25	1
6800.50	14	2	6	0.110	25	1
7231.12	3	2	4	0.362	25	13

C III:

Wavelength, Å	Multiplet no.	g_i	g_k	Transition probability A_{ki}, 10^8 s^{-1}	Accuracy, %	Source*
386.203	2 uv	1	3	25	50	8
†459.57	6 uv	9	15	95	50	9
977.026	1 uv	1	3	16.1	25	15
†1175.7	4 uv	9	9	13	50	1
1247.37	9 uv	3	1	12	50	1

λ (Å)	Mult.	g_i	g_k	A_{ki}	Acc.	Source
[11653]	29	3	1	0.157	25	1
11656.0	29	3	3	0.158	25	1
11677.0	25	7	5	0.0101	25	1
11747.5	24	3	5	0.202	25	1
[11778]	24	5	5	0.0375	25	1
11801.8	24	7	7	0.0266	25	1
11849.3	23	5	5	0.017	50	1
11863.0	23	3	3	0.029	50	1
11880.4	23	3	1	0.11	50	1
12551.0	30	1	3	0.0352	25	1
12582.3	30	3	5	0.0262	25	1
12602.6	30	5	3	0.0435	25	1
12614.8	30	5	5	0.078	25	1
16890	—	5	7	0.123	25	1
C II:						
†687.25	5 uv	6	10	28.0	25	13
†904.09	3 uv	6	6	41.6	25	13
†1010.2	7 uv	12	4	34.3	25	13
†1036.8	2 uv	6	2	22.2	25	13
†1323.9	11 uv	10	10	5.3	25	13
†1335.3	1 uv	6	10	2.65	25	10, 13
2509.11	14 uv	2	4	0.63	25	13, 14
2511.71	14 uv	4	4	0.126	25	13, 14
2512.03	14 uv	4	6	0.75	25	13, 14
2836.71	13 uv	2	4	0.359	25	13, 14
2837.60	13 uv	2	2	0.359	25	13, 14
†3876.7	33	28	36	2.66	25	1
3918.98	4	2	2	0.62	25	1, 14
3920.68	4	4	2	1.24	25	1, 14
4074.53	36	4	8	1.96	25	1
4076.00	36	8	10	2.28	25	1
†4267.2	6	10	14	2.46	25	1

λ (Å)	Mult.	g_i	g_k	A_{ki}	Acc.	Source
2296.89	8 uv	3	5	1.20	25	7, 16
3170.16	8	1	3	0.325	25	1
†3609.3	10	9	15	0.95	25	1
3703.52	12	3	3	0.82	25	1
†3887.1	15	15	21	1.81	25	1
4056.06	24	5	7	1.45	25	1
4122.05	17	3	5	1.04	25	1
4325.70	7	3	5	1.08	25	1
4388.24	14	7	5	0.224	25	1
†4516.5	9	9	3	1.66	25	1
4647.40	1	3	5	0.68	25	7
4663.53	5	3	1	0.84	25	1
4665.90	5	5	5	0.63	25	1
4673.91	5	5	3	0.347	25	1
5249.6	23	5	3	0.52	25	1
5253.55	4	3	3	0.194	25	1
5272.56	4	5	3	0.320	25	1
6727.1	3	1	3	0.149	25	1
6730.7	3	3	5	0.201	25	1
6744.2	3	5	7	0.266	25	1
C IV:						
†244.907	3 uv	2	6	22.3	10	4
†259.52	10 uv	6	10	27.6	10	17
289.143	9 uv	2	4	49.5	10	4
296.857	8 uv	2	2	5.27	10	4
296.951	8 uv	4	2	10.5	10	4
312.418	2 uv	2	4	45.7	10	1
312.455	2 uv	2	2	45.5	10	1
384.032	7 uv	2	4	148	10	1
419.525	6 uv	2	2	14.3	10	1
419.714	6 uv	4	2	28.5	10	1

* For references see pp. 7-208 and 7-209.

TABLE 7i-4. TRANSITION PROBABILITIES FOR ALLOWED LINES (Continued)

Carbon (Continued)

Wavelength, Å	Multiplet no.	Statistical weights g_i	Statistical weights g_k	Transition probability A_{ki}, 10^8 s^{-1}	Accuracy, %	Source*
1548.20	1 uv	2	4	2.65	3	1
1550.77	1 uv	2	4	2.63	3	1
†2524.40	14 uv	10	14	6.62	10	17
†2595.14	13 uv	10	6	0.673	10	17
2697.73	12 uv	2	2	1.17	10	17
2698.70	12 uv	4	2	2.33	10	17
†3936	2	2	6	0.330	10	17
5021	3	2	2	0.464	10	17
5023	3	4	2	0.930	10	17
5801.51	1	2	4	0.319	10	1
5812.14	1	2	2	0.316	10	1
C V:						
34.973	—	1	3	2550	3	1
40.270	—	1	3	8870	3	1
†186.72	—	9	15	142	10	5
197.02	—	3	5	124	10	5
†227.22	—	3	9	136	3	1
[247.31]	—	1	3	128	3	1
†248.71	—	9	15	425	3	1
267.26	—	3	5	396	3	1
†2273.9	—	3	9	0 565	3	1
[3540.8]	—	1	3	0 165	3	1

Nitrogen (Continued)

Wavelength, Å	Multiplet no.	Statistical weights g_i	Statistical weights g_k	Transition probability A_{ki}, 10^8 s^{-1}	Accuracy, %	Source*
6945.22	29	6	6	0.0149	25	1
7442.30	3	4	4	0.106	25	1
7468.31	3	6	4	0.161	25	1
8184.85	2	4	6	0.063	25	1
8188.01	2	2	4	0.092	25	1
8216.32	2	6	6	0.160	25	1
8223.12	2	4	2	0.202	25	1
8242.37	2	6	4	0.102	25	1
8590.01	8	2	2	0.190	25	1
8629.24	8	4	4	0.238	25	1
8680.27	1	6	8	0.191	25	1
8683.40	1	4	6	0.133	25	1
8686.16	1	2	4	0.079	25	1
8703.26	1	2	2	0.171	25	1
8711.71	1	4	4	0.101	25	1
8718.84	1	6	6	0.054	25	1
9028.92	15	2	2	0.255	10	1
9045.88	—	2	8	0.269	10	1
9060.72	15	2	4	0.257	10	1
9386.8†	7	2	4	0.183	25	1
9392.79	7	4	6	0.218	25	1
9822.75	19	6	6	0.0542	10	1
9863.33	19	8	8	0.101	10	1
10105.1	18	2	4	0.262	10	1
†10108.9	18	4	6	0.281	10	1

Nitrogen

λ (Å)	Source	g_i	g_k	A_{ki}	Accuracy	Source
N I:						
1134.17	2 uv	4	2	1.82	25	1n, 10
1134.42	2 uv	4	4	1.82	25	1n, 10
1134.98	2 uv	4	6	1.60	25	1n, 10
†1164.0	7 uv	10	10	0.343	25	1n, 10
†1167.9	6 uv	10	14	0.87	25	1n, 10
1199.55	1 uv	4	6	4.01	25	1n, 10
1200.22	1 uv	4	4	3.86	25	1n, 10
1200.71	1 uv	4	2	4.01	25	1n, 10
†1243.3	5 uv	10	10	3.35	25	1n, 10
†1310.7	13 uv	6	10	0.95	25	1n, 10
†1411.94	10 uv	6	10	0.379	25	1n, 10
1494.67	4 uv	4	2	3.65	25	1n, 10
†1743.6	9 uv	6	6	1.46	25	1n, 10
4099.95	10	2	4	0.034	50	1
4109.96	10	4	6	0.040	50	1
4214.73	5	4	6	0.022	50	1
4215.92	5	2	4	0.031	50	1
4230.35	5	6	4	0.033	50	1
4914.90	9	2	2	0.00759	10	1
4935.03	9	4	2	0.0158	10	1
[5197.8]	—	2	2	0.023	50	1
[5201.8]	—	2	4	0.023	50	1
5281.18	14	6	6	0.00282	25	1
5328.70	13	6	8	0.00254	25	1
5401.45	—	2	2	0.00369	25	1
5411.88	—	4	2	0.0075	25	1
6644.96	20	8	6	0.0311	25	1
6646.51	20	2	2	0.0194	25	1
6653.46	20	6	4	0.0244	25	1
6656.51	20	4	2	0.0193	25	1
10112.5	18	6	8	0.321	10	1
10114.6	18	8	10	0.374	10	1
10128.3	18	4	4	0.104	10	1
10147.3	18	6	6	0.0898	10	1
10164.8	18	8	8	0.0523	10	1
10500.3	28	2	4	0.0652	10	1
10507.0	28	4	6	0.132	10	1
10513.4	28	2	2	0.174	10	1
10520.6	28	4	4	0.162	10	1
10539.6	28	6	8	0.242	10	1
10549.6	28	6	6	0.126	10	1
10591.9	—	6	8	0.326	10	1
10644.0	—	4	2	0.107	10	1
10653.0	—	2	4	0.0532	10	1
10713.6	—	4	6	0.0376	10	1
10718.0	—	6	4	0.0564	10	1
10757.9	—	6	6	0.0868	10	1
11291.7	17	8	6	0.117	25	1
11294.2	17	2	2	0.0731	25	1
11313.9	17	6	4	0.0920	25	1
11997.9	37	4	4	0.054	25	1
12074.1	37	6	6	0.055	25	1
12186.9	27	6	6	0.054	25	1
[12330]	34	4	4	0.124	25	1
[12384]	34	4	6	0.123	25	1
12461.2	36	4	6	0.202	25	1
12467.8	36	6	8	0.217	25	1
N II:						
644.825	4 uv	3	3	30.2	25	18
645.167	4 uv	5	3	51	25	18
†671.48	3 uv	9	9	9.9	25	18

* For references see pp. 7-208 and 7-209.

TABLE 7i-4. TRANSITION PROBABILITIES FOR ALLOWED LINES (*Continued*)

Nitrogen (Continued)

Wavelength, Å	Multiplet no.	g_i	g_k	Transition probability A_{ki}, 10^8 s^{-1}	Accuracy, %	Source*
†916.34	2 uv	9	9	10.4	25	18
†1085.1	1 uv	9	15	3.56	25	10, 18
1886.82	14 uv	3	3	0.52	50	1
2206.10	15 uv	3	5	0.49	50	1
2461.30	23 uv	5	3	0.353	25	1
2709.82	22 uv	5	7	0.35	50	1
3006.86	18	3	3	0.54	25	1
3328.79	22	7	5	0.93	25	1
3330.30	22	3	1	1.11	25	1
3331.32	22	5	3	0.83	25	1
3437.16	13	3	1	2.40	25	1
3593.60	26	3	5	0.231	25	1
3609.09	26	3	3	0.228	25	1
3829.80	30	3	5	0.175	25	1
3838.39	30	5	5	0.52	25	1
3919.01	17	3	3	1.00	25	1
3995.00	12	3	5	1.58	25	1
4026.08	40	7	9	0.90	25	1
†4040.9	39	21	27	2.64	25	1
4124.08	65	5	5	0.276	25	1
4133.67	65	5	5	0.458	25	1
4145.76	65	7	5	0.64	25	1
4176.16	42	5	7	2.19	25	1
4227.75	33	5	3	1.06	25	1

Nitrogen (Continued)

Wavelength, Å	Multiplet no.	g_i	g_k	Transition probability A_{ki}, 10^8 s^{-1}	Accuracy, %	Source*
5940.25	28	3	3	0.235	25	1
5941.67	28	5	7	0.564	25	1
6167.82	36	9	7	0.333	25	1
6170.16	36	5	3	0.362	25	1
6173.40	36	7	5	0.320	25	1
6242.52	57	7	5	0.341	25	1
6340.57	46	7	5	0.258	25	1
6356.55	46	5	3	0.229	25	1
6357.57	46	3	1	0.304	25	1
6482.07	8	3	3	0.365	25	1
6504.61	45	7	7	0.052	25	1
6632.55	45	5	5	0.0404	25	1
6610.58	31	5	7	0.59	25	1
6629.80	41	5	3	0.283	25	1
6809.99	54	5	3	0.199	25	1
6834.09	54	3	3	0.118	25	1
6941.75	53	5	5	0.065	25	1
N III:						
685.513	3 uv	2	2	39.0	25	18
685.816	3 uv	4	4	48.8	25	18
†990.98	1 uv	6	10	4.20	25	18
1804.3	22 uv	2	2	2.26	25	1
1805.5	22 uv	4	2	4.51	25	1

Upper table

		gA	g_k	g_i	Ref.	λ(Å)
1	25	11.9	14	10	24 uv	†1885.25
1	25	11.0	14	10	27 uv	†1908.11
1	25	10.9	8	6	30 uv	2063.50
1	25	11.3	10	8	30 uv	2063.99
1	25	0.93	2	2	25 uv	2972.60
1	25	0.461	2	4	25 uv	[2977.3]
1	25	0.230	4	2	25 uv	[2978.8]
1	25	1.14	4	4	25 uv	2983.58
1	25	1.45	2	4	5	3365.79
1	25	1.22	6	6	5	3367.36
1	25	0.78	4	6	5	3374.06
1	25	0.209	4	2	4	3745.83
1	25	0.416	4	4	4	3754.62
1	25	0.61	4	6	4	3771.08
1	25	0.80	4	2	8	3934.41
1	25	0.96	6	4	8	3938.52
1	25	0.160	4	4	8	3942.78
1	25	0.96	4	2	1	4097.31
1	25	0.97	2	2	1	4103.37
1	25	0.84	4	2	6	4195.70
1	25	1.00	6	4	6	4200.62
1	25	0.165	4	4	6	4215.69
1	25	0.198	8	8	10	4348.36
1	25	0.70	8	6	3	4514.89
1	25	0.58	2	2	3	4518.18
1	25	0.372	4	4	3	4523.60
1	25	9.54	8	6	9	4861.33
1	25	0.152	6	6	9	4873.58
1	25	0.089	8	8	9	4884.14
1	25	0.181	4	2	14	6445.05
1	25	0.362	2	2	14	6450.78
1	25	0.304	6	4	14	6453.95

Lower table

		gA	g_k	g_i	Ref.	λ(Å)
1	25	2.14	21	15	48	†4239.4
1	25	1.30	3	3	15	4447.03
1	25	1.69	9	7	59	4530.40
1	25	0.76	9	7	58	4552.54
1	25	0.270	5	3	5	4601.48
1	25	0.340	3	1	5	4607.16
1	25	0.196	3	3	5	4613.87
1	25	0.90	1	3	5	4621.39
1	25	0.84	5	5	5	4630.54
1	25	0.466	3	5	5	4643.09
1	25	1.65	5	3	62	4677.93
1	25	0.269	3	3	20	4779.71
1	25	0.248	5	5	20	4788.13
1	25	0.313	7	7	20	4803.27
1	25	0.189	3	1	34	5104.45
1	25	0.139	7	5	69	5338.66
1	25	0.194	5	7	69	5340.20
1	25	0.275	7	7	69	5351.21
1	25	0.100	5	3	29	5478.13
1	25	0.167	3	5	29	5480.10
1	25	0.298	5	5	29	5495.70
1	25	0.198	5	3	63	5526.26
1	25	0.377	7	5	63	5530.27
1	25	0.327	5	5	63	5543.49
1	25	0.423	5	3	3	5666.64
1	25	0.310	3	1	3	5676.02
1	25	0.56	7	5	3	5679.56
1	25	0.231	3	3	3	5686.21
1	25	0.137	5	5	3	5710.76
1	25	0.315	3	1	28	5927.82
1	25	0.425	5	3	28	5931.79

* For references see pp. 7-208 and 7-209.

TABLE 7i-4. TRANSITION PROBABILITIES FOR ALLOWED LINES (Continued)

Wavelength, Å	Multiplet no.	Statistical weights g_i	Statistical weights g_k	Transition probability A_{ki} 10^8, s^{-1}	Accuracy, %	Source*
Nitrogen (Continued)						
6463.03	14	4	4	0.232	25	1
6466.86	14	6	8	0.432	25	1
6478.69	14	6	6	0.129	25	1
N IV:						
†225.17	6 uv	9	15	92	25	1
247.05	2 uv	1	3	110	50	9
†283.53	5 uv	9	15	264	25	9
335.050	10 uv	3	5	200	25	18
765.140	1 uv	1	3	20.5	25	15
921.982	3 uv	3	5	3.57	25	18
922.507	3 uv	1	3	4.82	25	18
923.045	3 uv	3	3	3.58	25	18
923.211	3 uv	5	5	10.7	25	18
923.669	3 uv	3	1	14.4	25	18
924.274	3 uv	5	3	5.9	25	18
1718.52	7 uv	3	5	3.23	25	19
3463.36	1	5	5	0.94	25	1
3478.69	1	3	5	1.09	25	9, 20
3482.98	1	3	3	1.09	25	9, 20
3484.90	1	3	1	1.07	25	9, 20
3747.66	8	3	5	1.06	25	1
4495	6	3	3	0.189	25	1
4528	6	5	3	0.305	25	1
[4685.4]	11	3	3	0.089	25	1
Nitrogen (Continued)						
[185.09]	—	3	5	825	3	1
1896.82	—	3	5	0.683	3	1
1907.34	—	3	3	0.672	3	1
1907.87	—	3	1	0.671	3	1
[2914.6]	—	1	3	0.206	3	1
Oxygen						
O I:						
1152.16	6 uv	5	5	5.3	25	21
1302.17	2 uv	5	3	3.14	25	21, 22
1304.87	2 uv	3	3	1.94	25	21, 22
1306.04	2 uv	1	3	0.61	25	21, 22
†3947.29	3	5	15	0.00326	25	1
†4368.30	5	3	9	0.0066	25	1
†5330.0	12	15	25	0.0197	25	1
5435.16	11	3	5	0.0061	25	1
5435.76	11	5	5	0.0102	25	1
5436.83	11	7	5	0.0142	25	1
†6046.4	22	9	3	0.0234	25	1
†6157.3	10	15	25	0.0701	10	1
†6259.6	50	21	27	0.063	25	1
6453.64	9	3	5	0.0142	10	1
6454.48	9	5	5	0.0237	10	1
6456.01	9	7	5	0.0331	10	1
6653.78	65	3	1	0.600	10	1

λ(Å)	Mult.	g_i	g_k	A	Acc.%	Ref.
4733	11	5	5	0.081	25	1
4752	11	7	7	0.102	25	1
5236	5	3	5	0.261	25	1
5245	5	5	7	0.345	25	1
5734	9	3	5	0.178	25	1
6383	2	1	3	0.193	25	1
7109.48	4	3	5	0.107	25	9
7123.10	4	5	7	0.142	25	9
N V:						
†162.562	3 uv	2	6	57.2	10	17
†186.13	6 uv	6	10	142	10	17
†209.28	2 uv	2	6	120	10	1
247.563	5 uv	2	4	357	10	1
266.192	4 uv	2	2	30.2	10	1
266.375	4 uv	4	2	60.6	10	1
1238.81	1 uv	2	4	3.38	3	1
1242.80	1 uv	2	2	3.36	3	1
†3161	2	6	6	3.06	10	17
†4335	3	2	2	0.368	10	17
4603.83	1	2	4	0.415	10	1
4619.9	1	2	2	0.411	10	1
†4751	5	6	10	0.958	10	17
†4933	7	10	14	1.62	10	1
†5273	4	6	2	1.40	10	17
N VI:						
24.898	—	1	3	5160	3	1
28.787	—	1	3	18100	3	1
†161.22	—	3	9	285	3	1
[173.34]	—	1	3	269	3	1
†173.92	—	9	15	876	3	1

λ(Å)	Mult.	g_i	g_k	A	Acc.%	Ref.
†7002.1	21	9	15	0.0325	25	1
7156.80	38	5	5	0.473	10	1
†7254.4	20	9	3	0.062	25	1
7471.36	55	5	3	0.0114	10	1
7473.23	55	5	5	0.102	10	1
7476.45	55	5	7	0.408	10	1
7477.21	55	3	3	0.170	10	1
7479.06	55	3	5	0.306	10	1
7480.66	55	1	3	0.226	10	1
7771.96	1	5	7	0.340	10	1
7774.18	1	5	5	0.340	10	1
7775.40	1	5	3	0.340	10	1
7886.31	64	3	5	0.370	10	1
7939.49	35	7	5	0.00165	25	1
7943.15	35	7	7	0.0417	25	1
7950.83	35	5	7	0.331	25	1
7952.18	35	3	5	0.313	25	1
7995.12	19	5	7	0.29	50	1
8227.64	34	5	3	0.0834	10	1
8232.99	34	3	3	0.261	10	1
8235.31	34	3	5	0.0432	10	1
†8446.5	4	3	9	0.280	10	1
8508.63	—	3	3	0.289	25	1
8820.45	37	5	7	0.261	25	1
†9263.9	8	15	25	0.419	25	1
†11287	—	9	15	0.235	25	1
11295.0	7	3	5	0.054	25	1
11297.5	7	5	5	0.091	25	1
11302.2	7	7	5	0.127	25	1
†13164	—	9	3	0.188	25	1

TABLE 7i-4. TRANSITION PROBABILITIES FOR ALLOWED LINES (Continued)

Oxygen (Continued)

Wavelength, Å	Multiplet no.	g_i	g_k	Transition probability A_{ki} 10^8, s^{-1}	Accuracy, %	Source*
O II:						
2733.34	20 uv	2	4	0.37	50	1
2747.46	20 uv	2	2	0.36	50	1
3122.62	14	6	6	0.278	25	1
3129.44	14	4	4	0.493	25	1
3134.32	14	2	2	0.77	25	1
3134.82	14	8	6	1.23	25	1
3138.44	14	6	4	0.96	25	1
3139.77	14	4	2	0.76	25	1
3277.69	23	4	6	0.259	25	1
3287.59	23	6	6	0.60	25	1
3290.13	23	2	4	0.356	25	1
3305.15	23	6	4	0.379	25	1
3306.60	23	4	2	0.70	25	1
3377.20	9	2	2	1.88	25	1
3390.25	9	2	4	1.86	25	1
3470.42	27	4	2	1.24	25	1
3470.81	27	6	4	1.12	25	1
3712.75	3	2	4	0.280	25	1
3727.33	3	4	4	0.59	25	1
3739.92	31	4	6	0.267	25	1
3749.49	3	6	4	0.90	25	1
3762.63	31	4	4	0.269	25	1
3777.60	31	4	2	0.252	25	1
4650.84	1	2	2	0.82	25	1
4661.64	1	4	4	0.52	25	1
4676.23	1	6	6	0.257	25	1
4861.03	57	2	4	0.366	25	1
4871.58	57	4	6	0.435	25	1
[4872.2]	57	4	4	0.073	25	1
4890.93	28	4	2	0.68	25	1
4906.88	28	4	4	0.68	25	1
4924.60	28	4	6	0.67	25	1
4941.12	33	2	4	0.83	25	1
4943.06	33	4	6	1.06	25	1
4955.78	33	4	4	0.256	25	1
5160.02	32	2	2	0.350	25	1
5176.00	32	4	2	0.171	25	1
5190.56	32	2	4	0.137	25	1
5206.73	32	4	4	0.391	25	1
6640.90	4	2	2	0.098	25	1
6721.35	4	4	4	0.189	25	1
6895.29	45	10	8	0.298	25	1
6906.54	45	8	6	0.272	25	1
6908.11	45	4	2	0.332	25	1
6910.75	45	6	4	0.267	25	1

O III:

λ	ref	g_i	g_k	A	Acc.	Ref.
2454.99	19 uv	3	1	4.00	25	1
2558.06	21 uv	7	5	1.16	25	1
2597.69	20 uv	5	3	0.97	25	1
2605.41	20 uv	3	3	0.58	25	1
2695.49	23 uv	3	5	2.09	25	1
2983.78	6	3	5	2.24	25	1
2996.51	10	3	3	0.51	25	1
3004.35	10	5	5	0.472	25	1
3017.63	10	7	7	0.59	25	1
3035.43	4	3	3	0.51	25	1
3043.02	4	3	1	2.03	25	1
3047.13	4	5	5	1.52	25	1
3059.30	4	5	3	0.84	25	1
3083.65	26	7	7	0.311	25	1
3084.63	26	7	5	0.248	25	1
3038.04	26	9	9	0.52	25	1
3315.73	12	3	1	1.39	25	1
3121.71	12	3	3	1.38	25	1
3132.86	12	3	5	1.36	25	1
3200.95	31	3	3	0.499	25	1
3207.12	31	5	5	0.460	25	1
3215.97	31	7	7	0.58	25	1
3260.98	8	5	7	1.84	25	1
3265.46	8	7	9	2.07	25	1
3267.31	8	3	5	1.73	25	1
3382.69	27	5	7	0.97	25	1
[3383.5]	27	5	3	0.363	25	1
3383.85	27	5	5	0.85	25	1
3384.95	27	7	9	1.45	25	1
3394.26	27	7	7	0.480	25	1

λ	ref	g_i	g_k	A	Acc.	Ref.
3803.14	34	4	4	0.55	25	1
3919.29	17	4	2	1.40	25	1
3945.05	6	2	4	0.217	25	1
3954.37	6	2	2	0.95	25	1
3973.26	6	4	4	1.27	25	1
3982.72	6	4	2	0.447	25	1
†4060.8	97	14	18	2.20	25	1
4072.16	10	6	8	1.70	25	1
4075.87	10	8	10	1.98	25	1
4078.86	10	4	4	0.55	25	1
4153.86	19	4	6	0.77	25	1
4169.54	19	6	4	0.157	25	1
4169.28	19	6	6	0.220	25	1
4185.46	36	6	8	2.43	25	1
†4253.9	101	18	22	2.63	25	1
†[4272.3]	—	2	6	1.08	25	1
4395.95	26	6	6	0.398	25	1
4414.91	5	4	6	1.15	25	1
4416.98	5	2	4	0.95	25	1
4443.05	35	6	6	0.57	25	1
[4443.7]	35	6	8	0.0212	25	1
[4447.7]	35	8	6	0.0282	25	1
4448.21	35	8	8	0.57	25	1
4452.38	5	4	4	0.154	25	1
4489.48	86	2	4	1.51	25	1
4491.25	86	4	6	1.81	25	1
4590.97	15	6	8	1.11	25	1
4602.11	93	6	6	1.70	25	1
4609.42	93	8	8	1.82	25	1
4638.85	1	6	4	0.422	25	1
4641.81	1	4	6	0.79	25	1
4649.14		6	8	1.04	25	1

TABLE 7i-4. TRANSITION PROBABILITIES FOR ALLOWED LINES (Continued)

Oxygen (Continued)

Wavelength, Å	Multiplet no.	g_i	g_k	Transition probability A_{ki}, 10^8 s^{-1}	Accuracy, %	Source*
[3395.5]	27	7	5	0.096	25	1
[3520.7]	24	1	3	0.493	25	1
[3530.7]	24	3	1	1.47	25	1
[3532.8]	24	3	3	0.367	25	1
[3534.3]	24	3	5	0.366	25	1
[3555.3]	24	5	3	0.60	25	1
3556.92	24	5	5	1.08	25	1
3638.70	35	5	7	1.40	25	1
3645.20	35	5	5	0.347	25	1
3646.84	35	3	5	1.04	25	1
3650.70	35	3	3	0.58	25	1
3653.00	35	1	3	0.77	25	1
3961.59	17	5	7	1.28	25	1
[4072.3]	23	1	3	0.52	25	1
4073.90	23	3	5	0.71	25	1
4081.10	23	5	7	0.94	25	1
4440.1	33	5	3	0.495	25	1
4447.82	33	5	5	0.492	25	1
4461.56	33	5	7	0.486	25	1
5268.06	19	1	3	0.311	25	1
5500.11	16	5	5	0.112	25	1
5592.37	5	3	3	0.328	25	1

Oxygen (Continued)

Wavelength, Å	Multiplet no.	g_i	g_k	Transition probability A_{ki}, 10^8 s^{-1}	Accuracy, %	Source*
4783.43	9	4	6	0.213	25	1
4794.22	9	4	4	0.161	25	1
4798.25	9	6	8	0.303	25	1
4813.07	9	6	6	0.090	25	1
5305.3	11	4	4	0.069	25	1
5362.4	11	6	6	0.069	25	1
O V:						
†192.85	5 uv	9	15	600	25	9
220.352	10 uv	3	5	450	25	9
629.732	1 uv	1	3	25.2	25	15
760.445	3 uv	5	5	16	50	1
1371.29	7 uv	3	5	7.4	25	24
3058.68	6	3	5	1.30	25	1
3239	5	3	3	0.342	25	1
3275.67	5	5	3	0.55	25	1
3717	8	5	5	0.109	25	1
3747	8	7	7	0.136	25	1
4135.9	11	3	3	0.261	25	1
4158.76	11	3	5	0.257	25	1
4554.28	7	3	5	0.233	25	1
5114	1	1	3	0.273	25	8
5343	13	1	3	0.304	25	1
5352	13	3	1	0.91	25	1
5376	13	3	3	0.223	25	1

λ	mult	g_i	g_k	A	acc	ref
O IV:						
787.710	1 uv	2	4	4.87	25	23
790.103	1 uv	4	4	0.97	25	23
790.203	1 uv	4	6	5.8	25	23
[2494.8]	5	2	2	1.02	25	1
[2511.4]	5	4	2	2.01	25	1
3063.46	1	2	4	4.48	25	1
3071.66	1	2	2	1.47	25	1
3194.75	7	6	6	0.194	25	1
3209.64	7	8	8	0.286	25	1
3348.08	4	2	4	1.03	25	1
3349.11	4	4	6	1.23	25	1
3354.31	8	4	2	0.69	25	1
3362.63	8	4	4	0.69	25	1
3375.50	8	4	6	0.68	25	1
3385.55	3	6	8	1.06	25	1
3390.37	3	2	2	0.88	25	1
3396.83	3	4	4	0.56	25	1
3411.76	2	4	6	1.15	25	1
3489.84	14	4	6	0.99	25	1
3560.42	12	4	6	1.08	25	1
3563.36	12	6	8	1.15	25	1
3729.03	6	6	8	0.69	25	1
3744.73	6	6	6	0.194	25	1
3758.45	6	8	8	0.112	25	1
3995.17	10	6	6	0.215	25	1
†4568	15	14	10	0.124	25	1
[4652.5]	13	2	2	0.301	25	1
[4685.4]	13	2	4	0.295	25	1
4772.57	9	2	4	0.128	25	1
4779.09	9	2	2	0.254	25	1
5417	13	3	5	0.218	25	1
5432	13	5	3	0.361	25	1
5473	13	5	5	0.64	25	1
5573	3	1	3	0.107	25	9
5582	3	3	5	0.145	25	9
5600	3	5	7	0.190	25	9
6329	14	5	7	0.136	25	1
6790	12	3	5	0.057	25	1
6830	12	5	7	0.075	25	1
†7438	17	3	9	0.295	25	8
O VI:						
†129.84	5 uv	6	10	292	10	17
†150.10	2 uv	2	6	259	10	1
172.935	4 uv	2	4	737	10	1
183.937	3 uv	2	2	56.7	10	1
184.117	3 uv	4	2	113	10	1
1031.95	1 uv	2	4	4.09	3	1
1037.63	1 uv	2	2	4.02	3	1
†3068	2	2	6	0.865	10	17
†3314	4	6	10	2.01	10	17
†3426	6	10	14	3.34	10	1
†3509	5	10	6	0.868	10	17
†3622	3	6	2	2.70	10	17
3811.35	1	2	4	0.513	10	1
3834.24	1	2	2	0.503	10	1
O VII:						
18.627	—	1	3	9370	3	1
21.602	—	1	3	33000	3	1
†120.331	—	3	9	533	3	1
[128.25]	—	1	3	504	3	1
†128.46	—	9	15	1620	3	1

* For references see pp. 7-208 and 7-209.

TABLE 7i-4. TRANSITION PROBABILITIES FOR ALLOWED LINES (Continued)

Wavelength, Å	Multiplet no.	Statistical weights g_i	Statistical weights g_k	Transition probability A_{ki}, 10^8 s^{-1}	Accuracy, %	Source*
Oxygen (Continued)						
[135.77]	—	3	5	1530	3	1
1623.29	—	3	5	0.805	3	1
1637.96	—	3	3	0.784	3	1
1639.58	—	3	1	0.781	3	1
[2475.4]	—	1	3	0.246	3	1
Fluorine						
F I:						
6239.64	3	6	4	0.29	50	1
6348.50	3	4	4	0.18	50	1
6413.66	3	2	4	0.090	50	1
6773.97	2	6	6	0.14	50	1
6834.26	2	4	4	0.24	50	1
6856.02	2	6	8	0.45	50	1
6870.22	2	2	2	0.38	50	1
6902.46	2	4	6	0.31	50	1
6909.82	2	2	4	0.18	50	1
6966.35	6	4	2	0.16	50	1
7037.45	6	4	4	0.38	50	1
7127.88	6	2	2	0.30	50	1
7202.37	6	2	4	0.072	50	1
7311.02	5	4	2	0.27	50	1
7331.95	1	6	4	0.17	50	1
7398.68	1	6	6	0.25	50	1
7425.64	1	4	2	0.30	50	1
7489.14	5	2	2	0.13	50	1
Fluorine (Continued)						
3039.75	3	4	6	2.56	25	1
3113.58	1	2	4	0.67	50	1
3115.67	1	4	6	1.1	50	1
3121.52	1	6	8	1.6	50	1
3124.76	1	2	2	1.3	50	1
3134.21	1	4	4	0.84	50	1
3142.78	4	2	4	1.16	25	1
3145.54	1	4	2	0.26	50	1
3146.96	1	6	6	0.47	50	1
3154.39	4	4	6	1.38	25	1
3156.11	4	4	4	0.230	25	1
3174.13	2	4	6	1.7	25	1
3174.73	2	2	4	1.4	50	1
3213.97	2	4	4	0.27	50	1
Neon						
Ne I:						
735.89	2 uv	1	3	6.6	25	1
743.70	1 uv	1	3	0.476	25	1
3454.19	2	3	1	0.085	25	1
3472.57	2	5	7	0.099	25	1
3520.47	7	3	1	0.073	25	1
5433.65	—	3	3	0.0029	50	26n
5852.49	6	3	1	0.706	10	25
5881.90	1	5	3	0.102	10	25

λ (Å)	Note	gᵢ	gₖ	gf	Acc.	Source
5939.32	—	3	5	0.0021	50	26n
5944.83	1	5	5	0.112	10	25
5975.53	1	3	5	0.0349	10	25
6030.00	3	3	3	0.0512	10	25
6046.13	↓	3	3	0.0024	50	26n
6064.54	—	1	3	0.0026	58	26n
6074.34	3	1	3	0.583	10	25
6096.16	3	5	3	0.179	10	25
6118.03	—	3	5	0.0065	50	26n
6128.45	3	3	3	0.0070	25	25
6143.06	1	5	5	0.285	10	25
6163.59	5	3	1	0.141	10	25
6217.28	1	3	5	0.0601	10	25
6266.50	5	3	1	0.254	10	25
6293.74	—	3	3	0.0069	50	26n
6304.79	3	5	3	0.0424	10	25
6313.69	—	1	3	0.0053	50	26n
6328.16	—	3	5	0.037	50	26n
6334.43	1	5	5	0.180	10	25
6351.86	—	3	1	0.0037	50	26n
6382.99	3	3	3	0.316	10	25
6402.25	1	7	5	0.506	10	25
6421.71	—	1	3	0.0033	50	26n
6506.53	3	5	3	0.298	10	25
6532.88	5	3	1	0.106	50	25
6598.95	6	3	3	0.225	10	25
6678.28	6	5	3	0.231	10	25
6717.04	6	3	3	0.217	10	25
6929.47	6	5	3	0.174	10	25
7032.41	1	3	5	0.253	10	25
7173.94	6	5	3	0.0321	10	25
7245.17	3	3	3	0.100	10	25
7552.24	1	4	6	0.10	50	1
7573.41	1	2	4	0.14	50	1
7754.70	1	4	6	0.35	50	4
7800.22	1	2	4	0.29	50	4
F II:						
3202.74	1	5	5	1.4	50	8
†3504.0	1	15	25	2.86	25	3
[3535.2]	1	3	1	2.1	50	6
3536.84	1	5	3	1.5	50	6
[3538.6]	1	3	3	0.51	50	6
3541.77	1	7	5	1.7	50	6
[3544.5]	1	5	5	0.31	50	6
†3641.7	1	21	21	0.147	25	11
3847.09	1	5	7	1.3	50	1
3849.99	1	5	5	1.3	50	1
3851.67	1	5	3	1.3	50	1
4024.73	1	3	5	1.2	50	2
4025.01	1	3	1	1.2	25	2
4025.50	1	3	3	1.2	50	2
†4103.4	1	9	15	2.05	50	4
4109.17	1	7	7	1.6	50	5
4116.55	1	5	5	1.2	50	5
[4117.1]	1	5	3	0.45	50	5
[4118.8]	1	3	5	0.27	50	5
4119.22	1	3	3	1.3	25	5
†4246.16	1	25	35	2.47	25	9
4299.18	1	5	7	1.7	50	7
†4446.9	1	15	21	2.35	25	10
F III:						
3034.54	1	6	6	0.184	25	3
3039.25	1	6	8	2.75	25	3

* For references see pp. 7-208 and 7-209.

TABLE 7i-4. TRANSITION PROBABILITIES FOR ALLOWED LINES (Continued)

Neon (Continued)

Wavelength, Å	Multiplet no.	g_i	g_k	Transition probability A_{ki}, 10^8 s^{-1}	Accuracy, %	Source*
7304.82	—	1	3	0.0030	50	26n
7438.90	5	1	3	0.0242	10	25
7488.87	—	3	5	0.349	25	1
8377.61	12	7	9	0.51	25	1
8495.36	18	5	7	0.357	25	1
8654.38	33	5	7	0.445	25	1
Ne II:						
[2858.0]	—	6	6	0.91	50	1
[2870.0]	—	6	6	0.11	50	1
[2873.0]	—	6	4	0.46	50	1
[2910.4]	—	2	4	0.43	50	1
[2925.7]	—	2	2	0.52	50	1
[2955.7]	4	6	4	1.2	50	1
3001.65	4	4	4	0.78	50	1
3034.48	8	6	8	3.1	50	1
3037.73	8	4	4	2.0	50	1
3045.58	8	2	2	2.5	50	1
3047.57	8	4	6	1.8	50	1
3054.69	8	2	4	0.93	50	1
3118.02	16	8	6	0.11	50	1
3169.30	16	6	4	0.17	50	1
3248.15	15	4	4	0.14	50	1
3255.39	23	6	4	0.12	50	1
3263.43	15	2	4	0.36	50	1

Neon (Continued)

Wavelength, Å	Multiplet no.	g_i	g_k	Transition probability A_{ki}, 10^8 s^{-1}	Accuracy, %	Source*
[4292.4]	57	10	10	0.20	50	1
[4346.9]	57	8	8	0.33	50	1
4379.50	56	8	8	0.20	50	1
4385.00	56	6	6	0.18	50	1
4391.94	57	8	10	2.2	50	1
4397.94	56	10	10	0.24	50	1
4409.30	57	6	8	2.0	50	1
4413.20	57	4	6	2.0	50	1
Ne III:						
2086.96	—	3	3	2.96	25	27
2087.44	—	5	3	0.99	25	27
2088.92	—	3	5	0.59	25	27
2089.43	—	5	5	2.73	25	27
2095.54	—	7	7	3.47	25	27
†2413.0	11 uv	9	15	4.87	25	27
2590.04	11 uv	5	7	1.69	25	27
2593.60	11 uv	5	5	1.69	25	27
2595.68	—	5	3	1.69	25	27
2610.03	—	7	9	2.01	25	27
2613.41	—	5	7	1.78	25	27
2615.87	—	3	5	1.68	25	27
†2678.2	12 uv	3	9	2.70	25	27

λ (Å)		gᵢ	gₖ	gA	Acc	Ref
3297.74	2	6	6	0.53	50	1
3323.75	7	4	4	1.56	25	1n, 27
3453.10	21	4	4	0.59	50	1
3456.68	28	2	4	1.0	50	1
3503.61	28	2	2	1.9	50	1
3551.52	24	2	4	0.055	50	1
3557.84	6	2	2	0.51	25	1n, 27
3561.23	31	4	6	0.11	50	1
3565.84	34	4	4	0.82	50	1
3568.53	9	6	8	1.14	25	1n, 27
3571.26	31	4	4	0.43	50	1
3590.47	32	4	6	0.087	50	1
3594.18	34	4	2	1.3	50	1
3612.35	26	2	4	0.22	50	1
3628.06	41	4	4	0.57	50	1
3632.75	33	4	4	0.090	50	1
3659.93	33	4	6	0.11	50	1
3664.09	1	6	4	0.51	25	1n, 27
3679.80	41	4	2	0.36	50	1
3694.22	1	6	6	0.73	25	1n, 27
3697.09	41	2	2	0.34	50	1
3701.81	40	4	6	0.25	50	1
3709.64	1	4	2	0.84	25	1n, 27
3713.09	5	4	6	1.19	25	1n, 27
3766.29	1	4	6	0.245	25	1n, 27
3800.02	39	4	4	0.35	50	1
3818.44	39	2	4	0.69	50	1
3829.77	39	4	6	0.88	50	1
4219.76	52	8	8	0.33	50	1
4231.60	52	6	6	0.22	50	1
4290.40	57	10	12	2.5	50	1

λ (Å)		gᵢ	gₖ	gA	Acc	Ref
Ne IV:						
541.124	1 uv	4	2	15.2	25	23
542.076	1 uv	4	4	15.2	25	23
543.884	1 uv	4	6	15.2	25	23
2018.44	—	4	4	3.7	50	1
2022.19	—	6	6	3.8	50	1
[2174.4]	—	2	4	1.3	50	1
[2176.1]	—	4	6	0.96	50	1
2203.88	—	6	6	2.2	50	1
[2206.4]	—	4	2	2.5	50	1
2220.81	—	6	4	1.4	50	1
2258.02	—	6	8	2.7	50	1
2262.08	—	6	6	2.7	50	1
2264.54	—	4	4	2.7	50	1
2285.79	—	6	8	2.8	50	1
2293.49	—	4	6	2.6	50	1
2350.84	—	2	4	1.0	50	1
2352.52	—	4	8	1.7	50	1
2357.96	—	6	8	2.5	50	1
2372.16	—	4	4	1.3	50	1
2384.95	—	6	6	0.72	50	1
Ne V:						
563.418	1 uv	1	3	6.7	25	23
569.759	1 uv	3	3	4.97	25	23
569.830	1 uv	5	5	8.9	25	23
572.106	1 uv	5	5	2.94	25	23
572.336	1 uv	5	7	11.7	25	23
2227.42	—	5	7	0.13	50	1
2232.41	—	7	9	0.20	50	1
2259.57	—	3	5	1.7	50	1
2263.39	—	1	3	1.2	50	1
2265.71	—	5	7	2.2	50	1

TABLE 7i-4. TRANSITION PROBABILITIES FOR ALLOWED LINES (Continued)

Neon (Continued)

Wavelength, Å	Multiplet no.	g_i	g_k	Transition probability A_{ki}, 10^8 s^{-1}	Accuracy, %	Source*
2282.61	—	3	3	0.89	50	1
2306.31	—	5	5	0.52	50	1
Ne VI:						
†122.62	—	6	10	1400	50	1
2042.38	—	2	4	2.73	25	1
2055.93	—	2	2	2.68	25	1
[2213.1]	—	2	4	1.54	25	1
Ne VIII:						
†88.11	—	2	6	853	10	1
†98.20	—	6	10	2760	10	1
†103.00	—	6	2	462	10	1
770.409	—	2	4	5.72	10	1
780.324	—	2	2	5.50	10	1
†[2860.1]	—	2	6	0.696	10	1
†[8454.3]	—	6	10	0.0214	10	1
Ne IX:						
[11.558]	—	1	3	24800	3	1
13.44	—	1	3	88700	3	1
†74.4	—	3	9	1460	10	1
[82.010]	—	3	5	4180	3	1
†[1297.5]	—	3	9	0.980	3	1
[1901.5]	—	1	3	0.329	3	1

Sodium (continued)

Wavelength, Å	Multiplet no.	g_i	g_k	Transition probability A_{ki}, 10^8 s^{-1}	Accuracy, %	Source*
22083.7	—	2	2	0.062	25	2
23348.4	—	2	4	0.056	25	2
[91380]	—	4	2	0.00156	25	2
Na II:						
300.151	4 uv	1	3	30	50	2
301.432	3 uv	1	3	9.5	50	2
372.069	2 uv	1	3	3.1	50	2
Na III:						
1752.65	—	6	6	3.3	50	2
1849.58	—	6	8	7.2	50	2
1856.73	—	4	6	5.1	50	2
1935.54	—	4	6	7.0	50	2
1939.32	—	6	8	7.6	50	2
1951.21	—	6	4	2.7	50	2
1965.04	—	8	10	8.8	50	2
[1976.4]	—	4	6	8.3	50	2
1985.58	—	4	6	1.7	50	2
1995.62	—	6	6	2.0	50	2
[2004.8]	—	2	4	4.6	50	2
[2011.9]	—	6	8	8.4	50	2
[2028.6]	—	8	8	1.7	50	2
[2036.9]	—	2	2	4.4	50	2
[2045.5]	—	6	6	1.1	50	2

Sodium

Na I:

λ (Å)	Mult.	gᵢ	gₖ	gf	Acc. (%)	Ref.
†2852.8	1 uv	2	6	0.0060	25	2
3302.37	2	2	4	0.0290	25	2
3302.98	2	2	2	0.0293	25	2
4494.18	15	2	4	0.0126	25	2
4664.81	12	2	4	0.0214	25	2
4747.94	11	2	2	0.0059	25	2
4751.82	11	4	2	0.0119	25	2
4978.54	9	2	4	0.0418	25	2
5148.84	8	4	2	0.0110	25	2
5153.40	8	4	2	0.0220	25	2
5682.63	6	2	4	0.109	25	2
5889.95	1	2	4	0.630	25	2
5895.92	1	2	2	0.628	25	2
6154.23	5	2	2	0.0241	25	2
6160.75	5	4	2	0.0482	25	2
8183.26	4	2	4	0.413	25	2
†8650.3	19	2	6	0.00231	25	2
†9465.94	24	10	14	0.0079	25	2
†9961.28	23	10	14	0.0127	25	2
10749.3	18	2	2	0.0074	25	2
†10834.9	22	10	14	0.0224	25	2
11381.5	3	2	2	0.084	25	2
11403.8	3	4	2	0.167	25	2
12311.5	—	2	4	0.0108	25	2
†12679.2	21	10	14	0.0471	25	2
14767.5	—	2	4	0.0217	25	2
16373.9	—	2	2	0.0058	25	2
16388.9	—	4	2	0.0115	25	2
†18465.3	—	10	14	0.140	25	2
22056.4	—	2	4	0.062	25	2

λ (Å)		gᵢ	gₖ	value	Acc. (%)	Ref.
[2067.4]	—	4	4	2.8	50	2
[2107.7]	—	2	2	2.1	50	2
[2151.2]	—	2	4	4.4	50	2
[2174.5]	—	4	6	5.3	50	2
[2180.8]	—	4	6	3.6	50	2
[2194.8]	—	4	4	3.7	50	2
[2222.8]	—	4	8	3.5	50	2
[2230.3]	—	6	4	3.7	50	2
[2232.2]	—	4	6	3.3	50	2
[2246.7]	—	4		2.4	50	2
[2278.5]	—	2	2	2.4	50	2
[2310.0]	—	4	2	2.3	50	2
[2367.3]	—	2	2	1.1	50	2
[2459.4]	—	4	6	3.0	50	2
[2468.9]	—	2	4	2.4	50	2
[2497.0]	—	6	6	1.7	50	2
Na IV:						
319.638	—	5	3	170	50	2
360.761	—	1	3	23	50	2
410.371	—	5	5	76	50	2
Na V:						
307.152	—	4	2	270	50	2
360.319	—	2	2	52	50	2
360.367	—	4	2	100	50	2
367.557	—	4	10	120	50	2
†445.14	—	6		11	50	2
459.897	—	4	2	31	50	2
461.051	—	4	4	31	50	2
†463.263	—	4	6	31	50	2
511.193	—	4	4	68	50	2

* For references see pp. **7**-208 and **7**-209.

TABLE 7i-4. TRANSITION PROBABILITIES FOR ALLOWED LINES (Continued)

Wavelength, Å	Multiplet no.	Statistical weights g_i	g_k	Transition probability A_{ki}, 10^8 s^{-1}	Accuracy, %	Source*
Magnesium						
Mg I:						
2025.82	2 uv	1	3	1.2	50	2
2736.54	9 uv	5	7	0.207	25	2
2776.69	6 uv	3	5	1.31	25	2
2778.27	6 uv	1	3	1.76	25	2
2781.42	6 uv	3	1	5.3	25	2
2782.97	6 uv	5	3	2.16	25	2
2846.72	5 uv	1	3	0.15	50	2
2938.47	3 uv	3	3	0.052	50	2
2942.00	3 uv	5	3	0.086	50	2
3091.07	5	1	3	0.313	25	2
3329.92	4	1	3	0.034	50	2
3332.15	4	3	3	0.10	50	2
3336.67	4	5	3	0.17	50	2
3829.35	3	1	3	0.940	10	2
4351.91	14	3	5	0.21	50	2
4702.99	11	3	5	0.16	50	2
5167.32	2	1	3	0.116	10	2
5172.68	2	3	3	0.346	10	2
5183.60	2	5	3	0.575	10	2
5528.40	9	3	5	0.14	50	2
†7657.8	22	3	9	0.0148	25	2
8806.76	7	3	5	0.14	50	2
8923.57	25	1	3	0.011	50	2
Magnesium (Continued)						
Mg III:						
186.510	4 uv	1	3	170	50	2
187.194	3 uv	1	3	100	50	2
231.730	2 uv	1	3	87	50	2
Mg IV:						
[1230.3]	—	6	4	4.1	50	2
[1245.2]	—	6	6	5.9	50	2
[1246.6]	—	2	4	3.4	50	2
[1253.7]	—	4	6	2.6	50	2
[1375.4]	—	4	4	4.5	50	2
1459.52	—	6	4	4.6	50	2
1490.41	—	4	4	2.8	50	2
[1525.2]	—	4	4	6.7	50	2
[1548.1]	—	4	6	6.4	50	2
1658.92	—	6	6	1.8	50	2
1680.02	—	4	4	3.1	50	2
1698.83	—	4	6	3.9	50	2
[1703.4]	—	2	4	2.4	50	2
1874.59	—	6	4	1.8	50	2
1893.87	—	6	6	2.8	50	2
1906.71	—	4	2	3.2	50	2
1946.20	—	4	6	1.1	50	2
1956.58	—	2	4	1.5	50	2

λ (Å)	I	g_1	g_2	value	Acc	Ref
9255.78	27	5	7	0.089	25	2
9414.96	38	15	21	0.022	25	2
†10811.1	37	15	21	0.0452	25	2
10953.3	35	1	3	0.025	50	2
11828.2	6	3	1	0.26	50	2
12083.7	26	5	7	0.170	25	2
†14877.6	—	15	21	0.105	25	2
†15031	—	3	9	0.139	25	2
17108.7	—	1	3	0.094	25	2
Mg II:						
†2660.8	4 uv	10	14	0.38	50	2
2790.77	3 uv	2	4	3.94	10	2
2795.53	1 uv	2	4	2.68	10	2
2802.70	1 uv	2	2	2.66	10	2
2928.63	2 uv	2	2	1.07	25	2
2936.51	2 uv	4	2	2.15	25	2
†3104.8	6	10	14	0.81	25	2
4384.64	10	2	4	0.14	50	2
4427.99	9	2	2	0.107	25	2
4433.99	9	4	2	0.214	25	2
4481.2	4	10	14	2.25	10	2
†5264.3	17	10	14	0.125	25	2
†6346.8	16	10	14	0.216	25	2
7877.05	8	2	4	0.66	25	2
8213.99	7	2	2	0.260	25	2
8234.64	7	4	2	0.52	25	2
9218.25	1	2	4	0.359	25	2
9244.27	1	2	2	0.356	25	2
9632.2	15	10	14	0.413	25	2
10951.8	3	4	2	0.166	25	2

λ (Å)	I	g_1	g_2	value	Acc	Ref
Mg V:						
276.581	—	5	3	200	50	2
312.311	—	1	3	27	50	2
351.089	—	5	3	50	50	2
352.202	—	3	1	120	50	2
353.094	—	5	5	88	50	2
353.300	—	3	3	29	50	2
354.223	—	1	1	40	50	2
355.326	—	3	5	29	50	2
Aluminum						
Al I:						
2145.56	—	2	4	0.233	25	2
2168.83	9 uv	2	4	0.306	25	2
2367.05	4 uv	2	4	0.71	25	2
2373.12	4 uv	4	6	0.85	25	2
2373.35	4 uv	4	4	0.14	50	2
2567.98	2 uv	2	4	0.221	25	2
2575.10	2 uv	4	6	0.264	25	2
2652.48	1 uv	2	2	0.133	25	2
2660.39	1 uv	2	2	0.264	25	2
3082.15	3	4	4	0.61	25	2
3944.01	1	2	2	0.493	25	2
3961.52	1	4	2	0.98	25	2
6696.02	5	2	4	0.0169	25	2
6698.67	5	2	2	0.0169	25	2
7835.31	10	4	6	0.057	50	2
8772.87	9	4	6	0.098	50	2
10873.0	12	4	2	0.011	50	2
10891.7	12	4	2	0.022	50	2
11253.2	8	2	6	0.166	25	2
13123.4	4	4	4	0.182	25	2

* For references see pp. 7-208 and 7-209.

TABLE 7i-4. TRANSITION PROBABILITIES FOR ALLOWED LINES (*Continued*)

Wavelength, Å	Multiplet no.	Statistical weights g_i	g_k	Transition probability A_{ki}, 10^8 s^{-1}	Accuracy, %	Source*
Aluminum (Continued)						
13150.8	4	2	2	0.181	25	2
16719.0	—	2	4	0.085	25	2
16750.6	—	4	6	0.101	25	2
16763.4	—	4	4	0.017	50	2
Al II:						
†1191.0	10 uv	9	15	1.7	50	2
1539.74	2 uv	3	5	8.8	50	2
1670.81	6 uv	1	3	14.6	10	2
1719.46	5 uv	1	3	6.79	10	2
1760.10	5 uv	3	5	3.30	25	2
1761.98	5 uv	1	3	4.38	25	2
1765.81	5 uv	3	1	13.1	25	2
1767.60	5 uv	5	3	5.4	25	2
1855.95	4 uv	1	3	0.832	10	2
1858.05	4 uv	3	3	2.48	10	2
1862.34	4 uv	5	3	4.12	10	2
†1908.7	—	9	9	8.1	50	2
1931.05	—	1	1	10.8	25	2
†1963.0	—	9	15	12	50	2
1989.85	8 uv	3	5	14.7	25	2
†2193.8	—	15	21	3.1	50	2
2816.19	—	3	1	3.83	25	2
†2996.8	14	9	15	0.11	50	2
3088.52	20	3	5	0.15	50	2
†3653.0	12	9	15	0.27	50	2
Aluminum (Continued)						
3713.10	4	4	2	2.27	25	2
3980.56	12	10	14	0.229	25	2
†4150.1	5	10	14	2.19	25	2
4357.24	9	2	4	0.070	50	2
4512.54	3	2	4	2.15	25	2
4903.71	11	10	14	0.351	25	2
5696.47	2	2	4	0.882	10	2
5722.65	2	2	2	0.870	10	2
Al IV:						
129.729	—	1	3	340	50	2
[130.37	—	1	3	630	50	2
160.073	—	1	3	170	50	2
Silicon						
Si I:						
1255.28	UV 41.12	1	3	3.1	50	28
1256.49	UV 41.12	3	3	9.5	50	28
1258.80	UV 41.12	5	3	15	50	28
1637.01	UV 104	5	5	4.9	50	28
1638.28	UV 104	5	5	5.0	50	28
1640.27	UV 104	5	7	5.2	50	28
1675.21	UV 23	5	5	0.45	50	28
1845.52	UV 10	1	3	0.99	50	28

Al II (continued)

λ (Å)	Mult.	g′	g″	gf	Acc. %	Ref.
1847.47	UV 10	3	5	1.4	50	28
1848.15	UV 10	3	3	0.65	50	28
1850.67	UV 10	5	7	1.9	50	28
1852.47	UV 10	5	5	0.42	50	28
1901.34	UV 57	5	7	0.80	50	28
1977.60	UV 7	1	3	0.18	50	28
1979.21	UV 7	3	1	0.51	50	28
1983.23	UV 7	3	5	0.14	50	28
1988.99	UV 7	5	5	0.41	50	28
2054.84	UV 103	5	7	1.3	50	28
2061.19	UV 103	5	5	1.4	50	28
2065.52	UV 103	5	3	1.5	50	28
2124.12	UV 48	5	7	2.4	50	28
2207.98	UV 3	1	3	0.25	50	2, 28
2210.89	UV 3	3	5	0.34	50	2, 28
2211.74	UV 3	3	3	0.19	50	2, 28
2216.67	UV 3	5	7	0.46	50	2, 28
2435.15	UV 45	5	5	0.28	50	28
2506.90	UV 1	3	5	0.415	25	2, 28
2514.32	UV 1	1	3	0.55	25	2, 28
2516.11	UV 1	5	5	1.22	25	2, 28
2519.20	UV 1	3	3	0.415	25	2, 28
2524.11	UV 1	3	1	1.62	25	2, 28
2528.51	UV 43	5	3	0.69	25	2, 28
2881.58	UV 3	1	1	1.75	25	2
3905.52	UV 2	3	5	0.145	25	2
4102.94		1	3	0.0016	50	
4782.99	11.06	5	3	0.018	50	12
4947.61	17.09	3	1	0.041	50	12
5006.06	17.08	3	5	0.025	50	12
5645.61	10	3	5	0.0044	50	12
5665.55	10	1	3	0.011	50	12

λ (Å)	Mult.	g′	g″	gf	Acc. %	Ref.
3703.22	18	3	5	0.38	50	2
3733.91	11	3	3	0.13	50	2
3738.00	11	5	3	0.21	50	2
3866.16	17	3	1	0.37	50	2
5593.23	16	3	5	2.3	50	2
5613.19	77	5	7	0.070	50	2
†5859.7	41	15	21	0.24	50	2
†6237.4	10	9	15	1.1	50	2
6335.74	22	5	3	0.14	50	2
6816.69	9	1	3	0.11	25	2
6823.48	9	3	3	0.34	25	2
6837.14	9	5	3	0.57	25	2
6917.93	75	5	7	0.16	50	2
6919.96	15	3	1	0.96	50	2
7042.06	3	3	5	0.59	25	2
7056.60	3	3	3	0.58	25	2
7063.64	3	3	1	0.58	25	2
7449.12	98	3	5	0.12	50	2
7471.41	21	5	7	0.94	50	2
7624.48	91	1	3	0.050	50	2
†8358.2	40	15	21	0.50	50	2
8640.70	4	1	3	0.286	25	2
Al III:						
1379.67	—	2	2	4.51	25	2
1384.14	—	4	2	8.9	25	2
1605.7		2	4	12.1	10	2
1854.72	1 uv	2	4	5.67	10	2
1862.78	1 uv	2	2	5.60	10	2
†1935.88	—	10	14	12.2	25	2
3612.35	1	4	2	1.48	25	2
3702.09	4	2	2	1.14	25	2

TABLE 7i-4. TRANSITION PROBABILITIES FOR ALLOWED LINES (Continued)

Silicon (Continued)

Wavelength, Å	Multiplet no.	g_i	g_k	Transition probability A_{ki}, 10^8 s^{-1}	Accuracy, %	Source*
1250.43	UV 13.05	6	6	35	50	28
1251.16	UV 8	6	4	19	50	2
1260.42	UV 4	2	4	25	50	2
1304.37	UV 3	2	2	3.6	50	2
1309.27	UV 3	4	2	7.0	50	2
1526.72	UV 2	2	2	3.73	25	2
1533.45	UV 2	4	2	7.4	25	2
†2072.4	UV 9	10	14	1.0	50	2
2500.93	UV 18	4	6	0.38	50	2
2904.28	UV 17	4	6	0.67	50	2
3203.87	7	2	4	0.39	50	2
3333.14	6	2	2	0.15	50	2
3339.82	6	4	2	0.30	50	2
3856.02	1	6	4	0.25	50	2
3862.60	1	4	2	0.28	50	2
4128.07	3	4	6	1.32	25	2
†4621.5	7.05	10	14	0.16	50	2
5041.03	5	2	4	0.98	50	2
†5466.6	7.03	10	14	0.26	50	2
5957.56	4	2	2	0.42	50	2
5978.93	4	4	2	0.81	50	2
6347.10	2	2	4	0.70	25	2
6371.36	2	2	2	0.69	25	2

Silicon (Continued)

Wavelength, Å	Multiplet no.	g_i	g_k	Transition probability A_{ki}, 10^8 s^{-1}	Accuracy, %	Source*
5684.48	11	5	3	0.039	50	12
5701.11	10	3	1	0.031	50	12
5708.40	10	5	5	0.025	50	12
5772.15	17	3	1	0.080	50	12
5780.38	9	1	3	0.011	50	12
5793.07	9	3	5	0.014	50	12
5797.86	9	5	7	0.014	50	12
5948.55	16	3	5	0.044	50	12
6721.85	38	3	5	0.034	50	2
6976.52	60	3	5	0.023	50	2
7003.57	60	5	7	0.024	50	2
7005.88	60	7	9	0.027	50	2
7680.27	36	3	5	0.062	50	2
7918.39	57	3	5	0.054	50	2
7932.35	57	5	7	0.054	50	2
7944.00	57	7	9	0.049	50	2
8093.24	34	3	3	0.012	50	12
9413.51	14	1	1	0.29	50	12
10288.9	6	3	3	0.027	50	2
10371.3	6	3	3	0.081	50	2
10585.1	6	5	3	0.19	50	12
10603.4	5	3	5	0.048	50	12
10661.0	5	1	3	0.089	50	12
10689.7	53	3	5	0.12	50	2
10694.3	53	5	7	0.12	50	2

λ(Å)	Mult.	g_i	g_k	Value	%	Ref.
10727.4	53	7	9	0.12	50	2
10749.4	5	3	3	0.10	50	12
10786.9	5	3	1	0.24	50	12
10827.1	5	5	5	0.19	50	12
10843.9	31	3	5	0.098	50	12
10869.5	13	3	5	0.24	50	12
10979.3	5	5	3	0.042	50	12
11984.2	4	3	5	0.15	50	12
11991.6	4	1	3	0.11	50	12
12031.5	4	5	7	0.18	50	12
12103.5	4	3	3	0.061	50	12
12270.7	4	5	5	0.033	50	12
15557.8	42.21	5	5	0.013	50	2
15584.4	42.21	3	3	0.020	50	2
15888.4	11.12	3	3	0.082	50	12
15960.0	42.21	7	5	0.070	50	2
16060.0	42.21	3	1	0.083	50	2
16094.8	42.21	5	3	0.060	50	2
Si II:						
989.867	UV 6	2	4	6.7	50	2
1190.42	UV 5	2	4	7.2	50	2
1193.28	UV 5	2	2	29	50	2
1194.50	UV 5	4	4	35	50	2
1197.39	UV 5	4	2	14	50	2
1223.91	UV 8.02	4	2	20	50	28
1227.60	UV 8.02	4	4	11	50	28
1229.39	UV 8.02	6	6	24	50	28
1246.74	UV 8.01	6	8	36	50	28
1248.43	UV 8	2	4	6.3	50	2
1250.09	UV 13.05	4	4	13	50	2
‑	‑	4	4	38	50	28

λ(Å)	Mult.	g_i	g_k	Value	%	Ref.
6818.45	7.20	2	4	0.11	50	2
7113.45	7.19	2	2	0.051	50	2
7125.84	7.19	4	2	0.098	50	2
7848.80	7.02	4	6	0.39	50	2
Si III:						
883.398	UV 27	5	7	63	50	2
994.787	UV 6	3	3	7.89	10	12
997.389	UV 6	5	3	13.1	10	12
1108.37	UV 5	1	3	16.2	10	12
1140.55	UV 32	1	3	22	50	12
1141.58	UV 32	3	5	30	50	12
1142.28	UV 32	3	3	16	50	12
1144.31	UV 32	5	7	39	50	2
1144.96	UV 32	5	5	9.7	50	2
1155.00	UV 31	1	3	7.5	50	12
1155.96	UV 31	3	1	22	50	2
1156.78	UV 31	3	3	5.2	50	2
1158.10	UV 31	3	5	5.5	50	2
1160.26	UV 31	5	3	9.1	50	2
1161.58	UV 31	5	5	16	50	2
1207.52	UV 22	5	5	19	50	2
1294.54	UV 4	3	5	5.62	10	2
1296.73	UV 4	1	3	7.46	10	2
1301.15	UV 4	3	1	22.2	10	2
1303.32	UV 4	5	3	9.18	10	2
1328.81	UV 48	1	3	27	50	2
1362.37	UV 38	3	1	11	10	2
1417.24	UV 9	3	1	26.0	25	2
1435.78	UV 61	5	7	21	50	2
1588.95	UV 59	5	3	11	50	2
1778.72	UV 35	7	9	4.4	50	2
1842.55	UV 20	5	3	2.61	25	2

TABLE 7i-4. TRANSITION PROBABILITIES FOR ALLOWED LINES (Continued)

Wavelength, Å	Multiplet no.	Statistical weights g_i	Statistical weights g_k	Transition probability A_{ki}, 10^8 s^{-1}	Accuracy, %	Source*
Silicon (Continued)						
†2449.48	UV 78	15	21	1.2	50	2
2528.47	UV 81	5	7	0.81	50	2
2546.09	UV 56	5	5	0.61	50	2
2559.21	UV 55	5	7	7.7	50	2
3233.95	6	3	3	1.3	50	2
3241.62	6	5	3	2.3	50	2
†3486.91	8.06	15	21	1.8	50	2
3590.47	7	3	5	3.9	50	2
3681.40	10.09	5	3	0.33	50	2
3791.41	5	1	3	2.0	50	2
4338.50	3	1	3	0.147	25	2
4341.40	46	3	1	1.8	50	2
4494.05	15	3	3	0.46	50	2
4552.62	2	3	5	1.26	25	2
4554.00	15	5	3	0.76	50	2
4567.82	2	3	3	1.25	25	2
4574.76	2	3	1	1.25	25	2
4619.66	13	3	5	0.33	50	2
4638.28	13	1	3	0.43	50	2
4665.87	13	3	3	0.32	50	2
4683.02	13	5	5	0.95	50	2
Silicon (Continued)						
2120.18	UV 18	2	2	3.0	25	2
2127.47	UV 18	4	2	6.0	25	2
†2287.04	UV 22	10	14	6.4	25	2
†2675.2	UV 25	14	10	0.280	25	2
†2723.81	UV 32	10	14	1.1	50	2
3149.56	2	2	4	4.02	25	2
3773.15	3	4	4	2.36	25	2
4088.85	1	2	4	1.56	10	2
4116.10	5	2	2	1.54	10	2
†4212.41	5	10	14	1.72	25	2
4314.10	4	2	2	1.08	25	2
4328.18	4	4	2	2.14	25	2
†4403.73	14	10	14	0.41	50	2
6667.56	3.02	4	4	1.14	25	2
†6998.36	12	10	14	0.55	25	2
7068.41	4.01	4	2	1.00	25	2
7630.50	9	2	2	0.440	25	2
7654.56	9	4	2	0.88	25	2
†8240.61	15	14	10	0.126	25	2
8957.25	3.01	2	4	0.421	25	2
9018.16	3.01	2	2	0.413	25	2

Phosphorus

P I (continued)

λ (Å)	Mult.	g_i	g_k	gA	%	Ref.
1774.99	1 uv	4	6	2.17	25	2
1782.87	1 uv	4	4	2.14	25	2
1787.68	1 uv	4	2	2.13	25	2
†1859.2	5 uv	10	10	2.81	25	2
2136.18	4 uv	6	4	2.83	25	2
2149.14	4 uv	4	2	3.18	25	2
2152.94	9 uv	2	4	0.485	25	2
2533.99	8 uv	4	4	0.200	25	2
2535.61	8 uv	4	4	0.95	25	2
2553.25	8 uv	2	2	0.71	25	2
2554.90	8 uv	2	2	0.300	25	2
8046.79	—	4	6	0.023	50	2
8090.08	—	8	4	0.020	50	2
8637.62	—	6	2	0.079	50	2
8741.54	—	2	4	0.091	50	2
9175.85	3	2	4	0.050	50	2
9304.88	3	4	4	0.096	50	2
9525.78	3	6	6	0.14	50	2
9563.45	2	4	4	0.081	50	2
9593.54	2	2	4	0.11	50	2
9750.73	2	4	2	0.22	50	2
9790.08	4	2	4	0.045	50	2
9796.79	2	6	6	0.18	50	2
9903.74	4	2	2	0.18	25	2
9976.65	2	6	4	0.11	25	2
10084.2	4	4	4	0.21	25	2
10204.7	4	4	2	0.083	10	2
10511.4	1	2	4	0.088	10	2
10529.5	1	4	6	0.15	25	2
10581.5	1	6	8	0.21	25	2

λ (Å)	Mult.	g_i	g_k	gA	%	Ref.
4663.80	13	3	1	1.3	50	2
4716.65	8.09	5	7	2.8	50	2
4730.52	13	5	3	0.52	50	2
5473.05	12.08	5	7	0.79	50	2
5490.11	12.08	3	3	0.33	50	2
5539.93	12.08	5	5	0.19	50	2
5696.50	8.17	5	3	0.20	50	2
5704.60	8.17	7	5	0.18	50	2
5716.29	8.17	9	7	0.19	50	2
5739.73	4	1	3	0.47	50	2
6169.84	22	5	7	0.12	50	2
6314.46	10.02	3	1	1.2	50	2
6521.49	17	3	5	0.32	50	2
6831.56	10.07	5	3	0.74	50	2
7612.36	10.01	3	5	1.1	50	2
8262.57	10.06	5	7	0.91	50	2
8265.64	10.06	5	5	0.23	50	2
8269.32	10.06	3	5	0.70	50	2
8341.93	44	3	5	0.26	50	2
9799.91	8.08	5	3	0.39	50	2

Si IV:

λ (Å)	Mult.	g_i	g_k	gA	%	Ref.
†645.759	UV 15	10	14	7.0	50	2
†749.941	UV 13	10	14	14.5	25	2
815.049	UV 4	2	2	12.3	25	2
818.129	UV 4	4	2	24.4	25	2
†1066.63	UV 11	10	14	39.1	25	2
1122.49	UV 3	2	4	22.2	25	2
1393.76	UV 1	2	4	9.20	10	2
1402.77	UV 1	2	2	9.03	10	2
†1533.22	UV 24	10	14	3.57	25	2
1727.38	UV 10	4	2	5.5	25	2

* For references see pp. **7**-208 and **7**-209.

TABLE 7i-4. TRANSITION PROBABILITIES FOR ALLOWED LINES (Continued)

Phosphorus (Continued)

Wavelength, Å	Multiplet no.	Statistical weights g_i	Statistical weights g_k	Transition probability A_{ki}, 10^8 s^{-1}	Accuracy, %	Source*
10596.9	1	2	2	0.17	50	2
10681.4	1	4	4	0.11	50	2
10813.0	1	6	6	0.060	50	2
P II:						
1301.87	2 uv	1	3	0.53	25	2
1304.47	2 uv	3	1	1.57	25	2
1304.68	2 uv	3	3	0.392	25	2
1305.48	2 uv	3	5	0.392	25	2
1309.87	2 uv	5	3	0.65	25	2
1310.70	2 uv	5	5	1.17	25	2
1535.90	1 uv	3	5	0.096	25	2
1542.29	1 uv	5	7	0.127	25	2
4385.35	—	3	3	0.40	50	2
4402.09	—	1	3	0.73	50	2
4414.28	—	3	5	0.18	50	2
4417.30	—	3	3	0.55	50	2
4420.71	—	3	1	1.6	50	2
4424.07	—	3	1	0.73	50	2
4463.00	—	5	5	0.54	50	2
4467.98	—	1	3	0.25	50	2
4475.26	—	5	7	1.3	50	2
4483.68	—	3	3	0.19	50	2
4499.24	—	5	7	1.4	50	2
4530.81	—	3	5	1.0	50	2

Phosphorus (Continued)

Wavelength, Å	Multiplet no.	Statistical weights g_i	Statistical weights g_k	Transition probability A_{ki}, 10^8 s^{-1}	Accuracy, %	Source*
5583.27	—	5	3	0.19	50	2
5588.34	—	3	5	0.15	50	2
5727.71	—	3	3	0.15	50	2
6024.18	—	3	5	0.51	50	2
6034.04	—	1	3	0.37	50	2
6043.12	—	5	7	0.68	50	2
6055.50	—	5	3	0.69	50	2
6087.82	—	3	3	0.27	50	2
6165.59	—	5	5	0.16	50	2
7735.06	—	1	3	0.11	50	2
7845.63	—	3	3	0.33	50	2
P III:						
3219.32	4	2	4	3.9	50	2
†3280.22	6	10	14	1.8	50	2
3717.63	10	2	4	0.34	50	2
3744.22	10	4	4	0.68	50	2
3802.08	10	6	4	0.97	50	2
3895.03	9	4	6	0.54	50	2
3904.79	9	2	4	0.75	50	2
3951.51	9	4	2	1.4	50	2
3957.64	9	6	6	1.2	50	2
3997.17	9	6	4	0.76	50	2
4057.39	1	4	4	0.10	50	2
4059.27	1	6	4	0.90	50	2

	Acc.	gA	g_k	g_i		λ (Å)
2	50	0.99	2	4	1	4080.04
2	50	1.5	4	2	3	4222.15
2	50	1.4	2	2	3	4246.68
2	50	0.11	10	14	7	†4587.91
						P IV:
2	25	5.5	3	1	4 uv	628.983
2	25	16.5	3	3	4 uv	629.914
2	25	27.3	3	5	4 uv	631.765
2	25	24.2	1	3	—	776.366
2	10	26.3	3	1	3 uv	823.181
2	50	48	5	3	—	846.999
2	50	66	7	5	—	849.764
2	50	84	5	3	5 uv	[855.05]
2	50	26	5	5	—	866.84
2	10	39.4	3	1	1 uv	950.662
2	25	29.0	3	1	—	963.993
2	25	7.7	5	3	2 uv	1025.58
2	25	10.1	3	1	2 uv	1028.13
2	25	29.9	1	3	2 uv	1033.14
2	25	12.4	3	5	2 uv	1035.54
2	50	18	9	15	—	†1090.0
2	25	32.4	1	3	—	1118.59
2	50	7.3	7	5	1	[1847.5]
2	25	2.13	5	3	1	3347.72
2	25	2.09	3	3	—	3364.44
2	50	2.1	1	3	1	3371.10
2	50	2.0	5	7	3	[3719.3]
2	50	1.8	3	5	3	3728.67
2	25	0.84	3	1	2	4249.57

	Acc.	gA	g_k	g_i		λ (Å)
2	50	0.31	3	5	—	4533.96
2	50	0.96	5	3	—	4554.83
2	50	0.96	1	3	—	4565.27
2	50	0.33	5	5	—	4582.17
2	50	1.7	7	5	—	4588.04
2	50	1.6	5	3	—	4559.86
2	50	1.9	9	7	—	4602.08
2	50	0.30	5	5	—	4626.70
2	50	0.97	3	3	—	4628.77
2	50	0.21	7	7	—	4658.31
2	50	0.11	5	5	—	4864.42
2	50	0.19	3	3	—	4927.20
2	50	0.63	3	1	—	4935.62
2	50	0.63	5	7	—	4943.53
2	50	0.78	1	3	—	4954.39
2	50	0.58	3	5	—	4969.71
2	50	0.40	5	3	—	5040.80
2	50	0.12	3	1	—	5152.23
2	50	0.35	3	3	—	5191.41
2	50	1.0	5	3	—	5253.52
2	50	0.55	3	5	—	5296.13
2	50	0.24	5	3	—	5316.07
2	50	0.32	3	1	—	5344.75
2	50	0.11	5	3	—	5378.20
2	50	0.23	3	3	—	5386.88
2	50	0.93	1	3	—	5409.72
2	50	0.69	5	5	—	5425.91
2	50	0.33	5	5	—	5450.74
2	50	0.15	3	1	—	5443.55
2	50	0.37	3	5	—	5499.73
2	50	0.11	3	3	—	5507.19
2	50	0.45	1	3	—	5541.14

TABLE 7i-4. TRANSITION PROBABILITIES FOR ALLOWED LINES (Continued)

Phosphorus (Continued)

Wavelength, Å	Multiplet no.	g_i	g_k	Transition probability A_{ki}, 10^8 s^{-1}	Accuracy, %	Source*
P V:						
542.567	—	2	2	25	25	2
544.914	—	4	2	49	25	2
†673.90	—	10	14	97	25	2
865.435	—	2	4	31.0	25	2
[997.53]	—	4	4	1.7	25	2
997.641	—	6	4	15	25	2
1000.36	—	4	2	16	25	2
1117.98	—	2	4	12.0	25	2
1128.00	—	2	2	11.6	25	2
[1379.7]	—	2	2	6.6	25	2
1385.11	—	4	2	13	25	2
[2424.3]	—	2	4	6.4	25	2
[2440.8]	—	4	3	7.4	25	2
[2441.1]	—	4	4	1.2	25	2
3175.16	1	2	4	2.34	25	2
3204.06	1	2	2	2.28	25	2

Sulfur

Wavelength, Å	Multiplet no.	g_i	g_k	Transition probability A_{ki}, 10^8 s^{-1}	Accuracy, %	Source*
S I:						
1295.66	9 uv	5	5	4.8	50	2
1296.17	9 uv	5	3	2.4	50	2
1302.34	9 uv	3	5	1.3	50	2
1302.87	9 uv	3	3	1.1	50	2
1303.11	9 uv	3	1	4.8	50	2

Sulfur (Continued)

Wavelength, Å	Multiplet no.	g_i	g_k	Transition probability A_{ki}, 10^8 s^{-1}	Accuracy, %	Source*
†8684.2	6	15	25	0.12	50	2
†9036.7	13	9	15	0.029	50	2
9212.91	1	5	7	0.30	50	2
9228.11	1	5	5	0.28	50	2
9237.49	1	5	3	0.28	50	2
10455.5	3	3	5	0.22	50	2
10456.8	3	3	1	0.22	50	2
10459.5	3	3	3	0.22	50	2
S II:						
1124.39	8 uv	2	4	0.84	50	2
1125.00	8 uv	4	4	3.1	50	2
1131.05	8 uv	2	2	2.7	50	2
1131.65	8 uv	2	2	1.1	50	2
1234.14	7 uv	4	4	0.048	50	2
1250.50	1 uv	4	2	0.46	25	2
1253.79	1 uv	4	4	0.42	25	2
1259.53	1 uv	4	6	0.34	25	2
3567.17	56	4	4	0.35	50	2
3616.92	56	6	6	0.36	50	2
3892.32	50	6	6	0.63	50	2
3933.29	55	6	8	2.0	50	2
4032.81	59	4	6	1.2	50	2
4142.29	44	2	4	1.7	50	2
4145.10	44	4	6	1.8	50	2

λ (Å)	Mult.	g_i	g_k	gf	Acc.	Ref.
4153.10	44	6	8	2.0	50	2
4162.70	44	8	10	2.3	50	2
4165.11	64	6	6	0.74	50	2
4259.18	66	6	8	1.5	50	2
4294.43	49	6	8	1.7	50	2
4463.58	43	8	6	0.53	50	2
4483.42	43	6	4	0.31	50	2
4552.38	40	4	2	1.3	50	2
4656.74	9	2	4	0.12	50	2
4716.23	9	4	4	0.23	50	2
4792.02	46	6	6	0.37	50	2
4815.52	9	6	4	0.64	50	2
4824.07	52	6	4	0.76	50	2
4885.63	15	2	4	0.13	50	2
4917.15	15	2	2	0.55	50	2
4924.08	7	4	6	0.22	50	2
4925.32	7	2	4	0.24	50	2
4942.47	7	2	4	0.15	50	2
4991.94	7	4	4	0.15	50	2
5009.54	7	4	2	0.70	50	2
5014.03	15	4	4	0.72	50	2
5027.19	1	4	4	0.26	50	2
5032.41	7	6	6	0.66	50	2
5047.28	15	4	2	0.32	50	2
5103.30	7	6	4	0.50	50	2
5142.33	1	2	2	0.19	50	2
†5208.0	39	10	10	0.79	50	2
5320.70	38	6	8	0.84	50	2
5400.67	61	4	4	0.40	50	2
5428.64	6	2	4	0.38	50	2
5432.77	6	4	6	0.61	50	2
5453.81	6	6	8	0.78	50	2

λ (Å)	Mult.	g_i	g_k	gf	Acc.	Ref.
1303.42	—	5	3	1.9	50	2
1305.89	9 uv	1	3	1.7	50	2
†1320.0	8 uv	9	15	0.94	50	2
1401.54	6 uv	5	3	0.91	50	2
1409.37	6 uv	3	3	0.50	50	2
1412.90	6 uv	1	3	0.16	50	2
†1429.1	5 uv	9	15	3.6	50	2
1448.25	12 uv	5	3	6.9	50	2
1474.01	3 uv	5	7	1.6	50	2
1474.39	3 uv	5	5	0.57	50	2
1483.04	3 uv	3	5	1.2	50	2
1483.23	3 uv	3	3	0.75	50	2
1485.61	4 uv	1	3	0.023	50	2
1487.15	3 uv	5	3	0.89	50	2
1666.69	11 uv	5	5	5.8	25	2
1687.49	13 uv	1	3	0.94	50	2
1782.26	2 uv	1	3	1.5	50	2
1807.34	2 uv	5	3	4.1	25	2
1820.36	2 uv	3	3	2.2	25	2
1826.26	2 uv	1	3	0.73	25	2
4694.13	2	5	7	0.0076	50	2
4695.45	2	5	5	0.0074	50	2
4696.25	2	5	3	0.0072	50	2
†5278.7	4	3	9	0.0038	50	2
6403.58	9	3	5	0.0057	50	2
6408.13	9	5	5	0.0095	50	2
6415.50	9	7	5	0.013	50	2
†6751.2	8	15	25	0.079	50	2
7679.60	7	3	5	0.012	50	2
7686.13	7	5	5	0.020	50	2
7696.73	7	7	5	0.028	50	2
†8451.6	14	9	3	0.050	50	2

* For references see pp. 7-208 and 7-209.

TABLE 7i-4. TRANSITION PROBABILITIES FOR ALLOWED LINES (Continued)

Wavelength, Å	Multiplet no.	Statistical weights g_i	g_k	Transition probability A_{ki}, 10^8 s^{-1}	Accuracy, %	Source*
Sulfur (Continued)						
5473.59	6	2	2	0.74	50	2
5509.67	6	4	4	0.39	50	2
5526.22	11	8	8	0.081	50	2
5536.77	11	4	6	0.066	50	2
5564.94	6	6	6	0.16	50	2
5578.85	11	6	6	0.074	50	2
5606.11	11	10	8	0.30	50	2
5616.63	11	4	4	0.083	50	2
5639.96	14	4	6	0.75	50	2
5646.98	14	2	4	0.68	50	2
5659.95	11	6	4	0.34	50	2
5664.73	11	4	2	0.38	50	2
6305.51	19	8	6	0.18	50	2
6312.68	26	6	4	0.20	50	2
7967.43	12	2	2	0.080	50	2
8314.73	12	4	2	0.16	50	2
S III:						
2460.50	17 uv	5	5	0.45	50	2
2489.59	17 uv	3	3	0.77	50	2
2496.24	17 uv	7	5	2.5	50	2
2499.08	17 uv	3	1	3.1	50	2
2508.15	17 uv	5	3	2.3	50	2
Sulfur (Continued)						
4332.71	4	1	3	0.64	50	2
4340.30	4	3	3	0.48	50	2
4361.53	4	5	5	0.28	50	2
S IV:						
551.17	—	2	2	20.6	25	2
3097.46	1	2	4	2.6	50	2
3117.75	1	2	2	2.5	50	2
S V:						
437.37	4 uv	1	3	11.2	25	2
438.19	4 uv	3	3	33.3	25	2
439.65	4 uv	5	3	55	10	2
658.262	3 uv	1	3	36.2	10	2
786.476	1 uv	1	3	52.5	10	2
849.241	2 uv	3	5	10.7	25	2
852.185	2 uv	1	3	14.1	25	2
857.872	2 uv	3	1	41.4	25	2
860.462	2 uv	5	3	17.1	25	2
S VI:						
†464.654	5 uv	10	14	202	25	2
706.480	3 uv	2	4	41.7	25	2
712.682	3 uv	4	6	48.5	25	2
712.844	3 uv	4	4	8.1	50	2
933.382	1 uv	2	4	16.3	25	2
944.517	1 uv	2	2	15.7	25	2

Chlorine

Cl I:

λ (Å)	Mult.	g	g	A	%	Code
1201.36	—	4	2	2.39	25	2
1335.72	2 uv	2	4	1.74	25	2
1347.24	2 uv	4	4	4.19	25	2
1351.66	2 uv	2	2	3.23	25	2
1363.45	2 uv	4	2	0.75	25	2
4323.35	9	4	4	0.011	50	2
4363.30	8	6	2	0.0067	50	2
4369.52	8	4	4	0.0070	50	2
4379.90	7	4	6	0.012	50	2
4438.48	6	6	6	0.014	50	2
4469.37	15	2	4	0.016	50	2
4475.31	7	6	4	0.0043	50	2
4526.20	15	4	4	0.041	50	2
4601.00	15	2	2	0.039	25	2
4661.22	15	4	2	0.010	50	2
4691.53	—	2	4	0.011	50	2
4976.62	—	4	4	0.0035	50	2
5099.80	—	2	2	0.0085	50	2
7256.63	5	4	6	0.19	50	2
7414.10	4	4	6	0.047	50	2
7547.06	5	4	4	0.13	50	2
7717.57	4	4	4	0.030	50	2
7744.94	5	4	2	0.065	50	2
7769.18	—	6	6	0.045	50	2
7821.35	—	8	6	0.095	50	2
7830.76	—	4	4	0.069	50	2
7878.22	3	6	6	0.018	50	2

λ (Å)	Mult.	g	g	A	%	Code
2636.88	19 uv	3	5	0.45	50	2
2665.40	19 uv	5	5	1.4	50	2
2680.47	19 uv	1	3	0.62	50	2
2691.68	19 uv	3	3	0.46	50	2
2702.76	19 uv	3	1	1.9	50	2
2718.88	16 uv	3	3	1.2	50	2
2721.40	19 uv	5	3	0.77	50	2
2726.82	20 uv	3	5	0.60	50	2
2731.10	16 uv	5	5	1.1	50	2
2741.01	16 uv	5	3	0.39	50	2
2756.89	16 uv	7	7	1.4	50	2
2775.25	16 uv	7	5	0.24	50	2
2785.49	20 uv	3	3	0.61	50	2
2856.02	15 uv	5	7	5.1	50	2
2863.53	15 uv	7	9	5.7	50	2
2872.00	15 uv	3	5	4.7	50	2
2904.31	15 uv	7	7	0.61	50	2
2950.23	18 uv	3	5	3.0	50	2
2964.80	18 uv	5	7	4.0	50	2
2985.98	18 uv	5	5	0.99	50	2
3662.01	6	3	3	0.64	50	2
3717.78	6	5	3	1.0	50	2
3778.90	5	3	5	0.44	50	2
3831.85	5	1	3	0.56	50	2
3837.80	5	3	3	0.42	50	2
3838.32	5	5	5	1.3	50	2
3860.64	5	3	1	1.6	50	2
3899.09	5	3	3	0.67	50	2
4253.59	4	5	7	1.2	50	2
4284.99	4	3	5	0.90	50	2

* For references see pp. 7-208 and 7-209.

TABLE 7i-4. TRANSITION PROBABILITIES FOR ALLOWED LINES (Continued)

Chlorine (Continued)

Wavelength, Å	Multiplet no.	Statistical weights g_i	Statistical weights g_k	Transition probability A_{ki}, 10^8 s^{-1}	Accuracy, %	Source*
7899.28	—	4	6	0.058	50	2
7915.09	—	2	2	0.061	50	2
7924.62	4	2	4	0.021	50	2
7935.00	—	6	8	0.046	50	2
7976.95	—	2	4	0.041	50	2
7997.80	3	4	4	0.021	50	2
8085.54	—	4	4	0.38	50	2
8086.67	—	6	6	0.40	50	2
8212.00	2	6	6	0.079	50	2
8333.29	2	4	4	0.16	50	2
8375.95	2	6	8	0.28	50	2
8428.25	2	2	2	0.24	50	2
8550.46	13	4	2	0.019	50	2
8575.25	2	2	4	0.12	50	2
8948.01	1	6	4	0.12	50	2
9073.15	12	4	2	0.19	50	2
9121.10	1	6	6	0.17	50	2
9191.67	1	4	2	0.21	50	2
9584.77	1	4	6	0.066	50	2
9592.20	11	4	6	0.24	50	2
9632.37	12	2	2	0.083	50	2
9702.35	1	2	4	0.091	50	2
9875.95	11	2	4	0.19	50	2

Chlorine (Continued)

Wavelength, Å	Multiplet no.	Statistical weights g_i	Statistical weights g_k	Transition probability A_{ki}, 10^8 s^{-1}	Accuracy, %	Source*
4130.86	60	5	5	0.41	50	2
4132.48	29	5	5	1.6	50	2
4133.66	60	3	3	0.45	50	2
4147.09	60	7	7	0.53	50	2
4208.03	43	5	5	1.1	50	2
4224.92	83	7	5	0.82	50	2
4241.38	24	5	5	0.60	50	2
4253.51	24	7	5	0.84	50	2
4261.22	66	5	3	0.83	50	2
4270.61	66	7	5	0.74	50	2
4276.51	66	9	7	0.76	50	2
4291.76	19	3	1	1.0	50	2
4304.07	19	3	3	0.25	50	2
4307.42	19	5	3	0.76	50	2
4336.26	19	5	5	0.15	50	2
4343.62	19	7	5	0.84	50	2
4399.14	46	3	3	1.3	50	2
4569.42	35	3	3	0.55	50	2
4768.68	40	3	5	0.77	50	2
4778.93	40	3	3	0.43	50	2
4785.44	40	5	5	0.26	50	2
4794.54	1	5	7	1.18	25	2
4810.06	1	5	5	1.13	25	2

λ(Å)	Ref	g	g	gf	Acc	
Cl II:						
1063.83	1 uv	5	3	0.482	25	2
1071.05	1 uv	5	5	0.85	25	2
1071.76	1 uv	3	3	0.285	25	2
1079.08	1 uv	3	5	0.277	25	2
2546.94	13 uv	3	5	0.58	50	2
2549.85	13 uv	5	7	0.76	50	2
2906.25	14 uv	3	3	0.86	50	2
3022.93	57	3	5	0.60	50	2
3231.75	73	7	5	0.12	50	2
3315.44	37	3	5	1.1	50	2
3329.12	37	5	7	1.5	50	2
3522.14	64	7	7	1.4	50	2
3568.04	78	5	5	1.2	50	2
3618.88	77	5	3	1.2	50	2
3639.19	77	3	3	0.72	50	2
3781.23	72	7	7	0.87	50	2
3798.80	62	5	7	1.6	50	2
3805.24	62	7	9	1.8	50	2
3809.51	62	3	5	1.5	50	2
3850.97	25	5	7	1.8	50	2
3854.75	84	3	5	2.2	50	2
3868.62	84	7	9	2.7	50	2
3883.80	55	3	5	0.33	50	2
3913.92	68	9	9	0.82	50	2
3916.70	68	7	7	0.74	50	2
3917.57	68	5	5	0.78	50	2
3954.21	82	5	5	1.1	50	2
3990.19	76	5	7	0.84	50	2
4020.06	76	3	5	0.62	50	2
4036.53	76	1	3	0.46	50	2
4811.57	74	5	7	0.34	50	2
4857.04	74	3	5	0.25	50	2
4896.77	17	7	9	0.88	50	2
4904.76	17	5	7	0.81	50	2
4907.17	39	3	3	0.32	50	2
4914.32	17	7	7	0.10	50	2
4917.72	17	3	5	0.75	50	2
4922.14	17	5	5	0.14	50	2
5068.10	16	5	7	0.097	50	2
5078.25	16	7	7	0.77	50	2
5098.34	16	3	5	0.13	50	2
5099.30	16	3	3	0.64	50	2
5103.04	16	5	5	0.59	50	2
5104.08	16	5	3	0.21	50	2
5113.36	16	7	5	0.13	50	2
5221.34	3	3	3	0.77	25	2
5392.12	28	5	7	0.89	50	2
5443.42	2	7	5	0.15	50	2
5444.25	2	5	5	0.095	50	2
5444.99	2	3	5	0.024	50	2
5456.27	2	5	3	0.084	50	2
5568.81	80	5	5	0.50	50	2
6094.65	26	5	3	0.53	50	2
Cl III:						
2253.07	15 uv	6	6	0.61	50	2
2268.95	15 uv	4	4	1.1	50	2
2278.34	15 uv	2	2	1.8	50	2
2283.93	15 uv	8	6	2.7	50	2
2298.51	19 uv	4	4	4.2	50	2

* For references see pp. 7-208 and 7-209.

TABLE 7i-4. TRANSITION PROBABILITIES FOR ALLOWED LINES (Continued)

Wavelength, Å	Multiplet no.	g_i	g_k	Transition probability A_{ki}, 10^8 s^{-1}	Accuracy, %	Source*
Chlorine (Continued)						
2340.64	19 uv	6	6	4.2	50	2
2370.37	24 uv	8	6	2.8	50	2
2403.32	17 uv	6	6	1.4	50	2
2416.42	17 uv	2	4	0.88	50	2
2484.27	13 uv	4	4	0.73	50	2
2486.91	21 uv	4	6	0.68	50	2
2504.23	13 uv	6	6	1.0	50	2
2510.92	13 uv	6	4	0.63	50	2
2519.45	13 uv	8	8	1.5	50	2
2531.76	22 uv	2	4	4.4	50	2
2532.48	22 uv	4	6	5.3	50	2
2577.13	18 uv	4	6	4.3	50	2
2580.67	18 uv	6	8	4.7	50	2
2601.16	12 uv	2	4	4.6	50	2
2603.59	12 uv	4	6	5.0	50	2
2609.50	12 uv	6	8	5.7	50	2
2616.97	12 uv	8	10	6.6	50	2
2618.78	12 uv	4	4	1.8	50	2
2624.71	23 uv	6	4	0.44	50	2
2651.19	12 uv	8	8	0.92	50	2
2661.65	16 uv	4	6	3.4	50	2
[2662.3]	16 uv	2	4	2.0	50	2
[2663.2]	16 uv	2	2	4.0	50	2
2665.54	16 uv	6	8	4.8	50	2
2669.6]	16 uv	4	4	2.6	50	2
Chlorine (Continued)						
[3071.4]	—	1	3	1.3	50	2
[3076.7]	—	5	7	2.3	50	2
[3106.0]	—	3	3	0.92	50	2
[3167.9]	—	5	5	0.52	50	2
Cl V:						
390.148	—	2	2	40.0	25	2
392.433	—	4	2	79	25	2
Argon						
A I:						
866.80	—	1	3	3.13	25	2
876.06	—	1	3	2.70	25	2
1048.22	—	1	3	5.1	25	2
1066.66	—	1	3	1.19	25	2
4044.42	—	3	5	0.00346	25	2
4158.59	—	5	5	0.0145	25	2
4181.88	—	1	3	0.0058	25	2
4198.32	—	3	1	0.0276	25	2
4200.67	—	5	7	0.0103	25	2
4259.36	—	3	1	0.0415	25	2
4266.29	—	3	5	0.00333	25	2
4272.17	—	3	3	0.0084	25	2
4300.10	—	3	5	0.00394	25	2
4333.56	—	3	5	0.0060	25	2
4335.34	—	3	3	0.00387	25	2

λ (Å)	Int.	g	g	gA	Acc.	Ref.
4345.17	—	3	3	0.00313	25	2
4510.73	—	3	1	0.0123	25	2
4887.95	—	3	3	0.014	50	2
4894.69	—	3	1	0.019	50	2
5151.39	—	3	1	0.0249	25	2
5162.29	—	3	3	0.0198	25	2
5187.75	—	3	5	0.0138	25	2
5495.87	—	7	9	0.0176	25	2
5558.70	—	3	5	0.0148	25	2
5606.73	—	3	3	0.0229	25	2
5650.70	—	3	1	0.0333	25	2
5882.62	—	3	1	0.0128	25	2
5888.58	—	7	5	0.0134	25	2
5912.09	—	3	3	0.0105	25	2
5928.81	—	5	3	0.011	50	2
5971.60	—	3	1	0.011	50	2
6032.13	—	7	9	0.0246	25	2
6043.22	—	5	7	0.0153	25	2
6105.64	—	5	5	0.0126	25	2
6416.31	—	3	5	0.0121	25	2
6752.84	—	3	5	0.0201	25	2
6871.29	—	3	3	0.0290	25	2
6965.43	—	5	3	0.067	25	2
7030.25	—	7	5	0.0278	25	2
7067.22	—	5	5	0.0395	25	2
7063.73	—	5	3	0.021	50	2
7158.83	—	3	1	0.022	50	2
7206.98	—	5	3	0.0258	25	2
7311.72	—	3	3	0.018	50	2
7316.01	—	3	3	0.010	50	2
7350.78	—	3	1	0.012	50	2
7372.12	—	7	9	0.020	50	2

λ (Å)	Int.	g	g	gA	Acc.	Ref.
2710.37	20 uv	4	6	3.5	50	2
2965.56	11 uv	6	4	2.7	50	2
2991.82	11 uv	4	2	3.0	50	2
3104.46	3	2	4	0.44	50	2
3139.34	3	4	4	0.86	50	2
3191.45	3	6	4	1.2	50	2
3244.44	6	2	4	0.41	50	2
3259.32	6	2	2	1.6	50	2
3283.41	2	4	6	0.68	50	2
3289.80	2	2	4	0.93	50	2
3320.57	6	4	4	1.9	50	2
3336.16	6	4	2	0.76	50	2
3387.60	2	2	4	0.93	50	2
3392.89	11	6	4	1.9	50	2
3393.45	11	6	6	1.9	50	2
3530.03	10	6	8	1.8	50	2
3560.68	10	4	6	1.7	50	2
3612.85	1	6	8	1.7	50	2
3622.69	1	4	6	1.2	50	2
	1	2	4	0.70	50	2
3656.95	1	2	2	1.4	50	2
3670.28	1	4	4	0.86	50	2
3682.05	1	6	6	0.48	50	2
3720.45	5	4	6	1.7	50	2
3748.81	5	2	4	1.3	50	2

Cl IV:

λ (Å)	Int.	g	g	gA	Acc.	Ref.
1532.19	—	7	5	6.3	50	2
1539.30	—	5	3	5.6	50	2
1617.43	—	5	5	3.5	50	2
[2782.4]	—	5	5	2.3	50	2
[3063.1]	—	3	5	1.7	50	2

* For references see pp. 7-208 and 7-209.

TABLE 7i-4. TRANSITION PROBABILITIES FOR ALLOWED LINES (Continued)

Argon (Continued)

Wavelength, Å	Multiplet no.	Statistical weights g_i	Statistical weights g_k	Transition probability A_{ki}, 10^8 s^{-1}	Accuracy, %	Source*
7383.98	—	3	5	0.087	25	2
7435.33	—	5	5	0.0094	50	2
7503.87	—	3	1	0.472	25	2
7514.65	—	3	1	0.430	25	2
7635.11	—	5	5	0.274	25	2
7723.76	—	5	3	0.057	25	2
7724.21	—	1	3	0.127	25	2
7948.18	—	1	3	0.196	25	2
8006.16	—	3	5	0.0468	25	2
8014.79	—	5	5	0.096	25	2
8103.69	—	3	3	0.277	25	2
8115.31	—	5	7	0.366	25	2
8264.52	—	3	3	0.168	25	2
8408.21	—	3	5	0.244	25	2
8424.65	—	3	5	0.233	25	2
8521.44	—	3	3	0.147	25	2
8605.78	—	5	5	0.0108	25	2
8667.94	—	1	3	0.0280	25	2
8761.69	—	3	5	0.0099	50	2
9122.97	—	5	3	0.212	25	2
9194.64	—	3	3	0.0198	25	2
9224.50	—	3	5	0.059	25	2
9291.53	—	3	1	0.0366	25	2

Argon (Continued)

Wavelength, Å	Multiplet no.	Statistical weights g_i	Statistical weights g_k	Transition probability A_{ki}, 10^8 s^{-1}	Accuracy, %	Source*
13499.2	—	5	3	0.027	50	2
13504.0	—	5	7	0.12	50	2
13573.6	—	3	1	0.051	50	2
13599.2	—	5	5	0.025	50	2
13622.4	—	3	5	0.082	50	2
13678.5	—	3	5	0.070	50	2
13825.7	—	5	5	0.033	50	2
14093.6	—	1	3	0.048	50	2
14596.3	—	5	12	0.053	50	2
14634.1	—	7	16	0.090	50	2
14786.3	—	5	12	0.0021	50	2
15046.4	—	1	3	0.058	50	2
15172.3	—	1	3	0.015	50	2
15302.3	—	7	16	0.054	50	2
15402.6	—	7	9	0.014	50	2
15899.9	—	3	5	0.077	50	2
15989.3	—	1	3	0.021	50	2
16436.9	—	3	5	0.059	50	2
16549.8	—	3	8	0.016	50	2
16940.4	—	5	5	0.028	50	2
23844.8	—	9	7	0.012	50	2
Ar II:						
718.091	4 uv	4	2	9.5	50	2
723.361	4 uv	4	4	23	50	2

λ (Å)	Aki	g	g′	Source	Acc. (%)	Code
725.550	19	2	2	4 uv	50	2
730.929	4.5	4	2	4 uv	50	2
919.782	1.41	2	4	1 uv	25	2
932.053	0.67	2	2	1 uv	25	2
3000.44	1.5	4	4	69	50	2
3028.91	2.3	4	2	—	50	2
3093.40	4.4	6	4	—	50	2
3139.02	1.0	6	6	47	50	2
3161.37	1.8	4	2	—	50	2
3169.67	0.82	6	4	47	50	2
3181.04	0.63	4	6	47	50	2
3194.23	0.24	4	6	46	50	2
3204.32	0.40	4	4	71	50	2
3236.81	0.52	4	2	83	50	2
3243.69	2.0	2	4	47	50	2
3249.80	2.0	4	2	47	50	2
3263.57	0.35	4	2	46	50	2
3273.32	0.37	2	4	71	50	2
3281.70	0.73	2	2	47	50	2
3293.64	1.7	4	4	83	50	2
3307.23	3.4	2	2	83	50	2
3350.93	1.5	6	6	109	50	2
3366.59	0.41	2	4	83	50	2
3376.44	1.5	8	8	109	50	2
3388.53	1.9	4	2	96	50	2
3429.62	0.22	6	8	107	50	2
3432.59	0.32	4	6	107	50	2
3454.10	0.45	4	6	44	50	2
3464.13	0.37	6	6	70	50	2
3476.75	1.34	6	6	44	25	2
3509.78	2.5	2	2	44	50	2

λ (Å)	Aki	g	g′	Source	Acc. (%)	Code
9657.78	0.060	3	3	—	25	2
9784.50	0.0161	5	3	—	25	2
10470.1	0.0117	3	1	—	25	2
10478.0	0.0274	3	3	—	25	2
10506.5	0.0158	12	5	—	25	2
10673.6	0.049	5	3	—	50	2
11393.7	0.0249	1	5	—	25	2
11441.8	0.0156	3	5	—	25	2
11668.7	0.0423	5	5	—	25	2
11719.5	0.0107	3	5	—	25	2
11943.5	0.046	8	3	—	50	2
12112.2	0.035	7	7	—	50	2
12139.8	0.051	3	3	—	50	2
12343.7	0.022	7	5	—	50	2
12356.8	0.0135	12	5	—	25	2
12402.9	0.12	3	3	—	50	2
12439.2	0.055	5	3	—	50	2
12456.1	0.10	3	5	—	50	2
12487.6	0.12	5	7	—	50	2
12702.4	0.080	3	3	—	50	2
12733.6	0.012	5	5	—	50	2
12746.3	0.022	3	3	—	50	2
12802.7	0.064	5	5	—	50	2
12933.3	0.11	1	3	—	50	2
12956.6	0.083	3	3	—	50	2
13008.5	0.10	3	5	—	50	2
13214.7	0.091	1	3	—	50	2
13231.4	0.046	3	3	—	50	2
13273.1	0.17	7	5	—	50	2
13313.4	0.15	5	3	—	50	2
13367.1	0.034	3	3	—	50	2
13406.6	0.065	20	9	—	50	2

* For references see pp. 7-208 and 7-209.

Table 7i-4. Transition Probabilities for Allowed Lines (Continued)

Wavelength, Å	Multiplet no.	Statistical weights g_i	Statistical weights g_k	Transition probability A_{ki}, 10^8 s^{-1}	Accuracy, %	Source*
Argon (Continued)						
3514.39	44	4	6	1.23	50	2
3520.00	56	6	6	0.80	50	2
3535.32	44	2	4	0.82	50	2
3548.52	56	4	4	1.1	50	2
3559.51	70	6	8	3.9	50	2
3561.03	106	8	10	4.0	50	2
3565.03	57	2	4	1.1	50	2
3576.61	56	6	8	2.77	25	2
3581.61	56	2	4	1.8	50	2
3582.36	56	4	6	3.72	25	2
3588.45	56	8	10	3.39	25	2
3600.22	115	4	4	2.2	50	2
3622.14	42	4	2	0.64	50	2
3639.83	116	4	6	1.4	50	2
3650.89	43	2	4	0.12	50	2
3655.28	82	4	6	0.23	50	2
3671.01	115	4	2	0.71	50	2
3678.27	42	6	4	0.25	50	2
3680.06	116	2	4	1.2	50	2
3718.21	131	4	6	2.0	50	2
3724.52	131	6	6	0.34	50	2
3729.31	10	6	4	0.60	50	2
Argon (Continued)						
4266.53	7	6	6	0.156	25	2
4275.16	77	2	4	0.26	50	2
4277.52	32	6	4	1.0	50	2
4337.07	113	2	4	0.34	50	2
4348.06	7	6	8	1.24	25	2
4352.20	1	2	2	0.228	25	2
4362.07	39	4	6	0.057	50	2
4370.75	39	4	4	0.65	25	2
4371.33	1	6	4	0.233	25	2
4375.95	17	4	2	0.200	25	2
4379.67	7	2	2	1.04	25	2
4400.10	1	4	4	0.164	25	2
4400.99	1	8	6	0.322	25	2
4426.01	7	4	6	0.83	25	2
4448.88	127	6	6	0.65	50	2
4481.81	39	6	6	0.494	25	2
4545.05	15	4	4	0.413	10	2
4564.42	85	4	2	0.29	50	2
4579.35	17	2	2	0.82	25	2
4589.90	31	4	6	0.82	25	2
4609.56	31	6	8	0.91	25	2
4657.89	15	4	2	0.81	25	2
4721.59	85	4	4	0.15	50	2

λ (Å)	Ref.	g	g	gA	%	code
4726.86	14	4	4	0.50	25	2
4735.91	6	6	4	0.58	25	2
4764.86	15	2	4	0.575	10	2
4806.02	6	6	6	0.79	25	2
4847.82	6	4	2	0.85	25	2
4865.92	85	4	6	0.15	50	2
4879.86	14	4	6	0.78	25	2
4933.21	6	4	4	0.143	25	2
4965.07	14	2	4	0.347	25	2
5009.33	6	4	6	0.147	25	2
5062.04	6	2	4	0.221	25	2
6638.23	20	6	4	0.129	25	2
6639.74	20	4	2	0.181	25	2
6643.72	20	10	8	0.167	25	2
6684.31	20	8	6	0.113	25	2
Ar III:						
871.099	1 uv	5	3	1.20	25	2
875.534	1 uv	3	1	2.81	25	2
878.728	1 uv	5	5	2.09	25	2
879.622	1 uv	3	3	0.69	25	2
883.179	1 uv	1	3	0.91	25	2
887.404	1 uv	3	5	0.68	25	2
3024.05	4	5	7	2.6	50	2
3054.82	4	3	5	1.9	50	2
3064.77	4	3	3	1.0	50	2
3078.15	4	1	3	1.4	50	2
3285.85	1	5	7	2.0	50	2
3301.88	1	5	5	2.0	50	2
3311.25	1	5	3	2.0	50	2
3336.13	3	7	9	2.0	50	2
3344.72	3	5	7	1.8	50	2

λ (Å)	Ref.	g	g	gA	%	code
3737.89	131	6	8	2.3	50	2
3763.50	54	8	6	0.14	50	2
3765.27	42	6	4	0.98	50	2
3770.52	42	2	4	0.41	50	2
3780.84	54	8	8	0.94	50	2
3796.60	129	4	6	0.25	50	2
3799.38	54	6	4	0.23	50	2
3803.17	129	4	6	1.5	50	2
3809.46	42	4	6	0.44	50	2
3825.68	128	6	4	0.76	50	2
3826.81	54	2	6	0.15	50	2
3841.52	54	6	2	0.27	50	2
3850.58	10	4	4	0.47	50	2
3868.52	90	4	6	1.9	50	2
3872.14	54	4	4	0.19	50	2
3880.34	54	2	2	0.22	50	2
3925.72	105	6	4	1.4	50	2
3928.63	10	2	4	0.30	25	2
3932.55	90	4	4	1.1	50	2
3946.10	105	8	6	1.4	50	2
3952.73	89	4	4	0.35	25	2
3979.36	90	4	2	1.3	50	2
4013.86	2	8	8	0.107	50	2
4033.82	52	4	2	0.98	50	2
4042.90	33	8	4	1.4	50	2
4072.01	33	6	6	0.57	25	2
4131.73	32	4	2	1.4	50	2
4156.09	52	4	4	0.39	50	2
4218.67	64	4	4	0.36	50	2
4222.64	77	4	2	0.69	50	2
4226.99	113	4	6	0.41	50	2
4228.16	8	4	6	0.130	25	2

TABLE 7i-4. TRANSITION PROBABILITIES FOR ALLOWED LINES (Continued)

Wavelength, Å	Multiplet no.	g_i	g_k	Transition probability A_{ki}, 10^8 s^{-1}	Accuracy, %	Source*
Argon (Continued)						
3358.49	3	3	5	1.6	50	2
3480.55	2	7	7	1.6	50	2
3499.67	2	3	3	1.3	50	2
3503.58	2	5	5	1.2	50	2
Ar IV:						
850.602	1 uv	4	6	2.67	25	2
2640.34	5 uv	6	6	2.2	50	2
2757.92	6 uv	6	8	2.8	50	2
2776.26	4 uv	2	4	1.1	50	2
2784.47	6 uv	4	6	2.5	50	2
2788.96	4 uv	4	6	1.9	50	2
2809.44	4 uv	6	8	2.6	50	2
Potassium						
K I:						
4044.15	3	2	4	0.0124	25	2
4047.21	3	2	2	0.0124	25	2
[6911.1]	—	2	2	0.0272	25	2
[6938.8]	—	4	2	0.054	25	2
7664.91	1	2	4	0.387	10	2
7698.98	1	2	2	0.382	10	2
[8904.1]	—	4	6	0.020	50	2
Potassium (Continued)						
K III:						
2550.02	8 uv	6	4	2.0	50	2
2635.11	8 uv	4	4	1.2	50	2
2689.90	8 uv	2	4	0.60	50	2
2938.45	7 uv	6	6	0.77	50	2
2986.20	7 uv	4	4	1.3	50	2
2992.24	7 uv	6	8	2.5	50	2
[3023.4]	7 uv	2	2	2.1	50	2
3052.07	7 uv	4	6	1.7	50	2
3056.84	7 uv	2	4	1.0	50	2
[3061.2]	5	4	2	0.88	50	2
3201.95	5	4	4	1.8	50	2
3209.34	5	2	2	1.5	50	2
3278.79	1	6	4	0.86	50	2
3289.06	4	4	6	2.0	50	2
3322.40	1	6	6	1.3	50	2
[3358.5]	1	4	2	1.5	50	2
3364.22	5	2	4	0.32	50	2
3421.83	4	2	4	1.5	50	2
3468.32	1	4	6	0.48	50	2
3513.88	1	2	4	0.65	50	2

Calcium

Ca I:

Wavelength (Å)	Ref.	gᵢ	gₖ	gA	Acc. (%)	
2200.73	7 uv	1	3	0.153	25	2
2398.56	5 uv	1	3	0.167	25	2
2994.96	17	1	5	0.367	25	2
2997.31	17	3	5	0.241	25	2
2999.64	17	3	3	0.279	25	2
3000.86	17	3	1	1.58	25	2
3006.86	17	5	5	0.75	25	2
3009.21	17	5	3	0.430	25	2
3150.75	15	5	7	0.086	50	2
†3220.5	13	9	15	0.15	50	2
3344.51	11	1	3	0.151	25	2
3487.60	10	5	3	0.078	50	2
3624.11	9	1	3	0.212	25	2
3870.48	26	3	5	0.072	50	2
3973.71	6	5	3	0.175	25	2
4092.63	25	3	5	0.11	50	2
4108.53	39	5	7	0.90	50	2
4226.73	5	1	1	2.18	10	2
4283.01	5	3	5	0.434	25	2
4289.36	5	5	3	0.60	25	2
4298.99	5	3	3	0.466	25	2
4302.53	5	5	5	1.36	25	2
4307.74	5	3	1	1.99	25	2
4318.65	5	3	3	0.74	25	2
4355.08	37	5	7	0.19	50	2
4425.44	4	1	3	0.468	25	2
4434.96	4	3	5	0.63	25	2
4435.69	4	3	3	0.356	25	2
4454.78	4	5	7	0.86	25	2
4455.89	4	5	5	0.208	25	2
9597.76	10	4	6	0.033	50	2
†11022.3	9	10	14	0.066	50	2
11690.2	6	2	4	0.220	25	2
11772.8	6	4	6	0.259	25	2
12432.2	5	2	2	0.079	25	2
12522.1	5	4	2	0.156	25	2
[12526]	—	2	4	0.0045	50	2
[12540]	—	2	2	0.0045	50	2
13377.9	—	6	4	0.0037	50	2
13397.1	—	4	2	0.0041	50	2
15168.4	—	4	6	0.15	50	2
†16963	—	10	14	0.0060	50	2
[17939]	—	2	2	0.0056	50	2
[18000]	—	4	2	0.011	50	2
†18627	—	10	14	0.0088	50	2
†21945	—	10	14	0.014	50	2
[27068]	—	2	4	0.046	50	2
[27185]	—	6	4	0.0025	50	2
[27206]	—	2	2	0.045	50	2
[27226]	—	4	2	0.0029	50	2
†31162	—	10	14	0.020	50	2
[31381]	—	6	4	0.014	50	2
[31591]	—	4	2	0.015	50	2
[36363]	—	2	2	0.016	50	2
[36613]	—	4	2	0.032	50	2
[37072]	—	2	4	0.029	50	2
[37333]	—	4	4	0.0057	50	2
[37348]	—	4	6	0.034	50	2
[62068]	—	6	4	0.0078	50	2
[62436]	—	4	2	0.0083	50	2

* For references see pp. **7**-208 and **7**-209.

TABLE 7i-4. TRANSITION PROBABILITIES FOR ALLOWED LINES (Continued)

Wavelength, Å	Multiplet no.	Statistical weights g_i	Statistical weights g_k	Transition probability A_{ki}, 10^8 s^{-1}	Accuracy, %	Source*
Calcium (Continued)						
4526.94	36	5	3	0.41	50	2
4578.55	23	3	5	0.176	25	2
4685.27	51	3	5	0.080	50	2
4878.13	35	5	7	0.188	25	2
5041.62	34	5	3	0.33	50	2
5188.85	49	3	5	0.40	50	2
5261.71	22	3	3	0.15	50	2
5262.24	22	3	1	0.60	50	2
5264.24	22	5	5	0.091	50	2
5265.56	22	5	3	0.44	50	2
5270.27	22	7	5	0.50	50	2
5581.97	21	5	7	0.060	50	2
5588.76	21	7	7	0.49	50	2
5590.12	21	3	5	0.083	50	2
5594.47	21	5	5	0.38	50	2
5598.49	21	3	3	0.43	50	2
5601.29	21	7	5	0.086	50	2
5602.85	21	5	3	0.14	50	2
5857.45	47	3	5	0.66	50	2
6102.72	3	1	3	0.077	25	2
Calcium (Continued)						
6122.22	3	3	3	0.231	25	2
6163.76	20	3	3	0.056	50	2
6166.44	20	3	1	0.22	50	2
6439.07	18	7	9	0.53	50	2
6449.81	19	5	5	0.090	50	2
6462.57	18	5	7	0.47	50	2
6471.66	18	7	7	0.059	50	2
6493.78	18	3	5	0.44	50	2
6499.65	18	5	5	0.081	50	2
Ca II:						
[1329.8]	—	6	6	0.459	25	2
[1368.4]	—	6	6	0.66	25	2
1433.1	7 uv	4	6	1.01	25	2
1553.5	6 uv	4	6	1.59	25	2
1807.74	11 uv	2	4	0.412	25	2
[1814.6]	11 uv	4	4	0.081	25	2
1815.04	11 uv	4	6	0.486	25	2
1838.08	4 uv	4	6	2.44	25	2
1843.6	10 uv	2	2	0.155	25	2
1851.10	10 uv	4	2	0.308	25	2

* For references see pp. 7-208 and 7-209.

TABLE 7i-4. TRANSITION PROBABILITIES FOR ALLOWED LINES (Continued)

Calcium (Continued)

Wavelength, Å	Multiplet no.	Statistical weights g_i	Statistical weights g_k	Transition probability A_{ki}, 10^8 s^{-1}	Accuracy, %	Source*
2103.24	9 uv	2	4	0.93	25	2
2112.76	9 uv	4	6	1.10	25	2
2113.19	9 uv	4	4	0.182	25	2
2131.43	3 uv	6	4	0.018	50	2
2132.25	3 uv	4	2	0.020	50	2
2197.79	8 uv	2	2	0.313	25	2
2208.61	8 uv	2	2	0.62	25	2
3158.87	4	2	4	3.05	25	2
3179.33	4	4	6	3.59	25	2
3181.28	4	4	4	0.60	25	2
3706.03	3	2	2	0.84	25	2
3736.90	3	4	2	1.65	25	2
3933.66	1	2	4	1.50	25	2
3968.47	1	2	2	1.46	25	2
4097.12	17	2	4	0.099	50	2
4109.83	17	4	6	0.12	50	2
4110.33	17	4	4	0.019	50	2

Calcium (Continued)

Wavelength, Å	Multiplet no.	Statistical weights g_i	Statistical weights g_k	Transition probability A_{ki}, 10^8 s^{-1}	Accuracy, %	Source*
5001.49	15	2	4	0.20	50	2
5019.98	15	4	6	0.23	50	2
5285.34	14	2	2	0.078	50	2
5307.30	14	4	2	0.15	50	2
8203.2	13	2	4	0.51	25	2
8250.2	13	4	6	0.61	25	2
8256.1	13	4	4	0.10	25	2
8498.02	2	4	4	0.0111	25	2
8542.09	2	6	4	0.099	25	2
8662.14	2	4	2	0.106	25	2
9856.7	12	2	2	0.19	50	2
9933.3	12	4	2	0.38	50	2
11836.4	5	2	4	0.23	50	2
11947.0	5	2	2	0.23	50	2

* For references see pp. 7-208 and 7-209.

TABLE 7i-5. TRANSITION PROBABILITIES FOR FORBIDDEN LINES

Wavelength, Å	Multiplet no.	Statistical weights g_i	Statistical weights g_k	Transition probability A_{ki}, s⁻¹	Accuracy, %	Source*
Hydrogen						
1420.4 MHz†	—	1	3	2.87(−15)	3	1
Carbon						
C I:						
4621.5	2F	3	1	2.60(−3)	25	1
4627.3	2F	5	1	1.9(−5)	50	1
8727.4	3F	5	1	5.0(−1)	25	1
9823.4	1F	3	5	7.8(−5)	25	1
9849.5	1F	5	5	2.31(−4)	25	1
C III:						
[2000.0]	—	1	3	1.42(−3)	25	1
Nitrogen						
N I:						
5198.5	1F	4	4	1.63(−5)	25	1
5200.7	1F	4	6	6.9(−6)	25	1
N II:						
3063.0	2F	3	1	3.40(−2)	25	1
3070.8	2F	5	1	1.6(−4)	50	1
5754.8	3F	5	1	1.08	25	1
6548.1	1F	3	5	1.03(−3)	25	1
6583.6	1F	5	5	3.04(−3)	25	1
N IV:						
[1573.4]	—	1	3	1.18(−2)	25	1
Oxygen						
O I:						
[2958.4]	2F	5	1	3.7(−4)	50	1
2972.3	2F	3	1	6.7(−2)	25	1
5577.35	3F	5	1	1.34	25	1
6300.23	1F	5	5	5.12(−3)	25	1
6363.88	1F	3	5	1.64(−3)	25	1

Wavelength, Å	Multiplet no.	Statistical weights g_i	Statistical weights g_k	Transition probability A_{ki}, s⁻¹	Accuracy, %	Source*
Neon (continued)						
3342.9	2F	5	1	2.80	25	1
3868.74	1F	5	5	1.70(−1)	25	1
3967.51	1F	3	5	5.2(−2)	25	1
Ne IV:						
[1608.8]	—	4	4	1.33	25	1
[1609.0]	—	4	2	5.3(−1)	25	1
[2438.6]	—	4	4	5.6(−3)	25	1
[2441.3]	—	4	6	5.9(−4)	50	1
4714.25	1F	6	4	4.01(−1)	25	1
4715.61	1F	6	2	1.10(−1)	25	1
4724.15	1F	4	4	4.37(−1)	25	1
4725.62	1F	4	2	3.89(−1)	25	1
Ne V:						
[1575.2]	—	3	1	4.20	25	1
[1592.7]	—	5	1	6.8(−3)	50	1
2972	2F	5	1	2.60	25	1
3345.9	1F	3	5	1.38(−1)	25	1
3425.8	1F	5	5	3.82(−1)	25	1
Sodium						
Na III:						
[73294]	—	4	2	4.56(−2)	3	2
Na IV:						
[1497.5]	—	5	1	1.2(−2)	50	2
[1522.7]	—	3	1	7.6	25	2
[2803.3]	—	5	1	3.5	25	2
3319.3	1F	5	5	5.6(−1)	25	2
3445.9	1F	3	3	1.67(−1)	10	2
[90391]	—	5	3	3.04(−2)	25	2
Na V:						
[1379.4]	—	4	4	4.3	25	2
[1380.2]	—	4	2	1.7	25	2

Spectrum / λ (Å)	Type			A	Acc.	Ref.
O II:						
[2100.4]	—	4	4	1.26(−2)	25	2
[2101.5]	—	6	4	1.2(−3)	50	2
4011.2	1F	4	6	9.0(−1)	25	2
4015.3	1F	4	4	1.30	25	2
4017.5	1F	6	2	1.3(−1)	50	2
4021.6	1F	4	4	9.3(−1)	25	2

Magnesium

Spectrum / λ (Å)	Type			A	Acc.	Ref.
Mg I:						
3848.91	—	3	3	1.8(+1)	50	2
3853.96	—	5	5	2.5(+1)	50	2
3854.97	—	5	3	5.3(+1)	50	2
Mg IV:						
[44911]	—	4	2	1.98(−1)	3	2
Mg V:						
[1286.8]	—	5	1	2.7(−2)	50	2
[1317.0]	—	3	1	2.3(+1)	25	2
[2416.8]	—	5	1	4.2	25	2
[2750.4]	—	5	5	1.90	25	2
[2892.0]	—	3	5	5.5(−1)	25	2
[56164]	—	5	3	1.27(−1)	10	2

Aluminum

Spectrum / λ (Å)	Type			A	Acc.	Ref.
Al II:						
[4451.6]	—	1	3	2.88(−3)	25	2
Al V:						
[29062]	—	4	2	7.31(−1)	3	2

Silicon

Spectrum / λ (Å)	Type			A	Acc.	Ref.
Si I:						
6526.78	1F	3	3	3.55(−2)	25	2
6589.61	1F	5	5	1.1(−3)	50	2
10991.4	2F	5	5	8.0(−1)	50	2
16068.3	0.01F	3	3	9.7(−4)	25	2
16454.5	0.01F	5	5	2.74(−3)	25	2

Spectrum / λ (Å)	Type			A	Acc.	Ref.
O II:						
3726.16	1F	4	4	1.70(−4)	25	1
3728.91	1F	4	6	4.84(−5)	25	1
7318.6	2F	6	2	6.1(−2)	25	1
7319.4	2F	6	4	1.15(−1)	25	1
7329.9	2F	4	2	1.00(−1)	25	1
7330.7	2F	4	4	6.1(−2)	25	1
O III:						
[2321.1]	—	3	1	2.30(−1)	25	1
[2331.6]	—	5	1	7.1(−4)	50	1
4363.21	2F	5	1	1.60	25	1
4958.91	1F	3	5	7.1(−3)	25	1
5006.84	1F	5	5	2.10(−2)	25	1
O V:						
[1304.2]	—	1	3	6.4(−2)	25	1

Fluorine

Spectrum / λ (Å)	Type			A	Acc.	Ref.
F II:						
[2225.5]	—	5	1	1.6(−3)	50	1
[2246.6]	2F	3	1	4.90(−1)	25	1
4157.5	2F	5	5	2.10	25	1
4789.5	1F	3	5	3.82(−2)	25	1
4869.3	1F	5	5	1.21(−2)	25	1
F III:						
[2930.0]	—	4	4	1.42(−3)	25	1
[2933.1]	—	4	6	1.31(−4)	50	1
F IV:						
[1875.5]	—	3	1	1.10	25	1
[1889.3]	—	5	1	2.3(−3)	50	1
3532.2	2F	5	5	2.10	25	1
3996.3	1F	3	5	3.42(−2)	25	1
4059.3	1F	5	5	9.8(−2)	25	1

Neon

Spectrum / λ (Å)	Type			A	Acc.	Ref.
Ne III:						
[1793.8]	—	5	1	5.1(−3)	50	1
[1814.8]	—	3	1	2.20	25	1

* For references see pp. 7-208 and 7-209.
† For this line the frequency in megahertz is listed.

TABLE 7i-5. TRANSITION PROBABILITIES FOR FORBIDDEN LINES (Continued)

Wavelength Å	Multiplet no.	g_i	g_k	Transition probability A_{ki}, s⁻¹	Accuracy %	Source*
Silicon (continued)						
Si III:						
[3314.7]	—	1	3	1.82(−2)	25	2
Phosphorus						
P I:						
5332.4	2F	4	4	1.08(−1)	25	2
5339.7	2F	4	2	4.26(−2)	25	2
8787.6	1F	4	6	2.0 (−4)	50	2
8799.1	1F	4	4	2.97(−4)	25	2
[13533]	—	4	4	7.5 (−2)	25	2
[13562]	—	6	4	1.13(−1)	25	2
[13580]	—	4	2	1.01(−1)	25	2
[13609]	—	6	2	5.3 (−2)	25	2
P II:						
4669.5	2F	3	1	2.20(−1)	25	2
4736.6	2F	5	1	6.3 (−3)	50	2
7869.5	3F	5	1	2.0	50	2
11483.2	1F	3	5	6.3 (−3)	25	2
11898.2	1F	5	5	1.70(−2)	25	2
P IV:						
[2681.7]	—	1	3	7.8 (−2)	25	2
Sulfur						
S I:						
4506.9	2F	5	1	7.3 (−3)	50	2
4589.26	2F	3	1	3.5 (−1)	25	2
7725.04	3F	5	1	1.78	25	2
10819.8	1F	5	5	2.77(−2)	25	2
11305.8	1F	3	5	8.0 (−3)	25	2
S II:						
4068.60	1F	4	4	3.41(−1)	25	2
4076.35	1F	4	2	1.34(−1)	25	2
Chlorine (continued)						
Cl IV:						
3118.3	2F	3	1	2.61	25	2
3203.3	2F	5	1	3.8 (−2)	50	2
5323.29	3F	5	1	3.2	50	2
7530.54	1F	3	5	8.0 (−2)	25	2
8045.63	1F	5	5	1.97(−1)	25	2
Cl V:						
[67000]	—	2	4	2.98(−2)	3	2
Argon						
Ar II:						
[69842]	—	4	2	5.26(−2)	3	2
Ar III:						
3005.1	2F	5	1	4.3 (−2)	50	2
3109.0	3F	3	1	4.02	25	2
5191.82	3F	5	1	3.10	25	2
7135.80	1F	5	5	3.35(−2)	25	2
7751.06	1F	3	5	8.3 (−2)	25	2
[89896]	—	5	3	3.08(−2)	10	2
Ar IV:						
[2853.6]	—	4	4	2.55	25	2
[2868.2]	—	4	4	9.7 (−1)	25	2
4711.33	1F	4	6	9.6 (−3)	50	2
4740.20	1F	4	4	7.7 (−2)	25	2
7170.62	2F	4	4	9.1 (−1)	25	2
7237.26	2F	6	4	6.7 (−1)	25	2
7262.76	2F	4	2	6.8 (−1)	25	2
7332.0	2F	6	2	1.22(−1)	25	2
Ar V:						
[2691.1]	—	3	1	6.8	25	2
[2786.1]	—	5	1	8.1 (−2)	50	2
4625.54	2F	5	1	3.8	50	2

(Columns: λ | Multiplet | g_i | g_k | value | % | code)

Sulfur / Chlorine

λ	Mult.	g_i	g_k	value	%	code
6716.42	2F	4	6	4.7 (−4)	50	2
6730.78	2F	4	4	4.3 (−4)	50	2
10284.3	3F	4	4	1.75(−1)	25	2
10317.7	3F	6	4	2.14(−1)	25	2
10336.0	3F	4	2	1.98(−1)	25	2
10369.7	3F	6	2	8.7 (−2)	25	2
S III:						
3721.8	2F	3	1	8.5 (−1)	25	2
3796.7	2F	5	1	1.6 (−2)	50	2
6312.1	3F	5	1	2.54	25	2
9069.4	1F	3	5	2.49(−2)	25	2
9532.1	1F	5	5	6.4 (−2)	25	2
S V:						
[2268.0]	—	1	3	2.36(−1)	25	2
Chlorine						
Cl II:						
3583.2	2F	5	1	1.8 (−2)	50	2
3675.0	2F	3	1	1.34	25	2
6152.9	3F	5	1	2.29	25	2
8579.5	1F	5	5	1.04(−1)	25	2
9125.8	1F	3	5	2.94(−2)	25	2
Cl III:						
3342.7	2F	4	4	9.6 (−1)	25	2
3353.4	2F	4	2	3.74(−1)	25	2
5517.66	1F	4	6	1.0 (−3)	50	2
5537.6	1F	4	4	7.1 (−3)	25	2
8433.7	3F	4	4	3.90(−1)	25	2
8481.6	3F	6	4	3.64(−1)	25	2
8501.8	3F	4	2	3.51(−1)	25	2
8550.5	3F	4	2	1.08(−1)	25	2

Potassium / Calcium

λ	Mult.	g_i	g_k	value	%	code
6435.10	1F	3	5	2.23(−1)	25	2
7005.67	1F	5	5	5.2 (−1)	25	2
[78905]	—	3	5	2.73(−2)	10	2
Potassium						
K III:						
[46240]	—	4	2	1.81(−1)	3	2
K IV:						
[2593.5]	—	5	1	8.6 (−2)	50	2
[2711.2]	2F	3	1	1.04(+1)	25	2
4510.9	1F	5	1	3.9	25	2
6101.83	1F	5	5	8.3 (−1)	25	2
6794.8	—	3	5	2.01(−1)	25	2
[59757]	—	5	3	1.05(−1)	10	2
K V:						
[2495.3]	—	4	4	6.5	25	2
[2515.3]	—	4	2	2.40	25	2
4122.63	1F	4	6	6.9 (−3)	50	2
4163.30	1F	4	4	1.11(−1)	25	2
6223.4	2F	4	4	2.26	25	2
6316.6	2F	6	4	1.46	25	2
6349.5	2F	4	2	1.50	25	2
6446.5	2F	6	4	1.9 (−1)	25	2
Calcium						
Ca IV:						
[32090]	—	4	2	5.43(−1)	3	2
Ca V:						
[2280.0]	—	5	1	1.6 (−1)	50	2
[2412.3]	—	3	1	2.4 (+1)	25	2
3996.3	2F	5	1	4.6	25	2
5309.18	1F	5	5	1.94	25	2
6086.92	1F	3	5	4.31(−1)	25	2
[41651]	—	5	3	3.11(−1)	10	2

* For references see pp. 7-208 and 7-209.
† For this line the frequency in megahertz is listed.

Section 8

NUCLEAR PHYSICS

J. B. MARION, Editor

The University of Maryland

CONTENTS

8a. Nuclear Constants and Calibrations

JERRY B. MARION

University of Maryland

This section collects the various nuclear quantities that are useful in designing or analyzing experiments in nuclear physics. For an extensive collection of graphs and tables, the reader is directed to Marion and Young [1].

8a-1. Nuclear Constants in the MeV System. A complete list of fundamental physical constants and derived quantities is to be found inside the front cover of this volume. For many nuclear physics calculations, however, it is convenient to have certain of these quantities already expressed in MeV energy units. The following list has been generated by using the fundamental constants of Taylor, Parker, and Langenberg [2].

$$m_0 c^2 = 0.5110043 \text{ MeV}$$
$$M_p c^2 = 938.2595 \text{ MeV}$$
$$M_n c^2 = 939.5529 \text{ MeV}$$
$$c^2 = 931.481 \text{ MeV/amu}$$
$$h = 4.135705 \times 10^{-21} \text{ MeV-sec}$$
$$\hbar = 6.582180 \times 10^{-22} \text{ MeV-sec}$$
$$\hbar c = 1.973288 \times 10^{-11} \text{ MeV-cm}$$
$$\hbar^2 c^2 = 389.387 \text{ MeV}^2\text{-barn}$$
$$e = 3.794703 \times 10^{-7} \text{ (MeV-cm)}^{\frac{1}{2}}$$
$$e^2 = 1.439977 \times 10^{-13} \text{ MeV-cm}$$
$$e/\hbar c = 1.923036 \times 10^4 \text{ (MeV-cm)}^{-\frac{1}{2}}$$
$$(e/\hbar c)^2 = 3.698066 \times 10^8 \text{ (MeV-cm)}^{-1}$$
$$\hbar/m_0 c^2 = 1.288087 \times 10^{-21} \text{ sec}$$

8a-2. Natural Units. If $\hbar = 1$ and $c = 1$, then

Mass, energy, and impulse are in units of cm^{-1}
Angular momentum is dimensionless
$e = 1/\sqrt{137}$
$1 \text{ MeV} = 0.506 \times 10^{11} \text{ cm}^{-1}$
If $\hbar = 1$, $c = 1$, and $m_0 = 1$, then
$1 \text{ sec} = 7.764 \times 10^{20}$ natural units
$1 \text{ cm} = 2.58 \times 10^{10}$ natural units
$1 \text{ MeV} = 1.96$ natural units

8a-3. Alpha-particle Calibration Energies. Listed in Table 8a-1 are the values for alpha-particle momenta and energies recommended by Wapstra [3]. The energies have been calculated for Wapstra's $B\rho$ values by using the expression

$$E = a(B\rho)^2 + b(B\rho)^4 + c(B\rho)^6$$

where a, b, and c are [1]

$$a = 48225.33 \times 10^{-12} \text{ keV (G-cm)}^{-2}$$
$$b = -311.98 \times 10^{-24} \text{ keV (G-cm)}^{-4}$$
$$c = 4.04 \times 10^{-36} \text{ keV (G-cm)}^{-6}$$

TABLE 8a-1. ALPHA-PARTICLE CALIBRATION ENERGIES

Source	$B\rho$, G-cm	Energy, keV
Po^{210}	$331,722 \pm 15$	5304.5 ± 0.5
Bi^{211}	$370,720 \pm 40$	6621.9 ± 1.4
Po^{211}	$393,190 \pm 50$	7448.1 ± 1.9
$Bi^{212}(ThC\ \alpha_0)*$	$354,326 \pm 20$	6049.6 ± 0.7
$Bi^{212}(ThC\ \alpha_1)*$	$355,475 \pm 20$	6088.9 ± 0.7
$Po^{212}(ThC')$	$427,060 \pm 20$	8785.0 ± 0.8
Bi^{214}	$338,170 \pm 70$	5510.9 ± 2.3
Po^{214}	$399,488 \pm 16$	7688.4 ± 0.6
Po^{215}	$391,490 \pm 40$	7383.9 ± 1.5
Po^{216}	$375,050 \pm 40$	6777.3 ± 1.5
Po^{218}	$352,870 \pm 70$	6000.1 ± 2.4
Rn^{219}	$376,160 \pm 40$	6817.5 ± 1.5
Rn^{220}	$361,260 \pm 60$	6288.5 ± 2.1
Rn^{222}	$337,410 \pm 70$	5486.2 ± 2.3
Ra^{223}	$349,010 \pm 50$	5869.6 ± 1.7
Ra^{224}	$343,450 \pm 40$	5684.2 ± 1.3
Ra^{226}	$314,990 \pm 80$	4781.8 ± 2.4
Th^{227}	$354,070 \pm 60$	6040.9 ± 2.0
Th^{228}	$335,570 \pm 60$	5426.6 ± 2.0
Th^{230}	$311,960 \pm 160$	4690.3 ± 4.8

* Intensity ratio: $\alpha_0/\alpha_1 = 2.57$.

8a-4. Gamma-ray Calibration Energies. Listed in Tables 8a-2 to 8a-4 are the weighted mean values of the energies of gamma rays frequently used as calibration standards [4]. (A more comprehensive list may be found in ref. 4.) Also, relative intensities are given for Co^{56} since the gamma rays from this nucleus span such a wide energy range and are therefore of great value for both energy and efficiency calibrations. Gamma rays from both radioactive sources and nuclear reactions are given.

8a-5. Accelerator-energy Calibration Points. In order to know with precision the energy of the beam from an accelerator, unless an absolute instrument of some type is available, the beam-analyzing system must be calibrated against some accurately known energy points. One method frequently used to calibrate such analyzers is to measure a number of gamma-ray resonances and neutron thresholds to establish several points of the energy scale. Listed in Tables 8a-5 to 8a-7 are a number of energy points suitable for calibration purposes. Only the weighted mean values are given; more complete details can be found elsewhere [5].

TABLE 8a-2. GAMMA RAYS FROM RADIOACTIVE SOURCES

Source	Energy, keV	Half life	Source	Energy, keV	Half life
m_0c^2.....	511.006 ± 0.002		Cs^{137}.....	661.635 ± 0.076	30 y
Be^7......	477.57 ± 0.05	53 d	Au^{198}....	411.795 ± 0.009	2.70 d
Na^{22}.....	1274.55 ± 0.04	2.60 y		569.62 ± 0.06	
Na^{24}.....	$\begin{cases} 1368.526 \pm 0.044 \\ 2753.92 \pm 0.12 \end{cases}$	15.0 h	Bi^{207}.....	$\begin{cases} 1063.44 \pm 0.09 \\ 1769.71 \pm 0.13 \end{cases}$	30 y
Cr^{51}....	320.080 ± 0.013	27.8 d		510.723 ± 0.020	
Mn^{54}....	834.81 ± 0.03	314 d	Tl^{208}..... (ThC'')..	$\begin{cases} 583.139 \pm 0.023 \\ 2614.47 \pm 0.10 \end{cases}$	(1.91 y)
Co^{60}.....	$\begin{cases} 1173.23 \pm 0.04 \\ 1332.49 \pm 0.04 \end{cases}$	5.26 y	Am^{241}....	$\begin{cases} 26.348 \pm 0.010 \\ 59.543 \pm 0.015 \end{cases}$	433 y
Zn^{65}.....	1115.40 ± 0.12	246 d			
Y^{88}......	$\begin{cases} 898.04 \pm 0.04 \\ 1836.13 \pm 0.04 \end{cases}$	106.6 d			

TABLE 8a-3. GAMMA RAYS FROM Co^{56}

Energy, keV	Relative intensity	Energy, keV	Relative intensity
733.79 ± 0.19	0.1 ± 0.05	2015.49 ± 0.20	2.93 ± 0.16
787.92 ± 0.15	0.40 ± 0.11	2035.03 ± 0.12	7.33 ± 0.30
846.76 ± 0.05	100	2113.00 ± 0.10	0.37 ± 0.08
977.47 ± 0.13	1.52 ± 0.16	2598.80 ± 0.12	16.77 ± 0.57
1037.97 ± 0.07	13.02 ± 0.35	3009.99 ± 0.24	0.84 ± 0.16
1175.26 ± 0.13	1.86 ± 0.23	3202.25 ± 0.19	3.15 ± 0.16
1238.34 ± 0.09	69.35 ± 1.47	3253.82 ± 0.15	7.70 ± 0.34
1360.35 ± 0.09	4.38 ± 0.16	3273.38 ± 0.18	1.55 ± 0.11
1771.57 ± 0.10	15.30 ± 0.53	3452.18 ± 0.22	0.88 ± 0.10
1964.88 ± 0.45	0.72 ± 0.08	3548.11 ± 0.25	0.18 ± 0.10

TABLE 8a-4. GAMMA RAYS FROM NUCLEAR REACTIONS

Nucleus	γ-ray energy, keV	Nucleus	γ-ray energy, keV
F^{17}.........	495.33 ± 0.10	C^{13}.........	4945.46 ± 0.17^e
F^{18}.........	658.75 ± 0.7^a	N^{14}.........	5104.87 ± 0.18
O^{17}.........	870.81 ± 0.22	O^{15}.........	5240.53 ± 0.52
B^{12}.........	953.10 ± 0.60	N^{15}.........	5268.9 ± 0.2
B^{12}.........	1673.52 ± 0.60	N^{15}.........	5297.9 ± 0.2^g
N^{14}.........	2312.68 ± 0.10^b	O^{16}.........	6129.3 ± 0.4
Be^{10}........	2589.9 ± 0.25^c	Be^{10}.......	6809.4 ± 0.4^e
N^{14}.........	2792.68 ± 0.15^d	O^{16}.........	7117.02 ± 0.49
Be^{10}........	3367.4 ± 0.2^e	Pb^{209}.......	7367.5 ± 1^e
C^{12}.........	4439.0 ± 0.2^f	N^{14}.........	9173 ± 1^h
		N^{15}.........	10829.2 ± 0.4^e

a From 1.70–1.04 MeV decay.
b Doppler shifted unless formed in $O^{14}(\beta^+)N^{14}$.
c From 5.96–3.37 MeV decay (thermal neutron capture).
d From 5.10–2.31 MeV decay.
e From thermal neutron capture.
f Doppler shifted unless formed in $B^{12}(\beta^-)C^{12}$.
g Doppler shifted unless formed in $O^{15}(\beta^-)N^{15}$ or by thermal neutron capture.
h Calculated from $C^{13}(p,\gamma)N^{14}$ resonance energy (1747.6 ± 0.9 keV) and 1964 masses; value given for observation at 0 deg to beam direction.

TABLE 8a-5. PROTON RESONANCE ENERGIES

Reaction	E_R, keV	Γ, keV
$F^{19}(p,\alpha\gamma)O^{16}$	340.46 ± 0.04	2.4 ± 0.2
$F^{19}(p,\alpha\gamma)O^{16}$	872.11 ± 0.20	4.7 ± 0.2
$Al^{27}(p,\gamma)Si^{28}$	991.90 ± 0.04	0.10 ± 0.02
$C^{13}(p,\gamma)N^{14}$	1747.6 ± 0.9	0.077 ± 0.012
$O^{16}(p,p)O^{16}$	12714 ± 8^a	<3
$C^{12}(p,p)C^{12}$	14233 ± 8	<1

[a] See ref. 6.

TABLE 8a-6. (p,n) THRESHOLD ENERGIES

Reaction	E_{th}, keV
$Li^7(p,n)Be^7$	1880.60 ± 0.07
$C^{13}(p,n)N^{13}$	3235.7 ± 0.7
$F^{19}(p,n)Ne^{19}$	4234.3 ± 0.8
$Al^{27}(p,n)Si^{27}$	5796.9 ± 3.8
$S^{34}(p,n)Cl^{34}$	6451.1 ± 4.5
$Ni^{60}(p,n)Cu^{60}$	7023.6 ± 3.9
$Fe^{54}(p,n)Co^{54}$	9202.7 ± 4.8
$Ni^{58}(p,n)Cu^{58}$	9515.2 ± 2.9
$C^{12}(p,n)N^{12}$	19684 ± 8

Other useful calibration points are:

$$O^{16}(d,n)F^{17} \quad E_{th} = 1829.2 \pm 0.6 \text{ keV}^a$$
$$Mg^{24}(\alpha,\gamma)Si^{28} \quad E_R = 3200 \pm 1 \text{ keV}^b$$

[a] See ref. 7.
[b] See ref. 8.

Threshold energies for various helium-ion-induced reactions can be calculated from the 1964 adjustment of atomic masses [9].

TABLE 8a-7. CALCULATED NEUTRON THRESHOLD ENERGIES

Reaction	E_{th}, MeV $\pm$ keV
$Li^6(He^3,n)B^8$	2.9650 ± 1.5
$Li^6(\alpha,n)B^9$	6.6239 ± 2.6
$Li^7(\alpha,n)B^{10}$	4.3843 ± 1.9
$C^{12}(He^3,n)O^{14}$	1.4366 ± 0.5
$C^{12}(\alpha,n)O^{15}$	11.3463 ± 1.7
$N^{14}(\alpha,n)F^{17}$	6.0888 ± 0.8
$N^{15}(\alpha,n)F^{18}$	8.1324 ± 1.5
$O^{16}(He^3,n)Ne^{18}$	3.7987 ± 5.7
$O^{16}(\alpha,n)Ne^{19}$	15.1761 ± 2.0

References

1. Marion, J. B., and F. C. Young: "Nuclear Reaction Analysis: Graphs and Tables," North-Holland, Publishing Company, Amsterdam, 1968.
2. Taylor, B. N., W. H. Parker, and D. N. Langenberg: *Rev. Mod. Phys.* **41**, 375 (1969).
3. Wapstra, A. H.: *Nucl. Phys.* **57**, 48 (1964).
4. Marion, J. B.: *Nucl. Data* **A4**, 301 (1968).
5. Marion, J. B.: *Rev. Mod. Phys.* **38**, 660 (1966).
6. Patterson, J. R., H. Winkler, and C. S. Zaidins: *Phys. Rev.* **163**, 1051 (1967).
7. Bondelid, R. O., J. W. Butler, and C. A. Kennedy: *Phys. Rev.* **120**, 889 (1960).
8. Rytz, A., H. Staub, H. Winkler, and F. Zamboni: *Nucl. Phys.* **43**, 229 (1963).
9. Mattauch, J. H. E., W. Thiele, and A. H. Wapstra: *Nucl. Phys.* **67**, 73 (1965).

8b. Properties of Nuclides

DAVID T. GOLDMAN

National Bureau of Standards

Table 8b-1 lists the properties that serve to characterize stable and radioactive nuclides. They are shown in order of ascending atomic number. The stable isotopes which constitute the naturally occurring elements are distinguished by **boldface** type. The literature was reviewed through December, 1968. In the interests of legibility and conciseness, no specific references are given with the table. Detailed information about the nuclides can be found in the list of general references that immediately precedes the table. The numbers cited here have been rounded off to the last significant figure before the uncertainty.

The first three columns of the table give for each element the atomic number, accepted symbol, and name. For the isotopes within each element, columns 4 and 5 give A, mass number, and N, number of neutrons. The atomic number is Z, and $A = Z + N$.

Column 6 lists the mass excess, or the difference between the actual weight of the nuclide and A, in milliatomic mass units (1 amu = 931.481 MeV). The scale is chosen such that the atomic mass of ^{12}C is exactly 12, and, hence, the mass excess of ^{12}C is identically zero. The values are taken mainly from the analysis of Mattauch, Thiele, and Wapstra [1], and from a later revision by Wapstra for $A \geq 212$ [2]. References 1 and 2 also contribute to column 12, the beta-decay Q values, the energy difference between the ground states of parent and daughter nuclei.

Two major sources of radioactive decay data are the products of the Nuclear Data Group at Oak Ridge National Laboratory [3] and the "Table of Isotopes" compiled at Lawrence Radiation Laboratory [4]. Except for more recently available data, these two references form the general source of most of the material given in columns 7, 8, 11, and 13. Other references to limited regions of the periodic table are the series prepared by Ajzenberg-Selove and Lauritsen [7] for $A \leq 20$ and Endt and Vander Leun [8] for $11 \leq Z \leq 21$.

The isotopic abundances of the naturally occurring isotopes are shown in **boldface** in column 8 and are those given by Fuller and Nier [5]. Columns 9 and 10 present the nuclear magnetic moments (in nuclear magnetons) and quadrupole moments (in barns), respectively. They are taken from an evaluation by Fuller and Cohen [6], which also provides data for the spins and parities of the nuclides (column 7).

Column 11 gives the decay particle, the most important decay energies in MeV and, in parentheses, the probability, in percent, of this decay energy in each transition. Column 13 gives similar information for the main gamma radiation.

Column 14 presents the 2,200-m/s (often called "thermal") neutron-absorption cross section. It is taken from the analysis of Goldman, et al. [9].

Much of the work presented here was initiated in the preparation of the "Chart of the Nuclides," the latest of which is the ninth edition [10].

References

1. Mattauch, J. H. E., W. Thiele, and A. H. Wapstra: 1964 Atomic Mass Table, *Nucl. Phys.* **67**, 1 (1965), as revised by N. B. Gove and A. H. Wapstra, to be published.
2. Wapstra, A. H.: 1967 Mass Table for $A = 212$, *Proc. 3d Intern. Conf. on Atomic Masses*, p. 153, R. C. Berber, ed., University of Manitoba Press, Winnipeg, 1967.
3. *Nuclear Data Sheets* published as sec. B of the journal *Nuclear Data*, K. Way, ed. This is a continuing series, with properties of the isotopes given in complete detail, which appears periodically. Before 1965 this series was published by the National Academy of Sciences.
4. Lederer, C. M., J. M. Hollander, and I. Perlman: "Table of the Isotopes," 6th ed., John Wiley & Sons, Inc., New York, 1967.
5. Fuller, G. H., and A. O. Nier: Appendix 2, "Relative Isotopic Abundances," *Nucl. Data Sheets*.
6. Fuller, G. H., and V. W. Cohen: *Nucl. Data* **A5**(5), 6 (1968).
7. Ajzenberg-Selove, F., and T. Lauritsen: Energy Levels of Light Nuclei $A = 11$–12, *Nucl. Phys.* **A114**, 1 (1968). Other references in this series are contained herein.
8. Endt, P. M., and C. Vander Leun: Energy Levels of $Z = 11$–21 Nuclei, *Nucl. Phys.* **A105**, 1 (1967).
9. Goldman, D. T., P. Aline, R. Sher, and J. R. Stehn: Twenty-two Hundred Meter per Second Neutron Absorption Cross Sections, submitted for publication.
10. Goldman, D. T., and J. R. Roesser: "Chart of the Nuclides," 9th ed., General Electric Co., 1966.

TABLE 8b-1. PROPERTIES OF NUCLIDES

(1) Atomic number Z	(2) Symbol	(3) Name	(4) Mass number A	(5) Number of neutrons N	(6) Mass excess, amu $\times 10^{-3}$	(7) Spin and parity	(8) % abundance or half life	(9) Magnetic moment, nuclear magnetons	(10) Quadrupole moment, barns	(11) Mode of decay, energy, and intensity, MeV (%)	(12) β-decay Q values, MeV	(13) Energy and intensity of γ-ray transitions, MeV (%)	(14) 2,200-m/s neutron-absorption cross section, barns
0	n	Neutron	1	1	8.6652	$\frac{1}{2}^+$	11 m	−1.9131		β-0.78			
1	H	Hydrogen	1	0	7.8252	$\frac{1}{2}^+$	99.985%	+2.79278					0.332
			2	1	14.1022	1^+	0.015%	+0.85742	+0.0028				0.00052
			3	2	16.0497	$\frac{1}{2}^+$	12.3 y	+2.9789		β-0.0186	0.01861		
2	He	Helium	3	1	16.0297	$\frac{1}{2}^+$	0.00013%	−2.1276					5327
			4	2	2.603	0^+	100%						0
			6	4	18.89	0^+	0.802 s			β-3.51(100)	3.51		
			8	6	34.	0^+	0.122 s			β-10, ···, n	10.7		
3	Li	Lithium	6	3	15.123	1^+	7.42%	+0.82202	−0.0008				950
			7	4	16.004	$\frac{3}{2}^-$	92.58%	+3.2564	−0.04			0.99(88)	0.037
			8	5	22.487	2^+	0.85 s	+1.6532		β-13, α1.6	16.0		
			9	6	26.80	$(\frac{3}{2}^-)$	0.172 s			β-11.0(75), 13.5(2), n0.7, ···	13.61		
4	Be	Beryllium	7	3	16.929	$\frac{3}{2}^-$	53.37 d	−1.1776	+0.05	ε(100)	0.862	0.478(10)	51,000
			9	5	12.183	$\frac{3}{2}^-$	100%						≤0.001
			10	6	13.534	0^+	2.7×10^6 y			β-0.555(100)	0.555		
			11	7	21.67	$(\frac{1}{2})^+$	13.6 s			β-11.5(61), 9.3(29), 4.7(6), 3.6(4)	11.61	2.12(32), 6.79(4), 5.86(2), 4.64(2), 7.97(2)	
			12	8	27.	(2^+)	0.011 s			β-, n			
5	B	Boron	8	3	24.609	$(2+)$	0.774 s			β+14, α21.6	18.0		
			10	5	12.9385	3^+	19.78%	+1.8007	+0.08				3836
			11	6	9.305	$\frac{3}{2}^-$	80.22%	+2.6885	+0.04				0.005
			12	7	14.354	1^+	0.0204 s	±1.002		β-13.37(98), 9.0(1), ···, α0.195(1.5)	13.370	4.43(1), ···	
			13	8	17.78		0.019 s			β-13.4(93), ···, n	13.437	3.67(7)	
6	C	Carbon	9	3	31.04		0.0127 s			β+ β8.2(60), 1.1(40), 2α			
			10	4	16.86	0^+	19.4 s			β+1.87(0.98), 0.85(1.6), 1.87		0.717(100), 1.023(1.6)	
			11	5	11.432	$\frac{3}{2}^-$	20.4 m	±1.03	±0.031	β+0.97(99+), ε(0.2)	1.98		
			12	6	0.	0^+	98.89%						0.0034

Z	Element	A	N	Mass excess	Jπ	Half‑life / abundance	μ	Q	Decay mode, energy (MeV)	Disintegration energy (MeV)	γ‑ray energy (MeV) (intensity %)	σ (barns)
		13	7	3.354	½−	**1.11%**	+0.7024					0.0009
		14	8	3.2420	0+	5730 y			β−0.156	0.1561		
		15	9	10.600	½+	2.4 s			β−4.51(68), 9.82(32)	9.77	5.299(68)	
		16	10	14.70	0+	0.74 s			β−, n	8.0		
7	N, Nitrogen	**12**	5	18.62	1+	0.0110 s	±0.46		β+16.38, ···, (100); 3α0.195(3)	17.36	4.43(2.4)	1.89
		13	6	5.738	½−	10.0 m	±0.3221		β+1.19(100)	2.22		0.000024
		14	7	3.0744	1+	**99.63%**	+0.4036					
		15	8	0.108	½−	**0.37%**	−0.2831					
		16	9	6.101	2−	7.2 s			β−4.3(68), 10.4(26), ···; α1.7(0.001)	10.422	6.13(69), 7.11(5), 2.75(1)	
		17	10	8.45	(½)−	4.16 s			β−4.1(95), 7.81(3), 8.68(2); n0.40(45), 1.2(45), 1.81(5)	8.68	0.87(3), 2.19(0.5)	
		18	11	14.25	(0,1,2)−	0.63 s			β−9.4	13.9	1.98(100), 0.82(59), 1.65(59), 2.47(41)	
		13	5	24.81	(3/2−)	0.0087 s			β+, p6.40(80), 6.97(20)	17.8		
		14	6	8.597	0+	71.0 s			β+1.81(99.4), 4.12(0.6)	5.1443	2.312(99)	
		15	7	3.070	½−	124 s	±0.7189		β+1.74(100)	2.760		
8	O, Oxygen	**16**	8	−5.0850	0+	**99.759%**						0.00178
		17	9	−0.867	5/2+	**0.037%**	−1.8937	−0.026				0.235
		18	10	−0.8400	0+	**0.204%**						0.00021
		19	11	3.578	5/2+	27 s			β−3.25(62), 4.60(38)	4.819	0.197(97), 1.37(59)	
		20	12	4.08	0+	14 s			β−	3.81	1.06(100)	
		17	8	2.096	5/2+	66 s	±4.722		β+1.74(100)	2.759		
		18	9	0.937	1+	109.7 m			β+0.635(97), ε(3)	0.655		
9	F, Fluorine	**19**	10	−1.595	½+	**100%**	+2.6288					0.0098
		20	11	−0.017	2+	11.4 s	+2.094		β−5.42(100)	7.030	1.63(100)	
		21	12	−0.049	5/2+	4.4 s			β−5.4(87), 4.0(13)	5.68	0.350(100), 1.38(13)	
		22	13	3.04	(3+)	4.0 s			β−11	12	1.28(100), 2.06(67)	
		17	7	17.7	(½−)	0.10 s			β+, p4.59, 3.80, 5.08, 6.95, ···			
		18	8	5.711	0+	1.5 s			β+3.42(93), ···	4.45	1.04(7)	
		19	9	1.881	½+	17.4 s	−1.887		β+2.22(100)	3.238		
10	Ne, Neon	**20**	10	−7.560	0+	**90.92%**						0.038
		21	11	−6.153	3/2+	**0.257%**	−0.6618	+0.09				
		22	12	−8.615	0+	**8.82%**						0.036
		23	13	−5.529	(5/2+)	37.6 s	−1.08		β−4.38(67), 3.95(32), 2.4(1), ···	4.380	0.439(33), 1.64(1), ···	
		24	14	−6.39	0+	3.38 m			β−1.99(92), 1.10(8)	2.47	0.472(100), 0.88(8)	

TABLE 8b-1. PROPERTIES OF NUCLIDES (*Continued*)

(1) Atomic number Z	(2) Symbol	(3) Name	(4) Mass number A	(5) Number of neutrons N	(6) Mass excess, amu $\times 10^{-4}$	(7) Spin and parity	(8) % abundance or half life	(9) Magnetic moment, nuclear magnetons	(10) Quadrupole moment, barns	(11) Mode of decay, energy, and intensity, MeV (%)	(12) β-decay Q values, MeV	(13) Energy and intensity of γ-ray transitions, MeV (%)	(14) 2,200-m/s neutron-absorption cross section, barns
11	Na	Sodium	20	9	7.4	....	0.402 s	....	....	$\beta^+11.25, 5.55, \ldots$; $\alpha 2.14, \ldots$	13.91	1.63, ...	
			21	10	−2.35	3/2+	22.8 s	+2.386		$\beta^+2.52(98), 2.17(2)$	3.54	0.35(2.2)	
			22	11	−5.563	3+	2.602 y	+1.746		$\beta^+0.544(90), 1.82(0.05),$ $\epsilon(9.5)$	2.843	1.274(99.5)	40,000
			23	12	−10.229	3/2+	**100%**	+2.2175	+0.14				0.534
			24	13	−9.036	4+	14.98 h	+1.690		$\beta^-1.39(99), \ldots$	5.515	2.754(99), 1.369(99), ...	
			25	14	−10.05	5/2+	60 s			$\beta^-3.8(65), 3.3(30), 2.2(5)$	3.83	0.98(15), 0.39(14), 0.58(14), 1.61(6)	
12	Mg	Magnesium	26	15	−8.1	2 or 3+	1.04 s			$\beta^-6.7(100), \ldots$	8.5	1.81(100)	
			20	8	19.		0.6 s						
			21	9	11.71		0.12 s			$\beta^+, p3.4, 4.0, 4.3, 4.8,$ $5.9, 6.45$	13.1		
			22	10	−0.41	0+	4.00 s			$\beta^+3.16(59), 3.23(36),$ $1.88(5)$	4.80	0.073(59), 0.583(100), 1.28(5)	
			23	11	−5.875	3/2+	12.0 s			$\beta^+3.04(91), 2.60(9)$	4.056	0.439(9)	
			24	12	−14.956	0+	**78.70%**						0.05
			25	13	−14.161	5/2+	**10.13%**	−0.8551	+0.22				0.18
			26	14	−17.407	0+	**11.17%**						0.030
			27	15	−15.657	1/2+	9.49 m			$\beta^-1.75(69), 1.59(31)$	2.61	0.84(70), 1.0(30), 0.18(1)	
			28	16	−16.121	0+	21.3 h			$\beta^-0.46(100)$	1.836	0.032(96), 1.35(70), 0.40(31), 0.95(29)	
13	Al	Aluminum	24m			1+	0.13 s			$\beta^+13.3(4.4), 11.9(1.9),$	14.32	IT0.439(93), 1.369(1.9)	
			24	11	0.5	4+	2.09 s			$\beta^+3.40(48), 4.42(41),$ $8.74(8), 6.8(3)$	13.88	1.37(40), 2.73(32), 4.22(15), 7.1(7) 5.4(3)	
			25	12	−9.568	5/2+	7.23 s			$\beta^+3.24(100)$			
			26m	13	....	0+	6.38 s						
			26	13	−12.???								

Z	Element	Sym	N	A	Mass excess	Jπ	Abundance / half‑life	μ	Q	Decay mode, energies (MeV) and %	E_max (MeV)	γ (MeV, %)	σ (b)
			14	27	−18.459	5/2+	**100%**	+3.6414	+0.15	......			0.232
			15	28	−18.088	3+	2.27 m			β⁻2.87(100)	4.634	1.779(100)	
			16	29	−19.552	5/2+	6.52 m			β⁻2.5(93), 1.5(7)	3.68	1.27(93), 2.43(7)	
			17	30	−17.1	(2,3)+	3.3 s			β⁻3.8(84), 5.1(16)	8.5	2.23(58), 3.51(42)	
14	Silicon	Si	11	25	4.1	(5/2+)	0.22 s			β⁺, p4.25, 1.95, 3.47, 2.31, 2.18, 0.97, ...	12.7		
			12	26	−7.66	0+	2.1 s			β⁺3.83(66), 3.00(34)	5.07	0.82(34)	
			13	27	−13.297	5/2+	4.17 s			β⁺3.8(100), ...	4.81		
			14	28	−23.071	0+	**92.21%**			......	......		0.16
			15	29	−23.504	1/2+	**4.70%**	−0.5553		......	......		0.28
			16	30	−26.228	0+	**3.09%**			......	......		0.10
			17	31	−24.636	3/2+	2.62 h			β⁻1.49(99.9), 0.21(0.1)	1.492	1.266(0.1)	
			18	32	−25.86	0+	650 y			β⁻0.21(100)	0.21		
15	Phosphorus	P	13	28	−8.2	(3)+	0.27 s			β⁺11.5(52), 6.96(16), 5.25(13), 3.94(13), 8.80(7)	14	1.78(100), 4.50(29), 3.04(8), 7.50(7), 2.84(6),	
			14	29	−18.19	1/2+	4.23 s			β⁺3.94(98), 2.68(1), ...	4.95	1.27(1)	
			15	30	−21.68	1+	2.50 m			β⁺3.24(99.5), 1.01(0.5)	4.24	2.23(0.5)	
			16	31	−26.235	1/2+	**100%**	+1.1317		......	......		0.19
			17	32	−26.091	1+	14.29 d	−0.2523		β⁻1.710(100)	1.710		
			18	33	−28.272	1/2+	25.3 d			β⁻0.248(100)	0.248		
			19	34	−26.7	1+	12.4 s			β⁻5.1(75), 3.2(25)	5.1	2.1(25), 4.0(0.2)	
16	Sulfur	S	13	29	−3.	(5/2+)	0.19 s			β⁺, p3.5, 3.73, 5.0, 5.2, 5.4, ...	~14	0.68(80), ...	
			14	30	−15.10	0+	1.25 s			β⁺4.43(80), 5.10(20), ...	6.13		
			15	31	−20.39	1/2+	2.61 s			β⁺4.39(99), ...	5.44	1.266(1)	
			16	32	−27.926	0+	**95.0%**			......	......		0.53
			17	33	−28.541	3/2+	**0.76%**	+0.6433	−0.055	......	......		0.18
			18	34	−32.130	0+	**4.22%**			......	......		0.02
			19	35	−30.967	3/2+	87.0 d	+1.00 or −1.07	+0.04	β⁻0.167(100)	0.167		
			20	36	−32.91	0+	**0.014%**			......	......		0.14
			21	37	−28.88	(7/2−)	5.06 m			β⁻1.6(94), 4.8(6), ...	4.8	3.11(94), 3.71(0.4)	
			22	38	−28.8	0+	2.87 h			β⁻1.1(95), 3.0(5)	3.0	1.88(95)	
17	Chlorine	Cl	15	32	−14.24	2+	0.30 s			β⁺9.5(60), 4.7(25), 7.5(14), 6.2(10), 11.6(1)	12.8	2.23(89), 4.77(25), 2.46(5), 1.65(5), 3.31(4),	
			16	33	−22.56	3/2+	2.52 s			β⁺4.5(99.7), ...	5.57	2.9(0.3)	

TABLE 8b-1. PROPERTIES OF NUCLIDES (Continued)

(1) Atomic number Z	(2) Symbol	(3) Name	(4) Mass number A	(5) Number of neutrons N	(6) Mass excess, amu $\times 10^{-3}$	(7) Spin and parity	(8) % abundance or half life	(9) Magnetic moment, nuclear magnetons	(10) Quadrupole moment, barns	(11) Mode of decay, energy, and intensity, MeV (%)	(12) β-decay Q values, MeV	(13) Energy and intensity of γ-ray transitions, MeV (%)	(14) 2,200-m/s neutron-absorption cross section, barns
17	Cl	Chlorine	34m			3+	32.2 m			β+2.5(28), 1.3(27), . . .	5.63	IT0.146(45), 2.13(41), 3.30(14), 4.12(0.4)	
			34	17	−26.250	0+	1.57 s			β+4.5(100)	5.48		
			35	18	−31.146	3/2+	75.53%	+0.82133	−0.079				44
			36	19	−31.693	2+	3.07 × 10⁵ y	+1.285	−0.017	β−0.71(98.3), ε(1.7), β+0.12(0.002)	0.712 1.14		
			37	20	−34.097	3/2+	24.47%	+0.68411	−0.062				0.43
			38m			5−	0.74 s						
			38	21	−32.00	2−	37.2 m			β−4.9(53), 1.1(36), 2.8(11)	4.9	2.17(47), 1.64(36), 3.81(0.02)	
			39	22	−31.99	3/2+	55.5 m			β−1.91(85), 2.18(8), 3.44(7)	3.44	1.27(50), 1.52(43), 0.25(42)	
			40	23	−29.6	(2−)	1.42 m			β−~7.5, ~3.2	7.5	1.45, 3.1, 0.33, 2.85, . . .	
18	Ar	Argon	33	15	−10.	(1/2+)	0.18 s			β+, p3.27, 2.17, 2.54, 3.93, . . .	12		
			34	16	−19.7	0+	0.9 s			β+5.05	6.06	0.67	
			35	17	−24.75	3/2+	1.80 s	+0.63		β+4.95(93), . . .	5.96	1.19(5), 1.73(2)	
			36	18	−32.453	0+	0.337%						6
			37	19	−33.223	3/2+	34.8 d	+0.95		ε(100)	0.814		
			38	20	−37.267	0+	0.063%						0.8
			39	21	−35.683	7/2−	269 y	−1.3		β−0.565(100)	0.565		500
			40	22	−37.616	0+	99.60%						0.65
			41	23	−35.500	7/2−	1.83 h			β−1.198(99), 2.489(0.8)	2.49	1.294(99), 1.67(0.05)	0.5
			42	24	−36.95	0+	33 y			β−	0.60		
			43	25			6 m			β−		0.976	
			44	26			14 m			β−		1.887	
19	K	Potassium	36	17	−18.6		0.27 s			β+9.9(70), 5.3(25)	12.9	1.97(89), 2.43(23), 2.21(19), 4.44(5)	
			37	18	−26.64	3/2+	1.23 s	+0.204		β+5.12(98), . . .	6.15	2.80(2)	
			38m			0+	0.95 s			β+5.03(100)	6.05		

Element	Z	A	N	Δ (MeV)	Spin	Abundance or half-life	μ	Q	Decay, particle energy (MeV)	Disint. energy	γ energy (MeV)	σ (b)
		38	19	−30.90	3+	7.68 m	+1.374		β⁺2.7(99.8), ...	5.93	2.17(99.8), 3.94(0.2)	
		39	20	−36.290	3/2+	**93.10%**	+0.3914	+0.055				2.2
		40	21	−36.000	4−	1.28×10⁹ y, **0.0118%**	−1.298	−0.07	β⁻1.31(89.4), ...; ε(10.6), β⁻(0.001)	1.314; 1.505	1.461(10.6)	70
		41	22	−38.173	3/2+	**6.88%**	+0.2149	+0.067				1.3
		42	23	−37.59	2−	12.36 h	−1.141		β⁻3.52(87), 2.00(18),	3.52	1.52(18), ...	
		43	24	−39.27	3/2+	21.8 h	±0.163		β⁻0.82(83), 0.46(10), 1.24(3), 1.82(1.3)	1.82	0.372(85), 0.616(60), 0.396(20), 0.593(13), 0.222(3), 1.02(2), 1.16(65), ...	
		44	25	−38.44	(2−)	22.0 m			β⁻5.2(35), 2(30), 4(9),	5.2	0.175, 1.71, ...	
		45	26	−39.3	3/2+	20 m	±0.173		β⁻2.1(70), 1.1(20), 4(10)	4.19	1.347(100), 3.70(31), 3.02(10), 2.27(9), 1.78(9), ...	
		46	27	−38.0	2(−)	115 s			β⁻6.3(50), ...	7.72	2.0, 2.6	
		47	28	−38.3	1/2+	17.5 s			β⁻4.1(99), 6(1)	6.65	1.57(20), ...	
Calcium Ca	20	37	17	−14.2		0.173 s			β⁺, p3.16, 1.74, 2.77, 1.98, 2.54, ...	11.56		
		38	18	−23.74	0+	0.5 s			β⁺5.59(79), 4.02(21),	6.6		
		39	19	−29.29	3/2+	0.88 s			β⁺5.48(100)	6.50		
		40	20	−37.408	0+	**96.97%**						0.41
		41	21	−37.721	7/2−	7.7×10⁴ y	−1.595		ε(100)	0.41		
		42	22	−41.372	0+	**0.64%**						0.70
		43	23	−41.223	7/2−	**0.145%**	−1.317					6.2
		44	24	−44.510	0+	**2.06%**						1.0
		45	25	−43.807	(7/2)−	162.7 d			β⁻0.256(100), ...	0.256	0.0124(0.002)	
		46	26	−46.31	0+	**0.0033%**						0.7
		47	27	−45.46	7/2−	4.56 d			β⁻0.67(83), 1.98(17)	1.98	1.297(75), 0.807(7), 0.489(7), ...	
		48	28	−47.47	0+	**0.18%**						1.1
		49	29	−44.33	3/2−	8.8 m			β⁻2.1(89), 1.0(10), ...	5.26	3.1(89), 4.06(10), 4.7(0.3), ...	
		50	30	−42.49	0+	9 s			β⁻3.1	4.97		
Scandium Sc	21	40	19	−22.03	4−	0.180 s			β⁺5.7(51), 9.58(20), 8.8(15), 7.5(15)	14.33	0.257, 0.072, 1.52, 1.59, 3.74(100), 0.78(46), 1.88(23), 2.05(23), 3.17(13), 1.13(12), ...	
		41	20	−30.75	7/2−	0.60 s			β⁺5.5(100)	6.5		

TABLE 8b-1. PROPERTIES OF NUCLIDES (*Continued*)

(1) Atomic number Z	(2) Symbol	(3) Name	(4) Mass number A	(5) Number of neutrons N	(6) Mass excess, amu $\times 10^{-3}$	(7) Spin and parity	(8) % abundance or half life	(9) Magnetic moment, nuclear magnetons	(10) Quadrupole moment, barns	(11) Mode of decay, energy, and intensity, MeV (%)	(12) β-decay Q values, MeV	(13) Energy and intensity of γ-ray transitions, MeV (%)	(14) 2,200-m/s neutron-absorption cross section, barns
21	Sc	Scandium	42m			7+	61 s			β+2.9(100)	6.96	0.438(100), 1.23(100), 1.52(100)	
			42	21	−34.47	0+	0.65 s			β+5.39(100)	6.43		
			43	22	−38.84	7/2−	3.94 h	+4.62	−0.26	β+1.20(78), 0.82(22)	2.22	0.374(22)	
			44m			6+	59 h	+3.88	−0.19	ε(1.3)		IT0.271(99), 1.02(1.3), 1.12(1.3), 1.16(1.3)	
			44	23	−40.594	2+	3.92 h	+2.56	+0.10	β+1.467(95), ε(5)	3.647	1.156(100), 1.5(0.8), 2.7(0.1)	
			45m			3/2+	0.30 s					IT0.0124(100)	
			45	24	−44.083	7/2−	100%	+4.7564	−0.22				25
			46m			7(+)	19 s					IT0.143	
			46	25	−44.829	4+	83.80 d	+3.03	+0.12	β−0.357(100), 1.48(0.004); β+(1.6×10^{-5})	2.367; 1.38	0.889(100), 1.220(100)	8
			47	26	−47.589	7/2−	3.35 d			β−0.44(73), 0.60(27)	0.600	0.159(73)	
			48	27	−47.78	6+	1.82 d	+5.34	−0.22	β−3.33(88), 3.51(12)	3.98	1.311(100), 0.983(100), 1.037(98), 0.175(9), 1.212(3)	
			49	28	−49.97	(7/2)−	57.5 m			β−2.01(100), . . .	2.01	1.78(0.03)	
			50m			(2+)	0.35 s					IT0.258	
			50	29	−47.82	(5+)	1.73 m			β−3.6, . . .	6.89	1.12(100), 1.56(100), 0.52(90)	
			51	30	−46.40	(7/2−)	12 s			β−5.04(55), 4.32(45)	6.52	1.45(55), 2.15(45)	
22	Ti	Titanium	41	19			0.088 s			β+, p4.81, 3.14, 3.80, 1.58, 4.27, 2.32, . . .	13		
			43	21	−31.48	7/2−	0.49 s			β+5.8(100)	6.8		
			44	22	−40.43	0+	48 y			ε(100)	0.16	0.0784(100), 0.0678(100)	
			45	23	−41.87	7/2−	3.09 h	±0.095	≈±0.02	β+1.04(85), . . ., ε(15)	2.06	0.719(0.4), . . .	
			46	24	−47.368	0+	7.93%						0.6

Z	Element	A	N	Mass excess	Spin	% abund. / half-life	μ	Q	Decay	E (MeV)	γ-ray energies (MeV)	σ (b)
		47	25	−48.232	5/2−	7.28%	−0.7883	+0.29				1.7
		48	26	−52.051	0+	73.94%						8.3
		49	27	−52.128	7/2−	5.51%	−1.1039	+0.24				1.9
		50	28	−55.216	0+	5.34%						0.14
		51	29	−53.40	3/2−	5.8 m			β−2.13(94), 1.50(6)	2.46	0.320(95), 0.928(5), 0.605(1)	
		52	30	53.11	(0+)	1.7 m			β−1.8	1.97	0.125, 0.17	
23	V Vanadium	46	23	−39.79	0+	0.426 s			β+6.0(100)	7.05		
		47	24	−45.10	3/2−	33 m			β+1.90(96), ε(4)	2.92	1.5(0.7), 1.8(5), ...	
		48	25	−47.741	4+	16.0 d	±4.5	±0.06	β+0.694(60), ε(40)	4.013	0.983(100), 1.312(98), 2.24(3), ...	
		49	26	−51.482	7/2−	330 d			ε(100)	0.605		
		50	27	−52.836	6+	6 × 10¹⁵ y 0.24%	+3.3470		ε(70)	2.22	1.55(70)	100
									β−(30)	1.03	0.783(30)	
		51	28	−56.036	7/2−	99.76%	+5.149	−0.05				4.8
		52	29	−55.22	3+	3.73 m			β−2.47(99), ...	3.97	1.43(100), ...	
		53	30	−55.67	(7/2−)	2.0 m			β−2.5(100)	~7	1.0(100)	
		54	31	−53.6	...	55 s			β−3.3(100)		2.21(100), 0.99(100), 0.84(100)	
24	Cr Chromium	46	22	...	...	1.1 s			β+			
		48	24	−46.0	0+	23 h			ε(100)	1.4	0.116(100), 0.31(100)	
		49	25	−48.73	(5/2)−	41.9 m	±0.48		β+1.54(65), 1.42(29), ...	2.56	0.090(30), 0.063(29), 0.153(4), ...	
		50	26	−53.951	0+	4.31%						16
		51	27	−55.229	7/2−	27.8 d	+0.94		ε(100)	0.752	0.320(9)	
		52	28	−59.490	0+	83.76%						0.76
		53	29	−59.349	3/2−	9.55%	−0.4743	+0.03				18.2
		54	30	−61.12	0+	2.38%						0.38
		55	31	−59.17	3/2−	3.5 m			β−2.50(99+)	2.59	1.52, 2.24	
		56	32	−59.4	0+	5.9 m			β−1.5(100)	1.6	0.083(100), 0.026(100)	
25	Mn Manganese	50(m)?			(0+)	2 m			β+(99), ε(1)		0.79, 1.11, ...	
		50	25	−45.76	0+	0.286 s			β+6.6	7.63		
		51	26	−51.79	5/2−	46 m	+3.57		β+2.17(97), ε(3)	3.19		
		52m		...	2+	21 m	+0.0076		β+2.63(92), ε(8)	4.708	1.435(100), ...	
		52	27	−54.43	6+	5.7 d	+3.05		ε(62), β+0.58(38)		1.435(100), 0.938(100), 0.747(99), ...	
		53	28	−58.71	7/2−	2 × 10⁶ y	+5.01		ε(100)	0.598		≈170
		54	29	−59.64	3+	313 d	+3.29		ε(100)	1.38	0.835(100)	<10
		55	30	−61.954	5/2−	100%	+3.444	+0.4				13.3

TABLE 8b-1. PROPERTIES OF NUCLIDES (Continued)

(1) Atomic number Z	(2) Symbol	(3) Name	(4) Mass number A	(5) Number of neutrons N	(6) Mass excess, amu $\times 10^{-3}$	(7) Spin and parity	(8) % abundance or half life	(9) Magnetic moment, nuclear magnetons	(10) Quadrupole moment, barns	(11) Mode of decay, energy, and intensity, MeV (%)	(12) β-decay Q values, MeV	(13) Energy and intensity of γ-ray transitions, MeV (%)	(14) 2,200-m/s neutron-absorption cross section, barns
25	Mn	Manganese	56	31	−61.093	3+	2.576 h	+3.218		$\beta^-2.85(53), 1.04(30), 0.75(16), \dots$	3.70	0.847(100), 1.81(30), 2.12(15), $\dots$	
			57	32	−61.9	(5/2−)	1.7 m			$\beta^-2.55(82), 1.1(18)$	2.7	0.117, 0.134, $\dots$	
			58	33	−60.2	$\dots$	1.1 m			β^-	~6.1	0.36, 0.41, 0.52, 0.57, 0.82, 1.0, 1.25, 1.4, 1.6, 2.2, 2.8	
26	Fe	Iron	52	26	−51.88	0+	8.2 h			$\beta^+0.8(56), \epsilon(44)$	2.37	0.165(100)	
			53	27	−54.69	(7/2)−	8.5 m			$\beta^+2.8(50), 2.4(38), \dots$	3.98	0.38(39)	
			54	28	−60.38	0+	5.82%			$\epsilon(2)$			2.8
			55	29	−61.705	3/2−	2.6 y			$\epsilon(100)$	0.232		
			56	30	−65.066	0+	91.66%						2.6
			57	31	−64.609	1/2−	2.19%	+0.0902					2.5
			58	32	−66.725	0+	0.33%						1.23
			59	33	−65.131	3/2−	45.1 d			$\beta^-0.467(54), 0.273(48),$	1.566	1.10(56), 1.29(44), 0.192(3), $\dots$	
			60	34	−66.95	0+	3×10^5 y			$\beta^-0.135(100)$	0.194	0.059(100)	
			61	35	−63.4	3/2(−)	6.07 m			$\beta^-2.62(39), 2.50(26), 2.78(18), \dots$	3.80	1.20(43), 1.025(43), 0.297(21), $\dots$	
27	Co	Cobalt	54m		$\dots$	(6,7+)	1.43 m			$\beta^+4.5(100)$		0.41(100), 1.14(100), 1.41(100)	
			54	27	−51.53	(0+)	0.194 s			$\beta^+(100)$	8.25		
			55	28	−57.99	7/2−	18.2 h			$\beta^+1.50(50), 1.03(30), 0.3(2), \dots \epsilon(25)$	3.46	0.93(80), 1.41(13), 0.48(12), $\dots$	
			56	29	−60.15	4+	77 d	±3.83		$\epsilon(80),$	4.57	0.845(100), 1.24(70), 2.60(16), 1.77(15), 1.038(14), 2.04(8), $\dots$	
			57	30	−63.711	7/2−	270 d	±4.62		$\beta^+1.46(19), 1.44(1)$	0.837	0.122(89), 0.0144(89), 0.136(11), $\dots$	
			58m		$\dots$	5+	9.1 h			$\epsilon(100)$		IT0.025	140,000

Element	A	N	Mass excess	Jπ	T½ / abundance	μ	Q	Decay modes	E	γ-rays	σ
	58	31	−64.248	2+	71.3 d	+4.03		ε(85) β⁺0.48(15), ...	2.309	0.8105(100), ...	1700
	59	32	−66.811	7/2−	**100%**	+4.62	+0.4				37.2
	60m		...	2+	10.47 m			β⁻1.55(0.2), ...		IT0.0588(99.8), 1.33(.2), ...	58
	60	33	−66.189	5+	5.26 y	+3.78		β⁻0.315(99.7), 0.67(0.2), 1.49(0.1)	2.82	1.33(100), 1.17(99.9)	2.0
	61	34	−67.56	7/2−	99 m			β⁻1.22(100)	1.29	0.07(100)	
	62		...	(1+, 2+)	1.55 m			β⁻			
	62	35	−66.05	(4+, 5+)	13.9 m			β⁻2.88(75), 0.88(25)	5.22	1.170(100), 1.163(82), 1.47(20), 1.74(20), 2.03(7)	
	63	36	−66.41	(5/2−, 7/2−)	52 s			β⁻3.6(100)	3.7	0.058(100)	
	64m		...	(4+)	28 s			β⁻		IT0.095(100)	
	64	37	−64.5	(1+)	<28 s				~7.0		
Ni Nickel 28	56	28	−57.88	0+	6.10 d			ε(100)	2.11	0.163(99), 0.812(84), 0.755(51), 0.276(34), 0.472(34), ...	
	57	29	−60.23	3/2−	36.0 h			ε(63), β⁺0.85(40), 0.72(6), ...	3.24	1.37(86), 0.127(14), 1.89(14), ...	
	58	30	−64.664	0+	**67.88%**						4.7
	59	31	−65.660	3/2−	7.5 × 10⁴ y			ε(100)	1.073		
	60	32	−69.220	0+	**26.23%**						2.8
	61	33	−68.950	3/2−	**1.19%**	−0.7487	+0.16				2.5
	62	34	−71.660	0+	**3.66%**						15
	63	35	−70.339	1/2−	92 y			β⁻0.0659(100)	0.0659		1.52
	64	36	−72.04	0+	**1.08%**						20
	65	37	−69.93	5/2−	2.54 h			β⁻2.14(58), 0.65(30), 1.02(10), ...	2.13	1.482(25), 1.115(16), 0.366(5), ...	
	66	38	−70.92	0+	55 h			β⁻0.20	0.20		
	67	39	−67.8	(−)	50 s			β⁻4.1(50), 2(30), 3.2(4), ...	4.1		
Cu Copper 29	58	29	−55.46		3.20 s			β⁺	8.57	0.90(50), 0.89(24), 1.26(22)	
	59	30	−60.50	3/2−	82 s			β⁺3.75(71), ... , ε(1)	4.80	1.31(10), 0.88(9), 0.34(4), 0.87(5), ...	
	60	31	−62.64	2+	23 m	+1.22		β⁺2.00(47), 3.00(14), 3.92(8), ... , ε(8)	6.13	1.33(87), 1.79(45), 0.83(19), ...	
	61	32	−66.54	3/2−	3.32 h	+2.13		β⁺1.22(52), 0.93(6), 0.56(3), ... , ε(37)	2.24	0.283(13), 0.656(10), 0.067(5), 1.19(4), ...	
	62	33	−67.43	1+	9.8 m	−0.38		β⁺2.92(94), 1.75(2), 0.87(2), ε	3.94	1.17(3), 0.875(1), ...	

TABLE 8b-1. PROPERTIES OF NUCLIDES (Continued)

(1) Atomic number Z	(2) Symbol	(3) Name	(4) Mass number A	(5) Number of neutrons N	(6) Mass excess, amu ×10⁻³	(7) Spin and parity	(8) % abundance or half life	(9) Magnetic moment, nuclear magnetons	(10) Quadrupole moment, barns	(11) Mode of decay, energy, and intensity, MeV (%)	(12) β-decay Q values, MeV	(13) Energy and intensity of γ-ray transitions, MeV (%)	(14) 2,200-m/s neutron-absorption cross section, barns
29	Cu	Copper	63	34	−70.410	$\tfrac{3}{2}-$	69.09%	+2.223	−0.180				4.5
			64	35	−70.243	1+	12.82 h	−0.216		ε(41), β⁺0.654(19), β⁻0.573(40)	1.677, 0.573	1.348(0.6)	<6,000
			65	36	−72.21	$\tfrac{3}{2}-$	30.91%	+2.382	−0.195				2.3
			66	37	−71.13	1+	5.1 m	±0.283		β⁻2.63(91), 1.6(9), ...	2.63	1.039(9), ...	130
			67	38	−72.24	$(\tfrac{3}{2}-)$	62 h			β⁻0.395(51), 0.484(28), 0.577(20)	0.576	0.185(45), 0.0933(35), 0.0913(6), ...	
			68	39	−70.23	(1.)+	31 s			β⁻3.5(75), 2.3(16), 2.7(4), ...	4.58	1.08(97), 1.26(15), ...	
			69	40	−70.8	$(\tfrac{3}{2})-$	3.0 m			β⁻2.6(79), ...	2.6	1.007(10), 0.834(6), 0.531(3), 0.649(1), ...	
30	Zn	Zinc	60	30	−55.18	0+	2.1 m			ε, β⁺			
			61	31	−60.8	$\tfrac{3}{2}-$	1.48 m			β⁺(99), ε(1)	5.91	0.47(11), 1.64(6), 0.97(3), 0.69(2)	
			62	32	−65.62	0+	9.15 h			ε(87), β⁺0.66(13)	1.68	0.041(36), 0.597(20), 0.506(13), ...	
			63	33	−66.79	$\tfrac{3}{2}-$	38.4 m	−0.282	+0.31	β⁺2.34(76), 1.69(10), 1.40(7), ..., ε(6)	3.36	0.670(11), 0.962(8), ...	
			64	34	−70.860	0+	48.89%						0.80
			65	35	−70.77	$\tfrac{5}{2}-$	244 d	+0.769	−0.026	ε(98), β⁺0.325(2)	1.353	1.115(51), ...	0.9
			66	36	−73.960	0+	27.81%						6
			67	37	−72.868	$\tfrac{5}{2}-$	4.11%	+0.8754	+0.17				1.08
			68	38	−75.152	0+	18.57%						
			69m			$\tfrac{9}{2}+$	13.9 h					IT0.439(100)	
			69	39	−73.46	$\tfrac{1}{2}-$	58 m			β⁻0.9(100)	0.92	0.439(100)	
			70	40	−74.67	0+	0.62%						0.099
			71m			$\tfrac{9}{2}+$	3.92 h			β⁻1.45(98), ...		0.387(94), 0.488(70), 0.620(65)	

Z	Sym	Element	A	N	Mass excess	Spin/parity	Half-life	μ	Q	Decay (MeV)	E (MeV)	γ energies (MeV)	σ
			71	41	−72.28	(1/2 −)	2.4 m			β−2.61(80), 2.10(14), ...	2.61	0.512(13), 0.923(3), ...	
			72	42	−73.14	0+	46.5 h			β−0.296(90), 0.25(10)	0.457	0.145(92), 0.193(8), 0.016(8), ...	
31	Ga	Gallium	63	32	−61.	(3/2 −)	33 s			β+	~5.5		
			64	33	−63.26	0+	2.6 m			β+2.89(33), 6.05(22), ..., ε(2)	7.08	0.992(46), 3.366(17), 0.809(14), 0.919(8), 2.375(8), ...	
			65	34	−67.27		15.2 m			β+2.12(53), 2.03(10), 2.24(7), ..., ε(10)	3.26	0.115(55), 0.061(15), 0.153(9), 0.752(7), ...	
			66	35	−68.39	0(+)	9.4 h			β+4.153(51), 0.94(3), ..., ε(42)	5.175	1.039(40), 2.748(25), 0.828(6), 2.190(6), ...	
			67	36	−71.78	3/2 −	78 h	+1.850	+0.22	ε(100)	1.00	0.0933(70), 0.1845(2), 0.300(15), 0.393(4), ...	
			68	37	−72.01	1+	68.3 m	±0.0117	+0.031	β+1.90(88), 0.82(1), ε(10)	2.92	1.078(3), ...	
			69	38	−74.420	3/2 −	60.4%	+2.016	+0.19				1.8
			70	39	−73.973	1+	21.1 m			β−1.65(99.5), ...	1.66	1.04(0.5), 0.173(0.2)	
			71	40	−75.294	3/2 −	39.6%	+2.562	+0.12				5
			72	41	−73.63	3−	14.1 h	−0.1322	+0.59	β−0.95(28), 0.66(21), 0.64(15), 3.15(10), 1.47(9), ...	4.000	0.834(95), 2.201(29), 0.630(23), 2.508(14), ...	
			73	42	−74.87	(3/2 −)	4.8 h			β−1.19(85), 0.40(9), 1.16(5), ...	1.55	0.296(87), 0.325(13), 0.742(7), ...	
			74	43	−72.92	(3−)	7.9 m			β−2.6(52), 4.3, ...	5.5	0.596(85), 2.35(49), 0.609(13), ...	
			75	44	−73.6		1.9 m			β−3.3	3.3	0.58, 0.36	
			76	45		(3−)	32 s			β−5.8, 5.3, 2.2, ...	~7	1.12, 0.560, ...	
			77	46			17 s			β−			
			78	47			~4 s			β−			
32	Ge	Germanium	65	33	−60.		1.5 m			β+5.5, 3.8, ...	6.5	0.67(3), 1.72(2)	
			66	34	−66.15	0+	2.4 h			β+1.3, 2.0, ..., ε	3.0	0.044(41), 0.382(29), 0.109(17), 0.273(12), 0.338(10), 0.065(8), ...	
			67	35	−67.6		19 m			β+3.24(51), 1.6(12), 2.3(8), ..., ε(6)	4.43	0.167(90), 0.915(10), 1.48(9), ...	
			68	36	−71.60	0+	275 d			ε(100)	~0.7		
			69	37	−72.034		39 h			β+1.21(32), ..., ε(63)	2.227	1.107(26), 0.574(12), 0.872(9), ...	
			70	38	−75.749	0+	20.52%						0.28
			71	39	−75.044	1/2 −	11 d	+0.546		ε(100)	0.235		

TABLE 8b-1. PROPERTIES OF NUCLIDES (Continued)

(1) Atomic number Z	(2) Symbol	(3) Name	(4) Mass number A	(5) Number of neutrons N	(6) Mass excess, amu ×10^-3	(7) Spin and parity	(8) % abundance or half life	(9) Magnetic moment, nuclear magnetons	(10) Quadrupole moment, barns	(11) Mode of decay, energy, and intensity, MeV (%)	(12) β-decay Q values, MeV	(13) Energy and intensity of γ-ray transitions, MeV (%)	(14) 2,200-m/s neutron-absorption cross section, barns
32	Ge	Germanium	**72**	40	−77.918	0+	**27.43%**						0.98
			73m			($\frac{1}{2}$−)	0.53 s					IT0.0533(99), 0.0670(1), 0.0135(99)	
			73	41	−76.536	$\frac{9}{2}$+	**7.76%**	−0.8792	−0.28				14
			74	42	−78.821	0+	36.54%						0.45
			75m		−77.12	($\frac{7}{2}$+)	46 s	±0.51				IT0.133(100)	
			75	43		($\frac{1}{2}$)(−)	83 m			β−1.19(87), 0.919(11), ...	1.189	0.2646(11), 0.199(1), ...	
			76	44	−78.595	0+	**7.76%**						
			77m			($\frac{1}{2}$)−	53 s			β−2.90(52), ...	2.76	IT0.159(26), 0.215(22), ...	
			77	45	−76.39	($\frac{7}{2}$+)	11.3 h			β−		0.263(50), 0.210(32), 0.215(28), 0.417(25), 0.558(18), ...	
33	As	Arsenic	78	46	−77.0	0+	1.5 h			β−0.70(100)	0.98	0.278(100)	0.23
			79	47	74.5		50 s			β−	4.3		
			68	35	−67.77		~7 m			β+	3.9	0.23	
			69	36	−69.07		15 m			β+2.9, ϵ	6.22		
			70	37		4(+)	52 m			β+2.144(74), 1.44(10), 2.89(6), ϵ(10)		1.040(80), 0.668(26), 0.595(24), 0.760(24), 1.14(24), 1.708(23), ...	
			71	38	−72.89	($\frac{5}{2}$)−	64 h			ϵ(66), β+0.812(33), ...	2.009	0.175(99), 1.10(3), ...	
			72	39	−73.24	2−	26 h			β+2.50(60), 3.34(18), 1.87(5), ϵ(10)	4.357	0.834(78), 2.201(20), 2.508(10), 0.630(8), ...	
			73	40	−76.17	($\frac{3}{2}$−)	76 d			ϵ(100)	0.341	0.0533(99), 0.0135(99), 0.0670(1)	
			74m			(5)	8 s					IT0.283(100)	

Element	A	N	Mass excess	J^π	Half-life / abundance	μ	Q	Decay	E	γ-rays (MeV)	σ
	74	41	−76.069	2−	17.7 d			ε(33), β+0.91(28), 1.51(4), β−1.36(19), 0.72(16)	2.56	0.596(58), 0.635(16), …	4.3
	75	42	−78.400	3/2(−)	**100%**	+1.439	+0.29		1.354		
	76	43	−77.602	2−	26.4 h	−0.905	±7	β−2.97(52), 2.41(29), 1.76(10), …	2.972	0.5593(41), 0.657(6), 1.216(5), …	
	77	44	−79.35	3/2−	38.8 h			β−0.686(97), …	0.686	0.239(2), …	
	78	45	−78.1	(2−)	1.5 h			β−4.1(25), 1.4(25), …	4.1	0.641(56), 0.695(21), 1.310(14), …	
	79	46	−79.1	(3/2)−	9.0 m			β−2.14(95), 1.70(2), 1.80(1.5), 1.25(1.5), …	2.24	0.36(2), 0.43(2), …	
	80	47	−77.0	1+	15 s			β−6.0(56), …	6.0	0.666(42), 1.64(2), 1.22(2), …	
	81	48	−77.9	(3/2)−	32 s			β−3.8(100)	3.8	0.655, 0.817	
	82	49		(1/2)−	15 s			β−			
	83	50			14 s			β−			
	84	51			6 s			β−			
	85	52			2.1 s			β−, n	~2.5	0.021, 0.032, 0.050, 0.113, 0.203, 0.427, …	
Selenium Se 34	70	36	−68.	0+	39 m			β+	4.4	0.15, 0.83, 0.87, 1.10, …	53
	71	37	−73.	(3/2,5/2)	4.5 m	+0.534		β+3.4	~0.6	0.046	85
	72	38		0+	8.5 d			ε		0.084, 0.254, 0.394, 0.402, 0.578, 1.08, …	42
	73m			(1/2)−	42 m			β+1.71, 1.63(85), ε(15)			0.4
	73	39	−73.23	(9/2+)	7.1 h		+1.0	β+1.32(70), …, ε(29)	2.74	0.360(99), 0.0669(65), 0.0757(50), …	
	74	40	−77.524	0+	**0.87%**			ε(100)	0.865		
	75	41	−77.471	5/2(+)	120 d	−1.02	+0.8			0.264(59), 0.136(57), 0.279(25), 0.121(17), …	
	76	42	−80.788	0+	**9.02%**						85
	77	43	−80.087	1−	**7.58%**						42
	78	44	−82.691	0+	**23.52%**				0.154		0.4
	79m				3.9 m					IT0.096(100)	
	79	45	−81.516		6.5 × 10⁴ y			β−0.15(100)			
	80	46	−83.475	0+	**49.82%**						0.61
	81m			(7/2+)	57 m			β−1.17(0.15), 1.05(0.05)		IT0.103(99.8)	

TABLE 8b-1. PROPERTIES OF NUCLIDES (Continued)

(1) Atomic number Z	(2) Symbol	(3) Name	(4) Mass number A	(5) Number of neutrons N	(6) Mass excess, amu ×10⁻³	(7) Spin and parity	(8) % abundance or half life	(9) Magnetic moment, nuclear magnetons	(10) Quadrupole moment, barns	(11) Mode of decay, energy, and intensity. MeV (%)	(12) β-decay Q values. MeV	(13) Energy and intensity of γ-ray transitions, MeV (%)	(14) 2,200-m/s neutron-absorption cross section, barns
34	Se	Selenium	81	47	−82.00	$(\tfrac{1}{2})-$	18 m			β⁻1.58(98), 1.03(1), 0.77(1), . . .	1.576	0.276(2), 0.290(1), 0.565(1), . . .	
			82	48	−83.29	0+	9.19%						0.05
			83m			$(\tfrac{1}{2})-$	69 s			β⁻3.5(50), 2.40(25), 1.50(25)		1.031(22), 0.356(16), 0.989(14), 0.676(13), . . .	
			83	49	−80.99	$(\tfrac{9}{2})+$	23 m			β⁻0.93, 1.51, . . .	3.58	0.356(73), 0.512(45), 0.226(34), 0.720(22), . . .	
			84	50	−81.5	0+	3.2 m			β⁻1.4(100)	1.8	0.408(100)	
			85	51			39 s			β⁻			
			86	52			16 s			β⁻			
			87	53			6 s			β⁻, n			
35	Br	Bromine	70?	35	−70.	0+, 1+	23 s			β⁺, p2.5	~12		
			74	39			36 m			β⁺4.7	5.7	0.630, 0.720, . . .	
			75	40	−74.26	$(\tfrac{3}{2}-)$	97 m			β⁺1.72(51), 1.1(10), 0.65(3), ε(30)	3.02	0.620, 0.285	
			76	41	−75.3	1(+)	16 h	±0.548	±0.25	β⁺3.58, 0.93, 1.80, ε		0.56, 0.65, . . .	
			77m			$\tfrac{9}{2}$	4.3 m					IT0.1076(100)	
			77	42	−78.62		56 h			ε(99), β⁺0.36(1)	1.371	0.239(27), 0.298(8), 0.578(7), . . .	
			78	43	−78.85	1+	6.4 m			β⁺2.52(82), . . . ε(7)	3.57	0.614(13)	
			79m			$\tfrac{9}{2}$	4.9 s					IT0.208(100)	
			79	44	−81.668	$\tfrac{3}{2}-$	50.54%	+2.106	+0.31				10.5
			80m			5−	4.42 h	+1.317	+0.71			IT0.048(100), 0.037(100)	
			80	45	−81.464	1−	17.6 m	+0.514	±0.18	β⁻2.02(84), 1.38(7), ε(6), β⁺0.866(3)	2.02	0.616(7), 0.666(1), . . .	
			81	46	−83.708	$\tfrac{3}{2}-$	49.46%	+2.270	+0.26		1.873		3.1
			82m			2−	6.1 m			β⁻1.66, 2.36		IT0.046(99.8), 0.777(0.2), . . .	

Z	Element	A	N	Mass excess	Jπ	T½ / abund.	μ (±1.626)	Q (±0.70)	Decay	γ-rays	E	σ
		82	47	−83.20	5−	35.4 h	±1.626	±0.70	β⁻0.44(98), ...	0.7768(83), 0.5543(70), 0.6191(43), ...	3.088	
		83	48	−84.83	(3/2)−	2.40 h			β⁻0.925(99), 0.395(1), ...	0.632(100), 0.093(100), 0.521(1)	0.967	
		84				6.0 m				0.88(75), 1.46(75), 0.44(68), 1.59(16)		
		84	49	−83.45		31.8 m			β⁻1.9(72), 0.8(20), ...	0.88(48), 1.9(10), 2.47(7), 1.01(6), ...	4.7	
		85	50	−84.5		3.0 m			β⁻2.5		2.8	
		86	51	−81.5		54 s			β⁻7.4(15), 5.6(15), ...	1.56, 1.36, 2.75, 5.44, ...	7.4	
		87	52	−80.		55 s			β⁻2.6(70), ..., n	1.44, 2.98, 2.56, 4.2, ...	6.5	
		88	53			16 s			β⁻, n			
		89	54			4.5 s			β⁻, n	0.76		
		90	55			1.6 s			β⁻, n			
36	Kr Krypton	74	38	−67.		16 m			β⁺3.1		4.1	
		75	39	−69.		5.5 m			β⁺, ε	0.316, 0.267, ...	~5	
		76	40	−74.	0+	14.3 h			ε(100)		~1.0	
		77	41	−75.5		1.2 h			β⁺1.85(40), 1.70(35), 1.87(3), ..., ε(19)	0.130(87), 0.146(42), 0.106(10), ...	2.99	
		78	42	−79.599	0+	0.35%						4.7
		79m			(7/2+)	55 s				IT0.127(100)		
		79	43	−79.93	(1/2−)	34.9 h			ε(91), β⁺0.604(7), ...	0.2613(11), 0.3974(10), 0.6059(10), ...	1.628	
		80	44	−83.62	0+	2.27%						14
		81m			1/2−	13 s				IT0.190(100)		
		81	45	−83.4	7/2+	2.1 × 10⁵ y			ε(100)		0.29	
		82	46	−86.518	0+	11.56%						37
		83m			1/2−	1.9 h				IT0.032(100), 0.0.3(100)		
		83	47	−85.869	9/2+	11.55%	−0.970	+0.26				170
		84	48	−88.495	0+	56.99%						0.13
		85m			1/2−	4.4 h			β⁻0.82(77)	IT0.1495(77), IT0.305(23)	0.67	
		85	49	−87.463	9/2+	10.76 y	±1.005	+0.43	β⁻0.66(99.6), ...	0.514(0.4)		
		86	50	−89.384	0+	17.37%						0.06
		87	51	−86.64	(5/2+)	76 m			β⁻3.5(55), 3.9(15), 1.3(14), 3.1(8), 1.5(6), ...	0.403(58), 2.57(14), 0.845(8), 2.01(3), 0.674(3), ...	3.89	
		88	52	−85.6	0+	2.80 h			β⁻0.52(68), 2.8(20), 0.9(12)	2.40(35), 0.19(35), 0.85(23)	2.8	

TABLE 8b-1. PROPERTIES OF NUCLIDES (Continued)

(1) Atomic number Z	(2) Symbol	(3) Name	(4) Mass number A	(5) Number of neutrons N	(6) Mass excess, amu ×10⁻³	(7) Spin and parity	(8) % abundance or half life	(9) Magnetic moment, nuclear magnetons	(10) Quadrupole moment, barns	(11) Mode of decay, energy, and intensity, MeV (%)	(12) β-decay Q values, MeV	(13) Energy and intensity of γ-ray transitions, MeV (%)	(14) 2,200-m/s neutron-absorption cross section, barns
36	Kr	Krypton	89	53	−82.19	....	3.2 m	....		β^-4.9, 4.6, 3.8, ...	5.2	0.221(25), 0.568(21), 1.533(11), 0.498(11), 1.472(10), 0.904(7), ...	
			90	54	−80.3	0+	33 s	....		β^-2.80(47), ...	4.56	0.120(65), 0.536(48), 1.11(48), ...	
			91	55	−77.	....	9 s	....		β^-~3.6	6.5	0.108(65), 0.509(25)	
			92	56	....	0+	2.0 s	....		β^-	....	0.14(50), ...	
			93	57	....	....	1.2 s	....		β^-	....		
			94	58	....	....	1.4 s	....		β^-	....		
			95	59	....	....	Short	....		β^-	....		
37	Rb	Rubidium	78	41	....	....	6.5 m	....			....	0.455, 0.664, 1.110, 1.943, 1.148. ...	
			79	42	−76.1	....	23 m	....		β^+	3.5	0.688, 0.183, 0.147, 0.143, 0.130, 0.622, ...	
			80	43	−77.	1+	34 s	....		β^+4.1(60), ...	5.8	0.616(40)	
			81m	..	....	9/2+	32 m	....		β^+1.4(50)	....	IT0.085(50)	
			81	44	−91.0	3/2−	4.7 h	+2.05		ϵ(68), β^+1.05(30), 0.575(2)	2.26	0.190(65), 0.446(22), 0.457(3), 0.538(2), ...	
			82m	..	....	5−	6.4 h	+1.643		β^+0.8(100)	....	0.7768(83), 0.5543(70), 0.619(41), 0.698(35), 1.044(31), 1.317(25), ...	
			82	45	−81.8	1+	1.3 m	....		β^+3.15(83), ... , ϵ(5)	4.17	0.7768(13), 1.384(1), ...	
			83	46	−85.	5/2−	83 d	+1.4		ϵ(100)	1.0	0.521(46), 0.530(30), ...	
			84 m	..	....	6+	20 m	....		ϵ?	....	IT0.464(52), 0.216(48), 0.250(48)	
			84	47	−85.619	2−	33.0 d	−1.32		ϵ(76), β^+0.80(11), 1.67(10), β^-0.89(3)	2.680, 0.886	0.883(73), ...	

A	N	Mass excess	J^π	Half-life	μ	Q	Decay	Energy	γ-rays	σ
85	48	−88.200	5/2−	72.15%	+1.3524	+0.26				0.45
86m			...	1.02 m					IT0.56(100)	
86	49	−88.822	2−	18.66 d	−1.691		β^-1.78(91), 0.70(9); ϵ(0.005)	1.78; 0.52	1.078(9)	
87	50	−90.814	3/2−	5×10^{10} y; 27.85%	+2.7500	+0.12	β^-0.274(100)	0.274		1.0
88	51	−88.7	2−	17.8 m	±0.51		β^-5.2(76), 2.5(14), 3.4(4), ...	5.2	1.863(21), 0.898(13), 2.68(2), ...	
89	52	−87.72	...	15.4 m			β^-1.26(38), 2.21(32), 4.49(18), 1.92(4), ...	4.49	1.03(60), 1.25(47), 2.19(17), 2.57(12), 0.659(10), 0.949(10), ...	
90			0+	4.3 m			β^-6.6, 5.8, 4.4, 2.2, ...	6.6	0.83(61), 4.34(18), 3.34(15), ...	
90	53	−85.2	(2−)	2.6 m			β^-4.6	5.7		
91	54	−84.	...	57 s			β^-	~7.9	0.58	
92	55	−81.	...	4.4 s			β^-, n	~7		
93	56	−78.	...	5.9 s			β^-, n			
94	57		...	2.7 s			β^-, n			
95	58		...	0.36 s			β^-, n			
96	59		...	0.23 s						
97	60		...	0.14 s						

Strontium Sr 38

A	N	Mass excess	J^π	Half-life	μ	Q	Decay	Energy	γ-rays	σ
80	42	−77.	0+	1.7 h			ϵ(100)	~4		
81	43	−81.	...	29 m			β^+	~0.6		
82	44	−82.3	0+	25 d			ϵ(100)			
83	45		7/2+	33 h			ϵ(80), β^+1.15(11), 0.81(7), ...	2.21	0.763(43), 0.385(35), 0.040(23), ...	
84	46	−86.570	0+	0.56%						0.88
85m			(1/2−)	70 m			ϵ(14)		IT0.007(85), 0.231(85), 0.150(14), 0.237(1)	
85	47	−87.06	(9/2+)	64 d			ϵ(100)	1.11	0.514(100)	
86	48	−90.724	0+	9.86%						0.8
87m			1/2−	2.83 h			ϵ(1)		IT0.388(99)	
87	49	−91.108	9/2+	7.02%	−1.093	+0.3				
88	50	−94.372	0+	82.56%						0.005
89	51	−92.53	5/2+	52 d			β^-1.46(99.9), ...	1.463	0.91(0.01)	0.5
90	52	−92.25	0+	28.1 y			β^-0.546(100)	0.546		0.9
91	53	−89.84	5/2+	9.67 h			β^-1.09(33), 1.36(29), 2.67(26), 0.62(7), 2.04(4)	2.67	1.025(30), 0.748(27), 0.645(15), 1.413(5), 0.93(3)	

Table 8b-1. Properties of Nuclides (Continued)

(1) Atomic number Z	(2) Symbol	(3) Name	(4) Mass number A	(5) Number of neutrons N	(6) Mass excess, amu ×10⁻³	(7) Spin and parity	(8) % abundance or half life	(9) Magnetic moment, nuclear magnetons	(10) Quadrupole moment, barns	(11) Mode of decay, energy, and intensity, MeV (%)	(12) β-decay Q values, MeV	(13) Energy and intensity of γ-ray transitions, MeV (%)	(14) 2,200-m/s neutron-absorption cross section, barns
38	Sr	Strontium	92	54	-89.0	0+	2.71 h			$\beta^-0.55(90), 1.5(10),$	1.9	1.37(90), 0.44(4), 0.23(3)	
			93	55	-85.8		8 m			$\beta^-2.9(65), 2.6(25),$ 3.9(14),...	4.3	0.60, 0.8, 1.2, ...	
			94	56	-84.6	0+	1.3 m			$\beta^-2.1$	3.5	1.42(100)	
			95	57	$-81.$		26 s			β^-	~5.7		
			96	58			4.0 s			β^-			
			97	59			~0.4 s			β^-			
39	Y	Yttrium	82	43			Short						
			83	44	$-78.$		7.5 m						
			84	45	-79.11		41 m			$\epsilon, \beta^+3.5, 2.9$	6.3	0.795(100), 0.982(100), 1.041(50), ...	
			85m			$(\tfrac{1}{2}-)$	2.8 h			$\beta^+1.54(50), 1.1, \ldots,$ $\epsilon(45)$		0.92(9), 0.503, 0.70, 0.77	
			85	46	-83.56	$(\tfrac{9}{2}+)$	4.9 h			$\beta^+2.24(55), 2.1(10),$ $1.1(4), \ldots, \epsilon(30)$	3.26	0.231(13), 2.16(9), 0.77(8), ...	
			86m			5–	48 m					IT0.102(100), 0.208(100)	
			86	47	-85.05	4–	14.6 h			$\epsilon(73), \beta^+1.2(11), \ldots$	5.27	1.077(82), 0.63(37), 1.16(35), 0.778(21), ...	
			87m			$(\tfrac{9}{2}+)$	14 h			$\beta^+(5)$	1.9	IT0.381(99)	
			87	48	-89.09	$(\tfrac{1}{2}-)$	80 h			$\epsilon(99.7), \beta^+0.7(0.3)$		0.483, 0.388	
			88	49	-90.49	$(4-)$	108 d			$\epsilon(99.8), \beta^+0.76(0.2)$	3.621	1.836(100), 0.898(91)	
			89m			$\tfrac{9}{2}+$	16 s					IT0.91(100)	
			89	50	-94.133	$\tfrac{1}{2}-$	100%	-0.1373					0.001
			90m			$(7+)$	3.1 h			$\beta^-(0.4)$		IT0.202(99.6), 0.483 (99.6)	
			90	51	-92.84	2–	64.2 h	-1.63	-0.15	$\beta^-2.27(99.8), \ldots$	2.27	1.75(0.2)	
			91m			$\tfrac{9}{2}+$	50 m	±0.164				IT0.551(100)	1.4
			91	52	-92.71	$\tfrac{1}{2}-$	58.8 d	0.164		$\beta^-1.545(99.7), \ldots$	1.545	1.21(0.3)	
			92	53	-91.07	2–	3.53 h			$\beta^-3.63(86), \ldots$	3.63	0.934(14), 1.40(5), ...	
			93	54	-90.45	$(\tfrac{1}{2}-)$	10.2 h			$\beta^-2.89(90), \ldots$	2.89	0.267(6), 0.94(2), 1.90(2), ...	
			94	55	-88.3		20.3 m			$\beta^-5.0(50), \ldots$	5.0	0.92(43), 0.56(6), ...	

Z	Element	Symbol	A	N	Δ (MeV)	Jπ	T½	μ	Particle energy, MeV (abundance)	Disint. energy (MeV)	γ-ray energy, MeV (%)
			95	56	−87.2	(1/2−)	10.7 m		β^-4.43(82), 1.31(5), 0.86(4), ...	4.43	0.953, 2.175, 1.323, 2.631, 3.576, ...; 0.7, 1.0, 1.5, ... (0.05, 1.0, 0.15, <4, 0.075)
			96	57	−84.		2.3 m		β^-3.5	~7	
			97	58	−82.		1.11 s		β^-	~6	
			99	60			~0.8 s		β^-		
40	Zirconium	Zr	81	41			~10 m				
			82	42		0+	9.5 m				
			83	43			5–10 m				
			84	44		0+	16 m				
			85	45			15 m		ϵ, β^+	~5	0.04
			85	45	−78.		1.4 h		ϵ, β^+		
			86	46	−84.	0+	16.5 h		ϵ(100)	~1	0.243(96), 0.028(20), 0.612(5) (0.05)
			87	47	−85.33		1.6 h		β^+2.10(83), ϵ(17)	3.50	0.394(97)
			88	48	−90.8	0+	85 d		ϵ(100)	~0.7	IT0.588(94), 1.51(6)
			89m	49		(1/2−)	4.18 m		ϵ(5), β^+0.89(1), ...		
			89	49	−91.091	(5/2+)	78.4 h		ϵ(78), β^+0.90(22)	2.834	0.909(99), 1.713(11), ...
			90m			5−	0.81 s				IT2.32(94), 0.133(6), 2.18(6)
			90	50	−95.300	0+	51.46%				
			91	51	−94.358	5/2+	11.23%	−1.303			
			92	52	−94.961	0+	17.11%				
			93	53	−93.552	5/2+	1.5×10⁶ y		β^-0.060(95), ...	0.063	
			94	54	−93.680	0+	17.40%				
			95	55	−91.965	5/2+	65 d		β^-0.362(55), 0.399(43), 0.88(1), ...	1.121	0.756(55), 0.723(43), 0.235(1)
			96	56	−91.718	0+	2.80%				
			97	57	−89.03		17.0 h		β^-1.93(86), 0.56(5), 1.42(3), ...	2.67	0.743(95), 0.508(5)
			98	58	−87.25	0+	31 s		β^-2.1	2.2	
			99	59	−84.		2.4 s		β^-	~4.5	
			100	60	−83.	0+	~1 s		β^-	~3	
			101	61	−78.		~3.3 s		β^-	~6.5	
41	Niobium	Nb	88	47	−82.		14 m		β^-3.2, ..., ϵ	~7.2	1.08, 1.06, 0.67, 0.40, 0.27, 0.075, 1.083
			89m			(1/2−)	42 m		β^+3.1, ϵ	3.9	0.588
			89	48	−86.9	(5/2+)	1.9 h		β^+2.9, ϵ		1.626, 3.577, 3.838, ...
			90m			(4)−	20 s				IT<0.003, 0.122
			90	49	−88.74	(8+)	14.6 h		β^-1.5(55), ϵ(45)	6.11	1.14(97), 2.32(82), 0.142(75), 2.18(14)
			91m			(1/2−)	62 d		ϵ(3)		IT0.1045(97), 1.21(3)
			91	50	−93.00	(5/2+)	Long		ϵ(100)	1.26	
			92m			2+	10.16 d		ϵ(99.9), β^+		0.934(99), 0.9(2), 1.8(1)

TABLE 8b-1. PROPERTIES OF NUCLIDES (*Continued*)

(1) Atomic number Z	(2) Symbol	(3) Name	(4) Mass number A	(5) Number of neutrons N	(6) Mass excess, amu $\times 10^{-3}$	(7) Spin and parity	(8) % abundance or half life	(9) Magnetic moment, nuclear magnetons	(10) Quadrupole moment, barns	(11) Mode of decay, energy, and intensity, MeV (%)	(12) β-decay Q values, MeV	(13) Energy and intensity of γ-ray transitions, MeV (%)	(14) 2,200-m/s neutron-absorption cross section, barns
41	Nb	Niobium	92	51	-92.79	(7+)	>350 y						
			93m	..	..	1/2-	13.6 y					IT0.0304(100)	
			93	52	-93.620	9/2+	**100%**	+6.167	-0.22				1.0
			94m	..	..	3+	6.3 m			β^-1.2(0.2), ...		IT0.0407(99+), 0.871(0.2)	11
			94	53	-92.70	6+	2.0×10^6 y			β^-0.5(100)	2.06	0.702(100), 0.871(100)	
			95m	..	..	(1/2-)	90 h					IT0.253(100)	<7
			95	54	-93.170	(9/2+)	35.1 d			β^-0.160(99), ...	0.925	0.765(99)	
			96	55	-91.90		23.4 h			β^-0.748(95), 0.50(5)	3.19	0.778(97), 0.569(59), 1.092(49), 0.459(28), 0.851(22), 1.200(21),	
			97m	..	..	(1/2-)	1.0 m					IT0.747(100)	
			97	56	-91.90	(9/2+)	72 m			β^-1.27(98), ...	1.93	0.658(98), 1.0(2)	
			98m	..	..		2.8 s			β^-4.3		0.780, 1.020, . . .	
			98	57	-89.6	(4,5+)	51 m			β^-2.32(38), 1.94(29), 1.42(25), 3.1(8)	4.6	0.780(100), 0.720(75), 1.44(10), 1.68(10), 1.93(8), 0.330(8)	
			99m	..	..		10 s			β^-		IT	
			99	58	-89.		2.4 m			β^-3.2, . . .	~3	0.100(1), 0.260(1)	
			100				3.0 m			β^-		0.53(100), 0.36(55), 0.45(40), 0.140(10), ...	
			100	59	-86.		11 m			β^-3.5(45), 3.1(45), 4.2(10)	~6	0.53(100), 0.62(60), 1.04(10), 1.15(10), 1.47(5)	
			101		-85.		1.0 m			β^-	~4		
42	Mo	Molybdenum	88	46	-77.		27 m			β^+2.5	~5	2.69	
			89	47	-81.		7 m			β^+4.9, 4.0	~6		
			90	48	-86.1	0+	5.7 h			ϵ(75), β^+1.09(25)	2.49	0.257(85), 0.122(71), 0.942(10), 0.445(9), 1.273(8), ...	

		Z	Mass excess	Spin-parity	Half-life / abundance	μ	±	Decay		Gamma rays	σ
91m	...	...	...	...	66 s	...	...	...	...	IT0.658(60), 1.21(20), 1.53(15)	
91		49	−88.23	(9/2+)	15.49 m	...	...	β+2.78(30), 3.99(5), ...	...	...	
92		50	−93.192	0+	15.84%	...	...	β+3.44(94), ε(6)	4.46	...	
92		...	...	(2,3,+)	6.9 h	...	...			IT0.685, 1.479, 0.264	<0.3
93m		51	−93.19	5/2+	3000 y	...	...	ε(100)	0.40	...	
93		52	−94.910	0+	9.04%						14.4
94		53	−94.163	5/2+	15.72%	−0.9133	±0.12	...	...	...	1.0
95		54	−95.326	0+	16.53%						1.9
96		55	−93.977	5/2+	9.46%	−0.9325	±1.1	...	...	...	0.14
97		56	−94.591	0+	23.78%						
98		57	−92.28	1/2+	66.7 h	...	...	β−1.23(82), 0.45(17), 0.86(1), ...	1.37	0.141(99), 0.740(14), 0.181(7), 0.778(4), 0.041(2), ...	
99		58	−92.522	0+	9.63%				2.82		0.20
100		59	−89.65	(5/2+)	14.6 m	...	...	β−0.74(38), 1.61(25), 0.84(13), 1.23(11), 2.23(10), ...		0.191(25), 1.02(25), 0.59(21), 2.08(16), 0.89(15), ...	
101		60	−90.	0+	11 m	...	...	β−1.2	1.2	...	
102		...	...	...	5 h	...	...	β−		0.54, 1.3, 1.7, 1.1, 1.5	
103		51	−86.	0+	62 s	...	...	β−	~4	0.070	
104		62	...	...	1.3 m	...	...	β−4.8, 2.2			
105		63	...	0+	40 s	...	...	β−			
106		64	...	...	9.5 s	...	...	β−			

43 Tc Technetium

		Z	Mass excess	Spin-parity	Half-life			Decay		Gamma rays	
92		49	−84.7	(8,9+)	4.4 m	...	...	β+4.1(92), ε(8)	7.9	1.54(100), 0.79(95), 0.33(90), 0.14(67), 0.24(30), 0.090(20)	
93m		...	...	(1/2−)	43 m	...	...	ε(18)		IT0.390(82), 2.66(18)	
93		50	−89.77	(9/2+)	2.7 h	...	...	ε(87), β+0.82(11), 0.68(2)	3.19	1.35(65), 1.49(33)	
94m		...	...	2+	53 m	...	...	β+2.47(63), ..., ε(34)		0.871(91), 1.53(6), 1.87(6), 2.74(4), 0.993(2), ...	
94		51	−90 337	6, 7+	293 m	...	...	ε(89), β+0.816(11)	4.26	0.849(100), 0.702(100), 0.871(100), ...	
95m		...	...	(1/2−)	61 d	...	...	ε(95), β+0.71(0.3), 0.49(0.1)		IT0.0389(4), 0.204(70), 0.584(36), 0.838(27), ...	
95		52	−92.34	(9/2+)	20.0 h	...	...	ε(100)	1.70	0.768(82), 0.84(11), 1.06(4)	
96m		...	...		52 m	...	...	β+2.0, 0.4(0.01)		IT0.0344(99+)	

TABLE 8b-1. PROPERTIES OF NUCLIDES (Continued)

(1) Atomic number Z	(2) Symbol	(3) Name	(4) Mass number A	(5) Number of neutrons N	(6) Mass excess, amu ×10⁻³	(7) Spin and parity	(8) % abundance or half life	(9) Magnetic moment, nuclear magnetons	(10) Quadrupole moment, barns	(11) Mode of decay, energy, and intensity, MeV (%)	(12) β-decay Q values, MeV	(13) Energy and intensity of γ-ray transitions, MeV (%)	(14) 2,200-m/s neutron-absorption cross section, barns
43	Tc	Technetium	96	53	−92.17		4.3 d			ϵ(100)	2.9	0.778(100), 0.851(100), 0.81(84), 1.12(16), 0.32(5)	
			97m			$(\tfrac{1}{2}-)$	90 d					IT0.0965(100)	
			97	54	−93.61	$(\tfrac{9}{2}+)$	2.6 × 10⁶ y			ϵ(100)	~0.3		
			98	55	−92.9	(7,6+)	1.5 × 10⁶ y			β^-0.30(100)	1.7	0.76(100), 0.66(100)	22
			99m			$\tfrac{1}{2}-$	6.0 h					IT0.002, 0.141	
			99	56	−93.751	$\tfrac{9}{2}+$	2.14 × 10⁵ y	+5.68	+0.3	β^-0.292(100)	0.292		
			100	57	−92.16	1+	17 s			β^-3.38, 2.88, 2.2	3.4	0.540, 0.60, 0.71, 0.81	
			101	58	−92.67	$(\tfrac{5}{2}+)$	14.0 m			β^-1.32(90), 1.07(10)	1.63	0.307(91), 0.545(8) · · ·	
			102			(1)+	5 s			β^-4.4		0.47, 0.63, 1.1, 2.19, 1.60,	
			102	59	−91.		4.5 m			β^-2.2	~4.5		
			103	60	−91.2		50 s			β^-2.2, 2.0	2.4	0.135, 0.21, 0.35	
			104	61	−89.		18 m			β^-4.6, 3.3, 2.4, 1.8, · · ·	~5.3	0.36, 0.53, 0.89, 1.6, 2.5, · · ·	
			105	62	−88.7		7.7 m			β^-3.4	3.4	0.159, 0.143, 0.108, 0.321	
			106	63	−86.		37 s			β^-	~6.5	· · ·	
			107	64			29 s			β^-			
44	Ru	Ruthenium	93	49	−83.		50 s			β^+	~6		
			94	50	−88.64		52 m			ϵ		0.37, 0.89	
			95	51	−89.59	$(\tfrac{5}{2}+)$	1.65 h			ϵ(85), β^+1.1(10), 1.33(3), 0.7(1)	1.58, 2.35	0.340(70), 1.09(21), 0.625(13), 1.43(5)	
			96	52	−92.402	0+	**5.51%**						0.21
			97	53	−92.4	$(\tfrac{5}{2}+)$	2.9 d			ϵ(100)	1.2	0.215(91), 0.324(8)	
			98	54	−94.711	0+	**1.87%**						<8
			99	55	−94.063	$\tfrac{5}{2}+$	**12.72%**	−0.63					9.9
			100	56	−95.782	0+	**12.62%**						9.7
			101	57	−94.423	$\tfrac{5}{2}+$	**17.07%**	−0.69					2.9

Element	Symbol	Z	A	N	Mass excess	Spin/parity	Half-life / abundance	Decay	μ	Disint. energy	Gamma energies	σ (barns)
			102	58	−95.652	0+	**31.61%**					1.44
			103	59	−93.67	(5/2+)	39.6 d	β−0.21(89), 0.1(7, 0.70(3), …		0.74	0.040(100), 0.497(88), 0.610(6), …	
			104	60	−94.572	0+	**18.58%**					0.47
			105	61	−92.25	(5/2+)	4.44 h	β−1.14, 1.08, 0.91, 1.9, 0.52		1.92	0.726(48), 0.470(18), 0.678(17), 0.317(11), 0.970(2), …	0.20
			106	62	−92.68	0+	357 d	β−0.0394(100)		0.0394	0.0394(100)	0.146
			107	63	−89.9		4.2 m	β−3.2, 2.1, 2.3, …		3.2	0.195(14), 0.86(7), 0.93(4), 1.03(4), 1.29(4), …	
			108	64	−90.0	0+	4.5 m	β−1.3(72), 1.1(28)		1.3	0.165(28)	
			109	65			35 s	β−				
Rhodium	Rh	45	95	50	−84.		5 m	ε, β+		~5	0.942, 1.36	
			96 m				1.5 m	ε, β+			IT?	
			96	51	−85.49		9 m	ε, β+		6.44	0.83, 0.63, 0.69	
			97	52	−88.6		32 m	β+2.47, 2.06, 1.8		3.49	0.08, 0.19, 0.43, 0.26, 0.86, 1.57, 2.54, …	
			98 m				3 m	β+2.5, 3.3			IT	
			98	53	−89.28		8.7 m	ε(90), β+0.74(10)		5.1	0.65(100)	
			99 m				4.7 h	ε, β+1.03, 0.71, 0.59, 0.44		2.05	0.34(70), 0.62(20), …	
			99	54	−91.86		16 d				0.528, 0.089, 0.353, 0.322, 0.618, …	
			100	55	−91.89	1, 2−	20 h	ε(93), β+2.62(3), 2.07(3), …		3.64	0.540(88), 2.37(39), 0.820(25), 1.55(23), 1.35(20), …	
			101 m			(9/2+)	4.5 d	ε(90)			IT0.157(10), 0.307(83), 0.545(6), …	
			101	56	−93.82	(1/2−)	3 y	ε(100)		0.56	0.127(88), 0.198(75), 0.325(11)	
			102				2.9 y	ε(100)			0.475(95), 0.63(54), 0.698(41), 1.05(41), 0.768(30), …	
			102	57	−93.16		206 d	ε(72), β+1.30(9), 0.83(3), β−1.15		2.32	0.475(57), 0.628(4), 1.103(3), …	
			103	58	−94.489	1/2−	**100%**		−0.0883	1.15		150
			104 m			5+	4.30 m	β−(0.18)			IT0.0319(99+), 0.0457(97), 0.051(47), …	800
			104	59	−93.337	1+	43 s	β−2.5(98), 1.9(1.9), … ε(0.5)		2.47	0.56(1.9), 1.24(0.1)	40

TABLE 8b-1. PROPERTIES OF NUCLIDES (*Continued*)

(1) Atomic number Z	(2) Symbol	(3) Name	(4) Mass number A	(5) Number of neutrons N	(6) Mass excess, amu ×10⁻³	(7) Spin and parity	(8) % abundance or half life	(9) Magnetic moment, nuclear magnetons	(10) Quadrupole moment, barns	(11) Mode of decay, energy, and intensity, MeV (%)	(12) β-decay Q values, MeV	(13) Energy and intensity of γ-ray transitions, MeV (%)	(14) 2,200-m/s neutron-absorption cross section, barns
45	Rh	Rhodium	105m	...	.....	$(\frac{1}{2}-)$	45 s				.....	IT0.129(100)	
			105	60	−94.31	$(\frac{7}{2}+)$	35.6 h			β^-0.565(75), 0.246(20),	0.565	0.319(19), 0.306(5)	18,700
			106m	...	.....	......	130 m			β^-0.92(66), 0.70(22), 1.31(3), ...	.....	0.512(88), 0.735(41), 0.451(35), 0.82(35), 1.04(34), ...	
			106	61	−92.72	1+	30.4 s			β^-3.54(79), 2.4(11), 3.1(8), ...	3.54	0.512(21), 0.623(11),	
			107	62	−93.25	$(\frac{5}{2})+$	22 m			β^-1.20(71), 1.12(8), 0.84(7), ...	1.51	0.305(73), 0.390(11), 0.68(3), ...	
			108	63	−91.3	1+	17 s			β^-4.5(50), ...	~4.5	0.434(43), 0.62(22), 0.51(10), ...	
			109m	...	.....	......	50 s				.....	IT0.11	
			109	64	−91.		30 s			β^-	~2.5	0.49, 0.31	
			110	65	−89.0	1+	5 s			β^-5.5, ...	5.5	0.374	
46	Pd	Palladium	98	52	−87.	0+	17 m			ε	~2	0.132?	
			99	53	−88.2		22 m			β^+2.0	3.4	0.140, 0.275, 0.420, 0.67	
			100	54	−91.	0+	4.0 d			ε(100)	~0.4	0.084(49), 0.074(34), 0.126(16), 0.159(4)	
			101	55	−91.69	$(\frac{5}{2}+)$	8.4 h			ε(97.5), β^+0.78(2), 0.49(0.5)	1.99	0.296(30), 0.590(24), 0.566(7), 0.723(5), ...	
			102	56	−94.39	0+	**0.96%**						
			103	57	−93.89	$(\frac{5}{2}+)$	17 d			ε(100)	0.56	0.040(100), ...	
			104	58	−95.99	0+	**10.97%**						
			105	59	−94.91	$\frac{5}{2}+$	**22.23%**	−0.642	+0.8				4.8
			106	60	−96.51	0+	**27.33%**						
			107	61	−94.871	$\frac{5}{2}+$	~7×10⁶ y			β^-0.035(100)	0.035		0.28
			108	62	−96.109	0+	**26.71%**						
			109m	...	.....	$(\frac{1}{2}+)$	4.7 m					IT0.188(60), 0.08(40), 0.11(40)	

Z	Element	A	N	Mass excess	J	Half-life	μ	Decay	Q	γ (MeV) (intensity)	σ
		109	63	−94.046	5/2+	13.47 h		β⁻1.03(99+), . . .	1.115	0.088(89), . . .	0.4
		110	64	−94.84	0+	11.81%					
		111m				5.5 h		β⁻(25)		IT0.17(75), 0.07	
		111	65	−92.33		22 h		β⁻2.2(100)	2.2	0.38, 0.58, . . .	
		112	66	−92.61	0+	21 h		β⁻0.28(100)	0.30	0.0185(20)	
		113	67			1.4 m		β⁻			
		114	68		0+	2.4 m		β⁻			
		115	69			40 s		β⁻			
		117	71			5 s		β⁻~4.5(73), ~4.4(27)			
		118	72		0+	3.1 s		β⁻			
47	Ag Silver	99	52	−81.7		2.8 m		β⁺3.32(53), 1.7(24), 2.45(23)	6.0	1.04(45), 1.70(40), 1.53(32), . . .	
		100	53	−84.		9 m		δ⁺ 5.8		0.728	
		101	54	−87.0		11 m		δ⁺2.73(44), 2.18(23), 1.56(17), . . . , ε	4.4	0.26, 1.16, 0.65, 0.67, 0.58, . . .	
		102m			2	7.7 m	+4	β⁺3.4, 4.06, 3.07, ε(13)		0.86, 0.73, 1.60, 0.558, . . .	
		102	55	−88.4	5	4 m		β⁺2.4(40), 2.9, 3.3, 3.6	5.56	0.55, 0.78, 1.27, 2.06	
		102				13 m		β⁺2.26(37), 1.90(13), 1.50(3), . . . , ε(45)		0.558(85), 0.727(65), 1.80(42), 2.07(20), . . .	
		103m			(1/2−)	5.7 s	+4.4	ε(50), β⁺1.7, 1.3	2.7	IT0.138	
		103	56	−91.0	7/2+	66 m		β⁺2.7(60), ε(10)		0.27(34), 0.12(26), 0.15(23), 0.24(10), 1.16(9), . . .	
		104m			2+	30 m	+3.7	ε, β⁺0.99	4.10	IT0.02(30), 0.556(70)	
		104	57	−91.58	5+	67 m	+4.0	ε(100), β⁺		0.556(84), 0.764(48), 0.854(30), 1.34(8), 1.62(8), . . .	
		105	58	−93.48	1/2−	40 d	±0.101	ε(100)	1.34	0.344(42), 0.280(32), 0.064(10), 0.443(10), 0.644(10), 1.088(4), . . .	
		106m			6+	8.4 d				0.512(20)	
		106	59	−93.32	1+	24.7 m	+2.9	β⁺1.96(50), . . . , ε(30)	2.97	0.115(55), 0.061(15), 0.153(9), 0.752(7), . . .	
		107m			7/2+	44.3 s		ε(90)		IT0.093(100)	
		107	60	−94.909	1/2−	51.82%	−0.1135				35
		108m			(6+)	>5 y		β⁻1.6(96), . . .	1.64	IT0.03(9), 0.08(9), 0.722(90), 0.614(90)	
		108	61	−94.047	1+	2.42 m	+2.80	ε(2), β⁺0.88(0.28)	1.92	0.632(2), 0.434(0.45)	

TABLE 8b-1. PROPERTIES OF NUCLIDES (*Continued*)

(1) Atomic number Z	(2) Symbol	(3) Name	(4) Mass number A	(5) Number of neutrons N	(6) Mass excess, amu $\times 10^{-3}$	(7) Spin and parity	(8) % abundance or half life	(9) Magnetic moment, nuclear magnetons	(10) Quadrupole moment, barns	(11) Mode of decay, energy, and intensity, MeV (%)	(12) β-decay Q values, MeV	(13) Energy and intensity of γ-ray transitions, MeV (%)	(14) 2,200-m/s neutron-absorption cross section, barns
47	Ag	Silver	109m			7/2+	39.8 s	±4.3				IT0.088(100)	
			109	62	-95.244	1/2-	**48.18%**	-0.1305					93
			110m			6+	253 d	+3.604		β⁻0.087(67), 0.53(31), · · ·		0.658(96), 0.885(71), 0.937(32), 0.764(213), 1.384(21), · · · ,	82
			110	63	-93.886	1+	24.4 s	+2.85		β⁻2.88(96), 2.2(4), · · · ε(0.3)	2.88 / 0.87	IT0.116(1) 0.658(4), · · ·	
			111m			(7/2+)-	74 s			β⁻1.05(93), 0.71(6), 0.80(1), · · ·	1.05	IT0.065(100)	
			111	64	-94.70	1/2-	7.5 d	-0.145				0.34(6), 0.25(1)	
			112	65	-92.94	2-	3.2 h	±0.054		β⁻3.96(54), 3.35(22), 1.96(10), · · ·	3.96	0.617(41), 1.40(5), 1.63(3), 2.11(3). 2.55(2), · · ·	
			113m			1/2-	1.2 m			β⁻ <2		0.14, 0.30, 0.39, 0.56, 0.70	
			113	66	-93.44		5.3 h	±0.159		β⁻	2.00	0.30(100), 0.67(17)	
			114	67	-91.		5 s			β⁻4.6	4.6	3.12(10), 0.58(5). · · ·	
			115m			(1/2-)	49 s			β⁻~3.2		0.57	
			115	68	-91.1		21 m			β⁻3.2, 3.0, 2.7, 2.5, 2.1, 1.0, · · ·	3.2	0.22(46), 0.28(13), 2.12(13), 0.14(12), 1.48(11), 0.36(11), · · ·	
			116	69	-89.		2.5 m			β⁻5.0, · · ·	~6.3	0.52(10), 0.70(2)	
			117	70			1.1 m			β⁻			
			118	71	-81.0		5.3 s			β⁻			
48	Cd	Cadmium	101	53	-85.		1.2 m			β⁺	5.5	0.118, 0.481, · · ·	
			102	54	-86.	0+	5.5 m			β⁺	~3	0.22, 0.63, 0.85	
			103	55			10 m			β⁺	~4.4	0.084	
			104	56	-90.	0+	57 m			ε(100)	~1.2		

Z	El	A	N	Δ (MeV)	J	t½ / abund.	μ	Q	Decay (MeV)	E (MeV)	γ-rays (MeV)	σ (b)
		105	57	−91.	5/2	55 m	−0.74	+0.5	β+1.69, ...	~2.8	0.31, 0.34, 0.35, 0.43, 0.61, 1.9, 2.0, 2.3	1.0
		106	58	−93.537	0+	**1.22%**						
		107	59	−93.388	5/2+	6.5 h	−0.6144	+0.8	ε(99.7), β+0.302(0.3)	1.417	0.093(100), ...	1.5
		108	60	−95.811	0+	**0.88%**						700
		109	61	−95.050	5/2+	453 d	−0.8270	+0.8	ε(100)	0.182	0.088(100)	0.1
		110	62	−96.990	0+	**12.39%**						
		111m			11/2−	48.6 m	−1.11	−1.0			IT0.150, 0.247	
		111	63	−95.814	1/2+	**12.75%**	−0.5943					0.06
		112	64	−97.238	0+	**24.07%**						
		113m			11/2−	14 y	−1.087	−0.8	β−0.58(100)		IT?	
		113	65	−95.592	1/2+	**12.26%**	−0.6217					20,000
		114	66	−96.637	0+	**28.86%**						0.34
		115m			11/2−	43 d	−1.040	−0.6	β−1.62(97), 0.68(1.6),		0.935(2), 1.29(1), ...	
		115	67	−94.57	1/2+	53.5 h	−0.6477		β−1.11(60), 0.58(31), ...	1.45	0.336(95), 0.526(26), 0.492(10), ...	
		116	68	−95.238	0+	**7.58%**						0.077
		117m				3.4 h			β−0.67, 0.41		0.273(18), 1.998(15), 1.24(11), 0.880(10), 1.433(10), 1.408(8), ...	
		117	69	−92.76	1/2+	2.4 h			β−0.65, 0.79, 2.23, ...	2.52	0.273(31), 1.303(19), 0.345(18), 1.577(17), 0.314(16), 0.897(7), ...	
		118	70	−93.08	0+	50 m			β−	~0.8	0.8?	
		119m	71	−90.4		2.7 m			β−3.5			
		119				10 m			β−3.5	3.5		
		121	73	−86.30		13 s			β−	3.5		
49	In	106	57	−89.6	2,3+	5.3 m			β+4.89, 2.7	6.7	0.53, 0.63, 0.86, 1.66, 0.99, ...	
	Indium	107	58	−90.3	(6,7+)	32 m			β+2.2, ..., ε	3.5	0.22(46), 0.32, 0.73, 0.84, 0.94, 1.05, 1.25, ...	
		108m			(9/2+)	39 m			ε3.50, 2.66, 2.28, ε	5.15	0.383, 0.633, 0.842	
		108	59			56 m			β+1.29, ε	5.11	0.633, 0.842, 0.872, 0.243, 0.150, ...	
		109m2			(1/2−)	0.21 s					IT0.68(100), 1.44(80)	
		109m1				1.3 m					IT0.658(100)	
		109	60	−92.88	9/2+	4.3 h	+5.53	+0.86	ε(94), β+0.79(6)	2.02	0.205, 0.28, 0.35, 0.65, 0.91	

TABLE 8b-1. PROPERTIES OF NUCLIDES (Continued)

(1) Atomic number Z	(2) Symbol	(3) Name	(4) Mass number A	(5) Number of neutrons N	(6) Mass excess, amu $\times 10^{-3}$	(7) Spin and parity	(8) % abundance or half life	(9) Magnetic moment, nuclear magnetons	(10) Quadrupole moment, barns	(11) Mode of decay, energy, and intensity, MeV (%)	(12) β-decay Q values, MeV	(13) Energy and intensity of γ-ray transitions, MeV (%)	(14) 2,200-m/s neutron-absorption cross section, barns
49	In	Indium	110m			7(+)	4.9 h	+10.4 or −10.7	−0.21 or +0.22	ϵ, β^+?		0.66, 0.91	
			110		−92.77	2+	67 m	+4.36	+0.36	$\beta^+2.20(71),\ldots,(29)$	3.93	0.658(99),...	
			111m				7.3 m					IT0.539	
			111		−94.93	9/2+	2.81 d	+5.53	+0.85	$\epsilon(100)$	0.83	0.247(94), 0.173(89),...	
			112m			4+	20.7 m			$\beta^+0.66(44)$	0.66	IT0.156(100)	
			112	63	−94.46	1+	14 m	+2.81	+0.089	$\epsilon(34), \beta^+1.56(21)$	2.59	0.617(6)	
			113m			1/2−	100 m	−0.210				IT0.393(100)	
			113	64	−95.91	9/2+	4.28%	+5.523	+0.82				10.7
			114m			5+	50.0 d	+4.7		$\epsilon(<0.02)$		IT0.1916(96.5), 0.724(3.5), 0.556(3.5)	
			114	65	−95.10	1+	72 s			$\beta^-1.986(98),\ldots, \epsilon(2), \beta^+0.4(0.004)$	1.986	1.30(0.2)	
			115m			1/2−	4.50 h	−0.244		$\beta^-0.83(5)$	1.44	IT0.335(95)	
			115	66	−96.13	9/2+	6×10^{14} y; 95.72%	5.534	+0.83	$\beta^-0.49(100)$	0.49		198
			116m2				2.16 s					IT0.164(100)	
			116m1			5+	54.0 m	+4.3		$\beta^-1.00(49), 0.87(40), 0.60(11),\ldots$		1.293(80), 1.09(53), 0.417(36), 2.111(20), 0.819(17), 1.508(11),...	
			116	67	−94.74	1+	14 s			$\beta^-3.3(99),\ldots$	3.27	1.293(1),...	
			117m				1.93 h	−0.2515		$\beta^-1.77(37), 1.62(16)$		IT0.314(47), 0.158(16)	
			117	68	−95.47	9/2+	44 m			$\beta^-0.74(100)$	1.47	0.56(100), 0.158(100)	
			118m2				8.5 s			$\beta^-1.8(1)$		IT0.138(99)	
			118m			(4,5+)	4.4 m			$\beta^-1.3(53), 2.0(32),\ldots$		1.23(97), 1.05(80), 0.69(41),...	
			118	69	−93.9	1+	5 s			$\beta^-4.2(80), 3.0(16),\ldots$	4.2	1.23(15)	
			119m			(1/2−)	18 m			$\beta^-2.7,\ldots,(95)$		0.023, 0.91, IT0.30(5)	

Z	El	Name	A	N	Δ	Jπ	T½ or abundance	μ	Decay	E (MeV)	γ-ray energies (MeV)	σ (b)
			119	70	−94.2	(9/2+)	2.4 m		β⁻1.6(100)	2.35	0.82(95), 0.73(5)	
			120m?			(1)+	3.2 s		β⁻~5.6(~85)		1.17(~15)	
			120	71	−92.	(4,5+)	46 s		β⁻2.2(41), 3.1(27), ...	5.6	1.171(100), 1.02(61), 1.28(14), 0.090(12), 0.71(12), 0.94(12), ...	
			121				3.1 m		β⁻3.7	3.4	0.94	
			121	72	−92.1		30 s		β⁻	~7	0.99, 1.14	
			122	73	−89.		8 s		β⁻5	4.4		
			123	74	−89.		36 s		β⁻4.4	7.4		
			124	75	−86.8		4 s		β⁻5	~2		
50	Sn	Tin	108	58	−88.	0+	9 m		ε			
			109n				1.5 m		β⁺		1.13, 0.99, 3.21	
			109	59	−8.9		18.1 m		ε(80), β⁺1.52(20)	~4	0.28, 0.42 IT	
			110	60	−92.13	0+	4.0 h		ε	0.59	1.12(50), 0.65(44), 0.33(26), 1.55(18), ...	
			111	61	−92.24		35 m		ε(70), β⁺1.51(30)	2.52	0.283(95)	
			112	62	−95.17	0+	0.96%				1.14(2), 0.75(1), 1.89(1), ...	1.15
			113m	63	−94.81	(7/2+)	20 m	±0.88	ε(9)		...	
			113			1/2+	115 d		ε(100)	0.76	IT0.079(91)	
			114	64	−97.24	0+	0.66%				0.393(100), 0.255(1)	
			115	65	−96.65	1/2+	0.35%	−0.918				
			116	66	−98.252	0+	14.30%					0.006
			117m	67	−97.039	11/2−	14.0 d				IT0.158(100)	
			117			1/2+	7.61%	−1.000				0.01
			118	68	−98.387	0+	24.03%					0.14
			119m			11/2−	~250 d				IT0.065(100), 0.024	
			119	69	−96.684	1/2+	8.58%	+1.046				
			120	70	−97.793	0+	32.85%					0.15
			121m			(11/2−)	76 y	±0.08				
			121	71	−95.762	3/2+	27 h	±0.70	β⁻0.35(100)	0.383	0.037(100)	
			122	72	−96.549	0+	4.72%		β⁻0.383(100)			
			123m			(3/2+)	40 m		β⁻1.26(100)		0.16(100)	
			123	73	−94.26	11/2−	129 d		β⁻1.42(99), ...	1.42	1.09(1), ...	
			124	74	−94.717	0+	5.94%					0.14
			125m			(3/2+)	9.7 m		β⁻2.04(97), ...		0.325(97), ...	
			125	75	−92.21	11/2−	9.6 d		β⁻2.34(95), ...	2.34	1.067(4), 0.822(1.5), 0.915(1.4), ...	
			126	76	−92.34	0+	~10⁵ y		β⁻	~0.3	0.060, 0.067, 0.092	
			127				4 m		β⁻2.7		0.495	

TABLE 8b-1. PROPERTIES OF NUCLIDES (Continued)

(1) Atomic number Z	(2) Symbol	(3) Name	(4) Mass number A	(5) Number of neutrons N	(6) Mass excess, amu $\times 10^{-3}$	(7) Spin and parity	(8) % abundance or half life	(9) Magnetic moment, nuclear magnetons	(10) Quadrupole moment, barns	(11) Mode of decay, energy, and intensity, MeV (%)	(12) β-decay Q values, MeV	(13) Energy and intensity of γ-ray transitions, MeV (%)	(14) 2,200-m/s neutron-absorption cross section, barns
50	Sn	Tin	127	77	−90.	....	2.1 h	....	....	β^-1.5	....	0.44, 0.49, 0.82, 1.10, 2.00, 2.32, · · ·	
			128	78	−89.5	0+	59 m	....	....	β^-0.08, 0.7	1.3	0.50(61), 0.57(22), 0.072(19), 0.044(7)	
			129m	79	....	....	2 m	....	....	β^-	....		
			129	78	....	....	9 m	....	....	β^-	....	1.15, · · ·	
			130	79	....	0+	2.6 m	....	....	β^-	....		
			131	80	....	....	1.3 m	....	....	β^-	....		
			132		−87.87	....	2.2 m	....	....	β^-	....		
51	Sb	Antimony	112	61	−87.87	....	0.9 m	....	....	ϵ, β^+2.43, 1.85	6.8	1.27	
			113	62	−90.63	....	6.7 m	....	....	β^+2.7, ϵ	3.90	0.32, 0.6–0.9, 1.03, 1.2	
			114	63	−91.12	....	3.3 m	....	....	β^+2.7, ϵ	5.7	0.9, 1.30	
			115	64	−93.40	$\frac{5}{2}$(+)	32 m	+3.46	−0.27	ϵ(67), β^+1.51(33)	3.03	0.499(100), 0.98(5), 1.24(5), 2.22(1)	
			116m		....	(8−)	60 m	....	....	ϵ(81), β^+1.16(19)	....	1.293(100), 0.96(75), 0.545(68), 0.406(36), 0.099(30), 0.140(30)	
			116	65	−93.42	3(+)	16 m	....	....	ϵ(72), β^+2.3, 1.5	4.5	1.293(85), 0.93(26), 2.23(14)	
			117	66	−95.16	$\frac{5}{2}$+	2.7 h	+2.67	−0.4	ϵ(98), β^+0.57(2)	1.75	0.158(87)	
			118m		....	(8)−	5.0 h	....	....	ϵ(99+)	....	1.049(100), 1.230(100), 0.254(93), 0.041(29)	
			118	67	−94.43	1+	3.5 m	±2.5	....	β^+2.7(75), · · · , ϵ(25)	3.7	1.23(3), · · ·	
			119	68	−96.07	($\frac{5}{2}$+)	38.3 h	+3.45	−0.30	ϵ(100)	0.58	0.024	
			120		....	(8−)	5.8 d	....	....	ϵ(100)	....	1.171(100), 1.03(99), 0.200(88), 0.090(81)	
			120	69	−94.92	1+	16.5 m	±2.3	....	ϵ(56), β^+1.7(44)	2.68	1.17(1)	
			121	70	−96.188	$\frac{5}{2}$+	57.25%	+3.359	−0.29	....	....	....	6.2
			122m		....	(8−)	4.2 m	....	....	....	....	IT(100), 0.061, 0.075, · · ·	

A	N	mass excess	spin/parity	half-life	μ	Q	decay	E	γ energies	σ
122	71	−94.820	2−	2.8 d	−1.90	+0.69		1.972	0.564(66), 0.686(3), 1.26(1), 1.14(1)	3.4
123	72	−95.780	7/2+	**42.75%**	+2.547	−0.37		1.61	....	
124m2				21 m					IT0.025(100)	
124m1				93 s					IT0.01(80), 0.505(20)	
124	73	−94.067	3−	60.2 d			β⁻1.19(20)	2.916	0.644(20), 0.603(20), 0.603(97), 1.691(50), 0.722(14), 2.091(7), 0.644(7), ...	6.5
125	74	−94.75	7/2+	2.75 y	±2.6		β⁻0.299(43), 0.126(28), 0.615(13), ...	0.76	0.427(31), 0.60(24), 0.634(11), 0.463(10), 0.176(6), ...	
126m		−92.7		19.1 m			β⁻1.9		0.41, 0.67, ...	
126	75			12.5 d			β⁻1.9	3.7	0.29, 0.41, 0.58, 0.69, 0.85, 0.99, ...	
127	76	−93.07	(7/2+)	92 h			β⁻0.90(35), 1.12(24), 0.81(17), ...	1.60	0.685(35), 0.473(22), 0.784(13), 0.253(9), 0.604(5), 0.543(3), ...	
128		−90.9		9 h			β⁻1		0.314, 0.53, 0.64, 0.75	
128	77			10 m			β⁻2.6	4.3	0.75(200), 0.32(83), 1.07(4)	
129	78	−90.81		4.3 h			β⁻0.58(32), 1.55(22), 1.82(12), 1.06(9), ...	2.4	0.916, 1.03, 0.683, 1.73, ...	
130	79	−88.		7 m			β⁻	~5	0.20, 0.82, 1.03, 1.16	
130	80	−88.		36 m			β⁻	~3	0.19, 0.33, 0.82, 0.94	
131	81	−85.		23 m			β⁻	~6	0.95(48), 0.64(37)	
132	82	−85.		2.1 u			β⁻	~4		
133				4.2 m			β⁻			
134				11 s			β⁻, n			
135				2 s						
107	55	−64.		2.2 s			α3.28			
108	56	−70.	0+	5.3 s			α3.1, β⁺, p2.6, 3.4, 3.7	~10		
109	57	−72.		4.2 s			β⁺, p	~7		
110 or 111	58 or 59			19 s			β⁺, p2.46, 2.67, 2.82	~9		
114	62	−88.		17 m			β⁺	~3		
115m				0.1 s					IT0.275	
115	63	−88.5		6.0 m			β⁺2.8, ... ε(20)	4.5	0.72(34), 1.28(32), 1.38(32), 1.08(24), 0.96(6), 1.58(6)	

52   Te   Tellurium

TABLE 8b-1. PROPERTIES OF NUCLIDES (*Continued*)

(1) Atomic number Z	(2) Symbol	(3) Name	(4) Mass number A	(5) Number of neutrons N	(6) Mass excess, amu $\times 10^{-3}$	(7) Spin and parity	(8) % abundance or half life	(9) Magnetic moment, nuclear magnetons	(10) Quadrupole moment, barns	(11) Mode of decay, energy, and intensity, MeV (%)	(12) β-decay Q values, MeV	(13) Energy and intensity of γ-ray transitions, MeV (%)	(14) 2,200-m/s neutron-absorption cross section, barns
52	Te	Tellurium	116	64	-91.7	0+	2.50 h			ε, β+0.44?	1.6	0.094(100)	
			117				1.9 h			β+(1.7)		IT0.03, 0.27	
			117m				0.1 s						
			117	65	-91.40	1/2+	61 m			ε(70), β+1.8(30)	3.50	0.72(65), 1.78(9), 0.93(6)	
			118	66	-94.	0+	6.0 d			ε(100)	~0.3	1.212(67), 0.153(62), 0.270(25), 1.137(7), 2.09(4),···	
			119m			1/2-	4.7 d			ε(100)		0.644(85), 0.70(11), 1.75(4),···	
			119	67	-93.60	1/2+	15.9 h	±0.25		ε(98), β+0.627(2), 0.57(0.3)	2.294		
			120	68	-95.98	0+	0.089%						2.3
			121m			11/2-	154 d			ε(10)		IT0.0818(90), 0.2122(90), 1.10(3)	
			121	69	-95.	1/2+	17 d			ε(100)	~1.	0.573(80), 0.508(18),···	
			122	70	-96.944	0+	2.46%						3.1
			123m			11/2-	117 d			ε(100)		IT0.088(100), 0.159(100)	
			123	71	-95.718	1/2+	~1.2 × 10^13 y; 0.87%	-0.7359			0.06		440
			124	72	-97.170	0+	4.61%						6.8
			125m			11/2-	58 d					IT0.1094(100),	
			125	73	-95.574	1/2+	6.99%	-0.8871				0.0355(100)	1.56
			126	74	-96.688	0+	18.71%						1.02
			127m			(11/2-)	109 d			β-(1)		IT0.0887(100), 0.417(0.3),···	
			127	75	-94.79	(3/2+)	9.3 h			β-0.69(99.7),···	0.69		
			128	76	-95.532	0+	31.79%						0.22
			129m			11/2-	34 d			β-1.60(30), 0.91(6),		IT0.1056(64), 0.696(6), 0.730,···	
			129	77	-93.40	3/2+	69 m			β-1.45(89), 1.00(9), 0.37(1),···	1.48	0.027(19), 0.460(15), 1.08(2),···	
			130	78	-93.768	0+	34.48%						0.22

53 I Iodine

A	Mass excess (MeV)	I^π	$T_{1/2}$	μ	Q	Decay	Decay energy (MeV)	γ-ray energies (MeV)
131m		$\tfrac{1}{2}-$	30 h					IT0.182(18), 0.78(60), 0.85(31), 1.127(13), 1.206(11), 0.336(9), ...
131	−91.45	$\tfrac{3}{2}+$	25 m			2.28	β^-0.42(43), 0.57(31), 0.22(4), 2.46(4), ...	0.150(68), 0.453(16), 1.147(6), 0.493(5), 0.603(4), ...
132	−91.46	0+	78 h			0.50	β^-2.14(62), 1.68(20), 1.35(15), ...	0.228(100), 0.050(16), 0.116(2), 0.112(2)
133m		$(1\tfrac{1}{2}-)$	54 m				β^-0.22(100)	IT0.334(13), 0.913(98), 0.647(33), 0.864(78), 0.915(18), ...
133	−89.0	$(\tfrac{3}{2}+)$	12.5 m			3.0	β^-1.3, 2.4(87)	0.312(70), 0.407(31), 1.333(11), 0.720(8), 0.787(7), 0.931(5), ...
134	−89.	0+	42 m			~1.4	β^-2.24(35), 2.65(26), 1.63(14), 1.24(9), 1.40(6), ...	0.204(21), 0.262(19), 0.17(16), 0.08(13)
135	−83.		29 s			~6	β^-	
136	−33.	0+	~33 s			~5.5	β^-	
115	−84.		1.3 m			~7	β^+	
116	−86.8		<0.5 m				β^+	
117			2.7 m				β^+, ϵ	
118m			8 m			4.3	β^+4.9(57), 4.3(43)	0.104
118	−88.		13 m				β^+5.5(51), ... , ϵ(46)	0.274, 0.325, 0.599, 0.605(100), 0.600(44), 0.612(19), IT0.104
119	−90.2		19.3 m			3.2	ϵ(46)	0.605(100), 1.15(13), ... , 0.258, 0.644, ... , 0.26, 0.78
120m			50 m				β^+2.22(57), 2.16(3), 1.84(2), ϵ(38)	
120	−90.		1.3 h			5.6	ϵ(95), β^+3.8, 3.1	0.56(100), 0.60(100), 0.61(31), ... , 0.56, 1.53, 2.46, 2.57, 0.64, 0.60...
121	−93.	$(\tfrac{5}{2}+)$	2.1 h			2.35	ϵ(54), β^+4.0, 2.1(46)	0.212(90), 0.32(6), 0.27(3)
122	−92.49	1+	3.5 m			4.14	ϵ(91), β^+1.2(9)	0.564(10), 0.69(1), 0.78(1)
123	−94.4	$\tfrac{5}{2}+$	13.3 h			1.2	β^+3.12(90), 2.6(10), ... , ϵ(10)	0.159(97), 0.53(2), ...
124	−93.78	2−	4.2 d			3.17	ϵ(75), β^+1.53(14), 2.14(11)	0.605(67), 0.73(14), 1.69(14), 0.644(12), 1.51(4), ...
125	−95.415	$\tfrac{5}{2}+$	59.9 d	+3.0	−0.89	0.149	ϵ(100)	0.0355(7)
126	−94.38	2−	13 d			2.150; 1.251	ϵ(55), β^+1.13(1), 0.46(0.3); β^-0.865(30), 1.25(9), 0.385(5)	3.667(33), 0.75(3), ...; 0.386(34), 0.48(4), ...

(^{125}I: 900)

TABLE 8b-1. PROPERTIES OF NUCLIDES (*Continued*)

(1) Atomic number Z	(2) Symbol	(3) Name	(4) Mass number A	(5) Number of neutrons N	(6) Mass excess, amu $\times 10^{-3}$	(7) Spin and parity	(8) % abundance or half life	(9) Magnetic moment, nuclear magnetons	(10) Quadrupole moment, barns	(11) Mode of decay, energy, and intensity, MeV (%)	(12) β-decay Q values, MeV	(13) Energy and intensity of γ-ray transitions, MeV (%)	(14) 2,200-m/s neutron-absorption cross section, barns
53	I	Iodine	127	...	−95.526	5/2+	100%	+2.808	−0.79	...	...	...	6.2
			128	...	−94.19	1+	25.0 m	...	...	β⁻2.12(79), 1.67(13), 1.13(2), ε(6), β⁺	2.12, 1.27	0.44(14), 0.528(1.4), ...	
			129	76	−95.013	7/2+	1.7 × 10⁷ y	+2.617	−0.55	β⁻0.150(100)	0.190	0.040(9)	28
			130m	...	...	...	9.0 m	...	...	β⁻2.5(13), 1.9(2)		IT(85)	
			130	77	−93.28	5(−)	12.3 h	...	...	β⁻1.04(52), 0.62(48), 1.7(0.4)	2.99	0.669(100), 0.536(99), 0.743(87), 0.419(35), 1.15(12)	18
			131	78	−93.873	7/2+	8.07 d	+2.74	−0.40	β⁻0.606(90), 0.33(7), 0.25(2), ...	0.970	0.364(82), 0.637(7), 0.284(5), 0.080(3), 0.723(2)	~0.7
			132	79	−92.00	4+	2.3 h	±3.08	±0.08	β⁻1.20(20), 2.16(18), 1.00(18), 0.80(16), 2.16(9), ...	3.56	0.67(144), 0.773(89), 0.955(22), 0.52(20), 1.40(14), 1.14(6), ...	
			133	80	−92.17	7/2+	20.9 h	+2.84	−0.26	β⁻1.27(85), 0.89(3), ...	1.80	0.529(90), 0.875(8), 1.30, ...	
			134	81	−90.15	(4,5+)	52 m			β⁻2.43(25), 1.2(23), 1.4(15), 2.2(12), ...	4.2	0.85(95), 0.89(65), 0.61(18), 0.55(8), 0.41(8), 1.79(5), ...	
			135	82	−89.94	7/2+	6.7 h			β⁻1.0, 0.5, 1.4	2.73	1.14(37), 1.28(34), 1.72(19), 1.46(12), 0.86(11), 1.80(11), ...	
			136	83	−85.3	(2−)	83 s			β⁻4.3(23), 5.6(15), 2.7(8), 7.0(5), ...	7.0	1.32(95), 2.3(19), 0.39(19), 0.27(18), 0.20(12), 2.63(10), ...	
			137	84	−82.	...	23 s			β⁻, n 0.6	~5		
			138	85	...	...	5.9 s			β⁻, n			
			139	86	...	...	2 s			β⁻, n			
			140	87	−75.	...	1.5 s			β⁻			
54	Xe	Xenon	115	61	−75.	...	19 s			ε	~7.5		
			116	62	−79.	0+	55 s			ε, β⁺	~4.5	0.104	

A	N	Mass excess	J^π	% or T½	μ	Q	Type of decay	E (MeV)	γ-ray energies (MeV)	σ (barns)
117	63	−80.		1.1 m			ε, β+	~6.0	0.117, 0.222, 0.295, ...	
118	64	−84.	0+	6 m			β+	~3	0.05, ...	
119	65	−84.8		6 m			β+	5.0	0.10, ...	
120	66	−88.	0+	41 m			ε	~2	0.055, 0.073, 0.176, 0.76	
121	67	−88.5		39 m			β+ 2.8	3.8	0.060, 0.090, 0.110, 0.158, 0.180, 0.345	
122	68	−91.	0+	20 h			ε(100)	~1		
123	69	−91.6		2.1 h			ε, β+ 1.51	2.7	0.149, 0.090, 0.110, 0.178, 0.329, 0.68, 1.10, ...	
124	70	−93.9	0+	0.096%						100
125m			(1/2−)	55 s					IT(?)0.075, 0.111	
125	71	−93.5	(1/2+)	17 h			ε(99), β+(1)	1.7	0.188, 0.243, 0.055, 0.454, ...	
126	72	−95.71	0+	0.090%						3
127m			(1/2−)	70 s					IT0.175, 0.125	
127	73	−94.8	(3/2+)	36.4 d			ε(100)	0.66	0.203(65), 0.172(22), 0.375(20), 0.145(4), ...	
128	74	−96.468	0+	1.92%						<8
129m			1/2−	8.0 d					IT0.197, 0.040	
129	75	−95.216	1/2+	26.44%	−0.7768					21
130	76	−96.491	0+	4.08%						<26
131m			11/2−	11.8 d					IT0.164, ...	
131	77	−94.915	3/2+	21.18%	+0.6908	−0.12				90
132	78	−95.843	0+	26.89%						0.4
133m			(11/2−)	2.26 d					IT0.232	
133	79	−94.19	(3/2+)	5.27 d			β− 0.346(99)	0.428	0.0810(37), ...	190
134	80	−94.602	0+	10.44%						0.25
135m			(11/2−)	15.6 m					IT0.527	
135	81	−92.86	(3/2+)	9.2 h			β− 0.91(97), 0.55(3)	1.16	0.250(97), 0.608(3), ...	2.64 × 10⁶
136	82	−92.779	0+	8.87%						0.17
137	83	−88.26		3.9 m			β− 4.1(67), 3.6(33)	4.1	0.455(33)	
138	84	−86.	0+	14 m			β− 2.4	2.8	0.26, 1.76, 2.02, 0.42, 0.16, 0.51	
139	85	−81.6		40 s			β− 4.6(31), 4.5(28), 4.8(20), ...	4.8	0.219(77), 0.175(29), 0.297(24), 0.290(10), 0.394(8), ...	
140	86	−79.	0+	14 s			β− 4.7		0.805, 1.413, 0.622, 1.315, 1.309, ...	
141	87			1.7 s			β−			
142	88			1.15 s			β−		1.325, 0.572, 0.657, 0.204, 0.645, 0.416, ...	

TABLE 8b-1. PROPERTIES OF NUCLIDES (Continued)

(1) Atomic number Z	(2) Symbol	(3) Name	(4) Mass number A	(5) Number of neutrons N	(6) Mass excess, amu × 10⁻³	(7) Spin and parity	(8) % abundance or half life	(9) Magnetic moment, nuclear magnetons	(10) Quadrupole moment, barns	(11) Mode of decay, energy, and intensity, MeV (%)	(12) β-decay Q values, MeV	(13) Energy and intensity of γ-ray transitions, MeV (%)	(14) 2,200-m/s neutron-absorption cross section, barns
54	Xe	Xenon	143	89			0.96 s			β^-			
			144	90			8.8 s			β^-			
55	Cs	Cesium	119	64			33 s						
			120	65			61 s						
			121	66			2 m						
			122?	67			0.5 s						
			123	68	$-87.$		5.6 m			$\beta^+2.6(50)$, 2.5(20), $\epsilon(30)$	~4	0.1	
			125	70	-90.3	$\tfrac{1}{2}$	45 m			$\epsilon(51)$, $\beta^+2.05(49)$	3.07	0.098	
			126	71	-90.4	1+	1.6 m	+1.41		$\beta^+3.8(54)$, $\cdots$, $\epsilon(15)$	4.8	0.112	
			127	72	-92.5	$\tfrac{1}{2}+$	6.2 h	+1.46		$\epsilon(97)$, $\beta^+1.07(1)$, 0.677(1), $\cdots$	2.1	0.385(38), 0.406(72), 0.125(10), 0.462, $\cdots$	
			128	73	-92.27	1+	3.8 m			$\beta^+2.9(37)$, 2.5(14), 1.9(4), $\epsilon(45)$	3.91	0.441(27), 0.97(1), 1.12(1), 0.528, 0.576	
			129	74	$-94.$	$\tfrac{1}{2}+$	32.4 h	(+)1.479		$\epsilon(100)$	~1.1	0.371(48), 0.410(30), 0.550(5), 0.320(4), 0.280(3), 0.040(2)	
			130	75	-93.24	1+	29.1 m	+1.37 or −1.45		$\epsilon(53)$, $\beta^+1.97(46)$, $\cdots$; $\beta^-0.442(2)$	2.99; 0.442	0.536(5), 0.586(1), 0.894(1), $\cdots$	
			131	76	-94.53	$\tfrac{5}{2}+$	9.70 d	+3.54	−0.57	$\epsilon(100)$	0.35		
			132	77	-93.59	2(−)	6.5 d	+2.22	+0.46	$\epsilon(97)$, $\beta^+0.40(0.6)$	2.10	0.668(99), 1.32(0.6), 1.14(0.5), 0.48(4)	
			133	78	-94.56	$\tfrac{7}{2}+$	**100%**	+2.578	−0.003				29.0
			134m			8−	2.90 h	+1.096		$\beta^-(2)$	1.3	IT0.128, $\cdots$	
			134	79	-93.30	4+	2.05 y	+2.990	+0.36	$\beta^-0.658(71)$, 0.089(27), 0.415(2)	2.06	0.605(98), 0.796(88), 0.569(16), 1.365(3), 1.168(2), 1.038(1), $\cdots$	134
			135m			$(\tfrac{19}{2}-)$	53 m					IT0.840(100), 0.78(100)	
			135	80	-94.10	$\tfrac{7}{2}+$	3 × 10⁶ y	+2.729	+0.044	$\beta^-0.21(100)$	0.210		8.7
			136	81	-92.71	5	13 d	+3.70		$\beta^-0.34(91)$, 0.68(7), 0.56(2)	2.54	0.818(100), 1.05(82), 0.340(53), 0.16(36), 1.25(20), 0.273(18), $\cdots$	
			137	82	-92.93	$\tfrac{7}{2}+$	30.0 y	+2.838	+0.045	$\beta^-0.514(93)$, 1.176(7)	1.176	0.662(93)	0.11

Z / El	A	N	Δ (MeV)	I π	Half-life / abundance	μ	Q	Decay (MeV)	Energy (MeV)	γ-ray energies (MeV)	σ
	138	83	−89.	3	32.2 m	±0.5		β−3.40, 2.68, ...	5.4	1.426(73), 1.01(25), 0.463(23), 2.21(18), 2.63(9), 0.55(8)	
	139	84	−86.7		9.1 m			β−4.0(90), 2.9, ...	4.0	1.28, 0.63, 1.65, 0.05, 0.80, ...	
	140	85	−83.2		65 s			β−5.1, 2.3	5.7	0.60, 0.91, 1.13, 1.63, 1.82, 2.10, 2.33, 2.72, 3.15, ...	
	141	86	−80.		24 s			β−	~5		
	142	87	−76.		2.3 s			β−7.6, n	~7		
	143	88			1.6 s			β−			
	144	89			1.06 s			β−			
56 Ba Barium	123	67			2 m			β+, ε			
	124	68			2.5 m			β+			
	125m				8 m						
	125	69	−86.		3 m			β+3.4	4.5		
	126	70	−88.	0+	97 m			ε(100)	~1.8	0.076, 0.083, 0.141, 0.056, ...	
	127m				10 m			β+			
	127	71	−89.		18 m			β+3.14	4.3	0.23(100), 0.70(33), ...	
	128	72	−92.	0+	2.4 d			ε(100)	~0.7	0.11, 0.07, 0.18, 0.20, 0.09, 0.134, 0.278, ...	
	129m				2.1 h			ε		0.182, ...	
	129	73	−91.		2.5 h			β+1.43, 1.24, 1.0, ...	2.45	0.21, 0.22, ...	
	130	74	−93.72	0+	0.101%						11
	131m				15 m			ε(100)		IT0.078, 0.11, ...	
	131	75	−93.10	(1/2+)	12 d			ε(100)	1.34	0.496(48), 0.124(28), 0.216(19), 0.373(13), 0.25(5), 0.60(3), ...	
	132	76	−94.95	0+	0.097%						8.5
	133m			11/2−	38.9 h					IT0.276(100), 0.0123	
	133	77	−94.01	3/2+	7.8 y			ε(100)	0.515	0.356(52), 0.080(31), 0.303(15), 0.384(7), 0.276(6), ...	
	134	78	−95.51	0+	2.42%						2
	135m			11/2−	28.7 h		+0.18			IT0.268(100)	
	135	79	−94.33	3/2+	5.59%	+0.8365					5.8
	136m			7−	0.32 s					IT0.164, 1.05, 0.82	
	136	80	−95.44	0+	7.81%						0.4
	137m			11/2−	2.558 m		+0.28			IT0.6610(100)	
	137	81	−94.18	3/2+	11.32%	+0.9357					5.1

TABLE 8b-1. PROPERTIES OF NUCLIDES (Continued)

(1) Atomic number Z	(2) Symbol	(3) Name	(4) Mass number A	(5) Number of neutrons N	(6) Mass excess, amu × 10^-3	(7) Spin and parity	(8) % abundance or half life	(9) Magnetic moment, nuclear magnetons	(10) Quadrupole moment, barns	(11) Mode of decay, energy, and intensity, MeV (%)	(12) β-decay Q values, MeV	(13) Energy and intensity of γ-ray transitions, MeV (%)	(14) 2,200-m/s neutron-absorption cross section, barns
56	Ba	Barium	**138**	82	−94.76	0+	**71.66%**					....	0.35
			139	83	−91.17	($\frac{7}{2}$−)	82.9 m			β−2.3(72), 2.2(23), . . .	2.3	0.166(23), 1.43(0.5), . . .	4
			140	84	−89.36	0+	12.8 d			β−1.02(52), 0.47(34), 0.59(11), . . .	1.05	0.537(34), 0.030(11), 0.163(6), 0.305(6), 0.438(5), . . .	1.6
			141	85	−85.8	....	18 m			β−3.0, 2.8, 2.6, 2.4, 2.0, . . .	3.0	0.191, 0.305, 0.277, 0.460, 0.344, 0.648, . . .	
			142	86	−83.5	0+	11 m			β−1.0, 1.7	2.2	0.255, 1.180, 0.227, 0.080, 0.905, 0.425, . . .	
57	La	Lanthanum	143	87	−79.	....	12 s			β−	~4		
			144	88	−77.	....	11.4 s			β−	~3		
			124	67		....	7 m						
			125	68		....	<1 m						
			126	69	−83.	....	1.0 m			β+, ε	~5	0.26	
			127	70		....	3.5 m			β+, ε			
			128m			....	0.6 s					0.75, 0.115	
			128	71	−84.	....	4.7 m			β+4.2, ε	~7	0.283, 0.479, 0.644, 0.602, . . .	
			129	72	−87.	....	10 m			β+, ε	~4	0.36, 0.45, 0.55, 0.72, 0.81, 0.91, 1.01, 1.19, 1.45, . . .	
			130	73	−88.	....	8.7 m			β+, ε	~6		
			131	74	−89.9	($\frac{1}{2}$+)	59 m			ε(72), β+1.41(7), 1.57(6), 1.94(4), 1.83(3), . . .	2.96	0.108(44), 0.417(19), 0.365(18), 0.285(15), 0.453(7), . . .	
			132	75	−89.9	2−	4.5 h			ε(64), β+3.20(13), 2.62(13), 3.66(10), 2.5(1)	4.7	0.464(74), 0.567(15), 0.663(9), 1.031(8), 0.540(7), 0.515(5), . . .	
			133	76	−92.	....	4.0 h			ε, β−~1.2	~2	0.29, 0.62, 0.86, 1.08, . . .	

Z	El	A	N	Mass excess	Spin	Half-life / abundance	μ	±	Decay	Q	γ	σ
		134	77	−91.52	1+	6.7 m			β⁺2.67(62), 2.07(2), ε(36)	3.7	0.605(6), …	8.8
		135	78	−93.2	(5/2+)	19.5 h			ε(100)	1.0	0.481(2), 0.87(0.2), 0.59(0.1), …	2.8
		136	79	−92.4	1+	9.9 m			ε(61), β⁺1.85(39), …	2.9	0.82(2.5), …	
		137	80	−94.	7/2+	6×10⁴ y			ε(100)	~0.5		
		138	81	−92.84	5−	1.05×10¹¹ y; **0.089%**	+3.707	±0.8	β⁻0.21(30)	1.78	1.426(70)	
		139	82	−93.60	7/2+	**99.911%**	+2.778	+0.22		1.01	0.80(30)	
		140	83	−90.47	3−	40.22 h			β⁻1.34(35), 1.24(19), 1.67(17), 2.17(8), …	3.767	1.596(96), 0.487(40), 0.329(20), 0.815(19), 0.923(10), 2.53(3), …	
		141	84	−89.07	2−	3.9 h			β⁻2.43(98), 0.9(2)	2.43	1.37(2)	
		142	85	−85.85		92 m			β⁻2.11(23), 1.98(19), 4.517(13), 1.79(11), 0.87(11), …	4.517	0.645(48), 2.41(15), 2.55(11), 0.898(9), 1.91(8), 2.99(5), …	
		143	86	−84.0	(7/2+)	14 m			β⁻3.30(75), …	3.30	0.625, 1.17, 1.98, 1.58, 2.56, 1.07	
		144	87	−80.		41 s			β⁻	~5.5	0.080, 0.32, 0.75	
58	Ce Cerium	129	71			~13 m			β⁺, ε		0.13	
		130	72		0+	30 m			ε, β⁺		IT0.231	
		131m	:			5 m						
		131	73	−85.		10 m			ε(89), β⁺~2.8	~5	0.170(20), 1.44(9), 0.396(9), 0.119(6), …	
		132	74	−88.	0+	4.2 h			ε	~1.4	0.182(82), 0.155(12), … IT0.097, 0.077, 0.088	
		133m				97 m			ε			
		133	75	−89.		5.4 h			ε(90), β⁺1.3(10)	~3.3	0.477, 0.131, 0.510, 0.278, 0.784, 0.618, …	
		134	76	−91.	0+	72 h			ε(100)	~0.5	0.265(100), 0.606(98), 0.300(56), 0.517(46), 0.783(22), 0.828(22), …	
		135	77	−91.		17.2 h			ε(99), β⁺0.81(1)	~2.3		
		136	78	−92.82	0+	**0.193%**					…	7.3
		137m	:		1/2−	34.4 h			ε(0.6)		IT0.255(99.4), 0.825(0.5), …	
		137	79	−92.	3/2+	9.0 h			ε(100)	1.2	0.446(2), …	

TABLE 8b-1. PROPERTIES OF NUCLIDES (*Continued*)

(1) Atomic number Z	(2) Symbol	(3) Name	(4) Mass number A	(5) Number of neutrons N	(6) Mass excess, amu ×10⁻³	(7) Spin and parity	(8) % abundance or half life	(9) Magnetic moment, nuclear magnetons	(10) Quadrupole moment, barns	(11) Mode of decay, energy, and intensity, MeV (%)	(12) β-decay Q values, MeV	(13) Energy and intensity of γ-ray transitions, MeV (%)	(14) 2,200-m/s neutron-absorption cross section, barns
58	Ce	Cerium	**138**	80	-93.97	0+	**0.250%**						1.1
			139m			$(\tfrac{11}{2}-)$	56 s					IT0.746(100)	
			139	81	-93.30	3/2+	140 d			ε(100)	0.27	0.1658(100)	0.58
			140	82	-94.52	0+	**98.48%**						29
			141	83	-91.68	7/2−	32.4 d	±0.9		β⁻0.437(70), 0.581(30)	0.581	0.1454(70)	0.95
			142	84	-90.70	0+	**11.07%**						6.0
			143	85	-87.59	3/2−	33.7 h			β⁻1.093(53), 1.387(29), 0.71(16), 0.50(2), 28(1)	1.445	0.293(49), 0.057(11), 0.722(7), 0.664(7), 0.491(2), ⋯	
			144	86	-86.31	0+	284 d			β⁻0.32(76), 0.18(19), 0.24(4), ⋯	0.32	0.134(17), 0.080(6), 0.041(1), 0.059(1), 0.034(1), ⋯	1.0
			145	87	-82.8		3.0 m			β⁻1.7(70), 2.1(4), ⋯	2.5	0.725(69), 0.063(19), 0.285(10), 0.435(6), ⋯	
			146	88	-81.3	0+	14.2 m			β⁻0.75(95), ⋯	1.08	0.317(53), 0.1335(16), 0.264(11), 0.1415(7), 0.210(6), 0.100(3), ⋯	
			147	89	$-78.$		70 s			β⁻	~3		
			148	90	$-76.$	0+	43 s			β⁻	~2		
59	Pr	Praseodymium	134	75	$-84.$		17 m			β⁺	~6	0.22, 0.30, 0.41, 0.64, 0.96	
			135	76	$-87.$		22 m			β⁺2.5	3.6	0.080, 0.22, 0.30	
			136	77	$-87.$		1.1 h			ε(67), β⁺2.0(33)	~5	0.17, ⋯	
			137	78	$-89.$		1.3 h			ε(73), β⁺1.7(27), ⋯	2.7	0.837(8), 0.434(5), 0.161(4), 1.087(2), ⋯	
			138m			(6,7,8−)	2.1 h			ε(77), β⁺1.65(21), ⋯		1.04(100), 0.79(100), 0.298(77), 0.40(9)	
			138	79	-89.21	(1+)	1.5 m			β⁺3.43(73), ε(27)	4.44	0.79, 0.69, ⋯	
			139	80	-91.03	(5/2+)	4.5 h			ε(91), β⁺1.09(9)	2.11	1.35(0.5), ⋯	
			140	81	-90.88	(1+)	3.39 m			β⁺2.32(50), ε(50)	3.34	1.596(0.3), ⋯	
			141	82	-92.30	5/2+	**100%**	+4.3	-0.07				11.5

Element (Z)	A	N	Δ (mass excess)	I	T½ / abundance	μ (±0.25)	Q (±0.03)	Decay mode, energy (MeV)	E	γ-rays, MeV (%)	σ
	142	83	−89.91	2−	19.2 h			β−2.16(96), 0.59(4), ε(0.01)	2.16	1.57(4)	18
	143	84	−89.14	7/2+	13.6 d			β−0.931	0.931		89
	144	85	−86.65	0−	17.3 m			β−2.996(98), 2.296(1), 0.806(1),...	2.996	0.696(1), 2.186(0.7),...	
	145	86	−85.45	(7/2+)	5.98 h			β−1.805(98),...	1.805	0.674(0.5), 0.749(0.4),...	
	146	87	−82.5	(3)−	24.2 m			β−1.1(40), 2.1(29), 3.6(11),...	4.08	0.454(49), 1.526(18), 0.736(8), 0.790(8), 1.378(6),...	
	147	88	−81.0		12 m			β−2.10(45), 1.45(35),...	2.7	0.565(29), 0.645(16), 1.26(11), 0.61, 1.18, 0.078,...	
	148	89	−78.		2.0 m			β−4.2	~5	0.31	
	149	90	−76.6		2.3 m			β−3.0	3.0	0.11, 0.14, 0.17, 0.30, 0.74	
Nd — Neodymium 60	135	75			12 m			β+		0.21, 0.44	
	136	76		0+	55 m			β+1.32, 2.97, 6		0.109, 0.553, 0.540, 0.576	
	138	78	−88.	0+	5.2 h			ε, β+1.0		0.73, 0.114, 0.983, 0.327, 1.03, 0.90,...	
	139m				5 h			ε, β+3.1		IT0.23	
	139	79	−88.		30 m			β+, ε		0.41	
	140	80	−90.4	0+	3.3 d			ε(100)	0.47		
	141m				63 s					IT0.755(100)	
	141	81	−90.36	3/2+	2.5 h			ε(97), β+0.79(3)	1.80	1.14(2), 1.30(1),...	
	142	82	−92.23	0+	27.11%						18.8
	143	83	−90.14	7/2−	12.17%	−1.08	−0.48				32.0
	144	84	−89.87	0+	2.1 × 10¹⁵ y; 23.85%			α1.83			4.0
	145	85	−87.39	7/2−	8.30%	−0.66	−0.25				48
	146	86	−86.85	0+	17.22%						1.4
	147	87	−83.87	5/2−	11.06 d	+0.59		β−0.807(81), 0.365(16), 0.21(2), 0.408(1),...	0.898	0.0911(27), 0.531(13), 0.440(1), 0.275(1),...	
	148	88	−83.07	0+	5.73%						2.5
	149	89	−79.85	5/2−	1.72 h			β−1.42(38), 1.02(30), 1.13(26), 1.55(6),...	1.67	0.210(27), 0.27(26), 0.114(18), 0.541(10), 0.424(9), 0.654(9),...	
	150	90	−79.09	0+	5.62%						1, 3
	151	91	−76.11		12 m			β−2.1, 1.8, 1.6, 1.2	2.46	0.118(40), 0.174(10), 1.180(9), 0.138(6), 0.086(5), 0.737(5),...	

TABLE 8b-1. PROPERTIES OF NUCLIDES (*Continued*)

(1) Atomic number Z	(2) Symbol	(3) Name	(4) Mass number A	(5) Number of neutrons N	(6) Mass excess, amu ×10⁻³	(7) Spin and parity	(8) % abundance or half life	(9) Magnetic moment, nuclear magnetons	(10) Quadrupole moment, barns	(11) Mode of decay, energy, and intensity, MeV (%)	(12) β-decay Q values, MeV	(13) Energy and intensity of γ-ray transitions, MeV (%)	(14) 2,200-m/s neutron-absorption cross section, barns
61	Pm	Promethium	139	78	$-83.$	....	~6 m	....		$\epsilon(65)$, β^+	~4.5	0.42, 0.77, 1.03	
			140	79	-84.0	....	5.8 m	....		$\beta^+2.6(57)$, $\epsilon(43)$	~6	0.195(13), 1.22, 0.89, 1.58, 0.63,...	
			141	80	-86.4	....	23 m	....		$\beta^+3.80(81)$, ...$\epsilon(19)$	3.7	1.572(0.2)	
			142	81	-87.2	1+	36 s	....		$\epsilon(100)$	4.82	0.742(46)	
			143	82	-89.00	$(\tfrac{5}{2}+)$	265 d	....		$\epsilon(100)$	1.07	0.618(100), 0.696(100), 0.476(43),...	
			144	83	-87.3	(5,6–)	363 d	....		$\epsilon(100)$	2.37	0.072(11), 0.067(8),...	
			145	84	-87.21	$(\tfrac{5}{2}+)$	17.7 y	....		$\epsilon(100)$	0.170	0.454(65), 0.736(21)	
			146	85	-85.26	....	5.5 y	....		$\epsilon(65)$, $\beta^-0.790(32)$,...	1.477, 1.537	0.747(34), 0.633(2),...	84.00
			147	86	-84.83	$\tfrac{7}{2}+$	2.6234 y	+2.7	+0.7	$\beta^-0.2247(100)$	0.2247		187
			148m	..		(6)–	43 d			$\beta^-0.397(51)$, 0.683(23), 0.495(17), 1.0(5)	....	0.550(92), 0.630(88), 0.915(20), 0.414(18), ...IT0.062(6),...	25,000
			148	87	-82.50	1–	5.4 d	+2.0	+0.2	$\beta^-2.465(50)$, 1.020(37), 1.922(13)	2.465	0.076(6)	2,000
			149	88	-81.63	$\tfrac{7}{2}+$	53.1 h	....		$\beta^-1.071(97)$, 0.785(3),...	1.071	0.550(28), 1.465(22), 0.912(15)	1,350
			150	89	-78.9	(1)	2.7 h	....		$\beta^-2.3(29)$, 1.8(22), 1.4(22), 3.1(12),...	3.4	0.334(71), 1.165(23), 1.33(22), 0.831(18), 0.88(12), 0.71(8),...	
			151	90	-78.76	$\tfrac{5}{2}+$	28 h	±1.6	±1.9	$\beta^-0.85(41)$, 1.19(10), 1.13(8), 1.02(7), 0.74(7),...	1.19	0.340(21), 0.17(18), 0.10(7), 0.275(6), 0.72(6), 0.07(5),...	<700
			152	91	-76.6	....	6 m	....		$\beta^-2.2$	~3.4	0.122, 0.245	
			153	92	-76.0	....	5.5 m	....		$\beta^-1.65$	1.8	0.09, 0.12, 0.18	
			154	93	$-73.$	....	2.5 m	....		$\beta^-2.5$	~4		

Element	Z	A	N	Atomic mass (excess)	Jπ	Abundance or half-life	μ	Q	Decay mode & energy (MeV)	Decay energy	γ energies (MeV)	σ (barns)
Samarium	62	140	78	...	...	14 m			β+			
		141	79	−82.	...	23 m			ε, β+	~4.4		
		142	80	−84.9	0+	72.5 m			ε(94), β+1.03(6)	2.05	0.195, 0.437, 0.780, 0.92, 1.48, ...	
		143m	:	...	(11/2−)	65 s					IT0.750(100)	
		143	81	−85.26	(3/2+)	8.83 m			β+2.47(50), ε(50)	3.49	1.055(3), 1.51(1), 1.17(1), ...	
		144	82	−87.93	0+	**3.09%**						~0.7
		145	83	−86.52	7/2−	340 d			ε(100)	0.638	0.0614(13), ...	~110
		146	84	−86.90	0+	1.00 × 10⁸ y			α2.50			
		147	85	−85.08	7/2−	1.07 × 10¹¹ y	−0.813	−0.20	α2.232			75
		148	86	−85.15	0+	**14.97%**						4.73
		149	87	−82.79	7/2−	**11.24%**	−0.670	+0.058				40,000
		150	88	−82.70	0+	**13.83%**						102
		151	89	−80.04	(5/2, 7/2−)	90 y			β⁻0.076(98)	0.076	0.0217(2)	15,000
		152	90	−80.24	0+	**26.72%**						210
		153	91	−77.88	3/2+	47 h	−0.022	+1.2	β⁻0.70(48), 0.63(32), 0.80(20), ...	0.801	0.103(28), 0.070(5), ...	
		154	92	−77.78	0+	**22.71%**						5.5
		155	93	−75.35	3/2(−)	23 m		±0.9	β⁻1.54(92), 1.4(6), ...	1.65	0.104(93), 0.246(5), ...	
		156	94	−74.46	0+	9.4 h			β⁻0.72(51), 0.43(44), ...	0.72	0.088(30), 0.204(20), 0.166(10), 0.25(5), 0.291(3)	
		157	95	...	...	0.5 m			β⁻		0.57	
Europium	63	142	79	−79.9	(1+)	1.2 m			β+, ε		1.03(1), 0.77(1)	
		143	80	−81.13	(5/2+)	2.6 m			β+4.0(100)	5.0	1.659, 0.820, 2.481	
		144	81	−83.61	1+	10.5 s			β+5.2, ...	6.33	0.894(67), 0.654(15), 1.658(14), 1.997(7), 0.543(5), ...	
		145	82		(5/2+)	5.8 d			ε(6), β+1.72(1.5), 0.79(5)	2.72		
		146	83	−82.74	(4−)	4.6 d			ε(96), β+1.467(3.5), 0.80(0.5), 2.108(0.2)	3.872	0.747(100), 0.633(40), 0.634(37), 0.704(9), 0.666(7), 1.53(6), ...	
		147	84	−83.18	(5/2+)	22 d			ε(99), β+0.747(0.6), 0.522(0.4), 0.548(0.3), α2.90(0.002)	1.767	0.1212(46), 0.1974(32), 0.678(12), 0.602(8), 0.799(7), ...	
		148	85	−81.81	(5−)	54 d			ε(100)	3.10	0.550(100), 0.630(72), 0.611(20), 0.553(17), 0.726(12), 0.414(8), ...	

TABLE 8b-1. PROPERTIES OF NUCLIDES (Continued)

(1) Atomic number Z	(2) Symbol	(3) Name	(4) Mass number A	(5) Number of neutrons N	(6) Mass excess, amu $\times 10^{-3}$	(7) Spin and parity	(8) % abundance or half life	(9) Magnetic moment, nuclear magnetons	(10) Quadrupole moment, barns	(11) Mode of decay, energy, and intensity, MeV (%)	(12) β-decay Q values, MeV	(13) Energy and intensity of γ-ray transitions, MeV (%)	(14) 2,200-m/s neutron-absorption cross section, barns
63	Eu	Europium	149	86	−82.	$(\frac{5}{2}+)$	106 d			ε(100)	~0.8	0.328(4), 0.277(3), . . .	
			150m	..	. . . .	(4,5−)	6.2 y			ε(100)	. . . .	0.334(96), 0.439(86), 0.584(60), 0.74(21), 1.049(9), 1.248(5), . . .	
			150	87	−80.21	(0,1−)	12.6 h			β^-1.01(90)	1.010	0.334(4), 0.406(3), . . .	
			151	88	−80.12	$\frac{5}{2}(+)$	**47.82%**	+3.464	+1.1	ε(9), β^+1.24(0.5)	2.25		8,000
			152m2	..	. . . .	8−	96 m			β^-1.87(73), 1.55(2), 0.56(2), ε(23),		IT0.0397, 0.0183, 0.0895, 0.842(13), 0.963(12), 0.122(8), 0.344(3), 1.32(1), 1.39(1)	
			152m1	..	. . . .	(0−)	9.3 h			β^+0.89(0.007), . . .			
			152	89	−78.22	3−	12 y	±1.924	±3.0	ε(72), β^+0.71(0.01), . . .	1.88	0.122(37), 1.408(22), 0.965(15), 1.113(14)	
			153	90	−78.74	$\frac{5}{2}+$	**52.18%**	+1.530	+2.8	β^-0.69(15), 1.48(9), 0.36(2), . . .	1.82	0.344(27), 0.779(14), . . .	450
			154	91	−76.99	3−	16 y	±2.000		β^-0.592(38), 0.270(26), 0.861(18), 1.866(11), . . .	1.978	0.123(38), 1.274(37), 1.004(31), 0.723(21), 0.873(12), 0.248(7), . . .	1,500
			155	92	−77.10	$(\frac{5}{2}+)$	1.81 y			β^-0.16(40), 0.143(34), 0.248(16), 0.188(10)	0.248	0.0865(40), 0.1053(34), 0.060(10), . . .	14,000
			156	93	−75.22	0+	15 d			β^-0.48(33), 2.45(31), 0.30(17), . . .	2.45	1.24(16), 1.15(14), 0.812(9), 0.089(8), 1.97(7), 0.723(6), . . .	
			157	94	−74.57	. . . . . .	15.2 h			β^-0.86(30), 1.28(30), 0.90(15), 0.66(15), 1.34(10), . . .	1.36	0.413(23), 0.064(19), 0.37(14), 0.623(6), 0.477(3), 0.055(3), . . .	
			158	95	−72.2	. . . . . .	46 m			β^-2.5(44), 1.6(36), 1.2(15), 3.4(5), . . .	3.4	0.080(60), 0.95(55), 0.90(20), 0.98(12), 1.19(10), 1.35(7), . . .	

Z	El.	A	N	Mass excess	I	Half-life / abundance	μ	Q	Radiations (%)	E (MeV)	γ-ray energies (MeV)	σ (barns)
		159	96	−70.77	....	18 m			β⁻ 2.57(25), 2.35(21), 1.90(21), 1.75(11), 1.50(11), 1.0(10)	2.63	0.07(42), 0.67(21), 0.09(18), 0.15(14), 0.73(10), ...	
		160	97	−68.	....	~2.5 m			β⁻ 3.6(100)	~4		
64	Gd (Gadolinium)	144	80	−77.	0+	4.5 m			β⁺	~4		
		145	81	−78.	(7/2−)	23 m			β⁺ 2.5(66), ε(34)	5.3	1.76, 1.882, 0.95, 1.04, 0.809, 0.330	
		146	82	−81.	0+	50 d			ε(100)	~1	0.1148(100), 0.1156(100), 0.1547(62)	
		147	83	−80.68	(7/2−)	38 h			ε(99.8), β⁺ 0.97(0.2), ...	2.33	0.2292(77), 0.3961(37), 0.370(16), 0.766(10), 1.068(9), 0.894(8), ...	
		148	84	−81.81	0+	93 y			α 3.18	0.01		
		149	85	−80.61	(7/2−)	9.3 d			ε(99+), α 3.01(0.0005)	~1.4	0.150(48), 0.299(26), 0.347(25), 0.750(11), 0.790(10), 0.94(5), ...	
		150	86	−81.30	0+	1.8 × 10⁶ y			α 2.80			
		151	87	−79.62	(7/2−)	120 d			ε(100), α 2.60	0.46	0.154(7), 0.244(7), 0.175(3), 0.0216(3), 0.308(1)	
		152	88	−80.18	0+	1.1 × 10¹⁴ y; **0.20%**			α 2.24			<125
		153	89	−78.52	3/2+	242 d			ε(100)	0.241	0.099(55), 0.070, ...	
		154	90	−79.11	0+	**2.15%**						102
		155	91	−77.36	3/2−	**14.73%**	−0.254	+1.3				61,000
		156	92	−77.86	0+	**20.47%**						8
		157	93	−77.03	3/2−	**15.68%**	−0.39	+1.5				254,000
		158	94	−75.88	0+	**24.87%**						3.5
		159	95	−73.59	3/2−	18.0 h			β⁻ 0.95(63), 0.89(24), 0.59(13), ...	0.95	0.058, 0.363, ...	
		160	96	−72.93	0+	**21.90%**						0.77
		161	97	−70.31	(5/2)−	3.6 m			β⁻ 1.56(87), 1.39(5), 1.66(4), 1.50(3), ...	1.95	0.361(63), 0.315(24), 0.102(18), 0.077(10), 0.284(6), ...	96,000
65	Tb (Terbium)	162	98	−69.0	....	10.4 m			β⁻	1.4	0.410, 0.428	
		147m		:	....	2.5 m			ε(47), β⁺(3)		IT(50), 0.799, 0.164	
		147	82	−76.	....	1.6 h			ε(95), β⁺(5)	~5	0.694(32), 0.139(24), 0.119(5)	
		148	83	−75.8	....	70 m			β⁺ 4.6, ...	5.6	1.12, 0.78	
		149m		:	....	4.3 m			ε, β⁺; α 3.99(0.025)		IT	

TABLE 8b-1. PROPERTIES OF NUCLIDES (*Continued*)

(1) Atomic number Z	(2) Symbol	(3) Name	(4) Mass number A	(5) Number of neutrons N	(6) Mass excess, amu ×10⁻³	(7) Spin and parity	(8) % abundance or half life	(9) Magnetic moment, nuclear magnetons	(10) Quadrupole moment, barns	(11) Mode of decay, energy, and intensity, MeV (%)	(12) β-decay Q values, MeV	(13) Energy and intensity of γ-ray transitions, MeV (%)	(14) 2,200-m/s neutron-absorption cross section, barns
65	Tb	Terbium	149	84	−76.62	.....	4.1 h			ε(77, α3.95(23)	3.70	0.166, 0.352, 0.388, 0.187, ...	
			150	85	−76.28	.....	3.1 h			ε, β+3.6(50), α3.49	4.67	0.64(100), 0.93(35), ...	
			151	86	−76.82	.....	18 h			ε(99+), α3.42(0.0005)	2.61	0.108(35), 0.252(35), 0.288(32), 0.18(18), 0.40, 0.60, ...	
			152m	.....	.....	.....	4.0 m					0.14, 0.23, ...	
			152	87	−76.08	.....	18 h			ε(88), β+2.80(4), 1.87(2), 2.46(1), ... ε, β+, α(0.002)	3.82	0.344, 0.432, 0.587, 0.271, 1.048, ...	
			153	88	−77.	(5/2 −)	55 h			α~3	~1.8	0.212(30), 0.11(12), 0.083(11), 0.17(9), 0.25, 0.33, 0.88, ...	
			154m	.....	.....	.....	8.5 h			ε(100)	....	0.123, 0.248, 0.53, 0.65, ...	
			154	89	−75.	.....	21 h			ε	~3.4	0.123, 0.248, 0.30, 0.35, 0.53, 2.5, ...	
			155	90	−76.46	.....	5.6 d			ε	0.85	0.087(37), 0.105(25), 0.180(8), 0.163(8), 0.262(7), 0.368(4)	
			156m	.....	.....	(0+)	5.5 h			ε, β~−0.14	....	IT0.088	
			156	91	−75.	(3−)	5.4 d			ε(100)	2.3	0.535(70), 0.199(40), 1.22(29), 0.089(7, 1.16(17), 1.42(15), ...	
			157	92	−75.96	(3/2+)	150 y			ε(100)	0.064	0.054(0.2)	
			158m	.....	.....	(2+)	11 s	±1.74	+2.7	.....	....	IT0.11(100)	
			158	93	−74.54	3(−)	1.2 × 10⁹ y			ε(87), ... β⁻0.86(12), ...	1.20 / 0.94	0.95(69), 0.099(14), 0.086(12), 0.182(10), 0.782(10), ...	

		A	N	Δ (mass excess)	J	half-life / abundance	μ (±)	Q	decay	E	γ-rays	σ
		159	94	−74.61	3/2+	**100%**	±1.99	+1.3	....		....	3.0
		160	95	−72.80	3−	72.1 d	±1.68	+3.0	β−0.56(44), 0.86(29), 0.46(10), ...	1.83	0.879(30), 0.299(27), 0.960(25), 1.178(15), 0.087(13), 1.27(7), ...	525
		161	96	−72.40	3/2+	6.9 d	−0.46	+2.3	β−0.51(64), 0.45(26), 0.58(10), ...	0.58	0.026(21), 0.049(19), 0.075(10), 0.057(5), ...	
		162	97	−70.5		7.5 m			β−1.25(90), 1.5(5), ...		0.260(80), 0.808(45), 0.888(35), 0.185(16), 0.882(15), 0.080(8), ...	
		163	98	−69.44		6.5 h	+0.64	+2.5	β−1.4(40), 1.65(30), 1.5(15), 1.1(15)	1.68	0.330(40), 0.025(30), 0.235(15), 0.510(15), ...	
		163	98	−69.44		19.5 m			β−0.80, 1.28, 1.40	1.68	0.390(31), 0.351(26), 0.494(23), 0.074(15), ...	
		164	99	−67.2		23 h			β−	3.3	...	
66	Dy (Dysprosium)	149	83	−72.	0+	~15 m			ε	~4		
		150	84	−74.		7.2 m			ε, β+ α4.23(18)	~2.0	0.39	
		151	85	−74.	0+	18 m			β+, ε, α4.06(6)	3.0	0.145	
		152	86	−75.21		2.4 h			ε, α3.65(0.05)	0.8	0.257	
		153	87	−74.17		6.4 h			ε, α3.48(0.003) α3.37	~2.1	0.081, 0.100, 0.255, ...	
		154m				13 h			α2.87			
		154	88	−75.53	(3/2−)	10.7 y			ε(98), β+0.85(2), ...		0.227(68), 0.664(3), 1.090(3), 1.000(3), 0.905(2), ...	
		155	89	−74.20		10.2 h				2.10		
		156	90	−75.67	0+	**0.052%**			ε(100)	1.36	0.326(91), ...	96
		157	91	−74.50	(3/2−)	8.1 h				0.365		
		158	92	−75.55	0+	**0.090%**			ε(100)		0.058(4), ...	
		159	93	−74.22	3/2−	144 d			ε(100)			
		160	94	−74.77	0+	**2.29%**	−0.46					55
		161	95	−73.03	5/2+	**18.83%**	+0.64	+2.3				585
		162	96	−73.16	0+	**25.53%**						200
		163	97	−71.23	5/2−	**24.97%**		+2.5				140
		164	98	−70.80	0+	**28.18%**	±0.50					2,600
		165m2			1/2−	32 s			IT		IT0.108(97), 0.514(2), ...	
		165m1			7/2+	1.26 m			β−0.89(2), 1.0(0.5);			
		165	99	−68.25	7/2+	139.2 m	±0.50		β−1.3(83), 1.2(15), 0.3(1), ...	1.30	0.095(4), ...	

Table 8b-1. Properties of Nuclides (Continued)

(1) Atomic number Z	(2) Symbol	(3) Name	(4) Mass number A	(5) Number of neutrons N	(6) Mass excess, amu $\times 10^{-3}$	(7) Spin and parity	(8) % abundance or half life	(9) Magnetic moment, nuclear magnetons	(10) Quadrupole moment, barns	(11) Mode of decay, energy, and intensity, MeV (%)	(12) β-decay Q values, MeV	(13) Energy and intensity of γ ray transitions, MeV (%)	(14) 2,200-m/s neutron-absorption cross section, barns
66	Dy	Dysprosium	166	100	−67.17	0+	81.5 h			β^-0.40(92), 0.48(5),...	0.482	0.082(12),...	
			167	101	−66.		4.4 m			β^-	~7		
67	Ho	Holmium	150	83	−66.		~20 s			ϵ, β^+			
			151		−68.		42 s			ϵ, β^+, α4.60(30)	~5		
			151	84			36 s			ϵ, β^+, α4.51(20)			
			152		−68.		52 s			ϵ, β^+, α4.45(19)	6.4		
			152	85	−68.4		2.4 m			ϵ, β^+, α4.38(30)	4.3		
			153	86	−69.60		9 m			ϵ, β^+, α3.95(0.1)	5.8		
			154	87	−69.35	1	12 m			β^+, α3.91	~3	0.335	
			155	88	−71.	5/2	50 m			β^+2.1, ϵ		0.092, 0.138, 0.117, 0.209, 0.243, 0.326	
			156	89	−70.	1	55 m			ϵ, β^+1.80, 1.3, 2.9	~5	0.138(100), 0.266(99), 0.367(23), 0.685, 0.89, 1.41,...	
			157	90	−72.	7/2	14 m			β^+, ϵ	~2.5	0.087, 0.152, 0.196, 0.227, 0.71, 1.20,...	
			158m			2−	29 m					IT0.067, 0.098, 0.218, 0.32,...	
			158	91	−71.29	5+	11 m			β^+1.32, ϵ	3.98	0.098, 0.218, 0.320, 0.52.	
			159m				6.9 s					IT0.206(100)	
			159	92	−72.	7/2	33 m			ϵ	~1.7	0.057, 0.080, 0.17, 0.25, 0.309,...	
			160m			2−	5.0 h			ϵ(34)		IT0.060(66)	
			160	93	−71.64	5+	2.6 m			ϵ(99.8), β^+0.57(0.2),...	2.92	0.729(50), 0.97(35), 0.880(26), 0.65(20), 0.197(20), 0.539(5),...	
			161m			1+	6 s					IT0.211(100)	
			161	94	−72.15	1−	2.5 h			ϵ(100)	0.82	0.026(23), 0.075(15), 0.176(2), 0.157(1),...	

Element	A	N	I^π	Mass excess	μ	Q	Abundance / $T_{1/2}$	Decay mode, energy	E	σ	γ energies
	162m		6−				68 m	ε(37)			IT0.0578, 0.383(63)
	162	95	1+	−70.83			15 m	ε(95), β+1.10, ...	2.16		0.081(8)
	163m		1/2+				1.1 s				IT0.30(100)
	163	96	7/2−	−71.23			33 y	ε	0.01		
	164m		6				38 m	β⁻0.97(35), 0.88(22)	0.97		IT0.046, 0.052, 0.037
	164	97	1	−69.73			25 m	ε(43)	1.11		0.091(5) 0.073(5)
	165	98	7/2−	−69.64	+4.12	+3.0	**100%**			65	0.184(90), 0.810(60), 0.711(58), 0.081(12), 0.412(12), 0.532(12), ...
	166m		(7−)				1.2 × 10³ y	β⁻0.07, ...			
	166	99	0−	−67.69		1.84	26.7 h	β⁻1.84(52), 1.76(47), 0.38(1), ...			0.081(5), 1.380(0.1), ...
	167	100	(7/2−)	−66.9		0.97	3.1 h	β⁻0.30(44), 0.62(21), 0.97(16), 0.89(14), ...			0.347(58), 0.079(14), 0.208(13), 0.084(9), ...
	168	101		−64.1		2.8	3.3 m	β⁻2.2			0.85
	169	102	(1/2−)	−63.1		2.2	4.6 m	β⁻1.20(75), 1.95(25)			0.065(60), 0.78(30), 0.075(18), 0.850(13), 0.150(13), 0.760(11), ...
	170	103		−60.5		3.7	45 s	β⁻3.1			
Erbium **Er** **68**	152	84	0+	−65.		~3	11 s	α4.80(90), ε, β+			0.43
	153	85		−65.		~5	36 s	α4.67(75), ε, β			
	154	86	0+	−67.		~2	4.5 m	α4.15			
	155	87		−66.6		~4	5.3 m	α4.01			
	157	89	3/2?	−68.		~4	24 m	β+, ε			0.117, 0.386, 1.32, 1.66, 1.82, 2.0
	158	90	0+	−70.		~1.5	2.3 h	ε, β+0.8			0.072, 0.067, 0.315, 0.387,
	159	91	3/2?	−69.		~3	36 m	ε, β+			
	160	92	0+	−71.		~0.8	29 h	ε			0.206, 0.627, 1.20, 1.80, 2.60, ...
	161	93	(3/2−)	−70.		2.0	3.1 h	ε(99), β+1.2(1)			
	162	94	0+	−71.17	+1.1	+3.9	**0.136%**			160	0.826(63), 0.211(9), 0.592(8), 1.17(8), 1.37(5), 0.305(3), ...
	163	95	5/2−	−69.92			75 m	ε(99.9), β+0.19(0.004), ...	1.21		0.44(0.06), 1.11(0.05), ...
	164	96	0+	−70.76	±0.65	±2.2	**1.56%**			13	
	165	97	5/2−	−69.21			10.3 h	ε(100)	0.37		

TABLE 8b-1. PROPERTIES OF NUCLIDES (Continued)

(1) Atomic number Z	(2) Symbol	(3) Name	(4) Mass number A	(5) Number of neutrons N	(6) Mass excess, amu ×10⁻³	(7) Spin and parity	(8) % abundance or half life	(9) Magnetic moment, nuclear magnetons	(10) Quadrupole moment, barns	(11) Mode of decay, energy, and intensity, MeV (%)	(12) β-decay Q values, MeV	(13) Energy and intensity of γ-ray transitions, MeV (%)	(14) 2,200-m/s neutron-absorption cross section, barns
68	Er	Erbium	**166**	98	−69.69	0+	**33.41%**						30
			167m			$\frac{1}{2}$−	2.3 s					IT0.2078(100)	
			167	99	−67.92	$\frac{7}{2}$+	**22.94%**	−0.564	+2.8				700
			168	100	−67.60	0+	**27.07%**						1.9
			169	101	−65.38	$\frac{1}{2}$−	9.4 d	+0.513		β⁻0.34(58), 0.33(42)	0.340	0.00842	
			170	102	−64.51	0+	**14.88%**						6
			171	103	−61.94	$\frac{5}{2}$−	7.6 h	±0.70	±2.4	β⁻1.065(91), 0.58(4), 1.49(2)	1 490	0.308(63, 0.296(28), 0.112(25) 0.124(9), . . .	
			172	104	−60.63	0+	49 h			β⁻0.36(44), 0.29(42), 0.9(10), . . .	0.9	0.407(4?), 0.610(40), . . .	
			173	105	−57.3		12 m			β⁻2.3(70), 1.8(30)	2.8	0.20(60), 0.40(30), 0.18(20), 0.52(20), 0.36(19), 0.16(18), . . .	
69	Tm	Thulium	153	84	−51.		1.6 s			α5.11	∼7		
			154				3.0 s			α5.04			
			154	85	−58.		5 s			α4.96	∼8		
			161	92	−66.2		30 m			ε(100)	3.5	0.0456, 0.0595, 0.1061, 0.1126, 0.0845, 0.147, 0.172, . . .	
			162			(3)	77 m			ε(100)	4.89	0.102(100), 0.236(13)	
			162	93	−66.0		22 m	±0.08		β⁺3.82, 2.3, . . . , ε	2.42	0.102	
			163	94	−67.33	$\frac{1}{2}$(+)	1.8 h			ε, β⁺1.1, 0.40		0.104(8), 0.24(5), 0.29(3), 0.24(3), . . .	
			164	95	−66.51	(1+)	1.9 m			ε(61), β⁺2.94(26), 2.85(13)	3.95	0.091(4), 0.356, 0.39, 0.77, 0.86, 1.16, 1.31, 1.67, . . .	
			165	96	−67.56	$\frac{1}{2}$+	30.1 h			ε(99+), β⁺0.30(0.007)	1.57	0.243(50), 0.297(35), 0.807(15), 0.34(10), 1.13(5), 0.70(2), . . .	

Z	Sym	Name	A	N	Δ (mass excess)	I^π	$T_{1/2}$ / abund.			Decay	E	γ	σ
			166	97	−66.42	2+	7.7 h	±4.5	±0.047	ε(98), β+1.94(1.5), 1.1(0.3)	3.04	0.0806(90), 0.184(20), 0.780(19), 2.054(16), 0.208(99), 0.057(4), 0.532(2)	115, 92
			167	98	−67.12	½+	9.6 d			ε(100)	0.75		
			168	99	−65.77		85 d			ε(100)	1.70	0.816(88), 0.198(77), 0.747(40), 0.447(27), 0.080(11), 0.917(4), …	4.5
			169	100	−65.76	½−	**100%**		−0.232				
			170	101	−64.17	1−	129 d	±0.57	±0.246	β−0.97(77), 0.885(23)	0.967	0.0843(23)	
			171	102	−63.54	½+	1.92 y		±0.229	β−0.098(98), 0.03(2)	0.098	0.0667	
			172	103	−61.39	2−	63.7 h			β−1.80(39), 1.88(23), 0.41(14), 0.28(10), 0.71(8), …	1.87	1.09(10), 1.39(9), 1.46(7), 1.53(7), 0.079(5), 1.61(5), 0.399(89), 0.465(8), …	
			173	104	−60.35		8.2 h			β−0.90(75), 0.86(23), 1.3(2)	1.32	0.066(1)	
			174	:			5.2 m			β−1.2(80), 0.7(20)	3.0	0.365(93), 0.99(89), 0.273(85), 0.176(67), 0.50(15)	
			174	105	−57.8		5.5 m			β−2.5(100)	2.5	0.511, 0.362, 0.943, 0.985, …	
			175	106	−56.1		15 m			β−0.90(30), 1.50(20), 1.87(20), …	2.5	0.19(100), 0.39(80), 0.29(70), 0.05(60), 1.05(45), 0.87(40), …	
			176	107	−52.9	0+	1.4 m			β−1.15(40), 2.00(40), 3.05(20)	4.1		
70	Yb	Ytterbium	154	84	−54.		0.39 s			α5.33			
			155	85	−54.		1.6 s			α5.21			
			162	92	−64.		~24 m			ε	≳2		
			163	93	−63.		11 m			ε	≳4		
			164	94	−65.		76 m			ε	≳1		
			165	95	−64.61	0+	10 m			ε(190), β+1.7(10)	2.75	0.0409	
			166	96	−66.14	0+	57.5 h			ε(100)	0.26	0.080(70), 0.069(10), 1.09(5), 0.118(4), …	
			167	97	−65.02	(5/2−)	18 m			ε(99+), β+0.65(0.3)	1.96	0.082(17)	
			168	98	−66.08	0+	**0.135%**			ε(100)		0.113(90), 0.176(15), …	3,200
			169	99	−64.78	(7/2+)	32 d				0.91	0.063(45), 0.198(35), 0.177(22), 0.110(19), 0.308(10), …	
			170	100	−65.21	0+	**3.03%**						9.4
			171	101	−63.65	½−	**14.31%**		+0.4919				50

TABLE 8b-1. PROPERTIES OF NUCLIDES (*Continued*)

(1) Atomic number Z	(2) Symbol	(3) Name	(4) Mass number A	(5) Number of neutrons N	(6) Mass excess, amu $\times 10^{-3}$	(7) Spin and parity	(8) % abundance or half life	(9) Magnetic moment, nuclear magnetons	(10) Quadrupole moment, barns	(11) Mode of decay, energy, and intensity, MeV (%)	(12) β-decay Q values, MeV	(13) Energy and intensity of γ-ray transitions, MeV (%)	(14) 2,200-m/s neutron-absorption cross section, barns
70	Yb	Ytterbium	**172**	102	−63.59	0+	**21.82%**						0.4
			173	103	−60.35	5/2−	**16.13%**	−0.6776	+3.0				19
			174	104	−61.11	0+	**31.84%**						65
			175	105	−58.70	7/2−	101 h			β⁻0.467(87), 0.073(11), 0.35(2), 0.22	0.467	0.396(6), 0.283(4), 0.114(2), ...	
			176m			(8−)	11 s					IT0.093, 0.38, 0.294, 0.187, 0.082(100)	
			176	106	−57.42	0+	**12.73%**						5.5
			177m			1/2−	6.5 s					IT0.228, 0.104(100)	
			177	107	−54.72	9/2+	1.9 h			β⁻1.40(60), 1.3, 0.17	1.40	0.151(16), 1.080(5), 0.122(3), 1.24(13), ...	
71	Lu	Lutetium	155	84	−46.	...	0.07 s			α5.63			
			156			...	0.23 s			α5.54			
			156	85	−47.	...	0.5 s			α5.41			
			167	96	−61.73	...	54 m						
			168	97	−61.40	(1−)	7.1 m			ε(98), β⁺1.5(1), 1.1(1)	3.1	0.030, 0.079, 0.239, 0.100, ...	
			169m			(1/2−)	2.7 m					IT0.029(100)	
			169	98	−62.34	7/2+	34 h			ε(95), β⁺1.20	4.4	0.99(13), 0.90(10), 0.087(7), ...	
			170m			4−	0.7 s					IT0.048, 0.044(100)	
			170	99	−61.51	0+	2.0 d			ε, β⁺1.2, 0.4	2.26	0.063, 0.024, 0.071, 0.075, 0.087, 0.092, 0.104, 0.111, ...	
			171m			(1/2−)	78 s					IT0.071(100)	
			171	100	−62.	7/2+	8.3 d			ε(99.8), β⁺2.39(0.2)	3.41	0.084, 0.194, ...; 0.741(68), 0.019(20), 0.668(14), 0.075(8), 0.842(7), ...	
			172m			(1−)	3.7 m			ε(99.9), β⁺	~1.4	IT0.0419(100)	

Element	A	N	Δ	Spin (π)	Half-life (abundance)	μ	Q	Decay mode (MeV, %)	E (MeV)	γ-ray energies (MeV, %)	σ (b)
	172	101	−61.	$(4-)$	6.70 d			ε	~2.5	1.093(60), 0.901(45), 0.182(26), 0.810(21), 0.912(21), 0.079(13), …	
	173	102	−61.02	$(7/2+)$	1.37 y			ε(100)	0.69	0.272(18), 0.079(14), 0.101(7), 0.17(5), …	
	174m			$6-$	150 d			ε		IT0.0591, 0.0671, 0.0447, 0.994, 0.63, 0.076, 0.176, …	
	174	103	−59.65	$1-$	3.6 y			ε(100)	1.4	1.24(9), 0.076(6), …	
	175	104	−59.20	$7/2+$	97.41%	+2.23	+5.6				21
	176m			$1(-)$	3.7 h	+0.318	−2.3	β^-1.22(60), 1.31(40)			
	176	105	−57.29	$7(-)$	4×10^{10} y 2.59%	+3.18	+8.0	β^-0.43(100)	1.02	0.088(1.0), 0.105, 1.14, 0.306(95), 0.203(85), 0.088(15)	2,050
	177m			$23/2-$	161 d			β^-(78)		IT0.228(22), 0.104(22), 0.208(62), 0.378(29), 0.113(23), 0.418(21), …	
	177	106	−56.22	$7/2+$	6.7 d	+2.24	+5.4	β^-0.497(86), 0.385(7), 0.175(7), 0.249(0.1)	0.497	0.113(14), 0.208(7), …	
	178				25 m			β^-1.50(100)		0.089, 0.214, 0.326, 0.427, …	
	178	107	−53.7		5 m			β^-2.25(100)	2.25	0.33, …	
	179	108	−52.7	$(5/2+)$	4.6 h			β^-1.34(87), 1.1(13)	1.34	0.213f(3), …	
	180	109	−49.9		2.5 m			β^-3.3	3.3		
72 Hf Hafnium	157	85	−42.		0.12 s			α5.68			
	158		−45.	0+	3 s			α5.27			
	168	96	−59.	0+	25 m			ε	~2	0.117, 0.098, 0.183, 0.160, 0.203, …	
	169	97	−58.7		1.5 h			ε, β^+~1.3	~5	0.12, …	
	170	98	−60.	0+	12.2 h			ε	~1	0.165, 0.120, 0.99, 1.28, …	
	171	99	−59.		11 h			ε	~3	0.662, 0.347, 0.470, 0.296, 0.122, 0.519, …	
	172	100	−60.	0+	5 y			ε	~0.4	0.024(22), 0.125(21), 0.082(10), …	
	173	101	−59.	$(1/2-)$	23.6 h			ε	~3	0.13(96), 0.30(62), 0.162(5), 0.898(2), 1.04(1), 1.20(0.4), …	

TABLE 8b-1. PROPERTIES OF NUCLIDES (*Continued*)

(1) Atomic number Z	(2) Symbol	(3) Name	(4) Mass number A	(5) Number of neutrons N	(6) Mass excess, amu $\times 10^{-3}$	(7) Spin and parity	(8) % abundance or half life	(9) Magnetic moment, nuclear magnetons	(10) Quadrupole moment, barns	(11) Mode of decay, energy, and intensity, MeV (%)	(12) β-decay Q values, MeV	(13) Energy and intensity of γ-ray transitions, MeV (%)	(14) 2,200-m/s neutron-absorption cross section, barns
72	Hf	Hafnium	**174**	102	−59.86	0+	2.0×10^{15} y **0.18%**			α2.50		....	390
			175	103	−58.55	$(\frac{5}{2}-)$	70 d			ε	0.59	0.343(85), 0.089(3), 0.433(1), ...	
			176	104	−58.57	0+	**5.20%**					....	15
			177m	...	...	$2\frac{3}{2}+$	1.1 s					IT0.208(81), 0.228(48), 0.378(37), 0.113(30), 0.418(27), 0.327(23), ...	
			177	105	−56.75	$\frac{7}{2}-$	**18.50%**	+0.61	+3			....	380
			178m	...	...	(8−)	4.3 s					IT0.427(97), 0.326(94), 0.214(75), 0.089(54), 0.093(14)	
			178	106	−56.28	0+	**27.14%**					....	75
			179m	...	...	$\frac{1}{2}-$	18.6 s					IT0.161(99), 0.217(99), 0.378(1)	
			179	107	−54.16	$\frac{9}{2}+$	**13.75%**	−0.47	+3			....	65
			180m	...	...	8−	5.5 h					IT0.333(93), 0.215(82), 444(80), 0.058(48), 501(17), 0.093(16)	
			180	108	−53.43	0+	**35.24%**					....	12.6
			181	109	−50 87	$(\frac{1}{2}-)$	42.4 d			β⁻0.41	1.023	0.482(81), 0.133(48), 0.346(13), ...	40
			182	110	−49.28	0+	9×10^6 y			β⁻	0.5	0.271	
			183m	...	...		91 d			β⁻2.2, 1.9		0.095, 0.250, 0.340, ...	
			183	111	−46.40		65 m			β⁻1.18(56), 1.54(34), 1.00(10)	2.0	0.784, 0.458, 0.073, ...	
73	Ta	Tantalum	172	99	−55.		44 m			β⁺, ε	~5	0.092, 0.208, 0.27, 1.1, 1.3, ...	

A	N	Δ	Jπ	T½ / abundance	μ	Q	σ	Decay	E(decay)	γ-ray energies (MeV)
173	100	−56.	....	3.7 h				ϵ	~3	0.0698, 0.172, 0.160, 0.181, 0.090, ...
174	101	−56.	....	1.2 h				ϵ, β^+	~4	0.090, 0.125, 0.060, 0.205, 0.28, 0.35, ...
175	102	−56.	....	10.5 h				ϵ	~2.3	0.0815, 0.0505, 0.267, 0.0705, 0.104, 0.163, 0.207, ...
176	103	−55.2	(1−)	8.0 h				ϵ	3.1	0.0884, 0.202, 0.147, 0.176, 0.158, ...
177	104	−55.51	(7/2+)	56.6 h				ϵ	1.17	0.113(6), 0.208(1), 1.058, 0.425, ...
178	:	....	....	2.1 h				ϵ	....	0.328(100), 0.427(97), 0.214(75), 0.089(54), 0.093(13), ...
178m	105	−54.2	1+	9.4 m				$\epsilon(99), \beta^+0.89, 0.80(1)$	1.9	0.093(35), 1.35(1), ...
179	106	−54.03	(7/2+)	~600 d				$\epsilon(100)$	0.119	....
180m	:	....	1(−)	8.1 h				$\epsilon(87), \beta^+0.71(10), 0.61(3)$	....	0.0933(27), 0.103(3)
180	107	−52.44	7/2+	**0.0123%**						....
181	108	−51.97	(8+)	**99.988%**	+2.36	+4.2	22			....
182m	:	....	....	16.5 m			8,200		....	IT 0.184(40), 0.147(40), 0.172(40), 0.319(5), ...
182	109	−49.83	3−	115 d				$\beta^-0.54(45), 0.25(31), 0.41(24), ...$	1.807	0.1001(69), 0.0677(52), 1.122(35), 1.222(28), 0.0847(24), 1.189(16), ...
183	110	−48.57	7/2(+)	5.0 d				$\beta^-0.615(87), ...$	1.068	0.0465(48), 0.1079(47), 0.2461(36), 0.0526(36), 0.0991(35), 0.1614(18), ...
184	111	−45.73	(5−)	8.7 h				$\beta^-1.170(80), 1.110(20)$	3.0	0.111(84), 0.413(79), 0.252(58), 0.922(33), 0.318(28), 0.793(16), ...
185	112	−44.44	7/2(+)	49 m				$\beta^-1.72(100)$	1.96	0.060, 0.175, 0.100, 0.24, ...
186	113	−41.4	....	10 m				$\beta^-2.27, 2.56$	3.9	0.20, 0.73, 0.51, 0.61, 0.13, 0.30, ...

TABLE 8b-1. PROPERTIES OF NUCLIDES (Continued)

(1) Atomic number Z	(2) Symbol	(3) Name	(4) Mass number A	(5) Number of neutrons N	(6) Mass excess, amu × 10^-3	(7) Spin and parity	(8) % abundance or half life	(9) Magnetic moment, nuclear magnetons	(10) Quadrupole moment, barns	(11) Mode of decay, energy, and intensity, MeV (%)	(12) β-decay Q values, MeV	(13) Energy and intensity of γ-ray transitions, MeV (%)	(14) 2,200-m/s neutron-absorption cross section, barns
74	W	Tungsten	160?	86	.....	.....	.....	.....	.....	α5.75	.....	.....	
			173	99	−52.	.....	16 m	.....	.....	ϵ	~4	.....	
			174	100	−54.	0+	31 m	.....	.....	ϵ	~6	.....	
			175	101	−53.	.....	34 m	.....	.....	ϵ	~3	.....	
			176	102	−54.	0+	2.3 h	.....	.....	ϵ(99.5), β^+(0.5)	~1	0.26, 0.80, 1.3, 1.6, ... 0.0949, 0.1002, 0.0613, 0.0336, ...	
			177	103	−54.	.....	135 m	.....	.....	ϵ	~2	0.0705, 0.0305, 0.1157, 0.1864, 0.1560, 0.427, ...	
			178	104	−54.	0+	21.5 d	.....	.....	ϵ(100)	~0.09	.....	
			179m	...	.....	(1/2−)	5.2 m	.....	.....	.....	.....	.....	
			179	105	−53.	(7/2−)	38 m	.....	.....	ϵ(100)	~1	0.031(22), ...	
			180	106	−53.3	0+	1.14%	.....	.....	.....	.....	.....	10
			181	107	−51.77	(9/2+)	140 d	.....	.....	ϵ(100)	0.19	0.006, 0.136, 0.152	21
			182	108	−51.75	0+	26.41%	.....	.....	.....	.....	.....	
			183m	...	.....	(9/2+)	5.3 s	.....	.....	.....	.....	IT0.105, 0.047, 0.161, 0.210	
			183	109	−49.73	1/2−	14.40%	+0.117	.....	.....	.....	.....	10.2
			184	110	−49.03	0+	30.64%	.....	.....	.....	.....	.....	1.8
			185m	...	.....	.....	1.6 m	.....	.....	.....	.....	IT0.131(100), 0.175, 0.060, 0.175, ... 0.125(0.02)	
			185	111	−46.53	3/2−	74 d	.....	.....	β−0.429(99.98), 0.30(0.02)	0.429	.....	
			186	112	−45.60	0+	28.41%	.....	.....	.....	.....	.....	38
			187	113	−42.80	3/2−	23.8 h	.....	.....	β−0.628(59), 1.315(19), 33(8), 0.71(7), 0.55(4), ...	1.311	0.1342(30), 0.686(29), 0.480(25), 0.0720(23), 0.618(7), 0.552(6), ...	90
			188	114	−41.47	0+	69 d	.....	.....	β−0.349(99), ...	0.349	0.0630(0.9), ...	
			189	115	−38.0	.....	11 m	.....	.....	β−2.0, 2.5	2.5	0.258, 0.417, 0.555, 0.855, 0.955, 0.178, ...	

Element	Z	A	N	Jπ	Mass excess	Half-life	μ	Q	Decay	Q / Emax	γ-ray energies	σ (b)
Rhenium (Re)	75	175	100		−48.	5 m			β+, ε	~6	0.240, 0.110	
		176	101		−48.	3 m			β+3.1	~4		
		177	102		−49.	17 m			ε	~5	0.106, 0.237	
		178	103		−49.	15 m				~3	0.290, 0.429, 0.415, 0.477, . . .	
		179	104		−50.	20 m						
		180	105		−49.2	2.4 m			ε(93), β+1.77(7)	2.8	0.902(100), 0.104(26), 0.826(11), . . .	
		181	106		−49.	18 h			ε	~1	0.365, 0.110, 0.065, 0.0435, 0.0197, 0.1103, . . .	
		182	107	(7+)		64 h			ε		0.1001(74), 0.2292(25), 1.428(19), 0.178(18), 0.169(12), 0.0679, . . .	
		182	107	(2+)	−48.68	13 h			ε(99.8), β+1.74(0.2)	2.86	0.1001(66), 0.0677(62), 1.122(38), 1.222(31), 0.0847(29), 1.189(18), . . .	
		183	108	(5/2+)	−49.14	70 d			ε		0.0465(78), 0.1623(48), 0.0526(25), 0.0991(15), 0.0847(12), 0.1079(11), . . .	
		184m		(8+)		165 d			ε(30)	0.59	IT0.083, 0.105(70), 0.1112(18), 0.904(11), 0.793(8), 0.895(6), . . .	
		184	109	(3−)	−47.	38 d			ε	1.62	0.1112(59), 0.904(41), 0.793(36), 0.895(19), 0.253(5), . . .	
		185	110	5/2+	−46.99	**37.07%**	+3.172	+2.7				110
		186	111	1(−)	−44.96	90 h	+1.73	~±0.4	β−1.07(74), 0.935(21), ε(5)	1.07, 0.52	0.1372(21), 0.1225(2), . . .	
		187	112	5/2+	−44.21	5 × 10¹⁰ y, **62.93%**	+3.204	+2.6	β−0.0026(100)	0.0026		75
		188m		(6)−		18.6 m					IT0.0024, 0.016, 0.0636, 0.0925, . . .	
		188	113	1−	−41.85	16.8 h	+1.78	~+0.4	β−2.13(80), 1.98(16)	2.13	0.155(16), . . .	
		189	114	(5/2+)	−40.73	24 h			β−1.01(50), 0.79(19), γ725(7), . . .	1.01	0.0309(21), 0.0696(15), 0.186(5), 0.245(5), 0.036(4), . . .	<2

TABLE 8b-1. PROPERTIES OF NUCLIDES (*Continued*)

(1) Atomic number Z	(2) Symbol	(3) Name	(4) Mass number A	(5) Number of neutrons N	(6) Mass excess, amu ×10⁻³	(7) Spin and parity	(8) % abundance or half life	(9) Magnetic moment, nuclear magnetons	(10) Quadrupole moment, barns	(11) Mode of decay, energy, and intensity, MeV (%)	(12) β-decay Q values, MeV	(13) Energy and intensity of γ-ray transitions, MeV (%)	(14) 2,200-m/s neutron-absorption cross section, barns
75	Re	Rhenium	190	115	−38.1		2.8 m			$\beta^-1.6$	3.2	0.191(10), 0.392(10), 0.57(10), 0.88(3)	
			191	116	−37.		10 m			$\beta^-1.8$	1.8		
			192	117			6 s			$\beta^-2.5$			
76	Os	Osmium	176	100	−44.	0+	3.6 m			ε	~8	0.20, 0.29, 0.37, 0.48, 0.57	
			178	102	−46.	0+	5 m				~7	0.776, 1.291, 0.857, . . .	
			179	103	−47.		8 m				~5		
			180	104	−47.		22 m			ε		0.880, . . .	
			181				2.7 m			ε(94), $\beta^+1.75(6)$		0.145, 0.118, . . .	
			181	105	−47.		105 m			ε	~3	0.239, 0.118, 1.061, 0.167, 0.0276, 0.510, 0.1802, 0.0555, . . .	
			182	106		0+	21 h			ε			
			183m			(½ −)	10 h			ε(54)		IT0.1707(46), 1.102(26), 1.108(22), 0.0673(6), 1.035(6)	
			183	107	−47.	(9⁄2 +)	12 h			ε(100)	~1	0.382(88), 0.1145(71), 0.168(18), 0.236(6), . . .	3,000
			184	108	−47.40	0+	0.018%					0.646(81), 0.879(7), 0.718(4), 0.1253(2), . . .	
			185	109	−45.89	(½ −)	94 d			ε(100)	1.015		
			186	110	−46.13	0+	1.59%						
			187	111	−44.21	½−	1.64%	+0.0643					
			188	112	−44.12	0+	13.3%						
			189	113	−41.82.	3⁄2−	16.1%	+0.6566	+0.8			. . .	
			190m			(10)−	9.9 m					IT0.616(99), 0.502(98), 0.361(94), 0.187(70)	0.0003
			190	114	−41.52	0+	26.4%						
			191m			(3⁄2−)	13 h					IT0.042(100)	13
			191	115	−39.03	(9⁄2−)	15 d			$\beta^-0.143$	0.310	0.042, 0.129, . . .	
			192	116	−38.49	0+	41.0%						2.0

930

77	Iridium	Ir	A	N	Mass excess	Spin	Half-life	Decay mode	Moment	Decay energy	γ-ray energies
			193	117	−35.82		31 h	β⁻1.132(70), 0.99(9), 0.67(7), 1.06(6), ...		1.132	0.460(4), 0.139(3), 0.558(2), 0.387(1), ...
			194	118	−34.77	0+	6.0 y	β⁻0.097(67), 0.054(33),		0.097	0.043(33), 0.082(0.05)
			195	119	−32.0		6.5 m	β⁻2		2.0	
			171	94	−29.		1 s	α5.91			
			172	95	−29.		2 s	α5.81			
			173	96	−31.		3 s	α5.66			
			174	97	−33.		4 s	α5.48			
			175	98	−35.		4.5 s	α5.39			
			176	99	−36.		8 s	α5.12			
			177	100	−38.		21 s	α5.01			
			182	105	−42.	(5−)	15 m	β⁺, ε		~6	0.13, 0.28
			183	106	−43.		58 m	ε		~3	0.238
			184	107	−42.3		3.2 h	ε, β⁺		4.7	0.264, 0.125, 0.391, 0.840, 0.96, 1.09
			185	108	−43.	(3/2+)	14 h	ε(100)		~2.5	0.0374(72), 0.0599(38), 0.0973(23), 0.1007(15), 0.2542(15), 0.090(11), ...
			186	109	−42.00	(7)	15 h	ε(98), β⁺1.94(2), ...		3.831	0.1372(95), 0.297(75), 0.4348(35), 0.773(9), 0.636(7), 0.933(5), ...
			187	110	−43.	(3/2+)	11 h	ε(100)		~1.5	0.0743(16), 0.0652(15), 0.0255(4), 0.178(3), 0.427(3), 0.611(3), ...
			188	111	−41.08	(2−)	41 h	ε(99), β⁺1.656(1), ...		2.833	0.1550(57), 0.633(22), 0.478(15), 2.217(12), 1.210(7), 0.635(6), ...
			189	112	−41.	(3/2+)	13.3 d	ε(100)		~0.5	0.0696(21), 0.2448(8), 0.0591(4), 0.0363(4), ...
			190m₂			(11−)	3.2 h	ε(94)			IT0.1487(6), 0.616(93), 0.502(92), 0.361(88), 0.187(66)
			190m₁			(7+)	1.2 h				IT0.0263(100)
			190	113	−39.3	(4+)	11 d	ε(100)		2.1	0.56(72), 0.187(51), 0.604(47), 0.518(39), 0.40(39), 0.37(39), ...
			191m			1/2−	4.9 s		+1.3		IT0.129, ...
			191	114	−39.36	3/2+	37.3%		+0.145		

TABLE 8b-1. PROPERTIES OF NUCLIDES (*Continued*)

(1) Atomic number Z	(2) Symbol	(3) Name	(4) Mass number A	(5) Number of neutrons N	(6) Mass excess, amu $\times 10^{-3}$	(7) Spin and parity	(8) % abundance or half life	(9) Magnetic moment, nuclear magnetons	(10) Quadrupole moment, barns	(11) Mode of decay, energy, and intensity, MeV (%)	(12) β-decay Q values, MeV	(13) Energy and intensity of γ-ray transitions, MeV (%)	(14) 2,200-m/s neutron-absorption cross section, barns
77	Ir	Iridium	$192m_2$	...	...	$9(+)$	>5 y	...	...	...	...	IT0.161(100)	...
			$192m_1$	...	...	$1(+)$	1.4 m	...	...	β^-1.2(0.008), 1.5(0.007),....	...	IT0.058(99+),.... 0.317(0.008),....	
			192	115	-37.36	$4(-)$	74.2 d	±1.8	...	β^-0.672(49), 0.53(42), 0.24(5),....	1.457	0.317(81), 0.468(49), 0.308(30), 0.296(29), 0.604(9), 0.612(6),....	1,100
			$193m$	...	...	$11/2-$	12 d	...	...	ϵ(4)	...	IT0.080(100)	
			193	116	-37.04	$3/2+$	**62.7%**	$+0.158$	$+1.2$	...	...	...	0.05
			$194m$	117	-34.88	...	171 d	...	...	β^-<0.25	1.2	0.33(100), 0.48(100),....	
			194			$1-$	17.4 h	...	...	β^-2.24(89), 1.92(5), 0.98(2), 1.62(1),....	2.24	0.328(10), 0.64(1),....	120
			$195m$				4.0 h	...	...	...	...	0.685, 0.433, 0.319, 0.365, ...	
			195	118	-34.19		2.8 h	...	...	β^-1.0, 0.6	1.0	0.099, 0.130	
			$196m$	119			1.4 h	...	...	β^-1.16(80), 0.4(5),....	...	0.647(100), 0.522(99), 0.394(95), 0.441(95), 0.356(94), 0.100(33)	
			196		-31.6		52 s	...	...	β^-3.20(80), 2.1(16),....	3.2	0.355(20), 0.779(11), 0.447(5),....	
			197	120	-30.5		7 m	...	...	β^-2.0(50), 1.5(50)	2.0	0.50(50)	
78	Pt	Platinum	173	95	$-22.$	...	Short	...	...	α6.19	~6		
			174	96	$-26.$	...	0.7 s	...	...	α6.03(80), $\epsilon+\beta^+$(20)			
			175	97	$-27.$	...	2.1 s	...	...	α5.95	~6		
			176	98	$-30.$	...	6 s	...	...	$\epsilon+\beta^+$(99), α5.74(1)	~7		
			177	99	$-31.$	...	6.8 s	...	...	$\epsilon+\beta^+$(99+), α5.51(0.3)	~6		
			178	100	$-33.$	...	21 s	...	...	$\epsilon+\beta^+$(99), α5.44(1.3), 5.28(0.07)	~5		
			179	101	$-34.$	...	33 s	...	...	$\epsilon+\beta^+$(99+), α5.15(0.1)	~6		
			180	102	$-36.$	...	50 s	...	...	$\epsilon+\beta^+$(99+), α5.14(0.3)	~4		

A	N	Mass excess	Spin	Half-life / Abundance	Decay	Energy	μ	γ energies	σ
181	103	−36.		51 s	ε + β⁺(99+), α5.02(0.0006)	~5		0.68, 1.72, 1.85	
182	104	−38.	0+	3 m	ε + β⁺(99+), α4.82(0.02)	~4		0.155, 0.190	
183	105	−38.		6 m	ε + β⁺(99+), α4.72(0.001)	~5		0.065, 0.140, 0.190	
184m	—			42 m	ε	~2.5		0.1064, 0.1101, 0.1795, 2.01, 0.1391, 0.1840, …	
184	106	−40.		20 m	ε, α4.47	~4		0.1976(41), 0.1951(41), …	
185	107	−39.		1.1 h	ε	~2		0.0547(11), 0.3816(9), …	
186	108	−40.		2.8 h	ε	~3		0.0419(5), 0.424(4), …	
187	109	−39.		2.3 h	ε			0.0821(50), 0.0942(33),	
188	110	−40.50	0+	10.2 d	ε	0.54		0.1138(17), 0.1411(17), 0.1867(9), 0.721(8), …	
189	111	−39.	(3/2−,7/2−)	11 h	ε	~2		0.539(9), 0.410(5), 0.082(3), 0.457(2), …	
190	112	−40.03	0+	5×10¹¹ y, 0.0127%	α3.18				<14
191	113	−38.29	(3/2−)	3.0 G	ε(100)	1.00		IT0.0126, 0.1355	
192	114	−38.92	0+	0.78%					<2
193m	—			4 3 d		0.06		IT0.1299, 0.031, 0.099,	
193	115	−36.96		16 y	ε				
194	116	−37.29	0+	32.9%					
195m	—			4.1 d				IT0.346, 0.053(97), 0.279(3)	
195	117	−35.19	1/2−	33.8%			+0.6060		24
196	118	−35.03	0+	25.3%					0.9
197m	—			80 n.	β−0.737(3)	0.75		0.077(100), 0.191(9), …	
197	119	−32.65	1/2−	18 h	β−0.67(90), 0.48(10)				
198	120	−32.11	0+	7.21%					4.0
199m	—			14.1 s		1.68		IT0.393, 0.032(100)	
199	121	−29.42		30 m	β−1.68(63), 1.14(14, 0.89(10), 0.95(8), …			0.540(24), 0.475(12), 0.197(9), 0.075(9), 0.715(3), 0.960(2), …	15
200	122	−29.	0+	11.5 ?	β−	~0.7		0.15, 0.23, 1.76	
201	123	−25.2		2.5 m	β−2.66, …	2.66			

TABLE 8b-1. PROPERTIES OF NUCLIDES (*Continued*)

(1) Atomic number Z	(2) Symbol	(3) Name	(4) Mass number A	(5) Number of neutrons N	(6) Mass excess, amu ×10⁻³	(7) Spin and parity	(8) % abundance or half life	(9) Magnetic moment nuclear magnetons	(10) Quadrupole moment, barns	(11) Mode of decay, energy, and intensity, MeV (%)	(12) β-decay Q values, MeV	(13) Energy and intensity of γ-ray transitions, MeV (%)	(14) 2,200-m/s neutron-absorption cross section, barns
79	Au	Gold	177	98	$-22.$	....	1.3 s	....	....	α6.12			
			178	99	$-23.$	....	2.6 s	....	....	α5.92			
			179	100	$-26.$	....	7.2 s	....	....	α5.85			
			181	102	$-30.$	....	11.5 s	....	....	α5.62(55), 5.48(45)			
			183	104	$-32.$	....	50 s	....	....	α5.34			
			184	105	$-32.$	....	1.0 m	....	....	ε, β^+	~7	0.163, 0.273, 0.362	
			185	106	$-34.$	....	4.3 m	....	....	ε, β^+, α5.07(0.01)	~5		
			186	107	$-34.$	....	12 m	....	....	ε	~6	0.16, 0.22, 0.30, 0.40	
			187	108	$-35.$	....	8 m	....	....	ε, α4.69	~4		
			188	109	$-35.$	....	8 m	....	....	ε	~5	0.25, 0.33, 0.63	
			189m	...	$-36.$	....	4.7 m	....	....	ε	....	0 168(80)	
			189	110	....	....	30 m	....	....	ε	~3	0.713, 0.448, 0.813, 0.168, ...	
			190	111	$-35.$	1−	39 m	±0.066	....	ε, β^+	~4.4	0.29, 0.60, ...	
			191	112	$-36.$	$\tfrac{3}{2}+$	3.2 h	±0.137	....	ε	~2.0	0.030, 0.048, 0.091, 0.278, 0.133, ...	
			192	113	-35.15	1(−)	5.0 h	±0.00785	....	ε(99), β^+2.19, 2.49	3.5	0.316, 0.157, 0.0451, 0.295, 0.105, 1.140, ...	
			193m	...	....	$\tfrac{1}{2}-$	3.9 s	....	....	ε(0.03)	....	IT0.0323, 0.258, 0.220, 0.291, ...	
			193	114	$-36.$	$\tfrac{3}{2}+$	16 h	±0.139	....	ε(100)	~1	0.18(11), 0.26(9), 0.114(5), 0.440(3), ...	
			194	115	-34.58	1−	39.5 h	±0.074	....	ε(97), β^+1.5(2), 1.2(1)	2.51	0.328(68), 0.294(12), 1.469(8), 1.887(4), 2.044(4), 1.596(3), ...	
			195m	...	....	$\tfrac{11}{2}-$	31 s	....	....			IT0.0567, 0.2615(97), 0.200(2), ...	
			195	116	-34.95	$\tfrac{3}{2}+$	183 d	±0.147	....	ε(100)	0.229	0.0985, 0.0308, ...	
			196m	...	....	12(−)	9.7 h	....	....		....	IT0.148(42), 0.188(32), 0.285(5), 0.316(5)	

El / Z	A	N	I(π)	Mass excess	Half-life	μ	Q	Disint. energy	Decay	γ-rays (MeV)	σ
	196	117	2−	−33.44	6.18 d	+0.58 or −0.62		1.48; 0.684	ε(94); β−0.259(6)	0.356(94), 0.333(25), 0.426(6), 1.091(0.2)	
	197m		11/2−		7.4 s					IT0.130, 0.279, ..., (100)	
	197	118	3/2+	−33.45	100%	+0.14486	+0.58				98.8
	198	119	2−	−31.769	2.697 d	+0.590		1.374	β−0.961(99), 0.29(1)	0.412(99), 0.676(1), 1.088(0.2)	25,800
	199	120	3/2+	−31.23	3.15 d			0.45	β−0.30(73), 0.25(2), 0.46(6)	0.1584(73), 0.2082(21)	
	200	121	(1−)	−29.3	48.4 m	+0.270		2.2	β−2.2(70), 0.7(25), ...	0.368(24), 1.227(23), 1.593(1), ...	30
	201	122		−28.1	26 m			1.5	β−1.5	0.54	
	202	123		−25.6	30 s			3.5	β−3.5(90), 3(10)	0.44(10), 0.52(0.3)	
	203	124		−24.	55 s			~2.5	β−1.9	0.69	
	204	125		−21.7	4 s			4.5	β−4.5	0.43	
80 Hg Mercury	179	99		−17.	3.5 s				α6.08		
	180	100		−21.	5.9 s				α5.96		
	181	101		−22.	3.6 s				α6.00, 5.91		
	182	102		−24.	10.5 s				α5.86		
	183	103		−25.	8.8 s				α5.91, 5.83		
	184	104		−27.	32 s				α5.54		
	185	105		−28.	52 s			~6	ε, α5.65, 5.57	0.12, 0.27, 0.35, 0.44	
	186	106		−30.	1.4 m			~3.5	ε, α5.11	0.175, 0.255, 0.400	
	187	107		−30.	3 m			~5	ε	0.140	
	188	108		−32.	3.3 m			~3	ε, α5.14	0.165, 0.240, 0.320, 0.500	
	189	109		−32.	9 m			~4	ε	0.0288, 0.143, 0.130, 0.155, ...	
	190	110	0+	−33.	20 m			~2	ε	0.26, ...	
	191	111		−33.	55 m	−1.063	−1.2		ε	0.274(100), 0.157(20), 0.114(10), ...	
	192	112	0+	−34.	4.8 h			~3	ε		
	193m		13/2+		10 h			~0.9	ε(82), β+1.17(2)	IT0.0395, 0.1012(16), 0.257, 0.218, 0.574, 0.220, ...	
	193	113	3/2−	−33.	4 h	−0.62	−2	2.34	ε	0.038, 0.187, 0.564, 0.762, 0.855, 1.040, ...	
	194	114	0+	−34.	1.2 y			0.05	ε(100)		
	195m		13/2(+)		40 h	−1.049	+1.3		ε(50)	IT0.123(52), 0.261(30), 0.560(8), ...	

TABLE 8b-1. PROPERTIES OF NUCLIDES (*Continued*)

(1) Atomic number Z	(2) Symbol	(3) Name	(4) Mass number A	(5) Number of neutrons N	(6) Mass excess, amu × 10⁻³	(7) Spin and parity	(8) % abundance or half life	(9) Magnetic moment, nuclear magnetons	(10) Quadrupole moment, barns	(11) Mode of decay, energy, and intensity, MeV (%)	(12) β-decay Q values, MeV	(13) Energy and intensity of γ-ray transitions, MeV (%)	(14) 2,200-m/s neutron-absorption cross section, barns
80	Hg	Mercury	195	115	−33.	$\frac{1}{2}(-)$	9.5 h	+0.538		ε(100)	~1.4	0.780(8), 0.062(7), 0.585(3), 0.207(2), 0.261(2), 1.11(1), ...	
			196	116	−34.18	0+	**0.146%**						3,100
			197m			1,3/2+	24 h	+1.032	+1.5	ε(6)		IT0.165, 0.134(94), 0.279(6), 0.130	
			197	117	−33.00	$\frac{1}{2}-$	65 h	+0.524		ε(100)	0.42	0.077(18), 0.191(2), 0.268, ...	
			198	118	−33.252	0+	**10.02%**						0.02
			199m			1,3/2+	43 m					IT0.375, 0.158(100)	
			199	119	−31.725	$\frac{1}{2}-$	**16.84%**	+0.5027					2,000
			200	120	−31.679	0+	**23.13%**						<60
			201	121	−29.696	$\frac{3}{2}-$	**13.22%**	−0.5567	+0.45				<60
			202	122	−29.358	0+	**29.80%**						4.9
			203	123	−27.125	$\frac{5}{2}-$	46.9 d	+0.84	± ≦13	β⁻0.214(100)	0.492	0.279(100)	
			204	124	−26.502	0+	**6.85%**						0.43
			205	125	−23.92	$(\frac{1}{2}-)$	5.5 m			β⁻1.5	1.5	0.203	
			206	126	−22.49	0+	8.2 m			β⁻1.3(59), 1.0(36), 0.65(5), ...	1.31	0.31, 0.65	
81	Tl	Thallium	191	110	−27.		10 m			ε, β+	~5		
			192	111	−27.		11 m			ε, β+	~6.3	0.424, 0.64	
			193m	112		$(\frac{9}{2}+)$	2.1 m					IT0.365	
			193		−29.	$(\frac{1}{2}+)$	23 m			ε, β+	~4.2	0.241, 0.252, 0.261, 0.309, 0.270, 0.330, ...	
			194m	113	−29.		32.8 m			ε(100)	~5.5	0.097, 0.748, 0.636, 0.427	
			194				33.0 m			ε(100)		0.427	
			195m	114	−30.	$(\frac{9}{2}-)$	3.5 s					IT0.099, 0.383(100)	
			195			$\frac{1}{2}+$	1.2 h			ε, β⁻~1.8	~3.2	0.0372, 0.242, 0.226, 0.562, ...	

Mass No.	N	J^{π}	Mass excess	Half-life	μ	Q	Decay	γ-ray energies (MeV)	σ
196m	..	(7+)	.....	1.41 h	.....	.....	$\epsilon(96)$	IT0.120(4), 0.084, 0.428, 0.636, 0.696	.....
196	115	(2−)	−29.	1.84 h	.....	~4.4	ϵ	0.428	
197m	: :	(9/2−)	.....	0.54 s	.....	.....		IT0.222, 0.385(100)	
197	116	1/2+	−30.	2.8 h	+1.55	~2.4	ϵ	0.152, 0.426, 0.134, ...	
198m	: :	7+	.....	1.87 h	.....	.....	$\epsilon(45)$	IT0.261, ... (55), 0.412(45), 0.586(35), 0.635(35), 0.283(30)	
198	117	2−	−29.5	5.3 h	±<0.002	3.5	$\epsilon(99)$, $\beta^{+}2.4$, 2.1, 14,	0.412(90), 0.65(40), 1.42(24), 1.20(21), 2.01(15), 2.45(5), ...	
199	118	1/2+	−30.	7.4 h	+1.59	~1.4	$\epsilon(100)$	0.455(14), 0.208(12), 0.247(9), 0.158(5)	
200	119	2−	−29.04	26.1 h	±≤0.15	2.454	$\epsilon(99.6)$, $\beta^{+}1.07(0.3)$, 1.44(0.06), ...	0.368(88), 1.21(35), 0.579(10), 0.829(8), 1.364(4), 1.517(4), ...	
201	120	1/2+	−29.25	73 h	+1.60	0.41	$\epsilon(100)$	0.167(8), 0.135(2), ...	
202	121	2−	−28.03	12.2 d	±≤0.15	1.24	$\epsilon(100)$	0.439(95), 0.522(0.1), 0.961(0.07)	
203	122	1/2+	−27.65	**29.50%**	+1.6115	.....			11.0
204	123	2−	−26.13	3.83 y	±0.089	0.763 / 0.34	$\beta^{-}0.763(98)$ / $\epsilon(2)$		22
205	124	1/2+	−25.56	**70.50%**	+1.6274	.....			0.5
206	125	0(−)	−23.88	4.19 m	.....	1.533	$\beta^{-}1.53$		
207m	: :	(11/2−)	.....	1.3 s	.....	.....		IT1.00, 0.35(100)	
207	126	(1/2+)	−22.55	4.79 m	.....	1.43	$\beta^{-}1.43(99)$	0.898(0.2)	
208	127	(5+)	−17.99	3.06 m	.....	4.994	$\beta^{-}1.80(49)$, 1.29(24), 1.52(23), 1.04(5), ...	2.614(100), 0.583(86), 0.511(23), 0.860(12)	
209	128	(1/2+)	−14.63	2.2 m	.....	3.98	$\beta^{-}1.99(100)$	0.45(100), 1.56(100), 0.12(50), ...	
210 (RaC'')	129	.....	−9.90	1.3 m	.....	5.50	$\beta^{-}1.9(56)$, 1.3(25), 2.3(19)	0.795(100), 0.296(80), 1.31(21), 1.21(17), 1.06(12), 2.43(9), ...	
Pb Lead 82									
194	112	0+	−25.	11 m	.....	~3.2	ϵ	0.204	
195	113	.....	−25.	17 m	.....	~4.6	ϵ	0.099, 0.383, 0.393	
196	114	0+	−27.	37 m	.....	~2.4	ϵ	0.253, 0.240, 0.367, 0.192,	
197m	: :	(13/2+)	.....	42 m	.....	.....	$\epsilon(80)$	IT0.234(20), 0.222(80), 0.387, 0.0849	

TABLE 8b-1. PROPERTIES OF NUCLIDES (*Continued*)

(1) Atomic number Z	(2) Symbol	(3) Name	(4) Mass number A	(5) Number of neutrons N	(6) Mass excess, amu × 10⁻³	(7) Spin and parity	(8) % abundance or half life	(9) Magnetic moment, nuclear magnetons	(10) Quadrupole moment, barns	(11) Mode of decay, energy, and intensity, MeV (%)	(12) β-decay Q values, MeV	(13) Energy and intensity of γ-ray transitions, MeV (%)	(14) 2,200-m/s neutron-absorption cross section, barns
82	Pb	Lead	197	115	$-26.$	$(\tfrac{3}{2}-)$	?			ϵ	~3.7	0.386	
			198	116	$-28.$	0+	2.4 h			$\epsilon(100)$	~1.5	0.38(40), 0.173(28), 0.290(16), 0.259(8), 0.865(6), 0.575(4), ...	
			199m	...	...	$(\tfrac{13}{2}+)$	12.2 m					IT0.424(100)	
			199	117	$-27.$	$(\tfrac{5}{2}-)$	90 m			$\epsilon, \beta^{+}2.8$	~3	0.367(80), 0.353(17), 0.720(10)	
			200	118	$-28.$	0+	21.5 h			ϵ	~0.7	0.1476, 0.1423, 0.2356, 0.1095, 0.2684, 0.2572, ...	
			201m	...	...		61 s			$\epsilon, \beta^{+}0.55$	~1.8	IT0.629(100)	
			201	119	$-27.$		9.4 h					0.331, 0.361, 0.406, 0.585, ...	
			202m	...	...	9-	3.62 h			$\epsilon(10)$		IT0.422(90), 0.961(90), 0.787(45), 0.658(35), 0.490(10), 0.460(8), ...	
			202	120	-27.97	0+	~3 × 10⁵ y			$\epsilon(100)$	0.05	IT0.825(92), 0.820(10), ...	
			203m	...	...	$(\tfrac{13}{2}+)$	6.1 s						
			203	121	-26.60	$(\tfrac{5}{2}-)$	52.1 h			$\epsilon(100)$	0.98	0.279(81), 0.401(5), 0.680(1)	
			204m	...	...	9-	66.9 m					IT0.90, 0.375, ...	
			204	122	-26.96	0+	1.48%			$\epsilon(100)$	0.04		0.66
			205	123	-25.52	$(\tfrac{5}{2}-)$	3 × 10⁷ y						0.30
			206	124	-25.53	0+	23.6%					IT1.064, 0.570(100)	
			207m	125	-24.097	$\tfrac{13}{2}+$	0.77 s						0.71
			207			$\tfrac{1}{2}-$	22.6%	+0.5895					
			208	126	-23.34	0+	52.3%						0.015
			209	127	-18.90	$(\tfrac{9}{2}+)$	3.30 h			$\beta^{-}0.635$	0.64		

Element	Z	A	N	Mass excess	Jπ	Half-life	μ	Q	Decay, energies (MeV)	Dis. energy	γ energies (MeV)
		210 (RaD)	128	−15.80	0+	22 y	...	...	β⁻0.017(81), 0.063(19); α3.72	0.063	0.0465
		211 (AcB)	129	−11.23	(9/2+)	36.1 m			β⁻1.37(92), 0.53(6), 0.95(1), 0.25(0.7), ...	1.37	0.405(3), 0.832(3), 0.427(2), 0.766(1), ...
		212 (ThB)	130	−8.11	0+	10.64 h			β⁻0.34(81), 0.58(14), ...	0.57	0.239(47), 0.300(3), ...
		213	131	−3.4	...	10.2 m			...	2	
		214 (RaB)	132	−0.16	...	26.8 m			β⁻0.69(42), 0.74(36), 1.03(6), ...	1.04	0.352(42), 0.295(25), 0.242(12), 0.053(11), ...
Bismuth	83	190	107			?			...		
Bi		191m				?			...		
		191	108	−13.		15 s			α6.46		
		192	109	−14.		40 s			α6.90		
		193m				3.2 s			α6.31		
		193	110	−16.		70 s			α6.06		
		194	111	−16.		85 s			α6.50		
		195m				60 s			α5.90		
		195	112	−19.		2.5 m			α5.61		
		196?	113	−19.		7.8 m			α6.10		
		197	114	−21.	9/2	8 m			α5.42	~8	
		198	115	−21.	7	11.9 m			ε, α5.8	~5.5	0.20, 0.32, 0.56, 1.06
		199	116	−22.		24.7 m			ε(9⁺), α5.81(0.05)	~7	
		200	117	−22.		35 m			ε(99⁺), α5.48(0.01)	~5	
		201m				59. m			ε	~6	
		201	118	−23.	9/2	1.8 h			ε(9⁺), α5.24(0.01)	~4	0.629
		202	119	−22.	5	95 m			ε	~5	0.422, 0.961
		203	120	−23.17	9/2	11.8 h	+4.59	−0.64	ε; β⁺1.35, 0.74	3.19	0.82(78), 1.87(35), 1.52(31), 1.034(16), 0.381(9), 0.186(6), 0.375, 0.0785, 0.176, 0.249, 0.216, 0.899
		204	121	−22.	6+	11.2 h	+4.25	−0.41	e(100)	~4.4	0.375, 0.0785, 0.176, 0.703(28), 1.766(27), 0.988(17), 0.57(14), 1.044(8), 1.864(6), ...
		205	122	−22.62	9/2	15.3 d	+5.5		ε(99.9), β⁺0.98(0.06)	2.70	0.249, 0.216, 0.899, 0.703(28), ...
		206	123	−21.61	6+	6.24 d	+4.56	−0.19	ε	3.65	0.803(99), 0.880(72), 0.516(46), 1.720(36), 0.538(34), 0.343(26), ...
		207	124	−21.52	(9/2−)	30 y			ε	2.40	0.570(98), 1.063(77), 1.771(9)

TABLE 8b-1. PROPERTIES OF NUCLIDES (Continued)

(1) Atomic number Z	(2) Symbol	(3) Name	(4) Mass number A	(5) Number of neutrons N	(6) Mass excess, amu ×10⁻³	(7) Spin and parity	(8) % abundance or half life	(9) Magnetic moment, nuclear magnetons	(10) Quadrupole moment, barns	(11) Mode of decay, energy, and intensity, MeV (%)	(12) β-decay Q values, MeV	(13) Energy and intensity of γ-ray transitions, MeV (%)	(14) 2,200-m/s neutron-absorption cross section, barns
83	Bi	Bismuth	208	125	-20.27	$(5+)$	3.7×10^5 y	......	......	$\epsilon(100)$	2.87	2.614(100)	0.034
			209	126	-19.60	$\frac{9}{2}-$	**100%**	$+4.080$	-0.35	......	......	......	0.054
			210m	...	......	......	3×10^4 y	......	......	$\alpha4.96(58), 4.92(36), 4.57(6), 4.43(0.4), \beta^-(0.4)$	......	0.266(45), 0.305(23), 0.344, 0.650, ...	
			210 (RaE)	127	-15.87	$1-$	5.01 d	±0.0442	±0.13	$\beta^-1.160(99+), \alpha4.65, 4.69$	1.160	......	
			211 (AcC)	128	-12.70	$\frac{9}{2}-$	2.15 m	......	......	$\alpha6.62(84), 6.28(16), ..., \beta^-(0.3)$	0.59	0.351(14)	
			212 (ThC)	129	-8.72	$1(-)$	60.6 m	......	......	$\beta^-2.25(54), 1.52(5), ..., \alpha6.05(25), 6.09(10), ...$	2.246	0.727(7), 1.620(2), 0.040(2), 0.785(1), ...	
			213	130	-5.62	$\frac{9}{2}-$	46 m	......	......	$\beta^-1.42(66), 1.02(31), ...\alpha5.87(2), 5.55$	1.42	0.440(32), ...	
			214 (RaC)	131	-1.27	$(1-)$	19.8 m	......	......	$\beta^-3.28(20), 1.51(18), 1.55(15), 1.88(9), ..., \alpha5.45, 5.511, ... (0.02)$	3.28	0.609(42), 1.764(17), 1.120(15), 0.769(6), 1.378(5), ...	
			215	132	1.8	$(\frac{9}{2}-)$	7.4 m	......	......	β^-	2.24		
84	Po	Polonium	192	108		......	0.5 s	......	......	$\alpha6.58$			
			193	109	$-8.$	......	4 s	......	......	$\alpha7.0$			
			194	110	$-10.$	$0+$	0.6 s	......	......	$\alpha6.85$			
			195m	...		......	2.0 s	......	......	$\alpha6.71$			
			195	111	$-11.$	......	4.5 s	......	......	$\alpha6.62$			
			196	112	$-13.$	$0+$	5 s	......	......	$\alpha6.52$			
			197m	...		......	27 s	......	......	$\alpha6.38$			
			197	113	$-14.$	......	56 s	......	......	$\alpha6.28$			
			198	114	$-16.$	$0+$	1.7 m	......	......	$\epsilon, \alpha6.18$			
			199m	...		......	4.1 m	......	......	$\epsilon(74), \alpha6.05(26)$	~4		

Nuclide	N	Mass excess	I π	T½	μ	Q	Decay mode (energy, %)	E (MeV)	γ-ray energy (MeV) (%)
199	115	−16.	.....	5.1 m	.....	.....	ε(97), α5.94(3)	~6	
200	116	−18.	0+	11.4 m	.....	.....	ε(88), α5.86(12)	~3.7	
201m	.....	.....	.....	9 m	.....	.....	ε(97), α5.78(3)	~5	
201	117	−18.	3/2	15.4 m	.....	.....	ε(99), α5.68(1)	~3	
202	118	−19.	0+	44 m	.....	.....	ε(98), α5.58(2)		
203m	.....	.....	.....	30 m	.....	.....	ε(99+), α5.38(0.1)	~4	
203	119	−19.	5/2(−)	42 m	.....	.....	ε(99+), α5.49(0.02)	~2.3	
204	120	−20.	0+	3.6 h	.....	.....	ε(99+), α5.38(0.7)	~3.4	
205	121	−19.	5/2−	1.8 h	~+0.26	+0.17	ε(99+), α5.22(0.07)		<0.03
206	122	−19.65	0+	8.8 d			ε(95), α5.22(5)	1.82	0.51(100), 1.02(85), 0.807(60), 0.338(40), 0.286(35)
207m	.....	.....	.....	2.8 s	.....	.....			IT0.82(100), 0.26(42), 0.31(40)
207	123	−18.39	5/2−	5.7 h	~+0.27	+0.28	ε(99), β+0.89(0.3), 1.14(0.2), α5.12(0.01)	2.91	0.95(84), 0.74(36), 0.41(13), 1.15(6), 0.25(5), 1.37(4), ...
208	124	−18.76	0+	2.93 y	.....	.....	α5.12(99+), 4.22(0.0002), ε(0.006)	1.41	0.285(0.006), ...
209	125	−17.56	1/2	103 y	+0.76	.....	α4.88(99), 4.62(0.5), ε(0.5)	1.89	0.91(0.5), 0.261(0.4)
210 (RaF)	126	−17.12	0+	138.4 d	.....	.....	α5.305(100)		
211m	.....	.....	(25/2+)	25 s	.....	.....	α7.28(91), 8.88(7), 8.06(1.7), 8.31(0.25)		0.570(92), 1.063(77)
211 (AcC′)	127	−13.34	(9/2+)	0.52 s	.....	.....	α7.45(99), 6.89(0.5)		0.57(0.5), 0.90(0.5)
212m	.....	.....	(18+)	45 s	.....	.....	α11.65(97), ...		2.61(3), 0.57(2)
212 (ThC′)	128	−11.135	0+	3.04 × 10⁻⁷ s	.....	.....	α8.78(100), ...		
213	129	−7.15	(9/2+)	4.2 × 10⁻⁶ s	.....	.....	α8.38(100), ...		
214 (RaC′)	130	−4.79	0+	1.62 × 10⁻⁴ s	.....	.....	α7.688(100), ...		
215 (AcA)	131	−0.55	(7/2, 9/2+)	1.78 × 10⁻³ s	.....	.....	α8.384(100), ... β⁻(0.0002)	0.740	0.44
216 (ThA)	132	1.92	0+	0.15 s	.....	.....	α6.777(100), ...		
217	133	6.	.....	<10 s	.....	.....	α6.54(>80), β⁻(<20)	~1.6	
218 (RaA)	134	9.01	0+	3.05 m	.....	.....	α6.000(99.98), ... β⁻(0.02)	0.27	

TABLE 8b-1. PROPERTIES OF NUCLIDES (*Continued*)

(1) Atomic number Z	(2) Symbol	(3) Name	(4) Mass number A	(5) Number of neutrons N	(6) Mass excess, amu $\times 10^{-3}$	(7) Spin and parity	(8) % abundance or half life	(9) Magnetic moment, nuclear magnetons	(10) Quadrupole moment, barns	(11) Mode of decay, energy, and intensity, MeV (%)	(12) β-decay Q values, MeV	(13) Energy and intensity of γ-ray transitions, MeV (%)	(14) 2,200-m/s neutron-absorption cross section, barns
85	At	Astatine	196	111	-4.	. . . .	0.3 s	. . . .	. . . .	α7.06			
			197	112	-6.	. . . .	0.4 s	. . . .	. . . .	α6.96			
			198m	. . .	-7.	. . . .	1.5 s	. . . .	. . . .	α6.85			
			198	113	-7.	. . . .	4.9 s	. . . .	. . . .	α6.75			
			199	114	-9.	. . . .	7.2 s	. . . .	. . . .	α6.64			
			200m	. . .	. . .	. . . .	4.3 s	. . . .	. . . .	α6.54			
			200	115	-9.	. . . .	42 s	. . . .	. . . .	α6.46, 6.41, ε			
			201	116	-11.	. . . .	1.5 m	. . . .	. . . .	α6.34, ε			
			202m	. . .	. . .	. . . .	2.6 m	. . . .	. . . .	α6.23			
			202	117	-11.	. . . .	3.0 m	. . . .	. . . .	ε(90), α6.13(10)	~7		
			203	118	-13.	. . . .	7.4 m	. . . .	. . . .	ε(86), α6.09(14)	~6		
			204	119	-13.	. . . .	9.3 m	. . . .	. . . .	ε(95), α5.95(5)	~7		
			205	120	-14.	. . . .	26 m	. . . .	. . . .	ε(82), α5.90(18)	~5		
			206	121	-14.	. . . .	32 m	. . . .	. . . .	α5.70(88), ε(12)	~6		
			207	122	-14.3	. . . .	1.8 h	. . . .	. . . .	ε(90), α5.76(10)	3.84		
			208	123	-13.	. . . .	1.6 h	. . . .	. . . .	ε(99.5), α5.65(0.5)	~4.9	0.685(99), 0.660(80), 0.177(20), . . .	
			209	124	-13.83	$(\frac{9}{2}-)$	5.5 h	. . . .	. . . .	ε(95), α5.65(5)	3.48	0.780(94), 0.545(62), 0.195(23)	
			210	125	-12.96	(5+)	8.3 h	. . . .	. . . .	ε(99.8), α5.36, 5.53, 5.44	3.87	1.80(100), 0.245(79), 1.483(48), 1.436(29), 1.599(14), . . .	
			211	126	-12.49	$\frac{9}{2}(-)$	7.21 h	. . . .	. . . .	ε(59), α5.87(41)	0.79	0.67	
			212m	. . .	. . .	. . . .	0.12 s	. . . .	. . . .	α7.82(80), 7.88(20)	. . . .	0.063	
			212	127	-9.26	. . . .	0.30 s	. . . .	. . . .	α7.66(80), 7.60(20)	. . . .	0.063	
			213	128	-7.1	. . . .	Short	. . . .	. . . .	α9.06			
			214	129	-3.67	. . . .	Short	. . . .	. . . .	α8.78(99.8), 8.45, 8.24			
			215	130	-1.34	. . . .	1×10^{-4} s	. . . .	. . . .	α8.00(99.95), 7.60(0.05)		0.40	

Radon, Rn, Z = 86

Z	Sym / Name	A	N	Δ (MeV)	I	T½	Major radiations (MeV)	γ-rays (MeV)	(MeV)	
		216	131	2.42		3×10^{-4} s	α7.81(97), 7.69(2), 7.61, 7.57, 7.48, 7.40, 7.33, 7.25			
		217	132	4.71		0.03 s	α7.066(99.9), 6.82, 6.62, 6.49, ...	0.260, 0.455, 0.595		
		218	133	8.71		~2 s	α6.70(90), 6.65(6), 6.76(3.4), β⁻(0.1)			
		219	134	11.3		0.9 m	α6.27(97), β⁻(3)		2.88	
86	Rn Radon	201	115	−4.		3 s	α6.27(97)		1.69	
		202	116	−6.	0+	13 s	β⁻(3)			
		203m	:			28 s	α6.77			
		203	117	−6.		45 s	α6.64			
		204	118	−8.	0+	75 s	α6.55			
		205	119	−8.		1.8 m	α6.50			
		206	120	−8.	0+	6.5 m	α6.42		~4	
		207	121	−9.		11 m	α6.26		~5	
		208	122	−10.	0+	23 m	α6.26(65), ε(35)		~3	
		209	123	−10.		30 m	ε(96), α6.15(4)		~4	
		210	124	−10.46	0+	2.42 h	ε(80), α6.14(20)		2.33	
		211	125	−9.38	(1/2−)	14.6 h	ε(83), α6.04(17)	0.680(74), 1.37(38), 0.445(29), 1.13(23), 0.946(21), 0.865(18)	2.89	
		212	126	−9.29	0+	25 m	α6.04(96), ε(4)			
		213	127	−6.11		0.019 s	ε(71), α5.78(17), 5.85(9), 5.62(0.5)			
		215	129	−1.25	0+	Short	α6.26			
		216	130	0.27	0+	4.5×10^{-6} s	α8.10			
		217	131	3.94		5.4×10^{-4} s	α8.6			
		218	132	5.61	0+	0.03 s	α8.05	0.61(0.2)		
		219	133	9.51	(3/2, 5/2+)	3.96 s	α7.74	0.272(9), 0.401(5), . . .		
		220 (Tn)	134	11.40	0+	55.6 s	α7.13(99.8), 6.54(0.2)	0.542(0.07)		<0.2
		221	135	15.		25 m	α6.817(81), 6.551(12), 6.423(7), 6.53, 6.31, 6.16, ...		~1.1	
		222 (Rn)	136	17.61	0+	3.824 d	α6.29(99.9), 5.75(0.07)	0.51		0.72
		223	137			43 m	β⁻(80), α(20)		1.85	
		224	138		0+	1.90 h	α5.486(99+), 4.98, 4.82		0.55	
		225				4.5 m	β⁻			
		226			0+	6 m	β⁻			

TABLE 8b-1. PROPERTIES OF NUCLIDES (Continued)

(1) Atomic number Z	(2) Symbol	(3) Name	(4) Mass number A	(5) Number of neutrons N	(6) Mass excess, amu $\times 10^{-3}$	(7) Spin and parity	(8) % abundance or half life	(9) Magnetic moment, nuclear magnetons	(10) Quadrupole moment, barns	(11) Mode of decay, energy, and intensity, MeV (%)	(12) β-decay Q values, MeV	(13) Energy and intensity of γ-ray transitions, MeV (%)	(14) 2,200-m/s neutron-absorption cross section, barns
87	Fr	Francium	203	116	2.	....	0.7 s	....	....	α7.13			
			204m	: 117	–.	....	2.2 s	....	....	α7.03			
			204	117	1.	....	3.3 s	....	....	α6.97			
			205	118	–0.8	....	3.7 s	....	....	α6.92			
			206	119	–1.	....	15.7 s	....	....	α6.79			
			207	120	–3.	....	15 s	....	....	α6.77			
			208	121	–3.	....	59 s	....	....	α6.65			
			209	122	–4.	....	52 s	....	....	α6.65			
			210	123	–4.	....	3.2 m	....	....	α6.57			
			211	124	–4.5	....	3.1 m	....	....	α6.53			
			212	125	–4.	....	19 m	....	....	ε(56), α6.39(17), 6.42(16), 6.35(11), ...	~5.1	0.039, 0.123, ...	
			213	126	–3.82	....	35 s	....	....	α6.77(99.5), ε(0.5)	2.14		
			214m	:	....	(9–)	3.4×10^{-8} s	....	....	α8.48(51), 8.55(46), 7.71(1), 8.05(1), ...			
			214	127	–1.02	....	5.0×10^{-8} s	....	....	α8.43(68), 7.45(10), 7.83(10), 8.36(7), ...			
			215	128	0.36	....	$\ll 10^{-8}$ s	....	....	α9.4			
			217	130	4.64	....	Short	....	....	α8.31			
			218	131	7.52	....	Short	....	....	α7.85(93), 7.55(5), 7.52(1), 7.71, 7.37			
			219	132	9.25	....	0.02 s	....	....	α7.30(98.4), 6.95(1), 7.14, 6.68, 6.72		0.163, 0.189, 0.352, 0.493, 0.530	
			220	133	12.32	....	27.5 s	....	....	α6.68(80), 6.63(14), 6.57(5), 6.52(2), 6.48, 6.43, 6.40			
			221	134	14.26	....	4.8 m	....	....	α6.34(82), 6.12(16), 6.24(1), 5.98(5), 5.94, 6.07, 5.78, ...		0.22(13), ...	
			222	135	17.57	....	14.8 m	....	....	β−(100)	2.028	0.050(40), 0.080(13), ...	
			223 (AcK)	136	19.76	....	22 m	....	....	β−1.10, ... (99+), α5.34(0.005)	1.16	0.234(4), ...	

Z	Element	A	N	Mass excess	J	Half-life	Particle radiations (MeV)	E	γ radiations (MeV)	σ
88	Ra Radium	224	137	23.		<2 m	β⁻			
		225	138			3.9 m	β⁻			
		226	139		0+	1.4 m	β⁻			
		206	118	5.	0+	0.4 s	α7.27			
		207	119	4.		1.3 s	α7.13			
		208	120	2.	0+	1.2 s	α7.13			
		209	121	2.		4.7 s	α7.01			
		210	122	0.5	0+	3.8 s	α7.02			
		211	123	0.8		15 s	α6.91			
		212	124	-0.3	0+	13 s	α6.87			
		213	125	0.1		2.7 m	α6.62(49), 6.73(45), 6.52(6)			
		214	126	-0.03		2.6 s	α7.14	~2.9		
		215	127	2.74		1.6 × 10⁻³ s	α8.70(96), 7.88(3), ...			
		216	128	3.49	0+	<10⁻³ s	α9.3			
		217	129	6.32		Short	α6.98			
		218	130	7.15	0+	Short	α8.39			
		219	131	10.08		Short	α8.0			
		220	132	11.03	0+	0.023 s	α7.45(99), ...		0.465(1)	
		221	133	13.93	0+	30 s	α6.61(35), 6.76(31), 6.67(21), 6.59(8), 6.58(3), 6.25(0.7), ...		0.152(13), 0.176(2), 0.320(0.7), ...	
		222	134	15.38	0+	38 s	α6.56(96), ...		0.325(4), ...	
		223 (AcX)	135	18.53	(½+)	11.4 d	α5.714(53), 5.605(24), 5.745(9), 5.538(9), 5.432(2), 5.870(1), ...		0.270(14), 0.154(5), 0.180(5), ...	130 / Fission 1.0
		224 (ThX)	136	20.20	0+	3.64 d	α5.684(95), 5.45(5), ...	0.387	0.241(4), ...	12.0
		225	137	23.63	(5/2)−	14.8 d	β⁻0.32(60), ...		0.04(30)	
		226 (Ra)	138	25.44	0+	1,600 y	α4.781(94), 4.598(5), ...		0.186(3), ...	20
		227	139	29.20		41 m	β⁻1.31	1.31	0.027, 0.30, 0.50	
		228 (MsTh₁)	140	31.10	0+	5.75 y	β⁻0.048(70), 0.024(30)	0.055	0.0067, 0.0263	36
89	Ac Actinium	229	141	35.		Short	β⁻	~2		
		230	142	37.		1 h	β⁻	~0.8		
		209	120	10.		0.19 s	α7.59			
		210	121	10.		0.35 s	α7.46			
		211	122	8.		0.25 s	α7.48			

TABLE 8b-1. PROPERTIES OF NUCLIDES (Continued)

(1) Atomic number Z	(2) Symbol	(3) Name	(4) Mass number A	(5) Number of neutrons N	(6) Mass excess, amu $\times 10^{-3}$	(7) Spin and parity	(8) % abundance or half life	(9) Magnetic moment, nuclear magnetons	(10) Quadrupole moment, barns	(11) Mode of decay, energy, and intensity, MeV (%)	(12) β-decay Q values, MeV	(13) Energy and intensity of γ-ray transitions, MeV (%)	(14) 2,200-m/s neutron-absorption cross section, barns
89	Ac	Actinium	212	123	8.		0.93 s			α7.38			
			213	124	7.		0.80 s			α7.36			
			214	125	7.		8.2 s			α7.21(52), 7.08(44), 7.00(4)			
			215	126	6.		0.17 s			α7.60			
			216	127	9.		3.9×10^{-4} s			α9.14			
			218	129	11.54		Short			α9.21			
			219	130	12.43		Short			α8.66			
			220	131	14.76		0.024 s			α7.64, 7.43			
			221	132	15.60		Short			α7.63(70), 7.42(30)			
			222	133	17.78	$(\frac{5}{2}-)$	4.2 s			α7.00(93), 6.96(6), ε(1)	2.237	0.082, 0.096, 0.12, ...	
			223	134	19.13		22 m			α6.65(42), 6.66(38), 6.56(13), 6.52(4), ... ε(1)			
			224	135	21.72		2.9 h			ε(90), α6.04(3), 6.20(3), 6.13(3), ...	0.563	0.216, 0.32, ...	
			225	136	23.23	$(\frac{3}{2}-)$	10.0 d			α5.83(52), 5.79(29), 5.72(10), 5.63(4), 5.72(3), ...	1.40	0.084, 0.088, 0.100, 0.150, 0.110, 0.063	
			226	137	26.10		29 h			β⁻0.89(46), 1.1(31) ε(23)	1.12, 0.63	0.072(30), 0.230, 0.253, 0.158, 0.185, 0.068	
			227 (Ac)	138	27.77	$\frac{3}{2}(-)$	21.8 y	+1.1	+1.7	β⁻0.044, ... (98.6), α4.95(0.7, 4.94(0.5), ...	0.044	0.009, 0.025, 0.015	810
			228 (MsTh₂)	139	31.04	3(+)	6.13 h			β⁻1.19(35), 1.76(20), 2.10(13), 0.62(6), ...	2.14	0.911(23), 0.966(20), 0.338(15), 0.0575, 0.129, 1.593, ...	
			229	140	33.		66 m			β⁻	1		
			230	141	36.6		<1 m			β⁻~2.2	2.9		
			231	142	38.6		15 m			β⁻2.1	2.1	0.185, 0.28, 0.29, 0.71	
90	Th	Thorium	213	123	13.		0.15			α7.69			
			214	124	12.		0.13 s			α7.68			

A	N	Mass excess	J^π	Half-life	μ	Q	Particle radiations (MeV)	Intensity	γ-rays (MeV)	σ
215	125	12.	....	1.2 s	....	....	α7.39(52), 7.52(40), 7.633(8)	....		
216	126	11.	....	0.03 s	....	....	α7.92			
217	127	13.	....	<0.0003 s	....	....	α9.25			
221	131	18.19	0+	0.0017 s	....	....	α 8.48			
222	132	18.48	0+	~0.0028 s	....	....	α7.98			
223	133	20.69	....	0.9 s	....	....	α7.55			
224	134	21.47	0+	1.05 s	....	....	α7.17(79), 6.99(19), 6.77(1), 6.70(0.4)	....	0.177(9), 0.410(1), ...	
225	135	23.96	(3/2+)	8.0 m	....	....	α6.48(41), 6.44(14), 6.50(13), 6.80(9), ...; e(5)	....	0.322(30), 0.246(5), 0.362(5), ...	
226	136	24.92	0+	30.9 m	....	....	α6.33(79), 6.22(19), 6.10(1), ...	0.680	0.111(5), 0.242(1), ...	Fission 150
227 (Rdac)	137	27.73	(3/2+)	18.5 d	....	....	α6.037(25), 5.976(23), 5.755(20), 5.707(8), 5.712(5), 5.699(4)	....	0.236(12), 0.050(10), 0.256(6), 0.330(3), ...	
228 (RdTh)	138	28.73	0+	1.913 y	....	....	α5.42(71), 5.34(28), ...	....	0.0845(16), 0.216(3), ...	120
229	139	31.78	5/2+	7340 y	+0.38	~4.6	α4.84(58), 4.90(11), 4.81(11), 4.97(10), 5.05(7)	....	0.20(10), 0.137(3), ...	Fission 32
230 (Io)	140	33.16	0+	8.0×10⁴ y	....	....	α4.68(76), 4.62(4), ...	0.387	0.07(0.6), 0.14, 0.18, 0.25	33
231 (UY)	141	36.32	(5/2+)	25.5 h	....	....	β^-0.30(46), 0.22(27), 0.14(20), 9.09(7)	....	0.084(10), 0.026(2), 0.090, ...	
232 (Th)	142	38.08	0+	1.41×10¹⁰ y; 100%	....	....	α4.01(76), 3.95(24)	....	0.059	7.4
233	143	41.60	....	22.2 m	....	....	β^-1.244(~85), 1.158, 1.073, 0.88, 0.79, 0.58	1.244	0.0660(3), 0.0292(2), 0.453(1), 0.171(0.7), ...	1,500
234 (UX₁)	144	43.64	0+	24.1 d	....	....	β^-0.19(81), 0.10(19)	0.263	0.093(4), 0.063(3), ...	Fission 15 / 1.8
235	145	....	....	<<10 m	....	....	β^-			
Pa — 91 Protactinium										
224	133	25.55	....	0.60 s	....	....	α			
225	134	26.12	....	2 s	....	....	α7.25			
226	135	27.88	....	1.8 m	....	....	α5.86(38), 6.82(34), 6.73(1); e(26)	2.77		
227	136	28.80	(5/2−)	38.3 m	....	....	α6.46(43), 6.41(13), 6.42(10), 6.40(8), ...; e(15)	1.000	0.065(6), 0.110(2), ...	

TABLE 8b-1. PROPERTIES OF NUCLIDES (Continued)

(1) Atomic number Z	(2) Symbol	(3) Name	(4) Mass number A	(5) Number of neutrons N	(6) Mass excess, amu ×10⁻³	(7) Spin and parity	(8) % abundance or half life	(9) Magnetic moment, nuclear magnetons	(10) Quadrupole moment, barns	(11) Mode of decay, energy, and intensity, MeV (%)	(12) β-decay Q values, MeV	(13) Energy and intensity of γ-ray transitions, MeV (%)	(14) 2,200-m/s neutron-absorption cross section, barns
91	Pa	Protactinium	228	137.	31.01	(3+)	26 h			ϵ(98)	2.11	0.95(93), 0.46(32), 0.33(18), 0.41(13), 0.20(9), 1.57(7), . . .	
			229	138	32.10	($\frac{5}{2}$−)	1.5 d			α6.08, 6.11, 6.12, 6.14, . . . , (2) ϵ(99+), α5.58(0.1), 5.62, 5.67, . . .	0.29	0.0424, . . .	
			230	139	34.56	(2−)	17.7 d			ϵ(90) β^-0.41(10), α5.34, . . . (0.003)	1.29	0.954(50), 0.91(24), 0.45(18), 0.51(8), . . .	Fission 1,500
			231	140	35.90	$\frac{3}{2}$−	3.24 × 10⁴ y	±1.98		α5.01(24), 5.02(23), 4.95(22), 4.73(11), 5.06(10)	0.56	0.027(6), 0.29(6), . . .	200 Fission 0.01
			232	141	38.59	−	1.31 d			β^-0.32(98), 1.19(0.8), 1.30(0.7)	1.34	0.87(51), 0.971(40), 0.150(12), 0.46(9), 0.57(8), 0.107(5), . . .	760
			233	142	40.27	$\frac{3}{2}$−	27.0 d	+3.4	−3.0	β^-0.26(55), 0.15(40), 0.57(5)	0.571	0.31(34), 0.300(6), 0.341(4), . . .	43
			234m (UX₂)			(0−)	1.17 m			β^-2.29(98). . . .		IT0.07(0.1), 0.043(2), 1.00(0.6). . . .	
			234 (UZ)	143	43.35	(4+)	6.75 h			β^-0.51(66), 0.23(14), 0.73(11), 1.02(7), 1.35(2)	2.22	0.90(70), 0.100(50), 0.126(26), 0.70(24), 0.56(15), 0.22(14), . . .	
			235	144	45.4		23.7 m			β^-1.4	1.4		
			236	145	49.0		12 m			β^-3.3	~3		
			237	146	52.22		39 m			β^-2.30(60), 1.3(30), 0.8(10)	2.30	0.46, 0.87, 0.92, 0.205, 0.090, 0.75, . . .	
92	U	Uranium	227	135	31.		1.3 m			α6.8			
			228	136	31.38	0+	9.2 m			α6.68(66), 6.59(29), . . . ϵ(5)	0.35	0.246, 0.187, 0.152	
			229	137	33.50	($\frac{3}{2}$+)	58 m			ϵ(80), 6.36(13), 6.33(4), 6.30(2), . . .	1.32		

Z	Element	A	N	Atomic mass	Spin	Half-life / abundance	μ	Q	Decay (MeV)	Disint. E	γ-ray energies (MeV)	Cross section (barns)
		230	138	33.94	0+	20.8 d			α5.89(67), 5.82(32),...		0.072(0.5), 0.231(0.2),...	Fission 25
		231	139	36.29	(5/2−)	4.3 d			ε(99+), α5.46(0.006)	0.36	0.026(12), 0.084(7), 0.220(1),...	Fission 400
		232	140	37.15	0+	72 y			α5.32(68), 5.27(32),... 5.14(0.3),...		0.058(0.2),...	150 Fission 75
		233	141	39.65	5/2+	1.62 × 10⁵ y	+0.54	+3.5	α4.82(83), 4.78(15),... 4.72(1.5),...		0.042, 0.097, 0.055, 0.029,	579 Fission 532
		234 (UII)	142	40.98	0+	2.47 × 10⁵ y / 0.0057%			α4.77(72), 4.72(28),...		0.22, 0.32,...	95
		235m			1/2+	26.1 m					0.053(0.2),...	
		235 (AcU)	143	43.94	1/2−	7.1 × 10⁸ y / 0.72%	−0.35	+4.1	α4.396(57), 4.366(18),... 4.216(5.7), 4.597(4.6), 4.415(4), 4.323(3),...		IT 0.000075 0.185(54), 0.143(11), 0.163(5), 0.204(5), 0.110(3),...	681 Fission 582
		236	144	45.59	0+	2.39 × 10⁷ y			α4.49,...		0.05	6.0
		237	145	48.75	1/2+	6.75 d			β−0.25(96), 0.10(4),...	0.517	0.060(36), 0.208(23), 0.026(2), 0.175(2), 0.33(1),...	480 Fission 2
		238 (UI)	146	50.82	0+	4.51 × 10⁹ y / 99.27%			α4.20(75), 4.15(25)		0.048	2,720
		239	147	54.33	(5/2+)	23.5 m			β−1.21(80), 1.28(20)	1.28	0.075(51), 0.044(4),...	36 Fission 14
		240	148	56.63	0+	14.1 h			β−0.36(100)	0.51	0.044	
93	Np Neptunium	228 or 227	135 or 134			60 s			SF			
		229	136	36.25		4.0 m			α6.89			
		230	137	37.79		4.6 m			α6.66			
		231	138	38.27		≈50 m			α6.29	2.7		
		232	139	40.		≈13 m			ε(99+)	1.1		
		233	140	41.		35 m			α5.54(0.001)			
		234	141	42.91	(0+)	4.40 d			ε(99+), β+0.8(0.05)	1.80	1.56(20), 1.53(12), 1.60(10), 1.44(7), 1.19(6), 1.57(6),...	Fission 900
		235	142	44.08	5/2+	410 d			ε(99+), α5.02, 5.10,... (0.002)	0.113		
		236m				>5 × 10³ y			β−?			2,800

TABLE 8b-1. PROPERTIES OF NUCLIDES (*Continued*)

(1) Atomic number Z	(2) Symbol	(3) Name	(4) Mass number A	(5) Number of neutrons N	(6) Mass excess, amu ×10⁻³	(7) Spin and parity	(8) % abundance or half life	(9) Magnetic moment, nuclear magnetons	(10) Quadrupole moment, barns	(11) Mode of decay, energy, and intensity, MeV (%)	(12) β-decay Q values, MeV	(13) Energy and intensity of γ-ray transitions, MeV (%)	(14) 2,200-m/s neutron-absorption cross section, barns
93	Np	Neptunium	237	144	48.19	$\frac{5}{2}+$	2.14×10^6 y	+3.3	—	α4.79(49), 4.77(26), 4.76(8), 4.64(7), 4.66(5), 4.87(3), ...	.....	0.029(14), 0.087(14), 0.145(1), ...	169
			238	145	50.97	2(+)	2.1 d	.....	.....	β^-1.25(45), 0.26(38), 0.20(15), ...	1.29	1.01(42), 0.044, ...	Fission 0.019
			239	146	52.95	$\frac{5}{2}+$	2.35 d	.....	.....	β^-0.437(48), 0.332(32), 0.393(7), 0.713(7), 0.654(4), ...	0.723	0.106(23), 0.278(14), 0.228(12), 0.209(4), ...	Fission 2,000
			240m	...	.....	1(−)	7.3 m	.....	.....	β^-2.18(41), 1.60(32), 1.30(10), 0.7(7), ...	2.1	0.56(21), 0.60(13), 0.92(3), 1.5(3), ...	
			240	147	56.08	.....	67 m	.....	.....	β^-0.89		0.56, 0.160, 0.245, 0.290, 1.00, 1.16, ...	63
94	Pu	Plutonium	241	...	58.3	.....	3.4 h	.....	.....	β^-	1.4		Fission 170
			241	148		.....	16 m	.....	.....	β^-1.4	1.1		
			232	138	11.17	0+	36 m	.....	.....	ϵ(98), α6.59(2)	2.0		
			233	139	42.99	.....	20 m	.....	.....	ϵ(99+), α6.31(0.1)	0.39		
			234	140	43.33	0+	9.0 h	.....	.....	ϵ(94), α6.20(4), 6.19(2), ...		0.048(0.3), 0.109(0.01)	
			235	141	45.28	.....	26 m	.....	.....	ϵ(99+), α5.86(0.003)	1.12		
			236	142	46.05	0+	2.85 y	.....	.....	α5.77(69), 5.72(31), 5.62(0.2), ...			
			237m	...		$(\frac{1}{2}+)$	0.18 s	.....	.....			IT0.145(100)	
			237	143	48.43	$(\frac{7}{2}-)$	45.6 d	.....	.....	ϵ(99+), α5.37, 5.56	0.22	0.060(5), 0.033, 0.044, ...	Fission 2,500
			238	144	49.58	0+	86 y	.....	.....	α5.50(72), 5.46(28), ...		0.043(0.04), ...	577 / Fission 16.5
			239	145	52.18	$\frac{1}{2}+$	2.44×10^4 y	+0.200	.....	α5.157(73), 5.144(15), 5.107(11), ...		0.052(0.02), 0.039(0.007), ...	1,005 / Fission 736
			240	146	53.84	0+	6,580 y	.....	.....	α5.17(76), 5.12(24), 5.02(0.1), ...		0.045, ...	290 / Fission 0.05

Z	Sym	Name	A	N	Mass excess	Spin	Half-life	μ	Q	Decay modes and energies (MeV)		γ-ray energies (MeV)	Thermal cross section (barns)
			241	147	56.87	5/2+	14.0 y	−0.73	+5.6	β^-0.0208(99+), α4.90(0.002), …	0.0208	……	1,371 / Fission 1,011
			242	148	58.77	0+	3.79×10^5 y			α4.90(76), 4.86(24)		……	20
			243	149	62.03	(7/2+)	4.98 h			β^-0.58(61), 0.49(38), …	0.59	0.084(21), 0.381(0.7), …	271 / Fission 196
			244	150	64.24	0+	8×10^7 y			α		…	1.8
			245	151	67.83		10.5 h			β^-0.93(48), 1.2, …	1.26	0.327(49), …	260
			246	152	70.13	0+	10.9 d			β^-0.15(90), 0.33(10)	0.37	0.044(30), 0.224(25), 0.180(10)	
95	Am	Americium	237	142	50.		~1.3 h			ε(99+), α6.02(0.005)	1.5	0.98(80), 1.35(76),	
			238	143	52.		1.9 h			ε(100)	~2.3	0.58(29), 0.36(12)	
			239	144	53.04	(5/2−)	12.1 h			ε(99+), α5.78(0.005)	0.81	0.228(18), 0.278(17), 0.209(5), 0.068, 0.057, 0.049, …	
			240	145	55.		51 h			ε(100)	1.4	1.00(77), 0.90(23), 1.40	
			241	146	56.85	5−	433 y	+1.59	+4.9	α5.49(86), 5.44(13), 5.39(1), 5.55(0.3), …		0.060(36), 0.026(3), …	787 / Fission 3.3
			242m2			5(−)	0.014 s			SF			
			242m1			1(−)	152 y	±0.382	±2.8	α5.21(0.4), …		IT0.048(99.5), …	Fission 2,100
			242	147	59.57	5/2−	16.0 h			β^-0.62(50), 0.66(34); ε(16)	0.66	0.042	
			243	148	61.39	(1−)	7.37×10^3 y	+1.4	+4.9	α5.28(88), 5.23(11), 5.18(1), …	0.75	0.045 / 0.075(61), 0.044(5), …	189
			244m			(6−)	26 m			β^-1.50(80), …, (20); ε(0.04)		0.043, …	
			244	149	64.31	(5/2+)	10.1 h			β^-0.387(100)	1.43	0.746(66), 0.900(25), 0.154(19), 0.099(5), …	Fission 2,300
			245	150	66.48	(2+)	2.05 h			β^-0.91(78), 0.65(17), 0.60(5)	0.91	0.253(20), 0.240(1), 0.296(1)	
			246m				39 m			β^-		0.680, 0.205, 0.154, 0.757, …	
			246	151	69.72		25.0 m			β^-1.31(79), 1.60(14), 2.10(7)	2.30	1.079(32), 0.799(29), 1.063(19), 1.037(14), 1.086(2), 0.834(2), …	
96	Cm	Curium	247	152	72.	0+	22 m			β^-	~1.6		
			238	142	53.03		2.5 h			ε(90), α6.51(10)	~0.9	0.285, 0.227	
			239	143	55.		2.9 h			ε(100)	1.7	0.188	

Table 8b-1. Properties of Nuclides (*Continued*)

(1) Atomic number Z	(2) Symbol	(3) Name	(4) Mass number A	(5) Number of neutrons N	(6) Mass excess, amu ×10⁻³	(7) Spin and parity	(8) % abundance or half life	(9) Magnetic moment, nuclear magnetons	(10) Quadrupole moment, barns	(11) Mode of decay, energy, and intensity, MeV (%)	(12) β-decay Q values, MeV	(13) Energy and intensity of γ-ray transitions, MeV (%)	(14) 2,200-m/s neutron-absorption cross section, barns
96	Cm	Curium	240	144	55.52	0+	26.8 d	....	....	α6.29(72), 6.25(28), 6.15(0.04)	....	....	
			241	145	57.68	($\frac{1}{2}$+)	35 d	....	....	ϵ(99), α5.94(0.7), 5.93(0.2), ...	0.77	0.475(95), 0.60, 0.145	
			242	146	58.86	0+	163 d	....	....	α6.12(73), 6.07(27), 5.97(0.03), ...	....	0.044(0.04), ...	20
			243	147	61.40	$\frac{5}{2}$+	32 y	....	....	α5.79(73), 5.74(12), 5.99(6), 6.06(6), 5.69(2), . . ., ϵ(0.3)	....	0.278(14), 0.228(12), 0.209(4), 0.106	950 Fission 700
			244	148	62.78	0+	18.0 y	....	....	α5.81(77), 5.77(23), . . .	0.007	0.043(0.02), . . .	14 Fission 1.0
			245	149	65.51	$\frac{7}{2}$+	9.3 × 10³ y	....	....	α5.36(87), 5.46(5), 5.31(4), 5.25(2), 5.53(2), . . .	....	0.173(14), 0.13(5)	2,250 Fission 250
			246	150	67.25	0+	5.5 × 10³ y	....	....	α5.39(81), 5.34(19)	....	....	8.4
			247	151	70.38	....	1.6 × 10⁷ y	....	....	α	....	....	288
			248	152	72.38	0+	4.7 × 10⁵ y	....	....	α5.08(73), 5.04(16), SF(11)	....	....	Fission 108 6
			249	153	75.99	0+	64 m	....	....	β^-0.9	0.9	....	
			250	154	78.	....	1.2 × 10⁴ y	....	....	SF	....	....	
97	Bk	Berkelium	243	146	63.02	($\frac{3}{2}$−)	4.6 h	....	....	ϵ(99+), α6.57(0.04), 6.54(0.03), 6.76(0.02), 6.21(0.02), . . .	1.51	0.76, 0.84, 0.95, . . .	1.6
			244	147	65.	....	4.4 h	....	....	ϵ(99+), α6.67(0.003), 6.62(0.003)	2.3	0.218, 0.892, 0.922, 0.188, 0.490, 0.334, . . .	
			245	148	66.39	($\frac{3}{2}$−)	4.98 d	....	....	ϵ(99+), α5.89(0.02), 6.15(0.02), 6.36(0.02), . . .	0.82	0.253(31), 0.381(4), . . ., 0.385(1), . . .	

Z	Element	A	N	Atomic mass excess	Jπ	Half-life	Decay mode, α energy (MeV)	Eβ, EC (MeV)	γ energy (MeV)	σ (barns)
		246	149	69.	...	1.8 d	ε(100)	~1.6	0.800(40), 1.082(4), 1.124(3), 0.834(3), 0.084(40), 0.27(30)	
		247	150	70.32	(9/2−)	1.4 × 10³ y	α5.52(58), 5.68(37), 5.31(5)	...	...	1,600
		248	...	...	(8−)	>9 y	?	0.7	...	
		248	151	73.	(0−)	16 h	β⁻0.65(70), ε(30)	0.6	...	
		249	152	75.00	7/2+	314 d	β⁻0.125(99+), α5.42(0.002), ...	0.125	0.32, ...	Fission 960
		250	153	78.34	(2−)	3.22 h	β⁻0.73(89), 1.76(11)	1.76	0.990(47), 1.032(39), 0.037, 0.94, 0.140, 0.184	
		251	154	81.	...	57 m	β⁻~0.5, ~1.0	~1.1	...	
98	Californium (Cf)	242	144	63.72	0+	3.3 m	α7.37	~2.2	...	
		243	145	65.	...	11 m	ε(90), α7.06, 7.17		...	
		244	146	65.99	0+	20 m	α7.21(75), 7.17(25)		...	
		245	147	68.07	...	44 m	ε(70), α7.13(30)	1.56	...	
		246	148	68.84	0+	36 h	α6.76(78), 6.72(22), 6.63(0.2), ...		0.042(0.01), 0.096(0.01), 0.146(0.004)	
		247	149	71.2	...	2.5 h	ε(100)	~0.7	0.295(1), 0.417, 0.460, ...	
		248	150	72.22	0+	350 d	α6.27(82), 6.22(18)		...	2,035 / Fission 300
		249	151	74.87	9/2−	360 y	α5.81(84), 5.76(4), 5.95(3), 5.91(3), 6.20(2), 5.85(1), ...		0.388(72), 0.333(15), 0.253(3), ...	
		250	152	76.43	0+	13 y	α6.03(83), 5.99(17), 5.89(0.3), ...		0.043	1,500
		251	153	79.61	1/2+	~800 y	α5.67(35), 5.85(28), 6.01(12), 5.63(5), ...		0.177(19), 0.224(7)	6,100 / Fission 4,000
		252	154	81.66	0+	2.65 y	α6.12(82), 6.08(15), ... SF(3.1)		0.043(0.01), 0.100, (0.01), 0.16	20
		253	155	85.17	(7/2+)	17.6 d	β⁻0.27(99+), α5.98(0.31), ...	0.27	...	1,500
		254	156	87.	0+	60.5 d	SF(99+), α5.84(0.2), ...		...	150
99	Einsteinium (Es)	245	146	71.	...	1.3 m	ε(83), α7.70(17)	~3	...	
		246	147	73.	...	7.3 m	ε(90), α7.33(10)	~4	...	
		247	148	73.62	...	5.0 m	ε(93), α7.33(7)	2.3	...	
		248	149	75.	...	25 m	ε(99+), α6.88(0.3)	~3	...	
		249	150	76.38	...	2 h	ε(99+), α6.77(0.13)	1.41	...	

TABLE 8b-1. PROPERTIES OF NUCLIDES (Continued)

(1) Atomic number Z	(2) Symbol	(3) Name	(4) Mass number A	(5) Number of neutrons N	(6) Mass excess, amu ×10⁻³	(7) Spin and parity	(8) % abundance or half life	(9) Magnetic moment, nuclear magnetons	(10) Quadrupole moment, barns	(11) Mode of decay, energy, and intensity, MeV (%)	(12) β-decay Q values, MeV	(13) Energy and intensity of γ-ray transitions, MeV (%)	(14) 2,200-m/s neutron-absorption cross section, barns
99	Es	Einsteinium	250	151	79.	...	8 h	...	...	ε(100)	~2		
			251	152	80.00	...	1.5 d	...	...	ε(99+), α6.48(0.53)	0.35		
			252	153	83.	(7+)	~140 d	...	...	α6.64(82), 6.58(13), 6.49(2), 6.26(1), ...		0.40(1), 0.074(0.3), 0.23(0.2), 0.28(0.2), ...	150
			253	154	84.85	($\frac{7}{2}$+)	20.47 d	...	...	α6.64(90), 6.60(7), 6.55(1), ...		0.387(0.05), 0.429, ...	
			254m	...		(2−)	39.3 h	...	...	β⁻0.48(75), 1.127(25)		0.69(38), 0.65(31), ...	Fission, 1,840
			254	155	88.05	(7+)	276 d	...	...	ε(0.08), 6.39, α6.44(93), 6.37(3), 6.42(2), 6.36(1), ...		0.063(2), 0.31(0.2), 0.27(0.1), 0.39(0.07), ...	Fission, 3,060
100	Fm	Fermium	255	156	90.	...	38.3 d	...	...	β⁻(91.5), α6.31(8.5), SF	~0.3	...	~40
			256	157	93.71	...	25 m	...	...	β⁻			
			257	158		...	<20 h	...	...	β⁻	1.8		
			244	144	75.31	...	0.0033 s	...	...	SF			
			245	145	77.	...	4.2 s	...	...	α8.15			
			246	146	77 19	...	1.6 s	...	...	α8.25			
			247	147	79.	...	30 s	...	...	α7.87(70), 793(30)			
			248	148		0+	0.6 m	...	...	α7.87(80), 7.83(20), SF(0.1)			
			249	149		...	~2.5 m	...	...	α7.9			

Z	Element	Symbol	A	N	Mass excess	Spin	Half-life	Major radiations, energies (MeV)		γ-ray energies (MeV)	
			250	150	79.54	0^+	30 m	α7.44			
			251	151	82.		7 h	ε(99), α6.90(1)	~1.5		
			252	152	82.49	0^+	23 h	α7.05			
			253	153	85.21		3 d	ε(80), α6.96(9), 6.91(2)	0.34	0.041(0.2), 0.098(0.03), 0.151	75
			254	154	86.88	0^+	3.24 h	α7.20(85), 7.16(14), 7.06(0.9), ..., SF(0.055)			
			255	155	89.98	$7/2^+$	20.1 h	α7.03(93), 6.97(5), 6.90(0.6), ..., SF		0.081(1.1), 0.059(0.9), ...	26
			256	156	91.81	0^+	2.0 h	SF(92), α6.92(8)			
			257	157	95.15	$(9/2^+)$	80 d	α6.52(94), 6.70(3), 6.4(2), ..., SF		0.242(10), 0.180(8), ...	
			258	158		0^+	<5 s	SF			
101	Mendelevium	Md	255	154	91.		27 m	ε(90), α7.34(10)	~1.1		
			256	155	94.		77 m	ε(97), α7.23(3)	~2.0		
			257	156	96.		4.5 h	ε(92), α7.08(8)	~0.4		
			258	157			54 d	α6.73, 6.78			
102	Nobelium	No	251	149	89.		0.8 s	α8.60(80), 8.68(20)			
			252	150	88.97	0^+	2.3 s	α8.41(70), SF(30)			
			253	151	91.		95 s	α8.01			
			254	152	90.98	0^+	55 s	α8.10			
			255	153	93.		180 s	α8.11			
			256	154	94.28	0^+	3.1 s	α8.43(99+), SF(0.5)			
			257	155	97.		23 s	α8.27(50), 8.23(50)			
103	Lawrencium	Lr	256	153			35 s	α8.42, ..., ε(?)			
			257	154	99.		0.7 s	α8.87, 8.81			
			258	155	100.		4.05	α8.6			
104	⋯	⋯	257	153			~4.5	α9.00, 8.95, 8.78, 8.70			
			258?	154			0.1 s	SF			
			259	155			~3 s	α8.77, 8.86			
105	⋯	⋯	260	155			1.6 s	α9.06(55), 9.10(75), 9.14(20)			

8c. Atomic Mass Formulas

PHILIP A. SEEGER

Los Alamos Scientific Laboratory

8c-1. Introduction. The nuclear or atomic mass is a direct measure of the total binding energy of the nucleus, and thus of the ground state of the nuclear Hamiltonian. If the Hamiltonian were known, the mass-law problem would be solved: it would be possible in principle to write the binding energy in terms of the atomic number Z and the mass number $A = N + Z$. Note that the mass is

$$M(Z,A) = A \cdot u + Z \cdot \Delta M_H + (A - Z) \cdot \Delta M_n - \frac{B(Z,A)}{c^2} \qquad (8\text{c-}1)$$

where u is the atomic mass unit $= 931.487$ MeV/c², $\Delta M_H = 7.82519$ mu is the mass excess of the hydrogen atom, $\Delta M_n = 8.66520$ mu is the neutron mass excess, and $B(Z,A)$ is the binding energy in MeV.

Even an incomplete nuclear theory can be used to predict the forms of some terms in the mass law. Weizäcker [1] pointed out that arbitrary multipliers could be used with such terms to gain insight both for the theory and the masses. His mass law, as simplified by Bethe and Bacher [2], has formed the basis for most subsequent studies. Many recent formulations and summaries are given in proceedings of topical conferences held at Vienna [3], Lysekil [4], and Winnipeg [5].

8c-2. Uses of the Mass Law. The complexity of a mass law depends on its intended use. For instance, in the calculation of nuclear kinematics, the mass number A is often a sufficient approximation, whereas for nuclear reaction theory quite sophisticated treatments are required. Uses may be classed as theoretical or experimental.

Theoretical uses include the comparison of calculated coefficients to values fitted to experimental data. Another use is an indirect determination of arbitrary constants in the theory: e.g., parameters in a proposed form of the nucleon-nucleon interaction can be found by calculating mass-law terms as functions of the interaction [6]. A third theoretical use is subtraction of the smoothly varying part of the mass law from the experimental data to isolate the small terms.

The mass law is used "experimentally" to estimate binding energies for use in other calculations or experiments. For "interpolation"—finding binding energies in the region of known data—the mass law should be discarded whenever practical in favor of tabulated experimental values. If it is necessary to use a mass law, a formula such as that of Zeldes et al. [7], which uses a large number of parameters to reproduce the experimental data as well as possible, may be used. Extrapolation to unknown masses requires the mass law. If the extrapolation is only a short distance from known data, and if only a few binding energies are needed, the values given by the mass law should be corrected by comparison of calculated and experimental data in the neighborhood, or a local extrapolation should be made. If a long extrapolation or a large number of calculations must be performed, a sophisticated mass-law formula must be used, and it should then be used for all binding energies in the problem, including known data. Two suitable formulas are those of Myers and Swiatecki [8] and of Seeger [9]; the latter is presented below.

8c-3. Terms in the Mass Law. Mathematically, the function $B(Z,A)$ can be expanded in terms of any two functions of Z and A which remain small over the ranges of Z and A to be considered. Since the binding energy per particle is nearly constant for $A > 10$, it is convenient to expand $B(Z,A)/A$. The range of nuclear force being short compared to the nuclear radius, a convenient expansion parameter is $1/R$. Constancy of nuclear density implies that $R \sim A^{\frac{1}{3}}$, and so the usual expansion parameter is $1/A^{\frac{1}{3}}$.

The distance of Z from the line of beta stability is a possible choice for the other parameter. In deriving terms from a model of the nucleus, however, the beta-stability line is not explicitly known, and it is more natural to expand about the symmetry line $Z = A/2$; from the statistical model [2] the form is $[(A - 2Z)/A]^2$.

There are some terms in the nuclear binding energy which it is not convenient to expand. The Coulomb force, for example, does not have a short range, and the Coulomb energy can be included explicitly if the charge distribution is assumed; for a uniformly charged sphere of radius $r_0 A^{\frac{1}{3}}$,

$$E_c = \frac{3Z^2 e^2}{5 r_0 A^{\frac{1}{3}}}$$

The binding energy can be expressed quite generally as

$$B(Z,A) = A \cdot f\left[\left(\frac{A - 2Z}{A} \right)^2, A^{-\frac{1}{3}} \right] - E_c \tag{8c-2}$$

where f is a power series in its arguments, and E_c represents Coulomb energy and any other terms which are not expanded. Although the original derivation was in terms of the liquid-drop model, the terms can be calculated analytically or numerically from any model. A calculation for infinite nuclear matter with $N = Z$ will yield the zero-order term in the expansion, the volume term αA. (Adjustable multipliers are denoted by Greek letters.) A mass law of this simple form, with one parameter determined by least-squares fit to known odd-A binding energies [10, 11] and the Coulomb energy derived from electron-scattering experiments [12], is illustrated in Fig. 8c-1a; it is clear from the figure that finite nuclei cannot be adequately represented by infinite nuclear matter.

The two first-order terms are the symmetry and the surface terms of the liquid drop: $-\beta(A - 2Z)^2/A - \gamma A^{\frac{2}{3}}$. These can be found for other models by calculating respectively infinite nuclear matter with $N \neq Z$ and semi-infinite matter with a plane surface. The negative signs indicate decreased binding energy. Inclusion of these terms completes the Weizsäcker formula [2]; the residual discrepancies following a least-squares fit to odd-A nuclides with $A > 40$ are shown in Fig. 8c-1b. The calculated binding energies are accurate to about $\frac{1}{4}$ percent; the error is greater than 1 percent for only nine of the lightest nuclides included.

Myers [6], using a nuclear force with constants determined by fitting to the four-parameter mass law above, has carried the expansion of Eq. (8c-2) to second-order terms: $A^{\frac{1}{3}}$, $(A - 2Z)^2/A^{\frac{4}{3}}$, and $(A - 2Z)^4/A^3$. The expressions become very complicated because the Coulomb force affects the density distribution of protons compared to neutrons. Only one of the second-order terms, the surface-symmetry term $\eta(A - 2Z)^2/A^{\frac{4}{3}}$, is commonly included in the mass law, and its effect is so weak that the coefficient is determined only poorly.

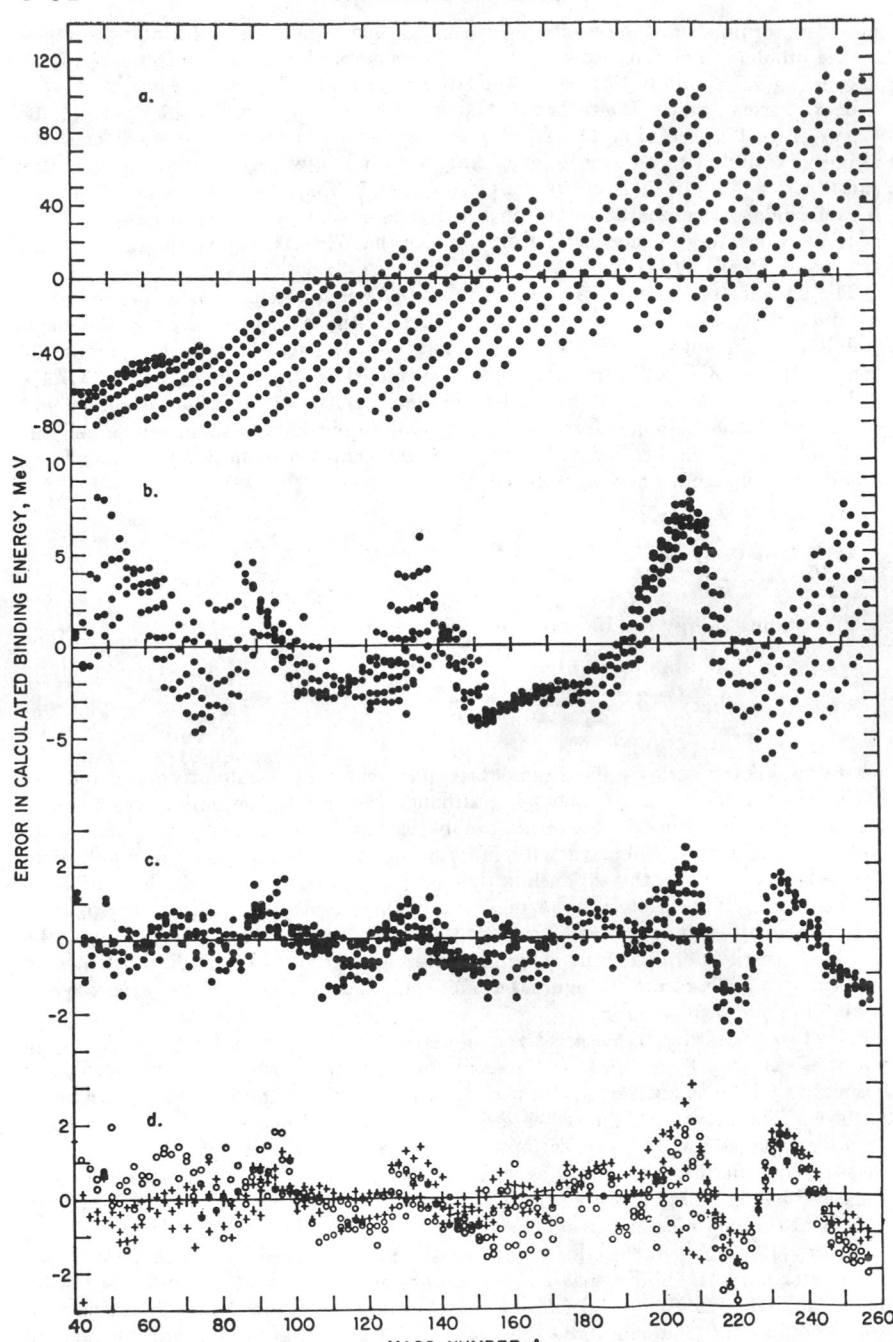

FIG. 8c-1. Errors of calculated binding energies versus mass number A: (a) for mass law with volume term and Coulomb energy only, fitted with 1 parameter to odd-A binding energies; (b) for 4-parameter liquid-drop mass law; (c) residual errors for odd-A nuclides, for Eq. (8c-3) fitted to 1,148 odd- and even-A nuclides; (d) same mass law, residual errors of even-even (+) and odd-odd (0) nuclides.

The expansion of Eq. (8c-2) is accurate only to the extent that the discrete levels occupied by nucleons can be represented by a smooth distribution, and the structure apparent in Fig. 8c-1b is due principally to the breaking down of this assumption. A correction term to the liquid-drop mass law can be constructed by comparing a single-particle-level diagram such as that of the Nilsson model to a smoothed average of the same levels. The method used is that of Strutinsky, extended by Tsang [13], who has shown that the results reach a limit which is independent of the details of the smoothing. The calculations [9] yield two functions $\delta U_N(N,\epsilon)$ and $\delta U_Z(Z,\epsilon)$, where ϵ is a measure of the spheroidal deformation of the nucleus. The coefficients of these functions in the mass law depend only on the radii of the neutron and proton distributions, $r_N A^{\frac{1}{3}}$ and $r_0 A^{\frac{1}{3}}$, respectively. The parameter r_N is new, but r_0 is the same radius constant which describes the proton charge distribution in the Coulomb energy.

Pairing correlation energy cannot be included in an average nuclear potential. It is calculated by applying the Bardeen-Cooper-Schrieffer (BCS) formalism to the single-particle levels, using as the average pairing matrix element $G \sim 1/A$. For a given value of the one adjustable parameter $G_N r_N{}^2 = G_P r_0{}^2$, the BCS ground-state energy for each particle number is found, and the difference in binding energy between it and the sum of the Nilsson levels is called $P_N(N,\epsilon)$ or $P_Z(Z,\epsilon)$ [9]. Since the presence of an unpaired particle decreases the binding energy of the BCS solution, the even-odd mass difference is calculated directly with no additional parameters. (A simple alternative phenomenological form for the even-odd difference is $\pm \delta/A^{\frac{1}{2}}$, where the $+$ sign is for even-even nuclides, $-$ for odd-odd, and the term is omitted for odd A. The least-squares determined value for δ is 10.6 $\pm$ 1.1 MeV.)

It is known that many nuclei, e.g., the rare earths and actinides, have nonspherical equilibrium shapes which are represented approximately in the Nilsson model by spheroids. The terms δU and P are explicit functions of the deformation parameter ϵ; the surface and Coulomb terms in the liquid-drop mass law can also be expanded in powers of ϵ. Then by maximizing total binding energy with respect to ϵ, the equilibrium deformation ϵ_0 is found; the results [9] agree qualitatively with experiment.

Several other small terms are included in the mass law. In the Coulomb energy there are an exchange term [2] and a correction for the diffuseness of the nuclear surface [8]. A first-order term in $(A - 2Z)/A$ seems to be required to represent extra binding of nuclei with $N = Z$; a rapidly decreasing exponential is used [8]. The binding of the atomic electrons [14] is included, although small, to prevent falsification of other terms. The complete formula is, in MeV,

$$B(Z,A) = \alpha A - \frac{\beta(A - 2Z)^2}{A} - \left[\gamma A^{\frac{2}{3}} - \frac{\eta(A - 2Z)^2}{A^{\frac{1}{3}}} \right] \left(1 + \frac{8}{45}\epsilon_0{}^2 + \frac{88}{2,835}\epsilon_0{}^3 \right)$$

$$- \frac{3}{5}e^2 \frac{Z^2}{r_0 A^{\frac{1}{3}}} \left(1 - \frac{0.76361}{Z^{\frac{2}{3}}} - \frac{2.453}{r_0{}^2 A^{\frac{2}{3}}} - \frac{4}{45}\epsilon_0{}^2 - \frac{92}{2,835}\epsilon_0{}^3 \right)$$

$$+ 7 \exp\left(-\frac{6|A - 2Z|}{A} \right) + 14.33 \times 10^{-6} Z^{2.39}$$

$$+ \delta U_N(A - Z, \epsilon_0) + \delta U_Z(Z, \epsilon_0) + P_N(A - Z, \epsilon_0) + P_Z(Z, \epsilon_0) \quad (8c\text{-}3)$$

The value used for $\frac{3}{5}e^2$ is 0.864 MeV-fm.

8c-4. Determination and Testing of Coefficients. The principal method used to determine coefficients is least-squares fitting to tables of experimentally derived binding energies. From a statistician's [15] point of view, this is not a valid procedure because there are correlations among the data of the mass table. Therefore Eq. (8c-3) has been fitted both to the mass table and to the raw experimental data. Other methods, e.g., fitting the Coulomb radius to a fission barrier [8], have also been used. In this mass law, the four parameters of the Nilsson model were chosen [9] by trial and error to reproduce known level structures as well as possible. The value for the BCS parameter was found by solving the problem with several values of the BCS parameter, iterating to find the solution which minimized the sum of residuals.

The least-squares solution fitting the remaining six parameters to 1,148 binding energies from the 1964 [10] and 1967 [11] mass tables is given in the second column of Table 8c-1, and the solution fitted to 552 mass-spectroscopic doublets [16] and 957

TABLE 8c-1. MASS-LAW COEFFICIENTS

Parameter	Fitted to mass table	Fitted to doublets and reactions
α, MeV.........	15.8089 ± 0.0170	15.8570 ± 0.0322
β, MeV.........	30.157 ± 0.142	31.402 ± 0.168
γ, MeV.........	20.230 ± 0.052	20.337 ± 0.105
η, MeV.........	47.66 ± 0.94	53.52 ± 0.92
r_0, fm...........	1.18729 ± 0.00229	1.17641 ± 0.00376
r_N, fm..........	1.2285 ± 0.0070	1.1983 ± 0.0078
$G_p r_0^2$, MeV-fm^2...	28.70	27.67
σ_1, MeV.........	0.805	1.916
σ_2, MeV.........	0.464	0.449

nuclear reaction Q values [11,16] is given in the third column. The standard deviation σ_1 is the fit to total binding energies, and σ_2 is the fit to the doublets and reaction energies.

The quoted errors in Table 8c-1 are the square roots of the diagonal elements of the error matrix adjusted to force χ^2 = degrees of freedom. For the first column they show only the relative uncertainties in the determination of the parameters; for the second column they are a more accurate estimate of statistical uncertainties. The values of the coefficients are slightly different from those in the "Winnipeg Proceedings" [9] because of the elimination of the free parameter in the Strutinsky smoothing and the addition of a neutron radius different from the proton radius. The residual errors of the calculated binding energies with the coefficients of the first column are shown for odd-A nuclides in Fig. 8c-1c, and for even-A in Fig. 8c-1d. The systematic errors remaining above $A = 200$ are due to undercorrection for the doubly closed shell at ^{208}Pb and to higher-order shapes of deformation [13]. Figure 8c-1d indicates systematic differences between even-even and odd-odd nuclides, demonstrating that the even-odd mass difference as calculated by BCS theory in this mass law is not accurate. Different forms for the A dependence of the pairing matrix element, e.g., $G \sim 1/A^{0.8}$, were found to give a qualitatively better fit to even-odd mass differences, but always with a considerably larger sum of residuals than the solution presented here.

The first test of a mass law is its ability to fit the known binding energies. In this case the rms deviation is 805 keV. Much better fits have been obtained, e.g., 168 keV by Zeldes [7], but generally by using a large number of phenomenological terms to represent shell, pairing, and deformation effects. Since the probability of successful extrapolation decreases with increased number of parameters, a "figure of merit" is sometimes applied which is the product of the rms deviation and the number of parameters. The mass law presented here has only 11 adjustable parameters, including all model parameters.

To test interpolation the data were placed in random order and divided into two groups of 574 binding energies each; to test extrapolation the 622 data with N less than the stability line were placed in one group and 526 neutron-rich data in another. A separate determination of the parameters was made for each of the four groups, and the sums of residuals were compared to the sum of residuals for the mass law fitted to all the data by applying the variance-ratio or F-distribution test.[1] The respective values of F obtained for the four groups are 1.04, 0.95, 1.03, and 1.01; since the F distribution for this number of degrees of freedom is approximately normal about 1.00 with standard deviation ± 0.07, the statistical test is well satisfied both for interpolation and for extrapolation from either side of the beta-stability line to the other. There is, of course, no guarantee that the extrapolation continues accurately beyond the known nuclides.

Wing [17] has developed and used tests for comprehensive comparison of various mass formulas—based on alpha-decay energies as a function of N, neutron-pair separation energies as a function of Z, local roughness of beta-decay energies, the beta-stability line and the steepness of the valley of beta stability, separation of even and odd mass surfaces, and delayed neutron and proton precursors. No mass laws tested to date satisfy all tests.

The ultimate test is use. Two examples of problems involving extrapolation to neutron-rich isotopes which have been used to compare formulas are the r process of nucleosynthesis [18] and the study of delayed neutron emitters [19]. In both these cases the mass law presented here behaved well.

8c-5. Table of Binding Energies. Binding energies for 2,827 nuclides calculated from Eq. (8c-3) with the coefficients of the second column of Table 8c-1 are given in Table 8c-2. From 16 to 34 isotopes of each element from $Z = 20$ to $Z = 114$ are given. The isotope nearest to Green's [20] approximation of the beta-stability line, $N - Z = 0.4A^2/(A + 200)$, is indicated by a star. The four data columns give, in MeV: the spherical liquid-drop part of the binding energy [first three lines of Eq. (8c-3), with $\epsilon = 0$]; the shell correction $\delta U_N(N,\epsilon_0) + \delta U_Z(Z,\epsilon_0)$; the BCS pairing energy $P_N(N,\epsilon_0) + P_Z(Z,\epsilon_0)$; and finally the total binding energy. The deformation energy of the liquid drop is the difference between the final column and the other three columns.

Particle separation energies (binding energies of the last particles) for nucleus (Z,A) can be found as follows:

$$S_p(Z,A) = B(Z,A) - B(Z - 1, A - 1)$$
$$S_n(Z,A) = B(Z,A) - B(Z, A - 1)$$
$$S_\alpha(Z,A) = B(Z,A) - B(Z - 2, A - 4) - 28.3 \text{ MeV}$$

The energy available for ground-state negative beta decay is

$$Q_\beta(Z,A) = B(Z + 1, A) - B(Z,A) + 0.8 \text{ MeV}$$

and the end point of the positron energy for decay to the ground state is

$$Q_{\beta+}(Z,A) = B(Z - 1, A) - B(Z,A) - 1.8 \text{ MeV}$$

References

1. von Weizsäcker, C. F.: *Z. Physik* **96**, 431 (1935).
2. Bethe, H. A., and R. F. Bacher: *Rev. Mod. Phys.* **8**, 82 (1936).
3. Johnson, W. H., Jr., ed.: *Proc. 2d Intern. Conf. on Nuclidic Masses*, Springer-Verlag OHG, Vienna, 1964; referenced as "Vienna Proceedings."
4. Forsling, W., C. J. Herrlander, and H. Ryde, eds.: "Nuclides Far Off the Stability Line," Almqvist and Wiksell, Stockholm, 1967; also *Arkiv Fysik* **36**; referenced as "Lysekil Proceedings."
5. Barber, R. C., ed.: *Proc. 3d Intern. Conf. on Atomic Masses*, University of Manitoba Press, Winnipeg, 1967; referenced as "Winnipeg Proceedings."

(*References continued on p.* 8-142.)

[1] Application of the F distribution is correctly described in report LA-3751 [9]; comments concerning it in the corresponding paper in the "Winnipeg Proceedings" are not accurate.

TABLE 8c-2. CALCULATED BINDING ENERGIES IN MeV

$Z = 20$: Calcium

Number of neutrons N	Mass number A	Liquid drop	Shell correction	BCS pairing energy	Total binding energy
20	40	339.1	−1.3	2.6	340.4
21	41	350.0	−2.7	2.1	349.3
22	42	360.3	−3.6	5.1	361.8
23*	43	369.9	−3.5	3.4	369.9
24	44	379.1	−3.3	5.6	381.3
25	45	387.7	−2.3	3.4	388.8
26	46	395.7	−1.1	4.7	399.3
27	47	403.2	1.0	1.9	406.1
28	48	410.3	3.2	1.9	415.4
29	49	416.8	1.4	1.8	420.0
30	50	422.9	0.	4.0	426.9
31	51	428.5	−0.9	2.6	430.2
32	52	433.7	−1.3	4.3	436.6
33	53	438.5	−2.8	3.3	438.7
34	54	442.8	−4.2	6.1	444.6
35	55	446.8	−4.9	4.3	446.2

$Z = 21$: Scandium

Number of neutrons N	Mass number A	Liquid drop	Shell correction	BCS pairing energy	Total binding energy
20	41	343.2	−2.5	1.2	341.8
21	42	356.8	−3.6	0.6	353.6
22	43	367.8	−4.1	3.1	366.7
23	44	378.2	−3.7	1.3	375.7
24*	45	388.1	−3.2	3.1	387.8
25	46	397.4	−1.9	0.9	396.0
26	47	406.1	−1.3	2.5	407.2
27	48	414.4	0.3	0.4	414.9
28	49	422.1	2.0	0.5	424.6
29	50	429.3	0.5	0.4	430.1
30	51	436.1	−0.6	2.2	437.7

$Z = 23$: Vanadium (Continued)

Number of neutrons N	Mass number A	Liquid drop	Shell correction	BCS pairing energy	Total binding energy
28	51	442.6	1.2	2.1	445.9
29	52	451.2	0.2	1.4	452.7
30	53	459.3	−1.6	4.0	461.6
31	54	466.9	−2.6	2.6	467.0
32	55	474.1	−3.0	4.3	475.4
33	56	480.9	−3.4	2.4	479.6
34	57	487.1	−5.4	5.6	487.4
35	58	493.0	−5.7	3.6	490.9
36	59	498.5	−6.1	5.6	498.0
37	60	503.6	−6.6	4.0	500.9
38	61	508.3	−6.4	5.6	507.5
39	62	512.7	−6.9	4.2	509.9
40	63	516.7	−6.8	6.1	515.9

$Z = 24$: Chromium

Number of neutrons N	Mass number A	Liquid drop	Shell correction	BCS pairing energy	Total binding energy
20	44	350.2	−3.0	5.2	352.5
21	45	366.0	−3.1	3.5	366.3
22	46	381.1	−5.1	7.5	383.5
23	47	395.4	−4.4	5.3	396.3
24	48	409.1	−4.9	8.0	412.2
25	49	420.5	−3.7	5.7	422.4
26	50	431.3	−2.8	7.2	435.7
27	51	441.6	0.9	3.0	445.3
28*	52	451.3	1.5	4.4	457.2
29	53	460.6	0.7	3.4	464.6
30	54	469.4	−1.6	6.4	474.2
31	55	477.7	−2.4	4.9	480.2
32	56	485.5	−2.9	6.6	489.2
33	57	492.9	−2.9	4.2	494.0

Columns: N (neutron number) | A (mass number) | M_1 | Δ | M_2 | ±

Z = 21: Scandium

N	A	M_1	Δ	M_2	±
31	52	442.4	−1.8	441.6	1.1
32	53	448.3	−2.4	448.8	3.0
33	54	453.7	−3.5	451.7	1.8
34	55	458.7	−4.2	458.2	3.9
35	56	463.3	−5.0	460.4	2.3
36	57	467.6	−5.4	466.3	4.4

Z = 22: Titanium

N	A	M_1	Δ	M_2	±
20	42	346.4	−3.3	347.9	4.8
21	43	360.7	−3.9	360.3	3.6
22	44	374.3	−5.3	376.1	7.0
23	45	385.5	−4.9	385.7	5.1
24	46	396.0	−4.9	398.5	7.4
25	47	406.0	−3.7	407.3	5.0
26*	48	415.5	−2.9	419.3	6.6
27	49	424.5	0.	427.5	3.2
28	50	432.9	1.2	438.1	4.0
29	51	440.8	0.1	444.2	3.4
30	52	448.2	−1.7	452.4	6.0
31	53	455.2	−2.6	457.1	4.5
32	54	461.7	−3.1	464.8	6.2
33	55	467.8	−3.7	468.3	4.4
34	56	473.5	−5.8	475.5	7.8
35	57	478.7	−6.2	478.3	5.8
36	58	483.6	−6.7	484.8	8.0
37	59	488.1	−6.9	487.2	6.0
38	60	492.2	−6.7	493.1	7.6

Z = 23: Vanadium

N	A	M_1	Δ	M_2	±
20	43	348.7	−3.3	348.3	2.8
21	44	363.8	−3.7	361.4	1.4
22	45	378.1	−5.1	377.9	4.8
23	46	391.8	−4.4	390.1	2.7
24	47	403.0	−4.5	403.6	5.0
25	48	413.7	−2.9	413.1	2.4
26	49	423.9	−2.6	425.7	4.4
27*	50	433.5	0.3	434.7	1.1

Z = 24: Chromium

N	A	M_1	Δ	M_2	±
34	58	499.8	−5.8	502.4	8.4
35	59	506.3	−6.0	506.5	6.3
36	60	512.4	−6.7	514.3	8.6
37	61	518.1	−6.5	517.8	6.2
38	62	523.4	−6.4	524.9	7.9
39	63	528.3	−6.7	527.9	6.3
40	64	532.9	−6.4	534.5	8.0
41	65	537.2	−6.1	537.1	6.1

Z = 25: Manganese

N	A	M_1	Δ	M_2	±
20	45	350.9	−1.9	351.6	2.6
21	46	367.4	−2.1	366.2	1.0
22	47	383.2	−3.9	384.0	4.7
23	48	398.2	−3.0	397.6	2.4
24	49	412.6	−3.4	414.2	5.0
25	50	426.3	−1.7	426.8	2.2
26	51	437.8	−1.6	440.7	4.5
27	52	448.7	1.8	450.9	0.6
28	53	459.1	2.4	463.5	1.9
29	54	469.1	1.6	471.6	1.0
30*	55	478.5	−0.5	481.8	3.8
31	56	487.4	−1.4	488.4	2.4
32	57	495.8	−1.8	498.1	4.1
33	58	503.9	−4.5	503.6	1.9
34	59	511.4	−4.6	512.6	5.7
35	60	518.5	−5.1	517.4	3.5
36	61	525.2	−5.4	525.7	5.6
37	62	531.5	−5.2	529.8	3.8
38	63	537.4	−5.7	537.5	5.4
39	64	543.0	−5.4	541.2	3.9
40	65	548.1	−5.2	548.3	5.7
41	66	552.9	−5.3	551.5	3.9
42	67	557.4	−4.9	558.1	6.0
43	68	561.5		561.0	4.4

Z = 26: Iron

N	A	M_1	Δ	M_2	±
21	47	368.0	−0.9	369.4	2.3
22	48	384.5	−2.6	387.9	6.1

TABLE 8c-2. CALCULATED BINDING ENERGIES IN MeV (Continued)

Number of neutrons N	Mass number A	Liquid drop	Shell correction	BCS pairing energy	Total binding energy
		Z = 26: Iron (Continued)			
23	49	400.2	-2.3	4.2	402.2
24	50	415.3	-2.5	6.6	419.4
25	51	429.6	-1.5	4.5	432.7
26	52	443.4	-0.4	5.8	448.8
27	53	455.0	2.9	1.9	459.6
28	54	466.0	3.8	3.1	472.9
29	55	476.6	2.8	2.3	481.5
30	56	486.6	0.7	5.1	492.4
31*	57	496.2	-0.1	3.6	499.7
32	58	505.2	-0.6	5.3	510.0
33	59	513.9	-1.0	3.2	516.0
34	60	522.0	-3.5	7.1	525.7
35	61	529.8	-4.0	5.2	531.0
36	62	537.1	-4.4	7.4	539.9
37	63	544.0	-4.4	5.1	544.7
38	64	550.4	-4.3	6.8	552.9
39	65	556.6	-4.6	5.1	557.1
40	66	562.3	-4.4	6.9	564.8
41	67	567.7	-4.1	5.1	568.6
42	68	572.7	-4.5	7.6	575.7
43	69	577.4	-4.1	5.8	579.1
44	70	581.7	-3.6	7.6	585.7
		Z = 27: Cobalt			
22	49	385.0	0.4	2.1	387.3
23	50	401.4	0.7	0.5	402.3
24	51	417.1	1.2	2.1	420.2
25	52	432.2	2.2	0.1	434.2
26	53	446.6	3.0	1.5	450.8
		Z = 28: Nickel (Continued)			
39	67	580.8	-0.7	2.4	582.5
40	68	587.7	-0.6	4.2	591.3
41	69	594.2	-0.9	2.8	596.1
42	70	600.3	-0.9	5.0	604.4
43	71	606.1	-0.5	3.3	608.9
44	72	611.5	0.	4.9	616.5
45	73	616.6	0.9	3.2	620.7
46	74	621.4	2.1	4.2	627.7
47	75	625.9	3.9	1.7	631.5
48	76	630.0	5.5	2.5	638.0
		Z = 29: Copper			
24	53	418.6	0.8	2.1	421.3
25	54	434.9	1.8	0.2	436.6
26	55	450.6	2.5	1.6	454.6
27	56	465.6	4.0	-0.3	469.0
28	57	480.0	5.5	-0.2	485.4
29	58	493.9	4.1	-0.3	497.5
30	59	505.8	3.0	1.4	510.1
31	60	517.1	1.9	0.3	519.3
32	61	528.0	1.3	2.2	531.4
33	62	538.4	0.2	1.0	539.4
34	63	548.4	-0.6	3.1	550.7
35*	64	557.9	-1.4	1.6	557.9
36	65	566.9	-1.7	3.6	568.6
37	66	575.6	-2.4	2.1	575.1
38	67	583.8	-2.6	4.0	585.1
39	68	591.6	-3.0	2.7	591.0
40	69	599.0	-3.1	4.6	600.3

N	M	δ	δ	M
27	460.4	4.3	−0.4	464.1
28	472.0	5.8	−0.2	477.6
29	483.2	4.5	−0.4	487.1
30	493.9	3.4	1.4	498.5
31	504.0	2.2	0.3	506.4
32*	513.7	1.5	2.3	517.4
33	523.0	0.5	1.0	524.2
34	531.7	−0.2	3.1	534.3
35	540.1	−1.0	1.5	540.3
36	548.0	−1.4	3.5	549.8
37	555.4	−2.1	2.0	555.1
38	562.5	−2.4	4.2	564.0
39	569.2	−2.8	2.8	568.8
40	575.5	−2.8	4.7	577.1
41	581.4	−2.6	2.9	581.4
42	587.0	−2.2	4.5	589.1
43	592.2	−1.5	2.4	593.0
44	597.1	−0.4	3.6	600.1
45	601.7	0.5	1.7	603.7

Z = 28: Nickel

N	M	δ	δ	M
23	401.8	1.6	1.5	404.9
24	418.2	1.7	3.5	423.5
25	433.9	2.6	1.6	438.1
26	449.0	3.8	2.7	455.5
27	463.4	5.9	−0.2	469.0
28	477.2	7.9	0.	485.1
29	489.0	6.2	−0.2	495.0
30	500.2	4.8	2.1	507.1
31	511.0	3.9	0.7	515.6
32	521.3	3.5	2.3	527.1
33	531.1	2.0	1.2	534.3
34*	540.5	0.6	4.1	545.2
35	549.4	−0.2	2.5	551.8
36	557.9	−0.7	4.6	561.8
37	565.9	−0.7	2.5	567.7
38	573.6	−0.5	4.0	577.1

N	M	δ	δ	M	A
41	605.8	2.9	−2.9	606.1	70
42	614.5	4.5	−2.5	612.7	71
43	619.6	2.5	−1.8	619.0	72
44	627.7	3.7	−0.8	625.0	73
45	632.4	1.9	0.	630.6	74
46	639.9	2.9	1.2	635.9	75
47	644.3	0.9	2.7	640.9	76
48	651.2	1.8	3.9	645.5	77
49	655.4	−0.3	5.9	649.9	78

Z = 30: Zinc

N	M	δ	δ	M	A
26	456.7	5.9	−0.6	451.4	56
27	471.6	2.3	2.3	467.1	57
28	488.7	3.2	3.4	482.1	58
29	501.4	2.6	2.2	496.6	59
30	516.1	5.2	0.4	510.5	60
31	525.8	3.8	−0.4	522.4	61
32	538.4	5.4	−0.9	533.9	62
33	546.9	3.6	−1.5	544.9	63
34	558.9	7.1	−3.7	555.4	64
35	566.7	5.4	−4.2	565.5	65
36	577.8	7.5	−4.7	575.1	66
37*	584.9	5.3	−4.6	584.3	67
38	595.4	6.9	−4.5	593.1	68
39	601.9	5.3	−4.8	601.4	69
40	611.8	7.0	−4.6	609.4	70
41	617.8	5.4	−4.5	617.0	71
42	627.1	7.7	−4.8	624.2	72
43	632.7	6.0	−4.3	631.0	73
44	641.3	7.7	−3.9	637.5	74
45	646.6	6.1	−3.2	643.7	75
46	654.6	7.1	−1.9	649.5	76
47	659.5	4.4	0.2	654.9	77
48	667.0	5.4	1.4	660.1	78
49	671.5	2.4	4.2	665.0	79
50	678.3	2.9	5.9	669.5	80
51	680.8	2.5	4.6	673.8	81

TABLE 8c-2. CALCULATED BINDING ENERGIES IN MeV (Continued)

Z = 31: Gallium

Number of neutrons N	Mass number A	Liquid drop	Shell correction	BCS pairing energy	Total binding energy
27	58	467.9	1.1	1.0	469.7
28	59	483.5	1.9	2.0	487.4
29	60	498.5	1.0	1.3	500.7
30	61	513.0	−0.9	3.8	515.9
31	62	526.9	−1.7	2.5	527.6
32	63	539.0	−2.2	4.1	540.8
33	64	550.5	−2.5	2.2	550.0
34	65	561.6	−4.5	5.4	562.5
35	66	572.3	−4.8	3.5	570.8
36	67	582.4	−5.2	5.5	582.6
37	68	592.2	−5.7	3.9	590.3
38*	69	601.5	−5.5	5.4	601.3
39	70	610.4	−6.0	4.0	608.4
40	71	618.9	−5.9	5.9	618.8
41	72	627.1	−5.7	4.2	625.4
42	73	634.8	−5.6	6.1	635.2
43	74	642.2	−5.3	4.5	641.3
44	75	649.2	−4.7	6.0	650.5
45	76	655.8	−4.3	4.7	656.2
46	77	662.1	−3.0	5.6	664.8
47	78	668.1	−0.9	3.0	670.2
48	79	673.8	0.2	4.2	678.1
49	80	679.1	3.0	1.2	683.2
50	81	684.2	4.5	1.8	690.5
51	82	688.9	3.3	1.4	693.5
52	83	693.4	1.5	3.4	698.3

Z = 32: Germanium

Number of neutrons N	Mass number A	Liquid drop	Shell correction	BCS pairing energy	Total binding energy
29	61	499.8	0.	3.6	503.2
30	62	514.8	−2.3	6.5	519.0

Z = 33: Arsenic (Continued)

Number of neutrons N	Mass number A	Liquid drop	Shell correction	BCS pairing energy	Total binding energy
41*	74	644.7	−5.2	3.4	641.9
42	75	653.4	−7.9	7.1	652.6
43	76	661.8	−7.4	5.4	659.8
44	77	669.8	−7.0	7.1	670.0
45	78	677.5	−6.3	5.5	676.8
46	79	684.8	−5.0	6.5	686.3
47	80	691.8	−2.9	3.8	692.7
48	81	698.5	−1.7	4.9	701.6
49	82	704.8	1.0	2.0	707.7
50	83	710.8	2.7	2.4	715.9
51	84	716.5	1.4	2.1	719.9
52	85	721.9	−0.3	4.0	725.6
53	86	727.0	−1.3	3.2	728.9
54	87	731.8	−2.2	4.4	734.0
55	88	736.4	−2.6	2.9	736.6
56	89	740.6	−3.1	3.9	741.4

Z = 34: Selenium

Number of neutrons N	Mass number A	Liquid drop	Shell correction	BCS pairing energy	Total binding energy
30	64	516.3	−3.2	6.0	519.1
31	65	531.9	−4.0	4.7	532.6
32	66	547.0	−4.5	6.6	548.8
33	67	561.5	−4.4	4.7	561.1
34	68	575.5	−4.8	6.4	576.4
35	69	587.8	−5.8	5.0	586.5
36	70	599.6	−5.9	5.0	599.8
37	71	611.0	−5.8	5.2	609.2
38	72	622.0	−6.0	7.0	621.9
39	73	632.5	−5.0	4.8	630.6
40	74	642.6	−5.9	6.9	642.5

650.7	5.1	−5.5	652.3	75	41
661.9	7.3	−6.3	661.5	76	42*
669.6	5.4	−5.5	670.4	77	43
680.3	7.8	−6.3	678.9	78	44
687.5	7.0	−6.5	687.1	79	45
697.6	7.9	−5.3	694.9	80	46
704.4	5.1	−3.0	702.4	81	47
713.8	6.3	−2.0	709.5	82	48
720.4	3.8	0.4	716.3	83	49
729.0	3.8	2.4	722.8	84	50
733.5	3.7	0.9	729.0	85	51
739.7	5.4	−0.6	734.8	86	52
743.4	4.6	−1.6	740.4	87	53
749.0	5.8	−2.4	745.7	88	54
752.1	4.4	−2.9	750.7	89	55
757.3	5.3	−3.4	755.4	90	56
759.8	4.9	−4.5	759.9	91	57

Z = 35: Bromine

547.3	4.7	−4.8	547.8	67	32
560.2	2.8	−4.2	562.9	68	33
576.1	4.4	−4.3	577.5	69	34
588.1	2.9	−4.4	591.6	70	35
601.9	4.6	−4.7	603.9	71	36
612.0	2.3	−3.4	615.9	72	37
625.1	4.5	−3.3	627.3	73	38
634.5	2.2	−4.2	638.4	74	39
646.8	4.1	−4.0	648.9	75	40
655.5	2.5	−4.4	659.1	76	41
667.1	4.6	−4.4	668.9	77	42
675.3	2.9	−6.3	678.3	78	43
686.4	5.7	−5.7	687.2	79	44*
694.1	4.0	−4.8	695.9	80	45
704.6	5.3	−3.2	704.2	81	46
712.0	3.2	−2.4	712.2	82	47
721.9	4.5	−0.1	719.8	83	48
729.0	2.1	1.7	727.1	84	49
738.0	2.3		734.0	85	50

531.3	5.0	−3.0	529.3	63	31
546.4	6.7	−3.5	543.2	64	32
556.1	4.4	−3.3	555.4	65	33
569.1	8.4	−6.3	567.0	66	34
578.1	5.6	−5.6	578.2	67	35
590.3	7.6	−6.0	588.9	68	36
598.6	6.2	−6.8	599.2	69	37
610.2	7.6	−6.4	609.1	70	38
617.8	5.9	−6.2	618.6	71	39
628.7	8.1	−6.8	627.6	72	40*
635.8	6.4	−6.6	636.3	73	41
646.1	8.3	−6.7	644.5	74	42
652.8	6.8	−6.4	652.4	75	43
662.5	8.7	−6.1	659.9	76	44
668.7	7.3	−5.6	667.1	77	45
677.8	8.3	−4.4	673.9	78	46
683.6	5.3	−2.0	680.4	79	47
692.1	6.6	−1.0	686.5	80	48
697.1	3.4	2.0	692.4	81	49
705.4	4.1	3.4	697.9	82	50
708.9	3.5	2.3	703.1	83	51
714.1	5.7	0.3	708.1	84	52
716.9	4.8	−0.6	712.7	85	53
721.6	6.1	−1.5	717.1	86	54

Z = 33: Arsenic

501.1	2.1	−1.2	500.3	62	29
517.5	4.6	−3.0	515.9	63	30
530.4	3.2	−3.7	530.9	64	31
546.1	4.8	−4.2	545.4	65	32
557.7	2.7	−3.8	559.5	66	33
571.2	6.5	−6.9	571.6	67	34
580.7	3.1	−5.1	583.4	68	35
593.6	5.1	−5.3	594.7	69	36
602.5	3.5	−5.4	605.6	70	37
614.6	5.3	−5.8	616.0	71	38
622.8	3.3	−5.2	626.0	72	39
634.2	5.3	−5.7	635.5	73	40

TABLE 8c-2. CALCULATED BINDING ENERGIES IN MeV (Continued)

Z = 35: Bromine (Continued)

Number of neutrons N	Mass number A	Liquid drop	Shell correction	BCS pairing energy	Total binding energy
51	86	740.6	0.4	2.1	743.0
52	87	747.0	−1.1	3.8	749.6
53	88	753.0	−1.1	2.8	753.8
54	89	758.7	−2.7	3.9	759.9
55	90	764.2	−3.4	2.8	763.5
56	91	769.4	−4.2	4.1	769.2
57	92	774.4	−2.6	2.2	772.1
58	93	779.0	−3.1	3.6	777.7
59	94	783.5	−1.9	1.9	780.7

Z = 36: Krypton

Number of neutrons N	Mass number A	Liquid drop	Shell correction	BCS pairing energy	Total binding energy
33	69	563.6	−5.0	4.9	562.4
34	70	578.7	−4.8	6.2	578.8
35	71	593.3	−4.3	4.2	591.3
36	72	607.4	−4.1	5.8	606.8
37	73	619.9	−2.9	3.4	617.4
38	74	631.8	−3.3	5.0	631.0
39	75	643.4	−2.8	3.3	640.9
40	76	654.4	−3.7	5.2	653.7
41	77	665.2	−3.5	3.6	662.9
42	78	675.4	−5.8	6.7	675.0
43	79	685.3	−6.0	5.4	683.7
44	80	694.8	−6.2	7.1	695.2
45*	81	703.9	−6.2	5.9	703.5
46	82	712.7	−5.9	7.6	714.5
47	83	721.1	−3.5	4.8	722.3
48	84	729.2	−3.0	6.4	732.7
49	85	737.0	−0.2	3.6	740.2
50	86	744.4	1.3	4.0	749.7

Z = 37: Rubidium (Continued)

Number of neutrons N	Mass number A	Liquid drop	Shell correction	BCS pairing energy	Total binding energy
61	98	821.2	0.1	0.6	818.2
62	99	825.7	0.1	1.7	823.8

Z = 38: Strontium

Number of neutrons N	Mass number A	Liquid drop	Shell correction	BCS pairing energy	Total binding energy
35	73	594.9	−1.9	2.6	592.6
36	74	610.0	−1.6	3.9	609.1
37	75	624.6	−0.8	1.8	622.0
38	76	638.7	−0.4	2.9	637.7
39	77	651.3	−0.8	1.8	648.6
40	78	663.3	−1.7	3.7	662.3
41	79	675.0	−1.8	2.3	672.5
42	80	686.2	−4.1	5.2	685.5
43	81	697.1	−3.9	3.8	695.1
44	82	707.5	−6.4	7.1	707.6
45	83	717.6	−6.1	5.5	716.7
46	84	727.3	−5.5	6.9	728.7
47	85	736.7	−3.7	4.6	737.5
48*	86	745.8	−2.7	5.8	748.8
49	87	754.5	−0.6	3.6	757.3
50	88	762.8	1.5	3.4	767.7
51	89	770.8	−0.1	3.4	774.1
52	90	778.5	−1.4	5.0	782.1
53	91	785.9	−2.3	4.1	787.7
54	92	793.0	−3.1	5.2	795.1
55	93	799.8	−3.8	4.0	800.0
56	94	806.3	−4.9	5.9	807.0
57	95	812.6	−2.6	3.2	811.3
58	96	818.5	−2.7	4.4	818.2
59	97	824.3	−0.5	1.9	822.6

(Table printed sideways. Continuation blocks and two labeled elements, Z = 37: Rubidium and Z = 39: Yttrium.)

(continuation)

A	N	M_a	Δ_a	Δ_b	M_b
87	51	751.5	0.2	3.6	755.2
88	52	758.3	−1.6	5.6	762.3
89	53	764.8	−2.4	4.6	766.9
90	54	771.0	−3.3	5.8	773.5
91	55	776.8	−3.7	4.4	777.4
92	56	782.5	−4.2	5.4	783.6
93	57	787.8	−5.4	5.1	787.0
94	58	793.0	−3.5	5.1	793.0
95	59	797.9	−1.5	2.9	796.4
96	60	802.4	−1.1	3.7	802.0

Z = 37: Rubidium

A	N	M_a	Δ_a	Δ_b	M_b
72	35	590.7	1.9	−2.8	594.5
73	36	606.7	3.1	−2.5	609.0
74	37	619.1	1.2	−1.8	623.2
75	38	633.1	2.4	−1.5	635.6
76	39	643.6	1.2	−1.9	647.7
77	40	656.8	2.9	−2.4	659.3
78	41	666.6	1.4	−2.4	670.5
79	42	679.0	3.5	−3.6	681.2
80	43	688.2	2.3	−3.7	691.6
81	44	700.1	5.3	−6.0	701.5
82	45	708.7	4.4	−6.8	711.1
83	46	720.2	5.6	−5.7	720.4
84	47*	728.6	3.2	−3.9	729.3
85	48	739.4	4.4	−2.9	737.9
86	49	747.4	2.0	−0.6	746.1
87	50	757.4	2.2	1.2	754.0
88	51	763.4	2.0	−0.1	761.6
89	52	770.9	3.7	−1.6	768.8
90	53	776.0	2.8	−2.5	775.7
91	54	783.0	3.9	−3.3	782.4
92	55	787.4	2.7	−3.9	788.7
93	56	794.0	4.3	−4.4	794.8
94	57	798.0	1.9	−2.5	800.7
95	58	804.5	2.9	−2.4	806.2
96	59	808.5	1.0	−0.9	811.5
97	60	814.5	1.7	−0.3	816.5

(continuation)

A	N	M_a	Δ_a	Δ_b	M_b
98	60	829.0	2.5	0.3	829.7
99	61	833.2	1.3	0.9	834.9
100	62	839.1	2.3	0.9	839.8
101	63	842.7	1.2	1.2	844.4
102	64	848.1	2.0	1.2	848.8

Z = 39: Yttrium

A	N	M_a	Δ_a	Δ_b	M_b
76	37	621.6	0.3	0.3	625.5
77	38	637.7	1.2	0.8	640.1
78	39	650.4	0.1	0.8	654.3
79	40	664.3	2.1	−0.6	666.8
80	41	675.1	0.7	−0.7	678.9
81	42	688.3	2.3	−1.7	690.6
82	43	698.4	1.9	−3.3	701.9
83	44	711.2	5.0	−5.6	712.8
84	45	720.9	2.7	−4.1	723.4
85	46	733.2	4.7	−4.8	733.6
86	47	742.5	3.0	−3.6	743.4
87	48	754.2	4.0	−0.9	752.9
88	49	763.2	2.2	0.9	762.1
89	50*	774.0	2.2	−0.4	770.9
90	51	781.0	2.1	−1.6	779.3
91	52	789.3	3.4	−2.3	787.5
92	53	795.4	2.6	−3.0	795.3
93	54	803.3	3.6	−4.1	802.9
94	55	808.6	2.8	−5.1	810.1
95	56	816.1	4.8	−2.4	817.0
96	57	821.0	1.7	−2.1	823.8
97	58	828.4	2.7	0.1	830.1
98	59	833.3	0.5	1.0	836.3
99	60	840.2	1.0	1.2	842.2
100	61	844.8	0.2	1.4	847.8
101	62	851.1	1.3	1.6	853.0
102	63	855.2	0.2	1.2	858.1
103	64	861.0	0.9	0.9	862.9
104	65	864.5	0.2		867.4
105	66	869.7	1.2		871.7

TABLE 8c-2. CALCULATED BINDING ENERGIES IN MeV (Continued)

Z = 40: Zirconium

Number of neutrons N	Mass number A	Liquid drop	Shell correction	BCS pairing energy	Total binding energy
38	78	640.5	−0.8	3.1	639.5
39	79	655.2	−1.3	2.0	652.6
40	80	669.3	−2.1	4.0	668.3
41	81	681.9	−2.1	2.5	679.5
42	82	694.1	−4.1	5.2	693.4
43	83	705.9	−3.8	3.7	704.0
44	84	717.3	−5.9	6.7	717.4
45	85	728.3	−6.0	5.4	727.5
46	86	739.0	−5.9	7.3	740.4
47	87	749.4	−3.9	4.8	750.2
48	88	759.3	−3.6	6.6	762.3
49	89	768.9	−0.8	3.8	771.7
50	90	778.2	0.6	4.3	783.1
51*	91	787.1	−0.4	3.8	790.4
52	92	795.7	−2.2	5.8	799.3
53	93	804.0	−2.9	4.7	805.8
54	94	811.9	−3.7	5.9	814.1
55	95	819.6	−4.3	4.5	819.8
56	96	827.0	−4.7	5.7	827.7
57	97	834.1	−2.5	3.1	832.9
58	98	840.9	−3.0	4.5	840.6
59	99	847.5	−0.9	2.2	845.8
60	100	853.8	−0.3	2.8	853.0
61	101	859.8	0.6	1.5	858.0
62	102	865.5	0.5	2.5	864.7
63	103	870.9	0.9	1.4	869.1
64	104	876.1	0.9	2.2	875.3
65	105	881.0	0.7	1.4	879.2
66	106	885.7	0.3	2.5	884.8
67	107	890.2	−0.1	1.7	888.2

Z = 42: Molybdenum (Continued)

Number of neutrons N	Mass number A	Liquid drop	Shell correction	BCS pairing energy	Total binding energy
46	88	747.9	−5.5	7.4	749.7
47	89	759.1	−3.2	4.6	760.4
48	90	770.0	−3.3	6.8	773.5
49	91	780.5	0.1	3.4	783.8
50	92	790.7	0.8	4.5	796.0
51	93	800.5	0.2	3.6	804.2
52	94	810.0	−1.9	6.0	814.0
53	95	819.1	−2.5	4.8	821.3
54	96	827.9	−3.4	6.0	830.5
55*	97	836.5	−3.7	4.4	837.2
56	98	844.7	−3.9	5.2	845.9
57	99	852.6	−4.4	4.2	851.9
58	100	860.3	−3.7	5.1	860.3
59	101	867.7	−3.1	2.3	866.1
60	102	874.8	−1.4	3.3	874.2
61	103	881.6	−0.2	1.7	879.8
62	104	888.1	−0.3	2.7	887.3
63	105	894.4	0.2	1.6	892.5
64	106	900.3	−0.1	2.6	899.5
65	107	906.1	0.	1.6	904.2
66	108	911.5	−0.3	2.6	910.6
67	109	916.7	−0.5	1.7	914.8
68	110	921.7	−1.0	2.9	920.9
69	111	926.5	−1.2	2.1	924.7
70	112	931.0	−1.4	3.0	930.3
71	113	935.3	−1.4	1.9	933.7

Z = 43: Technetium

Number of neutrons N	Mass number A	Liquid drop	Shell correction	BCS pairing energy	Total binding energy
42	85	699.8	−3.3	3.6	698.6
43	86	714.0	−3.0	1.9	711.6

Z = 41: Niobium

A	A				
40	81	670.2	−1.8	668.0	2.6
41	82	684.4	−1.9	680.8	1.1
42	83	697.0	−3.4	695.1	3.4
43	84	709.3	−3.4	706.2	2.1
44	85	721.2	−4.4	720.0	4.3
45	86	732.7	−3.4	730.6	2.3
46	87	743.8	−4.8	743.9	5.0
47	88	754.6	−3.4	754.1	3.1
48	89	765.0	−3.1	766.7	4.8
49	90	775.1	−0.5	776.6	2.1
50	91	784.8	0.6	788.3	2.9
51	92	794.2	−0.2	796.2	2.3
52	93	803.2	−1.9	805.4	4.2
53*	94	811.9	−2.4	812.4	3.0
54	95	820.3	−3.2	821.1	4.1
55	96	828.4	−4.0	827.3	3.0
56	97	836.2	−4.4	835.6	4.1
57	98	843.8	−2.2	841.4	1.6
58	99	851.0	−2.6	849.5	3.0
59	100	858.0	−0.7	855.1	0.9
60	101	864.7	−0.3	862.8	1.7
61	102	871.1	0.3	868.1	0.5
62	103	877.2	0.3	875.2	1.5
63	104	883.0	0.6	880.1	0.5
64	105	888.6	0.5	886.6	1.3
65	106	893.9	0.4	891.0	0.5
66	107	899.0	0.1	897.0	1.5
67	108	903.8	−0.3	900.8	0.7
68	109	908.4	−0.7	906.4	1.8
69	110	912.8	−1.0	909.8	1.0

Z = 42: Molybdenum

A	A				
42	84	699.2	−4.9	698.9	5.8
43	85	711.9	−4.8	710.5	4.4
44	86	724.3	−5.0	724.9	6.1
45	87	736.3	−5.1	736.0	5.0

A	A				
44	87	726.8	−4.3	726.4	4.4
45	88	739.2	−4.3	737.9	3.2
46	89	751.3	−6.1	752.2	7.0
47	90	763.0	−2.6	763.3	3.0
48	91	774.3	−2.9	776.8	5.4
49	92	785.3	0.7	787.6	1.7
50	93	795.9	1.4	800.2	3.0
51	94	806.1	0.9	808.9	2.0
52	95	816.0	−1.5	819.1	4.6
53	96	825.6	−2.2	826.9	3.4
54	97	834.8	−3.1	836.5	4.8
55*	98	843.8	−3.1	843.6	2.9
56	99	852.4	−3.4	852.7	3.7
57	100	860.8	−3.6	859.1	2.5
58	101	868.9	−2.3	868.0	2.9
59	102	876.7	−1.4	874.2	1.2
60	103	884.2	−1.5	882.7	2.4
61	104	891.4	−0.6	888.7	0.8
62	105	898.3	−0.7	896.6	1.8
63	106	905.0	−0.3	902.1	0.7
64	107	911.3	−0.6	909.5	1.8
65	108	917.4	−0.4	914.6	0.7
66	109	923.3	−0.6	921.5	1.6
67	110	928.9	−0.7	926.1	0.8
68	111	934.3	−1.1	932.5	1.9
69	112	939.4	−1.2	936.7	1.0
70	113	944.3	−1.2	942.7	1.9
71	114	949.0	−1.0	946.5	0.6
72	115	953.4	−0.8	952.1	1.3
73	116	957.7	−1.3	955.6	0.8

Z = 44: Ruthenium

A	A				
44	88	728.6	−4.6	730.1	6.3
45	89	741.5	−6.4	742.2	7.1
46	90	754.0	−5.1	756.8	7.9
47	91	766.2	−2.2	768.4	4.5
48	92	777.9	−2.0	782.4	6.4
49	93	789.3	1.6	793.5	2.7

TABLE 8c-2. CALCULATED BINDING ENERGIES IN MeV (Continued)

Number of neutrons N	Mass number A	Liquid drop	Shell correction	BCS pairing energy	Total binding energy
		Z = 44: Ruthenium (Continued)			
50	94	800.4	2.2	4.0	806.6
51	95	811.1	1.7	3.0	815.7
52	96	821.4	-0.6	5.6	826.4
53	97	831.4	-1.4	4.5	834.6
54	98	841.1	-2.3	5.8	844.6
55	99	850.5	-2.3	4.0	852.1
56	100	859.5	-3.1	5.2	861.6
57	101	868.3	-2.8	3.3	868.4
58*	102	876.8	-3.0	4.5	877.6
59	103	885.0	-2.6	3.0	884.0
60	104	892.8	-2.5	4.1	892.8
61	105	900.5	-1.1	2.0	899.0
62	106	907.8	-1.2	3.0	907.3
63	107	914.8	-0.6	1.7	913.2
64	108	921.6	-1.0	2.9	921.0
65	109	928.1	-0.6	1.6	926.4
66	110	934.3	-0.7	2.6	933.7
67	111	940.3	-0.8	1.7	938.7
68	112	946.1	-1.2	2.9	945.6
69	113	951.6	-1.0	1.8	950.2
70	114	956.9	-1.0	2.6	956.6
71	115	961.9	-0.8	1.4	960.8
72	116	966.7	-0.7	2.3	966.8
73	117	971.3	-0.9	1.6	970.7
74	118	975.7	-1.2	2.9	976.4
75	119	979.9	-1.2	2.1	980.1
		Z = 45: Rhodium			
45	90	743.2	-2.8	2.9	743.2
46	91	756.1	-2.3	4.5	758.3

Number of neutrons N	Mass number A	Liquid drop	Shell correction	BCS pairing energy	Total binding energy
		Z = 46: Palladium (Continued)			
52	98	830.2	1.9	4.4	836.5
53	99	841.0	1.0	3.5	845.5
54	100	851.6	0.1	4.7	856.4
55	101	861.8	-0.1	3.1	864.7
56	102	871.6	-0.8	4.2	875.1
57	103	881.2	-1.1	2.7	882.5
58	104	890.5	-4.0	6.2	892.6
59	105	899.4	-2.7	3.2	899.6
60	106	908.1	-3.3	4.7	909.1
61*	107	916.5	-2.5	3.0	915.8
62	108	924.6	-2.6	4.1	925.0
63	109	932.5	-1.6	2.1	931.4
64	110	940.0	-1.6	3.2	940.0
65	111	947.3	-1.1	1.7	946.1
66	112	954.3	-1.2	2.9	954.3
67	113	961.1	-1.2	1.7	960.0
68	114	967.6	-1.3	3.0	967.7
69	115	973.9	-0.8	1.6	973.0
70	116	979.9	-0.8	2.6	980.3
71	117	985.6	-0.5	1.4	985.2
72	118	991.2	-0.6	2.5	992.1
73	119	996.5	-0.3	1.5	996.8
74	120	1001.6	-0.7	3.0	1003.3
75	121	1006.5	-0.4	2.1	1007.7
76	122	1011.2	-1.5	4.6	1014.1
77	123	1015.7	-0.3	3.1	1018.4
78	124	1020.0	-0.1	4.7	1024.6
		Z = 47: Silver			
48	95	785.0	3.1	2.4	790.4
49	96	797.7	4.9	0.4	802.9

Z = 46: Palladium

N	A	M	Δ	Δ	M
47	92	768.7	−0.6	2.2	770.3
48	93	780.9	−0.3	4.0	784.7
49	94	792.8	2.6	1.0	796.3
50	95	804.2	3.5	2.1	809.8
51	96	815.4	2.8	1.3	819.4
52	97	826.1	0.9	3.4	830.4
53	98	836.6	0.3	2.2	839.1
54	99	846.7	−0.5	3.3	849.5
55	100	856.5	−1.1	2.1	857.4
56	101	865.9	−1.4	3.0	867.4
57	102	875.1	−1.9	1.9	874.6
58	103	884.0	−2.2	3.2	884.2
59*	104	892.6	−2.4	2.1	891.0
60	105	900.8	−2.6	3.3	900.3
61	106	908.9	−1.6	1.4	906.8
62	107	916.6	−1.6	2.4	915.5
63	108	924.0	−1.1	1.0	921.7
64	109	931.1	−1.4	2.3	929.9
65	110	938.1	−0.9	0.9	935.7
66	111	944.7	−1.0	1.8	943.4
67	112	951.1	−0.9	0.8	948.8
68	113	957.2	−1.2	2.1	956.0
69	114	963.1	−0.9	0.9	961.0
70	115	968.7	−0.9	1.7	967.8
71	116	974.1	−0.6	0.5	972.4
72	117	979.3	−0.5	1.4	978.8
73	118	984.3	−0.5	0.6	983.1
74	119	989.0	−0.7	1.8	989.2
75	120	993.6	−0.5	0.9	993.2
76	121	997.9	−0.9	2.4	999.1

N	A	M	Δ	Δ	M
47	93	770.6	−0.3	4.0	774.2
48	94	783.3	0.6	5.2	789.0
49	95	795.5	3.6	2.0	801.0
50	96	807.4	4.8	2.8	815.0
51	97	819.0	3.8	2.2	825.0

Z = 48: Cadmium

N	A	M	Δ	Δ	M
50	97	810.0	6.6	0.6	817.2
51	98	822.0	5.3	0.5	827.7
52	99	833.6	4.1	1.9	839.5
53	100	844.9	3.2	1.0	849.1
54	101	855.8	2.5	2.1	860.3
55	102	866.4	1.6	1.1	869.1
56	103	876.7	1.0	2.2	879.9
57	104	886.7	−0.2	1.6	887.8
58	105	896.3	−1.4	3.4	898.2
59	106	905.7	−2.0	2.3	905.6
60	107	914.8	−2.8	3.9	915.6
61	108	923.6	−2.5	2.3	922.6
62*	109	932.0	−3.0	3.8	932.1
63	110	940.3	−2.3	2.0	938.8
64	111	948.2	−2.2	3.0	947.9
65	112	955.9	−1.6	1.4	954.3
66	113	963.3	−1.9	2.7	962.9
67	114	970.4	−1.5	1.4	968.9
68	115	977.3	−1.6	2.6	977.0
69	116	983.9	−1.1	1.2	982.7
70	117	990.3	−1.0	2.2	990.4
71	118	996.5	−0.6	0.9	995.7
72	119	1002.4	−0.8	2.2	1003.0
73	120	1008.1	−0.2	1.0	1008.0
74	121	1013.5	−0.5	2.5	1015.0
75	122	1018.8	0.	1.4	1019.8
76	123	1023.8	0.1	2.8	1026.5
77	124	1028.6	0.9	1.7	1031.2
78	125	1033.3	1.7	2.7	1037.6
79	126	1037.7	3.0	1.4	1042.1

N	A	M	Δ	Δ	M
49	97	799.2	6.3	0.9	806.4
50	98	811.9	8.3	1.0	821.3
51	99	824.3	6.8	0.9	832.1
52	100	836.4	5.4	2.6	844.4
53	101	848.1	4.4	1.9	854.3

TABLE 8c-2. CALCULATED BINDING ENERGIES IN MeV (Continued)

Z = 48: Cadmium (Continued)

Number of neutrons N	Mass number A	Liquid drop	Shell correction	BCS pairing energy	Total binding energy
54	102	859.4	3.6	3.0	866.0
55	103	870.4	3.1	1.7	875.2
56	104	881.1	2.7	2.5	886.3
57	105	891.5	1.1	2.0	894.5
58	106	901.5	-0.6	4.5	905.5
59	107	911.3	-1.5	3.4	913.1
60	108	920.7	-2.9	5.6	923.5
61	109	929.9	-3.7	5.9	930.8
62	110	938.8	-4.1	5.9	940.6
63	111	947.4	-4.6	4.7	947.5
64*	112	955.7	-4.4	5.9	956.9
65	113	963.8	-3.4	3.7	963.5
66	114	971.5	-3.5	5.0	972.5
67	115	979.1	-2.5	3.0	978.8
68	116	986.3	-2.7	4.5	987.4
69	117	993.3	-1.8	2.6	993.4
70	118	1000.1	-2.1	4.2	1001.6
71	119	1006.6	-1.3	2.6	1007.2
72	120	1012.9	-1.8	4.3	1015.0
73	121	1018.9	-0.8	2.6	1020.3
74	122	1024.7	-1.3	4.6	1027.9
75	123	1030.3	0.	2.9	1033.0
76	124	1035.7	0.	4.6	1040.3
77	125	1040.9	1.9	2.5	1045.2
78	126	1045.9	3.1	3.2	1052.2
79	127	1050.6	4.5	1.9	1057.0
80	128	1055.2	6.0	2.3	1063.5
81	129	1059.5	7.8	0.9	1068.2
82	130	1063.7	9.6	1.0	1074.3

Z = 50: Tin (Continued)

Number of neutrons N	Mass number A	Liquid drop	Shell correction	BCS pairing energy	Total binding energy
54	104	864.7	6.7	2.0	873.5
55	105	876.6	6.2	0.7	883.4
56	106	888.0	5.8	1.5	895.4
57	107	899.2	4.2	0.9	904.3
58	108	910.1	2.6	3.5	916.1
59	109	920.6	1.5	2.5	924.6
60	110	930.8	0.3	4.6	935.8
61	111	940.8	-0.5	3.5	943.8
62	112	950.4	-1.1	5.0	954.4
63	113	959.8	-1.5	3.7	962.0
64	114	968.8	-1.6	5.0	972.2
65	115	977.6	-2.2	3.9	979.3
66	116	986.1	-2.7	5.7	989.2
67*	117	994.4	-2.8	4.4	996.0
68	118	1002.4	-3.1	6.2	1005.4
69	119	1010.1	-2.7	4.6	1012.0
70	120	1017.6	-2.7	6.2	1021.0
71	121	1024.9	-2.3	4.7	1027.3
72	122	1031.9	-1.5	5.7	1036.0
73	123	1038.6	-0.6	4.0	1042.0
74	124	1045.2	0.4	4.8	1050.4
75	125	1051.5	1.9	2.8	1056.1
76	126	1057.5	2.9	3.7	1064.1
77	127	1063.4	4.6	1.7	1069.7
78	128	1069.0	6.1	2.2	1077.4
79	129	1074.5	7.4	1.0	1082.9
80	130	1079.7	9.0	1.3	1090.0
81	131	1084.7	10.7	-0.1	1095.3

$Z = 49$: Indium

A	N	Mass	Δ	Mass	Δ
100	51	826.1	8.3	834.1	−0.2
101	52	838.5	6.9	846.9	1.5
102	53	850.6	5.9	857.2	0.8
103	54	862.4	5.0	869.3	1.9
104	55	873.8	4.5	878.9	0.6
105	56	884.9	4.1	890.5	1.4
106	57	895.7	2.5	899.0	0.9
107	58	906.1	0.9	910.4	3.4
108	59	916.3	−0.2	918.5	2.4
109	60	926.1	−1.4	929.2	4.5
110	61	935.7	−2.7	936.9	3.4
111	62	944.9	−3.2	947.1	4.9
112	63	953.9	−3.3	954.4	3.6
113	64	962.6	−3.8	964.2	4.9
114	65	971.0	−4.3	971.0	3.9
115	66*	979.1	−3.8	980.4	5.6
116	67	987.0	−3.9	986.9	3.9
117	68	994.7	−3.1	996.0	5.4
118	69	1002.0	−3.2	1002.2	3.5
119	70	1009.2	−2.6	1010.9	5.0
120	71	1016.0	−2.6	1016.8	3.5
121	72	1022.7	−1.6	1025.1	5.0
122	73	1029.1	−1.1	1030.8	3.4
123	74	1035.3	−0.6	1038.8	4.6
124	75	1041.0	1.3	1044.2	2.4
125	76	1047.0	3.1	1051.9	3.6
126	77	1052.5	4.5	1057.1	1.5
127	78	1057.8	5.9	1064.4	2.1
128	79	1062.9	7.4	1069.6	0.9
129	80	1067.8	9.1	1076.4	1.2
130	81	1072.5	10.9	1081.4	−0.2
131	82	1077.0	9.6	1087.8	−0.1
132	83	1081.3	8.4	1090.8	−0.2
133	84	1085.4		1094.8	1.0

$Z = 50$: Tin

A	N	Mass	Δ	Mass	Δ
102	52	840.1	8.6	850.2	1.6
103	53	852.6	7.6	861.0	0.9
132	82	1089.6	12.5	1102.1	0.
133	83	1094.2	11.2	1105.3	−0.1
134	84	1098.6	10.0	1109.7	1.1
135	85	1102.9	9.1	1112.6	0.6

$Z = 51$: Antimony

A	N	Mass	Δ	Mass	Δ
105	54	873.1	1.6	866.5	5.0
106	55	883.5	0.5	878.7	4.3
107	56	895.8	1.5	890.6	3.8
108	57	905.3	0.9	902.2	2.3
109	58	917.4	3.0	913.4	1.0
110	59	926.3	2.0	924.3	0.
111	60	937.8	3.8	935.0	−0.9
112	61	946.2	2.8	945.3	−1.7
113	62	957.3	4.3	955.3	−2.2
114	63	965.3	3.1	965.0	−2.5
115	64	975.9	4.8	974.5	−3.1
116	65	983.6	3.1	983.7	−2.8
117	66	993.7	4.5	992.5	−3.0
118	67	1001.1	2.8	1001.2	−2.4
119	68	1010.8	4.2	1009.5	−2.6
120	69*	1017.8	2.4	1017.6	−1.8
121	70	1027.1	3.8	1025.5	−1.8
122	71	1033.8	2.0	1033.1	−0.8
123	72	1042.7	3.4	1040.4	−0.9
124	73	1049.1	1.9	1047.5	0.4
125	74	1057.7	3.1	1054.4	1.4
126	75	1063.9	1.6	1061.0	1.9
127	76	1072.1	2.8	1067.5	3.1
128	77	1078.1	1.4	1073.7	4.3
129	78	1086.0	2.1	1079.6	5.6
130	79	1091.8	0.8	1085.4	7.2
131	80	1099.3	1.1	1091.0	8.8
132	81	1105.0	−0.2	1096.3	10.6
133	82	1112.6	−0.1	1101.5	9.4
134	83	1115.6	−0.1	1106.4	8.3
135	84	1120.3	0.8	1111.2	7.4
136	85	1123.5	0.3	1115.8	

TABLE 8c-2. CALCULATED BINDING ENERGIES IN MeV (Continued)

Number of neutrons N	Mass number A	Liquid drop	Shell correction	BCS pairing energy	Total binding energy
		Z = 51: Antimony (Continued)			
86	137	1120.2	6.7	1.1	1127.9
87	138	1124.4	6.0	0.4	1130.8
		Z = 52: Tellurium			
56	108	892.5	1.9	3.3	897.8
57	109	904.5	0.7	2.4	907.5
58	110	916.1	−1.3	5.3	920.1
59	111	927.4	−2.1	4.0	929.3
60	112	938.4	−3.5	6.3	941.3
61	113	949.1	−4.4	5.3	950.1
62	114	959.5	−4.6	6.6	961.5
63	115	969.6	−5.3	5.5	969.8
64	116	979.5	−5.2	6.5	980.8
65	117	989.0	−3.8	4.0	988.8
66	118	998.3	−4.1	5.5	999.3
67	119	1007.3	−2.9	3.3	1007.1
68	120	1016.0	−3.3	4.9	1017.1
69	121	1024.5	−2.2	2.9	1024.5
70*	122	1032.7	−2.7	4.6	1034.2
71*	123	1040.6	−1.6	2.8	1041.2
72	124	1048.3	−2.3	4.7	1050.5
73	125	1055.8	−1.6	3.3	1057.2
74	126	1063.0	−2.3	5.6	1066.2
75	127	1070.0	−0.6	3.4	1072.7
76	128	1076.7	−0.6	5.2	1081.4
77	129	1083.3	1.4	3.0	1087.6
78	130	1089.6	2.4	3.9	1095.9
79	131	1095.7	3.7	2.7	1102.1
80	132	1101.6	5.3	3.0	1109.9

Number of neutrons N	Mass number A	Liquid drop	Shell correction	BCS pairing energy	Total binding energy
		Z = 53: Iodine (Continued)			
84	137	1134.4	5.0	2.1	1141.6
85	138	1139.6	4.3	1.6	1145.4
86	139	1144.6	3.4	2.5	1150.5
87	140	1149.4	2.9	1.7	1153.9
88	141	1154.1	2.3	2.3	1158.7
89	142	1158.5	2.0	1.2	1161.7
90	143	1162.8	1.6	1.7	1166.1
		Z = 54: Xenon			
59	113	931.9	−3.4	3.6	931.9
60	114	943.6	−4.3	5.4	944.5
61	115	955.2	−3.3	3.3	954.3
62	116	966.3	−3.4	4.5	966.5
63	117	977.2	−2.6	2.7	975.9
64	118	987.8	−2.6	3.7	987.6
65	119	998.1	−2.1	2.3	996.6
66	120	1008.1	−2.3	3.4	1007.8
67	121	1017.8	−2.0	2.3	1016.4
68	122	1027.2	−2.2	3.5	1027.0
69	123	1036.4	−1.8	2.1	1035.2
70	124	1045.3	−1.7	3.1	1045.4
71	125	1053.9	−1.4	1.9	1053.2
72	126	1062.3	−1.6	3.1	1063.0
73*	127	1070.4	−1.1	2.0	1070.4
74*	128	1078.3	−1.5	3.5	1079.8
75	129	1086.0	−1.0	2.5	1087.0
76	130	1093.4	−2.2	5.1	1096.2
77	131	1100.6	−0.7	3.3	1103.2

The data below are rotated tabular entries. Each block lists, for each isotope, the mass number A, the neutron number N, and four computed mass-formula values.

(upper-left block)

A	N				
132	78	1112.1	4.8	-0.3	1107.6
133	79	1118.9	3.6	0.9	1114.4
134	80	1127.4	3.9	2.5	1120.9
135	81	1134.1	2.4	4.4	1127.3
136	82	1142.1	2.6	6.0	1133.4
137	83	1146.7	2.3	5.1	1139.3
138	84	1152.3	3.6	3.6	1145.0
139	85	1156.5	3.0	2.9	1150.6
140	86	1161.8	4.1	1.9	1155.9
141	87	1165.6	3.3	1.3	1161.0
142	88	1170.7	3.9	0.8	1166.0
143	89	1174.0	2.5	0.8	1170.7
144	90	1178.8	3.2	0.2	1175.3
145	91	1181.3	2.5	-0.8	1179.7
146	92	1186.1	5.2	-3.0	1183.9

‖ *Z = 55: Cesium*

A	N				
116	61	955.0	1.8	-2.2	957.4
117	62	967.5	2.8	-2.3	968.9
118	63	977.5	1.4	-1.7	980.2
119	64	989.5	2.7	-2.1	991.1
120	65	999.0	1.3	-1.6	1001.7
121	66	1010.4	2.3	-1.8	1012.1
122	67	1019.5	1.3	-1.7	1022.1
123	68	1030.4	2.5	-2.0	1031.9
124	69	1039.0	1.3	-1.7	1041.4
125	70	1049.4	2.2	-1.5	1050.7
126	71	1057.6	1.0	-1.4	1059.7
127	72	1067.5	1.9	-1.4	1068.4
128	73	1075.3	1.1	-1.5	1076.9
129	74	1084.9	2.2	-1.3	1085.1
130	75*	1092.4	1.4	-2.4	1093.1
131	76	1101.8	3.5	-1.6	1100.8
132	77	1109.1	2.5	-1.3	1108.4
133	78	1118.3	4.0	-0.1	1115.7
134	79	1125.5	2.8	1.5	1122.8
135	80	1134.3	3.1	3.4	1129.7
136	81	1141.3	1.6	4.9	1136.3
137	82	1149.6	1.9		1142.8

(lower-left block)

A	N				
133	81	1115.9	1.5	7.1	1107.3
134	82	1123.3	1.7	8.8	1112.8
135	83	1127.2	1.5	7.7	1118.0
136	84	1132.2	2.8	6.3	1123.1
137	85	1135.8	2.2	5.5	1128.0
138	86	1140.5	3.2	4.5	1132.7
139	87	1143.6	2.5	3.9	1137.2
140	88	1148.1	3.1	3.4	1141.6
141	89	1150.8	1.7	3.3	1145.7

Z = 53: Iodine

A	N				
110	57	906.2	-0.5	1.7	907.3
111	58	918.2	-2.2	4.2	920.2
112	59	930.0	-2.6	2.7	929.9
113	60	941.3	-3.6	4.5	942.2
114	61	952.5	-2.9	2.5	951.4
115	62	963.2	-4.0	4.4	963.2
116	63	973.7	-3.1	2.4	972.2
117	64	983.9	-3.3	3.6	983.5
118	65	993.9	-2.5	1.8	992.1
119	66	1003.5	-2.8	3.2	1002.9
120	67	1012.9	-2.2	1.7	1011.1
121	68	1021.9	-2.4	2.9	1021.4
122	69	1030.7	-1.8	1.5	1029.3
123	70	1039.3	-1.8	2.6	1039.2
124	71	1047.6	-1.3	1.3	1046.6
125	72*	1055.6	-1.7	2.8	1056.1
126	73	1063.4	-0.7	1.3	1063.2
127	74	1071.0	-1.2	2.9	1072.4
128	75	1078.3	-0.5	1.7	1079.2
129	76	1085.4	-0.9	3.6	1088.1
130	77	1092.3	0.4	2.2	1094.7
131	78	1098.9	1.1	3.3	1103.3
132	79	1105.4	2.2	2.2	1109.8
133	80	1111.6	3.9	2.4	1117.9
134	81	1117.6	5.7	1.0	1124.3
135	82	1123.4	7.3	1.2	1132.0
136	83	1129.0	6.3	1.0	1136.2

TABLE 8c-2. CALCULATED BINDING ENERGIES IN MeV (*Continued*)

Z = 55: Cesium (Continued)

Number of neutrons N	Mass number A	Liquid drop	Shell correction	BCS pairing energy	Total binding energy
83	138	1149.0	4.0	1.5	1154.5
84	139	1155.1	2.6	2.9	1160.5
85	140	1160.9	1.9	2.2	1165.0
86	141	1166.5	0.8	3.3	1170.7
87	142	1172.0	0.3	2.5	1174.8
88	143	1177.2	-0.3	3.2	1180.1
89	144	1182.3	-0.2	1.7	1183.8
90	145	1187.2	-0.8	2.4	1188.8
91	146	1191.9	-1.8	1.8	1191.7
92	147	1196.4	-4.1	4.5	1196.8
93	148	1200.9	-1.8	1.7	1199.6
94	149	1205.1	-1.5	2.3	1204.4

Z = 56: Barium

Number of neutrons N	Mass number A	Liquid drop	Shell correction	BCS pairing energy	Total binding energy
62	118	970.9	-1.7	3.3	970.2
63	119	982.5	-1.2	2.0	980.6
64	120	993.7	-1.6	3.3	992.9
65	121	1004.8	-1.2	2.0	1002.8
66	122	1015.4	-1.5	3.0	1014.5
67	123	1025.9	-1.5	2.1	1023.9
68	124	1036.0	-2.0	3.3	1035.1
69	125	1045.9	-1.8	2.3	1044.1
70	126	1055.4	-1.9	3.2	1054.8
71	127	1064.8	-1.7	2.0	1063.3
72	128	1073.8	-1.6	2.9	1073.5
73	129	1082.7	-1.8	2.2	1081.7
74	130	1091.2	-2.1	3.4	1091.5
75	131	1099.6	-2.0	2.6	1099.3
76	132	1107.6	-2.4	4.1	1108.9

Z = 57: Lanthanum (Continued)

Number of neutrons N	Mass number A	Liquid drop	Shell correction	BCS pairing energy	Total binding energy
81	138	1152.6	1.8	1.7	1156.1
82	139	1159.7	3.5	1.8	1165.0
83	140	1166.6	2.4	1.7	1170.6
84	141	1173.2	1.1	2.8	1177.2
85	142	1179.7	0.4	2.3	1182.4
86	143	1186.0	-0.5	3.2	1188.6
87	144	1192.0	-1.0	2.4	1193.4
88	145	1197.9	-1.6	3.1	1199.4
89	146	1203.6	-1.8	1.9	1203.7
90	147	1209.1	-2.2	2.4	1209.3
91	148	1214.5	-2.4	2.7	1213.0
92	149	1219.6	-2.3	2.7	1218.6
93	150	1224.6	-1.7	1.6	1222.6
94	151	1229.4	-1.3	2.2	1228.2
95	152	1234.1	-0.5	1.1	1232.1
96	153	1238.5	-0.2	1.6	1237.3
97	154	1242.8	0.4	0.6	1241.0

Z = 58: Cerium

Number of neutrons N	Mass number A	Liquid drop	Shell correction	BCS pairing energy	Total binding energy
66	124	1020.5	-0.1	2.3	1019.5
67	125	1031.6	-0.3	1.6	1029.7
68	126	1042.5	-0.7	2.6	1041.4
69	127	1053.0	-0.9	1.8	1051.1
70	128	1063.3	-1.1	2.7	1062.3
71	129	1073.3	-1.0	1.5	1071.4
72	130	1083.0	-0.8	2.2	1082.3
73	131	1092.5	-1.5	1.8	1091.0
74	132	1101.8	-2.0	3.0	1101.3
75	133	1110.7	-2.3	2.4	1109.7

77*	133	1115.5	−2.0	3.3	1116.6
78	134	1123.1	−2.1	5.0	1126.0
79	135	1130.6	−1.0	3.9	1133.5
80	136	1137.8	0.7	4.2	1142.6
81	137	1144.8	2.5	2.7	1150.0
82	138	1151.5	4.1	3.0	1158.6
83	139	1158.1	3.2	2.6	1163.9
84	140	1164.4	1.8	3.9	1170.1
85	141	1170.6	1.1	3.3	1175.0
86	142	1176.6	0.1	4.3	1180.9
87	143	1182.3	−0.4	3.4	1185.4
88	144	1187.9	−1.0	4.2	1191.0
89	145	1193.2	−1.0	2.8	1195.0
90	146	1198.4	−1.6	3.5	1200.3
91	147	1203.5	−2.7	3.1	1203.6
92	148	1208.3	−4.7	5.4	1208.9
93	149	1213.0	−2.0	2.7	1212.2
94	150	1217.5	1.7	3.2	1217.4
95	151	1221.9	−0.9	2.1	1220.9

76	134	1119.5	−2.8	3.9	1119.7
77	135	1128.0	−2.8	3.2	1127.9
78	136	1136.3	−3.3	5.0	1137.9
79	137	1144.3	−2.0	3.7	1146.1
80*	138	1152.2	−0.4	4.0	1155.8
81*	139	1159.8	1.2	2.7	1163.8
82	140	1167.2	3.1	2.7	1173.0
83	141	1174.4	1.8	2.7	1178.9
84	142	1181.4	0.7	3.7	1185.8
85	143	1188.2	−0.1	3.2	1191.3
86	144	1194.8	−1.0	4.1	1197.9
87	145	1201.1	−1.4	3.3	1203.0
88	146	1207.3	−2.1	4.0	1209.2
89	147	1213.3	−2.4	3.0	1213.9
90	148	1219.1	−2.6	3.3	1219.8
91	149	1224.8	−2.7	2.9	1223.9
92	150	1230.3	−2.3	3.5	1229.9
93	151	1235.6	−1.3	2.1	1234.2
94	152	1240.7	−0.8	2.6	1240.1
95	153	1245.6	−0.1	1.6	1244.4
96	154	1250.3	0.2	2.2	1249.9
97	155	1254.9	0.9	1.2	1253.9
98	156	1259.3	1.4	1.6	1259.1
99	157	1263.5	1.4	1.1	1262.8

Z = 57: *Lanthanum*

64	121	995.9	−0.6	1.9	994.1
65	122	1007.2	−0.5	0.9	1004.4
66	123	1018.3	−0.8	1.8	1016.4
67	124	1029.1	−1.0	1.2	1026.2
68	125	1039.5	−1.3	2.1	1037.7
69	126	1049.7	−1.4	1.3	1047.0
70	127	1059.7	−1.5	2.1	1057.9
71	128	1069.3	−1.4	1.0	1066.7
72	129	1078.7	−1.2	1.6	1077.2
73	130	1087.7	−1.7	1.1	1085.7
74	131	1096.8	−1.9	2.2	1095.7
75	132	1105.5	−2.0	1.5	1103.8
76	133	1113.9	−2.6	2.9	1113.5
77	134	1122.1	−2.2	2.1	1121.5
78*	135	1130.1	−2.8	4.0	1131.2
79*	136	1137.7	−1.6	2.8	1139.0
80	137	1145.3	0.	3.1	1148.4

Z = 59: *Praseodymium*

67	126	1033.7	0.4	0.5	1031.0
68	127	1044.8	0.1	1.5	1043.1
69	128	1055.7	−0.2	0.7	1053.1
70	129	1066.4	−0.3	1.6	1064.6
71	130	1076.7	−0.2	0.4	1074.0
72	131	1086.8	−0.2	1.0	1085.1
73	132	1096.6	−1.1	0.7	1094.0
74	133	1106.2	−1.4	1.8	1104.6
75	134	1115.5	−1.9	1.2	1113.2
76	135	1124.5	−2.4	2.4	1123.5
77	136	1133.4	−2.7	2.0	1131.9
78	137	1142.0	−3.9	4.0	1142.0

TABLE 8c-2. CALCULATED BINDING ENERGIES IN MeV (Continued)

Number of neutrons N	Mass number A	Liquid drop	Shell correction	BCS pairing energy	Total binding energy
		Z = 59: Praseodymium (Continued)			
79	138	1150.4	-2.7	2.8	1150.4
80	139	1158.5	-1.2	3.2	1160.4
81	140	1166.5	0.3	2.0	1168.7
82*	141	1174.2	2.1	2.0	1178.3
83	142	1181.7	0.9	2.0	1184.6
84	143	1189.0	-0.1	2.8	1191.8
85	144	1196.1	-0.7	2.3	1197.6
86	145	1203.0	-1.4	2.9	1204.5
87	146	1209.7	-1.9	2.3	1209.9
88	147	1216.2	-2.5	2.9	1216.5
89	148	1222.4	-3.2	2.3	1221.4
90	149	1228.6	-3.0	3.0	1227.7
91	150	1234.6	-2.0	1.6	1232.5
92	151	1240.4	-1.3	1.9	1238.9
93	152	1246.0	-0.5	0.8	1243.7
94	153	1251.3	0.	1.4	1249.9
95	154	1256.5	0.5	0.5	1254.5
96	155	1261.6	0.9	1.1	1260.4
97	156	1266.4	1.5	0.2	1264.7
98	157	1271.1	1.9	0.6	1270.2
99	158	1275.6	1.9	0.1	1274.2
100	159	1279.9	1.9	0.8	1279.2
		Z = 60: Neodymium			
69	129	1057.9	0.7	0.9	1056.2
70	130	1068.8	0.5	1.8	1068.0
71	131	1079.6	0.6	0.7	1077.8
72	132	1090.0	0.5	1.4	1089.1
73	133	1100.1	-0.7	1.2	1098.3
74	134	1110.0	-1.1	2.4	1109.2

Number of neutrons N	Mass number A	Liquid drop	Shell correction	BCS pairing energy	Total binding energy
		Z = 61: Promethium (Continued)			
79	140	1160.7	-3.1	2.8	1160.0
80	141	1169.4	-2.7	3.9	1170.6
81	142	1178.0	-0.9	2.5	1179.6
82	143	1186.4	0.7	2.7	1189.8
83	144	1194.5	-0.3	2.4	1196.7
84	145	1202.5	-1.6	3.6	1204.5
85	146	1210.2	-1.7	2.6	1210.9
86*	147	1217.7	-2.5	3.3	1218.4
87	148	1225.0	-2.6	2.5	1224.5
88	149	1232.1	-2.6	2.7	1231.7
89	150	1239.0	-2.5	1.8	1237.3
90	151	1245.8	-1.7	2.1	1244.5
91	152	1252.4	-0.7	0.8	1250.1
92	153	1258.7	0.1	1.2	1257.3
93	154	1264.9	0.7	0.3	1262.8
94	155	1270.9	1.2	1.0	1269.7
95	156	1276.7	1.6	0.3	1274.9
96	157	1282.3	2.1	0.7	1281.4
97	158	1287.7	2.6	0.	1286.4
98	159	1292.9	2.9	0.4	1292.4
99	160	1298.0	2.9	0.	1297.0
100	161	1302.8	3.0	0.6	1302.5
101	162	1307.6	3.1	0.1	1306.8
102	163	1312.1	3.2	0.5	1311.9
103	164	1316.5	3.4	-0.1	1315.9
104	165	1320.7	3.5	0.1	1320.5
		Z = 62: Samarium			
73	135	1105.4	0.2	1.0	1103.7
74	136	1116.0	-0.6	2.4	1115.2

N	A	M_1	δ_1	δ_2	M_2
75	137	1124.7	1.6	−1.1	1126.3
76	138	1135.7	2.4	−1.4	1136.3
77	139	1145.0	2.1	−2.0	1146.1
78	140	1155.8	3.8	−3.0	1155.6
79	141	1165.1	3.9	−3.4	1165.0
80	142	1176.1	5.2	−3.1	1174.1
81	143	1185.3	3.7	−1.3	1183.0
82	144	1195.9	4.0	0.3	1191.6
83	145	1203.1	3.6	−0.6	1200.1
84	146	1211.2	4.9	−2.0	1208.3
85	147	1218.0	3.8	−2.1	1216.4
86	148	1225.8	4.5	−2.8	1224.2
87	149	1232.1	3.5	−2.9	1231.8
88*	150	1239.6	3.7	−2.8	1239.2
89	151	1245.5	2.5	−2.3	1246.4
90	152	1253.0	2.7	−1.5	1253.5
91	153	1258.9	1.6	−0.7	1260.4
92	154	1266.4	2.0	0.0	1267.0
93	155	1272.2	1.1	0.7	1273.5
94	156	1279.3	1.6	1.1	1279.7
95	157	1284.8	0.9	1.6	1285.8
96	158	1291.5	1.4	2.0	1291.7
97	159	1296.8	0.5	2.6	1297.4
98	160	1303.1	0.9	3.0	1302.9
99	161	1308.0	0.5	3.0	1308.2
100	162	1313.8	1.1	3.0	1313.4
101	163	1318.3	0.6	3.1	1318.4
102	164	1323.7	0.4	3.3	1323.3
103	165	1327.9	0.7	3.5	1327.9
104	166	1332.9	0.5	3.5	1332.4
105	167	1336.5	1.2	2.8	1336.7
106	168	1341.1		2.5	1340.9

Z = 63: Europium

N	A	M_1	δ_1	δ_2	M_2
74	137	1116.8	1.5	0.1	1118.2
75	138	1126.5	0.8	−0.7	1128.8
76	139	1137.8	1.4	−0.8	1139.1
77	140	1147.2	0.9	−1.5	1149.2

N	A	M	δ_1	δ_2
75	135	1119.6	−1.7	1.9
76	136	1129.0	−2.4	3.2
77	137	1138.2	−2.8	2.8
78	138	1147.1	−4.0	5.0
79	139	1155.8	−3.8	4.6
80	140	1164.3	−2.2	4.9
81	141	1172.5	−0.4	3.4
82	142	1180.6	1.2	3.6
83	143	1188.4	0.3	3.3
84*	144	1196.0	−1.1	4.6
85	145	1203.4	−1.5	3.8
86	146	1210.6	−2.8	5.0
87	147	1217.6	−2.4	2.8
88	148	1224.4	−4.0	5.0
89	149	1231.0	−3.8	3.5
90	150	1237.5	−3.0	3.8
91	151	1243.8	−1.6	2.1
92	152	1249.8	−0.8	2.3
93	153	1255.7	0.1	1.1
94	154	1261.4	0.6	1.7
95	155	1266.9	1.2	0.9
96	156	1272.2	1.6	1.4
97	157	1277.3	2.2	0.5
98	158	1282.3	2.6	0.9
99	159	1287.0	2.6	0.5
100	160	1291.6	2.6	1.1
101	161	1296.1	2.8	0.5
102	162	1300.4	2.9	1.0

Z = 61: Promethium

N	A	M_1	δ_1	δ_2	M_2
71	132	1081.9	1.1	0.2	1079.5
72	133	1092.6	1.1	0.7	1091.1
73	134	1103.2	−0.6	0.2	1100.6
74	135	1113.3	−0.6	1.6	1111.7
75	136	1123.2	−1.5	0.8	1120.8
76	137	1133.0	−2.2	1.7	1131.5
77	138	1142.4	−3.0	1.4	1140.5
78	139	1151.6		2.9	1151.0

TABLE 8c-2. CALCULATED BINDING ENERGIES IN MeV (Continued)

Z = 63: Europium (Continued)

Number of neutrons N	Mass number A	Liquid drop	Shell correction	BCS pairing energy	Total binding energy
78	141	1159.1	−2.0	2.1	1158.2
79	142	1168.7	−3.0	2.4	1167.8
80	143	1178.1	−3.5	4.3	1178.9
81	144	1187.3	−1.6	2.8	1188.5
82	145	1196.3	0.	2.9	1199.3
83	146	1205.1	−0.9	2.7	1206.8
84	147	1213.6	−2.4	4.0	1215.3
85	148	1222.0	−1.8	2.5	1222.4
86	149	1230.1	−2.1	2.8	1230.5
87	150	1238.0	−2.5	2.1	1237.2
88	151	1245.8	−1.8	2.1	1245.1
89*	152	1253.3	−1.3	1.0	1251.5
90	153	1260.7	−0.7	1.5	1259.4
91	154	1267.8	0.	0.5	1265.7
92	155	1274.8	0.7	1.0	1273.5
93	156	1281.5	1.2	0.3	1279.7
94	157	1288.1	1.5	0.9	1287.1
95	158	1294.4	1.9	0.3	1293.0
96	159	1300.6	2.4	0.7	1300.0
97	160	1306.5	2.8	0.	1305.5
98	161	1312.3	3.2	0.4	1312.1
99	162	1317.9	3.2	0.	1317.2
100	163	1323.4	3.2	0.6	1323.4
101	164	1328.7	3.3	0.1	1328.2
102	165	1333.7	3.5	0.5	1333.8
103	166	1338.7	3.6	−0.1	1338.3
104	167	1343.4	3.7	0.1	1343.5
105	168	1348.0	3.0	0.6	1347.4
106	169	1352.5	2.7	0.7	1352.3
107	170	1356.8	2.3	0.2	1355.8

Z = 65: Terbium (Continued)

Number of neutrons N	Mass number A	Liquid drop	Shell correction	BCS pairing energy	Total binding energy
80	145	1184.6	−3.3	4.1	1185.3
81	146	1194.4	−1.7	2.9	1195.6
82	147	1204.1	−0.2	3.1	1206.9
83	148	1213.4	−1.0	2.7	1215.1
84	149	1222.6	−2.2	3.8	1224.2
85	150	1231.6	−1.0	1.9	1232.0
86	151	1240.3	−1.2	2.1	1240.7
87	152	1248.8	−1.9	1.7	1248.1
88	153	1257.2	−1.0	1.6	1256.6
89	154	1265.3	−0.3	0.6	1263.7
90	155	1273.3	0.	1.2	1272.3
91	156	1281.0	0.4	0.4	1279.2
92	157	1288.5	1.0	0.9	1287.5
93*	158	1295.8	1.3	0.3	1294.3
94	159	1302.9	1.7	0.9	1302.2
95	160	1309.8	2.0	0.3	1308.7
96	161	1316.6	2.4	0.8	1316.2
97	162	1323.1	2.9	0.	1322.3
98	163	1329.5	3.2	0.4	1329.4
99	164	1335.6	3.2	0.	1335.2
100	165	1341.6	3.3	0.6	1341.8
101	166	1347.4	3.4	0.1	1347.2
102	167	1353.1	3.5	0.5	1353.4
103	168	1358.6	3.7	−0.1	1358.4
104	169	1363.9	3.8	−0.1	1364.2
105	170	1369.0	3.1	−0.1	1368.6
106	171	1374.0	2.8	−0.6	1374.0
107	172	1378.8	2.4	0.2	1378.0
108	173	1383.5	2.2	0.7	1383.1
109	174	1387.9	1.6	0.3	1386.8

Z = 64: Gadolinium

N	A	M₁	Δ₁	Δ₂	M₂
76	140	1141.4	−0.6	1.9	1140.9
77	141	1151.8	−1.4	1.6	1150.7
78	142	1162.0	−2.2	3.1	1162.1
79	143	1171.9	−3.0	3.4	1171.9
80	144	1181.6	−3.5	5.2	1183.4
81	145	1191.2	−1.8	4.0	1193.3
82	146	1200.5	0.	3.9	1204.4
83	147	1209.5	−1.1	3.9	1212.2
84	148	1218.4	−2.3	4.9	1221.0
85	149	1227.0	−1.8	3.4	1228.4
86	150	1235.5	−2.1	3.8	1236.8
87	151	1243.7	−2.5	3.0	1243.9
88	152	1251.7	−2.0	3.1	1252.0
89	153	1259.6	−1.4	1.9	1258.6
90	154	1267.2	−0.5	2.1	1266.8
91*	155	1274.7	0.1	1.0	1273.4
92	156	1281.9	0.7	1.5	1281.4
93	157	1288.9	1.2	0.7	1287.8
94	158	1295.8	1.6	1.4	1295.5
95	159	1302.4	2.0	0.7	1301.6
96	160	1308.8	2.4	1.2	1308.9
97	161	1315.1	2.9	0.4	1314.7
98	162	1321.2	3.3	0.8	1321.6
99	163	1327.1	3.3	0.4	1327.0
100	164	1332.8	3.3	1.0	1333.4
101	165	1338.3	3.4	0.5	1338.5
102	166	1343.7	3.6	0.9	1344.4
103	167	1348.9	3.7	0.3	1349.2
104	168	1353.9	3.8	0.5	1354.6
105	169	1358.8	3.1	0.3	1358.8
106	170	1363.5	2.8	1.0	1363.9
107	171	1368.1	2.4	0.6	1367.7
108	172	1372.4	2.1	1.1	1372.5
109	173	1376.7	1.5	0.7	1376.0
110	175	1392.3	1.3	0.9	1391.6
111	176	1396.5	0.9	0.4	1395.0

Z = 65: Terbium

N	A	M₁	Δ₁	Δ₂	M₂
78	143	1164.4	−1.1	1.5	1163.6
79	144	1174.6	−2.3	2.0	1173.7

Z = 66: Dysprosium

N	A	M₁	Δ₁	Δ₂	M₂
80	146	1187.0	−3.4	5.3	1188.9
81	147	1197.2	−1.6	3.8	1199.4
82	148	1207.1	−0.1	4.2	1211.1
83	149	1216.8	−0.8	3.7	1219.6
84	150	1226.3	−2.3	5.0	1228.9
85	151	1235.5	−0.9	2.7	1237.0
86	152	1244.6	−1.0	2.9	1246.1
87	153	1253.4	−1.7	2.5	1253.7
88	154	1262.0	−1.2	2.6	1262.5
89	155	1270.4	−0.7	1.4	1269.7
90	156	1278.7	−0.1	1.9	1278.5
91	157	1286.7	0.4	0.9	1285.7
92	158	1294.6	0.9	1.5	1294.3
93	159	1302.1	1.3	0.7	1301.2
94	160	1309.5	1.7	1.4	1309.5
95*	161	1316.7	2.0	0.7	1316.2
96	162	1323.7	2.4	1.2	1324.0
97	163	1330.6	2.9	1.1	1330.3
98	164	1337.2	3.3	0.8	1337.7
99	165	1343.6	3.3	0.4	1343.7
100	166	1349.9	3.4	1.0	1350.7
101	167	1356.0	3.5	0.5	1356.3
102	168	1361.9	3.7	0.9	1362.8
103	169	1367.7	3.8	0.3	1368.0
104	170	1373.2	3.2	0.5	1374.1
105	171	1378.6	2.9	1.0	1378.8
106	172	1383.9	2.5	0.5	1384.4
107	173	1388.9	2.2	1.0	1388.8
108	174	1393.9	1.7	1.0	1394.1
109	175	1398.6	1.4	1.3	1398.1
110	176	1403.2	0.9	0.8	1403.2
111	177	1407.7	0.6	1.4	1406.8
112	178	1412.0	0.3	0.9	1411.6
113	179	1416.2			1415.1

Z = 67: Holmium

N	A	M₁	Δ₁	Δ₂	M₂
81	148	1199.4	−1.1	2.6	1200.9
82	149	1209.6	0.1	3.2	1212.9

TABLE 8c-2. CALCULATED BINDING ENERGIES IN MeV (Continued)

Z = 67: Holmium (Continued)

Number of neutrons N	Mass number A	Liquid drop	Shell correction	BCS pairing energy	Total binding energy
83	150	1219.6	-0.2	2.3	1221.7
84	151	1229.4	-2.0	3.9	1231.3
85	152	1239.0	-0.2	1.3	1239.7
86	153	1248.3	0.	1.3	1249.1
87	154	1257.4	-0.9	1.1	1257.0
88	155	1266.4	-0.6	1.6	1266.2
89	156	1275.1	-0.2	0.6	1273.8
90	157	1283.6	0.2	1.2	1283.0
91	158	1291.9	0.7	0.3	1290.5
92	159	1300.0	1.1	0.9	1299.4
93	160	1307.9	1.5	0.2	1306.6
94	161	1315.6	1.8	0.9	1315.2
95	162	1323.1	2.1	0.2	1322.1
96*	163	1330.4	2.5	0.8	1330.3
97	164	1337.5	2.9	0.	1336.9
98	165	1344.4	3.3	0.3	1344.6
99	166	1351.1	3.3	0.	1350.8
100	167	1357.7	3.3	0.6	1358.0
101	168	1364.0	3.4	0.1	1363.9
102	169	1370.2	3.5	0.5	1370.7
103	170	1376.2	3.7	-0.1	1376.2
104	171	1382.1	3.8	0.1	1382.5
105	172	1387.7	3.2	-0.1	1387.5
106	173	1393.2	2.9	-0.6	1393.4
107	174	1398.6	2.5	0.1	1398.0
108	175	1403.8	2.3	0.6	1403.6
109	176	1408.8	1.8	0.2	1407.8
110	177	1413.6	1.5	0.8	1413.2
111	178	1418.4	1.0	0.3	1417.1
112	179	1422.9	0.8	1.0	1422.1
113	180	1427.3	0.4	0.4	1425.8
114	181	1431.6	0.2	1.0	1430.6

Z = 69: Thulium (Continued)

Number of neutrons N	Mass number A	Liquid drop	Shell correction	BCS pairing energy	Total binding energy
87	156	1263.9	0.	0.7	1264.2
88	157	1273.4	-0.2	1.4	1273.8
89	158	1282.7	-0.3	0.8	1281.9
90	159	1291.8	-0.1	1.4	1291.5
91	160	1300.7	0.2	0.6	1299.5
92	161	1309.4	0.4	1.2	1308.8
93	162	1317.8	0.8	0.4	1316.6
94	163	1326.1	1.1	1.1	1325.6
95	164	1334.1	1.4	0.4	1333.1
96	165	1342.0	1.7	0.9	1341.7
97	166	1349.7	2.2	0.	1348.8
98	167	1357.1	2.6	0.4	1357.1
99	168	1364.4	2.6	0.7	1363.8
100*	169	1371.5	2.6	0.7	1371.6
101	170	1378.4	2.7	0.1	1378.0
102	171	1385.1	2.8	0.6	1385.3
103	172	1391.7	3.0	-0.1	1391.4
104	173	1398.0	3.1	0.2	1398.3
105	174	1404.2	2.7	-0.1	1403.9
106	175	1410.3	2.4	0.5	1410.3
107	176	1416.1	2.1	-0.1	1415.5
108	177	1421.8	1.9	0.6	1421.6
109	178	1427.3	1.4	0.1	1426.5
110	179	1432.7	1.2	0.7	1432.3
111	180	1438.0	0.9	0.2	1436.8
112	181	1443.0	0.7	0.8	1442.4
113	182	1448.0	0.3	0.3	1446.6
114	183	1452.7	0.	1.0	1452.0
115	184	1457.3	-0.2	0.4	1456.0
116	185	1461.8	-0.2	0.9	1461.1
117	186	1466.1	-0.3	0.3	1464.9
118	187	1470.3	-0.4	0.8	1469.9

Z = 68: Erbium

N	A	M	δ	δ	M
83	151	1221.9	0.2	3.2	1225.2
84	152	1232.0	-1.5	4.6	1235.2
85	153	1241.9	0.2	2.1	1243.9
86	154	1251.5	0.	2.3	1253.5
87	155	1260.9	-0.6	1.8	1261.7
88	156	1270.1	-0.7	2.4	1271.0
89	157	1279.1	-0.8	1.7	1278.8
90	158	1287.9	-0.3	2.0	1288.1
91	159	1296.6	-0.4	0.9	1295.8
92	160	1304.9	0.9	1.4	1304.9
93	161	1313.1	1.4	0.6	1312.4
94	162	1321.1	1.7	1.2	1321.2
95	163	1328.9	2.1	0.5	1328.4
96	164	1336.5	2.5	1.0	1336.8
97	165	1343.9	3.0	0.2	1343.7
98*	166	1351.0	3.4	0.5	1351.6
99	167	1358.0	3.4	0.1	1358.1
100	168	1364.8	3.4	0.8	1365.6
101	169	1371.5	3.5	0.2	1371.8
102	170	1377.9	3.6	0.7	1378.8
103	171	1384.2	3.8	0.	1384.6
104	172	1390.3	3.9	0.3	1391.2
105	173	1396.3	3.3	0.1	1396.5
106	174	1402.0	3.0	0.7	1402.6
107	175	1407.6	2.6	0.3	1407.5
108	176	1413.1	2.3	0.8	1413.4
109	177	1418.3	1.8	0.4	1417.9
110	178	1423.5	1.5	1.1	1423.5
111	179	1428.4	1.1	0.6	1427.8
112	180	1433.2	0.9	1.2	1433.1
113	181	1437.9	0.4	0.7	1437.0
114	182	1442.4	0.	1.5	1442.1
115	183	1446.8	-0.3	1.0	1445.8
116	184	1451.0	-0.6	1.7	1450.7

Z = 69: Thulium

N	A	M	δ	δ	M
85	154	1244.3	1.0	0.9	1245.7
86	155	1254.2	0.8	1.1	1255.7

Z = 70: Ytterbium

N	A	M	δ	δ	M
87	157	1266.3	0.4	1.5	1267.9
88	158	1276.1	0.2	1.9	1277.8
89	159	1285.7	-0.2	1.3	1286.0
90	160	1295.1	-0.4	2.1	1295.7
91	161	1304.3	-0.3	1.3	1303.8
92	162	1313.2	0.	2.0	1313.4
93	163	1322.0	0.4	1.1	1321.3
94	164	1330.5	0.7	1.7	1330.6
95	165	1338.9	1.1	1.0	1338.3
96	166	1347.0	1.3	1.5	1347.2
97	167	1355.0	1.8	0.7	1354.6
98	168	1362.7	2.3	1.0	1363.1
99	169	1370.2	2.2	0.7	1370.1
100	170	1377.6	2.2	1.3	1378.1
101	171	1384.8	2.3	1.3	1384.8
102*	172	1391.8	2.4	0.6	1392.4
103	173	1398.6	2.6	0.8	1398.7
104	174	1405.2	2.7	0.5	1405.9
105	175	1411.7	2.4	1.0	1411.7
106	176	1418.0	2.2	0.6	1418.5
107	177	1424.1	1.8	1.1	1423.9
108	178	1430.0	1.6	0.6	1430.4
109	179	1435.8	1.3	1.1	1435.5
110	180	1441.5	1.1	0.6	1441.6
111	181	1447.0	0.8	1.2	1446.4
112	182	1452.3	0.6	0.8	1452.3
113	183	1457.4	0.2	1.5	1456.8
114	184	1462.5	0.	0.9	1462.5
115	185	1467.3	-0.1	1.5	1466.8
116	186	1472.1	-0.2	0.8	1472.3
117	187	1476.6	-0.1	1.4	1476.4
118	188	1481.1	0.	1.0	1481.7
119	189	1485.4	0.	1.8	1485.9
120	190	1489.5	0.		1491.0

Z = 71: Lutetium

N	A	M	δ	δ	M
89	160	1288.2		0.8	1288.3
90	161	1297.9	-0.4	1.7	1298.3
91	162	1307.3	-0.5	1.0	1306.5

TABLE 8c-2. CALCULATED BINDING ENERGIES IN MeV (Continued)

Z = 71: Lutetium (Continued)

Number of neutrons N	Mass number A	Liquid drop	Shell correction	BCS pairing energy	Total binding energy
92	163	1316.6	-0.5	1.8	1316.3
93	164	1325.6	-0.2	0.9	1324.4
94	165	1334.4	0.	1.5	1334.9
95	166	1343.1	0.5	0.6	1341.9
96	167	1351.5	0.8	1.2	1351.0
97	168	1359.7	1.3	0.3	1358.6
98	169	1367.7	1.6	0.7	1367.4
99	170	1375.6	1.8	0.3	1374.7
100	171	1383.2	1.7	1.0	1383.0
101	172	1390.6	1.8	0.4	1389.9
102	173	1397.9	1.9	0.9	1397.8
103	174	1405.0	2.2	0.2	1404.4
104*	175	1411.9	2.2	0.5	1411.8
105	176	1418.6	1.9	0.1	1418.0
106	177	1425.2	1.7	0.6	1425.0
107	178	1431.5	1.4	0.2	1430.8
108	179	1437.7	1.2	0.7	1437.5
109	180	1443.8	0.9	0.2	1442.9
110	181	1449.7	0.9	0.7	1449.4
111	182	1455.4	0.6	0.3	1454.5
112	183	1461.0	0.4	0.9	1460.6
113	184	1466.4	0.	0.5	1465.4
114	185	1471.7	-0.2	1.2	1471.4
115	186	1476.8	-0.2	0.6	1476.0
116	187	1481.8	-0.2	1.1	1481.7
117	188	1486.6	0.	0.4	1486.1
118	189	1491.3	0.2	0.9	1491.7
119	190	1495.8	0.4	0.4	1496.1
120	191	1500.2	0.7	0.8	1501.5
121	192	1504.5	1.0	0.6	1505.9
122	193	1508.6	1.1	1.3	1510.8

Z = 73: Tantalum (Continued)

Number of neutrons N	Mass number A	Liquid drop	Shell correction	BCS pairing energy	Total binding energy
94	167	1340.7	-0.4	1.8	1340.7
95	168	1349.9	0.	0.9	1349.0
96	169	1358.9	0.3	1.3	1358.7
97	170	1367.7	0.7	0.5	1366.7
98	171	1376.2	1.0	0.9	1376.0
99	172	1384.6	1.1	0.4	1383.8
100	173	1392.8	1.0	1.2	1392.6
101	174	1400.8	1.1	0.6	1400.1
102	175	1408.6	1.1	1.2	1408.5
103	176	1416.2	1.4	0.3	1415.5
104	177	1423.6	1.5	0.7	1423.6
105	178	1430.9	1.4	0.3	1430.3
106	179	1437.9	1.3	0.8	1437.9
107*	180	1444.9	1.1	0.3	1444.3
108	181	1451.6	0.9	0.9	1451.6
109	182	1458.2	0.8	0.4	1457.6
110	183	1464.6	0.8	0.8	1464.6
111	184	1470.8	0.6	0.4	1470.2
112	185	1476.9	0.4	1.0	1477.0
113	186	1482.8	0.2	0.6	1482.3
114	187	1488.6	0.1	1.3	1488.9
115	188	1494.2	0.1	0.7	1494.0
116	189	1499.6	0.2	1.2	1500.3
117	190	1505.0	0.6	0.4	1505.3
118	191	1510.1	1.0	0.9	1511.5
119	192	1515.2	1.6	0.1	1516.5
120	193	1520.0	2.1	0.6	1522.5
121	194	1524.8	2.5	0.2	1527.4
122	195	1529.4	2.8	0.8	1532.9
123	196	1533.9	3.3	0.5	1537.6
124	197	1538.2	3.7	1.1	1542.9
125	198	1542.4	4.1	1.0	1547.5

Z = 72: Hafnium

N	A	value	err	value	err
90	162	1300.1	−0.6	1301.6	2.6
91	163	1309.9	−0.5	1310.0	1.7
92	164	1319.4	−0.7	1320.0	2.5
93	165	1328.7	−0.5	1328.3	1.6
94	166	1337.8	−0.2	1338.1	2.2
95	167	1346.8	0.2	1346.2	1.3
96	168	1355.4	0.5	1355.6	1.8
97	169	1363.9	0.9	1363.5	1.0
98	170	1372.2	1.3	1372.5	1.4
99	171	1380.4	1.4	1380.0	0.9
100	172	1388.2	1.3	1388.6	1.7
101	173	1396.0	1.4	1395.8	1.1
102	174	1403.5	1.4	1404.0	1.6
103	175	1410.9	1.7	1410.7	0.8
104	176	1418.0	1.8	1418.5	1.2
105*	177	1425.0	1.6	1425.0	0.8
106	178	1431.8	1.5	1432.3	1.2
107	179	1438.5	1.2	1438.3	0.8
108	180	1444.9	1.0	1445.3	1.4
109	181	1451.0	0.8	1451.0	0.9
110	182	1457.4	0.8	1457.8	1.3
111	183	1463.4	0.5	1463.1	0.9
112	184	1469.2	0.3	1469.6	1.6
113	185	1474.9	0.	1474.7	1.1
114	186	1480.4	−0.2	1480.9	1.9
115	187	1485.8	−0.2	1485.8	1.3
116	188	1491.0	−	1491.8	1.9
117	189	1496.0	0.1	1496.5	1.1
118	190	1501.0	0.4	1502.5	1.6
119	191	1505.8	0.9	1507.1	0.9
120	192	1510.4	0.9	1512.9	1.5
121	193	1514.9	1.6	1517.5	1.2
122	194	1519.3	1.8	1522.9	1.9
123	195	1523.5	2.3	1527.3	1.6

Z = 73: Tantalum

N	A	value	err	value	err
92	165	1321.7	−0.6	1322.2	1.9
93	166	1331.4	−0.5	1330.7	1.2

Z = 74: Tungsten

N	A	value	err	value	err
94	168	1343.1	−0.4	1343.9	2.3
95	169	1352.6	0.	1352.4	1.3
96	170	1361.8	0.3	1362.3	1.8
97	171	1370.9	0.6	1370.6	1.0
98	172	1379.7	0.9	1380.2	1.4
99	173	1388.4	1.0	1388.1	0.9
100	174	1396.8	0.9	1397.3	1.7
101	175	1405.1	1.1	1405.0	1.0
102	176	1413.2	1.0	1413.7	1.7
103	177	1421.1	1.3	1420.9	0.8
104	178	1428.7	1.4	1429.3	1.3
105	179	1436.3	1.4	1436.3	0.7
106	180	1443.6	1.3	1444.2	1.2
107	181	1450.8	1.2	1450.8	0.7
108	182	1457.7	1.0	1458.4	1.3
109*	183	1464.6	1.0	1464.7	0.7
110	184	1471.2	1.0	1472.0	1.2
111	185	1477.7	0.8	1477.9	0.7
112	186	1484.0	0.7	1484.9	1.4
113	187	1490.2	0.5	1490.6	0.9
114	188	1496.2	0.4	1497.4	1.6
115	189	1502.1	0.6	1502.9	1.0
116	190	1507.8	0.8	1509.5	1.5
117	191	1513.3	1.0	1514.8	0.9
118	192	1518.8	1.3	1521.4	1.6
119	193	1524.0	1.8	1526.6	0.9
120	194	1529.2	2.6	1532.5	1.2
121	195	1534.2	3.3	1538.1	0.7
122	196	1539.0	3.3	1544.0	1.7
123	197	1543.7	4.3	1548.9	1.5
124	198	1548.3	4.7	1554.5	1.3
125	199	1552.7	5.3	1559.4	1.4
126	200	1557.1	5.9	1564.4	1.2
127	201	1561.2	5.3	1567.7	1.2

Z = 75: Rhenium

N	A	value	err	value	err
96	171	1364.3	0.2	1364.6	1.4
97	172	1373.6	0.6	1373.0	0.4
98	173	1382.7	0.8	1382.9	1.0

TABLE 8c-2. CALCULATED BINDING ENERGIES IN MeV (Continued)

Z = 75: Rhenium (Continued)

Number of neutrons N	Mass number A	Liquid drop	Shell correction	BCS pairing energy	Total binding energy
99	174	1391.7	1.0	0.3	1391.0
100	175	1400.4	0.7	1.2	1400.5
101	176	1408.9	0.9	0.5	1408.3
102	177	1417.2	0.8	1.2	1417.4
103	178	1425.4	1.2	0.3	1424.9
104	179	1433.3	1.2	0.8	1433.6
105	180	1441.1	1.5	0.1	1440.8
106	181	1448.7	1.4	0.6	1449.1
107	182	1456.1	1.3	0.1	1455.9
108	183	1463.4	1.1	0.8	1463.8
109	184	1470.5	1.2	0.2	1470.4
110	185	1477.4	1.2	0.7	1478.0
111*	186	1484.1	1.0	0.2	1484.2
112	187	1490.7	0.8	1.1	1491.5
113	188	1497.1	0.7	0.6	1497.4
114	189	1503.4	0.4	1.4	1504.5
115	190	1509.5	0.6	0.7	1510.3
116	191	1515.4	0.9	1.4	1517.3
117	192	1521.2	1.3	0.7	1523.0
118	193	1526.9	2.1	1.0	1529.8
119	194	1532.4	2.9	0.1	1535.3
120	195	1537.8	3.8	0.4	1541.8
121	196	1543.0	4.4	−0.1	1547.2
122	197	1548.1	5.0	0.2	1553.3
123	198	1553.1	5.5	−0.1	1558.4
124	199	1557.9	6.3	−0.1	1564.2
125	200	1562.5	6.8	−0.1	1569.2
126	201	1567.1	7.5	−0.1	1574.5
127	202	1571.5	6.7	−0.1	1578.0
128	203	1575.8	6.0	0.5	1582.2
129	204	1579.9	5.5	0.2	1585.5

Z = 77: Iridium (Continued)

Number of neutrons N	Mass number A	Liquid drop	Shell correction	BCS pairing energy	Total binding energy
101	178	1415.0	0.4	0.9	1414.9
102	179	1423.9	0.3	1.6	1424.5
103	180	1432.5	0.6	0.7	1432.5
104	181	1441.0	0.7	1.3	1441.8
105	182	1449.3	1.2	0.3	1449.4
106	183	1457.4	1.4	0.7	1458.4
107	184	1465.4	1.4	0.2	1465.7
108	185	1473.2	1.3	1.0	1474.2
109	186	1480.8	1.4	0.2	1481.3
110	187	1488.2	1.6	0.3	1489.5
111	188	1495.4	1.4	0.8	1496.2
112	189	1502.5	0.7	1.5	1504.1
113	190	1509.4	0.9	0.9	1510.6
114*	191	1516.2	1.1	1.5	1518.4
115	192	1522.8	1.5	0.8	1524.7
116	193	1529.2	1.9	1.5	1532.4
117	194	1535.5	2.6	0.6	1538.5
118	195	1541.7	3.5	0.8	1545.9
119	196	1547.7	4.2	0.1	1551.9
120	197	1553.5	5.0	0.4	1558.9
121	198	1559.2	5.7	0.	1564.8
122	199	1564.8	6.2	0.4	1571.3
123	200	1570.2	6.9	−0.1	1577.0
124	201	1575.5	7.7	−0.1	1583.2
125	202	1580.6	8.3	−0.1	1588.8
126	203	1585.6	8.9	−0.1	1594.5
127	204	1590.5	8.1	−0.1	1598.5
128	205	1595.2	7.3	0.6	1603.1
129	206	1599.8	6.8	0.3	1606.8
130	207	1604.3	6.3	0.8	1611.3
131	208	1608.6	5.9	0.3	1614.8
132	209	1612.8	5.5	0.7	1619.0

Z = 76: Osmium

A	N	value	δ	value	δ
174	98	1385.2	1.0	1386.0	1.3
175	99	1394.5	1.2	1394.4	0.5
176	100	1403.4	0.9	1404.1	1.5
177	101	1412.2	1.1	1412.3	0.7
178	102	1420.8	1.0	1421.6	1.4
179	103	1429.2	1.3	1429.3	0.5
180	104	1437.4	1.4	1438.3	1.1
181	105	1445.5	1.8	1445.8	0.2
182	106	1453.3	1.8	1454.3	0.7
183	107	1461.0	1.7	1461.4	0.2
184	108	1468.5	1.5	1469.6	1.0
185	109	1475.9	1.6	1476.4	0.3
186	110	1483.0	1.5	1484.3	0.9
187	111	1490.0	1.3	1490.7	0.5
188	112*	1496.8	0.8	1498.4	1.6
189	113*	1503.5	0.7	1504.6	1.2
190	114	1510.0	0.8	1512.0	1.8
191	115	1516.4	1.1	1518.1	1.0
192	116	1522.5	1.5	1525.5	1.5
193	117	1528.6	2.4	1531.4	0.4
194	118	1534.5	3.5	1538.6	0.6
195	119	1540.3	4.0	1544.5	0.2
196	120	1545.9	4.6	1551.3	0.8
197	121	1551.4	5.3	1556.9	0.2
198	122	1556.7	6.0	1563.3	0.6
199	123	1561.9	6.8	1568.7	0.
200	124	1566.9	7.7	1574.8	0.1
201	125	1571.8	8.3	1580.1	0.
202	126	1576.6	9.0	1585.6	0.
203	127	1581.2	8.1	1589.3	0.
204	128	1585.7	7.3	1593.7	0.7
205	129	1590.1	6.6	1597.2	0.4
206	130	1594.4	6.0	1601.4	1.1
207	131	1598.5	5.5	1604.7	0.7

Z = 77: Iridium

A	N	value	δ	value	δ
176	99	1396.7	0.4	1396.6	0.8
177	100	1405.9	0.1	1406.6	1.8

Z = 78: Platinum

A	N	value	δ	value	δ
179	101	1417.3	−0.1	1417.6	1.3
180	102	1426.4	0.	1427.6	2.1
181	103	1435.4	0.2	1435.7	1.1
182	104	1444.1	0.4	1445.3	1.8
183	105	1452.7	0.9	1453.2	0.6
184	106	1461.1	1.2	1462.4	1.1
185	107	1469.3	1.3	1470.0	0.5
186	108	1477.3	1.2	1478.9	1.3
187	109	1485.1	1.5	1486.2	0.4
188	110	1492.8	1.7	1494.7	1.0
189	111	1500.3	1.5	1501.7	0.6
190	112	1507.7	1.4	1510.0	1.6
191	113	1514.8	1.6	1516.7	0.9
192	114	1521.8	1.8	1524.7	1.6
193	115	1528.7	2.2	1531.3	0.8
194	116*	1535.4	2.5	1539.1	1.5
195	117	1541.9	3.1	1545.4	0.7
196	118	1548.3	3.8	1553.0	1.1
197	119	1554.5	4.6	1559.2	0.2
198	120	1560.6	5.4	1566.5	0.5
199	121	1566.6	6.3	1572.6	0.1
200	122	1572.4	7.1	1579.4	0.7
201	123	1578.0	7.8	1585.2	0.1
202	124	1583.5	8.3	1591.8	0.4
203	125	1588.9	9.0	1597.5	0.3
204	126	1594.1	8.2	1603.5	0.4
205	127	1599.2	7.3	1607.7	0.2
206	128	1604.2	6.8	1612.6	1.1
207	129	1609.0	6.1	1616.5	0.7
208	130	1613.7	5.7	1621.2	1.3
209	131	1618.3	5.2	1624.9	0.9
210	132	1622.7	5.2	1629.4	1.4
211	133	1627.0	4.7	1632.8	0.6
212	134	1631.2		1637.1	1.2

Z = 79: Gold

A	N	value	δ	value	δ
182	103	1437.7	−0.7	1437.5	1.2
183	104	1446.7	−0.6	1447.4	1.9

TABLE 8c-2. CALCULATED BINDING ENERGIES IN MeV (Continued)

Z = 79: Gold (Continued)

Number of neutrons N	Mass number A	Liquid drop	Shell correction	BCS pairing energy	Total binding energy
105	184	1455.6	-0.2	0.8	1455.4
106	185	1464.2	0.1	1.4	1465.0
107	186	1472.7	0.3	0.6	1472.8
108	187	1480.9	0.3	1.5	1482.0
109	188	1489.1	0.6	0.5	1489.5
110	189	1497.0	0.8	1.2	1498.4
111	190	1504.7	0.8	0.7	1505.6
112	191	1512.3	0.7	1.7	1514.2
113	192	1519.7	1.0	0.9	1521.3
114	193	1527.0	1.4	1.6	1529.6
115	194	1534.1	1.8	0.9	1536.4
116	195	1541.0	2.3	1.5	1544.5
117	196	1547.8	2.9	0.6	1551.2
118*	197	1554.4	3.9	0.8	1559.0
119	198	1560.9	4.6	0.1	1565.5
120	199	1567.2	5.4	0.4	1573.0
121	200	1573.4	6.0	0.	1579.3
122	201	1579.5	6.6	0.4	1586.4
123	202	1585.3	7.3	-0.1	1592.5
124	203	1591.1	8.0	-0.1	1599.2
125	204	1596.7	8.6	-0.1	1605.2
126	205	1602.1	9.3	0.	1611.4
127	206	1607.5	8.4	-0.1	1615.8
128	207	1612.6	7.6	0.6	1620.9
129	208	1617.7	7.1	0.3	1625.1
130	209	1622.6	6.5	0.8	1630.0
131	210	1627.4	6.2	0.3	1633.9
132	211	1632.1	5.8	0.8	1638.6
133	212	1636.6	5.5	0.2	1642.3
134	213	1641.0	5.2	0.6	1646.8
135	214	1645.2	5.0	-0.1	1650.2
136	215	1649.4	4.9	0.1	1654.4

Z = 81: Thallium (Continued)

Number of neutrons N	Mass number A	Liquid drop	Shell correction	BCS pairing energy	Total binding energy
109	190	1495.4	-0.9	1.2	1495.5
110	191	1503.8	-1.2	2.6	1505.0
111	192	1512.1	-0.6	1.4	1512.7
112	193	1520.2	-0.8	2.8	1522.0
113	194	1528.1	0.	1.5	1529.5
114	195	1535.8	0.4	2.4	1538.6
115	196	1543.4	1.3	1.2	1545.9
116	197	1550.9	2.3	1.5	1554.7
117	198	1558.1	3.3	0.3	1561.8
118	199	1565.3	4.4	0.6	1570.2
119	200	1572.2	4.9	0.1	1577.2
120	201	1579.0	5.5	0.7	1585.2
121	202	1585.7	6.1	0.2	1592.0
122*	203	1592.2	6.8	0.6	1599.6
123	204	1598.5	7.7	-0.1	1606.1
124	205	1604.7	8.5	-0.1	1613.3
125	206	1610.8	9.1	-0.1	1619.8
126	207	1616.7	9.8	-0.1	1626.5
127	208	1622.5	8.9	-0.1	1631.3
128	209	1628.1	8.1	0.7	1636.9
129	210	1633.6	7.5	0.4	1641.5
130	211	1639.0	6.9	1.0	1646.8
131	212	1644.2	6.4	0.6	1651.2
132	213	1649.3	6.0	1.1	1656.4
133	214	1654.3	5.8	0.4	1660.5
134	215	1659.1	5.5	0.8	1665.4
135	216	1663.8	5.5	-0.1	1669.2
136	217	1668.4	5.4	0.1	1673.9
137	218	1672.9	4.1	0.	1676.9
138	219	1677.2	2.7	1.7	1681.6
139	220	1681.4	1.8	1.3	1684.5
140	221	1685.5	0.5	2.9	1688.9

Z = 80: Mercury

N	value	dev	A	value	dev
105	1457.9	1.2	185	1458.0	−0.6
106	1467.7	1.8	186	1466.9	−0.4
107	1475.8	1.0	187	1475.6	−0.1
108	1485.2	1.9	188	1484.1	−0.2
109	1493.0	0.9	189	1492.5	0.2
110	1502.1	1.7	190	1500.7	0.2
111	1509.6	1.1	191	1508.7	0.3
112	1518.5	2.1	192	1516.5	0.2
113	1525.8	1.2	193	1524.2	0.7
114	1534.5	2.0	194	1531.7	1.0
115	1541.6	1.1	195	1539.0	1.6
116	1550.0	1.6	196	1546.2	2.3
117	1556.9	0.6	197	1553.2	3.1
118	1565.1	0.8	198	1560.1	4.2
119	1571.8	0.3	199	1566.8	4.7
120*	1579.6	0.9	200	1573.4	5.3
121	1586.1	0.3	201	1579.8	6.0
122	1593.8	0.8	202	1586.0	6.7
123	1599.8	0.1	203	1592.2	7.5
124	1606.8	0.3	204	1598.1	8.3
125	1613.0	0.1	205	1604.0	8.9
126	1619.4	0.2	206	1609.7	9.6
127	1624.0	0.1	207	1615.2	8.7
128	1629.4	0.9	208	1620.6	7.9
129	1633.8	0.6	209	1625.9	7.3
130	1638.9	1.2	210	1631.0	6.7
131	1643.1	0.8	211	1636.0	6.2
132	1648.0	1.3	212	1640.9	5.8
133	1651.9	0.6	213	1645.7	5.6
134	1656.6	1.0	214	1650.3	5.3
135	1660.2	0.1	215	1654.8	5.3
136	1664.6	0.3	216	1659.1	5.2
137	1667.4	0.2	217	1663.4	3.9
138	1672.0	1.9	218	1667.5	2.5

Z = 81: Thallium

N	value	dev	A	value	dev
107	1477.7	1.4	188	1478.0	−1.4
108	1487.5	2.7	189	1486.8	−1.7

Z = 82: Lead

N	value	dev	A	value	dev
109	1498.1	2.9	191	1497.9	−2.5
110	1508.1	4.5	192	1506.6	−3.0
111	1516.0	2.6	193	1515.1	−1.7
112	1525.6	3.6	194	1523.4	−1.3
113	1533.3	1.9	195	1531.6	−0.2
114	1542.7	2.4	196	1539.6	0.8
115	1550.3	1.2	197	1547.4	1.6
116	1559.3	1.6	198	1555.1	2.6
117	1566.6	0.4	199	1562.6	3.6
118	1575.3	0.6	200	1569.9	4.8
119	1582.6	0.2	201	1577.1	5.2
120	1590.8	0.8	202	1584.2	5.7
121	1597.8	0.3	203	1591.1	6.4
122	1605.6	0.7	204	1597.8	7.1
123	1612.4	0.1	205	1604.4	8.0
124*	1619.8	0.1	206	1610.8	8.9
125	1626.5	0.	207	1617.1	9.5
126	1633.4	0.	208	1623.3	10.2
127	1638.5	0.	209	1629.3	9.3
128	1644.3	0.7	210	1635.1	8.4
129	1649.1	0.5	211	1640.8	7.8
130	1654.7	1.1	212	1646.4	7.2
131	1659.3	0.7	213	1651.9	6.7
132	1664.7	1.2	214	1657.2	6.3
133	1669.0	0.6	215	1662.4	6.0
134	1674.2	0.9	216	1667.4	5.8
135	1678.2	0.1	217	1672.4	5.8
136	1683.1	0.1	218	1677.2	5.8
137	1686.3	0.	219	1681.8	4.4
138	1691.3	1.8	220	1686.4	3.1
139	1694.3	1.4	221	1690.8	2.1
140	1698.9	3.0	222	1695.1	0.8
141	1701.8	2.4	223	1699.3	0.1
142	1706.2	3.8	224	1703.4	−1.0

Z = 83: Bismuth

N	value	dev	A	value	dev
111	1517.3	1.3	194	1517.7	−1.3
112	1527.0	2.4	195	1526.2	−1.4

TABLE 8c-2. CALCULATED BINDING ENERGIES IN MeV (Continued)

Z = 83: Bismuth (Continued)

Number of neutrons N	Mass number A	Liquid drop	Shell correction	BCS pairing energy	Total binding energy
113	196	1534.6	-0.6	1.2	1535.0
114	197	1542.9	-0.3	2.1	1544.6
115	198	1550.9	0.6	1.0	1552.4
116	199	1558.8	1.4	1.4	1561.5
117	200	1566.6	2.2	0.4	1569.1
118	201	1574.2	3.3	0.6	1578.0
119	202	1581.6	3.8	0.1	1585.5
120	203	1588.9	4.5	0.6	1593.9
121	204	1596.0	5.1	0.1	1601.2
122	205	1603.0	5.8	0.5	1609.2
123	206	1609.8	6.6	-0.1	1616.3
124	207	1616.4	7.4	-0.1	1623.9
125	208	1623.0	8.0	-0.1	1630.8
126*	209	1629.3	8.7	-0.	1637.9
127	210	1635.6	7.8	-0.1	1643.3
128	211	1641.7	7.0	0.6	1649.3
129	212	1647.6	6.5	0.3	1654.4
130	213	1653.4	5.9	0.8	1660.1
131	214	1659.1	5.5	0.4	1665.0
132	215	1664.6	5.1	0.9	1670.6
133	216	1670.0	4.8	0.3	1675.1
134	217	1675.3	4.5	0.7	1680.5
135	218	1680.4	4.4	-0.1	1684.8
136	219	1685.5	4.3	0.1	1689.8
137	220	1690.3	3.0	0.	1693.3
138	221	1695.1	1.8	1.6	1698.5
139	222	1699.7	1.3	1.3	1701.8
140	223	1704.3	-0.3	2.6	1706.6
141	224	1708.6	-0.9	2.0	1709.7
142	225	1712.9	-1.7	3.1	1714.3
143	226	1717.1	-2.0	2.3	1717.4
144	227	1721.1	-2.5	3.3	1721.8

Z = 85: Astatine (Continued)

Number of neutrons N	Mass number A	Liquid drop	Shell correction	BCS pairing energy	Total binding energy
116	201	1564.9	-0.7	1.9	1566.0
117	202	1573.2	0.1	1.0	1574.1
118	203	1581.2	0.9	1.3	1583.4
119	204	1589.1	1.5	0.8	1591.4
120	205	1596.8	2.1	1.3	1600.3
121	206	1604.4	2.8	0.8	1608.0
122	207	1611.9	3.4	1.3	1616.5
123	208	1619.1	4.2	0.7	1624.0
124	209	1626.3	4.9	1.0	1632.1
125	210	1633.2	5.4	0.8	1639.5
126	211	1640.1	6.1	0.9	1647.1
127	212	1646.8	5.3	0.8	1652.8
128	213	1653.3	4.5	1.5	1659.3
129*	214	1659.7	4.0	1.2	1664.8
130	215	1665.9	3.4	1.2	1671.1
131	216	1672.1	3.0	1.3	1676.3
132	217	1678.0	2.5	1.8	1682.4
133	218	1683.9	2.4	1.2	1687.4
134	219	1689.6	2.0	1.6	1693.2
135	220	1695.2	2.0	0.7	1697.9
136	221	1700.6	1.8	1.0	1703.4
137	222	1705.9	0.6	0.8	1707.3
138	223	1711.1	-0.8	2.6	1712.9
139	224	1716.2	-1.5	2.0	1716.6
140	225	1721.1	-2.7	3.4	1721.9
141	226	1725.9	-3.0	2.6	1725.5
142	227	1730.6	-3.8	3.7	1730.5
143	228	1735.2	-3.6	2.5	1734.0
144	229	1739.7	-3.6	3.0	1738.8
145	230	1744.1	-3.6	0.9	1742.4
146	231	1748.3	-1.9	1.3	1747.2
147	232	1752.5	-1.4	1.0	1750.8

Z = 84: Polonium

N	A	mass	Δ	mass	Δ
112	196	1528.5	−1.9	1529.1	2.7
113	197	1537.2	−1.3	1537.4	1.7
114	198	1545.6	−1.5	1547.1	3.1
115	199	1554.0	−0.5	1555.1	1.8
116	200	1562.1	−0.1	1564.6	2.6
117	201	1570.1	1.0	1572.4	1.4
118	202	1577.9	1.9	1581.6	1.8
119	203	1585.6	2.5	1589.3	1.2
120	204	1593.1	2.9	1598.0	1.9
121	205	1600.4	3.6	1605.5	1.4
122	206	1607.6	4.3	1613.7	1.8
123	207	1614.7	5.3	1621.0	1.0
124	208	1621.6	6.0	1628.9	1.3
125	209	1628.3	6.6	1636.1	1.1
126	210	1634.9	7.3	1643.4	1.2
127*	211	1641.4	6.5	1648.9	1.1
128	212	1647.7	5.6	1655.2	1.9
129	213	1653.9	5.0	1660.5	1.6
130	214	1659.9	4.3	1666.5	2.3
131	215	1665.8	3.9	1671.5	1.9
132	216	1671.6	3.4	1677.4	2.4
133	217	1677.2	3.3	1682.1	1.7
134	218	1682.7	3.0	1687.7	2.1
135	219	1688.0	3.1	1692.2	1.3
136	220	1693.3	2.9	1697.5	1.1
137	221	1698.4	1.7	1701.2	1.1
138	222	1703.3	0.2	1706.6	3.0
139	223	1708.2	−0.6	1710.0	2.4
140	224	1712.9	−2.0	1715.1	4.1
141	225	1717.5	−2.4	1718.4	3.3
142	226	1722.0	−3.5	1723.2	4.8
143	227	1726.4	−3.5	1726.4	3.6
144	228	1730.6	−4.3	1731.1	4.8
145	229	1734.7	−3.2	1734.2	2.8

Z = 85: Astatine

N	A	mass	Δ	mass	Δ
114	199	1548.0	−1.5	1548.2	2.0
115	200	1556.6	−1.1	1556.5	1.3

Z = 86: Radon

N	A	mass	Δ	mass	Δ
117	203	1575.8	−0.7	1576.6	1.7
118	204	1584.0	0.	1586.2	2.2
119	205	1592.2	0.5	1594.4	1.7
120	206	1600.1	0.9	1603.5	2.4
121	207	1608.0	1.7	1611.4	1.8
122	208	1615.6	2.1	1620.2	2.4
123	209	1623.1	3.1	1627.9	1.6
124	210	1630.5	3.7	1636.3	2.1
125	211	1637.7	4.3	1643.9	1.9
126	212	1644.7	4.9	1651.7	2.0
127	213	1651.6	4.2	1657.6	1.8
128	214	1658.4	3.3	1664.4	2.7
129	215	1665.0	2.8	1670.1	2.3
130	216	1671.5	2.1	1676.6	3.0
131*	217	1677.8	1.7	1682.0	2.6
132	218	1684.0	1.1	1688.3	3.1
133	219	1690.1	0.7	1693.5	2.3
134	220	1696.0	1.0	1699.6	2.8
135	221	1701.8	0.6	1704.5	1.7
136	222	1707.5	−0.4	1710.2	2.1
137	223	1713.0	−2.1	1714.3	1.7
138	224	1718.4	−2.6	1720.1	3.8
139	225	1723.7	−4.1	1724.1	3.0
140	226	1728.8	−4.0	1729.5	4.8
141	227	1733.9	−5.0	1733.3	3.5
142	228	1738.7	−4.0	1738.5	4.8
143	229	1743.6	−2.2	1742.3	2.9
144	230	1748.3	−1.9	1747.4	2.1
145	231	1752.9	−1.7	1751.4	1.5
146	232	1757.4	−1.5	1756.5	2.0
147	233	1761.7	−1.2	1760.5	1.8
148	234	1765.9		1765.3	2.2

Z = 87: Francium

N	A	mass	Δ	mass	Δ
119	206	1594.8	−0.2	1595.6	1.1
120	207	1603.0	0.5	1605.0	1.5
121	208	1611.0	1.2	1613.2	1.0
122	209	1618.9	1.4	1622.1	1.8

TABLE 8c-2. CALCULATED BINDING ENERGIES IN MeV (Continued)

Z = 87: Francium (Continued)

Number of neutrons N	Mass number A	Liquid drop	Shell correction	BCS pairing energy	Total binding energy
123	210	1626.7	2.3	1.1	1630.1
124	211	1634.2	2.8	1.6	1638.6
125	212	1641.7	3.3	1.5	1646.5
126	213	1648.9	4.0	1.6	1654.5
127	214	1656.1	3.3	1.3	1660.7
128	215	1663.1	2.4	2.2	1667.6
129	216	1669.9	1.9	1.8	1673.6
130	217	1676.6	1.2	2.5	1680.3
131	218	1683.2	0.9	2.0	1686.0
132	219	1689.6	0.4	2.5	1692.5
133*	220	1695.9	0.3	1.7	1697.9
134	221	1702.0	−0.1	2.3	1704.2
135	222	1708.0	0.1	1.2	1709.3
136	223	1713.9	−0.3	1.7	1715.3
137	224	1719.6	−1.3	1.3	1719.6
138	225	1725.2	−2.9	3.3	1725.6
139	226	1730.7	−3.3	2.4	1729.8
140	227	1736.1	−4.7	4.1	1735.4
141	228	1741.3	−4.4	2.7	1739.5
142	229	1746.4	−4.7	3.3	1744.9
143	230	1751.6	−2.4	1.0	1749.2
144	231	1756.5	−2.1	1.4	1754.7
145	232	1761.3	−2.0	1.1	1759.0
146	233	1765.9	−1.8	1.6	1764.3
147	234	1770.5	−1.6	1.3	1768.5
148	235	1774.9	−1.2	1.6	1773.5
149	236	1779.2	−0.4	0.8	1777.6
150	237	1783.3	−0.2	1.0	1782.3

Z = 88: Radium

Number of neutrons N	Mass number A	Liquid drop	Shell correction	BCS pairing energy	Total binding energy
121	209	1613.6	0.5	1.7	1615.9
122	210	1621.8	0.6	2.7	1625.0

Z = 89: Actinium (Continued)

Number of neutrons N	Mass number A	Liquid drop	Shell correction	BCS pairing energy	Total binding energy
129	218	1678.3	0.4	2.0	1680.8
130	219	1685.4	−0.3	2.8	1687.9
131	220	1692.4	−0.7	2.3	1694.1
132	221	1699.3	−1.2	2.9	1701.0
133	222	1706.0	−1.2	2.0	1706.9
134	223	1712.6	−1.7	2.7	1713.5
135	224	1719.0	−1.4	1.5	1719.1
136	225	1725.3	−1.8	1.5	1725.5
137*	226	1731.5	−2.7	1.5	1730.3
138	227	1737.5	−4.5	3.6	1736.7
139	228	1743.4	−4.7	2.6	1741.3
140	229	1749.2	−6.2	4.4	1747.4
141	230	1755.0	−3.2	1.3	1752.3
142	231	1760.6	−2.5	1.5	1758.4
143	232	1766.0	−2.2	0.9	1763.4
144	233	1771.3	−2.1	1.6	1769.4
145	234	1776.6	−1.8	1.2	1774.2
146	235	1781.6	−1.5	1.6	1779.9
147	236	1786.6	−0.9	0.9	1784.6
148	237	1791.3	−0.6	1.2	1790.1
149	238	1796.1	0.	0.7	1794.6
150	239	1800.6	0.2	0.9	1799.7
151	240	1805.0	0.3	0.7	1804.0
152	241	1809.3	0.3	0.8	1808.6
153	242	1813.6	0.1	0.7	1812.5
154	243	1817.6	−0.1	1.0	1816.9

Z = 90: Thorium

Number of neutrons N	Mass number A	Liquid drop	Shell correction	BCS pairing energy	Total binding energy
125	215	1650.9	1.3	2.5	1654.6
126	216	1658.8	1.9	2.6	1663.3

(continued — upper band)

N	A	Mass (I)	δ	δ	Mass (II)
127	217	1666.6	1.2	2.4	1670.2
128	218	1674.3	0.3	3.2	1677.8
129	219	1681.8	−0.1	2.7	1684.5
130	220	1689.2	−0.8	3.4	1691.8
131	221	1696.4	−1.0	2.8	1698.2
132	222	1703.5	−1.6	3.5	1705.3
133	223	1710.4	−1.5	2.6	1711.4
134	224	1717.2	−2.2	3.4	1718.3
135	225	1723.8	−2.0	2.3	1724.1
136	226	1730.3	−2.3	2.7	1730.7
137	227	1736.7	−3.3	2.3	1735.7
138	228	1743.0	−5.0	4.4	1742.3
139*	229	1749.1	−5.1	3.3	1747.2
140	230	1755.2	−3.6	2.8	1753.6
141	231	1761.2	−2.7	1.6	1758.9
142	232	1766.9	−2.0	1.8	1765.3
143	233	1772.6	−1.9	1.4	1770.6
144	234	1778.1	−1.7	2.0	1776.8
145	235	1783.5	−1.4	1.5	1781.8
146	236	1788.8	−1.0	1.9	1787.8
147	237	1793.9	−0.5	1.3	1792.7
148	238	1798.9	−0.2	1.6	1798.3
149	239	1803.8	0.3	1.1	1803.1
150	240	1808.5	0.6	1.2	1808.3
151	241	1813.2	0.7	1.1	1812.9
152	242	1817.7	0.5	1.2	1817.7
153	243	1822.1	0.3	1.1	1821.8
154	244	1826.4	−0.1	1.4	1826.3
155	245	1830.6	−0.5	1.3	1829.9
156	246	1834.6	−0.5	1.9	1834.4

Z = 91: Protactinium

N	A	Mass (I)	δ	δ	Mass (II)
128	219	1677.1	−0.2	2.6	1679.5
129	220	1684.9	−0.7	2.1	1686.4
130	221	1692.5	−1.4	2.9	1693.9
131	222	1699.9	−2.0	2.6	1700.5
132	223	1707.2	−2.3	3.0	1707.9
133	224	1714.3	−2.0	1.9	1714.2

(continued — lower band)

N	A	Mass (I)	δ	δ	Mass (II)	N
123	211	1629.7	1.5	2.0	1633.2	127
124	212	1637.5	2.0	2.5	1642.0	128
125	213	1645.2	2.5	2.4	1650.0	129
126	214	1652.7	3.1	2.5	1658.3	130
127	215	1660.1	2.5	2.2	1664.7	131
128	216	1667.3	1.5	3.1	1671.9	132
129	217	1674.3	1.1	2.6	1678.1	133
130	218	1681.2	0.4	3.4	1685.0	134
131	219	1688.0	0.0	2.9	1690.9	135
132	220	1694.7	−0.5	3.5	1697.6	136
133	221	1701.2	−0.5	2.6	1703.3	137
134	222	1707.5	−1.0	3.3	1709.7	138
135*	223	1713.7	−0.7	2.1	1715.1	139*
136	224	1719.8	−1.1	2.6	1721.3	140
137	225	1725.8	−2.0	2.1	1725.8	141
138	226	1731.6	−3.8	4.2	1732.0	142
139	227	1737.3	−4.0	3.1	1736.4	143
140	228	1742.9	−5.6	5.0	1742.3	144
141	229	1748.4	−4.5	3.0	1746.6	145
142	230	1753.8	−3.1	2.5	1752.4	146
143	231	1759.0	−2.3	1.5	1757.1	147
144	232	1764.1	−2.1	2.1	1762.8	148
145	233	1769.1	−2.0	1.8	1767.4	149
146	234	1774.0	−1.1	2.3	1772.9	150
147	235	1778.7	−1.4	1.8	1777.0	151
148	236	1783.3	−1.0	2.1	1782.6	152
149	237	1787.9	−0.3	1.4	1786.9	153
150	238	1792.2	−0.1	1.6	1791.8	154
151	239	1796.4	0.	1.4	1796.0	155
152	240	1800.5	0.1	1.6	1800.4	156

Z = 89: Actinium

N	A	Mass (I)	δ	δ	Mass (II)	N
123	212	1632.3	0.8	1.4	1634.5	128
124	213	1640.4	1.3	1.9	1643.5	129
125	214	1648.3	1.8	1.8	1651.8	130
126	215	1656.0	2.4	1.9	1660.3	131
127	216	1663.6	1.8	1.6	1667.0	132
128	217	1671.0	0.8	2.5	1674.4	133

Table 8c-2. Calculated Binding Energies in MeV (Continued)

Z = 91: Protactinium (Continued)

Number of neutrons N	Mass number A	Liquid drop	Shell correction	BCS pairing energy	Total binding energy
134	225	1721.3	-2.7	2.7	1721.3
135	226	1728.2	-2.6	1.7	1727.3
136	227	1734.9	-2.7	1.9	1734.2
137	228	1741.5	-4.0	1.8	1739.4
138	229	1748.0	-5.4	3.6	1746.2
139	230	1754.5	-3.5	1.5	1751.6
140	231	1760.7	-2.9	1.7	1758.4
141*	232	1766.9	-2.3	0.8	1764.0
142	233	1772.8	-1.7	1.0	1770.7
143	234	1778.7	-1.4	0.7	1776.2
144	235	1784.4	-1.2	1.2	1782.6
145	236	1790.0	-0.9	0.8	1787.9
146	237	1795.5	-0.6	1.1	1794.0
147	238	1800.8	-0.2	0.6	1799.2
148	239	1806.0	0.2	0.9	1805.0
149	240	1811.1	0.6	0.5	1810.0
150	241	1816.0	0.8	0.6	1815.4
151	242	1820.8	0.9	0.5	1820.2
152	243	1825.6	1.0	0.6	1825.2
153	244	1830.2	0.8	0.5	1829.4
154	245	1834.6	0.6	0.8	1834.1
155	246	1839.0	0.3	0.6	1838.0
156	247	1843.3	-0.1	1.2	1842.6
157	248	1847.4	-0.4	0.8	1846.3

Z = 92: Uranium

Number of neutrons N	Mass number A	Liquid drop	Shell correction	BCS pairing energy	Total binding energy
130	222	1695.3	-1.6	3.4	1697.1
131	223	1703.0	-1.6	2.7	1704.0
132	224	1710.5	-2.3	3.3	1711.5

Z = 93: Neptunium (Continued)

Number of neutrons N	Mass number A	Liquid drop	Shell correction	BCS pairing energy	Total binding energy
141	234	1776.9	-1.4	0.5	1774.4
142	235	1783.3	-0.8	0.8	1781.5
143	236	1789.5	-0.7	0.5	1787.5
144	237	1795.6	-0.4	1.0	1794.3
145*	238	1801.6	-0.2	0.6	1800.1
146	239	1807.5	0.2	0.9	1806.6
147	240	1813.2	0.6	0.4	1812.1
148	241	1818.8	1.0	0.6	1818.4
149	242	1824.3	1.3	0.3	1823.7
150	243	1829.6	1.6	0.4	1829.6
151	244	1834.9	1.6	0.3	1834.7
152	245	1840.0	1.7	0.4	1840.1
153	246	1845.0	1.5	0.3	1844.7
154	247	1849.8	1.3	0.6	1849.8
155	248	1854.6	1.0	0.4	1854.1
156	249	1859.2	0.7	0.9	1859.0
157	250	1863.8	0.4	0.6	1863.1
158	251	1868.2	0.2	1.1	1867.8
159	252	1872.5	0.1	0.7	1871.7
160	253	1876.7	0.1	1.0	1876.2
161	254	1880.8	0.1	0.4	1879.9

Z = 94: Plutonium

Number of neutrons N	Mass number A	Liquid drop	Shell correction	BCS pairing energy	Total binding energy
134	228	1731.2	-3.8	3.5	1730.9
135	229	1738.7	-4.0	2.9	1737.6
136	230	1746.0	-4.3	3.3	1745.0
137	231	1753.4	-3.0	1.7	1751.4
138	232	1760.6	-2.3	1.9	1759.0
139	233	1767.7	-2.0	1.4	1765.6

N	A	M	Δ	±	M
133	225	1717.8	−2.4	2.7	1718.1
134	226	1725.0	−2.9	3.3	1725.4
135	227	1732.1	−3.1	2.6	1731.6
136	228	1739.1	−3.0	2.6	1738.6
137	229	1745.9	−4.7	3.0	1744.1
138	230	1752.7	−3.9	3.0	1751.1
139	231	1759.3	−2.9	1.6	1757.0
140	232	1765.3	−2.3	1.9	1764.1
141	233	1772.1	−1.8	1.1	1769.9
142	234	1778.3	−1.1	1.3	1776.9
143*	235	1784.3	−0.8	1.0	1782.6
144	236	1790.2	−0.8	1.6	1789.2
145	237	1796.0	−0.5	1.2	1794.7
146	238	1801.7	−0.2	1.5	1801.0
147	239	1807.2	0.2	1.0	1806.4
148	240	1812.6	0.6	1.3	1812.4
149	241	1817.9	0.9	0.9	1817.6
150	242	1823.0	1.2	1.0	1823.2
151	243	1828.1	1.2	0.9	1828.2
152	244	1833.0	1.3	1.0	1833.4
153	245	1837.8	1.1	0.9	1837.8
154	246	1842.5	1.0	1.2	1842.7
155	247	1847.0	0.7	0.9	1846.8
156	248	1851.5	0.3	1.5	1851.6
157	249	1855.5	0.1	1.2	1855.4
158	250	1860.0	−0.1	1.6	1860.0
159	251	1864.2	−0.2	1.2	1863.7

Z = 93: Neptunium

N	A	M	Δ	±	M
132	225	1713.3	−2.2	2.3	1713.3
133	226	1720.9	−2.6	1.9	1720.1
134	227	1728.3	−3.1	2.4	1727.5
135	228	1735.6	−3.6	2.0	1734.0
136	229	1742.8	−3.8	2.2	1741.2
137	230	1749.9	−3.6	1.6	1747.3
138	231	1756.9	−2.7	1.6	1754.7
139	232	1763.7	−2.4	0.9	1760.9
140	233	1770.4	−1.9	1.3	1768.3

N	A	M	Δ	±	M
140	234	1774.5	−1.4	1.7	1773.2
141	235	1781.3	−0.9	0.9	1779.5
142	236	1787.8	−0.3	1.2	1786.9
143	237	1794.3	−0.2	0.8	1793.0
144	238	1800.6	0.	1.3	1800.0
145	239	1806.8	0.3	0.9	1806.0
146	240	1812.8	0.7	1.2	1812.7
147*	241	1818.7	1.0	0.7	1818.5
148	242	1824.5	1.4	1.0	1824.9
149	243	1830.2	1.7	0.6	1830.5
150	244	1835.8	2.0	0.7	1836.5
151	245	1841.2	2.2	0.6	1841.8
152	246	1846.5	1.9	0.7	1847.4
153	247	1851.7	1.8	0.6	1852.3
154	248	1856.8	1.5	0.9	1857.5
155	249	1861.7	1.2	0.7	1862.0
156	250	1866.6	0.8	0.9	1867.1
157	251	1871.3	0.6	1.4	1871.3
158	252	1875.9	0.5	1.0	1876.3
159	253	1880.4	0.4	1.4	1880.3
160	254	1884.8	0.5	0.8	1885.1
161	255	1889.1	0.6	1.0	1888.9
162	256	1893.3	0.3	0.8	1893.4
163	257	1897.7			1897.1

Z = 95: Americium

N	A	M	Δ	±	M
135	230	1741.3	−4.2	2.2	1739.2
136	231	1749.0	−2.7	1.6	1747.1
137	232	1756.6	−2.2	0.8	1754.0
138	233	1764.0	−1.8	1.5	1762.1
139	234	1771.2	−1.4	0.7	1768.8
140	235	1778.3	−0.8	1.1	1776.7
141	236	1785.2	−0.4	0.3	1783.2
142	237	1791.9	0.1	0.6	1790.8
143	238	1798.6	0.3	0.2	1797.2
144	239	1805.1	0.5	0.8	1804.4
145	240	1811.5	0.8	0.3	1810.6
146	241	1817.7	1.2	0.6	1817.5

TABLE 8c-2. CALCULATED BINDING ENERGIES IN MeV (Continued)

Z = 95: Americium (Continued)

Number of neutrons N	Mass number A	Liquid drop	Shell correction	BCS pairing energy	Total binding energy
147	242	1823.8	1.5	0.1	1823.4
148*	243	1829.8	1.9	0.4	1830.1
149	244	1835.7	2.2	0.	1835.8
150	245	1841.5	2.5	0.2	1842.1
151	246	1847.1	2.5	0.	1847.6
152	247	1852.6	2.6	0.1	1853.4
153	248	1858.0	2.3	0.	1858.4
154	249	1863.2	2.2	0.3	1863.9
155	250	1868.4	1.9	0.1	1868.5
156	251	1873.4	1.7	0.6	1873.8
157	252	1878.4	1.3	0.3	1878.2
158	253	1883.2	1.1	0.8	1883.3
159	254	1887.9	0.9	0.4	1887.6
160	255	1892.5	0.9	0.8	1892.5
161	256	1896.9	0.9	0.2	1896.5
162	257	1901.3	1.0	0.4	1901.2
163	258	1905.6	0.7	0.1	1905.0
164	259	1909.8	0.5	0.6	1909.4

Z = 96: Curium

Number of neutrons N	Mass number A	Liquid drop	Shell correction	BCS pairing energy	Total binding energy
138	234	1766.9	-1.2	1.8	1765.7
139	235	1774.3	-0.7	0.9	1772.7
140	236	1781.5	-0.2	1.3	1780.7
141	237	1788.6	0.2	0.5	1787.5
142	238	1795.6	0.7	0.8	1795.3
143	239	1802.4	0.8	0.5	1801.8
144	240	1809.1	1.1	1.0	1809.3
145	241	1815.7	1.3	0.5	1815.6
146	242	1822.2	1.7	0.8	1822.8

Z = 97: Berkelium (Continued)

Number of neutrons N	Mass number A	Liquid drop	Shell correction	BCS pairing energy	Total binding energy
157	254	1891.2	2.2	0.2	1891.8
158	255	1896.4	1.9	0.8	1897.3
159	256	1901.5	1.7	0.4	1901.8
160	257	1906.4	1.6	0.8	1907.1
161	258	1911.3	1.5	0.3	1911.5
162	259	1916.0	1.7	0.4	1916.6
163	260	1920.7	1.5	0.1	1920.7
164	261	1925.2	1.4	0.4	1925.5
165	262	1929.7	1.2	0.1	1929.4
166	263	1934.0	1.1	0.4	1934.0
167	264	1938.2	0.8	0.1	1937.6

Z = 98: Californium

Number of neutrons N	Mass number A	Liquid drop	Shell correction	BCS pairing energy	Total binding energy
142	240	1801.6	1.3	0.9	1802.0
143	241	1808.9	1.4	0.6	1809.0
144	242	1816.0	1.6	1.1	1816.8
145	243	1823.0	1.9	0.6	1823.6
146	244	1829.8	2.3	0.9	1831.1
147	245	1836.6	2.7	0.4	1837.6
148	246	1843.2	3.1	0.6	1844.9
149	247	1849.6	3.3	0.3	1851.2
150	248	1856.0	3.6	0.4	1858.0
151	249	1862.2	3.6	0.3	1864.1
152	250	1868.3	3.7	0.4	1870.5
153	251	1874.2	3.5	0.3	1876.1
154*	252	1880.1	3.3	0.6	1882.2
155	253	1885.8	3.0	0.4	1887.4
156	254	1891.4	2.8	0.8	1893.3
157	255	1896.9	2.5	0.6	1898.2

N	A	M₁	δ₁	M₂	δ₂
158	256	1903.9	1.1	1902.3	2.3
159	257	1908.6	0.7	1907.6	2.0
160	258	1914.1	1.1	1912.8	1.9
161	259	1918.6	0.6	1917.8	1.9
162	260	1923.9	0.9	1922.7	1.8
163	261	1928.2	0.4	1927.6	1.9
164	262	1933.3	0.7	1932.3	1.6
165	263	1937.3	0.4	1936.9	1.4
166	264	1942.1	0.7	1941.4	1.1
167	265	1945.9	0.5	1945.8	0.9
168	266	1950.4	0.9	1950.1	0.5
169	267	1954.0	0.6	1954.3	

Z = 99 : Einsteinium

N	A	M₁	δ₁	M₂	δ₂
144	243	1819.6	0.7	1818.8	2.0
145	244	1826.6	0.2	1826.0	2.3
146	245	1834.3	0.5	1833.0	2.8
147	246	1841.1		1840.0	3.1
148	247	1848.5	0.2	1846.7	3.5
149	248	1855.1	−0.1	1853.4	3.7
150	249	1862.1	−0.1	1859.9	4.0
151	250	1868.4		1866.3	4.0
152	251	1874.9	−0.1	1872.6	4.1
153	252	1880.7		1878.8	3.9
154	253	1887.0	−0.1	1884.9	3.7
155	254	1892.4	−0.3	1890.8	3.4
156*	255	1898.4		1896.6	3.2
157	256	1903.6	0.5	1902.3	2.9
158	257	1909.5	0.2	1907.9	2.7
159	258	1914.4	0.7	1913.3	2.5
160	259	1920.0	0.3	1918.7	2.4
161	260	1924.7	0.8	1923.9	2.3
162	261	1930.2	0.3	1929.0	2.3
163	262	1934.7	0.6	1934.1	2.3
164	263	1939.9		1939.0	2.4
165	264	1944.2	0.2	1943.8	2.1
166	265	1949.1	0.4	1948.5	1.8
167	266	1953.1	0.1	1953.1	1.5

N	A	M₁	δ₁	M₂	δ₂
147	243	1828.9	0.3	1828.5	2.1
148	244	1835.7	0.6	1834.7	2.5
149	245	1841.7	0.2	1840.8	2.7
150*	246	1848.1	0.4	1846.7	3.0
151	247	1853.8	0.2	1852.5	3.0
152	248	1859.8	0.3	1858.2	3.1
153	249	1865.0	0.2	1863.8	2.9
154	250	1870.7	0.5	1869.3	2.8
155	251	1875.6	0.3	1874.6	2.5
156	252	1881.0	0.8	1879.9	2.2
157	253	1885.6	0.5	1885.0	1.9
158	254	1890.9	1.0	1890.0	1.7
159	255	1895.3	0.7	1894.9	1.4
160	256	1900.4	1.0	1899.6	1.4
161	257	1904.6	0.4	1904.3	1.4
162	258	1909.5	0.6	1908.9	1.6
163	259	1913.5	0.4	1913.3	1.2
164	260	1918.1	0.8	1917.7	1.1
165	261	1921.8	0.4	1921.9	0.9

Z = 97 : Berkelium

N	A	M₁	δ₁	M₂	δ₂
140	237	1783.5	1.0	1784.4	0.1
141	238	1790.5	0.3	1791.7	0.5
142	239	1798.5	0.6	1798.8	0.9
143	240	1805.2	0.2	1805.9	1.1
144	241	1812.9	0.7	1812.8	1.3
145	242	1819.4	0.2	1819.6	1.6
146	243	1826.8	0.6	1826.2	2.0
147	244	1833.1	0.1	1832.8	2.3
148	245	1840.2	0.3	1839.1	2.7
149	246	1846.3	0.	1845.4	3.0
150	247	1853.0	0.1	1851.6	3.2
151	248	1858.9	0.	1857.6	3.3
152*	249	1865.0	0.1	1863.5	3.4
153	250	1870.4	0.	1869.2	3.1
154	251	1876.3	0.3	1874.9	3.0
155	252	1881.3	0.1	1880.5	2.7
156	253	1887.0	0.5	1885.9	2.5

TABLE 8c-2. CALCULATED BINDING ENERGIES IN MeV (Continued)

Z = 99: Einsteinium (Continued)

Number of neutrons N	Mass number A	Liquid drop	Shell correction	BCS pairing energy	Total binding energy
168	267	1957.6	1.4	0.5	1957.8
169	268	1961.9	0.9	0.2	1961.5
170	269	1966.3	0.7	0.7	1966.1
171	270	1971.0	1.7	0.	1969.6

Z = 100: Fermium

Number of neutrons N	Mass number A	Liquid drop	Shell correction	BCS pairing energy	Total binding energy
146	246	1835.8	3.3	0.5	1837.7
147	247	1842.9	3.6	0.1	1844.7
148	248	1849.9	4.0	0.3	1852.3
149	249	1856.8	4.2	0.1	1859.0
150	250	1863.5	4.5	0.1	1866.2
151	251	1870.1	4.5	0.	1872.7
152	252	1876.6	4.6	0.1	1879.4
153	253	1882.9	4.4	0.	1885.4
154	254	1889.2	4.2	0.3	1891.9
155	255	1895.3	3.9	0.1	1897.5
156	256	1901.3	3.7	0.5	1903.7
157	257	1907.2	3.4	0.2	1909.1
158*	258	1913.0	3.2	0.7	1915.1
159	259	1918.6	3.0	0.4	1920.2
160	260	1924.2	2.8	0.8	1926.1
161	261	1929.6	2.7	0.4	1931.0
162	262	1934.9	2.7	0.6	1936.6
163	263	1940.1	2.7	0.1	1941.3
164	264	1945.2	2.8	0.3	1946.7
165	265	1950.2	2.5	0.1	1951.1
166	266	1955.1	2.3	0.5	1956.2
167	267	1959.8	2.0	0.2	1960.4
168	268	1964.5	1.8	0.6	1965.3

Z = 102: Nobelium (Continued)

Number of neutrons N	Mass number A	Liquid drop	Shell correction	BCS pairing energy	Total binding energy
153	255	1889.9	4.4	0.1	1892.7
154	256	1896.5	4.2	0.4	1899.5
155	257	1903.0	3.9	0.2	1905.5
156	258	1909.4	3.7	0.6	1912.2
157	259	1915.7	3.4	0.4	1917.9
158	260	1921.8	3.2	0.9	1924.3
159	261	1927.8	3.0	0.5	1929.8
160	262	1933.8	2.9	0.9	1936.1
161	263	1939.6	3.0	0.3	1941.4
162*	264	1945.2	3.1	0.5	1947.4
163	265	1950.8	2.8	0.2	1952.4
164	266	1956.3	2.7	0.6	1958.2
165	267	1961.6	2.5	0.2	1963.0
166	268	1966.9	2.5	0.5	1968.5
167	269	1972.0	2.1	0.2	1973.0
168	270	1977.0	1.9	0.6	1978.3
169	271	1982.0	1.5	0.3	1982.6
170	272	1986.8	1.3	0.8	1987.6
171	273	1991.5	1.1	0.4	1991.7
172	274	1996.1	0.9	0.7	1996.6
173	275	2000.4	-0.1	0.7	2000.6
174	276	2004.8	-0.1	0.9	2005.5
175	277	2009.1	-0.6	1.2	2009.5
176	278	2013.3	-0.5	1.4	2013.9

Z = 103: Lawrencium

Number of neutrons N	Mass number A	Liquid drop	Shell correction	BCS pairing energy	Total binding energy
153	256	1892.8	4.3	-0.1	1895.5
154	257	1899.6	4.2	0.2	1902.5
155	258	1906.3	3.9	0.	1908.7

(continued)

N	A				
169	269	1969.0	1.3	1969.2	0.3
170	270	1973.5	1.1	1973.9	0.8
171	271	1978.5	2.2	1977.8	0.1
172	272	1982.5	2.3	1982.5	0.4
173	273	1987.0	2.2	1986.2	0.1

Z = 101: Mendelevium

N	A				
149	250	1859.7	4.1	1861.9	−0.1
150	251	1866.6	4.4	1869.3	−0.1
151	252	1873.4	4.4	1876.0	−0.1
152	253	1880.1	4.6	1882.9	−0.1
153	254	1886.6	4.3	1889.1	0.3
154	255	1893.1	4.2	1895.7	0.5
155	256	1899.4	3.9	1901.5	0.2
156	257	1905.6	3.7	1908.0	0.7
157	258	1911.7	3.4	1913.5	0.4
158	259	1917.6	3.1	1919.8	0.8
159	260	1923.4	2.9	1925.1	0.3
160*	261	1929.2	2.8	1931.1	0.5
161	262	1934.8	2.7	1936.2	0.2
162	263	1940.3	2.8	1942.0	0.4
163	264	1945.7	2.7	1946.9	0.8
164	265	1951.0	2.8	1952.5	0.2
165	266	1956.1	2.5	1957.1	0.4
166	267	1961.2	2.3	1962.4	0.4
167	268	1966.2	2.0	1966.8	0.1
168	269	1971.0	1.8	1971.8	0.5
169	270	1975.7	1.4	1975.9	0.2
170	271	1980.4	1.2	1980.8	0.7
171	272	1984.9	0.9	1984.7	0.2
172	273	1989.4	0.8	1989.4	0.6
173	274	1994.3	1.7	1993.2	0.1
174	275	1998.5	1.5	1997.9	0.7

Z = 102: Nobelium

N	A				
151	253	1876.3	4.5	1879.2	0.1
152	254	1883.2	4.6	1886.3	0.2

(continued)

N	A				
156	259	1912.9	3.7	1915.5	0.5
157	260	1919.3	3.4	1921.4	0.2
158	261	1925.6	3.1	1928.0	0.7
159	262	1931.8	3.0	1933.8	0.3
160	263	1937.9	3.0	1940.2	0.7
161	264	1943.9	3.0	1945.7	0.1
162	265	1949.8	3.1	1951.9	0.3
163	266	1955.5	2.9	1957.1	
164*	267	1961.2	2.7	1963.0	0.4
165	268	1966.7	2.5	1968.1	0.1
166	269	1972.2	2.5	1973.7	0.3
167	270	1977.5	2.1	1978.5	0.1
168	271	1982.7	1.9	1983.9	0.5
169	272	1987.8	1.6	1988.4	0.2
170	273	1992.8	1.3	1993.6	0.6
171	274	1997.7	1.1	1997.9	0.6
172	275	2002.5	0.9	2002.9	0.6
173	276	2007.0	0.4	2007.0	0.1
174	277	2011.5	0.5	2012.0	0.4
175	278	2016.0	0.	2016.2	0.4
176	279	2020.4	0.	2020.8	0.8
177	280	2024.7	−0.4	2024.8	0.7
178	281	2028.9	−0.2	2029.2	0.8

Z = 104

N	A				
155	259	1909.1	4.0	1911.8	0.2
156	260	1915.9	3.8	1918.8	0.6
157	261	1922.5	3.4	1924.9	0.4
158	262	1929.0	3.3	1931.8	0.8
159	263	1935.4	3.1	1937.7	0.5
160	264	1941.7	3.1	1944.3	0.8
161	265	1947.9	3.3	1950.0	0.4
162	266	1953.9	3.0	1956.4	0.1
163	267	1959.9	2.8	1961.8	0.6
164	268	1965.7	2.6	1967.9	0.5
165	269	1971.4	2.6	1973.1	0.2
166*	270	1977.0	2.2	1978.9	
167	271	1982.5		1983.9	

TABLE 8c-2. CALCULATED BINDING ENERGIES IN MeV (Continued)

Z = 104 (Continued)

Number of neutrons N	Mass number A	Liquid drop	Shell correction	BCS pairing energy	Total binding energy
168	272	1987.9	2.0	0.6	1989.5
169	273	1993.2	1.7	0.3	1994.2
170	274	1998.4	1.5	0.8	1999.6
171	275	2003.4	1.0	0.5	2004.0
172	276	2008.3	0.7	0.9	2009.4
173	277	2013.1	0.5	0.8	2014.0
174	278	2017.8	0.7	0.9	2019.1
175	279	2022.5	0.2	1.1	2023.5
176	280	2027.1	0.1	1.4	2028.3
177	281	2031.5	−0.1	1.3	2032.5
178	282	2035.9	−0.1	1.4	2037.1
179	283	2040.1	−0.7	1.7	2041.0
180	284	2044.3	−0.8	1.9	2045.4

Z = 105

Number of neutrons N	Mass number A	Liquid drop	Shell correction	BCS pairing energy	Total binding energy
157	262	1925.3	3.3	0.3	1927.6
158	263	1932.0	3.2	0.7	1934.6
159	264	1938.6	3.1	0.3	1940.7
160	265	1945.1	3.1	0.6	1947.5
161	266	1951.4	3.1	0.	1953.4
162	267	1957.7	3.3	0.2	1960.0
163	268	1963.8	3.0	0.	1965.7
164	269	1969.8	2.8	0.4	1971.9
165	270	1975.7	2.6	0.	1977.3
166	271	1981.5	2.6	0.3	1983.3
167	272	1987.2	2.3	0.1	1988.5
168*	273	1992.8	2.0	0.5	1994.3
169	274	1998.2	1.7	0.2	1999.5
170	275	2003.6	1.5	0.6	2004.7

Z = 107

Number of neutrons N	Mass number A	Liquid drop	Shell correction	BCS pairing energy	Total binding energy
162	269	1963.9	3.1	0.2	1966.3
163	270	1970.4	2.9	−0.1	1972.4
164	271	1976.8	2.9	0.2	1979.0
165	272	1983.0	2.6	0.	1984.8
166	273	1989.2	2.5	0.3	1991.2
167	274	1995.2	2.2	0.1	1996.7
168	275	2001.1	1.9	0.6	2002.9
169	276	2006.9	1.6	0.3	2008.2
170	277	2012.6	1.5	0.7	2014.3
171	278	2018.2	1.6	0.2	2019.6
172*	279	2023.7	1.6	0.5	2025.5
173	280	2029.0	1.7	0.2	2030.7
174	281	2034.3	2.1	0.3	2036.4
175	282	2039.5	1.8	0.3	2041.4
176	283	2044.6	1.8	0.5	2046.7
177	284	2049.6	1.6	0.3	2051.4
178	285	2054.5	1.6	0.5	2056.5
179	286	2059.3	1.1	0.7	2061.0
180	287	2063.9	1.1	0.9	2065.9
181	288	2068.5	0.2	1.5	2070.1
182	289	2073.0	−0.4	2.1	2074.8
183	290	2077.4	−0.4	1.8	2078.8
184	291	2081.7	−0.4	1.9	2083.2
185	292	2085.9	−0.7	1.6	2086.8

Z = 108

Number of neutrons N	Mass number A	Liquid drop	Shell correction	BCS pairing energy	Total binding energy
164	272	1979.6	2.9	0.5	1982.2
165	273	1986.1	2.7	0.2	1988.3
166	274	1992.4	2.7	0.4	1994.9

N	A	M	δ	δ	M'
167	275	2000.6	0.2	2.4	1998.6
168	276	2007.0	0.8	2.1	2004.7
169	277	2012.7	0.4	2.0	2010.7
170	278	2019.0	0.8	2.0	2016.5
171	279	2024.5	0.4	2.1	2022.3
172	280	2030.7	0.8	2.2	2028.0
173	281	2036.2	0.5	2.3	2033.5
174*	282	2042.0	0.6	2.6	2039.0
175	283	2047.2	0.6	2.4	2044.4
176	284	2052.7	0.9	2.3	2049.6
177	285	2057.6	0.8	2.2	2054.8
178	286	2062.9	1.2	2.0	2059.8
179	287	2067.7	1.3	1.6	2064.8
180	288	2072.7	1.8	1.3	2069.7
181	289	2077.2	1.8	1.0	2074.4
182	290	2082.0	2.4	0.6	2079.1
183	291	2086.2	2.1	0.4	2083.7
184	292	2090.8	2.3	0.4	2088.1
185	293	2094.6	2.0	0.1	2092.5
186	294	2099.1	3.0	-0.6	2096.5
187	295	2102.8	2.3	-0.6	2101.0

Z = 109

N	A	M	δ	δ	M'
166	275	1997.7	0.1	2.9	1995.2
167	276	2003.7	0.	2.6	2001.6
168	277	2010.3	0.4	2.5	2007.9
169	278	2016.2	0.1	2.4	2014.1
170	279	2022.7	0.4	2.5	2020.1
171	280	2028.4	0.	2.6	2026.1
172	281	2034.8	0.3	2.8	2031.9
173	282	2040.4	0.	3.0	2037.6
174	283	2046.5	0.1	3.3	2043.3
175	284	2051.8	0.1	3.1	2048.8
176*	285	2057.5	0.3	3.1	2054.3
177	286	2062.6	0.1	3.0	2059.6
178	287	2068.0	0.5	2.8	2064.8
179	288	2072.9	0.6	2.5	2070.0
180	289	2078.2	1.0	2.2	2075.0

N	A	M	δ	δ	M'
171	276	2009.4	0.3	1.1	2008.8
172	277	2014.9	0.6	1.0	2013.8
173	278	2019.6	0.3	0.9	2018.8
174	279	2025.0	0.4	1.1	2023.8
175	280	2029.5	0.4	0.7	2028.6
176	281	2034.5	0.8	0.6	2033.3
177	282	2038.8	0.5	0.5	2038.0
178	283	2043.6	0.7	0.6	2042.5
179	284	2047.6	0.9	0.	2046.9
180	285	2052.2	1.1	-0.1	2051.3
181	286	2056.1	1.8	-1.2	2055.5
182	287	2060.3	2.0	-1.4	2059.7

Z = 106

N	A	M	δ	δ	M'
160	266	1948.0	3.2	0.7	1950.8
161	267	1954.6	3.2	0.1	1956.8
162	268	1961.0	3.4	0.3	1963.6
163	269	1967.3	3.1	0.1	1969.5
164	270	1973.5	3.1	0.4	1975.9
165	271	1979.6	2.8	0.1	1981.5
166	272	1985.5	2.6	0.5	1987.7
167	273	1991.4	2.3	0.2	1993.0
168	274	1997.1	2.1	0.6	1999.0
169*	275	2002.7	1.7	0.4	2004.1
170	276	2008.2	1.3	1.0	2009.9
171	277	2013.6	1.1	0.6	2015.0
172	278	2018.9	1.2	1.0	2020.8
173	279	2024.1	1.2	0.7	2025.8
174	280	2029.2	1.5	0.8	2031.3
175	281	2034.2	1.2	0.8	2036.1
176	282	2039.2	1.2	1.1	2041.3
177	283	2044.0	1.0	0.9	2045.8
178	284	2048.7	0.4	1.2	2050.7
179	285	2053.3	0.3	1.4	2055.1
180	286	2057.8		1.7	2059.8
181	287	2062.2	-0.6	2.3	2063.9
182	288	2066.5	-0.9	2.7	2068.3
183	289	2070.8	-1.1	2.5	2072.2

TABLE 8c-2. CALCULATED BINDING ENERGIES IN MeV (Continued)

Z = 109 (Continued)

Number of neutrons N	Mass number A	Liquid drop	Shell correction	BCS pairing energy	Total binding energy
181	290	2079.9	1.9	1.1	2082.8
182	291	2084.7	1.5	1.6	2087.8
183	292	2089.5	1.3	1.3	2092.2
184	293	2094.1	1.3	1.5	2096.9
185	294	2098.7	1.0	1.2	2100.9
186	295	2103.1	0.3	2.2	2105.6
187	296	2107.5	0.3	1.6	2109.4
188	297	2111.8	-0.1	2.2	2113.8
189	298	2115.9	0.1	1.5	2117.5

Z = 110

Number of neutrons N	Mass number A	Liquid drop	Shell correction	BCS pairing energy	Total binding energy
168	278	2010.7	2.9	0.5	2013.7
169	279	2017.0	2.9	0.2	2019.8
170	280	2023.2	3.2	0.4	2026.6
171	281	2029.4	3.1	0.3	2032.6
172	282	2035.4	3.4	0.5	2039.2
173	283	2041.3	3.6	0.3	2045.1
174	284	2047.2	3.9	0.4	2051.3
175	285	2052.9	3.7	0.4	2056.9
176	286	2058.5	3.7	0.7	2062.8
177	287	2064.0	3.5	0.6	2068.1
178*	288	2069.4	3.4	1.1	2073.8
179	289	2074.7	2.8	1.4	2078.9
180	290	2079.9	3.0	1.4	2084.3
181	291	2085.0	2.8	1.4	2089.2
182	292	2090.0	2.5	1.8	2094.3
183	293	2094.9	2.4	1.6	2098.9
184	294	2099.7	2.4	1.7	2103.8
185	295	2104.4	2.0	1.5	2107.9

Z = 112 (Continued)

Number of neutrons N	Mass number A	Liquid drop	Shell correction	BCS pairing energy	Total binding energy
177	289	2071.6	5.1	0.5	2077.2
178	290	2077.4	5.0	0.9	2083.3
179	291	2083.0	5.0	0.8	2088.8
180	292	2088.6	5.3	0.8	2094.6
181	293	2094.0	5.1	0.7	2099.8
182*	294	2099.3	4.8	1.1	2105.3
183	295	2104.6	4.7	0.8	2110.2
184	296	2109.7	4.7	1.0	2115.4
185	297	2114.8	4.3	0.8	2119.9
186	298	2119.7	3.7	1.7	2125.1
187	299	2124.6	3.5	1.3	2129.4
188	300	2129.4	3.1	1.9	2134.3
189	301	2134.0	3.0	1.4	2138.5
190	302	2138.6	2.8	1.9	2143.3
191	303	2143.1	2.7	1.4	2147.2
192	304	2147.5	2.6	1.8	2151.9
193	305	2151.8	2.8	1.2	2155.7
194	306	2156.0	2.6	1.6	2160.2

Z = 113

Number of neutrons N	Mass number A	Liquid drop	Shell correction	BCS pairing energy	Total binding energy
175	288	2062.7	6.1	-0.1	2068.6
176	289	2068.8	6.2	0.1	2075.1
177	290	2074.8	6.2	0.2	2080.9
178	291	2080.8	6.3	-0.1	2087.2
179	292	2086.6	6.4	-0.1	2092.9
180	293	2092.3	6.6	-0.1	2098.8
181	294	2097.9	6.4	-0.1	2104.2
182	295	2103.4	6.2	-0.2	2109.8
183	296	2108.8	6.1	-0.1	2114.8

(Z = 110, continued)

N	A	M₁	Δ	δ	M₂
186	296	2109.1	1.3	2.4	2112.8
187	297	2113.6	1.2	1.9	2116.7
188	298	2118.0	0.8	2.6	2121.4
189	299	2122.4	0.8	2.0	2125.2
190	300	2126.6	0.6	2.4	2129.6
191	301	2130.8	0.6	1.9	2133.3

Z = 111

N	A	M₁	Δ	δ	M₂
171	282	2032.3	3.6	0.1	2035.9
172	283	2038.5	4.0	0.3	2042.7
173	284	2044.6	4.2	0.	2048.7
174	285	2050.6	4.4	0.2	2055.1
175	286	2056.5	4.4	0.1	2060.9
176	287	2062.3	4.3	0.3	2066.9
177	288	2068.0	4.3	0.2	2072.4
178	289	2073.6	4.1	0.5	2078.3
179	290	2079.1	4.2	0.4	2083.5
180*	291	2084.4	4.0	0.5	2089.1
181	292	2089.7	3.6	0.5	2094.2
182	293	2094.9	3.5	0.9	2099.5
183	294	2100.0	3.5	0.7	2104.2
184	295	2104.9	3.1	0.9	2109.3
185	296	2109.8	2.4	0.6	2113.5
186	297	2114.6	2.4	1.5	2118.6
187	298	2119.3	2.0	1.0	2122.7
188	299	2123.9	2.0	1.6	2127.5
189	300	2128.4	1.9	1.0	2131.5
190	301	2132.8	1.9	1.4	2136.1
191	302	2137.1	1.6	0.9	2139.9
192	303	2141.3		1.4	2144.4

Z = 112

N	A	M₁	Δ	δ	M₂
173	285	2047.5	4.8	0.4	2052.7
174	286	2053.7	5.0	0.6	2059.3
175	287	2059.8	5.1	0.4	2065.2
176	288	2065.8	5.1	0.7	2071.5

Z = 113

N	A	M₁	Δ	δ	M₂
184*	297	2114.2	6.0	0.1	2120.3
185	298	2119.4	5.5	0.	2124.9
186	299	2124.5	5.2	0.6	2130.3
187	300	2129.5	4.9	0.3	2134.7
188	301	2134.4	4.7	0.8	2139.9
189	302	2139.3	4.5	0.4	2144.2
190	303	2144.0	4.4	0.8	2149.1
191	304	2148.7	4.3	0.3	2153.3
192	305	2153.2	4.2	0.7	2158.1
193	306	2157.7	4.2	0.2	2162.1
194	307	2162.0	4.2	0.5	2166.7
195	308	2166.3	4.4	−0.1	2170.5
196	309	2170.5		0.	2174.9

Z = 114

N	A	M₁	Δ	δ	M₂
177	291	2077.7	7.4	0.	2085.1
178	292	2083.7	7.7	0.1	2091.6
179	293	2089.7	7.7	0.	2097.5
180	294	2095.6	7.9	0.	2103.6
181	295	2101.4	7.7	0.	2109.1
182	296	2107.1	7.5	0.3	2114.9
183	297	2112.7	7.4	0.	2120.1
184	298	2118.2	7.5	0.1	2125.7
185	299	2123.6	6.9	0.	2130.5
186*	300	2128.8	6.4	0.8	2136.0
187	301	2134.0	6.1	0.5	2140.6
188	302	2139.1	5.8	1.1	2145.9
189	303	2144.1	5.6	0.7	2150.4
190	304	2149.0	5.4	1.1	2155.5
191	305	2153.8	5.3	0.7	2159.8
192	306	2158.5	5.2	1.0	2164.8
193	307	2163.2	5.3	0.4	2168.9
194	308	2167.7	5.3	0.7	2173.7
195	309	2172.1	5.5	0.	2177.6
196	310	2176.5	5.8	0.	2182.3
197	311	2180.7	4.9	0.	2185.6
198	312	2184.9	4.1	1.2	2190.1

6. Myers, W. D.: "Winnipeg Proceedings," p. 61; W. D. Myers and W. J. Swiatecki: *Annals of Physics* **55**, 395 (1969).
7. Zeldes, N.: "Lysekil Proceedings," p. 361. Calculations are tabulated by N. Zeldes, A. Grill, and A. Simievic: *Mat. Fys. Skr. Dan. Vid. Selsk.* **3**(5), (1967).
8. Myers, W. D., and W. J. Swiatecki: "Lysekil Proceedings," p. 343; *Nucl. Phys.* **81** (1966). A table of calculated values is included in report UCRL-11980.
9. Seeger, P. A.: "Winnipeg Proceedings," p. 85; P. A. Seeger and R. C. Perisho, report LA-3751.
10. Mattauch, J. H. E., W. Thiele, and A. H. Wapstra: *Nucl. Phys.* **67**, 1 (1965).
11. Wapstra, A., C. Kurzeck, and A. Anisimoff: "Winnipeg Proceedings," p. 153.
12. Hahn, B., D. G. Ravenhall, and R. Hofstadter: *Phys. Rev.* **101**, 1131 (1956).
13. Nilsson, S. G., C. F. Tsang, A. Sobiczewski, Z. Szymański, S. Wycech, C. Gustafson, I. L. Lamm, P. Möller, and B. Nilsson: *Nucl. Phys.* **A131**, 1 (1969).
14. Foldy, L. L.: *Phys. Rev.* **83**, 397 (1951).
15. Breitenberger, E.: "Vienna Proceedings," p. 91.
16. Mattauch, J. H. E., W. Thiele, and A. H. Wapstra: *Nucl. Phys.* **67**, 73 (1965).
17. Wing, James: "Winnipeg Proceedings," p. 194.
18. Seeger, P. A.: "Lysekil Proceedings," p. 495: "Winnipeg Proceedings," p 85.
19. Talbert, W. L., Jr., A. B. Tucker, and G. M. Day: *Phys. Rev.* **177**, 1805 (1969).
20. Green, A. E. S.: "Nuclear Physics," Mc-Graw-Hill Book Company, New York, 1955.

8d. Passage of Charged Particles Through Matter[1]

HANS BICHSEL

University of Washington

8d-1. Introductory Note. This section presents some of the commonly used formulas and principal data on the passage of fast charged particles through matter. Because of space limitations, much useful material has been omitted. The bibliography includes mainly the newest available references. Most of the technical reports cited are available from the National Technical Information Service, Springfield, Virgina 22151. An extensive review of the field is found in Publication 1133 of the National Academy of Sciences–National Research Council (NA67). The Bibliography of Atomic and Molecular Processes (ORNL-AMPIC 13, UC-34-Physics for January–December, 1969), is published annually by the Atomic and Molecular Processes Information Center, Oak Ridge National Laboratory, Oak Ridge, Tennessee. It contains sections concerned with energy losses, ionization, particle range, etc. The Information Center at JILA (Joint Institute for Laboratory Astrophysics), University of Colorado, Boulder, Colorado 80302, also disseminates information of this nature.

A number of papers concerned with particles at the lowest energies considered in this article have appeared in the *Proceedings of an International Conference on Atomic Collisions and Penetration Studies with Energetic Ion Beams*, Chalk River, Ontario, September 18–21, 1967 (DA68), in the abstracts of the V International Conference of the Physics of Electronic and Atomic Collisions (FL67), and of the Sixth Inter-

[1] This work was supported by Public Health Service Grant CA-08150 from the National Cancer Institute and in part by the U.S. Atomic Energy Commission Contract AT(04-3)-136.

national Conference on the Physics of Electronic and Atomic Collisions, M.I.T., July 28–August 2 (1969), (The Massachusetts Institute of Technology Press, Cambridge, Massachusetts 02142). The seventh Conference took place in Amsterdam, July 26–30, 1971.

8d-2. Atomic Collision Cross Sections.   The following notation will be used.

The kinetic energy of particles will be denoted by T, the energy of a secondary electron (delta ray) by E or by W if expressed in atomic shell units [Eq. (8d-3)]. Thicknesses s are usually measured in g cm^{-2} ($s = \rho x$, x thickness in cm, ρ density in g cm^{-3}). The stopping power (usually called dE/dx) will then be denoted as $S = -dT/ds$.

Except for particles with very small or extremely large velocities v, the interaction between energetic charged particles (of charge ze) and matter leads mainly to the excitation and ionization of atoms or molecules (FA63). The probability for a collision leading to an atomic state of energy E_n is described by the collision cross section σ_n. Relatively little information is available about the details of σ_n (e.g., FC68, RU68, OL67, ES69). In energy-loss experiments, the quantities observed are usually averages over E_n and σ_n (e.g., the stopping power dT/ds is $\Sigma_n E_n \sigma_n$), and even a coarse approximation of σ_n will give satisfactory answers.

Frequently, the free-electron approximation is used for a description of σ_n. The energies E_n then are continuously distributed and are equal to the energy E of the electron after the collision. The collision cross section is differential with respect to E and is given, nonrelativistically, by (see, e.g., BI68)

$$nd\sigma' = \left(\frac{PZ}{\beta^2}\right) E^{-2}\, dE \tag{8d-1}$$

where $P = 2\pi z^2 mc^2 r_0{}^2 N_0/A = 0.15354 \times z^2/A$ MeV/(g/cm^2)

  z = charge number of incident particle

  $\beta = v/c$, velocity of incident particle relative to velocity of light [see Eq. (8d-37)]

  $r_0{}^2 = e^4/m^2 c^4 = 7.9408 \times 10^{-26}$ cm^2 (square of "classical electron radius")

  m = rest mass of electron, $mc^2 = 0.511004$ MeV

  N_0 = Avogadro's number = 6.02217×10^{23}

  e = electron charge

  E = energy of electron after collision

  A = atomic weight of stopping material, in grams

  Z = atomic number of stopping material

  n = number of electrons in a thickness $s = 1$ g/cm^2

Using the Born approximation, Bethe (BE30, IN71) has given the nonrelativistic quantum mechanical derivation of $d\sigma$ for bound electrons:

$$d\sigma = \frac{2P}{\beta^2} \sum_i J_i(\eta_i, W)\, dW \tag{8d-2}$$

where J_i is called the excitation function (WA56). Electron energies W and equivalent particle energies η_i are measured in atomic shell units:

$$W = \frac{E}{(Z - d_i)^2 R_y} \tag{8d-3}$$

$$\eta_i = \frac{mc^2\beta^2}{2R_y(Z - d_i)^2} = \frac{18{,}800\beta^2}{(Z - d_i)^2} \tag{8d-4}$$

where R_y = Rydberg = 13.60 eV

  d_i = electron defect, depending on the atomic shell i ($i = K, L, M, \ldots$ shell)

$$d_K = 0.3 \qquad d_L = 4.15$$

The excitation functions J_i have been evaluated, using hydrogenic wave functions, for the K, L, and M shells (WA51, WA52, WA56, BI67, KM66, KH68). Whereas J_K probably is reasonably correct for all Z, it appears that J_L is acceptable without modifications for $Z > 30$ only, and J_M will have to be recalculated with more realistic wave functions.

An appreciation of the difference between the two approximations, $d\sigma'$ and $d\sigma$, can be obtained from a study of a plot of $J_i W^2$ versus W (Fig. 8d-1). Further comments will be made later at appropriate places (see also BI69).

Generally, the Born approximation is valid for $\beta \gg z/137$ (protons with $\beta = 1/137$ have a kinetic energy of 25 keV). Some tests have been made for small-particle velocities: for protons incident on helium, the Bethe-Born approximation is valid for energies above 450 keV (TH67), while for the vacuum ultraviolet emission of hydrogen

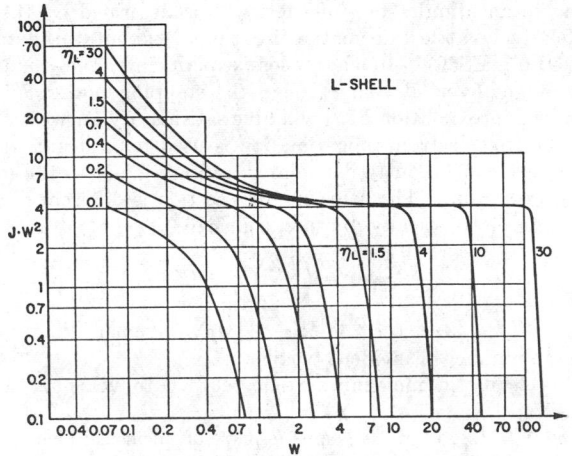

Fig. 8d-1. First Born approximation of the excitation function J for L-shell electrons relative to free-electron excitation function $J' = 1/W^2$. Plotted is $J/J' = JW^2$ as a function of δ-ray energy $W = E/[13.6 \text{ eV}(Z - 4.15)^2]$. The "ionization" energy $W_{min} \cong I_L$ is approximately 0.09 for Al, 0.17 for Pb. The matrix elements are calculated with hydrogenic wave functions. In Bohr's papers, the rise at small W is described as a resonance effect.

gas, produced by fast protons, it appears to be valid above 150 keV (DD68). Distributions in energy and angle of electrons ejected from atoms by fast protons are described in TO71.

Almost 100 papers concerning atomic and molecular excitation by electron impact alone are listed in the Bibliography of Atomic and Molecular Processes for January to June of 1968. In particular, the following may be of interest: ES69, KY68, OL67, VS68.

Measurements of the excitation of the inner shells with protons have been made: KP67, DK68; see KJ68 and ML58 for further references. At low energies it is necessary to take into account the Coulomb deflection of the incident particle to get reasonable agreement with the Born approximation (BL69). Similar corrections are necessary for incident electrons.

A large fraction of energy losses below 50 eV, in solids or liquids, are caused by the excitation of plasmons (volume and surface) (PO67), collective oscillations of electrons (SP63, CR66), discrete excited states, etc. Most of this information has been obtained from experiments performed with electrons, but similar results have to be expected for heavy particle interactions. While these small energy losses are of relatively little importance for the stopping power (about 30 percent is contributed to it by energy losses below 60 eV) and straggling, they may be very important for chemical and biological effects (where 1 or 2 eV may be sufficient to break a DNA molecule).

8d-3. Stopping Power for Heavy Charged Particles. Since the stopping power of heavy charged particles depends largely on the velocity and the charge of the particle, but not on its mass (IS67), the discussion of this section applies to all heavy charged particles, with the exceptions specified in Sec. 8d-6. The tables and data presented apply to protons and can be converted for other particles with the procedures described in Sec. 8d-6.

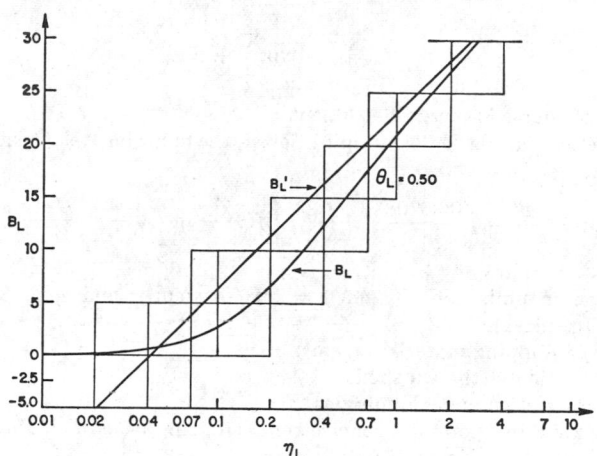

FIG. 8d-2. The stopping number $B = \int WJ\, dW$ for L-shell electrons in copper. Also given is the asymptotic expression $B' = S_L' \ln (2mv^2/I_L)$. The difference between the two functions is the shell correction C_L [Eq. (8d-8)]; it is a basic part of the quantum-mechanical theory.

The mean energy loss per unit path length is called the stopping power S. It is defined by

$$S = \frac{-dT}{ds} = \int W\, d\sigma = \frac{2P}{\beta^2} \sum_i B_i \tag{8d-5}$$

where the stopping number B_i is defined by

$$B_i = \int_{I_i}^{\infty} J_i W\, dW \tag{8d-6}$$

$I_i = W_{\min}$ is the energy required to lift an electron from the ith shell to the lowest unoccupied atomic level, and the integral includes a sum over the discrete atomic energy levels. For large velocities, Bethe (LB37) has derived the asymptotic expression

$$B' = \Sigma B_i' = Z \ln \frac{2mv^2}{I_{\rm av}} \tag{8d-7}$$

$I_{\rm av}$ is defined in Eq. (8d-17). B_L and B_L' are shown in Fig. 8d-2. The shell corrections are defined by

$$\sum_i C_i(\eta_i, Z) = B' - \sum_i B_i \tag{8d-8}$$

and are thus an integral part of the quantum-mechanical theory. For higher-order Born approximations they will presumably depend on the particle charge ze.

If S is calculated using the free-electron cross sections $d\sigma'$, an unphysical minimum energy $E_{\min} = I_e^2/2mv^2$ has to be used as the lower limit of the integral to get approximately the correct stopping power (e.g., p. 192, Eq. 8 in NA67):

$$S' = \frac{P}{\beta^2}\int_{E_{\min}}^{2mv^2}\frac{ZE}{E^2}\,dE = \frac{2PZ}{\beta^2}\ln\frac{2mv^2}{I_e} \tag{8d-9}$$

This choice of $E_{\min}$ is necessary to take into account the increase of J over J' at small energies W (see Fig. 8d-1) but it will not give exact agreement with the quantum mechanical theory. To achieve this, it is necessary to choose

$$I_e = I_{\mathrm{av}}\exp\sum_i\frac{C_i}{Z} \tag{8d-10}$$

where I_e now of course is energy-dependent.

For the practical calculation of stopping power, the following, relativistically correct formula is used:

$$S = \frac{-dT}{ds} = \left(\frac{0.30708}{\beta^2}\right)z^2\left(\frac{Z}{A}\right)\left[f(\beta) - \ln I_{\mathrm{av}} - \sum_i\frac{C_i}{Z} - \frac{\delta}{2}\right] \tag{8d-11}$$

Stopping power in units MeV/(g/cm²) = keV/(mg/cm²); and z, β, Z, and A are defined with Eq. (8d-1).

ρ = density of stopping material (g/cm³)

C_i = shell correction of the ith shell

δ = density correction at high energies

I_{av} = average excitation potential per electron of stopping atom (including low-velocity density effect), a constant by definition.

$$f(\beta) = \ln\left(\frac{2mc^2\beta^2}{1 - \beta^2}\right) - \beta^2 \tag{8d-11a}$$

β^2 and $f(\beta)$ are listed in Table 8d-1 as functions of the kinetic energy T of several particles. $f(\beta)$ is applicable for any charged particle of velocity $v = \beta c$ and mass

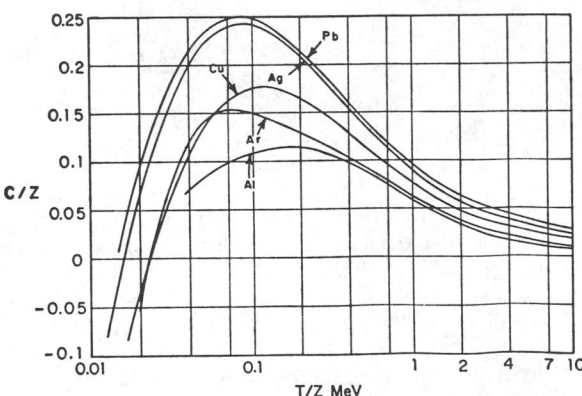

Fig. 8d-3. Practical shell correction C/Z for particles of charge $+1$. The abscissa is $T/Z = T_i/(m_rZ)$; see Eq. (8d-12). For $Z \leq 25$, Walske's, and for $Z > 25$, Bonderup's shell corrections are modified to fit experimental data for protons and deuterons. In this procedure. deviations from the first Born approximation are included in C/Z, and the shell corrections depend on the incident particle charge z. For $C/Z < -0.1$, the Bonderup corrections do not fit the data well.

$M \gg m$. If an ion of mass M_i and kinetic energy T_i is under consideration, its velocity can be found by looking up in Table 8d-1 the value of β corresponding to a proton energy

$$T = \frac{T_i}{m_r} \tag{8d-12}$$

where $m_r = M_i c^2/938.259$ MeV. In general it will be easiest to use existing tables, e.g., NS70. Due to the generalized approach used in NS70, large differences from experimental data are found, e.g., for alpha particle ranges in argon (50 percent at 1 MeV, no less than 4 percent up to 10 MeV). The shell corrections can be obtained from Fig. 8d-3, and I values from Fig. 8d-4.

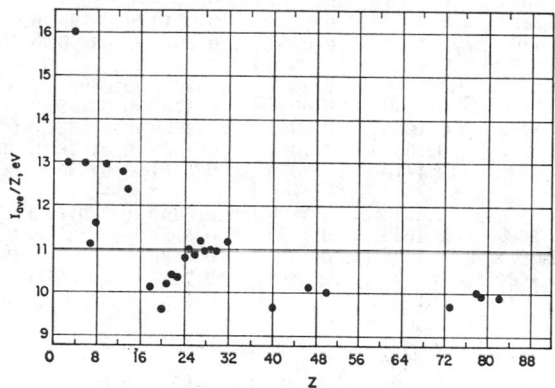

FIG. 8d-4. The mean excitation energy I_{av} for different elements. Given is I_{av}/Z versus Z. For H_2, $I_{\mathrm{av}} = 19.2$ eV, for He, $I_{\mathrm{av}} = 41.3$ eV, from α-particle measurements. The values represent the author's present opinion, and may change by several percent. The strong fluctuations found for neighboring elements are significant though.

For most metals the density effect δ is negligible for proton energies below 1000 MeV. For details see ST67, FA56, CF70, and page 69 of BK58. Experimental confirmation is found, e.g., in NM67, and BH67.

At low energies (proton energies of less than 0.5 MeV, alpha-particle energies below 2 MeV), the charged particle will not have its full charge (see Sec. 8d-6).

A list of values for S computed (BJ67) from Eq. (8d-11) is given in Table 8d-2. For emulsion, see BD63 and BA63. For the other materials, the I values given in Fig. 8d-5 were used. The shell corrections are discussed in Sec. 8d-5. The density effect is not used.

For proton energies of 0.05 to 12 MeV, the experimental stopping powers for many substances are given in Table 8d-3. Most of these numbers are read from the graphs of WH58, and the tables of AH67, AS68, and AV69. This seems to be the best way to average the experimental results, but see also MA68, OR68, WM67, JK68, SP70. The stopping cross section in eV-cm² per atom can be obtained by multiplying S with the factor $(A/N_0) \times 10^6$ (Avogadro's number N_0, atomic weight A).

For protons in other elements, interpolation for Z by the method of Lindhard and Scharff (LS53) can be used, but direct computation from Eq. (8d-11) is recommended. (A discussion of experimental results is found in BK67.)

The stopping power of compounds is within a few percent an additive function of the stopping power of the elements which make up the compound (Bragg rule, see, e.g., BI68 or BT68). Precise measurements at 300 MeV (TH52) have shown deviations of about 1 percent from additivity. At energies between 4 and 30 MeV energy-

NUCLEAR PHYSICS

TABLE 8d-1. RELATIVISTIC VELOCITY $\beta = v/c$, β^2, AND STOPPING-NUMBER FUNCTION $f(\beta)$ FOR HEAVY IONS AS A FUNCTION OF KINETIC ENERGY T

Kinetic energy T for					β	β^2	$f(\beta)$
Protons, MeV	Alphas, MeV	Pions, MeV	Muons, MeV	Electrons, keV			
0.50	1.9863	0.0744	0.0563	0.2723	0.032634	0.001065	6.9925
0.55	2.1849	0.0818	0.0619	0.2995	0.034225	0.001171	7.0877
0.60	2.3836	0.0893	0.0676	0.3268	0.035745	0.001278	7.1746
0.65	2.5822	0.0967	0.0732	0.3540	0.037204	0.001384	7.2546
0.70	2.7808	0.1041	0.0788	0.3812	0.038606	0.001490	7.3286
0.75	2.9795	0.1116	0.0845	0.4085	0.039960	0.001597	7.3975
0.80	3.1781	0.1190	0.0901	0.4357	0.041269	0.001703	7.4620
0.85	3.3767	0.1265	0.0957	0.4629	0.042537	0.001809	7.5225
0.90	3.5753	0.1339	0.1014	0.4902	0.043769	0.001916	7.5796
0.95	3.7740	0.1413	0.1070	0.5174	0.044966	0.002022	7.6336
1.00	3.9726	0.1488	0.1126	0.5446	0.046132	0.002128	7.6848
1.10	4.3699	0.1636	0.1239	0.5991	0.048380	0.002341	7.7800
1.20	4.7671	0.1785	0.1351	0.6536	0.050528	0.002553	7.8668
1.30	5.1644	0.1934	0.1464	0.7080	0.052587	0.002765	7.9467
1.40	5.5616	0.2083	0.1577	0.7625	0.054567	0.002978	8.0206
1.50	5.9589	0.2231	0.1689	0.8169	0.056478	0.003190	8.0895
1.60	6.3562	0.2380	0.1802	0.8714	0.058326	0.003402	8.1539
1.70	6.7534	0.2529	0.1914	0.9259	0.060116	0.003614	8.2143
1.80	7.1507	0.2678	0.2027	0.9803	0.061854	0.003826	8.2713
1.90	7.5479	0.2827	0.2140	1.0348	0.063544	0.004038	8.3252
2.00	7.9452	0.2975	0.2252	1.0893	0.065189	0.004250	8.3764
2.10	8.3425	0.3124	0.2365	1.1437	0.066794	0.004461	8.4250
2.20	8.7397	0.3273	0.2477	1.1982	0.068360	0.004673	8.4714
2.30	9.1370	0.3422	0.2590	1.2526	0.069891	0.004885	8.5157
2.40	9.5342	0.3570	0.2703	1.3071	0.071388	0.005096	8.5581
2.50	9.9315	0.3719	0.2815	1.3616	0.072855	0.005308	8.5987
2.60	10.3288	0.3868	0.2928	1.4160	0.074292	0.005519	8.6378
2.70	10.7260	0.4017	0.3041	1.4705	0.075701	0.005731	8.6754
2.80	11.1233	0.4165	0.3153	1.5250	0.077084	0.005942	8.7116
2.90	11.5205	0.4314	0.3266	1.5794	0.078442	0.006153	8.7465
3.00	11.9178	0.4463	0.3378	1.6339	0.079776	0.006364	8.7803
3.10	12.3151	0.4612	0.3491	1.6884	0.081089	0.006575	8.8129
3.20	12.7123	0.4760	0.3604	1.7428	0.082380	0.006786	8.8445
3.30	13.1096	0.4909	0.3716	1.7973	0.083650	0.006997	8.8751
3.40	13.5068	0.5058	0.3829	1.8517	0.084901	0.007208	8.9048
3.50	13.9041	0.5207	0.3941	1.9062	0.086134	0.007419	8.9336
3.60	14.3014	0.5356	0.4054	1.9607	0.087349	0.007630	8.9616
3.70	14.6986	0.5504	0.4167	2.0151	0.088547	0.007841	8.9889
3.80	15.0959	0.5653	0.4279	2.0696	0.089728	0.008051	9.0154
3.90	15.4931	0.5802	0.4392	2.1241	0.090894	0.008262	9.0412
4.00	15.8904	0.5951	0.4505	2.1785	0.092045	0.008472	9.0664
4.10	16.2877	0.6099	0.4617	2.2330	0.093181	0.008683	9.0909
4.20	16.6849	0.6248	0.4730	2.2874	0.094303	0.008893	9.1148
4.30	17.0822	0.6397	0.4842	2.3419	0.095411	0.009103	9.1382
4.40	17.4794	0.6546	0.4955	2.3964	0.096507	0.009314	9.1610

TABLE 8d-1. RELATIVISTIC VELOCITY $\beta = v/c$, β^2, AND STOPPING-NUMBER
FUNCTION $f(\beta)$ FOR HEAVY IONS AS A FUNCTION
OF KINETIC ENERGY T (Continued)

Kinetic energy T for					β	β^2	$f(\beta)$
Protons, MeV	Alphas, MeV	Pions, MeV	Muons, MeV	Electrons, keV			
4.50	17.8767	0.6694	0.5068	2.4508	0.097589	0.009524	9.1834
4.60	18.2740	0.6843	0.5180	2.5053	0.098660	0.009734	9.2052
4.70	18.6712	0.6992	0.5293	2.5598	0.099718	0.009944	9.2265
4.80	19.0685	0.7141	0.5405	2.6142	0.100766	0.010154	9.2474
4.90	19.4657	0.7289	0.5518	2.6687	0.101802	0.010364	9.2679
5.00	19.8630	0.7438	0.5631	2.7232	0.102827	0.010573	9.2879
5.50	21.8493	0.8182	0.6194	2.9955	0.107803	0.011622	9.3825
6.00	23.8356	0.8926	0.6757	3.2678	0.112552	0.012668	9.4687
6.50	25.8219	0.9670	0.7320	3.5401	0.117102	0.013713	9.5480
7.00	27.8082	1.0414	0.7883	3.8124	0.121474	0.014756	9.6213
7.50	29.7945	1.1157	0.8446	4.0847	0.125688	0.015797	9.6895
8.00	31.7808	1.1901	0.9009	4.3570	0.129758	0.016837	9.7533
8.50	33.7671	1.2645	0.9572	4.6294	0.133699	0.017875	9.8131
9.00	35.7534	1.3389	1.0135	4.9017	0.137521	0.018912	9.8695
9.50	37.7397	1.4133	1.0698	5.1740	0.141233	0.019947	9.9228
10.00	39.7260	1.4876	1.1261	5.4463	0.144845	0.020980	9.9733
10.50	41.7123	1.5620	1.1824	5.7186	0.148363	0.022012	10.0213
11.00	43.6986	1.6364	1.2387	5.9909	0.151795	0.023042	10.0671
11.50	45.6849	1.7108	1.2950	6.2632	0.155145	0.024070	10.1108
12.00	47.6712	1.7852	1.3514	6.5356	0.158420	0.025097	10.1526
12.50	49.6575	1.8596	1.4077	6.8079	0.161623	0.026122	10.1927
13.00	51.6438	1.9339	1.4640	7.0802	0.164759	0.027145	10.2311
13.50	53.6301	2.0083	1.5203	7.3525	0.167831	0.028167	10.2681
14.00	55.6164	2.0827	1.5766	7.6248	0.170844	0.029188	10.3037
14.50	57.6027	2.1571	1.6329	7.8971	0.173800	0.030206	10.3380
15.00	59.5890	2.2315	1.6892	8.1695	0.176701	0.031223	10.3712
15.50	61.5753	2.3059	1.7455	8.4418	0.179552	0.032239	10.4032
16.00	63.5616	2.3802	1.8018	8.7141	0.182353	0.033253	10.4342
16.50	65.5479	2.4546	1.8581	8.9864	0.185108	0.034265	10.4643
17.00	67.5342	2.5290	1.9144	9.2587	0.187818	0.035276	10.4934
17.50	69.5205	2.6034	1.9707	9.5310	0.190486	0.036285	10.5216
18.00	71.5068	2.6778	2.0270	9.8033	0.193112	0.037292	10.5490
18.50	73.4931	2.7522	2.0833	10.0757	0.195700	0.038298	10.5757
19.00	75.4794	2.8265	2.1396	10.3480	0.198249	0.039303	10.6016
19.50	77.4657	2.9009	2.1960	10.6203	0.200762	0.040306	10.6269
20.00	79.4520	2.9753	2.2523	10.8926	0.203241	0.041307	10.6514
21.00	83.4246	3.1241	2.3649	11.4372	0.208097	0.043305	10.6988
22.00	87.3972	3.2728	2.4775	11.9819	0.212829	0.045296	10.7438
23.00	91.3698	3.4216	2.5901	12.5265	0.217443	0.047281	10.7868
24.00	95.3424	3.5704	2.7027	13.0711	0.221947	0.049261	10.8279
25.00	99.3150	3.7191	2.8153	13.6158	0.226348	0.051234	10.8673
26.00	103.2876	3.8679	2.9279	14.1604	0.230652	0.053200	10.9051
27.00	107.2602	4.0167	3.0405	14.7050	0.234864	0.055161	10.9414
28.00	111.2328	4.1654	3.1532	15.2496	0.238989	0.057116	10.9763
29.00	115.2054	4.3142	3.2658	15.7943	0.243032	0.059064	11.0100

TABLE 8d-1. RELATIVISTIC VELOCITY $\beta = v/c$, β^2, AND STOPPING-NUMBER
FUNCTION $f(\beta)$ FOR HEAVY IONS AS A FUNCTION
OF KINETIC ENERGY T (*Continued*)

Kinetic energy T for					β	β^2	$f(\beta)$
Protons, MeV	Alphas, MeV	Pions, MeV	Muons, MeV	Electrons, keV			
30.00	119.1780	4.4629	3.3784	16.3389	0.246996	0.061007	11.0425
31.00	123.1506	4.6117	3.4910	16.8835	0.250885	0.062943	11.0738
32.00	127.1232	4.7605	3.6036	17.4282	0.254704	0.064874	11.1042
33.00	131.0958	4.9092	3.7162	17.9728	0.258454	0.066799	11.1335
34.00	135.0684	5.0580	3.8288	18.5174	0.262140	0.068717	11.1620
35.00	139.0410	5.2068	3.9414	19.0621	0.265763	0.070630	11.1896
36.00	143.0136	5.3555	4.0541	19.6067	0.269327	0.072537	11.2164
37.00	146.9862	5.5043	4.1667	20.1513	0.272833	0.074438	11.2424
38.00	150.9588	5.6531	4.2793	20.6959	0.276284	0.076333	11.2677
39.00	154.9314	5.8018	4.3919	21.2406	0.279683	0.078222	11.2923
40.00	158.9040	5.9506	4.5045	21.7852	0.283030	0.080106	11.3163
41.00	162.8766	6.0994	4.6171	22.3298	0.286328	0.081984	11.3396
42.00	166.8492	6.2481	4.7297	22.8745	0.289579	0.083856	11.3624
43.00	170.8218	6.3969	4.8424	23.4191	0.292784	0.085722	11.3845
44.00	174.7944	6.5457	4.9550	23.9637	0.295944	0.087583	11.4062
45.00	178.7670	6.6944	5.0676	24.5084	0.299062	0.089438	11.4273
46.00	182.7396	6.8432	5.1802	25.0530	0.302138	0.091287	11.4480
47.00	186.7122	6.9919	5.2928	25.5976	0.305173	0.093131	11.4682
48.00	190.6848	7.1407	5.4054	26.1422	0.308170	0.094969	11.4879
49.00	194.6574	7.2895	5.5180	26.6869	0.311129	0.096801	11.5072
50.00	198.6300	7.4382	5.6306	27.2315	0.314051	0.098628	11.5261
52.50	208.5615	7.8102	5.9122	28.5931	0.321203	0.103171	11.5716
55.00	218.4930	8.1821	6.1937	29.9547	0.328147	0.107680	11.6149
57.50	228.4245	8.5540	6.4752	31.3162	0.334896	0.112155	11.6562
60.00	238.3560	8.9259	6.7568	32.6778	0.341463	0.116597	11.6956
62.50	248.2875	9.2978	7.0383	34.0394	0.347858	0.121005	11.7333
65.00	258.2190	9.6697	7.3198	35.4010	0.354091	0.125380	11.7695
67.50	268.1505	10.0416	7.6014	36.7625	0.360170	0.129723	11.8041
70.00	278.0820	10.4135	7.8829	38.1241	0.366105	0.134033	11.8375
72.50	288.0135	10.7855	8.1644	39.4857	0.371903	0.138312	11.8696
75.00	297.9450	11.1574	8.4460	40.8473	0.377569	0.142558	11.9005
77.50	307.8765	11.5293	8.7275	42.2088	0.383111	0.146774	11.9304
80.00	317.8080	11.9012	9.0090	43.5704	0.388534	0.150958	11.9592
82.50	327.7395	12.2731	9.2906	44.9320	0.393843	0.155112	11.9871
85.00	337.6710	12.6450	9.5721	46.2936	0.399043	0.159236	12.0141
87.50	347.6025	13.0169	9.8536	47.6551	0.404140	0.163329	12.0403
90.00	357.5340	13.3888	10.1352	49.0167	0.409136	0.167392	12.0657
92.50	367.4655	13.7608	10.4167	50.3783	0.414036	0.171426	12.0903
95.00	377.3970	14.1327	10.6982	51.7399	0.418845	0.175431	12.1142
97.50	387.3285	14.5046	10.9798	53.1014	0.423564	0.179407	12.1375
100.00	397.2600	14.8765	11.2613	54.4630	0.428198	0.183354	12.1601
105.00	417.1230	15.6203	11.8243	57.1862	0.437222	0.191163	12.2036
110.00	436.9860	16.3641	12.3874	59.9093	0.445938	0.198860	12.2450
115.00	456.8490	17.1080	12.9505	62.6325	0.454366	0.206448	12.2844
120.00	476.7120	17.8518	13.5135	65.3556	0.462525	0.213929	12.3220

TABLE 8d-1. RELATIVISTIC VELOCITY $\beta = v/c$, β^2, AND STOPPING-NUMBER
FUNCTION $f(\beta)$ FOR HEAVY IONS AS A FUNCTION
OF KINETIC ENERGY T (Continued)

Kinetic energy T for					β	β^2	$f(\beta)$
Protons, MeV	Alphas, MeV	Pions, MeV	Muons, MeV	Electrons, keV			
125.00	496.5750	18.5956	14.0766	68.0788	0.470431	0.221305	12.3579
130.00	516.4380	19.3394	14.6397	70.8019	0.478098	0.228577	12.3923
135.00	536.3010	20.0833	15.2027	73.5251	0.485539	0.235748	12.4254
140.00	556.1640	20.8271	15.7658	76.2482	0.492767	0.242820	12.4572
145.00	576.0270	21.5709	16.3289	78.9714	0.499793	0.249793	12.4878
150.00	595.8900	22.3147	16.8919	81.6945	0.506627	0.256671	12.5173
155.00	615.7530	23.0586	17.4550	84.4177	0.513279	0.263455	12.5457
160.00	635.6161	23.8024	18.0181	87.1408	0.519756	0.270146	12.5733
165.00	655.4791	24.5462	18.5811	89.8640	0.526067	0.276747	12.5999
170.00	675.3421	25.2900	19.1442	92.5871	0.532220	0.283258	12.6257
175.00	695.2051	26.0339	19.7072	95.3103	0.538221	0.289682	12.6507
180.00	715.0681	26.7777	20.2703	98.0334	0.544077	0.296019	12.6749
185.00	734.9311	27.5215	20.8334	100.7566	0.549793	0.302273	12.6985
190.00	754.7941	28.2653	21.3964	103.4797	0.555377	0.308443	12.7214
195.00	774.6571	29.0092	21.9595	106.2029	0.560832	0.314532	12.7437
200.00	794.5201	29.7530	22.5226	108.9260	0.566163	0.320541	12.7655
205.00	814.3831	30.4968	23.0856	111.6492	0.571377	0.326471	12.7866
210.00	834.2461	31.2406	23.6487	114.3723	0.576476	0.332324	12.8073
215.00	854.1091	31.9845	24.2118	117.0955	0.581464	0.338101	12.8274
220.00	873.9721	32.7283	24.7748	119.8186	0.586347	0.343803	12.8471
225.00	893.8351	33.4721	25.3379	122.5418	0.591128	0.349432	12.8663
230.00	913.6981	34.2159	25.9010	125.2649	0.595809	0.354989	12.8851
235.00	933.5611	34.9597	26.4640	127.9881	0.600396	0.360475	12.9035
240.00	953.4241	35.7036	27.0271	130.7112	0.604889	0.365891	12.9215
245.00	973.2871	36.4474	27.5901	133.4344	0.609294	0.371239	12.9391
250.00	993.1501	37.1912	28.1532	136.1575	0.613611	0.376519	12.9564
255.00	1013.0131	37.9350	28.7163	138.8807	0.617845	0.381733	12.9734
260.00	1032.8761	38.6789	29.2793	141.6038	0.621998	0.386882	12.9900
265.00	1052.7391	39.4227	29.8424	144.3270	0.626073	0.391967	13.0063
270.00	1072.6021	40.1665	30.4055	147.0501	0.630070	0.396989	13.0223
275.00	1092.4651	40.9103	30.9685	149.7733	0.633994	0.401949	13.0380
280.00	1112.3281	41.6542	31.5316	152.4964	0.637846	0.406848	13.0534
285.00	1132.1911	42.3980	32.0947	155.2196	0.641628	0.411687	13.0686
290.00	1152.0541	43.1418	32.6577	157.9427	0.645342	0.416467	13.0835
295.00	1171.9171	43.8856	33.2208	160.6659	0.648991	0.421189	13.0982
300.00	1191.7801	44.6295	33.7838	163.3890	0.652575	0.425854	13.1126
310.00	1231.5061	46.1171	34.9100	168.8353	0.659558	0.435016	13.1409
320.00	1271.2321	47.6048	36.0361	174.2816	0.666304	0.443961	13.1682
330.00	1310.9581	49.0924	37.1622	179.7279	0.672826	0.452695	13.1948
340.00	1350.6841	50.5801	38.2884	185.1742	0.679135	0.461225	13.2206
350.00	1390.4101	52.0677	39.4145	190.6205	0.685242	0.469557	13.2458
360.00	1430.1361	53.5554	40.5406	196.0668	9.691156	0.477697	13.2703
370.00	1469.8621	55.0430	41.6667	201.5131	0.696887	0.485651	13.2942
380.00	1509.5881	56.5307	42.7929	206.9594	0.702442	0.493425	13.3176
390.00	1549.3141	58.0183	43.9190	212.4057	0.707830	0.501024	13.3404

TABLE 8d-1. RELATIVISTIC VELOCITY $\beta = v/c$, β^2, AND STOPPING-NUMBER
FUNCTION $f(\beta)$ FOR HEAVY IONS AS A FUNCTION
OF KINETIC ENERGY T (Continued)

Kinetic energy T for					β	β^2	$f(\beta)$
Protons, MeV	Alphas, MeV	Pions, MeV	Muons, MeV	Electrons, keV			
400.00	1589.0401	59.5060	45.0451	217.8520	0.713059	0.508453	13.3626
410.00	1628.7661	60.9936	46.1713	223.2983	0.718135	0.515717	13.3845
420.00	1668.4921	62.4813	47.2974	228.7446	0.723064	0.522822	13.4058
430.00	1708.2181	63.9689	48.4235	234.1909	0.727854	0.529772	13.4267
440.00	1747.9441	65.4566	49.5496	239.6372	0.732510	0.536570	13.4473
450.00	1787.6701	66.9442	50.6758	245.0835	0.737036	0.543223	13.4674
460.00	1827.3961	68.4319	51.8019	250.5298	0.741440	0.549733	13.4871
470.00	1867.1221	69.9195	52.9280	255.9761	0.745724	0.556105	13.5065
480.00	1906.8482	71.4071	54.0542	261.4224	0.749895	0.562342	13.5256
490.00	1946.5742	72.8948	55.1803	266.8687	0.753956	0.568450	13.5444
500.00	1986.3002	74.3824	56.3064	272.3150	0.757911	0.574430	13.5628
510.00	2026.0262	75.8701	57.4325	277.7613	0.761765	0.580286	13.5809
520.00	2065.7522	77.3577	58.5587	283.2076	0.765521	0.586023	13.5988
530.00	2105.4782	78.8454	59.6848	288.6539	0.769183	0.591643	13.6164
540.00	2145.2042	80.3330	60.8109	294.1002	0.772754	0.597149	13.6337
550.00	2184.9302	81.8207	61.9371	299.5465	0.776237	0.602545	13.6508
560.00	2224.6562	83.3083	63.0632	304.9928	0.779636	0.607832	13.6677
570.00	2264.3822	84.7960	64.1893	310.4391	0.782953	0.613015	13.6843
580.00	2304.1082	86.2836	65.3154	315.8854	0.786191	0.618096	13.7007
590.00	2343.8342	87.7713	66.4416	321.3317	0.789353	0.623078	13.7168
600.00	2383.5602	89.2589	67.5677	326.7780	0.792441	0.627963	13.7328
610.00	2423.2862	90.7466	68.6938	332.2243	0.795458	0.632753	13.7486
620.00	2463.0122	92.2342	69.8200	337.6706	0.798406	0.637451	13.7642
630.00	2502.7382	93.7219	70.9461	343.1169	0.801287	0.642060	13.7795
640.00	2542.4642	95.2095	72.0722	348.5632	0.804103	0.646582	13.7947
650.00	2582.1902	96.6972	73.1983	354.0095	0.806857	0.651018	13.8098
660.00	2621.9162	98.1848	74.3245	359.4558	0.809550	0.655372	13.8246
670.00	2661.6422	99.6725	75.4506	364.9021	0.812185	0.659644	13.8393
680.00	2701.3682	101.1601	76.5767	370.3484	0.814762	0.663837	13.8539
690.00	2741.0942	102.6478	77.7029	375.7947	0.817284	0.667954	13.8683
700.00	2780.8202	104.1354	78.8290	381.2410	0.819753	0.671995	13.8825
710.00	2820.5462	105.6231	79.9551	386.6873	0.822170	0.675963	13.8966
720.00	2860.2722	107.1107	81.0812	392.1336	0.824536	0.679859	13.9106
730.00	2899.9982	108.5984	82.2074	397.5799	0.826853	0.683686	13.9244
740.00	2939.7242	110.0860	83.3335	403.0262	0.829123	0.687444	13.9380
750.00	2979.4502	111.5737	84.4596	408.4725	0.831346	0.691136	13.9516
760.00	3019.1762	113.0613	85.5857	413.9188	0.833524	0.694763	13.9650
770.00	3058.9022	114.5490	86.7119	419.3651	0.835659	0.698326	13.9783
780.00	3098.6282	116.0366	87.8380	424.8114	0.837751	0.701827	13.9915
790.00	3138.3542	117.5243	88.9641	430.2577	0.839802	0.705268	14.0045
800.00	3178.0803	119.0119	90.0903	435.7040	0.841813	0.708649	14.0175
810.00	3217.8063	120.4996	91.2164	441.1503	0.843785	0.711973	14.0303
820.00	3257.5323	121.9872	92.3425	446.5966	0.845718	0.715239	14.0430
830.00	3297.2583	123.4749	93.4686	452.0429	0.847615	0.718451	14.0556
840.00	3336.9843	124.9625	94.5948	457.4892	0.849476	0.721609	14.0681

TABLE 8d-1. RELATIVISTIC VELOCITY $\beta = v/c$, β^2, AND STOPPING-NUMBER
FUNCTION $f(\beta)$ FOR HEAVY IONS AS A FUNCTION
OF KINETIC ENERGY T (*Continued*)

Protons, MeV	Alphas, MeV	Pions, MeV	Muons, MeV	Electrons, keV	β	β^2	$f(\beta)$
850.00	3376.7103	126.4502	95.7209	462.9355	0.851301	0.724714	14.0805
860.00	3416.4363	127.9378	96.8470	468.3818	0.853093	0.727767	14.0928
870.00	3456.1623	129.4255	97.9732	473.8281	0.854851	0.730770	14.1050
880.00	3495.8883	130.9131	99.0993	479.2744	0.856576	0.733723	14.1172
890.00	3535.6143	132.4008	100.2254	484.7207	0.858270	0.736628	14.1292
900.00	3575.3403	133.8884	101.3515	490.1670	0.859933	0.739485	14.1411
910.00	3615.0663	135.3761	102.4777	495.6133	0.861566	0.742297	14.1529
920.00	3654.7923	136.8637	103.6038	501.0596	0.863170	0.745063	14.1647
930.00	3694.5183	138.3514	104.7299	506.5059	0.864745	0.747785	14.1763
940.00	3734.2443	139.8390	105.8561	511.9522	0.866293	0.750463	14.1879
950.00	3773.9703	141.3266	106.9822	517.3985	0.867813	0.753099	14.1994
960.00	3813.6963	142.8143	108.1083	522.8448	0.869306	0.755694	14.2108
970.00	3853.4223	144.3019	109.2344	528.2911	0.870774	0.758248	14.2221
980.00	3893.1483	145.7896	110.3606	533.7374	0.872216	0.760762	14.2334
990.00	3932.8743	147.2772	111.4867	539.1837	0.873634	0.763237	14.2446
1000.00	3972.6003	148.7649	112.6128	544.6300	0.875028	0.765673	14.2556

dependent deviations up to 3 percent have been observed for Al_2O_3, SiO_2, and Lucite (TS67 and BT68). At small energies, energy-loss measurements (SZ65, BP71) have also shown deviations from the Bragg rule.

For the approximation with an analytic function, the expression

$$S = CT^\alpha$$

may be used over limited energy ranges; e.g., for protons with $5 < T < 20$ MeV in Ge, $C = 136.7$ and $\alpha = -0.7313$ will be accurate to better than 0.4 percent (see BI68 for other values). If particles of initial energy T are absorbed in a material of thickness s, the mean residual energy $\bar{T}_1$ of the particles can be calculated directly:

$$\bar{T}_1 = (C_R T^\gamma - s)^{1/\gamma}$$

where
$$C_R = (C\gamma)^{-1} \quad \text{and} \quad \gamma = 1 - \alpha.$$

If the stopping power is used to obtain $\bar{T}_1$, successive approximations must be calculated. The computer program of BJ67 produces the coefficients C, C_R, and α.

8d-4. Range-energy Relations. As long as fewer than about 20 percent of the particles are removed from the incident beam by nuclear reactions, the *median* projected range $R_m(T)$ is defined as the thickness of material through which one-half of the incident monoenergetic charged particles of energy T are transmitted (see page 203 of BI68).

The *mean* range of monoenergetic particles of kinetic energy T is defined by

$$R(T) = \int f(R) R \, dR \qquad (8d-13)$$

TABLE 8d-2. CALCULATED MASS STOPPING POWER S/ρ
IN MeV/(G/CM²) FOR PROTONS

T, MeV	I 64 eV Be	78 eV Graphite	66.6 eV Water	166 eV Al	320 eV Cu	475 eV Ag	820 eV Pb
10.0	37.720	40.875	46.641	33.776	27.169	23.213	17.620
10.5	36.252	39.303	44.840	32.531	26.218	22.435	17.068
11.0	34.904	37.858	43.185	31.385	25.341	21.714	16.556
11.5	33.662	36.525	41.666	30.325	24.528	21.045	16.079
12.0	32.513	35.292	40.254	29.343	23.773	20.422	15.633
12.5	31.448	34.147	38.944	28.429	23.069	19.840	15.216
13.0	30.456	33.082	37.724	27.577	22.409	19.294	14.823
13.5	29.531	32.087	36.586	26.779	21.790	18.781	14.454
14.0	28.666	31.156	35.521	26.032	21.209	18.299	14.105
14.5	27.855	30.283	34.522	25.330	20.663	17.844	13.775
15.0	27.094	29.463	33.583	24.669	20.148	17.415	13.463
15.5	26.376	28.690	32.700	24.045	19.662	17.009	13.167
16.0	25.700	27.960	31.865	23.456	19.202	16.625	12.885
16.5	25.061	27.271	31.077	22.898	18.764	16.259	12.618
17.0	24.456	26.618	30.331	22.369	18.348	15.910	12.363
17.5	23.882	25.999	29.623	21.866	17.953	15.579	12.120
18.0	23.337	25.411	28.951	21.389	17.577	15.263	11.888
18.5	22.820	24.852	28.312	20.934	17.218	14.961	11.666
19.0	22.327	24.320	27.703	20.500	16.876	14.673	11.454
19.5	21.857	23.812	27.123	20.086	16.549	14.398	11.251
20.0	21.409	23.327	26.569	19.690	16.237	14.134	11.056
21.0	20.571	22.421	25.534	18.949	15.651	13.639	10.688
22.0	19.802	21.590	24.584	18.268	15.110	13.181	10.348
23.0	19.095	20.824	23.710	17.640	14.609	12.756	10.032
24.0	18.442	20.117	22.902	17.059	14.145	12.362	9.738
25.0	17.837	19.462	22.153	16.519	13.714	11.995	9.464
26.0	17.275	18.852	21.457	16.017	13.312	11.653	9.207
27.0	16.750	18.284	20.808	15.548	12.936	11.333	8.965
28.0	16.261	17.753	20.202	15.109	12.585	11.033	8.738
29.0	15.802	17.256	19.634	14.697	12.254	10.750	8.524
30.0	15.372	16.789	19.101	14.310	11.943	10.483	8.323
31.0	14.967	16.349	18.600	13.946	11.648	10.230	8.132
32.0	14.586	15.935	18.127	13.602	11.370	9.992	7.952
33.0	14.225	15.544	17.681	13.276	11.107	9.766	7.780
34.0	13.885	15.174	17.258	12.969	10.857	9.553	7.617
35.0	13.562	14.823	16.859	12.677	10.620	9.349	7.461
36.0	13.256	14.491	16.479	12.399	10.395	9.156	7.313
37.0	12.965	14.175	16.119	12.135	10.181	8.972	7.172
38.0	12.689	13.874	15.775	11.884	9.977	8.797	7.037
39.0	12.425	13.587	15.449	11.645	9.782	8.629	6.908
40.0	12.174	13.314	15.137	11.416	9.596	8.469	6.785
41.0	11.934	13.053	14.839	11.198	9.418	8.315	6.667
42.0	11.704	12.804	14.555	10.989	9.248	8.167	6.554
43.0	11.485	12.565	14.282	10.788	9.085	8.025	6.445
44.0	11.275	12.336	14.022	10.597	8.928	7.889	6.340

TABLE 8d-2. CALCULATED MASS STOPPING POWER S/ρ
IN MeV/(G/CM²) FOR PROTONS (Continued)

T, MeV	I — 64 eV Be	78 eV Graphite	66.6 eV Water	166 eV Al	320 eV Cu	475 eV Ag	820 eV Pb
45.0	11.073	12.117	13.771	10.413	8.777	7.759	6.239
46.0	10.880	11.906	13.531	10.236	8.632	7.633	6.142
47.0	10.694	11.704	13.301	10.066	8.493	7.513	6.049
48.0	10.515	11.509	13.079	9.903	8.358	7.396	5.958
49.0	10.343	11.322	12.866	9.745	8.229	7.284	5.872
50.0	10.178	11.142	12.660	9.594	8.104	7.176	5.788
52.5	9.790	10.719	12.179	9.238	7.811	6.922	5.590
55.0	9.435	10.333	11.738	8.911	7.543	6.689	5.409
57.5	9.109	9.977	11.333	8.611	7.295	6.475	5.241
60.0	8.808	9.649	10.959	8.334	7.066	6.275	5.085
62.5	8.530	9.345	10.613	8.077	6.854	6.090	4.940
65.0	8.271	9.064	10.293	7.839	6.657	5.917	4.804
67.5	8.031	8.801	9.994	7.616	6.474	5.756	4.678
70.0	7.807	8.557	9.715	7.409	6.302	5.606	4.560
72.5	7.597	8.328	9.454	7.214	6.140	5.465	4.449
75.0	7.400	8.113	9.210	7.032	5.988	5.332	4.344
77.5	7.215	7.911	8.980	6.860	5.846	5.207	4.246
80.0	7.041	7.721	8.764	6.699	5.711	5.090	4.153
82.5	6.877	7.542	8.560	6.546	5.584	4.978	4.065
85.0	6.722	7.373	8.368	6.402	5.463	4.873	3.982
87.5	6.576	7.213	8.185	6.266	5.350	4.773	3.902
90.0	6.437	7.061	8.013	6.136	5.241	4.679	3.827
92.5	6.305	6.917	7.849	6.013	5.139	4.589	3.755
95.0	6.180	6.780	7.693	5.897	5.041	4.503	3.687
97.5	6.060	6.650	7.545	5.785	4.948	4.422	3.622
100.0	5.947	6.526	7.403	5.679	4.859	4.343	3.559
105.0	5.735	6.294	7.140	5.481	4.693	4.197	3.443
110.0	5.541	6.083	6.899	5.300	4.542	4.063	3.337
115.0	5.364	5.888	6.678	5.134	4.402	3.940	3.238
120.0	5.200	5.709	6.475	4.980	4.274	3.826	3.148
125.0	5.049	5.544	6.287	4.839	4.155	3.721	3.064
130.0	4.909	5.391	6.113	4.707	4.044	3.623	2.986
135.0	4.779	5.248	5.951	4.585	3.942	3.532	2.912
140.0	4.657	5.116	5.800	4.471	3.845	3.447	2.844
145.0	4.544	4.992	5.659	4.364	3.755	3.368	2.780
150.0	4.438	4.876	5.527	4.264	3.671	3.293	2.720
155.0	4.338	4.767	5.403	4.171	3.592	3.224	2.664
160.0	4.245	4.664	5.287	4.083	3.517	3.158	2.611
165.0	4.157	4.568	5.177	4.000	3.447	3.096	2.560
170.0	4.073	4.477	5.074	3.921	3.381	3.037	2.513
175.0	3.995	4.391	4.976	3.847	3.318	2.982	2.468
180.0	3.921	4.309	4.883	3.777	3.259	2.929	2.426
185.0	3.850	4.232	4.796	3.710	3.202	2.879	2.386
190.0	3.783	4.159	4.712	3.647	3.149	2.832	2.347
195.0	3.720	4.089	4.633	3.587	3.098	2.787	2.311

TABLE 8d-2. CALCULATED MASS STOPPING POWER S/ρ
IN MEV/(G/CM²) FOR PROTONS (*Continued*)

T, MeV \ I	64 eV Be	78 eV Graphite	66.6 eV Water	166 eV Al	320 eV Cu	475 eV Ag	820 eV Pb
200.0	3.659	4.023	4.558	3.530	3.049	2.744	2.276
205.0	3.601	3.960	4.486	3.475	3.003	2.703	2.243
210.0	3.547	3.900	4.418	3.424	2.959	2.664	2.211
215.0	3.494	3.842	4.353	3.374	2.917	2.626	2.181
220.0	3.444	3.787	4.290	3.326	2.877	2.590	2.152
225.0	3.396	3.735	4.230	3.281	2.838	2.556	2.125
230.0	3.350	3.684	4.173	3.238	2.801	2.523	2.098
235.0	3.306	3.636	4.118	3.196	2.766	2.492	2.072
240.0	3.264	3.590	4.066	3.156	2.732	2.461	2.048
245.0	3.223	3.546	4.015	3.117	2.699	2.432	2.024
250.0	3.184	3.503	3.967	3.081	2.668	2.404	2.001
255.0	3.147	3.462	3.920	3.045	2.638	2.378	1.980
260.0	3.111	3.422	3.876	3.011	2.609	2.352	1.959
265.0	3.076	3.384	3.832	2.978	2.581	2.327	1.938
270.0	3.043	3.348	3.791	2.946	2.554	2.303	1.919
275.0	3.010	3.312	3.751	2.916	2.528	2.280	1.900
280.0	2.979	3.278	3.712	2.886	2.503	2.258	1.882
285.0	2.949	3.245	3.675	2.858	2.479	2.236	1.864
290.0	2.920	3.214	3.639	2.830	2.456	2.215	1.847
295.0	2.892	3.183	3.604	2.804	2.433	2.195	1.831
300.0	2.865	3.153	3.570	2.778	2.411	2.176	1.815
310.0	2.814	3.097	3.506	2.729	2.370	2.139	1.785
320.0	2.766	3.044	3.446	2.683	2.331	2.104	1.757
330.0	2.720	2.994	3.389	2.640	2.294	2.071	1.730
340.0	2.678	2.947	3.336	2.600	2.260	2.041	1.705
350.0	2.637	2.903	3.286	2.562	2.227	2.012	1.681
360.0	2.599	2.862	3.239	2.526	2.196	1.984	1.659
370.0	2.564	2.822	3.194	2.492	2.168	1.958	1.638
380.0	2.530	2.785	3.152	2.459	2.140	1.934	1.618
390.0	2.497	2.750	3.112	2.429	2.114	1.911	1.599
400.0	2.467	2.717	3.074	2.400	2.089	1.889	1.581
410.0	2.438	2.685	3.038	2.372	2.066	1.868	1.564
420.0	2.410	2.655	3.004	2.346	2.044	1.848	1.548
430.0	2.384	2.626	2.971	2.321	2.022	1.829	1.533
440.0	2.359	2.598	2.940	2.297	2.002	1.811	1.518
450.0	2.335	2.572	2.910	2.275	1.983	1.794	1.504
460.0	2.313	2.547	2.882	2.253	1.964	1.778	1.491
470.0	2.291	2.524	2.855	2.232	1.947	1.762	1.478
480.0	2.270	2.501	2.829	2.213	1.930	1.747	1.466
490.0	2.250	2.479	2.804	2.194	1.914	1.733	1.454
500.0	2.231	2.458	2.780	2.176	1.898	1.719	1.443
510.0	2.213	2.438	2.758	2.158	1.884	1.706	1.432
520.0	2.195	2.419	2.736	2.142	1.870	1.693	1.422
530.0	2.179	2.400	2.715	2.126	1.856	1.681	1.412
540.0	2.162	2.383	2.695	2.111	1.843	1.670	1.403

TABLE 8d-2. CALCULATED MASS STOPPING POWER S/ρ
IN MeV/(G/CM^2) FOR PROTONS (*Continued*)

T, MeV	I 64 eV Be	78 eV Graphite	66.6 eV Water	166 eV Al	320 eV Cu	475 eV Ag	820 eV Pb
550.0	2.147	2.366	2.676	2.096	1.830	1.659	1.394
560.0	2.132	2.349	2.657	2.082	1.818	1.648	1.385
570.0	2.118	2.334	2.639	2.068	1.807	1.638	1.377
580.0	2.104	2.319	2.622	2.055	1.796	1.628	1.369
590.0	2.091	2.304	2.606	2.043	1.785	1.618	1.361
600.0	2.078	2.290	2.590	2.030	1.775	1.609	1.353
610.0	2.065	2.277	2.574	2.019	1.765	1.600	1.346
620.0	2.054	2.263	2.560	2.007	1.755	1.592	1.339
630.0	2.042	2.251	2.545	1.997	1.746	1.584	1.333
640.0	2.031	2.239	2.531	1.986	1.737	1.576	1.326
650.0	2.020	2.227	2.518	1.976	1.728	1.568	1.320
660.0	2.010	2.216	2.505	1.966	1.720	1.561	1.314
670.0	2.000	2.205	2.493	1.957	1.712	1.554	1.308
680.0	1.990	2.194	2.481	1.948	1.704	1.547	1.303
690.0	1.981	2.184	2.469	1.939	1.697	1.540	1.297
700.0	1.972	2.174	2.458	1.930	1.690	1.534	1.292
710.0	1.963	2.165	2.447	1.922	1.683	1.528	1.287
720.0	1.955	2.155	2.437	1.914	1.676	1.522	1.282
730.0	1.947	2.146	2.426	1.906	1.669	1.516	1.277
740.0	1.939	2.138	2.417	1.899	1.663	1.510	1.273
750.0	1.931	2.129	2.407	1.892	1.657	1.505	1.268
760.0	1.924	2.121	2.398	1.885	1.651	1.500	1.264
770.0	1.916	2.113	2.389	1.878	1.645	1.495	1.260
780.0	1.909	2.106	2.380	1.871	1.640	1.490	1.256
790.0	1.903	2.098	2.372	1.865	1.634	1.485	1.252
800.0	1.896	2.091	2.363	1.859	1.629	1.480	1.248
810.0	1.890	2.084	2.355	1.853	1.624	1.476	1.245
820.0	1.883	2.077	2.348	1.847	1.619	1.471	1.241
830.0	1.877	2.071	2.340	1.841	1.614	1.467	1.238
840.0	1.871	2.064	2.333	1.836	1.610	1.463	1.235
850.0	1.866	2.058	2.326	1.830	1.605	1.459	1.231
860.0	1.860	2.052	2.319	1.825	1.601	1.455	1.228
870.0	1.855	2.046	2.312	1.820	1.596	1.451	1.225
880.0	1.850	2.040	2.306	1.815	1.592	1.448	1.222
890.0	1.845	2.035	2.299	1.811	1.588	1.444	1.220
900.0	1.840	2.029	2.293	1.806	1.585	1.441	1.217
910.0	1.835	2.024	2.287	1.801	1.581	1.437	1.214
920.0	1.830	2.019	2.282	1.797	1.577	1.434	1.212
930.0	1.826	2.014	2.276	1.793	1.574	1.431	1.209
940.0	1.821	2.009	2.270	1.789	1.570	1.428	1.207
950.0	1.817	2.005	2.265	1.785	1.567	1.425	1.204
960.0	1.813	2.000	2.260	1.781	1.563	1.422	1.202
970.0	1.809	1.996	2.255	1.777	1.560	1.419	1.200
980.0	1.805	1.991	2.250	1.773	1.557	1.417	1.198
990.0	1.801	1.987	2.245	1.770	1.554	1.414	1.196
1000.0	1.797	1.983	2.240	1.766	1.551	1.412	1.193

where $f(R)$ is the experimentally measured distribution function (the "probability density" of the mathematicians) and can be determined quite readily in cloud or bubble chambers and in photographic emulsions (except for problems connected with the last bubble or grain). It is not a practical quantity for experiments in which the tracks of the particles cannot be followed. In particular, the *mean projected* range is difficult to determine experimentally because of the removal of particles from the beam due to nuclear reactions and multiple scattering.

At energies higher than a few MeV, the number of particles is sensibly reduced owing to nuclear reactions (KO64, BI60, and BA61), and appropriate corrections must be applied (see Sec. 8d-8 under Nuclear Reactions).

The quantity related to $R(T)$ which can be calculated from stopping-power theory is the theoretical mean range $R_t(T)$ in the continuous slowing-down approximation (csda):

$$R_t(T) = \int_{T_1}^{T} S^{-1} \, dT \tag{8d-14}$$

In principle, T_1 is the thermal energy of the particle. For small velocities the description of the stopping power given in Sec. 8d-6 under Very Low Velocity Particles can be used. If S is not known accurately at these energies, a more accurate result for $R_t(T)$ may be obtained when T_1 is chosen to be a higher energy (e.g., 1 MeV for protons), and an experimental value of $R(T_1)$ is added to the integral to take care of the low-energy contribution to the range. For experimental measurements it will be necessary to consider the detector threshold energy as the energy T_1 (BM57 and HP60).

A small difference between $R(T)$ and $R_t(T)$ is caused by the use of the csda approximation (LE52 and TT68). A simple relation that exists between the ranges for different particles is discussed in Sec. 8d-6.

Mean csda ranges for protons in several elements have been computed (BJ67) by numerical integration of the values of Tables 8d-2 and 8d-3. They are listed in Table 8d-4. Values for R (1 MeV) are obtained from BF60, MR67, and RY55. For other elements, the method of SU60 can be used to obtain range-energy relations. For other particles (mesons or heavier ions) see Sec. 8d-6. Extensive tabulations can be found in JA66, BJ69, BB67, and NO67. For high energies ($T > 1000$ MeV for protons) nuclear interactions absorb most of the particles, and range becomes a rather meaningless term.

While the straggling in pathlength can be represented approximately by a gaussian (see Sec. 8d-7), the asymmetry of multiple scattering (the zigzag path taken by a particle can only be longer than the foil thickness, see Sec. 8d-8), and the residual skewness of the electron-loss straggling causes an asymmetry in the range straggling. The *median* range therefore, is different from the *mean* range.

The total median range $R_m(T)$ (equal to the foil thickness), neglecting the straggling asymmetries, can be obtained from the computed mean pathlength $R_t(T)$ by the application of the multiple-scattering correction ΔR:

$$R_m(T) = R_t(T) - \Delta R$$

The relative correction of $\Delta R/R$ for several elements is plotted in Fig. 8d-5. Further discussion is given in BU60, BF61, BZ67, and TB68. No discussion of the relation of mean and median range seems to be available (see Sec. 8d-7).

8d-5. Shell Corrections and I Values. In principle, the stopping power S can be calculated theoretically using atomic collision cross sections [Eq. (8d-5)]. At present, no complete sets of cross sections for all shells are available, and the expression Eq. (8d-11) is used for the calculation of S. The unknown functions B_K, B_L, ... are then replaced by one unknown constant, $I = I_{av}$, and the unknown functions

TABLE 8d-3A. Low-energy Proton Stopping Power S in MeV/(g/cm²) for Several Substances: Accuracy 2 to 10%

T, MeV	H_2	He	Li	Be	C	N_2	O_2	Ne	Al	A	Ni	Cu	Kr	Ag	Sn	Xe	Au	Pb	Air	H_2O
0.01	...	...	...	...	440	...	...	...	260	...	100	70	...	...	...	...	22	...	...	...
0.02	...	...	...	...	560	...	...	...	360	...	145	160	...	...	...	...	44	...	...	...
0.03	...	...	...	...	640	...	...	...	410	...	177	190	...	...	...	...	60	...	...	...
0.04	...	...	...	...	700	...	...	...	440	...	200	200	...	...	...	...	75	...	...	...
0.05	3800	1050	...	690	720	750	600	350	460	480	220	210	270	...	...	240	85	...	730	890
0.10	3500	1090	750	700	710	780	610	440	440	480	260	220	290	...	...	230	105	122	730	910
0.15	2800	960	680	640	650	690	600	440	390	430	270	220	250	...	...	210	112	127	650	830
0.20	2300	830	610	570	580	610	540	420	340	380	260	220	220	...	...	192	119	127	580	740
0.25	1990	740	550	510	540	530	500	390	320	330	250	210	198	...	...	176	116	120	520	660
0.30	1740	660	500	460	490	480	450	360	310	300	230	200	182	...	...	163	110	113	480	600
0.35	1560	600	450	430	460	440	410	340	290	270	220	192	169	...	...	152	104	106	430	540
0.40	1410	550	420	390	430	400	380	320	280	250	210	183	159	151	142	143	98	100	410	500
0.45	1280	510	390	370	390	370	360	300	270	230	193	175	150	142	134	134	93	95	380	460
0.50	1180	480	360	350	370	350	340	290	250	220	182	168	143	134	127	127	88	90	350	430
0.55	1090	450	340	330	350	320	320	270	240	210	173	161	137	128	121	121	84	86	330	400
0.60	1020	420	320	310	330	310	300	260	230	200	165	155	132	122	115	115	81	83	310	380
0.70	910	380	290	280	290	280	270	230	210	184	151	144	123	113	107	106	75	77	280	340
0.80	810	340	260	260	270	250	250	210	197	171	141	135	116	105	100	98	70	71	260	310
0.90	740	310	240	240	250	240	230	198	185	160	133	128	109	99	94	92	66	67	240	290
1.00	680	290	230	220	230	220	220	185	173	150	126	121	104	94	89	87	63	63	220	260
1.1	630	270	220	210	220	210	200	174	163	142	120	114	99	89	85	82	60	60	210	260
1.2	590	250	210	198	200	194	192	164	155	134	115	109	94	85	81	78	57	58	198	230
1.3	550	240	200	187	192	185	182	156	147	127	110	104	90	82	78	75	54	55	186	220
1.4	520	220	195	179	183	176	173	149	140	121	106	99	87	79	75	72	53	53	177	210
1.5	500	210	188	170	175	168	165	144	134	116	101	95	84	75	71	69	51	52	168	197
1.6	470	200	184	161	167	160	157	137	129	112	97	92	81	72	68	66	49	50	160	188
1.8	430	183	173	148	154	148	144	127	119	103	91	85	76	68	65	62	47	47	147	172
2.0	390	168	164	137	143	137	134	119	111	95	85	80	72	64	61	59	44	45	136	159

TABLE 8d-3B. EXPERIMENTAL PROTON STOPPING POWER S IN $\mathrm{MeV}/(\mathrm{G}/\mathrm{CM}^2)$*

T, MeV	Be	Al	Ca	Sc	Ti	V	Cr	Mn	Fe	Co	Ni	Cu	Zn	Zr	Ag	Gd	Ta	Pt	Au
2.00	134.25	110.67	107.21	96.58	93.19	90.61	89.57	86.51	87.30	83.74	86.45	81.09	80.89	71.49	63.74	55.31	49.27	45.43	45.78
2.25	122.70	101.92	98.91	89.24	86.11	83.73	82.93	80.07	80.83	77.64	80.14	75.19	75.02	66.56	59.63	51.65	46.20	42.85	43.12
2.50	113.21	94.68	92.03	83.15	80.23	78.02	77.41	74.72	75.45	72.56	74.89	70.28	70.13	62.44	56.18	48.58	43.62	40.67	40.87
2.75	105.19	88.52	86.15	77.94	75.20	73.14	72.69	70.13	70.85	68.18	70.35	66.07	65.92	58.83	53.16	45.87	41.35	38.71	38.89
3.00	98.35	83.19	81.09	73.40	70.87	68.91	68.60	66.16	66.84	64.37	66.41	62.44	62.32	55.70	50.46	43.51	39.35	36.97	37.11
3.25	92.42	78.56	76.65	69.45	67.11	65.22	65.01	62.66	63.32	61.04	62.96	59.25	59.16	52.92	48.07	41.42	37.59	35.39	35.52
3.50	87.24	74.51	72.76	65.97	63.78	61.93	61.82	59.59	60.21	58.09	59.97	56.43	56.36	50.42	45.94	39.57	36.02	33.96	34.09
3.75	82.65	70.94	69.30	62.87	60.81	59.04	58.98	56.86	57.45	55.46	57.28	53.91	53.85	48.20	44.02	37.91	34.60	32.66	32.79
4.00	78.57	67.76	66.18	60.08	58.15	56.44	56.42	54.40	55.00	53.10	54.88	51.65	51.60	46.20	42.28	36.42	33.32	31.47	31.62
4.25	74.90	64.85	63.38	57.57	55.74	54.09	54.09	52.19	52.79	50.97	52.69	49.60	49.55	44.40	40.70	35.06	32.15	30.39	30.55
4.50	71.60	62.21	60.83	55.29	53.54	51.95	51.97	50.16	50.78	49.03	50.70	47.72	47.69	42.74	39.26	33.83	31.07	29.40	29.56
4.75	68.58	59.80	58.51	53.21	51.53	50.00	50.04	48.31	48.94	47.26	48.86	45.99	45.99	41.23	37.93	32.69	30.07	28.49	28.65
5.00	65.87	57.59	56.37	51.30	49.68	48.21	48.25	46.62	47.24	45.62	47.17	44.42	44.42	39.84	36.71	31.65	29.15	27.65	27.79
5.25	63.36	55.56	54.41	49.53	47.98	46.56	46.61	45.05	45.67	44.11	45.61	42.95	42.97	38.55	35.58	30.69	28.29	26.85	27.00
5.50	61.06	53.68	52.59	47.90	46.40	45.03	45.09	43.60	44.22	42.72	44.16	41.59	41.61	37.36	34.53	29.80	27.49	26.12	26.25
5.75	58.93	51.93	50.91	46.38	44.97	43.61	43.68	42.25	42.87	41.41	42.87	40.33	40.36	36.25	33.55	28.96	26.75	25.43	25.56
6.00	56.96	50.31	49.35	44.97	43.56	42.29	42.37	40.99	41.60	40.20	41.57	39.15	39.19	35.22	32.63	28.18	26.06	24.79	24.91
6.50	53.42	47.38	46.53	42.41	41.10	39.90	39.98	38.72	39.31	38.00	39.30	37.01	37.07	33.35	30.97	26.77	24.79	23.60	23.72
7.00	50.34	44.81	44.04	40.17	38.92	37.80	37.89	36.71	37.28	36.06	37.29	35.12	35.19	31.69	29.48	25.50	23.66	22.54	22.67
7.50	47.62	42.52	41.83	38.17	36.99	35.93	36.02	34.92	35.47	34.33	35.51	33.44	33.52	30.21	28.14	24.37	22.65	21.58	21.73
8.00	45.21	40.47	39.85	36.38	35.27	34.26	34.35	33.32	33.85	32.78	33.90	31.93	32.01	28.89	26.94	23.36	21.73	20.72	20.85
8.50	43.05	38.64	38.08	34.77	33.71	32.75	32.85	31.88	32.39	31.37	32.45	30.57	30.65	27.69	25.85	22.43	20.89	19.94	20.06
9.00	41.10	36.97	36.47	33.31	32.30	31.39	31.49	30.57	31.06	30.09	31.14	29.33	29.41	26.60	24.86	21.59	20.13	19.22	19.34
9.50	39.34	35.46	35.00	31.98	31.02	30.15	30.25	29.38	29.85	28.92	29.95	28.21	28.29	25.60	23.95	20.82	19.42	18.56	18.67
10.00	37.74	34.08	33.66	30.76	29.84	29.01	29.11	28.28	28.74	27.85	28.84	27.17	27.26	24.69	23.11	20.11	18.77	17.95	18.06
10.50	36.27	32.82	32.43	29.65	28.77	27.97	28.06	27.28	27.72	26.87	27.83	26.22	26.30	23.85	22.34	19.46	18.17	17.39	17.49
11.00	34.93	31.66	31.30	28.61	27.77	27.01	27.10	26.35	26.78	25.96	26.89	25.35	25.42	23.07	21.62	18.85	17.62	16.88	16.97
11.50	33.69	30.58	30.25	27.66	26.85	26.12	26.21	25.49	25.90	25.12	26.02	24.54	24.61	22.35	20.96	18.29	17.10	16.39	16.48
12.00	32.54	29.58	29.28	26.77	26.00	25.30	25.37	24.69	25.09	24.33	25.21	23.78	23.85	21.68	20.34	17.77	16.62	15.93	16.02

* From AH67, AS68, and AV69.

TABLE 8d-4. CALCULATED CSDA RANGES R IN G/CM^2 FOR PROTONS
OF KINETIC ENERGY T

T, MeV	Be	Graphite	Water	Al	Cu	Ag	Pb
1.0	0.0029	0.0039	0.0039	0.0042	0.0061	0.0080	0.0116
1.1	0.0034	0.0043	0.0043	0.0048	0.0070	0.0091	0.0133
1.2	0.0039	0.0048	0.0047	0.0054	0.0078	0.0103	0.0151
1.3	0.0044	0.0053	0.0051	0.0061	0.0088	0.0115	0.0169
1.4	0.0050	0.0059	0.0056	0.0068	0.0098	0.0128	0.0188
1.5	0.0055	0.0064	0.0061	0.0075	0.0108	0.0141	0.0208
1.6	0.0062	0.0070	0.0066	0.0083	0.0118	0.0154	0.0228
1.7	0.0068	0.0076	0.0071	0.0091	0.0129	0.0168	0.0248
1.8	0.0075	0.0083	0.0077	0.0099	0.0141	0.0183	0.0270
1.9	0.0082	0.0089	0.0083	0.0108	0.0153	0.0198	0.0291
2.0	0.0089	0.0096	0.0089	0.0117	0.0165	0.0213	0.0314
2.1	0.0097	0.0104	0.0095	0.0126	0.0178	0.0229	0.0336
2.2	0.0105	0.0111	0.0101	0.0136	0.0190	0.0245	0.0360
2.3	0.0113	0.0119	0.0108	0.0146	0.0204	0.0262	0.0384
2.4	0.0121	0.0127	0.0115	0.0156	0.0218	0.0279	0.0408
2.5	0.0130	0.0135	0.0122	0.0166	0.0232	0.0296	0.0433
2.6	0.0139	0.0143	0.0130	0.0177	0.0246	0.0314	0.0458
2.7	0.0148	0.0152	0.0137	0.0188	0.0261	0.0332	0.0484
2.8	0.0158	0.0161	0.0145	0.0200	0.0276	0.0351	0.0511
2.9	0.0167	0.0170	0.0153	0.0211	0.0291	0.0370	0.0538
3.0	0.0177	0.0180	0.0161	0.0223	0.0307	0.0390	0.0565
3.1	0.0188	0.0189	0.0170	0.0236	0.0323	0.0409	0.0593
3.2	0.0198	0.0199	0.0178	0.0248	0.0340	0.0430	0.0621
3.3	0.0209	0.0209	0.0187	0.0261	0.0357	0.0450	0.0650
3.4	0.0220	0.0220	0.0196	0.0274	0.0374	0.0471	0.0680
3.5	0.0232	0.0230	0.0205	0.0287	0.0392	0.0493	0.0709
3.6	0.0243	0.0241	0.0215	0.0301	0.0409	0.0515	0.0740
3.7	0.0255	0.0252	0.0225	0.0315	0.0428	0.0537	0.0771
3.8	0.0267	0.0263	0.0234	0.0329	0.0446	0.0559	0.0802
3.9	0.0279	0.0275	0.0245	0.0344	0.0465	0.0582	0.0834
4.0	0.0292	0.0287	0.0255	0.0358	0.0484	0.0605	0.0866
4.1	0.0305	0.0299	0.0265	0.0373	0.0504	0.0629	0.0899
4.2	0.0318	0.0311	0.0276	0.0389	0.0524	0.0653	0.0932
4.3	0.0331	0.0323	0.0287	0.0404	0.0544	0.0677	0.0965
4.4	0.0345	0.0336	0.0298	0.0420	0.0564	0.0702	0.0999
4.5	0.0359	0.0349	0.0309	0.0436	0.0585	0.0727	0.1034
4.6	0.0373	0.0362	0.0321	0.0453	0.0606	0.0753	0.1069
4.7	0.0387	0.0375	0.0332	0.0469	0.0628	0.0778	0.1104
4.8	0.0402	0.0389	0.0344	0.0486	0.0649	0.0805	0.1140
4.9	0.0416	0.0403	0.0356	0.0503	0.0672	0.0831	0.1176
5.0	0.0432	0.0417	0.0369	0.0521	0.0694	0.0858	0.1213
5.5	0.0510	0.0490	0.0433	0.0612	0.0810	0.0997	0.1403
6.0	0.0595	0.0569	0.0502	0.0709	0.0934	0.1145	0.1603
6.5	0.0686	0.0653	0.0575	0.0812	0.1066	0.1302	0.1814
7.0	0.0782	0.0742	0.0653	0.0922	0.1205	0.1466	0.2035
7.5	0.0884	0.0837	0.0736	0.1038	0.1351	0.1639	0.2267
8.0	0.0992	0.0937	0.0824	0.1160	0.1504	0.1820	0.2508
8.5	0.1106	0.1042	0.0915	0.1288	0.1664	0.2008	0.2759
9.0	0.1225	0.1152	0.1012	0.1421	0.1831	0.2205	0.3019
9.5	0.1349	0.1266	0.1112	0.1561	0.2005	0.2409	0.3289

TABLE 8d-4. CALCULATED CSDA RANGES R IN G/CM^2 FOR PROTONS
OF KINETIC ENERGY T (*Continued*)

T, MeV	Be	Graphite	Water	Al	Cu	Ag	Pb
10.0	0.1479	0.1386	0.1217	0.1706	0.2186	0.2620	0.3568
10.5	0.1614	0.1511	0.1327	0.1857	0.2373	0.2840	0.3856
11.0	0.1755	0.1641	0.1440	0.2013	0.2567	0.3066	0.4154
11.5	0.1901	0.1775	0.1558	0.2175	0.2768	0.3300	0.4460
12.0	0.2052	0.1915	0.1680	0.2343	0.2975	0.3541	0.4776
12.5	0.2208	0.2059	0.1807	0.2516	0.3188	0.3790	0.5100
13.0	0.2370	0.2207	0.1937	0.2695	0.3408	0.4045	0.5433
13.5	0.2536	0.2361	0.2072	0.2879	0.3635	0.4308	0.5775
14.0	0.2708	0.2519	0.2211	0.3068	0.3867	0.4578	0.6125
14.5	0.2885	0.2682	0.2353	0.3263	0.4106	0.4855	0.6484
15.0	0.3067	0.2849	0.2500	0.3463	0.4351	0.5138	0.6851
15.5	0.3254	0.3021	0.2651	0.3668	0.4602	0.5429	0.7226
16.0	0.3446	0.3198	0.2806	0.3879	0.4860	0.5726	0.7610
16.5	0.3643	0.3379	0.2965	0.4095	0.5123	0.6030	0.8002
17.0	0.3845	0.3564	0.3128	0.4316	0.5393	0.6341	0.8403
17.5	0.4052	0.3755	0.3295	0.4542	0.5668	0.6659	0.8811
18.0	0.4264	0.3949	0.3465	0.4773	0.5950	0.6983	0.9228
18.5	0.4481	0.4148	0.3640	0.5009	0.6237	0.7314	0.9652
19.0	0.4702	0.4351	0.3819	0.5251	0.6530	0.7651	1.0085
19.5	0.4929	0.4559	0.4001	0.5497	0.6830	0.7995	1.0525
20.0	0.5160	0.4771	0.4187	0.5748	0.7135	0.8346	1.0974
21.0	0.5637	0.5209	0.4571	0.6266	0.7762	0.9066	1.1894
22.0	0.6132	0.5663	0.4970	0.6804	0.8412	0.9812	1.2845
23.0	0.6646	0.6135	0.5385	0.7361	0.9086	1.0584	1.3826
24.0	0.7179	0.6624	0.5814	0.7937	0.9781	1.1380	1.4838
25.0	0.7731	0.7129	0.6258	0.8533	1.0499	1.2201	1.5880
26.0	0.8301	0.7651	0.6717	0.9148	1.1240	1.3047	1.6952
27.0	0.8889	0.8190	0.7190	0.9782	1.2002	1.3918	1.8052
28.0	0.9495	0.8745	0.7678	1.0434	1.2786	1.4812	1.9182
29.0	1.0119	0.9317	0.8180	1.1106	1.3591	1.5730	2.0341
30.0	1.0760	0.9904	0.8696	1.1795	1.4418	1.6672	2.1529
31.0	1.1420	1.0508	0.9227	1.2503	1.5266	1.7638	2.2744
32.0	1.2096	1.1127	0.9772	1.3229	1.6135	1.8627	2.3988
33.0	1.2791	1.1763	1.0330	1.3974	1.7025	1.9640	2.5259
34.0	1.3502	1.2414	1.0903	1.4736	1.7935	2.0675	2.6558
35.0	1.4231	1.3081	1.1489	1.5516	1.8867	2.1733	2.7885
36.0	1.4977	1.3763	1.2089	1.6313	1.9818	2.2814	2.9239
37.0	1.5740	1.4461	1.2703	1.7129	2.0791	2.3918	3.0620
38.0	1.6520	1.5174	1.3330	1.7961	2.1783	2.5043	3.2027
39.0	1.7316	1.5903	1.3971	1.8811	2.2795	2.6191	3.3462
40.0	1.8129	1.6646	1.4625	1.9679	2.3827	2.7361	3.4922
41.0	1.8959	1.7405	1.5292	2.0563	2.4879	2.8553	3.6409
42.0	1.9805	1.8178	1.5972	2.1465	2.5951	2.9766	3.7922
43.0	2.0668	1.8967	1.6666	2.2383	2.7042	3.1002	3.9461
44.0	2.1546	1.9770	1.7373	2.3319	2.8152	3.2258	4.1025
45.0	2.2441	2.0588	1.8092	2.4271	2.9282	3.3537	4.2615
46.0	2.3353	2.1421	1.8825	2.5239	3.0431	3.4836	4.4231
47.0	2.4280	2.2268	1.9570	2.6224	3.1599	3.6157	4.5871
48.0	2.5223	2.3129	2.0329	2.7226	3.2786	3.7498	4.7537
49.0	2.6182	2.4006	2.1099	2.8244	3.3992	3.8861	4.9228

Table 8d-4. Calculated csda Ranges R in g/cm^2 for Protons of Kinetic Energy T (Continued)

T, MeV	Be	Graphite	Water	Al	Cu	Ag	Pb
50.0	2.7156	2.4896	2.1883	2.9278	3.5216	4.0244	5.0943
52.5	2.9661	2.7184	2.3897	3.1934	3.8359	4.3792	5.5339
55.0	3.2263	2.9560	2.5988	3.4690	4.1617	4.7466	5.9886
57.5	3.4960	3.2023	2.8156	3.7545	4.4988	5.1266	6.4583
60.0	3.7752	3.4571	3.0400	4.0496	4.8470	5.5188	6.9426
62.5	4.0637	3.7204	3.2718	4.3544	5.2063	5.9233	7.4415
65.0	4.3614	3.9921	3.5111	4.6686	5.5764	6.3398	7.9547
67.5	4.6681	4.2720	3.7576	4.9922	5.9573	6.7682	8.4821
70.0	4.9839	4.5602	4.0113	5.3251	6.3487	7.2083	9.0235
72.5	5.3086	4.8563	4.2722	5.6671	6.7507	7.6601	9.5786
75.0	5.642	5.161	4.540	6.018	7.163	8.123	10.147
77.5	5.984	5.473	4.815	6.378	7.586	8.598	10.729
80.0	6.335	5.792	5.097	6.747	8.018	9.083	11.325
82.5	6.694	6.120	5.386	7.124	8.461	9.580	11.933
85.0	7.062	6.455	5.681	7.511	8.914	10.088	12.555
87.5	7.438	6.798	5.983	7.905	9.376	10.606	13.189
90.0	7.822	7.149	6.292	8.309	9.848	11.135	13.836
92.5	8.215	7.506	6.607	8.720	10.330	11.675	14.495
95.0	8.615	7.871	6.929	9.140	10.821	12.225	15.167
97.5	9.024	8.244	7.257	9.568	11.322	12.785	15.852
100.0	9.440	8.623	7.592	10.004	11.832	13.355	16.548
105.0	10.297	9.404	8.279	10.901	12.879	14.527	17.976
110.0	11.184	10.212	8.992	11.828	13.962	15.738	19.452
115.0	12.101	11.047	9.729	12.787	15.081	16.988	20.973
120.0	13.048	11.910	10.489	13.776	16.233	18.276	22.539
125.0	14.024	12.799	11.273	14.795	17.420	19.601	24.150
130.0	15.029	13.713	12.080	15.843	18.640	20.963	25.803
135.0	16.061	14.654	12.909	16.919	19.893	22.361	27.499
140.0	17.121	15.619	13.760	18.024	21.177	23.794	29.236
145.0	18.208	16.008	14.633	19.156	22.493	25.262	31.015
150.0	19.322	17.622	15.527	20.315	23.840	26.763	32.833
155.0	20.462	18.659	16.442	21.500	25.217	28.298	34.691
160.0	21.627	19.720	17.378	22.712	26.623	29.865	36.587
165.0	22.817	20.803	18.334	23.950	28.059	31.464	38.521
170.0	24.032	21.909	19.309	25.212	29.524	33.095	40.492
175.0	25.272	23.037	20.304	26.500	31.017	34.757	42.500
180.0	26.536	24.186	21.319	27.812	32.538	36.449	44.544
185.0	27.823	25.357	22.352	29.147	34.086	38.171	46.622
190.0	29.133	26.549	23.404	30.507	35.661	39.922	48.735
195.0	30.466	27.761	24.474	31.889	37.262	41.702	50.882
200.0	31.821	28.994	25.562	33.294	38.888	43.510	53.063
205.0	33.199	30.247	26.668	34.722	40.541	45.346	55.276
210.0	34.598	31.519	27.791	36.171	42.218	47.210	57.521
215.0	36.018	32.811	28.932	37.643	43.920	49.100	59.798
220.0	37.460	34.122	30.089	39.135	45.646	51.018	62.106
225.0	38.922	35.451	31.262	40.649	47.396	52.961	64.444
230.0	40.404	36.799	32.452	42.183	49.169	54.930	66.813
235.0	41.907	38.165	33.659	43.737	50.965	56.924	69.211
240.0	43.429	39.549	34.880	45.312	52.784	58.943	71.638
245.0	44.971	40.951	36.118	46.906	54.625	60.987	74.094

TABLE 8d-4. CALCULATED CSDA RANGES R IN G/CM^2 FOR PROTONS
OF KINETIC ENERGY T (Continued)

T, MeV	Be	Graphite	Water	Al	Cu	Ag	Pb
250.0	46.531	42.370	37.371	48.520	56.489	63.054	76.578
255.0	48.111	43.806	38.639	50.152	58.373	65.145	79.090
260.0	49.709	45.258	39.921	51.804	60.279	67.260	81.630
265.0	51.325	46.727	41.219	53.473	62.206	69.397	84.196
270.0	52.960	48.213	42.531	55.161	64.154	71.557	86.789
275.0	54.612	49.714	43.857	56.867	66.121	73.739	89.408
280.0	56.282	51.232	45.197	58.591	68.109	75.943	92.052
285.0	57.968	52.765	46.551	60.332	70.116	78.169	94.722
290.0	59.672	54.313	47.918	62.090	72.143	80.416	97.416
295.0	61.392	55.876	49.299	63.865	74.188	82.683	100.135
300.0	63.129	57.455	50.693	65.657	76.253	84.971	102.877
310.0	66.651	60.655	53.520	69.289	80.437	89.608	108.434
320.0	70.236	63.912	56.397	72.984	84.692	94.322	114.081
330.0	73.883	67.225	59.323	76.741	89.017	99.113	119.818
340.0	77.588	70.591	62.297	80.558	93.410	103.978	125.641
350.0	81.352	74.010	65.318	84.434	97.868	108.914	131.548
360.0	85.171	77.479	68.383	88.365	102.389	113.920	137.535
370.0	89.045	80.998	71.492	92.352	106.973	118.993	143.601
380.0	92.972	84.565	74.644	96.392	111.616	124.131	149.744
390.0	96.951	88.178	77.836	100.484	116.317	129.333	155.960
400.0	100.980	91.837	81.070	104.626	121.075	134.597	162.249
410.0	105.058	95.540	84.342	108.818	125.889	139.921	168.607
420.0	109.183	99.286	87.653	113.057	130.756	145.304	175.033
430.0	113.355	103.074	91.000	117.342	135.675	150.743	181.525
440.0	117.572	106.902	94.384	121.673	140.644	156.237	188.081
450.0	121.832	110.770	97.803	126.048	145.663	161.785	194.699
460.0	126.136	114.677	101.256	130.465	150.731	167.384	201.378
470.0	130.481	118.621	104.742	134.924	155.844	173.035	208.115
480.0	134.866	122.602	108.261	139.423	161.003	178.734	214.909
490.0	139.291	126.619	111.812	143.962	166.207	184.482	221.758
500.0	143.754	130.670	115.393	148.539	171.453	190.276	228.662
510.0	148.255	134.755	119.005	153.154	176.741	196.115	235.618
520.0	152.792	138.873	122.645	157.805	182.070	201.999	242.624
530.0	157.365	143.023	126.314	162.492	187.439	207.925	249.681
540.0	161.972	147.205	130.011	167.213	192.846	213.893	256.785
550.0	166.614	151.417	133.735	171.967	198.291	219.902	263.937
560.0	171.288	155.658	137.486	176.755	203.773	225.951	271.134
570.0	175.994	159.929	141.262	181.574	209.290	232.038	278.375
580.0	180.732	164.228	145.063	186.425	214.842	238.162	285.660
590.0	185.501	168.555	148.889	191.306	220.428	244.323	292.987
600.0	190.299	172.908	152.739	196.216	226.047	250.520	300.355
610.0	195.126	177.288	156.612	201.156	231.697	256.751	307.763
620.0	199.982	181.693	160.508	206.124	237.380	263.017	315.211
630.0	204.865	186.124	164.426	211.119	243.093	269.315	322.696
640.0	209.775	190.579	168.365	216.140	248.835	275.645	330.218
650.0	214.712	195.057	172.326	221.188	254.607	282.006	337.776
660.0	219.675	199.559	176.308	226.262	260.406	288.397	345.369
670.0	224.662	204.083	180.309	231.360	266.234	294.819	352.996
680.0	229.675	208.630	184.330	236.482	272.088	301.269	360.657
690.0	234.711	213.198	188.371	241.628	277.968	307.747	368.349

TABLE 8d-4. CALCULATED CSDA RANGES R IN G/CM^2 FOR PROTONS OF KINETIC ENERGY T (Continued)

T, MeV	Be	Graphite	Water	Al	Cu	Ag	Pb
700.0	239.770	217.787	192.430	246.797	283.874	314.253	376.074
710.0	244.853	222.396	196.507	251.989	289.805	320.785	383.829
720.0	249.957	227.026	200.603	257.202	295.760	327.344	391.614
730.0	255.084	231.676	204.715	262.437	301.739	333.928	399.428
740.0	260.231	236.344	208.845	267.693	307.741	340.537	407.270
750.0	265.400	241.031	212.991	272.969	313.765	347.170	415.140
760.0	270.589	245.737	217.154	278.265	319.812	353.827	423.037
770.0	275.797	250.460	221.332	283.581	325.880	360.507	430.960
780.0	281.025	255.201	225.526	288.916	331.968	367.209	438.909
790.0	286.272	259.959	229.736	294.269	338.078	373.933	446.882
800.0	291.537	264.733	233.960	299.640	344.207	380.679	454.880
810.0	296.821	269.524	238.198	305.029	350.355	387.445	462.902
820.0	302.122	274.330	242.451	310.435	356.523	394.232	470.946
830.0	307.440	279.152	246.717	315.858	362.709	401.039	479.013
840.0	312.775	283.989	250.997	321.297	368.913	407.864	487.102
850.0	318.127	288.841	255.290	326.752	375.134	414.709	495.212
860.0	323.494	293.708	259.596	332.223	381.373	421.572	503.343
870.0	328.878	298.588	263.914	337.710	387.628	428.454	511.494
880.0	334.276	303.483	268.245	343.211	393.900	435.352	519.665
890.0	339.690	308.391	272.588	348.727	400.188	442.268	527.855
900.0	345.119	313.312	276.943	354.257	406.492	449.200	536.064
910.0	350.562	318.246	281.309	359.802	412.810	456.149	544.291
920.0	356.019	323.193	285.686	365.359	419.144	463.113	552.536
930.0	361.489	328.151	290.075	370.930	425.492	470.093	560.798
940.0	366.973	333.123	294.474	376.514	431.854	477.088	569.077
950.0	372.470	338.105	298.883	382.111	438.230	484.098	577.373
960.0	377.981	343.100	303.303	387.720	444.619	491.122	585.684
970.0	383.503	348.105	307.733	393.341	451.022	498.160	594.012
980.0	389.038	353.122	312.173	398.974	457.437	505.211	602.354
990.0	394.585	358.149	316.622	404.619	463.865	512.276	610.711
1000.0	400.143	363.187	321.081	410.275	470.305	519.354	619.083

C_i, which are important only at small energies. If extensive experimental data are available, the shell corrections, $C/Z = \Sigma_i C_i/Z$, can be determined experimentally (AN69), together with the I value. Usually, experimental uncertainties and limited coverage in energy do not permit this approach. In a modification of an earlier approach (BI61), it is suggested now, that, for $8 \leq Z \leq 25$, Walske's shell corrections (WA52, WA56, BI67, KH68) be used in modified form:

$$\frac{C}{Z} = \frac{C_K + VC_L(H\beta^2)}{Z} \qquad (8d\text{-}15)$$

with parameters H, V, and I determined in a least-squares fit to experimental data. Similarly, for $Z \leq 8$, $C/Z = VC_K(H\beta^2)$. For $Z \geq 25$, Bonderup's shell corrections C_B (BO67) are used, also in a modified form:

$$\frac{C}{Z} = \frac{VC_B(Hv^2/v_0^2Z)}{Z} \qquad (8d\text{-}16)$$

Good fits to experimental data for protons and deuterons are obtained as long as $C_B \geq 0$. Values for H, V, and I may be found in BJ67. Typically, for $Z \geq 47$, $H = 0.755$, $V = 0.68$, and $I_{Ag} = 476$ eV, $I_{Au} = 780$ eV. For $Z = 29$, $H = 0.55$, $V = 0.61$, and $I_{Cu} = 319.5$ eV. These fits include effects due to the higher Born approximations and are therefore valid only for particles of charge $+e$.

It was found that the least-squares fits do not show singular and distinct minima. For experimental data covering a limited energy range, different local minima will give almost the same χ^2. This is fairly obvious from Eq. (8d-11): for a limited velocity range, an increase in I can be almost entirely compensated by a decrease in the shell corrections (BH69).

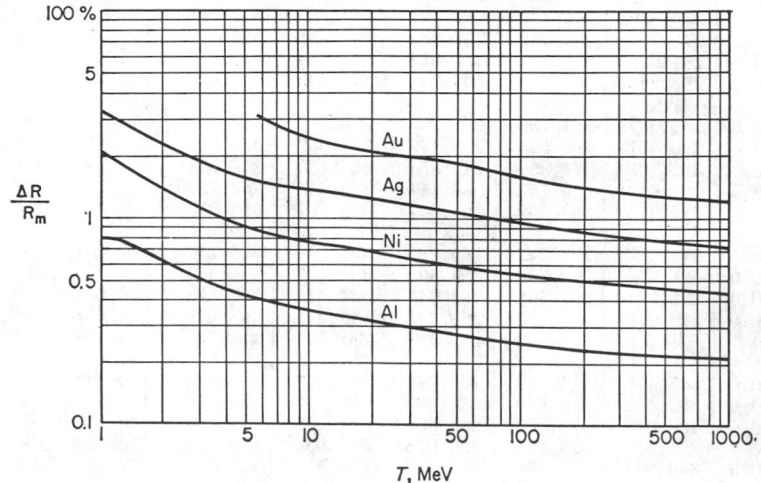

Fig. 8d-5. The fractional multiple-scattering correction for different elements as a function of proton energy T. The experimental median projected range R_m is related to the csda range R_0 through $R_0 = R_m + \Delta R$. Corrections due to nuclear diffraction scattering are neglected. Accuracy 10 to 20 percent.

Values of C/Z for protons and deuterons adopted in this section are given in Fig. 8d-4.

Although I values are properly defined by

$$\ln I_{av} \equiv \sum_n f_n \ln I_n \qquad (8d\text{-}17)$$

(DT68), only a few values for light elements have been calculated with this expression (BE66, WH33). They are not as accurate as the experimental values. The quotient $k = I/Z$ is expected to be a constant if I is evaluated using the Thomas-Fermi model (BL33). Figure 8d-5 shows a plot of the best available values of k. Both the rise of k for $20 \leq Z \leq 30$ and the oscillation for even and odd values are unexpected. The interpolation schemes suggested in the past (DT68) cannot be considered reliable, and further measurements appear to be very desirable. Recent data are given in VK69.

8d-6. Miscellaneous Effects. A difference in the ranges of positive and negative mesons has been observed (BD63,HL69). Similarly, Andersen, Simonsen, and Sørensen (AS69) found a difference between the stopping power of particles of charge one (p,d) and of charge two (He^3, He^4). This difference presumably is caused by effects due to higher Born approximations. In the further discussions of this section, these effects are implicitly included in the definition of z^*.

The first Born approximation used in the derivation of the collision cross sections [Eq. (8d-2)] is valid for $\beta \gg \beta_1 = z/137$. For particles with $\beta < \beta_1$, atom-atom collisions will contribute increasingly to the stopping process, and an approach based on the use of the Thomas-Fermi model of both the incident ions (with an effective charge $z^*e < ze$) and the absorber atoms has been fruitful (see Very Low Velocity Particles in this section).

The stopping power S_M for any particle of mass M, nuclear charge ze (values for different particles are given in Table 8d-5), and kinetic energy T can be calculated from the proton stopping power S_p with

$$S_M(T) = z^{*2}S_p(\tau) \tag{8d-18}$$

where $\tau = T/m_r$, and z^* is discussed under Charge-state Correction in this section. Similarly, a simple relation exists between the range R_M of the particle and the range R_p of a proton:

$$R_M(T) = \frac{m_r}{z^2} R_p(\tau) + m_r z^{\frac{1}{3}} C_z \left(\frac{\beta}{z}\right) \tag{8d-19}$$

where $m_r = Mc^2/938.259$ MeV, and the second term is called the range extension caused by the reduced charge z^*. C_z is a universal function for any ion in a specific substance. For emulsion, C_z is found in fig. 5 of HP60, and it is defined for any substance in eq. (7) of HP60 (see BB67 for data). Another approach can be used: Use Eq. (8d-19) to find the range difference $R_M(T) - R_M(T_1)$ and add $R_M(T_1)$ as defined under Very Low Velocity Particles to find $R_M(T)$.

In general, a numerical calculation for a specific case, using Eq. (8d-11) with appropriate effective charge z^* will be preferable to the use of Eq. (8d-19).

EXAMPLES

1. The mean range of 20-MeV muons in aluminum ($m_r = 0.1126$ from Table 8d-5)

$$R_\mu(20 \text{ MeV}) = 0.1126 \times R_p(177.6 \text{ MeV}) = 0.1126 \times 27.15 = 3.057 \text{ g/cm}^2$$

2. The mean range of 50-MeV alphas ($m_r = 3.9726$) in copper is

$$R_\alpha(50 \text{ MeV}) = \frac{3.9726}{4} \times R_p(12.602 \text{ MeV}) = 0.3219 \text{ g/cm}^2$$

where R_p is obtained from Table 8d-4, and C_z has been neglected. An extensive discussion for heavy ions is given in NO67 and NS70, with many graphs for different incident particles.

Charge-state Correction. For velocities $\beta < \beta_2 = 0.04z^{\frac{2}{3}}$ it is observed that the nuclear charge ze is not fully effective. A reduced effective charge z^*e is used in Eq. (8d-11) instead of the nuclear charge ze (RO60, NO63, NO67, HP60). If z^* is defined to give the correct observed stopping power, it is not equal to the mean charge per particle of a beam leaving an absorber (PB68, BG65). With an accuracy of about 5 percent, z^* can be obtained from

$$\frac{z^*}{z} = 1 - \exp\left(-1.316x + 0.1112x^2 - 0.0650x^3\right) \tag{8d-20}$$

where $x = 100\beta/z^{\frac{2}{3}}$. This expression is valid for $x > 0.27$. In gases, the values are several percent smaller (AR69). It should be noted that the approach described in the next section overlaps the range of validity of Eq. (8d-20).

TABLE 8d-5. PROPERTIES OF CHARGED PARTICLES*
(Electron Masses to Be Divided by 1,000)

Ion	z	Lifetime, nanosec	Charge, $10^{-19}\,C$	Mass 10^{-24} g	Mass amu	Mass MeV	m_r
Electron	1	Stable	±1.60219	0.910956	0.548593	511.004	0.544630
Muon	1	2198.3	±1.60219	0.188357	0.113432	105.6598	0.112613
Pion	1	26.04	±1.60219	0.248823	0.149846	139.578	0.148763
Kaon	1	12.35	±1.60219	0.880322	0.530147	493.82	0.526317
Sigma +	1	0.081	+1.60219	2.120318	1.276895	1189.40	1.267671
Sigma −	1	0.164	−1.60219	2.134436	1.285398	1197.32	1.276112
		Mass excess, MeV					
1 N	0	8.0714	0.	1.674920	1.0086652	939.553	1.0013786
1 H	1	7.2890	1.60219	1.672614	1.0072766	938.259	1.0000000
2 H	1	13.1359	1.60219	3.343569	2.0135536	1875.587	1.9990076
3 H	1	14.9500	1.60219	5.007334	3.0155011	2808.883	2.9937170
3 He	2	14.9313	3.20438	5.006390	3.0149325	2808.353	2.9931526
4 He	2	2.4248	3.20438	6.644626	4.0015059	3727.328	3.9725990
6 Li	3	14.0884	4.80658	9.985570	6.0134789	5601.443	5.9700375
7 Li	3	14.9073	4.80658	11.647561	7.0143581	6533.743	6.9636862
7 Be	4	15.7689	6.40877	11.648186	7.0147345	6534.093	6.9640599
9 Be	4	11.3505	6.40877	14.961372	9.0099911	8392.637	8.9449027
10 B	5	12.0522	8.01096	16.622243	10.0101958	9324.309	9.9378820
11 B	5	8.6677	8.01096	18.276741	11.0065623	10252.406	10.9270507
12 C	6	0.	9.61315	19.920910	11.9967084	11174.708	11.9100440
13 C	6	3.1246	9.61315	21.587011	13.0000629	12109.314	12.9061502
14 C	6	3.0198	9.61315	23.247356	13.9999504	13040.691	13.8988145
14 N	7	2.8637	11.21534	23.246166	13.9992342	13040.024	13.8981035
15 N	7	0.1004	11.21534	24.901771	14.9962676	13968.741	14.8879343
16 O	8	−4.7365	12.81753	26.552769	15.9905263	14894.875	15.8750105
17 O	8	−0.8077	12.81753	28.220304	16.9947441	15830.285	16.8719738
18 O	8	−0.7824	12.81753	29.880881	17.9947713	16761.791	17.8647767
19 F	9	−1.4860	14.41973	31.539247	18.9934674	17692.058	18.8562582
20 Ne	10	−7.0415	16.02192	33.188963	19.9869546	18617.472	19.8425685
21 Ne	10	−5.7299	16.02192	34.851833	20.9883627	19550.265	20.8367424
22 Ne	10	−8.0249	16.02192	36.508273	21.9858989	20479.451	21.8270724
23 Na	11	−9.5283	17.62411	38.165213	22.9837363	21408.918	22.8177014
24 Mg	12	−13.9333	19.22630	39.816981	23.9784587	22335.483	23.8052379
25 Mg	12	−13.1907	19.22630	41.478836	24.9792559	23267.707	24.7988053
26 Mg	12	−16.2142	19.22630	43.133977	25.9760100	24196.165	25.7883589
27 Al	13	−17.1961	20.82849	44.791847	26.9744073	25126.153	26.7795437
28 Si	14	−21.4899	22.43068	46.443813	27.9692490	26052.830	27.7671987
29 Si	14	−21.8936	22.43068	48.103625	28.9688156	26983.907	28.7595444
30 Si	14	−24.4394	22.43068	49.759617	29.9660826	27912.843	29.7496071
31 P	15	−24.4376	24.03288	51.419241	30.9655359	28843.815	30.7418404
32 S	16	−26.0127	25.63507	53.076053	31.9632963	29773.210	31.7323930
33 S	16	−26.5826	25.63507	54.735569	32.9626845	30704.121	32.7245615
34 S	16	−29.9335	25.63507	56.390126	33.9590871	31632.251	33.7137661
36 S	16	−30.6650	25.63507	59.709903	35.9583126	33494.492	35.6985491
35 Cl	17	−29.0145	27.23726	58.051385	34.9595251	32564.140	34.7069770
37 Cl	17	−31.7648	27.23726	61.367545	36.9565725	34424.353	36.6895976
36 Ar	18	−30.2316	28.83945	59.708836	35.9576699	33493.894	35.6979111
38 Ar	18	−34.7182	28.83945	63.021900	37.9528533	35352.369	37.6786813
40 Ar	18	−35.0383	28.83945	66.342392	39.9525096	37215.012	39.6638921
39 K	19	−33.8033	30.44164	64.683151	38.9532869	36284.254	38.6718877
40 K	19	−33.5333	30.44164	66.344164	39.9535768	37216.006	39.6649515
40 Ca	20	−34.8476	32.04383	66.340910	39.9516172	37214.180	39.6630060

* From refs. BB69, TP69, and MT65.

For ions with $21 \leq z \leq 39$, Hvelplund and Fastrup (HV68, FB68) have found a periodic dependence of the stopping cross section on z for a carbon absorber. Similar effects were found in WI68, NA69, and HA68. Fractional charges for carbon absorbers in CC67 agree with Eq. (8d-20) to better than 5 percent for most ions. The fluctuations for different absorbers found in table III of that reference could be due to shell corrections.

When available, experimental data should be used. Recent papers include:

Br and I ions in Be, C, Al, Ag, Au	MB66
O^{16} ions in Ag, Au; S^{32} ions in Au	AH68
S^{32}, Cl^{35}, Br^{79}, I^{127} ions in Mylar	PB68
O and Cl ions in C, Al, Ni, Ag, Au	BG65
I^{127} ions in C, Al, Ni, Ag, Au, UF_4	BN67
C, N, O, F, Ne in Be, C	CB68
$21 \leq z \leq 39$ in C	HV68
$3 \leq z \leq 13$ in Ar	AR69

Interesting results for charge-state populations (I^{127} in gas and solid) have been found by Moak et al. (ML68). Many references to earlier work are included.

Very Low Velocity Particles. At low velocities, $\beta \leq \beta_1 = z^{\frac{2}{3}}/137$, ions will carry a reduced charge, and for $\beta \ll \beta_0 = 1/137 = 0.0073$, they will be neutral. The collisions then will be between neutral atoms, and are commonly called *nuclear collisions* (LS63, OH63). Even for this case, energy loss to atomic electrons is still possible (LS63). From a Thomas-Fermi description of the atoms, it is expected that the following dimensionless parameters should result in universal range-energy curves:

Energy:
$$\epsilon = 32.53 \times T(\text{keV}) \, M_2/[zZ(M_1 + M_2) \sqrt{\zeta}] \qquad (8\text{d-}21)$$

Range:
$$\rho = 1.660 \times 10^5 \times R(\text{mg cm}^{-2}) \frac{M_1}{(M_1 + M_2)^2 \zeta} \qquad (8\text{d-}22)$$

where M_1 = atomic mass of incident particle
M_2 = atomic mass of absorber material
z = atomic number of incident particle (usually called Z_1)
Z = atomic number of absorber material (usually called Z_2)
$\zeta = z^{\frac{2}{3}} + Z^{\frac{2}{3}}$

It is found that the stopping power consists of contributions by electronic and nuclear stopping:

$$S = S_e + S_n \qquad (8\text{d-}23)$$

From (LS63),

$$S_e = k \sqrt{\epsilon} \qquad (8\text{d-}24)$$

where

$$k = 0.0793 \times \xi_e \sqrt{zZ} \frac{(M_1 + M_2)^{\frac{3}{2}}}{[(z^{\frac{2}{3}} + Z^{\frac{2}{3}})^{\frac{3}{4}} M_1^{\frac{3}{2}} M_2^{\frac{1}{2}}]} \qquad (8\text{d-}25)$$

and ξ_e is approximately given by $z^{\frac{1}{6}}$. This formula is valid for $\epsilon < 1000$.

The nuclear collision stopping power depends on the ion-atom potential (discussed, e.g., in NV66, KE68, LS63, LN68). From table I of SC66, the following analytic form has been derived (similar to an expression given in BS68):

$$\left(\frac{d\epsilon}{d\rho}\right)_n = \frac{0.5455 \ln \epsilon}{\epsilon(1 - 0.9988 \times \epsilon^{-1.5391})} \qquad (8\text{d-}26)$$

and

$$S_n = 1.96 \times 10^{-4} \frac{d\epsilon}{d\rho} M_2(M_1 + M_2) \frac{\sqrt{\zeta}}{zZM_1} \qquad (8\text{d-}27)$$

$$S_n \text{ in keV/(mg/cm}^2)$$

It is seen that $(d\epsilon/d\rho)_n$ is a universal function of ϵ, while S_e, through k, depends on zZ. It is therefore only possible to produce a universal range curve

$$\rho(\epsilon) = \int_0^\epsilon \frac{d\epsilon'}{(d\epsilon'/d\rho)_n}$$

for the nuclear collisions, and if the electronic collisions are of importance, different range curves will be obtained for different values of k.

Different quantities have been defined to describe the path taken by the particle: linear range (total pathlength), vector range (vector distance from point of incidence to stopping point), and projected range (projection of vector range onto direction of incidence). A particle will experience only few collisions: e.g., for $T = 12$ keV argon atoms in a germanium absorber, the mean collision number is ~ 6 (KE68). Both statistical and continuous methods have been used to calculate mean ranges.

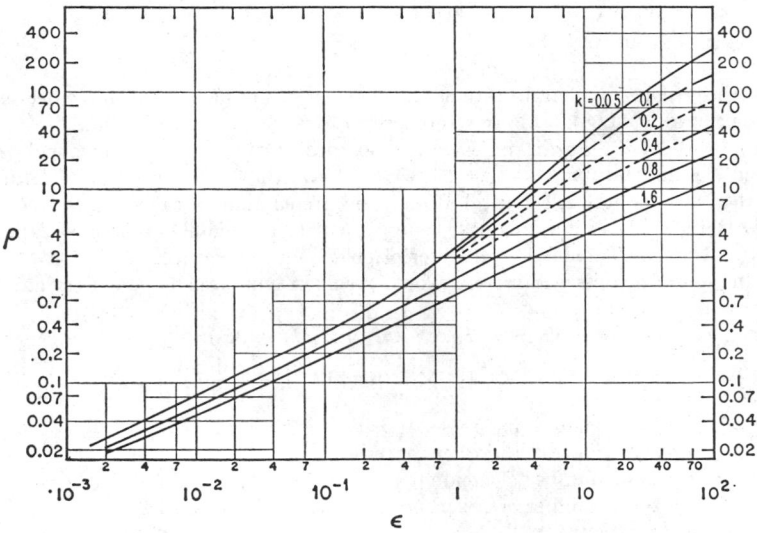

Fig. 8d-6. Range-energy relation for low-energy ions. The dimensionless paramenters ϵ (for the kinetic energy) and ρ (for the range) are defined in Eqs. (8d-21) and (8d-22). The parameter k [Eq. (8d-25)] is related to the low-energy electronic stopping power.

For $M_1 \geq M_2$, the ratio of mean projected range $\overline{R_p}$ and linear range R is approximately $R/\overline{R_p} \sim 1 + M_2/3M_1$ (LS63, MS65). A modification of this procedure is suggested in MS65, giving a better agreement with experiment for $\epsilon < 1$.

Using Eqs. (8d-14) and (8d-27), range-energy curves have been calculated (SC66) for different values of k, and are plotted in Fig. 8d-6. In general, the agreement between theory and experiment is satisfactory, with accuracies of about 20 percent: AG68, BL68, BS68, CA68, JD67, LS67. The use of logarithmic scales in the plots of experimental data tends to hide the differences. Usually, the value of k in Eq. (8d-24) is considered an adjustable parameter, and better agreement with the theory can then be achieved (e.g., CS68, CB68).

Moak and Brown (MB66) and Kahn and Forgue (KF67) have found deviations from the $\sqrt{\epsilon}$ behavior predicted by Eq. (8d-24) for $\epsilon \sim 200$. The deviations in k for light elements are not unexpected: the Thomas-Fermi model may not give a good approximation for $Z < 20$.

At higher values of ϵ (say, $\epsilon > 300$), the approach presented here overlaps with the Bethe theory using effective charges (see under Charge-state Correction in this section), and experimental data have to be consulted to find the more reliable approach. Useful data are found in AG69 for protons with $0.5 \leq T \leq 30$ keV in 10 materials.

Small Volumes. The energy losses discussed in Sec. 8d-3 are as experienced by the charged particles and are not directly related to the energy gained by the absorber material (see the discussion of LET in Sec. 8d-9).

EXAMPLES

1. For an energy $T = 50$ MeV, in a silicon detector of the transmission type thicker than 5 mg/cm^2 $\simeq$ 20 μm, in a vacuum, about 5 percent of all the protons will each knock out delta rays of mean energy 40 keV. The mean energy loss $\bar{\Delta}$ of all protons is reduced by 2 keV. The most probable energy loss Δ_p will be changed much less, though. Contrary to expectation, the spectrum of these delta ray losses is proportional to $E^{0.23}$.

2. In very small volumes (diameter of 1 μm or less of a material of density $\rho = 1$ g/cm^3, corresponding to the size of living cells), the energy lost by a particle of moderate or large energy is quite uncorrelated to the energy absorbed in the volume. Since the behavior of low-energy electrons is not well known (energies of less than 1 keV), and since the collision cross sections are not known for low-Z materials, calculations are extremely unreliable at present (KL68, EB70).

Channeling. In single crystals it is observed that energy loss depends on the direction of the particle path with respect to the crystal axes. A detailed discussion of various aspects of the problem is given by Lindhard (LI65). Other calculations are available in several of the experimental papers mentioned below and in BR68.

If particles travel parallel to a major axis of the lattice, some can move "in between" the atoms, reducing the number of collisions with small impact parameters (energy loss and straggling would then both be reduced; see AE67 and DM69) while others would move close to nuclear positions, increasing the effects. For a well-collimated beam with small multiple scattering, a fraction of the beam may keep away from atoms for long distances.

A number of experiments have recently been published: an especially instructive diagram is given in RS67, a study of 3- to 11-MeV protons in Si and Ge is of interest for the use of solid-state detectors (AE67). Other studies are described in DW68, DM69, ER67, RO69, SV68.

8d-7. Straggling of Heavy Particles. Particles, in passing through an absorber of thickness s, experience a random number of collisions with a wide range of possible energy transfers. The energy losses Δ of a monoenergetic beam of particles thus will fluctuate ("straggle") about the mean energy loss $\bar{\Delta} = sS$.

The straggling distribution function $f(\Delta)$ depends only slightly on the properties of the incident particle (β,z) and the material (Z,A,S). It is highly asymmetric for small $\bar{\Delta}$, reaching minimum asymmetry for $\bar{\Delta} \simeq 0.5T$.

Straggling theories frequently are based on the use of the moments μ_n of the distribution functions (SY48, TT68, PA69):

$$\mu_n = \int f(\Delta)\, \Delta^n\, d\Delta \qquad \mu_0 \equiv 1$$

Also used are the central moments C_n:

$$C_n = \int f(\Delta)\, (\Delta - \bar{\Delta})^n\, d\Delta \qquad \bar{\Delta} = \mu_1 \qquad\qquad (8d\text{-}28)$$

and the moments M_n of the primary collision cross section:

$$M_n = \int W^n\, d\sigma \qquad [d\sigma,\ \text{e.g., from Eq. (8d-2)}]$$

Thin Absorbers. Simple equations relate the moments (FA53):

$$\mu_1 = sM_1 \qquad C_3 = sM_3$$
$$C_2 = sM_2 \qquad C_4 = sM_4 + 3s^2M_2{}^2 \qquad \text{etc.}$$

If an experimental straggling function (e.g., NI61, KO68, MR68, AL69) is known to have negligible spurious contributions for large Δ (e.g., from slit edge scattering or delta ray escape losses) the comparison of C_n with M_n is much simpler than the comparison with theoretical straggling functions. There is no simple relation between the full-width-at-half-maximum (FWHM) and $\sigma = \sqrt{C_2}$ (SB67) except for a gaussian: FWHM $= 2.355\sigma$.

Landau (LA44), Symon (SY48), Vavilov (VA57), Bichsel (BJ70) and others have discussed straggling in thin absorbers. Most of these calculations are based on the use of the free electron collision spectrum. Thus:

$$M'_n = \int E^n d\sigma'_r = \frac{0.1535\, Z\, z^2}{\beta^2\, A} \frac{E_{\max}^{n-1}}{n-1} \left[1 - \beta^2 \frac{(n-1)}{n} \right] \qquad \text{(MeV)}^n\, \text{cm}^2/\text{g} \qquad (8d\text{-}29)$$

using the relativistic form of Eq. (8d-1):

$$d\sigma'_r = d\sigma' \left(1 - \beta^2 \frac{E}{E_{\max}} \right) \qquad \text{and} \qquad E_{\max} \simeq \frac{2\, mc^2\, \beta^2}{(1 - \beta^2)}$$

In particular (e.g., BO48):

$$M'_2 = \frac{(0.1569 Z z^2)}{A} \frac{1 - \beta^2/2}{1 - \beta^2} \qquad \text{(MeV)}^2\, \text{cm}^2/\text{g} \qquad (8d\text{-}30)$$

The Vavilov parameter

$$\kappa = 0.1503 Z z^2\, (1 - \beta^2)\, s/(A\beta^4) \qquad (8d\text{-}31)$$

is used customarily in the discussion of $f(\Delta)$. Extensive tables of $f(\Delta)$ according to Vavilov are given in SB67. It should be noted that the numerical convergence of the Vavilov calculation is unsatisfactory for $\xi = \kappa E_{\max} < 7\ \text{I}$ (HB68).

No complete discussions are available based on the use of Eq. (8d-2). An estimate of the effect can be obtained from Figs. 8d-7 and 8d-8. For the K-shell, the ratios are somewhat smaller (BI69), and they are not expected to be much different for the outer shells $(M, N, \ldots)$. Experimental data confirm this assumption (Fig. 7 of NI61).

Corrections to the Vavilov functions using Eq. (8d-2) for L-shell electrons are discussed in BJ70. The corrections are especially important for $\kappa < D_2 = M_2/M'_2 - 1$ (BL50, BK58: the quantity b^2 used in these papers is equal to $2\, D_2/\kappa$).

For applications in thin silicon detectors, see Fig. 8d-9 (taken from BI70).

Thick Absorbers. An extensive discussion for large energy losses is given by Symon (SY48), by Tschalär (TS67, TS68, TT68), and by Payne (PA69). For experimental results, see TM70. For moderate energy losses, Tschalär's results for heavy particles of initial kinetic energy T and residual mean energy T_1 can be approximated by the following expression for the second moment (accurate to about 2 percent):

$$C_2 = s M'_2 Q$$

where
$$Q = \left(\frac{T}{T_1} \right)^{\frac{1}{3}} \qquad \text{for} \quad \frac{B}{Z} \sim 2.3 \quad \text{and} \quad \frac{T_1}{T} > 0.4$$

$$= 0.99 \left(\frac{T}{T_1} \right)^{\frac{1}{2}} \qquad \frac{B}{Z} \sim 3.5 \qquad \frac{T_1}{T} > 0.4$$

$$= 0.985 \left(\frac{T}{T_1} \right)^{\frac{2}{3}} \qquad \frac{B}{Z} \sim 6.9 \qquad \frac{T_1}{T} > 0.6$$

where B is the stopping number, Eq. (8d-8), and

$$\frac{B}{Z} = f(\beta) - \ln I - \sum_i \frac{C_i}{Z}$$

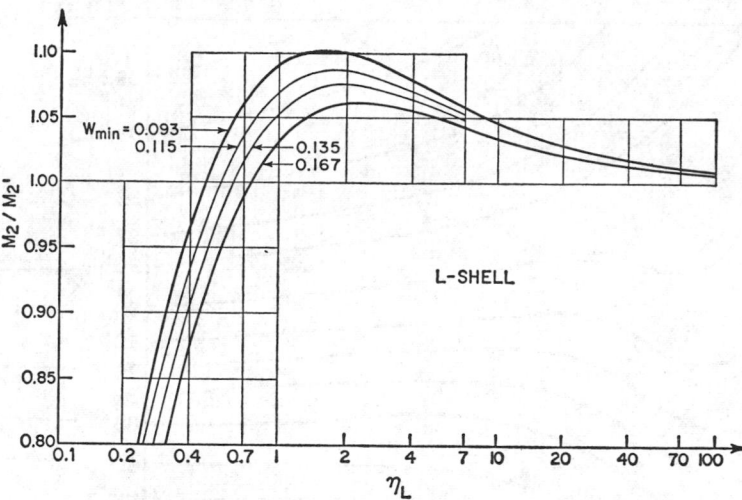

FIG. 8d-7. The ratio M_2/M_2' of the second moments of the quantum-mechanical [Eq. (8d-2)] and the free-electron cross sections [Eq. (8d-1)] for the L shell. The curves apply for silicon ($W_{min} = 0.093$), copper (0.115) silver (0.135), and lead (0.167).

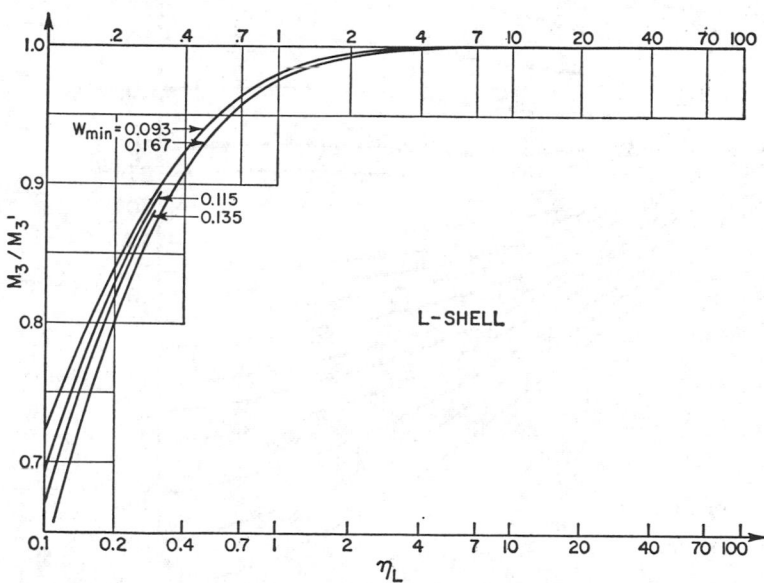

FIG. 8d-8. The ratio M_3/M_2' of the third moments for the L shell (see Fig. 8d-8 for the elements). Notice that the asymmetry (skewness) is reduced at lower energies.

For larger energy losses, TS68 should be consulted. For the asymmetry of the curves, the third moment should be studied. Tschalär uses the skewness parameter $\gamma_3' = C_3/C_2^{3/2}$ for this purpose. From his results it is found that the expression for thin absorbers, $\gamma_3' = sM_3'/(sM_2')^{3/2}$, is accurate to a few percent for $B/Z \sim 2.3$ and $T_1/T > 0.5$ and for $B/Z \sim 6$ and $T_1/T > 0.7$. It may be noted that the distribution func-

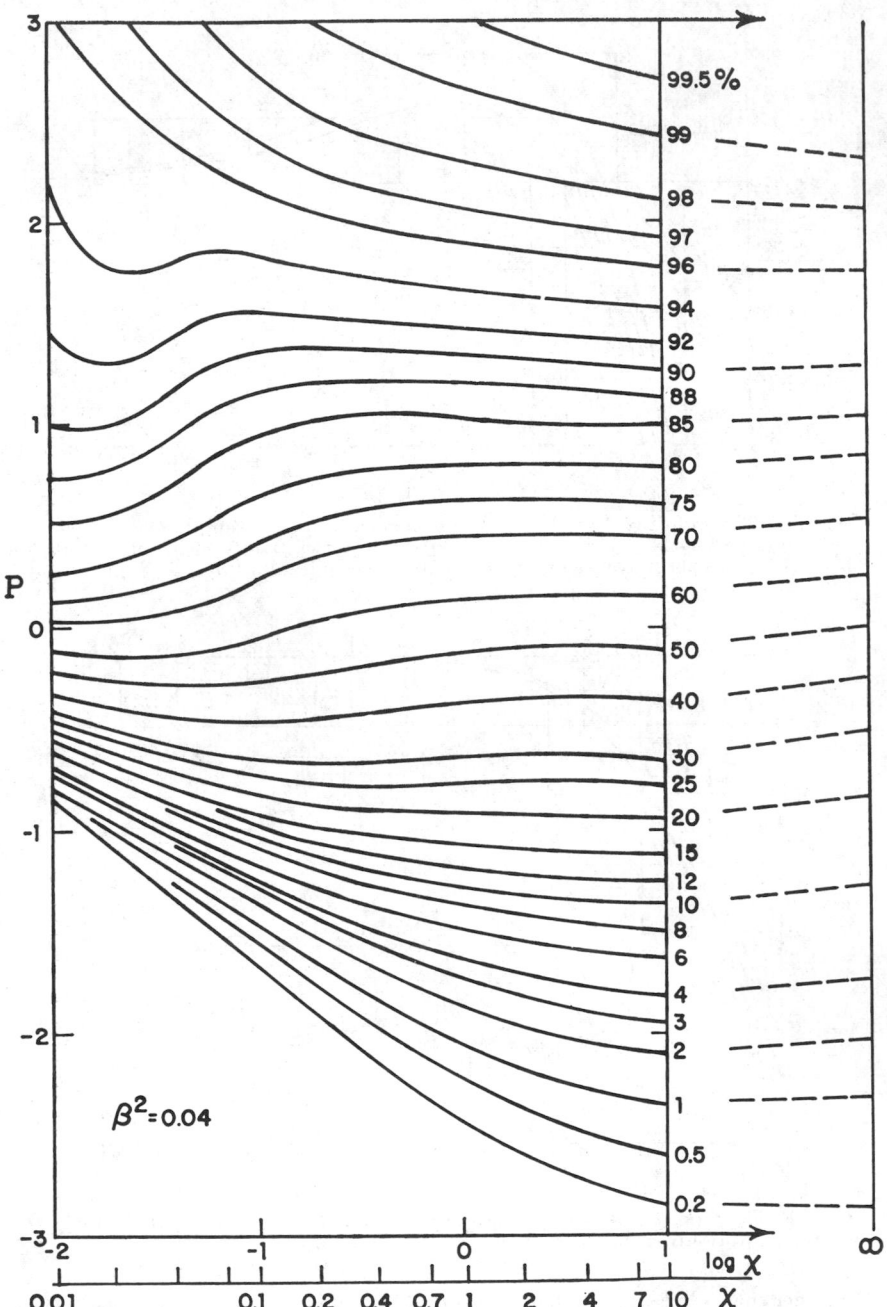

Fɪɢ. 8d-9. Contour lines for the straggling distribution function $\Phi\,[\Phi(\Delta_u) = \int_0^{\Delta_u} f(\Delta)\, ds$ where $f(\Delta)$ is the Vavilov function] in silicon for particles of velocity $\beta^2 = 0.04$ ($T \sim 20$ MeV for protons). The curves are similar for other velocities. The Vavilov theory has been

tions for the cases discussed above are approximately given by the Vavilov functions for the value $\kappa_v = 0.25\gamma_3{}^{-2}$ of the Vavilov parameter $\kappa_v = \xi/E_{\max}$ (SB67).

For the ranges R of particles with a mean value $\bar{R}$ the second central moment, also called the *mean-square fluctuation* σ^2 is defined by

$$\sigma^2 = \langle R^2 \rangle - \bar{R}^2 \tag{8d-32}$$

The distribution $f(R)$ is usually approximated by a gaussian:

$$f(R) \approx \frac{1}{\sigma \sqrt{2\pi}} \exp\left[-\frac{(R - \bar{R})^2}{2\sigma^2} \right] \tag{8d-33}$$

and the probability p of finding a particle with range between R and $R + dR$ is $p\,dR = f(R)\,dR$. The deviations from a gaussian are small, but not negligible. They are discussed in LE52 and TT68. Their influence on the Bragg curve has not been studied yet (VK69).

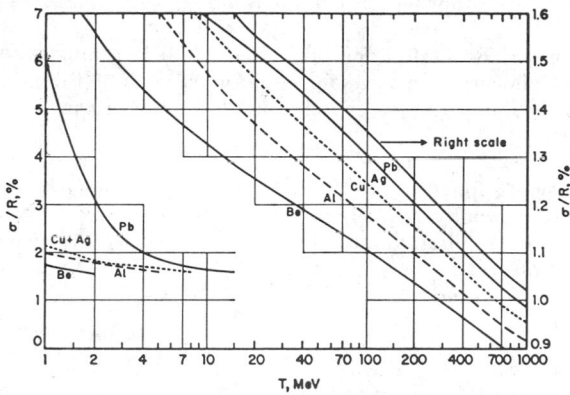

FIG. 8d-10. The range straggling parameter, σ/R (percent), for protons of kinetic energy T in different elements. σ/R is corrected for the quantum-mechanical effects (estimated from Fig. 8d-8).

The ratio of σ to the total mean range R is given in Fig. 8d-10 for protons in several elements. For other particles of mass M, the value can be calculated from

$$\frac{\sigma}{R}(T)_M = \sqrt{\frac{1}{m_r}} \frac{\sigma}{R}(\tau)_{\text{proton}} \tag{8d-34}$$

Estimates for the quantum-mechanical corrections have been incorporated in the calculations for Fig. 8d-10. The values of σ/R are considerably smaller than the

modified for the quantum-mechanical corrections. The Vavilov parameter is

$$\kappa_V = 7.49 \times 10^{-2} s z^2 (1 - \beta^2)/\beta^2$$

(for silicon; s in g cm^{-2}). Plotted is the energy loss p (dimensionless) which exceeds the energy loss of Φ percent of the incident particles. The actual energy loss is $\Delta = \bar{\Delta} + p\sigma$, where $\bar{\Delta}$ is the mean energy loss ($\bar{\Delta} = sS$), and σ is the standard deviation

$$\sigma^2 = 78{,}250 s z^2 (1 - \beta^2/2)/(1 - \beta^2)\ \text{keV}^2.$$

EXAMPLE: 40-MeV protons, $s = 0.02$ g cm^{-2}, $\beta^2 = 0.08$, $\kappa_V = 0.22$, $\bar{\Delta} = 0.02 \times 11.72 = 0.234$ MeV, $\sigma = 40$ keV. For $\Delta = \bar{\Delta}$, $p = 0$, and about 58 percent of all the protons lose less than 234 keV. The exact answer is 61.6 percent. On the other hand, for $\Phi = 96$ percent, $p \sim 2.0$, and $\Delta = 234 + 80$ keV $= 314$ keV. Thus. 4 percent of the protons lose more than 314 keV (the exact answer is 315 keV).

values calculated by Sternheimer (ST60), but they are still slightly larger than experimental values (BU60), which were evaluated neglecting the skewness of the range straggling curves. The observed straggling in range-energy measurements is composed of the energy-loss straggling, and an additional asymmetric contribution caused by the multiple-scattering process (BU60, BI60).

8d-8. Coulomb and Multiple Scattering, and Nuclear Interactions. *Coulomb Scattering.* The differential cross section for Coulomb scattering of a charged particle of kinetic energy T (in MeV), momentum p, velocity v, and charge ze by a nucleus of charge Ze and mass number A into the solid angle $2\pi \sin \theta \, d\theta$ is given by the Rutherford formula:

$$d\Phi(\theta) = \frac{2\pi e^4 z^2 Z(Z+1)}{4p^2 v^2 \sin^4(\theta/2)} \sin \theta \, d\theta$$

$$\cong \frac{0.814 z^2 Z(Z+1)}{T^2} \frac{\sin \theta \, d\theta}{\sin^4(\theta/2)} \times 10^{-26} \text{ cm}^2 \qquad (8d\text{-}35)$$

where θ is the angle of scattering from the incident direction. The above formula assumes that the mass of the incident particle is negligible compared with the mass of the nucleus.

Deviations from the Rutherford formula will occur at large angles as the particles begin to feel the influence of nuclear forces. An estimate of the minimum energy T_m for which a deviation can be expected at $\theta = 180$ deg can be obtained from

$$T = zZ(A+3)^{-\frac{1}{3}} \qquad \text{MeV} \qquad (8d\text{-}36)$$

A detailed discussion is found in EP61 and JA68. At small angles, the cross section will be smaller than given by Eq. (8d-35) because the atomic electrons will shield the nuclear charge. The Rutherford cross section is reduced by 10 percent at an angle θ_q given by (from MO47)

$$\theta_q = \theta_0(61.7 + 421\alpha^2)^{\frac{1}{2}}$$

and by 50 percent at

$$\theta_r = \theta_0(2.75 + 10.85\alpha^2)^{\frac{1}{2}}$$

where

$$\theta_0 = \frac{0.244 Z^{\frac{1}{3}}}{pc \text{ (MeV)}} \sim \frac{0.244 Z^{\frac{1}{3}}}{\sqrt{2M_0 c^2 T}}$$

and $\alpha = Zz/137\beta$. For large kinetic energies,

$pc = (T^2 + 2TM_0 c^2)^{\frac{1}{2}}$, and with $\zeta = T/M_0 c^2$,

$$\beta^2 = \zeta \frac{\zeta + 2}{(\zeta + 1)^2} \qquad (8d\text{-}37)$$

EXAMPLE. 10-MeV alpha particles in Au: from Table 8d-1, $\beta = 0.073$, $\alpha = 15.8$. $\theta_0 = 1.05/(74,600)^{\frac{1}{2}} = 3.84 \times 10^{-3}$ deg. Finally,

$$\theta_q = 3.84 \times 10^{-3}(61.7 + 105,000)^{\frac{1}{2}} = 1.25 \text{ deg.}$$

This reduction is of great importance in the derivation of the multiple-scattering formulas.

Multiple Scattering in Thin Absorbers. Multiple Coulomb scattering in thin foils will cause a parallel beam of particles to spread out into a cone. Recent discussions are found in HF68, SC63, and GD68. Moliere's theory (MO48, BE53, and MO55) is a small-angle approximation to the general problem (BR59, NS61, and TM59) which is in agreement with experimental results, with the possible exception of electrons in heavy elements and also possibly at small energies ($\beta^2 < 2 \times 10^{-3}$).

The characteristic quantity occurring in the theory is the angle θ_0, defined by $\theta_0 = \theta_1 B^{\frac{1}{2}}$ where

$$\theta_1^2 = 0.157 \frac{Z(Z+1)z^2}{A} \frac{s}{(pv)^2} \qquad (8d\text{-}38)$$

θ_1 is in radians; s is the foil thickness in g/cm², p the momentum, and v the velocity of the particle (pv in MeV); z, Z, and A have the same meaning as in Sec. 8d-2. B is defined in MO48; for practical purposes it can be obtained from MZ67 or from Table 8d-6, for particles with charge 1 with an accuracy of better than 5 percent. A few values are listed for $z > 1$. It is not obvious whether z^* or z should be used for a computation of the multiple scattering of heavy ions. The use of z^* is suggested. For $z \geq 6$ and $Z \geq 50$, all values $B(\beta,z)$ are larger than $0.98 \times B(\beta = 0, z = 1)$; and for $z \geq 6$ and $Z \geq 20$, all values $B(\beta,z) \geq 0.95 \times B(\beta = 0, z = 1)$, but less than $B(0,1)$.

TABLE 8d-6. B OF MOLIERE'S THEORY FOR $z = 1$, 2, AND 6, VARIABLE β, AND THICKNESS s*

Z	s, g/cm²	z = 1									z = 2		z = 6	
		$\beta^2 = 0$	0.005	0.01	0.02	0.05	0.1	0.2	0.5	1.0	0.1	1.0	0.1	1.0
3	10^{-3}	10.5	8.8	8.3	7.6	6.6	5.7	4.9	3.8	2.8	7.4	4.6		
	10^{-2}	13.0	11.5	10.8	10.2	9.2	8.5	7.7	6.6	5.7	10.0	7.4		
	10^{-1}	15.4	14.0	13.3	12.8	11.7	11.0	10.3	9.2	8.5	12.5	10.0		
	1	17.9	16.4	15.8	15.2	14.2	13.5	12.8	11.8	11.0	14.9	12.6		
10	10^{-3}	8.2	8.0	7.7	7.4	6.7	6.0	5.2	4.2	3.2	7.2	4.9	8.1	7.2
	10^{-2}	10.7	10.5	10.3	9.9	9.25	8.7	8.0	7.0	6.2	9.8	7.7	10.6	9.7
	10^{-1}	13.3	13.0	12.8	12.4	11.8	11.2	10.5	9.6	8.8	12.3	10.3	13.1	12.3
	1	15.7	15.4	15.2	14.8	14.3	13.7	13.1	12.1	11.4	14.8	12.8	15.5	14.7
20	10^{-3}	6.8	6.7	6.6	6.5	6.2	5.8	5.2	4.2	3.5	6.5	5.0	6.8	6.4
	10^{-2}	9.4	9.3	9.3	9.2	8.9	8.5	7.9	7.1	6.4	9.2	7.8	9.4	9.1
	10^{-1}	12.0	11.9	11.8	11.7	11.4	11.0	10.5	9.7	9.0	11.7	10.3	11.9	11.6
	1	14.4	14.4	14.3	14.2	13.9	13.5	13.1	12.2	11.5	14.2	12.8	14.4	14.2
50	10^{-3}	4.7	4.7	4.7	4.6	4.6	4.5	4.3	3.7	3.2	4.6	4.1	4.7	4.6
	10^{-2}	7.5	7.5	7.5	7.4	7.4	7.3	7.2	6.6	6.0	7.5	7.0	7.5	7.4
	10^{-3}	10.0	10.0	10.0	10.0	10.0	9.9	9.7	9.2	8.8	10.0	9.6	10.1	10.0
	1	12.5	12.5	12.5	12.5	12.5	12.4	12.2	11.8	11.3	12.6	12.1	12.5	12.5
100	10^{-3}	3.1	3.1	3.1	3.1	3.0	3.0	3.0	2.8	2.5	3.1	2.9	3.1	3.1
	10^{-2}	6.0	6.0	6.0	6.0	6.0	5.9	5.9	5.7	5.4	6.0	5.8	6.0	6.0
	10^{-1}	8.7	8.7	8.7	8.7	8.7	8.6	8.6	8.4	8.2	8.7	8.5	8.7	8.7
	1	11.2	11.2	11.2	11.2	11.2	11.1	11.1	10.9	10.7	11.2	11.0	11.2	11.2

* For any value of z at $\beta = 0$, B is the same as for $z = 1$. The theory is valid only for $B \gtrsim 4.5$. Linear interpolation for Z or β^2 will give sufficient accuracy. Logarithmic interpolation is required for s.

Moliere's theory modified by Nigam et al. (NS61) gives the distribution function $F(x)\, dx$ for the relative number of particles entering a cone of angle x and width dx. The reduced angle x is defined by

$$x = \frac{\theta}{\theta_0}$$

An extensive discussion of the problem is given in MZ67. Table 8d-7 giving $F(x)$ is obtained from MZ67.

Also of interest is the relative number N/N_0 of particles entering a cone of half angle α:

$$\frac{N}{N_0} = \int_0^{\alpha/\theta_0} f(x)x\, dx \tag{8d-39}$$

Values are given in Table 8d-8. For experimental tests of the theory, see BI58 MO58, LO67, BN66.

EXAMPLE. 2-MeV protons penetrating 3 mg/cm^2 of Ni foil: The average energy in the foil is 1.87 MeV. $\beta^2 \sim 3.96 \times 10^{-3}$ from Table 8d-1, $B \sim 7.7$ from Table 8d-6. $\theta_1^2 = 4.72 \times 10^{-4}$, $\theta_0 = 6.03 \times 10^{-2}$ rad = 3.46 deg. Thus, inside a cone of half angle 7 deg, all but about 6.3 percent of the protons will be found (see Table 8d-8).

TABLE 8d-7. MULTIPLE-SCATTERING DIFFERENTIAL DISTRIBUTION FUNCTION* $F(x)$

x	$B = 4$	5	6	7	8	9	10	12
0	1.0	1.0	1.0	1.0	1.0	1.0	1.0	1.0
0.2	0.94070	0.94546	0.94850	0.95058	0.95208	0.95321	0.95409	0.95537
0.4	0.78389	0.79992	0.81017	0.81721	0.82232	0.82616	0.82916	0.83351
0.6	0.58102	0.60800	0.62535	0.63731	0.64601	0.65259	0.65772	0.66520
0.8	0.38726	0.41889	0.43939	0.45363	0.46402	0.47192	0.47811	0.48716
1.0	0.23800	0.26632	0.28491	0.29793	0.30752	0.31486	0.32063	0.32913
1.2	0.14139	0.16116	0.17437	0.18377	0.19077	0.19616	0.20045	0.20681
1.4	0.08650	0.09681	0.10393	0.10911	0.11304	0.11612	0.11859	0.12231
1.6	0.05666	0.05986	0.06226	0.06410	0.06556	0.06673	0.06769	0.06918
1.8	0.03899	0.03840	0.03816	0.03807	0.03805	0.03806	0.03809	0.03817
2.0	0.02685	0.02506	0.02387	0.02303	0.02240	0.02192	0.02154	0.02097
2.2	0.01793	0.01628	0.01507	0.01416	0.01345	0.01288	0.01241	0.01170
2.4	0.01164	0.01048	0.00956	0.00883	0.00824	0.00775	0.00735	0.00673
2.6	0.00799	0.00716	0.00646	0.00589	0.00543	0.00504	0.00471	0.00419
2.8	0.00549	0.00489	0.00438	0.00396	0.00361	0.00332	0.00308	0.00269
3.0	0.00397	0.00349	0.00310	0.00277	0.00251	0.00229	0.00211	0.00182
3.2	0.00300	0.00259	0.00227	0.00202	0.00181	0.00164	0.00150	0.00128
3.4	0.00232	0.00198	0.00171	0.00151	0.00135	0.00122	0.00111	0.00094
3.6	0.00182	0.00154	0.00132	0.00116	0.00103	0.00093	0.00084	0.00071
3.8	0.00145	0.00121	0.00103	0.00090	0.00080	0.00072	0.00065	
4.0	0.00115	0.00096	0.00082	0.00071	0.00063	0.00056	0.00051	
4.2	0.00093	0.00077	0.00065	0.00057	0.00050	0.00045	0.00041	
4.4	0.00075	0.00062	0.00053	0.00046	0.00041	0.00037	0.00033	
4.6	0.00062	0.00051	0.00043	0.00038	0.00033	0.00030	0.00027	
4.8	0.00051	0.00042	0.00036	0.00031	0.00027	0.00025	0.00022	
5.0	0.00043	0.00035	0.00030	0.00026	0.00023	0.00021	0.00019	
5.2	0.00036	0.00030	0.00025	0.00022	0.00019	0.00018	0.00016	
5.4	0.00030	0.00025	0.00021	0.00019	0.00016	0.00015	0.00013	
5.6	0.00026	0.00021	0.00018	0.00016	0.00014	0.00013		
5.8	0.00022	0.00019	0.00016	0.00014	0.00012	0.00011		
6.0	0.00019	0.00016	0.00014	0.00012	0.00010	0.00010		

* From J. B. Marion and B. A. Zimmerman, *Nucl. Instr. Methods* **51**, 93 (1967).

Caution must be used with the incident particle of mass approximately equal to or larger than the mass of the scattering nucleus. In this case a considerable fraction of the energy can be lost to the recoil nucleus. This effect is, of course, not included in the fundamental energy-loss formula [Eq. (8d-11)].

Multiple Scattering in Thick Absorbers. For thick absorbers the mean energy correction due to multiple scattering has been calculated in TB68 for energy losses between $0.5T$ and $0.1T$, for $10 < T < 140$ MeV, for detector angles between 0.005 and 0.5 rad for protons in Al, Ag, and Au.

TABLE 8d-8. MULTIPLE-SCATTERING INTEGRAL DISTRIBUTION FUNCTION
(Given is the fraction of incident particles found inside a cone of half angle x.)

x	$B = 4$	5	6	7	8	9	10	12
0.2	0.04617	0.04431	0.04320	0.04247	0.04195	0.04153	0.04123	0.04078
0.4	0.16893	0.16330	0.15993	0.15773	0.15616	0.15485	0.15393	0.15253
0.6	0.33004	0.32259	0.31815	0.31523	0.31316	0.31132	0.31008	0.30814
0.8	0.48890	0.48427	0.48156	0.47981	0.47856	0.47716	0.47637	0.47496
1.0	0.61973	0.62202	0.62359	0.62473	0.62554	0.62555	0.62592	0.62614
1.2	0.71612	0.72641	0.73300	0.73759	0.74088	0.74266	0.74449	0.74676
1.4	0.78446	0.80062	0.81102	0.81829	0.82357	0.82679	0.82981	0.83380
1.6	0.83429	0.85269	0.86473	0.87324	0.87948	0.88340	0.88704	0.89194
1.8	0.87231	0.88987	0.90159	0.90998	0.91620	0.92011	0.92378	0.92875
2.0	0.90166	0.91679	0.92709	0.93457	0.94016	0.94358	0.94690	0.95136
2.2	0.92375	0.93623	0.94485	0.95118	0.95591	0.95868	0.96149	0.96519
2.4	0.93964	0.94997	0.95714	0.96242	0.96636	0.96849	0.97080	0.97375
2.6	0.95110	0.95983	0.96584	0.97026	0.97353	0.97513	0.97700	0.97928
2.8	0.95964	0.96714	0.97224	0.97596	0.97869	0.97985	0.98136	0.98308
3.0	0.96607	0.97259	0.97697	0.98014	0.98244	0.98325	0.98447	0.98575
3.2	0.97115	0.97684	0.98062	0.98334	0.98528	0.98581	0.98680	0.98772
3.4	0.97529	0.98024	0.98351	0.98584	0.98750	0.98779	0.98860	0.98924
3.6	0.97872	0.98302	0.98584	0.98786	0.98927	0.98938	0.99002	0.99043
3.8	0.98158	0.98531	0.98776	0.98950	0.99071	0.99066	0.99117	0.99140
4.0	0.98398	0.98722	0.98934	0.99086	0.99189	0.99172	0.99212	0.99224
4.2	0.98600	0.98882	0.99067	0.99199	0.99288	0.99260	0.99291	0.99296
4.4	0.98771	0.99018	0.99179	0.99295	0.99372	0.99334	0.99357	0.99359
4.6	0.98917	0.99134	0.99275	0.99377	0.99443	0.99397	0.99413	0.99413
4.8	0.99043	0.99233	0.99357	0.99447	0.99504	0.99452	0.99462	0.99461
5.0	0.99152	0.99320	0.99429	0.99508	0.99557	0.99500	0.99504	0.99503
5.2	0.99247	0.99395	0.99491	0.99561	0.99603	0.99541	0.99541	0.99541
5.4	0.99331	0.99461	0.99545	0.99607	0.99644	0.99578	0.99573	0.99574
5.6	0.99405	0.99519	0.99594	0.99648	0.99680	0.99610	0.99602	0.99604
5.8	0.99470	0.99571	0.99637	0.99685	0.99712	0.99639	0.99628	0.99631
6.0	0.99530	0.99618	0.99676	0.99719	0.99741	0.99666	0.99651	0.99655
7.0	0.99655	0.99720	0.99762	0.99793	0.99810	0.99755	0.99744	0.99747
8.0	0.99736	0.99785	0.99818	0.99842	0.99854	0.99812	0.99804	0.99806
9.0	0.99791	0.99830	0.99856	0.99875	0.99885	0.99852	0.99845	0.99847
10.0	0.99831	0.99863	0.99883	0.99899	0.99907	0.99880	0.99874	0.99876

The multiple-scattering correction for median ranges has been discussed in Sec. 8d-3.

Nuclear Interactions. Heavy charged particles will be removed from beams by nuclear interactions: the beam intensity will be attenuated exponentially

$$I = I_0 e^{-s\Sigma} \tag{8d-40}$$

where I is the flux density, and Σ is the macroscopic cross section $\Sigma = \sigma_t n$ (σ_t = total microscopic cross section, n = number of nuclei per g). For estimates, $\Sigma = 0.032/A^{\frac{1}{3}}$ cm^2/g may be used (A = atomic number of absorber).

8d-9. Electrons. While electrons in passing through matter will experience interactions similar to heavy-particle interactions, two basic differences are manifest:

1. In the collisions with atomic electrons, large energy losses can occur.
2. Electrons with energies of only a few hundred keV will show relativistic effects.

An extensive review of the theory is found in BK58, and extensive tabulations are contained in BS67. The derivation of the stopping-power formula is similar to the heavy-particle case. It will be assumed that after a collision by a negative electron, the electron with the higher velocity will be considered the primary. The mean collision loss in MeV cm^2 g^{-1} is given by BS67:

$$- \left(\frac{dT}{\rho \, ds} \right)_{\text{col}}^{\pm} = \frac{0.1535}{\beta^2} \frac{Z}{A} \left[\ln \frac{2(\tau + 2)}{(I/mc^2)^2} + F^{\pm}(\tau,\Delta) - \delta \right] \qquad (8d\text{-}41)$$

where for electrons $\Delta = \frac{1}{2}\tau$ and

$$F^- = -1 - \beta^2 + \ln \left[(\tau - \Delta)\Delta \right] + \frac{\tau}{\tau - \Delta} + \frac{\frac{1}{2}\Delta^2 + (2\tau + 1) \ln (1 - \Delta/\tau)}{(\tau + 1)^2} \qquad (8d\text{-}42)$$

and for positrons $\Delta = \tau$ and

$$F^+ = \ln (\tau\Delta) - \frac{\beta}{\tau} \left[\tau + \Delta - \frac{\frac{5}{4}\Delta^2}{\tau + 2} + \frac{(\tau + 1)(\tau + 3)\Delta - \frac{1}{3}\Delta^3}{(\tau + 2)^2} \right.$$
$$\left. - \frac{(\tau + 1)(\tau + 3)\frac{1}{4}\Delta^4 - \tau/3\Delta^3 + \frac{1}{4}\Delta^4}{(\tau + 2)^3} \right] \qquad (8d\text{-}43)$$

Here $\tau = T/mc^2$, δ is the density correction, and $mc^2 = 511,004$ eV; Δ is the maximum energy given to delta rays, divided by mc^2. The other symbols here have the same meaning as in Eq. (8d-11). In particular, the same I values are used as for the heavier particles.

The shell corrections are not included, because their contribution above 0.1 MeV amounts to less than 1 percent. If desired, the shell corrections discussed above (Fig. 8d-4) can be used to correct stopping-power values obtained from Eq. (8d-41). The differences between electrons and positrons have been studied by Rohrlich and Carlson (RC54).

The energy loss due to *bremsstrahlung* is important for electrons at relatively small energies. An estimate of the ratio r of the bremsstrahlung energy loss to $(dT/ds)_{\text{coll}}$ is given by

$$r \sim T \frac{Z + 1.2}{700} \qquad (T \text{ in MeV}) \qquad (8d\text{-}44)$$

At $T_c \sim 700/(Z + 1.2)$ MeV the two energy losses are equal. An important quantity is associated with the traversal of matter by electrons of energies above T_c; this is the distance X_0 in which an electron's energy is reduced to $1/e = 0.3679$ of its original value. X_0 is called the *radiation length* and is given in Table 8d-9, together with more accurate values of T_c. Recent experimental results are found in DR68, and for $T \leq 4$ keV in BC69.

Restricted Stopping Power (LET). Secondary radiation (delta rays or bremsstrahlung photons) may travel quite far from the track of a particle. An estimate of the energy deposited inside of a small cylinder around a track can be obtained by setting the quantity Δ in Eq. (8d-41) equal to the energy of delta rays capable of escaping from the volume of interest. Heavy particles produce relatively few delta rays of high energy [see Eq. (8d-1)], and the difference between LET and dT/ds is relatively small for energies below Mc^2 (see Small Volumes in Sec. 8d-6, however).

Practical Considerations for Stopping Power. Computed values of the electron stopping power are given for some elements in Fig. 8d-11. Extensive tables are found in BS67. For $T < 5$ MeV, $(dT/\rho \, ds)_{\text{coll}} \sim Z^{-\frac{1}{4}}$. This factor should be used for interpolation in Fig. 8d-11.

Straggling (discussed in detail in KM61) is much larger for electrons than for heavier particles (see, e.g., Fig. 12 in BI68 or Fig. 2 in BR64). The width at half

TABLE 8d-9. CRITICAL ENERGY T_c AND RADIATION LENGTH X_0 FOR
VARIOUS SUBSTANCES*

Substance	T_c, MeV	X_0, g cm^{-2}
Hydrogen............	340	58
Helium.............	220	85
Carbon.............	103	42.5
Nitrogen............	87	38
Oxygen.............	77	34.2
Aluminum..........	47	23.9
Argon..............	34.5	19.4
Iron...............	24	13.8
Copper.............	21.5	12.8
Lead...............	6.9	5.8
Air................	83	36.5
Water.............	93	35.9

* From H. A. Bethe and J. Ashkin, "Experimental Nuclear Physics," vol. 1, p. 166, John Wiley &
Sons, New York, 1952.

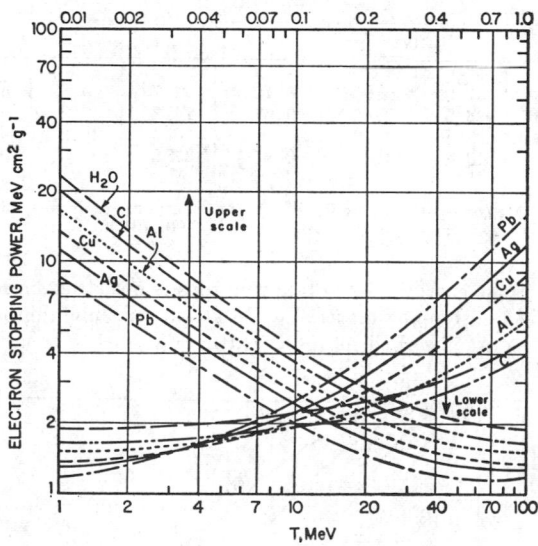

FIG. 8d-11. Calculated electron-mass stopping power S, including collision and radiation loss for different materials (BS67). The stopping power for NaI is within 1 percent of S for Ag.

maximum of a straggling distribution may amount to more than 50 percent of the mean energy loss. Multiple scattering (VV68) and backscattering contribute to the problem. Comparison of mean energy losses calculated from Eq. (8d-41) with experimental data (e.g., HU57, HA59, HR68) can be expected to be accurate to better than 10 percent only if a detailed study of straggling etc., has been made. A comparison of experiment and theory for 1- and 2-MeV electrons in silicon is found in SI67.

Electron Ranges and Energy Deposition in Thick Absorbers. For electrons traversing thick absorbers, lateral and backscattering will be very important, and electron distribution functions will extend over wide ranges in space, angle, and energy. A general treatment is found in BE63, KK68, RO68, SP55, and SP54. Practical results for many substances are given in SP59, KE66, BS67, LP57, and PE62 and KK68. Detailed investigations have been performed for 5- to 30-keV electrons (CT65), and for 40- to 160-keV electrons (GF59). For higher energies, see, e.g., BH58. Electron ranges calculated by the use of Eq. (8d-14) do not have a simple relation to any observed quantity: see Table 8d-10.

TABLE 8d-10. THE COMPARISON OF MAXIMUM ELECTRON RANGES R_{max}
WITH SPENCER'S X_{max}*

T, MeV	R_{max} exp.	csda range	Ratio	Spencer's X_{max}	R_{max} exp.	csda range	Ratio	Spencer's X_{max}
	Aluminum				Copper			
0.05	5.05	5.71	0.884	0.875	5.42	6.90	0.786	0.775
0.10	15.44	18.64	0.829	0.875	17.1	22.1	0.772	0.775
0.10+	14.4	17.3	0.832		16.1	20.7	0.778	
0.15	31.0	36.4	0.850	0.875	34.0	42.8	0.795	0.760
	Silver				Gold			
0.05	5.04	7.99	0.63	0.70	4.73	9.88	0.48	
0.10	15.6	25.2	0.62	0.67	14.3	30.3	0.47	0.57
0.10+	16.5	23.5	0.70		18.5	28.2	0.66	
0.15	30.2	48.4	0.62	0.65	27.6	57.5	0.48	

* Positrons of 0.1 MeV are indicated by 0.1+. Experimental ranges from GF59; csda ranges from BS67; X_{max} is the value at which $J(X)$ reaches a value of 0.001 (SP59).

The practical range-energy relation for electrons is not strongly dependent on the atomic number of the stopping material. Only that for aluminum is given. Mono-energetic electrons are absorbed as indicated in Fig. 8d-12, which serves to define

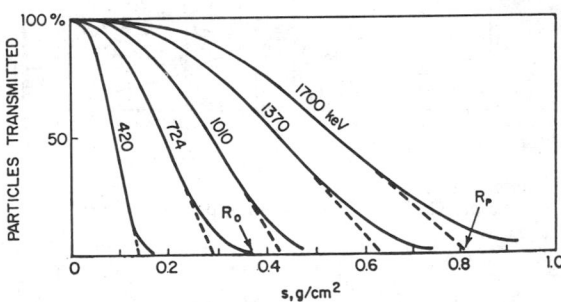

FIG. 8d-12. Absorption curve of monoenergetic electrons. R_p is defined as the extrapolated range, R_0 as the maximum range (BK58).

the "practical range" R_p and the "maximum range" R_0. The practical range, in aluminum is given by

$$R_p = 0.537T \left(1 - \frac{0.9815}{1 + 0.003123T} \right) \tag{8d-45}$$

R_p is in mg/cm², and T in keV, for the energy range 0.3 keV $\leq T \leq$ 20 MeV, with an accuracy of about ± 6 percent (KK68). A graph of this relation is given in Fig. 8d-13.

The formulas given above for monoenergetic electrons can be used for continuous beta-ray spectra where R_p and T_0 refer to the maximum beta-ray range and energy, respectively. For a discussion of the methods of determining the range from an absorption curve, see KP52.

For practical applications in which information on electron range and energy deposition is required, it appears best to use Spencer's calculations (SP59; see also BI68); but some information is found also in KK68.

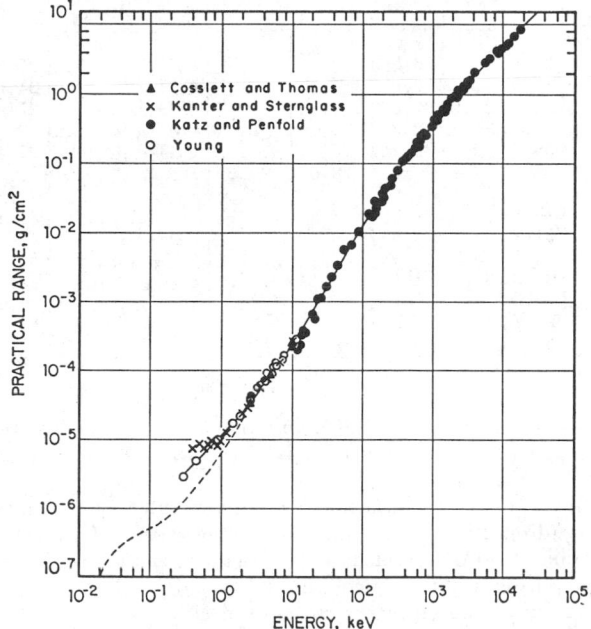

Fig. 8d-13. Practical range in aluminum versus electron energy (KK68). [*Coslett and Thomas* (CT65); *Kanter and Sternglass, Phys. Rev.* **126**, 620 (1962); *Katz and Penfold* (KP52); *Young, Phys. Rev.* **103**, 292 (1956).] The dotted line gives experimental data in air and collodion (CO69).

Unlike the case for heavy charged particles, determination of electron energies from transmission measurements is not accurate enough for most applications. Energies can be determined much more accurately by measurements with calibrated scintillation or solid-state detectors.

8d-10. Mean Energy for the Formation of an Ion Pair. *Gases.* The energy loss W of a charged particle per ion pair formed in the material traversed is nearly independent of the energy and type of particle for velocities $\beta^2 > 10^{-4}z$, as can be seen in Table 8d-11. For further values see MY68.

From the measurements of Phipps, Boring, and Lowry (PB64), the following approximate velocity dependence of W has been derived for ions in argon with $A < 40$:

$$W = \frac{0.119}{\beta} \quad \text{eV} \quad \text{for } \beta \leq 0.0043$$

TABLE 8d-11. AVERAGE ENERGY W IN eV FOR THE FORMATION OF AN ION PAIR FOR VARIOUS PARTICLES

Gas	Particle			Fission fragments		Gas	Particle		
	β	p	α	Light	Heavy		β	p	α
Kinetic energy $T =$	0.3	1	5	90	60	Kinetic energy $T =$	0.3	1	5
H_2	36.6	....	36.2	....	....	C_2H_4	26.3	....	28.03
He	41.5	....	46.0	....	....	C_2H_6	24.6	....	26.6
N_2	34.6	36.6	36.39	....	....	C_3H_8	27.8		
O_2	31.8	31.5	32.3	....	....	C_4H_{10}	23.0	....	24.8
Ne	36.2	28.6	35.7	....	....	C_6H_{14}	22.4		
Ar	26.2	26.4	26.3	28.0	29.5	BF_3	....	....	35.6
Kr	24.3	....	24.0	....	....	NH_3	34.8	....	30.5
Xe	21.9	....	22.8	....	....	C_2H_5OH	....	....	32.6
Air	33.7	36.0	34.98	....	....	CCl_2F_2	....	....	29.5
CO_2	32.9	34.9	34.1	....	....	SO_2	....	....	32.5
CH_4	27.3	....	29.1	....	....	H_2O	30.1	....	37.6
C_2H_2	25.7	....	27.3						

For more accurate values, PB64, BS65, and LH65 should be consulted. For Pb^{206} ions with $T = 103$ keV, measurements have been made by Cano (CA68); see also PL69.

Mixtures of gases do not follow a simple additivity rule for the value of W (MY68, BH54). A large drop in W of argon for small concentrations of C_2H_4 has been observed. For further details see MY68. Ionization fluctuations and the resolution of ionization chambers are discussed extensively in AK67.

Solids. A recent discussion of the response of NaI(Tl) to heavy ions is found in KA68, with references to earlier work.

The ionization in silicon and germanium has been studied extensively (see almost any issue of *IEEE Transactions on Nuclear Science*). The average energy ϵ for the generation of an electron-hole pair is much smaller than for gases. For Si, $\epsilon \sim 3.6$ eV, for Ge, $\epsilon \sim 2.96$ eV. For silicon, the following effects have been observed:

1. For low-energy electrons (produced with gamma rays), pulse heights, after correction for charge collection efficiency, are proportional to energy within 0.2 percent (ZM69).

2. For a change in temperature from 300 to 90 K, an increase of 4 percent in ϵ has been observed (PG68, KR71).

3. ϵ is about one percent smaller for alpha particles than for electrons (PG68).

4. For heavy ions, ϵ is energy-dependent at small energies (BB63, FK67, FS69, KA67, BB69, SA65), mainly due to "nuclear collisions" (LN63). The energy T'_M calculated from a measured ionization pulse should be increased by ΔT to obtain the correct kinetic energy T_M of the particle.

Until better information becomes available, $\Delta T \sim 4M$ (keV) ("ionization defect", BB63) can be used for $T_M \gg 6M$ (keV) (M = atomic mass of particle). For α particles, the upper curve in Fig. 10 of LN63 may be used; for protons, $\Delta T \sim 1 - 2$ keV (FS69).

Somewhat different results are given in RB69. Similar results have been obtained for germanium detectors (DB67, PR69). Several factors determine the resolution of solid-state detectors (BL67, AN67, TS67); some of the more important are:

1. Electronic noise and drift of amplifier system
2. Ballistic deficit
3. Pulse pileup
4. Recombination and trapping
5. Channeling (see Sec. 8d-7)
6. Absorption in surface layers
7. Statistics of the number N of electron-hole pairs produced.

Fano (FA47) has shown that the standard deviation of the mean number $\bar{N}$ is: $\Delta N^2 = \langle (N - \bar{N})^2 \rangle = F\bar{N}$, where $F \leq 1$. Bilger (BL67) found $F = 0.13$ for germanium. Alkhazov et al. (AK67) obtained $F \sim 0.1$ for silicon. The problem is also discussed in DF67, ZA70. PG70 give an upper limit $F \simeq 0.08$.

Energy-loss tables for p, d, t, He^3, He^4, and Li^6 with data useful for particle identifier systems are given in BT67 and SK67. Information about the straggling in thin silicon detectors is given in Fig. 8d-7.

References

AE67. Appleton, B. R., C. Erginsoy, and W. M. Gibson: *Phys. Rev.* **161**, 330 (1967).
AG68. Ait-Salem, M., H. Gerhardt, F. Gönnenwein, H. Hipp, and H. Paap: *Nucl. Instr. Methods* **60** 45 (1968).
AG69. Arkhipov, E. P., and Yu. V. Gott: *Sov. Phys.—JETP* **29**, 615 (1969) [*Zh. Eksp. Teor. Fiz.* **56**, 1146 (1969)].
AH67. Andersen, H. H., C. C. Hanke, H. Sørensen, and P. Vajda: *Phys. Rev.* **153**, 338 (1967).
AH68. Armitage, B. H., and B. W. Hooton: *Nucl. Instr. Methods* **58**, 29 (1968).
AK67. Alkhazov, G. D., A. P. Komar, and A. A. Vorob'ev: *Nucl. Instr. Methods* **48**, 1 (1967).
AL69. Aitken, D. W., W. L. Lakin, and H. R. Zulliger: *Phys. Rev.* **179**, 393 (1969).
AN67. Andersson-Lindstroem, G.: *Nucl. Instr. Methods* **56**, 309 (1967).
AN69. Andersen. H. H., H. Sørensen, and P. Vajda: *Phys. Rev.* **180**, 373 (1969).
AR69. Andreev, V. N., V. G. Nedopekin, and V. I. Rogov: *Sov. Phys.—JETP* **29**, 807 (1969) [*Zh. Eksp. Teor. Fiz.*, **56**, 1504 (1969)].
AS68. Andersen, H. H., C. C. Hanke, H. Simonsen, H. Sørensen, and P. Vajda: *Phys. Rev.* **175**, 389 (1968).
AS69. Andersen, H. H., H. Simonsen, and H. Sørensen: *Nucl. Phys.* **A125**, 171 (1969).
AV69. Andersen, H. H., H. Simonsen, H. Sørensen, and P. Vajda: *Phys. Rev.* **186**, 372 (1969).
BA61. Barkas, W. H.: *Phys. Rev.* **124**, 897 (1961).
BA63. Barkas, W. H.: "Nuclear Research Emulsions," Academic Press, Inc., New York, 1963.
BB63. Bilger, H., E. Baldinger, and W. Czaja: *Helv. Phys.* **36**, 405 (1963).
BB67. Barkas, W. H., and M. J. Berger: Paper 7 of NA67.
BB69. Barash-Schmidt, N., A. Barbaro-Galtieri, L. R. Price, A. H. Rosenfeld, P. Söding, C. G. Wohl, M. Roos, and G. Conforto: *Rev. Mod. Phys.* **41**, 109 (1969).
BC69. Bronshtein, I. M. and A. N. Brozdnichenko: *Sov. Phys.—Solid State* **11**, 140 (1969).
BD63. Barkas, W. H., J. N. Dyer, and H. H. Heckman: *Phys. Rev. Letters* **11**, 26 (1963).
BE30. Bethe, H.: *Ann. Phys.* **5**, 325 (1930).
BE53. Bethe, H.: *Phys. Rev.* **89**, 1256 (1953).
BE63. Berger, M. J.: "Methods in Computational Physics," vol. 1, p. 135, Alder, Fernbach, and Rotenberg, eds., Academic Press, Inc., New York, 1963.
BE66. Bell, R. J., and A. Dalgarno: *Proc. Phys. Soc. (London)* **89**, 55 (1966); **86**, 375 (1965).
BF60. Bichsel, H., and B. J. Farmer: *Bull. Am. Phys. Soc.* **5**, 263 (1960).
BF61. Barkas, W. H., and S. von Friesen: *Nuovo Cimento* (10) **19**, suppl. 1, p. 41 (1961).
BG65. Booth, W., and I. S. Grant: *Nucl. Phys.* **63**, 481 (1965).
BH54. Bortner, T. E., and G. S. Hurst: *Phys. Rev.* **93**, 1236 (1954).
BH58. Breuer, H., D. Harder, and W. Pohlit: *Z. Naturforsch.* **13a**, 567 (1958).
BH67. Bellamy, E. H., R. Hofstadter, W. L. Lakin, J. Cox, M. L. Perl, W. T. Toner, and T. F. Zipf: *Phys. Rev.* **164**, 417 (1967).

BH69. Bichsel, H., C. C. Hanke, and J. Buechner: USC-136-148, March, 1969; also *Mat. Fys. Medd. Dan. Vid. Selsk.* **38**, 3 (1970).
BI58. Bichsel, H.: *Phys. Rev.* **112**, 182 (1958).
BI60. Bichsel, H.: *Phys. Rev.* **120**, 1012 (1960).
BI61. Bichsel, H.: Univ. of Southern Calif. Physics Dept. Tech. Report No. 3 (June 1961).
BI67. Bichsel, H.: USC-136-120, April, 1967.
BI68. Bichsel, H.: "Radiation Dosimetry," vol. 1, chap. 4, F. H. Attix and Wm. R. Roesch, eds., Academic Press, Inc., New York, 1968.
BI69. Bichsel, H.: USC-136-147, January, 1969.
BI70. Bichsel, H.: *Nucl. Instr. Methods* **78**, 277 (1970).
BJ67. Bichsel, H.: *Univ. Calif. LRL Tech. Rept.* UCRL-17538, May 1967.
BJ69. Bichsel, H.: "Range Energy Tables," to be published.
BJ70. Bichsel, H.: *Phys. Rev.* **B1**, 2854 (1970).
BK58. Birkhoff, R. D.: "Encyclopedia of Physics," vol. 34, p. 34, Springer-Verlag OHG. Berlin, 1958.
BK67. Bichsel, H.: Paper 2 in NA67.
BL33. Bloch, F.: *Z. Phys.* **81**, 363 (1933).
BL50. Blunck, O., and S. Leisegang: *Z. Phys.* **128**, 500 (1950).
BL67. Bilger, H. R.: *Phys. Rev.* **163**, 238 (1967).
BL68. Bowman, W. W., F. M. Lanzafame, C. K. Kline, Yu-Wen Yu, and M. Blann: *Phys. Rev.* **165**, 485 (1968).
BL69. Brandt, W., and R. Laubert: *Phys. Rev.* **178**, 225 (1969).
BM57. Bichsel, H., R. F. Mozley, and W. A. Aron: *Phys. Rev.* **105**, 1788 (1957).
BN66. Bednyakov, A. A., V. S. Nikolaev, A. V. Rudchenko, and A. F. Tulinov: *Soviet Phys. JETP* **23**, 391 (1966).
BN67. Bridwell, L. B., L. C. Northcliffe, S. Datz, C. D. Moak, and H. O. Lutz: *Phys. Rev.* **159**, 276 (1967).
BO48. Bohr, N.: *Dan. Mat. Fys. Medd.* **18**(8), (1948); (2 ed., 1953).
BO67. Bonderup, E.: *Dan. Mat. Fys. Medd.* **35**(17), (1967).
BP71. Bourland, P. D., and D. Powers: *Phys. Rev.* (to be published, 1971).
BR59. Breitenberger, E.: *Proc. Roy. Soc. (London),* ser. A, **250**, 514 (1959).
BR64. Breuer, H.: *Z. Physik* **180**, 209 (1964).
BR68. Brice, D. K.: *Phys. Rev.* **165**, 475 (1968).
BS65. Boring, J. W., G. E. Strohl, and F. R. Woods: *Phys. Rev.* **140**, A1065 (1965).
BS67. Berger, M. J., and S. M. Seltzer: Paper 10 in NA67.
BS68. Biersack, J. P.: *Z. Physik* **211**, 495 (1968).
BT67. Bichsel, H., and C. Tschalär: *Nucl. Data* **A3**, 343 (1967), and UCRL-17663.
BT68. Bichsel, H., and C. Tschalär: *Phys. Rev.* **175**, 476 (1968).
BU60. Bichsel, H., and E. A. Uehling: *Phys. Rev.* **119**, 1670 (1960).
BZ67. Berger, M. J., and S. M. Seltzer: Paper 5 in NA67.
CA68. Cano, G. L.: *Phys. Rev.* **169**, 277 (1968).
CB68. Chu, W. K., P. D. Bourland, K. H. Wang, and D. Powers: *Phys. Rev.* **175**, 342 (1968); also *Phys. Rev.* **187**, 478 (1969).
CC67. Cumming, J. B., and V. P. Crespo: *Phys. Rev.* **161**, 287 (1967).
CF70. Crispin, A., and G. N. Fowler: *Rev. Mod. Phys.* **42**, 290 (1970).
CO69. Cole, A.: *Rad. Res.* **38**, 7 (1969).
CR66. Creuzberg, M.: *Z. Phys.* **196**, 433 (1966).
CS68. Croft, P. D., and K. Street, Jr.: *Phys. Rev.* **165**, 1375 (1968).
CT65. Coslett, V. E., and R. N. Thomas: *Brit. J. Appl. Phys.* **15**, 883, 1283 (1964); **16**, 779 (1965).
DA68. Davies, J. A.: *Can. J. Phys.* **46** (Mar. 15, 1968).
DB67. Donnelly, D. P., H. W. Baer, J. J. Reidy, and M. L. Wiedenbeck: *Nucl. Instr. Methods* **57**, 219 (1967).
DD68. Dahlberg, D. A., D. K. Anderson, and I. E. Dayton: *Phys. Rev.* **170**, 127 (1968).
DF67. Di Cola, G., and L. Farese: *Phys. Rev.* **162**, 690 (1967).
DK68. Der, R. C., T. M. Kavanagh, J. M. Khan, B. P. Curry, and R. J. Fortner: *Phys. Rev. Letters* **21**, 1731 (1968).
DM69. Datz, S., C. D. Moak, T. S. Noggle, B. R. Appleton, and H. O. Lutz: *Phys. Rev.* **179**, 315 (1969).
DR68. Dance, W. E., D. H. Rester, B. J. Farmer, J. H. Johnson, and L. L. Baggerly: *J. Appl. Phys.* **39**, 2881 (1968).
DT68. Dalton, P., and J. E. Turner: *Health Phys.* **15**, 257 (1968).
DW68. Davies, J. A., and J. L. Whitton: *Phys. Rev.* **165**, 345 (1968).
EB70. Ebert, H. G.: "Proceedings: Second Symposium on Microdosimetry (Stresa, Italy, Oct. 20–24, 1969)," Commission of the European Communities, EUR 4452 d-f-e, Centre for Information and Documentation, Brussels, January, 1970 ($16.-).

EP61. Eisberg, R. M., and C. E. Porter: *Rev. Mod. Phys.* **33**, 190 (1961).
ER67. Eriksson, L.: *Phys. Rev.* **161**, 235 (1967).
ES69. Ehrhardt, H., M. Schulz, T. Tekaat, and K. Willmann: *Phys. Rev. Letters* **22**, 89 (1969).
FA47. Fano, U.: *Phys. Rev.* **72**, 26 (1947).
FA53. Fano, U.: *Phys. Rev.* **92**, 328 (1953).
FA56. Fano, U.: *Phys. Rev.* **103**, 1202 (1956).
FA63. Fano, U.: *Ann. Rev. Nucl. Sci.* **13**, 1 (1963).   Reprinted in NA67.
FB68. Fastrup, B., A. Borup, and P. Hvelplund: *Can. J. Phys.* **46**, 489 (1968).
FC68. Fano, U., and J. W. Cooper: *Rev. Mod. Phys.* **40**, 441 (1968).
FK67. Forgue, V., and S. Kahn: *Nucl. Instr. Methods* **48**, 93 (1967).
FL67. Flaks, I. P.: *5th Intern. Conf. Phys. of Electron. and Atomic Collisions, Leningrad,* July 17–23, 1967, Publishing House, "Nauka," Leningrad.
FS69. Forcinal, G., P. Siffert, and A. Coche: *IEEE Trans. Nucl. Sci.* **NS**15(1), 475 (1969).
GD68. Gnedin, Yu. N., A. Z. Dolginov, and A. I. Tsygan: *Soviet Phys. JETP* **27**, 267 (1968).
GF59. Gubernator, K., and A. Flammersfeld: *Z. Phys.* **156**, 179 (1959).
HA59. Hall, H. E., A. O. Hanson, and D. Jamnik: *Phys. Rev.* **115**, 633 (1959).
HA68. Harrison, D. E., Jr.: *Appl. Phys. Letters* **13**, 277 (1968).
HB68. Hilbert, J. W., N. A. Baily and R. G. Lane: *Phys. Rev.* **168**, 290 (1968).
HF68. Hemmer, P. C., and I. E. Farquhar: *Phys. Rev.* **168**, 294 (1968).
HL69. Heckman, H. H., and P. J. Lindstroem: *Phys. Rev. Letters* **22**, 871 (1969).
HP60. Heckman, H. H., B. L. Perkins, W. G. Simon, F. M. Smith, and W. H. Barkas: *Phys. Rev.* **117**, 544 (1960).
HR68. Hara, E.: *Nucl. Instr. Methods* **65**, 85 (1968).
HU57. Hudson, A. M.: *Phys. Rev.* **105**, 1 (1957).
HV68. Hvelplund, P., and B. Fastrup: *Phys. Rev.* **165**, 408 (1968).
IN71. Inokuti, M.: *Rev. Mod. Phys.* (to be published, 1971).
IS67. Ishiwari, R., N. Shiomi, Y. Mori, T. Onata, and Y. Uemura: *Bull. Inst. Chem. Res. Kyoto Univ.* **45**, 379 (1967).
JA66. Janni, J. F.: AFWL-TR-65-150 (Sept. 1966).
JA68. Jackson, D. F., and C. G. Morgan: *Phys. Rev.* **175**, 1402 (1968).
JD67. Jespersgard, P., and J. A. Davies: *Can. J. Phys.* **45**, 2983 (1967).
JK68. Johnson, C. H., and R. L. Kernell: *Phys. Rev.* **169**, 974 (1968).
KA67. Krulisch, A. H., and R. C. Axtmann: *Nucl. Instr. Methods* **55**, 238 (1967).
KA68. Katz, R., and E. J. Kobetich: *Phys. Rev.* **170**, 397 (1968).
KE66. Kessaris, N. D.: *Phys. Rev.* **145**, 164 (1966).
KE68. Kessel'man, V. S.: *Soviet Phys. Semicond.* **2**, 76, (1968).
KF67. Kahn, S., and V. Forgue: *Phys. Rev.* **163**, 290 (1967).
KH68. Khandelwal, G. S.: *Nucl. Phys.* **A116**, 97 (1968).
KJ68. Khandelwal, G. S.: *Phys. Rev.* **167**, 136 (1968).
KK68. Kobetich, E. J., and R. Katz: *Phys. Rev.* **170**, 391 (1968).
KL68. Kellerer, A. M.: "Microdosimetry," GSF—Bericht B-1, Strahlenbiologisches Inst. der Univ. München. (November, 1968).
KM61. Knop, G., A. Minten, and B. Nellen: *Z. Phys.* **165**, 533 (1961).
KM66. Khandelwal, G. S., and E. Merzbacher: *Phys. Rev.* **144**, 349 (1966).
KO64. Koschmieder, L.: *Z. Naturforsch.* **19a**, 1414 (1964).
KO68. Kolata, J. J., T. M. Amos, and H. Bichsel: *Phys. Rev.* **176**, 484 (1968).
KP52. Katz, L., and A. S. Penfold: *Rev. Mod. Phys.* **24**, 28 (1952).
KP67. Khan, J. M., D. L. Potter, R. D. Worley, and H. P. Smith: *hys. Rev.* **163**, 81 (1967).
KR71. Key, J. R., and T. A. Rabson: *IEEE, Trans. Nucl. Sci.* **NS-18**, 184 (1971).
KY68. Kyle, H. L., and K. Omidvar: *Phys. Rev.* **176**, 164 (1968).
LA44. Landau, L.: *USSR J. Phys.* **8**, 201 (1944).
LB37. Livingston, M. S., and H. Bethe: *Rev. Mod. Phys.* **9**, 263 (1937).
LE52. Lewis, H. W.: *Phys. Rev.* **85**, 20 (1952).
LH65. Leimgruber, R., P. Huber, and E. Baumgartner: *Helv. Phys. Acta* **38**, 499 (1965).
LI65. Lindhard, Jens: *Mat. Fys. Medd. Dan. Vid. Selsk.* **34**, 14 (1965).
LN63. Lindhard, J., V. Nielsen, M. Scharff, and P. V. Thomsen: *Mat. Fys. Medd. Dan. Vid. Selsk* **33**, No. 10 (1963).
LN68. Lindhard, J., V. Nielsen, and M. Scharff: *Mat. Fys. Medd. Dan. Vid. Selsk.* **36**, 10 (1968).
LO67. Lassen, N. O., and A. Ohrt: *Mat. Fys. Medd. Dan. Vid. Selsk* **36**, 9 (1967).
LP57. Leiss, J. E., S. Penner, and C. S. Robinson: *Phys. Rev.* **107**, 1544 (1957).
LS53. Lindhard, J., and M. Scharff: *Mat. Fys. Medd. Dan. Vid. Selsk.* **27**, 15 (1953).
LS63. Lindhard, J., M. Scharff, and H. E. Schiøtt: *Mat. Fys. Medd. Dan. Vid. Selsk,* **33**, 14 (1963).

LS67. Lehmann, L., H. Spehl, and N. Wertz· *Nucl. Instr. Methods* **55**, 201 (1967).
MA68. Morton, A. H., D. A. Aldcroft, and M. F. Payne: *Phys. Rev.* **165**, 415 (1968).
MB66. Moak, C. D., and M. D. Brown: *Phys. Rev.* **149**, 244 (1966).
ML58. Merzbacher, E., and H. W. Lewis: "Encyclopedia of Physics," vol. 34, Springer-Verlag OHG, Berlin, 1958.
ML68. Moak, C. D., H. O. Lutz, L. B. Bridwell, L. C. Northcliffe, and S. Datz: *Phys. Rev.* **176**, 427 (1968).
MO47. Moliere, G.: *Z. Naturforsch.* **2A**, 133 (1947).
MO48. Moliere, G.: *Z. Naturforsch.* **3A**, 78 (1947).
MO55. Moliere, G.: *Z. Naturforsch.* **10A**, 177 (1955).
MO58. Mozley, R. F., R. C. Smith, and R. E. Taylor: *Phys. Rev.* **111**, 647 (1958).
MR67. Marcinkowski, A., H. Rzewuski, and Z. Werner: *Nucl. Instr. Methods* **57**, 338 (1967).
MR68. Maccabee, H. D., M. R. Raju, and C. A. Tobias: *Phys. Rev.* **165**, 469 (1968).
MS65. Morbitzer, L., and A. Scharmann: *Z. Phys.* **185**, 488 (1965).
MT65. Mattauch, J. H. E., W. Thiele, and A. H. Wapstra: *Nucl. Phys.* **67**, 1 (1965).
MY68. Myers, I. T.: "Radiation Dosimetry," vol. 1, chap. 7, Ionization, 2d ed., Academic Press, Inc., New York, 1968.
MZ67. Marion, J. B., and B. A. Zimmerman: *Nucl. Instr. Methods* **51**, 93 (1967).
NA67. *Natl. Acad. Sci.–Natl. Res. Council Publ.* 1133, U. Fano, ed., 2d printing, 1967.
NA69. Nakata, H.: *Can. J. Phys.* **47**, 2545 (1969).
NI61. Nielsen, L. P.: *Mat. Fys. Medd. Dan. Vid. Selsk.* **33**, No. 6 (1961).
NM67. Nicolleta, C. A., P. J. McNulty, and P. L. Jain: *Phys. Rev.* **164**, 1693 (1967).
NO63. Northcliffe, L. C.: *Ann. Rev. Nucl. Sci.* **13**, 67 (1963).   Reprinted in NA67.
NO67. Northcliffe, L. C.: Paper 8 in NA67.
NS61. Nigam, B. P., M. K. Sundaresan, and Ta-You Wu: *Phys. Rev.* **115**, 491 (1959).
NS70. Northcliffe, L. C., and R. F. Schilling: *Nuclear Data Tables* **A7**, 233 (1970).
NV66. Nichols, D. K., and V. A. J. Van Lint: *Solid State Phys.* **18**, 1 (1966).   "Advances in Research Applications," Academic Press, Inc., New York.
OH63. Oen, O. S., D. K. Holmes, and M. T. Robinson: *J. Appl. Phys.* **34**, 302 (1963).
OL67. Oldham, W. J. B.: *Phys. Rev.* **161**, 1 (1967).
OR68. Ormrod, J. H.: *Can. J. Phys.* **46**, 497 (1968).
PA69. Payne, M. G.: *Phys. Rev.* **185**, 611 (1969).
PB64. Phipps, J. A., J. W. Boring, and R. A. Lowry: *Phys. Rev.* **135**, A36 (1964).
PB68. Pierce, T. E., W. W. Bowman, and M. Blann: *Phys. Rev.* **172**, 287 (1968).
PE62. Perkins, J. F.: *Phys. Rev.* **126**, 1781 (1962).
PG68. Pehl, R. H., F. S. Goulding, D. A. Landis, and M. Lenzlinger: *Nucl. Instr. Methods* **59**, 45 (1968).
PG70. Pehl, R. H., and F. S. Goulding: *Nucl. Instr. Methods*, **81**, 329 (1970).
PL69. Pivovar, L. I., Yu. Z. Levchenko, A. N. Grigor'ev, and S. M. Khazan: *Sov. Phys.—JETP* **29**, 399 (1969) [*Zh. Eksp. Teor. Fiz.* **56**, 736 (1969)].
PO67. Powell, C. J.: *Health Phys.* **13**, 1265 (1967).
PR69. Palms, J. M., P. V. Rao, R. E. Wood: *IEEE Trans. Nucl. Sci.* **NS 16**(1), 36 (1969).
RB69. Ray, J. A., and C. F. Barnett: *IEEE Trans. Nucl. Sci.* **NS-16**(1), 82 (1969).
RC54. Rohrlich, F., and B. C. Carlson: *Phys. Rev.* **93**, 38 (1954).
RO60. Roll, R. G., and F. E. Steigert: *Nucl. Phys.* **17**, 54 (1960).
RO68. Roesch, Wm. C.: "Radiation Dosimetry," vol. 1, chap. 5, 2d ed., F. H. Attix and Wm. C. Rosech, eds., Academic Press, Inc., New York, 1968.
RO69. Robinson, M. T.: *Phys. Rev.* **179**, 327 (1969).
RS67. Remillieux, J., J. J. Samueli, and A. Sarazin: *J. Phys. Radium* **28**, 832 (1967).
RU68. Rudge, M. R. H.: *Rev. Mod. Phys.* **40**, 564 (1968).
RY55. Rybakov, B. V.: *Soviet Phys. JETP* **1**, 435 (1955).
SA65. Sattler, A. R.: *Phys. Rev.* **138**, A1815 (1965).
SB67. Seltzer, S. M., and M. J. Berger: Section 9 in NA67.
SC63. Scott, W. T.: *Rev. Mod. Phys.* **35**, 231 (1963).
SC66. Schiøtt, H. E.: *Mat. Fys. Medd. Dan. Vid. Selsk.* **35**, 9, (1966).
SI67. Singh, J. J.: *NASA Tech. Note* D-3927, May, 1967.
SK67. Skyrme, D. J.: *Nucl. Instr. Methods* **57**, 61 (1967).
SP54. Spencer, L. V., and U. Fano: *Phys. Rev.* **93**, 1172 (1954).
SP55. Spencer, L. V.: *Phys. Rev.* **98**, 1597 (1955).
SP59. Spencer, L. V.: *Natl. Bur. Standards (U.S.) Monograph* 1.
SP63. Swanson, N., and C. J. Powell: *Jour. Chem. Phys.* **39**, 630 (1963).
SP70. Swint, J. B., R. M. Prior, and J. J. Ramirez: *Nucl. Instr. & Meth.* **80**, 134 (1970).
ST60. Sternheimer, R. M.: *Phys. Rev.* **117**, 485 (1960).
ST67. Sternheimer, R. M.: *Phys. Rev.* **164**, 349 (1967).
SU60. Sternheimer, R. M.: *Phys. Rev.* **118**, 1045 (1960).

SV68. Sattler, A. R., and F. L. Vook: *Phys. Rev.* **175**, 526 (1968).
SY48. Symon, K. R.: Thesis, Harvard University, Cambridge, Mass., 1948.
SZ65. Sautter, C. A., and E. J. Zimmerman: *Phys. Rev.* **140**, A490 (1965).
TB68. Tschalär, C., and H. Bichsel: *Nucl. Instr. Methods* **62**, 208 (1968).
TH52. Thompson, T. J.: UCRL-1910. (Thesis, Univ. of California, Berkeley, 1952).
TH67. Thomas, E. W.: *Phys. Rev.* **164**, 151 (1967).
TM59. Ter-Mikayelian, M. L.: *Nucl. Phys.* **9**, 679 (1958–1959).
TM70. Tschalär, C., and H. D. Maccabee: *Phys. Rev.* **B1**, 2863 (1970).
TO71. Toburen, L. H.: *Phys. Rev.* **A3**, 216 (1971).
TP69. Taylor, B. N., W. H. Parker, and D. N. Langenberg: *Rev. Mod. Phys.* **41**, 375 (1969).
TS67. Tschalär, C.: Thesis, University of Southern California, Los Angeles, January 1967.
TS68. Tschalär, C.: *Nucl. Instr. Methods* **61**, 141 (1968).
TT68. Tschalär, C.: *Nucl. Instr. Methods* **64**, 237 (1968).
VA57. Vavilov, P. V.: *Soviet Phys. JETP* **5**, 749 (1957).
VK69. Vasilevskii, I. M., I. I. Karpov, V. I. Petrushkin, and Yu. D. Prokoshkin: *Sov. J. Nucl. Phys.* **9**, 583 (1969); [*Yad. Fiz.* **9**, 997 (1968)].
VS68. Vriens, L., J. A. Simpson, and S. R. Mielczarek: *Phys. Rev.* **165**, 7 (1968); **170**, 163 (1968).
VV68. Van Camp, K. J., and V. J. Vanhuyse: *Z. Phys.* **211**, 152 (1968).
WA51. Walske, M. C.: Thesis, Cornell University, 1951.
WA52. Walske, M. C.: *Phys. Rev.* **88**, 1283 (1952).
WA56. Walske, M. C.: *Phys. Rev.* **101**, 940 (1956).
WH58. Whaling, Ward: "Encyclopedia of Physics," vol. 34, p. 202 Springer-Verlag OHG, Berlin, 1958.
WH33. Wheeler, J. A.: *Phys. Rev.* **43**, 258 (1933).
WI68. Winterbon, K. B.: *Can. J. Phys.* **46**, 2429 (1968).
WM67. White, W., and R. M. Mueller: *J. Appl. Phys.* **38**, 3660 (1967).
ZA70. Zulliger, H. R., and D. W. Aitken: *IEEE Trans. Nucl. Sci.* NS-17, 187 (1970).
ZM69. Zulliger, H. R., L. M. Middleman, and D. W. Aitken: *IEEE Trans. Nucl. Sci.* **NS16**, 1, 47 (1969).

8e. Gamma Rays

ROBLEY D. EVANS

Massachusetts Institute of Technology

8e-1. Attenuation of Gamma Rays and X Rays. *The Photon.* Photons are classified according to their mode of origin, not their quantum energy. *Gamma rays* are the electromagnetic radiations which accompany nuclear transitions. *X rays* are the electromagnetic radiations which accompany electronic transitions, including the characteristic or fluorescent line spectra of X rays, the bremsstrahlung or continuous spectra of X rays, and the positron-negatron annihilation radiation. By extension, the X-ray category includes all photons originating outside an atomic nucleus and due to the transitions of other elementary particles, for example, proton bremsstrahlung, mu-mesonic X rays, and π°-decay photons. In this section we deal only with the interaction of photons with matter. These interactions are thought to be independent of the origin of the photon. Hence we use the term photon to refer here to both γ rays and X rays.

The quantum energy of a photon is $E = h\nu$, where ν is the frequency and $h = 4.135 \times 10^{-21}$ MeV sec is Planck's constant. The corresponding wavelength is $\lambda = c/\nu = hc/E = 0.012\ 397 \times 10^{-8}/E$ cm, when E is in MeV.

Competing Interactions. A photon can interact with matter by any one of several competing alternative mechanisms. For details, see refs. B3, E1, F3, G1, and G2. In each case the interaction is an all-or-nothing affair. The interaction can be with the entire atom (photoelectric effect and Rayleigh scattering) or with one electron in the atom (Compton effect and pair production in the field of an electron) or with the atomic nucleus (pair production, resonance elastic scattering, photodisintegration, and meson production). The probability for each of these many competing independent processes can be expressed as a collision cross section per atom, per electron, or per nucleus in the absorber. The sum of all these cross sections (corrected for coherence in some low-energy cases), normalized to a per atom basis, is then the probability that the incident photon will have an interaction of some kind while passing through a very thin absorber which contains one atom per cm² of area normal to the path of the incident photon.

Attenuation, Scattering, and Absorption. The total collision cross section per atom, when multipled by the number of atoms per cm³ of absorber, is then the *linear attenuation coefficient* μ_0 per centimeter of travel in the absorber. The fraction of incident photons which can pass through a thickness x of absorber whose density is ρ without having an interaction of any kind is $e^{-\mu_0 x}$ or $e^{-(\mu_0/\rho)(\rho x)}$, where ρ is the density of the medium and μ_0/ρ is the mass attenuation coefficient.

The absorption coefficient μ_a is a much more restricted concept than the attenuation coefficient μ_0. Attenuation can be by some purely elastic process, such as Rayleigh scattering or nuclear resonance (Mössbauer) scattering, in which the photon is merely deflected and does not give up any of its initial energy to the medium. Here only a scattering coefficient would be involved. However, in a photoelectric interaction, the entire energy of the incident photon is truly absorbed by an atom of the medium; there is no scattered residual photon. Here the attenuation of the primary radiation is due to complete absorption of the energy of the incident photon. The intermediate case of greatest importance is the Compton effect, in which some energy is absorbed and appears in the medium as kinetic energy of a Compton recoil electron while the balance of the incident energy is not absorbed but is present as a Compton scattered photon. *Scattering*, then, involves the deflection of incident photon energy, *absorption* involves the conversion of incident photon energy into the kinetic energy of a

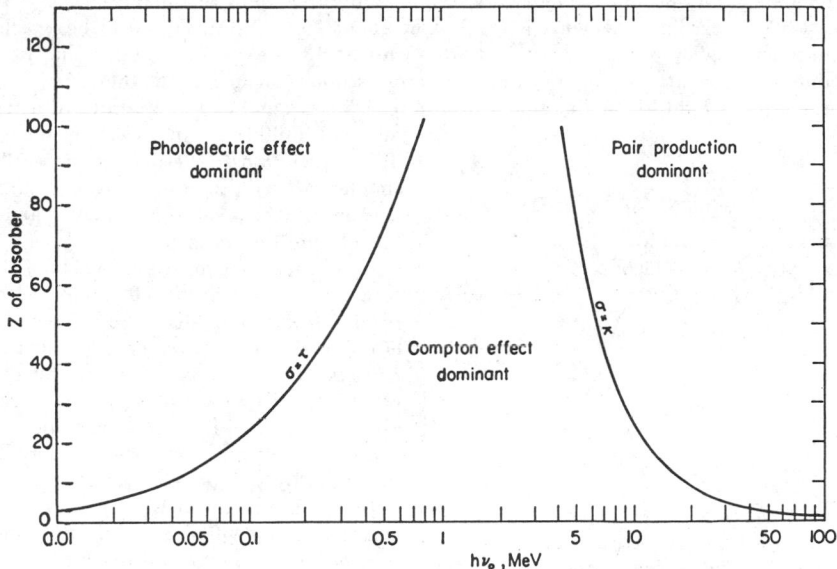

Fig. 8e-1. Relative importance of the three major types of γ-ray interaction. The lines show the values of Z and $h\nu$ for which the two neighboring effects are just equal. [*From Evans* (E1).]

charged particle (usually an electron; see Sec. 8e-5), and the *attenuation* coefficient μ_0 is the sum of the scattering coefficient μ_s and the absorption coefficient μ_a, or

$$\mu_0 = \mu_s + \mu_a \qquad (8e\text{-}1)$$

In most practical cases μ_s is simply the Compton scattering coefficient σ_s of Eqs. (8e-25) and (8e-34) (see page 720 of ref. E1).

Compton, Photoelectric, and Pair-production Competition. In the energy domain met most frequently, say 0.01 to 10 MeV, all but a few minor effects are due to only three of the many competing processes. These are the Compton effect, the photoelectric effect, and pair production. Figure 8e-1 provides a handy guide to the relative importance of these three principal interactions over broad ranges of energies $h\nu_0$ of the incident photons and atomic numbers Z of the attenuating material. For any Z the Compton effect predominates for photon energies between 0.8 and 4 MeV; for low-Z materials the Compton effect predominates over a much wider domain of photon energies. For moderately large Z, the photoelectric interaction is dominant at very small $h\nu_0$ and the pair-production interaction is dominant at large $h\nu_0$.

Nuclear Disintegrations by High-energy Photons. The "separation energy," or energy required to remove one neutron or one proton from a nucleus, differs from the average binding energy per nucleon and, for most stable middleweight or heavyweight nuclei, lies in the domain of about 6 to 16 MeV (see ref. E4 for a brief discussion). Above these appropriate individual threshold energies a nucleus can absorb a photon and emit a neutron in a (γ,n) reaction, or emit a proton in a (γ,p) reaction. Above the threshold energy, the cross section for these *photonuclear reactions* increases with increasing photon energy $h\nu$, reaches a maximum value, then decreases with further increases of $h\nu$. The shape of this peak is characteristic of resonance reactions and is called *the giant resonance.* Typically, the maximum value of this photonuclear cross section in most light elements is reached at about 20 to 25 MeV, and in heavy elements at smaller photon energies down to about 13 MeV in uranium. The peak value of the giant resonance in the (γ,n) cross section is typically of the order of 10 millibarns in light elements, and increases with mass number to the order of a few hundred millibarns in heavy elements. The (γ,p) cross section is generally smaller than the (γ,n) cross section because of the nuclear Coulomb barrier against proton emission.

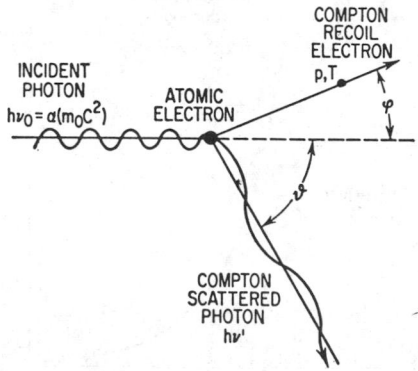

In all cases, the maximum value of the total cross section for all photonuclear reactions is smaller than about 5 percent of the total cross section for Compton and pair-production interactions. Thus in nearly all practical cases the energy absorbed from a high-energy photon beam by any medium is not materially increased by photonuclear interactions.

In very heavy nuclei, photon absorption can also induce fission. The threshold for *photofission* in various isotopes of Th, U, and Pu is between 5.0 and 5.5 MeV. The photofission cross section for U^{238} has a giant resonance shape, a peak cross section of 125 millibarns at 14 MeV, and

Fig. 8e-2. Trajectories in the scattering plane for the incident photon $h\nu_0$, the scattered photon $h\nu'$, and the scattering electron which acquires momentum p and kinetic energy T.

a full width at half maximum of 8.8 MeV. Again the cross section is very small compared with the cross section for pair production (19 barns) and Compton interactions (3 barns) in uranium at 14 MeV.

When nuclei absorb photons whose energy is above about 140 MeV, pi mesons can be produced in reactions such as $p(\gamma,\pi^+)n$. The cross section for *meson production* rises slowly with photon energy above about 140 MeV, reaching values of about 0.2 millibarn at 250 MeV for $H(\gamma,\pi^+)n$. For other elements, the cross section varies approximately as $Z^{\frac{2}{3}}$. For comparison, the pair-production cross section at 250 MeV is about $6Z^2$ barns for all elements, or more than 10,000 times larger than the meson production cross section.

8e-2. Compton Effect. The incident photon has an energy $h\nu_0$ and momentum $p = h\nu_0/c$. It spends these on a single individual atomic electron in a given collision. If the initial binding energy of the electron is small compared with $h\nu_0$, then the struck electron can be considered initially free and unbound. Conservation of momentum and energy leads to the usual Compton equations, and the cross sections are given by the Dirac electron theory as first applied to the Compton collision by Klein and Nishina. As in Fig. 8e-2, the electron recoils at an angle φ with kinetic energy T, and the remaining energy, $h\nu' = h\nu_0 - T$, is found in a Compton scattered photon emitted at an angle ϑ.

The *Compton shift* is the difference between the wavelength λ_0 (or the energy $h\nu_0$) of the incident photon and the wavelength λ' (or energy $h\nu'$) of the Compton scattered photon and is

$$\frac{1}{h\nu'} - \frac{1}{h\nu_0} = \frac{1}{m_0 c^2}(1 - \cos\vartheta) \qquad (8e\text{-}2)$$

or

$$\lambda' - \lambda_0 = \frac{h}{m_0 c}(1 - \cos\vartheta) \qquad (8e\text{-}3)$$

Note that the Compton shift in wavelength $(\lambda' - \lambda_0)$ in any particular direction is independent of the energy of the incident photon but that the Compton shift in energy $(h\nu_0 - h\nu')$ increases very strongly as $h\nu_0$ increases. The length $h/m_0 c = \lambda_c = 2.426 \times 10^{-10}$ cm is called the Compton wavelength for an electron. It is equal to the wavelength of a photon whose energy is just equal to the rest energy of the electron $m_0 c^2 = 0.5110$ MeV.

Writing the incident energy in terms of the dimensionless quantity

$$\alpha \equiv \frac{h\nu_0}{m_0 c^2} \qquad (8e\text{-}4)$$

the conservation laws give, for the energy of the Compton scattered photon,

$$\frac{\nu'}{\nu_0} = \frac{1}{1 + \alpha(1 - \cos\vartheta)} \qquad (8e\text{-}5)$$

$$(h\nu')_{\min} = m_0 c^2 \frac{\alpha}{1 + 2\alpha} = h\nu_0 \frac{1}{1 + 2\alpha} \qquad \text{for } \vartheta = 180 \text{ deg} \qquad (8e\text{-}6)$$

Figure 8e-3 gives $h\nu'$ vs. $h\nu_0$ for 10 values of ϑ. The curve for $\vartheta = 180$ deg gives $(h\nu')_{\min}$ and thus evaluates the backscatter peak and the energy separation between the Compton edge and the total energy peak in γ-ray scintillation spectroscopy (see Sec. 8e-6). The energy of the backscattered photon $(h\nu'_{\min})$ approaches its maximum value of $m_0 c^2/2 = 0.25$ MeV for high-energy incident photons, $\alpha \gg 1$.

The angle φ and the kinetic energy T of the Compton recoil electron are related to the photon scattering angle ϑ by

$$\cot\varphi = (1 + \alpha)\tan\frac{\vartheta}{2} \qquad (8e\text{-}7)$$

$$T = h\nu_0 - h\nu' = h\nu_0 \frac{\alpha(1 - \cos\vartheta)}{1 + \alpha(1 - \cos\vartheta)} \qquad (8e\text{-}8)$$

$$T_{\max} = h\nu_0 - (h\nu')_{\min} = h\nu_0 \frac{2\alpha}{1 + 2\alpha} \qquad \text{for } \vartheta = 180 \text{ deg} \qquad (8e\text{-}9)$$

Klein-Nishina Collision Differential Cross Section. The *collision* differential cross section $d(_e\sigma)$, where the subscript connotes per electron in the attenuator, refers to the *number* of collisions of a particular type. The *number* of photons which are scattered in a particular direction, as a fraction of the *number* of incident photons, is $d(_e\sigma)$ and has the dimensions

$$d(_e\sigma) = \frac{\text{number scattered [number/(sec} \cdot \text{electron)]}}{\text{number incident [number/(cm}^2 \cdot \text{sec)]}} = \frac{\text{cm}^2}{\text{electron}} \qquad (8e\text{-}10)$$

As first shown by Klein and Nishina, the collision differential cross section for unpolarized photons striking unbound, randomly oriented electrons is

$$d(_e\sigma) = \frac{r_0^2}{2} d\Omega \left(\frac{\nu'}{\nu_0}\right)^2 \left(\frac{\nu_0}{\nu'} + \frac{\nu'}{\nu_0} - \sin^2\vartheta\right) \qquad \frac{\text{cm}^2}{\text{electron}} \qquad (8e\text{-}11)$$

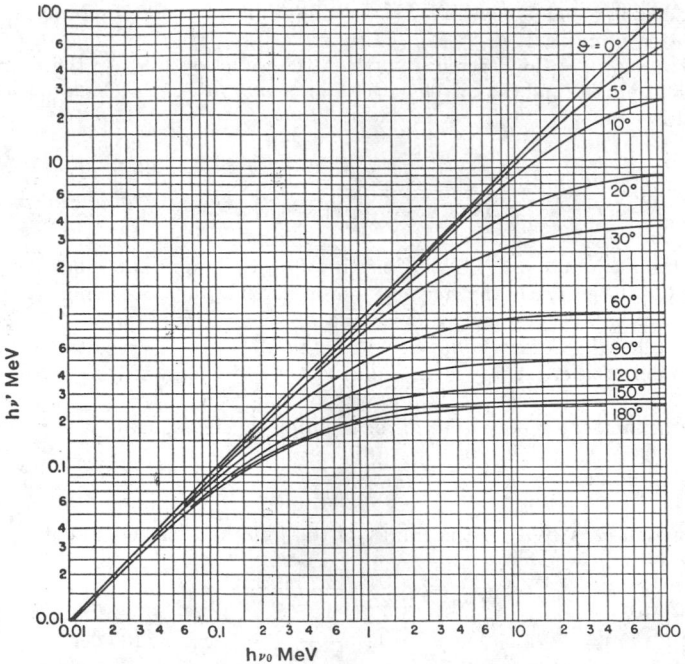

FIG. 8e-3. Dependence of the energy $h\nu'$ of the Compton-scattered photon on $h\nu_0$ and the photon scattering angle ϑ, from Eq. (8e-5). [*From Evans* (E2).]

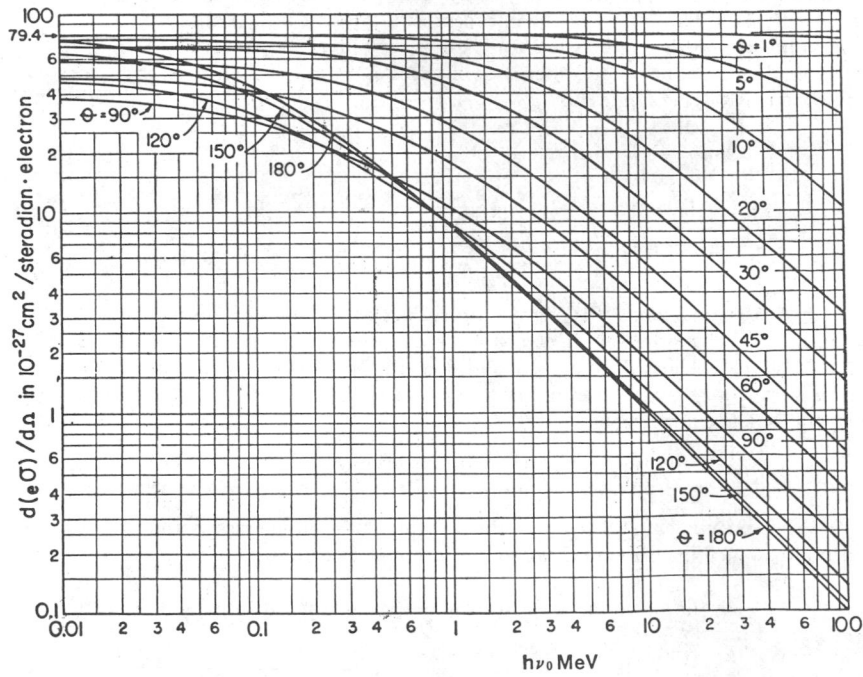

FIG. 8e-4. Collision differential cross section $d(_e\sigma)/d\Omega$ for the *number* of photons scattered per unit solid angle in the direction ϑ. [*From Evans* (E2).]

where the scattered photon $h\nu'$ goes into the solid angle $d\Omega$ steradians at mean angle ϑ and the classical radius of the electron r_0 has the value

$$r_0 \equiv \frac{e^2}{m_0 c^2} = 2.818 \times 10^{-13} \text{ cm} \tag{8e-12}$$

Substituting ν'/ν_0 in terms of ϑ and the incident photon energy $h\nu_0 = \alpha(m_0 c^2)$ gives the equivalent explicit relationship

$$d(_e\sigma) = r_0^2 \, d\Omega \left[\frac{1}{1 + \alpha(1 - \cos\vartheta)}\right]^2 \left(\frac{1 + \cos^2\vartheta}{2}\right)$$
$$\left\{1 + \frac{\alpha^2(1 - \cos\vartheta)^2}{(1 + \cos^2\vartheta)[1 + \alpha(1 - \cos\vartheta)]}\right\} \quad \frac{\text{cm}^2}{\text{electron}} \tag{8e-13}$$

Table 8e-1 and Fig. 8e-4 give numerical values of $d(_e\sigma)/d\Omega$ in 10^{-27} cm^2 (or millibarn)/ steradian per electron. The classical or Thomson differential cross section is

$$\frac{d(_e\sigma_{\text{Thom}})}{d\Omega} = r_0^2 \left(\frac{1 + \cos^2\vartheta}{2}\right) \quad \frac{\text{cm}^2}{\text{steradian} \cdot \text{electron}} \tag{8e-14}$$

TABLE 8e-1. COLLISION DIFFERENTIAL CROSS SECTION $d(_e\sigma)/d\Omega$ IN
10^{-27} CM^2/STERADIAN PER ELECTRON*
[From Eq. (8e-13)]

$h\nu_0$, MeV	ϑ										
	1°	5°	10°	20°	30°	45°	60°	90°	120°	150°	180°
0.01	79.4	79.1	78.1	74.6	69.1	59.0	48.6	38.3	46.4	64.7	73.6
0.04	79.4	79.0	78.0	74.0	68.0	57.5	45.5	34.4	40.4	53.5	60.2
0.1	79.4	79.0	77.7	73.0	66.0	53.7	41.0	29.3	31.3	39.2	43.2
0.2	79.4	78.8	77.3	71.3	62.8	48.2	35.0	22.9	23.0	27.2	29.3
0.4	79.4	78.6	76.4	68.2	57.2	41.0	28.5	16.8	15.7	17.2	17.9
1	79.3	77.9	73.7	60.2	45.0	27.7	17.7	10.4	8.80	8.45	8.35
2	79.3	76.8	69.7	50.1	33.0	18.3	11.5	6.80	5.30	4.73	4.54
4	79.2	74.6	62.8	37.3	21.7	11.5	7.15	4.09	2.98	2.50	2.39
10	78.9	68.6	48.4	21.2	11.0	5.6	3.48	1.86	1.28	1.05	0.98
20	78.4	60.4	34.2	12.5	6.3	2.93	1.86	0.97	0.66	0.535	0.498
40	77.6	48.6	22.0	7.00	3.40	1.56	0.97	0.492	0.336	0.270	0.254
100	75.0	30.5	10.5	3.08	1.45	0.64	0.40	0.204	0.135	0.108	0.102

* From R. D. Evans, Compton Effect, in "Handbuch der Physik," vol. XXXIV, pp. 218–298, S. Flügge, ed., Springer-Verlag, Berlin, 1958.

where $r_0^2 = 79.41 \times 10^{-27}$ cm^2 is the upper limit approached by this Klein-Nishina collision differential cross section at any ϑ as $\alpha = h\nu_0/m_0 c^2$ approaches zero and at any α as ϑ approaches zero.

Klein-Nishina Scattering Differential Cross Section. The *scattering* differential cross section $d(_e\sigma_s)$ refers to the amount of *energy* scattered in a particular direction; thus

$$d(_e\sigma_s) = \frac{\text{scattered energy per sec [MeV/(sec} \cdot \text{electron)]}}{\text{incident intensity [MeV/(cm}^2 \cdot \text{sec)]}} = \frac{\text{cm}^2}{\text{electron}} \tag{8e-15}$$

The scattered *energy* is the number of scattered photons times the quantum energy $h\nu'$ of each, and the incident intensity is the number of incident photons per unit area times the quantum energy $h\nu_0$ of each. Not all the energy $h\nu_0$ is scattered, but only the fraction $h\nu'/h\nu_0$. Therefore the *scattering* differential cross section for unpolarized radiation is

$$d(_e\sigma_s) = \frac{\nu'}{\nu_0}\, d(_e\sigma)$$

$$= \frac{r_0^2}{2}\, d\Omega \left(\frac{\nu'}{\nu_0}\right)^3 \left(\frac{\nu_0}{\nu'} + \frac{\nu'}{\nu_0} - \sin^2\vartheta\right) \qquad \frac{\text{cm}^2}{\text{electron}} \qquad (8\text{e-}16)$$

Tables and graphs of $d(_e\sigma_s)$ over the range $1\deg \le \vartheta \le 180\deg$ and 0.01 MeV $\le h\nu_0 \le 100$ MeV are available in ref. E2.

Angular Distribution of the Number of Scattered Photons. The total solid angle available per unit scattering angle is

$$\frac{d\Omega}{d\vartheta} = 2\pi \sin\vartheta \qquad (8\text{e-}17)$$

and approaches zero in the forward- and backward-scattering directions. The *number-vs.-angle* distribution of scattered photons is

$$\frac{d(_e\sigma)}{d\vartheta} = \frac{d(_e\sigma)}{d\Omega}\, 2\pi \sin\vartheta \qquad \frac{\text{cm}^2}{\text{electron}\cdot\text{radian}} \qquad (8\text{e-}18)$$

and has a forward maximum which is in the vicinity of $\vartheta = 20$ deg for $h\nu_0 = 3$ MeV and is at larger angles for smaller $h\nu_0$ (see Fig. 20 of ref. E2).

Angular Distribution of the Energy of Scattered Photons. The distribution of scattered photon *energy* in any angular interval, that is, between two cones of half angles ϑ and $\vartheta + d\vartheta$, is

$$\frac{d(_e\sigma_s)}{d\vartheta} = \frac{d(_e\sigma_s)}{d\Omega}\, 2\pi \sin\vartheta \qquad \frac{\text{cm}^2}{\text{electron}\cdot\text{radian}} \qquad (8\text{e-}19)$$

and is more sharply peaked than the number-vs.-angle distribution because of the variation of $h\nu'$ with ϑ (see fig. 24 of ref. E2).

Angular Distribution of Compton Recoil Electrons. The ionization which actuates many radiation detectors is due primarily to Compton recoil electrons produced in the detector or its walls and projected between $\varphi = 0$ and $\varphi = 90$ deg. The initial *number-vs.-angle* distribution of the recoil electrons is

$$\frac{d(_e\sigma)}{d\varphi} = \frac{d(_e\sigma)}{d\Omega} \left[\frac{2\pi(1 + \cos\vartheta)\sin\vartheta}{(1 + \alpha)\sin^2\varphi}\right] \qquad \frac{\text{cm}^2}{\text{electron}\cdot\text{radian}} \qquad (8\text{e-}20)$$

For photon energies below about 0.5 MeV this distribution has two maxima, in the neighborhood of 20 and 60 deg. At higher photon energies the wide-angle maximum disappears and the small-angle maximum occurs at smaller angles as $h\nu_0$ increases (see fig. 21 of ref. E2 and table III of ref. J1).

Energy Distribution of Compton Recoil Electrons. The *number-vs.-energy* spectrum of Compton electrons is

$$\frac{d(_e\sigma)}{dT} = \frac{d(_e\sigma)}{d\Omega}\, \frac{d\Omega}{d\vartheta}\, \frac{d\vartheta}{dT}$$

$$= \frac{d(_e\sigma)}{d\Omega}\, \frac{2\pi m_0 c^2}{(h\nu')^2} = \frac{d(_e\sigma)}{d\Omega}\, \frac{2\pi m_0 c^2}{(h\nu_0 - T)^2}$$

$$= \frac{\pi r_0^2}{\alpha^2 m_0 c^2} \left\{ 2 + \left(\frac{T}{h\nu_0 - T}\right)^2 \left[\frac{1}{\alpha^2} + \frac{h\nu_0 - T}{h\nu_0} - \frac{2}{\alpha}\frac{(h\nu_0 - T)}{T}\right]\right\}$$

$$\frac{\text{cm}^2}{\text{keV}\cdot\text{electron}} \qquad (8\text{e-}21)$$

where $r_0 = 2.818 \times 10^{-13}$ cm and $m_0c^2 = 511.0$ keV. The electron spectra for $h\nu_0 =$ 0.5 to 3.5 MeV, in steps of 0.5 MeV, are shown in Fig. 8e-5 (see also table II of ref. J1 and fig. VII of ref. N1). The pronounced number maximum which occurs just at the maximum electron energy $T_{\max}$ is called the *Compton edge* in γ-ray spectroscopy (see Sec. 8e-6).

Energy Distribution of Compton Scattered Photons. Each recoil electron has a companion scattered photon whose energy is $h\nu' = h\nu_0 - T$. Hence the energy spectrum

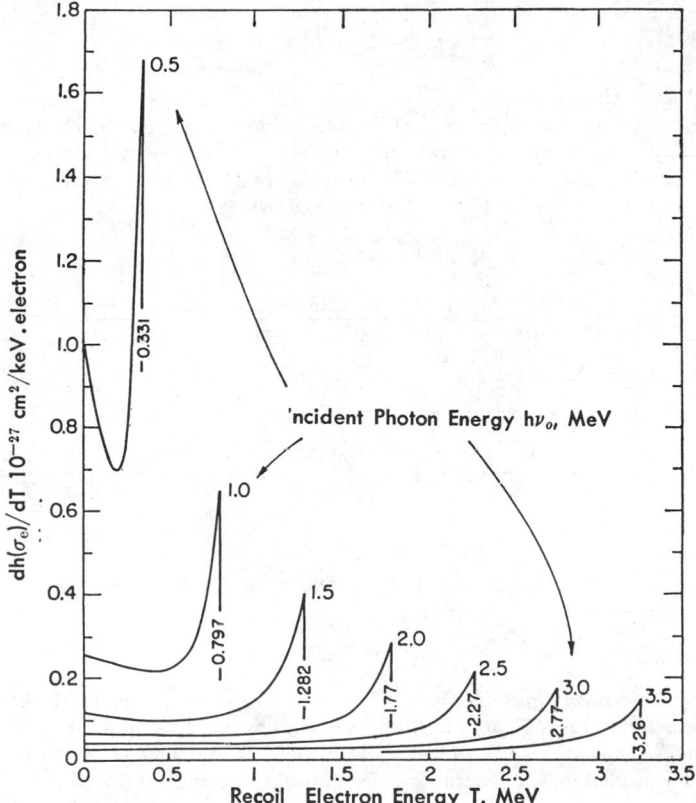

Fig. 8e-5. Number-vs.-energy distribution of Compton recoil electrons, for seven values of the incident photon energy $h\nu_0$, in 10^{-27} cm² (millibarn)/keV per free electron. The energy spectrum of scattered photons is obtained by transforming the energy scale from T to $h\nu_0 - T$ for each curve. [*From Evans* (E2).]

of scattered photons is complementary to the energy spectrum of recoil electrons and is given by replacing T by $(h\nu_0 - h\nu')$ and dT by $d(h\nu')$ in Eq. 8e-21 and in Fig. 8e-5.

Average Collision Cross Section. The average (or total) collision cross section $_e\sigma$ is the probability of any Compton interaction by one photon while passing normally through a material containing one electron per cm² and is given by

$$_e\sigma = \int_0^\pi \frac{d(_e\sigma)}{d\Omega} 2\pi \sin\vartheta \, d\vartheta = 2\pi r_0^2 \left\{ \frac{1+\alpha}{\alpha^2} \left[\frac{2(1+\alpha)}{1+2\alpha} - \frac{\ln(1+2\alpha)}{\alpha} \right] \right.$$
$$\left. + \frac{\ln(1+2\alpha)}{2\alpha} - \frac{1+3\alpha}{(1+2\alpha)^2} \right\} \quad \frac{\text{cm}^2}{\text{electron}} \quad (8\text{e-}22)$$

Numerical values are given in Table 8e-2 (see also fig. 27 of ref. E2 and fig. VIII of ref. N1). For small values of $\alpha \equiv h\nu_0/m_0c^2$, accuracy is best preserved by using the expansion

$$_e\sigma \simeq \frac{8}{3}\pi r_0^2 \left(1 - 2\alpha + \frac{26}{5}\alpha^2 - \frac{133}{10}\alpha^3 + \frac{1{,}144}{35}\alpha^4 - \frac{544}{7}\alpha^5 + \frac{3{,}784}{20}\alpha^6 - \cdots\right)$$
$$\frac{\text{cm}^2}{\text{electron}} \quad (8\text{e-}23)$$

while for $\alpha \gg 1$,

$$_e\sigma \simeq \pi r_0^2 \frac{1 + 2\ln 2\alpha}{2\alpha} \qquad \frac{\text{cm}^2}{\text{electron}} \qquad (8\text{e-}24)$$

is a good approximation.

Average Scattering Cross Section. The total *scattered energy* in photons of various energies $h\nu'$, scattered on the average by each electron per cm^2 of scattering material, is the *average scattering cross section* $_e\sigma_s$ multiplied by the incident energy expressed as (number of photons) $\times$ (energy $h\nu_0$ per photon), where

$$_e\sigma_s = \int_0^\pi \frac{d(_e\sigma_s)}{d\Omega} 2\pi \sin\vartheta \, d\vartheta$$
$$= \pi r_0^2 \left[\frac{\ln(1 + 2\alpha)}{\alpha^3} + \frac{2(1 + \alpha)(2\alpha^2 - 2\alpha - 1)}{\alpha^2(1 + 2\alpha)^2} + \frac{8\alpha^2}{3(1 + 2\alpha)^3}\right]$$
$$\frac{\text{cm}^2}{\text{electron}} \quad (8\text{e-}25)$$

For small α, use the expansion

$$_e\sigma_s = \frac{8}{3}\pi r_0^2 \left(1 - 3\alpha + \frac{94}{10}\alpha^2 - 28\alpha^3 + \frac{552}{7}\alpha^4 - 212\alpha^5 + \frac{1{,}648}{3}\alpha^6 - \cdots\right)$$
$$\frac{\text{cm}^2}{\text{electron}} \quad (8\text{e-}26)$$

Table 8e-2 gives numerical values of $_e\sigma_s$ as well as the average energy $(h\nu')_\text{av}$ per scattered photon, which is

$$(h\nu')_\text{av} = h\nu_0 \frac{_e\sigma_s}{_e\sigma} \qquad (8\text{e-}27)$$

Average Absorption Cross Section. The total *kinetic energy*, in recoil electrons of various kinetic energies T, produced on the average per electron per cm^2 of material is the *average absorption cross section* $_e\sigma_a$ multiplied by the incident energy expressed as (number of photons) $\times$ (energy $h\nu_0$ per photon), where

$$_e\sigma_a = {_e\sigma} - {_e\sigma_s} = 2\pi r_0^2 \left[\frac{2(1 + \alpha)^2}{\alpha^2(1 + 2\alpha)} - \frac{1 + 3\alpha}{(1 + 2\alpha)^2} - \frac{(1 + \alpha)(2\alpha^2 - 2\alpha - 1)}{\alpha^2(1 + 2\alpha)^2}\right.$$
$$\left. - \frac{4\alpha^2}{3(1 + 2\alpha)^3} - \left(\frac{1 + \alpha}{\alpha^3} - \frac{1}{2\alpha} + \frac{1}{2\alpha^3}\right)\ln(1 + 2\alpha)\right] \qquad \frac{\text{cm}^2}{\text{electron}} \quad (8\text{e-}28)$$

For small α, use the expansion

$$_e\sigma_a = \frac{8}{3}\pi r_0^2 \left(\alpha - \frac{42}{10}\alpha^2 + \frac{147}{10}\alpha^3 - \frac{1{,}616}{35}\alpha^4 + \frac{940}{7}\alpha^5 - \frac{7{,}752}{21}\alpha^6 + \cdots\right)$$
$$\frac{\text{cm}^2}{\text{electron}} \quad (8\text{e-}29)$$

Table 8e-2 gives numerical values of $_e\sigma_a$ as well as the average energy T_av of the Compton recoil electrons, which is

$$T_\text{av} = h\nu_0 - (h\nu')_\text{av} = h\nu_0 \frac{_e\sigma_a}{_e\sigma} \qquad (8\text{e-}30)$$

TABLE 8e-2. KLEIN-NISHINA CROSS SECTIONS FOR COMPTON INTERACTIONS IN 10^{-27} CM^2 (MILLIBARNS) PER FREE ELECTRON AND RELATED QUANTITIES*
[Calculated from the following equations: $_e\sigma$ (8e-22) and (8e-23), $_e\sigma_s$ (8e-25) and (8e-26), $_e\sigma_a$ (8e-28) and (8e-29), $(h\nu')_{av}$ (8e-27), T_{av} (8e-30), $T_{av}/h\nu_0$ (8e-30). Using $r_0 = 2.818 \times 10^{-13}$ cm and $m_0c^2 = 0.5110$ MeV]

Photon energy $h\nu_0$, MeV	Cross sections, 10^{-27} cm^2/electron			Scattered photon average energy $(h\nu')_{av}$, MeV	Recoil electron	
	Collision $_e\sigma$ †	Scattering $_e\sigma_s$ †	Absorption $_e\sigma_a$		Average energy $(T)_{av}$, MeV	Fraction of incident photon energy $(T)_{av}/h\nu_0$
0.010	640.5	628.5	12.0	0.0098	0.0002	0.0187
0.015	629.0	611.6	17.4	0.0146	0.0004	0.0277
0.020	618.0	595.7	22.3	0.0193	0.0007	0.0361
0.030	597.6	566.5	31.1	0.0284	0.0016	0.0520
0.040	578.7	540.1	38.6	0.0373	0.0027	0.0667
0.050	561.5	516.2	45.3	0.0460	0.0040	0.0807
0.060	545.7	494.5	51.2	0.0544	0.0056	0.0938
0.080	517.3	456.7	60.6	0.0706	0.0094	0.1171
0.100	492.8	424.8	68.0	0.0862	0.0138	0.1380
0.150	443.6	363.1	80.5	0.1228	0.0272	0.1815
0.200	406.5	318.6	87.9	0.1568	0.0432	0.2162
0.300	353.5	258.2	95.3	0.2191	0.0809	0.2696
0.400	316.7	218.6	98.1	0.276	0.124	0.3098
0.500	289.7	190.5	99.2	0.329	0.171	0.3424
0.600	267.5	169.2	98.3	0.379	0.221	0.3675
0.800	235.0	138.9	96.1	0.473	0.327	0.4089
1.00	211.2	118.3	92.9	0.560	0.440	0.4399
1.50	171.6	86.70	84.9	0.758	0.742	0.4948
2.00	146.4	68.67	77.7	0.939	1.061	0.5307
3.00	115.1	48.65	66.4	1.269	1.731	0.5769
4.00	95.98	37.73	58.25	1.57	2.428	0.6069
5.00	82.87	30.83	52.04	1.86	3.140	0.6280
6.00	73.23	26.07	47.16	2.14	3.864	0.6440
8.00	59.89	19.93	39.96	2.66	5.338	0.6672
10	50.99	16.14	34.85	3.16	6.835	0.6835
15	37.71	10.94	26.77	4.35	10.65	0.7099
20	30.25	8.272	21.98	5.47	14.53	0.7266
30	22.00	5.563	16.44	7.58	22.42	0.7473
40	17.46	4.191	13.27	9.6	30.4	0.7600
50	14.58	3.362	11.22	11.5	38.5	0.7695
60	12.54	2.807	9.733	13.4	46.6	0.7762
80	9.882	2.110	7.772	17.1	62.9	0.7865
100	8.199	1.690	6.509	20.6	79.4	0.7939

* From R. D. Evans, Compton Effect, in "Handbuch der Physik," vol. XXXIV, pp. 218-298. S. Flügge, ed., Springer-Verlag, Berlin, 1958.

† These numerical calculations for $_e\sigma$ and $_e\sigma_s$ were done on the IBM computer at the Massachusetts Institute of Technology under the direction of Mr. W. B. Thurston.

The fraction of the incident photon energy which is absorbed and appears as kinetic energy of Compton recoil electrons in the average of all Compton collisions is $T_{\text{av}}/h\nu_0 = {}_e\sigma_a/{}_e\sigma$. This fraction starts at zero for very low energy photons and increases monotonically with $h\nu_0$, as shown in the right-hand column of Table 8e-2.

The Compton collision (or attenuation) cross section is then the sum of the Compton scattering and absorption cross sections; that is,

$$
\begin{aligned}
{}_e\sigma &= \frac{(h\nu')_{\text{av}}}{h\nu_0} \, ({}_e\sigma) + \frac{T_{\text{av}}}{h\nu_0} \, ({}_e\sigma) \\
&= \frac{{}_e\sigma_s}{{}_e\sigma} \, ({}_e\sigma) + \frac{{}_e\sigma_a}{{}_e\sigma} \, ({}_e\sigma) \\
&= {}_e\sigma_s + {}_e\sigma_a
\end{aligned}
\qquad (8e\text{-}31)
$$

8e-3. Photoelectric Effect. The entire primary photon energy $h\nu$ is absorbed by the struck atom. One electron (usually from the K or L shell) is ejected with kinetic energy T, where

$$
T = h\nu - B_e
\qquad (8e\text{-}32)
$$

and B_e is binding energy of the electron before being ejected from the atom. Momentum is conserved by the backward recoil of the entire residual atom. The energy B_e is emitted promptly by the residual atom as characteristic X rays and Auger electrons from the filling of the vacancy in the inner shell. Because the entire atom participates in the interaction, photoelectric interactions are described by an *atomic* cross section ${}_a\tau$ cm²/atom. No single closed formula describes ${}_a\tau$ accurately over a wide range of $h\nu$. A crude but useful guide is

$$
{}_a\tau \simeq \text{const} \, \frac{Z^4}{(h\nu)^3}
\qquad (8e\text{-}33)
$$

The experimental and theoretical material has been summarized in refs. D1, D2, E1, G2, and H3. Numerical tables of blended theoretical and experimental "best" values of ${}_a\tau$ for 11 elements from 0.01 to 100 MeV are given in ref. H3. Mainly owing to differences in the interpretation of experimental results, H3 values of ${}_a\tau$ at small $h\nu_0$ (<0.1 MeV) and large Z are smaller than D2 and larger than G2 values. For large $h\nu_0$ (>1 MeV) and large Z, ${}_a\tau$ is smaller in H3 and D2 than in G2, owing to revisions in the theoretical values. For 1 MeV and large Z there is agreement among D2, G2, and H3. Photoelectric mass absorption coefficients for air, water, Al, Cu, NaI, and Pb and interpolation formulas are given in Sec. 8e-5.

8e-4. Pair Production by Photons. In the field of a charged particle, usually an atomic nucleus but also to some degree in the field of an atomic electron, a photon may be totally absorbed and a positron-negatron pair emitted. A minimum incident photon energy of $h\nu = 2m_0c^2 = 1.02$ MeV is required for pair production in the field of a nucleus, and a minimum of $h\nu = 4m_0c^2 = 2.04$ MeV in the field of an atomic electron.

The atomic cross section ${}_a\kappa$ for nuclear pair production increases with Z^2 (reduced somewhat at very large photon energies by electron screening of the nuclear field) and with the photon energy $h\nu$. The kinetic energies of the positron and the negatron pair electrons are continuously distributed, each from a minimum of zero up to a maximum of $h\nu - 2m_0c^2$. Tables are given for 11 elements in ref. H3, and for 24 elements and several mixtures in ref. G2.

Analytical expressions for ${}_a\kappa$ are complicated (see ref. H1). Tables are given for 11 elements in ref. H3, and for 24 elements and several mixtures in ref. G2. Graphs and interpolation formulas are given in ref. E1 and in Sec. 8e-5.

8e-5. Mass Attenuation and Absorption Coefficients for Photons in Narrow-beam Geometry. Linear attenuation coefficients, σ (Compton), τ (photo), and κ (pair), are the atomic cross sections (cm²/atom) multiplied by atoms/cm³ of material and have dimensions of cm^{-1}. Mass attenuation coefficients are the linear coefficients (cm^{-1}) divided by the density ρ (g/cm^3), thus σ/ρ, τ/ρ, κ/ρ, with dimensions of cm^2/g, and have the advantage of being independent of the actual density and physical state of the attenuator. Each mass attenuation coefficient for an element is the corresponding atomic cross section multiplied by the number of atoms per gram (Avogadro's number/atomic weight). Compton mass attenuation coefficients are nearly independent of Z because the number of electrons per gram varies only slightly among all elements except hydrogen.

The total *mass attenuation coefficient* μ_0/ρ is the sum of Compton absorption (σ_a/ρ), Compton scattering (σ_s/ρ), photoelectric attenuation (τ/ρ), and pair-production attenuation (κ/ρ),

$$\frac{\mu_0}{\rho} = \frac{\sigma_a}{\rho} + \frac{\sigma_s}{\rho} + \frac{\tau}{\rho} + \frac{\kappa}{\rho} = \frac{\sigma}{\rho} + \frac{\tau}{\rho} + \frac{\kappa}{\rho} \qquad (8e\text{-}34)$$

In some narrow-beam attenuation situations a portion of the coherent, elastic Rayleigh scattering may also be effective.

The *mass absorption coefficient* relates only to the actual absorption of photon energy. This is a two-step process involving, first, the conversion of photon energy to kinetic energy of secondary electrons and to rest energy of electron pairs and, second, the dissipation of this kinetic energy mainly by ionization and excitation of the atoms in the absorbing medium, but to a small extent also by bremsstrahlung from radiative collisions of the secondary electrons with atomic nuclei in the absorber. There is no ambiguity about the first step, but the variety of treatments of the second step has led to a confusing group of absorption coefficients, including the so-called "true-absorption," "real-absorption," "energy-absorption," "dose-absorption," and "energy-transfer-absorption" coefficients.

If a collimated beam containing n photons/cm^2 sec, each having energy $h\nu_0$ MeV, is normally incident on an absorber of thickness dx and density ρ, then the number dn of photons/cm^2 sec which will have collisions is

$$dn = n\mu_0 \, dx = n\left(\frac{\mu_0}{\rho}\right)(\rho \, dx) \qquad (8e\text{-}35)$$

The incident photon *intensity* is $I = nh\nu_0$ MeV/cm^2 sec, and the energy transferred from incident photons to secondary electrons, which is closely similar to the "absorbed dose rate" in MeV/g sec, can be written as

$$dI = nh\nu_0\left(\frac{\mu_{absn}}{\rho}\right)(\rho \, dx) \qquad \frac{\text{MeV}}{\text{cm}^2 \text{ sec}} \qquad (8e\text{-}36)$$

or

$$\frac{dI}{\rho \, dx} = I\left(\frac{\mu_{absn}}{\rho}\right) \qquad \frac{\text{MeV}}{\text{g sec}} \qquad (8e\text{-}37)$$

where dI is the change in the intensity of the photons, and the mass-absorption coefficient (μ_{absn}/ρ) can be defined in a variety of slightly different ways.

Most generally one can write

$$\frac{\mu_{absn}}{\rho} \equiv \left(\frac{\sigma}{\rho}\right)f_c + \left(\frac{\tau}{\rho}\right)f_\tau + \left(\frac{\kappa}{\rho}\right)f_\kappa \qquad (8e\text{-}38)$$

where the dimensionless factors f_c (Compton), f_τ (photo), and f_κ (pair) represent the fraction of the incident photon energy $h\nu_0$ which is considered to be absorbed in the medium from each type of interaction. The size of the "region of interest" relative to the mean free path of the secondary photons in the medium plays an important role in the choice of these dimensionless factors.

The situation is clearest for the Compton interaction. The energy of the Compton scattered photon is usually large enough to permit it to escape from a small "region of interest" without interacting. Then the photon energy transferred to electrons is simply the kinetic energy acquired by the Compton recoil electron. Then, by Eq. (8e-30)

$$f_c = \frac{T_{av}}{h\nu_0} = \frac{e\sigma_a}{e\sigma} = \frac{\sigma_a/\rho}{\sigma/\rho} \qquad (8e\text{-}39)$$

Hence

$$\left(\frac{\sigma}{\rho}\right)f_c = \frac{\sigma_a}{\rho} = \frac{\sigma}{\rho} - \frac{\sigma_s}{\rho} \qquad (8e\text{-}40)$$

or just the Compton mass absorption coefficient σ_a/ρ.

In the photoelectric interaction the kinetic energy of the ejected photoelectron is, by Eq (8e-32), $T = h\nu_0 - B_e$; hence one extreme estimate of f_τ is

$$(f_\tau)_1 = \frac{T}{h\nu_0} = \frac{h\nu_0 - B_e}{h\nu_0} \qquad (8e\text{-}41)$$

which would be valid if the binding energy B_e is not released as electron kinetic energy in the volume of interest. However the excitation energy B_e may be emitted as Auger electrons or as $K, L, M, \ldots$ X-ray photons. When it is emitted as Auger electrons, the energy B_e is locally present in the medium as kinetic energy of electrons and $f_\tau = 1$. When B_e is emitted as X rays, these photons are somewhat analogous to a Compton scattered photon and could be excluded from f_τ. If Φ is the average fluorescence yield [Φ increases with Z, rising from 0.01 for $Z = 10$ (Ne) to about 0.4 for $Z = 29$ (Cu) and 0.95 for $Z = 82$ (Pb)], then another estimate of f_τ is

$$(f_\tau)_2 = \frac{h\nu_0 - \Phi B_e}{h\nu_0} = 1 - \frac{\Phi B_e}{h\nu_0} \qquad (8e\text{-}42)$$

The correction term $\Phi B_e/h\nu_0$ is negligible for light elements because Φ is so small. For heavy elements, where Φ approaches unity, $\Phi B_e/h\nu_0$ is comparable to unity for photon energies near the absorption edge (K edge = 0.088 MeV in Pb), then decreases in importance as $h\nu_0$ increases. But the fluorescence radiation has a very short mean free path (for example, 0.06 cm in Pb) and is therefore reabsorbed very close to the emitting atom. The correction $\Phi B_e/h\nu_0$ is therefore justifiable only if the "volume of interest" for energy absorption is very small, for example, less than 1 mm³ in Pb. Therefore, in absorbers having an appreciable thickness the energy B_e is all reabsorbed in a small volume and the effective value of f_τ would be

$$(f_\tau)_3 = 1 \qquad (8e\text{-}43)$$

In the pair-production interaction, the total kinetic energy of the electron pair (or of the triplet in the case of pair production in the field of an atomic electron) is $h\nu_0 - 2m_0c^2$. Hence the fraction of $h\nu_0$ which appears at once as kinetic energy of secondary electrons is

$$(f_\kappa)_1 = \frac{h\nu_0 - 2m_0c^2}{h\nu_0} = 1 - \frac{2m_0c^2}{h\nu_0} = 1 - \frac{2}{\alpha} \qquad (8e\text{-}44)$$

The energy $2m_0c^2$ is reemitted as two 0.511-MeV annihilation photons at the point of annihilation of the positron member of the electron pair. This annihilation radiation is the analog of Compton scattered radiation. In Pb, it has a mean free path of about 0.6 cm. For absorbers whose dimensions are small, the annihilation radiation clearly plays the role of a scattered radiation and should be so treated as in $(f_\kappa)_1$. For larger absorbers it has been a common approximation to ignore this correction term for annihilation radiation and to take

$$(f_\kappa)_2 = 1 \qquad (8e\text{-}45)$$

This approximation usually introduces only a small change in μ_{absn}/ρ even for thin absorbers of heavy elements because at modest energies, say 1 to 3 MeV, the pair-production interaction is only a small component of the total absorption while at very large photon energies, where the pair production is predominant, the correction $2m_0c^2/h\nu_0$ becomes small. The magnitude of the correction is shown in Figs. 8e-6 through 8e-11.

For certain dosimetric applications in radiological physics (see chap. 1 of ref. A1 for details) a value of μ_{absn}/ρ is utilized which includes only the kinetic energy transferred to charged particles per unit mass of irradiated material in an infinitesimally small

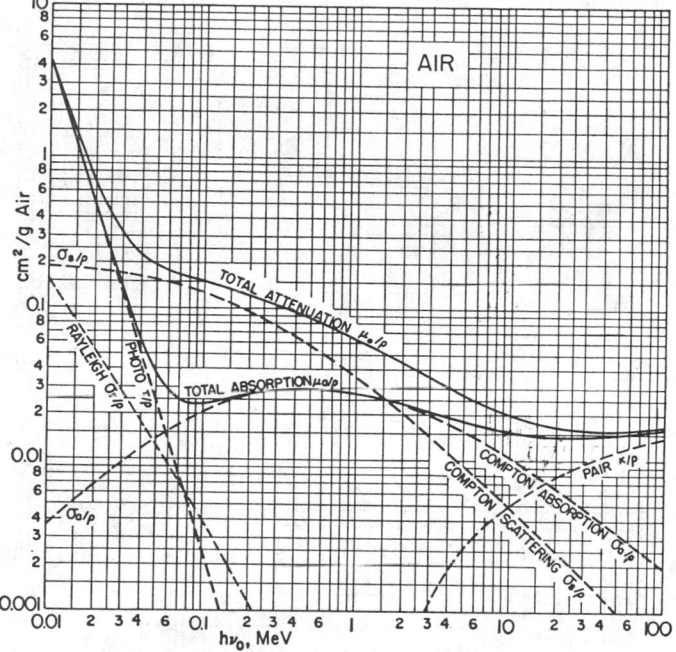

FIG. 8e-6. Mass attenuation coefficients for photons in "air" taken as 78.04 volume percent nitrogen, 21.02 volume percent oxygen, and 0.94 volume percent argon. At 0°C and 760 mm Hg pressure, the density of air is $\rho = 0.001293$ g/cm³. [From Evans (E1).]

region. This kinetic energy per unit mass has been named *kerma* (an acronym for *k*inetic *e*nergy *r*eleased in *ma*terial) by the ICRU (ref I1). The corresponding value of μ_{absn}/ρ received in 1962 (ref. I1) an "official" ICRU symbol and name (μ_K/ρ), the *mass energy-transfer coefficient*. For this particular special case of μ_{absn}/ρ, applicable under carefully specified conditions (ref. A1) to very small volumes of absorber, Eq. (8e-38) becomes

$$\frac{\mu_K}{\rho} = \frac{\sigma_a}{\rho} + \frac{\tau}{\rho}\left(1 - \frac{\Phi B_e}{h\nu_0}\right) + \frac{\kappa}{\rho}\left(1 - \frac{2m_0c^2}{h\nu_0}\right) \qquad (8e-46)$$

Tables of μ_K/ρ are given in refs. E4 and H3. For light elements and mixtures such as water, air, and Al, μ_K/ρ is substantially equal to μ_a/ρ of Eq. (8e-48) from 0.01 to 10 MeV.

For certain other dosimetric applications in radiological physics (chap. 1 of ref. A1), in which only the dissipation of electron kinetic energy by ionization and excitation is considered, the bremsstrahlung losses by each type of secondary electron are deducted in the individual terms of Eq. (8e-46) which make up μ_K/ρ. The bremsstrahlung spectrum is heavily weighted with soft photons whose mean free path in the absorber usually will be small. However, for very high energy photons in heavy elements some of the higher-energy bremsstrahlung photons may have greater mean free paths than the primary photons. When the mass energy-transfer coefficient (μ_K/ρ) is reduced to account for bremsstrahlung losses by the secondary Compton, photoelectric, and pair electrons, the resulting coefficient is now called officially (ref. I1) the *mass energy-absorption coefficient* (μ_{en}/ρ). Note that μ_{en}/ρ depends upon the material with which the photon interacts within the infinitesimal "region of interest," as does μ_K/ρ in Eq. (8e-46), but in addition μ_{en}/ρ depends upon the atomic

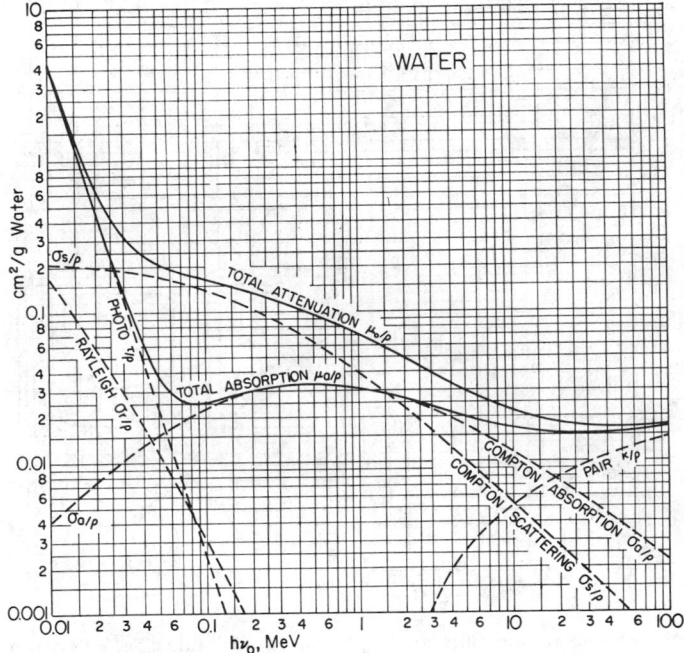

FIG. 8e-7. Mass attenuation coefficients for photons in water. [*From Evans* (E1).]

number of the material through which the secondary electron travels. Existing tables of μ_{en}/ρ are based on the assumption that the secondary electrons are absorbed in the same material in which they were formed. Bremsstrahlung losses by fast tertiary electrons produced in hard collisions by the Compton, photo, and pair electrons are usually ignored in computations of μ_{en}/ρ. Then for 2-MeV photons μ_{en}/ρ is less than μ_K/ρ by about 1 percent in Al, and about 7 to 8 percent in Pb or U. In order to use either μ_K/ρ or μ_{en}/ρ for computations of energy absorption or "absorbed dose," it is necessary to specify completely the spectrum of photon flux density, including all primary, secondary, tertiary, etc., photons, passing through the region of interest. The mass energy-absorption coefficient was first defined by Fano (ref. F2) in 1953, and the 1961 tables of μ_{en}/ρ by R. T. Berger (ref. B5) were called "mass energy-transfer" coefficients just before the name "mass energy-absorption coefficient" became official for μ_{en}/ρ in 1962 (ref. I1). Extensive tables of μ_{en}/ρ as well as μ_K/ρ are given in refs. E4 and H3.

We have noted that the mean free path of the secondary photons often is small. Hence for many practical cases involving absorbers of small to moderate size, we can select $(f_\tau)_3$ and $(f_\kappa)_1$ and write

$$\frac{\mu_{absn}}{\rho} = \frac{\sigma_a}{\rho} + \frac{\tau}{\rho} + \frac{\kappa}{\rho}\left(1 - \frac{2m_0c^2}{h\nu_0}\right) \qquad (8e\text{-}47)$$

or, using $(f_\kappa)_2$, the common but approximate expression, usually called the *mass absorption coefficient* μ_a/ρ, given by

$$\frac{\mu_a}{\rho} = \frac{\sigma_a}{\rho} + \frac{\tau}{\rho} + \frac{\kappa}{\rho} \qquad (8e\text{-}48)$$

The *mass absorption coefficient* (μ_a/ρ) is a slightly more general concept than either the mass energy-transfer coefficient (μ_K/ρ) or the mass energy-absorption coefficient (μ_{en}/ρ); both the latter are special cases of μ_a/ρ designed for the computation of radiation dose or dose rate in infinitesimally small volumes of absorbing material. Tables of μ_a/ρ as well as μ_K/ρ and μ_{en}/ρ for 18 elements from H to U over the energy range 0.01 to 10 MeV are given in ref. H3.

Figures 8e-6 through 8e-11 give the mass attenuation coefficients μ_0/ρ of Eq. (8e-34) and the mass absorption coefficients μ_a/ρ of Eq. (8e-48) and (dashed) μ_{absn}/ρ of Eq.

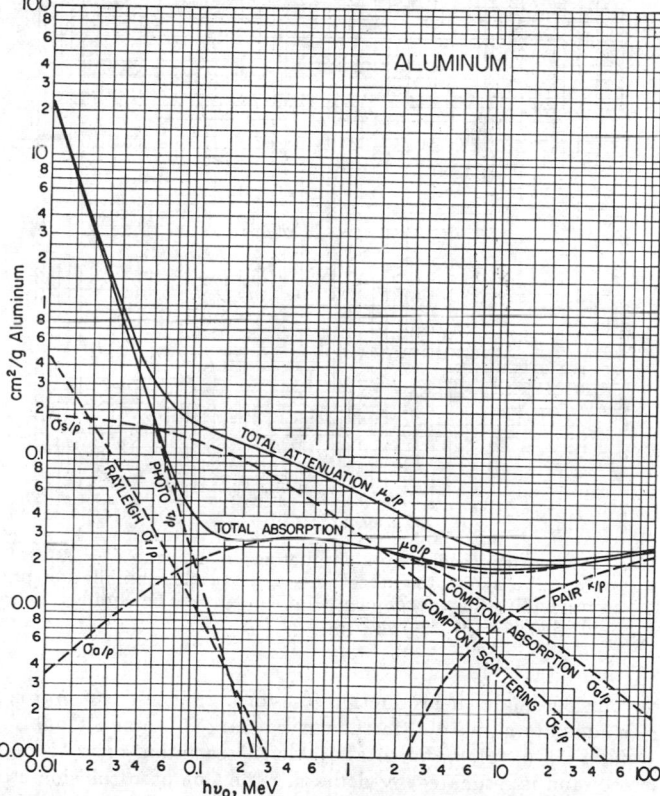

Fig. 8e-8. Mass attenuation coefficients for photons in aluminum ($Z = 13$). The dashed branch on the μ_a/ρ curve shows the effect of excluding annihilation photons [Eq. (8e-47)]. The corresponding linear coefficients for aluminum may be obtained by multiplying all curves by $\rho = 2.70$ g/cm³ Al. [*From Evans* (E1).]

(8e-47) for air, water, Al, Cu, NaI, and Pb from refs. E1 and E2 as computed from the tables in ref. W1 of the collision cross sections per atom for photoelectric interactions, pair-production interactions, and coherent scattering and from the tables for σ, σ_a, and σ_s in ref. E2. The pair-production coefficients include the effects of the atomic electrons. The curves marked "total absorption" are $(\mu_a/\rho) = (\sigma_a/\rho) + (\tau/\rho) + (\kappa/\rho)$, where σ_a, τ, and κ are the corresponding linear coefficients for Compton absorption,

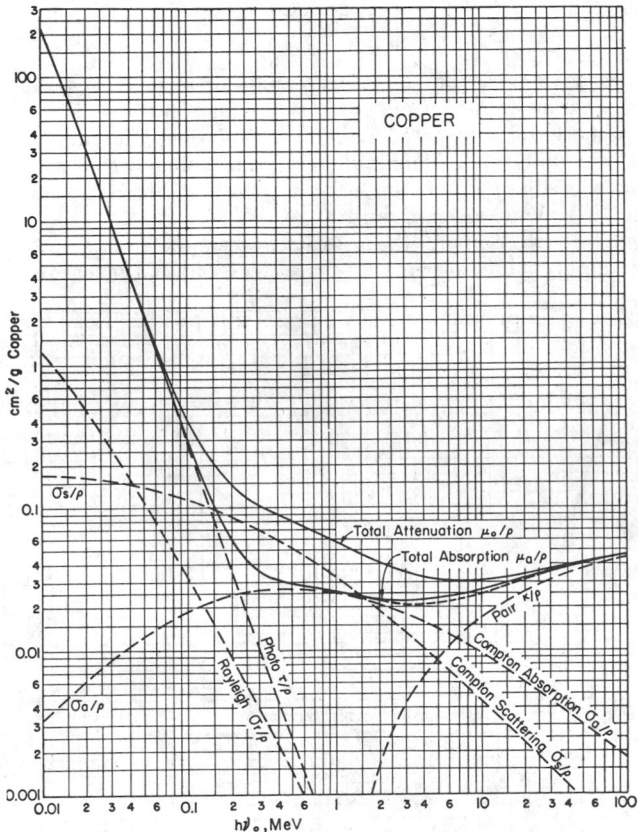

Fig. 8e-9. Mass attenuation coefficients for photons in copper ($Z = 29$). The dashed branch on the μ_a/ρ curve shows the effect of excluding the annihilation photons [Eq. (8e-47)]. The corresponding linear coefficients for copper may be obtained by multiplying all curves by $\rho = 8.92$ g/cm³ Cu. [From Evans (E2).]

photoelectric collisions, and pair-production collisions. When the Compton scattering coefficient σ_s is added to μ_a, we obtain the curves marked "total attenuation," which are $(\mu_0/\rho) = (\mu_a/\rho) + (\sigma_s/\rho)$. The total Rayleigh scattering cross section (σ_r/ρ) is shown separately and is estimated by deducting the Compton collision cross section $_e\sigma$ of ref. E2 from the total coherent cross sections of ref. W1. Because the Rayleigh scattering is elastic and is confined to small angles, it has not been included in μ_0/ρ.

A dashed branch on the mass-absorption-coefficient curves μ_a/ρ in the vicinity of 2 to 30 MeV shows the effect of correcting the κ/ρ contribution for annihilation radi-

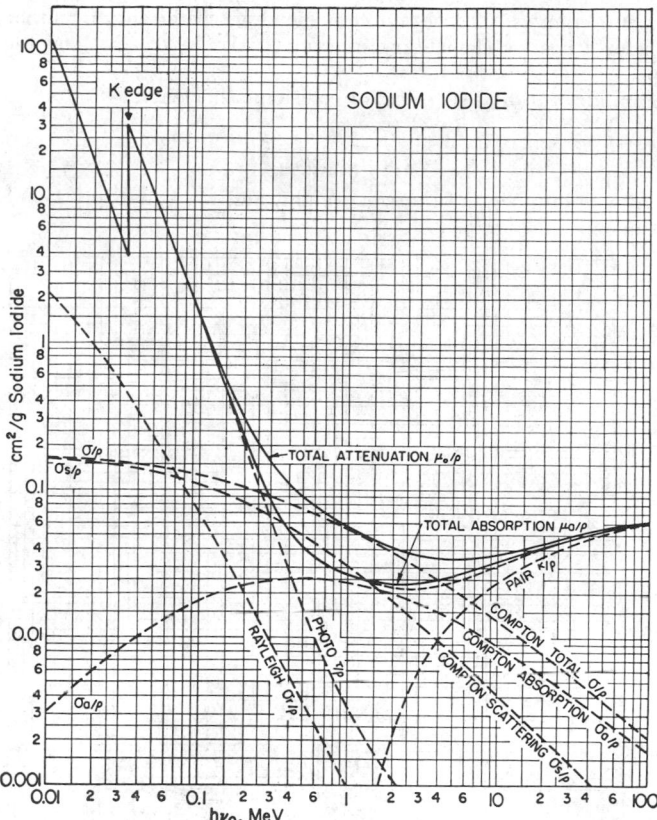

FIG. 8e-10. Mass attenuation coefficients for photons in pure NaI. The "Compton total" attenuation coefficient $(\sigma/\rho) = (\sigma_a/\rho) + (\sigma_s/\rho)$ is shown explicitly, because of its usefulness in predicting the behavior of NaI(Tl) scintillators. The 0.1 to 0.2 percent thallium activator in NaI(Tl) scintillators is ignored here. The dashed branch on the μ_a/ρ curve shows the effect of excluding annihilation photons [Eq. (8e-47)]. Linear attenuation coefficients for NaI may be obtained using $\rho = 3.67$ g/cm^3 NaI. [*From Evans* (E1).]

ation in accord with Eq. (8e-47). This correction has a maximum value of about 12 percent for Pb between 3 and 8 MeV and decreases with atomic number. The correction is too small (<3 percent) to be visible on the curves for water and air.

Mixtures and Compounds. An absorber whose bulk density is ρ and which is made up of a mixture of elements whose mass attenuation (or absorption) coefficients are (μ_1/ρ_1), (μ_2/ρ_2), . . . will have an over-all mass attenuation coefficient given by

$$\frac{\mu}{\rho} = \frac{\mu_1}{\rho_1} w_1 + \frac{\mu_2}{\rho_2} w_2 + \cdots \qquad (8e\text{-}49)$$

where w_1, w_2, . . . are the fractions by weight of the elements which make up the absorber. This relationship is valid when all the (μ/ρ)'s represent total attenuation coefficients, total absorption coefficients, or any one or more partial (Compton, photo, pair) effects. Because the chemical binding energies between atoms in a molecule are very small, chemical compounds are treated as mixtures.

Interpolation Formulas. To obtain corresponding coefficients for other elements, use a graph for a nearby elementary substance (Al, Cu, or Pb), and the following relationships:

1. Comptom absorption, scattering, or collision coefficients:

$$\frac{\sigma_1}{\rho_1} = \frac{\sigma_2}{\rho_2} \frac{A_2}{A_1} \frac{Z_1}{Z_2} \tag{8e-50}$$

where Z is atomic number, A is atomic weight, and the subscripts 1 and 2 refer to any two elements.

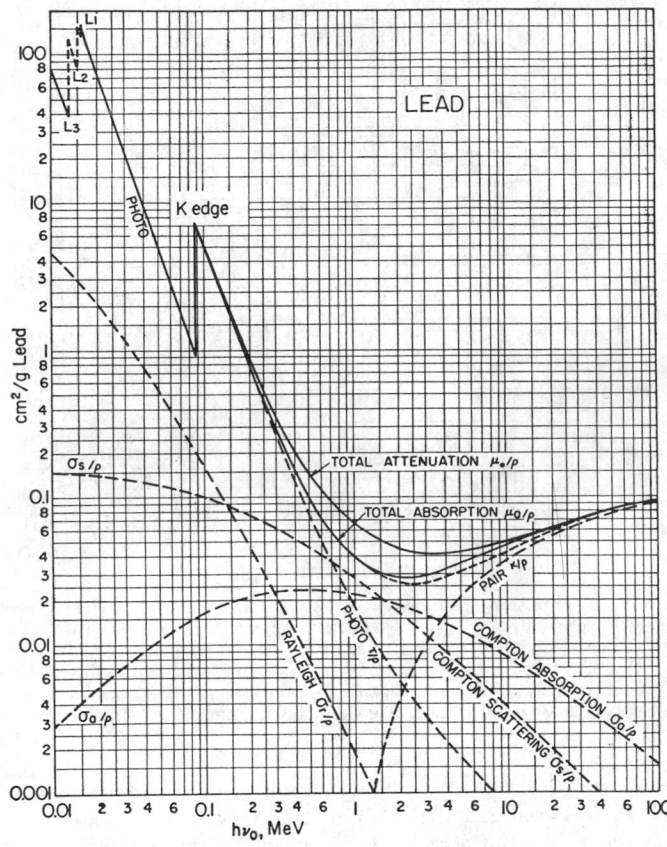

FIG. 8e-11. Mass attenuation coefficients for photons in lead. The dashed branch on the μ_a/ρ curve shows the effect of excluding annihilation photons [Eq. 8e-47)]. The corresponding linear coefficients for lead may be obtained using $\rho = 11.35$ g/cm³ Pb. [*From Evans* (E1).]

2. Photoelectric attenuation coefficients:

$$\frac{\tau_1}{\rho_1} = \frac{\tau_2}{\rho_2} \frac{A_2}{A_1} \left(\frac{Z_1}{Z_2}\right)^n \tag{8e-51}$$

where the exponent n is an empirical function of $h\nu$ as given in Fig. 8e-12.

3. Pair-production attenuation coefficients, screening neglected:

$$\frac{\kappa_1}{\rho_1} = \frac{\kappa_2}{\rho_2} \frac{A_2}{A_1} \left(\frac{Z_1}{Z_2}\right)^2 \tag{8e-52}$$

8e-6. Interpretation of Scintillation Spectrometer Pulse-height Spectra. Incident monoenergetic photons produce secondary electrons of many energies T. The scintillations from these electrons therefore give a continuous distribution of pulse heights. This distribution is further broadened by statistical fluctuations (see ref. B4) in the actual light-pulse output from every monoenergetic subgroup of electrons, such as the photoelectrons.

Figure 8e-13 is a typical pulse-height spectrum, characteristic of an NaI(Tl) scintillator in the size range of 2 by 2 to 4 by 4 in., irradiated by the 0.662-MeV γ rays emitted in the decay of Cs^{137}. Pulses in the *total energy peak* arise whenever the total energy $h\nu$ of one photon is absorbed in the scintillator and include a primary photoelectric event accompanied by absorption within the scintillator of the resulting X rays and Auger electrons, a primary Compton event accompanied by absorption also of the Compton scattered photon, and in the case of $h\nu > 1.02$ MeV a primary pair-production event followed by absorption of both quanta of annihilation radiation.

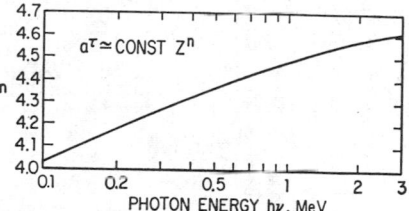

Escape peaks are present also at energies of $h\nu - 28$ keV when the K_α X ray of iodine is not absorbed and at $h\nu - 511$ keV and $h\nu - 1,022$ keV when one or both annihilation photons are not absorbed in the crystal.

The *resolution* of the spectrometer is usually described as the full width at half maximum of the total energy peak, divided by the total energy, and is commonly of the order of 8 percent for 0.662-MeV γ rays in 4- by 4-in. NaI(Tl)

Fig. 8e-12. Approximate variation of the photoelectric cross section $_a\tau$ cm^2/atom with Z^n, for various values of $h\nu$. For use with the interpolation formula [Eq. (8e-51)]. [By N. C. Rasmussen, from Evans (E1).]

crystals. For a given crystal and photomultiplier the width of the total energy peak increases roughly as $\sqrt{h\nu}$; hence the percent resolution varies roughly as $1/\sqrt{h\nu}$.

The Compton edge in Fig. 8e-13 is the high-energy end of the Compton recoil-electron distribution of Fig. 8e-5 and of Eq. (8e-21), broadened out somewhat by multiple Compton collisions in the crystal and by the same type of inherent statistical fluctuations which produce the width of the total energy peak (see ref. B4). The energy of the Compton edge is, by Eq. (8e-9),

$$T_{\max} = h\nu_0 \frac{2\alpha}{1 + 2\alpha} = h\nu_0 - (h\nu')_{\min} \qquad (8e-53)$$

The *backscatter peak* is an unwanted "ghost line" produced by photons which have been scattered into the scintillator from surrounding materials. These are mainly Compton large-angle backscattered photons; hence their energy and that of the backscatter peak which arises from their total absorption in the crystal are, by Eq. (8e-6), just slightly greater than

$$(h\nu')_{\min} = m_0 c^2 \frac{\alpha}{1 + 2\alpha} \qquad (8e-54)$$

Figure 8e-3 contains a plot of $h\nu'$ for $\vartheta = 180$ deg, which is $(h\nu')_{\min}$, and hence gives the minimum energy $(h\nu')_{\min}$ of the backscatter peak and also the energy separation $(h\nu_0 - T_{\max}) = (h\nu')_{\min}$ between the total energy peak $(h\nu_0)$ and the Compton edge $T_{\max}$. The center of gravity of the backscatter peak is usually at an energy greater than $(h\nu')_{\min}$ because photons which have been scattered through less than $\vartheta = 180$ deg are strongly involved. For many experimental situations the effective backscatter angle is between $\vartheta = 120$ and 150 deg, for which $h\nu'$ is also given in Fig. 8e-3.

The *intrinsic efficiency* (or interaction ratio) of a NaI(Tl) crystal is the total number of pulses under the entire pulse-height distribution curve (excluding the backscatter pulses) per primary photon incident on the crystal face. The intrinsic efficiency, as well as the pulse-height distribution, depends upon the geometry of source and crystal and has been calculated by Monte Carlo methods for several arrangements (see refs. B1, B2, and M1) using the collision cross sections and total mass attenuation coeffi-

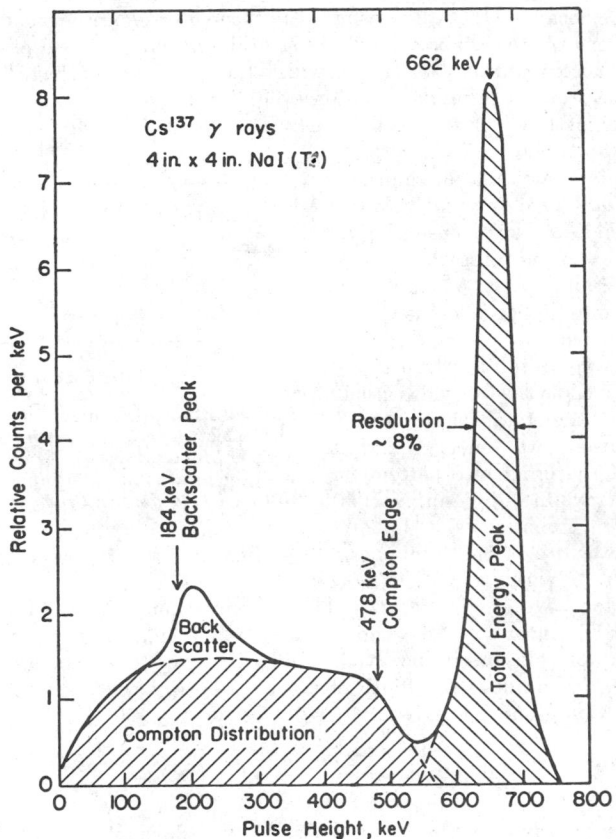

FIG. 8e-13. Typical pulse-height distribution in a 4 in. × 4 in. NaI(Tl) scintillator irradiated by 0.662-MeV γ rays from the decay of Cs^{137}. The resolution is about 8 percent of the energy of the total-energy peak. For $h\nu_0 = 662$ keV, $(h\nu')_{min} = 184$ keV, and $T_{max} = 478$ keV as marked.

cients for NaI as given in Fig. 8e-10. The minor effect of the 0.1 to 0.2 percent thallium content of NaI(Tl) is generally ignored. Calculations have been made by the same methods of the so-called *photofraction*, which is the ratio of the number of pulses under the total energy peak to the total number of pulses in the entire distribution (excluding the backscatter peak because it is not due to primary photons and depends on the experimental arrangement). Note that the photofraction includes pulses arising from all total absorption mechanisms and thus exceeds the effect of photoelectric collisions.

Figure 8e-14 gives illustrative values of the *intrinsic efficiency* and of the *photofraction* for some common sizes of NaI(Tl) crystals irradiated by monoenergetic photons from a point source placed along the extended cylindrical axis of the crystal, according to calculations at five γ-ray energies from 0.279 to 4.45 MeV by Miller, Reynolds, and Snow (ref. M1; for other geometries and crystal sizes see refs. M1, B1, B2).

Especially when a number of primary photon energies are present, the interpretation of a pulse-height distribution may become very complicated and dependent upon the preparation of a response matrix and an *inversion matrix* for the particular geometry and scintillator (see ref. H2).

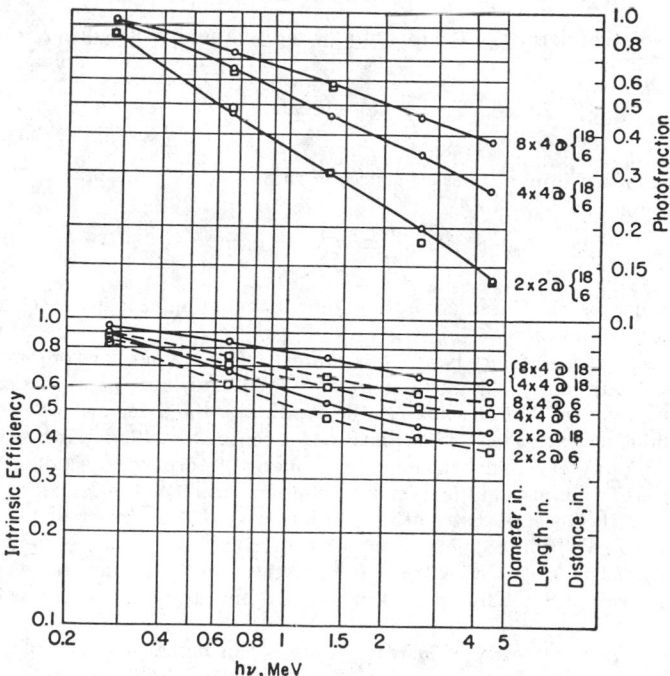

Fig. 8e-14. Illustrative calculated values of the *photofraction* (pulses in total-energy peak/ pulses in entire distribution) and the *intrinsic efficiency* (pulses per photon striking the crystal face) for point sources of γ rays located on the axis of a NaI(Tl) crystal at 18 in. (circles) or 6 in. (squares) from the crystal face, for crystals whose diameters and lengths in inches are shown opposite the curves.

Germanium semiconductor γ-ray detectors have much higher resolution than NaI(Tl) scintillators but thus far are available only in relatively small sizes and must be operated at greatly reduced temperatures. The individual interaction coefficients for photons in germanium ($Z = 32$) are not included in most primary tables, but can be computed by interpolation from nearby elements such as copper ($Z = 29$). The Compton, photoelectric, and pair-production coefficients for germanium, from 0.05 to 10 MeV, have been computed by Chapman (see ref. C1).

8e-7. Self-absorption in γ-ray Sources. There is appreciable self-absorption in substantially all practical γ-ray sources, due to interactions suffered by primary photons while emerging from within the source. The common types of γ-ray standards used in most laboratories may have from 1 to 10 percent self-absorption. The quantitative corrections for self-absorption depend strongly on the relative response of the detecting instruments to primary and to degraded photons (see refs. E1 and E3).

When the response of the detector is proportional to the γ-ray intensity (MeV/sec cm^2) and independent of the photon energy, when the distance between the source and detector is large compared with the dimensions of the source, and when the self-absorption is less than about 10 percent, a good approximation for the ratio of the net intensity I to the intensity I_0 in the absence of self-absorption is:

1. For radiation along the axis of a cylindrical source of length l:

$$\frac{I}{I_0} \simeq e^{-(\frac{1}{2})\mu_a l} \tag{8e-55}$$

2. For radiation normal to the axis of a cylindrical source of radius R:

$$\frac{I}{I_0} \simeq e^{-\left(\frac{8}{3\pi}\right)\mu_a R} \tag{8e-56}$$

3. For radiation from a spherical source of radius R:

$$\frac{I}{I_0} \simeq e^{-(\frac{3}{4})\mu_a R} \tag{8e-57}$$

In each case μ_a is the linear absorption (not attenuation) coefficient of the source material.

In the limiting case of very large γ-ray sources, whose dimensions are several mean free paths (1 mfp = $1/\mu_0$), the external γ-ray intensity is simply proportional to the solid angle subtended by the source at the position of the detector.

8e-8. Build-up Factor for Photons in Broad-beam Attenuation. In most practical situations in γ-ray and X-ray shielding or in energy absorption, a significant fraction of the scattered photons and the secondary photons can reach the detector. Accurate calculation of the spectral distribution and intensity of the scattered and secondary photons in these "poor geometry" or "broad-beam" situations is often extremely complicated. The status of extensive theoretical and experimental work at the National Bureau of Standards and elsewhere is admirably summarized as of 1958 in ref. F3.

The transmitted intensity of *primary* photons can be calculated easily, using the total attenuation coefficient μ_0, because *the primary radiation is always in good geometry*. The complicated contribution of scattered and secondary photons is described by the so-called *build-up factor*, which is the ratio of the observed effect to the effect produced only by the residual primary radiation. Thus there are different build-up factors for photon intensity or energy flux, photon number, dose rate in an air cavity, and energy absorption in the medium (see ref. G1). The build-up factor B for dose rate is defined as

$$B \equiv \frac{\text{total observed dose rate}}{\text{primary dose rate}}$$
$$= 1 + \frac{\text{dose rate due to scattered and secondary radiation}}{\text{dose rate due to primary radiation}} \tag{8e-58}$$

The ratio of the dose rate from secondaries dP_{sec} to dose rate from primaries dP_{prim}, and hence the build-up factor B, is found experimentally and theoretically (see ref. G1) to increase monotonically with the thickness r of absorbing material. Thus "equilibrium" between secondary and primary photons is never reached. To a reasonable approximation, especially for the dose rate measured inside an effectively infinite medium at a distance r from a point source in the medium, the increase of B with r can be represented by

$$B = 1 + \frac{dP_{\text{sec}}}{dP_{\text{prim}}} = 1 + a(\mu_0 r)^k \qquad (8e\text{-}59)$$

where $\mu_0 r$ is the distance r from the source measured in mean free paths of the primary radiation and k is a constant which depends upon the photon energy $h\nu$ of the photons

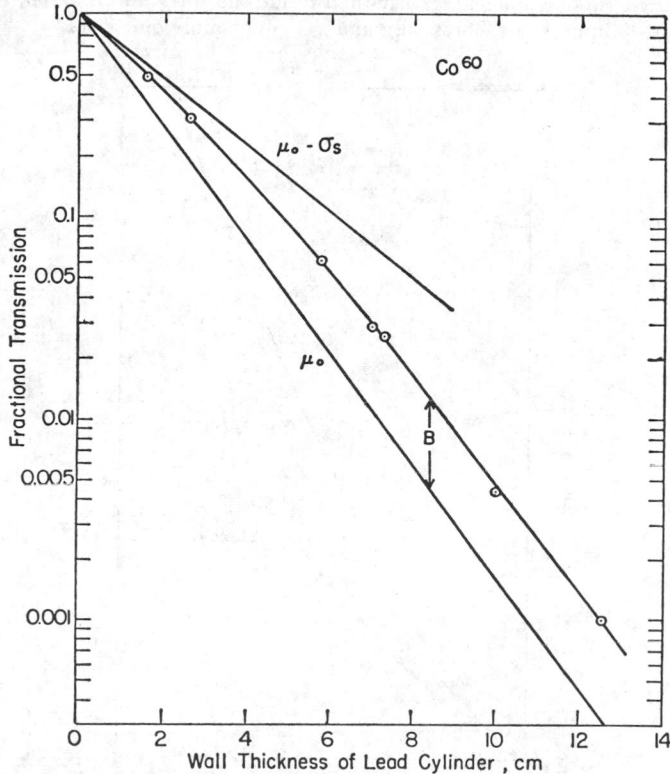

Fig. 8e-15. Transmission of Co^{60} γ rays through cylindrical Pb shields. Theoretical lines for the transmission of undeflected primary photons (μ_0) and for the absorption of primary energy ($\mu_0 - \sigma_s = \mu_a$) are shown for comparison. The ratio of the experimentally observed exposure dose rate outside the shield to the theoretical dose rate due to transmitted primary photons (μ_0) is the dose-rate build-up factor, marked B. [Data by A. Morrison, from Evans (E1).]

and the atomic number Z of the absorber (see refs. F1 and P1) and is of the order of unity. The proportionality constant a can be shown (see ref. E1) from conservation of energy to be

$$a = \frac{1}{\Gamma(k+1)} \frac{\sigma_s}{\mu_a} \qquad (8e\text{-}60)$$

where $\Gamma(k + 1)$ is the gamma function of $k + 1$; hence $\Gamma(k + 1) = k!$ whenever k is an integer. The linear Compton scattering coefficient σ_s, the linear absorption coefficient σ_a, and the linear attenuation coefficient $\mu_0 = \mu_a + \sigma_s$ are each evaluated for the *primary* photon energy in the medium.

An oversimplified but useful approximate theory (see ref. E1) for γ-ray energies in the neighborhood of 1 MeV and for low or medium Z results when k is taken as unity; then

$$B \simeq 1 + \frac{\sigma_s}{\mu_a} (\mu_0 r) \qquad\qquad (8e\text{-}61)$$

The effects of boundaries (see refs. F3 and E1) and hence the build-up factor for radiation measured outside a shield, rather than inside the shielding material, are much more difficult to approach theoretically and are now mainly empirical.

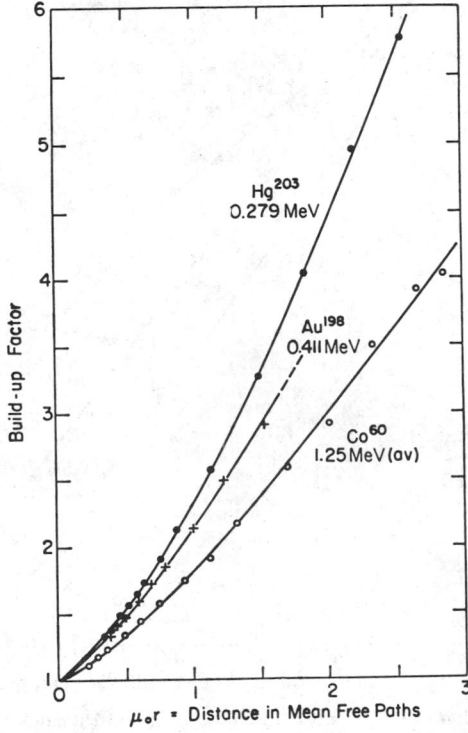

FIG. 8e-16. Build-up factors for point sources of γ rays of three energies measured in an essentially infinite water medium. [Data by M. A. Van Dilla, from Evans (E1).]

Figure 8e-15 illustrates the γ-ray dose rate measured outside a Pb shield surrounding a Co^{60} source and the physical meaning of the dose-rate build-up factor B.

Figure 8e-16 illustrates the monotonic increase of the dose-rate build-up factor inside a large water medium with distance from point sources of 1.25, 0.411, and 0.279-MeV γ rays.

8e-9. Gamma-ray Output of Radionuclides. Many radionuclides emit γ rays and can be used as photon sources. It is convenient to describe the source strength of a particular radionuclide in terms of the so-called *specific γ-ray constant*, Γ, which is defined as the exposure dose rate in milliroentgens (mR) per hour produced in air at a distance of one meter from a point source of one millicurie (mCi) of the radionuclide. The exposure dose rate has usually been taken as equivalent to $I(\mu_a/\rho)$, where I is the γ-ray intensity (MeV/cm² sec), and μ_a/ρ is the mass absorption coefficient of air for the particular photon energy involved [see Eqs. (8e-37) and (8e-48)]. Inserting the geometrical factors, and omitting any attenuation of the γ rays in 1 meter of air, the specific γ-ray constant becomes

$$\Gamma = 19.3 \sum_i n_i (h\nu)_i \left(\frac{\mu_a}{\rho}\right)_i, \qquad \frac{mR}{mCi \cdot hr} \text{ at 1 meter} \qquad (8e\text{-}62)$$

where n_i photons of energy $(h\nu)_i$ are emitted per disintegration, and the average energy to form one ion pair in air has been taken as $W = 34.0$ eV per ion pair. Because μ_a/ρ for air varies only slightly ($+15$ percent over the usual domain of nuclear γ-ray energies (see Fig. 8e-6), a convenient approximate rule is

$$\Gamma \approx \tfrac{1}{2} \Sigma(h\nu) \qquad mR/mCi \cdot hr \text{ at 1 meter} \qquad (8e\text{-}63)$$

where $\Sigma(h\nu)$ is the total photon energy in MeV emitted per disintegration.

TABLE 8e-3. THE SPECIFIC γ-RAY CONSTANT Γ IN MILLIROENTGENS PER HOUR PRODUCED AT 1 METER BY THE NUCLEAR γ RAYS AND THE ANNIHILATION RADIATION FROM 1 MILLICURIE OF THE RADIONUCLIDES LISTED*

Nuclide	Half period	mR hr·mCi at 1 meter	Nuclide	Half period	mR hr·mCi at 1 meter
Na²²	2.58 yr	1.23	As⁷⁶	26.5 hr	0.4
Na²⁴	15.0 hr	1.84	Br⁸²	35.3 hr	1.5
Al²⁸	2.30 min	0.85	I¹²⁸	25.0 min	0.017
Mn⁵²	5.7 days	1.85	I¹³⁰	12.5 hr	1.20
Mn⁵⁴	314 days	0.47	I¹³¹	8.05 days	0.23
Fe⁵⁹	45 days	0.62	Cs¹³⁷	30 yr	0.31
Co⁵⁸	71 days	0.54	Ta¹⁸²	115 days	0.6
Co⁶⁰	5.26 yr	1.29	Au¹⁹⁸	64.8 hr	0.24
Cu⁶⁴	12.9 hr	0.12	Ra²²⁶	1,620 yr	0.84†
Zn⁶⁵	245 days	0.26			

* Revised from Evans (p. 722 of ref. E1) using $W = 34$ eV per ion pair.
† With 0.5-mm Pt filtration. Γ for Ra²²⁶ is usually expressed in terms of mg instead of mCi, the accepted experimental value being 0.825 mR/hr·mg at 1 meter (ref. I2). One gram of Ra²²⁶ has 3.62×10^{10} disintegrations/sec; 1 Ci = 3.7×10^{10}/sec. Thus the experimental value can be expressed as 0.844 mR/hr·mCi, in good agreement with the tabulated value.

Table 8e-3 gives numerical values computed from Eq. (8e-62) for the specific γ-ray constant, or exposure dose rate produced by nuclear γ rays (plus the annihilation radiation in the case of positron emitters) at 1 meter from a 1-mCi point source of several common radionuclides. These values are in reasonable agreement with computed and experimental values compiled by Mann and others (ref. I2).

References

A1. Attix, F. H., and W. C. Roesch, eds.: "Radiation Dosimetry," vol. 1, 2d ed., Academic Press, Inc., New York, 1968. The current authoritative treatise on the fundamental principles of radiological physics; 8 chapters carefully edited and cross-referenced.
B1. Bell, P. R.: The Scintillation Method, in "Beta- and Gamma-ray Spectroscopy," pp. 133–164, Kai Siegbahn, ed., Interscience Publishers, Inc., New York, 1955. Review of basic principles and experimental methods. A permanently valuable intro-

duction to the later literature. The revised version of this article is J. H. Neiler, and P. R. Bell, in "Alpha-, Beta-, and Gamma-ray Spectroscopy," vol. 1, pp. 245–302, Kai Siegbahn, Ed., North-Holland Publishing Co., Amsterdam, 1965.

B2. Berger, M. J., and J. Doggett: Response Function of NaI(Tl) Scintillation Counters, *Rev. Sci. Instr.* **27**, 269 (1956), and *Natl. Bur. Standards J. Research* **56**, 355 (1956). Monte Carlo calculations of intrinsic efficiency and photofraction for several sizes (0.5 in. diameter by 0.5 in. long to 5 in. diameter by 9 in. long) NaI(Tl) crystals irradiated axially by collimated monoenergetic γ rays at six energies from 0.279 to 4.45 MeV.

B3. Bethe, H. A., and J. Ashkin: Penetration of Gamma Rays, in "Experimental Nuclear Physics," vol. I, pp. 304–349, E. Segre, ed., John Wiley & Sons, Inc., New York, 1953. Theory and graphs for Compton, photoelectric, and pair-production interactions.

B4. Breitenberger, E.: Scintillation Spectrometer Statistics, in "Progress in Nuclear Physics," vol. 4, pp. 56–94, O. R. Frisch, ed., Pergamon Press, London, 1955. Theory and some experimental results on line width and shape from scintillation counters, extensive bibliography.

B5. Berger, Rosemary T.: The X- or Gamma-ray Energy Absorption or Transfer Coefficient; Tabulations and Discussions, *Radiation Research* **15**, 1, 1961. Definitive summary of various types of absorption coefficients with tables of the mass energy $1 = 1$ absorption coefficients for 0.01-MeV to 10-MeV photons in water, air, and 14 elements from H to Cu.

C1. Chapman, G. T.: Gamma-ray Attenuation Coefficients for Germanium, *Nucl. Instr. Methods* **52**, 101 (1967). Tables and a graph of the Compton, photoelectric, and pair-production coefficients for germanium ($Z = 32$), computed by interpolation among five elements ($Z = 20, 26, 29, 42,$ and 50) in the tables by G. White Grodstein (ref. G2).

D1. Davisson, C. M., and R. D. Evans: Gamma-ray Absorption Coefficients, *Revs. Mod. Phys.* **24**, 79 (1952). Review of theory and experimental results through 1951 on Compton, photoelectric, and pair-production interactions, with graphs or tables for 24-elements.

D2. Davisson, C. M.: Interaction of Gamma Radiation With Matter, in "Alpha-, Beta- and Gamma-ray Spectroscopy," pp. 37–78, 827–843, K. Siegbahn, ed., North-Holland Publishing Company, Amsterdam, 1965. A thorough review of the theory, with tables for 21 elements from H to U and 4 compounds or mixtures, for Compton, Rayleigh, photoelectric, pair-production, and total cross sections from 0.01 to 100 MeV. The numerical tables are utilized and updated by those in refs. E4 and H3.

E1. Evans, R. D.: "The Atomic Nucleus," McGraw-Hill Book Company, New York, 1955. Textbook and reference book on nuclear physics. Chapters 23, 24, and 25 deal with interaction of photons with matter; theory, experiment, applications, graphs, and bibliography.

E2. Evans, R. D.: Compton Effect, in "Handbuch der Physik," vol. XXXIV, pp. 218–298, S. Flügge, ed., Springer-Verlag OHG, Berlin, 1958. Theory and experimental results on all aspects of the Compton interaction for free and bound electrons, polarized and unpolarized photons, with 42 figures and 8 tables.

E3. Evans, R. D., and R. O. Evans: Studies of Self-absorption in Gamma-ray Sources, *Rev. Mod. Phys.* **20**, 305 (1948). Theory and experimental results for self-absorption in cylindrical standard sources, particularly of radium.

E4. Evans, R. D.: X-ray and Gamma-ray Interactions, in "Radiation Dosimetry," vol. 1, pp. 93–155, 2d ed., F. H. Attix and W. C. Roesch, eds., Academic Press, Inc., New York, 1968. An amended and enlarged version of the present chapter, plus extensive new tables of mass attenuation coefficients μ_0/ρ, mass energy-transfer coefficients μ_K/ρ, and mass energy-absorption coefficients μ_{en}/ρ for 22 elements from H to U and for 11 mixtures or compounds, prepared by J. H. Hubbell as an extension of ref. H3.

F1. Fano, U.: Gamma-ray Attenuation. Analysis of Penetration, *Nucleonics* **11**(9), 55 (1953). Summary of the progress to 1953 of the NDA-NBS program of calculation and measurement of multiple scattering, deep penetration, and build-up factors in infinite homogeneous media. See Goldstein and Wilkins (ref. G1) and Fano, Spencer, and Berger (ref. F3) for later developments.

F2. Fano, U.: Gamma-ray Attenuation. Basic Processes, *Nucleonics* **11**(8), 8 (1953). Review of photon interactions and qualitative description of corrections to the mass absorption coefficient, for fluorescence radiation, bremsstrahlung, and rest energy of electron pairs, to obtain energy-absorption coefficient μ_{en}. Table of μ_{en} from 0.088 to 10 MeV for water, Al, Fe, and Pb.

F3. Fano, U., L. V. Spencer, and M. J. Berger: Penetration and Diffusion of X Rays, in "Handbuch der Physik," vol. XXXVIII/2, pp. 660–817, S. Flügge, ed., Springer-

Verlag OHG, Berlin, 1959. The definitive treatise on multiple scattering, build-up factors, and related quantities in homogeneous infinite media and in bounded media, with 57 figures and 23 tables.

G1. Goldstein, H., and J. E. Wilkins, Jr.: Calculations of the Penetration of Gamma Rays, *U.S. Atomic Energy Comm. Rept.* NYO-3075, 1954, 196 pp. Theory and results of calculations of spectra of multiply scattered photons, build-up factors, and related quantities for monoenergetic 0.5- to 10-MeV γ-ray sources in infinite homogeneous media, by the "moments method" of Spencer and Fano; 140 tables and 75 graphs.

G2. Grodstein, G. W.: X-ray Attenuation Coefficients from 10 keV to 100 MeV, *Natl. Bur. Standards Circ.* 583, 1957, 54 pp. Review of theory of photon interactions, with some experimental comparisons. Tables of theoretical atomic cross sections for Compton interactions with and without coherence terms, photoelectric effect, and pair production by nuclear and by electronic interactions for 24 elements and five mixtures. A widely used basic reference. Earlier editions of these tables appeared (see ref. W1) under the same title by G. R. White as *Natl. Bur. Standards Rept.* 1003, 1952, and on pp. 857–874 of "Beta- and Gamma-ray Spectroscopy," Kai Siegbahn, ed., Interscience Publishers, Inc., New York, 1955. A supplement to Circular 583, 1959, under the same title, by R. T. McGinnies, gives improved estimates of the photoelectric cross sections and total mass attenuation coefficients at low energies. Other more recent evaluations (refs. D2 and H3) have revised the photoelectric coefficients, see Sec. 8e-3, herein, for discussion.

H1. Heitler, W.: "The Quantum Theory of Radiation," 3d ed., 430 pp., Oxford University Press, New York, 1954. Standard treatise on theory of interaction of photons with matter. No detailed experimental treatment.

H2. Hubbell, J. H.: Response of a Large Sodium-iodide Detector to High Energy X-rays, *Rev. Sci. Instr.* **29,** 65 (1958). A clearly explained example of the calculation of a 28-row by 28-column response matrix and inversion matrix for the case of a 5-in.-diameter by 4-in.-long NaI(Tl) crystal irradiated axially and centrally by well-collimated photons from 0.01 to 8 MeV.

H3. Hubbell, J. H., and M. J. Berger: Photon Attenuation and Energy Absorption Coefficients, Tabulations and Discussions, *Natl. Bur. Standards (U.S.) Rept.* 8681, 2d ed., 1966, 118 pp. Also to be published in "Engineering Compendium on Radiation Shielding," R. G. Jaeger, ed., IAEA, Vienna. An invaluable basic reference. The most recent among the series of critical evaluations of photon coefficients, intended to improve and extend the earlier numerical values given in refs. B5, D1, D2, G2, F2, I2, and W1. The tables cover 22 elements from H to U and 4 compounds and mixtures, mostly from 0.01 to 10 MeV, some to 100 MeV. Coefficients for total attenuation (μ_0/ρ), three types of absorption (μ_a/ρ), (μ_K/ρ), and (μ_{en}/ρ), and for the Compton, Rayleigh, photoelectric, and pair-production components are tabulated. These tables are based on the same smoothed-input data used for the tables in ref. E4, where a different selection of elements and many more mixtures are tabulated.

I1. International Commission on Radiological Units and Measurements (ICRU): Radiation Quantities and Units, report 10a, *Natl. Bur. Standards (U.S.) Handbook* **84** (1962). First promulgation of a revised set of definitions and units for use in radiological physics: kerma, fluence, exposure, mass energy-transfer coefficients, and others.

I2. International Commission on Radiological Units and Measurements (ICRU): Report of the ICRU for 1959, *Natl. Bur. Standards (U.S.) Handbook* **78** (1961). Definitions (some now superseded) of quantities and units for radiological physics, clinical, and biological factors; physical aspects of dosimetry; and measurement of radioactivity for radiological use.

J1. Johns, H. E., D. V. Cormack, S. A. Denesuk, and G. F. Whitmore: Initial Distribution of Compton Electrons, *Can. J. Phys.* **30,** 556 (1952). Tables for 10-keV to 30-MeV photons of the angular distribution of Compton scattered photons, the energy distribution and angular distribution of Compton recoil electrons.

M1. Miller, W. F., J. Reynolds, and W. J. Snow: Efficiencies and Photofractions for Sodium-iodide Crystals, *Rev. Sci. Instr.* **28,** 717 (1957); **30,** 141 (1959); and *Argonne Natl. Lab. Rept.* ANL-5902, 1958. Results of Monte Carlo calculations for NaI(Tl) crystals from 2 to 32 in. diameter and 1 to 8 in. length for 0.279- to 4.45-MeV photons from a parallel beam or from a point source or a disk source at 0, 6, 12, 18, or 30 in. distance along the crystal axis.

N1. Nelms, A. T.: Graphs of the Compton Energy-angle Relationship and the Klein-Nishina Formula from 10 keV to 500 MeV, *Natl. Bur. Standards Circ.* 542, 1953, 89 pp. A succinct review of the Compton laws, the Klein-Nishina cross sections for unpolarized photons, and the incoherent scattering function, with 81 clear and accurate graphs.

P1. Price, B. T., C. C. Horton, and K. T. Spinney: "Radiation Shielding," Pergamon Press, London, 1957. Chapter 2 deals with broad-beam γ-ray attenuation and build-up factors, particularly with applications to the shielding of nuclear reactors.

W1. White, G. R.: X-ray Attenuation Coefficients from 10 keV to 100 MeV, *Natl. Bur. Standards Rept.* 1003, 1952. This mimeographed "first edition" of ref. G2 filled a great need and has been widely used. It contains on p. 17 a table of Compton "true absorption" coefficients $_e\sigma_a$ in which the low-energy values are incorrect but continue to be used by some workers (for example, ref. P1). Correct values of $_e\sigma_a$ and other Compton coefficients, over the same range of photon energies, are in ref. E2.

W2. Whyte, G. N.: "Principles of Radiation Dosimetry," John Wiley & Sons, Inc., New York, 1959, 124 pp. Physical principles and experimental methods for photon, electron, and neutron dosimetry.

8f. Neutrons

MURREY D. GOLDBERG

Brookhaven National Laboratory

JOHN A. HARVEY

Oak Ridge National Laboratory

Neutrons have been extensively used as probes to investigate the properties of nuclei ever since the discovery of the neutron in 1932. Its lack of charge gives it an advantage over charged particles in many experimental areas, although the inability to define its energy to the same precision as that of charged particles counters this advantage in other areas of nuclear structure studies. Neutrons are available in copious supply from many sources throughout an energy range extending from below 10^{-4} eV to above 100 MeV. An enormous number of data have been accumulated, covering a wide variety of nuclear interactions for all nuclides of the periodic table.

No attempt will be made to even summarize the data here. For a detailed compilation of neutron cross-section experimental data, the reader is referred to the many volumes of the report BNL-325 (1958, 1960, 1964–1966) for cross-section data at thermal energies, for parameters of compound-nucleus resonances due to neutron interactions, and for energy-dependent cross-section data through the entire energy range; and to the report BNL-400 (1970) for angular distribution data. (A complete bibliography of these reports is given in Sec. 8f-10.)

For the techniques involved in neutron physics measurements, the reader is referred to the two-volume work "Fast Neutron Physics," edited by J. B. Marion and J. L. Fowler and published by Interscience Publishers, Inc., New York (part I in 1960; part II in 1963); and to "Experimental Neutron Resonance Spectroscopy" edited by J. A. Harvey and published by Academic Press, Inc., New York, 1970.

This section contains selected topics which seem appropriate. A noticeable omission is a table of thermal neutron cross sections. Such a listing can be found in the Table of Isotopes by D. T. Goldman (Sec. 8b), and is not duplicated here.

8f-1. Neutron Properties. Where needed to obtain quantities for this table, the values of physical constants used were taken from B. N. Taylor, W. H. Parker, and D. N. Langenberg, *Rev. Mod. Phys.* **41**, 375 (1969):

Spin, $\hbar/2$

Statistics, Fermi-Dirac

Radioactive decay, half life = 11.0 ± 0.3 min; beta-decay energy = 782.45 ± 0.07 keV

Magnetic moment $\mu_n = -1.913159 \pm 0.000047$ nuclear magnetons

Neutron mass, $1.00866520 \pm 0.00000010$ mass units [unified scale, $C^{12} = 12$], $(1.674920 \pm 0.000011) \times 10^{-24}$ g, 939.5527 ± 0.0052 MeV

Compton wavelength:

$$\lambda_c = h/m_n c = (1.3196217 \pm 0.0000090) \times 10^{-13} \text{ cm}$$
$$\lambdabar_c = \lambda_c/2\pi = (2.100243 \pm 0.000014) \times 10^{-14} \text{ cm}$$

Nonrelativistic conversion formulas [E (eV) as function of T (K), v (m/sec), and λ (cm)]:

$$E = kT = 8.6171 \times 10^{-5}T$$
$$E = m_n v^2/2 = 5.22695 \times 10^{-9}v^2$$
$$E = h^2/2m_n\lambda^2 = 8.18015 \times 10^{-18}/\lambda^2$$

For a neutron velocity of 2,200 m/sec:

$$E = 0.0252984 \text{ eV}$$
$$T = 293.58 \text{ K}$$
$$\lambda = 1.79818 \times 10^{-8} \text{ cm}$$

Figures 8f-1 to 8f-3 provide graphs from which reasonably accurate values of the relationships between energy-dependent properties can be read. The formula for

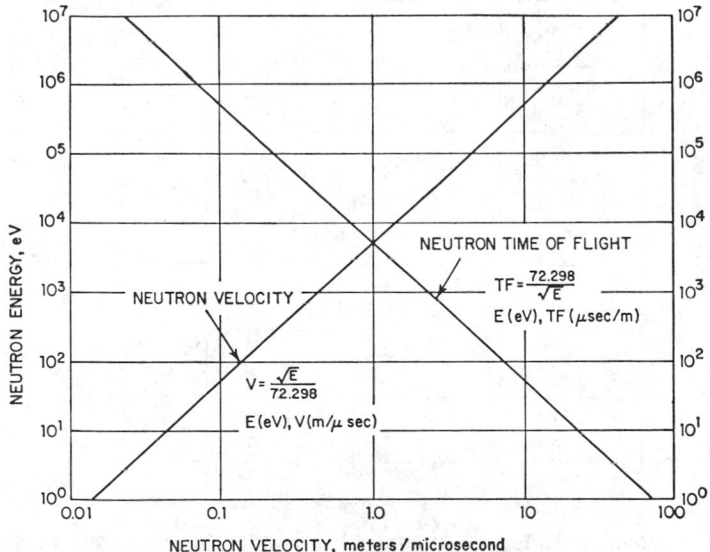

FIG. 8f-1. Variation of neutron velocity and neutron time of flight with energy for the neutron energy range 1 eV to 10 MeV.

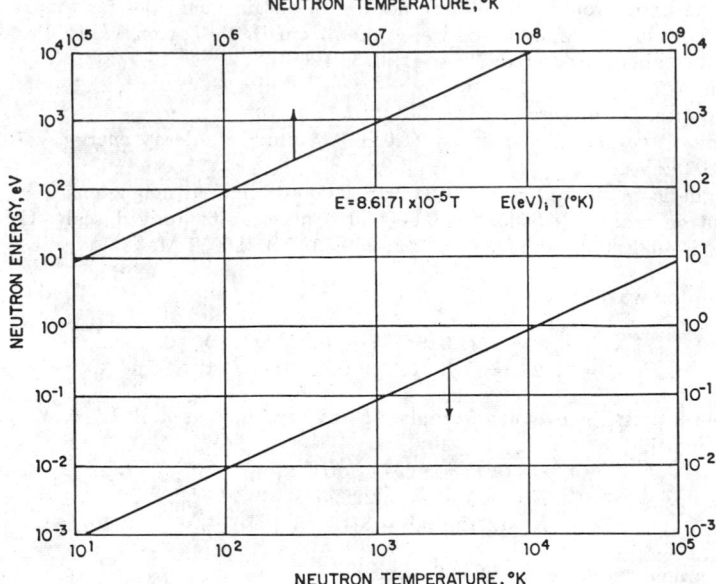

FIG. 8f-2. Variation of neutron temperature with energy for the neutron energy range 0.001 eV to 10 keV.

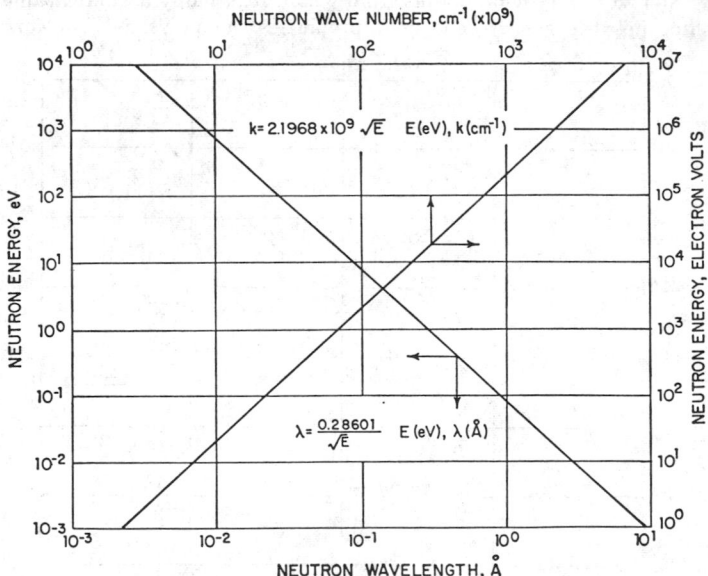

FIG. 8f-3. Variation of neutron wavelength with energy for the neutron energy range 0.001 eV to 10 keV and the variation of the neutron wave number with energy for the neutron energy range 1 eV to 10 MeV.

each relationship is given on the curve so that more accurate values can be obtained quickly when needed or can be calculated for values at energies outside the range included in the graph.

8f-2. Neutron Separation Energies. The compound nucleus $(A + 1)$ formed by the addition of a neutron to a target nuclide A at rest has an excitation energy equal to the neutron's separation energy B_n of the compound nucleus plus the neutron's kinetic energy E_n minus the compound nucleus' recoil energy $[\sim E_n/(A + 1)]$. The energy of the gamma radiation to the ground state of the residual nucleus will equal the excitation energy of the compound nucleus plus its kinetic energy minus the kinetic energy of the residual nucleus. For example, the energy of the ground-state gamma ray from the capture of *thermal* neutrons by ^{14}N will be the neutron separation energy of ^{15}N $(10,834.0 \pm 0.8$ keV$)$ minus the recoil energy of ^{15}N $\{\sim0.537[E_\gamma{}^2(\text{in MeV})/(A + 1)]$ keV $= 4.2$ keV$\}$ which equals 10,829.8 keV. For a photoneutron reaction, the energy of the gamma ray will have to exceed the neutron separation energy of the target nuclide by the recoil energy of the compound nucleus. For example, for the photodisintegration of the deuteron, the gamma ray's energy must exceed the neutron separation of deuterium $(2,224.55 \pm 0.07$ keV$)$ by the recoil energy 1.32 keV.

Neutron separation energies with standard deviations are listed in Table 8f-1 for the stable nuclides, for radioactive nuclides with half lives $>1,000$ years, for shorter-lived radioactive nuclides which have been produced in milligrams or larger quantities, and for these nuclides plus a neutron. The values for the neutron separation energies come principally from least-squares adjustments of experimental data by J. H. E. Mattauch, W. Thiele, and Λ. H. Wapstra, *Nuclear Physics* **67**, 32 (1965); by A. H. Wapstra, C. Kurzeck, and A. Anisimoff, *Proceedings of the Third International Conference on Atomic Masses*, August 28–September 1, 1967, edited by R. C. Barber, University of Manitoba Press; and from vols. I and II, sec. B of the *Nuclear Data Sheets*, Academic Press, Inc. Recent accurate values obtained from the (n,γ) reaction were taken from a Compendium of Thermal Neutron Capture γ-ray Measurements, part I, $Z \leq 46$, *Nuclear Data* **A3** (4–6), 367 (1967); part II, $Z = 47$ to $Z = 67$ (Ag to Ho), *Nuclear Data Tables* **A5** (1–2), 1 (1968); part III, $Z = 68$ to $Z = 94$ (Er to Pu), *Nuclear Data Tables* **A5** (4–5), 243 (1969); and from recent accurate unpublished (n,γ) data. For nuclides where no error is assigned, the neutron separation energies were estimated from nuclear systematics.

8f-3. Coherent Scattering Amplitudes. Since the wavelengths of thermal neutrons are of the order of one angstrom, their scattering interactions with matter exhibit such well-known optical phenomena as diffraction and refraction. A scattering interaction can be characterized by an *amplitude*, which represents the distance that the neutron wave is shifted by the scattering nucleus.

For a nucleus of nonzero spin, there will be two amplitudes, a_+ and a_-, corresponding to the two possible spin states of the compound nucleus, $I \pm \frac{1}{2}$, formed by the combination of a neutron (spin $\frac{1}{2}$) with a target nucleus (spin I). The compound states are not equally probable but are weighted in the ratio $(I + 1)$ to I. Hence, the *coherent amplitude*, which is the weighted mean of the two amplitudes, is given by

$$a_{\text{coh}} = \frac{I + 1}{2I + 1} a_+ + \frac{I}{2I + 1} a_-$$

The *coherent cross section*, neglecting interference with other nuclei, will then be given by $4\pi a_{\text{coh}}^2$. The coherent amplitude for an element with two or more isotopes can be obtained from

$$a_{\text{coh}} = \Sigma f_i a_i$$

TABLE 8f-1. TABLE OF NEUTRON SEPARATION ENERGIES

Atomic no. Z	Element	Mass no. A = N + Z	Number of neutrons N	Separation energy Bₙ, keV
1	H	2	1	2,224.55 ± 0.07
		3	2	6,257.4 ± 0.1
2	He	4	2	20,578.0 ± 0.4
		5	3	−958 ± 19
3	Li	6	3	5,662 ± 37
		7	4	7,252.5 ± 1.3
		8	5	2,032.5 ± 1.0
4	Be	9	5	1,665.1 ± 0.6
		10	6	6,811.8 ± 0.4
5	B	10	5	8,437.9 ± 1.4
		11	6	11,455.9 ± 0.5
		12	7	3,368.9 ± 1.4
6	C	12	6	18,719.8 ± 1.1
		13	7	4,946.5 ± 0.2
		14	8	8,176.2 ± 0.9
		15	9	1,218.0 ± 0.8
7	N	14	7	10,552.9 ± 1.1
		15	8	10,834.0 ± 0.8
		16	9	2,486.7 ± 3.5
8	O	16	8	15,667.9 ± 1.2
		17	9	4,142.5 ± 0.9
		18	10	8,046.2 ± 0.9
		19	11	3,956.3 ± 2.9
9	F	19	10	10,429.9 ± 1.0
		20	11	6,602 ± 1
10	Ne	20	10	16,865.0 ± 1.6
		21	11	6,759.8 ± 1.5
		22	12	10,366.5 ± 1.5
		23	13	5,194.9 ± 3.3
11	Na	23	12	12,417.6 ± 3.2
		24	13	6,959.3 ± 0.4
12	Mg	24	12	16,532.4 ± 3.2
20	Ca	48	28	9,940 ± 11
		49	29	5,144 ± 6
21	Sc	45	24	11,319 ± 6
		46	25	8,766.6 ± 3.9
22	Ti	46	24	13,192 ± 5
		47	25	8,875.4 ± 2.5
		48	26	11,628.0 ± 2.7
		49	27	8,146.0 ± 2.2
		50	28	10,944.4 ± 3.5
		51	29	6,379 ± 5
23	V	50	27	9,337 ± 5
		51	28	11,054.6 ± 3.9
		52	29	7,308.8 ± 4.4
24	Cr	50	26	12,930 ± 11
		51	27	9,261 ± 3
		52	28	12,034.9 ± 3.9
		53	29	7,941.4 ± 3.1
		54	30	9,719 ± 2
		55	31	6,254 ± 6
25	Mn	55	30	10,225 ± 6
		56	31	7,270.4 ± 3.1
26	Fe	54	28	13,396 ± 16
		55	29	9,299.5 ± 3.9
		56	30	11,203.3 ± 4.6
		57	31	7,646.5 ± 1.0
		58	32	10,043.1 ± 1.0
		59	33	6,586.3 ± 4.1
		60	34	8,864 ± 30
		61	35	5,730 ± 50
27	Co	59	32	10,461.1 ± 4.9
		60	33	7,491.6 ± 1.9
28	Ni	58	30	12,195 ± 16

Element	Z	A	N	Value	±
		59	31	8,999.2	0.6
		60	32	11,387.4	2.7
		61	33	7,817.4	2.9
		62	34	10,596.2	1.5
		63	35	6,835.7	2.8
		64	36	9,658.8	2.0
		65	37	6,099	5
Cu	29	63	34	10,840.5	4.5
		64	35	7,916.3	0.8
		65	36	9,911	4
		66	37	7,065.2	0.7
Zn	30	64	34	11,855.1	3.5
		65	35	7,979.2	0.8
		66	36	11,051	5
		67	37	7,052.4	0.7
		68	38	10,198.1	0.5
		69	39	6,482.2	0.9
		70	40	9,195	8
		71	41	6,050	50
Ga	31	69	38	10,324	4
		70	39	7,640	7
		71	40	9,311	7
		72	41	6,516	6
Ge	32	70	38	11,529.0	4.8
		71	39	7,413.2	4.8
		72	40	10,750.9	4.7
		73	41	6,785.2	1.5
		74	42	10,200.0	1.6
		75	43	6,489	19
		76	44	9,445	19
		77	45	5,986	32
As	33	75	42	10,244	5
		76	43	7,334	5
Se	34	74	40	12,072	19
		75	41	8,025	8
		76	42	11,161	8
		77	43	7,415	6
		78	44	10,492	5
		79	45	6,971.3	4.4

Element	Z	A	N	Value	±
		25	13	7,330	1
		26	14	11,098	1
		27	15	6,443	1
Al	13	27	14	13,056.7	2.6
		28	15	7,723	1
Si	14	28	14	17,175.4	3.5
		29	15	8,476	1
		30	16	10,617.2	4.3
		31	17	6,594	5
P	15	31	16	12,312	8
		32	17	7,936.5	11
S	16	32	16	15,092	1
		33	17	8,646	3.7
		34	18	11,422.3	2.9
		35	19	6,985.1	8
		36	20	9,879	70
		37	21	4,420	6
Cl	17	35	18	12,635	2
		36	19	8,583	4.1
		37	20	10,316.6	8
		38	21	6,110	17
Ar	18	36	18	15,252	2.5
		37	19	8,790.8	2.4
		38	20	11,838.7	6
		39	21	6,591	6
		40	22	9,872	2
		41	23	6,098	10
K	19	39	20	13,089	2.7
		40	21	7,801.5	0.8
		41	22	10,096.0	2
		42	23	7,535	23
Ca	20	40	20	5,619	1
		41	21	8,364	8
		42	22	11,471	4.8
		43	23	7,927.8	5
		44	24	11,135	1
		45	25	7,420.3	4.9
		46	26	10,401	10
		47	27	7,281	11

TABLE 8f-1. TABLE OF NEUTRON SEPARATION ENERGIES (*Continued*)

Atomic no. Z	Element	Mass no. $A = N + Z$	Number of neutrons N	Separation energy B_n, keV
34	Se	80	46	9,902.8 ± 4.9
		81	47	6,715 ± 6
		82	48	9,262 ± 9
35	Br	83	49	5,970 ± 50
		79	44	10,698 ± 6
		80	45	7,876.0 ± 4.4
		81	46	10,164 ± 6
36	Kr	82	47	7,601 ± 8
		78	42	12,010 ± 50
		79	43	8,340 ± 8
		80	44	11,520 ± 9
		81	45	7,850 ± 100
		82	46	10,980 ± 100
		83	47	7,467 ± 6
		84	48	10,519 ± 5
		85	49	7,122 ± 6
		86	50	9,848 ± 7
		87	51	5,511 ± 8
37	Rb	85	48	10,475 ± 6
		86	49	8,637 ± 9
		87	50	9,940 ± 7
		88	51	6,130 ± 90
38	Sr	84	46	11,580 ± 31
		85	47	8,482 ± 31
		86	48	11,522 ± 10
		87	49	8,437 ± 5
		88	50	11,113 ± 1
		89	51	6,393 ± 8
39	Y	89	50	11,477 ± 8
		90	51	6,857 ± 2
40	Zr	90	50	11,997 ± 6
		91	51	7,194 ± 5
46	Pd	106	60	9,544 ± 4
		107	61	6,532 ± 7
		108	62	9,227 ± 8
		109	63	6,150 ± 9
		110	64	8,807 ± 14
		111	65	5,740 ± 50
47	Ag	107	60	9,531 ± 9
		108	61	7,267 ± 1
		109	62	9,182 ± 9
		110	63	6,810 ± 1
48	Cd	106	58	10,870 ± 7
		107	59	7,929 ± 7
		108	60	10,334 ± 7
		109	61	7,381 ± 7
		110	62	9,856 ± 4.5
		111	63	6,975.4 ± 4.8
		112	64	9,399.6 ± 3.8
		113	65	6,538.2 ± 1
		114	66	9,039 ± 9
		115	67	6,143 ± 9
		116	68	8,694 ± 13
		117	69	5,764 ± 12
49	In	113	64	9,427 ± 12
		114	65	7,312 ± 10
		115	66	9,034 ± 23
		116	67	6,725 ± 5
50	Sn	112	62	11,080 ± 200
		113	63	7,744 ± 17
		114	64	10,320 ± 16
		115	65	7,537 ± 9
		116	66	9,563 ± 7
		117	67	6,941 ± 5

Left portion (Z = 40–46)

Z	El.	A	N	Energy (keV)	±
		92	52	8,634	2
		93	53	6,750	6
		94	54	8,198	6
		95	55	6,468	6
		96	56	7,838	7
		97	57	5,575	22
41	Nb	93	52	8,844	8
		94	53	7,229.5	1.5
		95	54	8,510	15
42	Mo	92	50	12,580	50
		93	51	8,053	13
		94	52	9,692	13
		95	53	7,373.8	3.8
		96	54	9,154.2	0.5
		97	55	6,816.1	3.4
		98	56	8,642.8	1.0
		99	57	5,918	9
		100	58	8,300	9
		101	59	5,390	19
43	Tc	97	54	9,450	
		98	55	7,350	
		99	56	8,880	200
		100	57	6,590	60
44	Ru	96	52	10,124	37
		97	53	8,040	
		98	54	10,250	
		99	55	7,469	5
		100	56	9,671	6
		101	57	6,806	6
		102	58	9,216	5
		103	59	6,248	20
		104	60	8,887	20
		105	61	5,976	16
45	Rh	103	58	9,312	9
		104	59	6,999.3	1.5
46	Pd	102	56	10,360	50
		103	57	7,608	23
		104	58	10,023	20
		105	59	7,091	12

Right portion (Z = 50–54)

Z	El.	A	N	Energy (keV)	±
		118	68	9,331.1	4.6
		119	69	6,484	3
		120	70	9,110.1	4.3
		121	71	6,181	6
		122	72	8,804	7
		123	73	5,957	5
		124	74	8,506	11
		125	75	5,732	5
51	Sb	121	70	9,250	8
		122	71	6,806.0	1.5
		123	72	8,975	7
		124	73	6,468	2
52	Te	120	68	10,283	24
		121	69	6,976	47
		122	70	10,058	45
		123	71	6,925	3
		124	72	9,425	2
		125	73	6,630	3
		126	74	9,117	5
		127	75	6,290	3
		128	76	8,754	10
		129	77	6,116	10
		130	78	8,385	10
		131	79	5,907	2
53	I	127	74	9,153	8
		128	75	6,826	3
		129	76	8,865	10
		130	77	6,498	31
54	Xe	124	70	10,500	350
		125	71	7,610	350
		126	72	10,240	
		127	73	7,200	
		128	74	9,640	
		129	75	6,913	7
		130	76	9,259	7
		131	77	6,603	6
		132	78	8,932	6
		133	79	6,531	36
		134	80	8,460	36

TABLE 8f-1. TABLE OF NEUTRON SEPARATION ENERGIES (*Continued*)

Atomic no. Z	Element	Mass no. A = N + Z	Number of neutrons N	Separation energy B_n, keV
54	Xe	135	81	6,560 ± 100
		136	82	7,880 ± 100
		137	83	4,460 ± 100
55	Cs	133	78	9,038 ± 27
		134	79	6,705 ± 15
		135	80	9,050 ± 110
		136	81	6,610 ± 70
56	Ba	130	74	10,260
		131	75	7,695 ± 12
		132	76	9,560 ± 280
		133	77	7,257 ± 42
		134	78	9,252 ± 16
		135	79	6,974 ± 3
		136	80	9,106.4 ± 0.8
		137	81	6,904 ± 4
		138	82	8,611.1 ± 0.8
		139	83	4,723.4 ± 0.7
57	La	138	81	7,260
		139	82	8,792 ± 20
		140	83	5,161.0 ± 1.0
58	Ce	136	78	9,990
		137	79	7,840
		138	80	9,470
		139	81	7,508 ± 26
		140	82	9,039 ± 48
		141	83	5,428.6 ± 0.6
		142	84	7,159 ± 8
		143	85	5,182 ± 10
59	Pr	141	82	9,386 ± 18
		142	83	5,843.6 ± 1.2
60	Nd	142	82	9,809 ± 9
		143	83	6,123 ± 2
		144	84	7,817.2 ± 1.8
		145	85	5,760.4 ± 1.9
66	Dy	160	94	8,590 ± 30
		161	95	6,448 ± 12
		162	96	8,193 ± 3
		163	97	6,270 ± 3
		164	98	7,654 ± 3
		165	99	5,715 ± 2
67	Ho	165	98	8,043 ± 36
		166	99	6,243 ± 3
		167	100	7,290 ± 100
68	Er	162	94	9,200
		163	95	6,840 ± 90
		164	96	8,795 ± 42
		165	97	6,645 ± 40
		166	98	8,549 ± 33
		167	99	6,436.2 ± 0.5
		168	100	7,771.2 ± 0.5
		169	101	5,997 ± 12
		170	102	7,190 ± 70
		171	103	5,676 ± 10
69	Tm	169	100	8,055 ± 35
		170	101	6,595 ± 2.5
70	Yb	168	98	8,980
		169	99	6,867.2 ± 0.5
		170	100	8,550
		171	101	6,616 ± 3
		172	102	8,023 ± 3
		173	103	6,365 ± 3
		174	104	7,465 ± 3
		175	105	5,819 ± 3
		176	106	6,640 ± 80
		177	107	5,565 ± 16
71	Lu	175	104	7,801 ± 42
		176	105	6,293.2 ± 1.2
		177	106	6,890 ± 2

Neutron binding energies for Hafnium through Platinum ($Z = 72$–78):

Element	Z	A	N	Binding energy	±
Hf	72	174	102	8,700	70
		175	103	6,910	3
		176	104	8,110	3
		177	105	6,370	3
		178	106	7,622	3
		179	107	6,098	21
		180	108	7,383	
		181	109	5,693	
Ta	73	181	108	7,640	
		182	109	6,063.0	0.8
		183	110	6,929	11
W	74	180	106	8,400	36
		181	107	6,947	21
		182	108	8,035	2
		183	109	6,191	4
		184	110	7,413	2
		185	111	5,754	43
		186	112	7,206	2
		187	113	5,467	5
Re	75	185	110	7,800	60
		186	111	6,178	
		187	112	7,290	
		188	113	5,872.2	1.5
Os	76	184	108	8,000	70
		185	109	6,820	70
		186	110	8,310	60
		187	111	6,220	19
		188	112	7,837	90
		189	113	6,000	
		190	114	7,793.5	1.5
		191	115	5,890	90
		192	116	7,630	80
		193	117	5,480	60
Ir	77	191	114	8,250	180
		192	115	6,145	9
		193	116	7,786	46
		194	117	6,103	27
Pt	78	190	112	8,680	
		191	113		
		192	114	8,360	

Neutron binding energies for Promethium through Dysprosium ($Z = 61$–66, continued from the preceding $Z = 60$ entries):

Element	Z	A	N	Binding energy	±
		146	86	7,570.2	2.1
		147	87	5,293.2	4.0
		148	88	7,333.7	4.8
		149	89	5,042	12
		150	90	7,332	11
		151	91	5,309	15
Pm	61	147	86	7,684	23
		148	87	5,904	11
Sm	62	144	82	10,616	36
		145	83	6,763	11
		146	84	8,411	19
		147	85	6,371	17
		148	86	8,140.0	1.2
		149	87	5,846.2	4.8
		150	88	7,985.2	0.8
		151	89	5,609	15
		152	90	8,224	16
		153	91	5,867.0	0.4
		154	92	7,904	12
		155	93	5,819	11
Eu	63	151	88	7,933	23
		152	89	6,305	4
		153	90	8,544	12
		154	91	6,439	2
Gd	64	152	88	8,480	13
		153	89	8,606	20
		154	90	6,456	9
		155	91	8,527	5
		156	92	6,347	5
		157	93	7,929.4	3.7
		158	94	6,031	27
		159	95	7,376	28
		160	96	5,650	70
		161	97	8,177	25
Tb	65	159	94	6,377	2
		160	95	9,890	
Dy	66	156	90	6,830	
		157	91	8,840	34
		158	92	6,851	
		159	93		

TABLE 8f-1. TABLE OF NEUTRON SEPARATION ENERGIES (*Continued*)

Atomic no. Z	Element	Mass no. A = N + Z	Number of neutrons N	Separation energy B_n, keV
78	Pt	193	115	6,288 ± 48
		194	116	8,384 ± 20
		195	117	6,126 ± 13
		196	118	7,920.9 ± 1.5
		197	119	5,854 ± 14
		198	120	7,561 ± 19
		199	121	5,570 ± 19
79	Au	197	118	8,084 ± 12
		198	119	6,513.2 ± 0.8
80	Hg	196	116	8,810
		197	117	6,637 ± 42
		198	118	8,634 ± 41
		199	119	6,652.8 ± 4.9
		200	120	8,028.8 ± 0.5
		201	121	6,226.5 ± 4.6
		202	122	7,755.1 ± 1.5
		203	123	5,987 ± 7
		204	124	7,499 ± 7
		205	125	5,540 ± 100
81	Tl	203	122	7,696 ± 23
		204	123	6,654 ± 2
		205	124	7,534 ± 7
		206	125	6,504 ± 3
82	Pb	202	120	8,870
		203	121	6,930 ± 39
		204	122	8,244 ± 12
		205	123	6,734.2 ± 1.5
		206	124	8,082 ± 6
		207	125	6,736.4 ± 1.5
		208	126	7,367.7 ± 1.5
		209	127	3,944 ± 9
83	Bi	208	125	6,867 ± 6
		209	126	7,453.6 ± 4.4
		210	127	4,599.7 ± 4.8
88	Ra	226	138	6,387 ± 6
		227	139	4,586 ± 21
90	Th	229	139	5,233 ± 12
		230	140	6,787 ± 4
		231	141	5,129 ± 4
		232	142	6,431 ± 4
		233	143	4,787 ± 5
91	Pa	231	140	6,803 ± 21
		232	141	5,567 ± 23
92	U	232	140	7,270 ± 50
		233	141	5,737 ± 12
		234	142	6,840 ± 4
		235	143	5,307 ± 4
		236	144	6,545 ± 2
		237	145	5,129 ± 4
		238	146	6,144 ± 4
		239	147	4,803.4 ± 2.4
93	Np	236	143	5,715 ± 16
		237	144	6,591 ± 16
94	Pu	238	144	5,486 ± 9
		239	145	7,002 ± 7
		240	145	5,657 ± 4
		241	146	6,533.7 ± 1.5
		242	147	5,243 ± 3
		243	148	6,305 ± 4
		244	149	5,034 ± 7
		245	150	6,018 ± 13
95	Am	241	151	4,750
		242	146	6,660
		243	147	5,535 ± 9
		244	148	6,376 ± 9
		245	149	5,355 ± 5
96	Cm	245	149	5,524 ± 4
		246	150	6,452 ± 4
		247	151	5,156 ± 8
		248	152	6,209 ± 8
		249	153	4,713 ± 6

where the f_i are the abundances ($\Sigma f_i = 1$) of the individual isotopes with coherent amplitudes a_i. The *index of refraction* for a noncapturing medium can be calculated from the coherent scattering amplitude as

$$n^2 = 1 - \frac{\lambda^2 N a_{\mathrm{coh}}}{\pi}$$

where λ is the neutron wavelength, and N is the number of nuclei per cm³.

Table 8f-2 lists the coherent scattering amplitudes for those elements, or particular isotopes, for which experimental values have been determined. The amplitude values are given in femtometers ($\equiv 10^{-13}$ cm) and are preceded by a positive or negative sign. The standard convention adopted here is that positive amplitude represents hard-sphere scattering, i.e., a phase shift of 180 deg. For two values the sign is omitted, since no explicit experimental assignment has been made, but both cases are probably positive.

8f-4. Recommended 2,200-m/sec Cross Sections for Fissile Isotopes. In Table 8f-3, which is taken from G. C. Hanna, C. H. Westcott, H. D. Lemmel, B. R. Leonard, Jr., J. S. Story, and P. M. Attree, *Atomic Energy Rev.* **7**, (4) 3 (1969), IAEA, Vienna, the results of a careful study of relevant experimental measurements are presented. A least-squares fitting procedure was used, and both direct and indirect measurements of the quantities listed were considered.

The quantities appearing in the table have the following meanings:

σ_a Absorption cross section: $\sigma(n,\gamma) + \sigma(n,f) = \sigma_{\mathrm{tot}} - \sigma_{\mathrm{scat}}$
σ_f Fission cross section: $\sigma(n,f)$
σ_γ Radiative capture cross section: $\sigma(n,\gamma)$
α Ratio: $\sigma(n,\gamma)/\sigma(n,f)$
η Number of neutrons produced per absorption event: prompt + delayed
$\bar{\nu}_{\mathrm{T}}$ Number of neutrons produced per fission event: prompt + delayed

The ν value for Cf²⁵², which is the standard used in the fissile ν measurements, was evaluated to be 3.765 ± 0.012.

8f-5. s-wave Neutron Strength Functions, Observed Resonance Spacings, and Average Radiation Widths. Neutron cross sections of most nuclides exhibit individual resonances in the energy region from 0.1 eV to 100 keV. These resonances correspond to excited states of the compound nucleus at an excitation energy just above the neutron separation energy. The resonances can be described by the following parameters:

E_0 Resonance energy
Γ Total width
Γ_n Neutron width
Γ_γ Radiation width
Γ_F Fission width
l Angular momentum of the neutron, s, p, d, etc.
J Spin of the compound nucleus

Since the parameters of over 10,000 resonances have now been measured, it is not possible to present a complete listing of these parameters in this section. The detailed parameters may be found in the many volumes of the report BNL-325 (1958, 1960, 1964–1966). However, for many purposes the average of the parameters for each nuclide are sufficient. Some of these averages for s-wave interactions which are predominant in the energy region $\lesssim 10$ keV are listed in Table 8f-4.

The s-wave neutron strength function for a nuclide is defined as $\overline{\Gamma}_n^\circ/D$. $\overline{\Gamma}_n^\circ$ is the average of the reduced neutron widths Γ_n° of s-wave resonances of the same spin and parity, where Γ_n° is equal to $\Gamma_n/\sqrt{E_0}$ (in eV). D is the average level spacing for resonances of the same spin and parity. s-wave strength functions can be determined from the parameters of resolved resonances or from the energy dependence of the

TABLE 8f-2. COHERENT SCATTERING AMPLITUDES

Atomic no. Z	Element	Mass no. A = N + Z	Coherent amplitude a_coh, fm	Atomic no. Z	Element	Mass no. A = N + Z	Coherent amplitude a_coh, fm
1	H	1	−3.719 ± 0.002	47	Ag	...	+6.1 ± 0.4
		2	+6.21 ± 0.04			107	+8.3 ± 0.5
		3	+4.7 ± 0.3			109	+4.3 ± 0.4
2	He	...	+3.0 ± 0.5	48	Cd	...	+3.32 ± 0.20
3	Li	...	−1.94 ± 0.05	49	In	...	+3.8 ± 0.1
		6	+1.8	50	Sn	...	+6.1 ± 0.4
		7	−2.1			116	+5.8 ± 0.1
4	Be	...	+7.74 ± 0.07			117	+6.4 ± 0.25
5	B	...	+5.40 ± 0.04			118	+5.8 ± 0.1
		10	+6.53 ± 0.35			119	+6.0 ± 0.25
		11	+6.1 ± 0.1			120	+6.4 ± 0.1
6	C	...	+6.656 ± 0.004			122	+5.5 ± 0.3
		13	+6.0 ± 0.8			124	+5.9 ± 0.2
7	N	...	+9.14 ± 0.10	51	Sb	...	+5.4 ± 0.1
8	O	...	+5.80 ± 0.05	52	Te	...	+5.6 ± 0.4
9	F	...	+5.6 ± 0.1			120	+5.3 ± 0.1
10	Ne	...	+4.60 ± 0.05			123	+5.8 ± 0.1
11	Na	...	+3.5 ± 0.1			124	+5.5 ± 0.7
12	Mg	...	+5.2 ± 0.1			125	+5.6
13	Al	...	+3.5 ± 0.3			130	5.7
14	Si	...	+4.1646 ± 0.0022	53	I	...	+5.2 ± 0.3
15	P	...	+5.1 ± 0.1	54	Xe	...	+4.62 ± 0.09
16	S	...	+3.1 ± 0.2	55	Cs	...	+4.9 ± 0.3
17	Cl	...	+9.9 ± 0.6	56	Ba	...	+5.3 ± 0.3
18	Ar	...	+1.89 ± 0.02	57	La	...	+8.3 ± 0.6
19	K	...	+3.70 ± 0.04	58	Ce	...	+4.84 ± 0.06
20	Ca	...	+4.88 ± 0.07			140	+4.7 ± 0.2
		40	+4.9 ± 0.2			142	+4.5 ± 0.4
		44	+1.8 ± 0.1	59	Pr	...	+4.4 ± 0.4
21	Sc	...	+11.8 ± 1.0	60	Nd	...	+7.2 ± 0.6
22	Ti	...	−3.5 ± 0.1			142	+7.7 ± 0.6
		46	+4.8 ± 0.2			144	+2.8 ± 0.6
		47	+3.3 ± 0.2			146	+8.7 ± 0.4

Z	Element	A	Amplitude
		48	−5.8 ± 0.3
		49	+0.8 ± 0.2
		50	+5.5 ± 0.3
23	V	⋯	−0.5 ± 0.1
24	Cr	⋯	+3.52 ± 0.06
		52	+4.9
25	Mn	⋯	−3.7 ± 0.2
26	Fe	⋯	+9.2 ± 0.2
		54	+4.2 ± 0.3
		56	+10.1 ± 0.1
		57	+2.3 ± 0.3
27	Co	⋯	+2.5 ± 0.2
28	Ni	⋯	+10.3 ± 0.3
		58	+14.4 ± 0.3
		60	+3.0 ± 0.06
		61	+7.60 ± 0.4
		62	−8.7 ± 0.07
		64	−0.38 ± 0.3
29	Cu	⋯	+7.7 ± 0.15
		63	+6.72 ± 0.2
		65	+11.1 ± 0.1
30	Zn	⋯	+5.7 ± 0.1
31	Ga	⋯	+7.2 ± 0.5
32	Ge	⋯	+8.4 ± 0.1
33	As	⋯	+6.4 ± 0.14
34	Se	⋯	+7.79 ± 0.5
35	Br	⋯	+6.7 ± 0.15
36	Kr	⋯	7.44 ± 0.1
37	Rb	⋯	+8.5 ± 0.1
		85	+8.3 ± 0.07
38	Sr	⋯	+6.83 ± 0.05
39	Y	⋯	+7.88 ± 0.1
40	Zr	⋯	+7.0 ± 0.2
41	Nb	⋯	+6.9 ± 0.2
42	Mo	⋯	+6.7 ± 0.2
43	Tc	99	+6.8 ± 0.3
44	Ru	⋯	+7.3
45	Rh	⋯	+6.0 ± 0.7
46	Pd	⋯	+5.9

Z	Element	A	Amplitude
61	Pm	⋯	
62	Sm	152	−5 ± 2
		154	+8 ± 2
63	Eu	⋯	+6.3 ± 2
64	Gd	⋯	+15 ± 2
65	Tb	⋯	+7.56 ± 0.20
66	Dy	⋯	+16.9 ± 0.4
		160	+6.7 ± 0.4
		161	+10.3 ± 0.4
		162	−1.4 ± 0.5
		163	+5.0 ± 0.4
		164	+49.4 ± 0.5
67	Ho	⋯	+8.5 ± 0.2
68	Er	⋯	+7.9 ± 0.4
69	Tm	⋯	+7.20 ± 0.06
70	Yb	⋯	+12.90 ± 0.07
71	Lu	⋯	+7.3 ± 0.2
72	Hf	⋯	+7.77 ± 0.14
73	Ta	⋯	+7.0 ± 0.2
74	W	⋯	+4.66 ± 0.5
75	Re	⋯	+9.2 ± 0.09
76	Os	⋯	+10.7
		189	+11.0
		190	+11.4
		192	+11.9
77	Ir	⋯	+10.6
78	Pt	⋯	+9.5 ± 0.6
79	Au	⋯	+7.6 ± 0.1
80	Hg	⋯	+12.68 ± 0.02
81	Tl	⋯	+8.9 ± 0.9
82	Pb	⋯	+9.34 ± 0.02
83	Bi	⋯	+8.5239 ± 0.0014
90	Th	⋯	+9.8 ± 0.1
92	U	⋯	+8.4 ± 0.2
		235	+9.8 ± 0.6
93	Np	⋯	+10.57 ± 0.06
94	Pu	239	+7.5 ± 0.3
		240	+3.8

TABLE 8f-3. RECOMMENDED 2,200-M/SEC CROSS SECTIONS FOR FISSILE ISOTOPES

Parameter	U^{233}	U^{235}	Pu^{239}	Pu^{241}
σ_a	577.6 ± 1.8	678.5 ± 1.9	1012.9 ± 4.1	1375.4 ± 8.6
σ_f	530.6 ± 1.9	580.2 ± 1.8	741.6 ± 3.1	1007.3 ± 7.2
σ_γ	47.0 ± 0.9	98.3 ± 1.1	271.3 ± 2.6	368.1 ± 7.8
α	0.0885 ± 0.0018	0.1694 ± 0.0021	0.3659 ± 0.0039	0.3654 ± 0.0090
η	2.2844 ± 0.0063	2.0719 ± 0.0060	2.1085 ± 0.0066	2.149 ± 0.014
$\bar{\nu}_T$	2.4866 ± 0.0069	2.4229 ± 0.0066	2.8799 ± 0.0090	2.934 ± 0.012

total cross section averaged over many resonances. The values with standard deviations listed in Table 8f-4 were taken principally from a compilation by K. K. Seth, *Nuclear Data* **A2**, 299 (Sept. 1966). Standard deviations are given for values where the errors are less than 50 percent of the value. A few spin-$\frac{3}{2}$ target nuclides have s-wave strength functions which are different for the two compound nucleus spin states 1 and 2. For these nuclides the s-wave strength function for the resonances with $J = 2$ is approximately twice that for $J = 1$ resonances.

The values for the observed resonance spacings with standard deviations for s-wave neutrons D_{obs} listed in Table 8f-4 were taken principally from a summary by J. E. Lynn, "The Theory of Neutron Resonance Reactions," Clarendon Press, Oxford, 1968. For zero-spin target nuclides, the average level spacing D for resonances with $J = \frac{1}{2}$ is equal to D_{obs}. For nonzero-spin target nuclides, D for the two spin states is greater than D_{obs} and may be computed from the formula

$$D_J = 2D_{obs} \frac{2I + 1}{2J + 1}$$

where I = spin of target nucleus

J = spin of compound nucleus

$J = I \pm \frac{1}{2}$ for s-wave resonances

For example, for $I = \frac{1}{2}$, $D_{J=0} = 4D_{obs}$ and $D_{J=1} = \frac{4}{3}D_{obs}$. Several of the values listed as lower limits were determined from the energy of only the lowest resonance.

The values of the average radiation widths with standard deviations for s-wave neutrons $\bar{\Gamma}_\gamma$ listed in Table 8f-4 were determined principally from the resonance data tabulated in BNL-325 (1958, 1960, 1964–1966). The standard deviation includes both the experimental errors in the measurements of the radiation widths of the individual resonances and the standard deviation arising from the width of the distribution of Γ_γ. For heavy nuclides the width of the distribution is $\sim$10 percent corresponding to a chi-squared distribution with 200 degrees of freedom, while for light nuclides the width is $\sim$30 percent which corresponds to $\sim$22 degrees of freedom.

Values denoted with an asterisk are based on data for only one or two resonances and sometimes are computed from thermal capture cross sections or resonance capture integrals. These values with an asterisk may not include the standard deviation arising from a poor sampling from the distribution of Γ_γ. Hence, the correct $\bar{\Gamma}_\gamma$ for these nuclides might be different from the value listed by $\sim$10 to $\sim$30 percent. A few nonzero-spin target nuclides have average radiation widths which are different for the two spin states of the compound nucleus. For these nuclides values are listed for both spin states.

8f-6. Infinite-dilution Resonance Integrals.[1] The neutron cross sections for most nuclides exhibit resonance structure. The incident neutron energy range in which

[1] We should like to express our appreciation to Dr. M. K. Drake, of the Gulf General Atomic Corporation (now at Brookhaven National Laboratory), for supplying the information contained in this section and in Table 8f-5.

the individual resonances can be observed varies from nuclide to nuclide; but for most of the heavier nuclides, this energy range begins at near-thermal neutron energies and extends to approximately 10 keV. The resonance integral is a quantity that is frequently used to characterize the magnitude of the neutron cross section for the resonance energy region. The resonance integral has been found to be particularly useful in characterizing the absorption cross section for materials used in reactor physics analysis.

When a resonance absorber is placed in a moderator at near-zero concentrations, the resonance absorption integral is not affected by energy self-shielding or by Doppler broadening. Under these conditions, the material absorbs neutrons in the slowing-down spectrum of the moderator. The resonance integral is expressed as

$$\text{R.I.} = \int_{E_c}^{E_{\max}} \sigma(E)\varphi(E)\ dE$$

where $\sigma(E)$ is the neutron cross section as a function of energy E. In the case where the absorber is at near-zero concentrations in a moderator, the weighting function $\varphi(E)$ (neutron flux distribution) is proportional to $1/E$ for neutron energies greater than a few tenths of an electron volt. The upper limit of the integral is generally taken as a few MeV. The lower-energy limit is generally taken as the cutoff, between where the neutron flux distribution can be treated as being $1/E$ and where a Maxwellian distribution can be used. In most experimental measurements of the resonance integral, E_c is determined by the type and thickness of the filter used to absorb the neutrons in the Maxwellian portion of the spectrum. Cadmium is the material generally used as the filter, and an appropriate thickness of the material results in a cutoff energy of 0.5 eV.

In Table 8f-5 are listed recommended infinite-dilution resonance integrals. The values given in this table have been taken from various sources. In most cases the recommended values have been taken from experimental integral measurements. In other cases the values have been obtained by integrating experimentally measured, differential cross-section data. Resonance integrals for several reaction mechanisms, i.e., (n,γ), $(n, \text{fission})$, and $(n, \text{absorption})$, are included. The particular reaction mechanism is given in the Reaction column. In certain cases the (n,γ) reaction produces two or more different states of the residual nucleus, and this information is also given in the Reaction column. The recommended resonance integrals (R.I.) at infinite dilution are given in the final column. In all cases the cutoff energy E_c has been taken as 0.5 eV, and the upper energy limit $E_{\max}$ has been taken to be 15 MeV. Also, the integrals given in Table 8f-5 contain the contribution from the $1/v$ part of the low-energy cross sections.

8f-7. Neutron Flux Standards. Because of the uncharged nature of the neutron, its direct detection is difficult. For many cross-section measurements a knowledge of the incident neutron flux, either absolute or relative, is required, and many techniques are employed to accomplish this end. For absolute flux measurements, techniques used include the production of known flux by means of source reactions (see Sec. 8f-9), the utilization of the reasonably well-known characteristics of the interactions of neutrons with protons (the n-p interaction), and the invocation of certain well-determined cross sections as standards for measurement of other lesser-known cross sections. The n-p interaction is, of course, only a special case of the last technique.

The total n-p cross section has been measured to high precision at a number of neutron energies. These data have been fitted by an analytical form based on effective range theory. The resulting equation, which gives the total cross section σ_τ in barns for an incident laboratory neutron energy E in MeV, seems to fit the high-

TABLE 8f-4. s-WAVE NEUTRON STRENGTH FUNCTIONS Γ_n°/D, s-WAVE OBSERVED RESONANCE SPACINGS D_{obs}, AND s-WAVE AVERAGE RADIATION WIDTHS Γ_γ

Target nucleus Z	A	$\Gamma_n^\circ/D \times 10^4$	D_{obs}, eV	Γ_γ, mV
9 F	19	0.7	$(6 \pm 3) \times 10^4$	
11 Na	23	0.8	$(2.7 \pm 1.4) \times 10^5$	
13 Al	27	0.9	$(4 \pm 2) \times 10^4$	
16 S	32	1.2	$(2.9 \pm 0.9) \times 10^5$	
17 Cl	35	0.65 ± 0.15	$(1.3 \pm 0.4) \times 10^4$	600 ± 100*
	37	0.30 ± 0.20	$(1.3 \pm 0.4) \times 10^4$	3000 ± 1000*
18 Ar	36	0.63 ± 0.20		
19 K	39	1.3 ± 0.4	$(1.0 \pm 0.3) \times 10^4$	500 ± 20*
20 Ca	40	3.1	$(4.9 \pm 1.0) \times 10^4$	
	42	2.2 ± 0.7	$(2.8 \pm 0.4) \times 10^4$	1800 ± 200*
	43	1.4 ± 0.6	$(3.3 \pm 0.6) \times 10^3$	
	44	2.7 ± 1.0	$(3.3 \pm 0.5) \times 10^4$	
	45	5.0 ± 0.9	$(1.3 \pm 0.4) \times 10^3$	
21 Sc	45			
22 Ti	46	2.6 ± 1.1	$(3.0 \pm 0.6) \times 10^4$	350 ± 110*
	47	2.9 ± 0.8	$(2.6 \pm 0.4) \times 10^3$	
	48	3.6 ± 1.4	$(2.2 \pm 0.4) \times 10^4$	
	49	3.0 ± 0.8	$(4.0 \pm 1.0) \times 10^3$	
	50	1.4	$(1.2 \pm 0.5) \times 10^5$	
23 V	50	4.5 ± 1.7	$(1.1 \pm 0.3) \times 10^3$	~1500*
	51	4.2 ± 1.0	$(3.6 \pm 0.9) \times 10^3$	2900 ± 900*
24 Cr	50	4.0 ± 1.0	$(1.6 \pm 0.2) \times 10^4$	
	52	2.5 ± 1.0	$(4.4 \pm 0.8) \times 10^4$	
	53	4.1 ± 1.0	$(3.0 \pm 0.7) \times 10^3$	
	54	2.2 ± 0.9	$(2.3 \pm 0.4) \times 10^4$	450 ± 40*
25 Mn	55	4.2 ± 1.7	$(2.1 \pm 0.8) \times 10^3$	
26 Fe	54	5.2 ± 1.7	$(2.5 \pm 0.4) \times 10^4$	1400 ± 200*
	56	1.6 ± 0.5	$(2.9 \pm 0.4) \times 10^4$	2000 ± 1000*
	57	3.0 ± 1.0	$(8 \pm 3) \times 10^3$	450 ± 50
27 Co	59	3.5 ± 1.0	960 ± 210	

Target nucleus Z	A	$\Gamma_n^\circ/D \times 10^4$	D_{obs}, eV	Γ_γ, mV
38 Sr	84	0.5	340 ± 90	
	86	0.65 ± 0.3	$(1.0 \pm 0.3) \times 10^3$	205 ± 20*
	87	0.5 ± 0.1	200 ± 60	
	88	0.30 ± 0.08	$(5.5 \pm 1.9) \times 10^4$	
39 Y	89	0.9	$(1.0 \pm 0.3) \times 10^3$	
40 Zr	90	0.4 ± 0.2	$(4.5 \pm 1.6) \times 10^3$	110 ± 20*
	91	0.65 ± 0.25	315 ± 85	
	92	1.4	$(1.2 \pm 0.4) \times 10^3$	
	94	0.8 ± 0.3	$(2.4 \pm 0.9) \times 10^3$	220 ± 50*
	96	1.5	$(1.0 \pm 0.3) \times 10^3$	170 ± 20
41 Nb	93	1.5 ± 0.7	70 ± 10	
42 Mo	92	0.4 ± 0.1	$(2.4 \pm 1.0) \times 10^3$	170 ± 30*
	94	0.6 ± 0.15	$(1.0 \pm 0.4) \times 10^3$	150 ± 30*
	95	1.0	110 ± 30	150 ± 30*
	96	0.55 ± 0.22	$(1.0 \pm 0.5) \times 10^3$	130 ± 30*
	97	0.8	90 ± 20	100 ± 40*
	98	0.6 ± 0.24	270 ± 80	120 ± 20
	100	0.8 ± 0.4	430 ± 150	
43 Tc	99	0.9	20 ± 5	
44 Ru	99	0.45 ± 0.18	120 ± 60	165 ± 20*
	101	0.3 ± 0.15	16 ± 5	200 ± 20
	102		>100	290 ± 50*
45 Rh	103	0.4 ± 0.2	26 ± 8	155 ± 5
46 Pd	105	0.43 ± 0.15	10 ± 3	155 ± 10
	108	0.35 ± 0.10	30 ± 14	100 ± 15*
47 Ag		0.45 ± 0.1		

Elements 28–37

Element	A	Col 1	Col 2	Col 3
28 Ni	58	2.4	$(2.7 \pm 0.5) \times 10^4$	
	60	3.0 ± 0.9	$(2.3 \pm 0.4) \times 10^4$	
	61	2.8 ± 0.7	$(2.4 \pm 0.6) \times 10^4$	
	62	2.9 ± 0.6	$(1.9 \pm 0.3) \times 10^4$	
	64	2.0 ± 0.8	$(2.8 \pm 0.4) \times 10^4$	800 ± 200*
29 Cu	63	2.5 ± 0.8	$(1.2 \pm 0.3) \times 10^3$	510 ± 50
	65	1.7 ± 0.7	$(1.7 \pm 0.3) \times 10^3$	340 ± 40
30 Zn	64	1.7 ± 0.8	$(2.6 \pm 0.9) \times 10^3$	300 ± 30*
	66	1.1 ± 0.6	$(5.0 \pm 1.3) \times 10^3$	200 ± 50*
	67	1.5	600 ± 300	400 ± 70*
	68	3.1	$(1.0 \pm 0.2) \times 10^4$	180 ± 30*
31 Ga	69	3.0 ± 0.8	340 ± 95	210 ± 50*
	71	1.2 ± 0.5	170 ± 63	350 ± 110*
32 Ge	70	1.5	$(1.7 \pm 0.3) \times 10^3$	
	72	1.8 ± 0.7	$(2.1 \pm 0.4) \times 10^3$	
	73	1.3 ± 0.5	$(8.5 \pm 1.0) \times 10^3$	
	74	2.0 ± 0.7	$(8 \pm 1) \times 10^3$	
	76	0.8		
33 As	75	{ 1.0 ± 0.3 / 2.5 ± 0.6† }	87 ± 14	290 ± 20
34 Se	74	1.3	370 ± 70	260 ± 50*
	76	2.6 ± 0.6	700 ± 150	230 ± 40*
	77	1.7 ± 0.8	100 ± 25	380 ± 30*
	78	1.0 ± 0.5	1000 ± 200	220 ± 50*
	80	1.4 ± 0.4	1200 ± 400	220 ± 50*
	82	1.0 ± 0.8	$(7 \pm 1) \times 10^3$	
35 Br	79	{ 0.9 ± 0.3 / 1.9 ± 0.4† }	57 ± 19	340 ± 20
	81	1.5 ± 0.6	51 ± 26	300 ± 30
36 Kr	80	1.3	530 ± 280	400 ± 90*
	83	0.24	200 ± 150	220 ± 60*
37 Rb	85	1.1 ± 0.4	180 ± 30	215 ± 20*
	87	2.0 ± 0.4	1600 ± 400	145 ± 30*

Elements 48–54

Element	A	Col 1	Col 2	Col 3
48 Cd	107	0.4 ± 0.1	30 ± 10	141 ± 4
	109	0.8 ± 0.3	13 ± 3	133 ± 3
	110	0.4 ± 0.15	> 45	130 ± 40*
	111	0.35	26	95 ± 20*
	112	0.45	200	90 ± 30*
	113	0.4	25 ± 5	110 ± 5*
	114	0.6	160 ± 50	150 ± 50*
49 In	113	0.6 ± 0.3	6.5 ± 2.0	70 ± 20*
	115	0.3 ± 0.15	6.7 ± 2.0	76 ± 5*
50 Sn	112	0.2 ± 0.1	25 ± 7	110 ± 30*
	114	0.5 ± 0.2	150 ± 60	70 ± 30*
	115	0.7	50 ± 30	70 ± 15
	116	0.3	150 ± 20	79 ± 14*
	117	0.26 ± 0.05	25 ± 5	70 ± 30*
	118	0.16 ± 0.03	180 ± 50	100 ± 50*
	119	0.4 ± 0.2	30 ± 8	130 ± 60*
	120	0.08 ± 0.03	200 ± 70	
	122	0.12 ± 0.06	400 ± 200	
	124	0.4	400 ± 200	
51 Sb	121	0.4 ± 0.1	14 ± 4	92 ± 4
	123	0.5	28 ± 12	90 ± 20
52 Te	122	0.6	130 ± 15	104 ± 3*
	123	0.5 ± 0.2	26 ± 5	150 ± 80*
	124	0.8 ± 0.2	147 ± 11	
	125	1.0 ± 0.2	38 ± 4	
	126	0.7	210 ± 20	
	128	0.49 ± 0.10	260 ± 30	
	130	0.30 ± 0.10	870 ± 90	
53 I	127	0.25 ± 0.10	13.0 ± 0.5	
	129	0.14 ± 0.04	31 ± 10	
54 Xe	129	0.6 ± 0.1	40 ± 15	120 ± 7*
	130	0.3		113 ± 8*
	131	1.0	31 ± 16	91 ± 1*
	135	2.0 ± 0.7		

* Based on data for only one or two resonances (sometimes computed from thermal-capture cross sections or resonance-capture integrals).

† $(J = 1)$ and $(J = 2)$.

TABLE 8f-4. s-WAVE NEUTRON STRENGTH FUNCTIONS $\Gamma_n°/D$, s-WAVE OBSERVED RESONANCE SPACINGS D_{obs}, AND s-WAVE AVERAGE RADIATION WIDTHS Γ_γ (Continued)

Target nucleus Z A	$\Gamma_n°/D \times 10^4$	D_{obs}, eV	Γ_γ, mV
55 Cs 133	0.7 ± 0.1	20 ± 1	120 ± 20*
56 Ba 130	3.2 ± 0.3	150 ± 50	
135	0.8 ±	35 ± 7	
136	......	(4 ± 2) × 10³	107 ± 5
137	0.34	400 ± 200	180 ± 20*
138	1.8 ± 0.6	(10 ± 4) × 10³	
57 La 138	0.7 ± 0.5	23 ± 7	99 ± 6*
139	0.8 ± 0.2	90 ± 20	80 ± 30*
58 Ce 140	1.0	(3 ± 1) × 10³	
142	1.2	(1.0 ± 0.2) × 10³	
59 Pr 141	2.1 ± 0.4	51 ± 16	82 ± 5
60 Nd 142	2.5 ± 0.5	1000 ± 250	
143	0.6 ± 0.3	38 ± 6	76 ± 11*
144	4.3 ± 1.4	520 ± 70	78 ± 12*
145	4.8 ± 2.0	19 ± 3	58 ± 8
146	3.0 ± 0.7	310 ± 40	55 ± 8*
148	4.5 ± 1.9	220 ± 20	96 ± 14
150	3.6 ± 1.1	230 ± 30	84 ± 12
61 Pm 147	2.0 ± 0.8	4.0 ± 1.0	65 ± 10*
62 Sm 147	4.2 ± 1.8	8.0 ± 1.7	61 ± 5*
149	2.3	2.4 ± 0.6	62 ± 2
151	4.5 ± 1.0	2.3 ± 0.4	
152	3.5 ± 1.9	53 ± 8	65 ± 5
154	2.7 ± 0.8	125 ± 20	75 ± 15
63 Eu 151	1.9 ± 0.7	1.04 ± 0.07	95 ± 4
153	2.9 ± 0.5	1.45 ± 0.12	94 ± 5
64 Gd	2.3 ± 0.7		
	2.2 ± 0.7		
	1.5 ± 0.3		

$\Gamma_n°/D \times 10^4$	Target nucleus Z A	D_{obs}, eV	Γ_γ, mV
1.6	179	4.5 ± 0.7	30 ± 5*
0.3	180	100 ± 40	56 ± 2
2.0 ± 0.5	73 Ta 180	2.0 ± 1.0	
1.7	181	4.4 ± 0.4	
2.7	182		
1.0 ± 0.5	74 W 180	20 ± 7	56 ± 2
2.6 ± 0.5	182	63 ± 10	74 ± 3
2.4 ± 0.5	183	15 ± 4	68 ± 6
2.6 ± 0.8	184	95 ± 14	62 ± 3
2.2 ± 0.5	186	92 ± 13	
2.4 ± 0.5	75 Re 185	2.8 ± 0.6	53 ± 2
2.1 ± 0.7	187	4.5 ± 1.0	54 ± 2
2.4 ± 0.5	76 Os 189	5.1 ± 1.2	90 ± 7
2.2 ± 0.5	77 Ir 191	3.1 ± 0.6	71 ± 2*
2.0 ± 0.5	193	8.2 ± 1.6	87 ± 1*
1.7 ± 0.3	78 Pt 192	150 ± 60	52 ± 8*
1.7	194	200 ± 100	70 ± 15*
1.4	195	12 ± 1	{ 93 ± 10 / 118 ± 6‡ }
{ 2.0 ± 0.7 / 1.9 ± 0.4‡ }	196		120 ± 20*
	198		150 ± 30*
{ 0.9 ± 0.4 / 2.3 ± 0.6† }	79 Au 197	16.8 ± 1.6	125 ± 4
1.4 ± 0.5	80 Hg 196	>50	220 ± 100*
1.0	198	90 ± 30	135 ± 15

	152	4.6 ± 1.8	15 ± 2	57 ± 15*
	154	2.4 ± 1.0	15 ± 2	63 ± 15*
	155	2.2 ± 0.3	1.9 ± 0.2	109 ± 2
	156	1.8 ± 0.6	47 ± 4	82 ± 12
	157	2.3 ± 0.3	5.6 ± 0.8	103 ± 4
	158	1.5 ± 0.5	85 ± 9	89 ± 13
	160	2.6 ± 1.0	170 ± 20	98 ± 15
65 Tb	159	1.9 ± 0.6	3.9 ± 0.6	87 ± 2
66 Dy		1.0		
	161	1.8 ± 0.4	2.9 ± 0.4	110 ± 10
	162	2.5 ± 0.9	72 ± 15	155 ± 15*
	163	1.7 ± 0.5	10 ± 2	110 ± 10
	164	1.2 ± 0.4	200 ± 50	55 ± 3*
67 Ho	165	1.9 ± 0.3	5.0 ± 1.0	77 ± 2
68 Er		1.9 ± 0.5		
	162	2.1 ± 0.7	6 ± 2	
	164	1.5 ± 0.5	20 ± 8	
	166	1.7 ± 0.5	49 ± 5	90 ± 10
	167	2.2 ± 0.4	4.3 ± 0.3	87 ± 3
	168	1.4 ± 0.4	130 ± 17	80 ± 10*
	170	1.4 ± 0.5	250 ± 50	
69 Tm	169	{1.4 ± 0.4 {1.3 ± 0.2‡	7.3 ± 0.5	85 ± 2
70 Yb		1.3		
	168	....	15 ± 10	70 ± 10*
	170	2.4 ± 0.5	37 ± 6	
	171	1.5 ± 0.3	6.3 ± 0.6	73 ± 5
	172	1.1 ± 0.3	62 ± 10	
	173	1.6 ± 0.4	7.5 ± 0.8	74 ± 6
	174	0.9 ± 0.4	230 ± 50	
	176	1.8 ± 0.6	190 ± 50	
71 Lu	175	1.9 ± 0.7	3.7 ± 0.7	72 ± 2*
	176	1.7 ± 0.7	2.3 ± 0.4	59.0 ± 0.2*
72 Hf		2.5 ± 0.5		
	174	2.8	16 ± 3	
	176	1.4 ± 0.5	32 ± 7	
	177	2.0	2.3 ± 0.3	64 ± 3*
	178	2.1 ± 0.7	60 ± 20	

	199	2.0	70 ± 20	{230 ± 30* {300 ± 15‡
	200	2.1	(2.2 ± 0.7) × 10³	
	201	1.4	100 ± 40	{460 ± 30* {290 ± 20*†
	202	....	(2.4 ± 1.3) × 10³	
81 Tl		1.2 ± 0.4		
	203	....	(2 ± 1) × 10³	640 ± 70*
	205	....	(1.0 ± 0.3) × 10⁴	
82 Pb	204	0.55 ± 0.2	(2.7 ± 0.5) × 10³	
	206	1.8	~5 × 10⁴	
	207	0.3	~8 × 10³	5500 ± 900*
	208	....	>3.5 × 10⁵	
83 Bi	209	0.5 ± 0.2	(3.5 ± 1.2) × 10³	<44*
88 Ra	226	....	....	29 ± 1*
90 Th	229	0.62 ± 0.25	0.60 ± 0.15	47 ± 4*
	230	1.3 ± 0.5	11 ± 3	26 ± 2*
	232	0.9 ± 0.1	17.5 ± 0.7	23.0 ± 1.5
91 Pa	231	0.85 ± 0.15	0.45 ± 0.07	44 ± 3
	233	1.9	0.8 ± 0.2	47 ± 4
92 U	232	1.4 ± 0.6	5.3 ± 1.0	
	233	0.9 ± 0.2	0.62 ± 0.05	45 ± 5
	234	1.2 ± 0.4	13 ± 2	25 ± 6*
	235	0.95 ± 0.10	0.53 ± 0.03	40 ± 3
	236	1.2 ± 0.3	15 ± 2	23.9 ± 1.0
	238	0.9 ± 0.1	17.7 ± 0.7	24.0 ± 1.1
93 Np	237	1.0 ± 0.15	0.57 ± 0.06	34 ± 2*
94 Pu	238	1.3 ± 0.2	13 ± 4	45 ± 10*
	239	1.2 ± 0.2	2.39 ± 0.12	41.6 ± 1.5
	240	1.05 ± 0.16	13.5 ± 1.0	22 ± 2
	241	1.10 ± 0.2	1.2 ± 0.2	40 ± 8
	242	0.9 ± 0.3	†15 ± 2	25 ± 3*
95 Am	241	1.1 ± 0.2	0.43 ± 0.07	41 ± 3
	243^m	1.4	0.6 ± 0.2	
	243	0.84 ± 0.25	1.4 ± 0.3	40 ± 3*
96 Cm	244	0.76 ± 0.30	13 ± 3	38 ± 3
	246	0.5	30 ± 15	35 ± 5*

* Based on data for only one or two resonances (sometimes computed from thermal-capture cross sections or resonance-capture integrals).
† (J = 1) and (J = 2).
‡ (J = 0) and (J = 1).

TABLE 8f-5. TABLE OF INFINITE-DILUTION RESONANCE INTEGRALS

Target nucleus Z	A	Reaction	Resonance integral, barns
1 H	1	n,γ	0.149 ± 0.001
	2	n,γ	$(0.26 \pm 0.02) \times 10^{-3}$
2 He		Abs.	$(3.3 \pm 1.0) \times 10^{-3}$
	3	n,p	$2{,}397 \pm 10$
3 Li		Abs.	32 ± 1
	6	n,α	430 ± 10
	6	n,γ	$(20 \pm 6) \times 10^{-3}$
	7	n,γ	$(17 \pm 3) \times 10^{-3}$
4 Be		n,γ	$(4.5 \pm 0.6) \times 10^{-3}$
5 B		Abs.	341 ± 1
	10	n,α	$1{,}725 \pm 6$
	10	n,γ	0.23 ± 0.1
	11	n,γ	$(25 \pm 10) \times 10^{-3}$
6 C		Abs.	$(1.3 \pm 0.2) \times 10^{-3}$
		n,γ	$1{,}03 \pm 0.03$
7 N	14	n,p	1.00 ± 0.03
	14	n,γ	$(34 \pm 4) \times 10^{-3}$
8 O		n,γ	$(0.08 \pm 0.02) \times 10^{-3}$
9 F	19	Abs.	0.25 ± 0.1
	19	n,γ	$(25 \pm 5) \times 10^{-3}$
10 Ne		n,γ	$(17 \pm 5) \times 10^{-3}$
11 Na	22	n,γ	$(2.0 \pm 0.3) \times 10^{5}$
	23	Abs.	0.314 ± 0.01
12 Mg		n,γ	0.35 ± 0.1
13 Al	27	Abs.	0.07 ± 0.02
	27	n,γ	0.28 ± 0.01
14 Si		Abs.	0.13 ± 0.01
		n,γ	0.54 ± 0.06
15 P	31	Abs.	0.09 ± 0.02
	31	n,γ	0.30 ± 0.05
16 S		Abs.	0.09 ± 0.01
		n,γ	1.3 ± 0.2
17 Cl		Abs.	0.4 ± 0.1
		n,γ	13.5 ± 1.0
		n,γ	13.0 ± 1.0

Target nucleus Z	A	Reaction	Resonance integral, barns
36 Kr	81	$(n,\gamma)^{82m}$Br(6.2 min)	34 ± 2
	81	$(n,\gamma)^{82z}$Br(36 hr)	7 ± 1
	83	n,γ	150 ± 50
	84	n,γ	5 ± 2
37 Rb	85	n,γ	14 ± 7
	85	n,γ	0.05 ± 0.02
	86	n,γ	5 ± 2
	87	n,γ	6 ± 2
38 Sr	88	n,γ	3 ± 1
	90	n,γ	14 ± 4
39 Y	89	n,γ	0.06 ± 0.03
40 Zr	90	n,γ	2 ± 1
	91	n,γ	0.9 ± 0.2
	92	n,γ	1.20 ± 0.1
	94	n,γ	0.25 ± 0.05
41 Nb	93	n,γ	7.3 ± 1.0
	94	n,γ	0.6 ± 0.2
42 Mo	92	n,γ	0.3 ± 0.1
	94	n,γ	3.5 ± 1.5
	95	n,γ	9.0 ± 0.5
	96	n,γ	160 ± 30
	97	n,γ	25 ± 2
	98	n,γ	0.53 ± 0.04
43 Tc	99	n,γ	0.95 ± 0.05
	100	n,γ	110 ± 5
44 Ru	96	n,γ	25 ± 1
	98	n,γ	16 ± 1
	99	n,γ	7.3 ± 0.5
	100	n,γ	4.8 ± 0.5
		n,γ	190 ± 20
		n,γ	60 ± 25
		n,γ	6 ± 1
		n,γ	10 ± 5
		n,γ	210 ± 20
		n,γ	12 ± 2

Element	A	Reaction	Cross section
18 Ar	35	n,γ	17.0 ± 1.0
	37	n,γ	0.30 ± 0.03
19 K	40	n,γ	0.41 ± 0.03
		Abs.	3.0 ± 0.5
	41	n,γ	2.0 ± 1.0
20 Ca		Abs.	0.96 ± 0.05
	45	n,γ	1.65 ± 0.2
		Abs.	0.45 ± 0.1
21 Sc	45	n,γ	11.0 ± 1.0
22 Ti		Abs.	3.0 ± 0.2
23 V		n,γ	2.9 ± 0.2
		Abs.	2.6 ± 0.2
24 Cr		n,γ	2.54 ± 0.25
		Abs.	1.6 ± 0.1
	50	n,γ	1.5 ± 0.1
	52	n,γ	7.5 ± 0.2
	53	n,γ	0.43 ± 0.04
25 Mn	55	n,γ	9.5 ± 0.5
26 Fe		n,γ	14.4 ± 0.4
27 Co	58m	n,γ	2.25 ± 0.2
	59	n,γ	$(2.5 \pm 1.0) \times 10^5$
28 Ni		n,γ	72 ± 3
		Abs.	2.8 ± 0.1
	63	n,γ	2.2 ± 0.1
29 Cu		Abs.	4.3 ± 0.4
	65	n,γ	4.2 ± 0.3
30 Zn		n,γ	5.1 ± 0.2
	64	n,γ	2.3 ± 0.2
	68	n,γ	1.6 ± 0.2
	68	$(n,y)\,^{69g}$Zn(51 min)	2.0 ± 0.4
	68	$(n,\gamma)\,^{69m}$Zn(13.8 hr)	1.8 ± 0.4
31 Ga	69	n,γ	0.21 ± 0.03
	71	n,γ	8.0 ± 2.0
33 As	75	n,γ	6.4 ± 2.0
34 Se		n,γ	10.5 ± 2.0
		n,γ	80 ± 5
35 Br	79	n,γ	12 ± 3
	81	n,γ	75 ± 20
		n,γ	110 ± 30
		n,γ	41 ± 2

Element	A	Reaction	Cross section
45 Rh	101	n,γ	85 ± 10
	102	n,γ	5 ± 1
	104	n,γ	6 ± 2
	106	Abs.	2.0 ± 0.6
	103	n,γ	82 ± 4
	103	$(n,\gamma)\,^{104m}$Rh	1,080 ± 40
	103	$(n,\gamma)\,^{104g}$Rh	1,160 ± 40
	105	n,γ	$(1.8 \pm 0.5) \times 10^4$
46 Pd	102	n,γ	95 ± 15
	104	n,γ	14 ± 4
	105	n,γ	20 ± 5
	106	n,γ	90 ± 20
	108	n,γ	16 ± 5
	110	n,γ	250 ± 30
47 Ag	107	n,γ	16 ± 5
	109	n,γ	760 ± 60
	109	$(n,\gamma)\,^{110m}$Ag	120 ± 15
	111	n,γ	50 ± 5
48 Cd	106	n,γ	1,460 ± 80
	108	n,γ	100 ± 20
	110	n,γ	68 ± 10
	111	n,γ	7 ± 3
	112	n,γ	8 ± 3
	113	n,γ	40 ± 5
	114	n,γ	45 ± 5
	116	n,γ	15 ± 5
49 In	113	n,γ	380 ± 20
	115	n,γ	23 ± 3
	115	$(n,\gamma)\,^{116m}$In(54 min)	2 ± 1
	115	$(n,\gamma)\,^{116g}$In(13 sec)	3,200 ± 100
50 Sn	112	n,γ	840 ± 60
	114	n,γ	2,650 ± 100
	115	n,γ	650 ± 30
	116	n,γ	3,300 ± 100
	117	n,γ	7.5 ± 1.0
	118	n,γ	30 ± 3
		n,γ	7.5 ± 2.0
		n,γ	3.8 ± 1.0
		n,γ	16 ± 2
		n,γ	17 ± 2
		n,γ	6.5 ± 1.0

TABLE 8f-5. TABLE OF INFINITE-DILUTION RESONANCE INTEGRALS (*Continued*)

Target nucleus Z	A	Reaction	Resonance integral, barns	Resonance integral, barns	Reaction	Target nucleus Z	A
	119	n, γ	5 ± 1	2,300 ± 300	n, γ	63 Eu	151
	120	n, γ	2.0 ± 0.5	3,300 ± 300	n, γ		153
	122	n, γ	1.5 ± 0.5	1,450 ± 200	n, γ		154
	124	n, γ	13 ± 2	950 ± 300	n, γ		155
51 Sb	121	n, γ	170 ± 15	6,000 ± 1,000	n, γ	64 Gd	154
	123	n, γ	200 ± 10	430 ± 40	n, γ		155
52 Te	120	n, γ	125 ± 10	500 ± 200	n, γ		156
	122	n, γ	54 ± 5	1,730 ± 200	n, γ		157
	123	n, γ	2.0 ± 0.5	110 ± 10	n, γ		158
	124	n, γ	65 ± 5	790 ± 50	n, γ		160
	125	n, γ	5,600 ± 500	70 ± 7	n, γ	65 Tb	159
	126	n, γ	4 ± 1	13 ± 2	n, γ	66 Dy	158
	128	n, γ	19 ± 2	400 ± 20	n, γ		160
	130	n, γ	10 ± 1	1,425 ± 200	n, γ		161
53 I	127	n, γ	1.5 ± 0.2	100 ± 50	n, γ		162
	129	n, γ	0.35 ± 0.05	1,100 ± 200	n, γ		163
	131	n, γ	154 ± 5	1,190 ± 150	n, γ		164
54 Xe	124	n, γ	35 ± 5	2,575 ± 300	n, γ	67 Ho	165
	128	n, γ	8 ± 4	1,650 ± 200	n, γ	68 Er	162
	129	n, γ	265 ± 50	380 ± 30	n, γ		164
	130	n, γ	3,600 ± 500	700 ± 30	n, γ		166
	131	n, γ	35 ± 5	780 ± 60	n, γ		167
	132	n, γ	300 ± 100	440 ± 40	n, γ		168
	133	n, γ	12 ± 3	90 ± 10	n, γ		170
	134	n, γ	840 ± 50	115 ± 10	n, γ	69 Tm	169
	135	n, γ	2.5 ± 0.5	3,150 ± 200	n, γ	70 Yb	168
	136	n, γ	160 ± 20	38 ± 4	n, γ		170
55 Cs	133	n, γ	2 ± 1	45 ± 4	n, γ		171
	133	$(n,\gamma)^{134m}$ Cs	7600 ± 500	1,550 ± 100	n, γ		172
	133	$(n,\gamma)^{134g}$ Cs	12 ± 2	185 ± 15	n, γ		173
	135	n, γ	35 ± 3	31,000 ± 3,000	n, γ		
	137	n, γ	385 ± 40	350 ± 40	n, γ		
		n, γ	420 ± 40	340 ± 30	n, γ		
		n, γ	60 ± 5	28 ± 3	n, γ		
		n, γ	1.0 ± 0.5	400 ± 40	n, γ		

Elements 56–62

Z, Element	A	Reaction	σ (barns)
56 Ba	130	n, γ	9 ± 2
	132	n, γ	15 ± 3
	134	n, γ	6 ± 3
	135	n, γ	15 ± 5
	136	n, γ	100 ± 20
	137	n, γ	18 ± 2
	138	n, γ	5 ± 1
	140	n, γ	0.25 ± 0.05
57 La	138	n, γ	14 ± 2
	139	n, γ	12.3 ± 1.0
	140	n, γ	330 ± 30
58 Ce	140	n, γ	12 ± 1
	142	n, γ	70 ± 5
	144	n, γ	0.45 ± 0.05
59 Pr	141	n, γ	2.5 ± 0.5
	143	n, γ	2.6 ± 0.3
	144	n, γ	18.3 ± 1.0
60 Nd	142	n, γ	190 ± 40
	143	n, γ	60,000 ± 30,000
	144	n, γ	45 ± 5
	145	n, γ	12 ± 2
	146	n, γ	115 ± 10
	148	n, γ	10 ± 2
	150	n, γ	260 ± 15
		n, γ	12 ± 2
		n, γ	18 ± 1
		n, γ	16 ± 1
61 Pm	147	n, γ	2,300 ± 400
	147	$(n,\gamma)^{148m}\mathrm{Pm}$	1,200 ± 300
	147	$(n,\gamma)^{148g}\mathrm{Pm}$	1,100 ± 300
	148m	n, γ	30,000 ± 10,000
	148g	n, γ	40,000 ± 10,000
62 Sm	144	n, γ	1,400 ± 150
	147	n, γ	10 ± 5
	148	n, γ	590 ± 20
	149	n, γ	20 ± 10
	150	n, γ	3,200 ± 100
	151	n, γ	350 ± 50
	152	n, γ	2,450 ± 300
	154	n, γ	3,100 ± 100
		n, γ	40 ± 20

Elements 71–80

Z, Element	A	Reaction	σ (barns)
	174	n, γ	36 ± 4
	176	n, γ	14 ± 2
71 Lu	175	n, γ	670 ± 70
	176	n, γ	660 ± 70
72 Hf	174	n, γ	950 ± 100
	176	n, γ	2,000 ± 200
	177	n, γ	450 ± 50
	178	n, γ	800 ± 80
	179	n, γ	7,300 ± 200
	180	n, γ	1,900 ± 200
73 Ta	180	n, γ	625 ± 50
	181	n, γ	45 ± 5
74 W	180	n, γ	740 ± 40
	182	n, γ	100 ± 30
	183	n, γ	740 ± 40
	184	n, γ	350 ± 30
	186	n, γ	10 ± 10
75 Re	185	n, γ	590 ± 10
	187	n, γ	380 ± 15
76 Os	187	n, γ	13 ± 2
	189	n, γ	490 ± 50
77 Ir	191	n, γ	850 ± 50
	193	n, γ	1,770 ± 100
78 Pt	190	n, γ	310 ± 20
	192	n, γ	210 ± 20
	194	n, γ	250 ± 30
	195	n, γ	760 ± 100
	196	n, γ	2,050 ± 150
	198	n, γ	3,100 ± 200
79 Au	197	n, γ	1,400 ± 100
80 Hg	196	n, γ	140 ± 10
	198	n, γ	80 ± 30
		n, γ	90 ± 10
		n, γ	14 ± 2
		n, γ	380 ± 20
		n, γ	5 ± 2
		n, γ	50 ± 5
		n, γ	1,565 ± 40
		n, γ	90 ± 10
		n, γ	1,350 ± 200
		n, γ	70 ± 10

TABLE 8f-5. TABLE OF INFINITE-DILUTION RESONANCE INTEGRALS (*Continued*)

Target nucleus Z	A	Reaction	Resonance integral, barns
81 Tl	199	n, γ	410 ± 50
	200	n, γ	20 ± 5
	201	n, γ	40 ± 10
	202	n, γ	6 ± 2
	203	n, γ	12 ± 2
	205	n, γ	40 ± 5
82 Pb	203	n, γ	0.7 ± 0.1
	204	n, γ	0.18 ± 0.02
	205	n, γ	2.7 ± 0.3
	206	n, γ	0.12 ± 0.02
	207	n, γ	0.45 ± 0.05
	208	n, γ	0.015 ± 0.005
83 Bi	209	n, γ	0.31 ± 0.03
89 Ac	227	n, γ	1,300 ± 50
90 Th	230	n, γ	1,050 ± 100
	232	n, γ	84 ± 2
	233	n, γ	500 ± 200
91 Pa	231	n, γ	510 ± 50
	232	n, γ	890 ± 50
92 U	232	n, γ	240 ± 40
	232	Fiss.	320 ± 40
	232	Abs.	560 ± 60
	233	n, γ	138 ± 8
	233	Fiss.	780 ± 20
	233	Abs.	918 ± 25
	234	n, γ	650 ± 50
	235	n, γ	144 ± 6
	235	Fiss.	280 ± 15
	235	Abs.	424 ± 20
	236	n, γ	400 ± 80
	238	n, γ	275 ± 10
93 Np	237	n, γ	850 ± 200
	237	Fiss.	7 ± 3
	237	Abs.	860 ± 200
	238	Fiss.	750 ± 250
	238	Abs.	1,500 ± 500
	239	n, γ	450 ± 300
94 Pu	238	n, γ	145 ± 8
	238	Fiss.	24 ± 4
	238	Abs.	169 ± 10
	239	n, γ	190 ± 10
	239	Fiss.	310 ± 15
	239	Abs.	500 ± 20
	240	n, γ	8,200 ± 600
	241	n, γ	160 ± 20
	241	Fiss.	590 ± 30
	241	Abs.	750 ± 40
	242	n, γ	1,200 ± 100
95 Am	241	n, γ	2,400 ± 800
	241	$(n,\gamma)^{242m}$ Am	300 ± 100
	241	$(n,\gamma)^{242g}$ Am	2,100 ± 700
	242m	Fiss.	20 ± 5
	243	Fiss.	1,600 ± 200
	243	n, γ	2,300 ± 400
	243	$(n,\gamma)^{244m}$ Am	2,200 ± 400
	243	$(n,\gamma)^{244g}$ Am	110 ± 20
96 Cm	244	n, γ	650 ± 60
	244	Fiss.	20 ± 4
	244	Abs.	670 ± 60
	245	n, γ	250 ± 50
	245	Fiss.	500 ± 100
	245	Abs.	750 ± 100
	246	n, γ	250 ± 50
	248	n, γ	350 ± 50
97 Bk	249	n, γ	1,200 ± 200
98 Cf	250	n, γ	5,300 ± 500
99 Es	251	n, γ	1,000 ± 100
	253	n, γ	3,500 ± 600

precision cross-section measurements to much better than 1 percent. It can be expressed as follows:

$$\sigma_T = 3\pi[1.206E + (-1.860 + 0.0941491E + 0.000130658E^2)^2]^{-1}$$
$$+ \pi[1.206E + (0.4223 + 0.1300E)^2]^{-1} \qquad 0 \leq E \leq 40 \text{ MeV}$$

For the angular distribution in n-p scattering, the apparent isotropy of the scattering in the center-of-mass system up to approximately 10 MeV leads to the equation for the laboratory differential cross section at a laboratory scattering angle θ:

$$\sigma(\theta) = \frac{\sigma_T}{\pi} \cos \theta \qquad 0 \leq E \lesssim 10 \text{ MeV}$$

For further discussion of these n-p relations, see the chapters by J. L. Gammel and by J. E. Perry, Jr., in "Fast Neutron Physics." The constants in the σ_T equation are those of Marion and Young in "Nuclear Reaction Analysis," published by North-Holland Publishing Company, Amsterdam, 1968.

The use of well-measured cross sections as standards against which newer measurements are made occurs for all energy neutrons from thermal to very high energies. For cross sections using a thermal spectrum, some much-used 2,200 m/sec standard cross section values are shown in Table 8f-6.

TABLE 8f-6. 2,200-M/SEC STANDARD VALUES FOR CROSS SECTIONS USING A THERMAL SPECTRUM

Target	Reaction	Cross section, barns
He^3	(n,p)	$5,327 \pm 10$
Li^6	(n,t)	950 ± 15
B^{10}	(n,α)	$3,836 \pm 9$
Co^{59}	(n,γ)	37.2 ± 0.3
Au^{197}	(n,γ)	98.8 ± 0.3
U^{235}	(n,f)	580.2 ± 1.8
Pu^{239}	(n,f)	741.6 ± 3.1

For neutron energies in the resonance region (1 eV $\lesssim E \lesssim$ 10 keV), capture or fission data are usually made absolute either by measurements of the integrated flux over a particular well-determined resonance (or group of resonances) in a standard material or by normalization to other measurements on resonances in the nucleus under study. In addition, some reaction cross sections whose variations are monotonic and well-defined are used as detectors and provide excellent relative standards and occasionally adequate absolute standards. Examples are the $He^3(n,p)T$, the $Li^6(n,t)He^4$, and the $B^{10}(n,\alpha)Li^7$ reactions, all of which vary as $1/\sqrt{E}$ below 10 keV.

For fast neutrons, flux measurements are made using a variety of total cross sections which are either constant or monotonically varying in the region of interest. (One, of course, is the n-p cross section, mentioned above.) Also various reaction cross sections can be used under the same conditions. The three reactions mentioned above for He^3, Li^6, and B^{10} continue their $1/\sqrt{E}$ behavior above 10 keV and are used up to energies where their energy variation is known to sufficient accuracy for the problem at hand. Some (n,γ) and (n,f) cross sections can be used to higher energies, though they are not usually known to the accuracies of the light-target reactions. Threshold reactions (see Sec. 8f-8) are useful for flux measurements on broad-energy-spectrum sources; and many (n,p), (n,α), and $(n,2n)$ cross sections, measured near 14 MeV, have served as standards for subsequent 14-MeV measurements.

Curves showing the variation of $H^1(n,n)H^1$ (the n-p total cross section), $He^3(n,p)T$, $Li^6(n,t)He^4$, and $B^{10}(n,\alpha)Li^7$ above 100 eV are shown in Fig 8f-4, reproduced from "Nuclear Reaction Analysis" by J B. Marion and F. C. Young. (We appreciate the permission granted by the authors and publisher to use this figure.)

8f-8. Threshold Reactions. Many nuclear reactions resulting from the bombardment of nuclei with neutrons have negative Q values and thus have an energetic threshold below which the reaction cannot occur. When this threshold energy is exceeded, it is found that the cross section rises rapidly, sometimes showing considerable structure superposed on this underlying increase. Beyond a certain energy, competition with other energetically possible reactions leads to a decreasing rise and an eventual fall in the cross section. When the resulting residual nuclei are radioactive and have convenient lifetimes and decay characteristics, the presence of the threshold offers a means of using the reaction as a rough spectrometer in a broad neutron spectrum.

Many reactions have been used as threshold detectors in experimental measurements. Some of the more common (n,p), (n,α), and $(n,2n)$ excitation functions used for this purpose are shown in Figs. 8f-5 to 8f-9. Most of the measurements were made with insufficient energy resolution to detect whatever fine structure might exist, but their use as detectors for broad-spectrum sources achieves the same integration effect.

For a detailed description of the characteristics and use of many threshold detectors, see chap. IV. C, by P. R. Byerly, Jr., in "Fast Neutron Physics," part I.

8f-9. Fast-neutron Source Reactions. The production of monoenergetic beams of fast neutrons ($E \gtrsim 1$ keV) is usually achieved by the interactions of light charged particles, using cyclotrons, cascade generators, or electrostatic accelerators to provide high-quality energy precision and variability. The energy of a neutron beam will depend upon the energy of the incident charged particles, the Q of the reaction, the masses of the incident and target nuclei, and the neutron emission angle. An energy "spread" of the neutron beam results from such considerations as the energy spread of the incident particles, the thickness of the target materials, and the angular spread caused by the geometric size of the observation apparatus. The kinematics of these reactions are well defined and are best described in chap. I. B, by J. Monahan, in "Fast Neutron Physics" part I.

The monoenergetic character of source reactions is only approximate, since complex-particle breakup or excitation of higher states in the product nucleus can contribute groups of neutrons with lower energies than those of the primary group. For the common source reactions the characteristics of the lower-energy neutrons are known well enough for appropriate corrections to be made, or else the energy separation from the primary group is sufficiently great for straightforward energy discrimination.

The four most used fast-neutron source reactions are the $Li^7(p,n)Be^7$, $T(p,n)He^3$, $D(d,n)He^3$, and $T(d,n)He^4$ reactions. With the high precision available on proton and deuteron energies with electrostatic generators, good-quality neutron beams can be produced with energies from a few keV up to approximately 30 MeV. Most measurements with these reactions are performed with neutrons emitted at 0 deg with respect to the incident-beam direction, since yields are higher and polarization effects are avoided. Exceptions to this rule include the use of back-angle beams near the threshold of the $Li^7(p,n)$ reaction in order to obtain the lowest possible energies, the use of beams at 90 deg from the $T(d,n)$ reaction with low-energy cascade generators in order to enhance the detection of the associated He^4 particle, and for polarization measurements when advantage is taken of the source-reaction polarization.

Figure 8f-10 shows the variation of the neutron energy with incident charged-particle energy at 0 deg for the four reactions. In addition, the energy of neutrons resulting from leaving Be^7 in its first excited state of 431 keV energy in the $Li^7(p,n)$ reaction is shown, as well as the 90 deg energy variation of the $T(d,n)$ reaction for low-

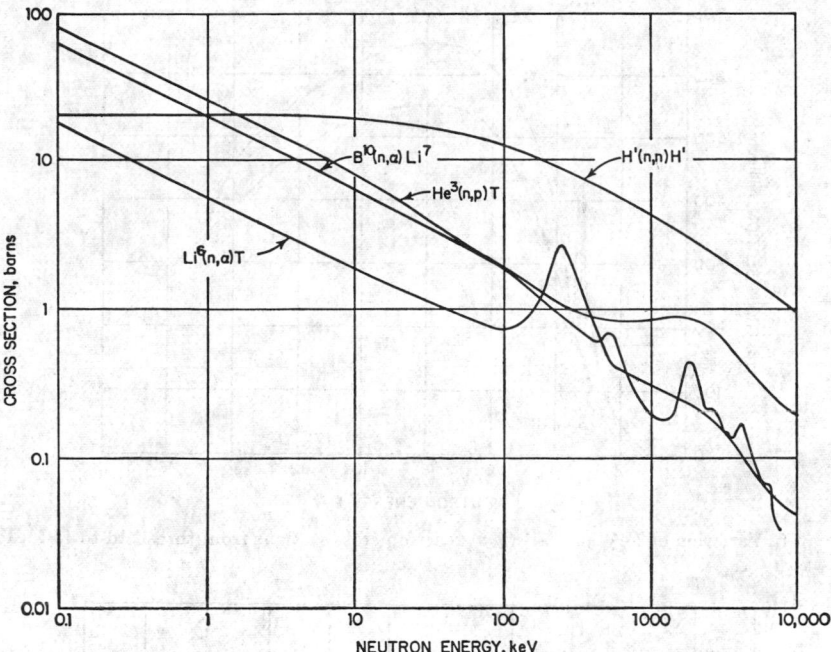

FIG. 8f-4. Variation of the cross section for $H^1(n,n)$, $He^3(n,p)$, $Li^6(n,t)$, and $B^{10}(n,\alpha)$ reactions for the neutron energy range 100 eV to 10 MeV.

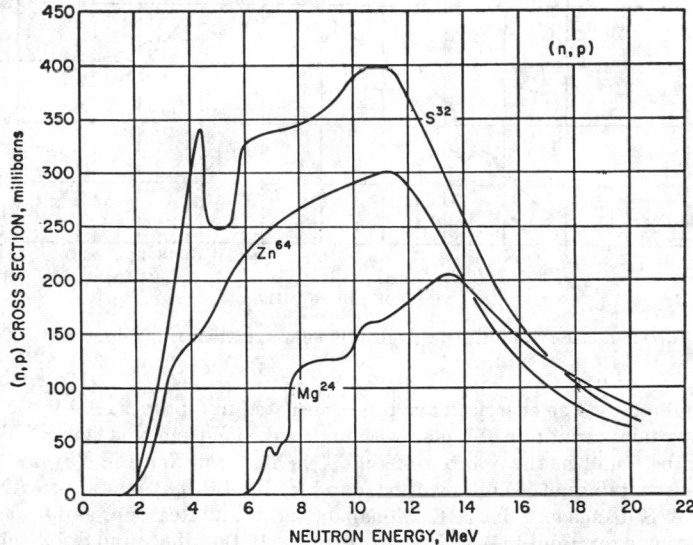

Fig. 8f-5. Variation of Mg^{24}, S^{32}, and $Zn^{64}(n,p)$ reaction cross sections from threshold to 20 MeV.

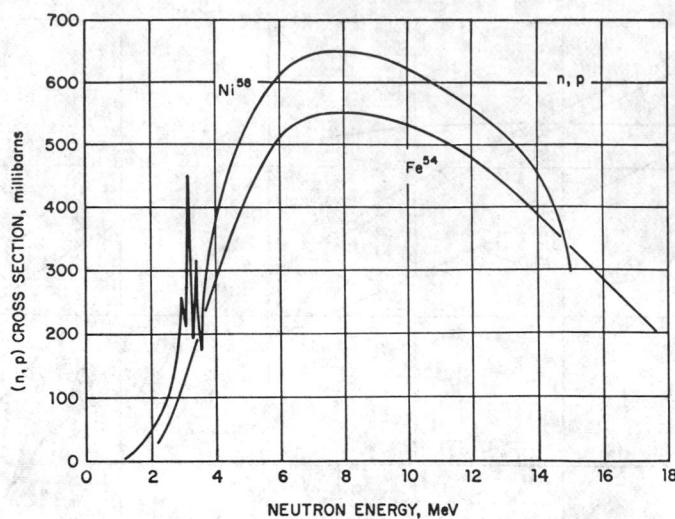

FIG. 8f-6. Variation of Fe^{54} and Ni^{58}(n,p) reaction cross sections from threshold to 15 MeV.

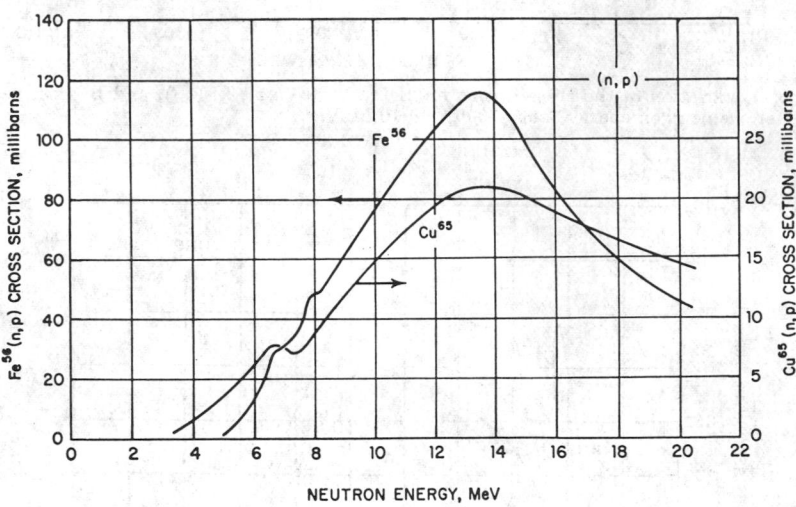

FIG. 8f-7. Variation of Fe^{56} and Cu^{65}(n,p) reaction cross sections from threshold to 20 MeV.

energy deuterons.   The energies have been calculated in a proper relativistic computation, but values cannot be obtained with high precision from this figure.   Tables of
values can be found in the appropriate chapters of "Fast Neutron Physics" part I.
More extensive tables of energies can be found for the Li^7(p,n) reaction in ANL-5219
(1954) by A. S. Langsdorf, Jr., J. E. Monahan, and W. A. Reardon; and for the three
hydrogen-source reactions in AECU-3118 (1956) by L. Blumberg and S. I. Schlesinger.

The Li^7(p,n) reaction has a Q value of -1.644 MeV with a corresponding threshold
energy of 1.881 MeV.   At a threshold energy of 2.378 MeV a second group of neutrons
becomes energetically possible, owing to reactions in which the product Be^7 nucleus is

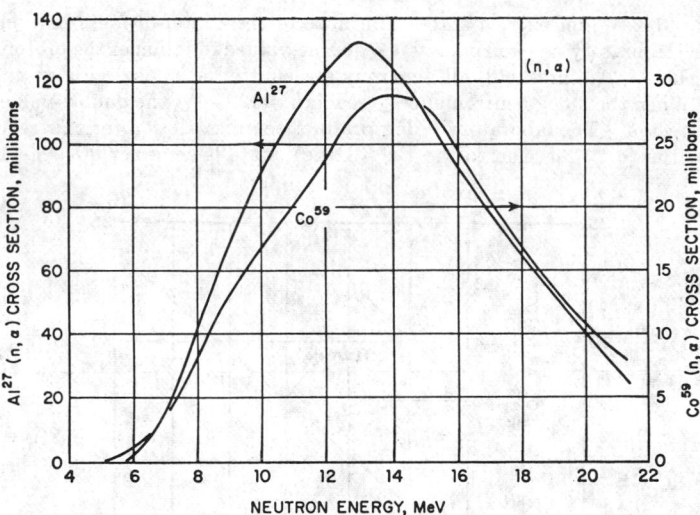

FIG. 8f-8. Variation of Al^{27} and $Co^{59}(n,\alpha)$ reaction cross sections from threshold to 21 MeV.

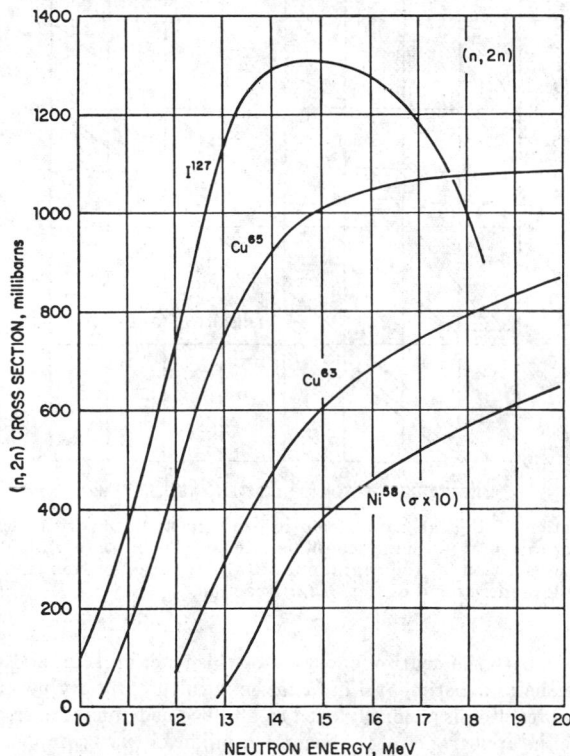

FIG. 8f-9. Variation of Ni^{58}, Cu^{63}, Cu^{65}, and $I^{127}(n,2n)$ reaction cross sections from threshold to 20 MeV. Note the multiplication of the true Ni^{58} cross section by 10 for plotting purposes.

left in its 431-keV first excited state. Because of the center-of-mass motion of the system, the laboratory neutron energy is a double-valued function of the proton energy near threshold. At threshold, all neutrons are emitted at 0 deg with an energy of 30 keV. When the 0 deg neutron energy reaches ≈ 120 keV, the double-valued character disappears. The laboratory 0 deg production cross section for this reaction as well as for the (p,n') reaction to the Be^7 431-keV state is shown in Fig. 8f-11. The

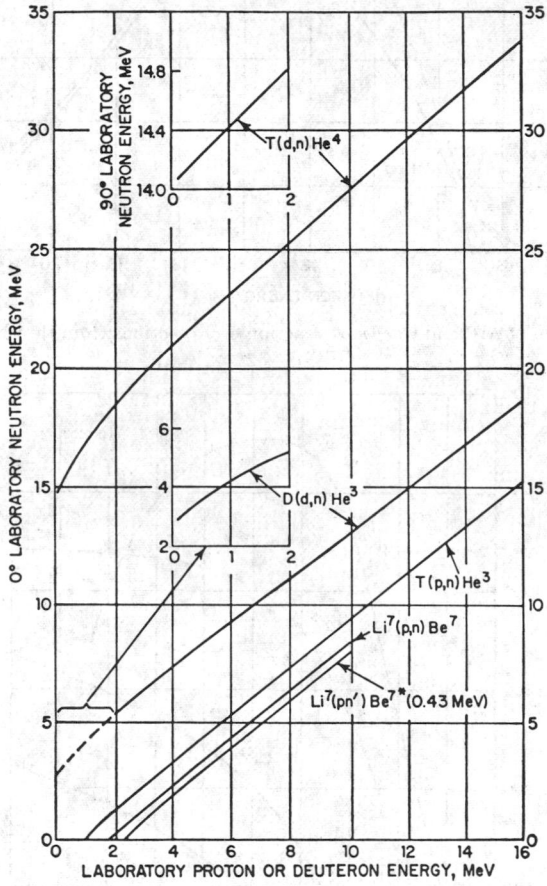

FIG. 8f-10. Variation of the laboratory neutron energy at 0 deg with incident charged-particle energy for the neutron source reactions $Li^7(p,n)$, $T(p,n)$, $D(d,n)$, and $T(d,n)$. Also shown is the same variation for the contaminant $Li^7(p,n')$ reaction and the neutron energy at 90 deg for the low-energy end of the $T(d,n)$ reaction.

latter curve, along with the neutron energies for this reaction (Fig. 8f-10), can be used to make corrections to experimental data taken with the primary neutron group.

The $T(p,n)He^3$ reaction is generally used to produce neutrons of energy greater than ≈ 700 keV, since the appearance of the second group in $Li^7(p,n)$ complicates use of that reaction to produce higher-energy neutrons. The Q value is -0.764 MeV, with a corresponding threshold energy of 64 keV. The laboratory 0 deg production cross section is shown in Fig. 8f-12. A possible contamination of the monoenergetic char-

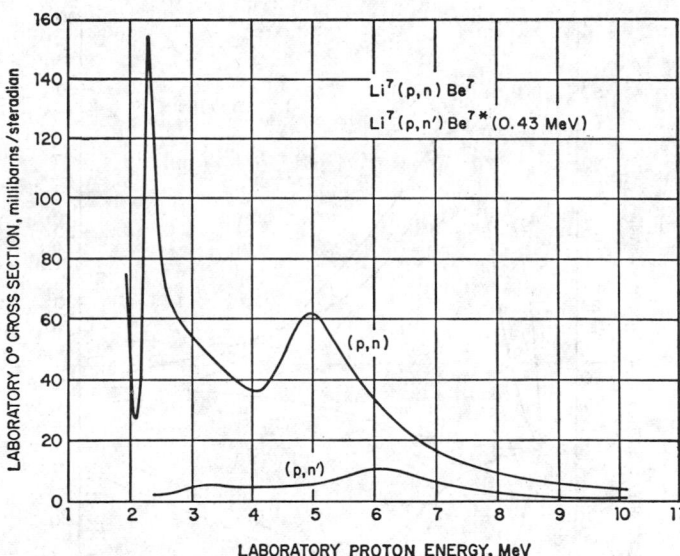

FIG. 8f-11. The laboratory 0-deg production cross section for the Li⁷(p,n) neutron source reaction and for the Li⁷(p,n') contaminant reaction.

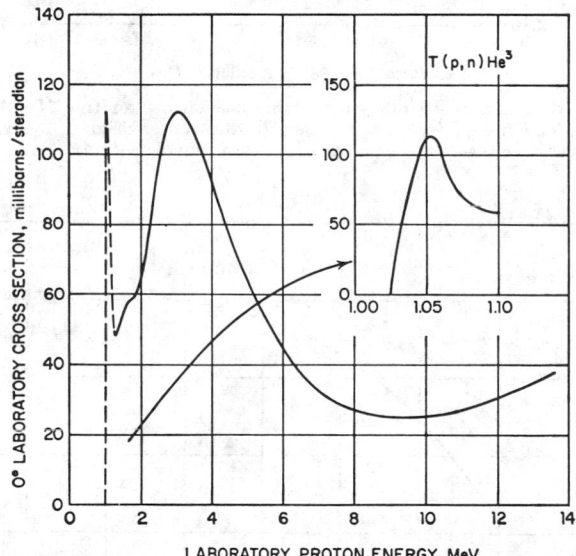

FIG. 8f-12. The laboratory 0-deg production cross section for the T(p,n) neutron source reaction.

acter of the beam would be the many-body breakup of the triton. The threshold for this reaction is at 8.34 MeV, but no breakup has been observed up to a proton energy of 13 MeV (0 deg breakup production cross section <5 millibarns/steradian). Measured differential cross section for the T(p,n) reaction can be transformed to the center-of-mass system and fitted by a sum of Legendre polynomials; i.e., $\sigma(\theta_{\mathrm{cm}}) =$

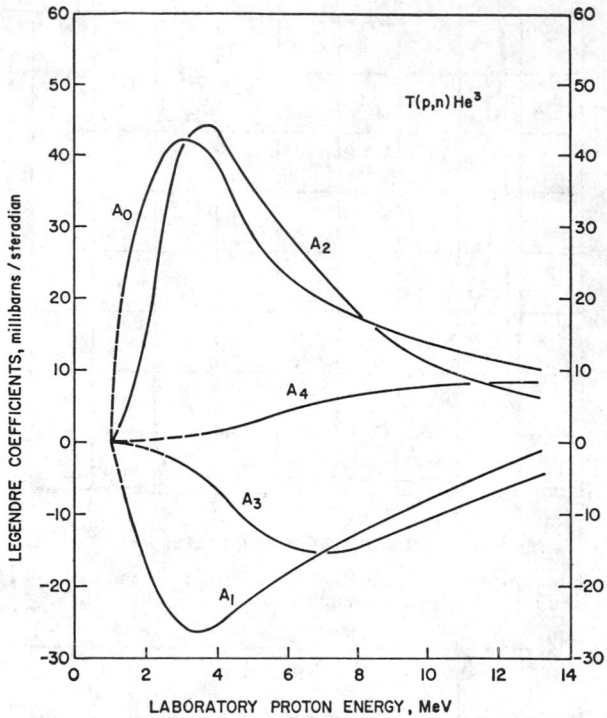

FIG. 8f-13. Variation of the Legendre-sum fitting coefficients for the $T(p,n)$ reaction with proton energy. The fitting is done for data in the center-of-mass system. The total integrated production cross section for the reaction is equal to $4\pi A_0{}^2$.

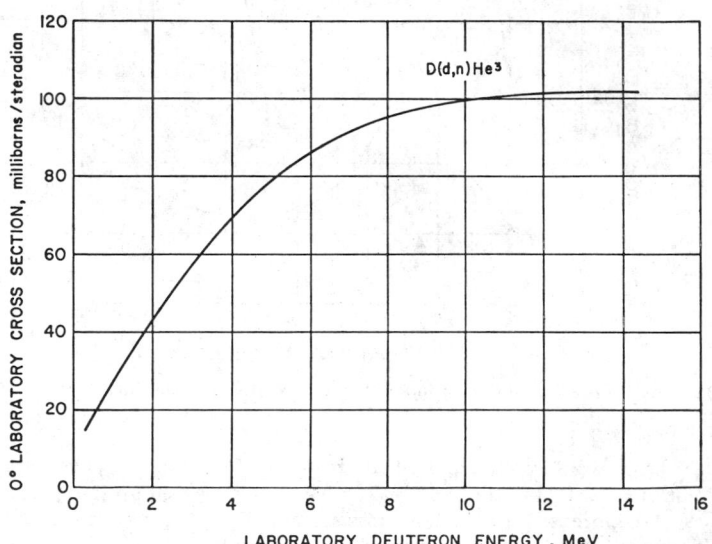

FIG. 8f-14. The laboratory 0-deg production cross section for the $D(d,n)$ neutron source reaction.

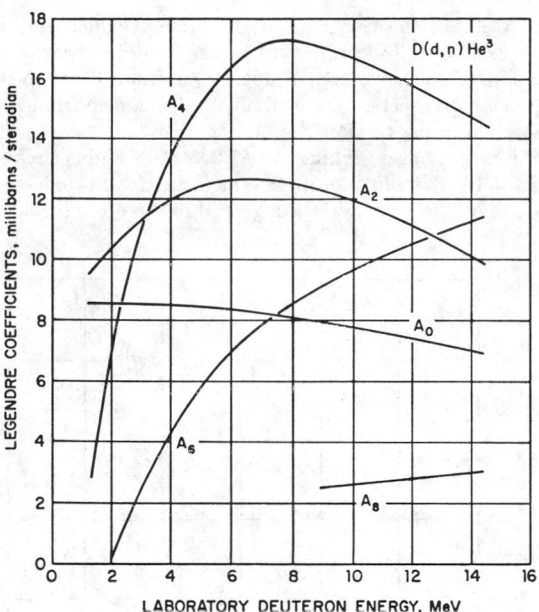

FIG. 8f-15. Variation of the Legendre-sum fitting coefficients for the D(d,n) reaction with deuteron energy. The fitting is done for data in the center-of-mass system. The total integrated production cross section for the reaction is equal to $4\pi A_0^2$.

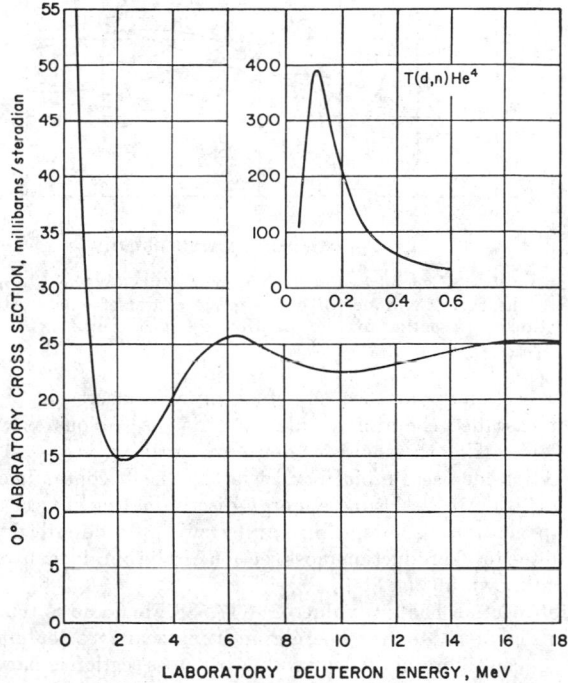

FIG. 8f-16. The laboratory 0-deg production cross section for the T(d,n) neutron source reaction.

$\Sigma A_n P_n (\cos \theta_{cm})$. A plot of the variation of the fitting coefficients with energy allows construction of angular distribution curves at any energy through the range shown. For this reaction, Fig. 8f-13 shows such a plot. One can also see the shape of the total integrated production cross section (as well as obtain values) from the variation of the A_0 coefficient, since this cross section is equal to $4\pi A_0^2$.

The $D(d,n)He^3$ reaction has a Q value of $+3.268$ MeV and is thus exoergic. As can be seen from Fig. 8f-10, the neutron energy at 0 deg for cascade-generator deuteron energies is ≈ 3 MeV. The laboratory 0 deg production cross section is shown in Fig.

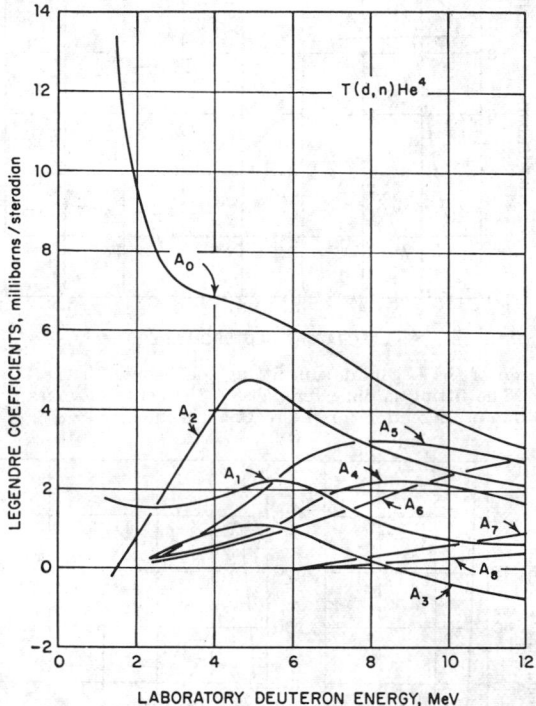

FIG. 8f-17. Variation of the Legendre-sum fitting coefficients for the $T(d,n)$ reaction with deuteron energy. The fitting is done for data in the center-of-mass system. The total integrated production cross section for the reaction is equal to $4\pi A_0^2$.

8f-14. The cross section approaches zero smoothly with decreasing deuteron energy, and so fluxes at cascade generator energies are not as copious as with the $T(d,n)$ reaction (see below). The threshold for deuteron breaking is at 4.45 MeV, and this reaction cross section increases rapidly and is a significant contaminant for deuteron energies above about 6 MeV. Usually energy-discriminating devices are adequate to separate breakup neutrons from the primary group. As described above, angular distributions and the total production cross section can be obtained, using the Legendre fitting coefficients shown in Fig. 8f-15.

The $T(d,n)He^4$ reaction has a Q value of $+17.588$ MeV and is thus exoergic. As can be seen from Fig. 8f-10, neutron energies at 0 deg near zero deuteron energy begin at near 15 MeV. The laboratory 0 deg production cross section is shown in Fig. 8f-16. The large production cross section at a deuteron energy of 110 keV, due to a resonance in He[5], makes this reaction a copious source of high-energy neutrons for deuteron energies available with cascade generators. As can be seen from the top insert in Fig. 8f-10, the energy varies less rapidly at 90 than at 0 deg, and many experiments have been done at this angle. The threshold for deuteron breakup is 3.71 MeV, but the large energy gap between the breakup and primary groups makes energy discrimination quite easy. As described above, angular distributions and the total production cross section can be obtained using the Legendre fitting coefficients shown in Fig. 8f-17.

8f-10. Cross-section Compilation Bibliography. The Sigma Center (now part of the National Neutron Cross Section Center) at Brookhaven National Laboratory has produced over the years compilations of measured neutron cross-section data and parametric information derived from these measurements. Below are listed still current volumes from this work.

1. Hughes, Donald J., and Robert B. Schwartz: "Neutron Cross Sections," BNL 325, 2d ed., July, 1958.
2. Hughes, D. J., B. A. Magurno, and M. K. Brussel: "Neutron Cross Sections," BNL 325, 2d ed., supplement 1, Jan. 1, 1960.
3. Stehn, John R., et al., "Neutron Cross Sections," BNL 325, 2d ed., supplement 2, vol. 1, [Z = 1–20], May, 1964.
4. Goldberg, Murrey D., et al., "Neutron Cross Sections," BNL 325, 2d ed., supplement 2, vol. IIA [Z = 21–40], February, 1966; *ibid.*, vol. IIB [Z = 41–60], May, 1966; *ibid.*, vol. IIC [Z = 61–87], August, 1966.
5. Stehn, John R., et al., "Neutron Cross Sections," BNL 325, 2d ed., supplement 2., vol. III [Z = 88–98], February, 1965.
6. Garber, Donald I., et al., "Angular Distributions in Neutron-induced Reactions," BNL 400, 3d ed., vol. I [Z = 1–20] and vol. II [Z = 21–94], 1970.

The world's experimental neutron cross-section data are stored at Brookhaven on magnetic tape and are available to those who request them.

8g. Nuclear Fission[1]

WALTER D. LOVELAND

Oregon State University

8g-1. The Probability of Fission. *Spontaneous Fission Half Lives.* Table 8g-1 lists the known half lives for decay by spontaneous fission from the ground states of various nuclei.

Fission Cross Sections. Tables 8g-2 and 8g-3 give the values of the fission cross section in barns for the thermal-neutron-induced and 14-MeV neutron-induced fission of various nuclei. Similarly, Figs. 8g-1 to 8g-3 show the energy variation of the fission cross section for proton-, alpha-particle-, and photon-induced fission, respectively. Moderate excitation-energy-induced fission may occur after the emission of 0, 1, 2, . . . neutrons; and thus the observed fission properties are a combination of the charac-

[1] Work supported in part by the U.S. Atomic Energy Commission.

teristics of fission of many different isotopes with different excitation energies. Figure 8g-4 shows the ratio of neutron width to fission width versus mass of the fissioning nucleus and is very useful in sorting out these situations involving "multiple-chance" fission.

8g-2. Fission Product Distributions. *Mass Distributions.* Table 8g-4 is the well-known "Katcoff table" of radiochemically measured fission yields for the thermal-

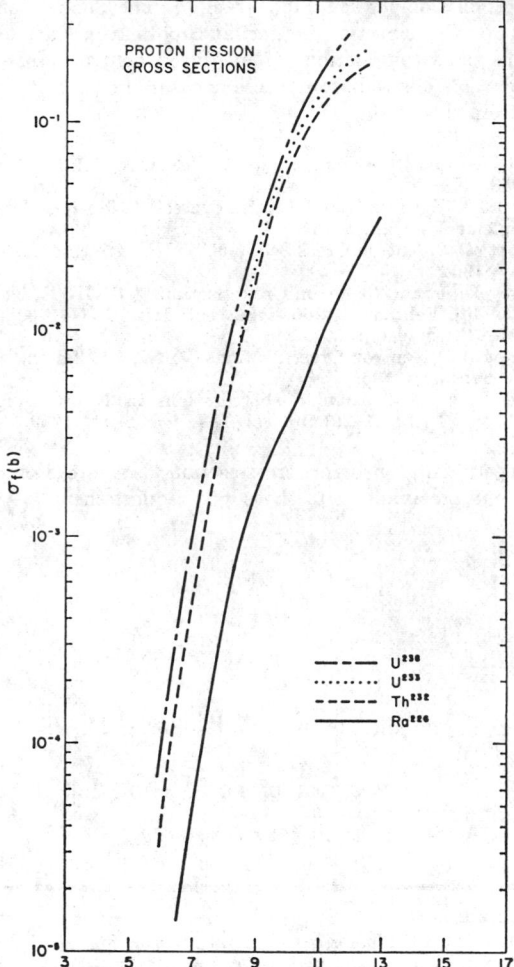

Fig. 8g-1. Energy variation of the fission cross section for proton-induced fission of various targets.

neutron-induced fission of U^{233}, U^{235}, and Pu^{239}. Figure 8g-5 summarizes the same information graphically. Figures 8g-6 and 8g-7 show similar mass-yield curves measured by physical techniques for a few representative cases of charged-particle-induced fission.

Charge Distributions. The most probably primary fragment charge Zp for fission fragments of mass A is shown in Fig. 8g-5 as a function of A for the thermal-neutron-

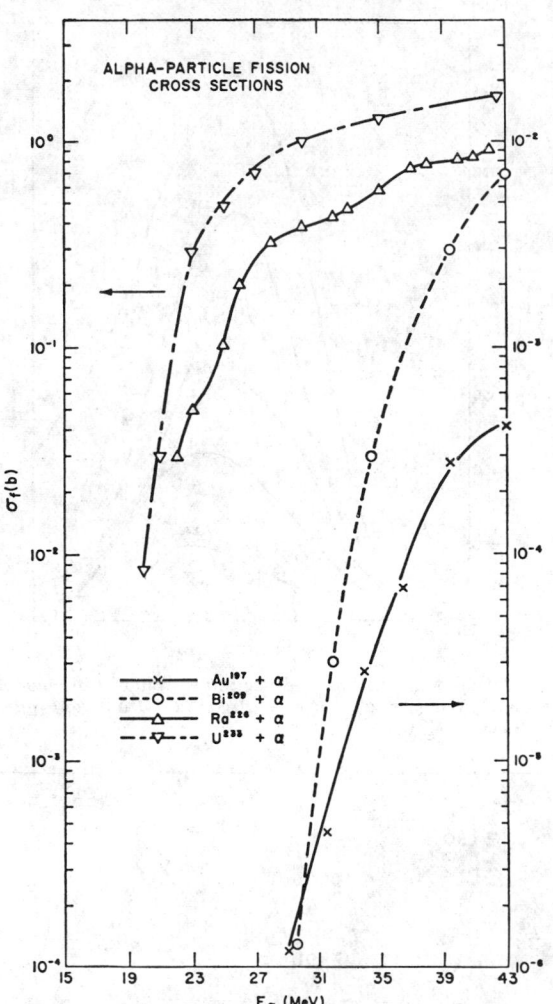

FIG. 8g-2. Energy variation of the fission cross section for α-particle-induced fission of various targets.

induced fission of U^{235}. The distribution of yields of fission products of charge Z is generally assumed to be gaussian for each fragment mass A and is given by

$$P(Z) = \frac{1}{\sqrt{c\pi}} \exp\left[-\frac{(Z - Zp)^2}{c}\right]$$

where c is an empirical constant. Wahl has found that a value of $c = 0.86$ fits a good deal of the data although there is no reason to expect c to have the same value for all mass numbers.

Kinetic Energy Release. Tables 8g-5 and 8g-6 show the average values of the fragment kinetic energies and masses prior to prompt neutron emission by the fragments for the thermal-neutron-induced fission of U^{235}, Pu^{239}, and Pu^{241} and the alpha-particle-

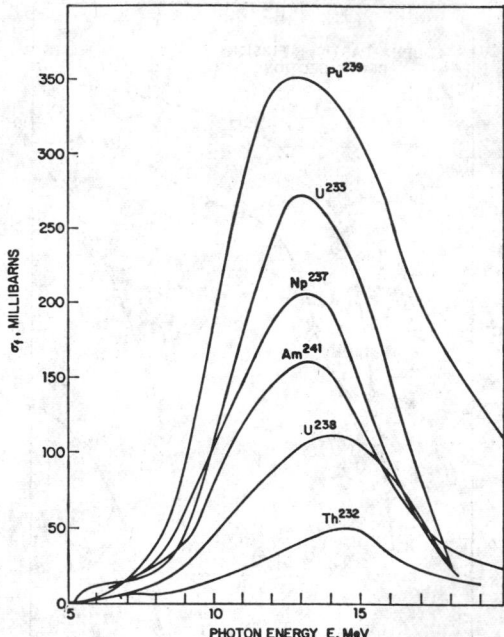

FIG. 8g-3. Energy variation of the fission cross section for photon-induced fission of various targets. [*L. Katz, A. P. Baerg, and F. Brown, Proc. 2d U. N. Conf. on Peaceful Uses of Atomic Energy* **15,** P/200 (1958).]

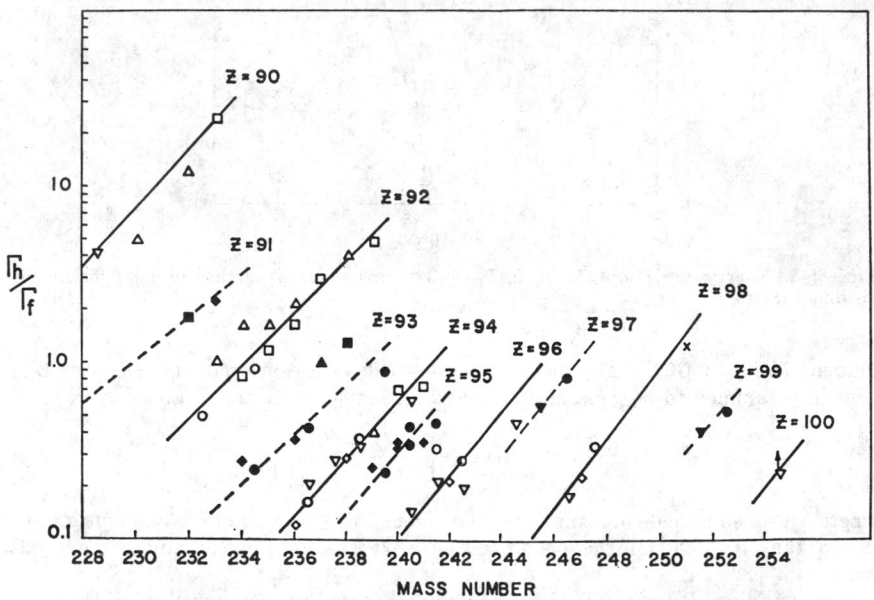

FIG. 8g-4. Neutron-width to fission-width ratios versus mass number of the fissioning nucleus. [*R. Vandenbosch and J. R. Huizenga, Proc. 2d U. N. Conf. on Peaceful Uses of Atomic Energy* **15,** 284 (1958).]

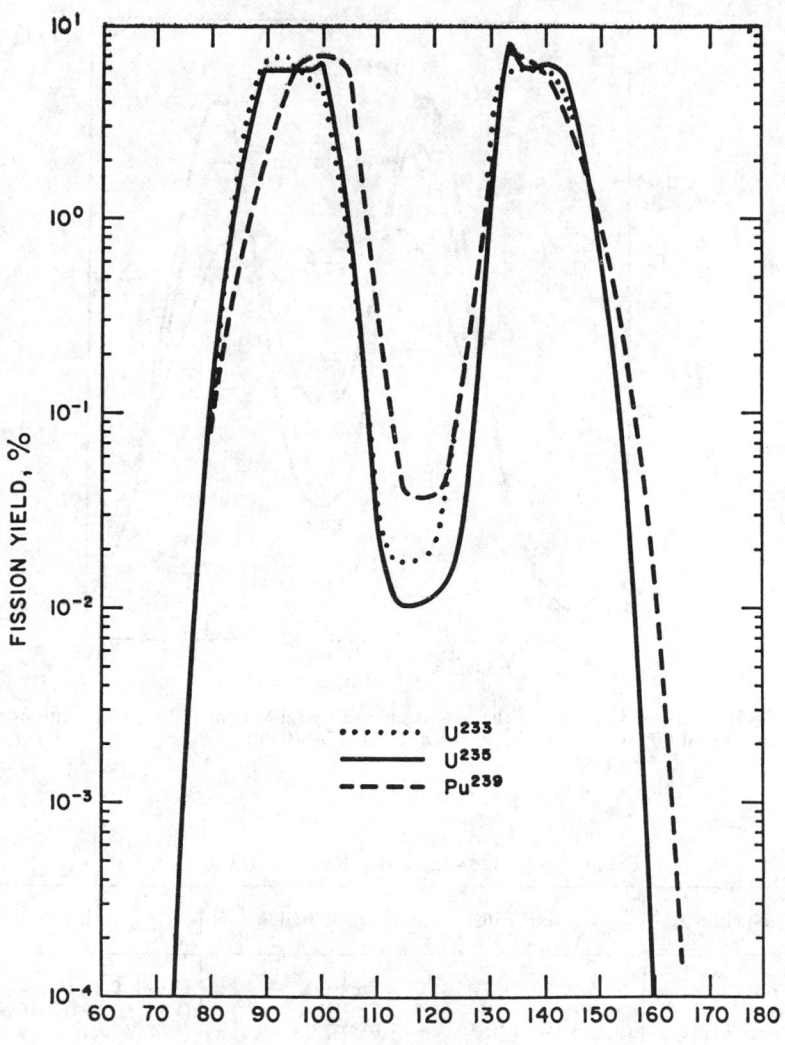

FIG. 8g-5. Fission-fragment mass distributions for the thermal-neutron-induced fission of U^{233}, U^{235}, and Pu^{239}.

induced fission of Th^{230}, Th^{232} and U^{233}, respectively. The variation in fragment kinetic energy with fragment mass is shown in Figs. 8g-9 to 8g-12 for these same fissioning nuclei.

Neutron Emission. Table 8g-7 shows the average number of prompt neutrons emitted per fission $\bar{\nu}$ for various nuclides. The distribution of neutron energies (as measured in the laboratory system) seems to be reasonably represented by a Maxwellian distribution of the form

$$N(E) = \left(\frac{2}{\pi^{\frac{1}{2}}T^{\frac{3}{2}}}\right) E^{\frac{1}{2}} e^{-E/T}$$

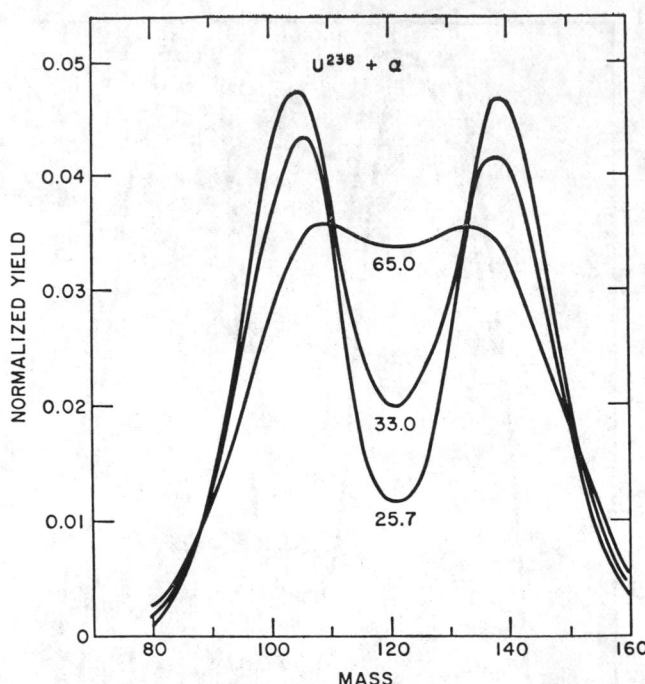

FIG. 8g-6. Mass-yield curves for the α-particle-induced fission of U^{238}. Yields are normalized to total of 200 percent. (*D. S. Burnett*, UCRL 11006.)

TABLE 8g-1. SPONTANEOUS FISSION HALF LIVES

Nuclide	Half Life	Nuclide	Half Life
Th^{230}	$\geq 1.5 \times 10^{17}$ y	Cm^{248}	$(4.6 \pm 0.5)(10^6)$ y
Th^{232}	$> 10^{21}$ y	Cm^{250}	$(1.13 \pm 0.05)(10^4)$ y
U^{232}	$(8 \pm 5.5)(10^{13})$ y	Bk^{249}	$(1.87 \pm 0.09)(10^9)$ y
U^{233}	$(1.2 \pm 0.3)(10^{17})$ y	Cf^{246}	(2.1 ± 0.3) y
U^{234}	1.6×10^{16} y	Cf^{248}	7×10^3 y
U^{235}	$(3.5 \pm 0.9)(10^{17})$ y	Cf^{249}	$(6.87 \pm 0.33)(10^{10})$ y
U^{236}	2×10^{16} y	Cf^{250}	$(1.66 \pm 0.08)(10^4)$ y
U^{238}	$(1.01 \pm 0.03)(10^{16})$ y	Cf^{252}	(8.55 ± 0.5) y
Np^{237}	$> 10^{18}$ y	Cf^{254}	(60.5 ± 0.2) d
Pu^{236}	3.5×10^9 y	Es^{253}	$(6.3 \pm 0.2)(10^5)$
Pu^{238}	$(5 \pm 0.6)(10^{10})$ y	Es^{254}	$> 2.5 \times 10^7$ y
Pu^{240}	$(1.340 \pm 0.015)(10^{11})$ y	Es^{255}	$2,440 \pm 140$ y
Pu^{242}	$(6.5 \pm 0.7)(10^{10})$ y	Fm^{252}	$> 3,000$ d
Pu^{244}	$(2.5 \pm 0.8)(10^{10})$ y	Fm^{254}	228 ± 1 d
Am^{241}	$(2.3 \pm 0.8)(10^{14})$ y	Fm^{255}	$(1.0 \pm 0.6)(10^4)$ y
Am^{243}	$(3.3 \pm 0.3)(10^{13})$ y	Fm^{256}	3 h
Cm^{240}	1.9×10^6 y	Fm^{257}	~ 100 y
Cm^{242}	7.2×10^6 y	No^{254}	~ 6 s
Cm^{244}	$(1.346 \pm 0.006)(10^7)$ y	No^{256}	8.2 ± 1.0 s
Cm^{246}	$(1.78 \pm 0.04)(10^7)$ y	Ku^{260}	(0.3 ± 0.1) s

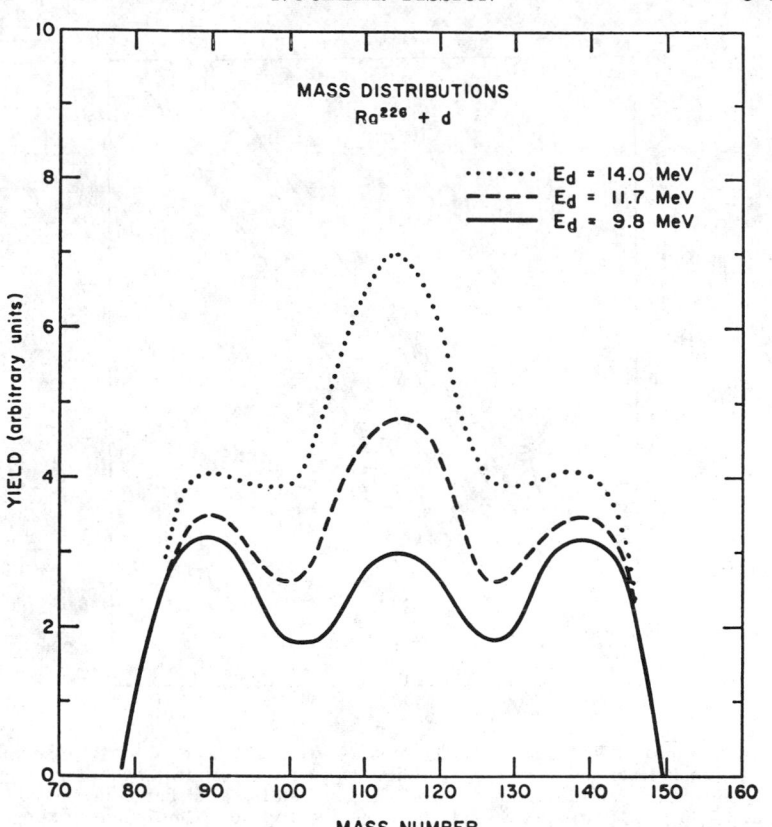

FIG. 8g-7. Mass distributions for the deuteron-induced fission of Ra^{226}. [*H. C. Britt, H. E. Wegner and J. Gursky, Phys. Rev.* **129**, 2239 (1963).]

TABLE 8g-2. THERMAL-NEUTRON-FISSION CROSS SECTIONS

Nuclide	$\sigma_f(b)$	Nuclide	$\sigma_f(b)$
Th^{229}	32 ± 3	Np^{234}	900 ± 300
Th^{230}	<0.001	$Np_2^{236}(5000\ y)$	2800 ± 800
Th^{232}	$(6 \pm 2)(10^{-5})$	Np^{237}	0.019 ± 0.003
Pa^{231}	0.010 ± 0.005	Pu^{238}	18.4 ± 0.9
U^{232}	72 ± 10	Pu^{239}	741 ± 4
U^{233}	524 ± 2	Pu^{240}	0.03 ± 0.045
U^{234}	<0.65	Pu^{241}	950 ± 30
U^{235}	577 ± 1	Am^{241}	3.13 ± 0.15
U^{238}	<0.5		

where T, the nuclear temperature, equals two-thirds the average energy, $\bar{E}$. Table 8g-8 gives some characteristics of fission neutron spectra while Table 8g-9 shows some *delayed* neutron yields from thermal-neutron-induced fission. The variation of the number of prompt neutrons emitted by a fragment of mass A with fragment mass is shown in Fig. 8g-13.

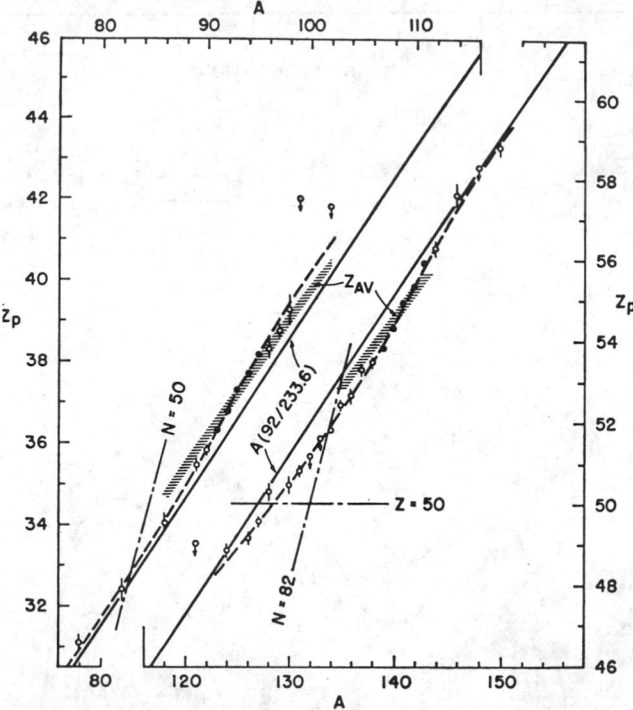

Fig. 8g-8. Empirical Zp values for fission products from U^{235} thermal-neutron fission. $\bullet Zp$ values obtained from gaussian charge distribution curves determined by two or more fractional yields. $\bigcirc Zp$ values estimated from the gaussian isobaric charge-distribution curve with $c = 0.86 \pm 0.15$ fitted to a single fractional yield value. Continuous lines represent the average charge density, $A(92/233.6)$. Broken lines represent an empirical Zp function derived from the points. [A. C. Wahl, "*Physics and Chemistry of Fission*," vol. I, 317 (1965).]

TABLE 8g-3. 14-MeV NEUTRON-FISSION CROSS SECTIONS

Nuclide	$\sigma_f(b)$
Bi^{209}	$(85 \pm 10)(10^{-9})$
Th^{230}	0.72 ± 0.15
Th^{232}	0.35 ± 0.03
U^{233}	2.25 ± 0.05
U^{235}	2.35 ± 0.100
U^{238}	1.23 ± 0.05
Np^{237}	2.5 ± 0.1
Pu^{239}	2.65 ± 0.10
Pu^{240}	2.4 ± 0.3
Pu^{241}	2.6 ± 0.1

Gamma-ray Emission and Beta Decay. The characteristics of the prompt gamma rays and beta particles emitted during the deexcitation of the fission products are given in Tables 8g-10 and 8g-11.

8g-3. Use of Semiconductor Radiation Detectors in Fission Studies. A great deal of the new and significant data in nuclear fission physics is due to the use of semi-

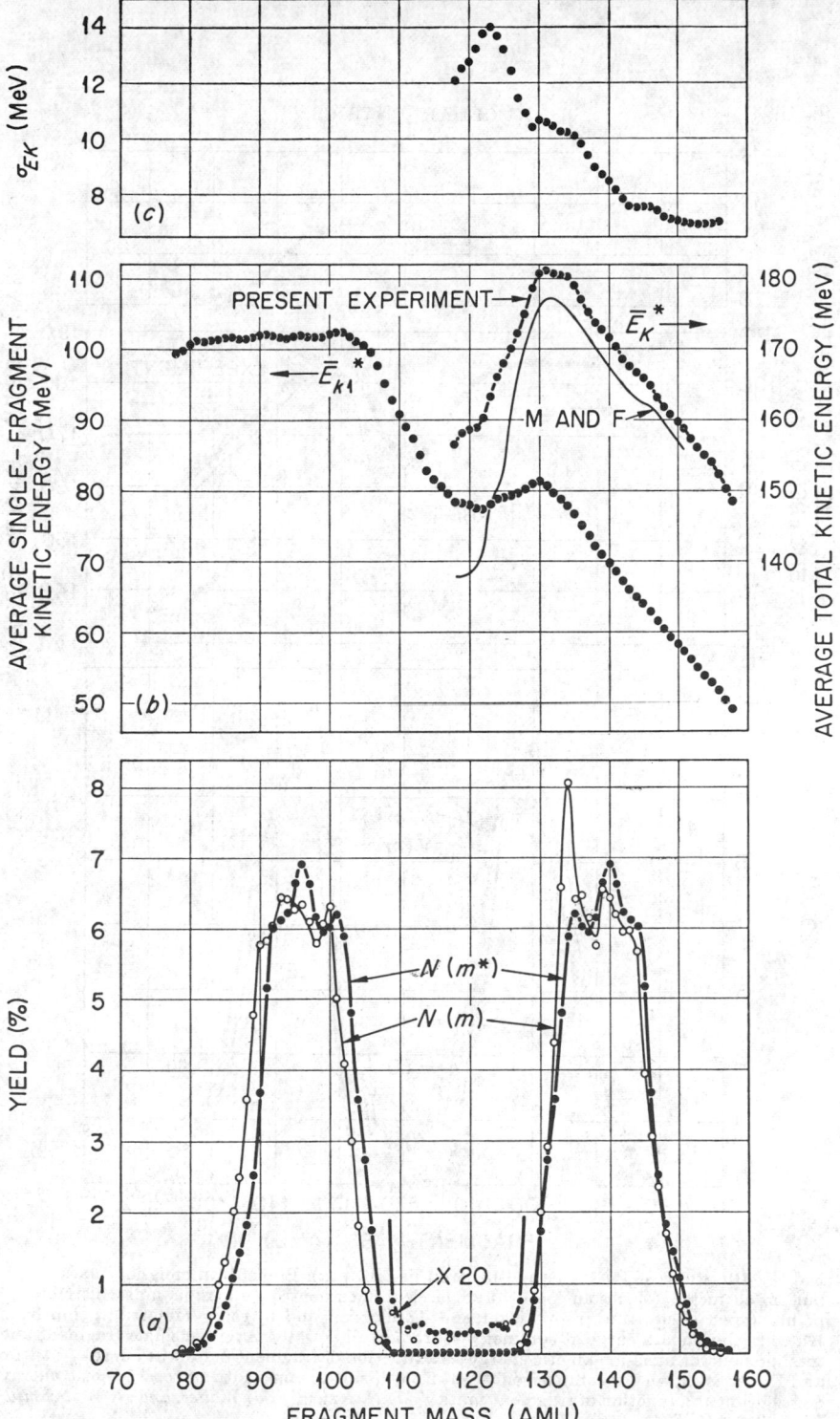

FIG. 8g-9. Study of fragment kinetic energy and mass for U^{239} thermal-neutron-induced fission. (a) Preneutron emission $N(m*)$ and postneutron emission $N(m)$ mass distributions. (b) Average single-fragment and total preneutron emission kinetic energy as a function of mass. The total kinetic energy curve of Milton and Fraser is shown for comparison. (c) Root-mean-square width of total kinetic energy distribution as a function of fragment mass. [*H. W. Schmitt, J. H. Neiler, and F. J. Walter, Phys. Rev.* **141**, 1140 (1966).]

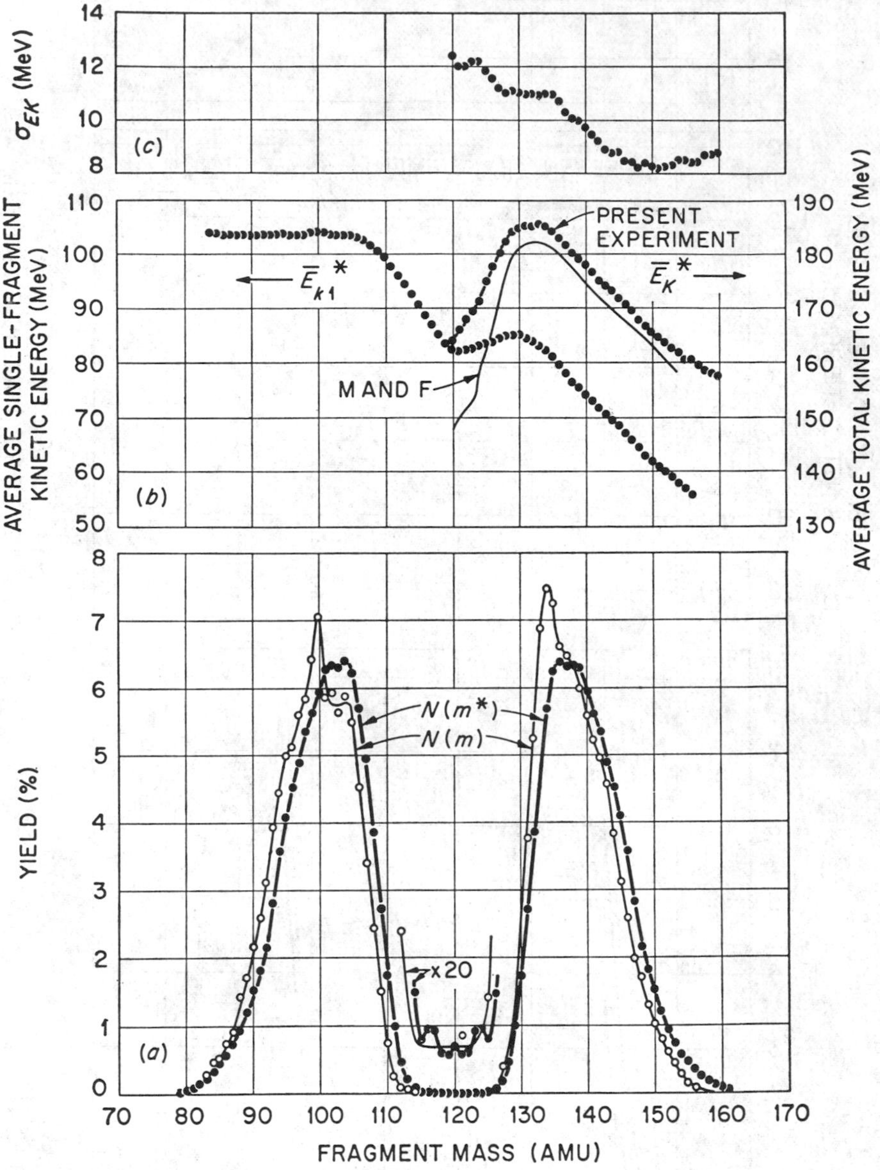

FIG. 8g-10. Study of Pu²³⁹ thermal-neutron fission. (a) Preneutron emission mass distribution corrected for resolution (closed circles); postneutron emission mass distribution points (open circles) are from Fickel and Tomkinson and in the symmetric region from Katcoff; the smooth curve at symmetry is from Walker. (b) Average single-fragment and total preneutron emission kinetic energy as a function of fragment mass; the curve of Milton and Fraser is shown for comparison. (c) Root-mean-square width of total kinetic energy distribution as a function of fragment mass. [*J. H. Neiler, F. J. Walter, and H. W. Schmitt, Phys. Rev.* **149**, 894 (1966).]

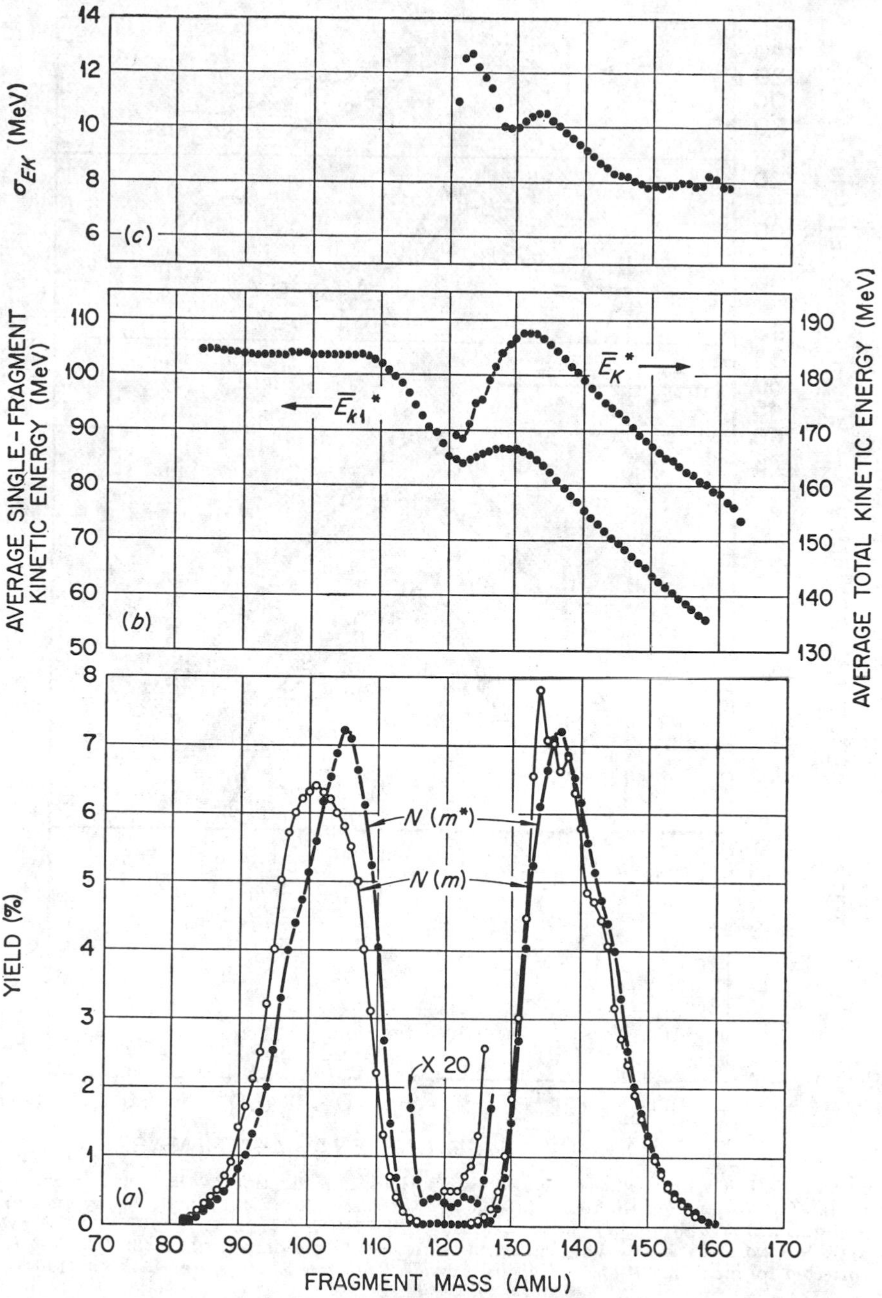

FIG. 8g-11. Study of Pu²⁴¹ thermal-neutron fission. (a) Preneutron emission mass distribution corrected for resolution (closed circles); the postneutron emission mass yields shown (open circles) are from Farrar *et al*. (b) Average single fragment and total preneutron emission kinetic energy as a function of fragment mass. (c) Root-mean-square width of total kinetic energy distribution as a function of fragment mass. [*J. H. Neiler, F. J. Walter, and H. W. Schmitt, Phys. Rev.* **149**, 894 (1966).]

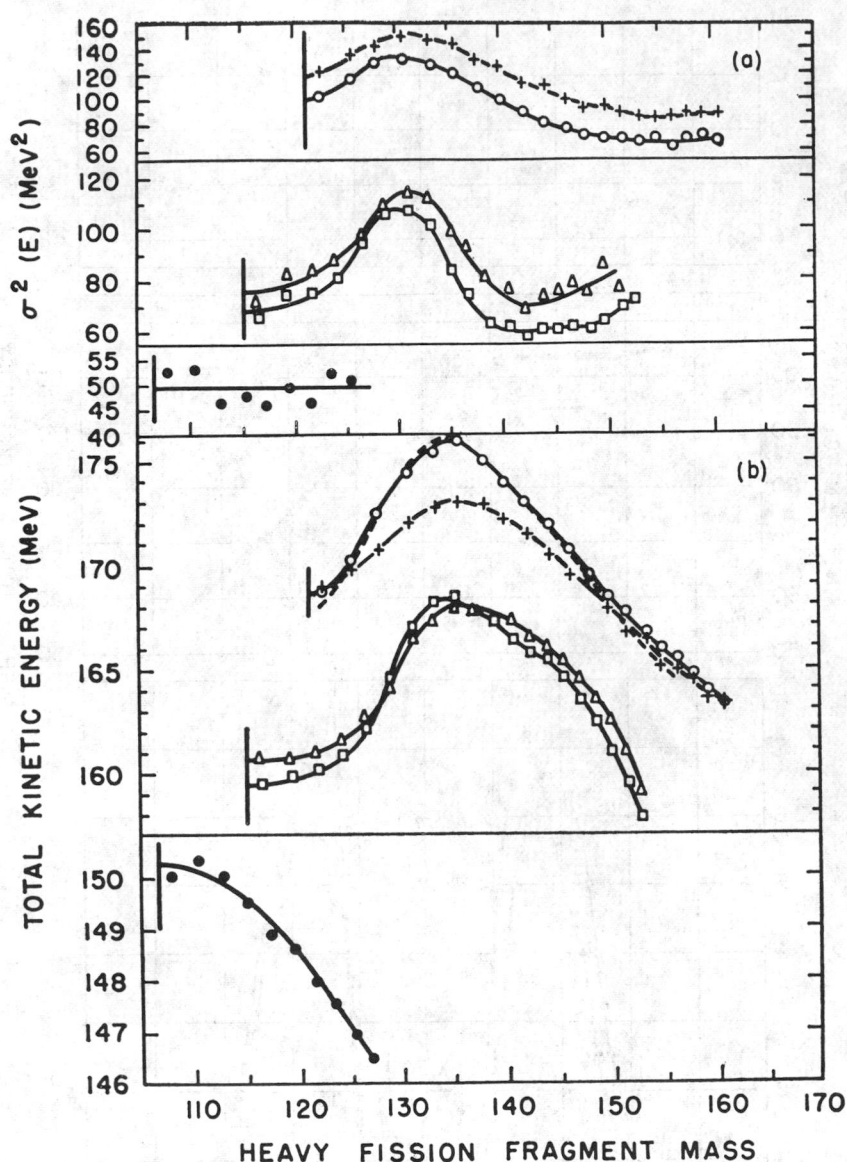

FIG. 8g-12. Initial total kinetic energy distributions as a function of the heavy fragment mass. (a) Variance of the total kinetic energy release. (b) Total kinetic energy release: ●Bi²⁰⁹ (42 MeV α,f); □Ra²²⁶ (30.8 MeV α,f); △Ra²²⁶ (38.7 MeV α,f); ○U²³⁸ (29.7 MeV α,f); + (42.0 MeV α,f). The dashed curve represents the data for U²³⁸ (29.7 MeV α,f) corrected for mass resolution. [*J. P. Unik and J. R. Huizenga, Phys. Rev.* **134,** B90 (1964).]

conductor radiation detectors rather than radiochemical techniques for the measurement of fission-fragment energies, masses, etc. Of particular importance in this regard has been the work of H. W. Schmitt and his coworkers[1] in formulating a mass-dependent energy calibration for these detectors (which corrects for the incomplete collection of the charge deposited by a heavy ion in the detector) and standards for the selection of good-quality detectors.

What one does to calibrate one's detectors in a given situation is to measure the fission-fragment pulse-height spectrum for a thin Cf^{252} or U^{235} source with the detectors. The fragment pulse-height spectrum is then used to define two points P_L and P_H, the midpoint of the light and heavy fragment peak, respectively, at three-fourths maximum. Then

$$E = (a + a'm)X + b + b'm$$

where E is the fragment kinetic energy, m is the mass, X is the pulse height, and a, a', b, b' are constants. The values of the constants are shown in Table 8g-12.

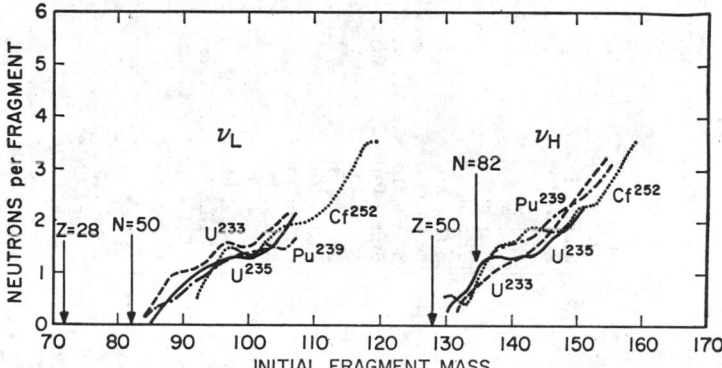

FIG. 8g-13. Neutron yields as a function of fragment mass derived from mass yield data. Also shown are the approximate initial fragment masses corresponding to various magic numbers based upon an unchanged charge-to-mass ratio (UCD) for the initial fragments. [J. Terrell, Phys. Rev. **127**, 880 (1962).]

Similarly, by measuring this pulse-height spectrum, one can define reasonable minimum characteristics for these heavy-ion detectors. Figure 8g-14 shows a typical Cf^{252} pulse-height spectrum with various shape parameters of the spectrum defined. Reasonable limits on these parameters are given in Table 8g-13. To get a feel for the importance of these parameters, note that a detector with $N_L/N_H = 1.25$ and $N_L/N_V = 2.73$ gave a factor of ~ 3 worse detector resolution than an acceptable detector.

8g-4. Cf^{252} Spontaneous Fission. Frequently, the measurement of fission-fragment energies, velocities, etc., is made relative to a primary standard, Cf^{252} spontaneous fission. This section presents the "best values" for the properties of Cf^{252} spontaneous fission as of June, 1968.

Fragment Kinetic Energies and Masses. Table 8g-14 summarizes the data of Whetstone concerning average fragment energies and masses. All quantities refer to preprompt neutron emission. Figure 8g-15 shows the variation in fragment kinetic energy with fragment mass.

Charge Distribution. The data on the most probable primary fragment charge Z_p as measured by K X-ray-fission coincidence measurements is given in Fig. 8g-16.

[1] *Phys. Rev.* **137**, B837 (1966); and *Nucl. Instr. Methods* **40**, 204 (1966).

TABLE 8g-4. THERMAL-NEUTRON-FISSION YIELDS (PERCENT) FROM U^{233}, U^{235}, AND Pu^{239}

Fission product	U^{233}	U^{235}	Pu^{239}	Fission product	U^{233}	U^{235}	Pu^{239}
47-hr Zn^{72}	.....	1.6×10^{-5}	1.2×10^{-4}	10.3-hr Y^{93}	.....	6.1	3.97
4.9-hr Ga^{73}	.....	1.1×10^{-4}		1.1×10^{6}-yr Zr^{93}	6.98	6.45	4.48
7.8-min Ga^{74}	.....	3.5×10^{-4}		Stable Zr^{94}	6.68	6.40	5.8
11.3-hr Ge^{77}	0.011	0.0031		65-day Zr^{95}	6.1	6.2	5.03
38.7-hr As^{77}	0.021	0.0083		Stable Mo^{95}	6.11	6.27	5.17
2.1-hr Ge^{78}	.....	0.020		Stable Zr^{96}	5.58	6.33	3.6×10^{-3}
91-min As^{78}	.....	0.020		23-hr Nb^{96}	6.5×10^{-3}	6.1×10^{-3}	5.5
9.0-min As^{79}	.....	0.056		17.0-hr Zr^{97}	.....	5.9	5.65
Total Br^{80}	3.9×10^{-4}	1.0×10^{-5}		Stable Mo^{97}	5.37	6.09	0.20
57-min Se^{81m}	.....	0.0084		52-min Nb^{98}	0.20	0.064	5.89
18.4-min Se^{81}	1.1×10^{-3}	0.14		Stable Mo^{98}	5.15	5.78	6.10
35.9-hr Br^{82}	.....	4×10^{-5}		66.5-hr Mo^{99}	4.80	6.06	7.10
25-min Se^{83}	.....	0.22	0.084	Stable Mo^{100}	4.41	6.30	5.91
2.4-hr Br^{83}	0.87	0.51		Stable Ru^{101}	2.91	5.0	5.99
Stable Kr^{83}	1.17	0.544	0.29	Stable Ru^{102}	2.22	4.1	5.67
6.0-min Br^{84}	.....	0.019		39.7-day Ru^{103}	1.8	3.0	5.93
31.8-min Br^{84}	.....	0.92		Stable Ru^{104}	0.94	1.8	
Stable Kr^{84}	1.95	1.00	0.47	4.45-hr Ru^{105}	.....	0.9	
39-sec Se^{85}	.....	~1.1		36-hr Rh^{105}	.....	.....	3.9
10.6-yr Kr^{85}	0.58	0.293	0.127	1.01-yr Ru^{106}	0.24	0.38	4.57
Stable Rb^{85}	2.51	1.30	0.539	22-min Rh^{107}	.....	0.19	
Stable Kr^{86}	3.27	2.02	0.76	13.4-hr Pd^{109}	0.044	0.030	1.40
18.6-day Rb^{86}	2.3×10^{-4}	2.9×10^{-5}	2.3×10^{-5}	7.6-day Ag^{111}	0.024	0.019	0.23
16-sec $Se^{(87)}$	.....	~2		21.0-hr Pd^{112}	0.016	0.010	0.12
5×10^{10}-yr Rb^{87}	4.56	2.49	0.92	43-day Cd^{115m}	0.0011	0.0007	0.0031
Stable Sr^{88}	5.37	3.57	1.42	53-hr Cd^{115}	0.020	0.0097	0.0038
50.5-day Sr^{89}	5.86	4.79	1.71	Total 115	0.021	0.0104	0.041
28-yr Sr^{90}	6.43	5.77	2.25	3.0-hr Cd^{117m}	.....	0.011	
9.7-hr Sr^{91}	5.57	5.81	2.43	27.5-hr Sn^{121}	0.018	0.015	0.043
58-day Y^{91}	5.1	~5.4	2.9	136-day Sn^{123}	.....	0.0013	
2.7-hr Sr^{92}	6.43	5.84	2.61	9.6-day Sn^{125}	0.052	0.013	0.071
Stable Zr^{92}	6.64	6.03	3.14	2.0-yr Sb^{125}	.....	0.021	
				91-hr Sb^{127}	0.60	0.13	0.39

TABLE 8g-4. THERMAL-NEUTRON-FISSION YIELDS (PERCENT) FROM U^{233}, U^{235}, AND Pu^{239} (Continued)

Fission product	U^{233}	U^{235}	Pu^{239}
105-day Te^{127m}	...	0.035	...
57-min Sn^{128}	...	0.37	...
25.0-min I^{128}	...	3×10^{-5}	...
37-day Te^{129m}	...	0.35	...
1.7×10^{7}-yr I^{129}	...	0.8	...
2.6-min Sn^{130}	...	2.0	...
12.6 hr I^{130}	...	5×10^{-4}	...
30-hr Te^{131m}	2.9	0.44	...
8.05-day I^{131}	3.39	~3.1	3.77
Stable Xe^{131}	4.4	2.93	3.78
77-hr Te^{132}	4.64	~4.7	5.1
Stable Xe^{132}	...	4.38	5.26
20.8-hr I^{133}	...	~6.9	5.2
5.27-day Xe^{133}	5.78	6.62	6.91
Stable Cs^{133}	...	6.59	6.91
52.5-min I^{134}	5.95	7.8	7.47
Stable Xe^{134}	5.5	8.06	5.7
6.7-hr I^{135}	...	6.1	7.17
9.2-hr Xe^{135}	...	6.3	2.1
2.6×10^{6}-hr Cs^{135}	6.03	6.41	6.63
86-sec I^{136}	1.8	3.1	0.11
Stable Xe^{136}	6.63	6.46	6.63
13-day Cs^{136}	0.12	0.0068	6.31
30-yr Cs^{137}	6.58	6.15	5.87
Stable Ba^{138}	...	5.74	5.4
83-min Ba^{139}	6.45	6.55	5.60
12.8-day Ba^{140}	5.4	6.35	5.7
Stable Ce^{140}	6.47	6.44	...
3.8-hr La^{141}	7.1	6.4	...

Fission product	U^{233}	U^{235}	Pu^{239}
33-day Ce^{141}	...	~6.0	5.1
Stable Pr^{141}	6.4	...	(4.5)*
Stable Ce^{142}	6.83	6.01	5.01
33-hr Ce^{143}	...	5.7	5.3
Stable Nd^{143}	5.99	6.03	4.57
280-day Ce^{144}	4.5	~6.0	3.79
5×10^{15}-yr Nd^{144}	4.61	5.62	3.93
Stable Nd^{145}	3.47	3.98	3.13
Stable Nd^{146}	2.63	3.07	2.60
11.1-day Nd^{147}	1.9	~2.7	2.2
2.6-yr Pm^{147}	1.98	...	1.94
1.3×10^{11}-yr Sm^{147}	...	2.36	2.07
Stable Nd^{148}	1.34	1.71	1.73
53.1-hr Pm^{149}	...	...	1.4
Stable Sm^{149}	0.76	1.13	1.32
Stable Nd^{150}	0.56	0.67	1.01
80-yr Sm^{151}	0.335	0.44	0.80
Stable Sm^{152}	0.220	0.281	0.62
47-hr Sm^{153}	0.11	0.15	0.37
Stable Eu^{153}	0.13	0.169	...
Stable Sm^{154}	0.045	0.077	0.29
24-min Sm^{155}	...	0.033	0.23
4-yr Eu^{155}	...	0.033	...
15.4-day Eu^{156}	0.011	0.014	0.11
15.4-hr Eu^{157}	...	0.0078	...
60-min Eu^{158}	...	0.002	...
18.0-hr Gd^{159}	...	0.00107	0.021
6.9-day Tb^{161}	...	7.6×10^{-5}	0.0039
82-hr Dy^{166}	...	...	6.8×10^{-5}

Reprinted from S. Katcoff, Nucleonics 18(11), 203, Copyright, 1960, McGraw-Hill Publishing Company, New York.

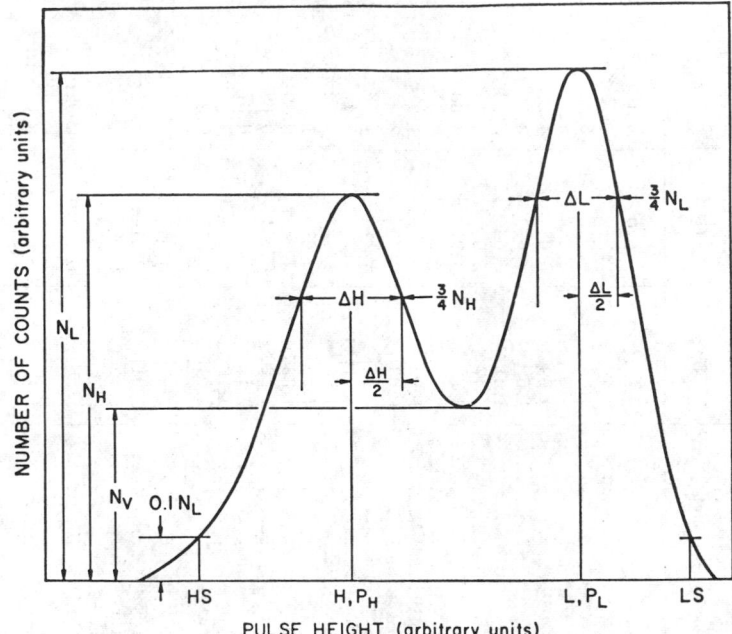

FIG. 8g-14. Shape parameters for Cf^{252} fission-fragment pulse-height distribution.

TABLE 8g-5. AVERAGE FRAGMENT ENERGIES AND MASSES FOR
THERMAL-NEUTRON-INDUCED FISSION

Target nucleus	U^{235}	Pu^{239}	Pu^{241}
Total kinetic energy E_K.....................	171.9 ± 1.4	177.7 ± 1.8	179.6 ± 1.8
σ_{E_K}..	10.9	11.09	11.46
Kinetic energy, light fragment E_L............	101.56	103.2 ± 1.0	103.2 ± 1.0
Kinetic energy, heavy fragment E_H..........	70.34	74.5 ± 0.8	76.3 ± 0.8
Mass, light fragment M_L..................	96.57	100.34	102.58
Mass, heavy fragment M_H..................	139.43	139.66	139.42
$\sigma_{ML} = \sigma_{MH}$...............................	5.36	6.01	5.71

All quantities above are average quantities prior to prompt neutron emission by the fragments.
Values are those of H. W. Schmitt, J. H. Neiler, and F. J. Walter, *Phys. Rev.* **141**, 1146 (1966); and
J. H. Neiler, F. J. Walter, and H. W. Schmitt, *Phys. Rev.* **149**, 894 (1966).

Neutron Distribution. The average number of prompt neutrons emitted in the
spontaneous fission of Cf^{252} is 3.771 ± 0.030. The properties of the neutron distribution in angle, energy, and number are shown in Figs. 8g-17 to 8g-20.

Gamma-ray Distribution. The gamma-ray yield as a function of fragment mass is
shown in Fig. 8g-21. The average number of photons per fission is 10.3, and the
average photon energy released per fission is 8.2 MeV.

Charged-particle Yields. The yield of charged particles emitted in Cf^{252} spontaneous
fission is given in Table 8g-15.

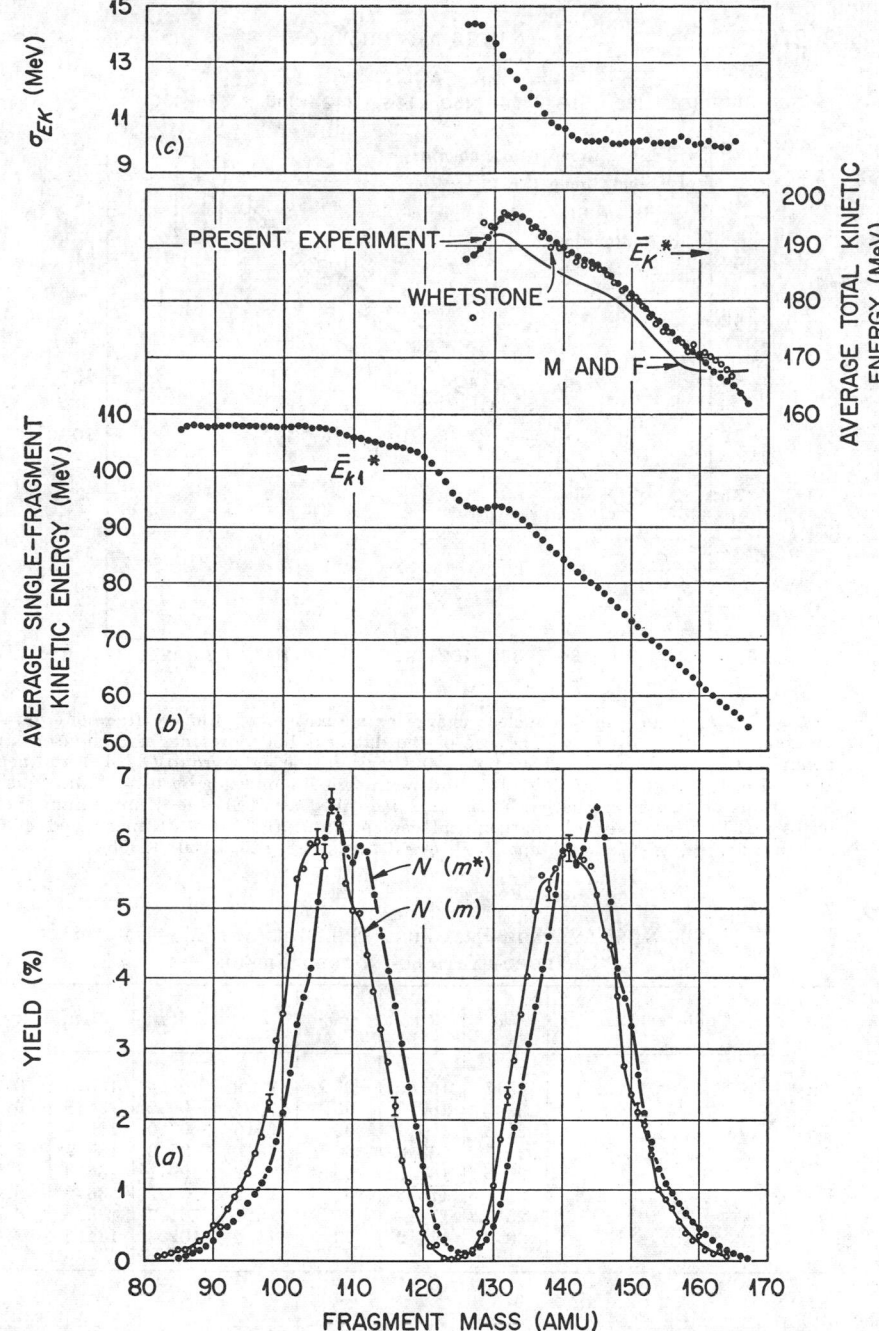

FIG. 8g-15. Study of Cf^{252} spontaneous fission. (a) Preneutron emission mass distribution $N(m*)$ corrected for mass resolution; the postneutron-emission mass distribution $N(m)$ is from Schmitt (1965). (b) Average single-fragment and total preneutron emission kinetic energy as a function of mass; the total kinetic energy curves of Whetstone and Milton and Fraser are shown for comparison. (c) Root-mean-square width of total kinetic energy distribution as a function of fragment mass. [H. W. Schmitt, J. H. Neiler, and F. J. Walter, Phys. Rev. **141**, 1146 (1966).]

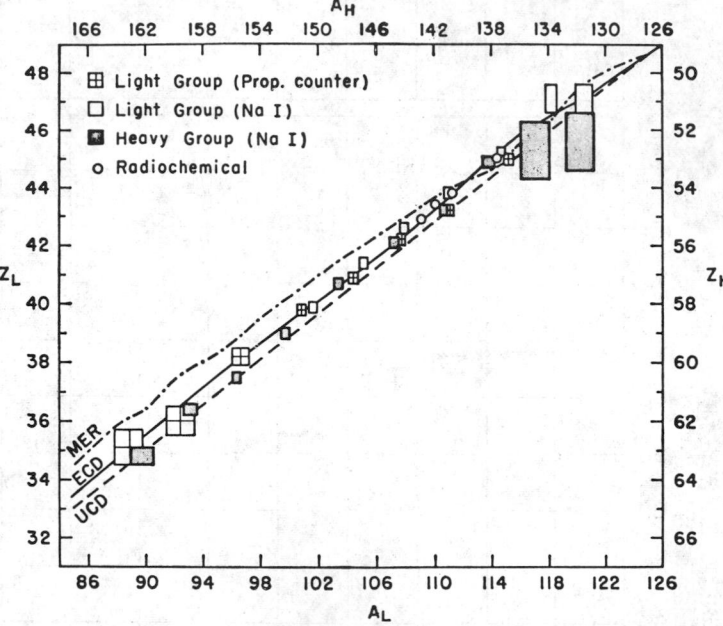

FIG. 8g-16. Average primary nuclear charge as a function of primary fragment mass in spontaneous fission of Cf^{252}. The size of the data symbol represents estimated errors in determinations of Z and A. The charge and mass of the light group (Z_L, A_L) and heavy group (Z_H, A_H) fragments are folded around symmetric fission ($Z = 49$, $A = 126$). Curves for various postulates of charge division are identified as MER ($\cdot\!\!-\!\!\cdot$), maximum energy release; ECD (———), equal charge displacement; and UCD (- - -), unchanged charge distribution. [*L. E. Glendenin and J. P. Unik, Phys. Rev.* **140**, B1301 (1965).]

TABLE 8g-6. AVERAGE FRAGMENT ENERGIES AND MASSES FOR
CHARGED-PARTICLE-INDUCED FISSION

Target and reaction	E_α	E_K	E_L	E_H	M_L	M_H	σ_M
$Th^{230} + \alpha$...............	25.7	167.5	97.0	70.5	98.4	135.6	8.7
	29.7	166.0	95.6	70.4	99.2	134.8	9.1
$Th^{232} + \alpha$...............	21.8	169.1	99.5	69.6	97.0	139.0	7.9
	25.7	168.2	98.1	70.1	98.3	137.7	8.2
	29.5	167.0	96.9	70.2	99.1	136.9	8.8
$U^{233} + \alpha$...............	21.8	176.3	101.7	74.6	100.2	136.8	8.4
	25.7	174.9	99.8	75.1	101.7	135.3	8.8
	29.7	174.2	98.9	75.4	102.4	134.6	9.0

TABLE 8g-7. AVERAGE NUMBER OF PROMPT NEUTRONS EMITTED
PER FISSION FOR VARIOUS NUCLIDES

Fissioning nucleus	Bondarenko (1958)[a]	Leachman (1958)[a]	Recent values
Spontaneous Fission			
U^{238}	2.30 ± 0.20		1.97 ± 0.07^b
Pu^{236}	2.17 ± 0.20		
Pu^{238}	2.28 ± 0.10		
Pu^{240}	2.23 ± 0.05	$\begin{cases}2.26 \pm 0.05 \\ 2.22 \pm 0.11\end{cases}$	$\begin{cases}2.154 \pm 0.028^b \\ 2.189 \pm 0.026^c\end{cases}$
Pu^{242}	2.28 ± 0.13	2.18 ± 0.09	
Cm^{242}	2.59 ± 0.11		
Cm^{244}	2.82 ± 0.09		
Bk^{249}	3.72 ± 0.16		
Cf^{246}	2.92 ± 0.19		
Cf^{252}	3.84 ± 0.12		$\begin{cases}3.771 \pm 0.031^c \\ 3.799 \pm 0.034^d \\ 3.704 \pm 0.015^e\end{cases}$
Cf^{254}	3.90 ± 0.14		
Fm^{254}	4.05 ± 0.19		
Thermal Neutron Fission			
Th^{230}	2.13 ± 0.03		
U^{234}	2.52 ± 0.03	$\begin{cases}2.54 \pm 0.04 \\ 2.55 \pm 0.05\end{cases}$	2.473 ± 0.026^c
U^{236}	2.47 ± 0.03	$\begin{cases}2.47 \pm 0.05 \\ 2.46 \pm 0.03\end{cases}$	$\begin{cases}2.425 \pm 0.020^c \\ 2.369 \pm 0.015^e \\ 2.417 \pm 0.015^f\end{cases}$
Pu^{240}		$\begin{cases}2.88 \pm 0.04 \\ 2.95 \pm 0.06\end{cases}$	2.831 ± 0.028^c
Pu^{242}		3.03 ± 0.06	$\begin{cases}3.14 \pm 0.06^g \\ 2.96 \pm 0.08^h\end{cases}$
Am^{242}	3.14 ± 0.04		

* From J. Gindler and J. R. Huizenga, Nuclear Fission, in "Nuclear Chemistry," vol. II, L. Yaffe, ed., Academic Press, Inc., New York, 1968.
[a] I. I. Bondarenko, B. D. Kuzminov, L. S. Kutsayeva, L. I. Prokhorova, and G. N. Smirenkin, *Proc. U.N. Intern. Conf. Peaceful Uses At. Energy (Geneva)* **15**, 353 (1958). R. B. Leachman, *ibid.*, 229.
[b] Asplund-Nilsson, H. Condé, and N. Starfelt, *Nucl. Sci. Eng.* **15**, 213 (1963).
[c] J. C. Hopkins and B. C. Diven, *Nucl. Phys.* **48**, 433 (1963).
[d] I. Asplund-Nilsson, H. Condé, and N. Starfelt, *Nucl. Sci. Eng.* **16**, 124 (1963).
[e] D. W. Colvin and M. G. Sowerby, "Physics and Chemistry of Fission," vol. II, p. 25, IAEA, Vienna, 1965.
[f] H. Condé and M. Holmberg, *ibid*, p. 57.
[g] G. de Saussure and E. G. Silver, *Nucl. Sci. Eng.* **5**, 49 (1959).
[h] A. H. Jaffey, C. T. Hibdon, and R. Sjoblom, *J. Nucl. Energy*, pt. A, **11**, 21 (1959).

TABLE 8g-8. CHARACTERISTICS OF FISSION NEUTRON SPECTRA*

Fissile nuclide	Average energy E, MeV	Maxwellian temperature, $T = 2\bar{E}/3$, MeV
$U^{233} + n_{th}$	$1.98 \pm 0.05\dagger$	1.32 ± 0.03
$U^{235} + n_{th}$	$1.95 \pm 0.05\dagger$	1.30 ± 0.03
$Pu^{239} + n_{th}$	$2.03 \pm 0.05\dagger$	1.35 ± 0.03
$Pu^{241} + n_{th}$	$2.002 \pm 0.051\ddagger$	1.335 ± 0.034
Cf^{252}	$2.15 \pm 0.08\dagger$	1.43 ± 0.05

* From J. Gindler and J. R. Huizenga, Nuclear Fission, in "Nuclear Chemistry," vol. II, L. Yaffe, ed., Academic Press, Inc., New York, 1968.
† J. Terrell, *Phys. Rev.* **127**, 880 (1967).
‡ A. B. Smith, R. Sjoblom, and J. H. Roberts, Phys. Rev. **123**, 2140 (1961).

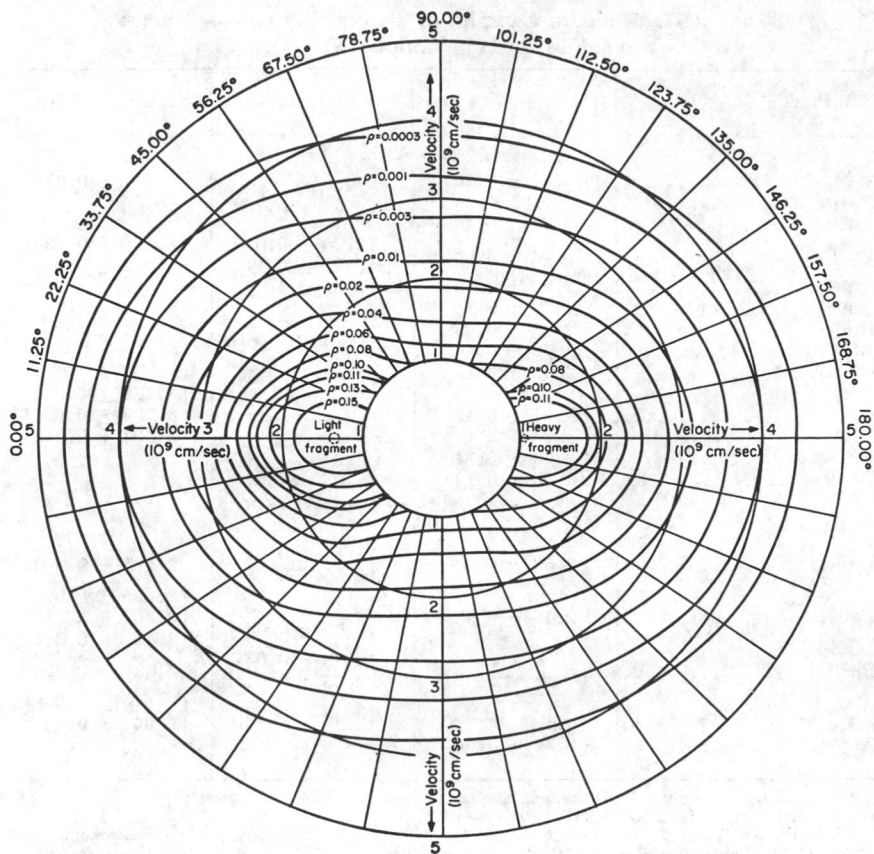

Fig. 8g-17. Contour diagram in polar coordinates of observed neutron density distribution $p(V,\phi)$ as a function of neutron velocity and angle. The contour lines are lines of constant neutron density. The average velocities of the light and heavy fragments are also shown. [*H. R. Bowman, S. G. Thompson, J. C. D. Milton, and W. J. Swiatecki, Phys. Rev.* **126**, 2120 (1962); **129**, 2133 (1963).]

TABLE 8g-9. ABSOLUTE YIELDS OF DELAYED NEUTRONS FROM THERMAL-NEUTRON-INDUCED FISSION*

Target fissile nuclide	Delayed neutrons per fission
U^{233}	0.0066 ± 0.0003†
U^{235}	0.0158 ± 0.0005†
Pu^{239}	0.0061 ± 0.0003†
Pu^{241}	0.0154 ± 0.0015‡

* From J. Gindler and J. R. Huizenga, Nuclear Fission, in "Nuclear Chemistry," vol. II, L. Yaffe' ed., Academic Press, Inc., New York, 1968.
† G. R. Keepin, T. F. Wimett, and R. K. Zeigler, *Phys. Rev.* **107**, 1044 (1957).
‡ S. A. Cox, *Phys. Rev.* **123**, 1735 (1961).

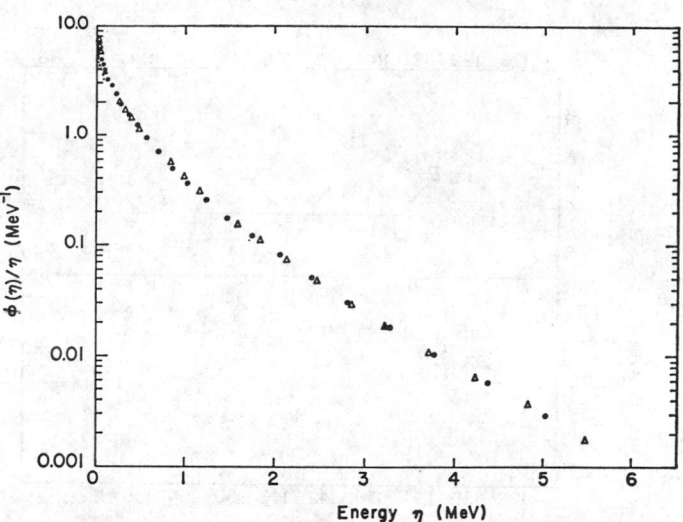

FIG. 8g-18. The center-of-mass neutron spectrum $\phi(\eta)$ divided by η, the neutron energy in the center-of-mass system. The dots represent neutrons emitted in the direction of the light fragments; the triangles represent neutrons emitted in the direction of the heavy fragments. The curve for the light fragments was reduced by the factor 1.16, the ratio of the number of neutrons from light fragments to the number from heavy fragments if all neutrons are emitted from moving fragments. See Fig. 8g-17 for reference.

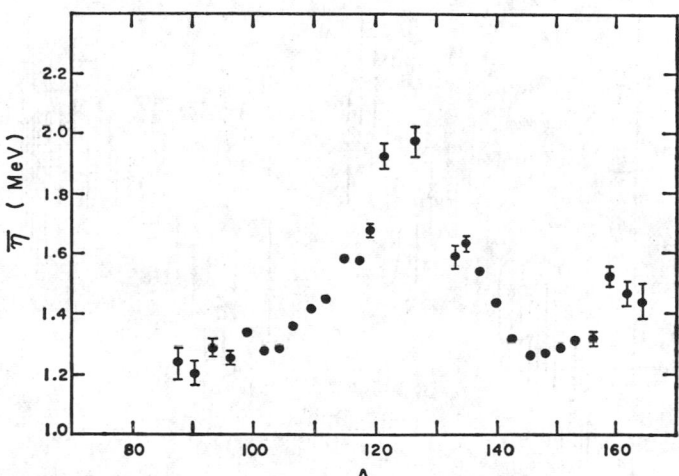

FIG. 8g-19. The average center-of-mass neutron kinetic energy as a function of fragment mass, corrected for mass resolution. For reference, see Fig. 8g-17.

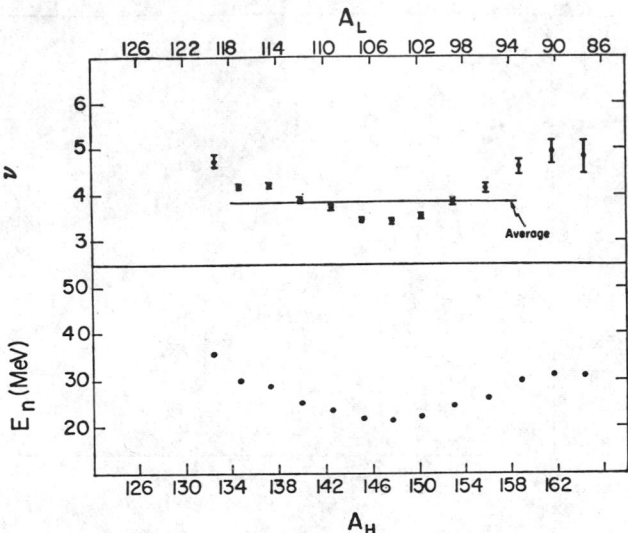

FIG. 8g-20. Total number of neutrons ν and total energy E_u appearing in the form of neutrons as a function of fragment mass. For reference, see Fig. 8g-17.

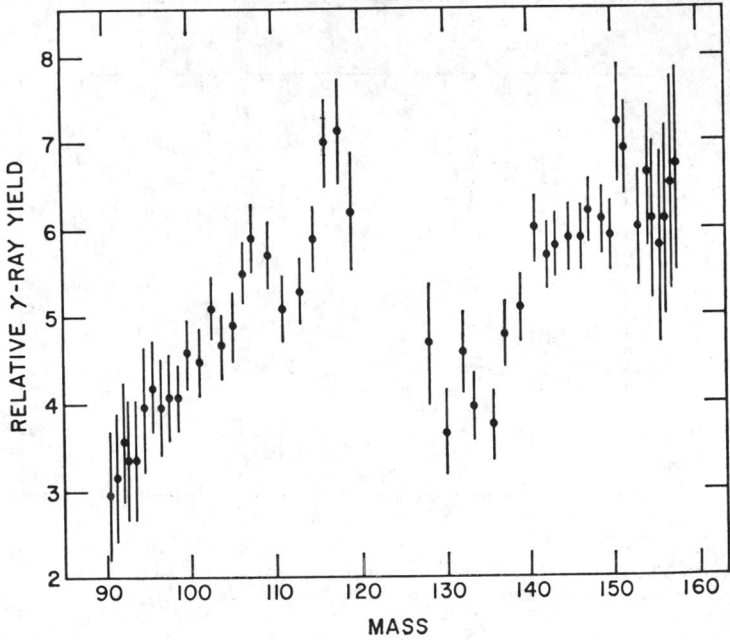

FIG. 8g-21. The γ-ray yield as a function of mass.   [S. A. E. Johansson, Nucl. Phys. **60**, 378 (1964).]

TABLE 8g-10. CHARACTERISTICS OF PROMPT GAMMA
RADIATIONS EMITTED IN FISSION*

Fissioning nuclide	Average no. of photons per fission	Average photon energy released per fission
$U^{235} + n_{th}{}^{a}$	7.5	7.46
$U^{235} + n_{th}{}^{b}$	7.2 ± 0.8	7.4 ± 0.8
$U^{235} + n_{th}{}^{c}$	7.93 ± 0.48	9.51 ± 0.23
Cf^{252d}	10	9
Cf^{252e}	10.3	8.2

* From J. Gindler and J. R. Huizenga, Nuclear Fission, in "Nuclear Chemistry," vol. II., L. Yaffe, ed., Academic Press, Inc., New York, 1968.
[a] J. Francis and R. Gamble, *Oak Ridge Nat. Lab. Rept.* ORNL-1879 (unpublished).
[b] F. C. Maienschein, R. W. Peele, W. Zohel, and T. A. Love, *Proc. U.N. Intern. Conf. Peaceful Uses At. Energy (Geneva)* **15**, 366 ($0.3 \le E\gamma \le 10$ MeV), (1958).
[c] F. E. W. Rau, *Ann. Physik* **10**, 252 (1963).
[d] H. R. Bowman and S. G. Thompson, *Proc. U.N. Intern. Conf. Peaceful Uses At. Energy (Geneva)* **15**, 212 (1958).
[e] A. Smith, P. Fields, A. Friedman, S. Cox, and R. Sjoblom, *ibid*, **15**, 392 (1958).

TABLE 8g-11. AVERAGE NUMBER AND ENERGY OF BETA DECAYS
PER FISSION FOR U^{235} AND U^{233} THERMAL NEUTRON FISSION*

U^{235}		U^{233}
N_β	E_β	N_β
6.6 ± 0.9^{a}		
6.9 ± 0.4^{b}	8.1 ± 0.4^{b}	
6.6 ± 0.2^{c}		
5.93 ± 0.2^{d}	....	5.25 ± 0.2^{d}
6.10^{f}	....	5.27^{d}
	7.6 ± 0.5^{e}	

* From J. Gindler and J. R. Huizenga, Nuclear Fission, in "Nuclear Chemistry," vol. II, L. Yaffe, ed., Academic Press, Inc., New York, 1968.
[a] G. Alzmann, *Nukleonik* **3**, 295 (1961).
[b] P. Armbruster and H. Meister, *Z. Physik* **170**, 274 (1962).
[c] P. Armbruster, D. Hovestadt, H. Meister, and H. J. Specht, *Nucl. Phys.* **54**, 586 (1964).
[d] H. J. Specht and H. Seyfarth, "Physics and Chemistry of Fission," vol. II, p. 253, IAEA, Vienna, 1965.
[e] J. F. Perkins and R. W. King, *Nucl. Sci. Eng.* **3**, 726 (1958).
[f] Calculated value.

TABLE 8g-12. CALIBRATION CONSTANTS FOR HEAVY-ION DETECTORS

Cf^{252} $\qquad\qquad\qquad\qquad$ U^{235}

$$a = \frac{24.0203}{P_L - P_H} \qquad\qquad a = \frac{30.9734}{P_L - P_H}$$

$$a' = \frac{0.03574}{P_L - P_H} \qquad\qquad a' = \frac{0.04596}{P_L - P_H}$$

$$b = 89.6083 - aP_L \qquad\qquad b = 87.8626 - aP_L$$

$$b' = 0.1370 - a'P_L \qquad\qquad b' = 0.1345 - a'P_L$$

TABLE 8g-13. LIMITS ON SPECTRUM SHAPE PARAMETERS

Spectrum characteristic	Cf^{252}	U^{235}
N_L/N_V	>2.85	~ 19
N_H/N_V	~ 2.2	~ 12.5
N_L/N_V	1.30	$1.49-1.55$
$\Delta L/(L - H)$	$\lesssim 0.38$	0.22
$\Delta H/(L - H)$	$\lesssim 0.45$	0.35
$(H - HS)/(L - H)$	$\lesssim 0.70$	0.38
$(LS - L)/(L - H)$	$\lesssim 0.49$	0.27
$(LS - HS)/(L - H)$	$\lesssim 2.18$	~ 1.66

TABLE 8g-14. AVERAGE FRAGMENT ENERGIES AND MASSES
FOR Cf^{252} SPONTANEOUS FISSION

$\langle V_H \rangle$	1.036 cm/nsec	$\langle E_L \rangle$	105.71 MeV
$\langle V_L \rangle$	1.375 cm/nsec	σ_{EH}	8.43 MeV
$\sigma(V_H)$	0.0789 cm/nsec	σ_{EL}	5.61 MeV
$\sigma(V_L)$	0.0650 cm/nsec	$\langle E_K \rangle$	185.7 MeV
$\langle M_H \rangle$	143.61 amu	σ_{EK}	11.0 MeV
$\langle M_L \rangle$	108.39 amu	$\langle R_A \rangle$	1.334
$\sigma_{ML} = \sigma_{MH}$	6.72 amu	σ_{RA}	0.137
$\langle E_H \rangle$	80.01 MeV		

All quantities are preneutron emission [S. L. Whetstone, *Phys. Rev.* **131**, 1232 (1963)].

TABLE 8g-15. CHARGED-PARTICLE YIELDS FROM Cf^{252} SPONTANEOUS FISSION

Particle	Yield, particles/fission	Particle	Yield, particles/fission
p	$(5.1 \pm 0.5)(10^{-5})$	He^6	$(7.8 \pm 1.6)(10^{-5})$
d	$(2.0 \pm 0.1)(10^{-5})$	He^8	$(5.9 \pm 1.6)(10^{-6})$
t	$(1.90 \pm 0.06)(10^{-4})$	He^{10}	$(3 \pm 3)(10^{-7})$
He^3	$\lesssim 2.9 \times 10^{-5}$	Li	$(3.9 \pm 2.0)(10^{-6})$
α	$(3.27 \pm 0.10)(10^{-3})$	Be	$>3 \times 10^{-7}$

From S. L. Whetstone and T. D. Thomas, *Phys. Rev.* **154**, 1174 (1967).

8g-5. Bibliography. For a comprehensive summary of nuclear fission data, the reader is referred to the following excellent sources:

1. Hyde, E. K.: "The Nuclear Properties of the Heavy Elements," vol. III, Prentice-Hall, Inc., Englewood Cliffs, N.J., 1964.
2. "Physics and Chemistry of Fission," vols. I and II, (IAEA, Vienna, 1965).
3. Gindler, J., and J. R. Huizenga: Nuclear Fission, in "Nuclear Chemistry," vol. II, L. Yaffe, ed., Academic, Press, Inc., New York, 1968.

8h. Elementary Particles and Interactions[1]

ARTHUR H. ROSENFELD

University of California, Berkeley

GEORGE A. SNOW

University of Maryland

Matter as we know it is built up of a large number of particles (called "elementary" partly on account of our ignorance about them), which interact with one another via four kinds of forces. No theory has been completely successful in predicting or explaining the number of different particles that exist or many of their intrinsic properties, although a substantial amount of order has emerged in recent years.

In Sec. 8h-1 the four basic interactions between particles are described.

Section 8h-2 contains a general description of the different kinds of particles: some of which are called "stable," others unstable "resonances."

Section 8h-3 describes the conservation laws obeyed by the forces in nature which then lead to quantum numbers used to classify elementary particles.

Section 8h-4 contains a table of "elementary" particles and their intrinsic properties.

Section 8h-5 briefly describes the SU₃ classification of particles, called the eightfold way.

Section 8h-6 suggests some further reading.

8h-1. The Four Basic Interactions. All the physical phenomena and all the states of matter observed in the universe are apparently manifestations of one or more of four basic interactions between particles. These four kinds of forces differ enormously in strength and in range. In order of increasing strength, they are: gravitation, the "weak" interaction, electromagnetism, and the "strong" interaction. The basic properties of these four forces are summarized in Table 8h-1.

TABLE 8h-1. THE FOUR BASIC FORCES AND THEIR CHARACTERISTICS

Force	Acts on:	Strength	Range	Examples
Gravity	Mass or energy	$\sim 10^{-38}$	$\infty\,(\sim 1/r^2)$	Solar system
Weak	Leptons, hadrons	$\sim 10^{-14}$	$<10^{-14}$ cm	Radioactivity
Electromagnetic	Charged particles	$\frac{1}{137}$	$\infty\,(\sim 1/r^2)$	Atoms, molecules, . . .
Strong	Hadrons	~ 1	$\simeq 10^{-13}$ cm	Nuclei

The *force of gravity* acts between all objects that have mass or energy. With large aggregates of matter, the force of gravity even at large distances can be dominating—for example, the force between earth and sun or between moon and earth. However, at the scale of atomic or subatomic particles it is by far the weakest of the four forces

[1] Supported in part by the U.S. Atomic Energy Commission.

8-277

If one introduces a dimensionless constant to characterize the strength of each interaction and assigns a strength of 1 to the strongest interaction (not surprisingly called the *strong interaction*), then for gravity that constant is about 10^{-38}. The range of the gravitational force extends essentially to infinity decreasing inversely as the square of the distance between the masses ($\sim 1/r^2$).

The *weak interaction* is indeed weak relative to the electromagnetic or strong interactions since its characteristic dimensionless strength constant is $\sim 10^{-14}$. The weak interaction is responsible for natural and artificial radioactivity and for reactions amongst the lightest elementary particles, called *leptons* (i.e., neutrinos, electrons, and muons). It also gives rise to interactions between leptons and all other more massive elementary particles. The characteristic range of the weak interaction is known to be very small, less than 10^{-14} cm, but its actual size has not yet been measured. (Higher-energy accelerators are needed to examine these interactions at such small distances.) The fact that the range of the weak interaction is so small means that, unlike gravity, its influence over macroscopic distances is completely negligible.

The *electromagnetic interaction* is the force between electrically charged particles. Just as for gravity, the range of this force extends to infinity ($\sim 1/r^2$). Despite the fact that the intrinsic strength of electromagnetism is much larger than that of gravitation ($e^2/\hbar c = \frac{1}{137}$, where e is the basic unit of charge, $\hbar$ is Planck's constant divided by 2π, and c is the velocity of light), massive aggregates of matter (e.g., raindrops, or the earth) tend to have zero electric charge (one can neutralize charge but not mass!) so that the electric forces cancel. The electromagnetic force is the one that keeps electrons and positively charged nuclei bound to each other in atoms, molecules, and crystals, so that all of chemistry and biology is governed by the laws of electromagnetism.

The *strong interaction* manifests itself as the force that binds neutrons and protons together in nuclei, but it has many other forms as well. All particles that experience the strong interaction are called *hadrons*. These include *baryons* like the proton and the neutron, as well as *mesons*, described in the next section. Hadrons and leptons are mutually exclusive in that only the former participate in the strong interaction, whereas both sets of particles participate in the three other interactions—electromagnetic, weak, and gravitational. The range of the strong interaction is $\sim 10^{-13}$ cm, so that they dominate all other forces only at very small subatomic distances, despite their great strength.

8h-2. Types of Particles. *Gravitons.* When the gravitational field that is generated by any particle is quantized, the theory predicts the existence of gravitons. They have zero rest mass and intrinsic spin $2\hbar$, and interact with other particles through the gravitational interaction only. This interaction is extremely weak. The first report of the experimental detection of gravitational waves was made by J. Weber in 1969 [1].

Photons (γ). These are the quanta of the electromagnetic field. They have zero rest mass and intrinsic spin $1\hbar$, and are emitted and absorbed exclusively by the charge or current of other particles via the electromagnetic interaction.

Leptons. These are relatively light particles (rest mass either zero or small) of spin $\frac{1}{2}\hbar$, whose interactions with each other and with all other known particles (other than photons and gravitons) are "weak." They include two kinds of neutrinos ν_e and ν_μ with zero electric charge, the electron e^-, the negative muon μ^-, and the antiparticles[1] of these four particles denoted as antineutrinos $\bar{\nu}_e$ and $\bar{\nu}_\mu$, the positron e^+, and the positive muon μ^+. Whenever leptons are produced, transformed into each other, or annihilated by the weak interaction, the number of leptons minus the number of antileptons is preserved, e.g. $n \rightarrow pe^-\bar{\nu}_e$ in ordinary neutron radioactivity. This property of the weak interaction gives rise to a conservation law for leptons that will be discussed further in Sec 8h-3.

[1] Except for a few special particles (e.g., γ, π^0, η) every particle has a distinct antiparticle counterpart.

Hadrons. These are particles that interact with each other via the strong interaction. They also experience electromagnetic, weak, and gravitational interactions. The strongest interaction that a particle experiences determines its classification. Hadrons are divided into two classes: mesons and baryons.

MESONS. The lightest mesons are particles with *zero spin*, intermediate in mass between leptons and protons, that interact strongly with each other and with all the baryons. These spin-0 mesons are designated as pi mesons (π^-, π^0, π^+) [the superscript denotes electric charge], η mesons (0 charge), and K mesons (K^0, K^+) and their antiparticles ($\bar{K}^0$, K^-). Unlike the leptons and the baryons, the antiparticle of a meson can be itself as is the case of the π^0 and η mesons. The π^- meson is the antiparticle of the π^+ meson. These eight mesons are "stable with respect to disintegration via the strong interactions," but they are unstable with respect to disintegrations into leptons, photons, or lighter mesons via the weak and electromagnetic interactions. Table 8h-3 in Sec. 8h-4 lists these decays in detail.

There are many heavier mesons, all with *integral spin* in units of $\hbar$, including a family of eight mesons of spin 1 called "vector" mesons. These heavier mesons play an important role in all phenomena involving the strong interaction, but the number and variety of them keeps us from discussing them in detail here. They differ in one important way from the lighter-spin zero mesons: namely, they are unstable with respect to disintegration into the lighter mesons via the strong interactions. As a result, the mean lives of vector mesons and other still heavier mesons are very short, $\sim 10^{-22}$ sec, as compared to mean lives $\sim 10^{-8}$ to 10^{-10} sec for spin-0 mesons that decay only via the weak interaction. The heavier mesons are sometimes called *resonant* states, that is states of the constituent particles into which they disintegrate. For example, the ρ^+ meson, a vector meson that decays via the reaction $\rho^+ \to \pi^+ + \pi^0$, can be considered to be a resonant state of the $\pi^+\pi^0$ system with total angular momentum equal to one. Particles and resonant states are two different names for the same thing. In fact, there is no profound difference between particles that are stable with respect to the strong interactions and particles or resonant states such as the vector mesons that are not stable; rather instability or stability simply depends upon whether or not there exist appropriate hadronic states of smaller mass into which the particle can decay without violating the laws of physics.

BARYONS. These are strongly interacting particles of half-integral spin ($\frac{1}{2}\hbar$, $\frac{3}{2}\hbar$, ...) of mass greater than or equal to the mass of the proton. The lightest spin-$\frac{1}{2}$ group consists of the *nucleons* N (proton p and neutron n) and the *hyperons* Y called a lambda Λ (0 charge), three sigmas Σ^+, Σ^0, Σ^- and two xis Ξ^-, Ξ^0. These hyperons are heavier than nucleons, and stable with respect to strong interactions, but unstable via the weak interaction. Corresponding to each baryon there exists an *antibaryon* which has identical mass and spin as the baryon but opposite charge and magnetic moment. These antibaryons are called antinucleons $\bar{N}(\bar{p}, \bar{n})$ and antihyperons $\bar{Y}(\bar{\Lambda}, \bar{\Sigma}^+, \bar{\Sigma}^0, \bar{\Sigma}^-, \bar{\Xi}^0, \bar{\Xi}^-)$. All these antibaryons have been produced and observed at high-energy accelerators. In the processes of production or annihilation of antibaryons in high-energy collisions, an antibaryon is always produced in a pair together with a baryon. The kinetic energy of motion of the incoming beam particle is transformed into rest-mass energy of a baryon-antibaryon pair. Just as the weak interaction preserves the number of leptons minus antileptons in any reaction, the strong interactions preserve the number of baryons minus antibaryons. In fact, this property of conserving the number of baryons minus antibaryons holds not just for the strong interaction but for all interactions, and is responsible for the stability of the proton[1] and all stable nuclei, and hence for the stability of matter as we know it.

The forces that hold nucleons together to form nuclei are due to exchanges of mesons

---

[1] If it were not for this conservation law, protons would decay by reactions like $p \to e^+\gamma$, $e^+\pi^0$, ... see Sec. 8h-3.

between pairs of nucleons. The short range r_0 of the nuclear force is related to the mass M of the exchanged mesons via the Heisenberg uncertainty principle, $r_0 \sim \hbar/Mc \sim 10^{-13}$ cm. Beside the spin-$\frac{1}{2}$ baryons mentioned above, there are a host of higher-mass baryon resonant states or particles with spins $\frac{3}{2}\hbar$, $\frac{5}{2}\hbar$, and higher. None of these states is stable with respect to disintegration via the strong interaction except for the Ω^- hyperon, of spin $\frac{3}{2}\hbar$, which decays via the weak interaction, and to which we shall return later. Again we shall not discuss in detail the properties of such higher-mass unstable baryonic states.

8h-3. Conservation Laws and Quantum Numbers. Our understanding of the interactions in the physical world rests heavily upon the discovery of the existence of conservation laws in physics. Some of these conservation laws are found to have universal validity for all interactions, whereas others are only approximate, holding for one kind of interaction but not for another. In either case, the elementary particles are labeled by quantum numbers and continuous parameters, such as mass, that denote how much of each kind of conserved quantity the particular particle has. In addition to such additive quantum numbers, there exist multiplicative quantum numbers, such as parity that are related to discrete symmetries of particle interactions, which are conserved for some of the known interactions but not for others.

In order to decipher the quantum numbers assigned to the elementary particles listed in Table 8h-3 of Sec. 8h-4, the conservation laws and their associated quantum numbers are listed and briefly discussed below in order of decreasing generality.

Conservation of Energy and Momentum (Four-momentum) and the Concept of Rest Mass. The total energy E and the total linear momentum $\mathbf{p}$ are conserved in any reaction. Within the framework of Einstein's special theory of relativity these two conservation laws can be combined into the law of conservation of four-momentum $(E, c\mathbf{p})$. The fact that the quantity $E^2 - (c\mathbf{p})^2 \equiv (mc^2)^2$ is an invariant with respect to Lorentz transformations (transformations to different constant-velocity reference frames) makes it convenient to assign a rest mass m to each elementary particle. The name *rest mass* is used because its mass is just m in a Lorentz frame in which the particle is at rest. In such a frame, the total energy of the particle is given by $E = mc^2$, so that it is common to measure the mass in units of energy such as MeV (million electron volts, where 1 MeV $= 1.6 \times 10^{-6}$ ergs). Conservation of energy and momentum implies that if particle A of rest mass m_A is at rest and spontaneously disintegrates into two other particles B and C of rest masses m_B and m_C, then $m_A c^2 = E_B + E_C$, and $0 = \mathbf{p}_B + \mathbf{p}_C$, so that the invariant mass of particles $B + C$ defined as $c^{-2}[(E_B + E_C)^2 - (\mathbf{p}_B + \mathbf{p}_C)^2 c^2]^{\frac{1}{2}}$ equals m_A.

Conservation of Angular Momentum, the Spin-quantum Number, and Statistics. The fact that total angular momentum is conserved in *all* reactions is profoundly related to the local isotropy, or nondirectionality, of space. Total angular momentum $\mathbf{J}$ is found to be quantized so that $\mathbf{J}^2 = J(J + 1)\hbar^2$, where J, the total angular-momentum quantum number, can take on only half-integral or integral values. For a given particle $\mathbf{J}$ is the sum of $\mathbf{l} + \mathbf{S}$ where $\mathbf{l}$ is the orbital angular momentum of the particle about some axis, and $\mathbf{S}$ is the intrinsic-spin angular momentum of the particle. If the particle is at rest, $\mathbf{l} = 0$ and $\mathbf{J} = \mathbf{S}$, so that the intrinsic spin of the particle is identical to the total angular momentum of that particle in its rest system. Table 8h-3 of Sec. 8h-4 denotes the intrinsic spin of each particle by the symbol J. As mentioned earlier, all baryons and leptons have half-integral spin, while all mesons have integral spin. When one considers a system containing more than one indistinguishable particle, further restrictions occur in the "statistics" of these identical particles, that is, in the ways of combining these into quantum-mechanical states. Identical half-integral spin particles (e.g., e^-e^- or $\mu^-\mu^-$) obey Fermi-Dirac statistics, which means that the allowed states must be antisymmetric with respect to the interchange of any two such particles. (The Pauli principle for electrons in atoms is a famous example of this relationship.) On the other hand, integral-spin particles obey Bose-Einstein statistics, which imply that the allowed states must be symmetric with respect to the interchange of identical particles.   (Planck's law for black-body radiation is a famous illustration of this relationship.)   As a result of this connection between spin and statistics, half-integral spin particles are often called *fermions;* and integral spin particles, *bosons.*

Conservation of Electric Charge. One of the oldest and best-established conservation laws is that for electric charge.   (This was first clearly stated and demonstrated by Benjamin Franklin.)   The total amount of charge Q is found to be conserved in all reactions.   In addition, charge is quantized in units of $+e$ or $-e$, where e is the magnitude of charge on the electron.   The total charge in any state is obtained by simply adding algebraically the charges on each of the particles in that state.   All the particles listed in Table 8h-3 of Sec. 8h-4 have charges 0 or $\pm e$.   On the other hand, there do exist higher-energy excited states of nucleons that are doubly charged such as N^{++} which decays to $p + \pi^+$.

Conservation of Baryon Number. If one assigns an additive quantum number $B = +1$ to each baryon, $B = -1$ to each antibaryon, and $B = 0$ to all other particles (leptons, mesons); then all reactions conserve the total baryon number.   For ordinary nuclei the total baryon quantum number is identical with the atomic mass number. This conservation law implies that the mode of production of antibaryons must be in the form of baryon-antibaryon pairs, e.g., $p + p \rightarrow p + p + \bar{p} + p$.   Even more significant for the stability of the universe is the fact that baryon conservation forbids baryons from decaying into leptons, via reactions such as $p \rightarrow e^+ + \gamma$, which otherwise would be allowed by the laws of conservation of energy, momentum, angular momentum, and charge.   When antimatter such as an antiproton $\bar{p}$, which has charge $-e$, interacts with a proton or a neutron, the final products are mesons: e.g., $\bar{p} + n \rightarrow \pi^+ + \pi^- + \pi^- + \pi^0$.   This is an example of "annihilation" of matter by antimatter.

In Table 8h-3 of Sec. 8h-4 the antibaryons are not listed separately from the baryons, since all their properties can be deduced directly from the properties of the baryons.   The relationships are same mass, same intrinsic spin, opposite charge, opposite magnetic moment, same lifetime, and same proportion of decay modes, where the emitted particles in each decay mode are also converted from particles to antiparticles.   This symmetry between particle and antiparticle is related to a symmetry of the relativistic quantum theory of strong, electromagnetic, and weak interactions under the operation of TCP. T, C, and P stand for time reversal, particle-antiparticle interchange, and spatial reflection, respectively.   As far as we know at present, all interactions are invariant under the simultaneous action of the TCP operator, but as we shall discuss below, this is not true for each of these operators separately.

Conservation of Lepton Number and Muon Number. If one assigns an additive quantum number $l = +1$ to each lepton $(e^-, \mu^-, \nu_e, \nu_\mu)$, $l = -1$ to each antilepton $(e^+, \mu^+, \bar{\nu}_e, \bar{\nu}_\mu)$, and $l = 0$ to all other particles, then all observed reactions conserve the total lepton number.   For example, energetic neutrinos produced by π^+ decays, $\pi^+ \rightarrow \mu^+ + \nu_\mu$, are observed to interact with nucleons via the reaction $\nu_\mu + n \rightarrow$
$l=0 \quad -1 \quad +1$ $\qquad\qquad\qquad\qquad\qquad\qquad\qquad\qquad\qquad\qquad\qquad\qquad\qquad l=+1 \quad 0$
$p + \mu^-$ but *not* via the reaction $\nu_\mu + p \rightarrow n + \mu^+$.   The muon number can be
$0 \quad +1 \qquad\qquad\qquad\qquad\qquad\qquad\qquad\quad l=+1 \quad 0 \quad 0 \quad -1$
introduced to distinguish muons from electrons by assigning a muon number $+1$ to μ^-, ν_μ, -1 to $\mu^+, \bar{\nu}_\mu$, and 0 to all other particles, including $e^\pm, \nu_e, \bar{\nu}_e$.   Conservation of muon number can be invoked to "explain" the observed absence of decays such as $\mu^+ \rightarrow e^+ + \gamma$ or $\mu^\pm \rightarrow e^\pm + e^+ + e^-$.   The usual mode of disintegration of a muon is $\mu^+ \rightarrow e^+ + \nu_e + \bar{\nu}_\mu$ or $\mu^- \rightarrow e^- + \bar{\nu}_e + \nu_\mu$.   These decays conserve total lepton number and total muon number.

The five conservation laws described so far in this section are absolute in the sense that no violations of any of them have been observed at our present level of sensitivity of experiment. The conservation laws discussed in the remainder of this section are valid only for subsets of the four kinds of interactions but not for all of them.

Invariance under Spatial Inversion P and Particle-Antiparticle Conjugation C. Parity, denoted by the symbol P, refers to the symmetry of a wave function describing a given state under the operation of spatial inversion, $(x \rightarrow -x, y \rightarrow -y, z \rightarrow -z)$. P can be either plus or minus one. Experimentally, parity is conserved in all strong and electromagnetic interactions but not in the weak interaction. The parity of a given state is determined by multiplying the intrinsic parities of its constituent particles with the parity due to any orbital angular momentum between these particles. For example, the parity of a wave function describing two identical particles with relative orbital angular momentum l is $(-1)^l$.

Particle-antiparticle conjugation C, often called *charge conjugation*, refers to the operation of transforming a particle into its antiparticle. Again the strong and electromagnetic interactions are invariant under such a transformation, but the weak interaction is not. This means that all the properties of strong reactions such as $p + p \rightarrow p + p$, or electromagnetic decays of excited atoms, (atom)$^* \rightarrow$ (atom) $+ \gamma$, are exactly the same as the corresponding antiparticle reactions $\bar{p} + \bar{p} \rightarrow \bar{p} + \bar{p}$ or $\overline{(atom)}^* \rightarrow \overline{(atom)} + \gamma$. However, the weak decay of a Λ hyperon, $\Lambda \rightarrow p + \pi^-$, is not identical in all respects to the decay $\bar{\Lambda} \rightarrow \bar{p} + \pi^+$. Note that under the operation C, baryons go to antibaryons, and leptons go to antileptons, but mesons and photons go to mesons and photons. Antimesons belong to the same species as mesons, so that an anti-π^- meson is just a π^+ meson. Even more remarkable, an anti-π^0 meson is itself, a π^0 meson. An antiphoton is "minus" itself. Neutral mesons have the possibility of being their own antiparticles, whereas neutral baryons or neutral leptons do not because of the reversal of baryon number and lepton number, respectively. When a neutral particle is its own antiparticle, it is assigned a C quantum number, listed in Table 8h-3 of Sec. 8h-4, which may be either $+1$ or -1. However, not all neutral mesons are their own antiparticles.

Until a few years ago it was thought that the weak interaction was invariant to the combined operation of particle-antiparticle conjugation and spatial inversion, i.e., CP. However, at the level of $\sim 10^{-3}$ of the usual weak amplitude, the weak decay $K_L^0 \rightarrow \pi^+ + \pi^-$ has been observed. This process violated CP invariance. Hence the only discrete symmetry invariance of the weak interaction that appears to be exact is TCP invariance, not P or C or CP separately.

We turn now to the conservation laws for the quantities I spin and hypercharge. The former is conserved for the strong interactions only, the latter for both strong and electromagnetic interactions.

The two concepts, I spin and hypercharge, are closely tied to each other and have played an essential part in the current classification scheme of all the hadrons.

Conservation of I spin and Hypercharge, and the Gell-Mann–Nishijima Classification Scheme. The concept of I spin, or "isotopic spin," is introduced in order to describe the charge multiplet structure of hadrons. The particles in a given multiplet have identical spin and parity, and almost identical mass, but different charges. Examples of such multiplets are the (n,p) doublet and the (π^-, π^0, π^+) triplet. One defines an I-spin vector operator $\mathbf{I}$ with three components analogous to the angular-momentum operator $\mathbf{J}$. $\mathbf{I}$ operates in a new *internal* coordinate space of the hadrons, distinct from ordinary space. In exact analogy to J in real space, the eigenvalues of the operator $\mathbf{I}^2$ are $I(I + 1)$, where I is called the total I-spin quantum number. The number of different charge states in a multiplet I is $2I + 1$. For nucleons and π mesons, the charge state is related to the third component I_3 of the I-spin vector by the relation

$$Q = I_3 + \frac{B}{2} \tag{8h-1}$$

where Q is the charge in units of $+e$, and B is the baryon number [$+1$ for (n,p), 0 for π mesons]. The formula implies that $I_3 = +\frac{1}{2}$ and $-\frac{1}{2}$ for proton and neutron; $I_3 = +1$, 0, -1 for π^+, π^0, π^- mesons, respectively. The neutron and proton are seen to form an $I = \frac{1}{2}$ multiplet in isotopic spin space, while the π^+, π^0, π^- mesons form an $I = 1$ multiplet. The I-spin formalism was found to be useful because the strong interactions conserve total I spin ($I_{\text{final}} = I_{\text{initial}}$, or $\Delta I = 0$). This property of the strong interaction is called *charge independence*. The total I spin for a system of more than one particle is built up out of the component I spins of each particle by rules of addition that are identical to the rules of addition for total angular momentum. For example, with the help of Table 8h-2 it can be seen that a (p,p) system is in a pure $I = 1$ state, whereas an (n,p) system is in a $50:50$ mixture of $I = 0$ and $I = 1$ states.

Charge independence means that the strong interaction for the (p,p) system is identical to the strong interaction for that *part* of the (n,p) system that is in the $I = 1$ state. By using the I-spin formalism, one can extend the concept of identical particles from protons alone or neutrons alone to both together, that is, to nucleons. Just as a proton can have its spin angular-momentum quantum number S_Z equal to $+\frac{1}{2}$ or $-\frac{1}{2}$ (in units of $\hbar$), a "nucleon" can have the third component of its I spin, I_3, equal to $+\frac{1}{2}$ or $-\frac{1}{2}$, corresponding to a proton or a neutron, respectively. Neither the electromagnetic nor the weak interaction conserves I or I_3, but instead each changes the hadronic I and I_3 quantum numbers in a definite way.

When the K mesons and hyperons were first discovered, they too were found to cluster in charge multiplets, as listed in Sec. 8h-2. However, these K mesons and hyperons presented a puzzle in that they were produced copiously via the strong interactions in collisions between pions and nucleons; yet they had a very low transition probability, characteristic of the weak interactions, for decay back to pions and nucleons. As a result, they were called "strange" particles. Gell-Mann and Nishijima independently showed how to classify the K mesons and hyperons and their interactions by introducing a new quantum number S for "strangeness." The "ordinary" particles like the π mesons and nucleons were assigned $S = 0$, whereas the new strange particles were assigned nonzero integral values of S. Actually it is more common now to use the hypercharge quantum number Y which is simply related to strangeness S by the equation $Y = B + S$, where $B =$ baryon number. The new idea was to generalize the relation between charge Q and I_3, given for pions and nucleons by Eq. (8h-1), to

$$Q = I_3 + \frac{Y}{2} \tag{8h-2}$$

They also added the hypothesis that all strong and electromagnetic reactions conserved hypercharge while the weak interactions, which were responsible for the *decays* of the strange particles, did not conserve hypercharge. Table 8h-2 lists the values of I, I_3, and Y for the eight pseudoscalar mesons ($J^P = 0^-$) and for the eight spin-$\frac{1}{2}$ baryons.

The hypercharge quantum numbers for antiparticles are obtained from those for particles by letting $Y \to -Y$. Furthermore, since $Q \to -Q$, Eq. (8h-2) implies that $I_3 \to -I_3$ when particle $\to$ antiparticle. The total I spin is the same for a particle and its antiparticle since the multiplicity of charge states does not change. So the hypercharge Y is a simple additive quantum number just as ordinary charge or baryon number.

The Gell-Mann–Nishijima classification scheme for all reactions involving only mesons and baryons can be simply codified by the following selection rules:

    1. Strong interactions: $\Delta Q = \Delta B = \Delta Y = \Delta I = 0$
    2. Electromagnetic interactions: $\Delta Q = \Delta B = \Delta Y = 0$
    3. Weak interactions: $\Delta Q = \Delta B = 0$, $\Delta Y = \pm 1$

TABLE 8h-2. I-SPIN AND HYPERCHARGE QUANTUM NUMBERS
FOR STABLE MESONS AND BARYONS

Mesons	π^+	π^0	π^-	K^+	K^0	$\bar{K}^0$	K^-	η
I	1	1	1	$\frac{1}{2}$	$\frac{1}{2}$	$\frac{1}{2}$	$\frac{1}{2}$	0
I_3	$+1$	0	-1	$+\frac{1}{2}$	$-\frac{1}{2}$	$+\frac{1}{2}$	$-\frac{1}{2}$	0
$Y(B=0)^a$	0	0	0	$+1$	$+1$	-1	-1	0

Baryons	p	n	Λ	Σ^+	Σ^0	Σ^-	Ξ^0	Ξ^-
I	$\frac{1}{2}$	$\frac{1}{2}$	0	1	1	1	$\frac{1}{2}$	$\frac{1}{2}$
I_3	$+\frac{1}{2}$	$-\frac{1}{2}$	0	$+1$	0	-1	$+\frac{1}{2}$	$-\frac{1}{2}$
$Y(B=+1)^a$	$+1$	$+1$	0	0	0	0	-1	-1

$^a Q = I_3 + Y/2.$

[The symbol ΔQ means $(Q_{\text{initial}} - Q_{\text{final}})$ for any reaction.] The leptons are not included in the Gell-Mann–Nishijima scheme. The quantum number of I spin and hypercharge Y are not assigned to them since they do not participate in the strong interactions. A photon has $Y = 0$ but no definite I spin. As a consequence of these rules, the strong interactions allow reactions of the type

$$\pi^- + p \rightarrow \Lambda + K^0 \tag{8h-3}$$
$$K^- + p \rightarrow \Xi^- + K^+ \tag{8h-4}$$
$$\bar{p} + p \rightarrow \Lambda + \bar{\Lambda} \tag{8h-5}$$

but forbid reactions of the type

$$\pi^- + p \rightarrow \Sigma^+ + K^- \quad \text{(violates } Y\text{)} \tag{8h-6}$$
$$n + p \rightarrow \Lambda + p \quad \text{(violated } Y \text{ and } I\text{)} \tag{8h-7}$$
$$\Lambda \rightarrow p + \pi^- \quad \text{(violates } Y \text{ and } I\text{)} \tag{8h-8}$$

The weak interaction is responsible for the decays of most hyperons and K mesons. The final products may be nonleptonic, e.g., $\Lambda \rightarrow p\pi^-$ or $K^+ \rightarrow \pi^+\pi^0$, or leptonic, e.g., $\Lambda \rightarrow pe^-\bar{\nu}_e$ or $K^+ \rightarrow \mu^+\nu_\mu$. For three neutral particles, π^0, η, and Σ^0, the predominant modes of decay are electromagnetic, not weak, e.g., $\Sigma^0 \rightarrow \Lambda + \gamma$ or $\pi^0 \rightarrow \gamma + \gamma$. The weak decays of strange particles seem to obey other selection rules. For example, in leptonic decays of hadrons the selection rule $\delta S = \delta Q$ holds, where δS and δQ denote the change in strangeness and in charge of strongly interacting particles: e.g., $\Sigma^- \rightarrow n + l^- + \bar{\nu}$ is allowed ($\delta S = \delta Q = +1$), but $\Sigma^+ \rightarrow n + l^+ + \bar{\nu}$ is forbidden ($\delta S = +1$, $\delta Q = -1$).

The neutral K mesons provide a unique and fascinating application of the ideas described above when coupled with ordinary quantum mechanics. The K^0 meson is a particle with $Y = S = +1$, its antiparticle denoted by $\bar{K}^0$ has opposite hypercharge $Y = S = -1$. Once either particle is produced in a strong reaction, it is observed to decay with two lifetimes, not one. One calls the short-lived particle a K_S^0 and the longer-lived particle a K_L^0. Each of these particles has different linear combinations of K^0 and $\bar{K}^0$ mesons, approximately $50:50$ mixtures in either case. By the general TCP theorem, the K^0 and $\bar{K}^0$ particles have identical mass, but the K_S^0 and K_L^0 differ slightly in mass, in fact, by the incredibly small amount of 2 parts in 10^{14}. The remarkable $K^0 - \bar{K}^0$ system is also the only one to date in which the weak interaction has been found not to conserve the value of the operator CP, in particular in both the leptonic and nonleptonic decays of K_L^0.

8h-4. Properties of Elementary Particles. A detailed list of the intrinsic properties of the stable elementary particles is given in Table 8h-3. By "stable" we mean stable with respect to the strong interactions, but not with respect to the weak or electromagnetic interactions. This table is reproduced essentially verbatim from the latest edition of the annual Review of Particle Properties, that is printed each January by *Reviews of Modern Physics* [2]. This article contains all the references to the experimental measurements that have gone into the data in Table 8h-3. Since the journal *Reviews of Modern Physics* is readily available, we have chosen not to reproduce the hundreds of references here. That article also contains a detailed listing of the properties of the known unstable particles as well.

In Table 8h-3 the first column lists the symbol for each particle. The second lists four quantum numbers: the I-spin, I, angular momentum, and parity (where appropriate) in the symbol J^P, and the C quantum number if applicable. The third column lists the mass in MeV, the mass^2 in GeV^2, and the mass difference where it has been measured directly for members of the same charge multiplet. The next column has the mean life in seconds along with the mean distance for decay $c\tau$ in centimeters. The remaining columns contain a list of the partial decay modes for each unstable particle, the fraction of the total decay probability for each decay mode, and the unique momentum (two body decay) or maximum momentum (three or more bodies) of a secondary particle in the rest system of the decaying particle. These detailed properties are given for each particle, but not for its antiparticle, since the TCP theorem implies that the properties of the antiparticle are identical to those of each particle except for the appropriate quantum-number transformations (see page 8-283).

8h-5. SU$_3$ Classification of Hadrons—The Eightfold Way. *Supermultiplets.* We have already shortened our table of hadrons by grouping them into I-spin multiplets. Thus, it was pointed out in Sec. 8h-3 that the neutron n and proton p both belonged to the I-spin doublet called the nucleon N, and the π mesons, which can appear with three electric charges Q, form an I-spin triplet (π^-,π^0,π^+). But we have so far treated the different multiplets as independent and "elementary."

Now we proceed to point out that particle physicists further group these multiplets into "supermultiplets," of 1, 8, or 10 particles; so that, in fact, all the mesons in Table 8h-3 are said to belong to the $J^P = 0^-$ octet, and all the baryons except the Ω^- belong to the $J^P = \frac{1}{2}^+$ octet.

Typical supermultiplets are illustrated in Fig. 8h-1. Each dot represents a particle, plotted in a space where electric charge Q increases to the right and hypercharge Y increases upward. (More precisely, $x = Q - Q_{\mathrm{av}} = I_3, y = Y$, with I_3 and Y defined in Sec. 8h-3.) The eight baryons with $J^P = \frac{1}{2}^+$ are arranged at the upper left. Here the nucleon doublet N, (with $Y = +1$) contributes two dots (n and p), the Σ triplet ($Y = 0$) adds three ($\Sigma^-,\Sigma^0,\Sigma^+$), the Λ singlet one, and the $\Xi(Y = -1)$ adds the other two. We shall comment below on the symmetry of the hexagon thus created, but first we continue empirically.

The next array also turns out to form a hexagon. This hexagon consists of the eight $J^P = 0^-$ mesons which happen all to be stable (against strong decay, Sec. 8h-2), and hence are listed in Table 8h-3 along with the eight stable baryons.

The next array (also hexagonal!) is made of $J^P = 1^-$ mesons which happen all to be unstable, and so they are called "resonances" and are omitted from Table 8h-3. (A table of $\sim$50 resonant multiplets can be found in ref. 2.) Several other meson octets are now known.

Finally, Fig. 8h-1 shows a triangular "decuplet" of the 10 baryons with $J^P = \frac{3}{2}^+$. Nine of these are resonances; one is the stable Ω^- baryon.

Quarks. The SU$_3$ explanation of the hexagons and triangles is also sketched in Fig. 8h-1. In 1961 Gell-Mann and Ne'eman independently pointed out that these supermultiplets of 10, 8, 1 would be built up out of a single supermultiplet of 3 "primitive"

TABLE 8h-3. INTRINSIC PROPERTIES OF STABLE ELEMENTARY PARTICLES: JANUARY, 1970

Particle	$I^G(J^P)C$	Mass, MeV / Mass², GeV²	Mean life, sec / $c\tau$, cm	Partial mode	Fraction[a]	P or $P_{max,}$[b] MeV/c
γ	$0, 1(1^-)-$	$0(<2\times10^{-21})$	Stable	Stable		
ν	$\nu_e, J=\frac{1}{2}$ / ν_μ	$0(<60 \text{ eV})$ / $0(<1.6)$	Stable	Stable		
e	$J=\frac{1}{2}$	0.511006 ± 0.000002	Stable ($>2\times10^{21}\,y$)	Stable		
μ	$J=\frac{1}{2}$	105.659 ± 0.002 $m^2 = 0.0112$ $m_\mu - m_{\pi^\pm} = -33.920 \pm 0.013$	$(2.1983 \pm 0.0008) \times 10^{-6}$ $c\tau = 6.592 \times 10^4$	$e\nu\bar\nu$ $e\gamma\gamma$ $3e$ $e\gamma$	100 $(<1.6\ )10^{-5}$ $(<1.3\ )10^{-7}$ $(<2\ )10^{-8}$	53 53 53 53
$\pi^\pm$	$1^-(0^-)$	139.578 ± 0.013 $m^2 = 0.0195$	$(2.603 \pm 0.006) \times 10^{-8}$, $S = 2.0^*$ $c\tau = 781$ $(\tau^+ - \tau^-)/\bar\tau = (0.05 \pm 0.07)\%$ (test of CPT)	$\mu\nu$ $e\nu$ $\mu\nu\gamma$ $\pi^0 e\nu$ $e\nu\gamma$	100 % $(\ 1.24 \pm 0.03\ )10^{-4}$ [c]$(\ 1.24 \pm 0.25\ )10^{-4}$ $(\ 1.02 \pm 0.07\ )10^{-8}$ [c]$(\ 3.0 \pm 0.5\ )10^{-8}$	30 70 30 5 70
π^0	$1^-(0^-)+$	134.975 $m^2 = 0.0177$ $m_{\pi^\pm} - m_{\pi^0} = 4.6041 \pm 0.0037$	$(0.89 \pm 0.18) \times 10^{-16}$, $S = 1.6^*$ $c\tau = 2.67 \times 10^{-6}$	$\gamma\gamma$ $\gamma e^+ e^-$ $\gamma\gamma\gamma$ $e^+e^+e^-e^-$	$(\ 98.83 \pm 0.04\ )\%$ $(\ 1.17 \pm 0.04\ )\%$ $(<5\ )10^{-6}$ [d]$(\ 3.47\ )10^{-5}$	67 67 67 67
$K^\pm$	$\frac{1}{2}(0^-)$	493.82 ± 0.11 $m^2 = 0.244$	$(1.235 \pm 0.004) \times 10^{-8}$, $S = 1.8^*$ $c\tau = 370$ $(\tau^+ - \tau^-)/\bar\tau = (0.09 \pm 0.12)\%$	$\mu\nu$ π^0 $\pi^-\pi^+$ $\pi^0\pi^0$	$(\ 63.77 \pm 0.29\ )\%\ S = 1.1^*$ $(\ 20.93 \pm 0.30\ )\%\ S = 1.2^*$ $(\ 5.57 \pm 0.04\ )\%\ S = 1.2^*$ $(\ 1.70 \pm 0.05\ )\%$	236 205 126 133

Particle	J^P	Properties	Decay mode	Fraction		Ref.
		(test of CPT) $S = 1.3*$	$\mu\pi^0\nu$	(3.18 ± 0.11)	%$S = 2.0*$	215
			$e\pi^0\nu$	(4.85 ± 0.07)	%$S = 1.2*$	228
			$\pi\pi^\mp e^\pm\nu$	(3.3 ± 0.3)	10^{-5}	203
			$\pi\pi^\pm\mu^\mp\nu$	(<7)	10^{-7}	151
			$\pi\pi^\mp\mu^\pm\nu$	(0.9 ± 0.4)	10^{-5}	151
			$\pi\pi^\mp\mu^\pm\nu$	(<3)	10^{-6}	247
			$e\nu$	(1.2 ± 0.3)	10^{-4}	205
			$\pi\pi^0\gamma$	(<1.9)	10^{-5}	126
			$\pi e\nu\gamma$	c(10 ± 4)	10^{-4}	227
			πe^+e^-	c(6 ± 4)	10^{-5}	227
			$\pi\mu^+\mu^-$	c(<0.4)	10^{-6}	172
			$\pi\gamma\gamma$	(<2.4)	10^{-6}	227
				(<1.1)	10^{-4}	
K^0	$\frac{1}{2}(0^-)$	$m_{K^\pm} - m_{K^0} = -3.94 \pm 0.13$		50 % K_{short}, 50 % K_{long}		
K^0_S	$\frac{1}{2}(0^-)$	497.76 ± 0.16 $S = 1.5*$ $m^2 = 0.248$ $(0.862 \pm 0.006) \times 10^{-10}$, $S = 1.2*$ $c\tau = 2.59$	$\pi^+\pi^-$	(68.7 ± 0.6)	%$S = 1.6*$	206
			$\pi^0\pi^0$	(31.3)	%	209
			$\mu^+\mu^-$	(<3.1)	10^{-7}	225
			e^+e^-	(<2.2)	10^{-7}	249
			$\pi^+\pi^-\gamma$	c(3.3 ± 1.2)	10^{-3}	206
K^0_L	$\frac{1}{2}(0^-)$	$m_{K_S} - m_{K_L} = -(0.469 \pm 0.015) \times \dfrac{1}{\tau_S}$ $(5.38 \pm 0.19) \times 10^{-8}$ $c\tau = 1614$ $S = 1.6*$ $\dfrac{\Gamma(K_S \to \pi^+\pi^-\pi^0)}{\Gamma(K_L \to \pi^+\pi^-\pi^0)} < 0.45$ (test of CP)	$\pi^0\pi^0\pi^0$	(21.5 ± 0.7)	%$S = 1.2*$	139
			$\pi^+\pi^-\pi^0$	(12.6 ± 0.3)	%	133
			$\pi\mu\nu$	(26.8 ± 0.7)	%$S = 1.2*$	216
			$\pi e\nu$	(38.8 ± 0.8)	%$S = 1.2*$	229
			$\pi^+\pi^-$	(0.157 ± 0.005)	%	206
			$\pi^0\pi^0$	c(0.121 ± 0.029)	%$S = 1.5*$	209
			$\pi^+\pi^-\gamma$	(<0.4)	10^{-3}	206
			$\gamma\gamma$	(5.2 ± 0.5)	$10^{-4}\ S = 1.6*$	249
			$e\mu$	(<0.6)	10^{-5}	238
			$\mu^+\mu^-$	(<1.5)	10^{-6}	225
			e^+e^-	(<1.7)	10^{-5}	249
η	$0^+(0^-)^+$	548.8 ± 0.6 $m^2 = 0.301$ $\Gamma = (2.63 \pm 0.64)$ keV Neutral decays 71.5 %	$\gamma\gamma$	(38.2 ± 2.1)	%	274
			$\pi^0\gamma\gamma$	(2.0 ± 2.8)	% } $S = 1.4*$	258
			$3\pi^0$	e(31.4 ± 2.7)	%	179

TABLE 8h-3. INTRINSIC PROPERTIES OF STABLE ELEMENTARY PARTICLES: JANUARY, 1970 (Continued)

Particle	$I^G(J^P)C$	Mass, MeV / Mass², GeV²	Mean life, sec / $c\tau$, cm	Decays — Partial mode	Decays — Fraction[a]	P or P_{max},[b] MeV/c
η			Charged decays 28.5 %	$\begin{cases}\pi^+\pi^-\pi^0\\ \pi^+\pi^-\gamma\\ \pi^0 e^+e^-\\ \pi^+\pi^- e^+e^-\end{cases}$	(23.0 ± 1.1)% S = 1.2* (5.4 ± 0.5)% (<0.01)% (0.1 ± 0.1)%	174 236 258 236
p	$\tfrac{1}{2}(\tfrac{1}{2}^+)$	938.256 ± 0.005 $m^2 = 0.880$	stable ($>2 \times 10^{28}\ y$)			
n	$\tfrac{1}{2}(\tfrac{1}{2}^+)$	939.550 ± 0.005 $m^2 = 0.882$ $m_p - m_n = -1.2933 \pm 0.0001$	$^e(0.932 \pm 0.014) \times 10^3$ $c\tau = 2.80 \times 10^{13}$	$pe^-\nu$	100 %	1
Λ	$0(\tfrac{1}{2}^+)$	1115.60 ± 0.08 S = 1.3* $m^2 = 1.245$	$(2.51 \pm 0.03) \times 10^{-10}$, S = 1.3* $c\tau = 7.54$	$p\pi^-$ $n\pi^0$ $pe\nu$ $p\mu\nu$	(65.3 ± 1.3)% 34.7)% 0.85 ± 0.07)10^{-3} 1.35 ± 0.60)10^{-3} S = 1.3*	100 104 163 131
Σ^+	$1(\tfrac{1}{2}^+)$	1189.40 ± 0.19 S = 1.7* $m^2 = 1.412$ $m_{\Sigma^+} - m_{\Sigma^-} = -7.92 \pm 0.13$	$(0.802 \pm 0.007) \times 10^{-10}$ $c\tau = 2.41$ $\dfrac{\Gamma(\Sigma^+ \to l^+\nu\nu)}{\Gamma(\Sigma^- \to l^-\nu\nu)} = {}<0.03$	$p\pi^0$ $n\pi^+$ $p\gamma$ $n\pi^+\gamma$ $\Lambda e^+\nu$ $\leftarrow \begin{cases} n\mu^+\nu \\ ne^+\nu \end{cases}$	(51.7 ± 0.8)% 48.3)% 1.16 ± 0.17)10^{-3} 1.3 ± 0.3)10^{-4} S = 1.4* 2.02 ± 0.47)10^{-5} <1.1)10^{-5} <0.7)10^{-5}	189 185 225 185 72 202 224
Σ^0	$1(\tfrac{1}{2}^+)$	1192.46 ± 0.12 S = 1.2* $m^2 = 1.422$	$<1.0 \times 10^{-14}$ $c\tau < 3 \times 10^{-4}$	$\Lambda\gamma$ Λe^+e^-	100 % $^d($ 5.45 ±)10^{-3}	75

	$I(J^P)$	Mass, lifetime	Decay mode	Fraction		p_{max}
Σ^-	$1(\frac{1}{2}^+)$	1197.32 ± 0.11 $S = 1.3*$ $m^2 = 1.434$ $m_{\Sigma^0} - m_{\Sigma^-} = 4.86 \pm 0.07$			%	
		$^e(1.49 \pm 0.03) \times 10^{-10}$, $S = 2.1*$ $c\tau = 4.47$	$n\pi^-$	100		193
			$ne^-\nu$	1.06 ± 0.05	$)10^{-3}$	230
			$n\mu^-\nu$	0.45 ± 0.04	$)10^{-3}$	210
			$\Lambda e^-\nu$	0.60 ± 0.06	$)10^{-4}$	79
			$n\pi^-\gamma$	$^c(1.0 \pm 0.2$	$)10^{-4}$	193
Ξ^0	$\frac{1}{2}(\frac{1}{2}^+)^f$	1314.7 ± 0.7 $m^2 = 1.728$ $m_{\Xi^0} - m_{\Xi^-} = -6.5 \pm 0.7$			%	
		$(3.03 \pm 0.18) \times 10^{-10}$ $c\tau = 9.10$	$\Lambda\pi^0$	100		135
			$p\pi^-$	$(<0.9$	$)10^{-3}$	299
			$pe^-\nu$	$(<1.3$	$)10^{-3}$	323
			$\Sigma^+e^-\nu$	$(<1.5$	$)10^{-3}$	119
			$\Sigma^-e^+\nu$	$(<1.5$	$)10^{-3}$	112
			$\Sigma^+\mu^-\nu$	$(<1.5$	$)10^{-3}$	64
			$\Sigma^-\mu^+\nu$	$(<1.5$	$)10^{-3}$	49
			$p\mu^-\nu$	$(<1.3$	$)10^{-3}$	309
Ξ^-	$\frac{1}{2}(\frac{1}{2}^+)^f$	1321.25 ± 0.18 $m^2 = 1.746$			%	
		$(1.66 \pm 0.04) \times 10^{-10}$, $S = 1.1*$ $c\tau = 4.98$	$\Lambda\pi^-$	100		139
			$\Lambda e^-\nu$	$^g(0.67 \pm 0.23$	$)10^{-3}$	190
			$\Sigma^0 e^-\nu$	$(<0.5$	$)10^{-3}$	122
			$\Lambda\mu^-\nu$	$(<1.3$	$)10^{-3}$	163
			$\Sigma^0\mu^-\nu$	$(<0.5$	$)\%$	70
			$n\pi^-$	$(<1.1$	$)10^{-3}$	303
			$ne^-\nu$	$(<1.0$	$)\%$	327
Ω^-	$0(\frac{3}{2}^+)^f$	1672.5 ± 0.5 $m^2 = 2.797$				
		$1.3^{+0.4}_{-0.3} \times 10^{-10}$ $c\tau = 3.9$	$\Xi^0\pi^-$ $\Xi^-\pi^0$ ΛK^-	Total of 28 events seen		293 289 210

From Review of Particle Properties, UCRL-8030. N. Barash-Schmidt, A. Barbaro-Galtieri, C. Bricman, S. E. Derenzo, L. R. Price, A. Rittenberg, Matts Roos, A. H. Rosenfeld, Paul Söding, and C. G. Wohl. (Closing date for data: Nov. 1, 1969.)
Quantities in italics have changed by more than one (old) standard deviation since January, 1969.

* S = Scale factor = $\sqrt{\chi^2/(N-1)}$, where $N \approx$ number of experiments. S should be ≈ 1. If $S > 1$, we have enlarged the error of the mean, δx, i.e., $\delta x \to S \, \delta x$. This convention is still inadequate, since if $S \gg 1$, the experiments are probably inconsistent, and therefore the real uncertainty is probably even greater than $S \, \delta x$. See text and ideogram in data card listings, UCRL-8030.
a Quoted upper limits correspond to a 90% confidence level.
b In decays with more than two bodies, P_{max} is the maximum momentum that any particle can have.
c See data card listings, (UCRL-8030) for energy limits used in measuring this branching ratio.
d Theoretical value; see also data card listings, UCRL-8030.
e See note in data card listings, UCRL-8030.
f Predicted from SU_3.
g Assumes rate for $\Xi^- \to \Sigma^0 e^- \nu$ small compared with $\Xi^- \to \Lambda e^- \nu$.

particles called quarks and an antimultiplet (antiquarks). (It is not known whether quarks exist in nature or only as a mathematical explanation.) The quarks "exist" as an I-spin doublet (such as n and p), and a singlet (such as Λ). In their simplest form they would have surprising *fractional* quantum numbers, $B = \frac{1}{3}$, $Q = -\frac{1}{3}$ or $+\frac{2}{3}$, etc. Mesons are then tightly bound states of quark + antiquark $(q\bar{q})$; baryons "contain" three quarks (qqq) held together by the strong interaction.

The mathematics of how three primitive objects can be combined into larger groups is called *group theory*, and the particular combination that correctly explains nature is called, in group theory, SU_3; hence the title for this section: "SU_3 Classification."

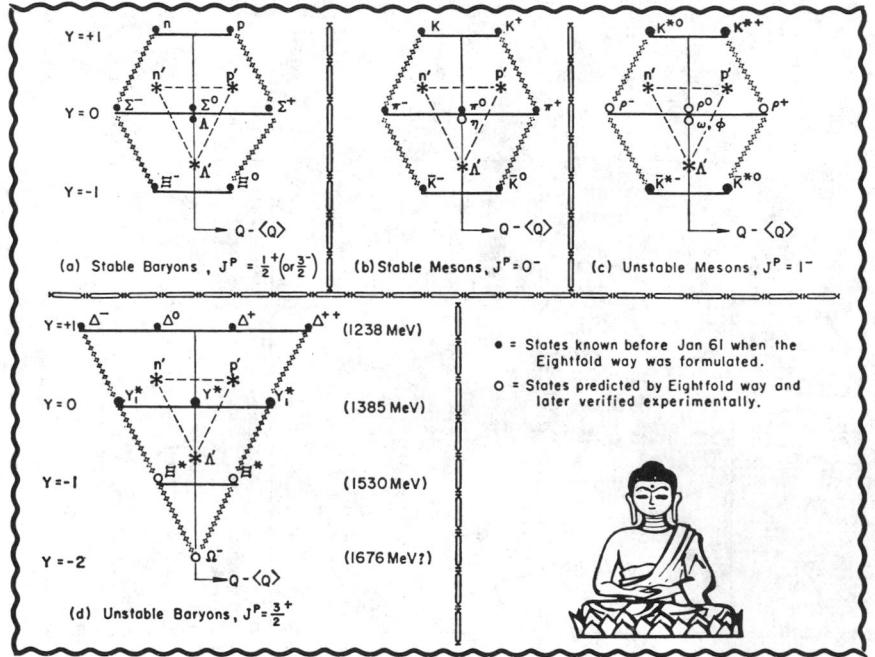

FIG. 8h-1. The asterisks labeled n', p', and Λ' are a possible set of primitive particles called "quarks," from which the mesons and baryons can be formed.

The algebraic rules of SU_3 explain much more than the size of multiplets—they also explain quite well the masses and decay modes of particles and resonances (see any textbook on particle physics [3]).

8h-6. Further Reading. There are many textbooks on particle physics. A sample of them are listed in ref. 3. Two fairly recent and complete books are those by Gasiorowicz [3] and by Frazer [3]. Many excellent semipopular articles can be found in the *Scientific American* [4], and more technical review articles in the *Annual Review of Nuclear Science* [5]. A mild apology to the reader—this text is rather compact and not too easy to read; two articles which cover much of the same material but in a more leisurely fashion have been written by Ne'eman [6] and Rosenfeld [7]. A more extended but nonmathematical discussion of the subject can be found in a readable book by Ford [8].

Acknowledgement. We wish to thank Dr. LeRoy Price of the Berkeley Particle Data Group for his help and criticism.

References

1. Weber, J.: *Phys. Rev. Letters* **22**, 1320 (1969).
2. Particle Data Group: *Rev. Mod. Phys.* **41**, 109 (1969).
3.   (i) Gasiorowicz, S.: "Elementary Particle Physics," John Wiley & Sons, Inc., New York, 1966.
  (ii) Frazer, W.: "Elementary Particles," Prentice-Hall, Inc., Englewood Cliffs, N.J., 1966.
  (iii) Bernstein, J.: "Elementary Particles and Their Currents," W. H. Freeman and Company, San Francisco, 1968.
  (iv) Gell-Mann, M., and Y. Ne'eman: "The Eightfold Way," W. A. Benjamin, Inc., New York, 1964.
  (v) Källen, G.: "Elementary Particle Physics," Addison-Wesley Press, Inc., Cambridge, Mass., 1964.
  (vi) Sakurai, J. J.: "Invariance Principles and Elementary Particles," Princeton University Press, Princeton, N.J., 1964.
  (vii) Adair, R. K., and E. C. Fowler: "Strange Particles," Interscience Publishers, a division of John Wiley & Sons, Inc., New York, N.Y., 1963.
  (viii) Levi-Setti, R.: "Elementary Particles," University of Chicago, 1963.
  (ix) Yang, C. N.: "Elementary Particles," Princeton University Press, Princeton, N.J., 1962.
  (x) Williams, W. S. C.: "An Introduction to Elementary Particles," Academic Press, Inc., New York, 1961.
4. *Scientific American* Articles:
  (i) The Overthrow of Parity, P. Morrison, April, 1957.
  (ii) Pions, R. Marshak, January, 1957.
  (iii) Elementary Particles, Gell-Mann and Rosenfeld, July, 1957.
  (iv) The Weak Interactions, S. B. Treiman, March, 1959.
  (v) Two Neutrino Experiment, L. Lederman, January, 1962.
  (vi) Strongly Interacting Particles, Chew, Gell-Mann, and Rosenfeld, February, 1964.
  (vii) The Omega-Minus Experiment, W. B. Fowler and N. P. Samios, October, 1964.
  (viii) Violations of Symmetry in Physics, E. P. Wigner, December, 1965.
5.   (i) Lee, T. D., and C. S. Wu: *Ann. Rev. Nucl. Sci.* **15**, 381(1965); **16**, 471(1966).
  (ii) Tripp, R. D.: *ibid.* **15**, 325 (1965).
  (iii) Feinberg, G., and L. M. Lederman: *ibid.* **13**, 431 (1963).
6. Ne'eman, Y.: "Science Year (the World Book Science Annual), 1968.
7. Rosenfeld, A. H.: Elementary Guide, UCRL-11 100 (unpublished).
8. Ford, K. W.: "The World of Elementary Particles," Blaisdell Publishing Company, a division of Ginn and Company, Waltham, Mass., 1963.

8i. Health Physics[1]

KARL Z. MORGAN AND JAMES E. TURNER

Oak Ridge National Laboratory

8i-1. Introduction. The practice of health physics utilizes knowledge gained in all sciences to furnish an understanding of the mechanisms of radiation damage and to provide adequate and reasonable limits for exposure, measurements of exposure, and the specification of conditions and procedures to ensure protection. It embodies

[1] Work sponsored by the U.S. Atomic Energy Commission under contract with Union Carbide Corporation.

the application of many scientific and technical disciplines, i.e., physics, biology, chemistry, engineering, etc., to the end that any situation involving possible radiation hazard to man can be analyzed correctly, and suitable steps can be taken to prevent harm to man or to his environment. Health physics involves research, engineering, educational, and applied activities. It deals with the scattering and loss of energy of ionizing radiation and the damage produced by the passage of this radiation through matter. Thus, in addition to applied activities, there are many health-physics research and engineering problems such as (1) shielding, (2) dosimetry, (3) studies of physical parameters relating to dosimetry (e.g., stopping power, attachment coefficient, energy to produce an ion pair, etc.), (4) radioactive-waste disposal, (5) studies of human exposures, (6) determination of permissible exposure values, (7) studies of effects of ionizing radiation on the environment, etc. A health physicist is a person engaged in and dedicated to a study and practice of problems of providing radiation protection. He is concerned with obtaining an understanding of mechanisms of radiation damage and with the development and implementation of instruments, methods, and procedures so that he can determine the existence of hazardous ionizing radiation and provide protection to man and his environment from its unwarranted deleterious effects.

8i-2. Definition of Units and Terms Used in Health Physics.[1] *Absorbed Dose:* the amount of energy imparted by ionizing radiation to a sample of matter per unit mass. The unit of absorbed dose is the rad ($= 100$ ergs/g).

Absorption Coefficient $(\mu - \sigma_s)$: the difference between the attenuation coefficient and that for Compton scattering. This quantity, which is used to a good approximation to describe photon energy absorption, excludes the part of the original photon energy that escapes as a degraded photon from the site of interaction. The dimension of $\mu - \sigma_s$ is reciprocal distance (e.g., cm^{-1}).

Activity: the disintegration rate of a radionuclide. The unit of activity is the curie (1 Ci corresponds to 3.7×10^{10} disintegrations/sec).

Attenuation Coefficient μ (*Macroscopic Cross Section*): the probability of interaction per unit distance traveled. The dimension of μ is reciprocal distance (e.g., cm^{-1}). EXAMPLE: The relative number of photons of a given energy that do not experience an interaction in traveling a distance x is $e^{-\mu x}$. In terms of σ_t, the total microscopic cross section (e.g., cm^2), $\mu = N\sigma_t$, where N = number of electrons per unit volume.

The *Bragg-Gray principle* and applications of it are used as the basis of many measurements of ionizing radiation. According to this principle the energy absorbed per unit mass $(dE/dm)_b$ in a given medium b is related to the ionization in a small gas-filled cavity in that medium by the expression

$$\left(\frac{dE}{dm}\right)_b = P_b W_g J_g \tag{8i-1}$$

Here P_b is the relative mass stopping power of the medium b with respect to the gas g, W_g is the average energy required to produce an ion pair in the gas, and J_g, the quantity that is usually determined experimentally, is the number of ion pairs produced per unit mass of the gas in the cavity. It should be emphasized that, in order for this principle to hold always, the gas cavity must be small compared with the range of the ionizing particles, and both W_g and P_b must be independent of the energy of the radiation. When the walls of the chamber and the gas are made of the same material, e.g., air- or tissue-equivalent substances, the Bragg-Gray principle also applies with a cavity large compared with the range of the ionizing particles.

---

[1] More detailed discussions are given in Radiation Quantities and Units, report 10a of the International Commission of Radiological Units and Measurements, *Natl. Bur. Standards Handbook* 84, 1962.

Curie, Ci: unit of activity: 1 Ci = 3.7 × 10¹⁰ disintegrations/sec. The millicurie (1 mCi = 10^{-3} Ci), microcurie (1 μCi = 10^{-6} Ci), nanocurie (1 nCi = 10^{-9} Ci), and picocurie (1 pCi = 10^{-12} Ci) are often used.

Dose Equivalent: defined for purposes of radiation protection as the product of the absorbed dose and relevant modifying factors, such as those for radiation quality (e.g., QF, which relates to LET, and the *H* factor, which relates to the damage from internally deposited, bone-seeking radionuclides relative to that of radium). The unit of dose equivalent is the rem. Dose equivalents from different sources of radiation are additive in protection work.

Exposure: the amount of charge (of either sign) produced in air by X- or gamma-ray photons per unit mass of air. The unit of exposure is the roentgen (= 2.58 × 10^{-4} coul/kg).

Fluence: the ratio of the number of particles or photons that enter a small, imaginary test sphere placed in a radiation field and the cross-sectional area of the sphere. The dimension of fluence is the square of reciprocal distance (e.g., cm^{-2}).

Flux Density: fluence per unit time. The dimensions of flux density are reciprocal area times reciprocal time (e.g., $cm^{-2}\ sec^{-1}$).

Linear Energy Transfer, LET: linear rate of energy loss along the track of a particle. LET is often expressed in keV/μm or MeV/cm (1 keV/μm = 10 MeV/cm).

Mass-absorption Coefficient, $(\mu - \sigma_s)/\rho$: the quotient of the absorption coefficient and the density of a material. Dimensions are area times reciprocal mass (e.g., cm^2/g).

Mass-attenuation Coefficient, μ/ρ: the quotient of the attenuation coefficient and the density of a material. Dimensions are area times reciprocal mass (e.g., cm^2/g).

Mass Stopping Power, P/ρ: the stopping power of a material divided by its density. Often expressed in MeV/(g/cm^2) or ergs/(g/cm^2). See Sec. 8i-3.1 and Fig. 8i-1.

Quality Factor, QF: numerical linear-energy-transfer-dependent factor, depending on the kind of incident radiation and its energy. The product of QF and absorbed dose gives the dose equivalent used for purposes of radiation protection. See tables given in Sec. 8i-13.

Rad: Unit of absorbed dose: 1 rad = 100 ergs/g.

Relative Biological Effectiveness, RBE: the biological effectiveness of any type and energy of ionizing radiation in producing a specific biological effect (e.g., a certain incidence or degree of leukemia, anemia, sterility, carcinomas, cataracts, shortening of life-span, etc.) relative to damage produced by X rays, having an energy of about 200 keV or a linear energy transfer in water of about 3 keV/μm delivered at a rate of about 10 rads/min. Gamma radiation from ^{60}Co is often used as the reference standard. The RBE is given frequently as an average value in the common energy range of a particular type of ion and/or throughout the medium under study.

Rem: roentgen-equivalent-man. Unit of dose equivalent.

Roentgen, R: unit of exposure: 1 R = 2.58 × 10^{-4} coul/kg. This quantity is numerically the same as that implied by the older definition of the roentgen as that quantity of X or gamma radiation that produces 1 esu of charge of either sign per 0.001293 g of dry air (1 cc at 0°C and 760 mm Hg).

Stopping Power, P or $-dE/dx$: mean rate of energy loss of a charged particle per unit distance traveled. The stopping power of a medium for a given particle is numerically equal to LET. Often expressed in ergs/cm or MeV/cm. (See mass stopping power.)

Specific Ionization S: the average number of ion pairs produced per unit distance along the track of a particle.

W Value: the average energy needed to produce an ion pair. (See page 8-296.)

8i-3. Useful Data and Equations. *Stopping Powers.* Figure 8i-1 shows the stopping powers of water for a number of particles. These values differ only slightly

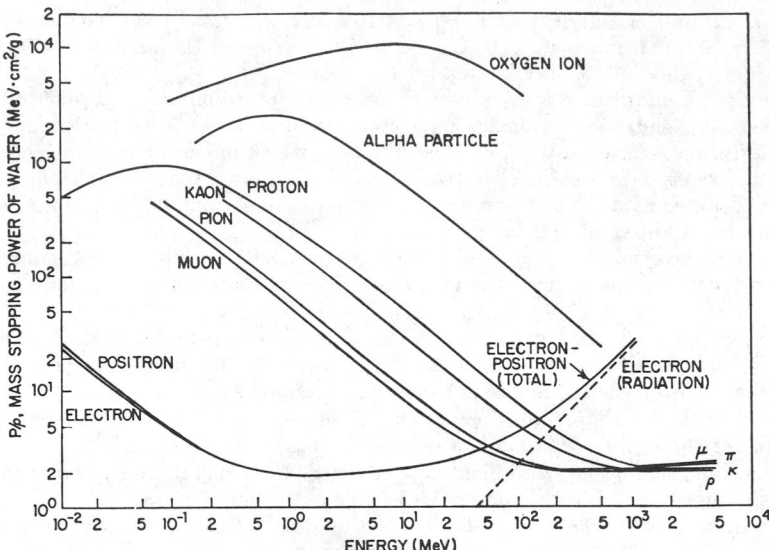

FIG. 8i-1. Mass stopping power of water for several particles.   The dashed curve shows the contribution of radiation (bremsstrahlung) to the mass stopping power for electrons.

from those of muscle or other soft tissue.   These curves are based on the following sources:

Barkas, W. H., and M. J. Berger: Studies in Penetration of Charged Particles, U. Fano ed., *Natl. Acad. Sci.–Natl. Res. Council Publ.* 1133, 1964.

Berger, M. J., and S. M. Seltzer: Studies in Penetration of Charged Particles, U. Fano, ed. See above.   We are grateful to Dr. Berger for furnishing the positron data.

Bichsel, H.: "American Institute of Physics Handbook, 3d ed., D. W. Gray, ed., McGraw-Hill Book Company, New York, 1972.   (This volume.)

Neufeld, J., and W. S. Snyder: in "Selected Topics in Radiation Dosimetry," International Atomic Energy Agency, Vienna, 1961.

Steward, P. G.: Stopping Power and Range for any Nucleus in the Specific Energy Interval 0.01–500 MeV/AMU in any Nongaseous Material, *LRL Rept.* UCRL-18127, 1968.

Whaling, W.: "Encyclopedia of Physics," vol. 34(2), p. 214, Springer-Verlag OHG, Berlin, 1958.

Stopping powers of a number of materials for different charged particles can be calculated over a wide range of energies from the information given by H. Bichsel in Sec. 8d of this Handbook.

Average stopping power $\bar{P}$ (MeV/cm) of particle of energy E (MeV) over its range R (cm):

$$\bar{P} = \frac{E}{R} \tag{8i-2}$$

Average mass stopping power (MeV cm^2/g) is $\bar{P}/\rho$, where ρ is the density (g/cm^3) of the medium traversed.

Ranges.   The mean ranges of electrons, protons, and alpha particles in water, muscle, bone, and lead are shown in Fig. 8i-2.   The ranges of these particles in air are given in Fig. 8i-3.   These figures are based on the following sources:

Barkas, W. H., and M. J. Berger: *ibid.*

Berger, M. J., and S. M. Selzer: *ibid.*

Bethe, H. A., and J. Ashkin: "Experimental Nuclear Physics," vol. I, E. Segrè, ed., John Wiley & Sons, Inc., New York, 1953.

Evans, R. D.: "The Atomic Nucleus," McGraw-Hill Book Company, New York, 1955.
Snyder, W. S., and J. Neufeld: On the Energy Dissipation of Moving Ions in Tissue, *Oak Ridge Natl. Lab. Rept.* ORNL-1083, Oak Ridge, Tenn., 1951.
Steward, P. G.: *ibid.*

These mean ranges have been calculated at high energies without allowance for nuclear cascades, i.e., absorption of a proton by a nucleus.

Except at low velocities, where capture and loss of electrons by a moving ion occurs, the ranges of other heavy particles (e.g., muons, pions, deuterons, tritons) can be found from the range-energy curves given in Fig. 8i-2, since energy loss depends in a

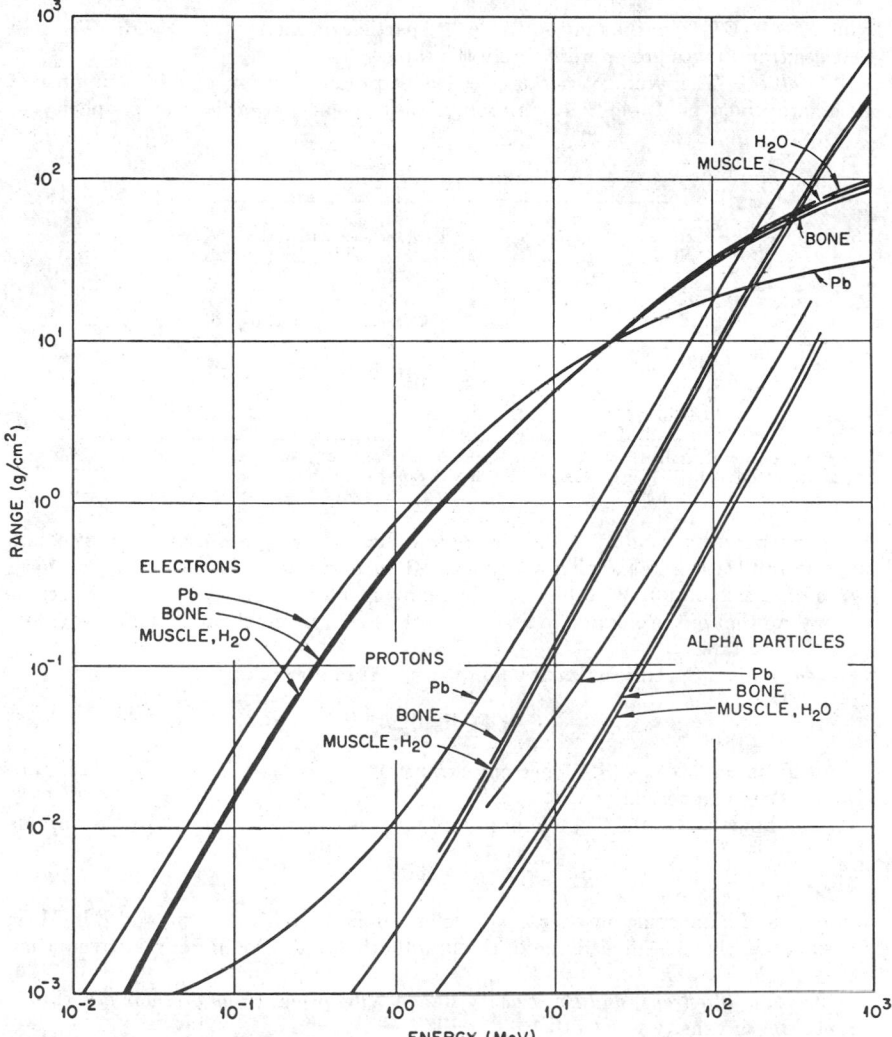

FIG. 8i-2. Mean ranges of electrons, protons, and α particles in water, muscle, bone, and lead. See text for determining ranges of other heavy particles. Ranges in other materials can be approximated from the water and lead curves by interpolating on the basis of average atomic number.

known way on charge and velocity. For example, the ranges $R_1(v)$ and $R_2(v)$ of two heavy particles, moving with the same speed v in a medium and having charges z_1 and z_2 and masses M_1 and M_2, are related by the equation

$$R_1(v) = \left(\frac{z_2}{z_1}\right)^2 \frac{M_1}{M_2} R_2(v) \tag{8i-3}$$

Nonrelativistically, the range $R_1(E)$ of one particle at energy E is given by

$$R_1(E) = \left(\frac{z_2}{z_1}\right)^2 \frac{M_1}{M_2} R_2\left(\frac{M_2 E}{M_1}\right) \tag{8i-4}$$

where $R_2(M_2 E/M_1)$ is the range of the other particle at energy $(M_2/M_1)E$. Electron and positron ranges are approximately the same.

W Values. The average energies needed to produce an ion pair in a number of gases are given in Table 8i-1. Although these values, regarded for simplicity as

TABLE 8i-1. W VALUES IN eV FOR SEVERAL GASES*

He	42	CO_2	34
Ne	37	CH_4	28
Ar	26	C_2H_2	27
Kr	24	C_2H_4	27
Xe	22	C_2H_6	26
H_2	36	C_3H_8	26
N_2	36	C_4H_{10}	26
O_2	31	BF_3	36
Air	35		

* Based on data summarized by L. W. Cochran in chap. 5, "Principles of Radiation Protection," K. Z. Morgan and J. E. Turner, eds., John Wiley & Sons, Inc., New York, 1967.

being independent of the type and energy of radiation, are appropriate in most health physics applications (e.g., with X rays, radiation from radioactive sources, and most types of accelerators), W values for slow-moving heavy ions (i.e., at ion velocities lower than that of the electron in first Bohr orbit) may be much larger than the values given in the table.

Alpha Rays. Specific ionization (ion pairs/cm) in air:[1]

$$S \approx 11 \times 10^4 E^{-0.74} \tag{8i-5}$$

where E is in MeV. (<10 percent error for alpha energies $2 \leq E \leq 50$). See Table 8i-2 for numerical values.

Limited portions of the range R–energy E curve in Fig. 8i-3 can be fit by the formula

$$R = AE^k \tag{8i-6}$$

where A and k are constant over a particular portion. With R in cm and E in MeV, for example, the measured range in the neighborhood of 5 to 10 MeV is given accurately by $R = 0.31E^{1.5}$.

Beta Rays, Electrons, and Positrons. Range (cm) of electrons of energy E (MeV) in medium of density ρ (g/cm³):[1]

$$R \approx \frac{1}{\rho} [0.54E - 0.13(1 - e^{-4E})] \tag{8i-7}$$

[1] Empirical formula developed by K. Z. Morgan.

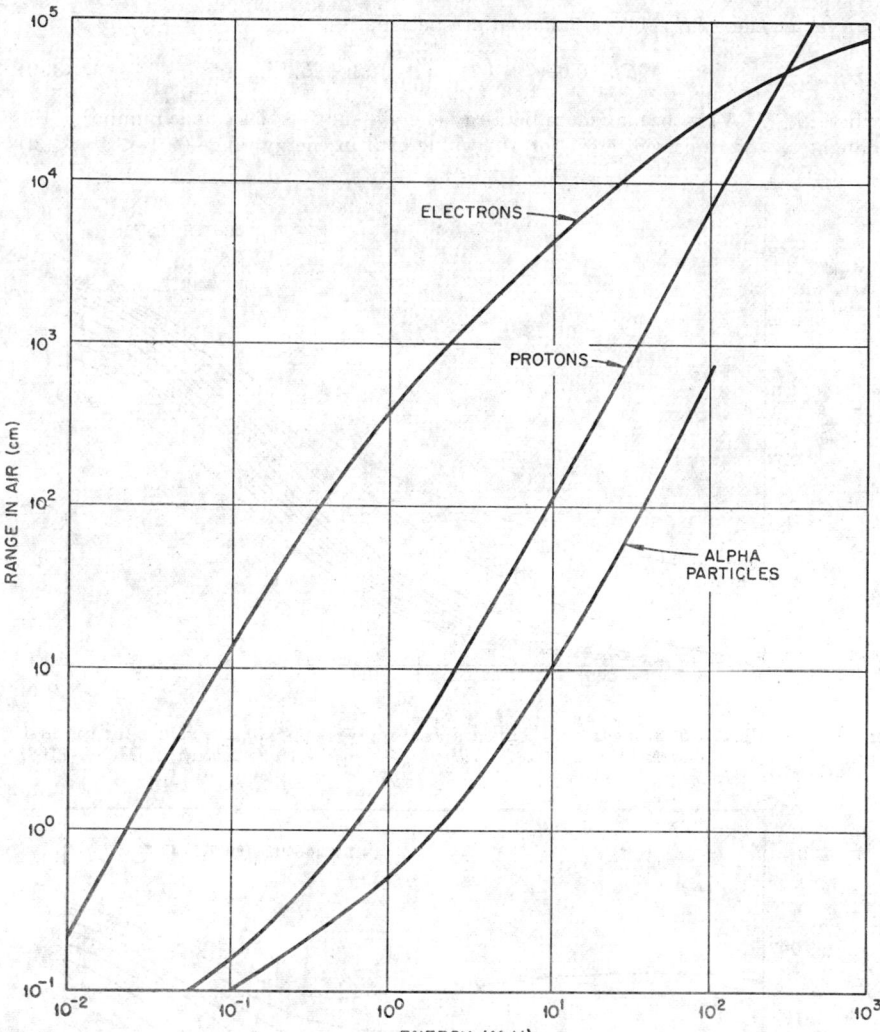

FIG. 8i-3. Mean ranges of electrons, protons, and α particles in air ($\rho = 0.001293$ g/cm³). See text for determining ranges of other heavy particles.

(For water, this formula is in error by not more than ± 25 percent for electron energies $0.01 \leq E \leq 30$.)

In the interval E (MeV) ≈ 1 to ≈ 20 the range R (g/cm²) is given by[1]

$$R \approx 0.530E - 0.106 \tag{8i-8}$$

(For water, this formula is in error by not more than ± 15 percent for electron energies $1 \leq E \leq 20$.)

[1] Katz L. and A. S. Penfold, *Rev. Mod. Phys.* **24**, 28 (1952).

Average energy $\bar{E}$ (MeV) of allowed β^- spectrum:[1]

$$\bar{E} \approx 0.099 E_m \left(1 - \frac{Z^{0.5}}{50}\right) (3 + E_m{}^{0.6}) \tag{8i-9}$$

where E_m (MeV) is the maximum beta-ray energy, and Z is the atomic number of the daughter. (<10 percent error for any value of Z in energy range $0.01 < E_m < 3$.)

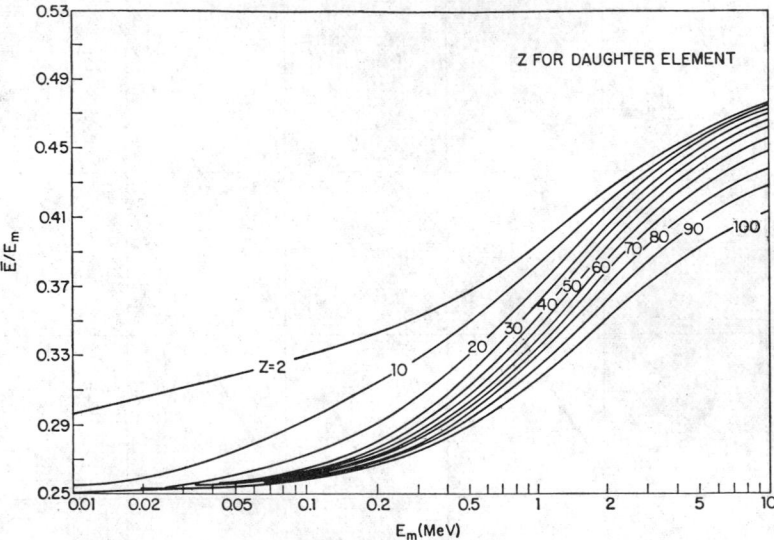

FIG. 8i-4a. Allowed β^- transitions. This graph also gives approximate values for the first forbidden nonunique transitions. [*L. T. Dillman, Oak Ridge Natl. Lab. Rept.* ORNL-4168, pp. 233ff., *Oak Ridge, Tenn.* (1967).]

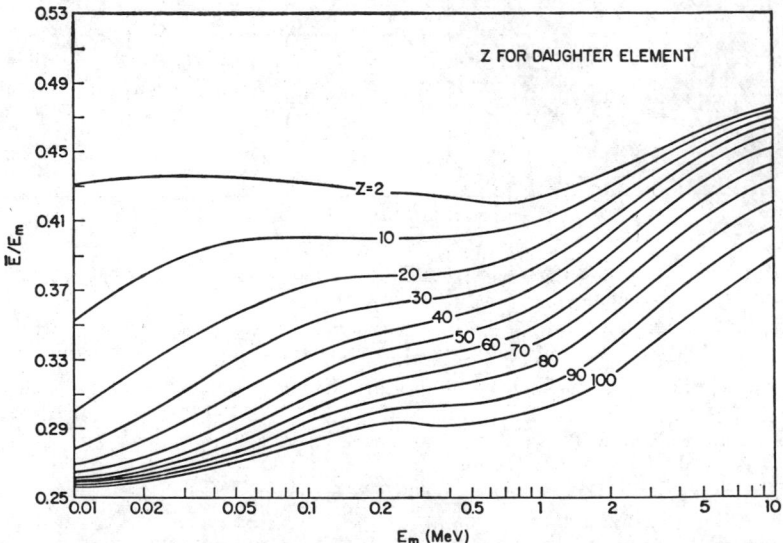

FIG. 8i-4b. First forbidden unique β^- transitions. This graph also gives approximate values for the second forbidden nonunique transitions. [*L. T. Dillman, ibid.*]

[1] Empirical formula developed by K. Z. Morgan.

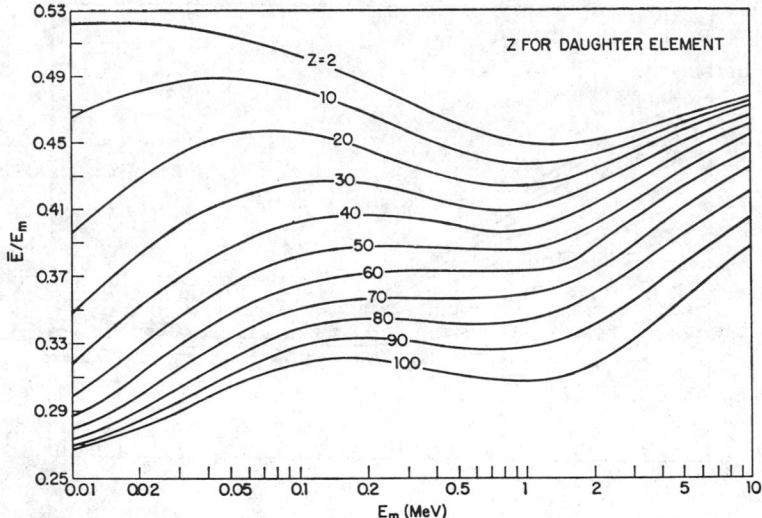

FIG. 8i-4c. Second forbidden unique β^- transitions. [L. T. Dillman, ibid.]

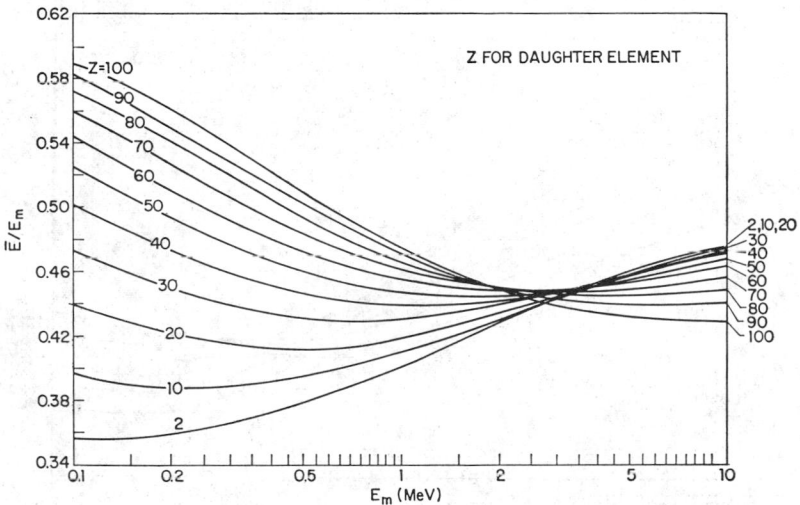

FIG. 8i-4d. Allowed β^+ transitions. This graph also gives approximate values for the first forbidden nonunique transitions. [L. T. Dillman, ibid.]

$\bar{E}$ has been calculated for electrons for a wide range of values of E_m and Z by James, Steel, and Story[1] and for both electrons and positrons by Dillman.[2] Figures 8i-4a through 8i-4f show values of $\bar{E}/E_m$ as obtained by Dillman. Note that in these figures values of Z are the atomic numbers of the daughter elements. The ratio $\bar{E}/E_m$ varies from about 0.25 to 0.63. As a rule of thumb, it is sometimes assumed that $\bar{E}/E_m \approx \frac{1}{3}$.

[1] M. F. James, B. G. Steel, and J. S. Story, Average Electron Energy in Beta Decay, U.K. Atomic Energy Authority Rept. AERE-M 640, Harwell, Berkshire, England, 1960.
[2] L. T. Dillman, see Oak Ridge Natl. Lab. Rept. ORNL-4168, pp. 233 ff., Oak Ridge, Tenn., 1967.

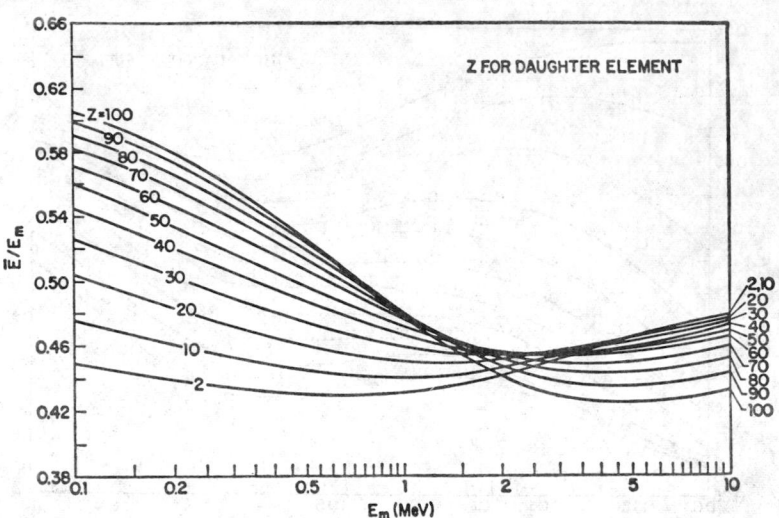

FIG. 8i-4e. First forbidden unique β^+ transitions.   This graph also gives approximate values for the second forbidden nonunique transitions.   [L. T. Dillman, ibid.]

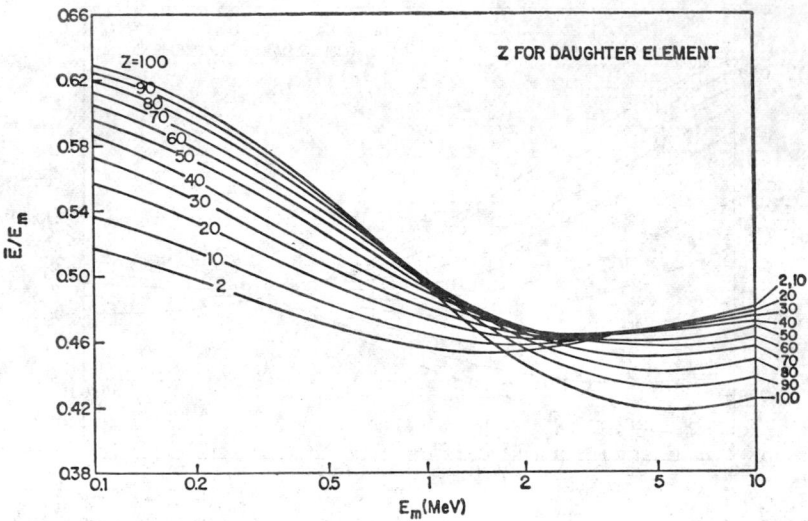

FIG. 8i 4f. Second forbidden unique β^+ transitions.   [L. T. Dillman, ibid.]

Apparent absorption coefficient (cm²/g) for beta particle of energy E (MeV) in tissue:[1]

$$\mu \approx 20E^{-1.49} \qquad (8\text{i-}10)$$

($<$10 percent error for $0.1 \leq E \leq 3$.)

Specific ionization (ion pairs/cm) for β^- with allowed spectrum and maximum energy E_m (MeV) in air:[1]

$$S \approx 33 + 63E_m^{-0.9} \qquad (8\text{i-}11)$$

($<$10 percent error for $0.05 \leq E \leq 2$.)   See Table 8i-2.

[1] Empirical formula developed by K. Z. Morgan.

TABLE 8i-2. SPECIFIC IONIZATION AND RANGES OF ALPHA AND BETA PARTICLES

Particle energy, MeV	Specific ionization in air,[a] ion pair/cm			Ranges, cm			
				Alpha particle		Beta particle	
	Alpha particle[b]	Electron[c,d]	Beta particle[e]	Air[a,f]	Soft tissue[g,h]	Air[a,c]	Soft tissue[c]
0.01	......	750	2,100	......	......	0.22	0.00025
0.05	31,000	220	900	0.06	......	3.8	0.0043
0.1	39,000	140	540	0.10	......	13	0.014
0.4	61,000	72	175	0.26	......	110	0.13
0.6	68,000	66	130	0.34	......	200	0.23
0.8	71,000	64	110	0.43	......	290	0.33
1.0	71,000	63	97	0.52	0.0007	380	0.43
1.2	69,000	63	89	0.60	0.0008	470	0.54
1.5	62,000	63	80	0.74	0.0009	610	0.70
2.0	52,000	64	73	1.01	0.0012	840	0.96
3.0	39,000	66	69	1.67	0.0020	1,300	1.5
4.0	33,000	68	70	2.50	0.0030	1,700	2.0
5.0	29,000	70	71	3.52	0.0041	2,100	2.4
6.0	25,000	71	71	4.67	0.0054	2,500	2.9
7.0	23,000	72	72	5.96	0.0068	2,900	3.4
8.0	20,000	73	73	7.36	0.0084	3,300	3.8
10	17,000	75	75	10.5	0.012	4,000	4.7
20	9,900	81	81	34	0.037	7,300	8.6
50	......	88	.....	170	0.17	15,000	18
100	......	93	.....	590	0.58	25,000	30
400	......	99	.....	......	7.0	53,000	65
1,000	......	100	.....	......	......	77,000	96

[a] 15°C, 760 mm Hg.
[b] M. S. Livingston and H. A. Bethe, *Rev. Mod. Phys.* **9**, 270 (1937). The value $W = 35.5$ eV/ion pair was used for α particles in air.
[c] M. J. Berger and S. M. Seltzer, Studies in Penetration of Charged Particles, U. Fano, ed., *Natl. Acad. Sci.–Natl. Res. Council Publ.* 1133, 1964.
[d] The value $W = 34$ eV/ion pair was used for electrons in air.
[e] Calculated on the basis of electron values by R. D. Birkhoff, Oak Ridge National Laboratory. The specific ionization in this column is given for the mixture of β-particle energies having the Fermi distribution for which column 1 shows the maximum energy.
[f] H. A. Bethe and J. Ashkin in "Experimental Nuclear Physics," vol. I, p. 180, E. Segrè, ed., John Wiley & Sons, Inc., New York, 1953.
[g] W. S. Snyder and J. Neufeld, *Oak Ridge Natl. Lab. Rept.* ORNL-1083, Oak Ridge, Tenn., 1951.
[h] P. G. Steward, *Univ. Calif. Rept.* UCRL-18127, Berkeley, 1968.

Specific ionization (ion pairs/cm) of electrons of energy E (MeV) in air:[1]

$$S \approx 55 + 9E^{-1} \tag{8i-12}$$

(<10 percent error for $0.01 \leq E \leq 2$.)   See Table 8i-2.

Stopping power in MeV/(g/cm^2) at depth x cm in a medium of density ρ (g/cm^3) from particles of average initial energy $\bar{E}$:

$$P = \frac{\mu \bar{E} e^{-\mu x}}{\rho} \tag{8i-13}$$

[1] Empirical formula developed by K. Z. Morgan.

Ratio of radiative (bremsstrahlung) and ionization energy-loss rates of an electron with total energy E (MeV) moving in a medium with atomic number Z:

$$\frac{(-dE/dx) \text{ radiation}}{(-dE/dx) \text{ ionization}} \approx \frac{EZ}{800} \qquad (8\text{i-}14)$$

Fraction of energy of an incident electron of kinetic energy E (MeV) converted into bremsstrahlung in a thick target of atomic number Z:

$$f \approx 0.0007ZE \qquad (8\text{i-}15)$$

($E \lesssim$ several MeV.)

X and Gamma Rays. Figure 8i-5 shows the relative importance of the photoelectric effect, Compton scattering, and pair production in absorbers of different atomic number Z. With $Z = 50$, for example, the figure shows that the photoelectric and Compton attenuation coefficients are equal at a photon energy of about 0.3 MeV; and the Compton and pair-production coefficients, at about 6 MeV.

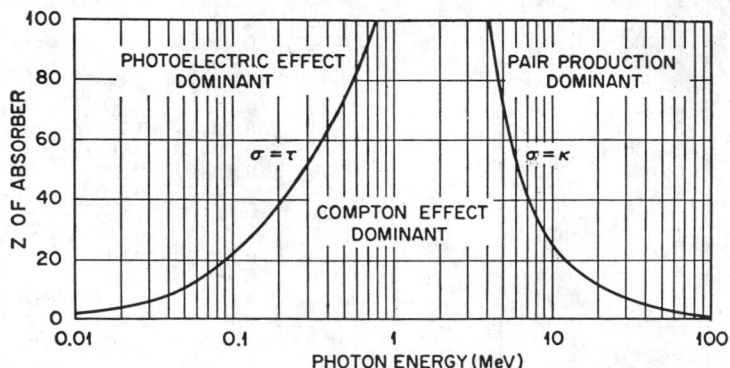

Fig. 8i-5. Relative importance of the three major types of γ-ray interaction. The lines show the values of atomic number Z and photon energy for which the two neighboring effects are just equal. [*From R. D. Evans, "The Atomic Nucleus," McGraw-Hill Book Company, New York*, 1955.)

At low photon energies $h\nu$, the photoelectric cross section for a given Z varies as $(h\nu)^{-\frac{7}{2}}$. At a given (low) photon energy, the cross section for the K-shell photoelectric interaction, which is the dominant process, is proportional to Z^5. The Compton cross section at a given energy is almost independent of Z, and the attenuation coefficient is proportional to the number of electrons per unit volume. The mass attenuation coefficient is thus roughly the same for any absorber in the Compton region. The average fraction of the energy of the incident photon imparted to a Compton electron increases with photon energy. Pair production can occur only when $h\nu \geq 1.02$ MeV. The pair-production cross section at energies of several MeV and greater increases approximately as Z^2 and very slowly with $h\nu$.

Figures 8i-6 through 8i-9 give the mass-attenuation and mass-absorption coefficients of a number of materials. These figures are taken from "Principles of Radiation Protection," K. Z. Morgan and J. E. Turner, eds., John Wiley & Sons, Inc., New York, 1967.

Neutrons. Classification by energies:[1]

THERMAL. In thermal equilibrium with their surroundings, neutrons have a Maxwellian velocity distributiom. At 23°C the most probable speed is 2,200 me-

---

[1] Measurement of Absorbed Dose of Neutrons and of Mixtures of Neutrons and Gamma Rays, *Natl. Bur. Standards Handbook* 75, 1961.

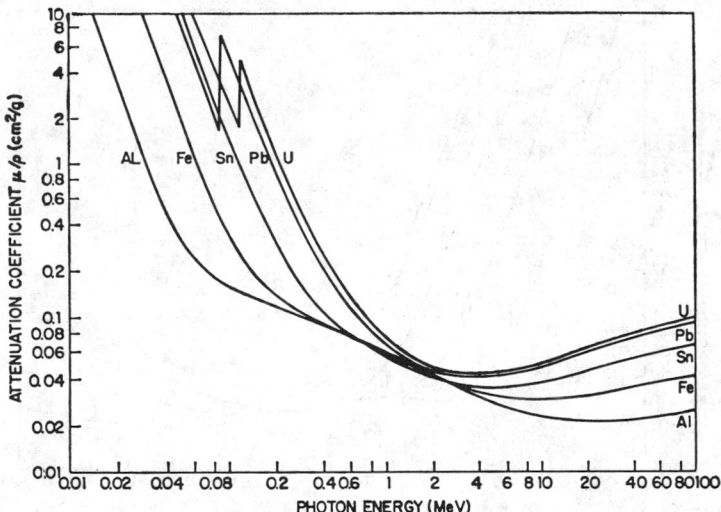

FIG. 8i-6. Mass-attenuation coefficients for various elements. (*D. Z. Morgan and J. E. Turner*, eds., *"Principles of Radiation Protection,"* John Wiley & Sons., Inc., New York, 1967.)

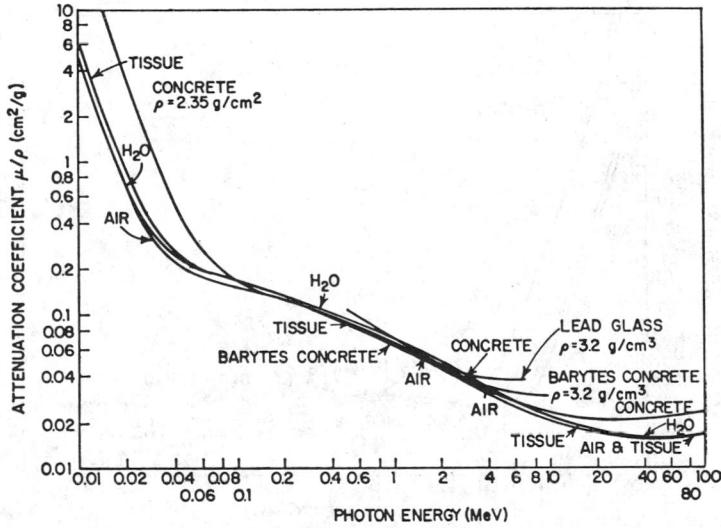

FIG. 8i-7. Mass-attenuation coefficients for various materials. (*D. Z. Morgan and J. E. Turner, ibid.*)

ters/sec; the most probably energy is 0.025 eV. The most important interactions are those involving neutron capture, such as (n,γ) and (n,p) reactions. In tissue the reaction $_1^1H(n,\gamma)_1^2H$ produces a 2.2-MeV gamma ray and the $_7^{14}N(n,p)_6^{14}C$ reaction produces a 0.6-MeV proton. The reaction $_5^{10}B(n,\alpha)_3^7Li$, which releases 2.8 MeV of energy, is utilized in a number of low-energy neutron detectors. Many reaction cross sections are inversely proportional to the neutron velocity (1/v law).

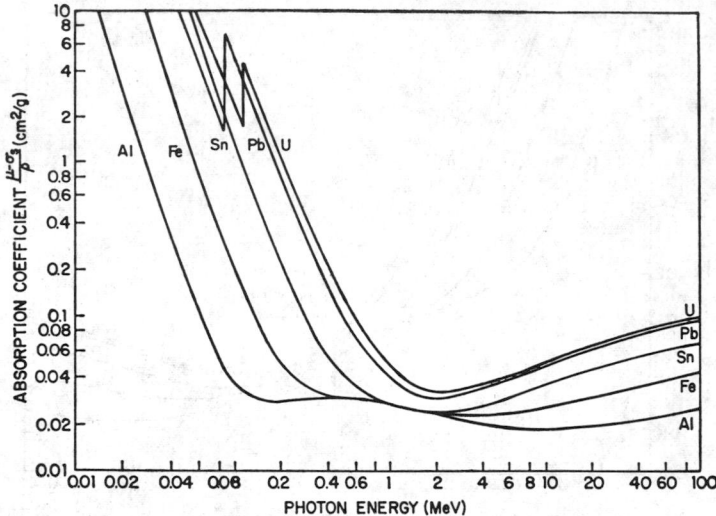

FIG. 8i-8. Mass-absorption coefficients for various elements.　(*D. Z. Morgan and J. E. Turner, ibid.*)

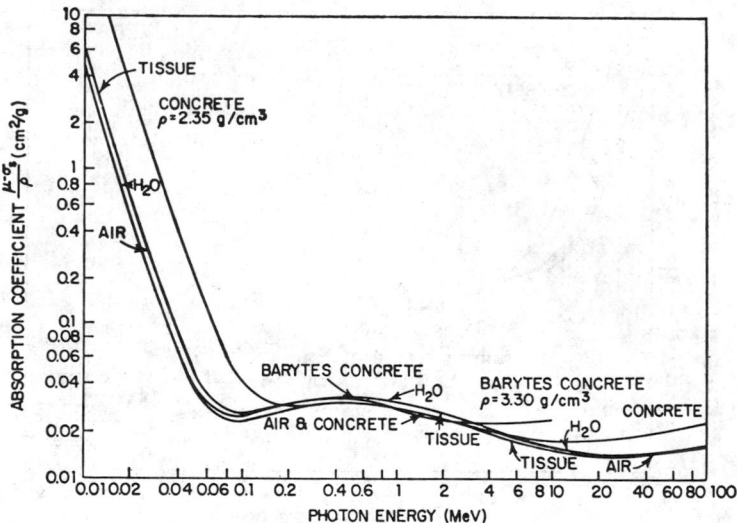

FIG. 8i-9. Mass-absorption coefficients for various materials.　(*D. Z. Morgan and J. E. Turner, ibid.*)

INTERMEDIATE ENERGIES (0.5 eV to 10 keV). Neutrons with energies above thermal are sometimes called *epithermal*. Reaction cross sections are characterized by resonance structure.

FAST NEUTRONS (10 keV to 14 MeV). Elastic scattering is usually the most important interaction in the lower portion of this energy range. Inelastic scattering becomes important at the higher energies ($\gtrsim$10MeV). The average energy lost by a neutron (mass m and energy E) scattered elastically and isotropically (in the center-of-mass coordinate system) by a nucleus of mass M is

$$\overline{\Delta E} = \frac{2mME}{(m + M)^2} \tag{8i-16}$$

This expression describes neutron scattering from hydrogen from a few eV to $\sim$14 MeV. With heavier elements this formula is restricted to lower energies. In soft tissue, which is rich in hydrogen, collisions with hydrogen produce $\gtrsim$85 percent of the first-collision absorbed dose with neutron energies $E \leq 10$ MeV and $\sim$76 percent with $E \lesssim 14$ MeV.[1]

HIGHER ENERGIES ($>$14 MeV). Inelastic processes, involving ejection of secondary particles, are important with nuclei other than hydrogen in this energy range. Further discussion is included in the next paragraph.

High-energy Protons and Neutrons (50 MeV to 2 GeV). A high-energy proton passing through tissue deposits energy by ionizing atoms and, like a high-energy neutron, by means of nuclear reactions. Entering a nucleus, a high-energy nucleon can cause a cascade in which a number of secondary nucleons and pions are emitted. The residual nucleus generally returns to its ground state by evaporating additional nucleons and/or heavier nuclear fragments and by emitting gamma rays. As a rough estimate, the total attenuation coefficient of soft tissue for a high-energy nucleon is $\approx$0.01 cm^{-1}. Approximately 30 percent of the nucleons incident normally on a soft-tissue slab have a nuclear reaction in penetrating to a depth of 30 cm, roughly the thickness of the human torso.

Exposure-to-fluence Conversion in Air

$$1 \text{ roentgen} = \frac{2.08 \times 10^3 \, W}{E(\mu - \sigma_s)_a} \text{ photons/cm}^2 \approx \frac{2.15 \times 10^9}{E} \text{ photons/cm}^2 \tag{8i-17}$$

Here W (eV/ion pair) is the average energy needed to produce an ion pair in air, E (MeV) is the photon energy, and $(\mu - \sigma_s)_a$ is the absorption coefficient of air (cm^{-1}). The final, approximate equation is not in error by more than $\pm$13 percent in the range $0.07 \leq E \leq 2.0$. The value $W = 34$ eV was used for secondary electrons produced by the photons.

Values in Air Equivalent to an Exposure of 1 Roentgen

2.58 $\times$ 10^{-4} coul/kg (definition of the roentgen)
773 esu/g
1.61 $\times$ 10^{12} ion pairs/g
5.47 $\times$ 10^{13} eV/g
87.7 ergs/g = 0.8770 rad } (These values are based on $W = 34$ eV/ion pair for
2.10 $\times$ 10^{-6} cal/g(20°C) secondary electrons.)

Absorbed Dose-to-Fluence Conversion in Soft Tissue. For photons,

$$1 \text{ rad} = \frac{6.24 \times 10^7 \rho_t}{E(\mu - \sigma_s)_t} \text{ photons/cm}^2 \approx \frac{2.17 \times 10^9}{E} \text{ photons/cm}^2 \tag{8i-18}$$

For charged particles,

$$1 \text{ rad} = \frac{6.24 \times 10^{13} \rho_a}{SWP} \text{ particles/cm}^2 \tag{8i-19}$$

Here $\rho_t = 1$ g/cm^3 and $\rho_a = 0.001293$ g/cm^3 are the densities of tissue and air, E (MeV) is the photon energy, $(\mu - \sigma_s)_t$ is the absorption coefficient of tissue (cm^{-1}), S is the average specific ionization in air (ion pairs/cm), W (eV) is the average energy needed to produce an ion pair in air, and P is the mass stopping power of tissue

[1] *Natl. Bur. Standards Handbook* 75, *ibid.*

relative to air. The approximate equation for photons is in error by no more than ± 13 percent in the range $0.07 \leq E \leq 2.0$ MeV. For most purposes one can assume that $W = 35$ eV. For more accurate work, the value W should be taken as 34 eV for electrons from X or gamma radiation or for beta particles, and 35.5 eV for alpha particles. As noted on page 8-296, however, W values for slowly moving heavy ions are often larger, and one should consult the literature.

Values in Soft Tissue Equivalent to an Absorbed Dose of 1 Rad.

2.94×10^{-4} coul/kg

882 esu/g $\left.\rule{0pt}{3em}\right\}$ (These values are based on an assumed value $W = 34$ eV/ion pair.)

1.84×10^{12} ion pairs/g

6.24×10^{13} eV/g

100 ergs/g (definition of the rad)

2.39×10^{-6} cal/g

Quality Factors. X rays, gamma rays, electrons, and positrons: QF = 1. Values for heavy charged particles and neutrons are shown in Tables 8i-3 and 8i-4.

TABLE 8i-3. LET–QUALITY FACTOR RELATIONSHIP FOR
HEAVY CHARGED PARTICLES*

Average LET, keV/μm in water†	QF
≤ 3.5	1
3.5–7.0	1–2
7.0–23	2–5
23–53	5–10
53–175	10–20

* Report of the RBE Committee to the ICRP and ICRU, *Health Phys.* **9**, 357 (1963).
† 1 keV/μm = 10 MeV/cm.

TABLE 8i-4. NEUTRON QUALITY FACTORS*

Neutron energy, MeV	QF	Neutron energy, MeV	QF
Thermal	3	10	6.5
10^{-4}	2	50	5
0.02	5	400	3.5
0.1	8	1000	$\{$ ~10 at surface / ~2.5 at depth 20–30 cm
0.5	10		
2.5	8	2000	$\{$ ~10 at surface / ~2.5 at depth 20–30 cm
5.0	7		

* Values for energies through 10 MeV are given in *Natl. Bur. Standards Handbook* 63, 1957; values at 50 and 400 MeV are based on work of J. Neufeld, W. S. Snyder, J. E. Turner, and H. A. Wright, *Health Phys.* **12**, 227 (1966); values at 1000 and 2000 MeV are based on calculations of H. A. Wright, V. E. Anderson, J. E. Turner, J. Neufeld, and W. S. Snyder, **16**, 13 (1969) *Health Phys.*

Estimates of quality factors for high-energy (50-MeV to 2-GeV) protons and neutrons have been made by applying the values shown in Table 8i-3 to detailed studies of nuclear interactions. For protons, QF $\cong$ 1.5 in the region from 50 to 600 MeV[1,2] and rises to about 2 at 2 GeV.[2] For neutrons, QF decreases from ~5 at 50 MeV to ~3.5 at 400 MeV.[1] Owing to the build-up of secondary protons (low

[1] J. Neufeld, W. S. Snyder, J. E. Turner, and H. A. Wright, *Health Phys.* **12**, 227 (1966).
[2] H. A. Wright, V. E. Anderson, J. E. Turner, J. Neufeld, and W. S. Snyder, *Health Phys.* **16**, 13 (1969).

LET), the neutron quality factor decreases with increasing depth, this effect becoming more pronounced at higher neutron energies. At 2 GeV, the neutron quality factor decreases from ∼10 at the surface of a tissue phantom to ∼2.5 at a depth of 20 to 30 cm.[1]

8i-4. External Dose Equations

1. Exposure from a 1-curie point source in time t (hr) at distance r (cm):

$$\epsilon_\gamma = \frac{1.5 \sum_i (\mu - \sigma_s)_i E_i e^{-\mu_i x} f_i B_i t 10^8}{r^2} \qquad \text{roentgens} \qquad (8\text{i-}20)$$

where μ_i = attenuation coefficient (cm^{-1}) of medium between source and point of measurement, f_i = fraction of emitted photons having energy E_i, $(\mu - \sigma_s)_i$ = absorption coefficient (cm^{-1}) of air for photons of energy E_i. Figures 8i-6 through 8i-9 can be used to obtain values of μ and $\mu - \sigma_s$. The term B_i is the build-up factor due to the scattered radiation of energy E_i. Its value[2,3] depends upon the width of the beam and the distance, volume, and atomic number of the scattering medium. When the medium is air ($x = r$) and ϵ_γ is measured at not too great a distance r from the point source (i.e., a few meters), then

$$e^{-\mu_i x} B_i \approx 1$$

Some values of ϵ_γ and corresponding absorbed doses are given as a function of E_i in Table 8i-5.

TABLE 8i-5. THEORETICAL VALUES OF EXPOSURE AND ABSORBED DOSE IN AN INFINITESIMAL VOLUME 1 METER FROM A 1-CURIE SOURCE*

Energy, MeV	Exposure ϵ_γ, R/hr	Absorbed dose in tissue, rads/hr	Energy, MeV	Exposure ϵ_γ, R/hr	Absorbed dose in tissue, rads/hr
0.02	0.19	0.19	2.0	0.89	0.85
0.04	0.060	0.051	4.0	1.5	1.4
0.06	0.041	0.035	6.0	2.0	1.8
0.08	0.039	0.037	8.0	2.5	2.3
0.10	0.045	0.042	10	2.9	2.7
0.20	0.10	0.098	20	5.4	4.8
0.40	0.22	0.22	40	11	8.8
0.60	0.34	0.33	60	16	14
0.80	0.43	0.42	100	29	24
1.0	0.52	0.51			

* W. S. Snyder and J. L. Powell, *Oak Ridge Natl. Lab. Rept.* ORNL-421, March, 1950. These values do not include contributions due to air scattering and absorption. Inclusion of absorption by air reduces the exposure rate at 1 meter by 8 percent at a photon energy of 0.02 MeV, by 3 percent at 0.04 MeV, and by one percent at 0.06 MeV. The reduction is insignificant at higher energies.

2. Absorbed dose in a small volume surrounded by a radioactive material uniformly distributed in an infinite medium of density ρ_m (g/cm^3) in t days:

$$D = \frac{51 \, C \bar{E} P_t t}{\rho_m} \qquad \text{rads} \qquad (8\text{i-}21)$$

[1] H. A. Wright, V. E. Anderson, J. E. Turner, J. Neufeld, and W. S. Snyder, *ibid.*
[2] H. Goldstein and J. E. Wilkins, Report NYO-3075, Office Tech. Serv. Rept., U.S. Department of Commerce, 1954.
[3] T. Rockwell III, ed., "Reactor Shielding Design Manual," D. Van Nostrand Company, Inc., Princeton, N.J., 1956.

The medium contains (a constant) activity C (μCi/cm^3) of a radionuclide that emits per disintegration an average energy $\bar{E}$ (MeV). When the dose is in tissue, P_t is the mass stopping power of tissue relative to the medium.

3. First-collision absorbed dose in a small volume of tissue in t days from thermal neutrons with a flux density N neutrons/(sec)(cm^2) and energy E MeV due to the (n,p) reaction with nitrogen:

$$D_{N_t} = 1.38 \times 10^{-3} \, Ntf_N \sigma_N E = 1.8 \times 10^{-6} \, Nt \qquad \text{rads} \qquad (8\text{i-}22)$$

There are f_N atoms of nitrogen per gram of tissue having a thermal-neutron absorption cross section σ_N cm^2. In this case the ${}^1_1\text{H}(n,\gamma){}^2_1\text{D}$ reaction accounts for more energy loss than the ${}^{14}_7\text{N}(n,p){}^{14}_6\text{C}$ reaction but is less significant at the peak of the dose equivalent (at about 3 mm depth in tissue) because of the large QF that must be applied to the proton energy.[1]

4. First-collision absorbed dose to tissue in t days from fast neutrons:

$$D_{N_f} = 1.38 \times 10^{-3} \, NEt \sum_i f_i \sigma_i e_i \qquad \text{rads}$$

$$f_i = \text{atoms of } i\text{th type per g of tissue} = \frac{6.02 \times 10^{23} F_i}{A_i} \qquad (8\text{i-}23)$$

$$e_i = \text{fraction of energy lost per collision} = \frac{2M_i m}{(M_i + m)^2}$$

Here σ_i is the cross section (cm^2) for neutrons of energy E (MeV), F_i is the fraction by weight of the ith element in tissue, A_i is the atomic weight of the ith element, M_i is the mass of the atoms of the ith element, and m is the mass of the neutron.

5. Absorbed dose from ionizing particles (α, β, p, etc.) in t days:

$$D_i = 1.38 \times 10^{-9} NPt \qquad \text{rads} \qquad (8\text{i-}24)$$

Here P (eV $\times$ cm^2/g) is the stopping power of the medium, and N is the flux density of ionizing particles (cm^{-2} sec^{-1}).

6. Exposure from a flux density N/photons/(sec/cm^2) of X or gamma radiation in t days:

$$D_x = 1.22 NEt(\mu - \sigma_s)_a \qquad \text{roentgens} \qquad (8\text{i-}25)$$

when the photons of energy E (MeV) have an absorption coefficient $(\mu - \sigma_s)_a$ (cm^{-1}) in air.

7. Absorbed dose from a flux density N/photons/(sec/cm^2) of X or gamma radiation in t days:

$$D_x = \frac{1.38 \times 10^{-3} NEt(\mu - \sigma_s)_m}{\rho_m} \qquad \text{rads} \qquad (8\text{i-}26)$$

when the photons of energy E (MeV) have an absorption coefficient $(\mu - \sigma_s)_m$ (cm^{-1}) in medium m of density ρ_m (g/cm^3).

[1] Snyder, W. S. *Nucleonics* **6**(2), 46 (1950).

8i-5. Internal Dose Equations[1]

1. Effective half life of a radionuclide:

$$T = \frac{T_r T_b}{T_r + T_b} \tag{8i-27}$$

in which T_r is the radioactive half life, and T_b is the biological half life of the radionuclide in a body organ.

2. Maximum permissible body burden:

$$q = \frac{2.8 \times 10^{-3} mR}{f_2 \Sigma E(\text{QF}) H} \quad \mu\text{Ci} \tag{8i-28}$$

in which q (μCi) in the total body delivers a dose equivalent rate of R (rem/week) to the critical body organ of mass m(g). The relative hazard factor H is taken as 5 for alpha, beta, and atomic-recoil components of energy emitted by radioisotopes for which the bone is the critical organ, with the exception of the case when the parent element of the chain is an isotope of radium, in which case H is 1. The term f_2 is the fraction in the critical organ of that in the total body, E is the average energy (MeV) absorbed in the organ per disintegration, and QF is the quality factor of the radiation. (See Table 8i-3: QF = 1 for β^+, β^-, e^-, X, and gamma radiation; 10 for alpha; and 20 for atomic recoils. In the special case when $E_{\max} \leq 0.03$ MeV for β^+, β^-, or e^-, QF = 1.7.) The critical body organ is the organ receiving the radioisotope that results in the greatest body damage.

In the case of alpha- or beta-emitting radioisotopes for which the bone is the critical organ, use is made of the long-standing generally accepted value of $q = 0.1$ μCi for ^{226}Ra by making a comparison on an effective energy basis with ^{226}Ra by means of the equation

$$q = \frac{11}{f_2 \Sigma E(\text{QF}) H} \quad \mu\text{Ci} \tag{8i-29}$$

3. Maximum permissible concentration in air (MPC)$_a$ and water (MPC)$_w$:

$$(\text{MPC})_a = \frac{10^{-7} q f_2}{T f_a (1 - e^{-0.693t/T})} \quad \mu\text{Ci/cc} \tag{8i-30}$$

$$(\text{MPC})_w = \frac{9.2 \times 10^{-4} q f_2}{T f_w (1 - e^{-0.693t/T})} \quad \mu\text{Ci/cc} \tag{8i-31}$$

Here (MPC)$_a$ and (MPC)$_w$ are given in μCi/cc of air and water, respectively, that will result in a maximum permissible burden $q f_2$ in the critical organ after an occupational exposure, 40 hr/week, 50 weeks/year, for a time t days equal to 50 years. f_a and f_w are the fractions that arrive in the critical organ from inhalation and ingestion, respectively. T is the effective half life in days in the critical organ.

---

[1] For detailed information on maximum permissible exposure levels, refer to "Maximum Permissible Body Burdens and Maximum Permissible Concentrations of Radionuclides in Air and in Water for Occupational Exposure," Handbook 69, Superintendent of Documents, Washington, D.C.; ICRP Publ. 2, Report of Committee II on Permissible Dose for Internal Radiation, 1959, Health Phys. 3, 1 (1960); and Recommendations of the International Commission on Radiological Protection, ICRP Publ. 6, Pergamon Press, New York, 1962; K. Z. Morgan and M. R. Ford, Developments in Internal Dose Determinations, Nucleonics 12(6), 32–39 (June, 1954); K. Z. Morgan, W. S. Snyder, and M. R. Ford, Maximum Permissible Concentrations of Radioisotopes in Air and Water for Short Period Exposure, in "Peaceful Uses of Atomic Energy," vol. 13, United Nations, 1956; and "Background Material for the Development of Radiation Protection Standards," Federal Radiation Council, Superintendent of Documents, Washington, D.C., May, 1960; September, 1961; July, 1964; and May, 1965.

In the case of an inert gas,

$$(MPC)_a = \frac{0.024 R \rho_a P_a}{\Sigma E(QF) H P_t} \qquad \mu Ci/cc \tag{8i-32}$$

and when the maximum permissible equivalent rate R is 0.1 rem/week,

$$(MPC)_a = \frac{2.6 \times 10^{-6}}{\Sigma E(QF) H} \qquad \mu Ci/cc \tag{8i-33}$$

In these equations R = dose-equivalent rate (rem/week), ρ_a = density of air ($= 0.001293$ g/cc), P_a/P_t = stopping power of air relative to tissue ($= 1/1.13$ for beta, X, and gamma radiation).

4. Dose equivalent delivered to the critical body organ after a single intake:

$$D = \frac{74 \Sigma E(QF) H f I_0 T}{m} (1 - e^{-0.693t/T}) \qquad rem \tag{8i-34}$$

in which D is the dose equivalent (rem) delivered to the critical organ of mass m (g) in time t (days) when I_0 (μCi) is taken into the body in a single event, and the fraction f is deposited in the critical organ.

5. Dose equivalent in t days to the body organ containing a constant burden of $q f_2$ (μCi):

$$D = \frac{51 q f_2 t \Sigma E(QF) H}{m} \qquad rem \tag{8i-35}$$

6. Dose equivalent to lower large intestine from single intake I_0 (μCi):

$$D_{GI} = \frac{0.13 f I_0 \Sigma E(QF) H}{G} \qquad rem \tag{8i-36}$$

Except for radionuclides of very short radioactive half life, the lower large intestine receives the largest dose of any portion of the GI tract. In this equation G is given by the relation

$$G = \frac{0.693 (h_1 - h_0)}{(e^{-0.693 h_0/T_r} - e^{-0.693 h_1/T_r}) T_r} \tag{8i-37}$$

Usually h_0/T_r and h_1/T_r are small, and in such cases $G = 1$. In these equations f = fraction of ingested or inhaled radionuclide going to the lower large intestine, h_0 = time of arrival in lower large intestine ($= 13/24$ days), h_1 = time of departure from lower large intestine ($= 31/24$ days).

7. Dose equivalent in t weeks at distance r (cm) from a microscopic radioactive particle lodged in the body in which $Q = \mu Ci$ of particle, S_a = specific ionization of beta radiation in air (ion pair/cm), W_a = energy to produce ion pair (eV/ip), P_t = stopping power tissue relative to air, μ_t = coefficient of attenuation of beta radiation (cm^{-1}), and T_r = radioactive half life of radionuclide (weeks):

$$D_p = \frac{4.1 \times 10^{-5} T_r Q S_a W_a P_t (QF) H e^{-\mu_t r} (1 - e^{-0.693t/T_r})}{r^2 \rho_a} \qquad rem \tag{8i-38}$$

It should be noted that $S_a W_a/\rho_a$ eV/(g/cm^2) is the stopping power of air and $S_a W_a P_t/\rho_a = P$ eV/(g/cm^2) is the stopping power of the tissue. Values of stopping power for water may be found from Fig. 1, and average values over the range of the particle may be found for various media from Figs. 8i-2 and 8i-3 by using the relation $P = \bar{E}/R$, in which $\bar{E}$ is the average energy (eV), and R is the range (g/cm^2).

8. Average dose equivalent in time t (weeks) in a sphere of radius r (cm) resulting from a microscopic radioactive particle lodged in tissue at the center of the sphere (in this case, $r <$ range of beta rays):

$$D_p = \frac{1.23 \times 10^{-4} T_r Q S_a W_a P_t (\mathrm{QF}) H (1 - e^{-\mu_t r})(1 - e^{-0.693t/T_r})}{r^3 \mu_t \rho_a} \qquad \text{rem} \qquad (8\text{i-}39)$$

9. Average dose equivalent in time t (weeks) in a sphere of radius r (cm) $\geq$ range of alpha or beta radiation resulting from a microscopic radioactive particle lodged in tissue at the center of the sphere:

$$D_p = \frac{1.23 \times 10^2 T_r Q \Sigma E (\mathrm{QF}) H (1 - e^{-0.693t/T_r})}{r^3} \qquad \text{rem} \qquad (8\text{i-}40)$$

10. Dose equivalent from fallout from a nuclear detonation:

$$D_f = 5 t_0 \dot{D}_0 \left[1 - \left(\frac{t_0}{t_1} \right)^{0.2} \right] \qquad \text{rem} \qquad (8\text{i-}41)$$

In this case D_f is the dose equivalent during time $t_1 - t_0$, and $\dot{D}_0$ is the dose-equivalent rate measured at time t_0 since the nuclear detonation. It is assumed that $\dot{D}_0 \propto t^{-1.2}$.

8i-6. Radiation Protection Guides and Standards. One should avoid unnecessary exposure to ionizing radiation. The Federal Radiation Council has stated[1] that ". . . the establishment of radiation protection standards involves a balancing of the benefits to be derived from the controlled use of radiation and atomic energy against the risk of radiation exposure." Table 8i-6 summarizes some of the 1968 values of

TABLE 8i-6. PERMISSIBLE DOSE EQUIVALENT TO BODY ORGANS OF WORKERS
AS RECOMMENDED (1968) BY THE NATIONAL COUNCIL
ON RADIATION PROTECTION

Organs	Maximum permissible dose equivalent in any 13-week period, rem	Annual maximum permissible dose equivalent, rem
Red bone marrow, total body, head, trunk, gonads, lenses of the eyes*..	3	5
Skin, thyroid, bone................	10(15)*	30
Feet, ankles, hands, forearms.......	25(38)*	75
Other single organs................	5(8)*	15

* The values recommended by the ICRP are identical to those recommended by the NCRP with exception of the ICRP values given in parentheses. Also, the ICRP now includes the lenses of the eyes with the "other single organs."

maximum permissible dose in general use for occupational exposure. These amounts, permitted when necessary in order to carry out operations, are independent of an individual's exposure for medical reasons or his exposure to natural background radiation. The recommendations of various organizations differ in some respects from one another, and specific recommendations may be different from the values given in Table 8i-6. Generally, values one-tenth those for occupational exposure are applied to exposure of individuals in uncontrolled areas or to the critical segments

[1] Report No. 2, Washington, D.C., 1961.

of this population. Further information and details are given in the following publications:

Recommendations of the International Commission on Radiological Protection (ICRP), Pergamon Press, London:
  Publ. 1, 1959.
  Publ. 2, 1959; and *Health Phys.* **3**, 1 (1960). The bibliography of ICRP Publ. 2 appears only in the latter.
  Publ. 3, 1960.
  Publ. 4, 1963.
  Publ. 5, 1964.
  Publ. 6, 1964.
  Publ. 7, 1965.
  Publ. 8, 1966.
  Publ. 9, 1966.
  Health Phys. **12**, 129 (1966).
  "Principles of Radiation Protection," K. Z. Morgan and J. E. Turner, eds., John Wiley & Sons, Inc., New York, 1967.
Recommendations of the National Council on Radiation Protection and Measurements (NCRP):
  National Bureau of Standards Handbook 52, 1953.
  National Bureau of Standards Handbook 59, supplements, Jan. 8, 1957, and Apr. 15, 1958.
  National Bureau of Standards Handbook 59, 1959.
  Radiology **75**, 122 (1960).
Federal Radiation Council, Washington, D.C.:
  Rept. 1, 1960.
  Rept. 2, 1961.
  Rept. 5, 1964.
  Rept. 7, 1965.
  Federal Register, p. 6953, May 22, 1965.

Perhaps the most important recent development in radiation protection standards is the concept of maximum permissible dose commitment. Obviously it would not be proper to permit an employee to accumulate a maximum permissible body burden of a radionuclide with a long effective half life during, for example, one year of employment because for a long period thereafter he could be permitted no further external or internal exposure. To avoid this, the International Commission on Radiological Protection set the limit at an annual permissible dose commitment which is defined as the dose resulting from a body intake of a radionuclide by a person occupationally exposed for one year at the maximum permissible concentration, MPC, of a radionuclide. It can be shown that if the period of body intake of the radionuclide is τ years, the critical body-organ dose equivalent integrated over $50 + \tau/2$ years as a result of a dose commitment would be equal numerically to the annual permissible dose-equivalent value R_{50} as given in Table 8i-6 for the critical organ multiplied by τ. Table 8i-7[1] summarizes values of maximum permissible dose commitment for the three categories of occupational exposure: (A) routine application, (B) single exposure or quarterly dose, and (C) planned special exposures. In this table R_{50} corresponds to the dose rate reached in the critical body organ after occupational exposure at the MPC of a radionuclide for 50 years.

Time-average neutron flux densities that deliver 100 mrem in a 40-hr work week are given in Table 8i-8. Additional flux densities have been calculated for neutron energies up to 400 MeV.[2] For normally incident beams it is estimated that the maximum permissible flux density decreases smoothly from about 15 neutrons/(sec/cm²) at 60 MeV to about 10 neutrons/(sec/cm²) at 400 MeV.

---

[1] Taken from "Principles of Radiation Protection," K. Z. Morgan and J. E. Turner, eds., John Wiley & Sons, Inc., New York, 1967.
[2] J. Neufeld, W. S. Snyder, J. E. Turner, and H. A. Wright, *Health Phys.* **12**, 227 (1966).

TABLE 8i-7. OCCUPATIONAL EXPOSURE LEVELS* RECOMMENDED BY ICRP

Categories of exposure	Maximum dose equivalent for external exposure and/or internal exposure to radionuclides with a short effective half life		Dose equivalent integrated over a 50-year period resulting from a permissible quarterly dose equivalent or from a planned special exposure	Maximum permissible intake of any radionuclide for single, quarterly, or planned exposure corresponds to an intake:	
	rem in 13 weeks	rem in 1 year ‡	rem in 50 years†	For 13 weeks at:	At the MPC for:
A. Maximum permissible dose equivalent for routine application	$R_{50}/4$	R_{50}	$R_{50}/4$	$1 \times$ MPC	$\frac{1}{4}$ year
B. Maximum permissible dose equivalent for single exposures or for exposures on a quarterly basis	$R_{50}/2$‡	$2R_{50}$‡ to total body, gonads, or red bone marrow if $5(N-18)$ is not exceeded. R_{50}‡ to all other organs	$R_{50}/2$	$2 \times$ MPC	$\frac{1}{2}$ year
C. Planned special exposures§	$2R_{50}$ committed in any single event to any body organ		$2R_{50}$	$8 \times$ MPC	2 years
D. Summation of all planned special exposures in a lifetime	$5R_{50}$ committed in a lifetime to any body organ from planned special exposures		$5R_{50}$	$20 \times$ MPC	5 years

* Exposure levels are given in terms of the annual permissible dose equivalent R_{50}. Values of R_{50} for individual body organs in various radiosensitivity organ groups are as follows:

Group 1 for gonads, total body, and red bone marrow: $R_{50} = 5$, $R_{50}/2 \to 3$, $2R_{50} \to 12$, and $5R_{50} = 25$ rem.

Group 2 for thyroid, skin, and bone: $R_{50} = 30$, $R_{50}/2 = 15$, $2R_{50} = 60$, and $5R_{50} = 150$ rem.

Group 3 for hands, forearms, feet, and ankles: $R_{50} = 75$, $R_{50}/2 \to 38$, $2R_{50} = 150$, and $5R_{50} = 375$ rem.

Group 4 for all other body organs (including the lenses of the eyes): $R_{50} = 15$, $R_{50}/2 \to 8$, $2R_{50} = 30$, and $5R_{50} = 75$ rem.

† Actually, over a period of $50 + \tau/2$ years for external and/or internal exposures to radionuclides of any effective half life, τ is period of intake of a radionuclide.

‡ The ICRP states that it would be undesirable to repeat this quarterly dose equivalent of $R_{50}/2$ at close intervals, but to provide flexibility, it is permitted on infrequent occasions to receive $2R_{50}$ in a year to the gonads, total body, and red bone marrow, provided the accumulated dose equivalent of $5(N - 18)$ rem is not exceeded. Ordinarily, the annual dose equivalent R_{50} should not be exceeded for exposure to any of the body organs.

§ Planned special exposures are not permitted if a single exposure in excess of $R_{50}/2$ has been received in the previous 12 months or if at any time the worker has received an abnormal exposure in excess of $5R_{50}$. Planned special exposures are not permitted to women of reproductive capacity. They are not permitted to gonads, total body, or red bone marrow if as a consequence $5(N - 18)$ rem is exceeded.

Maximum permissible concentrations of specific radionuclides in water and in air are recommended by the International Commission on Radiological Protection, Publication 2, *Health Phys.* **3,** 1 (1960). Concentrations of unidentified radionuclides in water and air are given in Tables 8i-9 and 8i-10.

TABLE 8i-8. MAXIMUM PERMISSIBLE NEUTRON FLUX DENSITIES
AS GIVEN BY THE NATIONAL COUNCIL ON RADIATION PROTECTION*

Neutron energy, MeV	Average flux density, $cm^{-2} sec^{-1}$	Neutron energy, MeV	Average flux density, $cm^{-2} sec^{-1}$
Thermal	670	2.5	20
10^{-4}	500	5.0	18
0.02	280	7.5	17
0.1	80	10	17
0.5	30	10–30	10
1.0	18		

* *Natl. Bur. Standards Handbook* 63, 1957.

TABLE 8i-9. MAXIMUM PERMISSIBLE CONCENTRATION OF UNIDENTIFIED
RADIONUCLIDES IN WATER, $(MPCU)_w$ VALUES* FOR
CONTINUOUS OCCUPATIONAL EXPOSURE

Limitations μCi/cc *of water*†

If no one of the radionuclides ^{90}Sr, ^{126}I, ^{129}I, ^{131}I, ^{210}Pb, ^{210}Po, ^{211}At, ^{223}Ra, ^{224}Ra, ^{226}Ra, ^{228}Ra, ^{227}Ac, ^{230}Th, ^{231}Pa, ^{232}Th, and Th-nat is present, then the $(MPCU)_w$ is.. 3×10^{-5}
If no one of the radionuclides ^{90}Sr, ^{129}I, ^{210}Pb, ^{210}Po, ^{223}Ra, ^{226}Ra, ^{228}Ra, ^{231}Pa, and Th-nat is present, then the $(MPCU)_w$ is......................... 2×10^{-5}
If no one of the radionuclides ^{90}Sr, ^{129}I, ^{210}Pb, ^{226}Ra, and ^{228}Ra is present, then the $(MPCU)_w$ is.. 7×10^{-6}
If neither ^{226}Ra nor ^{228}Ra is present, then the $(MPCU)_w$ is.............. 10^{-6}
If no analysis of the water is made, then the $(MPCU)_w$ is............... 10^{-7}

* Each $(MPCU)_w$ value is the smallest $(MPC)_w$ value of the National Council on Radiation Protection—*Natl. Bur. Standards Handbook* 69, June 5, 1959, or *ICRP Publ.* 2, 1959, for radionuclides other than those listed opposite the value. Thus these $(MPCU)_w$ values are permissible levels for continuous occupational exposure (168 hr/week) for any radionuclide or mixture of radionuclides where the indicated isotopes are not present [i.e., where the concentration of the radionuclide in water is small compared with the $(MPC)_w$ value for this radionuclide]. The $(MPCU)_w$ may be much smaller than the more exact maximum permissible concentration of the material, but the determination of this $(MPC)_w$ often requires expensive and time-consuming identification of the radionuclides present and the concentration of each.
† Use one-tenth of these values for interim application in the neighborhood of an atomic-energy plant or other controlled area.

Regulations for the Shipment of Radioactive Materials. The reader should refer to official publications[1] for detailed information on the shipment of radioactive materials.

[1] Robley D. Evans, Chairman of the Subcommittee on Shipment of Radioactive Substances, "Physical, Biological and Administrative Problems Associated with the Transportation of Radioactive Substances." ICC shipping regulations are given in Title 49, parts 71 to 78, of the Code of Federal Regulations; Civil Aeronautics Board regulations are given in part 49 of the Civil Air Regulations, "Transportation of Explosives and Other Dangerous Articles"; regulations of the United States Coast Guard are given in the Federal Register, July 17, 1952, pp. 6460ff.; regulations governing the transportation of radioactive materials in the United States mails are given in the U.S. Postal Manual, parts 124.24 and 125.24.

TABLE 8i-10. MAXIMUM PERMISSIBLE CONCENTRATION OF UNIDENTIFIED
RADIONUCLIDES IN AIR, $(MPCU)_a$ VALUES* FOR
CONTINUOUS OCCUPATIONAL EXPOSURE

Limitations $\mu Ci/cc$ *of air†*

If there are no α-emitting radionuclides, and if no one of the β-emitting radionuclides ^{90}Sr, ^{129}I, ^{210}Pb, ^{227}Ac, ^{228}Ra, ^{230}Pa, ^{241}Pu, and ^{249}Bk is present, then the $(MPCU)_a$ is.. 10^{-9}

If there are no α-emitting radionuclides, and if no one of the β-emitting radionuclides ^{210}Pb, ^{227}Ac, ^{228}Ra, and ^{241}Pu is present, then the $(MPCU)_a$ is.... 10^{-10}

If there are no α-emitting radionuclides, and if the β-emitting radionuclide ^{227}Ac is not present, then the $(MPCU)_a$ is............................. 10^{-11}

If no one of the radionuclides ^{227}Ac, ^{230}Th, ^{231}Pa, ^{232}Th, Th-nat, ^{238}Pu, ^{239}Pu, ^{240}Pu, ^{242}Pu, and ^{249}Cf is present, then the $(MPCU)_a$ is................. 10^{-12}

If no one of the radionuclides ^{231}Pa, Th-nat, ^{239}Pu, ^{240}Pu, ^{242}Pu, and ^{249}Cf is present, then the $(MPCU)_a$ is....................................... 7×10^{-13}

If no analysis of the air is made, then the $(MPCU)_a$ is.................... 4×10^{-13}

* Each $(MPCU)_a$ value is the smallest $(MPC)_a$ value of the National Council on Radiation Protection—*Natl. Bur. Standards Handbook* 69, June 5, 1959, or of *Intern. Comm. Radiol. Protec. Publ.* 2, 1959, for radionuclides other than those listed opposite the value. Thus these $(MPCU)_a$ values are permissible levels for continuous occupational exposure (168 hr/week) for any radionuclide or mixture of radionuclides where the indicated isotopes are not present [i.e., where the concentration of the radionuclide in air is small compared with the $(MPC)_a$ value for this radionuclide]. The $(MPCU)_a$ value may be much smaller than the more exact maximum permissible concentration of the material, but the determination of this $(MPC)_a$ often requires expensive and time-consuming identification of the radionuclides present and the concentration of each.

† Use one-tenth of these values for interim application in the neighborhood of an atomic-energy plant or other controlled area.

General limitations for the shipment of radioisotopes are:

1. A package must not be less than 4 in. in its smallest outside dimension.

2. A single package must not contain more than 2 curies (2.7 curies of less dangerous radioisotopes) or as noted below.[1]

3. The surface of the package must contain no significant contamination.

4. The dose rate at any accessible surface must not exceed 200 mR/hr (or equivalent in mrem/hr).

5. The dose rate at 1 m must not exceed 10 mR/hr.

6. Shipments of radioactive materials by rail and motor express, air, and boat fall into five categories (groups I, II, III, IV, and exempt). Only exempt shipments may be made by mail.

[1] Not more than 300 curies of solid cesium-137, cobalt-60, gold-198, or iridium-192 may be packed in one outside container for shipment by rail freight, rail express, or highway except by special arrangements and under conditions approved by the Bureau of Explosives.

8j. Particle Accelerators

MARTIN P. REISER

University of Maryland

8j-1. Introduction. Particle accelerators are devices designed to impart high kinetic energies to charged particles such as electrons, protons, and positive or negative ions of atoms and molecules. In the broadest sense this definition includes any device in which particle acceleration occurs, e.g., electron and ion guns, X-ray machines, cathode-ray tubes, etc. However, the term *particle accelerator* is commonly used only for machines which are capable of generating particle energies in excess of 1 MeV. By definition, 1 electron volt (eV) is the kinetic energy acquired by a particle with charge $e = 1.602 \times 10^{-19}C$ in falling freely across an electric potential difference of 1 volt: $1 \text{ eV} = 1.602 \times 10^{-19}J$. Kinetic energies of 1 MeV and higher are necessary for the particles to penetrate and probe the structure of the nucleus of the atom or to achieve nuclear disintegration. At particle energies above 150 MeV mesons are created, whereas in the range of a few GeV[1] and higher, other members of the family of elementary particles, such as antiprotons, hyperons, etc., are produced in collisions of the accelerated particles with matter.

The genesis of particle accelerators and the construction of the various types of devices now existing form an exciting and interesting chapter of modern science and technology. The first successful accelerator was a small two-stage linear accelerator (linac) built by R. Wideröe [1] in 1928. It was followed in 1931 and 1932 by the invention of the electrostatic accelerator by R. J. Van de Graaff [2], the cyclotron by E. O. Lawrence [3], and the voltage multiplier by Cockroft and Walton [4], the last mentioned being the first device used for artificial nuclear disintegration. Other important milestones in the history of particle accelerators are: the invention of the betatron in 1928 by Wideröe [1] and the successful construction of the first machine of this kind by Kerst [5] in 1940 and 1941; the invention of the synchrotron principle [6] by McMillan and Veksler in 1945; the proposal of strong focusing by Christofilos in 1950 and, independently, by Courant, Livingston and Snyder in 1952 [7]; and the proposal of sector-focusing by L. H. Thomas in 1938 [8]; which led to the development of the isochronous cyclotron after the principle was tested in 1949 to 1956 at the Lawrence Radiation Laboratory [9].

During the four decades of accelerator development the energies of accelerated particles have increased on the average by roughly a factor of 10 every six years [10], the largest machines presently operating being the 33-GeV synchrotrons at Brookhaven National Laboratory and the European Nuclear Research Center (CERN) at Geneva, and the 70-GeV synchrotron at the Serpukhov in the USSR. A 500-GeV synchrotron is being built at the National Accelerator Laboratory, Batavia, Illinois.

The following review of the various types of accelerators and their design principles

---

[1] GeV = 1 billion electron volts = 10^9 eV; in the United States the term BeV is frequently used for 10^9 eV.

is organized in a topical rather than chronological order. After a general survey of the fundamental concepts and a classification of the various types of accelerators some useful general formulas are presented. This is followed by a review of the basic theory and mode of operation of each type of accelerator. Only the fundamental equations and major results of the theory are discussed. Important relations are presented in a numerical form suitable for calculations in practical units.

For a comprehensive and detailed account of particle accelerators several books are available; also one volume of the "Encyclopedia of Physics" is devoted to this topic. In addition, the proceedings of several international and national conferences on particle accelerators present a collection of valuable review and technical papers describing the developments during the last two decades. They are listed in the bibliography at the end of this section.

8j-2. Acceleration Principles and Types of Accelerators. The simplest method of acceleration is to let the particles cross a gap between two electrodes, one of which is at a high electrostatic potential with respect to the other. Electrical breakdown sets an intrinsic upper limit to the potential differences that can be achieved between electrodes, and, consequently, such single-gap or potential-drop accelerators are feasible only at low energies corresponding to potentials of less than about 10 MV. To get above this limit the energy must be accumulated in many steps by directing the beam through a series of gaps (or by multiple traversal of one gap), in which case time-varying electric fields must be employed. The "electrodes" are formed by conducting tubes which are separated by small gaps and connected to a rf power source. After passing through a gap, the particles travel through the field-free interior region of a tube. By the time they enter the gap on the other end of the tube, the electric field has reversed its polarity and the particles are accelerated again. As the velocity increases, the tube sections between gaps have to be longer to assure that the particles are in synchronism with the applied rf fields in the gaps. This is the principle of the linear accelerator (linac).

Since the length of a linear accelerator increases with increasing energy, a limit is set by the sheer physical size of a machine (the largest linac so far is 2 miles long!). This drawback is to a large extent avoided in circular or cyclic accelerators where a magnetic field is employed to force the particles into cyclic orbits, during which they pass many times through one or several rf acceleration gaps. The classical machine of this type is the cyclotron, and all circular accelerators are based on this principle.

Figure 8j-1 presents a classification of particle accelerators which divides the various types of machines into three main groups: d-c accelerators, linear accelerators, and circular accelerators. The first group, where only machines capable of acceleration to energies above 1 MeV are mentioned, can be subdivided into voltage multipliers and electrostatic accelerators. Linear accelerators can be grouped into proton or heavy-ion linacs and electron linacs.

Circular accelerators are divided into two branches, one representing the betatron, and the second branch comprising all the other circular machines. The latter are grouped according to the type of magnetic field employed for focusing the particles. In conventional cyclotrons, microtrons, and synchrocyclotrons, the magnetic field is axially symmetric, $B = B(r)$, and a constant negative gradient, $dB/dr < 0$, provides the focusing forces for the particles. In the isochronous cyclotron B is a function of both radius r and azimuth angle ϕ; the radial gradient of the average field $\bar{B}(r)$ is positive, i.e., $d\bar{B}/dr > 0$, and focusing is provided by the azimuthal variation (sector focusing). The constant-gradient synchrotrons use a negative field gradient for focusing like conventional cyclotrons; but, in addition the magnetic field is pulsed, $B = B(r,t)$, the time dependence being programmed to keep the orbit radius constant during acceleration. Finally, in the AG synchrotrons the magnetic ring consists of sectors with alternating gradients: a sector with $dB/dr < 0$ is followed by one where

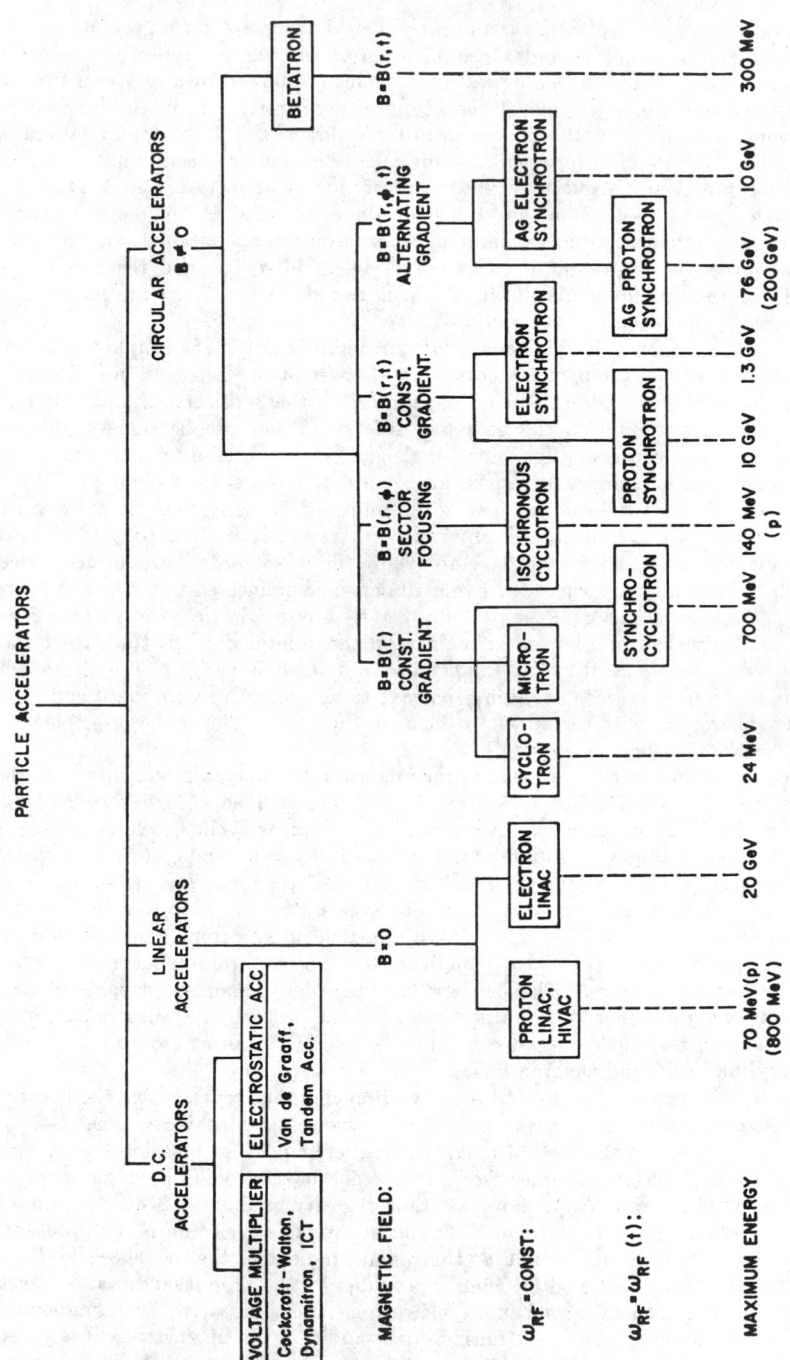

FIG. 8j-1. Classification of particle accelerators.

$dB/dr > 0$, which results in an overall strong focusing effect similar to the combination of focusing and defocusing lenses in optics.

Within each of these four groups further distinction is made between machines where the frequency of the accelerating rf fields is constant and those where it is varying with time.

Below each type of linear and circular accelerator the maximum energy achieved in an existing machine is given; figures in parenthesis indicate design goals for machines presently planned or under construction.

8j-3. General Relations and Beam Characteristics. In most accelerators particle velocities v reach values where relativistic effects are significant. This section presents a number of useful general relations in relativistic form, followed by general parameters and definitions which are commonly used to characterize the properties of an accelerator beam. Practical units (m, kg, sec, V, A, etc.) are used except where noted.

The relativistic mass increase is given by

$$\frac{m}{m_0} = \frac{E}{E_0} = \gamma = (1 - \beta^2)^{-\frac{1}{2}} = 1 + \frac{E_k}{E_0} \quad (8j\text{-}1)$$

where $E_0 = m_0 c^2$ = rest energy
$E = E_0 + E_k$ = total energy
E_k = kinetic energy,
$\beta = v/c$

It follows that

$$\gamma^2 = \frac{1}{1 - \beta^2} \quad (8j\text{-}2)$$

$$\beta^2 = \frac{\gamma^2 - 1}{\gamma^2} \quad (8j\text{-}3)$$

$$\beta = \frac{(E^2 - E_0^2)^{\frac{1}{2}}}{E} = \frac{[(E_k/E_0)^2 + 2E_k/E_0]^{\frac{1}{2}}}{1 + E_k/E_0} \quad (8j\text{-}4)$$

Figure 8j-2 shows β versus kinetic energy for several particles.

The relativistic relation between momentum and energy is

$$p = mv = \frac{1}{c} (E^2 - E_0^2)^{\frac{1}{2}} \quad (8j\text{-}5)$$

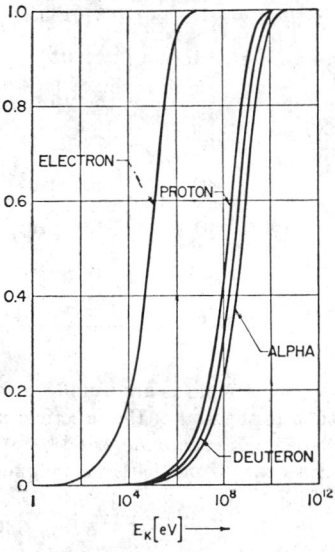

Fig. 8j-2. v/c of electrons, protons, deuterons, and α particles as a function of kinetic energy.

The orbit radius of charged particles with momentum p and charge q in a magnetic field of strength B (Wb/m²) along the orbit is determined by the relation

$$R = \frac{p}{qB} = \frac{1}{qBc} (E^2 - E_0^2)^{\frac{1}{2}} = \frac{E_0}{qBc} \left[\left(\frac{E_k}{E_0} \right)^2 + 2 \frac{E_k}{E_0} \right]^{\frac{1}{2}} \quad (8j\text{-}6)$$

The radian or cyclotron frequency is

$$\omega = \frac{qB}{m} = \frac{qBc^2}{E} = \frac{qB}{m_0\gamma} = \frac{qBc^2}{E_0\gamma} \quad (8j\text{-}7)$$

with the revolution period being $T = 2\pi/\omega$.

With $E_e = 0.511$ MeV being the rest energy of an electron, the rest energy of an ion can be calculated to good approximation as follows:

$$E_0 = MA - Z \cdot E_e (\text{MeV}) \quad (8j\text{-}8)$$

where $A = 931.481$ MeV represents the atomic mass unit based on $^{12}C = 12$ (exactly); M is the mass number of the element, Z the number of electrons removed from the atomic shell (the binding energy of the removed electrons can be neglected). The rest energy of a number of ions of light elements which are of practical interest for accelerators are given in Table 8j-1.

TABLE 8j-1. REST ENERGIES OF SOME ISOTOPES AND IONS

Isotope	M, amu	Rest energy, MeV	Ion	Rest energy, MeV
^1H	1.007825	938.7698	$^1H^+$	938.2588
^2H	2.01410	1876.0959	$^2H^+$	1875.5849
^3He	3.01603	2809.3746	$^3He^+$	2808.8636
			$^3He^{2+}$	2808.3526
^4He	4.0026	3728.3459	$^4He^+$	3727.8348
			$^4He^{2+}$	3727.3238
^6Li	6.01512	5602.96999	$^6Li^+$	5602.458999
			$^6Li^{3+}$	5601.43698
^{12}C	12.0	11,177.772	$^{12}C^{3+}$	11,176.23898
			$^{12}C^{6+}$	11,174.70597
^{14}N	14.00307	13,043.59365	$^{14}N^+$	13,043.08265
			$^{14}N^{7+}$	13,040.01662
^{16}O	15.99491	14,898.9548	$^{16}O^+$	14,898.4438
			$^{16}O^{8+}$	14,894.8667
^{20}Ne	19.99244	18,622.578	$^{20}Ne^+$	18,622.06699
			$^{20}Ne^{10+}$	18,617.46796

Magnetic fields are usually measured in units of gauss (G) or kilogauss (kG) rather than in the unit of the practical international system which is variously denoted with webers per square meter (Wb/m²) or tesla, where 1 Wb/m² = 10 kG. The "rigidity" of a charged particle in a magnetic field can be expressed by the formula

$$RB = 3.33564 \times 10^{-2} \frac{(2E_0E_k + E_k^2)^{\frac{1}{2}}}{Z} \qquad (8j\text{-}9)$$

Solving for the kinetic energy yields

$$E_k = (E_0^2 + 898.755 \, R^2B^2Z^2)^{\frac{1}{2}} - E_0 \qquad (8j\text{-}10)$$

where R is in meters, B in kG, and the energies in MeV.

Figure 8j-3 shows a plot of the rigidity versus kinetic energy for electrons, protons, and deuterons (on a logarithmic scale). The interesting feature is that at extremely relativistic energies (10 GeV and above) the rigidity is the same for all three particles. Also note that in this region RB increases to good approximation linearly with E_k, as follows from Eq. (8j-9) if $E_k \gg E_0$. At low, nonrelativistic energies $(E_k < E_0)RB$ is proportional to $\sqrt{E_k}$, and one can use the formula

$$RB = 1.44 \frac{(ME_k)^{\frac{1}{2}}}{Z} \qquad (8j\text{-}11)$$

where again R is in m, B in kG, and E_k in MeV.

For the orbital frequency $f = 2\pi/\omega$ of the particles one obtains the relations

$$f = \frac{1.43041 \times 10^3 \times Z \times B}{\gamma E_0} = \frac{1.53563 \times Z \times B}{M\gamma} \qquad (8j\text{-}12)$$

where f is in MHz, B in kG, and E_0 in MeV. Specifically,

$$f\,(\text{MHz}) = \frac{1.52454B\,(\text{kG})}{\gamma} \quad \text{for protons} \qquad (8\text{j-}13)$$

and
$$f\,(\text{GHz}) = \frac{2.79922B\,(\text{kG})}{\gamma} \quad \text{for electrons} \qquad (8\text{j-}14)$$

The characteristics and quality of a particle beam obtained from an accelerator is usually described in terms of intensity, kinetic energy, time structure, energy spread, and phase-space area or emittance. The various parameters associated with this description of beam properties are explained in Figs. 8j-4 and 8j-5. All beams from linacs, cyclotrons, and synchrotrons are composed of short bursts spaced at intervals determined by the oscillation period $T = 2\pi/\omega_{\text{rf}}$ of the accelerating r-f (or microwave) fields. The width of the individual bursts τ (which may be defined as full width at

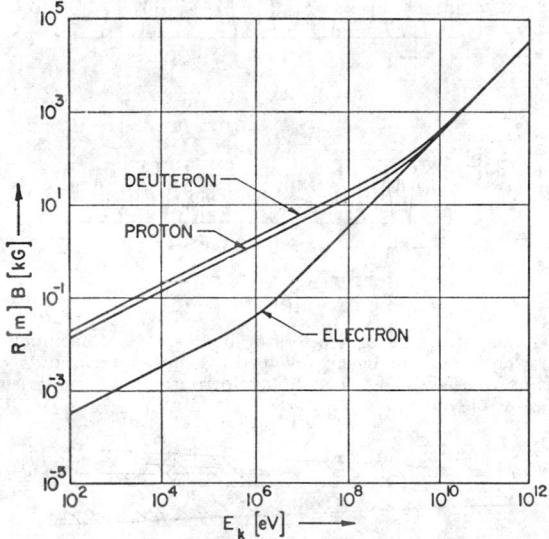

FIG. 8j-3. Rigidity RB of electrons, protons, and deuterons versus kinetic energy.

half maximum) is usually much smaller than T, and the ratio $\delta = \tau/T$ is known as the *(microscopic) duty factor*. In cyclotrons the "pulse train" of ion bursts is continuous (Fig. 8j-4b). Synchrocyclotrons and synchrotrons, where ω_{rf} is modulated, and linacs operating in pulsed schemes to avoid excessive power problems, generate beams of a pulsed nature, as shown in Fig. 8j-4c. The repetition period T_m is determined by the modulation frequency or repetition rate. The width τ_m of the macroscopic pulses, which may contain many thousands of "microbursts," defines what is known as the *macroscopic duty* factor $\delta_m = \tau_m/T_m$. If I_0 denotes peak current, then the time-averaged current is roughly $I = I_0\delta$ in Fig. 8j-4b and $I = I_0\delta\delta_m$ in Fig. 8j-4c.

The state and history of a beam during the acceleration process can be described, in principle, by the space and momentum coordinates of all the particles in six-dimensional phase space and the phase-space particle density as a function of time. If the mutual interaction of the particles can be neglected, Liouville's theorem states that the volume occupied by the beam in phase space remains constant, even though the shape of this volume may change considerably during acceleration. Let x denote the distance from the center line or axis of the beam, and p_x the transverse momentum of a particle in the x direction and a given time, as illustrated in Fig. 8j-5. If the external electric and magnetic forces acting on the particles in the three directions are not coupled, or in the absence of such forces, the area occupied by the beam in two-dimensional phase space, $\int dp_x \, dx$, remains a constant. Under ideal conditions this

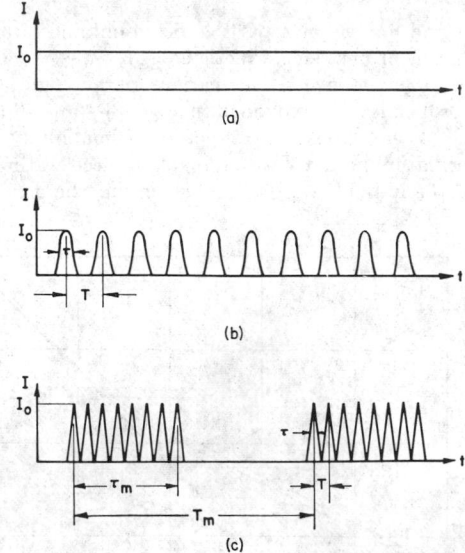

FIG. 8j-4. Typical time structure of accelerator beams. (a) Continuous beam from d-c accelerators. (b) Continuous rf beam (cyclotron); microscopic duty factor $\delta = \tau/T$. (c) Pulsed rf beam (pulsed linacs, synchrocyclotrons, etc.); macroscopic duty factor $\delta_m = \tau_m/T_m$; microscopic duty factor $\delta = \tau/T$.

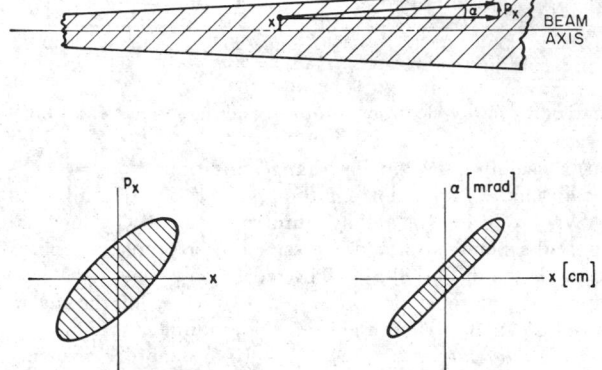

FIG. 8j-5. Transverse phase space or emittance of accelerator beam.

area has the shape of an ellipse, as depicted in Fig. 8j-5. However, in practice this shape may be quite irregular and "filamented," which is an indication that nonlinear effects and other factors have caused a deterioration of beam quality. For practical purposes it is more convenient to express the beam quality in terms of the angle of divergence α; in most cases this angle is very small so that $\alpha = p_x/p$, where p is the total momentum of the particles. The area filled by the beam in $\alpha - x$ "phase space" is called the *emittance* and usually measured in cm mrad (see Fig. 8j-5) If the beam has no axial symmetry, the emittance in the other transverse direction must also be given to fully describe the beam quality. In contrast to $\int dp_x \, dx$, the emittance $\int d\alpha \, dx$ is not constant but is inversely proportional to the momentum. For comparison of different accelerators or beams of different energy it is, therefore, better to define the emittance in terms of $\text{cm} \cdot \text{mrad MeV}^{-\frac{1}{2}}$.

The energy spread of an accelerator beam is commonly defined as $\Delta E/E$, where ΔE represents the full width at half maximum of the intensity-versus-energy curve.

8j-4. D-C Accelerators. *Voltage Multipliers.* Voltage-multiplying circuits had been developed by Schenkel in 1919 and Greinacher in 1921 [11]. Cockroft and

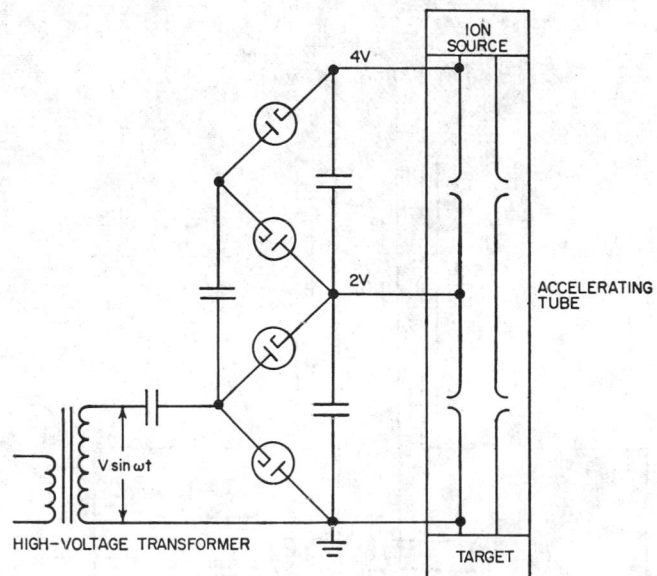

Fig. 8j-6. Schematic diagram of Cockcroft-Walton accelerator.

Walton adapted the Greinacher scheme for their famous accelerator which achieved the first nuclear disintegration in 1932 [4]. Accelerators based on this principle of voltage multiplication are now commonly called Cockroft-Walton machines. The circuit, which achieves voltage multiplication through a cascade-rectification scheme, is shown in Fig. 8j-6. The ac or r-f voltage across the secondary winding of the high-voltage transformer is applied to a multistage combination of capacitors and rectifiers. With no load current the rectified output voltage is equal to the total number of capacitors times the peak voltage of the transformer. Cockroft-Walton used a capacitor-rectifier system with a voltage-multiplication factor of 4 and an output voltage of about 700 kV.

Under load conditions the voltage is reduced below the ideal maximum by an amount which is proportional to the current and inversely proportional to the frequency input voltage.

Voltage-multiplier systems of the Cockroft-Walton type are used in many laboratories, as high-voltage power supplies, generators for intense neutron beams, or injectors for linear accelerators. A special high-frequency version of the Cockroft-Walton is the Dynamitron, which was developed by Radiation Dynamics, Inc. and generates voltages up to 3 MV with beam currents up to 10 mA.

Another type of voltage-multiplication accelerator is the insulating-core transformer (ICT) developed by the High-Voltage Engineering Corporation. This machine is basically a transformer with a magnetic circuit consisting of insulated segments. The outputs of the secondary windings of each segment are rectified and connected in series, yielding direct voltages up to 4 MV with maximum beam currents in the range of 10 mA.

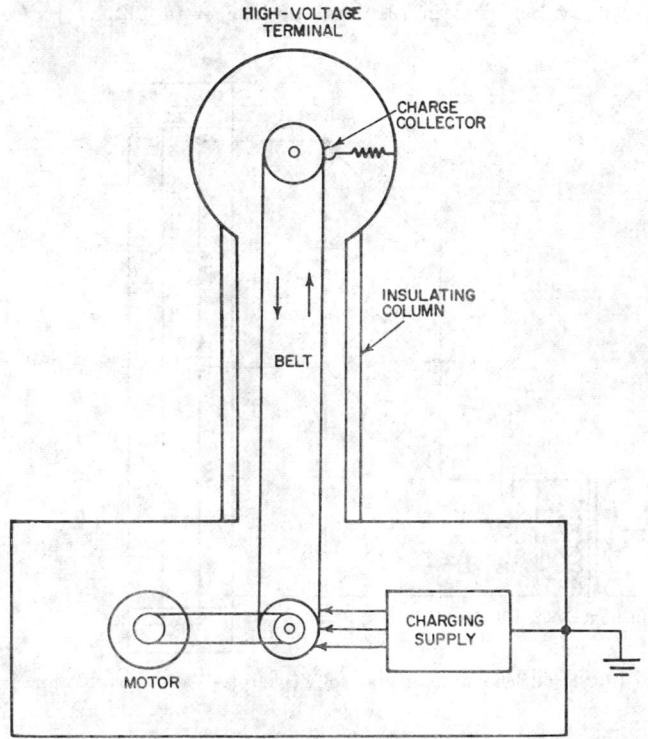

FIG. 8j-7. Schematic diagram of Van de Graaff accelerator.

The Electrostatic Accelerator. The most successful dc accelerator is the electrostatic generator developed by R. J. Van de Graaff in 1930. The "Van de Graaff," as it is commonly called, is relatively simple in concept but involves a sophisticated technology in coping with the numerous engineering problems, primarily the fundamental limitations due to electrical breakdown. The principle of operation is illustrated schematically in Fig. 8j-7. The main structure is a large metal sphere, called the high-voltage terminal, which is supported by an insulating column and charged to high potentials through a moving belt. At the grounded end an electric power supply with 20 to 30 kV voltage produces a corona discharge from a row of sharp metallic needles pointed at the motor-driven belt. A charge of the desired polarity is sprayed on the belt, which carries it to the high-voltage terminal where it is removed by a collecting device. As the charge deposited on the sphere is increased, the potential rises until the charging current is balanced by the accelerated ion beam and leak currents to ground (or until breakdown occurs).

The electron gun or ion source is housed inside the terminal, and the beam is accelerated to ground through the accelerating tube which may be in the same insulating column as the charging belt or, as in early models, is located in a separate column.

Early Van de Graaff generators operated in air where voltages up to 2 MV could be achieved. However, at higher voltages where air-insulated machines become unreasonably large in size, pressurized tanks with gas mixtures of nitrogen and carbon dioxide are used for insulation. With gas pressures of several hundred pounds per square inch, voltages as high as 8 to 10 MV have been achieved in such pressurized models. Pressure-tank Van de Graaffs have been built as vertical as well as horizontal machines and for acceleration of electrons as well as ions.

In recent years a multistage electrostatic generator, the so-called "tandem accelerator," has been developed by the High Voltage Engineering Company. In the simplest model, the two-stage tandem, the positive high-voltage terminal is in the center of a horizontal cylindrical pressure tank with accelerating columns on either side. Both ion source and target are at ground potential. Negative ions produced in a special source are injected into one of the accelerating tubes and accelerated to the terminal in the center. Inside the terminal they pass through a gas cell where two or more electrons are stripped, leaving the ions in a positive-charge state. These positive ions are then accelerated again through the second acceleration tube and arrive at the other end at ground potential with an energy twice that available from a conventional single-stage Van de Graaff of the same terminal voltage. By adding another two-stage tandem with opposite terminal potential, even higher final energies can be achieved. In one version the particles are injected as neutral atoms, unaffected by the electric field, until they reach the negative terminal where an electron is added in a special gas canal. Then acceleration occurs to the other end at ground, and from there into the positive terminal of the second tank. At this point electrons are stripped, and the positive ions are further accelerated to ground, the final energy being three times that of a single-stage generator. In yet another version, the positive ions are returned to the first stage through bending magnets and accelerated to the negative terminal where they hit the target with an energy equivalent to four times the terminal voltage. With terminal voltages of 8 MV, energies between 20 and 30 MeV (for single-charged particles) can thus be obtained from such multistage tandems.

Many hundreds of electrostatic accelerators have been built throughout the world. Most of them are used as high-precision tools for nuclear reaction studies and nuclear spectroscopy at low energies.

8j-5. Linear Accelerators. *Proton and Heavy-ion Linacs.* The concept of the type of linear accelerator originally proposed by Wideröe is easily explained with the help of Fig. 8j-8a. The accelerating structure consists of a straight-line array of hollow cylindrical metal tubes of increasing length through which the beam is traveling. The tube sections are separated by small gaps, and alternate tubes are connected to opposite terminals of an ac generator; thus a time-varying sinusoidal electric field is produced across the gaps. To obtain continuous acceleration the electric potential of the tubes must change polarity while the particles coast through the field-free interior of the so-called drift tubes. As the particle velocity v increases, with the frequency f of the supplied alternating voltage being fixed, the length L of the tube sections has to be increased to maintain "resonance" between particles and electric field. If L_n is the length of the nth gap-plus-tube section, τ the period, and λ the wavelength of the ac field, then

$$L_n = v_n \frac{\tau}{2} = \frac{v_n}{2f} = v_n \frac{\lambda}{2c} = \frac{\beta_n \lambda}{2} \qquad (8j\text{-}15)$$

Under nonrelativistic conditions $v_n \approx (2E_{kn}/m_0)^{\frac{1}{2}}$, and if the energy is gained in equal increments of ΔE_k, we can write

$$L_n \approx \frac{\lambda}{2c}\left(\frac{2E_{kn}}{E_0}\right)^{\frac{1}{2}} = \frac{\lambda}{2}\left(\frac{2n\,\Delta E_k}{E_0}\right)^{\frac{1}{2}} \qquad (8j\text{-}16)$$

This shows that the length of the sections must increase roughly as the square root of the section number n. If, specifically, n denotes the last section, then the final kinetic energy, $E_{kn} = (2/\lambda^2)L_n^2 E_0$ is fixed, being determined only by particle rest mass, wavelength λ, and the geometry. Thus the linear accelerator does not permit a variation of energy, as is possible with dc machines. In the Wideröe-type linac the wavelength λ of the applied field is large compared to the length of a tube section L_n ($\beta_n \ll 1$, $\lambda \gg L_n$), so that each drift tube is essentially at a (spatially) constant

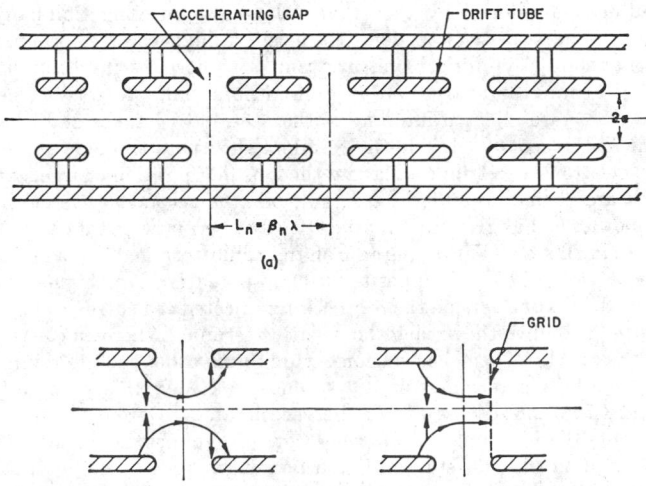

Fig. 8j-8. Linear accelerator. (a) Acceleration system of a proton linac. (b) Electric force lines in the accelerating gap. (c) Accelerating gap with focusing grid.

potential. On the other hand, the total length of a linac is proportional to λ, which implies that such an accelerator becomes excessively large if it is designed for low-frequency operation.

All modern linacs differ from the original Wideröe scheme in that they are designed as short-wave cylindrical resonant cavities or waveguides in either a standing-wave or a traveling-wave mode of operation. The mode of operation depends on the particle velocities. Electrons are injected at sufficiently high energies that v is equal to the speed of light, which permits the use of traveling waves, as will be described in the next section. With protons and heavier ions, on the other hand, velocities usually remain far below c, and a standing-wave mode has to be employed. Because of this fundamental difference, electron linacs generally operate at frequencies of several thousand MHz while proton accelerators usually work with frequencies in the range of 100 to 200 MHz.

The present form of the standing-wave linac for protons and heavy ions was first worked out by L. W. Alvarez in 1945 and built at Berkeley in 1946 and 1947 [12]. In this design the cylindrical tank enclosing the drift tubes forms a resonant cavity which is excited in the TM_{010} mode with the wavelength of the longitudinal electric field given by $\lambda = 2.61a$ (where $2a$ = diameter of tank). The nodes of the standing field waves are at the center of the drift tubes; the maxima at the gaps. Thus the electric field has the same direction in each gap at any given instant of time. If the particles are to be continuously accelerated, the drift time inside the tubes between consecutive gap crossings must be equal to a full rf cycle. The resonance condition for an Alvarez-type linac is therefore

$$L_n = v_n \tau = \beta_n \lambda \qquad (8j\text{-}17)$$

The energy gain per gap, neglecting the finite transit time, is given approximately by

$$\Delta E_k = E_z g \cos \phi \qquad (8j\text{-}18)$$

where E_z is the average electric field on the axis, g the gap width, and ϕ the phase of the field at the moment of particle transit. A particle is synchronous with the rf field if it crosses each gap at the same phase ϕ_s. A fundamental requirement for a resonance accelerator like the linac is the existence of phase stability which assures that nonsynchronous particles (ϕ different from ϕ_s) are not lost during the acceleration process. Simple consideration shows that this is achieved only if ϕ_s lies in the phase interval where E_z is increasing, i.e., $-\pi/2 < \phi_s < 0$. In this case particles crossing the gaps at phases different from ϕ_s are forced into phase oscillations about ϕ_s, provided the starting phase is within certain limits.

Unfortunately this principle of phase stability is incompatible with the requirements of transverse focusing. The electric fields in the gaps constitute electric lenses which are focusing only if the transverse field components in the entrance region (see Fig. 8j-8b) are stronger than the defocusing components in the exit half of the gap, i.e., if $0 < \phi < \pi/2$. This dilemma was solved in earlier designs by the use of grids (Fig. 8j-8c), reducing the defocusing field components, and/or the use of solenoidal magnetic lenses incorporated in the drift tubes. The latest linacs, however, generally employ magnetic quadrupole lenses as proposed by Blewett [13], which provide the most effective means of focusing resulting in only negligible beam loss.

One of the greatest drawbacks of any linear accelerator is the high power required, which is in the range of megawatts. It is therefore necessary to operate linacs in a pulsed scheme with relatively low (macroscopic) duty factor to keep power losses at a manageable level.

The largest existing proton linac in the world is the 100-MeV injector for the Serpukhov 75-GeV synchrotron. (The 70-MeV linac at Minnesota which was the largest machine in the past, has recently been shut down.) A 800-MeV proton machine is being built at Los Alamos.

Electron Linacs. In the electron linac the electron velocity is practically equal to the speed of light, which permits operating the cavity as a waveguide in a traveling-wave mode. No drift tubes are necessary since the electrons are riding on the crest of the wave, being continuously accelerated to full energy. However, the phase velocity of a traveling wave in an empty waveguide is greater than the speed of light, and to reduce it to the value c, it is necessary to load the cavity with disk-shaped irises spaced at intervals of $\lambda/4$.

The first successfully operating electron linacs were developed in 1945 to 1947 by W. W. Hansen and collaborators at Stanford and D. W. Fry and coworkers at the Telecommunications Research Establishment in England [14].

In contrast to the proton linac, transverse focusing poses no great problem in electron linacs. This is due to the fact that, at the much higher frequencies of the electron linacs, the azimuthal magnet field B_θ resulting from the time-varying electric field produces a focusing force vB_θ which is comparable to the defocusing qE_r term.

The net defocusing force is proportional to $1 - \beta^2$ and thus goes to zero as β approaches 1.

The largest electron accelerator in the world is the 2-mile linac at Stanford which is designed for electrons of 20-GeV energy.

8j-6. The Conventional Cyclotron. The cyclotron, invented by Lawrence in 1930, was the first successful circular accelerator. It is based on the fact that a magnetic field B forces charged particles into circular orbits with angular frequency $\omega = qB/m$ (Eq. 8j-7) and orbit radius $R = v/\omega$ (Eq. 8j-6). During each revolution the particles

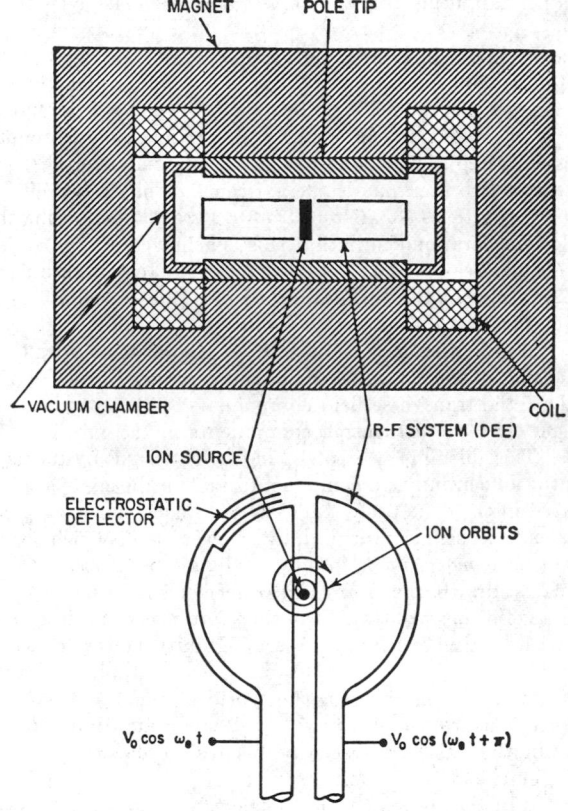

Fig. 8j-9. Cyclotron.

pass through an acceleration gap across which a r-f voltage $V = V_m \cos \omega_e t$ is maintained. When the radio frequency is in "resonance" with the circulating ions, i.e., when $\omega_e = \omega$, continuous acceleration occurs, and the ions travel on an expanding spiraling orbit from the center of the magnetic field ($R = 0$) to some maximum energy and radius determined by the size of the pole shoes of the magnet. The r-f system consists of a large pillbox-type structure split into two halves, which are shaped like a D and therefore called "dees" (see Fig. 8j-9). Each dee is part of a quarter-wave resonance system. The two resonators are oscillating in a push-pull mode (180 deg out of phase); i.e., $V_1 = V_0 \cos \omega_e t$, $V_2 = V_0 \cos (\omega_e t + \pi)$, and the peak voltage across the acceleration gap is $V_m = 2V_0$. The dees are located in the gap between the poles of an electromagnet and are enclosed by a vacuum chamber. The ions are produced in a low-pressure gas-discharge tube at the center of the magnetic field. Ion source design and technology have been considerably improved over the years; the "open-arc source" of the early days was replaced by the "hooded" structure, and today most cyclotrons use the "chimney"-type source with reflector developed by R. S. Livingston and R. J. Jones at Oak Ridge [15].

The magnetic field in the conventional cyclotron must decrease slightly with radius to produce the required force component toward the median plane, which serves to focus the beam during the many revolutions from center to maximum radius. The equation of motion for the z direction (perpendicular to the median plane) is

$$\ddot{z} = \frac{q}{m} v B_r \tag{8j-19}$$

Using the linear term of the Taylor expansion, $B_r = (\partial B_r/\partial z)z$, and div $B = 0$, or $\partial B_r/\partial z = \partial B_z/\partial r$, we can write

$$\ddot{z} = \frac{q}{m} v \frac{dB}{dr} z = \frac{qB}{m} \frac{\omega}{r} \frac{r}{B} \frac{dB}{dr} z \tag{8j-20}$$

where $B = B_z(r)$ is the field in the median plane. Introducing the "field index" $n = -(r/B)dB/dr$, and azimuth angle $\theta = \omega t$, Eq. (8j-20) can be written in the form

$$\frac{d^2z}{d\theta^2} + v_z^2 z = 0 \tag{8j-21}$$

where

$$v_z^2 = n \tag{8j-22}$$

If $n > 0$, or $dB/dr < 0$, the particles perform stable oscillations about the median plane. These oscillations are known as *betatron oscillations* because they were first investigated in connection with the betatron [16]. The parameter v_z measures the number of betatron oscillations per revolution.

A similar equation can be derived for the radial motion, yielding for the radial betatron frequency the relation

$$v_r^2 = 1 - n \tag{8j-23}$$

with the stability condition $n < 1$.

Thus simultaneous stability in both vertical and radial directions can exist only if

$$0 < n < 1 \tag{8j-24}$$

The focusing requirement of $dB/dr < 0$ implies that the orbital frequency $\omega = qB/m$ of ions is not a constant, but decreases with radius. As a result, the resonance condition $\omega_e = \omega$ is violated, particles get out of step with the r-f voltage, and after a certain number of turns the phase slip is large enough so that deceleration occurs. This dilemma is still further enhanced by the relativistic mass increase which, unfortunately, goes in the same direction as a radially decreasing field. In practice, the electric frequency is set to correspond with the orbital frequency at some intermediate radius r_c. Inside this radius the phase shift per turn, $\Delta\phi = 2\pi(\omega_e - \omega)/\omega$, is negative ($\omega > \omega_e$); the ion phase ϕ with respect to the radio frequency reaches a minimum $\phi_{min} < 0$ at r_c, then increases in the region $r > r_c$ (where $\Delta\phi > 0$), goes through the peak-voltage phase $\phi = 0$, and reaches a maximum value ϕ_{max} close to $\pi/2$ at the final radius. At this point, just before deceleration would occur, the particles have reached the maximum energy attainable and enter an electrostatic deflector which extracts them out of the magnet for bombardment of an external target.

The maximum energy attainable in this type of cyclotron depends on the number of revolutions, which is inversely proportional to the peak dee voltage. The largest conventional machine is the 86-in. cyclotron at Oak Ridge National Laboratory, which accelerates protons to 24 MeV (with a dee-to-ground voltage of 250 kV, or

$V_m = 500$ kV!). With the development of the sector-focusing cyclotrons, conventional cyclotrons are no longer built, and many existing machines are being converted or shut down.

8j-7. The Microtron. Cyclotrons are only capable of accelerating protons or heavy ions since the orbital frequencies of these particles are low enough to permit the use of quarter-wave resonators where the wavelength λ is substantially larger than the magnet pole diameter. Electron frequencies for magnetic fields between 10 and 20 kG are in the range of several thousand MHz, and quarter-wave resonance systems are impractical (λ is too small). Besides, the relativistic mass increase of electrons begins at much lower energies (1 percent increase at 5 keV!) than that for protons.

The microtron, or electron cyclotron, first proposed by Veksler [6] is a cyclotron-type device for the acceleration of electrons. It employs a small microwave cavity near the periphery of the magnetic field, through which the electron beam passes once per revolution. The orbits form a family of circles of increasing radius with a common tangent at the point where they intersect the cavity gap. Resonance acceleration occurs if the electrons cross the gap at the same voltage phase in each revolution. The rotation period of the electrons is given by

$$\tau = \frac{2\pi}{\omega} = \frac{2\pi E}{eBc^2} = \frac{2\pi E_0}{eBc^2}\,\gamma = \tau_0\gamma \qquad (8j\text{-}25)$$

Resonance exists if the electron rotation period on the first revolution τ_1 and the difference $\Delta\tau = \tau_{n+1} - \tau_n$ between consecutive orbits are each equal to some integral multiple of the r-f period τ_{rf}, i.e.,

$$\tau_1 = \tau_0\gamma_1 = \tau_0\left(1 + \frac{E_{k1}}{E_0}\right) = k\tau_{rf} \qquad (8j\text{-}26)$$

$$\Delta\tau = \tau_{n+1} - \tau_n = \tau_0\Delta\gamma = \tau_0\frac{\Delta E_k}{E_0} = m\tau_{rf} \qquad (8j\text{-}27)$$

where k and m are positive integers.

If $E_{k1} = \Delta E_k$, we find by elimination of τ_{rf} the relation

$$\frac{\Delta E_k}{E_0} = \frac{m}{k - m} \qquad (8j\text{-}28)$$

In contrast to the cyclotron, the energy gain in a microtron cannot be arbitrary but must occur in fractions of the rest energy E_0 as determined by the choice of the integers k and m in Eq. (8j-28). Note also that k must be larger than m (minimum k value is 2) in order for ΔE_k to be finite and positive. For $k = 2$, $m = 1$ the energy gain must be equal to the rest energy.

It should be pointed out that the magnetic field in a microtron is practically uniform, $2\pi E_0/eBc^2 = \tau_0 = $ const, which implies that there is little or no vertical focusing. However, if the number of orbits is not too large, this is not a serious problem. Microtrons for electron energies between 1 and 30 MeV have been built at many places. However, very little is known about operating experience and performance characteristics.

8j-8. The Synchrocyclotron. In 1945 McMillan and Veksler independently proposed the synchrotron and synchrocyclotron [6] which made it possible to get beyond the energy limits of the conventional cyclotron. The two basic ingredients in this new accelerator concept are (1) the modulation of the electric frequency (and in the synchrotron also the magnetic field) with time to achieve synchronism between radio frequency and circulating particles; (2) the existence of phase stability which assures the continuous acceleration of nonsynchronous particles within certain limits.

The synchrocyclotron employs a cyclotron-type r-f system with frequency ω_e modulated by the use of a rotating capacitor, tuning fork, or other means, such that ω_e is a function of time, decreasing in synchronism with the orbital frequency of the ions. After a group of ions is accelerated to full energy, the radio frequency returns to its starting value and begins another cycle of acceleration. The major drawback of this scheme is that beam intensities are down by a factor of 10^2 to 10^4 compared to those of the fixed-frequency cyclotrons. On the other hand, the substantial increase in particle energy more than outweighs this disadvantage. Many synchrocyclotrons were built throughout the world, the largest machines producing protons of more than 600 MeV.

The time variation of the electric frequency in a synchrocyclotron is determined by the rate of change of the orbital frequency of the synchronous particle:

$$\frac{d\omega_s}{dt} = -\frac{\omega_s{}^2 KqV \cos \phi_s}{2\pi E_s} \tag{8j-29}$$

where

$$K = 1 + \frac{n}{(1-n)\beta_s{}^2} \tag{8j-30}$$

and $qV \cos \phi_s$ is the energy gain per turn, and E_s the total energy of the synchronous particle. A particle which passes the acceleration gap at a phase ϕ different from ϕ_s will gain a different amount of energy, and, therefore, its orbital radius will be slightly different from that of the synchronous particle. If $\Delta p/p$ is the fractional difference in momentum between nonsynchronous and synchronous particles, then the corresponding difference in revolution time is given by the relation

$$\frac{\Delta T}{T} = \frac{1}{\alpha} - \frac{1}{\gamma^2}\frac{\Delta p}{p} \tag{8j-31}$$

where $\alpha = 1 - n$, $\gamma = E/E_0$.

In synchrocyclotrons the values of α and γ are such that $\Delta T/T$ has the same sign as $\Delta p/p$. This implies that phase stability exists only in the phase interval $0 < \phi_s < \pi/2$ where the voltage falls: A particle crossing the gap at a phase $\phi > \phi_s$ gains less energy, i.e., $\Delta p < 0$; as a result has a shorter revolution time than the synchronous particle; and arrives, therefore, earlier at the next gap crossing. A similar argument can be made if $\phi < \phi_s$. In both cases the phase ϕ oscillates about the synchronous phase ϕ_s. The differential equation for these phase oscillations is

$$\frac{d}{dt}\left(\frac{E_s}{\omega_s{}^2 K}\frac{d\phi}{dt}\right) = \frac{qV}{2\pi}(\cos \phi - \cos \phi_s) \tag{8j-32}$$

Ion capture takes place only during a small time interval Δt at the beginning of each modulation cycle. With a few simplifying assumptions Bohm and Foldy derived the expression [17]

$$\Delta t = \frac{4}{\omega_s}\sqrt{\frac{\pi E_s}{KqV}}\frac{L(\phi_0,\phi_s)}{\cos \phi_s} \tag{8j-33}$$

where $L(\phi_0,\phi_s)$ is a function of starting phase ϕ_0 and synchronous phase ϕ_s. As was first pointed out by McKenzie [18], the beam current that can be accelerated in a synchrocyclotron is space-charge limited. If $I_{sp.ch.}$ denotes the maximum (direct) current that can pass through the available beam space within the dees under space-charge conditions, the captured average current in a synchrocyclotron is given by

$$I = I_{sp.ch.}\frac{\Delta t}{T_m}\frac{\Delta\phi_0}{2\pi} = I_{sp.ch.}\,\Delta t f_m\frac{\Delta\phi_0}{2\pi} \tag{8j-34}$$

$\Delta\phi_0/2\pi$ is the microscopic, $\Delta t/T_m$ the macroscopic duty factor (see Fig. 8j-4). $I_{sp.ch.}$ is roughly proportional to the voltage V and the square of the vertical focusing frequency, $v_z{}^2$ in the center [10,20]. Since the repetition rate f_m is proportional to V, and Δt is proportional to $V^{-\frac{1}{2}}$, the beam current in a synchrocyclotron is in this crude approximation proportional to $V^{\frac{3}{2}}$ or some similar power of V. In all existing synchrocyclotrons the dee voltage is very small (5 to 20 kV) to minimize r-f power losses. The low voltage also necessitates the use of an open-arc source since ions would not be able to clear the chimney-type structure of the type of ion source used in fixed-frequency machines.

All these factors explain the very low internal beam currents (down by a factor 10^2 to 10^3 compared to fixed-frequency cyclotrons), poor beam quality, and poor extraction efficiency (a few percent compared with typically 40 to 90 percent in FF cyclotrons). After the successful development of sector-focusing cyclotrons, several synchrocyclotrons are being modified and improved to remain competitive with the new type of machines. All these synchrocyclotron conversion programs (a survey is given by Blosser in [21]) involve an increase of the dee voltage and thus the repetition rate, the installation of a "chimney"-type ion source used in other cyclotrons, and an improvement of the vertical focusing in the center through use of magnetic bumps or sectors.

8j-9. Sector-focusing (Isochronous) Cyclotrons. In 1938 L. Thomas had shown in a theoretical study that it should be possible to build a cyclotron with constant ion frequency ω by employing a magnetic field which varies sinusoidally with azimuth angle. The average magnetic field increases with radius to compensate the relativistic mass increase, thus keeping $\omega = qB/m$ a constant, while at the same time vertical focusing is provided by the azimuthal field variation (called "flutter"). Because of World War II and the invention of the synchrotron, this idea was not acted upon until 1950 when a group at the Lawrence Radiation Laboratory began a study and built an electron model which proved the feasibility of the new cyclotron concept [9]. Similar studies were soon started at other places in the United States and Europe, and since then a large number of sector-focusing cyclotrons have been built and are now in operation.

Details of the theory, design, and performance characteristics of sector-focusing cyclotrons can be found in J. R. Richardson's monography [22] and in the proceedings of several international conferences [23].

Most sector-focusing cyclotrons employ a wedge-shaped rather than a sinusoidal variation in azimuth. Besides, in most cases the pole-shoe sectors or "hills" are spiral-shaped rather than straight, which provides additional focusing, as was first proposed by the MURA group (Midwestern Universitities Research Association) in 1955 [24]. In this general case the median-plane magnetic field is of the form

$$B(r,\theta) = \bar{B}(r)\left[1 + \sum_n f_n(r)\cos n\,(\theta - \phi_n(r))\right](n = N, 2N, 3N, \text{etc.}) \quad (8j\text{-}35)$$

The number of sectors or periods N is 3 or 4 in most existing cyclotrons. The average magnetic field $\bar{B}(r)$ increases with radius according to the relativistic mass change:

$$\bar{B}(r) = B_0\left(1 + \frac{E_k}{E_0}\right) = B_0(1 - \beta^2)^{-\frac{1}{2}} = B_0\left(1 - \frac{r\omega_0{}^2}{c}\right)^{-\frac{1}{2}} \quad (8j\text{-}36)$$

Then

$$\omega = \omega_0 = \frac{q\dot{B}_0}{m_0} = \text{const} \quad (8j\text{-}37)$$

Calculation of the betatron frequencies for such a sector field leads to rather complicated analytical expressions. (For high accuracy, numerical orbit integration by computer is required.) Neglecting a number of less important terms, first-order theory gives the following approximate results:

$$v_r{}^2 = 1 + k \quad (8j\text{-}38)$$

$$v_z{}^2 = -k + \frac{N^2}{N^2 - 1}F(1 + 2\tan^2\alpha) \quad (8j\text{-}39)$$

where

$$k = \frac{r}{\bar{B}}\frac{d\bar{B}}{dr} = -n \qquad F \approx \frac{1}{2}\sum_n f_n{}^2 \quad (8j\text{-}40)$$

α is the (effective) spiral angle defined by

$$\tan \alpha = r \, d\phi/dr \qquad (8j\text{-}41)$$

More accurate formulas are given in the literature ([22,24] for example).

Equation (8j-38) for the radial frequency is identical with Eq. (8j-23) except that in this case $\nu_r \geq 1$ as k is positive. With respect to the vertical frequency (Eq. 8j-39), the spiral angle α and flutter amplitude F must be large enough to compensate for the defocusing average field and, in addition, provide a net focusing effect such that $\nu_z > 0$ (in most cases ν_z is between 0.1 and 0.2). At small radii, sector focusing ceases to be effective since the azimuthal field amplitude, measured by $F(r)$, goes to zero as $(r/g)^N$, where g is the magnet gap width, and N the number of sectors. To achieve good focusing at small radii, the number of sectors should be small, i.e. three or four (fields with fewer than three sectors are unstable for the radial motion). The problem can be further alleviated by utilizing electric focusing through careful programming of the particles' phase history with respect to the radio frequency [25] and, if necessary, employing a small magnetic bump with negative k. Improved central-region design (source position, beam optics, space-charge compensation, defining slits, etc.) is one of the main reasons for the excellent beam quality in sector-focusing cyclotrons [26]. It is also possible to control the pulse width to a certain extent and achieve microscopic duty factors of 10 to 20 percent and higher or get narrow nanosecond pulses for time-of-flight experiments by employing phase-selection slits in the center.

Sector-focusing cyclotrons are limited in energy by resonances in the radial motion which arise whenever the betatron frequency ν_r passes through certain critical values. Under the condition of isochronism,

$$k = \gamma^2 - 1 = (1 + E_k/E_0)^2 - 1 \qquad (8j\text{-}42)$$

and hence, approximately,

$$\nu_r = \gamma = 1 + \frac{E_k}{E_0} \qquad (8j\text{-}43)$$

Thus ν_r starts at unity and increases linearly with energy. According to the theory of resonances in sector fields, a stop band occurs in the radial motion whenever $\nu_r = N/2$, where N is the number of sectors. A two-sector field is therefore intrinsically unstable. According to Eq. (8j-43), in a three-sector cyclotron the stop band $\nu_r = \frac{3}{2}$ occurs at a proton energy of 469 MeV, while $N = 4$ ($\nu_r = 2$) leads to a limit of 938 MeV. If terms neglected in Eqs. (8j-38) and (8j-43) are taken into account, the stop-band energy limits are found to be considerably lower than these values. The resonance problem as well as practical considerations such as achieving the desired field shapes put an upper limit for isochronous cyclotrons at a proton energy of about 800 to 1000 MeV.

The largest sector-focusing machine built so far is the isochronous cyclotron at the University of Maryland, which is capable of accelerating protons to a maximum energy of 140 MeV. At energies above about 200 MeV, new design concepts must be invoked. Several projects in this category are presently under study or construction: A 500-MeV sector-focusing "meson factory" designed as a ring accelerator with a 70-MeV isochronous cyclotron as injector, is presently under construction at Zurich, while a somewhat similar ring machine for 200-MeV protons is being designed at Indiana University. The concept of a separated-orbit cyclotron (SOC) has been studied at Oak Ridge National Laboratory, and a negative-hydrogen cyclotron (H^-) for 500 MeV, called TRIUMF, is being built at Vancouver, Canada.

One of the outstanding features of most existing isochronous cyclotrons is the variability of energy and the possibility of accelerating different types of particles. The r-f system can be tuned over a wide range of frequencies, and the desired magnet-field profiles at various excitation levels are achieved by means of a system of trimming coils.

Extraction of the beam out of the cyclotron [27] is generally accomplished by inducing a coherent radial oscillation at the $v_r = 1$ resonance, which occurs at the transition from the isochronous field to the fringe field. In traversing the resonance, the radial amplitude, and thus separation between consecutive turns, is increased sufficiently so that the beam can enter an electrostatic deflector (or a combination of electric and magnetic deflector channels) which bends it into the external beam pipe. Extraction efficiencies of 40 to 90 percent have been achieved in existing machines for proton currents between 10 and 100 μA. The energy spread of the extracted beam is in the range of 0.1 to 0.3 percent while the emittance in radial and vertical direction is typically between 10 and 30 mm mrad. Sector-focusing cyclotrons are, therefore, excellent tools for nuclear-structure physics in the intermediate-energy range of 10 to 200 MeV for light nuclei.

8j-10. Constant-gradient Synchrotons. For acceleration of protons to energies above 1 GeV, linacs and cyclotrons are impractical, as the size of such machines would become prohibitively large. The only type of accelerator that has been capable so far of generating protons in the billion-volt energy range is the synchrotron, which is based on the principle of phase-stable synchronous acceleration proposed by Veksler and McMillan. Fundamentally the synchrotron is an extension of the synchrocyclotron, the main difference being that the orbit radius is kept constant, and the guiding magnetic field is provided by a number of individual magnets placed along the orbit. The particles are first preaccelerated in a Van de Graaff, Cockcroft-Walton, or linac, and then injected into the synchrotron ring. To keep the orbit radius constant in the synchrotron, the magnets are pulsed such that $B = B(t)$ increases from a minimum value at injection to the maximum given by the final energy of the particles. Orbit stability is provided by constant-gradient focusing as in cyclotrons. Magnets and pole shoes have to be designed carefully to keep the field index $n = -(r/B)dB/dr$ within acceptable limits over the entire range of variation of the magnetic field.

The orbital frequency of the particles is determined by the radius of curvature R in the magnets and the length l of the straight drift section between the magnets. With N straight sections, the circumference of an orbit is $L = 2\pi R + Nl$, and substituting Eq. (8j-4) for $v = \beta c$, one gets

$$\omega = \frac{2\pi}{T} = \frac{2\pi v}{L} = \frac{2\pi c}{L} \left(\frac{E^2 - E_0{}^2}{E^2} \right)^{\frac{1}{2}} \tag{8j-44}$$

or, in view of Eq. (8j-6)

$$\omega = \frac{2\pi c}{L} \frac{BRqc}{[(BRqc)^2 + E_0{}^2]^{\frac{1}{2}}} \tag{8j-45}$$

The particles are accelerated by r-f resonators located in the straight sections between magnets. From Eq. (8j-6) the rate of energy increase dE/dt is determined by dB/dt:

$$\frac{dE}{dt} = \left(\frac{E^2 - E_0{}^2}{E^2} \right)^{\frac{1}{2}} qRc \frac{dB}{dt} \tag{8j-46}$$

The corresponding energy gain per turn $\Delta E = qV \cos \phi = (\omega/2\pi)dE/dt$, is then obtained from Eqs. (8j-44) and (8j-46), and is given by

$$\Delta E = L \, qR \frac{dB}{dt} \tag{8j-47}$$

Since accurate timing of the magnet pulse is exceedingly difficult at such high power levels, no predetermined time schedule can be set up for the variation of B, ΔE, and ω_{rf} with time. Instead, ω_{rf} and ΔE are controlled electronically to follow the rate of change of the magnetic field. A pickup loop in the magnetic field supplies a signal proportional to dB/dt, from which B is obtained at any given time through electronic integration. A computer solves Eq. (8j-45), and the values for ω and V are sent to the control circuits of the r-f oscillators. The required frequency bandwidth of the oscillators is porportional to the range of velocities between injection and full energy; it is the smaller, the higher the energy of the preaccelerator.

As in the synchrocyclotron, phase stability in the constant-gradient synchrotron is obtained when the phase of the synchronous particle lies in the interval of decreasing voltage amplitude.

Electron synchrotrons differ from proton machines in several aspects. Because of the smaller rest mass, electron velocities at energies above a few MeV are essentially equal to the speed of light. The orbital frequency is thus higher than for the proton machines but practically constant, so that frequency modulation is unnecessary. In addition, energy losses due to electromagnetic radiation of the accelerated electrons are substantially higher than for the protons where they are practically negligible. However, these losses are automatically compensated for by the mechanism of phase stability: A decrease in momentum due to radiation losses causes a shrinkage of the orbit radius, so that the electron arrives earlier at the acceleration gap and thus gains additional energy which compensates for these losses.

Historically the electron synchrotron preceded the proton synchrotron by several years. The largest constant-gradient electron machine is the 1.3-GeV synchrotron at Cornell University, while the largest proton synchrotron with constant gradient is the 10-GeV accelerator at Dubna.

The focusing forces in constant-gradient synchrotrons are inherently weak, and, consequently, the amplitudes of the betatron oscillations are relatively large. This necessitates the use of magnets with large gap dimensions to contain the beam, and makes an accelerator of this kind prohibitively expensive if the energy exceeds more than a few GeV. (The magnets for the 10-GeV "Synchrophasotron" at Dubna weigh 36,000 tons!) The invention of the alternating-gradient or strong-focusing principle was, therefore, a major breakthrough in high-energy accelerator design. Alternating-gradient synchrotrons can be built with smaller magnets and have better beam quality and higher beam intensities than constant-gradient machines.

8j-11. Alternating-gradient Synchrotrons. The principle of strong focusing was independently discovered first in 1949 by Christofilos, whose work was not published then, and shortly after that by Courant, Livingston, and Snyder in 1952. This new concept is most easily understood in terms of its well-known optical analog, the combination of focusing and defocusing lenses. If two lenses of focal lengths f_1 and f_2 are combined, with a separation d between them, the focal length F of this system is given by

$$\frac{1}{F} = \frac{1}{f_1} + \frac{1}{f_2} - \frac{d}{f_1 f_2} \qquad (8j\text{-}48)$$

In the special case of a converging and diverging lens of equal, but opposite, strength, one has $f_2 = -f_1$, and hence

$$F = \frac{f_1{}^2}{d} \qquad (8j\text{-}49)$$

The focal length of such a two-lens system is thus always positive (focusing). The application of this idea to synchrotrons implies the combination of strongly focusing and defocusing magnets. According to the theory of betatron oscillations, Eqs. (8j-20) to (8j-24), a magnet with negative gradient, $dB/dr < 0$, is focusing vertically while defocusing radially if $n > 1$. A radially increasing field ($n < 0$), on the other hand, focuses the particles only in the radial direction and is defocusing with respect to the vertical motion. The alternating-gradient synchrotron ring consists of a succession of magnets arranged in such a way that a magnet with large positive gradient is followed by one with a negative gradient of equal strength. The absolute values of n are typically in the range of 200 to 300, as compared to 0.5 in the conventional weak focusing machines. Consequently, the frequencies of the corresponding radial and vertical oscillations are between one and two orders of magnitude larger than in constant-gradient accelerators.

The strong focusing forces reduce the required beam space and the size of the magnets, and thus result in substantial reduction of costs and in improvement of beam quality.

With regard to synchrotron oscillations and phase stability, the alternating-gradient machines are distinctly different from the weak focusing accelerators. The theory shows that the parameter α in Eq. (8j-31) is always greater than 1, and hence $1/\alpha < 1$, in contrast to the constant-gradient machines. At low energies, where $\gamma^2 < \alpha$, an increase in momentum causes a decrease in revolution time. This implies that stability exists if the synchronous particle crosses the accelerating gaps when the voltage is rising. As γ increases, a critical transition energy occurs where $\gamma^2 = \alpha$. Above that energy ($\gamma^2 > \alpha$), particles behave as in the synchrocyclotron and constant-gradient synchrotron; i.e., the synchronous phase must be in a region of falling voltage. This means that in AG synchrotrons provisions must be made to shift the phase of the accelerating voltage at the point where the particles pass through the transition energy. If the injection energy is, however, higher than the transition energy, this difficulty can be avoided.

A major problem in the design of strong focusing synchrotrons is the existence of resonances which occur whenever the values of the betatron frequencies are integers or integral fractions. The operating point must be carefully chosen, taking into account the effects of misalignments and space-charge forces. The electric and magnetic self-fields of the circulating beam produce a net defocusing force which is equivalent to an effective change of the field index n given by

$$\Delta n = \frac{qQR}{4\pi^2\epsilon_0 a^2 E} \frac{1 - \beta^2}{\beta^2} = \frac{qQR}{4\pi^2\epsilon_0 a^2 E_0} \frac{1}{\gamma(\gamma^2 - 1)} \qquad (8j\text{-}50)$$

where Q is the total charge, and R is the major and a the minor radius of the toroidal ring of circulating beam.

If Δn denotes the maximum tolerable change in field index (to stay away from a resonance or avoid defocusing), then the maximum number of particles N_{lim} which can be contained in the ring is given by

$$N_{\text{lim}} = 2.18 \times 10^{15} \frac{a^2}{R} E_0 \gamma(\gamma^2 - 1) \Delta n \qquad (8j\text{-}51)$$

where E_0 is in MeV, a and R in m. (A detailed analysis of space-charge effects, including image effects in surrounding walls, was made by Laslett [28].) For electrons at extremely relativistic energies ($\gamma \gg 1$) the total current $I_{\text{lim}} = qN_{\text{lim}}v/2\pi R$ contained in the ring is given by the relation

$$I = 8,500 \left(\frac{a}{R}\right)^2 \gamma^3 \Delta n \qquad (8j\text{-}52)$$

By proper choice of the n values and careful alignment of the magnets it was possible to overcome the difficulties imposed by resonances and space-charge effects. Several AG synchrotrons are now operating successfully. The presently largest proton accelerator in the world is the 70-GeV alternating-gradient synchrotron at Serpukhov (U.S.S.R.) which went into operation during 1968. In second and third place follow the 33-GeV AGS at Brookhaven (U.S.A.) and the 28-GeV proton synchrotron at CERN, the European Nuclear Research Center at Geneva, Switzerland.

The largest alternating-gradient synchrotron for the acceleration of electrons is the 10-GeV accelerator built at Cornell University. Other large electron machines are the 7-GeV synchrotron (DESY) at Hamburg, Germany, which began operation in 1964; the 6-GeV machine at Cambridge, Massachusetts, operating since 1962; and a 6.5-MeV accelerator in the Soviet Union.

Although synchrotron radiation losses put an upper limit in the range of 10 GeV to electron synchrotrons, proton machines with energies up to 1000 GeV appear to be within the reach of technical feasibility. Preliminary studies of a 1000-GeV accelerator have been carried out in the United States and are in progress in the U.S.S.R. At the National Accelerator Laboratory at Batavia, Illinois, a 200-GeV alternating-gradient synchrotron is being constructed, with provisions to extend the energy to 500 GeV at a later time. Acceleration in this project will take place in three stages: A 200-MeV linac will inject the beam into a 10-GeV booster synchrotron (diameter of 150 m), from which the particles are steered into the main ring (diameter of 2,000 m) for acceleration to full energy.

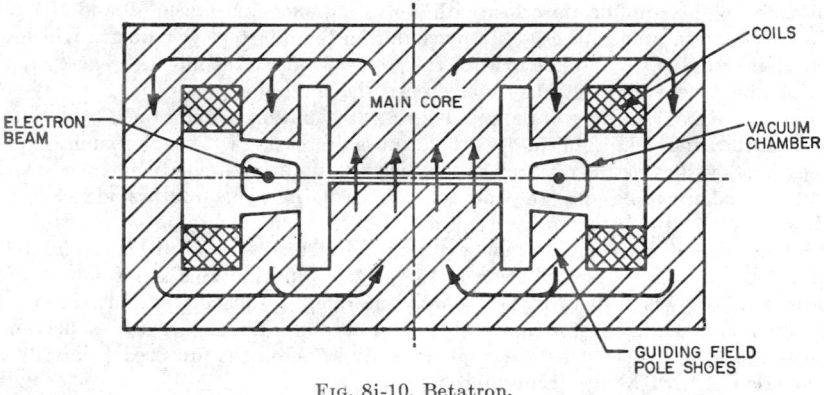

FIG. 8j-10. Betatron.

8j-12. Betatrons. The betatron differs from other circular accelerators in that the electromotive force for accelerating the particles is generated by the time variation of the magnetic flux. It is only suitable for the acceleration of electrons. The magnet structure of a betatron (Fig. 8j-10) resembles that of a cyclotron; the major difference is in the design and shape of the core part with the pole shoes. As in a synchrotron, the orbit radius of the circulating electron beam is kept constant throughout the acceleration process. This implies that the increase in energy due to the changing magnetic flux linked by the circulating electrons must be precisely in step with the increase of the magnetic field strength at the orbit radius. The accelerating electric field E along the circular orbit is determined by Maxwell's second equation

$$\int \mathbf{E} \cdot \mathbf{dL} = -\frac{\partial}{\partial t} \int \mathbf{B} \cdot \mathbf{dS} \qquad (8j\text{-}53)$$

In cylindrical coordinates, if $\mathbf{B} = B\mathbf{a}_z$ and increasing in time, $\mathbf{E} = -E\mathbf{a}_\phi$, and thus

$$2\pi R E - \frac{\partial}{\partial t} \int_0^R 2\pi B R \, dR = \pi R^2 \frac{d\bar{B}}{dt} \qquad (8j\text{-}54)$$

$\bar{B}$ is the average magnetic field inside the circular orbit with constant radius R. The rate of change of momentum of the electron is given by

$$\frac{d}{dt}(mv) = eE = \frac{d}{dt}(eBR) = eR\frac{dB}{dt} \qquad (8j\text{-}55)$$

Elimination of E from Eqs. (8j-54) and (8j-55) gives the fundamental betatron relation

$$\frac{d\bar{B}}{dt} = 2\frac{dB}{dt} \tag{8j-56}$$

which says that the change in the space-averaged field inside the orbit $\bar{B}$ must equal twice the change in the field at the orbit $B(R)$. If both the average core field and the field at the orbit are zero when the acceleration process starts, as is usually the case, integration of (8j-56) gives

$$\bar{B} = 2B \tag{8j-57}$$

The average core field must thus be twice as high as the field at the orbit ("two-to-one" rule) which explains the shape of the magnet core and pole shoes in Fig. 8j-10.

The magnet is driven with an ac power supply which generates a sinusoidally varying current at a frequency in the range of 30 to 60 Hz. To minimize eddy currents the magnet structure is laminated. The electrons are injected from an electron gun close to the equilibrium orbit, with a starting energy between 10 and 100 keV. The acceleration process then takes place during the quarter cycle during which the field rises from the value (close to zero) that corresponds to the injection energy to the peak value, where the electrons have reached the maximum energy.

Radial and axial stability of the beam during acceleration is maintained by constant-gradient focusing; i.e., the field near the orbit is decreasing with radius such that the index n has values between 0 and 1. In fact, the resulting oscillations are known as *betatron oscillations* because the theory of gradient focusing was first developed in connection with the betatron by Kerst and Serber.

At the end of each acceleration cycle the electron beam is displaced from the equilibrium orbit by a perturbation in the magnetic field. This is accomplished by additional coils which disturb the "two-to-one rule," resulting in an increase of the orbit radius and thereby forcing the beam to hit the internal target or deflecting it out of the magnetic field for external use. Most betatrons are used primarily for production of hard X rays from internal targets.

Electromagnetic radiation emitted by the circulating electrons sets an upper-energy limit to betatron-type acceleration. In the relativistic electron-energy range above a few MeV, the rate of energy loss due to radiation is proportional to the fourth power of the kinetic energy and inversely proportional to the orbit radius,

$$\Delta E_{\text{rad}} = 8.8 \times 10^{-8}\frac{E_k{}^4}{R} \tag{8j-58}$$

where ΔE_{rad} is in electron volts per revolution, E_k in MeV, and R in meters.

The betatron was invented by Wideröe, but the first successful machine was built by Kerst. Today a large number of betatrons are in operation in hospitals, for industrial applications as well as for scientific use. The largest betatron is the 300-MeV machine at the University of Illinois, Urbana.

8j-13. New Developments. Accelerator technology is advancing at a rapid rate in many areas. New design concepts have been proposed or are being investigated, and in all likelihood new types of accelerators will be built in the future. It is impossible to survey all these developments, but below a few examples will be discussed briefly to illustrate major present trends.

Heavy-ion Accelerators. In principle all the existing types of accelerators with the exception of the betatron and microtron are capable of accelerating ions of heavy elements. The main problem in practically every instance is that a high charge state is either required to facilitate acceleration or desired to obtain a sufficiently high energy per nucleon. Most ion sources, however, which are utilizing a gas discharge produce ions with only a few electrons removed (typically 1 to 5). However, to accelerate heavy ions $(M > 20)$ with low charge state $(Z < 4)$ in a cyclotron, for example, the wavelength of the r-f system would have to be impractically large, or operation at a very high harmonic, $\omega_{rf}/\omega = N \gg 1$, would be necessary, which again is not feasible. There are basically two solutions to this problem: One is to develop new types of ion sources which yield higher charge states; the other approach is to accelerate ions with low charge state to some intermediate energy, then remove more electrons by stripping in a foil or gas cell, and accelerate further. Thus a negatively charged heavy ion can be injected into a tandem where stripping takes place in the positive-voltage terminal, followed by several steps of acceleration and stripping until the ions with various charge states and energies arrive at ground potential. If desired, one ion component can then be injected into a cyclotron for acceleration to even higher energies. Similar possibilities exist with a multistage linear accelerator or combination of linac and synchrotron. Various schemes of this kind are discussed in the Proceedings of the 1969 Accelerator Conference in Washington, D.C.

High-energy Cyclotrons. Several sector-focusing cyclotron projects in the 200- to 500-MeV range are under construction (Indiana, Zurich, Vancouver), and should come into operation in the 1970s. In addition, the improvement of existing synchrocyclotrons is of great interest as currents in such converted machines should be close to those achieved in isochronous cyclotrons. The 600-MeV synchrocyclotron at CERN is being improved by a change of the rf system (higher dee voltage and repetition rate) and of the ion source and central region. The 385-MeV synchrocyclotron of Columbia University, New York, is being converted into a 500-MeV machine by changing rf voltage, repetition rate, and central region, as in the CERN case, but also adding sector focusing in the magnetic field. These modifications should increase internal beam currents by a factor of 10 to 20 and external beams by 100.

The Collective-ion or Electron-ring Accelerator (ERA). First proposed by Veksler in 1956 [29], the ERA involves an entirely new acceleration concept which holds great promise for the acceleration of protons to superhigh energies. The basic idea involves the formation of a relativistic high-density electron ring (typically 10^{13} to 10^{14} particles, major radius 5 cm, minor radius 1 mm, energy 20 to 25 MeV) in a strong magnetic field. After formation of the ring, gas is admitted, the ions formed by collisions with the electrons are trapped in the deep potential well of the electron cluster, and the ring with ions is subsequently accelerated to high energies. Since the ions travel with the same speed as the electrons, their final kinetic energy is substantially larger than that of the electrons. If $M_i c^2$ is the rest energy of the ions, E_{e0} the total energy of the ring electrons before and E_{ef} after acceleration, the final total ion energy is given by $E_{if} = (E_{ef}/E_{e0})M_i c^2$. Thus to obtain a proton energy of 1 GeV, requires $E_{ef} \approx 2E_{e0}$, and if the initial energy of the ring electrons is $E_{e0} = 25$ MeV, an additional amount of 25 MeV must be added by acceleration of the ring. If the energy is gained at a rate of 40 keV/cm, the accelerator needs only a length of a little more than 6 m to produce the 1-GeV protons. The size of a multi-GeV proton accelerator would therefore be substantially smaller than that of a synchrotron, which explains the attractiveness of the electron-ring accelerator concept. At the same time the ERA holds great promise also as an accelerator for heavy ions. The various design problems and prospects of the ERA are discussed in the proceedings of a symposium in Berkeley [30]. Compressed electron rings in a pulsed magnetic field were obtained during 1968 in experiments at Dubna, Berkeley [31], and the University of Maryland]32]. A promising alternative to a pulsed system is the formation of the electron ring in a static magnetic field [33]. For further information on the ERA see the article by D. Keefe in the journal *Particle Accelerators* [34].

Other interesting developments in the accelerator field, such as *storage rings, superconducting linacs,* and the *racetrack microtron,* are reviewed in the proceedings of the latest accelerator conferences listed in the general bibliography.

General Bibliography

1. Livingood, John J.: "Principles of Cyclic Particle Accelerators," D. Van Nostrand Company, Inc., Princeton, N.J., 1961.
2. Livingston, M. S., and J. P. Blewett: "Particle Accelerators," McGraw-Hill Book Company, New York, 1962.
3. Kollath, R.: "Particle Accelerators," Sir Isaac Pitman & Sons, Ltd., London, 1962.
4. Kolomensky, A. A., and A. N. Lebedev: "Theory of Cyclic Accelerators," North-Holland Publishing Company, Amsterdam, 1966.
5. Persico, E., E. Ferrari, and S. E. Segre: "Principles of Particle Accelerators," W. A. Benjamin, Inc., New York, 1968.
6. Nuclear Instrumentation I, "Encyclopedia of Physics," vol. 44, Springer Verlag OHG, Berlin, 1959.
7. Proc. CERN Symp. High Energy Accelerators and Pion Phys. 1, E. Regenstreif, ed., CERN, Geneva, 1956.
8. Proc. Intern. Conf. High Energy Accelerators and Instrumentation, L. Kowarski, ed. CERN, Geneva, 1959.
9. Proc. Intern. Conf. High Energy Accelerators, M. H. Blewett, ed., Brookhaven, 1961.
10. Proc. Intern. Conf. High Energy Accelerators, Dubna, 1963, A. A. Kolomensky, chief ed., Atomizdat, 1964.
11. Proc. Intern. Conf. High Energy Accelerators, Frascati, 1965.
12. Proc. 6th Intern. Conf. High Energy Accelerators, Cambridge, Mass., 1967.
13. First National Particle Accelerator Conference, IEEE NS-12(3), June, 1965.
14. U.S. National Particle Accelerator Conference, IEEE N5-14(3), June, 1967.
15. 1969 Particle Accelerator Conference, IEEE NS-16(3), June, 1969.

References

1. Wideröe, R.: On a New Principle for Production of High Potentials, Arch. Elektrotech. 21, 387–406 (1928).
2. Van de Graaff, R. J.: A 1,500,000 Volt Electrostatic Generator, Phys. Rev. 38, 1919 (1931).
3. Lawrence, E. O., and N. E. Edlefsen: On the Production of High Speed Protons, Science 72, 376–377 (1930). Lawrence, E. O., and M. S. Livingston: The Production of High Speed Light Ions without the Use of High Voltages, Phys. Rev. 40, 19–35 (1932).
4. Cockcroft, J. D. and E. T. S. Walton: Experiments with High Velocity Positive Ions, Proc. Roy. Soc. (London), ser. A, 136, 619–630 (1932).
5. Kerst, D. W.: The Acceleration of Electrons by Magnetic Induction, Phys. Rev. 60, 47–53 (1941).
6. McMillan, E. M.: The Synchrotron: a Proposed High Energy Particle Accelerator, Phys. Rev. 68, 143–144 (1945). Veksler, V.: A New Method of Acceleration of Relativistic Particles, J. Phys. (U.S.S.R.) 9, 153–158 (1945).
7. Christofilos, N.: Focusing System for Ions and Electrons, U.S. Patent 2,736,799 (filed March 10, 1950, issued Feb. 28, 1956). E. D. Courant, M. S. Livingston, and H. S. Snyder: The Strong-Focusing Synchrotron: a New High Energy Accelerator, Phys. Rev. 88, 1190–1196 (1952).
8. Thomas, L. H.: The Paths of Ions in the Cyclotron, Phys. Rev. 54, 580–588 (1938).
9. Kelly, E. L., P. V. Pyle, R. L. Thornton, J. R. Richardson, and B. T. Wright: Two Electron Models of a Constant Frequency Relativistic Cyclotron, Rev. Sci. Instr. 27, 493–503 (1956).
10. Livingston, M. S., and J. P. Blewett: "Particle Accelerators," McGraw-Hill Book Company, New York, 1962.
11. Schenkel, M.: Eine neue Schaltung fur die Erzeugung hoher Gleichspannungen, Elektrotech. Z. 40, 333–334 (1919). H. Greinacher: Uber eine neue Methode, Wechselstrom mittels elektrischer Ventile und Kondensatoren in hochgespannten Gleichstrom zu verwandeln, Z. Physik 4, 195–205 (1921).
12. Alvarez, L. W.: The Design of a Proton Linear Accelerator, Phys. Rev. 70, 799–800 (1946). L. W. Alvarez, H. Bradner, J. V. Frank, H. Gordon, J. D. Gow, L. C. Marshall, F. Oppenheimer, W. K. H. Panofsky, C. Richman, and J. R. Woodyard: Berkeley Proton Linear Accelerator, Rev. Sci. Instr. 26, 111–133 (1955).
13. Blewett, J. P.: Radial Focusing in the Linear Accelerator, Phys. Rev. 88, 1197–1199 (1952).
14. Ginzton, E. L., W. W. Hansen, and W. R. Kennedy: "Linear Electron Accelerator," Rev. Sci. Instr. 19, 89–108 (1948). D. W. Fry, R. B. R. Shersby-Harvie, L. B. Mullet, and W. Walkinshaw: Traveling Wave Linear Accelerator for Electrons, Nature 160, 351–352 (1947).

15. Livingston, R. S., and R. J. Jones: High Intensity Ion Source for Cyclotrons, *Rev. Sci. Instr.* **25**, 552–557 (1954).
16. Kerst, D. W., and R. Serber: Electronic Orbits in the Induction Accelerator, *Phys. Rev.* **60**, 53–58 (1941).
17. Bohm D., and L. Foldy: Theory of the Synchro-cyclotron, *Phys. Rev.* **72**, 649–661 (1947).
18. McKenzie, K. R.: Space Charge Limits and Cyclotron Beam Enhancement, *Nucl. Instr. Methods* **31**, 139–146 (1964).
19. Blosser, H. G., and M. M. Gordon: Performance Estimates for Injector Cyclotrons, *Nucl. Instr. Methods* **13**, 101 (1961).
20. Reiser, M.: Space Charge Effects and Current Limitations in Cyclotrons, *IEEE Trans. Nucl. Sci.* **NS-13**(4), 171–178 (1966).
21. Blosser, H. G.: Synchrocyclotron Improvement Programs, *IEEE Trans. Nucl. Sci.* **NS-16**(3), June, 1969.
22. Richardson, J. R.: Sector Focusing Cyclotrons, in "Progress in Nuclear Techniques and Instrumentation," vol. I, North-Holland Publishing Company, Amsterdam, 1965.
23. Conference on Sector Focused Cyclotrons, Sea Island, Ga., February, 1959, *Natl. Acad. Sci. Publ.* 656, 1959. Proceedings of the International Conference on Sector-focused Cyclotrons, Los Angeles, Calif., April, 1962, in *Nucl. Instr. Methods* **18, 19** (1962). Proceedings of the International Conference on Sector-focused Cyclotrons and Meson Factories, Geneva, Switzerland, April, 1963, *CERN Rept.* 63-19, May 29, 1963. International Conference on Isochronous Cyclotrons, Gatlinburg, Tenn., May, 1966, in *IEEE Trans.* **NS-13**(4), (1966).
24. Symon, K. R., D. W. Kerst, L. W. Jones, and K. M. Terwilliger: Fixed-field Alternating Gradient Accelerators, *Phys. Rev.* **98**, 1152–1153 (1955). K. R. Symon, D. W. Kerst, L. W. Jones, L. J. Laslett, and K. M. Terwilliger: Fixed-field Alternating Gradient Particle Accelerators, *Phys. Rev.* **103**, 1837–1859 (1956).
25. Smith, W. I. B.: Improved Focusing near the Cyclotron Source, *Nucl. Instr. Methods* **9**, 49–54 (1960). M. Reiser: Ion Capture and Initial Orbits in the Karlsruhe Isochronous Cyclotron, *Nucl. Instr. Methods* **13**, 55–69 (1961).
26. Reiser, M.: Central Orbit Program for a Variable Energy Multi-particle Cyclotron, *Nucl. Instr. Methods* **18, 19**, 370–377 (1962). H. G. Blosser: Problems and Performance in the Cyclotron Central Region, *IEEE Trans. Nucl. Sci.* **NS-13**(4), 1–14 (1966).
27. The various extraction methods are discussed in papers by Gordon, Kim, Hagedorn and Kramer, Paul and Wright in *IEEE Trans.* **NS-13**(4), 48–83 (1966).
28. Laslett, L. J.: On Intensity Limitations Imposed by Transverse Space Charge Effects in Circular Particle Accelerators, *Proc. 1963 Summer Study on Storage Rings, Accelerators, and Experimentation at Super-high Energies*, BNL 7534, 1963.
29. Veksler, V. I.: *Proc. CERN Symp. on High Energy Accelerators*, p. 80, 1956.
30. Proceedings of the Symposium on Electron Ring Accelerators, *LRL Rept.* UCRL-18103, February, 1968.
31. Keefe, D., et al.: Experiments on Forming Intense Rings of Electrons Suitable for the Acceleration of Ions, *Phys. Rev. Letters* **22**, 558–561 (1969).
32. Trivelpiece, A. W., R. E. Pechacek, and C. A. Kapetanakos: *Phys. Rev. Letters* **21**, 1436 (1968).
33. Berg, R. E., Hogil Kim, M. P. Reiser, and G. T. Zorn: Possibilities of Forming a Compressed Electron Ring in a Static Magnetic Field, *Phys. Rev. Letters* **22**, 419–421 (1969). See also papers by Laslett and Sessler, Christofilos, Berg, et al. in Proceedings of 1969 Accelerator Conference, Washington, D.C., *IEEE Trans.* **NS-16** (3) 1969.
34. Keefe, D.: Research on the Electron Ring Accelerator, *Particle Accelerators* **1**, 1–13 (1970).

Section 9

SOLID-STATE PHYSICS

H. P. R. FREDERIKSE, Editor

The National Bureau of Standards

CONTENTS

9a. Crystallographic Properties

J. D. H. DONNAY[1]

The Johns Hopkins University

W. P. MASON AND E. A. WOOD[2]

Bell Telephone Laboratories, Inc.

9a-1. Crystal System, Space Group, Cell Content, Lattice Constants, Structure Type. These data are presented for all the chemical elements (Table 9a-2) and for certain selected compounds (Table 9a-3).

In each table the first column contains the *chemical formula*, with mention of the polymorphic form, if necessary, and of the temperature, if known, at which the lattice constants have been determined.

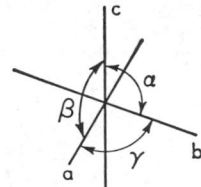

FIG. 9a-1. Coordinate axes (= "crystallographic axes").

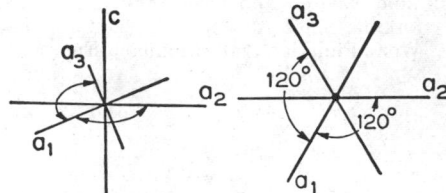

FIG. 9a-2. Coordinate axes for the hexagonal system (can also be used for the rhombohedral system).

The *crystal system*, listed in column 2, is based on the point symmetry of the lattice[3] of the crystal structure. It is given by the initial letter of its name (see Table 9a-1). The coordinate axes x, y, z are taken along three concurrent cell edges that form a right-handed system (a, b, c in Fig. 9a-1; a_1, a_2, c in Fig. 9a-2). Symmetry governs the relative values of the unit lengths a, b, c and of the interaxial angles α, β, γ. The symmetry requirements entail a specialization of the lattice constants (Table 9a-1) and a corresponding reduction in the number of values that must be listed in the tables of data.

The *space group* is given (Tables 9a-2 and 9a-3, column 3) in both Schoenflies and and Hermann-Mauguin notations. The symbols of the 32 crystal point groups, needed for comparison with the space-group symbols, will be found in Table 9a-1, where the

---

[1] Crystallographic Data.

[2] Tensor Properties of Crystals.

[3] "Lattice" *s.s.*: triperiodic assemblage of points, the termini of the vectors $\mathbf{L}(uvw) = u\mathbf{a} + v\mathbf{b} + w\mathbf{c}$, where u, v, w take all integral values—the geometrical expression of a translation group, described by a repeating parallelepiped ("cell") whose edges are preferably chosen along symmetry axes of the lattice.

Hermann-Mauguin symbol is given for every orientation and the Schoenflies symbol follows between parentheses.

A Hermann-Mauguin *point-group symbol* states what symmetry a specified discontinuous vectorial property possesses along certain directions of the crystal. These directions are those of the symmetry axes of the lattice (Table 9a-1, column 2). They are grouped in sets of equivalent directions, some being chosen as cell edges as shown in Table 9a-1 (column 3). An Arabic numeral represents a rotation axis of symmetry along one direction (*examples:* any 2 in 222, the 3 in 3*m*) or along each direction of a set (the 2 in $\bar{4}2m$, either 2 in 622, the 3 in *m3m*). Surmounted by a bar the numeral indicates a rotatory-inversion axis. *Example:* the $\bar{4}$ axis stands for a cyclic group in which the first power of the symmetry operation is a 90 deg rotation followed by an inversion through a point[1] on the axis—the fixed point in the point group. The $\bar{1}$ axis

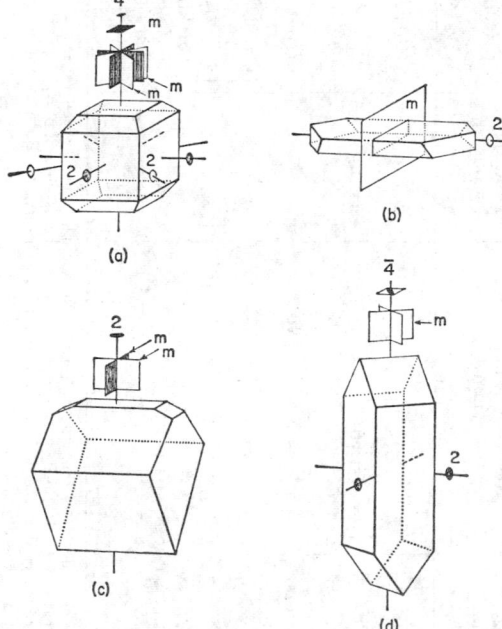

Fig. 9a-3. Examples of macroscopic crystal symmetry (point groups): (*a*) tetragonal, 4/*mmm*; (*b*) monoclinic, 2/*m*; (*c*) orthorhombic, *mm*; (*d*) tetragonal, $\bar{4}2m$. (*After W. P. Mason and E. A. Wood.*)

is not defined in direction: it symbolizes the center of symmetry. A mirror plane of symmetry, designated *m*, is perpendicular to the direction it describes. *Example:* in 6/*mmm* the first *m* is perpendicular to the *c* axis, the second *m* and the third *m* represent three mirrors each that are perpendicular to a_1, a_2, a_3 and the bisectors, respectively. The following point groups contain the center of symmetry: *mmm* (= 2/*m* 2/*m* 2/*m*), *m3m* (= 4/*m* $\bar{3}$ 2/*m*), *m3* (= 2/*m* $\bar{3}$), $\bar{3}$, and *N*/*m*, where *N* is an even number. Figure 9a-3 illustrates the assemblages of symmetry elements in four selected point groups, which express the morphological symmetry of well-formed crystals.

A Hermann-Mauguin *space-group symbol* begins with a capital letter that tells the lattice mode: primitive (*P*), body-centered (*I*), one-face-centered (*C, A,* or *B*), all-face-centered (*F*), rhombohedral (*R*). Additional symmetry elements appear. In a

---

[1] Note that this point is *not* a center of symmetry.

TABLE 9a-1. THE SEVEN CRYSTAL SYSTEMS BASED ON THE POINT SYMMETRY OF THE LATTICE

Name	Lattice symmetry (holohedry)	Symmetry directions[a]	Restrictions imposed by symmetry	Tabulated values	Merohedries (subgroups of the lattice symmetry in each system[e])
Anorthic (= triclinic)	$\bar{1}(C_i)$	None	None	$a, b, c, \alpha, \beta, \gamma$	$1(C_1)$;
Monoclinic	$2/m(C_{2h})$	b	$\gamma = \alpha = 90°$	a, b, c, β	$m(C_s),\ 2(C_2)$
Orthorhombic (= rhombic)	$mmm(D_{2h})$	$(a)(b)(c)$	$\alpha = \beta = \gamma = 90°$	a, b, c	$mm2[2mm][m2m](C_{2v}),\ 222(D_2)^f$;
Tetragonal	$4/mmm(D_{4h})$	$(c)(ab)$(bisectors)[c]	$b = a,\ \alpha = \beta = \gamma = 90°$	a, c	$\bar{4}2m[\bar{4}m2](D_{2d}),\ 4mm(C_{4v}),\ 422(D_4)$; $4/m(C_{4h}),\ \bar{4}(S_4),\ 4(C_4)$;
Hexagonal	$6/mmm(D_{6h})$	$(c)(a_1a_2a_3)$(bisec.)[c]	$b = a,\ \alpha = \beta = 90°,\ \gamma = 120°$	a, c	$\bar{6}m2[\bar{6}2m](D_{3h}),\ 6mm(C_{6v}),\ 622(D_6)$; $6/m(C_{6h}),\ \bar{6}(C_{3h}),\ 6(C_6)$; $3m1[31m](D_{3d}),\ 3m1[31m1](C_{3v}),\ 321[312](D_3)$; $3(C_{3i}),\ 3(C_3)^g$;
Rhombohedral[b]	$3m(D_{3d})$	$(c)(a_1a_2a_3)$ .... (abc)(b.d.)(f.d.)[d]	$b = a,\ \alpha = \beta = 90°,\ \gamma = 120°$ $b = c = a_{rh},\ \beta = \gamma = \alpha$	a, c a_{rh}, α	$3m(C_{3v}),\ 32(D_3);\ 3(C_{3i}),\ 3(C_3)^g$;
Cubic (= isometric)	$m3m(O_h)$		$b = c = a,\ \alpha = \beta = \gamma = 90°$	a	$\bar{4}3m(T_d),\ 432(O);\ m3(T_h),\ 23(T)$

[a] Symmetry directions in the lattice (in the same sequence as in the Hermann-Mauguin symbol), and how cell edges are chosen from them.

[b] A rhombohedral lattice can be described by means of a triple cell, which has the same shape as the cell of a hexagonal lattice (a,c) but has additional points at $\frac{1}{3}\frac{2}{3}\frac{2}{3}$ and $\frac{2}{3}\frac{1}{3}\frac{1}{3}$ (fractions of a_1, a_2, c, respectively), or by means of the primitive rhombohedral cell (a_{rh},α). Both descriptions are given in Tables 9a-2 and 9a-3.

[c] Bisectors of the angles between the axes: $a0b$, $a0\bar{b}(T)$ and $a_1 0\bar{a}_3$, $\bar{a}_3 0 a_2$, $a_1 0 \bar{a}_1(H)$. (Fig. 9a-2)

[d] The four body diagonals (b.d.) and the six face diagonals (f.d.) of the cube.

[e] Semicolons separate the 11 Laue classes.

[f] Alternate orientations, shown between brackets, are needed in space-group symbols.

[g] All five *trigonal* point groups appear in both hexagonal (H) and rhombohedral (R) systems.

TABLE 9a-2. CRYSTALLOGRAPHIC DATA FOR THE ELEMENTS

Formula (temp., °C, for the lattice constants given)	Crystal system	Space group[a]	Z	Lattice constants,[b] $a, b, c; \alpha, \beta, \gamma$	Structure type
Ar (extrapolated, 0K)	Cub.	O_h^5–$Fm3m$	4	5.3109 ± 0.0001	A1 (f.c.c.)
Ag (25)	Cub.	O_h^5–$Fm3m$	4	c4.08641 Å*	A1 (f.c.c.)
Al (24.8)	Cub.	O_h^5–$Fm3m$	4	4.04964	A1 (f.c.c.)
As (26)	Rhdr.	D_{3d}^5–$R\bar{3}m$	6	3.760, 10.548	A7
			2	(a_{rh} 4.132, α 54°07½')	
Au (25)	Cub.	O_h^5–$Fm3m$	4	4.0786	A1 (f.c.c.)
B	Tetr.	D_{4h}^{12}–$P4_2/nmn$	50	8.74, 5.07	
Ba (26)	Cub.	O_h^9–$Im3m$	2	5.025	A2 (b.c.c.)
Be (18)	Hex.	D_{6h}^4–$P6_3/mmc$	2	2.2854, 3.5807	A3 (h.c.p.)
Be (630)(stable 500–700)	Hex.	....	ca. 60	7.1, 10.8	
Bi (25)	Rhdr.	D_{3d}^5–$R\bar{3}m$	6	4.546, 11.860	A7 (As)
			2	(a_{rh} 4.745, α 57°14½')	
Br₂ (−150)	Orth.	D_{2h}^{18}–$Bmab$	4	6.68, 8.74, 4.49	A14 (I_2)
C (diamond)(26)	Cub.	O_h^7–$Fd3m$	8	3.5667	A4 (diamond)
C (graphite)(14.6)	Hex.	D_{6h}^4–$P6_3/mmc$	4	2.4612 ± 0.0001, 6.7079 ± 0.0007	A9
C (graphite)	Rhdr.	D_{3d}^5–$R\bar{3}m$	6	2.461, 10.064	
			2	(a_{rh} 3.642, α 39.49°)	
Ca (α)(electrolytic)	Cub.	O_h^5–$Fm3m$	4	5.582 ± 0.004	A1 (f.c.c.)
Ca (γ)(above 464)	Cub.	O_h^9–$Im3m$	2	4.477 ± 0.007	A2 (b.c.c.)
Cd (26)	Hex.	D_{6h}^4–$P6_3/mmc$	2	2.9793, 5.6181	A3 (h.c.p.)
Ce (γ)	Cub.	O_h^5–$Fm3m$	4	5.150 ± 0.002	A1 (f.c.c.)
Ce (β)	Hex.	D_{6h}^4–$P6_3/mmc$	2	3.66, 5.97	A3 (h.c.p.)
Ce (α)	Cub.	O_h^5–$Fm3m$	4	5.1612 ± 0.0005	A1 (f.c.c.)
Ce (δ)(stable above 730)	Cub.	O_h^9–$Im3m$	2	4.11	A2 (b.c.c.)
Cl₂ (−185)	Tetr.	D_{4h}^{16}–$P4_2/ncm$	8	8.58, 6.13	A 18
Co (α)(20)	Hex.	D_{6h}^4–$P6_3/mmc$	2	2.5074, 4.0699	A3 (h.c.p.)
Co (20)	Cub.	O_h^5–$Fm3m$	4	3.5442	A1 (f.c.c.)
Cr	Cub.	T_d^3–$\bar{4}3m$	58	8.735	A12 (αMn)
Cr (25)	Cub.	O_h^9–$Im3m$	2	2.8845 ± 0.0005	A2 (b.c.c.)
Cs (−100)	Cub.	O_h^9–$Im3m$	2	6.09	A2 (b.c.c.)
Cu (25)	Cub.	O_h^5–$Fm3m$	4	3.61509 ± 0.00004	A1 (f.c.c.)
Dy (99.8% pure)	Hex.	D_{6h}^4–$P6_3/mmc$	2	3.5903 ± 0.0001, 5.6475 ± 0.0002	A3 (h.c.p.)
Er (99.8% pure)	Hex.	D_{6h}^4–$P6_3/mmc$	2	3.5588 ± 0.0003, 5.5874 ± 0.0003	A3 (h.c.p.)
Eu (98–99% pure)	Cub.	O_h^9–$Im3m$	2	4.606 ± 0.001	A2 (b.c.c.)
Fe (α)(20)(stable to 900)	Cub.	O_h^9–$Im3m$	2	2.86645 ± 0.00001	A2 (b.c.c.)
Fe (γ)(stable 900–1400)	Cub.	O_h^5–$Fm3m$	4	3.64	A1 (f.c.c.)
Fe (δ)(stable above 1400)	Cub.	O_h^9–$Im3m$	2	2.94	A2 (b.c.c.)
Ga (unstable form)(−16.3)	Orth.	D_{2h}–$Amam$	4	3.17, 8.13, 2.90, all ± 0.03	
Ga (stable form)	Orth.	D_{2h}^{18}–$Abam$	8	4.524, 7.661, 4.523	A11
Gd (99.7% pure)	Hex.	D_{6h}^4–$P6_3/mmc$	2	3.6360 ± 0.0009, 5.7826 ± 0.0006	A3 (h.c.p.)
Ge (24.6)	Cub.	O_h^7–$Fd3m$	8	5.657764 ± 0.000010 Å*	A4 (diamond)
H₂ (above 1.30 K)	Hex.	D_{6h}^4–$P6_3/mmc$	2	3.761 ± 0.007, 6.105 ± 0.011	A3 (h.c.p.)
He³ (3.48 K, 163 atm)	Hex.	D_{6h}^4–$P6_3/mmc$	2	3.501, 5.721	A3 (h.c.p.)
He⁴ (3.95 K, 129 atm)	Hex.	D_{6h}^4–$P6_3/mmc$	2	3.470, 5.540	A3 (h.c.p.)
Hf (26)	Hex.	D_{6h}–$P6_3/mmc$	2	3.1967, 5.0578	A3 (h.c.p.)
Hg (5 K)	Rhdr.	D_{3d}^5–$R\bar{3}m$	3	3.457, 6.664	} A10
			1	(a_{rh} 2.9863, α 70° 44.6')	
Ho (99.4% pure)	Hex.	D_{6h}^4–$P6_3/mmc$	2	3.5773 ± 0.0001, 5.6158 ± 0.0002	A3 (h.c.p.)
I₂ (26 ± 1)	Orth.	D_{2h}^{18}–$Bmab$	4	7.271, 9.803, 4.792	A14
In (20)	Tetr.	D_{4h}^{17}–$I4/mmm$	2	3.2517, 4.9459	A6
Ir (26)	Cub.	O_h^5–$Fm3m$	4	3.8394	A1 (f.c.c.)
K (20)	Cub.	O_h^9–$Im3m$	2	5.344 ± 0.005	A2 (b.c.c.)
Kr (−252.5)	Cub.	O_h^5–$Fm3m$	4	5.60	A1 (f.c.c.)
Kr (89 K)	Cub.	O_h^5–$Fm3m$	4	5.709 ± 0.014	A1 (f.c.c.)
La (α)(99.8% pure)	Hex.	....	4	3.770 ± 0.002, 12.159 ± 0.008	
La (β)(stable above ca. 260)	Cub.	O_h^5–$Fm3m$	4	.307 ± 0.002 (99.6% pure)	A1 (f.c.c.)
La (γ)(stable above 864 C)	Cub.	O_h^9–$Im3m$	2	4.26	A2 (b.c.c.)
Li⁷ (20)	Cub.	O_h^9–$Im3m$	2	{ 3.5092 ± 0.0006	} A2 (b.c.c.)
Li⁶ (20)				3.5107 ± 0.0009	
Li (−195)	Cub.	O_h^9–$Im3m$	2	3.491 ± 0.002	A2 (b.c.c.)
Li (−195)	Hex.	D_{6h}^4–$P6_3/mmc$	2	3.111 ± 0.001, 5.093 ± 0.009	A3 (h.c.p.)
Li (−195)	Cub.	O_h^5–$Fm3m$	4	4.404 ± 0.030 (induced by deformation)	A1 (f.c.c.)
Lu (99.9% pure)	Hex.	D_{6h}^4–$P6_3/mmc$	2	3.5031 ± 0.0004, 5.5509 ± 0.0004	A3 (h.c.p.)
Mg (25°)(99.995% pure)	Hex.	D_{6h}^4–$P6_3/mmc$	2	3.20939 ± 0.00003, 5.21053 ± 0.00005 Å*	A3 (h.c.p.)

TABLE 9a-2. CRYSTALLOGRAPHIC DATA FOR THE ELEMENTS (*Continued*)

Formula (temp., °C, for the lattice constants given)	Crystal system	Space group[a]	Z	Lattice constants,[b] $a, b, c; \alpha, \beta, \gamma$	Structure type
Mn (α)	Cub.	T_d^3–$I\bar{4}3m$	58	8.911 ± 2	A12 (type)
Mn (β)(stable 705–1100)	Cub.	O_h^6–$Fm3c$	160	12.61	A13 (type)
Mn (γ)	Tetr.	D_{4h}^{17}–$I4/mmm$	2	2.671 ± 0.002, 3.533 ± 2	
Mo (25)	Cub.	O_h^9–$Im3m$	2	3.1472	A2 (b.c.c.)
Mo	Cub.	O_h^5–$Fm3m$	4	4.16 (precipitated in vacuo)	A1 (f.c.c.)
N₂ (α)(20 K)(stable below 35.6 K)	Cub.	T_h^6–$Pa3$	4	5.661 ± 0.008	
N₂ (β)(50 K)(stable 35.6–63.1 K)	Hex.	$P6_3/mmc$	2	3.93 ± 0.16, 6.50 ± 51	
Na	Cub.	O_h^9–$Im3m$	2	4.2906 ± 0.0005	A2 (b.c.c.)
Na (−195)	Cub.	O_h^5–$Fm3m$	4	5.350 (induced by deformation at −253°)	A1 (f.c.c.)
Nb (20)(H₂-free)	Cub.	O_h^9–$Im3m$	2	3.3008 ± 0.0003	A2 (b.c.c.)
Nd (α)(99.8% pure)	Hex.		4	3.6579 ± 0.0003, 11.7992 ± 0.0005	
Nd (β)(stable above 862)	Cub.	O_h^9–$Im3m$	2	4.13	A2 (b.c.c.)
Ne (3 K)	Cub.	O_h^5–$Fm3m$	4	4.446 ± 0.009	A1 (f.c.c.)
Ni (25)(99.99% pure)	Cub.	O_h^5–$Fm3m$	4	3.52394 ± 0.00008	A1 (f.c.c.)
O₂ (α)(stable below 23.5 K)					
O₂ (β)(stable 23.5–43.4 K)					
O₂ (γ)(50 K)	Cub.	O_h^3–$Pm3n$	8	6.83 ± 0.05	
Os (26)	Hex.	D_{6h}^4–$P6_3/mmc$	2	2.7341, 4.3197	A3 (h.c.p.)
P₄ (white)(−30)	Cub.	T_d^2–$I\bar{4}3m$, or O^5–$I43$?	56	18.51 ± 0.03	
P (black)(22)	Orth.	D_{2h}^{18}–$Abam$	8	4.3763 ± 0.0005, 10.478 ± 0.001, 3.3136 ± 0.0005	A17
P (red)	Mon.			7.34 (pseudocubic), 11.31 (cubic)	
Pb (25)(99.999% pure)	Cub.	O_h^5–$Fm3m$	4	4.9505	A1 (f.c.c.)
Pd (25)	Cub.	O_h^5–$Fm3m$	4	3.8898	A1 (f.c.c.)
Po (α)(10 ± 10)[d]	Cub.	O_h^1–$Pm3m$		3.345 ± 0.002	
Po (β)(75 ± 15)[d]	Rhdr.	D_{3d}^5–$R\bar{3}m$ {		2.222, 2.151 (a_{rh} 3.359 ± 0.002, 98° 13′ ± 3′)	
Pr (α)(99.9% pure)	Hex.		4	3.6725 ± 0.0007, 11.8354 ± 0.0012	
Pr (β)	Cub.	O_h^5–$Fm3m$	4	5.161 ± 0.002	A1 (f.c.c.)
Pt (25)	Cub.	O_h^5–$Fm3m$	4	3.9231	A1 (f.c.c.)
Pu (α)(21)(stable 122 ± 2)	Mon.	C_{2h}^2–$P2_1/m$	16	10.973 ± 0.001, 4.8244 ± 0.0005, 6.1835 ± 0.0005, 101.81°	
Pu (β)(190)(stable to 206 ± 3)	Mon.	C_{2h}^1–$I2/m$	34	9.284, 10.463, 7.859, 92.13° ± 0.03°	
Pu (γ)(235)[d] (stable to 319 ± 5)	Orth.	D_{2h}^{24}–$Fddd$	8	5.7682 ± 0.0004, 10.162 ± 0.002, 3.1587 ± 0.0004	
Pu (δ)(320)(stable to 451 ± 4)	Cub.	O_h^5–$Fm3m$	4	4.6370	A1 (f.c.c.)
Pu (δ')(477)(stable to 476 ± 5)	Tetr.	D_{4h}^{17}–$I4/mmm$	2	3.339 ± 0.003, 4.446 ± 0.007	
Pu (ϵ)(490)(liquid at 640 ± 2)	Cub.	O_h^9–$Im3m$	2	3.6361 ± 4(99.97% pure)	A2 (b.c.c.)
Rb (19)	Cub.	O_h^9–$Im3m$	2	5.709	A2 (b.c.c.)
Re (20)	Hex.	D_{6h}^4–$P6_3/mmc$	2	2.7608 ± 0.0004, 4.4582 ± 0.0003	A3 (h.c.p.)
Rh (18)	Cub.	O_h^5–$Fm3m$	4	3.8043 ± 3	A1 (f.c.c.)
Ru (20)	Hex.	D_{6h}^4–$P6_3/mmc$	2	2.70389, 4.28168	A3 (h.c.p.)
S(β)(103)	Mon.	C_{2h}^5–$P2_1/c$	48	11.04, 10.98, 10.92, 96°44′	
S (γ)	Mon.	C_{2h}^5–$P2/n$	32	8.54, 13.08, 8.25, 112°53′	
S₈ (α)(24.8)	Orth.	D_{2h}^{24}–$Fddd$	16	12.86654 ± 0.00010,24.48694 ± 0.00026, 10.46500 ± 0.00009 Å*	
S (unstable)	Rhdr.	C_{3i}^2–$R\bar{3}$ {	18 6	10.9, 4.27: a_{rh} 6.46, α 115°18′	
Sb (26)	Rhdr.	D_{3d}^5–$R\bar{3}m$ {	6 2	4.307, 11.273 (a_{rh} 4.506, α 57°06′)	A7
Sc (99.6% pure)	Hex.	D_{6h}^4–$P6_3/mmc$	2	3.3090 ± 0.0001, 5.2733 ± 0.0016	A3 (h.c.p.)
Sc	Cub.	O_h^5–$Fm3m$	4	4.541 ± 5	A1 (f.c.c.)
Se (26)	Hex.	D_3^4–$P3_121$ or D_3^6–$P3_221$	3	4.3662, 4.9536	A8 (type)
Se (α)	Mon.	C_{2h}^5–$P2_1/n$	32	11.61, 9.07, 9.05, ±0.01 90°46′ ± 5′	
Se (β)	Mon.	C_{2h}^5–$P2_1/a$	32	12.85, 8.07, 9.31, ±0.01, 93°08′ ± 5′	
Si (25)	Cub.	O_h^7–$Fd3m$	8	5.43072 ± 0.00005	A4 (diamond)
Si (dense form)	Cub.	T_h^7–$Ia3$	16	6.64 ± 0.01	
Sm	Rhdr.	D_{3d}^5–$R\bar{3}m$ {	9 3	3.629 ± 0.002, 26.20 ± 0.01 (a_{rh} 8.982, α23.31°)	
Sn (α, gray)(25)	Cub.	O_h^7–$Fd3m$	8	6.489 (stable to 13.2°)	A4 (diamond)
Sn (β, white)(25	Tetr.	D_{4h}^{19}–$I4_1/amd$	4	5.8318 ± 0.0003, 3.1819 ± .0003	A5

Table 9a-2. Crystallographic Data for the Elements (*Continued*)

Formula (temp., °C, for the lattice constants given)	Crystal system	Space group[a]	Z	Lattice constants,[b] $a, b, c; \alpha, \beta, \gamma$	Structure type
Sr (α)(25)(stable below 215 ± 10)	Cub.	O_h^5–$Fm3m$	4	6.0849 ± 0.0005	A1 (f.c.c.)
Sr (β)(248°)(stable 215–605)....	Hex.	D_{6h}^4–$P6_3/mmc$	2	4.32 ± 0.01, 7.06 ± 0.01	A3 (h.c.p.)
Sr (γ)(614°)(stable above 605 ± 10)	Cub.	O_h^9–$Im3m$	2	4.85 ± 0.01	A2 (b.c.c.)
Ta (25)	Cub.	O_h^9–$Im3m$	2	3.3058	A2 (b.c.c.)
Tb (99.9% pure)	Hex.	D_{6h}^4–$P6_3/mmc$	2	3.6010 ± 0.0003, 5.6936 ± 0.0002	A3 (h.c.p.)
Tc	Hex.	D_{6h}^4–$P6_3/mmc$	2	2.741 ± 0.001, 4.397 ± 0.001	A3 (h.c.p.)
Te (25)	Hex.	D_3^4–$P3_121$ or D_3^6–$P3_221$	3	4.4570, 5.9290	A8 (Se)
Th (α)(stable below 1400)(25)	Cub.	O_h^5–$Fm3m$	4	5.0847 ± 0.0002	A1 (f.c.c.)
Th (β)(1450)[d]	Cub.	O_h^9–$Im3m$	2	4.11 ± 0.01	A2 (b.c.c.)
Ti (α)(25)	Hex.	D_{6h}^4–$P6_3/mmc$	2	2.950, 4.686	A3 (h.c.p.)
Ti (β)(900°)(stable above 882 ± 20)	Cub.	O_h^9–$Im3m$	2	3.33	A2 (b.c.c.)
Tl (α)(18)(99.995% pure)	Hex.	D_{6h}^4–$P6_3/mmc$	2	3.4496 ± 0.0002, 5.5137 ± 0.0004	A3 (h.c.p.)
Tl (β)(262)(stable above 230)	Cub.	O_h^9–$Im3m$	2	3.882 ± 0.001 (99.995% pure)	A2 (b.c.c.)
Tm (99.9% pure)	Hex.	D_{6h}^4–$P6_3/mmc$	2	3.5375 ± 0.0001, 5.55146 ± 0.0004	A3 (h.c.p.)
U (α)(24.6)	Orth.	D_{2h}^{17}–$Amam$	4	4.95572 ± 0.00006, 5.87007 ± 0.00006, 2.85373 ± 0.00005 Å*	A20
U (β)(stable 660–760)	Tetr.	C_{4v}^4–$P4_2nm$ or D_{2d}^8–$P\bar{4}n2$	30	10.52, 5.57	
U (γ)(stable, 760 to mp)(room temp.)	Cub.	O_h^9–$Im3m$	2	3.474 ± 0.005	A2 (b.c.c.)
U (γ)(800)	Cub.	O_h^9–$Im3m$	2	3.49	A2 (b.c.c.)
V (25)	Cub.	O_h^9–$Im3m$	2	3.0399 ± 0.0003	A2 (b.c.c.)
W (α)(25)	Cub.	O_h^9–$Im3m$	2	[c]3.16517 Å*	A2 (b.c.c.)
W (β)[e] (transforms irreversibly to α above 700)	Cub.	O_h^3–$Pm3n$	8	5.048 ± 0.003	A15
Xe (75.0 K ⊥ 0.5)	Cub.	O_h^5–$Fm3m$	4	6.2023 ± 0.0010	A1 (f.c.c.)
Y	Hex.	D_{6h}^4–$P6_3/mmc$	2	3.6515 ± 0.0002, 5.7474 ± 0.0004	A3 (h.c.p.)
Yb (99.9% pure)	Cub.	O_h^5–$Fm3m$	4	5.4862 ± 0.0004	A1 (f.c.c.)
Zn (17)	Hex.	D_{6h}^4–$P6_3/mmc$	2	2.6589, 4.9349	A3 (h.c.p.)
Zr (α)(25)	Hex.	D_{6h}^4–$P6_3/mmc$	2	3.232, 5.147	A3 (h.c.p.)
Zr (β)(stable above 870)(979)...	Cub.	O_h^9–$Im3m$	2	3.616 ± 0.002	A2 (b.c.c.)

[a] Schoenflies symbol followed by Hermann-Mauguin symbol.

[b] For nearly all the substances listed, the unit of length is the Å based on the 1948 values of the X-ray wavelengths [*Acta Cryst.* **1**, 48 (1948)]. Pre-1949 values were published in kX units (1 kX = 1,000 X, where X is the unit used by Siegbahn to express his numerical values of X-ray wavelengths); they have been converted to Å by means of the relation 1 kX = 1.00202 Å. For highly accurate determinations, the original values have been converted to Bearden's redefined angstrom unit (1964), which is identified by an asterisk (Å*); in the formula 1 kX = ΛÅ* the conversion factor Λ is the ratio λ_B/λ_S of the numerical values used by Bearden and Siegbahn for the wavelength of the X-ray line employed: Λ is equal to 1.002,061 for NiKα_1; 1.002,058 for CuKα_1; etc. Neither the Å nor the Å* is a metric unit. The Å used in the tables is equal to 10^{-10} m within 1 in 25,000; the Å*, within a few ppm. See J. W. M. Dumond, *Proc. Natl. Acad. Sci.* **45**, 1052 (1959); J. A. Bearden, "X-ray Wavelengths," U.S. Atomic Energy Commission, Oak Ridge, Tenn., 1964.

[c] H. E. Swanson et al., *NBS Monograph* 25(4), p. 3, 1966.

[d] W. B. Pearson, "A Handbook of Lattice Spacings and Structures of Metals and Alloys," Pergamon Press, New York, 1958–1967.

[e] For controversy on the existence of $W(\beta)$, see R. L. Moss and I. Woodward, *Acta Cryst.* **12**, 255 (1959).

screw axis the numeral carries a subscript (*example*, 6_3); this means that the rotation is accompanied by a glide parallel to the axis and equal to a fraction of the corresponding lattice translation (in 6_3 the glide that accompanies a rotation of 360 deg/6 amounts to 3/6 of **c**). In a *glide plane* the reflection is combined with a glide parallel to the plane and equal to half a lattice translation; the glide plane is designated by a letter that identifies this lattice translation as being: a cell edge (*a*, *b*, or *c*),[1] the diagonal of the unit parallelogram (*mesh*) of the net in the plane (*n*), or half the diagonal if the mesh is centered (*d*). To read the point-group symmetry of a crystal from its space-group symbol: skip the capital letter, replace every small letter by *m*, and omit all subscripts (*example*: $P2_1/a$ gives $2/m$).

[1] In $R3c$ and $R\bar{3}c$ the *c* is that of the hexagonal description (a,c).

TABLE 9a-3. CRYSTALLOGRAPHIC DATA FOR SELECTED COMPOUNDS

Formula (temp., °C, for the lattice constants given)	Crystal system	Space group[a]	Z	Lattice constants,[b] $a, b, c; \alpha, \beta, \gamma$	Structure type
AgCl (26)	Cub.	O_h^5–$Fm3m$	4	5.5491	B_1 (NaCl)
AgBr (26)	Cub.	O_h^5–$Fm3m$	4	5.7745	B1 (NaCl)
Al₂O₃ (α)(corundum)(24.7)	Rhdr.	D_{3d}^6–$R\bar{3}c$	6	4.759216 ± 0.000027, 12.99127 ± 0.00024	D51
			2	(a_{rh} 5.12863; 55°17.36′)	
Al₂O₃ (β)	Hex.	D_{6h}^4–$P6_3/mmc$	12	5.57, 22.60	D56
BaTiO₃	Tetr.	D_{4h}^1–$P4/mmm$	1	3.9939, 4.0346	Deformed G5
BaTiO₃ (200)	Cub.	O_h^1–$Pm3m$	1	4.0121 ± 0.0005	G5
CaCO₃ (calcite)(18)	Rhdr.	D_{3d}^6–$R\bar{3}c$	6	4.9898 ± 0.0003, 17.060 ± 0.005	G1
			2	(a_{rh} 6.3748; 46°04.8′)	
CaCO₃ (aragonite)	Orth.	D_{2h}^{16}–$Pnam$	4	5.741, 7.968, 4.959	G2 (type)
CaF₂ (fluorite)	Cub.	O_h^5–$Fm3m$	4	5.46398 ± 0.00007 Å*	C1 (type)
CdI₂ (type 2H)(25)	Hex.	D_{3d}^3–$P\bar{3}m1$	1	4.24, 6.835	C6
CoFe₂O₄	Cub.	O_h^7–$Fd3m$	8	8.429 ± 0.005	H11 (spinel)[c]
COOK·(CHOH)₂·COONa·4H₂O (Rochelle salt)	Orth.	D_2^3–$P2_12_12$	4	11.93, 14.33, 6.18	
CsBr (α)(25)	Cub.	O_h^1–$Pm3m$	1	4.2953	B2 (CsCl)
CsBr (β)	Cub.	O_h^5–$Fm3m$	4	7.23 ± 0.02	B1 (NaCl)
CsCl (25)	Cub.	O_h^1–$Pm3m$	1	4.10 ± 0.02	B2 (type)
CsCl (β)(500)(stable above 460)	Cub.	O_h^5–$Fm3m$	4	7.09 ± 0.02	B1 (NaCl)
CsI (α)(26)	Cub.	O_h^1–$Pm3m$	1	4.5679	B2 (CsCl)
CsI (β)	Cub.	O_h^7–$Fm3m$	4	7.66 ± 0.02	B1 (NaCl)
Fe₃O₄ (26)(iron ferrite or magnetite)	Cub.	O_h^7–$Fd3m$	8	8.396	H11 (spinel)[c]
KBr (25 ± 0.2)	Cub.	O_h^5–$Fm3m$	4	6.5982 ± 0.0002	B1 (NaCl)
KCl (25)	Cub.	O_h^5–$Fm3m$	4	6.29294 ± 0.00008	B1 (NaCl)
KF (26)	Cub.	O_h^5–$Fm3m$	4	5.347	B1 (NaCl)
KI (25)	Cub.	O_h^5–$Fm3m$	4	7.06555 ± 0.00015	B1 (NaCl)
LiBr (26)	Cub.	O_h^5–$Fm3m$	4	5.5013	B1 (NaCl)
LiCl (25)	Cub.	O_h^5–$Fm3m$	4	5.13988 ± 0.00004	B1 (NaCl)
LiF (25 ± 0.2)	Cub.	O_h^5–$Fm3m$	4	4.0262 ± 0.0001	B1 (NaCl)
KH₂PO₄ (26)	Tetr.	D_{2d}^{12}–$I\bar{4}2d$	4	7.448, 6.977	H22
MgAl₂O₄ (spinel)	Cub.	O_h^7–$Fd3m$	8	8.0800	H11 (type)
MgFe₂O₄	Cub.	O_h^7–$Fd3m$	8	8.359 ± 0.005	H11 (spinel)[c]
MgO (25)	Cub.	O_h^5–$Fm3m$	4	4.213	B1 (NaCl)
MnFe₂O₄ (20)	Cub.	O_h^7–$Fd3m$	8	8.499	H11 (spinel)[c]
NaBr (25)	Cub.	O_h^5–$Fm3m$	4	5.9737	B1 (NaCl)
NaBrO₃	Cub.	T^4–$P2_13$	4	6.72	G3
NaCl (26)	Cub.	O_h^5–$Fm3m$	4	5.6402; also reported; 5.64009 ± 0.00003 (25°)	B1 (type)
NaClO₃	Cub.	T^4–$P2_13$	4	6.568 ± 0.001	G3
NaF (25)	Cub.	O_h^5–$Fm3m$	4	4.6342	B1 (NaCl)
NH₄Br (250)(stable above 137.8)	Cub.	O_h^5–$Fm3m$	4	6.91	B1 (NaCl)
NH₄Br (18)(stable below 137.8)	Cub.	O_h^1–$Pm3m$	1	4.059	B2 (CsCl)
NH₄Br (γ)(−100)	Tetr.	D_{2d}^7–$P\bar{4}b2$	1	4.257, 4.043, ±0.004	B25
NH₄Cl (250)(stable above 184.3)	Cub.	O_h^5–$Fm3m$	4	6.547	B1 (NaCl)
NH₄Cl (26)(stable below 184.3)	Cub.	O_h^1–$Pm3m$	1	3.8756	B2 (CsCl)
NiAs (pure)	Hex.	D_{6h}^4–$P6_3/mmc$	2	3.638, 5.059	B8
NiFe₂O₄ (25)	Cub.	O_h^7–$Fd3m$	8	8.339	H11 (spinel)[c]
NH₄H₂PO₄	Tetr.	D_{2d}^{12}–$I\bar{4}2d$	4	7.499, 7.548	H22
RbF	Cub.	O_h^5–$Fm3m$	4	5.64 ± 0.02	B1 (NaCl)
RbF (~12 kb)	Cub.	O_h^1–$Pm3m$	1	3.29	B2 (CsCl)
RbCl (α)(−190)	Cub.	O_h^1–$Pm3m$	1	3.749	B2 (CsCl)
RbCl (β)(20)	Cub.	O_h^5–$Fm3m$	4	6.548; also 6.5810(27°)	B1 (NaCl)
TlCl (26)	Cub.	O_h^1–$Pm3m$	1	3.8421	B2 (CsCl)
TlBr (25)	Cub.	O_h^1–$Pm3m$	1	3.9850	B2 (CsCl)
TlI	Cub.	O_h^1–$Pm3m$	1	4.206	B2 (CsCl)
TlI (25)	Orth.	D_{2h}^{17}–$Amam$	4	5.251, 12.92, 4.582	B33
SiC (α)(type 2H)	Hex.	C_{6v}^4–$P6_3mc$	2	3.076 ± 0.001, 5.048 ± 0.001	(many types)
SiC (β)(25)	Cub.	T_d^2–$P\bar{4}3m$	4	4.3597	B3, ZnS
SiO₂ (low quartz stable up to 573 ± 1)(25)	Hex.	D_3^4–$P3_121$ or D_3^6–$P3_221$	3	4.91343 ± 0.00001, 5.40506 ± 0.00003 Å*	C8, α
SiO₂ (high quartz)(stable 573–870)	Hex.	D_6^4–$P6_222$ or D_6^5–$P6_422$	3	5.01, 5.47	C8, β
SiO₂ (high tridymite)(stable 870–1470)	Hex.	D_{6h}^4–$P6_3/mmc$	4	5.04, .24 (determined outside stability range?)	C10

TABLE 9a-3. CRYSTALLOGRAPHIC DATA FOR SELECTED COMPOUNDS (*Continued*)

Formula (temp., °C, for the lattice constants given)	Crystal system	Space group[a]	Z	Lattice constants,[b] $a, b, c; \alpha, \beta, \gamma$	Structure type
SiO₂ (low tridymite)	Hex.	....	864	30.08, 49.08	
SiO₂ (high cristobalite(stable 1470–1710)	Cub.	$T_6{}^4$–$P2_13$	8	7.1473 (at 1300°)	
SiO₂ (coesite)(high pressure)	Mon.	$C_{2h}{}^6$–$C2/c$	16	7.17, 12.38, 7.17, 120°	
ZnO (18)	Hex.	$C_{6v}{}^4$–$P6_3mc$	2	3.2427 ± 0.0001, 5.1948 ± 0.0003	B4
ZrSiO₄ (zircon)	Tetr.	$D_{4h}{}^{19}$–$I4_1/amd$	4	6.58, 5.93	H0₃
ZnS (sphalerite,blende) (contains 0.16 wt % Fe)	Cub.	$T_d{}^2$–$F\bar{4}3m$	4	5.423 ± 0.006	B3
ZnS (wurtzite)(type 2H)	Hex.	$C_{6v}{}^4$–$P6_3mc$	2	3.8231, 6.2613	B4

[a] Schoenflies symbol followed by Hermann-Mauguin symbol.

[b] For nearly all the substances listed, the unit of length is the Å based on the 1948 values of the X-ray wavelengths *Acta Cryst.* **1**, 48 (1948) . Pre-1949 values were published in kX units (1 kX = 1,000 X, where X is the unit used by Siegbahn to express his numerical values of X-ray wavelengths); they have been converted to Å by means of the relation 1 k**X** = 1.00202 Å. For highly accurate determinations, the original values have been converted to Bearden's redefined angstrom unit (1964), which is identified by an asterisk (Å*); in the formula 1 kX = AÅ* the conversion factor A is the ratio λ_B/λ_S of the numerical values used by Bearden and Siegbahn for the wavelength of the X-ray line employed: A is equal to 1.002,061 for NiKα₁; 1.002,058 for CuKα₁; etc. Neither the Å nor the Å* is a metric unit. The Å used in the tables is equal to 10^{-10} m within 1 in 25,000; the Å*, within a few ppm. See J. W. M. Dumond, *Proc. Natl. Acad. Sci.* **45**, 1052 (1959); J. A. Bearden, "X-ray Wavelengths," U.S. Atomic Energy Commission, Oak Ridge, Tenn., 1964.

[c] "Spinel" and "inverse spinel" both belong to type H11.

The value of Z (Tables 9a-2 and 9a-3, column 4) is the number of formula units in the cell, the formula being expressed as in column 1, and the cell as in column 5.

The cell given in column 5 is chosen according to the conventions of the second edition of "Crystal data" (ref. 1). The numerical values have been updated from the manuscript of the forthcoming third edition. The substances known to crystallize in any given space group are listed in ref. 2.

The symbols in column 6 identify structure types in *Strukturbericht* (refs. 4 and 5): A1, cubic close packed (= face-centered cubic, f.c.c.); A2, body-centered cubic (b.c.c.); A3, hexagonal close packed (h.c.p.); A4, diamond-type structures; etc. In many cases the prototype of the structure is added between parentheses: A7(As), A8(Se), etc.

Illustrative References

1. Donnay, J. D. H., Gabrielle Donnay, E. G. Cox, Olga Kennard, and M. V. King: Crystal Data, Determinative Tables, 2d ed., *Am. Cryst. Assoc. Monograph* 5, Polycrystal Book Service, Box 11567, Pittsburgh, Pa., 1963.
2. Nowacki, W.: Crystal Data, Systematic Tables, 2d ed., *Am. Cryst. Assoc. Monograph* 6, Polycrystal Book Service, Pittsburgh, Pa., 1967.
3. Palache, C., H. Berman, and C. Frondel: "The System of Mineralogy," 7th ed., vols. 1–3, John Wiley & Sons, Inc., New York; Chapman & Hall, Ltd., London, 1944, 1951, 1962.
4. Strunz, H.: "Mineralogische Tabellen," 5th ed., Akademische Verlagsgesellschaft Geest & Portig K.-G., Leipzig, 1969.
5. "Structure Reports," the continuation of "Strukturbericht," published for the International Union of Crystallography by N. V. A. Oosthoek's, Utrecht, Netherlands. "Strukturbericht," a digest of crystal-structure literature from 1913 through 1939, published in conjunction with *Z. Krist.*
6. Wyckoff, R. W. G.: "Crystal Structures," 2d ed., vols. 1–5, Interscience Publishers, a division of John Wiley & Sons, Inc., New York, 1963–1966.
7. Pearson, W. B.: "A Handbook of Lattice Spacings and Structures of Metals and Alloys," Pergamon Press, New York, 1958–1967.
8. Barrett, C. S.: "Structure of Metals," 3d ed., McGraw-Hill Book Company, New York, 1966.
9. Lipson, H., and C. A. Taylor: "Fourier Transforms and X-ray Diffraction," G. Bell & Sons, Ltd., London, 1958.

9a-2. Effect of Symmetries on Tensor Properties for Crystals.[1] The point-group symmetries inherent in the 32 crystal classes have an effect on the types of relations that can exist between electric, magnetic, thermal, optical, and elastic stress variables and their corresponding strain variables. Such relations are usually expressed in tensor form and for cartesian coordinates can be expressed in terms of simple cartesian tensors. With such tensors, the transformations considered are restricted to be orthogonal transformations from one cartesian system to another. By this restriction, we are avoiding many complications which arise in the case of oblique coordinates or nonplanar coordinate surfaces.

The simplest relations exist between scalar quantities and *first-rank tensors* (vectors). Examples are given by the pyroelectric effect, the electrocaloric effect, the heat of polarization, and the field due to increase in thermal energy, which satisfy the equations

$$D_n = p_n \, \Delta T \qquad \Delta S = p_i E_i \qquad \Delta S = q_n D_n \qquad E_i = -q_i \, \Delta T \qquad (9a\text{-}1)$$

where the symbols and their meaning are given in Table 9a-8. First-rank tensors can also arise from a contracted third-rank tensor. An example of interest is the piezoelectric displacement due to a hydrostatic pressure:

$$D_n = d_{nkk} T_{kk} \qquad (9a\text{-}2)$$

Table 9a-4 shows the resulting constants for the various crystallographic symmetries. Only those classes which have unique polar axes have constants different from zero.

TABLE 9a-4. FIRST-RANK TENSORS

$1; q_1, q_2, q_3 \qquad 2; 0, q_2, 0 \ (y = \text{unique axis})$

$m; q_1, 0, q_3 \ (m \perp y) \qquad mm2, 4, 4mm, 3, 3m, 6, 6mm; 0, 0, q_3$

All other classes have a zero result. All classes giving a positive result have a unique polar axis.

Second-rank tensors can arise as a relation between two vectors or as a relation between a scalar and a quantity expressed by a second-rank tensor. Examples of the first case are electric permittivity, dielectric impermeability, magnetic permeability and magnetic impermeability, electric conductivity and resistivity, thermal conductivity and resistivity, and Thomson thermoelectricity (ref. 2). Equations for these effects are given by (9a-3) with a glossary of terms in Table 9a-8:

$$
\begin{aligned}
D_i &= \epsilon_{ij} E_j & E_i &= \beta_{ij} D_j & B_i &= \mu_{ij} H_j \\
H_i &= \beta_{ij} B_j & I_i &= \sigma_{ij} E_j & E_i &= \rho_{ij} I_j \\
h_i &= -k_{ij}\left(\frac{\partial T}{\partial x_j}\right) & \frac{\partial T}{\partial x_i} &= -r_{ij} h_j & \frac{\partial \bar{\mu}}{\partial x_i} &= -\Sigma_{ik}\frac{\partial T}{\partial x_k}
\end{aligned}
\qquad (9a\text{-}3)
$$

Examples of the second case for which the tensors arise as a relation between a scalar and quantities expressed as second-rank tensors are thermal expansions, stresses due to temperature changes, strain for a hydrostatic stress, and Peltier thermoelectric coefficients. These relations are given by Eq. (9a-4):

$$S_{ij} = \alpha_{ij} \, \Delta T \qquad T_{ij} = -\lambda_{ij} \, \Delta T \qquad S_{ij} = s_{ijkk} T_{kk} \qquad \Pi_{ik} = \frac{T}{e} \Sigma_{ik} \qquad (9a\text{-}4)$$

All the second-rank tensors are symmetric except the thermoelectric tensors. Table 9a-5 shows the terms for the various crystal symmetries and the changes caused by the relations $\alpha_{ij} = \alpha_{ji}$.

[1] This subsection was originally contributed by W. P. Mason and E. A. Wood, Bell Telephone Laboratories, Inc.

Third-rank tensors have been employed in expressing the direct and inverse piezo-electric effect with four different forms depending on the sets of variable used. They have also been employed in defining the electrooptical effect and the Hall effect. These relations are given by Eq. (9a-5):

$$D_n = d_{nij}T_{ij} \qquad S_{ij} = d_{mij}E_m \qquad T_{kl} = -e_{mkl}E_m \qquad D_n = e_{nij}S_{ij}$$
$$T_{kl} = -h_{nkl}D_n \qquad E_m = -h_{mij}S_{ij} \qquad S_{ij} = g_{nij}D_n \qquad E_m = -g_{mij}T_{ij} \qquad (9a-5)$$
$$E_m = D_n(\beta_{mn}^S + \gamma_{mno}^S D_0) \qquad E_i = \epsilon_{ijk}I_j(R_{km}H_m)$$

In all third-rank tensors two pairs of indices can be interchanged, for example *ij* in

TABLE 9a-5. SECOND-RANK TENSORS

Triclinic 1, $\bar{1}$ 9 constants	π_{11} π_{12} π_{13} π_{21} π_{22} π_{23} π_{31} π_{32} π_{33}	If symmetric $\alpha_{12} = \alpha_{21}$; $\alpha_{13} = \alpha_{31}$ $\alpha_{23} = \alpha_{32}$; 6 constants
Monoclinic 2, *m*, 2/*m* 5 constants	π_{11} 0 π_{13} 0 π_{22} 0 π_{31} 0 π_{33}	If symmetric $\alpha_{13} = \alpha_{31}$ 4 constants
Trigonal, tetragonal, hexagonal 3, $\bar{3}$, 4, $\bar{4}$, 4/*m* 6, $\bar{6}$, 6/*m* 3 constants	π_{11} π_{12} 0 $-\pi_{12}$ π_{11} 0 0 0 π_{33}	If symmetric $\alpha_{12} = 0$ 2 constants
Trigonal, tetragonal, hexagonal 32, 3*m*, $\bar{3}m$ 422, 4*mm*, $\bar{4}2m$, 4/*mmm* 622, 6*mm*, $\bar{6}m2$, 6/*mmm* 2 constants	π_{11} 0 0 0 π_{11} 0 0 0 π_{33}	If symmetric the same number of constants
Cubic or isotropic 23, *m*3, $\bar{4}3m$ 432, *m*3*m* 1 constant	π_{11} 0 0 0 π_{11} 0 0 0 π_{11}	Same for symmetric tensor

d_{nij}, since T_{ij} is a symmetric tensor with $T_{ij} = T_{ji}$. Hence it is usual to replace the two indices by a single one according to the convention

$$11 = 1 \quad 22 = 2 \quad 33 = 3 \quad 23 = 32 = 4 \quad 13 = 31 = 5 \quad 12 = 21 = 6 \quad (9a-6)$$

Table 9a-6 gives the resulting third-rank tensors for the various crystal symmetries.

All the *fourth-rank tensors* in general use express relations between two second-rank tensors such as stress and strain or between a second-rank tensor and the product of two vectors. Examples are elasticity equations, photoelastic relations, magneto-strictive and electrostrictive relations, magnetoresistance effects, and piezoresistance effects, expressed by the equations

$$S_{ij} = s_{ijkl}T_{kl} \qquad T_{kl} = c_{ijkl}S_{ij} \qquad E_m = D_n(\beta_{mn}{}^s + m_{ijmn}S_{ij})$$
$$S_{ij} = M_{ijkl}B_kB_l \qquad S_{ij} = q_{ijkl}D_kD_l \qquad E_i = \alpha_{ijkl}I_jH_kH_l \qquad (9a-7)$$
$$E_i = (\rho_{ij} + \pi_{ijkl}T_{kl})I_j$$

Except in the case of ferroelectric or ferromagnetic crystals (ref. 13), it is generally believed that $T_{ij} = T_{ji}$, so that the compliance tensor s_{ijkl} and the elastic stiffness tensor c_{ijkl} would indicate 36 independent constants. On account of Maxwell-type

TABLE 9a-6. THIRD-RANK TENSORS

Class							Note	Class							Note
1	e_{11}	e_{12}	e_{13}	e_{14}	e_{15}	e_{16}	; $\bar{1}=0$;	**2**	0	0	0	e_{14}	0	e_{16}	
	e_{21}	e_{22}	e_{23}	e_{24}	e_{25}	e_{26}			e_{21}	e_{22}	e_{23}	0	e_{25}	0	
	e_{31}	e_{32}	e_{33}	e_{34}	e_{35}	e_{36}			0	0	0	e_{34}	0	e_{36}	
m	e_{11}	e_{12}	e_{13}	0	e_{15}	0	; $2/m=0$;	**222**	0	0	0	e_{14}	0	0	
	0	0	0	e_{24}	0	e_{26}			0	0	0	0	e_{25}	0	
	e_{31}	e_{32}	e_{33}	0	e_{35}	0			0	0	0	0	0	e_{36}	
mm2	0	0	0	0	e_{15}	0	; $mmm=0$;	**$\bar{4}$**	0	0	0	e_{14}	e_{15}	0	
	0	0	0	e_{24}	0	0			0	0	0	e_{15}	$-e_{14}$	0	
	e_{31}	e_{32}	e_{33}	0	0	0			e_{31}	$-e_{31}$	0	0	0	e_{36}	
4,6	0	0	0	e_{14}	e_{15}	0	; $4/m=0$;	**$\bar{4}2m$**	0	0	0	e_{14}	0	0	; $4/mmm=0$
	0	0	0	e_{15}	$-e_{14}$	0			0	0	0	0	e_{14}	0	
	e_{31}	e_{31}	e_{33}	0	0	0			0	0	0	0	0	e_{36}	
422 622	0	0	0	e_{14}	0	0	;	**4mm 6mm** Transverse isotropy	0	0	0	0	e_{15}	0	
	0	0	0	0	$-e_{14}$	0			0	0	0	e_{15}	0	0	
	0	0	0	0	0	0			e_{31}	e_{31}	e_{33}	0	0	0	
3	e_{11}	$-e_{11}$	0	e_{14}	e_{15}	$-e_{22}$	;	**32**	e_{11}	$-e_{11}$	0	e_{14}	0	0	; $\bar{3}=0$
	$-e_{22}$	e_{22}	0	e_{15}	$-e_{14}$	$-e_{11}$			0	0	0	0	$-e_{14}$	$-e_{11}$	
	e_{31}	e_{31}	e_{33}	0	0	0			0	0	0	0	0	0	
$\bar{6}$	e_{11}	$-e_{11}$	0	0	0	$-e_{22}$	;	**3m**	0	0	0	0	e_{15}	$-e_{22}$	; $\bar{3}m=0$
	$-e_{22}$	e_{22}	0	0	0	$-e_{11}$			$-e_{22}$	e_{22}	0	e_{15}	0	0	
	0	0	0	0	0	0			e_{31}	e_{31}	e_{33}	0	0	0	
$\bar{6}m2$	e_{11}	$-e_{11}$	0	0	0	0	;	**23 $\bar{4}3m$**	0	0	0	e_{14}	0	0	$6/mmm,\ 6/m=0$
	0	0	0	0	0	$-e_{11}$			0	0	0	0	e_{14}	0	; $m3,\ m3m,\ 432=0$
	0	0	0	0	0	0			0	0	0	0	0	e_{14}	

For all systems with a center of symmetry, the third rank tensors vanish.

TABLE 9a-7. FOURTH-RANK TENSORS

(Type M_{ijkl}, $i \to j$, $k \to l$; Type c_{ijkl}, $i \to j$, $k \to l$, $ij \to kl$; Type K_{ijkl}, $i \to j \to k \to l$)

Group I — Triclinic $1, \bar{1}$ — 36 constants

M_{11}	M_{12}	M_{13}	M_{14}	M_{15}	M_{16}
M_{21}	M_{22}	M_{23}	M_{24}	M_{25}	M_{26}
M_{31}	M_{32}	M_{33}	M_{34}	M_{35}	M_{36}
M_{41}	M_{42}	M_{43}	M_{44}	M_{45}	M_{46}
M_{51}	M_{52}	M_{53}	M_{54}	M_{55}	M_{56}
M_{61}	M_{62}	M_{63}	M_{64}	M_{65}	M_{66}

c constants the same except $c_{ab} = c_{ba}$, resulting in 21 constants. K constants the same as c except $K_{44} = K_{23}$, $K_{55} = K_{13}$, $K_{66} = K_{12}$, $K_{46} = K_{25}$, $K_{56} = K_{14}$, $K_{45} = K_{36}$, resulting in 15 constants.

Group II — Monoclinic $2, m, 2/m$ — 20 constants — y = unique axis

M_{11}	M_{12}	M_{13}	0	M_{15}	0
M_{21}	M_{22}	M_{23}	0	M_{25}	0
M_{31}	M_{32}	M_{33}	0	M_{35}	0
0	0	0	M_{44}	0	M_{46}
M_{51}	M_{52}	M_{53}	0	M_{55}	0
0	0	0	M_{64}	0	M_{66}

c constants the same except $c_{ab} = c_{ba}$, resulting in 13 constants. K constants the same as c, except $K_{44} = K_{23}$, $K_{55} = K_{13}$, $K_{66} = K_{12}$, $K_{46} = K_{25}$, resulting in 9 constants.

Group III — Orthorhombic $mm2, 222, mmm$ — 12 constants

M_{11}	M_{12}	M_{13}	0	0	0
M_{21}	M_{22}	M_{23}	0	0	0
M_{31}	M_{32}	M_{33}	0	0	0
0	0	0	M_{44}	0	0
0	0	0	0	M_{55}	0
0	0	0	0	0	M_{66}

c constants the same except $c_{ab} = c_{ba}$, resulting in 9 constants. K constants the same as c except $K_{44} = K_{23}$, $K_{55} = K_{13}$, $K_{66} = K_{12}$, resulting in 6 constants.

Group IV — Trigonal $3, \bar{3}$ — 12 constants

M_{11}	M_{12}	M_{13}	M_{14}	$-M_{25}$	$2M_{62}$
M_{12}	M_{11}	M_{13}	$-M_{14}$	M_{25}	$-2M_{62}$
M_{31}	M_{31}	M_{33}	0	0	0
M_{41}	$-M_{41}$	0	M_{44}	M_{45}	$2M_{52}$
$-M_{52}$	M_{52}	0	$-M_{45}$	M_{44}	$2M_{41}$
$-M_{62}$	M_{62}	0	$-M_{25}$	M_{14}	$M_{11} - M_{12}$

c constants the same, except $c_{ab} = c_{ba}$ and $c_{62} = c_{45} = 0$. $c_{46} = c_{25}, c_{56} = c_{14}$, resulting in 7 constants. K constants the same as c constants except $K_{44} = K_{23}$, resulting in 6 constants.

Group V — Trigonal $3m, 32, \bar{3}m$ — 8 constants

M_{11}	M_{12}	M_{13}	M_{14}	0	0
M_{12}	M_{11}	M_{13}	$-M_{14}$	0	0
M_{31}	M_{31}	M_{33}	0	0	0
M_{41}	$-M_{41}$	0	M_{44}	0	0
0	0	0	0	M_{44}	$2M_{41}$
0	0	0	0	M_{14}	$M_{11} - M_{12}$

c constants the same except $c_{ab} = c_{ba}$, $2c_{41} = c_{14} = c_{56}$. 6 constants. K constants the same as the c constants except $K_{44} = K_{23}$. 5 constants.

Group VI — Tetragonal $4, \bar{4}, 4/m$ — 10 constants

M_{11}	M_{12}	M_{13}	0	0	M_{16}
M_{12}	M_{11}	M_{13}	0	0	$-M_{16}$
M_{31}	M_{31}	M_{33}	0	0	0
0	0	0	M_{44}	M_{45}	0
0	0	0	$-M_{45}$	M_{44}	0
M_{61}	$-M_{61}$	0	0	0	M_{66}

c constants the same except $c_{13} = c_{31}$, $c_{16} = c_{61}$. $c_{45} = 0$. 7 constants. K constants the same except $K_{44} = K_{23}$, $K_{66} = K_{12}$. 5 constants.

TABLE 9a-7. FOURTH-RANK TENSORS (Continued)

Group VII
Tetragonal
$4mm$, $\bar{4}2m$, 422, $4/mmm$
7 constants

$$
\begin{vmatrix}
M_{11} & M_{12} & M_{13} & 0 & 0 & 0 \\
M_{12} & M_{11} & M_{13} & 0 & 0 & 0 \\
M_{31} & M_{31} & M_{33} & 0 & 0 & 0 \\
0 & 0 & 0 & M_{44} & 0 & 0 \\
0 & 0 & 0 & 0 & M_{44} & 0 \\
0 & 0 & 0 & 0 & 0 & M_{66}
\end{vmatrix}
$$

c constants the same except $c_{13} = c_{31}$. 6 constants.
K constants the same except $K_{44} = K_{23}$, $K_{66} = K_{12}$. 4 constants.

Group VIII
Hexagonal
$\bar{6}$, 6, $6/m$
8 constants

$$
\begin{vmatrix}
M_{11} & M_{12} & M_{13} & 0 & 0 & 2M_{62} \\
M_{12} & M_{11} & M_{13} & 0 & 0 & -2M_{62} \\
M_{31} & M_{31} & M_{33} & 0 & 0 & 0 \\
0 & 0 & 0 & M_{44} & M_{45} & 0 \\
0 & 0 & 0 & -M_{45} & M_{44} & 0 \\
M_{62} & -M_{62} & 0 & 0 & 0 & M_{66}
\end{vmatrix}
$$

c constants the same except $c_{13} = c_{31}$, $c_{61} = 0$, $c_{45} = 0$. 5 constants.
K constants the same as c constants except $K_{44} = K_{23}$. 4 constants.

Group IX
Hexagonal
$\bar{6}m2$, 622, $6mm$, $6/mmm$
6 constants

$$
\begin{vmatrix}
M_{11} & M_{12} & M_{13} & 0 & 0 & 0 \\
M_{12} & M_{11} & M_{13} & 0 & 0 & 0 \\
M_{31} & M_{31} & M_{33} & 0 & 0 & 0 \\
0 & 0 & 0 & M_{44} & 0 & 0 \\
0 & 0 & 0 & 0 & M_{44} & 0 \\
0 & 0 & 0 & 0 & 0 & M_{11} - M_{12}
\end{vmatrix}
$$

c constants the same except $c_{13} = c_{31}$. 5 constants.
K constants the same as c constants, except $K_{44} = K_{23}$. 4 constants.

Group X
Cubic
23, $m3$
4 constants

$$
\begin{vmatrix}
M_{11} & M_{12} & M_{13} & 0 & 0 & 0 \\
M_{13} & M_{11} & M_{12} & 0 & 0 & 0 \\
M_{12} & M_{13} & M_{11} & 0 & 0 & 0 \\
0 & 0 & 0 & M_{44} & 0 & 0 \\
0 & 0 & 0 & 0 & M_{44} & 0 \\
0 & 0 & 0 & 0 & 0 & M_{44}
\end{vmatrix}
$$

c constants the same except $c_{12} = c_{13}$. 3 constants.
K constants the same as c constants except $K_{44} = K_{12}$. 2 constants.

Group XI
Cubic
$\bar{4}3m$, 432, $m3m$
3 constants

$$
\begin{vmatrix}
M_{11} & M_{12} & M_{12} & 0 & 0 & 0 \\
M_{12} & M_{11} & M_{12} & 0 & 0 & 0 \\
M_{12} & M_{12} & M_{11} & 0 & 0 & 0 \\
0 & 0 & 0 & M_{44} & 0 & 0 \\
0 & 0 & 0 & 0 & M_{44} & 0 \\
0 & 0 & 0 & 0 & 0 & M_{44}
\end{vmatrix}
$$

c constants the same. 3 constants.
K constants the same as c constants except $K_{44} = K_{12}$. 2 constants.

Group XII
Isotropic
2 constants

$$
\begin{vmatrix}
M_{11} & M_{12} & M_{12} & 0 & 0 & 0 \\
M_{12} & M_{11} & M_{12} & 0 & 0 & 0 \\
M_{12} & M_{12} & M_{11} & 0 & 0 & 0 \\
0 & 0 & 0 & M_{11} - M_{12} & 0 & 0 \\
0 & 0 & 0 & 0 & M_{11} - M_{12} & 0 \\
0 & 0 & 0 & 0 & 0 & M_{11} - M_{12}
\end{vmatrix}
$$

c and K constants the same. 2 constants.

relations, one can interchange the ij with the kl moduli, and this reduces the number to 21. When it is not permissible to interchange ij with kl as in the magnetostrictive equations,

$$S_{ij} = M_{ijkl}B_k B_l \qquad (9a\text{-}8)$$

there are 36 possible constants. Table 9a-7 for fourth-rank tensors shows how the crystal symmetries affect the number and relations among the independent constants. Type c relations indicated are for the case that ij can be interchanged with kl. A

TABLE 9a-8. GLOSSARY OF TENSOR TERMS

Symbol	Meaning	Symbol	Meaning
ΔQ	Increment of heat	r_{ij}	Thermal resistive constants
ΔT	Increment of temperature	R_{kl}	Hall-effect constants
ΔS	Increment of entropy	s_{ijkl}	Compliance constants
B_i	Magnetic flux density	S_{ij}	Strain components
c_{ijkl}	Elastic stiffness constants	T	Absolute temperature
D	Electric displacements	T_{kl}	Stress components
D_o	Electric displacement at optical frequencies	x_i	Length variable
		α_{ij}	Temperature-expansion coefficients
d_{nij}	Piezoelectric constants		
e	Electronic charge	α_{ijkl}	Magnetoresistive constants
e_{mkl}	Piezoelectric constants	β_{ij}	Dielectric or magnetic impermeabilities
E	Electric fields		
g_{nij}	Piezoelectric constants	ν_{mno}	Electrooptic constants
h_i	Flow of heat per unit area	ϵ_{ij}	Dielectric constants
h_{nkl}	Piezoelectric constants	ϵ_{ijk}	Rotation tensor (see ref. 4, p. 393)
H_j	Magnetic fields		
I_i	Electric current densities	λ_{ij}	Temperature coefficients of stress at constant volume
k_{ij}	Thermal conductivities		
m_{ijmn}	Photoelastic constants	$\bar{\mu}$	Electrochemical potential
M_{ijkl}	Magnetostrictive constants	μ_{ij}	Magnetic permeability constants
p_n, p_i	Pyroelectric or pyromagnetic constants	π_{ijkl}	Piezoresistive constants
		Π_{ik}	Peltier thermoelectric coefficients
P_i	Polarization	ρ_{ij}	Electrical resistivity constants
q_n, q_i	Pyroelectric or pyromagnetic constants	σ_{ij}	Electrical conductivity constants
		Σ_{ik}	Thermoelectric coefficients (Thomson)
q_{ijkl}	Electrostrictive constants		

third type of symmetry for fourth-rank tensors occurs when all the indices i, j, k, and l are interchangeable. Such a case occurs when the elastic moduli satisfy the Cauchy relationship. This is denoted by type-K symmetry in Table 9a-7.

Table 9a-8 shows the symbols used in the above equations and their meaning.

References for Section 9a-2

1. Mason, W. P.: "Piezoelectric Crystals and Their Application to Ultrasonics," D. Van Nostrand Company, Inc., Princeton, N.J., 1640.
2. Nye, J. F.: "Physical Properties of Crystals," Oxford University Press, New York, 1957.
3. Huntington, H. B.: The Elastic Constants of Crystals, *Solid State Phys.* **7** (1958).
4. Mason, W. P.: "Physical Acoustics and the Properties of Solids," D. Van Nostrand Company, Inc, Princeton, N.J. 1958.

9b. Structure, Melting Point, Density, and Energy Gap of Simple Inorganic Compounds

H. P. R. FREDERIKSE

The National Bureau of Standards

Table 9b-1 lists the following properties of inorganic compounds:
Crystal structure (see also Sec. 9a)
Space group (see also Sec. 9a)
Melting point (see also Secs. 4d and 4j)
Density (see also Secs. 2b, 3f, and 4c)
Energy gap (for definition see Sec. 9c-1)
The compounds are listed not alphabetically but according to the location of the constituent elements in the periodic table (see Sec. 7b). The bulk of the table presents data on binaries; a few ternaries are also listed. Compounds are listed in groups beginning with the constituent elements from the first column and the seventh column and successively progressing toward the middle of the periodic system as follows (Roman numerals refer to columns):

IA-VII	IIB-VI
IA-VI	IIB-V
IA-V	IIIB-VI
IB-VII	IIIB-V
IB-VI	IVB-VII
IB-V	IVB-VI
IB-IV	IVB-V
IIA-VII	IVB-IV
IIA-VI	VB-VI
IIA-V	Transition metal oxides, sulfides, etc.
IIA-IV	Transition metal phosphides, arsenides, etc.
IIB-VII	Ternaries
	Noble gas compounds

With a few exceptions only those compounds have been listed for which at least one of the four properties has been measured. The list of compounds is, of course, far from complete; the cutoff is by necessity somewhat arbitrary.

There is often some disagreement among authors or sources. For an evaluation of the reliability of a particular figure one should go back to the original literature.

For further information the reader is referred to the references at the end of the table.

Abbreviations

cub	cubic	d	decomposes
tetr	tetragonal	b.p.	boiling point
hex	hexagonal	tr	transition (the compound listed is stable *below* the
orth	orthorhombic		transition temperature)
mon	monoclinic	liq	liquid
tricl	triclinic	s	sublimes
rhomb	rhombohedral	ign	ignites
Z	zinc blende	calc	calculated
W	wurtzite	met	metallic (conduction)
per	perovskite		

TABLE 9b-1. CLASSIFICATION AND PROPERTIES OF INORGANIC COMPOUNDS

Compound	Structure	Space group	Melting point, °C	Density, g/cm³	Energy gap, eV
IA-VII (alkali halides (ref. 7):					
LiF.........	cub (NaCl)	$Fm3m$	870	2.601	~12
LiCl.........	cub (NaCl)	$Fm3m$	614	2.06_8	~10
LiBr.........	cub (NaCl)	$Fm3m$	547	3.46_4	~8.5
LiI.........	cub (NaCl)	$Fm3m$	446	4.06_1	≥ 5.9
NaF.........	cub (NaCl)	$Fm3m$	992	2.79	≥ 10.5
NaCl.........	cub	$Fm3m$	800	2.16_4	8.5
NaBr.........	cub (NaCl)	$Fm3m$	755	3.210	7.7
NaI.........	cub (NaCl)	$Fm3m$	651	3.665	≥ 5.8
KF.........	cub (NaCl)	$Fm3m$	880	2.505	10.9
KCl.........	cub (NaCl)	$Fm3m$	790	1.9917	8.5
KBr.........	cub (NaCl)	$Fm3m$	730	2.754	7.8
KI.........	cub (NaCl)	$Fm3m$	723	3.114	≥ 6.2
RbF.........	cub (NaCl)	$Fm3m$	760	2.88	10.4
RbCl.........	cub (NaCl)	$Fm3m$	715	2.76	8.2
RbBr.........	cub (NaCl)	$Fm3m$	682	3.35	7.7
RbI.........	cub (NaCl)	$Fm3m$	642	3.55	≥ 6.1
CsF.........	cub (NaCl)	$Fm3m$	683	3.58_6	10.0
CsCl.........	cub	$Pm3m$	tr 460	3.988	≥ 8.0
CsCl (β)......	cub (NaCl)	$Fm3m$	646	3.54 (calc.)	≥ 7.5
CsBr.........	cub (CsCl)	$Pm3m$	636	4.43_3	7.0–8.0
CsI.......	cub (CsCl)	$Pm3m$	621	4.51	≥ 6.3
IA-VI:					
Li₂O.........	cub (CaF₂)	$Fm3m$	>1700	2.01_3	
Li₂S.........	cub (CaF₂)	$Fm3m$	....	1.66	
Li₂Se.........	cub (CaF₂)	$Fm3m$	....	2.91	
Li₂Te.........	cub (CaF₂).....	$Fm3m$	....	3.24	
Na₂O.........	cub (CaF₂)	$Fm3m$	s	2.27	
Na₂S.........	cub (CaF₂)	$Fm3m$	950	1.85_6	
Na₂Se.......	cub (CaF₂)	$Fm3m$	>875	2.58	
Na₂Te.........	cub (CaF₂)	$Fm3m$	....	2.90	
K₂O.........	cub (CuF₂)	$Fm3m$	....	2.32	
K₂S.........	cub (CaF₂)	$Fm3m$	471	1.80_5	
IA-V:					
Li₃N.........	hex	$P6/mmm$	840	....	2–3
NaN₃.........	orth (?)	....	tr 19		
NaN₃.........	hex	$R32$ or $R\bar{3}m$	d 340	1.853	
KN₃.........	tetr	$I4/mcm$	350	2.038	
Rb₃N.........	tetr	$I4/mcm$	....	2.788	
Li₃P.........	hex (β Al₂O₃)	$P6_3/mmc$	....	1.43	
Na₃P.........	hex	$P6_3/mmc$	d	1.74 (calc)	
Li₃As.........	hex	$P6_3/mmc$	....	2.42 (calc)	
Na₃As.......	hex	$P6_3/mmc$	....	2.328	
K₃As.........	hex	$P6_3/mmc$	....	2.14 (calc)	
Li₃Sb.........	hex	$P6_3/mmc$	>950	2.96 (calc)	
NaSb.........	mon	$P2_1/n$	465	4.03 (calc)	~0.8
Na₃Sb.........	hex (β Al₂O₃)	$P6_3/mmc$	856	2.67 (calc)	
KSb.........	....	....	605	....	0.9
K₃Sb.........	hex (β Al₂O₃)	$P6_3/mmc$	812	2.35 (calc)	0.8
Cs₃Sb.........	cub	$Fd3m$	....	5.01 (calc)	0.8
Na₃Bi.......	hex	$P6_3/mmc$	773	3.70 (calc)	
K₃Bi.........	hex	$P6_3/mmc$	....	2.98 (calc)	
Cs₃Bi.........	cub	$Fd3m$	....	5.01 (calc)	0.5–0.6
IB-VII:					
CuCl (1)......	cub (Z)	$F\bar{4}3m$	tr 407	4.136	3.31
CuCl (2)......	hex (W)	$P6_3mc$	422		
CuBr (1)......	cub (Z)	$F\bar{4}3m$	tr 382	4.72	2.98
CuBr (2)......	hex (W)	$P6_3mc$	488		
CuI.........	cub (Z)	$F\bar{4}3m$	605	5.667	3.06
AgF.........	cub (NaCl)	$Fm3m$	435	5.85_2	
AgCl.........	cub (NaCl)	$Fm3m$	455	....	3.0
AgBr.........	cub (NaCl)	$Fm3m$	430	....	2.9
AgI (1)......	cub (Z)	$F\bar{4}3m$	....	6.0	2.8
AgI (2).......	hex (W)	$P6_3mc$	558	5.68	
AgI (α) (146–558°C)......	cub				

TABLE 9b-1. CLASSIFICATION AND PROPERTIES OF
INORGANIC COMPOUNDS (*Continued*)

Compound	Structure	Space group	Melting point, °C	Density, g/cm³	Energy gap, eV
IB-VII (*Cont.*):					
AuCl.........	....	....	tr 170 ($\rightarrow$ AuCl₃)	7.4	
AuBr.........	....	....	d 115	7.9	
AuI..........	....	$P4_2/n$	d 120	8.25	
IB-VI:					
CuO..........	mon	$A2/a$	d	6.40	~1.95
Cu₂O.........	cub	$Pn3m$	1236	6.0	2.2
CuS..........	hex	$P6_3/mmc$	tr 103	4.681	
Cu₁.₈S........	cub (NaCl?)	$Fm3m(?)$	....	5.6 (170°C)	
Cu₂S (α)......	orth	$Cmma$ $Cm2a$ } (?) $C2ma$	tr 105	5.8	
Cu₂S (β)......	cub (CaF₂)	$Fm3m(?)$	1100	5.6	
CuSe.........	hex	$P6_3/mmc$	....	5.99	
Cu₂Se (β).....	cub (CaF₂)	$Fm3m$	1148	6.75	
Cu₂Te.......	hex	$P6/mmm$	1125	7.41 (calc)	
AgO..........	cub	?	d >100	7.44	
Ag₂O.........	cub (Cu₂O)	$Pn3m$	d 300	7.14₃	
Ag₂S (β)......	mon	$P2_1/n$	tr 175	7.32₆	~1.3
Ag₂S (α)......	cub (CsCl)	$Pm3m$	825	7.3₇	met
Ag₂Se (β).....	mon	$P2_1/n$	....	....	~0.075
Ag₂Se (α)....	cub (CsCl)	$Pm3m$	897	8.187	met (?)
Ag₂Te (α).....	mon	$P2_1/n$	955	8.350	0.17
AuTe₂........	mon	$C2/m$	464	9.31 (calc)	
IB-V:					
Cu₃N.........	cub	?	d 300	6.12 (calc)	
Cu₃P.........	hex	?	....	7.15	
Cu₃As........	hex	$P\bar{3}c$	830	7.85	
Cu₂Sb.......	tetr	$P4/nmm$	585		
Cu₃Sb.......	hex	?	687		
Ag₃Sb........	orth	?	559(?)	9.74	
AuSb₂........	cub (FeS₂)	$Pa3$	460(?)	9.98	
Au₂Bi........	cub (spinel)	$Fd3m$	373	15.46	
IB-IV:					
AuSn.........	hex (NiAs)	$P6_3/mmc$	418	11.6	
Au₂Pb........	cub	$Fd3m$ } ? $F4_132$			
IIA-VII:					
BeF₂.........	tetr	....	800	2.01	
BeCl₂........	orth	$Ibam$	s 405	1.90	
BeBr₂........	....	....	s 488	3.46₅	
BeI₂.........	....	....	480	4.36 (calc)	
MgF₂........	tetr (SnO₂)	$P4/mnm$	1263	3.148	~11
MgBr₂.......	hex (CdI₂)	$P\bar{3}m1$	711	3.72	
MgCl₂.......	hex (CdI₂)	$P\bar{3}m1$	714	2.32	
MgI₂........	hex (CdI₂)	$P\bar{3}m1$	d	4.43	
CaF₂.........	cub	$Fm3m$	1418	3.18	~10
CaCl₂........	orth	$Pnnm$	782	2.22	
CaBr₂........	....	....	760	3.35₃	
CaI₂.........	hex	$P\bar{3}m$	575	3.95₆	
SrF₂.........	cub (CaF₂)	$Fm3m$	1400	4.18	
SrCl₂........	cub (CaF₂)	$Fm3m$	875	3.05₂	
SrBr₂........	orth	$Pbnm$	643	4.21₆	
SrI₂.........	....	....	402	4.54₉	
BaF₂.........	cub (CaF₂)	$Fm3m$	1320	4.893	
BaCl₂ (1).....	mon	?	tr 925	3.85₆	
BaCl₂ (2).....	cub (CaF₂)	$Fm3m$	962		
BaBr₂........	orth	$Pnam$	850	4.886	
BaI₂.........	orth	$Pnam$	740	5.236	
IIA-VI:					
BeO..........	hex (W)	$P6_3mc$	2550	3.01–3.09	
BeS..........	cub (Z)	$F\bar{4}3m$	....	2.36	
BeSe........	cub (Z)	$F\bar{4}3m$	....	4.32 (calc)	
BeTe........	cub (Z)	$F\bar{4}3m$	....	5.09 (calc)	
MgO........	cub (NaCl)	$Fm3m$	2800	3.65	7.3

TABLE 9b-1. CLASSIFICATION AND PROPERTIES OF
INORGANIC COMPOUNDS (*Continued*)

Compound	Structure	Space group	Melting point, °C	Density, g/cm³	Energy gap, eV
IIA-VI (*Cont.*):					
MgS	cub (NaCl)	$Fm3m$	d	2.82	
MgSe	cub (NaCl)	$Fm3m$			
MgTe	hex (W)	$P6mc$	....	3.86 (calc)	
CaO	cub (NaCl)	$Fm3m$	2600	2.62	6–7
CaS	cub (NaCl)	$Fm3m$	....	2.80	
CaSe	cub (NaCl)	$Fm3m$			
CaTe	cub (NaCl)	$Fm3m$	....	7.59_3	
SrO	cub (NaCl)	$Fm3m$	2415	3.9–4.8	~6
SrS	cub (NaCl)	$Fm3m$	....	3.7	
SrSe	cub (NaCl)	$Fm3m$	....	4.53 (calc)	~2
SrTe	cub (NaCl)	$Fm3m$	....	....	~2
BaO	cub (NaCl)	$Fm3m$	1923	4.7–5.7	~4.8
BaS	cub (NaCl)	$Fm3m$	....	4.25	
BaSe	cub (NaCl)	$Fm3m$			
BaTe	cub (NaCl)	$Fm3m$			
IIA-V:					
Be₃N₂	cub (Tl₂O₃)	$Ia3$	~2200	2.70_9	
Be₃P₂	cub (Tl₂O₃)	$Ia3$	....	2.23_4	
Mg₃N₂	cub (Tl₂O₃)	$Ia3$	d 1500	2.71	
Mg₃P₂	cub (Tl₂O₃)	$Ia3$	....	2.05_5	
Mg₃As₂	cub (Tl₂O₃)	$Ia3$	800	3.148	
Mg₃Sb₂	hex	$P\bar{3}m$	930	4.00	0.82
Mg₃Bi₂	hex	$P\bar{3}m$	715	5.94	met (?)
Ca₃N₂	cub (Tl₂O₃)	$Ia3$	1195	2.63	
Ca₃P₂	cub (Tl₂O₃) (?)	$Ia3$ (?)	>1600	2.51	
Ca₃As₂	cub (Tl₂O₃) (?)	$Ia3$ (?)	d	2.50	
Ca₃Sb₂	cub (Tl₂O₃) (?)	$Ia3$ (?)			
Ca₃Bi₂	cub (Tl₂O₃) (?)	$Ia3$ (?)	928		
IIA-IV:					
Be₂C	cub (CaF₂)	$Fm3m$	d >2100	1.9	
Mg₂Si	cub (CaF₂)	$Fm3m$	1102	1.88	0.77
Mg₂Ge	cub (CaF₂)	$Fm3m$	1115	3.09	0.6–0.7
Mg₂Sn	cub (CaF₂)	$Fm3m$	778	3.591	0.3
Mg₂Pb	cub (CaF₂)	$Fm3m$	550	3.29	met (?)
Ca₂C	tetr				
Ca₂Si	tetr	....	920	....	1.9
CaSi₂	hex	$R\bar{3}m$	1220	2.456	
Ca₂Ge	orth	$Pnam$			
Ca₂Sn	tetr	....	1122	....	0.9
Ca₂Pb	....	....	1150	....	0.4–0.5
IIB-VII:					
ZnF₂	tetr (SnO₂)	$P4/mnm$	872	4.84	
ZnCl₂	hex (CdCl₂)	$R\bar{3}m$	262	2.91	
ZnBr₂	hex (CdCl₂) (?)	$R\bar{3}m$ (?)	394	4.21_9	
ZnI₂	hex (CdCl₂)	$R\bar{3}m$	446	4.696	
CdF₂	cub (CaF₂)	$Fm3m$	1110	6.64	>6.0
CdCl₂	hex (rhomb)	$R\bar{3}m$	568	4.01_7	
CdBr₂	hex (rhomb)	$R\bar{3}m$	568	5.19_2	
CdI₂	hex (W)	$P6mc$	387	5.4–5.6	
HgF₂	cub (FeS₂)	$Pa3$	d 645	8.95	
Hg₂F₂	cub	$I4/mmm$	570	8.73	
HgCl₂	orth	$Pmnb$	277	5.6	
Hg₂Cl₂	tetr	$I4/mmm$	s 400	6.47	
HgBr₂	orth	$Bb2m$	241	6.05_3	
Hg₂Br₂	tetr	$I4/mmm$	s 345	7.307	
HgI₂	tetr	$P4/nmc$	tr 126	6.28	
HgI₂	orth	....	259	6.27_1	
Hg₂I₂	tetr	$I4/mmm$	s 140	7.70	
IIB-VI (refs. 11, 12):					
ZnO	hex (W)	$P6mc$	1975	5.7	3.436
ZnS (β)	cub (Z)	$F\bar{4}3m$	tr 1020	4.10_2	3.84
ZnS (α)	hex (W)	$P6mc$	1850 (150 atm)	4.08	3.91
ZnSe	cub (Z)	$F\bar{4}3m$	~1500	5.65	2.83
ZnTe	cub (Z)	$F\bar{4}3m$	1238	5.54–6.39	2.39

TABLE 9b-1. CLASSIFICATION AND PROPERTIES OF
INORGANIC COMPOUNDS (*Continued*)

Compound	Structure	Space group	Melting point, °C	Density, g/cm^3	Energy gap, eV
IIB-VI (refs. 11, 12) (*Cont.*):					
CdO.........	cub (NaCl)	$Fm3m$	s 1559	8.15	2.2 (?)
CdS (β)......	cub (Z)	$F\bar{4}3m$	s 685	4.87	2.5
CdS (α)......	hex (W)	$P6mc$	1750 (100 atm)	4.82	2.582
CdSe........	hex (W)	$P6mc$	>1258	5.81	1.84
CdTe........	cub (Z)	$F\bar{4}3m$	1098	6.20	1.607
HgO.........	orth	$Pmnn$	d 100	11.23	
HgS (α)......	hex	$P3_121$	tr 386	8.176	
HgS (β)......	cub (Z)	$F\bar{4}3m$	s 583	7.65	2.5
HgSe........	cub (Z)	$F\bar{4}3m$	798	8.24 (calc)	met
HgTe........	cub (Z)	$F\bar{4}3m$	670	8.12 (calc)	met
IIB-V (ref. 13):					
Zn$_3$N$_2$.......	cub	$Ia3$	....	6.4 (calc)	
Zn$_3$P$_2$ (1).....	cub	$Pn3m$	>420	4.678 (calc)	
Zn$_3$P$_2$ (2).....	tetr	$P4/nmc$	....	4.54 (calc)	
Zn$_3$As$_2$ (1)....	cub	$Pn3m$	1015	5.578	1.0
Zn$_3$As$_2$ (2)....	tetr	$P4/nmc$	....	4.21–4.76	
ZnSb........	orth	$Pbcn$	544	6.383	0.56
Cd$_3$N$_2$.......	cub	$Ia3$	....		
Cd$_3$P$_2$ (1).....	cub	$Pn3m$	....	5.95$_6$ (calc)	0.6
Cd$_3$P$_2$ (2).....	tetr	$P4/nmc$	....	5.956	
Cd$_3$As$_2$ (1)....	cub	$Pn3m$	....	6.21	
Cd$_3$As$_2$ (2)....	tetr	$P4/nmc$	721	4.25	0.13
Cd$_3$Sb$_2$........	mon	....	421		
CdSb.........	orth	$Pbcn$	456	6.92	0.48
IIIB-VII:					
B$_2$F$_4$..........	mon	$P2_1/n$	$-$ 56	1.92 (calc)	
BCl$_3$..........	hex	$P6_3$	$-$107	1.80 (calc)	
BBr$_3$.........	hex	?	$-$ 46	3.41 (calc)	
BI$_3$..........	hex	$P6_3$	43	3.197 (calc)	
AlF$_3$..........	hex	$R32$	1040	3.197 (calc)	
AlCl$_3$.........	mon	$A2/m$	s 178	2.48 (calc)	
AlBr$_3$........	mon	$P2_1/a$	97.5	3.205	
GaF$_3$.........	hex	$R\bar{3}c$	>1000		
GaCl$_2$........	orth	$Pcnn$	170.5	2.74 ?	
GaI$_3$.........	orth	$Amma$	212		
InF$_3$.........	hex	$R\bar{3}c$	1170		
InCl$_2$........	orth	?	235	3.64	
InCl$_3$........	mon	$C2/m$	s 400		
InBr.........	orth	$Amam$	220	4.96	
InI..........	orth	$Amam$	351	5.39 (calc)	
TlF..........	orth	$Fmmm$	b.p. 300	8.23	
TlCl..........	cub	$Pm3m$	430	7.02	3.41
TlCl$_3$.........	mon	$C2/m$	25		
TlBr.........	cub	$Pm3m$	460	7.54	3.02
TlI	orth	$Amam$	tr 175		
TlI	cub	$Pm3m$	440	7.45 (calc)	
IIIB-VI)ref. 8):					
B$_2$O$_3$..........	cub or hex (?)	$P3_1$	294	2.44	
B$_2$S$_5$..........	cub or hex (?)	....	390	1.85	
B$_2$S$_3$..........	cub or hex (?)	....	310	1.55	
B$_2$Se$_3$.........	cub or hex (?)	....	d		
Al$_2$O$_3$ (α)......	hex (Fe$_2$O$_3$)	$R\bar{3}c$	2050	3.99	8.3
Al$_2$O$_3$ (β)......	hex (NiAs)	$P6_3/mmc$	2040	3.30	
Al$_2$O$_3$ (γ)......	cub (spinel)	$P4_13$	tr to α	3.619	
Al$_2$S$_3$........	hex	....	1118	2.32	4.1
Al$_2$Se$_3$........	hex	$P6mc$	....	3.21	3.1
Al$_2$Te$_3$........	hex	$P6mc$	900	4.54	2.5
Ga$_2$O$_3$........	hex (Fe$_2$O$_3$)	$R3c$	1740	6.44 (calc)	4.4
Ga$_2$O.........	....	....	>660	4.77	
Ga$_2$S$_3$ (β).....	hex (W)	$P6mc$	1255	3.67 (calc)	~2.5
Ga$_2$S$_3$ (α).....	cub (Z)........	$F\bar{4}3m$	tr 550	3.63	2.85
GaS..........	hex	$P6/mmc$	965	3.86	~2.9
Ga$_2$S........	....	....	>800	4.18	
Ga$_2$Se$_3$ (β).....	hex (W)	$P6mc$	1020	4.92	

TABLE 9b-1. CLASSIFICATION AND PROPERTIES OF
INORGANIC COMPOUNDS (*Continued*)

Compound	Structure	Space group	Melting point, °C	Density, g/cm³	Energy gap, eV
IIIB–VI (ref. 8) (*Cont.*):					
Ga_2Se_3 (α).....	cub (Z)	$F\bar{4}3m$	....	....	~1.9
GaSe.........	hex	$P6_3/mmc$	960	5.03	2.04
Ga_2Se.........	....	....	....	5.02	
Ga_2Te_3 (β)....	hex (W)	$P6mc$	790		
Ga_2Te_3 (α)....	cub (Z)	$F\bar{4}3m$	tr 670	5.57	1.2 or 1.5
GaTe.........	....	....	824	5.44	1.7
In_2O_3.........	cub (Tl_2O_3)	$Ia3$	d 850	7.18	~2.8
InO.........					
In_2O.........	....	....	2 650–700 (in vac)	6.99	
In_2S_3.........	cub	$Fd3m$	1050	4.63	~2.0
InS.........	....	....	692	5.18	
In_2S.........	....	....	653	5.87	
In_2Se_3 (β)....	....	....	890		
In_2Se (α).....	....	....	tr 196	5.48	1.2
InSe.........	hex	$P6_3/mmc$	660	5.55	1.05
In_2Se.........	....	....	....	6.17	
In_2Te_3 (α).....	cub (Z)	$F\bar{4}3m$	667	5.75	~1.0
InTe.........	tetr	$I4/mcm$	696	6.29	
In_2Te.........	....	....	460	6.47	
Tl_2O_3.........	cub	$Ia3$	717	10.19	
Tl_2O.........	....	....	300		
Tl_2S_3.........	....	....	260		
TlS.........	tetr	$I4/mcm$	....	7.62	
Tl_2S.........	hex	$R\bar{3}$ or $R3$	448	8.0	~1.0
Tl_2Se_2.........					
TlSe.........	tetr	$I4/mcm$	....	5.175	0.57
Tl_2Se.........	....	....	398		
Tl_2Te_3.........	....	....	428		
IIIB–V (refs. 9, 10):					
BN.........	cub (Z)	$F\bar{4}3m$	s 3000	2.20	4.6
BP.........	cub (Z)	$F\bar{4}3m$	ign 200		
AlN.........	hex (W)	$P6_3mc$	>2200	3.26	~3.3
AlP.........	cub (Z)	$F\bar{4}3m$	....	2.424 (calc)	2.5
AlAs.........	cub (Z)	$F\bar{4}3m$	1600	3.598	2.3
AlSb.........	cub (Z)	$F\bar{4}3m$	1060	4.34	1.55
GaN.........	hex (W)	$P6mc$	....	6.10	
GaP.........	cub (Z)........	$F\bar{4}3m$	~1350	....	2.35
GaAs.........	cub (Z)	$F\bar{4}3m$	1280	....	1.35
GaSb.........	cub (Z)	$F\bar{4}3m$	728	....	0.7
InN.........	hex (W)	$P6mc$	....	6.88	2.4
InP.........	cub (Z)	$F\bar{4}3m$	1055	....	1.3
InAs.........	cub (Z)........	$F\bar{4}3m$	942	....	0.35
InSb.........	cub (Z)	$F\bar{4}3m$	525	....	0.17
InBi.........	tetr	....	110	....	met
TlSb.........	cub (CsCl)	$Pm3m$			
TlBi.........	cub (CsCl)	$Pm3m$	230		
IVB–VII:					
CBr_4 (α)......	mon	....	tr 47	3.42	
CBr_4 (β)......	cub	$P\bar{4}3m$	90		
CI_4.........	cub	$P\bar{4}3m$	d 171	4.32	
$SiBr_4$.........	....	....	5	2.81₄	
SiI_4.........	cub (FeS_2?)	$Pa3$	120.5		
$GeBr_2$.........	....	....	122		
$GeBr_4$.........	....	....	26.1	3.13₂	
GeI_4.........	cub (FeS_2)	$Pa3$	144.0	4.32₂	
$SnCl_2$.........	orth	....	247	3.9	
$SnCl_4$.........	....	....	−33	2.23 (liq)	
$SnBr_2$.........	orth	....	232	5.12	
$SnBr_4$.........	orth	....	31	3.34 (liq)	
SnI_2.........	mon	....	320	5.21	
SnI_4.........	cub (FeS_2)	$Pa3$	145	4.46	

TABLE 9b-1. CLASSIFICATION AND PROPERTIES OF
INORGANIC COMPOUNDS (*Continued*)

Compound	Structure	Space group	Melting point, °C	Density, g/cm³	Energy gap, eV
IVB-VII (*Cont.*):					
PbF₂.........	orth	*Pnam*	tr 200	8.37	
PbF₂........	cub (CaF₂)	*Fm3m*	822	7.66	~5.0
PbCl₂........	orth	?	501	5.85	
PbCl₄........	....	....	−15	3.18	
PbBr₂........	orth	*Pnam*	373	6.71 (calc)	
PbI..........	....	....	d 300		
PbI₂..........	hex	*P3̄m1*	402	6.18	2.57
IVB-VI:					
SiO₂:					
α-Cristo-balite.....	pseudocub.	*P2₁2₁2₁*	....	2.30	
β-Cristo-balite.....	cub	*P2₁3*	1728	2.32	
α-Quartz....	hex	*P3₂21* or *P3₁21*	tr 600	2.66	
β-Quartz....	hex	*P6₂22* or *P6₄22*			
α-Tridymite.	orth	....	1680	2.3	
β-Tridymite.	hex	*P6₃/mmc*			
Fused silica.	tetr				
SiS₂..........	orth	*Ibam*	1090	2.02	
SiS..........	....	....	....	1.85	
GeO₂.........	hex	*P3₂21*	1115	4.7	
GeS₂.........	orth	*Fdd2*	~800	3.01	
GeS..........	orth	*P6nm*	625	4.01	1.8
GeSe........	orth	*P6nm*	780	....	1.0
GeSe₂........	orth	....	707	4.56	
GeTe........	cub (NaCl)	*Fm3m*	725		
SnO₂........	tetr	*P4/mnm*	d 1127	7.0	
SnO..........	tetr (PbO)	*P4/nmm*	d 700–950	6.45	
SnS₂..........	hex (W)	*P6mc* (?)	d	4.5	2.3
SnS..........	orth	*P6nm*	880	5.08	~1.1
SnSe₂........	....	....	650	5.0	~1.0
SnSe........	orth	....	860	6.18	1.3
SnTe........	cub (NaCl)	*Fm3m*	800	6.48	0.3
PbO₂........	tetr (SnO₂)	*P4/mnm*	d 290	9.33–9.44	
PbO (red).....	tetr	*P4/nmm*	888	9.13	~2.6
PbO (yellow)..	orth	*Pca2*	....	9.52	
Pb₂O.........	cub (Cu₂O)	*Pn3m*	d	8.35	
PbS..........	cub (NaCl)	*Fm3m*	1114	7.5	0.37
PbSe........	cub (NaCl)	*Fm3m*	1065	8.1–8.2	0.27
PbTe........	cub (NaCl)	*Fm3m*	905	8.16	0.33
IVB-V:					
SnAs........	cub (NaCl)	*Fm3m*	600		
Sn₃As₂.......	orth	....	585		
SnSb........	cub (NaCl)	*Fm3m*	425		
IVB-IV:					
SiC..........	hex (W)	*Pb₃mc*	~2700	3.17–3.22	3.1
SiC (carborundum).......	cub (Z)	*F4̄3m*	....	3.21₆	2.86
VB-VI (ref. 8):					
As₂O₃ (1).....	mon	*P2₁n*	315	4.14	
As₂O₃ (2).....	cub	*Fd3m*	s 193	3.874	
As₂S₃.........	mon	*P2₁n*	300	3.43	
As₂Se₃........	hex (rhomb)	*R3̄m* (?)	360	4.75	1.3
As₂Te₃........	mon	*P2₁n* (?)	362	....	1.0–1.2
Sb₂O₃ (1).....	cub	*Fd3m*	656	5.1–5.8	
Sb₂O₃ (2).....	orth	*Pnaa*	656	~5.7	
Sb₂S₃........	orth	*Pbnm*	550	4.64	1.7
Sb₂Se₃........	orth	*Pbnm*	611	5.8	1.2–1.35
Sb₂Te₃........	hex (rhomb)	*R3̄m* (?)	629	....	0.3
Bi₂O₃ (1)......	orth	....	820	8.9	
Bi₂O₃ (α)......	cub	*Pn3m*	tr 704	8.2	
Bi₂O₃ (β)......	tetr	*P4̄b2*	....	9.14	
Bi₂S₃........	orth	*Pbnm*	850 (?)	7.39	1.1–1.3
Bi₂Se₃........	hex (rhomb)	*R3̄m* (?)	710	6.82	0.35
Bi₂Te₃........	hex (rhomb)	*R3̄m*	580	7.65	0.15

TABLE 9b-1. CLASSIFICATION AND PROPERTIES OF
INORGANIC COMPOUNDS (*Continued*)

Compound	Structure	Space group	Melting point, °C	Density, g/cm³	Energy gap, eV
Transition Metal Oxides, Sulfides, Selenides, and Tellurides					
Sc_2O_3..........	cub (Tl_2O_3)	$Ia3$	....	3.86_1	
TiO_2 (rutile).....	tetr (SnO_2)	$P4/mnm$	1835	4.283	3.05
Ti_2O_3..........	hex (Fe_2O_3)	$R\bar{3}c$	2130	4.6	
TiO...........	cub (NaCl)	$Fm3m$	1750	4.93	met
V_2O_5...........	orth	$Pnm2$	700–800	3.57_7	
V_2O_4...........	tetr (SnO_2)	$P4/mnm$	1967	4.4	
V_2O_3...........	hex (Fe_2O_3)	$R\bar{3}c$	1970	4.78	
Cr_2O_3...........	hex (Fe_2O_3)	$R\bar{3}c$	1990	5.215	
MnO_2...........	tetr (SnO_2)	$P4/mnm$	d 535	5.02_6	
Mn_2O_3..........	cub (Tl_2O_3)	$Ia3$	d 1080	4.5 or 4.8	
MnO...........	cub (NaCl)	$Fm3m$	1650	5.4	
Mn_3O_4..........	tetr	$I4/amd$	1705	4.8	
MnS_2...........	cub (FeS_2)	$Pa3$	d	3.46	
MnS...........	cub (NaCl)	$Fm3m$	d	3.95	
$MnSe_2$...........	cub (FeS_2)	$Pa3$	....	....	0.15
$MnSe$...........	cub (NaCl)	$Fm3m$	....	5.59	2.5
$MnTe_2$...........	cub (FeS_2)	$Pa3$	....	6.15 (calc)	
$MnTe$.........	hex (NiAs)	$P6_3/mmc$			
Fe_2O_3 (hematite)	hex	$R\bar{3}c$	1565	5.24	
Fe_2O_3 (γ).......	cub	$P4_33$ or $P2_13$	....	4.59	
FeO...........	cub (NaCl)	$Fm3m$	1420	5.7	
Fe_3O_4 (magnetite)	cub (spinel)	$Fd3m$	d 1538	5.17	
FeS...........	hex (NiAs)	$P6_3/mmc$	1193	4.84	
FeS_2 (1) (pyrite).	cub	$Pa3$	1171	5.005	1.2
S_2 (2)........	orth	$Pnnm$	tr 450	4.92	
Se............	hex (NiAs)	$P6_3/mmc$			
$FeSe_2$...........	cub (FeS_2)	$Pa3$	....	5.0	
CoO...........	cub (NaCl)	$Fm3m$	1935	5.7–6.7	
CoS...........	hex (NiAs)	$P6_3/mmc$	>1116	5.45	
CoS_2...........	cub (FeS_2)	$Pa3$	....	4.27	
Co_3S_4..........	cub (spinel)	$Fd3m$	....	4.86	
$CoSe$...........	hex (NiAs)	$P6_3/mmc$	....	7.65	
$CoTe$...........	hex (NiAs)	$P6_3/mmc$			
NiO...........	cub (NaCl)	$Fm3m$	2090	6.7–6.9	~4.0
NiS (millerite)...	hex	....	797	5.41	
NiS (β)........	hex (NiAs)	$P6_3/mmc$	797	4.6	
NiS_2............	cub (FeS_2)	$Pa3$	....	4.3 (calc)	
$NiSe$...........	hex (NiAs)	$P6_3/mmc$	....	8.46	
$NiTe$...........	hex (NiAs)	$P6_3/mmc$			
ZrO_2 (1).......	cub (CaF_2)	$Fm3m$	2715	5.35	
ZrO_2 (2).......	mon	$P2_1/a$	2700	5.82	
Nb_2O_5..........	orth	....	1520	4.5–4.6	
Nb_2O_3..........	....	....	1772		
MoO_2...........	tetr (SnO_2)	$P4/mnm$	....	6.44–6.47	
MoO_3...........	orth	$Pbnm$	795	4.5	
MoS_2 (molybde- nite)..........	hex (NiAs)	$P6_3/mmc$	1185	4.92	~1.0
Ta_2O_5...........	orth	....	d 1470	8.73_5	
WO_2............	tetr (SnO_2)	$P4/mnm$	ign	12.11	
WO_3............	tricl	....	1470	7.16	
ThO_2...........	cub (CaF_2)	$Fm3m$	3050	9.87	
UO_2...........	cub (CaF_2)	$Fm3m$	2227	10.9	
Transition Metal Phosphides, Arsenides, etc.					
$CrAs$...........	hex	....	....	6.35	
$CrSb$...........	hex (NiAs)	$P6_3/mmc$	1100		
MnP...........	orth	$Pnam$	1190	5.49	
$MnAs$...........	orth	$Pnam$	d 400	~6.2	
$MnSb$..........	hex (NiAs)	$P6_3/mmc$	809		
Mn_2Sb..........	tetr	$P4/nmm$	948		
$MnBi$..........	hex (NiAs)	$P6_3/mmc$			
$MnSi$...........	cub (FeSi)	$P2_13$	1280	5.9	
FeP...........	orth	$Pnam$	>1000	5.2 or 6.07	
$FeAs$ (η)........	hex (NiAs)	$P6_3/mmc$	1020	7.83	

TABLE 9b-1. CLASSIFICATION AND PROPERTIES OF
INORGANIC COMPOUNDS (*Continued*)

Compound	Structure	Space group	Melting point, °C	Density, g/cm³	Energy gap, eV
		Transition Metal Phosphides, Arsenides, etc. (*Cont.*):			
FeSb............	hex (NiAs)	$P6_3/mmc$	~1000		
FeSi............	cub	$P2_13$	1410	6.21	
NiAs............	hex	$P6_3/mmc$	968	7.72	
NiSb............	hex	$P6_3/mmc$	1158	7.54	
NiSi............	cub (FeSi)	$P2_13$	1000		
		Ternaries (refs. 17, 18)			
CuFeS₂ (chalcopyrite)........	tetr	$I\bar{4}2d$	1085	4.1–4.3	0.53
CuAlS₂.........	tetr	$I\bar{4}2d$	....	3.45	2.5
CuInS₂.........	tetr	$I\bar{4}2d$	950	4.71	1.2
CuInSe₂........	tetr	$I\bar{4}2d$	990	5.65	0.92
CuInTe₂........	tetr	$I\bar{4}2d$	790	6.00	0.95
CuTlS₂.........	tetr	$I\bar{4}2d$	....	6.07	
AgInS₂.........	tetr	$I\bar{4}2d$	850	4.97	1.9
AgInSe₂........	tetr	$I\bar{4}2d$	....	5.80	1.18
AgInTe₂........	tetr	$I\bar{4}2d$	....	6.08	0.96
ZnSiAs₂........	tetr	$I\bar{4}2d$	....	....	2.1
ZnGeP₂.........	tetr	$I\bar{4}2d$	....	4.04	2.2
CdGeP₂.........	tetr	$I\bar{4}2d$	....	....	1.8
ZnGeAs₂........	tetr	$I\bar{4}2d$	....	5.26	>0.6
Cu₃SbS₃........	cub	$I\bar{4}3m$	550	4.4–5.1	~1.0
Cu₃AsS₃........	cub	$I\bar{4}3m$	640	~4.5	~1.0
Ag₃SbS₃........	hex	$R3c$	....	5.85	~1.9
Ag₃AsS₃........	hex	$R3c$	....	5.69 (calc)	~2.0
AgSbS₂.........	mon	$A2/a$	....	5.2–5.3	
AgSbSe₂........	cub	....	611	6.64	~0.7
AgSbTe₂........	cub	....	555	7.12	~0.6
MgAl₂O₄ (spinel).	cub	$Fd3m$	2135	3.57	
ZnFe₂O₄........	cub (spinel)	$Fd3m$	1590	5.29	
CuFe₂O₄........	cub (spinel)	$Fd3m$	....	5.42	
NiFe₂O₄........	cub (spinel)	$Fd3m$	....	5.268	
MnFe₂O₄........	cub (spinel)	$Fd3m$	....	4.52	
ZnAl₂S₄........	cub (spinel)	$Fd3m$	....	3.30	
CaIn₂S₄........	cub (spinel)	$Fd3m$	....	4.10	
HgIn₂S₄........	cub (spinel)	$Fd3m$	....	5.79	
CaTiO₃ (perovskite)...	cub	$Pm3m$	1915	~4.0	3.7
BaTiO₃ (1)......	cub (per)	$Pm3m$	1618	~6.0	
BaTiO₃ (2)......	tetr	$P4/mmm$	tr 120	6.02 (calc)	3.5
SrTiO₃.........	cub (per)	$Pm3m$	1910	5.11 (calc)	3.4
PbTiO₃ (2)......	tetr	$P4/mmm$	tr 490	7.94 (calc)	
FeTiO₃.........	hex	$R\bar{3}$	1470	4.4–4.9	
PbZrO₃ (2)......	tetr	$P4/mmm$	tr 233	8.10 (calc)	
KNbO₃ (1)......	cub (per)	$Pm3m$	1039	4.634 (calc)	
KNbO₃ (2)......	orth	....	tr 434		
KTaO₃ (1)......	cub (per)	$Pm3m$	1357	7.022 (calc)	3.5
NaNbO₃ (1).....	cub (per)	$Pm3m$	1450	4.609 (calc)	
LaMnO₃........	pseudo cub (distorted per)	....	....	6.89 (calc)	
NaClO₃........	cub (FeSi)	$P2_13$	248	2.49	
NaBrO₃........	cub (FeSi)	$P2_13$	381	3.254	
NaIO₃.........	orth	$Pnma$	....	~4.26	
KClO₃.........	mon	$P2_1/m$	368	2.32	
		Noble gas compounds (ref. 19):			
XeF₂...........	tetr	$I4/mmm$	140	4.32 (calc)	
XeF₄...........	mon	$P2_1/n$	~114	4.04 (calc)	
XeF₆...........	....	....	46		
XeO₃..........	orth	$P2_12_12_1$	....	4.55	

References to Table 9b-1

General References

1. "Handbook of Chemistry and Physics," 46th ed., Chemical Rubber Publishing Company, Cleveland, Ohio, 1966.
2. "Lange's Handbook of Chemistry," McGraw-Hill Book Company, New York, 1952.
3. *NBS Circ.* 500.
4. Donnay, J. D. H.: "Crystal Data," 2d ed., American Crystallographic Association, 1963.
5. Wyckoff, R. W. G.: "Crystal Structures," 2d ed., vols. 1–3, Interscience Publishers, a division of John Wiley & Sons, Inc., New York, 1963–1965.
6. Hansen, M. and Anderko, K.: "Constitution of Binary Alloys," 2d ed., McGraw-Hill Book Company, New York, 1958.

Specific References

7. Eby, Teegaarden, and Dutton: *Phys. Rev.* **116**, 1099 (1959) (energy gap).
8. Aigrain, P., and M. Balkanski: "Selected Constants of Semiconductors," Pergamon Press, New York, 1961.
9. Hannay, N. B., ed.: "Semiconductors," Reinhold Publishing Corporation, New York, 1959.
10. Willardson, R. K., and A. C. Beer, eds.: "Semiconductors and Semimetals," vols. 1–3, Academic Press, Inc., New York, 1966–1968.
11. Reynolds, D. C., et al.: *Phys. Stat. Solidi* **9**, 645 (1965); **12**, 3 (1965).
12. Harman, T. C.: "Proceedings International Conference on II-VI Compounds" D. G. Thomas, ed., W. A. Benjamin, Inc., New York, 1966.
13. Turner, W. J., et al: *Phys. Rev.* **121**, 759 (1961).
14. Morin, F. J.: ref. 9, p. 600.
15. Levin, E. M.: H. F. McMurdy, and F. P. Hall: "Phase Diagrams for Ceramists," vols. 1 and 2, American Ceramic Society, Columbus, Ohio, 1956, 1959. (Oxides, melting points)
16. Hutson, A. R.: ref. 9, p. 541.
17. Hahn, Harry, et al.: *Z. Anorg. Allgem. Chem.* **271**, 153 (1953); *ibid.* **279**, 241 (1955). (Chalcopyrites: structure)
18. Winkler, U.: *Helv. Phys. Acta* **28**, 633 (1955). (Appendix 2: selected semiconductors: energy gaps)
19. Hyman, Herbert H., ed.: "Noble Gas Compounds," University of Chicago Press, Chicago, 1963.

9c. Electronic Properties of Solids

H. P. R. FREDERIKSE[1]

The National Bureau of Standards

J. C. SLATER[2]

University of Florida

DEFINITIONS AND FORMULAS

9c-1. Energy-band Theory of Solids (refs. 1, 2, and 3). According to quantum theory an electron bound to an atom can exist in only a limited number of discrete energy states. A large number of noninteracting identical atoms will all have the same set of allowed discrete energy states. If, now, these atoms are brought closer together and finally to their actual distances in a solid, they begin to interact and the energy levels will split. In a periodic array of atoms (crystalline solid), the allowed states tend to cluster into continuous groups of energy levels called *energy bands*. These energy bands may or may not overlap. The solid may also consist of two, three, or more kinds of different atoms (compounds).

Metal: A material in which the highest occupied energy band is only partly filled. The resistivity of metals *increases* with temperature; the temperature dependence is close to linear except at low temperatures.

Semiconductor (refs. 1, 2, and 3): A material in which the highest occupied energy band (*valence band*) is completely filled at absolute zero. The *energy gap* between the valence band and the next higher band (*conduction band*) is between zero and 4 or 5 eV. The resistivity *decreases* in certain temperature ranges exponentially with increasing temperature.

Insulator: A material in which the highest occupied energy band is completely filled. The difference between insulators and semiconductors is only gradual. Materials with energy gaps larger than 4 or 5 eV are usually called insulators. The resistivity of pure insulators at room temperature is extremely high. At elevated temperature ionic conduction often dominates electronic conduction.

Effective Mass (refs. 1, 2, and 3). Near the top or the bottom of a band the energy is generally a quadratic function of the wave vectors, so that by analogy with the expression $\varepsilon = p^2/2m = \hbar^2 k^2/2m$ for free electrons we can define an effective mass m^* such that $\partial^2 \varepsilon/\partial k^2 = \hbar^2/m^*$ (p = momentum, k = wavevector, $\hbar$ = Planck's constant $\times 1/2\pi$). The effective mass of electrons is positive. Near the top of a band m^* is negative, so that the motion corresponds to that of a positive charge (*hole*).

9c-2. Distribution Function, Fermi Energy, etc. The probability that a given state of energy ε is occupied is given by

$$f = \frac{1}{e^{\frac{\varepsilon - \varepsilon_F}{kT}} + 1}$$

[1] Definitions and Formulas.
[2] Bibliography of Energy Band Calculations.

This is called the *Fermi-Dirac distribution function*. $\mathcal{E}_F$ is the *Fermi energy*. At absolute zero $\mathcal{E}_F{}^0$ has the significance of a cutoff energy. All states with energy less than $\mathcal{E}_F{}^0$ are occupied, and all states with energy greater than $\mathcal{E}_F{}^0$ are vacant. The distribution is called *degenerate* when $\mathcal{E}_F \gg kT$ and *nondegenerate* when $\mathcal{E}_F \ll kT$. In the latter case the distribution function becomes

$$f = e^{\frac{\mathcal{E}_F - \mathcal{E}}{kT}} = A \cdot e^{-\frac{\mathcal{E}}{kT}}$$

This is known as the *Maxwell-Boltzmann* or *classical distribution function*. The *density of states* (or number of states with energy $\mathcal{E}$) per unit volume is given by

$$g(\mathcal{E}) = \frac{4\pi(2m^*)^{\frac{3}{2}}}{h^3} \mathcal{E}^{\frac{1}{2}}$$

(for spherical energy surfaces). The Fermi energy or Fermi level is determined by the total number of electrons per unit volume (n_0). One calculates for Fermi-Dirac statistics:

$$\mathcal{E}_F \cong \mathcal{E}_F{}^0 \left[1 - \frac{\pi^2}{12} \left(\frac{kT}{\mathcal{E}_F{}^0} \right)^2 + \cdots \right]$$

where

$$\mathcal{E}_F{}^0 = \frac{h^2}{2m^*} \left(\frac{3n_0}{8\pi} \right)^{\frac{2}{3}}$$

and for Maxwell-Boltzmann statistics:

$$\mathcal{E}_F = kT \ln \frac{n_0 h^3}{2(2\pi m^* kT)^{\frac{3}{2}}}$$

9c-3. Transport Properties. *Electrical Conductivity.* In a solid where ohmic conduction occurs, the current density **J** is given by

$$\mathbf{J} = \sigma \mathbf{E}$$

where σ is the conductivity and **E** the applied electric field. In a homogeneous isothermal crystal σ is a tensor having the symmetry of the crystal.

Mobility. The drift mobility of charge carriers is defined as the drift velocity per unit applied electric field (v_D/E). The relation to the *collision time* τ_c is given by

$$\mu^{(D)} = \frac{e\tau_c}{m^*}$$

Hall Effect. When a magnetic field is applied to a conductor carrying a current density **J**, an electric field $\mathbf{E}_H$ (Hall field) is developed given by the relation

$$\mathbf{E}_H = R\mathbf{J} \times \mathbf{B}$$

R is called the *Hall coefficient* and **B** is the magnetic induction. When the current density is in the length direction of the sample (J_x) and the field in the z direction, the Hall coefficient (for electrons *or* holes) is

$$R = \mp \frac{r}{ne} = \mp \frac{\mu}{\sigma}$$

where n = carriers/cm^3
   e = 1.6×10^{-19} coul
   μ = cm^2/volt-sec
   σ = (ohm-cm)$^{-1}$
   r = a scattering factor of the order of 1

Hence $R = cm^3/coul.$ μ is called the *Hall mobility* and is usually somewhat different from the drift mobility.

Magnetoresistance.   The resistance of a metal or semiconductor is altered by the presence of a magnetic field.   The relative change in resistance is

$$\frac{\Delta\rho}{\rho} = \frac{aB^2}{1 + \mu^2B^2}$$

The theory for a single isotropic energy band gives no change in resistance for metals. For semiconductors (with one type of carrier scattered by acoustical lattice vibrations) one finds at low fields that

$$a = 0.38\mu^2 \times 10^{-16}$$

where the mobility μ is measured in $cm^2/volt\text{-}sec$ and B in oersteds.

Seebeck Effect (Thermoelectric Power).   If two different conductors are joined together at both ends and the two junctions kept at different temperatures, an electromotive force is set up which is proportional to the temperature difference (for small ΔT).   The thermoelectromotive force per degree centigrade is called the *thermoelectric power* (Q).

For metals:
$$Q = \frac{\pi^2 k^2 T}{3e} \left(\frac{\partial \log \sigma(\mathcal{E})}{\partial \mathcal{E}} \right)_{\mathcal{E} = \mathcal{E}_F}$$

where $\sigma(\mathcal{E})$ is the electrical conductance due to charge carriers of energy $\mathcal{E}$.

For semiconductors see Sec. 9e-4.

Thomson Effect.   When an electric current J passes between two points of a homogeneous conductor, with a temperature difference ΔT existing between these points, an amount of heat $\sigma_T J \, \Delta T$ is emitted or absorbed in addition to the Joule heat.   The parameter σ_T is called the *Thomson coefficient.*

Peltier Effect.   If two conductors are joined together and kept at a constant temperature while a current J passes through the junction, heat is generated or absorbed at the junction in addition to the Joule heat.   The *Peltier coefficient* Π_{12} is defined so that the heat emitted or absorbed per second at the junction is $\Pi_{12}J$.

Kelvin Relations

$$Q_{12} = \Pi_{12}T$$
$$T\frac{dQ_{12}}{dT} = \sigma_{T_1} - \sigma_{T_2}$$

Nernst Effect (ref. 4).   If a temperature gradient is maintained in an electronic conductor ($J = 0$) in the presence of a transverse magnetic field, a transverse electric field develops which is given by

$$\mathbf{E}_t = Q_N \, \nabla T \times \mathbf{B}$$

Q_N is called the isothermal *Nernst coefficient.*   For semiconductors (one type of carrier, classical statistics, and acoustical lattice scattering):

$$Q_N = -\frac{3\pi}{16}\frac{k}{e}\mu$$

Ettinghausen Effect (ref. 4).   If a temperature difference is maintained across an electronic conductor perpendicular to a current of density J in the presence of a magnetic field, a transverse temperature gradient is established given by

$$\nabla_t T = PJ \times \mathbf{B}$$

P is called the *Ettinghausen coefficient.*   The Ettinghausen coefficient P, the Nernst coefficient Q_N, and the thermal conductivity κ are related by the expression

$$\kappa P = T Q_N$$

Righi-Leduc Effect (ref. 4). If a temperature difference is maintained in an electronic conductor in the presence of a magnetic field in which $J = 0$, a transverse temperature gradient is established:

$$\nabla_t T = SB \times \nabla T$$

S is called the *Righi-Leduc coefficient*.

Thermal Conductivity. If a temperature difference is maintained across a solid, the heat transported per unit time and unit cross-sectional area is

$$q = \kappa \nabla T$$

where κ is the *thermal conductivity*. The thermal conductivity of an electronic conductor can be written as the sum of two compounds; κ_l is due to heat transport via the lattice, and κ_e stems from the electronic heat transport:

$$\kappa = \kappa_l + \kappa_e = \kappa_l + L\sigma T$$

where σ is the electrical conductivity and L the *Lorenz number* or *Wiedemann-Franz ratio*. For degenerate free electrons,

$$L = \frac{\pi^2}{3}\left(\frac{k}{e}\right)^2 = 2.45 \times 10^{-8}\frac{\text{watt-ohm}}{\text{deg}^2}$$

For nondegenerate free electrons and acoustical lattice scattering,

$$L = 2\left(\frac{k}{e}\right)^2$$

Thermionic Emission. The current density of electrons emitted from a metal at temperature T is

$$J = AT^2 e^{-\phi/kT}$$

This is the *Richardson-Dushman equation*. ϕ is the *work function*, $A = 4\pi mek^2/h^3 = 120 \text{ amp/cm}^2/\text{deg}^2$.

9c-4. Specific Heat. The specific heat of an electronic conductor consists of two terms

$$C_v = \gamma T + BT^3$$

where the first term is the electronic and the second the lattice contribution. The former can usually be observed only at very low temperatures.
For degenerate free electrons:

$$\gamma = \frac{\pi^2 k^2}{\varepsilon_F} \text{ ergs/deg}^2 \text{ per electron}$$

For nondegenerate free electrons:

$$\gamma = \frac{3}{2}\frac{k}{T} \text{ ergs/deg}^2 \text{ per electron}$$

9c-5. Magnetic Properties of Electrons. *Cyclotron Resonance.* Current carriers in a solid when accelerated by a microwave electric field perpendicular to an externally applied static magnetic field H will spiral about the magnetic field. For sufficiently large mean free path l or collision time τ—the condition is $\omega_c \tau > 1$—a resonance absorption is observed for a frequency

$$\omega_c = \frac{eH}{m^*c}$$

where c is the velocity of light. This technique provides a direct measurement of the effective mass electrons (or holes) m^*.

Magnetic Susceptibility of Charge Carriers. Charge carriers contribute a diamagnetic effect through their translational motion and a paramagnetic effect due to their spin. For nondegenerate conductors (semiconductors),

$$\chi_e = \frac{n\mu_B{}^2}{kT}\left(1 - \frac{m^2}{3m^{*2}}\right)$$

where n is the concentration of free carriers and μ_B the Bohr magneton. If m^* is small (Ge), the susceptibility is mainly diamagnetic. If m^* is large (TiO$_2$), the paramagnetic effect dominates.

For degenerate conductors (metals, semimetals, and impure semiconductors) at low temperature,

$$\chi_e = \frac{3n\mu_B{}^2}{2\varepsilon_F}\left(1 - \frac{m^2}{3m^{*2}}\right)$$
$$= \frac{4m^*\mu^2}{h^2}(3\pi^2 n)^{\frac{1}{3}}\left(1 - \frac{m^2}{3m^{*2}}\right)$$

Transition metals have a large m^*, and consequently they show a high magnetic susceptibility (*Pauli paramagnetism*); semimetals with small m^* (e.g., Bi) have a diamagnetic susceptibility.

Knight Shift. Polarization of conduction electrons will produce a shift in frequency at which nuclear magnetic resonance absorption will occur for a given type of nucleus in a metal relative to a particular nonmetallic solid.

9c-6. Optical Properties of Electrons. *Optical Absorption.* Electromagnetic radiation of wavelengths in the ultraviolet, visible, or infrared region will be absorbed by a semiconductor or metal through the excitation of electrons and phonons.

As far as the electronic excitation is concerned, three mechanisms can be distinguished:

1. Electronic transition between different energy bands
2. Electronic transitions within an energy band ("free carrier absorption")
3. Electronic transitions between a localized state of an imperfection and an energy band

The absorption coefficient α is deduced from the measured transmission by means of the following expression:

$$T = \frac{I}{I_0} = \frac{(1 - R^2)e^{-\alpha d}}{1 - R^2 e^{-2\alpha d}}$$

where I_0 = incident light intensity
  I = transmitted light intensity
  d = thickness of sample
  R = reflectivity = $[(n-1)^2 + k^2]/[(n+1)^2 + k^2]$
  n = refractive index
  k = extinction coefficient = $\alpha \times$ wavelength$/4\pi$

Photoconductivity: An increase of electrical conductivity under illumination due to excitation of electrons or holes into conducting states.

The resulting current is given by

$$J = \frac{eIV\mu\tau}{L^2}$$

where I = absorbed light intensity
  V = applied voltage
  μ = carrier mobility
  τ = carrier lifetime
  L = length of sample

Photovoltaic Effect: The generation of a voltage (due to optical excitations) when a semiconductor is illuminated at the electrodes or at internal barriers or p-n junctions.

Carrier Lifetime: The length of time that an electron (hole) spends in conducting states before being captured by a hole (electron) or imperfection. The decay of excess carriers follows the law

$$\frac{dn}{dt} = \frac{n_0 - n}{\tau}$$

where n_0 is the equilibrium density of carriers and τ the *carrier lifetime*.

Exciton: A bound electron-hole pair in an insulator or a semiconductor. The exciton energy levels are states in the forbidden energy gap, below the conduction band. The exciton may move through the crystal, transporting energy but no electrical charge, because it is neutral.

Specific References

1. Herman, F.: *Proc. IRE* **43**, 1703 (1955).
2. Lax, B.: *Rev. Mod. Phys.* **30**, 122 (1958).
3. Brooks, H.: *Advan. Electron. Electron Phys.* **7**, 120 (1955).
4. Scanlon, W. W.: "Methods of Experimental Physics," vol. 6B, p. 166, L. Marton, ed., Academic Press, Inc., New York, 1959.

General References

I. Condon, E. U., and H. Odishaw: "Handbook of Physics," McGraw-Hill Book Company, New York, 1958.
II. Seitz, F.: "The Modern Theory of Solids," McGraw-Hill Book Company, New York, 1940.
III. Wilson, A. H.: "Theory of Metals," 2d ed., Cambridge University Press, London, 1954.
IV. Kittel, C.: "Introduction to Solid State Physics," 3d ed., John Wiley & Sons, Inc., New York, 1966.
V. Shockley, W.: "Electrons and Holes in Semiconductors," D. Van Nostrand Company, Inc., Princeton, N.J., 1950.
VI. Van Vleck, J. H.: "The Theory of Electric and Magnetic Susceptibilities," Oxford University Press, New York, 1932.
VII. Wannier, G. H.: "Elements of Solid State Theory," Cambridge University Press, London, 1959.
VIII. Ziman, J. M.: "Electrons and Phonons," Oxford University Press, New York, 1960.

BIBLIOGRAPHY OF ENERGY BAND CALCULATIONS[1]

This bibliography contains most of the principal papers dealing with detailed energy-band calculations. The references are arranged chronologically under each type of crystal.

9c-7. Alkali Metals

Wigner, E., and F. Seitz: *Phys. Rev.* **43**, 804 (1933); **46**, 509 (1934). Sodium, cellular.
Slater, J. C.: *Phys. Rev.* **45**, 794 (1934). Sodium, cellular.
Millman, J.: *Phys. Rev.* **47**, 286 (1935). Lithium, cellular.
Seitz, F.: *Phys. Rev.* **47**, 400 (1935). Lithium, cellular.
Gombas, P.: *Z. Physik* **113**, 150 (1939). Na, K, Rb, Cs, pseudopotential.

[1] Updated version of bibliography in J. C. Slater, "Quantum Theory of Molecules and Solids," vol. 2, pp. 300–305, McGraw-Hill Book Company, New York, 1965.

von der Lage, F., and H. Bethe: *Phys. Rev.* **71**, 612 (1947).   Sodium, cellular.
Sternheimer, R.: *Phys. Rev.* **78**, 235 (1950).   Cesium, cellular.
Silverman, R. A.: *Phys. Rev.* **85**, 227 1952).   Lithium, **k · p.**
Parmenter, R. H.: *Phys. Rev.* **86**, 552 (1952).   Lithium, OPW.
Howarth, D., and H. Jones: *Proc. Phys. Soc. (London)*, ser. A, **65**, 355 (1952).   Sodium, cellular.
Kohn, W., and J. Rostoker: *Phys. Rev.* **94**, 1111 (1954).   Lithium, Green's function.
Schiff, B.: *Proc. Phys. Soc. (London)*, ser. A, **67**, 2 (1954).   Lithium, cellular.
Callaway, J.: *Phys. Rev.* **103**, 1219 (1956).   Potassium, OPW and cellular.
Miasek, M.: *Bull. Acad. Polon. Sci.*, Cl. III, **4**, 453 (1956).   Sodium, variation.
Callaway, J., and E. L. Haase: *Phys. Rev.* **108**, 217 (1957).   Cesium, OPW and cellular.
Brown, E., and J. A. Krumhansl: *Phys. Rev.* **109**, 30 (1958).   Lithium, modified OPW.
Glasser, M. L., and J. Callaway: *Phys. Rev.* **109**, 1541 (1958).   Lithium, OPW.
Callaway, J.: *Phys. Rev.* **112**, 322 (1958).   Sodium, pseudopotential.
Callaway, J., and D. F. Morgan, Jr.: *Phys. Rev.* **112**, 334 (1958).   Rubidium, cellular.
Callaway, J.: *Phys. Rev.* **112**, 1061 (1958).   Cesium, cellular, terms in k^2.
Cohen, M. H., and V. Heine: *Advan. Phys.* **7**, 395 (1958).   Alkali metals.
Bassani, F.: *J. Phys. Chem. Solids* **8**, 375, 379 (1959).   Sodium in diamond lattice, OPW.
Bassani, F., and V. Celli: *Nuovo Cimento* **11**, 805 (1959).   Lithium in diamond lattice, OPW.
Callaway, J.: *Phys. Rev.* **119**, 1012 (1960).   Potassium, cellular, terms in k^2.
Cornwell, J. F., and E. P. Wohlfarth: *Nature* **186**, 379 (1960).   Lithium, pseudopotential.
Callaway, J.: *Phys. Rev.* **123**, 1255 (1961).   Sodium, cellular, terms in k^2.
Callaway, J.: *Phys. Rev.* **124**, 1824 (1961).   Lithium, OPW, and comparison of various results.
Cornwell, J.: *Proc. Roy. Soc. (London)*, ser. A, **261**:551 (1961).   Alkali metals and beryllium, pseudopotential.
Callaway, J., and W. Kohn: *Phys. Rev.* **127**, 1913 (1962).   Lithium, cellular, terms in k^2.
Ham, F. S.: *Phys. Rev.* **128**, 82 (1962).   Alkali metals, Green's function.
Ham, F. S.: *Phys. Rev.* **128**, 2524 (1962).   Alkali metals, Green's function.
Kenney, J. F.: *Quart. Progr. Rept.* 53, p. 38, Solid-state and Molecular Theory Group, MIT, July 15, 1964.   Lithium, sodium, APW.
De Leener, M., and A. Bellemans: *J. Chem. Phys.* **43**, 3075 (1965).   Alkali metals, free-electron approximation, cohesion.
Meyer, A., and W. H. Young: *Phys. Rev.* **139**, A401 (1965).   Lithium, pseudopotential.
Lafon, E. E., and C. C. Lin: *Phys. Rev.* **152**, 579 (1966).   Lithium, tight-binding.
Kenney, J. F.: *Quart. Progr. Rept.* 66, Solid-State and Molecular Theory Group, MIT, Oct. 15, 1967.   Rubidium, cesium, APW.

9c-8. Divalent and Trivalent Elements

Manning, M. F., and H. Krutter: *Phys. Rev.* **51**, 761 (1937).   Calcium, cellular.
Herring, C., and A. G. Hill: *Phys. Rev.* **58**, 132 (1940).   Beryllium, OPW.
Matyas, Z.: *Phil. Mag.* **30**, 429 (1948).   Aluminum, tight-binding.
Jones, H.: *Phil. Mag.* **41**, 663 (1950).   Magnesium, nearly free electrons.
Donovan, B.: *Phil. Mag.* **43**, 868 (1952).   Beryllium, cellular.
Trlifaj, M.: *Czech. J. Phys.* **1**, 110 (1952).   Magnesium, APW.
Antoncik, E.: *Czech. J. Phys.*, **2**, 18 (1953).   Aluminum, APW.
Heine, V.: *Proc. Roy. Soc. (London)*, ser. A, **240**, 340, 354, 361 (1957).   Aluminum, OPW.
Harrison, W. A.: *Phys. Rev.* **116**, 555 (1959); **118**, 1182 (1960).   Aluminum, nearly free electrons, pseudopotential.
Segall, B.: *Phys. Rev.* **124**, 1797 (1961).   Aluminum, Green's function.
Cornwell, J. F.: *Proc. Roy. Soc. (London)*, ser. A, **261**, 551 (1961).   Beryllium, pseudopotential.
Falicov, L. M.: *Phil. Trans. Roy. Soc. London*, ser. A, **255**, 55 (1962).   Magnesium, OPW.
Harrison, W. A.: *Phys. Rev.* **126**, 497 (1962); **129**, 2503, 2512 (1963).   Zinc, pseudopotential.
Segall, B.: *Phys. Rev.* **131**, 121 (1963).   Aluminum, Green's function.
Loucks, T. L., and P. H. Cutler: *Phys. Rev.* **133**, A819 (1964).   Beryllium, OPW.
Loucks, T. L., and P. H. Cutler: *Phys. Rev.* **134**, A1618 (1964).   Beryllium, OPW.
Terrall, J. H.: *Phys. Letters* **8**, 149 (1964).   Beryllium, APW.
Kimball, J. C., R. W. Stark, and F. M. Mueller: *Phys. Rev.* **162**, 600 (1967).   Magnesium, pseudopotential.
Vasvari, B., A. E. O. Animalu, and V. Heine: *Phys. Rev.* **154**, 535 (1967).   Calcium, strontium, barium under pressure, pseudopotential.
Snow, E. C.: *Phys. Rev.* **158**, 683 (1967).   Aluminum, APW.

9c-9. Diamond, Silicon, Germanium, 3-5 Compounds

Kimball, G. E.: *J. Chem. Phys.* **3**, 560 (1935). Diamond, cellular.
Hund, F., and B. Mrowka: *Sächsische Akad. Wiss. Leipzig* **87**, 185, 325 (1935). Diamond, cellular.
Mullancy, J. F.: *Phys. Rev.* **66**, 326 (1944). Silicon, cellular.
Morita, A.: *Sci. Rept. Tohoku Univ.*, **33**:92 (1949). Diamond, tight-binding.
Holmes, D. K.: *Phys. Rev.* **87**, 782 (1952). Silicon, cellular.
Herman, F.: *Phys. Rev.* **88**, 1210 (1952). Diamond, OPW.
Hall, G. G.: *Phil. Mag.* **43**, 338 (1952); *Phys. Rev.* **90**, 317 (1953). Diamond, equivalent orbital tight-binding.
Herman, F., and J. Callaway: *Phys. Rev.* **89**, 518 (1953). Germanium, OPW.
Yamaka, E., and T. Sugita: *Phys. Rev.* **90**, 992 (1953). Silicon, cellular.
Herman, F.: *Phys. Rev.* **93**, 1214 (1954). Diamond and germanium, OPW.
Herman, F.: *Physica* **20**, 801 (1954). Germanium, OPW.
Jenkins, D. P.: *Physica* **20**, 967 (1954). Silicon, cellular.
Bell, D. G., R. Hensman, D. P. Jenkins, and L. Pincherle: *Proc. Phys. Soc. (London)*, ser. A, **67**, 562 (1954). Silicon, cellular.
Herman, F.: *Proc. Inst. Radio Engrs.* **43**, 1703 (1955). Silicon and germanium, OPW.
Herman, F.: *J. Electron.* **1**, 103 (1955). General discus ion, OPW and experiment.
Woodruff, T. O.: *Phys. Rev.* **98**, 1741 (1955); **103**, 1159 (1956). Silicon, OPW.
Jenkins, D. P.: *Proc. Phys. Soc. (London)*, ser. A, **69**, 548 (1956). Silicon, cellular.
Kobayasi, S.: *J. Phys. Soc. Japan* **11**, 175 (1956); **13**, 261 (1958). Carborundum, tight-binding, OPW.
Bassani, F.: *Phys. Rev.* **108**, 263 (1957). Silicon, OPW, tight-binding, interpolation.
Callaway, J.: *J. Electronics* **2**, 330 (1957). GaAs, perturbation of Ge.
Kane, E. O.: *J. Phys. Chem. Solids* **1**, 82 (1957). Germanium, silicon, $\mathbf{k} \cdot \mathbf{p}$.
Kane, E. O.: *J. Phys. Chem. Solids* **1**, 249 (1957). InSb, $\mathbf{k} \cdot \mathbf{p}$.
Hall, G. G.: *Phil. Mag.* **3**, 429 (1958). Diamond, silicon, germanium, equivalent orbital tight-binding.
Morita, A.: *Progr. Theoret. Phys. (Kyoto)* **19**, 534 (1958). Diamond type, semilocalized combination of orbitals.
Kleinman, L., and J. C. Phillips: *Phys. Rev.* **116**, 880 (1959). Diamond, pseudopotential.
Phillips. J. C.: *J. Phys. Chem. Solids* **8**, 369, 379 (1959). Silicon, germanium, pseudopotential.
Segall, B.: *J. Phys. Chem. Solids* **8**, 371, 379 (1959). Germanium, Green's function.
Gubanov, A. I., and A. A. Nranyan: *Fiz. Tverd. Tela* **1**, 1044 (1959). 3-5 compounds, tight-binding and equivalent orbitals.
Bassani, F.: *Nuovo Cimento* **13**, 244 (1959). Silicon, tight-binding.
Kleinman, L., and J. C. Phillips: *Phys. Rev.* **117**, 460 (1960). BN, pseudopotential.
Kleinman, L., and J. C. Phillips: *Phys. Rev.* **118**, 1153 (1960). Silicon, pseudopotential.
Nranyan, A. A.: *Fiz. Tverd. Tela* **2**, 1650 (1960). 3-5 compounds, tight-binding equivalent orbitals.
Gashimzade, F. M., and V. E. Khartsiev: *Fiz. Tverd. Tela* **3**, 1453 (1961). Silicon, germanium, GaAs, OPW.
Phillips. J. C.: *Phys. Rev.* **125**, 1931 (1962). Silicon and germanium, general discussion.
Braunstein, R., and E. O. Kane: *J. Phys. Chem. Solids* **23**, 1423 (1962). 3-5 compounds.
Coulson, C. A., L. B. Redei, and D. Stocker: *Proc. Roy. Soc. (London)*, ser. A, **270**, 357 (1962). 3-5 compounds, OPW.
Redei, L. B.: *Proc. Roy. Soc. (London)*, ser. A, **270**, 373, 383 (1962). Diamond, OPW.
Stocker, D.: *Proc. Roy. Soc. (London)*, ser. A, **270**, 397 (1962). 3-5 compounds, OPW.
Bassani, F., and M. Yoshimine: *Phys. Rev.* **130**, 20 (1963). Group 4 elements and 3-5 compounds, OPW.
Braunstein, R.: *Phys. Rev.* **130**, 869 (1963). Germanium-silicon alloys.
Bassani, F., and L. Liu: *Phys. Rev.* **132**, 2047 (1963). Gray tin.
Cohan, N. V., D. Pugh, and R. H. Tredgold: *Proc Phys. Soc. (London)* **82**, 65 (1963). Diamond, tight-binding, equivalent orbitals.
Brust, D.: *Phy³. Rev.* **134**, A1337 (1964). Germanium and silicon, pseudopotential.
Herman, F.: "Proceedings International Conference on the Physics of Semiconductors," M. Hulin, ed., Dunod, Paris, 1964. Diamond-type crystals, OPW.
Kreher, K.: *Fortschr. Physik* **12**, 489 (1964). Gallium arsenide.
Cardona, M., F. H. Pollak, and J. G. Broerman: *Phys. Letters* **19**, 276 (1965). Gallium arsenide, spin-orbit splitting.
Chow, P. C., and L. Liu: *Phys. Rev.* **140**, A1817 (1965). 3-5 compounds, relativistic effect, perturbation.

Doggett, G.: *Proc. Phys. Soc. (London)* **86**, 393 (1965). Diamond, excited states, tight-binding, equivalent orbitals.
Harrison, W. A.: *Physica* **31**, 1692 (1965). Silicon, pseudopotential.
Nakayima, M.: *J. Phys. Soc. Japan* **20**, 56 (1965). Semiconductors; effect of homogeneous deformation on band structure.
Shindo, K., A. Morita, and H. Kamimura: *J. Phys. Soc. Japan* **20**, 2054 (1965). Crystals with zincblende and wurtzite structure, spin-orbit coupling.
Cardona, M., and F. H. Pollak: *Phys. Rev.* **142**, 530 (1966). Germanium and silicon, k · p method.
Cohen, M. L., and T. K. Bergstresser: *Phys. Rev.* **141**, 789 (1966). 14 semiconductors, pseudopotential.
Doggett, G.: *Phys. Chem. Solids* **27**, 99 (1966). Spin-orbit splitting, 4-4 and 3-5 compounds, tight-binding.
Herman, F., R. L. Kortum, C. D. Kuglin, and R. A. Short: "Proceedings International Conference on the Physics of Semiconductors," Kyoto, 1966. Diamond-type crystals, OPW.
Herman, F., R. L. Kortum, C. D. Kuglin, and R. A. Short: "Quantum Theory of Atoms, Molecules, and the Solid State," Academic Press, Inc., New York, 1966. Silicon, germanium, gray tin, OPW.
Kane, E. O.: *Phys. Rev.* **146**, 558 (1966). Silicon, pseudopotential.
Keown, R.: *Phys. Rev.* **150**, 568 (1966). Diamond, APW.
Pollak, F. H., and M. Cardona: *Phys. Chem. Solids* **27**, 423 (1966). Germanium and gallium arsenide, k · p method.
Saslow, W., T. K. Bergstresser, and M. L. Cohen: *Phys. Rev. Letters* **16**, 354 (1966). Diamond, pseudopotential.
Dresselhaus, G., and M. S. Dresselhaus: *Phys. Rev.* **160**, 649 (1967). Silicon and germanium, tight-binding.
Wiff, D. R., and R. Keown: *J. Chem. Phys.* **47**, 3113 (1967). Boron nitride, APW.

9c-10. Transition and Other Elements with f.c.c., b.c.c. or Hexagonal Structure

Krutter, H. M.: *Phys. Rev.* **48**, 664 (1935). Copper, cellular.
Tibbs, S. R.: *Proc. Cambridge Phil. Soc.* **34**, 89 (1938). Copper, silver, cellular.
Chodorow, M. I.: *Phys. Rev.* **55**, 675 (1939). Copper, APW.
Manning, M. F., and M. I. Chodorow: *Phys. Rev.* **56**, 787 (1939). Tungsten, cellular.
Manning, M. F.: *Phys. Rev.* **63**, 190 (1943). Iron, cellular.
Greene, J. B., and M. F. Manning: *Phys. Rev.* **63**, 203 (1943). Iron, fcc, cellular.
Fletcher, G. C., and E. P. Wohlfarth: *Phil. Mag.* **42**, 106 (1951). Nickel, tight-binding.
Fletcher, G. C.: *Proc. Phys. Soc. (London)*, ser. A, **65**, 192 (1952). Nickel, tight-binding.
Howarth, D. J.: *Proc. Roy. Soc. (London)*, ser. A, **220**, 513 (1953). Copper, cellular.
Koster, G. F.: *Phys. Rev.* **98**, 901 (1955). Nickel, tight-binding.
Howarth, D. J.: *Phys. Rev.* **99**, 469 (1955). Copper, APW.
Callaway, J.: *Phys. Rev.* **99**, 500 (1955). Iron, OPW.
Schiff, B.: *Proc. Phys. Soc. (London)*, ser. A, **68**, 686 (1955); **69**, 185 (1956). Titanium, cellular.
Fukuchi, M.: *Progr. Theoret. Phys.* **16**, 222 (1956). Copper, OPW.
Altmann, S. L., and N. V. Cohan: *Proc. Phys. Soc. (London)* **71**, 383 (1958). Titanium, cellular.
Altmann, S. L.: *Proc. Roy. Soc. (London)*, ser. A, **244**, 141, 153 (1958). Zirconium, cellular.
Stern, F.: *Phys. Rev.* **116**, 1399 (1959). Iron, tight-binding.
Belding, E. F.: *Phil. Mag.* **4**, 1145 (1959). Cr, Fe, Ni, tight-binding.
Wood, J. H.: *Phys. Rev.* **117**, 714 (1960). Iron, APW.
Segall, B.: *Phys. Rev. Letters* **7**, 154 (1961). Copper, Green's function.
Burdick, G. A.: *Phys. Rev. Letters* **7**, 156 (1961). Copper, APW.
Asdente, M., and J. Friedel: *Phys. Rev.* **124**, 384 (1961); **126**, 2262 (1962). Chromium, tight-binding, 4s omitted.
Knox, R. S., and F. Bassani: *Phys. Rev.* **124**, 652 (1961). Argon, tight-binding and OPW.
Cornwell, J. F.: *Phil. Mag.* **6**, 727 (1961). Noble metals, pseudopotential.
Segall, B.: *Phys. Rev.*, **125**:109 (1962). Copper, Green's function.
Wood, J. H.: *Phys. Rev.* **126**, 517 (1962). Iron, APW.
Asdente, M.: *Phys. Rev.*, **127**, 1949 (1962). Chromium, tight-binding.
Altmann, S. L., and C. J. Bradley: *Phys. Letters* **1**, 336 (1962). Zirconium, cellular.
Cornwell, J. F., and E. P. Wohlfarth: *J. Phys. Soc. Japan* **17** (suppl. B-1), 32 (1962). Iron.
Glasser, M. L.: *Rev. Mex. Fis.* **11**, 31 (1962). Silver.
Lomer, W. M.: *Proc. Phys. Soc. (London)* **80**, 489 (1962). Chromium, general discussion.
Burdick, G. A.: *Phys. Rev.* **129**, 138 (1963). Copper, APW.

Yamashita, J., M. Fukuchi, and S. Wakoh: *J. Phys. Soc. Japan* **18**, 999 (1963). Nickel, tight-binding and Green's function.

Mattheiss, L. F.: *Bull. Am. Phys. Soc.*, ser. II, **8**, 222 (1962). 3d transition elements to Cu, Zn, APW.

Fowler, W. B.: *Phys. Rev.* **132**, 1594 (1963). Krypton, tight-binding and OPW.

Mattheiss, L. F.: *Phys. Rev.* **133**, A1399 (1964). Argon, APW.

Mattheiss, L. F.: *Phys. Rev.* **134**, A970 (1964). Iron transition series, APW.

Mattheiss, L. F., and R. E. Watson: *Phys. Rev. Letters* **13**, 526 (1964). Tungsten, spin-orbit parameters.

Wakoh, S., and J. Yamashita: *J. Phys. Soc. Japan* **19**, 1342 (1964). Nickel, KKR and APW.

Abate, E., and M. Asdente: *Phys. Rev.* **140**, A1303 (1965). Iron, tight-binding.

Altmann, S. L., and C. J. Bradley: *Proc. Phys. Soc. (London)* **86**, 915 (1965). Hexagonal metals, cellular.

Dimmock, J. O., A. J. Freeman, and R. E. Watson: *J. Appl. Phys.* **36**, 1142 (1965). Gadolinium, APW.

Gandelman, G. M.: *Zh. Eksperim. i. Teor. Fiz.* **48**, 758 (1965). Argon, transition to metallic state under pressure, statistical method.

Harrison, W. A.: *Phys. Rev.* **139**, A179 (1965). Lead, pseudopotential.

Hodges, L., and H. Ehrenreich: *Phys. Letters* **16**, 203 (1965). Ferromagnetic nickel, pseudopotential.

Katsuki, S., and M. Tsuji: *J. Phys. Soc. Japan* **20**, 1136 (1965). Cadmium, pseudopotential.

Lomer, W. N.: "Proceedings International Conference on Magnetism," p. 127, London, 1965. Chromium, magnetic properties.

Loucks, T. L.: *Phys. Rev. Letters* **14**, 693 (1965). Tungsten, APW.

Loucks, T. L.: *Phys. Rev.* **139**, A1181 (1965). Chromium, molybdenum, and tungsten, APW.

Loucks, T. L.: *Phys. Rev.* **139**, A1333 (1965); 143, 506 (1966). Tungsten, APW.

Mattheiss, L. F.: *Phys. Rev.* **138**, A112 (1965). V$_3$X type compounds, APW.

Mattheiss, L. F.: *Phys. Rev.* **139**, A1893 (1965). Tungsten, APW.

Nagamiya, T., K. Motizuki, and K. Yamasaki: "Proceedings International Conference on Magnetism," p. 195, London, 1965. Chromium, spin-density waves.

Wakoh, S.: *J. Phys. Soc. Japan* **20**, 1894 (1965). Copper and nickel, APW and KKR.

Beeby, J. L.: *Phys. Rev.* **141**, 781 (1966). Transition metals, ferromagnetism.

Chatterjee, S., and S. K. Sen: *Proc. Phys. Soc. (London)* **87**, 779 (1966). Silver, OPW.

Freeman, A. J., J. O. Dimmock, and R. E. Watson: "Quantum Theory of Atoms, Molecules, and the Solid State," Academic Press, Inc., New York, 1966. Rare earths, APW.

Freeman, A. J., A. M. Furdyna, and J. O. Dimmock: *J. Appl. Phys.* **37**, 1256 (1966). Palladium, APW.

Hermanson, J., and J. C. Phillips: *Phys. Rev.* **150**, 652 (1966). Excitons, pseudopotential.

Hermanson, J.: *Phys Rev.* **150**, 660 (1966). Rare-gas solid, excitons, pseudopotential.

Hodges, L., H. Ehrenreich, and N. D. Lang: *Phys. Rev.* **152**, 505 (1966). Noble and transition metals, interpolation method.

Keeton, S. C., and T. L. Loucks: *Phys. Rev.* **146**, 429 (1966). Thorium, actinium, and lutecium, APW.

Loucks, T. L.: *Phys. Rev.* **144**, 504 (1966). Yttrium, APW.

Snow, E. C., J. T. Waber, and A. C. Switendick: *J. Appl. Phys.* **37**, 1342 (1966). Nickel, APW.

Spicer, W. E.: *J. Appl. Phys.* **37**, 947 (1966). Copper, nickel, silver, and iron, density of states from experiment.

Switendick, A. C.: *J. Appl. Phys.* **37**, 1022 (1966). Chromium, APW.

Williams, R. W., T. L. Loucks, and A. R. Mackintosh: *Phys. Rev. Letters* **16**, 168 (1966). Rare earth metals, APW and experiment.

Yamashita, J., S. Wakoh, and S. Asano: "Quantum Theory of Atoms, Molecules, and the Solid State," p. 497, Academic Press, Inc., New York, 1966. Iron, nickel, chromium, CoFe, KKR.

Asdente, M., and M. Delitala: *Phys. Rev.* **163**, 497 (1967). Iron, tight-binding.

Connolly, J. W. D.: *Phys. Rev.* **159**, 415 (1967). Nickel, APW.

Deegan, R. A., and W. D. Twose: *Phys. Rev.* **164**, 993 (1967). Niobium, OPW.

De Cicco, P. D., and A. Kitz: *Phys. Rev.* **162**, 486 (1967). Iron, APW.

Falicov, L. M., and M. J. Zuckermann: *Phys. Rev.* **160**, 372 (1967). Antiferromagnetic metals, pseudopotential.

Faulkner, J. S., H. L. Davis, and H. W. Joy: *Phys. Rev.* **161**, 656 (1967). Copper, KKR.

Heine, V.: *Phys. Rev.* **153**, 673 (1967). Transition metals, pseudopotential.

Loucks, T. L.: *Phys. Rev.* **159**, 544 (1967). Zirconium, APW.

Mueller, F. M.: *Phys. Rev.* **153**, 659 (1967). Noble metals, interpolation method.

Mueller, F. M., and J. C. Phillips: *Phys. Rev.* **157**, 600 (1967). Copper, interpolation method.
Snow, E. C., and J. T. Waber: *Phys. Rev.* **157**, 570 (1967). Copper, APW.
Sokoloff, J. B.: *Phys. Rev.* **161**, 540 (1967). Magnetic impurities in copper.
Stark, R. W., and L. M. Falicov: *Phys. Rev. Letters* **19**, 795 (1967). Zinc and cadmium, pseudopotential.
Tsui, D. C.: *Phys. Rev.* **164**, 669 (1967). Nickel, comparison with experiment.

9c-11. Graphite

Wallace, P. R.: *Phys. Rev.* **71**, 622 (1947); **72**, 258 (1947). Tight-binding.
Coulson, C. A.: *Nature* **159**, 265 (1947). Tight-binding.
Coulson, C. A., and R. Taylor: *Proc. Phys. Soc. (London)*, ser. A, **65**, 815 (1952). Tight-binding.
Carter, J. L., and J. A. Krumhansl: *J. Chem. Phys.* **21**, 2238 (1953). Tight-binding.
Ariyama, K., and S. Mase: *Progr. Theoret. Phys.* **12**, 244 (1954). Tight-binding.
Lomer, W. M.: *Proc. Roy. Soc. (London)*, ser. A, **227**, 330 (1955). Tight-binding.
Johnston, D. F.: *Proc. Roy. Soc. (London)*, ser. A., **227**, 349 (1955); **237**, 48 (1956). Tight-binding.
McClure, J. W.: *Phys. Rev.* **108**, 612 (1957). k · p.
Yamazaki, M.: *J. Chem. Phys.* **26**, 930 (1957). Tight-binding.
Corbato, F.: "Proceedings 1957 Carbon Conference," p. 173, Pergamon Press, New York. Tight-binding.
Slonczewski, J. C., and P. R. Weiss: *Phys. Rev.* **109**, 272 (1958). k · p.
Haering, R. R.: *Can. J. Phys.* **36**, 352 (1958). Tight-binding.
Mase, S.: *J. Phys. Soc. Japan* **13**, 563 (1958). Tight-binding.
Peacock, T. E., and R. McWeeny: *Proc. Phys. Soc. (London)* **74**, 385 (1959). Tight-binding.
Barriol, J.: *J. Chim. Phys.* **57**, 837 (1960); J. Barriol and J. Metzger, *J. Chim. Phys.* **57**, 848 (1960). Tight-binding.
Anno, T., and C. A. Coulson: *Proc. Roy. Soc. (London)*, ser. A, **264**, 165 (1961). Tight-binding (semiempirical).
Dresselhaus, G., and M. S. Dresselhaus: *Phys. Rev.* **140**, A401 (1965). Graphite, spin-orbit interaction, perturbation method.
Linderberg, J.: *Arkiv Fysik* **30**, 557 (1965). Graphite.

9c-12. Elements with Other Crystal Structures

Jones, H.: *Proc. Roy. Soc. (London)*, ser. A, **147**, 396 (1934). Bismuth, nearly free electrons.
Morita, A.: *Sci. Rept. Tohoku Univ.* **33**, 144 (1949). Bismuth, tight-binding.
Reitz, J. R.: *Phys. Rev.* **105**, 1233 (1957). Selenium, tellurium, tight-binding.
Gaspar, R.: *Acta Phys. Hung.* **7**, 289 (1957). Selenium and tellurium, tight-binding.
Ridley, E. C.: *Proc. Roy. Soc. (London)*, ser. A, **247**, 199 (1958). Uranium, cellular.
Mase, S.: *J. Phys. Soc. Japan* **13**, 434 (1958); **14**, 584 (1959). Bismuth, tight-binding.
de Carvalho, A. P.: *Compt. Rend.* **248**, 778 (1959). Tellurium, tight-binding.
Harrison, W. A.: *J. Phys. Chem. Solids* **17**, 171 (1960). Bismuth, pseudopotential.
Miasek, M.: *Bull. Acad. Polon. Sci., Ser. Sci. Math. Astron. Phys.* **8**, 9 (1960). White tin, OPW.
Bergson, G.: *Arkiv Kemi*, **16**, 315 (1960). Sulfur, tight-binding.
Behrens, E.: *Z. Physik*, **161**, 279 (1961). Bismuth, tight-binding.
Behrens, E.: *Z. Physik*, **163**, 140 (1961). Selenium, tight-binding.
Miasek, M.: *Phys. Rev.* **130**, 11 (1963). White tin, OPW.
Wood, J. H.: *Bull. Am. Phys. Soc.*, ser. II, **8**, 222 (1963). Gallium, APW.
Golin, S.: *Phys. Rev.* **140**, A993 (1965). Arsenic, OPW.
Beissner, R. E.: *Phys. Rev.* **145**, 479 (1966). Tellurium, pseudopotential.
Falicov, L. M., and P. J. Lin: *Phys. Rev.* **141**, 562 (1966). Antimony, pseudopotential.
Lin, P. J., and J. C. Phillips: *Phys. Rev.* **147**, 469 (1966). Antimony, pseudopotential.

9c-13. Compounds, Other Than 3-5

Jones, H.: *Proc. Roy. Soc. (London)*, ser. A, **144**, 225 (1934). Alloys, γ-phase, nearly free electrons.
Slater, J. C., and W. Shockley: *Phys. Rev.* **50**, 705 (1936). Sodium chloride, general discussion.
Shockley, W.: *Phys. Rev.* **50**, 754 (1936). Sodium chloride, cellular.
Ewing, D. H., and F. Seitz: *Phys. Rev.* **50**, 760 (1936). LiF and LiH, cellular.
Tibbs, S. R.: *Trans. Faraday Soc.* **35**, 1471 (1939). Sodium chloride, cellular.

Morita, A., and C. Horie: *Sci. Rept. Tohoku Univ.* **36**, 259 (1952). Barium oxide, tight-binding.

Bell, D. G., D. M. Hum, L. Pincherle, D. W. Sciama, and P. M. Woodward: *Proc. Roy. Soc. (London)*, ser. A, **217**, 71 (1953). PbS, cellular.

Casella, R. C.: *Phys. Rev.* **104**, 1260 (1956). Sodium chloride, tight-binding.

Yamazaki, M.: *J. Chem. Phys.* **27**, 746 (1957). Boron carbide, tight-binding.

Kucher, T. I.: *Zh. cksperim. i Tcor. Fiz.*, **34**:394 (1958); **35**:1049 (1958). NaCl, tight-binding.

Birman, J. L.: *Phys. Rev.* **109**, 810 (1958). ZnS, cellular.

Shakin, C., and J. Birman: *Phys. Rev.* **109**, 818 (1958). ZnS, cellular.

Howland, L. P.: *Phys. Rev.* **109**, 1927 (1958). Potassium chloride, tight-binding.

Tolpygo, K. B., and O. F. Tomasevich: *Ukr. Fiz. Zh.* **3**, 145 (1958). Sodium chloride, tight-binding.

Birman, J. L.: *Phys. Rev.* **115**, 1493 (1959). ZnS, tight-binding.

Birman, J. L.: *J. Phys. Chem. Solids* **8**, 35 (1959). ZnS, general discussion.

O'Sullivan, W.: *J. Chem. Phys.* **30**, 379 (1959). BeO, tight-binding.

Kudinov, E. K.: *Fiz. Tverd. Tela* **1**, 1851 (1959). Bi_2Te_3, tight-binding.

Flodmark, S.: *Arkiv Fiz.* **14**, 513 (1959); **18**, 49 (1960). Type BM_6, tight-binding.

Balkanski, M., and J. des Cloizeaux: *J. Phys. Radium* **21**, 825 (1960); *Abhandl. Deut. Akad. Wiss. Berlin, Kl. Math. Phys. Tech.* **76** (1960). CdS, spin-orbit interaction.

Kucher, T. I., and K. B. Tolpygo: *Fiz. Tverd. Tela* **2**, 2301 (1960). Sodium chloride, tight-binding.

Tolpygo, K. B., and O. F. Tomasevich: *Fiz. Tverd. Tela* **2**, 3110 (1960). Sodium chloride, tight-binding.

Kudinov, E. K.: *Fiz. Tverd. Tela* **3**, 317 (1961). Bi_2Te_3, tight-binding.

Zhilich, A. G., and V. P. Makarov: *Fiz. Tverd. Tela*, **3**:585 (1961). Cuprous oxide, Green's function.

Wood, V. E., and J. R. Reitz: *J. Phys. Chem. Solids* **23**, 229 (1962). Cesium gold, cellular.

Gashimzade, F. M., and V. E. Khartsiev: *Fiz. Tverd. Tela* **4**, 434 (1962). SnS-type compounds, OPW.

Evseev, Z. Ya., and K. B. Tolpygo: *Fiz. Tverd. Tela* **4**, 3644 (1962). Sodium chloride, tight-binding.

Johnson, L. E., J. B. Conklin, and G. W. Pratt, Jr.: *Phys. Rev. Letters* **11**, 538 (1963). PbTe, relativistic APW.

Evseev, Z. Ya.: *Fiz. Tverd. Tela* **5**, 2345 (1963). Sodium chloride, tight-binding.

Lee, P. M., and L. Pincherle: *Proc. Phys. Soc. (London)* **81**, 461 (1963). Bismuth telluride, APW.

Yamashita, J.: *J. Phys. Soc. Japan* **18**, 1010 (1963). TiO and NiO. Tight-binding.

Mackintosh, A. R.: *J. Chem. Phys.*, **38**, 1991 (1963). Tungsten bronzes.

Beleznay, F., and G. Biczo: *J. Chem. Phys.* **41**, 2351 (1964). DNA, Hückel approximation.

Harman, T. C., et al.: *Solid State Commun.* **2**, 305 (1964). HgTe and HgTe-CdTe alloys.

Kahn, A. H., and A. J. Leyendecker: *Phys. Rev.* **135**, A1321 (1964). Strontium titanate, tight binding.

Kahn, A. H., H. P. R. Frederikse, and J. H. Becker: From "Transition Metal Compounds," p. 53, Gordon and Breach, Science Publishers, Inc., New York, 1964. $SrTiO_3$ and TiO_2.

Ladik, J., and K. Appel: *J. Chem. Phys.* **40**, 2470 (1964). Polynucleotides, Hückel approximation.

Pratt, G. W., Jr., and L. G. Ferreira: "Proceedings Internal Conference on Physics of Semiconductors," M. Hulin, ed., Dunod, Paris, 1964. PbTe, $k \cdot p$ method.

Sandrock, R., and J. Treusch: *Z. Naturforsch.* **19a**, 844 (1964). Chalcopyrite structure, $k \cdot p$.

Conklin, J. B., Jr., L. E. Johnson, and G. W. Pratt, Jr.: *Phys. Rev.* **137**, A1282 (1965). PbTe, relativistic APW.

Ern, V., and A. C. Switendick: *Phys. Rev.* **137**, A1927 (1965). TiC, TiN, TiO, APW.

Frei, V., and B. Velicky: *Czech. J. Phys.* **B15**, 43 (1965). CdSb, symmetry and pseudopotential.

Gorzkowski, W.: *Phys. Stat. Solidi* **11**, K131 (1965). HgTe, $k \cdot p$.

Hassan, S. S. A. Z.: *Proc. Phys. Soc. (London)* **85**, 783 (1965). Sodium-chloride type, plane-wave approximation.

Johnson, K. H., and H. Amar: *Phys. Rev.* **139**, A760 (1965). Ordered beta brass, KKR.

Ladik, J., and G. Biczo: *J. Chem. Phys.* **42**, 1658 (1965). DNA, Hückel approximation.

Miyakawa, T., and S. Oyama: *Mem. Defense Acad. Japan* **5**, 161 (1965). NaCl-type crystal, plane wave method.

Scop, P. M.: *Phys. Rev.* **139**, A934 (1965). AgCl. AgBr, APW.

Amar, H., K. H. Johnson, and K. P. Wang: *Phys. Rev.* **148**, 672 (1966). Beta-phase alloys, KKR.

Dahl, J. P., and A. C. Switendick: *Phys. Chem. Solids* **27**, 931 (1966). Cuprous oxide, APW.

Onodera, Y., M. Okazaki, and T. Inui: *J. Phys Soc. Japan* **21**, 1816 (1966). Potassium iodide, relativistic Green's function.

Oyama, S., and T. Miyakawa: *J. Phys. Soc. Japan* **21**, 868 (1966). KCl, plane wave method.

Yamashita, J., S. Wakoh, and S. Asano: *J. Phys. Soc. Japan* **21**, 53 (1966) CoFe superlattice, KKR.

Amar, H., K. H. Johnson, and C. B. Sommers: *Phys. Rev.* **153**, 655 (1967). Beta-brass, KKR.

Arlinghaus, F. J.: *Phys. Rev.* **157**, 491 (1967). Beta-brass, APW.

Bergstresser, T. K., and M. L. Cohen: *Phys. Rev.* **164**, 1069 (1967) CdSe, CdS, and ZnS pseudopotential.

Cho, S. J.: *Phys. Rev.* **157**, 632 (1967). EuS, APW.

De Cicco, P. D.: *Phys. Rev.* **153**, 931 (1967). KCl, APW.

Eckelt, P., O. Madelung, and J. Treusch: *Phys. Rev. Letters* **18**, 656 (1967). ZnS, KKR.

Euwema, R. N., T. C. Collins, D. G. Shankland, and J. S. De Witt: *Phys. Rev.* **162**, 710 (1967). CdS, OPW.

Gray, D., and E. Brown: *Phys. Rev.* **16** , 567 (1967). Cu₃Au, OPW.

Kunz, A. B.: *Phys. Rev.* **159**, 738 (1967 . Alkali halides, spin-orb t effects.

9d. Properties of Metals

JULIUS BABISKIN

U.S. Naval Research Laboratory

J. ROBERT ANDERSON

University of Maryland

9d-1. Electrical Resistivity and Hall Coefficient. The temperature-dependent ideal resistivity values ρ_i for very pure metals are listed in Table 9d-1 at 0 and 22°C where these ρ_i values are closely equal to the measured resistivity values ρ of pure metals. ρ_i was obtained either by subtracting ρ_0, the residual resistivity at very low temperatures due to impurities and imperfections, from ρ or by choosing the lowest reported values of ρ for high-purity metals. The ratio of the resistivity at 100,000 kg/cm² (ρ_p) to that at zero pressure (ρ) at 20°C and the Hall coefficient (R) at 20°C are also listed in Table 9d-1.

9d-2. Ideal Electrical Resistivity at Low Temperatures. *Matthiessen's rule* states that the *measured resistivity* ρ at a given temperature T is composed of the temperature-dependent *ideal resistivity* ρ_i due to electron scattering by lattice vibrations and the temperature-independent *residual resistivity* ρ_0 caused by impurities and imperfections; that is, $\rho = \rho_i + \rho_0$. At higher temperatures, $\rho_i \propto T$ for $T \gtrsim 0.25\theta$, where θ is the Debye characteristic temperature. At very low temperatures, $\rho_i \propto T^n$ where $n = 5$ for a free-electron metal. For many transition metals, $n \cong 2$ to 3 at low T owing to electron-electron interactions. Tables 9d-2a and 9d-2b list values for ρ_i at various temperatures below 273 K, while Table 9d-1 lists values for ρ_i at

TABLE 9d-1. SOME ELECTRICAL PROPERTIES OF PURE METALS

Metal	ρ_i,* microhm-cm 0°C	ρ_i,* microhm-cm 22°C	ρ_p/ρ,† at 100,000 kg/cm²	R,‡ cm³/coul × 10⁴
Aluminum	2.50, 2.44[a]	2.74	0.770	−0.30
Antimony	37.6	41.3	0.605	
Arsenic	26	29	0.928	
Barium	36	39	2.618	
Beryllium	2.71	3.25	0.876	+2.44
Bismuth	105	116	0.474	
Cadmium	6.73	7.27	0.658	+0.60
Calcium	3.08[b]	3.35[b]	4.399	
Cerium	79, 76.7[c]	81	....	+0.181
Cesium	18.0	19.96	5.33	−7.8
Chromium	12.1[d]	12.9[d]	0.558	
Cobalt	5.15[d]	5.80[d]	0.951	−1.33
Copper	1.55[d], 1.545[a]	1.70[d]	0.866	−0.55
Dysprosium	87.5, 56[c]	90	....	−1.3
Erbium	77	81	....	−0.34
Europium	86	89		
Gadolinium	127.5	134.1	....	−0.95
Gallium	13.65	14.85		
Gold	2.01[d]	2.20[d]	0.816	−0.72
Hafnium	28.0[d]	30.6[d]		
Holmium	74.5	77.7		
Indium	8.0	8.75	0.493	−0.07
Iridium	4.65[d]	5.07[d]	0.886	
Iron	8.7[d]	9.8[d]	0.841	+0.245
Lanthanum	75, 62.4[c]	79	....	−0.8
Lead	19.3, 19.2[a]	21.0	0.487	+0.09
Lithium	8.494	9.32	1.704	−1.7
Lutecium	49	53		
Magnesium	3.94	4.30	0.767	−0.94
Manganese	136[d], 91[e]	136[d]	....	−0.93
Mercury (liq.)	94.1	95.9	0.555	
Molybdenum	4.84[d]	5.33[d]	0.892	+1.26
Neodymium	56.5	59	....	+0.97
Neptunium	116	118.5		
Nickel	6.20[d]	7.04[d]	0.858	−0.611
Niobium	13.5[d]	14.5[d]	0.894	
Osmium	8.35[d]	9.13[d]		
Palladium	9.70[d]	10.55[d]	0.847	−0.68
Platinum	9.59[d]	10.42[d]	.861	−0.24
Plutonium	144	143		
Polonium	42	46		
Potassium	6.447, 6.1[e]	7.19	0.596	−4.2
Praseodymium	64	67	....	−0.71
Rhenium	16.9[d]	18.6[d]		
Rhodium	4.36[d]	4.78[d]	0.872	370
Rubidium	11.25	12.51	2.95	
Ruthenium	6.69[d]	7.37[d]		
Samarium	95, 88[c]	99		
Scandium	42.9	46.8		
Silver	1.47[d]	1.61[d]	0.802	−0.84
Sodium	4.289	4.75	0.479	−2.5
Strontium	19.8	21.5	1.810	
Tantalum	12.1[d]	13.1[d]	0.882	+1.01
Terbium	109	111		
Thallium	15	16.4	0.265	+0.24
Thorium	14.0, 13[e]	15.	0.821	
Thulium	58	62		

TABLE 9d-1. SOME ELECTRICAL PROPERTIES OF PURE METALS (*Continued*)

Metal	ρ_i,* microhm-cm 0°C	ρ_i,* microhm-cm 22°C	ρ_p/ρ,† at 100,000 kg/cm²	R,‡ cm³/coul × 10⁴
Tin.............	10.1	11.0	0.548	−0.04
Titanium.........	39.0^d	43.1^d	0.916	
Tungsten.........	4.82^d	5.33^d	0.895	+1.18
Uranium.........	24.1	25.7	0.724	
Vanadium........	18.3^d	19.9^d	0.878	
Ytterbium........	25.5	26.4		
Yttrium..........	53.7	58.5		
Zinc.............	5.45	5.92	0.679	+0.33
Zirconium.......	38.6^d	42.4^d	0.9836	

* Unless otherwise indicated, most of the ρ_i values were taken from G. T. Meaden, "Electrical Resistance of Metals," Plenum Press, Plenum Publishing Corporation, New York, 1965.

† ρ_p/ρ taken from P. W. Bridgeman, *Proc. Am. Acad. Arts Sci.* **81**, 165 (1952).

‡ R taken from J. Bardeen, "Handbook of Physics," pp. 4–74, E. U. Condon and H. Odishaw, eds., McGraw-Hill Book Company, New York, 1958.

^{a} L. A. Hall, "Survey of Electrical Resistivity Measurements on 16 Pure Metals in the Temperature Range 0 to 273°K," *NBS Tech. Note* 365, February, 1968.

^{b} F. X. Kayser and S. D. Soderquist, *J. Phys. Chem. Solids*, **28**, 2343 (1967).

^{c} J. A. Gibson et al.: "The Properties of Rare Earth Metals and Compounds," Battelle Memorial Institute, Columbus, Ohio, 1959.

^{d} G. K. White and S. B. Woods, *Phil. Trans. Roy. Soc. London*, ser. A, **251**, 273 (1959).

^{e} R. B. Stewart and V. J. Johnson, eds., A Compendium of the Properties of Materials at Low Temperatures (Phase II), *WADD Tech. Rept.* 60–56, part IV, chap. 6, Wright-Patterson Air Force Base, Ohio: Aeronautical Systems Division, Air Force Systems Command, December, 1961.

273 K (0°C) and at 295 K (22°C). Table 9d-2a lists ρ_i, θ, and n for the noble metals (Group IB) and the transition metals (Groups IVA, VA, VIA, VIIA, and VIIIA). Table 9d-2b lists ρ_i and θ for the remaining groups of metals other than the noble and the transition metals.

9d-3. Electronic Structure of Metals. The metals listed in Table 9d-3 are divided into three groups, simple metals, transition metals, and semimetals. The reference list is not complete, but the numbers next to the element names refer to recent papers which contain fairly complete references. A recent review article [1] gives rather complete references to de Haas–van Alphen effect studies up to 1968.

In the first column under the name of the metal are given the lattice constants in angstroms and the crystal structure. Values of the lattice constants are given at low temperatures, approximately 4.2 K, where these are available. In some cases these have been estimated from low-temperature thermal expansion data. Where low-temperature data are not available, room-temperature (R.T.) values are listed. One useful reference is Pearson's compilation [2].

In the next four columns the Fermi surface description is given. For most metals the identifications are based upon band structure calculations, and in some cases the descriptions are extremely tentative. The letters in the description refer to symmetry points in the Brillouin zone following the standard convention as given, for example, by Koster [3]. The names of the parts of Fermi surface are taken from the appropriate references. e and h refer to electrons and holes, respectively. In the majority of cases the type of carrier has been determined from band structure calculations rather than from actual experiments. The magnetic field direction is given in column 6 and refers to the normal to an extremal Fermi surface cross section. The frequencies given in column 7 were obtained from de Haas–van Alphen effect measurements. When

TABLE 9d-2a. IDEAL ELECTRICAL RESISTIVITIES IN MICROHM-CM OF PURE METALS AT LOW TEMPERATURES*

T, K	Group IB			Group IVA			Group VA			Group VIA			Group VIIA		Group VIIIA								
	Cu	Ag	Au	Ti	Zr	Hf	V	Nb	Ta	Cr	Mo	W	Mn	Re	Fe	Ru	Os	Co	Rh	Ir	Ni	Pd	Pt
250	1.40	1.34	1.83	34.8	34.6	25.3	16.6_5	12.3	11.0	10.9_5	4.3_2	4.3_2	133	15.2	7.5_5	5.9_6	7.50	4.50	3.9_0	4.1_9	5.4_0	8.8_5	8.70
220	1.20	1.16	1.60	29.3	29.4	21.7	14.5	10.8	9.6	9.0_5	3.6_4	3.6_6	131	12.9_5	6.2	5.0_2	6.4_5	3.72	3.3_1	3.5_0	4.3_5	7.6_5	7.54
200	1.06	1.04	1.44	25.7	26.1	19.3	12.9	9.8	8.6	7.7_5	3.1_8	3.2_2	131	11.4_5	5.3	4.3_8	5.7_0	3.2_3	2.9_2	3.2_0	3.7_2	6.9_0	6.76
180	0.92	0.92	1.28	22.1	22.6	16.9	11.2	8.7	7.6_5	6.4	2.7_4	2.7_8	130	9.9_5	4.4_0	3.7_5	5.0_0	2.7_5	2.5_2	2.8_0	3.1_0	6.0_6	5.97
160	0.77_5	0.79_5	1.12	18.5	19.3	14.5	9.5	7.5_5	6.6_5	5.2	2.2_7	2.3_3	127	8.4	3.5_5	3.1_0	4.2	2.26	2.1_2	2.3_8	2.5_5	5.1_9	5.18
140	0.63_5	0.67_5	0.95_5	14.8	16.0	12.2	7.7_5	6.4	5.6	3.9	1.8_2	1.8_8	125	6.9	2.7_3	2.4_8	3.50	1.7_8	1.7_1	1.9_6	1.9_7	4.3	4.37_5
120	0.49_0	0.54_0	0.79_0	11.2	12.8	9.9	6.0	5.2	4.6	2.6_5	1.36	1.44	123	5.3_5	1.9_5	1.8_5	2.70	1.3_2	1.2_8	1.5_5	1.4_6	3.4_6	3.56_5
100	0.35_0	0.42_0	0.63_0	7.9	9.5_5	7.6	4.3	3.9_5	3.5_5	1.6_2	0.92	1.02	121	3.9_5	1.24	1.2_5	1.90	0.91	0.8_9	1.1_0	1.0_6	2.6_0	2.74_2
90	0.28_6	0.35_5	0.54_5	6.3_5	7.9_0	6.5	3.5_5†	3.30	3.0_5	1.18	0.714	0.82_0	120	3.2	0.92	0.91	1.50	0.72	0.69_5	0.90	0.75	2.1_7	2.32_6
80	0.21_5	0.29_0	0.46_0	4.8_5	6.4	5.4	2.6_5	2.6_8	2.5_0	0.8_1	0.51_5	0.60_0	121	2.5_5	0.64	0.64	1.10	0.54	0.51	0.72	0.55	1.7_2	1.90_9
70	0.15_3	0.23_0	0.38	3.5	4.9_0	4.3	1.9_0	2.0_7	1.9_6	0.5_2	0.354	0.42_5	122	1.8_6	0.42	0.43	0.79	0.38	0.34	0.53	0.38	1.3_0	1.49_7
60	0.095	0.17	0.29	2.3†	3.5_0	3.2	1.2_7	1.5	1.4_3	0.3_0	0.21_6†	0.27_1	122	1.2_7	0.25	0.24	0.50	0.25	0.20_4	0.35	0.24_5	0.92	1.09_4
50	0.050	0.11	0.20	1.4	2.2_5	2.1†	0.7_5	0.9_7	0.9_5	0.16_5	0.11_3	0.15_1	117	0.7_7	0.13_5†	0.10_5	0.26	0.14_5	0.10_5	0.20	0.15	0.58	0.719
40	0.022	0.058	0.12	0.6_5	1.2_0	1.2_5	0.3_8	0.56†	0.5_4	0.07_8†	0.04_7	0.06_7†	105	0.3_7	0.06_0	0.03_6†	0.11†	0.07_5†	0.043	0.10	0.07_3	0.32	0.396
30	0.0063†	0.020	0.050	0.2_5	0.4†	0.5_1	0.1_4	0.25	0.2†	0.02_9	0.01_2	0.02_2	82	0.11	0.02_2	0.01_6	0.02_8	0.02_7	0.011_5†	0.03_2†	0.03_0†	0.13	0.160
25	0.0025	0.010_7†	0.027	0.07_5	0.23_5	0.2_6	0.07_6	0.1_5	0.1_2	0.015_5	0.004_6	0.011_5	65	0.04_7†	0.012_5	0.00_5	0.01_2	0.01_4	0.004_9	0.014_5	0.017	0.074	0.0837
20	0.0008	0.0038	0.0125†	0.02_0	0.09_0	0.10_5	0.03_7	0.08	0.05_1	0.007_2	0.004	0.005_6	46	0.016_5	0.00_7			0.006_6	0.001_8	0.005_0	0.009	0.036†	0.0359
15	0.0001_7	0.0011	0.0037		0.02_5	0.02_7	0.01_4	0.03_5	0.01_7	0.002_7		0.002_2	28	0.004_5				0.002_7		0.001_3	0.004_5	0.014_5	0.0116†
10	......	0.0002	0.0006			0.005			0.003_2				12†					0.001_1				0.004	0.0029
θ, K	310	229	185	360	250	210	390	250	230	480	380	315	410	280	400	500	400	380	350	290	390	295	225
n	5.1	4.7	5.1	5.3	4.5	4.7	3.4	2.7	3.8	3.2	5.1	4.0	2.0	4.6	3.3	4.7	4.7	3.3	4.6	4.7	3.1	3.2	3.7

* Data taken from G. K. White and S. B. Woods: *Phil. Trans. Roy. Soc. London*, ser. A, **251**, 273 (1959).
† Values for ρ_i at which $\rho_i \lesssim \rho_0$ (or $\rho \lesssim 2\rho_0$).

TABLE 9d-2b. IDEAL ELECTRICAL RESISTIVITIES IN MICROHM-CM OF PURE METALS AT LOW TEMPERATURES*

Metal	20 K	50 K	80 K	100 K	150 K	200 K	250 K	θ, K†
Group IA:								
Li	0.015	0.27	0.995	1.714	3.708	5.704	7.613	370
Na	0.0165	0.317	0.805	1.145	1.994	2.874	3.821	158
K	0.1074	0.719	1.389	1.836	3.005	4.281	5.720	90
Rb	0.433	1.57_3	2.70_0	3.46_1	5.46_6	7.64_8	10.01	52
Cs	0.882	2.65_5	4.42_4	5.63_7	8.78_0	12.22	16.06	54‡
Group IIA:								
Be	0.0004	0.0077	0.038_9	0.090_7	0.436	1.15_1	2.15_6	1160
Mg	0.008_5	0.1_2	0.55	0.89	......	2.6	......	400
Ca	0.03	0.25	0.57	0.87	1.52	2.14	2.74	230
Sr	0.48	2.5	4.6	6.3	......	14.1	......	147
Ba	0.73	3.5	7.8	10.7	......	25	......	110
Group IIIA:								
Sc	0.16	2.9	8	$11._2$	$20._7$	$29._8$	$38._8$	
Y	0.3_6	4.8	$11._2$	$15._4$	$26._6$	$37._6$	$48._6$	214‡
Group IIB:								
Zn	0.052	0.49	1.16	1.6_2	2.7_2	3.8_4	4.9_3	310
Cd	0.13	0.87	1.7	2.3	3.6	4.9	6.2	188
Hg	1.24	3.9_5	6.6_3	8.6	13.3	18.4	92.2	80
Group IIIB:								
Al	0.0006	0.05	0.25	0.47	1.0_6	1.6_5	2.24	428
Ga	0.09	......	2.7_3	3.9_5	6.8	9.5_5	12.3	320
In	0.16	0.9_4	1.8_0	2.3_8	3.8_4	5.4_3	7.1_5	108
Tl	0.42	2.0	3.6	4.7	7.5	10.3_5	13.5	87
Group IVB:								
Sn	0.10	0.9_5	2.1_5	2.9_5	4.9_5	7.0	9.1	178
Pb	0.56	2.76	4.97	6.5	10.2	$13._9$	$17._6$	110
Group VB:								
As	0.2_9	$1._9$	$4._5$	$6._4$	......	......	......	285‡
Sb	0.4_2	3.2	7.2	10.0	17.9	25.9	34.0	207
Bi	5.8	19	30	37	55	74	96	119
Rare-earth metals:								
La	3.3	17	29	36	$49._5$	61	71	142
Pr	......	$8._5$	23	36	46	54	61	74‡
Nd	$8._3$	17	25	$29._5$	$38._5$	46	53	
Sm	14	33	52	64	73	82	91	
Eu	$8._5$	33	61	78	75	78	83	
Gd	1.0	$12._5$	$29._7$	$41._2$	$69._0$	$95._6$	$119._0$	152‡
Tb	0.9_5	$12._5$	27	38	64	93	108	158‡
Dy	1.1	$11._8$	$26._6$	$40._3$	$72._5$	81	85	140
Ho	3.4	15	31	43	56	64	71	
Er	$4._5$	24	39	42	52	63	73	
Tm	$2._1$	21	$25._5$	29	38	46	55	
Yb	$1._3$	$6._3$	$10._8$	$13._8$	$17._9$	$21._5$	$24._4$	
Lu	0.7_3	$6._0$	$11._9$	16	26	36	45	166‡
Actinide metals:								
Th	0.19	1.67	3.34	4.4	7.2	9.9_4	12.7	170
U	0.52	4.54	7.4	9.4	14.0	18.3	22.3	200
Np	1.91	24.2	49.8	63.1	87.3	102.5	112.7	
Pu	20	116	153	156	153	148	145	

* Except for calcium, the ρ_i values were taken from G. T. Meaden, "Electrical Resistance of Metals," Plenum Press, Plenum Publishing Corporation, New York, 1965. The ρ_i values for calcium were taken from F. X. Kayser and S. D. Soderquist, *J. Phys. Chem. Solids,* **28**, 2343 (1967).

† Unless otherwise indicated, most of the θ values were taken from G. T. Furukawa and T. B. Douglass, "American Institute of Physics Handbook," 2d ed., pp. 4–61, D. E. Gray, ed., McGraw-Hill Book Company, New York, 1963.

‡ These θ values were taken from F. J. Blatt: "Physics of Electronic Conduction in Solids," pp. 48–49, McGraw-Hill Book Company, New York, 1968.

extremal areas A were given in angstroms or atomic units, the conversion to frequency F was made, using the following relations:

$$F \text{ (gauss)} = A \text{ (a.u.}^{-2}) \times 3.741 \times 10^8$$
$$F \text{ (gauss)} = A \text{ (angstroms}^{-2}) \times 1.04728 \times 10^8$$

An $\sim$ sign is used for values estimated from graphs. Error estimates are not given here, but can be obtained from the references. If no reference is indicated for a specific measurement, the first reference given for that element is implied.

In columns 8 and 9 are given cyclotron mass values obtained from de Haas–van Alphen effect and cyclotron resonance measurements. No attempt has been made to give a complete listing of the values obtained from cyclotron resonance even though more accurate measurements are usually obtained by this technique.

In the final column are listed other experiments that have been performed on these metals, using the following abbreviations:

ASE	Anomalous skin effect	MA	Magnetoacoustic
CR	Cyclotron resonance	MT	Magnetothermal
GM	Galvanomagnetic	PA	Positive annihilation
H	Helicons	SE	Size effect
KE	Kohn effect		

Descriptions of these experiments can be found by referring to the references given in this table.

TABLE 9d-3. ELECTRONIC STRUCTURE OF METALS

Metal	Band	Description	Carrier	Orbit description	Magnetic field direction	F (in 10^6 Gauss)	Mass values, m^*/m de Haas–van Alphen	Cyclotron resonance	Other experiments
I. SIMPLE METALS									
1. Aluminum [4] f.c.c. $a = 4.0236$ [5]	1	Full							
	2	Large closed surface centered at Γ	h	....	[110]	436.6 [6, 7]	......	......	GM, MA, SE, ASE, KE, H
					[111]	411 [7]	1.3 [6, 7]	1.3 [8]	
					[100]	680 [7]			
	3	Multiply connected surface of [110] arms	e	Central sections through K	[110]	2.86	0.130		
					[111]	3.44	0.150	0.161 [8]	
					[100]	3.89	0.180	0.183 [8]	
				Minimum arm cross sections	[110]	~0.26	~0.09		
					[100]	0.28	0.091		
				Arm joints	[111]	0.36	0.102		
					[100]	0.466	0.118		
					[110]	~0.51			
2. Beryllium [9] h.c.p. $a = 3.5814$ $c = 2.2828$	1 and 2	6-cornered coronet	h	Neck	[11$\bar2$0]	0.109	0.0196		GM, PA, MT
				B_1 belly	[11$\bar2$0]	12.4	0.25		
				B_2 belly	[10$\bar1$0]	14.82	0.34		
				Inner circle	[0001]	381 [10]			
	3 and 4	Cigars	e	Waist central	[0001]	9.42	0.164		
				Waist noncentral	[0001]	9.72	0.174		
				Long section	[10$\bar1$0]	53.5			
					[11$\bar2$0]	53.1			
3. Cadmium [11] h.c.p. $a = 2.9684$ [12] $c = 5.5261$ [12]	1 and 2	Pinched-off monster	h	....	[0001]	5.98 [14]			GM, MA, CR, SE
					[11$\bar2$0]	12.7 [15]			
					[10$\bar1$0]	11.4 [15]			
		Band 1 caps at H							
		Band 2 undulating cylinder along K–H		AHL plane	[0001]	~6.4 [14]			
				Belly	[0001]	~61 [14]			
	3 and 4	Lens-shaped centered at Γ	e		[0001]	196 [15]		1.23 [16]	
					[11$\bar2$0]	64 [15]		~0.59 [16]	
					[10$\bar1$0]	63 [15]		~0.59 [16]	
4. Calcium [17] f.c.c. $a = 5.57$ (R.T.)		Data from polycrystalline samples only		"Hyperboloidal surface"		3.3	0.35		
						13	0.62		
						17.6	0.65		

Metal	No.	Surface description		Region	[hkl]		1.25–1.40	[ref]	Methods
5. Cesium [18], [106] b.c.c. a = 6.045	1	Free-electron-like spherical surface	e	....	[110]	136.4	1.25–1.40	0.0973– 0.896 [20]	GM, MA, SE
					[100]	~139			
					[111]	~140			
					[100]	0.135			
						0.495			
						0.855			
						23.5			
						56.7			
6. Gallium [19]	....	De Haas–Van Alphen frequencies and masses have not been completely correlated with band structure	...	....	[010]	0.345		0.0513– 0.728 [20]	
						0.725			
						19.2			
						22.5			
						30.0			
						63.5			
					[001]	0.20		0.063– 0.772 [20]	
						0.22			
						0.765			
						8.3			
						13			
						20.5			
7. Indium [21] f.c.t. a = 4.5557 c = 4.9342	1	Full	h						GM, MA, SE
	2	Large closed surface centered at Γ		Central	[110]	295 [23]	~1.2 [22]	1.17 [23]	
				Noncentral	[110]	339 [23]	~1.3 [22]	1.34 [23]	
				Central	[011]	332 [23]	....	2.07 [23]	
				Central	[001]	476 [23]	~1.5 [22]	1.54 [23]	
					[111]	317 [23]	0.204	0.202 [23]	
	3	Rings of [110] arms	e	Arm cross section	[110]	4.59	0.36	0.36 [23]	
				Centered at K	[011]	8.25		0.27 [23]	
					[100]	6.05			
				Arm junction	[011]	0.092	0.20	0.18 [23]	
					[110]	0.148	0.18		
						0.140			
8. Lead [24] f.c.c. a = 4.90	1 2	Full Large closed surface centered at Γ	h	....	[110]	159 [26]	1.09 [25]	1.12 [27]	GM, MA, SE, ASE, KE
					[111]	156 [26]	1.11 [25]	1.15 [27]	
					[100]	204 [28]	1.47 [25]	1.58 [27]	
	3	Multiply connected surface of [110] arms	e	Arm cross sections centered at K	[110]	18.1 [26]	0.51 [25]	0.56 [27]	
					[100]	24 [26]	0.70 [25]	0.75 [27]	
					[111]	22.4 [26]	0.65 [25]	0.70 [27]	
				Junction of arms centered at W	[100]	51.3 [26]	1.20 [25]	1.23 [27]	
				Inside four arms	[100]	36.0 [26]	0.87 [25]		

TABLE 9d-3. ELECTRONIC STRUCTURE OF METALS (Continued)

Metal	Band	Description	Carrier	Orbit description	Magnetic field direction	F (in 10^6 Gauss)	de Haas-van Alphen	Cyclotron resonance	Other experiments
I. SIMPLE METALS									
9. Magnesium [29] h.c.p. $a = 3.20$ (R.T.) $c = 5.20$	1	Cap	h	....	[0001]	1.18	....	....	GM, MA, CR, ASE
	2	Monster	h	Necks tilted ~28.7° from ΓA zone line in (10Ī0) Waists	[0001]	0.804	0.11		
					[10Ī0]	1.92			
					[11Ž0]	1.53			
					[11Ž0]	2.70			
					[10Ī0]	3.16			
					[0001]	2.24			
	3	Cigar	e	....	[10Ī0]	11.7	0.138		
					[11Ž0]	10.7	0.162		
					[0001]	115	0.10		
		Lens	e	....	[10Ī0]	27.2	0.42		
					[11Ž0]	27.16	0.42		
					[0001]	13.9	0.49		
	3 and 4	Magnetic breakdown couples 3d-band butterfly and 4th-band pockets	e	....	[10Ī0]	8.64	0.32		
					[11Ž0]	7.78			
10. Mercury [30, 31] $a, b = 2.9863$ $c = 70°44.6'$ Rhomb	1	Multiply connected cylinders parallel to [001]	h		[011]	0.94	0.20		GM, CR, PA
					[2Ī1]	1.07	0.23		
					[100]	0.735	0.16		
					[111]	1.06			
					[10Ī]	1.34	0.15		
					[2Ī1]	19.3			
				Larger orbits	[100]	21.5			
	2	Surfaces centered at L	e		[10Ī]	15.8	0.90		
					[100]	34.0			
					[10Ī]	32.0			
					[111]	34.5			
					[2Ī1]	40.0			
					[1Ī0]	32.2			
11. Potassium[b] [32], [106] $a = 5.225^a$ [33] b.c.c.	1	Nearly spherical, centered at Γ	e		[100]	182.7 [34]	1.18–1.25 [34]	1.21 [35]	GM, MA, SE, H
					[110]	182.4 [34]			
					[123]	182.4 [34]			

Element	Zone	Shape	Type	Section	Direction[b]	Area[b]	[c]		References
12. Sodium[b] [36] a = 4.225 [33] b.c.c.	1	Nearly spherical, centered at Γ	e	....	Arbitrary	281.8[b]	1.24 [34]	....	GM, CR, PA, H
13. Rubidium[b] [34] a = 5.585 [33] b.c.c.	1	Nearly spherical, centered at Γ	e	....	Average of several directions	160.3	1.28		CR, SE
14. Thallium[b] [37], [111], [112] a = 3.438 [38] c = 5.478 h.c.p.	1 and 2	Full	h	....			....		
	3	Crown		Central	[1120]	93.5 [39]			GM, MA
				Central	[1010]	98.9 [39]			
					[0001]	209 [39]			
	4	Hexagonal network	e	Central arm	[1120]	27.4 [39]			
				Noncentral arm	[1010]	37.6 [39]			
				Central	[0001]	218 [39]			
15. Tin[b] [41], [105] a = 5.80 [42] c = 3.15 b.c.t.	1 and 2	Full	h	....	[0001]	1.8 [40]	0.25 [40]	....	GM, MA, CR, SE
	3	Dumbbells centered at X	h	Central	[001]	1.72	0.16 [43]		
				Noncentral	[001]	3.25			
					[100]	15.8			
					[110]	16.7			
	4	Multiply connected intersecting tubes centered at Γ	h	....	[001]	112			
					[001]	103			
					[100]	32.9	0.56 [43]		
					[110]	25.6	0.51 [43]		
		Crossed convex lens-shaped reentrant region centered at Γ	e	....	[001]	34.1			
	5	Multiply connected tilted tubes with alternate top-up and top-down pear-shaped pieces	e	Large part of pear	[001]	68.1	0.57 [43]		
				Smallest cross section of pear	[001]	63.2			
				tion of pear inside of tilted tube network	[001]	52.8			
				Pear section	[110]	80.4	0.55 [43]		
					[110]	67.7			
				Tilted tubes	[100]	20.6			
					[001]	20.9			
	6	Molar-shaped surface centered at Γ	e	....	[110]	4.45	0.31 [43]		
					[100]	5.87			
					[100]	4.54			

a Low-temperature lattice constant may be in error owing to strained samples [32].
b The possibility of a martensitic transformation at the low temperatures makes interpretation difficult.

TABLE 9d-3. ELECTRONIC STRUCTURE OF METALS (Continued)

Mass values, m^*/m are given in the two columns "de Haas–van Alphen" and "Cyclotron resonance."

Metal	Band	Description	Carrier	Orbit description	Magnetic field direction	F (in 10^6 Gauss)	de Haas–van Alphen	Cyclotron resonance	Other experiments
I. SIMPLE METALS (Cont.):									
16. Zinc [44, 45] $a = 2.651$ $c = 4.838$ h.c.p.	2	Monster	h	Arms minimum	28° from [0001]	4.44 [46]			GM, MA, CR, SE, MT
				4-arm orbit	[0001]	5.13 [46]	0.44 [44]		
					$[11\bar{2}0]$	26.6 [46]			
					$[10\bar{1}0]$	33.0			
				Waists	$[11\bar{2}0]$	0.446 [46]	~0.11		
					$[10\bar{1}0]$	1.11 [46]	0.13		
	3	Cigar	e		[0001]	0.0157 [46]	0.0075		
					$[11\bar{2}0]$	0.265 [46]	0.09		
		Lens	e		$[10\bar{1}0]$	73.5 [46]	~0.54 [44]	~0.55 [47, 16]	
					$[10\bar{1}0]$	73.5 [46]	~0.59 [44]	~0.57 [47, 16]	
					[0001]			1.20 [47, 16]	
1a. Barium [108] $a = 5.000$ [118] bcc				α	[111]	3.29	0.37		
				β	[111]	4.56	0.42		
				γ	[100]	19.5	0.92		
II. TRANSITION METALS									
1. Chromium [48, 103] b.c.c.^{c}		De Haas–van Alphen data not tabulated. See [48]. Preliminary results				~0.2–40			GM
					$[11\bar{2}3]$	~4.9			
					[0001]	~3.8			
1a. Cobalt [104,110] $a = 2.5071$ $c = 4.0686$ (R.T.) h.c.p.									
2. Copper [49] $a = 3.603$ [50] f.c.c.	1	Sphere with necks touching [111]	e	Neck	[111]	21.77 [97]	0.45 [51]		GM, MA, CR, ASE, PA, MT
		Brillouin zone faces		Belly	[111]	581.4 [97]	1.5 [50]		
					[100]	599.8 [97]			
				Dog's bone	[110]	251.4 [97]	1.4 [51]		
				4-cornered rosette	[100]	246.2 [97]	~1.3 [51]		
3. Gold [49] $a = 4.065$ [50] f.c.c.	1	Sphere with necks touching [111]	e	Neck	[111]	15.32	~0.29 [47]		GM, MA, CR
		Brillouin zone faces		Belly	[111]	449.3	~1.1 [52]		
					[100]	485.0 [97]	~1.1 [52]		
				Dog's bone	[110]	193.8 [97]	~1.0 [52]		
				4-cornered rosette	[100]	200.3 [97]	~1.1 [50]		
3a. Iridium [109,116] $a = 3.8387$ (R.T.) f.c.c.			h		[100]	37.8			
					[100]	55.3			
					[111]	46.3			
					[110]	41.9			
			e		[100]	205			

Metal	Surface	h/e	Feature	Direction	F	m^*	Notes
4. Iron [53, 54] $a = 2.86$ b.c.c.	Ellipsoids			[100]	3.84	0.35	GM, PA
				[010]	4.08		
				[111]	4.11		
				[110]	3.89		
				[101]	4.10		
	Surface (1) centered at Γ			[100]	23.8	0.58	
				[111]	28.0		
	Surface (2) centered at Γ			[110]	369	2.6	
				[111]	347	2.6	
	Octahedron centered at H			[111]	154	1.7	
	Surface centered at Γ (minority)			[111]	51.8	2.8	
5. Lutetium [96]	?			[111]	11.3		
	Preliminary results			[0001]	3.8		
				[100]	23		
6. Molybdenum [55] [98, 100] $a = 3.147$ (R.T.) b.c.c.	Ellipsoids centered at N	h		[001]	31	0.34	GM, MA
				[110]	26		
				[011]	~29		
				[101]	~39		
				[111]	24		
	Octahedron centered at H	h		[100]	31	~1.0 [56]	
				[110]	154 [56]		
				[111]	~116		
	Jack centered at Γ	e	Neck	[100]	110	0.8–0.95 [56]	
				[110]	~12	~0.43	
				[111]	~32		
				[100]	36.5		
				[111]	80 [56]		
	Lenses lying along ΓH	e	Waists	[100]	~5.3	~0.44	
				[001]	~8.4		
				[110]	~5.0		
				[101]	~5.8		
				[111]	5.5		
7. Nickel [57, 58] $a = 3.5172$ [5, 59] f.c.c.	Pockets centered at X	h		[100]	10.12[d]	1.0	
				[010]	24.9	1.9	
				[110]	21.9	1.4	
				[101]	14.2	1.35	
				[111]	15.8		
	Necks centered at L	e		[110]	3.84	0.36	
				[111]	2.68	0.25	

[c] Linearly polarized spin-density wave $Q\|[001]$.
[d] The dimensions of the hole pockets depend on the orientation of the magnetic field. Thus the de Haas–van Alphen frequencies cannot be referred to a simple rigid ellipsoidal model. See Hodge et al. [58].

TABLE 9d-3. ELECTRONIC STRUCTURE OF METALS (*Continued*)

Metal	Fermi surface nomenclature Band	Fermi surface nomenclature Description	Carrier	Orbit description	Magnetic field direction	F (in 10^6 Gauss)	Mass values, m^*/m de Haas–van Alphen	Mass values, m^*/m Cyclotron resonance	Other experiments
II. TRANSITION METALS (*Cont.*): 8. Niobium [60], [62], [99] a = 3.29 (R.T.) b.c.c.	......	Ellipsoids centered at N	...	......	~[111]	63–86 [62]	~1.0 [61]	....	GM
							1.28 [62]		
		Jungle gym	...	Minimum-arm cross section	[100]	14.5 [62]			
9. Osmium [63] a = 2.7304 (R.T.) c = 4.3097 h.c.p.	......	Similar to Ruthenium	...	......	[0001]	~210		....	GM
					[0001]	~150			
					[0001]	~3			
10. Palladium [64, 65] a = 3.884 f.c.c.	......	Closed surface centered at Γ	e	......	[100]	275	2.0		
					[111]	309	2.31		
						244	1.95		
		Ellipsoids centered at X	h	......	[100]	8.95	1.05		
					[001]	5.71	0.625		
					[110]	8.95	1.05		
					[101]	6.84	0.770		
					[111]	7.49	0.862		
		Open surface	h	......	[100]	27	2.37		
					[110]		6.2		
11. Platinum [66], [107] a = 3.907 [5] f.c.c.	......	Closed surface centered at Γ	e	......	[100]	290 [65]	2.44 [65]	....	GM
					[110]	324 [65]	3.16 [65]		
					[111]	260 [65]	2.06 [65]		
		Ellipsoids centered at X	h	......	[001̄]	1.11 [67]			
					[111]	1.7 [67]	0.363 [65]		
					[100]	1.45 [65]			
		Open surface	h	......	[110]	27.9 [65]	1.53 [65]		
					[101]	81.6 [65]	3.32 [65]		
						68.1 [65]	3.62 [65]		
12. Rhenium [68] a = 4.447 [69] c = 2.758 [69] h.c.p.	5	Ellipsoids centered at L	h	......	[0001]	4.56			
					[101̄0]	0.77			
					[112̄0]	2.63			
	6	Dumbbells centered at L	h	......	[0001]	7.6			
					[101̄0]	15.6			
					[101̄0]	16.2			
				Rotated 60°	[112̄0]	14.3			
	7	Closed surface centered at L	h	Rotated 60°	[112̄0]	13.6			
					[101̄0]	15.5			
					[101̄0]	79.7			
				Rotated 60°	[101̄0]	64.8			
				Rotated 60°	[112̄0]	68			

Metal	Orbit	Surface	Type	Orbit section	Direction			Method
13. Rhodium [70, 71] $a = 3.8044$ (R.T.) f.c.c.	8	Open cylindrical surface	e	....	~[1120]	90		GM, MA, CR, ASE, MT
	7 and 8	....		....	[100]	300	0.20	
		Ellipsoids along Γ–L			[110]	3.34	0.14	
					[101]	2.69	0.22	
					[111]	5.07	0.12	
		Ellipsoids along Γ–X			[111]	2.32	0.30	
					[100]	4.25	0.35	
					[001]	15.6	0.42	
					[110]	24.6	0.43	
					[101]	17.5	~0.5	
		Closed surface along Γ–X (tentative)			[111]	26.3	0.43	
					[110]	18.9	1.2	
		Closed surface centered at Γ (tentative)			[111]	48.5		
					[111]	180	1.65	
14. Ruthenium [72]* $a = 2.69844$ (R.T.) $c = 4.27305$ h.c.p.	....	Closed surface probably centered at Γ	h	....	[0001]	~8		
					[1100]	~15		
					[1120]	~15		
		Two closed surfaces centered at Γ	e	....	[0001]	160–210		
					[1010]	130–190		
					[1120]	130–200		
		Surface centered on line M–L		....	[0001]	~20		
					[1010]	~7.5		
					[1120]	~8.0		
		Surface centered at L		....	[1010]	~3.5		
					[1120]	~4		
15. Silver [73] $a = 4.069$ [50] f.c.c.	....	Sphere with necks touching [111]	e	Neck	[111]	8.921 [49]	~0.4 [74]	GM
				Belly	[111]	460.0 [49]	~0.9 [74]	
		Brillouin zone faces		....	[100]	474.6 [97]		
				Dog's bone	[110]	201.6 [97]	~1.0 [74]	
				4-cornered rosette	[100]	196.3 [97]		
16. Tantalum [62], [75] $a = 3.30$ (R.T.) b.c.c.	....	Distorted ellipsoids centered at N		Minimum-arm cross section	[100]	45–63 [62]	~0.8 [76] (1.1–1.25) [77]	
		Jungle gym		....	[100]	~29 [62]		
17. Thorium [78] $a = 5.084$ (R.T.) f.c.c.		Butterfly-shaped pieces along [110]	e		[100]	10.0 [79]	0.66 [79]	
					[100]	11.9 [79]	0.58 [79]	
					[110]	2.0 [79]		
					[110]	9.6 [79]	0.58 [79]	
					[111]	11.7 [79]		

* Identification of orbits tentative.

TABLE 9d-3. ELECTRONIC STRUCTURE OF METALS (Continued)

Metal	Fermi surface nomenclature		Carrier	Orbit description	Magnetic field direction	F (in 10⁶ Gauss)	Mass values, m^*/m		Other experiments
	Band	Description					de Haas–van Alphen	Cyclotron resonance	
II. TRANSITION METALS (Cont.):									
18. Tungsten [80, 81] $a = 3.162$ [80] b.c.c.	......	Quasi-spherical surface centered at Γ	h	......	[110]	24.8 [79]			GM, MA, CR, SE
					[111]	24.8 [79]			
					[100]′	22.1 [79]			
		Dumbbell-shaped pieces centered at L with axes along [111]	h	......	[110]	19.9 [79]	0.75 [79]		
					[111]	22.5 [79]	0.58 [79]		
					[111]	10.9 [79]			
		Ellipsoids centered at N	h		[110]	6.87	0.27	0.27 [82]	
					[101]	8.03	0.32	0.32 [82]	
					[011]	9.22	0.36	0.36 [82]	
					[111]	7.03	0.287 [55]		
					[111]	7.66	0.287 [55]		
					[100]	5.93	0.28 [55]		
					[001]	8.54	0.37		
		Octahedron centered at H	h	......	[100]	143.5	0.93	0.23 [82]	
					[111]	98.8	0.60	0.33 [82]	
					[110]	106.9	0.67	1.05 [82]	
		Jack centered at Γ	e	Necks	[100]	6.12	0.25	0.57 [82]	
					[110]	23.03		0.67 [82]	
				Ball of jack	[100]	19.5	0.75 [55]	0.54 [82]	
					[110]	21.81	0.58	0.55–0.58 [82]	
				Central orbit around jack body	[112]	22.84	0.60		
				Two-ball orbit	[111]	69.1	0.9	0.83 [82]	
				Four-ball orbit	[100]	63.8		2.86 [82]	
					[110]	178.4		1.83 [82]	
					[110]	120.4			
19. Vanadium [96] $a = 3.0259$ b.c.c.	......	Preliminary results	...	Ellipsoids at N	[110]	66.75	1.78~		
					[110]	55.5	2.3		
					[110]	52.7			
					[100]	60.3			
					[100]	~52			
20. Ytterbium [17], [83] $a = 5.486$ [84] f.c.c.	......	......	...		[110]	~0.17			
					[112]	~0.17			
					[111]	1.4–1.8			
21. Zirconium [85], [114], [115] $a = 3.23$ $c = 5.146$ (R.T.) h.c.p.	3	Surface centered at Γ	h	......	[0001]	~34.5 [86]			
					[1120]	~37 [86]			
					[1010]	~36 [86]			
	4		h		[0001]	~50 [86]			

	5	Multiply connected	Surface description		Cross section	Direction				Reference
III. SEMIMETALS						[0001] [0001] [0001]	~110 [86] ~77 [86] ~59 [86]			
1. Antimony [87] $a = 4.3007$ $c = 11.222$ [88] (hexagonal axes) trigonal	......	e	Closed pockets centered at L	e	Minimum cross section	[0001]	87.7°g		......	GM, MA, CR, MT
					Maximum cross section	[0001]	174.0°g	0.68		
								4.35		
			Six equivalent ellipsoidal pockets centered on the mirror plane	h	Minimum cross section	Binary	53.0°g	3.6	0.084	
					Maximum cross section		148.8°g	0.613	0.069	
					Binary			1.98		
2. Arsenic [89, 90] trigonal	......	e	Three closed centrosymmetric pockets tilted 86.4° from trigonal	e	Minimum cross section		86.4°g	2.16	0.163	MA, CR
					Maximum cross section		−9.0°g	2.13	0.130	
					Binary	......				
			Six pockets connected by long thin necks	h	Neck minimum cross section	......	−9.6°g	9.58		
					Neck	Trigonal		7.68		
					Principal pockets minimum cross section		37.25°g	0.0258		
								0.028		
								1.49		
3. Bismuth [91], [113] $a = 4.53$ $c = 11.797$ (hexagonal axes) trigonal	......	e	Ellipsoids with one axis parallel to the binary axis and the other two tilted ~6° from the trigonal and bisectrix axes, respectively	e	......	Binary		0.189	0.14 [92]	GM, CR, MA, MT, H, SE, ASE
						Binary		0.0139	......	
						Bisectrix		0.0240	0.009 [92]	
						Bisectrix		0.012	0.11 [92]	
						Trigonal		0.084		
			Ellipsoid	h	......	Binary		0.223	......	
						Bisectrix		0.223		
						Trigonal		0.0635		
4. Carbon (graphite) [93] $a = 2.46$ $c = 6.70$ hex	......	......	Ellipsoidal surface centered at K along K–H	h^A	......	‖c		0.0625 [94]	0.067	GM, MA, CR
						⊥c		0.77 [94]	0.057 [94] 0.68 [94]	
			Ellipsoidal surfaces along K–H	e^h	......	‖c		0.046 [94]	0.039 [94]	
						⊥c		0.67 [94]	0.47 [94]	
			Caps at ends of electron surfaces above (minority carriers)	e^h	......	‖c		0.0074 [94]	0.0023 [94]	
						⊥c		0.067 [94]	0.017 [94]	

f Frequencies associated with dumbbell and quasi-sphere merge at [100].
g Tilt from trigonal in trigonal-bisectrix plane.
h Experiments on pyrolytic graphite [95] suggest that the surfaces attributed to holes should be attributed to electrons and vice versa.

References for Section 9d and Table 9d-3

1. Gold, A. V.: "Solid State Physics," vol. 1, pp. 120–126, J. F. Cochran and R. Haering, eds, Gordon and Breach, Science Publishers, Inc., New York, 1968.
2. Pearson, W. B.: "Lattice Spacings and Structure of Metals and Alloys," Pergamon Press, New York, 1958.
3. Koster, G. F.: *Solid State Phys.* **5**, 173–256 (1957).
4. Larson, C. O., and W. L. Gordon: *Phys. Rev.* **156**, 703 (1967).
5. Armstrong, R. W.: Private communication.
6. Priestley, M. G.: *Phil. Mag.* **7**, 1205 (1962).
7. Anderson, J. R., and S. Lane: *Phys. Rev.* **B2**, 298 (1970).
8. Spong, F. W., and A. F. Kip: *Phys. Rev.* **137**, A431 (1965).
9. Tripp, J. H., W. L. Gordon, P. M. Everett, and R. W. Stark: *Phys. Letters* **26A**, 98 (1967).
10. Watts, B. R.: *Proc. Roy. Soc (London)*, ser. A, **282**, 521 (1964).
11. Alekseyevsky, N. E., and V. S. Yegorov: *Zh. Eksperim. i Teor. Fiz.* **55**, 1153 (1968).
12. Jones, R. C., R. G. Goodrich, and L. M. Falicov: *Phys. Rev.* **174**, 672 (1968).
13. Naberezhnykh, V. P., A. A. Mar'Yakhin, and V. L. Mel'Nik: *Soviet Phys.—JETP* **25**, 403 (1967).
14. Tsui, D. C., and R. W. Stark: *Phys. Rev. Letters* **16**, 19 (1966).
15. Grassie, A. D. C.: *Phil. Mag.* **9**, 847 (1964).
16. Shaw, M. P., T. G. Eck, and D. A. Zych: *Phys. Rev.* **142**, 406 (1966).
17. Condon, J. H., and J. A. Marcus: *Phys. Rev.* **134**,A 446 (1964).
18. Okumura, K., and I. M. Templeton: *Proc. Roy. Soc. (London)*, ser. A, **287**, 89 (1965).
19. Goldstein, A., and S. Foner: *Phys. Rev.* **146**, 442 (1966).
20. Moore, T. W.: *Phys. Rev.* **165**, 864 (1968).
21. Hughes, A. J., and J. P. G. Shepherd: *Journal of Physics C (Solid State Physics)* **2**, 661 (1969).
22. O'Sullivan, W. J., J. E. Schirber, and J. R. Anderson: *Phys. Letters* **27A**, 144 (1968)
23. Mina, R. T., and M. S. Khaikin: *Soviet Phys—JETP* **24**, 42 (1966).
24. Anderson, J. R., and A. V. Gold: *Phys. Rev.* **139**, A1459 (1965).
25. Phillips, R. A., and A. V. Gold: *Phys. Rev.*, **178**, 932 (1969).
26. Anderson, J. R., W. J. O'Sullivan, and J. E. Schirber: To be published.
27. Mina, R. T., and M. S. Khaikin: *Soviet Phys.—JETP* **18**, 896 (1964).
28. Anderson, J. R., and D. C. Hines: *Phys. Rev.* **B2**, 4752 (1970).
29. Stark, R. W.: *Phys. Rev.* **162**, 589 (1967).
30. Brandt, G. B., and J. A. Rayne: *Phys. Rev.* **148**, 644 (1966).
31. Dishman, J. M., and J. A. Rayne: *Phys. Rev.* **166**, 728 (1968).
32. Thomas, R. L., and G. Turner: *Phys. Rev.* **176**, 768 (1968).
33. Barrett, C. S.: *Acta Cryst.* **9**, 671 (1956).
34. Shoenberg, D., and P. J. Stiles: *Proc. Roy. Soc. (London)*, ser. A, **281**, 62 (1964).
35. Grimes, C. C., and A. F. Kip: *Phys. Rev.* **132**, 1991 (1963).
36. Lee, M. J. G.: *Proc. Roy. Soc. (London)*, ser. A., **295**, 440 (1966).
37. Aleksandrov, B. N.: *Soviet Phys. JETP—26*, 508 (1968).
38. Barrett, C. S.: *Phys. Rev.* **110**, 1071 (1968).
39. Priestley, M. G.: *Phys. Rev.* **148**, 580 (1966).
40. Anderson, J. R., J. E. Schirber, and D. Stone: *Grenoble High Pressure Conference Proceedings* **188**, 131 (1970).
41. Craven, J. E., and R. W. Stark: *Phys. Rev.* **168**, 849 (1968).
42. Statleu, M. D., and A. R. de Vrooman: *Phys. Stat. Solidi* **23**, 675, 683 (1967).
43. Vaughan, R. W., and D. D. Elleman: *Bull. APS* **13**, 1454 (1968).
44. Ventsel, V. A., A. I. Likhter, and A. V. Rudnex: *Soviet Phys—JETP* **26**, 73 (1968).
45. Ventsel, V. A.: *Zh. Eksperim. i Teor. Fiz.* **55**, 1191 (1968).
46. Higgins, R. J., J. A. Marcus, and D. H. Whitmore: *Phys. Rev.* **137A**, 1172 (1965).
47. Shaw, M. P., P. I. Sampath, and T. G. Eck: *Phys. Rev.* **142**, 399 (1966).
48. Graebner, J., and J. A. Marcus: *Phys. Rev.* **175**, 659 (1968).
49. Jan, J. P., and I. M. Templeton: *Phys. Rev.* **161**, 556 (1967).
50. Shoenberg, D.: *Phil. Trans. Roy. Soc. London*, ser. A, **255**, 85 (1966).
51. Joseph, A. S., A. C. Thorsen, E. Gertner, and L. E. Valby: *Phys. Rev.* **148**, 569 (1966).
52. Joseph, A. S., A. C. Thorsen, and F. A. Blum: *Phys. Rev.* **140**, A2046 (1965).
53. Panousis, P. T.: *USAEC Rept.* IS-T-175, 1967; and to be published.
54. Gold, A. V.: *J. Appl. Phys.* **39**, 768 (1968).
55. Sparlin, D. M., and J. A. Marcus: *Phys. Rev.* **144**, 484 (1966).
56. Meyers, A., and G. Leaver: "Proceedings 10th Conference on Low Temperature Physics," vol. 3, p. 290, Viniti Publishing House, Moscow, 1967.
57. Tsui, D. C.: *Phys. Rev.* **164**, 669 (1967).

58. Hodges, L., D. R. Stone, and A. V. Gold: *Phys. Rev. Letters* **19**, 655 (1967).
59. Heumann, T.: *Naturwissenschaften* **32**, 296 (1944).
60. Fawcett, E., W. A. Reed, and R. R. Soden: *Phys. Rev.* **159**, 513 (1967).
61. Thorsen, A. C., and T. G. Berlincourt: *Phys. Rev. Letters* **7**, 244 (1961).
62. Halloran, M., J. H. Condon, J. E. Graebner, J. E. Kunzler, and F. S. L. Hsu, *Phys. Rev.* **1B**, 366 (1970).
63. Kamm, G. N., and J. R. Anderson, *Phys. Rev.* **B2**, 2944 (1970).
64. Vuillemin, J.: *Phys. Rev.* **144**, 396 (1966).
65. Windmiller, L. R., J. B. Ketterson, and S. Hornfeldt, *J. Appl. Phys.* **40**, 1291 (1969).
66. Windmiller, L. R., and J. B. Ketterson: *Phys. Rev. Letters* **21**, 1076 (1968).
67. Stafleu, M. D., and A. R. DeVroomen: *Phys. Letters* **19**, 81 (1965).
68. Thorsen, A. C., A. S. Joseph, and L. E. Valby: *Phys. Rev.* **150**, 523, (1966).
69. Matteiss, L. F.: *Phys. Rev.* **151**, 450 (1966).
70. Coleridge, P. T.: *Proc. Roy. Soc. (London)*, ser. A, **295**, 458 (1966).
71. Ketterson, J. B., L. R. Windmiller, and S. Hornfeldt: *Phys. Letters* **26A**, 115 (1968).
72. Coleridge, P. T., *Phys. Letters* **22**, 367 (1966) and *Journal of Low Temperature Physics* **1**, 577 (1969).
73. Lewis, P. E., and P. M. Lee: *Phys. Rev.* **175**, 795 (1968).
74. Joseph, A. S., and A. C. Thorsen: *Phys. Rev.* **A138**, 1159 (1965).
75. Fawcett, E., W. A. Reed, and R. R. Soden: *Phys. Rev.* **159**, 533 (1967).
76. Thorsen, A. C., and T. G. Berlincourt: *Phys. Rev. Letters* **7**, 244 (1961).
77. Condon, J. H.: *Bull. Am. Phys. Soc.* **11**, 170 (1966).
78. Thorsen, A. C., A. S. Joseph, and L. E. Valby: *Phys. Rev.* **162**, 574 (1967).
79. Boyle, D. J.: *USAEC Rept.* IS-T-261, 1968; D. J. Boyle and A. V. Gold: *Phys. Rev. Letters* **22**, 461 (1969).
80. Girvan, R. F., A. V. Gold, and R. A. Phillips: *J. Phys. Chem. Solids* **29**, 1485 (1968).
81. Girvan, R. F.: *USAEC Rept.* IS-T-103, 1966.
82. Walsh, W. M., Jr.: In "Solid State Physics, vol. 1, p. 160, J. F. Cochran and R. Haering, eds., Gordon and Breach, Science Publishers, Inc., New York, 1968.
83. Tanuma, S., Y. Ishizawa, H. Nagasawa, and T. Sugawava: *Phys. Letters* **25A**, 669 (1967).
84. Gschneidner, K. A., Jr.: "Rare Earth Alloys," D. Van Nostrand Company, Inc., Princeton, N.J., 1961.
85. Loucks, T. L.: *Phys. Rev.* **159**, 544 (1967).
86. Thorsen, A. C., and A. S. Joseph: *Phys. Rev.* **131**, 2078 (1963).
87. Windmiller, L. R.: *Phys. Rev.* **149**, 472 (1966).
88. Barrett, C. S., P. Cucka, and K. Haefner: *Acta Cryst.* **16**, 451 (1963).
89. Vanderkooy, J., and W. R. Datars: *Phys. Rev.* **156**, 671 (1967).
90. Priestley, M. G., L. R. Windmiller, J. B. Ketterson, and Y. Eckstein: *Phys. Rev.* **154**, 671 (1967).
91. Bhargava, R. N.: *Phys. Rev.* **156**, 785 (1967).
92. Kao, Y. H.: *Phys. Rev.* **129**, 1122 (1963)
93. McClure, J. W., and W. J. Spry: *Phys. Rev.* **165**, 809 (1968).
94. Soule, D. E.: *IBM J. Res. Develop.* **8**, 268 (1964).
95. Schroeder, P. P., M. S. Dresselhaus, and A. Javan: *Phys. Rev. Letters* **20**, 1292 (1968).
96. Phillips, R. A.: Private communication.
97. O'Sullivan, W. J., A. C. Switendick, and J. E. Schirber: *Phys. Rev.* **1B**, 1443 (1970).
98. Boiko, V. V., V. A. Gasparov, I. G. Gverdtsiteli, *Soviet Phys. JETP* **29**, 267 (1969).
99. Scott, G. B., and M. Springford, *Proc. Roy. Soc. (London)* **A320**, 115 (1970).
100. Leaver, G., and A. Myers, *Phil. Mag.* **19**, 465 (1969).
101. Henmann, R., *Phys. Stat. Sol.* **25**, 661 (1968).
102. Cucka, P., and C. S. Barrett, *Acta Cryst.* **15**, 865 (1962).
103. Arko, A. J., J. A. Marcus, and W. A. Reed, *Phys. Rev.* **185**, 901 (1969).
104. Reed, W. A., and E. Fawcett, *Proc. of the Int'l Conf. on Magnetism (Inst. of Phys. and Phys. Soc., London)* 120 (1964).
105. Vaughan, R. W., D. D. Elleman, and D. G. McDonald, *J. Phys. Chem. Solids* **31**, 117 (1970).
106. Glinski, R., and I. M. Templeton, *Jour. of Low Temp. Phys.* **1**, 223 (1969).
107. Ketterson, J. B., and L. R. Windmiller, *Phys. Rev.* **2B**, 4813 (1970).
108. McEwen, K. A., *Phys. Letters* **30A**, 77 (1969).
109. Volkenshtein, N. V., V. A. Vovosydov, V. E. Startsev, *Soviet Phys. JETP* **31**, 862 (1970).
110. Anderson, J. R., and D. R. Stone—private communication.
111. Ishizawa, Y., and A. E. Dixon, *Bull. Am. Phys. Soc.* **16**, 82 (1971).
112. Capocci, F. A., P. M. Holtham, D. Parsons, and M. G. Priestley, *Jour. of Physics C (Solid State Physics)* **3**, 2081 (1970).

113. Brown, Rodney D., III, *Phys. Rev.* **B2**, 928 (1970).
114. Everett, P. M., *Bull. Am. Phys. Soc.* **16**, 336 (1971).
115. Schirber, J. E., *Phys. Letters* **33A**, 172 (1970).
116. Grodski, J. J., and A. E. Dixon, *Bull. Am. Phys. Soc.* **16**, 82 (1971).
117. Tanuma, S., W. R. Datars, H. Doi, and A. Dunsworth, *Solid State Comm.* **8,** 1107 (1970).
118. Barrett, C. S., *J. Chem. Phys.* **25**, 1123 (1956).

9e. Properties of Semiconductors

H. P. R. FREDERIKSE

The National Bureau of Standards

9e-1. Introduction. This chapter contains a number of numerical values of semiconductor parameters collected from the literature up to November, 1968. Our knowledge is, however, still very uneven. In some cases very detailed information about band structure, transport properties, etc., is available, while for other semiconductors even the energy gap has not yet been determined unambiguously. The data in this chapter are therefore limited to a few groups: elemental semiconductors, III-V compounds, some II-IV compounds, some II-VI compounds, the lead compounds, and a few others.

For definitions of electronic parameters and some simple formulas the reader is referred to Sec. 9c.

9e-2. Band Structure. In order to illustrate concepts like anisotropic effective mass, spin-orbit splitting, etc., the electronic energy band structure of Ge and Si will be briefly discussed (refs. 1 to 5).

Germanium. CONDUCTION BAND. The constant-energy surfaces near the bottom of the conduction band consist of four symmetrically equivalent ellipsoids, whose lengths are oriented along the $\langle 111 \rangle$ directions. The four minima (valleys) are located at the edge of the Brillouin zone. Each of these ellipsoids can be described by the following energy-momentum relation:

$$\varepsilon = \frac{p_x{}^2 + p_y{}^2}{2m_t} + \frac{p_z{}^2}{2m_l}$$

where m_t and m_l are the transverse and longitudinal effective masses, respectively.

The effective mass m^* in any particular direction of momentum space is given by

$$\frac{1}{m^{*2}} = \frac{\cos^2 \theta}{m_t{}^2} + \frac{\sin^2 \theta}{m_t m_l}$$

where θ is the angle between the chosen direction and the longitudinal axis of the ellipsoid.

Experiments (ref. 2) indicate that

$$m_l = (1.59 \pm 0.03)m_0$$
$$m_t = (0.082 \pm 0.0003)m_0$$

At the center of the zone (Γ_2'),

$$m^* \approx 0.034m_0$$

The density-of-state effective mass $m^{(N)}$ (or average effective mass), which is obtained from conductivity and Hall-effect experiments, is given by

$$m^{(N)} = (m_l m_t^2)^{\frac{1}{3}} \nu^{\frac{2}{3}}$$

where ν is the number of ellipsoids. The value of this parameter for the conduction band of germanium is $m_e^{(N)} = 0.55m_0$.

VALENCE BAND. The energy-band structure at the top of the valence band consists of three nearly spherical (warped) surfaces. The maxima are located at the center of the zone $[\vec{k} = 0]$. Two of the surfaces are degenerate at the center point. The third is somewhat lower in energy owing to *spin-orbit interaction*. (The spin-orbit interaction results from coupling between the magnetic dipole fields of the spin and the orbital motion of an electron.) Close to the band edge the energies of the two degenerate surfaces V_1 and V_2 can be described by

$$\varepsilon_{1,2} = -\frac{1}{2m_0} \{Ap^2 \pm [B^2p^4 + C^2(p_x^2 p_y^2 + p_y^2 p_z^2 + p_z^2 p_x^2)]^{\frac{1}{2}}\}$$

and that of the split-off band V_3 by

$$\varepsilon_3 = -\Delta - \frac{1}{2m_0} A p^2$$

where Δ is the spin-orbit splitting energy. The plus sign in the expression for $\varepsilon_{1,2}$ refers to light holes, and the minus sign to heavy holes. In Ge the parameters have the following values:

$$\Delta = 0.29 \text{ ev} \qquad m_{V_1} = 0.34m_0$$
$$A = -13.3 \pm 0.2 \qquad m_{V_2} = 0.043m_0$$
$$B = \pm 8.6 \pm 0.1 \qquad m_{V_3} = 0.08m_0$$
$$C = \pm 12.5 \pm 0.2$$

For further details see Fig. 9e-1.

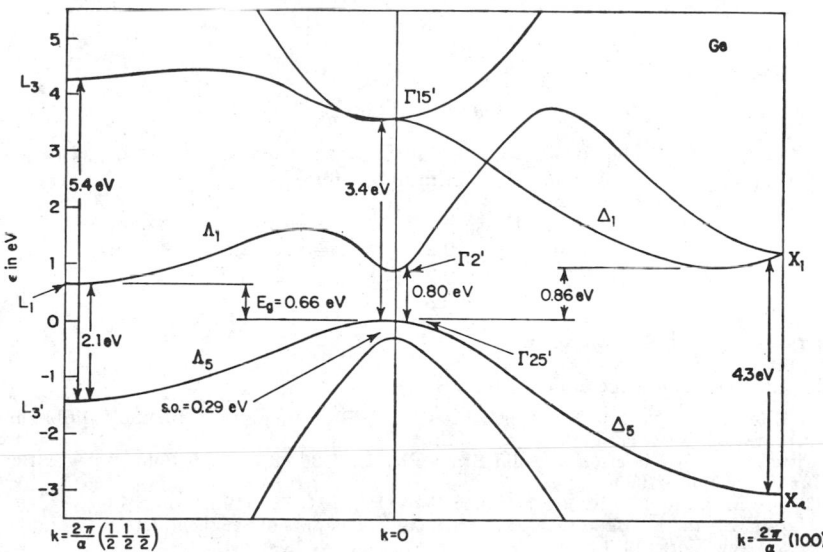

FIG. 9e-1. The band structure of germanium near the band gap. Energy as a function of wave vector k for the $\langle 111 \rangle$ and $\langle 100 \rangle$ directions. (*After Charles Kittel, "Introduction to Solid State Physics," 3rd ed., John Wiley & Sons, Inc., New York, 1966.*)

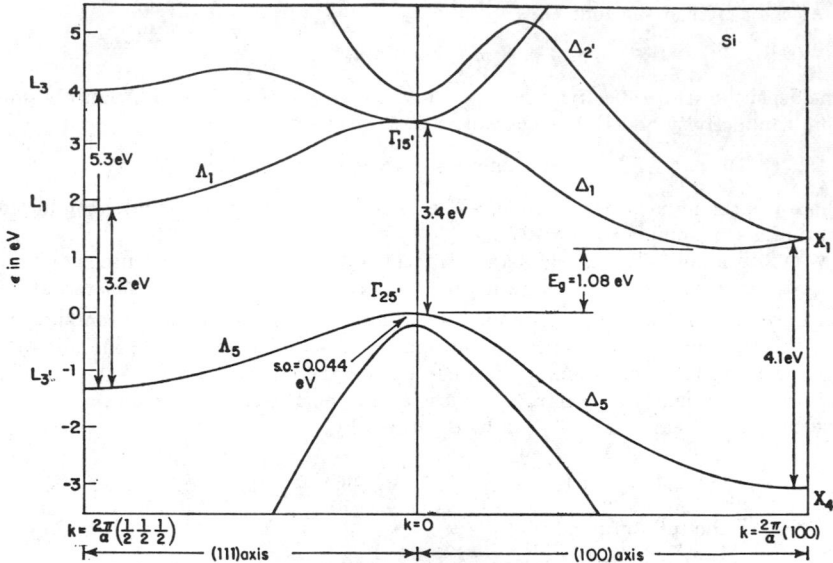

FIG. 9e-2. The band structure of silicon near the band gap. Energy as a function of wave vector k for the $\langle 111 \rangle$ and $\langle 100 \rangle$ directions. (*After Charles Kittel, "Introduction to Solid State Physics," 3d ed., John Wiley & Sons, Inc., New York, 1966.*)

Silicon. CONDUCTION BAND. The constant-energy surfaces near the bottom of the conduction band consist of three symmetrically equivalent ellipsoids whose length axes are oriented along $\langle 100 \rangle$ directions. The six minima (valleys) are inside the Brillouin zone.

The effective electron mass values are

$$m_l = (0.98 \pm 0.04)m_0$$
$$m_t = (0.19 \pm 0.01)m_0$$

The density-of-states effective mass is

$$m_e{}^{(N)} = 1.1m_0$$

VALENCE BAND. The constant-energy surfaces are similar to those of germanium at the top of the valence band. The parameters for Si are

$\Delta = 0.044$ eV	$m_{V_1} = 0.52m_0$
$A = -4.0 \pm 0.2$	$m_{V_2} = 0.16m_0$
$B = \pm 1.1 \pm 0.2$	$m_{V_3} = 0.25m_0$
$C = \pm 4.1 \pm 0.5$	

For further details see Fig. 9e-2.

References on Band Structure of Ge and Si:

1. Geballe, T. H.: "Semiconductors," p. 313, N. B. Hannay, ed., Reinhold Publishing Corporation, New York, 1959.
2. Kittel, C.: "Introduction to Solid State Physics," 3d ed., p. 316, John Wiley & Sons, Inc., New York, 1966.
3. Cohen, M. L., and T. K. Bergstresser: *Phys. Rev.* **141**, 789 (1966).
4. Aigrain, P., and M. Balkanski, eds.: "Selected Constants of Semiconductors," Pergamon Press, New York, 1961.
5. Putley, E. H.: "The Hall Effect and Related Phenomena," Butterworth & Co. (Publishers), Ltd., London, 1960.

III-V Compounds. The bandstructures of the III-V compounds are similar to those of Ge and Si. However, most of the III-V compounds have the maximum of the valence band and the minimum of the conduction band at the center of the Brillouin zone ($k = 0$). Values of the energy gap E_g, effective masses (electron mass m_e^*, light and heavy hole masses m_{lh}^* and m_{hh}^*, mass of the "split-off" valence band m_{sh}^*, free-electron mass m_0), and spin-orbit splitting $\Delta_{\text{s.o.}}$ are listed in the first seven columns of Table 9e-1.

TABLE 9e-1. CHARACTERISTICS OF III-V COMPOUNDS

Compound	ϵ_g at $T = 0$ K, eV		Effective masses				$\Delta_{\text{s.o.}}$ eV	Mobilities, cm²/volt-sec at 300 K		Impurity activation energy, eV
	Direct	Indirect	m_e^*/m_0	m_{lh}^*/m_0	m_{hh}^*/m_0	m_{sh}^*/m_0		μ_e	μ_h	
InSb	0.23⁶	....	0.015	0.021	0.39	0.11(c)	~0.9	78,000	750	0.007 (Zn,p)
InAs	0.36	....	0.026	0.025	0.41	0.08³	0.43	33,000	460	
InP	1.29	....	0.073	0.078(c)	0.4	0.15(c)	~0.2	4,600	150	
InN	2.4									
GaSb	0.81⁴	....	0.047	0.06	0.3	0.14(c)	~0.8	4,000	1,400	0.024 (Zn,p)
GaAs	1.52	1.85 (X)†	0.07	0.12	0.68	0.20	0.34	8,800	400	
GaP	2.88 (Γ)†	2.32 (X)†	0.13(c)	0.14(c)	0.86(c)	0.24(c)	0.13	300	100	
GaN	3.3									
AlSb	2.20 (Γ)†	1.65 (X)†	0.11(c)	0.11(c)	0.9	0.22(c)	0.75	200	550	0.07 (Te,n) 0.16 (Se,n)
AlN	4.6									
BP	6.0 (?)	2.0								
Ref	1, 2		3				4	5		6, 7

† At 77 K.
(c) = calculated.

Other Compounds and Elemental Semiconductors. Somewhat less is known about the band structure of other semiconducting compounds and elements. Values for energy gaps, effective masses, and mobilities can be found in Table 9e-2.

Temperature and Pressure Dependence of the Energy Gap. For most semiconductors the energy gap decreases with increasing temperature. Exceptions are the lead compounds. The change is nearly linear with temperature except at low temperatures. The energy gap also changes with pressure. The thermodynamic relationship between the two is

$$\left(\frac{\partial \mathcal{E}_g}{\partial T}\right)_P = \left(\frac{\partial \mathcal{E}_g}{\partial T}\right)_V + \left(\frac{\partial V}{\partial T}\right)_P \left(\frac{\partial P}{\partial V}\right)_T \left(\frac{\partial \mathcal{E}_g}{\partial P}\right)_T$$

Temperature and pressure coefficients for a few materials are listed in Table 9e-3.

References for Table 9e-1

1. Willardson, R. K., and A. C. Beer, eds.: "Semiconductors and Semimetals," vol. 1, p. 7, Academic Press, Inc., New York, 1966–1968.
2. InN: Ormont, B. F.: *Zh. Neorgan. Khim.* **4**, 2176 (1959); transl.: *Russ. J. Inorg. Chem.* **4**, 988 (1959).
   GaN: Grimmeis, H. G., et al.: *Z. Naturforsch.* **159**, 799 (1960).
   AlN: Ivey, H. F.: *Advan. Electron Electron. Phys.*, suppl. 1, 169 (1963).
   BP: Archer, R. J., et al.: *Phys. Rev. Letters* **12**, 538 (1964).
3. Reference 1, vol. 2, p. 151.
4. Reference 1, vol. 2, p. 141.
5. Reference 1, vol. 1, p. 16.
6. Hannay, N. B., ed.: "Semiconductors," p. 389, Reinhold Publishing Corporation, New York, 1959.
7. AlSb: Turner, W. J., and W. E. Reese: *Phys. Rev.* **117**, 1003 (1960).

TABLE 9e-2. CHARACTERISTICS OF SEVERAL ELEMENTAL AND COMPOUND SEMICONDUCTORS

Compound	Structure	ϵ_g, eV ($T = 0$ K)	Effective masses		Mobilities at 300 K, cm²/volt-sec		References
			m_e/m_0	m_h/m_0	μ_e	μ_h	
Diamond	cub	5.4	0.2	0.7(lh), 2.12(hh) 1.06(sh)	1,800	1,200	1–5
Graphite	hex	0.0	0.03–0.04	……	4×10^5	1.5×10^5 (lh)	6–8
Se	trig	1.8 ($T = 300$ K)	……	……	~10	……	8, 9
Te	trig	0.33	$m_e^{long} = 0.37 m_0$, $m_e^{trans} = 0.32 m_0$	0.06–0.07	1,700	1,200	8, 9
α-Sn	cub	0.0†	0.02	$m_{lh} = 0.08 m_0$, $m_{hh} = 0.12 m_0$	2,500	2,400	8, 10a
SiC	hex	3.0	$m_{eh} = 0.04 m_0$	$m_{lh} = 0.06 m_0$, $m_{hh} = 0.3 m_0$	100	50	8
	cub	1.9	0.46	……	……	……	8
Mg₃Si	antifluorite	0.78	……	……	400	60	10b
Mg₃Ge	antifluorite	0.57	……	……	500	100	8, 11
Mg₃Sn	antifluorite	0.185 (indirect) 0.35 (direct)	……	……	320	260	9, 12, 13
Mg₃Sb₂	hex	0.82	0.32‡	0.57‡	19	82	14
Zn₃As₂	tetr	0.93	……	……	……	10	15
ZnAs₂	monocl	0.90(‖c), 0.93(⊥c)	……	……	……	50	15
ZnSb	orth	0.56	$m_e^{long} = 0.175 m_0$, $m_e^{trans} = 0.146 m_0$	……	340	10	9
Cd₃As₂	tetr	<0.02	0.04	……	15,000	……	16
CdAs₂	tetr	1.00(‖c), 1.04(⊥c)	$m_e^{long} = 0.15 m_0$, $m_e^{trans} = 0.58 m_0$	……	$\mu^c = 400$, $\mu^a = 100$	……	15, 17
CdSb	orth	0.5–0.6	$m_e^{long} = 0.14 m_0$, $m_e^{trans} = 0.16 m_0$	$m_h^{long} = 0.094 m_0$, $m_h^{trans} = 0.35 m_0$	250	300–700	8
PbS	f.c.c.	0.29	$m_e^{long} = 0.104 m_0$, $m_e^{trans} = 0.08 m_0$	$m_h^{long} = 0.105 m_0$, $m_h^{trans} = 0.075 m_0$	550	600	18a, 18b, 9
PbSe	f.c.c.	0.17	$m_e^{long} = 0.068 m_0$, $m_e^{trans} = 0.041 m_0$	$m_h^{long} = 0.084 m_0$, $m_h^{trans} = 0.0465 m_0$	1,020	930	18a, 18c, 9
PbTe	f.c.c.	0.19	$m_e^{long} = 0.23 m_0$, $m_e^{trans} = 0.024 m_0$	$m_h^{long} = 0.381 m_0$, $m_h^{trans} = 0.031 m_0$	1,620	750	18a, 18d, 9

							Ref.
ZnO	hex W§	3.436			100 [20]		19 (E_g, m^*)
ZnS	W§	3.910	$m_\perp = 0.27$, $m_\parallel = 0.28$	1.8	140 [21]		
ZnS	Z§	3.84	0.38				20–22
ZnSe	W§	2.795					
ZnSe	Z§	2.83	0.1	0.6	100 [9]		23, 24 (μ)
ZnTe	Z§	2.39					
CdS	W§	2.582	$m_t = 0.171m_0$, $m_l = 0.15m_0$	$m_\perp = 5m_0$, $m_\parallel = 0.7m_0$	210 [8]		
CdSe	W§	1.840		$m_\perp = 0.45$, $m_\parallel \geq m_0$	600 [22]		
CdTe	Z§	1.607	0.13		900 [23]	75 [24]	}25
HgSe	Z§	−0.24¶		0.17‡			
HgTe	Z§	−0.30¶	0.027†	0.35‡			
As₂Se₃	amorph.	1.6					
As₂Te₃	monocl.	1.0					
Sb₂S₃	orthorh.	~1.7			170	80	26
Sb₂Se₃	orthorh.	~1.2			15	45	
Bi₂Te₃	rhomb.	0.3				270	
Bi₂S₃	orthorh.	1.3					
Bi₂Se₃	rhomb.	0.35			600		
Bi₂Te₃	rhomb.	0.2	0.45	0.51	1,250	515	26, 27
Cu₂O	cub	2.172	0.5	0.5		100	28a, 28b, 8'
TiO₂	tetr.	∥3.03; ⊥3.04			1.0($\parallel c$); 0.2($\perp c$)		9, 29
SrTiO₃	cub	3.4	$m_t = 1.5m_0$, $m_l = 6.0m_0$		5		30, 31

† At $k = 0$.
‡ Density-of-states mass.
§ W = wurzite; Z = zincblende.
¶ HgSe and HgTe are semimetals. The "energy gap" quoted is $E(\Gamma_6) - E(\Gamma_8)$ which is negative.

References for Table 9e-2

1. Mitchell, E. J. W.: *J. Phys. Chem. Solids* **8**, 444 (1958).
2. Clark, C. D.: *Proc. Roy. Soc. (London)*, ser. A, **277**, 312 (1964).
3. Rauch, C. J.: *Phys. Rev. Letters* **7**, 83 (1961).
4. Rauch, C. J.: "Proceedings International Conference on the Physics of Semiconductors," p. 276, Exeter, Institute of Physics, L ndon, 1962.
5. Redfield, A. G.: *Phys. Rev.* **94**, 526 (1954).
6. Soule, D. E., and J. W. McClure: *J. Phys. Chem. Solids* **8**, 29 (1959).
7. Soule, D. E.: *Phys. Rev.* **112**, 698 (1958).
8. Putley, E. H.: "The Hall Effect and Related Phenomena," Butterworth & Co. (Publishers), Ltd., London, 1960.
9. Aigrain, P., and M. Balkanski, eds.: "Selected Constants of Semiconductors," Pergamon Press, New York, 1961.
10a. Groves, S., and W. Paul: "Proceedings of the International Conference on the Physics of Semiconductors," p. 41, M. Hulin, ed., Dunod, Paris, 1964.
10b. Morris, R. G., R. D. Redin, and G. C. Danielson: *Phys. Rev.* **109**, 1909 (1958).
11. Lott, L., and D. Lynch: *Phys. Rev.* **141**, 681 (1966).
12. Lipson, H. G., and A. Kahan: *Phys. Rev.* **133**, 800 (1964).
13. Lawson, W. D., et al.: *J. Electron.* **1**, 203 (1955).
14. Busch, G., et al.: *Helv. Phys. Acta* **27**, 249 (1954).
15. Turner, W. J., et al.: *Phys. Rev.* **121**, 759 (1961).
16. Haidemenakis, E. D., et al.: *J. Phys. Soc. Japan* **21**, 189 (1966).
17. Turner, W. J., et al.: *J. Appl. Phys.* **32**, 2241 (1961).
18a. Mitchell, E. D., et al.: "Proceedings International Conference on the Physics of Semiconductors," p. 325, M. Hulin, ed., Dunod, Paris, 1964.
18b. Cuff, K. F., et al.: ref. 18a, p. 677.
18c. Berman, S.: *Phys. Rev.* **158**, 723 (1967).
18d. Numata, H., and Y. Uemara: *J. Phys. Soc. Japan* **19**, 2140 (1964).
19. Reynolds, D. C., et al., *Phys. Stat. Solidi* **12**, 3 (1965).
2). Thomas, D. G.: *J. Phys. Chem. Solids* **10**, 47 (1959).
21. Aven, M., and C. A. Mead: *Appl. Phys. Letters* **7**, 8 (1965).
22. Heinz, D. M., and E. Banks: *J. Chem. Phys.* **24**, 391 (1956).
23. Segall, B., et al.: *Phys. Rev.* **129**, 2471 (1963).
24. Yamada, S.: *J. Phys. Soc. Japan* **17**, 645 (1962).
25. Harman, T. C.: "Proceedings of the International Conference on II-V Compounds," p. 982, D. G. Thomas, ed., W. A. Benjamin, Inc., New York, 1967.
26. Black, et al.: *J. Phys. Chem. Solids* **2**, 240 (1957).
27. Drabble, J. R.: *Proc. Phys. Soc. (London)* **71**, 430 (1958); *ibid.*, **72**, 380 (1958).
28a. Knox, R. S.: *Solid State Phys.* suppl. 5, 53 (1963).
28b. Gross, E. F.: *J. Phys. Chem. Solids* **8**, 172 (1959).
29. Frederikse, H. P. R.: *J. Appl. Phys.* **32** (suppl.), 2211 (1961).
30. Cohen, M. I., and R. F. Blunt: *Phys. Rev.* **168**, 929 (1968).
31. Frederikse, H. P. R., et al.: *J. Phys. Soc. Japan* **21**, (suppl.), 32 (1966 .

TABLE 9e-3. TEMPERATURE AND PRESSURE DEPENDENCE OF THE ENERGY GAP

	$(\delta\epsilon g/\delta T)_P$, eV/deg $\times 10^4$	Ref.	$(\delta\epsilon g/\delta P)_T$, eV m^2/kg $\times 10^2$	Ref.
C..................	....	...	$< 1.0\ (X)$	3
Si.................	$-\ 2.3\ (X)$	1	$-\ 1.5\ (X)$	3
Ge.................	$-\ 3.7$	1	$5.0\ (L)$	3
α-Sn..............	....	1	$5.0\ (L)$	3
Se.................	-14	2	-20.0	2
Te.................	$-\ 0.3$	2	-19.0	2
SiC................	$-\ 3.3$	1		
InSb...............	$-\ 2.9\ (\Gamma)$	1	$(14.2),\ 15.5\ (\Gamma)$	1
InAs..............	$-\ 3.3\ (\Gamma)$	1	$(4.8,5.5)8.5$	1
InP...............	$-\ 4.6\ (\Gamma)$	1	$4.6,\ 8.4$	
GaSb..............	$-\ 4.1\ (\Gamma)$	1	$12.0,\ (16.0)$	1
GaAs..............	$-\ 5.0\ (\Gamma)$	1	$-8.7\ (X)\ 9.4$	1
GaP...............	$-\ 5.4\ (\Gamma)$	1	10.7	1
AlSb..............	$-\ 3.5\ (X?)$	1	$-\ 1.6\ (X)$	
ZnO (W)...........	$-\ 9.5\ (\Gamma)$	1	$0.6\ (\Gamma)$	1
ZnS (W)...........	$-\ 3.8\ (\Gamma)$	1	$9.0\ (\Gamma)$	1
ZnSe (W)	$-\ 7.2\ (\Gamma)$	1	$6.0\ (\Gamma)$	1
CdS (W)...........	$-\ 5.0\ (\Gamma)$	1	$+\ 3.3\ (\Gamma)$	1
CdSe (W)..........	$-\ 4.6\ (\Gamma)$	1	....	1
PbS...............	$+\ 3.7\ (L)$	1	$6.9\ (L)$	1
PbSe..............	$\sim 4.0\ (L)$	1		
PbTe..............	$\sim 4.0\ (L)$	1		

References for Table 9e-3

1. Aigrain, P., and M. Balkanski, eds.: "Selected Constants of Semiconductors," Pergamon Press, New York, 1961.
2. Caldwell, R. S., and H. Y. Fan: *Phys. Rev.* **114,** 664 (1959).
3. Willardson, R. K., and A. C. Beer, eds.: "Semiconductors and Semimetals," vol. 1, p. 200, Academic Press, Inc., New York, 1966–1968.

9e-3. Hall Coefficient and Mobility. The Hall effect is sometimes described by the "Hall angle" θ which is the angle through which the equipotential planes in a long rectangular specimen carrying a current are tilted when a magnetic induction B is applied normal to the direction of current flow. For conduction by electrons or holes only

$$\theta = \mu B$$

For a semiconductor with spherical energy surfaces containing only electrons which are nondegenerate and are scattered by acoustical lattice modes

$$R = \frac{-3\pi}{8}\frac{1}{nec} = \frac{-1.18}{nec}$$

If the scattering is due to ionized impurities, then

$$R = \frac{-315\pi}{512}\frac{1}{nec} = \frac{-1.93}{nec}$$

In case of degeneracy

$$R = -\frac{1}{nec} = \frac{-6.25 \times 10^{18}}{n} \qquad \text{cm}^3/\text{coul}$$

For hole conduction n becomes p (= density of holes) and the minus sign changes to a plus sign. The conductivity is given by

$$\sigma = ne\mu$$

where μ is the drift mobility of electrons or holes.

The *Hall mobility* is given by $\mu^{(H)} = R\sigma$ in the extrinsic range (only electrons or only holes). If both electrons and holes are present, the expressions for the Hall coefficient and the conductivity become

$$R = \frac{-n\mu_e^2 + p\mu_h^2}{(n\mu_e + p\mu_h)^2} \frac{3\pi}{8ec}$$

and

$$\sigma = e(n\mu_e + p\mu_h)$$

when scattering is by acoustical modes.

For intrinsic conduction $n = p$. The diffusion constants D_e and D_h for electrons and holes are related to the mobilities μ_e and μ_h by the *Einstein relation:*

$$D = \frac{kT}{e}\mu$$

The magnitude and temperature dependence of the mobility depend on the scattering mechanism. On the basis of a simplified theoretical model the temperature dependence is as follows:

Acoustical lattice scattering $\mu \sim T^{-\frac{3}{2}}$

Optical lattice scattering $\mu \sim (e^{\theta_l/T} - 1)$

Ionized impurity scattering $\mu \sim T^{\frac{3}{2}}$

Neutral impurity scattering $\mu \sim$ temperature-independent

Scattering by (edge) dislocations $\mu \sim T$

For details see ref. 1 and 3.

In reality the temperature dependence often deviates from the above proportionalities. The drift mobilities for Ge and Si in the lattice scattering range are given by (ref. 2)

Ge: $\mu_e^{(D)} = 4.90 \times 10^7 T^{-1.66}$ (100 K $< T <$ 280 K)

$\mu_h^{(D)} = 1.05 \times 10^9 T^{-2.33}$ (100 K $< T <$ 290 K)

Si: $\mu_e^{(D)} = 4.0 \times 10^9 T^{-2.6}$

$\mu_h^{(D)} = 2.5 \times 10^8 T^{-2.3}$

The Hall mobilities at room temperatures are (ref. 3)

Ge: $\mu_e^{(H)} = 4,500$ cm²/volt-sec

$\mu_h^{(H)} = 3,500$ cm²/volt-sec

Si: $\mu_e^{(H)} = 1,300$ cm²/volt-sec

$\mu_h^{(H)} = 500$ cm²/volt-sec

The $R\sigma$ product and the drift mobility of holes in germanium differ considerably; this is because of a small percentage of "fast" holes.

Hall mobilities of other semiconductors are listed in Tables 9e-1 and 9e-2.

References

1. Blatt, F. J.: *Solid State Phys.* **4**, 199 (1957).
2. Geballe, T. H.: "Semiconductors," p. 349, N. B. Hannay, ed., Reinhold Publishing Corporation, New York, 1959.
3. Putley, E. H.: "The Hall Effect and Related Phenomena," Butterworth & Co. (Publishers), Ltd., London, 1960.

9e-4. Thermoelectric Power. The thermoelectric power of semiconductors Q [measured in (micro) volts per degree] depends on the temperature, the number of carriers, the statistics, and the scattering mechanism. Some of the most common formulas (neglecting phonon-drag effects, ref. 2) are given below:

Extrinsic Range (one type of isotropic carriers, classical statistics, ref. 1):

$$Q = \pm \frac{k}{e}\left(r + \frac{\mathcal{E}_F}{kT}\right) = \pm \frac{k}{e}\left[r - \ln\frac{nh^3}{2(2\pi m^{(N)}kT)^{\frac{3}{2}}}\right]$$

where $+$ refers to p-type, $-$ to n-type semiconductors.

$r = 2$ for acoustical lattice scattering

$r = 4$ for ionized impurity scattering [actually (ref. 2) $r = 3.2$]

$r = 3$ for polar scattering

$r = 2.5$ for neutral impurity scattering

n = concentration of carriers, cm^{-3}

$m^{(N)}$ = density-of-states effective mass

ε_F = Fermi energy measured from bottom of conduction band or top of valence band (positive in both cases), eV

k = Boltzmann constant, eV/deg

Transition Range (electrons and holes, classical statistics, ref. 1):

$$Q = -\frac{k}{e(nc + p)} \left[r(nc - p) - nc \ln \frac{nh^3}{2(2\pi m_e^{(N)}kT)^{\frac{3}{2}}} + p \ln \frac{ph^3}{2(2\pi m_h^{(N)}kT)^{\frac{3}{2}}} \right]$$

where n = concentration of electrons

p = concentration of holes

c = mobility ratio = μ_e/μ_h

Intrinsic Range (classical statistics, ref. 1):

$$Q = -\frac{k}{e}\frac{c-1}{c+1}\left(\frac{\varepsilon_g}{2kT} + r + \frac{3}{4}\frac{c+1}{c-1}\ln\frac{m_e^{(N)}}{m_h^{(N)}} \right)$$

where $\varepsilon_g = \varepsilon_0 + aT$ is the energy gap at temperature T.

A typical example of the temperature dependence of the thermoelectric power in n- and p-type semiconductors is shown in Fig. 9e-3. Phonon drag effects appear at lower temperatures.

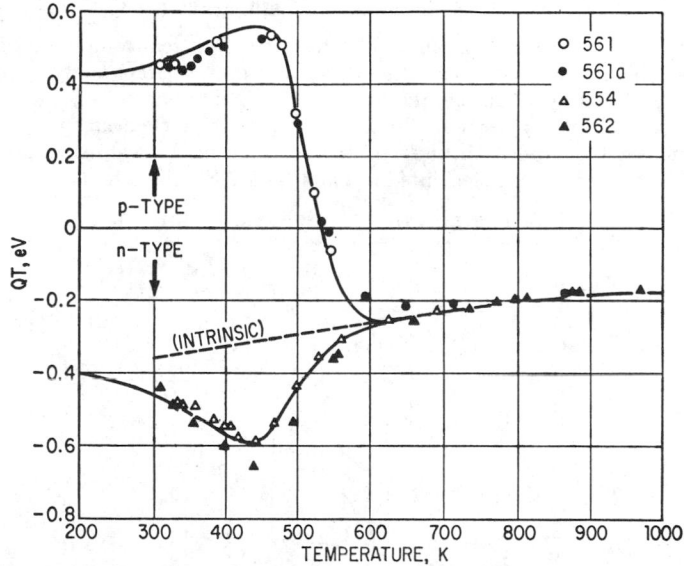

FIG. 9e-3. The thermoelectric power of silicon. Measured points and calculated curves for samples containing 8×10^{14} excess boron atoms per cm^3 (p-type) and 3×10^{14} excess phosphorus atoms per cm^3 (n-type). [*After Geballe and Hull, Phys. Rev.* **98**, 940 (1955).]

References

1. Johnson, V. A.: "Progress in Semiconductors," vol. I, p. 63, A. F. Gibson, ed., John Wiley & Sons, Inc., New York, 1956.
2. Herring, C.: *Phys. Rev.* **96**, 1163 (1954).

9e-5. Thermal Conductivity of Semiconductors. Values of the thermal conductivity for a few semiconductors are listed in Table 9e-4.

TABLE 9e-4. THERMAL CONDUCTIVITY OF SEMICONDUCTORS AT 300 K

	κ_L, watts/cm deg	Ref.		κ_L, watts/cm deg	Ref.
Diamond..........	1.51	1	AlP.............	0.8	2
Si...............	1.412	2	Te...............	0.063	1
Ge..............	0.606	2	PbS............	0.024	3
InSb.............	0.166	2	PbSe...........	0.017	3
InAs.............	0.273	2	PbTe............	0.022	3
InP..............	0.680	2	CdS.............	0.16	4
GaSb............	0.390	2	As$_2$Te$_3$...........	0.025	3
GaAs............	0.455	2	Bi$_2$Te$_3$...........	0.025	3
GaP.............	0.77	2	Mg$_2$Sn...........	0.10	3
AlSb.............	0.57	2	TiO$_2$ (c axis)......	$\sim$0.10	5
AlAs............	0.9	2			

References for Table 9e-4

1. Winkler, U.: *Helv. Phys. Acta* **28**, 633 (1955).
2. Willardson, R. K., and A. C. Beer, eds.: "Semiconductors and Semimetals," vol. 2, p. 3, Academic Press, Inc., New York, 1966–1968.
3. Aigrain, P., and M. Balkanski, eds.: "Selected Constants of Semiconductors," Pergamon Press, New York, 1961.
4. "American Institute of Physics Handbook," 2d ed., p. 4–94, McGraw-Hill Book Company, New York, 1963.
5. Thurber, W. R., and A. J. H. Mante: *Phys. Rev.* **139**, A1655 (1965).

9e-6. Impurities in Semiconductors. Purification of semiconductors by zone melting or by pulling a crystal from the melt is based on the fact that in most cases impurities tend to stay in the liquid rather than go into the solid. The essential parameter in this process is the *distribution coefficient k*, defined as the ratio of the concentrations of the impurity in the solid to that in the liquid host material. Distribution coefficients of certain elements in Si and Ge are compiled in Table 9e-5; these

TABLE 9e-5. DISTRIBUTION COEFFICIENTS IN SI AND GE†

Element	k (in Si)	k (in Ge)
Cu............	....	1.5×10^{-5}
Ag............	....	10^{-4}–10^{-6}
Au...........	3×10^{-5}	3×10^{-5}
Zn............	....	10^{-2}
B.............	6.8×10^{-1}	$\sim$20
Al............	1.6×10^{-3}	10^{-1}
Ga...........	4×10^{-3}	10^{-1}
In............	3×10^{-4}	1.1×10^{-3}
Tl............	....	4×10^{-5}
P.............	4×10^{-2}	1.2×10^{-1}
As............	7×10^{-2}	4×10^{-2}
Sb............	1.8×10^{-2}	3×10^{-3}
Fe............	....	2×10^{-6}
Co............	....	2×10^{-6}
Ni............	....	5×10^{-6}
Pt............	10^{-5}	2×10^{-6}

† From W. Crawford Dunlap, "Introduction to Semiconductors," John Wiley & Sons, Inc., New York, 1957. (Dunlap uses for k the term segregation coefficient S.)

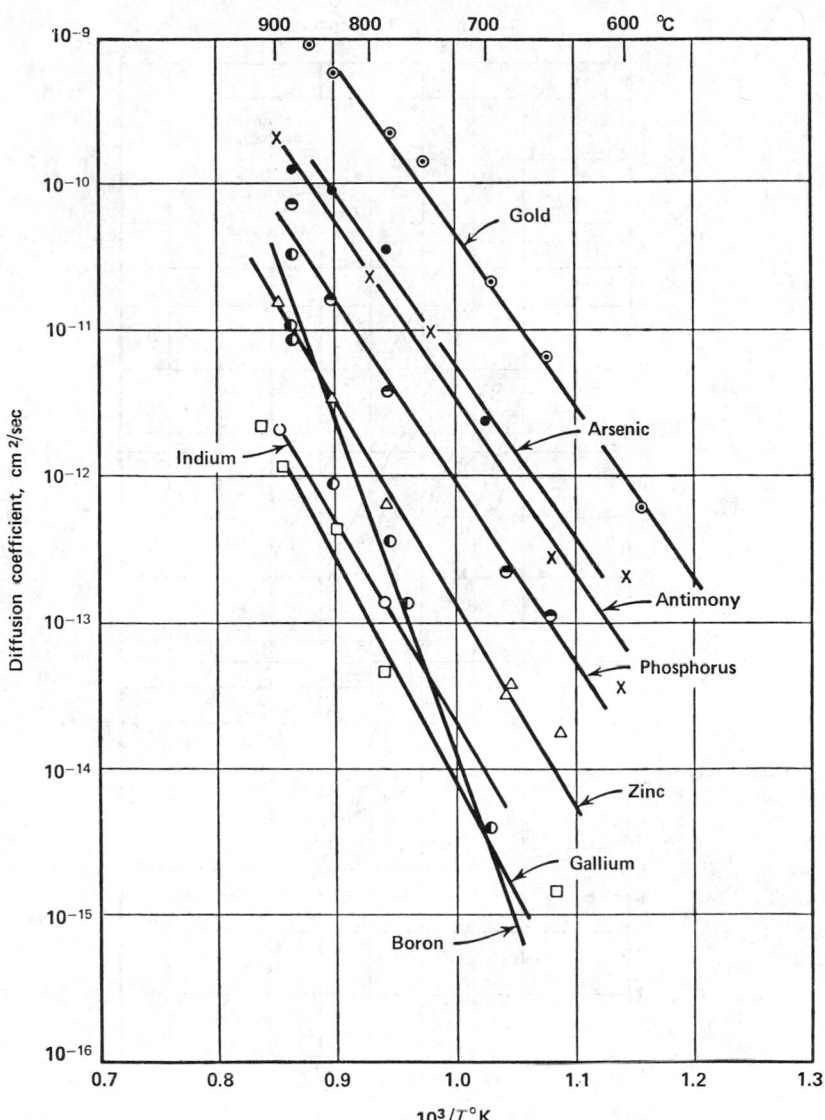

FIG. 9e-4. Diffusion coefficients of some impurities into germanium. (*After W. Crawford Dunlap, Jr., "An Introduction to Semi-conductors," p. 252, John Wiley & Sons, Inc., New York*, 1957.)

values refer to small impurity concentrations and equilibrium conditions (i.e., negligibly small gradients of concentration and temperature in the system).

Diffusion Coefficient. The "doping" of semiconductors as well as the preparation of *p-n* barriers depends greatly on the rate of diffusion. Extensive information exists about the diffusion coefficient *D* of impurities in Si and Ge. For other materials the data are scant. Approximate values of *D* are given in Figs. 9e-4 and 9e-5 and Table 9e-6. The magnitude of *D* depends somewhat on crystal perfection and

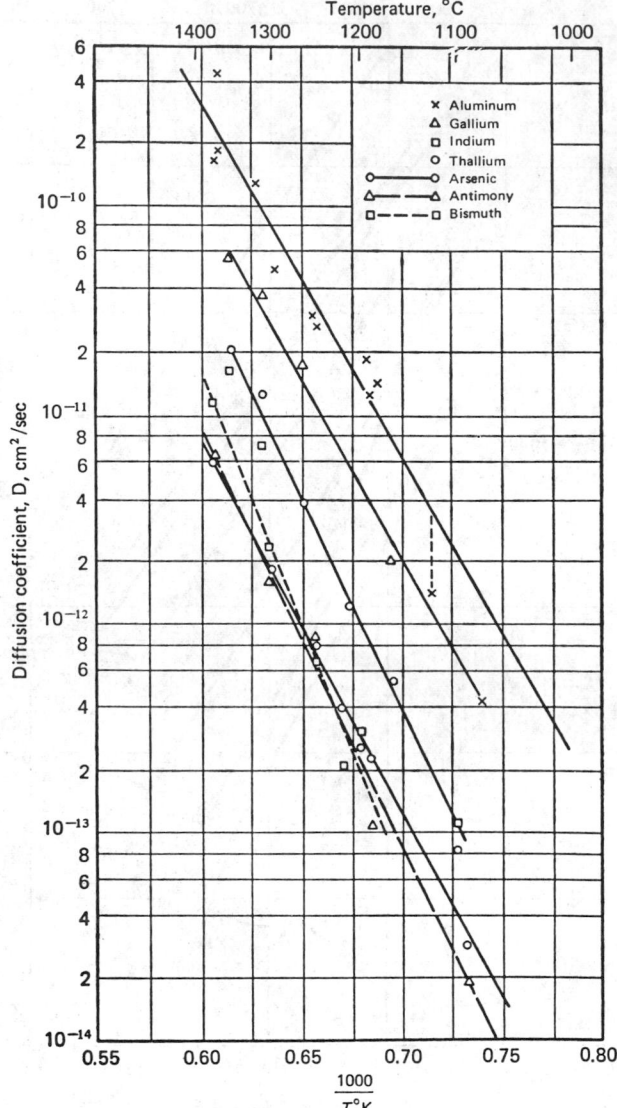

FIG. 9e-5. Diffusion coefficients of some impurities into silicon. [*After Fuller and Ditzen-burger, J. Appl. Phys.* **27,** 544 (1956).]

impurity content. The activation energies of diffusion are usually of the order of a few tenths to a few electron volts.

Activation Energies of Carriers from Donors and Acceptors. Chemical additives or physical imperfections (vacancies, interstitials, dislocations) create localized energy states which can release or capture (trap) free carriers. The locations of these energy levels for certain impurities in Si and Ge are indicated in Figs. 9e-6 and 9e-7. The activation energies are given for low concentrations of impurities; at high concentration the activation energies decrease.

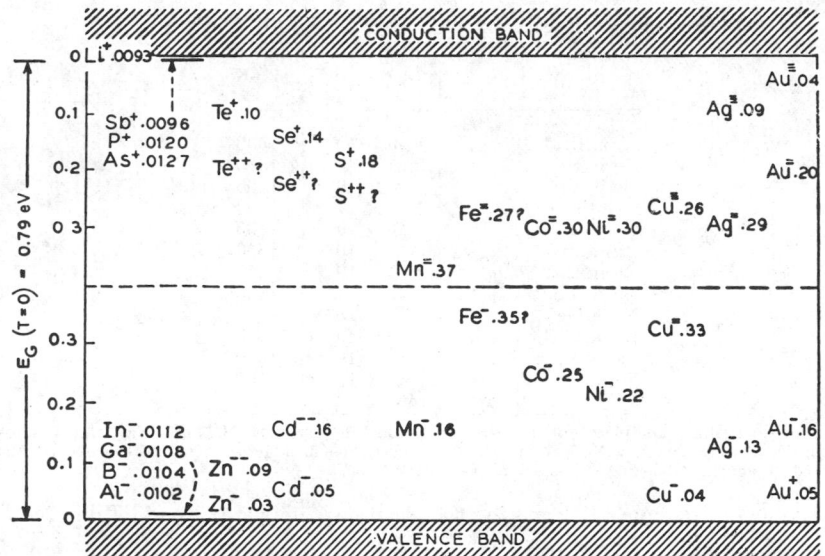

FIG. 9e-6. Energy levels of donors and acceptors between the valence and conduction bands of Ge. The ionization energy indicated (in eV) is required to produce what is presumed to be the ionic state shown and is measured from the nearest band edge, i.e., from the conduction band in the upper half of the diagram and from the valence band in the lower half. (*After T. H. Geballe, "Semiconductors," chap. 8, N. B. Hannay, ed., Reinhold Publishing Corporation, New York, 1959.*)

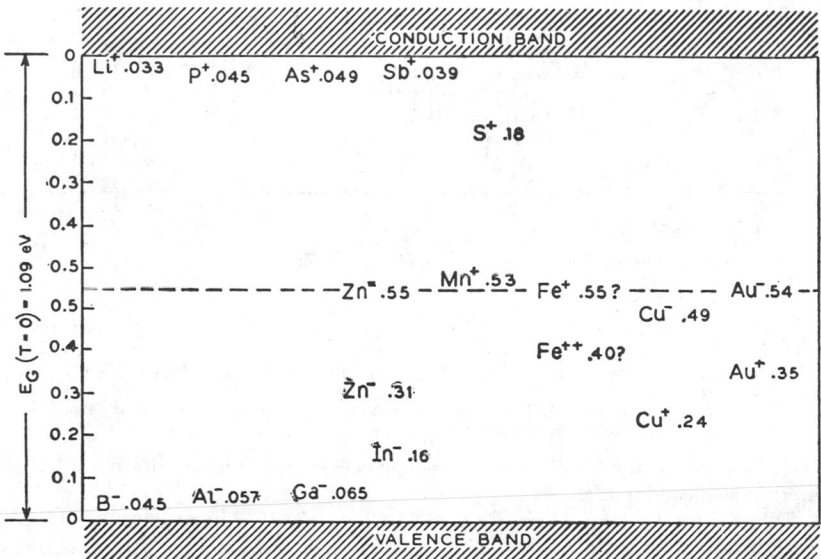

FIG. 9e-7. Energy levels of donors and acceptors between the valence and conduction bands of Si. (*After T. H. Geballe, "Semiconductors," chap. 8, N. B. Hannay, ed., Reinhold Publishing Corporation, New York, 1959.*)

TABLE 9e-6. DIFFUSION COEFFICIENTS AT TEMPERATURE T†

	D, cm²/sec	T, K
H in Si	2.4×10^{-4}	1500
He in Si	2.6×10^{-5}	1500
He in Ge	4×10^{-6}	1100
Li in Si	1.3×10^{-5}	1500
Li in Ge	1.3×10^{-5}	1100
Cu in Si	10^{-5}	1500
Cu in Ge	10^{-5}	1100
Cu in PbS	6×10^{-5}	800
Ni in Ge	10^{-5}	1100
Ni in PbS	2×10^{-5}	800
Fe in Si	1×10^{-6}	1500
Fe in Ge	2×10^{-6}	1100

† Sources: H. Reiss and C. S. Fuller, "Semiconductors," pp. 234, 244, N. B. Hannay, ed., Reinhold Publishing Corporation, New York, 1959; D. G. Thomas, "Semiconductors," pp. 289–290, N. B. Hannay, ed., Reinhold Publishing Corporation, New York, 1959.

Data on several impurity levels in other semiconductors are given in Table 9e-7.

TABLE 9e-7. ACTIVATION ENERGY OF IMPURITIES IN SEVERAL SEMICONDUCTORS

	Donor	$\mathcal{E}_d$, eV	Acceptor	$\mathcal{E}_a$, eV	Ref.
α-Sn	Sb	0.004	Mg	0.001	1
			Al	0.005	1
InSb	Te	<0.001	Zn	0.007	2
GaSb	....	......	Zn	0.024	3
AlSb	Te	0.068	...	.....	4
	Se	0.16	...	.....	4
GdS	Cl, Ga S-vacancy	0.03	...	.....	5 5
PbS	Cu	0.02	...	.....	6
	Ni	0.03	...	.....	6
ZnO	H, Zn (Interst.)	0.05	...	.....	7
SiC	N	0.085	Al	0.275	8

References for Table 9e-7

1. Busch, G., and E. Mooser: *Helv. Phys. Acta* **26,** 611, 697 (1953).
2. H. Hrostowski et al.: *Phys. Rev.* 100, 1672 (1955).
3. Welker, H.: *Physica* **20,** 893 (1954).
4. Turner, W. J., and W. E. Reese: *Phys. Rev.* **117,** 1003 (1960).
5. Kröger et al.: *Z. Physik. Chem.* **203,** 1 (1954).
6. Scanlon, W. W.: *Solid State Phys.* **9,** 124 (1959).
7. Hutson, A. R.: "Semiconductors," p. 582, N. B. Hannay, ed., Reinhold Publishing Corporation, New York, 1959.
8. Lely, J. A., and F. A. Kröger: "Halbleiter und Phosphore," p. 525, Schön and Welker, eds., Friedrich Vieweg und Sohn, Brunswick, Germany, 1958.

9e-7. Charge-carrier Lifetime. Several mechanisms affect the lifetime of charge carriers in semiconductors.

1. Direct recombination of an electron in the conduction band with a hole in the valence band. The radiative recombination lifetime τ_r associated with this process is given by

$$\tau_r = \frac{n_i}{2R}$$

where n_i is the intrinsic carrier concentration and R is the rate of radiative recombination.[1]

2. Indirect recombination (via recombination centers). The electron (hole) is captured by an impurity center, which subsequently captures a hole (electron). In most semiconductors the lifetime is limited by this process. The impurity centers can be either volume or surface states.

3. Auger process. An electron recombines with a hole, and the emitted energy is absorbed by a third carrier.

4. Trapping process. An electron (hole) is captured by a center (trap) but is thermally reexcited into the conduction (valence) band before it can recombine with a hole (electron). Surface states often act as traps.

Except for direct recombination, lifetimes pertain to electrons *or* holes. The minority-carrier lifetimes can also be expressed in terms of the capture cross sections of a center (A_n for electron capture by empty centers and A_p for hole capture by filled centers). The lifetime and the cross section are related by the following expression:

$$\tau_{n,p} = \frac{1}{N_t \langle v A_{n,p} \rangle}$$

where N_t is the density of centers and v is the thermal velocity of electrons or holes.

Data on some observed lifetimes τ_o and calculated radiative lifetimes τ_c at 300 K are listed in Table 9e-8.

TABLE 9e-8. CARRIER LIFETIMES

	τ_o, sec[*]	τ_c, sec	Ref.
Ge..............	10^{-2}	0.3	1
Si..............	10^{-4}	3.5	1
PbS.............	20×10^{-6}	63×10^{-6}	2
PbSe............	....	80×10^{-6}	2
PbTe............	....	6.8×10^{-6}	2
GaAs...........	$\tau_{maj} > 10^{-6}$	....	3
	$\tau_{min} \ll 10^{-6}$		3
GaSb...........	10^{-6}	....	3
InP.............	$\tau_e \approx 10^{-6}$	....	3
	$\tau_h \approx 2 \times 10^{-3}$		3
InAs...........	$10^{-7} - 10^{-8}$	....	3
InSb...........	10^{-7}	....	3

[*] Largest values observed in single crystals. Films or specially treated samples may show much longer, or much shorter, "effective" lifetimes.

References for Table 9e-8

1. Petritz, R. L.: In "Photoconductivity Conference," p. 62, R. G. Breckenridge et al., eds., John Wiley & Sons, Inc., New York, 1956.
2. Scanlon, W. W.: *Solid State Phys.* **9**, 117, 136 (1959).
3. Willardson, R. K., and A. C. Beer, eds.: "Semiconductors and Semimetals," vol. 2, p. 205, Academic Press, Inc., New York, 1966–1968.

9e-8. Lattice Properties. Most semiconductors show an optical absorption and an anomalous dispersion in the far-infrared region. This effect is rather small in covalent semiconductors like Ge and Si; it increases, however, with increasing polar character. On the basis of an expression derived by Born and Huang (ref. 1) it is possible to calculate an *effective charge* e^* which describes the deviation from homo-

[1] W. Van Roosbroeck and W. Shockley, *Phys. Rev.* **94**, 1558 (1954).

polarity. The optical lattice vibrations are divided into transverse and longitudinal modes with frequencies ω_T and ω_L.

The acoustical lattice vibrations are related to the *elastic constants* C_{lm}; these modes also determine the lattice specific heat, which is described by the *Debye temperature* θ. The phonon dispersion (frequency ω vs. wave vector q) has been measured for a number of cubic crystals, using the technique of inelastic neutron scattering (ref. 2). An example (Ge) is shown in Fig. 9e-8.

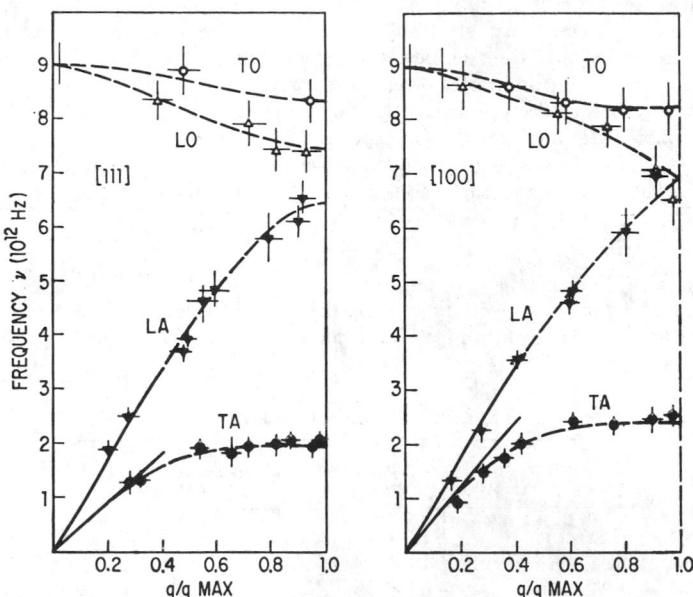

FIG. 9e-8. Frequency–wave-number relations for the $<100>$ $<111>$ directions in germanium. [*After Brockhouse and Iyengar, J. Phys. Chem. Solids* **8**, 400 1959.]

Data on ω_T, ω_L, e^*, θ_0, and elastic constants for a number of semiconductors are assembled in Table 9e-9.

References

1. Born, M., and V. Huang: "Dynamical Theory of Lattices," p. 114, Oxford University Press, New York, 1954.
2. Brockhouse, B. N., and P. K. Iyengar: *Phys. Rev.* **111**, 747 (1958).

9e-9. Refractive Index and Dielectric Constant. The *dielectric constant* in ionic lattices depends on frequency. Neglecting dissipative forces this dependence is given by

$$\epsilon(\omega) = \epsilon_\infty + \frac{\epsilon_{st} - \epsilon_\infty}{1 - (\omega/\omega_T)^2}$$

where ω_T is the lattice vibration frequency for long-wavelength transverse waves. For high frequencies $\omega \gg \omega_T$: $\epsilon = \epsilon_\infty$, the optical dielectric constant; for low frequencies $\epsilon = \epsilon_{st}$, the static dielectric constant.

The optical dielectric constant is related to the *refractive index* by the following expression:

$$n^2 = \epsilon_\infty$$

For nonpolar materials $\epsilon_{st} = \epsilon_\infty$. The dielectric constants of Ge and Si are 16 and 11.8, respectively. Values of n and ϵ_{st} for other semiconductors are given in Table 9e-9.

TABLE 9e-9. LATTICE PROPERTIES OF SEMICONDUCTORS

Material	ω_T, cm⁻¹	ω_L, cm⁻¹	Ref.	C_{11}	C_{12}	C_{44}	θ	Ref.	ϵ_s	n	e^*	Ref.
				10^{11} N/m² at 300 K								
Diamond	1,333		1a	10.76	1.25	5.76	2240	1c	5.5	2.4	0	1d
Si	518		1a	1.656	0.639	0.796	645	4	11.7	3.44	0	1d
Ge	309		1a	1.288	0.483	0.671	374	4	15.8	3.97		1d
InSb	180	191.3	1b	0.672	0.367	0.302	203	3	17.88	3.96	0.42	2
InAs	218.9	243.3	1b	0.833	0.453	0.396	247	3	14.55	3.44	0.56	2
InP	303.7	345	1b				321	3	12.37	3.1	0.66	2
GaSb	230.5	240.3	1b	0.885	0.404	0.433	266	3	15.69	3.8	0.33	2
GaAs	268.2	290.5	1b	1.188	0.538	0.594	344	3	13.13	3.3	0.51	2
GaP	366.3	401.9	1b						10.18	2.9	0.58	2
AlSb	318.8	339.6	1b	0.894	0.442	0.415	292	3	11.2	3.14	0.48	2
BN	820	835	6a						7.1	2.1	1.14	5
ZnS(Z)	339	298	6a	1.046	0.653	0.461	315	6b	8.3	2.24	0.48	6a
CdS(W)	261	295		0.907; $C_{13}=0.510$	0.581; $C_{33}=0.938$	0.150		6b	$\epsilon_{11}=9.4$; $\epsilon_{33}=10.3$	2.29		
PbS	238	223	9	1.27	0.298	0.248	227	11,14	205	17.4		8
PbSe	65	44	8	1.02	0.38	0.25	138	12	280	23.6		8
PbTe	31	110	10	1.08	0.077	0.134	125	14	400	5.63	0.55	5
TiO₂	125–167, 400–533	360–400, 770–823	15	0.273; $C_{13}=0.149$	0.176; $C_{33}=0.484$	0.125; $C_{66}=0.194$	~758	13,14; 16	173 (c); 89 (a)			15

References for Table 9e-9

1*a*. Cowley, R. A.: *Proc. Phys. Soc. (London)* **88**(*II*), 463 (1966).
1*b*. Hass, Marvin: in "Semiconductors and Semimetals," vol. 1, p. 7, R. K. Willardson and A. C. Beer, eds., Academic Press, Inc., New York, 1966–1968.
1*c*. McSkimin, H. J., and W. L. Bond: *Phys. Rev.* **105**, 116 (1957).
1*d*. Aigrain, P., and M. Balkanski, eds.: "Selected Constants of Semiconductors," Pergamon Press, New York, 1961.
2. Hass, Marvin: In "Semiconductors and Semimetals," vol. 1, p. 14, R. K. Willardson and A. C. Beer, eds., Academic Press, Inc., New York, 1966.
3. Drabble, J. R.: in "Semiconductors and Semimetals," vol. 2, p. 110, R. K. Willardson and A. C. Beer, eds., Academic Press, Inc., New York, 1967.
4. McSkimin, H. J.: *J. Appl. Phys.* **24**, 988 (1953).
5. Burstein, E.: In "Lattice Vibrations," p. 315, R. Wallis, ed., Pergamon Press, New York, 1965.
6*a*. Reynolds, D. C., et al.: *Phys. Stat. Solidi* **12**, 3 (1965).
6*b*. Berlincourt, D., et al.: *Phys. Rev.* **129**, 1009 (1963).
7. Zemel, J. N., et al.: *Phys. Rev.* **140**, A330 (1965).
8. Burstein, E., et al.: "Proceedings of the International Conference on the Physics of Semiconductors," p. 1065, M. Hulin, ed., Dunod, Paris, 1964.
9. Elcombe, M. M.: *Proc. Roy. Soc. (London)*, ser. A, **300**, 210 (1967).
10. Cochran, W., et al.: *Proc. Roy. Soc. (London)* ser. A, **293**, 433 (1966).
11. Bhagavantam, S., and T. S. Rao: *Nature* **168**, 42 (1951).
12. Ramachandran, G. N., and N. A. Wooster: *Acta Cryst.* **4**, 335 (1951).
13. Houston, B., et al.: *J. Appl. Phys.* **39**, 3913 (1968).
14. Parkinson, D. H., and J. E. Quarrington: *Proc. Phys. Soc. (London)*, ser. A, **67**, 569 (1954).
15. Cronemeyer, D. C.: *Phys. Rev.* **112**, 800 (1958).
16. Values corrected by F. Birch from work by Verma, R. K.: *J. Geophys. Res.* **65**, 757 (1960).

9f. Properties of Ionic Crystals

R. J. FRIAUF[1]

University of Kansas

K. F. YOUNG[2]

The National Bureau of Standards

W. J. MERZ[3]

RCA Laboratories Zurich, Switzerland

IONIC CONDUCTIVITY AND DIFFUSION IN IONIC CRYSTALS

9f-1. Ionic Conductivity and Diffusion. These phenomena are ascribed to the presence of ionic defects—vacancies where ions are missing from normally occupied positions and ions in interstitial positions in the structure. *Schottky defects* are com-

[1] Ionic Conductivity and Diffusion.
[2] Dielectric Constants of Inorganic Crystals.
[3] Piezoelectric and Pyroelectric Properties; Ferroelectric and Antiferroelectric Properties.

binations of cation and anion vacancies, as in the alkali halides and alkaline earth oxides. *Frenkel defects* are combinations of vacancies and interstitial ions, for cations as in the silver halides, or for anions as in the alkaline earth halides. At high temperatures the defects exist in thermodynamic equilibrium in the crystal; for Schottky defects in MX crystals, for example, the concentration or mole fraction increases with temperature according to (ref. 23)

$$x = x_0 \exp\left(-\frac{1}{2}\frac{h_f}{kT}\right) \qquad x_0 = \exp\left(\frac{1}{2}\frac{s_f}{k}\right) \tag{9f-1}$$

where h_f and s_f are the enthalpy and entropy of formation of a pair of defects. At lower temperatures the mole fraction is usually controlled by the presence of aliovalent impurities.

The random jumping of a defect gives rise to a *microscopic diffusion coefficient* for the defect of

$$d = d_0 \exp\left(-\frac{\Delta h}{kT}\right) \qquad d_0 = \tfrac{1}{6}\nu a^2 \exp\left(\frac{\Delta s}{k}\right) \tag{9f-2}$$

where ν is an attempt frequency, a is the jump distance, and Δh and Δs are the activation enthalpy and entropy for the jump. (The factor $\frac{1}{6}$ is appropriate for a cubic lattice.) In an electric field there is also a drift *mobility*

$$\mu = \mu_0 \exp\left(-\frac{\Delta h}{kT}\right) \qquad \mu_0 = \left(\frac{q}{kT}\right) d_0 \tag{9f-3}$$

Here μ_0 has been obtained from d_0 with the *microscopic Einstein relation*

$$\frac{d}{\mu} = \frac{kT}{q} \tag{9f-4}$$

The conversion factor is $k/e = 0.862 \times 10^{-4}$ volt/K with d in cm²/sec and μ in cm²/volt-sec. Equations (9f-1) to (9f-3) are used to express the observed conductivity and diffusion coefficients in the following sections.

Ionic crystals covered in these tables include halides, simple inorganic radicals (such as nitrates and azides), binary oxides, and the other chalcogenides (sulfides, selenides, and tellurides). Excluded from consideration are III-V compounds, ternary oxides (such as spinels and perovskites), and glasses and zeolites. Conductivity and self-diffusion coefficients are given for pure crystals only, but some information from experiments on doped crystals is contained in Table 9f-2. The effect of high pressure on conductivity and data for mixed electronic and ionic conductors are also presented. Space limitations prevent any consideration of the extensive recent literature on dielectric and anelastic relaxation, thermoelectric phenomena, and effects of radiation and plastic deformation on conductivity and diffusion. Similarly the diffusion of all foreign ions is excluded because of the proliferation of results. Many of the excluded topics are discussed in some of the books and review articles given in the general references.

9f-2. Conductivity for Ionic Conductors. The conductivity can be determined by passage of direct current through the sample if sufficient precautions are taken. More recently, however, most measurements have been made with current pulses of the order of 10^{-2} to 10^{-3} sec duration or alternating currents at frequencies of 1 to 10 kHz, in order to avoid large polarization effects at the electrodes.

In most cases a plot of log σ vs. $1/T$ is approximately a straight line, at least for a limited temperature range, allowing an empirical representation of the data as

$$\sigma = \sigma_0 \exp\left(-\frac{W}{kT}\right) \tag{9f-5}$$

The parameters σ_0 and W are listed in Table 9f-1. The conductivity at the melting temperature has been calculated from Eq. (9f-5) if it is not given in the references.

The values for σ_0 and W are not always so accurate as the number of significant figures would indicate. With good single-crystal or polycrystalline samples of high purity a careful worker can reproduce results within a few percent, but data from different laboratories may differ by 5 to 10 percent, and discrepancies of 50 percent are not uncommon. Hence W may be reliable to a few percent in favorable circumstances or to perhaps 10 percent in less favorable cases, and a discrepancy of 50 percent in σ_0, which is very sensitive to the choice of W, is not surprising. For this reason several representative sets of data, if available, have been given for each substance.

9f-3. Concentration and Mobility of Defects in Ionic Crystals. The conductivity of a crystal containing several types of defects is

$$\sigma = N\Sigma_j q_j x_j \mu_j \qquad (9f\text{-}6)$$

where N is the number of molecules per unit volume of the perfect crystal, and q_j is the magnitude of the charge of the jth defect. If only one type of defect makes an appreciable contribution to the conductivity, the use of Eqs. (9f-1) and (9f-3) gives the observed form of Eq. (9f-5). In the *intrinsic* region for temperatures near the melting point $W_{\text{intr}} = \frac{1}{2}h_f + \Delta h$ and $\sigma_0 = Nqx_0\mu_0$, and in the *extrinsic* region for lower temperatures $W_{\text{extr}} = \Delta h$ and $\sigma_0 = Nqc\mu_0$ since x is maintained constant at the impurity concentration c. This simple explanation corresponds to the frequent observation of two different temperature ranges with different slopes in the plot of log σ vs. $1/T$, especially for the initial observations on a substance, and the two slopes are often combined to obtain h_f and Δh from the expressions for W_{intr} and W_{extr}. This is presumably the extent of the analysis when only activation enthalpies are given in Table 9f-2. Recent work has shown, however, that such an analysis is at best only tentative because of contributions of other types of ions, association and precipitation of impurities, and overlapping of the different temperature regions.

In early work many transport number determinations were made by electrolysis in order to identify the ions carrying the current. When only one type of ion contributes to the conductivity, these experiments have, in fact, verified Faraday's laws of mass transport to an accuracy of 1 percent. When several types of ions, or both ions and electrons, however, make appreciable contributions, such experiments have not given very reliable results, presumably because of experimental difficulties at the electrodes and at the interfaces between the several samples involved. Hence only a handful of these experiments have been reported in the last ten years, and no separate table of results is provided.

In a few recent investigations of alkali halides an attempt has been made to separate cation and anion contributions to the conductivity by fitting a sum of two terms of the form of Eq. (9f-5) to the observed total conductivity, as indicated by Eq. (9f-6), and some results are given in this form in Table 9f-1. Often measurements of tracer diffusion coefficients allow evaluation of ionic transport numbers, but even these may not be completely unambiguous if vacancy pairs contribute noticeably to diffusion (ref. Ne1). The most reliable results are obtained from analysis of measurements on crystals intentionally doped with aliovalent impurities, with due account taken of mass-action laws, association of charged defects and impurities, and long-range Debye-Hückel interactions (ref. Be1). Most of the results in Table 9f-2 have been obtained in this way.

The temperature dependences of x and μ are given by Eqs. (9f-1) and (9f-3). It should be observed, however, that μ_0 contains a factor $1/T$, which is also carried over into σ_0. For this reason d_0 is listed rather than μ_0 in Table 9f-2; the conversion is obtained immediately from Eq. (9f-4). When the factor of $1/T$ is not explicitly removed from μ_0 or σ_0, the apparent activation energy is smaller than the correct value by kT, which is of the order of 0.05 to 0.15 eV for temperatures from 300 to 1500°C.

9f-4. Effect of Pressure on Conductivity. When the effect of high pressure is taken into consideration, Eqs. (9f-1) and (9f-3) are modified to give

$$x = x_0 \exp \left(- \frac{h_f + P v_f}{2kT} \right) \qquad \mu = \mu_0 \exp \left(- \frac{\Delta h + P \, \Delta v}{kT} \right) \qquad \text{(9f-7)}$$

where v_f is the change in volume of the crystal when a pair of Schottky defects is formed, and Δv is the activation volume when a defect moves from one position to another. If only one type of defect contributes appreciably to the conductivity, the pressure dependence of the conductivity is given by

$$\sigma = \sigma_0 \exp \left(- \frac{P \, \Delta V}{kT} \right) \qquad \text{(9f-8)}$$

where $\Delta V_{\text{intr}} = \frac{1}{2} v_f + \Delta v$, for instance, in the intrinsic range.

The pressure dependence of the original data is expressed by a *pressure coefficient*

$$\alpha = - \left(\frac{\partial \ln \sigma}{\partial P} \right)_T \qquad \text{(9f-9)}$$

The corresponding *free volume* from Eq. (9f-8) is

$$\Delta V = RT\alpha = 82.0 \times T \times \alpha$$

with ΔV in cm³/mole and α in atm⁻¹. Values of pressure coefficients and free volumes are given for a number of substances in Table 9f-3.

9f-5. Mixed Electronic and Ionic Conductors. Many ionic crystals have an appreciable electronic conductivity in addition to their ionic conductivity. Exclusive ionic conductivity occurs for nearly all halides (the cuprous halides being the only noteworthy exception) and for crystals with simple inorganic radicals. Beryllia also has mainly ionic conductivity, but the other alkaline earth oxides show progressively larger amounts of electronic conductivity, especially at higher temperatures. The only other predominantly ionic conductors are crystals with the fluorite structure such as calcia-stabilized zirconia and even sodium sulfide, perhaps some rare-earth-type trioxides such as scandia and neodymia, and a new class of complex sulfides typified by Ag_3SI. Appreciable, but not exclusive, ionic conductivity is displayed by the cuprous halides, some simple metal oxides such as alumina and tetragonal zirconia, and most rare-earth oxides such as ceria and dysprosia. Traces of ionic conductivity (a few percent) are present in the copper and silver chalcogenides. Electronic conductivity (by electrons or holes) is dominant in transition-metal oxides such as Cr_2O_3, and in all other divalent chalcogenides such as ZnO and PbS.

It should be clear that a fairly complicated situation exists when both electronic and ionic defects are present to an appreciable extent in a crystal. The treatment of the various interactions (refs. 21 and 22) shows that the defect structure may be profoundly influenced by the atmosphere surrounding the crystal or by deviations from stoichiometry of the crystal. Hence conductivity results are practically meaningless unless these conditions are specified, and similar remarks apply to diffusion. Fortunately much more attention has been devoted in recent years to control and measurement of the environment, and this information is provided where pertinent in Tables 9f-5 and 9f-6 in one of three ways: saturation of one constituent by contact with the metal or high vapor pressure of a volatile component, measurement of the oxygen partial pressure, or determination of the deviation from stoichiometry.

Several experimental techniques have been used to distinguish between electronic and ionic conductivity (ref. 56). (1) The earliest was direct determination of mass transport by electrolysis, but this has often been unreliable (ref. He1) and is seldom used at present.   (2) Polarization effects are often observed; namely, the a-c conductivity at moderately high frequencies like 100 kHz is considerably less than the d-c conductivity.   The simplest assumption is that the a-c value is due to the electronic conductivity only, whereas the d-c value represents the total conductivity (ref. Ve3). Despite the appeal of this interpretation the results are usually ambiguous, and much clarification is needed to make this method reliable (ref. Mc1).   (3) If the potential drop between the electrodes is kept below the decomposition voltage of the sample, it may be assumed that an ionic current cannot flow to the electrode, and the remaining current is then ascribed to electronic conductivity (refs. Wa5, Wa6).   This method appears to be fairly reliable in some cases, but note must be taken of the range of chemical potentials occurring in such experiments.   (4) If the conductivity is completely ionic, an emf that can be calculated from thermodynamic data should be established when the ends of the sample are at different chemical potentials (ref. Wa2), and this has been amply verified for calcia-stabilized zirconia, for instance.   If some electronic conductivity is also present, part of the emf is effectively short-circuited out, and hence the reduction of the observed emf below the thermodynamic value gives an indication of the amount of electronic transport (ref. Sc6).   This method is the most commonly used, especially for oxides, and appears to give a reliable estimate of the average transport number if care is taken to establish a well-defined chemical potential at each end of the sample and to ensure thermodynamic equilibrium.   An unfortunate aspect of this method is that it does not distinguish between electrons and holes for the electronic part of the conductivity, or between different types of ions for the ionic conductivity, but often other information is available.   (5) The amount of ionic conductivity can be calculated from tracer diffusion coefficients with the Einstein relation if the charge on the defect and the correlation factor are known (see Sec. 9f-7). Since the last two items require a rather detailed knowledge of the diffusion mechanism, this approach is most often useful to establish an order of magnitude, especially when the ionic conductivity is very much smaller than the electronic part.

Table 9f-4 gives in most cases the *total conductivity*, which can often be determined more accurately than the transport numbers.   Table 9f-5 gives the *ionic transport numbers*, which are defined as the fraction of the total current carried by ions.   The two tables should be used together to obtain an estimate of the magnitude and nature of the conductivity for a particular substance.   Substances have been listed only when there is some information about the ionic part of the conductivity; thus the numerous articles dealing solely with semiconducting behavior in ionic crystals such as ZnO and CdS are not included.

9f-6. Diffusion.   The tracer diffusion coefficient for an ion which can diffuse by means of several types of defects is

$$D_T = \Sigma_j f_j x_j d_j \tag{9f-10}$$

where f_j is the correlation factor (see Sec. 9f-7) and x_j and d_j are given by Eqs. (9f-1) and (9f-2).   When only a single mechanism is important, the temperature dependence is given by

$$D_T = D_0 \exp\left(-\frac{W}{kT}\right) \tag{9f-11}$$

and this form is usually used to represent experimental results.   Empirically determined values of D_0 and W are given in Table 9f-6.

In the intrinsic region the parameters in Eq. (9f-11) are given by

$$D_0 = \tfrac{1}{6}\nu a^2 x_0 f \exp\left(\frac{\tfrac{1}{2}s_f + \Delta s}{k}\right) \quad \text{and} \quad W = \tfrac{1}{2}h_f + \Delta h \tag{9f-12}$$

Theoretical estimates indicate that W should be several electron volts, as observed, and that $\frac{1}{2}s_f + \Delta s$ should be at most a few entropy units, leading to a value of $D_0 \sim 10^{-3}$ to $10 \text{ cm}^2/\text{sec}$. When an appreciably different value of D_0 is obtained empirically, it is usually an indication that some disturbing influence, such as impurities or grain-boundary diffusion, is dominating over the assumed thermodynamic equilibrium for volume diffusion.

Since the temperature dependence is the same for all types of defects, indirect methods must be used to distinguish a particular type of defect; this has been done with considerable success in many instances, as indicated in Table 9f-6. Some of these methods are (1) determination of the influence of aliovalent impurities in doped crystals, (2) study of correlation effects as described in Sec. 9f-7, and (3) observation of the effect of varying the stoichiometry or ambient pressure of one of the constituents of the crystal.

Among experimental methods for measuring diffusion coefficients with radioactive or isotopic tracers, *sectioning* is the most direct and reliable. *Surface counting, gaseous exchange,* and *solution exchange* are more sensitive but sometimes less reliable. Other methods of detection involve changes in *optical absorption, X-ray emission,* and *semi-conducting properties* or observation of *additive coloration* or *electrotransport.* The line width in *nuclear magnetic resonance* allows a determination of the temperature dependence and an estimate of the magnitude of diffusion for stable nuclei. The rate of *oxidation* and *sintering* processes can also be used to evaluate diffusion coefficients when the process is sufficiently well understood.

The remarks concerning the accuracy of the results for ionic conductivity apply here with even more need for caution. For most halides pure single crystals are available, the melting points are not excessively high, and the influence of the surrounding atmosphere is often unimportant (ref. 50); hence in favorable cases an accuracy approaching that for the conductivity may be realized. For the usually semiconducting and often refractory chalcogenides, however, the situation is much less favorable. The high melting temperatures and difficulties of obtaining pure materials suggest that very few intrinsic properties have yet been observed for these substances (ref. 54). Further-more the influence of grain boundaries is just beginning to be investigated, and yet a number of measurements have been made on sintered or pressed powder samples with porosities up to 5 or 10 percent. Finally the defect structure is strongly influenced by any excess or deficit of the constituents, as discussed in Sec. 9f-5. The data in Table 9f-6 may nonetheless be useful both as a survey of existing experimental efforts and as a stimulus to better understanding.

9f-7. Correlation Effects in Diffusion. Both the ionic conductivity and diffusion of a charged defect are caused by the jumping of the defect through the crystal, and the connection of these two phenomena is given by the microscopic Einstein relation in Eq. (9f-4). If a single type of defect is responsible for all the observed conductivity and diffusion, Eqs. (9f-6) and (9f-10) may be combined (without the correlation factor) to give a *macroscopic Einstein relation* that defines $D_{\text{conductivity}}$.

$$D_{\text{conductivity}} = \left(\frac{kT}{Nq^2}\right)\sigma \qquad (9f\text{-}13)$$

In many instances this relationship is at least approximately satisfied, but there are four ways in which deviations may occur. (1) There may be another contribution to the conductivity, such as an electronic part or another type of ionic defect. (2) There may be neutral complexes of defects, such as vacancy pairs in the alkali halides, which contribute to the diffusion but not to the conductivity. (3) In the diffusion of tracers there are correlations in the random-walk motion of tracer atoms that lead to correlation factors, as first described by Bardeen and Herring (ref. Ba2). (4) In interstitialcy mechanisms there are also different displacements for the tracer atom and for the

charge of the defect.   This displacement effect is usually included with the genuine
correlation effects to give an overall correlation factor for interstitialcy mechanisms.
   The *experimental correlation factor* is defined by

$$f = \frac{D_{\text{tracer}}}{D_{\text{conductivity}}}$$

Theoretical correlation factors may be calculated by considering the geometry of the
diffusion mechanism and of the lattice (see refs. below).   Comparison of experimental
and theoretical values will then often point to a particular mechanism for diffusion.
Experimental and theoretical correlation factors are presented in Table 9f-7.

Guide to references on theoretical correlation factors.
   General treatment: 24, Ba2, Co2, Co3, Ho10
   Vacancy mechanisms: Ba2, Co2, Fr1, Sc8
   Interstitial and interstitialcy mechanisms: Co3, Fr2, Mc2
   Vacancy pairs and impurity complexes: Co2, Ho10, Le1, Li1
   Anisotropic lattices: Co2, Gh1, Hu1, Hu2, Md1, Mu3
   Disordered lattices: Ri2, Yo3
   Diffusion by nuclear magnetic resonance: Ei2, St5
   Isotope effects: Le2, Th1

TABLE 9f-1. CONDUCTIVITY FOR IONIC CONDUCTORS
[The conductivity is given as $\sigma = \sigma_0 \exp(-W/kT)$.]

Substance	Form	T_m, °C	$\sigma(T_m)$, (ohm-cm)$^{-1}$	T range, °C	σ_0, (ohm-cm)$^{-1}$	W, eV	Specific reference	Other references
ALKALI HALIDES								
LiH..................	sc	688	4×10^{-2}	480–630	4×10^7	1.72	Pr1	
				240–480	1	0.53	Pr1	
LiD..................	sc	......	....	480–630	1×10^7	1.72	Pr1	
LiF..................	sc	842	2.4×10^{-3}	540–720	$6 \times 10^9/T$	2.07	Ja6	Be7, Be8,
				340–540	$5 \times 10/T$	0.70	Ja6	Ha19, Le5,
	sc	......	1.5×10^{-3}	560–750	$1.6 \times 10^9/T$	1.99	Ba7	St7
				330–560	$4.5 \times 10^{10}/T$	1.65	Ba7	
LiCl..................	sc	606	9×10^{-3}	480–570	2.5×10^6	1.47	Ha19	Le5
	pc	......	1.8×10^{-3}	400–550	2.5×10^5	1.42	Gi1	
				30–350	1.2	0.59	Gi1	
LiBr..................	sc	550	1.8×10^{-2}	440–540	1.4×10^6	1.29	Ha19	
	pc	......	1.4×10^{-2}	350–500	4.2×10^5	1.22	Gi1	
				30–300	3.3	0.56	Gi1	
	sc	......	....	160–360	8×10^{-2}	0.43	Al3	
LiI..................	sc	452	5×10^{-2}	340–420	9.6×10^5	1.05	Ha19	
	pc	......	7×10^{-2}	250–350	1.8×10^5	0.92	Gi1	
				30–150	1.4×10^{-1}	0.36	Gi1	
NaF..................	pc	992	3×10^{-3}	330–990	1.3×10^3	1.42	Ph1	Le5
NaCl..................	sc	800	1.0×10^{-3}	520–740	$4.7 \times 10^8/T$	1.86ᵃ	Fu3	Bi2, Br5,
				520–740	$1.2 \times 10^9/T$	2.07ᵃ	Fu3	Do6, Dr2,
			1.2×10^{-3}	720–800	$2.4 \times 10^{10}/T$	2.19	Ne1	Et1, Ja4,
				550–650	$9.2 \times 10^8/T$	1.92	Ne1	Ka4, Ko1,
				275–425	$3/T$	0.65	Ne1	La4, Ma6, Le1
NaBr..................	sc	755	1.2×10^{-3}	450–700	$2.1 \times 10^8/T$	1.68	Ma6	Le5, Ph1,
				300–450	$3.5 \times 10^2/T$	0.84	Ma6	Sc2
	sc	......	2.1×10^{-3}	610–730	$2.3 \times 10^8/T$	1.64	Ho8	
				490–570	$3 \times 10^2/T$	0.80	Ho8	
NaI..................	pc	661	2.1×10^{-3}	350–600	8.1×10^3	1.23	Ph1	Le5
				170–350	6×10^{-2}	0.60	Ph1	
KF..................	sc	846	6×10^{-4}	660–790	2×10^7	2.34	Ka1	Le5
				400–500	4×10^{-1}	1.02	Ka1	
KCl..................	sc	768	2.4×10^{-4}	570–750	$4.1 \times 10^6/T$	1.66ᶜ	Fu4	As3, Be2,
				570–750	$5.6 \times 10^{10}/T$	2.36ᵃ	Fu4	Bi4, Gr1,
	sc	......	1.4×10^{-4}	340–640	$2.3 \times 10^8/T$	1.90	Mi4	He4, Le5,
	sc	......	4.5×10^{-5}	480–680	$5.9 \times 10^7/T$	1.88	Al4	Me2, Ph2, Pe1, Wa4
KBr..................	sc	728	1.3×10^{-4}	300–700	$3.1 \times 10^8/T$	1.91ᶜ	Da9	Gr4, Le5,
				300–700	$7.1 \times 10^9/T$	2.21ᵃ	Da9	Pe1, Ph2
	sc	......	2.1×10^{-4}	440–680	1.1×10^6	1.93	Ro4	
	sc	......	1.0×10^{-4}	560–680	7.9×10^5	1.97	Ho9	
KI..................	sc	680	1.0×10^{-3}	430–600	$1.4 \times 10^7/T$	1.69ᶜ	Pe1	Bi4, Ec1,
				430–600	$1.6 \times 10^{12}/T$	2.31ᵃ	Pe1	He4, Le5,
	sc	......	2.1×10^{-4}	450–650	1.6×10^6	1.87	Ka1	Ph2
RbCl..................	sc	717	1.3×10^{-5}	550–700	$3.6 \times 10^6/T$	1.58ᶜ	Fu3	Le5, Pi1
				550–700	$8.8 \times 10^{11}/T$	2.55ᵃ	Fu3	
RbBr..................	sc	681	3.4×10^{-5}	....	1.8×10^6	2.03	Le5	
CsF..................	pc	684	1.1×10^{-3}	550–660	1.6×10^5	1.55	Ha6	
				330–550	2	0.85	Ha6	
CsCl (α)..............	sc	636	6.1×10^{-5}	480–610	$1.0 \times 10^8/T$	1.67	Ar3	Ar2
	pc	......	7×10^{-5}	470–580	1	0.95	Ha6	
CsCl (β)..............	sc	tr. 469	....	250–480	$8.0 \times 10^7/T$	1.33	Ar3	Ha6, Ha9,
	sc	......	....	150–460	1.0×10^2	1.05	Mo7	He4, Ho6
CsBr..................	sc	636	2.6×10^{-3}	475–590	2.5×10^5	1.44	Ly1	
				300–475	2.5×10^4	1.28	Ly1	
	pc	......	4×10^{-4}	340–620	1×10^3	1.15	Ha6	
CsI..................	sc	621	1.9×10^{-3}	480–595	2.2×10^5	1.43	Ly1	Be10, Ec1,
				300–480	1.4×10^4	1.25	Ly1	Ha6
	sc	......	2.1×10^{-3}	300–550	1.1×10^5	1.37	Ho7	
OTHER MONOVALENT HALIDES								
NH₄Cl.................	sc	......	....	40–170	4.4×10^5	1.15	He4	
NH₄Br.................	pp	......	....	70–150	2.6	0.83	He4	
NH₄I..................	pp	......	....	0–130	1.9×10^7	1.23	He4	

TABLE 9f-1. CONDUCTIVITY FOR IONIC CONDUCTORS (*Continued*)

Substance	Form	$T_m,$ °C	$\sigma(T_m),$ (ohm-cm)$^{-1}$	T range, °C	$\sigma_0,$ (ohm-cm)$^{-1}$	W, eV	Specific reference	Other references
OTHER MONOVALENT HALIDES (*Cont.*)								
AgCl.................	sc	455	3.5×10^{-2}	160–380	$3.9 \times 10^7/T$	0.99^c	Ab1	Ab2, Co4,
				160–380	$5.0 \times 10^6/T$	0.78^i	Ab1	Sh6, Wa9
	sc	......	4.6×10^{-2}	50–250	$2.8 \times 10^7/T$	0.97^c	Mu1	
				50–250	$7.1 \times 10^6/T$	0.78^i	Mu1	
	sc	......	1.3×10^{-1}	225–400	2.4×10^5	0.93	Eb1	
AgBr.................	sc	422	1.3×10^{-1}	20–180	$6.3 \times 10^7/T$	0.87^c	Mu1	Ku3, Lu1,
				20–180	$4.9 \times 10^6/T$	0.68^i	Mu1	Mi1, Te1,
	sc	......	7.0×10^{-1}	345–410	1.1×10^8	1.13	Fr2	Wa6, Wa10
				250–345	1.4×10^6	0.89	Fr2	
				140–250	2.1×10^4	0.70	Fr2	
AgI (α).........	pc	555	2.7	220–530	5.5	0.051	Kv5	Li3
	pc	......	2.7	145–555	9.2	0.064^x	Bi3	
AgI (β).........	sc^c	tr. 146	....	85–145	6×10^7	0.97	La2	Li3, Mr1
				15–85	5×10^2	0.61	La2	
	sc^a	......	....	90–145	3×10^7	0.93	La2	
				20–90	1.3	0.40	La2	
TlCl.................	sc	427	6.6×10^{-3}	375–425	$8.8 \times 10^6/T$	0.87	Fr4	Ha10, Le5,
				325–380	$2.3 \times 10^6/T$	0.80	Fr4	Ph2, Sa1
				235–330	$1.15 \times 10^6/T$	0.76	Fr4	
	sc	......	5.7×10^{-3}	200–350	$8.6 \times 10^5/T$	0.75	Ja8	
TlBr.................	pc	458	1.0×10^{-3}	295–420	3.8×10^4	1.10	He3	Le5, Mo6,
				175–295	4.0×10^2	0.80	He3	Sa1
	pc	......	2.0×10^{-3}	150–400	2.2×10^2	0.73	Ph2	
TlI (α).........	pc	438	1.4×10^{-3}	163–400	4.2×10	0.63	Ph2	
TlI (β).........	pc	tr. 163	....	90–163	2.5×10^{-3}	0.41	Ph2	Sa1
MIXED HALIDES								
Na$_2$CdCl$_4$........	pc	......	....	230–350	2.8×10^3	0.86	Ja7	
KHF$_2$ (β)......	pc	239	....	195–225	1.1×10^5	0.91	Da6	
KAg$_4$I$_5$.........	pc	253	3.4×10^{-1}	40–250	$3.1 \times 10^3/T$	0.13	Br2	Ow1
K$_2$BaCl$_4$........	pc	662	....	500–635	1.6×10^6	1.54	Kr2	Sc10
K$_2$BaBr$_4$........	pc	......	....	430–600	7.1×10^4	1.33	Ja7	
RbAg$_4$I$_5$........	pc	228	8.6×10^{-1}	20–220	$1.1 \times 10^4/T$	0.14	Br3	Ow1
(NH$_4$)$_2$SnCl$_6$....	sc			20–180	4.0×10^{-8}	0.30	He4	
NH$_4$PF$_6$.........	pp	......	....	20–160	7.7×10^3	1.03	He4	
Cu$_2$HgI$_4$ (α)......	pp			67–80	3.3×10^3	0.59	Su2	Ja2
Cu$_2$HgI$_4$ (β)......	pp	tr. 67		15–66	2.2×10^{-1}	0.39	Su2	
Ag$_2$HgI$_4$ (α)......	pp			50–80	7.5×10^3	0.44	Su2	Ne2
	pp			50–84	1.8×10^6	0.61	We3	
Ag$_2$HgI$_4$ (β)......	pp	tr. 50		10–50	1.4×10^6	0.71	Su2	Ma5
	pp	......		22–35	9.9×10^4	0.65	Ne2	
POLYVALENT HALIDES								
CaF$_2$..............	sc	1418	2.9	620–980	$1.1 \times 10^{10}/T$	2.13	Ba6	Ar4, As1,
	sc	......	2.7	560–1000	$5.6 \times 10^9/T$	2.04	Ur1	Ch2, Ni1,
				200–560	3×10^{-1}	0.80	Ur1	So3
SrF$_2$..............	sc	1190	1.6×10^{-1}	700–1010	$5.6 \times 10^9/T$	2.14	Ba6	Bo3, Cr1,
	sc	......	5×10^{-2}	510–800	2.6×10^4	1.65	Ar4	Ni1
				50–510	2×10^{-1}	0.9	Ar4	
SrCl$_2$.............	sc	873	5.0×10^{-2}	380–660	$1.8 \times 10^7/T$	1.25	Ba6	
	sc	......	2	650–780	$3 \times 10^{17}/T$	3.2	Ho5	
				400–500	$3 \times 10/T$	0.4	Ho5	
BaF$_2$..............	sc	1287	2.2	620–900	$6.8 \times 10^8/T$	1.64	Ba8	Ba6
				300–620	$7.2 \times 10^7/T$	1.49	Ba8	
			1.3×10^{-1}	50–800	1.3×10^3	1.24	Ar4	
BaCl$_2$.............	pp	950	6×10^{-4}	310–760	3.5×10^{-2}	0.43	Ja3	
BaBr$_2$.............	pc	847	1.4×10^{-3}	390–750	1.0×10^{-1}	0.41	Ja7	
CdCl$_2$.............	pc	568	1.1×10^{-1}	260–520	1.6×10^5	1.03	Ja7	Bi6
HgI$_2$ (α).........	pc	......	....	127–150	1.1×10^7	0.43	Ja2	
HgI$_2$ (β).........	pc	tr. 126	....	92–125	3.8×10^{14}	0.96	Ja2	
PbCl$_2$.............	sc	500	4.0×10^{-3}	340–440	2.5×10^3	0.89	De8	Sc9
				200–340	3.5×10^{-2}	0.30	De8	
	pc	......	....	175–350	8.7	0.40	Si1	
PbBr$_2$.............	sc	373	2.2×10^{-4}	230–330	$4.9 \times 10^4/T$	0.71	Ve1	
				6C–140	$4.8/T$	0.36	Ve1	
PbI$_2$.............	sc	402	4.0×10^{-4}	210–270	2×10^5	1.30	Da8	
				170–210	2×10^{-1}	0.71	Da8	
	sc^c	......	1.2×10^{-5}	270–400	2.1×10^4	1.24	Se7	
	sc^a	......	9×10^{-5}	180–370	6×10^{-2}	0.38	Se7	

TABLE 9f-1. CONDUCTIVITY FOR IONIC CONDUCTORS (*Continued*)

Substance	Form	T_m, °C	$\sigma(T_m)$, (ohm-cm)$^{-1}$	T range, °C	σ_0, (ohm-cm)$^{-1}$	W, eV	Specific reference	Other references
POLYVALENT HALIDES (*Cont.*)								
AlCl$_3$	pc	189	2.4×10^{-6}	150–189	3.8×10^{12}	1.76	Bi5	
	pc	......	1.1×10^{-5}	160–189	2×10^4	0.85	Se11	
GaCl$_3$	pc	78	1.5×10^{-6}	60– 70	2.5×10^{12}	1.21	Gr3	
GaBr$_3$	pc	122	1.5×10^{-6}	80–100	3.4×10	0.52	Gr3	
GaI$_3$	pc	211	6×10^{-5}	190–205	6.7×10^{18}	2.23	Gr3	
LaF$_3$	sc	1490	....	160–560	$3/T$	0.084	Sh3	
				20– 80	$2 \times 10^5/T$	0.46	Sh3	
SIMPLE INORGANIC RADICALS								
Li$_2$SO$_4$	pc	860	3.0	575–800	1.2×10^2	0.36	Kv2	Kv1, Kv3
LiN$_3$	pp	......	....	172–280	4.3×10^7	1.87	Ga1	Ja1
				90–170	5.3×10^{-2}	0.53	Ga1	
NaNO$_2$	sc^e	tr. 163	....	163–190	1.4×10^5	1.15	As2	
				130–160	2.5×10^{12}	1.85	As2	
	sc^f		....	163–200	1.4×10	0.64	So1	
		tr. 163	....	40–163	1.0×10	0.68	So1	
NaNO$_3$	sc^c	306	5×10^{-6}	240–300	4.0×10^{13}	2.17	Ra2	Bi7
				20–230	1.3×10	0.94	Ra2	
	sc^c	......	5×10^{-5}	250–280	3×10^{22}	3.08	Ma7	
				40–240	3.6×10	0.87	Ma7	
NaBrO$_3$	sc	dec 315	....	182–298	1.6×10^{-3}	1.1	Ra1	
				50–162	9×10^{-6}	0.46	Ra1	
Na$_2$C$_2$	pp	dec 797	....	190–270	1.3×10^4	0.97	An3	
NaN$_3$	sc^c	......	....	270–330	2.5×10^7	1.82	To2	Ja1
				100–270	1.0×10^{-2}	0.87	To2	
	sc^d	......	....	230–400	1.0×10^{-1}	0.87	To2	
				130–230	1.0×10^{-4}	0.56	To2	
KNO$_3$	sc^c	334	....	130–165	1.0×10^2	0.80	As2	Cl1
		tr. 130	....	90–130	2.3	0.80	As2	
KCN	sc	634	1.0×10^{-4}	320–630	2.8×10^3	1.34	Lo2	
				160–320	3.4×10^{-1}	0.93	Lo2	
KCNS	sc	175	3×10^{-6}	120–175	3.1×10^{18}	2.13	Lo1	Pl1
				20–120	4.4×10^7	1.30	Lo1	
KN$_3$	pp	......	....	40–160	2.0×10^7	1.36	To1	
				120–230	4×10^4	1.30	Ja1	
CsNO$_3$	pc	404	2.0×10^{-3}	270–390	1.3×10^4	1.05	Bi8	
				220–270	2×10^2	0.74	Bi8	
	sc	......	....	160–210	6×10^{-5}	0.20	So2	
		tr. 154	....	35–130	8	0.72	So2	
NH$_4$NO$_3$	pp	......	....	125–143	3.8×10^4	0.68	Br6	
		tr. 125	....	48–124	3.1×10^5	0.96	Br6	
		tr. 84	....	48– 84	4.8×10	0.75	Br6	
NH$_4$ClO$_4$	pp	......	....	50–110	3.5×10^5	1.15	Zi2	
CuSO$_4$	pp	dec 770	6×10^{-4}	500–570	2.8×10^7	2.21	Ja3	
AgNO$_3$	pc	208	5×10^{-1}	206–208	....	......	Ce2	
	pc	......	....	164–207	1×10^{12}	1.46	Da7	
Ag$_2$SO$_3$	pp	dec 100	....	40– 80	5×10^2	0.80	Za1	
Ag$_2$SO$_4$ (α)	pp	656	6×10^{-2}	430–600	3.6×10^2	0.70	Ha12	Ja3
Ag$_2$SO$_4$ (β)	...	tr. 425	....	200–410	8.0×10^{-1}	0.42	Ha12	
Ag$_2$C$_2$O$_4$	pp	......	....	40–100	8.3×10^{-2}	0.70	Bo2	
	pp	......	....	....	1.0×10^{-6}	0.46	Fi1	
AgN$_3$	pp	......	....	180–250	2	0.46	Gr2	Ba10
	pp	......	....	25–110	3.3×10^4	0.82	Za2	
Ca(N$_3$)$_2$	pp	......	....	80–110	4.8	0.94	To1	
	pp	......	....	20–100	2.8×10^{-10}	0.23	Ja1	
Sr(N$_3$)$_2$	pp	......	....	30–110	2.0×10^{-11}	0.22	Ja1	
Ba(N$_3$)$_2$	sc^d	......	....	80–130	3×10^{-6}	0.56	To3	Ja1
	sc^c	......	....	70–110	2×10^{-7}	0.41	To3	
ZnSO$_4$	pp	761	1.4×10^{-6}	504–761	2.4	1.28	Ja3	
PbSO$_4$	pp	1080	8×10^{-4}	625–810	5.1×10^2	1.56	Ja3	

Form of Sample:

sc single crystal
pc polycrystalline
pp pressed powder
tr. transition
dec decomposes

Crystal Directions:

sc^a $\|a$ axis
sc^c $\|c$ axis
sc^d $\perp$ (100) plane
sc^f $\perp$ (001) plane
cs^f $\|$ ferroelectric axis

Activation Energy Notes:

^{a}Anion vacancy contribution
^{c}Cation vacancy contribution
^{f}Interstitial cation contribution
$^x \sigma = \sigma_0 \exp(-W/kT)[1 + \exp(-W/kT)]^{-1}$

TABLE 9f-2. CONCENTRATION AND MOBILITY OF DEFECTS IN IONIC CRYSTALS

Substance	T range, °C	x_0	h, eV	Defect	d_0, cm²/sec	Δh, eV	Defect	Method	Specific reference	Other references
ALKALI HALIDES										
LiH	240–630		2.38	Schottky	1.6×10^{-1}	0.53	V_{Li}'	con	Pr1	Be9
LiF	400–700		2.42	Schottky	$5.2\times10^{-5}\times T$	0.73	V_{Li}'	con	Ba7	Ei1,
	460–840	9.2×10	2.68	Schottky		0.65	V_{Li}'	con	Ha19	Ja6,
	500–800	5.0×10^{2}		Schottky		1.0	V_F'	nmr	St7	St4
LiCl	310–570	1.7×10^{3}	2.12	Schottky	$2.8\times10^{-5}\times T$	0.41	V_{Li}'	con	Ha19	
LiBr	270–540	8.1×10^{3}	1.80	Schottky	$3.9\times10^{-5}\times T$	0.39	V_{Li}'	nmr	Ha19	
	160–360		2.0	Schottky		0.43	V_{Li}'	con	Al3	
LiI	180–420	5.0×10^{2}	1.34	Schottky	$5.6\times10^{-5}\times T$	0.38	V_{Li}'	nmr	Ha19	
NaF	250–650	2.2×10	3.0	Schottky	$\geq 2.0\times10^{-3}$	0.52	V_{Na}'	con	Pe2	Ma2
NaCl	250–720	3.4	2.12	Schottky	4.1×10^{-1}	0.80	V_{Na}'	con	Dr1	Bi2, Ch3,
	250–790		2.17	Schottky	3.7×10^{-1}	0.66	V_{Na}'	con	Al5	Et1, Ja5,
	250–550	8		V_{Na}'	2.0×10^{2}	1.15	V_{Cl}'	con	Al5	Ka4, Ki3,
	400–550		0.95	$(V_{Na}V_{Cl})^{x}$		1.25	Cl into $(V_{Na}V_{Cl})^{x}$	dis	Da4	Pi1, St10
NaBr	630–760	3.0×10^{4}	1.30	Schottky	1.7×10^{-2}	0.80	V_{Na}'	die	Ec3	
	490–730	8.3		Schottky	7.8×10^{-1}	1.02	V_{Na}'	con	Ho8	Ma6, Sc2
KF	400–790	3.5×10	2.64	Schottky	4.8×10^{-1}	0.84	V_K'	con	Ka1	Al4, Bi4
KCl	400–740		2.22	Schottky	1.1×10^{-1}	0.71	V_K'	con	Dr1	Ch3, Ka1,
	270–640	1.5	2.26	Schottky	9.1	1.04	V_K'	con	Be1	Ma12, Pel,
	270–640	4.4×10	2.31	Schottky	1.1	0.95	V_{Cl}'	dif	Be1	Wa4
	560–760	~ 1	0.7	$(V_KV_{Cl})^{x}$		1.04	Cl into $(V_KV_{Cl})^{x}$	dis	Fu2	Ka3
KBr	250–350	4.5×10^{5}	2.53	Schottky	1.1×10^{-3}	0.66	V_K'	die	Ka2	Ec2
	640–770	1.4×10^{2}		Schottky	9.1×10^{-2}	0.87	V_{Br}'	con	Sa2	Ch3, Ka1,
	440–680			Schottky	7.4×10^{-1}	0.83	V_K'	con	Ro4	Ma12, Pel
	440–680			$(V_KV_{Br})^{x}$	7.9×10^{-1}	1.19	Br into $(V_KV_{Br})^{x}$	con	Ro4	
	410–680	7.7×10	2.40	Schottky	2.3×10^{-2}	1.21	V_K'	die	Ho9	
KI	600–730	5.5×10^{5}	1.41	Schottky		0.7	V_K'	con	Ec3	Ch3, Pel
	350–560		2.00	Schottky		0.6		con	Ec1	
RbI	230–260		2.1	Schottky		0.34	V_{Cl}'	con	Ch3	
CsCl (α)	480–610		1.86	Schottky	0.7	0.58	V_{Cl}'	dif	Ar2	Ar2, Mo7
CsCl (β)	280–460		2.0	Schottky	0.6	0.27	V_{Cs}'	dif	Ha9	
	280–460				0.34	0.67	V_{Cl}'	cdf	Ha9	
CsBr	320–550			Schottky	0.58	0.3	V_{Cs}'	cdf	Ly1	Be10, Ly1
	320–550				0.27		V_{Br}'	dif	Ly1	
CsI	300–500		2.14	Schottky	0.67		V_{Cs}'	dif	Ho7	
	300–500				0.3		V_I'	con	Ho7	
OTHER HALIDES										
AgCl	160–380	1.6×10^{2}	1.44	Ag Frenkel	5.7×10^{-3}	0.27	V_{Ag}'	con	Ab1	Ab2, Eb1,
	160–380				7.2×10^{-4}	0.055	$Ag_i^{\cdot}$	con	Ab1	Mu1
	300–440				2.1×10^{-4}	0.008	collin. instlcy.	dif	Wel	
	300–440				7.5×10^{-4}	0.13	noncoll. instlcy.	dif	Wel	
	20–60		0.82	$(V_{Ag}V_{Cl})^{x}$			Cl into $(V_{Ag}V_{Cl})^{x}$	con	La6	
AgBr	200–300	5.3×10^{2}	1.13	Ag Frenkel	1.5×10^{-2}	0.30	V_{Ag}'	con	Te1	Ku3, Mu1
	200–300				1.2×10^{-4}	0.17	$Ag_i^{\cdot}$	con	Te1	
	200–330				2.5×10^{-4}	0.058	collin. instlcy.	dif	Wel	
	300–400	4×10^{-2}	1.62	Schottky	3.0×10^{-2}	0.27	noncoll. instlcy.	dif	Kr1	Fr2

Crystal	T, °C	(A)	(B)	Reaction	Defect	val 1	val 2	Method	Ref.	Ref.
AgI (α)	150–250			Ag disorder				con	Li3	
AgI (β)	200–140		0.69	Ag Frenkel	$Ag^{\cdot}$	0.057		con	Li3	Mr1
TlCl	200–330	5.5×10	1.36	Schottky	$Ag;$	0.14		spc	Ja8	Ch8, Fr4
	280–400				$V_{Tl}^{\cdot}$	0.44		spc	Ja8	
CaF₂	640–920	6.0×10^2	2.81	F Frenkel	F_i'	0.105	2.2×10^{-2}	con	Url	
	250–650				F_i'	1.64	9.9×10^{-4}	con	Url	
	230–980				$V_F^{\cdot}$	0.60	1.2×10^3	con		
							$9 \times 10^{-3} \times T$			
SrF₂	370–1010	6×10^2	2.4	F Frenkel	$V_F^{\cdot}$	0.9		con	Ba6	Bo3, Cr1
SrCl₂	180–660	1×10^3	2.3	F Frenkel	$V_F^{\cdot}$ or F_i'	1.0		con	Ba6	Ba6, Ma2,
BaF₂	350–650	2.1×10	1.8	Cl Frenkel	$V_{Cl}^{\cdot}$	0.4	4×10^{-2}	con	Ba6	Ni1
			1.86	F Frenkel	F_i'	0.56	5.5×10^{-2}	con	Ba8	De8
PbCl₂	150–500	4.6×10^{-1}	1.16	Schottky	F_i'	0.79	3.5×10^{-1}	con	Ba9	
	160–310			$V_{Pb}'' + 2V_{Cl}^{\cdot}$	V_{Pb}''	0.60	$1.4 \times 10^{-6} \times T$	con	Sc9	
PbBr₂	60–330		1.44	$V_{Pb}'' + 2V_{Br}^{\cdot}$	V_{Cl}'	0.12	$2.6 \times 10^{-7} \times T$	con	Sc9	
LaCl₃	20–440	5.8×10	0.28	$V_{La}^{x} + 3V_F^{x}$	$V_{Br}^{\cdot}$	0.29		con	Vel	
					$V_F^{\cdot}$	0.42		thx	Sh3	
SIMPLE INORGANIC RADICALS										
NaNO₃	120–300		2.46	Na Frenkel	$Na^{\cdot}$	0.94		con	Ra2	
NaN₃	100–330	7×10^{-1}	1.90	Schottky	$V_{Na}^{\cdot}$	0.87	$8 \times 10^{-5} \times T$	con	To2	
KNO₃	130–330			K Frenkel	$V_K^{\cdot}$	0.40		con	Cl1	
CsNO₃	220–390		0.62			0.74		con	Bi8	
Ag₂C₂O₄	50–110			Ag Frenkel	$V_{Ag}^{\cdot}$	0.82	$7 \times 10^{-7} \times T$	con	Bo2	
OXIDES										
SrO	1040–1340		2.0	$\tfrac{1}{2}O_{2(g)} \rightarrow O_i' + h^{\cdot}$	O_i'	0.6		P1	Co8	
	790–1100		3.0	$V_O^{\cdot\cdot} \rightarrow O_i'$	O_i'	0.6		P2	Co8	
ZnO	900–1025			Zn Frenkel	$Zn_i^{\cdot}$	1.78		dif	Se2	
CdO	630–850	1.2×10^2	2.78	$2V_O^{\cdot\cdot} + Cd_{Cd}'' + 2Cd_i'$	$V_O^{\cdot\cdot}$	2.7	$3.3 \times 10^{-6} \times T$	dif	Ha17	
Al₂O₃	1670–1900		6.5	$2V_{Al}''' + 3V_O^{\cdot\cdot}$				dif	Oi1	
			6.8		V_{Al}'''	2.2		dif	Pa1	
TiO₂	1000–1500		9.6	$Ti_i^{\cdot\cdot\cdot} + 3e' + O_{2(g)}$				sem	Bl2	
Nb₂O₅ (α)	800–1160		2.8	$V_{Cr}''' + e' + \tfrac{1}{2}O_{2(g)}$	$V_O^{\cdot}$	1.2		con	Ch4	
Cr₂O₃	1040–1550		3.9	$V_{Cr}''' + Cr_i^{\cdot} + h^{\cdot}$	V_{Cr}''' or $Cr_i^{\cdot\cdot}$	1.3		cdf	Ha1	
UO₂	320–850		3.1	$V_O^{\cdot\cdot} + O_i$	O_i'' inst/cy	1.3		dif	Be4	
	100–800				V_U'''' or $U_i^{\cdot\cdot}$	2.3		con	Na1	
SULFIDES AND TELLURIDES										
Na₂S	350–800		1.77	Na Frenkel	$V_{Na}^{\cdot}$ or $Na_i^{\cdot}$	0.76		cdf	Mo2	
	350–800				$V_S^{\cdot}$	1.17		cdf	Mo2	Ch3
Cu₂S (β)	130–210	8×10^{-4}	0	V_{Cu}'	V_{Cu}'		$3.6 \times 10^{-3} \times T$	sem	Yo1	
CdTe	700–1000	1.7×10^{-4}	1.04	$V_{Cd}' + Cd_i^{\cdot}$		0.26		sem	De5	

Method:

cdf	Ionic conductivity and diffusion
con	Ionic conductivity in doped samples
die	Dielectric relaxation
dif	Diffusion with tracers
dis	Charged dislocations
nmr	Nuclear magnetic resonance
sem	Semiconducting properties
the	Ionic thermoelectric effect
thx	Thermal expansion
spc	Space charge polarization

P1	$P_{O_2} = 10^{-3}$ to 1 atm
P2	$P_{O_2} = 10^{-12}$ to 10^{-5} atm
P3	$P_{O_2} = 10^{-5}$ to 1 atm

TABLE 9f-3. EFFECT OF PRESSURE ON CONDUCTIVITY

I. PRESSURE COEFFICIENT $\alpha = -(\partial \ln \sigma / \partial P)_T$

Substance	Form	T range, °C	P range, kiloatm	$\alpha_0 \times 10^4$, atm^{-1}	Comments	Specific reference	Other references
NaCl.......	sc	600–700	0– 5	3.7	intrinsic	Bi2	
	sc	220–510	0–10	1.33	extrinsic	Pi1	
KCl........	sc	550–700	0– 5	5.1	intrinsic	Bi4	Ta12
	sc	220–490	0–10	1.46	extrinsic	Pi1	
KI.........	sc	400–630	0– 5	5.3	intrinsic	Bi4	
RbCl.......	sc	300	0– 5	4.0	extrinsic	Pi1	
CuBr (β)....	pc	380	0– 0.15	0.29	Cu electrodes	Bi1	
CuBr (γ)....	pc	250–380	0– 0.15	1.2	Cu electrodes	Bi1	Ne3
AgCl.......	sc	144–336	0– 8	1.3 to 2.6	intrinsic	Ab2	Sh7
	sc	200–350	0– 1.5	3.6 to 4.5	Ag diffusion	Mu6	
AgBr.......	sc	202–406	0– 2	1.2 to 2.1	intrinsic	Ku3	Sc3
	pc	250–350	0– 0.15	3.3 to 3.9	....	Wa1	
AgI (α).....	pc	191	0– 0.15	0.37	....	Wa1	Li3
		400	0– 0.15	0.15	....	Wa1	
AgI (β).....	pc	110–130	0– 0.03	−1.9 to −12	....	Li3	Pa8, Ri4,
	pc	110–135	0– 0.15	−2.2 to − 2.8	....	Wa1	Sc7
	pp	90	0– 2.0	−7.1	Ag diffusion	Mu5	
	pc	20	3– 7	5.8 to 5.4	High P phases	Ne3	
Ag$_2$HgI$_4$ (α).	pp	65– 85	0– 4	−2.1	....	We3	
		65– 85	4– 8	5.0 to 4.3	High P phases	We3	
KNO$_3$......	pp	200–320	0– 0.5	2.0 to 2.8	....	Cl2	
ZrO$_2$ (monocl.) .	pp	600–800	10–30	−1.1	....	Wh1	
		1000	0– 5	−3.9	....	Wh1	

II. FREE VOLUMES

Substance	T range, °C	P range, kiloatm	$v_{formation}$, cm^3/mole	Defect	Δv_{motion}, cm^3/mole	Defect	Specific reference	Other references
NaCl....	400–700	0– 5	43	Schottky	9.5	V_{Na}'	Bi4	Bi2, Sh8
	220–510	0–10	....	....	7.7	V_{Na}'	Pi1	
KCl.....	400–700	0– 5	67	Schottky	10	V_K'	Bi4	Ta12
	220–490	0–10	....	....	7.0	V_K'	Pi1	
	685, 745	7–17	....	....	11.7	$V_{Cl}\cdot$	Ra3	
AgCl....	30–350	0– 8	16.7	Ag Frenkel	4.7	V_{Ag}'	Ab2	
	30–350	0– 8	....	....	3.2	$Ag_i\cdot$	Ab2	
AgBr....	200–290	0– 8	16	Ag Frenkel	7.4	V_{Ag}'	Ku3	
	200–290	0– 8	....	....	2.6	$Ag_i\cdot$	Ku3	
	350–410	0– 5	43	Schottky	20	$V_{Br}\cdot$	Ta10	
KNO$_3$...	200–330	0– 0.5	8 to 11	K Frenkel	4 to 9	V_K'	Cl2	

TABLE 9f-4. CONDUCTIVITY FOR MIXED CONDUCTORS
Total conductivity $\sigma = \sigma_0 \exp(-W/kT)$

Substance	Form	T_m, °C	T range, °C	σ_0, (ohm-cm)$^{-1}$	W, eV	Environment	Specific reference	Other references
HALIDES								
CuCl	sc	426	100– 400	6.3×10^6	1.06	Cu	Hs1	Tu2, Wa7
CuBr (α)	pc	491	470– 491	6.5	0.039	Cu	Bi3	
CuBr (β)	pc	tr. 470	379– 450	2.0×10	0.21	Cu	Bi1	Wa7
CuBr (γ)	pc	tr. 379	230– 379	2.8×10^{10}	1.47	Cu	Bi1	Wa3, Wa7
CuI (α)	pc	602	402– 440	8	0.20	Cu	Bi1	Wa7
CuI (β)	pc	tr. 402	370– 400	2×10^7	1.09	Cu	Bi1	Wa7
CuI (γ)	pc	tr. 369	330– 369	1.8×10^{10}	1.52	Cu	Bi1	Wa7, We4
Ag$_2$HgI$_4$ (β)	pp	tr. 50	27– 48	7.5×10^8	0.89	Ag	We3	
OXIDES						PO_2 (atm)		
BeO	pc	....	1100–1300	$1.6 \times 10^5/T$	2.52	air	De3	Br1, Pr3
			600–1100	$2.8 \times 10^{-2}/T$	0.69	air	De3	
	sp	....	1380–1600	$4.7 \times 10^6/T$	3.12	air	Cl4	
MgO	sc	....	930–1500	1.3^i	2.00^i	O$_2$, air	Mi3	Bu1, Da5
	sc	....	770–1300	1.8×10^3	2.7	10^{-8}	Mi2	
	sc	....	400– 750	1.3×10^3	2.8	10^{-7}	Le6	
CaO	sc	2550	1000–1300	2.9×10	1.73	$\sim 10^{-4}$	Gu2	Pa2
	sc	....	770–1150	8×10^7	3.5	vacuum	Su3	
SrO	sp	....	1040–1340	2.8×10^2	1.6	1	Co8	Su3
				$\sigma \propto P_{O_2}^{\frac{1}{4}}$	....	10^{-3} to 1	Co8	
			790–1100	2.0×10^2	2.1	10^{-12} to 10^{-5}	Co8	
BaO	sc	....	440– 733	4×10^{-2}	0.5	vacuum	Do2	
Al$_2$O$_3$	sc, pc	....	750–1320	8×10^{-3}	1.9	air	Ma9	Ch1, Da3, Fl1, Ha7, Ha8, Pe4
	sc	....	1625–1725	5.1×10^9	5.5	10^{-5}	Pa5	
			1300–1600	9.1×10	2.62	10^{-5}	Pa5	
Sc$_2$O$_3$	pp	....	800–1300	....	1.70	air	No1	
Y$_2$O$_3$	sp	....	1200–1600	8×10^2	1.94	10^{-1}	Ta8	No1
				$\sigma \propto P_{O_2}^{\frac{1}{16}}$	....	10^{-1} to 10^{-7}	Ta8	
La$_2$O$_3$ (hex)	pp	2315	400– 700	4.6×10^6	0.67	air	Me1	No1
(cub)	....	tr. 590	350– 580	1.5×10^{10}	1.05	air	Me1	
ZrO$_2$ (tetr)	sc	....	1500–1780	2.1×10^6	3.3	air	An2	Mc1, Ve2
			1150–1340	6.7×10^{-2}	0.74	air	An2	
ZrO$_2$:CaO (15%)	sp	....	100–1100	1.4×10^2	1.13	10^{-9}	Ve3	Di1, Jo1, Ki1, St9
ZrO$_2$:YO$_{1.5}$ (20%)	sp	....	700–1350	1.8×10^2	0.83	air	St9	
HfO$_2$	sp	....	1000–1600	9.3×10	1.45	1	Ro1	
				$\sigma \propto P_{O_2}^{0.19}$	....	1 to 10^{-4}	Ro1	
HfO$_2$:CaO (12%)	sp	....	800–2000	1.8×10^3	1.43	max. cond.	Jo1	
HfO$_2$:YO$_{1.5}$ (16%)	sp	....	900–1600	8×10^2	1.12	max. cond.	Be11	
CeO$_2$	sc	....	130–1000	1.5×10^3	1.28	air	Vi1	No1
ThO$_2$	sp	3050	800–1100	1.1×10	1.41	10^{10} to 10^{-22}	La3	Da1
ThO$_2$:YO$_{1.5}$ (15%)	sp	....	800–1100	3.1×10	0.92	10^{-5} to 10^{-22}	La3	
Pr$_2$O$_3$ (hex)	pp	....	720– 850	....	0.86	dry H$_2$	Me1	No1
(cub)	pp	tr. 780	700– 780	....	0.84	dry H$_2$	Me1	
Nd$_2$O$_3$ (hex)	pp	....	400– 700	....	0.75	air	Me1	No1
(cub)	pp	tr. 550	400– 540	....	1.26	air	Me1	
Sm$_2$O$_3$	pp	....	800–1300	....	1.17	air	No1	
Eu$_2$O$_3$	pp	....	800–1300	....	1.24	air	No1	
Gd$_2$O$_3$	pp	....	800–1300	....	1.36	air	No1	
Tb$_4$O$_7$	pp	....	800– 900	....	0.40	air	No1	
Dy$_2$O$_3$	pp	....	800–1300	....	1.39	air	No1	
Er$_2$O$_3$	pp	....	800–1300	....	1.40	air	No1	
Yb$_2$O$_3$	pp	....	800–1300	....	1.53	air	No1	
SULFIDES, SELENIDES, AND TELLURIDES								
Na$_2$S	sp	1169	520– 800	3.4×10^7	1.64	H$_2$	Mo2	
			350– 520	8.0×10	0.75	H$_2$	Mo2	
Cu$_2$S (β)	....	tr. 470	110– 470	$9 \times 10^4/T$	0.24	....	Mi5	Ha16, **We2**
	pc	....	400	$\sigma_i = 2.4$	....	Pt, Cu	Yo2	Yo1
Cu$_2$Se	pc	....	580– 750	3×10^i	0.17^i	Cu$_{1.96}$ Se	Ce1	
Ag$_2$S (α)	pc	835	180– 300	$2.9 \times 10^4/T^i$	0.11^i	....	Ok1	He1
Ag$_2$S (β)	....	tr. 179	130– 160	1.3×10^8	0.71	Ag	Mi6	Ri1
			130– 170	1.4×10^6	0.69	S	Mi6	
Ag$_3$SBr	sp	....	100– 250	8^i	0.24^i	Ag	Re4	
Ag$_3$SI (α)	sp	700	235– 400	2	0.04	Ag	Re4	Ta1
Ag$_3$SI (β)	sp	tr. 235	0– 80	2.9	0.14	Ag	Ta4	Ta5
Ag$_2$Se (α)	pc	....	130– 300	$1.7 \times 10^4/T^i$	0.10^i	....	Ok1	Mi7
Ag$_2$Se (β)	pc	tr. 133	100– 130	3.4×10^i	0.39^i	Ag	Mi9	
Ag$_2$Te (α)	....	....	165– 225	9×10^4	0.31	....	Mi8	
Ag$_2$Te (β)	pc	....	80– 140	6×10^4	0.56	....	Mi9	Ta6

Form of Sample:
  sc   single crystal
  pc   polycrystalline
  pp   pressed powder
  sp   sintered powder

Conductivity:
  ^{i}Ionic portion of conductivity.

TABLE 9f-5. TRANSPORT NUMBERS FOR MIXED CONDUCTORS

Substance	T, °C	Transport numbers	Form	Environment	Specific reference	Other references
HALIDES		$t_c = 1 - t_e$				
CuCl	150	[1]0.008 [2]0.04	[1]pc [2]pc	Cu, Cu₂O	[1]Ma3 [2]Tu2	Wa7
	250	0.027 0.85				
	350	0.51 0.99				
CuBr (α)	470–491	1.00	pc	Cu	Tu3	Wa7
CuBr (β)	390–445	1.00	pc	Cu	Tu3	Wa7
CuBr (γ)	100	[1]0.004 [2]0.01 [3]0.005	[1]pc [2]pc [3]pc	Cu, Cu₂O	[1]Ma3 [2]Tu3 [3]Ki4	
	200	0.042 0.10 0.032				
	300	0.26 0.87 0.36				
	390	0.25 1.00 0.65				
CuI (α)	402–500	1.00	pc	Cu	Tu1	Wa7
CuI (β)	375–400	0.99	pc	Cu, Cu₂O	Tu1	Wa7
CuI (γ)	200	[1]0.003 [2]0.00	[1]pc [2]pc	Cu, Cu₂O	[1]Ma3 [2]Tu1	Wa7, We4
	300	0.004 0.25		Cu		
Ag₂HgI₄ (α)	50–84	0.97	pp	Ag, Pt	We3	
Ag₂HgI₄ (β)	27–48	0.4	pp	Ag, Pt	We3	
Ag₂SO₄ (α)	602	$t_e = 3 \times 10^{-3}$	…	$P_{O_2} = 10$ to 700 torr	Ha12	
	446	$t_h = 1 \times 10^{-4}$	…	$P_{O_2} = 0.1$ to 700 torr	Ha12	
OXIDES		$t_c = 1 - t_e$	…	P_{O_2} (atm)		
Cu₂O	800–1000	(2 to 5) $\times 10^{-4}$	pp	Cu	Gul	
BeO	1000–1200	[2]1.00 [3]0.80 to 0.96	[1]sp [2]pp [3]pp	10 to 10⁻³	[1]Cl3 [2]Pa2 [3]Pa2	
	1200–1300	1.00		O₂, air		
	1400–1700	[1]0.94		O₂, CO – CO₂		
MgO	900	[1]1.0 [3]0.57	[1]sc [2]sc [3]pp	O₂, air	[1]Mi3 [2]Sc6 [3]Pa2	Sc4
	1100	0.9 0.91		10⁻⁴		
	1300	[2]0.87 0.5 0.63		O₂, CO – CO₂		

Compound	Temp (°C)	$t_i = 1 - t_e - t_h$	P_{O_2} (atm) / condition	Method	Ref.	Ref.
CaO	900		O_2, air	[1]pp	[1]Pa2	
	1100	[1]0.52 [2]0.57 [3]0.02	O_2, CO – CO_2	[2]pp	[2]Pa2	
SrO	1300	0.61 0.91 0.02	~10^{-4}	[3]sc	[3]Gu2	
	1000	[1]0.03 0.63	O_2, air	[1]pp	[1]Pa2	
BaO	1150	0.08 [2]0.68	O_2, CO – CO_2	[2]pp	[2]Pa2	
	200–700	0.0 0.89	vacuum	sc	Do2	
		(= $1 - t_e - t_h$)	P_{O_2} (atm)			
Al$_2$O$_3$	800–1300	0.9 to 0.0	O_2, air	sc	Ma9	
	1100–1300	1.0 to 0.0	1 to 10^{-13}	sp	Sc6	
Sc$_2$O$_3$	800–1000	1.0	10^{-12}, 10^{-17}	sp	Sc6	
Y$_2$O$_3$	700–800	0.3 to 0.15	10^{-15}	sp	Ta8	
	1200–1600	0.00	10^{-1} to 10^{-17}	sp	Ta8	
La$_2$O$_3$	825	≥ 0.5	6×10^{-5} to 3×10^{-22}	sp	Ta11	
	400–1000	~1	air	pp	Me1	
		$t_a = 1 - t_e - t_h$	P_{O_2} (atm)			
ZrO$_2$ (tetr)	1300–1600	0.9 to 0.4	10^{-5} to 10^{-11}	sp	Ve2	
	1140–1340	0.6 to 0.2	10^{-11} to 10^{-14}	sp	Mc1	Ta7
ZrO$_2$ (monocl)	990	4.5×10^{-3}	0.4	sc	Ma1	Ve3
ZrO$_2$:CaO (15%)	1000	≥ 0.99	1 to 10^{-15}	sp	St1	
	800	[1]$t_h = 10^{-2.6}$ to $10^{-3.3}$ [2]$t_e = 10^{-6.6}$ to $10^{-4.9}$	1 to 10^{-5}	[1]sp	Pa6	
	1000	$t_h = 10^{-1.7}$ to $10^{-3.1}$ $t_e = 10^{-5.0}$ to $10^{-2.7}$	10^{-10} to 10^{-20}	[2]sp	Pa6	
ZrO$_2$:YO$_{1.5}$ (20%)	900–1100	$t_e \leq 0.01$	air or H_2	sp	Br4	Sm1, Tr1
HfO$_2$	1000	0.71 to 0.94	10^{-6} to 10^{-15}	sp	Ro1	St1
	1000	0.01 to 0.03	1 to 10^{-3}		Ro1	
	1400	0.06 to 0.13	1 to 10^{-3}		Ro1	
	1000–1500	0.00	1 to 10^{-18}		Ta9	
HfO$_2$:YO$_{1.5}$ (16%)	800–1050	1.0	10^{-10} to 10^{-15}	sp	Be1	
Nb$_2$O$_5$(α)	1000–1200	0.03	1 to 10^{-5}	sp	El1	Vol

TABLE 9f-5. TRANSPORT NUMBERS FOR MIXED CONDUCTORS (*Continued*)

Substance	T, °C	Transport numbers	Form	Environment P_{O_2} (atm)	Specific reference	Other references
	....	$t_c = 1 - t_e - t_h$	....			
Cr₂O₃...........	980–1550	5.7×10^{-6} to 1.2×10^{-3}	sp	N₂	Ha1	
MnO...........	1010	1×10^{-5}	sp	Mn	Bo1	
FeO...........	720–1020	1.4×10^{-5} to 2.6×10^{-4}	pc	Fe₀.₈₈₇O	De6	Ha1
	....	$t_a = 1 - t_e - t_h$	....	P_{O_2} (atm)		
CeO₂...........	450–1300	$0 < t_a < 1$	sp	air	Ho4	No2, Vi2
CeO₂:LaO₁.₅ (30%)........	1000–1100	1.00	sp	air	Ne4	Ta3
ThO₂...........	1000	0.06 to 0.93	sp	1 to 10^{-9}	La3	Da1, St1
ThO₂...........	1000	1.00 to 0.96	sp	10^{-12} to 10^{-21}	La3	
ThO₂:YO₁.₅ (15%).......	1000	≧0.99	sp	10^{-5} to 10^{-24}	St1	La3, Su1, Wi1
	800	[1]$t_h = 10^{-2.7}$ to $10^{-3.2}$ [2]$t_e = 10^{-5.7}$	[1]sp	1 to 10^{-10}	Pa6	
UO₂...........	1000	$t_h = 10^{-4.7}$ to $10^{-2.7}$ $t_e = 10^{-4.4}$	[2]sp	10^{-5} to 10^{-25}	Pa6	
	900–1100	4×10^{-6} to 1.2×10^{-4}	sp	UO₂.₀₀₀₃	Do3	Ii1

	Temp. (°C)	$t_i = 1 - t_e - t_h$	Form	P_{O_2} (atm)	References	
Nd₂O₃	800–1000	1.00	pp	$10^{-12}, 10^{-17}$	Sc6	Hi2, Ku4, Wa8, We2
Sm₂O₃	800–1000	1.00	pp	$10^{-12}, 10^{-17}$	Sc6	
Gd₂O₃	800–1100	$0 < t_i < 1$			Ra4	
Dy₂O₃	800–1100	$0 < t_i < 1$			Ra4	
Yb₂O₃	800–1000	1.00	pp	$10^{-12}, 10^{-17}$	Sc6	

	Temp. (°C)	$t_c = 1 - t_e$	Form	P_{O_2} (atm)	References	
SULFIDES, SELENIDES, AND TELLURIDES						
Na₂S	350–800	1.00	sp		Mo2	
Cu₂S	134–207 400	0.05 to 0.15 0.86	pc	Pt, Cu	Yo1 Yo2	
Cu₂Se	580–750	0.06 to 0.11	pc	Cu₁.₉₆Se	Ce1	
Cu₂Te	335–410	1×10^{-4}			Re3	
Ag₂S (α)	130–200	$t_c = 10^{-2}$ to 10^{-3} $t_a = 10^{-8}$		Ag	Wa2	Bu3
Ag₂S (β)	130–170	0.011 0.59		S	Mi6 Mi6	Hel
Ag₃SBr	100–250	0.98 to 0.93	sp	Ag	Re4	Ta1, Ta2
Ag₃SI (α)	300	0.95	sp	Ag	Re4 Re4	Ta4
Ag₃SI (β)	100–200	0.995 to 0.986	sp	Ag	Mi7	Bu2
Ag₂Se (α)	220	2×10^{-3}		Ag	Mi9	
Ag₂Se (β)	124	1×10^{-7}	pc	Ag	Yo2	
Ag₂Te (α)	270	0.010	pc	Ag	Mi9	Mi8
Ag₂Te (β)	100	4×10^{-7}		Ag	Yo2	Ta6
ZnSe	200–400	0.08 to 0.13	sc	Fe – FeO, Cu – Cu₂O	Ki4	

Form of Sample:
sc single crystals
pc polycrystalline
pp pressed powder
sp sintered powder

Transport Numbers:
t_i Ionic
t_c Cation
t_a Anion
t_e Electron
t_h Hole

TABLE 9f-6. DIFFUSION IN IONIC CRYSTALS

[The diffusion coefficient is given as $D = D_0 \exp\,(-W/kT)$.]

Substance	Form	Isotope	T range, °C	D_0, cm²/sec	W, eV	Defect	Method	Environment	Comments	Specific reference	Other references
ALKALI HALIDES											
LiF	sc	^{7}Li	560–770	2.3	1.81	V_{Li}'	nmr			Ei1	Ma2, St6
			360–560	4.5×10^{-7}	0.71	V_{Li}'	nmr			Ei1	
NaF	pw	^{19}F	600–790	6.1×10	2.2	V_F	nmr			Ei1	St7
		^{23}Na	550–650	1.6×10	2.0	V_{Na}'	nmr			Pe2	Ma2
NaCl			250–400	2.3×10^{-7}	0.52	V_{Na}'	nmr			Pe2	
	sc	^{22}Na	600–720	2.9×10	1.97	V_{Na}'	sct			Be5	Ba4, Do6,
	sc	^{22}Na	670–770	1.8×10	2.10	V_{Na}'	sct		single vacancy	Ne1	Ei3, La5,
		^{22}Na	670–770	1.1×10^3	2.35	$(V_{Na}V_{Cl})^x$	sct		vacancy pair	Ne1	Ma6
	sc	^{36}Cl	520–745	2.2	2.07	V_{Cl}	sct		single vacancy	La4	Ba5, La5
			520–745	9.9×10^2	2.50	$(V_{Na}V_{Cl})^x$	sct		vacancy pair	Fu3	
NaBr	sc	^{24}Na	425–700	6.7×10^{-1}	1.53	V_{Na}'	sct			Ma6	
	sc	^{82}Br	450–690	5.0×10^{-2}	2.02	V_{Br}	sct			Sc2	Do1
KF	pp	^{42}K	580–840	2	1.78	V_K'	sct			La5	Ar6, Wi3
KCl	sc	^{42}K	450–760	4×10^{-2}	1.48	V_K'	sct			La5	Ba3, La5,
	sc	^{36}Cl	560–760	3.6×10	2.10	V_{Cl}	sct		single vacancy	Fu2	Ra3
			560–760	8.6×10^3	2.65	$(V_K V_{Cl})^x$	sct		vacancy pair	Fu2	
KBr	sc	^{42}K	470–730	1×10^4	1.26	V_K'	sct			La5	Do1, La5
	sc	^{82}Br	400–700	3×10^4	2.61	V_{Br}	gsx			Da10	La5
	sc	^{42}K	430–690	1.2×10^{-5}	0.64	V_K'	sct			No6	La5
KI	sc	^{131}I	430–690	3.3×10	1.12	V_I'	sct			No6	Ma4
RbBr	sc	^{86}Rb	600–760	3.1	1.99	V_{Rb}'	sct			Ar1	
CsF	pc	^{137}Cs	480–640	1×10^{-1}	1.67	$\{V_{Cs}'\}$	sct			La5	
CsCl (α)	pp	^{137}Cs	465–620	7×10^{-1}	1.39	$\{V_{Cs}'\}$	sct			La5	
	pp	^{36}Cl	465–620	2.4×10	1.56	$\{V_{Cl}'\}$	sct			La5	
CsCl (β)	sc	^{137}Cs	280–460	1.5	1.53	V_{Cs}'	sfc			Ha9	La5
	sc	^{36}Cl	280–460	1.5×10	1.27	V_{Cl}'	sfc			Ha9	Ho6, La5
CsBr	sc	^{134}Cs	320–550	3.9	1.54	V_{Cs}'	sct			Ly1	
	sc	^{82}Br	415–530	1.4×10	1.42	V_{Br}'	sct			Ly1	Ho7, Kl5
CsI	sc	^{134}Cs	320–550	2.1	1.53	V_{Cs}'	sct			Ly1	Kl5
	sc	^{131}I	410–540		1.37	V_I'	sct			Ly1	
OTHER MONOVALENT HALIDES											
CuI (α,β)	pp	^{131}I	370–500	7.5	1.48	$\{V_I'\}$	sfc			No3	
CuI (γ)	pp	^{131}I	350–370	5.0×10^4	1.96	$\{V_I'\}$	sfc			No3	
AgCl	sc	^{110}Ag	380–440	6.5×10^2	1.13	$V_{Ag}' + Ag_i^{.}$	sct	intersitialcy		We1	Co4, Re1
			300–380	1.8	0.92	$V_{Ag}' + Ag_i^{.}$	sct	intersitialcy		We1	
	sc	^{36}Cl	300–450	8.5×10	1.57	V_{Cl}	sct			La1	Co4, No5

Compound	Type	Isotope	Temp. range	D	Q	Defect	Method	Notes	Ref.	Ref.
AgBr	sc	^{110}Ag	345– 410	6.8×10^2	1.10	$V_{Ag}' + Ag\cdot$	set	interstitialcy	Fr2	Mil, Mu4, Se6, St2, St3
			250– 345	5.7×10	0.97	$V_{Ag}' + Ag$	set	interstitialcy	Fr2	
			140– 250	1.3	0.79	$V_{Ag}' + Ag_i$	set	interstitialcy	Fr2	
	sc	^{32}Br	370– 415	9.4×10^3	2.53	$V_{Br}\cdot$	set		Ta10	Ba1, Zi1
			332– 370	9.4×10^4	1.93	$V_{Br}\cdot$	set		Ta10	
AgI (α)	pp	^{110}Ag	145– 220	1.6×10^{-4}	0.097	V_{Ag}'	set		Jo3	Jo3, Jo4
	pp	^{131}I	145– 540	4.4×10^{-4}	0.70	$\{V_I\cdot\}$	sfc	cation disorder	No4	
AgI (β)	pp	^{115}Ag	20– 145	4.4×10^{-4}	0.62	$\{V_{Ag} + Ag\cdot\}$	slx		Jo3	Mu5, Zi1
	sc	^{121}I	80– 140	4.7×10^{-10}	0.29	$\{V_I\cdot\}$	set		La2	
		^{121}I	50– 140	$D = (8.2 \text{ to } 3.4) \times 10^{-4}$					La2	
AgI (γ)	pp	^{110}Ag	20– 147	5×10^{-4}	0.37	$\{V_{Ag} + Ag\cdot\}$	slx		Jo3	
AgHgI$_4$ (α)	pc	^{111}Ag	40– 110	1.8	0.47	V_{Ag}'	sfc	cation disorder	Zi1	
		^{209}Hg	50– 160		0.89	V_{Hg}''	sfc	cation disorder	Zi1	
TlCl	sc	^{204}Tl	290– 390	6.2×10^{-1}	1.10	V_{Tl}'	set		Fr4	Fr3
	sc	^{36}Cl	270– 420	3.1×10^{-2}	0.77	$V_{Cl}\cdot$	set		Fr4	
POLYVALENT HALIDES										
CaF$_2$	sc	^{44}Ca	800–1250	1.3×10^2	3.75	$\{V_{Ca}''\}$	set		Ma11	Sh9
	sc	^{18}F	670– 950	1.9×10	1.91	$V_F' + F_i'$	sfc		Ma10	
		^{18}F	360– 800	1.1×10^{-4}	0.91	V_F'	sfc		Ma10	
CaCl$_2$	sc	^{90}Sr	700	$D = 3 \times 10^{-10}$		$\{V_{Ca}''\}$	sfc		Ho5	
		^{36}Cl	800	$D = 3 \times 10^{-7}$		$V_{Cl}\cdot$ or Cl_i'	sfc		Ho5	
PbCl$_2$	pp	^{212}Pb	180– 270	1.2×10^2	1.65	$\{V_{Pb}''\}$	sfc		He5	He5
PbI$_2$	sc	^{212}Pb	260– 320	2.0×10^5	1.37	$\{V_{Pb}''\}$	sfc	(see next table)	He5	
LaF$_3$	sc	^{19}F	0– 230	7.1×10^{-4}	0.50	$\{V_F'\}$	nmr		Se7 Lu3	Le3
SIMPLE INORGANIC RADICALS										
Li$_2$SO$_4$	pc	^{6}Li	640– 790	1.9×10^{-3}	0.34	V_{Li}'	set	cation disorder	Kv4	Ku5, Lu2
LiOH	sc	^{6}Li	25	$D = 7.3 \times 10^{-11}$					Ku5	Li4
As$_2$SO$_4$ (α)	sp	^{111}Ag	430– 600	2.5	1.15	$\{V_{Ag} + Ag\cdot\}$	sfc		Jo2	Ja3
Ag$_2$SO$_4$ (β)	sp	^{111}Ag	100– 430	6.7×10^{-5}	0.58	$\{V_{Ag} + Ag\cdot\}$	sfc		Jo2	
CaCO$_3$	sc	^{13}C	606– 848	4.5×10^{-4}	2.51		gsx		Ha13	
MONOVALENT OXIDES								P_{O_2} (atm)		
Cu$_2$O	pc	^{64}Cu	800–1050	4.4×10^{-2}	1.57	V_{Cu}'	set	1.3×10^{-3}	Mo3	Ca4, Sh4
Ag$_2$O	sc	^{18}O	1020–1120	5.4×10^{-4}	1.70	$O_i\cdot\cdot$	gsx	1.3×10^{-1}	E52	Mo4, Si1
	pp	^{110}Ag	20– 160	5.4×10^{-8}	0.30	$\{V_{Ag}'\}$	sfc		Ro7	
DIVALENT OXIDES								P_{O_2} (atm)		
BeO	sc	^{7}Be	1720–1960	1.2×10^{-6}	1.56	V_{Be}''	set	Ar	Au2	Au1, Au3
	pc	^{7}Be	1500–1760	1.3×10^{-3}	2.78	V_{Be}''	set	Ar	Au2	
	sp	^{7}Be	1100–1800	3.2×10^{-3}	2.73	V_{Be}''	set	vacuum	De3	De2
	sc	^{18}O	1500–2130	5.9×10^{-5}	2.12	V_{Be}''	set	$D_i/D_a = 1.3$	Co7	
MgO	sc	^{28}Mg	1400–1600	3.0×10^{-5}	2.97		gsx	4.1×10^{-1} air	Ho1	Au2
	sc	^{18}O	1400–1600	2.5×10^{-1}	3.42		set		Li13	H3, Se4
	sc	^{18}O	1000–1150	4.3×10^{-5}	3.56	$O_i\cdot\cdot$	gsx	$(1 - 120) \times 10^{-2}$	Ro6	Au2
		^{18}O	750–1000	4.8×10^{-14}	1.31	$O_i\cdot\cdot$	gsx	$(1 - 120) \times 10^{-2}$	Ro6	H3, Se4
	pc	^{18}O	1650	$D < 10^{-14}$			ard	1.6×10^{-1}	Ho2	Oi2

TABLE 9f-6. DIFFUSION IN IONIC CRYSTALS (Continued)

Substance	Form	Isotope	T range, °C	D_0, cm²/sec	W, eV	Defect	Method	Environment P_{O_2} (atm)	Comments	Specific reference	Other references
CaO	sc	^{45}Ca	1000–1400	8.8×10^{-3}	1.50	$[V_{Ca}'']$	set	1.3×10^{-4}	……	Gu2	Li6
BaO	sc	^{140}Ba	1080–1230	1×10^{09}	11	$[Ba_x^z]$	set	……	……	Re2	Bel2, De1
				1×10^{01}	12	$[V_{Ba}'']$	set	……	……	Re2	
			330–1080	1×10^{-0}	0.44	$[Ba_x^z]$	set	……	……	Re2	
				3×10^{-18}	0.3	$[V_{Ba}'']$	set	……	……	Re2	
ZnO	sc	O	800–1300	2.5×10^{03}	2.8	V_O^z	ade	Ba(g)	……	Sp2	Le4, Li5,
	pw	^{65}Zn	720–840	4×10^{-1}	3.32	$Zn_i\cdot$	gsx	Zn(g)	$D \propto P z_{Zn}^{0.65}$	Se5	Pa4, Ro2,
	sc	^{65}Zn	940–1025	5.0	3.25	……	sfc	……	……	Mu2	Ro3, Se1,
			850–940	3.0×10^{-9}	0.87	$[Zn_i\cdot]$	sfc	……	……	Mu2	Se2, Se4,
	sc	^{65}Zn	1000–1250	1.3×10^{-6}	1.9	……	set	……	……	Mo5	
	sc	^{18}O	1100–1300	6.5×10^{11}	7.15	[disloc]	gsx	O_2	……	Mo5	Sp1
CdO	sc	^{18}O	630–855	3.8×10^{6}	3.99	$V_O\cdot$	gsx	1 (to 10⁻¹)	$D \propto P{o_2}^{\frac{1}{2}}$	Ha17	Ha14, Ha15
SnO₂	sp	^{119}Sn	980–1380	1×10^{8}	5.14	……	sfc	1 (to 10⁻³)	$D \propto P{o_2}^{-\delta}$	Li10	Si4
PbO (α)	pp	^{210}Pb	600–680	4×10^{9}	3.5	……	sfc	air	……	Da2	Li8
	pw	^{18}O	500–650	5.4×10^{-6}	0.93	……	gsx	O_2	……	Th3	
PbO (β)	pc	^{212}Pb	200–460	1.6×10^{-11}	0.56	……	sfc	air	……	Li9	
POLYVALENT OXIDES								P_{O_2} (atm)			
Al₂O₃	sp	^{26}Al	1670–1905	2.8×10	4.95	V_{Al}'''	set	air	……	Pa1	Co1, He6
	sc	^{18}O	1500–1780	1.9×10^{3}	6.6	$V_O\cdot$	gsx	2.0×10^{-2}	……	Oi1	
			1200–1620	6.3×10^{-8}	2.5	……	gsx	2.0×10^{-2}	……	Oi1	
In₂O₃	pf	In	308–407	7.8×10^{-3}	1.35	$In_i\cdot\cdot\cdot$	oxy	5×10^{-4}	on InSb	Ro5	
Bi₂O₃	pp	^{210}Bi	720–780	4.5×10^{-1}	2.00	V_{Bi}'''	sfc	air	……	Pa3	
			600–700	4.3×10^{-4}	0.90	V_{Bi}'''	sfc	air	……	Pa3	
Y₂O₃	sp	^{91}Y	1400–1800	2.4×10^{-4}	1.90	……	set	vacuum	……	Be6	
	sp	O	1000–1500	7.2	2.54	……	oxy	air	……	Wi2	
TiO₂	sc	^{18}O	710–1300	2.0×10^{-3}	2.60	$V_O\cdot$	gsx	$(2-6) \times 10^{-1}$	$D_\perp/D_\parallel = 1.6$	Ha18	Ha16
ZrO₂	sc	^{18}O	800–1000	9.7×10^{-3}	2.43	$V_O\cdot$	gsx	4×10^{-1}	……	Ma1	De4, Do4
ZrO₂:CaO (16%)	pf	^{95}Zr	300–390	3.5×10^{-4}	1.24	V_{Zr}''''	set	1	……	Sm2	Mo1
	pc	^{46}Ca	1700–2150	4.4×10^{-1}	4.01	V_{Ca}''	set	H_2	……	Rh1	Mo1
			1700–2100	1.8×10^{-2}	4.35	$V_O\cdot$	set	H_2	……	Rh1	
ZrO₂:CaO (14%)	sc	^{18}O	780–1100	……	1.35	$V_O\cdot$	set	air	O disorder	Si3	
Nb₂O₅ (α)	sc	^{18}O	850–1200	1.2×10^{-3}	2.14	$V_O\cdot$	gsx	1 (to 10⁻²)	$D \propto P{o_2}^{-\frac{1}{4}}$	Ch6	Ch4, Do5,
	sc	^{18}O	850, 900	$D_\parallel/D_\perp = 60, 190$	……	O_2	ard	$10^{-7}, 10^{-2}$	[010] axis	Sh1	Sh2
Nb₂O₅ (γ)	pf	O	540–840	1.0	1.85	$[V_O\cdot\cdot]$	oxy	1 to 10⁻³		Sb2	

Material	type	Tracer	T (°C)	D	Q	Defect	Method	P_{O_2} (atm) / conditions	Remarks	Refs
TRANSITION METAL OXIDES										
Cr₂O₃	sp	⁵¹Cr	1040–1550	1.4×10^{-7}	2.64	V_{Cr}'' or $Cr_i^{\cdot\cdot\cdot}$	sfc	N₂		Ha1, Fe1,Li11
	sc	⁵¹Cr	1300	$D \propto (P_{H_2O}/P_{H_2})^{0.4}$		V_{Cr}''	sfc	$P_{H_2O}/P_{H_2}=2$ to 18		Wa1
	pc	¹⁸O	1100–1450	7.4×10^{-7}	4.38		gsx	1.6×10^{-1}		Ha3, Ha5
MnO	sp	⁵⁴Mn	900–1150	7.4×10^{-7}	0.79	V_{Mn}''	set	10^{-12} (to 10^{-14})	$D \propto P_{O_2}^{4}$	Bo1
FeO	pc	⁵⁹Fe	700–1120	1.1×10^{-2}	1.31	V_{Fe}''	set	$Fe_{0.917}O$		De6, De7,He2
	sp	⁵⁵Fe	700–1000	1.3×10^{6}	1.31	V_{Fe}''	set	$Fe_{0.921}O$		Ca2, Hi1
Fe₂O₃ (α)	pc	⁵⁹Fe	950–1050	2.0	4.35	$\{V_O^{\cdot\cdot}\}$	gsx	air		Iz2, Li7
Fe₃O₄	sp	¹⁸O	900–1250	5.2	3.38	V_{Fe}''	sfc	1.6×10^{-1}		Ha4, Ki2
	pc	⁵⁵Fe	750–1000	6×10^{5}	2.38		sfc	$Fe_{2.99}O_4$		Hi1, Iz1
	sc	⁵⁵Fe	850–1075	$D \propto (P_{CO_2}/P_{CO})^{0.3}$	3.64	V_{Fe}''	set	Ar		Kl1
	sp	⁵⁵Fe	1115	3.2×10^{-14}			sfc	$P_{CO_2}/P_{CO}=10$ to $10^{3.5}$		Se5
CoO	sc	¹⁸O	300–550	2.2×10^{-3}	0.74	V_{Co}''	gsx	$H_2O(g)$		Ca5, Ca3,Pr2
	sp	⁶⁰Co	1010–1340	9×10	1.50		set	1	$D \propto P_{O_2}^{0.3}$	Ca2
	sc	¹⁸O	1150–1500	2.4×10^{22}	4.2	V_{Ni}''	gsx	2.1×10^{-1}		Ch5, Ho3,Th2
Co₃O₄	pc	⁶³Ni	830–860	1.8×10^{-3}	7.6	V_{Ni}''	set	air		Th2
NiO	sc	⁶³Ni	1000–1470	4.4×10^{-4}	1.98	$\{O_i''\}$	sfc	air		Ch7, Kl2
	sc	¹⁸O	1000–1500	6.2×10^{-4}	1.92		gsx	7×10^{-2}		Sh5, Li12
RARE EARTH OXIDES										
Pr₂O₃	sc	¹⁸O	700–990	4.5×10^{-2}	2.49	$V_O^{\cdot\cdot}$	oxy	vacuum		Ok2
Nd₂O₃	pw	O	700–1000	1.3×10^{-4}	1.82		gsx	4×10^{-2} to 4×10^{-1}		Ku2
Er₂O₃	pc	¹⁸O	700–1000	6.0×10^{-6}	1.34		gsx	4×10^{-2} to 4×10^{-1}		St8
UO₂	pc	O	850–1250	1.2	0.93		oxy	air		St8, Be4,Li2
	pw	²³³U	1450–1700	$D = 1.6\times10^{-11}\times y^{1.9}$	2.07	$\{V_U''''\}$, UO_{2+x}	sec	H₂		Wi2, Li14,Ya1
	sc	²³⁵U	1500	1.2×10^{3}	3.04	O_i''	sfc	$y = 0.007$ to 0.17	interstitialcy	Al1, Ar7,Be3
	sp	²³⁵U	550–850	2.1×10^{-3}	2.83	O_i''	gsx	$UO_{2.002}$	interstitialcy	Ma8, Do3,Th4
	pw	¹⁸O	320–500		1.29		gsx	$UO_{2.063}$		Be4
MONOVALENT CHALCOGENIDES										
Na₂S	sp	²²Na	520–700	8.3×10^{2}	1.66		sfc	H₂		Be4, We2
	sp	³⁵S	400–520	1.6×10^{-3}	0.77		sfc	H₂		Mo2
			420–800	3.8×10^{-4}	1.77		sfc	H₂		Mo2
CuS	pc	⁶⁴Cu	140–450	2	<0.2	Cu disorder		no effect		Mo2
Cu₂Se	pc	Cu	580–750	9	0.27	Ag disorder	smc	$Cu_{0.96}Se$		Pa7
Ag₂S (α)	pc	¹¹⁰Ag	200–400	2.8×10^{-4}	0.15	$V_S^{\cdot}$	set	no effect		Ce1, Ce1
	pc	³⁵S	650–1000	2.4×10^{-4}	1.04	$\{Ag_i^{\cdot}\}$	set			Al2, Al2
Ag₂S (β)	pw	¹¹⁰Ag	95–175	6.3×10^{-3}	0.45		set			Is1, Is1
	pw	¹¹⁰Ag	25–70	9.3×10^{-3}	0.40		slx			Al1
	pc	³⁵S	120–141	2.4×10^{-1}	1.07		slx			Pe3
AgSbS₂	pc	Ag	400	$D = 4\times10^{-7}$		elt	S_2	no effect		Pe3
		Sb	400	$D = 3\times10^{-11}$		oxy	S_2			Ri3
Ag₂Se	pc	¹¹⁰Ag	150–280	2.1×10^{-3}	0.12	Ag disorder	set	vacuum		Ri3, Ok1
Tl₂Se	pc	²⁰⁴Tl	150–300	1.2×10^{-3}	0.61		set	vacuum		Ok1, Ja3,Sh4
	pc	⁷⁵Se	150–300	2.2×10^{-3}	0.58		set	vacuum		Ak1, Ak1

TABLE 9f-6. DIFFUSION IN IONIC CRYSTALS (Continued)

Substance	Form	Isotope	T range, °C	D_0, cm²/sec	W, eV	Defect	Method	Environment P (atm)	Comments	Specific reference	Other references
DIVALENT CHALCOGENIDES											
ZnS	sc	^{65}Zn	1030–1075	1×10^6	6.5	$[V_{Zn}'']$	gsx	$P_{Zn}=1$	....	Se3	
	sc		940–1030	1.5×10^4	3.25	$[Zn_i^{\cdot\cdot}]$	gsx	$P_{Zn}=1$	....	Se3	
	sc	^{35}S	700–890	2.9×10^4	3.4	....	sct		....	B11	
	sc	^{35}S	740–1100	8×10^{-5}	2.2	....	ard	$P_S=5\times10^{-1}$	....	Go1	
ZnSe	sc	^{75}Se	1000–1150	2.3×10^{-1}	2.7	Se_i^x	sct	P_{Se_2} max	....	Re5	Wo2
ZnTe	sc	^{65}Zn	780–950	1.4×10	2.69	V_{Zn}^x or Zn_i^x	sct	no effect	....	Re5	
	sc	^{123}Te	780–950	1.9×10^4	3.78	Te_i^x	sct	Te(g)	or $(V_{Zn}V_{Te})^z$	Re5	
CdS	sc	^{115}Cd	700–1130	3	2.0	V_{Cd}^x	sct	Cd sat.	....	Wo1	
CdSe	sc	^{75}Se	700–1000	1.3×10^8	4.43	Se_i^x	sct	$P_{Cd}=10^{-1}$ (to 10^{-7})	$D \propto P_{Cd}^{-1}$	Wo2	
	sc	^{75}Se	700–1000	2.6×10^{-3}	1.55	Se_i^x	sct	P_{Se_2} max	....	Wo2	
CdTe	sc	^{109}Cd	660–920	3.3×10^2	2.67	$V_{Cd}^x + Cd_i^x$	sct	Cd sat.	....	Bo6	An1, Sel
	sc		640–850	1.6×10	2.44	$V_{Cd}^x + Cd_i^x$	sct	Te sat.	....	Bo6	
	sc	^{123}Te	660–900	8.5×10^{-7}	1.42	Te_i^x	sct	Cd sat.	....	Bo6	Se9
	sc	^{123}Te	510–780	1.7×10^{-4}	1.38	Te_i^x	sct	Te sat.	....	Bo6	
PbS	sc	^{210}Pb	500–800	8.6×10^{-5}	1.5	$V_{Pb}^x + Pb_i^x$	sct	Pb 10^{18} cm^{-3}	....	Si2	
	sc		500–800	2.6×10^{-5}	1.4	Pb_i^x	sct	S excess	....	Si2	
	sc		500–800	5.5×10^{-7}	1.0	V_{Pb}^x	sct	stoich.	....	Si2	
	sc	^{35}S	500–750	6.8×10^{-5}	1.38	$((V_{Pb}V_S)^z)^x$	sct	Pb 10^{18} cm^{-3}	....	Se10	
	sc		500–750	1.9×10^{-6}	1.16	$\{S_i^x\}$	sct	S 10^{18} cm^{-3}	....	Se10	
	sc		500–750	4.6×10^{-6}	1.2	$\{S_i^x\}$	sct		....	Se10	
PbSe	sc	^{210}Pb	400–800	5.0×10^{-6}	0.83	$V_{Pb}^x + Pb_i^x$	sct	vacuum	....	Se8	Bo5
	sc	^{75}Se	650–800	2.1×10^{-5}	1.2	V_{Se}^x	sct	vacuum	....	Bo5	
PbTe	sc	^{210}Pb	310–820	3×10^{-3}	1.51	Pb_i^x	sct		....	Ca1	Bo4, Go2
	sc	^{127}Te	500–800	2.7×10^{-6}	0.75	V_{Te}^x	sct	vacuum	....	Bo5	Go2
BiSe	pc	^{75}Se	100–200	2.5×10^{-9}	0.67	....	sct	vacuum	....	Ku1	
Bi₂Se₃	pc	^{75}Se	100–200	8.5×10^{-9}	2.17	....	sct	vacuum	....	Ku1	
FeS	pc	Fe	350–700	3.2×10^{-2}	0.97	V_{Fe}^x	sct	Fe$_{0.85}$S	....	Co6	Co5, Me3
	sc	S	900–1060	1×10^{33}	10.4	....	....	Fe$_{0.85}$S	....	Co6	Co5
CoTe (γ)	pc	^{60}Co	400–800	$D = 1.5 \times 10^{-7}$	....	....	sct		D at 805°	Ar5	
	pc	^{125}Te	400–800	$D = 1.6 \times 10^{-9}$	....	....	sct		D at 410°	Ar5	
NiS	sc	^{63}Ni	725–880	1.1×10^{-2}	1.11	$V_{Ni}^z + Ni_i^z$	sct	Ni$_{0.97}$S	∥ c axis	Kl3	Fu1, Kl4
			725–880	8.5×10^{-3}	1.11	$V_{Ni}^z + Ni_i^z$	sct	Ni$_{0.97}$S	⊥ c axis	Kl3	
	sc	^{35}S	800–880	2.5×10^2	2.84	V_S^z	sct	Ni$_{0.97}$S	∥ c axis	Kl3	Kl4
	sc		800–880	2.2×10^6	3.79	V_S^z	sct	Ni$_{0.97}$S	⊥ c axis	Kl3	

Form of Sample
sc   single crystal
pc   polycrystalline
pf   polycrystalline film
pw   powder
pp   pressed powder
sp   sintered powder

Defect
V_{Ba}'' Ba ion vacancy
V_{Zn}^z Zn vacancy, neutral
V_{Br}^z Br ion vacancy
$(V_{Na}V_{Cl})^x$ Vacancy pair, neutral
Ag_i^z Interstitial Ag ion
Te_i^x Interstitial Te atom
{ }   Tentative assignment

Method
s:t   sectioning
sfc   surface counting
gsx   gaseous exchange
slx   solution exchange
adc   additive coloration
ard   autoradiography
elt   electrotransport
nmr   nuclear magnetic resonance
oxy   oxidation or weight loss
sme   semiconducting properties

TABLE 9f-7. CORRELATION EFFECTS IN DIFFUSION

[The correlation factor is $f = D_{\text{tracer}}/D_{\text{conductivity}}$.]

Substance	Isotope	T range, °C	f_{expt}	Specific reference	Other references	Defect and lattice	f_{theor}	Reference
NaCl...........	22Na + 36Cl	580- 680	0.9 -1.0	Do6	Ma6	$V_{Na}' + V_{Cl}'$, fcc	0.781	Ba2, Co2
	22Na + 36Cl	640- 790	0.85-1.00	Ne1				
NaBr...........	24Na	450- 700	1.0	Ma6	....	V_{Na}', fcc	0.781	Ba2, Co2
KCl.............	42K	500- 700	1.0	As3	Wi3	V_K', fcc	0.781	Ba2, Co2
CsCl (β)........	137Cs + 36Cl	280- 460	1.4 -1.5	Ha9	....	$V_{Cs}' + V_{Cl}'$, sc	0.653	Ba2, Co2
CsBr...........	134Cs + 82Br	330- 530	0.83-0.88	Ly1	....	$V_{Cs}' + V_{Br}'$, sc	0.653	Ba2, Co2
CsI............	134Cs + 131I	330- 530	0.68-0.86	Ly1	....	$V_{Cs}' + V_I'$, sc	0.653	Ba2, Co2
AgCl...........	110Ag	300- 440	0.48-0.54	We1	Co4, Mu6	$V_{Ag}' + Ag_i\cdot$, fcc	0.33-0.78	Co3, Mc2
AgCl:Cd.......	110Ag	130- 230	0.78-0.74	Gr1	Co4, Lu4	V_{Ag}', fcc	0.781	Ba2, Co2
AgBr...........	110Ag	140- 410	0.47-0.65	Fr2	Mi1, Mu4	$V_{Ag}' + Ag_i\cdot$, fcc	0.33-0.78	Co3, Fr2
AgBr:Cd.......	110Ag	100- 200	0.80	Mi1	St2	V_{Ag}', fcc	0.781	Ba2, Co2
AgI (β).........	110Ag	90- 110	6 -11	Mu5	....	{Ag ring}		
AgI.............	110Ag	90- 110	2 -8	Mu5	....	High P phase		
TlCl...........	204Tl + 36Cl	290- 300	0.75 0.70	Fr4	....	$V_{Tl}' + V_{Cl}'$, sc	0.653	Ba2, Co2
CaCl2...........	36Cl	650- 700	0.6	Ho5	....	$V_{Cl}\cdot$ or Cl_i', sc		
		200- 500	2.0 -2.6	Ho5				
Li2SO4..........	6Li	600- 800	0.9 -0.7	Kv4	Kv2	Li disorder		
BeO............	7Be	1100-1250	0.4 -0.8	De3	....	V_{Be}'', hcp	0.783	Co2, Mu3
	7Be	1550	0.8	Cl5				
ZrO2:CaO (15%)	18O	780-1100	0.5 -0.8	Si3	Ki1	$V_O\cdot\cdot$, sc	0.653	Ba2, Co2
Na2S............	22Na	400- 700	1.0	Mo2	....	V_{Na}', sc	0.653	Ba2, Co2
Ag2S (α)........	110Ag	180- 280	0.26-0.30	Ok1	....	Ag disorder	0.3	Yo3
	110Ag	200- 400	0.27-0.38	Ri1	....	Ag disorder	0.5	Ri2
Ag2Se (α).......	110Ag	140- 280	0.33-0.40	Ok1	....	Ag disorder	0.3 -0.5	Ri2, Yo3

References for Tables 9f-1 through 9f-7

The literature has been surveyed to about the middle of 1968. Most of the references cover only the period 1958 to 1968; some earlier articles have been included, especially when more recent work is not available. The *specific references* in the tables provide the most complete or reliable information, but the *other references* also contain either appreciable data or extensive discussion.

Collections of Data and Bibliographies

Halides: 9

Oxides: 1, 2, 4, 5, 7

General: 3, 6, 8

1. Berard, M. F.: "Diffusion in Ceramic Systems: A Selected Bibliography," Ames Laboratory, Iowa State University, 1962.
2. Cumming, P. A., and P. J. Harrop: *U.K. At. Energy Authority Res. Group Bibliog.* AERE-BIB 143, 22 pp., 1965.
3. "Diffusion Data," Diffusion Information Center, Cleveland, Ohio.
4. Dragoo, A. L.: *J. Res. NBS,* **72A,** 157 (1967).
5. Harrop, P. J.: *J. Mater. Sci.* **3,** 206 (1968).
6. Landolt-Boernstein: "Zahlenwerte und Funktionen aus Physik, Chemie, Astronomie, Geophysik, und Technik," 6th ed., Band II, Teil 6, Springer-Verlag OHG, Berlin, 1959.
7. O'Keeffe, M.: In "Sintering and Related Phenomena," p. 57, G. C. Kuczynski, N. A. Horton, and C. F. Gibbon, eds., Gordon and Breach, Science Publishers, Inc., New York, 1967.
8. Touloukian, Y. S., ed.: "Thermophysical Properties Research Literature: A Retrieval Guide," 2d ed., Plenum Press, Plenum Publishing Corporation, New York, 1967.
9. Sueptitz, P. and J. Teltow: *Phys. Stat. Solidi* **23,** 9 (1967).

Books and Monographs

10. Adda, Y. and J. Philibert: "La Diffusion dans les Solides," Presses Univ. France, Paris, 1966.
11. Boltaks, B. I.: "Diffusion in Semiconductors," Infosearch Ltd., London, 1961.

12. Franklin, A. D., ed.: "Calculation of the Properties of Vacancies and Interstitials," *NBS Misc. Publ.* 287, 1966.
13. Girifalco, L. A.: "Atomic Migration in Solids," Blaisdell Publishing Company, a division of Ginn and Company, Waltham, Mass., 1964.
14. Gruber, B., ed.: "Theory of Crystal Defects," Academic Press, Inc., New York, 1966.
15. Hasiguti, R. R., ed.: "Lattice Defects and Their Interactions," Gordon and Breach, Science Publishers, Inc., New York, 1967.
16. Hauffe, K.: "Oxydation von Metallen und Metallegierungen," Springer-Verlag OHG, Berlin, 1956; "Oxidation of Metals," Plenum Press, Plenum Publishing Corporation, New York, 1965.
17. Howard, R. E., and A. B. Lidiard: "Matter Transport in Solids, *Rept. Progr. Phys.* **27**, 161 (1964).
18. Inokuchi, H.: "Electrical Conduction in Solids," Routledge & Kegan Paul, Ltd., London, 1965.
19. Jost, W.: "Diffusion in Solids, Liquids, and Gases," Academic Press, Inc., New York, 1952.
20. Kingery, W. D., ed.: *Kinetics High-Temp. Processes, Conf., Dedham, Mass.*, p. 37, 1959.
21. Kröger, F. A., and H. J. Vink: *Solid State Phys.* **3**, 310 (1956).
22. Kröger, F. A.: "The Chemistry of Imperfect Crystals," North-Holland Publishing Company, Amsterdam, 1964.
23. Lidiard, A. B.: Ionic Conductivity, in "Handbuch der Physik," vol. XX, p. 246, S. Flügge, ed., Springer-Verlag OHG, Berlin, 1957.
24. Manning, J. R.: "Diffusion Kinetics for Atoms in Crystals," D. Van Nostrand Company, Inc., Princeton, N.J., 1968.
25. Murin, A. N., and B. G. Luré: "Diffuziya Mechenykh Atomov i Provodimost v Ionnykh Kristallakh (Diffusion of Labeled Atoms and Conductivity in Ionic Crystals), Izd. Leningradsk Univ., Leningrad, 1967.
26. Pick, H.: "Struktur von Stoerstellen in Alkalihalogenid-kristallen," Springer-Verlag OHG, Berlin, 1966.
27. Schmalzried, H.: *Progr. Solid State Chem.* **2**, 265 (1965).
28. Shewmon, P. G.: "Diffusion in Solids," McGraw-Hill Book Company, New York, 1963.
29. Smith, A. C., J. F. Janak, and R. B. Adler, "Electric Conduction in Solids," McGraw-Hill Book Company, New York, 1967.
30. Stasiw, O.: "Elektronen-und Ionenprozesse in Ionenkristallen," Springer-Verlag OHG, Berlin, 1959.
31. Van Bueren, H. G.: "Imperfections in Crystals," 2d ed., North-Holland Publishing Company, Amsterdam, 1961.
32. Wachtman, J. B., Jr., and A. D. Franklin, eds.: "Mass Transport in Oxides," *NBS Spec. Publ.* 296, 1968.

Review Articles

Conduction processes, ionic and electronic: 53, 56, 61, 63, 64, 68
Correlation and isotope effects: 35, 42, 43, 44
Dielectric and anelastic relaxation: 33, 46, 47
Diffusion, general discussion: 54, 55, 58, 59, 60, 62, 66, 67
Diffusion and ionic conductivity: 36, 38, 48, 50, 57, 69
Diffusion of divalent ions and inert gases: 37, 39, 45, 48
Point defects, formation and general properties: 34, 40, 49, 51, 52
Pressure effects on ionic conductivity: 41, 65

Halides

33. Cole, R. H.: *Progr. Dielectrics* **3**, 1 (1961).
34. Curien, H.: *J. Phys. Radium* **24**, 543 (1963).
35. Friauf, R. J.: *J. Appl. Phys.* **33**, 494 (1962).
36. Friauf, R. J.: *J. Phys. Chem.* **66**, 2380 (1962).
37. Friauf, R. J.: *J. Phys. Chem. Solids* **30**, 429 (1969).
38. Haven, Y.: *Proc. Brit. Ceram. Soc.* **1**, 93 (1964).
39. Kelly, R., and C. Jech: *Proc. Brit. Ceram. Soc.* **9**, 243 (1967).
40. Lawson, A. W.: *J. Appl. Phys.* **33**, 466 (1962).
41. Lazarus, D.: *Progr. Very High Pressure Res., Proc. Intern. Conf., Bolton Landing, Lake George, N.Y.*, p. 46, 1961.
42. LeClaire, A. D.: *High Temp. Technol., Proc. Int. Symp.*, Pacific Grove, Calif., p. 255, 1964.
43. Lidiard, A. B.: *Proc. Intern. Symp. Reactivity Solids, 4th, Amsterdam*, p. 52, 1961.
44. Lidiard, A. B.: Interaction Radiation Solids, *Proc. Intern. Summer School*, Mol, Belg., p. 804, 1964.

45. Matzke, H. J.: *Can. J. Phys.* **46**, 621 (1968).
46. Meakins, R. J.: *Progr. Dielectrics* **3**, 151 (1961).
47. Nowick, A. S.: *Advan. Phys.* **16**, 1 (1967).
48. Seitz, F.: Interaction Radiation Solids, *Proc. Intern. Summer School*, Mol, Belg., p. 362, 1964.
49. Shlichta, P. J.: *Geol. Soc. Am. Spec. Paper* 88, p. 597, 1968.
50. Slifkin, L.: In "Mass Transport in Oxides," J. B. Wachtman, Jr., and A. D. Franklin, eds., *NBS Spec. Publ.* 296, p. 1, 1968.
51. Smakula, A.: *Mol. Designing Mater. Devices*, 69 (1965).
52. Smoluchowski, R.: Interaction Radiation Solids, *Proc. Intern. Summer School*, Mol, Belg., p. 378, 1964.

Oxides

53. Anthony, A. M.: *Journees Intern. Combust. Conversion Energie (Paris)*, 719 (1964).
54. Birchenall, C. E.: In "Mass Transport in Oxides," J. B. Wachtman, Jr., and A. D. Franklin, eds., *NBS Spec. Publ.* 296, p. 119, 1968.
55. Garbunov, N. S., and V. I. Izvekov: *Usp. Fiz. Nauk* **72**, 273 (1960).
56. Heyne, L.: in "Mass Transport in Oxides," J. B. Wachtman, Jr., and A. D. Franklin, eds., *NBS Spec. Publ.* 296, p. 149, 1968.
57. Hirano, K.: *Yogyo Kyokai Shi* **74**, 215 (1966).
58. Kingery, W. D., ed.: *Kinetics High-Temp. Processes, Conf., Dedham, Mass.*, p. 37, 1959.
59. Lindner, R.: *Inst. Intern. Chim. Colvay, Conseil Chim.*, 10^e, Brussels, p. 459, 1956.
60. Lindner, R.: *Proc. U.N. Intern. Conf. Peaceful Uses At. Energy*, 2d, Geneva, **20**, 116 (1958).
61. Mitoff, S. P.: *Progr. Ceram. Soc.* **4**, 217 (1966).
62. Moore, W. J.: *Radioisotopes Sci. Res, Proc. Intern. Conf.* **1**, 528 (1957).
63. Volger, J.: *Progr. Semicond.* **4**, 205 (1960).

General

64. Krogh-Moe J.: "Selected Topics High Temp. Chem.," p. 79, 1966.
65. Lacam, A., and M. Lallemand: *J. Phys. Radium* **25**, 402 (1964).
66. Philibert, J.: *J. Phys. Radium* **24**, 417 (1963).
67. Philibert, J.: *Silicates Ind.* **28**, 449 (1963).
68. Wagner, C.: *Proc. 7th Meeting Intern. Comm. Electrochem. Thermodyn. Kinet.*, p. 361, 1957.
69. Wagner, C.: *Mol. Designing Mater. Devices*, 122 (1965).

Articles

Ab1. Abbink, H. C., and D. A. Martin: *J. Phys. Chem. Solids* **27**, 205 (1966).
Ab2. Abey, A. E., and C. T. Tomizuka: *J. Phys. Chem. Solids* **27**, 1149 (1966).
Ak1. Akhundov, G. A., and G. B. Abdullaev: *Soviet Phys. Doklady* **3**, 390 (1958).
Al1. Alcock, C. B., R. J. Hawkins, A. W. Hills, and P. McNamara: *Thermodynamics, Proc. Symp., Vienna* **2**, 57 (1966).
Al2. Allen, R. L., and W. J. Moore: *J. Phys. Chem.* **63**, 223 (1959).
Al3. Allen, R. R., and M. J. Weber: *J. Chem. Phys.* **38**, 2970 (1963).
Al4. Allnatt, A. R., and P. W. M. Jacobs: *Trans. Faraday Soc.* **58**, 116 (1962).
Al5. Allnatt, A. R., and P. Pantelis: *Solid State Commun.* **6**, 309 (1968).
An1. Anderson, J. S., and J. R. Richards: *J. Chem. Soc.* **1946**, 537.
An2. Anthony, A. M., A. Guillot, and P. Nicolau: *Compt. Rend.* **B262**, 896 (1966).
An3. Antropoff, A. V., and J. Fr. Mueller: *Z. Anorg. Allgem. Chem.* **204**, 305 (1932).
Ar1. Arai, G., and J. G. Mullen: *Phys. Rev.* **143**, 663 (1966).
Ar2. Arends, J., and H. Nijboer: *Solid State Commun.* **5**, 163 (1967).
Ar3. Arends, J., and H. Nijboer: *Phys. Stat. Solidi* **26**, 537 (1968).
Ar4. Arkhangel'skaya, V. A., B. G. Mikheev, T. I. Nikitinskaya, and M. S. Tyutin: *Soviet Phys.—Solid State* **9**, 539 (1967).
Ar5. Arkharov, V. I., S. M. Klotsman, A. N. Timofeev, and I. Sh. Trakhtenberg: *Phys. Metals Metallog.* **14**, (1) 62 (1962).
Ar6. Arnikar, H. J., and M. Chemla: *Compt. Rend.* **242**, 2132 (1956).
Ar7. Aronson, S., R. B. Roof, Jr., and J. Belle: *J. Chem. Phys.* **27**, 137 (1957).
As1. Asadi, P.: *Phys. Stat. Solidi* **20**, K55, K59 (1967).
As2. Asao, Y., I. Yoshida, R. Ando, and S. Sawada: *J. Phys. Soc. Japan* **17**, 442 (1962).
As3. Aschner, J. F.: *Phys. Rev.* **94**, 771 (1954).
Au1. Austermann, S. B.: In *Kinetics High-Temp. Processes, Conf., Dedham, Mass.*, p. 66, W. D. Kingery, ed., 1959.
Au2. Austerman, S. B.: *J. Nucl. Mater* **14**, 248 (1964).

Au3. Austerman, S. B., and J. W. Wagner: *J. Am. Ceram. Soc.* **49**, 94 (1966).
Ba1. Baranovskii, V. L., B. G. Lur'e, and A. M. Murin: *Dokl. Akad. Nauk SSSR* **105**, 1188 (1955).
Ba2. Bardeen, J., and C. Herring: In "Atom Movements," p. 87, J. H. Holloman, ed., American Society for Metals, Cleveland, 1951; also in "Imperfections in Nearly Perfect Crystals," W. Shockley, ed., p. 261, John Wiley & Sons, Inc., New York, 1952.
Ba3. Barr, L. W., I. M. Hoodless, J. A. Morrison, and R. Rudman: *Trans. Faraday Soc.* **56**, 679 (1960).
Ba4. Barr, L. W., and A. D. LeClaire: *Proc. Brit. Ceram. Soc.* **1**, 109 (1964).
Ba5. Barr, L. W., J. A. Morrison, and P. A. Schroeder: *J. Appl. Phys.* **36**, 624 (1965).
Ba6. Barsis, E., and A. Taylor: *J. Chem. Phys.* **45**, 1154 (1966).
Ba7. Barsis, E., E. Lilley, and A. Taylor: *Proc. Brit. Ceram. Soc.* **9**, 203 (1967).
Ba8. Barsis, E., and A. Taylor: *J. Chem. Phys.* **48**, 4357 (1968).
Ba9. Barsis, E., and A. Taylor: *J. Chem. Phys.* **48**, 4362 (1968).
Ba10. Bartlett, B. E., F. C. Tompkins, and D. A. Young: *Proc. Roy. Soc. (London)*, ser. A, **246**, 206 (1958).
Be1. Beaumont, J. H., and P. W. M. Jacobs: *J. Chem. Phys.* **45**, 1496 (1966).
Be2. Beaumont, J. H., and P. W. M. Jacobs: *Phys. Stat. Solidi* **17**, K45 (1966).
Be3. Belle, J., and A. B. Auskern: In *Kinetics High-Temp. Processes, Conf., Dedham, Mass.*, p. 44, W. D. Kingery, ed., 1959; also A. B. Auskern and J. Belle: *J. Chem. Phys.* **28**, 171 (1958).
Be4. Belle, J., A. B. Auskern, W. A. Bostrom, and F. S. Susko: *Proc. Intern. Symp. Reactivity Solids, 4th, Amsterdam*, 452 (1961).
Be5. Bénière, F., and F. Chemla: *Compt. Rend.* **C266**, 660 (1968).
Be6. Berard, M. F., and D. R. Wilder: *J. Appl. Phys.* **34**, 2318 (1963).
Be7. Bergé, P.: *Bull. Soc. Franc. Mineral. Crist.* **83**, 57 (1960).
Be8. Bergé, P., M. Benveniste, G. Blanc, and M. Dubois: *Compt. Rend.* **258**, 5839 (1964).
Be9. Bergé, P., C. Gago, G. Blanc, M. Adam-Benveniste, and M. Dubois: *J. Phys. Radium* **27**, 295 (1966).
Be10. Besson, H., D. Chauvy, and J. Rossel: *Helv. Phys. Acta* **35**, 211 (1962).
Be11. Besson, J., C. Deportes, and G. Robert: *Compt. Rend.* **C262**, 527 (1966).
Be12. Bever, R. S.: *J. Appl. Phys.* **24**, 1008 (1953).
Bi1. Biermann, W., and H. J. Oel: *Z. Physik. Chem. (Frankfurt)* **17**, 163 (1958).
Bi2. Biermann, W.: *Z. Physik. Chem. (Frankfurt)* **25**, 90 (1960).
Bi3. Biermann, W., and W. Jost: *Z. Physik. Chem. (Frankfurt)* **25**, 139 (1960).
Bi4. Biermann, W.: *Z. Physik. Chem. (Frankfurt)* **25**, 253 (1960).
Bi5. Biltz, W., and A. Voigt: *Z. Anorg. Allgem. Chem.* **126**, 39 (1923).
Bi6. Biltz, W., and W. Klemm: *Z. Physik. Chem.* **110**, 318 (1924).
Bi7. Bizouard, M., and P. Cerisier: *Compt. Rend.* **B262**, 1 (1966).
Bi8. Bizouard, M., P. Cerisier, and J. Pantaloni: *Compt. Rend.* **C264**, 144 (1967).
Bl1. Blount, G. H., G. A. Marlor, and R. H. Bube: *J. Appl. Phys.* **38**, 3795 (1967).
Bl2. Blumenthal, R. N., J. Baukus, and W. M. Hirthe: *J. Electrochem. Soc.* **114**, 172 (1967).
Bo1. Bocquet, J. P., M. Kawahara, and P. Lacombe: *Compt. Rend.* **C265**, 1318 (1967).
Bo2. Boldyrev, V. V., Yu A. Zakharov, V. M. Lykhin, and L. A. Votinova: *Kinetics Catalysis* **4**, 587 (1963).
Bo3. Bollman, W., P. Goerlich, H. Karras, and H. Mothes: In ref. 9.
Bo4. Boltaks, B. I., and Yu N. Mokhov: *Soviet Phys.—Tech. Phys.* **1**, 2366 (1957).
Bo5. Boltaks, B. I., and Yu N. Mokhov: *Soviet Phys.—Tech. Phys.* **3**, 974 (1958).
Bo6. Borsenberger, P. M., and D. A. Stevenson: *J. Phys. Chem. Solids* **29**, 1277 (1968).
Br1. Bradhurst, D. H., and H. J. de Bruin: *J. Nucl. Mater.* **24**, 261 (1967).
Br2. Bradley, J. N., and P. D. Greene: *Trans. Faraday Soc.* **62**, 2069 (1966).
Br3. Bradley, J. N., and P. D. Greene: *Trans. Faraday Soc.* **63**, 424 (1967).
Br4. Bray, D. T., and U. Merten: *J. Electrochem. Soc.* **111**, 447 (1964).
Br5. Brown, N., and I. M. Hoodless: *J. Phys. Chem. Solids* **28**, 2297 (1967).
Br6. Brown, R. N., and A. C. McLaren: *Proc. Roy. Soc. (London)* **266**, 329 (1962).
Bu1. Budnikov, P. P., and V. K. Yanovskii: *J. Appl. Chem. U.S.S.R.* **37**, 1249 (1964).
Bu2. Busch, G., and P. Junod: *Helv. Phys. Acta* **30**, 470 (1957).
Bu3. Busch, G., and P. Junod: *Helv. Phys. Acta* **31**, 567 (1958).
Ca1. Card, F. E.: Thesis, Syracuse University, 1957.
Ca2. Carter, R. E., and F. D. Richardson: *Trans. Met. Soc. AIME* **200**, 1244 (1954).
Ca3. Carter, R. E., F. D. Richardson, and C. Wagner: *Trans. Met. Soc. AIME* **203**, 336 (1955).
Ca4. Castellan, G. W., and W. J. Moore: *J. Chem. Phys.* **17**, 41 (1949).
Ca5. Castle, J. E., and P. L. Surman: *J. Phys. Chem.* **71**, 4255 (1967).

Ce1. Celustka, B., and Z. Ogorelec: *J. Phys. Chem. Solids* **27**, 957 (1966).
Ce2. Cerisier, P., and M. Bizouard: *Compt. Rend.* **261**, 5100 (1965).
Ch1. Champion, J. A.: *Brit. J. Appl. Phys.* **15**, 633 (1964).
Ch2. Champion, J. A.: *Brit. J. Appl. Phys.* **16**, 805 (1965).
Ch3. Chang, R.: *Proc. Brit. Ceram. Soc.* **9**, 193 (1967).
Ch4. Chen, W. K., and R. A. Swalin: *J. Phys. Chem. Solids* **27**, 57 (1966).
Ch5. Chen, W. K., and R. A. Jackson: *Am. Ceram. Soc. Bull.* **46**, 357 (1967).
Ch6. Chen, W. K., and R. A. Jackson: *J. Chem. Phys.* **47**, 1144 (1967).
Ch7. Choi, J. S., and W. J. Moore: *J. Phys. Chem.* **66**, 1308 (1962).
Ch8. Christy, R. W., and H. S. Dobbs: *J. Chem. Phys.* **46**, 722 (1967).
Cl1. Cleaver, B.: *Z. Physik. Chem. (Frankfurt)* **45**, 346 (1965).
Cl2. Cleaver, B.: *Z. Physik. Chem. (Frankfurt)* **45**, 359 (1965).
Cl3. Cline, C. F., J. Carlberg, and W. Newkirk: *J. Am. Ceram. Soc.* **50**, 55 (1967).
Cl4. Cline, C. F., H. W. Newkirk, and R. R. Vandervoort: *J. Am. Ceram. Soc.* **50**, 221 (1967).
Cl5. Cline, C. F., H. W. Newkirk, R. H. Condit, and Y. Hashimoto: In "Mass Transport in Oxides," J. B. Wachtman, Jr., and A. D. Franklin, eds., *NBS Spec. Publ.* 296, p. 177, 1968.
Co1. Coble, R. L.: *J. Am. Ceram. Soc.* **41**, 55 (1958).
Co2. Compaan, K., and Y. Haven: *Trans. Faraday Soc.* **52**, 786 (1956).
Co3. Compaan, K., and Y. Haven: *Trans. Faraday Soc.* **54**, 1498 (1958).
Co4. Compton, W. D.: *Phys. Rev.* **101**, 1209 (1956); W. D. Compton and R. J. Maurer: *J. Phys. Chem. Solids* **1**, 191 (1956).
Co5. Condit, R. H., and C. E. Birchenall: *U.S. Dept. Commerce Office Tech. Serv. PB Rept.* **147**, 772, 1960.
Co6. Condit, R. H.: Thesis, Princeton University, 1961.
Co7. Condit, R. H., and Y. Hashimoto: *J. Am. Ceram. Soc.* **50**, 425 (1967).
Co8. Copeland, W. D., and R. A. Swalin: *J. Phys. Chem. Solids* **29**, 313 (1968).
Cr1. Croatto, V., and M. Bruno: *Gazz. Chim. Ital.* **78**, 95 (1948).
Da1. Danforth, W. E., and J. H. Bodine: *J. Franklin Inst.* **260**, 467 (1955).
Da2. Dasgupta, A. K., D. N. Sitharamarao, and G. D. Palkan: *Nature* **207**, 628 (1965).
Da3. Dasgupta, S.: *Brit. J. Appl. Phys.* **17**, 267 (1966).
Da4. Davidge, R. W.: *Phys. Stat. Solidi* **3**, 1851 (1963).
Da5. Davies, M. O.: *J. Chem. Phys.* **38**, 2047 (1963).
Da6. Davis, M. L., and E. F. Westrum, Jr.: *J. Phys. Chem.* **65**, 338 (1961).
Da7. Davis, W. J., S. E. Rogers, and A. R. Ubbelohde: *Proc. Roy. Soc. (London)*, ser. A, **220**, 14 (1953).
Da8. Dawood, R. I., and A. J. Forty: *Phil. Mag.* **7**, 1633 (1962).
Da9. Dawson, D. K., and L. W. Barr *Phys. Rev. Letters* **19**, 844 (1967).
Da10. Dawson, D. K., and L. W. Barr: *Proc. Brit. Ceram. Soc.* **9**, 171 (1967).
De1. Debiesse, J., and G. Neyret: *Vide* **6**, 1098 (1951).
De2. DeBruin, H. J., and G. M. Watson: *J. Nucl. Mater.* **14**, 239 (1964).
De3. DeBruin, H. J., G. M. Watson, and G. M. Blood: *J. Appl. Phys.* **37**, 4543 (1966).
De4. Debuigne, J., and P. Lehr: *Compt. Rend.* **256**, 1113 (1963).
De5. de Nobel, D.: *Philips Res. Rept.* **14**, 361, 430 (1959).
De6. Desmarescaux, P., J. P. Bocquet, and P. Lacombe: *Bull. Soc. Chim. France* **1965**, 1106.
De7. Desmarescaux, P.: *Publ. Sci. Tech. Min. Air (France)* **434**, (1967).
De8. de Vries, K. J., and J. F. van Santen: *Physica* **29**, 482 (1963).
Di1. Dixon, J. M., L. D. Lagrange, U. Merten, C. F. Miller, and J. T. Porter: *J. Electrochem. Soc.* **110**, 276 (1963).
Do1. Dobrovinska, O. R., V. I. Solunskii, and A. G. Shakhova: *Ukr. Fiz. Zh.* **12**, 868 (1967).
Do2. Dolloff, R. T.: *J. Appl. Phys.* **27**, 1418 (1956).
Do3. Dornelas, W., and P. Lacombe: *Compt. Rend.* **C265**, 359 (1967).
Do4. Douglas, D. L.: Corrosion Reactor Material, *Proc. Conf. Salzburg, Austria* **2**, 224 (1962).
Do5. Douglas, D. L.: Corrosion Reactor Material, *Proc. Conf. Salzburg, Austria* **2**, 233 (1962).
Do6. Downing, H. L., Jr., and R. J. Friauf: *J. Phys. Chem. Solids* **31**, 845 (1970).
Dr1. Dreyfus, R. W., and A. S. Nowick: *J. Appl. Phys.* **33**, 473 (1962).
Dr2. Dreyfus, R. W., and A. S. Nowick: *Phys. Rev.* **126**, 1367 (1962).
Eb1. Ebert, I., and J. Teltow: *Ann. Physik* **15**, 268 (1955).
Eb2. Ebisuzaki, Y.: Thesis, Indiana University, 1963.
Ec1. Ecklin, D., C. Nadler, and J. Rossel: *Helv. Phys. Acta* **37**, 692 (1964).
Ec2. Economu, N. A.: *Phys. Rev.* **135**, A1020 (1964).

Ec3. Economu, N. A., and P. V. Sastry: *Phys. Stat. Solidi* **6**, 135 (1964).
Ei1. Eisenstadt, M.: *Phys. Rev.* **132**, 630 (1963).
Ei2. Eisenstadt, M., and A. G. Redfield: *Phys. Rev.* **132**, 635 (1963).
Ei3. Eisenstadt, M.: *Phys. Rev.* **133**, A191 (1964).
El1. Elo, R., R. A. Swalin, and W. K. Chen: *J. Phys. Chem. Solids* **28**, 1625 (1967).
Et1. Etzel, H. W., and R. J. Maurer: *J. Chem. Phys.* **18**, 1003 (1950).
Fe1. Fedorchenko, I. M., and Yu B. Ermolovich: *Ukr. Khim. Zh.* **26**, 429 (1960).
Fi1. Finch, A., P. W. Jacobs, and F. Tompkins: *J. Chem. Soc.* **1954**, 2053.
Fl1. Floris, J. V.: *J. Am. Ceram. Soc.* **43**, 262 (1960).
Fr1. Franklin, A. D.: *J Res. NBS (Phys. and Chem.)* **69A**, 301 (1965).
Fr2. Friauf, R. J.: *Phys. Rev.* **105**, 843 (1957).
Fr3. Friauf, R. J.: *J. Phys. Chem. Solids* **18**, 203 (1961).
Fr4. Friauf, R. J.: *Z. Naturforsch.*, to be published.
Fu1. Fueki, K., Y. Oguri, and T. Mukaibo: *Bull. Chem. Soc. Japan* **41**, 569 (1968).
Fu2. Fuller, R. G.: *Phys. Rev.* **142**, 524 (1966).
Fu3. Fuller, R. G., and M. H. Reilly: *Phys. Rev. Letters* **19**, 113 (1967).
Fu4. Fuller, R. G., M. H. Reilly, C. L. Marquardt, and J. C. Wells, Jr.: *Phys. Rev. Letters* **20**, 662 (1968).
Ga1. Gallais, F., and E. Masdupuy: *Compt. Rend.* **227**, 635 (1948).
Gh1. Ghate, P. B.: *Phys. Rev.* **133**, A1167 (1964).
Gi1. Ginnings, D. C., and T. E. Phipps: *J. Am. Chem. Soc.* **52**, 1340 (1930).
Go1. Gobrecht, H., H. Nelkowski, J. W. Baars, and M. Weigt: *Solid State Commun.* **5**, 777 (1967).
Go2. Gomez, M. P.: Thesis, Stanford University, 1965.
Gr1. Gracey, J. P., and R. J. Friauf: *J. Phys. Chem. Solids* **30**, 421 (1969).
Gr2. Greener, E. H., G. A. Fehr, and W. M. Hirthe: *J. Chem. Phys.* **38**, 133 (1963).
Gr3. Greenwood, N. N., and I. J., Worrall: *J. Inorg. Nucl. Chem.* **3**, 357 (1957).
Gr4. Gruendig, H.: *Z. Phys.* **158**, 577 (1960).
Gu1. Gundermann, J., K. Hauffe, and C. Wagner: *Z. Physik. Chem.* **B37**, 148 (1937); J. Gundermann and C. Wagner: *ibid.*, 155, 157.
Gu2. Gupta, Y. P., and L. J. Weirick: *J. Phys. Chem. Solids* **28**, 811 (1967).
Ha1. Hagel, W. C., and A. U. Seybolt: *J. Electrochem. Soc.* **108**, 1146 (1961).
Ha2. Hagel, W. C.: *J. Electrochem. Soc.* **110**, 63C (1963).
Ha3. Hagel, W. C.: *J. Am. Ceram. Soc.* **48**, 70 (1965).
Ha4. Hagel, W. C.: *Trans. Met. Soc. AIME* **236**, 179 (1966).
Ha5. Hagel, W. C., P. J. Jorgensen, and D. S. Tomalin: *J. Am. Ceram. Soc.* **49**, 23 (1966).
Ha6. Harpur, W. W., R. L. Moss, and A. R. Ubbelohde: *Proc. Roy. Soc. (London)*, ser. A, **232**, 196 (1955).
Ha7. Harrop, P. J., and R. H. Creamer: *Brit. J. Appl. Phys.* **14**, 335 (1963).
Ha8. Harrop, P. J.: *Brit. J. Appl. Phys.* **16**, 729 (1965).
Ha9. Harvey, P. J., and I. M. Hoodless: *Phil. Mag.* **16**, 543 (1967).
Ha10. Hauffe, K., and A. L. Griessbach-Vierk: *Z. Electrochem.* **57**, 248 (1953).
Ha11. Hauffe, K., and H. Pfeiffer: *Z. Metallk.* **44**, 27 (1953).
Ha12. Hauffe, K., and D. Hoeffgen: *Z. Physik. Chem. (Frankfurt)* **49**, 94 (1966).
Ha13. Haul, R. A. W., and L. H. Stein: *Trans. Faraday Soc.* **51**, 1280 (1955).
Ha14. Haul, R., and D. Just: *Naturwissenschaften* **45**, 435 (1958).
Ha15. Haul, R., D. Just, and G. Duembgen: *Proc. Intern. Symp. Reactivity Solids, 4th, Amsterdam*, p. 65, 1961.
Ha16. Haul, R., and G. Duembgen: *Z. Elektrochem.* **66**, 636 (1962).
Ha17. Haul, R., and D. Just: *J. Appl. Phys.* **33**, 487 (1962).
Ha18. Haul, R., and G. Duembgen: *J. Phys. Chem. Solids* **26**, 1 (1965).
Ha19. Haven, Y.: *Rec. Trav. Chim.* **69**, 1471 (1950).
He1. Hebb, M. H.: *J. Chem. Phys.* **20**, 185 (1952).
He2. Hembree, P. L.: Thesis, Northwestern University, 1967.
He3. Hermann, P.: *Z. Physik. Chem. (Leipzig)* **227**, 338 (1964).
He4. Herrington, T. M., and L. A. K. Staveley: *J. Phys. Chem. Solids* **25**, 921 (1964).
He5. Hevesy, G., and W. Seith: *Z. Physik* **56**, 790 (1929).
He6. Hewson, C. W., and W. D. Kingery: *J. Am. Ceram. Soc.* **50**, 218 (1967).
Hi1. Himmel, L., R. F. Mehl, and C. E. Birchenall: *Trans. Met. Soc. AIME* **197**, 827 (1953).
Hi2. Hirahara, E.: *J. Phys. Soc. Japan* **6**, 422 (1951).
Hi3. Hirashima, M.: *J. Phys. Soc. Japan* **10**, 1055 (1955).
Ho1. Holt, J. B.: *J. Nucl. Mater.* **11**, 107 (1964).
Ho2. Holt, J. B., and R. H. Condit: *Mater. Sci. Res.* **3**, 13 (1966).
Ho3. Holt, J. B.: *Proc. Brit. Ceram. Soc.* **9**, 157 (1967).
Ho4. Holverson, E. L., and C. J. Kevane: *J. Chem. Phys.* **44**, 3692 (1966).

Ho5. Hood, G. M.. and J. A. Morrison: *J. Appl. Phys.* **38**, 4796 (1967).
Ho6. Hoodless, I. M., and J. A. Morrison: *J. Phys. Chem.* **66**, 557 (1962).
Ho7. Hoodless, I. M., and B. D. McNicol: *Phil. Mag.* **17**, 1223 (1968).
Ho8. Hoshino, H., and M. Shimoji: *J. Phys. Chem. Solids* **28**, 1169 (1967).
Ho9. Hoshino, H., and M. Shimoji: *J. Phys. Chem. Solids* **29**, 1431 (1968).
Ho10. Howard, R. E.: *Phys. Rev.* **144**, 650 (1966).
Hs1. Hsueh, Y. W., and R. W. Christy: *J. Chem. Phys.* **39**, 3519 (1963).
Hu1. Huntington, H. B., and P. B. Ghate: *Phys. Rev. Letters* **8**, 421 (1962).
Hu2. Huntington, H. B., P. B. Ghate, and J. H. Rosolowski: *J. Appl. Phys.* **35**, 3027 (1964).
Ii1. Iida, S.: *Japan. J. Appl. Phys.* **6**, 77 (1967).
Is1. Ishiguro, M., F. Oda, and T. Fujino: *Mem. Inst. Sci. Ind. Research Osaka Univ.* **10**, 1 (1953).
Iz1. Izvekov, V. I.: *Inzh. Fiz. Zh., Akad. Nauk Belorussk. SSR* **1**, 64 (1958).
Iz2. Izvekov, V. I., N. S. Garbunov, and A. A. Babad-Zakhryakin: *Phys. Metals Metallog.* **14**, 30 (1962).
Ja1. Jacobs, P. W. M., and F. C. Tompkins: *J. Chem. Phys.* **23**, 1445 (1955).
Ja2. Jaffray, J.: *J. Rech. Centre Natl. Rech. Sci., Lab Bellevue (Paris)* **39**, 125 (1957).
Ja3. Jagitsch, R.: *Trans. Chalmers Univ. Technol. Gothenburg* **11**, 1 (1942).
Ja4. Jain, S. C., and S. L. Dahake: *Phys. Letters (Netherlands)* **3**, 308 (1963).
Ja5. Jain, S. C., and S. L. Dahake: *Indian J. Pure Appl. Phys.* **2**, 71 (1964).
Ja6. Jain, S. C., and G. D. Sootha: *Phys. Stat. Solidi* **22**, 505 (1967).
Ja7. Jander, W.: *Z. Anorg. Allgem. Chem.* **199**, 306 (1931).
Ja8. Jackson, B. J. H., and D. A. Young: *Trans. Faraday Soc.* **63**, 2246 (1967).
Jo1. Johansen, H. A., and J. G. Cleary: *J. Electrochem. Soc.* **111**, 100 (1964).
Jo2. Johansson, G., and R. Lindner: *Acta Chem. Scand.* **4**, 782 (1950).
Jo3. Jordan, P., and M. Pochon: *Helv. Phys. Acta* **30**, 33 (1957).
Jo4. Jost, W., and J. Noelting: *Z. Physik. Chem. (Frankfurt)* **7**, 383 (1956); Jost, W., **and** H. J. Oel: *Discussions Faraday Soc.* **23**, 137 (1957).
Ka1. Kalbitzer, S.: *Z. Naturforsch.* **17a**, 1071 (1962).
Ka2. Kanzaki, H., K. Kido, and T. Ninomiya: *J. Appl. Phys.* **33**, 482 (1962).
Ka3. Kanzaki, H., K. Kido, and S. Ohzora: *J. Phys. Soc. Japan* **18**, suppl. 3, 115 (1963).
Ka4. Kanzaki, H., K. Kido, S. Tamura, and S. Oki: *J. Phys. Soc. Japan* **20**, 2305 (1965).
Ki1. Kingery, W. D., J. Pappis, M. E. Doty, and D. C. Hill: *J. Am. Ceram. Soc.* **42**, 393 (1959).
Ki2. Kingery, W. D., D. C. Hill, and R. P. Nelson: *J. Am. Ceram. Soc.* **43**, 473 (1960).
Ki3. Kirk, D. L., and P. L. Pratt: *Proc. Brit. Ceram. Soc.* **9**, 215 (1967).
Ki4. Kirovskaya, I. A., L. G. Maidanovskaya, and V. D. Zhukova: *Inorg. Mater.* **3**, 260 (1967).
Kl1. Klotsman, S. M., A. N. Timofeev, and I. Sh. Trakhtenberg: *Phys. Metals Metallog.* **10**(5), 93 (1960).
Kl2. Klotsman, S. M., A. N. Timofeev, and I. Sh. Trakhtenberg: *Phys. Metals Metallog.* **14**(3), 91 (1962).
Kl3. Klotsman, S. M., A. N. Timofeev, and I. Sh. Trakhtenberg: *Phys. Metals Metallog.* **16**(5), 92 (1963).
Kl4. Klotsman, S. M., A. N. Timofeev, and I. Sh. Trakhtenberg: *Phys. Metals Metallog.* **17**(1), 119 (1964).
Kl5. Klotsman, S. M., I. P. Polikarkova, A. N. Timofeev, and I. Sh. Trakhtenberg: *Soviet Phys.—Solid State* **9**, 1956 (1967).
Ko1. Kobayashi, K., and T. Tomiki: *J. Phys. Soc. Japan* **15**, 1982 (1960).
Kr1. Kröger, F. A.: *J. Phys. Chem. Solids* **26**, 901 (1965).
Kr2. Krogh-Moe, J., M. Vikan, and C. Krohn: *Acta Chem. Scand.* **21**, 309 (1967).
Ku1. Kuliev, A. A., and G. B. Abdullaev: *Soviet Phys.—Solid State* **1**, 545 (1959).
Ku2. Kuntz, U. E., and L. Eyring: In *Kinetics High-Temp. Processes, Dedham, Mass.*, p. 50, W. D. Kingery, ed., 1959.
Ku3. Kurnick, S.: *J. Chem. Phys.* **20**, 218 (1952).
Ku4. Kushida, T.: *J. Sci. Hiroshima Univ.*, ser. A, **14**, 147 (1950).
Ku5. Kuznets, E. D., and L. M. Yakimenko: *J. Appl. Chem. U.S.S.R.* **40**, 754 (1967).
Kv1. Kvist, A., and A. Lunden: *Z. Naturforsch.* **19a**, 1058 (1964).
Kv2. Kvist, A., and A. Lunden: *Z. Naturforsch.* **20a**, 235 (1965).
Kv3. Kvist, A.: *Z. Naturforsch.* **21a**, 487 (1966).
Kv4. Kvist, A., and U. Trolle: *Z. Naturforsch.* **22a**, 213 (1967).
Kv5. Kvist, A., and A. M. Josefson: *Z. Naturforsch.* **23a**, 625 (1968).
La1. Lakatos, E., and K. H. Lieser: *Z. Physik. Chem. (Frankfurt)* **48**, 213 (1966).
La2. Lakatos, E., and K. H. Lieser: *Z. Physik. Chem. (Frankfurt)* **48**, 228 (1966).
La3. Lasker, M. F., and R. A. Rapp: *Z. Physik. Chem. (Frankfurt)* **49**, 198 (1966).

La4. Laurance, N.: *Phys. Rev.* **120**, 57 (1960).
La5. Laurent, J. F., and J. Bénard: *J. Phys. Chem. Solids* **3**, 7 (1957); **7**, 218 (1958); J. F. Laurent: *Ann. Chim. (Paris)* **3**, 712 (1958); J. Bénard, and J. F. Laurent: *Radioisotopes Sci. Res. Proc. Intern. Conf. Paris* **1**, 577 (1957).
La6. Layer, H., M. G. Miller, and L. Slifkin: *J. Appl. Phys.* **33**, 478 (1962); H. Layer and L. Slifkin: *J. Phys. Chem* **66**, 2396 (1962); M. N. Kabler, H. Layer, M. G. Miller, and L. Slifkin: *Mater. Sci. Res.* **1**, 82 (1963).
Le1. LeClaire, A. D., and A. B. Lidiard: *Phil. Mag.* **1**, 518 (1956).
Le2. LeClaire, A. D.: *Phil. Mag.* **14**, 1271 (1966).
Le3. Lee, K., and A. Sher: *Phys. Rev. Letters* **14**, 1027 (1965).
Le4. Lee, V. J., and G. Parravano: *J. Appl. Phys.* **30**, 1735 (1959).
Le5. Lehfeldt, W.: *Z. Physik* **85**, 717 (1933).
Le6. Lewis, T. J., and A. J. Wright: *Brit. J. Appl. Phys.* **1**, 441 (1968).
Li1. Lidiard, A. B.: *Phil. Mag.* **46**, 815, 1218 (1955).
Li2. Lidiard, A. B.: *J. Nucl. Mater.* **19**, 106 (1966).
Li3. Lieser, K. H.: *Z. Physik. Chem. (Frankfurt)* **9**, 302, 308 (1956).
Li4. Lindner, R.: *Z. Elektrochem.* **54**, 430 (1950).
Li5. Lindner, R., D. Campbell, and A. Akerstroem: *Acta Chem. Scand.* **6**, 457 (1952).
Li6. Lindner, R., St. Austruemdal, and A. Akerstroem: *Acta Chem. Scand.* **6**, 468 (1952).
Li7. Lindner, R., A. G. Verduch, and A. Akerstroem: *Arkiv. Kemi* **4**, 381 (1952).
Li8. Lindner, R.: *Arkiv Kemi* **4**, 385 (1952).
Li9. Lindner, R., and H. N. Terem: *Arkiv Kemi* **7**, 273 (1954).
Li10. Lindner, R., and O. Enqvist: *Arkiv Kemi* **9**, 471 (1956).
Li11. Lindner, R., and A. Akerstroem: *Z. Physik. Chem. (Frankfurt)* **6**, 162 (1956).
Li12. Lindner, R., and A. Akerstroem: *Discussions Faraday Soc.* **23**, 133 (1957).
Li13. Lindner, R., and G. D. Parfitt: *J. Chem. Phys.* **26**, 182 (1957).
Li14. Lindner, R., and F. Schmitz: *Z. Naturforsch.* **16a**, 1373 (1961).
Lo1. Lomelin, J. M., and T. J. Neubert: *J. Phys. Chem.* **67**, 1115 (1963).
Lo2. Lothian, T. A., and T. J. Neubert: *J. Chem. Phys.* **47**, 3092 (1967).
Lu1. Luckey, G., and W. West: *J. Chem. Phys.* **24**, 879 (1956).
Lu2. Lunden, A.: *Z. Naturforsch.* **17a**, 142 (1962).
Lu3. Lundin, A. G., S. P. Gabuda, and A. I. Lifshits: *Soviet Phys.—Solid State* **9**, 273 (1967).
Lu4. Lur'e, B. G., A. N. Murin, and I. V. Murin: *Soviet Phys.—Solid State* **8**, 2991 (1967).
Ly1. Lynch, D. W.: *Phys. Rev.* **118**, 468 (1960).
Ma1. Madeyski, A., and W. W. Smeltzer: *Mater. Res. Bull.* **3**, 369 (1968).
Ma2. Mahendroo, P. P., and A. W. Nolle: *Phys. Rev.* **126**, 125 (1962).
Ma3. Maidanovskaya, L. G., I. A. Kirovskaya, and G. L. Lobanova: *Inorg. Mater.* **3**, 839 (1967).
Ma4. Makarov, L. L., B. G. Lur'e, and V. N. Malyshev: *Soviet Phys.—Solid State* **2**, 79 (1960).
Ma5. Manakin, B. A., I. A. Voloshchenko, and V. A. Kolesnikov: *Russ. J. Phys. Chem.* **41**, 1861 (1967).
Ma6. Mapother, D., H. N. Crookes, and R. J. Maurer: *J. Chem. Phys.* **18**, 1231 (1950).
Ma7. Mariani, E., J. Eckstein, and E. Rubinova: *Czech. J. Phys.* **17**, 552 (1967).
Ma8. Marin, J. F., H. Michaud, and P. Contamin: *Compt. Rend.* **C264**, 1633 (1967).
Ma9. Matsumura, T.: *Can. J. Phys.* **44**, 1685 (1966).
Ma10. Matzke, H. J.: *J. Nucl. Mater.* **11**, 344 (1964).
Ma11. Matzke, H. J., and R. Lindner: *Z. Naturforsch.* **19a**, 1178 (1964).
Ma12. Maycock, J. N.: *J. Appl. Phys.* **35**, 1512 (1964).
Mc1. McClaine, L. A., and C. P. Coppel: *J. Electrochem. Soc.* **113**, 80 (1966).
Mc2. McCombie, C. W., and A. B. Lidiard: *Phys. Rev.* **101**, 1210 (1956).
Md1. Mdivani, O. M.: *Tr. Tbilissk. Gos. Univ.* **103**, 151 (1965).
Me1. Mehrotra, P. N., G. V. Chandrasekhar, C. N. R. Rao, and E. C. Subbaro: *Trans. Faraday Soc.* **62**, 3586 (1966).
Me2. Melik-Gaikazyan, I. Ya., L. I. Roschina, M. I. Ignat'eva, and E. P. Kurakina: *Izv. Vysshikh Uchebn. Zavedenii Fiz.* **10**, 141 (1967).
Me3. Meussner, R. A., and C. E. Birchenall: *Corrosion* **13**, 677 (1957).
Mi1. Miller, A. S., and R. J. Maurer: *J. Phys. Chem. Solids* **4**, 196 (1958).
Mi2. Mitoff, S. P.: *J. Chem. Phys.* **31**, 1261 (1959).
Mi3. Mitoff, S. P.: *J. Chem. Phys.* **36**, 1383 (1962).
Mi4. Miyata, T., R. Sano, and T. Tomiki: *J. Phys. Soc. Japan* **20**, 638 (1965).
Mi5. Miyatani, S. Y., and Y. Suzuki: *J. Phys. Soc. Japan* **8**, 680 (1953).
Mi6. Miyatani, S.: *J. Phys. Soc. Japan* **10**, 786 (1955).
Mi7. Miyatani, S.: *J. Phys. Soc. Japan* **13**, 317 (1958).
Mi8. Miyatani, S.: *J. Phys. Soc. Japan* **13**, 341 (1958).

Mi9. Miyatani, S.: *J. Phys. Soc. Japan* **14**, 996 (1959).
Mo1. Moebius, H. H., H. Witzmann, and D. Gerlach: *Z. Chem.* **4**, 154 (1964).
Mo2. Moebius, H. H., H. Witzmann, and R. Hartung: *Z. Physik. Chem.* (*Leipzig*) **227**, 40 (1964).
Mo3. Moore, W. J., and B. Selikson: *J. Chem. Phys.* **19**, 1539 (1951); **20**, 927 (1952).
Mo4. Moore, W. J., Y. Ebisuzaki, and J. A. Sluss: *J. Phys. Chem.* **62**, 1438 (1958).
Mo5. Moore, W. J., and E. I. Williams: *Discussions Faraday Soc.* **28**, 86 (1959).
Mo6. Morkel, A., and H. Schmalzried: *J. Chem. Phys.* **36**, 3101 (1962).
Mo7. Morlin, Z.: *Acta Phys. Acad. Sci. Hung.* **21**, 137 (1966).
Mr1. Mrgudich, J. N.: *J. Electrochem. Soc.* **107**, 475 (1960).
Mu1. Mueller, P.: *Phys. Stat. Solidi* **9**, K193 (1965); **12**, 775 (1965).
Mu2. Muennich, F.: *Naturwissenschaften* **42**, 340 (1955).
Mu3. Mullen, J. G.: *Phys. Rev.* **124**, 1723 (1961); *Phys. Rev. Letters* **9**, 383 (1962).
Mu4. Murin, A. N., B. G. Lur'e, and N. A. Lebedev: *Soviet Phys.—Solid State* **2**, 2324 (1960).
Mu5. Murin, A. N., B. G. Lur'e, and Yu P. Tarlakov: *Soviet Phys.—Solid State* **3**, 2395 (1961).
Mu6. Murin, A. N., B. G. Lur'e, and I. V. Murin: *Soviet Phys.—Solid State* **9**, 1840 (1968).
Na1. Nagels, P., W. van Lierde, R. de Batist, M. Denayer, L. de Jonghe, and R. Gevers: *Thermodynamics, Proc. Symp., Vienna* **2**, 311 (1966).
Ne1. Nelson, V. C., and R. J. Friauf: *J. Phys. Chem. Solids* **31**, 825 (1970).
Ne2. Neubert, T. J., and G. M. Nichols: *J. Am. Chem. Soc.* **80**, 2619 (1958).
Ne3. Neuhaus, A., and E. Hinze: *Ber. Bunsenges. Physik. Chem.* **70**, 1073 (1966).
Ne4. Neuimin, A. D., and S. F. Pal'guev: *Dokl. Phys. Chem.* **143**, 315 (1962).
Ni1. Nikitinskaya, T. I., E. V. Suntsov, and M. S. Tyutin: *Soviet Phys.—Solid State* **9**, 1656 (1967).
No1. Noddack, W., and H. Walch: *Z. Elektrochem.* **63**, 269 (1959).
No2. Noddack, W., and H. Walch: *Z. Physik. Chem.* (*Leipzig*) **211**, 194 (1959).
No3. Noelting, J.: *Z. Physik. Chem.* (*Frankfurt*) **19**, 118 (1959).
No4. Noelting, J.: *Z. Physik. Chem.* (*Frankfurt*) **32**, 154 (1962).
No5. Noelting, J.: *Z. Physik. Chem.* (*Frankfurt*) **38**, 154 (1963).
No6. Noyer, F., and J. F. Laurent: *Compt. Rend.* **242**, 3068 (1956).
Oi1. Oishi, Y., and W. D. Kingery: *J. Chem. Phys.* **33**, 480 (1960).
Oi2. Oishi, Y., and W. D. Kingery: *J. Chem. Phys.* **33**, 905 (1960).
Ok1. Okazaki, H.: *J. Phys. Soc. Japan* **23**, 355 (1967).
Ok2. O'Keefe, M., and W. J. Moore: *J. Phys. Chem.* **65**, 1438, 2277 (1961).
Ow1. Owens, B. B., and G. R. Argue: *Science* **157**, 308 (1967).
Pa1. Paladino, A. E., and W. D. Kingery: *J. Chem. Phys.* **37**, 957 (1962).
Pa2. Pal'guev, S. F., and A. D. Neuimin: *Soviet Phys.—Solid State* **4**, 629 (1962).
Pa3. Palkar, G. D., D. N. Sitharamararo, and A. K. Dasgupta: *Trans. Faraday Soc.* **59**, 2634 (1963).
Pa4. Panasyuk, G. P., M. N. Danchevskaya, and N. I. Kobozev: *Russ. J. Phys. Chem.* **41**, 354 (1967).
Pa5. Pappis, J., and W. D. Kingery: *J. Am. Ceram. Soc.* **44**, 459 (1961).
Pa6. Patterson, J. W., E. C. Bogren, and R. A. Rapp: *J. Electrochem. Soc.* **114**, 752 (1967).
Pa7. Pavyluchenko, M. M., I. I. Pokrovskii, and A. S. Tikhonov: *Dokl. Akad. Nauk. Belorussk. S.S.R.* **9**, 235 (1965).
Pa8. Payne, R. T., and A. W. Lawson: *J. Chem. Phys.* **34**, 2201 (1961).
Pe1. Pershits, Ya. N., and E. V. Pavlov: *Soviet Phys.—Solid State* **10**, 1125 (1968).
Pe2. Persyn, G. A., and A. W. Nolle: *Phys. Rev.* **140**, A1610 (1965).
Pe3. Peschanski, D.: *J. Chim. Phys.* **47**, 933 (1950).
Pe4. Peters, D. W., L. Feinstein, and C. Peltzer: *J. Chem. Phys.* **42**, 2345 (1965).
Ph1. Phipps, T. E., W. D. Lansing, and T. G. Cooke: *J. Am. Chem. Soc.* **48**, 112 (1926).
Ph2. Phipps, T. E., and E. G. Partridge: *J. Am. Chem. Soc.* **51**, 1331 (1929).
Pi1. Pierce, C. B.: *Phys. Rev.* **123**, 744 (1961).
Pl1. Plester, D. W., S. E. Rogers, and A. R. Ubbelohde: *Proc. Roy. Soc.* (*London*) **235**, 469 (1956).
Pr1. Pretzel, F. E., D. T. Vier, E. G. Szklarz, and W. B. Lewis: *USAEC Rept. LA2463* (1960).
Pr2. Price, J. B., and J. B. Wagner, Jr.: *Z. Physik. Chem.* (*Frankfurt*) **49**, 257 (1966).
Pr3. Pryor, A. W.: *J. Nucl. Mater.* **14**, 258 (1964).
Ra1. Ramasastry, C., and Y. V. G. S. Murti: *J. Phys. Chem. Solids* **24**, 1384 (1963).
Ra2. Ramasastry, C., and Y. V. G. S. Murti: *Proc. Roy. Soc.* (*London*), ser. A, **305**, 441 (1968).
Ra3. Rapoport, E.: Thesis, University of Maryland, 1964.

Ra4. Rapp, R. A.: *USAEC Rept.* COO-1440-3, 1967.
Re1. Reade, R. F., and D. S. Martin: *J. Appl. Phys.* **31**, 1965 (1960).
Re2. Redington, R. W.: *Phys. Rev.* **87**, 1066 (1952).
Re3. Reinhold, H., and H. Braueninger: *Z. Physik. Chem.* **B41**, 397 (1938).
Re4. Reuter, B., and K. Hardel: *Ber. Bunsenges. Physik. Chem.* **70**, 82 (1966).
Re5. Reynolds, R. A.: Thesis, Stanford University, 1966.
Rh1. Rhodes, W. H., and R. E. Carter: *J. Am. Ceram. Soc.* **49**, 244 (1966).
Ri1. Rickert, H.: *Z. Physik. Chem. (Frankfurt)* **23**, 355 (1960).
Ri2. Rickert, H.: *Z. Physik. Chem. (Frankfurt)* **24**, 418 (1960).
Ri3. Rickert, H., and C. Wagner: *Z. Elektrochem.* **64**, 793 (1960).
Ri4. Riggleman, B. M., and H. G. Drickamer: *J. Chem. Phys.* **38**, 2721 (1963).
Ro1. Robert, G., C. Deportes, and J. Besson: *J. Chim. Phys.* **64**, 1275 (1967).
Ro2. Roberts, J. P., and C. Wheeler: *Phil. Mag.* **2**, 708 (1957).
Ro3. Roberts, J. P., and C. Wheeler: *Trans. Faraday Soc.* **56**, 570 (1960).
Ro4. Rolfe, J.: *Can. J. Phys.* **42**, 2195 (1964).
Ro5. Rosenberg, A. J., and M. C. Lavine: *J. Phys. Chem.* **64**, 1135, 1143 (1960).
Ro6. Rovner, L. H.: Thesis, Cornell University, 1966.
Ro7. Rozenblyum, N. D., N. C. Bubyreva, V. I. Bukhareva, and G. Z. Kazakevich: *Russ. J. Phys. Chem.* **40**, 1324 (1966).
Sa1. Samara, G. A.: *Phys. Rev.* **165**, 959 (1968).
Sa2. Sastry, P. V., and T. V. Srinivasan: *Phys. Rev.* **132**, 2445 (1963).
Sc1. Scanlon, W. W., and R. F. Brebrick: *Physica* **20**, 1090 (1954).
Sc2. Schamp, H. W., and E. Katz: *Phys. Rev.* **94**, 828 (1954).
Sc3. Schmalzried, H.: *Z. Physik. Chem. (Frankfurt)* **22**, 199 (1959).
Sc4. Schmalzried, H.: *J. Chem. Phys.* **33**, 940 (1960).
Sc5. Schmalzried, H.: *Z. Physik. Chem. (Frankfurt)* **31**, 184 (1962).
Sc6. Schmalzried, H.: *Z. Physik. Chem. (Frankfurt)* **38**, 87 (1963).
Sc7. Schock, R. N., and S. Katz: *J. Chem. Phys.* **48**, 2094 (1968).
Sc8. Scholten P. C., and K. J. Mysels: *Trans. Faraday Soc.* **56**, 994 (1960).
Sc9. Schwab, G. M., and G. Eulitz: *Z. Physik. Chem. (Frankfurt)* **55**, 179 (1967).
Sc10. Scott, K. T., and L. L. Wassell: *Proc. Brit. Ceram. Soc.* **7**, 375 (1967).
Se1. Secco, E. A., and W. J. Moore: *J. Chem. Phys.* **23**, 1170 (1955).
Se2. Secco, E. A., and W. J. Moore: *J. Chem. Phys.* **26**, 942 (1957).
Se3. Secco, E. A.: *J. Chem. Phys.* **29**, 406 (1958).
Se4. Secco, E. A.: *Discussions Faraday Soc.* **28**, 94 (1959).
Se5. Secco, E. A.: *Can. J. Chem.* **39**, 1544 (1961).
Se6. Seifert, G.: *Z. Physik.* **161**, 132 (1961).
Se7. Seith, W.: *Z. Elektrochem.* **39**, 538 (1933).
Se8. Seltzer, M. S., and J. B. Wagner, Jr.: *J. Chem. Phys.* **36**, 130 (1962).
Se9. Seltzer, M. S., and J. B. Wagner, Jr.: *J. Phys. Chem. Solids* **24**, 1525 (1963).
Se10. Seltzer, M. S., and J. B. Wagner, Jr.: *J. Phys. Chem. Solids* **26**, 233 (1965).
Se11. Semenenko, K. N., and T. N. Naumova: *Russ. J. Inorg. Chem.* **9**, 718 (1964).
Sh1. Sheasby, J. S., and B. Cox: *J. Less-Common Metals* **15**, 129 (1968).
Sh2. Sheasby, J. S., W. W. Smeltzer, and A. E. Jenkins: *J. Electrochem. Soc.* **115**, 338 (1968).
Sh3. Sher, A., R. Solomon, K. Lee, and M. W. Muller: *Phys. Rev.* **144**, 593 (1966).
Sh4. Shim, M. T.: Thesis, Indiana University, 1957.
Sh5. Shim, M. T., and W. J. Moore: *J. Chem. Phys.* **26**, 802 (1957).
Sh6. Shimizu, K.: *Rev. Phys. Chem. Japan* **30**, 1 (1960).
Sh7. Shimizu, K.: *Rev. Phys. Chem. Japan* **30**, 73 (1960).
Sh8. Shimizu, K.: *Rev. Phys. Chem. Japan* **31**, 67 (1962).
Sh9. Short, J. M., and R. Roy: *J. Phys. Chem.* **68**, 3077 (1964).
Si1. Simkovich, G.: *J. Phys. Chem. Solids* **24**, 213 (1963).
Si2. Simkovich, G., and J. B. Wagner, Jr.: *J. Chem. Phys.* **38**, 1368 (1963).
Si3. Simpson, L. A., and R. E. Carter: *J. Am. Ceram. Soc.* **49**, 139 (1966).
Si4. Sinclair, W. R., and T. C. Loomis: In *Kinetics High-Temp. Processes, Conf., Dedham, Mass.*, p. 58, W. D. Kingery, ed., 1959.
Sl1. Sluss, J. A., Jr.: Thesis, Indiana University, 1962.
Sm1. Smith, A. W., F. W. Meszaros, and C. D. Amata: *J. Am. Ceram. Soc.* **49**, 240 (1966).
Sm2. Smith, T.: *J. Electrochem. Soc.* **112**, 560 (1965).
So1. Sonin, A. S., and I. S. Zheludev: *Kristallografiya* **8**, 57 (1963).
So2. Sonin, A. S., and I. S. Zheludev: *Kristallografiya* **8**, 285 (1963).
So3. Southgate, P. D.: *J. Phys. Chem. Solids* **27**, 1623 (1966).
Sp1. Spicar, E.: Thesis, Stuttgart, 1956.
Sp2. Sproull, R. L., R. S. Bever, and G. G. Libowitz: *Phys. Rev.* **92**, 77 (1953).
St1. Steele, B. C. H., and C. B. Alcock: *Trans. Met. Soc. AIME* **233**, 1359 (1965).

St2. Steiger, R.: *Chimia* **18**, 306 (1964).
St3. Steiger, R., K. Boustany, and Ch. G. Boissonnas: *Helv. Chim. Acta* **49**, 787 (1966).
St4. Stoebe, T. G., T. O. Ogurtani, and R. A. Huggins: *Phys. Rev.* **134**, 963 (1964).
St5. Stoebe, T. G., T. O. Ogurtani, and R. A. Huggins: *Phys. Stat. Solidi* **12**, 649 (1965).
St6. Stoebe, T. G., and R. A. Huggins: *J. Mater. Sci.* **1**, 117 (1966).
St7. Stoebe, T. G., and P. L. Pratt: *Proc. Brit. Ceram. Soc.* **9**, 181 (1967).
St8. Stone, G. D., G. R. Weber, and L. Eyring: in "Mass Transport in Oxides," J. B. Wachtman, Jr., and A. D. Franklin, eds., *NBS Spec. Publ.* 296, p. 179, 1968.
St9. Strickler, D. W., and W. G. Carlson: *J. Am. Ceram. Soc.* **47**, 122 (1964).
St10. Strumane, R., and R. de Batist: *Phys. Stat. Solidi* **6**, 817 (1964).
Su1. Subbarao, E. C., P. H. Sutter, and J. Hrizo: *J. Am. Ceram. Soc.* **48**, 443 (1965).
Su2. Suchow, L., and G. R. Pond: *J. Am. Chem. Soc.* **75**, 5242 (1953).
Su3. Surplice, N. A.: *Brit. J. Appl. Phys.* **17**, 175 (1966).
Ta1. Takahashi, T., and O. Yamamoto: *Denki Kagaku* **32**, 610 (1964); **33**, 346 (1965).
Ta2. Takahashi, T., and O. Yamamoto: *Denki Kagaku* **33**, 733 (1965).
Ta3. Takahashi, T., and H. Iwahara: *Denki Kagaku* **34**, 254 (1966).
Ta4. Takahashi, T., and O. Yamamoto: *Electrochim. Acta* **11**, 779 (1966).
Ta5. Takahashi, T., O. Yamamoto, K. Tsukada, and A. Baba: *Denki Kagaku* **35**, 32 (1967).
Ta6. Takahashi, T., K. Kuwabara, and O. Yamamoto: *Denki Kagaku* **35**, 682 (1967).
Ta7. Tallan, N. M., R. W. Vest, and H. C. Graham: *Mater. Sci. Res.* **2**, 33 (1965).
Ta8. Tallan, N. M., and R. W. Vest: *J. Am. Ceram. Soc.* **49**, 401 (1966).
Ta9. Tallan, N. M., W. C. Tripp, and R. W. Vest: *J. Am. Ceram. Soc.* **50**, 279 (1967).
Ta10. Tannhauser, D. S.: *J. Phys. Chem. Solids* **5**, 224 (1958).
Ta11. Tare, V. B., and H. Schmalzried: *Z. Physik. Chem. (Frankfurt)* **43**, 30 (1964).
Ta12. Taylor, W. H., W. B. Daniels, B. S. H. Royce, and R. Smoluchowski: *J. Phys. Chem. Solids* **27**, 39 (1966).
Te1. Teltow, J.: *Ann. Physik* **5**, 63, 71 (1949).
Th1. Tharmalingam, K., and A. B. Lidiard: *Phil. Mag.* **4**, 899 (1959).
Th2. Thompson, B. A.: Thesis, Rensselaer Polytechnic Institute, 1962.
Th3. Thompson, B. A., and R. L. Strong: *J. Phys. Chem.* **67**, 594 (1963).
Th4. Thorn, R. J., and G. H. Winslow: *J. Chem. Phys.* **44**, 2822 (1966).
To1. Tompkins, F. C., and D. A. Young: *Discussions Faraday Soc.* **23**, 202 (1957).
To2. Torkar, K., and G. W. Herzog: *Monatsh. Chem.* **97**, 765 (1966).
To3. Torkar, K., and H. T. Spath: *Monatsh. Chem.* **98**, 2382 (1967).
Tr1. Tretjakov, J., and H. Schmalzried: *Ber. Brunsenges. Phys. Chem.* **69**, 396 (1965).
Tu1. Tubandt, C., E. Rindtorff, and W. Jost: *Z. Anorg. Allgem. Chem.* **165**, 195 (1927).
Tu2. Tubandt, C., and M. Baudouin: In "Landolt-Boernstein Physikalisch-Chemische Tabellen," 5th ed., Springer-Verlag OHG, Berlin, 1931.
Tu3. Tubandt, C., and J. Geiler: In "Landolt-Boernstein Physikalisch-Chemische Tabellen," 5th ed., Springer-Verlag OHG, Berlin, 1931.
Ue1. Ueda, A., Y. Asano, N. Nishimaki, K. Kojima, and M. Ishiguro: *Mem. Inst. Sci. Ind. Res. Osaka Univ.* **17**, 89 (1960).
Ur1. Ure, R. W., Jr.: *J. Chem. Phys.* **26**, 1363 (1957).
Ve1. Verwey, J. F., and J. Schoonman: *Physica* **35**, 386 (1967).
Ve2. Vest, R. W., and N. M. Tallan: *J. Am. Ceram. Soc.* **48**, 472 (1965).
Ve3. Vest, R. W., and N. M. Tallan: *J. Appl. Phys.* **36**, 543 (1965).
Vi1. Vinokurov, I. V., Z. N. Zonn, and V. A. Ioffe: *Inorg. Mater.* **3**, 901 (1967).
Vo1. Volchenkova, Z. S., and S. F. Pal'guev: *Tr. Inst. Electrokhim., Akad. Nauk S.S.S.R., Ural'sk. Filial* **5**, 133 (1964).
Wa1. Wagener, K.: *Z. Physik. Chem. (Frankfurt)* **23**, 305 (1960).
Wa2. Wagner, C.: *Z. Physik. Chem.* **B21**, 25 (1933); **B23**, 469 (1933).
Wa3. Wagner, C.: *J. Chem. Phys.* **18**, 62 (1950).
Wa4. Wagner, C., and P. Hantleman: *J. Chem. Phys.* **18**, 72 (1950).
Wa5. Wagner, C.: *Proc. Intern. Comm. Electrochem. Thermodyn. Kinet.*, 7th meeting, 1955, p. 361, Butterworth Scientific Publications, London, 1957.
Wa6. Wagner, C.: *Z. Elektrochem.* **63**, 1027 (1959).
Wa7. Wagner, J. B., and C. Wagner: *J. Chem. Phys.* **26**, 1597 (1957).
Wa8. Wagner, J. B., and C. Wagner: *J. Chem. Phys.* **26**, 1602 (1957).
Wa9. Wakabayashi, H.: *J. Phys. Soc. Japan* **15**, 2000 (1960).
Wa10. Wakabayashi, H.: *J. Phys. Soc. Japan* **17**, 292 (1962).
We1. Weber, M. D., and R. J. Friauf: *J. Phys. Chem. Solids* **30**, 407 (1969).
We2. Wehefritz, V.: *Z. Physik. Chem. (Frankfurt)* **26**, 339 (1960).
We3. Weil, R., and A. W. Lawson: *J. Chem. Phys.* **41**, 832 (1964).
We4. Weiss, K.: *Z. Physik. Chem. (Frankfurt)* **12**, 68 (1957).
Wh1. Whitney, E. D.: *J. Electrochem. Soc.* **112**, 91 (1965).

Wi1. Wimmer, J. M., L. R. Bidwell, and N. M. Tallan: *J. Am. Ceram. Soc.* **50**, 198 (1967).
Wi2. Wirkus, C. D., M. F. Berard, and D. R. Wilder: *J. Am. Ceram. Soc.* **50**, 113 (1967).
Wi3. Witt, H.: *Z. Physik* **134**, 117 (1953).
Wo1. Woodbury, H. H.: *Phys. Rev.* **134**, A492 (1964).
Wo2. Woodbury, H. H., and R. B. Hall: *Phys. Rev.* **157**, 641 (1967).
Ya1. Yajima, S., H. Furuya, and T. Hirai: *J. Nucl. Mater.* **20**, 162 (1966).
Yo1. Yokota, I.: *J. Phys. Soc. Japan* **8**, 595 (1953).
Yo2. Yokota, I., and S. Miyatani: *Japan. J. Appl. Phys.* **1**, 144 (1962).
Yo3. Yokota, I.: *J. Phys. Soc. Japan* **21**, 420 (1966).
Za1. Zakharov, Yu A., and G. G. Savel'ev: *Kinetics Catalysis* **5**, 307 (1964).
Za2. Zakharov, Yu A., and A. A. Kabanov: *Russ. J. Phys. Chem.* **38**, 1567 (1964).
Zi1. Zimen, K. E., G. Johansson, and M. Hillert: *J. Chem. Soc.*, suppl. 2, S392 (1949).
Zi2. Zirkind, P., and E. S. Freeman: *Nature* **199**, 1280 (1963).

DIELECTRIC PROPERTIES

9f-8. Dielectric Constants of Crystalline Solids. The dielectric constants of inorganic compounds are listed in Table 9f-8; the chemical formula is given in the second column. The column headed t, °C gives the temperature of the measurements in degrees Celsius; the column headed ν, Hz gives the frequency of the measurement in Hz; the column headed ϵ/ϵ_v gives the dielectric constant (relative capacitance); and the final column gives the reference to the source of the information.

Discrepancies in the dielectric constant of the order of 10 percent are frequently found in the literature.

TABLE 9f-8. INORGANIC SOLIDS—CRYSTALLINE

Name	Formula	t, °C	ν, Hz	ϵ/ϵ_v	Ref.
Alums:					
Ammonium alum..............	$Al(NH_4)(SO_4)_2 \cdot 12H_2O$	r.t.	10^{12}	6	34
Cesium alum.................	$CsAl(SO_4)_2 \cdot 12H_2O$	....	10^{12}	5.0	34
Potassium alum..............	$KAl(SO_4)_2 \cdot 12H_2O$	....	aud.	6.5	15
Rubidium alum..............	$RbAl(SO_4)_2 \cdot 12H_2O$	....	10^{12}	5.1	34
Rubidium chrome alum.......	$RbCr(SO_4)_2 \cdot 12H_2O$	....	10^{12}	5.0	34
Aluminum antimonide..........	AlSb	....	....	11.2	66
Aluminum oxide................	Al_2O_3	25	10^2 to 8×10^9	$9.34\perp$	58
				$11.54\|$	38
Aluminum fluosilicate (topaz)....	$Al_2(F_2SiO_4)$	24	7×10^3	6.62a	64
				6.58b	
				6.95c	
Aluminum phosphate...........	$AlPO_4$	r.t.	10^3	6.05	37
Ammonium bromide............	NH_4Br	r.t.	10^{12}	7.3	34
Ammonium chloride...........	NH_4Cl	r.t.	2×10^6	6.96	28
Ammonium iodide (CsCl structure)	NH_4I	r.t.	6.15×10^3	7.3	23
Ammonium iodide (NaCl structure)	NH_4I	....	....	9.8	23
Ammonium sulfate.............	$(NH_4)_2SO_4$	20	10^3	9.8	61
Ammonium uranyl oxalate........	$(NH_4)_2UO_2(C_2O_4)_2$	....	....	8.14	20
Ammonium uranyl oxalate trihydrate.........................	$(NH_4)_2UO_2(C_2O_4)_2 \cdot 3H_2O$	....	....	6.06	20
Ammonium tartrate............	$(NH_4)_2(C_4H_4O_6)$	r.t.	10^3	6.45	37
Antimonous sesquioxide.........	Sb_2O_3	....	$1.5 - 2 \times 10^3$	12.8	24
Barium carbonate..............	$BaCO_3$	18	2×10^5	8.53	51
Barium chloride................	$BaCl_2$	....	....	9.81	30
Barium chloride dihydrate.......	$BaCl_2 \cdot 2H_2O$	....	....	9.00	30
Barium fluoride................	BaF_2	....	2×10^6	7.33	28
Barium formate................	$Ba(COOH)_2$	r.t.	10^3	7.9	37
Barium nitrate.................	$Ba(NO_3)_2$	19	2×10^5	4.95	51
Barium oxide..................	BaO	-25 to 60	$60\sim$ to 6×10^7	34	5
Barium peroxide..............	BaO_2	r.t.	2×10^6	10.7	24
Barium stannate..............	$BaSnO_3$	25	25×10^5	18	6
Barium sulfate................	$BaSO_4$	15	10^6	11.4	
Barium sulfide.................	BaS	....	7.25×10^6	19.230	52
Beryllium aluminum silicate (beryl)	$Be_3Al_2(Si_6O_{18})$	24	7×10^3	5.95	64
				6.86	64
Beryllium carbonate............	$BeCO_3$	18	2×10^5	9.7	51
Beryllium oxide................	BeO	18	2×10^6	7.35	28
Bismuth trioxide..............	Bi_2O_3	r.t.	2×10^6	18.2	24
Bismuth titanate..............	$Bi_4Ti_3O_{12}$	r.t.	10^3	135–220	59
Boron nitride..................	BN	....	....	7.1	66
Cadmium arsenide..............	Cd_3As_2	-269	....	$18.5\|$	36
Cadmium sulfide.............	CdS	....	....	9.4(11)	43
				10.3(33)	43
Cadmium fluoride.............	CdF_2	27	10^5–10^7	8.33 ± 0.08	67
Cadmium bromide..............	CdBr	20	5×10^5	8.6	18
Cadmium malonate............	$Cd(C_3H_2O_4)$	20	5×10^5	4.5	18
Calcium carbonate.............	$CaCO_3$	....	10^8	$8.5\perp$	48
				$8.0\|$	
Calcium fluoride................	CaF_2	....	10^5	6.76	29, 34, 45, 46
			aud.	6.85	15, 44, 56, 40
Calcium nitrate................	$Ca(NO_3)_2$	19	2×10^5	6.54	51
Calcium oxide.................	CaO	10	2×10^6	11.8	28
Calcium sulfide................	CaS	....	7.25×10^6	6.699	52
Calcium sulfate (gypsum)........	$CaSO_4 \cdot 2H_2O$	....	....	5.10a	17
				5.24b	
				10.30c	
Cadmium telluride..............	CdTe	24	....	10.60 ± 0.15	35
		-196	10^4	9.65	4
Ceric oxide....................	CeO_2	r.t.	2×10^6	7.0	24
Cesium bromide................	CsBr	....	2×10^6	6.51	28
Cesium carbonate..............	Cs_2CO_3	18	2×10^5	6.53	51
Cesium chloride	CsCl	19	2×10^5	6.34	51
Cesium iodide.................	CsI	25	10^6	5.65	28, 27
Cesium nitrate.................	$CsNO_3$	r.t.	5×10^5	9.4a	8
				8.3c	

TABLE 9f-8. INORGANIC SOLIDS—CRYSTALLINE (Continued)

Name	Formula	t, °C	ν, Hz	ϵ/ϵ_v	Ref.
Chromic sesquioxide............	Cr_2O_3	25.5	10^3	13.3a	21
				11.9c	
Cupric oxide....................	CuO	r.t.	2×10^6	18.1	24
Cupric sulfate pentahydrate......	$CuSO_4 \cdot 5H_2O$	....	....	6.60	30
Cuprous bromide................	CuBr	20	5×10^5	8.0	18
Cuprous chloride................	CuCl	....	10^3	9.8	3
Cuprous oxide..................	Cu_2O	r.t.	10^5	7.60 ± 0.06	40.1
Dextrose sodium bromide........	$C_6H_{12}O_6 \cdot NaBr$	....	10^3	4.0	37
Diamond.......................	C	....	....	5.5	49
Europium sulfide................	EuS	-193	$5 \times (10^2-10^5)$	13.10 ± 0.04	10
Ferrous oxide..................	FeO	r.t.	2×10^6	14.2	24
Gallium arsenide...............	GaAs	....	....	13.13	66
Gallium antimonide.............	GaSb	....	....	15.69	66
Gallium phosphide..............	GaP	....	....	10.18	66
Hexamine cobalt (III) chloride....	$[Co(NH_3)_6]Cl_3$	....	10^6	7.31	39
Ice I (1 bar)...................	H_2O	-23.4	....	97.5	11
(1600 bar)................				100.8	
Indium antimonide..............	InSb	....	....	17.88	66
Indium arsenide................	InAs	....	....	14.55	66
Indium phosphide..............	InP	....	....	12.37	66
Iodic acid.....................	HIO_3	....	10^3	7.5	37
Iodine........................	I_2	....	5×10^4	6a	53
			to	3b	
			5×10^7	40c	
Lead acetate...................	$Pb(C_2H_3O_2)_2$	17–22	10^6	2.6	
Lead bromide..................	$PbBr_2$	20	$0.5-3 \times 10^6$	>30	18
Lead carbonate................	$PbCO_3$	15	10^3	18.6	
Lead chloride..................	$PbCl_2$	20	$0.5-3 \times 10^6$	33.5	18
Lead fluoride..................	PbF_2	r.t.	....	26.3	7
Lead iodide....................	PbI_2	20	$0.5-3 \times 10^6$	20.8	18
Lead molybdate (wulfenite).......	$PbMoO_4$	....	3×10^8	26.8$\parallel$	49
Lead nitrate...................	$Pb(NO_3)_2$	....	$0.5-3 \times 10^6$	16.8	18
Lead oxide....................	PbO	r.t.	2×10^6	25.9	24
Lead selenide..................	PbSe	....	....	280	66
Lead sulfate...................	$PbSO_4$	17–22	10^6	14.3	
Lead sulfide...................	PbS	....	....	205	66
Lead telluride.................	PbTe	....	....	400	66
Lithium bromide................	LiBr	....	2×10^6	12.1	28
Lithium chloride................	LiCl	....	2×10^6	11.05	18
Lithium carbonate..............	Li_2CO_3	18	2×10^5	4.9	51
Lithium fluoride................	LiF	20	10^6	9.27	28
		25	10^2-10^7	9.00	63
		80	10^2-10^7	9.11	63
Lithium iodide..................	LiI	....	2×10^6	11.03	28
Lithium sulfate monohydrate.....	$Li_2SO_4 \cdot H_2O$	....	10^3	5.6	37
Lithium trisodium chromate......	$LiNa_3CrO_4 \cdot 6H_2O$	....	10^3	8.0	37
Lithium trisodium molybdate.....	$LiNa_3MoO_4 \cdot 6H_2O$	....	10^3	8.1	37
Magnesium borate monochloride..	$Mg_3B_7O_{13}Cl$	r.t.	5×10^5	14.1a	54
Magnesium carbonate...........	$MgCO_3$	18	2×10^5	8.1	51
Magnesium malonate...........	$Mg(C_2H_2O_4)$	20	5×10^5	5.8	18
Magnesium mica (phlogopite).....	$(K,H)_3Mg_3Al(SiO_4)_3$	r.t.	5×10^4	275	1
Magnesium oxalate.............	$Mg(C_2O_4)$	20	5×10^5	5.2	18
Magnesium oxide...............	MgO	25	10^2-10^8	9.65	63
Magnesium sulfate.............	$MgSO_4$	20	5×10^5	8.2	18
Magnesium sulfate heptahydrate..	$MgSO_4 \cdot 7H_2O$	....	....	5.46	30
Manganese dioxide.............	MnO_2	25	....	10^4	68
Manganese monoxide...........	MnO	r.t.	$10^6, 273 \times 10^5$	18.0 ± 0.5	14
Mercuric chloride...............	$HgCl_2$	....	10^{12}	6.5	34
Mercurous chloride.............	HgCl	....	10^{12}	14.0$\perp$	34
Mercurous selenide.............	HgSe	r.t.	10^4-10^6	25.6	32
Mercurous sulfide..............	HgS	r.t.	10^4-10^6	30.6	32
Mica—ruby, muscovite..........		26	$10^2-3 \times 10^3$	5.4	63
Mica—Canadian................		25	10^2-10^4	6.9$\perp$	63
		25	10^4	7.3$\parallel$	63
Nickel sulfate hexahydrate.......	$NiSO_4 \cdot 6H_2O$	....	10^3	6.2	37
Phosphorus, red................	P	....	10^8	4.1	
yellow.....................		....	10^8	3.6	49

TABLE 9f-8. INORGANIC SOLIDS—CRYSTALLINE (*Continued*)

Name	Formula	t, °C	ν, Hz	ϵ/ϵ_v	Ref.
Potassium bromate............	$KBrO_3$	r.t.	2×10^6	7.3	57
Potassium bromide............	KBr	r.t.	2×10^6	4.78	28
Potassium carbonate...........	K_2CO_3	18	2×10^5	4.96	51
Potassium chlorate............	$KClO_3$	r.t.	2×10^6	5.1	57
Potassium chloride............	KCl	29.5	10^6	4.64	27
		80	10^6	4.80	27
Potassium chromate...........	K_2CrO_4	....	6×10^7	7.3	
Potassium cyanide.............	KCN	r.t.	2×10^6	6.15	57
Potassium dihydrogen arsenate....	KH_2AsO_4	r.t.	2×10^6	31	57
Potassium dihydrogen phosphate..	KH_2PO_4	....	10^3	46	37
Potassium fluoride............	KF	....	2×10^6	6.05	28
Potassium iodate..............	KIO_3	r.t.	2×10^6	16.85	57
Potassium iodide..............	KI	....	2×10^6	4.94	28
Potassium nitrate.............	KNO_3	20	2×10^5	4.37	51
Potassium perchlorate..........	$KClO_4$	r.t.	2×10^6	5.9	57
Potassium orthophosphate.......	K_3PO_4	r.t.	2×10^6	7.75	57
Potassium monohydrogen ortho-phosphate.................	K_2HPO_4	r.t.	2×10^6	9.05	57
Potassium dihydrogen orthophosphate.....................	KH_2PO_4	r.t.	2×10^6	>31	57
Potassium sulfate.............	K_2SO_4	r.t.	2×10^6	6.4	57
Potassium tantalate–niobate (KTN)....................	$KTa_{0.66}Nb_{0.34}O_3$	−1	10^4	6,000	12
		0	....	34,000	
		20	....	6,000	
Potassium thiocyanate..........	KSCN	r.t.	2×10^6	7.9	57
Potassium thionates:					
Potassium trithionate........	$K_2S_3O_6$	20	1.8×10^6	5.7	50
Potassium tetrathionate.......	$K_2S_4O_6$	20	1.8×10^6	5.5	50
Potassium pentathionate......	$K_2S_5O_6 \cdot H_2O$	20	1.8×10^6	7.8	50
Potassium hexathionate.......	$K_2S_6O_6$	20	1.8×10^6	7.8	50
Rubidium bromide (NaCl structure)......................	RbBr	r.t.	....	4.9	25
Rubidium bromide (CsCl structure)......................	RbBr	r.t.	....	6.5	25
Rubidium carbonate...........	Rb_2CO_3	19	2×10^5	6.73	51
Rubidium chloride.............	RbCl	....	2×10^6	5.0	28
Rubidium fluoride.............	RbF	....	2×10^6	5.91	28
Rubidium iodide..............	RbI	....	2×10^6	5.0	28
Rubidium indium sulfate........	$RbIn(SO_4)_2$	....	....	6.85	19
Rubidium nitrate..............	$RbNO_3$	130–215	10^6	20–380	16
		215–265	10^6	30	16
Selenium.....................	Se	25	3×10^8	11.0	63
		25	3×10^9	10.4	63
		25	2×10^{10}	7.5	63
Selenium, amorphous...........	Se	25	$10^2–10^{1C}$	6.00	63
Silicon monoxide..............	SiO	r.t.	10^3	5.8	22
Silicon dioxide (α-quartz).......	SiO_2	r.t.	....	4.5$\perp$	
				4.6$\parallel$	9
Silicon carbide.................	SiC	....	10^5	10.2$\parallel$	26
Silicon nitride................	Si_3N_4	r.t.	10^3	4.2	22
Silver bromide................	AgBr	....	2×10^6	13.1	28, 18
Silver chloride................	AgCl	....	2×10^6	12.3	28, 18
Silver cyanide................	AgCN	....	10^6	5.6	
Silver nitrate.................	$AgNO_3$	20	5×10^5	9.0	18
Sodium ammonium tartrate tetra-hydrate...................	$NaNH_4(C_4H_4O_6) \cdot 4H_2O$	....	10^3	9.0	37
Sodium bromide...............	NaBr	....	2×10^6	5.99	28
Sodium carbonate.............	Na_2CO_3	18	2×10^5	8.75	51
Sodium carbonate decahydrate....	$Na_2CO_3 \cdot 10H_2O$		6×10^7	5.3	30
Sodium chlorate...............	$NaClO_3$	....	....	5.28	31
Sodium chloride...............	NaCl	20	2×10^6	5.62	28
		25	$10^2–10^7$	5.9	63
		85	$10^4–10^7$	5.98	63
Sodium cyanide...............	NaCN	20	10^5	7.55	58
Sodium fluoride...............	NaF	19	2×10^6	6.0	28

SOLID-STATE PHYSICS

TABLE 9f-8. INORGANIC SOLIDS—CRYSTALLINE (*Continued*)

Name	Formula	t, °C	ν, Hz	ϵ/ϵ_v	Ref.
Sodium iodide...................	NaI	....	2×10^6	6.60	28
Sodium nitrate.................	NaNO$_3$	19	2×10^5	6.85	51
Sodium nitrite.................	NaNO$_2$	r.t.	5×10^5	6.8a	55
		r.t.	5×10^5	6.4b	
		r.t.	5×10^5	7.8c	
Sodium perchlorate.............	NaClO$_4$	....	10^3	5.76	37
Sodium sulfate.................	Na$_2$SO$_4$	....	....	7.90	30
Sodium sulfate decahydrate......	Na$_2$SO$_4$·10H$_2$O	....	....	5.0	30
Sodium uranyl oxalate..........	Na$_2$UO$_2$(C$_2$O$_4$)$_2$	....	....	5.18	20
Stannic dioxide.................	SnO$_2$	r.t.	10^4–10^{10}	$9.0 \pm 0.5\parallel$	60
		r.t.	10^4–10^{10}	$14 \pm 2\perp$	60
Strontium carbonate............	SrCO$_3$	18	2×10^5	8.85	51
Strontium chloride.............	SrCl$_2$	....	....	9.19	30
Strontium chloride hexahydrate...	SrCl$_2$·6H$_2$O	....	....	8.52	30
Strontium fluoride.............	SrF$_2$	....	2×10^6	7.69	28
Strontium formate dihydrate.....	Sr(COOH)$_2$·2H$_2$O	....	10^3	6.1	37
Strontium nitrate..............	Sr(NO$_3$)$_2$	19	2×10^5	5.33	51
Strontium oxide................	SrO	....	2×10^6	13.3	28
Strontium sulfide..............	SrS	....	7.25×10^6	11.310	52
Strontium titanate.............	SrTiO$_3$	25	10^3	332	65
		−195	10^3	2,080	65
Sulfur (100)....................	S	25	10^2–10^3	3.75	63
(010).........................		25	10^2–10^3	3.95	63
(001).........................		25	10^2–10^3	4.44	63
Sublimed......................		25	10^2–10^3	3.69	63
Tantalum pentoxide (α).........	Ta$_2$O$_5$	−196	10^3	30$\perp$	42
		−196	10^3	65$\parallel$	42
Tantalum pentoxide (β).........	Ta$_2$O$_5$	19	10^3	24	42
Thallous bromide...............	TlBr	25	10^3–10^7	30.3	63
Thallous chloride..............	TlCl	....	2×10^6	31.9	28
Thallous iodide (orthorhombic)....	TlI	20	10^4	21.2 ± 0.2	47
(cubic)........................	TlI	20	10^4	29.6 ± 0.5	47
(orthorhombic)................	TlI	193	10^7	37.3	63
Thallous nitrate...............	TlNO$_3$	20	5×10^5	16.5	18
			27–37×10^9	13.5	33
Thallous sulfate...............	Tl$_2$SO$_4$	20	5×10^5	25.5	18
Thorium dioxide................	ThO$_2$	....	3×10^5	18.9 ± 0.4	2
		r.t.	2×10^6	10.6	24
Tin antimonide.................	SnSb	r.t.	10^4–10^6	147	32
Titanium dioxide (rutile)........	TiO$_2$	r.t.	....	89 a	41
		r.t.	....	173 c	41
Uranium dioxide................	UO$_2$	....	....	24	2
Ytterbium sesquioxide..........	Yb$_2$O$_3$	r.t.	10^3	5.0	22
Zinc malonate.................	Zn(C$_3$H$_2$O$_4$)	20	5×10^5	5.6	
Zinc monoxide..................	ZnO	....	....	8.14	13
Zinc selenide..................	ZnSe	25	10^4	9.12	4
Zinc sulfide...................	ZnS	25	10^4	8.37	4
Zinc telluride.................	ZnTe	25	10^4	10.10	4
Zirconium dioxide.............	ZrO$_2$	r.t.	2×10^6	12.5	24

References for Table 9f-8

1. Afanaev, Popova, and Metsik: *Izv. Vysshikh Uchebn. Zavedenii Fiz.* **1962**, (6), 64.
2. Axe and Pettit: *Phys. Rev.* **151**, 676 (1966).
3. Belyaev, Belikova, Dobrzhanskii, Netesov, and Schaldin: *Fiz. Tverd. Tela* **6**, 2526–2528 (1964).
4. Berlincourt, Jaffe, and Shiozawa: *Phys. Rev.* **129**, 1009 (1963).
5. Bever and Sproull: *Phys. Rev.* **83**, 801 (1951).
6. Bosman and Havinga: *Phys. Rev.* **129**, 1593 (1963).
7. Bosomworth: *Phys. Rev.* **157**, 709 (1967).
8. Brown and Koenig: *Phys. Letters* **2**, 309 (1962).
9. Cady: "Piezoelectricity," McGraw-Hill Book Company, New York, 1946.
10. Campbell and Lawson: *J. Phys. Chem. Solids*, **30**, 775–776, 1969.
11. Chan, Davidson, and Whalley: *J. Chem. Phys.* **43**, 2376 (1965).
12. Chen, Geusic, Kurtz, Skinner, and Wemple: *J. Appl. Phys.* **37**, 388 (1966).

13. Collins and Kleinman: *J. Phys. Chem. Solids*, **11**, 190–194 (1959).
14. Crevecoeur: Private communication.
15. Curie: *Ann. Chim. Phys.* **17**, 385 (1889).
16. Dantsiger and Fesenko: *Soviet Phys.—Cryst.* **10**, 272 (1965).
17. Demau: *J. Phys. Radium* **24**, 284 (1963).
18. Eucken and Buchner: *Z. Physik. Chem.* **27**(B), 321 (1934).
19. Ezuchevskaya, Syrkin, and Deichman: *Zh. Neorg. Khim.* **9**, 1495 (1964).
20. Ezuchevskaya, Syrkin, and Shchelokov: *Zh. Neorg. Khim.* **9**, 1758–1759 (1964).
21. Fang and Brower: *Phys. Rev.* **129**, 1561 (1963).
22. Feldman and Hacskaylo: *Rev. Sci. Instr.* **33**, 1459 (1962).
23. Gibbs and Hill: *Phil. Mag.* **9**, 367–375 (1964).
24. Guntherschultze and Keller: *Z. Physik* **75**, 78 (1932).
25. Havinga and Bosman: *Phys. Rev.* **140**, A292 (1965).
26. Hofman, Lely, and Volger: *Physica* **23**, 236 (1957).
27. Højendahl: *Z. Physik. Chem.* **20**(B), 54 (1933).
28. Højendahl: *Kgl. Danske Videnskab. Selskab Mat.-Fys. Medd.* **16**, 1–132 (1938).
29. Jaeger: *Ann. Physik* **53**, 409 (1917).
30. Kamiyoshi and Miyamoto: *Sci. Rept. Res.* Inst. Tohoku Univ., ser. A, **2**, 370 (1950).
31. Kiriyama: *Science (Japan)* **17**, 239 (1947).
32. Kir'yashkina, Popov, Bilenko, and Kir'yashkina: *Soviet Phys.* **2**, 69–73 (1957).
33. LeFevre and Ritchie: *J. Chem. Soc.* **1963**, 4933.
34. Liebisch and Rubens: *Sitzber. Preuss. Akad. Wiss. Physik-Math. Kl.* **1919**, 876.
35. Lorimer and Spitzer: *J. Appl. Phys.* **36**, 1841 (1965).
36. Malone and Ferguson: *J. Chem. Phys.* **2**, 99 (1934).
37. Mason: "Piezoelectric Crystals and Their Application to Ultrasonics," D. Van Nostrand Company, Inc., Princeton, N.J., 1950.
38. Morgan and Lowry: *J. Phys. Chem.* **34**, 2385 (1930).
39. Nakano, Satuka, and Saruwatari: *Nippon Kagaku Zasshi* **84**, 902–909 (1963).
40. Naragamo Rao: *Proc. Indian Acad. Sci.* **30A**, 82 (1949).
40.1. Noguet: *J. de Phys.*, **31**, 393 (1970).
41. Parker: *Phys. Rev.* **124**, 1719 (1961).
42. Pavlovic: *J. Chem. Phys.* **40**, 951–956 (1964).
43. Reynolds, et al.: *Phys. Stat. Solidi* **12**, 3 (1965).
44. Romich and Nowak: *Sitzber. Akad. Wiss. Wien, Math.-Naturw. Kl.* **7011**, 380 (1875).
45. Rubens: *Sitzber. Preuss. Akad. Wiss., Phys.-Math. Kl.* **1915**, I, 4.
46. Rubens: *Z. Physik* **1**, 11 (1920).
47. Samara: *Phys. Rev.* **165**, 959 (1968).
48. Schmidt: *Ann. Physik* **9**, 919 (1902).
49. Schmidt: *Ann. Physik* **11**, 114 (1903).
50. Schmidt and Sand: *J. Inorg. Nucl. Chem.* **26**, 1189–1190 (1964).
51. Schupp: *Z. Physik* **75**, 84 (1932).
52. Sharma and Gupta: *Indian J. Phys.* **37**, 33 (1963).
53. Simhony: *J. Phys. Chem. Solids* **24**, 1297–1300 (1963).
54. Sonin and Zheludev: *Kristallografiya* **8**, 283 (1963).
55. Sonin and Zheludev: *Kristallografiya* **8**, 285 (1963).
56. Starke: *Ann. Physik* **60**, 629 (1897).
57. Steulmann: *Z. Physik* **77**, 114 (1932).
58. Tables of Dielectric Materials, vol. 6, *MIT Lab. for Insul. Res. Tech. Rept.* 126, June, 1958.
59. Tambovtsev, Skorikov, and Zheludev: *Kristallografiya* **8**, 889–893 (1963).
60. van Daal: *J. Appl. Phys.* **39**, 4467 (1968).
61. Unruh: *Phys. Letters* **17**, 8–9 (1965).
62. Voigt: "Lehrbuch der Kristallphysik," p. 459.
63. Von Hippel: "Dielectric Materials and Applications," John Wiley & Sons, Inc., New York, 1954.
64. Wappler: *Z. Phys. Chem.* **228**, 33 (1965).
65. Weaver: *J. Phys. Chem. Solids* **11**, 274 (1959).
66. Willardson and Beer: "Semiconductors and Semimetals," vol. 1, p. 14, Academic Press, Inc., New York.
67. Young and Frederikse: *J. Appl. Phys.*, July, 1969.
68. Yousef and Farag: *Physica* **31**, 706 (1965).

9f-9. Piezoelectric and Pyroelectric Constants

TABLE 9f-9. PIEZOELECTRIC STRAIN CONSTANTS*

Substance	Formula	d_{11}	d_{14}	d_{25}	d_{36}	Ref.
1. Aluminum phosphate..............	$AlPO_4$	±3.3	±1.5	......	......	29*
		+1.4	Small	......	......	12*
2. Ammonium dihydrogen arsenate......	$NH_4H_2AsO_4$		+41	......	+31	12*
3. Ammonium dihydrogen phosphate.....	$NH_4H_2PO_4$	......	−1.5	......	+48	12*
			+1.5	......	−45.6	40
			+1.7	......	+49	29*
4. Ammonium ditartrate...............	$NH_4HC_4H_4O_6$	......	−1.6	7.0	−0.4	35
5. Barium formate....................	$Ba(HCOO)_2$	......	±4.0	±2.7	±4.7	29*
6. Benzil...........................	$C_{14}H_{10}O_2$	+8.0	....	......	......	42
7. Benzophenon......................	$(C_6H_5)_2CO$	......	+12.3	+2.0	+20.3	45
8. Beryllium sulfate tetrahydrate........	$BeSO_4{:}4H_2O$	......	7	......	......	12*
9. Cadmiumtelluride..................	$CdTe\ (-196°C)$	......	+1.7	......	......	9*
10. Cesium tartrate....................	$Cs_2C_4H_4O_6$	2.7	0.17	......	......	12*
11. Deutero ammonium dideuterium phosphate.................	$ND_4D_2PO_4$		10		75	31
12. Dextrose plus sodium bromide........	$C_6H_{12}O_6 + 2NaBr$	−3.7	−1.8	......	......	29*
13. Dextrose plus sodium chloride........	$C_6H_{12}O_6 + 2NaCl$	−7.0	+0.3	......	......	29*
14. Dextrose plus sodium iodide..........	$C_6H_{12}O_6 + 2NaI$	−3.8	+0.7	......	......	29*
15. Galliumarsenide...................	$GaAs$	......	+2.6	......	......	15
16. Heavy rochelle salt.................	$KNaC_4D_2H_2O_6{\cdot}4D_2O$	......	Very large (see Table 9f-10)	−73	+13.3	30
17. Hexamethylentetramine..............	$(CH_2)_6N_4$	......	+17.5	......	......	18*
18. Iodic acid.........................	HIO_3	......	±18.9	±15.3	±23.5	29*
19. Lithium ammonium tartrate monohydrate.....................	$LiNH_4C_4H_4O_6{\cdot}H_2O$	......	±4.4	±6.5	±4.9	29*
			7.7	−5.3	6.8	35
20. Lithium potassium tartrate monohydrate.....................	$LiKC_4H_4O_6{\cdot}H_2O$	......	+3.2	+11.2	±7.6	29*
			+2.0	−9.4	+6.6	35
			2.1	10.0	6.8	12*
21. Magnesium sulfate heptahydrate......	$MgSO_4{\cdot}7H_2O$	......	−2.1	−2.7	−3.8	40
22. Mercury sulfide...................	HgS	+19.1	~1.7	......	......	11*
23. Nickel sulfate heptahydrate..........	$NiSO_4{\cdot}7H_2O$	......	−2.0	−2.9	−3.2	40
24. Nickel sulfate hexahydrate...........	$NiSO_4{\cdot}6H_2O$	......	−5.3	......	......	40
			±6.0	......	......	29*
25. Patchouli camphor..................	$C_{15}H_{26}O$	+0.05	....	......	......	42
26. Potassium dideuterium phosphate.....	KD_2PO_4	......	....	......	+51.7	2*
					+58	37
27. Potassium dihydrogen arsenate.......	KH_2AsO_4	......	+23.5	......	+22	12*
			26.6	......	22.4	33
28. Potassium dihydrogen phosphate......	KH_2PO_4	......	+1.3	......	−20.9	40
			1.4	......	23	29*
			+1.3	......	+21	12*
29. Potassium ditartrate...............	$KHC_4H_4O_6$	......	−4.3	3.4	−1.0	35
30. Potassium dithionate................	$K_2S_2O_6$.............	1.4	2.0	......	......	12*
31. Quartz............................	SiO_2	+2.31	−0.73	......	......	4
		+2.3	−0.67	......	......	13, 40
		−2.25	+0.85	......	......	29*
32. Rochelle salt......................	$KNaC_4H_4O_6{\cdot}4H_2O$	......	Very large (see Table 9f-10)	−56	+11.8	29*
				−53	+11.7	13
33. Rubidium dihydrogen phosphate......	RbH_2PO_4	......	4.5	......	37	12*
34. Rubidium tartrate..................	$Rb_2C_4H_4O_6$	+2.7	....	......	......	42
35. Selenium..........................	Se	65	....	......	......	17
36. Sodium ammonium tartrate tetrahydrate........................	$NaNH_4C_4H_4O_6{\cdot}4H_2O$	......	+18.7	−49.8	+9.4	28
			±19	±31.7	±10.3	29
37. Sodium bromate...................	$NaBrO_3$	......	−2.6	......	......	4
			−2.4	......	......	40
			+2.7	......	......	29*
38. Sodium chlorate....................	$NaClO_3$	......	−1.75	......	......	4, 40
			+2.0	......	......	29*
39. Strontium formate dihydrate.........	$Sr(HCOO)_2{\cdot}2H_2O$	......	±8.5	±11.5	±2.3	29*
40. Zinc selenide......................	$ZnSe$	......	+1.1	......	......	9*
41. Zinc sulfide (zincblende)........	ZnS	......	−3.2	......	......	24
			+3.2	......	......	9*
42. Zinc telluride......................	$ZnTe$	......	+0.9	......	......	9*
43. Zinc sulfate heptahydrate............	$ZnSO_4{\cdot}7H_2O$	......	−1.9	−3.5	−3.1	40

TABLE 9f-9. PIEZOELECTRIC STRAIN CONSTANTS* *(Continued)*

Substance	Formula	d_{15}	d_{22}	d_{24}	d_{31}	d_{32}	d_{33}	Ref.
44. Aluminium nitride	AlN						5.0	23*
45. Ammonium pentaborate tetrahydrate	$NH_4B_5O_8 \cdot 4H_2O$	+6.7		+13	-1.9	-6.6	+6.9	16*
46. Antimony sulfoiodide	SbSI				~150		1,300	10*
47. Barium antimonyl tartrate	$Ba(SbO)_2$ $(C_4H_4O_6)_2 \cdot H_2O$						+3.7	42*
48. Barium titanate	$BaTiO_3$	392			-37		84	14*
					-34.5		85.6	7*
49. Barium titanate ceramic	$BaTiO_3$	270			-79		191	7*
50. Beryllium oxide	BeO				-0.12		+0.24	1*
51. Boracite							+0.6	39
52. Cadmium selenide	CdSe	-10.5			-3.9		+7.8	9*
53. Cadmium sulfide	CdS	-14.3			-3.7		+10.7	22*
54. Cesium nitrate	$CsNO_3$						+0.5	38
55. Lithium gallium oxide	$LiGaO_2$	+5.9		+5.1	-2.8	-2.4		11*
56. Lithium niobate	$LiNbO_3$	+74	+20.8		-0.86		+16.2	43
57. Lithium trisodium chromate hexahydrate	$LiNa_3(CrO_4)_2 \cdot 6H_2O$		±2.9					29*
58. Lithium trisodium molybdate hexahydrate	$LiNa_3(MoO_4)_2 \cdot 6H_2O$		±2.5		±1.3		±1.9	29*
59. Potassium lithium sulfate	$KLiSO_4$	+0.9			-2.35		+5.2	12*
60. Potassium pentaborate tetrahydrate	$KB_5O_8 \cdot 4H_2O$	+1.7		+9.5	<1	-5.4	+5.6	16*
61. Resorcinol	$C_6H_4(OH)_2$	+18.0		+18.4	-4.1	-4.3	+5.6	26
62. Sodium calcium aluminosilicate	$(Na_2Ca)(AlSiO_4)_5$ $CO_3(H_2O)_{0-3}$	+9.0			+0.7		+4.3	27
63. Sodium lithium sulfate	$NaLiSO_4$		0.85		+0.01		+0.3	12*
64. Sodium nitrite	$NaNO_2$	+9.3		-20.2	-1.2	-2.8	+1.7	19
65. Terpine monohydrate	$C_{10}H_{18}(OH)_2 \cdot H_2O$	+4.3		+5.8	-2.2	+3.5	+2.2	36
66. Tourmaline	variable	+3.7	-0.23		+0.25		+1.9	34
		-3.6	-0.33		-0.34		-1.8	29*
67. Zinc oxide	ZnO	-10 to -13			-4.7		+12	22*

Substance	Formula	d_{14}	d_{16}	d_{21}	d_{22}	d_{23}	d_{25}	d_{34}	d_{36}	Ref.
68. Anthracene	$C_{14}H_{10}$			-0.07	+0.16	-0.08				44
69. Cane sugar	$C_{12}H_{22}O_{11}$	+1.2	-2.4	+1.5	-3.4	+0.7	-0.9	-4.2	+0.4	20
70. Diammonium tartrate	$(NH_4)_2C_4H_4O_6$	+3.1	-2.8	+5.9	-8.7	+0.6	-2.0	-4.7	+1.9	40
		+3.3	+1.7	-6.7	+8.6	-0.6	+2.4	+1.8	+2.0	12*
71. Dipotassium tartrate hemihydrate	$K_2C_4H_4O_6 \cdot \frac{1}{2}H_2O$	+7.9	+3.5	-0.8	+4.5	-5.3	-6.5	-12.3	-23.2	6*
72. Ethylene diamine tartrate	$C_6H_{14}O_6N_2$	-10.0	-12.2	+10.1	+2.2	-11.3	-18.0	-17.0	-18.4	3*
73. Guanidine tartrate	$C_5H_{11}O_6N_3$			+2.6		-3.9	+3.3			12*
74. Lithium sulfate monohydrate	$Li_2SO_4 \cdot H_2O$	+0.76	-2.0	-3.6	+16.3	+1.7	-5.0	-2.1	-4.2	5*
75. Rhamnose	$C_6H_{12}O_5 \cdot H_2O$	+0.7	+5.0	+2.7	-3.0	-5.0	+12.2	-12.0	+1.1	32
76. Sorbitol hexacetate	$C_{18}H_{26}O_{12}$	1.4	23	0.5	-8	0.8		2	2	12
77. Tartaric acid	$C_4H_6O_6$	-13.2	+9.5	+2.0	-2.2	+0.65	+1.3	+9.3	-8.0	41
		+8.0	+5.3	-0.8	-2.2	-2.1	+0.4	-10.8	+11.7	29*
78. Triglycine sulfate	$(CH_2NH_2COOH)_3$	2.8	-4.6	23.6	7.9	25.3	24.3	-3.2	2.8	25
	H_2SO_4			23	22.0	84.0				21

Substance	Formula	d_{11}	d_{12}	d_{13}	d_{15}	d_{24}	d_{26}	d_{31}	d_{32}	d_{33}	d_{35}	Ref.
79. Lithium trihydrogen selenite	$LiH_3(SeO_3)_2$	+23.2	-22.3	-12.1	-46.6	-12.8	-14.6	-18.4	+5.5	+19.9	+53.2	8*

ADDENDUM

Substance	Formula	d_{14}	Ref.
80. Aluminum antimonide	AlSb	1.64	46
81. Indium antimonide	InSb	2.35	46
82. Indium arsenide	InAs	1.14	46
83. Gallium antimonide	GaSb	2.9	46

* According to the standards on piezoelectric crystals of the IRE [*Proc. IRE* **37**, 1378 (1949)] we define the piezoelectric strain constants $d_{ik} = (\partial D_i)/(\partial T_k)_E$ or $d_{ik} = (\partial S_k)/(\partial E_i)_T$, where $i = 1, 2, 3$, and $k = 1$ to 6; T_k = stress; S_k = strain; D_i = electric displacement; and E_i = electric field. The units are coulomb/newton or meter/volt (rationalized mks). The listed numbers have to be multiplied by 10^{-12}. In all cases marked by an asterisk the IRE convention of tension being a positive stress has been followed. For the other values quoted, the convention used is somewhat uncertain, although pressure is usually taken as positive.

References for Table 9f-9

1. Austermann, S. B., D. A. Berlincourt, and H. A. Krueger: *J. Appl. Phys.* **34,** 339 (1963).
2. Bantle, W.: *Helv. Phys. Acta* **18,** 245 (1945).
3. Bechmann, R.: *Proc. Phys. Soc.* **63B,** 577 (1950).
4. Bechmann, R.: *Proc. Phys. Soc.* **64B,** 323 (1951).
5. Bechmann, R.: *Proc. Phys. Soc.* **65B,** 375 (1952).
6. Bechmann, R., and S. Ayers: *Proc. Phys. Soc.* **67B,** 422 (1954).
7. Berlincourt, D., and H. Jaffe: *Phys. Rev.* **111,** 143 (1958).
8. Berlincourt, D., and W. R. Cook: *Bull. Am. Phys. Soc.* **6,** 140 (1961).
9. Berlincourt D., H. Jaffe, and L. R. Shiozawa: *Phys. Rev.* **129,** 1009 (1963).
10. Berlincourt, D., H. Jaffe, W. J. Merz, and R. Nitsche: *Appl. Phys. Letters* **4,** 61 (1964).
11. Berlincourt, D. A.: Clevite Corporation, Cleveland, February, 1965.
12. Brush Laboratories Company, Cleveland, Ohio: Reports.
13. Cady, W.: "Piezoelectricity," McGraw-Hill Book Company, New York, 1946.
14. Caspari, M. E., and W. J. Merz: *Phys. Rev.* **80,** 1082 (1950).
15. Charlson, E. J., and G. Mott: *Proc. IRE* **51,** 1239 (1963).
16. Cook, W. R., and H. Jaffe: *Acta Cryst.* **10,** 705 (1957).
17. Gobrecht, H., H. Hamisch, and A. Tausend: *Z. Physik* **148,** 209 (1957).
18. Gockel, H.: *Phys. Z.* **37,** 657 (1936).
19. Hamano, K., K. Negishi, M. Marutake and S. Nomura: Japan. *J. Appl. Phys.* **2,** 83 (1963).
20. Holman, W. F.: *Ann. Phys.* **29,** 160 (1909).
21. Husimi, K., K. Kataoka: *J. Phys. Soc.* Japan **14,** 105 (1959).
22. Hutson, A. R.: *Phys. Rev. Letters* **4,** 505 (1960).
23. Jaffe, H. and D. Berlincourt: *Proc. IEEE* **53,** 1372 (1965).
24. Knol, K. S.: *Koninkl. Akad. Amst.* **35,** 99 (1932).
25. Konstaninova, U. P., I. M. Sil'vestrova and K. S. Aleksandrov: *Translation Soviet Phys.—Cryst.* **4,** 63 (1960).
26. Koptsik, V. A.: *Translation Soviet Phys.—Cryst.* **4,** 197 (1960).
27. Koptsik, V. A., and I. B. Kobyakov: *Translation Soviet Phys.—Cryst.* **4,** 201 (1960).
28. Mandell, W.: *Proc. Roy. Soc (London)* **121,** 130 (1928).
29. Mason, W. P.: "Piezoelectric Crystals and Their Application to Ultrasonics," D. Van Nostrand Company, Inc., Princeton, N.J., 1950.
30. Mason, W. P., and A. N. Holden: *Phys. Rev.* **57,** 54 (1940).
31. Mason, W. P., and B. T. Matthias: *Phys. Rev.* **88,** 477 (1952).
32. Meyer, G.: Dissertation, Göttingen, 1937.
33. Niemiec, T.: *Phys. Rev.* **75,** 215 (1949).
34. Riecke, E., and W. Voigt: *Wied. Ann. Phys.* **45,** 523 (1892).
35. Scholz, H.: Dissertation, Göttingen, 1941.
36. Sil'vestrova, I. M., K. S. Aleksandrov, and A. A. Chumakov: *Soviet Phys.—Cryst.* **3,** 388 (1958).
37. Sliker, T. R. and S. R. Burlage: *J. Appl. Phys.* **34,** 1837 (1963).
38. Sonin, A. S., and I. S. Zheludev: *Soviet Phys.—Cryst.* **8,** 219 (1963).
39. Sonin, A. S., and I. S. Zheludev: *Soviet Phys.—Cryst.* **8,** 217 (1963).
40. Spitzer, F.: Dissertation, Göttingen, 1938.
41. Tamaru, T.: *Phys. Z.* **6,** 379 (1905).
42. Van der Veen: *Z. Krist.* **51,** 545 (1913).
43. Yamada, T., N. Niizeki, and H. Toyoda: *Jap. J. Appl. Phys.* **6,** 151 (1967).
44. Zheludev, I. S., and V. M. Fridkin: *Soviet Phys.—Cryst.* **3,** 319 (1958).
45. Zheludev, I. S., and M. M. Tagieva: *Soviet Phys.—Cryst.* **7,** 473 (1963).
46. Arlt, G., and P. Quadflieg: *Phys. Stat. Solidi* **25,** 323 (1968).

TABLE 9f-10. TEMPERATURE DEPENDENCE OF SOME PIEZOELECTRIC STRAIN CONSTANTS
($\times 10^{-12}$ in coulomb/newton or meter/volt)

Substance	Formula		123 K	153 K	193 K	243 K	258 K	273 K	293 K	307 K	321 K	343 K	390 K	423 K	473 K	Ref.
Ammonium dihydrogen phosphate	$NH_4H_2PO_4$	d_{36}		-83	-69	-55	-51	-48	-46							6
Barium titanate	$BaTiO_3$	d_{31}							-37	-38	-39	-42	-170	0	0	4
Barium titanate ceramics	$BaTiO_3$	d_{31}				-60	-65	-75	-85	-69	-64	-55				2
Heavy rochelle salt	$KNaC_4H_2D_2O_6 \cdot 4D_2O$	d_{14}			50	29	195	140	193	2,800	213					3
Potassium dihydrogen phosphate	KH_2PO_4	d_{36}	20,000	135			26	23	21							1
Rochelle salt	$KNaC_4H_4O_6 \cdot 4H_2O$	d_{14}			5.5	96	700	765	780	250						7
						22	1,200	2,250	2,700	250						8
						150	2,1C0	350	740	320	120					2
Sodium bromate	$NaBrO_3$	d_{14}							2.7	2.8	3.0	3.2	3.7	4.1	5.1	5
Sodium chlorate	$NaClO_3$	d_{14}							2.0	2.2	2.3	2.5	3.1	3.8	5.4	5

References for Table 9f-10

1. Bantle, W., and C. Caflisch: *Helv. Phys. Acta* **16**, 235 (1943).
2. Brush Laboratories Company, Cleveland, Ohio: Reports.
3. Cady, W.: "Piezoelectricity," McGraw-Hill Book Company, New York, 1946.
4. Caspari, M. E., and W. J. Merz: *Phys. Rev.* **80**, 1082 (1950).
5. Mason, W. P.: "Piezoelectric Crystals and Their Application to Ultrasonics," D. Van Nostrand Company, Inc., Princeton, N.J., 1950.
6. Matthias, B., W. Merz, and P. Scherrer: *Helv. Phys. Acta* **20**, 273 (1947).
7. Valasek, J.: *Science* **65**, 235 (1927).
8. Valasek, J.: *Phys. Rev.* **20**, 639 (1922).

TABLE 9f-11. TEMPERATURE COEFFICIENT OF SOME PIEZOELECTRIC STRAIN
CONSTANTS AT ROOM TEMPERATURE

Substance	Formula	α_{11}	α_{14}	α_{25}	α_{36}	Ref.
Iodic acid..............	HIO_3		$+3.5$	-3.5	-0.9	3
Lithium ammonium tartrate monohydrate.....	$LiNH_4C_4H_4O_6 \cdot H_2O$		$+39$	-50	$+31$	3
Quartz.................	SiO_2	~ -10 -2.15				2 1
Rochelle salt...........	$NaKC_4H_4O_6 \cdot 4H_2O$		12.9 See Table 9f-10.	$+49$	$+10.9$	4
Sodium ammonium tartrate tetrahydrate.....	$NaNH_4C_4H_4O_6 \cdot 4H_2O$		$+2.1$	-19	$+12.1$	3
Strontium formate dihydrate................	$Sr(HCOO)_2 \cdot 2H_2O$		-8	-3.8	-14.7	3

Temperature coefficient α is defined as $\alpha_{ik} = \dfrac{1}{d_{ik}} \dfrac{\partial d_{ik}}{\partial \theta}$ where θ is the temperature. α_{ik} is measured in degrees^{-1}. The listed numbers have to be multiplied by 10^{-4}.

References for Table 9f-11

1. Bechmann, R.: *Proc. Phys. Soc. (London)*, ser. B, **64**, 323 (1951).
2. Cady, W.: "Piezoelectricity," McGraw-Hill Book Company, New York, 1946.
3. Mason, W. P.: "Piezoelectric Crystals and Their Application," D. Van Nostrand Company, Inc., Princeton, N.J., 1950.
4. J., Valasek: *Science* **65**, 235 (1927).

TABLE 9f-12. TABLE OF PYROELECTRIC CONSTANTS*

Substance	Formula	p	References
1. Barium titanate (ceramic)............	$BaTiO_3$	20	5
2. Calamine.........................	$2ZnO \cdot SiO_2 \cdot H_2O$	2.0	2, 6
3. Cane sugar.......................	$C_{12}H_{22}O_{11}$	0.18	4
4. Colmanite.......................	$CaB_3O_4(OH)_3 \cdot H_2O$	0.03–0.3	3
5. Diammonium tartrate...............	$(NH_4)_2C_4H_4O_6$	0.95	1, 4
6. Dipotassium tartrate hemihydrate.....	$K_2C_4H_4O_6 \cdot \frac{1}{2}H_2O$	2.0	1, 4
7. Lithium selenate monohydrate.......	$Li_2SeO_4 \cdot H_2O$	5.7	1, 4
8. Lithium sodium sulfate..............	$LiNaSO_4$	0.75	1, 4
9. Lithium sulfate monohydrate........	$Li_2SO_4 \cdot H_2O$	7.7	1, 4
10. Lithium trisodium selenate hexahydrate	$LiNa_3(SeO_4)_2 \cdot 6H_2O$	1.8	1, 4
11. Potassium lithium sulfate...........	$KLiSO_4$	1.6	1, 4
12. Resorcinol.......................	$C_6H_6O_2$	2.6	4
13. Rhamnose.......................	$C_6H_{12}O_5 \cdot H_2O$	1.2, 0.17	4
14. Scolecite........................	$CaAl_2Si_3O_{10} \cdot 3H_2O$	0.33	4
15. Strontium ditartrate tetrahydrate.....	$Sr(HC_4H_4O_6)_2 \cdot 4H_2O$	0.24, 2.7	1, 6
16. Tartaric acid.....................	$C_4H_6O_6$	2.5	4
17. Tourmaline......................	Variable	0.35–0.44	1

* The pyroelectric constant is defined as $p = (\partial D/\partial \theta)_{E=0}$, where D is the electric displacement, and θ the temperature. The units for p are coulomb/meter^2 degree (rationalized mks). The listed numbers have to be multiplied by 10^{-5}. They include both the true and the false pyroeffect.

References for Table 9f-12

1. Ackermann, A. W.: *Ann. Physik* **46**, 197 (1915).
2. Curie and Curie: *Compt. Rend.* **91**, 383 (1880).
3. Davisson, J. W.: *Acta Cryst.* **9**, 9 (1956).
4. Hayashi, F.: Dissertation, Göttingen, 1912.
5. Perls, T. A., T. J. Diesel and W. J. Dobrov: *J. Appl. Phys.* **29**, 1297 (1958).
6. Van der Veen, A.: Thesis, Delft, 1911.

TABLE 9f-13. TEMPERATURE DEPENDENCE OF SOME PYROELECTRIC CONSTANTS*
($\times 10^{-5}$ in coulomb/meter^2 degree)

Substance	Formula	23 K	88 K	198 K	253 K	274 K	293 K	352 K	372 K	408 K	488 K	578 K	648 K
Diammonium tartrate	$(NH_4)_2C_4H_4O_6$	0.05	0.10	0.39	0.80	0.86	0.95	1.14					
Lithium selenate monohydrate	$Li_2SeO_4 \cdot H_2O$	0.31	0.77	3.28	4.82	5.33	5.70	6.45					
Lithium sodium sulfate	$LiNaSO_4$	0.04	0.10	0.29	0.54	0.68	0.75	0.91					
Lithium sulfate monohydrate	$Li_2SO_4 \cdot H_2O$	0.40	1.27	4.07	6.15	6.82	7.75	9.0					
Lithium trisodium selenate hexahydrate	$LiNa_3(SeO_4)_2 \cdot 6H_2O$	0.12	0.31	0.98	1.53	1.69	1.80	2.12					
Potassium lithium sulfate	$KLiSO_4$	....	0.23	0.83	1.36	1.50	1.61	1.78					
Dipotassium tartrate hemihydrate	$K_2C_4H_4O_6 \cdot \frac{1}{2}H_2O$	0.13	0.33	1.10	1.70	1.87	1.99	2.30					
Strontium ditartrate tetrahydrate	$Sr(HC_4H_4O_6)_2 \cdot 4H_2O$	0.013	0.04	0.15	0.21	0.23	0.24	0.27					
Tourmaline Yellow-green		0.027	0.097	0.32	0.40	0.41	0.43	0.44	0.45	0.46	0.50	0.56	0.62
Rose-red		0.027	0.10	0.33	0.41	0.42	0.44	0.47	0.48	0.49	0.51	0.58	0.65
Blue-green		0.013	0.047	0.22	0.31	0.34	0.35	0.39	0.40	0.41	0.42	0.46	0.51

* W. Ackermann, *Ann. Physik* **46**, 197 (1915).

TABLE 9f-14. FERROELECTRIC CRYSTALS

Substance	Formula	Structure at room temp.	Ferroelectric axis	Curie point, K	Max spont. polarization, coulombs/meter^2	Small-signal dielectric const at room temp.			Ref.
						ϵ_a/ϵ_v	ϵ_b/ϵ_c	ϵ_c/ϵ_v	
Heavy rochelle salt	$KNaC_4H_2D_2O_6\cdot4D_2O$	monocl.	a	308 upper, 251 lower	0.37×10^{-2} at 279 K	2,300 at θ up	9.4	9.8	17, 32, 49
Lithium ammonium tartrate monohydrate	$LiNH_4C_4H_4O_6\cdot H_2O$	orthorh.	b	106	0.21×10^{-2} at $T \ll \theta$	7.2	8.0	6.9	53, 58
Lithium thallium tartrate monohydrate	$LiTlC_4H_4O_6\cdot H_2O$	orthorh.	a	10	0.14×10^{-2} at $T \ll \theta$				53
Rochelle salt	$KNaC_4H_4O_6\cdot4H_2O$	monocl.	a	297 upper, 255 lower	0.24×10^{-2} at 276 K	4,000 at θ up	10.0	9.6	17
Cesium dideuterium arsenate	CsD_2AsO_4	tetrag.	c	212					86
Cesium dihydrogen arsenate	CsH_2AsO_4	tetrag.	c	143					26, 86
Potassium dideuterium arsenate	KD_2AsO_4	tetrag.	c	162					86
Potassium dideuterium phosphate	KD_2PO_4	tetrag.	c	213	4.8×10^{-2} at θ	88	88	90	4, 86
Potassium dihydrogen arsenate	KH_2AsO_4	tetrag.	c	97	5.0×10^{-2} at θ	62	62	22	16, 86
Potassium dihydrogen phosphate	KH_2PO_4	tetrag.	c	123	4.95×10^{-2} at $T \ll \theta$	42	42	21	16, 86
Rubidium dideuterium arsenate	RhD_2AsO_4	tetrag.	c	178					86
Rubidium dideuterium phosphate	RbD_2PO_4	tetrag.	c	218					54
Rubidium dihydrogen arsenate	RbH_2AsO_4	tetrag.	c	111					50, 86
Rubidium dihydrogen phosphate	RbH_2PO_4	tetrag.	c	146	5.6×10^{-2} at $T \ll \theta$	35	35	22	50, 86
Guanidine aluminum sulfate hexahydrate	$C(NH_2)_3Al(SO_4)_2\cdot6H_2O$	trigonal	c	none	0.35×10^{-2} at 296 K	5	5	6	34, 35
Guanidine chromium sulfate hexahydrate	$C(NH_2)_3Cr(SO_4)_2\cdot6H_2O$	trigonal	c	none	0.37×10^{-2} at 296 K	5	5	6	34, 35
Guanidine gallium sulfate hexahydrate	$C(NH_2)_3Ga(SO_4)_2\cdot6H_2O$	trigonal	c	none	0.37×10^{-2} at 296 K	5	5	6	34, 35
Guanidine vanadium sulfate hexahydrate	$C(NH_2)_3V(SO_4)_2\cdot6H_2O$	trigonal	c	none	0.36×10^{-2} at 296 K				77
Guanidine aluminum selenate hexahydrate	$C(NH_2)_3Al(SeO_4)_2$	trigonal	c	none	0.45×10^{-2} at 296 K	5	5	6	34, 35
Guanidine chromium selenate hexahydrate	$C(NH_2)_3Cr(SeO_4)_2\cdot6H_2O$	trigonal	c	none	0.47×10^{-2} at 296 K	5	5	6	34, 35
Guanidine gallium selenate hexahydrate	$C(NH_2)_3Ga(SeO_4)_2\cdot6H_2O$	trigonal	c	none	0.47×10^{-2} at 296 K	5	5	6	34, 35
Deuteroguanidine aluminum sulfate hexahydrate	$C(ND_2)_3Al(SO_4)_2\cdot6D_2O$	trigonal	c	none	0.35×10^{-2} at 296 K	9	9	9	64
Methyl ammonium aluminum alum	$(CH_3NH_3)Al(SO_4)_2\cdot12H_2O$	cubic	cubic 111	176	1.0×10^{-2} at 166 K				74
Methyl ammonium chromium alum	$(CH_3NH_3)Cr(SO_4)_2\cdot12H_2O$	cubic	cubic 111	164	1.0×10^{-2} at 162 K				74
Methyl ammonium gallium alum	$(CH_3NH_3)Ga(SO_4)_2\cdot12H_2O$	cubic	cubic 111	171					74
Methyl ammonium indium alum	$(CH_3NH_3)In(SO_4)_2\cdot12H_2O$	cubic	cubic 111	164	1.2×10^{-2} at 162 K				74
Methyl ammonium iron alum	$(CH_3NH_3)Fe(SO_4)_2\cdot12H_2O$	cubic	cubic 111	169	1.3×10^{-2} at 167 K				74
Methyl ammonium vanadium alum	$(CH_3NH_3)V(SO_4)_2\cdot12H_2O$	cubic	cubic 111	157	0.9×10^{-2} at 155 K				74
Methyl ammonium aluminum (selenate) alum	$(CH_3NH_3)Al(SeO_4)_2\cdot12H_2O$	cubic	cubic 111	216	1.2×10^{-2} at 214 K				74
Deuterated methyl ammonium aluminum alum	$(CH_3ND_3)Al(SO_4)_2\cdot12D_2O$	cubic	cubic 111	177	1.0×10^{-2} at 175 K				74
Ammonium indium alum	$NH_4In(SO_4)_2\cdot12H_2O$	cubic	cubic 111	127	1.2×10^{-2} at 125 K				74
Ammonium iron alum	$NH_4Fe(SO_4)_2\cdot12H_2O$	cubic	cubic 111	88	0.4×10^{-2} at 86 K				74
Ammonium vanadium alum	$NH_4V(SO_4)_2\cdot12H_2O$	cubic	cubic 111	116	1.0×10^{-2} at 114 K				74
Deuterated ammonium iron alum	$ND_4Fe(SO_4)_2\cdot12D_2O$	cubic	cubic 111	88	0.4×10^{-2} at 86 K				74
Urea chromium alum	$(CONH_2NH_2)Cr(SO_4)_2\cdot12H_2O$	cubic	cubic 111	160	0.1×10^{-2} $T \ll \theta$				64
Ammonium bisulfate	$(NH_4)HSO_4$	monocl.	c	upper 270 lower 154	0.8×10^{-2} at 200 K			16	71
Ammonium sodium sulfate dihydrate	$NH_4NaSO_4\cdot2H_2O$								

Substance	Formula	System	Direction	Transition temp. (K)	Spontaneous polarization				Refs.
Ammonium lithium sulfate	NH$_4$LiSO$_4$	orthorh.	c	224	0.25–0.45 × 10^{-2} at 215 K	10	9	9	37, 55, 62
Ammonium sulfate	(NH$_4$)$_2$SO$_4$	orthorh.	b	176	0.19 × 10^{-2} at 163 K	9	10	9	37, 55, 62
Ammonium fluoberyllate	(NH$_4$)$_2$BeF$_4$	orthorh.	c	224	0.45 × 10^{-2} at 215 K	10	9	9	37
Deuteroammonium sulfate	(ND$_4$)$_2$SO$_4$	orthorh.	b	179	0.19 × 10^{-2} at 163 K	9	10	9	37
Deuteroammonium fluoberyllate	(ND$_4$)$_2$BeF$_4$	cubic	cubic 111	89	0.3 × 10^{-2} at R.T.	9	9		45
Diammonium dicadmium sulfate	(NH$_4$)$_2$Cd$_2$(SO$_4$)$_3$	cubic		none		29	13	30	72
Lithium trihydrogen selenite	LiH$_3$(SeO$_3$)$_2$	monocl.	c	258	15 × 10^{-2} at 83 K	7	8	10	73
Rubidium bisulfate	RbHSO$_4$	monocl.	c	271	0.65 × 10^{-2} at 178 K				7
Sodium trideuterium selenite	NaD$_3$(SeO$_3$)$_2$	monocl.	a	194		~100			7, 72
Sodium trihydrogen selenite	NaH$_3$(SeO$_3$)$_2$	monocl.	b			~100			
Deuterotriglycine sulfate	(ND$_2$CH$_2$COOD)$_2$SO$_4$	monocl.	b	333	1.2 × 10^{-2} at θ	6.6	8.1	7.4	68
Diglycine manganous chloride dihydrate	(NH$_2$CH$_2$COOH)$_2$·MnCl$_2$·2H$_2$O	monocl.	b	328	1.3 × 10^{-2} at 300 K		14		69
Diglycine nitrate	(NH$_2$CH$_2$COOH)$_2$·HNO$_3$	monocl. (101)		196	0.6 × 10^{-2} at 196 K				67
Glycine silver nitrate	NH$_2$CH$_2$COOH·AgNO$_3$	monocl.	b	218	0.55 × 10^{-2} at 78 K	9	15	4	36
Triglycine fluoberyllate	(CH$_2$NH$_2$COOH)$_3$H$_3$BeF$_4$	monocl.	b	343	3.2 × 10^{-2} at R.T.				21, 36, 56
Triglycine selenate	(CH$_2$NH$_2$COOH)$_3$H$_2$SeO$_4$	monocl.	b	295–300		9	50	4	36, 56
Triglycine sulfate	(CH$_2$NH$_2$COOH)$_3$H$_2$SO$_4$	monocl.	b	321	2.2 × 10^{-2} at R.T.	9	50	4	60
Antimony sulfur bromide	SbSBr	orthorh.	c	93					23
Antimony sulfur iodide	SbSI	orthorh.	c	294	~25 × 10^{-2} at 273 K	25	25	5 × 10^4	60
Bismuth sulfur bromide	BiSBr	orthorh.	c	103					60
Bismuth sulfur iodide	BiSI	orthorh.	c	113					38
Hydrogen bromide	HBr			89.7	0.4 × 10^{-2} at 78 K				38
Hydrogen chloride	HCl	trig.		98.4	1.2 × 10^{-2} at 90 K				63
Germanium telluride	GeTe			670	small				66
Ammonium monochloracetate	(CH$_2$ClCOO)NH$_4$	monocl.	b	~123	~0.1 × 10^{-2} at 78 K	~5 along ferroelectric axis			40
Ammonium hydrogen dimonochloroacetate	(CH$_2$ClCOO)$_2$H·NH$_4$	monocl.	102 axis	128	0.18 × 10^{-2} at 78 K				46
Boracite	Mg$_3$B$_7$O$_{13}$Cl			265	small		~14		2
Nickel-iodine boracite	Ni$_3$B$_7$O$_{13}$I			64			~20		30
Colmanite	CaB$_3$O$_4$(OH)$_3$·H$_2$O	cubic		271	0.5 × 10^{-2} at 235 K			14	57
Dicalcium strontium propionate	Ca$_2$Sr(CH$_2$CH$_2$COO)$_6$	monocl.	b	282	0.3 × 10^{-2} at 228 K				70
Lithium hydrazinium sulfate	Li(N$_2$H$_5$)SO$_4$	t-trag.	c	none	0.3 × 10^{-2} at 300 K				95, 96
Potassium ferrocyanide trihydrate	K$_4$Fe(CN)$_6$·3H$_2$O	orthorh.	c	251	1.4 × 10^{-2} at 193 K				96
Potassium rutenocyanide trihydrate	K$_4$Ru(CN)$_6$·3H$_2$O	monocl.		259	1.4 × 10^{-2} at 210 K				96
Potassium osmocyanide trihydrate	K$_4$Os(CN)$_6$·3H$_2$O	monocl.		271	3.5 × 10^{-2} at 227 K				96
Potassium manganocyanide trihydrate	K$_4$Mn(CN)$_6$·3H$_2$O	monocl.	(010) plane	233					22, 24
Tetramethyl ammonium trichloromercurate	((CH$_3$)$_4$N)HgCl$_3$	monocl.	(010) plane	none	1.2 × 10^{-2} at R.T.	50 along ferroelectric axis			24, 22
Tetramethyl ammonium tribromomercurate	((CH$_3$)$_4$N)HgBr$_3$	monocl.	(010) plane	none	1 × 10^{-2} at R.T.				24
Tetramethyl ammonium triiodomercurate	((CH$_3$)$_4$N)HgI$_3$	monocl.	(010) plane	none	1.7 × 10^{-2} at R.T.				24
Tetramethyl phosphonium tribromomercurate	((CH$_3$)$_4$P)HgBr$_3$	monocl.		none	3 × 10^{-2} at R.T.				29, 85
Thiourea	(NH$_2$)$_2$CS	orthorh.	b	169	2.5 × 10^{-2} at 169 K	3	32	3	48
Trisarcosine calcium chloride	(CH$_3$NHCH$_2$COOH)$_3$CaCl$_2$	orthorh.	b	127	0.27 × 10^{-2} at 78 K		5		18
Aluminum metaniobate (ceramics)	AlNb$_3$O$_9$			493			~170		18
Barium metaniobate (ceramics)	BaNb$_2$O$_6$	tetrag.		~343			~130		59, 81, 94
Barium titanate	BaTiO$_3$		c	393	26 × 10^{-2} at 296 K	~5,000	~5,000	~160	75, 90
Bismuth titanate	Bi$_4$Ti$_3$O$_{12}$	orthorh.	c	916–948	3.5 × 10^{-2} at R.T.			~160	19, 44, 81
Cadmium niobate (ceramics)	CdNb$_2$O$_7$	cubic		188		~310	~310	112	25, 31
Lead metaniobate	Pb(NbO$_2$)$_2$	orthorh.	a + b	843	6 × 10^{-2} at 93 K (111)	200	200	600	

TABLE 9f-14. FERROELECTRIC CRYSTALS (Continued)

Substance	Formula	Structure at room temp.	Ferroelectric axis	Curie point, K	Max spont. polarization, coulombs/meter²	Small-signal dielectric const at room temp. ϵ_a/ϵ_v	ϵ_b/ϵ_v	ϵ_c/ϵ_v	Ref.
Lead metatantalate (ceramics)	$Pb(TaO_3)_2$	orthorh.	...	423	...	...	~1,200	...	25
				513			~300		42
Lead titanate (ceramics)	$PbTiO_3$	tetragonal	c	763	...	...	~50	...	80
Lithium niobate	$LiNbO_3$	trigonal	c	1468	71×10^{-2} at R.T.	84	84	30	1, 51, 97
Lithium tantalate	$LiTaO_3$	trigonal	c	891–938	50×10^{-2} at R.T.	51	51	45	1, 51, 76, 97
Manganese metatantalate (ceramics)	$MnTaO_4$	...	...	...	...	...	~750	...	18
Potassium iodate	KIO_3	trigonal	c	485	0.3×10^{-2} at R.T.	...	20	20	33
Potassium niobate	$KNbO_3$	orthorh.	c	691	26×10^{-2} at 691 K	...	~500	~500	81, 89
Potassium nitrate	KNO_3	orthorh.	c	397	8.0×10^{-2} at 393 K	5	5	5	79
Potassium tantalate	$KTaO_3$	cubic	c	13	...	...	~500	~500	39
Rare-earth manganates	$ErMnO_3$	hexag.	c	...	...	...	...	...	6
	$HoMnO_3$	hexag.	c						6
	$LuMnO_3$	hexag.	c						6
	$TmMnO_3$	hexag.	c						6
	$YbMnO_3$	hexag.	c	973–998	$4–5 \times 10^{-2}$ at R.T.	...	~20	~20	9, 41
Yttrium manganate	$YMnO_3$	hexag.	c	933	$4–5 \times 10^{-2}$ at R.T.	...	...	~20	9, 41
Rare-earth molybdates	$Eu_2(MoO_4)_3$	orthorh.	c	434	0.14×10^{-2} at R.T.	...	...	9.5	13
	$Gd_2(MoO_4)_3$	orthorh.	c	432	0.17×10^{-2} at R.T.	...	...	10	13
	$Sm_2(MoO_4)_3$	orthorh.	c	463	0.24×10^{-2} at R.T.	...	...	12	13
	$Tb_2(MoO_4)_3$	orthorh.	c	430	0.18×10^{-2} at R.T.	...	...	11	13
Sodium niobate	$NaNbO_3$	orthorh.	c	<64	...	76	76	~670	20
Sodium nitrate	$NaNO_2$	orthorh.	b	437	8.6×10^{-2} at R.T.	7.5	8	5	61
Tungsten trioxide	WO_3	monocl.			probably ferroelectric				52
Mixed niobates									
(ceramic)	Ba_2BiNbO_6	rhomb.		633					93
	$Sr_{0.75}Ba_{0.25}Nb_2O_6$	tetrag.	c	333	14×10^{-2} at R.T.		3,400	450	5, 47, 97
	$Sr_{0.5}Ba_{0.5}Nb_2O_6$	tetrag.	c	403				450	5, 47
	$Sr_{0.25}Ba_{0.75}Nb_2O_6$	tetrag.	c	472	25×10^{-2} at R.T.			118	5, 47, 97
	$Pb_{0.4}Ba_{0.3}Sr_{0.3}Nb_2O_6$	tetrag.	c	415					78
	$PbO = 2.5Nb_2O_5$	orthorh.		908				300	78
(ceramic)	$Ba_3CuNb_2O_9$	tetrag.	c	653			~50		93
(ceramic)	$BaBi_2Nb_2O_9$	orthorh.		473			~280		87
(ceramic)	$PbBi_2Nb_2O_9$	orthorh.		833			~170		82, 87
(ceramic)	$SrBi_2Nb_2O_9$	orthorh.		713			~190		87
(ceramic)	$Sr_3CuNb_2O_9$	tetrag.		663					93
(ceramic)	$PbSc_{\frac{1}{2}}Cr_{\frac{1}{2}}Nb_{\frac{1}{2}}O_3$			330			~1,270		88
	$PbFe_{\frac{1}{2}}Nb_{\frac{1}{2}}O_3$	rhomb.	c	393			~1,200		14
	$PbMg_{\frac{1}{3}}Nb_{\frac{2}{3}}O_3$	cubic	c	265	24×10^{-2} at 77 K		~10,000		8, 12

Substance	Form	Crystal system	axis	T_c (K)	P_s	ε_a	ε_c	ε_{max}	Ref.
$PbNi_{1/3}Nb_{2/3}O_3$		cubic	o	148			~1,600		8
$K_3Li_2Nb_5O_{15}$		tetrag.	c	693	25×10^{-2} at R.T.	309		100	97, 11, 92
$NaSr_2Nb_5O_{15}$		orthorh.	c	543				2,000	92
$Na_3Li_2Nb_5O_{15}$		orthorh.	c	>1073				100	92
$K_6NaLi_3Nb_5O_{15}$		tetrag.	c	429				500	92
$KSr_2Nb_5O_{15}$		tetrag.	c	650–665	8×10^{-2} at R.T.	1,200	1,200	1,000	28, 97
$KBa_2Nb_5O_{15}$		tetrag.	c	651					28, 15
$KPb_2Nb_5O_{15}$		orthorh.	c	833–858					28
$NaBa_2Nb_5O_{15}$		tetrag.	c	779	40×10^{-2} at R.T.	242	242	51	27, 28, 78, 97
$NaPb_2Nb_5O_{15}$		tetrag.	c	412					28
$RbSr_2Nb_5O_{15}$		tetrag.	c	714					28
$RbPb_2Nb_5O_{15}$		tetrag.	c	303					28
$K_3Li_2Ta_{3.5}Nb_{1.5}O_{15}$			c					30,000	11
Mixed tantalate									
Ba_2BiTaO_6	(ceramic)	rhomb.		673					93
$Ba_3CuTa_2O_9$	(ceramic)	tetrag.	c	743					93
$Sr_2CuTa_2O_9$	(ceramic)	tetrag.	c	1523					93
$BaBi_2Ta_2O_9$	(ceramic)	orthorh.		383			~400		87
$PbBi_2Ta_2O_9$	(ceramic)	orthorh.		703			~180		87
$SrBi_2Ta_2O_9$	(ceramic)	orthorh.		608	5.8×10^{-2} at R.T.		~180		87
$PbFe_{1/2}Ta_{1/2}O_3$	(ceramic)			243					83
Mixed titanates									
$BiK(TiO_3)_2$		tetrag.	c	643				500	14, 43
$BiNa(TiO_3)_2$		tetrag.	c	593	8×10^{-2} at 389 K			700	14, 43
$BaBi_4Ti_4O_{15}$		orthorh.		668			~150		87
$PbBi_4Ti_4O_{15}$		orthorh.		843			~220		87
$SrBi_4Ti_4O_{15}$		orthorh.		928			~190		87
$Na_{0.5}Bi_{4.5}Ti_4O_{15}$		orthorh.		823			~200		87
$K_{0.5}Bi_{4.5}Ti_4O_{15}$		orthorh.		602			~140		87
$Ba_2Bi_4Ti_5O_{18}$	(ceramic)	tetrag.	c	583	2×10^{-2} at R.T.		~400	400	3
$Pb_2Bi_4Ti_5O_{18}$	(ceramic)			558	6×10^{-2} at 508 K		~280		87
$Sr_2Bi_4Ti_5O_{18}$	(ceramic)				3.5×10^{-2} at 528 K				87
Mixed tungstates									
Ba_2CuWO_6	(ceramic)	tetrag.	c	1473					93
Sr_2CuWO_6	(ceramic)	tetrag.	c	1193					93
Ba_3BiWO_9	(ceramic)	rhombo.		723					93
$PbSc_{2/3}W_{1/3}O_3$	(ceramic)			261			~2,000		93
$PbLi_{1/3}Nb_{1/3}W_{1/3}O_3$	(ceramic)			233					93
$PbLi_{1/4}Fe_{1/4}W_{1/2}O_3$	(ceramic)			203					93
$PbLi_{1/4}Yb_{1/4}W_{1/2}O_3$	(ceramic)			243			~400		93
$PbY_{1/2}Ti_{1/4}W_{1/4}O_3$	(ceramic)			323					93
$PbFe_{2/3}W_{1/3}O_3$	(ceramic)			178			~2,000		84
$PbCo_{2/3}W_{1/3}O_3$	(ceramic)	rhombo.		77			~250		10

References for Table 9f-14

1. Abrahams, S. C., and J. L. Bernstein: *J. Phys. Chem. Solids* **28**, 1685 (1967).
2. Ascher, E., H. Rieder, H. Schmid, and H. Stössel: *J. Appl. Phys.* **37**, 1404 (1966).
3. Aurivillius, B., and P. H. Fang: *Phys. Rev.* **126**, 893 (1962).
4. Bantle, W.: *Helv. Phys. Acta.* **15**, 373 (1942).
5. Ballman, A. A., and H. Brown: *J. Cryst. Growth* **1**, 311 (1967).
6. Bertaut, E. F., and F. Lissalde: *Solid State Commun.* **5**, 173 (1967).
7. Blinc, R., A. Jovanovic, A. Levstik, and A. Prelesnik: *J. Phys. Chem. Solids* **26**, 1359 (1965).
8. Bokov, V. A., and I. E. Mylnikova: *Soviet Phys.—Solid State* **3**, 613 (1961).
9. Bokov, V. A., G. A. Smolenskii, S. A. Kizhaev, and I. E. Mylnikova: *Soviet Phys.—Solid State* **5**, 2646 (1964).
10. Bokov, V. A., S. A. Kizhaev, I. E. Mylnikova and A. G. Tutov: *Soviet Phys.—Solid State* **6**, 2419 (1965).
11. Bonner, W. A., W. H. Grodkiewicz, and L. G. Van Uitert: *J. Cryst. Growth* **1**, 318 (1967).
12. Bonner, W. A., E. F. Dearborn, J. E. Geusic, H. M. Marcos, and L. G. van Uitert: *Appl. Phys. Letters* **10**, 163 (1967).
13. Borchardt, H. J., and P. E. Bierstedt: *J. Appl. Phys.* **38**, 2057 (1967); and *Appl. Phys. Letters* **8**, 50 (1966).
14. Buhrer, C. F.: *J. Chem. Phys.* **36**, 798 (1962).
15. Burns, G., D. F. O'Kane, E. A. Giess, and B. A. Scott: *Solid State Commun.* **6**, 223 (1968).
16. Busch, G.: *Helv. Phys. Acta* **11**, 269 (1938).
17. Cady, W. G.: "Piezoelectricity," McGraw-Hill Book Company, New York, 1946.
18. Coates, R. V. and H. F. Kay: *Phil. Mag.* **3**, 1449 (1958).
19. Cook, W. R. and Hans Jaffe: *Phys. Rev.* **88**, 1426 (1952).
20. Cross, L. E.: *Phil. Mag.* **46**, 453 (1955).
21. Fatuzzo, E., and R. Nitsche: *Z. Elektrochem.* **63**, 970 (1959).
22. Fatuzzo, E., and R. Nitsche: *Phys. Rev.* **117**, 936 (1960).
23. Fatuzzo, E., G. Harbeke, W. J. Merz, R. Nitsche, H. Roetschi, and W. Ruppel: *Phys. Rev.* **127**, 2036 (1962).
24. Fatuzzo, E., R. Nitsche, H. Roetschi, and S. Zingg: *Phys. Rev.* **125**, 514 (1962).
25. Francombe, M. H., and B. Lewis: *Acta Cryst.* **11**, 696 (1958).
26. Frazer, B. C. and R. Pepinsky: *Phys. Rev.* **91**, 212 (1953).
27. Geusic, J. E., H. J. Levinskin, J. J. Rubin, S. Singh, and L. G. Van Uitert: *Appl. Phys. Letters* **11**, 269 (1967).
28. Giess, E. A., G. Burns, D. F. O'Kane, and A. W. Smith: *Appl. Phys. Letters* **11**, 233 (1967); and private communication.
29. Goldsmith, G. J. and J. G. White: *J. Chem. Phys.* **31**, 1175 (1959).
30. Goldsmith, G. J.: *Bull. Am. Phys. Soc.* II(1), 322 (1956).
31. Goodman, G.: *J. Am. Ceram. Soc:* **36**, 368 (1953).
32. Hablützel, J.: *Helv. Phys. Acta* **12**, 489 (1939).
33. Herlach, F.: *Helv. Phys. Acta* **34**, 305 (1961).
34. Holden, A. N., B. T. Matthias, W. J. Merz, and J. P. Remeika: *Phys. Rev.* **98**, 546 (1955).
35. Holden, A. N., W. J. Merz, J. P. Remeika, and B. T. Matthias: *Phys. Rev.* **101**, 962 (1956).
36. Hoshino, S., T. Mitsui, F. Jona, and R. Pepinsky: *Phys. Rev.* **107**, 1255 (1957).
37. Hoshino, S., K. Vedam, Y. Okaya, and R. Pepinsky: *Phys. Rev.* **112**, 405 (1958).
38. Hoshino, S., K. Shimaoka, and N. Niimura: *Phys. Rev. Letters* **19**, 1286 (1967).
39. Hulm, J. K., B. T. Matthias, and E. A. Long: *Phys. Rev.* **79**, 885 (1950).
40. Ichikawa, M., and T. Mitsui: *Phys. Rev.* **152**, 495 (1966).
41. Ismailzade I. G., and S. A. Kizhaev: *Soviet Phys.—Solid State* **7**, 236 (1965).
42. Isupov, V. A.: *Translation: Soviet Phys.—Solid State* **1**, 216 (1959).
43. Isupov, V. A., P. L. Strelets, I. A. Serova, N. D. Yataenko, and T. M. Shirobokikh, *Soviet Phys.—Solid State* **6**, 615 (1964).
44. Jona, F., G. Shirane, and R. Pepinsky: *Phys. Rev.* **98**, 903 (1955).
45. Jona, F., and R. Pepinsky: *Phys. Rev.* **103**, 1126 (1956).
46. LeCorre: *J. Phys. Radium* **18**, 629 (1957).
47. Lenzo, P. V., E. G. Spencer and A. A. Ballman: *Appl. Phys. Letters,* **11**, 23 (1967).
48. Makita Y.: *J. Phys. Soc. Japan* **20**, 2073 (1965).
49. Mason, W. P., and A. N. Holden: *Phys. Rev.* **57**, 54 (1940).
50. Matthias, B., W. Merz, and P. Scherrer: *Helv. Phys. Acta* **20**, 273 (1947).
51. Matthias, B. T., and J. P. Remeika: *Phys. Rev.* **76**, 1886 (1949).
52. Matthias, B. T.: *Phys. Rev.* **76**, 430 (1949).
53. Matthias, B. T., and J. K. Hulm: *Phys. Rev.* **82**, 108 (1951).

54. Matthias, B. T.: *Phys. Rev.* **85**, 723 (1952).
55. Matthias, B. T., and J. P. Remeika: *Phys. Rev.* **103**, 262 (1956).
56. Matthias, B. T., C. E. Miller, and J. P. Remeika: *Phys. Rev.* **104**, 849 (1956).
57. Matthias, B. T., and J. P. Remeika: *Phys. Rev.* **107**, 1727 (1957).
58. Merz, W. J.: *Phys. Rev.* **82**, 562 (1951).
59. Merz, W. J.: *Phys. Rev.* **91**, 513 (1953).
60. Nitsche, R., H. Roetschi, and P. Wild: *Appl. Phys. Letters* **4**, 210 (1964).
61. Nomura, S.: *J. Phys. Soc. Japan* **16**, 2440 (1961).
62. Okaya, Y., K. Vedam, and R. Pepinsky: *Acta Cryst.* **11**, 307 (1958).
63. Pawley, G. S., W. Cochran, R. W. Cowley, and G. Dolling: *Phys. Rev. Letters* **17**, 753 (1966).
64. Pepinsky, R., F. Jona and G. Shirane: *Phys. Rev.* **102**, 1181 (1956).
65. Pepinsky R., and F. Jona: *Phys. Rev.* **105**, 344 (1957).
66. Pepinsky, R., Y. Okaya, and T. Mitsui: *Acta Cryst.* **10**, 600 (1957).
67. Pepinsky, R., Y. Okaya, D. P. Eastman, and F. Mitsui: *Phys. Rev.* **107**, 1538 (1957).
68. Pepinsky, R., K. Vedam, and Y. Okaya: *Phys. Rev.* **110**, 1309 (1958).
69. Pepinsky, R., K. Vedam, S. Hoshino, and K. Okaya: *Phys. Rev.* **111**, 430 (1958).
70. Pepinsky, R., K. Vedam, Y. Okaya, and S. Hoshino: *Phys. Rev.* **111**, 1467 (1958).
71. Pepinsky, R., K. Vedam, S. Hoshino, and Y. Okaya: *Phys. Rev.* **111**, 1508 (1958).
72. Pepinsky, R., and K. Vedam: *Phys. Rev.* **114**, 1217 (1959).
73. Pepinsky, R., and K. Vedam: *Phys. Rev.* **117**, 1502 (1960).
74. Pepinsky, R.: Private communication.
75. Pulvari, Ch., and A. S. de la Paz: *J. Appl. Phys.* **37**, 1754 (1966).
76. Razbirin, B. S.: *Soviet Phys.—Solid State* **6**, 254 (1964).
77. Remeika, J. P., and W. J. Merz: *Phys. Rev.* **102**, 295 (1956).
78. Rubin, J. J., L. G. Van Uitert, and H. J. Levinstein: *J. Cryst. Growth* **1**, 315 (1967).
79. Sawada, S., and S. Nomura: *J. Phys. Soc. Japan* **16**, 2486 (1961).
80. Shirane, G., and S. Hashino: *J. Phys. Soc. Japan* **6**, 265 (1951).
81. Shirane, G., F. Jona, and R. Pepinsky: *Proc. IRE* **43**, 1738 (1955).
82. Smolenskii, G. A., V. A. Isupov, and A. I. Agranoskaya: *Soviet Phys.—Solid State* **1**, 149 (1959).
83. Smolenskii, G. A., A. I. Agranovskaya, and V. A. Isupov: *Soviet Phys.—Solid State* **1**, 907 (1959).
84. Smolenskii, G. A., and V. A. Bakov: *J. Appl. Phys.* **35**, 915 (1964).
85. Solomon, A. L.: *Phys. Rev.* **104**, 1191 (1956).
86. Stephenson, C.: *J. Chem. Phys.* **21**, 1110 (1953).
87. Subbarao, E. C.: *J. Phys. Chem. Solids* **23**, 665 (1962).
88. Tomashpolskii, Y. Y., and Y. N. Venertsev: *Soviet Phys.—Solid State* **7**, 412 (1965).
89. Triebwasser, S.: *Phys. Rev.* **101**, 993 (1956).
90. Van Uitert, L. G., and L. Egerton: *J. Appl. Phys.* **32**, 959 (1961).
91. Van Uitert, L. G., S. Singh, H. J. Levinstein, J. E. Geusic, and W. A. Bonner: *Appl. Phys. Letters* **11**, 161 (1967).
92. Van Uitert, L. G., H. J. Levinstein, J. J. Rubin, C. D. Capio, E. F. Dearborn, and W. A. Bonner: *Mater. Res. Bull.* **3**, 47 (1968).
93. Venevtsev, N., A. G. Kapyshev, A. S. Viskov, V. M. Lebedev, V. M. Petrov, and G. S. Zhdanov: *Proc. Intern. Meeting Ferroelectricity*, Prague, **1**, 261 (1966).
94. von Hippel, A.: *Rev. Mod. Phys.* **22**, 221 (1950).
95. Waku, W., H. Hirabayashi, H. Toyoda, and H. Iwasaki: *J. Phys. Soc. Japan* **14**, 973 (1959).
96. Waku, S., K. Masuno, T. Tanake, and H. Iwasaki: *J. Phys. Soc. Japan*, **15**, 1185 (1960).
97. Wemple, S. H., M. Di Domenico, and I. Camlibel: *Appl. Phys. Letters* **12**, 209 (1968).

TABLE 9f-15. ANTIFERROELECTRIC CRYSTALS

Substance	Formula	Structure at room temperature	Transition temperature, K	Small-signal dielectric const. at room temperature			Ref.
				ϵ_a/ϵ_v	ϵ_b/ϵ_v	ϵ_c/ϵ_v	
Ammonium chromium sulfate	$NH_4Cr(SO_4)_2 \cdot 12H_2O$	....	82	...	....	....	22
Ammonium dihydrogen arsenate	$NH_4H_2AsO_4$	tetrag.	216	75	75	12	5,
Ammonium dihydrogen phosphate	$NH_4H_2PO_4$	tetrag.	148	56	56	15, 5	5, 30
Ammonium paraperiodate	$(NH_4)_2H_3IO_6$	trigonal	251	143	143	180	2, 6
Barium bismuth molybdate (ceramic)	$Ba_3Bi_2MoO_9$	monocl.	773	...	....	....	29
Barium bismuth vanadate (ceramic)	Ba_2BiVO_6	monocl.	593	...	~25	....	29
Barium hafnate (ceramic)	$BaHfO_3$	cubic	488	...	~90	....	19
Cesium trihydrogen selenite	$CsH_3(SeO_3)_2$	triclinic	145	80	60	30	14
Copper formate tetrahydrate	$Cu(HCOO)_2 \cdot 4H_2O$	monocl.	234	20	400	....	21
Deuteroammonium dideuterium arsenate	$ND_4D_2AsO_4$	orthorh.	304	...	....	....	9
Deuteroammonium dideuterium phosphate	$ND_4D_2PO_4$	tetragonal	243	73	73	22, 5	15, 18
Deuteroammonium paraperiodate	$(ND_4)_2D_3IO_6$	trigonal	266	...	....	....	1
Deuterosilver paraperiodate	$Ag_2D_3IO_6$	trigonal	279	...	....	....	11
Lead cadmium niobium tungsten oxide (ceramic)	$PbCd_{\frac{1}{4}}Nb_{\frac{2}{3}}W_{\frac{1}{3}}O_3$	....	768	...	~200	....	28, 29
Lead cobalt tungstate	Pb_2CoWO_6	rhomb.	305	...	....	~250	4
Lead hafnate (ceramic)	$PbHfO_3$	tetragonal	488	...	~90	....	24
Lead magnesium tungstate (ceramic)	Pb_2MgWO_6	....	311	...	~140	....	27
Lead manganese tungsten oxide (ceramic)	Pb_2MnWO_6	....	423	...	~150	....	29
Lead metaniobate	$PbTa_2O_6$	....	343	...	~860	....	7
Lead nickel tungsten oxide (ceramic)	Pb_2NiWO_6	cubic	290	...	~50	....	20
Lead silicate (ceramic)	Pb_4SiO_6	....	428	...	~23	....	13
Lead vanadite (ceramic)	$Pb_3V_2O_8$	....	373	...	~30	....	12
Lead zirconate (ceramic)	$PbZrO_3$	orthorh.	506	...	~80	....	25, 26
Silver paraperiodate (powder)	$Ag_2H_3IO_6$	trigonal	227	...	~57	....	2, 10, 11
Sodium niobate	$NaNbO_3$	orthorh.	911	76	76	670	8, 16
Sodium tantalate	$NaTaO_3$	orthorh.	....	...	....	....	16
Sodium trideuterium selenite	$NaD_3(SeO_3)_2$	monocl.	271	100	....	....	3
Tungsten trioxide	WO_3	triclinic	1013	...	....	....	17, 23

References for Table 9f-15

1. Aboav, D., H. Gränicher, and W. Petter: *Helv. Phys. Acta* **28**, 299 (1955).
2. Baertschi, P.: *Helv. Phys. Acta* **18**, 267 (1945).
3. Blinc, R., A. Jovanovic, A. Levstik, and A. Prelesnik: *J. Phys. Chem. Solids* **26**, 1359 (1965).
4. Bokov, V. A., S. A. Kizhaev, I. E. Mylnikova, and A. G. Tutov: *Soviet Phys.—Solid State* **6**, 2419 (1965).
5. Busch, G.: *Helv. Phys. Acta* **11**, 269 (1938).
6. Busch, G., W. Känzig, and W. M. Meier: *Helv. Phys. Acta* **26**, 385 (1953).
7. Coates, R. V., and H. F. Kay: *Phil. Mag.* **3**, 1449 (1958).
8. Cross, L. E.: *Phil. Mag.* **46**, 453 (1955).
9. Frazer, B. C.: *Phys. Rev.* **91**, 246 (1953).
10. Gränicher, H., W. M. Meier, and W. Petter: *Helv. Phys. Acta* **27**, 216 (1954).
11. Herlach, F., B. Aboav, H. Gränicher, and W. Petter: *Helv. Phys. Acta* **30**, 252 (1957).
12. Isupov, V. A., N. N. Krainik, I. D. Fridberg, and I. E. Zelenkova: *Soviet Phys.—Solid State* **7**, 844 (1965).
13. Isupov, V. A.: *Soviet Phys.—Solid State* **7**, 1789 (1966).
14. Makita, Y.: *J. Phys. Soc. Japan* **20**, 1567 (1965).

15. Mason, W. P., and B. T. Matthias: *Phys. Rev.* **88**, 477 (1952).
16. Matthias, B. T., and J. P. Remeika: *Phys. Rev.* **75**, 1711 (1949).
17. Matthias, B. T., and E. A. Wood: *Phys. Rev.* **84**, 1255 (1951).
18. Matthias, B. T.: *Phys. Rev.* **85**, 141 (1952).
19. Megaw, H. D.: "Ferroelectricity in Crystals," Methuen & Co., Ltd., London, 1957.
20. Nomura, S., T. Nakagawa, O. Fukunaga, and S. Saito: *J. Phys. Soc. Japan* **24**, 957 (1968).
21. Okada, K.: *Phys. Rev. Letters* **15**, 252 (1965).
22. Pepinsky, R., and F. Jona: *Phys. Rev.* **105**, 344 (1957).
23. Sawada, S., and G. C. Danielson: *Phys. Rev.* **113**, 1005, 1008 (1952).
24. Shirane, G., and R. Pepinsky: *Phys. Rev.* **91**, 812 (1953).
25. Shirane, G., E. Sawaguchi, and Y. Takagi: *J. Phys. Soc. Japan* **6**, 208 (1951).
26. Shirane, G., E. Sawaguchi, and Y. Takagi: *Phys. Rev.* **84**, 476 (1951).
27. Smolenskii, G. A., A. E. Agranovskaya, and V. A. Isupov: *Soviet Phys.—Solid State* **1**, 907 (1959).
28. Tomashpolskii, Y. Y., and Y. N. Venevtsev: *Soviet Phys.—Solid State* **7**, 412 (1965).
29. Venetsev, N., A. G. Kapyshev, A. S. Viscov, V. M. Lebedev, V. M. Petrov, and G. S. Zhdanov: *Proc. Intern. Meeting Ferroelectricity*, Prague, **1**, 261 (1966).
30. Wood, E. A., W. J. Merz, and B. T. Matthias: *Phys. Rev.* **87**, 544 (1952).

9g. Properties of Superconductors

B. W. ROBERTS

General Electric Research and Development Center

9g-1. Introduction. The historically first observed and most distinctive property of a superconductive body is the near total loss of resistance at a critical temperature T_c characteristic of each material. Figure 9g-1a illustrates schematically two types of possible transitions. The sharp vertical discontinuity in resistance is indicative of that found for a single crystal of a very pure element or one of a few well-annealed alloy compositions. The broad transition, illustrated by broken lines, suggests the transition shape seen for materials that are inhomogeneous and contain unusual strain distributions. Careful testing of the resistivity limits for superconductors shows that it is less than 4×10^{-23} ohm-cm while the lowest resistivity observed in metals is of the order of 10^{-13} ohm-cm. If one compares the resistivity of a superconductive body to that of copper at room temperature, the superconductive body is at least 10^{17} times less resistive.

The temperature interval ΔT_c, over which the transition between the normal and superconductive states takes place, may be of the order of as little as 2×10^{-5} K *or* several kelvins in width, depending upon the material state. The narrow transition width was attained in 99.9999 percent pure gallium single crystals.

A *type I superconductor* below T_c, as exemplified by a pure metal, exhibits perfect diamagnetism and excludes a magnetic field up to some critical field H_c, whereupon it reverts to the normal state as shown in the H-T diagram of Fig. 9g-1b.

The difference in entropy near absolute zero between the superconductive and normal states relates directly to the electronic specific heat γ:

$$(S_s - S_n)_{T \to 0} = -\gamma T$$

High-field Superconductivity.   The discovery of the large current-carrying capability of Nb_3Sn and other similar alloys has led to an extensive study of the physical properties of these alloys.   In brief, a high-field superconductor, or *type II superconductor*, passes from the perfect diamagnetic state at low magnetic fields to a mixed state and finally to a sheathed state before attaining the normal resistive state of the metal. The magnetization of a typical high-field superconductor is shown in Fig. 9g-1c.   The magnetic field values separating the four stages are given as H_{c1}, H_{c2}, and H_{c3}.   The superconductive state below H_{c1} is perfectly diamagnetic, identical to the state of most pure metals of the "soft" or type I.   Between H_{c1} and H_{c2} a "mixed superconductive

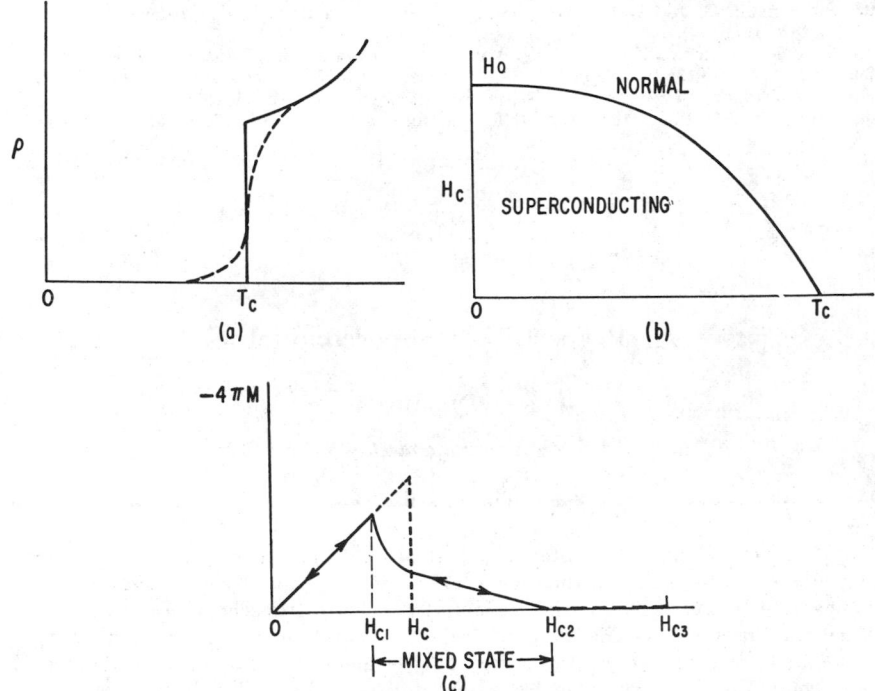

FIG. 9g-1. Physical properties of superconductors.   (*a*) Resistivity versus temperature for a pure and perfect lattice (solid line).   Impure and/or imperfect lattice (dashed line). (*b*) Magnetic field temperature dependence for type I or "soft" superconductors.   (*c*) Schematic magnetization curve for "hard" or type II superconductors.

state" is found in which fluxons (a minimal unit of magnetic flux) create lines of normal superconductor in a superconductive matrix.   The volume of the normal state is proportional to $-4\pi M$ in the "mixed state" region.   Thus at H_{c2} the fluxon density has become so great as to drive the interior volume of the superconductive body completely normal.   Between H_{c2} and H_{c3} the superconductor has a sheath of current-carrying superconductive material at the body surface, and above H_{c3} the normal state exists along with "fluctuations."   With several types of careful measurement, it is possible to determine H_{c1}, H_{c2} and H_{c3}.   Table 9g-3 contains some of the available data on high-field superconductive materials.

High-field superconductive phenomena are also related to specimen dimension and configuration.   For instance, the *type I superconductor*, Hg, has entirely different magnetization behavior in high magnetic fields when contained in the very fine set of filamentary tunnels found in an unprocessed Vycor glass. The great majority of superconductive materials are type II. The elements in very pure form and a very few precisely stoichiometric and well-annealed compounds are type I with the possible exceptions of vanadium and niobium.

Metallurgical Aspects. The sensitivity of superconductive properties to the material state is most pronounced and has been used in a reverse sense to study and specify the detailed state of alloys. The mechanical state, the homogeneity, and the presence of impurity atoms and other electron-scattering centers are all capable of controlling the critical temperature and the current-carrying capabilities in high magnetic fields. Well-annealed specimens tend to show sharper transitions than those that are strained or inhomogeneous. This sensitivity to mechanical state underlines a general problem in the tabulation of properties for superconductive materials. The occasional divergent values of the critical temperature and of the critical fields quoted for a type II superconductor may lie in the variation in sample preparation. Critical temperature of materials studied early in the history of superconductivity must be evaluated in light of the probable metallurgical state of the material as well as the availability of less pure starting elements. It has been noted that recent work has given extended consideration to the metallurgical aspects of sample preparation.

9g-2. Notes Concerning Data. Table 9g-1 lists the elements and some of their superconductive properties. The data have been selected generally from recent studies in which sample purity and perfection appear to have been considered.

Table 9g-2 is a general and selected listing of superconductive materials. All compositions are denoted on an atomic basis, i.e., AB, AB_2 or AB_3 for compounds, unless otherwise noted. Solid solutions or odd compositions may be denoted as A_xB_{1-x}, or A_xB. A series of three or more alloys is indicated as A_xB_{1-x} or by actual indication of the atomic fraction range such as $A_{0-0.6}B_{1-0.4}$. The critical temperature of such a series of alloys is denoted by a range of values or possibly the maximum value.

The selection of the critical temperature from a transition observed in the effective permeability or the change in resistance, or possibly the incremental changes in frequency observed by certain techniques, is not often obvious from the literature. Most authors choose the midpoint of such curves as the probable critical temperature of the idealized material, and others will choose the highest temperature at which a deviation from the normal state property is observed. And in view of the previous discussions concerning the variability of the superconductive properties as a function of purity and other metallurgical aspects, it is recommended that appropriate literature be checked to determine the most probable critical temperature or critical field of a given alloy.

Table 9g-3 lists high magnetic field superconductors.

References to the data presented in this section, plus additional entries on superconductive materials as well as those materials specifically tested and found nonsuperconductive to some low temperature, may be found in the following publications by B. W. Roberts:

Superconductive Materials and Some of Their Properties, in "Progress in Cryogenics," vol. IV, pp. 160–231, Heywood and Co., London, 1964.

Superconductive Materials and Some of Their Properties, *NBS Tech. Note* 408, 1966.

or a successor report in preparation to the one above. (NBS Tech. Note 482, 1969).

TABLE 9g-1. PROPERTIES OF SUPERCONDUCTIVE ELEMENTS

Element	T_c, K		H_0, oersteds		θ_D*, K	γ, mJ mole^{-1} K^{-2}[†]
	Calorimetric	Magnetic	Calorimetric	Magnetic		
Al...........	1.183	1.196	104	99	420	1.36
Cd...........	0.518	0.56	29.6	30	209	0.688
Ga...........	1.087	1.091	59	51	317, 324.7	0.601, 0.596
Ga (β)......	.....	6.2				
Ga (γ).......	.....	7.62				
Hg (α).......	4.16	4.154	380	410.9	87, 71.9	1.81
Hg (β).......	.....	3.949	....	339	93	1.37
In...........	3.407	3.4035	282.7	293	109	1.66
Ir............	.....	0.14	....	19	420	3.2
La (α).......	4.80	4.9	....	....	142	10.0
La (β).......	5.91	6.06	....	1,600	132	6.7
Mo...........	0.917	0.92	95	98	460	1.83
Nb...........	9.17	9.26	1,944	1,980	277	7.79
Os...........	.....	0.655	....	65	500	2.35
Pa...........	.....	1.4				
Pb...........	7.23	7.193	....	803	96.3	3.0
Re...........	1.699	1.698	188	198	415	2.35
Ru...........	.....	0.49	....	66	550	3.0
Sb	.....	2.6–2.7				
Sn...........	3.722	3.722	303	305.50	195	1.74
Ta...........	4.39	4.483	780	830	258	6.0
Tc...........	.....	8.22, 7.92				
Th...........	.....	1.368	131	162	168	4.65
Ti...........	0.42	0.39	56	100	425	3.32
Tl...........	2.38	2.39	176.5	171	78.5	1.47
U (α)........	.....	0.68, 0.23	....	....	206	12.2
U (pseudo-γ) .	.....	1.80 (extrapolated value)				
V............	5.37	5.30	1,310	1,020	399	9.8
W...........	.....	0.012	....	1.07	550	3.0
Zn...........	0.852	0.875	51.8	53	309	0.66
Zr............	.....	0.546	....	47	290	2.78
Zr (ω)........	.....	0.65				

THIN FILMS FORMED AT VARIOUS TEMPERATURES

Element	T_c, K (magnetic)	H_0, oersteds (magnetic)	Element	T_c, K (magnetic)
Al...............	1.3–3.7		Mo..............	~5
Be...............	~6, ~8.4	$H_{c2} \gg 11,000$	Nb..............	6.5–9.4
Bi...............	~6.0		Re..............	1.9–~7
Ga...............	8.4, 7.2		Sn..............	4.6–4.7, 4.1
In...............	3.94–4.25, 3.7		Ti..............	1.3 max.
La...............	5.00–6.74		W..............	1.7–4.1

THIN FILMS FORMED UNDER HIGH PRESSURE

Element	T_c, K (magnetic)	Pressure	Element	T_c, K (magnetic)	Pressure
Bi II........	3.916	25,000 atm	Se II.....	6.75, 6.95	~130 kb
	3.90	25,200 atm			
	3.86	26,800 atm	Si........	7.1	120–130 kb
Bi III.......	7.25	27,000–28,400 atm	Te........	~3.3	~56,000 atm
Ce...........	1.7	50 kb	Tl (f.c.c.) .	1.45	35 kb
Ge...........	4.85–5.4	~120 kb	Tl (h.c.p.)	1.95	35 kb

* For another data set see K. Mendelssohn, "Cryophysics," p. 178, Interscience Publishers, Inc., New York, 1960.

[†] D. H. Parkinson, *Rept. Progr. Phys.* **21**, 226 (1958). See also F. Heiniger, E. Bucher, and J. Müller, Low Temperature Specific Heat of Transition Metals and Alloys, *Phys. Kondens. Materie* **5**, 243–284 (1966).

TABLE 9g-2. SELECTED SUPERCONDUCTIVE COMPOUNDS AND
ALLOYS WITH CRITICAL TEMPERATURES, CRITICAL FIELDS,
AND CRYSTAL STRUCTURE TYPE*

Substance	T_c, K	H_0, oersteds	Crystal structure type†
$Ag_xAl_yZn_{1-x-y}$	0.5–0.845		
$Ag_7BF_4O_8$	0.15	....	Cubic
$AgBi_2$	3.0–2.78		
$Ag_7F_{0.25}N_{0.75}O_{10.25}$	0.85–0.90		
Ag_7FO_8	0.3	....	Cubic
Ag_2F	0.066	2.5	
$Ag_{0.8-0.3}Ga_{0.2-0.7}$	6.5–8		
Ag_4Ge	0.85	....	Hex., c.p.
$Ag_{0.438}Hg_{0.562}$	0.64	....	$D8_2$
$AgIn_2$	~2.4	....	$C16$
$Ag_{0.1}In_{0.9}Te$ $(n = 1.40 \times 10^{22})$‡	1.20–1.89	....	$B1$
$Ag_{0.2}In_{0.8}Te$ $(n = 1.07 \times 10^{22})$	0.77–1.00		$B1$
AgLa (9.5 kb)	1.2	....	$B2$
Ag_7NO_{11}	1.04	57	Cubic
Ag_xPb_{1-x}	7.2 max.		
Ag_xSn_{1-x} (film)	2.0–3.8		
Ag_xSn_{1-x}	1.5–3.7		
$AgTe_3$	2.6	....	Cubic
$AgTh_2$	2.26	....	$C16$
$Ag_{0.03}Tl_{0.97}$	2.67		
$Ag_{0.94}Tl_{0.06}$	2.32		
Ag_xZn_{1-x}	0.5–0.845		
Al (film)	1.3–2.31		
Al (1 to 21 katm)	1.170–0.687	....	$A1$
$AlAu_4$	0.4–0.7	....	Like $A13$
Al_2CMo_3	10.0	....	$A13$
Al_2CMo_3	9.8–10.2	1700	$A13$ + trace 2d phase
Al_2CaSi	5.8		
$Al_{0.131}Cr_{0.088}V_{0.781}$	1.46	....	Cubic
$AlGe_2$	1.75		
$Al_{0.5}Ge_{0.5}Nb$	12.6	....	$A15$
$Al_{\sim0.8}Ge_{\sim0.2}Nb_3$	20.7	....	$A15$
$AlLa_3$	5.57	....	$D0_{19}$
Al_2La	3.23	....	$C15$
Al_3Mg_2	0.84	....	Cubic, f.c.
$AlMo_3$	0.58	....	$A15$
$AlMo_6Pd$	2.1		
AlN	1.55	....	$B4$
Al_2NNb_3	1.3	....	$A13$
$AlNb_3$	18.0	....	$A15$
Al_xNb_{1-x}	<4.2–13.5	....	$D8_b$
Al_xNb_{1-x}	12–17.5	....	$A15$
$Al_{0.27}Nb_{0.73-0.48}V_{0-0.25}$	14.5–17.5	....	$A15$
$AlNb_xV_{1-x}$	<4.2–13.5		
AlOs	0.39	....	$B2$
Al_3O_s	5.90		
AlPb (films)	1.2–7		
Al_2Pt	0.48–0.55	....	$C1$
Al_3Re_{24}	3.35	....	$A12$
Al_3Th	0.75	....	$D0_{19}$

* See Sec. 9g-2, Notes concerning Data.
† See Key Table, p. 9-145.
‡ n = number of normal carriers per cubic centimeter for semiconductor superconductors.

TABLE 9g-2. SELECTED SUPERCONDUCTIVE COMPOUNDS AND
ALLOYS WITH CRITICAL TEMPERATURES, CRITICAL FIELDS,
AND CRYSTAL STRUCTURE TYPE* (*Continued*)

Substance	T_c, K	H_0, oersteds	Crystal structure type†
$Al_xTi_yV_{1-x-y}$	2.05–3.62	....	Cubic
$Al_{0.108}V_{0.892}$	1.82	....	Cubic
Al_xZn_{1-x}	0.5–0.845		
$AlZr_3$	0.73	....	$L1_2$
$AsBiPb$	9.0		
$AsBiPbSb$	9.0		
$As_{0.33}InTe_{0.67}$ ($n = 1.24 \times 10^{22}$)	0.85–1.15	....	$B1$
$As_{0.5}InTe_{0.5}$ ($n = 0.97 \times 10^{22}$)	0.44–0.62	....	$B1$
$As_{0.50}Ni_{0.06}Pd_{0.44}$	1.39		
$AsPb$	8.4	....	$C2$
$AsPd_2$ (low-temperature phase)	0.60	....	Hexagonal
$AsPd_2$ (high-temperature phase)	1.70	....	$C22$
$AsPd_5$	0.46	....	Complex
$AsRh$	0.58	....	$B31$
$AsRh_{1.4-1.6}$	<0.03–0.56	....	Hexagonal
$AsSn$	4.10		
$AsSn$ ($n = 2.14 \times 10^{22}$)	3.41–3.65	....	$B1$
$As_{\sim2}Sn_{\sim3}$	3.5–3.6, 1.21–1.17		
As_3Sn_4 ($n = 0.56 \times 10^{22}$)	1.16–1.19	....	Rhombohedral
Au_5Ba	0.4–0.7	....	$D2_d$
$AuBe$	2.64	....	$B20$
Au_2Bi	1.80	....	$C15$
Au_5Ca	0.34–0.38	....	$C15_b$
$AuGa$	1.2	....	$B31$
$Au_{0.40-0.92}Ge_{0.60-0.08}$	<0.32–1.63	....	Complex
$AuIn$	0.4–0.6	....	Complex
$AuLu$	<0.35	....	$B2$
$AuNb_3$	11.5	....	$A15$
$AuNb_3$	1.2	....	$A2$
$Au_{0-0.3}Nb_{1-0.7}$	1.1–11.0		
$Au_{0.02-0.98}Nb_3Rh_{0.98-0.02}$	2.53–10.9	....	$A15$
$AuNb_{3(1-x)}V_{3x}$	1.5–11.0	....	$A15$
$AuPb_2$	3.15		
$AuPb_2$ (film)	4.3		
$AuPb_3$	4.40		
$AuPb_3$ (film)	4.25		
Au_2Pb	1.18, 6–7	....	$C15$
$AuSb_2$	0.58	....	$C2$
$AuSn$	1.25	....	$B8_1$
Au_xSn_{1-x} (film)	2.0–3.8		
Au_5Sn	0.7–1.1	....	$A3$
Au_3Te_5	1.62	....	Cubic
$AuTh_2$	3.08	....	$C16$
$AuTl$	1.92		
AuV_3	0.74	....	$A15$
Au_xZn_{1-x}	0.50–0.845		
$AuZn_3$	1.21	....	Cubic
Au_xZr_y	1.7–2.8	....	$A3$
$AuZr_3$	0.92	....	$A15$
$BCMo_2$	5.4	....	Orthorhombic
$B_{0.03}C_{0.51}Mo_{0.47}$	12.5		
$BCMo_2$	5.3–7.0	....	Orthorhombic
BHf	3.1	....	Cubic
B_6La	5.7		

TABLE 9g-2. SELECTED SUPERCONDUCTIVE COMPOUNDS AND
ALLOYS WITH CRITICAL TEMPERATURES, CRITICAL FIELDS,
AND CRYSTAL STRUCTURE TYPE* (*Continued*)

Substance	T_c, K	H_0, oersteds	Crystal structure type†
$B_{12}Lu$	0.48		
BMo	0.5 (extrapolated)		
BMo_2	4.74	....	$C16$
BNb	8.25	....	B_f
BRe_2	2.80, 4.6		
$B_{0.3}Ru_{0.7}$	2.58	....	$D10_2$
$B_{12}Sc$	0.39		
BTa	4.0	....	B_f
B_6Th	0.74		
BW_2	3.1	....	$C16$
B_6Y	6.5–7.1		
$B_{12}Y$	4.7		
BZr	3.4	....	Cubic
$B_{12}Zr$	5.82		
$BaBi_3$	5.69	740	Tetragonal
$Ba_xO_3Sr_{1-x}Ti$ ($n = 4.2 - 11 \times 10^{19}$)	<0.1–0.55		
$Ba_{0.13}O_3W$	1.9	....	Tetragonal
$Ba_{0.14}O_3W$	<1.25–2.2	....	Hexagonal
$BaRh_2$	6.0	....	$C15$
$Be_{22}Mo$	2.51	....	Cubic, like $Be_{22}Re$
$Be_8Nb_5Zr_2$	5.2		
$Be_{0.98-0.92}Re_{0.02-0.08}$ (quenched)	9.5–9.75	....	Cubic
$Be_{0.957}Re_{0.043}$	9.62	....	Cubic, like $Be_{22}Re$
$BeTc$	5.21	....	Cubic
$Be_{22}W$	4.12	....	Cubic, like $Be_{22}Re$
$Be_{13}W$	4.1	....	Tetragonal
Bi_3Ca	2.0		
$Bi_{0.5}Cd_{0.13}Pb_{0.25}Sn_{0.12}$ (weight fractions)	8.2		
$BiCo$	0.42–0.49		
Bi_2Cs	4.75	....	$C15$
Bi_xCu_{1-x} (electrodeposited)	2.2		
$BiCu$	1.33–1.40		
$Bi_{0.019}In_{0.981}$	3.86		
$Bi_{0.05}In_{0.95}$	4.65	....	α phase
$Bi_{0.10}In_{0.90}$	5.05	....	α phase
$Bi_{0.15-0.30}In_{0.85-0.70}$	5.3–5.4	....	α and β phases
$Bi_{0.34-0.48}In_{0.66-0.52}$	4.0–4.1		
Bi_3In_5	4.1		
$BiIn_2$	5.65		β phase
Bi_2Ir	1.7–2.3		
Bi_2Ir (quenched)	3.0–3.96		
BiK	3.6		
Bi_2K	3.58	....	$C15$
$BiLi$	2.47	....	$L1_0$, α phase
$Bi_{4-9}Mg$	0.7–~1.0		
Bi_3Mo	3–3.7		
$BiNa$	2.25	....	$L1_0$
$BiNb_3$ (high pressure and temperature)	3.05	....	$A15$
$BiNi$	4.25	....	$B8_1$
Bi_3Ni	4.06	....	Orthorhombic
$Bi_{1-0}Pb_{0-1}$	7.26–9.14		
$Bi_{1-0}Pb_{0-1}$ (film)	7.25–8.67		

TABLE 9g-2.   SELECTED SUPERCONDUCTIVE COMPOUNDS AND
ALLOYS WITH CRITICAL TEMPERATURES, CRITICAL FIELDS,
AND CRYSTAL STRUCTURE TYPE* (Continued)

Substance	T_c, K	H_0, oersteds	Crystal structure type†
$Bi_{0.05-0.40}Pb_{0.95-0.60}$	7.35–8.4	....	Hexagonal, c.p. to ϵ phase
$BiPbSb$	8.9		
$Bi_{0.5}Pb_{0.31}Sn_{0.19}$ (weight fractions)	8.5		
$Bi_{0.5}Pb_{0.25}Sn_{0.25}$	8.5		
$BiPd_2$	4.0		
$Bi_{0.4}Pd_{0.6}$	3.7–4	....	Hexagonal, ordered
$PiPd$	3.7	....	Orthorhombic
Bi_2Pd	1.70	....	Monoclinic, α phase
Bi_2Pd	4.25	....	Tetragonal, β phase
$BiPdSe$	1.0	....	$C2$
$BiPdTe$	1.2	....	$C2$
$BiPt$	1.21	....	$B8_1$
$BiPtSe$	1.45	....	$C2$
$BiPtTe$	1.15	....	$C2$
Bi_2Pt	0.155	10	Hexagonal
Bi_2Rb	4.25	....	$C15$
$BiRe_2$	1.9–2.2		
$BiRh$	2.06	....	$B8_1$
Bi_3Rh	3.2	....	Orthorhombic, like NiB_3
Bi_4Rh	2.7	....	Hexagonal
Bi_3Sn	3.6–3.8		
$BiSn$	3.8		
Bi_xSn_y	3.85–4.18		
Bi_3Sr	5.62	530	$L1_2$
Bi_3Te	0.75–1.0		
Bi_5Tl_3	6.4	>400	
$Bi_{0.26}Tl_{0.74}$	4.4	....	Cubic, disordered
$Bi_{0.26}Tl_{0.74}$	4.15	....	$L1_2$, ordered?
Bi_2Y_3	2.25		
Bi_3Zn	0.8–0.9		
$Bi_{0.3}Zr_{0.7}$	1.51		
$BiZr_3$	2.4–2.8		
CCs_x	0.020–0.135	....	Hexagonal
C_8K (gold)	0.55		
$CGaMo_2$	3.7–4.1	....	Hexagonal, H phase
$CHf_{0.5}Mo_{0.5}$	3.4	....	$B1$
$CHf_{0.3}Mo_{0.7}$	5.5	....	$B1$
$CHf_{0.25}Mo_{0.75}$	6.6	....	$B1$
$CHf_{0.7}Nb_{0.3}$	6.1	....	$B1$
$CHf_{0.6}Nb_{0.4}$	4.5	....	$B1$
$CHf_{0.5}Nb_{0.5}$	4.8	....	$B1$
$CHf_{0.4}Nb_{0.6}$	5.6	....	$B1$
$CHf_{0.25}Nb_{0.75}$	7.0	....	$B1$
$CHf_{0.2}Nb_{0.8}$	7.8	....	$B1$
$CHf_{0.9-0.1}Ta_{0.1-0.9}$	5.0–9.0	....	$B1$
CK (excess K)	0.55	....	Hexagonal
C_8K	0.39	....	Hexagonal
$C_{0.40-0.44}Mo_{0.60-0.56}$	9–13		
CMo	6.5, 9.26		
CMo_2	12.2	....	Orthorhombic
$C_{0.44}Mo_{0.56}$	1.3	....	$B1$
$C_{0.5}Mo_xNb_{1-x}$	10.8–12.5	....	$B1$
$C_{0.6}Mo_{4.8}Si_3$	7.6	....	$D8_8$
$CMo_{0.2}Ta_{0.8}$	7.5	....	$B1$
$CMo_{0.5}Ta_{0.5}$	7.7	....	$B1$
$CMo_{0.75}Ta_{0.25}$	8.5	....	$B1$

TABLE 9g-2. SELECTED SUPERCONDUCTIVE COMPOUNDS AND
ALLOYS WITH CRITICAL TEMPERATURES, CRITICAL FIELDS,
AND CRYSTAL STRUCTURE TYPE* (*Continued*)

Substance	T_c, K	H_0, oersteds	Crystal structure type†
$CMo_{0.8}Ta_{0.2}$	8.7	....	$B1$
$CMo_{0.85}Ta_{0.15}$	8.9	....	$B1$
CMo_xTi_{1-x}	10.2 max.	....	$B1$
$CMo_{0.83}Ti_{0.17}$	10.2	....	$B1$
CMo_xV_{1-x}	2.9–9.3	....	$B1$
CMo_xZr_{1-x}	3.8–9.5	....	$B1$
$C_{0.1-0.9}N_{0.9-0.1}Nb$	8.5–17.9		
$C_{0-0.38}N_{1-0.62}Ta$	10.0–11.3		
CNb (whiskers)	7.5–10.5		
$C_{0.984}Nb$	9.8	....	$B1$
CNb (extrapolated)	~14		
$C_{0.7-1.0}Nb_{0.3-0}$	6–11	....	$B1$
CNb_2	9.1		
CNb_xTa_{1-x}	8.2–13.9		
CNb_xTi_{1-x}	<4.2–8.8	....	$B1$
$CNb_{0.6-0.9}W_{0.4-0.1}$	12.5–11.6	....	$B1$
$CNb_{0.1-0.9}Zr_{0.9-0.1}$	4.2–8.4		$B1$
CRb_x (gold)	0.023–0.151	....	Hexagonal
$CRe_{0.01-0.08}W$	1.3–5.0		
$CRe_{0.06}W$	5.0		
CTa	~11 (extrapolated)		
$C_{0.987}Ta$	9.7		
$C_{0.845-0.987}Ta$	2.04–9.7		
CTa (film)	5.09	....	$B1$
CTa_2	3.26	....	L_3'
$CTa_{0.4}Ti_{0.6}$	4.8	....	$B1$
$CTa_{1-0.4}W_{0-0.6}$	8.5–10.5	....	$B1$
$CTa_{0.2-0.9}Zr_{0.8-0.1}$	4.6–8.3	....	$B1$
CTc (excess C)	3.85	....	Cubic
$CTi_{0.5-0.7}W_{0.5-0.3}$	6.7–2.1	....	$B1$
CW	1.0		
CW_2	2.74	....	L_3'
CW_2	5.2	....	Cubic, f.c.
$CaIr_2$	6.15	....	$C15$
$Ca_xO_3Sr_{1-x}Ti$ $(n = 3.7 - 11.0 \times 10^{19})$	<0.1–0.55		
$Ca_{0.1}O_3W$	1.4–3.4	....	Hexagonal
$CaPb$	7.0		
$CaRh_2$	6.40	....	$C15$
$Cd_{0.3-0.5}Hg_{0.7-0.5}$	1.70–1.92		
$CdHg$	1.77, 2.15	....	Tetragonal
$Cd_{0.0075-0.05}In_{1-x}$	3.24–3.36	....	Tetragonal
$Cd_{0.97}Pb_{0.03}$	4.2		
$CdSn$	3.65	>266	
$Cd_{0.17}Tl_{0.83}$	2.3		
$Cd_{0.18}Tl_{0.82}$	2.54		
$CeCo_2$	0.84	....	$C15$
$CeCo_{1.67}Ni_{0.33}$	0.46	....	$C15$
$CeCo_{1.67}Rh_{0.33}$	0.47	....	$C15$
$Ce_xGd_{1-x}Ru_2$	3.2–5.2	....	$C15$
$CeIr_3$	3.34		
$CeIr_5$	1.82		
$Ce_{0.005}La_{0.995}$	4.6		
Ce_xLa_{1-x}	1.3–6.3		
$Ce_xPr_{1-x}Ru_2$	1.4–5.3	....	$C15$
Ce_xPt_{1-x}	0.7–1.55		
$CeRu_2$	6.0	....	$C15$
$Co_xFe_{1-x}Si_2$	1.4 max.	....	$C1$

TABLE 9g-2. SELECTED SUPERCONDUCTIVE COMPOUNDS AND
ALLOYS WITH CRITICAL TEMPERATURES, CRITICAL FIELDS,
AND CRYSTAL STRUCTURE TYPE* (*Continued*)

Substance	T_c, K	H_0, oersteds	Crystal structure type†
$CoHf_2$	0.56	....	$E9_3$
$CoLa_3$	4.28		
$CoLu_3$	~0.35		
$Co_{0-0.01}Mo_{0.8}Re_{0.2}$	2–10		
$Co_{0.02-0.10}Nb_3Rh_{0.98-0.90}$	2.28–1.90	....	$A15$
$Co_xNi_{1-x}Si_2$	1.4 max.		$C1$
$Co_{0.5}Rh_{0.5}Si_2$	2.5		
$Co_xRh_{1-x}Si_2$	3.65 max.		
$Co{\sim}_{0.3}Sc{\sim}_{0.7}$	~0.35		
$CoSi_2$	1.40, 1.22	105	$C1$
Co_3Th_7	1.83	....	$D10_2$
Co_xTi_{1-x}	2.8 max.	....	Co in α-Ti
Co_xTi_{1-x}	3.8 max.	....	Co in β-Ti
$CoTi_2$	3.44	....	$E9_3$
$CoTi$	0.71	....	$A2$
CoU	1.7	....	$B2$, distorted
CoU_6	2.29	....	$D2_c$
$Co_{0.28}Y_{0.72}$	0.34		
CoY_3	<0.34		
$CoZr_2$	6.3	....	$C16$
$Co_{0.1}Zr_{0.9}$	3.9	....	$A3$
$Cr_{0.6}Ir_{0.4}$	0.4	....	Hexagonal, c.p.
$Cr_{0.65}Ir_{0.35}$	0.59	....	Hexagonal, c.p.
$Cr_{0.7}Ir_{0.3}$	0.76	....	Hexagonal, c.p.
$Cr_{0.72}Ir_{0.28}$	0.83		
Cr_3Ir	0.45	....	$A15$
$Cr_{0-0.1}NB_{1-0.9}$	4.6–9.2	....	$A2$
$Cr_{0.80}Os_{0.20}$	2.5	....	Cubic
Cr_xRe_{1-x}	1.2–5.2		
$Cr_{0.40}Re_{0.60}$	2.15	....	$D8_b$
$Cr_{0.8-0.6}Rh_{0.2-0.4}$	0.5–1.10	....	$A3$
Cr_3Ru (annealed)	3.3	....	$A15$
Cr_2Ru	2.02	....	$D8_b$
$Cr_{0.1-0.5}Ru_{0.9-0.5}$	0.34–1.65	....	$A3$
Cr_xTi_{1-x}	3.6 max.	....	Cr in α-Ti
Cr_xTi_{1-x}	4.2 max.	....	Cr in β-Ti
$Cr_{0.1}Ti_{0.3}V_{0.6}$	5.6	1360	
$Cr_{0.0175}U_{0.9825}$	0.75	....	β phase
$Cs_{0.32}O_3W$	1.12	....	Hexagonal
$Cu_{0.15}In_{0.85}$ (film)	3.75		
$Cu_{0.04-0.08}In_{1-x}$	4.4		
$CuLa$	5.85		
Cu_xPb_{1-x}	5.7–7.7		
CuS	1.62	....	$B18$
CuS_2	1.48–1.53	....	$C18$
$CuSSe$	1.5–2.0	....	$C18$
$CuSe_2$	2.3–2.43	....	$C18$
$CuSeTe$	1.6–2.0	....	$C18$
Cu_xSn_{1-x}	3.2–3.7		
Cu_xSn_{1-x} (film) (made at 10K)	3.6–7		
Cu_xSn_{1-x} (film) (made at 300K)	2.8–3.7		
$CuTe_2$	<1.25–1.3	....	$C18$
$CuTh_2$	3.49	....	$C16$
$Cu_{0-0.027}V$	3.9–5.3	....	$A2$
Cu_xZn_{1-x}	0.5–0.845		
$D_{0.13}Nb$	9.12		

TABLE 9g-2. SELECTED SUPERCONDUCTIVE COMPOUNDS AND
ALLOYS WITH CRITICAL TEMPERATURES, CRITICAL FIELDS,
AND CRYSTAL STRUCTURE TYPE* (Continued)

Substance	T_c, K	H_0, oersteds	Crystal structure type†
Er_xLa_{1-x}	1.4–6.3		
$Fe_{0-0.04}Mo_{0.8}Re_{0.2}$	1–10		
$Fe_{0.05}Ni_{0.05}Zr_{0.90}$	~3.9		
Fe_3Th_7	1.86	....	$D10$
Fe_xTi_{1-x}	3.2 max.	....	Fe in α-Ti
Fe_xTi_{1-x}	3.7 max.	....	Fe in β-Ti
$Fe_xTi_{0.6}V_{1-x}$	6.8 max.		
FeU_6	3.86	....	$D2_c$
$Fe_{0.1}Zr_{0.9}$	1.0	....	$A3$
$Ga_{0.5}Ge_{0.5}Nb_3$	7.3	....	$A15$
$GaLa_3$	5.84		
Ga_2Mo	9.5		
$GaMo_3$	0.76	....	$A15$
Ga_4Mo	9.8		
GaN (black)	5.85	....	$B4$
$GaNb_3$	14.5	....	$A15$
$Ga_xNb_3Sn_{1-x}$	14–18.37	....	$A15$
$Ga_{0.7}Pt_{0.3}$	2.9	....	$C1$
$GaPt$	1.74	....	$B20$
$GaSb$ (120 kb, 77K, annealed)	4.24	....	$A5$
$Ga_{0-1}Sn_{1-0}$ (quenched)	3.47–4.18		
$Ga_{0-1}Sn_{1-0}$ (annealed)	2.6–3.85		
Ga_5V_2	3.55	....	Tetragonal, Mn_2Hg_5 type
GaV_3	16.8	....	$A15$
$GaV_{2.1-3.5}$	6.3–14.45	....	$A15$
$GaV_{4.5}$	9.15		
Ga_3Zr	1.38		
Gd_xLa_{1-x}	<1.0–5.5		
$Gd_xOs_2Y_{1-x}$	1.4–4.7		
$Gd_xRu_2Th_{1-x}$	3.6 max.	....	$C15$
$GeIr$	4.7	....	$B31$
Ge_2La	1.49, 2.2	....	Orthorhombic, distorted $ThSi_2$ type
$GeMo_3$	1.43	....	$A15$
$GeNb_2$	1.9		
$GeNb_3$ (quenched)	6–17	....	$A15$
$Ge_{0.29}Nb_{0.71}$	6	....	$A15$
$Ge_xNb_3Sn_{1-x}$	17.6–18.0	....	$A15$
$Ge_{0.5}Nb_3Sn_{0.5}$	11.3		
$GePt$	0.40	....	$B31$
Ge_3Rh_5	2.12	....	Orthorhombic related to $InNi_2$
Ge_2Sc	1.3		
Ge_3Te_4 ($n = 1.06 \times 10^{22}$)	1.55–1.80	....	Rhombohedral
Ge_xTe_{1-x} ($n = 8.5 - 64 \times 10^{20}$)	0.07–0.41	....	$B1$
GeV_3	6.01	....	$A15$
Ge_2Y	3.80	....	C_c
$Ge_{1.62}Y$	2.4		
$H_{0.33}Nb_{0.67}$	7.28	....	Cubic, b.c.
$H_{0.1}Nb_{0.9}$	7.38	....	Cubic, b.c.
$H_{0.05}Nb_{0.95}$	7.83	....	Cubic, b.c.
$H_{0.12}Ta_{0.88}$	2.81	....	Cubic, b.c.
$H_{0.08}Ta_{0.92}$	3.26	....	Cubic, b.c.
$H_{0.04}Ta_{0.96}$	3.62	....	Cubic, b.c.

TABLE 9g-2. SELECTED SUPERCONDUCTIVE COMPOUNDS AND
ALLOYS WITH CRITICAL TEMPERATURES, CRITICAL FIELDS,
AND CRYSTAL STRUCTURE TYPE* (*Continued*)

Substance	T_c, K	H_0, oersteds	Crystal structure type†
$HfN_{0.989}$	6.6	....	$B1$
$Hf_{0-0.5}Nb_{1-0.5}$	8.3–9.5	....	$A2$
$Hf_{0.75}Nb_{0.25}$	>4.2		
$HfOs_2$	2.69	....	$C14$
$HfRe_2$	4.80	....	$C14$
$Hf_{0.14}Re_{0.86}$	5.86	....	$A12$
$Hf_{0.99-0.96}Rh_{0.01-0.04}$	0.85–1.51		
$Hf_{0-0.55}Ta_{1-0.45}$	4.4–6.5	....	$A2$
HfV_2	8.9–9.6	....	$C15$
Hg_xIn_{1-x}	3.14–4.55		
$HgIn$	3.81		
Hg_2K	1.20	....	Orthorhombic
Hg_3K	3.18		
Hg_4K	3.27		
Hg_8K	3.42		
Hg_3Li	1.7	....	Hexagonal
Hg_2Na	1.62	....	Hexagonal
Hg_4Na	3.05		
Hg_xPb_{1-x}	4.14–7.26		
$HgSn$	4.2		
Hg_xTl_{1-x}	2.30–4.109		
Hg_5Tl_2	3.86		
Ho_xLa_{1-x}	1.3–6.3		
$InLa_3$	9.83, 10.4	....	$L1_2$
$InLa_3$ (0–35 kb)	9.75–10.55		
$In_{1-0.86}Mg_{0-0.14}$	3.395–3.363	272.4–259.2	
$InNb_3$			
(high pressure and temperature)	4–8, 9.2	....	$A15$
$In_{0-0.3}Nb_3Sn_{1-0.7}$	18.0–18.19	....	$A15$
$In_{0.5}Nb_3Zr_{0.5}$	6.4		
$In_{0.11}O_3W$	<1.25–2.8	....	Hexagonal
$In_{0.95-0.85}Pb_{0.05-0.15}$	3.6–5.05		
$In_{0.98-0.91}Pb_{0.02-0.09}$	3.45–4.2		
$InPb$	6.65		
$InPd$	0.7	....	$B2$
$InSb$			
(quenched from 170 kb into liquid N_2)	4.8	....	Like $A5$
$(InSb)_{0.95-0.10}Sn_{0.05-0.90}$ (various heat treatments)	3.8–5.1		
$(InSb)_{0-0.07}Sn_{1-0.93}$	3.67–3.74		
In_3Sn	~5.5		
In_xSn_{1-x}	3.4–7.3		
$In_{0.82-1}Te$ ($n = 0.83$–1.71×10^{22})	1.02–3.45	....	$B1$
$In_{1.000}Te_{1.002}$	3.5–3.7	....	$B1$
In_3Te_4 ($n = 0.47 \times 10^{22}$)	1.15–1.25	....	Rhombohedral
In_xTl_{1-x}	2.7–3.374	252–284	
$In_{0.8}Tl_{0.2}$	3.223	252	
$In_{0.62}Tl_{0.38}$	2.760		
$In_{0.78-0.69}Tl_{0.22-0.31}$	3.18–3.32	....	Tetragonal
$In_{0.69-0.62}Tl_{0.31-0.38}$	2.98–3.3	....	Cubic, f.c.
Ir_2La	0.48	....	$C15$
Ir_3La	2.32	....	$D10_2$
Ir_3La_7	2.24	....	$D10_2$

TABLE 9g-2. SELECTED SUPERCONDUCTIVE COMPOUNDS AND
ALLOYS WITH CRITICAL TEMPERATURES, CRITICAL FIELDS,
AND CRYSTAL STRUCTURE TYPE* (*Continued*)

Substance	T_c, K	H_0, oersteds	Crystal structure type†
Ir_5La	2.13		
Ir_2Lu	2.47	....	$C15$
Ir_3Lu	2.89	....	$C15$
$IrMo$	<1.0	....	$A3$
$IrMo_3$	8.8	....	$A15$
$IrMo_3$	6.8	....	$D8_b$
$IrNb_3$	1.9	....	$A15$
$Ir_{0.4}Nb_{0.6}$	9.8	....	$D8_b$
$Ir_{0.37}Nb_{0.63}$	2.32		$D8_b$
$IrNb$	7.9		$D8_b$
$Ir_{0.02}Nb_3Rh_{0.98}$	2.43		$A15$
$Ir_{0.05}Nb_3Rh_{0.95}$	2.38	....	$A15$
$Ir_{0.287}O_{0.14}Ti_{0.573}$	5.5	....	$E9_3$
$Ir_{0.265}O_{0.035}Ti_{0.65}$	2.30	....	$E9_3$
Ir_xOs_{1-x}	0.3–0.98 (max.)–0.6		
$IrOsY$	2.6	....	$C15$
$Ir_{1.5}Os_{0.5}$	2.4	....	$C14$
Ir_2Sc	2.07	....	$C15$
$Ir_{2.5}Sc$	2.46	....	$C15$
$IrSn_2$	0.65–0.78	....	$C1$
Ir_2Sr	5.70	....	$C15$
$Ir_{0.5}Te_{0.5}$	~3		
$IrTe_3$	1.18	....	$C2$
$IrTh$	<0.37	....	B_f
Ir_2Th	6.50	....	$C15$
Ir_3Th	4.71		
Ir_3Th_7	1.52	....	$D10_2$
Ir_5Th	3.93	....	$D2_d$
$IrTi_3$	5.40	....	$A15$
IrV_2	1.39	....	$A15$
IrW_3	3.82		
$Ir_{0.28}W_{0.72}$	4.49		
Ir_2Y	2.18, 1.38	....	$C15$
$Ir_{0.69}Y_{0.31}$	1.98, 1.44	....	$C15$
$Ir_{0.70}Y_{0.30}$	2.16	....	$C15$
Ir_2Y	1.09	....	$C15$
Ir_2Y_3	1.61		
Ir_xY_{1-x}	0.3–3.7		
Ir_2Zr	4.10	....	$C15$
$Ir_{0.1}Zr_{0.9}$	5.5	....	$A3$
$K_{0.27-0.31}O_3W$	0.50	....	Hexagonal
$K_{0.40-0.57}O_3W$	1.5	....	Tetragonal
$La_{0.55}Lu_{0.45}$	2.2	....	Hexagonal, La type
$La_{0.8}Lu_{0.2}$	3.4	....	Hexagonal, La Type
$LaMg_2$	1.05	....	$C15$
LaN	1.35		
La_xNd_{1-x}	1.4–6.3		
$LaOs_2$	6.5	....	$C15$
$LaPt_2$	0.46	....	$C15$
$La_{0.28}Pt_{0.72}$	0.54	....	$C15$
$LaRh_3$	2.60		
$LaRh_5$	1.62		
La_7Rh_3	2.58	....	$D10_2$
$LaRu_2$	1.63	....	$C15$
La_3S_4	6.5	....	$D7_3$
La_3Se_4	8.6	....	$D7_3$
$LaSi_2$	2.3	....	C_c

TABLE 9g-2. SELECTED SUPERCONDUCTIVE COMPOUNDS AND
ALLOYS WITH CRITICAL TEMPERATURES, CRITICAL FIELDS,
AND CRYSTAL STRUCTURE TYPE* (Continued)

Substance	T_c, K	H_0, oersteds	Crystal structure type†
La_xY_{1-x}	1.7–5.4		
LaZn	1.04	....	$B2$
LiPb	7.2		
$LuOs_2$	3.49	....	$C14$
$Lu_{0.275}Rh_{0.725}$	1.27	....	$C15$
$LuRh_5$	0.49		
$LuRu_2$	0.86	....	$C14$
$Mg\sim_{0.47}Tl\sim_{0.53}$	2.75	220	$B2$
Mg_2Nb	5.6		
Mn_xTi_{1-x}	2.3 max	....	Mn in α-Ti
Mn_xTi_{1-x}	1.1–3.0	....	Mn in β-Ti
MnU_6	2.32	....	$D2_c$
MoN	12	....	Hexagonal
Mo_2N	5.0	....	Cubic, f.c.
Mo_xNb_{1-x}	0.016–9.2		
Mo_3Os	7.2	....	$A15$
$Mo_{0.62}Os_{0.38}$	5.65	....	$D8_b$
Mo_3P	5.31	....	DO_e
$Mo_{0.5}Pd_{0.5}$	3.52	....	$A3$
Mo_3Re	10.0		
Mo_xRe_{1-x}	1.2–12.2		
$MoRe_3$	9.25, 9.89	....	$A12$
$Mo_{0.42}Re_{0.58}$	6.35	....	$D8_b$
$Mo_{0.52}Re_{0.48}$	11.1		
$Mo_{0.57}Re_{0.43}$	14.0		
$Mo\sim_{0.60}Re_{0.395}$	10.6		
MoRh	1.97	....	$A3$
Mo_xRh_{1-x}	1.5–8.2	....	Cubic, b.c.
MoRu	9.5–10.5	....	$A3$
$Mo_{0.61}Ru_{0.39}$	7.18	....	$D8_b$
$Mo_{0.2}Ru_{0.8}$	1.66	....	$A3$
Mo_3Sb_4	2.1		
Mo_3Si	1.30	....	$A15$
$MoSi_{0.7}$	1.34		
Mo_xSiV_{3-x}	4.54–16.0	....	$A15$
Mo_xTc_{1-x}	10.8–15.8		
$Mo_{0.16}Ti_{0.84}$	4.18, 4.25	<985	
$Mo_{0.913}Ti_{0.087}$	2.95		
$Mo_{0.04}Ti_{0.96}$	2.0	....	Cubic
$Mo_{0.025}Ti_{0.975}$	1.8		
Mo_xU_{1-x}	0.7–2.1		
Mo_xV_{1-x}	0–$\sim$5.3		
Mo_2Zr	4.27–4.75	....	$C15$
NNb (whiskers)	10–14.5		
NNb (diffusion wires)	16.10		
NNb (film)	6–9	....	$B1$
$N_{0.988}Nb$	14.9	....	$B1$
$N_{0.824-0.988}Nb$	14.4–15.3	....	$B1$
$N_{0.70-0.795}Nb$	11.3–12.9	....	Cubic and tetragonal
NNb_xO_y	13.5–17.0	....	$B1$
NNb_xO_y	6.0–11		
$N_{100-42\ w/o}Nb_{0-58\ w/o}Ti$	15–16.8		
$N_{100-75\ w/o}Nb_{0-25\ w/o}Zr$	12.5–16.35		
NNb_xZr_{1-x}	9.8–13.8	....	$B1$
$N_{0.93}Nb_{0.85}Zr_{0.15}$	13.8	....	$B1$
$N_xO_yTi_z$	2.9–5.6	....	Cubic
$N_xO_yV_z$	5.8–8.2	....	Cubic

TABLE 9g-2. SELECTED SUPERCONDUCTIVE COMPOUNDS AND
ALLOYS WITH CRITICAL TEMPERATURES, CRITICAL FIELDS,
AND CRYSTAL STRUCTURE TYPE* (*Continued*)

Substance	T_c, K	H_0, oersteds	Crystal structure type †
$N_{0.34}Re$	4–5	....	Cubic, f.c.
NTa	12–14	....	$B1$
	–(extrapolated)		
NTa (film)	4.84	....	$B1$
$N_{0.6-0.987}Ti$	<1.17–5.8	....	$B1$
$N_{0.82-0.99}V$	2.9–7.9	....	$B1$
NZr	9.8	....	$B1$
$N_{0.906-0.984}Zr$	3.0–9.5	....	$B1$
$Na_{0.28-0.35}O_3W$	0.56	....	Tetragonal
$Na_{0.28}Pb_{0.72}$	7.2		
NbO	1.25		
$NbOs_2$	2.52	....	$A12$
Nb_3Os	1.05	....	$A15$
$Nb_{0.6}Os_{0.4}$	1.89, 1.78	....	$D8_b$
$Nb_3Os_{0.02-0.10}Rh_{0.98-0.90}$	2.42–2.30	....	$A15$
$Nb_{0.6}Pd_{0.4}$	1.60	....	$D8_f$ plus cubic
$Nb_3Pd_{0.02-0.10}Rh_{0.98-0.90}$	2.49–2.55	....	$A15$
$Nb_{0.62}Pt_{0.38}$	4.21	....	$D8_b$
Nb_3Pt	10.9	....	$A15$
Nb_5Pt_3	3.73	....	$D8_b$
$Nb_3Pt_{0.02-0.98}Rh_{0.98-0.02}$	2.52–9.6	....	$A15$
$Nb_{0.38-0.18}Re_{0.62-0.82}$	2.43–9.70	....	$A12$
Nb_3Rh	2.64	....	$A15$
$Nb_{0.60}Rh_{0.40}$	4.21	....	$D8_b$ plus other
$Nb_3Rh_{0.98-0.90}Ru_{0.02-0.10}$	2.42–2.44	....	$A15$
Nb_xRu_{1-x}	1.2–4.8	....	
NbS_2	6.1–6.3	....	Hexagonal, $NbSe_2$ type
NbS_2	5.0–5.5	....	Hexagonal/three-layer type
$Nb_3Sb_{0-0.7}Sn_{1-0.3}$	6.8–18	....	$A15$
$NbSe_2$	5.15–5.62	....	Hexagonal, NbS_2 type
$Nb_{1-1.05}Se_2$	2.2–7.0	....	Hexagonal, NbS_2 type
Nb_3Si	1.5	....	$L1_2$
Nb_3SiSnV_3	4.0		
Nb_3Sn	18.05		$A15$
$Nb_{0.8}Sn_{0.2}$	18.18, 18.5	....	$A15$
Nb_xSn_{1-x} (film)	2.6–18.5		
$NbSn_2$	2.60	620	Orthorhombic
Nb_3Sn_2	16.6		Tetragonal
$NbSnTa_2$	10.8	....	$A15$
Nb_2SnTa	16.4	....	$A15$
$Nb_{2.5}SnTa_{0.5}$	17.6	....	$A15$
$Nb_{2.75}SnTa_{0.25}$	17.8	....	$A15$
$Nb_{3x}SnTa_{3(1-x)}$	6.0–18.0		
$NbSnTaV$	6.2	....	$A15$
$Nb_2SnTa_{0.5}V_{0.5}$	12.2	....	$A15$
$NbSnV_2$	5.5	....	$A15$
Nb_2SnV	9.8	....	$A15$
$Nb_{2.5}SnV_{0.5}$	14.2	....	$A15$
Nb_xTa_{1-x}	4.4–9.2	....	$A2$
$NbTc_3$	10.5	....	$A12$
Nb_xTi_{1-x}	0.6–9.8		
$Nb_{0.6}Ti_{0.4}$	9.8		
Nb_xU_{1-x}	1.95 max.		
$Nb_{0.88}V_{0.12}$	5.7	....	$A2$
$Nb_{0.75}Zr_{0.25}$	10.8		

TABLE 9g-2. SELECTED SUPERCONDUCTIVE COMPOUNDS AND
ALLOYS WITH CRITICAL TEMPERATURES, CRITICAL FIELDS,
AND CRYSTAL STRUCTURE TYPE* (Continued)

Substance	T_c, K	H_0, oersteds	Crystal structure type †
$Nb_{0.66}Zr_{0.33}$	10.8		
$Ni_{0.3}Th_{0.7}$	1.98	....	$D10_2$
$NiZr_2$	1.52		
$Ni_{0.1}Zr_{0.9}$	1.5	....	$A3$
$O_4Rb_{0.27-0.29}W$	1.98	....	Hexagonal
O_3SrTi ($n = 1.7-12.0 \times 10^{19}$)	0.12–0.37		
O_3SrTi ($n = 10^{18}-10^{21}$)	0.05–0.47		
O_3SrTi ($n = \sim 10^{20}$)	0.47		
OTi	0.58		
$O_3Sr_{0.08}W$	2–4	....	Hexagonal
$O_3Tl_{0.30}W$	2.0–2.14	....	Hexagonal
OV_3Zr_3	7.5	....	$E9_3$
OW_3 (film)	3.35, 1.1	....	$A15$
$OsReY$	2.0	....	$C14$
Os_2Sc	4.6	....	$C14$
$OsTa$	1.95	....	$A12$
Os_3Th_7	1.51	....	$D10_2$
Os_xW_{1-x}	0.9–4.1		
OsW_3	~ 3		
Os_2Y	4.7	....	$C14$
Os_2Zr	3.0	....	$C14$
Os_xZr_{1-x}	1.50–5.6		
PPb	7.8		
$PPd_{3.0-3.2}$	$<0.35-0.7$	....	$D0_{11}$
P_3Pd_7 (high temperature)	1.0	....	Rhombohedral
P_3Pd_7 (low temperature)	0.70	....	Complex
PRh	1.22		
PRh_2	1.3	....	$C1$
PW_3	2.26	....	$D0_e$
Pb_2Pd	2.95	....	$C16$
Pb_4Pt	2.80	....	Related to $C16$
Pb_2Rh	2.66	....	$C16$
$PbSb$	6.6		
$PbTe$ (plus 0.1 w/o Pb)	5.19		
$PbTe$ (plus 0.1 w/o Tl)	5.24–5.27		
$PbTl_{0.27}$	6.43	756	
$PbTl_{0.17}$	6.73	796	
$PbTl_{0.12}$	6.88	849	
$PbTl_{0.075}$	6.98	880	
$PbTl_{0.04}$	7.06	864	
$Pb_{1-0.26}Tl_{0-0.74}$	7.20–3.68		
$PbTl_2$	3.75–4.1		
Pb_3Zr_5	4.60	....	$D8_8$
$PbZr_3$	0.76	....	$A15$
$Pd_{0.9}Pt_{0.1}Te_2$	1.65	....	$C6$
$Pd_{0.05}Ru_{0.05}Zr_{0.9}$	~ 9		
$Pd_{2.2}S$ (quenched)	1.63	....	Cubic
$PdSb_2$	1.25	....	$C2$
$PdSb$	1.50	....	$B8_1$
$PdSbSe$	1.0	....	$C2$
$PdSbTe$	1.2	....	$C2$
Pd_4Se	0.42	....	Tetragonal
$Pd_{6-7}Se$	0.66	....	Like Pd_4Te
$Pd_{2.8}Se$	2.3		
Pd_xSe_{1-x}	2.5 max.		
$PdSi$	0.93	....	$B31$
$PdSn$	0.41	....	$B31$

TABLE 9g-2. SELECTED SUPERCONDUCTIVE COMPOUNDS AND
ALLOYS WITH CRITICAL TEMPERATURES, CRITICAL FIELDS,
AND CRYSTAL STRUCTURE TYPE* (Continued)

Substance	T_c, K	H_0, oersteds	Crystal structure type †
$PdSn_2$	3.34		
Pd_2Sn	0.41	....	$C37$
Pd_3Sn_2	0.47–0.64	....	$B8_2$
$PdTe$	2.3, 3.85	....	$B8_1$
$PdTe_{1.02-1.08}$	2.56–1.88	....	$B8_1$
$PdTe_2$	1.69	....	$C6$
$PdTe_{2.1}$	1.89	....	$C6$
$PdTe_{2.3}$	1.85	....	$C6$
$Pd_{1.1}Te$	4.07	....	$B8_1$
$PdTh_2$	0.85	....	$C16$
$Pd_{0.1}Zr_{0.9}$	7.5	....	$A3$
$PtSb$	2.1	....	$B8_1$
$PtSi$	0.88	....	$B31$
$PtSn$	0.37	....	$B8_1$
$PtTe$	0.59	....	Orthorhombic
$PtTh$	0.44	....	B_f
Pt_3Th_7	0.98	....	$D10_2$
Pt_5Th	3.13		
$PtTi_3$	0.58	....	$A15$
$Pt_{0.02}U_{0.98}$	0.87	....	β phase
$PtV_{2.5}$	1.36	....	$A15$
PtV_3	2.87–3.20	....	$A15$
$PtV_{3.5}$	1.26	....	$A15$
$Pt_{0.5}W_{0.5}$	1.45	....	$A1$
Pt_xW_{1-x}	0.4–2.7		
Pt_2Y_3	0.90		
Pt_2Y	1.57, 1.70	....	$C15$
Pt_3Y_7	0.82	....	$D10_2$
$PtZr$	3.0	....	$A3$
$Re_{0.64}Ta_{0.36}$	1.46	....	$A12$
$Re_{24}Ti_5$	6.60	....	$A12$
Re_xTi_{1-x}	6.6 max.		
$Re_{0.76}V_{0.24}$	4.52	....	$D8_b$
$Re_{0.92}V_{0.08}$	6.8	....	$A3$
$Re_{0.6}W_{0.4}$	6.0		
$Re_{0.5}W_{0.5}$	5.12	....	$D8_b$
Re_2Y	1.83	....	$C14$
Re_2Zr	5.9	....	$C14$
Re_6Zr	7.40	....	$A12$
$Rh_{17}S_{15}$	5.8	....	Cubic
$Rh_{\sim0.24}Sc_{\sim0.76}$	0.88, 0.92		
Rh_xSe_{1-x}	6.0 max.		
Rh_2Sr	6.2	....	$C15$
$Rh_{0.4}Ta_{0.6}$	2.35	....	$D8_b$
$RhTe_2$	1.51	....	$C2$
$Rh_{0.67}Te_{0.33}$	0.49		
Rh_xTe_{1-x}	1.51 max.		
$RhTh$	0.36	....	B_f
Rh_3Th_7	2.15	....	$D10_2$
Rh_5Th	1.07		
Rh_xTi_{1-x}	2.25–3.95		
$Rh_{0.02}U_{0.98}$	0.96		
RhV_3	0.38	....	$A15$
RhW	~3.4	....	$A3$
RhY_3	0.65		
Rh_2Y_3	1.48		
Rh_3Y	1.07	....	$C15$
Rh_5Y	0.56		
$RhZr_2$	10.8	....	$C16$
$Rh_{0.005}Zr$ (annealed)	5.8		

TABLE 9g-2. SELECTED SUPERCONDUCTIVE COMPOUNDS AND
ALLOYS WITH CRITICAL TEMPERATURES, CRITICAL FIELDS,
AND CRYSTAL STRUCTURE TYPE* (Continued)

Substance	T_c, K	H_0, oersteds	Crystal structure type †
$Rh_{0-0.45}Zr_{1-0.55}$	2.1–10.8		
$Rh_{0.1}Zr_{0.9}$	9.0	....	Hexagonal, c.p.
Ru_2Sc	1.67	....	$C14$
Ru_2Th	3.56	....	$C15$
$RuTi$	1.07	....	$B2$
$Ru_{0.05}Ti_{0.95}$	2.5		
$Ru_{0.1}Ti_{0.9}$	3.5		
$Ru_xTi_{0.6}V_y$	6.6 max.		
$Ru_{0.45}V_{0.55}$	4.0	....	$B2$
RuW	7.5	....	$A3$
Ru_2Y	1.52	....	$C14$
Ru_2Zr	1.84	....	$C14$
$Ru_{0.1}Zr_{0.9}$	5.7	....	$A3$
$SbSn$	1.30–1.42, 1.42–2.37	....	$B1$ or distorted $B1$
$SbTi_3$	5.8	....	$A15$
Sb_2Tl_7	5.2		
$Sb_{0.01-0.03}V_{0.99-0.97}$	3.76–2.63	....	$A2$
SbV_3	0.80	....	$A15$
Si_2Th	3.2	....	C_c, α phase
Si_2Th	2.4	....	$C32$, β phase
SiV_3	17.1	....	$A15$
$Si_{0.9}V_3Al_{0.1}$	14.05	....	$A15$
$Si_{0.9}V_3B_{0.1}$	15.8	....	$A15$
$Si_{0.9}V_3C_{0.1}$	16.4	....	$A15$
$SiV_{2.7}Cr_{0.3}$	11.3	....	$A15$
$Si_{0.9}V_3Ge_{0.1}$	14.0	....	$A15$
$SiV_{2.7}Mo_{0.3}$	11.7	....	$A15$
$SiV_{2.7}Nb_{0.3}$	12.8	....	$A15$
$SiV_{2.7}Ru_{0.3}$	2.9	....	$A15$
$SiV_{2.7}Ti_{0.3}$	10.9	....	$A15$
$SiV_{2.7}Zr_{0.3}$	13.2	....	$A15$
Si_2W_3	2.8, 2.84		
$Sn_{0.174-0.104}Ta_{0.826-0.896}$	6.5–<4.2	....	$A15$
$SnTa_3$	8.35	....	$A15$, highly ordered
$SnTa_3$	6.2	....	$A15$, partially ordered
$SnTaV_2$	2.8	....	$A15$
$SnTa_2V$	3.7	....	$A15$
Sn_xTe_{1-x} ($n = 10.5$–20×10^{20})	0.07–0.22	....	$B1$
Sn_xTl_{1-x}	2.37–5.2		
SnV_3	3.8	....	$A15$
$Sn_{0.02-0.057}V_{0.98-0.943}$	2.87–~1.6	....	$A2$
$Ta_{0.025}Ti_{0.975}$	1.3	....	Hexagonal
$Ta_{0.05}Ti_{0.95}$	2.9	....	Hexagonal
$Ta_{0.05-0.75}V_{0.095-0.25}$	4.30–2.65	....	$A2$
$Ta_{0.8-1}W_{0.2-0}$	1.2–4.4	....	$A2$
$Tc_{0.1-0.4}W_{0.9-0.6}$	1.25–7.18	....	Cubic
$Tc_{0.50}W_{0.50}$	7.52	....	α plus σ
$Tc_{0.60}W_{0.40}$	7.88	....	σ plus α
Tc_6Zr	9.7	....	$A12$
$Th_{0-0.55}Y_{1-0.45}$	1.2–1.8		
$Ti_{0.70}V_{0.30}$	6.14	....	Cubic
Ti_xV_{1-x}	0.2–7.5		
$Ti_{0.5}Zr_{0.5}$ (annealed)	1.23		
$Ti_{0.5}Zr_{0.5}$ (quenched)	2.0		
V_2Zr	8.80	....	$C15$
$V_{0.26}Zr_{0.74}$	≈5.9		
W_2Zr	2.16	....	$C15$

Key to Crystal Structure Types Found in Table 9g-2

"Struck- turbericht" type*	Example	Class	"Struck- turbericht" type*	Example	Class
$A1$	Cu	Cubic f.c.	$C15$	Cu_2Mg	Cubic, f.c.
$A2$	W	Cubic, b.c.	$C15_b$	$AuBe_5$	Cubic
$A3$	Mg	Hexagonal, close packed	$C16$	$CuAl_2$	Tetragonal, b.c.
			$C18$	FeS_2	Orthorhombic
$A4$	Diamond	Cubic, f.c.	$C22$	Fe_2P	Trigonal
$A5$	White Sn	Tetragonal, b.c.	$C23$	$PbCl_2$	Orthorhombic
$A6$	In	Tetragonal, b.c. (f.c. cell usu- ally used)	$C32$	AlB_2	Hexagonal
			$C36$	$MgNi_2$	Hexagonal
			$C37$	Co_2Si	Orthorhombic
$A7$	As	Rhombohedral	$C49$	$ZrSi_2$	Orthorhombic
$A8$	Se	Trigonal	$C54$	$TiSi_2$	Orthorhombic
$A10$	Hg	Rhombohedral	C_c	Si_2Th	Tetragonal, b.c.
$A12$	α-Mn	Cubic, b.c.	$D0_3$	BiF_3	Cubic, f.c.
$A13$	β-Mn	Cubic	$D0_{11}$	Fe_3C	Orthorhombic
$A15$	"β-W," (WO_3)	Cubic	$D0_{18}$	Na_3As	Hexagonal
$B1$	NaCl	Cubic, f.c.	$D0_{19}$	Ni_3Sn	Hexagonal
$B2$	CsCl	Cubic	$D0_{20}$	$NiAl_3$	Orthorhombic
$B3$	ZnS	Cubic	$D0_{22}$	$TiAl_3$	Tetragonal
$B4$	ZnS	Hexagonal	$D0_e$	Ni_3P	Tetragonal, b.c.
$B8_1$	NiAs	Hexagonal	$D1_3$	Al_4Ba	Tetragonal, b.c.
$B8_2$	Ni_2In	Hexagonal	$D1_e$	$PtSn_4$	Orthorhombic
$B10$	PbO	Tetragonal	$D2_1$	CaB_6	Cubic
$B11$	γ-CuTi	Tetragonal	$D2_c$	MnU_6	Tetragonal, b.c.
$B17$	PtS	Tetragonal	$D2_d$	$CaZn_5$	Hexagonal
$B18$	CuS	Hexagonal	$D5_2$	La_2O_3	Trigonal
$B20$	FeSi	Cubic	$D5_8$	Sb_2S_3	Orthorhombic
$B27$	FeB	Orthorhombic	$D7_3$	Th_3P_4	Cubic, b.c.
$B31$	MnP	Orthorhombic	$D7_b$	Ta_3B_4	Orthorhombic
$B32$	NaTl	Cubic, f.c.	$D8_1$	Fe_3Zn_{10}	Cubic, b.c.
$B34$	PdS	Tetragonal	$D8_2$	Cu_5Zn_8	Cubic, b.c.
B_f	δ-CrB	Orthorhombic	$D8_3$	Cu_9Al_4	Cubic
B_g	MoB	Tetragonal, b.c.	$D8_8$	Mn_5Si_3	Hexagonal
B_h	WC	Hexagonal	$D8_b$	CrFe	Tetragonal
B_i	γ'-MoC	Hexagonal	$D8_i$	Mo_2B_5	Rhombohedral
$C1$	CaF_2	Cubic, f.c.	$D10_2$	Fe_3Th_7	Hexagonal
$C1_b$	MgAgAs	Cubic, f.c.	$E2_1$	$CaTiO_3$	Cubic
$C2$	FeS_2	Cubic	$E9_3$	Fe_3W_3C	Cubic, f.c.
$C6$	CdI_2	Trigonal	$L1_0$	CuAu	Tetragonal
$C11_b$	$MoSi_2$	Tetragonal, b.c.	$L1_2$	Cu_3Au	Cubic
$C12$	$CaSi_2$	Rhombohedral	$L2_b'$	ThH_2	Tetragonal, b.c.
$C14$	$MgZn_2$	Hexagonal	$L3'$	Fe_2N	Hexagonal

* See W. B. Pearson, "A Handbook of Lattice Spacings and Structures of Metals," vol. 1, p. 79; vol. 2, p. 3; Pergamon Press, New York, 1958, 1967.

TABLE 9g-3. HIGH CRITICAL MAGNETIC FIELD SUPERCONDUCTIVE COMPOUNDS
AND ALLOYS WITH CRITICAL TEMPERATURES, H_{c1}, H_{c2}, H_{c3}, AND
THE TEMPERATURE OF FIELD OBSERVATIONS, T_{obs}

Substance	T_c, K	H_{c1}, kg	H_{c2}, kg	H_{c3}, kg	T_{obs}, K*
Al_2CMo_3	9.8–10.2	0.091	156		1.2
$AlNb_3$		0.375		. . . . . .	
$Ba_xO_3Sr_{1-x}Ti$	<0.1–0.55	0.0039 max.			
$Bi_{0.5}Cd_{0.1}Pb_{0.27}Sn_{0.13}$			>24		3.06
Bi_xPb_{1-x}	7.35–8.4	0.122 max.	~30 max.		4.2
$Bi_{0.56}Pb_{0.44}$	8.8		15	. . . . . .	4.2
$Bi_{7.5w/o}Pb_{92.5w/o}$			2.32		
$Bi_{0.099}Pb_{0.901}$		0.29	2.8		
$Bi_{0.02}Pb_{0.98}$		0.46	0.73		
$Bi_{0.53}Pb_{0.32}Sn_{0.16}$		. . . .	>25		3.06
$Bi_{1-0.99}Sn_{0-0.07}$		. . . .	0–0.032		3.7
Bi_5Tl_3	6.4	. . . .	>5.56		3.35
C_8K (excess K)	0.55	. . . .	0.160 ($H \perp c$)	. . . . . .	0.32
C_8K	0.39	. . . .	0.730 ($H \parallel c$)	. . . . . .	0.32
			0.025 ($H \perp c$)	. . . . . .	0.32
			0.250 ($H \parallel c$)	. . . . . .	0.32
$C_{0.44}Mo_{0.56}$	12.5–13.5	0.087	98.5	. . . . . .	1.2
CNb	8–10	0.12	16.9	. . . . . .	4.2
$CNb_{0.4}Ta_{0.6}$	10–13.6	0.19	14.1	. . . . . .	1.2
CTa	9–11.4	0.22	4.6	. . . . . .	1.2
$Ca_xO_3Sr_{1-x}Ti$	<0.1–0.55	0.002–0.004		. . . . . .	
$Cd_{0.1}Hg_{0.9}$ (by weight)		0.23	0.34		2.04
$Cd_{0.05}Hg_{0.95}$		0.28	0.31	. . . . . .	2.16
$Cr_{0.10}Ti_{0.30}V_{0.60}$	5.6	0.071	84.4	. . . . . .	0
GaN	5.85	0.725	. . . .		4.2
Ga_xNb_{1-x}		. . . .	>28	. . . . . .	4.2
$GaSb$ (annealed)	4.24	. . . .	2.64	. . . . . .	3.5
$GaV_{1.95}$	5.3		73 §		
$GaV_{2.1-3.5}$	6.3–14.45		230–300‡		0
GaV_3		0.4	350 §	. . . . . .	0
			500‡	. . . . . .	
$GaV_{4.5}$	9.15	. . . .	121†		0
Hf_xNb_y		. . . .	>52–>102	. . . . . .	1.2
Hf_xTa_y		. . . .	>28–>86	. . . . . .	1.2
$Hg_{0.05}Pb_{0.95}$		0.235	2.3	. . . . . .	
$Hg_{0.101}Pb_{0.899}$		0.23	4.3		4.2
$Hg_{0.15}Pb_{0.85}$	~6.75	. . . .	>13		2.93
$In_{0.98}Pb_{0.02}$	3.45	0.1	. . . .	0.12	2.76
$In_{0.96}Pb_{0.04}$	3.68	0.1	0.12	0.25	2.94
$In_{0.94}Pb_{0.06}$	3.90	0.095	0.18	0.35	3.12
$In_{0.913}Pb_{0.087}$	4.2	~0.17	0.55	2.65	
$In_{0.316}Pb_{0.684}$		0.155	3.7	. . . . . .	4.2
$In_{0.17}Pb_{0.83}$		. . . .	2.8	5.5	4.2
$In_{1.000}Te_{1.002}$	3.5–3.7	. . . .	1.2†	. . . . . .	0
$In_{0.95}Tl_{0.05}$		0.263	0.263	. . . . . .	3.3
$In_{0.90}Tl_{0.10}$		0.257	0.257	. . . . . .	3.25
$In_{0.83}Tl_{0.17}$		0.242	0.39	. . . . . .	3.21
$In_{0.75}Tl_{0.25}$		0.216	0.50	. . . . . .	3.16
LaN	1.35	0.45	. . . .		0.76
La_3S_4	6.5	≈0.15	>25	. . . . . .	1.3
La_3Se_4	8.6	≈0.2	>25	. . . . . .	1.25
$Mo_{0.52}Re_{0.48}$	11.1	. . . .	14–21	22–33	4.2
			18–28	37–43	1.3
$Mo_{0.6\pm0.05}Re_{0.395}$	10.6	. . . .	14–20	20–37	4.2
			19–26	26–37	1.3
$Mo~_{0.5}Tc~_{0.5}$			~75†	. . . . . .	0
$Mo_{0.16}Ti_{0.84}$	4.18	0.028	98.7†	. . . . . .	0
			36–38	. . . . . .	3.0
$Mo_{0.913}Ti_{0.087}$	2.95	0.060	~15	. . . . . .	4.2
$Mo_{0.1-0.3}U_{0.9-0.7}$	1.85–2.06		>25	. . . . . .	
$Mo_{0.17}Zr_{0.83}$			~30		
$N_{(12.8w/o)}Nb$	15.2		>9.5		13.2
NNb (wires)	16.1		153†	. . . . . .	0
			132	. . . . . .	4.2
			95	. . . . . .	8
			53	. . . . . .	12
NNb_xO_{1-x}	13.5–17.0		~38		
NNb_xZr_{1-x}	9.8–13.8		4–>130		4.2
$N_{0.9}Nb_{0.85}Zr_{0.15}$	13.8		>130		4.2
$Na_{0.086}Pb_{0.914}$		0.19	6.0	. . . . . .	
$Na_{0.016}Pb_{0.984}$		0.28	2.05		
Nb	9.15		2.020		1.4
			1.710		4.2
Nb		0.4–1.1	3–5.5		4.2
Nb (unstrained)		1.1–1.8	3.40	6–9.1	4.2
Nb (strained)		1.25–1.92	3.44	6.0–8.7	4.2
Nb (cold-drawn wire)		2.48	4.10	≈10	4.2
Nb (film)		. . . .	>25	. . . . . .	4.2
$NbSc$		. . . .	>30		

TABLE 9g-3. HIGH CRITICAL MAGNETIC FIELD SUPERCONDUCTIVE COMPOUNDS
AND ALLOYS WITH CRITICAL TEMPERATURES, H_{c1}, H_{c2}, H_{c3}, AND
THE TEMPERATURE OF FIELD OBSERVATIONS, T_{obs} (Continued)

Substance	T_c, K	H_{c1}, kg	H_{c2}, kg	H_{c3}, kg	T_{obs}, K*
Nb_3Sn	....	0.170	221	......	4.2
			70	......	14.15
			54	......	15
			34	......	16
			17	......	17
$Nb_{0.1}Ta_{0.9}$	....	0.084	0.154	......	4.195
$Nb_{0.2}Ta_{0.8}$		....	10	......	4.2
$Nb_{0.65-0.73}Ta_{0.02-0.10}Zr_{0.25}$	....		>70->90	......	4.2
Nb_xTi_{1-x}		....	148 max.	......	1.2
			120 max.		4.2
$Nb_{0.222}U_{0.778}$		1.98	23	......	1.2
Nb_xZr_{1-x}			127 max.	......	1.2
			94 max.	......	4.2
O_3SrTi	0.43	0.0049†	0.504†	......	0
O_3SrTi	0.33	0.00195†	0.420†	......	0
$PbSb_{1w/o}$ (quenched)		....	>1.5	......	4.2
$PbSb_{1w/o}$ (annealed)		....	>0.7	......	4.2
$PbSb_{2.8w/o}$ (quenched)		....	>2.3	......	4.2
$PbSb_{2.8w/o}$ (annealed)		....	>0.7	......	4.2
$Pb_{0.871}Sn_{0.129}$		0.45	1.1		
$Pb_{0.965}Sn_{0.035}$		0.53	0.56		
$Pb_{1-0.26}Tl_{0-0.74}$	7.20-3.68	....	2-6.9†	......	0
$PbTl_{0.17}$	6.73	....	4.5†	......	0
$Re_{0.26}W_{0.74}$			>30		
$Sb_{0.93}Sn_{0.07}$			0.12	......	3.7
SiV_3	17.0	0.55	156§		
Sn_xTe_{1-x}		0.00043-0.00236	0.005-0.0775	......	0.012-0.079
Ta (99.95%)		0.425	1.850	......	1.3
		0.325	1.425	......	2.27
		0.275	1.175	......	2.66
		0.090	0.375	......	3.72
$Ta_{0.5}Nb_{0.5}$		....	3.55		4.2
$Ta_{0.65-0}Ti_{0.35-1}$	4.4-7.8	....	>14-138		1.2
$Ta_{0.5}Ti_{0.5}$		....	138	......	1.2
Te	~3.3	0.25†			0
Tc_xW_{1-x}	5.75-7.88		8-44		4.2
Ti		....		2.7	4.2
$Ti_{0.75}V_{0.25}$	5.3	0.029†	199†	......	0
$Ti_{0.775}V_{0.225}$	4.7	0.024†	172†	......	0
$Ti_{0.615}V_{0.385}$	7.07	0.050	~34	......	4.2
$Ti_{0.516}V_{0.484}$	7.20	0.062	~28	......	4.2
$Ti_{0.415}V_{0.585}$	7.49	0.078	~25	......	4.2
$Ti_{0.12}V_{0.88}$		....	17.3	28.1	4.2
$Ti_{0.09}V_{0.91}$		....	14.3	16.4	4.2
$Ti_{0.06}V_{0.94}$		....	8.2	12.7	4.2
$Ti_{0.03}V_{0.97}$		....	3.8	6.8	4.2
Ti_xV_{1-x}		....	108 max.	......	1.2
V	5.31	~0.8	~3.4	......	1.79
		~0.75	~3.15	......	2
		~0.45	~2.2	......	3
		~0.30	~1.2	......	4
$V_{0.26}Zr_{0.74}$	≈5.9	0.238	....	......	1.05
		0.227	....	......	1.78
		0.185	....	......	3.04
		0.165	....	......	3.5
W (film)	1.7-1.1	....	>34	......	1

* Temperature of critical field measurement.
† Extrapolated.
‡ Linear extrapolation.
§ Parabolic extrapolation.

9h. Color Centers and Dislocations

C. C. KLICK

U.S. Naval Research Laboratory

9h-1. General Properties of Color Centers. Color centers are imperfections in transparent solids that give rise to optical absorption. Most of these centers are associated with crystalline defects, but centers arising from the incorporation of chemical impurities are frequently also considered to be color centers. Work on this subject has progressed farthest in the alkali halides; and only these materials will be discussed here. Related centers appear in most transparent solids, but their atomic identification is uncertain in many cases.

The conditions under which each center appears will be discussed in more detail below for the various centers. The most common treatments (ref. 1) are exposure to ionizing radiation such as X rays, heating in the alkali metal vapor which leads to centers with trapped electrons, and heating in halogen vapor which leads to centers with trapped holes.

It is frequently useful to relate *the number of centers* to the strength of the absorption band produced. If the absorption band is gaussian in shape, then an approximate relation is (ref. 2)

$$N_0 f = 8.7 \times 10^{16} \frac{n}{(n^2 + 2)^2} \alpha_m W \qquad (9\text{h-}1)$$

where N_0 is the concentration of centers per cubic centimeter, f is the oscillator strength, n is the index of refraction of the material at the wavelength of the absorption band, α_m is the absorption coefficient at the maximum of the band in reciprocal centimeters, and W is the width of the absorption band in electron volts at an absorption coefficient one-half that of the maximum. If it is possible to measure the concentration N_0 by some method such as chemical analysis or magnetic susceptibility, then the oscillator strength can be obtained. Knowledge of this factor for a particular center allows the determination of the density of that center from optical measurements alone and also gives a measure of the degree to which the optical transition is an allowed one. If the curve is Lorentzian in shape, then the constant in Eq. (9h-1) is 12.9. This form of the equation, often called Smakula's equation, is used in much of the older work, but the Gauss curve is a better (refs. 3 and 4), but not perfect, fit to the observed bands. Oscillator strengths given here will be in terms of Eq. (9h-1); they can be converted to Smakula's equation by multiplying by 1.48.

9h-2. F-center and Other Trapped Electron Centers. The most widely investigated color center is the *F-center* now known *to consist of an electron trapped at a negative ion vacancy.* If an alkali halide crystal is heated in the vapor of the alkali metal for several hours and then quenched to room temperature, the F-band appears. To the short-wavelength side of the F-band there also appear several weak absorption bands which have been designated K-, L_1-, L_2-, and L_3-bands. It is believed that these are more highly excited states of the F-center and show a dependence of wavelength on lattice constant which is similar to that of the F-band (ref. 5).

If the F-band is bleached with light at low temperatures ($-150°C$ for KBr, for example) a new broad band grows to the long-wavelength side of the F-band. This absorption is due to the *F'-center* and *arises from an F-center that has captured an additional electron* (ref. 6).

Irradiation with light in the F-band at room temperature causes the F-band to decrease and produces the *M-band* which arises from *a pair of F-centers*, and then the R_1- and R_2-bands arising from *a cluster of three F-centers*.

The peak position of the absorption of these centers at room temperature is given in Table 9h-1. Wavelengths are given throughout this article in millimicrons ($m\mu$).

TABLE 9h-1. WAVELENGTH OF ABSORPTION OF ELECTRON TRAP CENTERS
(In millimicrons)

	L_3 −180°C	L_2 −180°C	L_1 −180°C	K −180°C	F 20°C	R_1 20°C	R_2 20°C	M 20°C	Width at half maximum of F-band, eV 20°C
LiF......	...	...	...	...	250	313	380	444	0.82
LiCl.....	...	...	...	...	385	...	580	650	0.62
NaF.....	...	...	...	...	341	...	415	505	0.62
NaCl....	...	...	...	...	458	545	596	725	0.47
NaBr....	...	...	...	...	540	...	...	...	0.52
NaI.....	...	...	...	...	588				
KF......	...	...	...	...	455	...	570	...	0.41
KCl.....	251	288	344	457	556	658	727	825	0.35
KBr.....	276	316	374	525	625	735	790	892	0.345
KI......	326	382	447	585	689	...	...	...	0.345
RbCl....	279	335	402	523	609	...	...	...	0.31
RbBr....	300	362	435	593	694	805	859	957	0.28
RbI.....	338	413	506	646	756	...	...	...	0.35
CsCl....	...	...	...	...	605				
CsBr....	...	...	...	...	680				

The values are somewhat approximate, since different workers report results varying by as much as 20 mμ. Also given in Table 9h-1 are the widths at half maximum of the F-band at room temperature.

It has been noted that the wavelengths of the absorption bands vary with the distance a between nearest neighbors of the alkali halides. Equations (sometimes called Ivey relations) governing these bands are as follows (ref. 7):

F-center: $\lambda_{abs} = 703a^{1.84}$

R_1-center $\lambda_{abs} = 816a^{1.84}$

R_2-center: $\lambda_{abs} = 884a^{1.84}$

M-center: $\lambda_{abs} = 1{,}400a^{1.56}$

Both λ and a are in angstroms.

The variation of the maximum of the F-center absorption band as a function of temperature is shown in Fig. 9h-1 (ref. 16). The width W at half maximum of the F-band absorption also varies with temperature and fits an equation of the form

$$W = W_0 \coth \left(\frac{h\nu}{2kT} \right)^{\frac{1}{2}}$$

                                                           (9h-2)

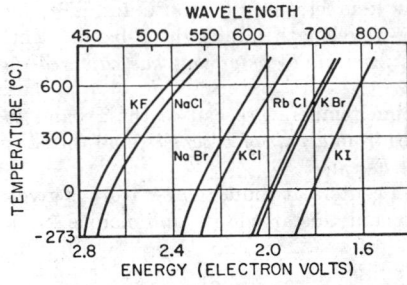

FIG. 9h-1. Effect of temperature on the position of the F-band maximum.

where W_0 is the width at absolute zero, h is Planck's constant, ν is a frequency related to the lattice vibrations of the solid, k is Boltzmann's constant, and T is the absolute temperature. At low temperatures (less than about 25° absolute) W is a constant; at high temperatures (above room temperature) W increases with the square root of the temperature. Table 9h-2 lists the values of W_0 and ν which give the best fit to experiment (ref. 17). W and W_0 are given in units of electron volts.

Direct measurements of oscillator strengths have been made for some of the F-centers using chemical techniques, electron-spin resonance measurements, and measurements

TABLE 9h-2. HALF WIDTH OF F-BAND AS A FUNCTION OF TEMPERATURE
[Constants for Eq. (9h-2)]

	W_0, eV	ν, Hz
LiF	0.43	4.1×10^{12}
NaCl	0.29	4.4×10^{12}
KCl	0.18	2.6×10^{12}
KBr	0.20	2.6×10^{12}
KI	0.18	3.6×10^{12}

of paramagnetic susceptibility. The agreement among these various methods is relatively poor. Oscillator strengths of the following values have been reported:

F-center in NaCl: 0.5[8], 0.58[9], 0.5[10], 0.57[11]
F-center in KCl: 0.55[8], 0.54[12], 0.57[9], 0.44[10], 0.78[13], 0.61[11]
F-center in KBr: 0.47[10], 0.57[11]
F-center in KI: 0.31[10]

From measurements on the growth of the M-band as the F-band is bleached by light, values of 0.2 (refs. 14 and 15) have been obtained for the oscillator strength of the M-band in KCl.

The effect of pressure on the position of the F-center maximum has been measured (ref. 18) in the range up to 50,000 atm. Data obtained at room temperature for the F-band in NaCl, KCl, and CsBr are shown in Fig. 9h-2. In the case of KCl the sharp

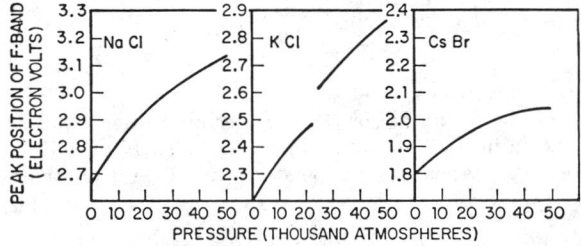

FIG. 9h-2. Effect of pressure on the position of the F-band maximum.

break in the curve occurs as the lattice changes from the normal NaCl type to that of CsCl at high pressures.

When color centers are excited, the stored energy may be released as luminescence. If only F-centers are present and measurements are made at liquid-nitrogen temperatures or below, F-center luminescence is observed (ref. 19). When M-centers are present, then excitation in both the F-band and M-band gives rise to M-center emission. Finally, when the R-centers arise, only emission characteristic of these centers can be observed. In general only the M-center emission can be seen on exciting at room temperature.

From measurements of the polarization of luminescence emission as a function of the polarization of the exciting light, the symmetry of the centers can be obtained. The F-center is found to be isotropic, but the M- and R-centers are asymmetrical and have their major axis along the $\langle 110 \rangle$ directions. Table 9h-3 gives the luminescence

TABLE 9h-3. LUMINESCENCE OF ELECTRON TRAP CENTERS AT 77 K

	F-center		M-center			R-center		
	Peak position, mμ	Half width, eV	Peak position, mμ	Half width, eV	P	Peak position, mμ	Half width, eV	P
LiF.......	.....	....	670	....	0.60			
NaCl.....	1,200	0.31	1,070	0.20	0.60	1,180	0.19	0.20
KCl.....	1,010	0.20	1,080	0.21	0.60	1,240	0.15	0.20
KBr.....	1,320	0.22						
KI........	1,470	0.16						
RbCl.....	1,120	0.23						

peak positions and half widths for various centers and gives also the polarization P of the luminescence of the M- and R-centers (ref. 20). This polarization is measured with the exciting light, polarizer, sample, analyzer, and detector in line. If the polarizer is set parallel to a $\langle 110 \rangle$ direction, then the luminescence measured with the analyzer parallel to the polarizer is $I_{\parallel}$ and with the analyzer crossed is $I_{\perp}$. The polarization P is defined as

$$P = \frac{I_{\parallel} - I_{\perp}}{I_{\parallel} + I_{\perp}}$$

For simple dipoles along the $\langle 110 \rangle$ direction, the value of P should be 0.66.

Electron-spin resonance has been observed for the F-center in many of the alkali halides. Table 9h-4 gives the g-values of the resonance (compared with a value of 2.023 for the free electron) and the width ΔH of the resonance absorption. This value is expressed in gauss for measurements in which the magnetic field is approximately 3,000 gauss and the frequency is approximately 9,000 MHz.

9h-3. Hole Trap Centers. A group of centers exists in the alkali halides characterized by having trapped a hole. Transfer of an electron from an electron trap center to one of these hole trap centers destroys them both. These centers do not follow an Ivey relation. Although some of them are formed by additive coloration at high temperature, the best-understood ones are formed at low temperatures by X-raying and show very detailed paramagnetic spectra. The peak position of the absorption bands for some of these centers is given in Table 9h-5.

The H-center is formed by X-raying at liquid-helium temperatures. From a study of its detailed paramagnetic resonance spectrum (ref. 25), the *H-center is found to be three halide ions and a halogen atom squeezed into the position normally occupied by three halide ions along a face diagonal.* The center bleaches thermally at 56 K in KCl.

X-raying at 77 K produces the V_1-center, an H-center near an impurity. Also at 77 K a weak absorption band, called the X_2^--band, is formed. The number of these centers is much larger if small amounts of Ag, Tl, or Pb are in the crystals (ref. 26). From a study of the paramagnetic resonance spectra (ref. 27) it is concluded that the X_2^--*center consists of a hole trapped between two halogen ions which have been displaced toward each other slightly from their equilibrium positions.*

TABLE 9h-4. ELECTRON-SPIN RESONANCE OF THE F-CENTER

Material	g-value	ΔH	Ref.
LiF.............	2.003	120	21
NaF............	2.002	50	21
NaCl...........	1.987	180	22
KCl............	1.995	61	22
KBr............	1.980	162	22
KI.............	1.971	200	23
RbCl...........	2	400	24
RbBr...........	2	380	24
RbI............	2	640	24

TABLE 9h-5. WAVELENGTH OF ABSORPTION OF HOLE TRAP CENTERS
(In millimicrons)

	H 4 K	V_1 77 K	X_2^- 77 K	V_2 300 K	V_3 300 K
LiF.............	...	...	348		
NaCl..........	330	345	...	223	210
KCl...........	335	356	365	230	212
KBr...........	380	410	385	265	231
KI.............	...	...	404		

The V_2- *and* V_3-*centers* are formed in alkali halides by heating them in halogen vapor and quenching to room temperature.

9h-4. Perturbed Lattice Transitions. Two bands, the α- and β-bands, have been found in alkali halides near the edge of the fundamental absorption band upon X-raying at liquid-nitrogen temperatures (ref. 28) (Table 9h-6). The strength of the α-band is correlated with the presence of negative ion vacancies and the β-band with F-centers. It is believed that both bands arise from transitions similar to those in the fundamental band of pure crystals but modified by the proximity of the various imperfections. The oscillator strength of the β-band is approximately unity, and that of the α-band somewhat less.

9h-5. Colloid Centers. Colloid centers are formed in crystals that have been colored by heating in alkali vapor and are then held at temperatures between 300 and 500°C. An absorption band to the long-wavelength side of the F-band appears. As the temperature increases over this range, the F-band intensity increases, the colloid band decreases, and its peak position shifts to longer wavelength. It is believed that these bands are due to colloid metal particles of from 10 to 50 Å in diameter.

9h-6. Impurity Absorption Bands. Alkali halide crystals containing hydrogen show an absorption band known as the U-band. The *U-center consists of a hydride ion substituting for a normal halide ion.* Irradiation with light in the U-band produces the F-center and a new center, the *U_2-center, due to interstitial hydrogen atoms. Interstitial hydrogen ions give rise to the U_1-center* (ref. 29). The absorption peak positions of the U-bands are given in Table 9h-7. For KCl the U_2-band occurs at 236 mμ and the U_1-band is a broad band near 275 mμ.

TABLE 9h-6. ABSORPTION OF α- AND β-BANDS
(Absorption peaks in millimicrons)

	α	β
NaF	131	127
NaCl	173	168
NaBr	199	
KCl	178	170
KBr	201	192
KI	238	226
RbBr	205	196
RbI	240	229

TABLE 9h-7. ABSORPTION BANDS FROM U-CENTERS, HYDROXIDE CENTERS, Z_1- AND Z_2-CENTERS
(In millimicrons)

	U-band	OH⁻-band	Z_1 band	Z_2 band
NaCl	192	185	505	512
NaBr	210			
KCl	214	204	590	635
KBr	228	214		
KI	244			
RbCl	229			
RbBr	242			

Incorporation of OH⁻ in alkali halides gives rise to the *hydroxide center* absorption bands shown in Table 9h-7, which follow an Ivey relation (refs. 30 and 31). The presence of these bands influences the amount of F-center coloration by X rays at room temperature and the formation of colloids (ref. 32).

Z-centers are formed from additively colored crystals that are doped with divalent impurities such as strontium, barium, or calcium (ref. 33). The Z_1-*center* can be formed by irradiating an additively colored crystal in the F-band at room temperature. The Z_2-*center* is formed by heating a crystal containing F- and Z_1-centers to approximately 100°C. Positions of these absorption bands are given in Table 9h-7. The Z_2-center in KCl is luminescent and emits at 1,140 mμ; the emission does not appear to be polarized (ref. 34).

The addition of heavy metal ions to the alkali halides produces absorption and emission bands largely characteristic of the ions. Table 9h-8 shows the optical properties of the centers due to incorporation of Tl, Pb, Ag, and Cu.

TABLE 9h-8. ABSORPTION AND EMISSION BANDS DUE TO Tl, Pb, Ag, AND Cu
(In millimicrons)

	Tl abs.	Tl emis.	Pb abs.	Pb emis.	Ag abs.	Ag emis.	Cu abs.	Cu emis.
NaCl...............	199	288	193	318	210	249	255	358
	254	...	274	384				
				453				
NaBr..............	216	295	220	...	219	263	259	365
	267	308	...	...	...	...	...	438
NaI................	234							
	293							
KCl...............	195	250	196	346	215	272	265	396
	247	305	273					
		475						
KBr..............	210	318	223	...	...	...	265	393
	261	350	302					
KI................	236	415	265					
	287							
RbCl.............	195	315	198					
	245							
RbBr.............	212							
	259							
RbI...............	240							
	286							
CsCl..............	196							
	248							
CsBr..............	214							
	263							
CsI...............	241							
	269							
	299							

References for Secs. 9h-1 through 9h-6

There are a series of survey articles and books on color centers each of which gives an excellent summary of the field at the time of its publication. They are listed below in chronological order.

Pohl, R. W.: *Proc. Phys. Soc.* **49**, (extra part), 3 (1937).
Mott, N. F., and R. W. Gurney: "Electronic Processes in Ionic Crystals," Oxford University Press, New York, 1940.
Seitz, F.: *Revs. Mod. Phys.* **18**, 384 (1946).
Przibram, K.: "Verfarbung und Lumineszenz," Springer-Verlag OHG, Berlin, 1953; "Irradiation Colours and Luminescence," Pergamon Press, Ltd., 1956.
Seitz, F.: *Revs. Mod. Phys.* **26**, 7 (1954).
Stöckmann, F.: In "Landolt-Börnstein," 6th ed., vol. 1, pt. 4 entitled Kristalle, p. 981, Springer-Verlag, Berlin, 1955.
Schulman, J. H., and W. D. Compton: "Color Centers," Pergamon Press, Ltd., 1963.
Markham, J. J.: "F-Centers in Alkali Halides," Academic Press, Inc., New York, 1966.
Fowler, W. B. ed.: "Physics of Color Centers," Academic Press, Inc., New York, 1968.
 1. Schulman, J. H., and H. W. Etzel: In "Methods of Experimental Physics," vol. 6, p. 324, Academic Press, Inc., New York, 1959.
 2. Dexter, D. L.: *Phys. Rev.* **101**, 48 (1956).
 3. Hesketh, R. V., and E. E. Schneider: *Phys. Rev.* **95**, 837 (1954).
 4. Markham, J. J.: *Rev. Mod. Phys.* **31**, 956 (1959).
 5. Lüty, F.: *Z. Phys.*, **160**, 1 (1960).
 6. Pick, H.: *Ann. Physik* **31**, 365 (1938).
 7. Ivey, H. F.: *Phys. Rev.* **72**, 341 (1947).
 8. Pick, H.: *Ann. Physik* **31**, (5), 365 (1938).
 9. Silsbee, R. H.: *Phys. Rev.* **103**, 1675 (1956).
10. Rauch, C. J., and C. V. Heer: *Phys. Rev.* **105**, 914 (1957).
11. Doyle, W. T.: *Phys. Rev.* **111**, 1072 (1958).

12. Kleinschrod, F. G.: *Ann. Physik* **27** (5), 97 (1936).
13. Scott, A. B., and M. E. Hills: *J. Chem. Phys.* **28**, 24 (1958).
14. Hirai, M.: *J. Phys. Soc. Japan* **14**, 1400 (1959).
15. Okura, H.: *J. Phys. Soc. Japan* **12**, 1313 (1957).
16. Mollwo, E.: *Z. Physik* **85**, 56 (1933).
17. Russell, G. A., and C. C. Klick: *Phys. Rev.* **101**, 1473 (1956).
18. Maisch, W. G., and H. G. Drickamer: *J. Phys. Chem. Solids* **5**, 328 (1958).
19. Botden, Th. P. J., C. Z. van Doorn, and Y. Haven: *Philips Res. Repts.* **9**, 469 (1954).
20. Lambe, J., and W. Dale Compton: *Phys. Rev.* **106**, 684 (1957).
21. Lord, N. W.: *Phys. Rev.* **105**, 756 (1957).
22. Kip, A. F., C. Kittel, R. A. Levy, and A. M. Portis: *Phys. Rev.* **91**, 1066 (1953).
23. Noble, G. A.: *Bull. Am. Phys. Soc.*, ser. 2, **3**, 178 (1958).
24. Wolf, H. C., and K. H. Hausser: *Naturwissenschaften* **23**, 646 (1959).
25. Känzig, W., and T. O. Woodruff: *J. Phys. Chem. Solids* **9**, 70 (1959).
26. Delbecq, C. J., B. Smaller, and P. H. Yuster: *Phys. Rev.* **111**, 1235 (1958).
27. Castner, T., and W. Känzig: *J. Phys. Chem. Solids* **3**, 178 (1957).
28. Delbecq, C. J., P. Pringsheim, and P. H. Yuster: *J. Chem. Phys.* **19**, 574 (1951); **20**, 746 (1952).
29. Delbecq, C. J., B. Smaller, and P. H. Yuster: *Phys. Rev.* **104**, 599 (1956).
30. Rolfe, J.: *Phys. Rev. Letters* **1**, 56 (1958).
31. Etzel, H. W., and D. A. Patterson: *Phys. Rev.* **112**, 1112 (1958).
32. Etzel, H. W.: *Phys. Rev.* **118**, 1150 (1960).
33. Pick, H.: *Ann. Physik* **35**, 73 (1939); *Z. Physik* **114**, 127 (1939).
34. West, E. J., and W. Dale Compton: *Phys. Rev.* **108**, 576 (1957).

9h-7. Dislocations. There are two important simple types of dislocations in crystals: the *edge dislocation* and the *screw dislocation*. Figure 9h-3 illustrates an edge dislocation. In this case one portion of the crystal has partially slipped with respect to the other. The plane along which the slip has occurred is plane *ABCD*. If the slip has been one atom distance, then all the atoms are once again in order except for those along the line *AD*. This line, along which the crystal is badly distorted, is

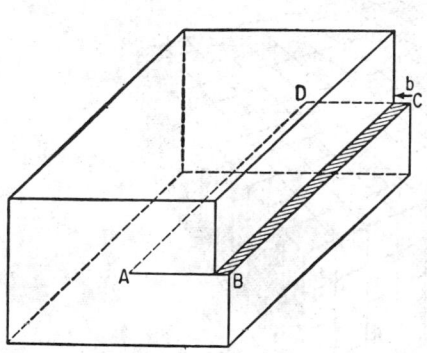

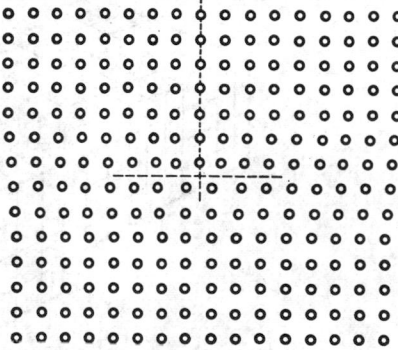

FIG. 9h-3. An edge dislocation. (*After W. T. Read, "Dislocations in Crystals," McGraw-Hill Book Company, New York, 1953.*)

FIG. 9h-4. View of the surface of a crystal with an edge dislocation. (*Reprinted with permission from W. Shockley, J. H. Hollomon, R. Maurer, and F. Seitz: "Imperfections in Nearly Perfect Crystals," John Wiley & Sons, Inc., New York, 1952.*)

the dislocation. The direction of displacement of the atoms due to the formation of the dislocation is *b*, the *Burgers vector*, and for an edge dislocation *b* is always perpendicular to the dislocation line. Figure 9h-4 shows an end view of the crystal where the dislocation line comes through to the surface at *A*. The *slip plane* is represented by the horizontal line. It is seen that the edge dislocation can also be thought of as consisting of the partial introduction of an extra vertical plane of atoms. The end of this plane is the dislocation line.

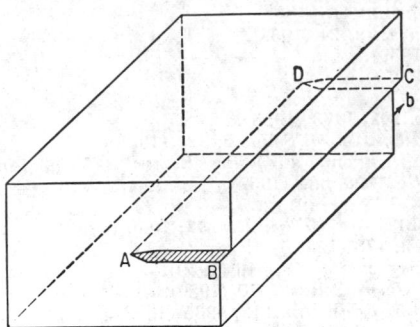

FIG. 9h-5. A screw dislocation. (*After W. T. Read, "Dislocations in Crystals," McGraw-Hill Book Company, New York, 1953.*)

A screw dislocation is illustrated in Fig. 9h-5, and the surface of the crystal through which it appears is shown in Fig. 9h-6. Here the Burgers vector b is parallel to the dislocation line.

More general forms of dislocation lines are possible. A ring, for instance, consists of portions which are pure edge and pure screw dislocations connected by segments which have both edge and screw character.

The Burgers vector can be obt ̣ ̣ned in general by making a circuit around a dislocation. Starting in an undistorted part of the crystal we might proceed by counting up l atoms, then left r atoms, then down l atoms, and right r atoms. If this bounded surface does not contain a dislocation, one arrives at the starting point. If a dislocation line does pass through the surface, the circuit will not close on the origin. The vector necessary to close the circuit is the Burgers vector.

The large amount of slip observed along single planes necessitates that there be a source for the creation of many dislocations within a strained crystal. One such model is the *Frank-Read source* illustrated in Fig. 9h-7. The line is a dislocation pinned at both ends by the presence of an impurity atom, for instance. Under stress

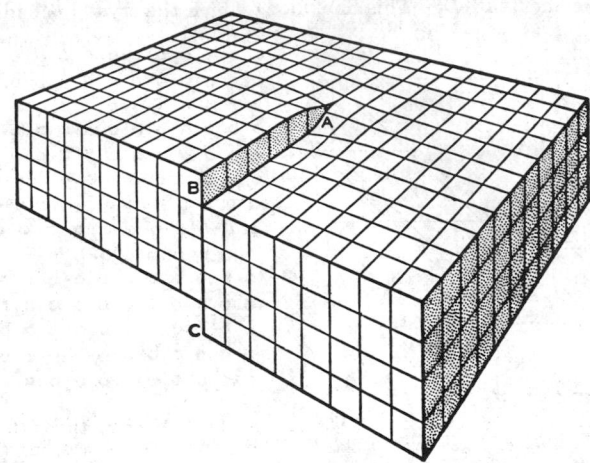

FIG. 9h-6. View of the surface of a crystal with a screw dislocation. (*Reprinted with permission from W. T. Read, "Dislocations in Crystals," McGraw-Hill Book Company, New York, 1953.*)

the line bows out and finally curls back on itself to touch. One loop continues on; the other returns to the original configuration from which the process may be repeated.

A *jog* in a dislocation is a sidewise step, usually of one atomic distance, of a dislocation line.

Dislocations have been made visible in silver halides and alkali halides by appropriate treatment. It is found that they generally appear to form a hexagonal network. A convenient way to determine the number of dislocations in a crystal is to

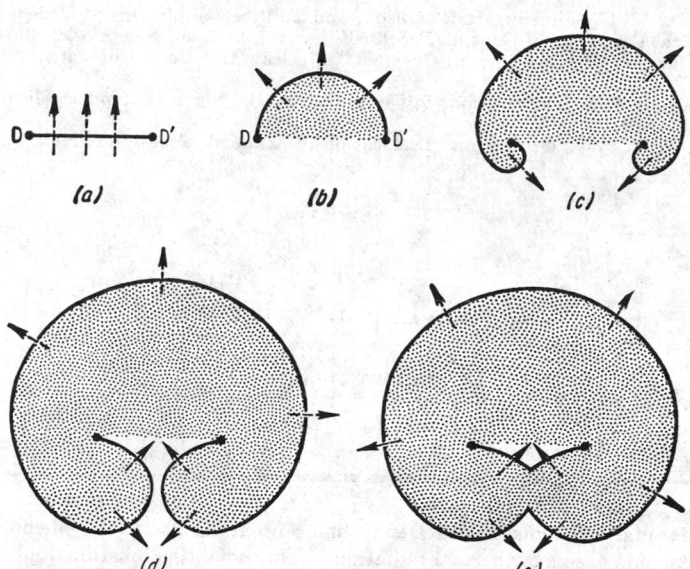

FIG. 9h-7. A Frank-Read source for dislocations. Steps *a* to *c* show the formation of a dislocation loop under a shear stress. (*Reprinted with permission from W. T. Read, "Dislocations in Crystals," McGraw-Hill Book Company, New York, 1953.*)

etch the surface and count the ends of dislocation lines which appear as etch pits The density of dislocation lines is given as those passing through a square centimeter. Normal crystals have approximately 10^5 to 10^9 dislocations per square centimeter, and this may be reduced in very carefully prepared crystals to a few hundred or less.

Etching also allows the motion of a single dislocation to be followed as stress is applied and thus permits studies of the mobility of dislocations.

Dislocations have made it possible to understand the growth of crystals under conditions of very low supersaturation. If a screw dislocation intersects the surface, as in Fig. 9h-6, then the atoms can easily continue to build on the spiral. When examined carefully with the microscope, a great many crystals show this spiral growth pattern.

Another problem solved by dislocations is that of the plastic flow of metals which occurs at stresses less by a factor of 10^4 than those calculated for a pure crystal. The relatively easy motion of dislocation lines has now been used to explain a large variety of mechanical properties.

In recent years it has been possible to obtain dislocation-free crystals. These are usually in the form of thin small needles called *whiskers*. It can be shown experimentally that these whiskers have the mechanical properties expected of the pure materials. Whiskers of many materials have been prepared, including Fe, Cu, CdS, *p*-toluidine, and the potassium halides. While most of the whiskers are straight, a commonly observed defect is a sharp bend, or *kink*, in the crystal.

References for Sec. 9h-7

A number of detailed expositions of dislocations are available.

Cohen, M., ed.: "Dislocations in Metals," American Institute of Mining and Metallurgical Engineers, New York, 1954.

Cottrell, A. H.: "Dislocations in Plastic Flow in Crystals," Oxford University Press, London, 1953.

Fisher, J. C., W. G. Johnston, R. Thomson, and T. Vreeland, Jr., eds.: "Dislocations and Mechanical Properties of Crystals," John Wiley & Sons, Inc., New York, 1957.
Read, W. T., Jr.: "Dislocations in Crystals," McGraw-Hill Book Company, New York, 1953.
Van Buerew, H. G.: "Imperfections in Crystals," 2d ed., North Holland Publishing Company, Amsterdam, 1961.
Verma, A. R.: "Crystal Growth and Dislocations," Academic Press, Inc., New York, 1953.

9i. Luminescence

FERD WILLIAMS

University of Delaware

9i-1. General Phenomena and Definitions. *Luminescence* is the phenomenon of light emission in excess of thermal radiation. This definition must be qualified, however, in order to exclude the Raman effect, Compton and Rayleigh scattering, and Cherenkov emission; and this is achieved by limiting luminescence to phenomena involving a time delay of emission after excitation which is long compared with the period of the emitted radiation, λ/c or approximately 10^{-14} sec. Luminescent emission involves optical transitions between electronic states characteristic of the radiating material. For ordinary luminescence the emission occurs by the Einstein spontaneous transition probability and is therefore incoherent; at high-excitation intensities the emission for *laser* materials occurs predominately by the Einstein induced transition probability and is therefore coherent. Although most investigations of luminescence have been concerned with visible emission and the term implies luminous or visible radiation, the same basic processes may yield infrared or ultraviolet radiation. Therefore, luminescence is also applied to such emission as is in excess of thermal radiation. Solid materials which luminesce when suitably excited are called *phosphors*. In many crystalline phosphors the luminescent emission originates in impurity systems called *activators*.

The general phenomenon of luminescence has been subdivided on the basis of the duration of the emission following excitation. Most investigators (refs. 1, 2, and 3) have made this subdivision by considering the mechanisms responsible for the afterglow. When the excitation is removed, there is invariably an exponential afterglow which depends on the lifetime of the emitting state of the activator. This spontaneous afterglow is called *fluorescence*. The time constant for the fluorescent emission may be as short as 10^{-8} or as long as 10^{-1} sec, depending on the phosphor, particularly on the identity of the activator. Frequently, there is an additional component to the afterglow which decays more slowly and with more complex kinetics. This component is called *phosphorescence*. In some phosphors, phosphorescence is attributed to metastable states of the activator; in others, to electron or hole traps spatially removed from the activator. Because thermal activation of the metastable activator or trap is prerequisite to emission, phosphorescence is strongly dependent on temperature. Phosphorescence may persist for times as short as milliseconds or as long as days or longer. During continuous excitation both the fluorescent and phosphorescent mechanisms contribute to the luminescent emission in proportions depending on the steady-state kinetics of these processes. Some authors (ref. 4) have chosen to define fluorescence as luminescent emission during excitation, and phosphorescence as emission after the excitation has been removed.

Luminescence has also been subdivided according to type of excitation. This subdivision is indicated by a prefix. For example, *photoluminescence* refers to luminescence excited by photons, electromagnetic radiation in the visible, ultraviolet, or infrared. *Cathodoluminescence* involves excitation by cathode rays, energetic electrons impinging on the phosphor. *Electroluminescence* involves the excitation of luminescence as a result of the existence of an applied electrical potential difference in the phosphor. This phenomenon may occur by several different mechanisms: excitation involving the injection of minority electronic charge carriers in semiconducting phosphors is designated *injection* electroluminescence; excitation involving the impact excitation of activators by electrons which have been accelerated to high kinetic energies in high applied fields is designated *collision-excitation* electroluminescence. Mechanical excitation such as grinding is termed *triboluminescence*. The conversion of the energy of a chemical reaction into luminescent emission is *chemiluminescence*, whereas the excitation of luminescence by biological processes is *bioluminescence*.

Thermoluminescence, however, does not refer to thermal excitation but rather to the thermal stimulation of luminescence. The phenomenon is essentially phosphorescence measured during conditions of increasing temperature. The stimulation of luminescence by the visible or infrared is described as *optical stimulation*.

Double prefixes are used to describe more complex luminescent phenomena. The first prefix refers to the control of the luminescence which has been excited in accordance with the mechanism described in the second prefix. For example, *electrophotoluminescence* refers to the modulation of photoluminescence by an applied electric field, whereas *photoelectroluminescence* refers to the control of electroluminescence by incident photons.

9i-2. The Representation of Absorption and Emission Spectra. Most luminescent solids exhibit broad, bell-shaped absorption bands near the fundamental absorption edge and emission bands corresponding to smaller transition energies. The emission bands of many luminescent solids are also broad and structureless; however, the emission of some, particularly of solid-state lasers, are narrow bands. The photoluminescent excitation spectrum usually coincides with the absorption bands. The excitation and emission spectra are determined by the characteristics of the activator systems. The energies of the discrete, localized states of the activator system are a function of the nuclear coordinates of the crystal. Since the atoms which are near the activator interact differently, depending on the electronic states of the activator, the transition energy between these discrete states depends on the internuclear coordinates of these atoms. This effect combined with the Franck-Condon principle accounts for the Stokes' shift of the emission to smaller transition energies compared with absorption in the activator system and for the breadth of the absorption and emission bands. This is of course for systems in which the transition for emission is the inverse of the transition for photoexcitation except for differences in interaction with the lattice for excited and unexcited activator. For systems in which there is a nonradiative transition between quite different excited electronic states of the activator following photoexcitation, the Stokes' shift originates mainly from the difference in energies of these excited states, and the emission may be narrow bands.

We shall first consider the origin and representation of the spectra of phosphors with broad and structureless absorption and emission bands. For simple activator systems, describable in terms of a single internuclear coordinate q, *the probability of the transition* involving a photon of wave number $\bar{\nu}$ per wave-number interval $P_{\bar{\nu}}$ has been shown to be (ref. 5):

$$P_{\bar{\nu}} = M^2 \left(\frac{K}{2\pi kT} \right)^{\frac{1}{2}} \exp \left[\frac{-Kq^2}{2k\theta \coth (\theta/T)} \right] \frac{dq}{d\bar{\nu}} \qquad (9i\text{-}1)$$

Table 9i-1. Activator Absorption and Emission Bands of Alkali Halide Phosphors[*]

Phosphor	Absorption maxima, $\bar{\nu}_0 \times 10^{-4}$ cm^{-1}				Emission maxima, $\bar{\nu}_0 \times 10^{-4}$ cm^{-1}		
NaCl:Tl	5.02	4.7	3.94	.....	3.35	.....	2.80[b]
KCl:Tl	5.10	4.8[e]	4.05	4 04[a,b]	3.27	2.62[b,c]	2.08[b]
RbCl:Tl	5.13	4.8	4.08	......	3.16	2.89[b]	
CsCl:Tl	5.10	4.8	4.03				
NaBr:Tl	4.63	....	3.74	......	3.38[d]		
KBr:Tl	4.76	....	3.83	......	3.13[b]	2.80[b]	
RbBr:Tl	4.72	....	3.86				
CsBr:Tl	4.67	....	3.80				
NaI:Tl	4.27	....	3.41	......	2.42		
KI:Tl	4.24	....	3.48	3.22[a]	2.39		
RbI:Tl	4.17	....	3.50		2.30		
CsI:Tl	4.15	3.7	3.34	......	1.68		
NaCl:Pb	5.18	....	3.65	......	3.13	2.67	2.17
KCl:Pb	5.10	....	3.66	......	2.89		
RbCl:Pb	5.05	....	3.68				
NaBr:Pb	4.54	....	3.29				
KBr:Pb	4.44	....	3.30				
NaCl:Sn	4.25[d]	....	3.51	......	2.25		
KCl:Sn	4.31	....	3.56[d]	......	2.01		
RbCl:Sn	.....	....	.....	......	1.93		
NaBr:Sn				......	2.15		
KBr:Sn	3.86[d]	....	3.28[d]	......	1.93		
RbBr:Sn					1.84		
LiI:Sn	.....	....	.....		1.89		
NaI:Sn					1.85		
KI:Sn	3.47[d]	....	2.85[d]	......	1.78		
RbI:Sn	.....		.....		1.73		
NaCl:Ga	4.60	....	3.75	......	2.44	......	1.99
KCl:Ga	4.61	....	.....	......	.....	......	1.99
KBr:Ga	4.42	....	3.64[e]	......	.....	......	1.78
NaCl:In	4.18	....	3.40	......	2.41		
KCl:In	4.35[d]	....	3.51[d]	......	2.30		
NaBr:In	4.00						
KBr:In	4.10	....	3.39[d]	......	2.28		
KI:In	3.82						
NaCl:Ge	4.30[d]	....	3.74	......	2.12[e]		
KCl:Ge	4.54[d]	....	.....		1.85		
KBr:Ge	3.98[d]	....	.....		1.75		
NaCl:Cu	3.94	....	.....		2.81		
KCl:Cu	3.88[e]	....	.....		2.53		
NaBr:Cu	3.86						
KBr:Cu	3.77	....	.....		2.52		
KI:Cu	3.75	....	.....		2.48	......	1.91
NaCl:Ag	4.76						
KCl:Ag	4.55[e]						
NaBr:Ag	4.56						
KI:Ag	3.70[d]						
NaBr:Eu	.....	....	.....	......	2.30		
KBr:Eu	4.00[d]	....	3.03[d]	......	2.36		
KI:Eu	.....	....	.....	......	2.28		

[*] The data of this table were assembled by P. D. Johnson, General Electric Research Laboratory.
[a] Excitation in large $\bar{\nu}$ absorption band.
[b] Observed at low temperature.
[c] Observed with high activator concentration.
[d] Multiple band.
[e] Disagreement in literature.

where M is the electronic part of the matrix element for the transition, K is the force constant for the displacement q from equilibrium in the initial state, and $k\theta$ is the zero-point energy for the vibration associated with displacement q. If we make the harmonic approximation, that is, that M is independent of q, and also if we assume that the force constants for the initial and final states are equal, then $P_{\bar{\nu}}$ is found to be gaussian in $\bar{\nu}$.

TABLE 9i-2. PHOTOLUMINESCENT SPECTRA OF ZINC SULFIDE PHOSPHORS*

Phosphor	Band	$\bar{\nu}_0 \times 10^{-4}$ cm^{-1}	$h \times 10^{-4}$ cm^{-1}	Comments
Hex. ZnS:Ag,Cl.	Silver blue	2.29	0.12	
Hex. ZnS:Cu,Cl.	Copper blue	2.22	0.15	Band skewed to small $\bar{\nu}$ because of copper green
Hex. ZnS:I.....	Self-activated blue	2.19	0.19	
Hex. ZnS:Cu,I..	Copper blue	2.19	0.16	Copper blue and self-activated blue resolvable at low temperature
Hex. ZnS.......	Self-activated blue	2.19	0.12	
Cub. ZnS:Ag,Cl.	Silver blue	2.21	0.14	
Cub. ZnS:Cu,Cl.	Copper blue	2.18	0.14	Band skewed to small $\bar{\nu}$ because of copper green
Cub. ZnS.......	Self-activated blue	2.13	0.18	
Cub. ZnS:Al....	Self-activated blue	2.09	0.19	
Hex. ZnS:Cu,Cl.	Copper green	1.92	0.12	
Cub. ZnS:Cu,Cl.	Copper green	1.88	0.12	
Hex. ZnS:Ag,In.	Silver red	1.56	0.23	Spectrum at 77 K
Cub. ZnS:Cu....	Copper red	1.49	0.24	
Cub. ZnS:Cu....	Copper red	1.39	0.20	Spectrum at 77 K
Hex. ZnS:Au,Ib.	Gold infrared	1.20	0.22	Spectrum at 77 K

* Spectra at 298 K unless otherwise indicated. The author is indebted to J. S. Prener, General Electric Research Laboratory, for evaluating these data.

This is not found to be precisely the situation experimentally, and actually *the radiant energy* of wave number $\bar{\nu}$ per unit wave-number interval $E_{\bar{\nu}}$ was empirically found to be the more accurately gaussian in $\bar{\nu}$ (refs. 6 to 9)

$$E_{\bar{\nu}} = b \exp\left[-\frac{(\bar{\nu} - \bar{\nu}_0)^2}{1.44h^2} \right] \qquad (9\text{i-}2)$$

where $2h$ is the half width of the band with maximum at $\bar{\nu}_0$. $P_{\bar{\nu}}$ and $E_{\bar{\nu}}$ are, of course, related as follows: $E_{\bar{\nu}} = hc\bar{\nu}P_{\bar{\nu}}$.

The parameters $\bar{\nu}_0$ and h of Eq. (9i-2) are used to characterize the emission spectra in Tables 9i-1 to 9i-6. This permits the most precise representation of the spectra with two parameters and also permits the straightforward transformation to and from the wavelength scale. The latter is of some importance, because many of the spectral data on phosphors are given as radiant energy of wavelength λ per unit wavelength interval E_λ. The relation between $E_{\tilde\nu}$ and E_λ is as follows: $E_{\tilde\nu} = -\lambda^2 E_\lambda$. The maximum at λ_0 in the plot of E_λ vs. λ does not coincide with the maximum $\tilde\nu_0$ in the

TABLE 9i-3. EMISSION SPECTRA AND QUANTUM EFFICIENCIES OF FLUORESCENT-LAMP PHOSPHORS*

Phosphor	$\tilde\nu_0 \times 10^{-4}$ cm^{-1}	$h \times 10^{-4}$ cm^{-1}	$\eta\dagger$	Comments
$(Ca,Zn)_3(PO_4)_2:Tl$	3.18	0.17	0.9	
$BaSi_2O_5:Pb$	2.88	0.18		
$Ca_3(PO_4)_2:Ce$	2.74	0.17	0.7	
$CaWO_4$	2.21	0.29	0.7	
$Ca_3(PO_4)_2:Cu,Sn$	2.07	0.18		
$Ca_3(PO_4)_2:Cu$	2.05	0.18		
$3Ca_3(PO_4)_2 \cdot Ca(F,Cl)_2:Sb,Mn$	2.04	0.28	0.8–0.95	Evidence of additional Mn emission at small $\tilde\nu$; relative intensities depend on Sb and Mn concentrations; $\tilde\nu_0$ depends slightly on F/Cl composition
	1.69	0.11		
$MgWO_4$	2.01	0.29	0.9	
$3Sr_3(PO_4)_2 \cdot Ca(F,Cl)_2:Sn,Mn$	1.93	0.28	....	Relative intensities of Sb and Mn bands depend on Sb and Mn concentration
	1.76	0.10		
$Zn_2SiO_4:Mn$ (willemite)	1.89	0.09	0.8	Slightly skewed to small $\tilde\nu$
$CaSiO_3:Pb,Mn$	2.80	0.33	....	$R = 0.42$ for overlapping Mn bands; R depends on Mn concentrations
	1.74	0.10		
	1.60	0.10		
$(Sr,Mg)_3(PO_4)_2:Sn$	1.58	0.13		
$Cd_2B_2O_5:Mn$	1.58	0.11	0.7	

* Excitation predominantly by 2,537-A radiation. The spectra, from which the $\tilde\nu_0$ and h were derived, were provided by H. C. Froehlich and F. J. Studer of Large Lamp Engineering of General Electric.
† J. Tregellas-Williams, *J. Electrochem. Soc.* **105**, 173 (1958).

plot $E_{\tilde\nu}$ **vs.** $\tilde\nu$, **but rather the** latter occurs at slightly smaller wave number

$$\tilde\nu_0 = 1/(\lambda_0 + \Delta\lambda)$$

where $\Delta\lambda = 1.44h^2\lambda^3$. The parameters $\tilde\nu_0$ and h for each band were determined, in accordance with Eq. (9i-2), by minimizing the sum of squared deviations in fitting the logarithm of $E_{\tilde\nu}$ to a quadratic function of $\tilde\nu$. For multiple-band emission arising from a single activator, a third parameter R is also given in Tables 9i-3 and 9i-4.

TABLE 9i-6. EMISSION SPECTRA OF ELECTROLUMINESCENT PHOSPHORS

Phosphor*	Emission maximum, $\bar{\nu}_0 \times 10^{-4}\,\mathrm{cm}^{-1}$	Width, $h \times 10^{-4}\,\mathrm{cm}^{-1}$	Temperature, K
ZnS:Cu, Br.............	2.16	0.14	300
ZnS:Cu, Cl.............	1.91	0.15	300
ZnS:Cu, Cl, Mn........	1.69	0.09	300
GaP:Zn, S.............	1.79	0.02	20
	1.73	0.04	300
GaP:Cd, O.............	1.49	0.07	20
	1.43	0.09	300
SiC:B, N†.............	1.73	0.19	300

* Electroluminescence of ZnS phosphors occurs by collision-excitation mechanism; for GaP and SiC, by minority-carrier injection.

† 6H polymorph.

For materials with narrow-band emission such as some of those in Table 9i-6 and all of those in Tables 9i-7 to 9i-9 we shall be concerned with somewhat different characteristics. Some of these materials are becoming important as sources of ordinary luminescent emission, however, their most striking characteristics become evident under the conditions for the generation of coherent emission. Under these conditions the narrow emission is important because in order to satisfy the requirement that the induced probability be large compared to the spontaneous probability, $\rho B > A$, where ρ is the radiation density per unit frequency or wave number, the same flux of radiant energy in a narrow frequency or wave-number interval yields a lower threshold power for laser action. The conditions for the generation of coherent optical radiation were first predicted (ref. 10) and then first observed (ref. 11) for ruby, $Al_2O_3:Cr^{3+}$. Since then, many solid-state lasers have been reported and reviewed (refs. 12 and 13). Some characteristics are tabulated in Tables 9i-7 to 9i-9.

Consistent with the accuracies of both the original data and the analyses, the values of the parameters given in Tables 9i-1 to 9i-9 are for most materials reliable to plus or minus 1 in the last digit given.

9i-3. Tabulated Spectra, Efficiencies, and Afterglow Characteristics. In Table 9i-1, which describes the absorption and emission bands of *alkali halide phosphors*, only the maxima $\bar{\nu}_0$ in the plots of α and $E_{\bar{\nu}}$ vs. $\bar{\nu}$ are given. Except for a few materials such as KCl:Tl, the data for these materials are not sufficiently complete to resolve the overlapping bands and attribute accurate half widths to each band. The value of the parameter h for these phosphors is a monotonic function of $\bar{\nu}_0$, so that the correction $\Delta\lambda$ involved in the transformation E_λ to $E_{\bar{\nu}}$ can be made within the accuracy of the data. The bands described in Table 9i-1, unless otherwise indicated, describe the spectra with quite low activator concentrations, e.g., 10^{-4} atomic fraction or less. At higher concentrations, additional bands at lower $\bar{\nu}$ frequently appear.

In Table 9i-2 the parameters $\bar{\nu}_0$ and h for the photoluminescent emission of *zinc sulfide phosphors* are given. For the same activator, e.g., copper, silver, or gold, a number of different emission bands are observed whose relative intensities depend on coactivator concentration and on the atmosphere during crystallization. The coactivator is an additional impurity which is necessary for luminescence in sulfide phosphors but which does not have the pronounced effect on the emission spectrum that the activator has. In this class of phosphors the coactivators have been identified, according to semiconductivity terminology, as donors; the activators, as acceptors (refs. 14 and 15). In semiconducting phosphors in general, radiative deexcitation has frequently been found to occur via donor-acceptor pairs (ref. 16). The substitution of

TABLE 9i-7. LASER MATERIALS AND TRANSITIONS*

Material	Transition	Emission, $\bar{\nu}_0 \times 10^{-4}$ cm^{-1}	Terminal state, cm^{-1}	Excitation, $\nu \times 10^{-4}$ cm^{-1}
Al_2O_3: 0.05 % Cr^{3+}	$^2E(\bar{E}) \rightarrow {}^4A_2$	1.4422	0	2.38–3.12
	$^2E(\bar{A}) \rightarrow {}^4A_2$	1.4432	0	1.66–2.00
Al_2O_3: 0.5 % Cr^{3+}	$^2E(\bar{A}) \rightarrow {}^4A_2$	1.4432	0	
	pair lines	1.4267	100	
	pair lines	1.4202	100	
	pair lines	1.3037	100	
MgF_2: 1 % Co^{2+}	$^4T_2 \rightarrow {}^4T_1$	0.5714	1,087	1.81–2.27
		0.5546	1,256	1.33–1.66
ZnF_2: 1 % Co^{2+}	....	0.3829		
MgF_2: 1 % Ni^{2+}	$^3T_2 \rightarrow {}^3A_2$	0.6165	340	2.08–2.63
				1.01–1.47
				0.64–0.91
$CaWO_4$: 0.5 % Pr^{3+}	$^1G_4 \rightarrow {}^3H_4$	0.9552	377	2.04–2.33
LaF_3: 1 % Pr^{3+}	$^3P_0 \rightarrow {}^3H_6$	1.6708	~4,200	2.08–2.33
CaF_2: 1 % Nd^{3+}	$^4F_{\frac{3}{2}} \rightarrow {}^4I_{\frac{11}{2}}$	0.9560	~2,000	
$CaMoO_4$: 1.8 % Nd^{3+}	$^4F_{\frac{3}{2}} \rightarrow {}^4I_{\frac{11}{2}}$	0.9425	~2,000	
$CaWO_4$: 1 % Nd^{3+}	$^4F_{\frac{3}{2}} \rightarrow {}^4I_{\frac{9}{2}}$	1.0934	471	1.67–1.75
	$^4F_{\frac{3}{2}} \rightarrow {}^4I_{\frac{11}{2}}$	0.9389	2,016	1.32–1.35
	$^4F_{\frac{3}{2}} \rightarrow {}^4I_{\frac{13}{2}}$	0.7467	4,004	
Glass: Nd^{3+}	$^4F_{\frac{3}{2}} \rightarrow {}^4I_{\frac{11}{2}}$	0.94	1,950	
LaF_3: 1 % Nd^{3+}	$^4F_{\frac{3}{2}} \rightarrow {}^4I_{\frac{13}{2}}$	0.9404	2,187	
$Y_3Al_5O_{12}$: Nd^{3+}	$^4F_{\frac{3}{2}} \rightarrow {}^4I_{\frac{11}{2}}$	0.9423	2,001	
		0.9391	2,111	
CaF_2: 0.01 % Sm^{2+}	$^5D_0 \rightarrow {}^7F_1$	1.4118	263	2.22–2.50
SrF_2: 0.01 % Sm^{2+}	$^5D_0 \rightarrow {}^7F_1$	1.4353	263	1.47–1.72
Y_2O_3: 5 % Eu^{3+}	$^5D_0 \rightarrow {}^7F_2$	1.6358	859	3.57–5.00
				2.13–2.17
				1.85–1.92
CaF_2: 0.01 % Dy^{2+}	$^5I_7 \rightarrow {}^5I_8$	0.4239	30	
CaF_2: Ho^{3+}	$^5S_2 \rightarrow {}^5I_8$	1.8142	~370	
$CaWO_4$: 0.5 % Ho^{3+}	$^5I_7 \rightarrow {}^5I_8$	0.4887	250	2.17–2.27
$Y_3Al_5O_{12}$: Ho^{3+}	$^5I_7 \rightarrow {}^5I_8$	0.4769	462	2.17–2.27
		0.4781	......	0.85–0.88
		0.4711	518	0.51–0.53
Glass: Ho^{3+}	$^5I_7 \rightarrow {}^5I_8$	0.5128		
$Ca(NbO_3)_2$: Er^{3+}	$^4I_{\frac{13}{2}} \rightarrow {}^4I_{\frac{15}{2}}$	0.6211		
$CaWO_4$: 1 % Er^{3+}	$^4I_{\frac{13}{2}} \rightarrow {}^4I_{\frac{15}{2}}$	0.6203	375	3.57–5.00·
$Y_3Al_5O_{12}$: Er^{3+}	$^4I_{\frac{15}{2}} \rightarrow {}^4I_{\frac{15}{2}}$	0.6023	525	3.57–5.00
		0.6078	......	2.12–2.17
				1.85–1.92
CaF_2: 0.01 % Tm^{3+}	$^2F_{\frac{5}{2}} \rightarrow {}^2F_{\frac{7}{2}}$	0.8960	0	2.94–3.57
				2.17–2.56
				1.59–1.89
$CaWO_4$: Tm^{3+}	$^3H_4 \rightarrow {}^4H_6$	0.5232	325	2.08–2.17
				0.82–0.86
		0.5219	325	0.55–0.59
Er_2O_3: Tm^{3+}	$^3H_4 \rightarrow {}^3H_6$	0.5170		
$Y_3Al_5O_{12}$: Tm^{3+}	$^3H_4 \rightarrow {}^3H_6$	0.5309	240	2.08–2.17
				0.82–0.86
		0.4967	582	0.55–0.59
$Y_3Al_5O_{12}$: Yb^{3+}	$^2F_{\frac{5}{2}} \rightarrow {}^2F_{\frac{7}{2}}$	0.9712	623	1.00–1.11
Glass (1): Yb^{3+}	$^2F_{\frac{5}{2}} \rightarrow {}^2F_{\frac{7}{2}}$	0.9900	400	
Glass (2): Yb^{3+}	$^2F_{\frac{5}{2}} \rightarrow {}^2F_{\frac{7}{2}}$	0.9823		
CaF_2: 0.05 % U^{3+}	$^4I_{\frac{11}{2}} \rightarrow {}^4I_{\frac{9}{2}}$	0.3982	505	0.77–0.83
		0.3827	609	
SrF_2: U^{3+}	$^4I_{\frac{11}{2}} \rightarrow {}^4I_{\frac{9}{2}}$	0.4154	334	

* Modified from Z. J. Kiss and R. J. Pressley, *Proc. IEEE*, **54**, 1238 (1966), and L. G. Van Uitert, "Luminescence of Inorganic Solids," Goldberg, ed., pp. 520–523, Academic Press, Inc., New York, 1966.

TABLE 9i-8. SENSITIZED LASER MATERIALS AND TRANSITIONS*

Material†	Transition	Emission, $\bar{\nu}_0 \times 10^{-4}$ cm^{-1}	Terminal state, cm^{-1}
$Y_3Al_5O_{12}$: Yb^{3+}, Er^{3+}, Tm^{3+}, Ho^{3+}......	$^5I_7 \to {}^5I_8$	0.4697	530
		0.4710	520
		0.4765	460
$Er_3Y_3Al_{10}O_{24}$: $Er^{3+} \to Ho^{3+}$...........	$^5I_7 \to {}^5I_8$	0.4766	462
$Y_3Al_5O_{12}$: Cr^{3+}, Ho^{3+}................	$^5I_7 \to {}^5I_8$	0.4767	462
		0.4711	510
Glass: Yb^{3+}, Ho^{3+}....................	$^5I_7 \to {}^5I_8$	0.481	$\sim$300
$CaMoO_4$: Er^{3+}, Ho^{3+}...............	$^5I_7 \to {}^5I_8$	0.4821	
		0.4829	$\sim$250
		0.4864	
$Er_3Y_3Al_{10}O_{24}$: $Er^{3+} \to Tm^{3+}$..........	$^3H_4 \to {}^3H_6$	0.4965	582
		0.5307	240
		0.5319	228
$Y_3Al_5O_{12}$: Yb^{3+}, Tm^{3+}..............	$^3H_4 \to {}^3H_6$	0.4966	580
		0.5308	
$Y_3Al_5O_{12}$: Cr^{3+}, Tm^{3+}..............	$^3H_4 \to {}^3H_6$	0.4967	582
		0.4952	$\sim$600
Er_2O_3: $Er^{3+} \to Tm^{3+}$................	$^3H_4 \to {}^3H_6$	0.5170	
$CaMoO_4$: Er^{3+}, Tm^{3+}...............	$^3H_4 \to {}^4H_6$	0.5231	$\sim$325
		0.5246	
Glass: $Yb^{3+} \to Er^{3+}$................	$^4I_{\frac{13}{2}} \to {}^4I_{\frac{15}{2}}$	0.6482	
Glass (silicate): UO_2^{2+}, Nd^{3+}.........	$^4F_{\frac{3}{2}} \to {}^4I_{\frac{11}{2}}$	0.94	$\sim$2,000
Phosphate glass: Mn^{2+}, Nd^{3+}.........	$^4F_{\frac{3}{2}} \to {}^4I_{\frac{11}{2}}$	0.94	$\sim$2,000
$Y_3Al_5O_{12}$: Cr^{3+}, Nd^{3+}..............	$^4F_{\frac{3}{2}} \to {}^4I_{\frac{11}{2}}$	0.9423	2,001
Glass: Nd^{3+}, Yb^{3+}..................	$^2F_{\frac{5}{2}} \to {}^2F_{\frac{7}{2}}$	0.982	
Silicate glass: UO_2^{2+}, Yb^{3+}...........	$^2F_{\frac{5}{2}} \to {}^2I_{\frac{7}{2}}$	0.985	$\sim$400

* Modified from L. G. Van Uitert, "Luminescence of Inorganic Solids," Goldberg, ed., pp. 520–523, Academic Press, Inc., New York, 1966. The excitation is in the characteristic bands of sensitizer or laser dopant.

| The last dopant in the formula is the activator with the lasing transition.

TABLE 9i-9. SEMICONDUCTOR LASERS*

Material	Emission, $\bar{\nu}_0 \times 10^{-4}$ cm^{-1}	Excitation†	Material	Emission, $\bar{\nu}_0 \times 10^{-4}$ cm^{-1}	Excitation†
$Pb_{0.73}Sn_{0.27}Te$.....	0.036	PL	$In(P,As)$.........	1.13	EL
$Pb_{0.8}Sn_{0.2}Te$......	0.038	PL	$GaAs$............	1.18	EL, CR, PL
$PbSe$............	0.118	EL, CR	$GaAs_xP_{1-x}$........	1.14–1.69‡	EL
$PbTe$............	0.154	EL, CR, PL	$CdTe$............	1.25	CR
$InSb$............	0.19	EL, CR, PL	$CdSe$............	1.45	CR
PbS.............	0.233	EL, CR	$GaSe$............	1.67	CR
$Cd_{0.3}Hg_{0.7}Te$......	0.25	PL	CdS.............	2.02	CR, PL
Te..............	0.27	CR	$ZnSe$............	2.17	CR
$InAs$............	0.31	EL, CR, PL	ZnO.............	2.67	CR
$GaSb$............	0.62	EL, CR	ZnS.............	3.06	CR
InP.............	1.11	EL			

* Modified from R. Rediker, *Proc. 1966 Intern. Conf. Luminescence*, edited by G. Szigeti (Hungarian Academy of Sciences, Budapest, 1968) p. 1756; M. R. Lorenz and M. H. Pilkuhn, *IEEE* Spectrum **4**, 87 (1967); D. C. Reynolds, *Trans. Met. Soc. AIME*, **239**, 300 (1967).

† CR = cathode ray, EL = electric injection, PL = photoexcitation.

‡ Depends on x in formula $GaAs_xP_{1-x}$.

cadmium sulfide for part of the zinc sulfide shifts the emission band to smaller $\bar{\nu}$ in accordance with the reduction in band gap of the material. The zinc sulfide phosphors activated with phosphorus, arsenic, and antimony are not included, since their emission spectra are not unambiguously described in the literature. The photoluminescence described in Table 9i-2 is efficiently excited by 3650-Å ultraviolet radiation.

The *fluorescent-lamp phosphors* described in Table 9i-3 have probably had their emission spectra most accurately measured. In some cases the parameters $\bar{\nu}$ and h are derived from data taken on experimental lamps; in other cases, on the phosphors directly. In either case, the excitation is predominantly by 2537-Å ultraviolet radiation. The calcium halophosphate phosphors merit additional comment. The emission spectrum consists of two broad and structureless bands. The intensity of the blue band depends on the antimony activator concentration; that of the orange band on the manganese activator concentration. In addition, $\bar{\nu}_0$ for the manganese emission shifts to smaller $\bar{\nu}$ with increasing chloro- to fluorophosphate content, with a maximum shift of approximately 2 percent of $\bar{\nu}_0$. The antimony emission shifts only 1 percent of $\bar{\nu}_0$ with complete chloride substitution. In contrast to most phosphors activated with divalent manganese, there is evidence for a second, less intense manganese band at smaller $\bar{\nu}$ (ref. 8). The quantum efficiencies shown in Table 9i-3 are based on the analysis by Tregellas-Williams (ref. 17).

Another class of photoluminescent phosphors are those activated by tetravalent manganese. Their emission is characterized by a fine structure consisting of bands approximately 100 cm^{-1} in width separated by approximately 200 cm^{-1}. This material and the magnesium arsenate maintain high photoluminescent (ref. 9) efficiencies to 650 K. The high-temperature stability combined with their red emission leads to their use for color correction in the medium-pressure mercury lamp. Rare-earth phosphors related to those in Tables 9i-7 and 9i-8 are becoming important in discharge lamps.

In principle, the *cathodoluminescent screens* described in Table 9i-4 are not confined to particular phosphors; however, the screen specifications are set up so that in practice only a particular phosphor or combination of phosphors meets the specifications. For screens composed of two or more phosphors, either as a mixture or as two distinct layers, the contribution of each phosphor to the emission spectrum may be slightly altered because of absorption by the other components. This accounts for some of the small discrepancies between the parameters given in Table 9i-4 which are based on the screen performance characteristics and the parameters based on data obtained on the separate phosphors. A single time constant does not describe the afterglow of the sulfide phosphors, because the afterglow is hyperbolic, rather than exponential, and is markedly dependent on current density. This arises from a broad distribution in energy of trapping states whose occupational probabilities depend on density of excitation.

Both organic and inorganic phosphors are used in *scintillation counters*. The emission spectra of the organic phosphors consist of a series of narrow bands compared with the broad, structureless emission bands of the inorganic phosphors. The persistences of the organic phosphors are shorter, whereas the efficiencies of the inorganic phosphors are greater for particles which produce large ionization densities. The characteristics of the phosphors commonly used in scintillation counters are given in Table 9i-5.

Electroluminescent materials are of two classes: those which operate by the collision-excitation mechanism, and those which operate by minority-charge carrier injection. Zinc sulfide is the principal representative of the first class; silicon carbide and gallium phosphide are the most important members of the second class. The first are used as particles in a dielectric matrix with an applied a-c field; the second, as single crystals with either a-c or d-c fields. The parameters describing the emission of these materials are given in Table 9i-6.

9i-4. Formulas for Some Luminescent Characteristics. In addition to Eqs. (9i-1) and (9i-2), which are the theoretical and empirical formulas, respectively, for the optical spectra of activator systems, there are a number of other simple formulas for interpreting or describing luminescent phenomena.

There are, of course, the *formulas for the afterglow.* The simplest is the exponential decay of the emission intensity,

$$I = I_0 \exp(-At) \tag{9i-3}$$

where A is the spontaneous transition probability or the reciprocal of the *lifetime* τ shown in Tables 9i-4 and 9i-5. At high temperatures the lifetime of the emitting state decreases because of nonradiative deexcitation, and under these conditions the afterglow is of the form (ref. 18):

$$I = I_0 \exp\left\{-\left[A + s \exp\left(-\frac{\epsilon}{kT}\right)\right]t\right\} \tag{9i-4}$$

where s is the frequency factor, and ϵ is the activation energy for the nonradiative process. Concurrently, under these conditions the luminescent efficiency decreases with increasing temperature (ref. 19):

$$\eta = \frac{1}{1 + (s/A) \exp(-\epsilon/kT)} \tag{9i-5}$$

The luminescent efficiency is also dependent on activator concentration, and for activators distributed at random lattice sites and capable of efficient luminescence if no other activators occupy the z nearest-neighbor sites, the efficiency is of the form (ref. 20):

$$\eta = \frac{c(1-c)^z}{c + \kappa(1-c)} \tag{9i-6}$$

where c is the atomic fraction of activator impurity, and κ is the ratio of capture cross sections for nonradiative and radiative processes.

The afterglow observed for zinc sulfide phosphor is fundamentally quite complex, since a distribution of electron traps and retrappings are apparently involved. Empirically, the following hyperbolic form has been extensively used:

$$I = I_0(1 + \alpha t)^{-n} \tag{9i-7}$$

Simple bimolecular recombination leads to an equation of his form with $n = 2$; however, experimentally $n \approx 1$ is usually found.

The distribution of trapping states is more clearly evident from thermoluminescent measurements. For traps which empty by first-order kinetics without retrapping and in accord with the activation energy ϵ' and frequency factor s', the thermoluminescent intensity at temperature T is (refs. 21 and 22):

$$I = n_0 s' \exp\left(-\frac{\epsilon'}{kT}\right) \exp\left[-\frac{s'}{dT/dt} \int_{T_0}^{T} \exp\left(-\frac{\epsilon'}{kT}\right) dT\right] \tag{9i-8}$$

where n_0 is the initial concentration of occupied trapping states, T_0 is the initial temperature, and dT/dt is the rate of temperature increase. Equations which include retrapping have also been formulated.

As noted earlier, radiative recombination of electrons and positive holes at donor-acceptor pairs is an important luminescent process in semiconductors. The transition energy for this emission at the ith pair with interimpurity distance R_i is as follows, neglecting overlap of the effective mass functions for donor electron and acceptor hole (ref. 16):

$$hc\bar{\nu}_i = E_g - (E_A + E_D) + \frac{e^2}{KR_i} \tag{9i-9}$$

where E_g is the band gap; E_A and E_D are the absolute values of the ionization energies of separated acceptor and donor, respectively; and K is the dielectric constant.

Various formulas have been proposed to describe the voltage dependence of the brightness of phosphors which electroluminesce by the collision-excitation mechanism. The most successful for present electroluminescent cells are of the following form (refs. 23 and 24):

$$B = V^m \exp\left(-\frac{d}{V^{\frac{1}{2}}}\right) \qquad (9\text{i-}10)$$

where $m = 0, 2$ have been used; V is the applied voltage; and d is a constant.

For injection electroluminescence involving a p-n junction the dependence of current J on applied voltage is

$$J = J_0 \exp\frac{eV}{\beta kT} \qquad (9\text{i-}11)$$

where β is found to be of the order of 1 or 2.

The solid-state laser acts as a resonance cavity with an amplifying medium. The threshold power for oscillation is (ref. 12):

$$P = \frac{3h^2c^2(1 - R)\Delta\bar{\nu}}{8\pi^2\tau Ml} \qquad (9\text{i-}12)$$

where R is the product of reflectivities of the end mirrors, τ is the lifetime of the excited state, M is the dipole matrix element of the transition, and l is the length of the cavity. As noted earlier, the threshold power is less for emission bands of narrower width $\Delta\bar{\nu}$. In Eq. 9i-12, h and c are respectively Planck's constant and the velocity of light.

References

Sources Referred to in Text

1. Pringsheim, P.: "Fluorescence and Phosphorescence," pp. 2–5, 290–297, Interscience Publishers, Inc., New York, 1949.
2. Kröger, F. A.: "Some Aspects of the Luminescence of Solids," p. 36, American Elsevier Publishing Company, Inc., New York, 1948.
3. Curie, D.: "Luminescence of Crystals," pp. 2–5, translated by G. F. J. Garlick, Methuen & Co., Ltd., London; John Wiley & Sons, Inc., New York, 1963.
4. Garlick, G. F. J.: In "Handbuch der Physik," vol. XXVI (2), Springer-Verlag OHG, Berlin, 1958.
5. Williams, F. E., and M. H. Hebb: *Phys. Rev.* **84**, 1181 (1951).
6. Henderson, S. T.: *Proc. Roy. Soc. (London)*, ser. A, **173**, 323 (1939).
7. Lord, Rees, and Wise: *Proc. Phys. Soc. (London)* **59**, 473 (1947).
8. Butler, K. H.: *J. Electrochem. Soc.* **93**, 143 (1948); **97**, 265 (1950).
9. Brinkman, H., and C. C. Vlam: *Physica* **14**, 650 (1948).
10. Schawlow, A. L., and C. J. Townes: *Phys. Rev.* **112**, 1940 (1958).
11. Maiman, T. H.: *Nature* **187**, 493 (1960).
12. Kiss, Z. J., and R. J. Pressley: *Proc. IEEE*, **54**, 1236 (1966).
13. Mak, A. A., Y. A. Anan'ev, and B. A. Ermakov: *Soviet Phys.* **92**, 419 (1968).
14. Bowers, R., and N. T. Melamed: *Phys. Rev.* **99**, 1781 (1955).
15. Prener, J. S., and F. Williams: *Phys. Rev.* **103**, 342 (1956).
16. Williams, F.: *Phys. Stat. Solidi* **25**, 493 (1968).
17. Tregellas-Williams, J.: *J. Electrochem. Soc.* **105**, 173 (1958).
18. Kröger, Hoogenstraaten, Bottema, and Botden: *Physica* **14**, 81 (1948).
19. Gurney, R. W., and N. F. Mott: *Trans. Faraday Soc.* **35**, 71 (1939).
20. Johnson, P. D., and F. Williams: *J. Chem. Phys.* **18**, 1477 (1950).
21. Urbach, F.: *Wien. Ber.* **139**(IIA), 363 (1930).
22. Randall, J. T., and M. H. F. Wilkins: *Proc. Roy. Soc. (London)*, ser. A, **184**, 367 (1945).
23. Destriau, G.: *Phil. Mag.* **38**, 700 (1947).
24. Curie, D.: *J. Phys. Radium* **13**, 317 (1952).

Monographs and Textbooks

Adirowitsch, E. I.: "Einige Fragen zur Theorie der Lumineszenz der Kristalle," Akademie-Verlag GmbH, Berlin, 1953. Originally published in Russian in 1950, this textbook

TABLE 9i-4. EMISSION SPECTRA AND AFTERGLOW OF CATHODOLUMINESCENT SCREENS*

Screen	Phosphor	$\bar{\nu}_0 \times 10^{-4}$ cm^{-1}	$h \times 10^{-4}$ cm^{-1}	Afterglow τ, sec	Comments
P1......	$Zn_2SiO_4:Mn$(rhbhd.)†	1.89	0.09	1.1×10^{-2}	
P2......	$ZnS:Cu,Ag$ (hex.)†	1.85	0.14	Medium, intensity-dependent hyperbolic	Ag blue only about $\frac{1}{10}$ of Cu green band
		2.22	0.15		
P3......	$(Zn,Be)_2SiO_4:Mn$	1.84	0.09	6×10^{-3}	Slightly concave upward exponential afterglow; $R = 0.30$
		1.64	0.11		
P4......	$ZnS:Ag$ (hex.)†	2.22	0.15	Medium short, hyperbolic	White screen composed of two phosphors
	$(Zn,Cd)S:Ag$ (hex.)†	1.76	0.16		
P5......	$CaWO_4$	2.21	0.29	1.1×10^{-5}	
P6......	$ZnS:Ag$ (hex.)	2.22	0.15	Short hyperbolic	Same as P4, except more $(ZnCd)S/ZnS$ for less blue white
	$(Zn,Cd)S:Ag$ (hex.)	1.76	0.16		
P7......	$ZnS:Ag$ (hex.)	2.26	0.17	Blue flash followed by slow yellow hyperbolic afterglow	Two-layer screen: electrons excite $ZnS:Ag$, whose emission excites $(ZnCd)S:Cu$
	$(Zn,Cd)S:Cu$	1.78	0.15		
P11‡....	$ZnS:Ag,Ni$ (cub.)	2.11	0.19	Short hyperbolic, current dependent	
P12.....	$(Zn,Mg)F_2:Mn$	1.68	0.11	8.5×10^{-2}	Tends to burn at high currents
P13.....	$MgSiO_3:Mn$	1.54	0.10	1.7×10^{-2}	
P14.....	$ZnS:Ag$ (hex.)	2.26	0.17	Blue flash followed by yellow-orange afterglow	Same as P7, except that higher Cd content in $(Zn,Cd)S:Cu$ yields orange afterglow
	$(Zn,Cd)S:Cu$ (hex.)†	1.65	0.16		
P15.....	ZnO	3.15	0.12	Ultraviolet 2×10^{-8}	
		2.42	0.24	Visible 4×10^{-7}	
P16.....	$(Ca,Mg)SiO_3:Ce$	2.60	0.20	4×10^{-8}	
P17.....	ZnO	2.21	0.15	10^{-6}	Similar to P7, except ZnO substituted for $ZnS:Ag$
	$(Zn,Cd)S:Cu$	1.78	0.15	Medium hyperbolic	
P18.....	$(Ca,Mg)SiO_3:Ti$	2.34	0.40	2×10^{-5}	
	$(Zn,Be)SiO_4:Mn$	1.84	0.09		
		1.64	0.11	6×10^{-3}	$R = 0.71$
P19.....	$KMgF_3:Mn$	1.68	0.09	7.5×10^{-2}	Tends to burn at higher currents
P20.....	$(Zn,Cd)S:Ag$	1.76	0.16	Medium short, hyperbolic	
P21.....	$MgF_2:Mn$	1.66	0.08	8×10^{-2}	Also weak blue band present; tends to burn at high currents
P22§....	$ZnS:Ag$	2.22	0.15	Short hyperbolic	Three-color screen
	$Zn_2SiO_4:Mn$	1.89	0.09	1.1×10^{-2}	
	$Zn_3(PO_4):Mn$	1.56	0.09	1.2×10^{-2}	
P23.....	$ZnS:Ag$	2.22	0.15	Medium short, hyperbolic afterglow	Similar to P4
	$(Zn,Cd)S:Ag$	1.73	0.16		
P24.....	ZnO	1.93	0.21	1.4×10^{-7}	No ultraviolet emission as in P15
P25.....	$CaSiO_3:Pb,Mn$	2.80	0.33		
		1.74	0.10		
		1.60	0.10	1.6×10^{-2}	$R = 0.42$ for Mn bands
P26.....	$ZnF_2:Mn$	1.67	0.09	8.3×10^{-2}	Tends to burn at high currents
P27.....	$Zn_3(PO_4)_2:Mn$	1.56	0.09	1.2×10^{-2}	
P28.....	$(Zn,Cd)S:Cu,Ag$	....	....	Long hyperbolic	Multiple emission bands in yellow green, two-phosphor screen

TABLE 9i-4. EMISSION SPECTRA AND AFTERGLOW OF CATHODOLUMINESCENT SCREENS*
(*Continued*)

Screen	Phosphor	$\bar{\nu}_0 \times 10^{-4}$ cm^{-1}	$h \times 10^{-4}$ cm^{-1}	Afterglow τ, sec	Comments
P29.....	ZnS:Cu,Ag CaWO$_4$	....	....	Medium persistence	Two-component screen composed of P2 and P5

* Based on "Optical Characteristics of Cathode Ray Tube Screens," compiled by the Joint Electron Tube Engineering Council, Oct. 1, 1959.
† Available as standard phosphors from the National Bureau of Standards.
‡ P8 and P9 are obsolete; P10 is the dark trace KCl screen which is not a luminescent material. P11 is also available as a transparent screen.
§ The blue and red components are currently for most P22 screens (Zn,Cd)S:Ag and Eu-doped vanadate, respectively.

This parameter R is the ratio of the parameter b of Eq. (9i-2) for the band with the large $\bar{\nu}$ to b with the smaller $\bar{\nu}_0$.

The parameters describing the absorption spectra as a function of wave number and wavelength are simply related. Because the absorption coefficient α is the probability of absorption of a photon of wave number $\bar{\nu}$ or of wavelength λ per centimeter of absorbing medium, the maximum at λ_0 in the plot of $\alpha(\lambda)$ vs. λ coincides with the maximum $\bar{\nu}_0$ in the plot of $\alpha(\bar{\nu})$ vs. $\bar{\nu}$; that is, $\bar{\nu}_0 = 1/\lambda_0$.

TABLE 9i-5. EMISSION, PULSE HEIGHT, AND DECAY TIME OF
SCINTILLATION PHOSPHORS

Phosphor	$\bar{\nu}_0 \times 10^{-4}$ cm^{-1}	Approximate energy conversion efficiency η_E, % for β-rays	Decay time τ, sec	Comments
Anthracene.....	2.4, 2.3, 2.2	5	3×10^{-8}	Several overlapping bands
Naphthalene...	3.16, 3.12, 3.07, 3.02, 2.97, 2.92	1	8×10^{-8}	Series of narrow overlapping bands; sublimes
Stilbene........	2.5, 2.4	3	8×10^{-9}	Several overlapping bands
Terphenyl......	2.6, 2.5, 2.3	2	1.2×10^{-8}	Several overlapping bands
NaI:Tl........	2.42	15	2.5×10^{-7}	Very hygroscopic
KI:Tl.........	2.39	2	10^{-6}	
CsI:Tl.........	1.68	9	$\sim 10^{-5}$	Hygroscopic
ZnS:Ag........	2.29	25	$\sim 10^{-5}$	$h = 1.2 \times 10^3$ cm^{-1}, also efficient for α-particles
CaWO$_4$........	2.21	5	1.1×10^{-5}	$h = 2.9 \times 10^3$ cm^{-1}, stable

discusses the luminescence of crystals using thermodynamic, kinetic, and quantum-mechanical methods.

Aven, M., and J. S. Prener, eds.: "Physics and Chemistry of II-IV Compounds," North-Holland Publishing Company, Amsterdam: John Wiley & Sons, Inc., New York, 1967. Authoritative review of electrical and optical phenomena in II-VI semiconductors and luminescent materials.

Curie, D.: "Luminescence Cristalline,' Dunod, Paris, 1960. Translated into English by G. F. J. Garlick, "Luminescence in Crystals," Methuen & Co., Ltd., London; John Wiley & Sons, Inc., New York, 1963. A compact textbook emphasizing theoretical work on luminescence in crystals.

Förster, T.: "Fluoreszenz organischer Verbindungen," Vandenhoeck and Ruprecht, Gottingen, 1951. A textbook on the fundamentals of luminescence in organic materials.

Garlick, G. F. J.: "Luminescent Materials," Oxford University Press, New York, 1949. An elementary textbook emphasizing the fundamental processes in phosphors.

Goldberg, P., ed.: "Luminescence of Inorganic Solids," Academic Press, Inc., New York, 1966. Comprehensive modern review with each chapter by different authority.

Ivey, H. F.: "Electroluminescence and Related Effects," Academic Press, Inc., New York, 1963. Review of luminescence involving electric fields.

Kröger, F. A.: "Some Aspects of the Luminescence of Solids," American Elsevier Publishing Company, Inc., New York, 1948. A monograph on inorganic phosphors classified by activators.

Kröger, F. A.: "Chemistry of Imperfect Crystals," North-Holland Publishing Company, Amsterdam; John Wiley & Sons, Inc., New York, 1964. Comprehensive treatise on defects and imperfections in luminescent materials and semiconductors.

Pringsheim, P.: "Fluorescence and Phosphorescence," Interscience Publishers, Inc., New York, 1949. A descriptive encyclopedic reference text on the luminescence of gases, liquids, and solids.

Review Articles

Curie, D.: Theories of Electroluminescence, in "Progress in Semiconductors," vol. 2, p. 251, Heywood and Co., London, 1957.

Destriau, G., and H. F. Ivey: Electroluminescence and Related Topics, *Proc. IRE* **43**, 1911 (1955).

Garlick, G. F. J.: Cathodoluminescence, *Advan. Electron.* **2**, 151 (1950).

Garlick, G. F. J.: Luminescence, in "Handbuch der Physik," vol. XXVI(2), Springer-Verlag OHG, Berlin, 1958.

Klick, C. C., and J. H. Schulman: Luminescence in Solids, *Solid State Phys.* **5**, 97 (1957).

Piper, W. W., and F. Williams: Electroluminescence, *Solid State Phys.* **6**, 95 (1958).

Williams, F.: Solid State Luminescence, *Advan. Electron.* **5**, 137 (1953).

Williams, F.: *Phys. Stat. Solidi*, **25**, 493 (1968).

Reports of Symposia

Fonda, G. R., and F. Seitz, eds.: "Solid Luminescent Materials," John Wiley & Sons, Inc., New York, 1948. Papers of the symposium at Cornell University in 1946.

Kallmann, H. P., and G. M. Spruch: "Luminescence of Inorganic and Organic Materials," John Wiley & Sons, Inc., New York, 1961. Papers and discussion of the symposium at New York University in 1961.

Luminescence, *Brit. J. Appl. Phys.* **54** (1955). Papers and discussion of the symposium at Cambridge, England, in 1954.

La Luminescence des corps cristallins anorganiques, *J. Phys. Radium* **17**, 609 (1956). Papers and discussion of the symposium at Paris in 1956.

Riehl, N., and Kallmann, H. P.: "International Symposium on Luminescence: The Physics and Chemistry of Scintillators," K. Thiemig, Munich 1966. Papers of symposium on luminescence emphasizing scintillators in Munich in 1965.

Thomas, D. G., ed.: "II-VI Semiconducting Compounds," W. A. Benjamin, Inc., New York, 1967. Paper and discussion of symposium at Brown University in 1967 on II-VI semiconductors and luminescent crystals.

Proceedings of 1966 Internaational Conference on Luminescence, edited by G. Szigeti (Hungarian Academy of Sciences, Budapest, 1968). Papers and discussion of conference in Budapest in 1966.

Proceedings of 1969 International Conference on Luminescence, edited by F. Williams (North Holland Publishing Co., Amsterdam, 1970). Paper and discussion, with indices, of conference at the University of Delaware in 1969.

9j. Work Function and Secondary Emission

GEORGE A. HAAS

U.S. Naval Research Laboratory

9j-1. Work Function Measurements. The work function ϕ of a substance is given as the difference in energy between the Fermi level (or electrochemical potential) of a solid and the electrostatic surface potential just outside. This is what is generally measured in thermionic (Th) and contact potential (C.P.D.) measurements. The photoelectric "work function" (P.E.), however, is normally taken as the measure of the photoelectric threshold and represents the energy difference between the level of the highest-lying electron at room temperature and the electrostatic surface potential. Although the highest-energy electron of a metal at room temperature is very near the Fermi level, this is not necessarily true for semiconductors. (This fact must be kept in mind when one is comparing photoelectric work functions of semiconductors with those derived from thermionic or contact potential techniques.) Table 9j-1 lists the work functions of elements as determined by *thermionic, photoelectric,* and *contact potential* difference methods, while Table 9j-2 gives the thermionic work functions of various compounds used in electron emitter applications.

The thermionic work function ϕ as a rule is obtained by analyzing the emitted current by means of the Richardson equation:

$$J = 120T^2 e^{-\phi/kT}$$

where J in the current density in amp/cm², T is the absolute temperature, k is $8.62 \cdot 10^{-5}$ eV/deg, and ϕ is given in electron volts. ϕ is usually not constant in the temperature range of measurement but for most substances can be expressed in terms of a linear temperature dependence $\phi = \phi_0 + \alpha T$, where ϕ_0 is the temperature-independent component of the work function, and α is the temperature coefficient.

In many thermionic measurements, unfortunately, the work function quoted and subsequently recorded in review articles is the value obtained from the slope of a Richardson plot (log J/T^2 vs. $1/T$), which is just ϕ_0. If there is an appreciable temperature dependence, this value can be highly erroneous since these measurements are obtained at fairly high temperatures (i.e., 1000 to 2000 K). Consequently, unless the temperature dependence can be determined from the published data by some other method (e.g., from the intercept of the Richardson plot or "Richardson A value") the results are not included in this review. Furthermore, since the linear approximation to the temperature dependence very likely does not hold for all temperatures, the temperature range in which the measurements were made is also included.

Except where specifically noted, the photoelectric work function measurements are carried out at room temperature and therefore require no additional information regarding temperature range. The same is also true of contact potential difference measurements which are normally obtained at room temperature by measuring the difference in electrostatic surface potential between a substance having a known work function value or "standard" (e.g., W, Hg, Ag, etc.), and the substance to be measured.

TABLE 9j-1. WORK FUNCTIONS OF THE ELEMENTS

Element	Work function eV	Technique	Notes	Reference	Year
Ag.........	$4.31 + 0.1 \times 10^{-4}T$ 1160–1200 K	Th	1	1	1953
	4.32 1230 K	Th	2	2	1956
	4.5	P.E.	......	3	1950
	4.3	P.E.	......	4	1953
	4.32	C.P.D.	3, 4	5	1951
	4.3	C.P.D.	4, 5	6	1964
	4.44	C.P.D.	4, 6	6	1964
	4.29	C.P.D.	4, 7	6	1964
Al..........	4.36	P.E.	......	7	1936
	4.08	P.E.	......	8	1936
	4.2	P.E.	......	9	1944
	4.24	C.P.D.	3, 4	5	1951
	4.19	C.P.D.	4, 5	10	1957
	4.18	C.P.D.	5, 8	11	1966
Au.........	$4.25 + 0.15 \times 10^{-4}T$ 1160–1280 K	Th	9	1	1953
	5.1	P.E.	9	12	1961
	5.4	P.E.	9	13	1966
	5.45	C.P.D.	9, 10	14	1966
	5.22	C.P.D.	5, 9, 11	15	1966
	5.4	C.P.D.	5, 9, 10	13	1966
As..........	4.66	P.E.	......	16	1949
	4.72	P.E.	......	17	1949
Ba.........	$2.3 + 5 \times 10^{-4}T$ 1000–1300 K	Th	2	18	1965
	2.49	P.E.	......	19	1939
	2.48	P.E.	......	20	1940
	2.42	C.P.D.	3	21	1935
	2.35	C.P.D.	5	22	1941
	2.66	C.P.D.	6	23	1952
	2.5	C.P.D.	......	24	1963
B..........	4.4–4.6	P.E.	......	25	1948
Be.........	3.67 920–1180 K	Th	......	26	1966
	3.3	P.E.	......	27	1934
	3.92	P.E.	......	28	1937
	3.89	C.P.D.	12	29	1963
Bi..........	4.31	P.E.	......	30	1936
	4.46	P.E.	......	31	1941
	4.34	P.E.	......	17	1949
C...........	$4.39 + 1.7 \times 10^{-4}T$ 1300–2200 K	Th	......	32	1947
	$4.6 + 0.6 \times 10^{-4}T$ 1490–1670 K	Th	1	33	1952
	4.81	P.E.	......	34	1926
Ca..........	3.2	P.E.	......	35	1932
	2.7	P.E	......	36	1936
	3.21	P.E	......	37	1937
Cd..........	4.07	P.E.	......	38	1931
	4.099	P.E.	......	9	1944
	4.0	C.P.D.	13	39	1953
	4.22	C.P.D.	4, 6	40	1955
Ce..........	$2.48 + 1.8 \times 10^{-4}T$ 1060–1450 K	Th	2, 4	41	1926
	2.84	P.E.	......	35	1932

TABLE 9j-1. WORK FUNCTIONS OF THE ELEMENTS (*Continued*)

Element	Work function eV	Technique	Notes	Reference	Year
Co..........	$4.4 + 0.9 \times 10^{-4}T$ $\simeq$1200–1450 K	Th	......	42	1942
	$4.41 + 0.6 \times 10^{-4}T$ 1410–1590 K	Th	1	43	1952
	4.12	P.E.	14	44	1931
	4.25	P.E.	15	44	1931
Cr..........	$4.58 + 0.6 \times 10^{-4}T$ 1450–1600 K	Th	1	43	1952
	$\simeq$3.9 1100–1400 K	Th	......	26	1966
	4.4	P.E.	......	12	1961
Cs..........	$\simeq$1.86 ~500 K	Th	......	45	1966
	2.14	P.E.	......	46	1964
	1.84	C.P.D.	3, 4	6	1964
Cu..........	4.5 1160–1280 K	Th	1	1	1953
	4.6 ~1350 K	Th	......	2	1956
	4.4 1100–1300 K	Th	......	26	1966
	4.76	P.E.	......	47	1934
	4.86⟨111⟩	P.E.	......	48	1935
	5.61⟨110⟩	P.E.	......	48	1935
	4.60	C.P.D.	3, 4	5	1951
	4.51	C.P.D.	4, 5	10	1957
Er..........	$2.97 + 0.65 \times 10^{-4}T$ 1150–1500 K	Th	......	48a	1967
Fe..........	$\phi_\beta = 4.48 + 1.3 \times 10^{-4}T$	Th	......	42	1942
	$\phi_\gamma = 4.21 + 3.75 \times 10^{-4}T$ $\simeq$1200–1450 K	Th	......	42	1942
	$4.31 + 0.6 \times 10^{-4}$ 1410–1610 K	Th	1	43	1952
	4.5 1200–1500 K	Th	......	26	1966
	$\phi_\beta = 4.62$	P.E.	......	49	1953
	$\phi_\gamma = 4.68$	P.E.	......	49	1953
	4.16	C.P.D.	4, 5	10	1957
Ga..........	3.8	C.P.D.	16	50	1938
Ge..........	$\simeq$3.5 ~900 K	Th	......	2	1956
	4.11⟨111⟩	P.E.	......	51	1959
Hf..........	$3.6 + 1.4 \times 10^{-4}T$ 1250–1820 K	Th	......	52	1957
	3.85 ~1000–1700 K	Th	2	53	1962
Hg..........	4.52	P.E.	......	54	1931
	4.5	P.E.	......	55	1934
Ir..........	$5.3 + 0.2 \times 10^{-4}T$ 1700–2200 K	Th	......	56	1951
	$5.4 - 0.3 \times 10^{-4}T$ 1590–2320 K	Th	......	57	1956
	$\simeq$5.28 1300–2000 K	Th	......	58	1966
	4.57	C.P.D.	16	50	1938
K..........	2.24	P.E.	......	59	1932
	2.26	P.E.	......	60	1937

TABLE 9j-1. WORK FUNCTIONS OF THE ELEMENTS (*Continued*)

Element	Work function cV	Technique	Notes	Reference	Year
La...........	$\simeq$3.0 1200–1500 K	Th	2	41	1926
Li...........	2.42	P.E.	....	47	1934
	1.4	C.P.D.	16	50	1938
Mg..........	3.66	P.E.	....	61	1964
	3.12	C.P.D.	13	62	1951
Mn..........	$3.83 + 1.1 \times 10^{-4}T$ 1370–1520 K	Th	1	43	1952
	3.76	P.E.	....	47	1934
Mo..........	$4.38-0.25 \times 10^{-4}T$ 1410–2110 K	Th	....	52	1957
	$4.33 + 0.1 \times 10^{-4}T$ $\sim$1300–1900 K	Th	....	63	1962
	4.25 1600 K	Th	....	64	1966
	$4.33-1.52 \times 10^{-4}T$ 1200–2000 K	Th	....	58	1966
	4.41	P.E.	....	63	1962
	4.20	C.P.D.	4	6	1964
Na..........	2.06	P.E.	....	47	1934
	2.29	P.E.	....	28	1937
	2.28	P.E.	....	20	1940
Nb..........	4.0 1400–2100 K	Th	....	65	1964
	4.3 2200 K	Th	....	66	1964
	$\simeq$4.19 1050–2100 K	Th	....	58	1966
	4.33$\langle$110$\rangle$	P.E.	....	67	1963
	4.55$\langle$335$\rangle$	P.E.	....	67	1963
	4.66$\langle$111$\rangle$	P.E.	....	67	1963
	4.38	C.P.D.	4	68	1964
Nd..........	$\simeq$2.95 1150–1450 K	Th	2	41	1926
Ni...........	$5.24 + 0.75 \times 10^{-4}T$ $\simeq$1100 K	Th	....	69	1949
	4.5 1410–1610 K	Th	1	43	1952
	4.41 1170–1250 K	Th	....	26	1966
	$6.27-1.0 \times 10^{-3}T$ 1380–1500 K	Th	....	26	1966
	5.05 $T = 623$ K	P.E.	....	69	1949
	5.2 $T = 1108$ K	P.E.	....	69	1949
	4.73	C.P.D.	4, 5	10	1957
	5.22	C.P.D.	10	14	1966
Os...........	$5.43-3.9 \times 10^{-4}T$ 1413–1640 K	Th	....	58	1966
	$\simeq$5.17 $\sim$1500 K(?)	Th	....	70	1967
	5.93	P.E.	....	71	1966
Pb...........	3.97	P.E.	....	72	1928
	3.49	C.P.D.	13	62	1951
	3.83	C.P.D.	4, 5	73	1956
Pd..........	4.64	C.P.D.	13	39	1953
Pr...........	$2.57 + 1.5 \times 10^{-4}T$ 1120–1410 K	Th	2	41	1926

TABLE 9j-1. WORK FUNCTIONS OF THE ELEMENTS (*Continued*)

Element	Work function eV	Technique	Notes	Reference	Year
Pt............	5.3–5.5 $\sim$1600–1900 K(?)	Th	....	74	1950
	$5.03 + 4.2 \times 10^{-4}T$ 1620–1946 K	Th	....	26	1966
	4.82	C.P.D.	13	39	1953
Rb...........	2.09	P.E.	....	59	1932
Re...........	$4.85 + 0.6 \times 10^{-4}T$ 1470–2150 K	Th	....	75	1963
	$4.7 + 0.75 \times 10^{-4}T$ 1820–2860 K	Th	....	76	1963
	4.8 1600–2200 K	Th	....	65	1964
	4.96 1325–2250 K	Th	....	58	1966
Rh...........	$4.8 + 1.1 \times 10^{-4}T$ $\sim$1500–1900 K(?)	Th	....	77	1938
	$4.9 + 0.2 \times 10^{-4}T$ 1550–1950 K	Th	....	56	1951
	4.92	P.E.	....	78	1931
Ru...........	4.52	C.P.D.	16	50	1938
Sb............	4.01	P.E.	....	79	1937
	4.6	P.E.	....	17	1949
	4.14	C.P.D.	16	50	1938
	4.1	C.P.D.	3, 4	6	1964
Sc............	$3.13 + 0.8 \times 10^{-4}T$ 1150–1500 K	Th	....	48a	1967
Se............	5.11	P.E.	....	47	1934
	4.42	C.P.D.	16	50	1938
Si............	$3.59 + 2.3 \times 10^{-4}T$ 1250–1700 K	Th	....	32	1947
	$4.02 + 2.6 \times 10^{-4}T$ 1373–1623 K	Th	....	80	1953
	5.4$\langle$111$\rangle$	P.E.	....	81	1962
	3.95 (*n* type)	C.P.D.	17	82	1947
	4.2 (*p* type)	C.P.D.	17	82	1947
	4.5 (*n* or *p*)	C.P.D.	18	83	1949
Sm...........	$\sim$3.15 1150–1600 K	Th	2	41	1926
Sn............	3.62	P.E.	....	72	1928
	4.21 liquid	P.E.	....	84	1929
	4.38 α phase	P.E.	....	84	1929
	4.50 β phase	P.E.	....	84	1929
	4.21	C.P.D.	13	85	1952
	4.42	C.P.D.	19	86	1963
Sr............	$2.3 + 0.5 \times 10^{-4}T$ 850–950 K	Th	20	87	1955
	2.24	P.E.	....	47	1934
	2.74	P.E.	....	88	1938
Ta...........	4.35 300–1860 K	Th	21	89	1957
	4.25 1100–2200 K	Th	....	58	1966
	$4.33 + 0.25 \times 10^{-4}T$ 1700–2230 K	Th	....	90	1966
	4.1	P.E.	....	35	1932
	4.05	P.E.	....	91	1935
	4.22	C.P.D.	4	6	1964

TABLE 9j-1. WORK FUNCTIONS OF THE ELEMENTS (*Continued*)

Element	Work function eV	Technique	Notes	Reference	Year
Te	5.0	P.E.	....	92	1953
	4.7	C.P.D.	16	50	1938
Th	$3.38 + 0.45 \times 10^{-4}T$ 1250–1800 K	Th	....	93	1926
	$\simeq 3.4$	P.E.	....	35	1932
	3.66	P.E.	....	47	1934
	3.46	C.P.D.	16	50	1938
	3.71	C.P.D.	7	94	1962
	3.44	C.P.D.	3	94	1962
Ti	$3.95 + 0.85 \times 10^{-4}T$ 1370–1520 K	Th	1	43	1952
	4.0 1300–1600 K	Th	....	65	1964
	4.45	P.E.	....	95	1954
	4.14	C.P.D.	16	50	1938
Tl	3.68	P.E.	....	96	1935
	3.84	C.P.D.	16	50	1938
U	$3.0 + 2.7 \times 10^{-4}T$ 1250–1400 K	Th	20	96	1959
	3.47 1250–1400 K	Th	22	96	1959
	$2.9 + 2.3 \times 10^{-4}T$ 1020–2000 K	Th	20	76	1963
	$\simeq 3.55$ 1000–1500 K	Th	22	97	1967
	3.47 (α)	P.E.	....	98	1962
	3.52 (β)	P.E.	....	98	1962
	3.39 (γ)	P.E.	....	98	1962
	3.65 (α)	P.E.	....	99	1967
	3.59 (β)	P.E.	....	99	1967
	3.45 (γ)	P.E.	....	99	1967
	3.19	C.P.D.	3, 4	100	1962
	3.63 (α)	C.P.D.	4	97	1967
	3.58 (β)	C.P.D.	4	97	1967
	3.53 (γ)	C.P.D.	4	97	1967
V	$4.12 + 0.75 \times 10^{-4}T$ 1410–1540 K	Th	1	43	1952
	3.77	P.E.	....	47	1934
	4.44	C.P.D.	16	50	1938
W	$4.52 + 0.6 \times 10^{-4}T$ 1350–2200 K	Th	....	101	1950
	$4.5 + 0.15 \times 10^{-4}T$ 1820–2940 K	Th	....	76	1963
	$4.58 + 0.15 \times 10^{-4}T$ 2100–2600 K	Th	....	90	1966
	4.52 1150–2200 K	Th	....	58	1966
	4.6	P.E.	....	35	1932
	4.565	P.E.	....	102	1935
	4.49	P.E.	....	103	1948
	4.55	C.P.D.	4, 5	10	1957
Y	$2.95 + 0.2 \times 10^{-4}T$ 1150–1400 K	Th	....	48a	1967
Zn	4.26	P.E.	....	104	1940
	4.307	P.E.	....	9	1944
	4.11	C.P.D.	4	105	1940
	4.22	C.P.D.	13	85	1952

TABLE 9j-1. WORK FUNCTIONS OF THE ELEMENTS (*Continued*)

Element	Work function eV	Technique	Notes	Reference	Year
Zr...........	3.78 $T = ?$	Th	....	106	1951
	3.73	P.E.	....	35	1932
	4.33	P.E.	....	95	1954
	3.60	C.P.D.	16	50	1938

However, since the resulting work function values depend on the accuracy with which the work function of the standard is known, the substance used for the standard and its assumed work function are also included.

Some of the earlier work function measurements quoted in other reference works are omitted here, especially where an appreciable number of results on the same subject have been recently published employing more refined experimental techniques. That is not to say, however, that "recentness" is synonomous with "cleanliness"; rather, the most recent studies involving techniques such as low-energy electron diffraction and Auger spectrum analysis serve to show how contaminated "clean" surfaces really are. Consequently, the values listed here merely serve to indicate the measured values of the work functions as they are presently limited by experimental refinement of measuring methods. For some substances which can be easily cleaned they are quite accurate; for others they will most certainly be considered outdated by the next review.

Notes for Table 9j-1

1. Effusion method.
2. Deduced from published results.
3. Film on W substrate.
4. Assuming ϕ for bulk W is 4.54.
5. Film on glass substrate.
6. Film on Ta substrate.
7. Bulk.
8. Assuming ϕ of Au is 5.22.
9. No Hg contamination.
10. Absolute work function value using field-emission-retarding potential method.
11. Assuming ϕ of Al $\simeq$ 4.2.
12. From breakdown voltage of metal insulator–metal junction, assuming ϕ of Al = 4.08.
13. Assuming ϕ of Ag = 4.31.
14. Hexagonal crystal structure.
15. Face-centered-cubic structure.
16. Assuming ϕ of Hg = 4.52.
17. Using various oxides, e.g., ϕ of PbO = 4.06.
18. Absolute work function value using photoemission-retarding potential method.
19. From breakdown of metal insulator–metal junction, assuming ϕ of Al = 4.08.
20. Monolayer on W substrate.
21. Also used contact potential shifts of two <211> Ta crystals to get low-temperature values of ϕ.
22. 15 layers.

References for Table 9j-1

1. Jain, S. C., and K. S. Krishnan: *Proc. Roy. Soc.* (*London*), ser. A, **217,** 451 (1953).
2. Bolshov, V. G.: *J. Tech. Phys. U.S.S.R.* **26,** 1123 (1956).
3. Blackmer, L., and H. Farnsworth: *Phys. Rev.* **77,** 826 (1950).
4. Gilleo, M.: *Phys. Rev.* **91,** 534 (1953).
5. Mitchell, E., and J. Mitchell: *Proc. Roy. Soc.* (*London*), ser. A, **210,** 70 (1951).
6. Hopkins, B. J., and J. C. Riviere: *Brit. J. Appl. Phys.* **15,** 941 (1964).
7. Gaviola, E., and J. Strong: *Phys. Rev.* **49,** 446 (1936).
8. Brady, J., and V. Jacobsmayer: *Phys. Rev.* **49,** 670 (1936).
9. Suhrmann, R., and J. Pietrzyk: *Z. Physik* **122,** 600 (1944).
10. Riviere, J. C.: *Proc. Phys. Soc.* (*London*) **70,** 676 (1957).
11. Huber, E. E., Jr., and C. T. Kirk, Jr.: *Surface Sci.* **5,** 447 (1966).
12. Katrich, G. A., and O. G. Sarbei: *Soviet Phys.—Solid State* **3,** 1181 (1961).

13. Sachtler, W. M. H., H. J. G. Dorgelo, and A. A. Holscher: *Surface Sci.* **5**, 221 (1966).
14. Holscher, A. A.: *Surface Sci.* **4**, 89 (1966).
15. Huber, E. E., Jr.: *Appl. Phys. Letters* **8**, 169 (1966).
16. Taft, E., and L. Apker: *Phys. Rev.* **75**, 1181 (1949).
17. Apker, L., E. Taft, and J. Dickey: *Phys. Rev.* **76**, 270 (1949).
18. Dyubua, B. Ch., and L. A. Stepanov: *Radio Eng. Electron. Phys.* **7**, 1878 (1965).
19. Cashman, R., and E. Bassoe: *Phys. Rev.* **55**, 63 (1939).
20. Maurer, R.: *Phys. Rev.* **57**, 653 (1940).
21. Anderson, P. A.: *Phys. Rev.* **47**, 958 (1935).
22. Anderson, P. A.: *Phys. Rev.* **59**, 1034 (1941).
23. Anderson, P. A.: *Phys. Rev.* **88**, 655 (1952).
24. Gavrilyuk, V. M., and V. K. Medvedev: *Soviet Phys.—Solid State* **4**, 1737 (1963).
25. Apker, L., E. Taft, and J. Dickey: *Phys. Rev.* **74**, 1462 (1948).
26. Wilson, R. G.: *J. Appl. Phys.* **37**, 2261 (1966).
27. Suhrmann, R., and A. A. Schallamach: *Z. Physik* **91**, 775 (1934).
28. Mann, M., and L. Du Bridge: *Phys. Rev.* **51**, 120 (1937).
29. Simmons, J. C.: *Phys. Rev. Letters* **10**, 10 (1963).
30. Rentschler, H. C., and D. E. Henry: *J. Opt. Soc. Am.* **26**, 30 (1936).
31. Weber, A., and C. Eisel: *Phys. Rev.* **59**, 473A (1941).
32. Braun, A., and G. Busch: *Helv. Phys. Acta.* **20**, 33 (1947).
33. Jain, S. C., and K. S. Krishnan: *Proc. Roy. Soc. (London)*, ser. A, **213**, 143 (1952).
34. Roy, S.: *Proc. Roy. Soc. (London)*, ser. A, **112**, 599 (1926).
35. Rentschler, H. C., D. E. Henry, and K. Smith: *Rev. Sci. Instr.* **3**, 794 (1932).
36. Jamison, N., and R. Cashman: *Phys. Rev.* **50**, 624 (1936).
37. Lieben, I.: *Phys. Rev.* **51**, 642 (1937).
38. Bomke, H.: *Ann. Phys.* **10**, 579 (1931).
39. Giner, J., and E. Lange: *Naturwissenschaften* **40**, 506 (1953).
40. Anderson, P. A.: *Phys. Rev.* **98**, 1739 (1955).
41. Schumacher, E. E., and J. E. Harris: *J. Am. Chem. Soc.* **48**, 3108 (1926).
42. Whalin, H.: *Phys. Rev.* **61**, 509 (1942).
43. Jain, S. C., and K. S. Krishnan: *Proc. Roy. Soc. (London)*, ser. A, **215**, 431 (1952).
44. Cardwell, A.: *Phys. Rev.* **38**, 2033 (1931).
45. Wilson, R. G.: *J. Appl. Phys.* **37**, 4125 (1966).
46. Boutry, G., R. Evard, and J. Richard: *Compt. Rend.* **258**, 143 (1964).
47. Schulze, R.: *Z. Physik* **92**, 212 (1934).
48. Underwood, N.: *Phys. Rev.* **47**, 502 (1935).
48a. Savitskiy, E. M., V. F. Terekhova, and E. V. Maslova: *Radio Eng. and Electron. Phys.* **7**, 1233 (1967).
49. Cardwell, A.: *Phys. Rev.* **92**, 554 (1953).
50. Klein, O., and E. Lange: *Z. Elektrochem* **44**, 542 (1938).
51. Haneman, D.: *J. Phys. Chem Solids* **11**, 205 (1959).
52. Hagstrum, H. D.: *J. Appl. Phys.* **28**, 323 (1957).
53. Dyubua, B. Ch., A. I. Pekarev, B. N. Popov, and M. A. Tylkina: *Radio Eng. El ctron. Phys.* **9**, 1463 (1962).
54. Roller, D., W. Jordan, and C. Woodward: *Phys. Rev.* **38**, 396 (1931).
55. Cassel, H., and A. Schneider: *Naturwissenschaften* **22**, 464 (1934).
56. Weinreich, O.: *Phys. Rev.* **82**, 573 (1951).
57. Goldwater, D. L., and W. E. Danforth: *Phys. Rev.* **103**, 871 (1956).
58. Wilson, R. G.: *J. Appl. Phys.* **37**, 3170 (1966).
59. Brady, J.: *Phys. Rev.* **41**, 613 (1932).
60. Mayer, H.: *Ann. Phys.* **29**, 129 (1937).
61. Roger, G.: *Compt. Rend.* **258**, 1458 (1964).
62. Fianda, F., and E. Lange: *Z. Elektrochem.* **55**, 237 (1951).
63. Jaklevic, R., and D. Juenker: *J. Appl. Phys.* **33**, 562 (1962).
64. Dyubua, B. Ch., O. K. Kultashev, and L. V. Gorshkova: *Soviet Phys.—Solid State* **8**, 882 (1966).
65. Dyubua, B. Ch., O. K. Kultashev, and I. A. Tsyganova: *Radio Eng. Electron. Phys.* **11**, 1716 (1964).
66. Mastkevich, T. L., T. V. Krachino, A. P. Kazansten, and L. S. Markova: *Zh. Tekhn. Fiz.* **34**, 2021 (1964).
67. Milton, O.: Doctoral Thesis, Brown University, 1963. Cited in *Ref. Dissert. Abstr.* **24**, 3808 (1964).
68. Hopkins, B. J., and K. J. Ross: *Brit. J. Appl. Phys.* **15**, 89 (1964).
69. Cardwell, A.: *Phys. Rev.* **76**, 125 (1949).
70. Houston, J. M.: *Proc. 27th Ann. Phys. Electron. Conf.*, MIT, Cambridge, Mass., p. 95, 1967.

71. Van der Velder, G. J. M.: As cited in P. Zalm and A. J. A. Van Stratum, *Philips Tech. Rev.* **27**, 69 (1966).
72. Lurisky, P., and S. Prilesaev: *Z. Phys.* **49**, 236 (1928).
73. Anderson, P. A., and A. L. Hunt: *Phys. Rev.* **102**, 367 (1956).
74. Ertel, A.: *Phys. Rev.* **78**, 353 (1950).
75. Anderson, J., E. W. Danforth, and A. I. Williams: *J. Appl. Phys.* **34**, 2260 (1963).
76. Haas, G. A., and J. T. Jensen: *J. Appl. Phys.* **34**, 3451 (1963).
77. Whalin, H., and L. Whitney: *J. Chem. Phys.* **6**, 594 (1938).
78. Dixon, E.: *Phys. Rev.* **37**, 60 (1931).
79. Middel, V.: *Z. Physik* **105**, 358 (1937).
80. Esaki, L.: *J. Phys. Soc. Japan* **8**, 347 (1953).
81. Van Laar, J., and J. Scheer: *Philips Res. Rept.* **17**, 101 (1962).
82. Meyerhof, W.: *Phys. Rev.* **71**, 727 (1947).
83. Smith, A.: *Phys. Rev.* **75**, 953 (1949).
84. Goetz, A.: *Phys. Rev.* **33**, 373 (1929).
85. Hirschberg, R., and E. Lange: *Naturwissenschaften* **39**, 131 (1952).
86. Simmons, J. G.: *Phys. Rev. Letters* **10**, 10 (1963).
87. Moore, G. E., and H. W. Allison: *J. Chem. Phys.* **23**, 1609 (1955).
88. Cashman, R., and E. Bassoe: *Phys. Rev.* **53**, 919A (1938).
89. Shelton, H.: *Phys. Rev.* **107**, 1535 (1957).
90. Zandberg, E. Ya., and A. Ya. Tontegode: *Soviet Phys.—Tech. Phys.* **10**, 1162 (1966).
91. Cashman, R., and W. Huxford: *Phys. Rev.* **48**, 734 (1935).
92. Taft, E., and L. Apker: *J. Opt. Soc. Am.* **43**, 81 (1953).
93. Zwikker, C.: *Proc. Roy Acad. Sci. Amsterdam* **29**, 792 (1926).
94. Riviere, J. C.: *Proc. Phys. Soc. (London)* **80**, 124 (1962).
95. Malamud, H., and A. Krumbein: *J. Appl. Phys.* **25**, 591 (1954).
96. Rauh, E. G., and R. J. Thorn: *J. Chem. Phys.* **31**, 1481 (1959).
97. Barry, D. E., B. J. Hopkins, and A. J. Sargood: *Surface Sci.* **7**, 365 (1967).
98. Fry, R., and A. Cardwell: *Phys. Rev.* **125**, 471 (1962).
99. Lea, C., and C. H. B. Mee: *Surface Sci.* **8**, 417 (1967).
100. Riviere, J. C.: *Proc. Phys. Soc. (London)* **80**, 116 (1962).
101. Nichols, M. H.: **78**, 158 (1950).
102. Krüger, F., and G. Stabenow: *Ann. Phys.* **23**, 713 (1935).
103. Apker, L., E. Taft, and J. Dickey: *Phys. Rev.* **73**, 46 (1948).
104. Klug, W., and H. Steyskal: *Z. Physik* **116**, 415 (1940).
105. Anderson, P. A.: *Phys. Rev.* **57**, 122 (1940).
106. Wahl, A.: *Phys. Rev.* **82**, 574 (1951).

TABLE 9j-2. WORK FUNCTIONS FOR VARIOUS COMPOUNDS

Emissive material	Base metal	Work function	Temperature range, K	Reference	Year
Oxides of:					
Ba............	Pt	$1.4 + 7 \times 10^{-4}T$	700–950	1	1965
Sr............	Pt	$1.6 + 7 \times 10^{-4}T$	1000–1180	1	1965
Ca............	Pt	$1.86 + 7 \times 10^{-4}T$	1080–1380	1	1965
Th............	W, Ta, Mo	$\sim\!2.6 + 2.4 \times 10^{-4}T$	$\sim\!1300$–2200	2, 3	1947–8
	Re, Nb	$2.37 + 4 \times 10^{-4}T$	1300–1700	4	1962
Hf............	W	$2.82 + 4.72 \times 10^{-4}T$	1300–1950	5	1959
Zr............	....	3.9	2000	6	1953
Ti............	....	3.7	2000	6	1953
Sc............	W	$3.66 + 2.2 \times 10^{-4}T$	1300–2000	5	1959
Y............	Re, Nb	$2.9 + 1.6 \times 10^{-4}T$	1300–1700	4	1962
La............	Ni	$3.0 + .075 \times 10^{-4}T$	$\sim\!1000$–1100	7	1959
Ce............	....	$2.3 + 4.1 \times 10^{-4}T$	$\sim\!1800$–2000	6	1953
Pr............	W	$2.8 + 5.4 \times 10^{-4}T$	$\sim\!1400$–1600	8	1958
Nd............	W	$2.3 + 4.1 \times 10^{-4}T$	$\sim\!1400$–1600	8	1958
Sm............	W	$2.8 + 5.1 \times 10^{-4}T$	$\sim\!1700$–1900	8	1958
Eu............	W	$2.6 + 6 \times 10^{-4}T$	$\sim\!1400$–1600	8	1958
Gd............	W	$2.1 + 4.5 \times 10^{-4}T$	1100–1600	8	1958
Tb............	W	$2.1 + 5.4 \times 10^{-4}T$	1400–1600	8	1958
Dy............	W	$2.1 + 4.16 \times 10^{-4}T$	1100–1600	8	1958
Ho............	W	$2.3 + 5.1 \times 10^{-4}T$	1350–1950	8	1958
Er............	W	$2.4 + 4.35 \times 10^{-4}T$	$\sim\!1700$–1900	8	1958
Yb............	W	$2.7 + 2.85 \times 10^{-4}T$	1500–1900	8	1958
Lu............	W	$2.3 + 6 \times 10^{-4}T$	$\sim\!1500$–1700	8	1958
U............	W	$\sim\!3.1 + 2.1 \times 10^{-4}T$	1350–2100	9	1963
Hexaborides of:					
Ca............	None	$2.86 + 3.3 \times 10^{-4}T$	1100–1400	10	1951
Sr............	None	$2.67 + 5.82 \times 10^{-4}T$	1250–1459	10	1951
Ba............	None	$3.45 + 1.74 \times 10^{-4}T$	1250–1650	10	1951
Th............	None	$2.92 + 4.72 \times 10^{-4}T$	1150–1600	10	1951
Y............	Ta	$2.2 + 1.8 \times 10^{-4}T$	....	11	1958
La............	None	$2.66 + 1.22 \times 10^{-4}T$	1080–1300	10	1951
Ce............	None	$2.59 + 3.0 \times 10^{-4}T$	1140–1370	10	1951
Pr............	Ta	$3.46 - 7.9 \times 10^{-4}T$	....	11	1958
Nd............	Ta	$3.97 - 1.08 \times 10^{-4}T$	....	11	1958
Gd............	Ta	$2.05 + 4.28 \times 10^{-4}T$	....	11	1958
Dy............	Ta	$3.53 + 1.37 \times 10^{-4}T$	....	11	1958
Ho............	Ta	$3.42 + 1.86 \times 10^{-4}T$	....	11	1958
Er............	Ta	$3.37 + 2.15 \times 10^{-4}T$	....	11	1958
Yb............	Ta	$3.13 + 3.32 \times 10^{-4}T$	....	11	1958
Lu............	Ta	$3.0 + 5.0 \times 10^{-4}T$	....	11	1958
Tm............	W	$2.75 + 3.3 \times 10^{-4}T$	1100–1800	12	1964
Carbides of:					
Ta............	W	$3.14 + 5.2 \times 10^{-4}T$	1400–2300	13	1951
	Ta	$3.17 + 5.4 \times 10^{-4}T$	1600–2250	14	1962
Th............	Ta	$3.2 + 0.16 \times 10^{-4}T$	1500–2000	6	1953
	W	$3.5 - 1.32 \times 10^{-4}T$	1400–2300	13	1951
Ti............	W	$3.35 + 1.35 \times 10^{-4}T$	1400–2300	13	1951
Zr............	W	$2.18 + 5.1 \times 10^{-4}T$	1173–2073	13	1951
	W	$3.24 + 2.0 \times 10^{-4}T$	1300–2000	15	1962
U............	W	$2.94 + 1.1 \times 10^{-4}T$	1200–2100	16	1960
Si............	Ta	$3.5 + 0.5 \times 10^{-4}T$	....	6	1953
Hf............	W	$3.42 + 1.75 \times 10^{-4}T$	1300–2000	15	1962
V............	W	3.85	1300–2000	15	1962
Mo............	None	$5.36 - 4 \times 10^{-4}T$	1400–1650	17	1967

TABLE 9j-2. WORK FUNCTIONS FOR VARIOUS COMPOUNDS (*Continued*)

Emissive material	Base metal	Work function	Temperature range, K	Reference	Year
Borides of:					
Ta(mono)	W	$2.89 + 2.14 \times 10^{-4}T$	1400–2100	13	1951
Ta(di)	W, Ta, C	~ 2.85	1600–1800	18	1962
Ta(di)	None	4.26	1100–1550	17	1967
Zr(mono)	W	$4.48 - 4.76 \times 10^{-4}T$	1500–2100	13	1951
Ti(di)	None	$4.6 - 1.17 \times 10^{-4}T$	1350–1850	19	1957
V(di)	None	$4.2 + 6.1 \times 10^{-4}T$	1450–1800	19	1957
Nb(di)	None	$4.0 + 3.65 \times 10^{-4}T$	1500–2000	19	1957
Cr(di)	None	$4.1 + 3.55 \times 10^{-4}T$	1450–1900	19	1957
Mo(di)	None	$3.85 + 4.9 \times 10^{-4}T$	1550–2050	19	1957
Hf(di)	HfC	~ 3.7	2240	20	1963
U(di)	W	$\sim 3.3 + .2 \times 10^{-4}T$	1300–2100	9	1963
U(tetra)	W	$\sim 3.4 - .8 \times 10^{-4}T$	1300–2100	9	1963
Y(tetra)	Ta	$3.2 + 2.5 \times 10^{-4}T$	1530–1890	21	1966
Gd(tetra)	Ta	$2.5 + 4.0 \times 10^{-4}T$	1560–1750	21	1966
Silicides of:					
Re(di)	W	$4.02 - 2.67 \times 10^{-4}T$	1200–1400	22	1966
Nb(di)	W	$4.34 - 5.25 \times 10^{-4}T$	1300–1700	22	1962
Zr(di)	W	$3.95 - 5.0 \times 10^{-4}T$	1200–1900	22	1962
V(di)	W	$3.26 - 7.5 \times 10^{-4}T$	1100–1600	22	1962
Cr(mono)	W	$3.49 - .58 \times 10^{-4}T$	1200–1400	22	1962
Cr(di)	W	$3.78 - 1.2 \times 10^{-4}T$	1200–1450	22	1962
U(di)	W	$3.0 + 2.1 \times 10^{-4}T$	1300–2000	9	1963
U(tri)	W	$3.22 + 1.1 \times 10^{-4}T$	1300–2000	9	1963
W(di)	None	~ 4.62	1350–1610	17	1967
Ta(di)	None	~ 4.71	1180–1600	17	1967
Mo(di)	None	~ 4.73	~ 1500	17	1967
Nitrides of:					
Zr	W	3.96	2000	13	1951
B	W	4.0	2000	13	1951
Ti	W	3.96	2000	13	1951
U	W	$\sim 3.1 + 2.14 \times 10^{-4}T$	1350–2100	9	1963

References for Table 9j-2

1. Beyner, K. S., and B. P. Nikonov: *Radio Eng. Electron. Phys.* **3**, 408 (1965).
2. Wright, D. A.: *Nature* **160**, 129 (1947).
3. Hanley, T. E.: *J. Appl. Phys.* **19**, 583 (1948).
4. Kaganovich, M. V., and R. A. Makarova: *Radio Eng. Electron. Phys.* **9**, 1475 (1962).
5. Bondarenko, B. V., and B. M. Tsarev: *Radiotekhn. i Elektron.* **4**, 1060 (1959).
6. Wright, D. A.: *Proc. Inst. Elec. Engrs.* (*London*), pt.C, **100**, 125 (1953).
7. Thermionic Emission Studies, *RCA Quart. Rept.* 3, Contract DA36-039-SC-78155, June 30, 1959.
8. Kulvarskaya, B. S., V. B. Marchenko, and G. V. Stepanov; *Radiotekhn. i Elektron.* **3**, 1005 (1958).
9. Haas, G. A., and J. T. Jensen, Jr.: *J. Appl. Phys.* **34**, 3451 (1963).
10. Lafferty, J. M.: *J. Appl. Phys.* **22**, 299 (1951).
11. Kudintseva, G. A., and B. M. Tsarev: *Radiotekhn. i Elektron.* **3**, 428 (1958).
12. Yermakov, S. V.: *Radio Eng. Electron. Phys.* **1**, 142 (1964).
13. Goldwater, D. L., and R. E. Haddad: *J. Appl. Phys.* **22**, 70 (1951).
14. Eckstein, B. H., and R. Forman: *J. Appl. Phys.* **35**, 82 (1962).
15. Bondarenko, B. V., and S. V. Yermakov: *Radio Eng. Electron. Phys.* **12**, 1953 (1962).
16. Haas, G. A., and J. T. Jensen, Jr.: *J. Appl. Phys.* **31**, 1231 (1960).
17. Wilson, R. G., and W. E. McKee: *J. Appl. Phys.* **38**, 1716 (1967).
18. Matskevich, T. L., T. V. Krachino, and A. P. Kazantsev: *Radio Eng. Electron. Phys.* **11**, 1833 (November, 1962).
19. Steinitz, R.: Final Report: Research on Thermionic Emission of Borides, AF 18(600)-1504, Div. File 20-20, American Electro Metal Division of Firth Sterling, Yonkers, N.Y., 1957.

20. Matskevich, T. L., A. P. Kazantsev, and T. V. Karchino: *Soviet Phys.—Tech. Phys.* **7**(10), 934 (1963).
21. Manelis, R. M., L. P. Grishina, and A. D. Runov: *Radio Eng. Electron. Phys.* **11**, 1855 (1966).
22. Yermakov, S. V., and B. M. Tsarev: *Radio Engr. Electron. Phys.* **7**(12), 1956 (1962).

9j-2. Secondary Emission Measurements. The secondary emission ratio δ of a material is the number of secondary electrons emitted on the average, per incident primary electron. Secondary electron emission involves three processes: (1) the excitation of electrons in the solid by the incident primary electron beam, (2) the diffusion of the excited electrons to the surface, and (3) the escape of the excited electrons through the surface barrier. The secondary emission ratio of a given surface depends on the primary electron energy, the angle of incidence of the primary electrons with respect to the surface, bulk properties of the material, and most important, the surface conditions. In so far as practicable, yield measurements given here have been selected to be representative of clean surfaces at normal incidence.

The secondary emission ratio increases with primary energy until it reaches a maximum denoted by δ_m in Table 9j-3. The primary energy at which this maximum occurs is listed as $E_{p,\mathrm{max}}$. When the primary energy is raised beyond the maximum, electrons are excited increasingly deeper within the material so that many of them are unable to diffuse to the surface. The secondary emission ratio therefore slowly decreases as the primary energy is increased beyond $E_{p,\mathrm{max}}$. The voltages for which the secondary emission ratio crosses the $\delta = 1$ line are called the *first* and *second crossover*, respectively, and are designated as E_I and E_II in the table.

In general, lower secondary emission ratios are obtained from metals than from insulators. Roughening the surface of a material or increasing its porosity will further decrease the secondary emission ratio since the secondary electrons intercept other parts of the roughened and porous surface and are prevented from escaping into the vacuum. A good example of such a low secondary emission surface is carbon deposited as soot.

The highest secondary emission yields have been previously obtained from alkali halides and various oxides, such as NaBr or MgO. However, certain solid-state concepts have been recently employed to substantially increase the secondary emission ratio. One approach using Cs-treated GaP depends on bending the energy bands near the surface so that the effective electron barrier for escaping secondaries is lowered. The resulting secondary emission ratio for this surface represents an order of magnitude increase over that previously achieved from "natural"-appearing compounds.

A number of reviews of secondary emission have recently been published.[1] Table 9j-3 is taken from the review by Whetten except for a few additions.

[1] See, for example:

R. Kollath, "Handbuch der Physik," vol 21, pp. 232–233, Elektronen-Emission Gasentladungen, Springer-Verlag OHG, Berlin, 1956.

O. Hachenberg and W. Brauer, Secondary Emission from Solids, *Advan. Electron. Electron Phys.* **11**, 413–499 (1959).

H. Bruining, "Physics and Applications of Secondary Electron Emission," Pergamon Press, New York, 1954.

A. J. Dekker, *Solid State Phys.* **6**, 251 (1958).

N. R. Whetten, "Methods of Experimental Physics," vol 4, part A, pp. 69–84, Academic Press, Inc., New York, 1967.

SOLID-STATE PHYSICS

TABLE 9j-3. SECONDARY EMISSION PARAMETERS

Elements	δ_{max}	$E_{p,max}$	E_I	E_{II}	References
Ag....................	1.5	800	200	>2000	1, 2, 3
Al....................	1.0	300	300	300	2
Au....................	1.4	800	150	>2000	1, 2, 3
B.....................	1.2	150	50	600	5
Ba....................	0.8	400	None	None	2
Be....................	0.5	200	None	None	2, 6, 7, 4
Bi....................	1.2	550	...	......	8, 9
C (diamond)...........	2.8	750	...	>5000	10
(graphite)...........	1.0	300	300	300	11
(soot)...............	0.45	500	None	None	11
Cd....................	1.1	450	300	700	12, 4
Co....................	1.2	600	200	......	13, 14
Cs....................	0.7	400	None	None	2, 15
Cu....................	1.3	600	200	1500	1, 12, 2
Fe....................	1.3	400	120	1400	14, 3, 16
Ga....................	1.55	500	75	......	17
Ge....................	1.15	500	150	900	5, 18, 9
Hg....................	1.3	600	350	>1200	17
K.....................	0.7	200	None	None	19, 20
Li....................	0.5	85	None	None	2
Mg....................	0.95	300	None	None	15, 2
Mo....................	1.25	375	150	1200	1, 21, 3, 22, 16
Na....................	0.82	300	None	None	24
Nb....................	1.2	375	150	1050	1, 3
Ni....................	1.3	550	150	>1500	1, 14, 13, 21, 16
Pb....................	1.1	500	250	1000	8, 17
Pd....................	>1.3	>250	120	......	24
Pt....................	1.8	700	350	3000	3
Rb....................	0.9	350	None	None	19
Sb....................	1.3	600	250	2000	25
Si....................	1.1	250	125	500	5
Sn....................	1.35	500	...	......	8, 23
Ta....................	1.3	600	250	>2000	1
Th....................	1.1	800	...	......	2
Ti....................	0.9	280	None	None	11
Tl....................	1.7	650	70	>1500	9
W.....................	1.4	650	250	>1500	26, 1, 27, 3
Zr....................	1.1	350	...	......	11

TABLE 9j-3. SECONDARY EMISSION PARAMETERS (*Continued*)

Compounds	δ_{max}	$E_{p,max}$	Reference
Alkali halides:			
CsCl...................	6.5	....	28
KBr(crystal)...................	14	1800	29, 30
KCl(crystal)...................	12	1600	31, 32
(layer)................	7.5	1200	31, 33
KI(crystal).................	10	1600	29, 32, 30
(layer)................	5.6	....	28
LiF(crystal)................	8.5	....	34
(layer)................	5.6	700	28
NaBr(crystal).................	24	1800	35, 32
(layer)................	6.3	....	28
NaCl(crystal).................	14	1200	36, 30, 29, 32, 34
(layer)................	6.8	600	28, 37
NaF(crystal)................	14	1200	34
(layer)................	5.7	....	28
NaI(crystal).................	19	1300	34
(layer)................	5.5	....	28
RbCl(layer)................	5.8	....	28
Oxides:			
Ag$_2$O........................	1.0	....	38
Al$_2$O$_3$(layer)................	2–9	....	39, 40, 36, 2
BaO(layer)...................	2.3–4.8	400	40, 2
BeO....................	3.4	2000	40
CaO....................	2.2	500	40
Cu$_2$O......................	1.2	400	28, 41
MgO(crystal)................	20–25	1500	42, 43, 44, 45
(layer)................	3–15	400–1500	40, 46, 47, 42
MoO$_2$....................	1.2	....	38
SiO$_2$(quartz)................	2.1–4	400	39, 40
SnO$_2$....................	3.2	640	48
Sulfides:			
MoS$_2$....................	1.1	....	28
PbS.....................	1.2	500	31
WS$_2$......................	1.0	....	28
ZnS.....................	1.8	350	49
Others:			
BaF$_2$(layer)................	4.5	....	28
CaF$_2$(layer)................	3.2	....	28
BiCs$_3$....................	6	1000	31
BiCs....................	1.9	1000	31
GeCs.....................	7	700	31
Rb$_3$Sb....................	7.1	450	31
SbCs$_3$....................	6	700	31, 50
Mica(natural)................	2.4	350	39
(synthetic)................	2.9	300	51
Glasses....................	2–3	300–450	39, 52
MgF$_2$....................	4.0	400	51
Talc....................	2.3	300	51
Teflon....................	1.7	600	53
ZnSiO$_4$....................	2.6	300	53
GaP + Cs....................	120*	2500*	54
Phosphors:			
P$_1$....................	2.7	750	55
P$_2$....................	3.4	750	55
P$_3$....................	3.9	1000	55
P$_4$....................	3.7	700	55

* Maximum voltage measured was 2500 volts, δ was still rising.

References for Table 9j-3

1. Warnecke, R.: *J. Phys. Radium* **7**, 270 (1936).
2. Bruining, H., and J. H. deBoer: *Physica* **5**, 17 (1938).
3. Kollath, R.: *Z. Physik* **38**, 202 (1937).
4. Suhrmann, R., and W. Kundt: *Z. Physik* **121**, 118 (1943).
5. Koller, L. R., and J. S. Burgess: *Phys. Rev.* **70**, 571 (1946).
6. Kollath, R.: *Ann. Physik* **33**, 285 (1938).
7. Schneider, E. G.: *Phys. Rev.* **54**, 185 (1938).
8. Morozov, P. M.: *Zh. Eksperim. i Teor. Fiz.* **11**, 410 (1941).
9. Gobrecht, H., and F. Speer: *Z. Physik* **135**, 602 (1953).
10. Johnson, J. B.: *Phys. Rev.* **92**, 843 (1953).
11. Bruining, H.: *Philips Tech. Rev.* **3**, 80 (1938).
12. Suhrmann, R., and W. Kundt: *Z. Physik* **120**, 363 (1943).
13. Wooldridge, D. E.: *Phys. Rev.* **56**, 1062 (1939).
14. Treloar, L. R. G., and D. H. Landon: *Proc. Phys. Soc. (London)*, ser. B, **50**, 625 (1938).
15. Klebnikov, N. S.: *Zh. Tekhn. Fiz.* **5**, 593 (1938).
16. Petry, R. L.: *Phys. Rev.* **26**, 346 (1925).
17. Brophy, J. J.: *Phys. Rev.* **83**, 534 (1951).
18. Johnson, J. B., and K. G. McKay: *Phys. Rev.* **93**, 668 (1954).
19. Afanasjewa, A., and P. W. Timofeew: *Zh. Tekhn. Fiz.* **4**, 953 (1937).
20. Joffe, M. S., and I. V. Nechlaev: *Zh. Eksperim. i Teor. Fiz.* **11**, 93 (1941).
21. Blankenfeld, G.: *Ann. Physik* **9**, 48 (1951).
22. Copeland, P. L.: *J. Franklin Inst.* **215**, 593 (1933).
23. Woods, J.: *Proc. Phys. Soc (London)*, ser. B, **67**, 843 (1954).
24. Farnsworth, H. E.: *Phys. Rev.* **25**, 41 (1925).
25. Kollath, R.: in "Handbuch der Physik," vol. 21, p. 232, S. Flügge, ed., Springer-Verlag OHG, Berlin, 1956.
26. Petry, R. L.: *Phys. Rev.* **28**, 362 (1926).
27. Coomes, E. A.: *Phys. Rev.* **55**, 519 (1939).
28. Bruining, H., and J. H. deBoer: *Physica* **6**, 834 (1939).
29. Dobretzov, D. N., and A. S. Titkow: *Dokl. Akad. Nauk S.S.S.R.* **100**, 33 (1955).
30. Shulman, A. R., and B. P. Dementyev: *Zh. Tekhn. Fiz.* **25**, 2256 (1955).
31. Hachenberg, O., and W. Brauer: *Advan. Electron. Electron Phys.* **11**, 438 (1959).
32. Whetten, N. Rey: *Bull. Am. Phys. Soc.*, ser. II, **5**, 347 (1960).
33. Knoll, M., O. Hachenberg, and J. Randmer: *Z. Physik* **122**, 137 (1944).
34. Dobretzov, D. N., and T. L. Matskevich: *Zh. Tekhn. Fiz.* **27**, 734 (1957).
35. Matskevich, T. L.: *Zh. Tekhn. Fiz.* **26**, 2399 (1956).
36. Shulman, A. R., W. L. Makedonsky, and J. D. Yaroshetsky: *Zh. Tekhn. Fiz.* **23**, 1152 (1953).
37. Vudinsky, M. M.: *Zh. Tekhn. Fiz.* **9**, 271 (1939).
38. Afanasjewa, A., P. Timofeew, and A. Ignaton: *Phys. Z. Sowjet*, **10**, 831 (1936).
39. Salow, H.: *Z. Tech. Phys.* **21**, 8 (1940).
40. Geyer, K. H.: *Ann. Phys.* **42**, 241 (1942).
41. Gornij, N. B.: *Zh. Eksperim. i Teor. Fiz.* **26**, 79 (1954).
42. Whetten, N. Rey, and A. B. Laponsky: *J. Appl. Phys.* **30**, 432 (1959).
43. Johnson, J. B., and K. G. McKay: *Phys. Rev.* **91**, 582 (1953).
44. Lye, R. G.: *Phys. Rev.* **99**, 1647 (1955).
45. Whetten, N. Rey, and A. B. Laponsky: *J. Appl. Phys.* **28**, 515 (1957).
46. Wargo, P., B. V. Haxby, and W. G. Shepherd: *J. Appl. Phys.* **27**, 1311 (1956).
47. Rappaport, P.: *J. Appl. Phys.* **25**, 288 (1954).
48. Mendenhall, H. E.: *Phys. Rev.* **72**, 532 (1947).
49. Gornij, N. B.: *Zh. Eksperim. i Teor. Fiz.* **26**, 88 (1954).
50. Morgulis, N. D., and B. I. Djatlowitskaja: *Zh. Tekhn. Fiz.* **10**, 657 (1940).
51. Stambach, G. L.: *NRL Rept.* 5029, Washington, D.C., November 19, 1957.
52. Mueller, C. W.: *J. Appl. Phys.* **16**, 453 (1945).
53. Stambach, G. L.: Report of NRL Progress, 50 pp., Washington, D.C., August, 1959.
54. Simon, R. E., and B. F. Williams: *IEEE Trans. Nucl. Sci.* **NS 15**(3), 167 (1968).
55. White, W. A.: *Overseas Res. Rept.* 83/51, PB 101420, Office of Technical Services, Washington, D.C.

Index

References involving tables are shown by the letter (T); references involving figures are shown by the letter (f).

A

Abbe number, definition of, **6**-7
Abbe's sine condition and lenses, **6**-99
Aberration in lenses, **6**-97, **6**-99
Absolute units, electrical (T), **5**-108 to **5**-109
Absolute viscosity:
  definition of, **2**-234, **2**-235
  units and conversion factors for (T), **2**-233
  of various gases (T), **2**-242 to **2**-243
Absorbed dose and health physics, definition of, **8**-292
Absorbers of charged particles (f's), **8**-171 to **8**-176
Absorptance:
  definition of, **6**-2
  formulas for intensity of, **6**-121
  spectral, of thermal-control materials (T), **6**-207
Absorption:
  definition of, **6**-2
  vs. frequency for various acoustic materials (f's), **3**-145
  of gamma rays, **8**-190 to **8**-191
  self-, in gamma-ray sources, **8**-211 to **8**-212
  and sounds in fluids, **3**-55 to **3**-65
  of various crystals and glasses, **6**-58 to **6**-94
Absorption bands:
  of alkali halide phosphors (T), **9**-160
  of alkali halides (T's), **9**-153, **9**-154
  of electron trap centers (T), **9**-149
  from various crystal centers (T), **9**-153
  for various substances (T), **9**-154
Absorption coefficients:
  acoustical: of seats and audience (T), **3**-147
  for various building materials (T), **3**-146

Absorption coefficients (*Cont.*):
  vs. frequency: for crystal quartz (T)(f), **6**-295, **6**-297
  for fused quartz (T)(f), **6**-300
  for Irtran VI (T)(f), **6**-303, **6**-304
  for sapphire (T)(f), **6**-296, **6**-298
  of Teflon (f), **6**-304
  and health physics, definition of, **8**-292
  for photons, definition of, **8**-201
  of several semiconductors (f), **6**-62
  vs. wave number: for GaAs (T)(f), **6**-306, **6**-308
  for germanium (T)(f), **6**-298, **6**-299
  for Irtran VI (T)(f), **6**-306, **6**-307
  for silicon (T)(f), **6**-298, **6**-299
Absorption cross section, definition of, **8**-198
Absorption curves for monoenergetic electrons (f), **8**-182
Absorption edges, short- and long-wavelength, of several crystals (f's), **6**-61
Absorption filters, **6**-170
Absorption lines, rotational, for water vapor (T), **6**-288 to **6**-290
Absorption loss, definition of acoustic, **3**-11
Absorption measure and sound in fluids, **3**-57 to **3**-58
Absorption spectrum:
  definition of, **6**-2
  and luminescence (T), **9**-159, **9**-160
  X-ray, of elements (T), **7**-96, **7**-120 to **7**-157
Absorption wavelength:
  of electron trap centers (T), **9**-149
  of hole trap centers (T), **9**-152
Abundance:
  of elements, **7**-6 to **7**-8
  of nuclides, percent (T), **8**-8 to **8**-91
Acceleration:
  angular, **2**-9
  definition of, **2**-4, **3**-2
  of gravity (T), **2**-91

1

W